2014

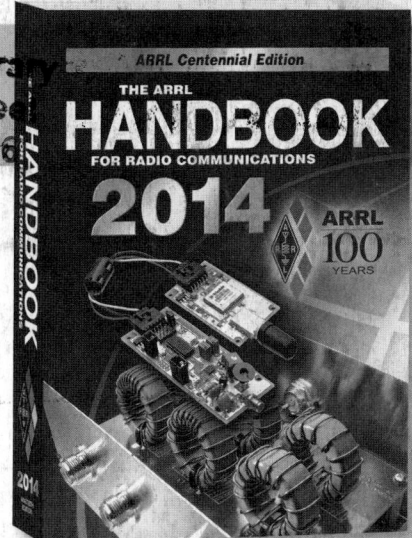

The ARRL Handbook

For Radio Communications

Ninety-First Edition

Published by:
ARRL

the national association for Amateur Radio™
Newington, CT 06111 USA

ARRL
100
YEARS

About the Cover: The blue toroid cores belong to an HF triplexer described in the Transmission Lines chapter by George Cutsogeorge, W2VJN, of International Radio. The availability of professional-quality filter design software, such as the programs provided by Tonne Software on the *Handbook's* CD-ROM, have enabled amateurs to design and build such advanced equipment. Above and to the left of the triplexer you see the balloon payload package by Bill Brown, WB8ELK. It transmits position and sensor information from high-altitude platforms, such as balloons, via the APRS network. The use of Amateur Radio in support of experiments and vehicle tracking is a rapidly growing activity. The new Handbook chapter on Telemetry and Navigation provides supporting information for building and using these systems.

D1065981

Editor
H. Ward Silver, NØAX

Contributing Editors
Steven R. Ford, WB8IMY
Mark J. Wilson, K1RO

Editorial Assistant
Maty Weinberg, KB1EIB

Technical Consultants
Michael E. Gruber, W1MG
Edward F. Hare, Jr., W1RFI
Zachary H.J. Lau, W1VT

Cover Design
Sue Fagan, KB1OKW
Bob Inderbitzen, NQ1R

Production
Michelle Bloom, WB1ENT
Nancy G. Hallas, W1NCY
Judi Kutzko
Jodi Morin, KA1JPA
David F. Pingree, N1NAS

Additional Contributors to the 2014 Edition

Alan Applegate, KØBG
Alan Bloom, N1AL
Bill Brown, WB8ELK
Pat Bunsold, WA6MHZ
Isidor Buchmann
Rick Campbell, KK7B
Paul Christensen, W9AC
Jim Colville, W7RY
Jeff Coval, ACØSC
George Cutsogeorge, W2VJN
Tim Duffy, K3LR
Courtney Duncan, N5BF
Joe Eisenberg, KØNEB
Dick Frey, K4XU
Joel Hallas, W1ZR
Rick Hilding, K6VVA
Riley Hollingsworth, K4ZDH
Scott Honaker, N7SS
Ken Humbertson, WØKAH

Dave Jones, KB4YZ
Matt Kastigar, WØXEU
Greg Lapin, N9GL
Rick Lindquist, WW1ME
Pete Loveall, AE5PL
Jim Lux, W6RMK
Carl Luetzelschwab, K9LA
Earl McCune, WA6SUH
Steve Morris, K7LXC
Tom O'Hara, W6ORG
Gary Pearce, KN4AQ
David Rowe, VK5DGR
Kai Siwiak, KE4PT
John Stanley, K4ERO
Jim Stewart
Rich Strand, KL7RA
Paul Verhage, KD4STH
Paul Wade, W1GHZ
Mel Whitten, KØPFX

Contents

A more detailed Table of Contents is included at the beginning of each chapter.

Foreword

Welcome to the 91st edition — the 2014 *ARRL Handbook*! In this ARRL centennial year, you'll find the book contains up-to-date information on the latest amateur technologies, along with solid and reliable electronic information that has been an integral part of the hobby through the decades. You'll particularly enjoy a special graphic-filled *Handbook* history written specially for this edition by one of the ARRL's favorite authors and editors, Joel Hallas, W1ZR. Complemented by a CD-ROM that is full of supplemental articles and projects, this edition has plenty for every ham.

We've updated many of the chapters — three-quarters of them have new or rewritten sections to keep them up-to-date. And there is an addition to the *Handbook* family — a completely new chapter on **Telemetry and Navigation**! Written by experts Paul Verhage, KD4STH, and Bill Brown, WB8ELK, this chapter is going to grow along with the rapid increase in the use of Amateur Radio in partnership with scientific experiments on land, sea, and in the atmosphere and even in space.

Key updates include the new material on batteries and battery charging in the **Power Sources** chapter — renamed to reflect the broad range of ways in which we get power for our radios. The **Oscillators and Synthesizers** chapter has been updated by industry expert Earl McCune, WA6SUH. To help out on the repair bench, the **Troubleshooting and Maintenance** chapter has been updated and reorganized to address the current needs of amateurs.

Speaking of that CD-ROM, more and more material is added to this valuable resource every year, including the full text and graphics from the current edition as convenient and searchable PDF files. This year, we've made a special effort to make sure you'll want to open the envelope inside the back cover and install it on your computer right away. There are more projects than ever on the CD-ROM, full construction details to complete projects in the printed book, and supporting articles that extend the technical material of the regular chapters.

Here are some highlights of what you'll find on this year's CD-ROM:

- W1ZR's annual transceiver model review organized by type and frequency coverage — if you are considering purchasing a radio, this is a great place to start!

- Historical development of amateur receivers and transmitters — W1ZR explains how our radio designs came to be as various challenges and needs were met.

- Jim Tonne, W4ENE, has contributed the latest amateur versions of his professional filter design software, *ELSIE*, along with several other filter and filter-related programs. We are once again very fortunate to include such high-quality software tools.

- Chapters providing operating guidance and station information on **Space Communications** by Joe Taylor, K1JT, and Steve Ford, WB8IMY; **Digital Communications** by Steve Ford, WB8IMY; and **Image Communications** by Tom O'Hara, W6ORG, and Dave Jones, KB4YZ

Projects have been added throughout the *Handbook* but the new ones on the transmit side are sure to catch your eye:

- The Everyham's Amplifier — a simple and flexible design by John Stanley, K4ERO, that is within reach of beginning builders who would like to add some output power to their signal.

- QSK Controllers for Amplifiers — one is an analog design by Jim Colville, W7RY, that can be adapted to many common linear amplifiers. The companion Arduino-based design by Paul Christensen, W9AC, provides the ultimate in flexibility.

- VHF signal sources by long-time amateur design guru Rick Campbell, KK7B, are a great way to start your 6 meter or 2 meter project.

- In addition, on the CD-ROM, you'll find a set of VHF/UHF amplifier designs from the pages of *QST* and *Ham Radio* magazine.

Even after more than 90 editions, the *ARRL Handbook* continues to pursue its mission of providing a technical reference for hams of all interests and abilities. Whether you choose to pursue public service communications, engage in technical experimentation, hone your operating skills, or just explore the magic of wireless communication, "the *Handbook*" has a place in your library and at your workbench.

David Sumner, K1ZZ
Chief Executive Officer
Newington, Connecticut
August 2013

The Amateur's Code

The Radio Amateur is:

CONSIDERATE...never knowingly operates in such a way as to lessen the pleasure of others.

LOYAL...offers loyalty, encouragement and support to other amateurs, local clubs, and the American Radio Relay League, through which Amateur Radio in the United States is represented nationally and internationally.

PROGRESSIVE...with knowledge abreast of science, a well-built and efficient station and operation above reproach.

FRIENDLY...slow and patient operating when requested; friendly advice and counsel to the beginner; kindly assistance, cooperation and consideration for the interests of others. These are the hallmarks of the amateur spirit.

BALANCED...radio is an avocation, never interfering with duties owed to family, job, school or community.

PATRIOTIC...station and skill always ready for service to country and community.

—*The original Amateur's Code was written by Paul M. Segal, W9EEA, in 1928.*

Common Schematic Symbols Used in Circuit Diagrams

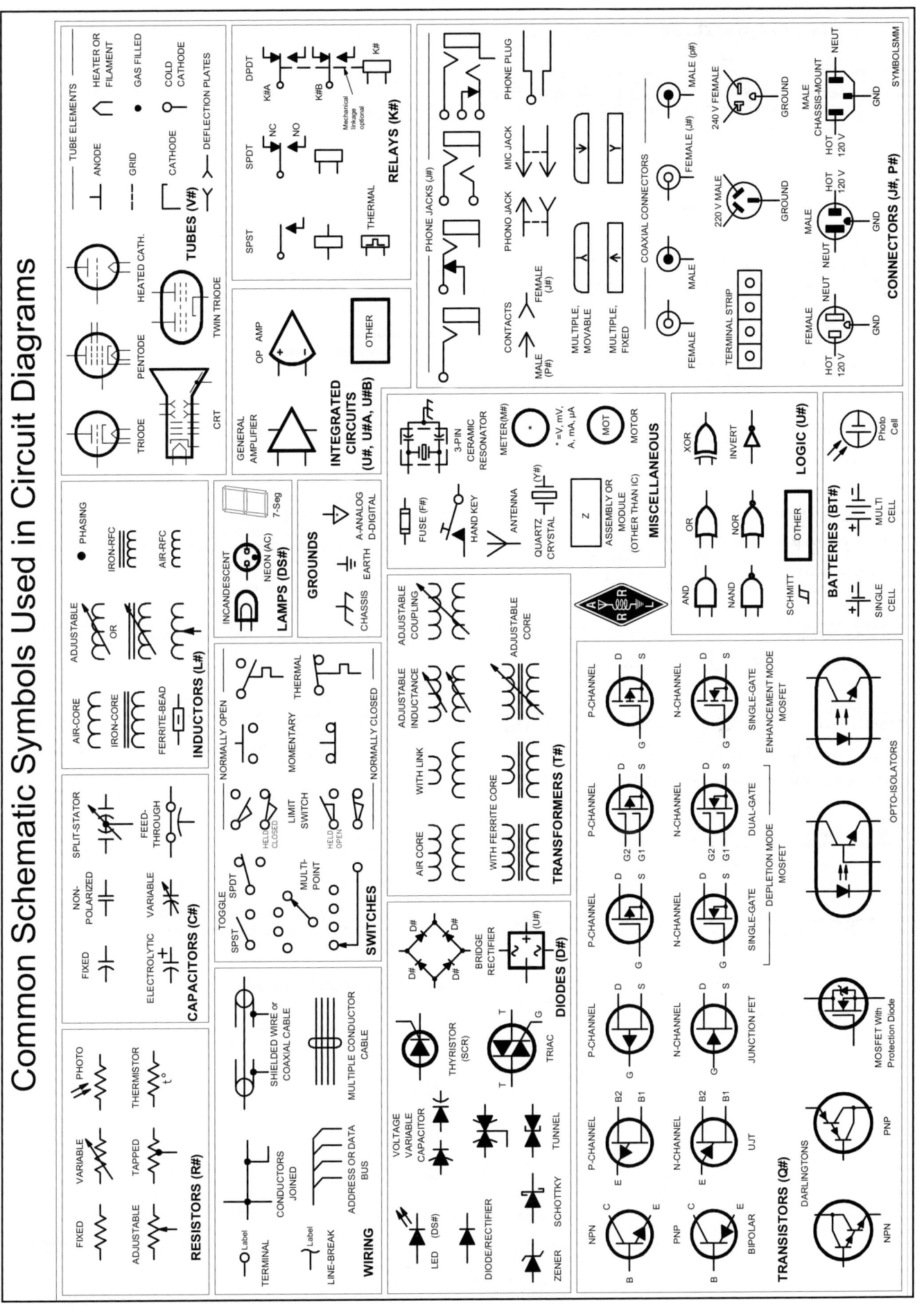

ARRL Member Services

 **Get Involved**
www.arrl.org/get-involved

 **Join or Renew**
www.arrl.org/join

 **Donate**
www.arrl.org/donate

 **Shop**
www.arrl.org/shop

Membership Benefits

Your ARRL membership includes *QST* magazine, plus dozens of other services and resources to help you **Get Started**, **Get Involved** and **Get On the Air**. ARRL members enjoy Amateur Radio to the fullest!

Members-Only Web Services

Create an online ARRL Member Profile, and get access to ARRL members-only Web services. Visit **www.arrl.org/myARRL** to register.

- **QST Digital Edition – www.arrl.org/qst**
  All ARRL members can access the online digital edition of *QST*. Enjoy enhanced content, convenient access and a more interactive experience. An app for *iOS* devices is also available.

- **QST Archive and Periodicals Search – www.arrl.org/qst**
  Browse ARRL's extensive online *QST* archive. A searchable index for *QEX* and *NCJ* is also available.

- **Free E-Newsletters**
  Subscribe to a variety of ARRL e-newsletters and e-mail announcements: ham radio news, radio clubs, public service, contesting and more!

- **Product Review Archive – www.arrl.org/qst**
  Search for, and download, *QST* Product Reviews published from 1980 to present.

- **E-Mail Forwarding Service**
  E-mail sent to your **arrl.net** address will be forwarded to any e-mail account you specify.

- **Customized ARRL.org home page**
  Customize your home page to see local ham radio events, clubs and news.

- **ARRL Member Directory**
  Connect with other ARRL members via a searchable online Member Directory. Share profiles, photos and more with members who have similar interests.

ARRL Technical Information Service — www.arrl.org/tis

Get answers on a variety of technical and operating topics through ARRL's Technical Information Service. ARRL Lab experts and technical volunteers can help you overcome hurdles and answer all your questions.

ARRL as an Advocate — www.arrl.org/regulatory-advocacy

ARRL supports legislation and regulatory measures that preserve and protect access to Amateur Radio Service frequencies. Members may contact the **ARRL Regulatory Information Branch** for information on FCC rules; problems with antenna, tower and zoning restrictions; and reciprocal licensing procedures for international travelers.

ARRL Group Benefit Programs* — www.arrl.org/benefits

- **ARRL "Special Risk" Ham Radio Equipment Insurance Plan**
  Insurance is available to protect you from loss or damage to your station, antennas and mobile equipment by lightning, theft, accident, fire, flood, tornado, and other natural disasters.

- **The ARRL Visa Signature® Card**
  Every purchase supports ARRL programs and services.

- **MetLife® Auto, Home, Renters, Boaters, Fire Insurance and Banking Products**
  ARRL members may qualify for up to a 10% discount on home or auto insurance.

  * ARRL Group Benefit Programs are offered by third parties through contractual arrangements with ARRL. The programs and coverage are available in the US only. Other restrictions may apply.

Programs

★ **New! ARRL Centennial 2014 — ARRL2014.org**
Second Century Campaign for the ARRL Endowment – **www.arrl.org/scc**
National Centennial Convention, July 17-20, 2014 – **ARRL2014.org**

Public Service — www.arrl.org/public-service
Amateur Radio Emergency Service® – **www.arrl.org/ares**
Emergency Communications Training – **www.arrl.org/emcomm-training**

Radiosport
Awards – **www.arrl.org/awards**
Contests – **www.arrl.org/contests**
QSL Service – **www.arrl.org/qsl**
Logbook of the World – **www.arrl.org/lotw**

Community
Radio Clubs (ARRL-affiliated clubs) – **www.arrl.org/clubs**
Hamfests and Conventions – **www.arrl.org/hamfests**
ARRL Field Organization – **www.arrl.org/field-organization**

Licensing, Education and Training
Find a License Exam Session – **www.arrl.org/exam**
Find a Licensing Class – **www.arrl.org/class**
ARRL Continuing Education Program – **www.arrl.org/courses-training**
Books, Software and Operating Resources – **www.arrl.org/shop**

Quick Links and Resources
QST – ARRL members' journal – **www.arrl.org/qst**
QEX – A Forum for Communications Experimenters – **www.arrl.org/qex**
NCJ – National Contest Journal – **www.arrl.org/ncj**
Support for Instructors – **www.arrl.org/instructors**
Support for Teachers – **www.arrl.org/teachers**
ARRL Volunteer Examiner Coordinator (ARRL VEC) – **www.arrl.org/vec**
Public and Media Relations – **www.arrl.org/media**
Forms and Media Warehouse – **www.arrl.org/forms**
FCC License Renewal – **www.arrl.org/fcc**
Foundation, Grants and Scholarships – **www.arrl.org/arrl-foundation**

Interested in Becoming a New Ham?

www.arrl.org/newham · newham@arrl.org · 1-800-326-3942 (US)

Contact Us

ARRL, the national association for Amateur Radio®
225 Main Street, Newington, CT 06111-1494 USA
Tel 1-860-594-0200, Mon-Fri 8 AM to 5 PM ET (except holidays)
FAX 1-860-594-0259, e-mail **hqinfo@arrl.org**,
website – **www.arrl.org/contact-arrl**

 Facebook
www.facebook.com/ARRL.org

 Follow us on Twitter
twitter.com/arrl · **twitter.com/w1aw** · **twitter.com/arrl_pr**
twitter.com/arrl_youth · **twitter.com/arrl_emcomm**
twitter.com/arrl_dxcc

 YouTube
www.youtube.com/ARRLHQ

The American Radio Relay League, Inc.

The American Radio Relay League, Inc. is a noncommercial association of radio amateurs, organized for the promotion of interest in Amateur Radio communication and experimentation, for the establishment of networks to provide communication in the event of disasters or other emergencies, for the advancement of the radio art and of the public welfare, for the representation of the radio amateur in legislative matters, and for the maintenance of fraternalism and a high standard of conduct.

ARRL is an incorporated association without capital stock chartered under the laws of the State of Connecticut, and is an exempt organization under Section 501(c)(3) of the Internal Revenue Code of 1986. Its affairs are governed by a Board of Directors, whose voting members are elected every three years by the general membership. The officers are elected or appointed by the directors. The League is noncommercial, and no one

with a pervasive and continuing conflict of interest is eligible for membership on its Board.

"Of, by, and for the radio amateur," the ARRL numbers within its ranks the vast majority of active amateurs in the nation and has a proud history of achievement as the standard-bearer in amateur affairs.

A *bona fide* interest in Amateur Radio is the only essential qualification of membership; an Amateur Radio license is not a prerequisite, although full voting membership is granted only to licensed amateurs in the US.

Membership inquiries and general correspondence should be addressed to the administrative headquarters: ARRL, 225 Main Street, Newington, Connecticut 06111-1494.

About the ARRL

The seed for Amateur Radio was planted in the 1890s, when Guglielmo Marconi began his experiments in wireless telegraphy. Soon he was joined by dozens, then hundreds, of others who were enthusiastic about sending and receiving messages through the air—some with a commercial interest, but others solely out of a love for this new communications medium. The United States government began licensing Amateur Radio operators in 1912.

By 1914, there were thousands of Amateur Radio operators—hams—in the United States. Hiram Percy Maxim, a leading Hartford, Connecticut inventor and industrialist, saw the need for an organization to band together this fledgling group of radio experimenters. In May 1914 he founded the American Radio Relay League (ARRL) to meet that need.

Today ARRL, with approximately 155,000 members, is the largest organization of radio amateurs in the United States. The ARRL is a not-for-profit organization that:
- promotes interest in Amateur Radio communications and experimentation
- represents US radio amateurs in legislative matters, and
- maintains fraternalism and a high standard of conduct among Amateur Radio operators.

At ARRL headquarters in the Hartford suburb of Newington, the staff helps serve the needs of members. ARRL is also International Secretariat for the International Amateur Radio Union, which is made up of similar societies in 150 countries around the world.

ARRL publishes the monthly journal *QST* and an interactive digital version of *QST*, as well as newsletters and many publications covering all aspects of Amateur Radio. Its headquarters station, W1AW, transmits bulletins of interest to radio amateurs and Morse code practice sessions. The ARRL also coordinates an extensive field organization, which includes volunteers who provide technical information and other support services for radio amateurs as well as communications for public-service activities. In addition, ARRL represents US amateurs with the Federal Communications Commission and other government agencies in the US and abroad.

Membership in ARRL means much more than receiving *QST* each month. In addition to the services already described, ARRL offers membership services on a personal level, such as the Technical Information Service—where members can get answers by phone, email or the ARRL website, to all their technical and operating questions.

Full ARRL membership (available only to licensed radio amateurs) gives you a voice in how the affairs of the organization are governed. ARRL policy is set by a Board of Directors (one from each of 15 Divisions). Each year, one-third of the ARRL Board of Directors stands for election by the full members they represent. The day-to-day operation of ARRL HQ is managed by an Executive Vice President and his staff.

No matter what aspect of Amateur Radio attracts you, ARRL membership is relevant and important. There would be no Amateur Radio as we know it today were it not for the ARRL. We would be happy to welcome you as a member! (An Amateur Radio license is not required for Associate Membership.) For more information about ARRL and answers to any questions you may have about Amateur Radio, write or call:

ARRL—the national association for Amateur Radio®
225 Main Street
Newington CT 06111-1494
Voice: 860-594-0200
Fax: 860-594-0259
E-mail: **hq@arrl.org**
Internet: **www.arrl.org**

Prospective new amateurs call (toll-free):
800-32-NEW HAM (800-326-3942)
You can also contact us via e-mail at **newham@arrl.org**
or check out the ARRL website at **www.arrl.org**

US Amateur Radio Bands

Published by:

ARRL *The national association for* AMATEUR RADIO®

www.arrl.org

225 Main Street, Newington, CT USA 06111-1494

Effective Date March 5, 2012

US AMATEUR POWER LIMITS

FCC 97.313 An amateur station must use the minimum transmitter power necessary to carry out the desired communications. (b) No station may transmit with a transmitter power exceeding 1.5 kW PEP.

160 Meters (1.8 MHz)

Avoid interference to radiolocation operations from 1.900 to 2.000 MHz

1.800 — 1.900 — 2.000 MHz — E,A,G

80 Meters (3.5 MHz)

3.500 — 3.600 — 3.700 — 3.800 — 4.000 MHz

3.525 3.600 N,T (*200 W*)

E / A / G

60 Meters (5.3 MHz)

2.8 kHz

5330.5 5346.5 5357.0 5371.5 5403.5 kHz

E,A,G (*100 W*)

General, Advanced, and Amateur Extra licensees may operate on these five channels on a secondary basis with a maximum effective radiated output of 100 W PEP. Permitted operating modes include upper sideband voice (USB), CW, RTTY, PSK31 and other digital modes such as PACTOR III as defined by the FCC Report and Order of November 18, 2011. USB is limited to 2.8 kHz centered on 5332, 5348, 5358.5, 5373 and 5405 kHz. CW and digital emissions must be centered 1.5 kHz above the channel frequencies indicated above. Only one signal at a time is permitted on any channel.

40 Meters (7 MHz)

7.000 — 7.125 — 7.300 MHz

7.025 7.125 N,T (*200 W*)

E / A / G

Phone and Image modes are permitted between 7.075 and 7.100 MHz for FCC licensed stations in ITU Regions 1 and 3 and by FCC licensed stations in ITU Region 2 West of 130 degrees West longitude or South of 20 degrees North latitude. See Sections 97.305(c) and 97.307(f)(11). Novice and Technician licensees outside ITU Region 2 may use CW only between 7.025 and 7.075 MHz and between 7.100 and 7.125 MHz. 7.200 to 7.300 MHz is not available outside ITU Region 2. See Section 97.301(e). These

30 Meters (10.1 MHz)

Avoid interference to fixed services outside the US.

200 Watts PEP

10.100 — 10.150 MHz — E,A,G

20 Meters (14 MHz)

14.000 — 14.150 — 14.350 MHz

14.025 14.150 14.225
14.175

E / A / G

17 Meters (18 MHz)

18.068 — 18.110 — 18.168 MHz — E,A,G

15 Meters (21 MHz)

21.000 — 21.200 — 21.450 MHz

21.025 21.200 21.275
21.225

E / A / G / N,T (*200 W*)

12 Meters (24 MHz)

24.890 — 24.930 — 24.990 MHz — E,A,G

10 Meters (28 MHz)

28.000 — 28.300 — 28.500 — 29.700 MHz

N,T (*200 W*)

E / A / G

6 Meters (50 MHz)

50.0 — 50.1 — 54.0 MHz — E,A,G,T

2 Meters (144 MHz)

144.0 — 144.1 — 148.0 MHz — E,A,G,T

1.25 Meters (222 MHz)

219.0 220.0 — 222.0 — 225.0 MHz

E,A,G,T / N (*25 W*)

70 cm (420 MHz)*

420.0 — 450.0 MHz — E,A,G,T

33 cm (902 MHz)*

902.0 — 928.0 MHz — E,A,G,T

23 cm (1240 MHz)*

1240 — 1270 — 1295 1300 MHz

E,A,G,T / N (*5 W*)

*Geographical and power restrictions may apply to all bands above 420 MHz. See *The ARRL Operating Manual* for information about your area.

All licensees except Novices are authorized all modes on the following frequencies:

2300-2310 MHz	10.0-10.5 GHz *	122.25-123.0 GHz
2390-2450 MHz	24.0-24.25 GHz	134-141 GHz
3300-3500 MHz	47.0-47.2 GHz	241-250 GHz
5650-5925 MHz	76.0-81.0 GHz	All above 275 GHz

* No pulse emissions

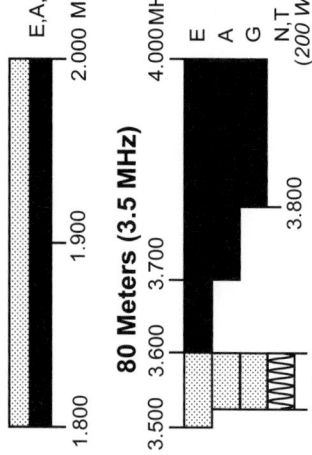

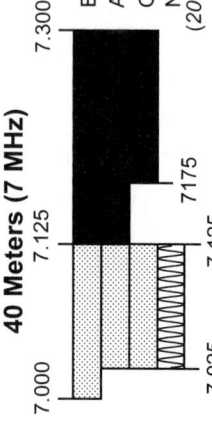
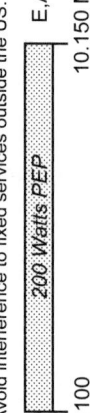
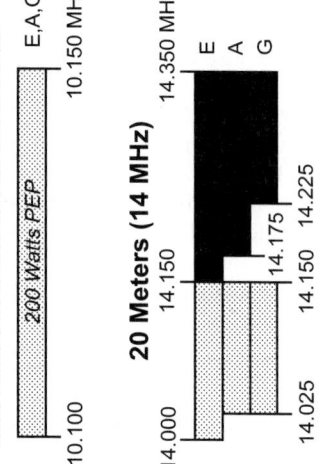

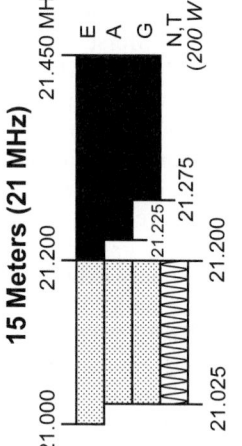

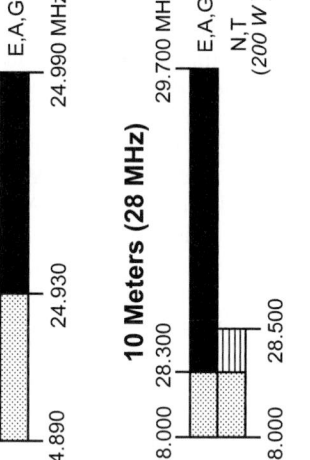
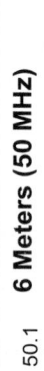

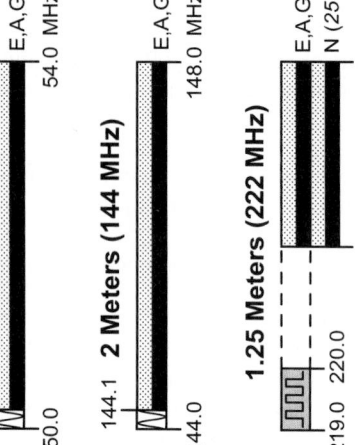
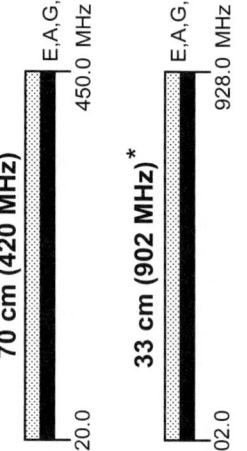
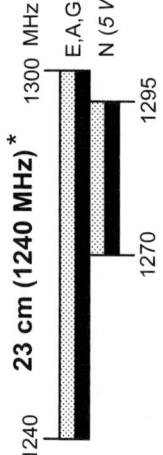

ARRL Handbook CD-ROM Contents

On the CD-ROM included with this book you'll find this entire edition of the *Handbook*, including text, drawings, tables, illustrations and photographs — many in color. Using Adobe *Reader*, you can view and print the text of the book, zoom in and out on pages, and copy selected parts of pages to the clipboard. A powerful search engine helps you find topics of interest. Also included is supplemental information and articles, PC board template packages, construction details for many projects, and companion software mentioned throughout. A README file is included on the CD-ROM for more information. The CD-ROM is included in protective envelope attached inside the back cover of the book.

Supplemental Files for Each Chapter

The CD-ROM provides supplemental information for most chapters of this book. This includes articles from *QST* and other sources, material from previous editions of the *ARRL Handbook*, tables and figures in support of the chapter material, and files that contain PC board layout and other design information to build and test the projects provided in the chapters. The supplemental information is arranged in folders, and a list of relevant CD-ROM content is included at the beginning of each chapter in this book.

Companion Software

TubeCalculator, a *Windows* application by Bentley Chan and John Stanley, K4ERO, accompanies the tube type RF power amplifier discussion in the **RF Power Amplifiers** chapter.

The following *Windows* programs by Tonne Software (**www.tonnesoftware.com**) are provided by Jim Tonne, W4ENE.

ClassEinstall204.exe — Designs single-ended Class E RF amplifiers.

DiplexerInstall209.exe — Designs both high-pass/low-pass and band-pass/band-stop types of diplexer circuits.

HelicalInstall206.exe — Designs and analyzes helical-resonator bandpass filters for the VHF and UHF frequency ranges.

JJSmithInstall210.exe — A graphics-intensive transmission-line calculator based on the Smith chart.

LCinstall257.exe — The free student edition of *Elsie,* a lumped-element filter design and analysis program.

MeterBasicInstall304.exe — Designs and prints professional-quality analog meter scales on your printer. The full-featured version of *Meter* is available from Tonne Software.

OptLowpassInstall203.exe — Designs and analyzes very efficient transmitter output low-pass filters.

PIELinstall217.exe — Designs and analyzes pi-L networks for transmitter output.

QuadNetInstall203.exe — Designs and analyzes active quadrature ("90-degree") networks for use in SSB transmitters and receivers.

SVCFilterInstall212.exe — Standard-value component routine to design low-pass and high-pass filters and delivers exact-values as well as nearest-5% values.

The ARRL
Radio Amateur's Handbook
— From Its Beginning

We will take a journey through time and space, looking over the ARRL *Handbooks* from their first edition to the modern era. After looking at the first offering, we will look in some detail at the changes on a decade by decade basis, highlighting the changes over each interval, with focus on the equipment in use and described in each volume. By the time of the first handbook, the art had already advanced from the days of spark and crystal sets to the use of vacuum tubes for both transmitting and receiving equipment.

In the early days of radio, before the ARRL published its first "handbook," there were attempts by others to fill the space. Some were reviewed in *QST*, but most were found to be lacking in technical content or clarity.[1] This might explain why the League decided to enter the handbook business, although they didn't get there until 1926. — *Joel R. Hallas, W1ZR, QST Contributing Editor*

[1]S. Kruse, 1OA, "Book Review — A. F. Collins, The Radio Amateur's Handbook," *QST*, Feb 1923, p 70.

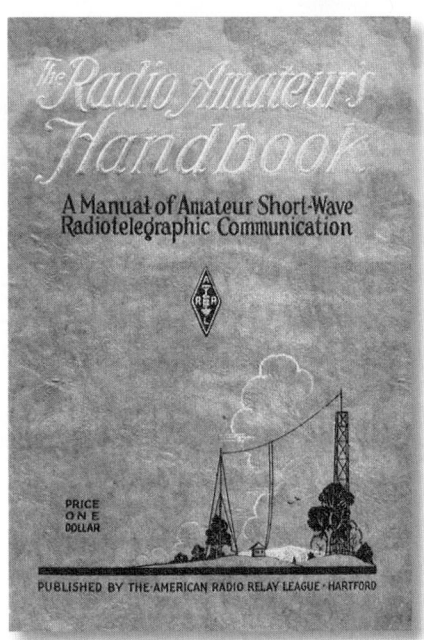

1926 Edition

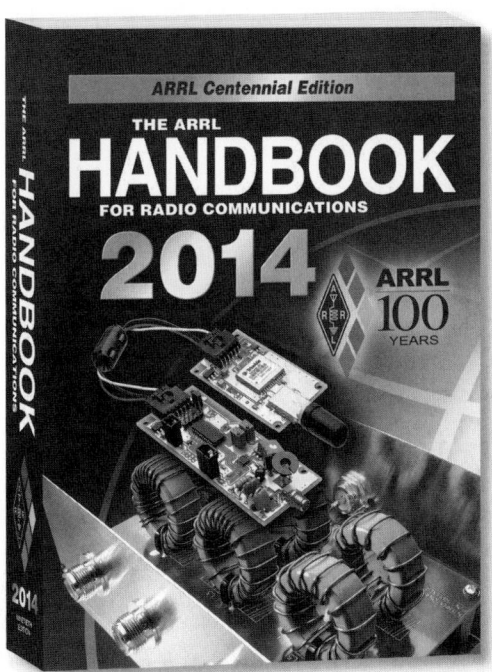

2014 Edition

ARRL Publishes the First *The Radio Amateur's Handbook* — The 1926 Edition

The first ARRL-published edition of *The Radio Amateur's Handbook* was announced by then *QST* Technical Editor Kenneth B. Warner, 1BHW, in *QST* of October 1926. Warner stated: "At last we have the honor of announcing *The Radio Amateur's Handbook*, the A.R.R.L. handbook we have been dreaming about for several years… The book has been written by Mr. F. E. Handy, A.R.R.L. Communication Manager, eminently qualified for the job not only because of his sound engineering knowledge but perhaps more particularly because this business of actually operating amateur stations is the subject about which his department deals daily and he knows it inside and out…."[2]

This volume was quite comprehensive for its time, as shown in the table of contents (see Figure 1926-1), with the emphasis spread among underlying technology, building a station, operating a station and a bit about the ARRL's activities. A total of 5000 copies of the first edition of the *Handbook* were produced.

The book opens to a view of typical full-size stations of the era (see Figure 1926-2). Note the use of open construction without attention to operator safety (no safety rules back then) and no attempt at shielding (that would change with the popularity of

[2]K.B. Warner, 1BHW, "Our *Handbook* Announced," *QST*, Oct 1926, p 8.

CONTENTS

vii

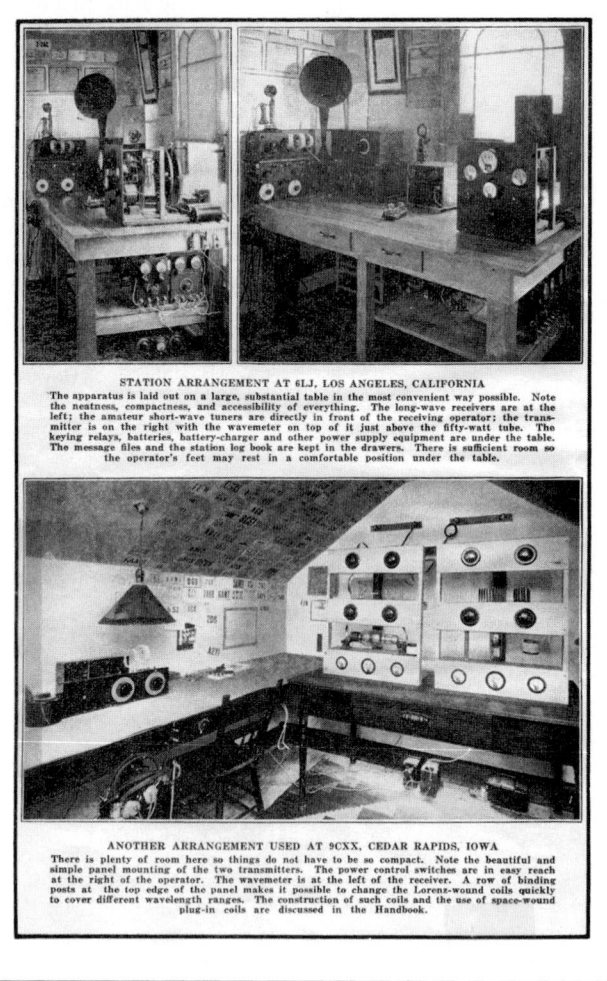

Figure 1926-1 — The Table of Contents of the First Edition of ARRL's *Radio Amateur's Handbook*, circa 1926.

STATION ARRANGEMENT AT 6LJ, LOS ANGELES, CALIFORNIA

The apparatus is laid out on a large, substantial table in the most convenient way possible. Note the neatness, compactness, and accessibility of everything. The long-wave receivers are at the left; the amateur short-wave tuners are directly in front of the receiving operator; the transmitter is on the right with the wavemeter on top of it just above the fifty-watt tube. The keying relays, batteries, battery-charger and other power supply equipment are under the table. The message files and the station log book are kept in the drawers. There is sufficient room so the operator's feet may rest in a comfortable position under the table.

ANOTHER ARRANGEMENT USED AT 9CXX, CEDAR RAPIDS, IOWA

There is plenty of room here so things do not have to be so compact. Note the beautiful and simple panel mounting of the two transmitters. The power control switches are in easy reach at the right of the operator. The wavemeter is at the left of the receiver. A row of binding posts at the top edge of the panel makes it possible to change the Lorenz-wound coils quickly to cover different wavelength ranges. The construction of such coils and the use of space-wound plug-in coils are discussed in the Handbook.

Figure 1926-2 — View of typical full size Amateur Radio stations of the era. From the inside cover of the 1926 ARRL *Handbook*.

over-the-air TV reception in the 1950s). The cover page (Figure 1926-3) is all business.

Construction projects included a breadboard long wave receiver (see Figure 1926-4), a shortwave regenerative receiver (Figure 1926-5) and both tuned (Figure 1926-6) and crystal controlled (Figure 1926-7) transmitters. At this time the regulations regarding exact frequency band allocations and technical standards were still a few years away with the consequence that many transmitters were shown operating in a "self rectifying" mode with raw ac on the plates. There was also attention given to various wire antenna arrangements.

I found the advertisement section at the rear to be fascinating. While some manufacturers advertised over many years, others came and went quickly.

Interestingly, Vibroplex ads were in all handbooks that I examined, often with ads for the same equipment year after year (see Figure 1926-8). Of course, Horace Martin started making his famous speed keys before there was radio, and the company continues to this day.

Vibroplex was not the only key manufacturer in this edition — Bunnel, developer of the famous sideswiper key, was also represented (see Figure 1926-9). Other common names included National, then making radio parts (see Figure 1926-10) with radio receivers some years away. While primarily focused on the amateur side of radio science, one ad stands out as a predecessor of many who would build on their interest in radio to move from amateur to professional status. Joseph E. Smith of The National Radio Institute (not related to National Radio) in their ad (see Figure 1926-11) even seems to promise salaries to be achieved through their studies.

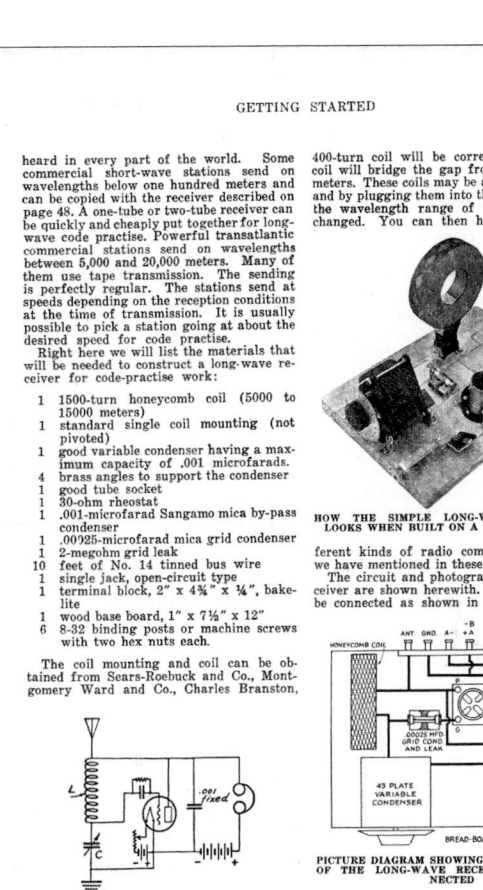

Figure 1926-4 — Construction projects were featured in even the first *Handbook*. Here is a one-tube regenerative long wave receiver you could build on a breadboard.

The
RADIO AMATEUR'S HANDBOOK

A MANUAL *of* AMATEUR SHORT-WAVE RADIOTELEGRAPHIC COMMUNICATION

BY
FRANCIS EDWARD HANDY
COMMUNICATIONS MANAGER. AMERICAN RADIO RELAY LEAGUE

FIRST EDITION

HARTFORD. CONN.
THE AMERICAN RADIO RELAY LEAGUE
1926

Figure 1926-3 — The 1926 ARRL *Handbook* title page — all business here.

All distributed capacity in coils does not come from the capacity between turns. Running leads through the finished coil adds to the distributed capacitance, especially if the lead happens to be one of the coil leads itself. If a lead *must* be run through a coil by all means run it through the center as this adds least distributed capacitance. Running the lead near one side of the coil raises the effective resistance of the coil considerably especially at the shorter wavelengths.

A simple way of making quite good space-wound coils is to wind the wire on a cardboard form, spacing it with a thread, string, or another wire which is taken off after the winding is completed Cardboard tubing should be first prepared by waterproofing with a celluloid varnish and binder. Wire should never be wound on the form until the varnish is dry. This prevents the varnish from soaking into the insulation which raises the dielectric constant making a poorer coil with higher effective resistance and distributed capacitance. A "Quaker Oats box" form is as good as they come electrically. Don't think that all coils must be of three inch diameter just because those mentioned here and all those you have seen are of that diameter. This is just a convenient size and good shortwave coils performing beautifully can be made 1½ and 2 inches in diameter. A thick form should be provided with longitudinal supporting strips to keep the wire off the form to reduce dielectric losses.

Here is a table showing the wavelength ranges that can be covered using 3½" diameter Lorenz coils of the following numbers of turns and a tuning condenser having the maximum and minimum capacity values specified. The wavelength ranges were determined by means of an oscillator (described in the appendix) and wavemeter (the next instrument we will build for the station after our receiver is in operation).

Tuning condenser capacity range: 20 to 380 micromicrofarads. A smaller maximum capacity (150 μμf) is recommended to give less crowd'ng on the amateur ranges.

Turns	Wavelength meters
55	200 to 625
24	85 to 273
10	40 to 126
5	24 to 70
2	11 to 35

Some space wound coils checked by the writer were three inches in diameter, wound on skeleton forms, of No. 18 B. and S. wire spaced the width of the wire itself. The tuning condenser had a range of from 15 to 100 micromicrofarads.

Turns	Wavelength (meters)
19	58 to 115
8	30 to 62
3	11 to 31

Although not allowing much overlap these ranges cover the three lower amateur bands nicely in the center of the dial. The confirmed traffic man will wish to use a still smaller condenser with a much larger number of coils (eight or ten) so that the tuning will not be critical and so that all stations will be spread out well over the dial.

BUILDING THE RECEIVER

A receiver built on a "breadboard" is very effective but picks up dust and dirt after a while and gets noisy. A panel-mounted set is about as cheap. The cabinet can be added later to protect the parts. Photographs show both "breadboard" and panel layouts. The sets can be wired from the photographs and the diagram.

Select every part carefully if you would make your receiver a permanent invest-

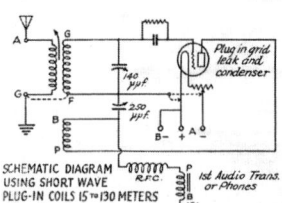

SCHEMATIC DIAGRAM USING SHORT WAVE PLUG-IN COILS 15 TO 130 METERS

THE CIRCUIT

Chosen from the several possible circuits, a fixed tickler and a "throttle" condenser for regeneration control are used. Smooth regeneration control is effected with this arrangement, while at the same time the tuning is but slightly changed by movements of the regeneration condenser.

ment. Do not patronize the local cut price or "gyp" store if you are after the best results obtainable for your money. "Bootleg" parts are cheap it is true—but standard parts and tubes from reputable manufacturers are not so likely to prove defective after a few weeks of use. Buy well and your set will still be giving satisfaction when the friend who was taken in by a misguided idea of saving is paying for expensive replacements.

In putting the "breadboard" receiver together lay out the apparatus as it is shown

---

master oscillator is not shielded to keep the coupling between them low so that there will be little feedback to endanger the crystal. The chokes are wound on small dowels using fine wire. The sockets shown are provided for two UX210s and two UV203As. A good mounting for the crystal is shown in the photograph. When a set is shielded, it is important that the coils be

Electrical measurements are based on the use of one or more calibrated instruments especially made for their application. These instruments vary in construction depending on whether they are for use with direct current, alternating current at commercial frequencies (25 or 60 cycles), or for radio measurements where you are dealing with frequencies of millions of cycles.

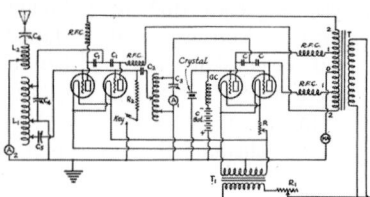

COMPLETE CIRCUIT OF A 40-METER A.C. CRYSTAL-CONTROLLED SENDING SET

T—Tapped plate transformer. 400 volts between 0 and 1. 1500 volts between 0 and 2.
T1—250-watt filament transformer with a 12-volt secondary.
R—2-ohm rheostat to adjust filaments of crystal tubes.
R1—Primary rheostat to adjust all filaments to supply voltage conditions.
RFC—200 turns No. 30 D.C.C. on half-inch wooden dowels.
GC—140 turns No. 26 D.C.C. on half-inch dowel, tapped every 20th turn after the 80th so that the "best" adjustment can be used.
C—2,000 μμf. Sangamo blocking condensers.
C1—2,000 μμf. mica blocking condensers (7500 volt breakdown).
C2—400 μμf. fixed mica grid condenser (4,000 volt breakdown).
C3—Double-spaced Cardwell receiving condenser, resultant max. capacity 250 μμf.
C4, C5, and C6—National or Cardwell 100-μμf. transmitting condensers.
C—BAT—40 volts.
MA—0-500 milliampere D.C. ammeter.
A1-A2—0-5 ampere scale thermo-ammeter.
L-L1—14 turns flatwise-wound ⅜" brass strip spaced ⅜". Diameter 4½ inches.
L2—6 turns No. 12 wire hinged on hard rubber form.

small to keep the magnetic field from spreading out and inducing heavy R.F. currents in the shielding. Such eddy currents will cause a serious power loss if the set is not very carefully planned and constructed. An un-neutralized set working on amplified harmonics of the crystal is best for the inexperienced builder to attempt.

ELECTRICAL MEASUREMENTS

The simple measurements of resistance (D.C.) and impedance (A.C.) have been described in the explanation of how to find the inductance of the filter choke. The proper use of voltmeter and ammeter followed by a substitution of the scale readings in Ohm's Law makes the determining of many circuit constants possible.

Two of the most important quantities to be measured are *current* and *voltage*. The instrument for measuring the rate of current flow is called an *ammeter* because the unit of current is taken as the "ampere". The unit of potential difference is the "volt" from which we name the instrument for measuring electrical pressure a *voltmeter*. Some instruments may be used for one kind of current only; others are suitable for both D.C. and A.C. Meters are built on various principles, each of which has a field of application and certain advantages and disadvantages. Most commercial instruments are built ruggedly and compactly. A pivoted movement carries a pointer moving over a scale which is calibrated directly in terms of volts, amperes, watts and so on, depending on the construction and circuit of

---

Figure 1926-5 — This shortwave regenerative receiver was another construction project in the 1926 *Handbook*.

Figure 1926-6 — Typical of its day, the 1926 *Handbook* featured a self-excited transmitter project.

capacitance (between turns) and no lumped capacity may be required. The trouble with this sort of thing is that when trying to adjust the set it is impossible to change one thing at a time as we desire—the controls are interlocking. The plate turns and the wavelength will be changed at the same time. Considerable inconvenience will result. Sometimes coils or parts of a circuit will give trouble by putting oscillations in the circuit at their natural period. Certain adjustments will appear to be stable but the tube later will "flop over" to another frequency of some particular part of the circuit. This emphasizes the desirability of using one tube rather than several in parallel on the short waves, as we thus get rid of troubles that may occur from "parasitic" oscillations between the different tubes and their leads. On the ultra-high frequencies we can make use of these oscillations in actual transmission but they are not well suited to practical communication and we will not spend any time on them.

PLANNING THE TRANSMITTER

To design and build a tube set for transmitting on short waves is quite simple. The choice of apparatus to fit the pocketbook is probably the most difficult task. Tube transmission and the use of shorter and shorter wavelengths have simplified this problem. The more powerful the set built, the more consistent the results obtained with least effort and care. The range of the transmitter in miles will not differ greatly with the power used. Low-powered transmitters using receiving tubes often give as good results in "DX" as more powerful sets. The atmospheric conditions, the wavelength used and the time of day all have greater effects on the "DX" worked than the power input.

Our transmitter may be built "breadboard" style or the apparatus can be mounted on a panel. To keep the expense down, a wooden panel or baseboard should be used. Dry oak or maple are as good as anything else we can buy for this.

Contrary to the general superstition, a panel-mounted set with the parts spaced sensibly and wired correctly will give much superior results to the hit-or-miss layout frequently seen with leads running helter-skelter over everything in sight. The apparatus can be neatly arranged in either style of mounting.

CIRCUITS

The choice of a transmitting circuit is not of great importance. There is no excuse for most of the variations from the standard and simple circuits that are used. The principle of feeding back r.f. voltage

from the plate to the grid circuit to produce continuous oscillations is the same in all vacuum-tube oscillator circuits. The fundamental circuits are the Hartley and Colpitts circuits, named for the prominent investigators who first used them. In both circuits an inductance and capacitance determine the wavelength. In the Hartley circuit the filament of the tube is connected to the middle of the coil and the

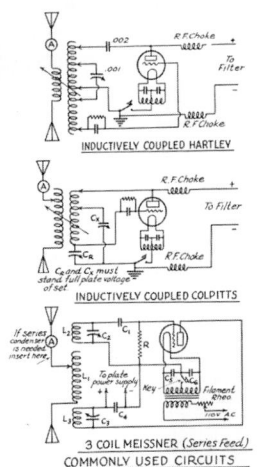

INDUCTIVELY COUPLED HARTLEY

INDUCTIVELY COUPLED COLPITTS

3 COIL MEISSNER (Series Feed)
COMMONLY USED CIRCUITS

plate and grid connections are made to the extreme ends of the coil. In the Colpitts circuit the filament of the tube is connected between two condensers and the plate and grid connections are made to the other condenser terminals. In one arrangement the tube gets its feedback inductively, in the other the capacitance voltage drops take care of the feedback.

The Hartley circuit has the tuning condenser across some of the turns both in the grid and in the plate part of the coil. When the condenser is entirely across plate turns, the circuit is a "tuned plate" circuit. When it is across the grid turns we have a "tuned grid" circuit.

Figure 1926-7 — A fairly revolutionary transmitter project had its frequency set by a piezoelectric quartz crystal. The extra stability came at a price — a separate crystal was required for each frequency.

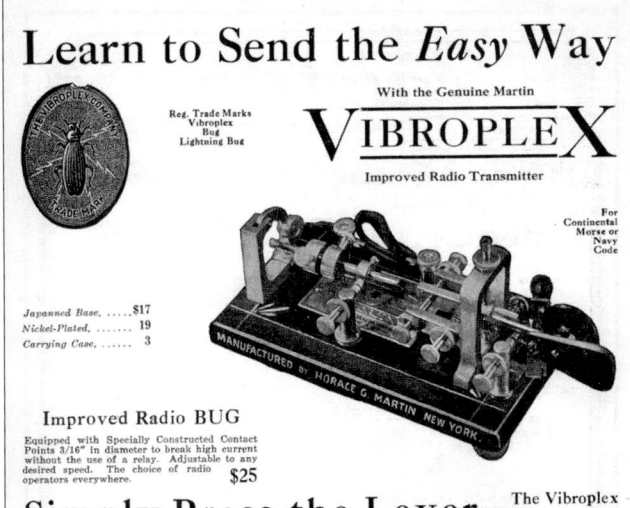
Figure 1926-8 — This 1926 ad from the Vibroplex Corporation was similar to those shown for 75 years.

Figure 1926-9 — Bunnell, developer of the famous sideswiper key, also advertised in the 1926 edition.

Figure 1926-10 — National Radio, then making radio parts, with complete radio receivers some years away, had an ad in the first edition of the *Handbook*.

Figure 1926-11 — Even in this first edition, there was interest in promoting Amateur Radio as a path to a career.

The Beginning of the Next Decade — The 1930 Edition

It's not obvious that the plan of the ARRL for the production of *The Radio Amateur's Handbook* was to make it an annual affair. The edition released in 1930 was, in fact, the seventh edition (see Figure 1930-1) — not the fifth, as would be the case if there were one per year. While the edition's lead photo (see Figure 1930-2) showed equipment at the ARRL headquarters station that looked somewhat similar to that of the first edition, in fact, there were many significant changes in Amateur Radio between these editions.

The changes in the late 1920s were driven more by regulatory actions than by technology. In 1926, radio was rather loosely regulated by the US Department of Commerce. Most "regulations" were in the form of loose understandings between various commercial and amateur operating groups. This included frequency allocations (amateurs were then permitted to use any frequency above 1500 kilocycles [now kHz]) as well as technical standards, which were not yet well defined.

The US Federal Radio Commission (FRC), the forerunner of today's more general FCC, was established by congress and signed into law by President Calvin Coolidge on February 23, 1927. Over a few years, the FRC defined strict frequency allocations for different services, including the Amateur Radio Service. The new allocations established bands with a similarity to today's bands, except that they were just those harmonically related as shown in Table 1. The original order described the bands in terms of the actual wavelength, as well as

Table 1
Amateur Radio Bands Established by the FRC in 1928

Band (meters)	Meters	Kilocycles (Frequency in KHz)
160	150.0 to 200.0	2000 to 1500
80	75.0 to 85.7	4000 – 3500
40	37.5 to 42.8	8000 – 7000
20	18.7 to 21.4	16,000 – 14,000
10	9.99 to 10.71	30,000 – 28,000
5	4.69 to 5.35	64,000 – 56,000
0.75	0.7477 to 0.7496	400,000 – 401,000

Table 2
Amateur Radio Bands for Voice Established by the FRC in 1928

Band (meters)	Meters	Kilocycles (Frequency in KHz)
160	150.0 to 175.0	2000 to 1715
80	84.5 to 85.7	3500 – 3550
5	4.69 to 5.35	64,000 – 56,000

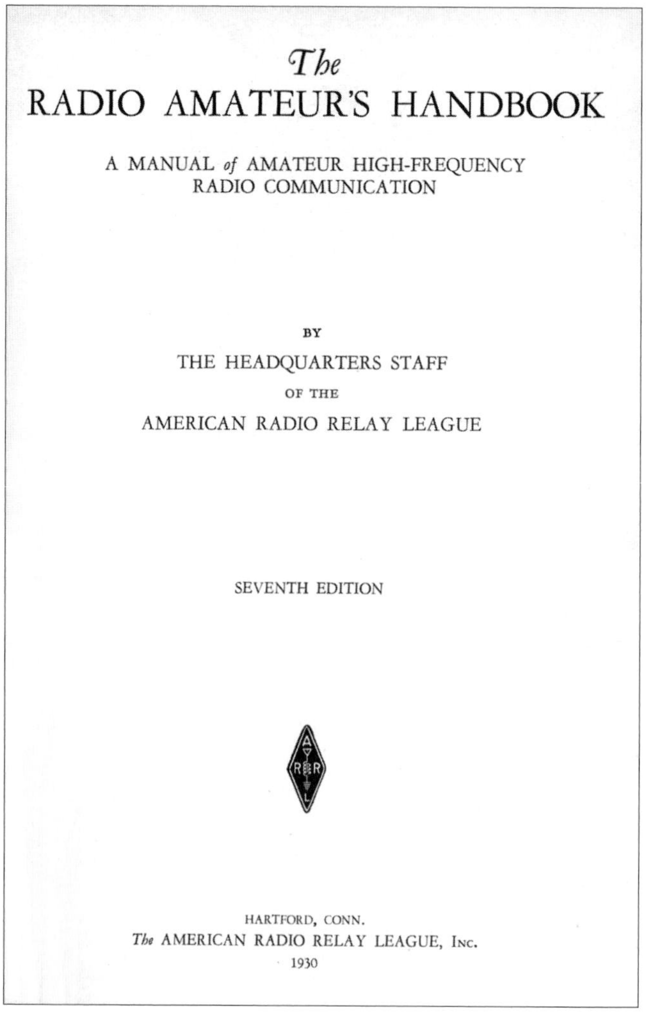

The
RADIO AMATEUR'S HANDBOOK

A MANUAL *of* AMATEUR HIGH-FREQUENCY
RADIO COMMUNICATION

BY

THE HEADQUARTERS STAFF

OF THE

AMERICAN RADIO RELAY LEAGUE

SEVENTH EDITION

HARTFORD, CONN.
The AMERICAN RADIO RELAY LEAGUE, INC.
· 1930

Figure 1930-1 — Title sheet of the seventh edition of *The Radio Amateur's Handbook.*

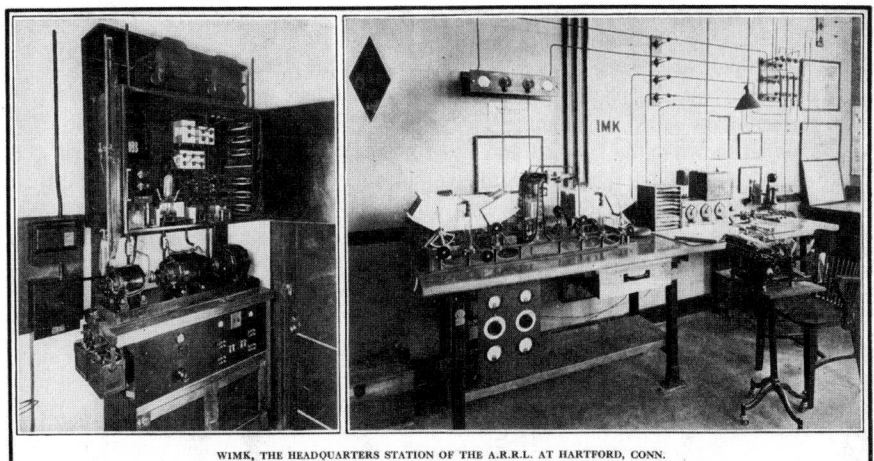

WIMK, THE HEADQUARTERS STATION OF THE A.R.R.L. AT HARTFORD, CONN.

The power supplies and the operating position are shown. Note the neatness and accessibility of every feature of the station arrangement. High voltage d. c. to the main and auxiliary transmitters is obtained from a motor-generator and a mercury-arc rectifier and filter. Fuses, relays, batteries, charging equipment and the like are all in the power-supply room. The receiver is in front of the operator, key and controls at his right, and the message file box at his left. Ample space is provided for the monitor, frequency meter, and station log when not in use. Two-wire voltage (Zeppelin) feed is used to separate antennas for the different transmitters and there is a separate receiving antenna to facilitate "break-in" work. The Official Broadcasts to A.R.R.L. Members are sent simultaneously on 3575 and 7150 kc. The main transmitter unit has interchangeable coils of heavy tubing with compression-type threaded brass couplings and may be shifted quickly to any of the different amateur frequency bands. WIMK is a busy station but always ready for a call from any "ham." Operating schedules are published regularly in QST.

Figure 1930-2 — The headquarters station of the ARRL, then in Hartford, Connecticut, was shown on the cover page of the seventh edition.

Figure 1930-3 — The National Radio ad still shows a sample of their fine parts, but now ads a complete receiver, the SW-5.

the frequency (radios of the day were more likely to be calibrated in wavelength than frequency). I have also shown the usual band descriptor that we use today. Table 2 shows the frequencies on which radiotelephony was allowed — not very many, at least initially.[3]

In addition to the frequency allocations, the FRC 1928 announcement included new technical standards. These now required amateur stations to use "loose coupling to the antenna system," typically inductive rather than conductive coupling, to reduce harmonic output and "key impacts." Additionally, plate supply modulation (self rectifying circuits) and spark transmitters were no longer allowed (interestingly, some ship transmitters continued to use spark for many years).

The new regulations were reflected in the details of the *Handbook* construction projects, although a cursory look might not have revealed the difference, since most of the technology did not change a lot. Even the 1926 edition did not include any spark transmitters, since most amateurs now realized the benefits and additional efficiency of CW transmitters made with vacuum tubes.

Again, the ads in the back of the 1930 edition provided a glimpse of what amateurs were buying. By this time, National Radio had moved from being a parts supplier to being a manufacturer of receivers for amateur use, including the SW-5 "thrill box," highlighted in their ad (see Figure 1930-3).

[3]K. B. Warner, 1BHW, "Recent Changes in Radio Law and Regulations," *QST*, May 1928, pp 14-15.

A Decade of Advancement — The 1940 Edition

Incremental advancement in technology occurred broadly in the decade preceding the 1940 edition. While the basic underlying physics remained the same, circuitry using vacuum tubes moved forward in large steps. By 1940, advanced amateur stations had moved from regenerative receivers on breadboards to bandswitching commercial superhets. Transmitters advanced from one or two tube self-excited transmitters on planks or breadboards to high power equipment in professional racks. This is perhaps best illustrated by comparing the cover leaf photos of the 1930 version of the ARRL headquarters station to that shown in the 1940 edition (see Figure 1940-1). This station, now the "Maxim Memorial Station, W1AW" was located at its present site in Newington, Connecticut, looking a lot like it would when I first visited it in 1956.

The cover page (Figure 1940-2) of the 1940 edition looks much like the previous ones, again, not identifying the publication year, but the edition — this one the 17th. Note that there was exactly one edition per year between 1930 and 1940, now firmly making the *Handbook* an annual affair.

By 1940, voice operation had been expanded into all HF bands, except 40 meters. Amateurs were allowed to operate on all frequencies above 110 Mc (MHz) and gravitated to the harmonics of lower frequency bands with operation on 2½ meters (112 MHz) and 1¼ meters (224 MHz).

Equipment projects were considerably

THE OPERATING ROOM AT THE MAXIM MEMORIAL STATION, W1AW, A.R.R.L HEADQUARTERS

Separate 1-kw. transmitters are installed for each band. Voice transmissions on 1806, 3950, 14,237 kc. and 28,600 follow simultaneous telegraph messages to all amateurs sent on 1762.5, 3825, 7280, 14,254 kc. and 28,600 kc. at 7.30 and 11 p.m. CST. Operators "Hal" Bubb (seated) and "Geo" Hart (standing) are always ready for a call from any amateur.

Figure 1940-1 — The ARRL headquarters Maxim Memorial Station, W1AW, was located at the site of the present ARRL headquarters (and station) in Newington, Connecticut. It is interesting to compare 1940's equipment to that of the 1930 headquarters station.

advanced over the decade. A snapshot of a receiver (see Figure 1940-3) shows a six-tube superhet that provides a significant improvement in performance over the regenerative receivers of the previous decade. As with most *Handbook* receivers, this set used plug-in coils, rather than bandswitching, in order to minimize construction complexity. While performance didn't suffer, operating convenience did. Most commercial superhet receivers described in the ad section (see Figure 1940-4) went a step beyond, with the exception of the famous National HRO (see Figure 1940-5), and

provided single switch bandswitching.

HF transmitters moved forward in a similar manner, many making use of metal chassis and panels and crystal frequency control. Bandswitching was not yet feasible in transmitter power stages due to the large tank inductors employed. Some used switching in low-level stages, but for many the marginal benefit was not worth the complexity. Figures 1940-6 and -7 illustrate one transmitter project, a medium power HF rig, from this edition. The ad from Hallicrafters (Figure 1940-8) not only showed a commercial bandswitching

receiver, but also a companion desktop transmitter.

With the additional VHF bands, operation on the (then called) "ultrahighs" gained in popularity. Early equipment for 2½ and 1¼ meters tended to be of the modulated oscillator variety and were described in their own chapter of the 1940 *Handbook* (see Figure 1940-9).

Ads from many manufacturers for complete radios, parts, instruction and accessories (see Figure 1940-10) continued to provide a window into what could be accomplished with enough resources.

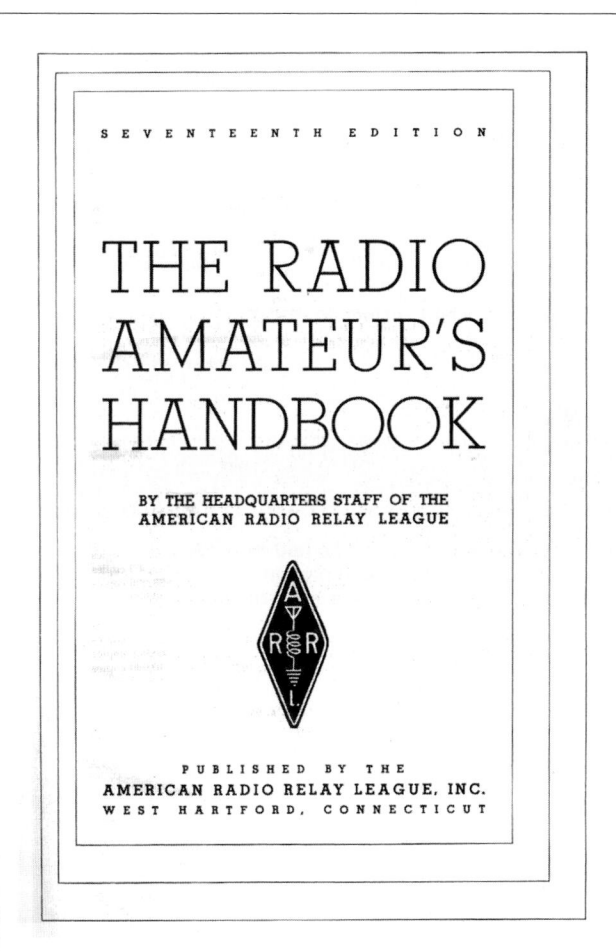

Figure 1940-2 — The cover page of the 1940 edition looks much like the previous ones, again, not identifying the publication year, but the edition — this one the 17th.

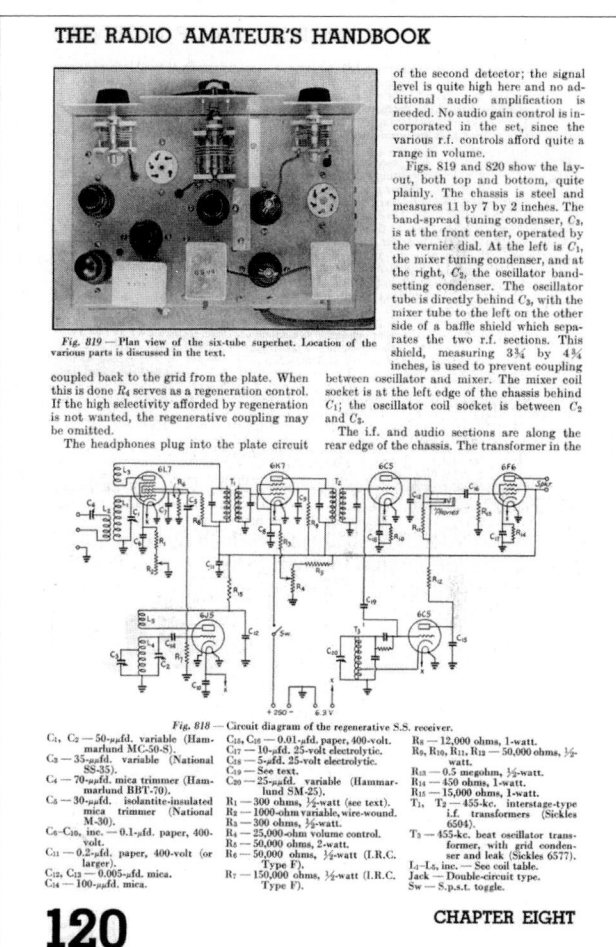

Figure 1940-3 — This 1940 *Handbook* construction project resulted in a modern HF ham band superhet using plug-in coils to change bands.

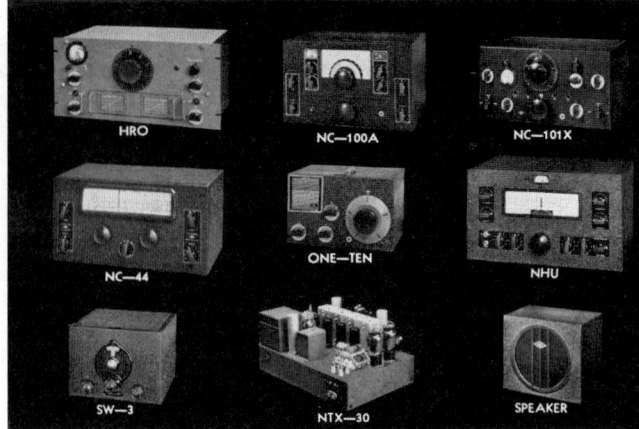
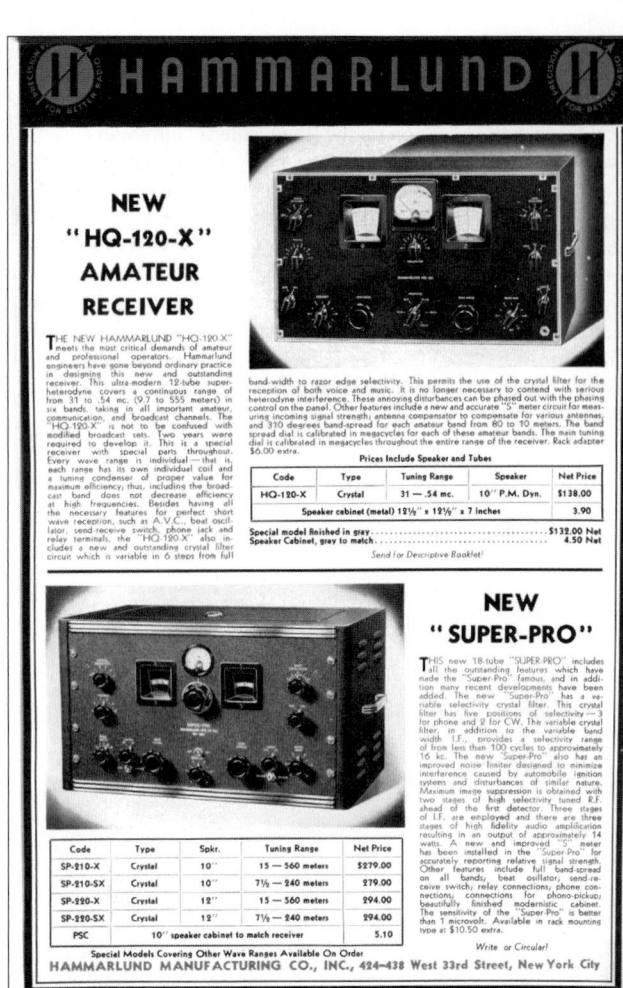
Figure 1940-4 — These receivers, advertised by Hammarlund, set the standard for modern receiver technology of the day — the architecture remaining popular for decades to come.

Figure 1940-5 — National Radio had expanded its receiver line to include bandswitching and plug-in coil superhets, but still offered the popular SW-3 regenerative receiver from years past.

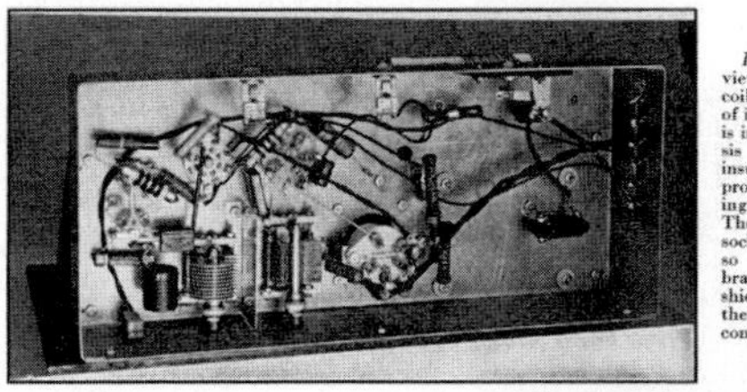

Fig. 1021 — Bottom view. The cathode-circuit coil may be seen to the left of its tuning condenser. C_3 is insulated from the chassis by four button-type insulators and the shaft provided with an insulating flexible shaft coupling. The 5-prong amplifier-tube socket is lowered an inch or so below the chassis on brackets. A small baffle shield is placed between the two small variable condensers.

No output-coupling arrangement is indicated in the diagram, this being left to the preferences of the constructor. There is ample room on the forms for a link. (Bib. 3).

● 100-TO-175-WATT TRANSMITTER OR EXCITER

The circuit of this unit is shown in Fig. 1023. The tube line-up consists of a 6L6 tetrode crystal oscillator, a 6L6 frequency doubler and a final amplifier which, in this case, employs a type HY51Z. The arrangement is suitable, however, for almost any triode-amplifier tube operating at plate voltages between 750 and 1000 volts in which the plate connection is at the top of the tube and the grid terminal is in the base.

Output at either the crystal frequency or the second harmonic is readily obtainable. The complication of neutralizing the second stage when operating at the crystal frequency is eliminated by cutting this stage out of use. This is accomplished by means of a "dummy" plug-in form which serves as a low-loss switch. Capacities suitable for coupling the final-amplifier grid to the output of either the oscillator or the doubler are mounted inside the "dummy" plug-in forms and connected as shown in the insert in the circuit diagram.

Most of the constructional details will be evident from an inspection of the photographs of Figs. 1022 and 1024. The coils for the oscillator and frequency doubler are wound on Hammarlund 1½-inch diameter plug-in forms, while those for the final amplifier are wound on National XR-10A ceramic forms which plug into the XB-15 jack base mounted on the chassis. All tank condensers are mounted underneath the chassis. The final-amplifier tank condenser C_3 is mounted by means of angle brackets on four ⅝-inch cone insulators which bring the shaft 1⅞ inches above the lower edge of the chassis and level with the shafts of the other two tank condensers which are shaft-hole mounted on the front edge of the chassis. The shaft of C_3 is fitted with an insulated flexible coupling and a bearing is set in the front edge of the chassis for the extension. Large clearance holes are cut in the panel for the shaft bushing of C_3 and the mounting nuts of the other two condensers. The dial plates are held in place by cementing them to the panel with Duco cement.

The socket of the final-amplifier tube is set

Fig. 1022 — The 100–175-watt transmitter-exciter. Controls from left to right are for the oscillator, doubler and final amplifier. The chassis is 17 in. by 8 in. by 3 in. and the Presdwood panel 19 in. by 8¾ in. The controls are 1⅞ in. above the bottom of the panel and the two outer controls 2 in. in from the edges of the chassis or 3 in. from the panel edges. The crystal-oscillator tube and plate coil are to the left, the doubler plate coil to the left of the 6L6 doubler and the change-over plug-in form at the rear.

Figure 1940-6 — A 1940 edition medium power transmitter construction project.

about an inch below the surface of the chassis on long machine screws to bring the plate terminal down closer to the tank-coil terminal. A pair of fibre lug strips supports the voltage-divider resistances for oscillator and doubler screen voltages. Other resistances and chokes are self-supported.

Connections between the final tank coil and condenser are made through feed-through insulators set in the chassis. The neutralizing condenser, which may be seen in front of the final tube socket, is mounted on spacers. A clearance hole in the chassis permits the shaft to protrude a half-inch or so above the chassis so that it may be adjusted with a screw driver.

All terminals for external connections, excepting that for the positive 1000-volt connection, are of the pin-jack type. The strips are mounted on small angle pieces behind a slot cut in the rear edge of the chassis. Insulated pin jacks are used to make connections and leave no exposed metal contacts. Separate connections are provided for meter and key connections as shown in the diagram.

When working at the crystal frequency, the "dummy" unit with connections shown in the

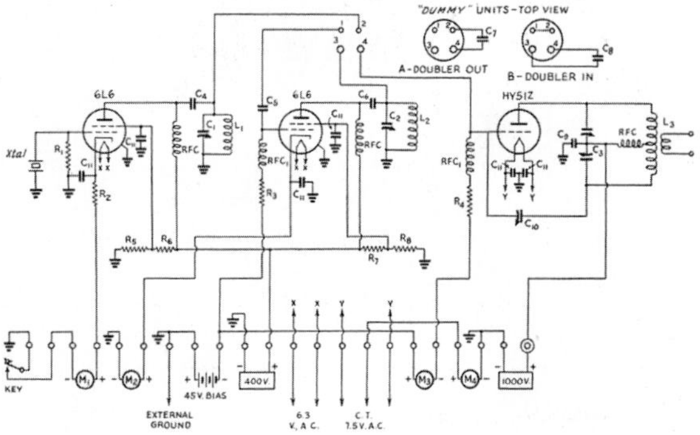

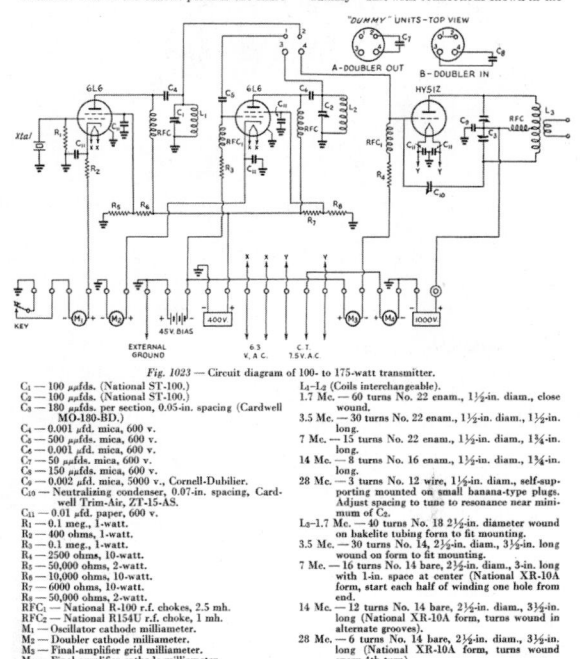

Fig. 1023 — Circuit diagram of 100- to 175-watt transmitter.

C_1 — 100 μμfds. (National ST-100.)
C_2 — 100 μμfds. (National ST-100.)
C_3 — 180 μμfds. per section, 0.05-in. spacing (Cardwell MO-180-BD.)
C_4 — 0.001 μfd. mica, 600 v.
C_5 — 500 μμfds. mica, 600 v.
C_6 — 0.001 μfd. mica, 600 v.
C_7 — 50 μμfds. mica, 600 v.
C_8 — 150 μμfds. mica, 600 v.
C_9 — 0.002 μfd. mica, 5000 v., Cornell-Dubilier.
C_{10} — Neutralizing condenser, 0.07-in. spacing, Cardwell Trim-Air, ZT-15-AS.
C_{11} — 0.01 μfd. paper, 600 v.
R_1 — 0.1 meg., 1-watt.
R_2 — 400 ohms, 1-watt.
R_3 — 0.1 meg., 1-watt.
R_4 — 2500 ohms, 10-watt.
R_5 — 50,000 ohms, 2-watt.
R_6 — 10,000 ohms, 10-watt.
R_7 — 6000 ohms, 10-watt.
R_8 — 50,000 ohms, 2-watt.
RFC_1 — National R-100 r.f. chokes, 2.5 mh.
RFC_2 — National R154U r.f. choke, 1 mh.
M_1 — Oscillator cathode milliameter.
M_2 — Doubler cathode milliameter.
M_3 — Final-amplifier grid milliameter.
M_4 — Final-amplifier cathode milliameter.

L_1–L_2 (Coils interchangeable).
1.7 Mc. — 60 turns No. 22 enam., 1½-in. diam., close wound.
3.5 Mc. — 30 turns No. 22 enam., 1½-in. diam., 1½-in. long.
7 Mc. — 15 turns No. 22 enam., 1½-in. diam., 1¾-in. long.
14 Mc. — 8 turns No. 16 enam., 1½-in. diam., 1¾-in. long.
28 Mc. — 3 turns No. 12 wire, 1½-in. diam., self-supporting mounted on small banana-type plugs. Adjust spacing to tune to resonance near minimum of C_2.
L_3–1.7 Mc. — 40 turns No. 18 2½-in. diameter wound on bakelite tubing form to fit mounting.
3.5 Mc. — 30 turns No. 14, 2½-in. diam., 3½-in. long wound on form to fit mounting.
7 Mc. — 16 turns No. 14 bare, 2½-in. diam., 3-in. long with 1-in. space at center (National XR-10A form, start each half of winding one hole from end.
14 Mc. — 12 turns No. 14 bare, 2½-in. diam., 3½-in. long (National XR-10A form, turns wound in alternate grooves).
28 Mc. — 6 turns No. 14 bare, 2½-in. diam., 3½-in. long (National XR-10A form, turns wound every 4th turn).

Figure 1940-7 — The second page of the project in Figure 1940-6, showing the schematic diagram.

Figure 1940-8 — Hallicrafters not only provided receivers, but also had a matching desktop transmitter.

bly will not be enough, and suppressors should be installed at each spark plug. Generator "whine" can be eliminated by winding up a choke consisting of a few turns of heavy wire and connecting it in series with the hot lead at the generator with a ¼-µfd. condenser be-tween the far side of the choke and the frame, in addition to the usual condenser directly across the generator terminals (Bib. 1).

● **A COMPLETE 56-MC. SUPERHET**

If a communications receiver or all-wave broadcast receiver is not available to be used as an i.f. amplifier for a 56-Mc. converter, a relatively-simple complete superhet may be constructed according to Fig. 2713. Photo-graphs of a receiver using the circuit are shown in Figs. 2712 and 2714. The circuit includes an 1851 r.f. stage, 954 mixer, 955 high-frequency oscillator, a 1600-kc. i.f. amplifier, regenerative second detector, and a pentode audio output. The regenerative second detector may be oper-ated in the oscillating condition for the recep-tion of c.w. signals, or just below oscillation for 'phone reception of weak signals when maxi-mum amplification is needed.

The receiver is constructed on a chassis

Fig. 2712 — An end view of the 56-Mc. superhet. Details of the construction of the r.f. end of the receiver are apparent from this photograph.

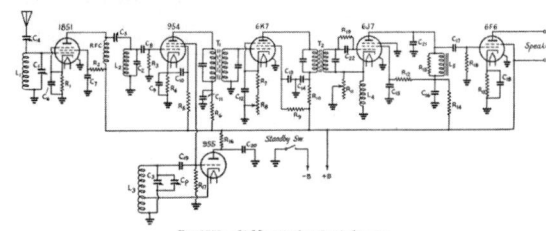

Fig. 2713 — 56-Mc. superhet circuit diagram.

C₃, C₂ — 10-µµfd. (Cardwell ZR-10-AS).
C₃ — 15-µµfd. (Cardwell ZR-15-AS).
C₄, C₅ — 3-35-µµfd. (Isolantite) padders.
Cₚ — 50-µµfd. air padder (Ham-marlund APC-50).
C₆, C₇, C₈, C₉ — 0.01-µfd., 400-volt tubular.
C₈, C₁₉, C₂₂ — 100-µµfd. midget mica.
C₁₁, C₁₄ — 0.05-µfd., 400-volt tubu-lar.
C₁₂, C₁₃, C₁₅, C₁₆ — 0.1-µfd., 400-volt tubular.
C₁₇ — 0.02-µfd., 400-volt tubular.
C₁₈ — 10-µfd., 25-volt tubular.
C₂₀ — 0.001-µfd. midget mica.

C₂₁ — 0.002-µfd., 400-volt tubular.
R₁ — 150 ohms, ½-watt.
R₂ — 60,000 ohms, ½-watt.
R₃ — 1 meg., ½-watt.
R₄, R₁₀ — 2000 ohms, ½-watt.
R₅, R₁₂, R₁₆ — 100,000 ohms, ½-watt.
R₆ — 2000 ohms, ½-watt.
R₇ — 300 ohms, ½-watt.
R₈ — 50,000-ohm potentiometer.
R₉, R₁₇ — 50,000 ohms, ½-watt.
R₁₁ — 1000-ohm potentiometer.
R₁₃ — 250,000 ohms, ½-watt.
R₁₄ — 25,000 ohms, ½-watt.
R₁₅ — 450 ohms, 10-watt.
R₁₈ — 500,000 ohms, ½-watt.
R₁₉ — 5 meg., ½-watt.

L₁ — 8 turns No. 14, ½" diameter, winding length 1½".
L₂ — 9 turns No. 14, ½" diameter, winding length 1½".
L₃ — 4 turns No. 14, ½" diameter, winding length ½" (cath-ode tap ½ turn from ground end).
L₄ — 30 turns No. 24, close-wound on ½" form.
L₅ — 1080-henry plate impedance (Thordarson T-29C27).
RFC — 2½-mh. r.f. choke (Na-tional R100).
T₁ — 1600-kc. iron core i.f. (Meiss-ner No. 16-8091).
T₂ — 1600-kc. iron core i.f. (Meiss-ner No. 16-8099).

CHAPTER TWENTY-SEVEN

387

Figure 1940-9 — The 1940 edition described VHF construction projects as well as those for HF.

Figure 1940-10 — Vibroplex continued to advertise its popular "Bugs."

The big news between the 1940 and 1950 editions was, of course, World War II, which was declared on December 7, 1941. All Amateur Radio operation was suspended "until further notice." While there were some exceptions for specific amateurs operating in support of local security and Civil Defense, including a special War Emergency Radio Service that mainly used the 2½ meter band, hams were essentially off the air until bands were gradually brought back starting after the end of hostilities.

One might have expected that this would have resulted in a suspension of *Handbook* publication, but that was not the case. *QST* and *The Radio Amateur's Handbook* continued unabated throughout the period, in fact there was even an extra edition published in 1942 for the war effort. This "Special Defense Edition" (see Figure 1950-1) was designed as a training manual for use by various military technical schools. At 288 pages, it was about one third the size of the

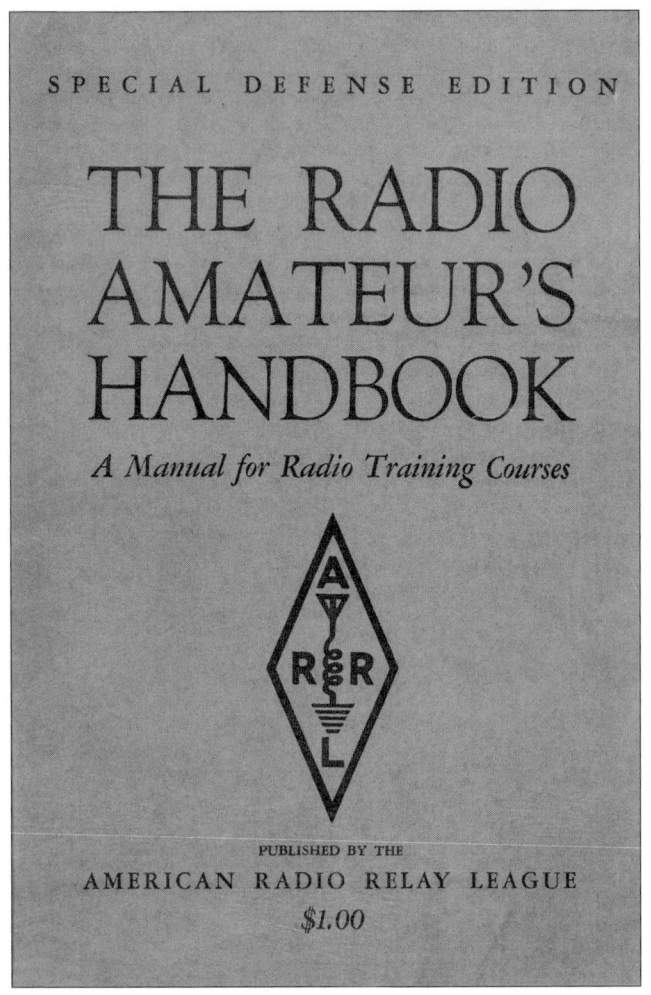

Figure 1950-1 — The cover of the 1942 Special Defense Edition of The Radio Amateur's *Handbook*.

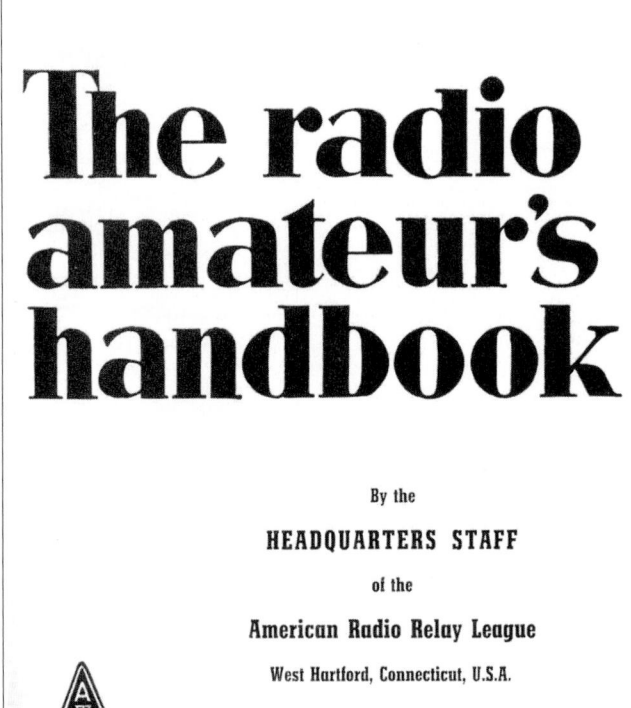

Figure 1950-2 — The cover page of the 1950 edition looks much like the previous ones, but a bit more modern and note that the year is explicitly shown.

regular editions, accomplished by eliminating most construction projects, the advertising section and material very specific to Amateur Radio.

The latter half of the 1940s found the amateur bands gradually returned to amateur use, but with some changes. Perhaps the most significant was the change in VHF allocations to accommodate the newly defined VHF television service. The 2½ and 5 meter bands were within the TV allocation, resulting in the shift to our current 2 and 6 meter bands.

The 1950 edition cover page (see Figure 1950-2) sported a new more modern look and now firmly indicated the year of publication, in addition to the edition identifier. Changes in the approach to technology employed reflected two different changes in the environment:

◊ The rapid development of radio science during the war years was reflected in Amateur Radio. The single signal effect, described by Lamb in the previous decade, has been improved by coupling it with the high selectivity of single crystal IF filters as shown in Figure 1950-3, a figure that will remain in *Handbooks* for decades. In addition to the adaptation of technology, the years following the war included an

---

106   **CHAPTER 5**

varies with signal strength, being less on strong signals, and the selectivity varies.

Crystal Filters

The most satisfactory method of obtaining high selectivity is by the use of a piezoelectric quartz crystal as a selective filter in the i.f. amplifier. Compared to a good tuned circuit, the Q of such a crystal is extremely high. The dimensions of the crystal are made such that it is resonant at the desired intermediate frequency. It is then used as a selective coupler between i.f. stages.

Fig. 5-23 gives a typical crystal-filter resonance curve. For single-signal reception, the audio-frequency image can be reduced by a factor of 1000 or more. Besides practically eliminating the a.f. image, the high selectivity of the crystal filter provides great discrimination against signals very close to the desired signal and, by reducing the band-width, reduces the response of the receiver to noise.

Crystal-Filter Circuits; Phasing

Several crystal-filter circuits are shown in Fig. 5-24. Those at A and B are practically identical in performance, although differing in details. The crystal is connected in a bridge circuit, with the secondary side of T_1, the input transformer, balanced to ground either through a pair of condensers, C-C (A), or by a center-tap on the secondary, L_2 (B). The bridge is completed by the crystal and the *phasing condenser*, C_3, which has a maximum capacity somewhat larger than the capacity of the crystal in its holder. When C_3 is set to balance the crystal-holder capacity, the resonance curve of the crystal circuit is practically symmetrical; the crystal acts as a series-resonant circuit of very high Q and thus allows signals of the desired frequency to be fed through C_3 to L_3L_4, the output transformer. Without C_3, the holder capacity (with the crystal acting as a dielectric) would pass undesired signals.

The phasing control has an additional function besides neutralization of the crystal-holder capacity. The holder capacity becomes a part of the crystal circuit and causes it to act as a parallel-tuned resonant circuit at a frequency slightly higher than its series-resonant frequency. Signals at the parallel-resonant frequency thus are prevented from reaching the output circuit. The phasing control, by varying the effect of the holder capacity, permits shifting the parallel-resonant frequency over a considerable range, providing adjustable rejection of interfering signals. The effect of rejection is illustrated in Fig. 5-23.

Additional I.F. Selectivity

Most commercial communications receivers do not have sufficient selectivity for amateur use, and their performance can be greatly improved by adding additional selectivity. One popular method is to couple a BC-453 aircraft receiver (war surplus, tuning range 190 to 550 kc.) to the tail end of the 465-kc. i.f. amplifier in the communications receiver and use the resultant output of the BC-453. The aircraft receiver uses an 85-kc. i.f. amplifier that is quite sharp — 6.5 kc. wide at −60 db. — and it helps tremendously in separating 'phone signals and in backing up crystal filters for improved c.w. reception. (See *QST*, January, 1948, page 40.)

If a BC-453 is not available, it is still a simple matter to enjoy the benefits of improved selectivity. It is only necessary to heterodyne to a lower frequency the 465-kc. signal existing in the receiver i.f. amplifier and then rectify it after passing it through the sharp low-frequency amplifier. The Hammarlund Company and the J. W. Miller Company both offer 50-kc. transformers for this application.

QST references on high i.f. selectivity include: McLaughlin, "Selectable Single Sideband," April, 1948; Githens, "C.W. Receiver," Aug., 1948.

● RADIO-FREQUENCY AMPLIFIERS

While selectivity to reduce audio-frequency images can be built into the i.f. amplifier, discrimination against radio-frequency images can only be obtained in circuits ahead of the first detector. These tuned circuits and their associated vacuum tubes are called **radio-frequency amplifiers**. For top performance of a communications receiver on frequencies above 7 Mc., it is mandatory that it have one or two stages of r.f. amplification, for image rejection and improved sensitivity.

Receivers with an i.f. of 455 kc. can be expected to have some r.f. image response at a signal frequency of 14 Mc. and higher if only one stage of r.f. amplification is used. (Regen-

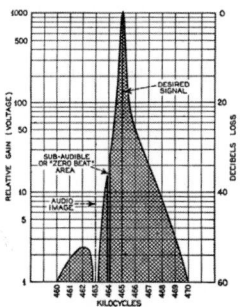

Fig. 5-23 — Graphical representation of single-signal selectivity. The shaded area indicates the over-all bandwidth, or region in which response is obtainable.

---

118   **CHAPTER 5**

An Amateur-Band Eight-Tube Superheterodyne

An advanced type of amateur receiver incorporating one r.f. amplifier stage, variable i.f. selectivity and audio noise limiting is shown in Figs. 5-33, 5-35 and 5-36. As can be seen from the circuit in Fig. 5-34, a 6SG7 pentode is used for the tuned r.f. stage ahead of the 6K8 converter. An antenna compensator, C_4, controlled from the panel, allows one to trim up the r.f. stage when using different antennas that might modify the tracking. The cathode bias resistor of the r.f. stage is made as low as possible consistent with the tube ratings, to keep the gain and hence the signal-to-noise ratio of the stage high. The oscillator portion of the 6K8 mixer is tuned to the high-frequency side of the signal except on the 28-Mc. band, the usual custom nowadays in communications receivers. The oscillator tuning condenser, C_{17}, is of higher capacity than the r.f. and mixer tuning condensers, in the interest of better oscillator stability.

The i.f. amplifier is tuned to 455 kc., and the first stage is made regenerative by soldering a short length of wire to the plate terminal of the socket and running it near the grid terminal, as indicated by C_{C1} in the diagram. Regeneration is controlled by reducing the gain of the tube, and R_{12}, a variable cathode-bias control, serves this function. The second i.f. stage uses a 6K7, selected because high gain is not necessary at this point.

Manual gain-control voltage is applied to the r.f. and second i.f. stages. It is not applied to the mixer because it might pull the oscillator frequency, and it is not tied in with the first i.f. amplifier because it would interlock with the regeneration control used for controlling the selectivity. However, the a.v.c. voltage is applied to the r.f. and both i.f. stages, with the result that the selectivity of the regenerative stage decreases with loud signals and gives a measure of automatic selectivity control. Using a negative-voltage power supply for the manual gain control is more expensive than the familiar cathode control, but it allows a wide range of control with less dissipation in the components. The a.v.c. is of the delayed type, the a.v.c. diode being biased about 1½ volts by the cathode resistor of the diode-triode detector-audio stage.

The second-detector-and-first-audio is the usual diode-triode combination and uses a 6SQ7. A 1N34 crystal diode is used as a noise limiter, and is left in the circuit all of the time. As is common with this type of circuit, it has little or no effect when the b.f.o. is on, but it is of considerable help to 'phone reception on the bands where automobile ignition is a factor. The constructor can satisfy himself on its operation when first building the receiver and working on it out of the case. By leaving one end of the 1N34 floating and touching it to the proper point in the circuit, a marked drop in ignition noise will be noted.

The b.f.o. is capacity-coupled to the detector by soldering one end of an insulated wire to the a.v.c. diode plate and wrapping several turns of the wire around the b.f.o. grid lead. This capacity is designated C_{C2} in the diagram. The wire was connected to the a.v.c. diode plate lead only for wiring convenience — the a.v.c. coupling condenser, C_{32}, passing the b.f.o. voltage without introducing appreciable attenuation.

Headphone output is obtained from the plate circuit of the 6SQ7 at J_1, and loudspeaker output is available from the 6F6 audio-amplifier stage. High-impedance or crystal headphones are recommended for maximum headphone output.

Fig. 5-33 — An amateur-band eight-tube receiver. The knobs on the left control audio volume (upper) and b.f.o. pitch, and the two on the right handle r.f. and i.f. gain (upper) and i.f. regeneration. The knob to the left of the large tuning knob is fastened to the M.A.N.-A.V.C.-B.F.O switch, and the one on the right is for the antenna trimmer. The toggle switch under the dial throws high negative bias on the r.f. stage during transmission periods.

Figure 1950-3 — The single signal effect, described by Lamb in the previous decade, has been improved by coupling it with the high selectivity of single crystal IF filters.

Figure 1950-4 — This receiver construction project, see schematic in Figure 1950-5, is based on the same architecture as the receiver shown in Figure 1940-3. While an additional RF and IF stage were added, there's not much new here.

abundance of low cost surplus radio equipment that could be put to use with varying levels of effort. In many cases this equipment made the decision to buy new commercial gear, or to build your own difficult. Still, the *Handbook* described receiver construction projects (see Figures 1950-4 and 5) that, while elegant looking, arguably were not very different from those shown in the 1940 edition.

◊ The introduction and increasing

popularity of broadcast television service as the decade drew to a close had a major impact on Amateur Radio, as it would for decades to come. Harmonic and spurious radiation in the VHF range that had been largely unnoticed for years suddenly resulted in non-amateurs becoming very aware of ham radio in a negative way. This gave rise to major changes in transmitter design and construction methods, including shielding, bypassing, filtering and the use of coaxial

cable — available as surplus following the war. This can be seen in Figure 1950-6, a transmitter not very different from those of the previous decade, but with attention paid to shielding and bypassing to "reduce the generation and transmission of harmonics."

Other developments were shown in the 1950 edition, including an early description of single-sideband suppressed-carrier transmission (see Figure 1950-7). This was to become known as "SSB" in a few years

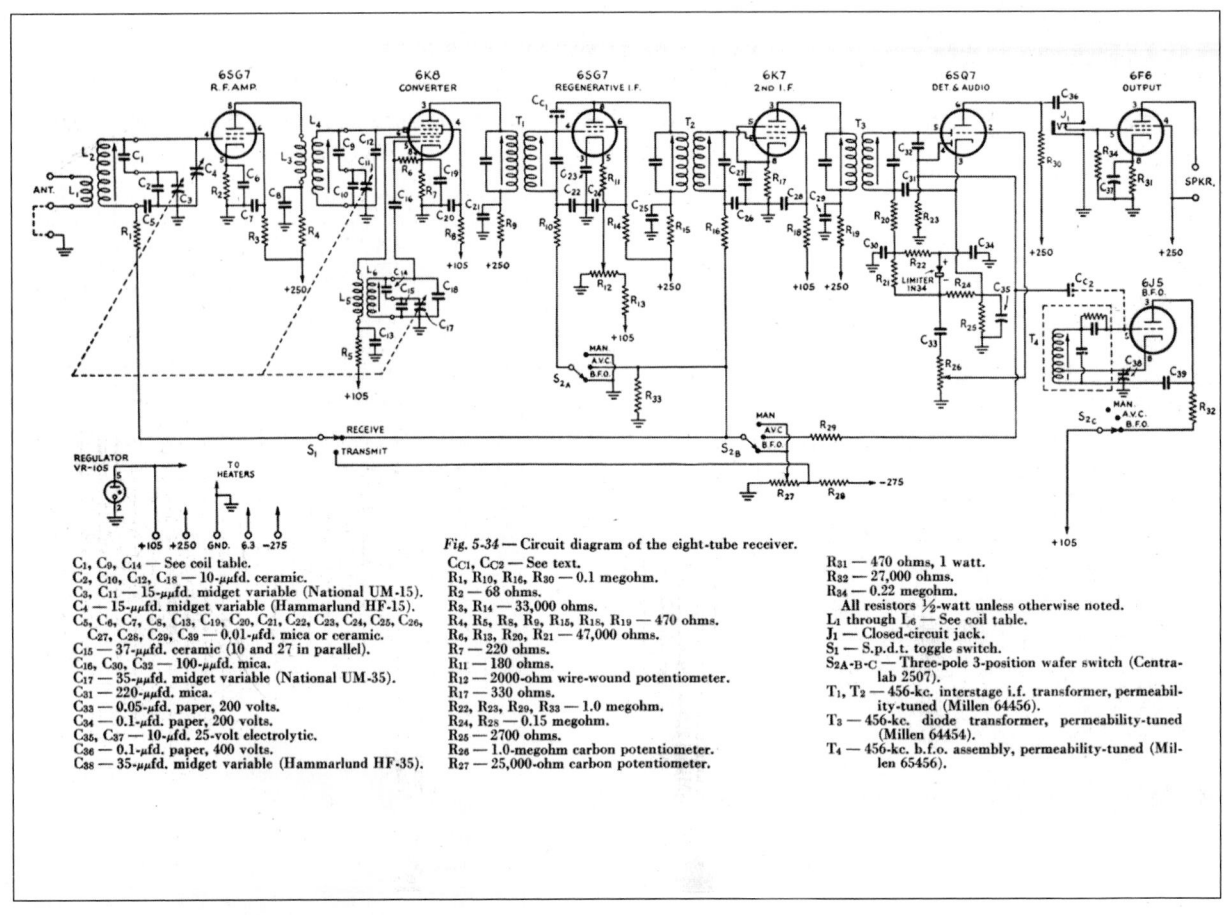

Figure 1950-5 — Schematic diagram of the receiver project shown in Figure 1950-4.

and, while the most popular HF voice mode today, took some years to displace (double-sideband full-carrier) AM from predominance in *Handbook* projects.

The ad section featured the newest of the "postwar" communication equipment. While perhaps shinier and prettier, these radios (see Figures 1950-8 and -9) were not terribly different from those of the previous generation — with the exception of the

introduction of the Collins 75A-1 (see Figure 1950-10) receiver. The 75A-1 featured down conversion to a tunable first IF stage, an architecture that would define the best of the SSB equipment of the upcoming generation. Vibroplex (see Figure 1950-11) continued to offer the same bugs as in previous decades, although the ad has a more modern look. Interestingly, the prices that dipped significantly between 1926 and

1940, returned to around their 1926 levels.

In 1939, the *Handbook* editors had concluded that there was enough material about antennas available that the *Handbook* could not do it justice. Thus a new book, *The ARRL Antenna Book*, was launched. Figure 1950-12 shows the cover of the first edition while Figure 1950-13 shows the cover of subsequent editions which maintained the familiar appearance for decades.

HIGH-FREQUENCY TRANSMITTERS   183

A 500-Watt Link-Coupled All-Band Transmitter

In the design of the transmitter shown in Figs. 6-58 through 6-69, an attempt has been made to incorporate means by which harmonic radiation and transmission may be minimized. In addition to the use of thorough shielding and power-line filtering, link coupling is used throughout.

Through the use of plug-in coils, the transmitter may be operated up to 21 Mc. with 1.75-Mc. crystals, and to 28 Mc. with either 3.5- or 7-Mc. crystals. With VFO input, it will go to 7 Mc. with 1.75-Mc. VFO output, to 21 Mc. with 3.5-Mc. VFO output, and to 28 Mc. with 7-Mc. VFO output.

The design of the push-pull triode final amplifier is suitable for any of the usual triodes with plate-cap connection, operating at plate voltages up to 1500 with plate modulation and a plate-voltage/total-plate-current ratio of 5 to 1 or greater.

The transmitter is made up in two sections mounted in a simple shielding enclosure consisting of a wood-strip frame covered with copper screening. The exciter unit is provided with pull handles and is designed to slide out for coil changing. As the unit is returned to the

enclosure, the power-supply connections are automatically made at the rear through a series of plugs which fit into jacks set along the side of a 3 × 4 × 17-inch chassis fastened permanently at the rear. This chassis also encloses and shields the harmonic-filter components for all power-supply leads.

The second section above includes the push-pull final amplifier and an antenna tuner. The top cover is hinged to provide access to the output-stage and antenna tank coils. The meters for the amplifier stage are set in a separate panel between the two main sections.

Circuit Details

Referring first to the circuit diagram of the exciter section shown in Fig. 6-60, either the built-in Pierce crystal oscillator or an external VFO may be used to feed a 6L6 stage which is operated as a doubler, as a tripler, or, when necessary, as a buffer amplifier. This stage feeds a push-push 807 driver stage that may be operated either as a doubler, or as a self-neutralized straight-through amplifier by opening S_2 which controls the heater of one of the 807s. This inactive tube then becomes the

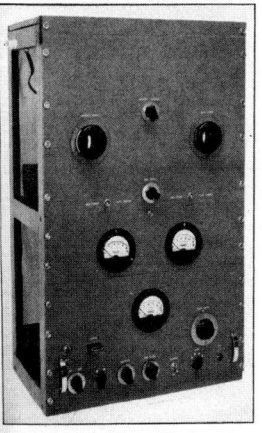 

Fig. 6-58 — A complete 500-watt all-band transmitter including antenna tuner. The exciter unit at the bottom slides out for coil changing. The panel screws on this unit are dummies cemented in place. The top of the screened enclosure is hinged to permit changing coils in the final amplifier and antenna tuner.

Fig. 6-59 — Rear view of the completed 500-watt all-band transmitter with the back screening panel removed. The rectangular enclosed unit to the rear of the exciter contains the v.h.f. power-lead filters. The two matching boxes above enclose the amplifier-stage milliammeters.

Figure 1950-6 — This 1950 edition medium power transmitter construction project includes shielding and filtering to reduce dreaded TVI.

RADIOTELEPHONY   309

Single-Sideband Transmission

The most recent development in amateur radiotelephony is the introduction of practical single-sideband suppressed-carrier transmission. This system has tremendous potentialities for increasing the effectiveness of 'phone transmission and for reducing interference. Because only one of the two sidebands normally produced in modulation is transmitted, the channel width is immediately cut in half. However, when only one sideband is transmitted the carrier — which is essential in double-sideband transmission — no longer is necessary; it can be supplied without too much difficulty at the receiver. With the carrier eliminated there is a great saving in power at the transmitter — or, from another viewpoint, a great increase in effective power output. Assuming that the same final-amplifier tube or tubes are used either for normal AM or for single-sideband, carrier suppressed, it can be shown that the use of SSB gives an effective gain of at least 9 db. over AM — equivalent to increasing the transmitter power 8 times. Eliminating the carrier also eliminates the heterodyne interference that wrecks so much communication in congested 'phone bands.

Two basic systems for generating SSB signals are shown in Fig. 9-59. One involves the use of a bandpass filter having sufficient selectivity to pass one sideband and reject the other. Filters having such characteristics can only be constructed for relatively low frequencies, and most filters used by amateurs are designed to work somewhere between 10 and 20 kc. Good sideband filtering can be done at frequencies as high as 100 kc. by using multiple-crystal filters. The low-frequency oscillator output is combined with the audio output of a speech amplifier in a "balanced modulator" — one in which the carrier is "neutralized" out, and only the upper and lower sidebands appear in the output. One of the sidebands is passed by the filter and the other rejected, so that an SSB signal is fed to the mixer. The signal is there mixed with the output of a high-frequency r.f. oscillator to produce the desired output frequency. For additional amplification a linear r.f. amplifier (Class A or Class B) must be used. When the SSB signal is generated at 10 or 20 kc., it is generally first heterodyned to somewhere around 500 kc. and then to the operating frequency. This simplifies the problem of rejecting the "image" frequencies resulting from the heterodyne process.

The second system is based on the phase relationships between the carrier and sidebands in a modulated signal. As shown in the diagram, the audio signal is split into two components that are identical except for a phase difference of 90 degrees. The output of the r.f. oscillator (which may be at the operating frequency, if desired) is likewise split into two

separate components having a 90-degree phase difference. One r.f. and one audio component are combined in each of two separate balanced modulators. The carrier is suppressed in the modulators, and the relative phases of the sidebands are such that one sideband is balanced out and the other is accentuated in the combined output. If the output from the balanced modulators is high enough, such an SSB exciter can work directly into the antenna, or the power level can be increased in a linear amplifier following the exciter.

Which is the better method of generating an SSB signal, the filter or the phasing method, is a controversial question. Properly adjusted, either system is capable of good results. Arguments in favor of the filter system are that it is

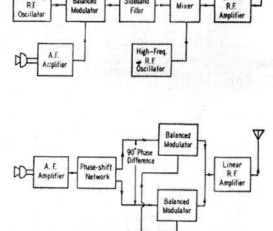

Fig. 9-59 — Two basic systems for generating single-sideband suppressed-carrier signals.

somewhat easier to adjust without an oscilloscope, since it requires only a receiver and a v.t.v.m. for alignment, and it is more likely to remain in adjustment over a long period of time. The chief argument against it, from the amateur viewpoint, is that it requires quite a few stages and at least two frequency conversions after modulation. The phasing system requires fewer stages and can be designed to require no frequency conversions, but its alignment and adjustment are often considered to be a little "trickier" than that of the filter system. This probably stems from lack of familiarity with the system rather than any actual difficulty. In most cases the phasing system will cost less to apply to an existing transmitter.

Figure 1950-7 — Single sideband transmission rated a few pages, but it would be some years before it would become mainstream for amateur use.

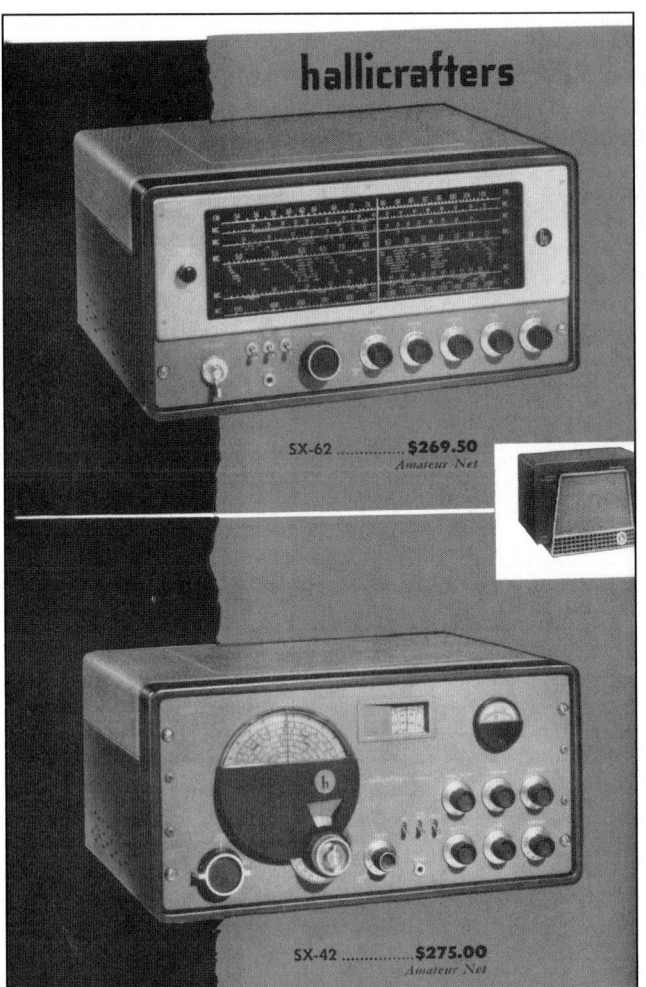

Figure 1950-8 — Hallicrafters receivers in 1950 had a definite postwar look. These receivers covered not only HF, but also VHF through 6 meters and the new FM broadcast band at 3 meters.

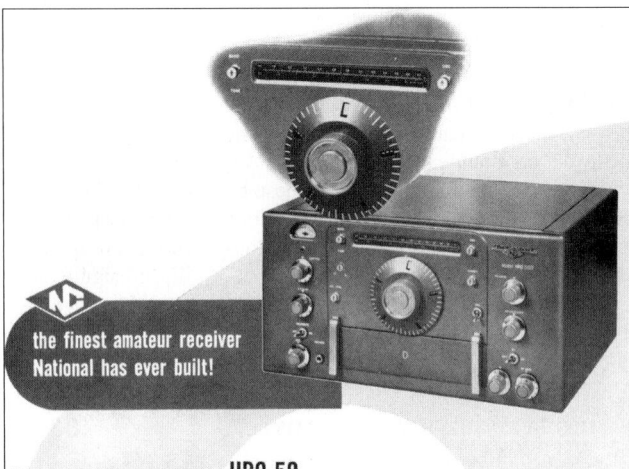

the finest amateur receiver National has ever built!

THE NEW *DIRECT READING* HRO-50

Now, National presents a great new HRO receiver after more than three years of designing, development and testing. Retaining all the world-famous, performance-proved HRO features, this superb receiver — the finest National has ever made — now incorporates no less than 14 advanced-design innovations. Exhaustive comparative tests indicate the new HRO-50, by far the most modern and versatile in its field, will set an entirely new standard of performance for communication receivers.

14 ALL NEW FEATURES

1. Direct frequency reading linear scale with a single range in view at a time. 2. Provisions for using 100/1000 kcs. crystal calibrator unit, switched from panel. 3. Variable front-of-panel antenna trimmer. 4. Built-in power supply with heat resistant barrier. 5. Front-of-panel oscillator compensation control. 6. B.F.O. switch separated from B.F.O. frequency control. 7. Provision for incorporation of NFM adapter inside receiver, switched from front panel. 8. Dimmer control for dial and meter illumination. 9. Miniature tubes in front end and high frequency oscillator. 10. Speaker matching transformer built into receiver with 8 and 500 -ohm output terminals. 11. High frequency and beat frequency oscillator circuits not disabled when receiver in "send" position. 12. High-fidelity push-pull audio amplifier, 8 watts undistorted output. 13. Tip jack for phono input. 14. Accessory socket for Select-o-Ject (see page 4).

TUBE COMPLEMENT:

1st RF, 6BA6; 2nd RF 6BA6; Mixer, 6BE6; HF oscillator 6C4; voltage regulator OB2; 1st I.F., 6K7; 2nd I.F., 6K7; Det./AVC, 6H6; B.F. Oscillator, 6J7; Noise Limiter, 6H6; 1st Audio, 6SJ7; phase inverter/ "S"-meter amp. 6SN7; Push-pull audio, 2-6V6; Rectifier, 5V4G; accessory crystal calibrator, 6AQ5; NFM adapter I.F. amplifier, 6SK7; Ratio detector, 6H6. Freq. range: 50 kc.-420 kc., 480 kc.-35 mc Coils AA, B, C, and D furnished covering standard amateur 160-10 meter bands.

National
® EST. 1914
NATIONAL COMPANY, Inc
MALDEN, MASSACHUSETTS

4

Figure 1950-9 — National Radio's premier postwar offering was the HRO-50 — an evolution of the prewar HRO that included miniature tubes, an internal power supply, push-pull (hi-fi) audio and a direct reading frequency scale. These were nice features, but it was now a two-person lift!

Figure 1950-10 — This ad by Collins radio doesn't really highlight the fact that the 75A-1 architecture was radical for its time, and would set the stage for the Collins SSB radios that would appear in the next decade.

Figure 1950-11 — Vibroplex continued to advertise its popular "Bugs." This ad has a more modern layout than earlier ones, and the prices are back to 1926 levels.

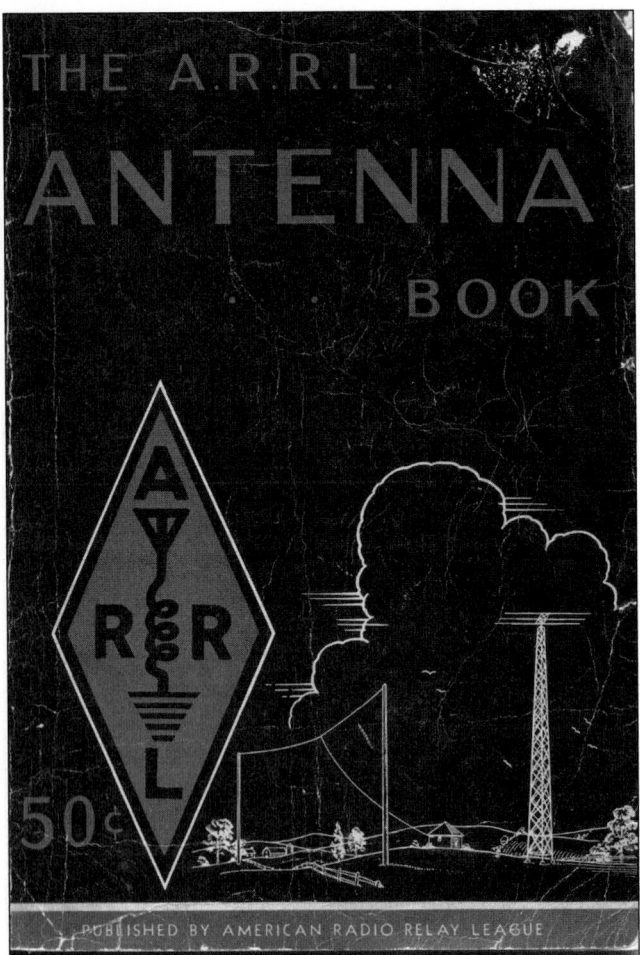

Figure 1950-12 — In 1939, the *Handbook* editors concluded that there was enough material about antennas available that a new reference text was needed, creating the *ARRL Antenna Book*.

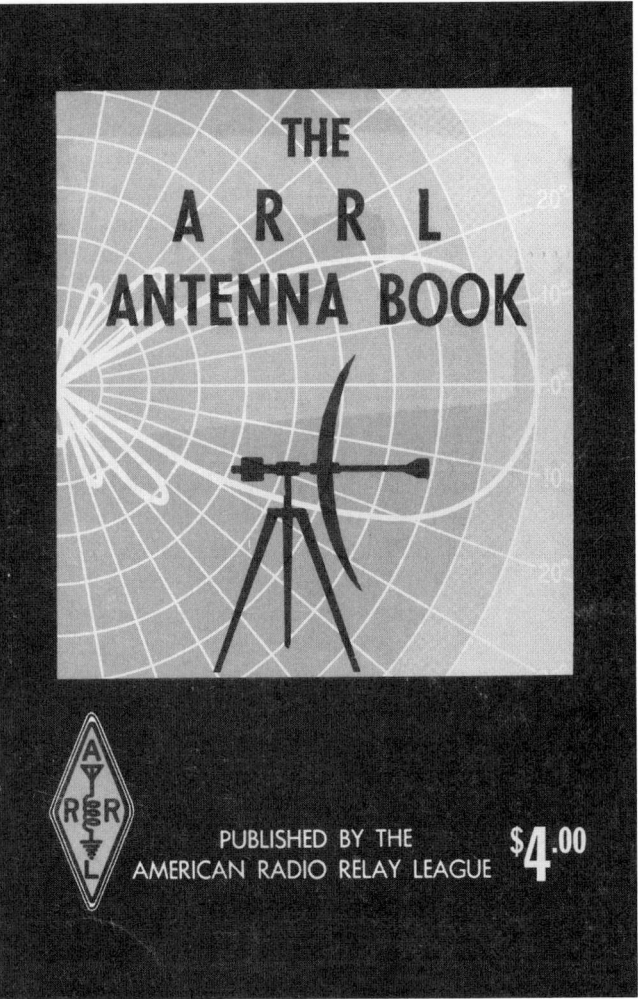

Figure 1950-13 — By 1949, the amount of material in the *Antenna Book* had increased substantially and the cover assumed the familiar look that subsequent editions maintained for decades.

A Decade of Advancement and Change — The 1960 Edition

The advancements in technology and engineering that were starting to emerge in the 1950s were solidified in the period building toward 1960, although not evident in the cover plate (see Figure 1960-1). While complete solid state equipment is some years away, transistor circuitry was showing up in projects where it was readily applied (see Figure 1960-2). In this figure, the mostly vacuum tube receiver includes a two transistor 100 kHz calibrator. Also note the two crystal IF band-pass filter. This arrangement was chosen with an ear toward SSB reception, in preference to the sharper single crystal filters of earlier year designs.

Figures 1960-3 and -4 illustrate a *Handbook* transmitter project that, while superficially similar to earlier CW transmitter projects, is of a significantly different design. In place of the plug-in link-coupled transmitter coils of earlier generations, this transmitter uses a pi-network output circuit that allows band-switching with a single switch section on the output circuit. It is now possible to have a tightly RF-sealed transmitter that provides instant bandswitching across the HF

THE RADIO AMATEUR'S HANDBOOK

By the HEADQUARTERS STAFF
of the
AMERICAN RADIO RELAY LEAGUE
WEST HARTFORD, CONN., U.S.A.

1960

Thirty-seventh Edition

Figure 1960-1 — The cover plate of the 1960 *Handbook* offers no clues to the changes over the past decade.

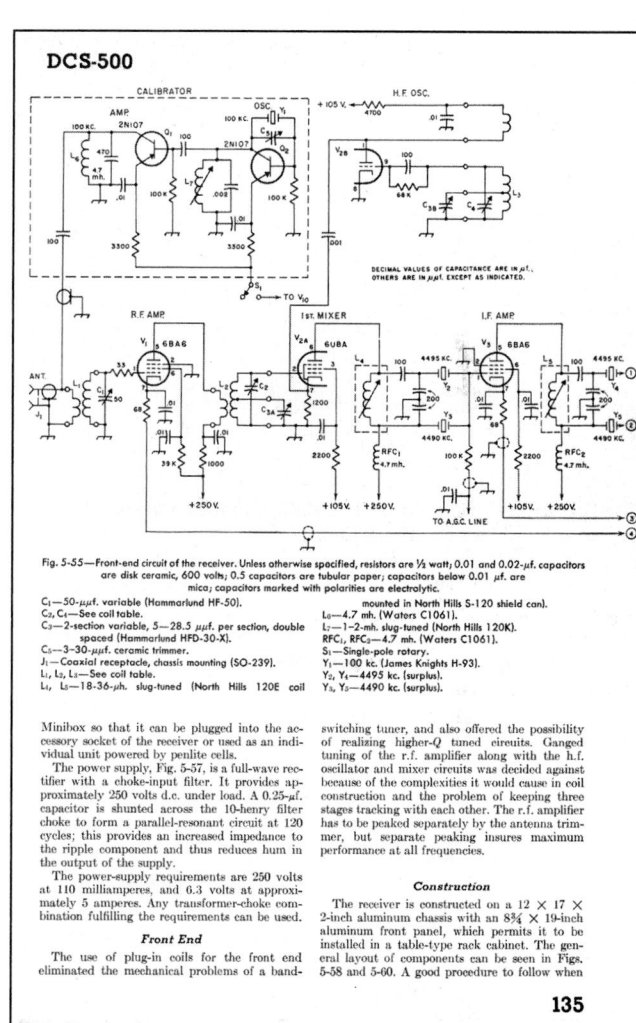

Figure 1960-2 — Schematic of a 1960 *Handbook* receiver project that makes use of transistors in the calibrator circuit. It also features a band-pass crystal filter appropriate for SSB reception.

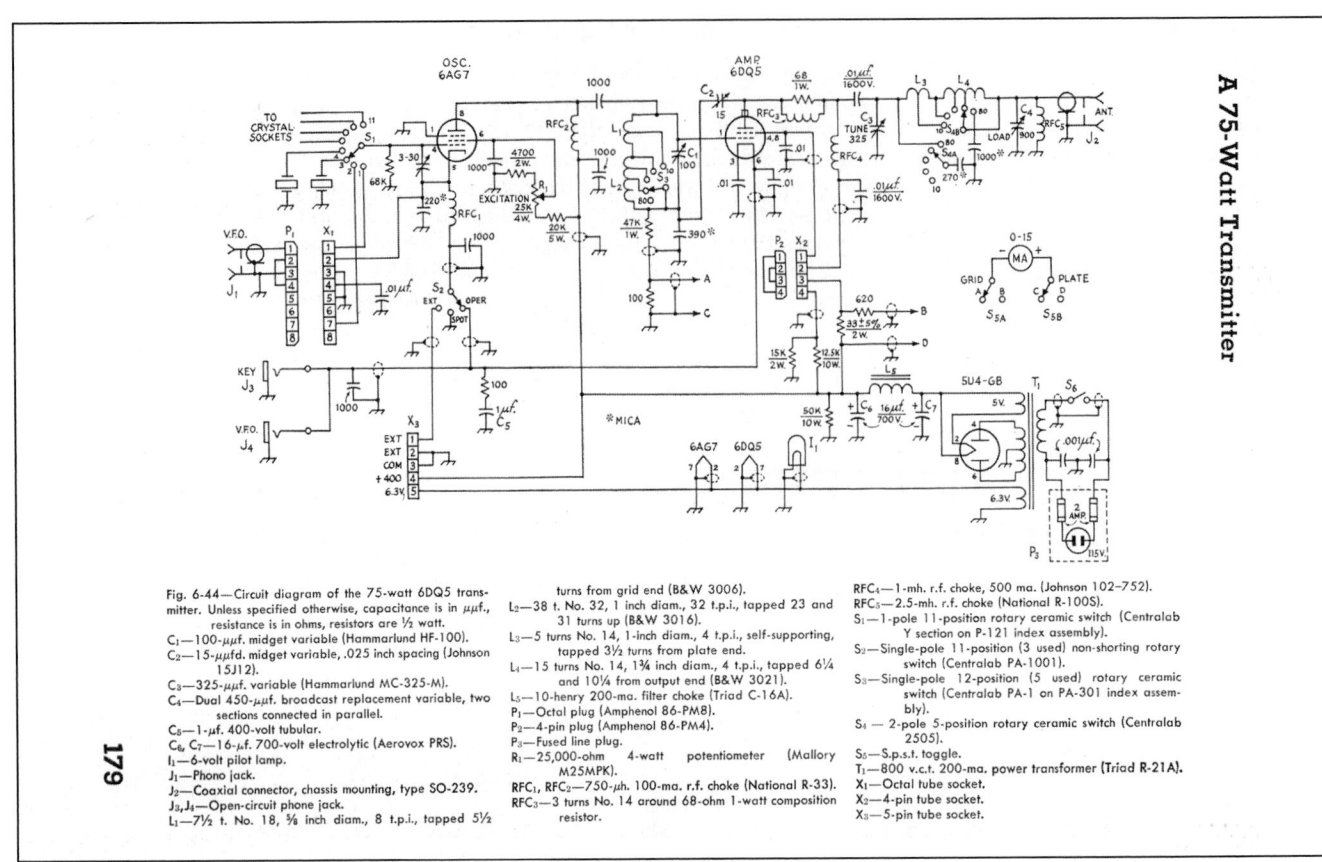

A 75-Watt Transmitter

Fig. 6-44—Circuit diagram of the 75-watt 6DQ5 transmitter. Unless specified otherwise, capacitance is in μμf., resistance is in ohms, resistors are ½ watt.

C₁—100-μμf. midget variable (Hammarlund HF-100).
C₂—15-μμfd. midget variable, .025 inch spacing (Johnson 15J12).
C₃—325-μμf. variable (Hammarlund MC-325-M).
C₄—Dual 450-μμf. broadcast replacement variable, two sections connected in parallel.
C₅—1-μf. 400-volt tubular.
C₆, C₇—16-μf. 700-volt electrolytic (Aerovox PRS).
I₁—6-volt pilot lamp.
J₁—Phono jack.
J₂—Coaxial connector, chassis mounting, type SO-239.
J₃, J₄—Open-circuit phone jack.
L₁—7½ t. No. 18, ⅝ inch diam., 8 t.p.i., tapped 5½

turns from grid end (B&W 3006).
L₂—38 t. No. 32, 1 inch diam., 32 t.p.i., tapped 23 and 31 turns up (B&W 3016).
L₃—5 turns No. 14, 1-inch diam., 4 t.p.i., self-supporting, tapped 3½ turns from plate end.
L₄—15 turns No. 14, 1¾ inch diam., 4 t.p.i., tapped 6¼ and 10¼ from output end (B&W 3021).
L₅—10-henry 200-ma. filter choke (Triad C-16A).
P₁—Octal plug (Amphenol 86-PM8).
P₂—4-pin plug (Amphenol 86-PM4).
P₃—Fused line plug.
R₁—25,000-ohm 4-watt potentiometer (Mallory M25MPK).
RFC₁, RFC₂—750-μh. 100-ma. r.f. choke (National R-33).
RFC₃—3 turns No. 14 around 68-ohm 1-watt composition resistor.

RFC₄—1-mh. r.f. choke, 500 ma. (Johnson 102-752).
RFC₅—2.5-mh. r.f. choke (National R-100S).
S₁—1-pole 11-position rotary ceramic switch (Centralab Y section on P-121 index assembly).
S₂—Single-pole 11-position (3 used) non-shorting rotary switch (Centralab PA-1001).
S₃—Single-pole 12-position (5 used) rotary ceramic switch (Centralab PA-1 on PA-301 index assembly).
S₄—2-pole 5-position rotary ceramic switch (Centralab 2505).
S₅—S.p.s.t. toggle.
T₁—800 v.c.t. 200-ma. power transformer (Triad R-21A).
X₁—Octal tube socket.
X₂—4-pin tube socket.
X₃—5-pin tube socket.

179

Figure 1960-3 — Schematic of a 1960 *Handbook* transmitter project that features a pi-network output circuit that allows band-switching and helps with harmonic reduction.

spectrum. The pi-network has other advantages — it is naturally low pass in response and it is unbalanced, and so it was a perfect fit to coax-fed antenna systems. Both features tended to reduce harmonic radiation and TVI.

The advertising section illustrated the diversity of technology available during the period. Collins Radio was at the head of the pack with its compact 100 W PEP HF SSB transceiver that could be used in a mobile setup, as shown in Figure 1960-5. This transceiver made use of the architectural design of the 75A-1 receiver

shown in Figure 1950-10.

Other equipment manufacturers offered a mix of technologies. The Heath Company (see Figure 1960-6) offered AM transmitters, such as their TX-1 Apache with an optional phasing type SSB exciter. Their companion multimode receiver, the RX-1 would be announced soon. E. F. Johnson offered a wide range of high quality AM and CW transmitters (see Figure 1960-7), including an SSB adapter similar to the one provided by Heath. Hallicrafters offered a modern heterodyning transmitter, the

HT-32, along with an SSB oriented receiver, the SX-101, and high power linear amplifier as shown in Figure 1960-8.

National Radio was still offering a full line of receivers, including a further update to the venerable HRO. Added to the line was an amateur band only receiver, the NC-300, a direct competitor to the Hallicrafters SX-101. A careful look at the Vibroplex ad (see Figure 1960-10) will reveal an entirely new product — the Vibrokeyer, "paddles" for an electronic keyer — a competitor to their mechanical bug.

6 – HIGH-FREQUENCY TRANSMITTERS

A 75-Watt 6DQ5 Transmitter

The transmitter shown in Fig. 6-43 is designed to satisfy the requirements of either a Novice or General class licensee. As described here it is capable of running the full 75 watts limit in the 80-, 40- and 15-meter Novice bands, with band-switching, crystal switching and other operating features. The General license holder can use the transmitter in any band 80 through 10 meters, and he can add v.f.o. control or amplitude modulation at any time without modifying the 6DQ5 transmitter. Crystal switching is a convenience for rapidly shifting frequency within a band to dodge QRM, and a SPOT position on the operate switch permits identifying one's frequency relative to others in a band. An accessory socket, X_3, furnishes a convenient point for borrowing power for a v.f.o. or for controlling the oscillator by an external switch.

Referring to Fig. 6-44, the circuit diagram of the transmitter, the crystal selector switch, S_1, is used to choose the desired crystal. For crystal-controlled operation crystals would be plugged in pins 1 and 3 and 5 and 7 of socket X_1. Similar sockets (not shown in the diagram) are used to hold the other crystals. When v.f.o. operation is desired, the v.f.o. output is connected to J_1, the plug P_1 is inserted in socket X_1, and the former 6AG7 crystal oscillator stage becomes an amplifier or multiplier stage when switch S_1 is turned to position 1.

Since the output of the 6AG7 stage will vary considerably with the bands in use, an excitation control, R_1, is included to allow for proper adjustment of the drive to the 6DQ5 amplifier. The 6DQ5, a highly sensitive tube, is neutralized to avoid oscillation; the small variable capacitor C_2 and the 390-$\mu\mu f$. mica capacitor form the neutralizing circuit. Screen or screen and plate modulation power can be introduced at socket X_2; for radiotelegraph operation these connections are completed by P_2. Grid or plate current of the 6DQ5 can be read by proper positioning of S_5: the 0–15 milliammeter reads 0–15 ma. in the grid-current position and 0–300 ma. in the plate-current position.

The transmitter is keyed at J_3, and a key-click filter (100-ohm resistor and C_8) is included to give substantially click-free keying. The v.f.o. jack, J_4, allows a v.f.o. to be keyed along with the transmitter for full break-in operation.

Construction

A 10 × 17 × 3-inch aluminum chassis is used as the base of the transmitter, with a standard 8¾-inch aluminum relay rack panel held in place by the bushings of the pilot light, excitation control and other components common to the chassis and panel. The panel was cut down to 17 inches in length so that the unit would take a minimum of room on the operating table. A good idea of the relative location of the parts can be obtained from the photographs. The support for the r.f. portion housing is made by fastening strips of 1-inch aluminum angle stock (Reynolds aluminum, available in many hardware stores) to the panel and to a sheet of aluminum 9½ inches long that is held to the rear chassis apron by screws and the key jack, J_3. A piece of aluminum angle must also be cut to mount on the chassis and hold the cane-metal (Reynolds aluminum) housing. Fig. 6-45 shows the three clearance holes for the screws that hold this latter angle to the chassis after the cane metal is in place. Build the can-metal housing as though the holes weren't there and the box has to hold water; this will minimize electrical leakage and the chances for TVI. To insure good electrical contact between panel and angle stock, remove the paint where necessary by heavy applications of varnish remover, with the rest of the panel

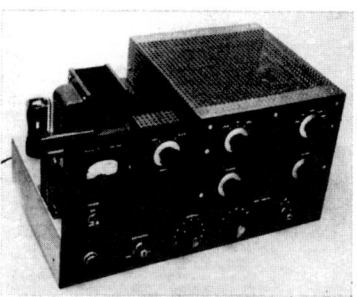

Fig. 6-43 — This 75-watt crystal-controlled transmitter has provision for the addition of v.f.o. control. A 6AG7 oscillator drives a 6DQ5 amplifier on 80 through 15 meters.

As a precaution against electrical shock, the meter switch, to the immediate right of the meter, is protected by a cane-metal housing. The switch to the right of the meter switch handles the spot-operate function, and the switch at the far top right is the plate-circuit band switch.

Along the bottom, from left to right: pilot light, excitation control, crystal switch, grid circuit band switch, and grid circuit tuning.

178

Figure 1960-4 — Front panel view of the 1960 *Handbook* transmitter project shown in Figure 1960-3. This tight design is fully enclosed and features front panel bandswitching.

Figure 1960-5 — This Collins Radio ad featuring their KWM-2 SSB transceiver was at the leading edge of Amateur Radio technology. The architecture closely followed that of the 75A-1 receiver shown in Figure 1950-10.

Figure 1960-6 — The Heath Company offered AM transmitters, such as their TX-1 Apache with an optional phasing type SSB exciter.

Figure 1960-7 — E. F. Johnson offered a wide range of high quality AM and CW transmitters, including an SSB adapter similar to the one provided by Heath.

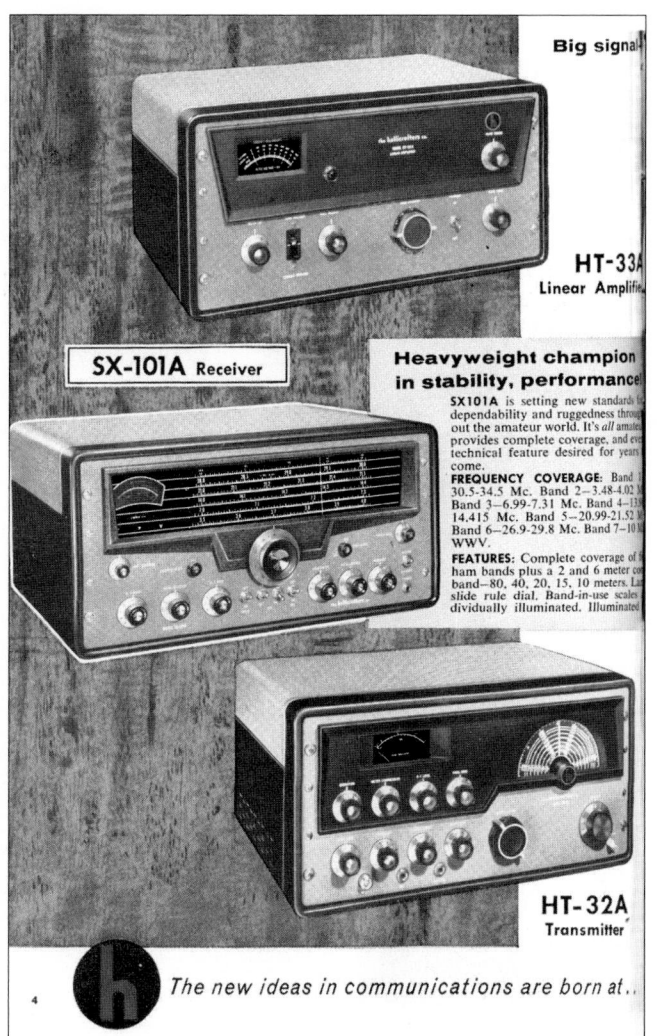

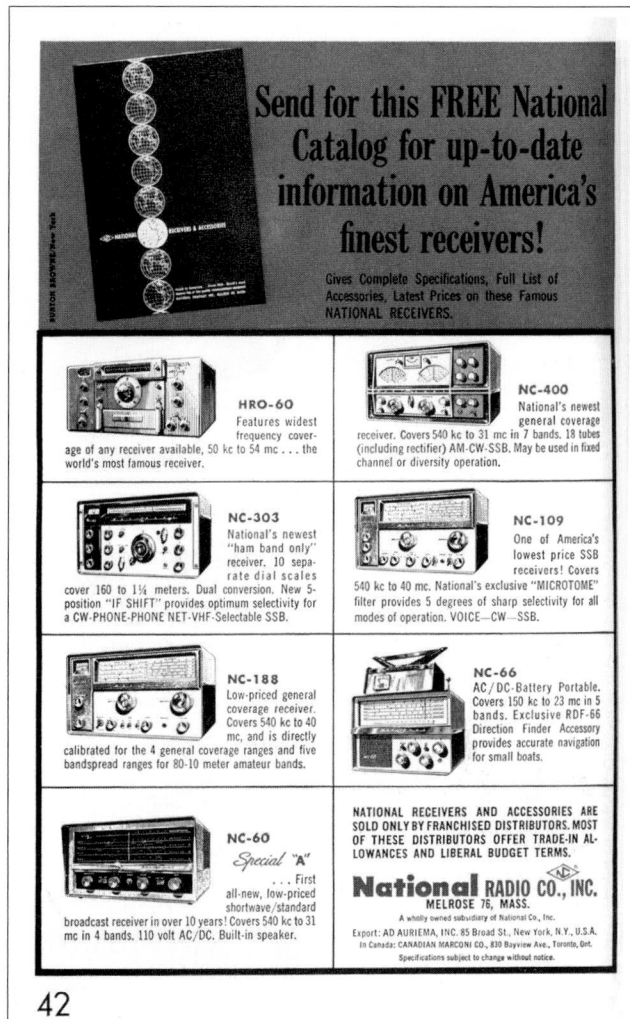

Figure 1960-8 — Hallicrafters offered a modern heterodyning transmitter, the HT-32, along with an SSB oriented receiver, the SX-101, and the HT-33 high power linear amplifier.

Figure 1960-9 — National Radio still offered the latest of the HRO series, the HRO-60 featuring double conversion, but they also would supply an amateur band only receiver designed for SSB. The new NC-300 was similar to the SX-101 from Hallicrafters.

Figure 1960-10 — The Vibroplex ad featured a new product — the Vibrokeyer, designed to be used with the newly popular electronic keyers.

The Decade of Solid State and the SSB Transceiver — The 1970 Edition

Between the 1960 and 1970 Handbooks (see Figure 1970-1), two major game changers went from being interesting technology to becoming mainstream Amateur Radio. The first was the acceptance of solid state devices as appropriate for most sections of radio equipment — although RF power stages would remain largely in the

domain of vacuum tubes for some time. This trend can be seen in a solid state receiver project (see Figure 1970-2) and a hybrid (solid state except for the transmit driver and power amplifier stages) SSB transmitter (see Figure 1970-3). There were still tube projects galore, since many amateurs felt most comfortable with the earlier

technology. The beginner transmitter shown in Figure 1970-4 could have appeared in much earlier editions.

The second advance was the emergence of single sideband telephony, and especially SSB transceivers, as the primary HF voice mode and equipment configuration. While SSB was introduced in a general way in past

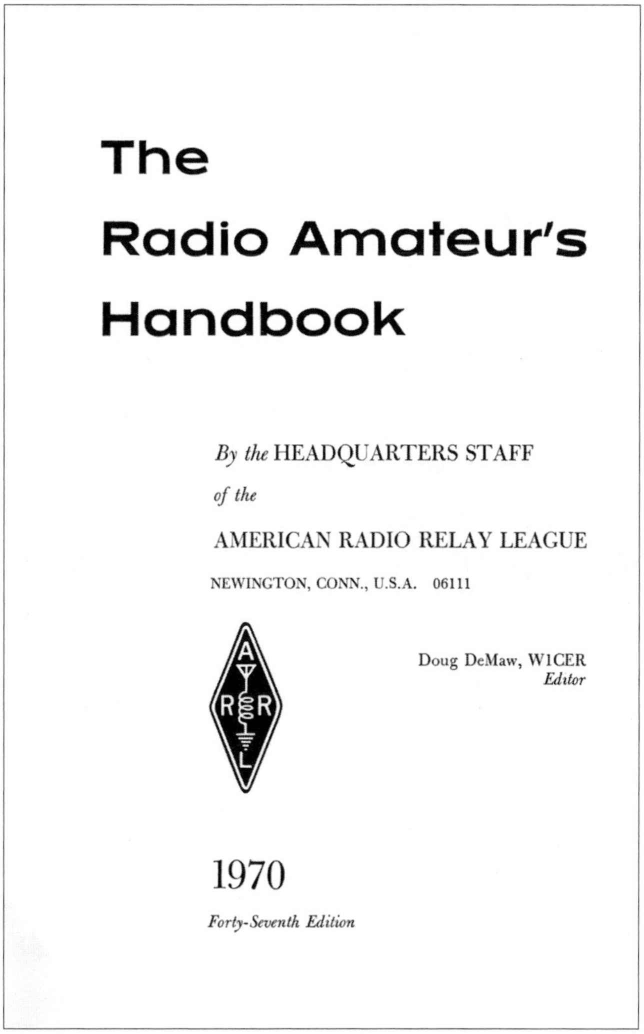

The

Radio Amateur's

Handbook

By the HEADQUARTERS STAFF

of the

AMERICAN RADIO RELAY LEAGUE

NEWINGTON, CONN., U.S.A.   06111

Doug DeMaw, W1CER
Editor

1970

Forty-Seventh Edition

Figure 1970-1 — The front matter on the 1970 *Handbook* had a more modern look than in the past.

decades, it was the emergence of the one box transceiver as shown previously in Figure 1960-5, that made the big difference. During this decade, the benefits of a transceiver, not only for SSB, but for CW as well, became apparent. Companies such as Collins Radio (see Figure 1970-5), the Heath Company (see Figure 1970-6), R.L.

Drake (see Figure 1970-7) and others provided their versions of both transceivers and separate transmitters and receivers that could be locked together as if they were transceivers.

Not everything changed. The Vibroplex ad (Figure 1970-8) maintained its

consistency over yet another decade. In recognition of the amount of operating, in contrast to technical information available, or needed by operators, *The Radio Amateur's Operating Manual* (see Figure 1970-9) was introduced in 1966 as something of a spinoff from the *Handbook*.

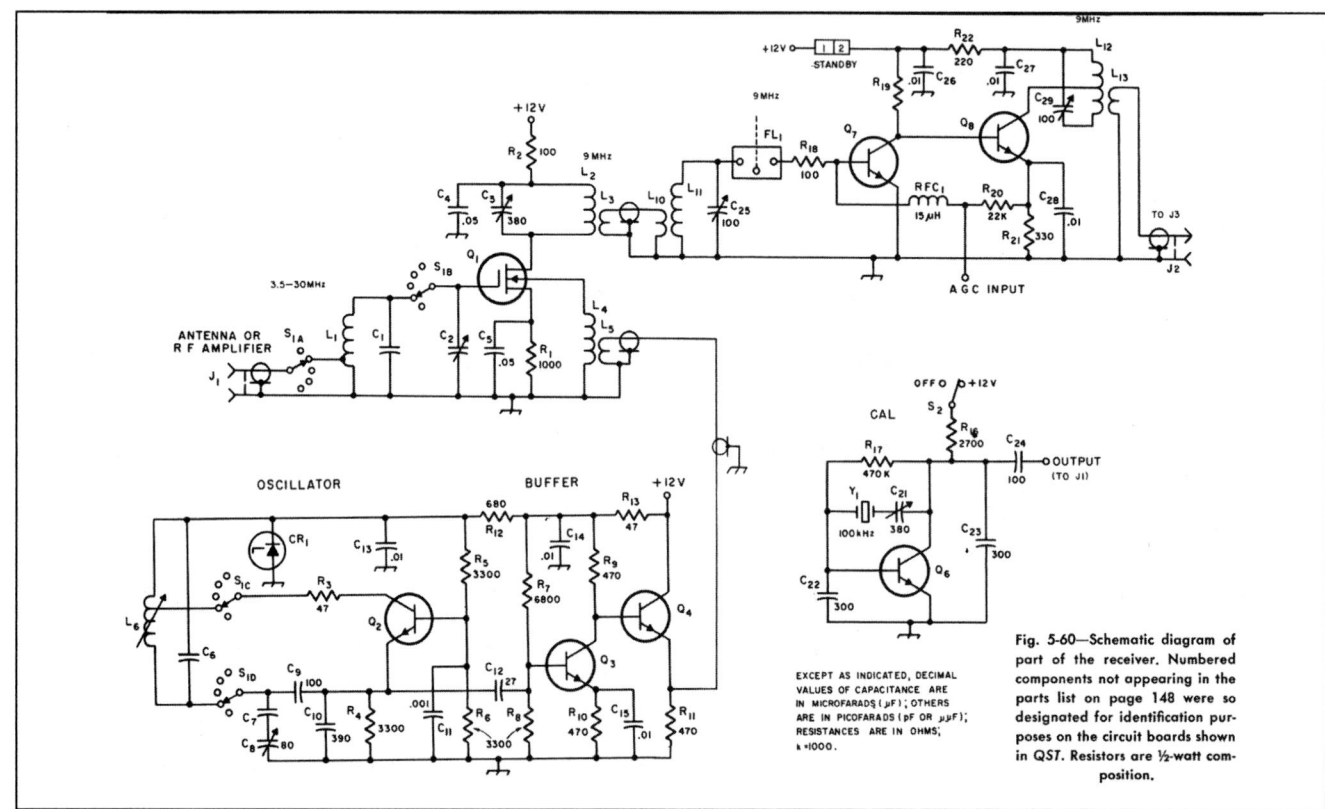

EXCEPT AS INDICATED, DECIMAL VALUES OF CAPACITANCE ARE IN MICROFARADS (μF); OTHERS ARE IN PICOFARADS (pF OR μμF); RESISTANCES ARE IN OHMS; k =1000.

Fig. 5-60—Schematic diagram of part of the receiver. Numbered components not appearing in the parts list on page 148 were so designated for identification purposes on the circuit boards shown in QST. Resistors are ½-watt composition.

Figure 1970-2 — This 1970 *Handbook* receiver project is entirely solid state.

A 175-WATT MECHANICAL-FILTER EXCITER

This 3.5 to 4.0-MHz. s.s.b./c.w. transmitter can be used by itself, or it can be used to drive any of the amplifiers described in Chapter 6. It will drive most commercially-built amplifiers also. The power output from this unit, while maintaining an acceptable IMD level (intermodulation distortion) is 100 watts, p.e.p.

Block diagrams have been added to each schematic illustration to help the reader understand how the circuit operates. The power supply, "A 650-Volt General-Purpose Supply," is shown in Chapter 12. Information on building the modular solid-state v.f.o. is given in Chapter 5 ("A General-Purpose V.F.O."). This transmitter was designed to be used with these two units.

This exciter has effective a.l.c., which helps to maintain a high average talk-power level. Grid-block keying is used for c.w. The keying is shaped to provide a clean, clickless note. If low-power a.m. operation is desired, carrier can be inserted for this purpose. The power input to the p.a. must be limited to approximately 25 watts if this is done, and the output signal will be *single-sideband a.m.*

Circuit Information

In the circuit of Fig. 9-31, output from a hi-impedance microphone is amplified by V_1 and fed to a twin-triode balanced-modulator, V_2. The 455-kHz. carrier is generated by V_{3A} and routed to V_2. The double-sideband suppressed-carrier a.m. signal from V_2 is next passed through FL_1 where it becomes a s.s.b.

suppressed-carrier signal; the sideband (upper or lower) depends upon the crystal selected at V_{3A}. Output from FL_1 is amplified by V_4, whose actual gain at a given instant is dependent upon the level of the minus-polarity a.l.c. voltage supplied to its grid; the lower the a.l.c. voltage, the higher will be the gain of V_4. V_{3B} is used as a cathode-follower to supply carrier (455 kHz.) to V_5 for tuneup, c.w., or a.m. operation, thus bypassing the mechanical filter and balanced-modulator with some of the signal. The carrier-insertion level is controlled by R_3, S_2, a part of R_3, opens that branch of the circuit during s.s.b. operation to minimize carrier leak-through from V_{3B} to V_5.

A transistorized v.f.o. is used to beat a 3045 to 3545-kHz. signal against the 455-kHz. s.s.b. signal at V_5. The *sum* frequency from the mixer provides the desired 3.5 to 4.0-MHz. transmitter output frequency. Output from the v.f.o. is filtered by L_4, L_5, and their related network capacitors. The filtering keeps spurious output from the v.f.o. from reaching the balanced mixer and generating unwanted frequencies. A vacuum-tube buffer stage, V_6, is used between the v.f.o. and V_5 to reduce "pulling" and to transform the v.f.o.'s low output impedance to a higher impedance for feeding the grid of the balanced mixer. L_6 is broadbanded (no parallel capacitor) to assure fairly constant mixer injection across the entire tuning range of the v.f.o.

The mixer tuned circuit, L_1-C_{2A}, is connected to the grid of the driver stage, V_6. The plate cir-

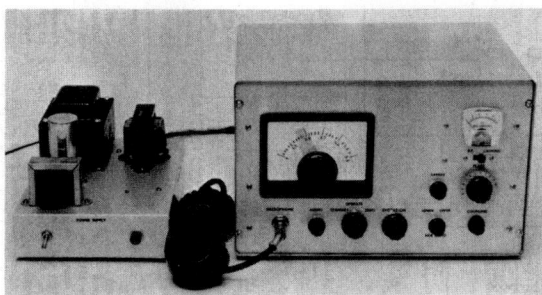

The exciter is housed in a home-made cabinet. Several commercial cabinets are available to the builder, many of which are similar in size and style to this one. An LMB type W-2J would be a good choice, and could be ventilated. Black decals are used for identifying the controls on the satin-finish aluminum panel. The panel was soaked in a lye bath to get the dull finish, then sprayed with clear lacquer.

Figure 1970-3 — The 1970 *Handbook* featured this compact 75 meter, 175 W SSB exciter/transmitter. This was a "hybrid" design, similar to some commercial equipment of the day — transistors were used in low level stages, while tubes remained in the transmitter power driver and output sections.

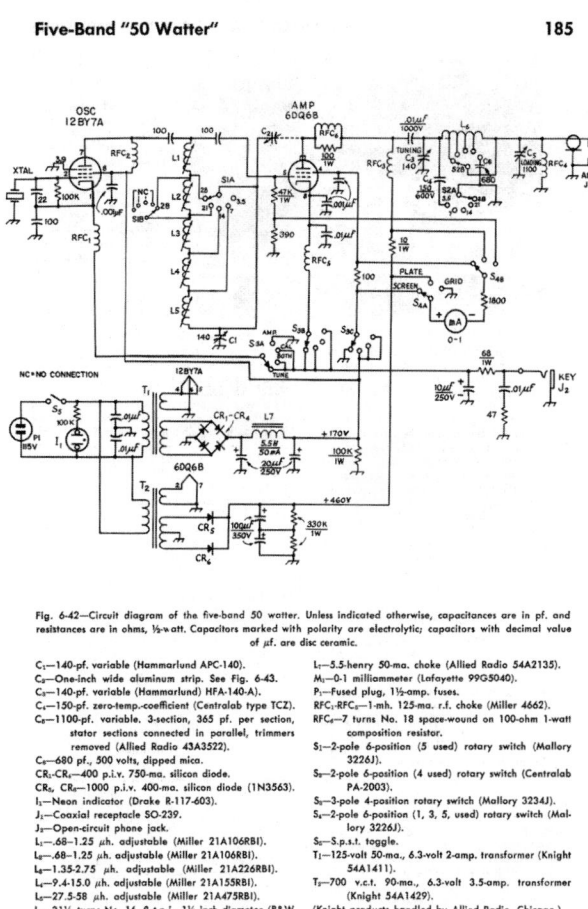

Fig. 6-42—Circuit diagram of the five-band 50 watter. Unless indicated otherwise, capacitances are in pf. and resistances are in ohms, ½-watt. Capacitors marked with polarity are electrolytic; capacitors with decimal value of μf. are disc ceramic.

C_1—140-pf. variable (Hammarlund APC-140).
C_2—One-inch wide aluminum strip. See Fig. 6-43.
C_3—140-pf. variable (Hammarlund HFA-140-A).
C_4—150-pf. zero-temp.-coefficient (Centralab type TCZ).
C_5—1100-pf. variable. 3-section, 365 pf. per section, stator sections connected in parallel, trimmers removed (Allied Radio 43A3522).
C_6—680 pf., 500 volts, dipped mica.
CR_1-CR_4—400 p.i.v. 750-ma. silicon diode.
CR_5, CR_6—1000 p.i.v. 400-ma. silicon diode (1N3563).
I_1—Neon indicator (Drake R-117-603).
J_1—Coaxial receptacle SO-239.
J_2—Open-circuit phone jack.
L_1—.68–1.25 μh. adjustable (Miller 21A106RBI).
L_2—.68–1.25 μh. adjustable (Miller 21A106RBI).
L_3—1.35–2.75 μh. adjustable (Miller 21A226RBI).
L_4—9.4–15.0 μh. adjustable (Miller 21A155RBI).
L_5—27.5–58 μh. adjustable (Miller 21A475RBI).
L_6—31½ turns No. 16, 8 t.p.i., 1¼ inch diameter (B&W 3018) tapped from C_5 end: 2¾, 6¼, 10¾, 22¾ turns.

L_7—5.5-henry 50-ma. choke (Allied Radio 54A2135).
M_1—0-1 milliammeter (Lafayette 99G5040).
P_1—Fused plug, 1½-amp. fuses.
RFC_1-RFC_5—1-mh. 125-ma. r.f. choke (Miller 4662).
RFC_6—7 turns No. 18 space-wound on 100-ohm 1-watt composition resistor.
S_1—2-pole 6-position (5 used) rotary switch (Mallory 3226J).
S_2—2-pole 6-position (4 used) rotary switch (Centralab PA-2003).
S_3—3-pole 4-position rotary switch (Mallory 3234J).
S_4—2-pole 6-position (1, 3, 5, used) rotary switch (Mallory 3226J).
S_5—S.p.s.t. toggle.
T_1—125-volt 50-ma., 6.3-volt 2-amp. transformer (Knight 54A1411).
T_2—700 v.c.t. 90-ma., 6.3-volt 3.5-amp. transformer (Knight 54A1429).
(Knight products handled by Allied Radio, Chicago.)

Figure 1970-4 — Not all projects were modernized — this beginner transmitter looks a lot like Novice transmitters from previous decades.

32S-3 Transmitter

The 32S-3 operates in SSB or CW modes with 175 watts PEP input from 3.4 to 5.0 MHz and from 6.5 to 30.0 MHz. Supplied crystals cover the 80-, 40-, 20-, and 15-meter bands, and 200 kHz for two more crystals. Features include mechanical filter sideband generation, permeability-tuned VFO and crystal-controlled HF oscillator.

75S-3B,-3C Receivers

The finest amateur receivers available for SSB, CW or RTTY. The 75S-3B provides SSB, CW and AM reception from 3.4 to 5.0 MHz and from 6.5 to 30.0 MHz. Crystals furnished with the 75S-3B cover the 80-, 40-, 20-, and 15-meter bands and 200 kHz of the 10-meter bands. The 75S-3C has provisions for 14 additional switch-selected crystals. Both units have crystal-controlled first oscillator, permeability-tuned VFO, and 2.1 kHz mechanical filter. Simple patch-cord connections permit operation with the 32S-3 as a transceiver.

312B-4 Speaker Console

The 312B-4 is a unitized control for the S-Line or KWM-2. It houses a speaker, RF directional wattmeter, and switches for station control functions.

30S-1 Linear Amplifier

Collins linear amplifiers can be driven by the KWM-1, KWM-2, 32S-3 or equivalent. The 30S-1 is a completely self-contained, single tube, grounded grid linear amplifier providing full legal power input for SSB, CW or RTTY. It may be used on any frequency between 3.4 and 30.0 MHz. A special comparator tuning circuit allows tune-up at low power. Select linear amplifier or exciter output by front panel push-button. Antenna relay included.

30L-1 Linear Amplifier

A compact linear amplifier for table-top or console operation, the 30L-1 provides 1 kw PEP input on SSB and 1 kw average on CW. It has a self-contained power supply, RF inverse feedback, and automatic load control. The 30L-1 features instant warm-up and self-contained antenna relay.

KWM-2,-2A Transceivers

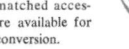

The KWM-2 and -2A SSB transceivers give unmatched flexibility for fixed-station or mobile operation. They cover the same bands as the S-Line with 175 watts PEP input SSB or 160 watts CW. Both have provision for two extra crystals, and 14 more can be added to the KWM-2A. Both have filter-type SSB generation, permeability-tuned oscillator, crystal controlled HF double conversion oscillator, VOX and anti-trip circuits, automatic load control and RF inverse feedback. Either unit will drive the 30S-1 or 30L-1 linear amplifiers. Optional matched accessories are available for mobile conversion.

COLLINS

Collins Radio Company • Cedar Rapids, Iowa

Figure 1970-5 — Collins Radio's 1970 ad features a full complement of gear built in the image of its successful KWM-2 showcased in Figure 1960-5.

Figure 1970-6 — The Heath Company took hold of the Collins idea and offered kits to build similar looking compact SSB gear.

17

Figure 1970-7 — The R. L. Drake Company, a relative newcomer to the scene, offered a line of similar and highly regarded transceivers and separate units.

Figure 1970-8 — Vibroplex continued to produce their fine line of semi-automatic keys and paddles.

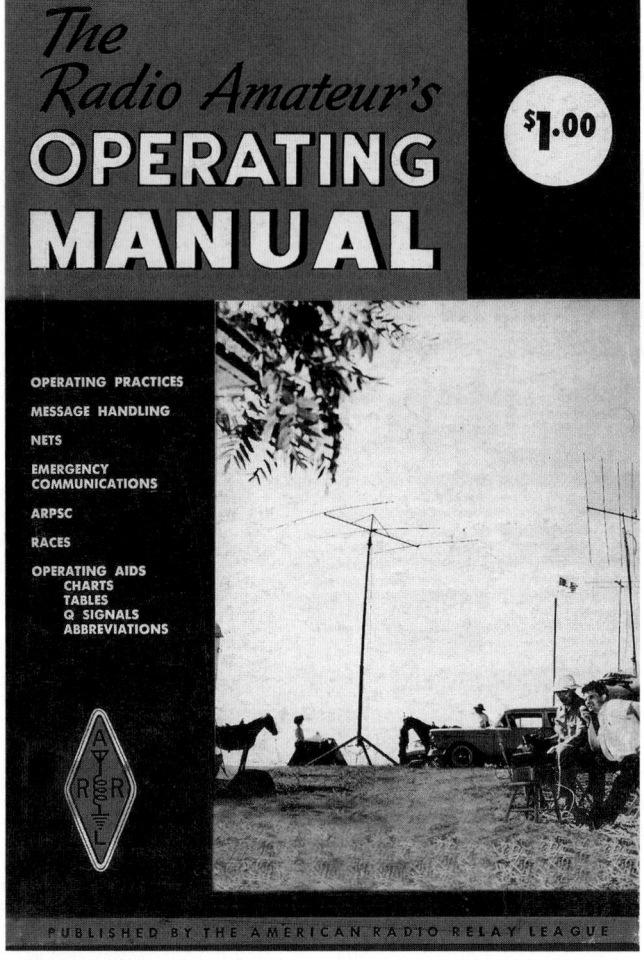

Figure 1970-9 — *The Radio Amateur's Operating Manual* **was introduced in 1966.**

FM Becomes the Mode on VHF — The 1980 Edition

The decade leading toward the 1980 *Handbook* saw a structural change — there was no longer an advertising section at the rear! Readers would need to look through the pages of *QST* to keep up with the advance of "Bug" technology from Vibroplex. The cover page (see Figure 1980-1) continued with an updated look.

During the decade between the 1970 and 1980 *Handbooks*, there were continued advancements in the application of solid state to mainstream projects as shown in a high

performance miniature solid state receiver (see Figures 1980-2 and -3) and an all solid state transmitter (see Figure 1980-4). While low power transmitters were headed toward solid state, a lack of available devices for typical HF power levels (100 W and higher) would keep vacuum tubes in the picture for a while longer.

One big change in the decade was the emergence of FM as the mode of choice for local voice communications. The genesis of this resulted from a change in bandwidth

requirements for commercial and government public service VHF users that resulted in a large amount of surplus equipment becoming available. By the 1970s, this equipment had been supplemented by equipment designed especially for amateurs, generally providing more channel choices than in the modified commercial gear. This was all accompanied by a new "FM and Repeaters" section in the *Handbooks* of the period.

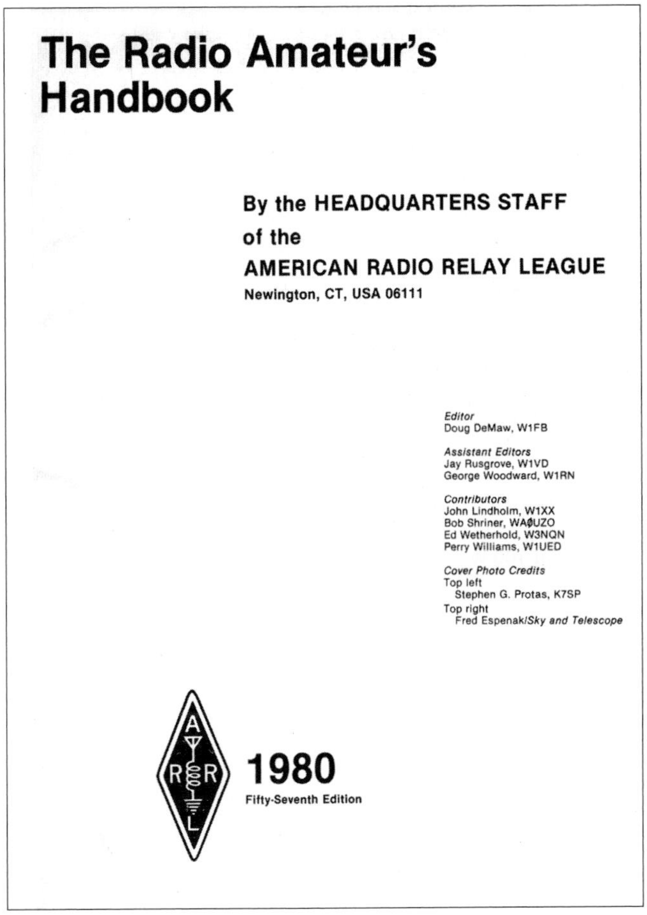

The Radio Amateur's Handbook

By the HEADQUARTERS STAFF
of the
AMERICAN RADIO RELAY LEAGUE
Newington, CT, USA 06111

Editor
Doug DeMaw, W1FB

Assistant Editors
Jay Rusgrove, W1VD
George Woodward, W1RN

Contributors
John Lindholm, W1XX
Bob Shriner, WA0UZO
Ed Wetherhold, W3NQN
Perry Williams, W1UED

Cover Photo Credits
Top left
Stephen G. Protas, K7SP
Top right
Fred Espenak/Sky and Telescope

1980
Fifty-Seventh Edition

Figure 1980-1 — The front matter in the 1980 *Handbook* had a more modern look than in the past.

Fig. 58 — Photograph of the miniature 5-band superheterodyne receiver. This layout arrangement was developed by WA0UZO of Circuit Board Specialists.

part of the tuned circuit at 50Ω and is a center-tapped broadband transformer.

Mixer and Filter

Two 40673 or 3N211 MOSFETs are used as a singly balanced active mixer (Q1 and Q2). Local-oscillator injection is applied to the paralleled gates no. 2 of Q1 and Q2. Output from the mixer is by means of T2, a broadband trifilar-wound pot-core transformer. A 10-kΩ resistor is bridged across the mixer drains to limit the signal swing — an aid to mixer IMD. FL2 filters the LO voltage to keep harmonic energy to the mixer at a low level. The 68-pF capacitor between FL2 and the mixer gates can be changed in value to obtain the 3-volt pk-pk LO voltage which is specified.

An MPF102 or 2N4416 JFET is used as a post-mixer amplifier (Q3). The drain load resistor is chosen to match the filter impedance. A 2-kΩ value is specified for use with the Collins F455FD-04 filter at FL1. If a different filter is used, such as an ssb type, the drain load resistor may have to be changed to match the impedance of the filter. Also, the end resonating capacitors at FL1 may need to be a different value (see manufacturer's data sheets).

I-F Amplifier

A two-stage CA3028A i-f amplifier is used following the i-f filter, FL1. Series regulation is applied to the operating voltage of this circuit in order to provide 9.1 volts. Automated gain control is developed by the agc strip and supplied to pin 4 of the CA3028As through Q7. Maximum gain occurs when the agc voltage is

+ 9. Minimum gain is at the + 2-volt agc level.

Product Detector and Audio Channel

A singly balanced product detector is fed from the low-Z secondary winding of T4. BFO injection is supplied by Q5, which is shown with a 454.3-kHz crystal for use with the Collins cw filter specified for FL1. If a Collins ssb i-f filter is used at FL1, the BFO crystal will have to be of a frequency that falls 20 dB down on the filter curve, upper or lower sideband. The manufacturer specifies which frequencies are correct for the filter being used. For upper and lower sideband operation it will be necessary to use two BFO crystals. Thus, a selector switch will be needed between the gate of Q5 and the two crystals.

RC filtering is used at the output of the product detector to keep the BFO and i-f leakage from reaching the audio preamplifier, Q4. Any JFET will suffice at Q4, such as a 2N4416.

Q17 functions as an audio muting switch during transmit. It is actuated via J2 by grounding the Q17 base line through an external set of relay contacts or bipolar-transistor dc switch.

U4 amplifies the audio to speaker level. LC filtering is used at the output of U4 to suppress unwanted hf oscillations which could interfere with overall receiver performance (spurious responses). The MC1306P IC is designed for low-voltage operation. Therefore, the 3-terminal regulator, U5, has been included in the circuit.

AGC System

Audio-derived agc is used in this

receiver. Output is sampled from the product detector, then amplified by means of U3. D3 and D4 rectify the amplified audio and drive a cascaded dc amplifier consisting of Q6 and Q7. The agc time constant is set by a 1-μF capacitor in parallel with a 1-MΩ resistor. The time constant can be varied to suit the operator by changing the value of the charging capacitor in that network. Values of less than 1 μF will shorten the discharge time and vice versa. D5 serves as a gating diode to prevent loss of the agc drive voltage through the i-f gain control, R3.

The agc strip drives the S meter, M1, through Q7. Agc voltage for U1 and U2 is obtained at the emitter of Q7. R2 should be set so that maximum receiver input signal provides + 9 volts at pin 4 or U1 and U2.

Local Oscillator

Q13 operates as a highly stable series-tuned Colpitts oscillator. Polystyrene fixed-value capacitors are used to ensure stability and offer drift compensation which corrects for the positive drift of the core material in L12. Silver-mica capacitors can be substituted with a possible degradation in long-term stability. C2 is the main-tuning capacitor. C4 is optional. If it is used it should be panel-mounted to permit dial calibration from band to band. This will be necessary because the converter crystals may not be precisely on the specified frequencies.

Q14 serves as a source-follower buffer stage. The signal level is built up by means of Q15, a Class A broadband amplifier. Sufficient LO output is available at 50Ω to swing the no. 2 gates of the balanced mixer to 3 volts pk-pk. The added power developed by Q15 is necessary to provide the required injection voltage across the 50-Ω load presented by harmonic filter FL2. A 40673 can be used in place of the 3N211 specified at Q13. Similarly, a 2N4416 will be satisfactory at Q14.

Front-End Converters

The same circuit-board pattern is used for the 40, 20 and 15 meter converters. In order to obtain a 200-kHz bandwidth on 80 meters, FL3 is used. Also, no rf amplifier stage is necessary on 80 meters. Therefore, the pc-board pattern is different from that for the other hf bands.

The converters of Fig. 60 are designed for a broadband i-f output of 1.8 to 2.0 MHz. They are selected for the desired operating band by means of S1. When this switch is placed in the 160-meter position the converters are bypassed to permit routing the antenna directly to the mixer of the main frame for reception on 160 meters.

Q8 performs as a low-gain, common-gate rf amplifier. The source tuned circuit is peaked for the center of the band segment of interest. It is broad enough in response to require no additional tuning.

Receiving Systems 8-36

Figure 1980-2 — This 1980 *Handbook* miniature receiver project is entirely solid state, but also incorporates features designed to limit IMD and improve dynamic range — issues with some early solid state receivers.

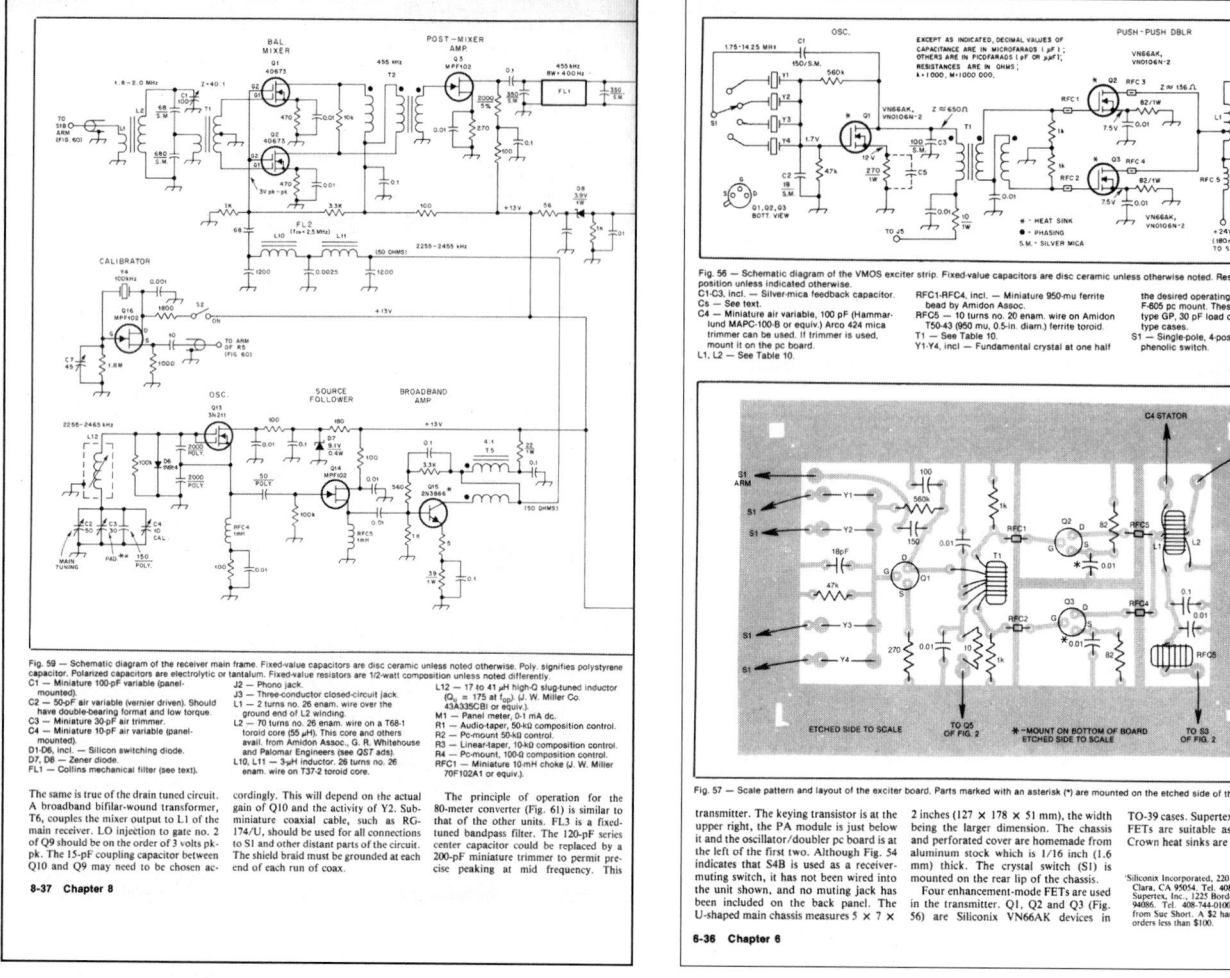

Fig. 59 — Schematic diagram of the receiver main frame. Fixed-value capacitors are disc ceramic unless noted otherwise. Poly. signifies polystyrene capacitor. Polarized capacitors are electrolytic or tantalum. Fixed-value resistors are 1/2-watt composition unless noted differently.

C1 — Miniature 100-pF variable (panel-mounted).
C2 — 50-pF air variable (vernier driven). Should have double-bearing format and low torque.
C3 — Miniature 30-pF air trimmer.
C4 — Miniature 10-pF air variable (panel-mounted).
D1-D6, incl. — Silicon switching diode.
D7, D8 — Zener diode.
FL1 — Collins mechanical filter (see text).

J2 — Phono jack.
J3 — Three-conductor closed-circuit jack.
L1 — 2 turns no. 26 enam. wire over the ground end of L2 winding.
L2 — 70 turns no. 26 enam. wire on a T68-1 toroid core (55 μH). This core and others avail. from Amidon Assoc., G. R. Whitehouse and Palomar Engineers (see QST ads).
L10, L11 — 3-μH inductor. 28 turns no. 26 enam. wire on T37-2 toroid core.

L12 — 17 to 41 μH high-Q slug-tuned inductor (Q_u = 175 at f_op). (J. W. Miller Co. 43A335CBI or equiv.).
M1 — Panel meter, 0-1 mA dc.
R1 — Audio-taper, 50-kΩ composition control.
R2 — Pc-mount 50-kΩ control.
R3 — Linear-taper, 10-kΩ composition control.
R4 — Pc-mount, 100-Ω composition control.
RFC1 — Miniature 10-mH choke (J. W. Miller 70F102A1 or equiv.).

8-37 Chapter 8

The same is true of the drain tuned circuit. A broadband bifilar-wound transformer, T6, couples the mixer output to L1 of the main receiver. LO injection to gate no. 2 of Q9 should be on the order of 3 volts pk-pk. The 15-pF coupling capacitor between Q10 and Q9 may need to be chosen ac-

cordingly. This will depend on the actual gain of Q10 and the activity of Y2. Subminiature coaxial cable, such as RG-174/U, should be used for all connections to S1 and other distant parts of the circuit. The shield braid must be grounded at each end of each run of coax.

The principle of operation for the 80-meter converter (Fig. 61) is similar to that of the other units. FL3 is a fixed-tuned bandpass filter. The 120-pF series center capacitor could be replaced by a 200-pF miniature trimmer to permit precise peaking at mid frequency. This

Fig. 56 — Schematic diagram of the VMOS exciter strip. Fixed-value capacitors are disc ceramic unless otherwise noted. Resistors are 1/2-W composition unless indicated otherwise.

C1-C3, incl. — Silver-mica feedback capacitor.
Cs — See text.
C4 — Miniature air variable, 100 pF (Hammarlund MAPC-100-B or equiv.) Arco 424 mica trimmer can be used. If trimmer is used, mount it on the pc board.
L1, L2 — See Table 10.

RFC1-RFC4, incl. — Miniature 950-mu ferrite bead by Amidon Assoc.
RFC5 — 10 turns no. 20 enam. wire on Amidon T50-43 (950 mu, 0.5-in. diam.) ferrite toroid.
T1 — See Table 10.
Y1-Y4, incl. — Fundamental crystal at one half

the desired operating frequency. Sockets ar F-605 pc mount. These and the crystals are type GP, 30 pF load capacitance in HG40 type cases.
S1 — Single-pole, 4-position, single-wafer phenolic switch.

Fig. 57 — Scale pattern and layout of the exciter board. Parts marked with an asterisk (*) are mounted on the etched side of the board.

transmitter. The keying transistor is at the upper right, the PA module is just below it and the oscillator/doubler pc board is at the left of the first two. Although Fig. 54 indicates that S4B is used as a receiver-muting switch, it has not been wired into the unit shown, and no muting jack has been included on the back panel. The U-shaped main chassis measures 5 × 7 ×

2 inches (127 × 178 × 51 mm), the width being the larger dimension. The chassis and perforated cover are homemade from aluminum stock which is 1/16 inch (1.6 mm) thick. The crystal switch (S1) is mounted on the rear lip of the chassis.

Four enhancement-mode FETs are used in the transmitter. Q1, Q2 and Q3 (Fig. 56) are Siliconix VN66AK devices in

TO-39 cases. Supertex VN0106N-2 VMOS FETs are suitable as direct substitutes. Crown heat sinks are required on all three.

Siliconix Incorporated, 2201 Laurelwood Rd., Santa Clara, CA 95054. Tel. 408-988-8000.
Supertex, Inc., 1225 Bordeaux Dr., Sunnyvale, CA 94086. Tel. 408-744-0100. Order VMOS devices from Sue Short. A $2 handling fee is required for orders less than $100.

6-36 Chapter 6

Figure 1980-3 — The schematic of part of the receiver of Figure 1980-2 shows the modern mixer and IF post mixer amplifier.

Figure 1980-4 — This MOSFET transmitter puts out 0.5 W. While low power transmitters have moved to solid state, the typical 100 W HF transceiver still uses tubes in its final stages.

We Embrace Computers and Space Communications — The 1990 Edition

During the 1980s, the *Handbook* quietly assumed a subtly different name. Instead of the longstanding title *The Radio Amateur's Handbook*, it was now *The ARRL Handbook for the Radio Amateur*. It also increased dramatically in page count and number of chapters and took on a different cover appearance, as shown in Figures 1990-1 and 1990-2.

The 1980s saw the beginning of the popularity of personal computers throughout the developed world. Amateur Radio was quick to embrace the new technology — first in off-line applications such as log keeping and contest scoring — then in radio system control and finally in signal processing in various forms.

Projects now added large numbers of integrated circuits to the mix, as well as digital logic and their families of integrated circuits.

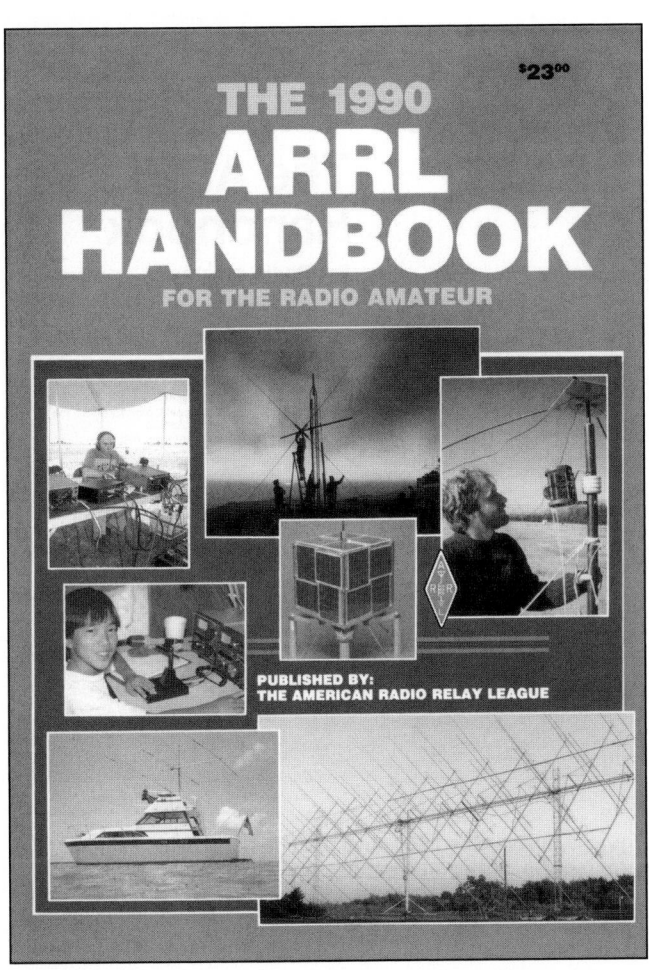

Figure 1990-1 — The cover of the 1990 *Handbook* has a completely different look than in the past. In addition, during the past decade, the size changed to approximately 8½ × 11 and the title has been slightly revised.

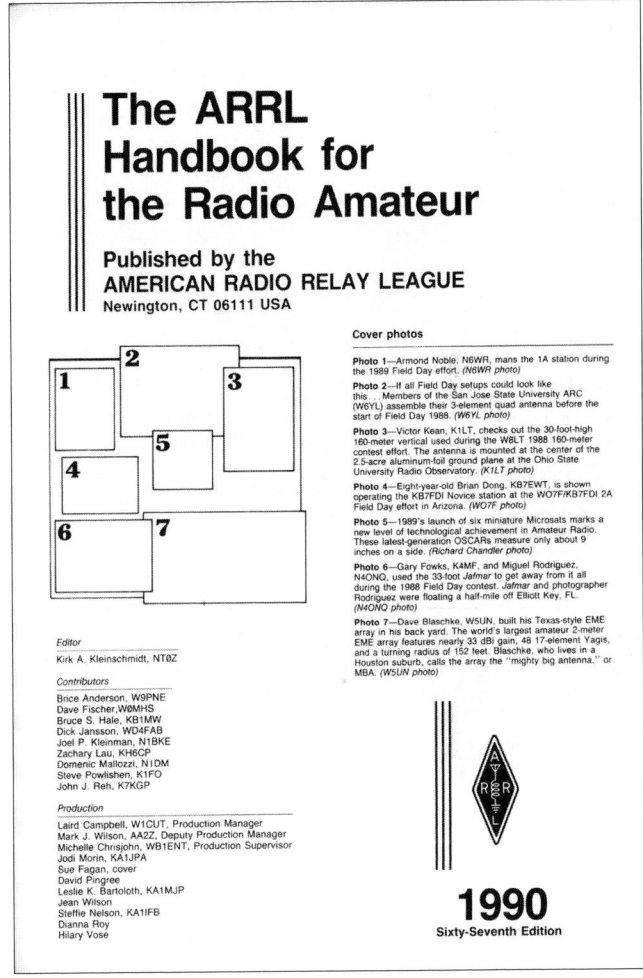

Figure 1990-2 — The 1990 cover has so much going on that it takes a navigation screen inside to keep track of it all.

The End of a Century — The 2000 Edition

During the 1990s, the *Handbook* continued the successful cover appearance, as shown in Figures 2000-1 and 2000-2.

The 1990s saw the expansion of the applications of integrated circuits. PC boards were being designed and produced using computer software and Amateur Radio was quick to embrace and integrate the technology of the ubiquitous Internet. The *Handbook* was becoming as much a reference book for electronic and radio engineering theory, as well as a source for projects, usually with much more design information than in previous generations.

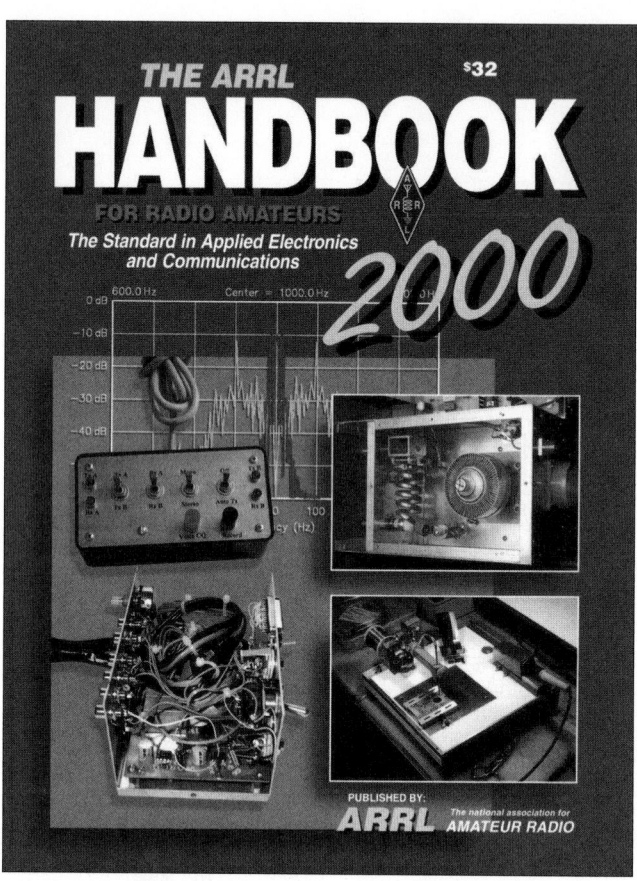

Figure 2000-1 — The cover of the 2000 *Handbook* builds on the look over the past decade.

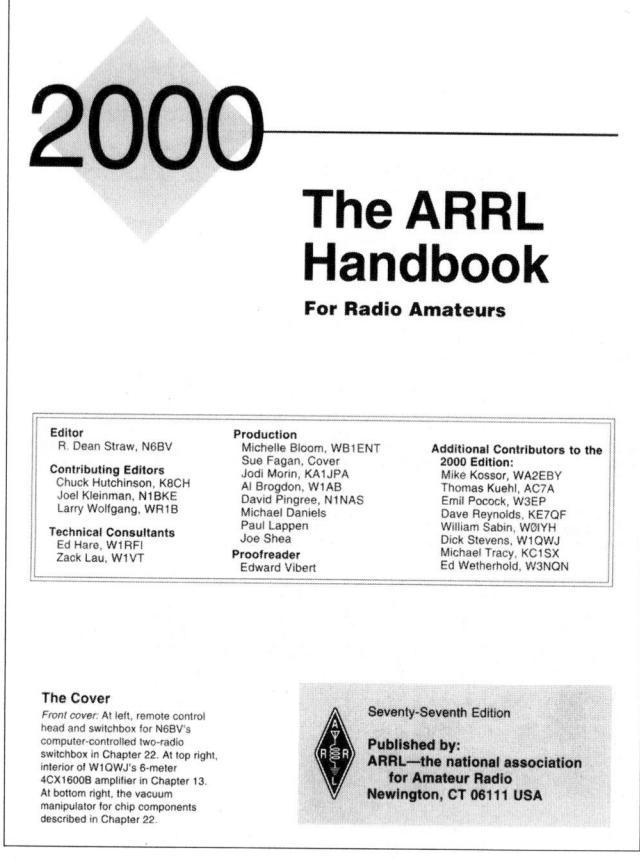

Figure 2000-2 — The 2000 cover can be described in a short paragraph — perhaps an improvement over 1990.

The Beginning of the Current Century — The 2010 Edition

During the 2000s, the *Handbook* continued the successful cover appearance, as shown in Figures 2010-1 and 2010-2. Note the subtle shift to a new name, now it is *The ARRL Handbook for Radio Communications*. This is perhaps in recognition of the fact we've known for a long time — the *Handbook* is not just for hams. Visit the office of a radio or telecommunications engineer and it's an even bet that a copy of the *Handbook* will be found on a bookshelf there. This is as true today as it was when I started my career as a radar engineer in 1969!

The 1990s saw the expansion of the applications of integrated circuits. PC boards were being designed and produced using computer software and Amateur Radio was quick to embrace and integrate the technology of the ubiquitous Internet.

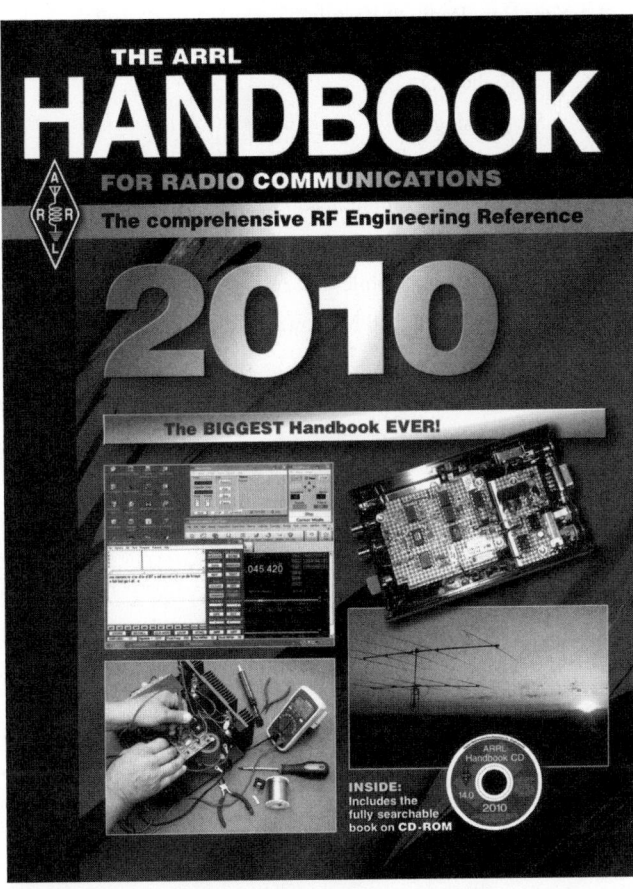

Figure 2010-1 — The cover of the 2010 *Handbook* builds on the look over the past decade. Note the change to a more general title — the *Handbook* isn't just for hams anymore — of course it never was.

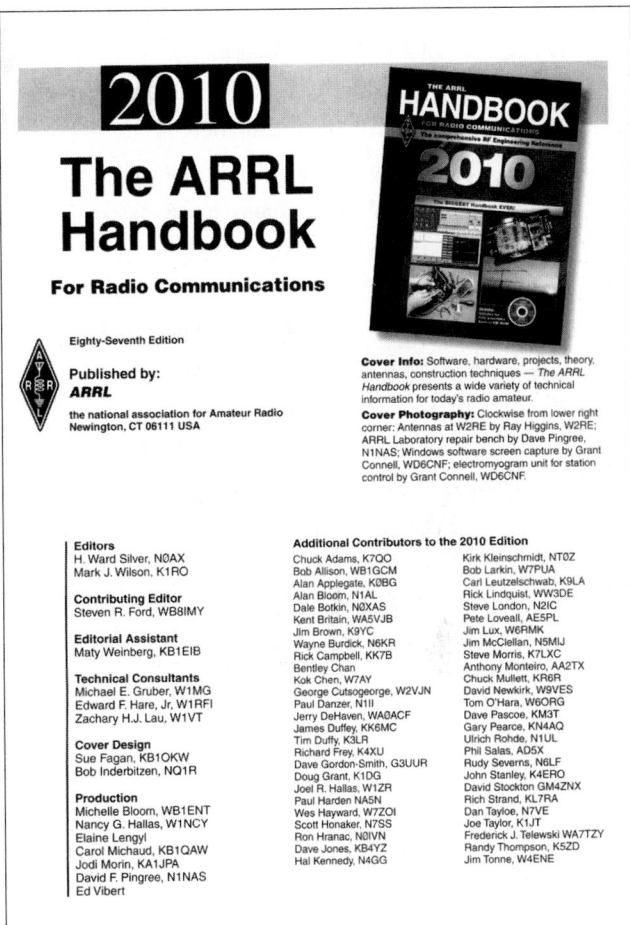

Figure 2010-2 — Look at the list of contributors to the 2010 edition — quite a difference from the single name on the inside of the 1926 edition! This major new edition featured lot of material by a lot of contributors — W1ZR is even on the list!

Contents

Chapter 1

What is Amateur (Ham) Radio?

For more than a century, a growing group of federally licensed radio hobbyists known as Amateur Radio — or "ham radio" — operators has had a front-row seat as radio and electronics have broadened our horizons and touched virtually all of our lives. Hams *pioneered* personal communication, even in the days before the telephone and household electricity were commonplace and the Internet not yet conceived. The word "radio" — or "wireless" — still evokes awe.

Today we enjoy wireless amenities that range from the ubiquitous mobile phone to sophisticated smartphones, tablet devices, and diminutive netbook PCs that go just about anywhere. Personal communication is the goal. The original "personal wireless" communication, Amateur Radio remains vital and active today. In this chapter, Rick Lindquist, WW1ME provides an overview of Amateur Radio activities and licensing requirements.

1.1 Do-It-Yourself Wireless

Amateur (or "ham") Radio operators have at their fingertips the ability to directly contact fascinating people they may never meet who live in distant places they'll never visit. They do this without any external infrastructure, such as a cell phone network or the Internet, sometimes using simple, inexpensive — often homemade — equipment and antennas. Since the earliest years of wireless communication, these radio experimenters, largely self-taught, developed and refined the means to contact one another without wires connecting them.

As a radio amateur, you can meet new friends, win awards, exchange "QSLs" (the ham's business card), challenge yourself and others in on-the-air competitions, educate yourself about radio technology, contribute to your community, travel, promote international goodwill, and continue the century-old wireless communication tradition. Your station is yours and yours alone, and it's independent of any other communication network.

Let's take a closer look.

AN AMAZING CENTURY OF HAM RADIO

Today we think of "wireless" as a relatively modern term that applies to a wide variety of electronic devices, but it's actually been around for more than a century. Wireless communication was a goal of early experimenters in the late 19th century and early 20th century. Equipment and methods for early wireless often were crude and rudimentary — a simple crystal radio (primitive "solid state" technology) to listen, and a spark gap transmitter to send Morse code, coupled with what was then called an "aerial." Little to no ready-made equipment was available, and parts for these early radio do-it-yourselfers were expensive and hard to obtain. On a good night, their transmissions might even span 50 miles! In the early 20th century, when not everyone had a telephone and calling long-distance was pricey, ham radio was, in more contemporary terms, "really cool technology."

In 1914, just two years after these early hams were required to hold licenses from the federal government, inventor and industrialist Hiram Percy Maxim, 1AW, and radio enthusiast Clarence Tuska, 1WD, established the American Radio Relay League (ARRL) to bring these US radio hobbyists under one tent to serve their common interests. These two founding fathers of ham radio and their peers would be awestruck to see how the world of Amateur Radio and wireless technology has expanded and evolved in the intervening century.

While Maxim and Tuska were not the first hams, the organization they founded, the ARRL — the national association for Amateur Radio — has championed and sustained these radio pioneers and their successors. Now 100 years down the road — light years in terms of radio science and technology — Amateur Radio continues to adapt to the times. While many traditions continue, today's ham radio is not the ham radio of yesteryear.

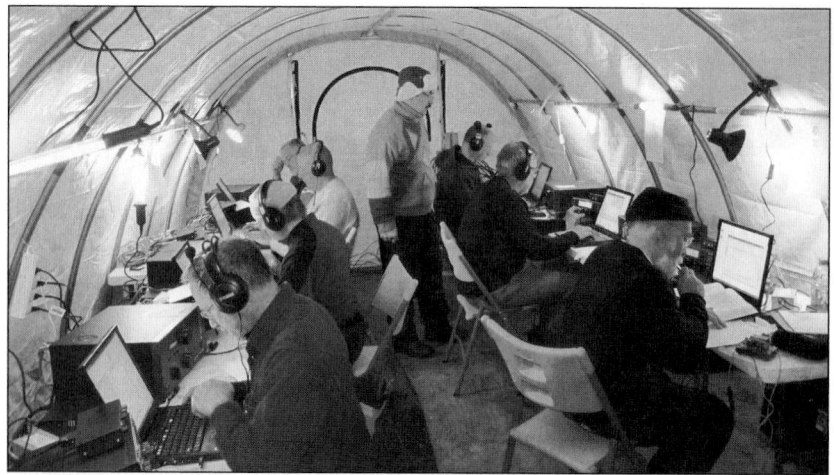

Fig 1.1 — Hams travel around the world, putting rare and unusual locations "on the air" to make thousands of contacts. The VP8ORK team traveled to Antarctica's South Orkney Islands by sea, operating from tents as shown here. Team members such as Nodir Tursoon-Zadeh, EY8MM came from as far away as Tajikistan to participate. [Lew Sayre, W7EW, photo]

Fig 1.4 — Kristina Whitley displays her excitement after making a contact at W4FOS, the Chesapeake Center for Science and Technology High School Amateur Radio Club station in Chesapeake, Virginia, during the 2011-2012 School Club Roundup. [Photo courtesy Richard Siff, W4BUE]

Fig 1.2 — Observers from the ARRL and other International Amateur Radio Union member-societies around the world have a role in preparing Amateur Radio positions for the World Radiocommunication Conference in Geneva, such as this one in 2012. At these gatherings, held every 3 or 4 years, countries agree on international radio regulations, including amateur allocations. [Carter Craigie, N3AO, photo]

Fig 1.3 — Rock musician Joe Walsh, WB6ACU (The Eagles) enjoys Amateur Radio history and vintage gear. During a visit to W1AW, he got to see ARRL co-founder Hiram Percy Maxim's "Old Betsy" rotary spark-gap transmitter (left) and a rack-mounted transmitter and a receiver (top) from an earlier incarnation of W1AW. [Joel Kleinman, N1BKE (SK), photo]

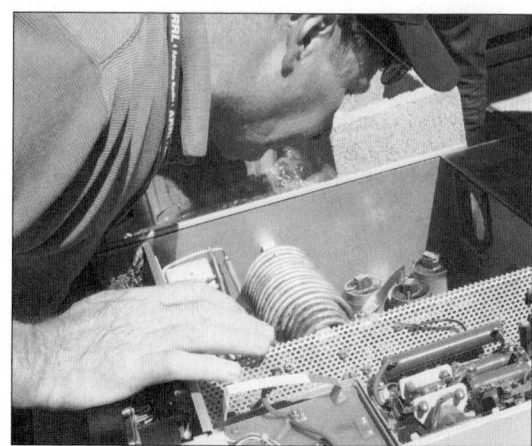

Fig 1.5 — Thrifty hams like Jon Zaimes, AA1K seen here in the flea market at Dayton Hamvention®, are always looking for a bargain and perhaps a "fixer-upper." [Rick Lindquist, WW1ME, photo]

THE ORIGINAL PERSONAL COMMUNICATION

In this age of multiple sophisticated communication platforms, it's not uncommon for people to ask, "Ham radio? Do they still *do* that?" Yes, "they" do. But, given the proliferation of communication alternatives, the larger question may be, *Why*?

Ham radio is a do-it-yourself (DIY) technological and social medium — *personal* communication with no bills, minutes or data plans. It's personal communication that's "off the grid," a wireless service you can rely on when other services aren't available.

It doesn't cost a lot to get into Amateur Radio and participation is open and accessible to everyone. Hams are mothers, fathers, and children of all ages, ethnic backgrounds, physical abilities, and walks of life who belong to a unique worldwide community of licensed radio hobbyists. Some are even well-known celebrities. All find joy and excitement by experiencing radio communication and electronics on a very personal level across a spectrum of activities.

The Federal Communications Commission (FCC) grants licenses in the Amateur Radio Service. With an emphasis on "service," the FCC has laid down five essential principles underlying Amateur Radio's fundamental purpose (see sidebar "Ham Radio's Rules of the Airwaves"). These recognize ham radio's value to the public as a "voluntary noncommercial communication service, particularly with respect to providing emergency communications." The service also exists to continue and expand Amateur Radio's "proven ability" to advance the state of the radio art, as well as both technical and communication skills. Further, the FCC says, the Amateur Radio Service should help to expand the "existing reservoir of trained operators, technicians, and electronics experts," and continue and extend the radio amateur's "unique ability to enhance international goodwill."

HAMS ARE EVERYWHERE

Spotting a radio amateur can be easy. The driver of that car sporting an "odd-looking" antenna may be a ham equipped for mobile operation. Your neighbor on the next block with the wires strung between trees or, perhaps, a tower supporting what looks like a very large television antenna probably is one too.

Modern technology continues to make ham radio more accessible to all, including those living on tight budgets or facing physical challenges. People lacking mobility may find the world of Amateur Radio a rewarding place to find lasting friendships — on the next block, in the next state, or around the globe.

Hams are ambassadors. For many radio

Ham Radio's Rules of the Airwaves

International and national radio regulations govern the operational and technical standards of all radio stations. The International Telecommunication Union (ITU) governs telecommunication on the international level and broadly defines radio services through the international *Radio Regulations*. In the US, the Federal Communication Commission (FCC) is the federal agency that administers and oversees the operation of nongovernmental and nonmilitary stations — including Amateur Radio. Title 47 of the *US Code of Federal Regulations* governs telecommunication. The Amateur Radio Service is governed by Part 97.

Experimentation has always been the backbone of Amateur Radio, and the Amateur Service rules provide a framework within which hams enjoy wide latitude to experiment in accordance with the "basis and purpose" of the service. The rules should be viewed as vehicles to promote healthy activity and growth, not as constraints that lead to stagnation. The FCC's rules governing Amateur Radio recognize five aspects, paraphrased below, in the Basis and Purpose of the Amateur Service.

- Amateur Radio's value to the public, particularly with respect to providing emergency communication support
- Amateur Radio's proven ability to contribute to the advancement of the radio art
- Encouraging and improving the Amateur Service through rules that help advance communication and technical skills
- Maintaining and expanding the Amateur Service as a source of trained operators, technicians and electronics experts
- Continuing and extending the radio amateur's unique ability to enhance international goodwill

The Amateur Radio Service rules, Part 97, are in six sections: General Provisions, Station Operation Standards, Special Operations, Technical Standards, Providing Emergency Communication and Qualifying Examination Systems. Part 97 is available in its entirety on the ARRL and FCC websites (see the Resources section at the end of this chapter for further information).

Hams on the Front Lines

Over the years, the military and the electronics industry have often drawn on the ingenuity of radio amateurs to improve designs or solve problems. Hams provided the keystone for the development of modern military communication equipment, for example. In the 1950s, the Air Force needed to convert its long-range communication from Morse code to voice, and jet bombers had no room for skilled radio operators. At the time, hams already were experimenting with and discovering the advantages of single sideband (SSB) voice equipment. With SSB, hams were greatly extending the distances they could transmit.

Air Force Generals Curtis LeMay and Francis "Butch" Griswold, both radio amateurs, hatched an experiment that used ham radio equipment at the Strategic Air Command headquarters. Using an SSB station in an aircraft flying around the world, LeMay and Griswold were able to stay in touch with Offutt Air Force Base in Nebraska from around the globe. The easy modification of this ham radio equipment to meet military requirements saved the government millions of dollars in research costs.

More recent technological experimentation has focused on such techniques as software defined radio (SDR). This amazing approach enables electronic circuit designers to employ software to replace more costly — and bulkier — hardware components. It's no coincidence or surprise that radio amateurs have been among those investigators doing the ground-level research and experimentation to bring this technology from the laboratory to the marketplace. Transceivers built on the SDR model now are making inroads within the Amateur Radio community and represent the likely wave of the future in equipment design.

Affirming the relationship between Amateur Radio and cutting-edge technology, Howard Schmidt, W7HAS, was White House Cybersecurity Coordinator from 2009 to 2012. An ARRL member, Schmidt is one of the world's leading authorities on computer security, with some 40 years of experience in government, business and law enforcement. Schmidt credits ham radio with helping to launch his career. "Building … computers to support my ham radio hobby gave me the technical skills that I needed to … start doing computer crime investigations and work on the early stages of computer forensics, in turn enabling me to start working on cybersecurity issues," he says. Hams are often found in industry and the military as technology presses ahead.

amateurs, a relaxing evening at home is having a two-way radio conversation with a friend in Frankfort, Kentucky or Frankfurt, Germany. Unlike any other hobby, Amateur Radio recognizes no international or political boundaries, and it brings the world together in friendship.

1.1.1 Making it Happen

A major feature of Amateur Radio's 100-year heritage has been the ham's ability to make do with what's at hand to get on the air. It is in the pursuit of such hands-on, do-it-yourself activities that this *Handbook* often comes into play, especially as electronic components are more plentiful today and circuit designs increasingly complex and creative.

Amateur Radio has always been about what its participants bring *to* it and what they make *of* it. Even today many enthusiasts enjoy making their own radio communication gear. Hams contact each other using equipment they've bought or built, or a combination of the two, over a wide range of the radio spectrum. The methods hams use to keep in touch range from the venerable Morse code — no longer a licensing requirement, by the way — to voice, modern digital (ie, computer-coded) modes, and even television.

The hybridization of Amateur Radio and computers and the Internet continues to blossom, as hams invent ever more creative ways to exploit this technology and make it an essential station component. Today it's possible for a ham to control a transmitting and receiving station via the Internet using nothing more than a laptop or smartphone — even if that station is thousands of miles distant. The wonder of software defined radio (SDR) techniques has even made it possible to create *virtual* radio communication gear. SDRs require a minimum of physical components; sophisticated computer software does the heavy lifting!

1.1.2 Your Ham Radio Comfort Zone

Amateur Radio offers such a wide range of activities that everyone can find a comfortable niche. As one of the few truly *international* hobbies, ham radio makes it possible to communicate with other similarly licensed

aficionados all over the world. On-the-air competition called contesting or "radiosport" — just to pick one activity many hams enjoy — helps participants to improve their skills and stations. Further, ham radio offers opportunities to serve the public by supporting communication in disasters and emergencies, and it remains a platform for sometimes cutting-edge scientific experimentation. Many of those who got into ham radio at a young age credit that involvement with later success in technological careers.

Ham radio's horizon extends into space. The International Space Station boasts a ham radio station, and most ISS crew members are Amateur Radio licensees. Thanks to the program Amateur Radio on the International Space Station (ARISS), suitably equipped hams can talk directly with NASA astronauts in space. Hams also contact each other through Earth-orbiting satellites designed and built by other radio amateurs, and they even bounce radio signals off the Moon and back to other hams on Earth.

Hams talk with one another from vehicles, while hiking or biking in the mountains, from remote camp sites, or while boating. Through a plethora of activities, hams learn a great deal, establish lifelong friendships and,

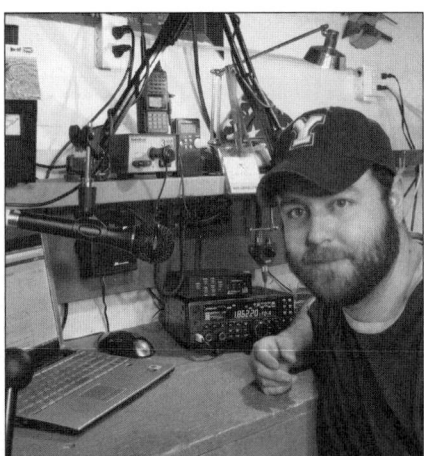

Fig 1.6 — ARRL Member Jordan Johns, KF7LUA of Idaho was the top scorer in the 7th call area in the 2011 ARRL April Rookie Roundup, a phone contest for newcomers to Amateur Radio. [Photo courtesy Jordan Johns, KF7LUA]

perhaps most important, have a *lot* of fun. Along the way, radio amateurs often contribute a genius that propels technological innovation.

Most likely you're already a ham or at least have experimented with radio and electronics yourself and are thinking about getting your ham license. This *Handbook* is an invaluable resource that reveals and explains the "mysteries" governing electronics in general and in radio — or wireless — communication in particular, especially as they pertain to Amateur Radio.

1.1.3 What's in it for Me?

As a community *of* communities, Amateur Radio can be whatever *you* want it to be. Whether you are looking for relaxation, excitement, enjoyment or a way to stretch your mental (and physical) horizons, Amateur Radio can provide it — even for those with time and money constraints. However it happens, communication between individuals is at the core of nearly all ham radio activities. In its most basic form, ham radio is two people saying "Hello!" to each other over the air, perhaps using inexpensive handheld transceivers or even homemade gear. In "Hamspeak," a two-way, on-the-air communication is known as a "QSO" — an old radiotelegraph, or Morse code, abbreviation often pronounced "CUE-so."

Ham radio can also be a group activity. Hams with common interests often gather on the airwaves to share their thoughts and even pictures. These get-togethers are called "nets" or "roundtables," depending on their formality. When hams meet on the air for an extended on-the-air conversation, they sometimes call it "ragchewing."

Nets form when like-minded hams gather on the air on a regular schedule. Nets often provide an on-the-air venue to find other hams with similar interests both inside and outside of Amateur Radio. Topics may be as diverse as vintage radio, chess, gardening, rock climbing, railroads, computer programming, teaching or an interest in certain types of radio equipment. Faith-based groups and scattered friends and families may also organize nets. You can find your special interest in *The ARRL Net Directory* on the ARRL website (**www.arrl.org/arrl-net-directory**).

1.2 Joining the Ham Radio Community

Morse code has been a major player in Amateur Radio's legacy, although it was a roadblock to some. Today it's no longer necessary to learn the code to become an Amateur Radio licensee. You still must hold a license granted by the Federal Communications Commission (FCC) to operate an Amateur Radio station in the United States, in any of its territories and possessions or from any vessel or aircraft registered in the US. There are no age or citizenship requirements to obtain a US Amateur Radio license, and the cost is minimal, sometimes free. Ham radio exams are regularly passed by children not yet in their teens!

The FCC offers three classes — or levels — of Amateur Radio license. From the easiest to the most difficult, they are Technician, General and Amateur Extra Class. Applicants must take and pass a multiple-choice written examination for each license. Official question pools for all ham radio license classes are publically available. The higher you climb the ladder, the more challenging the test and the more generous the operating privileges. To reach the top — Amateur Extra — you must pass the examinations for all three license classes.

1.2.1 Moving Through the Ranks

Most people start out in Amateur Radio by getting a Technician Class license or "ticket," as a ham license is sometimes called. Obtaining a Technician license requires passing a 35 question multiple-choice exam. The test covers FCC rules and regulations governing the airwaves, courteous operating procedures and techniques and some basic electronics. The privileges earned give Technicians plenty of room to explore and activities to try. For some, the Technician is the only ham license they'll ever want or need.

Technicians enjoy a wide, but somewhat limited, range of voice and digital radio operating privileges. These include access to some "high frequency" (HF or short-wave) frequency "bands" or segments of the radio spectrum. Depending upon license class, hams have access to up to 10 distinct HF bands in the range from 1.8 MHz to 29.7 MHz, where most direct international communication happens. (Frequency and wavelength terms are explained in the **Electrical Fundamentals** chapter.) Technicians also have all amateur privileges in the VHF-UHF and microwave spectrum, though, which allow operation on widely available FM voice repeaters. A repeater greatly extends the communication range of low-power, handheld radios or mobile stations too far apart to communicate with each other directly. The "Tech ticket" is a great introduction to the fun and excitement of ham radio and to the ways of the hobby.

By upgrading to General Class, a Technician licensee can earn additional operating privileges, such as access to all of the Amateur Radio HF bands. Upgrading to General entails passing another 35 question multiple-choice exam. In addition to Technician privileges, Generals enjoy worldwide communication using voice, digital, image and television techniques.

Reaching the top rung of the Amateur Radio ladder — Amateur Extra Class — means passing a more demanding 50 question examination. Amateur Extra licensees enjoy privileges on all frequency bands and communication modes available to hams. The exam may be challenging, but many hams consider it well worth the effort!

1.2.2 Study Aids

You can prepare for the exam on your own, with a group of friends or by taking a class sponsored by a ham radio club in your area. The ARRL offers materials and lesson plans for hams wishing to teach Amateur Radio licensing classes. Anyone can set up license classes. Many Amateur Radio clubs hold periodic classes, usually for the Technician license. The ARRL supports Registered Amateur Radio Instructors, but registration is not necessary to conduct a class. Check the ARRL website, **www.arrl.org**, for classes, clubs or volunteer examiners (VEs) in your area (more on VEs below).

Help is available at every step. The ARRL publishes study materials for all license classes. Visit the ARRL website or contact the ARRL's New Ham Desk for more information on how to get started. The Resources section at the end of this chapter includes an address and telephone number. The ARRL can help you find ham radio clubs in your area as well as ARRL-registered instructors and local Volunteer Examiner teams. Additional information on the ARRL website includes frequencies hams can use, popular operating activities, and how to order the latest ARRL study guide.

For newcomers seeking to obtain a Technician license, *The ARRL Ham Radio License Manual* includes the complete, up-to-date question pool, with the correct answers and clear explanations. The manual assumes no prior electronics background. It delves into the details behind the questions and answers, so you will *understand* the material, rather than simply memorize the correct answers. It includes a CD-ROM with software to help you practice for the exam.

If you already have some electronics background or just want brief explanations of the material, you might find *ARRL's Tech Q&A* manual a more appropriate choice. It also includes the entire Technician question pool to help you prepare.

When you are ready to upgrade to a General Class license, *The ARRL General Class License Manual* or *ARRL's General Q&A* can help you prepare. In like fashion, *The ARRL Extra Class License Manual* and *ARRL's Extra Q&A* will guide your study efforts for the Amateur Extra Class license. Check the ARRL website for detailed information on these and other license study options.

1.2.3 Taking the Test

While the FCC grants US Amateur Radio licenses, volunteer examiners (VEs) now administer all Amateur Radio testing. Other countries have adopted similar systems. Ham radio clubs schedule regular exam sessions, so you shouldn't have to wait long or travel far once you're ready. Exam sessions often are available on weekends (frequently at ham radio gatherings called "hamfests") or evenings. Most volunteer examiner teams charge a small fee to recover the cost of administering the test and handling the FCC paperwork.

The ARRL is a Volunteer Examiner Coordinator (VEC) and supports the largest VE program in the nation. More information about the VE program is available on the ARRL website.

The questions for each 35 or 50 question test come from a large "question pool" that's specific to each license class. All three question pools — Technician, General and Amateur Extra — are available to the public in study

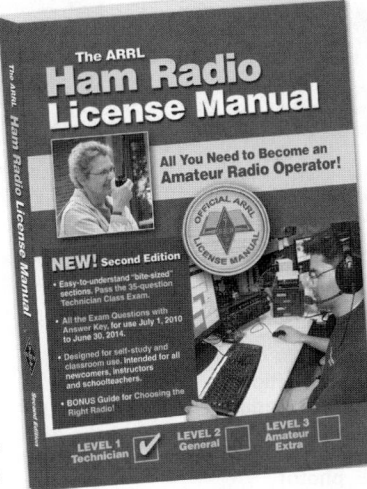

Fig 1.7 — ARRL's *Ham Radio License Manual* contains all the information you need to study for your Technician license.

guides and on the Internet. If you're studying, make sure you're working with the latest version, since question pools are updated on a set schedule. The Resources section at the end of this chapter has more information on where to find the question pools.

1.2.4 Your Ham Radio Mentor

Ham radio operators often learn the ropes from a mentor. In ham radio parlance, such an experienced ham willing to help newcomers is called an "Elmer." This individual teaches newcomers about Amateur Radio, often on a one-to-one basis. Your local ham radio club may be able to pair you up with an Elmer who will be there for you as you study, buy your first radio and set up your station — which many hams call their "radio shack" or "ham shack," a term held over from the days when ham stations often were in small buildings separate from the owner's residence. Elmers also are pleased and proud to help you with your first on-the-air contacts.

Elmers who belong to the international Courage Handi-Hams organization (**www. handiham.org**) focus on making study materials and ham radio station operation accessible to those with physical disabilities. Local Handi-Hams assist such prospective radio amateurs in getting licensed, and the Handi-Ham System may lend basic radio gear to get the new ham on the air.

1.2.5 Your Ham Radio Identity

Ham radio operators know and recognize each other by a unique call sign (some hams shorten this to simply "call") that the FCC

ARRL — the national association for Amateur Radio®

The American Radio Relay League (ARRL) is the internationally recognized society representing Amateur Radio in the US. Since its founding in 1914, the ARRL — the national association for Amateur Radio — has grown and evolved along with Amateur Radio. ARRL Headquarters and the Maxim Memorial Station W1AW are in Newington, Connecticut, near Hartford. Through its dedicated volunteers and a professional staff, the ARRL promotes the advancement of the Amateur Service in the US and around the world.

The ARRL is a nonprofit, educational and scientific organization dedicated to the promotion and protection of the many privileges that ham radio operators enjoy. Of, by and for the radio amateur, ARRL numbers some 160,000 members — the vast majority of active amateurs in North America. Licensees can become Full Members, while unlicensed persons are eligible to become Associate Members with all membership privileges except for voting in ARRL elections. Anyone with a genuine interest in Amateur Radio belongs in the ARRL.

The ARRL volunteer corps is called the Field Organization. Working at the state and local level, these individuals tackle ARRL's goals to further Amateur Radio. They organize emergency communication in times of disaster and work with agencies such as American Red Cross and Citizen Corps. Other volunteers keep state and local government officials abreast of the good that hams do at the state and local level.

When you join ARRL, you add your voice to those who are most involved with ham radio. The most prominent benefit of ARRL membership is its monthly journal *QST*, the premiere Amateur Radio magazine issued monthly in print and digital form. *QST* contains stories you'll want to read, articles on projects to build, announcements of upcoming contests and activities, reviews of new equipment, reports on the role hams play in emergencies and much more.

Being an ARRL member is far more than a subscription to *QST*. The ARRL represents your interests before the FCC and Congress, sponsors operating events throughout the year and offers membership services at a personal level. These include:
- low-cost ham equipment insurance
- the Volunteer Examiner program
- the Technical Information Service (which answers your questions about Amateur Radio technical topics)
- the QSL Service (which lets you exchange postcards with hams in other countries to confirm your contacts with them)

For answers to any questions about Amateur Radio, e-mail, call or write ARRL Headquarters. See the Resources section at the end of this chapter for contact information.

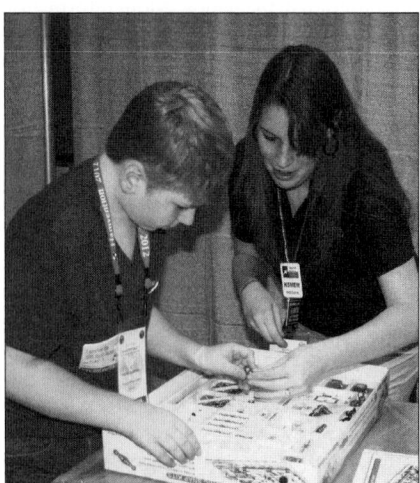

Fig 1.8 — At the Dayton Hamvention, ARRL Youth Lounge volunteer Megan, K5MEM assists prospective ham Travis Glidewell in learning electronics by doing. [Rick Lindquist, WW1ME, photo]

Fig 1.9 — This iconic brick building houses W1AW, the station operated by the ARRL in Newington, Connecticut, and known around the world. W1AW memorializes Hiram Percy Maxim, one of the founders of the ARRL. Visitors are welcome and often operate the station. [Rick Lindquist, WW1ME, photo]

issues when you get your license. Your call sign not only identifies your station on the air, it's an individual ham radio identity, and many hams become better known by their call signs than by their names! Although the FCC still issues hard-copy licenses, once your license and call sign grant appear online in the FCC's active Amateur Radio Service database, you have permission to operate.

A call sign also identifies the issuing country. US call signs, for example, begin with W, K, N or A followed by some combination of letters and one numeral. Each combination is different. One well-known ham radio call sign is W1AW, assigned to the Hiram Percy Maxim Memorial Station at ARRL Headquarters in Newington, Connecticut.

FCC-assigned call signs come in several flavors, with the shortest — and typically most desirable — combinations available only to Amateur Extra Class licensees. The FCC routinely assigns initial call signs to new Technician licensees in the longest format. These call signs start with two letters, a numeral from 0 to 9, and three more letters. The first part of a call sign including the numeral is called a *prefix*. The part following the numeral is called a *suffix* and is unique to a specific licensee. Typical prefixes in Canada are VE and VA, while the common prefix in Mexico is XE.

At one time, the numeral indicated a US station's geographical region — 1 for New England, 6 for California and 9 or Ø (zero) for the Midwest. The FCC has made ham radio call signs portable, however — just like telephone numbers. So a call sign with "1" following the prefix may belong to a ham located in Florida.

You don't have to keep the call sign the FCC assigns. For a modest fee, the FCC's *vanity call sign* program permits a ham to select a new personalized call sign from among the database of certain unassigned call signs, based on the applicant's license class.

Fig 1.10 — Most hams don't have this much "metal" in the air and rely on far more modest antennas. These impressive arrays are just a part of the W1AW "antenna farm" at ARRL Headquarters in Connecticut. [Rick Lindquist, WW1ME, photo]

1.3 Your Ham Radio Station

Amateur Radio costs as much or as little as your budget and enthusiasm dictate. Most hams set up home-based stations. By tradition the room or place where a station is located is your "ham shack," but many thrifty hams carry their stations with them in the form of relatively inexpensive handheld transceivers, some quite compact. Without requiring any antenna beyond the one attached to the radio itself, such radios can accompany you when you're out and about or traveling.

On the other end of the scale, radio amateurs serious about radio contesting or DXing (contacting distant stations in other countries) often invest in the latest equipment and extensive antenna systems. Most hams fall somewhere between these extremes. They have a modest equipment complement, simple wire antennas between two trees and maybe a small "beam" (directional antenna) on a backyard tower or mast. Whatever your investment level, you'll be able to talk around the world.

Fig 1.11 — Hams enjoy building and comparing antennas — all the way up to the microwave bands! In this photo from the 2010 Microwave Update convention, Kerry Banke, N6IZW (right) measures the performance of a 5.7 GHz dish built by Michelle Thompson, W5NYV. [Paul Wade, W1GHZ, photo]

Fig 1.12 — This very impressive and well-equipped shack of Jeff Blaine, AC0C belies the fact that his entire all-HF-band antenna system is hidden from view in the attic of his house. [Wei Qi Blaine, photo]

1.3.1 How Much Does It Cost?

Early radio amateurs generally built their own gear, mainly out of necessity; there were no well-stocked radio emporiums in the early 20th century. Constructing ham radio equipment from kits became popular in the mid-20th century, and several manufacturers still provide parts kits and circuit boards to make it even easier to build equipment yourself. A lot of hams still enjoy designing and building their own equipment (called "homebrewing") and the resulting cost savings. Many of these amateurs proudly stand at the forefront of technology and keep up with advances that may be applied within or even outside the hobby. Indeed, the projects you'll find in this *Handbook* provide a wide variety of equipment and accessories that make ham radio more convenient and enjoyable.

Today's radio amateurs most often start out using off-the-shelf commercially-made transceivers purchased new or on the used

Hamfests

Amateur Radio's broader social world extends beyond making on-the-air acquaintances. Regular ham radio gatherings, usually called "hamfests," offer opportunities to meet other hams in person — called "an eyeball contact" in ham parlance. Hams also enjoy buying, selling and trading ham radio equipment and accessories in the hamfest "flea market." *Every* ham loves a bargain. Other hamfest visitors take advantage of classroom sessions or forums to learn more about particular aspects of the hobby.

Hamfests are great places to get good deals on gear — some vendors offer substantial hamfest discounts — and to expand your knowledge. Thousands of radio amateurs from the US and around the world gather each spring at Dayton Hamvention® in Ohio. This truly international event epitomizes the goodwill that exists among the world's Amateur Radio enthusiasts.

The annual Dayton Hamvention attracts upward of 20,000 hams from around the world to meet and greet, learn, buy, sell and trade.

market, perhaps at a ham radio flea market or hamfest or through an Internet auction or classified ad site. An abundance of ham gear is readily available, and there's something out there to meet your needs within your budget. Used VHF or VHF-UHF (or "dual-band") handheld transceivers often are available for $100 or so, and the entrance of Chinese manufacturers into the ham radio market has resulted in a significant reduction in the price of some types of brand new gear.

Those interested in HF work can get in on the ground floor with a used, but serviceable, transceiver in the $200 to $500 range, and an excellent selection of new transceivers is available in the $500 to $1500 range. "Flea-power" CW (Morse code) transceivers covering single bands sell new for less than $100 in kit form; a new four-band CW transceiver manufactured in China and marketed by a US ham radio manufacturer is available for less than $300. An HF antenna such as a simple backyard dipole suspended from available trees is both inexpensive and effective. You'll find some great equipment choices advertised in ARRL's monthly journal, *QST*. In addition to advertising new ham gear, *QST* includes comprehensive equipment reviews.

More elaborate antennas and various accessories can add appreciably to the cost of your station, but less-expensive alternatives are available, including building your own.

1.3.2 Computers and Ham Radio

The marriage of ham radio and computers is solid and longstanding. Radio amateurs have discovered that interconnecting their PCs with their ham stations not only makes operating more convenient but can open the door to additional activities on the ham bands. Most radio amateurs now have a computer in the shack, often one that's dedicated to ham radio tasks. Software is available for many ham radio applications, from record keeping to antenna and circuit design.

Probably the most common use for a computer in the ham shack is logging — keeping a record of — contacts. This is especially true for contesters, where speed and accuracy are paramount. While there's no longer a legal requirement to maintain a detailed logbook of your on-the-air activities, many hams still keep one, even for casual operating, and computer logging can make the task less tedious (see sidebar "Keeping a Log").

Many computer logging applications also let you control many or most of your radio's functions, such as frequency or band selection, without having to leave your logging program. It's also possible to control various accessories, such as antenna rotators or selection switches, by computer.

Via your computer's soundcard, you can enjoy *digital modes* with nothing more than a couple of simple connections, operating software (often free) and an interface. RTTY (radioteletype) and PSK31 are two of the most popular HF "keyboard-to-keyboard" digital modes. PSK31 lets you communicate over great distances with a very modest ham station, typically at extremely low power levels.

Computers also can alert you to DX activity on the bands, help you practice taking Amateur Radio license examinations or improve your Morse code abilities. Many ham radio organizations, interest groups and even individuals maintain websites too.

Fig 1.13 — Computers often are front and center at ham radio stations today. Here, Rick Lindquist, WW1ME uses free software to log contacts during the ARRL 160 Meter Contest. The program also controls his radio. [Rick Lindquist, WW1ME, photo]

Keeping a Log

Keeping a log — on paper or using your computer — of your on-air activity is optional, but there are some important reasons for doing so. These include:

Awards tracking — A log lets you track contacts required for DXCC, WAS and other awards. Some computer logging programs do this automatically, so you can see how well you are progressing toward your goal.

An operating diary — A log book is a good place to record general information about your station. For example, you may want to note comparisons between different antennas or pieces of equipment based on reports from other stations. Your log is also a logical place to record new acquisitions (complete with serial numbers in case your gear is ever stolen). You can track other events as well, including the names and call signs of visiting operators, license upgrades, contests, propagation and so forth.

Legal protection — Good record keeping can help you protect yourself if you are ever accused of intentional interference or ever have a problem with unauthorized use of your call sign.

Paper or Computer?

Many hams, even some with computers, keep "hard copy" log books. A paper log is low tech; it doesn't consume power, it's flexible and can never suffer a hard-drive crash! Preprinted log sheets are available, or you can create your own customized log sheets in no time using word processing or publishing software.

On the other hand, computer logging offers many advantages, especially for contesters, DXers and those chasing awards. For example, a computerized log can instantly indicate whether you need a particular station for DXCC or WAS. Contesters use computer logs to manage contact data during a contest and to weed out duplicate contacts in advance. Most major contest sponsors prefer to receive computer log files, and some do not accept paper logs at all. Computer logs can also tell you at a glance how far along you are toward certain awards and even print QSL labels. And of course computer logs make it easy to submit your contacts to ARRL's online Logbook of The World.

Several of the most popular computer logging programs (and regular updates) are available at no cost, while others are available for a small fee. You also can purchase logging software from commercial vendors. Some are general-purpose programs, while others are optimized for contesting, DXing or other activities. Check the ads in *QST* and compare capabilities and requirements before you choose.

1.4 Getting on the Air

Amateur Radio is a *social* activity as well as a technical pursuit. It's a way to make new friends and acquaintances on the air that you may later meet in person. Some ham radio relationships last a lifetime, even though the individuals sometimes never meet face to face. Ham radio can be the glue that keeps high school and college friends in touch through the years.

Amateur Radio also can cement relationships between radio amateurs of different nationalities and cultures, leading to greater international goodwill and understanding — something that's especially beneficial in this era of heightened cultural tensions and misperceptions. When you become an Amateur Radio operator, you become a "world citizen." In return you can learn about the lives of the radio amateurs you contact in other countries.

"What do hams say to each other?" you might wonder. When they meet for the first time on the air, hams exchange the same sorts of pleasantries that anyone might when meeting face to face. Ham radio operators exchange name and location (abbreviated "QTH" by hams) and — specific to hamming — radio signal reports indicating how well they're hearing (or "copying") each other over the air. This name/location/signal report pattern is typical, regardless of radio mode. With these preliminaries out of the way, ham radio conversations often focus on equipment or may extend to other interests.

Although English is arguably the most common language on the ham bands (even spoken by hams whose first language may be something other than English), English speakers can make a favorable impression on hams in foreign countries if they can speak a little of the other person's language — even if it's as simple as *danke*, *gracias* or *arigato*.

Fig 1.14 — Fourteen year old Tommy James, KJ4SWI in Georgia, makes lots of contacts with this modest station and a simple wire antenna. [Thomas James, W4TBJ, photo]

1.4.1 Voice Modes

We've mentioned the use of voice (or "phone," short for "radiotelephone") and Morse code (or CW) on the amateur bands. Although more hams are embracing digital modes every day, phone and CW by far remain the most popular Amateur Radio communication modes. Ham voice modes are amplitude modulation (AM), which includes the narrower-bandwidth single sideband (SSB), and frequency modulation (FM). For the most part, SSB is heard on HF, while FM is the typical voice mode employed on VHF, UHF and microwave bands.

The great majority of ham radio HF phone operators use SSB (subdivided further into upper sideband and lower sideband), but a few still enjoy and experiment with the heritage "full-carrier AM." Once the primary ham radio voice mode, this type of AM still is heard on the standard broadcast band (530 to 1710 kHz). Today's AM buffs appreciate its warm, rich audio quality, and the simplicity of circuit design encourages restoring or modifying vintage radios or building from scratch. For more information about AM operation, visit **www.arrl.org/am-phone-operating-and-activities**.

1.4.2 Morse Code

Morse code was the very first radio transmission mode, although it wasn't long before early experimenters figured out how to transmit the human voice and even music over the airwaves. Morse is also the original digital mode; the message is transmitted by turning a radio signal on and off (a "1" and a "0" in digital terms) in a prescribed pattern to represent individual letters, numerals and characters. This pattern is the International Morse Code, sometimes called the "radio code," which varies in many respects from the original Morse-Vail Code (or "American Morse") used by 19th century railroad telegraphers. Leaning on longstanding tradition, hams often refer to Morse transmissions as "CW," after an archaic definition for "continuous wave" which described the type of radio wave involved.

Federal regulations once required that prospective radio amateurs be proficient in sending and receiving Morse code in order to operate on "worldwide" (ie, shortwave or HF) ham bands. Although this is no longer the case for any class of Amateur Radio license in the US, many hams still embrace CW as a favorite mode and use it routinely. Hams typically send Morse code signals by manipulating a manual telegraph key, a "semi-automatic" key (called a "bug") or a CW "paddle" and an electronic keyer that forms

the dots and dashes. Most hams decipher Morse code "by ear," either writing down the letters, numerals and characters as they come through the receiver's headphones or speaker or simply reading it in their heads. Some use one of the available computer-based or stand-alone accessories that can translate CW into plain text without the need to learn the code.

Hams who enjoy CW cite its narrow bandwidth — a CW signal takes up very little of the radio spectrum — simpler equipment and the ability of a CW signal to "get through" noise and interference with minimal transmitting power. CW is a common low-power (QRP) mode.

1.4.3 FM Repeaters

Hams often make their first contacts on local voice repeaters, although in recent years the nearly ubiquitous cell phone has reduced the popularity of — and even the need for — repeaters for everyday ham communication. Repeaters can greatly extend the useful range of a typical handheld FM transceiver much in the same way a cell tower retransmits your voice or text messages, and they carry the vast majority of VHF/UHF traffic, making local and even regional mobile commu-

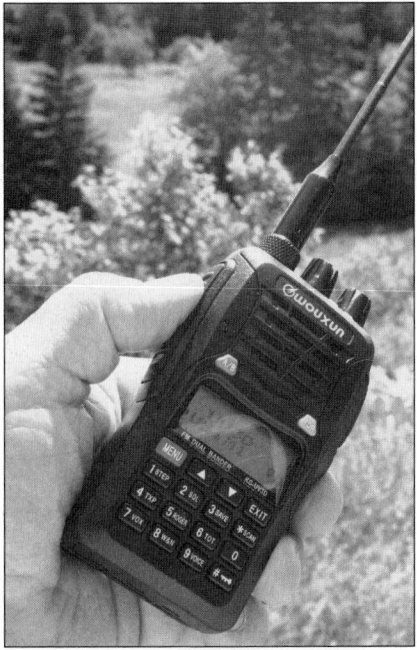

Fig 1.15 — Many thrifty hams carry a radio with them in the form of relatively inexpensive handheld transceivers such as this one, which feature long battery life and require no antenna beyond the one attached to the radio itself. Made in China, this unit cost slightly more than $100. [Rick Lindquist, WW1ME, photo]

nication possible for many hams. Located on hilltops, tall buildings or other high structures, repeaters strengthen signals and retransmit them. This provides communication over much farther distances than would be possible when operating point to point or "direct." The wider coverage can be especially important if the repeater is ever pressed into service during an emergency.

Typically, hams use repeaters for brief contacts, although socializing and "ragchewing" are routine on some "machines," as repeaters are often called. All repeater users give priority to emergency communications. Most repeaters are maintained by clubs or groups of hams. If you use a particular repeater frequently, you should join and support the repeater organization. Some hams set up their own repeaters as a service to the community.

The best way to learn the customs of a particular repeater is to listen for a while before transmitting. Most repeaters are *open*, meaning that any amateur may use the repeater, although repeaters typically require users to transmit an access tone (which you can select on any modern FM transceiver). A few repeaters are *closed*, meaning that usage is restricted to members of the club or group that own and operate the repeater. Some repeaters have *autopatch* capability that allows amateurs to make telephone calls through the repeater. The *ARRL Repeater Directory* shows repeater locations, frequencies, capabilities and whether the repeater is open or closed.

1.4.4 Digital Modes

Radioteletype (RTTY — often pronounced "Ritty") is a venerable data communication mode that remains in wide use today among radio amateurs. While RTTY does not support the features of newer computer-based data modes, it is well suited for keyboard-to-keyboard chats with other stations. It also is the most popular mode for worldwide digital contests and remains in common use among DXers and DXpeditions. RTTY was originally designed for use with mechanical

teleprinters, predating personal computers by several decades. Today, Amateur Radio RTTY uses soundcard-equipped computers and dedicated RTTY software.

An HF digital mode for general domestic and DX contacts is PSK31, a keyboard-to-keyboard mode that's especially effective for very low power (QRP) communication. It's easy to transmit and receive PSK31 using a PC and software — typically free — that operates via a PC soundcard. Most digital modes use the same basic PC/soundcard setup and new modes are developed on a regular basis.

One of the more recent software-based, soundcard modes is JT65 by Nobel laureate and radio amateur Joe Taylor, K1JT, who developed the software for this mode as part of *WSJT* (*Weak Signal communications by K1JT*). See **www.arrl.org/files/file/18JT65.pdf**. JT65 uses 64 audio tones to transmit information. Contacts in this mode, which often take place on the HF bands, are a bit different than typical two-way conversations. Stations take turns during synchronized 60-second transmissions cycles that alternate between odd and even minutes on the clock, until concluding the contact.

JT65 rivals Twitter in its brevity. Stations can transmit only about 13 characters each time, but the data are repeated each cycle to maximize information transfer. Successful JT65 contacts can take place under adverse radio conditions and using extremely low signal levels; a transmitter output power of 20 W is considered high power in the JT65 world.

Another digital mode, packet radio, is much less popular in today's ham radio world than it was in the 1990s. Packet's most important applications today include networking and *unattended* operation. The most common uses are the worldwide DX station spotting network, the Automatic Packet/Position Reporting System (APRS) and regional or local general-purpose networks.

APRS

Developed by Bob Bruninga, WB4APR, APRS or the Automatic Packet/Position

Reporting System (**www.aprs.org**) is used for tracking stations or objects in motion or in fixed positions and for exchanging data. It uses the *unconnected* packet radio mode to graphically indicate objects on maps displayed on a computer monitor. Using unconnected packets permits all stations to receive each transmitted APRS packet on a one-to-all basis rather than the one-to-one basis required by connected packets.

As with other packet transmissions, APRS data are relayed through stations called *digipeaters* (digital repeaters). Unlike standard packet radio, APRS stations use generic digipeater paths, so the operator needs no prior knowledge of the network. In addition, the Internet is an integral part of the system that is used for collecting and disseminating current APRS data in real time.

Virtually all VHF APRS activity occurs on 2 meters, specifically on 144.39 MHz, the recognized APRS operating channel in the US and Canada. On UHF, you'll find APRS activity on 445.925 MHz.

Many groups and individuals that participate in public service and disaster communications find APRS a useful tool. Others use it to view real-time weather reports. APRS also is useful to track roving (i.e., mobile) operators in certain ham radio operating events and supports the exchange of short messages.

1.4.5 Image Communication

Users of current technology often enjoy sharing photos or even talking face-to-face. While not as sophisticated, several ham radio communication modes allow the exchange of still or moving images over the air. Advances in technology have brought the price of image transmission equipment within reach of the average ham's budget. This has caused a surge of interest in image communication.

Amateur TV (ATV) is full-motion video over the air, sometimes called "fast-scan TV." Amateur Radio communication takes on an exciting, new dimension when you can actually *see* the person you're communicating with. In addition, ATV has proved to be very useful in emergency and disaster communication situations. Amateur groups in some areas have set up ATV repeaters, allowing lower-power stations to communicate over a fairly wide area. Since this is a wide-bandwidth mode, operation is limited to the UHF bands (70 cm and higher).

Digital ATV folds nicely into a newer Amateur Radio technological initiative called high-speed multimedia (HSMM) radio. The ham bands above 50 MHz can support computer-to-computer communication at speeds high enough to support multimedia applications — voice, data and image. One approach

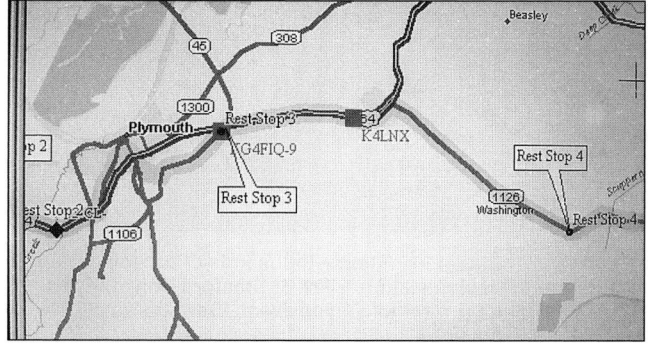

Fig 1.16 — APRS is a popular ham radio digital mode that allows tracking of mobile stations transmitting beacons via packet radio as they travel.

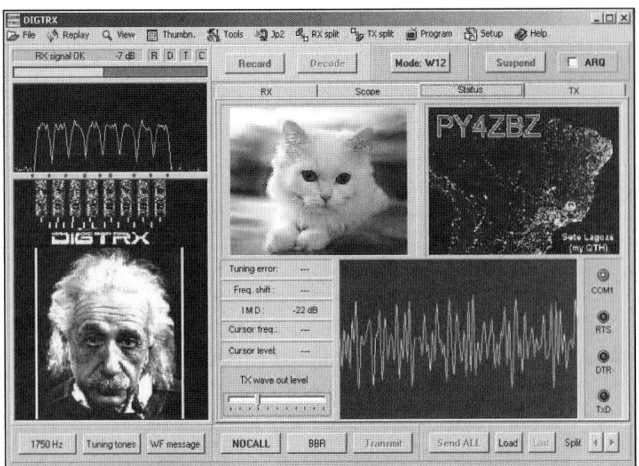

Fig 1.17 — Slow-scan television (SSTV) allows hams to exchange pictures over the air without wide-bandwidth live video equipment.

adapts IEEE 802 technologies, particularly 802.11b, operating on specific Amateur Radio frequencies in the 2400-2450 MHz band.

SSTV or "slow-scan TV" is an older, narrow-bandwidth image mode that remains popular in Amateur Radio. Instead of full-motion video, SSTV enthusiasts exchange photographs and other static images. Individual SSTV pictures take anywhere from 8 seconds to about 2 minutes to send, depending on the transmission method. These days most SSTV operation is done in color, using computers and soundcards in conjunction with software that's often free. Images are converted into a series of audio tones representing brightness level and colors. Since SSTV is a narrow-band mode, it is popular on HF on the same frequencies used for voice operation.

1.5 Your Ham Radio "Lifestyle"

After some on-the-air experience, many Amateur Radio enthusiasts focus on a particular mode or operating style and may identify themselves primarily as contesters, DXers, CW operators or VHF-UHFers. Others center their operating on such activities as specialized or experimental modes, mobile ham radio, very low-power operating (known as "QRP") and radio direction finding (RDF).

1.5.1 Ham Radio Contesting — Radiosport

Ham radio contesting, often called "radiosport," continues to grow in worldwide popularity. Hardly a weekend goes by when there isn't a ham radio contest of some sort. These on-the-air competitions range from regional operating events with a few hundred participants to national and worldwide competitions with thousands of stations on the air at the same time, attempting to communicate with one another for points.

Objectives vary from one event to another, but ham radio contests typically involve trying to contact — or "work" — as many *other* contest participants on the air within a specified period. In each contact, participants exchange certain information, often a signal report and a location, as the contest's rules dictate. A lot of contest scoring schemes place a premium on two-way contacts with stations in certain countries, states or zones. Top scorers in the various entry categories usually get certificates, but a few events offer sponsored plaques and trophies. Competition can be fierce among individual contesters and among contest clubs.

There are contests for nearly every mode

and operating preference available to Amateur Radio — voice, Morse code and digital modes. Some members of the contesting community are earnest competitors who constantly tweak their stations and skills to better their scores. Others take a more casual approach. All have lots of fun.

In the ARRL International DX Contest, for example, participants try to contact as many DX (foreign) stations as possible over the course of a weekend. Experienced hams with top-notch stations easily contact 1000 or more stations in more than 100 different countries in a single weekend, but even operators with more modest stations can make lots of contacts.

Other popular contests include state QSO parties, where the goal is to contact stations in as many of the sponsoring state's counties. ARRL November Sweepstakes (SS) is a high-energy US-and-Canadian contest that attracts thousands of operators each fall. One weekend is dedicated to CW, another to voice. VHF, UHF and microwave contests focus on making contacts using our highest-frequency bands. Digital-mode contests have gained in popularity in recent years, thanks to computer soundcards, radios that offer digital-mode capabilities, and often-free software.

You can find information on contests each month in ARRL's monthly membership

Fig 1.18 — Ham radio contesting or "radiosport" attracts thousands of participants all over the world. These operators used club station W6YX at Stanford University to compete in the 2012 ARRL RTTY Roundup, a radioteletype event. [Sawson Taheri, KG6NUB, photo]

journal *QST;* the contest calendar on the ARRL website also provides up-to-date information on upcoming operating events. The ARRL's bimonthly publication *National Contest Journal* (*NCJ*) focuses on topics of particular interest to contesting novices and veterans alike. For timely contest news and information, check "The ARRL Contest Update" e-newsletter at **www.arrl.org/contest-update-issues**, available every other week via e-mail and on the ARRL website.

ARRL FIELD DAY

An emergency communication training exercise with some elements of a contest, ARRL Field Day (FD) prompts thousands of participants outdoors the field on the fourth full weekend of June. Portable gear in tow, hams take to the hills, forests, campsites, parking lots and even emergency operations centers or vans to take part. Tracing its origins to the 1930s, Field Day started out as a way to publicly demonstrate ham radio's ability to operate "in the field" and "off the grid." The goal is not only to make lots of contacts but to operate successfully under the sorts of conditions that could prevail in the aftermath of a disaster or emergency.

Most stations are set up outdoors and use emergency power sources, such as generators, solar panels, wind turbines and occasionally even water wheels. Creativity reigns when it comes to power sources! Over the years, Field Day's contest-like nature has led to plenty of good-natured competition among clubs and groups. Field Day stations range from simple to elaborate. If a real disaster were to strike, stations such as these could be set up quickly wherever needed, without having to rely on commercial power.

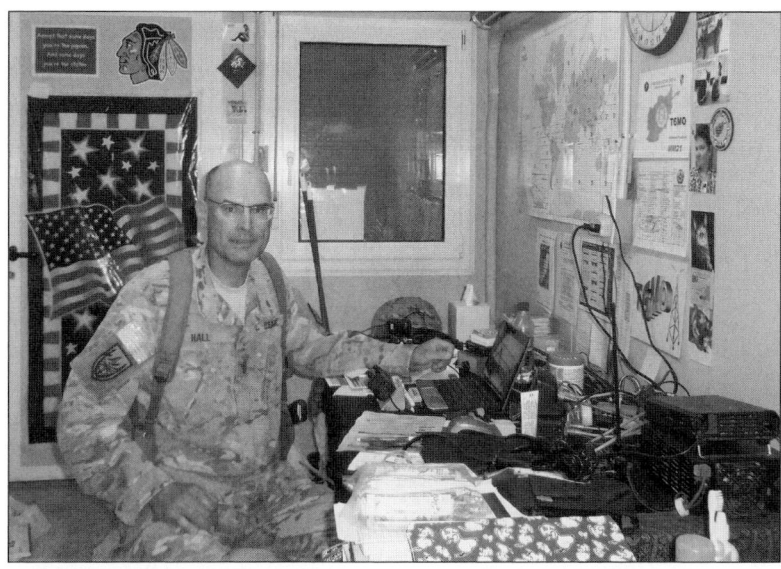

Fig 1.20 — Eric Hall, K9GY and T6MO, deployed with the US military in Afghanistan in 2012, spent some of his down time enjoying ham radio. [Eric Hall, K9GY/T6MO, photo]

1.5.2 Chasing DX

People unfamiliar with ham radio often ask, "How far can you talk?" Well, "talking far" is what chasing "DX" is all about. DX stations are those in distant places around the world. Chasing DX is a time-honored ham radio tradition. Hams who focus on contacting stations in far-flung and rare locations are called "DXers." Ham radio pioneers a century ago often competed in terms of how far *they* could talk; spanning the Atlantic via ham radio in the early 1920s was a stupendous accomplishment in its day. DXers often have as a goal attaining DX Century Club (DXCC) membership, earning a place on the vaunted

DXCC "Honor Roll" or entering the annual ARRL DX Challenge.

Working DX does not necessarily require an expensive radio and huge antenna system. It's possible to work a lot of DX all over the world with very low power and/or with modest antennas, including wires hung from trees or mobile (ie, vehicle-mounted) antennas.

Some hams specialize in certain ham bands to work DX, such as 160 meters, where DXing can be challenging due to the low operating frequency involved as well as frequent noise and infrequent DX propagation. Others prefer "the high bands," such as 20, 15 and 10 meters, where DX typically is more common and, in fact, typically abounds in times of favorable propagation.

DXPEDITIONS

DXers who have run out of new countries to work sometimes couple a love of ham radio and travel to *become* the DX! "DXpeditions" are journeys by hams to "rare" countries having few or no hams, where they set up a station (or stations), often making thousands of contacts in the space of a few days. They not only have a great time but can promote international goodwill.

Some DXpeditions are huge productions. Early in 2011, an international team activated Antarctica's South Orkney Islands, which have no indigenous population. South Orkney Islands count as a separate country (or "entity") for the DXCC award. In the space of about two weeks, the team completed nearly 64,000 two-way contacts with other hams around the world from their remote encampment. Members of the worldwide Amateur Radio community dig deep into their pockets to fund DXpeditions such as this one, which

Fig 1.19 — Tens of thousands of hams participate in the 24 hour ARRL Field Day each June. It's the largest Amateur Radio event on the planet! Here, Tom Thompson, WØIVJ works the nightshift at WØC. [Photo courtesy Tom Thompson, WØIVJ]

Fig 1.21 — These representatives from the Qatar Amateur Radio Society, A71DR (left) and A71AD, added an international flavor to ARRL Expo at the 2012 Dayton Hamvention®. [Rick Lindquist, WW1ME, photo]

QSL Cards

Long before the Internet and e-mail, hams began the custom of exchanging postcards that became known as QSL cards or simply QSLs. "QSL" is another radiotelegraph, or Morse code, abbreviation that means "I confirm receipt of your transmission." A QSL card contains information to verify that a two-way contact took place. Exchanging QSL cards can enhance your ham radio enjoyment and even lead to a regular correspondence.

Hams still take great pride in having distinctive QSL cards to exchange following a contact, although today, thanks to the Internet, electronic means exist, such as ARRL's Logbook of The World (see sidebar, "Logbook of The World") to confirm contacts.

DX stations, especially those in very rare places, are often inundated with QSL cards and requests from US hams. To ease the cost and administrative burden, most DX QSLs travel via QSL bureaus, which ship cards in bulk, then sort and distribute them on the receiving end. The Outgoing QSL Service is available to ARRL members at nominal cost. The incoming QSL bureaus are available to all amateurs. Bureau instructions and addresses are on the ARRL website.

Logbook of The World

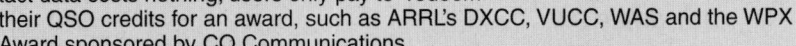

Instead of exchanging and collecting QSL cards, more and more radio amateurs are taking advantage of the ARRL's Logbook of The World to confirm contacts for award credit. LoTW is a world repository of individual radio contact records submitted by users. When both participants in a radio contact submit matching QSO records to LoTW, the result is a virtual QSL that each ham can apply toward ARRL award credit. Uploading contact data costs nothing; users only pay to "redeem" their QSO credits for an award, such as ARRL's DXCC, VUCC, WAS and the WPX Award sponsored by CQ Communications.

Once signed up as a Logbook user, you can submit new contact records whenever you wish. Your submissions are matched against those of other Logbook users. Whenever a match occurs, you receive instant credit for the contact.

To minimize the chance of fraudulent submissions, all LoTW QSO records are digitally "signed" by the licensee, who must hold an LoTW certificate. LoTW began operation in 2003 and within 10 years had more than 50,000 users and nearly half a billion QSO records in the system. Visit the Logbook of The World website, **www.arrl.org/logbook-of-the-world**, to learn more.

---

cost upward of $350,000. The 2012 HKØNA Malpelo Island DXpedition off the coast of Colombia logged more than 195,000 contacts on 160 through 6 meters, breaking the record for a "non-hotel, non-fly-in" DXpedition!

Most DXpeditions are smaller affairs in which one or two operators may combine a vacation with some on-air fun. Activity often peaks in conjunction with major DX contests. If you don't want to drag your radio and antennas along, fully equipped DX stations sometimes are available to rent in more-frequented locations, such as Hawaii and the Caribbean islands.

DX SPOTS AND NETS

The beginning DXer can get a good jump on DXCC by frequenting the DX spotting sites on the Internet. A DX spotting website is essentially an Internet clearing house of reports — or "spots" — posted by other DXers of stations actually heard or worked. The DX Summit website, **www.dxsummit.fi**, hosted in Finland is a popular one. Users around the world post spots in real time. Each lists the call sign and frequency of the DX station as well as the call sign of the station that posted the spot. Knowing where the DX station is being heard can tell you if you're likely to hear the DX station at *your* location.

DX nets offer another DX gateway. On DX

nets, a net control station tracks which DX stations have checked into the net, then allows individual operators on frequency to try working one of the DX stations. This permits weaker stations to be heard instead of their signals being covered up by stations calling in a "pileup."

1.5.3 Operating Awards

Earning awards that reflect Amateur Radio operating accomplishments is a time-honored tradition. Literally hundreds of operating awards are available to suit your level of activity and sense of accomplishment.

WORKED ALL CONTINENTS

The Worked All Continents (WAC) certificate is a good starting point for newcomers. Sponsored by the International Amateur Radio Union (IARU), earning WAC requires working and confirming contacts with one station on each of six continents (excluding Antarctica).

WORKED ALL STATES

Hams who can confirm two-way ham radio contacts with stations in each of the 50 United States can apply for the popular ARRL's Worked All States (WAS) award. Those who enjoy operating different bands and a seek a greater challenge may attempt the ARRL's 5-Band WAS (5BWAS) award by confirming contacts with all 50 US states on each of the 80, 40, 20, 15 and 10 meter bands.

A twist on the WAS award is the ARRL's Triple Play Award. Introduced in 2009, it was an instant hit. To earn the Triple Play Award, an amateur must contact other amateurs in each of the 50 US states using voice, Morse code and a digital mode, such as RTTY or

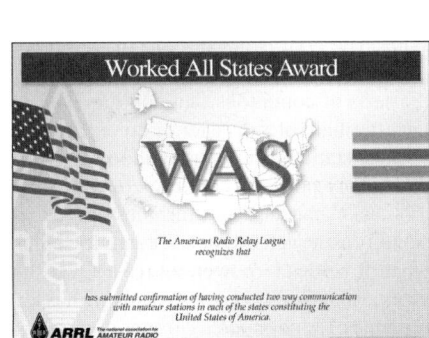

Fig 1.22 — The ARRL Worked All States (WAS) certificate is awarded for confirming contacts with hams in all 50 states. Confirmations may be made by traditional paper QSL cards, or electronically through ARRL's Logbook of The World online database.

PSK31. All qualifying contacts must be confirmed via the ARRL's Logbook of the World (LoTW — see sidebar, "Logbook of The World). The Triple Play Award is available to hams worldwide.

DX CENTURY CLUB

The most prestigious and popular DX award is the DX Century Club (DXCC), sponsored by the ARRL. Earning DXCC is quite a challenge. You must confirm two-way contact with stations in 100 countries (or "entities," as they're known in the DXCC program). Hams with very simple stations have earned DXCC. Operating in various DX contests when stations all over the world are looking for contact is a good way to combine DXing and contesting and to get a leg up on earning DXCC. There's also a 5-Band DXCC

(5BDXCC) for earning DXCC on each of five bands, 80, 40, 20, 15 and 10 meters.

Top-rung DX enthusiasts have been challenging themselves and each other through the ARRL DXCC Challenge. This ongoing activity involves confirming contacts with DXCC entities on all bands from 160 through 6 meters.

VHF/UHF CENTURY CLUB

Hams who operate on the VHF and UHF bands have a "century club" of their own, the VHF/UHF Century Club. Instead of working 100 DXCC entities, participants earn awards for making two-way contacts with a specified number of Maidenhead $2° × 1°$ grid locators or "grid squares," as they're more commonly known. Grid squares are designated by a combination of two letters and two numbers and represent a specific area on the globe. For operations on 6 meters, 2 meters and satellite, operators must contact 100 individual grid squares. More information is on the ARRL website, **www.arrl.org/awards/vucc**.

1.5.4 Satellite Communication

Amateur Radio established its initial foothold in space in 1961, with the launch of the OSCAR 1 satellite (OSCAR is an acronym for Orbiting Satellite Carrying Amateur Radio). Since then, amateurs have launched dozens of satellites, most of the low-earth orbit (LEO) variety and a small number in

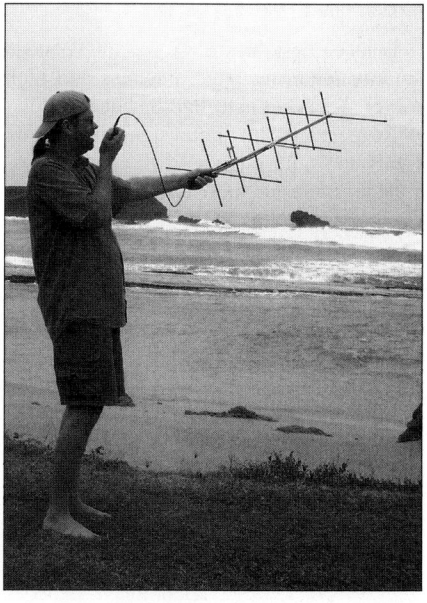

Fig 1.23 — Hams can use simple antennas and hand-held radios to contact amateur satellites, even the International Space Station. Sean Kutzko, KX9X took his satellite gear along during a vacation to Puerto Rico, making contacts via the AO-27 orbiting "bird." [Ward Silver, (NØAX, photo]

the high-earth orbit category. The history of Amateur Radio satellites and information on which ones are in operation is available on the AMSAT website, **www.amsat.org**.

Amateurs have pioneered several developments in the satellite industry, including low-earth orbit communication "birds" and PACSATs — orbiting packet-radio bulletin board systems — and CubeSats which are standard-sized miniature satellites constructed by student teams around the world. Operating awards, are available from ARRL and other organizations specifically for satellite operation.

Satellite operation is neither complex nor difficult; it's possible to work through some satellites with nothing more than a dual-band (VHF/UHF) handheld radio and perhaps a small portable antenna. More serious satellite work requires some specialized equipment. You may be able to work several Amateur Radio satellites (OSCARs) with the equipment that's now in your shack!

AMATEUR RADIO IN SPACE

In 1983, ham-astronaut Owen Garriott, W5LFL, operated the first ham radio station in space, communicating with earthbound hams from the space shuttle *Columbia.* NASA subsequently made Amateur Radio a part of many shuttle missions and later a permanent presence aboard the International Space Station. The Russian *Mir* space station also was equipped with ham radio, and since those early days, dozens of astronauts and cosmonauts from several countries have communicated from their orbital perches with hams on earth. NASA has promoted ham activity aboard spacecraft because of its proven educational and public relations value — and because ham radio could be used for backup communication in a pinch.

The Amateur Radio on the International Space Station (ARISS) program, **www.arrl. org/amateur-radio-on-the-international-space-station** and **www.ariss.rac.ca**, is the international organization for ham radio in space. ARISS is a consortium of the ARRL, AMSAT and NASA. The all-volunteer program seeks to inspire students worldwide to pursue careers in science, technology, engineering and math by making available opportunities to speak with on-orbit ISS crew members via ham radio. The ARISS International Working Group includes representatives of nine countries, including the US, several European nations and Japan, Russia and Canada. ARISS-provided ham gear on two ISS modules makes possible analog voice and digital communication between earthbound hams and the ISS crew as well as with the onboard ISS digipeater packet radio mailbox and APRS digipeater.

Most ISS crew members are Amateur Radio licensees. During scheduled contacts

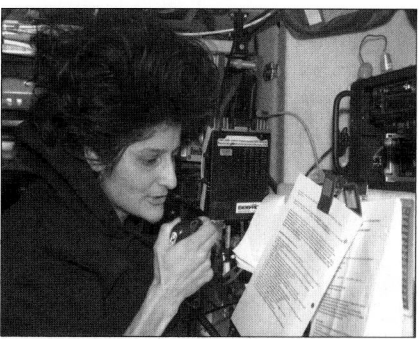

Fig 1.24 — At ham station NA1SS aboard the International Space Station, NASA Astronaut Sunitra Williams, KD5PLB answers questions posed by students on Earth during an ARISS school contact. [NASA photo]

with demonstration stations set up in schools around the world, students are able to ask the astronauts questions about their time in space and life aboard the space station. These voice contacts typically take place using VHF FM (2 meter) equipment. Some Amateur Radio satellites have been launched from the ISS.

1.5.5 QRP: Low-Power Operating

A very active segment of the Amateur Radio community enjoys operating with minimal transmitting power. They call themselves "QRP operators" or "QRPers" after the Morse code abbreviation for "I shall decrease transmitter power." According to the FCC rules regulating Amateur Radio, "An amateur station must use the minimum transmitter power necessary to carry out the desired communications." The FCC allows most hams to transmit or "run" up to 1500 W (watts), and many hams run 100 W. QRPers, however, typically use 5 W or less — sometimes *far* less (one ham achieved WAS while running 2 milliwatts — that's two-thousandths of a watt!).

Operating QRP can be challenging. Other stations may not hear your signal as easily, so patience becomes a real virtue, both for the low-power enthusiast and the station on the other end of the contact. What their stations lack in transmitting power QRPers make up for with effective antennas and skillful operating, and they make contacts around the world. This operating style has become so popular that many contests now include an entry category for stations running 5 W or less output power.

One of the best reasons to operate QRP is that low-power equipment typically is lightweight and less expensive. Many QRP operators enjoy designing and building their own "flea-power" transceivers, and various organizations support low-power operating

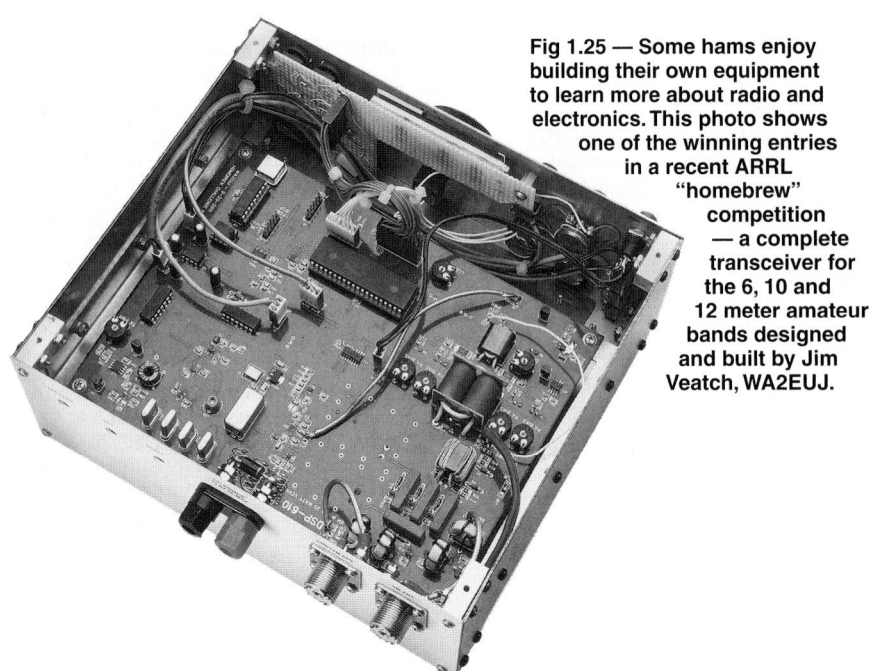

Fig 1.25 — Some hams enjoy building their own equipment to learn more about radio and electronics. This photo shows one of the winning entries in a recent ARRL "homebrew" competition — a complete transceiver for the 6, 10 and 12 meter amateur bands designed and built by Jim Veatch, WA2EUJ.

Fig 1.26 — Portable or "mountaintopping" operation is very popular on the VHF and UHF bands. In this photo Bob Witte, KØNR has set up a station atop Mt Herman, near Monument, Colorado. [Joyce Witte, KØJJW. photo]

by offering kits, circuits and advice. A few commercial manufacturers also market QRP equipment and kits. The QRP Amateur Radio Club International (**www.qrparci.org**) is perhaps the oldest organization to advance and promote QRP as a ham radio way of life. In addition to sponsoring various operating events throughout the year, QRP ARCI publishes *QRP Quarterly*, which includes articles of interest to both QRP operators and the broader ham radio community.

1.5.6 Operating Mobile

Many hams enjoy operating on the fly — usually from a car or truck but sometimes from a boat, a motorcycle, a bicycle and even while on foot (sometimes called "pedestrian mobile" or "manpack radio")! Operating radio gear installed in a motor vehicle is the most common form of "mobile," and manufacturers today offer a wide range of ham radio gear, including antennas, designed for such work. A mobile station can be as simple as a basic VHF or dual-band VHF/UHF radio and a little antenna attached magnetically to the roof, or as complex as an HF station and a more substantial antenna system. Some mobile stations are very sophisticated, with capabilities that rival those of many fixed stations.

While most hams who operate mobile use FM or SSB, a significant number operate CW while on the road. It takes a bit of practice, in part because the operator must learn to understand (or "copy") Morse code without having to write it down.

Hams on bicycle treks or hikes carry along lightweight radio gear. A lot of cyclists or hikers pack a small ham radio transceiver and wire antenna along with their sleeping bag, food and water.

1.5.7 VHF, UHF and Microwave Operating

Hams use many modes and techniques to extend the range of their line-of-sight VHF, UHF and microwave signals. Those who explore the potential of VHF/UHF communication often are called "weak-signal" operators to differentiate them from FM operators who communicate locally — although the signals involved often are not really *weak*.

These enthusiasts and experimenters probe the limits of propagation in the upper reaches of the Amateur Radio spectrum, often with the goal of discovering just how far they can communicate. They use directional antennas (beams or parabolic dishes) and very sensitive receivers. In some instances, they also employ considerable transmitter output power. As a result of their efforts, distance records are broken almost yearly. On 2 meters, for example, conversations between stations hundreds and even thousands of miles apart are not uncommon even though the stations are far beyond "line of sight" separation. Maximum distances decrease as frequencies increase, but communication regularly can span several hundred miles, even at microwave frequencies.

Weak-signal operators for many years depended on SSB and CW, but computer/sound card-based digital modes are now part of their arsenal. These modes use state-of-the-art digital signal processing (DSP) software for transmitting and receiving very weak signals that can be well below levels that the human ear can detect.

MOONBOUNCE (EME)

EME (Earth-Moon-Earth) communication, commonly called "moonbounce," fascinates many amateurs. The concept is simple: Use the Moon as a passive reflector of VHF and UHF signals. Considering the total path of some 500,000 miles, EME is the ultimate DX — at least to date. The first two-way amateur EME contacts took place in 1952.

In its earliest days EME was a CW mode activity requiring large antennas and high power. Advances in technology, such as low-noise receivers and digital signal processing (DSP) tools, have made EME contacts possible for more and more amateurs with modest stations.

METEOR SCATTER

Years ago hams discovered they could bounce signals off the ionized trails of vaporized matter that follow meteors entering Earth's atmosphere. Such trails often can reflect VHF radio signals for several seconds, during which stations can exchange extremely brief reports. During meteor showers, the ionized region becomes large enough — and lasts long enough — to sustain short contacts. It's exciting to hear a signal from hundreds of miles away pop out of the noise for a brief period!

Amateurs experimenting with meteor-scatter propagation use transmitter powers of 100 W or more and beam antennas. Most contacts are made using SSB, CW or digital modes. Although most SSB and CW QSOs

Fig 1.27 — Since antennas for the VHF, UHF and microwave bands are small enough for travel, "rover" stations are popular in competitions for these bands. Here, John D'Ausilio, W1RT adjusts his rover's antennas during the ARRL September VHF QSO Party. [Andy Zwirko, K1RA, photo]

Yet many amateurs still would rather repair and adjust their own equipment and covet the days when this was simpler and easier. This is but one reason behind the surge in vintage radio collecting and operating. Others enjoy vintage gear for its lower cost and wider availability, the novelty of operating older gear on today's ham bands, and for its rarity and antique value. Many of these radios are affectionately called "boat anchors" by vintage radio aficionados, since early radio gear tends to be relatively large and heavy.

Some enthusiasts enjoy the challenge of collecting and restoring older radios, sometimes striving to bring the equipment back to its original factory condition. Other vintage radio enthusiasts may have a parallel interest in conventional AM voice transmission. These activities take vintage radio fans back to an era when it was much more common for amateurs to build their own station equipment.

are made during annual meteor showers, digital mode contacts are possible any day of the year.

Nobel laureate astrophysicist and radio amateur Joe Taylor, K1JT, has developed open-source *Windows* and *Linux* software (under the *WSJT* umbrella) for weak-signal digital communication via meteor scatter, moonbounce and similar techniques on VHF and UHF, as well as for HF skywave propagation. Taylor says on his *WSJT* home page (**physics.princeton.edu/pulsar/K1JT**) that his program "can decode fraction-of-a-second signals reflected from ionized meteor trails and steady signals 10 dB below the audible threshold."

1.5.8 Vintage Radio

Many, if not most, veteran radio amateurs have a nostalgic streak, and this extends toward vintage radio gear. Present-day commercial Amateur Radio equipment has reached a level of complexity that often requires specialized test and troubleshooting equipment to repair or align. Modern component manufacturing technology such as surface-mount devices (SMDs) has become so commonplace that a modular approach to equipment repair is commonplace; rather than troubleshoot and replace a defective component, many manufacturers now prefer to swap out an entire module.

1.5.9 Radio Direction Finding (DF)

DFing is the art of locating a signal or noise source by tracking it with portable receivers and directional antennas. Direction finding is not only fun, it has a practical side. Hams who are proficient at DFing have been instrumental in hunting down signals from illegal jammers and malfunctioning transmitters in addition to locating noise (interference) sources. Because DFing only involves receiving, it does not require a ham ticket, however.

"Fox hunting" — also called "T-hunting,"

Fig 1.28 — Hams enjoy restoring and using old radio equipment. Rod Bunn, KA6ROD restored his Heathkit transceiver and station monitor to use during the annual ARRL Straight Key Night special operating event.

Fig 1.29 — Vintage vacuum tubes such as these were among the wares on sale in the Dayton Hamvention 2012 flea market, helping to satisfy the ham radio community's appetite for nostalgia. Flea markets are good sources for parts needed to refurbish older ham gear that many enjoy collecting. [Rick Lindquist, WW1ME, photo]

"radio-orienteering" or "bunny hunting" — is ham radio's answer to hide-and-seek. One player is designated the fox; he or she hides a transmitter, and the other players attempt to find it. Rules vary, but the fox must generally place the transmitter within certain boundaries and transmit at specific intervals.

Fox hunts differ around the world. American fox hunts often employ teams of fox hunters cruising in vehicles over a wide area. European and other fox hunters restrict their events to smaller areas and conduct fox hunts on foot. *Radiosport* competitions typically follow the European model and attract hundreds or more competitors.

DF techniques can come into play when tracking down sources of interference on the ham bands — intentional or inadvertent — or when there's a suspected "pirate" (unlicensed ham station) in the area.

1.6 Public Service

Providing communication in support of public service at no cost forms part of the Basis and Purpose of the Amateur Radio Service and has been a traditional responsibility of Amateur Radio from the start. Today, this most often involves ham radio's volunteer efforts during disasters and emergencies.

When Hurricane Sandy struck New York City and the Middle Atlantic States in the fall of 2012, the Amateur Radio community from Maine to the Carolinas responded to requests for assistance, activated local nets and supported the operations of the Hurricane Watch Net and the VoIP Hurricane Net. Hams volunteered around the clock to bridge the gap in the wake of downed utility lines to provide communication for evacuation efforts, as well as to link hospitals experiencing communications breakdowns, shelters, emergency operations centers and non-government relief agencies, such as The American Red Cross and The Salvation Army, which has its own Amateur Radio contingent, The Salvation Army Team Emergency Radio Network (SATERN). Radio amateurs also assisted after the storm by helping officials to assess damage. Many hams also are part of SKYWARN, which helps to identify and track severe weather activity via ham radio and coordinates its efforts with the National Weather Service.

Public service can take less dramatic forms: Hams also step forward to provide communication for walkathons, marathons, bike races, parades and other community events. The Boston and New York City marathons are two major events that welcome Amateur Radio assistance.

1.6.1 Public Service Communication

The ability to provide communication during disasters is a major justification for Amateur Radio's existence. Government officials on all levels and the general public have come to recognize that Amateur Radio works when other communications networks are unavailable. Despite the proliferation of cell phones and other personal communication devices, Amateur Radio continues to prove its value, since it can operate without

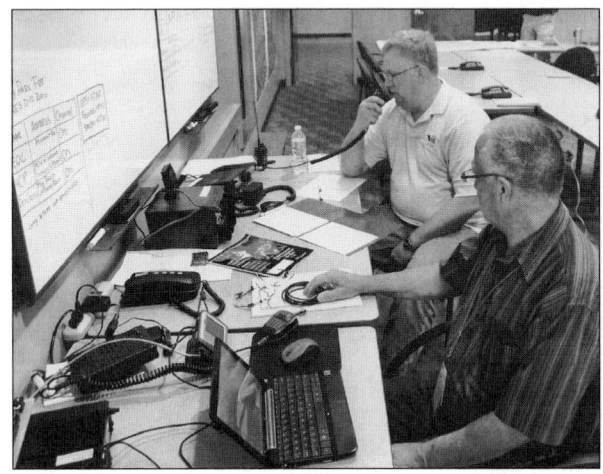

Fig 1.30 — Radio amateurs in Colorado helped to support disaster communications as wildfires raged across that state in 2012. Here, Larry Arave, W7LRY and John Cook, WA7NZE volunteer at the Larimer County Emergency Operations Center during the High Park Fire. [Rob Strieby, W0FT, photo]

an existing infrastructure. Ham radio doesn't need the mobile telephone network or the Internet.

Battery-powered equipment allows hams to provide essential communication even when power is knocked out. If need be, hams can make and install antennas on the spot from available materials. In the wake of hurricanes, forest fires, earthquakes and other natural disasters that cripple or compromise normal communications, hams may be called upon to handle thousands of messages in and out of the stricken region. The work that hams do during crisis situations cultivates good relations with neighbors and with local governments.

Amateur Radio operators have a long tradition of operating from back-up power sources. Through events such as Field Day, hams cultivate the ability to set up communication posts wherever they are needed. Moreover, Amateur Radio can provide computer networks (with over-the-air links as needed) and other services, such as video, that no other service can deploy on the fly, even on a wide scale.

1.6.2 Public Service Communication Organizations

Should a disaster or emergency arise, volunteer teams of amateurs from disaster

response organizations cooperate with first responders such as police and fire personnel, and with the Red Cross and medical personnel to provide or supplement communication. Hams sometimes are called upon to fill the communication gap between agencies whose radio systems are incompatible with one another.

ARES AND RACES

Ham radio disaster response activities typically take place under the umbrella of the Amateur Radio Emergency Service (ARES®), sponsored by the ARRL, and the Radio Amateur Civil Emergency Service (RACES), administered by the Federal Emergency Management Agency (FEMA). RACES works with government agencies to maintain civil preparedness and provide communication in times of civil emergency. RACES is activated at the request of a local, state or federal official. To remain at the ready, hams affiliated with emergency communication teams assess their systems and themselves through regularly scheduled nets and simulated emergency tests (SETS).

ARES and RACES organizations frequently work hand-in-hand. Amateurs serious about disaster response communication typically are active in both groups or may carry dual ARES/RACES membership. FCC rules make it possible for ARES and RACES to use many of the same frequencies, so an

ARES group also enrolled in RACES can work within either organization as circumstances dictate.

MILITARY AUXILIARY RADIO SYSTEM (MARS)

MARS is administered by the US armed forces. Its primary mission is to provide adjunct communication support to military and federal government homeland defense operations on a local, national and international basis. MARS volunteers joined Amateur Radio organizations to assist medical and humanitarian relief efforts in the wake of the devastating 2010 earthquake in Haiti. MARS has existed in one form or another since 1925.

There are three branches of MARS: Army MARS, Navy/Marine Corps MARS and Air Force MARS. Each branch has its own membership requirements, although all three branches require members to hold a valid US Amateur Radio license and to be at least 18 years old (in some cases, amateurs who are 17 years old may join with the signature of a parent or legal guardian).

MARS operation takes place on frequencies adjacent to the amateur bands and usually consists of nets scheduled to handle traffic or to handle administrative tasks. Various MARS branches also maintain repeaters or packet radio networks.

1.6.3 Public Service and Traffic Nets

The ARRL came into existence to coordinate and promote the formation of message-handling nets, so public service and traffic nets are part of a tradition that dates back almost to the dawn of Amateur Radio. In those early days, nets were needed to communicate over distances greater than a few miles. From their origination point, messages (also called "traffic") leapfrogged from amateur station to amateur station to their destination — thus the word "relay" in American Radio Relay League. It still works that way today, although individual stations typically have a much greater range.

Some nets and stations are typically only active in emergencies. These include Amateur Radio station WX4NHC at the National Hurricane Center, the Hurricane Watch Net, SKYWARN (weather observers), The Salvation Army Team Emergency Radio Network (SATERN), The Waterway Net and the VoIP (voice over Internet protocol) SKYWARN/Hurricane Net.

THE NATIONAL TRAFFIC SYSTEM (NTS)

The National Traffic System (NTS) exists to pass formal written messages from any point in the US to any other point. Messages, which follow a standard format called a "Radiogram," are relayed from one ham to another, using a variety of modes, including voice, Morse code, RTTY or packet. An NTS operator who lives near the recipient typically delivers the message by telephone.

During disasters or emergencies, radiograms communicate information critical to saving lives or property or to inquire about the health or welfare of disaster victims. At such times, the NTS works in concert with the Amateur Radio Emergency Service (ARES) and other emergency and disaster-relief organizations, such as the American Red Cross and The Salvation Army.

The NTS oversees many existing traffic nets, which meet daily. Most nets are local or regional. Handling routine message traffic such as birthday and holiday greetings keeps NTS participants prepared for emergencies.

1.7 Ham Radio in the Classroom

Amateur Radio is a terrific teaching tool! Many individuals began their path towards careers in electronics and wireless communication thanks to their experiences with Amateur Radio as children and teenagers.

Amateur Radio complements any school curriculum and gives students a chance to make a direct and immediate connection with their studies. For example, the math and science used in Amateur Radio apply equally in the classroom. Even geography takes on a new meaning when students are able to contact other countries around the globe and even to speak with the people who live in them!

Local volunteers are important to establishing an active Amateur Radio presence in schools. An HF or satellite station or even a VHF or UHF handheld transceiver tuned to the local repeater can prove an exciting and educational experience for pupils and volunteers alike.

Fig 1.31 — Project Blue Horizon was a balloon launch that served as a final class project for a team of Cornell University students. The balloon carried an instrumentation and control package that operated via an Amateur Radio control link, traveling more than 1100 miles in just over 30 hours. Shown here getting ready to release the balloon are (left to right) team members Matt Howells, NS3FD; Steve Orzechowski, KC2UFK; Angela Bratt, KC2UFL; John Ceccherelli, N2XE; and Dan Rondeau, KC2UFJ. [Photo courtesy John Ceccherelli, N2XE]

Thanks to the Amateur Radio on the International Space Station (ARISS) program, amateurs all over the nation have made it possible for students to speak directly with astronauts in space via ham radio.

1.7.1 ARRL Amateur Radio Education & Technology Program

Through the ARRL Amateur Radio Education & Technology Program (ETP), Amateur Radio can become a valuable resource for the classroom teacher. The goal of the ETP is "to build a foundation of wireless technology literacy to US teachers and students." Launched in 2000 the program continues to offer resources to schools, including ham radio equipment, at no cost, thanks to the support of donors in the Amateur Radio community. The ETP emphasizes the integration of technology, math, science, geography, writing, speaking and social responsibility within a global society. Applying Amateur Radio as

Fig 1.32 — Sixth-grade teacher Kaci Heins, KF7RCV (left in flight suit) stands by while one of her students in Flagstaff, Arizona, speaks with astronaut Joe Acaba, KE5DAR aboard the ISS during an Amateur Radio on the International Space Station (ARISS) school contact. Heins got her own ham ticket while preparing her class for the event. (Courtesy Kacy Heins, KF7RCV)

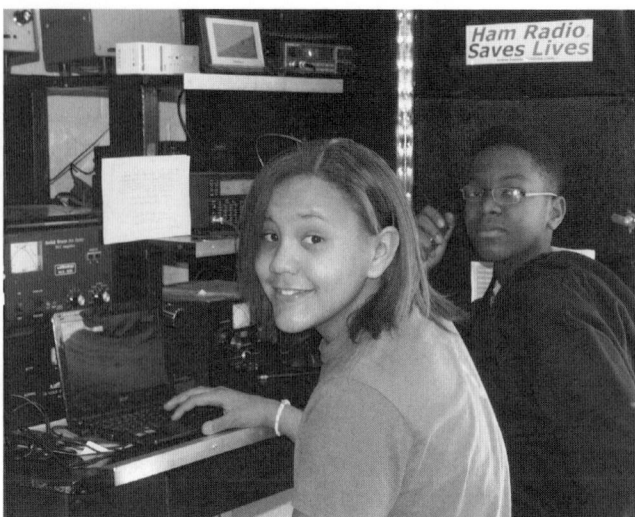

Fig 1.33 — At Eisenhower Middle School in Lawton, Oklahoma, Jada, KF5TAT (left) and Kerson, KF5TAQ sit at the Viking Radio Club station in the classroom of teacher Clifton Harper, KE5YZB. The school received a ham radio equipment grant through the ARRL Foundation and in 2012 Harper attended an ARRL Education & Technology Program Teacher's Institute on Wireless Technology at ARRL Headquarters. [Pamely Harper, KF5JXO, photo]

part of the class curriculum offers students a new dimension to learning. Each summer the ETP sponsors Teachers Institute on Wireless Technology sessions for educators, to enable them to make the most effective use of the ETP in their schools.

Amateur Radio emphasizes self-challenge, the value of lifelong learning and the importance of public service. From a more practical standpoint, future employers will be looking for candidates who are familiar not only with computers but with the sorts of wireless communication concepts used in Amateur Radio.

The ETP offers a range of resources to encourage educators. These include publications related to the use of technology in wireless communications; workshops, tips and ideas for teaching wireless technology in schools, community groups and clubs; and lesson plans and projects to help provide authentic, hands-on technological experiences for students.

Schools interested in incorporating Amateur Radio into their curricula, using it as an enrichment program or as a club activity may apply to become Project schools. See **www.arrl.org/education-technology-program** for more information on the ARRL Education & Technology Program.

1.8 Resources

ARRL—the National Association for Amateur Radio
225 Main St
Newington, CT 06111-1494
860-594-0200
Fax: 860-594-0259
e-mail: **hq@arrl.org**
Prospective hams call 1-800-32 NEW HAM (1-800-326-3942)
www.arrl.org
Membership organization of US ham radio operators and those interested in ham radio. Publishes study guides for all Amateur Radio license classes, a monthly journal, *QST*, and many books on Amateur Radio and electronics.

Amateur Radio Service Rules & Regulations — FCC Part 97
Available on the ARRL website: **www. arrl.org/part-97-amateur-radio**
Available as a PDF file on the FCC website: **www.fcc.gov/Bureaus/Engineering_ Technology/Documents/cfr/1998/47cfr97. pdf**

AMSAT NA (The Radio Amateur Satellite Corporation Inc)
850 Sligo Ave — Suite 600
Silver Spring, MD 20910
888-322-6728 or 301-589-6062
www.amsat.org
Membership organization for those interested in Amateur Radio satellites.

Courage Handi-Ham System
3915 Golden Valley Rd
Golden Valley, MN 55422
763-520-0512 or 866-426-3442
www.handiham.org
Provides assistance to persons with disabilities who want to earn a ham radio license or set up a station.

OMIK Amateur Radio Association
www.omikradio.org
OMIK, an ARRL affiliated club, is the largest predominately African-American Amateur Radio organization in the US. It promotes fellowship and Amateur Radio advancement and offers scholarships and other financial assistance for college-bound youth.

The ARRL Ham Radio License Manual
www.arrl.org/ham-radio-license-manual
Complete introduction to ham radio, including the exam question pool, complete explanations of the subjects on the exams. Tips on buying equipment, setting up a station and more.

The ARRL's Tech Q&A
www.arrl.org/shop/
Contains all of the questions in the Technician Class question pool, with correct answers highlighted and explained in plain English. Includes many helpful diagrams.

General Information and Other Study Material
The ARRL website (**www.arrl.org**) carries a wealth of information for anyone interested in getting started in Amateur Radio. For complete information on all options available for study material, check out the "Welcome to the World of Ham Radio" page, **www.arrl.org/what-is-ham-radio** and its associated links. You can also use the ARRL website to search for clubs, classes and Amateur Radio exam sessions near you.

1.9 Glossary

AM (Amplitude modulation) — The oldest voice operating mode still found on the amateur bands. The most common HF voice mode, SSB, is actually a narrower-bandwidth variation of AM.

Amateur Radio — A radiocommunication service for the purpose of self-training, intercommunication and technical investigation carried out by licensed individuals interested in radio technique solely with a personal aim and without pecuniary interest. (*Pecuniary* means payment of any type, whether money or goods.) Also called "ham radio."

Amateur Radio operator — A person holding an FCC license to operate a radio station in the Amateur Radio Service.

Amateur Radio station — A station licensed by the FCC in the Amateur Radio Service, including necessary equipment.

Amateur (Radio) Service — A radiocommunication service for the purpose of self-training, intercommunication and technical investigations carried out by licensed individuals interested in radio technique solely with a personal aim and without pecuniary interest.

AMSAT (Radio Amateur Satellite Corporation) — An international membership organization that designs, builds and promotes the use of Amateur Radio satellites, which are called "OSCARs."

APRS—Automatic Packet/Position Reporting System, a marriage of an application of the Global Positioning System and Amateur Radio to relay position and tracking information.

ARES (Amateur Radio Emergency Service) — An ARRL program for radio amateurs who participate in emergency communication.

ARISS — An acronym for Amateur Radio on the International Space Station. NASA, ARRL and AMSAT cooperate in managing the ARISS Program.

ARRL — The national association for Amateur Radio in the US; the US member-society in the *IARU* (International Amateur Radio Union).

ATV (Amateur television) — An Amateur Radio operating mode for sharing real-time video.

Band — A range of frequencies in the radio spectrum, usually designated by approximate wavelength in meters. For example, 7.0 to 7.3 MHz (megahertz) is the 40 meter amateur band. Hams are authorized to transmit on many different bands.

Bandwidth — In general, the width of a transmitted signal in terms of occupied spectrum. FCC definition: "The width of a frequency band outside of which the mean power of the transmitted signal is attenuated at least 26 dB below the mean power of the transmitted signal within the band."

Beacon — An amateur station transmitting communication for the purposes of observation of propagation and reception or other related experimental activities.

Beam antenna — A ham radio antenna having directional characteristics to enhance the transmitted signal in one direction at the expense of others. A "rotary beam" can be pointed in any direction.

Broadcasting — Transmissions intended for reception by the general public, either direct or relayed. Amateur Radio licensees are not permitted to engage in broadcasting.

Call sign — A series of unique letters and

numerals that the FCC assigns to an individual who has earned an Amateur Radio license.

Contact — A two-way communication between Amateur Radio operators.

Contest — A competitive Amateur Radio operating activity in which hams use their stations to contact the most stations within a designated time period.

Courage Handi-Ham System — Membership organization for ham radio enthusiasts with various physical disabilities and abilities.

CW — A synonym for radiotelegraphy (ie, Morse code by radio). CW is an abbreviation for "continuous wave," a term used in the early years of wireless.

Digital communication — Computer-based communication modes such as RTTY, PSK31, packet and other radio transmissions that employ an accepted digital code to convey intelligence or data.

Dipole antenna — Typically, a wire antenna with a feed line connected to its center and having two legs. Dipoles most often are used on the high-frequency (HF) amateur bands.

DSP (digital signal processing) — Technology that allows software to replace electronic circuitry.

DX — A ham radio abbreviation that refers to distant stations, typically those in other countries.

DXCC — DX Century Club, a popular ARRL award earned for contacting Amateur Radio operators in 100 different countries or "entities."

DXpedition — A trip to a location — perhaps an uninhabited island or other geographical or political entity — which has few, if any, Amateur Radio operators, thus making a contact rare.

Elmer — A traditional term for a person who enjoys helping newcomers get started in ham radio; a mentor.

Emergency communication — Amateur Radio communication during a disaster or emergency that support or supplants traditional means of telecommunication.

FCC (Federal Communications Commission) — The government agency that regulates Amateur Radio in the US.

Field Day — A popular, annual Amateur Radio activity sponsored by the ARRL during which hams set up radio stations, often outdoors, using emergency power sources to simulate emergency operation.

Field Organization — A cadre of ARRL volunteers who perform various services for the Amateur Radio community at the state and local level.

FM (Frequency modulation) — A method of transmitting voice and the mode commonly used on ham radio repeaters.

Fox hunt — A competitive radio direction-finding activity in which participants track down the source of a transmitted signal.

Fast-scan television — A mode of operation that Amateur Radio operators can use to exchange live TV images from their stations. Also called *ATV (Amateur Television)*.

Ham band — A range of frequencies in the radio spectrum on which ham radio communication is authorized.

Ham radio — Another name for Amateur Radio.

Ham radio operator — A radio operator holding a license granted by the FCC to operate on Amateur Radio frequencies.

HF (high frequency) — The radio frequencies from 3 to 30 MHz.

HSMM (high-speed multimedia) — A digital radio communication technique using spread spectrum modes primarily on UHF to simultaneously send and receive video, voice, text, and data.

IARU (International Amateur Radio Union) — The international organization made up of national Amateur Radio organizations or societies such as the ARRL.

Image — Facsimile and television signals.

International Morse Code — A digital code in which alphanumeric characters are represented by a defined set of short and long transmission elements — called "dots and dashes" or "dits and dahs" — that many Amateur Radio operators use to communicate.

ITU (International Telecommunication Union) — An agency of the United Nations that allocates the radio spectrum among the various radio services.

MARS — Military Auxiliary Radio System, a volunteer auxiliary communication program that supports the US Department of Defense and federal government homeland defense activities. Most MARS operators are Amateur Radio operators.

Mode — A type of ham radio communication, such as frequency modulation (FM voice), slow-scan television (SSTV), SSB (single sideband voice), CW (Morse code), or digital (eg, PSK-31 or JT65).

Morse code — A communication mode characterized by on/off keying of a radio signal to convey intelligence. Hams use the International Morse Code.

Net — An on-the-air meeting of hams at a set time, day and radio frequency, usually for a specific purpose.

Packet radio — A computer-to-computer radio communication mode in which information is encapsulated in short groups of data called packets. These packets contain addressing and error-detection information.

Phone — Emissions carrying speech or other sound information, such as FM, SSB or AM.

Public service — Activities involving Amateur Radio that hams perform to benefit their communities.

QRP — An abbreviation for low power.

QSL bureau — A system for sending and receiving Amateur Radio verification or "QSL" cards.

QSL cards — Cards that provide written confirmation of a communication between two hams.

QST — The monthly journal of the ARRL. *QST* means "calling all radio amateurs."

RACES (Radio Amateur Civil Emergency Service) — A radio service that uses amateur stations for civil defense communication during periods of local, regional or national civil emergencies.

RF (Radio frequency) — Electromagnetic radiation in the form of radio waves.

Radio (or Ham) shack — The room where Amateur Radio operators keep their station.

Radiotelegraphy — See **Morse code**.

Receiver — A device that converts radio signals into a form that can be heard or viewed.

Repeater — A typically unattended amateur station, usually located on a mountaintop, hilltop or tall building, that automatically and simultaneously receives and retransmits the signals of other stations on a different channel or channels for greater range.

RTTY (radio teletype) — Narrow-band direct-printing radioteletype that uses a digital code.

Space station — An amateur station located more than 50 km above Earth's surface.

SSB (Single sideband) — A common mode of voice of Amateur Radio voice transmission.

SSTV (Slow-scan television) — An operating mode ham radio operators use to exchange still pictures from their stations.

SWL (Shortwave listener) — A person who enjoys listening to shortwave radio broadcasts or Amateur Radio conversations. (A *BCL* is someone who listens for distant AM broadcast stations. Some SWLs also are BCLs.)

TIS (Technical Information Service) — A service of the ARRL that helps hams solve technical problems (**www.arrl.org/tis**).

Transceiver — A radio transmitter and receiver combined in one unit.

Transmitter — A device that produces radio-frequency (RF) signals.

UHF (Ultra-high frequency) — The radio frequencies from 300 to 3000 MHz.

VE (Volunteer Examiner) — An Amateur Radio operator who is qualified to administer Amateur Radio licensing examinations.

VHF (Very-high frequency) — The radio frequency range from 30 to 300 MHz.

WAS (Worked All States) — An ARRL award that is earned when an Amateur Radio operator confirms two-way radio contact with other stations in all 50 US states.

Wavelength — A means of designating a frequency band, such as the 80 meter band.

Work — To contact another ham.

Contents

Chapter 2

Electrical Fundamentals

Building on the work of numerous earlier contributors (most recently Roger Taylor, K9ALD's section on DC Circuits and Resistance), this chapter has been updated by Ward Silver, NØAX. Kai Siwiak, KE4PT, updated the material on Q and on resonant circuits. Look for the section "Radio Mathematics," available on the *Handbook* CD-ROM for a list of online math resources and sections on some of the mathematical techniques used in radio and electronics. Glossaries follow the sections on dc and ac theory, as well.

Chapter 2 — CD-ROM Content

Supplemental Articles

- "Radio Mathematics" — supplemental information about math used in radio and a list of online resources and tutorials about common mathematics
- "Hands-On Radio: Laying Down the Laws" by Ward Silver, NØAX
- "Hands-On Radio: Putting the Laws to Work" by Ward Silver, NØAX
- "Hands-On Radio: Kirchoff's Laws" by Ward Silver, NØAX
- "Hands-On Radio: Thevenin Equivalents" by Ward Silver, NØAX

Projects

- "Thermistors in Homebrew Projects" by Bill Sabin, WØIYH
- "Thermistor Based Temperature Controller" by Bill Sabin, WØIYH

2.1 Introduction to Electricity

The *atom* is the primary building block of matter and is made up of a *nucleus*, containing *protons* and *neutrons*, surrounded by *electrons*. Protons have a positive electrical charge, electrons a negative charge, and neutrons have no electrical charge. An *element* (or *chemical element*) is a type of atom that has a specific number of protons, the element's *atomic number*. Each different element, such as iron, oxygen, silicon, or bromine has a distinct chemical and physical identity determined primarily by the number of protons. A *molecule* is two or more atoms bonded together and acting as a single particle.

Unless modified by chemical, mechanical, or electrical processes, all atoms are electrically neutral because they have the same number of electrons as protons. If an atom loses electrons, it has more protons than electrons and thus has a net positive charge. If an atom gains electrons, it has more electrons than protons and a net negative charge. Atoms or molecules with a positive or negative charge are called *ions*. Electrons not bound to any atom, or *free electrons*, can also be considered as ions because they have a negative charge.

2.1.1 Electric Charge, Voltage and Current

Any piece of matter that has a net positive or negative electrical charge is said to be *electrically charged*. An electrical force exists between electrically charged particles, pushing charges of the same type apart (like charges repel each other) and pulling opposite charges together (opposite charges attract). This is the *electromotive force* (or EMF), also referred to as *voltage* or *potential*. The higher the EMF's voltage, the stronger is its force on an electrical charge.

Under most conditions, the number of positive and negative charges in any volume of space is very close to balanced and so the region has no net charge. When there are extra positive ions in one region and extra negative ions (or electrons) in another region, the resulting EMF attracts the charges toward each other. The direction of the force, from the positive region to the negative region, is called its *polarity*. Because EMF results from an imbalance of charge between two regions, its voltage is always measured between two points, with positive voltage defined as being in the direction from the positively-charged to the negatively-charged region.

If there is no path by which electric charge can move in response to an EMF (called a *conducting path*), the charges cannot move together and so remain separated. If a conducting path is available, then the electrons or ions will flow along the path, neutralizing the net imbalance of charge. The movement of electrical charge is called *electric current*. Materials through which current flows easily are called *conductors*. Most metals, such as copper or aluminum are good conductors. Materials in which it is difficult for current to flow are *insulators*. *Semiconductors*, such as silicon or germanium, are materials with much poorer conductivity than metals. Semiconductors can be chemically altered to acquire properties that make them useful in solid-state devices such as diodes, transistors and integrated circuits.

Voltage differences can be created in a variety of ways. For example, chemical ions can be physically separated to form a battery. The resulting charge imbalance creates a voltage difference at the battery terminals so that if a conductor is connected to both terminals at once, electrons flow between the terminals and gradually eliminate the charge imbalance, discharging the battery's stored energy. Mechanical means such as friction (static electricity, lightning) and moving conductors in a magnetic field (generators) can also produce voltages. Devices or systems that produce voltage are called *voltage sources*.

2.1.2 Electronic and Conventional Current

Electrons move in the direction of positive voltage — this is called *electronic current*. *Conventional current* takes the other point of view — of positive charges moving in the direction of negative voltage. Conventional current was the original model for electricity and results from an arbitrary decision made by Benjamin Franklin in the 18th century when the nature of electricity and atoms was still unknown. It can be imagined as electrons flowing "backward" and is completely equivalent to electronic current.

Conventional current is used in nearly all electronic literature and is the standard used in this book. The direction of conventional current direction establishes the polarity for most electronics calculations and circuit diagrams. The arrows in the drawing symbols for transistors point in the direction of conventional current, for example.

2.1.3 Units of Measurement

To measure electrical quantities, certain definitions have been adopted. Charge is measured in *coulombs* (C) and represented by q in equations. One coulomb is equal to 6.25×10^{18} electrons (or protons). Current, the flow of charge, is measured in *amperes* (A) and represented by i or I in equations. One ampere represents one coulomb of charge flowing past a point in one second and so amperes can also be expressed as coulombs per second. Electromotive force is measured in *volts* (V) and represented by e, E, v, or V in equations. One volt is defined as the electromotive force between two points required to cause one ampere of current to do one *joule* (measure of energy) of work in flowing between the points. Voltage can also be expressed as joules per coulomb.

2.1.4 Series and Parallel Circuits

A *circuit* is any conducting path through which current can flow between two points of that have different voltages. An *open circuit* is a circuit in which a desired conducting path is interrupted, such as by a broken wire or a switch. A *short circuit* is a circuit in which a conducting path allows current to flow directly between the two points at different voltages.

The two fundamental types of circuits are shown in **Fig 2.1**. Part A shows a *series circuit* in which there is only one current path. The current in this circuit flows from the voltage source's positive terminal (the symbol for a battery is shown with its voltage polarity shown as + and –) in the direction shown by the arrow through three *resistors* (electronic components discussed below) and back to

Schematic Diagrams

The drawing in Fig 2.1 is a *schematic diagram*. Schematics are used to show the electrical connections in a circuit without requiring a drawing of the actual components or wires, called a *pictorial diagram*. Pictorials are fine for very simple circuits like these, but quickly become too detailed and complex for everyday circuits. Schematics use lines and dots to represent the conducting paths and connections between them. Individual electrical devices and electronic components are represented by *schematic symbols* such as the resistors shown here. A set of the most common schematic symbols is provided in the **Component Data and References** chapter. You will find additional information on reading and drawing schematic diagrams in the ARRL Web site Technology section at **www.arrl.org/circuit-construction**.

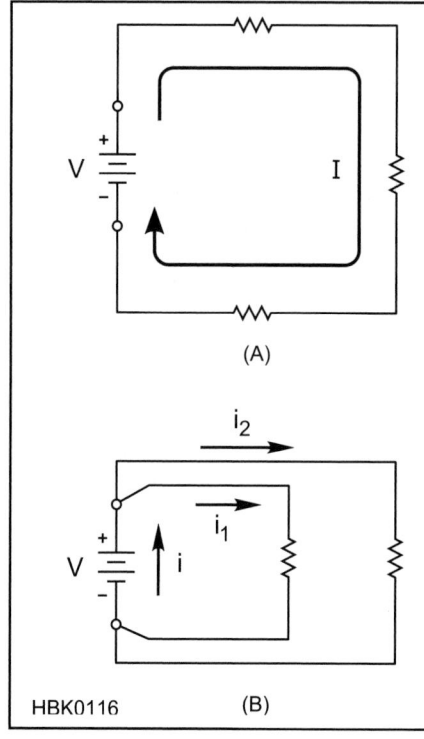

Fig 2.1 — A series circuit (A) has the same current through all components. Parallel circuits (B) apply the same voltage to all components.

the battery's negative terminal. Current is the same at every point in a series circuit.

Part B shows a *parallel circuit* in which there are multiple paths for the current to take. One terminal of both resistors is connected to the battery's positive terminal. The other terminal of both resistors is connected to the

battery's negative terminal. Current flowing out of the battery's positive terminal divides into smaller currents that flow through the individual resistors and then recombine at the battery's negative terminal. All of the components in a parallel circuit experience the same voltage. All circuits are made up of series and parallel combinations of components and sources of voltage and current.

2.1.5 Glossary — DC and Basic Electricity

Alternating current (ac) — A flow of charged particles through a conductor, first in one direction, then in the other direction.

Ampere — A measure of flow of charged particles per unit of time. One ampere (A) represents one coulomb of charge flowing past a point in one second.

Atom — The smallest particle of matter that makes up a distinct chemical element. Atoms consist of protons and neutrons in the central region called the nucleus, with electrons surrounding the nucleus.

Circuit — Conducting path between two points of different voltage. In a *series circuit*, there is only one current path. In a *parallel circuit*, there are multiple current paths.

Conductance (G) — The reciprocal of resistance, measured in siemens (S).

Conductor — Material in which electrons or ions can move easily.

Conventional Current — Current defined as the flow of positive charges in the direction of positive to negative voltage. Conventional current flows in the opposite direction of electronic current,

When Is E a V and V an E?

Beginners in electronics are often confused about the interchange of V and E to refer to voltage in a circuit. When should each be used? Unfortunately, there is no convention. Nevertheless, in ham radio E is usually used when referring to an electric field or the electromotive force around a circuit. E is also commonly used to state Ohm's Law: I = E / R. V is used when describing the difference in voltage between two points in a circuit or the terminal voltage of a power supply or battery. V is always used when referring to units of volts.

the flow of negative charges (electrons) from negative to positive voltage.

Coulomb — A unit of measure of a quantity of electrically charged particles. One coulomb (C) is equal to 6.25×10^{18} electrons.

Current (I) — The movement of electrical charge, measured in amperes and represented by *i* in equations.

Direct current (dc) — A flow of charged particles through a conductor in one direction only.

Electronic current — see **Conventional Current**

EMF — Electromotive Force is the term used to define the force of attraction or repulsion between electrically-charged regions. Also called *voltage* or *potential*.

Energy — Capability of doing work. It is usually measured in electrical terms as the number of watts of power consumed during a specific period of time, such as watt-seconds or kilowatt-hours.

Insulator — Material in which it is difficult for electrons or ions to move.

Ion — Atom or molecule with a positive or negative electrical charge.

Joule — Measure of a quantity of energy. One joule is defined as one newton (a measure of force) acting over a distance of one meter.

Ohm — Unit of resistance. One ohm is defined as the resistance that will allow one ampere of current when one volt of EMF is impressed across the resistance.

Polarity — The direction of EMF or voltage, from positive to negative.

Potential — See **EMF**.

Power — Power is the rate at which work is done. One watt of power is equal to one volt of EMF causing a current of one ampere through a resistor.

Resistance (R) — Opposition to current by conversion into other forms of energy, such as heat, measured in ohms (Ω).

Volt, voltage — See **EMF**.

Voltage source — Device or system that creates a voltage difference at its terminals.

2.2 Resistance and Conductance

2.2.1 Resistance

Any conductor connected to points at different voltages will allow current to pass between the points. No conductor is perfect or lossless, however, at least not at normal temperatures. The moving electrons collide with the atoms making up the conductor and lose some of their energy by causing the atoms to vibrate, which is observed externally as heat. The property of energy loss due to interactions between moving charges and the atoms of the conductor is called *resistance*. The amount of resistance to current is measured in *ohms* (Ω) and is represented by *r* or R in equations.

Suppose we have two conductors of the same size and shape, but of different materials. Because all materials have different internal structures, the amount of energy lost by current flowing through the material is also different. The material's ability to impede current flow is its *resistivity*. Numerically, the resis-

tivity of a material is given by the resistance, in ohms, of a cube of the material measuring one centimeter on each edge. The symbol for resistivity is the Greek letter rho, ρ.

The longer a conductor's physical path, the higher the resistance of that conductor. For direct current and low-frequency alternating currents (up to a few thousand hertz) the conductor's resistance is inversely proportional to the cross-sectional area of the conductor. Given two conductors of the same material and having the same length, but differing in cross-sectional area, the one with the larger area (for example, a thicker wire or sheet) will have the lower resistance.

One of the best conductors is copper, and it is frequently convenient to compare the resistance of a material under consideration with that of a copper conductor of the same size and shape. **Table 2.1** gives the ratio of the resistivity of various conductors to the resistivity of copper.

2.2.2 Conductance

The reciprocal of resistance (1/R) is *conductance*. It is usually represented by the symbol G. A circuit having high conductance has low resistance, and vice versa. In radio work, the term is used chiefly in connection with electron-tube and field-effect transistor characteristics. The units of conductance are siemens (S). A resistance of 1 Ω has a conductance of 1 S, a resistance of 1000 Ω has a conductance of 0.001 S, and so on. A unit frequently used in regards to vacuum tubes and the field-effect transistor is the μS or one millionth of a siemens. It is the conductance of a 1-MΩ resistance. Siemens have replaced the obsolete unit *mho* (abbreviated as an upside-down Ω symbol).

2.2.3 Ohm's Law

The amount of current that will flow through a conductor when a given EMF is

Table 2.1
Relative Resistivity of Metals

Material	Resistivity Compared to Copper
Aluminum (pure)	1.60
Brass	3.7-4.90
Cadmium	4.40
Chromium	8.10
Copper (hard-drawn)	1.03
Copper (annealed)	1.00
Gold	1.40
Iron (pure)	5.68
Lead	12.80
Nickel	5.10
Phosphor bronze	2.8-5.40
Silver	0.94
Steel	7.6-12.70
Tin	6.70
Zinc	3.40

The Origin of Unit Names

Many units of measure carry names that honor scientists who made important discoveries in or advanced the state of scientific knowledge of electrical and radio phenomena. For example, Georg Ohm (1787-1854) discovered the relationship between current, voltage and resistance that now bears his name as Ohm's Law and as the unit of resistance, the ohm. The following table lists the most common electrical units and the scientists for whom they are named. You can find more information on these and other notable scientists in encyclopedia entries on the units that bear their names.

Electrical Units and Their Namesakes

Unit	Measures	Named for
Ampere	Current	Andree Ampere 1775 -1836
Coulomb	Charge	Charles Coulomb 1736-1806
Farad	Capacitance	Michael Faraday 1791-1867
Henry	Inductance	Joseph Henry 1797-1878
Hertz	Frequency	Heinrich Hertz 1857-1894
Ohm	Resistance	Georg Simon Ohm 1787-1854
Watt	Power	James Watt 1736-1819
Volt	Voltage	Alessandro Volta 1745-1827

applied will vary with the resistance of the conductor. The lower the resistance, the greater the current for a given EMF. One ohm is defined as the amount of resistance that allows one ampere of current to flow between two points that have a potential difference of one volt. This proportional relationship is known as *Ohm's Law*:

$$R = \frac{E}{I} \qquad (1)$$

where

R = resistance in ohms,
E = potential or EMF in volts and
I = current in amperes.

Transposing the equation gives the other common expressions of Ohm's Law as:

$$E = I \times R \qquad (2)$$

and

$$I = \frac{E}{R} \qquad (3)$$

All three forms of the equation are used often in electronics and radio work. You must remember that the quantities are in volts, ohms and amperes; other units cannot be used in the equations without first being converted. For example, if the current is in milliamperes you must first change it to the equivalent fraction of an ampere before substituting the value into the equations.

The following examples illustrate the use of Ohm's Law in the simple circuit of **Fig 2.2**. If 150 V is applied to a circuit and the current is measured as 2.5 A, what is the resistance of the circuit? In this case R is the unknown, so we will use equation 1:

$$R = \frac{E}{I} = \frac{150\ V}{2.5\ A} = 60\ \Omega$$

No conversion of units was necessary because the voltage and current were given in volts and amperes.

If the current through a 20,000-Ω resistance is 150 mA, what is the voltage? To find voltage, use equation 2. Convert the current from milliamperes to amperes by dividing by 1000 mA / A (or multiplying by 10^{-3} A / mA) so that 150 mA becomes 0.150 A. (Notice the conversion factor of 1000 does not limit the number of significant figures in the calculated answer.)

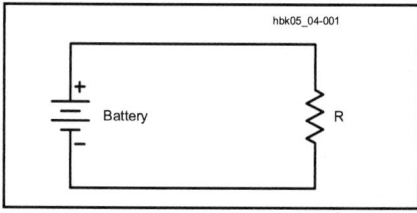

Fig 2.2 — A simple circuit consisting of a battery and a resistor.

$$I = \frac{150\ mA}{1000\ \frac{mA}{A}} = 0.150\ A$$

Then:

$$E = 0.150\ A \times 20{,}000\ \Omega = 3000\ V$$

In a final example, how much current will flow if 250 V is applied to a 5000-Ω resistor? Since I is unknown,

$$I = \frac{E}{R} = \frac{250\ V}{5000\ \Omega} = 0.05\ A$$

It is more convenient to express the current in mA, and 0.05 A × 1000 mA / A = 50 mA.

It is important to note that Ohm's Law applies in any portion of a circuit as well as to the circuit as a whole. No matter how many resistors are connected together or how they are connected together, the relationship between the resistor's value, the voltage across the resistor, and the current through the resistor still follows Ohm's Law.

2.2.4 Resistance of Wires

The problem of determining the resistance of a round wire of given diameter and length — or the converse, finding a suitable size and length of wire to provide a desired amount of resistance — can easily be solved with the help of the copper wire table given in the chapter on **Component Data and References**. This table gives the resistance, in ohms per 1000 ft, of each standard wire size. For example, suppose you need a resistance of 3.5 Ω, and some #28 AWG wire is on hand. The wire table shows that #28 AWG wire has a resistance of 65.31 Ω / 1000 ft. Since the desired resistance is 3.5 Ω, the required length of wire is:

$$Length = \frac{R_{DESIRED}}{\frac{R_{WIRE}}{1000\ ft}} = \frac{3.5\ \Omega}{\frac{66.17\ \Omega}{1000\ ft}} \qquad (4)$$

$$= \frac{3.5\ \Omega \times 1000\ ft}{65.31\ \Omega} = 55\ ft$$

As another example, suppose that the resistance of wire in a radio's power cable must not exceed 0.05 Ω and that the length of wire required for making the connections totals 14 ft. Then:

$$\frac{R_{WIRE}}{1000\ ft} < \frac{R_{MAXIMUM}}{Length} = \frac{0.05\ \Omega}{14.0\ ft} \qquad (5)$$

$$= 3.57 \times 10^{-3}\ \frac{\Omega}{ft} \times \frac{1000\ ft}{1000\ ft}$$

$$\frac{R_{WIRE}}{1000\ ft} < \frac{3.57\ \Omega}{1000\ ft}$$

Find the value of R_{WIRE} / 1000 ft that is less than the calculated value. The wire table shows that #15 AWG is the smallest size having a resistance less than this value. (The resistance of #15 AWG wire is given as 3.1810 Ω / 1000 ft.) Select any wire size larger than this for the connections in your circuit, to ensure that the total wire resistance will be less than 0.05 Ω.

When the wire in question is not made of copper, the resistance values in the wire table should be multiplied by the ratios shown in Table 2.1 to obtain the resulting resistance. If the wire in the first example were made from nickel instead of copper, the length required for 3.5 Ω would be:

$$Length = \dfrac{R_{DESIRED}}{\dfrac{R_{WIRE}}{1000 \text{ ft}}}$$ (6)

$$= \dfrac{3.5\,\Omega}{\dfrac{65.31\,\Omega}{1000 \text{ ft}} \times 5.1} = \dfrac{3.5\,\Omega \times 1000 \text{ ft}}{66.17\,\Omega \times 5.1}$$

$$Length = \dfrac{3500 \text{ ft}}{337.5} = 10.5 \text{ ft}$$

2.2.5 Temperature Effects

The resistance of a conductor changes with its temperature. The resistance of practically every metallic conductor increases with increasing temperature. Carbon, however, acts in the opposite way; its resistance decreases when its temperature rises. It is seldom necessary to consider temperature in making resistance calculations for amateur work. The temperature effect is important when it is necessary to maintain a constant resistance under all conditions, however. Special materials that have little or no change in resistance over a wide temperature range are used in that case.

2.2.6 Resistors

A package of material exhibiting a certain amount of resistance and made into a single unit is called a *resistor*. (See the **Component Data and References** chapter for information on resistor value marking conventions.) The size and construction of resistors having the same value of resistance in ohms may vary considerably based on how much power they are intended to dissipate, how much voltage is expected to be applied to them, and so forth (see **Fig 2.3**).

TYPES OF RESISTORS

Resistors are made in several different ways: carbon composition, metal oxide, carbon film, metal film and wirewound. In some circuits, the resistor value may be critical. In this case, precision resistors are used. These are typically wirewound or carbon-film devices whose values are carefully controlled during manufacture. In addition, special material or construction techniques may be used to provide temperature compensation, so the value does not change (or changes in a precise manner) as the resistor temperature changes.

Carbon composition resistors are simply small cylinders of carbon mixed with various binding agents to produce any desired resistance. The most common sizes of "carbon comp" resistors are ½- and ¼-W resistors. They are moderately stable from 0 to 60 °C (their resistance increases above and below this temperature range). They can absorb short overloads better than film-type resistors, but they are relatively noisy, and have relatively wide tolerances. Because carbon composition resistors tend to be affected by humidity and other environmental factors and because they are difficult to manufacture in surface-mount packages, they have largely been replaced by film-type resistors.

Metal-oxide resistors are similar to carbon composition resistors in that the resistance is supplied by a cylinder of metal oxide. Metal-oxide resistors have replaced carbon composition resistors in higher power applications because they are more stable and can operate at higher temperatures.

Wirewound resistors are made from wire, which is cut to the proper length and wound on a coil form (usually ceramic). They are capable of handling high power; their values are very stable, and they are manufactured to close tolerances.

Metal-film resistors are made by depositing a thin film of aluminum, tungsten or other metal on an insulating substrate. Their resistances are controlled by careful adjustments of the width, length and depth of the film. As a result, they have very tight tolerances. They are used extensively in surface-mount technology. As might be expected, their power handling capability is somewhat limited. They also produce very little electrical noise.

Carbon-film resistors use a film of carbon mixed with other materials instead of metal. They are not quite as stable as other film resistors and have wider tolerances than metal-film resistors, but they are still as good as (or better than) carbon composition resistors.

THERMAL CONSIDERATIONS FOR RESISTORS

Current through a resistance causes the conductor to become heated; the higher the resistance and the larger the current, the greater the amount of heat developed. Resistors intended for carrying large currents must be physically large so the heat can be radiated quickly to the surrounding air or some type of heat sinking material. If the resistor does not dissipate the heat quickly, it may get hot enough to melt or burn.

The amount of heat a resistor can safely dissipate depends on the material, surface area and design. Typical resistors used in amateur electronics (⅛ to 2-W resistors) dissipate heat primarily through the surface area of the case, with some heat also being carried away through the connecting leads. Wirewound resistors are usually used for higher power levels. Some have finned cases for better convection cooling and/or metal cases for better conductive cooling.

The major departure of resistors from ideal behavior at low-frequencies is their *temperature coefficient* (TC). (See the chapter on **RF Techniques** for a discussion of the behavior of

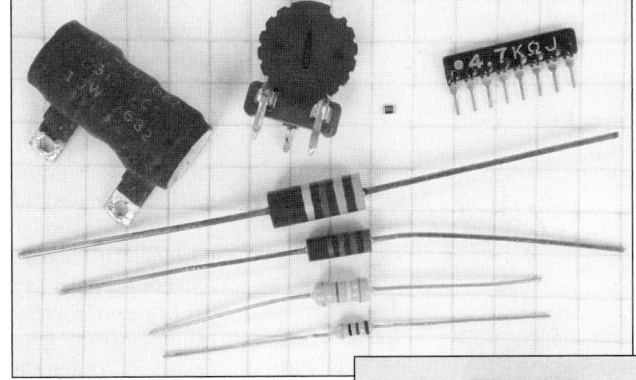

Fig 2.3 — Examples of various resistors. At the top left is a small 10-W wirewound resistor. A single in-line package (SIP) of resistors is at the top right. At the top center is a small PC-board-mount variable resistor. A tiny surface-mount (chip) resistor is also shown at the top. Below the variable resistor is a 1-W carbon composition resistor and then a ½-W composition unit. The dog-bone-shaped resistors at the bottom are ½-W and ¼-W film resistors. The ¼-inch-ruled graph paper background provides a size comparison. The inset photo shows the chip resistor with a penny for size comparison.

Table 2.2

Temperature Coefficients for Various Resistor Compositions

1 PPM = 1 part per million = 0.0001%

Type	TC (PPM/°C)
Wire wound	±(30 - 50)
Metal Film	±(100 - 200)
Carbon Film	+350 to −800
Carbon composition	±800

resistors at high frequencies.) The resistivity of most materials changes with temperature, and typical TC values for resistor materials are given in **Table 2.2**. TC values are usually expressed in parts-per-million (PPM) for each degree (centigrade) change from some nominal temperature, usually room temperature (77 °F or 27 °C). A positive TC indicates an increase in resistance with increasing temperature while a negative TC indicates a decreasing resistance. For example, if a 1000-Ω resistor with a TC of +300 PPM/°C is heated to 50 °C, the *change* in resistance is $300 \times (50 - 27) = 6900$ PPM, yielding a new resistance of

$$1000 \left(1 + \frac{6900}{1,000,000}\right) = 1006.9 \ \Omega$$

Carbon-film resistors are unique among the major resistor families because they alone have a negative temperature coefficient. They are often used to "offset" the thermal effects of the other components.

If the temperature increase is small (less than 30-40 °C), the resistance change with temperature is nondestructive — the resistor will return to normal when the temperature returns to its nominal value. Resistors that get too hot to touch, however, may be permanently damaged even if they appear normal. For this reason, be conservative when specifying power ratings for resistors. It's common to specify a resistor rated at 200% to 400% of the expected dissipation.

POTENTIOMETERS

Potentiometer (pronounced po-ten-tchee-AH-meh-tur) is a formal name for a variable resistor and the common name for these components is "pots." A typical potentiometer consists of a circular pattern of resistive material, usually a carbon compound similar to that used in carbon composition resistors, with a wiper contact on a shaft moving across the material. For higher power applications, the resistive material may be wire, wound around a core, like a wirewound resistor.

As the wiper moves along the material, more resistance is introduced between the wiper and one of the fixed contacts on the material. A potentiometer may be used to control current, voltage or resistance in a circuit. **Fig 2.4** shows several different types of potentiometers. **Fig 2.5** shows the schematic symbol for a potentiometer and how changing the position of the shaft changes the resistance between its three terminals. The figure shows a *panel pot*, designed to be

Fig 2.4 — This photo shows examples of different styles of potentiometers. The ¼-inch-ruled graph paper background provides a size comparison.

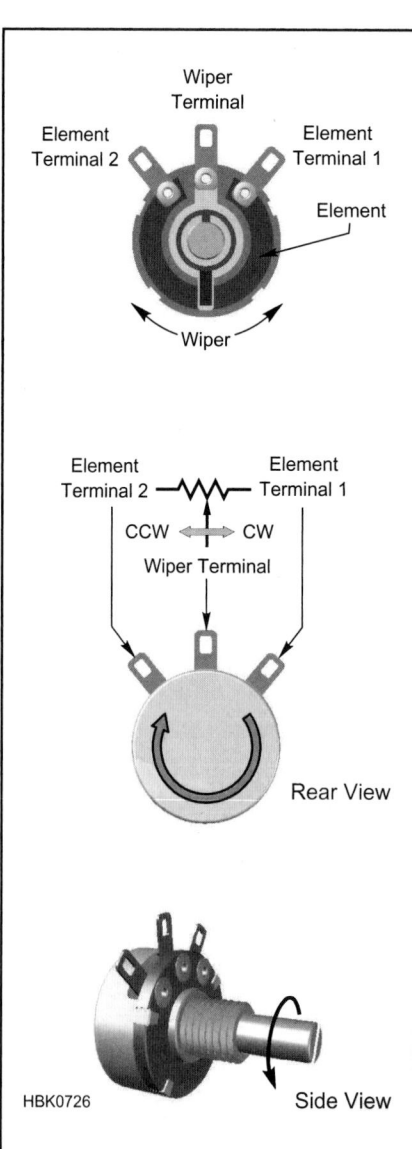

Fig 2.5 — Typical potentiometer construction and schematic symbol. Rotation on the shaft moves the wiper along the element, changing the resistance between the wiper terminal and the element terminals. Moving the wiper closer to an element terminal

mounted on an equipment panel and adjusted by an operator. The small rectangular *trimmer* potentiometers in Fig 2.5 are adjusted with a screwdriver and have wire terminals.

Typical specifications for a potentiometer include element resistance, power dissipation, voltage and current ratings, number of turns (or degrees) the shaft can rotate, type and size of shaft, mounting arrangements and resistance "taper."

Not all potentiometers have a *linear* taper. That is, the change in resistance may not be the same for a given number of degrees of shaft rotation along different portions of the resistive material. These are called *nonlinear tapers*

A typical use of a potentiometer with a nonlinear taper is as a volume control in an audio amplifier. Since the human ear has a logarithmic response to sound, a volume control must change the amplifier output much more near one end of the potentiometer than the other (for a given amount of rotation) so that the "perceived" change in volume is about the same for a similar change in the control's position. This is commonly called an *audio taper* or *log taper* as the change in resistance per degree of rotation attempts to match the response of the human ear. Tapers can be designed to match almost any desired control function for a given application. Linear and audio tapers are the most common tapers.

2.3 Basic Circuit Principles

Circuits are composed of *nodes* and *branches*. A node is any point in the circuit at which current can divide between conducting paths. For example, in the parallel circuit of **Fig 2.6**, the node is represented by the schematic dot. A branch is any unique conducting path between nodes. A series of branches that make a complete current path, such as the series circuit of Fig 2.1A, is called a *loop*.

Very few actual circuits are as simple as those shown in Fig 2.1. However, all circuits, no matter how complex, are constructed of combinations of these series and parallel circuits. We will now use these simple circuits of resistors and batteries to illustrate two fundamental rules for voltage and current, known as *Kirchoff's Laws*.

2.3.1 Kirchhoff's Current Law

Kirchhoff's Current Law (KCL) states, *"The sum of all currents flowing into a node and all currents flowing out of a node is equal to zero."* KCL is stated mathematically as:

$$(I_{in1} + I_{in2} + \cdots) - (I_{out1} + I_{out2} + \cdots) = 0$$
$$(7A)$$

The dots indicate that as many currents as necessary may be added.

Another way of stating KCL is that the sum of all currents flowing into a node must balance the sum of all currents flowing out of a node as shown in equation 7B:

$$(I_{in1} + I_{in2} + \cdots) = (I_{out1} + I_{out2} + \cdots) \quad (7B)$$

Equations 7A and 7B are mathematically equivalent.

KCL is illustrated by the following example. Suppose three resistors (5.0 kΩ, 20.0 kΩ and 8.0 kΩ) are connected in parallel as shown in Fig 2.6. The same voltage, 250 V, is applied to all three resistors. The current through R1 is I1, I2 is the current through R2 and I3 is the current through R3.

The current in each can be found from Ohm's Law, as shown below. For convenience, we can use resistance in kΩ, which gives current in milliamperes.

$$I1 = \frac{E}{R1} = \frac{250\ V}{5.0\ k\Omega} = 50.0\ mA$$

$$I2 = \frac{E}{R2} = \frac{250\ V}{20.0\ k\Omega} = 12.5\ mA$$

$$I3 = \frac{E}{R3} = \frac{250\ V}{8.0\ k\Omega} = 31.2\ mA$$

Notice that the branch currents are inversely proportional to the resistances. The 20,000-Ω resistor has a value four times larger than the 5000-Ω resistor, and has a current

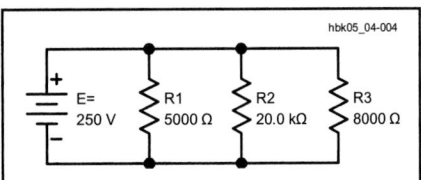

Fig 2.6 — An example of resistors in parallel.

one-quarter as large. If a resistor has a value twice as large as another, it will have half as much current through it when they are connected in parallel.

Using the balancing form of KCL (equation 7B) the current that must be supplied by the battery is therefore:

$$I_{BATT} = I1 + I2 + I3$$

$$I_{BATT} = 50.0\ mA + 12.5\ mA + 31.2\ mA$$

$$I_{BATT} = 93.7\ mA$$

2.3.2 Resistors in Parallel

In a circuit made up of resistances in parallel, the resistors can be represented as a single *equivalent* resistance that has the same value as the parallel combination of resistors. In a parallel circuit, the equivalent resistance is less than that of the lowest resistance value present. This is because the total current is always greater than the current in any individual resistor. The formula for finding the equivalent resistance of resistances in parallel is:

$$R_{EQUIV} = \frac{1}{\dfrac{1}{R1} + \dfrac{1}{R2} + \dfrac{1}{R3} + \dfrac{1}{R4} \cdots} \quad (8)$$

where the dots indicate that any number of parallel resistors can be combined by the same method. In the example of the previous section, the equivalent resistance is:

$$R = \frac{1}{\dfrac{1}{5000\ \Omega} + \dfrac{1}{20\ k\Omega} + \dfrac{1}{8000\ \Omega}} = 2.67\ k\Omega$$

The notation "//" (two slashes) is frequently used to indicate "in parallel with." Using that notation, the preceding example would be given as "5000 Ω // 20 kΩ // 8000 Ω."

For only two resistances in parallel (a very common case) the formula can be reduced to the much simpler (and easier to remember):

$$R_{EQUIV} = \frac{R1 \times R2}{R1 + R2} \quad (9)$$

Example: If a 500-Ω resistor is connected in parallel with one of 1200 Ω, what is the total resistance?

$$R = \frac{R1 \times R2}{R1 + R2} = \frac{500\ \Omega \times 1200\ \Omega}{500\ \Omega + 1200\ \Omega}$$

$$R = \frac{600,000\ \Omega^2}{1700\ \Omega} = 353\ \Omega$$

Any number of parallel resistors can be combined two at a time by using equation 9 until all have been combined into a single equivalent. This is a bit easier than using equation 8 to do the conversion in a single step.

CURRENT DIVIDERS

Resistors connected in parallel form a circuit called a *resistive current divider*. For any number of resistors connected in parallel (R1, R2, R3, ... R4), the current through one of the resistors, R_n, is equal to the sum of all resistor currents multiplied by the ratio of the equivalent of all parallel resistors *except* R_n to the sum of R_n and the equivalent value.

$$I_n = I_{TOT} (R_{EQ}/R_n + R_{EQ})$$

For example, in a circuit with three parallel resistors; R1, R2, and R3, the current through R2 is equal to:

$$I_2 = I \frac{R1//R3}{R2 + R1//R3}$$

where I is the total current through all the resistors and // indicates the equivalent parallel value. If I = 100 mA, R1 = 100 Ω, R2 = 50 Ω, and R3 = 200 Ω:

$$100\ mA \frac{(100//200)}{(50 + 100//200)} = 100 \frac{66.7}{116.7} = 57.2\ mA$$

2.3.3 Kirchhoff's Voltage Law

Kirchhoff's Voltage Law (KVL) states, *"The sum of the voltages around a closed current loop is zero."* Where KCL is somewhat intuitive, KVL is not as easy to visualize. In the circuit of **Fig 2.7**, KVL requires that the battery's voltage must be balanced exactly by the voltages that appear across the three resistors in the circuit. If it were not, the "extra" voltage would create an infinite current with no limiting resistance, just as KCL prevents

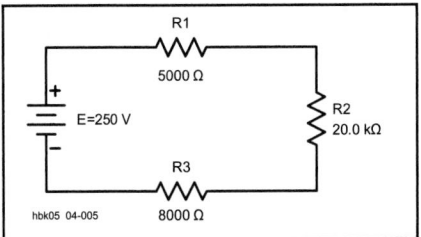

Fig 2.7 — An example of resistors in series.

charge from "building up" at a circuit node. KVL is stated mathematically as:

$$E_1 + E_2 + E_3 + \cdots = 0 \qquad (10)$$

where each E represents a voltage encountered by current as it flows around the circuit loop.

This is best illustrated with an example. Although the current is the same in all three of the resistances in the example of Fig 2.7, the total voltage divides between them, just as current divides between resistors connected in parallel. The voltage appearing across each resistor (the *voltage drop*) can be found from Ohm's Law. (Voltage across a resistance is often referred to as a "drop" or "I-R drop" because the value of the voltage "drops" by the amount $E = I \times R$.)

For the purpose of KVL, it is common to assume that if current flows *into* the more positive terminal of a component the voltage is treated as positive in the KVL equation. If the current flows *out* of a positive terminal, the voltage is treated as negative in the KVL. Positive voltages represent components that consume or "sink" power, such as resistors. Negative voltages represent components that produce or "source" power, such as batteries. This allows the KVL equation to be written in a balancing form, as well:

$$(E_{source1} + E_{source2} + \cdots) =$$
$$(E_{sink1} + E_{sink2} + \cdots)$$

All of the voltages are treated as positive in this form, with the power sources (current flowing *out* of the more positive terminal) on one side and the power sinks (current flowing *into* the more positive terminal) on the other side.

Note that it doesn't matter what a component terminal's *absolute* voltage is with respect to ground, only which terminal of the component is more positive than the other. If one side of a resistor is at +1000 V and the other at +998 V, current flowing into the first terminal and out of the second experiences a +2 V voltage drop. Similarly, current supplied by a 9 V battery with its positive terminal at –100 V and its negative terminal at –108.5 V still counts for KVL as an 8.5 V power source. Also note that current can flow *into* a battery's positive terminal, such as during recharging, making the battery a power sink, just like a resistor.

Here's an example showing how KVL works: In Fig 2.7, if the voltage across R1 is E1, that across R2 is E2 and that across R3 is E3, then:

$$-250 + I \times R1 + I \times R2 + I \times R3 = 0$$

This equation can be simplified to:

$$-250 + I(R1 + R2 + R3) =$$
$$-250 + I(33,000\ \Omega) = 0$$

Solving for I gives I = 250 / 33000 = 0.00758 A = 7.58 mA. This allows us to calculate the value of the voltage across each resistor:

E1 = I × R1 = 0.00758 A × 5000 Ω = 37.9 V

E2 = I × R2 = 0.00758 A × 20,000 Ω = 152 V

E3 = I × R3 = 0.00758 A × 8000 Ω = 60.6 V

Verifying that the sum of E1, E2, and E3 does indeed equal the battery voltage of 250 V ignoring rounding errors.

$$E_{TOTAL} = E1 + E2 + E3$$
$$E_{TOTAL} = 37.9\ V + 152\ V + 60.6\ V$$
$$E_{TOTAL} = 250\ V$$

2.3.4 Resistors in Series

The previous example illustrated that in a circuit with a number of resistances connected in series, the equivalent resistance of the circuit is the sum of the individual resistances. If these are numbered R1, R2, R3 and so on, then:

$$R_{EQUIV} = R1 + R2 + R3 + R4... \qquad (11)$$

Example: Suppose that three resistors are connected to a source of voltage as shown in Fig 2.7. The voltage is 250 V, R1 is 5.0 kΩ, R2 is 20.0 kΩ and R3 is 8.0 kΩ. The total resistance is then

$$R_{EQUIV} = R1 + R2 + R3$$
$$R_{EQUIV} = 5.0\ k\Omega + 20.0\ k\Omega + 8.0\ k\Omega$$
$$R_{EQUIV} = 33.0\ k\Omega$$

The current in the circuit is then

$$I = \frac{E}{R} = \frac{250\ V}{33.0\ k\Omega} = 7.58\ mA$$

VOLTAGE DIVIDERS

Notice that the voltage drop across each resistor in the KVL example is directly proportional to the resistance. The value of the 20,000 Ω resistor is four times larger than the 5000 Ω resistor, and the voltage drop across the 20,000 Ω resistor is four times larger. A resistor that has a value twice as large as another will have twice the voltage drop across it when they are connected in series.

Resistors in series without any other connections form a *resistive voltage divider*. (Other types of components can form voltage dividers, too.) The voltage across any specific resistor in the divider, R_n, is equal to the voltage across the entire string of resistors multiplied by the ratio of R_n to the sum of all resistors in the string.

For example, in the circuit of Fig 2.7, the voltage across the 5000 Ω resistor is:

$$E1 = 250\ \frac{5000}{5000 + 20,000 + 8000} = 37.9\ V$$

This is a more convenient method than calculating the current through the resistor and using Ohm's Law.

Voltage dividers can be used as a source of voltage. As long as the device connected to the output of the divider has a much higher resistance than the resistors in the divider, there will be little effect on the divider output voltage. For example, for a voltage divider with a voltage of E = 15 V and two resistors of R1 = 5 kΩ and R2 = 10 kΩ, the voltage across R2 will be 10 V measured on a high-impedance voltmeter because the measurement draws very little current from the divider. However, if the measuring device or load across R2 draws significant current, it will increase the amount of current drawn through the divider and change the output voltage.

A good rule of thumb is that the load across a voltage divider's output should be at least ten times the value of the highest resistor in the divider to stay reasonably close to the required voltage. As the load resistance gets closer to the value of the divider resistors, the current drawn by the load affects the voltage division and causes changes from the desired value.

Potentiometers (variable resistors described previously) are often used as adjustable voltage dividers and this is how they got their name. Potential is an older name for voltage and a "potential-meter" is a device that can "meter" or adjust potential, thus potentiometer.

2.3.5 Conductances in Series and Parallel

Since conductance is the reciprocal of resistance, G = 1/R, the formulas for combining resistors in series and in parallel can be converted to use conductance by substituting 1/G for R. Substituting 1/G into equation 11 shows that conductances in series are combined this way:

$$G = \frac{1}{\dfrac{1}{G1} + \dfrac{1}{G2} + \dfrac{1}{G3} + \dfrac{1}{G4}...}$$

and two conductances in series may be combined in a manner similar to equation 9:

$$G_{EQUIV} = \frac{G1 \times G2}{G1 + G2}$$

Substituting 1/G into equation 8 shows that conductances in parallel are combined this way:

$$G_{TOTAL} = G1 + G2 + G3 + G4...$$

This also shows that when faced with a large number of parallel resistances, converting them to conductances may make the math a little easier to deal with.

2.3.6 Equivalent Circuits

A circuit may have resistances both in parallel and in series, as shown in **Fig 2.8A**. In order to analyze the behavior of such a circuit, *equivalent circuits* are created and combined by using the equations for combining resistors in series and resistors in parallel. Each separate combination of resistors, series or parallel, can be reduced to a single equivalent resistor. The resulting combinations can be reduced still further until only a single resistor remains.

The method for analyzing the circuit of Fig 2.8A is as follows: Combine R2 and R3 to create the equivalent single resistor, R_{EQ} whose value is equal to R2 and R3 in parallel.

$$R_{EQ} = \frac{R2 \times R3}{R2 + R3} = \frac{20,000\,\Omega \times 8000\,\Omega}{20,000\,\Omega + 8000\,\Omega}$$

$$= \frac{1.60 \times 10^8\,\Omega^2}{28,000\,\Omega} = 5710\,\Omega = 5.71\text{ k}\Omega$$

This resistance in series with R1 then forms a simple series circuit, as shown in Fig 2.8B. These two resistances can then be combined into a single equivalent resistance, R_{TOTAL}, for the entire circuit:

$$R_{TOTAL} = R1 + R_{EQ} = 5.0\text{ k}\Omega + 5.71\text{ k}\Omega$$

$$R_{TOTAL} = 10.71\text{ k}\Omega$$

The battery current is then:

$$I = \frac{E}{R} = \frac{250\text{ V}}{10.71\text{ k}\Omega} = 23.3\text{ mA}$$

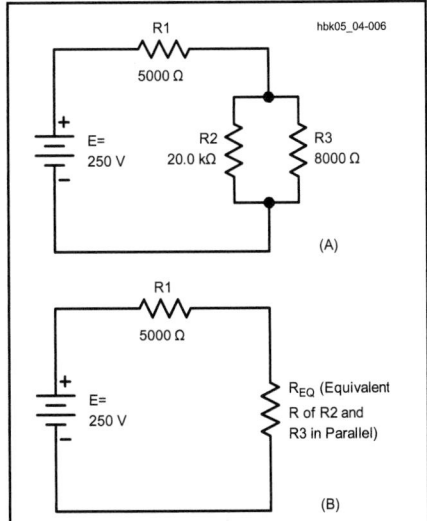

Fig 2.8 — At A, an example of resistors in series-parallel. The equivalent circuit is shown at B.

The voltage drops across R1 and R_{EQ} are:

$$E1 = I \times R1 = 23.3\text{ mA} \times 5.0\text{ k}\Omega = 117\text{ V}$$

$$E2 = I \times R_{EQ} = 23.3\text{ mA} \times 5.71\text{ k}\Omega = 133\text{ V}$$

These two voltage drops total 250 V, as described by Kirchhoff's Voltage Law. E2 appears across both R2 and R3 so,

$$I2 = \frac{E2}{R2} = \frac{133\text{ V}}{20.0\text{ k}\Omega} = 6.65\text{ mA}$$

$$I3 = \frac{E3}{R3} = \frac{133\text{ V}}{8.0\text{ k}\Omega} = 16.6\text{ mA}$$

where
  I2 = current through R2 and
  I3 = current through R3.

The sum of I2 and I3 is equal to 23.3 mA, conforming to Kirchhoff's Current Law.

2.3.7 Voltage and Current Sources

In designing circuits and describing the behavior of electronic components, it is often useful to use *ideal sources*. The two most common types of ideal sources are the *voltage source* and the *current source*, symbols for which are shown in **Fig 2.9**. These sources are considered ideal because no matter what circuit is connected to their terminals, they continue to supply the specified amount of voltage or current. Practical voltage and current sources can approximate the behavior of an ideal source over certain ranges, but are limited in the amount of power they can supply and so under excessive load, their output will drop.

Voltage sources are defined as having zero *internal impedance*, where impedance is a more general form of resistance as described in the sections of this chapter dealing with alternating current. A short circuit across an ideal voltage source would result in the source providing an infinite amount of current. Practical voltage sources have non-zero internal impedance and this also limits the amount of power they can supply. For example, placing a

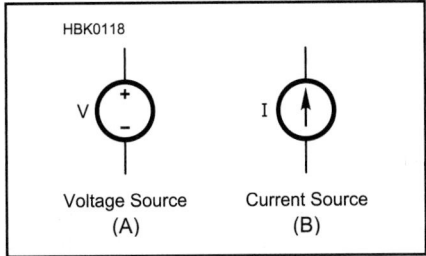

Fig 2.9 — Voltage sources (A) and current sources (B) are examples of ideal energy sources.

short circuit across the terminals of a practical voltage source such as 1.5 V dry-cell battery may produce a current of several amperes, but the battery's internal impedance acts to limit the amount of current produced in accordance with Ohm's Law — as if the resistor in Fig 2.2 was inside of or internal to the battery.

Current sources are defined to have infinite internal impedance. This means that no matter what is connected to the terminals of an ideal current source, it will supply the same amount of current. An open circuit across the terminal of an ideal current source will result in the source generating an infinite voltage at its terminals. Practical current sources will raise their voltage until the internal power supply limits are reached and then reduce output current.

2.3.8 Thevenin's Theorem and Thevenin Equivalents

Thevenin's Voltage Theorem (usually just referred to as "Thevenin's Theorem") is a useful tool for simplifying electrical circuits or *networks* (the formal name for circuits) by allowing circuit designers to replace a circuit with a simpler equivalent circuit. Thevenin's Theorem states, "*Any two-terminal network made up of resistors and voltage or current sources can be replaced by an equivalent network made up of a single voltage source and a series resistor.*"

Thevenin's Theorem can be readily applied to the circuit of Fig 2.8A, to find the current through R3. In this example, illustrated in **Fig 2.10**, the circuit is redrawn to show R1 and R2 forming a voltage divider, with R3 as the load (Fig 2.10A). The current drawn by the load (R3) is simply the voltage across R3, divided by its resistance. Unfortunately, the value of R2 affects the voltage across R3, just as the presence of R3 affects the voltage appearing across R2. Some means of separating the two is needed; hence the *Thevenin-equivalent circuit* is constructed, replacing everything connected to terminals A and B with a single voltage source (the *Thevenin-equivalent voltage, E_{THEV}*) and series resistor (the *Thevenin-equivalent resistance, R_{TH}*).

The first step of creating the Thevenin-equivalent of the circuit is to determine its *open-circuit voltage,* measured when there is no load current drawn from either terminal A or B. Without a load connected between A and B, the total current through the circuit is (from Ohm's Law):

$$I = \frac{E}{R1 + R2} \tag{12}$$

and the voltage between terminals A and B (E_{AB}) is:

$$E_{AB} = I \times R2 \tag{13}$$

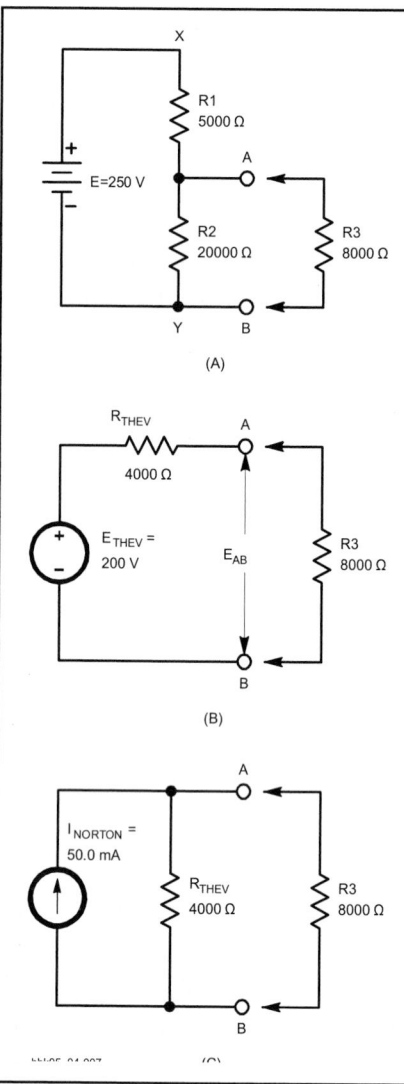

Fig 2.10 — Equivalent circuits for the circuit shown in Fig 2.8. A shows the circuit to be replaced by an equivalent circuit from the perspective of the resistor (R3 load). B shows the Thevenin-equivalent circuit, with a resistor and a voltage source in series. C shows the Norton-equivalent circuit, with a resistor and current source in parallel.

By substituting equation 12 into equation 13, we have an expression for E_{AB} in which all values are known:

$$E_{AB} = \frac{R2}{R1 + R2} \times E \tag{14}$$

Using the values in our example, this becomes:

$$E_{AB} = \frac{20.0\ k}{25.0\ k} \times 250\ V = 200\ V$$

when nothing is connected to terminals A or B.

E_{THEV} is equal to E_{AB} with no current drawn.

The equivalent resistance between terminals A and B is R_{THEV}. R_{THEV} is calculated as the equivalent circuit at terminals A and B with all sources, voltage or current, replaced by their internal impedances. The ideal voltage source, by definition, has zero internal resistance and is replaced by a short circuit. The ideal current source has infinite internal impedance and is replaced by an open circuit.

Assuming the battery to be a close approximation of an ideal source, replace it with a short circuit between points X and Y in the circuit of Fig 2.10A. R1 and R2 are then effectively placed in parallel, as viewed from terminals A and B. R_{THEV} is then:

$$R_{THEV} = \frac{R1 \times R2}{R1 + R2} \tag{15}$$

$$R_{THEV} = \frac{5000\ \Omega \times 20,000\ \Omega}{5000\ \Omega + 20,000\ \Omega}$$

$$R_{THEV} = \frac{1.0 \times 10^8\ \Omega^2}{25,000\ \Omega} = 4000\ \Omega$$

This gives the Thevenin-equivalent circuit as shown in Fig 2.10B. The circuits of Figs 2.10A and 2.10B are completely equivalent from the perspective of R3, so the circuit becomes a simple series circuit.

Once R3 is connected to terminals A and B, there will be current through R_{THEV}, causing a voltage drop across R_{THEV} and reducing E_{AB}. The current through R3 is equal to

$$I3 = \frac{E_{THEV}}{R_{TOTAL}} = \frac{E_{THEV}}{R_{THEV} + R3} \tag{16}$$

Substituting the values from our example:

$$I3 = \frac{200\ V}{4000\ \Omega + 8000\ \Omega} = 16.7\ mA$$

This agrees with the value calculated earlier.

The Thevenin-equivalent circuit of an ideal voltage source in series with a resistance is a good model for a real voltage source with non-zero internal resistance. Using this more realistic model, the maximum current that a real voltage source can deliver is seen to be

$$I_{sc} = \frac{E_{THEV}}{R_{THEV}}$$

and the maximum output voltage is $V_{oc} = E_{THEV}$.

Sinusoidal voltage or current sources can be modeled in much the same way, keeping in mind that the internal impedance, Z_{THEV},

for such a source may not be purely resistive, but may have a reactive component that varies with frequency.

2.3.9 Norton's Theorem and Norton Equivalents

Norton's Theorem is another method of creating an equivalent circuit. Norton's Theorem states, "*Any two-terminal network made up of resistors and current or voltage sources can be replaced by an equivalent network made up of a single current source and a parallel resistor.*" Norton's Theorem is to current sources what Thevenin's Theorem is to voltage sources. In fact, the Thevenin-resistance calculated previously is also the *Norton-equivalent resistance*.

The circuit just analyzed by means of Thevenin's Theorem can be analyzed just as easily by Norton's Theorem. The equivalent Norton circuit is shown in Fig 2.10C. The short circuit current of the equivalent circuit's current source, I_{NORTON}, is the current through terminals A and B with the load (R3) replaced by a short circuit. In the case of the voltage divider shown in Fig 2.10A, the short circuit completely bypasses R2 and the current is:

$$I_{AB} = \frac{E}{R1} \tag{17}$$

Substituting the values from our example, we have:

$$I_{AB} = \frac{E}{R1} = \frac{250\ V}{5000\ \Omega} = 50.0\ mA$$

The resulting Norton-equivalent circuit consists of a 50.0-mA current source placed in parallel with a 4000-Ω resistor. When R3 is connected to terminals A and B, one-third of the supply current flows through R3 and the remainder through R_{THEV}. This gives a current through R3 of 16.7 mA, again agreeing with previous conclusions.

A Norton-equivalent circuit can be transformed into a Thevenin-equivalent circuit and vice versa. The equivalent resistor, R_{THEV}, is the same in both cases; it is placed in series with the voltage source in the case of a Thevenin-equivalent circuit and in parallel with the current source in the case of a Norton-equivalent circuit. The voltage for the Thevenin-equivalent source is equal to the open-circuit voltage appearing across the resistor in the Norton-equivalent circuit. The current for a Norton-equivalent source is equal to the short circuit current provided by the Thevenin source. A Norton-equivalent circuit is a good model for a real current source that has a less-than infinite internal impedance.

2.4 Power and Energy

Regardless of how voltage is generated, energy must be supplied if current is drawn from the voltage source. The energy supplied may be in the form of chemical energy or mechanical energy. This energy is measured in joules (J). One joule is defined from classical physics as the amount of energy or *work* done when a force of one newton (a measure of force) is applied to an object that is moved one meter in the direction of the force.

Power is another important concept and measures the rate at which energy is generated or used. One *watt* (W) of power is defined as the generation (or use) of one joule of energy (or work) per second.

One watt is also defined as one volt of EMF causing one ampere of current to flow through a resistance. Thus,

$$P = I \times E \qquad (18)$$

where
  P = power in watts
  I = current in amperes
  E = EMF in volts.

(This discussion pertains only to direct current in resistive circuits. See the AC Theory and Reactance section of this chapter for a discussion about power in ac circuits, including reactive circuits.)

Common fractional and multiple units for power are the milliwatt (mW, one thousandth of a watt) and the kilowatt (kW, 1000 W).

Example: The plate voltage on a transmitting vacuum tube is 2000 V and the plate current is 350 mA. (The current must be changed to amperes before substitution in the formula, and so is 0.350 A.) Then:

$$P = I \times E = 2000 \text{ V} \times 0.350 \text{ A} = 700 \text{ W}$$

Power may be expressed in *horsepower* (hp) instead of watts, using the following conversion factor:

1 horsepower = 746 W

This conversion factor is especially useful if you are working with a system that converts electrical energy into mechanical energy, and vice versa, since mechanical power is often expressed in horsepower in the US. In metric countries, mechanical power is usually expressed in watts. All countries use the metric power unit of watts in electrical systems, however. The value 746 W/hp assumes lossless conversion between mechanical and electrical power; practical efficiency is taken up shortly.

2.4.1 Energy

When you buy electricity from a power company, you pay for electrical energy, not power. What you pay for is the work that the electrical energy does for you, not the rate at which that work is done. Like energy, work is equal to power multiplied by time. The common unit for measuring electrical energy is the *watt-hour* (Wh), which means that a power of one watt has been used for one hour. That is:

$$Wh = P \, t \qquad (19)$$

where
  Wh = energy in watt-hours
  P = power in watts
  t = time in hours.

Actually, the watt-hour is a fairly small energy unit, so the power company bills you for *kilowatt-hours* (kWh) of energy used. Another energy unit that is sometimes useful is the *watt-second* (Ws), which is equal to joules.

It is important to realize, both for calculation purposes and for efficient use of power resources, that a small amount of power used for a long time can eventually result in a power bill that is just as large as if a large amount of power had been used for a very short time.

A common use of energy units in radio is in specifying the energy content of a battery. Battery energy is rated in *ampere-hours* (Ah) or *milliampere-hours* (mAh). While the multiplication of amperes and hours does not result in units of energy, the calculation assumes the result is multiplied by a specified (and constant) battery voltage. For example, a rechargeable NiMH battery rated to store 2000 mAh of energy is assumed to supply that energy at a terminal voltage of 1.5 V. Thus, after converting 2000 mA to 2 A, the actual energy stored is:

$$\text{Energy} = 1.5 \text{ V} \times 2 \text{ A} \times 1 \text{ hour} = 3 \text{ Wh}$$

Another common energy unit associated with batteries is *energy density*, with units of Ah per unit of volume or weight.

One practical application of energy units is to estimate how long a radio (such as a hand-held unit) will operate from a certain battery. For example, suppose a fully charged battery stores 900 mAh of energy and that the radio draws 30 mA on receive. A simple calculation indicates that the radio will be able receive 900 mAh / 30 mA = 30 hours with this battery, assuming 100% efficiency. You shouldn't expect to get the full 900 mAh out of the battery because the battery's voltage will drop as it is discharged, usually causing the equipment it powers to shut down before the last fraction of charge is used. Any time spent transmitting will also reduce the time the battery will last. The **Power Sources** chapter includes additional information about batteries and their charge/discharge cycles.

2.4.2 Generalized Definition of Resistance

Electrical energy is not always turned into heat. The energy used in running a motor, for example, is converted to mechanical motion. The energy supplied to a radio transmitter is largely converted into radio waves. Energy applied to a loudspeaker is changed into sound waves. In each case, the energy is converted to other forms and can be completely accounted for. None of the energy just disappears! These are examples of the Law of Conservation of Energy. When a device converts energy from one form to another, we often say it *dissipates* the energy, or power. (Power is energy divided by time.) Of course the device doesn't really "use up" the energy, or make it disappear, it just converts it to another form. Proper operation of electrical devices often requires that the power be supplied at a specific ratio of voltage to current. These features are characteristics of resistance, so it can be said that any device that "dissipates power" has a definite value of resistance.

This concept of resistance as something that absorbs power at a definite voltage-to-current ratio is very useful; it permits substituting a simple resistance for the load or power-consuming part of the device receiving power, often with considerable simplification of calculations. Of course, every electrical device has some resistance of its own in the more narrow sense, so a part of the energy supplied to it is converted to heat in that resistance even though the major part of the energy may be converted to another form.

2.4.3 Efficiency

In devices such as motors and transmitters, the objective is to convert the supplied energy (or power) into some form other than heat. In such cases, power converted to heat is considered to be a loss because it is not useful power. The efficiency of a device is the useful power output (in its converted form) divided by the power input to the device. In a transmitter, for example, the objective is to convert power from a dc source into ac power at some radio frequency. The ratio of the RF power output to the dc input is the *efficiency* (*Eff* or η) of the transmitter. That is:

$$Eff = \frac{P_O}{P_I} \qquad (20)$$

where
  Eff = efficiency (as a decimal)
  P_O = power output (W)
  P_I = power input (W).

Example: If the dc input to the transmitter is 100 W, and the RF power output is 60 W, the efficiency is:

$$\text{Eff} = \frac{P_O}{P_I} = \frac{60 \text{ W}}{100 \text{ W}} = 0.6$$

Efficiency is usually expressed as a percentage — that is, it expresses what percent of the input power will be available as useful output. To calculate percent efficiency, multiply the value from equation 20 by 100%. The efficiency in the example above is 60%.

Suppose a mobile transmitter has an RF power output of 100 W with 52% efficiency at 13.8 V. The vehicle's alternator system charges the battery at a rate of 5.0 A at this voltage. Assuming an alternator efficiency of 68%, how much horsepower must the engine produce to operate the transmitter and charge the battery? Solution: To charge the battery, the alternator must produce 13.8 V × 5.0 A = 69 W. The transmitter dc input power is 100 W / 0.52 = 190 W. Therefore, the total electrical power required from the alternator is 190 + 69 = 259 W. The engine load then is:

$$P_I = \frac{P_O}{\text{Eff}} = \frac{259 \text{ W}}{0.68} = 381 \text{ W}$$

We can convert this to horsepower using the conversion factor given earlier to convert between horsepower and watts:

$$\frac{381 \text{ W}}{746 \text{ W/hp}} = 0.51 \text{ horsepower (hp)}$$

2.4.4 Ohm's Law and Power Formulas

Electrical power in a resistance is turned into heat. The greater the power, the more rapidly the heat is generated. By substituting the Ohm's Law equivalent for E and I, the following formulas are obtained for power:

$$P = \frac{E^2}{R} \tag{21}$$

and

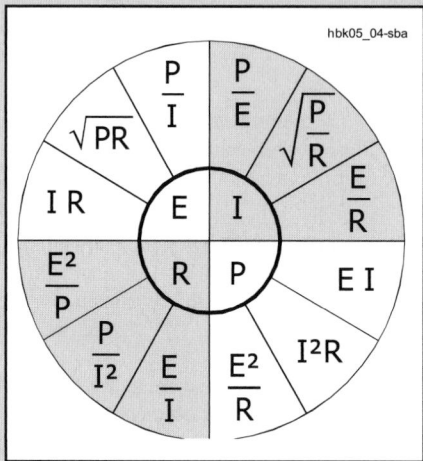
$$P = I^2 \times R \tag{22}$$

These formulas are useful in power calculations when the resistance and either the current or voltage (but not both) are known.

Example: How much power will be converted to heat in a 4000-Ω resistor if the potential applied to it is 200 V? From equation 21,

$$P = \frac{E^2}{R} = \frac{40,000 \text{ V}^2}{4000 \text{ }\Omega} = 10.0 \text{ W}$$

As another example, suppose a current of 20 mA flows through a 300-Ω resistor. Then:

$$P = I^2 \times R = 0.020^2 \text{ A}^2 \times 300 \text{ }\Omega$$

$$P = 0.00040 \text{ A}^2 \times 300 \text{ }\Omega$$

$$P = 0.12 \text{ W}$$

Note that the current was changed from milliamperes to amperes before substitution in the formula.

Resistors for radio work are made in many sizes, the smallest being rated to safely operate at power levels of about 1/16 W. The largest resistors commonly used in amateur equipment are rated at about 100 W. Large resistors, such as those used in dummy-load antennas, are often cooled with oil to increase their power-handling capability.

2.5 Circuit Control Components

2.5.1 Switches

Switches are used to allow or interrupt a current flowing in a particular circuit. Most switches are mechanical devices, although the same effect may be achieved with solid-state devices.

Switches come in many different forms and a wide variety of ratings. The most important ratings are the *voltage-handling* and *current-handling* capabilities. The voltage rating usually includes both the *breakdown voltage rating* and the *interrupt voltage rat-*ing. The breakdown rating is the maximum voltage that the switch can withstand when it is open before the voltage will arc between the switch's terminals. The interrupt voltage rating is the maximum amount of voltage that the switch can interrupt without arcing. Normally, the interrupt voltage rating is the lower value, and therefore the one given for (and printed on) the switch.

Switches typically found in the home are usually rated for 125 V ac and 15 to 20 A. Switches in cars are usually rated for 12 V dc and several amperes. The breakdown voltage rating of a switch primarily depends on the insulating material surrounding the contacts and the separation between the contacts. Plastic or phenolic material normally provides both structural support and insulation. Ceramic material may be used to provide better insulation, particularly in rotary (wafer) switches.

A switch's current rating includes both the *current-carrying capacity* and the *interrupt capability*. The current-carrying capacity of the switch depends on the contact material and size, and on the pressure exerted to keep the contacts closed. It is primarily determined from the allowable contact temperature rise. On larger ac switches and most dc switches, the interrupt capability is usually lower than the current carrying value.

Most power switches are rated for alternating current use. Because ac current goes through zero twice in each cycle, switches can successfully interrupt much higher alternating currents than direct currents without arcing. A switch that has a 10-A ac current rating may arc and damage the contacts if used to turn off more than an ampere or two of dc.

Switches are normally designated by the number of *poles* (circuits controlled) and *throws* or *positions* (circuit path choices). The simplest switch is the on-off switch, which is a single-pole, single-throw (SPST) switch as shown in **Fig 2.11A**. The off position does not direct the current to another circuit. The next step would be to change the current path to another path. This would be a single-pole, double-throw (SPDT) switch as shown in Fig 2.11B. Adding an off position would give a single-pole, double-throw, center-off switch as shown in Fig 2.11C.

Several such switches can be "ganged" to or actuated by the same mechanical activator to provide double-pole, triple-pole or even more, separate control paths all activated at once. Switches can be activated in a variety of ways. The most common methods include lever or toggle, push-button and rotary switches. Samples of these are shown in **Fig 2.12**. Most switches stay in position once set, but some are spring-loaded so they only stay in the desired position while held there. These are called *momentary* switches.

Rotary/wafer switches can provide very complex switching patterns. Several poles (separate circuits) can be included on each wafer. Many wafers may be stacked on the same shaft. Not only may many different circuits be controlled at once, but by wiring different poles/positions on different wafers together, a high degree of circuit switching logic can be developed. Such switches can select different paths as they are turned and

can also "short" together successive contacts to connect numbers of components or paths.

Rotary switches can also be designed to either break one contact before making another (*break-before-make*), or to short two contacts together before disconnecting the first one (*make-before-break*) to eliminate arcing or perform certain logic functions. The two types of switches are generally not interchangeable and may cause damage if inappropriately substituted for one another during circuit construction or repair. When buying rotary switches from a surplus or flea-market vendor, check to be sure the type of switch is correct.

Microswitches are designed to be actuated by the operation of machine components, opening or closing of a door, or some other mechanical movement. Instead of a handle or button-type actuator that would be used by a human, microswitches have levers or buttons more suitable for being actuated as part of an enclosure or machine.

In choosing a switch for a particular task, consideration should be given to function, voltage and current ratings, ease of use, availability and cost. If a switch is to be operated

frequently, a better-quality switch is usually less costly over the long run. If signal noise or contact corrosion is a potential problem (usually in low-current signal applications), it is best to get gold-plated contacts. Gold does not oxidize or corrode, thus providing surer contact, which can be particularly important at very low signal levels. Gold plating will not hold up under high-current-interrupt applications, however.

2.5.2 Fuses and Circuit Breakers

Fuses self-destruct to protect circuit wiring or equipment. The fuse *element* that melts or *blows* is a carefully shaped piece of soft metal, usually mounted in a cartridge of some kind. The element is designed to safely carry a given amount of current and to melt at a current value that is a certain percentage above the rated value.

The most important fuse rating is the *nominal current rating* that it will safely carry for an indefinite period without blowing. A fuse's melting current depends on the type of material, the shape of the element and the

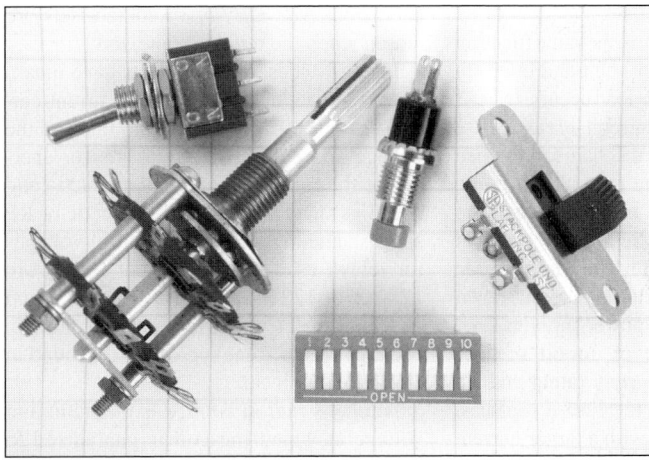

Fig 2.12 — This photo shows examples of various styles of switches. The ¼-inch-ruled graph paper background provides for size comparison.

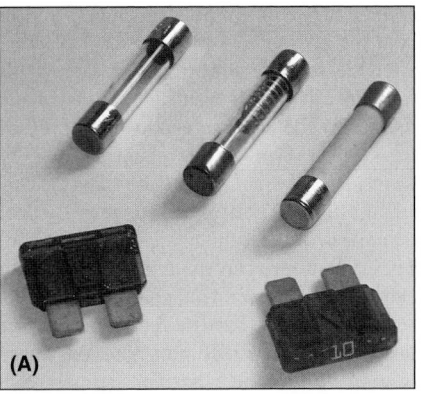

(A)

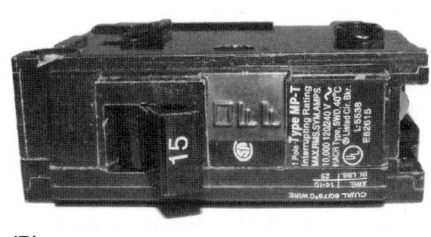

(B)

Fig 2.13 — These photos show examples of various styles of fuses. Cartridge-type fuses (A, top) can use glass or ceramic construction. The center fuse is a slow-blow type. Automotive blade-type fuses (A, bottom) are common for low-voltage dc use. A typical home circuit breaker for ac wiring is shown at B.

SPST
(A)

SPDT
(B)

SPDT, Center Off

hbk05_04-009   (C)

Fig 2.11 — Schematic diagrams of various types of switches. A is an SPST, B is an SPDT, and C is an SPDT switch with a center-off position.

heat dissipation capability of the cartridge and holder, among other factors.

Next most important are the timing characteristics, or how quickly the fuse element blows under a given current overload. Some fuses (*slow-blow*) are designed to carry an overload for a short period of time. They typically are used in motor-starting and power-supply circuits that have a large inrush current when first started. Other fuses are designed to blow very quickly to protect delicate instruments and solid-state circuits.

A fuse also has a voltage rating, both a value in volts and whether it is expected to be used in ac or dc circuits. The voltage rating is the amount of voltage an open fuse can withstand without arcing. While you should never substitute a fuse with a higher current rating than the one it replaces, you may use a fuse with a higher voltage rating.

Fig 2.13A shows typical cartridge-style cylindrical fuses likely to be encountered in ac-powered radio and test equipment. Automotive style fuses, shown in the lower half of Fig 2.13A, have become widely used in low-voltage dc power wiring of amateur stations. These are called "blade" fuses. Rated for vehicle-level voltages, automotive blade fuses should never be used in ac line-powered circuits.

Circuit breakers perform the same function as fuses — they open a circuit and interrupt current flow when an overload occurs. Instead of a melting element, circuit breakers use spring-loaded magnetic mechanisms to open a switch when excessive current is present. Once the overload has been corrected, the circuit-breaker can be reset. Circuit breakers are generally used by amateurs in home ac wiring (a typical ac circuit breaker is shown in Fig 2.13B) and in dc power supplies.

A replacement fuse or circuit breaker should have the same current rating and the same characteristics as the fuse it replaces. Never substitute a fuse with a larger current rating. You may cause permanent damage (maybe even a fire) to wiring or circuit elements by allowing larger currents to flow when there is an internal problem in equipment. (Additional discussion of fuses and circuit breakers is provided in the chapter on **Safety**.)

Fuses blow and circuit breakers open for several reasons. The most obvious reason is that a problem develops in the circuit, causing too much current to flow. In this case, the circuit problem needs to be fixed. A fuse can fail from being cycled on and off near its current rating. The repeated thermal stress causes metal fatigue and eventually the fuse blows. A fuse can also blow because of a momentary power surge, or even by rapidly turning equipment with a large inrush current on and off several times. In these cases it is only necessary to replace the fuse with

the same type and value.

Panel-mount fuse holders should be wired with the hot lead of an ac power circuit (the black wire of an ac power cord) connected to the end terminal, and the ring terminal is connected to the power switch or circuit inside the chassis. This removes voltage from the fuse as it is removed from the fuse holder. This also locates the line connection at the far end of the fuse holder where it is not easily accessible.

2.5.3 Relays And Solenoids

Relays are switches controlled by an electrical signal. *Electromechanical relays* consist of a electromagnetic *coil* and a moving *armature* attracted by the coil's magnetic field when energized by current flowing in the coil. Movement of the armature pushes the switch contacts together or apart. Many sets of contacts can be connected to the same armature, allowing many circuits to be controlled by a single signal. In this manner, the signal voltage that energizes the coil can control circuits carrying large voltages and/or currents.

Relays have two positions or *states* — energized and *de-energized*. Sets of contacts called *normally-closed* (NC) are closed when the relay is de-energized and open when it is energized. *Normally-open* contact sets are closed when the relay is energized.

Like switches, relay contacts have breakdown voltage, interrupting, and current-carrying ratings. These are not the same as the voltage and current requirements for energizing the relay's coil. Relay contacts (and housings) may be designed for ac, dc or RF signals. The most common control voltages for relays used in amateur equipment are 12 V dc or 120 V ac. Relays with 6, 24, and 28 V dc, and 24 V ac coils are also common. **Fig 2.14** shows some typical relays found in amateur equipment.

A relay's p*ull-in voltage* is the minimum voltage at which the coil is guaranteed to cause the armature to move and change the relay's state. *Hold-in voltage* is the minimum voltage at which the relay is guaranteed to hold the armature in the energized position after the relay is activated. A relay's pull-in voltage is higher than its hold-in voltage due to magnetic hysteresis of the coil (see the section on magnetic materials below). *Current-sensing relays* activate when the current through the coil exceeds a specific value, regardless of the voltage applied to the coil. They are used when the control signal is a current rather than a voltage.

Latching relays have two coils; each moves the armature to a different position where it remains until the other coil is energized. These relays are often used in portable and low-power equipment so that the contact configuration can be maintained without the need to continuously supply power to the relay.

Reed relays have no armature. The contacts are attached to magnetic strips or "reeds" in a glass or plastic tube, surrounded by a coil. The reeds move together or apart when current is applied to the coil, opening or closing contacts. Reed relays can open and close very quickly and are often used in transmit-receive switching circuits.

Solid-state relays (SSR) use transistors in-

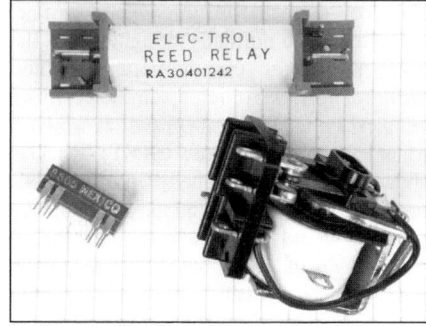

(A)

(B)

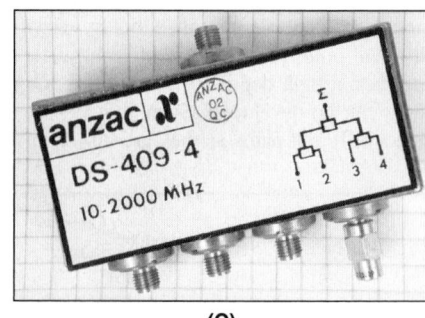

(C)

Fig 2.14 — These photos show examples of various styles and sizes of relays. Photo A shows a large reed relay, and a small reed relay in a package the size of a DIP IC. The contacts and coil can clearly be seen in the open-frame relay. Photo B shows a relay inside a plastic case. Photo C shows a four-position coaxial relay-switch combination with SMA connectors. The ¼-inch-ruled graph paper background provides a size comparison.

stead of mechanical contacts and electronic circuits instead of magnetic coils. They are designed as substitutes for electromechanical relays in power and control circuits and are not used in low-level ac or dc circuits.

Coaxial relays have an armature and contacts designed to handle RF signals. The signal path in coaxial relays maintains a specific characteristic impedance for use in RF systems. Coaxial connectors are used for the RF circuits. Coaxial relays are typically used to control antenna system configurations or to switch a transceiver between a linear amplifier and an antenna.

A *solenoid* is very similar to a relay, except that instead of the moving armature actuating switch contacts, the solenoid moves a lever or rod to actuate some mechanical device. Solenoids are not commonly used in radio equipment, but may be encountered in related systems or devices.

2.6 AC Theory and Waveforms

2.6.1 AC in Circuits

A circuit is a complete conductive path for current to flow from a source, through a load and back to the source. If the source permits the current to flow in only one direction, the current is *dc* or *direct current*. If the source permits the current to change direction, the current is *ac* or *alternating current*. **Fig 2.15** illustrates the two types of circuits. Circuit A shows the source as a battery, a typical dc source. Circuit B shows the more abstract voltage source symbol to indicate ac. In an ac circuit, both the current and the voltage reverse direction. For nearly all ac signals in electronics and radio, the reversal is *periodic*, meaning that the change in direction occurs on a regular basis. The rate of reversal may range from a few times per second to many billions per second.

Graphs of current or voltage, such as Fig 2.15, begin with a horizontal axis that represents time. The vertical axis represents the amplitude of the current or the voltage, whichever is graphed. Distance above the zero line indicates larger positive amplitude; distance below the zero line means larger negative amplitude. Positive and negative only designate the opposing directions in which current may flow in an alternating current circuit or the opposing *polarities* of an ac voltage.

If the current and voltage never change direction, then from one perspective, we have a dc circuit, even if the level of dc constantly changes. **Fig 2.16** shows a current that is always positive with respect to 0. It varies periodically in amplitude, however. Whatever the shape of the variations, the current can be called *pulsating dc*. If the current periodically reaches 0, it can be called *intermittent dc*. From another perspective, we may look at intermittent and pulsating dc as a combination of an ac and a dc current. Special circuits can separate the two currents into ac and dc *components* for separate analysis or use. There are circuits that combine ac and dc currents and voltages, as well.

2.6.2 AC Waveforms

A *waveform* is the pattern of amplitudes reached by an ac voltage or current as measured over time. The combination of ac and dc voltages and currents results in *complex waveforms*. **Fig 2.17** shows two ac waveforms fairly close in frequency and their combination. **Fig 2.18** shows two ac waveforms dissimilar in both frequency and wavelength, along with the resultant combined waveform. Note the similarities (and the differences) between the resultant waveform in Fig 2.18 and the combined ac-dc waveform in Fig 2.16.

Alternating currents may take on many useful waveforms. **Fig 2.19** shows a few that are commonly used in electronic and radio circuits. The *sine wave* is both mathematically and practically the foundation of all other forms of ac; the other forms can usually be reduced to (and even constructed from) a particular collection of sine waves. The *square wave* is vital to digital electronics. The *triangular* and *ramp waves* — the latter sometimes called a *sawtooth* waveform — are especially useful in timing circuits. The individual repeating patterns that make up these *periodic waveforms* are called *cycles*. Waveforms that do not consist of repetitive patterns are called *aperiodic* or *irregular* waveforms. Speech is

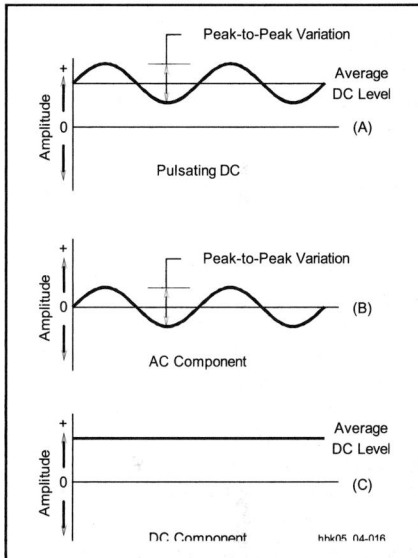

Fig 2.16 — A pulsating dc current (A) and its resolution into an ac component (B) and a dc component (C).

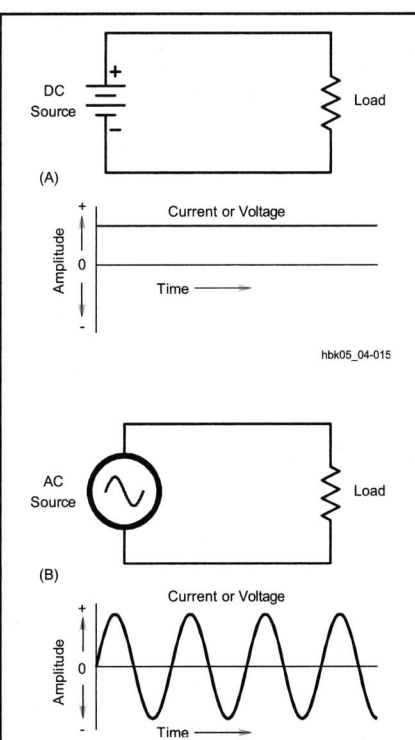

Fig 2.15 — Basic circuits for direct and alternating currents. With each circuit is a graph of the current, constant for the dc circuit, but periodically changing direction in the ac circuit.

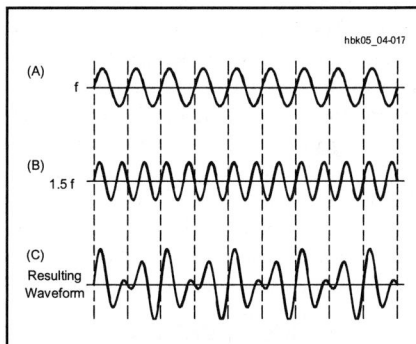

Fig 2.17 — Two ac waveforms of similar frequencies (f1 = 1.5 f2) and amplitudes form a composite wave. Note the points where the positive peaks of the two waves combine to create high composite peaks at a frequency that is the difference between f1 and f2. The beat note frequency is 1.5f – f = 0.5f and is visible in the drawing.

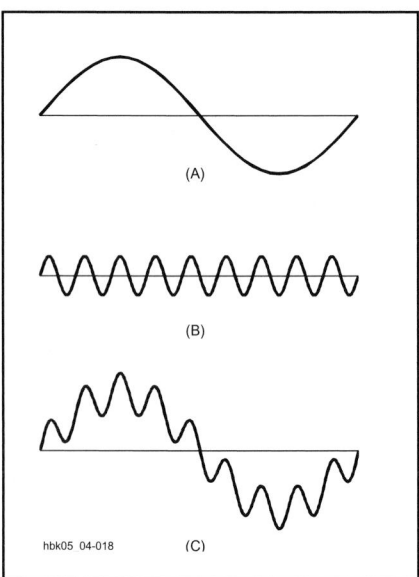

Fig 2.18 — Two ac waveforms of widely different frequencies and amplitudes form a composite wave in which one wave appears to ride upon the other.

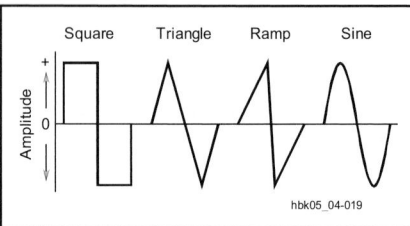

Fig 2.19 — Some common ac waveforms: square, triangle, ramp and sine.

a good example of an irregular waveform, as is noise or static.

There are numerous ways to generate alternating currents: with an ac power generator (an *alternator*), with a *transducer* (for example, a microphone) or with an electronic circuit (for example, an RF oscillator). The basis of the sine wave is circular or cyclical motion, which underlies the most usual methods of generating alternating current. The circular motion of the ac generator may be physical or mechanical, as in an alternator. Currents in the resonant circuit of an oscillator may also produce sine waves as repetitive electrical cycles without mechanical motion.

SINE WAVES AND CYCLICAL MOTION

The relationship between circular motion and the sine wave is illustrated in **Fig 2.20** as the correlation between the amplitude of an ac current (or voltage) and the relative positions of a point making circular rotations through one complete revolution of 360º. Following the height of the point above the horizontal

axis, its height is zero at point 1. It rises to a maximum at a position 90º from position 1, which is position 3. At position 4, 180º from position 1, the point's height falls back to zero. Then the point's height begins to increase again, but below the horizontal axis, where it reaches a maximum at position 5.

An alternative way of visualizing the relationship between circular motion and sine waves is to use a yardstick and a light source, such as a video projector. Hold the yardstick horizontally between the light source and the wall so that it points directly at the light source. This is defined as the 0° position and the length of the yardstick's shadow is zero. Now rotate the yardstick, keeping the end farthest from the projector stationary, until the yardstick is vertical. This is the 90° position and the shadow's length is a maximum in the direction defined as positive. Continue to rotate the yardstick until it reaches horizontal again. This is the 180° position and the shadow once again has zero length. Keep rotating the yardstick in the same direction until it is vertical and the shadow is at maximum, but now below the stationary end of the yardstick. This is the 270° position and the direction of the shadow is defined to be negative. Another 90° of rotation brings the pointer back to its original position at 0° and the shadow's length to zero. If the length of the shadow is plotted against the amount of rotation of the yardstick, the result is a sine wave.

The corresponding rise and fall of a *sinusoidal* or sine wave current (or voltage) along a linear time line produces the curve accompanying the circle in Fig 2.20. The curve is sinusoidal because its amplitude varies as the sine of the angle made by the circular movement with respect to the zero position. This is often referred to as the *sine function*, written sin(angle). The sine of 90° is 1, and 90° is also the position of maximum current (along with 270°, but with the opposite polarity). The sine of 45° (point 2) is 0.707, and the value of current at point 2, the 45° position of rotation, is 0.707 times the maximum current. Both ac current and voltage sine waves vary in the same way.

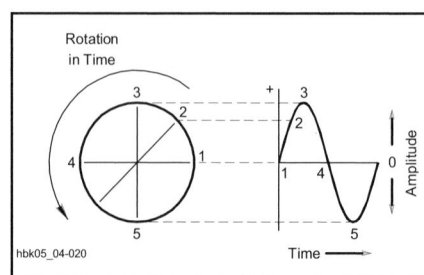

Fig 2.20 — The relationship of circular motion and the resultant graph of ac current or voltage. The curve is sinusoidal, a sine wave.

FREQUENCY AND PERIOD

With a continuously rotating generator, alternating current or voltage will pass through many equal cycles over time. This is a *periodic waveform*, composed of many identical cycles. An arbitrary point on any one cycle can be used as a marker of position on a periodic waveform. For this discussion, the positive peak of the waveform will work as an unambiguous marker. The number of times per second that the current (or voltage) reaches this positive peak in any one second is called the *frequency* of the waveform. In other words, frequency expresses the *rate* at which current (or voltage) cycles occur. The unit of frequency is *cycles per second*, or *hertz*—abbreviated Hz (after the 19th century radio-phenomena pioneer, Heinrich Hertz).

The length of any cycle in units of time is the *period* of the cycle, as measured between equivalent points on succeeding cycles. Mathematically, the period is the inverse of frequency. That is,

$$\text{Frequency (f) in Hz} = \frac{1}{\text{Period (T) in seconds}}$$

(23)

and

$$\text{Period (T) in seconds} = \frac{1}{\text{Frequency (f) in Hz}}$$

(24)

Example: What is the period of a 400-hertz ac current?

$$T = \frac{1}{f} = \frac{1}{400 \text{ Hz}} = 0.00250 \text{ s} = 2.5 \text{ ms}$$

The frequency of alternating currents used in radio circuits varies from a few hertz, or cycles per second, to thousands of millions of hertz. Likewise, the period of alternating currents amateurs use ranges from significant fractions of a second down to nanoseconds or smaller. In order to express compactly the units of frequency, time and almost everything else in electronics, a standard system of prefixes is used. In magnitudes of 1000 or 10^3, frequency is measurable in hertz, in kilohertz (1000 hertz or kHz), in megahertz (1 million hertz or MHz), gigahertz (1 billion hertz or GHz — pronounced "gee-gah" with a hard 'g' as in 'golf') and even in terahertz (1 trillion hertz or THz). For units smaller than unity, as in the measurement of period, the basic unit can be milliseconds (1 thousandth of a second or ms), microseconds (1 millionth of a second or μs), nanoseconds (1 billionth of a second or ns) and picoseconds (1 trillionth of a second or ps).

It is common for complex ac signals to contain a series of sine waves with frequencies related by integer multiples of some lowest or *fundamental* frequency. Sine waves with the higher frequency are called harmonics and are said to be *harmonically related* to the fundamental. For example, if a complex waveform is made up of sine waves with frequencies of 10, 20 and 30 kHz, 10 kHz is the fundamental and the other two are harmonics.

The uses of ac in radio circuits are many and varied. Most can be cataloged by reference to ac frequency ranges used in circuits. For example, the frequency of ac power used in the home, office and factory is ordinarily 60 Hz in the United States and Canada. In Great Britain and much of Europe, ac power is 50 Hz. Sonic and ultrasonic applications of ac run from about 20 Hz (audio) up to several MHz.

Radio circuits include both power- and sonic-frequency-range applications. Radio communication and other electronics work, however, require ac circuits capable of operation with frequencies up to the GHz range. Some of the applications include signal sources for transmitters (and for circuits inside receivers); industrial induction heating; diathermy; microwaves for cooking, radar and communication; remote control of appliances, lighting, model planes and boats and other equipment; and radio direction finding and guidance.

PHASE

When drawing the sine-wave curve of an ac voltage or current, the horizontal axis represents time. This type of drawing in which the amplitude of a waveform is shown compared to time is called the *time domain*. Events to the right take place later; events to the left occur earlier. Although time is measurable in parts of a second, it is more convenient to treat each cycle as a complete time unit divided into 360°. The conventional starting point for counting degrees in a sinusoidal waveform such as ac is the *zero point* at which the voltage or current begins the positive *half cycle*. The essential elements of an ac cycle appear in **Fig 2.21**.

The advantage of treating the ac cycle in this way is that many calculations and measurements can be taken and recorded in a manner that is independent of frequency. The positive peak voltage or current occurs at 90° into the cycle. Relative to the starting point, 90° is the *phase* of the ac at that point. Phase is the position within an ac cycle expressed in degrees or *radians*. Thus, a complete description of an ac voltage or current involves reference to three properties: frequency, amplitude and phase.

Phase relationships also permit the comparison of two ac voltages or currents at the same frequency. If the zero point of two signals with the frequency occur at the same time, there is zero phase difference between the signals and they are said to be *in phase*. **Fig 2.22** illustrates two waveforms with a constant phase difference. Since B crosses

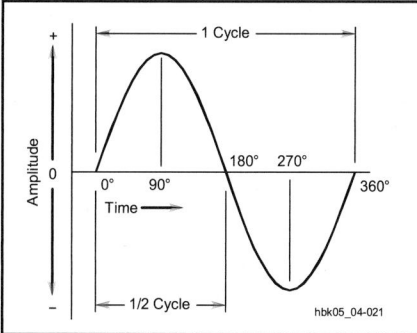

Fig 2.21 — An ac cycle is divided into 360° that are used as a measure of time or phase.

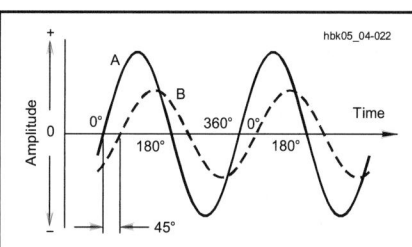

Fig 2.22 — When two waves of the same frequency start their cycles at slightly different times, the time difference or phase difference is measured in degrees. In this drawing, wave B starts 45° (one-eighth cycle) later than wave A, and so lags 45° behind A.

the zero point in the positive direction after A has already done so, there is a *phase difference* between the two waves. In the example, B *lags* A by 45°, or A *leads* B by 45°. If A and B occur in the same circuit, their composite waveform will also be a sine wave at an intermediate phase angle relative to each. Adding any number of sine waves of the same frequency always results in a sine wave at that frequency. Adding sine waves of different frequencies, as in Fig 2.17, creates a complex waveform with a *beat frequency* that is the difference between the two sine waves.

Fig 2.22 might equally apply to a voltage and a current measured in the same ac circuit. Either A or B might represent the voltage; that is, in some instances voltage will lead the current and in others voltage will lag the current.

Two important special cases appear in **Fig 2.23**. In Part A, line B lags 90° behind line A. Its cycle begins exactly one quarter cycle later than the A cycle. When one wave is passing through zero, the other just reaches its maximum value. In this example, the two sine waves are said to be *in quadrature*.

In Part B, lines A and B are 180° *out of phase*, sometimes called *anti-phase*. In this case, it does not matter which one is considered to lead or lag. Line B is always positive

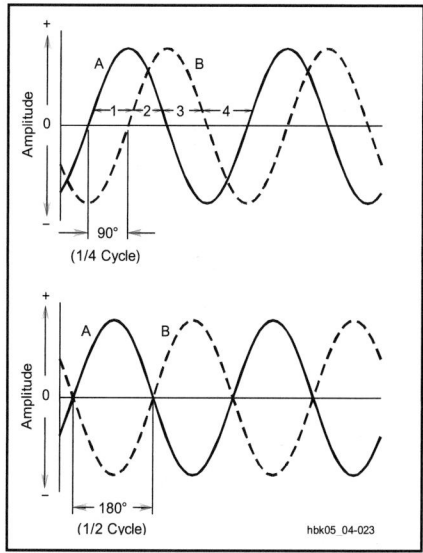

Fig 2.23 — Two important special cases of phase difference: In the upper drawing, the phase difference between A and B is 90°; in the lower drawing, the phase difference is 180°.

while line A is negative, and vice versa. If the two waveforms are of two voltages or two currents in the same circuit and if they have the same amplitude, they will cancel each other completely.

Phase also has other meanings. It is important to distinguish between *polarity*, meaning the sense in which positive and negative voltage or current relate to circuit operation, and phase, which is a function of time or position in a waveform. It is quite possible for two signals to have opposite polarities, but still be in phase, for example. In a multi-phase ac power system, "phase" refers to one of the distinct voltage waveforms generated by the utility.

Degrees and Radians

While most electronic and radio mathematics use degrees as a measure of phase, you will occasionally encounter *radians*. Radians are used because they are more convenient mathematically in certain equations and computations. Radians are used for angular measurements when the *angular frequency* (ω) is being used. There are 2π radians in a circle, just as there are 360°, so one radian = $360/2\pi$ ≈ 57.3°. Angular frequency is measured in radians/second (not revolutions/second or cycles/second) so ω = $2\pi f$. In this *Handbook*, unless specifically noted otherwise, the convention will be to use degrees in all calculations that require an angle.

2.6.3 Electromagnetic Energy

Alternating currents are often loosely classified as audio frequency (AF) and radio frequency (RF). Although these designations are handy, they actually represent something other than electrical energy: They designate special forms of energy that we find useful.

Audio or *sonic* energy is the energy imparted by the mechanical movement of a medium, which can be air, metal, water or even the human body. Sound that humans can hear normally requires the movement of air between 20 Hz and 20 kHz, although the human ear loses its ability to detect the extremes of this range as we age. Some animals, such as elephants, can detect air vibrations well below 20 Hz, while others, such as dogs and cats, can detect air vibrations well above 20 kHz.

Electrical circuits do not directly produce air vibrations. Sound production requires a *transducer*, a device to transform one form of energy into another form of energy; in this case electrical energy into sonic energy. The speaker and the microphone are the most common audio transducers. There are numerous ultrasonic transducers for various applications.

RF energy occurs at frequencies for which it is practical to generate and detect *electromagnetic* or *RF waves* that exist independently of the movement of electrical charge, such as a radio signal. Like sonic energy, a transducer — an antenna — is required to convert the electrical energy in a circuit to electromagnetic waves. In a physical circuit, such as a wire, electromagnetic energy exists as both electromagnetic waves and the physical movement of electrical charge.

Electromagnetic waves have been generated and detected in many forms with frequencies from below 1 Hz to above 10^{12} GHz, including at the higher frequencies infrared, visible, and ultraviolet light, and a number of energy forms of greatest interest to physicists and astronomers. **Table 2.3** provides a brief glimpse at the total spectrum of electromagnetic energy. The *radio spectrum* is generally considered to begin around 3 kHz and end at infrared light.

Despite the close relationship between electromagnetic energy and waves, it remains important to distinguish the two. To a circuit producing or amplifying a 15-kHz alternating current, the ultimate transformation and use of the electrical energy may make no difference to the circuit's operation. By choosing the right transducer, one can produce either a sonic wave or an electromagnetic wave — or both. Such is a common problem of video monitors and switching power supplies; forces created by the ac currents cause electronic parts both to vibrate audibly *and* to radiate electromagnetic energy.

All electromagnetic energy has one thing in common: it travels, or *propagates,* at the speed of light, abbreviated *c*. This speed is approximately 300,000,000 (or 3×10^8) meters per second in a vacuum. Electromagnetic-energy waves have a length uniquely associated with each possible frequency. The *wavelength* (λ) is the speed of propagation divided by the frequency (f) in hertz.

$$f\,(\text{Hz}) = \frac{3.0 \times 10^8 \left(\frac{\text{m}}{\text{s}} \right)}{\lambda\,(\text{m})} \qquad (25)$$

and

$$\lambda\,(\text{m}) = \frac{3.0 \times 10^8 \left(\frac{\text{m}}{\text{s}} \right)}{f\,(\text{Hz})} \qquad (26)$$

Example: What is the frequency of an RF wave with wavelength of 80 meters?

$$f\,(\text{Hz}) = \frac{3.0 \times 10^8 \left(\frac{\text{m}}{\text{s}} \right)}{\lambda\,(\text{m})}$$

$$= \frac{3.0 \times 10^8 \left(\frac{\text{m}}{\text{s}} \right)}{80.0\,\text{m}}$$

$$= 3.75 \times 10^6\ \text{Hz}$$

This is 3.750 MHz or 3750 kHz, a frequency in the middle of the ham band known as "80 meters."

A similar equation is used to calculate the wavelength of a sound wave in air, substituting the speed of sound instead of the speed of light in the numerator. The speed of propagation of the mechanical movement of air that we call sound varies considerably with air temperature and altitude. The speed of sound at sea level is about 331 m/s at 0 °C and 344 m/s at 20 °C.

To calculate the frequency of an electromagnetic wave directly in kilohertz, change the speed constant to 300,000 (3×10^5) km/s.

$$f\,(\text{kHz}) = \frac{3.0 \times 10^5 \left(\frac{\text{km}}{\text{s}} \right)}{\lambda\,(\text{m})} \qquad (27)$$

and

$$\lambda\,(\text{m}) = \frac{3.0 \times 10^5 \left(\frac{\text{km}}{\text{s}} \right)}{f\,(\text{kHz})} \qquad (28)$$

For frequencies in megahertz, change the speed constant to 300 (3×10^2) Mm/s.

$$f\,(\text{MHz}) = \frac{300 \left(\frac{\text{Mm}}{\text{s}} \right)}{\lambda\,(\text{m})} \qquad (29)$$

and

$$\lambda\,(\text{m}) = \frac{300 \left(\frac{\text{Mm}}{\text{s}} \right)}{f\,(\text{MHz})} \qquad (30)$$

Stated as it is usually remembered and used, "wavelength in meters equals 300 divided by frequency in megahertz." Assuming the proper units for the speed of light constant simplify the equation.

Example: What is the wavelength of an RF wave whose frequency is 4.0 MHz?

$$\lambda(\text{m}) = \frac{300}{f\,(\text{MHz})} = \frac{300}{4.0} = 75\ \text{m}$$

At higher frequencies, circuit elements with lengths that are a significant fraction of a wavelength can act like transducers. This property can be useful, but it can also cause problems for circuit operations. Therefore, wavelength calculations are of some importance in designing ac circuits for those frequencies.

Within the part of the electromagnetic-energy spectrum of most interest to radio applications, frequencies have been classified into groups and given names. **Table 2.4** provides a reference list of these classifications. To a significant degree, the frequencies within each group exhibit similar properties, both in circuits and as RF waves. For example, HF or high frequency waves, with frequencies from

---

Table 2.3
Key Regions of the Electromagnetic Energy Spectrum

Region Name	Frequency Range		
Radio frequencies	3.0×10^3 Hz	to	3.0×10^{11} Hz
Infrared	3.0×10^{11} Hz	to	4.3×10^{14} Hz
Visible light	4.3×10^{14} Hz	to	7.5×10^{14} Hz
Ultraviolet	7.5×10^{14} Hz	to	6.0×10^{16} Hz
X-rays	6.0×10^{16} Hz	to	3.0×10^{19} Hz
Gamma rays	3.0×10^{19} Hz	to	5.0×10^{20} Hz
Cosmic rays	5.0×10^{20} Hz	to	8.0×10^{21} Hz

Note: The range of radio frequencies can also be written as 3 kHz to 300 GHz

Table 2.4

Classification of the Radio Frequency Spectrum

Abbreviation	Classification	Frequency Range			
VLF	Very low frequencies	3	to	30	kHz
LF	Low frequencies	30	to	300	kHz
MF	Medium frequencies	300	to	3000	kHz
HF	High frequencies	3	to	30	MHz
VHF	Very high frequencies	30	to	300	MHz
UHF	Ultrahigh frequencies	300	to	3000	MHz
SHF	Superhigh frequencies	3	to	30	GHz
EHF	Extremely high frequencies	30	to	300	GHz

3 to 30 MHz, all exhibit *skip* or ionospheric refraction that permits regular long-range radio communications. This property also applies occasionally both to MF (medium frequency) and to VHF (very high frequency) waves, as well.

2.6.4 Measuring AC Voltage, Current and Power

Measuring the voltage or current in a dc circuit is straightforward, as **Fig 2.24A** demonstrates. Since the current flows in only one direction, the voltage and current have constant values until the resistor values are changed.

Fig 2.24B illustrates a perplexing problem encountered when measuring voltages and currents in ac circuits — the current and voltage continuously change direction and value. Which values are meaningful? How are measurements performed? In fact, there are several methods of measuring sine-wave voltage and current in ac circuits with each method providing different information about the waveform.

INSTANTANEOUS VOLTAGE AND CURRENT

By far, the most common waveform associated with ac of any frequency is the sine wave. Unless otherwise noted, it is safe to assume that measurements of ac voltage or current are of a sinusoidal waveform. **Fig 2.25** shows a sine wave representing a voltage or current of some arbitrary frequency and amplitude. The *instantaneous* voltage (or current) is the value at one instant in time. If a series of instantaneous values are plotted against time, the resulting graph will show the waveform.

In the sine wave of Fig 2.25, the instantaneous value of the waveform at any point in time is a function of three factors: the maximum value of voltage (or current) along the curve (point B, E_{max}), the frequency of the wave (f), and the time elapsed from the preceding positive-going zero crossing (t) in seconds or fractions of a second. Thus,

$$E_{inst} = E_{max} \sin (ft) \qquad (31)$$

assuming all sine calculations are done in degrees. (See the sidebar "Degrees and Radians".) If the sine calculation is done in radians, substitute $2\pi ft$ for ft in equation 31.

If the point's phase is known — the position along the waveform — the instantaneous voltage at that point can be calculated directly as:

$$E_{inst} = E_{max} \sin \theta \qquad (32)$$

where θ is the number of degrees of phase difference from the beginning of the cycle.

Example: What is the instantaneous value of voltage at point D in Fig 2.25, if the maximum voltage value is 120 V and point D's phase is 60.0°?

$$E_{inst} = 120 \text{ V} \times \sin 60°$$

$$= 120 \times 0.866 = 104 \text{ V}$$

PEAK AND PEAK-TO-PEAK VOLTAGE AND CURRENT

The most important of an ac waveform's instantaneous values are the maximum or *peak values* reached on each positive and negative half cycle. In Fig 2.25, points B and C represent the positive and negative peaks. Peak values (indicated by a "pk" or "p" subscript) are especially important with respect to component ratings, which the voltage or current in a circuit must not exceed without danger of component failure.

The *peak power* in an ac circuit is the

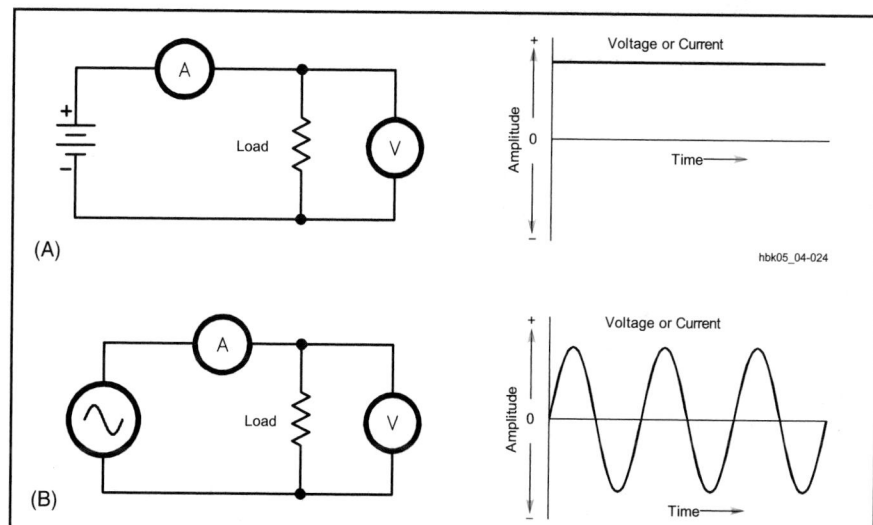

(A)

(B)

Fig 2.24 — Voltage and current measurements in dc and ac circuits.

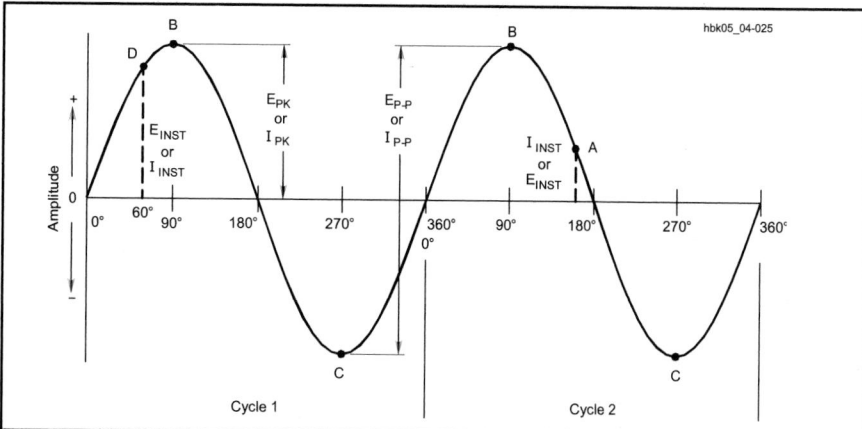

Fig 2.25 — Two cycles of a sine wave to illustrate instantaneous, peak, and peak-to-peak ac voltage and current values.

product of the peak voltage and the peak current, or

$$P_{pk} = E_{pk} \times I_{pk} \tag{33}$$

The span from points B to C in Fig 2.25 represents the largest difference in value of the sine wave. Designated the *peak-to-peak* value (indicated by a "P-P" subscript), this span is equal to twice the peak value of the waveform. Thus, peak-to-peak voltage is:

$$E_{P-P} = 2\,E_{pk} \tag{34}$$

Amplifying devices often specify their input limits in terms of peak-to-peak voltages. Operational amplifiers, which have almost unlimited gain, often require input-level limiting to prevent the output signals from distorting if they exceed the peak-to-peak output rating of the devices.

RMS VALUES OF VOLTAGES AND CURRENTS

The *root mean square* or *RMS* values of voltage and current are the most common values encountered in electronics. Sometimes called *effective* values, the RMS value of an ac voltage or current is the value of a dc voltage or current that would cause a resistor to dissipate the same average amount of power as the ac waveform. This measurement became widely used in the early days of electrification when both ac and dc power utility power were in use. Even today, the values of the ac line voltage available from an electrical power outlet are given as RMS values. Unless otherwise specified, unlabeled ac voltage and current values found in most electronics literature are normally RMS values.

The RMS values of voltage and current get their name from the mathematical method used to derive their value relative to peak voltage and current. Start by *squaring* the individual values of all the instantaneous values of voltage or current during an entire single cycle of ac. Take the average (*mean*) of these squares (this is done by computing an integral of the waveform) and then find the square *root* of that average. This procedure produces the RMS value of voltage or current for all waveforms, sinusoidal or not. (The waveform is assumed to be periodic.)

For the remainder of this discussion, remember that we are assuming that the waveform in question is a sine wave. The simple formulas and conversion factors in this section are generally *not* true for non-sinusoidal waveforms such as square or triangle waves (see the sidebar, "Measuring Non-Sinusoidal Waveforms"). The following formulas are true *only* if the waveform is a sine wave and the circuit is *linear* — that is, raising or lowering the voltage will raise or lower the current proportionally. If those conditions are true, the following conversion factors have been computed and can be used without any additional mathematics.

For a sine wave to produce heat equivalent to a dc waveform the peak ac power required is twice the dc power. Therefore, the average ac power equivalent to a corresponding dc power is half the peak ac power.

$$P_{ave} = \frac{P_{pk}}{2} \tag{35}$$

A sine wave's RMS voltage and current values needed to arrive at average ac power are related to their peak values by the conversion factors:

$$E_{RMS} = \frac{E_{pk}}{\sqrt{2}} = \frac{E_{pk}}{1.414} = E_{pk} \times 0.707 \tag{36}$$

$$I_{RMS} = \frac{I_{pk}}{\sqrt{2}} = \frac{I_{pk}}{1.414} = I_{pk} \times 0.707 \tag{37}$$

RMS voltages and currents are what is displayed by most volt and ammeters.

If the RMS voltage is the peak voltage divided by $\sqrt{2}$, then the peak voltage must be the RMS voltage multiplied by $\sqrt{2}$, or

$$E_{pk} = E_{RMS} \times 1.414 \tag{38}$$

$$I_{pk} = I_{RMS} \times 1.414 \tag{39}$$

Example: What is the peak voltage and the peak-to-peak voltage at the usual household ac outlet, if the RMS voltage is 120 V?

$$E_{pk} = 120\ \text{V} \times 1.414 = 170\ \text{V}$$

$$E_{P-P} = 2 \times 170\ \text{V} = 340\ \text{V}$$

In the time domain of a sine wave, the instantaneous values of voltage and current correspond to the RMS values at the 45°, 135°, 225° and 315° points along the cycle shown in **Fig 2.26**. (The sine of 45° is approximately 0.707.) The instantaneous value of voltage or current is greater than the RMS value for half the cycle and less than the RMS value for half the cycle.

Since circuit specifications will most commonly list only RMS voltage and current values, these relationships are important in finding the peak voltages or currents that will stress components.

Example: What is the peak voltage on a capacitor if the RMS voltage of a sinusoidal waveform signal across it is 300 V ac?

$$E_{pk} = 300\ \text{V} \times 1.414 = 424\ \text{V}$$

The capacitor must be able to withstand this higher voltage, plus a safety margin. (The capacitor must also be rated for ac use because of the continually reversing polarity and ac current flow.) In power supplies that convert ac to dc and use capacitive input filters, the output voltage will approach the peak value of the ac voltage rather than the RMS value. (See the **Power Sources** chapter for more information on specifying components in this application.)

AVERAGE VALUES OF AC VOLTAGE AND CURRENT

Certain kinds of circuits respond to the *average* voltage or current (not power) of an ac waveform. Among these circuits are electrodynamic meter movements and power supplies that convert ac to dc and use heavily inductive ("choke") input filters, both of which work with the pulsating dc output of a full-wave rectifier. The average value of each ac half cycle is the *mean* of all the instantaneous values in that half cycle. (The average value of a sine wave or any symmetric ac waveform over an entire cycle is zero!) Related to the peak values of voltage and current, average values for each half-cycle

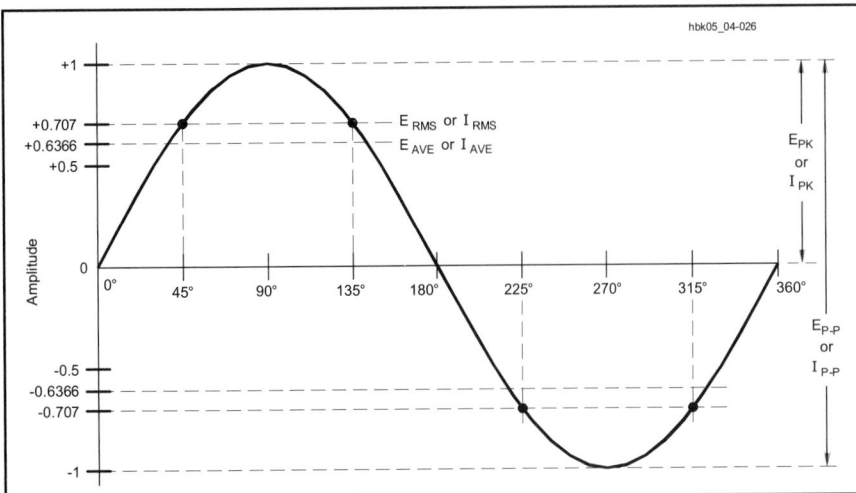

Fig 2.26 — The relationships between RMS, average, peak, and peak-to-peak values of ac voltage and current.

Measuring Nonsinusoidal Waveforms

Making measurements of ac waveforms is covered in more detail in the **Test Equipment and Measurements** chapter. However, this is a good point in the discussion to reinforce the dependence of RMS and values on the nature of the waveform being measured.

Analog meters and other types of instrumentation that display RMS values may only be calibrated for sine waves, those being the most common type of ac waveform. Using that instrumentation to accurately measure waveforms other than sine waves — such as speech, intermittent sine waves (such as CW from a transmitter), square waves, triangle waves or noise — requires the use of *calibration factors* or the measurement may not be valid.

To make calibrated, reliable measurements of the RMS or value of these waveforms requires the use of *true-RMS* instruments. These devices may use a balancing approach to a known dc value or if they are microprocessor-based, may actually perform the full root-mean-square calculation on the waveform. Be sure you know the characteristics of your test instruments if an accurate RMS value is important.

of a sine wave are $2/\pi$ (or 0.6366) times the peak value.

$$E_{ave} = 0.6366\,E_{pk} \tag{40}$$

$$I_{ave} = 0.6366\,I_{pk} \tag{41}$$

For convenience, **Table 2.5** summarizes the relationships between all of the common ac values. All of these relationships apply only to sine waves.

COMPLEX WAVEFORMS AND PEAK-ENVELOPE VALUES

Complex waveforms, as shown earlier in Fig 2.18, differ from sine waves. The amplitude of the peak voltage may vary signifi-

cantly from one cycle to the next, for example. Therefore, other amplitude measures are required, especially for accurate measurement of voltage and power with transmitted speech or data waveforms.

An SSB waveform (either speech or data) contains an RF ac waveform with a frequency many times that of the audio-frequency ac waveform with which it is combined. Therefore, the resultant *composite* waveform appears as an amplitude envelope superimposed upon the RF waveform as illustrated by **Fig 2.27**. For a complex waveform such as this, the *peak envelope voltage* (PEV) is the maximum or peak value of voltage anywhere in the waveform.

Peak envelope voltage is used in the calculation of *peak envelope power* (PEP). The Federal Communications Commission (FCC) sets the maximum power levels for amateur transmitters in terms of peak envelope power. PEP is the *average* power supplied to the antenna transmission line by the transmitter during one RF cycle at the crest of the modulation envelope, taken under normal operating conditions. That is, the average power for the RF cycle during which PEV occurs.

Since calculation of PEP requires the average power of the cycle, and the deviation of the modulated RF waveform from a sine wave is very small, the error incurred by using the conversion factors for sine waves is insignificant. Multiply PEV by 0.707 to obtain an RMS value. Then calculate PEP by using the square of the voltage divided by the load resistance.

$$PEP = \frac{(PEV \times 0.707)^2}{R} \tag{42}$$

Example: What is the PEP of a transmitter's

output with a PEV of 100 V into a 50-ohm load?

$$PEP = \frac{(100 \times 0.707)^2}{R} = \frac{(70.7)^2}{50} = 100\ W$$

2.6.5 – Glossary — AC Theory and Reactance

Admittance (Y) — The reciprocal of impedance, measured in siemens (S).

Capacitance (C) — The ability to store electrical energy in an electrostatic field, measured in farads (F). A device with capacitance is a capacitor.

Flux density (B) — The number of magnetic-force lines per unit area, measured in gauss.

Frequency (f) — The rate of change of an ac voltage or current, measured in cycles per second, or hertz (Hz).

Fundamental — The lowest frequency in a series of sine waves whose frequencies have an integer relationship.

Harmonic — A sine wave whose frequency is an integer multiple of a fundamental frequency.

Impedance (Z) — The complex combination of resistance and reactance, measured in ohms (Ω).

Inductance (L) — The ability to store electrical energy in a magnetic field, measured in henrys (H). A device, such as a coil of wire, with inductance is an inductor.

Peak (voltage or current) — The maximum value relative to zero that an ac voltage or current attains during any cycle.

Peak-to-peak (voltage or current) — The value of the total swing of an ac voltage

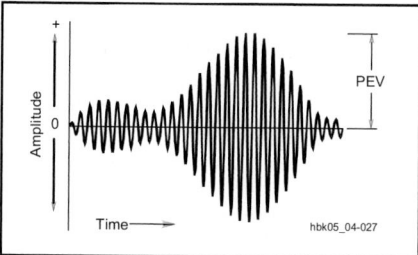

Fig 2.27 — The peak envelope voltage (PEV) for a composite waveform.

Table 2.5

Conversion Factors for Sinusoidal AC Voltage or Current

From	To	Multiply By
Peak	Peak-to-Peak	2
Peak-to-Peak	Peak	0.5
Peak	RMS	$1/\sqrt{2}$ or 0.707
RMS	Peak	$\sqrt{2}$ or 1.414
Peak-to-Peak	RMS	$1/(2 \times \sqrt{2})$ or 0.35355
RMS	Peak-to-Peak	$2 \times \sqrt{2}$ or 2.828
Peak	Average	$2/\pi$ or 0.6366
Average	Peak	$\pi/2$ or 1.5708
RMS	Average	$(2 \times \sqrt{2})/\pi$ or 0.90
Average	RMS	$\pi/(2 \times \sqrt{2})$ or 1.11

Note: These conversion factors apply only to continuous pure sine waves.

or current from its peak negative value to its peak positive value, ordinarily twice the value of the peak voltage or current.

Period (T) — The duration of one ac voltage or current cycle, measured in seconds (s).

Permeability (μ) — The ratio of the magnetic flux density of an iron, ferrite, or similar core in an electromagnet compared to the magnetic flux density of an air core, when the current through the electromagnet is held constant.

Power (P) — The rate of electrical-energy use, measured in watts (W).

Q (quality factor) — The ratio of energy stored in a reactive component (capacitor or inductor) to the energy dissipated, equal to the reactance divided by the resistance.

Reactance (X) — Opposition to alternating current by storage in an electrical field (by a capacitor) or in a magnetic field (by an inductor), measured in ohms (Ω).

Resonance — Ordinarily, the condition in an ac circuit containing both capacitive and inductive reactance in which the reactances are equal.

RMS (voltage or current) — Literally, "root mean square," the square root of the average of the squares of the instantaneous values for one cycle of a waveform. A dc voltage or current that will produce the same heating effect as the waveform. For a sine wave, the RMS value is equal to 0.707 times the peak value of ac voltage or current.

Susceptance (B) — The reciprocal of reactance, measured in siemens (S).

Time constant (τ) — The time required for the voltage in an RC circuit or the current in an RL circuit to rise from zero to approximately 63.2% of its maximum value or to fall from its maximum value 63.2% toward zero.

Toroid — Literally, any donut-shaped solid; most commonly referring to ferrite or powdered-iron cores supporting inductors and transformers.

Transducer — Any device that converts one form of energy to another; for example an antenna, which converts electrical energy to electromagnetic energy or a speaker, which converts electrical energy to sonic energy.

Transformer — A device consisting of at least two coupled inductors capable of transferring energy through mutual inductance.

2.7 Capacitance and Capacitors

It is possible to build up and hold an electrical charge in an *electrostatic field*. This phenomenon is called *capacitance*, and the devices that exhibit capacitance are called *capacitors*. (Old articles and texts use the obsolete term *condenser*.) **Fig 2.28** shows several schematic symbols for capacitors. Part A shows a fixed capacitor; one that has a single value of capacitance. Part B shows the symbol for variable capacitors; these are adjustable over a range of values. If the capacitor is of a type that is *polarized*, meaning that dc voltages must be applied with a specific polarity, the straight line in the symbol should be connected to the most positive voltage, while the curved line goes to the more negative voltage, which is often ground. For clarity,

the positive terminal of a polarized capacitor symbol is usually marked with a + symbol. The symbol for *non-polarized* capacitors may be two straight lines or the + symbol may be omitted. When in doubt, consult the capacitor's specifications or the circuits parts list.

2.7.1 Electrostatic Fields and Energy

An *electrostatic field* is created wherever a voltage exists between two points, such as two opposite electric charges or regions that contain different amounts of charge. The field causes electric charges (such as electrons or ions) in the field to feel a force in the direction of the field. If the charges are not free to move, as in an insulator, they store the field's energy as *potential energy*, just as a weight held in place by a surface stores gravitational energy. If the charges are free to move, the field's stored energy is converted to *kinetic energy* of motion just as if the weight is released to fall in a gravitational field.

The field is represented by *lines of force* that show the direction of the force felt by the electric charge. Each electric charge is surrounded by an electric field. The lines of force of the field begin on the charge and extend away from charge into space. The lines of force can terminate on another charge (such as lines of force between a proton and an electron) or they can extend to infinity.

The strength of the electrostatic field is measured in *volts per meter* (V/m). Stronger fields cause the moving charges to accelerate more strongly (just as stronger gravity causes weights to fall faster) and stores more energy in fixed charges. The stronger the field in V/m, the more force an electric charge in the field will feel. The strength of the electric field

diminishes with the square of the distance from its source, the electric charge.

2.7.2 The Capacitor

Suppose two flat metal plates are placed close to each other (but not touching) and are connected to a battery through a switch, as illustrated in **Fig 2.29A**. At the instant the switch is closed, electrons are attracted from the upper plate to the positive terminal of the battery, while the same quantity is repelled from the negative battery terminal and pushed into the lower plate. This imbalance of charge creates a voltage between the plates. Eventually, enough electrons move into one plate and out of the other to make the voltage between the plates the same as the battery voltage.

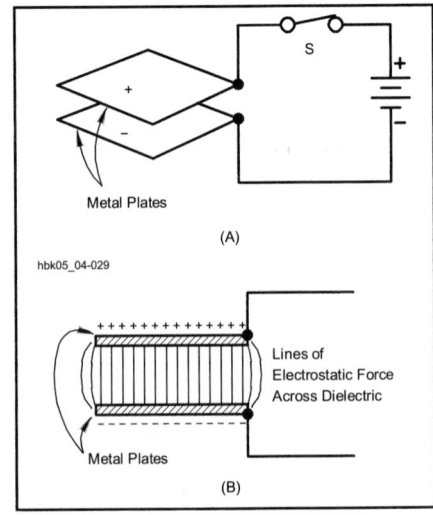

Fig 2.29 — A simple capacitor showing the basic charging arrangement at A, and the retention of the charge due to the electrostatic field at B.

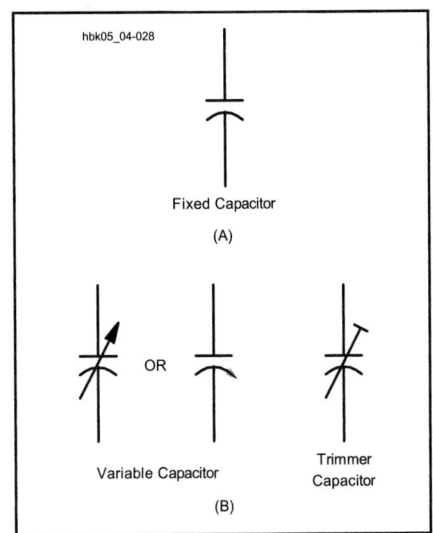

Fig 2.28 — Schematic symbol for a fixed capacitor is shown at A. The symbols for a variable capacitor are shown at B.

At this point, the voltage between the plates opposes further movement of electrons and no further current flow occurs.

If the switch is opened after the plates have been charged in this way, the top plate is left with a deficiency of electrons and the bottom plate with an excess. Since there is no current path between the two, the plates remain *charged* despite the fact that the battery no longer is connected to a source of voltage. As illustrated in Fig 2.29B, the separated charges create an electrostatic field between the plates. The electrostatic field contains the energy that was expended by the battery in causing the electrons to flow off of or onto the plates. These two plates create a *capacitor*, a device that has the property of storing electrical energy in an electric field, a property called *capacitance*.

The amount of electric charge that is held on the capacitor plates is proportional to the applied voltage and to the capacitance of the capacitor:

$$Q = CV \tag{43}$$

where

Q = charge in coulombs,
C = capacitance in farads (F), and
V = electrical potential in volts. (The symbol E is also commonly used instead of V in this and the following equation.)

The energy stored in a capacitor is also a function of voltage and capacitance:

$$W = \frac{V^2 C}{2} \tag{44}$$

where W = energy in joules (J) or watt-seconds.

If a wire is simultaneously touched to the two plates (short circuiting them), the voltage between the plates causes the excess electrons on the bottom plate to flow through the wire to the upper plate, restoring electrical neutrality. The plates are then *discharged*.

Fig 2.30 illustrates the voltage and current in the circuit, first, at the moment the switch is closed to charge the capacitor and, second, at the moment the shorting switch is closed to discharge the capacitor. Note that the periods of charge and discharge are very short, but that they are not zero. This finite charging and discharging time can be controlled and that will prove useful in the creation of timing circuits.

During the time the electrons are moving — that is, while the capacitor is being charged or discharged — a current flows in the circuit even though the circuit apparently is broken by the gap between the capacitor plates. The current flows only during the time of charge and discharge, however, and this time is usually very short. There is no continuous flow of direct current through a capacitor.

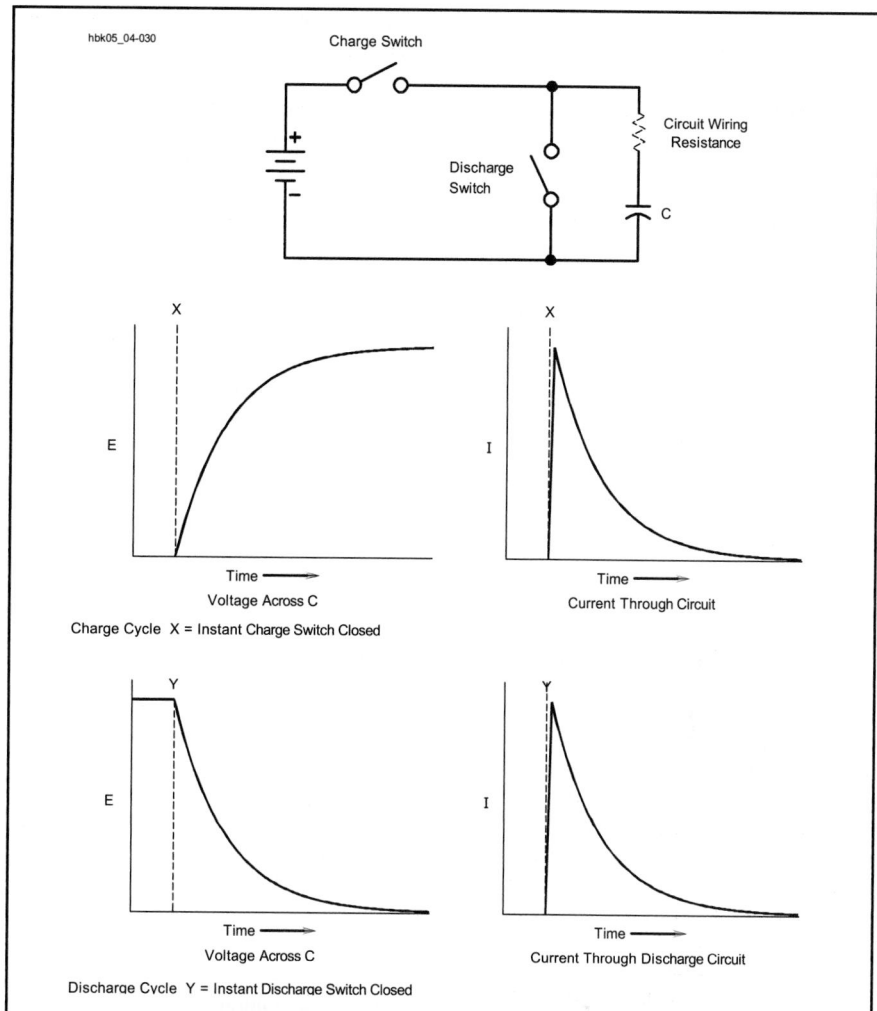

Fig 2.30 — The flow of current during the charge and discharge of a capacitor. The charging graphs assume that the charge switch is closed and the discharge switch is open. The discharging graphs assume just the opposite.

Although dc cannot pass through a capacitor, alternating current can. At the same time one plate is charged positively by the positive excursion of the alternating current, the other plate is being charged negatively at the same rate. (Remember that conventional current is shown as the flow of positive charge, equal to and opposite the actual flow of electrons.) The reverse process occurs during the second half of the cycle as the changing polarity of the applied voltage causes the flow of charge to change direction, as well. The continual flow into and out of the capacitor caused by ac voltage appears as an ac current, although with a phase difference between the voltage and current flow as described below.

UNITS OF CAPACITANCE

The basic unit of capacitance, the ability to store electrical energy in an electrostatic field, is the *farad*. This unit is generally too large for practical radio circuits, although capacitors of several farads in value are used in place of small batteries or as a power supply filter for automotive electronics. Capacitance encountered in radio and electronic circuits is usually measured in microfarads (abbreviated µF), nanofarads (abbreviated nF) or picofarads (pF). The microfarad is one millionth of a farad (10^{-6} F), the nanofarad is one thousandth of a microfarad (10^{-9} F) and the picofarad is one millionth of a microfarad (10^{-12} F). Old articles and texts use the obsolete term micromicrofarad (µµF) in place of picofarad.

CAPACITOR CONSTRUCTION

An ideal capacitor is a pair of parallel metal plates separated by an insulating or *dielectric* layer, ideally a vacuum. The capacitance of a vacuum-dielectric capacitor is given by

$$C = \frac{A \, \varepsilon_r \, \varepsilon_0}{d} \tag{45}$$

where

C = capacitance, in farads
A = area of plates, in cm^2
d = spacing of the plates in cm

ε_r = dielectric constant of the insulating material

ε_0 = permittivity of free space, 8.85×10^{-14} F/cm.

The actual capacitance of such a parallel-plate capacitor is somewhat higher due to *end effect* caused by the electric field that exists just outside the edges of the plates.

The *larger* the plate area and the *smaller* the spacing between the plates, the *greater* the amount of energy that can be stored for a given voltage, and the *greater* the capacitance. The more general name for the capacitor's plates is *electrodes*. However, amateur radio literature generally refers to a capacitor's electrodes as plates and that is the convention in this text.

The amount of capacitance also depends on the material used as insulating material between the plates; capacitance is smallest with air or a vacuum as the insulator. Substituting other insulating materials for air may greatly increase the capacitance.

The ratio of the capacitance with a material other than a vacuum or air between the plates to the capacitance of the same capacitor with air insulation is called the *dielectric constant*, or K, of that particular insulating material. The dielectric constants of a number of materials commonly used as dielectrics in capacitors are given in **Table 2.6**. For example, if polystyrene is substituted for air in a capacitor, the capacitance will be 2.6 times greater.

In practice, capacitors often have more than two plates, with alternating plates being connected in parallel to form two sets, as shown in **Fig 2.31**. This practice makes it possible to obtain a fairly large capacitance in a small space, since several plates of smaller

individual area can be stacked to form the equivalent of a single large plate of the same total area. Also, all plates except the two on the ends of the stack are exposed to plates of the other group on both sides, and so are twice as effective in increasing the capacitance.

The formula for calculating capacitance from these physical properties is:

$$C = \frac{0.2248 \text{ K A} (n-1)}{d} \qquad (46)$$

where

C = capacitance in pF,
K = dielectric constant of material between plates,
A = area of one side of one plate in square inches,
d = separation of plate surfaces in inches, and
n = number of plates.

If the area (A) is in square centimeters and the separation (d) is in centimeters, then the

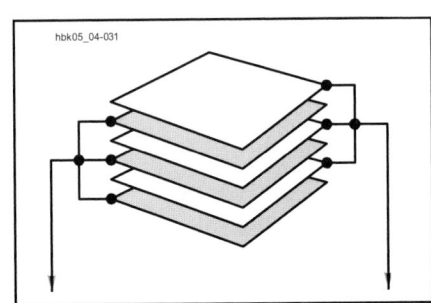

Fig 2.31 — A multiple-plate capacitor. Alternate plates are connected to each other, increasing the total area available for storing charge.

formula for capacitance becomes

$$C = \frac{0.0885 \text{ K A} (n-1)}{d} \qquad (47)$$

If the plates in one group do not have the same area as the plates in the other, use the area of the smaller plates.

Example: What is the capacitance of two copper plates, each 1.50 square inches in area, separated by a distance of 0.00500 inch, if the dielectric is air?

$$C = \frac{0.2248 \text{ K A} (n-1)}{d}$$

$$C = \frac{0.2248 \times 1 \times 1.50 \, (2-1)}{0.00500}$$

$$C = 67.4 \text{ pF}$$

What is the capacitance if the dielectric is mineral oil? (See Table 2.6 for the appropriate dielectric constant.)

$$C = \frac{0.2248 \times 2.23 \times 1.50 \, (2-1)}{0.00500}$$

$$C = 150.3 \text{ pF}$$

2.7.3 Capacitors in Series and Parallel

When a number of capacitors are connected in parallel, as in **Fig 2.32A**, the total capacitance of the group is equal to the sum of the individual capacitances:

$$C_{total} = C1 + C2 + C3 + C4 + \ldots + C_n \qquad (48)$$

When two or more capacitors are con-

Table 2.6
Relative Dielectric Constants of Common Capacitor Dielectric Materials

Material	Dielectric Constant (k)	(O)rganic or (I)norganic
Vacuum	1 (by definition)	I
Air	1.0006	I
Ruby mica	6.5 - 8.7	I
Glass (flint)	10	I
Barium titanate (class I)	5 - 450	I
Barium titanate (class II)	200 - 12000	I
Kraft paper	≈ 2.6	O
Mineral Oil	≈ 2.23	O
Castor Oil	≈ 4.7	O
Halowax	≈ 5.2	O
Chlorinated diphenyl	≈ 5.3	O
Polyisobutylene	≈ 2.2	O
Polytetrafluoroethylene	≈ 2.1	O
Polyethylene terephthalate	≈ 3	O
Polystyrene	≈ 2.6	O
Polycarbonate	≈ 3.1	O
Aluminum oxide	≈ 8.4	I
Tantalum pentoxide	≈ 28	I
Niobium oxide	≈ 40	I
Titanium dioxide	≈ 80	I

(Adapted from: Charles A. Harper, *Handbook of Components for Electronics*, p 8-7.)

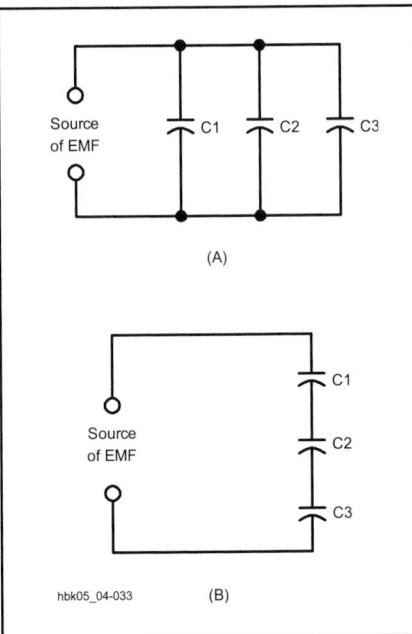

Fig 2.32 — Capacitors in parallel are shown at A, and in series at B.

nected in series, as in Fig 2.32B, the total capacitance is less than that of the smallest capacitor in the group. The rule for finding the capacitance of a number of series-connected capacitors is the same as that for finding the resistance of a number of parallel-connected resistors.

$$C_{total} = \cfrac{1}{\cfrac{1}{C1} + \cfrac{1}{C2} + \cfrac{1}{C3} + ... + \cfrac{1}{C_n}} \qquad (49)$$

For only two capacitors in series, the formula becomes:

$$C_{total} = \frac{C1 \times C2}{C1 + C2} \qquad (50)$$

The same units must be used throughout; that is, all capacitances must be expressed in µF, nF or pF, etc. Different units cannot be combined in the same equation.

Capacitors are often connected in parallel to obtain a larger total capacitance than is available in one unit. The voltage rating of capacitors connected in parallel is the lowest voltage rating of any of the capacitors.

When capacitors are connected in series, the applied voltage is divided between them according to Kirchhoff's Voltage Law. The situation is much the same as when resistors are in series and there is a voltage drop across each. The voltage that appears across each series-connected capacitor is inversely proportional to its capacitance, as compared with the capacitance of the whole group. (This assumes ideal capacitors.)

Example: Three capacitors having capacitances of 1, 2 and 4 µF, respectively, are connected in series as shown in **Fig 2.33**. The voltage across the entire series is 2000 V. What is the total capacitance? (Since this is a calculation using theoretical values to illustrate a technique, we will not follow the rules of significant figures for the calculations.)

$$C_{total} = \cfrac{1}{\cfrac{1}{C1} + \cfrac{1}{C2} + \cfrac{1}{C3}}$$

$$= \cfrac{1}{\cfrac{1}{1\,\mu F} + \cfrac{1}{2\,\mu F} + \cfrac{1}{4\,\mu F}}$$

$$= \cfrac{1}{\cfrac{7}{4\,\mu F}} = \frac{4\,\mu F}{7} = 0.5714\,\mu F$$

The voltage across each capacitor is proportional to the total capacitance divided by the capacitance of the capacitor in question. So the voltage across C1 is:

$$E1 = \frac{0.5714\,\mu F}{1\,\mu F} \times 2000\ V = 1143\ V$$

Similarly, the voltages across C2 and C3 are:

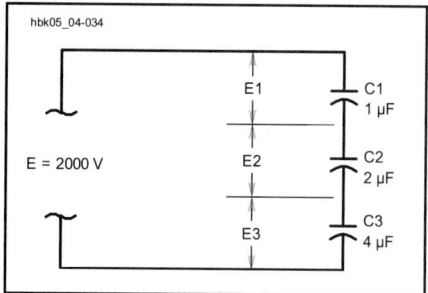

Fig 2.33 — An example of capacitors connected in series. The text shows how to find the voltage drops, E1 through E3.

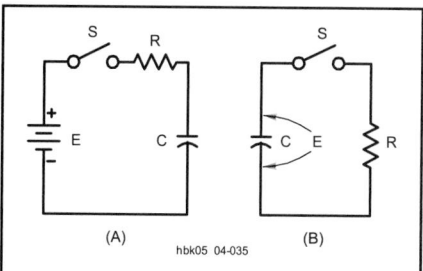

Fig 2.34 — An RC circuit. The series resistance delays the process of charging (A) and discharging (B) when the switch, S, is closed.

$$E2 = \frac{0.5714\,\mu F}{2\,\mu F} \times 2000\ V = 571\ V$$

and

$$E3 = \frac{0.5714\,\mu F}{4\,\mu F} \times 2000\ V = 286\ V$$

The sum of these three voltages equals 2000 V, the applied voltage.

Capacitors may be connected in series to enable the group to withstand a larger voltage than any individual capacitor is rated to withstand. The trade-off is a decrease in the total capacitance. As shown by the previous example, the applied voltage does not divide equally between the capacitors except when all the capacitances are precisely the same. Use care to ensure that the voltage rating of any capacitor in the group is not exceeded. If you use capacitors in series to withstand a higher voltage, you should also connect an "equalizing resistor" across each capacitor as described in the **Power Sources** chapter.

2.7.4 RC Time Constant

Connecting a dc voltage source directly to the terminals of a capacitor charges the capacitor to the full source voltage almost instantaneously. Any resistance added to the

circuit (as R in **Fig 2.34A**) limits the current, lengthening the time required for the voltage between the capacitor plates to build up to the source-voltage value. During this charging period, the current flowing from the source into the capacitor gradually decreases from its initial value. The increasing voltage stored in the capacitor's electric field offers increasing opposition to the steady source voltage.

While it is being charged, the voltage between the capacitor terminals is an exponential function of time, and is given by:

$$V(t) = E\left(1 - e^{-\frac{t}{RC}}\right) \qquad (51)$$

where
  $V(t)$ = capacitor voltage at time t,
  E = power source potential in volts,
  t = time in seconds after initiation of charging current,
  e = natural logarithmic base = 2.718,
  R = circuit resistance in ohms, and
  C = capacitance in farads.

(References that explain exponential equations, e, and other mathematical topics are found in the "Radio Mathematics" article on this book's CD-ROM.)

If t = RC, the above equation becomes:

$$V(RC) = E(1 - e^{-1}) \approx 0.632\ E \qquad (52)$$

The product of R in ohms times C in farads is called the *time constant* (also called the *RC time constant*) of the circuit and is the time in seconds required to charge the capacitor to 63.2% of the applied voltage. (The lower-case Greek letter tau, τ, is often used to represent the time constant in electronics circuits.) After two time constants (t = 2τ) the capacitor charges another 63.2% of the difference between the capacitor voltage at one time constant and the applied voltage, for a total charge of 86.5%. After three time constants the capacitor reaches 95% of the applied voltage, and so on, as illustrated in the curve of **Fig 2.35A**. After five time constants, a capacitor is considered fully charged, having reached 99.24% of the applied voltage. Theoretically, the charging process is never really finished, but eventually the charging current drops to an immeasurably small value and the voltage is effectively constant.

If a charged capacitor is discharged through a resistor, as in Fig 2.34B, the same time constant applies to the decay of the capacitor voltage. A direct short circuit applied between the capacitor terminals would discharge the capacitor almost instantly. The resistor, R, limits the current, so a capacitor discharging through a resistance exhibits the same time-constant characteristics (calculated in the same way as above) as a charging capacitor. The voltage, as a function of time while the capacitor is being discharged, is given by:

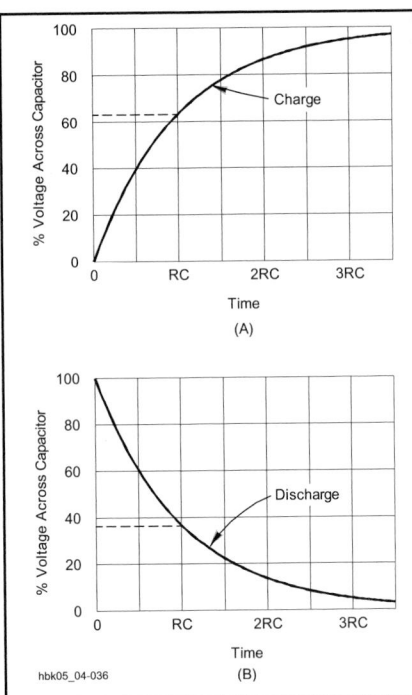

Fig 2.35 — At A, the curve shows how the voltage across a capacitor rises, with time, when charged through a resistor. The curve at B shows the way in which the voltage decreases across a capacitor when discharging through the same resistance. For practical purposes, a capacitor may be considered charged or discharged after five RC periods.

$$V(t) = E\left(e^{-\frac{t}{RC}}\right) \qquad (53)$$

where t = time in seconds after initiation of discharge and E is the fully-charged capacitor voltage prior to beginning discharge.

Again, by letting t = RC, the time constant of a discharging capacitor represents a decrease in the voltage across the capacitor of about 63.2%. After five time-constants, the capacitor is considered fully discharged, since the voltage has dropped to less than 1% of the full-charge voltage. **Fig 2.35B** is a graph of the discharging capacitor voltage in terms of time constants.

Time constant calculations have many uses in radio work. The following examples are all derived from practical-circuit applications.

Example 1: A 100-μF capacitor in a high-voltage power supply is shunted by a 100-kΩ resistor. What is the minimum time before the capacitor may be considered fully discharged? Since full discharge is approximately five RC periods,

$$t = 5 \times RC = 5 \times 100 \times 10^3 \ \Omega \times 100 \times 10^{-6} \ F$$

$$= 50,000 \times 10^{-3}$$

$$t = 50.0 \ s$$

Note: Although waiting almost a minute for the capacitor to discharge seems safe in this high-voltage circuit, never rely solely on capacitor-discharging resistors (often called *bleeder resistors*). Be certain the power source is removed and the capacitors are totally discharged before touching any circuit components. (See the **Power Sources** chapter for more information on bleeder resistors.)

Example 2: Smooth CW keying without clicks requires approximately 5 ms (0.005 s) of delay in both the rising and falling edges of the waveform, relative to full charging and discharging of a capacitor in the circuit. What typical values might a builder choose for an RC delay circuit in a keyed voltage line? Since full charge and discharge require 5 RC periods,

$$RC = \frac{t}{5} = \frac{0.005 \ s}{5} = 0.001 \ s$$

Any combination of resistor and capacitor whose values, when multiplied together, equal 0.001 would do the job. A typical capacitor might be 0.05 μF. In that case, the necessary resistor would be:

$$R = \frac{0.001 \ s}{0.05 \times 10^{-6} \ F}$$

$$= 0.02 \times 10^6 \ \Omega = 20,000 \ \Omega = 20 \ k\Omega$$

In practice, a builder would use the calculated value as a starting point. The final value would be selected by monitoring the waveform on an oscilloscope.

Example 3: Many modern integrated circuit (IC) devices use RC circuits to control their timing. To match their internal circuitry, they may use a specified threshold voltage as the trigger level. For example, a certain IC uses a trigger level of 0.667 of the supply voltage. What value of capacitor and resistor would be required for a 4.5-second timing period?

First we will solve equation 51 for the time constant, RC. The threshold voltage is 0.667 times the supply voltage, so we use this value for V(t).

$$V(t) = E\left(1 - e^{-\frac{t}{RC}}\right)$$

$$0.667 \ E = E\left(1 - e^{-\frac{t}{RC}}\right)$$

$$e^{-\frac{t}{RC}} = 1 - 0.667$$

$$\ln\left(e^{-\frac{t}{RC}}\right) = \ln(0.333)$$

$$-\frac{t}{RC} = -1.10$$

We want to find a capacitor and resistor combination that will produce a 4.5 s timing period, so we substitute that value for t.

$$RC = \frac{4.5 \ s}{1.10} = 4.1 \ s$$

If we select a value of 10 μF, we can solve for R.

$$R = \frac{4.1 \ s}{10 \times 10^{-6} \ F} = 0.41 \times 10^6 \ \Omega = 410 \ k\Omega$$

A 1% tolerance resistor and capacitor will give good results. You could also use a variable resistor and an accurate method of measuring time to set the circuit to a 4.5 s period.

As the examples suggest, RC circuits have numerous applications in electronics. The number of applications is growing steadily, especially with the introduction of integrated circuits controlled by part or all of a capacitor charge or discharge cycle.

2.7.5 Alternating Current in Capacitance

Whereas a capacitor in a dc circuit will appear as an open circuit except for the brief charge and discharge periods, the same capacitor in an ac circuit will both pass and limit current. A capacitor in an ac circuit does not handle electrical energy like a resistor, however. Instead of converting the energy to heat and dissipating it, capacitors store electrical energy when the applied voltage is greater than that across the capacitor and return it to the circuit when the opposite is true.

In **Fig 2.36** a sine-wave ac voltage having a maximum value of 100 V is applied to a capacitor. In the period OA, the applied voltage increases from 0 to 38, storing energy in the capacitor; at the end of this period the capacitor is charged to that voltage. In interval AB the voltage increases to 71; that is, by an additional 33 V. During this interval a smaller quantity of charge has been added than in OA, because the voltage rise during interval AB is smaller. Consequently the average current during interval AB is smaller than during OA. In the third interval, BC, the voltage rises from 71 to 92, an increase of 21 V. This is less than the voltage increase during AB, so the quantity of charge added is less; in other words, the average current during interval

RC Timesaver

When calculating time constants, it is handy to remember that if R is in units of MΩ and C is in units of μF, the result of R × C will be in seconds. Expressed as an equation: MΩ × μF = seconds

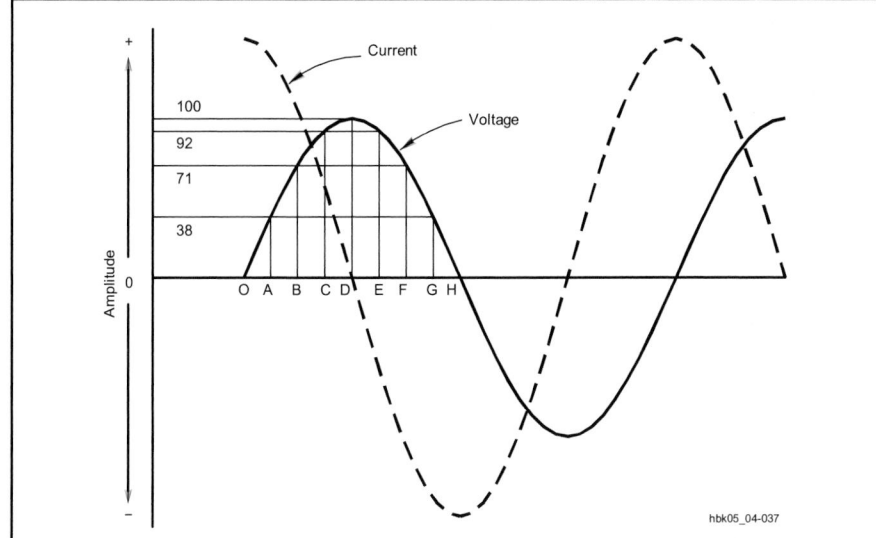

Fig 2.36 — Voltage and current phase relationships when an alternating current is applied to a capacitor.

where
X_C = capacitive reactance in ohms,
f = frequency in hertz,
C = capacitance in farads
π = 3.1416

Note: In many references and texts, angular frequency $\omega=2\pi f$ is used and equation 54 would read

$$X_C = \frac{1}{\omega C}$$

Example: What is the reactance of a capacitor of 470 pF (0.000470 μF) at a frequency of 7.15 MHz?

$$X_C = \frac{1}{2\pi f C}$$

$$= \frac{1}{2\pi \times 7.15\ \text{MHz} \times 0.000470\ \mu\text{F}}$$

$$= \frac{1\,\Omega}{0.0211} = 47.4\ \Omega$$

Example: What is the reactance of the same capacitor, 470 pF (0.000470 μF), at a frequency of 14.29 MHz?

$$X_C = \frac{1}{2\pi f C}$$

$$= \frac{1}{2\pi \times 14.3\ \text{MHz} \times 0.000470\ \mu\text{F}}$$

$$= \frac{1\,\Omega}{0.0422} = 23.7\ \Omega$$

BC is still smaller. In the fourth interval, CD, the voltage increases only 8 V; the charge added is smaller than in any preceding interval and therefore the current also is smaller.

By dividing the first quarter-cycle into a very large number of such intervals, it can be shown that the current charging the capacitor has the shape of a sine wave, just as the applied voltage does. The current is largest at the beginning of the cycle and becomes zero at the maximum value of the voltage, so there is a phase difference of 90° between the voltage and the current. During the first quarter-cycle the current is flowing in the original (positive) direction through the circuit as indicated by the dashed line in Fig 2.36, since the capacitor is being charged. The increasing capacitor voltage indicates that energy is being stored in the capacitor.

In the second quarter-cycle — that is, in the time from D to H — the voltage applied to the capacitor decreases. During this time the capacitor loses charge, returning the stored energy to the circuit. Applying the same reasoning, it is evident that the current is small in interval DE and continues to increase during each succeeding interval. The current is flowing *against* the applied voltage, however, because the capacitor is returning energy to (discharging into) the circuit. The current thus flows in the *negative* direction during this quarter-cycle.

The third and fourth quarter-cycles repeat the events of the first and second, respectively, although the polarity of the applied voltage has reversed, and so the current changes to correspond. In other words, an alternating current flows in the circuit because of the alternate charging and discharging of the capacitance. As shown in Fig 2.36, the current starts its cycle 90° before the voltage, so the

current in a capacitor *leads* the applied voltage by 90°. You might find it helpful to remember the word "ICE" as a mnemonic because the current (I) in a capacitor (C) comes before voltage (E). (see the sidebar "ELI the ICE man" in the section on inductors.) We can also turn this statement around, to say the voltage in a capacitor *lags* the current by 90°.

2.7.6 Capacitive Reactance and Susceptance

The quantity of electric charge that can be placed on a capacitor is proportional to the applied voltage and the capacitance. If the applied voltage is ac, this amount of charge moves back and forth in the circuit once each cycle. Therefore, the rate of movement of charge (the current) is proportional to voltage, capacitance and frequency. Stated in another way, capacitor current is proportional to capacitance for a given applied voltage and frequency.

When the effects of capacitance and frequency are considered together, they form a quantity called *reactance* that relates voltage and current in a capacitor, similar to the role of resistance in Ohm's Law. Because the reactance is created by a capacitor, it is called *capacitive reactance*. The units for reactance are ohms, just as in the case of resistance. Although the units of reactance are ohms, there is no power dissipated in reactance. The energy stored in the capacitor during one portion of the cycle is simply returned to the circuit in the next.

The formula for calculating the reactance of a capacitor at a given frequency is:

$$X_C = \frac{1}{2\pi f C} \tag{54}$$

Current in a capacitor is directly related to the rate of change of the capacitor voltage. The maximum rate of change of voltage in a sine wave increases directly with the frequency, even if its peak voltage remains fixed. Therefore, the maximum current in the capacitor must also increase directly with

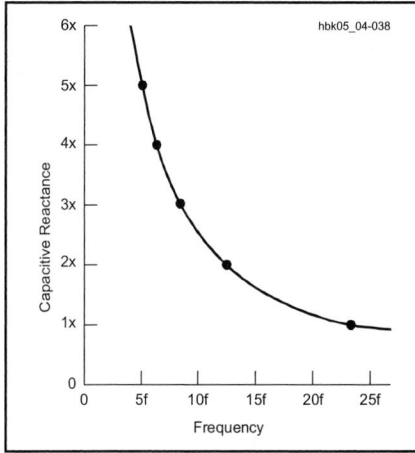

Fig 2.37 — A graph showing the general relationship of reactance to frequency for a fixed value of capacitance.

Capacitive Reactance Timesaver

The fundamental units for frequency and capacitance (hertz and farads) are too cumbersome for practical use in radio circuits. If the capacitance is specified in microfarads (µF) and the frequency is in megahertz (MHz), however, the reactance calculated from equation 54 is in units of ohms (Ω).

Table 2.7

Typical Temperature Coefficients and Leakage Resistances for Various Capacitor Constructions

Type	TC @ 20°C (PPM/°C)	DC Leakage Resistance (Ω)
Ceramic Disc	±300(NP0)	> 10 M
	+150/−1500(GP)	> 10 M
Mica	−20 to +100	> 100,000 M
Polyester	±500	> 10 M
Tantalum Electrolytic	±1500	> 10 MΩ
Small Al Electrolytic(≈ 100 µF)	−20,000	500 k - 1 M
Large Al Electrolytic(≈ 10 mF)	−100,000	10 k
Vacuum (glass)	+100	≈ ∞
Vacuum (ceramic)	+50	≈ ∞

frequency. Since, if voltage is fixed, an increase in current is equivalent to a decrease in reactance, the reactance of any capacitor decreases proportionally as the frequency increases. **Fig 2.37** illustrates the decrease in reactance of an arbitrary-value capacitor with respect to increasing frequency. The only limitation on the application of the graph is the physical construction of the capacitor, which may favor low-frequency uses or high-frequency applications.

CAPACITIVE SUSCEPTANCE

Just as conductance is sometimes the most useful way of expressing a resistance's ability to conduct current, the same is true for capacitors and ac current. This ability is called *susceptance* (abbreviated B). The units of susceptance are siemens (S), the same as that of conductance and admittance.

Susceptance in a capacitor is *capacitive susceptance*, abbreviated B_C. In an ideal capacitor with no losses, susceptance is simply the reciprocal of reactance. Hence,

$$B_C = \frac{1}{X_C}$$

where

X_C is the capacitive reactance, and
B_C is the capacitive susceptance.

2.7.7 Characteristics of Capacitors

The ideal capacitor does not conduct any current at dc, dissipates none of the energy stored in it, has the same value at all temperatures, and operates with any amount of voltage applied across it or ac current flowing through it. Practical capacitors deviate considerably from that ideal and have imperfections that must be considered when selecting capacitors and designing circuits that use capacitors. (The characteristics of capacitors at high frequencies is discussed in the **RF Techniques** chapter.)

LEAKAGE RESISTANCE

If we use anything other than a vacuum for the insulating layer, even air, two imperfections are created. Because there are atoms between the plates, some electrons will be available to create a current between the plates when a dc voltage is applied. The magnitude of this *leakage current* will depend on the insulator quality, and the current is usually very small. Leakage current can be modeled by a resistance R_L in parallel with the capacitance (in an ideal capacitor, R_L is infinite). **Table 2.7** shows typical dc leakage resistances for different dielectric materials. Leakage also generally increases with increasing temperature.

CAPACITOR LOSSES

When an ac current flows through the capacitor (even at low frequencies), capacitors dissipate some of the energy stored in the dielectric due to the electromagnetic properties of dielectric materials. This loss can be thought of as a resistance in series with the capacitor and it is often specified in the manufacturer's data for the capacitor as *effective (or equivalent) series resistance* (ESR).

Loss can also be specified as the capacitor's *loss angle*, θ. (Some literature uses δ for loss angle.) Loss angle is the angle between X_C (the reactance of the capacitor without any loss) and the impedance of the capacitor (impedance is discussed later in this chapter) made up of the combination of ESR and X_C. Increasing loss increases loss angle. The loss angle is usually quite small, and is zero for an ideal capacitor.

Dissipation Factor (DF) or *loss tangent* = tan θ = ESR / X_C and is the ratio of loss resistance to reactance. The loss angle of a given capacitor is relatively constant over frequency, meaning that ESR = (tan θ) / 2πfC goes down as frequency goes up. For this reason, ESR must be specified at a given frequency.

TOLERANCE AND TEMPERATURE COEFFICIENT

As with resistors, capacitor values vary in production, and most capacitors have a toler-ance rating either printed on them or listed on a data sheet. Typical capacitor tolerances and the labeling of tolerance are described in the chapter on **Component Data and References**.

Because the materials that make up a capacitor exhibit mechanical changes with temperature, capacitance also varies with temperature. This change in capacitance with temperature is the capacitor's *temperature coefficient* or *tempco (TC)*. The lower a capacitor's TC, the less its value changes with temperature.

TC is important to consider when constructing a circuit that will carry high power levels, operate in an environment far from room temperature, or must operate consistently at different temperatures. Typical temperature coefficients for several capacitor types are given in Table 2.7. (Capacitor temperature coefficient behaviors are listed in the **Component Data and References** chapter.)

VOLTAGE RATINGS AND BREAKDOWN

When voltage is applied to the plates of a capacitor, force is exerted on the atoms and molecules of the dielectric by the electrostatic field between the plates. If the voltage is high enough, the atoms of the dielectric will ionize (one or more of the electrons will be pulled away from the atom), causing a large dc current to flow discharging the capacitor. This is *dielectric breakdown,* and it is generally destructive to the capacitor because it creates punctures or defects in solid dielectrics that provide permanent low-resistance current paths between the plates. (*Self-healing* dielectrics have the ability to seal off this type of damage.) With most gas dielectrics such as air, once the voltage is removed, the arc ceases and the capacitor is ready for use again.

The *breakdown voltage* of a dielectric depends on the chemical composition and thickness of the dielectric. Breakdown voltage is not directly proportional to the thickness;

doubling the thickness does not quite double the breakdown voltage. A thick dielectric must be used to withstand high voltages. Since capacitance is inversely proportional to dielectric thickness (plate spacing) for a given plate area, a high-voltage capacitor must have more plate area than a low-voltage one of the same capacitance. High-voltage, high-capacitance capacitors are therefore physically large.

Dielectric strength is specified in terms of a *dielectric withstanding voltage* (DWV), given in volts per mil (0.001 inch) at a specified temperature. Taking into account the design temperature range of a capacitor and a safety margin, manufacturers specify *dc working voltage* (dcwv) to express the maximum safe limits of dc voltage across a capacitor to prevent dielectric breakdown.

For use with ac voltages, the peak value of ac voltage should not exceed the dc working voltage, unless otherwise specified in component ratings. In other words, the RMS value of sine-wave ac waveforms should be 0.707 times the dcwv value, or lower. With many types of capacitors, further derating is required as the operating frequency increases. An additional safety margin is good practice.

Dielectric breakdown in a gas or air dielectric capacitor occurs as a spark or arc between the plates. Spark voltages are generally given with the units *kilovolts per centimeter*. For air, the spark voltage or V_s may range from more than 120 kV/cm for gaps as narrow as 0.006 cm down to 28 kV/cm for gaps as wide as 10 cm. In addition, a large number of variables enter into the actual breakdown voltage in a real situation. Among the variables are the plate shape, the gap distance, the air pressure or density, the voltage, impurities in the air (or any other dielectric material) and the nature of the external circuit (with air, for instance, the humidity affects conduction on the surface of the capacitor plate).

Dielectric breakdown occurs at a lower voltage between pointed or sharp-edged surfaces than between rounded and polished surfaces. Consequently, the breakdown voltage between metal plates of any given spacing in air can be increased by buffing the edges of the plates. If the plates are damaged so they are no longer smooth and polished, they may have to be polished or the capacitor replaced.

2.7.8 Capacitor Types and Uses

Quite a variety of capacitors are used in radio circuits, differing considerably in physical size, construction and capacitance. Some of the different types are shown in **Fig 2.38** and many other types and packages are available. (See the **Component Data and References** chapter for illustrations of capacitor types and labeling conventions.)

The dielectric determines many properties of the capacitor, although the construction of the plates strongly affects the capacitor's ac performance and some dc parameters. Various materials are used for different reasons such as working voltage and current, availability, cost, and desired capacitance range.

Fixed capacitors having a single, non-adjustable value of capacitance can also be made with metal plates and with air as the dielectric, but are usually constructed from strips of metal foil with a thin, solid or liquid dielectric sandwiched between, so that a relatively large capacitance can be obtained in a small package. Solid dielectrics commonly used in fixed capacitors are plastic films, mica, paper and special ceramics. Two typical types of fixed capacitor construction are shown in **Fig 2.39**.

For capacitors with wire leads, there are

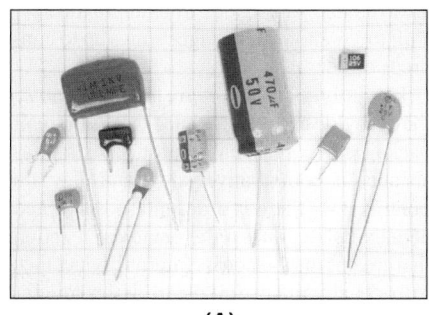

(A)

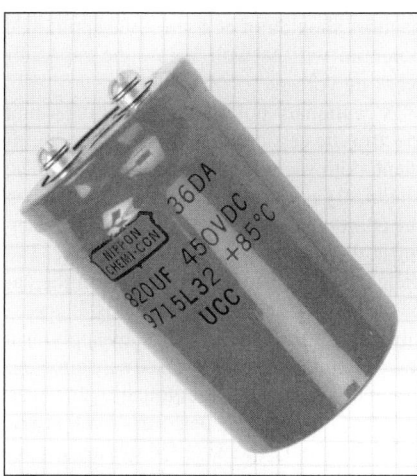

(B)

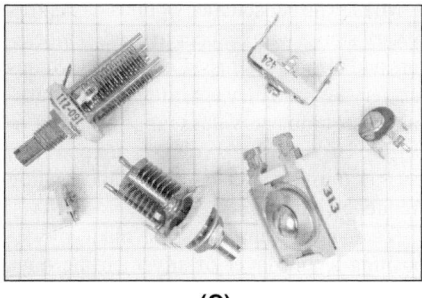

(C)

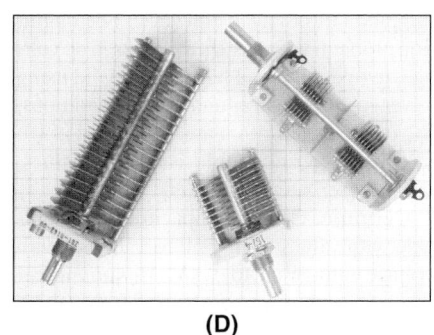

(D)

(E)

Fig 2.38 — Fixed-value capacitors are shown in parts A and B. Aluminum electrolytic capacitors are pictured near the center of photo A. The small tear-drop units to the left of center are tantalum electrolytic capacitors. The rectangular units are silvered-mica, polystyrene film and monolithic ceramic. At the right edge is a disc-ceramic capacitor and near the top right corner is a surface-mount capacitor. B shows a large "computer-grade" electrolytic. These have very low equivalent series resistance (ESR) and are often used as filter capacitors in switch-mode power supplies, and in series-strings for high-voltage supplies of RF power amplifiers. Parts C and D show a variety of variable capacitors, including air variable capacitors and mica compression units. Part E shows a vacuum variable capacitor such as is sometimes used in high-power amplifier circuits. The ¼-inch-ruled graph paper backgrounds provide size comparisons.

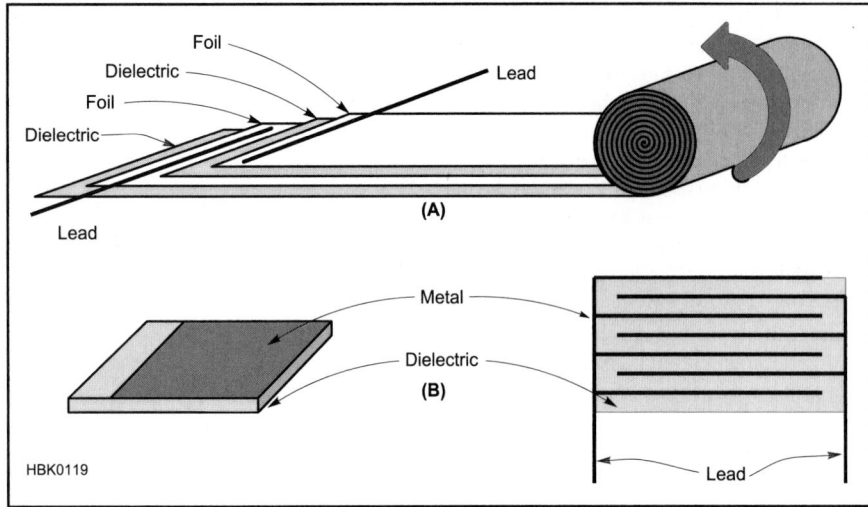

Fig 2.39 — Two common types of capacitor construction. A shows the roll method for film capacitors with axial leads. B shows the alternating layer method for ceramic capacitors. Axial leads are shown in A and radial leads in B.

two basic types of lead orientation; *axial* (shown in Fig 2.39A) in which the leads are aligned with the long axis of the capacitor body and *radial* (shown in Fig 2.39B) in which the leads are at right angles to the capacitor's length or width.

Vacuum. Both fixed and variable vacuum capacitors are available. They are rated by their maximum working voltages (3 to 60 kV) and currents. Losses are specified as negligible for most applications. The high working voltage and low losses make vacuum capacitors widely used in transmitting applications. Vacuum capacitors are also unaffected by humidity, moisture, contamination, or dust, unlike air-dielectric capacitors discussed next. This allows them to be used in environments for which air-dielectric capacitors would be unsuitable.

Air. Since K ≈ 1 for air, air-dielectric capacitors are large when compared to those of the same value using other dielectrics. Their capacitance is very stable over a wide temperature range, leakage losses are low, and therefore a high Q can be obtained. They also can withstand high voltages. Values range from a few tens to hundreds of pF.

For these reasons (and ease of construction) most variable capacitors in tuning circuits are *air-variable* capacitors made with one set of plates movable with respect to the other set to vary the area of overlap and thus the capacitance. A *transmitting-variable* capacitor has heavy plates far enough apart to withstand the high voltages and currents encountered in a transmitter. (Air variable capacitors with more closely-spaced plates are often referred to as *receiving-variables*.)

Plastic film. Capacitors with plastic film (such as polystyrene, polyethylene or Mylar) dielectrics are widely used in bypassing and coupling applications up to several mega-

hertz. They have high leakage resistances (even at high temperatures) and low TCs. Values range from tens of pF to 1 μF. Plastic-film variable capacitors are also available.

Most film capacitors are not polarized; however, the body of the capacitor is usually marked with a color band at one end. The band indicates the terminal that is connected to the outermost plate of the capacitor. This terminal should be connected to the side of the circuit at the lower potential as a safety precaution.

Mica. The capacitance of mica capacitors is very stable with respect to time, temperature and electrical stress. Leakage and losses are very low and they are often used in transmitting equipment. Values range from 1 pF to 0.1 μF. High working voltages are possible, but they must be derated severely as operating frequency increases.

Silver-mica capacitors are made by depositing a thin layer of silver on the mica dielectric. This makes the value even more stable, but it presents the possibility of silver migration through the dielectric. The migration problem worsens with increased dc voltage, temperature and humidity. Avoid using silver-mica capacitors under such conditions. Silver-mica capacitors are often used in RF circuits requiring stable capacitor values, such as oscillators and filters.

Ceramic. Ceramic capacitors are available with values from 1 pF to 1 μF and with voltage ratings up to 1 kV. *Monolithic ceramic* capacitors are constructed from a stack of thin ceramic layers with a metal coating on one side. The layer is then compressed with alternating metal coatings connected together to form the capacitor's plates. The high dielectric constant makes these capacitors physically small for their capacitance, but their value is not as stable and their dielectric properties vary with temperature, applied

voltage and operating frequency. They also exhibit piezoelectric behavior. Use them only in coupling and bypass roles. *Disc ceramic* capacitors are made similarly to monolithic ceramic capacitors but with a lower dielectric constant so they are larger and tend to have higher voltage ratings. Ceramic capacitors are useful into the VHF and UHF ranges.

Transmitting ceramic capacitors are made, like transmitting air-variables, with heavy plates and high-voltage ratings. They are relatively large, but very stable and have nearly as low losses as mica capacitors at HF.

Electrolytic. Electrolytic capacitors are constructed with plates made of aluminum-foil strips and a semi-liquid conducting chemical compound between them. They are sometimes called *aluminum electrolytics*. The actual dielectric is a very thin film of insulating material that forms on one set of plates through electrochemical action when a dc voltage is applied to the capacitor. The capacitance of an electrolytic capacitor is very large compared to capacitors having other dielectrics, because the dielectric film is so thin — much thinner than is practical with a solid dielectric. Electrolytic capacitors are available with values from approximately 1 μF to 1 F and with voltage ratings up to hundreds of volts.

Electrolytic capacitors are popular because they provide high capacitance values in small packages at a reasonable cost. Leakage resistance is comparatively low and they are polarized — there is a definite positive and negative plate, due to the chemical reaction that creates the dielectric. Internal inductance restricts aluminum-foil electrolytics to low-frequency applications such as power-supply filtering and bypassing in audio circuits. To maintain the dielectric film, electrolytic capacitors should not be used if the applied dc potential will be well below the capacitor working voltage.

A cautionary note is warranted regarding electrolytic capacitors found in older equipment, both vacuum tube and solid-state. The chemical paste in electrolytics dries out when the component is heated and with age, causing high losses and reduced capacitance. The dielectric film also disappears when the capacitor is not used for long periods. It is possible to "reform" the dielectric by applying a low voltage to an old or unused capacitor and gradually increasing the voltage. However, old electrolytics rarely perform as well as new units. To avoid expensive failures and circuit damage, it is recommended that electrolytic capacitors in old equipment be replaced if they have not been in regular use for more than ten years.

Tantalum. Related to the electrolytic capacitor, *tantalum* capacitors substitute a *slug* of extremely porous tantalum (a rare-earth metallic element) for the aluminum-foil strips as one plate. As in the electrolytic capacitor, the dielectric is an oxide film that forms on the surface of the tantalum. The slug is immersed in a liquid compound contained in a metal can that serves as the other plate. Tantalum capacitors are commonly used with values from 0.1 to several hundred µF and voltage ratings of less than 100 V. Tantalum capacitors are smaller, lighter and more stable, with less leakage and inductance than their aluminum-foil electrolytic counterparts but their cost is higher.

Paper. Paper capacitors are generally not used in new designs and are largely encountered in older equipment; capacitances from 500 pF to 50 µF are available. High working voltages are possible, but paper-dielectric capacitors have low leakage resistances and tolerances are no better than 10 to 20%.

Trimming capacitors. Small-value variable capacitors are often referred to as *trimmers* because they are used for fine-tuning or frequency adjustments, called *trimming*. Trimmers have dielectrics of Teflon, air, or ceramic and generally have values of less than 100 pF. *Compression trimmers* have higher values of up to 1000 pF and are constructed with mica dielectrics.

Oil-filled. Oil-filled capacitors use special high-strength dielectric oils to achieve voltage ratings of several kV. Values of up to 100 µF are commonly used in high-voltage applications such as high-voltage power supplies and energy storage. (See the chapter on **Power Sources** for additional information about the use of oil-filled and electrolytic capacitors.)

2.8 Inductance and Inductors

A second way to store electrical energy is in a *magnetic field*. This phenomenon is called *inductance*, and the devices that exhibit inductance are called *inductors*. Inductance is derived from some basic underlying magnetic properties.

2.8.1 Magnetic Fields and Magnetic Energy Storage

MAGNETIC FLUX

As an electric field surrounds an electric charge, magnetic fields surround *magnets*. You are probably familiar with metallic bar, disc, or horseshoe-shaped magnets. **Fig 2.40** shows a bar magnet, but particles of matter as small as an atom can also be magnets.

Fig 2.40 also shows the magnet surrounded by lines of force called *magnetic flux*, representing a *magnetic field*. (More accurately, a *magnetostatic field*, since the field is not changing.) Similar to those of an electric field, magnetic lines of force (or *flux lines*) show the direction in which a magnet would feel a force in the field.

There is no "magnetic charge" comparable to positive and negative electric charges. All magnets and magnetic fields have a polarity, represented as *poles*, and every magnet — from atoms to bar magnets — possesses both a *north* and *south pole*. The size of the source of the magnetism makes no difference. The north pole of a magnet is defined as the one attracted to the Earth's north magnetic pole. (Confusingly, this definition means the Earth's North Magnetic Pole is magnetically a south pole!) Like conventional current, the direction of magnetic lines of force was assigned arbitrarily by early scientists as pointing *from* the magnet's north pole *to* the south pole.

An electric field is *open* — that is, its lines of force have one end on an electric charge and can extend to infinity. A magnetic field is *closed* because all magnetic lines of force form a loop passing through a magnet's north and south poles.

Magnetic fields exist around two types of materials; *permanent magnets* and *electromagnets*. Permanent magnets consist of *ferromagnetic* and *ferrimagnetic* materials whose atoms are or can be aligned so as to produce a magnetic field. Ferro- or ferrimagnetic materials are strongly attracted to magnets. They can be *magnetized*, meaning to be made magnetic, by the application of a magnetic field. Lodestone, magnetite, and ferrites are examples of ferrimagnetic materials. Iron, nickel, cobalt, Alnico alloys and other materials are ferromagnetic. Magnetic materials with high *retentivity* form permanent magnets because they retain their magnetic properties for long periods. Other materials, such as soft iron, yield temporary magnets that lose their magnetic properties rapidly.

Paramagnetic substances are very weakly attracted to a magnet and include materials such as platinum, aluminum, and oxygen. *Diamagnetic* substances, such as copper, carbon, and water, are weakly repelled by a magnet.

The second type of magnet is an electrical conductor with a current flowing through it. As shown in **Fig 2.41**, moving electrons are surrounded by a closed magnetic field, illustrated as the circular lines of force around the wire lying in planes perpendicular to the current's motion. The magnetic needle of a compass placed near a wire carrying direct current will be deflected as its poles respond to the forces created by the magnetic field around the wire.

If the wire is coiled into a *solenoid* as shown in **Fig 2.42**, the magnetic field greatly intensifies. This occurs as the magnetic fields from each successive turn in the coil add together

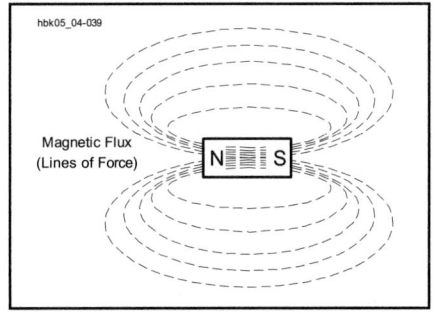

Fig 2.40 — The magnetic field and poles of a permanent magnet. The magnetic field direction is from the north to the south pole.

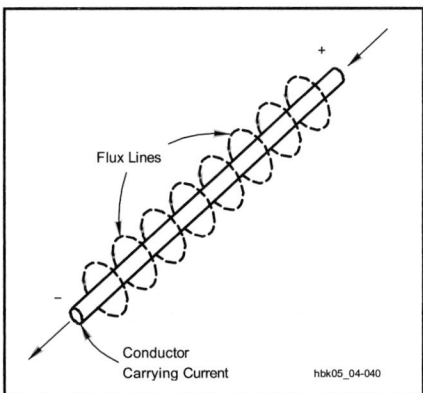

Fig 2.41 — The magnetic field around a conductor carrying an electrical current. If the thumb of your right hand points in the direction of the conventional current (plus to minus), your fingers curl in the direction of the magnetic field around the wire.

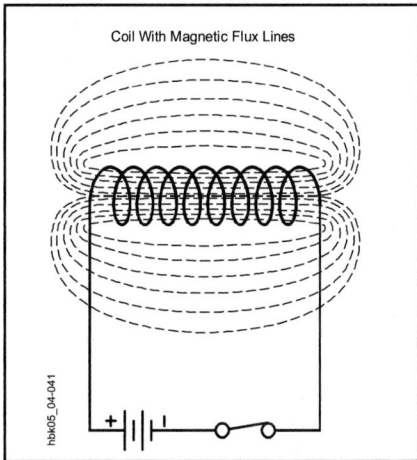

Fig 2.42 — Cross section of an inductor showing its flux lines and overall magnetic field.

The Right-hand Rule

How do you remember which way the magnetic field around a current is pointing? Luckily, there is a simple method, called the *right-hand rule*. Make your right hand into a fist, then extend your thumb, as in **Fig 2.A2**. If your thumb is pointing in the direction of conventional current flow, then your fingers curl in the same direction as the magnetic field. (If you are dealing with electronic current, use your left hand, instead!)

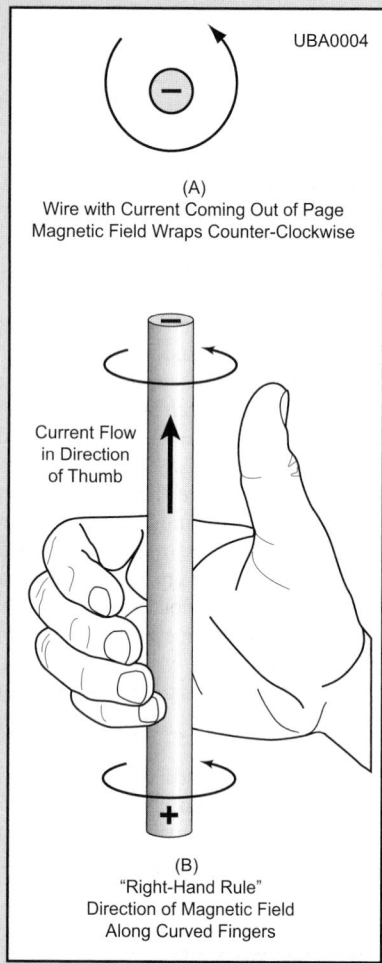

UBA0004

(A)
Wire with Current Coming Out of Page
Magnetic Field Wraps Counter-Clockwise

Current Flow
in Direction
of Thumb

(B)
"Right-Hand Rule"
Direction of Magnetic Field
Along Curved Fingers

Fig 2.A2 — Use the right-hand rule to determine magnetic field direction from the direction of current flow.

because the current in each turn is flowing in the same direction.

Note that the resulting *electromagnet* has magnetic properties identical in principle to those of a permanent magnet, including poles and lines of force or flux. The strength of the magnetic field depends on several factors: the number and shape of turns of the coil, the magnetic properties of the materials surrounding the coil (both inside and out), the length of the coil and the amplitude of the current.

Magnetic fields and electric current have a

Table 2.8
Magnetic Quantities

Value	Symbol	MKS	cgs
Magnetic Flux	lines	Weber, Wb = V-s	Maxwell, Mx = 10^{-8} Wb
Magnetic Flux Density	B	Tesla, T = Wb/m^2	Gauss, G = Mx/ cm^2
			T = 10,000 G
Magnetomotive Force	[T]	Amp-turn = A	Gilbert, Gb = 0.79577 A
Magnetic Field Strength	H	A / meter	Oersted, Oe = Gb/cm = 79.58 A/m

Magnetic Circuit Analogies

Electric Circuit	Magnetic Circuit
voltage drop V	Hl magnetovoltage drop
voltage source V	nl magnetomotive force
current I	psi = BA magnetic flux

Note – Magnetic circuit analogies as described by Shen and Kong, *Applied Electromagnetism*

special two-way relationship: voltage causing an electrical current (moving charges) in a conductor will produce a magnetic field and a moving magnetic field will create an electrical field (voltage) that produces current in a conductor. This is the principle behind motors and generators, converting mechanical energy into electrical energy and vice-versa.

MAGNETIC FLUX

Magnetic flux is measured in the SI unit (International System of Units) of the weber, which is a volt-second (Wb = Vs). In the *centimeter-gram-second* (*cgs*) metric system units, magnetic flux is measured in maxwells (1 Mx = 10^{-8} Wb). The volt-second is used because of the relationship described in the previous paragraph: 1 volt of electromotive force will be created in a loop of wire in which magnetic flux through the loop changes at the rate of 1 weber per second. The relationship between current and magnetic fields is one of motion and change.

Magnetic field intensity, known as *flux density*, decreases with the square of the distance from the source, either a magnet or current. Flux density (*B*) is represented in gauss (G), where one gauss is equivalent to one line of force (1 Mx) per square centimeter of area measured perpendicularly to the direction of the field (G = Mx / cm^2). The Earth's magnetic field at the surface is approximately one gauss. The gauss is a *cgs* unit. In SI units, flux density is represented by the tesla (T), which is one weber per square meter (T = Wb/m^2 and 1T = 10,000 G).

Magnetomotive Force and Field Strength

The magnetizing or *magnetomotive force* (ℑ) that produces a flux or total magnetic field is measured in gilberts (Gb). Magnetomotive force is analogous to electromotive force in that it produces the magnetic field. The SI unit of magnetomotive force is the ampere-turn, abbreviated A, just like the ampere. (1 Gb = 0.79577 A)

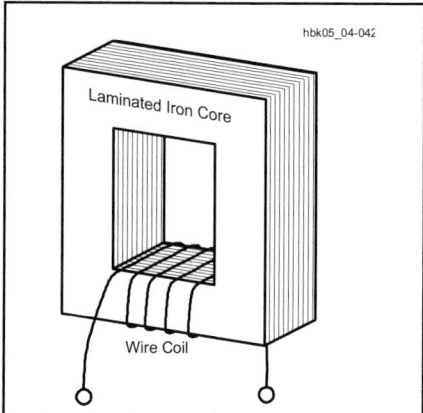

hbk05_04-042

Laminated Iron Core

Wire Coil

Fig 2.43 — A coil of wire wound around a laminated iron core.

$$\Im = \frac{10\,N\,I}{4\pi} \qquad (55)$$

where

ℑ = magnetomotive strength in gilberts,
N = number of turns in the coil creating the field,
I = dc current in amperes in the coil, and
π = 3.1416.

MAGNETIC FIELD STRENGTH

The magnetic field strength (*H*) measured in oersteds (Oe) produced by any particular magnetomotive force (measured in gilberts) is given by:

$$H = \frac{\Im}{\ell} = \frac{10\,N\,I}{4\,\pi\,\ell} \qquad (56)$$

where

H = magnetic field strength in oersteds, and
ℓ = mean magnetic path length in centimeters.

The *mean magnetic path length* is the average length of the lines of magnetic flux. If the inductor is wound on a closed core as shown in the next section, *l* is approximately the average of the inner and outer circumfer-

Table 2.9
Properties of Some High-Permeability Materials

Material	Approximate Percent Composition					Maximum Permeability
	Fe	Ni	Co	Mo	Other	
Iron	99.91	—	—	—	—	5000
Purified Iron	99.95	—	—	—	—	180,000
4% silicon-iron	96	—	—	—	4 Si	7000
45 Permalloy	54.7	45	—	—	0.3 Mn	25,000
Hipernik	50	50	—	—	—	70,000
78 Permalloy	21.2	78.5	—	—	0.3 Mn	100,000
4-79 Permalloy	16.7	79	—	—	0.3 Mn	100,000
Supermalloy	15.7	79	—	5	0.3 Mn	800,000
Permendur	49.7	—	50	—	0.3 Mn	5000
2V Permendur	49	—	49	—	2 V	4500
Hiperco	64	—	34	—	2 Cr	10,000
2-81 Permalloy*	17	81	—	2	—	130
Carbonyl iron*	99.9	—	—	—	—	132
Ferroxcube III**	$(MnFe_2O_4 + ZnFe_2O_4)$		1500			

Note: all materials in sheet form except * (insulated powder) and ** (sintered powder).
(Reference: L. Ridenour, ed., *Modern Physics for the Engineer*, p 119.)

ences of the core. The SI unit of magnetic field strength is the ampere-turn per meter. (1 Oe = 79.58 A/m)

2.8.2 Magnetic Core Properties

PERMEABILITY

The nature of the material within the coil of an electromagnet, where the lines of force are most concentrated, has the greatest effect upon the magnetic field established by the coil. All core materials are compared relatively to air. The ratio of flux density produced by a given material compared to the flux density produced by an air core is the *permeability* (μ) of the material. Air and non-magnetic materials have a permeability of one.

Suppose the coil in **Fig 2.43** is wound on an iron core having a cross-sectional area of 2 square inches. When a certain current is sent through the coil, it is found that there are 80,000 lines of force in the core. Since the area is 2 square inches, the magnetic flux density is 40,000 lines per square inch. Now suppose that the iron core is removed and the same current is maintained in the coil. Also suppose the flux density without the iron core is found to be 50 lines per square inch. The ratio of these flux densities, iron core to air, is 40,000 / 50 or 800. This ratio is the core's permeability.

Permeabilities as high as 10^6 have been attained. The three most common types of materials used in magnetic cores are these:

A. stacks of thin steel laminations (for power and audio applications, see the discussion on eddy currents below);

B. various ferrite compounds (for cores shaped as rods, toroids, beads and numerous

other forms); and

C. powdered iron (shaped as slugs, toroids and other forms for RF inductors).

The permeability of silicon-steel power-transformer cores approaches 5000 in high-quality units. Powdered-iron cores used in RF tuned circuits range in permeability from 3 to about 35, while ferrites of nickel-zinc and manganese-zinc range from 20 to 15,000. Not all materials have permeabilities higher than air. Brass has a permeability of less than one. A brass core inserted into a coil will decrease the magnetic field compared to an air core.

Table 2.9 lists some common magnetic materials, their composition and their permeabilities. Core materials are often frequency sensitive, exhibiting excessive losses outside the frequency band of intended use. (Ferrite materials are discussed separately in a later section of the chapter on **RF Techniques**.)

As a measure of the ease with which a magnetic field may be established in a material as compared with air, permeability (μ) corresponds roughly to electrical conductivity. Higher permeability means that it is easier to establish a magnetic field in the material. Permeability is given as:

$$\mu = \frac{B}{H} \tag{57}$$

where
  B is the flux density in gauss, and
  H is the magnetic field strength in oersteds.

RELUCTANCE

That a force (the magnetomotive force) is required to produce a given magnetic field strength implies that there is some opposition to be overcome. This opposition to the

creation of a magnetic field is called *reluctance*. Reluctance ($\Re$) is the reciprocal of permeability and corresponds roughly to resistance in an electrical circuit. Carrying the electrical resistance analogy a bit further, the magnetic equivalent of Ohm's Law relates reluctance, magnetomotive force, and flux density: $\Re = \Im / B$.

SATURATION

Unlike electrical conductivity, which is independent of other electrical parameters, the permeability of a magnetic material varies with the flux density. At low flux densities (or with an air core), increasing the current through the coil will cause a proportionate increase in flux. This occurs because the current passing through the coil forces the atoms of the iron (or other material) to line up, just like many small compass needles. The magnetic field that results from the atomic alignment is *much* larger than that produced by the current with no core. As more and more atoms align, the magnetic flux density also increases.

At very high flux densities, increasing the current beyond a certain point may cause no appreciable change in the flux because all of the atoms are aligned. At this point, the core is said to be *saturated*. Saturation causes a rapid decrease in permeability, because it decreases the ratio of flux lines to those obtainable with the same current using an air core. **Fig 2.44** displays a typical permeability curve, showing the region of saturation. The saturation point varies with the makeup of different magnetic materials. Air and other nonmagnetic materials do not saturate.

HYSTERESIS

Retentivity in magnetic core materials is caused by atoms retaining their alignment from an applied magnetizing force. Retentivity is desirable if the goal is to create a

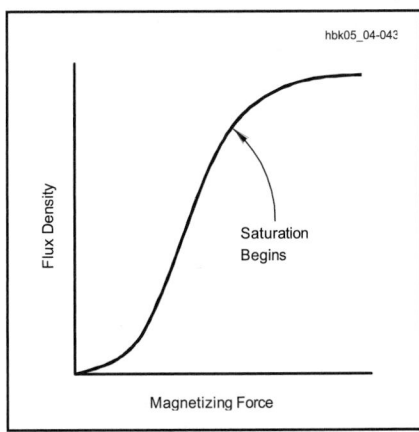

Fig 2.44 — A typical permeability curve for a magnetic core, showing the point where saturation begins.

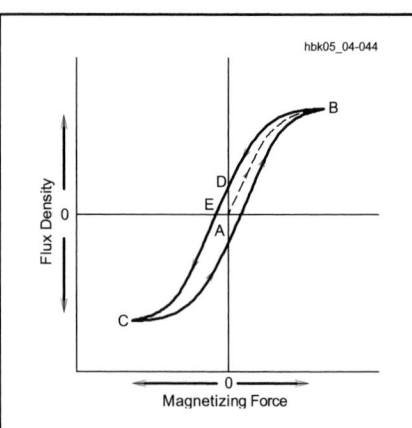

Fig 2.45 — A typical hysteresis curve for a magnetic core, showing the additional energy needed to overcome residual flux.

permanent magnet. In an electronic circuit, however, the changes caused by retentivity cause the properties of the core material to depend on the history of how the magnetizing force was applied.

Fig 2.45 illustrates the change of flux density (*B*) with a changing magnetizing force (*H*). From starting point A, with no flux in the core, the flux reaches point B at the maximum magnetizing force. As the force decreases, so too does the flux, but it does not reach zero simultaneously with the force at point D. As the force continues in the opposite direction, it brings the flux density to point C. As the force decreases to zero, the flux once more lags behind. This occurs because some of the atoms in core retain their alignment, even after the external magnetizing force is removed. This creates *residual flux* that is present even with no applied magnetizing force. This is the property of *hysteresis*.

In effect, a *coercive force* is necessary to reverse or overcome the residual magnetism retained by the core material. If a circuit carries a large ac current (that is, equal to or larger than saturation), the path shown in Fig 2.45 will be retraced with every cycle and the reversing force each time. The result is a power loss to the magnetic circuit, which appears as heat in the core material. Air cores are immune to hysteresis effects and losses.

2.8.3 Inductance and Direct Current

In an electrical circuit, any element whose operation is based on the transfer of energy into and out of magnetic fields is called an *inductor* for reasons to be explained shortly. **Fig 2.46** shows schematic-diagram symbols and photographs of a few representative inductors. The photograph shows an air-core inductor, a slug-tuned (variable-core) inductor with a nonmagnetic core and an inductor

---

Rate of Change

The symbol Δ represents change in the following variable, so that ΔI represents "change in current" and Δt "change in time". A rate of change per unit of time is often expressed in this manner. When the amount of time over which the change is measured becomes very small, the letter *d* replaces Δ in both the numerator and denominator to indicate infinitesimal changes. This notation is used in the derivation and presentation of the functions that describe the behavior of electric circuits.

---

with a magnetic (iron) core. Inductors are often called *coils* because of their construction.

As explained above, when current flows through any conductor — even a straight wire — a magnetic field is created. The transfer of energy to the magnetic field represents work performed by the source of the voltage. Power is required for doing work, and since power is equal to current multiplied by voltage, there must be a voltage drop across the inductor while energy is being stored in the field. This voltage drop, exclusive of any voltage drop caused by resistance in the conductor, is the result of an opposing voltage created in the conductor while the magnetic field is building up to its final value. Once the field becomes constant, the *induced voltage* or *back-voltage* disappears, because no further energy is being stored. Back voltage is analogous to the opposition to current flow in a capacitor from the increasing capacitor voltage.

The induced voltage opposes the voltage of the source, preventing the current from rising rapidly when voltage is applied. **Fig 2.47A** illustrates the situation of energizing an inductor or magnetic circuit, showing the relative amplitudes of induced voltage and the delayed rise in current to its full value.

The amplitude of the induced voltage is proportional to the rate at which the current changes (and consequently, the rate at which the magnetic field changes) and to a constant associated with the inductor itself, *inductance* (*L*). (*Self-inductance* is sometimes used to distinguish between mutual inductance as described below.) The basic unit of inductance is the *henry* (abbreviated H).

$$V = L\frac{\Delta I}{\Delta t} \tag{58}$$

where

    V is the induced voltage in volts,
    L is the inductance in henries, and
    $\Delta I/\Delta t$ is the rate of change of the current in amperes per second.

An inductance of 1 H generates an induced voltage of one volt when the inducing current is varying at a rate of one ampere per second.

The energy stored in the magnetic field of an inductor is given by the formula:

$$W = \frac{I^2 L}{2} \tag{59}$$

where

    W = energy in joules,
    I = current in amperes, and
    L = inductance in henrys.

This formula corresponds to the energy-storage formula for capacitors: energy storage is a function of current squared. Inductance is proportional to the amount of energy stored in an inductor's magnetic field for a given amount of current. The magnetic field strength, H, is proportional to the number of

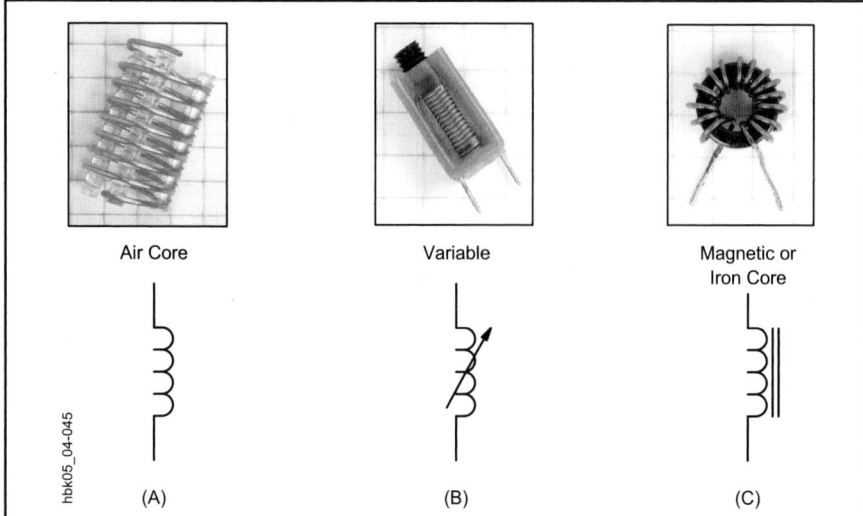

Fig 2.46 — Photos and schematic symbols for representative inductors. A, an air-core inductor; B, a variable inductor with a nonmagnetic slug and C, an inductor with a toroidal magnetic core. The ¼-inch-ruled graph paper background provides a size comparison.

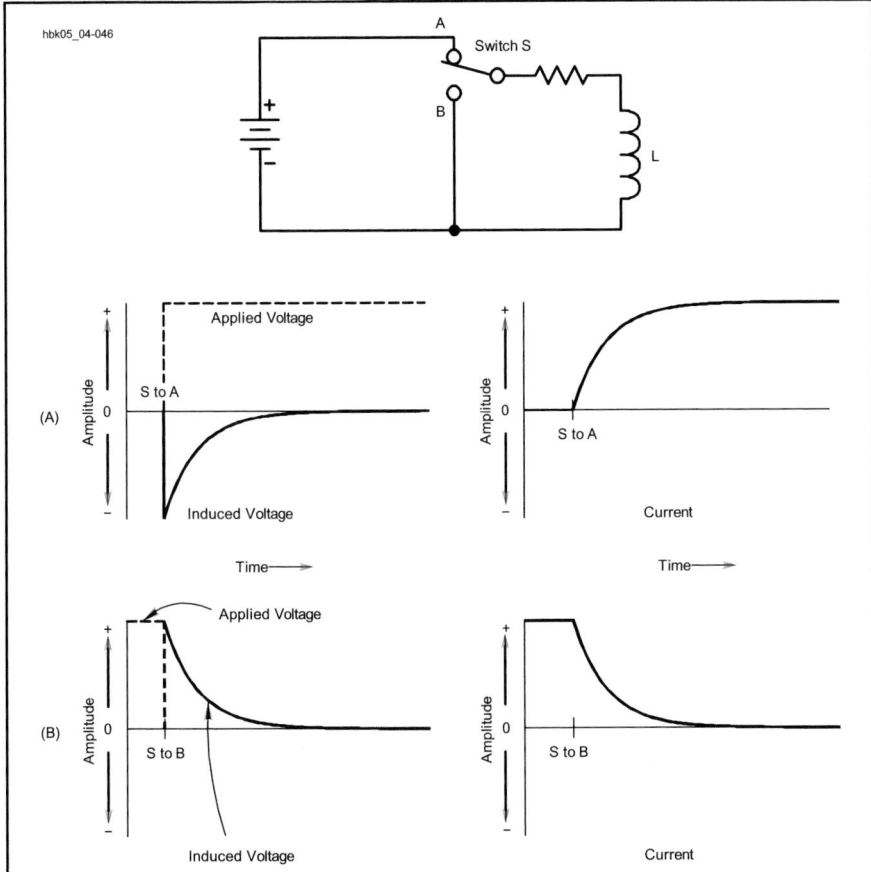

Fig 2.47 — Inductive circuit showing the generation of induced voltage and the rise of current when voltage is applied to an inductor at A, and the decay of current as the coil shorted at B.

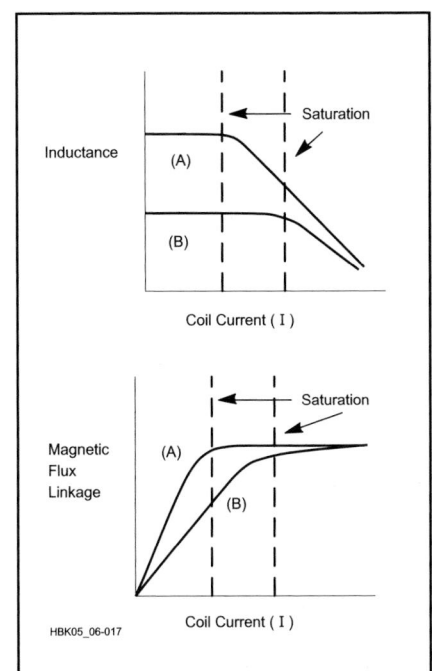

Fig 2.48 — Magnetic flux linkage and inductance plotted versus coil current for (A) a typical iron-core inductor. As the flux linkage Nφ in the coil saturates, the inductance begins to decrease since inductance = flux linkage / current. The curves marked B show the effect of adding an air gap to the core. The current-handling capability has increased, but at the expense of reduced inductance.

turns in the inductor's winding, N, (see equation 56) and for a given amount of current, to the value of μ for the core. Thus, inductance is directly proportional to both N and μ.

The polarity of the induced voltage is always such as to oppose any change in the circuit current. (This is why the term "back" is used, as in back-voltage or *back-EMF* for this reason.) This means that when the current in the circuit is increasing, work is being done against the induced voltage by storing energy in the magnetic field. Likewise, if the current in the circuit tends to decrease, the stored energy of the field returns to the circuit, and adds to the energy being supplied by the voltage source. The net effect of storing and releasing energy is that inductors oppose changes in current just as capacitors oppose changes in voltage. This phenomenon tends to keep the current flowing even though the applied voltage may be decreasing or be removed entirely. Fig 2.47B illustrates the decreasing but continuing flow of current caused by the induced voltage after the source voltage is removed from the circuit.

Inductance depends on the physical configuration of the inductor. All conductors, even straight wires, have inductance. Coiling a conductor increases its inductance. In

effect, the growing (or shrinking) magnetic field of each turn produces magnetic lines of force that — in their expansion (or contraction) — intercept the other turns of the coil, inducing a voltage in every other turn. (Recall the two-way relationship between a changing magnetic field and the voltage it creates in a conductor.) The mutuality of the effect, called *magnetic flux linkage* (ψ), multiplies the ability of the coiled conductor to store magnetic energy.

A coil of many turns will have more inductance than one of few turns, if both coils are otherwise physically similar. Furthermore, if an inductor is placed around a magnetic core, its inductance will increase in proportion to the permeability of that core, if the circuit current is below the point at which the core saturates.

In various aspects of radio work, inductors may take values ranging from a fraction of a nanohenry (nH) through millihenrys (mH) up to about 20 H.

EFFECTS OF SATURATION

An important concept for using inductors is that as long as the coil current remains below saturation, the inductance of the coil is essentially constant. **Fig 2.48** shows graphs of magnetic flux linkage (ψ) and inductance (L)

vs. current (I) for a typical iron-core inductor both saturated and non-saturated. These quantities are related by the equation

$$\psi = N\phi = LI \qquad (60)$$

where

ψ = the flux linkage
N = number of turns,
φ = flux density in webers
L = inductance in henrys, and
I = current in amperes.

In the lower graph, a line drawn from any point on the curve to the (0,0) point will show the effective inductance, L = Nφ / I, at that current. These results are plotted on the upper graph.

Note that below saturation, the inductance is constant because both ψ and I are increasing at a steady rate. Once the saturation current is reached, the inductance decreases because ψ does not increase anymore (except for the tiny additional magnetic field the current itself provides).

One common method of increasing the saturation current level is to cut a small air gap in the core (see **Fig 2.49**). This gap forces the flux lines to travel through air for a short dis-

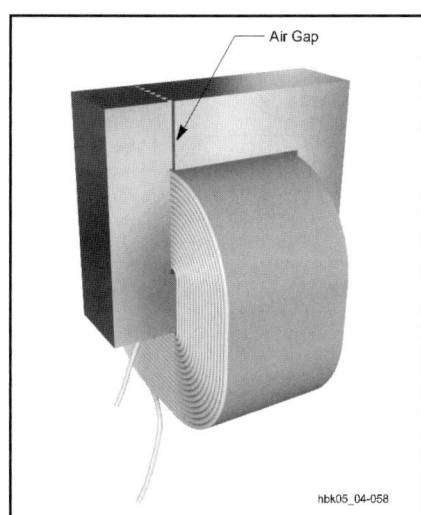

Fig 2.49 — Typical construction of a magnetic-core inductor. The air gap greatly reduces core saturation at the expense of reducing inductance. The insulating laminations between the core layers help to minimize eddy currents, as well.

tance, reducing the permeability of the core. Since the saturation flux linkage of the core is unchanged, this method works by requiring a higher current to achieve saturation. The price that is paid is a reduced inductance below saturation. The curves in Fig 2.48B show the result of an air gap added to that inductor.

Manufacturer's data sheets for magnetic cores usually specify the saturation flux density. Saturation flux density (ϕ) in gauss can be calculated for ac and dc currents from the following equations:

$$\phi_{ac} = \frac{3.49 \, V}{fNA}$$

$$\phi_{dc} = \frac{NIA_L}{10A}$$

where
   V = RMS ac voltage
   f = frequency, in MHz
   N = number of turns
   A = equivalent area of the magnetic path
      in square inches (from the data sheet)
   I = dc current, in amperes, and
   A_L = inductance index (also from the data
      sheet).

2.8.4 Mutual Inductance and Magnetic Coupling

MUTUAL INDUCTANCE

When two inductors are arranged with their axes aligned as shown in **Fig 2.50**, current flowing in through inductor 1 creates a magnetic field that intercepts inductor 2. Consequently, a voltage will be induced in inductor 2 whenever the field strength of inductor 1

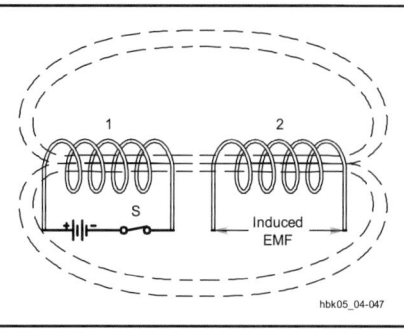

Fig 2.50 — Mutual inductance: When S is closed, current flows through coil number 1, setting up a magnetic field that induces a voltage in the turns of coil number 2.

is changing. This induced voltage is similar to the voltage of self-induction, but since it appears in the second inductor because of current flowing in the first, it is a mutual effect and results from the *mutual inductance* between the two inductors.

When all the flux set up by one coil intercepts all the turns of the other coil, the mutual inductance has its maximum possible value. If only a small part of the flux set up by one coil intercepts the turns of the other, the mutual inductance is relatively small. Two inductors having mutual inductance are said to be *coupled*.

The ratio of actual mutual inductance to the maximum possible value that could theoretically be obtained with two given inductors is called the *coefficient of coupling* between the inductors. It is expressed as a percentage or as a value between 0 and 1. Inductors that have nearly the maximum possible mutual inductance (coefficient = 1 or 100%) are said to be closely, or tightly, coupled. If the mutual inductance is relatively small the inductors are said to be loosely coupled. The degree of coupling depends upon the physical spacing between the inductors and how they are placed with respect to each other. Maximum coupling exists when they have a common or parallel axis and are as close together as possible (for example, one wound over the other). The coupling is least when the inductors are far apart or are placed so their axes are at right angles.

The maximum possible coefficient of coupling is closely approached when the two inductors are wound on a closed iron core. The coefficient with air-core inductors may run as high as 0.6 or 0.7 if one inductor is wound over the other, but will be much less if the two inductors are separated. Although unity coupling is suggested by Fig 2.50, such coupling is possible only when the inductors are wound on a closed magnetic core.

Coupling between inductors can be minimized by using separate closed magnetic cores for each. Since an inductor's magnetic field

is contained almost entirely in a closed core, two inductors with separate closed cores, such as the toroidal inductor in Fig 2.46 C, can be placed close together in almost any relative orientation without coupling.

UNWANTED COUPLING

The inductance of a short length of straight wire is small, but it may not be negligible. (In free-space, round wire has an inductance on the order of 1 µH/m, but this is affected by wire diameter and the total circuit's physical configuration.) Appreciable voltage may be induced in even a few inches of wire carrying ac by changing magnetic fields with a frequency on the order of 100 MHz or higher. At much lower frequencies or at dc, the inductance of the same wire might be ignored because the induced voltage would be very small.

There are many phenomena, both natural and man-made, that create sufficiently strong or rapidly-changing magnetic fields to induce voltages in conductors. Many of them create brief but intense pulses of energy called *transients* or "spikes." The magnetic fields from these transients intercept wires leading into and out of — and wires wholly within — electronic equipment, inducing unwanted voltages by mutual coupling.

Lightning is a powerful natural source of magnetically-coupled transients. Strong transients can also be generated by sudden changes in current in nearby circuits or wiring. High-speed digital signals and pulses can also induce voltages in adjacent conductors.

2.8.5 Inductances in Series and Parallel

When two or more inductors are connected in series (**Fig 2.51A**), the total inductance is equal to the sum of the individual inductances, provided that the inductors are sufficiently

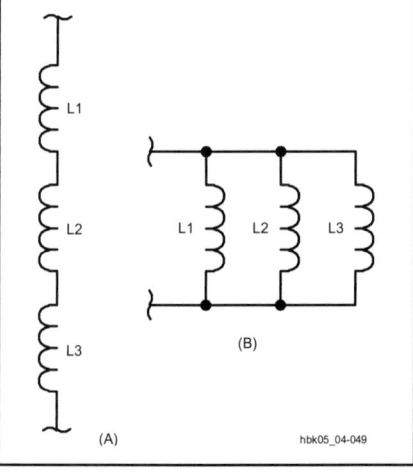

Fig 2.51 — Part A shows inductances in series, and Part B shows inductances in parallel.

separated so that there is no coupling between them (see the preceding section). That is:

$$L_{total} = L1 + L2 + L3 \ldots + L_n \qquad (61)$$

If inductors are connected in parallel (Fig 2.51B), again assuming no mutual coupling, the total inductance is given by:

$$L_{total} = \cfrac{1}{\cfrac{1}{L1} + \cfrac{1}{L2} + \cfrac{1}{L3} + \ldots + \cfrac{1}{L_n}} \qquad (62)$$

For only two inductors in parallel, the formula becomes:

$$L_{total} = \frac{L1 \times L2}{L1 + L2} \qquad (63)$$

Thus, the rules for combining inductances in series and parallel are the same as those for resistances, assuming there is no coupling between the inductors. When there is coupling between the inductors, the formulas given above will not yield correct results.

2.8.6 RL Time Constant

As with capacitors, the time dependence of inductor current is a significant property. A comparable situation to an RC circuit exists when resistance and inductance are connected in series. In **Fig 2.52**, first consider the case in which R is zero. Closing S1 sends a current through the circuit. The instantaneous transition from no current to a finite value, however small, represents a rapid change in current, and an opposing voltage is induced in L. The value of the opposing voltage is almost equal to the applied voltage, so the resulting initial current is very small.

The opposing voltage is created by change in the inductor current and would cease to exist if the current did not continue to increase. With no resistance in the circuit, the current would increase forever, always growing just fast enough to keep the self-induced opposing voltage just below the applied voltage.

When resistance in the circuit limits the current, the opposing voltage induced in L must only equal the difference between E and the drop across R, because that is the voltage actually applied to L. This difference becomes smaller as the current approaches its final value, limited by Ohm's Law to I = E/R. Theoretically, the opposing voltage never quite disappears, and so the current never quite reaches the Ohm's Law limit. In practical terms, the difference eventually becomes insignificant, just as described above for capacitors charging to an applied voltage through a resistor.

The inductor current at any time after the switch in Fig 2.52 has been closed, can be found from:

$$I(t) = \frac{E}{R}\left(1 - e^{-\frac{tR}{L}}\right) \qquad (64)$$

where

I(t) = current in amperes at time t,
E = power source potential in volts,
t = time in seconds after application of voltage,
e = natural logarithmic base = 2.718,
R = circuit resistance in ohms, and
L = inductance in henrys.

(References that explain exponential equations, e, and other mathematical topics are found in the "Radio Mathematics" article on this book's CD-ROM.) The term E/R in this equation represents the dc value of I, or the value of I(t) when t becomes very large; this is the *steady-state value* of I. If t = L/R, the above equation becomes:

$$V(L/R) = \frac{E}{R}(1 - e^{-1}) \approx 0.632\frac{E}{R} \qquad (65)$$

The time in seconds required for the current to build up to 63.2% of the maximum value is called the *time constant* (also the *RL time constant*), and is equal to L/R, where L is in henrys and R is in ohms. (Time constants are also discussed in the section on RC circuits above.) After each time interval equal to this constant, current increases by an additional 63.2% closer to the final value of E/R. This behavior is graphed in Fig 2.52. As is the case with capacitors, after five time constants the current is considered to have reached its maximum value. As with capacitors, we often use the lower-case Greek tau (τ) to represent the time constant.

Example: If a circuit has an inductor of 5.0 mH in series with a resistor of 10 Ω, how long will it take for the current in the circuit to reach full value after power is applied?

Since achieving maximum current takes approximately five time constants,

$$t = \frac{5\,L}{R} = \frac{5 \times 5.0 \times 10^{-3}\ H}{10\ \Omega}$$

$$= 2.5 \times 10^{-3}\ \text{seconds} = 2.5\ ms$$

Note that if the inductance is increased to 5.0 H, the required time increases by a factor of 1000 to 2.5 seconds. Since the circuit resistance didn't change, the final current is the same for both cases in this example. Increasing inductance increases the time required to reach full current.

Zero resistance would prevent the circuit from ever achieving full current. All practical inductors have some resistance in the wire making up the inductor.

An inductor cannot be discharged in the simple circuit of Fig 2.52 because the magnetic field ceases to exist or "collapses" as soon as the current ceases. Opening S1 does not leave the inductor charged in the way that a capacitor would remain charged. Energy storage in a capacitor depends on the separated charges staying in place. Energy storage in an inductor depends on the charges continuing to move as current.

The energy stored in the inductor's magnetic field attempts to return instantly to the circuit when S1 is opened. The rapidly changing (collapsing) field in the inductor causes a very large voltage to be induced across the inductor. Because the change in current is now in the opposite direction, the induced voltage also reverses polarity. This induced voltage (called *inductive kick-back*) is usually many times larger than the originally applied

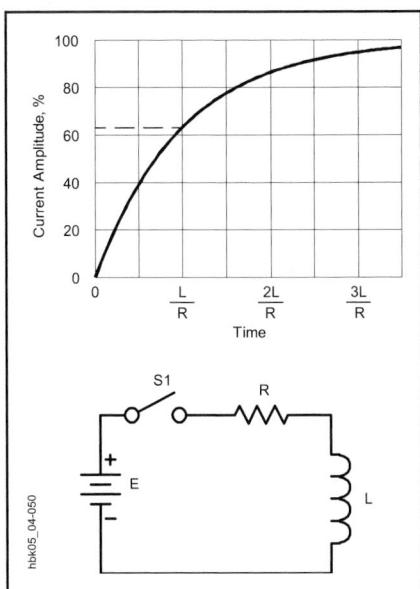

Fig 2.52 — Time constant of an RL circuit being energized.

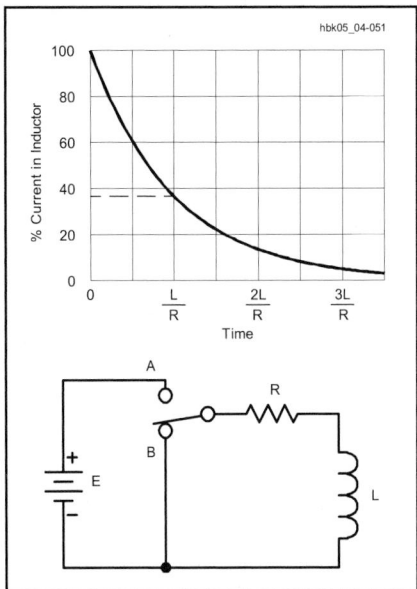

Fig 2.53 — Time constant of an RL circuit being de-energized. This is a theoretical model only, since a mechanical switch cannot change state instantaneously.

voltage, because the induced voltage is proportional to the rate at which the field changes.

The common result of opening the switch in such a circuit is that a spark or arc forms at the switch contacts during the instant the switch opens. When the inductance is large and the current in the circuit is high, large amounts of energy are released in a very short time. It is not at all unusual for the switch contacts to burn or melt under such circumstances.

The spark or arc at the opened switch can be reduced or suppressed by connecting a suitable capacitor and resistor in series across the contacts to absorb the energy non-destructively. Such an RC combination is called a *snubber network*. The current rating for a switch may be significantly reduced if it is used in an inductive circuit.

Transistor switches connected to and controlling inductors, such as relays and solenoids, also require protection from the high kick-back voltages. In most cases, a small power diode connected across the relay coil so that it does not conduct current when the inductor is energized (called a *kick-back diode*) will protect the transistor.

If the excitation is removed without breaking the circuit, as shown in **Fig 2.53**, the current will decay according to the formula:

$$I(t) = \frac{E}{R}\left(e^{-\frac{tR}{L}}\right) \qquad (66)$$

where t = time in seconds after removal of the source voltage.

After one time constant the current will decay by 63.2% of its steady-state value. (It will decay to 36.8% of the steady-state value.) The graph in Fig 2.53 shows the current-decay waveform to be identical to the voltage-discharge waveform of a capacitor. Be careful about applying the terms *charge* and *discharge* to an inductive circuit, however. These terms refer to energy storage in an electric field. An inductor stores energy in a magnetic field and the usual method of referring to the process is *energize* and *de-energize* (although it is not always followed).

2.8.7 Alternating Current in Inductors

For reasons similar to those that cause a phase difference between current and voltage in a capacitor, when an alternating voltage is applied to an ideal inductance with no resistance, the current is 90° out of phase with the applied voltage. In the case of an inductor, however, the current *lags* 90° behind the voltage as shown in **Fig 2.54**, the opposite of the capacitor current-voltage relationship. (Here again, we can also say the voltage across an inductor *leads* the current by 90°.) Interpreting Fig 2.54 begins with understanding that

the cause for current lag in an inductor is the opposing voltage that is induced in the inductor and that the amplitude of the opposing voltage is proportional to the rate at which the inductor current changes.

In time segment OA, when the applied voltage is at its positive maximum, the rate at which the current is changing is also the highest, a 38% change. This means that the opposing voltage is also maximum, allowing the least current to flow. In the segment AB, as a result of the decrease in the applied voltage, current changes by only 33% inducing a smaller opposing voltage. The process continues in time segments BC and CD, the latter producing only an 8% rise in current as the applied and induced opposing voltage approach zero.

In segment DE, the applied voltage changes polarity, causing current to begin to decrease, returning stored energy to the circuit from the inductor's magnetic field. As the current rate of change is now negative (decreasing) the induced opposing voltage also changes

polarity. Current flow is still in the original direction (positive), but is decreasing as less energy is stored in the inductor.

As the applied voltage continues to increase negatively, the current — although still positive — continues to decrease in value, reaching zero as the applied voltage reaches its negative maximum. The energy once stored in the inductor has now been completely returned to the circuit. The negative half-cycle then continues just as the positive half-cycle.

Similarly to the capacitive circuit discussed earlier, by dividing the cycle into a large number of intervals, it can be shown that the current and voltage are both sine waves, although with a difference in phase.

Compare Fig 2.54 with Fig 2.36. Whereas in a pure capacitive circuit, the current *leads* the voltage by 90°, in a pure inductive circuit, the current *lags* the voltage by 90°. These phenomena are especially important in circuits that combine inductors and capacitors. Remember that the phase difference between voltage and current in both types of circuits is a result of energy being stored and released as voltage across a capacitor and as current in an inductor.

EDDY CURRENT

Since magnetic core material is usually conductive, the changing magnetic field produced by an ac current in an inductor also induces a voltage in the core. This voltage causes a current to flow in the core. This *eddy current* (so-named because it moves in a closed path, similarly to eddy currents in water) serves no useful purpose and results in energy being dissipated as heat from the core's resistance. Eddy currents are a particular problem in inductors with iron cores. Cores made of thin strips of magnetic material, called *laminations*, are used to reduce

ELI the ICE Man

If you have difficulty remembering the phase relationships between voltage and current with inductors and capacitors, you may find it helpful to think of the phrase, "ELI the ICE man." This will remind you that voltage across an inductor leads the current through it, because the E comes before (leads) I, with an L between them, as you read from left to right. (The letter L represents inductance.) Similarly, I comes before (leads) E with a C between them.

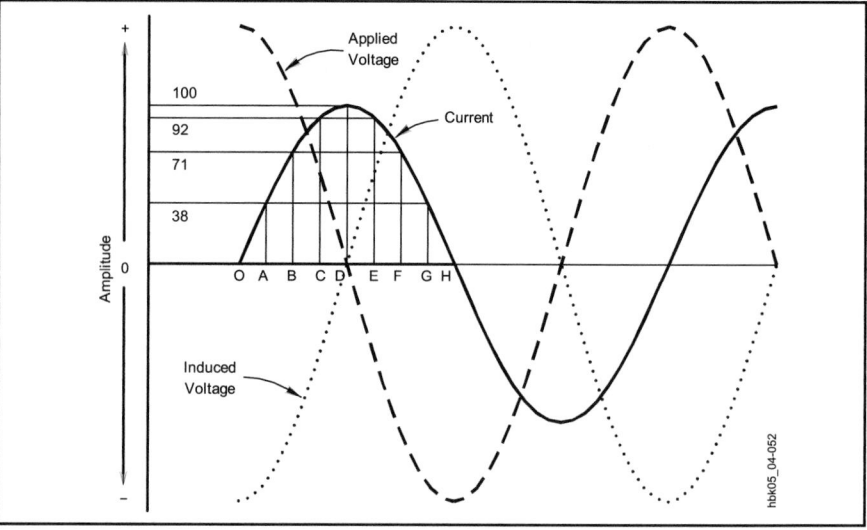

Fig 2.54 — Phase relationships between voltage and current when an alternating current is applied to an inductance.

eddy currents. (See also the section on Practical Inductors elsewhere in the chapter.)

2.8.8 Inductive Reactance and Susceptance

The amount of current that can be created in an inductor is proportional to the applied voltage but inversely proportional to the inductance because of the induced opposing voltage. If the applied voltage is ac, the rate of change of the current varies directly with the frequency and this rate of change also determines the amplitude of the induced or reverse voltage. Hence, the opposition to the flow of current increases proportionally to frequency. Stated in another way, inductor current is inversely proportional to inductance for a given applied voltage and frequency.

The combined effect of inductance and frequency is called *inductive reactance*, which — like capacitive reactance — is expressed in ohms. As with capacitive reactance, no power is dissipated in inductive reactance. The energy stored in the inductor during one portion of the cycle is returned to the circuit in the next portion.

The formula for inductive reactance is:

$$X_L = 2 \pi f L \qquad (67)$$

where
X_L = inductive reactance in ohms,
f = frequency in hertz,

Inductive Reactance Timesaver

Similarly to the calculation of capacitive reactance, if inductance is specified in microhenrys (µH) and the frequency is in megahertz (MHz), the reactance calculated from equation 66 is in units of ohms (Ω). The same is true for the combination of mH and kHz.

L = inductance in henrys, and
π = 3.1416.
(If $\omega = 2 \pi f$, then $X_L = \omega L$.)

Example: What is the reactance of an inductor having an inductance of 8.00 H at a frequency of 120 Hz?

$$X_L = 2 \pi f L$$
$$= 6.2832 \times 120 \text{ Hz} \times 8.0 \text{ H}$$
$$= 6030 \ \Omega$$

Example: What is the reactance of a 15.0-microhenry inductor at a frequency of 14.0 MHz?

$$X_L = 2 \pi f L$$
$$= 6.2832 \times 14.0 \text{ MHz} \times 15.0 \ \mu\text{H}$$
$$= 1320 \ \Omega$$

The resistance of the wire used to wind the inductor has no effect on the reactance, but simply acts as a separate resistor connected in series with the inductor.

Example: What is the reactance of the same inductor at a frequency of 7.0 MHz?

$$X_L = 2 \pi f L$$
$$= 6.2832 \times 7.0 \text{ MHz} \times 15.0 \ \mu\text{H}$$
$$= 660 \ \Omega$$

The direct relationship between frequency and reactance in inductors, combined with the inverse relationship between reactance and frequency in the case of capacitors, will be of fundamental importance in creating resonant circuits.

INDUCTIVE SUSCEPTANCE

As a measure of the ability of an inductor to limit the flow of ac in a circuit, inductive reactance is similar to capacitive reactance in having a corresponding *susceptance*, or ability to pass ac current in a circuit. In an ideal inductor with no resistive losses — that is, no energy lost as heat — susceptance is simply the reciprocal of reactance.

$$B = \frac{1}{X_L} \qquad (68)$$

where
X_L = reactance, and
B = susceptance.

The unit of susceptance for both inductors and capacitors is the *siemens*, abbreviated S.

2.9 Working with Reactance

2.9.1 Ohm's Law for Reactance

Only ac circuits containing capacitance or inductance (or both) have reactance. Despite the fact that the voltage in such circuits is 90° out of phase with the current, circuit reactance does oppose the flow of ac current in a manner that corresponds to resistance. That is, in a capacitor or inductor, reactance is equal to the ratio of ac voltage to ac current and the equations relating voltage, current and reactance take the familiar form of Ohm's Law:

$$E = I X \qquad (69)$$

$$I = \frac{E}{X} \qquad (70)$$

$$X = \frac{E}{I} \qquad (71)$$

where
E = RMS ac voltage in volts,
I = RMS ac current in amperes, and
X = inductive or capacitive reactance in ohms.

Example: What is the voltage across a capacitor of 200 pF at 7.15 MHz, if the current

through the capacitor is 50 mA?

Since the reactance of the capacitor is a function of both frequency and capacitance, first calculate the reactance:

$$X_C = \frac{1}{2 \pi f C}$$
$$= \frac{1}{2 \times 3.1416 \times 7.15 \times 10^6 \text{ Hz} \times 200 \times 10^{-12} \text{ F}}$$
$$= \frac{10^6 \ \Omega}{8980} = 111 \ \Omega$$

Next, use equation 69:

$$E = I \times X_C = 0.050 \text{ A} \times 111 \ \Omega = 5.6 \text{ V}$$

Example: What is the current through an 8.0-H inductor at 120 Hz, if 420 V is applied?

$$X_L = 2 \pi f L$$
$$= 2 \times 3.1416 \times 120 \text{ Hz} \times 8.0 \text{ H}$$
$$= 6030 \ \Omega$$

$$I = E / X_L = 420 / 6030 = 69.6 \text{ mA}$$

Fig 2.55 charts the reactances of capacitors

from 1 pF to 100 µF, and the reactances of inductors from 0.1 µH to 10 H, for frequencies between 100 Hz and 100 MHz. Approximate values of reactance can be read or interpolated from the chart. The formulas will produce more exact values, however. (The chart can also be used to find the frequency at which an inductor and capacitor have equal reactances, creating resonance as described in the section "At and Near Resonance" below.)

Although both inductive and capacitive reactance oppose the flow of ac current, the two types of reactance differ. With capacitive reactance, the current *leads* the voltage by 90°, whereas with inductive reactance, the current *lags* the voltage by 90°. The convention for charting the two types of reactance appears in **Fig 2.56**. On this graph, inductive reactance is plotted along the +90° vertical line, while capacitive reactance is plotted along the –90° vertical line. This convention of assigning a positive value to inductive reactance and a negative value to capacitive reactance results from the mathematics used for working with impedance as described elsewhere in this chapter.

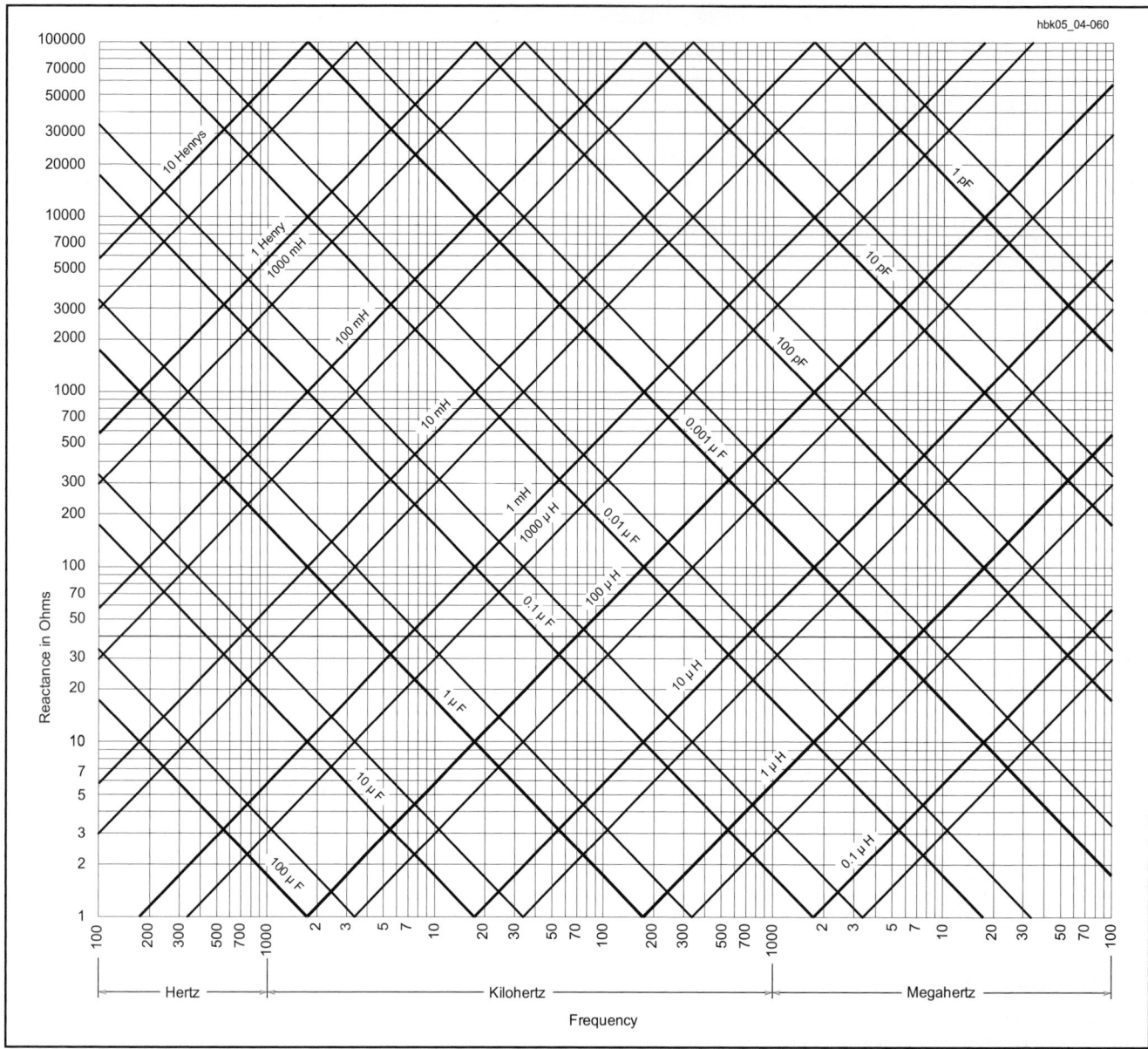

Fig 2.55 — Inductive and capacitive reactance vs frequency. Heavy lines represent multiples of 10, intermediate lines multiples of 5. For example, the light line between 10 μH and 100 μH represents 50 μH; the light line between 0.1 μF and 1 μF represents 0.5 μF, and so on. Other values can be extrapolated from the chart. For example, the reactance of 10 H at 60 Hz can be found by taking the reactance of 10 H at 600 Hz and dividing by 10 for the 10 times decrease in frequency. (Originally from Terman, *Radio Engineer's Handbook*. See references.)

In summary:

• conductance is reciprocal of resistance,

• susceptance is reciprocal of reactance, and

• admittance is reciprocal of impedance.

2.9.2 Reactances in Series and Parallel

If a circuit contains two reactances of the same type, whether in series or in parallel, the resulting reactance can be determined by applying the same rules as for resistances in series and in parallel. Series reactance is given by the formula

$$X_{total} = X1 + X2 + X3 \dots + X_n \qquad (72)$$

Example: Two noninteracting inductances are in series. Each has a value of 4.0 μH, and the operating frequency is 3.8 MHz. What is the resulting reactance?

The reactance of each inductor is:

$$X_L = 2 \pi f L$$

$$= 2 \times 3.1416 \times 3.8 \times 10^6 \text{ Hz} \times 4 \times 10^{-6} \text{ H}$$

$$= 96 \, \Omega$$

$$X_{total} = X1 + X2 = 96 \, \Omega + 96 \, \Omega = 192 \, \Omega$$

We might also calculate the total reactance by first adding the inductances:

$$L_{total} = L1 + L2 = 4.0 \text{ μH} + 4.0 \text{ μH} = 8.0 \text{ μH}$$

$$X_{total} = 2 \pi f L$$

$$= 2 \times 3.1416 \times 3.8 \times 10^6 \text{ Hz} \times 8.0 \times 10^{-6} \text{ H}$$

$$= 191 \, \Omega$$

(The fact that the last digit differs by one illustrates the uncertainty of the calculation caused by the limited precision of the measured values in the problem, and differences

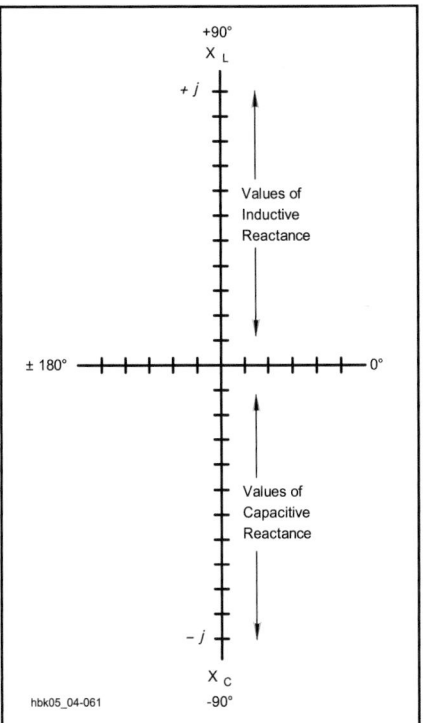

Fig 2.56 — **The conventional method of plotting reactances on the vertical axis of a graph, using the upward or "plus" direction for inductive reactance and the downward or "minus" direction for capacitive reactance. The horizontal axis will be used for resistance in later examples.**

caused by rounding off the calculated values. This also shows why it is important to follow the rules for significant figures.)

Example: Two noninteracting capacitors are in series. One has a value of 10.0 pF, the other of 20.0 pF. What is the resulting reactance in a circuit operating at 28.0 MHz?

$$X_{C1} = \frac{1}{2 \pi f C}$$

$$= \frac{1}{2 \times 3.1416 \times 28.0 \times 10^6 \text{ Hz} \times 10.0 \times 10^{-12} \text{ F}}$$

$$= \frac{10^6 \ \Omega}{1760} = 568 \ \Omega$$

$$X_{C2} = \frac{1}{2 \pi f C}$$

$$= \frac{1}{2 \times 3.1416 \times 28.0 \times 10^6 \text{ Hz} \times 20.0 \times 10^{-12} \text{ F}}$$

$$= \frac{10^6 \ \Omega}{3520} = 284 \ \Omega$$

$$X_{total} = X_{C1} + X_{C2} = 568 \ \Omega + 284 \ \Omega = 852 \ \Omega$$

Alternatively, combining the series capaci-

tors first, the total capacitance is 6.67×10^{-12} F or 6.67 pF. Then:

$$X_{total} = \frac{1}{2 \pi f C}$$

$$= \frac{1}{2 \times 3.1416 \times 28.0 \times 10^6 \text{ Hz} \times 6.67 \times 10^{-12} \text{ F}}$$

$$= \frac{10^6 \ \Omega}{1170} = 855 \ \Omega$$

(Within the uncertainty of the measured values and the rounding of values in the calculations, this is the same result as the 852 Ω we obtained with the first method.)

This example serves to remind us that *series capacitance* is not calculated in the manner used by other series resistance and inductance, but *series capacitive reactance* does follow the simple addition formula.

For reactances of the same type in parallel, the general formula is:

$$X_{total} = \frac{1}{\dfrac{1}{X1} + \dfrac{1}{X2} + \dfrac{1}{X3} + ... + \dfrac{1}{X_n}} \tag{73}$$

or, for exactly two reactances in parallel

$$X_{total} = \frac{X1 \times X2}{X1 + X2} \tag{74}$$

Example: Place the capacitors in the last example (10.0 pF and 20.0 pF) in parallel in the 28.0 MHz circuit. What is the resultant reactance?

$$X_{total} = \frac{X1 \times X2}{X1 + X2}$$

$$= \frac{568 \ \Omega \times 284 \ \Omega}{568 \ \Omega + 284 \ \Omega} = 189 \ \Omega$$

Alternatively, two capacitors in parallel can be combined by adding their capacitances.

$$C_{total} = C_1 + C_2 = 10.0 \text{ pF} + 20.0 \text{ pF} = 30 \text{ pF}$$

$$X_C = \frac{1}{2 \pi f C}$$

$$= \frac{1}{2 \times 3.1416 \times 28.0 \times 10^6 \text{ Hz} \times 30 \times 10^{-12} \text{ F}}$$

$$= \frac{10^6 \ \Omega}{5280} = 189 \ \Omega$$

Example: Place the series inductors above (4.0 µH each) in parallel in a 3.8-MHz circuit. What is the resultant reactance?

$$X_{total} = \frac{X_{L1} \times X_{L2}}{X_{L1} + X_{L2}}$$

$$= \frac{96 \ \Omega \times 96 \ \Omega}{96 \ \Omega + 96 \ \Omega} = 48 \ \Omega$$

Of course, a number (n) of equal reactances (or resistances) in parallel yields a reactance that is the value of one of them divided by n, or:

$$X_{total} = \frac{X}{n} = \frac{96 \ \Omega}{2} = 48 \ \Omega$$

All of these calculations apply only to reactances of the same type; that is, all capacitive or all inductive. Mixing types of reactances requires a different approach.

UNLIKE REACTANCES IN SERIES

When combining unlike reactances — that is, combinations of inductive and capacitive reactance — in series, it is necessary to take into account that the voltage-to-current phase relationships differ for the different types of reactance. **Fig 2.57** shows a series circuit with both types of reactance. Since the reactances are in series, the current must be the same in both. The voltage across each circuit element differs in phase, however. The voltage E_L *leads* the current by 90°, and the voltage E_C *lags* the current by 90°. Therefore, E_L and E_C have opposite polarities and cancel each other in whole or in part. The line E in Fig 2.57 approximates the resulting voltage, which is the *difference* between E_L and E_C.

Since for a constant current the reactance is directly proportional to the voltage, the net reactance is still the sum of the individual reactances as in equation 72. Because induc-

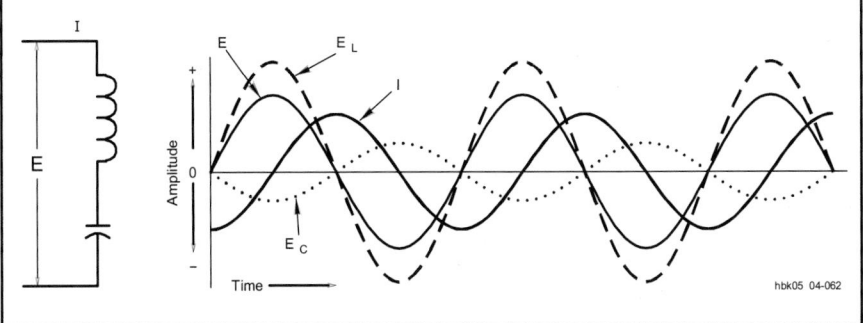

Fig 2.57 — **A series circuit containing both inductive and capacitive components, together with representative voltage and current relationships.**

tive reactance is considered to be positive and capacitive reactance negative, the resulting reactance can be either positive (inductive) or negative (capacitive) or even zero (no reactance).

Another common method of adding the unlike reactances is to use the absolute values of the reactances and subtract capacitive reactance from inductive reactance:

$$X_{total} = X_L - X_C \qquad (75)$$

The convention of using absolute values for the reactances and building the sense of positive and negative into the formula is the preferred method used by hams and will be used in all of the remaining formulas in this chapter. Nevertheless, before using any formulas that include reactance, determine whether this convention is followed before assuming that the absolute values are to be used.

Example: Using Fig 2.57 as a visual aid, let $X_C = 20.0 \ \Omega$ and $X_L = 80.0 \ \Omega$. What is the resulting reactance?

$$X_{total} = X_L - X_C$$

$$= 80.0 \ \Omega - 20.0 \ \Omega = +60.0 \ \Omega$$

Since the result is a positive value, reactance is inductive. Had the result been a negative number, the reactance would have been capacitive.

When reactance types are mixed in a series circuit, the resulting reactance is always smaller than the larger of the two reactances. Likewise, the resulting voltage across the series combination of reactances is always smaller than the larger of the two voltages across individual reactances.

Every series circuit of mixed reactance types with more than two circuit elements can be reduced to this simple circuit by combining all the reactances into one inductive and one capacitive reactance. If the circuit has more than one capacitor or more than one inductor in the overall series string, first use the formulas given earlier to determine the total series inductance alone and the total series capacitance alone (or their respective reactances). Then combine the resulting single capacitive

reactance and single inductive reactance as shown in this section.

UNLIKE REACTANCES IN PARALLEL

The situation of parallel reactances of mixed type appears in **Fig 2.58**. Since the elements are in parallel, the voltage is common to both reactive components. The current through the capacitor, I_C, *leads* the voltage by 90°, and the current through the inductor, I_L, *lags* the voltage by 90°. In this case, it is the currents that are 180° out of phase and thus cancel each other in whole or in part. The total current is the difference between the individual currents, as indicated by the line I in Fig 2.58.

Since reactance is the ratio of voltage to current, the total reactance in the circuit is:

$$X_{total} = \frac{E}{I_L - I_C} \qquad (76)$$

In the drawing, I_C is larger than I_L, and the resulting differential current retains the phase of I_C. Therefore, the overall reactance, X_{total}, is for capacitive in this case. The total reactance of the circuit will be smaller than the larger of the individual reactances, because the total current is smaller than the larger of the two individual currents.

In parallel circuits, reactance and current are inversely proportional to each other for a constant voltage and equation 74 can be used, carrying the positive and negative signs:

$$X_{total} = \frac{X_L \times (-X_C)}{X_L - X_C} = \frac{-X_L \times X_C}{X_L - X_C} \qquad (77)$$

As with the series formula for mixed reactances, follow the convention of using absolute values for the reactances, since the minus signs in the formula account for capacitive reactance being negative. If the solution yields a negative number, the resulting reactance is capacitive, and if the solution is positive, then the reactance is inductive.

Example: Using Fig 2.58 as a visual aid, place a capacitive reactance of $10.0 \ \Omega$ in parallel with an inductive reactance of $40.0 \ \Omega$. What is the resulting reactance?

$$X_{total} = \frac{-X_L \times X_C}{X_L - X_C}$$

$$= \frac{-40.0 \ \Omega \times 10.0 \ \Omega}{40.0 \ \Omega - 10.0 \ \Omega}$$

$$= \frac{-400 \ \Omega}{30.0 \ \Omega} = -13.3 \ \Omega$$

The reactance is capacitive, as indicated by the negative solution. Moreover, the resultant reactance is always smaller than the larger of the two individual reactances.

As with the case of series reactances, if each leg of a parallel circuit contains more than one reactance, first simplify each leg to a single reactance. If the reactances are of the same type in each leg, the series reactance formulas for reactances of the same type will apply. If the reactances are of different types, then use the formulas shown above for mixed series reactances to simplify the leg to a single value and type of reactance.

2.9.3 At and Near Resonance

Any series or parallel circuit in which the values of the two unlike reactances are equal is said to be *resonant*. For any given inductance or capacitance, it is theoretically possible to find a value of the opposite reactance type to produce a resonant circuit for any desired frequency.

When a series circuit like the one shown in Fig 2.57 is resonant, the voltages E_C and E_L are equal and cancel; their sum is zero. This is a *series-resonant* circuit. Since the reactance of the circuit is proportional to the sum of

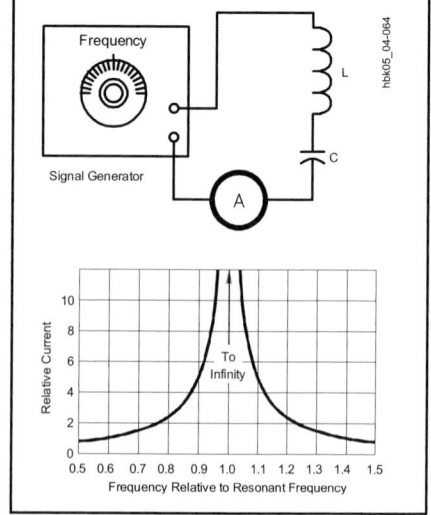

Fig 2.59 — The relative generator current with a fixed voltage in a series circuit containing inductive and capacitive reactances as the frequency approaches and departs from resonance.

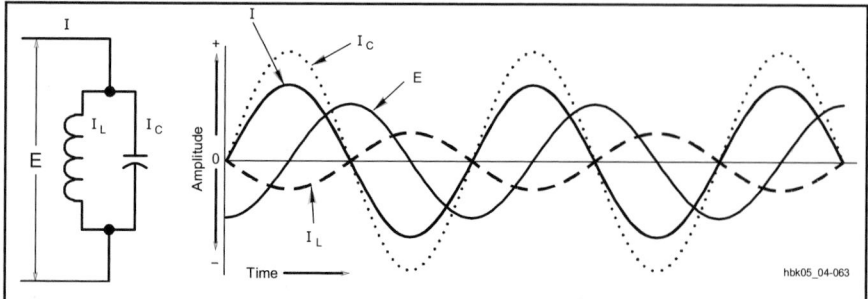

Fig 2.58 — A parallel circuit containing both inductive and capacitive components, together with representative voltage and current relationships.

these voltages, the net reactance also goes to zero. Theoretically, the current, as shown in **Fig 2.59**, can become infinite. In fact, it is limited only by losses in the components and other resistances that would exist in a real circuit of this type. As the frequency of operation moves slightly off resonance and the reactances no longer cancel completely, the net reactance climbs as shown in the figure. Similarly, away from resonance the current drops to a level determined by the net reactance.

In a *parallel-resonant* circuit of the type in Fig 2.58, the current I_L and I_C are equal and cancel to zero. Since the reactance is inversely proportional to the current, as the current ap-

proaches zero, the reactance becomes infinite. As with series circuits, component losses and other resistances in the circuit prevent the current from reaching zero. **Fig 2.60** shows the theoretical current curve near and at resonance for a purely reactive parallel-resonant circuit. Note that in both Fig 2.59 and Fig 2.60, the departure of current from the resonance value is close to, but not quite, symmetrical above and below the resonant frequency.

Example: What is the reactance of a series L-C circuit consisting of a 56.04-pF capacitor and an 8.967-µH inductor at 7.00, 7.10 and 7.20 MHz? Using the formulas from earlier in this chapter, we calculate a table of values:

Frequency (MHz)	X_L (Ω)	X_C (Ω)	X_{total} (Ω)
7.000	394.4	405.7	−11.3
7.100	400.0	400.0	0
7.200	405.7	394.4	11.3

The exercise shows the manner in which the reactance rises rapidly as the frequency moves above and below resonance. Note that in a series-resonant circuit, the reactance at frequencies below resonance is capacitive, and above resonance, it is inductive. **Fig 2.61** displays this fact graphically. In a parallel-resonant circuit, where the reactance becomes infinite at resonance, the opposite condition exists: above resonance, the reactance is capacitive and below resonance it is inductive, as shown in **Fig 2.62**. Of course, all graphs and calculations in this section are theoretical and presume a purely reactive circuit. Real circuits are never purely reactive; they contain some resistance that modifies their performance considerably. Real resonant circuits will be discussed later in this chapter.

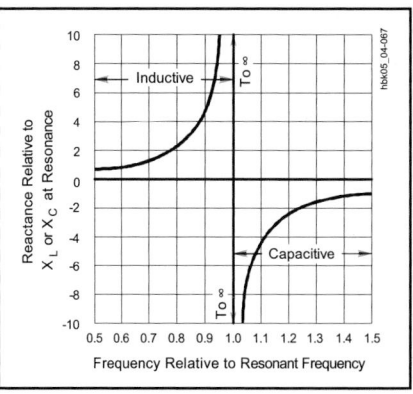

Fig 2.62 — The transition from inductive to capacitive reactance in a parallel-resonant circuit as the frequency passes resonance.

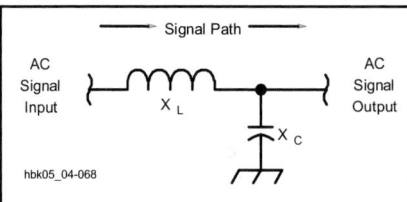

Fig 2.63 — A signal path with a series inductor and a shunt capacitor. The circuit presents different reactances to an ac signal and to its harmonics.

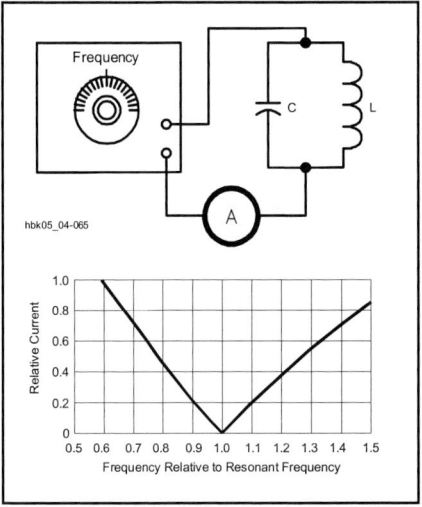

Fig 2.60 — The relative generator current with a fixed voltage in a parallel circuit containing inductive and capacitive reactances as the frequency approaches and departs from resonance. (The circulating current through the parallel inductor and capacitor is a maximum at resonance.)

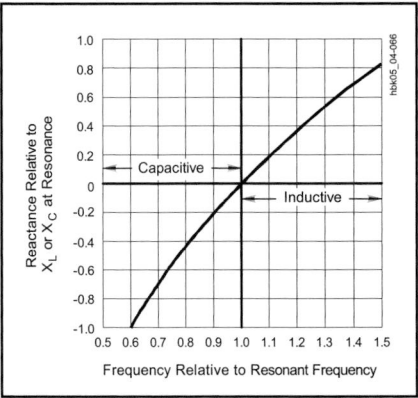

Fig 2.61 — The transition from capacitive to inductive reactance in a series-resonant circuit as the frequency passes resonance.

2.9.4 Reactance and Complex Waveforms

All of the formulas and relationships shown in this section apply to alternating current in the form of regular sine waves. Complex wave shapes complicate the reactive situation considerably. A complex or nonsinusoidal wave can be treated as a sine wave of some fundamental frequency and a series of harmonic frequencies whose amplitudes depend on the original wave shape. When such a complex wave — or collection of sine waves — is applied to a reactive circuit, the current through the circuit will not have the same wave shape as the applied voltage. The difference results because the reactance of an inductor and capacitor depend in part on the applied frequency.

For the second-harmonic component of the complex wave, the reactance of the inductor is twice and the reactance of the capacitor is half their respective values at the fundamental frequency. A third-harmonic component

produces inductive reactances that are triple and capacitive reactances that are one-third those at the fundamental frequency. Thus, the overall circuit reactance is different for each harmonic component.

The frequency sensitivity of a reactive circuit to various components of a complex wave shape creates both difficulties and opportunities. On the one hand, calculating the circuit reactance in the presence of highly variable as well as complex waveforms, such as speech, is difficult at best. On the other hand, the frequency sensitivity of reactive components and circuits lays the foundation for filtering, that is, for separating signals of different frequencies or acting upon them differently. For example, suppose a coil is in the series path of a signal and a capacitor is connected from the signal line to ground, as represented in **Fig 2.63**. The reactance of the coil to the second harmonic of the signal will be twice that at the fundamental frequency and oppose more effectively the flow of harmonic current. Likewise, the reactance of the capacitor to the harmonic will be half that to the fundamental, allowing the harmonic an easier current path away from the signal line toward ground. See the **RF and AF Filters** chapter for detailed information on filter theory and construction.

2.10 Impedance

When a circuit contains both resistance and reactance, the combined opposition to current is called *impedance*. Symbolized by the letter Z, impedance is a more general term than either resistance or reactance. Frequently, the term is used even for circuits containing only resistance or reactance. Qualifications such as "resistive impedance" are sometimes added to indicate that a circuit has only resistance, however.

The reactance and resistance comprising an impedance may be connected either in series or in parallel, as shown in **Fig 2.64**. In these circuits, the reactance is shown as a box to indicate that it may be either inductive or capacitive. In the series circuit at A, the current is the same in both elements, with (generally) different voltages appearing across the resistance and reactance. In the parallel circuit at B, the same voltage is applied to both elements, but different currents may flow in the two branches.

In a resistance, the current is in phase with the applied voltage, while in a reactance it is 90° out of phase with the voltage. Thus, the phase relationship between current and voltage in the circuit as a whole may be anything between zero and 90°, depending on the relative amounts of resistance and reactance.

As shown in Fig 2.56 in the preceding section, reactance is graphed on the vertical (Y) axis to record the phase difference between the voltage and the current. **Fig 2.65** adds resistance to the graph. Since the voltage is in phase with the current, resistance is recorded on the horizontal axis, using the positive or right side of the scale.

2.10.1 Calculating Z From R and X in Series Circuits

Impedance is the complex combination of resistance and reactance. Since there is a 90° phase difference between resistance and reactance (whether inductive or capacitive), simply adding the two values does not correspond to what actually happens in a circuit and will not give the correct result. Therefore, expressions like "Z = R ± X" can be misleading, because they suggest simple addition. As a result, impedance is often expressed "Z = R ± jX."

In pure mathematics, "*i*" indicates an imaginary number. Because *i* represents current in electronics, we use the letter "*j*" for the same mathematical operator, although there is nothing imaginary about what it represents in electronics. (References to explain imaginary numbers, rectangular coordinates, polar coordinates and how to work with them are provided in the "Radio Mathematics" article on this book's CD-ROM.) With respect to resistance and reactance, the letter *j* is normally assigned to those figures on the vertical axis, 90° out of phase with the horizontal axis. The actual function of *j* is to indicate that calculating impedance from resistance and reactance requires *vector addition*.

A *vector* is a value with both magnitude and direction, such as velocity; "10 meters/sec to the north". Impedance also has a "direction" as described below. In vector addition, the result of combining two values with a 90° phase difference is a quantity different from the simple *algebraic addition* of the two values. The result will have a phase difference intermediate between 0° and 90°.

RECTANGULAR FORM OF IMPEDANCE

Because this form for impedances, Z = R ± jX, can be plotted on a graph using rectangular coordinates, this is the *rectangular form* of impedance. The rectangular coordinate system in which one axis represents real number and the other axis imaginary numbers is called the *complex plane* and impedance with both real (R) and imaginary (X) components is called *complex impedance*. Unless specifically noted otherwise, assume that "impedance" means "complex impedance" and that both R and X may be present.

Consider **Fig 2.66**, a series circuit consisting of an inductive reactance and a resistance. As given, the inductive reactance is 100 Ω and the resistance is 50 Ω. Using *rectangular coordinates*, the impedance becomes

$$Z = R + jX \qquad (78)$$

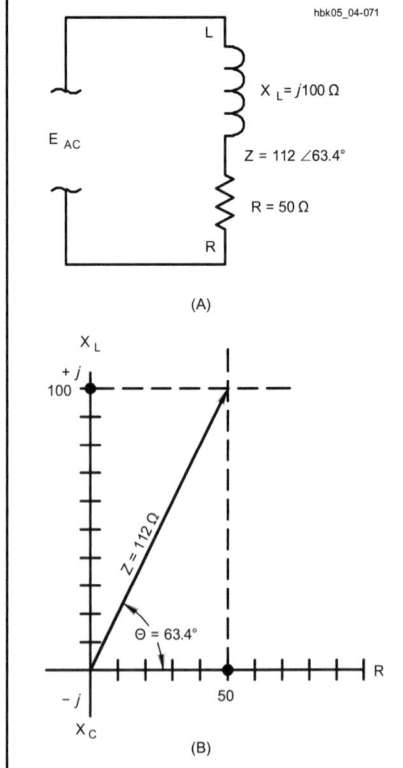

Fig 2.66 — A series circuit consisting of an inductive reactance of 100 Ω and a resistance of 50 Ω. At B, the graph plots the resistance, reactance, and impedance.

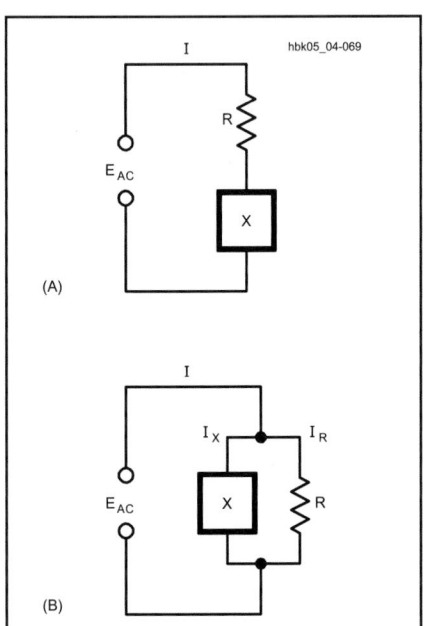

Fig 2.64 — Series and parallel circuits containing resistance and reactance.

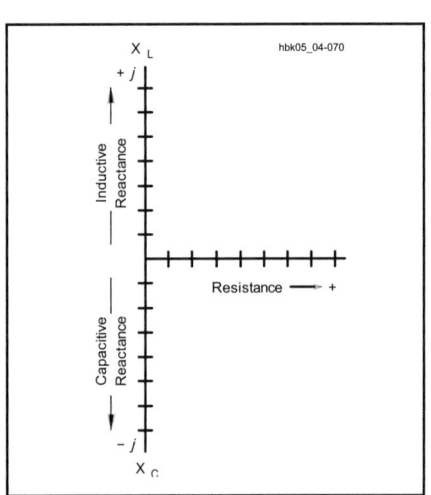

Fig 2.65 — The conventional method of charting impedances on a graph, using the vertical axis for reactance (the upward or "plus" direction for inductive reactance and the downward or "minus" direction for capacitive reactance), and using the horizontal axis for resistance.

where

Z = the impedance in ohms,
R = the resistance in ohms, and
X = the reactance in ohms.

In the present example,

$$Z = 50 + j100 \ \Omega$$

This point is located at the tip of the arrow drawn on the graph where the dashed lines cross.

POLAR FORM OF IMPEDANCE AND PHASE ANGLE

As the graph in Fig 2.66 shows, the impedance that results from combining R and X can also be represented by a line completing a right triangle whose sides are the resistance and reactance. The point at the end of the line — the complex impedance — can be described by how far it is from the origin of the graph where the axes cross (the *magnitude* of the impedance, |Z|) and the angle made by the line with the horizontal axis representing 0° (the *phase angle* of the impedance, θ). This is the *polar form* of impedance and it is written in the form

$$Z = |Z| \angle \theta \tag{79}$$

Occasionally, θ may be given in radians. The convention in this handbook is to use degrees unless specifically noted otherwise.

The length of the hypotenuse of the right triangle represents the magnitude of the impedance and can be calculated using the formula for calculating the hypotenuse of a right triangle, in which the square of the hypotenuse equals the sum of the squares of the two sides:

$$|Z| = \sqrt{R^2 + X^2} \tag{80}$$

In this example:

$$|Z| = \sqrt{(50 \ \Omega)^2 + (100 \ \Omega)^2}$$

$$= \sqrt{2500 \ \Omega^2 + 10,000 \ \Omega^2}$$

$$= \sqrt{12,500 \ \Omega^2} = 112 \ \Omega$$

The magnitude of the impedance that results from combining 50 Ω of resistance with 100 Ω of inductive reactance is 112 Ω. From trigonometry, the tangent of the phase angle is the side opposite the angle (X) divided by the side adjacent to the angle (R), or

$$\tan \theta = \frac{X}{R} \tag{81}$$

where

X = the reactance, and
R = the resistance.

Find the angle by taking the inverse tangent, or arctan:

$$\theta = \arctan \frac{X}{R} \tag{82}$$

Calculators sometimes label the inverse tangent key as "tan⁻¹". Remember to be sure your calculator is set to use the right angular units, either degrees or radians.

In the example shown in Fig 2.66,

$$\theta = \arctan \frac{100 \ \Omega}{50 \ \Omega} = \arctan 2.0 = 63.4°$$

Using the information just calculated, the complex impedance in polar form is:

$$Z = 112 \ \Omega \angle 63.4°$$

This is stated verbally as "112 ohms at an angle of 63 point 4 degrees."

POLAR TO RECTANGULAR CONVERSION

The expressions R ± jX and |Z| ∠θ both provide the same information, but in two different forms. The procedure just given permits conversion from rectangular coordinates into polar coordinates. The reverse procedure is also important. **Fig 2.67** shows an imped-

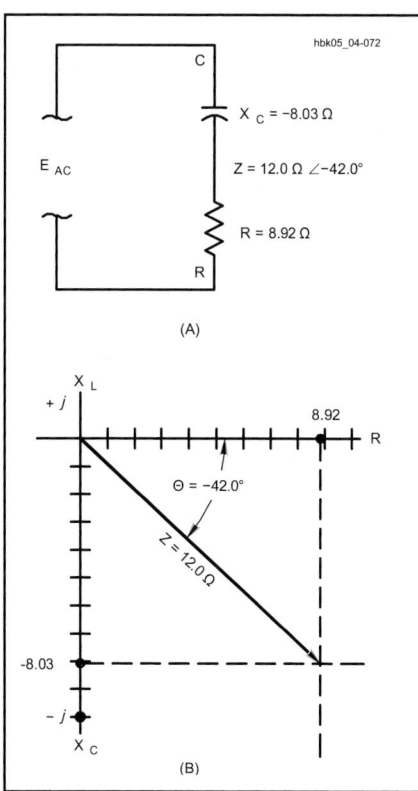

Fig 2.67 — A series circuit consisting of a capacitive reactance and a resistance: the impedance is given as 12.0 Ω at a phase angle θ of –42 degrees. At B, the graph plots the resistance, reactance, and impedance.

ance composed of a capacitive reactance and a resistance. Since capacitive reactance appears as a negative value, the impedance will be at a negative phase angle, in this case, 12.0 Ω at a phase angle of – 42.0° or Z = |12.0 Ω | ∠ – 42.0°.

Remember that the impedance forms a triangle with the values of X and R from the rectangular coordinates. The reactance axis forms the side opposite the angle θ.

$$\sin \theta = \frac{\text{side opposite}}{\text{hypotenuse}} = \frac{X}{|Z|}$$

Solving this equation for reactance, we have:

$$X = |Z| \times \sin \theta \ (\text{ohms}) \tag{83}$$

Likewise, the resistance forms the side adjacent to the angle.

$$\cos \theta = \frac{\text{side adjacent}}{\text{hypotenuse}} = \frac{R}{|Z|}$$

Solving for resistance, we have:

$$R = |Z| \times \cos \theta \ (\text{ohms}) \tag{84}$$

Then from our example:

$$X = 12.0 \ \Omega \times \sin (-42°)$$
$$= 12.0 \ \Omega \times -0.669 = -8.03 \ \Omega$$

$$R = 12.0 \ \Omega \times \cos (-42°)$$
$$= 12.0 \ \Omega \times 0.743 = 8.92 \ \Omega$$

Since X is a negative value, it is plotted on the lower vertical axis, as shown in Fig 2.67, indicating capacitive reactance. In rectangular form, Z = 8.92 Ω – j8.03 Ω.

In performing impedance and related calculations with complex circuits, rectangular coordinates are most useful when formulas require the addition or subtraction of values. Polar notation is most useful for multiplying and dividing complex numbers. (See the section on "Radio Math" on this book's CD-ROM for references dealing with the mathematics of complex numbers.)

All of the examples shown so far in this section presume a value of reactance that contributes to the circuit impedance. Reactance is a function of frequency, however, and many impedance calculations may begin with a value of capacitance or inductance and an operating frequency. In terms of these values, equation 80 can be expressed in either of two forms, depending on whether the reactance is inductive (equation 85) or capacitive (equation 86):

$$|Z| = \sqrt{R^2 + (2 \pi f L)^2} \tag{85}$$

$$|Z| = \sqrt{R^2 + \left(\frac{1}{2 \pi f C}\right)^2} \tag{86}$$

Example: What is the impedance of a circuit like Fig 2.66 with a resistance of 100 Ω and a 7.00-µH inductor operating at a frequency of 7.00 MHz? Using equation 85,

$$|Z| = \sqrt{R^2 + (2\,\pi\,f\,L)^2}$$

$$= \sqrt{\frac{(100\,\Omega)^2 + (2\,\pi \times 7.0 \times 10^{-6}\,\text{H}}{\times 7.0 \times 10^6\,\text{Hz})^2}}$$

$$= \sqrt{10,000\,\Omega^2 + (308\,\Omega)^2}$$

$$= \sqrt{10,000\,\Omega^2 + 94,900\,\Omega^2}$$

$$= \sqrt{104,900\,\Omega^2} = 323.9\,\Omega$$

Since 308 Ω is the value of inductive reactance of the 7.00-µH coil at 7.00 MHz, the phase angle calculation proceeds as given in the earlier example (equation 82):

$$\theta = \arctan\frac{X}{R} = \arctan\left(\frac{308.0\,\Omega}{100.0\,\Omega}\right)$$

$$= \arctan(3.08) = 72.0°$$

Since the reactance is inductive, the phase angle is positive.

2.10.2 Calculating Z From R and X in Parallel Circuits

In a parallel circuit containing reactance and resistance, such as shown in **Fig 2.68**, calculation of the resultant impedance from the values of R and X does not proceed by direct combination as for series circuits. The general formula for parallel circuits is:

$$|Z| = \frac{RX}{\sqrt{R^2 + X^2}} \qquad (87)$$

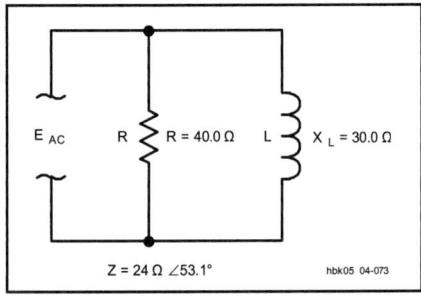

Z = 24 Ω ∠53.1°   hbk05 04-073

Fig 2.68 — A parallel circuit containing an inductive reactance of 30.0 Ω and a resistor of 40.0 Ω. No graph is given, since parallel impedances can not be manipulated graphically in the simple way of series impedances.

where the formula uses the absolute (unsigned) reactance value. The phase angle for the parallel circuit is given by:

$$\theta = \arctan\left(\frac{R}{X}\right) \qquad (88)$$

The sign of θ has the same meaning in both series and parallel circuits: if the parallel reactance is capacitive, then θ is a negative angle, and if the parallel reactance is inductive, then θ is a positive angle.

Example: An inductor with a reactance of 30.0 Ω is in parallel with a resistor of 40.0 Ω. What is the resulting impedance and phase angle?

$$|Z| = \frac{RX}{\sqrt{R^2 + X^2}} = \frac{30.0\,\Omega \times 40.0\,\Omega}{\sqrt{(30.0\,\Omega)^2 + (40.0\,\Omega)^2}}$$

$$= \frac{1200\,\Omega^2}{\sqrt{900\,\Omega^2 + 1600\,\Omega^2}} = \frac{1200\,\Omega^2}{\sqrt{2500\,\Omega^2}}$$

$$= \frac{1200\,\Omega^2}{50.0\,\Omega} = 24.0\,\Omega$$

$$\theta = \arctan\left(\frac{R}{X}\right) = \arctan\left(\frac{40.0\,\Omega}{30.0\,\Omega}\right)$$

$$\theta = \arctan(1.33) = 53.1°$$

Since the parallel reactance is inductive, the resultant angle is positive.

Example: A capacitor with a reactance of 16.0 Ω is in parallel with a resistor of 12.0 Ω. What is the resulting impedance and phase angle? (Remember that capacitive reactance is negative when used in calculations.)

$$|Z| = \frac{RX}{\sqrt{R^2 + X^2}} = \frac{-16.0\,\Omega \times 12.0\,\Omega}{\sqrt{(-16.0\,\Omega)^2 + (12.0\,\Omega)^2}}$$

$$= \frac{-192\,\Omega^2}{\sqrt{256\,\Omega^2 + 144\,\Omega^2}} = \frac{-192\,\Omega^2}{\sqrt{400\,\Omega^2}}$$

$$= \frac{-192\,\Omega^2}{20.0\,\Omega} = -9.60\,\Omega$$

$$\theta = \arctan\left(\frac{R}{X}\right) = \arctan\left(\frac{12.0\,\Omega}{-16.0\,\Omega}\right)$$

$$\theta = \arctan(-0.750) = -36.9°$$

Because the parallel reactance is capacitive and the reactance negative, the resultant phase angle is negative.

2.10.3 Admittance

Just as the inverse of resistance is conductance (G) and the inverse of reactance is susceptance (B), so, too, impedance has an inverse: admittance (Y), measured in

siemens (S). Thus,

$$Y = \frac{1}{Z} \qquad (89)$$

Since resistance, reactance and impedance are inversely proportional to the current (Z = E / I), conductance, susceptance and admittance are directly proportional to current. That is,

$$Y = \frac{I}{E} \qquad (90)$$

Admittance can be expressed in rectangular and polar forms, just like impedance,

$$Y = G \pm jB = |Y|\angle\theta \qquad (91)$$

The phase angle for admittance has the same sign convention as for impedance; if the susceptance component is inductive, the phase angle is positive, and if the susceptive component is capacitive, the phase angle is negative.

One handy use for admittance is in simplifying parallel circuit impedance calculations. Similarly to the rules stated previously for combining conductance, the admittance of a parallel combination of reactance and resistance is the vector addition of susceptance and conductance. In other words, for parallel circuits:

$$|Y| = \sqrt{G^2 + B^2} \qquad (92)$$

where

|Y| = magnitude of the admittance in siemens,

G = conductance or 1 / R in siemens, and

B = susceptance or 1 / X in siemens.

Example: An inductor with a reactance of 30.0 Ω is in parallel with a resistor of 40.0 Ω. What is the resulting impedance and phase angle? The susceptance is 1 / 30.0 Ω = 0.0333 S and the conductance is 1 / 40.0 Ω = 0.0250 S.

$$Y = \sqrt{(0.0333\,\text{S})^2 + (0.0250\,\text{S})^2}$$

$$Y = \sqrt{0.00173\,\text{S}^2} = 0.0417\,\text{S}$$

$$Z = \frac{1}{Y} = \frac{1}{0.0417\,\text{S}} = 24.0\,\Omega$$

The phase angle in terms of conductance and susceptance is:

$$\theta = \arctan\left(\frac{B}{G}\right) \qquad (93)$$

In this example,

$$\theta = \arctan\left(\frac{0.0333\,\text{S}}{0.0250\,\text{S}}\right) = \arctan(1.33) = 53.1°$$

Again, since the reactive component is inductive, the phase angle is positive. For a capacitively reactive parallel circuit, the phase angle would have been negative. Compare these results with the example calculation of the impedance for the same circuit earlier in the section.

Conversion between resistance, reactance and impedance and conductance, susceptance and admittance is very useful in working with complex circuits and in impedance matching of antennas and transmission lines. There are many on-line calculators that can perform these operations and many programmable calculators and suites of mathematical computer software have these functions built-in. Knowing when and how to use them, however, demands some understanding of the fundamental strategies shown here.

2.10.4 More than Two Elements in Series or Parallel

When a circuit contains several resistances or several reactances in series, simplify the circuit before attempting to calculate the impedance. Resistances in series add, just as in a purely resistive circuit. Series reactances of the same kind — that is, all capacitive or all inductive — also add, just as in a purely reactive circuit. The goal is to produce a single value of resistance and a single value of reactance that can be used in the impedance calculation.

Fig 2.69 illustrates a more difficult case in which a circuit contains two different reactive elements in series, along with a series resistance. The series combination of X_C and X_L reduce to a single value using the same rules of combination discussed in the section on purely reactive components. As Fig 2.69B demonstrates, the resultant reactance is the difference between the two series reactances.

For parallel circuits with multiple resistances or multiple reactances of the same type, use the rules of parallel combination to reduce the resistive and reactive components to single elements. Where two or more reactive components of different types appear in the same circuit, they can be combined using formulas shown earlier for pure reactances. As **Fig 2.70** suggests, however, they can also be combined as susceptances. Parallel susceptances of different types add, with attention to their differing signs. The resulting single susceptance can then be combined with the conductance to arrive at the overall circuit admittance whose inverse is the final circuit impedance.

2.10.5 Equivalent Series and Parallel Circuits

The two circuits shown in Fig 2.64 are equivalent if the same current flows when a given voltage of the same frequency is ap-

plied, and if the phase angle between voltage and current is the same in both cases. It is possible, in fact, to transform any given series circuit into an equivalent parallel circuit, and vice versa.

A series RX circuit can be converted into its parallel equivalent by means of the formulas:

$$R_P = \frac{R_S^2 + X_S^2}{R_S} \tag{94}$$

$$X_P = \frac{R_S^2 + X_S^2}{X_S} \tag{95}$$

where the subscripts P and S represent

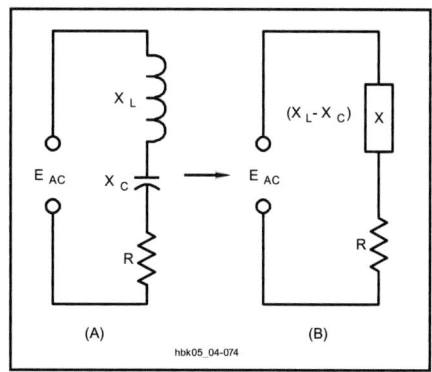

Fig 2.69 — A series impedance containing mixed capacitive and inductive reactances can be reduced to a single reactance plus resistance by combining the reactances algebraically.

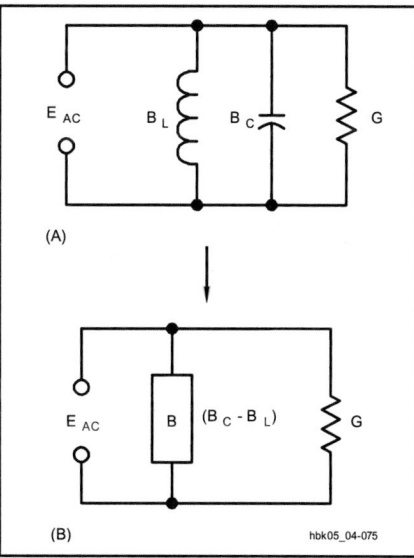

Fig 2.70 — A parallel impedance containing mixed capacitive and inductive reactances can be reduced to a single reactance plus resistance using formulas shown earlier in the chapter. By converting reactances to susceptances, as shown in A, you can combine the susceptances algebraically into a single susceptance, as shown in B.

the parallel- and series-equivalent values, respectively. If the parallel values are known, the equivalent series circuit can be found from:

$$R_S = \frac{R_P X_P^2}{R_P^2 + X_P^2} \tag{96}$$

and

$$X_S = \frac{R_P^2 X_P}{R_P^2 + X_P^2} \tag{97}$$

Example: Let the series circuit in Fig 2.64 have a series reactance of −50.0 Ω (indicating a capacitive reactance) and a resistance of 50.0 Ω. What are the values of the equivalent parallel circuit?

$$R_P = \frac{R_S^2 + X_S^2}{R_S} = \frac{(50.0\ \Omega)^2 + (-50.0\ \Omega)^2}{50.0\ \Omega}$$

$$= \frac{2500\ \Omega^2 + 2500\ \Omega^2}{50.0\ \Omega} = \frac{5000\ \Omega^2}{50\ \Omega} = 100\ \Omega$$

$$X_P = \frac{R_S^2 + X_S^2}{X_S} = \frac{(50.0\ \Omega)^2 + (-50.0\ \Omega)^2}{-50.0\ \Omega}$$

$$= \frac{2500\ \Omega^2 + 2500\ \Omega^2}{-50.0\ \Omega} = \frac{5000\ \Omega^2}{-50\ \Omega} = -100\ \Omega$$

In the parallel circuit of Fig 2.64, a capacitive reactance of 100 Ω and a resistance of 100 Ω to be equivalent would the series circuit.

2.10.6 Ohm's Law for Impedance

Ohm's Law applies to circuits containing impedance just as readily as to circuits having resistance or reactance only. The formulas are:

$$E = I\,Z \tag{98}$$

$$I = \frac{E}{Z} \tag{99}$$

$$Z = \frac{E}{I} \tag{100}$$

where
  E = voltage in volts,
  I = current in amperes, and
  Z = impedance in ohms.

Z must now be understood to be a complex number, consisting of resistive and reactive components. If Z is complex, then so are E and I, with a magnitude and phase angle. The rules of complex mathematics are then applied and the variables are written in boldface type as $\mathbf{Z}$, $\mathbf{E}$, and $\mathbf{I}$, or an arrow is added above them to indicate that they are complex, such as,

$$\vec{E} = \vec{I}\,\vec{Z}$$

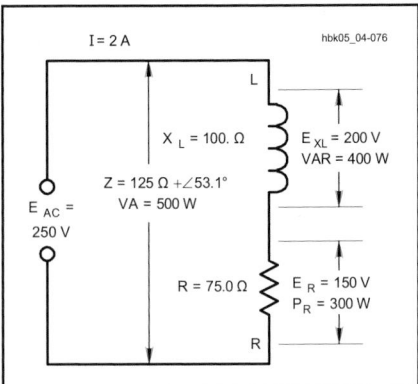

Fig 2.71 — A series circuit consisting of an inductive reactance of 100 Ω and a resistance of 75.0 Ω. Also shown is the applied voltage, voltage drops across the circuit elements, and the current.

If only the magnitude of impedance, voltage, and currents are important, however, then the magnitudes of the three variables can be combined in the familiar ways without regard to the phase angle. In this case E and I are assumed to be RMS values (or some other steady-state value such as peak, peak-to-peak, or average). **Fig 2.71** shows a simple circuit consisting of a resistance of 75.0 Ω and a reactance of 100 Ω in series. From the series-impedance formula previously given, the impedance is

$$Z = \sqrt{R^2 + X_L^2} = \sqrt{(75.0\ \Omega)^2 + (100\ \Omega)^2}$$

$$= \sqrt{5630\ \Omega^2 + 10,000\ \Omega^2}$$

$$= \sqrt{15,600\ \Omega^2} = 125\ \Omega$$

If the applied voltage is 250 V, then

$$I = \frac{E}{Z} = \frac{250\ V}{125\ \Omega} = 2.0\ A$$

This current flows through both the resistance and reactance, so the voltage drops are:

$$E_R = I\ R = 2.0\ A \times 75.0\ \Omega = 150\ V$$

$$E_{XL} = I\ X_L = 2.0\ A \times 100\ \Omega = 200\ V$$

Illustrating one problem of working only with RMS values, the simple arithmetical sum of these two drops, 350 V, is greater than the applied voltage because the two voltages are 90° out of phase. When phase is taken into account,

$$E = \sqrt{(150\ V)^2 + (200\ V)^2}$$

$$= \sqrt{22,500\ V^2 + 40,000\ V^2}$$

$$= \sqrt{62,500\ V^2} = 250\ V$$

2.10.7 Reactive Power and Power Factor

Although purely reactive circuits, whether simple or complex, show a measurable ac voltage and current, we cannot simply multiply the two together to arrive at power. Power is the rate at which energy is consumed by a circuit, and purely reactive circuits do not consume energy. The charge placed on a capacitor during part of an ac cycle is returned to the circuit during the next part of a cycle. Likewise, the energy stored in the magnetic field of an inductor returns to the circuit as the field collapses later in the ac cycle. A reactive circuit simply cycles and recycles energy into and out of the reactive components. If a purely reactive circuit were possible in reality, it would consume no energy at all.

In reactive circuits, circulation of energy accounts for seemingly odd phenomena. For example, in a series circuit with capacitance and inductance, the voltages across the components may exceed the supply voltage. That condition can exist because, while energy is being stored by the inductor, the capacitor is returning energy to the circuit from its previously charged state, and vice versa. In a parallel circuit with inductive and capacitive branches, the current circulating through the components may exceed the current drawn from the source. Again, the phenomenon occurs because the inductor's collapsing magnetic field supplies current to the capacitor, and the discharging capacitor provides current to the inductor.

To distinguish between the non-dissipated energy circulating in a purely reactive circuit and the dissipated or *real power* in a resistive circuit, the unit of *reactive power* is called the *volt-ampere reactive*, or *VAR*. The term watt is not used and sometimes reactive power is called "wattless" power. VAR has only limited use in radio circuits. Formulas similar to those for resistive power are used to calculate VAR:

$$VAR = I \times E \qquad (101)$$

$$VAR = I^2 \times X \qquad (102)$$

$$VAR = \frac{E^2}{X} \qquad (103)$$

where E and I are RMS values of voltage and current.

Real, or dissipated, power is measured in watts. *Apparent power* is the product of the voltage across and the current through an impedance. To distinguish apparent power from real power, apparent power is measured in *volt-amperes (VA)*.

In the circuit of Fig 2.71, an applied voltage of 250 V results in a current of 2.00 A, giving an apparent power of 250 V × 2.00 A = 500 W. Only the resistance actually consumes power, however. The real power dissipated by the resistance is:

$$E = I^2\ R = (2.0\ A)^2 \times 75.0\ V = 300\ W$$

and the reactive power is:

$$VAR = I^2 \times X_L = (2.0\ A)^2 \times 100\ \Omega = 400\ VA$$

The ratio of real power to the apparent power is called the circuit's *power factor* (PF).

$$PF = \frac{P_{consumed}}{P_{apparent}} = \frac{R}{Z} \qquad (104)$$

Power factor is frequently expressed as a percentage. The power factor of a purely resistive circuit is 100% or 1, while the power factor of a pure reactance is zero. In the example of Fig 2.71 the power factor would be 300 W / 500 W = 0.600 or 60%.

Apparent power has no direct relationship to the power actually dissipated unless the power factor of the circuit is known.

$$P = \text{Apparent Power} \times \text{power factor} \qquad (105)$$

An equivalent definition of power factor is:

$$PF = \cos\theta \qquad (106)$$

AC Component Summary

	Resistor	Capacitor	Inductor
Basic Unit Units Commonly Used	ohm (Ω)	farad (F) microfarads (μF) picofarads (pF)	henry (H) millihenrys (mH) microhenrys (μH)
Time constant	(None)	RC	L/R
Voltage-Current Phase	In phase	Current leads voltage Voltage lags current	Voltage leads current Current lags voltage
Resistance or Reactance Change with increasing frequency	Resistance No	$X_C = 1 / 2\pi fC$ Reactance decreases	$X_L = 2\pi fL$ Reactance increases
Q of circuit	Not defined	X_C / R	X_L / R

where θ is the phase angle of the circuit impedance.

Since the phase angle in the example equals:

$$\theta = \arctan\left(\frac{X}{R}\right) = \arctan\left(\frac{100\,\Omega}{75.0\,\Omega}\right)$$

$$\theta = \arctan(1.33) = 53.1°$$

and the power factor is:

$$PF = \cos 53.1° = 0.600$$

as the earlier calculation confirms.

Since power factor is always rendered as a positive number, the value must be followed by the words "leading" or "lagging" to identify the phase of the voltage with respect to the current. Specifying the numerical power factor is not always sufficient. For example, many dc-to-ac power inverters can safely operate loads having a large net reactance of one sign but only a small reactance of the opposite sign. Hence, the final calculation of the power factor in this example would be reported as "0.600, leading."

2.11 Quality Factor (Q) of Components

Components that store energy, such as capacitors and inductors, may be compared in terms of *quality factor* or *Q factor*, abbreviated *Q*. The concept of *Q* originated in 1914 (then dubbed *K*) and first appeared in print in 1923 when Kenneth S. Johnson used it to represent the ratio of reactance to resistance as a "figure of merit" for inductors in US patent 1,628,983. For a series or parallel representation of a reactive circuit element:

$$Q = \frac{X_S}{R_S} = \frac{R_P}{X_P} \qquad (107)$$

where for series-connected reactance and its series loss resistance (such as an inductor)

Q = quality factor (no units),
X_S = series reactance of the component (in ohms), and
R_S = the sum of all series resistances associated with the energy losses in the component (in ohms).

For a parallel connected reactance and its parallel loss resistance (such as a capacitor)

Q = quality factor (no units),
X_P = parallel-connected reactance of the component (in ohms), and
R_P = the total parallel resistance associated with the energy losses in the component (in ohms).

Several exactly equivalent formulas for Q may be seen in **Table 2.10**. In Table 2.10, equation [a] most naturally represents the Q of an inductor, while equation [b] is useful for a capacitor. Both representations are equivalent to equation [c] which relates the energy storage to energy losses in inductors and capacitors. Note that in a series circuit representation, the series resistance is proportional to energy loss, and the series reactance is proportional to stored energy. In a parallel circuit, however, the reciprocal of the resistance is proportional to the lost energy and the reciprocal of the reactance is proportional to the stored energy. The Q of a tuned circuit may be found by measuring the upper and lower frequencies where the resistance

Table 2.10

Equivalent Formulas for Expressing Q and Their Uses

[a]	$Q = \dfrac{\text{Series Reactance}}{\text{Series Resistance}}$	Johnson's historical definition of Q for inductors, used for series circuits
[b]	$Q = \dfrac{\text{Parallel Resistance}}{\text{Parallel Reactance}}$	Parallel equivalent circuit definition of Q, useful for capacitors.
[c]	$Q = \dfrac{2\pi \times \text{Stored energy}}{\text{Energy lost in one cycle}}$	Fundamental energy definition, useful with antennas, reactive components, and mechanical systems.
[d]	$Q = \dfrac{\sqrt{f_U f_L}}{f_U - f_L} = \dfrac{\text{Frequency}}{\text{Bandwidth}}$	Bandwidth formula for simple resonant circuits. Impedance Z = R + jX, and f_U is the upper frequency where R = X, and f_L is the lower frequency where R = −X, and $f_U − f_L$ represents the −3 dB bandwidth.

equals the magnitude of the reactance, and applying equation [d]. The geometrical mean frequency is $f = \sqrt{f_U f_L}$ and may be replaced by the center frequency for high-Q circuits. In circuits having several reactive components, such as the tuned circuits in Fig 2.84, the circuit Q is the parallel combination of the individual Q factors. For example:

$$Q = 1 / \left(\frac{1}{Q_C} + \frac{1}{Q_L}\right)$$

where Q_C is the capacitor Q (sometimes specified by a manufacturer) and the inductor Q is Q_L.

The Q of capacitors is ordinarily high. Good quality ceramic capacitors and mica capacitors may have Q values of 1200 or more. Microwave capacitors can have poor Q values — 10 or less at 10 GHz and higher frequencies because X_C will be low. Capacitors are subject to predominantly dielectric losses which are modeled as a parallel loss resistance across the capacitive reactance. Capacitors also have a series loss resistance associated with the conductor leads and capacitor plates, but this loss is often small enough to ignore.

Inductors are subject to several types of

electrical energy losses such as wire resistance (including skin effect) and core losses. All electrical conductors have some resistance through which electrical energy is lost as heat. Wire conductors suffer additional ac losses because alternating current tends to flow on the conductor surface due to the skin effect discussed in the chapter on **RF Techniques**. If the inductor's core is iron, ferrite or brass, the core will introduce additional losses of energy. Note that core losses for inductors are modeled as a resistor in parallel with the inductor (analogous to capacitor dielectric losses). The specific details of these losses are discussed in connection with each type of core material.

The sum of all core losses may be depicted by showing an equivalent series connected resistor with the inductor (as in Figs 2.52 and 2.53), although there is no separate component represented by the resistor symbol. As a result of inherent energy losses, inductor Q rarely approaches capacitor Q in a circuit where both components work together. Although many circuits call for the highest Q inductor obtainable, other circuits may call for a specific Q, even a very low one.

2.12 Practical Inductors

Various facets of radio circuits make use of inductors ranging from the tiny up to the massive. Small values of inductance, such as those inductors in **Fig 2.72A**, serve mostly in RF circuits. They may be self-supporting, air-core or air-wound inductors or the winding may be supported by nonmagnetic strips or a form. Phenolic, certain plastics and ceramics are the most common *coil forms* for air-core inductors. These inductors range in value from a few hundred μH for medium- and high-frequency circuits down to tenths of a μH at VHF and UHF.

The most common inductor in small-signal RF circuits is the *encapsulated inductor*. These components look a lot like carbon-composition or film resistors and are often marked with colored paint stripes to indicate value. (The chapter **Component Data and References** contains information on inductor color codes and marking schemes.) These inductors have values from less than 1 μH to a few mH. They cannot handle much current without saturating or over-heating.

It is possible to make solenoid inductors variable by inserting a moveable *slug* in the center of the inductor. (Slug-tuned inductors normally have a ceramic, plastic or phenolic insulating form between the conductive slug and the inductor winding.) If the slug material is magnetic, such as powdered iron, the inductance increases as the slug is moved into the center of the inductor. If the slug is brass or some other nonmagnetic material,

inserting the slug will reduce the inductor's inductance.

An alternative to air-core inductors for RF work are *toroidal* inductors (or *toroids*) wound on powdered-iron or ferrite cores. The availability of many types and sizes of powdered-iron cores has made these inductors popular for low-power fixed-value service. The toroidal shape concentrates the inductor's field nearly completely inside the inductor, eliminating the need in many cases for other forms of shielding to limit the interaction of the inductor's magnetic field with the fields of other inductors. (Ferrite core materials are discussed here and in the chapter on **RF Techniques**.)

Fig 2.72B shows samples of inductors in the millihenry (mH) range. Among these inductors are multi-section RF chokes designed to block RF currents from passing beyond them to other parts of circuits. Low-frequency radio work may also use inductors in this range of values, sometimes wound with *litz wire*. Litz wire is a special version of stranded wire, with each strand insulated from the others, and is used to minimize losses associated with skin effect.

For audio filters, toroidal inductors with values up to 100 mH are useful. Resembling powdered-iron-core RF toroids, these inductors are wound on ferrite or molybdenum-permalloy cores having much higher permeabilities.

Audio and power-supply inductors appear

in Fig 2.72C. Lower values of these iron-core inductors, in the range of a few henrys, are useful as audio-frequency chokes. Larger values up to about 20 H may be found in power supplies, as choke filters, to suppress 120-Hz ripple. Although some of these inductors are open frame, most have iron covers to confine the powerful magnetic fields they produce.

Although builders and experimenters rarely construct their own capacitors, inductor fabrication is common. In fact, it is often necessary, since commercially available units may be unavailable or expensive. Even if available, they may consist of inductor stock to be trimmed to the required value. Core materials and wire for winding both solenoid and toroidal inductors are readily available. The following information includes fundamental formulas and design examples for calculating practical inductors, along with additional data on the theoretical limits in the use of some materials.

2.12.1 Air-Core Inductors

Many circuits require air-core inductors using just one layer of wire. The approximate inductance of a single-layer air-core inductor may be calculated from the simplified formula:

$$L\,(\mu H) = \frac{d^2\,n^2}{18d + 40\ell} \tag{108}$$

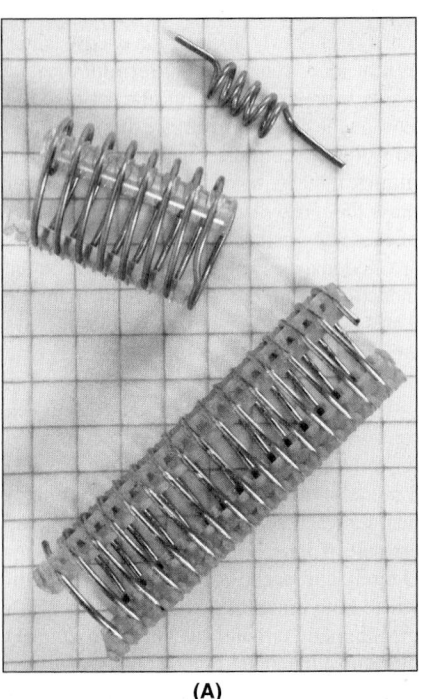

(A)

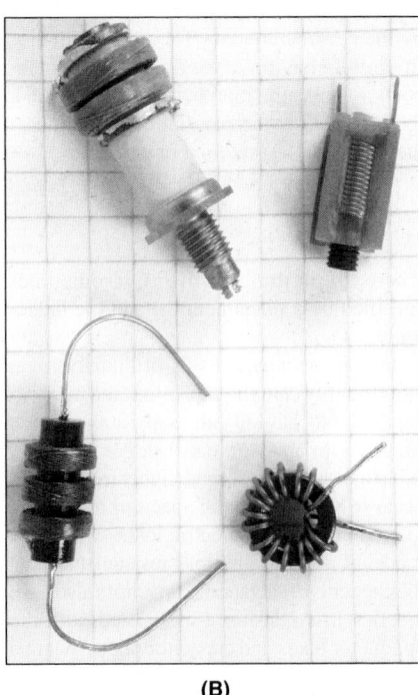

(B)

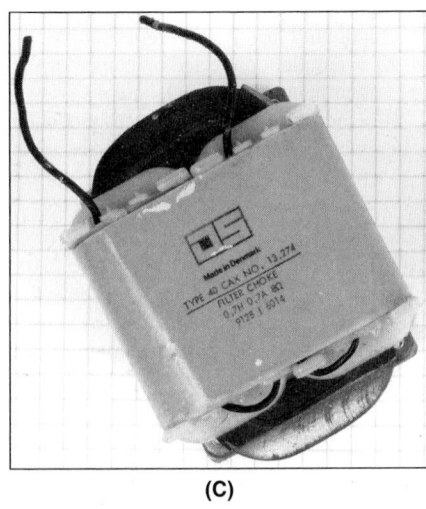

(C)

Fig 2.72 — Part A shows small-value air-wound inductors. Part B shows some inductors with values in the range of a few millihenrys and C shows a large inductor such as might be used in audio circuits or as power-supply chokes. The ¼-inch-ruled graph paper background provides a size comparison.

where

L = inductance in microhenrys,
d = inductor diameter in inches (from wire center to wire center),
ℓ = inductor length in inches, and
n = number of turns.

The notation is illustrated in **Fig 2.73**. This formula is a close approximation for inductors having a length equal to or greater than 0.4d. (Note: Inductance varies as the square of the turns. If the number of turns is doubled, the inductance is quadrupled. This relationship

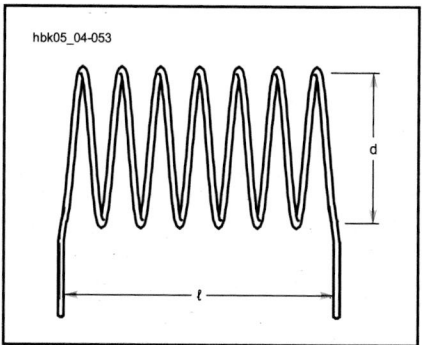

Fig 2.73 — Coil dimensions used in the inductance formula for air-core inductors.

is inherent in the equation, but is often overlooked. For example, to double the inductance, add additional turns equal to 1.4 times the original number of turns, or 40% more turns.)

Example: What is the inductance of an inductor if the inductor has 48 turns wound at 32 turns per inch and a diameter of ¾ inch? In this case, d = 0.75, ℓ = 48/32 = 1.5 and n = 48.

$$L\,(\mu H) = \frac{0.75^2 \times 48^2}{(18 \times 0.75)+(40 \times 1.5)}$$

$$= \frac{1300}{74} = 18\ \mu H$$

To calculate the number of turns of a single-layer inductor for a required value of inductance, the formula becomes:

$$\sqrt{L\,(18\,d\quad 40\quad)} \qquad (109)$$

Example: Suppose an inductance of 10.0 μH is required. The form on which the inductor is to be wound has a diameter of one inch and is long enough to accommodate an inductor of 1¼ inches. Then d = 1.00 inch, ℓ = 1.25 inches and L = 10.0. Substituting:

$$n = \frac{\sqrt{10.0\,[\,(18\times 1.0)+(40\times 1.25)\,]}}{1}$$

$$= \sqrt{680} = 26.1\ \text{turns}$$

A 26-turn inductor would be close enough in practical work. Since the inductor will be 1.25 inches long, the number of turns per inch will be 26.1 / 1.25 = 20.9. Consulting the wire table in the **Component Data and References** chapter, we find that #17 AWG enameled wire (or anything smaller) can be used. The proper inductance is obtained by winding the required number of turns on the form and then adjusting the spacing between the turns to make a uniformly spaced inductor 1.25 inches long.

Most inductance formulas lose accuracy when applied to small inductors (such as are used in VHF work and in low-pass filters built for reducing harmonic interference to televisions) because the conductor thickness is no longer negligible in comparison with the size of the inductor. **Fig 2.74** shows the measured inductance of VHF inductors and may be used as a basis for circuit design. Two curves are given; curve A is for inductors wound to an inside diameter of ½ inch; curve B is for inductors of ¾-inch inside diameter. In both curves, the wire size is #12 AWG and the winding pitch is eight turns to the inch (⅛-inch turn spacing). The inductance values include leads ½-inch long.

Machine-wound inductors with the preset diameters and turns per inch are available in many radio stores, under the trade names of B&W Miniductor and Airdux. Information on using such coil stock is provided in the **Component Data and References** chapter to simplify the process of designing high-quality inductors for most HF applications.

Forming a wire into a solenoid increases its inductance, but also introduces distributed capacitance. Since each turn is at a slightly different ac potential, each pair of turns effectively forms a capacitor in parallel with part of the inductor. (See the chapter on **RF Techniques** for information on the effects of these and other factors that affect the behavior of the "ideal" inductors discussed in this chapter.)

Moreover, the Q of air-core inductors is, in part, a function of the inductor shape, specifically its ratio of length to diameter. Q tends to be highest when these dimensions are nearly equal. With wire properly sized to the current carried by the inductor, and with high-caliber construction, air-core inductors can achieve Q above 200.

For a large collection of formulas useful in constructing air-core inductors of many configurations, see the "Circuit Elements" section in Terman's *Radio Engineers' Handbook*.

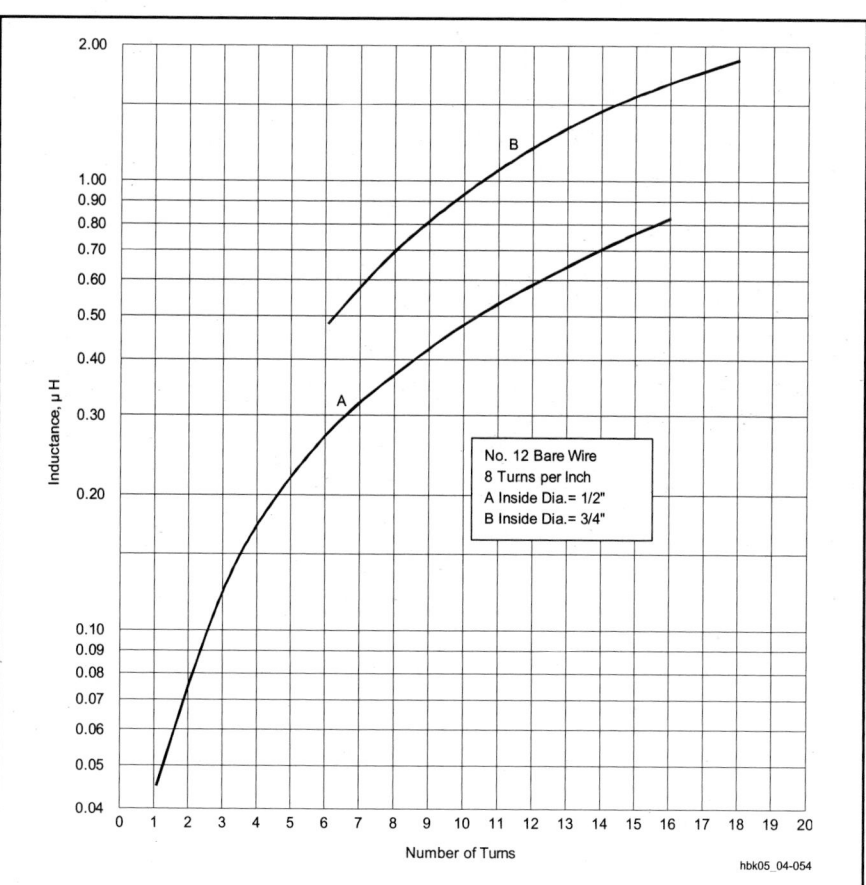

No. 12 Bare Wire
8 Turns per Inch
A Inside Dia.= 1/2"
B Inside Dia.= 3/4"

Fig 2.74 — Measured inductance of coils wound with #12 bare wire, eight turns to the inch. The values include half-inch leads.

2.12.2 Straight-Wire Inductance

At low frequencies the inductance of a straight, round, nonmagnetic wire in free space is given by:

$$L = 0.00508\, b \left\{ \left[\ln\left(\frac{2b}{a}\right) \right] - 0.75 \right\} \quad (110)$$

where
- L = inductance in μH,
- a = wire radius in inches,
- b = wire length in inches, and
- ln = natural logarithm = 2.303 × common logarithm (base 10).

If the dimensions are expressed in millimeters instead of inches, the equation may still be used, except replace the 0.00508 value with 0.0002.

Skin effect reduces the inductance at VHF and above. As the frequency approaches infinity, the 0.75 constant within the brackets approaches unity. As a practical matter, skin effect will not reduce the inductance by more than a few percent.

Example: What is the inductance of a wire that is 0.1575 inch in diameter and 3.9370 inches long? For the calculations, a = 0.0787 inch (radius) and b = 3.9370 inch.

$$L = 0.00508\, b \left\{ \left[\ln\left(\frac{2b}{a}\right) \right] - 0.75 \right\}$$

$$= 0.00508\,(3.9370) \times$$

$$\left\{ \left[\ln\left(\frac{2 \times 3.9370}{0.0787}\right) \right] - 0.75 \right\}$$

$$= 0.020 \times [\ln(100) - 0.75]$$

$$= 0.020 \times (4.60 - 0.75)$$

$$= 0.020 \times 3.85 = 0.077 \text{ μH}$$

Fig 2.75 is a graph of the inductance for wires of various radii as a function of length.

A VHF or UHF tank circuit can be fabricated from a wire parallel to a ground plane, with one end grounded. A formula for the inductance of such an arrangement is given in **Fig 2.76**.

Example: What is the inductance of a wire 3.9370 inches long and 0.0787 inch in radius, suspended 1.5748 inch above a ground plane? (The inductance is measured between the free end and the ground plane, and the formula includes the inductance of the 1.5748-inch grounding link.) To demonstrate the use of the formula in Fig 2.76, begin by evaluating these quantities:

$$b + \sqrt{b^2 + a^2}$$

$$= 3.9370 + \sqrt{3.9370^2 + 0.0787^2}$$

$$= 3.9370 + 3.94 = 7.88$$

$$b + \sqrt{b^2 + 4(h^2)}$$

$$= 3.9370 + \sqrt{3.9370^2 + 4(1.5748^2)}$$

$$= 3.9370 + \sqrt{15.50 + 4(2.480)}$$

$$= 3.9370 + \sqrt{15.50 + 9.920}$$

$$= 3.9370 + 5.0418 = 8.9788$$

$$\frac{2h}{a} = \frac{2 \times 1.5748}{0.0787} = 40.0$$

$$\frac{b}{a} = \frac{3.9370}{4} = 0.98425$$

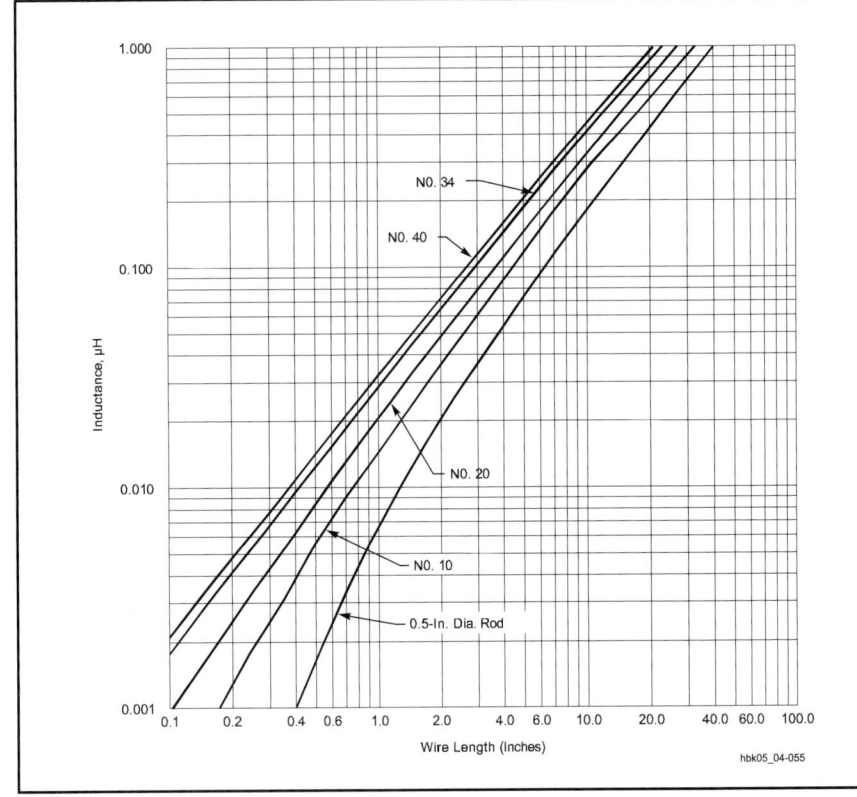

Fig 2.75 — Inductance of various conductor sizes as straight wires.

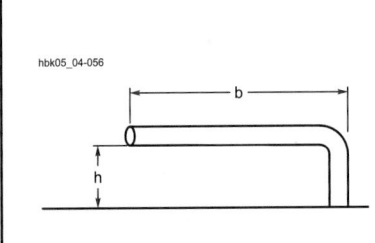

$$L = 0.0117b \left\{ \log_{10}\left[\frac{2h}{a}\left(\frac{b + \sqrt{b^2 + a^2}}{b + \sqrt{b^2 + 4h^2}} \right) \right] \right\} + 0.00508\left(\sqrt{b^2 + 4h^2} - \sqrt{b^2 + a^2} + \frac{b}{4} - 2h + a \right)$$

where
- L = inductance in μH
- a = wire radius in inches
- b = wire length parallel to ground plane in inches
- h = wire height above ground plane in inches

Fig 2.76 — Equation for determining the inductance of a wire parallel to a ground plane, with one end grounded. If the dimensions are in millimeters, the numerical coefficients become 0.0004605 for the first term and 0.0002 for the second term.

Substituting these values into the formula yields:

$$L = 0.0117 \times 3.9370$$

$$\left\{ \log_{10} \left[40.0 \times \left(\frac{7.88}{8.9788} \right) \right] \right\}$$

$$+ 0.00508 \times (5.0418 - 3.94 + 0.98425 - 3.1496 + 0.0787)$$

$$L = 0.0662 \ \mu H$$

Another conductor configuration that is frequently used is a flat strip over a ground plane. This arrangement has lower skin-effect loss at high frequencies than round wire because it has a higher surface-area to volume ratio. The inductance of such a strip can be found from the formula in **Fig 2.77**.

2.12.3 Iron-Core Inductors

If the permeability of an iron core in an inductor is 800, then the inductance of any given air-wound inductor is increased 800 times by inserting the iron core. The inductance will be proportional to the magnetic flux through the inductor, other things being equal. The inductance of an iron-core inductor is highly dependent on the current flowing in the inductor, in contrast to an air-core inductor, where the inductance is independent of current because air does not saturate.

Iron-core inductors such as the one sketched in Fig 2.49 are used chiefly in power-supply equipment. They usually have direct current flowing through the winding, and any variation in inductance with current is usually undesirable. Inductance variations may be overcome by keeping the flux density below the saturation point of the iron. Opening the core so there is a small air gap will achieve this goal, as discussed in the earlier section on inductors. The reluctance or magnetic resistance introduced by such a gap is very large compared with that of the iron, even though the gap is only a small fraction of an inch. Therefore, the gap — rather than the iron — controls the flux density. Air gaps in iron cores reduce the inductance, but they hold the value practically constant regardless of the current magnitude.

When alternating current flows through an inductor wound on an iron core, a voltage is induced. Since iron is a conductor, eddy currents also flow in the core as discussed earlier. Eddy currents represent lost power because they flow through the resistance of the iron and generate heat. Losses caused by eddy currents can be reduced by laminating the core (cutting the core into thin strips). These strips or laminations are then insulated from each other by painting them with some insulating material such as varnish or shellac. Eddy current losses add to hysteresis losses, which are

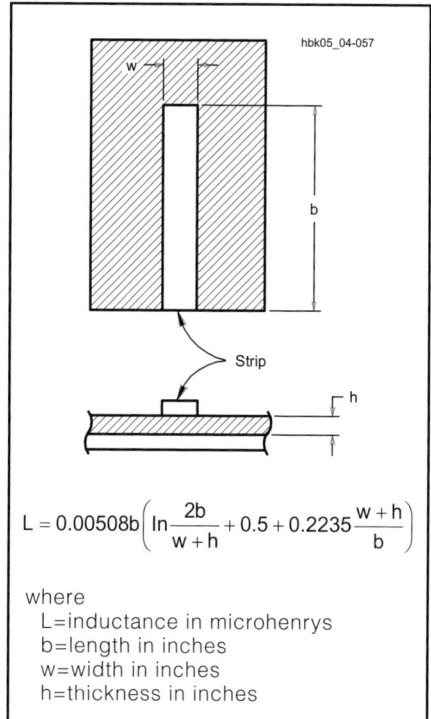

$$L = 0.00508b \left(\ln \frac{2b}{w+h} + 0.5 + 0.2235 \frac{w+h}{b} \right)$$

where
    L=inductance in microhenrys
    b=length in inches
    w=width in inches
    h=thickness in inches

Fig 2.77 — Equation for determining the inductance of a flat strip inductor.

also significant in iron-core inductors.

Eddy-current and hysteresis losses in iron increase rapidly as the frequency of the alternating current increases. For this reason, ordinary iron cores can be used only at power-line and audio frequencies — up to approximately 15000 Hz. Even then, a very good grade of iron or steel is necessary for the core to perform well at the higher audio frequencies. Laminated iron cores become completely useless at radio frequencies because of eddy current and hysteresis losses.

2.12.4 Slug-Tuned Inductors

For RF work, the losses in iron cores can be reduced to a more useful level by grinding the iron into a powder and then mixing it with a binder of insulating material in such a way that the individual iron particles are insulated from each other. Using this approach, cores can be made that function satisfactorily even into the VHF range. Because a large part of the magnetic path is through a nonmagnetic material (the binder), the permeability of the powdered iron core is low compared with the values for solid iron cores used at power-line frequencies.

The slug is usually shaped in the form of a cylinder that fits inside the insulating form on which the inductor is wound. Despite the fact that the major portion of the magnetic path for the flux is in air, the slug is quite effective in increasing the inductor inductance. By pushing (or screwing) the slug in and out

of the inductor, the inductance can be varied over a considerable range.

2.12.5 Powdered-Iron Toroidal Inductors

For fixed-value inductors intended for use at HF and VHF, the powdered-iron toroidal core has become the standard in low- and medium-power circuits. **Fig 2.78** shows the general outlines of a toroidal inductor on a magnetic core.

Manufacturers offer a wide variety of core materials, or *mixes*, to create conductor cores that will perform over a desired frequency range with a reasonable permeability. Permeabilities for powdered-iron cores fall in the range of 3 to 35 for various mixes. In addition, core sizes are available in the range of 0.125-inch outside diameter (OD) up to 1.06-inch OD, with larger sizes to 5-inch OD available in certain mixes. The range of sizes permits the builder to construct single-layer inductors for almost any value using wire sized to meet the circuit current demands.

The use of powdered iron in a binder reduces core losses usually associated with iron, while the permeability of the core permits a reduction in the wire length and associated resistance in forming an inductor of a given inductance. Therefore, powdered-iron-core toroidal inductors can achieve Q well above 100, often approaching or exceeding 200 within the frequency range specified for a given core. Moreover, these inductors are considered *self-shielding* since most of the magnetic flux is within the core, a fact that simplifies circuit design and construction.

Each powdered-iron core has an *inductance factor* or *index* A_L determined by the manufacturer. Amidon Corp. is the most common supplier of cores for amateurs (see the Component Data and References chapter) and specifies A_L in μH per 100 turns-squared. The following calculations are based on the Amidon specification. Other manufacturers specify A_L in μH per turn-squared.

To calculate the inductance of a powdered-iron toroidal inductor using the Amidon convention for A_L when the number of turns and the core material are known:

$$L = \frac{A_L \times N^2}{10,000} \qquad (111)$$

where
    L = the inductance in μH,
    A_L = the inductance index in μH per 100 turns-squared, and
    N = the number of turns.
The builder must then ensure that the core is capable of holding the calculated number of turns of wire of the required wire size.

Example: What is the inductance of a 60-turn inductor on a core with an A_L of 55? This

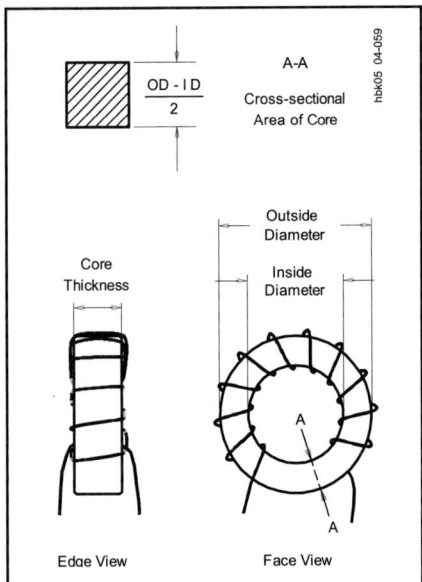

Fig 2.78 — A typical toroidal inductor wound on a powdered-iron or ferrite core. Some key physical dimensions are noted. Equally important are the core material, its permeability, its intended range of operating frequencies, and its A_L value. This is an 11-turn toroid.

A_L value was selected from manufacturer's information about a 0.8-inch OD core with an initial permeability of 10. This particular core is intended for use in the range of 2 to 30 MHz. See the **Component Data and References** chapter for more detailed data on the range of available cores.

$$L = \frac{A_L \times N^2}{10,000} = \frac{55 \times 60^2}{10,000}$$

$$= \frac{198,000}{10,000} = 19.8 \,\mu H$$

To calculate the number of turns needed for a particular inductance, use the formula:

$$N = 100\sqrt{\frac{L}{A_L}} \tag{112}$$

Example: How many turns are needed for a 12.0-µH inductor if the A_L for the selected core is 49?

$$N = 100\sqrt{\frac{L}{A_L}} = 100\sqrt{\frac{12.0}{49}}$$

$$N = 100\sqrt{0.245} = 100 \times 0.495 = 49.5 \text{ turns}$$

If the value is critical, experimenting with 49-turn and 50-turn inductors is in order, especially since core characteristics may vary slightly from batch to batch. Count turns by each pass of the wire through the center of the core. (*A straight wire through a toroidal*

core amounts to a one-turn inductor.) Fine adjustment of the inductance may be possible by spreading or compressing inductor turns.

The power-handling ability of toroidal cores depends on many variables, which include the cross-sectional area through the core, the core material, the numbers of turns in the inductor, the applied voltage and the operating frequency. Although powdered-iron cores can withstand dc flux densities up to 5000 gauss without saturating, ac flux densities from sine waves above certain limits can overheat cores. Manufacturers provide guideline limits for ac flux densities to avoid overheating. The limits range from 150 gauss at 1 MHz to 30 gauss at 28 MHz, although the curve is not linear. To calculate the maximum anticipated flux density for a particular inductor, use the formula:

$$B_{max} = \frac{E_{RMS} \times 10^8}{4.44 \times A_e \times N \times f} \tag{113}$$

where

B_{max} = the maximum flux density in gauss,
E_{RMS} = the voltage across the inductor,
A_e = the cross-sectional area of the core in square centimeters,
N = the number of turns in the inductor, and
f = the operating frequency in Hz.

Example: What is the maximum ac flux density for an inductor of 15 turns if the frequency is 7.0 MHz, the RMS voltage is 25 V and the cross-sectional area of the core is 0.133 cm²?

$$B_{max} = \frac{E_{RMS} \times 10^8}{4.44 \times A_e \times N \times f}$$

$$= \frac{25 \times 10^8}{4.44 \times 0.133 \times 15 \times 7.0 \times 10^6}$$

$$= \frac{25 \times 10^8}{62 \times 10^6} = 40 \text{ gauss}$$

Since the recommended limit for cores operated at 7 MHz is 57 gauss, this inductor is well within guidelines.

2.12.6 Ferrite Toroidal Inductors

Although nearly identical in general appearance to powdered-iron cores, ferrite cores differ in a number of important characteristics. Composed of nickel-zinc ferrites for lower permeability ranges and of manganese-zinc ferrites for higher permeabilities, these cores span a permeability range from 20 to above 10000. Nickel-zinc cores with permeabilities from 20 to 800 are useful in high-Q applications, but function more commonly in amateur applications as RF chokes. They are also useful in wide-band transformers, discussed in the chapter **RF Techniques**.

Ferrite cores are often unpainted, unlike powdered-iron toroids. Ferrite toroids and rods often have sharp edges, while powdered-iron toroids usually have rounded edges.

Because of their higher permeabilities, the A_L values for ferrite cores are higher than for powdered-iron cores. Amidon Corp. is the most common supplier of cores for amateurs (see the Component Data and References chapter) and specifies A_L in mH per 1000 turns-squared. Other manufacturers specify A_L in nH per turns-squared.

To calculate the inductance of a ferrite toroidal inductor using the Amidon convention for A_L when the number of turns and the core material are known:

$$L = \frac{A_L \times N^2}{1,000,000} \tag{114}$$

where

L = the inductance in mH,
A_L = the inductance index in mH per 1000 turns-squared, and
N = the number of turns.

The builder must then ensure that the core is capable of holding the calculated number of turns of wire of the required wire size.

Example: What is the inductance of a 60-turn inductor on a core with an A_L of 523? (See the chapter **Component Data and References** for more detailed data on the range of available cores.)

$$L = \frac{A_L \times N^2}{1,000,000} = \frac{523 \times 60^2}{1,000,000}$$

$$= \frac{1.88 \times 10^6}{1 \times 10^6} = 1.88 \text{ mH}$$

To calculate the number of turns needed for a particular inductance, use the formula:

$$N = 1000\sqrt{\frac{L}{A_L}} \tag{115}$$

Example: How many turns are needed for a 1.2-mH inductor if the A_L for the selected core is 150?

$$N = 1000\sqrt{\frac{L}{A_L}} = 1000\sqrt{\frac{1.2}{150}}$$

$$N = 1000\sqrt{0.008} = 1000 \times 0.089 = 89 \text{ turns}$$

For inductors carrying both dc and ac currents, the upper saturation limit for most ferrites is a flux density of 2000 gauss, with power calculations identical to those used for powdered-iron cores. More detailed information is available on specific cores and manufacturers in the **Component Data and References** chapter.

2.13 Resonant Circuits

A circuit containing both an inductor and a capacitor — and therefore, both inductive and capacitive reactance — is often called a *tuned circuit* or a *resonant circuit*. For any such circuit, there is a particular frequency at which the inductive and capacitive reactances are the same, that is, $X_L = X_C$. For most purposes, this is the *resonant frequency* of the circuit. At the resonant frequency — or at *resonance*, for short:

$$X_L = 2\pi f L = X_C = \frac{1}{2\pi f C}$$

By solving for f, we can find the resonant frequency of any combination of inductor and capacitor from the formula:

$$f = \frac{1}{2\pi\sqrt{L C}} \quad (116)$$

where
f = frequency in hertz (Hz),
L = inductance in henrys (H),
C = capacitance in farads (F), and
π = 3.1416.

For most high-frequency (HF) radio work, smaller units of inductance and capacitance and larger units of frequency are more convenient. The basic formula becomes:

$$f = \frac{10^3}{2\pi\sqrt{L C}} \quad (117)$$

where
f = frequency in megahertz (MHz),
L = inductance in microhenrys (µH),
C = capacitance in picofarads (pF), and
π = 3.1416.

Example: What is the resonant frequency of a circuit containing an inductor of 5.0 µH and a capacitor of 35 pF?

$$f = \frac{10^3}{2\pi\sqrt{L C}} = \frac{10^3}{6.2832\sqrt{5.0\times35}}$$

$$= \frac{10^3}{83} = 12 \text{ MHz}$$

To find the matching component (inductor or capacitor) when the frequency and one component is known (capacitor or inductor) for general HF work, use the formula:

$$f^2 = \frac{1}{4\pi^2 L C} \quad (118)$$

where f, L and C are in basic units. For HF work in terms of MHz, µH and pF, the basic relationship rearranges to these handy formulas:

$$L = \frac{25,330}{f^2 C} \quad (119)$$

$$C = \frac{25,330}{f^2 L} \quad (120)$$

where
f = frequency in MHz,
L = inductance in µH, and
C = capacitance in pF

For most radio work, these formulas will permit calculations of frequency and component values well within the limits of component tolerances.

Example: What value of capacitance is needed to create a resonant circuit at 21.1 MHz, if the inductor is 2.00 µH?

$$C = \frac{25,330}{f^2 L} = \frac{25,330}{(21.1^2\times2.0)}$$

$$= \frac{25,330}{890} = 28.5 \text{ pF}$$

Fig 2.55 can also be used if an approximate answer is acceptable. From the horizontal axis, find the vertical line closest to the desired resonant frequency. Every pair of diagonals that cross on that vertical line represent a combination of inductance and capacitance that will resonate at that frequency. For example, if the desired frequency is 10 MHz, the pair of diagonals representing 5 µH and 50 pF cross quite close to that frequency. Interpolating between the given diagonals will provide more resolution — remember that all three sets of lines are spaced logarithmically.

Resonant circuits have other properties of importance, in addition to the resonant frequency, however. These include impedance, voltage drop across components in series-resonant circuits, circulating current in parallel-resonant circuits, and bandwidth. These properties determine such factors as the selectivity of a tuned circuit and the component ratings for circuits handling significant amounts of power. Although the basic determination of the tuned-circuit resonant frequency ignored any resistance in the circuit, that resistance will play a vital role in the circuit's other characteristics.

2.13.1 Series-Resonant Circuits

Fig 2.79 presents a basic schematic diagram of a *series-resonant circuit*. Although most schematic diagrams of radio circuits would show only the inductor and the capacitor, resistance is always present in such circuits. The most notable resistance is associated with the series resistance losses in the inductor at HF. The dominant losses in the capacitor may be modeled as a parallel resistance (not shown), but these losses are low enough at HF to be ignored. The current meter shown in the circuit is a reminder that in series circuits, the same current flows through all elements.

At resonance, the reactance of the capacitor cancels the reactance of the inductor. The voltage and current are in phase with each other, and the impedance of the circuit is determined solely by the resistance. The actual current through the circuit at resonance, and for frequencies near resonance, is determined by the formula:

$$I = \frac{E}{Z}$$

$$= \frac{E}{\sqrt{R^2 + \left[2\pi f L - \frac{1}{(2\pi f C)}\right]^2}} \quad (121)$$

where all values are in basic units.

At resonance, the reactive factor in the formula is zero (the bracketed expression under the square root symbol). As the frequency is shifted above or below the resonant frequency without altering component values, however, the reactive factor becomes significant, and the value of the current becomes smaller than at resonance. At frequencies far from resonance, the reactive components become dominant, and the resistance no longer significantly affects the current amplitude.

The exact curve created by recording the current as the frequency changes depends on the ratio of reactance to resistance. When the reactance of either the coil or capacitor is of the same order of magnitude as the resistance, the current decreases rather slowly as the fre-

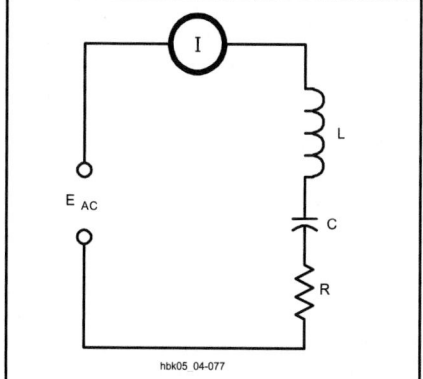

Fig 2.79 — A series circuit containing L, C, and R is resonant at the applied frequency when the reactance of C is equal to the reactance of L. The I in the circle is the schematic symbol for an ammeter.

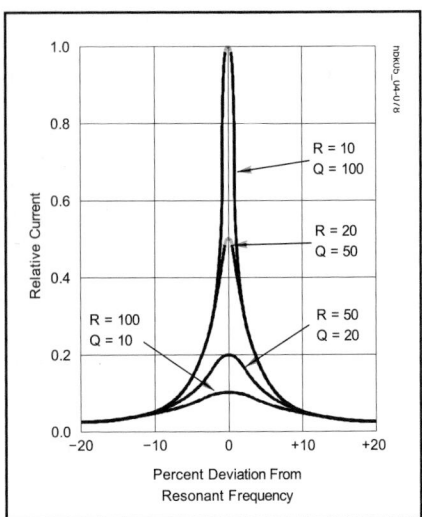

Fig 2.80 — Relative current in series-resonant circuits with various values of series resistance and Q. (An arbitrary maximum value of 1.0 represents current at resonance.) The reactance at resonance for all curves is 1000 Ω. Note that the current is hardly affected by the resistance in the circuit at frequencies more than 10% away from the resonant frequency.

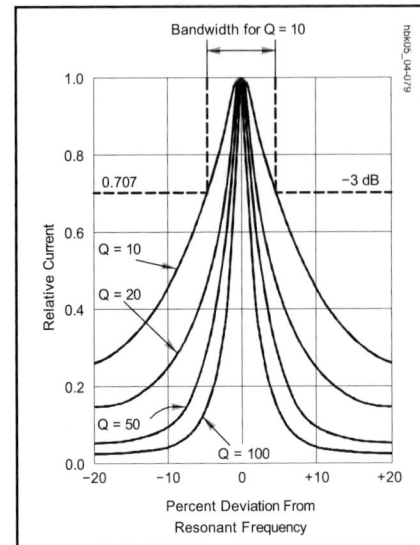

Fig 2.81 — Relative current in series-resonant circuits having different values of Q_U. The current at resonance is normalized to the same level for all curves in order to show the rate of change of decrease in current for each value of Q_U. The half-power points are shown to indicate relative bandwidth of the response for each curve. The bandwidth is indicated for a circuit with a Q_U of 10.

quency is moved in either direction away from resonance. Such a curve is said to be *broad*. Conversely, when the reactance is considerably larger than the resistance, the current decreases rapidly as the frequency moves away from resonance, and the circuit is said to be *sharp*. A sharp circuit will respond a great deal more readily to the resonant frequency than to frequencies quite close to resonance; a broad circuit will respond almost equally well to a group or band of frequencies centered around the resonant frequency.

Both types of resonance curves are useful. A sharp circuit gives good selectivity — the ability to respond strongly (in terms of current amplitude) at one desired frequency and to discriminate against others. A broad circuit is used when the apparatus must give about the same response over a band of frequencies, rather than at a single frequency alone.

Fig 2.80 presents a family of curves, showing the decrease in current as the frequency deviates from resonance. In each case, the inductive and capacitive reactances are assumed to be 1000 Ω. The maximum current, shown as a relative value on the graph, occurs with the lowest resistance, while the lowest peak current occurs with the highest resistance. Equally important, the rate at which the current decreases from its maximum value also changes with the ratio of reactance to resistance. It decreases most rapidly when the ratio is high and most slowly when the ratio is low.

UNLOADED Q

As noted in equation [a] of Table 2.10 earlier in this chapter, Q is the ratio of series reactance representing 2π times the stored energy (equation [c] in Table 2.10) to series resistance or consumed energy. Since both terms of the ratio are measured in ohms, Q has no units and is known as the *quality factor* (and less frequently, the *figure of merit* or the *multiplying factor*). The series resistive losses of the coil often dominate the energy consumption in HF series-resonant circuits, so the inductor Q largely determines the resonant-circuit Q. Since this value of Q is independent of any external load to which the circuit might transfer power, it is called the *unloaded Q* or Q_U of the circuit.

Example: What is the unloaded Q of a series-resonant circuit with a series loss resistance of 5 Ω and inductive and capacitive components having a reactance of 500 Ω each? With a reactance of 50 Ω each?

$$Q_{U1} = \frac{X1}{R} = \frac{500\ \Omega}{5\ \Omega} = 100$$

$$Q_{U2} = \frac{X2}{R} = \frac{50\ \Omega}{5\ \Omega} = 10$$

BANDWIDTH

Fig 2.81 is an alternative way of drawing the family of curves that relate current to frequency for a series-resonant circuit. By

assuming that the peak current of each curve is the same, the rate of change of current for various values of Q_U and the associated ratios of reactance to resistance are more easily compared. From the curves, it is evident that the lower Q_U circuits pass current across a greater *bandwidth* of frequencies than the circuits with a higher Q_U. For the purpose of comparing tuned circuits, bandwidth is often defined as the frequency spread between the two frequencies at which the current amplitude decreases to 0.707 (or $1/\sqrt{2}$) times the maximum value. Since the power consumed by the resistance, R, is proportional to the square of the current, the power at these points is half the maximum power at resonance, assuming that R is constant for the calculations. The half-power, or −3 dB, points are marked on Fig 2.81.

For Q values of 10 or greater, the curves shown in Fig 2.81 are approximately symmetrical. On this assumption, bandwidth (BW) can be easily calculated by inverting equation [d] in Table 2.10, and approximating the geometrical mean −3 dB frequency by f:

$$BW = \frac{f}{Q_U} \tag{122}$$

where BW and f are in the same units, that is, in Hz, kHz or MHz.

Example: What is the 3 dB bandwidth of a series-resonant circuit operating at 14 MHz with a Q_U of 100?

$$BW = \frac{f}{Q_U} = \frac{14\ MHz}{100} = 0.14\ MHz$$

$$= 140\ kHz$$

The relationship between Q_U, f and BW provides a means of determining the value of circuit Q when inductor losses may be difficult to measure. By constructing the series-resonant circuit and measuring the current as the frequency varies above and below resonance, the half-power points can be determined. Then:

$$Q_U = \frac{f}{BW} \tag{123}$$

Example: What is the Q_U of a series-resonant circuit operating at 3.75 MHz, if the −3 dB bandwidth is 375 kHz?

$$Q_U = \frac{f}{BW} = \frac{3.75\ MHz}{0.375\ MHz} = 10.0$$

Table 2.11 provides some simple formulas for estimating the maximum current and phase angle for various bandwidths, if both f and Q_U are known.

Table 2.11
The Selectivity of Resonant Circuits

Approximate percentage of current at resonance[1] or of impedance at resonance[2]	Bandwidth (between half-power or –3 dB points on response curve)	Series circuit current phase angle (degrees)
95	f / 3Q	18.5
90	f / 2Q	26.5
70.7	f / Q	45
44.7	2f / Q	63.5
24.2	4f / Q	76
12.4	8f / Q	83

[1]For a series resonant circuit
[2]For a parallel resonant circuit

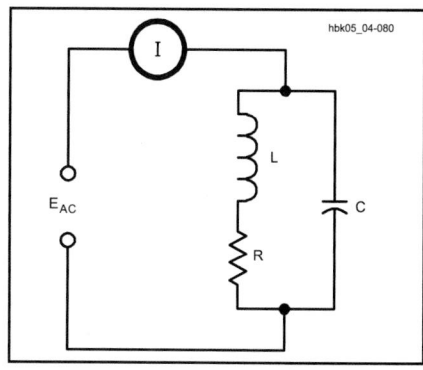

Fig 2.82 — A typical parallel-resonant circuit, with the resistance shown in series with the inductive leg of the circuit. Below a Q_U of 10, resonance definitions may lead to three separate frequencies which converge at higher Q_U levels. See text.

VOLTAGE DROP ACROSS COMPONENTS

The voltage drop across the coil and across the capacitor in a series-resonant circuit are each proportional to the reactance of the component for a given current (since E = I X). These voltages may be many times the applied voltage for a high-Q circuit. In fact, at resonance, the voltage drop is:

$$E_X = Q_U E_{AC} \qquad (124)$$

where

E_X = the voltage across the reactive component,
Q_U = the circuit unloaded Q, and
E_{AC} = the applied voltage in Fig 2.79.

(Note that the voltage drop across the inductor is the vector sum of the voltages across the resistance and the reactance; however, for Q greater than 10, the error created by using this is not ordinarily significant.) Since the calculated value of E_X is the RMS voltage, the peak voltage will be higher by a factor of 1.414. Antenna couplers and other high-Q circuits handling significant power may experience arcing from high values of E_X, even though the source voltage to the circuit is well within component ratings.

CAPACITOR LOSSES

Although capacitor energy losses tend to be insignificant compared to inductor losses up to about 30 MHz, the losses may affect circuit Q in the VHF range. Leakage resistance, principally in the solid dielectric that forms the insulating support for the capacitor plates, appears as a resistance in parallel with the capacitor plates. Instead of forming a series resistance, capacitor leakage usually forms a parallel resistance with the capacitive reactance. If the leakage resistance of a capacitor is significant enough to affect the Q of a series-resonant circuit, the parallel resistance (R_P) may be converted to an equivalent series resistance (R_S) before adding it to the inductor's resistance.

$$R_S = \frac{X_C^2}{R_P} = \frac{1}{R_P \times (2\,\pi\,f\,C)^2} \qquad (125)$$

Example: A 10.0 pF capacitor has a leakage resistance of 10,000 Ω at 50.0 MHz. What is the equivalent series resistance?

$$R_S = \frac{1}{R_P \times (2\,\pi\,f\,C)^2}$$

$$= \frac{1}{1.0 \times 10^4 \times (6.283 \times 50.0 \times 10^6 \times 10.0 \times 10^{-12})^2}$$

$$= \frac{1}{1.0 \times 10^4 \times 9.87 \times 10^{-6}}$$

$$= \frac{1}{0.0987} = 10.1\,\Omega$$

In calculating the impedance, current and bandwidth for a series-resonant circuit in which this capacitor might be used, the series-equivalent resistance of the unit is added to the loss resistance of the coil. Since inductor losses tend to increase with frequency because of skin effect in conductors, and capacitor dielectric losses also tend to increase with frequency, the combined losses in the capacitor and the inductor can seriously reduce circuit Q.

2.13.2 Parallel-Resonant Circuits

Although series-resonant circuits are common, the vast majority of resonant circuits used in radio work are *parallel-resonant circuits*. **Fig 2.82** represents a typical HF parallel-resonant circuit. As is the case for series-resonant circuits, the inductor is the chief source of resistive losses (that is, the parallel loss resistance across the capacitor is not shown), and these losses appear in series with the coil. Because current through parallel-resonant circuits is lowest at resonance, and impedance is highest, they are sometimes called *antiresonant* circuits. (You may encounter the old terms *acceptor* and *rejector*

referring to series- and parallel-resonant circuits, respectively.)

Because the conditions in the two legs of the parallel circuit in Fig 2.82 are not the same — the resistance is shown in only one of the legs — all of the conditions by which series resonance is determined do not occur simultaneously in a parallel-resonant circuit. **Fig 2.83** graphically illustrates the situation by showing the currents through the two components. (Currents are drawn in the manner of complex impedances shown previously to show the phase angle for each current.) When the inductive and capacitive reactances are identical, the condition defined for series resonance is met as shown at point (a). The impedance of the inductive leg is composed of both X_L and R, which yields an impedance greater than X_C and that is not 180° out of phase with X_C. The resultant current is greater than the minimum possible value and is not in phase with the voltage.

By altering the value of the inductor slightly (and holding the Q constant), a new frequency can be obtained at which the current reaches its minimum. When parallel circuits are tuned using a current meter as an indicator, this point (b) is ordinarily used as an indication of resonance. The current "dip" indicates a condition of maximum impedance and is sometimes called the *antiresonant* point or *maximum impedance resonance* to distinguish it from the condition at which $X_C = X_L$. Maximum impedance is achieved at this point by vector addition of X_C, X_L and R, however, and the result is a current somewhat out of phase with the voltage.

Point (c) in the figure represents the *unity-power-factor* resonant point. Adjusting the inductor value and hence its reactance (while holding Q constant) produces a new resonant frequency at which the resultant current is in phase with the voltage. The new value of

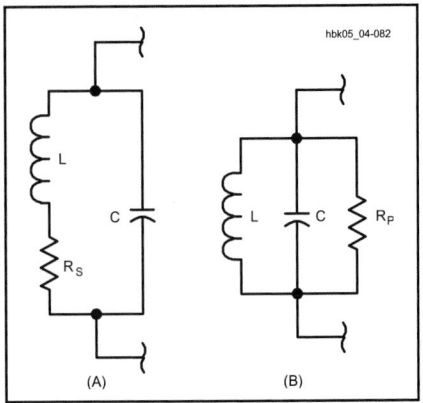

Fig 2.84 — Series and parallel equivalents when both circuits are resonant. The series resistance, R_S in A, is replaced by the parallel resistance, R_p in B, and vice versa. $R_P = X_L^2 / R_S$.

inductive reactance is the value required for a parallel-equivalent inductor and its parallel-equivalent resistor (calculated according to the formulas in the last section) to just cancel the capacitive reactance. The value of the parallel-equivalent inductor is always smaller than the actual inductor in series with the resistor and has a proportionally smaller reactance. (The parallel-equivalent resistor, conversely, will always be larger than the

coil-loss resistor shown in series with the inductor.) The result is a resonant frequency slightly different from the one for minimum current and the one for $X_L = X_C$.

The points shown in the graph in Fig 2.83 represent only one of many possible situations, and the relative positions of the three resonant points do not hold for all possible cases. Moreover, specific circuit designs can draw some of the resonant points together, for example, compensating for the resistance of the coil by retuning the capacitor. The differences among these resonances are significant for circuit Q below 10, where the inductor's series resistance is a significant percentage of the reactance. Above a Q of 10, the three points converge to within a percent of the frequency and the differences between them can be ignored for practical calculations. Tuning for minimum current will not introduce a sufficiently large phase angle between voltage and current to create circuit difficulties.

PARALLEL CIRCUITS OF MODERATE TO HIGH Q

The resonant frequencies defined above converge in parallel-resonant circuits with Q higher than about 10. Therefore, a single set of formulas will sufficiently approximate circuit performance for accurate predictions. Indeed, above a Q of 10, the performance

of a parallel circuit appears in many ways to be simply the inverse of the performance of a series-resonant circuit using the same components.

Accurate analysis of a parallel-resonant circuit requires the substitution of a parallel-equivalent resistor for the actual inductor-loss series resistor, as shown in **Fig 2.84**. Sometimes called the *dynamic resistance* of the parallel-resonant circuit, the parallel-equivalent resistor value will increase with circuit Q, that is, as the series resistance value decreases. To calculate the approximate parallel-equivalent resistance, use the formula:

$$R_P = \frac{X_L^{\,2}}{R_S} = \frac{(2\,\pi\,f\,L)^2}{R_S} = Q_U X_L \quad (126)$$

for $R_S \ll X_C \ll R_P$ and $X_P \approx X_S$ in equations 94 through 97.

Example: What is the parallel-equivalent resistance for a coil with an inductive reactance of 350 Ω and a series resistance of 5.0 Ω at resonance?

$$R_P = \frac{X_L^{\,2}}{R_S} = \frac{(350\,\Omega)^2}{5.0\,\Omega}$$

$$= \frac{122{,}500\,\Omega^2}{5.0\,\Omega} = 24{,}500\,\Omega$$

Since the coil Q_U remains the inductor's reactance divided by its series resistance, the coil Q_U is 70. Multiplying Q_U by the reactance also provides the approximate parallel-equivalent resistance of the coil series resistance.

At resonance, where $X_L = X_C$, R_P defines the impedance of the parallel-resonant circuit. The reactances just equal each other, leaving the voltage and current in phase with each other. In other words, the circuit shows only the parallel resistance. Therefore, equation 126 can be rewritten as:

$$Z = \frac{X_L^{\,2}}{R_S} = \frac{(2\,\pi\,f\,L)^2}{R_S} = Q_U X_L \quad (127)$$

In this example, the circuit impedance at resonance is 24,500 Ω.

At frequencies below resonance the current through the inductor is larger than that through the capacitor, because the reactance of the coil is smaller and that of the capacitor is larger than at resonance. There is only partial cancellation of the two reactive currents, and the total current therefore is larger than the current taken by the resistance alone. At frequencies above resonance the situation is reversed and more current flows through the capacitor than through the inductor, so the total current again increases.

The current at resonance, being determined wholly by R_P, will be small if R_P is large, and large if R_P is small. **Fig 2.85** illustrates the relative current flows through a parallel-tuned circuit as the frequency is moved from below

Fig 2.83 — Resonant conditions for a low- Q_U parallel circuit. Resonance may be defined as (a) $X_L = X_C$ (b) minimum current flow and maximum impedance or (c) voltage and current in phase with each other. With the circuit of Fig 2.82 and a Q_U of less than 10, these three definitions may represent three distinct frequencies.

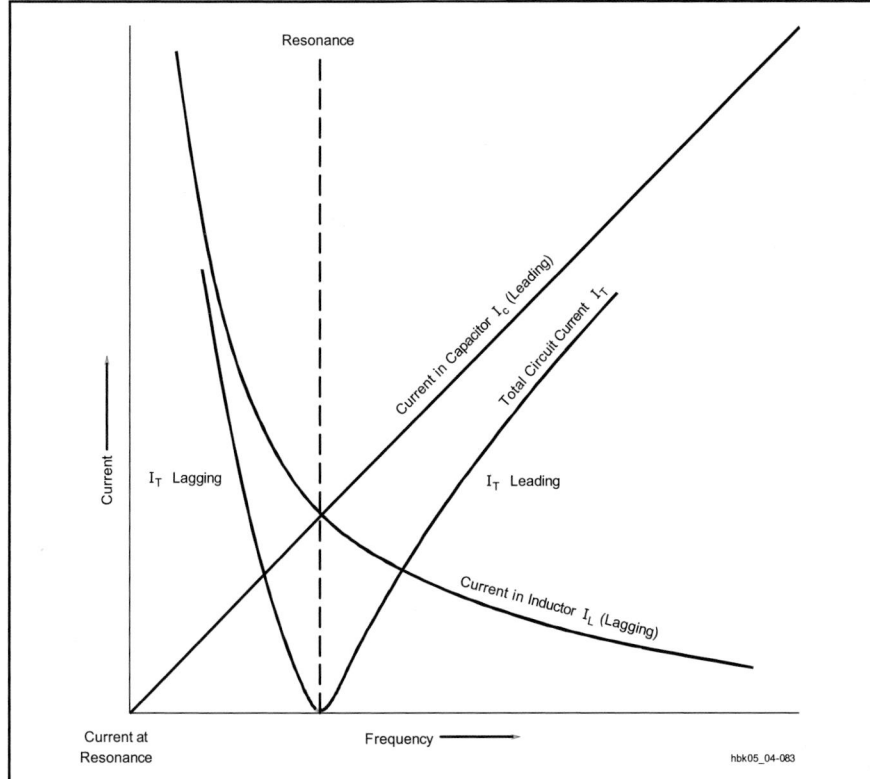

Fig 2.85 — The currents in a parallel-resonant circuit as the frequency moves through resonance. Below resonance, the current lags the voltage; above resonance the current leads the voltage. The base line represents the current level at resonance, which depends on the impedance of the circuit at that frequency.

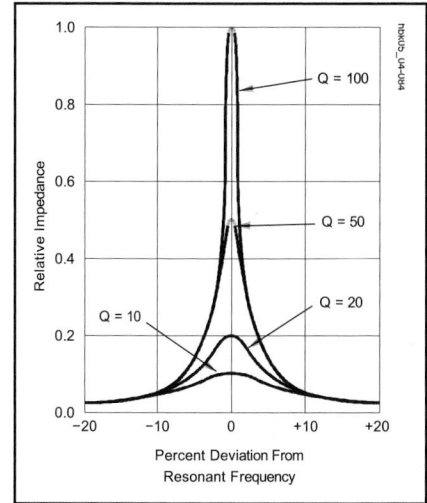

Fig 2.86 — Relative impedance of parallel-resonant circuits with different values of Q_U. The curves are similar to the series-resonant circuit current level curves of Fig 2.80. The effect of Q_U on impedance is most pronounced within 10% of the resonance frequency.

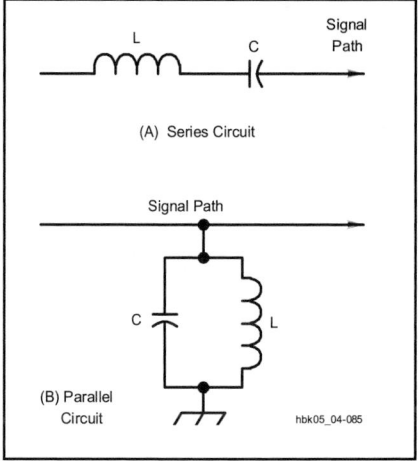

Fig 2.87 — Series- and parallel-resonant circuits configured to perform the same theoretical task: passing signals in a narrow band of frequencies along the signal path. A real design example would consider many other factors.

resonance to above resonance. The base line represents the minimum current level for the particular circuit. The actual current at any frequency off resonance is simply the vector sum of the currents through the parallel equivalent resistance and through the reactive components.

To obtain the impedance of a parallel-tuned circuit either at or off the resonant frequency, apply the general formula:

$$Z = \frac{Z_C Z_L}{Z_S} \qquad (128)$$

where

Z = overall circuit impedance

Z_C = impedance of the capacitive leg (usually, the reactance of the capacitor),

Z_L = impedance of the inductive leg (the vector sum of the coil's reactance and resistance), and

Z_S = series impedance of the capacitor-inductor combination as derived from the denominator of equation 121.

After using vector calculations to obtain Z_L and Z_S, converting all the values to polar form — as described earlier in this chapter — will ease the final calculation. Of course, each impedance may be derived from the resistance and the application of the basic

reactance formulas on the values of the inductor and capacitor at the frequency of interest.

Since the current rises away from resonance, the parallel-resonant-circuit impedance must fall. It also becomes complex, resulting in an ever-greater phase difference between the voltage and the current. The rate at which the impedance falls is a function of Q_U. Fig 2.86 presents a family of curves showing the impedance drop from resonance for circuit Q ranging from 10 to 100. The curve family for parallel-circuit impedance is essentially the same as the curve family for series-circuit current.

As with series-resonant circuits, the higher the Q of a parallel-tuned circuit, the sharper will be the response peak. Likewise, the lower the Q, the wider the band of frequencies to which the circuit responds. Using the half-power (–3 dB) points as a comparative measure of circuit performance, equations 122 and 123 apply equally to parallel-tuned circuits. That is, BW = f / Q_U and Q_U = f / BW, where the resonant frequency and the bandwidth are in the same units. As a handy reminder, Table 2.12 summarizes the performance of parallel-resonant circuits at high and low Q and above and below resonant frequency.

It is possible to use either series- or parallel-

resonant circuits to do the same work in many circuits, thus giving the designer considerable flexibility. Fig 2.87 illustrates this general principle by showing a series-resonant circuit in the signal path and a parallel-resonant circuit shunted from the signal path to ground. Assume both circuits are resonant at the same frequency, f, and have the same Q. The series-resonant circuit at A has its lowest impedance at f, permitting the maximum possible current to flow along the signal path. At all other frequencies, the impedance is greater and the current at those frequencies is less. The circuit passes the desired signal and tends to impede

Table 2.12

The Performance of Parallel-Resonant Circuits

A. High- and Low-Q Circuits (in relative terms)

Characteristic	High-Q Circuit	Low-Q Circuit
Selectivity	high	low
Bandwidth	narrow	wide
Impedance	high	low
Total current	low	high
Circulating current	high	low

B. Off-Resonance Performance for Constant Values of Inductance and Capacitance

Characteristic	Above Resonance	Below Resonance
Inductive reactance	increases	decreases
Capacitive reactance	decreases	increases
Circuit resistance	unchanged*	unchanged*
Relative impedance	decreases	decreases
Total current	increases	increases
Circulating current	decreases	decreases
Circuit impedence	capacitive	inductive

*This is true for frequencies near resonance. At distant frequencies, skin effect may alter the resistive losses of the inductor.

signals at undesired frequencies. The parallel circuit at B provides the highest impedance at resonance, f, making the signal path the lowest impedance path for the signal. At frequencies off resonance, the parallel-resonant circuit presents a lower impedance, thus presenting signals with a path to ground and away from the signal path. In theory, the effects will be the same relative to a signal current on the signal path. In actual circuit design exercises, of course, many other variables will enter the design picture to make one circuit preferable to the other.

CIRCULATING CURRENT

In a parallel-resonant circuit, the source voltage is the same for all the circuit elements. The current in each element, however, is a function of the element's reactance. **Fig 2.88** redraws the parallel-resonant circuit to indicate the total current and the current circulating between the coil and the capacitor. The current drawn from the source may be low, because the overall circuit impedance is high. The current through the individual elements may be high, however, because there is little resistive loss as the current circulates through the inductor and capacitor. For parallel-resonant circuits with an unloaded Q of 10 or greater, this *circulating current* is approximately:

$$I_C = Q_U I_T \qquad (129)$$

where
I_C = circulating current in A, mA or µA,
Q_U = unloaded circuit Q, and
I_T = total current in the same units as I_C.

Example: A parallel-resonant circuit permits an ac or RF total current of 30 mA and has a Q of 100. What is the circulating current through the elements?

$$I_X = Q_U I = 100 \times 30 \text{ mA} = 3000 \text{ mA} = 3 \text{ A}$$

Circulating currents in high-Q parallel-tuned circuits can reach a level that causes component heating and power loss. Therefore, components should be rated for the anticipated circulating currents, and not just the total current.

LOADED Q

In many resonant-circuit applications, the only power lost is that dissipated in the resistance of the circuit itself. At frequencies below 30 MHz, most of this resistance is in the coil. Within limits, increasing the number

of turns in the coil increases the reactance faster than it raises the resistance, so coils for circuits in which the Q must be high are made with relatively large inductances for the frequency.

When the circuit delivers energy to a load (as in the case of the resonant circuits used in transmitters), the energy consumed in the circuit itself is usually negligible compared with that consumed by the load. The equivalent of such a circuit is shown in **Fig 2.89**, where the parallel resistor, R_L, represents the load to which power is delivered. If the power dissipated in the load is at least 10 times as great as the power lost in the inductor and capacitor, the parallel impedance of the resonant circuit itself will be so high compared with the resistance of the load that for all practical purposes the impedance of the combined circuit is equal to the load impedance. Under these conditions, the load resistance replaces the circuit impedance in calculating Q. The Q of a parallel-resonant circuit loaded by a resistive impedance is:

$$Q_L = \frac{R_L}{X} \qquad (130)$$

where
Q_L = circuit loaded Q,
R_L = parallel load resistance in ohms, and
X = reactance in ohms of either the inductor or the capacitor.

Example: A resistive load of 3000 Ω is connected across a resonant circuit in which the inductive and capacitive reactances are each 250 Ω. What is the circuit Q?

$$Q_L = \frac{R_L}{X} = \frac{3000 \ \Omega}{250 \ \Omega} = 12$$

The effective Q of a circuit loaded by a parallel resistance increases when the reactances are decreased. A circuit loaded with a relatively low resistance (a few thousand ohms) must have low-reactance elements

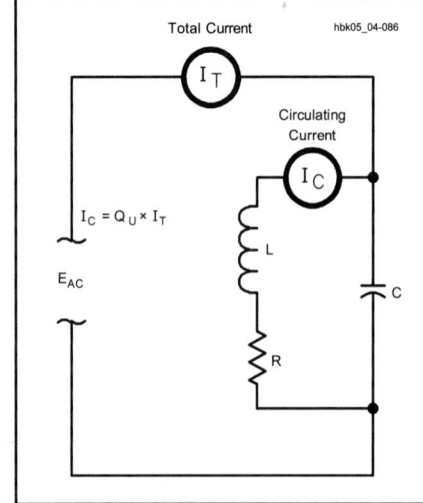

Fig 2.88 — A parallel-resonant circuit redrawn to illustrate both the total current and the circulating current.

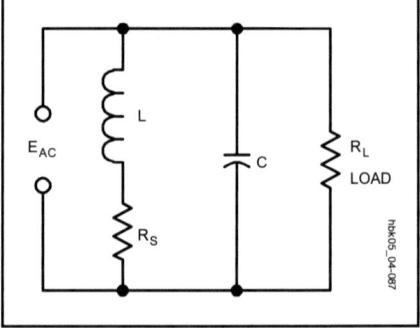

Fig 2.89 — A loaded parallel-resonant circuit, showing both the inductor-loss resistance and the load, R_L. If smaller than the inductor resistance, R_L will control the loaded Q of the circuit (Q_L).

(large capacitance and small inductance) to have reasonably high Q. Many power-handling circuits, such as the output networks of transmitters, are designed by first choosing a loaded Q for the circuit and then determining component values. See the chapter on **RF Power Amplifiers** for more details.

Parallel load resistors are sometimes added to parallel-resonant circuits to lower the circuit Q and increase the circuit bandwidth. By using a high-Q circuit and adding a parallel resistor, designers can tailor the circuit response to their needs. Since the parallel resistor consumes power, such techniques ordinarily apply to receiver and similar low-power circuits, however.

Example: Specifications call for a parallel-resonant circuit with a bandwidth of 400 kHz at 14.0 MHz. The circuit at hand has a Q_U of 70.0 and its components have reactances of 350 Ω each. What is the parallel load resistor that will increase the bandwidth to the specified value? The bandwidth of the existing circuit is:

$$BW = \frac{f}{Q_U} = \frac{14.0 \text{ MHz}}{70.0} = 0.200 \text{ MHz}$$

$$= 200 \text{ kHz}$$

The desired bandwidth, 400 kHz, requires a circuit with a Q of:

$$Q = \frac{f}{BW} = \frac{14.0 \text{ MHz}}{0.400 \text{ MHz}} = 35.0$$

Since the desired Q is half the original value, halving the resonant impedance or parallel-resistance value of the circuit is in order. The present impedance of the circuit is:

$$Z = Q_U X_L = 70.0 \times 350 \text{ } \Omega = 24500 \text{ } \Omega$$

The desired impedance is:

$$Z = Q_U X_L = 35.0 \times 350 \text{ } \Omega$$

$$= 12250 \text{ } \Omega = 12.25 \text{ k}\Omega$$

or half the present impedance.

A parallel resistor of 24,500 Ω, or the near-

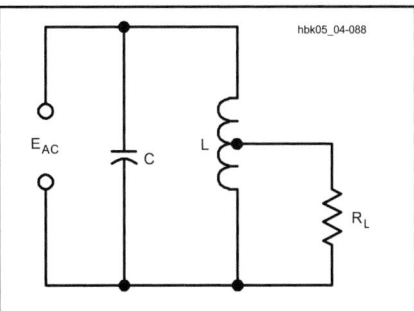

Fig 2.90 — A parallel-resonant circuit with a tapped inductor to effect an impedance match. Although the impedance presented to E_{AC} is very high, the impedance at the connection of the load, R_L, is lower.

est lower value (to guarantee sufficient bandwidth), will produce the required reduction in Q and bandwidth increase. Although this example simplifies the situation encountered in real design cases by ignoring such factors as the shape of the band-pass curve, it illustrates the interaction of the ingredients that determine the performance of parallel-resonant circuits.

IMPEDANCE TRANSFORMATION

An important application of the parallel-resonant circuit is as an impedance matching device. Circuits and antennas often need to be connected to other circuits or feed lines that do not have the same impedance. To transfer power effectively requires a circuit that will convert or "transform" the impedances so that each connected device or system can operate properly.

Fig 2.90 shows such a situation where the source, E_{AC}, operates at a high impedance, but the load, R_L, operates at a low impedance. The technique of impedance transformation shown in the figure is to connect the parallel-resonant circuit, which has a high impedance, across the source, but connect the load across only a portion of the coil. (This is called *tapping the coil* and the connection point is a *tap*.) The coil acts as an *autotransformer*, described in the following section, with the magnetic field of the coil shared between what are ef-

fectively two coils in series, the upper coil having many turns and the lower coil fewer turns. Energy stored in the field induces larger voltages in the many-turn coil than it does in the fewer-turn coil, "stepping down" the input voltage so that energy can be extracted by the load at the required lower voltage-to-current ratio (which is impedance). The correct tap point on the coil usually has to be experimentally determined, but the technique is very effective.

When the load resistance has a very low value (say below 100 Ω) it may be connected in series in the resonant circuit (such as R_S in Fig 2.84A, for example), in which case the series L-R circuit can be transformed to an equivalent parallel L-R circuit as previously described. If the Q is at least 10, the equivalent parallel impedance is:

$$Z_R = \frac{X^2}{R_L} \qquad (131)$$

where

Z_R = resistive parallel impedance at resonance,

X = reactance (in ohms) of either the coil or the capacitor, and

R_L = load resistance inserted in series.

If the Q is lower than 10, the reactance will have to be adjusted somewhat — for the reasons given in the discussion of low-Q parallel resonant circuits — to obtain a resistive impedance of the desired value.

These same techniques work in either "direction" — with a high-impedance source and low-impedance load or vice versa. Using a parallel-resonant circuit for this application does have some disadvantages. For instance, the common connection between the input and the output provides no dc isolation. Also, the common ground is sometimes troublesome with regard to ground-loop currents. Consequently, a circuit with only mutual magnetic coupling is often preferable. With the advent of ferrites, constructing impedance transformers that are both broadband and permit operation well up into the VHF portion of the spectrum has become relatively easy. The basic principles of broadband impedance transformers appear in the **RF Techniques** chapter.

2.14 Transformers

When the ac source current flows through every turn of an inductor, the generation of a counter-voltage and the storage of energy during each half cycle is said to be by virtue of self-inductance. If another inductor — not connected to the source of the original current — is positioned so the magnetic field of the first inductor intercepts the turns of the second inductor, coupling the two inductors and creating mutual inductance as described earlier, a voltage will be induced and current will flow in the second inductor. A load such as a resistor may be connected across the second inductor to consume the energy transferred magnetically from the first inductor.

Fig 2.91 illustrates a pair of coupled inductors, showing an ac energy source connected to one, called the *primary inductor*, and a load connected to the other, called the *secondary inductor*. If the inductors are wound tightly on a magnetic core so that nearly all or magnetic flux from the first inductor intersects with the turns of the second inductor, the pair is said to be *tightly coupled*. Inductors not sharing a common core and separated by a distance would be *loosely coupled*.

The signal source for the primary inductor may be household ac power lines, audio or other waveforms at low frequencies, or RF currents. The load may be a device needing power, a speaker converting electrical energy into sonic energy, an antenna using RF energy for communications or a particular circuit set up to process a signal from a preceding circuit. The uses of magnetically coupled energy in electronics are innumerable.

Mutual inductance (M) between inductors is measured in henrys. Two inductors have a mutual inductance of 1 H under the following conditions: as the primary inductor current changes at a rate of 1 A/s, the voltage across the secondary inductor is 1 V. The level of mutual inductance varies with many factors: the size and shape of the inductors, their relative positions and distance from each other, and the permeability of the inductor core material and of the space between them.

If the self-inductance values of two induc-

tors are known (self-inductance is used in this section to distinguish it from the mutual inductance), it is possible to derive the mutual inductance by way of a simple experiment schematically represented in **Fig 2.92**. Without altering the physical setting or position of two inductors, measure the total coupled inductance, L_C, of the series-connected inductors with their windings complementing each other and again with their windings opposing each other. Since, for the two inductors, $L_C = L1 + L2 + 2M$, in the complementary case, and $L_O = L1 + L2 - 2M$ for the opposing case,

$$M = \frac{L_C - L_O}{4} \qquad (132)$$

The ratio of magnetic flux set up by the secondary inductor to the flux set up by the primary inductor is a measure of the extent to which two inductors are coupled, compared to the maximum possible coupling between them. This ratio is the *coefficient of coupling* (k) and is always less than 1. If k were to equal 1, the two inductors would have the maximum possible mutual coupling. Thus:

$$M = k \sqrt{L1\,L2} \qquad (133)$$

where

    M = mutual inductance in henrys,
    L1 and L2 = individual coupled inductors, each in henrys, and
    k = the coefficient of coupling.

Using the experiment above, it is possible to solve equation 133 for k with reasonable accuracy.

Any two inductors having mutual inductance comprise a *transformer* having a *primary winding* or inductor and a *secondary winding* or inductor. The word "winding" is generally dropped so that transformers are said to have "primaries" and "secondaries." **Fig 2.93** provides a pictorial representation of a typical iron-core transformer, along with the schematic symbols for both iron-core and air-core transformers. Conventionally, the term *transformer* is most commonly applied to coupled inductors having a magnetic core material, while coupled air-wound inductors are not called by that name. They

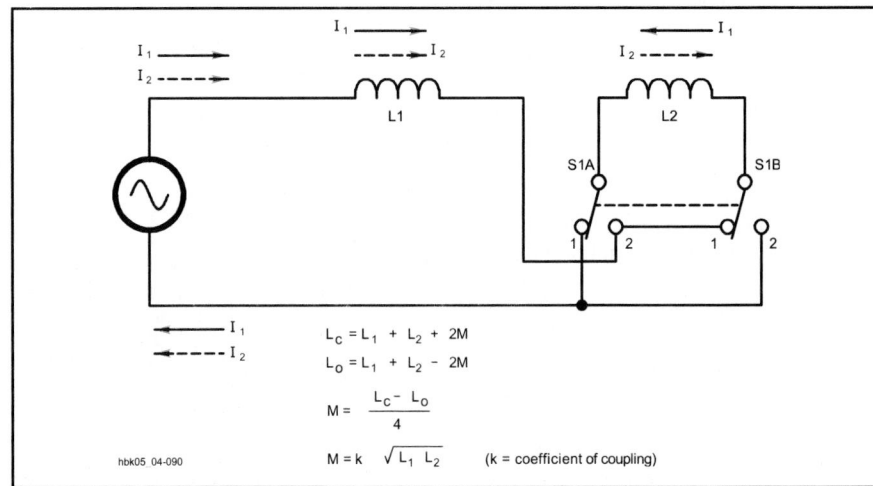

Fig 2.92 — An experimental setup for determining mutual inductance. Measure the inductance with the switch in each position and use the formula in the text to determine the mutual inductance.

$$L_C = L_1 + L_2 + 2M$$
$$L_O = L_1 + L_2 - 2M$$
$$M = \frac{L_C - L_O}{4}$$
$$M = k \sqrt{L_1 L_2} \quad (k = \text{coefficient of coupling})$$

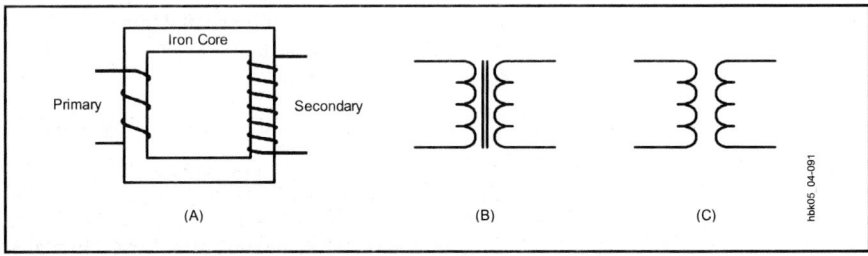

Fig 2.93 — A transformer. A is a pictorial diagram. Power is transferred from the primary coil to the secondary by means of the magnetic field. B is a schematic diagram of an iron-core transformer, and C is an air-core transformer.

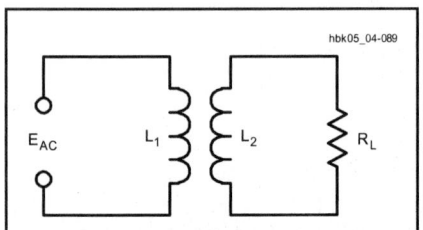

Fig 2.91 — A basic transformer: two inductors — one connected to an ac energy source, the other to a load — with coupled magnetic fields.

are still transformers, however.

We normally think of transformers as ac devices, since mutual inductance only occurs when magnetic fields are changing. A transformer connected to a dc source will exhibit mutual inductance only at the instants of closing and opening the primary circuit, or on the rising and falling edges of dc pulses, because only then does the primary winding have a changing field. There are three principle uses of transformers: to physically isolate the primary circuit from the secondary circuit, to transform voltages and currents from one level to another, and to transform circuit impedances from one level to another. These functions are not mutually exclusive and have many variations.

2.14.1 Basic Transformer Principles

The primary and secondary windings of a transformer may be wound on a core of magnetic material. The permeability of the magnetic material increases the inductance of the windings so a relatively small number of turns may be used to induce a given voltage value with a small current. A closed core having a continuous magnetic path, such as that shown in Fig 2.93, also tends to ensure that practically all of the field set up by the current in the primary winding will intercept or "cut" the turns of the secondary winding.

For power transformers and impedance-matching transformers used at audio frequencies, cores made of soft iron strips or sheets called *laminations* are most common and generally very efficient. At higher frequencies, ferrite or powdered-iron cores are more frequently used. This section deals with basic transformer operation at audio and power frequencies. RF transformer operation is discussed in the chapter on **RF Techniques**.

The following principles presume a coefficient of coupling (k) of 1, that is, a perfect transformer. The value k = 1 indicates that all the turns of both windings link with all the magnetic flux lines, so that the voltage induced per turn is the same with both windings. This condition makes the induced voltage independent of the inductance of the primary and secondary inductors. Iron-core transformers for low frequencies closely approach this ideal condition. **Fig 2.94** illustrates the conditions for transformer action.

VOLTAGE RATIO

For a given varying magnetic field, the voltage induced in an inductor within the field is proportional to the number of turns in the inductor. When the two windings of a transformer are in the same field (which is the case when both are wound on the same closed core), it follows that the induced voltages will

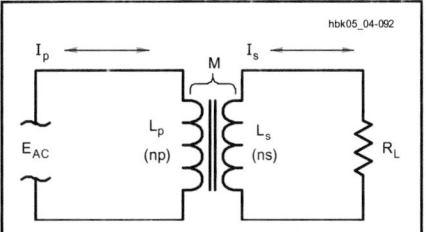

Fig 2.94 — The conditions for transformer action: two coils that exhibit mutual inductance, an ac power source, and a load. The magnetic field set up by the energy in the primary circuit transfers energy to the secondary for use by the load, resulting in a secondary voltage and current.

be proportional to the number of turns in each winding. In the primary, the induced voltage practically equals, and opposes, the applied voltage, as described earlier. Hence:

$$E_S = E_P \left(\frac{N_S}{N_P} \right) \tag{134}$$

where

E_S = secondary voltage,
E_P = primary applied voltage,
N_S = number of turns on secondary, and
N_P = number of turns on primary.

Example: A transformer has a primary with 400 turns and a secondary with 2800 turns, and a voltage of 120 V is applied to the primary. What voltage appears across the secondary winding?

$$E_S = 120 \text{ V} \left(\frac{2800}{400} \right) = 120 \text{ V} \times 7 = 840 \text{ V}$$

(Notice that the number of turns is taken as a known value rather than a measured quantity, so they do not limit the significant figures in the calculation.) Also, if 840 V is applied to the 2800-turn winding (which then becomes the primary), the output voltage from the 400-turn winding will be 120 V.

Either winding of a transformer can be used as the primary, provided the winding has enough turns (enough inductance) to induce a voltage equal to the applied voltage without requiring an excessive current. The windings must also have insulation with a voltage rating sufficient for the voltages applied or created. Transformers are called *step-up* or *step-down* transformers depending on whether the secondary voltage is higher or lower than the primary voltage, respectively.

CURRENT OR AMPERE-TURNS RATIO

The current in the primary when no current is taken from the secondary is called the *magnetizing current* of the transformer. An ideal transformer, with no internal losses, would consume no power, since the current

through the primary inductor would be 90° out of phase with the voltage. In any properly designed transformer, the power consumed by the transformer when the secondary is open (not delivering power) is only the amount necessary to overcome the losses in the iron core and in the resistance of the wire with which the primary is wound.

When power is transferred from the secondary winding to a load, the secondary current creates a magnetic field that opposes the field established by the primary current. For the induced voltage in the primary to equal the applied voltage, the original magnetizing field must be maintained. Therefore, enough additional current must flow in the primary to create a field exactly equal and opposite to the field set up by the secondary current, leaving the original magnetizing field.

In practical transformer calculations it may be assumed that the entire primary current is caused by the secondary load. This is justifiable because the magnetizing current should be very small in comparison with the primary load current at rated power output.

If the magnetic fields set up by the primary and secondary currents are to be equal, the number of ampere-turns must be equal in each winding. (See the previous discussion of magnetic fields and magnetic flux density.) Thus, primary current multiplied by the primary turns must equal the secondary current multiplied by the secondary turns.

$$I_P N_P = I_S N_S$$

and

$$I_P = I_S \left(\frac{N_S}{N_P} \right) \tag{135}$$

where

I_P = primary current,
I_S = secondary current,
N_P = number of turns in the primary winding, and
N_S = number of turns in the secondary winding.

Example: Suppose the secondary of the transformer in the previous example is delivering a current of 0.20 A to a load. What will be the primary current?

$$I_P = 0.20 \text{ A} \left(\frac{2800}{400} \right) = 0.20 \text{ A} \times 7 = 1.4 \text{ A}$$

Although the secondary voltage is higher than the primary voltage, the secondary current is lower than the primary current, and by the same ratio. The secondary current in an ideal transformer is 180° out of phase with the primary current, since the field in the secondary just offsets the field in the primary. The phase relationship between the currents in the windings holds true no matter what the phase difference between the current and the voltage of the secondary. In fact, the phase difference, if any, between voltage and current in the secondary winding will be *reflected* back to the primary as an identical phase difference.

Power Ratio

A transformer cannot create power; it can only transfer it and change the voltage and current ratios. Hence, the power taken from the secondary cannot exceed that taken by the primary from the applied voltage source. There is always some power loss in the resistance of the windings and in the iron core, so in all practical cases the power taken from the source will exceed that taken from the secondary.

$$P_O = \eta P_I \qquad (136)$$

where

P_O = power output from secondary,
P_I = power input to primary, and
η = efficiency factor.

The efficiency, η, is always less than 1 and is commonly expressed as a percentage: if η is 0.65, for instance, the efficiency is 65%.

Example: A transformer has an efficiency of 85.0% at its full-load output of 150 W. What is the power input to the primary at full secondary load?

$$P_I = \frac{P_O}{\eta} = \frac{150 \text{ W}}{0.850} = 176 \text{ W}$$

A transformer is usually designed to have the highest efficiency at the power output for which it is rated. The efficiency decreases with either lower or higher outputs. On the other hand, the losses in the transformer are relatively small at low output but increase as more power is taken. The amount of power that the transformer can handle is determined by its own losses, because these losses heat the wire and core. There is a limit to the temperature rise that can be tolerated, because too high a temperature can either melt the wire or cause the insulation to break down. A transformer can be operated at reduced output, even though the efficiency is low, because the actual loss will be low under such conditions. The full-load efficiency of small power transformers such as are used in radio receivers and transmitters usually lies between about 60 and 90%, depending on the size and design.

IMPEDANCE RATIO

In an ideal transformer — one without losses or leakage inductance (see Transformer Losses) — the primary power, $P_P = E_P I_P$, and secondary power, $P_S = E_S I_S$, are equal. The relationships between primary and secondary voltage and current are also known (equations 134 and 135). Since impedance is the ratio of voltage to current, $Z = E/I$, the impedances represented in each winding are related as follows:

$$Z_P = Z_S \left(\frac{N_P}{N_S} \right)^2 \qquad (137)$$

where

Z_P = impedance at the primary terminals from the power source,
Z_S = impedance of load connected to secondary, and
N_P/N_S = turns ratio, primary to secondary.

A load of any given impedance connected to the transformer secondary will thus be transformed to a different value at the primary terminals. The impedance transformation is proportional to the square of the primary-to-secondary turns ratio. (Take care to use the primary-to-secondary turns ratio, since the secondary-to-primary ratio is more commonly used to determine the voltage transformation ratio.)

The term *looking into* is often used to mean the conditions observed from an external perspective at the terminals specified. For example, "impedance looking into" the transformer primary means the impedance measured externally to the transformer at the terminals of the primary winding.

Example: A transformer has a primary-to-secondary turns ratio of 0.6 (the primary has six-tenths as many turns as the secondary) and a load of 3000 Ω is connected to the secondary. What is the impedance looking into the primary of the transformer?

$$Z_P = 3000 \ \Omega \times (0.6)^2 = 3000 \ \Omega \times 0.36$$

$$Z_P = 1080 \ \Omega$$

By choosing the proper turns ratio, the impedance of a fixed load can be transformed to any desired value, within practical limits. If transformer losses can be neglected, the transformed (reflected) impedance has the same phase angle as the actual load impedance. Thus, if the load is a pure resistance, the load presented by the primary to the power source will also be a pure resistance. If the load impedance is complex, that is, if the load current and voltage are out of phase with each other, then the primary voltage and current will have the same phase angle.

Many devices or circuits require a specific value of load resistance (or impedance) for optimum operation. The impedance of the actual load that is to dissipate the power may be quite different from the impedance of the source device or circuit, so an *impedance-matching transformer* is used to change the actual load into an impedance of the desired value. The turns ratio required is:

$$\frac{N_P}{N_S} = \sqrt{\frac{Z_P}{Z_S}} \qquad (138)$$

where

N_P / N_S = required turns ratio, primary to secondary,

Z_P = primary impedance required, and
Z_S = impedance of load connected to secondary.

Example: A transistor audio amplifier requires a load of 150 Ω for optimum performance, and is to be connected to a loudspeaker having an impedance of 4.0 Ω. What primary-to-secondary turns ratio is required in the coupling transformer?

$$\frac{N_P}{N_S} = \sqrt{\frac{Z_P}{Z_S}} = \frac{N_P}{N_S} \sqrt{\frac{150 \ \Omega}{4.0 \ \Omega}} = \sqrt{38} = 6.2$$

The primary therefore must have 6.2 times as many turns as the secondary.

These relationships may be used in practical circuits even though they are based on an ideal transformer. Aside from the normal design requirements of reasonably low internal losses and low leakage reactance, the only other requirement is that the primary have enough inductance to operate with low magnetizing current at the voltage applied to the primary.

The primary terminal impedance of an iron-core transformer is determined wholly by the load connected to the secondary and by the turns ratio. If the characteristics of the transformer have an appreciable effect on the impedance presented to the power source, the transformer is either poorly designed or is not suited to the voltage and frequency at which it is being used. Most transformers will operate quite well at voltages from slightly above to well below the design figure.

TRANSFORMER LOSSES

In practice, none of the formulas given so far provides truly exact results, although they afford reasonable approximations. Transformers in reality are not simply two coupled inductors, but a network of resistances and reactances, most of which appear in **Fig 2.95**. Since only the terminals numbered 1 through 4 are accessible to the user, transformer ratings and specifications take into account the additional losses created by these complexities.

In a practical transformer not all of the magnetic flux is common to both windings, although in well-designed transformers the amount of flux that cuts one winding and not the other is only a small percentage of the total flux. This *leakage flux* causes a voltage by self-induction in the winding creating the flux. The effect is the same as if a small *leakage inductance* existed independently of the main windings. Leakage inductance acts in exactly the same way as inductance inserted in series with the winding. It has, therefore, a certain reactance, depending on the amount of leakage inductance and the frequency. This reactance is called *leakage reactance* and is shown as X_{L1} and X_{L2} in Fig 2.95.

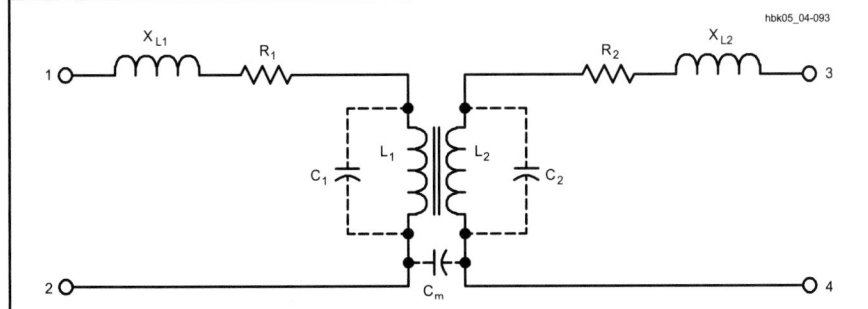

Fig 2.95 — A transformer as a network of resistances, inductances and capacitances. Only L1 and L2 contribute to the transfer of energy.

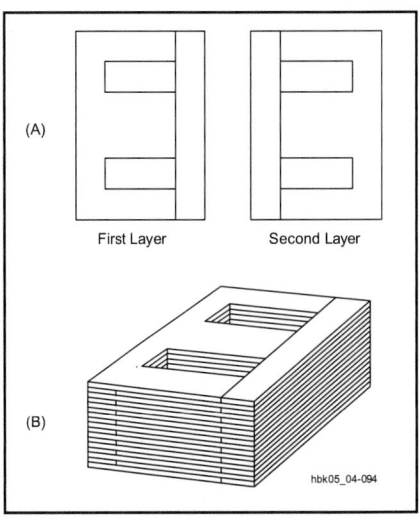

Fig 2.96 — A typical transformer iron core. The E and I pieces alternate direction in successive layers to improve the magnetic path while attenuating eddy currents in the core.

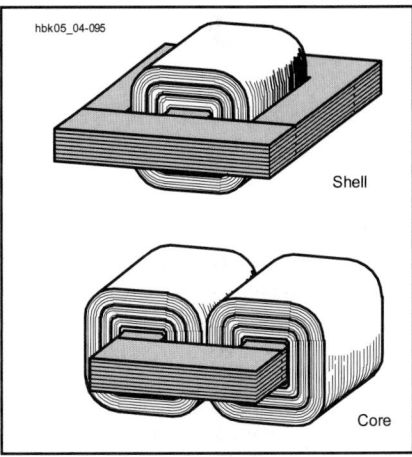

Fig 2.97 — Two common transformer constructions: shell and core.

Current flowing through the leakage reactance causes a voltage drop. This voltage drop increases with increasing current (or frequency); hence, it increases as more power is taken from the secondary. The resistances of the transformer windings, R1 and R2, also cause voltage drops when there is current. Although these voltage drops are not in phase with those caused by leakage reactance, together they result in a lower secondary voltage under load than is indicated by the transformer turns ratio. Thus, in a practical transformer, the greater the secondary current, the smaller the secondary terminal voltage becomes.

At ac line frequencies (50 or 60 Hz), the voltage at the secondary, with a reasonably well-designed iron-core transformer, should not drop more than about 10% from open-circuit conditions to full load. The voltage drop may be considerably more than this in a transformer operating at audio frequencies, because the leakage reactance increases with frequency.

In addition to wire resistances and leakage reactances, certain unwanted or "stray" capacitances occur in transformers. The wire forming the separate turns of the windings acts as the plates of a small capacitor, creating a capacitance between turns and between the windings. This *distributed capacitance* appears in Fig 2.95 as C1, C2, and C_M. Moreover, transformer windings can exhibit capacitance relative to nearby metal, for example, the chassis, the shield and even the core. When current flows through a winding, each turn has a slightly different voltage than its adjacent turns. This voltage causes a small current to flow in these *interwinding* and *winding-to-winding capacitances*.

Although stray capacitances are of little concern with power and audio transformers, they become important as the frequency increases. In transformers for RF use, the stray capacitance can resonate with either the leakage reactance or, at lower frequencies, with the winding reactances, L1 or L2, especially under very light or zero loads. In the frequency region around resonance, transformers no longer exhibit the properties formulated above or the impedance properties to be described below.

Iron-core transformers also experience losses within the core itself. *Hysteresis losses* include the energy required to overcome the retentivity of the core's magnetic material. Circulating currents through the core's resistance are *eddy currents*, which form part of the total core losses. These losses, which add to the required magnetizing current, are equivalent to adding a resistance in parallel with L1 in Fig 2.95.

CORE CONSTRUCTION

Audio and power transformers usually employ silicon steel as the core material. With permeabilities of 5000 or greater, these cores saturate at flux densities approaching 10^5 (Mx) per square inch of cross section. The cores consist of thin insulated laminations to break up potential eddy current paths.

Each core layer consists of an "E" and an "I" piece butted together, as represented in **Fig 2.96**. The butt point leaves a small gap. Each layer is reversed from the adjacent layers so that each gap is next to a continuous magnetic path so that the effect of the gaps is minimized. This is different from the air-gapped inductor of Fig 2.49 in which the air gap is maintained for all layers of laminations.

Two core shapes are in common use, as shown in **Fig 2.97**. In the shell type, both windings are placed on the inner leg, while in the core type the primary and secondary windings may be placed on separate legs, if desired. This is sometimes done when it is necessary to minimize capacitance between the primary and secondary, or when one of the windings must operate at very high voltage.

The number of turns required in the primary for a given applied voltage is determined by the size, shape and type of core material used, as well as the frequency. The number of turns required is inversely proportional to the cross-sectional area of the core. As a rough indication, windings of small power transformers frequently have about six to eight turns per volt on a core of 1-square-inch cross section

and have a magnetic path 10 or 12 inches in length. A longer path or smaller cross section requires more turns per volt, and vice versa.

In most transformers the windings are wound in layers, with a thin sheet of treated-paper insulation between each layer. Thicker insulation is used between adjacent windings and between the first winding and the core.

SHIELDING

Because magnetic lines of force are continuous loops, shielding requires a complete path for the lines of force of the leakage flux. The high-permeability of iron cores tends to concentrate the field, but additional shielding is often needed. As depicted in **Fig 2.98**, enclosing the transformer in a good magnetic material can restrict virtually all of the mag-

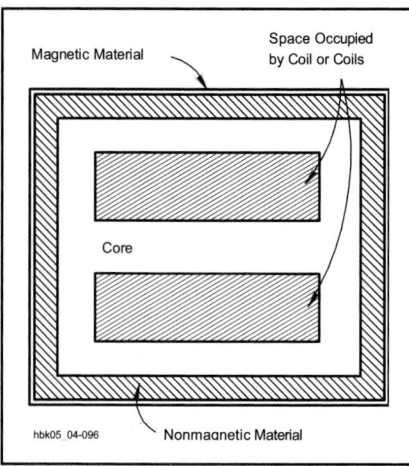

Fig 2.98 — A shielded transformer cross-section: the core plus an outer shield of magnetic material contain nearly all of the magnetic field.

netic field in the outer case. The nonmagnetic material between the case and the core creates a region of high reluctance, attenuating the field before it reaches the case.

2.14.2 Autotransformers

The transformer principle can be used with only one winding instead of two, as shown in **Fig 2.99A**. The principles that relate voltage, current and impedance to the turns ratio also apply equally well. A one-winding transformer is called an *auto-transformer*. The current in the common section (A) of the winding is the difference between the line (primary) and the load (secondary) currents, since these currents are out of phase. Hence, if the line and load currents are nearly equal, the common section of the winding may be wound with comparatively small wire. The line and load currents will be equal only when the primary

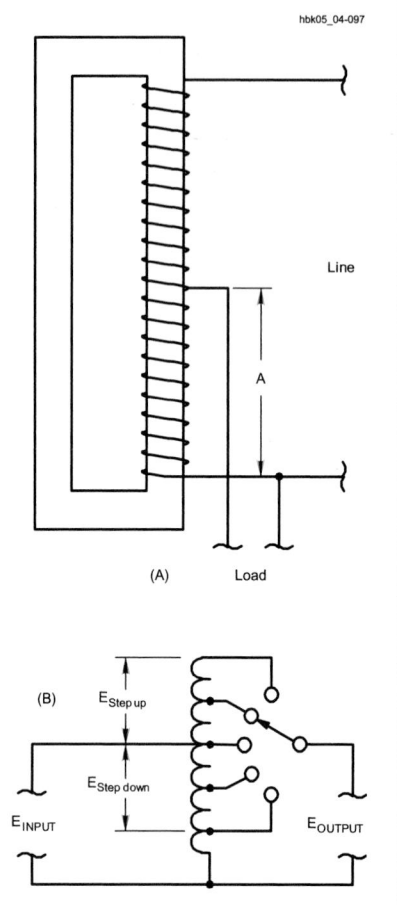

(line) and secondary (load) voltages are not very different.

Autotransformers are used chiefly for boosting or reducing the power-line voltage by relatively small amounts. Fig 2.99B illustrates the principle schematically with a switched, stepped autotransformer. Continu-

Fig 2.99 — The autotransformer is based on the transformer, but uses only one winding. The pictorial diagram at A shows the typical construction of an autotransformer. The schematic diagram at B demonstrates the use of an autotransformer to step up or step down ac voltage, usually to compensate for excessive or deficient line voltage.

ously variable autotransformers are commercially available under a variety of trade names; Variac and Powerstat are typical examples.

Technically, tapped air-core inductors, such as the one in the network in Fig 2.90 at the close of the discussion of resonant circuits, are also autotransformers. The voltage from the tap to the bottom of the winding is less than the voltage across the entire winding. Likewise, the impedance of the tapped part of the winding is less than the impedance of the entire winding. Because in this case, leakage reactances are great and the coefficient of coupling is quite low, the relationships that are true in a perfect transformer grow quite unreliable in predicting the exact values. For this reason, tapped inductors are rarely referred to as transformers. The stepped-down situation in Fig 2.90 is better approximated — at or close to resonance — by the formula

$$R_P = \frac{R_L\,X_{COM}^{\,2}}{X_L} \tag{139}$$

where
   R_P = tuned-circuit parallel-resonant impedance,
   R_L = load resistance tapped across part of the winding,
   X_{COM} = reactance of the portion of the winding common to both the resonant circuit and the load tap, and
   X_L = reactance of the entire winding.

The result is approximate and applies only to circuits with a Q of 10 or greater.

2.15 Heat Management

While not strictly an electrical fundamental, managing the heat generated by electronic circuits is important in nearly all types of radio equipment. Thus, the topic is included in this chapter. Information on the devices and circuits discussed in this section may be found in other chapters.

Any actual energized circuit consumes electric power because any such circuit contains components that convert electricity into other forms of energy. This dissipated power appears in many forms. For example, a loudspeaker converts electrical energy into sound, the motion of air molecules. An antenna (or a light bulb) converts electricity into electromagnetic radiation. Charging a

battery converts electrical energy into chemical energy (which is then converted back to electrical energy upon discharge). But the most common transformation by far is the conversion, through some form of resistance, of electricity into heat.

Sometimes the power lost to heat serves a useful purpose — toasters and hair dryers come to mind. But most of the time, this heat represents a power loss that is to be minimized wherever possible or at least taken into account. Since all real circuits contain resistance, even those circuits (such as a loudspeaker) whose primary purpose is to convert electricity to some *other* form of energy also convert some part of their input power to heat.

Often, such losses are negligible, but sometimes they are not.

If unintended heat generation becomes significant, the involved components will get warm. Problems arise when the temperature increase affects circuit operation by either

• causing the component to fail, by explosion, melting, or other catastrophic event, or, more subtly,

• causing a slight change in the properties of the component, such as through a temperature coefficient (TC).

In the first case, we can design conservatively, ensuring that components are rated to safely handle two, three or more times the maximum power we expect them to dissipate.

In the second case, we can specify components with low TCs, or we can design the circuit to minimize the effect of any one component. Occasionally we even exploit temperature effects (for example, using a resistor, capacitor or diode as a temperature sensor). Let's look more closely at the two main categories of thermal effects.

Not surprisingly, heat dissipation (more correctly, the efficient removal of generated heat) becomes important in medium- to high-power circuits: power supplies, transmitting circuits and so on. While these are not the only examples where elevated temperatures and related failures are of concern, the techniques we will discuss here are applicable to all circuits.

2.15.1 Thermal Resistance

The transfer of heat energy, and thus the change in temperature, between two ends of a block of material is governed by the following heat flow equation and illustrated in **Fig 2.100**):

$$P = \frac{kA}{L}\Delta T = \frac{\Delta T}{\theta} \qquad (140)$$

where

P = power (in the form of heat) conducted between the two points

k = *thermal conductivity*, measured in W/(m °C), of the material between the two points, which may be steel, silicon, copper, PC board material and so on

L = length of the block

A = area of the block, and

ΔT = *difference* in temperature between the two points.

Thermal conductivities of various common materials at room temperature are given in **Table 2.13**.

The form of equation 140 is the same as the variation of Ohm's Law relating current flow to the ratio of a difference in potential to resistance; I = E/R. In this case, what's flowing is heat (P), the difference in potential is a temperature difference (ΔT), and what's resisting the flow of heat is the *thermal resistance*;

$$\theta = \frac{L}{kA} \qquad (141)$$

with units of °C/W. (The units of resistance are equivalent to V/A.) The analogy is so apt that the same principles and methods apply to heat flow problems as circuit problems. The following correspondences hold:

• Thermal conductivity W/(m °C) ↔ Electrical conductivity (S/m).

• Thermal resistance (°C/W) ↔ Electrical resistance (Ω).

• Thermal current (heat flow) (W) ↔ Electrical current (A).

Table 2.13

Thermal Conductivities of Various Materials

Gases at 0 °C, Others at 25 °C; from *Physics*, by Halliday and Resnick, 3rd Ed.

Material	k in units of W/m°C
Aluminum	200
Brass	110
Copper	390
Lead	35
Silver	410
Steel	46
Silicon	150
Air	0.024
Glass	0.8
Wood	0.08

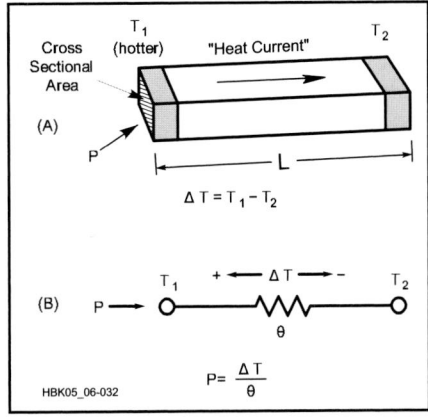

Fig 2.100 — Physical and "circuit" models for the heat-flow equation.

• Thermal potential (T) ↔ Electrical potential (V).

• Heat source ↔ Power source.

For example, calculate the temperature of a 2-inch (0.05 m) long piece of #12 copper wire at the end that is being heated by a 25 W (input power) soldering iron, and whose other end is clamped to a large metal vise (assumed to be an infinite heat sink), if the ambient temperature is 25 °C (77 °F).

First, calculate the thermal resistance of the copper wire (diameter of #12 wire is 2.052 mm, cross-sectional area is 3.31×10^{-6} m²)

$$\theta = \frac{L}{k\,A} = \frac{(0.05\ \text{m})}{(390\ \text{W}/(\text{m}\ ^{\circ}\text{C})) \times (3.31 \times 10^{-6}\ \text{m}^2)}$$

$$= 38.7\ ^{\circ}\text{C/W}$$

Then, rearranging the heat flow equation above yields (after assuming the heat energy actually transferred to the wire is around 10 W)

$$\Delta T = P\ \theta = (10\ \text{W}) \times (38.7\ ^{\circ}\text{C}/\text{W}) = 387\ ^{\circ}\text{C}$$

So the wire temperature at the hot end is 25 C + ΔT = 412 °C (or 774 °F). If this sounds

a little high, remember that this is for the steady state condition, where you've been holding the iron to the wire for a long time.

From this example, you can see that things can get very hot even with application of moderate power levels. For this reason, circuits that generate sufficient heat to alter, not necessarily damage, the components must employ some method of cooling, either active or passive. Passive methods include heat sinks or careful component layout for good ventilation. Active methods include forced air (fans) or some sort of liquid cooling (in some high-power transmitters).

2.15.2 Heat Sink Selection and Use

The purpose of a heat sink is to provide a high-power component with a large surface area through which to dissipate heat. To use the models above, it provides a low thermal-resistance path to a cooler temperature, thus allowing the hot component to conduct a large "thermal current" away from itself.

Power supplies probably represent one of the most common high-power circuits amateurs are likely to encounter. Everyone has certainly noticed that power supplies get warm or even hot if not ventilated properly. Performing the thermal design for a properly cooled power supply is a very well-defined process and is a good illustration of heat-flow concepts.

This material was originally prepared by ARRL Technical Advisor Dick Jansson, KD1K, during the design of a 28-V, 10-A power supply. (The chapter **Analog Basics** discusses semiconductor diodes, rectifiers, and transistors. The **Power Sources** chapter has more information on power supply design.) An outline of the design procedure shows the logic applied:

1. Determine the expected power dissipation (P_{in}).

2. Identify the requirements for the dissipating elements (maximum component temperature).

3. Estimate heat-sink requirements.

4. Rework the electronic device (if necessary) to meet the thermal requirements.

5. Select the heat exchanger (from heat sink data sheets).

The first step is to estimate the filtered, unregulated supply voltage under full load. Since the transformer secondary output is 32 V ac (RMS) and feeds a full-wave bridge rectifier, let's estimate 40 V as the filtered dc output at a 10-A load.

The next step is to determine our critical components and estimate their power dissipations. In a regulated power supply, the pass transistors are responsible for nearly all the power lost to heat. Under full load, and allowing for some small voltage drops in the

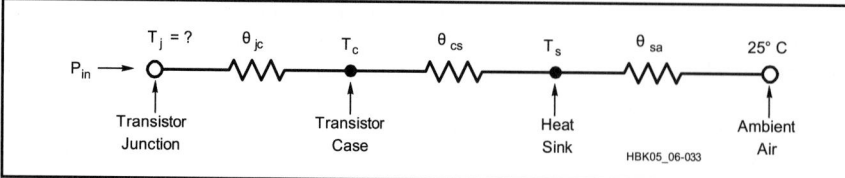

Fig 2.101 — Resistive model of thermal conduction in a power transistor and associated heat sink. See text for calculations.

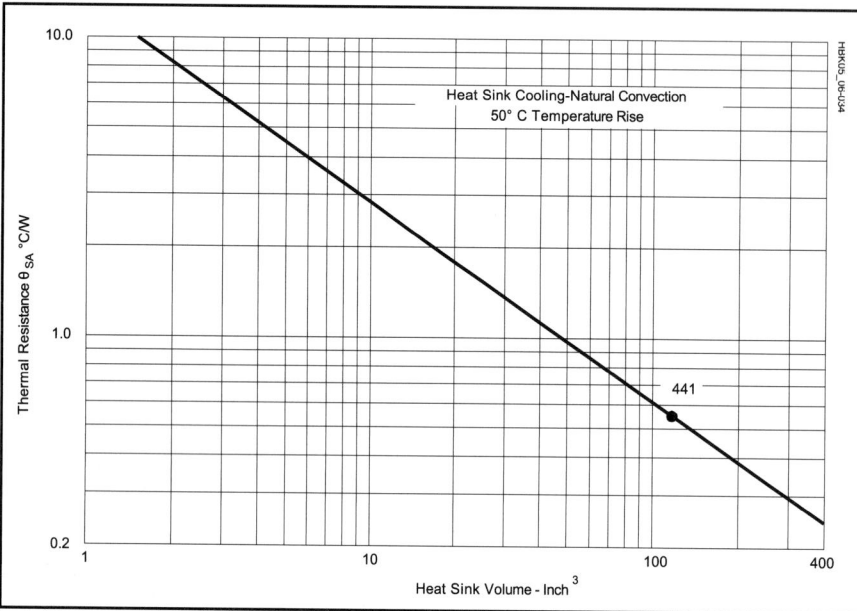

Fig 2.102 — Thermal resistance vs heat-sink volume for natural convection cooling and 50 °C temperature rise. The graph is based on engineering data from Wakefield Thermal Solutions, Inc.

power-transistor emitter circuitry, the output of the series pass transistors is about 29 V for a delivered 28 V under a 10-A load. With an unregulated input voltage of 40 V, the total energy heat dissipated in the pass transistors is (40 V − 29 V) × 10 A = 110 W. The heat sink for this power supply must be able to handle that amount of dissipation and still keep the transistor junctions below the specified safe operating temperature limits. It is a good rule of thumb to select a transistor that has a maximum power dissipation of twice the desired output power.

Now, consider the ratings of the pass transistors to be used. This supply calls for 2N3055s as pass transistors. The data sheet shows that a 2N3055 is rated for 15-A service and 115-W dissipation. But the design uses *four* in parallel. Why? Here we must look past the big, bold type at the top of the data sheet to such subtle characteristics as the junction-to-case thermal resistance, q_{jc}, and the maximum allowable junction temperature, T_j.

The 2N3055 data sheet shows $\theta_{jc} = 1.52$ °C/W, and a maximum allowable case (and junction) temperature of 220 °C. While it seems that one 2N3055 could barely, on paper

at least, handle the electrical requirements — at what temperature would it operate?

To answer that, we must model the entire "thermal circuit" of operation, starting with the transistor junction on one end and ending at some point with the ambient air. A reasonable model is shown in **Fig 2.101**. The ambient air is considered here as an infinite heat sink; that is, its temperature is assumed to be a constant 25 °C (77 °F). θ_{jc} is the thermal resistance from the transistor junction to its case. θ_{cs} is the resistance of the mounting interface between the transistor case and the heat sink. θ_{sa} is the thermal resistance between the heat sink and the ambient air. In this "circuit," the generation of heat (the "thermal current source") occurs in the transistor at P_{in}.

Proper mounting of most TO-3 package power transistors such as the 2N3055 requires that they have an electrical insulator between the transistor case and the heat sink. However, this electrical insulator must at the same time exhibit a low thermal resistance. To achieve a quality mounting, use thin polyimid or mica formed washers and a suitable thermal compound to exclude air from the interstitial space. "Thermal greases" are commonly

available for this function. Any silicone grease may be used, but filled silicone oils made specifically for this purpose are better.

Using such techniques, a conservatively high value for θ_{cs} is 0.50 °C/W. Lower values are possible, but the techniques needed to achieve them are expensive and not generally available to the average amateur. Furthermore, this value of θ_{cs} is already much lower than θ_{jc}, which cannot be lowered without going to a somewhat more exotic pass transistor.

Finally, we need an estimate of θ_{sa}. **Fig 2.102** shows the relationship of heat-sink volume to thermal resistance for natural-convection cooling. This relationship presumes the use of suitably spaced fins (0.35 inch or greater) and provides a "rough order-of-magnitude" value for sizing a heat sink. For a first calculation, let's assume a heat sink of roughly 6 × 4 × 2 inch (48 cubic inches). From Fig 2.102, this yields a θ_{sa} of about 1 °C/W.

Returning to Fig 2.101, we can now calculate the approximate temperature increase of a single 2N3055:

$$\delta T = P \, \theta_{total}$$

$$= 110 \text{ W} \times (1.52 \text{ °C / W} + 0.5 \text{ °C / W} + 1.0 \text{ °C / W})$$

$$= 332 \text{ °C}$$

Given the ambient temperature of 25 °C, this puts the junction temperature T_j of the 2N3055 at 25 + 332 = 357 °C! This is clearly too high, so let's work backward from the air end and calculate just how many transistors we need to handle the heat.

First, putting more 2N3055s in parallel means that we will have the thermal model illustrated in **Fig 2.103**, with several identical θ_{jc} and θ_{cs} in parallel, all funneled through the same θ_{as} (we have one heat sink).

Keeping in mind the physical size of the project, we could comfortably fit a heat sink of approximately 120 cubic inches (6 × 5 × 4 inches), well within the range of commercially available heat sinks. Furthermore, this application can use a heat sink where only "wire access" to the transistor connections is required. This allows the selection of a more efficient design. In contrast, RF designs require the transistor mounting surface to be completely exposed so that the PC board can be mounted close to the transistors to minimize parasitics. Looking at Fig 2.102, we see that a 120-cubic-inch heat sink yields a θ_{sa} of 0.55 °C/W. This means that the temperature of the heat sink when dissipating 110 W will be 25 °C + (110 W × 0.55 °C/W) = 85.5 °C.

Industrial experience has shown that silicon transistors suffer substantial failure when junctions are operated at highly elevated temperatures. Most commercial and military specifications will usually not permit design junction temperatures to exceed 125 °C. To arrive at a safe figure for our maximum

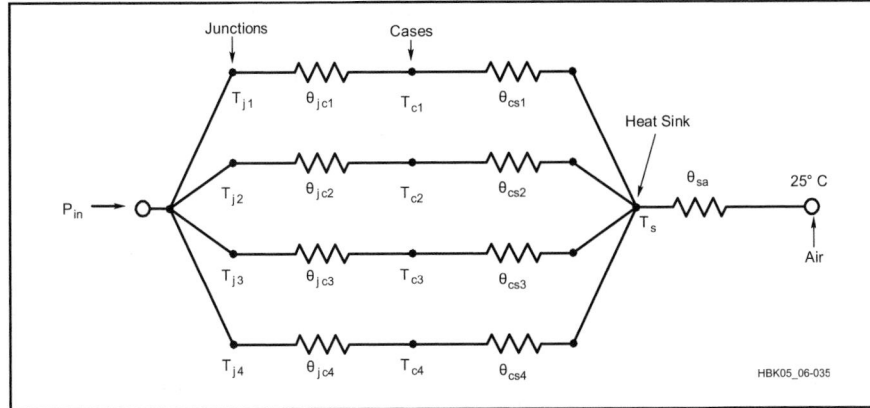

Fig 2.103 — Thermal model for multiple power transistors mounted on a common heat sink.

allowed T_j, we must consider the intended use of the power supply. If we are using it in a 100% duty-cycle transmitting application such as RTTY or FM, the circuit will be dissipating 110 W continuously. For a lighter duty-cycle load such as CW or SSB, the "key-down" temperature can be slightly higher as long as the average is less than 125 °C. In this intermittent type of service, a good conservative figure to use is $T_j = 150$ °C.

Given this scenario, the temperature rise across each transistor can be $150 - 85.5 = 64.5$ °C. Now, referencing Fig 2.103, remembering the total θ for each 2N3055 is $1.52 + 0.5 = 2.02$ °C/W, we can calculate the maximum power each 2N3055 can safely dissipate:

$$P = \frac{\delta T}{\theta} = \frac{64.5 \text{ °C}}{2.02 \text{ °C / W}} = 31.9 \text{ W}$$

Thus, for 110 W full load, we need four 2N3055s to meet the thermal requirements of the design. Now comes the big question:

What is the "right" heat sink to use? We have already established its requirements: it must be capable of dissipating 110 W, and have a θ_{sa} of 0.55 °C/W (see above).

A quick consultation with several manufacturer's catalogs reveals that Wakefield Thermal Solutions, Inc. model nos. 441 and 435 heat sinks meet the needs of this application. A Thermalloy model no. 6441 is suitable as well. Data published in the catalogs of these manufacturers show that in natural-convection service, the expected temperature rise for 100 W dissipation would be just under 60 °C, an almost perfect fit for this application. Moreover, the no. 441 heat sink can easily mount four TO-3-style 2N3055 transistors as shown in **Fig 2.104**. Remember: heat sinks should be mounted with the fins and transistor mounting area vertical to promote convection cooling.

The design procedure just described is applicable to any circuit where heat buildup is a potential problem. By using the thermal-resistance model, we can easily calculate whether or not an external means of cooling is necessary, and if so, how to choose it. Aside from heat sinks, forced air cooling (fans) is another common method. In commercial transceivers, heat sinks with forced-air cooling are common.

2.15.3 Transistor Derating

Maximum ratings for power transistors are usually based on a case temperature of 25 °C. These ratings will decrease with increasing operating temperature. Manufacturer's data sheets usually specify a *derating* figure or curve that indicates how the maximum ratings change per degree rise in temperature. If such information is not available (or even if it is!), it is a good rule of thumb to select a power transistor with a maximum power dissipation of at least twice the desired output power.

2.15.4 Rectifiers

Diodes are physically quite small, and they operate at high current densities. As a result their heat-handling capabilities are somewhat limited. Normally, this is not a problem in high-voltage, low-current supplies in which rectifiers in axial-lead DO-type packages are used. (See the **Component Data and References** chapter for information on device packages.) The use of high-current (2 A or greater) rectifiers at or near their maximum ratings, however, requires some form of heat sinking. The average power dissipated by a rectifier is

$$P = I_{AVG} \times V_F \qquad (142)$$

where
  I_{AVG} is the average current, and
  V_F is the forward voltage drop.

Average current must account for the conduction duty cycle and the forward voltage drop must be determined at the average current level.

Rectifiers intended for such high-current applications are available in a variety of packages suitable for mounting to flat surfaces. Frequently, mounting the rectifier on the main chassis (directly, or with thin mica insulating washers) will suffice. If the diode is insulated from the chassis, thin layers of thermal compound or thermal insulating washers should be used to ensure good heat conduction. Large, high-current rectifiers often require special heat sinks to maintain a safe operating temperature. Forced-air cooling is sometimes used as a further aid.

2.15.5 RF Heating

RF current often causes component heating problems where the same level of dc current may not. An example is the tank circuit of

Fig 2.104 — A Wakefield 441 heat sink with four 2N3055 transistors mounted.

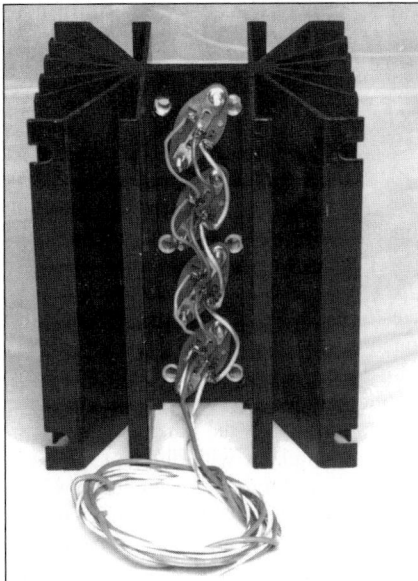

an RF oscillator. If several small capacitors are connected in parallel to achieve a desired capacitance, skin effect will be reduced and the total surface area available for heat dissipation will be increased, thus significantly reducing the RF heating effects as compared to a single large capacitor. This technique can be applied to any similar situation; the general idea is to divide the heating among as many components as possible.

2.15.6 Forced-Air and Water Cooling

In Amateur Radio today, forced-air cooling is most commonly found in vacuum-tube circuits or in power supplies built in small enclosures, such as those in solid-state transceivers or computers. Fans or blowers are commonly specified in cubic feet per minute (CFM). While the nomenclature and specifications differ from those used for heat sinks, the idea remains the same: to offer a low thermal resistance between the inside of the enclosure and the (ambient) exterior.

For forced air cooling, we basically use the "one resistor" thermal model of Fig 2.100. The important quantity to be determined is heat generation, P_{in}. For a power supply, this can be easily estimated as the difference between the input power, measured at the transformer primary, and the output power at full load. For variable-voltage supplies, the worst-case output condition is minimum voltage with maximum current. A discussion of forced-air cooling for vacuum tube equipment appears in the **RF Power Amplifiers** chapter.

Dust build-up is a common problem for forced-air cooling systems, even with powerful blowers and fans. If air intake grills and vents are not kept clean and free of lint and debris, air flow can be significantly reduced, leading to excessive equipment temperature and premature failure. Cleaning of air passageways should be included in regular equipment maintenance for good performance and maximum equipment life.

Water cooling systems are much less common in amateur equipment, used primarily for high duty cycle operating, such as RTTY, and at frequencies where the efficiency of the amplifier is relatively low, such as UHF and microwaves. In these situations, water cooling is used because water can absorb and transfer more than 3000 times as much heat as the same volume of air!

The main disadvantage of water cooling is that it requires pumps, hoses, and reservoirs whereas a fan or blower is all that is required for forced-air cooling. For high-voltage circuits, using water cooling also requires special insulation techniques and materials to allow water to circulate in close contact with the heat source while remaining electrically isolated.

Nevertheless, the technique can be effec-

tive. The increased availability of inexpensive materials designed for home sprinkler and other low-pressure water distribution systems make water-cooling less difficult to implement. It is recommended that the interested reader review articles and projects in the amateur literature to observe the successful implementation of water cooling systems.

2.15.7 Heat Pipe Cooling

A heat pipe is a device containing a *working fluid* in a reservoir where heat is absorbed and a channel to a second reservoir where heat is dissipated. Heat pipes work by *evaporative cooling*. The heat-absorbing reservoir is placed in thermal contact with the heat source which transfers heat to the working fluid, usually a liquid substance with a boiling point just above room temperature. The working fluid vaporizes and the resulting vapor pressure pushes the hot vapor through the channel to the cooling reservoir.

In the cooling reservoir, the working fluid gives up its heat of vaporization, returning to the fluid state. The cooled fluid then flows back through the channel to the heat-absorbing reservoir where the process is repeated.

Heat pipes require no fans or pumps — movement of the working fluid is driven entirely from the temperature difference between the two reservoirs. The higher the temperature difference between the absorbing and dissipating reservoirs, the more effective the heat pump becomes, up to the limit of the dissipating reservoir to dissipate heat.

The principle application for heat pipes is for space operations. Heat pipe applications in terrestrial applications are limited due to the gravity gradient sensitivity of heat pipes. At present, the only amateur equipment making use of heat pipes are computers and certain amplifier modules. Nevertheless, as more general-purpose products become available, this technique will become more common.

2.15.8 Thermoelectric Cooling

Thermoelectric cooling makes use of the *Peltier effect* to create heat flow across the junction of two different types of materials. This process is related to the *thermoelectric effect* by which thermocouples generate voltages based on the temperature of a similar junction. A *thermoelectric cooler* or *TEC* (also known as a *Peltier cooler*) requires only a source of dc power to cause one side of the device to cool and the other side to warm. TECs are available with different sizes and power ratings for different applications.

TECs are not available with sufficient heat transfer capabilities that they can be used in high-power applications, such as RF ampli-

fiers. However, they can be useful in lowering the temperature of sensitive receiver circuits, such as preamplifiers used at UHF and microwave frequencies, or imaging devices, such as charge-coupled devices (CCDs). Satellites use TECs as *radiative coolers* that dissipate heat directly as thermal or infrared radiation. TECs are also found in some computing equipment where they are used to remove heat from microprocessors and other large integrated circuits.

2.15.9 Temperature Compensation

Aside from catastrophic failure, temperature changes may also adversely affect circuits if the temperature coefficient (TC) of one or more components is too large. If the resultant change is not too critical, adequate temperature stability can often be achieved simply by using higher-precision components with low TCs (such as NP0/C0G capacitors or metal-film resistors). For applications where this is impractical or impossible (such as many solid-state circuits), we can minimize temperature sensitivity by *compensation* or *matching* — using temperature coefficients to our advantage.

Compensation is accomplished in one of two ways. If we wish to keep a certain circuit quantity constant, we can interconnect pairs of components that have equal but opposite TCs. For example, a resistor with a negative TC can be placed in series with a positive TC resistor to keep the total resistance constant. Conversely, if the important point is to keep the *difference* between two quantities constant, we can use components with the *same* TC so that the pair "tracks." That is, they both change by the same amount with temperature.

An example of this is a Zener reference circuit. Since a diode is strongly affected by operating temperature, circuits that use diodes or transistors to generate stable reference voltages must use some form of temperature compensation. Since, for a constant current, a reverse-biased PN junction has a negative voltage TC while a forward-biased junction has a positive voltage TC, a good way to temperature-compensate a Zener reference diode is to place one or more forward-biased diodes in series with it.

2.15.10 Thermistors

Thermistors can be used to control temperature and improve circuit behavior or protect against excessive temperatures, hot or cold. Circuit temperature variations can affect gain, distortion or control functions like receiver AGC or transmitter ALC. Thermistors can be used in circuits that compensate for temperature changes.

A *thermistor* is a small bit of intrinsic (no N or P doping) metal-oxide semiconductor compound material between two wire leads. As temperature increases, the number of liberated hole/electron pairs increases exponentially, causing the resistance to decrease exponentially. You can see this in the resistance equation:

$$R(T) = R(T0)e^{-\beta(1/T0-1/T)} \qquad (143)$$

where T is some temperature in Kelvins and T0 is a reference temperature, usually 298 K (25°C), at which the manufacturer specifies R(T0).

The constant β is experimentally determined by measuring resistance at various temperatures and finding the value of β that best agrees with the measurements. A simple way to get an approximate value of β is to make two measurements, one at room temperature, say 25 °C (298 K) and one at 100 °C (373 K) in boiling water. Suppose the resistances are 10 kΩ and 938 Ω. Eq 143 is solved for β

$$\beta = \frac{\ln\left(\frac{R(T)}{R(T0)}\right)}{\frac{1}{T} - \frac{1}{T0}} = \frac{\ln\left(\frac{938}{1000}\right)}{\frac{1}{373} - \frac{1}{298}} = 3507 \qquad (144)$$

With the behavior of the thermistor known — either by equation or calibration table — its change in resistance can be used to create an electronic circuit whose behavior is controlled by temperature in a known fashion. Such a circuit can be used for controlling temperature or detecting specific temperatures.

See "Termistors in Homebrew Projects" by Bill Sabin, WØIYH, and "Thermistor Based Temperature Controller" by Bill Sabin, WØIYH on this book's CD-ROM for practical projects using thermistors.

2.16 References and Bibliography

BOOKS

Alexander and Sadiku, *Fundamentals of Electric Circuits* (McGraw-Hill)

Banzhaf, W., WB1ANE, *Understanding Basic Electronics*, 2nd ed (ARRL)

Glover, T., *Pocket Ref* (Sequoia Publishing)

Grover, F., Inductance Calculations (Dover Publications)

Horowitz and Hill, *The Art of Electronics* (Cambridge University Press)

Kaiser, C., *The Resistor Handbook* (Saddleman Press, 1998)

Kaiser, C., *The Capacitor Handbook* (Saddleman Press, 1995)

Kaiser, C., *The Inductor Handbook* (Saddleman Press, 1996)

Kaiser, C., *The Diode Handbook* (Saddleman Press, 1999)

Kaiser, C., *The Transistor Handbook* (Saddleman Press, 1999)

Kaplan, S., *Wiley Electrical and Electronics Dictionary* (Wiley Press)

Orr, W., *Radio Handbook* (Newnes)

Terman, F., *Radio Engineer's Handbook* (McGraw-Hill)

MAGAZINE AND JOURNAL ARTICLES

ARRL Lab Staff, "Lab Notes — Capacitor Basics," *QST,* Jan 1997, pp 85-86

B. Bergeron, NU1N, "Under the Hood II: Resistors," *QST,* Nov 1993, pp 41-44

B. Bergeron, NU1N, "Under the Hood III: Capacitors," *QST,* Jan 1994, pp 45-48

B. Bergeron, NU1N, "Under the Hood IV: Inductors," *QST,* Mar 1994, pp 37-40

B. Bergeron, NU1N, "Under the Hood: Lamps, Indicators, and Displays," *QST,* Sep 1994, pp 34-37

J. McClanahan, W4JBM, "Understanding and Testing Capacitor ESR," *QST,* Sep 2003, pp 30-32

W. Silver, NØAX, "Experiment #24 — Heat Management," *QST,* Jan 2005, pp 64-65

W. Silver, NØAX, "Experiment #33 — The Transformer," *QST,* Oct 2005, pp 62-63

W. Silver, NØAX, "Experiment #62 — About Resistors," *QST,* Mar 2008, pp 66-67

W. Silver, NØAX, "Experiment #63 — About Capacitors," *QST,* Apr 2008, pp 70-71

W. Silver, NØAX, "Experiment #74 — Resonant Circuits," *QST,* Mar 2009, pp 72-73

J. Smith, K8ZOA, "Carbon Composition, Carbon Film and Metal Oxide Film Resistors," *QEX,* Mar/Apr 2008, pp 46-57

Contents

Chapter 3 — CD-ROM Content

Supplemental Articles

- "Hands-On Radio: The Common Emitter Amplifier" by Ward Silver, NØAX
- "Hands-On Radio: The Emitter-Follower Amplifier" by Ward Silver, NØAX
- "Hands-On Radio: The Common Base Amplifier" by Ward Silver, NØAX
- "Hands-On Radio: Field Effect Transistors" by Ward Silver, NØAX
- "Hands-On Radio: Basic Operational Amplifiers" by Ward Silver, NØAX
- Cathode Ray Tubes
- Large Signal Transistor Operation

Tools and Data

- *LTSpice* Simulation Files for Chapter 3
- Frequency Response Spreadsheet

Analog Basics

The first section of this chapter discusses a variety of techniques for working with analog signals, called signal processing. These basic functions are combined to produce useful systems that become radios, instruments, audio recorders and so on.

Subsequent sections present several types of active components and circuits that can be used to manipulate analog signals. (An active electronic component is one that requires a power source to function; passive components such as resistors, capacitors and inductors are described in the **Electrical Fundamentals** chapter.) The chapter concludes with a discussion of analog-digital conversion by which analog signals are converted into digital signals and vice versa.

Material in this chapter was updated by Ward Silver, NØAX, and Courtney Duncan, N5BF, building on material from previous editions by Greg Lapin, N9GL, and Leonard Kay, K1NU. The section "Poles and Zeroes" was adapted from material contributed by David Stockton, GM4ZNX and Fred Telewski, WA7TZY.

3.1 Analog Signal Processing

The term *analog signal* refers to voltages, currents and waves that make up ac radio and audio signals, dc measurements, even power. The essential characteristic of an analog signal is that the information or energy it carries is continuously variable. Even small variations of an analog signal affect its value or the information it carries. This stands in contrast to *digital signals* (described in the **Digital Basics** chapter) that have values only within well-defined and separate ranges called *states*. To be sure, at the fundamental level all circuits and signals are analog: Digital signals are created by designing circuits that restrict the values of analog signals to those discrete states.

Analog signal processing involves various electronic stages to perform functions on analog signals such as amplifying, filtering, modulation and demodulation. A piece of electronic equipment, such as a radio, is constructed by combining a number of these circuits. How these stages interact with each other and how they affect the signal individually and in tandem is the subject of sections later in the chapter.

3.1.1 Linearity

The premier properties of analog signals are *superposition* and *scaling*. Superposition is the property by which signals are combined, whether in a circuit, in a piece of wire, or even in air, as the sum of the individual signals. This is to say that at any one point in time, the voltage of the combined signal is the sum of the voltages of the original signals at the same time. In a *linear system* any number of signals will add in this way to give a single combined signal. (Mathematically, this is a *linear combination*.) For this reason, analog signals and components are often referred to as *linear signals* or *linear components*. A linear system whose characteristics do not change, such as a resistive voltage divider, is called *time-invariant*. If the system changes with time, it is *time-varying*. The variations may be random, intermittent (such as being adjusted by an operator) or periodic.

One of the more important features of superposition, for the purposes of signal processing, is that signals that have been combined by superposition can be separated back into the original signals. This is what allows multiple signals that have been received by an antenna to be separated back into individual signals by a receiver.

3.1.2 Linear Operations

Any operation that modifies a signal and obeys the rules of superposition and scaling is a *linear operation*. The following sections explain the basic linear operations from which linear systems are made.

AMPLIFICATION AND ATTENUATION

Amplification and *attenuation* scale signals to be larger and smaller, respectively. The operation of *scaling* is the same as multiplying the signal at each point in time by a constant value; if the constant is greater than one then the signal is amplified, if less than one then the signal is attenuated.

An *amplifier* is a circuit that increases the amplitude of a signal. Schematically, a generic amplifier is signified by a triangular symbol, its input along the left face and its output at the point on the right (see **Fig 3.1**). The linear amplifier multiplies every value of a signal by a constant value. Amplifier gain is often expressed as a multiplication factor (× 5, for example).

$$\text{Gain} = \frac{V_o}{V_i} \qquad (1)$$

where V_o is the output voltage from an amplifier when an input voltage, V_i, is applied. (Gain is often expressed in decibels (dB) as explained in the chapter on **Electrical Fundamentals**.)

Certain types of amplifiers for which currents are the input and output signals have *current gain*. Most amplifiers have both voltage and current gain. *Power gain* is the ratio of output power to input power.

Ideal linear amplifiers have the same gain for all parts of a signal. Thus, a gain of 10 changes 10 V to 100 V, 1 V to 10 V and –1 V to –10 V. (Gain can also be less than one.) The ability of an amplifier to change a signal's level is limited by the amplifier's *dynamic range*, however. An amplifier's dynamic range is the range of signal levels over which the amplifier produces the required gain without distortion. Dynamic range is limited for small signals by noise, distortion and other nonlinearities.

Dynamic range is limited for large signals because an amplifier can only produce output voltages (and currents) that are within the range of its power supply. (Power-supply voltages are also called the *rails* of a circuit.) As the amplified output approaches one of the rails, the output can not exceed a given voltage near the rail and the operation of the amplifier becomes nonlinear as described below in the section on Clipping and Rectification.

Another similar limitation on amplifier linearity is called *slew rate*. Applied to an amplifier, this term describes the maximum rate at which a signal can change levels and still be accurately amplified in a particular device. *Input slew rate* is the maximum rate of change to which the amplifier can react linearly. *Output slew rate* refers to the maximum rate at which the amplifier's output circuit can change. Slew rate is an important concept, because there is

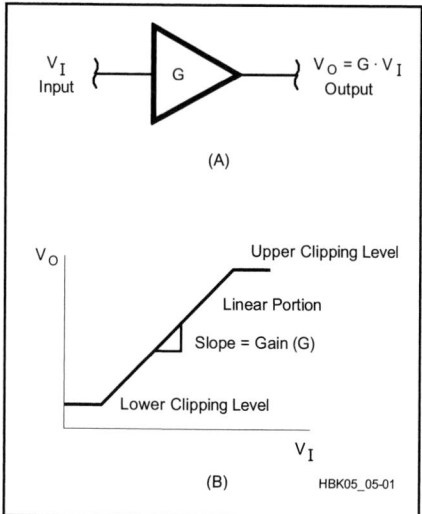

Fig 3.1 — Generic amplifier. (A) Symbol. For the linear amplifier, gain is the constant value, G, and the output voltage is equal to the input voltage times G; (B) Transfer function, input voltage along the x-axis is converted to the output voltage along the y-axis. The linear portion of the response is where the plot is diagonal; its slope is equal to the gain, G. Above and below this range are the clipping limits, where the response is not linear and the output signal is clipped.

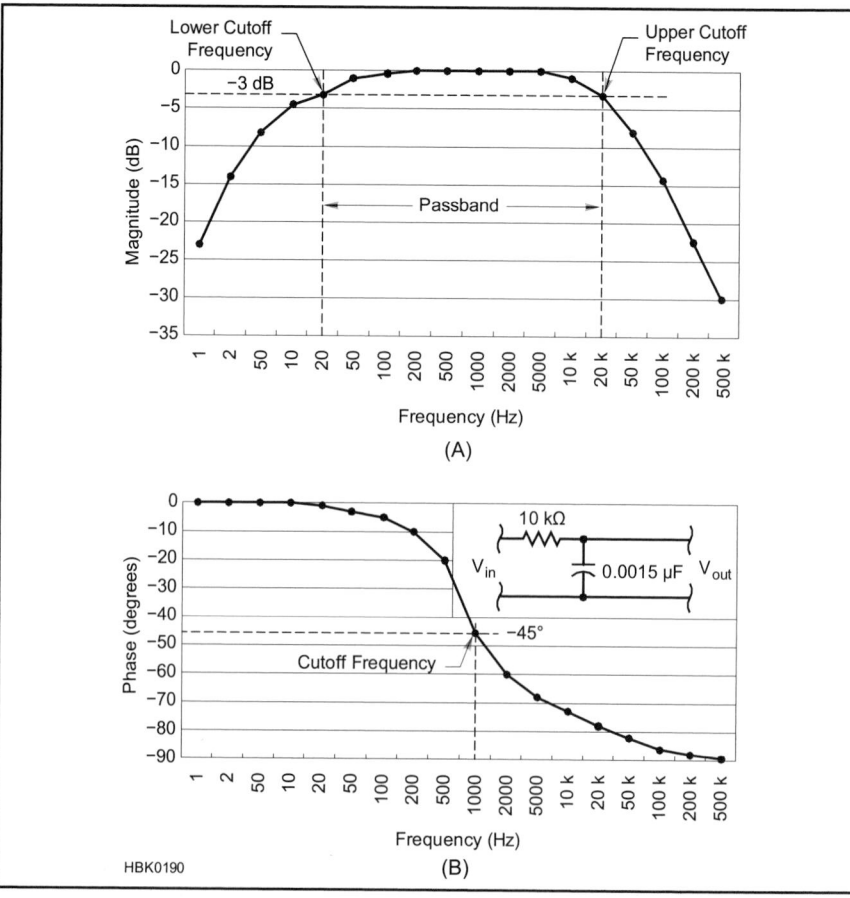

Fig 3.2 — Bode plot of (A) band-pass filter magnitude response and (B) an RC low-pass filter phase response.

a direct correlation between a signal level's rate of change and the frequency content of that signal. The amplifier's ability to react to or reproduce that rate of change affects its frequency response and dynamic range.

An *attenuator* is a circuit that reduces the amplitude of a signal. Attenuators can be constructed from passive circuits, such as the attenuators built using resistors, described in the chapter on **Test Equipment and Measurements**. Active attenuator circuits include amplifiers whose gain is less than one or circuits with adjustable resistance in the signal path, such as a PIN diode attenuator or amplifier with gain is controlled by an external voltage.

FREQUENCY RESPONSE AND BODE PLOTS

Another important characteristic of a circuit is its *frequency response*, a description of how it modifies a signal of any frequency. Frequency response can be stated in the form of a mathematical equation called a *transfer function*, but it is more conveniently presented as a graph of gain vs frequency. The ratio of output amplitude to input amplitude is often called the circuit's *magnitude* or *amplitude response*. Plotting the circuit's magnitude response in dB versus frequency on a logarithmic scale, such as in **Fig 3.2A**, is called a *Bode plot* (after Henrik Wade Bode). The combination of decibel and log-frequency scales is used because the behavior of most circuits depends on ratios of amplitude and frequency and thus appears linear on a graph in which both the vertical and horizontal scales are logarithmic.

Most circuits also affect a signal's phase along with its amplitude. This is called *phase shift*. A plot of phase shift from the circuit's input to its output is called the *phase response*, seen in Fig 3.2B. Positive phase greater than 0° indicates that the output signal *leads* the input signal, while *lagging* phase shift has a negative phase. The combination of an amplitude and phase response plot gives a good picture of what effect the circuit has on signals passing through it.

TRANSFER CHARACTERISTICS

Transfer characteristics are the ratio of an output parameter to an input parameter, such as output current divided by input current, h_{FE}. There are different families of transfer characteristics, designated by letters such as h, s, y or z. Each family compares parameters in specific ways that are useful in certain design or analysis methods. The most common transfer characteristics used in radio are the h-parameter family (used in transistor models) and the s-parameter family (used in RF design, particularly at VHF and above). See the **RF Techniques** chapter for more discussion of transfer characteristics.

COMPLEX FREQUENCY

We are accustomed to thinking of frequency as a real number — so many cycles per second — but frequency can also be a complex number, s, with both a real part, designated by σ, and an imaginary part, designated by $j\omega$. (ω is also equal to $2\pi f$.) The resulting complex frequency is written as $s = \sigma + j\omega$. At the lower left of **Fig 3.3** a pair of real and imaginary axes are used to plot values of s. This is called the *s-plane*. Complex frequency is used in Laplace transforms, a mathematical technique used for circuit and signal analysis. (Thorough treatments of the application of complex frequency can be found in college-level textbooks on circuit and signal analysis.)

When complex frequency is used, a sinusoidal signal is described by Ae^{st}, where A is the amplitude of the signal and t is time. Because s is complex, $Ae^{st} = Ae^{(\sigma+j\omega)t} = A(e^{\sigma t})(e^{j\omega t})$. The two exponential terms describe independent characteristics of the signal. The second term, $e^{j\omega t}$, is the sine wave with which we are familiar and that has frequency f, where $f = \omega/2\pi$. The first term, $e^{\sigma t}$, represents the rate at which the signal increases or decreases. If σ is negative, the exponential term decreases with time and the signal gets smaller and smaller. If σ is positive, the signal gets larger and larger. If $\sigma = 0$, the exponential term equals 1, a constant, and the signal amplitude stays the same.

Complex frequency is very useful in describing a circuit's stability. If the response to an input signal is at a frequency on the right-hand side of the s-plane for which $\sigma > 0$, the system is *unstable* and the output signal will get larger until it is limited by the circuit's power supply or some other mechanism. If the response is on the left-hand side of the s-plane, the system is *stable* and the response to the input signal will eventually die out. The larger the absolute value of σ, the faster the response changes. If the response is precisely on the $j\omega$ axis where $\sigma = 0$, the response will persist indefinitely.

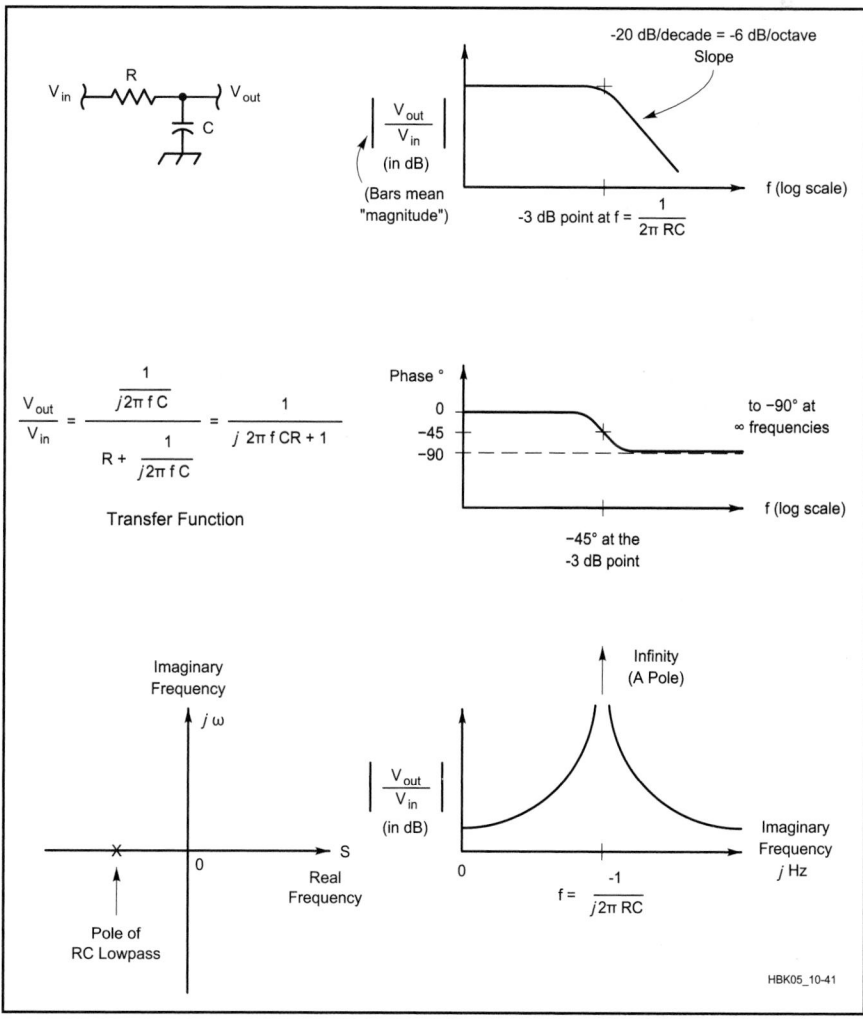

Fig 3.3 — The transfer function for a circuit describes both the magnitude and phase response of a circuit. The RC circuit shown at the upper left has a pole at f = 2πRC, the filter's −3 dB or cutoff frequency, at which the phase response is a 45° lagging phase shift. Poles cause an infinite response on the imaginary frequency axis.

In Fig 3.3 the equation for the simple RC-circuit's transfer function is shown at the left of the figure. It describes the circuit's behavior at real-world frequencies as well as imaginary frequencies whose values contain j. Because complex numbers are used for f, the transfer function describes the circuit's phase response, as well as amplitude. At one such frequency, $f = -j/2\pi RC$, the denominator of the transfer function is zero, and the gain is infinite! Infinite gain is a pretty amazing thing to achieve with a passive circuit — but because this can only happen at an imaginary frequency, it does not happen in the real world.

The practical effects of complex frequency can be experienced in a narrow CW crystal or LC filter. The poles of such a filter are just to the left of the $j\omega$ axis, so the input signal causes the filters to "ring", or output a damped sine wave along with the desired signal. Similarly, the complex frequency of an oscillator's output at power-up must have $\sigma > 0$ or the oscillation would never start! The output amplitude continues to grow until limiting takes place, reducing gain until $\sigma = 1$ for a steady output.

POLES AND ZEROES

Frequencies that cause the transfer function to become infinite are called *poles*. This is shown at the bottom right of Fig 3.3 in the graph of the circuit's amplitude response for imaginary frequencies shown on the horizontal axis. (The pole causes the graph to extend up "as a pole under a tent," thus the name.) Similarly, circuits can have *zeroes* which occur at imaginary frequencies that cause the transfer function to be zero, a less imaginative name, but quite descriptive.

A circuit can also have poles and zeroes at frequencies of zero and infinity. For example, the circuit in Fig 3.3 has a zero at infinity because the capacitor's reactance is zero at infinity and the transfer function is zero, as well. If the resistor and capacitor were exchanged, so that the capacitor was in series with the output, then at zero frequency (dc), the output would be zero because the capacitor's reactance was infinite, creating a zero.

Complex circuits can have multiple poles or multiple zeroes at the same frequency. Poles and zeroes can also occur at frequencies that are combinations of real and imaginary numbers. The poles and zeroes of a circuit form a pattern in the complex plane that corresponds to certain types of circuit behavior. (The relationships between the pole-zero pattern and circuit behavior is beyond the scope of this book, but are covered in textbooks on circuit theory.)

What is a Pole?

Poles cause a specific change in the circuit's amplitude and phase response for real-world frequencies, even though we can't experience imaginary frequencies directly. A pole is associated with a bend in a magnitude response plot that changes the slope of the response downward with increasing frequency by 6 dB per octave (20 dB per decade; an octave is a 2:1 frequency ratio, a decade is a 10:1 frequency ratio).

There are four ways to identify the existence and frequency of a pole as shown in Fig 3.3:

1. For a downward bend in the magnitude versus frequency plot, the pole is at the −3 dB frequency for a single pole. If the bend causes a change in slope of more than 6 dB/octave, there must be multiple poles at this frequency.

2. A 90° lagging change on a phase versus frequency plot, where the lag increases with frequency. The pole is at the point of 45° added lag on the S-shaped transition. Multiple poles will add their phase lags, as above.

3. On a circuit diagram, a single pole looks like a simple RC low-pass filter. The pole is at the −3 dB frequency (f = $1/2\pi RC$ Hz). Any other circuit with the same response has a pole at the same frequency.

4. In an equation for the transfer function of a circuit, a pole is a theoretical value of frequency that would result in infinite gain. This is clearly impossible, but as the value of frequency will either be absolute zero, or will have an imaginary component, it is impossible to make an actual real-world signal at a pole frequency.

For example, comparing the amplitude responses at top and bottom of Fig 3.3 shows that the frequency of the pole is equal to the circuit's −3 dB cutoff frequency ($1/2\pi fC$) multiplied by j, which is also the frequency at which the circuit causes a −45° (lagging) phase shift from input to output.

What Is a Zero?

A zero is the complement of a pole. In math, it is a frequency at which the transfer function equation of a circuit is zero. This is not impossible in the real world (unlike the pole), so zeroes can be found at real-number frequencies as well as complex-number frequencies.

Each zero is associated with an *upward* bend of 6 dB per octave in a magnitude response. Similarly to a pole, the frequency of the zero is at the +3 dB point. Each zero is associated with a transition on a phase-versus-frequency plot that reduces the lag by 90°. The zero is at the 45° leading phase point. Multiple zeroes add their phase shifts just as poles do.

In a circuit, a zero creates gain that increases with frequency forever above the zero frequency. This requires active circuitry that would inevitably run out of gain at some frequency, which implies one or more poles up there. In real-world circuits, zeroes are usu-

ally not found by themselves, making the magnitude response go up, but rather paired with a pole of a different frequency, resulting in the magnitude response having a slope between two frequencies but flat above and below them.

Real-world circuit zeroes are only found accompanied by a greater or equal number of poles. Consider a classic RC high-pass filter, such as if the resistor and capacitor in Fig 3.3 were exchanged. The response of such a circuit increases at 6 dB per octave from 0 Hz (so there must be a zero at 0 Hz) and then levels off at $1/2\pi RC$ Hz. This leveling off is due to the presence of a pole adding its 6 dB-per-octave roll-off to cancel the 6 dB-per-octave roll-up of the zero. The transfer function for such as circuit would equal zero at zero frequency and infinity at the imaginary pole frequency.

FEEDBACK AND OSCILLATION

The *stability* of an amplifier refers to its ability to provide gain to a signal without tending to oscillate. For example, an amplifier just on the verge of oscillating is not generally considered to be "stable." If the output of an amplifier is fed back to the input, the feedback can affect the amplifier stability. If the amplified output is added to the input, the output of the sum will be larger. This larger output, in turn, is also fed back. As this process continues, the amplifier output will continue to rise until the amplifier cannot go any higher (clamps). Such *positive feedback* increases the amplifier gain, and is called *regeneration*. (The chapter on **Oscillators and Synthesizers** includes a discussion of positive feedback.)

Most practical amplifiers have some intrinsic and unavoidable feedback either as part of the circuit or in the amplifying device(s) itself. To improve the stability of an amplifier, *negative feedback* can be added to counteract any unwanted positive feedback. Negative feedback is often combined with a phase-shift *compensation* network to improve the amplifier stability.

Although negative feedback reduces amplifier or stage gain, the advantages of *stable* gain, freedom from unwanted oscillations and the reduction of distortion are often key design objectives and advantages of using negative feedback.

The design of feedback networks depends on the desired result. For amplifiers, which should not oscillate, the feedback network is customized to give the desired frequency response without loss of stability. For oscillators, the feedback network is designed to create a steady oscillation at the desired frequency.

SUMMING

In a linear system, nature does most of

the work for us when it comes to adding signals; placing two signals together naturally causes them to add according to the principle of superposition. When processing signals, we would like to control the summing operation so the signals do not distort or combine in a nonlinear way. If two signals come from separate stages and they are connected together directly, the circuitry of the stages may interact, causing distortion of either or both signals.

Summing amplifiers generally use a resistor in series with each stage, so the resistors connect to the common input of the following stage. This provides some *isolation* between the output circuits of each stage. **Fig 3.4** illustrates the resistors connecting to a summing amplifier. Ideally, any time we wanted to combine signals (for example, combining an audio signal with a sub-audible tone in a 2 meter FM transmitter prior to modulating the RF signal) we could use a summing amplifier.

FILTERING

A *filter* is a common linear stage in radio equipment. Filters are characterized by their ability to selectively attenuate certain frequencies in the filter's *stop band*, while passing or amplifying other frequencies in the *passband*. If the filter's passband extends to or near dc, it is a *low-pass* filter, and if to infinity (or at least very high frequencies for the circuitry involved), it is a *high-pass* filter. Filters that pass a range of frequencies are *band-pass* filters. *All-pass* filters are designed to affect only the phase of a signal without changing the signal amplitude. The range of frequencies between a band-pass circuit's low-pass and high-pass regions is its *mid-band*.

Fig 3.2A is the amplitude response for a typical band-pass audio filter. It shows that the input signal is passed to the output with no loss (0 dB) between 200 Hz and 5 kHz. This is the filter's *mid-band response*. Above and below those frequencies the response of the filter begins to drop. By 20 Hz and 20 kHz, the amplitude response has been reduced to one-

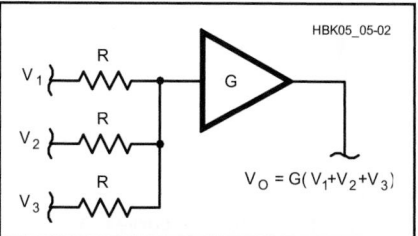

Fig 3.4 — Summing amplifier. The output voltage is equal to the sum of the input voltages times the amplifier gain, G. As long as the resistance values, R, are equal and the amplifier input impedance is much higher, the actual value of R does not affect the output signal.

half of (–3 dB) the mid-band response. These points are called the circuit's *cutoff* or *corner* or *half-power frequencies*. The range between the cutoff frequencies is the filter's passband. Outside the filter's passband, the amplitude response drops to 1/200th of (–23 dB) mid-band response at 1 Hz and only 1/1000th (–30 dB) at 500 kHz. The steepness of the change in response with frequency is the filter's *roll-off* and it is usually specified in dB/octave (an octave is a doubling or halving of frequency) or dB/decade (a decade is a change to 10 times or 1/10th frequency).

Fig 3.2B represents the phase response of a different filter — the simple RC low-pass filter shown at the upper right. As frequency increases, the reactance of the capacitor becomes smaller, causing most of the input signal to appear across the fixed-value resistor instead. At low frequencies, the capacitor has little effect on phase shift. As the signal frequency rises, however, there is more and more phase shift until at the cutoff frequency, there is 45° of lagging phase shift, plotted as a negative number. Phase shift then gradually approaches 90°.

Practical passive and active filters are described in the **RF and AF Filters** chapter. Filters implemented by digital computation (*digital filters*) are discussed in the chapter on **DSP and Software Radio Design**. Filters at RF may also be created by using transmission lines as described in the **Transmission Lines** chapter. All practical amplifiers are in effect either low-pass filters or band-pass filters, because their magnitude response decreases as the frequency increases beyond their gain-bandwidth products.

AMPLITUDE MODULATION/ DEMODULATION

Voice and data signals can be transmitted over the air by using *amplitude modulation* (AM) to combine them with higher frequency *carrier* signals (see the **Modulation** chapter). The process of amplitude modulation can be mathematically described as the multiplication (product) of the voice signal and the carrier signal. Multiplication is a linear process — amplitude modulation by the sum of two audio signals produces the same signal as the sum of amplitude modulation by each audio signal individually. Another aspect of the linear behavior of amplitude modulation is that amplitude-modulated signals can be demodulated exactly to their original form. Amplitude demodulation is the converse of amplitude modulation, and is represented as division, also a linear operation.

In the linear model of amplitude modulation, the signal that performs the modulation (such as the audio signal in an AM transmitter) is shifted in frequency by multiplying it with the carrier. The modulated waveform is considered to be a linear function of the signal.

The carrier is considered to be part of a time-varying linear system and not a second signal.

A curious trait of amplitude modulation (and demodulation) is that it can be performed nonlinearly, as well. Accurate analog multipliers (and dividers) are difficult and expensive to fabricate, so less-expensive nonlinear methods are often employed. Each nonlinear form of amplitude modulation generates the desired linear combination of signals, called *products*, in addition to other unwanted products that must then be removed. Both linear and nonlinear modulators and demodulators are discussed in the chapter on **Mixers, Modulators, and Demodulators**.

3.1.3 Nonlinear Operations

All signal processing doesn't have to be linear. Any time that we treat various signal levels differently, the operation is called *nonlinear*. This is not to say that all signals must be treated the same for a circuit to be linear. High-frequency signals are attenuated in a low-pass filter while low-frequency signals are not, yet the filter can be linear. The distinction is that the amount of attenuation at different frequencies is always the same, regardless of the amplitude of the signals passing through the filter.

What if we do not want to treat all voltage levels the same way? This is commonly desired in analog signal processing for clipping, rectification, compression, modulation and switching.

CLIPPING AND RECTIFICATION

Clipping is the process of limiting the range of signal voltages passing through a circuit (in other words, *clipping* those voltages outside the desired range from the signals). There are a number of reasons why we would like to do this. As shown in Fig 3.1, clipping is the process of limiting the positive and negative peaks of a signal. (Clipping is also called *clamping*.)

Clamping might be used to prevent a large audio signal from causing excessive deviation in an FM transmitter that would interfere with communications on adjacent channels. Clipping circuits are also used to protect sensitive inputs from excessive voltages. Clipping distorts the signal, changing it so that the original signal waveform is lost.

Another kind of clipping results in *rectification*. A *rectifier* circuit clips off all voltages of one polarity (positive or negative) and passes only voltages of the other polarity, thus changing ac to pulsating dc (see the **Power Sources** chapter). Another use of rectification is in a *peak detection* circuit that measures the peak value of a waveform. Only one polarity of the ac voltage needs to be measured and so a rectifier clips the unwanted polarity.

Another type of clipping occurs when an amplifier is intentionally operated with so much gain that the input signals result in an output that is clipped at the limits of its power supply voltages (or some other designated voltages) . The amplifier is said to be driven into *limiting* and an amplifier designed for this behavior is called a *limiter*. Limiters are used in FM receivers to amplify the signal until all amplitude variations in the signal are removed and the only characteristic of the original signal that remains is the frequency.

LOGARITHMIC AMPLIFICATION

It is sometimes desirable to amplify a signal logarithmically, which means amplifying low levels more than high levels. This type of amplification is often called *signal compression*. Speech compression is sometimes used in audio amplifiers that feed modulators. The voice signal is compressed into a small range of amplitudes, allowing more voice energy to be transmitted without overmodulation (see the **Modulation** chapter).

3.2 Analog Devices

There are several different kinds of components that can be used to build circuits for analog signal processing. Bipolar semiconductors, field-effect semiconductors and integrated circuits comprise a wide spectrum of active devices used in analog signal processing. (Vacuum tubes are discussed in the chapter on **RF Power Amplifiers**, their primary application in Amateur Radio.) Several different devices can perform the same function, each with its own advantages and disadvantages based on the physical characteristics of each type of device.

Understanding the specific characteristics of each device allows you to make educated decisions about which device would be best for a particular purpose when designing analog circuitry, or understanding why an existing circuit was designed in a particular way.

3.2.1 Terminology

A similar terminology is used when describing active electronic devices. The letter V or v stands for voltages and I or i for currents. Capital letters are often used to denote dc or bias values (bias is discussed later in this chapter). Lower-case often denotes instantaneous or ac values.

Voltages generally have two subscripts indicating the terminals between which the voltage is measured (V_{BE} is the dc voltage between the base and the emitter of a bipolar transistor). Currents have a single subscript indicating the terminal into which the current flows (I_C is the dc current into the collector of a bipolar transistor). If the current flows out of the device, it is generally treated as a negative value.

Resistance is designated with the letter R or r, and impedance with the letter Z or z. For example, r_{DS} is resistance between drain and source of an FET and Z_i is input impedance. For some parameters, values differ for dc and ac signals. This is indicated by using capital letters in the subscripts for dc and lower-case subscripts for ac. For example, the common-emitter dc current gain for a bipolar transistor is designated as h_{FE}, and h_{fe} is the ac current

gain. (See the section on transistor amplifiers later in this chapter for a discussion of the common-emitter circuit.) Qualifiers are sometimes added to the subscripts to indicate certain operating modes of the device. SS for saturation, BR for breakdown, ON and OFF are all commonly used.

Power supply voltages have two subscripts that are the same, indicating the terminal to which the voltage is applied. V_{DD} would represent the power supply voltage applied to the drain of a field-effect transistor.

Since integrated circuits are collections of semiconductor components, the abbreviations for the type of semiconductor used also apply to the integrated circuit. For example, V_{CC} is a power supply voltage for an integrated circuit made with bipolar transistor technology in which voltage is applied to transistor collectors.

3.2.2 Gain and Transconductance

The operation of an amplifier is specified by its *gain*. Gain in this sense is defined as the change (Δ) in the output parameter divided by the corresponding change in the input parameter. If a particular device measures its input and output as currents, the gain is called a *current gain*. If the input and output are voltages, the amplifier is defined by its *voltage gain*. *Power gain* is often used, as well. Gain is technically unit-less, but is often given in V/V. Decibels are often used to specify gain, particularly power gain.

If an amplifier's input is a voltage and the output is a current, the ratio of the change in output current to the change in input voltage is called *transconductance, g_m*.

$$g_m = \frac{\Delta I_o}{\Delta V_i} \tag{2}$$

Transconductance has the same units as conductance and admittance, Siemens (S), but is only used to describe the operation of active devices, such as transistors or vacuum tubes.

3.2.3 Characteristic Curves

Analog devices are described most completely with their *characteristic curves*. The characteristic curve is a plot of the interrelationships between two or three variables. The vertical (y) axis parameter is the output, or

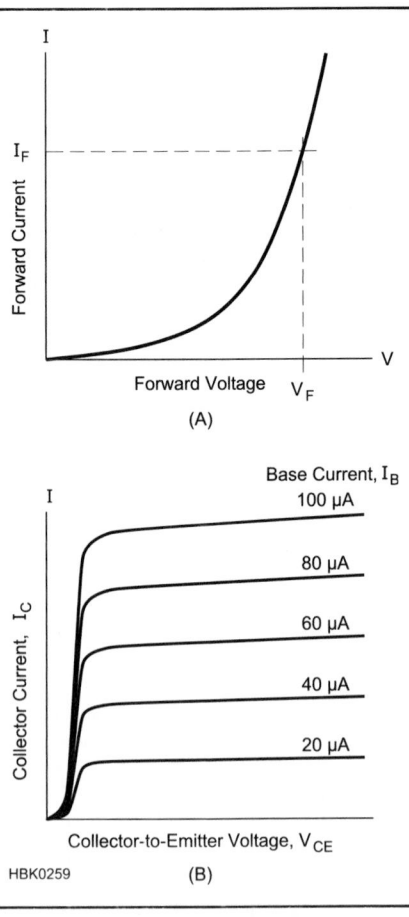

Fig 3.5 — Characteristic curves. A forward voltage vs forward current characteristic curve for a semiconductor diode is shown at (A). (B) shows a set of characteristic curves for a bipolar transistor in which the collector current vs collector-to-emitter voltage curve is plotted for five different values of base current.

result of the device being operated with an input parameter on the horizontal (x) axis. Often the output is the result of two input values. The first input parameter is represented along the x-axis and the second input parameter by several curves, each for a different value.

Almost all devices of concern are nonlinear over a wide range of operating parameters. We are often interested in using a device only in the region that approximates a linear response. Characteristic curves are used to graphically describe a device's operation in both its linear and nonlinear regions.

Fig 3.5A shows the characteristic curve for a semiconductor diode with the y-axis showing the forward current, I_F, flowing through the diode and the x-axis showing forward voltage, V_F, across the diode. This curve shows the relationship between current and voltage in the diode when it is conducting current. Characteristic curves showing voltage and current in two-terminal devices such as diodes are often called *I-V curves*. Characteristic curves may include all four quadrants of operation in which both axes include positive and negative values. It is also common for different scales to be used in the different quadrants, so inspect the legend for the curves carefully.

The parameters plotted in a characteristic curve depend on how the device will be used so that the applicable design values can be obtained from the characteristic curve. The slope of the curve is often important because it relates changes in output to changes in input. To determine the slope of the curve, two closely-spaced points along that portion of the curve are selected, each defined by its location along the x and y axes. If the two points are defined by (x_1, y_1) and (x_2, y_2), the slope, m, of the curve (which can be a gain, a resistance or a conductance, for example) is calculated as:

$$m = \frac{\Delta y}{\Delta x} = \frac{y_1 - y_2}{x_1 - x_2} \qquad (3)$$

It is important to pick points that are close together or the slope will not reflect the actual behavior of the device. A device whose characteristic curve is not a straight line will not have a linear response to inputs because the slope changes with the value of the input parameter.

For a device in which three parameters interact, such as a transistor, sets of characteristic curves can be drawn. Fig 3.5B shows a set of characteristic curves for a bipolar transistor where collector current, I_C, is shown on the y axis and collector-to-emitter voltage, V_{CE}, is shown on the x axis. Because the amount of collector current also depends on base current, I_B, the curve is repeated several times for different values of I_B. From this set of curves, an amplifier circuit using this transistor can be designed to have specific values of gain.

BIASING

The operation of an analog signal-processing device is greatly affected by which portion of the characteristic curve is used to do the processing. The device's *bias point* is its set of operating parameters when no input signal is applied. The bias point is also known as the *quiescent point* or *Q-point*. By changing the bias point, the circuit designer can affect the relationship between the input and output signal. The bias point can also be considered as a dc offset of the input signal. Devices that perform analog signal processing require appropriate input signal biasing.

As an example, consider the characteristic curve shown in **Fig 3.6**. (The exact types of device and circuit are unimportant.) The characteristic curve shows the relationship between an input voltage and an output current. Increasing input voltage results in an increase in output current so that an input signal is reproduced at the output. The characteristic curve is linear in the middle, but is quite nonlinear in its upper and lower regions.

In the circuit described by the figure, bias points are established by adding one of the three voltages, V_1, V_2 or V_3 to the input signal. Bias voltage V_1 results in an output current of I_1 when no input signal is present. This is shown as Bias Point 1 on the characteristic curve. When an input signal is applied, the input voltage varies around V_1 and the output current varies around I_1 as shown. If the dc

value of the output current is subtracted, a reproduction of the input signal is the result.

If Bias Point 2 is chosen, we can see that the input voltage is reproduced as a changing output current with the same shape. In this case, the device is operating linearly. If either Bias Point 1 or Bias Point 3 is chosen, however, the shape of the output signal is distorted because the characteristic curve of the device is nonlinear in this region. Either the increasing portion of the input signal results in more variation than the decreasing portion (Bias Point 1) or vice versa (Bias Point 3). Proper biasing is crucial to ensure that a device operates linearly.

3.2.4 Manufacturer's Data Sheets

Manufacturer's data sheets list device characteristics, along with the specifics of the part type (polarity, semiconductor type), identity of the pins and leads (*pinouts*), and the typical use (such as small signal, RF, switching or power amplifier). The pin identification is important because, although common package pinouts are normally used, there are exceptions. Manufacturers may differ slightly in the values reported, but certain basic parameters are listed. Different batches of the same devices are rarely identical, so manufacturers specify the guaranteed limits for the parameters of their device. There are

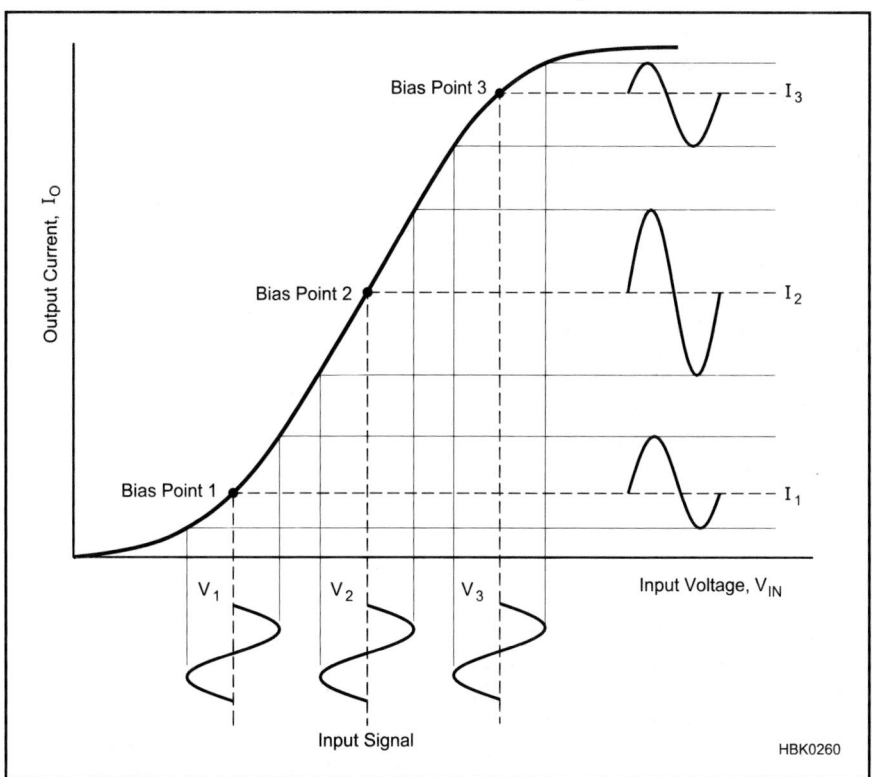

Fig 3.6 — Effect of biasing. An input signal may be reproduced linearly or nonlinearly depending on the choice of bias points.

usually three values listed in the data sheet for each parameter: guaranteed minimum value, the guaranteed maximum value, and/or the typical value.

Another section of the data sheet lists ABSOLUTE MAXIMUM RATINGS, beyond which device damage may result. For example, the parameters listed in the ABSOLUTE MAXIMUM RATINGS section for a solid-state device are typically voltages, continuous currents, total device power dissipation (P_D) and operating- and storage-temperature ranges.

Rather than plotting the characteristic curves for each device, the manufacturer often selects key operating parameters that describe the device operation for the configurations and parameter ranges that are most commonly used. For example, a bipolar transistor data sheet might include an OPERATING PARAMETERS section. Parameters are listed in an OFF CHARACTERISTICS subsection and an ON CHARACTERISTICS subsection that describe the conduction properties of the device for dc voltages. The SMALL-SIGNAL CHARACTERISTICS section might contain a minimum Gain-Bandwidth Product (f_T or GBW), maximum output capacitance, maximum input capacitance, and the range of the transfer parameters applicable to a given device. Finally, the SWITCHING CHARACTERISTICS section might list absolute maximum ratings for Delay Time (t_d), Rise Time (t_r), Storage Time (t_s), and Fall Time (t_f). Other types of devices list characteristics important to operation of that specific device.

When selecting equivalent parts for replacement of specified devices, the data sheet provides the necessary information to tell if a given part will perform the functions of another. Lists of cross-references and substitution guides generally only specify devices that have nearly identical parameters. There are usually a large number of additional devices that can be chosen as replacements. Knowledge of the circuit requirements adds even more to the list of possible replacements. The device parameters should be compared individually to make sure that the replacement part meets or exceeds the parameter values of the original part required by the circuit. Be aware that in some applications a far superior part may fail as a replacement, however. A transistor with too much gain could easily oscillate if there were insufficient negative feedback to ensure stability.

3.2.5 Physical Electronics of Semiconductors

In a conductor, such as a metal, some of the outer, or valence, electrons of each atom are free to move about between atoms. These free electrons are the constituents of electrical current. In a good conductor, the concentration of these free electrons is very high, on the order

of 10^{22} electrons/cm^3. In an insulator, nearly all the electrons are tightly held by their atoms and the concentration of free electrons is very small — on the order of 10 electrons/cm^3.

Between the classes of materials considered to be conductors and insulators is a class of elements called semiconductors, materials with conductivity much poorer than metals and much better than insulators. (In electronics, "semiconductor" means a device made from semiconductor elements that have been chemically manipulated as described below, leading to interesting properties that create useful applications.)

Semiconductor atoms (silicon, Si, is the most widely used) share their valence electrons in a chemical bond that holds adjacent atoms together, forming a three-dimensional lattice that gives the material its physical characteristics. A lattice of pure semiconductor material (one type of atom or molecule) can form a crystal, in which the lattice structure and orientation is preserved throughout the material. Monocrystalline or "single-crystal" is the type of material used in electronic semiconductor devices. Polycrystalline material is made of up many smaller crystals with their own individual lattice orientations.

Crystals of pure semiconductor material are called intrinsic semiconductors. When energy, generally in the form of heat, is added to a semiconductor crystal lattice, some electrons are liberated from their bonds and move freely throughout the lattice. The bond that loses an electron is then unbalanced and the space that the electron came from is referred to as a hole. In these materials the number of free electrons is equal to the number of holes.

Electrons from adjacent bonds can leave their positions to fill the holes, thus leaving behind a hole in their old location. As a consequence of the electron moving, two opposite movements can be said to occur: negatively charged electrons move from bond to bond in one direction and positively charged holes move from bond to bond in the opposite direction. Both of these movements represent forms of electrical current, but this is very different from the current in a conductor. While a conductor has free electrons that flow independently from the bonds of the crystalline lattice, the current in a pure semiconductor is constrained to move from bond to bond.

Impurities can be added to intrinsic semiconductors (by a process called doping) to enhance the formation of electrons or holes and thus improve conductivity. These materials are extrinsic semiconductors. Since the additional electrons and holes can move, their movement is current and they are called carriers. The type of carrier that predominates in the material is called the majority carrier. In N-type material the majority carriers are electrons and in P-type material, holes.

There are two types of impurities that can

be added: a donor impurity with five valence electrons donates free electrons to the crystalline structure; this is called an N-type impurity, for the negative charge of the majority carriers. Some examples of donor impurities are antimony (Sb), phosphorus (P) and arsenic (As). N-type extrinsic semiconductors have more electrons and fewer holes than intrinsic semiconductors. Acceptor impurities with three valence electrons accept free electrons from the lattice, adding holes to the overall structure. These are called P-type impurities, for the positive charge of the majority carriers; some examples are boron (B), gallium (Ga) and indium (In).

It is important to note that even though N-type and P-type material have different numbers of holes and free electrons than intrinsic material, they are still electrically neutral. When an electron leaves an atom, the positively-charged atom that remains in place in the crystal lattice electrically balances the roaming free electron. Similarly, an atom gaining an electron acquires a negative charge that balances the positively-charged atom it left. At no time does the material acquire a net electrical charge, positive or negative.

Compound semiconductor material can be formed by combining equal amounts of N-type and P-type impurity materials. Some examples of this include gallium-arsenide (GaAs), gallium-phosphate (GaP) and indium-phosphide (InP). To make an N-type compound semiconductor, a slightly higher amount of N-type material is used in the mixture. A P-type compound semiconductor has a little more P-type material in the mixture.

Impurities are introduced into intrinsic semiconductors by diffusion, the same physical process that lets you smell cookies baking from several rooms away. (Molecules diffuse through air much faster than through solids.) Rates of diffusion are proportional to temperature, so semiconductors are doped with impurities at high temperature to save time. Once the doped semiconductor material is cooled, the rate of diffusion of the impurities is so low that they are essentially immobile for many years to come. If an electronic device made from a structure of N- and P-type materials is raised to a high temperature, such as by excessive current, the impurities can again migrate and the internal structure of the device may be destroyed. The maximum operating temperature for semiconductor devices is specified at a level low enough to limit additional impurity diffusion.

The conductivity of an extrinsic semiconductor depends on the charge density (in other words, the concentration of free electrons in N-type, and holes in P-type, semiconductor material). As the energy in the semiconductor increases, the charge density also increases. This is the basis of how all semiconductor devices operate: the major difference is the way in which the energy level is increased. Variations include: The *transistor*, where conductivity is altered by injecting current into the device via a wire; the *thermistor*, where the level of heat in the device is detected by its conductivity, and the *photoconductor*, where light energy that is absorbed by the semiconductor material increases the conductivity.

3.2.6 The PN Semiconductor Junction

If a piece of N-type semiconductor material is placed against a piece of P-type semiconductor material, the location at which they join is called a *PN junction*. The junction has characteristics that make it possible to develop diodes and transistors. The action of the junction is best described by a diode operating as a rectifier.

Initially, when the two types of semiconductor material are placed in contact, each type of material will have only its majority carriers: P-type will have only holes and N-type will have only free electrons. The presence of the positive charges (holes) in the P-type material attracts free electrons from the N-type material immediately across the junction. The opposite is true in the N-type material.

These attractions lead to diffusion of some of the majority carriers across the junction, which combine with and neutralize the majority carriers immediately on the other side (a process called *recombination*). As distance from the junction increases, the attraction quickly becomes too small to cause the carriers to move. The region close to the junction is then *depleted* of carriers, and so is named the *depletion region* (also the *space-charge region* or the *transition region*). The width of the depletion region is very small, on the order of 0.5 μm.

If the N-type material (the *cathode*) is placed at a more negative voltage than the P-type material (the *anode*), current will pass through the junction because electrons are attracted from the lower potential to the higher potential and holes are attracted in the opposite direction. This *forward bias* forces the majority carriers toward the junction where recombination occurs with the opposite type of majority carrier. The source of voltage supplies replacement electrons to the N-type material and removes electrons from the P-type material so that the majority carriers are continually replenished. Thus, the net effect is a *forward current* flowing through the semiconductor, across the PN junction. The *forward resistance* of a diode conducting current is typically very low and varies with the amount of forward current.

When the polarity is reversed, majority carriers are attracted away from the junction, not toward it. Very little current flows across the PN junction — called *reverse leak-*

age current — in this case. Allowing only unidirectional current flow is what allows a semiconductor diode to act as rectifier.

3.2.7 Junction Semiconductors

Semiconductor devices that operate using the principles of a PN junction are called *junction semiconductors*. These devices can have one or several junctions. The properties of junction semiconductors can be tightly controlled by the characteristics of the materials used and the size and shape of the junctions.

SEMICONDUCTOR DIODES

Diodes are commonly made of silicon and occasionally germanium. Although they act similarly, they have slightly different characteristics. The *junction threshold voltage*, or *junction barrier voltage*, is the forward bias voltage (V_F) at which current begins to pass through the device. This voltage is different for the two kinds of diodes. In the diode response curve of **Fig 3.7**, V_F corresponds to the voltage at which the positive portion of the curve begins to rise sharply from the x-axis. Most silicon diodes have a junction threshold voltage of about 0.7 V, while the voltage for germanium diodes is typically 0.3 V. Reverse leakage current is much lower for silicon diodes than for germanium diodes.

The characteristic curve for a semiconductor diode junction is given by the following equation (slightly simplified) called the *Fundamental Diode Equation* because it describes the behavior of all semiconductor PN junctions.

$$I = I_S \left(e^{\frac{V}{\eta V_t}} - 1 \right) \qquad (4)$$

where
I = diode current
V = diode voltage
I_s = reverse-bias saturation current
$V_t = kT/q$, the thermal equivalent of voltage (about 25 mV at room temperature)
η = emission coefficient.

The value of I_s varies with the type of semiconductor material. η also varies from 1 to 2 with the type of material and the method of fabrication. (η is close to 2 for silicon at normal current levels, decreasing to 1 at high currents.) This curve is shown in **Fig 3.8B**.

The obvious differences between Fig 3.8A and B are that the semiconductor diode has a finite *turn-on* voltage — it requires a small but nonzero forward bias voltage before it begins conducting. Furthermore, once conducting, the diode voltage continues to increase very slowly with increasing current, unlike a true short circuit. Finally, when the applied voltage is negative, the reverse current is not exactly

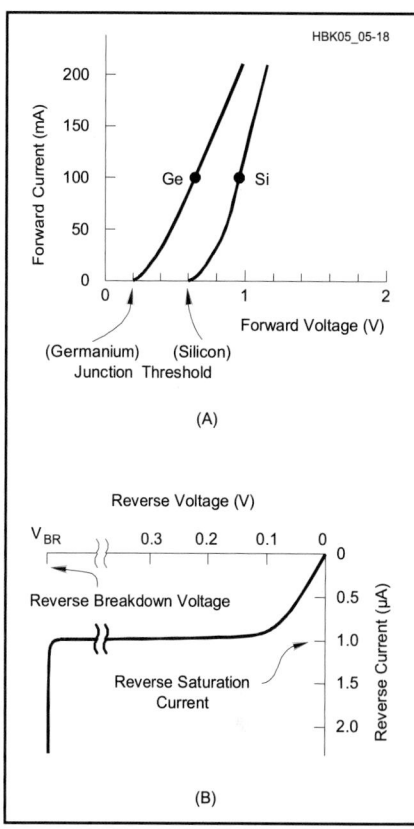

Fig 3.7 — Semiconductor diode (PN junction) characteristic curve. (A) Forward- biased (anode voltage higher than cathode) response for Germanium (Ge) and Silicon (Si) devices. Each curve breaks away from the x-axis at its junction threshold voltage. The slope of each curve is its forward resistance. (B) Reverse-biased response. Very small reverse current increases until it reaches the reverse saturation current (I_0). The reverse current increases suddenly and drastically when the reverse voltage reaches the reverse breakdown voltage, V_{BR}.

zero but very small (microamperes). The reverse current flow rapidly reaches a level that varies little with the reverse bias voltage. This is the *reverse-bias saturation current*, I_s.

For bias (dc) circuit calculations, a useful model for the diode that takes these two effects into account is shown by the artificial I-V curve in Fig 3.8C. This model neglects the negligible reverse bias current I_s.

When converted into an equivalent circuit, the model in Fig 3.8C yields the circuit in Fig 3.8D. The ideal voltage source V_a represents the turn-on voltage and R_f represents the effective resistance caused by the small increase in diode voltage as the diode current increases. The turn-on voltage is material-dependent: approximately 0.3 V for germanium diodes and 0.7 for silicon. R_f is typically on the order of 10 Ω, but it can vary according to the specific component. R_f can often be completely neglected in comparison to the other

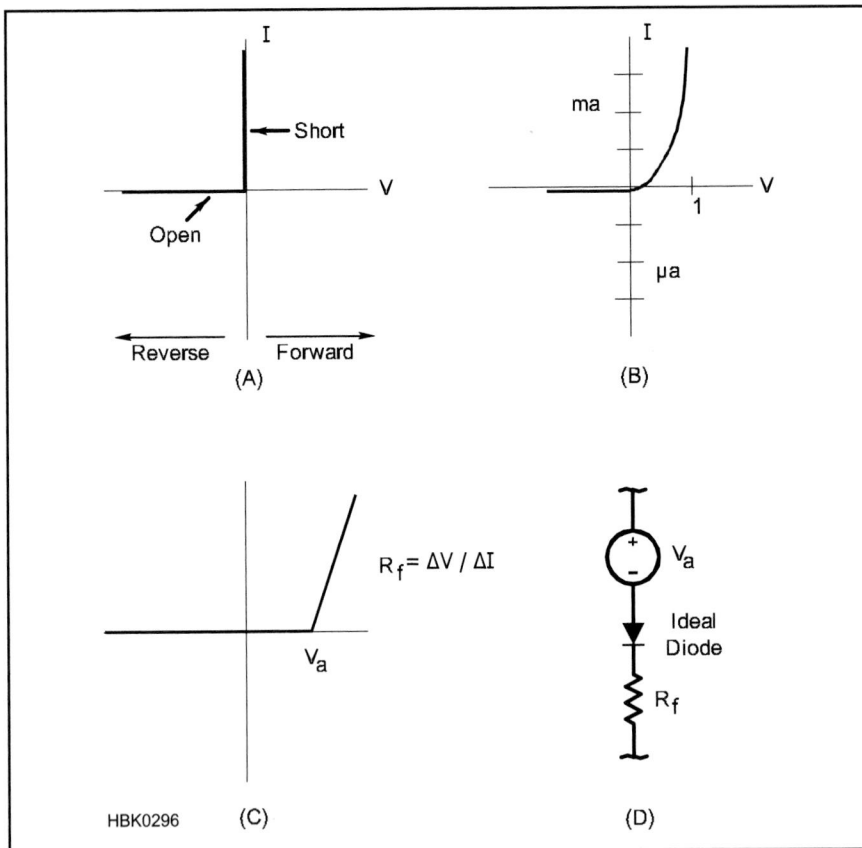

Figure 3.8 — Circuit models for rectifying switches (diodes). A: I-V curve of the ideal rectifier. B: I-V curve of a typical semiconductor diode showing the typical small leakage current in the reverse direction. Note the different scales for forward and reverse current. C shows a simplified diode I-V curve for dc-circuit calculations (at a much larger scale than B). D is an equivalent circuit for C.

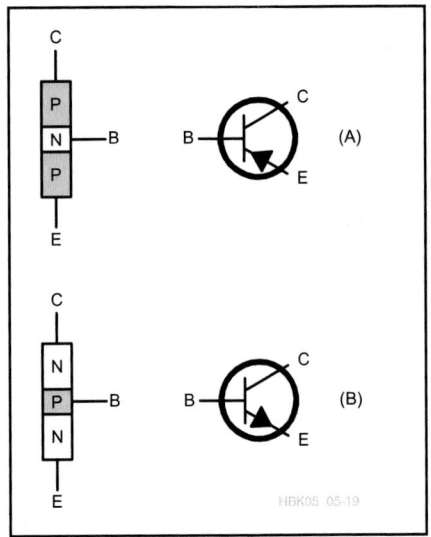

Fig 3.9 — Bipolar transistors. (A) A layer of N-type semiconductor sandwiched between two layers of P-type semiconductor makes a PNP device. The schematic symbol has three leads: collector (C), base (B) and emitter (E), with the arrow pointing in toward the base. (B) A layer of P-type semiconductor sandwiched between two layers of N-type semiconductor makes an NPN device. The schematic symbol has three leads: collector (C), base (B) and emitter (E), with the arrow pointing out away from the base.

resistances in the circuit. This very common simplification leaves only a pure *voltage drop* for the diode model.

BIPOLAR TRANSISTOR

A bipolar transistor is formed when two PN junctions are placed next to each other. If N-type material is surrounded by P-type material, the result is a PNP transistor. Alternatively, if P-type material is in the middle of two layers of N-type material, the NPN transistor is formed (**Fig 3.9**).

Physically, we can think of the transistor as two PN junctions back-to-back, such as two diodes connected at their *anodes* (the positive terminal) for an NPN transistor or two diodes connected at their *cathodes* (the negative terminal) for a PNP transistor. The connection point is the base of the transistor. (You can't actually make a transistor this way — this is a representation for illustration only.)

A transistor conducts when the base-emitter junction is forward biased and the base-collector is reverse biased. Under these conditions, the emitter region emits majority carriers into the base region, where they be-

come minority carriers because the materials of the emitter and base regions have opposite polarity. The excess minority carriers in the base are then attracted across the very thin base to the base-collector junction, where they are collected and are once again considered majority carriers before they can flow to the base terminal.

The flow of majority carriers from emitter to collector can be modified by the application of a bias current to the base terminal. If the bias current causes majority carriers to be injected into the base material (electrons flowing into an N-type base or out of a P-type base) the emitter-collector current increases. In this way, a transistor allows a small base current to control a much larger collector current.

As in a semiconductor diode, the forward biased base-emitter junction has a threshold voltage (V_{BE}) that must be exceeded before the emitter current increases. As the base-emitter current continues to increase, the point is reached at which further increases in base-emitter current cause no additional change in collector current. This is the condition of

saturation. Conversely, when base-emitter current is reduced to the point at which collector current ceases to flow, that is the situation of *cutoff.*

THYRISTORS

Thyristors are semiconductors made with four or more alternating layers of P- and N-type semiconductor material. In a four-layer thyristor, when the anode is at a higher potential than the cathode, the first and third junctions are forward biased and the center junction reverse biased. In this state, there is little current, just as in the reverse-biased diode. The different types of thyristor have different ways in which they turn on to conduct current and in how they turn off to interrupt current flow.

PNPN Diode

The simplest thyristor is a PNPN (usually pronounced like *pinpin*) diode with three junctions (see **Fig 3.10**). As the forward bias voltage is increased, the current through the device increases slowly until the *breakover (or firing) voltage*, V_{BO}, is reached and the flow of current abruptly increases. The PNPN diode is often considered to be a switch that is off below V_{BO} and on above it.

Bilateral Diode Switch (Diac)

A semiconductor device similar to two

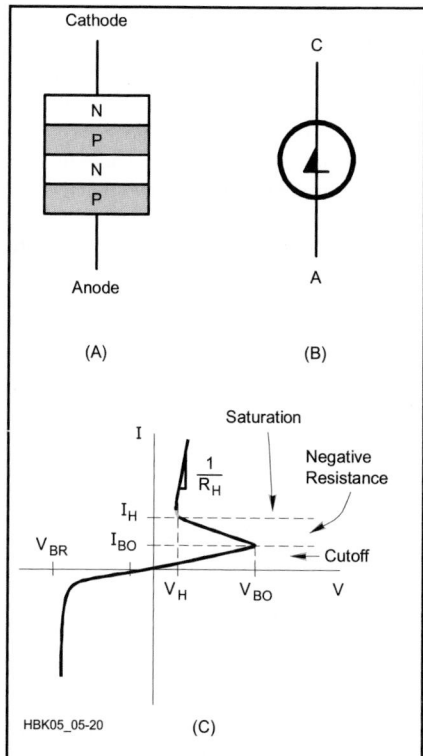

Fig 3.10 — PNPN diode. (A) Alternating layers of P-type and N-type semiconductor. (B) Schematic symbol with cathode (C) and anode (A) leads. (C) I-V curve. Reverse-biased response is the same as normal PN junction diodes. Forward biased response acts as a hysteresis switch. Resistance is very high until the bias voltage reaches V_{BO} (where the center junction breaks over) and exceeds the cutoff current, I_{BO}. The device exhibits a negative resistance when the current increases as the bias voltage decreases until a voltage of V_H and saturation current of I_H is reached. After this, the resistance is very low, with large increases in current for small voltage increases.

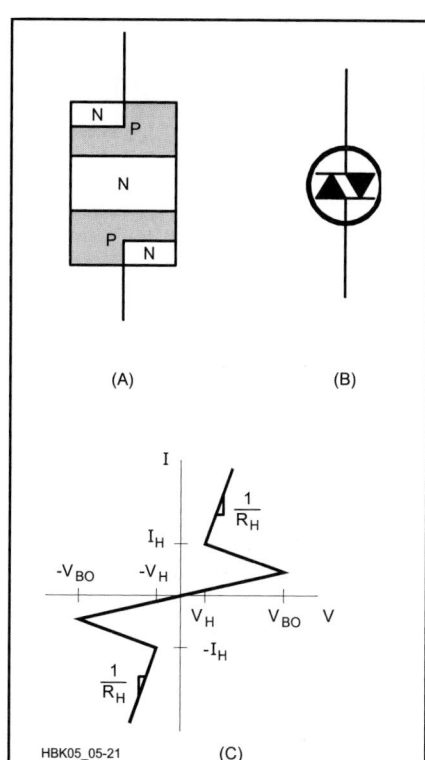

Fig 3.11 — Bilateral switch. (A) Alternating layers of P-type and N-type semiconductor. (B) Schematic symbol. (C) I-V curve. The right-hand side of the curve is identical to the PNPN diode response in Fig 3.10. The device responds identically for both forward and reverse bias so the left-hand side of the curve is symmetrical with the right-hand side.

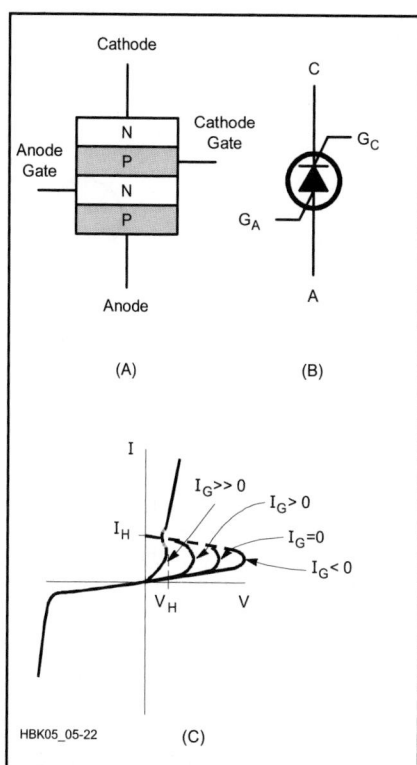

Fig 3.12 — SCR. (A) Alternating layers of P-type and N-type semiconductor. This is similar to a PNPN diode with gate terminals attached to the interior layers. (B) Schematic symbol with anode (A), cathode (C), anode gate (G_A) and cathode gate (G_C). Many devices are constructed without G_A. (C) I-V curve with different responses for various gate currents. $I_G = 0$ has a similar response to the PNPN diode.

PNPN diodes facing in opposite directions and attached in parallel is the *bilateral diode switch* or *diac*. This device has the characteristic curve of the PNPN diode for both positive and negative bias voltages. Its construction, schematic symbol and characteristic curve are shown in **Fig 3.11**.

Silicon Controlled Rectifier (SCR)

Another device with four alternate layers of P-type and N-type semiconductor is the *silicon controlled rectifier (SCR)*. (Some sources refer to an SCR as a thyristor, as well.) In addition to the connections to the outer two layers, two other terminals can be brought out for the inner two layers. The connection to the P-type material near the cathode is called the *cathode gate* and the N-type material near

the anode is called the *anode gate*. In nearly all commercially available SCRs, only the cathode gate is connected (**Fig 3.12**).

Like the PNPN diode switch, the SCR is used to abruptly start conducting when the voltage exceeds a given level. By biasing the gate terminal appropriately, the breakover voltage can be adjusted.

Triac

A five-layered semiconductor whose operation is similar to a bidirectional SCR is the *triac* (**Fig 3.13**). This is also similar to a bidirectional diode switch with a bias control gate. The gate terminal of the triac can control both positive and negative breakover voltages and the devices can pass both polarities of voltage.

Thyristor Applications

The SCR is highly efficient and is used in power control applications. SCRs are available that can handle currents of greater than

100 A and voltage differentials of greater than 1000 V, yet can be switched with gate currents of less than 50 mA. Because of their high current-handling capability, SCRs are used as "crowbars" in power supply circuits, to short the output to ground and blow a fuse when an overvoltage condition exists.

SCRs and triacs are often used to control ac power sources. A sine wave with a given RMS value can be switched on and off at preset points during the cycle to decrease the RMS voltage. When conduction is delayed until after the peak (as **Fig 3.14** shows) the peak-to-peak voltage is reduced. If conduction starts before the peak, the RMS voltage is reduced, but the peak-to-peak value remains the same. This method is used to operate light dimmers and 240 V ac to 120 V ac converters. The sharp switching transients created when these devices turn on are common sources of RF interference. (See the chapter on **RF Interference** for information on dealing with interference from thyristors.)

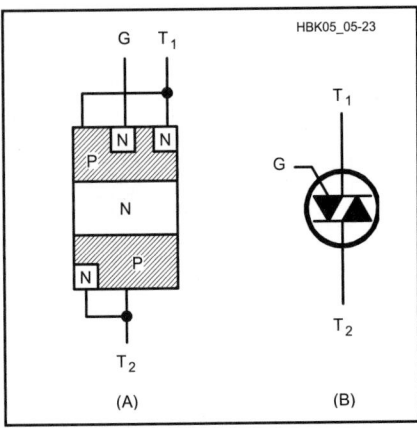

Fig 3.13 — Triac. (A) Alternating layers of P-type and N-type semiconductor. This behaves as two SCR devices facing in opposite directions with the anode of one connected to the cathode of the other and the cathode gates connected together. (B) Schematic symbol.

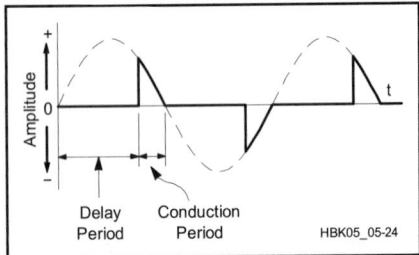

Fig 3.14 — Triac operation on sine wave. The dashed line is the original sine wave and the solid line is the portion that conducts through the triac. The relative delay and conduction period times are controlled by the amount or timing of gate current, I_G. The response of an SCR is the same as this for positive voltages (above the x-axis) and with no conduction for negative voltages.

3.2.8 Field-Effect Transistors (FET)

The *field-effect transistor (FET)* controls the current between two points but does so differently than the bipolar transistor. The FET operates by the effects of an electric field on the flow of electrons through a single type of semiconductor material. This is why the FET is sometimes called a *unipolar* transistor. Unlike bipolar semiconductors that can be arranged in many configurations to provide diodes, transistors, photoelectric devices, temperature sensitive devices and so on, the field effect technique is usually only used to make transistors, although FETs are also available as special-purpose diodes, for use as constant current sources.

FET devices are constructed on a *substrate* of doped semiconductor material. The channel is formed within the substrate and has the

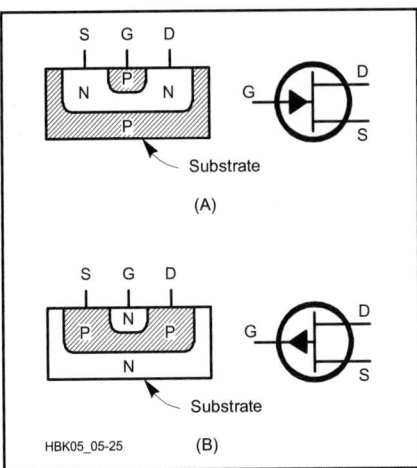

Fig 3.15 — JFET devices with terminals labeled: source (S), gate (G) and drain (D). A) Pictorial of N-type channel embedded in P-type substrate and schematic symbol. B) P-channel embedded in N-type substrate and schematic symbol.

opposite polarity (a P-channel FET has N-type substrate). Most FETs are constructed with silicon.

Within the FET, current moves in a *channel* as shown in **Fig 3.15**. The channel is made of either N-type or P-type semiconductor material; an FET is specified as either an N-channel or P-channel device. Current flows from the *source* terminal (where majority carriers are injected) to the *drain* terminal (where majority carriers are removed). A *gate* terminal generates an electric field that controls the current in the channel.

In N-channel devices, the drain potential must be higher than that of the source ($V_{DS} > 0$) for electrons (the majority carriers) to flow in channel. In P-channel devices, the flow of holes requires that $V_{DS} < 0$. The polarity of the electric field that controls current in the channel is determined by the majority carriers of the channel, ordinarily positive for P-channel FETs and negative for N-channel FETs.

Variations of FET technology are based on different ways of generating the electric field. In all of these, however, electrons at the gate are used only for their charge in order to create an electric field around the channel. There is a minimal flow of electrons through the gate. This leads to a very high dc input resistance in devices that use FETs for their input circuitry. There may be quite a bit of capacitance between the gate and the other FET terminals, however, causing the input impedance to be quite low at high frequencies.

The current through an FET only has to pass through a single type of semiconductor material. Depending on the type of material and the construction of the FET, drain-source resistance when the FET is conduct-

ing ($r_{DS(ON)}$) may be anywhere from a few hundred ohms to much less than an ohm. The output impedance of devices made with FETs is generally quite low. If a gate bias voltage is added to operate the transistor near cutoff, the circuit output impedance may be much higher.

In order to achieve a higher gain-bandwidth product, other materials have been used. Gallium-arsenide (GaAs) has *electron mobility* and *drift velocity* (both are measures of how easily electrons are able to move through the crystal lattice) far higher than the standard doped silicon. Amplifiers designed with GaAsFET devices operate at much higher frequencies and with a lower noise factor at VHF and UHF than those made with silicon FETs (although silicon FETs have improved dramatically in recent years).

JFET

One of two basic types of FET, the *junction FET (JFET)* gate material is made of the opposite polarity semiconductor to the channel material (for a P-channel FET the gate is made of N-type semiconductor material). The gate-channel junction is similar to a diode's PN junction with the gate material in direct contact with the channel. JFETs are used with the junction reverse-biased, since any current in the gate is undesirable. The reverse bias of the junction creates an electric field that "pinches" the channel. Since the magnitude of the electric field is proportional to the reverse-bias voltage, the current in the channel is reduced for higher reverse gate bias voltages. When current in the channel is completely halted by the electric field, this is called *pinch-off* and it is analogous to cutoff in a bipolar transistor. The channel in a JFET is at its maximum conductivity when the gate and source voltages are equal ($V_{GS} = 0$).

Because the gate-channel junction in a JFET is similar to a bipolar junction diode, this junction must never be forward biased; otherwise large currents will pass through the gate and into the channel. For an N-channel JFET, the gate must always be at a lower potential than the source ($V_{GS} < 0$). The prohibited condition is for $V_{GS} > 0$. For P-channel JFETs these conditions are reversed (in normal operation $V_{GS} > 0$ and the prohibited condition is for $V_{GS} < 0$).

MOSFET

Placing an insulating layer between the gate and the channel allows for a wider range of control (gate) voltages and further decreases the gate current (and thus increases the device input resistance). The insulator is typically made of an oxide (such as silicon dioxide, SiO_2). This type of device is called a *metal-oxide-semiconductor FET (MOSFET)* or *insulated-gate FET (IGFET)*.

The substrate is often connected to the source internally. The insulated gate is on the opposite side of the channel from the substrate (see **Fig 3.16**). The bias voltage on the gate terminal either attracts or repels the majority carriers of the substrate across its PN-junction with the channel. This narrows (*depletes*) or widens (*enhances*) the channel, respectively, as V_{GS} changes polarity. For example, in the N-channel enhancement-mode MOSFET, positive gate voltages with respect to the substrate and the source ($V_{GS} > 0$) repel holes from the channel into the substrate, thereby widening the channel and decreasing channel resistance. Conversely, $V_{GS} < 0$ causes holes to be attracted from the substrate, narrowing the channel and increasing the channel resistance. Once again, the polarities discussed in this example are reversed for P-channel devices. The common abbreviation for an N-channel MOSFET is *NMOS*, and for a P-channel MOSFET, *PMOS*.

Because of the insulating layer next to the gate, input resistance of a MOSFET is usually greater than 10^{12} Ω (a million megohms). Since MOSFETs can both deplete the channel, like the JFET, and also enhance it, the construction of MOSFET devices differs based on the channel size in the quiescent state, $V_{GS} = 0$.

A *depletion mode* device (also called a *normally-on MOSFET*) has a channel in the quiescent state that gets smaller as a reverse bias is applied; this device conducts current with no bias applied (see Fig 3.16A and B). An *enhancement mode* device (also called a *normally-off MOSFET*) is built without a channel and does not conduct current when $V_{GS} = 0$; increasing forward bias forms a temporary channel that conducts current (see Fig 3.16C and D).

Complementary Metal Oxide Semiconductors (CMOS)

Power dissipation in a circuit can be reduced to very small levels (on the order of a few nanowatts) by using MOSFET devices in complementary pairs (CMOS). Each amplifier is constructed of a series circuit of MOSFET devices, as in **Fig 3.17**. The gates are tied together for the input signal, as are the drains for the output signal. In saturation and cutoff, only one of the devices conducts. The current drawn by the circuit under no load is equal to the OFF leakage current of either device and the voltage drop across the pair is equal to V_{DD}, so the steady-state power used by the circuit is always equal to $V_{DD} \times I_{D(OFF)}$. Power is only consumed during the switching process, so for ac signals, power consumption is proportional to frequency.

CMOS circuitry could be built with discrete components, but the number of extra parts and the need for the complementary components to be matched has made that an unusual design technique. The low power consumption and ease of fabrication has made CMOS the most

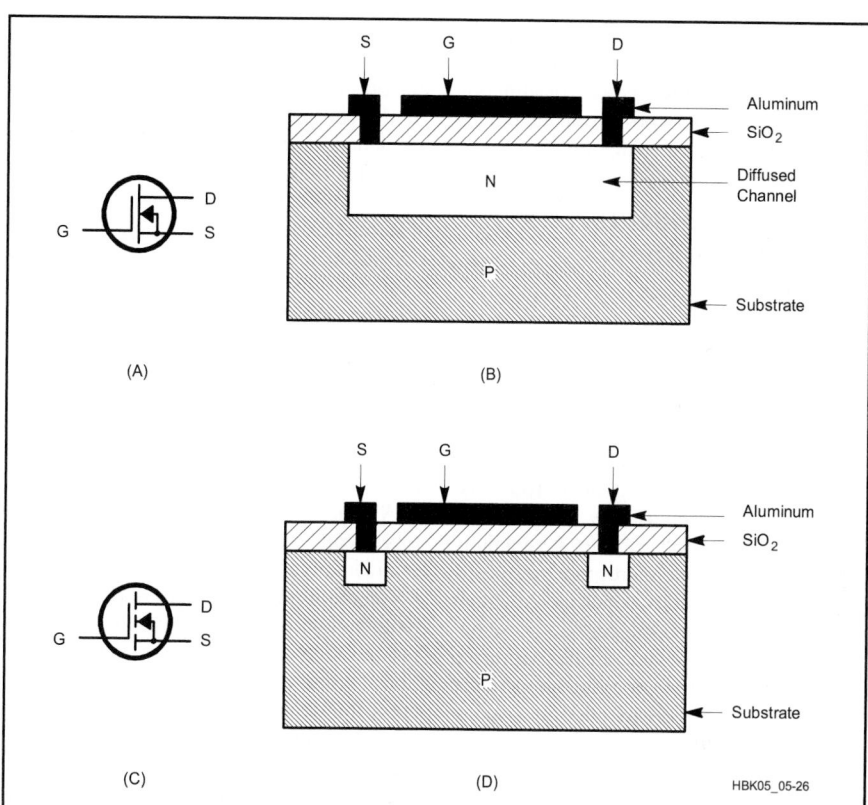

Fig 3.16 — MOSFET devices with terminals labeled: source (S), gate (G) and drain (D). N-channel devices are pictured. P-channel devices have the arrows reversed in the schematic symbols and the opposite type semiconductor material for each of the layers. (A) N-channel depletion mode device schematic symbol and (B) pictorial of P-type substrate, diffused N-type channel, SiO$_2$ insulating layer and aluminum gate region and source and drain connections. The substrate is connected to the source internally. A negative gate potential narrows the channel. (C) N-channel enhancement mode device schematic and (D) pictorial of P-type substrate, N-type source and drain wells, SiO$_2$ insulating layer and aluminum gate region and source and drain connections. Positive gate potential forms a channel between the two N-type wells by repelling the P-carriers away from the channel region in the substrate.

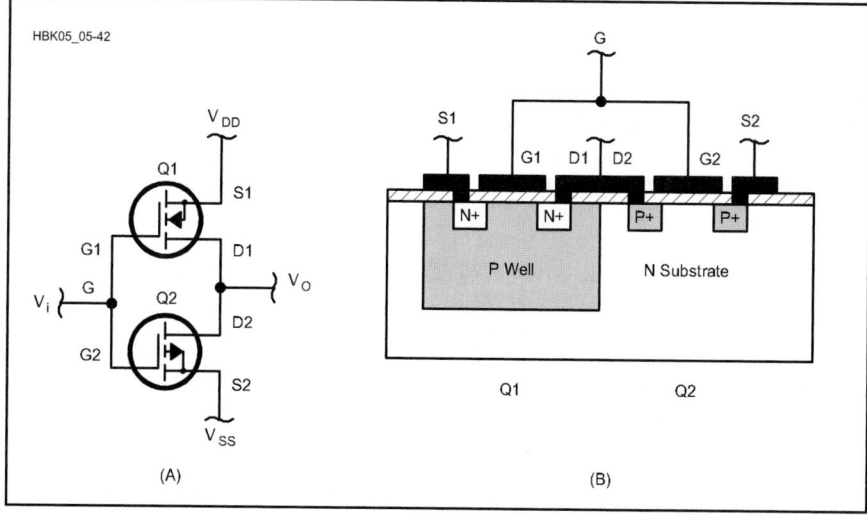

Fig 3.17 — Complementary metal oxide semiconductor (CMOS). (A) CMOS device is made from a pair of enhancement mode MOS transistors. The upper is an N-channel device, and the lower is a P-channel device. When one transistor is biased on, the other is biased off; therefore, there is minimal current from V_{DD} to ground. (B) Implementation of a CMOS pair as an integrated circuit.

common of all IC technologies. Although CMOS is most commonly used in digital integrated circuitry, its low power consumption has also been put to work by manufacturers of analog ICs, as well as digital ICs.

3.2.9 Semiconductor Temperature Effects

The number of excess holes and electrons in semiconductor material is increased as the temperature of a semiconductor increases. Since the conductivity of a semiconductor is related to the number of excess carriers, this also increases with temperature. With respect to resistance, semiconductors have a negative temperature coefficient. The resistance of silicon *decreases* by about 8% per °C and by about 6% per °C for germanium. Semiconductor temperature properties are the opposite of most metals, which *increase* their resistance by about 0.4% per °C. These opposing temperature characteristics permit the design of circuits with opposite temperature coefficients that cancel each other out, making a temperature insensitive circuit.

Semiconductor devices can experience an effect called *thermal runaway* as the current causes an increase in temperature. (This is primarily an issue with bipolar transistors.) The increased temperature decreases resistance and may lead to a further increase in current (depending on the circuit) that leads to an additional temperature increase. This sequence of events can continue until the semiconductor destroys itself, so circuit design must include measures that compensate for the effects of temperature.

Semiconductor Failure Caused by Heat

There are several common failure modes for semiconductors that are related to heat. The semiconductor material is connected to the outside world through metallic *bonding leads*. The point at which the lead and the semiconductor are connected is a common place for the semiconductor device to fail. As the device heats up and cools down, the materials expand and contract. The rate of expansion and contraction of semiconductor material is different from that of metal. Over many cycles of heating and cooling the bond between the semiconductor and the metal can break. Some experts have suggested that the lifetime of semiconductor equipment can be extended by leaving the devices powered on all the time, but this requires removal of the heat generated during normal operation.

A common failure mode of semiconductors is caused by the heat generated during semiconductor use. If the temperatures of the PN junctions remain at high enough levels for long enough periods of time, the impurities resume their diffusion across the PN junctions. When enough of the impurity atoms cross the depletion region, majority carrier recombination stops functioning properly and the semiconductor device fails permanently.

Excessive temperature can also cause failure anywhere in the semiconductor from heat generation within any current-carrying conductor, such as an FET channel or the bonding leads. Integrated circuits with more than one output may have power dissipation limits that depend on how many of the outputs are active at one time. The high temperature can cause localized melting or cracking of the semiconductor material, causing a permanent failure.

Another heat-driven failure mode, usually not fatal to the semiconductor, is excessive leakage current or a shift in operating point that causes the circuit to operate improperly. This is a particular problem in complex integrated circuits — analog and digital — dissipating significant amounts of heat under normal operating conditions. Computer microprocessors are a good example, often requiring their own cooling systems. Once the device cools, normal operation is usually restored.

To reduce the risk of thermal failures, the designer must comply with the limits stated in the manufacturer's data sheet, devising an adequate heat removal system. (Thermal issues are discussed in the **Electrical Fundamentals** chapter.)

3.2.10 Safe Operating Area (SOA)

Devices intended for use in circuits handling high currents or voltages are specified to have a *safe operating area (SOA)*. This refers to the area drawn on the device's characteristic curve containing combinations of voltage and current that the device can be expected to control without damage under specific conditions. The SOA combines a number of limits — voltage, current, power, temperature and various breakdown mechanisms — in order to simplify the design of protective circuitry. The SOA is also specified to apply to specific durations of use — steady-state, long pulses, short pulses and so forth. The device may have separate SOAs for resistive and inductive loads.

You may also encounter two specialized types of SOA for turning the device on and off. *Reverse bias safe operating area (RB-SOA)* applies when the device is turning off. *Forward bias safe operating* area (*FBSOA*) applies when turning the device on. These SOAs are used because the high rate-of-change of current and voltage places additional stresses on the semiconductor.

3.3 Practical Semiconductors

3.3.1 Semiconductor Diodes

Although many types of semiconductor diodes are available, they share many common characteristics. The different types of diodes have been developed to optimize particular characteristics for one type of application. You will find many examples of diode applications throughout this book.

The diode symbol is shown in **Fig 3.18**. Forward current flows in the direction from anode to cathode, in the direction of the arrow. Reverse current flows from cathode to anode. (Current is considered to be conventional current as described in the **Electrical Fundamentals** chapter.) The anode of a semiconductor junction diode is made of P-type material and the cathode is made of N-type material, as indicated in Fig 3.18. Most diodes are marked with a band on the cathode end (Fig 3.18).

DIODE RATINGS

Five major characteristics distinguish standard junction diodes from one another: current handling capacity, maximum voltage rating, response speed, reverse leakage current and junction forward voltage. Each of these characteristics can be manipulated during manufacture to produce special purpose diodes.

Current Capacity

The ideal diode would have zero resistance in the forward direction and infinite resistance in the reverse direction. This is not the case for actual devices, which behave as shown in the plot of a diode response in Fig 3.7. Note that the scales of the two graphs are drastically different. The inverse of the slope of the line (the change in voltage between two points on a straight portion of the line divided by the corresponding change in current) on the upper right is the resistance of the diode in the forward direction, R_F.

The range of voltages is small and the range of currents is large since the forward resistance is very small (in this example, about 2Ω). Nevertheless, this resistance causes heat dissipation according to $P = I_F{}^2 \times R_F$.

In addition, there is a forward voltage, V_F, whenever the forward current is flowing. This

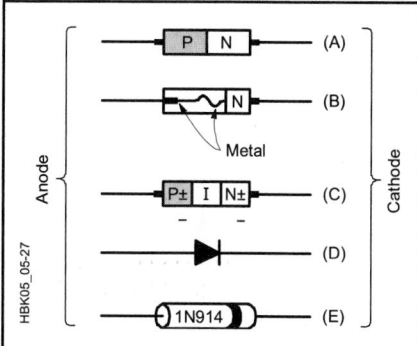

Fig 3.18 — Practical semiconductor diodes. All devices are aligned with anode on the left and cathode on the right. (A) Standard PN junction diode. (B) Point-contact or "cat's whisker" diode. (C) PIN diode formed with heavily doped P-type (P+), undoped (intrinsic) and heavily doped N-type (N+) semiconductor material. (D) Diode schematic symbol. (E) Diode package with marking stripe on the cathode end.

also results in heat dissipation as $P = I \times V_F$. In power applications where the average forward current is high, heating from forward resistance and the forward voltage drop can be significant. Since forward current determines the amount of heat dissipation, the diode's power rating is stated as a *maximum average current*. Exceeding the current rating in a diode will cause excessive heating that leads to PN junction failure as described earlier.

Peak Inverse Voltage (PIV)

In Fig 3.7, the lower left portion of the curve illustrates a much higher resistance that increases from tens of kilohms to thousands of megohms as the reverse voltage gets larger, and then decreases to near zero (a nearly vertical line) very suddenly. This sudden change occurs because the diode enters *reverse breakdown* or when the reverse voltage becomes high enough to push current across the junction. The voltage at which this occurs is the *reverse breakdown voltage*. Unless the current is so large that the diode fails from overheating, breakdown is not destructive and the diode will again behave normally when the bias is removed. The maximum reverse voltage that the diode can withstand under normal use is the *peak inverse voltage (PIV)* rating. A related effect is *avalanche breakdown* in which the voltage across a device is greater than its ability to control or block current flow.

Response Speed

The speed of a diode's response to a change in voltage polarity limits the frequency of ac current that the diode can rectify. The diode response in Fig 3.7 shows how that diode will act at dc. As the frequency increases, the diode

may not be able to turn current on and off as fast as the changing polarity of the signal.

Diode response speed mainly depends on *charge storage* in the depletion region. When forward current is flowing, electrons and holes fill the region near the junction to recombine. When the applied voltage reverses, these excess charges move away from the junction so that no recombination can take place. As reverse bias empties the depletion region of excess charge, it begins to act like a small capacitor formed by the regions containing majority carriers on either side of the junction and the depletion region acting as the dielectric. This *junction capacitance* is inversely proportional to the width of the depletion region and directly proportional to the cross-sectional surface area of the junction.

The effect of junction capacitance is to allow current to flow for a short period after the applied voltage changes from positive to negative. To halt current flow requires that the junction capacitance be charged. Charging this capacitance takes some time; a few µs for regular rectifier diodes and a few hundred nanoseconds for *fast-recovery* diodes. This is the diode's *charge-storage time*. The amount of time required for current flow to cease is the diode's *recovery time*.

Reverse Leakage Current

Because the depletion region is very thin, reverse bias causes a small amount of reverse leakage or reverse saturation current to flow from cathode to anode. This is typically 1 µA or less until reverse breakdown voltage is reached. Silicon diodes have lower reverse leakage currents than diodes made from other materials with higher carrier mobility, such as germanium.

The reverse saturation current I_s is not constant but is affected by temperature, with higher temperatures increasing the mobility of the majority carriers so that more of them cross the depletion region for a given amount of reverse bias. For silicon diodes (and transistors) near room temperature, I_s increases by a factor of 2 every 4.8 °C. This means that for every 4.8 °C rise in temperature, either the current doubles (if the voltage across it is constant), or if the current is held constant by other resistances in the circuit, the diode voltage will *decrease* by $V_T \times \ln 2$ = 18 mV. For germanium, the current doubles every 8 °C and for gallium-arsenide (GaAs), 3.7 °C. This dependence is highly reproducible and may actually be exploited to produce temperature-measuring circuits.

While the change resulting from a rise of several degrees may be tolerable in a circuit design, that from 20 or 30 degrees may not. Therefore it's a good idea with diodes, just as with other components, to specify power ratings conservatively (2 to 4 times margin) to prevent self-heating.

While component derating does reduce self-heating effects, circuits must be designed for the expected operating environment. For example, mobile radios may face temperatures from –20° to +140 °F (–29° to 60 °C).

Forward Voltage

The amount of voltage required to cause majority carriers to enter the depletion region and recombine, creating full current flow, is called a diode's *forward voltage*, V_F. It depends on the type of material used to create the junction and the amount of current. For silicon diodes at normal currents, $V_F = 0.7$ V, and for germanium diodes, $V_F = 0.3$ V. As you saw earlier, V_F also affects power dissipation in the diode.

POINT-CONTACT DIODES

One way to decrease charge storage time in the depletion region is to form a metal-semiconductor junction for which the depletion is very thin. This can be accomplished with a *point-contact diode*, where a thin piece of aluminum wire, often called a *whisker*, is placed in contact with one face of a piece of lightly doped N-type material. In fact, the original diodes used for detecting radio signals ("cat's whisker diodes") were made with a steel wire in contact with a crystal of impure lead (galena). Point-contact diodes have high response speed, but poor PIV and current-handling ratings. The 1N34 germanium point-contact diode is the best-known example of point-contact diode still in common use.

SCHOTTKY DIODES

An improvement to point-contact diodes, the *hot-carrier diode* is similar to a point-contact diode, but with more ideal characteristics attained by using more efficient metals, such as platinum and gold, that act to lower forward resistance and increase PIV. This type of contact is known as a *Schottky barrier*, and diodes made this way are called *Schottky diodes*. The junctions of Schottky diodes, being smaller, store less charge and as a result, have shorter switching times and junction capacitances than standard PN-junction diodes. Their forward voltage is also lower, typically 0.3 to 0.4 V. In most other respects they behave similarly to PN diodes.

PIN DIODES

The PIN diode, shown in Fig 3.18C is a *slow response* diode that is capable of passing RF and microwave signals when it is forward biased. This device is constructed with a layer of intrinsic (undoped) semiconductor placed between very highly doped P-type and N-type material (called P+-type and N+-type material to indicate the extra amount of doping), creating a *PIN junction*. These devices provide very effective switches for RF signals and are often used in transmit-receive switches

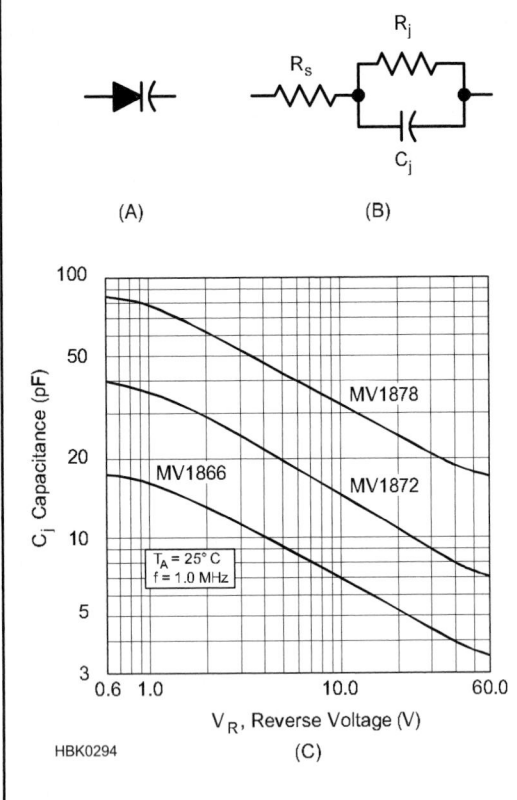

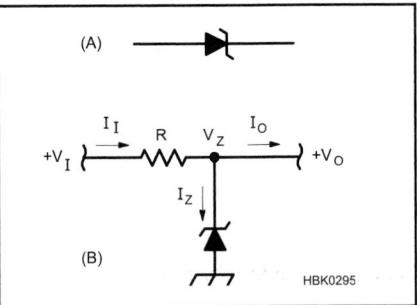

Fig 3.19 — Varactor diode. (A) Schematic symbol. (B) Equivalent circuit of the reverse biased varactor diode. R_S is the junction resistance, R_J is the leakage resistance and C_J is the junction capacitance, which is a function of the magnitude of the reverse bias voltage. (C) Plot of junction capacitance, C_J, as a function of reverse voltage, V_R, for three different varactor devices. Both axes are plotted on a logarithmic scale.

Fig 3.20 — Zener diode. (A) Schematic symbol. (B) Basic voltage regulating circuit. V_Z is the Zener reverse breakdown voltage. Above V_Z, the diode draws current until $V_I - I_I R = V_Z$. The circuit design should select R so that when the maximum current is drawn, $R < (V_I - V_Z) / I_O$. The diode should be capable of passing the same current when there is no output current drawn.

in transceivers and amplifiers. The majority carriers in PIN diodes have longer than normal lifetimes before recombination, resulting in a slow switching process that causes them to act more like resistors than diodes at high radio frequencies. The amount of resistance can be controlled by the amount of forward bias applied to the PIN diode and this allows them to act as current-controlled attenuators. (For additional discussion of PIN diodes and projects in which they are used, see the chapters on **Transmitters and Transceivers**, **RF Power Amplifiers**, and **Test Equipment and Measurements**.)

VARACTOR DIODES

Junction capacitance can be used as a circuit element by controlling the reverse bias voltage across the junction, creating a small variable capacitor. Junction capacitances are small, on the order of pF. As the reverse bias voltage on a diode increases, the width of the depletion region increases, decreasing its capacitance. A *varactor* (also known by the trade name Varicap diode) is a diode with a junction specially formulated to have a relatively large range of capacitance values for a modest range of reverse bias voltages (**Fig 3.19**).

As the reverse bias applied to a diode changes, the width of the depletion layer, and therefore the capacitance, also changes. The diode junction capacitance (C_j) under a reverse bias of V volts is given by

$$C_j = \frac{C_{j0}}{\sqrt{V_{on} - V}} \qquad (5)$$

where C_{j0} = measured capacitance with zero applied voltage.

Note that the quantity under the radical is a large *positive* quantity for reverse bias. As seen from the equation, for large reverse biases C_j is inversely proportional to the square root of the voltage.

Although special forms of varactors are available from manufacturers, other types of diodes may be used as inexpensive varactor diodes, but the relationship between reverse voltage and capacitance is not always reliable.

When designing with varactor diodes, the reverse bias voltage must be absolutely free of noise since any variations in the bias voltage will cause changes in capacitance. For example, if the varactor is used to tune an oscillator, unwanted frequency shifts or instability will result if the reverse bias voltage is noisy. It is possible to frequency modulate a signal by adding the audio signal to the reverse bias on a varactor diode used in the carrier oscillator. (For examples of the use of varactors in oscillators and modulators, see the chapters on **Mixers, Modulators, and Demodulators** and **Oscillators and Synthesizers**.)

ZENER DIODES

When the PIV of a reverse-biased diode is exceeded, the diode begins to conduct current

as it does when it is forward biased. This current will not destroy the diode if it is limited to less than the device's maximum allowable value. By using heavy levels of doping during manufacture, a diode's PIV can be precisely controlled to be at a specific level, called the *Zener voltage*, creating a type of voltage reference. These diodes are called Zener diodes after their inventor, American physicist Clarence Zener.

When the Zener voltage is reached, the reverse voltage across the Zener diode remains constant even as the current through it changes. With an appropriate series current-limiting resistor, the Zener diode provides an accurate voltage reference (see **Fig 3.20**).

Zener diodes are rated by their reverse-breakdown voltage and their power-handling capacity, where $P = V_Z \times I_Z$. Since the same current must always pass though the resistor to drop the source voltage down to the reference voltage, with that current divided between the Zener diode and the load, this type of power source is very wasteful of current.

The Zener diode does make an excellent and efficient voltage reference in a larger voltage regulating circuit where the load current is provided from another device whose voltage is set by the reference. (See the **Power Sources** chapter for more information about using Zener diodes as voltage regulators.) When operating in the breakdown region, Zener diodes can be modeled as a simple voltage source.

The primary sources of error in Zener-diode-derived voltages are the variation with load current and the variation due to heat. Temperature-compensated Zener diodes are available with temperature coefficients as low as 5 parts per million per °C. If this is unacceptable, voltage reference integrated circuits based on Zener diodes have been developed that include additional circuitry to counteract temperature effects.

A variation of Zener diodes, *transient voltage suppressor* (*TVS*) diodes are designed to dissipate the energy in short-duration, high-voltage transients that would otherwise damage equipment or circuits. TVS diodes have large junction cross-sections so that they can handle large currents without damage. These diodes are also known as TransZorbs. Since the polarity of the transient can be positive, negative, or both, transient protection circuits can be designed with two devices connected with opposite polarities.

RECTIFIERS

The most common application of a diode is to perform rectification; that is, permitting current flow in only one direction. Power rectification converts ac current into pulsating dc current. There are three basic forms of power rectification using semiconductor diodes: half wave (1 diode), full-wave center-tapped (2 diodes) and full-wave bridge (4 diodes). These applications are shown in **Fig 3.21A**, B and C and are more fully described in the **Power Sources** chapter.

The most important diode parameters to consider for power rectification are the PIV and current ratings. The peak negative voltages that are blocked by the diode must be smaller in magnitude than the PIV and the peak current through the diode when it is forward biased must be less than the maximum average forward current.

Rectification is also used at much lower current levels in modulation and demodulation and other types of analog signal processing circuits. For these applications, the diode's response speed and junction forward voltage are the most important ratings.

3.3.2 Bipolar Junction Transistors (BJT)

The bipolar transistor is a *current-controlled device* with three basic terminals; *emitter*, *collector* and *base*. The current between the emitter and the collector is controlled by the current between the base and emitter. The convention when discussing transistor operation is that the three currents into the device are positive (I_c into the collector, I_b into the base and I_e into the emitter). Kirchhoff's Current Law (see the **Electrical Fundamentals** chapter) applies to transistors just as it does to passive electrical networks: the total current entering the device must be zero. Thus, the relationship between the currents into a transistor can be generalized as

$$I_c + I_b + I_e = 0 \qquad (6)$$

which can be rearranged as necessary. For

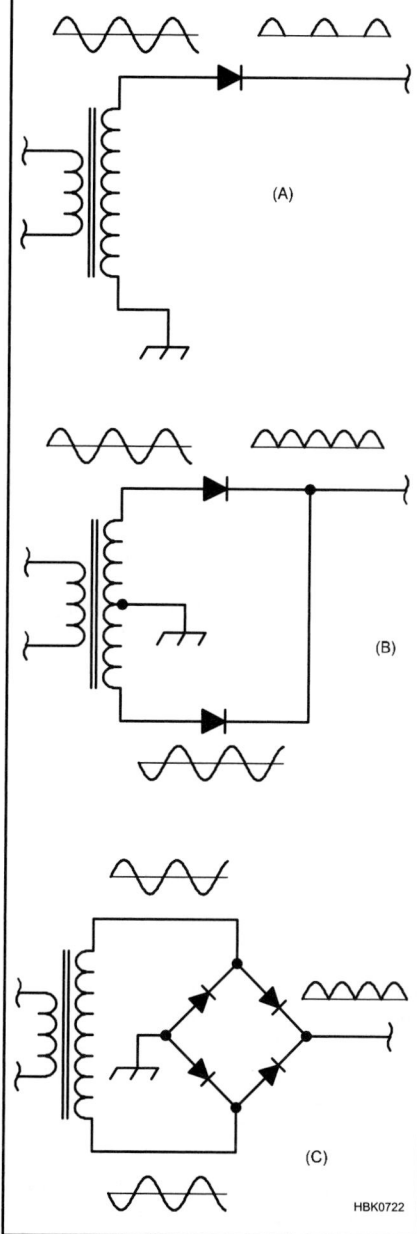

Fig 3.21 — Diode rectifier circuits. (A) Half wave rectifier circuit. Only when the ac voltage is positive does current pass through the diode. Current flows only during half of the cycle. (B) Full-wave center-tapped rectifier circuit. Center-tap on the transformer secondary is grounded and the two ends of the secondary are 180° out of phase. (C) Full-wave bridge rectifier circuit. In each half of the cycle two diodes conduct.

example, if we are interested in the emitter current,

$$I_e = -\left(I_c + I_b\right) \qquad (7)$$

The back-to-back diode model shown in Fig 3.9 is appropriate for visualization of transistor construction. In actual transistors,

however, the relative sizes of the collector, base and emitter regions differ. A common transistor configuration that spans a distance of 3 mm between the collector and emitter contacts typically has a base region that is only 25 μm across.

The operation of the bipolar transistor is described graphically by characteristic curves as shown in **Fig 3.22**. These are similar to the I-V characteristic curves for the two-terminal devices described in the preceding sections. The parameters shown by the curves depend on the type of circuit in which they are measured, such as common emitter or common collector. The output characteristic shows a set of curves for either collector or emitter current versus collector-emitter voltage at various values of input current (either base or emitter). The input characteristic shows the voltage between the input and common terminals (such as base-emitter) versus the input current for different values of output voltage.

CURRENT GAIN

Two parameters describe the relationships between the three transistor currents at low frequencies:

$$\alpha \approx -\frac{\Delta I_C}{\Delta I_E} \approx 1 \qquad (8)$$

$$\beta = \frac{\Delta I_C}{\Delta I_B} \qquad (9)$$

The relationship between α and β is defined as

$$\alpha = -\frac{\beta}{1+\beta} \qquad (10)$$

Another designation for β is often used: h_{FE}, the *forward dc current gain*. (The "h" refers to "h parameters," a set of transfer parameters for describing a two-port network and described in more detail in the **RF Techniques** chapter.) The symbol, h_{fe}, in which the subscript is in lower case, is used for the forward current gain of ac signals.

OPERATING REGIONS

Current conduction between collector and emitter is described by *regions* of the transistor's characteristic curves in Fig 3.22. (References such as *common-emitter* or *common-base* refer to the configuration of the circuit in which the parameter is measured.) The transistor is in its *active* or *linear region* when the base-collector junction is reverse biased and the base-emitter junction is forward biased. The slope of the output current, I_O, versus the output voltage, V_O, is virtually flat, indicating that the output current is nearly independent of the output voltage. In this region, the output circuit of the transistor can be modeled as a constant-current source controlled by

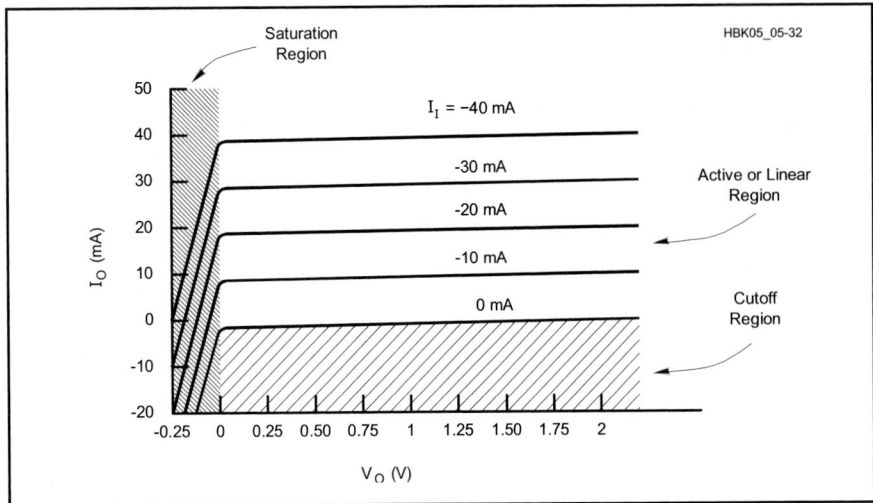

Fig 3.22 — Transistor response curve output characteristics. The x-axis is the output voltage, and the y-axis is the output current. Different curves are plotted for various values of input current. The three regions of the transistor are its cutoff region, where no current flows in any terminal, its active region, where the output current is nearly independent of the output voltage and there is a linear relationship between the input current and the output current, and the saturation region, where the output current has large changes for small changes in output voltage.

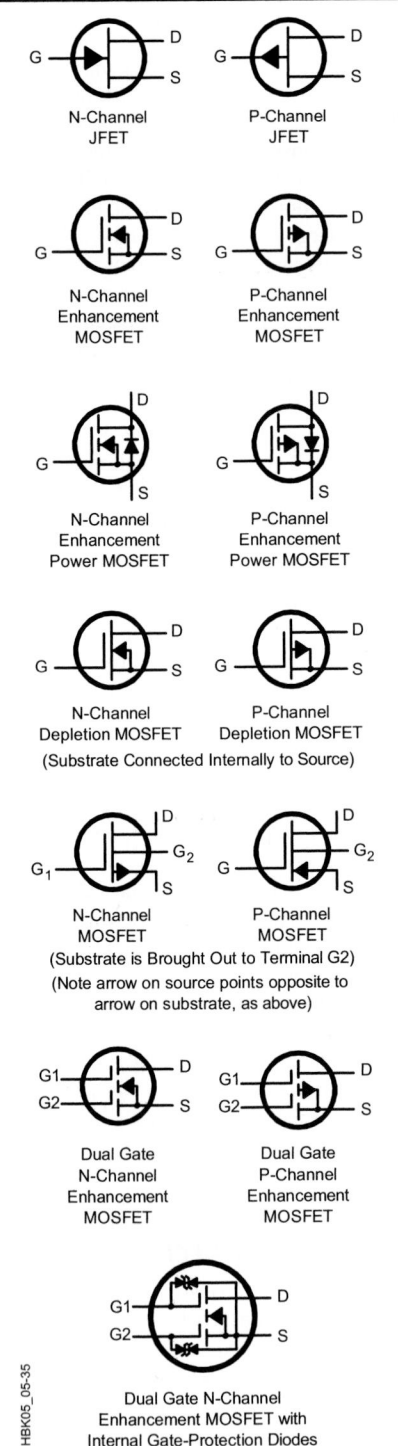

Fig 3.23—FET schematic symbols.

the input current. The slight slope that does exist is due to base-width modulation (known as the *Early effect*).

When both the junctions in the transistor are forward biased, the transistor is said to be in its *saturation region*. In this region, V_O is nearly zero and large changes in I_O occur for very small changes in V_O. Both junctions in the transistor are reverse-biased in the *cutoff region*. Under this condition, there is very little current in the output, only the nanoamperes or microamperes that result from the very small leakage across the input-to-output junction. Finally, if V_O is increased to very high values, avalanche breakdown begins as in a PN-junction diode and output current increases rapidly. This is the *breakdown region*, not shown in Fig 3.22.

These descriptions of junction conditions are the basis for the use of transistors. Various configurations of the transistor in circuitry make use of the properties of the junctions to serve different purposes in analog signal processing.

OPERATING PARAMETERS

A typical general-purpose bipolar-transistor data sheet lists important device specifications. Parameters listed in the ABSOLUTE MAXIMUM RATINGS section are the three junction voltages (V_{CEO}, V_{CBO} and V_{EBO}), the continuous collector current (I_C), the total device power dissipation (P_D) and the operating and storage temperature range. Exceeding any of these parameters is likely to cause the transistor to be destroyed. (The "O" in the suffixes of the junction voltages indicates that the remaining terminal is not connected, or open.)

In the OPERATING PARAMETERS section, three guaranteed minimum junction breakdown voltages are listed $V_{(BR)CEO}$, $V_{(BR)CBO}$ and $V_{(BR)EBO}$. Exceeding these voltages is likely to cause the transistor to enter avalanche breakdown, but if current is limited, permanent damage may not result.

Under ON CHARACTERISTICS are the guaranteed minimum dc current gain (β or h_{FE}), guaranteed maximum collector-emitter saturation voltage, $V_{CE(SAT)}$, and the guaranteed maximum base-emitter on voltage, $V_{BE(ON)}$. Two guaranteed maximum collector cutoff currents, I_{CEO} and I_{CBO}, are listed under OFF CHARACTERISTICS.

The next section is SMALL-SIGNAL CHARACTERISTICS, where the guaranteed minimum current gain-bandwidth product, BW or f_T, the guaranteed maximum output capacitance, C_{obo}, the guaranteed maximum input capacitance, C_{ibo}, the guaranteed range of input impedance, h_{ie}, the small-signal current gain, h_{fe}, the guaranteed maximum voltage feedback ratio, h_{re} and output admittance, h_{oe} are listed.

Finally, the SWITCHING CHARACTERISTICS section lists absolute maximum ratings for delay time, t_d; rise time, t_r; storage time, t_s; and fall time, t_f.

3.3.3 Field-Effect Transistors (FET)

FET devices are controlled by the voltage level of the input rather than the input current, as in the bipolar transistor. FETs have three basic terminals, the *gate*, the *source* and the *drain*. They are analogous to bipolar transistor terminals: the gate to the base, the source

to the emitter, and the drain to the collector. Symbols for the various forms of FET devices are pictured in **Fig 3.23**.

The FET gate has a very high impedance, so the input can be modeled as an open circuit. The voltage between gate and source, V_{GS}, controls the resistance of the drain-source

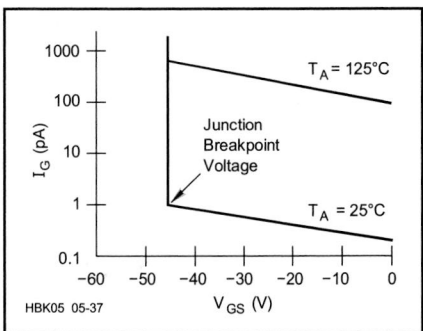

Fig 3.24 — JFET input leakage curves for common source amplifier configuration. Input voltage (V_{GS}) on the x-axis versus input current (I_G) on the y-axis, with two curves plotted for different operating temperatures, 25 °C and 125 °C. Input current increases greatly when the gate voltage exceeds the junction breakpoint voltage.

channel, r_{DS}, and so the output of the FET is modeled as a current source, whose output current is controlled by the input voltage.

The action of the FET channel is so nearly ideal that, as long as the JFET gate does not become forward biased and inject current from the base into the channel, the drain and source currents are virtually identical. For JFETs the *gate leakage current*, I_G, is a function of V_{GS} and this is often expressed with an *input curve* (see **Fig 3.24**). The point at which there is a significant increase in I_G is called the *junction breakpoint voltage*. Because the gate of MOSFETs is insulated from the channel, gate leakage current is insignificant in these devices.

The dc channel resistance, r_{DS}, is specified in data sheets to be less than a maximum value when the device is biased on ($r_{DS(on)}$). When the gate voltage is maximum ($V_{GS} = 0$ for a JFET), $r_{DS(on)}$ is minimum. This describes the effectiveness of the device as an analog switch. Channel resistance is approximately the same for ac and dc signals until at high frequencies the capacitive reactances inherent in the FET structure become significant.

FETs also have strong similarities to vacuum tubes in that input voltage between the grid and cathode controls an output current between the plate and cathode. (See the chapter on **RF Power Amplifiers** for more information on vacuum tubes.)

FORWARD TRANSCONDUCTANCE

The change in FET drain current caused by a change in gate-to-source voltage is called *forward transconductance*, g_m.

$$g_m = \frac{\Delta I_{DS}}{\Delta V_{GS}}$$

or

$$\Delta I_{DS} = g_m \Delta V_{GS} \qquad (11)$$

The input voltage, V_{GS}, is measured between the FET gate and source and drain current, I_{DS}, flows from drain to source. Analogous to a bipolar transistor's current gain, the units of transconductance are Siemens (S) because it is the ratio of current to voltage. (Both g_m and g_{fs} are used interchangeably to indicate transconductance. Some sources specify g_{fs} as the *common-source forward transconductance*. This chapter uses g_m, the most common convention in the reference literature.)

OPERATING REGIONS

The most useful relationships for FETs are the output and transconductance response characteristic curves in **Fig 3.25**. (References such as *common-source* or *common-gate* refer

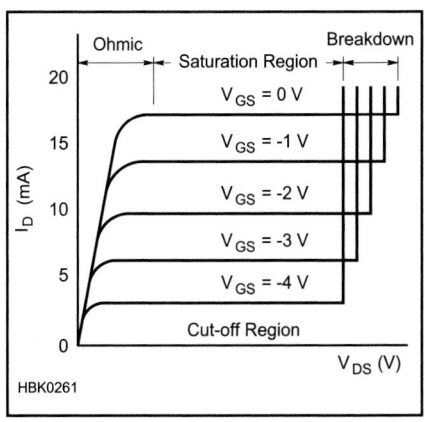

Fig 3.26 — JFET operating regions. At the left, I_D is increasing rapidly with V_{GS} and the JFET can be treated as resistance (R_{DS}) controlled by V_{GS}. In the saturation region, drain current, I_D, is relatively independent of V_{GS}. As V_{DS} increases further, avalanche breakdown begins and I_D increases rapidly.

to the configuration of the circuit in which the parameter is measured.) Transconductance curves relate the drain current, I_D, to gate-to-source voltage, V_{GS}, at various drain-source voltages, V_{DS}. The FET's forward transconductance, g_m, is the slope of the lines in the forward transconductance curve. The same parameters are interrelated in a different way in the output characteristic, in which I_D is shown versus V_{DS} for different values of V_{GS}.

Like the bipolar transistor, FET operation can be characterized by regions. The *ohmic region* is shown at the left of the FET output characteristic curve in **Fig 3.26** where I_D is increasing nearly linearly with V_{DS} and the FET is acting like a resistance controlled by V_{GS}. As V_{DS} continues to increase, I_D saturates and becomes nearly constant. This is the FET's *saturation region* in which the channel of the FET can be modeled as a constant-current source. V_{DS} can become so large that V_{GS} no longer controls the conduction of the device and avalanche breakdown occurs as in bipolar transistors and PN-junction diodes. This is the *breakdown region*, shown in Fig 3.26 where the curves for I_D break sharply upward. If V_{GS} is less than V_P, so that transconductance is zero, the FET is in the *cutoff region*.

OPERATING PARAMETERS

A typical FET data sheet gives ABSOLUTE MAXIMUM RATINGS for V_{DS}, V_{DG}, V_{GS} and I_D, along with the usual device dissipation (P_D) and storage temperature range. Exceeding these limits usually results in destruction of the FET.

Under OPERATING PARAMTERS the OFF CHARACTERISTICS list the gate-source breakdown voltage, $V_{GS(BR)}$, the reverse gate current, I_{GSS} and the gate-source cutoff voltage,

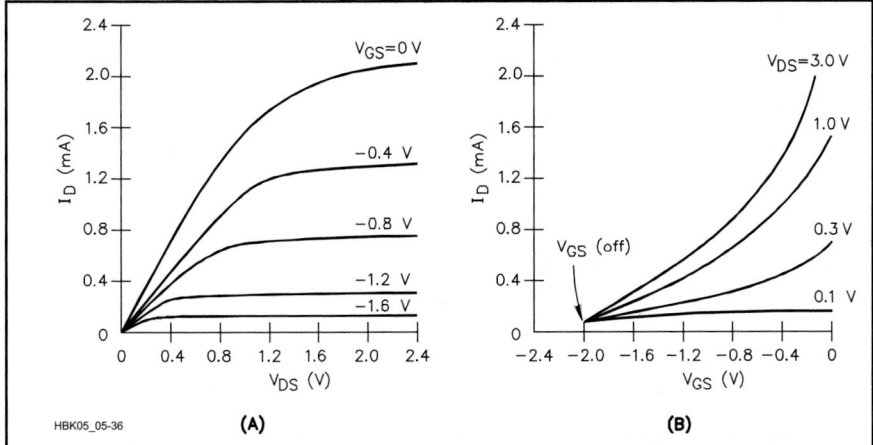

Fig 3.25 — JFET output and transconductance response curves for common source amplifier configuration. (A) Output voltage (V_{DS}) on the x-axis versus output current (I_D) on the y-axis, with different curves plotted for various values of input voltage (V_{GS}). (B) Transconductance curve with the same three variables rearranged: V_{GS} on the x-axis, I_D on the y-axis and curves plotted for different values of V_{DS}.

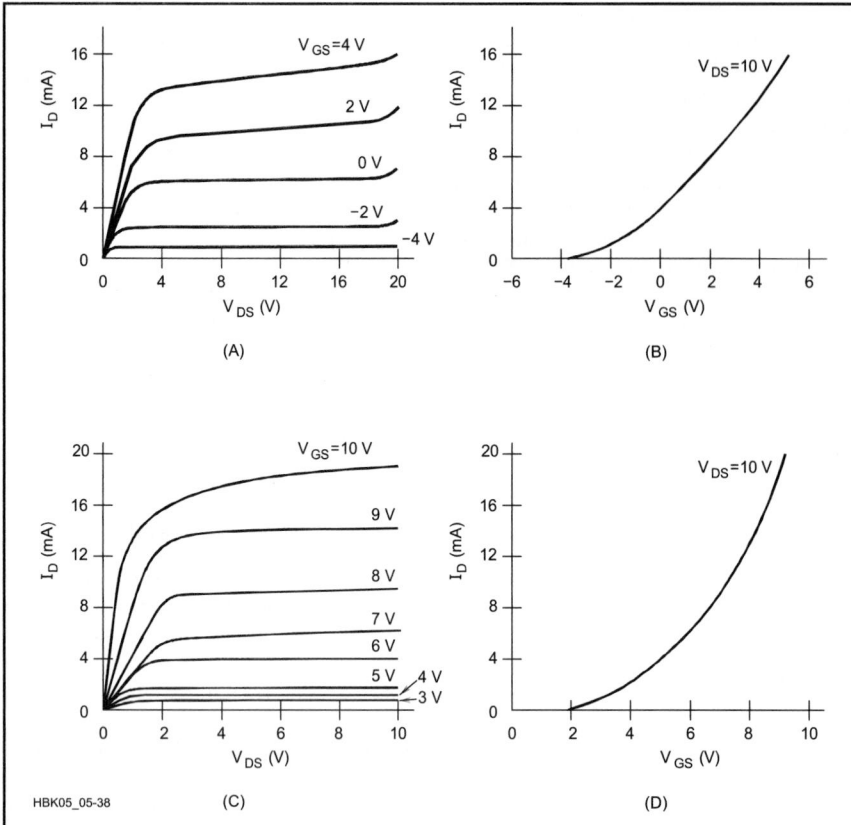

Fig 3.27 — MOSFET output [(A) and (C)] and transconductance [(B) and (D)] response curves. Plots (A) and (B) are for an N-channel depletion mode device. Note that V_{GS} varies from negative to positive values. Plots (C) and (D) are for an N-channel enhancement mode device. V_{GS} has only positive values.

$V_{GS(OFF)}$. Exceeding $V_{GS(BR)}$ will not permanently damage the device if current is limited. The primary ON CHARACTERISTIC parameters are the channel resistance, r_{DS}, and the zero-gate-voltage drain current (I_{DSS}). An FET's dc channel resistance, r_{DS}, is specified in data sheets to be less than a maximum value when the device is biased on ($r_{DS(on)}$). For ac signals, $r_{ds(on)}$ is not necessarily the same as $r_{DS(on)}$, but it is not very different as long as the frequency is not so high that capacitive reactance in the FET becomes significant.

The SMALL SIGNAL CHARACTERISTICS include the forward transfer admittance, y_{fs}, the output admittance, y_{os}, the static drain-source on resistance, $r_{ds(on)}$ and various capacitances such as input capacitance, C_{iss}, reverse transfer capacitance, C_{rss}, the drain-substrate capacitance, $C_{d(sub)}$. FUNCTIONAL CHARACTERISTICS include the noise figure, NF, and the common source power gain, G_{ps}.

MOSFETS

As described earlier, the MOSFET's gate is insulated from the channel by a thin layer of nonconductive oxide, doing away with any appreciable gate leakage current. Because of this isolation of the gate, MOSFETs do not need input and reverse transconductance curves. Their output curves (**Fig 3.27**) are similar to those of the JFET. The gate acts as a small capacitance between the gate and both the source and drain.

The output and transconductance curves in Fig 3.27A and 3.27B show that the depletion-mode N-channel MOSFET's transconductance is positive at $V_{GS} = 0$, like that of the N-channel JFET. Unlike the JFET, however, increasing V_{GS} does not forward-bias the gate-source junction and so the device can be operated with $V_{GS} > 0$.

In the enhancement-mode MOSFET, transconductance is zero at $V_{GS} = 0$. As V_{GS} is increased, the MOSFET enters the ohmic region. If V_{GS} increases further, the saturation region is reached and the MOSFET is said to be *fully-on*, with r_{DS} at its minimum value. The behavior of the enhancement-mode MOSFET is similar to that of the bipolar transistor in this regard.

The relatively flat regions in the MOSFET output curves are often used to provide a constant current source. As is plotted in these curves, the drain current, I_D, changes very little as the drain-source voltage, V_{DS}, varies in this portion of the curve. Thus, for a fixed gate-source voltage, V_{GS}, the drain current can be considered to be constant over a wide range of drain-source voltages.

Multiple gate MOSFETs are also available. Due to the insulating layer, the two gates are isolated from each other and allow two signals to control the channel simultaneously with virtually no loading of one signal by the other. A common application of this type of device is an automatic gain control (AGC) amplifier. The signal is applied to one gate and a rectified, low-pass filtered form of the output (the AGC voltage) is fed back to the other gate. Another common application is as a mixer in which the two input signals are applied to the pair of gates.

MOSFET Gate Protection

The MOSFET is constructed with a very thin layer of SiO_2 for the gate insulator. This layer is extremely thin in order to improve the transconductance of the device but this makes it susceptible to damage from high voltage levels, such as *electrostatic discharge* (ESD) from static electricity. If enough charge accumulates on the gate terminal, it can *punch through* the gate insulator and destroy it. The insulation of the gate terminal is so good that virtually none of this potential is eased by leakage of the charge into the device. While this condition makes for nearly ideal input impedance (approaching infinity), it puts the device at risk of destruction from even such seemingly innocuous electrical sources as static electrical discharges from handling.

Some MOSFET devices contain an internal Zener diode with its cathode connected to the gate and its anode to the substrate. If the voltage at the gate rises to a damaging level the Zener diode junction conducts, bleeding excess charges off to the substrate. When voltages are within normal operating limits the Zener has little effect on the signal at the gate, although it may decrease the input impedance of the MOSFET.

This solution will not work for all MOSFETs. The Zener diode must always be reverse biased to be effective. In the enhancement-mode MOSFET, $V_{GS} > 0$ for all valid uses of the part, keeping the Zener reverse biased. In depletion mode devices however, V_{GS} can be both positive and negative; when negative, a gate-protection Zener diode would be forward biased and the MOSFET gate would not be driven properly. In some depletion mode MOSFETs, back-to-back Zener diodes are used to protect the gate.

MOSFET devices are at greatest risk of damage from static electricity when they are out of circuit. Even though an electrostatic discharge is capable of delivering little energy, it can generate thousands of volts and high peak currents. When storing MOSFETs, the leads should be placed into conductive foam. When working with MOSFETs, it is a good idea to minimize static by wearing a grounded wrist strap and working on a grounded workbench or mat. A humidifier may help to decrease the static electricity in the air. Before inserting a MOSFET into a circuit board it helps to first touch the device leads with your hand and then touch the circuit board. This serves to equalize the excess charge so that little excess charge flows when the device is inserted into the circuit board.

Power MOSFETs

Power MOSFETs are designed for use as switches, with extremely low values of $r_{DS(on)}$; values of 50 milliohms (mΩ) are common. The largest devices of this type can switch tens of amps of current with V_{DS} voltage ratings of hundreds of volts. The **Component Data and References** chapter includes a table of Power FET ratings. The schematic symbol for power MOSFETs (see Fig 3.23) includes a *body diode* that allows the FET to conduct in the reverse direction, regardless of V_{GS}. This is useful in many high-power switching applications. Power MOSFETs used for RF amplifiers are discussed in more detail in the **RF Power Amplifiers** chapter.

While the maximum ratings for current and voltage are high, the devices cannot withstand both high drain current and high drain-to-source voltage at the same time because of the power dissipated; $P = V_{DS} \times I_D$. It is important to drive the gate of a power MOSFET such that the device is fully on or fully off so that either V_{DS} or I_D is at or close to zero. When switching, the device should spend as little time as possible in the linear region where both current and voltage are nonzero because their product (P) can be substantial. This is not a big problem if switching only takes place occasionally, but if the switching is repetitive (such as in a switching power supply) care should be taken to drive the gate properly and remove excess heat from the device.

Because the gate of a power MOSFET is capacitive (up to several hundred pF for large devices), charging and discharging the gate quickly results in short current peaks of more than 100 mA. Whatever circuit is used to drive the gate of a power MOSFET must be able to handle that current level, such as an integrated circuit designed for driving the capacitive load an FET gate presents.

The gate of a power MOSFET should not be left open or connected to a high-impedance circuit. Use a pull-down or pull-up resistor connected between the gate and the appropriate power supply to ensure that the gate is placed at the right voltage when not being driven by the gate drive circuit.

GaAsFETs

FETs made from gallium-arsenide (GaAs) material are used at UHF and microwave frequencies because they have gain at these frequencies and add little noise to the signal. The reason GaAsFETs have gain at these frequencies is the high mobility of the electrons in GaAs material. Because the electrons are more mobile than in silicon, they respond to the gate-source input signal more quickly and strongly than silicon FETs, providing gain at higher frequencies (f_T is directly proportional to electron mobility). The higher electron mobility also reduces thermally-generated noise generated in the FET, making the GaAsFET especially suitable for weak-signal preamps.

Because electron mobility is always higher than hole mobility, N-type material is used in GaAsFETs to maximize high-frequency gain. Since P-type material is not used to make a gate-channel junction, a metal Schottky junction is formed by depositing metal directly on the surface of the channel. This type of device is also called a *MESFET* (metal-semiconductor field-effect transistor).

3.3.4 Optical Semiconductors

In addition to electrical energy and heat energy, light energy also affects the behavior of semiconductor materials. If a device is made to allow photons of light to strike the surface of the semiconductor material, the energy absorbed by electrons disrupts the bonds between atoms, creating free electrons and holes. This increases the conductivity of the material (*photoconductivity*). The photon can also transfer enough energy to an electron to allow it to cross a PN junction's depletion region as current flow through the semiconductor (*photoelectricity*).

PHOTOCONDUCTORS

In commercial *photoconductors* (also called *photoresistors*) the resistance can change by as much as several kilohms for a light intensity change of 100 ft-candles. The most common material used in photoconductors is cadmium sulfide (CdS), with a resistance range of more than 2 MΩ in total darkness to less than 10 Ω in bright light. Other materials used in photoconductors respond best at specific colors. Lead sulfide (PbS) is most sensitive to infrared light and selenium (Se) works best in the blue end of the visible spectrum.

PHOTODIODES

A similar effect is used in some diodes and transistors so that their operation can be controlled by light instead of electrical current biasing. These devices, shown in **Fig 3.28**, are called *photodiodes* and *phototransistors*. The flow of minority carriers across the reverse biased PN junction is increased by light falling on the doped semiconductor material. In the dark, the junction acts the same as any reverse biased PN junction, with a very low current, I_{SC}, (on the order of 10 μA) that is nearly independent of reverse voltage. The presence of light not only increases the current but also provides a resistance-like relationship

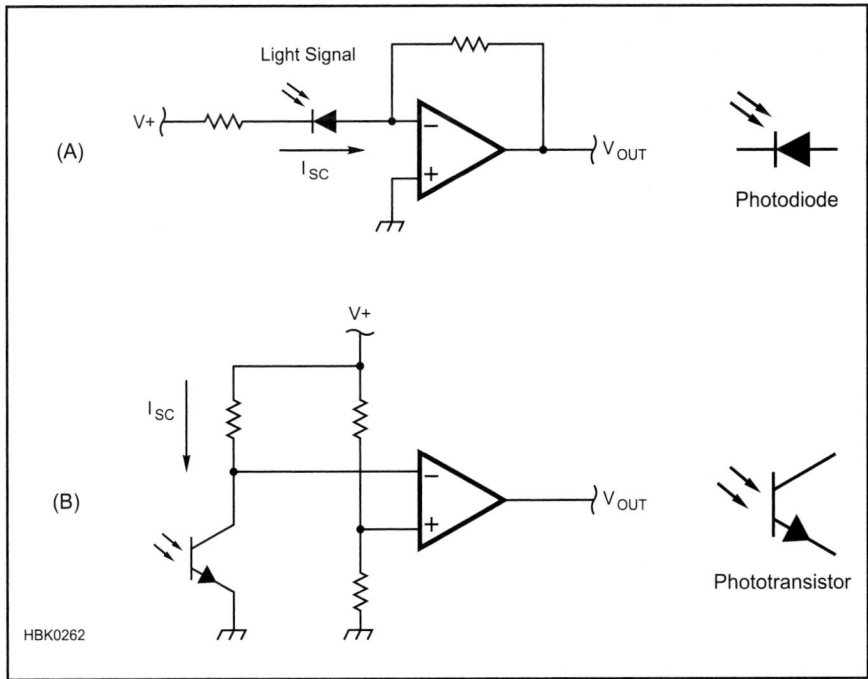

Fig 3.28 — The photodiode (A) is used to detect light. An amplifier circuit changes the variations in photodiode current to a change in output voltage. At (B), a phototransistor conducts current when its base is illuminated. This causes the voltage at the collector to change causing the amplifier's output to switch between ON and OFF.

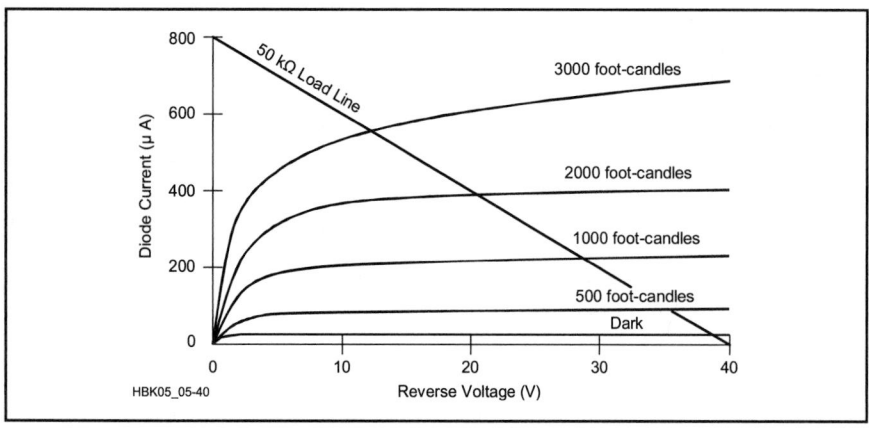

Fig 3.29 — Photodiode I-V curve. Reverse voltage is plotted on the x-axis and current through diode is plotted on the y-axis. Various response lines are plotted for different illumination. Except for the zero illumination line, the response does not pass through the origin since there is current generated at the PN junction by the light energy. A load line is shown for a 50-kΩ resistor in series with the photodiode.

(reverse current increases as reverse voltage increases). See **Fig 3.29** for the characteristic response of a photodiode. Even with no reverse voltage applied, the presence of light causes a small reverse current, as indicated by the points at which the lines in Fig 3.29 intersect the left side of the graph.

Photoconductors and photodiodes are generally used to produce light-related analog signals that require further processing. For example, a photodiode is used to detect infrared light signals from remote control devices as in Fig 3.28A. The light falling on the reverse-biased photodiode causes a change in I_{SC} that is detected as a change in output voltage.

Light falling on the phototransistor acts as base current to control a larger current between the collector and emitter. Thus the phototransistor acts as an amplifier whose input signal is light and whose output is current. Phototransistors are more sensitive to light than the other devices. Phototransistors have lots of light-to-current gain, but photodiodes normally have less noise, so they make more sensitive detectors. The phototransistor in Fig 3.28B is being used as a detector. Light falling on the phototransistor causes collector current to flow, dropping the collector voltage below the voltage at the amplifier's + input and causing a change in V_{OUT}.

PHOTOVOLTAIC CELLS

When illuminated, the reverse-biased photodiode has a reverse current caused by excess minority carriers. As the reverse voltage is reduced, the potential barrier to the forward flow of majority carriers is also reduced. Since light energy leads to the generation of both majority and minority carriers, when the resistance to the flow of majority carriers is decreased these carriers form a forward current. The voltage at which the forward current equals the reverse current is called the *photovoltaic potential* of the junction. If the illuminated PN junction is not connected to a load, a voltage equal to the photovoltaic potential can be measured across it as the *terminal voltage*, V_T, or *open-circuit voltage*, V_{OC}.

Devices that use light from the sun to produce electricity in this way are called *photovoltaic (PV)* or *solar cells* or *solar batteries*. The symbol for a photovoltaic cell is shown in **Fig 3.30A**. The electrical equivalent circuit of the cell is shown in Fig 3.30B. The cell is basically a large, flat diode junction exposed to light. Metal electrodes on each side of the junction collect the current generated.

When illuminated, the cell acts like a current source, with some of the current flowing through a diode (made of the same material as the cell), a shunt resistance for leakage current and a series resistor that represents the resistance of the cell. Two quantities define the electrical characteristics of common silicon photovoltaic cells. These are an open-circuit voltage, V_{OC} of 0.5 to 0.6 V and the output *short-circuit current*, I_{SC} as above, that depends on the area of the cell exposed to light and the degree of illumination. A measure of the cell's effectiveness at converting light into current is the *conversion efficiency*. Typical silicon solar cells have a conversion efficiency of 10 to 15% although special cells with stacked junctions or using special light-absorbing materials have shown efficiencies as high as 40%.

Solar cells are primarily made from single-crystal slices of silicon, similar to diodes and transistors, but with a much greater area. *Polycrystalline silicon* and *thin-film* cells are less expensive, but have lower conversion efficiency. Technology is advancing rapidly in the field of photovoltaic energy and there are a number of different types of materials and fabrication techniques that have promise in surpassing the effectiveness of the single-junction silicon cells.

Solar cells are assembled into arrays called *solar panels*, shown in Fig 3.30C. Cells are connected in series so that the combined output voltage is a more useful voltage, such as 12 V. Several strings of cells are then connected in parallel to increase the available output current. Solar panels are available with output powers from a few watts to hundreds of watts. Note that unlike batteries, strings of solar cells can be connected directly in parallel because they act as sources of constant current instead

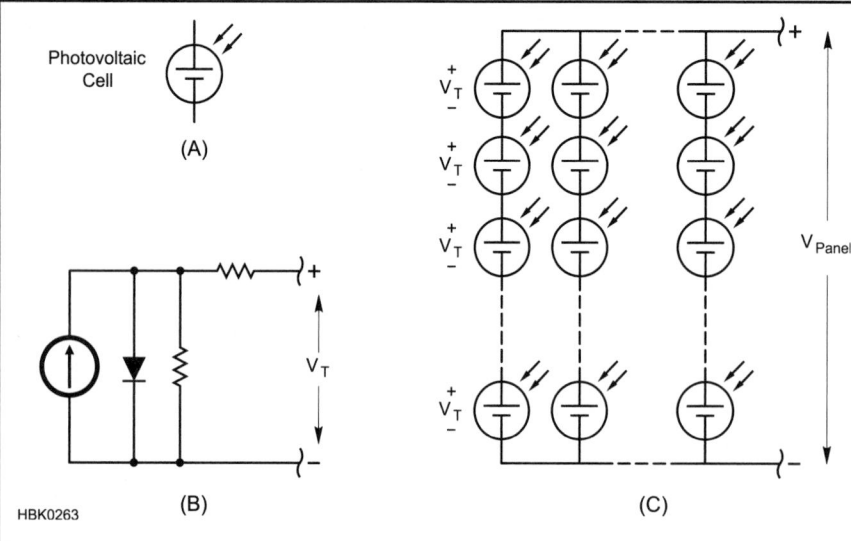

Fig 3.30 — A photovoltaic cell's symbol (A) is similar to a battery. Electrically, the cell can be modeled as the equivalent circuit at (B). Solar panels (C) consist of arrays of cells connected to supply power at a convenient voltage.

of voltage. (More information on the use of solar panels for powering radio equipment can be found in the chapter on **Power Sources**.)

LIGHT EMITTING DIODES AND LASER DIODES

In the photodiode, energy from light falling on the semiconductor material is absorbed to create additional electron-hole pairs. When the electrons and holes recombine, the same amount of energy is given off. In normal diodes the energy from recombination of carriers is given off as heat. In certain forms of semiconductor material, the recombination energy is given off as light with a mechanism called *electroluminescence*. Unlike the incandescent light bulb, electroluminescence is a cold (non-thermal) light source that typically operates with low voltages and currents (such as 1.5 V and 10 mA). Devices made for this purpose are called *light emitting diodes (LEDs)*. They have the advantages of low power requirements, fast switching times (on the order of 10 ns) and narrow spectra (relatively pure color).

The LED emits light when it is forward biased and excess carriers are present. As the carriers recombine, light is produced with a color that depends on the properties of the semiconductor material used. Gallium-arsenide (GaAs) generates light in the infrared region, gallium-phosphide (GaP) gives off red light when doped with oxygen or green light when doped with nitrogen. Orange light is attained with a mixture of GaAs and GaP (GaAsP). Silicon-carbide (SiC) creates a blue LED.

White LEDs are made by coating the inside of the LED lens with a white-light emitting phosphor and illuminating the phosphor with light from a single-color LED. White LEDs are currently approaching the cost of cold-florescent (CFL) bulbs and will eventually displace CFL technology for lighting, just as CFL is replacing the incandescent bulb.

The LED, shown in **Fig 3.31**, is very simple to use. It is connected across a voltage source

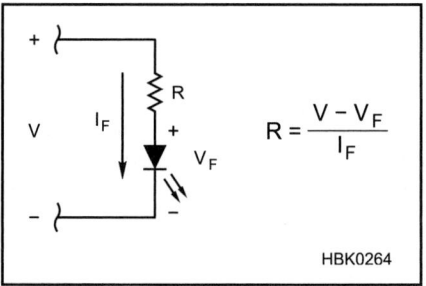

Fig 3.31 — A light-emitting diode (LED) emits light when conducting forward current. A series current-limiting resistor is used to set the current through the LED according to the equation.

(V) with a series resistor (R) that limits the current to the desired level (I_F) for the amount of light to be generated.

$$R = \frac{V - V_F}{I_F} \tag{12}$$

where V_F is the forward voltage of the LED.

The cathode lead is connected to the lower potential, and is specially marked as shown in the manufacturer's data sheet. LEDs may be connected in series for additional light, with the same current flowing in all of the diodes. Diodes connected in parallel without current-limiting resistors for each diode are likely to share the current unequally, thus the series connection is preferred.

The laser diode operates by similar principles to the LED except that all of the light produced is *monochromatic* (of the same color and wavelength) and it is *coherent*, meaning that all of the light waves emitted by the device are in phase. Laser diodes generally require higher current than an LED and will not emit light until the *lasing current* level is reached. Because the light is monochromatic and coherent, laser diodes can be used for applications requiring precise illumination and modulation, such as high-speed data links, and in data storage media such as CD-ROM and DVD. LEDs are not used for high-speed or high-frequency analog modulation because of recovery time limitations, just as in regular rectifiers.

OPTOISOLATORS

An interesting combination of optoelectronic components proves very useful in many analog signal processing applications. An *optoisolator* consists of an LED optically coupled to a phototransistor, usually in an enclosed package (see **Fig 3.32**). The optoisolator, as its name suggests, isolates different circuits from each other. Typically, isolation resistance is on the order of 10^{11} Ω and isolation capacitance is less than 1 pF. Maximum voltage isolation varies from 1000 to 10,000 V ac. The most common optoisolators are available in 6-pin DIP packages.

Optoisolators are primarily used for voltage level shifting and signal isolation. Voltage level shifting allows signals (usually digital signals) to pass between circuits operating at greatly different voltages. The isolation has two purposes: to protect circuitry (and operators) from excessive voltages and to isolate noisy circuitry from noise-sensitive circuitry.

Optoisolators also cannot transfer signals with high power levels. The power rating of the LED in a 4N25 device is 120 mW. Optoisolators have a limited frequency response due to the high capacitance of the LED. A typical bandwidth for the 4N25 series is 300 kHz. Optoisolators with bandwidths of several MHz are available, but are

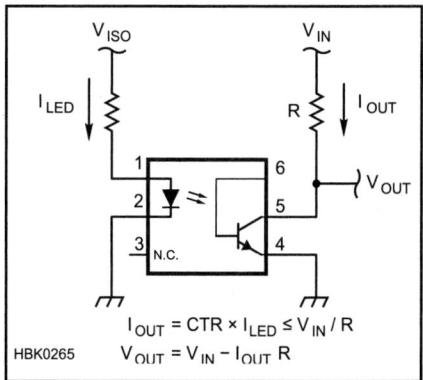

$$I_{OUT} = CTR \times I_{LED} \le V_{IN} / R$$
$$V_{OUT} = V_{IN} - I_{OUT} R$$

Fig 3.32 — The optoisolator consists of an LED (input) that illuminates the base of a phototransistor (output). The phototransistor then conducts current in the output circuit. CTR is the optoisolator's current transfer ratio.

somewhat expensive.

As an example of voltage level shifting, an optoisolator can be used to allow a low-voltage, solid-state electronic Morse code keyer to activate a vacuum-tube grid-block keying circuit that operates at a high negative voltage (typically about –100 V) but low current. No common ground is required between the two pieces of equipment.

Optoisolators can act as input protection for circuits that are exposed to high voltages or transients. For example, a short 1000-V transient that can destroy a semiconductor circuit will only saturate the LED in the optoisolator, preventing damage to the circuit. The worst that will happen is the LED in the optoisolator will be destroyed, but that is usually quite a bit less expensive than the circuit it is protecting.

Optoisolators are also useful for isolating different ground systems. The input and output signals are totally isolated from each other, even with respect to the references for each signal. A common application for optoisolators is when a computer is used to control radio equipment. The computer signal, and even its ground reference, typically contains considerable wide-band noise caused by the digital circuitry. The best way to keep this noise out of the radio is to isolate both the signal and its reference; this is easily done with an optoisolator.

The design of circuits with optoisolators is not greatly different from the design of circuits with LEDs and with transistors. On the input side, the LED is forward-biased and driven with a series current-limiting resistor whose value limits current to less than the maximum value for the device (for example, 60 mA is the maximum LED current for a 4N25). This is identical to designing with standalone LEDs.

On the output side, instead of current gain for a transistor, the optoisolator's *current transfer ratio* (*CTR*) is used. CTR is a ratio given in percent between the amount of current through the LED to the output transistor's maximum available collector current. For example, if an optoisolator's CTR = 25%, then an LED current of 20 mA results in the output transistor being able to conduct up to $20 \times 0.25 = 5$ mA of current in its collector circuit.

If the optoisolator is to be used for an analog signal, the input signal must be appropriately dc shifted so that the LED is always forward biased. A phototransistor with all three leads available for connection (as in Fig 3.32) is required. The base lead is used for biasing, allowing the optical signal to create variations above and below the transistor's operating point. The collector and emitter leads are used as they would be in any transistor amplifier circuit. (There are also linear optoisolators that include built-in linearizing circuitry.) The use of linear optoisolators is not common.

FIBER OPTICS

An interesting variation on the optoisolator is the *fiber-optic* connection. Like the optoisolator, the input signal is used to drive an LED or laser diode that produces modulated light (usually light pulses). The light is transmitted in a fiber optic cable, an extruded glass fiber that efficiently carries light over long distances and around fairly sharp bends. The signal is recovered by a photo detector (photoresistor, photodiode or phototransistor). Because the fiber optic cable is nonconductive, the transmitting and receiving systems are electrically isolated.

Fiber optic cables generally have far less loss than coaxial cable transmission lines. They do not leak RF energy, nor do they pick up electrical noise. Fiber optic cables are virtually immune to electromagnetic interference! Special forms of LEDs and phototransistors are available with the appropriate optical couplers for connecting to fiber optic cables. These devices are typically designed for higher frequency operation with gigahertz bandwidth.

3.3.5 Linear Integrated Circuits

If you look inside a transistor, the actual size of the semiconductor is quite small compared to the size of the packaging. For most semiconductors, the packaging takes considerably more space than the actual semiconductor device. Thus, an obvious way to reduce the physical size of circuitry is to combine more of the circuit inside a single package.

HYBRID INTEGRATED CIRCUITS

It is easy to imagine placing several small semiconductor chips in the same package. This is known as *hybrid circuitry*, a technology in which several semiconductor chips are placed in the same package and miniature wires are connected between them to make complete circuits.

Hybrid circuits miniaturize analog electronic circuits by replacing much of the packaging that is inherent in discrete electronics. The term *discrete* refers to the use of individual components to make a circuit, each in its own package. The individual components are attached together on a small circuit board or with bonding wires.

Once widespread in electronics, hybrid ICs have largely been displaced by fully integrated devices with all the components on the same piece of semiconductor material. Manufacturers often use hybrids when small size is needed, but there is insufficient volume to justify the expense of a custom IC.

A current application for hybrid circuitry is UHF and microwave amplifiers — they are in wide use by the mobile phone industry. For example, the Motorola MW4IC915N Wideband Integrated Power Amplifier is a complete 15-W transmitting module. Its TO-272 package is only about 1 inch long by 3/8-inch wide. This particular device is designed for use between 750 and 1000 MHz and can be adapted for use on the amateur 902 MHz band. Other devices available as hybrid circuits include oscillators, signal processors, preamplifiers and so forth. Surplus hybrids can be hard to adapt to amateur use unless they are clearly identified with manufacturing identification such that a data sheet can be obtained.

MONOLITHIC INTEGRATED CIRCUITS

In order to build entire circuits on a single piece of semiconductor, it must be possible to fabricate resistors and capacitors, as well as transistors and diodes. Only then can the entire circuit be created on one piece of silicon called a *monolithic integrated circuit*.

An integrated circuit (IC) or "chip" is fabricated in layers. An example of a semiconductor circuit schematic and its implementation in an IC is pictured in **Fig 3.33**. The base layer of the circuit, the *substrate*, is made of P-type semiconductor material. Although less common, the polarity of the substrate can also be N-type material. Since the mobility of electrons is about three times higher than that of holes, bipolar transistors made with N-type collectors and FETs made with N-type channels are capable of higher speeds and power handling. Thus, P-type substrates are far more common. For devices with N-type substrates, all polarities in the ensuing discussion would be reversed.

Other substrates have been used, one of the most successful of which is the *silicon-on-sapphire* (SOS) construction that has been used to increase the bandwidth of integrated circuitry. Its relatively high manufacturing cost has impeded its use, however, except for the demanding military and aerospace applications.

On top of the P-type substrate is a thin layer of N-type material in which the active and passive components are built. Impurities are diffused into this layer to form the appropriate component at each location. To prevent random diffusion of impurities into the N-layer, its upper surface must be protected. This is done by covering the N-layer with a layer of silicon dioxide (SiO_2). Wherever diffusion of impurities is desired, the SiO_2 is etched away. The precision of placing the components on the semiconductor material depends mainly on the fineness of the etching. The fourth layer of an IC is made of aluminum (copper is used in some high-speed digital ICs) and is used to make the interconnections between the components.

Different components are made in a single piece of semiconductor material by first diffusing a high concentration of acceptor impurities into the layer of N-type material. This process creates P-type semiconductor — often referred to as P+-type semiconductor because of its high concentration of acceptor atoms — that isolates regions of N-type material. Each of these regions is then further processed to form single components.

A component is produced by the diffusion of a lesser concentration of acceptor atoms into the middle of each isolation region. This results in an N-type *isolation well* that contains P-type material, is surrounded on its sides by P+-type material and has P-type material (substrate) below it. The cross sectional view in Fig 3.33B illustrates the various layers. Connections to the metal layer are often made by diffusing high concentrations of donor atoms into small regions of the N-type well and the P-type material in the well. The material in these small regions is N+-type and facilitates electron flow between the metal contact and the semiconductor. In some configurations, it is necessary to connect the metal directly to the P-type material in the well.

Fabricating Resistors and Capacitors

An isolation well can be made into a resistor by making two contacts into the P-type semiconductor in the well. Resistance is inversely proportional to the cross-sectional area of the well. An alternate type of resistor that can be integrated in a semiconductor circuit is a *thin-film resistor*, where a metallic film is deposited on the SiO_2 layer, masked on its upper surface by more SiO_2 and then etched to make the desired geometry, thus adjusting the resistance.

There are two ways to form capacitors in a semiconductor. One is to make use of the PN junction between the N-type well and the P-type material that fills it. Much like a varactor diode, when this junction is reverse biased

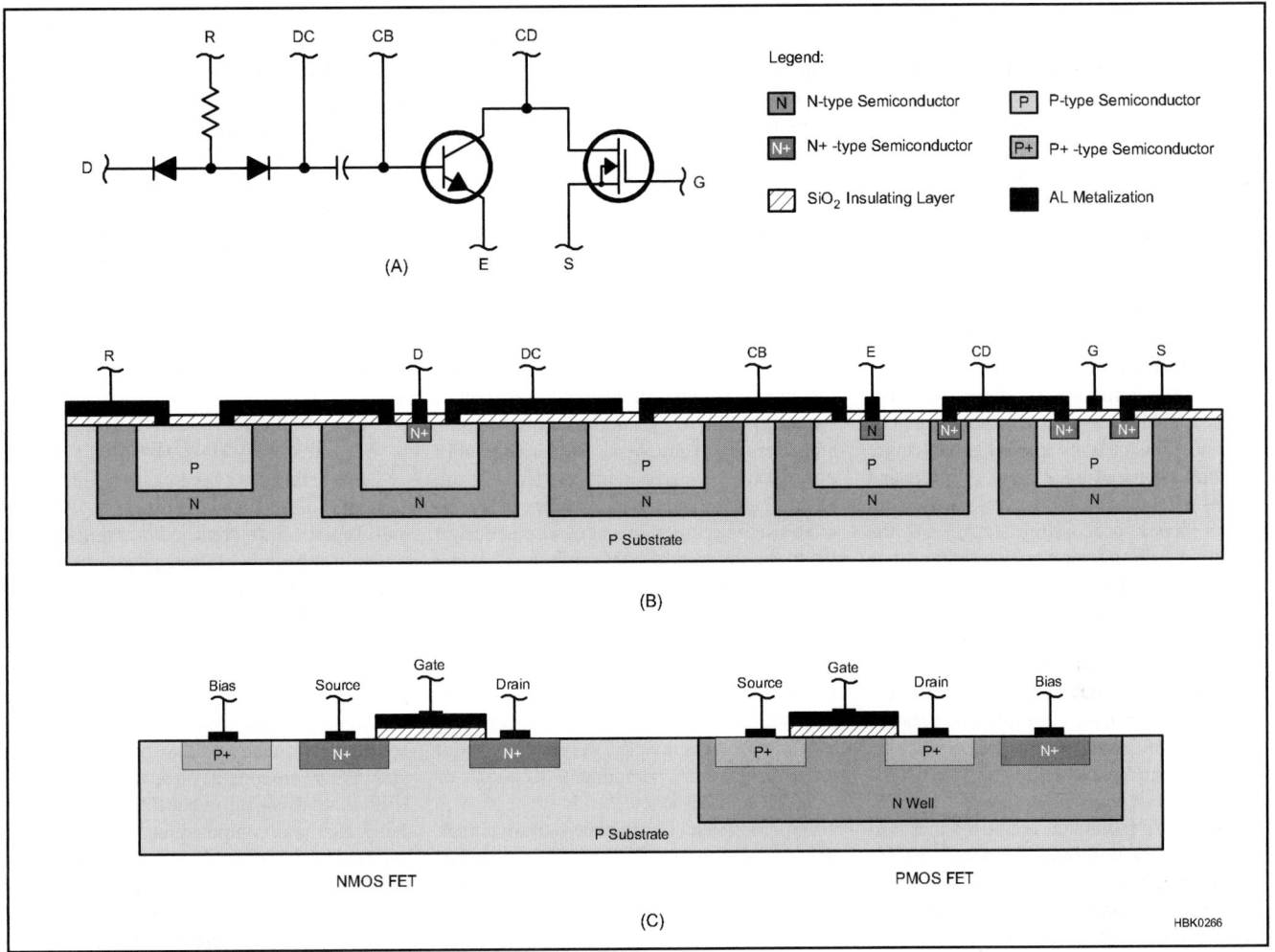

Fig 3.33 — Integrated circuit layout. (A) Circuit containing two diodes, a resistor, a capacitor, an NPN transistor and an N-channel MOSFET. Labeled leads are D for diode, R for resistor, DC for diode-capacitor, E for emitter, S for source, CD for collector-drain and G for gate. (B) Integrated circuit that is identical to circuit in (A). Same leads are labeled for comparison. Circuit is built on a P-type semiconductor substrate with N-type wells diffused into it. An insulating layer of SiO₂ is above the semiconductor and is etched away where aluminum metal contacts are made with the semiconductor. Most metal-to-semiconductor contacts are made with heavily doped N-type material (N⁺-type semiconductor).

a capacitance results. Since a bias voltage is required, this type of capacitor is polarized, like an electrolytic capacitor. Nonpolarized capacitors can also be formed in an integrated circuit by using thin film technology. In this case, a very high concentration of donor ions is diffused into the well, creating an N⁺-type region. A thin metallic film is deposited over the SiO₂ layer covering the well and the capacitance is created between the metallic film and the well. The value of the capacitance is adjusted by varying the thickness of the SiO₂ layer and the cross-sectional size of the well. This type of thin film capacitor is also known as a metal oxide semiconductor (MOS) capacitor.

Unlike resistors and capacitors, it is very difficult to create inductors in integrated circuits. Generally, RF circuits that need inductance require external inductors to be connected to the IC. In some cases, particu-larly at lower frequencies, the behavior of an inductor can be mimicked by an amplifier circuit. In many cases the appropriate design of IC amplifiers can reduce or eliminate the need for external inductors.

Fabricating Diodes and Transistors

The simplest form of diode is generated by connecting to an N⁺-type connection point in the well for the cathode and to the P-type well material for the anode. Diodes are often con-verted from NPN transistor configurations. Integrated circuit diodes made this way can either short the collector to the base or leave the collector unconnected. The base contact is the anode and the emitter contact is the cathode.

Transistors are created in integrated cir-cuitry in much the same way that they are fabricated in their discrete forms. The NPN transistor is the easiest to make since the wall of the well, made of N-type semiconduc-tor, forms the collector, the P-type material in the well forms the base and a small region of N⁺-type material formed in the center of the well becomes the emitter. A PNP transis-tor is made by diffusing donor ions into the P-type semiconductor in the well to make a pattern with P-type material in the center (emitter) surrounded by a ring of N-type ma-terial that connects all the way down to the well material (base), and this is surrounded by another ring of P-type material (collec-tor). This configuration results in a large base width separating the emitter and collector, causing these devices to have much lower current gain than the NPN form. This is one reason why integrated circuitry is designed to use many more NPN transistors than PNP transistors.

FETs can also be fabricated in IC form as shown in Fig 3.33C. Due to its many functional advantages, the MOSFET is the most common form used for digital ICs. MOSFETs are made in a semiconductor chip much the same way as MOS capacitors, described earlier. In addition to the signal processing advantages offered by MOSFETs over other transistors, the MOSFET device can be fabricated in 5% of the physical space required for bipolar transistors. CMOS ICs can contain 20 times more circuitry than bipolar ICs with the same chip size, making the devices more powerful and less expensive than those based on bipolar technology. CMOS is the most popular form of integrated circuit.

The final configuration of the switching circuit is CMOS as described in a previous section of this chapter. CMOS gates require two FETs, one of each form (NMOS and PMOS as shown in the figure). NMOS requires fewer processing steps, and the individual FETs have lower on-resistance than PMOS. The fabrication of NMOS FETs is the same as for individual semiconductors; P+ wells form the source and drain in a P-type substrate. A metal gate electrode is formed on top of an insulating SiO_2 layer so that the channel forms in the P-type substrate between the source and drain. For the PMOS FET, the process is similar, but begins with an N-type well in the P-type substrate.

MOSFETs fabricated in this manner also have bias (B) terminals connected to the positive power supply to prevent destructive *latch-up*. This can occur in CMOS gates because the two MOSFETs form a *parasitic SCR*. If the SCR mode is triggered and both transistors conduct at the same time, large currents can flow through the FET and destroy the IC unless power is removed. Just as discrete MOSFETs are at risk of gate destruction, IC chips made with MOSFET devices have a similar risk. They should be treated with the same care to protect them from static electricity as discrete MOSFETs.

While CMOS is the most widely used technology, integrated circuits need not be made exclusively with MOSFETs or bipolar transistors. It is common to find IC chips designed with both technologies, taking advantage of the strengths of each.

INTEGRATED CIRCUIT ADVANTAGES

The primary advantages of using integrated circuits as opposed to discrete components are the greatly decreased physical size of the circuit and improved reliability. In fact, studies show that failure rate of electronic circuitry is most closely related to the number of interconnections between components. Thus, using integrated circuits not only reduces volume, but makes the equipment more reliable.

The amount of circuitry that can be placed onto a single semiconductor chip is a function of two factors: the size of the chip and the size of the individual features that can be created on the semiconductor wafer. Since the invention of the monolithic IC in the mid-1960s, feature size limits have dropped below 100 nanometers ($\frac{1}{10}$th of a millionth of a meter) as of 2009. Currently, it is not unusual to find chips with more than one million transistors on them.

In addition to size and reliability of the ICs themselves, the relative properties of the devices on a single chip are very predictable. Since adjacent components on a semiconductor chip are made simultaneously (the entire N-type layer is grown at once; a single diffusion pass isolates all the wells and another pass fills them), the characteristics of identically formed components on a single chip of silicon are nearly identical. Even if the exact characteristics of the components are unknown, very often in analog circuit design the major concern is how components interact. For instance, push-pull amplifiers require perfectly matched transistors, and the gain of many amplifier configurations is governed by the ratio between two resistors and not their absolute values of resistance. With closely matched components on the single substrate, this type of design works very well without requiring external components to adjust or "trim" IC performance.

Integrated circuits often have an advantage over discrete circuits in their temperature behavior. The variation of performance of the components on an integrated circuit due to heat is no better than that of discrete components. While a discrete circuit may be exposed to a wide range of temperature changes, the entire semiconductor chip generally changes temperature by the same amount; there are fewer "hot spots" and "cold spots." Thus, integrated circuits can be designed to better compensate for temperature changes.

A designer of analog devices implemented with integrated circuitry has more freedom to include additional components that could improve the stability and performance of the implementation. The inclusion of components that could cause a prohibitive increase in the size, cost or complexity of a discrete circuit would have very little effect on any of these factors in an integrated circuit.

Once an integrated circuit is designed and laid out, the cost of making copies of it is very small, often only pennies per chip. Integrated circuitry is responsible for the incredible increase in performance with a corresponding decrease in price of electronics. While this trend is most obvious in digital computers, analog circuitry has also benefited from this technology.

The advent of integrated circuitry has also improved the design of high frequency circuitry, particularly the ubiquitous mobile phone and other wireless devices. One problem in the design and layout of RF equipment is the radiation and reception of spurious signals. As frequencies increase and wavelengths approach the dimensions of the wires in a circuit board, the interconnections act as efficient antennas. The dimensions of the circuitry within an IC are orders of magnitude smaller than in discrete circuitry, thus greatly decreasing this problem and permitting the processing of much higher frequencies with fewer problems of interstage interference. Another related advantage of the smaller interconnections in an IC is the lower inherent inductance of the wires, and lower stray capacitance between components and traces.

INTEGRATED CIRCUIT DISADVANTAGES

Despite the many advantages of integrated circuitry, disadvantages also exist. ICs have not replaced discrete components, even vacuum tubes, in some applications. There are some tasks that ICs cannot perform, even though the list of these continues to decrease over time as IC technology improves.

Although the high concentration of components on an IC chip is considered to be an advantage of that technology, it also leads to a major limitation. Heat generated in the individual components on the IC chip is often difficult to dissipate. Since there are so many heat generating components so close together, the heat can build up and destroy the circuitry. It is this limitation that currently causes many power amplifiers of more than 50 W output to be designed with discrete components.

Integrated circuits, despite their short interconnection lengths and lower stray inductance, do not have as high a frequency response as similar circuits built with appropriate discrete components. (There are exceptions to this generalization, of course. As described previously, monolithic microwave integrated circuits — MMICs — are available for operation to 10 GHz.) The physical architecture of an integrated circuit is the cause of this limitation. Since the substrate and the walls of the isolation wells are made of opposite types of semiconductor material, the PN junction between them must be reverse biased to prevent current from passing into the substrate. Like any other reverse-biased PN junction, a capacitance is created at the junction and this limits the frequency response of the devices on the IC. This situation has improved over the years as isolation wells have gotten smaller, thus decreasing the capacitance between the well and the substrate, and techniques have been developed to decrease the PN junction capacitance at the substrate. One such technique has been to create an N+-type layer between the well and the substrate, which decreases the capacitance of the PN junction as seen by the well. As a result, analog ICs are now available with gain-bandwidth products over 1 GHz.

A major impediment to the introduction

of new integrated circuits, particularly with special applications, is the very high cost of development of new designs for full custom ICs. The masking cost alone for a designed and tested integrated circuit can exceed $100,000 so these devices must sell in high volume to recoup the development costs. Over the past decade, however, IC design tools and fabrication services have become available that greatly reduce the cost of IC development by using predefined circuit layouts and functional blocks. These *application-specific integrated circuits (ASICs)* and *programmable gate arrays (PGA)* are now routinely created and used for even limited production runs. While still well beyond the reach of the individual amateur's resources, the ASIC or PGA is widely used in nearly all consumer electronics and in many pieces of radio equipment.

The drawback of the ASIC and PGA is that servicing and repairing the equipment at the integrated circuit level is almost impossible for the individual without access to the manufacturer's inventory of parts and proprietary information. Nevertheless, just as earlier amateurs moved beyond replacing individual components to diagnosing ICs, today's amateurs can troubleshoot to the module or circuit-board level, treating them as "components" in their own right.

3.3.6 Comparison of Semiconductor Devices for Analog Applications

Analog signal processing deals with changing a signal to a desired form. The three primary types of devices — bipolar transistors, field-effect transistors and integrated circuits — perform similar functions, each with specific advantages and disadvantages. The vacuum tube, once the dominant signal processing component, is relegated to high-power amplifier and display applications and is found only in the **RF Power Amplifiers** chapter of this *Handbook. Cathode-ray tubes* (CRTs) are covered in a supplement on the *Handbook* CD-ROM.

Bipolar transistors, when treated properly, can have virtually unlimited life spans. They are relatively small and, if they do not handle high currents, do not generate much heat. They make excellent high-frequency amplifiers. Compared to MOSFET devices they are less susceptible to damage from electrostatic discharge. RF amplifiers designed with bipolar transistors in their final amplifiers generally include circuitry to protect the transistors from the high voltages generated by reflections under high SWR conditions. Bipolar transistors and ICs, like all semiconductors, are susceptible to damage from power and lightning transients.

There are many performance advantages to FET devices, particularly MOSFETs. The extremely low gate currents allow the design of analog stages with nearly infinite input resistance. Signal distortion due to loading is minimized in this way. FETs are less expensive to fabricate in ICs and so are gradually replacing bipolar transistors in many IC applications.

The current trend in electronics is portability. Transceivers are decreasing in size and in their power requirements. Integrated circuitry has played a large part in this trend. Extremely large circuits have been designed with microscopic proportions, including combinations of analog and digital circuitry that previously required multiple devices. *Charge-coupled devices* (CCD) imaging technology has replaced vidicon tubes in video cameras and film cameras of all types. *Liquid crystal displays* (LCDs) in laptop computers and standalone computer displays have largely displaced the bulky and power-hungry cathode-ray tube (CRT), although it is still found in analog oscilloscopes and some types of video display equipment.

An important consideration in the use of analog components is the future availability of parts. At an ever increasing rate, as new components are developed to replace older technology, the older components are discontinued by the manufacturers and become unavailable for future use. ASIC technology, as mentioned earlier, brings the power of custom electronics to the radio, but also makes it nearly impossible to repair at the level of the IC, even if the problem is known. If field repair and service at the component level are to be performed, it is important to use standard ICs wherever possible. Even so, when demand for a particular component drops, a manufacturer will discontinue its production. This happens on an ever-decreasing timeline.

A further consideration is the trend toward digital signal processing and software-defined radio systems. (See the chapter on **DSP and Software Radio Design**.) More and more analog functions are being performed by microprocessors and the analog signals converted to digital at higher and higher frequencies. There will always be a need for analog circuits, but the balance point between analog and digital is shifting towards the latter. In future years, radio and test equipment will consist of a powerful, general-purpose digital signal processor, surrounded by the necessary analog circuitry to convert the signals to digital form and supply the processor with power.

3.4 Analog Systems

Many kinds of electronic equipment are developed by combining basic analog signal processing circuits, often treating them as independent functional blocks. This section describes several topics associated with building analog systems from multiple blocks. Although not all basic electronic functions are discussed here, the concepts associated with combining them can be applied generally.

An analog circuit can contain any number of discrete components (or may be implemented as an IC). Since our main concern is the effect that circuitry has on a signal, we often describe the circuit by its actions rather than by its specific components. A *black box* is a circuit that can be described entirely by the behavior of its interfaces with other blocks and circuitry. When circuits are combined in such as way as to perform sequential operations on a signal, the individual circuits are called *stages*.

The most general way of referring to a circuit is as a *network*. Two basic properties of analog networks are of principal concern: the effect that the network has on an analog signal and the interaction that the network has with the circuitry surrounding it. Interfaces between the network and the rest of the network are called *ports*.

Many analog circuits are analyzed as *two-port networks* with an input and an output port. The signal is fed into the input port, is modified inside the network and then exits from the output port. (See the chapter on **RF Techniques** for more information on two-port networks.)

3.4.1 Transfer Functions

The specific way in which the analog circuit modifies the signal can be described mathematically as a *transfer function*. The mathematical operation that combines a signal with a transfer function is pictured symbolically in **Fig 3.34**. The transfer function, h(t) or h(f), describes the circuit's modification of the input signal in the time domain where all values are functions of time, such as a(t) or b(t), or in the frequency domain where all values are functions of frequency, such as a(f) or b(f). The mathematical operation by which h(t) operates on a(t) is called *convolution* and is represented as a dot, as in a(t) • h(t) = b(t). In the frequency domain, the transfer function multiplies the input, as in a(f) × h(f) = b(f).

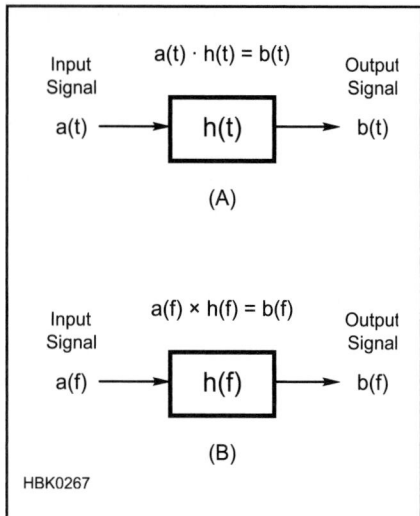

$$a(t) \cdot h(t) = b(t)$$

Input Signal → a(t) → [h(t)] → b(t) → Output Signal

(A)

$$a(f) \times h(f) = b(f)$$

Input Signal → a(f) → [h(f)] → b(f) → Output Signal

(B)

HBK0267

Fig 3.34 — Linear function blocks and transfer functions. The transfer function can be expressed in the time domain (A) or in the frequency domain (B). The transfer function describes how the input signal a(t) or a(f) is transformed into the output signal b(t) or b(f).

While it is not necessary to understand transfer functions mathematically to work with analog circuits, it is useful to realize that they describe how a signal interacts with other signals in an electronic system. In general, the output signal of an analog system depends not only on the input signal at the same time, but also on past values of the input signal. This is a very important concept and is the basis of such essential functions as analog filtering.

3.4.2 Cascading Stages

If an analog circuit can be described with a transfer function, a combination of analog circuits can also be described similarly. This description of the combined circuits depends upon the relationship between the transfer functions of the parts and that of the combined circuits. In many cases this relationship allows us to predict the behavior of large and complex circuits from what we know about the parts of which they are made. This aids in the design and analysis of analog circuits.

When two analog circuits are cascaded (the output signal of one stage becomes the input signal to the next stage) their transfer functions are combined. The mechanism of the combination depends on the interaction between the stages. The ideal case is the functions of the stages are completely independent. In other words, when the action of a stage is unchanged, regardless of the characteristics of any stages connected to its input or output.

Just as the signal entering the first stage is modified by the action of the first transfer function, the ideal cascading of analog circuits results in changes produced only by the individual transfer functions. For any number of stages that are cascaded, the combination of their transfer functions results in a new transfer function. The signal that enters the circuit is changed by the composite transfer function to produce the signal that exits in the cascaded circuits.

While each stage in a series may use feedback within itself, feedback around more than one stage may create a function — and resultant performance — different from any of the included stages. Examples include oscillation or negative feedback.

3.4.3 Amplifier Frequency Response

At higher frequencies a typical amplifier acts as a low-pass filter, decreasing amplification with increasing frequency. Signals within a range of frequencies are amplified consistently but outside that range the amplification changes. At high gains many amplifiers work properly only over a small range of frequencies. The combination of gain and frequency response is often expressed as a *gain-bandwidth product*. For many amplifiers, gain times bandwidth is approximately constant. As gain increases, bandwidth decreases, and vice versa.

Performance at lower frequencies depends on whether the amplifier is *dc- or ac-coupled*. Coupling refers to the transfer of signals between circuits. A dc-coupled amplifier amplifies signals at all frequencies down to dc. An ac-coupled amplifier acts as a high-pass filter, decreasing amplification as the frequency decreases toward dc. Ac-coupled circuits usually use capacitors to allow ac signals to flow between stages while blocking the dc bias voltages of the circuit.

3.4.4 Interstage Loading and Impedance Matching

Every two-port network can be further defined by its input and output impedance. The input impedance is the opposition to current, as a function of frequency, seen when looking into the input port of the network. Likewise, the output impedance is similarly defined when looking back into a network through its output port.

If the transfer function of a stage changes when it is cascaded with another stage, we say that the second stage has *loaded* the first stage. This often occurs when an appreciable amount of current passes from one stage to the next. Interstage loading is related to the relative output impedance of a stage and the input impedance of the stage that is cascaded after it.

In some applications, the goal is to transfer a maximum amount of power from the output of the stage to a load connected to the output. In this case, the output impedance of the stage is *matched* or transformed to that of the load (or vice versa). This allows the stage to operate at its optimum voltage and current levels. In an RF amplifier, the impedance at the input of the transmission line feeding an antenna is transformed by means of a matching network to produce the resistance the amplifier needs in order to efficiently produce RF power.

In contrast, it is the goal of most analog signal processing circuitry to modify a signal rather than to deliver large amounts of energy. Thus, an impedance-matched condition may not be required. Instead, current between stages can be minimized by using mismatched impedances. Ideally, if the output impedance of a network is very low and the input impedance of the following stage is very high, very little current will pass between the stages, and interstage loading will be negligible.

3.4.5 Noise

Generally we are only interested in specific man-made signals. Nature allows many signals to combine, however, so the desired signal becomes combined with many other unwanted signals, both man-made and naturally occurring. The broadest definition of noise is any signal that is not the one in which we are interested. One of the goals of signal processing is to separate desired signals from noise.

One form of noise that occurs naturally and must be dealt with in low-level processing circuits is called *thermal noise*, or *Johnson noise*. Thermal noise is produced by random motion of free electrons in conductors and semiconductors. This motion increases as temperature increases, hence the name. This kind of noise is present at all frequencies and is proportional to temperature. Naturally occurring noise can be reduced either by decreasing the circuit's bandwidth or by reducing the temperature in the system. Thermal noise voltage and current vary with the circuit impedance and follow Ohm's Law. Low-noise-amplifier-design techniques are based on these relationships.

Analog signal processing stages are characterized in part by the noise they add to a signal. A distinction is made between enhancing existing noise (such as amplifying it) and adding new noise. The noise added by analog signal processing is commonly quantified by the *noise factor, f*. Noise factor is the ratio of the total output noise power (thermal noise plus noise added by the stage) to the amplifier input noise power when the termination is at the standard temperature of 290 K (17 °C). When the noise factor is expressed in dB, we often call it *noise figure, NF*.

NF is calculated as:

$$NF = 10 \log \frac{P_{NO}}{A \, P_{N \, TH}} \qquad (13)$$

where

P_{NO} = total noise output power,

A = amplification gain

$P_{N \, TH}$ = input thermal noise power.

Noise factor can also be calculated as the difference between the input and output *signal-to-noise ratios* (SNR), with SNR expressed in dB.

In a system of many cascaded signal processing stages, such as a communications receiver, each stage contributes to the total noise of the system. The noise factor of the first stage dominates the noise factor of the entire system because noise added at the first stage is then multiplied by each following stage. Noise added by later stages is not multiplied to the same degree and so is a smaller contribution to the overall noise at the output.

Designers try to optimize system noise factor by using a first stage with a minimum possible noise factor and maximum possible gain. (Caution: A circuit that overloads is often as useless as one that generates too much noise.) See the **RF Techniques** chapter for a more complete discussion on noise. Circuit overload is discussed in the **Receivers** chapter.

3.4.6 Buffering

It is often necessary to isolate the stages of an analog circuit. This isolation reduces the loading, coupling and feedback between stages. It is often necessary to connect circuits that operate at different impedance levels be-tween stages. An intervening stage, a type of amplifier called a *buffer*, is often used for this purpose.

Buffers can have high values of amplification but this is unusual. A buffer used for impedance transformation generally has a low or unity gain. In some circuits, notably power amplifiers, the desired goal is to deliver a maximum amount of power to the output device (such as a speaker or an antenna). Matching the amplifier output impedance to the output-device impedance provides maximum power transfer. A buffer amplifier may be just the circuit for this type of application. Such amplifier circuits must be carefully designed to avoid distortion. Combinations of buffer stages can also be effective at isolating the stages from each other and making impedance transformations, as well.

3.5 Amplifiers

By far, the most common type of analog circuit is the amplifier. The basic component of most electronics — the transistor — is an amplifier in which a small input signal controls a larger signal. Transistor circuits are designed to use the amplifying characteristics of transistors in order to create useful signal processing functions, regardless of whether the input signal is amplified at the output.

3.5.1 Amplifier Configurations

Amplifier configurations are described by the *common* part of the device. The word "common" is used to describe the connection of a lead directly to a reference that is used by both the input and output ports of the circuit. The most common reference is ground, but positive and negative power sources are also valid references.

The type of circuit reference used depends on the type of device (transistor [NPN or PNP] or FET [P-channel or N-channel]), which lead is chosen as common, and the range of signal levels. Once a common lead is chosen, the other two leads are used for signal input and output. Based on the biasing conditions, there is only one way to select these leads. Thus, there are three possible amplifier configurations for each type of three-lead device. (Vacuum tube amplifiers are discussed in the chapter on **RF Power Amplifiers**.)

DC power sources are usually constructed so that ac signals at the output terminals are bypassed to ground through a very low impedance. This allows the power source to be treated as an *ac ground*, even though it may be supplying dc voltages to the circuit. When a circuit is being analyzed for its ac behavior, ac grounds are usually treated as ground, since dc bias is ignored in the ac analysis. Thus, a transistor's collector can be considered the "common" part of the circuit, even though in actual operation, a dc voltage is applied to it.

Fig 3.35 shows the three basic types of bipolar transistor amplifiers: the common-base, common-emitter, and common-collector. The common terminal is shown connected to ground, although as mentioned earlier, a dc bias voltage may be present. Each type of amplifier is described in the following sections. Following the description of the amplifier, additional discussion of biasing transistors and their operation at high frequencies and for large signals is presented.

3.5.2 Transistor Amplifiers

Creating a useful transistor amplifier depends on using an appropriate model for the transistor itself, choosing the right configuration of the amplifier, using the design equations for that configuration and insuring that the amplifier operates properly at different temperatures. This section follows that sequence, first introducing simple transistor models and then extending that knowledge to the point of design guidelines for common circuits that use bipolar and FETs.

DEVICE MODELS AND CLASSES

Semiconductor circuit design is based on equivalent circuits that describe the physics of the devices. These circuits, made up of voltage and current sources and passive components such as resistors, capacitors and inductors, are called models. A complete model that describes a transistor exactly over a wide frequency range is a fairly complex circuit. As a result, simpler models are used in specific circumstances. For example, the *small-signal model* works well when the device is operated close to some nominal set of characteristics such that current and voltage interact fairly linearly. The *large-signal model* is used when the device is operated so that it enters its saturation or cut-off regions, for example.

Different frequency ranges also require different models. The *low-frequency models* used in this chapter can be used to develop circuits for dc, audio and very low RF applications. At higher frequencies, small capacitances and inductances that can be ignored at low frequencies begin to have significant effects on device behavior, such as gain or impedance. In addition, the physical structure of the device also becomes significant as gain begins to drop or phase shifts between input and output signals start to grow. In this region, *high-frequency models* are used.

Amplifiers are also grouped by their *operating class* that describes the way in which the input signal is amplified. There are several classes of analog amplifiers; A, B, AB, AB1, AB2 and C.

The analog class designators specify over how much of the input cycle the active device is conducting current. A class-A amplifier's active device conducts current for 100 percent of the input signal cycle, such as shown in Fig 3.6. A class-B amplifier conducts during one-half of the input cycle, class-AB, AB1, and AB2 some fraction between 50 and 100 percent of the input cycle, and class-C for less than 50 percent of the input signal cycle.

Digital amplifiers, in which the active device is operated as a switch that is either fully-on or fully-off, similarly to switchmode power supplies, are also grouped by classes beginning with the letter D and beyond. Each different class uses a different method of con-

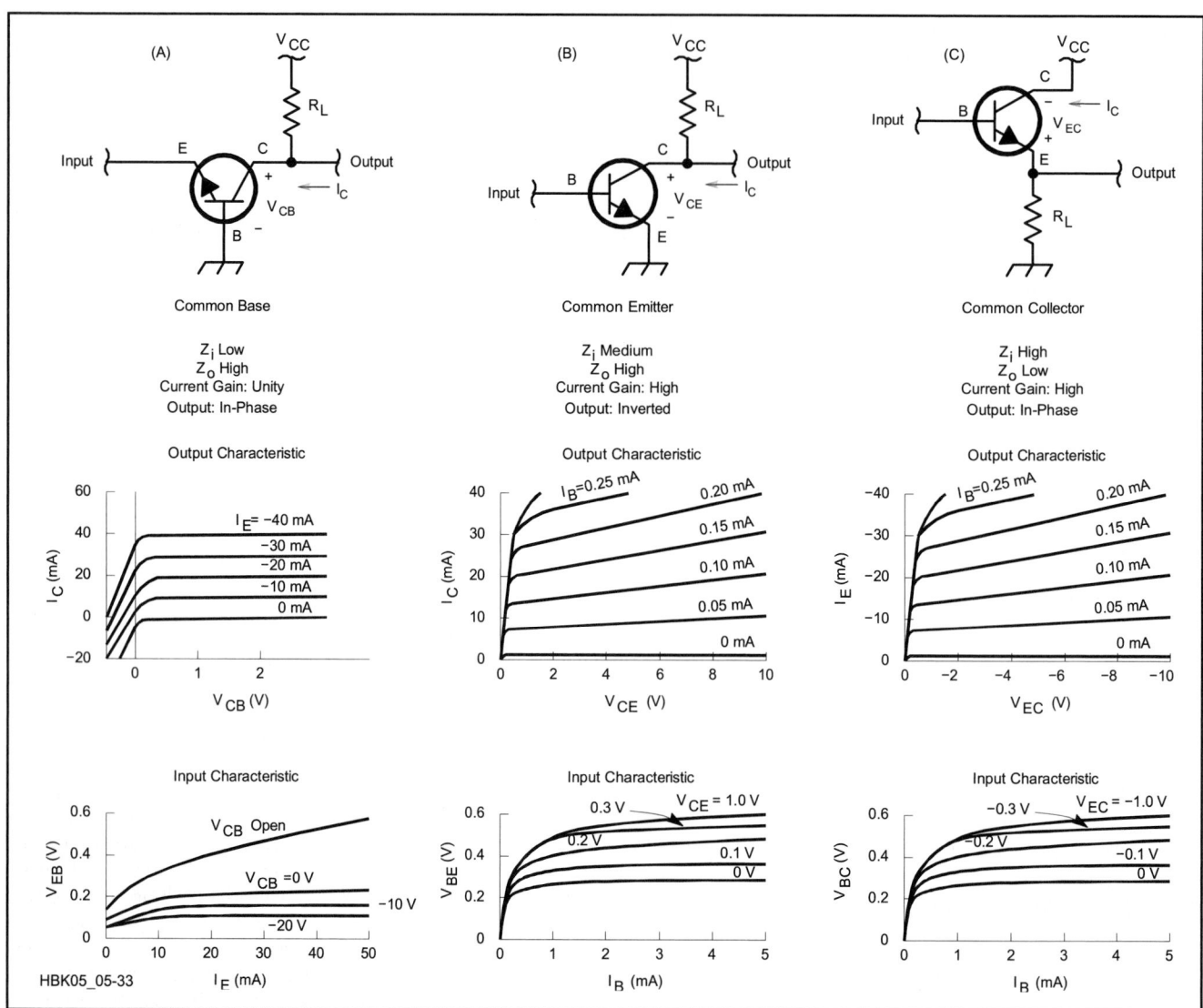

Fig 3.35 — The three configurations of bipolar transistor amplifiers. Each has a table of its relative impedance and current gain. The output characteristic curve is plotted for each, with the output voltage along the x-axis, the output current along the y-axis and various curves plotted for different values of input current. The input characteristic curve is plotted for each configuration with input current along the x-axis, input voltage along the y-axis and various curves plotted for different values of output voltage. (A) Common base configuration with input terminal at the emitter and output terminal at the collector. (B) Common emitter configuration with input terminal at the base and output terminal at the collector. (C) Common collector with input terminal at the base and output terminal at the emitter.

verting the switch's output waveform to the desired RF waveform.

Amplifier classes, models and their use at high-frequencies are discussed in more detail in the chapter on **RF Techniques**. In addition, the use of models for circuit simulation is discussed at length in the **Computer-Aided Circuit Design** chapter.

3.5.3 Bipolar Transistor Amplifiers

In this discussion, we will focus on simple models for bipolar transistors (BJTs). This discussion is centered on NPN BJTs but applies equally well to PNP BJTs if the bias voltage and current polarities are reversed.

This section assumes the small-signal, low-frequency models for the transistors.

SMALL-SIGNAL BJT MODEL

The transistor is usually considered as a *current-controlled* device in which the base current controls the collector current:

$$I_c = \beta I_b \qquad (14)$$

where
I_c = collector current
I_b = base current
β = *common-emitter current gain*, beta.

(The term "common-emitter" refers to the type of transistor circuit described below in which the transistor operates with base cur-

rent as its input and collector circuit as its output.) Current is positive if it flows *into* a device terminal.

The transistor can also be treated as a voltage-controlled device in which the transistor's emitter current, I_e, is controlled by the base-emitter voltage, V_{be}:

$$I_c = I_{es} \left[e^{(qV_{be}/kT)} - 1 \right] \approx I_{es} e^{(qV_{be}/kT)} \qquad (15)$$

where
q = electronic charge
k = Boltzmann's constant
T = temperature in degrees Kelvin (K)
I_{es} = emitter saturation current, typically 1×10^{-13} A.

The subscripts for voltages indicate the direction of positive voltage, so that V_{be} indicates positive is from the base to the emitter. It is simpler to design circuits using the current-controlled device, but accounting for the transistor's behavior with temperature requires an understanding of the voltage-controlled model.

Transistors are usually driven by both biasing and signal voltages. Equations 14 and 15 apply to both transistor dc biasing and signal design. Both of these equations are approximations of the more complex behavior exhibited by actual transistors. Equation 15 applies to a simplification of the first *Ebers-Moll model* in Reference 1. More sophisticated models for BJTs are described by Getreu in Reference 2. Small-signal models treat only the signal components. We will consider bias later.

The next step is to use these basic equations to design circuits. We will begin with small-signal amplifier design and the limits of where the techniques can be applied. Later, we'll discuss large-signal amplifier design and the distortion that arises from operating the transistor in regions where the relationship between the input and output signals is nonlinear.

Common-Emitter Model

Fig 3.36 shows a BJT amplifier connected in the common-emitter configuration. (The emitter, shown connected to ground, is common to both the input circuit with the voltage source and the output circuit with the transistor's collector.) The performance of this circuit is adequately described by equation 14. **Fig 3.37** shows the most common of all transistor small-signal models, a controlled current source with emitter resistance.

There are two variations of the model shown in the figure. Fig 3.37B shows the base as a direct connection to the junction of a current-controlled current source ($I_c = \beta I_b$) and a resistance, r_e, the *dynamic emitter resistance* representing the change in V_{be} with I_e. This resistance also changes with emitter current:

$$r_e = \frac{kT}{qI_c} \approx \frac{26}{I_e} \qquad (16)$$

where I_e is the dc bias current in milliamperes.

The simplified approximation only applies at a typical ambient temperature of 300 K because r_e increases with temperature. In Fig 3.37A, the emitter resistance has been moved to the base connection, where it has the value $(\beta+1)r_e$. These models are electrically equivalent.

The transistor's output resistance (the Thevenin or Norton equivalent resistance between the collector and the grounded emitter) is infinite because of the current source.

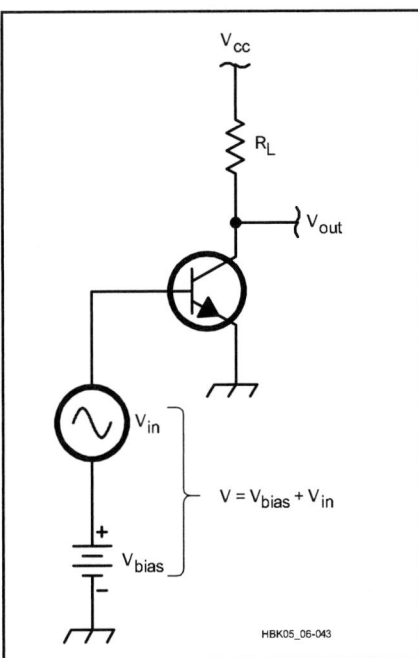

Fig 3.36 — Bipolar transistor with voltage bias and input signal.

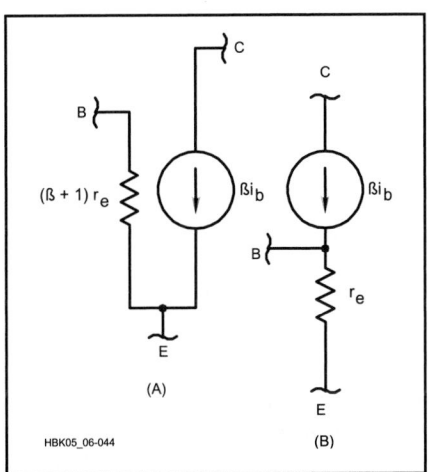

Fig 3.37 — Simplified low-frequency model for the bipolar transistor, a "beta generator with emitter resistance." $r_e = 26 / I_e$ (mA dc).

This is a good approximation for most silicon transistors at low frequencies (well below the transistor's gain-bandwidth product, F_T) and will be used for the design examples that follow.

As frequency increases, the capacitance inherent in BJT construction becomes significant and the *hybrid-pi model* shown in **Fig 3.38** is used, adding C_π in parallel with the input resistance. In this model the transfer parameter h_{ie} often represents the input impedance, shown here as a resistance at low frequencies.

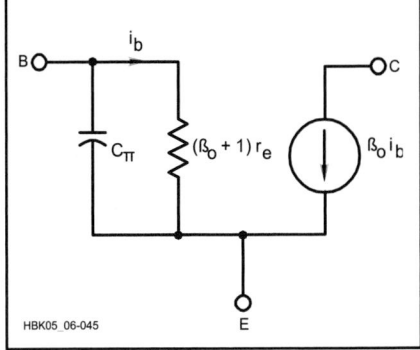

Fig 3.38 — The hybrid-pi model for the bipolar transistor.

THREE BASIC BJT AMPLIFIERS

Fig 3.39 shows a small-signal model applied to the three basic bipolar junction transistor (BJT) amplifier circuits: *common-emitter* (CE), *common-base* (CB) and *common-collector* (CC), more commonly known as the *emitter-follower* (EF). As defined earlier, the word "common" indicates that the referenced terminal is part of both the input and output circuits.

In these simple models, transistors in both the CE and CB configurations have infinite output resistance because the collector current source is in series with the output current. (The amplifier circuit's output impedance must include the effects of R_L.) The transistor connected in the EF configuration, on the other hand, has a finite output resistance because the current source is connected in parallel with the base circuit's equivalent resistance. Calculating the EF amplifier's output resistance requires including the input voltage source, V_s, and its impedance.

The three transistor amplifier configurations are shown as simple circuits in Fig 3.35. Each circuit includes the basic characteristics of the amplifier and characteristic curves for a typical transistor in each configuration. Two sets of characteristic curves are presented: one describing the input behavior and the other describing the output behavior in each amplifier configuration. The different transistor amplifier configurations have different gains, input and output impedances and phase relationships between the input and output signals.

Examining the performance needs of the amplifier (engineers refer to these as the circuit's *performance requirements*) determines which of the three circuits is appropriate. Then, once the amplifier configuration is chosen, the equations that describe the circuit's behavior are used to turn the performance requirements into actual circuit component values.

This text presents design information for the CE amplifier in some detail, then summarizes designs for the CC and CB ampli-

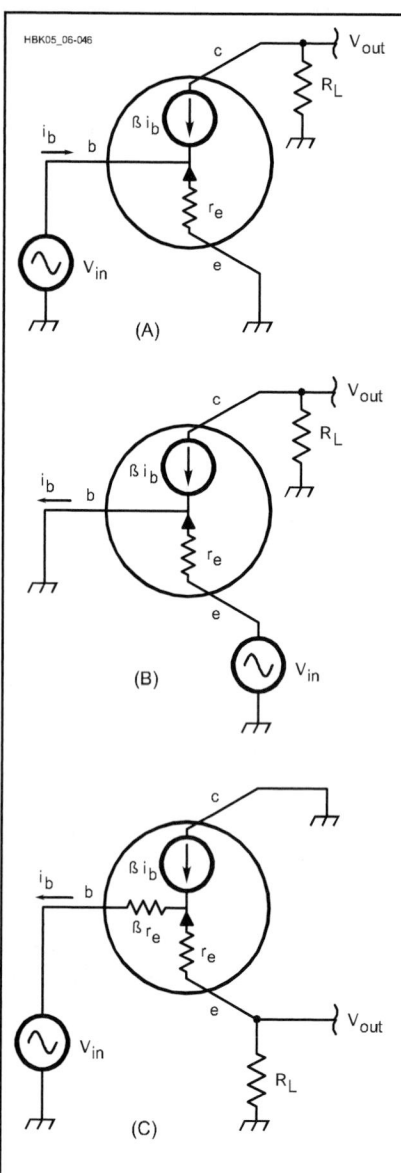

Fig 3.39 — Application of small-signal models for analysis of (A) the CE amplifier, (B) the CB and (C) the EF (CC) bipolar junction transistor amplifiers.

fiers. Detailed design analysis for all three amplifiers is described in the texts listed in the reference section for this chapter. All of the analysis in the following sections assume the small-signal, low-frequency model and ignore the effects of the coupling capacitors. High-frequency considerations are discussed in the **RF Techniques** chapter and some advanced discussion of biasing and large signal behavior of BJT amplifiers is available on the companion CD-ROM.

LOAD LINES AND Q-POINT

The characteristic curves in Fig 3.35 show that the transistor can operate with an infinite number of combinations of current (collector, emitter and base) and voltage (collector-

emitter, collector-base or emitter-collector). The particular combination at which the amplifier is operating is its *operating point*. The operating point is controlled by the selection of component values that make up the amplifier circuit so that it has the proper combination of gain, linearity and so forth. The result is that the operating point is restricted to a set of points that fall along a *load line*. The operating point with no input signal applied is the circuit's *quiescent point* or *Q-point*. As the input signal varies, the operating point moves along the load line, but returns to the Q-point when the input signal is removed.

Fig 3.40 shows the load line and Q-point for an amplifier drawn on a transistor's set of characteristic curves for the CE amplifier circuit. The two end-points of the load line correspond to transistor saturation (I_{Csat} on the I_C current axis) and cutoff (V_{CC} on the V_{CE} voltage axis).

When a transistor is in saturation, further increases in base current do not cause a further increase in collector current. In the CE amplifier, this means that V_{CE} is very close to zero and I_C is at a maximum. In the circuit of Fig 3.35B, imagine a short circuit across the collector-to-emitter so that all of V_{CC} appears across R_L. Increasing base current will not result in any additional collector current. At cutoff, base current is so small that V_{CE} is at a maximum because no collector current is flowing and further reductions in base current cause no additional increase in V_{CE}.

In this simple circuit, $V_{CE} = V_{CC} - I_C R_L$ and the relationship between I_C and V_{CE} is a straight line between saturation and cutoff. This is the circuit's *load line* and it has a slope of $R_L = (V_{CC} - V_{CE}) / I_C$. No matter what value of base current is flowing in the transistor, the resulting combination of I_C and V_{CE} will be somewhere on the load line.

With no input signal to this simple circuit, the transistor is at cutoff where $I_C = 0$ and $V_{CE} = V_{CC}$. As the input signal increases so that base current gets larger, the operating point begins to move along the load line to the left, so that I_C increases and the voltage drop across the load, $I_C R_L$, increases, reducing V_{CE}. Eventually, the input signal will cause enough base current to flow that saturation is reached, where $V_{CE} \approx 0$ (typically 0.1 to 0.3 V for silicon transistors) and $I_C \approx V_{CC} / R_L$. If R_L is made smaller, the load line will become steeper and if R_L increases, the load line's slope is reduced.

This simple circuit cannot reproduce negative input signals because the transistor is already in cutoff with no input signal. In addition, the shape and spacing of the characteristic curves show that the transistor responds nonlinearly when close to saturation and cutoff (the nonlinear regions) than it does in the middle of the curves (the linear or active region). Biasing is required so that the

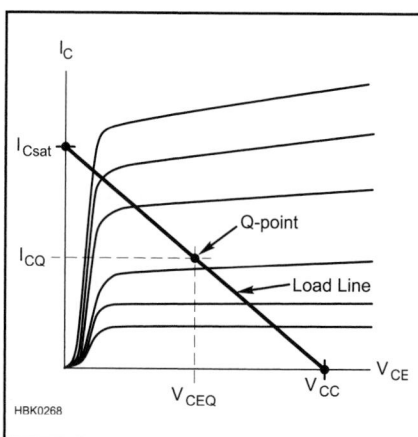

Fig 3.40 — A load line. A circuit's load line shows all of the possible operating points with the specific component values chosen. If there is no input signal, the operating point is the quiescent or Q-point.

circuit does not operate in nonlinear regions, distorting the signal as shown in Fig 3.6.

If the circuit behaves differently for ac signals than for dc signals, a separate *ac load line* can be drawn as discussed below in the section "AC Performance" for the common-emitter amplifier. For example, in the preceding circuit, if R_L is replaced by a circuit that includes inductive or capacitive reactance, ac collector current will result in a different voltage drop across the circuit than will dc collector current. This causes the slope of the ac load line to be different than that of the dc load line.

The ac load line's slope will also vary with frequency, although it is generally treated as constant over the range of frequencies for which the circuit is designed to operate. The ac and dc load lines intersect at the circuit's Q-point because the circuit's ac and dc operation is the same if the ac input signal is zero.

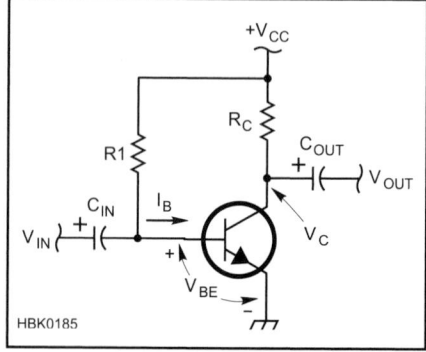

Fig 3.41 — Fixed-bias is the simplest common-emitter (CE) amplifier circuit.

COMMON-EMITTER AMPLIFIER

The *common-emitter amplifier (CE)* is the most common amplifier configuration of all — found in analog and digital circuits, from dc through microwaves, made of discrete components and fabricated in ICs. If you understand the CE amplifier, you've made a good start in electronics.

The CE amplifier is used when modest voltage gain is required along with an *input impedance* (the load presented to the circuit supplying the signal to be amplified) of a few hundred to a few kΩ. The current gain of the CE amplifier is the transistor's current gain, β.

The simplest practical CE amplifier circuit is shown in **Fig 3.41**. This circuit includes both coupling and biasing components. The capacitors at the input (C_{IN}) and output (C_{OUT}) block the flow of dc current to the load or to the circuit driving the amplifier. This is an ac-coupled design. These capacitors also cause the gain at very low frequencies to be reduced — gain at dc is zero, for example, because dc input current is blocked by C_{IN}. Resistor R_1 provides a path for bias current to flow into the base, offsetting the collector current from zero and establishing the Q-point for the circuit.

As the input signal swings positive, more current flows into the transistor's base through C_{IN}, causing more current to flow from the collector to emitter as shown by equation 14. This causes more voltage drop across R_L and so the voltage at the collector also drops. The reverse is true when the input signal swings negative. Thus, the output from the CE amplifier is inverted from its input.

Kirchoff's Voltage Law (KVL, see the **Electrical Fundamentals** chapter) is used to analyze the circuit. We'll start with the collector circuit and treat the power supply as a voltage source.

$$V_{cc} = I_c R_c + V_{ce}$$

We can determine the circuit's voltage gain, A_V, from the variation in output voltage caused by variations in input voltage. The output voltage from the circuit at the transistor collector is

$$V_c = V_{CC} - I_c R_c = V_{CC} - \beta I_B R_C \qquad (17)$$

It is also necessary to determine how base current varies with input voltage. Using the transistor's equivalent circuit of Fig 3.37A,

$$I_B = \frac{V_B}{(\beta + 1) r_e}$$

so that

$$V_c = V_{CC} - V_B \frac{\beta}{\beta + 1} \times \frac{R_C}{r_e} \qquad (18)$$

We can now determine the circuit's *voltage gain*, the variation in output voltage, ΔV_C, due to variations in input voltage, ΔV_B. Since V_{CC} is constant and β is much greater than 1 in our model:

$$A_V \approx -\frac{R_C}{r_e} \qquad (19)$$

Because r_e is quite small (typically a few ohms, see equation 16), A_V for this circuit can be quite high.

The circuit load line's end-points are $V_{CE} = V_{CC}$ and $I_C = V_{CC} / R_C$. The circuit's Q-point is determined by the collector resistor, R_C, and resistor R_1 that causes bias current to flow into the base. To determine the Q-point, again use KVL starting at the power source and assuming that $V_{BE} = 0.7$ V for a silicon transistor's PN junction when forward-biased.

$$V_{CC} - I_B R_1 = V_B = V_{BE} = 0.7 \text{ V}$$

so

$$I_B = \frac{V_{CC} - 0.7V}{R_1} \qquad (20)$$

And the Q-point is therefore

$$V_{CEQ} = V_{CC} - \beta I_B R_C \qquad (21a)$$

and

$$I_{CQ} = \beta I_B \qquad (21b)$$

The actual V_{BE} of silicon transistors will vary from 0.6-0.75 V, depending on the level of base current, but 0.7 V is a good compromise value and widely used in small-signal, low-frequency design. Use 0.6 V for very low-power amplifiers and 0.75 V (or more) for high-current switch circuits.

This simple *fixed-bias* circuit is a good introduction to basic amplifiers, but is not entirely practical because the bias current will change due to the change of V_{BE} with temperature, leading to thermal instability. In addition, the high voltage gain can lead to instability due to positive feedback at high frequencies.

To stabilize the dc bias, **Fig 3.42** adds R_E, a technique called *emitter degeneration* because the extra emitter resistance creates negative feedback: as base current rises, so does V_E, the voltage drop across R_E. This reduces the base-emitter voltage and lowers base current. The benefit of emitter degeneration comes from stabilizing the circuit's dc behavior with temperature, but there is a reduction in gain because of the increased resistance in the emitter circuit. Ignoring the effect of R_L for the moment,

$$A_V \approx -\frac{R_C}{R_E} \qquad (22)$$

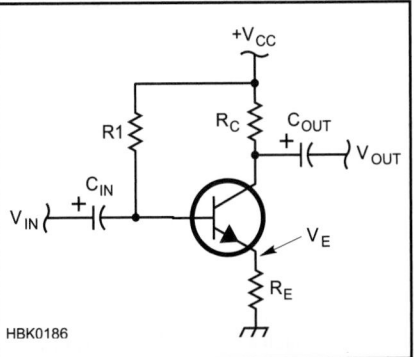

Fig 3.42 — Emitter degeneration. Adding R_E produces negative feedback to stabilize the bias point against changes due to temperature. As the bias current increases, the voltage drop across R_E also increases and causes a decrease in V_{BE}. This reduces bias current and stabilizes the operating point.

In effect, the load resistor is now split between R_C and R_E, with part of the output voltage appearing across each because the changing current flows through both resistors. While somewhat lower than with the emitter connected directly to ground, voltage gain becomes easy to control because it is the ratio of two resistances.

Biasing the CE Amplifier

Fig 3.43 adds R_1 and R_2 from a voltage divider that controls bias current by fixing the base voltage at:

$$V_B = V_{CC} \frac{R_2}{R_1 + R_2}$$

Since

$$V_B = V_{BE} + (I_B + I_C) R_E = 0.7 \text{ V} + (\beta + 1) I_B R_E$$

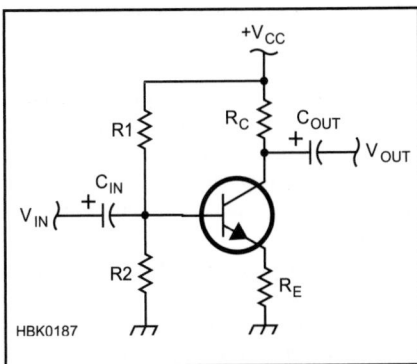

Fig 3.43 — Self-bias. R1 and R2 form a voltage divider to stabilize V_B and bias current. A good rule of thumb is for current flow through R1 and R2 to be 10 times the desired bias current. This stabilizes bias against changes in transistor parameters and component values.

base current is

$$I_B = \frac{V_B - 0.7\ V}{(\beta + 1)\ R_E} \qquad (23a)$$

and Q-point collector current becomes for high values of β

$$I_{CQ} = \beta I_B \approx \frac{V_{CC}\dfrac{R_2}{R_1 + R_2} - 0.7}{R_E} \qquad (23b)$$

This is referred to as *self-bias* in which the Q-point is much less sensitive to variations in temperature that affect β and V_{BE}.

A good rule-of-thumb for determining the sum of R_1 and R_2 is that the current flowing through the voltage divider, $V_{CC}/(R_1+R_2)$, should be at least 10 times the bias current, I_B. This keeps V_B relatively constant even with small changes in transistor parameters and temperature.

Q-point V_{CEQ} must now also account for the voltage drop across both R_C and R_E,

$$V_{CEQ} \approx V_{CC} - \beta I_B (R_C + R_E) \qquad (24)$$

More sophisticated techniques for designing the bias networks of bipolar transistor circuits are described in reference texts listed at the end of this chapter.

Input and Output Impedance

With R_E in the circuit, the small changes in input current, I_B, when multiplied by the transistor's current gain, β, cause a large voltage change across R_E equal to $\beta I_B R_E$. This is the same voltage drop as if I_B was flowing through a resistance equal to βR_E. Thus, the effect of β on impedance at the base is to multiply the emitter resistance, R_E by β, as well. At the transistor's base,

$$Z_B \approx (\beta + 1)\ R_E$$

The input source doesn't just drive the base, of course, it also has to drive the combination of R1 and R2, the biasing resistors. From an ac point of view, both R1 and R2 can be considered as connected to ac ground and they can be treated as if they were connected in parallel. When R1//R2 are considered along with the transistor base impedance, Z_B, the impedance presented to the input signal source is:

$$Z_{IN} = R1\ //\ R2\ //\ (\beta + 1)\ R_E \qquad (25)$$

where // designates "in parallel with."

For both versions of the CE amplifier, the collector output impedance is high enough that

$$Z_{OUT} \approx R_C \qquad (26)$$

CE Amplifier Design Example

The general process depends on the cir-

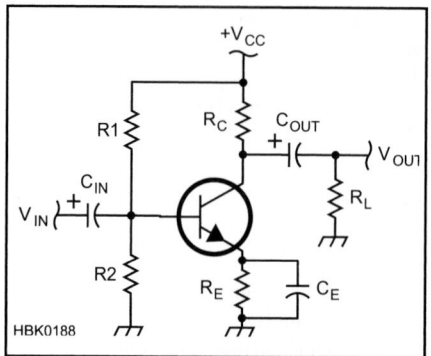

Fig 3.44 — Emitter bypass. Adding C_E allows ac currents to flow "around" R_E, returning ac gain to the value for the fixed-bias circuit while allowing R_E to stabilize the dc operating point.

cuit's primary performance requirements, including voltage gain, impedances, power consumption and so on. The most common situation in which a specific voltage gain is required and the circuit's Q-point has been selected based on the transistor to be used, and using the circuit of Fig 3.43, is as follows:

1) Start by determining the circuit's design constraints and assumptions: power supply $V_{CC} = 12$ V, transistor $\beta = 150$ and $V_{BE} = 0.7$ V. State the circuit's design requirements: $|A_V| = 5$, Q-point of $I_{CQ} = 4$ mA and $V_{CEQ} = 5$ V. (A $V_{CEQ} \approx \frac{1}{2} V_{CC}$ allows a wide swing in output voltage with the least distortion.)

2) Determine the values of R_C and R_E using equation 24: $R_C + R_E = (V_{CC} - V_{CEQ})/ I_{CQ} = 1.75$ kΩ

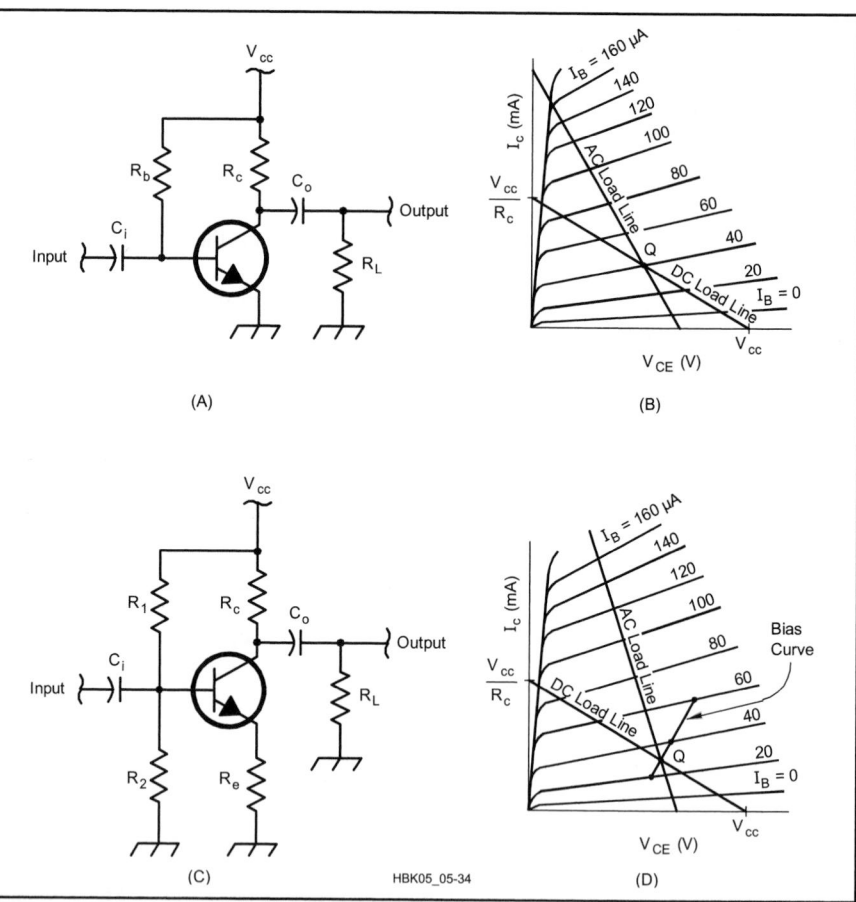

Fig 3.45 — Amplifier biasing and ac and dc load lines. (A) Fixed bias. Input signal is ac coupled through C_i. The output has a voltage that is equal to $V_{CC} - I_C \times R_C$. This signal is ac coupled to the load, R_L, through C_O. For dc signals, the entire output voltage is based on the value of R_C. For ac signals, the output voltage is based on the value of R_C in parallel with R_L. (B) Characteristic curve for the transistor amplifier pictured in (A). The slope of the dc load line is equal to $-1 / R_C$. For ac signals, the slope of the ac load line is equal to $-1 / (R_C // R_L)$. The quiescent-point, Q, is based on the base bias current with no input signal applied and the point where this characteristic line crosses the dc load line. The ac load line must also pass through point Q. (C) Self-bias. Similar to fixed bias circuit with the base bias resistor split into two: R1 connected to V_{CC} and R2 connected to ground. Also an emitter bias resistor, R_E, is included to compensate for changing device characteristics. (D) This is similar to the characteristic curve plotted in (B) but with an additional "bias curve" that shows how the base bias current varies as the device characteristics change with temperature. The operating point, Q, moves along this line and the load lines continue to intersect it as it changes. If C_E was added as in Fig 3.44, the slope of the ac load line would increase further.

3) $A_V = -5$, so from equation 22: $R_C = 5 R_E$, thus $6 R_E = 1.75$ kΩ and $R_E = 270$ Ω

4) Use equation 14 to determine the base bias current, $I_B = I_{CQ}/\beta = 27$ μA. By the rule of thumb, current through R_1 and $R_2 = 10 I_B = 270$ μA

5) Use equation 23 to find the voltage across $R_2 = V_B = V_{BE} + I_C R_E = 0.7 + 4$ mA (0.27 kΩ) = 1.8 V. Thus, $R_2 = 1.8$ V / 270 μA = 6.7 kΩ

6) The voltage across $R_1 = V_{CC} - V_{R2} = 12 - 1.8 = 10.2$ V and $R_1 = 10.2$ V / 270 μA = 37.8 kΩ

Use the nearest standard values ($R_E = 270$ Ω, $R_1 = 39$ kΩ, $R_2 = 6.8$ kΩ) and circuit behavior will be close to that predicted.

AC Performance

To achieve high gains for ac signals while maintaining dc bias stability, the *emitter-by-pass capacitor*, C_E, is added in **Fig 3.44** to provide a low impedance path for ac signals around R_E. In addition, a more accurate formula for ac gain includes the effect of adding R_L through the dc blocking capacitor at the collector. In this circuit, the ac voltage gain is

$$A_V \approx - \frac{R_C \,/\!/\, R_L}{r_e} \qquad (27)$$

Because of the different signal paths for ac and dc signals, the ac performance of the circuit is different than its dc performance. This is illustrated in **Fig 3.45** by the intersecting load lines labeled "AC Load Line" and "DC Load Line." The load lines intersect at the Q-point because at that point dc performance is the same as ac performance if no ac signal is present.

The equation for ac voltage gain assumes that the reactances of C_{IN}, C_{OUT}, and C_E are small enough to be neglected (less than one-tenth that of the components to which they are connected at the frequency of interest). At low frequencies, where the capacitor reactances become increasingly large, voltage gain is reduced. Neglecting C_{IN} and C_{OUT}, the low-frequency 3 dB point of the amplifier, f_L, occurs where $X_{CE} = 0.414 r_e$,

$$f_L = \frac{2.42}{2 \pi r_e C_E} \qquad (28)$$

This increases the emitter circuit impedance such that A_V is lowered to 0.707 of its midband value, lowering gain by 3 dB. (This ignores the effects of C_{IN} and C_{OUT}, which will also affect the low-frequency performance of the circuit.)

The ac input impedance of this version of the CE amplifier is lower because the effect of R_E on ac signals is removed by the bypass capacitor. This leaves only the internal emitter resistance, r_e, to be multiplied by the current gain,

$$Z_{IN} \approx R1 \,/\!/\, R2 \,/\!/\, \beta r_e \qquad (29)$$

and

$$Z_{OUT} \approx R_C \qquad (30)$$

again neglecting the reactance of the three capacitors.

The power gain, A_P, for the emitter-by-passed CE amplifier is the ratio of output power, V_O^2/Z_{OUT}, to input power, V_I^2/Z_{IN}. Since $V_O = V_I A_V$,

$$A_P = A_V^2 \frac{R1 \,/\!/\, R2 \,/\!/\, \beta r_e}{R_C} \qquad (31)$$

COMMON-COLLECTOR (EMITTER-FOLLOWER) AMPLIFIER

The common-collector (CC) amplifier in **Fig 3.46** is also known as the *emitter-follower* (EF) because the emitter voltage "follows" the input voltage. In fact, the amplifier has no voltage gain (voltage gain ≈ 1), but is used as a buffer amplifier to isolate sensitive circuits such as oscillators or to drive low-impedance loads, such as coaxial cables. As in the CE amplifier, the current gain of the emitter-follower is the transistor's current gain, β. It has relatively high input impedance with low output impedance and good power gain.

The collector of the transistor is connected directly to the power supply without a resistor and the output signal is created by the voltage drop across the emitter resistor. There is no 180° phase shift as seen in the CE amplifier; the output voltage follows the input signal with 0° phase shift because increases in the input signal cause increases in emitter current and the voltage drop across the emitter resistor.

The EF amplifier has high input impedance: following the same reasoning as for the CE amplifier with an unbypassed emitter

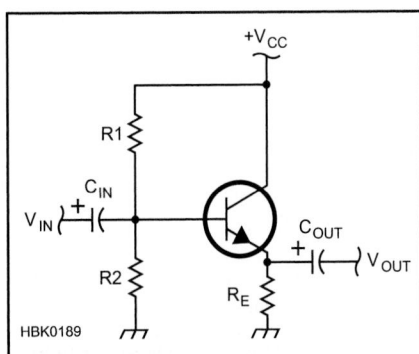

Fig 3.46 — Emitter follower (EF) amplifier. The voltage gain of the EF amplifier is unity. The amplifier has high input impedance and low output impedance, making it a good choice for use as a buffer amplifier.

resistor as described by equation 25,

$$Z_{IN} = R1 \,/\!/\, R2 \,/\!/\, (\beta + 1) R_E \qquad (32)$$

The impedance at the EF amplifier's output consists of the emitter resistance, R_E, in parallel with the series combination of the internal emitter resistance, r_e, the parallel combination of biasing resistors R1 and R2, and the internal impedance of the source providing the input signal. In this case, current gain acts to *reduce* the effect of the input circuit's impedance on output impedance:

$$Z_{OUT} = \left[\frac{R_S \,/\!/\, R1 \,/\!/\, R2}{(\beta + 1)} \right] /\!/ R_E \qquad (33)$$

In practice, with transistor β of 100 or more, $Z_{OUT} \approx R_S/\beta$. However, if a very high impedance source is used, such as an crystal microphone element or photodetector, the effects of the biasing and emitter resistors must be considered.

Because the voltage gain of the EF amplifier is unity, the power gain is simply the ratio of input impedance to output impedance,

$$A_P \approx \frac{R1 \,/\!/\, R2 \,/\!/\, (\beta + 1) R_E}{R_E} \qquad (34)$$

EF Amplifier Design Example

The following procedure is similar to the design procedure in the preceding section for the CE amplifier, except $A_V = 1$.

1) Start by determining the circuit's design constraints and assumptions: $V_{cc} = 12$ V (the power supply voltage), a transistor's β of 150 and $V_{BE} = 0.7$ V. State the circuit's design requirements: Q-point of $I_{CQ} = 5$ mA and $V_{CEQ} = 6$ V.

2) $R_E = (V_{CC} - V_{CEQ})/I_{CQ} = 1.2$ kΩ

3) Base current, $I_B = I_{CQ}/\beta = 33$ μA

4) Current through R1 and R2 = 10 I_B = 330 μA (10 I_B rule of thumb as with the CE amplifier)

5) Voltage across R2 = $V_{BE} + I_C R_E = 0.7 + 5$ mA (1.2 kΩ) = 6.7 V and R2 = 6.7 V / 330 μA = 20.3 kΩ (use the standard value 22 kΩ)

6) Voltage across R1 = $V_{CC} - 6.7$ V = 5.3 V

7) R1 = 5.3 V / 330 μA = 16.1 kΩ (use 16 kΩ)

8) $Z_{IN} = R1 \,/\!/\, R2 \,/\!/\, R_E(\beta + 1) \approx 8.5$ kΩ

COMMON-BASE AMPLIFIER

The common-base (CB) amplifier of **Fig 3.47** is used where low input impedance is needed, such as for a receiver preamp with a coaxial feed line as the input signal source. Complementary to the EF amplifier, the CB amplifier has unity current gain and high output impedance.

Fig 3.47A shows the CB circuit as it is usually drawn, without the bias circuit resistors connected and with the transistor symbol

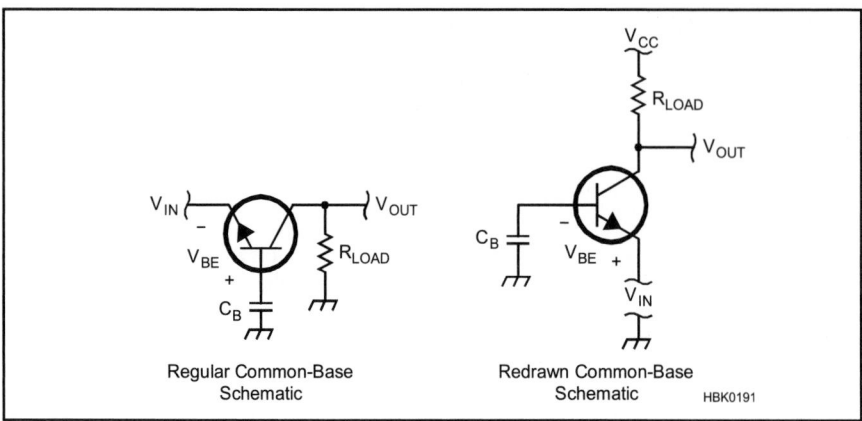

Regular Common-Base
Schematic

Redrawn Common-Base
Schematic

HBK0191

Fig 3.47 — The common-base (CB) amplifier is often drawn in an unfamiliar style (A), but is more easily understood when drawn similarly to the CE and EF amplifiers (B). The input signal to the CB amplifier is applied to the emitter instead of the base.

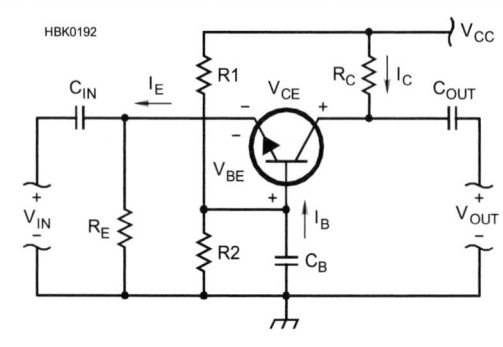

Fig 3.48 — A practical common-base (CB) amplifier. The current gain of the CB amplifier is unity. It has low input impedance and high output impedance, resulting in high voltage gain. The CB amplifier is used to amplify signals from low-impedance sources, such as coaxial cables.

turned on its side from the usual orientation so that the emitter faces the input. In order to better understand the amplifier's function, Fig 3.47B reorients the circuit in a more familiar style. We can now clearly see that the input has just moved from the base circuit to the emitter circuit.

Placing the input in the emitter circuit allows it to cause changes in the base-emitter current as for the CE and EF amplifiers, except that for the CB amplifier a positive change in input amplitude reduces base current by lowering V_{BE} and raising V_C. As a result, the CB amplifier is noninverting, just like the EF, with output and input signals in-phase.

A practical circuit for the CB amplifier is shown in **Fig 3.48**. From a dc point of view (replace the capacitors with open circuits), all of the same resistors are there as in the CE amplifier. The input capacitor, C_{IN}, allows the dc emitter current to bypass the ac input signal source and C_B places the base at ac ground while allowing a dc voltage for biasing. (All voltages and currents are labeled to aid in understanding the different orientation of the circuit.)

The CB amplifier's current gain,

$$A_I = \frac{i_C}{i_E} = \frac{\beta}{\beta + 1} \qquad (35)$$

is relatively independent of input and output impedance, providing excellent isolation between the input and output circuits. Output impedance does not affect input impedance, allowing the CB amplifier to maintain stable input impedance, even with a changing load.

Following reasoning similar to that for the CE and EF amplifiers for the effect of current gain on R_E, we find that input impedance for the CB amp is

$$Z_{IN} = R_E \mathbin{//} (\beta + 1) r_e \qquad (36)$$

The output impedance for the CB amplifier is approximately

$$Z_{OUT} = R_C \mathbin{//} \frac{1}{h_{oe}} \approx R_C \qquad (37)$$

where h_{oe} is the transistor's collector output admittance. The reciprocal of h_{oe} is in the range of 100 kΩ at low frequencies.

Voltage gain for the CB amplifier is

$$A_V \approx \frac{R_C \mathbin{//} R_L}{r_e} \qquad (38)$$

As a result, the usual function of the CB amplifier is to convert input current from a low-impedance source into output voltage at a higher impedance.

Power gain for the CB amplifier is approximately the ratio of output to input impedance,

$$A_P \approx \frac{R_C}{R_E \mathbin{//} (\beta + 1) r_e} \qquad (39)$$

CB Amplifier Design Example

Because of its usual function as a current-to-voltage converter, the design process for the CB amplifier begins with selecting R_E and A_V, assuming that R_L is known.

1) Start by determining the circuit's design constraints and assumptions: $V_{cc} = 12$ V (the power supply voltage), a transistor's β of 150 and $V_{BE} = 0.7$ V. State the circuit's design requirements: $R_E = 50$ Ω, $R_L = 1$ kΩ, $I_{CQ} = 5$ mA, $V_{CEQ} = 6$ V.

2) Base current, $I_B = I_{CQ}/\beta = 33$ μA

3) Current through R1 and R2 = 10 I_B = 330 μA (10 I_B rule of thumb as with the CE amplifier)

4) Voltage across R2 = $V_{BE} + I_C R_E$ = 0.7 + 5 mA (1.2 kΩ) = 6.7 V and R2 = 6.7 V / 330 μA = 20.3 kΩ (use the standard value 22 kΩ)

5) Voltage across R1 = $V_{CC} - 6.7$ V = 5.3 V

6) R1 = 5.3 V / 330 μA = 16.1 kΩ (use 16 kΩ)

7) $R_C = (V_{CC} - I_{CQ} R_E - V_{CEQ}) / I_{CQ}$ = (12 − 0.25 − 5) / 5 mA = 1.35 kΩ (use 1.5 kΩ)

8) $A_V = (1.5$ kΩ // 1 kΩ$) / (26 / I_E) = 115$

3.5.4 FET Amplifiers

The field-effect transistor (FET) is widely used in radio and RF applications. There are many types of FETs, with JFETs (junction FET) and MOSFETs (metal-oxide-semiconductor FET) being the most common types. In this section we will discuss JFETs, with the understanding that the use of MOSFETs is similar. (This discussion is based on N-channel JFETs, but the same discussion applies to P-channel devices if the bias voltages and currents are reversed.)

SMALL-SIGNAL FET MODEL

While bipolar transistors are most commonly viewed as current-controlled devices, the JFET, however, is purely a voltage-controlled device — at least at low frequencies. The input gate is treated as a reverse-biased diode junction with virtually no current flow. As with the bipolar transistor amplifier circuits, the circuits in this section are very basic and more thorough treatments of FET amplifier design can be found in the references at the end of the chapter.

The operation of an N-channel JFET for both biasing and signal amplification can be

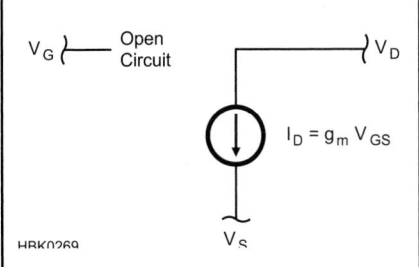

Fig 3.49 — Small-signal FET model. The FET can be modeled as a voltage-controlled current source in its saturation region. The gate is treated as an open-circuit due to the reverse-biased gate-channel junction.

characterized by the following equation:

$$I_D = I_{DSS} \left(\frac{V_P - V_{GS}}{V_P} \right)^2 \qquad (40)$$

where

I_{DSS} = drain saturation current
V_{GS} = the gate-source voltage
V_P = the pinch-off voltage.

I_{DSS} is the maximum current that will flow between the drain and source for a given value of drain-to-source voltage, V_{DS}. Note that the FET is a *square-law* device in which output current is proportional to the square of an input voltage. (The bipolar transistor's output current is an exponential function of input current.)

Also note that V_{GS} in this equation has the opposite sense of the bipolar transistor's V_{BE}. For this device, as V_{GS} increases (making the source more positive than the gate), drain current decreases until at V_P the channel is completely "pinched-off" and no drain current flows at all. This equation applies only if V_{GS} is between 0 and V_P. JFETs are seldom used with the gate-to-channel diode forward-biased ($V_{GS} < 0$).

None of the terms in Equation 40 depend explicitly on temperature. Thus, the FET is relatively free of the thermal instability exhibited by the bipolar transistor. As temperature increases, the overall effect on the JFET is to reduce drain current and to stabilize operation.

The small-signal model used for the JFET is shown in **Fig 3.49**. The drain-source channel is treated as a current source whose output is controlled by the gate-to-source voltage so that $I_D = g_m V_{GS}$. The high input impedance allows the input to be modeled as an open circuit (at low frequencies). This simplifies circuit modeling considerably as biasing of the FET gate can be done by a simple voltage divider without having to consider the effects of bias current flowing in the JFET itself.

The FET has characteristic curves as shown in Fig 3.25 that are similar to those of a bipolar

transistor. The output characteristic curves are similar to those of the bipolar transistor, with the horizontal axis showing V_{DS} instead of V_{CE} and the vertical axis showing I_D instead of I_C. Load lines, both ac and dc, can be developed and drawn on the output characteristic curves in the same way as for bipolar transistors.

The set of characteristic curves in Fig 3.25 are called *transconductance response curves* and they show the relationship between input voltage (V_{GS}), output current (I_D) and output voltage (V_{DS}). The output characteristic curves show I_D and V_{DS} for different values of V_{GS} and are similar to the BJT output characteristic curve. The input characteristic curves show I_D versus V_{GS} for different values of V_{DS}.

MOSFETs act in much the same way as JFETs when used in an amplifier. They have a higher input impedance, due to the insulation between the gate and channel. The insulated gate also means that they can be operated with the polarity of V_{GS} such that a JFET's gate-channel junction would be forward biased, beyond V_P. Refer to the discussion of depletion- and enhancement-mode MOSFETs in the previous section on Practical Semiconductors.

THREE BASIC FET AMPLIFIERS

Just as for bipolar transistor amplifiers, there are three basic configurations of amplifiers using FETs; the *common-source* (CS)(corresponding to the common-emitter), *common-drain* (CD) or *source-follower* (corresponding to the emitter-follower) and the *common-gate* (CG) (corresponding to the common base). Simple circuits and design methods are presented here for each, assuming low-frequency operation and a simple, voltage-controlled current-source model for the FET. Discussion of the FET amplifier at high frequencies is available in the **RF Techniques** chapter and an advanced discussion of biasing FET amplifiers and their large-signal behavior is contained on the companion CD-ROM.

COMMON-SOURCE AMPLIFIER

The basic circuit for a common-source FET amplifier is shown in **Fig 3.50**. In the ohmic region (see the previous discussion on FET characteristics), the FET can be treated as a variable resistance as shown in **Fig 3.50A** where V_{GS} effectively varies the resistance between drain and source. However, most FET amplifiers are designed to operate in the saturation region and the model of Fig 3.49 is used in the circuit of **Fig 3.50B** in which,

$$I_D = g_m V_{GS} \qquad (41)$$

where g_m is the FET's forward transconductance.

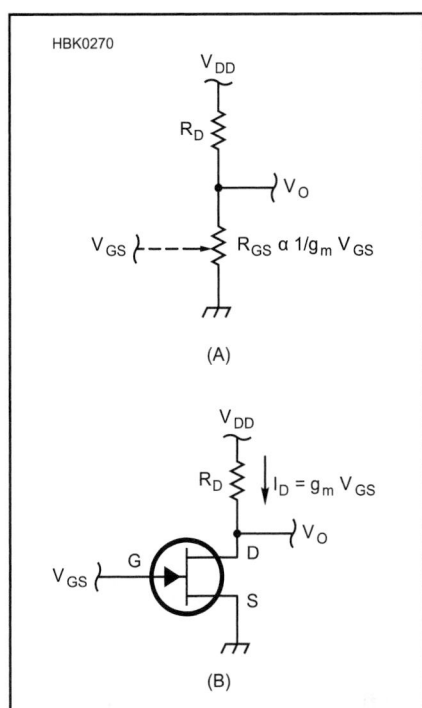

Fig 3.50 — In the ohmic region (A), the FET acts like a variable resistance, R_{DS}, with a value controlled by V_{GS}. The alpha symbol (α) means "is proportional to". In the saturation region (B), the drain-source channel of the FET can be treated like a current source with $I_D = g_m V_{GS}$.

If V_O is measured at the drain terminal (just as the common-emitter output voltage is measured at the collector), then

$$\Delta V_O = -g_m \Delta V_{GS} R_D$$

The minus sign results from the output voltage decreasing as drain current and the voltage drop across R_D increases, just as in the CE amplifier. Like the CE amplifier, the input and output voltages are thus 180° out of phase. Voltage gain of the CS amplifier in terms of transconductance and the drain resistance is:

$$A_V = -g_m R_D \qquad (42)$$

As long as $V_{GS} < 0$, this simple CS amplifier's input impedance at low frequencies is that of a reverse-biased diode — nearly infinite with a very small leakage current. Output impedance of the CS amplifier is approximately R_D because the FET drain-to-source channel acts like a current source with very high impedance.

$$Z_{IN} = \infty \text{ and } Z_{OUT} \approx R_D \qquad (43)$$

As with the BJT, biasing is required to create a Q-point for the amplifier that allows reproduction of ac signals. The practical circuit

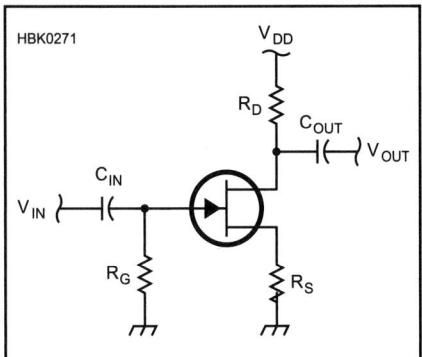

Fig 3.51 — Common-source (CS) amplifier with self-bias.

of Fig 3.50B is used to allow control of V_{GS} bias. A load line is drawn on the JFET output characteristic curves, just as for a bipolar transistor circuit. One end point of the load line is at $V_{DS} = V_{DD}$ and the other at $I_{DS} = V_{DD} / R_D$. The Q-point for the CS amplifier at I_{DQ} and V_{DSQ} is thus determined by the dc value of V_{GS}.

The practical JFET CS amplifier shown in **Fig 3.51** uses self-biasing in which the voltage developed across the source resistor, R_S, raises V_S above ground by $I_D R_S$ volts and $V_{GS} = -I_D R_S$ since there is no dc drop across R_G. This is also called *source degeneration*. The presence of R_S changes the equation of voltage gain to

$$A_V = -\frac{g_m R_D}{1 + g_m R_S} \approx -\frac{R_D}{R_S} \qquad (44)$$

The value of R_S is obtained by substituting $V_{GS} = I_D R_S$ into Equation 40 and solving for R_S as follows:

$$R_S = \frac{V_P}{I_{DQ}} \left(1 - \sqrt{\frac{I_{DQ}}{I_{DSS}}}\right) \qquad (45)$$

Once R_S is known, the equation for voltage gain can be used to find R_D.

The input impedance for the circuit of Fig 3.51 is essentially R_G. Since the gate of the JFET is often ac coupled to the input source through a dc blocking capacitor, C_{IN}, a value of 100 kΩ to 1 MΩ is often used for R_G to provide a path to ground for gate leakage current. If R_G is omitted in an ac-coupled JFET amplifier, a dc voltage can build up on the gate from leakage current or static electricity, affecting the channel conductivity.

$$Z_{IN} = R_G \qquad (46)$$

Because of the high impedance of the drain-source channel in the saturation region, the output impedance of the circuit is:

$$Z_{OUT} \approx R_D \qquad (47)$$

Designing the Common-Source Amplifier

The design of the CS amplifier begins with selection of a Q-point $I_{DQ} < I_{DSS}$. Because of variations in V_P and I_{DSS} from JFET to JFET, it may be necessary to select devices individually to obtain the desired performance.

1) Start by determining the circuit's design constraints and assumptions: $V_{DD} = 12$ V (the power supply voltage) and the JFET has an I_{DSS} of 35 mA and a V_P of –3.0V, typical of small-signal JFETs. State the circuit's design requirements: $|A_V| = 10$ and $I_{DQ} = 10$ mA.

2) Use equation 45 to determine $R_S = 139$ Ω

3) Since $|A_V| = 10$, $R_D = 10 R_S = 1390$ Ω.

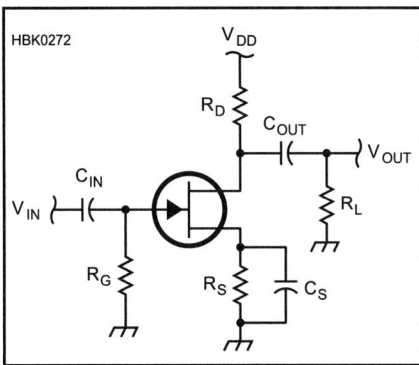

Fig 3.52 — Common-source amplifier with source bypass capacitor, C_S, to increase voltage gain without affecting the circuit's dc performance.

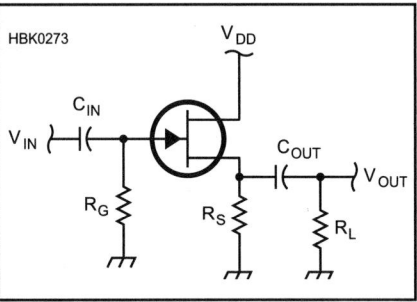

Fig 3.53 — Similar to the EF amplifier, the common-drain (CD) amplifier has a voltage gain of unity, but makes a good buffer with high input and low output impedances.

Use standard values for $R_S = 150$ Ω and $R_D = 1.5$ kΩ.

AC Performance

As with the CE bipolar transistor amplifier, a bypass capacitor can be used to increase ac gain while leaving dc bias conditions unchanged as shown in **Fig 3.52**. In the case of the CS amplifier, a *source bypass* capacitor is placed across R_S and the load, R_L, connected through a dc blocking capacitor. In this circuit voltage gain becomes:

$$A_V = -g_m (R_D // R_L) \qquad (48)$$

Assuming C_{IN} and C_{OUT} are large enough to ignore their effects, the low-frequency cutoff frequency of the amplifier, f_L, is approximately where $X_{CS} = 0.707 (R_D // R_L)$,

$$f_L = \frac{1.414}{2\pi (R_D // R_L) C_S} \qquad (49)$$

as this reduces A_V to 0.707 of its mid-band value, resulting in a 3 dB drop in output amplitude.

The low-frequency ac input and output impedances of the CS amplifier remain

$$Z_{IN} = R_G \text{ and } Z_{OUT} \approx R_D \qquad (50)$$

COMMON-DRAIN (SOURCE-FOLLOWER) AMPLIFIER

The *common-drain* amplifier in **Fig 3.53** is also known as a *source-follower* (SF) because the voltage gain of the amplifier is unity, similar to the emitter follower (EF) bipolar transistor amplifier. The SF amplifier is used primarily as a buffer stage and to drive low-impedance loads.

At low frequencies, the input impedance of the SF amplifier remains nearly infinite. The SF amplifier's output impedance is the source resistance, R_S, in parallel with the impedance of the controlled current source, $1/g_m$.

$$Z_{OUT} = R_S // \frac{1}{g_m}$$
$$= \frac{R_S}{g_m R_S + 1} \approx \frac{1}{g_m} \text{ for } g_m R_S \gg 1 \qquad (51)$$

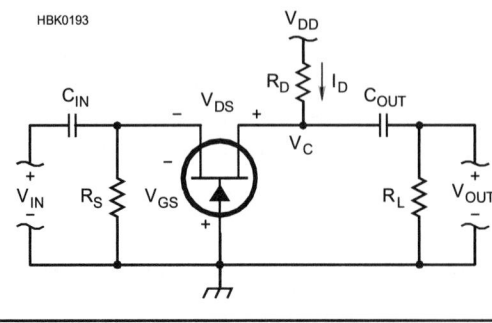

Fig 3.54 — FET common-gate (CG) amplifiers are often used as preamplifiers because of their high voltage gain and low input impedance. With the proper choice of transistor and quiescent-point current, the input impedance can match coaxial cable impedances directly.

Design of the SF amplifier follows essentially the same process as the CS amplifier, with $R_D = 0$.

THE COMMON-GATE AMPLIFIER

The *common-gate* amplifier in **Fig 3.54** has similar properties to the bipolar transistor common-base (CB) amplifier; unity current gain, high voltage gain, low input impedance and high output impedance. (Refer to the discussion of the CB amplifier regarding placement of the input and how the circuit schematic is drawn.) It is used as a voltage amplifier, particularly for low-impedance sources, such as coaxial cable inputs.

The CG amplifier's voltage gain is

$$A_V = g_m(R_D \ // \ R_L) \tag{52}$$

The output impedance of the CG amplifier is very high, we must take into account the output resistance of the controlled current source, r_o. This is analogous to the appearance of h_{oe} in the equation for output impedance of the bipolar transistor CG amplifier.

$$Z_O \approx r_o \, (g_m R_S + 1) \, // \, R_D \tag{53}$$

The CG amplifier input impedance is approximately

$$Z_I = R_S \, // \, \frac{1}{g_m} \tag{54}$$

Because the input impedance is quite low, the cascode circuit described later in the section on buffers is often used to present a higher-impedance input to the signal source.

Occasionally, the value of R_S must be fixed in order to provide a specific value of input impedance. Solving equation 40 for I_{DQ} results in the following equation:

$$I_{DQ} = $$
$$\frac{V_P}{2R_S{}^2 I_{DSS}} \left(V_P + \sqrt{V_P{}^2 - 4R_S I_{DSS} V_P} \right) - \frac{V_P}{R_S} \tag{55}$$

Designing the Common-Gate Amplifier

Follow the procedure for designing a CS amplifier, except determine the value of R_D as shown in equation 52 for voltage gain above.

1) Start by determining the circuit's design constraints and assumptions: $V_{DD} = 12$ V (the power supply voltage) and the JFET has a g_m of 15 mA/V, an I_{DSS} of 60 mA and $V_P = -6$ V. State the circuit's design requirements: $A_V = 10$, $R_L = 1$ kΩ and $R_S = 50$ Ω.

2) Solve equation 52 for R_D: $10 = 0.015 \times R_D//R_L$, so $R_D//R_L = 667$ Ω. $R_D = 667 \, R_L / (R_L - 667) = 2$ kΩ.

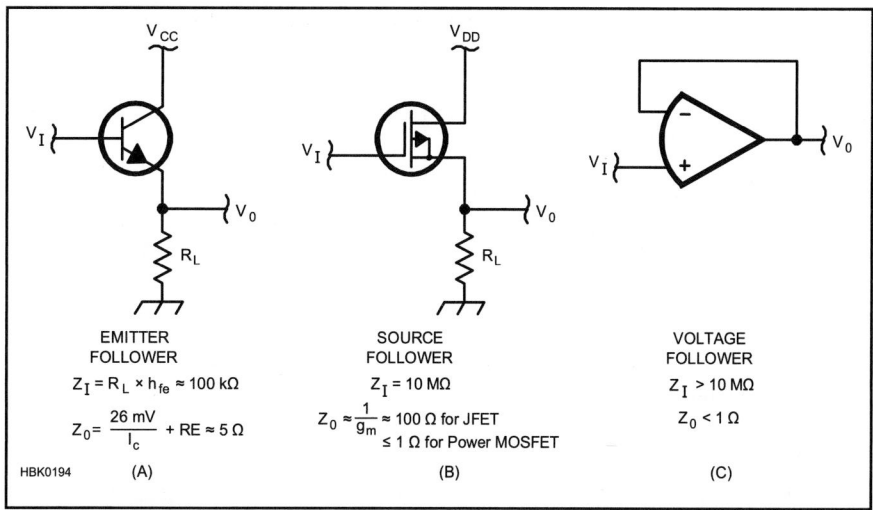

Fig 3.55 — Common buffer stages and some typical input (Z_I) and output (Z_O) impedances. (A) Emitter follower, made with an NPN bipolar transistor; (B) Source follower, made with an FET; and (C) Voltage follower, made with an operational amplifier. All of these buffers are terminated with a load resistance, R_L, and have an output voltage that is approximately equal to the input voltage (gain ≈ 1).

3) I_{DQ} is determined from equation 55: $I_{DQ} = 10$ mA. If I_{DQ} places the Q-point in the ohmic region, reduce A_V and repeat the calculations.

3.5.5 Buffer Amplifiers

Fig 3.55 shows common forms of buffers with low-impedance outputs: the *emitter follower* using a bipolar transistor, the *source follower* using a field-effect transistor and the *voltage follower*, using an operational amplifier. (The operational amplifier is discussed later in this chapter.) These circuits are called "followers" because the output "follows" the input very closely with approximately the same voltage and little phase shift between the input and output signals.

3.5.6 Cascaded Buffers

THE DARLINGTON PAIR

Buffer stages that are made with single active devices can be more effective if cascaded. Two types of such buffers are in common use. The *Darlington pair* is a cascade of two transistors connected as emitter followers as shown in **Fig 3.56**. The current gain of the Darlington pair is the product of the current gains for the two transistors, $\beta_1 \times \beta_2$.

What makes the Darlington pair so useful is that its input impedance is equal to the load impedance times the current gain, effectively multiplying the load impedance;

$$Z_I = Z_{LOAD} \times \beta_1 \times \beta_2 \tag{56}$$

For example, if a typical bipolar transistor has $\beta = 100$ and $Z_{LOAD} = 15$ kΩ, a pair of these transistors in the Darlington-pair configuration would have:

$$Z_I = 15 \text{ k}\Omega \times 100 \times 100 = 150 \text{ M}\Omega$$

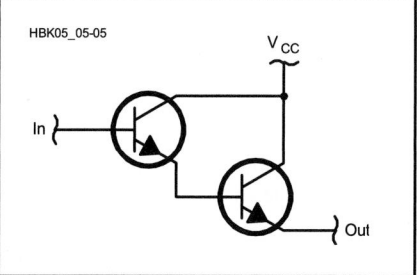

Fig 3.56 — Darlington pair made with two emitter followers. Input impedance, Z_I, is far higher than for a single transistor and output impedance, Z_O, is nearly the same as for a single transistor. DC biasing has been omitted for simplicity.

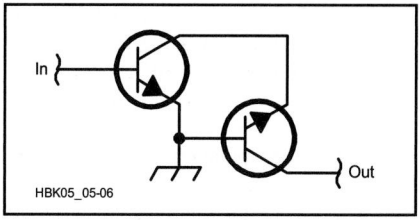

Fig 3.57 — Cascode buffer made with two NPN bipolar transistors has a medium input impedance and high output impedance. DC biasing has been omitted for simplicity.

This impedance places almost no load on the circuit connected to the Darlington pair's input. The shunt capacitance at the input of real transistors can lower the actual impedance as the frequency increases.

Drawbacks of the Darlington pair include lower bandwidth and switching speed. The extremely high dc gain makes biasing very sensitive to temperature and component tolerances. For these reasons, the circuit is usually used as a switch and not as a linear amplifier.

CASCODE AMPLIFIERS

A common-emitter amplifier followed by a common-base amplifier is called a *cascode buffer*, shown in its simplest form in **Fig 3.57**. (Biasing and dc blocking components are omitted for simplicity — replace the transistors with the practical circuits described earlier.) Cascode stages using FETs follow a common-source amplifier with a common-gate configuration. The input impedance and current gain of the cascode amplifier are approximately the same as those of the first stage. The output impedance of the common-base or –gate stage is much higher than that of the common-emitter or common-source amplifier. The power gain of the cascode amplifier is the product of the input stage current gain and the output stage voltage gain.

As an example, a typical cascode buffer made with BJTs has moderate input impedance ($Z_{IN} = 1$ kΩ), high current gain ($h_{fe} = 50$), and high output impedance ($Z_{OUT} = 1$ MΩ). Cascode amplifiers have excellent input/output isolation (very low unwanted internal feedback), resulting in high gain with good stability. Because of its excellent isolation, the cascode amplifier has little effect on external tuning components. Cascode circuits are often used in tuned amplifier designs for these reasons.

3.5.7 Using the Transistor as a Switch

When designing amplifiers, the goal was to make the transistor's output a replica of its input, requiring that the transistor stay within its linear region, conducting some current at all times. A switch circuit has completely different properties — its output current is either zero or some maximum value. **Fig 3.58** shows both a bipolar and metal-oxide semiconductor field-effect transistor (or MOSFET) switch circuit. Unlike the linear amplifier circuits, there are no bias resistors in either circuit. When using the bipolar transistor as a switch, it should operate in saturation or in cutoff. Similarly, an FET switch should be either fully-on or fully-off. The figure shows the waveforms associated with both types of switch circuits.

This discussion is written with power con-

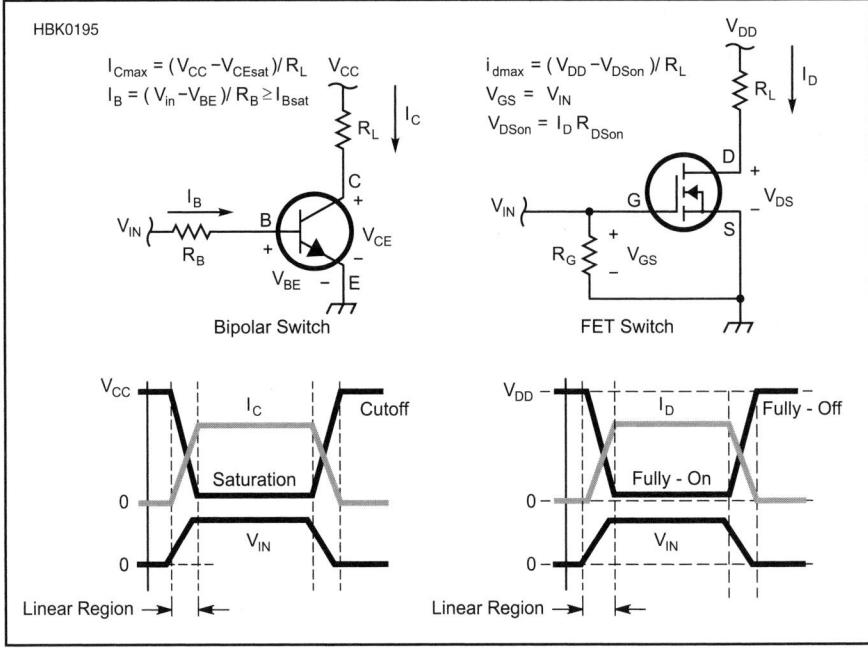

Fig 3.58 — A pair of transistor driver circuits using a bipolar transistor and a MOSFET. The input and output signals show the linear, cutoff and saturation regions.

trol in mind, such as to drive a relay or motor or lamp. The concepts, however, are equally applicable to the much lower-power circuits that control logic-level signals. The switch should behave just the same — switch between on and off quickly and completely — whether large or small.

DESIGNING SWITCHING CIRCUITS

First, select a transistor that can handle the load current and dissipate whatever power is dissipated as heat. Second, be sure that the input signal source can supply an adequate input signal to drive the transistor to the required states, both on and off. Both of these conditions must be met to insure reliable driver operation.

To choose the proper transistor, the load current and supply voltage must both be known. Supply voltage may be steady, but sometimes varies widely. For example, a car's 12 V power bus may vary from 9 to 18 V, depending on battery condition. The transistor must withstand the maximum supply voltage, V_{MAX}, when off. The load resistance, R_L, must also be known. The maximum steady-state current the switch must handle is:

$$I_{MAX} = \frac{V_{MAX}}{R_L} \tag{57}$$

If you are using a bipolar transistor, calculate how much base current is required to drive the transistor at this level of collector current. You'll need to inspect the transistor's data sheet because β decreases as collector current increases, so use a value for β speci-

fied at a collector current at or above I_{MAX}.

$$I_B = \frac{I_{MAX}}{\beta} \tag{58}$$

Now inspect the transistor's data sheet values for V_{CEsat} and make sure that this value of I_B is sufficient to drive the transistor fully into saturation at a collector current of I_{MAX}. Increase I_B if necessary — this is I_{Bsat}. The transistor must be fully saturated to minimize heating when conducting load current.

Using the *minimum* value for the input voltage, calculate the value of R_B:

$$R_B = \frac{V_{IN(min)} - V_{BE}}{I_{Bsat}} \tag{59}$$

The minimum value of input voltage must be used to accommodate the *worst-case* combination of circuit voltages and currents.

Designing with a MOSFET is a little easier because the manufacturer usually specifies the value V_{GS} must have for the transistor to be fully on, $V_{GS(on)}$. The MOSFET's gate, being insulated from the conducting channel, acts like a small capacitor of a few hundred pF and draws very little dc current. R_G in Fig 3.58 is required if the input voltage source does not actively drive its output to zero volts when off, such as a switch connected to a positive voltage. The MOSFET won't turn off reliably if its gate is allowed to "float." R_G pulls the gate voltage to zero when the input is open-circuited.

Power dissipation is the next design hurdle. Even if the transistors are turned completely on, they will still dissipate some heat. Just as for a resistor, for a bipolar transistor switch the power dissipation is:

$$P_D = V_{CE}I_C = V_{CE(sat)} I_{MAX}$$

where $V_{CE(sat)}$ is the collector-to-emitter voltage when the transistor is saturated.

Power dissipation in a MOSFET switch is:

$$P_D = V_{DS}I_D = R_{DS(on)} I_{MAX}^2 \qquad (60)$$

$R_{DS(on)}$ is the resistance of the channel from drain to source when the MOSFET is on. MOSFETs are available with very low on-resistance, but still dissipate a fair amount of power when driving a heavy load. The transistor's data sheet will contain $R_{DS(on)}$ specification and the V_{GS} required for it to be reached.

Power dissipation is why a switching transistor needs to be kept out of its linear region. When turned completely off or on, either current through the transistor or voltage across it are low, also keeping the product of voltage and current (the power to be dissipated) low. As the waveform diagrams in Fig 3.58 show, while in the linear region, both voltage and current have significant values and so the transistor is generating heat when changing from off to on and vice versa. It's important to make the transition through the linear region quickly to keep the transistor cool.

The worst-case amount of power dissipated during each on-off transition is approximately

$$P_{transition} = \frac{1}{4}V_{MAX} I_{MAX}$$

assuming that the voltage and current increase and decrease linearly. If the circuit turns on and off at a rate of f, the total average power dissipation due to switching states is:

$$P_D = \frac{f}{2}V_{MAX} I_{MAX} \qquad (61)$$

since there are two on-off transitions per switching cycle. This power must be added to the power dissipated when the switch is conducting current.

Once you have calculated the power the switch must dissipate, you must check to see whether the transistor can withstand it. The manufacturer of the transistor will specify a *free-air dissipation* that assumes no heat-sink and room temperature air circulating freely around the transistor. This rating should be at least 50% higher than your calculated power dissipation. If not, you must either use a larger transistor or provide some means of getting rid of the heat, such as heat sink. Methods of

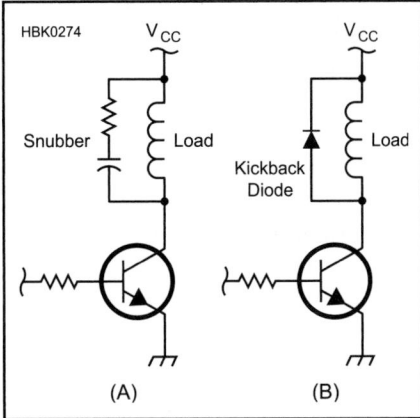

Fig 3.59 — The snubber RC circuit at (A) absorbs energy from transients with fast rise- and fall-times. At (B) a kickback diode protects the switching device when current is interrupted in the inductive load, causing a voltage transient, by conducting the energy back to the power source.

dissipating heat are discussed in the **Electrical Fundamentals** chapter.

INDUCTIVE AND CAPACITIVE LOADS

Voltage transients for inductive loads, such as solenoids or relays can easily reach dozens of times the power supply voltage when load current is suddenly interrupted. To protect the transistor, the voltage transient must be clamped or its energy dissipated. Where switching is frequent, a series-RC *snubber* circuit (see **Fig 3.59A**) is connected across the load to dissipate the transient's energy as heat. The most common method is to employ a "kickback" diode that is reverse-biased when the load is energized as shown in Fig 3.59B. When the load current is interrupted, the diode routes the energy back to the power supply, clamping the voltage at the power supply voltage plus the diode's forward voltage drop.

Capacitive loads such as heavily filtered power inputs may temporarily act like short circuits when the load is energized or de-energized. The surge current is only limited by the internal resistance of the load capacitance. The transistor will have to handle the temporary overloads without being damaged or overheating. The usual solution is to select a transistor with an I_{MAX} rating greater to the surge current. Sometimes a small current-limiting resistor can be placed in series with the load to reduce the peak surge current at the

expense of dissipating power continuously when the load is drawing current.

HIGH-SIDE AND LOW-SIDE SWITCHING

The switching circuits shown in Fig 3.58 are *low-side switches*. This means the switch is connected between the load and ground. A *high-side switch* is connected between the power source and the load. The same concerns for power dissipation apply, but the methods of driving the switch change because of the voltage of the emitter or source of the switching device will be at or near the power supply voltage when the switch is on.

To drive an NPN bipolar transistor or an N-channel MOSFET in a high-side circuit requires the switch input signal to be at least $V_{BE(sat)}$ or $V_{GS(on)}$ *above* the voltage supplied to the load. If the load expects to see the full power supply voltage, the switch input signal will have to be *greater* than the power supply voltage. A small step-up or boost dc-to-dc converter is often used to supply the extra voltage needed for the driver circuit.

One alternate method of high-side switching is to use a PNP bipolar transistor as the switching transistor. A small input transistor turns the main PNP transistor on by controlling the larger transistor's base current. Similarly, a P-channel MOSFET could also be employed with a bipolar transistor or FET acting as its driver. P-type material generally does not have the same high conductivity as N-type material and so these devices dissipate somewhat more power than N-type devices under the same load conditions.

3.5.8 Choosing a Transistor

With all the choices for transistors — Web sites and catalogs can list hundreds — selecting a suitable transistor can be intimidating. Start by determining the maximum voltage (V_{CEO} or $V_{DS(MAX)}$), current (I_{MAX}) and power dissipation ($P_{D(MAX)}$) the transistor must handle. Determine what dc current gain, β, or transconductance, g_m, is required. Then determine the highest frequency at which full gain is required and multiply it by either the voltage or current gain to obtain f_T or h_{fe}. This will reduce the number of choices dramatically.

The chapter on **Component Data and References** has tables of parameters for popular transistors that tend to be the lowest-cost and most available parts, as well. You will find that a handful of part types satisfy the majority of your building needs. Only in very special applications will you need to choose a corresponding special part.

3.6 Operational Amplifiers

An *operational amplifier*, or *op amp*, is one of the most useful linear devices. While it is possible to build an op amp with discrete components, and early versions were, the symmetry of the circuit demanded for high performance requires a close match of many components. It is more effective, and much easier, to implement as an integrated circuit. (The term "operational" comes from the op amp's origin in analog computers where it was used to implement mathematical operations.)

The op amp's performance approaches that of an ideal analog circuit building block: an infinite input impedance (Z_i), a zero output impedance (Z_o) and an open loop voltage gain (A_v) of infinity. Obviously, practical op amps do not meet these specifications, but they do come closer than most other types of amplifiers. These attributes allow the circuit designer to implement many different functions with an op amp and only a few external components.

3.6.1 Characteristics of Practical Op-Amps

An op amp has three signal terminals (see **Fig 3.60**). There are two input terminals, the *noninverting input* marked with a + sign and the *inverting input* marked with a – sign. Voltages applied to the noninverting input cause the op amp output voltage to change with the same polarity.

The output of the amplifier is a single terminal with the output voltage referenced to the external circuit's reference voltage. Usually, that reference is ground, but the op amp's internal circuitry allows all voltages to *float*, that is, to be referenced to any arbitrary voltage between the op amp's power supply voltages. The reference can be negative, ground or positive. For example, an op amp powered from a single power supply voltage amplifies just as well if the circuit reference voltage is halfway between ground and the supply voltage.

GAIN-BANDWIDTH PRODUCT AND COMPENSATION

An ideal op amp would have infinite frequency response, but just as transistors have an f_T that marks their upper frequency limit, the op amp has a *gain-bandwidth product* (GBW or BW). GBW represents the maximum product of gain and frequency available to any signal or circuit: voltage gain × frequency = GBW. If an op-amp with a GBW of 10 MHz is connected as a ×50 voltage amplifier, the maximum frequency at which that gain could be guaranteed is GBW / gain = 10 MHz / 50 = 200 kHz. GBW is an important consideration in high-performance filters and signal processing circuits whose design equations require high-gain at the frequencies over which they operate.

Older operational amplifiers, such as the LM301, have an additional two connections for *compensation*. To keep the amplifier from oscillating at very high gains it is often necessary to place a capacitor across the compensation terminals. This also decreases the frequency response of the op amp but increases its stability by making sure that the output signal can not have the right phase to create positive feedback at its inputs. Most modern op amps are internally compensated and do not have separate pins to add compensation capacitance. Additional compensation can be created by connecting a capacitor between the op amp output and the inverting input.

CMRR AND PSRR

One of the major advantages of using an op amp is its very high *common mode rejection ratio* (CMRR). *Common mode* signals are those that appear equally at all terminals. For example, if both conductors of an audio cable pick up a few tenths of a volt of 60 Hz signal from a nearby power transformer, that 60 Hz signal is a common-mode signal to whatever device the cable is connected. Since the op amp only responds to *differences* in voltage at its inputs, it can ignore or reject common mode signals. CMRR is a measure of how well the op amp rejects the common mode signal. High CMRR results from the symmetry between the circuit halves. CMRR is important when designing circuits that process low-level signals, such as microphone audio or the mV-level dc signals from sensors or thermocouples.

The rejection of power-supply imbalance is also an important op amp parameter. Shifts in power supply voltage and noise or ripple on the power supply voltages are coupled directly to the op amp's internal circuitry. The op amp's ability to ignore those disturbances is expressed by the *power supply rejection ratio* (PSRR). A high PSRR means that the op amp circuit will continue to perform well even if the power supply is imbalanced or noisy.

INPUT AND OUTPUT VOLTAGE LIMITS

The op amp is capable of accepting and amplifying signals at levels limited by the power supply voltages, also called *rails*. The difference in voltages between the two rails limits the range of signal voltages that can be processed. The voltages can be symmetrical positive and negative voltages (±12 V), a positive voltage and ground, ground and a negative voltage or any two different, stable voltages.

In most op amps the signal levels that can be handled are one or two diode forward voltage drops (0.7 V to 1.4 V) away from each rail. Thus, if an op amp has 15 V connected as its upper rail (usually denoted V+) and ground connected as its lower rail (V–), input signals can be amplified to be as high as 13.6 V and as low as 1.4 V in most amplifiers. Any values that would be amplified beyond those limits are clamped (output voltages that should be 1.4 V or less appear as 1.4 V and those that should be 13.6 V or more appear as 13.6 V). This clamping action was illustrated in Fig 3.1.

"Rail-to-rail" op amps have been developed to handle signal levels within a few tens of mV of rails (for example, the MAX406, from Maxim Integrated Products processes signals to within 10 mV of the power supply voltages). Rail-to-rail op-amps are often used in battery-powered products to allow the circuits to operate from low battery voltages for as long as possible.

INPUT BIAS AND OFFSET

The inputs of an op amp, while very high impedance, still allow some input current to flow. This is the *input bias current* and it is in the range of nA in modern op amps. Slight asymmetries in the op amp's internal circuitry result in a slight offset in the op amp's out-put voltage, even with the input terminals shorted together. The amount of voltage difference between the op amp's inputs required to cause the output voltage to be exactly zero is the *input offset voltage*, generally a few mV or less. Some op amps, such as the LM741, have special terminals to which a potentiometer can be connected to *null* the offset by correcting the internal imbalance. Introduction of a small dc correction voltage to the noninverting terminal is sometimes used to apply an offset voltage that counteracts the internal mismatch and

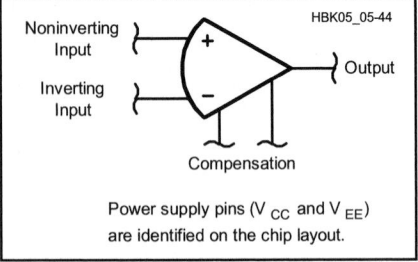

Fig 3.60 — Operational amplifier schematic symbol. The terminal marked with a + sign is the noninverting input. The terminal marked with a – sign is the inverting input. The output is to the right. On some op amps, external compensation is needed and leads are provided, pictured here below the device. Usually, the power supply leads are not shown on the op amp itself but are specified in the data sheet.

centers the signal in the rail-to-rail range.

DC offset is an important consideration in op amps for two reasons. Actual op amps have a slight mismatch between the inverting and noninverting terminals that can become a substantial dc offset in the output, depending on the amplifier gain. The op amp output voltage must not be too close to the clamping limits or distortion will occur.

A TYPICAL OP AMP

As an example of typical values for these parameters, one of today's garden-variety op amps, the TL084, which contains both JFET and bipolar transistors, has a guaranteed minimum CMRR of 80 dB, an input bias current guaranteed to be below 200 pA (1 pA = 1 millionth of a µA) and a gain-bandwidth product of 3 MHz. Its input offset voltage is 3 mV. CMRR and PSRR are 86 dB, meaning that an unwanted signal or power supply imbalance of 1 V will only result in a 2.5 nV change at the op amp's output! All this for 33 cents even purchased in single quantities and there are four op-amps per package — that's a lot of performance.

3.6.2 Basic Op Amp Circuits

If a signal is connected to the input terminals of an op amp without any other circuitry attached, it will be amplified at the device's *open-loop gain* (typically 200,000 for the TL084 at dc and low frequencies, or 106 dB). This will quickly saturate the output at the power supply rails. Such large gains are rarely used. In most applications, negative feedback is used to limit the circuit gain by providing a feedback path from the output terminal to the inverting input terminal. The resulting *closed-loop gain* of the circuit depends solely on the values of the passive components used to form the loop (usually resistors and, for frequency-selective circuits, capacitors). The higher the op-amp's open-loop gain, the closer the circuit's actual gain will approach that predicted from the component values. Note that the gain of the op amp itself has not changed — it is the configuration of the external components that determines the overall gain of the circuit. Some examples of different circuit configurations that manipulate the closed-loop gain follow.

INVERTING AND NONINVERTING AMPLIFIERS

The op amp is often used in either an *inverting* or a *noninverting* amplifier circuit as shown in **Fig 3.61**. (Inversion means that the output signal is inverted from the input signal about the circuit's voltage reference as described below.) The amount of amplification is determined by the two resistors: the feedback resistor, R_f, and the input resistor, R_i.

In the noninverting configuration shown in

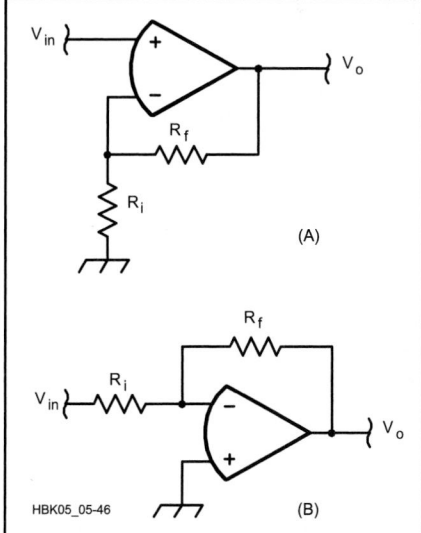

Fig 3.61 — Operational amplifier circuits. (A) Noninverting configuration. (B) Inverting configuration.

Fig 3.61A, the input signal is connected to the op-amp's noninverting input. The feedback resistor is connected between the output and the inverting input terminal. The inverting input terminal is connected to R_i, which is connected to ground (or the circuit reference voltage).

This circuit illustrates how op amp circuits use negative feedback, the high open-loop gain of the op amp itself, and the high input impedance of the op amp inputs to create a stable circuit with a fixed gain. The signal applied to the noninverting input causes the output voltage of the op-amp to change with the same polarity. That is, a positive input signal causes a positive change in the op amp's output voltage. This voltage causes current to flow in the voltage divider formed by R_f and R_i. Because the current into the inverting input is so low, the current through R_f is the same as R_i.

The voltage at the *summing junction*, the connection point for the two resistors and the inverting terminal, V_{INV}, is:

$$V_{INV} = V_O \left(\frac{R_i}{R_i + R_f} \right) \qquad (62)$$

The op amp's output voltage will continue to rise until the *loop error signal*, the difference in voltage between the inverting and noninverting inputs, is close to zero. At this point, the voltage at the inverting terminal is approximately equal to the voltage at the noninverting terminal, V_{in}, so that $V_{INV} = V_{in}$. Substituting in equation 62, the gain of this circuit is:

$$\frac{V_O}{V_{in}} = \left(1 + \frac{R_f}{R_i} \right) \qquad (63)$$

where
V_O = the output voltage
V_{in} = the input voltage.

The higher the op amp's open-loop gain, the closer will be the voltages at the inverting and noninverting terminals when the circuit is balanced and the more closely the circuit's closed-loop gain will equal that of Equation 63. So the negative feedback creates an electronic balancing act with the op amp increasing its output voltage so that the input error signal is as small as possible.

In the inverting configuration of Fig 3.61B, the input signal (V_{in}) is connected through R_i to the inverting terminal. The feedback resistor is again connected between the inverting terminal and the output. The noninverting terminal is connected to ground (or the circuit reference voltage). In this configuration the feedback action results in the output voltage changing to whatever value is needed such that the current through R_i is balanced by an equal and opposite current through R_f. The gain of this circuit is:

$$\frac{V_O}{V_{in}} = -\frac{R_f}{R_{in}} \qquad (64)$$

where V_{in} represents the voltage input to R_{in}.

For the remainder of this section, "ground" or "zero voltage" should be understood to be the circuit reference voltage. That voltage may not be "earth ground potential." For example, if a single positive supply of 12 V is used, 6 V may be used as the circuit reference voltage. The circuit reference voltage is a fixed dc voltage that can be considered to be an ac ground because of the reference source's extremely low ac impedance.

The negative sign in equation 64 indicates that the signal is inverted. For ac signals, inversion represents a 180° phase shift. The gain of the noninverting configuration can vary from a minimum of 1 to the maximum of which the op amp is capable, as indicated by A_v for dc signals, or the gain-bandwidth product for ac signals. The gain of the inverting configuration can vary from a minimum of 0 (gains from 0 to 1 attenuate the signal while gains of 1 and higher amplify the signal) to the maximum of which the device is capable.

The inverting amplifier configuration results in a special condition at the op amp's inverting input called *virtual ground*. Because the op amp's high open-loop gain drives the two inputs to be very close together, if the noninverting input is at ground potential, the inverting input will be very close to ground as well and the op amp's output will change with the input signal to maintain the inverting input at ground. Measured with a voltmeter, the input appears to be grounded, but it is merely maintained at ground potential by the action of the op amp and the feedback loop. This point in the circuit may not be connected to any other ground connection or circuit point because the resulting additional current flow will upset the balance of the circuit.

The *voltage follower* or *unity-gain buffer* circuit of **Fig 3.62** is commonly used as a buffer stage. The voltage follower has the input connected directly to the noninverting terminal and the output connected directly to the inverting terminal. This configuration has unity gain because the circuit is balanced when the output and input voltages are the same (error voltage equals zero). It also provides the maximum possible input impedance and the minimum possible output impedance of which the device is capable.

Differential and Difference Amplifier

A *differential amplifier* is a special application of an operational amplifier (see **Fig 3.63**). It amplifies the difference between two analog signals and is very useful to cancel noise under certain conditions. For instance, if an analog signal and a reference signal travel over the same cable they may pick up noise, and it is likely that both signals will have the same amount of noise. When the differential amplifier subtracts them, the signal will be unchanged but the noise will be completely removed, within the limits of the CMRR. The equation for differential amplifier operation is

$$V_O = \frac{R_f}{R_i}\left[\frac{1}{\frac{R_n}{R_g}+1}\left(\frac{R_i}{R_f}+1\right)V_n - V_i\right] \quad (65)$$

which, if the ratios R_i/R_f and R_n/R_g are equal, simplifies to:

$$V_O = \frac{R_f}{R_i}(V_n - V_1) \quad (66)$$

Note that the differential amplifier response is identical to the inverting amplifier response (equation 64) if the voltage applied to the noninverting terminal is equal to zero. If the voltage applied to the inverting terminal (V_i) is zero, the analysis is a little more complicated but it is possible to derive the noninverting amplifier response (equation 62)

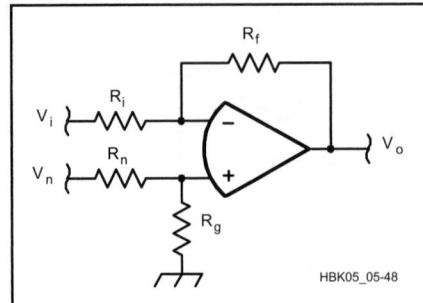

Fig 3.63 — Difference amplifier. This operational amplifier circuit amplifies the difference between the two input signals.

from the differential amplifier response by taking into account the influence of R_n and R_g. If all four resistors have the *same* value the *difference amplifier* is created and V_O is just the difference of the two voltages.

$$V_O = V_n - V_1 \quad (67)$$

Instrumentation Amplifier

Just as the symmetry of the transistors making up an op amp leads to a device with high values of Z_i, A_v and CMRR and a low value of Z_o, a symmetric combination of op amps is used to further improve these parameters. This circuit, shown in **Fig 3.64** is called an *instrumentation amplifier*. It has three parts; each of the two inputs is connected to a noninverting buffer amplifier with a gain of 1 + R2/R1. The outputs of these buffer amplifiers are then connected to a differential amplifier with a gain of R4/R3. V2 is the circuit's inverting input and V1 the noninverting input.

The three amplifier modules are usually all part of the same integrated circuit. This means that they have essentially the same temperature and the internal transistors and resistors are very well matched. This causes the subtle gain and tracking errors caused by temperature differences and mismatched components

between individual op amps to be cancelled out or dramatically reduced. In addition, the external resistors using the same designators (R2, R3, R4) are carefully matched as well, sometimes being part of a single integrated *resistor pack*. The result is a circuit with better performance than any single-amplifier circuit over a wider temperature range.

Summing Amplifier

The high input impedance of an op amp makes it ideal for use as a *summing amplifier*. In either the inverting or noninverting configuration, the single input signal can be replaced by multiple input signals that are connected together through series resistors, as shown in **Fig 3.65**. For the inverting summing amplifier, the gain of each input signal can be calculated individually using equation 64 and, because of the superposition property of linear circuits, the output is the sum of each input signal multiplied by its gain. In the noninverting configuration, the output is the gain times the weighted sum of the m different input signals:

$$V_O = V_{n1}\frac{R_{p1}}{R_1 + R_{p1}} + V_{n2}\frac{R_{p2}}{R_2 + R_{p2}} + \dots$$

$$+ V_{nm}\frac{R_{pm}}{R_m + R_{pm}} \quad (68)$$

where R_{pm} is the parallel resistance of all m resistors excluding R_m. For example, with three signals being summed, R_{p1} is the parallel combination of R_2 and R_3.

Comparators

A *voltage comparator* is another special form of op amp circuit, shown in **Fig 3.66**. It has two analog signals as its inputs and its output is either TRUE or FALSE depending on whether the noninverting or inverting signal voltage is higher, respectively. Thus, it "compares" the input voltages. TRUE generally corresponds to a positive output voltage and

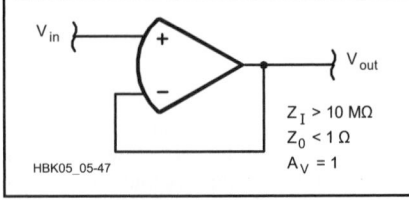

Fig 3.62 — Voltage follower. This operational amplifier circuit makes a nearly ideal buffer with a voltage gain of about one, and with extremely high input impedance and extremely low output impedance.

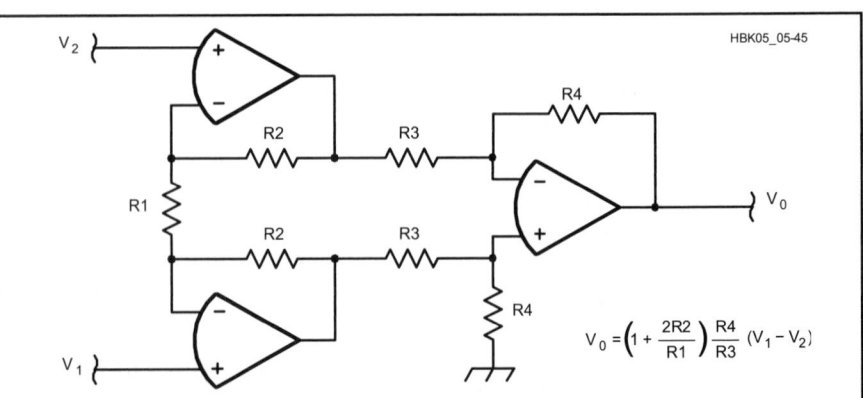

Fig 3.64 — Operational amplifiers arranged as an instrumentation amplifier. The balanced and cascaded series of op amps work together to perform differential amplification with good common-mode rejection and very high input impedance (no load resistor required) on both the inverting (V_1) and noninverting (V_2) inputs.

FALSE to a negative or zero voltage. The circuit in Fig 3.66 uses external resistors to generate a reference voltage, called the *setpoint*, to which the input signal is compared. A comparator can also compare two variable voltages.

A standard operational amplifier can be made to act as a comparator by connecting the two input voltages to the noninverting and inverting inputs with no input or feedback resistors. If the voltage of the noninverting input is higher than that of the inverting input, the output voltage will be driven to the positive clamping limit. If the inverting input is at a higher potential than the noninverting input, the output voltage will be driven to the negative clamping limit. If the comparator is comparing an unknown voltage to a known voltage, the known voltage is called the *setpoint* and the comparator output indicates whether the unknown voltage is above or below the setpoint.

An op amp that has been intended for use as a comparator, such as the LM311, is optimized to respond quickly to the input signals. In addition, comparators often have *open-collector outputs* that use an external *pull-up* resistor, R_{OUT}, connected to a positive power supply voltage. When the comparator output is TRUE, the output transistor is turned off and the pull-up resistor "pulls up" the output voltage to the positive power supply voltage. When the comparator output is FALSE, the transistor is driven into the saturation and the output voltage is the transistor's $V_{CE(sat)}$.

Hysteresis

Comparator circuits also use *hysteresis* to prevent "chatter" — the output of the comparator switching rapidly back and forth when the input voltage is at or close to the setpoint voltage. There may be noise on the input signal, as shown in **Fig 3.67A**, that causes the input voltage to cross the setpoint threshold repeatedly. The rapid switching of the output can be confusing to the circuits monitoring the comparator output.

Hysteresis is a form of positive feedback that "moves" the setpoint by a few mV in the direction *opposite* to that in which the input signal crossed the setpoint threshold. As shown in Fig 3.67B, the slight shift in the setpoint tends to hold the comparator output in the new state and prevents switching back to the old state. **Fig 3.68** shows how the output of the comparator is fed back to the positive input through resistor R3, adding or subtracting a small amount of current from the divider and shifting the setpoint.

Some applications of a voltage comparator

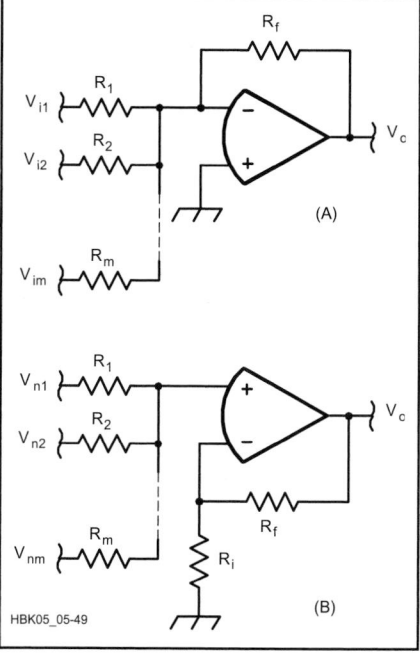

Fig 3.65 — Summing operational amplifier circuits. (A) Inverting configuration. (B) Noninverting configuration.

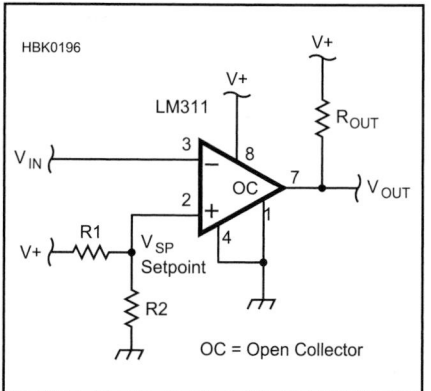

Fig 3.66 — A comparator circuit in which the output voltage is low when voltage at the inverting input is higher than the setpoint voltage at the noninverting input.

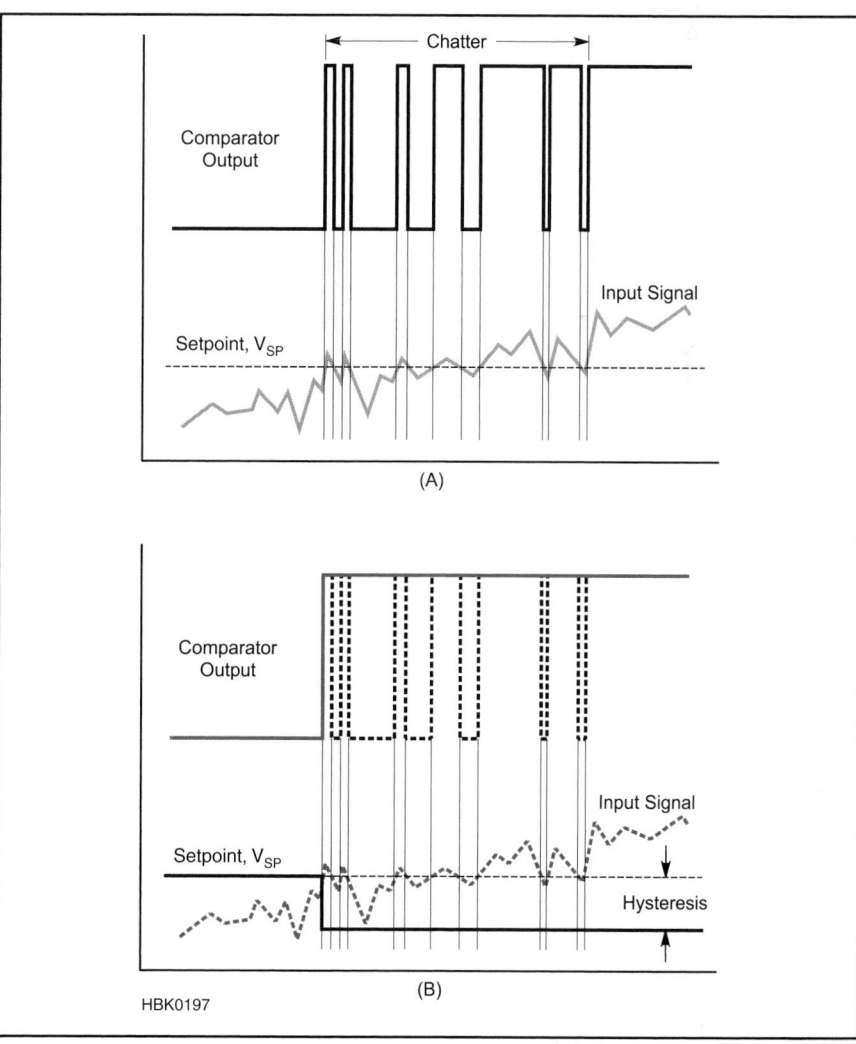

Fig 3.67 — Chatter (A) is caused by noise when the input signal is close to the setpoint. Chatter can also be caused by voltage shifts that occur when a heavy load is turned on and off. Hysteresis (B) shifts the setpoint a small amount by using positive feedback in which the output pulls the setpoint farther away from the input signal after switching.

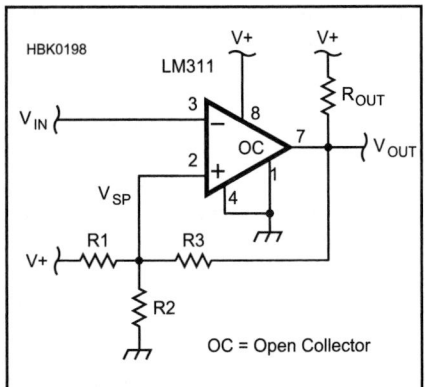

Fig 3.68 — Comparator circuit with hysteresis. R3 causes a shift in the comparator setpoint by allowing more current to flow through R1 when the comparator output is low.

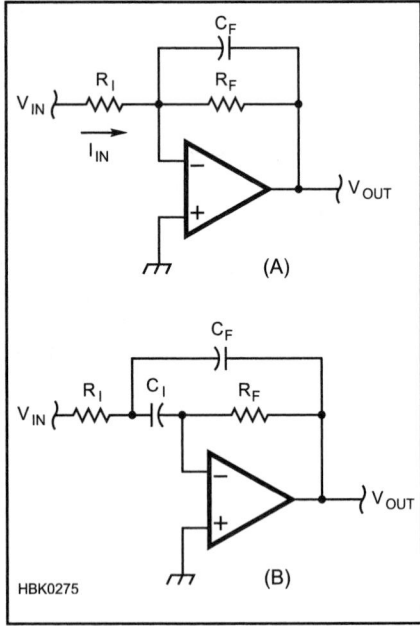

Fig 3.69 — Op amp active filters. The circuit at (A) has a low-pass response identical to an RC filter. The −3 dB frequency occurs when the reactance of C_F equals R_F. The band-pass filter at (B) is a multiple-feedback filter.

are a zero crossing detector, a signal squarer (which turns other cyclical wave forms into square waves) and a peak detector. An amateur station application: Circuits that monitor the CI-V band data output voltage from ICOM HF radios use a series of comparators to sense the level of the voltage and indicate on which band the radio is operating.

FILTERS

One of the most important type of op amp circuits is the *active filter*. Two examples of op amp filter circuits are shown in **Fig 3.69.** The simple noninverting low-pass filter in Fig 3.69A has the same response as a passive single-pole RC low-pass filter, but unlike the passive filter, the op amp filter circuit has a very high input impedance and a very low output impedance so that the filter's frequency and voltage response are relatively unaffected by the circuits connected to the filter input and output. This circuit is a low-pass filter because the reactance of the feedback capacitor decreases with frequency, requiring less output voltage to balance the voltages of the inverting and noninverting inputs.

The *multiple-feedback* circuit in Fig 3.69B results in a band-pass response while using only resistors and capacitors. This circuit is just one of many different types of active fil-

ters. Active filters are discussed in the **RF and AF Filters** chapter.

RECTIFIERS AND PEAK DETECTORS

The high open-loop gain of the op amp can also be used to simulate the I-V characteristics of an ideal diode. A *precision rectifier* circuit is shown in **Fig 3.70** along with the I-V characteristics of a real (dashed lines) and ideal (solid line) diode. The high gain of the op amp compensates for the V_F forward voltage drop of the real diode in its feedback loop with an output voltage equal to the input voltage plus V_F. Remember that the op amp's output increases until its input voltages are balanced. When the input voltage is negative, which would reverse-bias the diode, the op amp's output can't balance the input because the diode blocks any current flow through the feedback loop. The resistor at the output holds the voltage at zero until the input voltage is positive once again. Precision half-wave and full-wave rectifier circuits are shown in **Fig 3.71** and their operation is described in many reference texts.

One application of the precision rectifier circuit useful in radio is the *peak detector*, shown in **Fig 3.72**. A precision rectifier is used to charge the output capacitor which

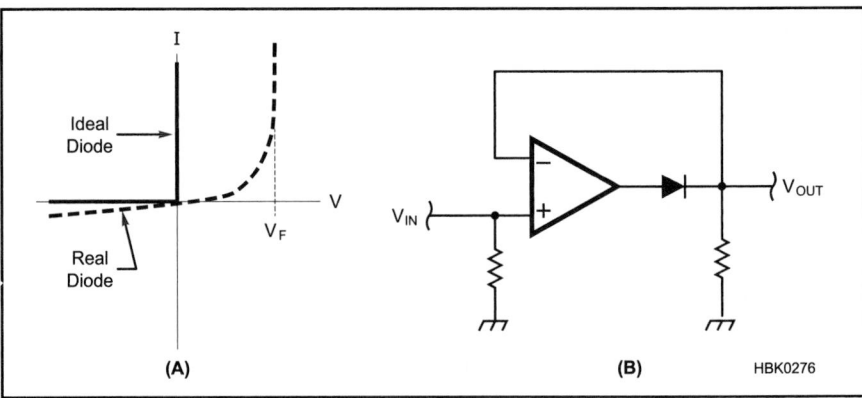

Fig 3.70 — Ideal and real diode I-V characteristics are shown at (A). The op amp precision rectifier circuit is shown at (B).

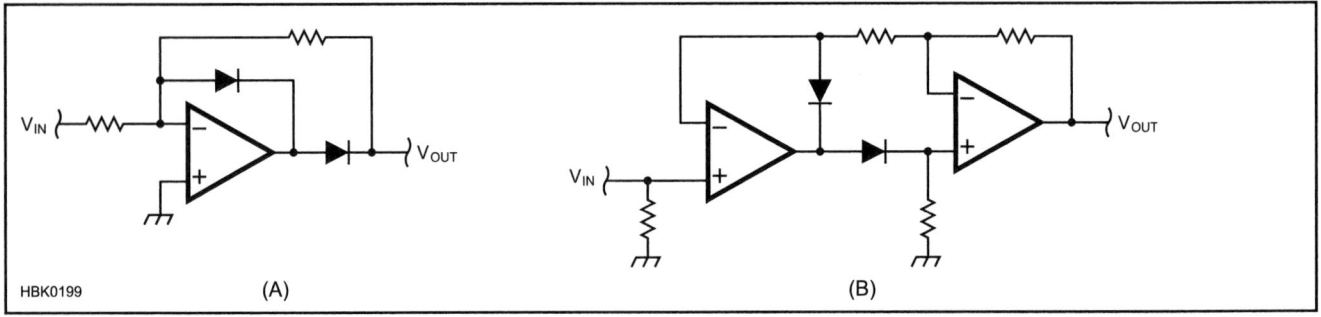

Fig 3.71 — Half-wave precision rectifier (A). The extra diode at the output of the op amp prevents the op amp from saturating on negative half-cycles and improves response time. The precision full-wave rectifier circuit at (B) reproduces both halves of the input waveform.

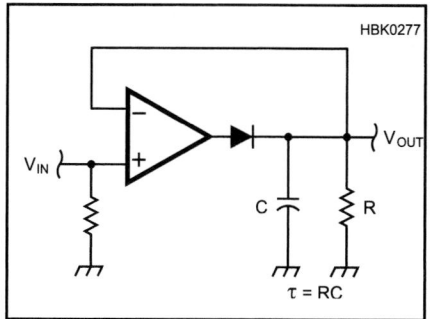

Fig 3.72 — Peak detector. Coupling a precision diode with a capacitor to store charge creates a peak detector. The capacitor will charge to the peak value of the input voltage. R discharges the capacitor with a time constant of $\tau = RC$ and can be omitted if it is desired for the output voltage to remain nearly constant.

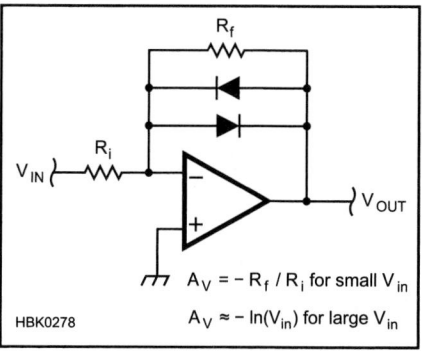

Fig 3.73 — Log amplifier. At low voltages, the gain of the circuit is $-R_f/R_i$, but as the diodes begin to conduct for higher-voltage signals, the gain changes to $-\ln(V_{in})$ in because of the diode's exponential current as described in the Fundamental Diode Equation.

holds the peak voltage. The output resistor sets the time constant at which the capacitor discharges. The resistor can also be replaced by a transistor acting as a switch to reset the detector. This circuit is used in AGC loops, spectrum analyzers, and other instruments that measure the peak value of ac waveforms.

LOG AMPLIFIER

There are a number of applications in radio in which it is useful for the gain of an amplifier to be higher for small input signals than for large input signals. For example, an audio compressor circuit is used to reduce the variations in a speech signal's amplitude so that the average power output of an AM or SSB transmitter is increased. A *log amplifier* circuit whose gain for large signals is proportional to the logarithm of the input signal's amplitude is shown in **Fig 3.73**. The log amp circuit is used in compressors and limiter circuits.

At signal levels that are too small to cause significant current flow through the diodes, the gain is set as in a regular inverting amplifier, $A_V = -R_f/R_i$. As the signal level increases, however, more current flows through the diode according to the Fundamental Diode Equation (equation 4) given earlier in this chapter. That means the op amp output voltage has to increase less (lower gain) to cause enough current to flow through R_i such that the input voltages balance. The larger the input voltage, the more the diode conducts and the lower the gain of the circuit. Since the

diode's current is exponential in response to voltage, the gain of the circuit for large input signals is logarithmic.

Voltage-Current Converters

Another pair of useful op amp circuits convert voltage into current and current into voltage. These are frequently used to convert currents from sensors and detectors into voltages that are easier to measure. **Fig 3.74A** shows a voltage-to-current converter in which the output current is actually the current in the feedback loop. Because the op amp's high open-loop gain insures that its input voltages are equal, the current $I_{R1} = V_{IN}/R1$. Certainly, this could also be achieved with a resistor and Ohm's Law, but the op amp circuit's high input impedance means there is little interaction between the input voltage source and the output current.

Going the other way, Fig 3.74B is a current-to-voltage converter. The op amp's output will change so that the current through the feedback resistor, R1, exactly equals the input current, keeping the inverting terminal at ground potential. The output voltage, $V_O = I_{IN}$ R1. Again, this could be done with just a resistor, but the op amp provides isolation between the source of input current and the output voltage. Fig 3.74C shows an application of a current-to-voltage converter in which the small currents from a photodiode are turned into voltage. This circuit can be used as a detector for amplitude modulated light pulses or waveforms.

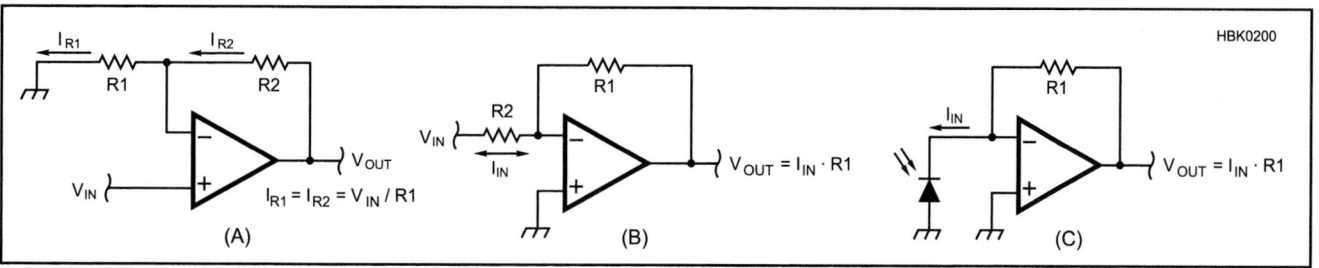

Fig 3.74 — Voltage-current converters. The current through R1 in (A) equals $V_{in}/R1$ because the op amp keeps both input terminals at approximately the same voltage. At (B), input current is balanced by the op amp, resulting in $V_{OUT} = I_{IN}R1$. Current through a photodiode (C) can be converted into a voltage in this way.

3.7 Analog-Digital Conversion

While radio signals are definitely analog entities, much of the electronics associated with radio is digital, operating on binary data representing the radio signals and the information they carry. The interface between the two worlds — analog and digital — is a key element of radio communications systems and the function is performed by analog-digital converters.

This section presents an overview of the different types of analog-digital converters and their key specifications and behaviors. The chapter on **Digital Basics** presents information on interfacing converters to digital circuitry and associated issues. The chapter on **DSP and Software Radio Design** discusses specific applications of analog-digital conversion technology in radio communications systems.

3.7.1 Basic Conversion Properties

Analog-digital conversion consists of taking data in one form, such as digital binary data or an analog ac RF waveform, and creating an equivalent representation of it in the opposite domain. Converters that create a digital representation of analog voltages or currents are called *analog-to-digital converters* (ADC), *analog/digital converters, A/D converters* or *A-to-D converters*. Similarly, converters that create analog voltages or currents from digital quantities are called *digital-to-analog converters* (DAC), *digital/analog converters, D/A converters* or *D-to-A converters*. The word "conversion" in this first section on the properties of converting information between the analog and digital domains will apply equally to analog-to-digital or digital-to-analog conversion.

Converters are typically implemented as integrated circuits that include all of the necessary interfaces and sub-systems to perform the entire conversion process. Schematic symbols for ADCs and DACs are shown in **Fig 3.75**.

Fig 3.76 shows two different representations of the same physical phenomenon; an analog voltage changing from 0 to 1 V. In the analog world, the voltage is continuous and can be represented by any real number between 0 and 1. In the digital world, the number of possible values that can represent any phenomenon is limited by the number of bits contained in each value.

In Fig 3.76, there are only four two-bit digital values 00, 01, 10, and 11, each corresponding to the analog voltage being within a specific range of voltages. If the analog voltage is anywhere in the range 0 to 0.25 V, the digital value representing the analog voltage will be 00, no matter whether the voltage is 0.0001 or 0.24999 V. The range 0.25 to 0.5 V is represented by the digital value 01, and so forth.

The process of converting a continuous range of possible values to a limited number of discrete values is called *digitization* and each discrete value is called a *code* or a *quantization code*. The number of possible codes that can represent an analog quantity is 2^N, where N is the number of bits in the code. A two-bit number can have four codes as shown in Fig 3.76, a four-bit number can have sixteen codes, an eight-bit number 256 codes, and so forth. Assuming that the smallest change in code values is one bit, that value is called the *least significant bit* (LSB) regardless of its position in the format used to represent digital numbers.

RESOLUTION AND RANGE

The *resolution* or *step size* of the conversion is the smallest change in the analog value that the conversion can represent. The *range* of the conversion is the total span of analog values that the conversion can process. The maximum value in the range is called the *full-scale* (F.S.) value. In Fig 3.76, the conversion range is 1 V. The resolution of the conversion is

$$resolution = \frac{range}{2^N}$$

In Fig 3.76, the conversion resolution is $\frac{1}{4} \times 1\,V = 0.25\,V$ in the figure. If each code had four bits instead, it would have a resolution of $\frac{1}{2}^4 \times 1\,V = 0.083\,V$. Conversion range does not necessarily have zero as one end point. For example, a conversion range of 5 V may span 0 to 5 V, –5 to 0 V, 10 to 15 V, –2.5 to 2.5 V, and so on.

Analog-digital conversion can have a range that is *unipolar* or *bipolar*. Unipolar means a conversion range that is entirely positive or negative, usually referring to voltage. Bipolar means the range can take on both positive and negative values.

Confusingly, the format in which the bits are organized is also called a *code*. The binary code represents digital values as binary numbers with the least significant bit on the right. *Binary-coded-decimal* (BCD) is a code in which groups of four bits represent individual decimal values of 0-9. In the *hexadecimal* code, groups of four bits represent decimal values of 0-15. There are many such codes. Be careful in interpreting the word "code" to be sure the correct meaning is used or understood.

To avoid having to know the conversion range to specify resolution, *percentage resolution* is used instead.

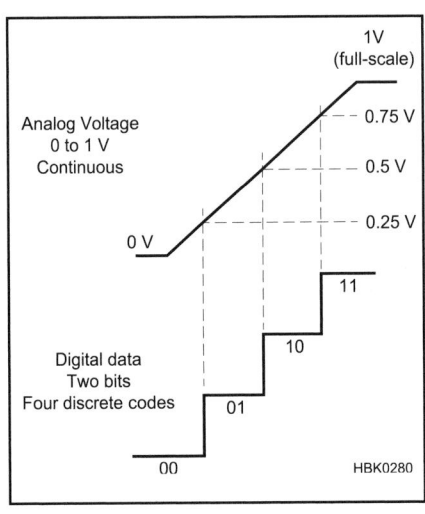

Fig 3.76 — The analog voltage varies continuously between 0 and 1 V, but the two-bit digital system only has four values to represent the analog voltage, so representation of the analog voltage is coarse.

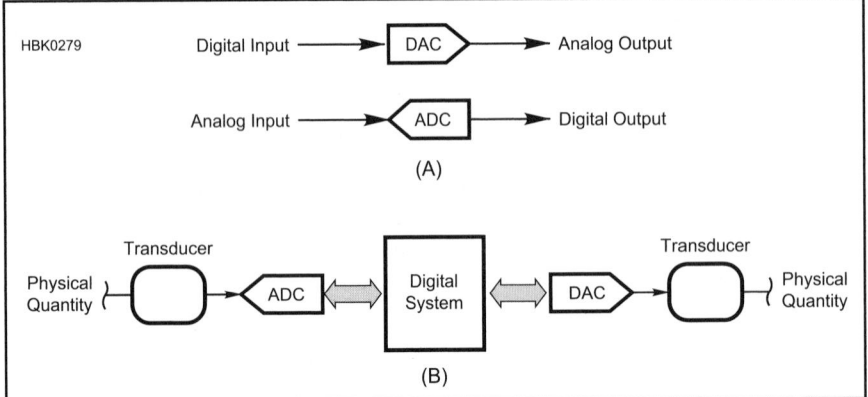

Fig 3.75 — Schematic symbols (A) for digital-to-analog converters (DAC) and analog-to-digital converters (ADC). The general block diagram of a system (B) that digitizes an analog signal, operates on it as digital data, then converts it back to analog form.

$$\% \text{ resolution} = \frac{\text{resolution}}{\text{full scale}} \times 100\% \qquad (69)$$

In Fig 3.76, the conversion's percentage resolution is

$$\% \text{ resolution} = \frac{0.25 \text{ V}}{1.0 \text{ V}} \times 100\% = 25\%$$

Because each code represents a range of possible analog values, the limited number of available codes creates *quantization error*. This is the maximum variation in analog values that can be represented by the same code. In Fig 3.76, any value from 0.25 through 0.50 V could be represented by the same code: 01. The quantization error in this case is 0.25 V. Quantization error can also be specified as % full-scale by substituting the value of error for resolution in equation 69.

Resolution can also be defined by the number of bits in the conversion. The higher the number of bits, the smaller the resolution as demonstrated above. Since many converters have variable ranges set by external components or voltages, referring to percent resolution or as a number of bits is preferred. The conversions between percent resolution and number of bits are as follows:

$$\% \text{ resolution} = \frac{1}{2^N} \times 100\% \qquad (70)$$

and

$$N = \frac{\log\left(\dfrac{100\%}{\% \text{ resolution}}\right)}{\log 2} \qquad (71)$$

Quantization error can also be specified as a number of least significant bits (LSB) where each bit is equivalent to the conversion's resolution.

When applied to receivers, dynamic range is more useful than percent resolution. For each additional bit of resolution, the resolution becomes two times greater, or 6.02 dB.

$$\text{Dynamic range (dB)} = N \times 6.02 \text{ dB} \qquad (72)$$

For example, a 16-bit conversion has a dynamic range of $16 \times 6.02 = 96.32$ dB. This is the conversion's theoretical dynamic range with no noise present in the system. If noise is present, some number of the smallest codes will contain only noise, reducing the dynamic range available to represent a signal. The noise inherent in a particular conversion process can be specified as an analog value (mV, μA and so on) or as a number of bits, meaning the number of codes that only represent noise. For example, if a conversion has five bits of noise, any value represented by the smallest five codes is considered to be noise. The conversion's *effective number of bits* (ENOB) describes the number of bits available to contain information about the signal.

ACCURACY

A companion to resolution, *accuracy* refers to the ability of the converter to either assign the correct code to an analog value or create the true analog value from a specific code. As with resolution, it is most convenient to refer to accuracy as either a percentage of full scale or in bits. *Full-scale error* is the maximum deviation of the code's value or the analog quantity's value as a percentage of the full scale value. If a converter's accuracy is given as 0.02% F.S. and the conversion range is 5 V, the conversion can be in error by as much as $0.02\% \times 5$ V = 1 mV from the correct or expected value. *Offset* has the same meaning in conversion as it does in analog electronics — a consistent shift in the value of the conversion from the ideal value.

Linearity error represents the maximum deviation in step size from the ideal step size. This is also called *integral nonlinearity* (INL). In the converter of Fig 3.76, ideal step size is 0.25 V. If the linearity error for the conversion was given as 0.05% F.S., any actual step size could be in error by as much as $0.05\% \times 5$ V = 2.5 mV. The amount of error is based on the full-scale value, not the step-size value. *Differential nonlinearity* is a measure of how much any two adjacent step sizes deviate from the ideal step size. Errors can be represented as a number of bits, usually assumed to be least significant bits, or LSB, with one bit representing the same range as the conversion resolution.

Accuracy and resolution are particularly important in conversions for radio applications because they represent distortion in the signal being processed. An analog receiver that distorts the received signal creates undesired spurious signals that interfere with the desired signal. While the process is not exactly the same, distortion created by the signal conversion circuitry of a digital receiver also causes degradation in performance.

DISTORTION AND NOISE

Distortion and noise in a conversion are characterized by several parameters all related to linearity and accuracy. THD+N (*Total Harmonic Distortion + Noise*) is a measure of how much distortion and noise is introduced by the conversion. THD+N can be specified in percent or in dB. Smaller values are better. SINAD (*Signal to Noise and Distortion Ratio*) is related to THD+N, generally specified along with a desired signal level to show what signal level is required to achieve a certain level of SINAD or the highest signal level at which a certain level of SINAD can be maintained. *Spurious-free Dynamic Range* (SFDR) is the difference between the amplitude of the desired signal and the highest unwanted signal. SFDR is generally specified in dB and a higher number indicates better performance.

CONVERSION RATE AND BANDWIDTH

Another important parameter of the conversion is the *conversion rate* or its reciprocal, *conversion speed*. A code that represents an analog value at a specific time is called a *sample*, so conversion rate, f_S, is specified in *samples per second* (sps) and conversion speed as some period of time per sample, such as 1 μsec/s. Because of the mechanics by which conversion is performed, conversion speed can also be specified as a number of cycles of clock signal used by the digital system performing the conversion. Conversion rate then depends on the frequency of the clock. Conversion speed may be smaller than the reciprocal of conversion rate if the system controlling the conversion introduces a delay between conversions. For example, if a conversion can take place in 1 μs, but the system only performs a conversion once per ms, the conversion rate is 1 ksps, not the reciprocal of 1 μs/s = 1 Msps.

The time accuracy of the digital clock that controls the conversion process can also affect the accuracy of the conversion. An error in long-term frequency will cause all conversions to have the same frequency error. Short-term variations in clock period are called *jitter* and they add noise to the conversion.

According to the *Nyquist Sampling Theorem*, a conversion must occur at a rate twice the highest frequency present in the analog signal. This minimum rate is the *Nyquist rate* and the maximum frequency allowed in the analog signal is the *Nyquist frequency*. In this way, the converter *bandwidth* is limited to one-half the conversion rate.

Referring to the process of converting analog signals to digital samples, if a lower rate is used, called *undersampling*, false signals called *aliases* will be created in the digital representation of the input signal at frequencies related to the difference between the Nyquist sampling rate and f_S. This is called *aliasing*. Sampling faster than the Nyquist rate is called *oversampling*.

Because conversions occur at some maximum rate, there is always the possibility of signals greater than the Nyquist frequency being present in an analog signal undergoing conversion or that is being created from digital values. These signals would result in aliases and must be removed by *band-limiting filters* that remove them prior to conversion. The mechanics of the sampling process are discussed further in the chapter **DSP and Software Radio Design**.

3.7.2 Analog-to-Digital Converters (ADC)

There are a number of methods by which the conversion from an analog quantity to a set of digital samples can be performed. Each has its strong points — simplicity, speed,

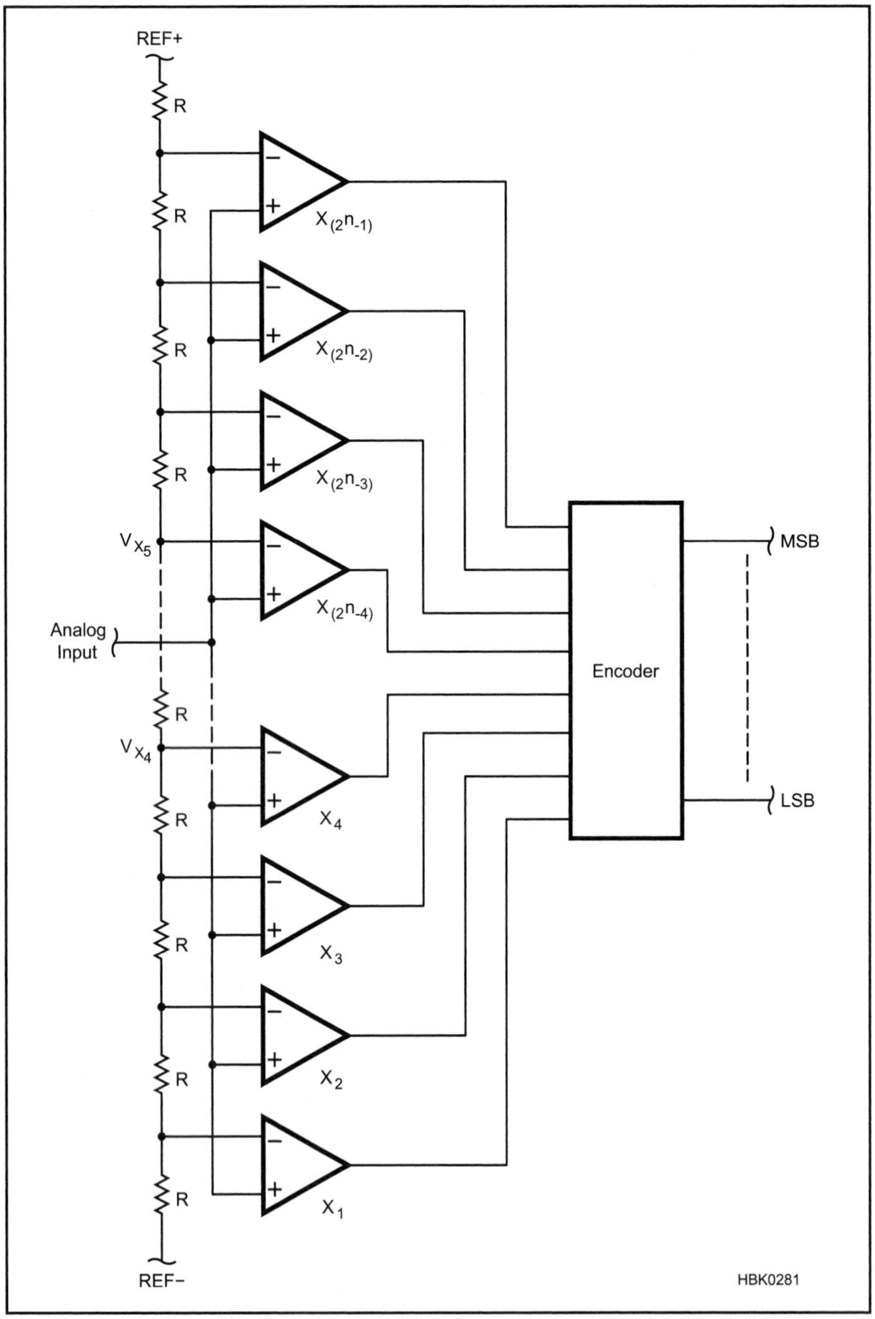

Fig 3.77 — The comparators of the flash converter are always switching state depending on the input signal's voltage. The decoder section converts the array of converter output to a single digital word.

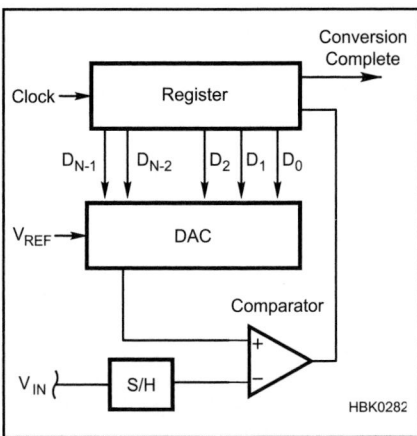

Fig 3.78 — The successive-approximation converter creates a digital word as it varies the DAC signal in order to keep the comparator's noninverting terminal close to the input voltage. A sample-and-hold circuit (S/H) holds the input signal steady while the measurement is being made.

available as quickly as the comparators can respond and the priority encoder can create the output code.

Flash converters are the fastest of all ADCs (conversion speeds can be in the ns range) but do not have high resolution because of the number of comparators and reference voltages required.

SUCCESSIVE-APPROXIMATION CONVERTER

The *successive-approximation converter* (SAC) is one of the most widely-used types of converters. As shown in **Fig 3.78**, it uses a single comparator and DAC (digital-to-analog converter) to arrive at the value of the input voltage by comparing it to successive analog values generated by the DAC. This type of converter offers a good compromise of conversion speed and resolution.

The DAC control logic begins a conversion by setting the output of the DAC to ½ of the conversion range. If the DAC output is greater than the analog input value, the output of the comparator is 0 and the most significant bit of the digital value is set to 0. The DAC output then either increases or decreases by ¼ of the range, depending on whether the value of the first comparison was 1 or 0. One test is made for each bit in digital output code and the result accumulated in a storage register. The process is then repeated, forming a series of approximations (thus the name of the converter), until a test has been made for all bits in the code.

While the digital circuitry to implement the SAC is somewhat complex, it is less expensive to build and calibrate than the array of comparators and precision resistors of the flash converter. Each conversion also takes

resolution, accuracy — all affect the decision of which method to use for a particular application. In order to pick the right type of ADC, it is important to decide which of these criteria most strongly affect the performance of your application.

FLASH CONVERTER

The simplest type of ADC is the *flash converter*, shown in **Fig 3.77**, also called a *direct-conversion ADC*. It continually generates a digital representation of the analog signal at its input. The flash converter uses an array of comparators that compare the amplitude of the input signal to a set of reference voltages. There is one reference voltage for each step.

The outputs of the comparator array represent a digital value in which each bit indicates whether the input signal is greater (1) or less than (0) the reference voltage for that comparator. A digital logic *priority encoder* then converts the array of bits into a digital output code. Each successive conversion is a known and fixed number of clock cycles. The higher the number of bits in the output code, the longer the conversion takes, all other things being equal.

DUAL-SLOPE INTEGRATING CONVERTERS

The *dual-slope integrating ADC* is shown in **Fig 3.79**. It makes a conversion by measuring the time it takes for a capacitor to charge and discharge to a voltage level proportional to the analog quantity.

A constant-current source charges a capacitor (external to the converter IC) until the comparator output indicates that the capacitor voltage is equal to the analog input signal. The capacitor is then discharged by the constant-current source until the capacitor is discharged. This process is repeated continuously. The frequency of the charge-discharge cycle is determined by the values of R and C in Fig 3.79. The frequency is then measured by frequency counter circuitry and converted to the digital output code.

Dual-slope ADCs are low-cost and relatively immune to noise and temperature variations. Due to the slow speed of the conversion (measured in tens of ms) these converters are generally only used in test instruments, such as multimeters.

DELTA-ENCODED CONVERTERS

Instead of charging and discharging a capacitor from 0 V to the level of the input signal, and then back to 0 V, the *delta-encoded ADC* in **Fig 3.80** continually compares the output of a DAC to the input signal using a comparator. Whenever the signal changes, the DAC is adjusted until its output is equal to the input signal. Digital counter circuits keep track of the DAC value and generate the digital output code. Delta-encoded counters are available with wide conversion ranges and high resolution.

SIGMA-DELTA CONVERTERS

The *sigma-delta converter* also uses a DAC and a comparator in a feedback loop to generate a digital signal as shown in **Fig 3.81**. An integrator stores the sum of the input signal and the DAC output. (This is "sigma" or sum in the converter's name.) The comparator output indicates whether the integrator output is above or below the reference voltage and that signal is used to adjust the DAC's output so that the integrator output stays close to the reference voltage. (This is the "delta" in the name.) The stream of 0s and 1s from the comparator forms a high-speed digital bit stream that is digitally-filtered to form the output code.

Sigma-delta converters are used where high resolution (16 to 24 bits) is required at low sampling rates of a few kHz. The digital filtering in the converter also reduces the

need for external band-limiting filters on the input signal.

ADC SUBSYSTEMS
Sample-and-Hold

ADCs that use a sequence of operations to create the digital output code must have a means of holding the input signal steady while the measurements are being made. This function is performed by the circuit of **Fig 3.82**. A high input-impedance buffer drives the external storage capacitor, C_{HOLD}, so that its voltage is the same as the input signal. Another high input-impedance buffer

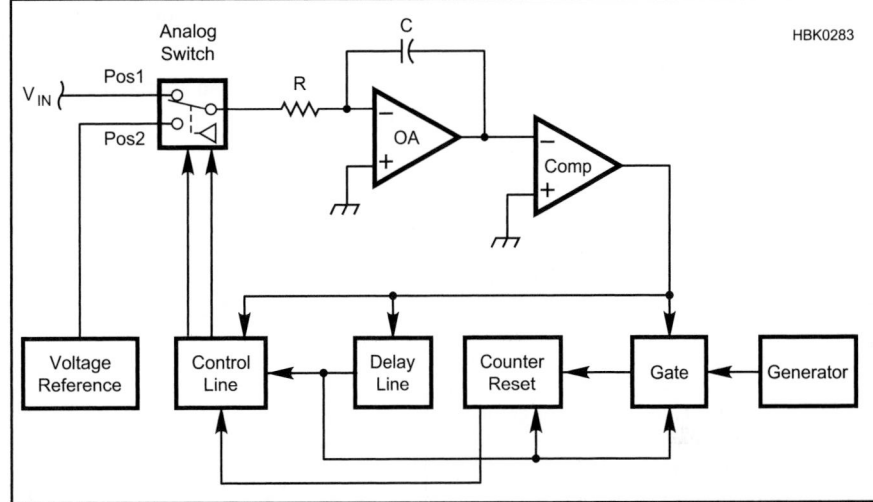

Fig 3.79 — Dual-slope integrating converter. By using a constant-current source to continually charge a capacitor to a known reference voltage then discharge it, the resulting frequency is directly proportional to the resistor value.

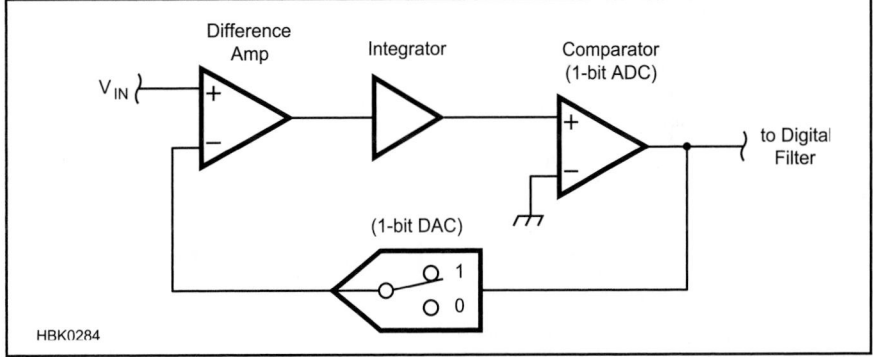

Fig 3.80 — Delta-encoded converter. The 1-bit DAC is operated in such a way that the bit stream out of the comparator represents the value of the input voltage.

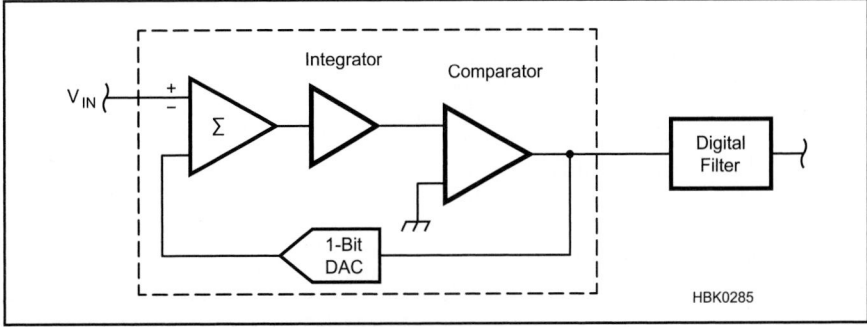

Fig 3.81 — Sigma-delta converter. Similar to the delta-encoded converter (Fig 3.80), the converter runs much faster than the output samples and uses a digital filter to derive the actual output value.

is used to provide a replica of the voltage on C_{HOLD} to the conversion circuitry.

When a conversion is started, a digital control signal opens the input switch, closes the output switch, and the capacitor's voltage is measured by the converter. It is important that the capacitor used for C_{HOLD} have low *leakage* so that while the measurement is being made, the voltage stays constant for the few ms required. This is of particular important in high-precision conversion.

Single-Ended and Differential Inputs

The input of most ADCs is *single-ended*, in which the input signal is measured between the input pin and a common ground. Shown in **Fig 3.83A**, this is acceptable for most applications, but if the voltage to be measured is small or is the difference between two non-zero voltages, an ADC with *differential inputs* should be used as in Fig 3.83B. Differential inputs are also useful when measuring current as the voltage across a small resistor in series with the current. In that case, neither side of the resistor is likely to be at ground, so a differential input is very useful. Differential inputs also help avoid the issue of noise contamination as discussed below.

Input Buffering and Filtering

The input impedance of most ADCs is high enough that the source of the input signal is unaffected. However, to protect the ADC input and reduce loading on the input source, an external buffer stage can be used. **Fig 3.84** shows a typical buffer arrangement with clamping diodes to protect against electrostatic discharge (ESD) and an RC-filter to prevent RF signals from affecting the input signal. In addition, to attenuate higher-frequency signals that might cause aliases, the input filters can also act as band-limiting filters.

Analog and Digital "Ground"

By definition, ADCs straddle the analog and digital domain. In principle, the signals remain separate and isolated from each other. In practice, however, voltages and currents from the analog and digital circuitry can be mixed together. This can result in the contamination of an analog signal with components of digital signals, and rarely, vice versa. Mostly, this is a problem when trying to measure small voltages in the presence of large power or RF signals.

The usual problem is that currents from high-speed digital circuitry find their way into analog signal paths and create transients and other artifacts that affect the measurement of the analog signal. Thus, it is important to have separate current paths for the two types of signals. The manufacturer of the converter will provide guidance for the proper use of the converter either in the device's data sheet or as application notes. Look for separate pins on

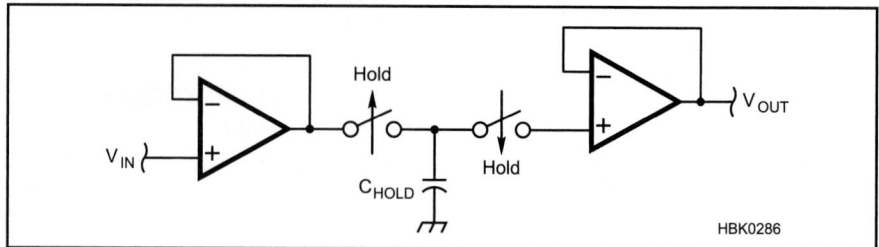

Fig 3.82 — Sample-and-hold (S/H). An input buffer isolates the sampled voltage from the input signal by charging the capacitor C_{HOLD} to that voltage with the input switch closed and the output switch open. When a measurement is being taken, the input switch is open to prevent the input signal from changing the capacitor voltage, and the output switch is closed so that the output buffer can generate a steady voltage at its output.

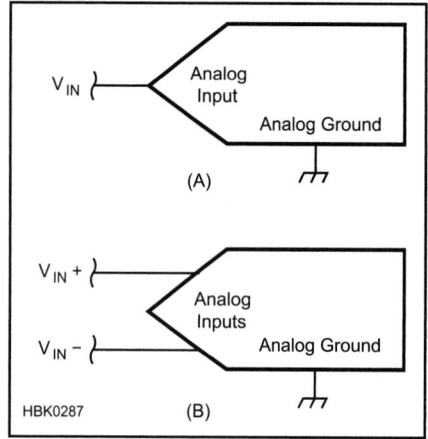

Fig 3.83 — Single-ended ADC inputs have a single active line and a ground or return line (A). Single-ended ADC input are often susceptible to noise and common mode signals or any kind of disturbance on their ground rails. In (B), the differential inputs used help the circuit "ignore" offsets and shifts in the input signal.

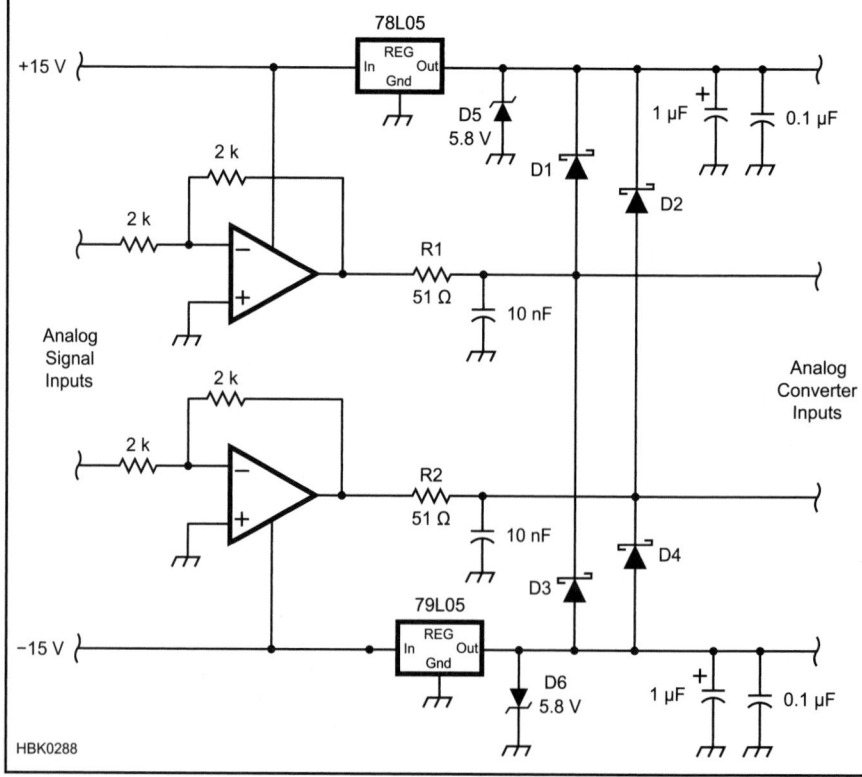

Fig 3.84 — Typical ADC input buffer-filter circuit. Unity-gain voltage followers help isolate the ADC from the input source. RC filters following the buffers act as band-limiting filters to prevent aliasing. Zener diodes are used to clamp the transient voltage and route the energy of transient into the power supply system.

the converter, such as "AGND" or "DGND" that indicate how the two types of signal return paths should be connected.

3.7.3 Digital-to-Analog Converters (DAC)

Converting a digital value to an analog quantity is considerably simpler than the reverse, but there are several issues primarily associated with DACs that affect the selection of a particular converter.

As each new digital value is converted to analog, the output of the DAC makes an abrupt *step change*. Even if very small, the response of the DAC output does not respond perfectly or instantaneously. *Settling time* is the amount of time required for the DAC's output to stabilize within a certain amount of the final value. It is specified by the manufacturer and can be degraded if the load connected to the DAC is too heavy or if it is highly reactive. In these cases, using a buffer amplifier is recommended.

Monotonicity is another aspect of characterizing the DAC's accuracy. A DAC is monotonic if in increasing the digital input value linearly across the conversion range, the output of the DAC increases with every step. Because of errors in the internal conversion circuitry, it is possible for there to be some steps that are too small or too large, leading to output values that seem "out of order." These are usually quite small, but if used in a precision application, monotonicity is important.

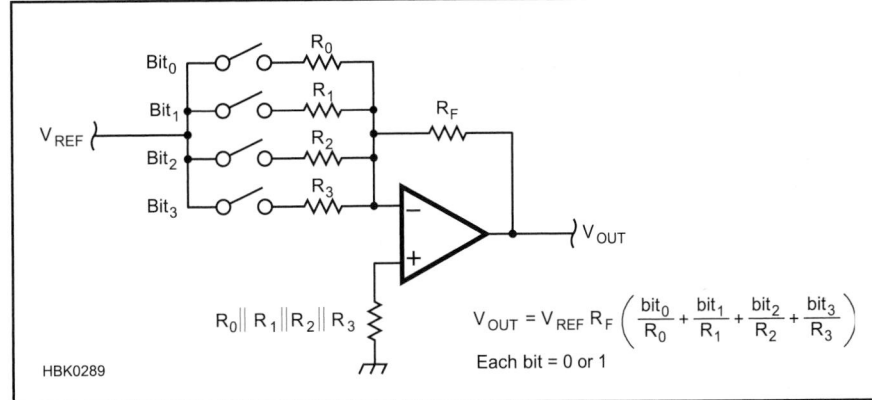

Fig 3.85 — Summing DAC. The output voltage is the inverted, weighted sum of the inputs to each summing resistor at the input. Digital data at the input controls the current into the summing resistors and thus, the output voltage.

$$V_{OUT} = V_{REF} \, R_F \left(\frac{bit_0}{R_0} + \frac{bit_1}{R_1} + \frac{bit_2}{R_2} + \frac{bit_3}{R_3} \right)$$

Each bit = 0 or 1

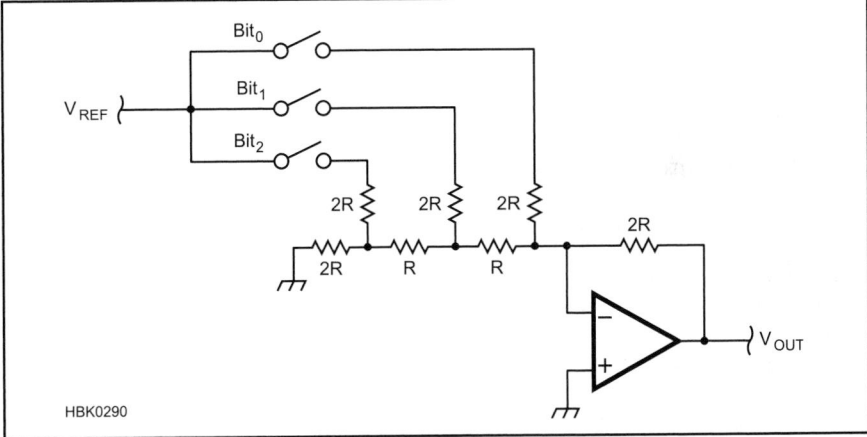

Fig 3.86 — R-2R Ladder DAC. This is the most common form of DAC because all of the resistor values are similar, making it easier to manufacture. The similarity in resistor values also means that there will be less variation of the comparator with temperature and other effects that affect all resistors similarly.

SUMMING DAC

A *summing DAC*, shown in **Fig 3.85**, is a summing amplifier with all of the inputs connected to a single reference voltage through switches. The digital value to be converted controls which switches are closed. The larger the digital value, the more switches are closed. Higher current causes the summing amplifier's output voltage to be higher, as well. Most summing DACs have a digital signal interface to hold the digital value between successive conversions, but not all, so be sure before you select a particular DAC.

The input resistors are *binary weighted* so that the summing network resistors representing the more significant bit values inject more current into the op amp's summing junction. Each resistor differs from its neighboring resistors in the amount of current it injects into the summing node by a factor of two, recreating the effect of each digital bit in the output voltage. At high resolutions, this becomes a problem because of the wide spread in resistor values — a 12-bit DAC would require a spread of 2048 between the largest and smallest resistor values. Summing DACs are generally only available with low resolution for that reason.

The *current output DAC* functions identically to the summing DAC, but does not have an op amp to convert current in the digitally-controlled resistor network to voltage. It consists only of the resistor network, so an external current-to-voltage circuit (discussed in the previous section on op amps) is required to change the current to a voltage. In some applications, the conversion to voltage is not required or it is already provided by some other circuit.

R-2R LADDER DAC

The summing and current output DACs both used binary weighted resistors to convert the binary digital value into the analog output. The practical limitation of this design is the large difference in value between the smallest and largest resistor. For example, in a 12-bit DAC, the smallest and largest resistors differ by a factor of 2048 (which is 2^{12-1}). This can be difficult to fabricate in an IC without

expensive *trimming* processes that adjust each resistor to the correct value and that maintain the powers-of-two relationship over wide temperature variations. For DACs with high-resolutions of 8 bits or more, the R-2R ladder DAC of **Fig 3.86** is a better design.

If the resistances are fairly close in value, the problems of manufacturing are greatly reduced. By using the *R-2R ladder* shown in the figure, the same method of varying current injected into an op amp circuit's summing junction can be used with resistors of only two values, R and 2R. In fact, since the op amp feedback resistor is also one of the IC resistors, the absolute value of the resistance R is unimportant, as long as the ratio of R:2R is maintained. This simplifies manufacturing greatly and is an example of IC design being based on ratios instead of absolute values. For this reason, most DACs use the R-2R ladder design and the performance differences lie mostly in their speed and accuracy.

3.7.4 Choosing a Converter

From the point of view of performance, choosing a converter, either an ADC or a DAC, comes down to resolution, accuracy and speed. Begin by determining the percent resolution or the dynamic range of the converter. Use the equations in the preceding section to determine the number of bits the converter must have. Select from converters with the next highest number of bits. For example, if you determine that you need 7 bits of resolution, use an 8-bit converter.

Next, consider accuracy. If the converter is needed for test instrumentation, you'll need to perform an *error budget* on the instrument's conversion processes, include errors in the analog circuitry. Once you have calculated percent errors, you can determine the requirements for FS error, offset error,

and nonlinearities. If an ADC is going to be used for receiving applications, the spur-free dynamic range may be more important than high precision. (The chapter on **DSP and Software Radio Design** goes into more detail about the performance requirements for ADCs used in these applications.)

The remaining performance criterion is the speed and rate at which the converter can operate. Conversions should be able to be made at a minimum of twice the highest frequency of signal you wish to reproduce. (The signal is assumed to be a sinusoid. If a complex waveform is to be converted, you must account for the higher frequency signals that create the nonsinusoidal shapes.) If the converter will be running near its maximum rate, be sure that the associated digital interface, supporting circuitry,

and software can support the required data rates, too!

Having established the conversion performance requirements, the next step is to consider cost, amount of associated circuitry, power requirements, and so forth. For example, a self-contained ADC is easier to use and takes up less PC board space, but is not as flexible as one that allows the designer to use an external voltage reference to set the conversion range. Many DACs with "current output" in their description are actually R-2R DACs with additional analog circuitry to provide the current output. DACs are available with both current and voltage outputs, as well. Other considerations, such as the nature of the required digital interface, as discussed in the next section, can also affect the selection of the converter.

3.8 Miscellaneous Analog ICs

The three main advantages of designing a circuit into an IC are to take advantage of the matched characteristics of its components, to make highly complex circuitry more economical, and to miniaturize the circuit and reduce power consumption. As circuits standardize and become widely used, they are often converted from discrete components to integrated circuits. Along with the op amp described earlier, there are many such classes of linear ICs.

3.8.1 Transistor and Driver Arrays

The most basic form of linear integrated circuit and one of the first to be implemented is the component array. The most common of these are the resistor, diode and transistor arrays. Though capacitor arrays are also possible, they are used less often. Component arrays usually provide space saving but this is not the major advantage of these devices. They are the least densely packed of the integrated circuits because each device requires a separate off-chip connection. While it may be possible to place over a million transistors on a single semiconductor chip, individual access to these would require a total of three million pins and this is beyond the limits of practicability. More commonly, resistor and diode arrays contain from five to 16 individual devices and transistor arrays contain from three to six individual transistors. The advantage of these arrays is the very close matching of component values within the array. In a circuit that needs matched components, the component array is often a good method of

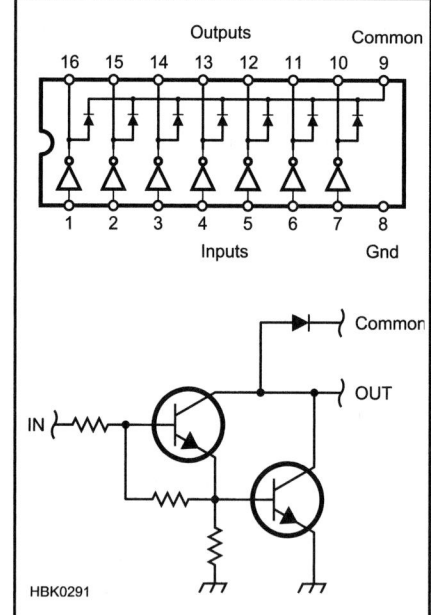

Fig 3.87 — Typical ULN2000-series driver array configuration and internal circuit. The use of driver array ICs is very popular as an interface between microprocessor or other low-power digital circuits and loads such as relays, solenoids or lamps.

obtaining this feature. The components within an array can be internally combined for special functions, such as termination resistors, diode bridges and Darlington pair transistors. A nearly infinite number of possibilities exists for these combinations of components and many of these are available in arrays.

Driver arrays, such as the ULN2000-series devices shown in Fig 3.87 are very useful in creating an interface between low-power circuits such as microprocessors and higher-power loads and indicators. Each driver consists of a Darlington pair switching circuit as described earlier in this chapter. There are different versions with different types and arrangements of resistors and diodes.

Many manufacturers offer driver arrays. They are available with built-in kickback diodes to allow them to drive inductive loads, such as relays, and are heavy enough to source or sink current levels up to 1 A. (All of the drivers in the array can not operate at full load at the same time, however. Read the data sheet carefully to determine what limitations on current and power dissipation may exist.)

3.8.2 Voltage Regulators and References

One of the most popular linear ICs is the voltage regulator. There are two basic types, the three-terminal regulator and the regulator controller. Examples of both are described in the **Power Sources** chapter.

The three-terminal regulator (input, ground, output) is a single package designed to perform all of the voltage regulation functions. The output voltage can be fixed, as in the 7800-series of regulators, or variable, as in the LM317 regulator. It contains a voltage reference, comparator circuits, current and temperature sensing protective circuits, and the main pass element. These ICs are usually contained in the same packages as power transistors and the same techniques of thermal management are used to remove excess heat.

Regulator controllers, such as the popular 723 device, contain all of the control and voltage reference circuitry, but require external components for the main pass element, current sensing, and to configure some of their control functions.

Voltage references such as the Linear Technology LT1635 are special semiconductor diodes that have a precisely controlled I-V characteristic. A buffer amplifier isolates the sensitive diode and provides a low output impedance for the voltage signal. Voltage references are used as part of power regulators and by analog-digital converter circuits.

3.8.3 Timers (Multivibrators)

A *multivibrator* is a circuit that oscillates between two states, usually with a square wave or pulse train output. The frequency of oscillation is accurately controlled with the addition of appropriate values of external resistance and capacitance. The most common multivibrator in use today is the 555 timer IC (NE555 by Signetics [now Philips] or LM555 by National Semiconductor). This very simple eight-pin device has a frequency range from less than one hertz to several hundred kilohertz. Such a device can also be used in *monostable* operation, where an input pulse generates an output pulse of a different duration, or in *a stable* or *free-running* operation, where the device oscillates continuously. Other applications of a multivibrator include a frequency divider, a delay line, a pulse width modulator and a pulse position modulator. (These can be found in the IC's data sheet or in the reference listed at the end of this chapter.)

Fig 3.88 shows the basic components of a 555. Connected between power input (V_{cc}) and ground, the three resistors labeled "R" at the top left of the figure form a *voltage divider* that divides V_{CC} into two equal steps—one at ⅔ V_{CC} and one at ⅓ V_{CC}. These serve as reference voltages for the rest of the circuit.

Connected to the reference voltages are blocks labeled *trigger comparator* and *threshold comparator*. (Comparators were discussed in a preceding section.) The trigger comparator in the 555 is wired so that its output is high whenever the trigger input is *less* than ⅓ V_{CC} and vice versa. Similarly, the threshold comparator output is high whenever the threshold input is *greater* than ⅔ V_{CC}. These two outputs control a digital *flip-flop* circuit. (Flip-flops are discussed in the **Digital Basics** chapter.)

The flip-flop output, *Q,* changes to high or low when the state of its *set* and *reset* input changes. The Q output stays high or low (it *latches* or *toggles*) until the opposite input changes. When the set input changes from low to high, Q goes low. When reset changes from low to high, Q goes high. The flip-flop

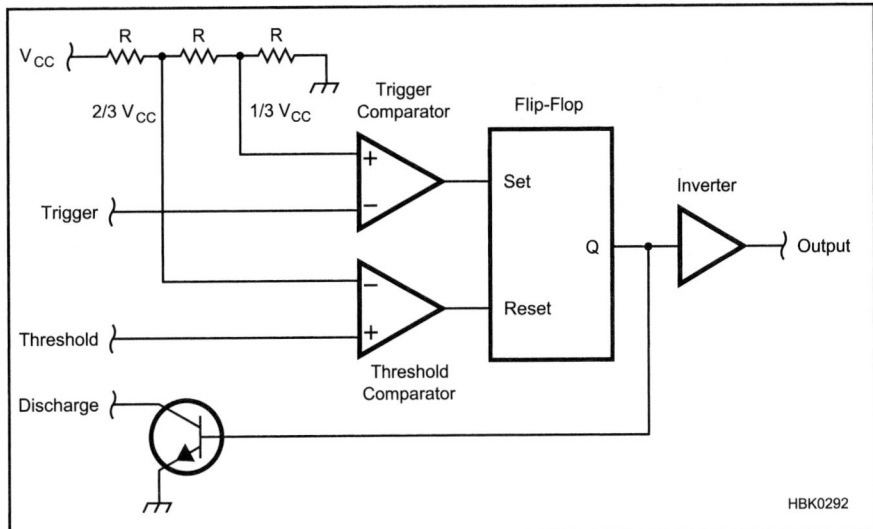

Fig 3.88 — Internal NE555 timer components. This simple array of components combine to make one of the most popular analog ICs. The 555 timer IC uses ratios of internal resistors to generate a precise voltage reference for generating time intervals based on charging and discharging a capacitor.

ignores any other changes. An inverter makes the 555 output high when Q is low and vice versa — this makes the timer circuit easier to interface with external circuits.

The transistor connected to Q acts as a switch. When Q is high, the transistor is on and acts as a closed switch connected to ground. When Q is low, the transistor is off and the switch is open. These simple building blocks — voltage divider, comparator, flip-flop and switch — build a surprising number of useful circuits.

THE MONOSTABLE OR "ONE-SHOT" TIMER

The simplest 555 circuit is the monostable circuit. This configuration will output one fixed-length pulse when triggered by an input pulse. **Fig 3.89** shows the connections for this circuit.

Starting with capacitor C discharged, the flip-flop output, Q, is high, which keeps the discharge transistor turned on and the voltage on C below ⅔ V_{CC}. The circuit is in its stable state, waiting for a trigger pulse.

When the voltage at the trigger input drops below ⅓ V_{CC}, the trigger comparator output changes from low to high, which causes Q to toggle to the low state. This turns off the transistor (opens the switch) and allows C to begin charging toward V_{CC}.

When C reaches ⅔ V_{CC}, this causes the threshold comparator to switch its output from low to high and that resets the flip-flop. Q returns high, turning on the transistor and discharging C. The circuit has returned to its stable state. The output pulse length for the monostable configuration is:

$$T = 1.1 \, R \, C_1 \tag{73}$$

Notice that the timing is independent of the absolute value of V_{CC} — the output pulse width is the same with a 5 V supply as it is with a 15 V supply. This is because the 555 design is based on ratios and not absolute voltage levels.

THE ASTABLE MULTIVIBRATOR

The complement to the monostable circuit is the astable circuit in **Fig 3.90**. Pins 2, 6 and 7 are configured differently and timing resistor is now split into two resistors, R1 and R2.

Start from the same state as the monostable circuit, with C completely discharged. The monostable circuit requires a trigger pulse to initiate the timing cycle. In the astable circuit, the trigger input is connected directly to the capacitor, so if the capacitor is discharged, then the trigger comparator output must be high. Q is low, turning off the discharge transistor, which allows C to immediately begin charging.

C charges toward V_{CC}, but now through the combination of R1 and R2. As the capacitor voltage passes ⅔ V_{CC}, the threshold comparator output changes from low to high, resetting Q to high. This turns on the discharge transistor and the capacitor starts to discharge through R2. When the capacitor is discharged below ⅓ V_{CC}, the trigger comparator changes from high to low and the cycle begins again, automatically. This happens over and over, causing a train of pulses at the output while C charges and discharges between ⅓ and ⅔ V_{CC} as seen in the figure.

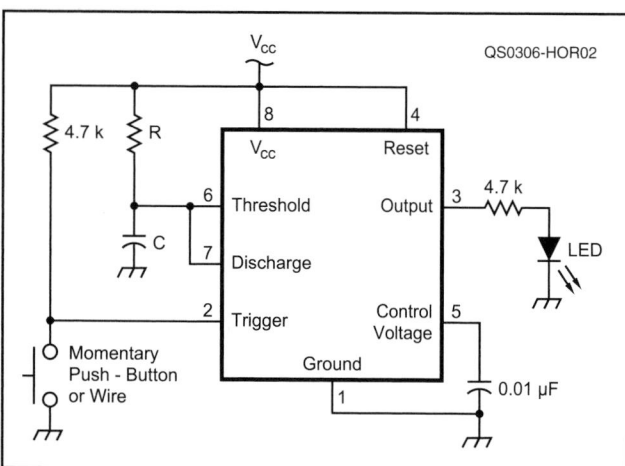

Fig 3.89 — Monostable timer. The timing capacitor is discharged until a trigger pulse initiates the charging process and turns the output on. When the capacitor has charged to 2/3 V_{CC}, the output is turned off, the capacitor is discharged and the timer awaits the next trigger pulse.

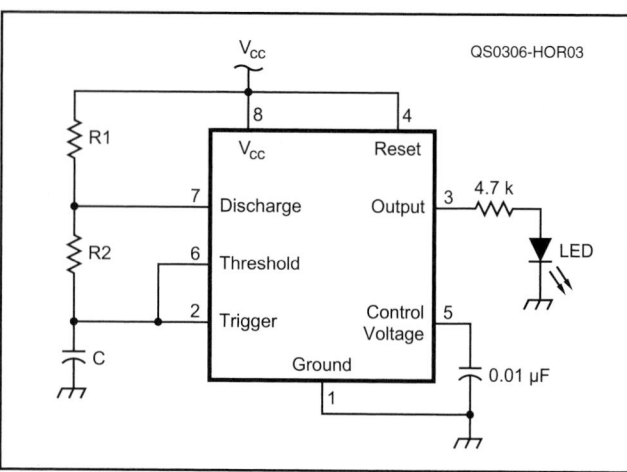

Fig 3.90 — Astable timer. If the capacitor discharge process initiates the next charge cycle, the timer will output a pulse train continuously.

The total time it takes for one complete cycle is the charge time, T_c, plus the discharge time, T_d:

$$T = T_c + T_d = 0.693\,(R_1 + R_2)\,C + 0.693\,R_2 C$$
$$= 0.693\,(R_1 + 2R_2)\,C \qquad (74)$$

and the output frequency is:

$$f = \frac{1}{T} = \frac{1.443}{(R_1 + 2\,R_2)\,C} \qquad (75)$$

When using the 555 in an application in or around radios, it is important to block any RF signals from the IC power supply or timing control inputs. Any unwanted signal present on these inputs, especially the Control Voltage input, will upset the timer's operation and cause it to operate improperly. The usual practice is to use a 0.01 µF bypass capacitor (shown on pin 5 in both Fig 3.89 and 3.90) to bypass ac signals such as noise or RF to ground. Abrupt changes in V_{CC} will also cause changes in timing and these may be prevented by connecting filter capacitors at the V_{CC} input to ground.

3.8.4 Analog Switches and Multiplexers

Arrays of analog switches, such as the Maxim MAX312-series, allow routing of audio through lower frequency RF signals without mechanical switches. There are several types of switch arrays. Independent switches have isolated inputs and outputs and are turned on and off independently. Both SPST and SPDT configurations are available. Multiple switches can be wired with common control signals to implement multiple-pole configurations.

Use of analog switches at RF through microwave frequencies requires devices specifically designed for those frequencies. The Analog Devices ADG901 is a switch usable to 2.5 GHz. It absorbs the signal when off, acting as a terminating load. The ADG902 instead reflects the signal as an open circuit when off. Arrays of three switches called "tee-switches" are used when very high isolation between the input and output is required.

Multiplexers or "muxes" are arrays of SPST switches configured to act as a multiposition switch that connects one of four to sixteen input signals to a single output. Demultiplexers ("demuxes") have a single input and multiple outputs. Multiplexer ICs are available as single N-to-1 switches (the MAX4617 is an 8-to-1 mux) or as groups of N-to-1 switches (the MAX4618 is a dual 4-to-1 mux).

Crosspoint switch arrays are arranged so that any of four to sixteen signal inputs can be connected to any of four to sixteen output signal lines. The Analog Devices AD8108 is an 8-by-8 crosspoint switch with eight inputs and eight outputs. These arrays are used when it is necessary to switch multiple signal sources among multiple signal receivers. They are most commonly used in telecommunications.

All analog switches use FET technology as the switching element. To switch ac signals, most analog switches require both positive and negative voltage power supplies. An alternative is to use a single power supply voltage and ground, but bias all inputs and output at one-half the power supply voltage. This requires dc blocking capacitors in all signal paths, both input and output, and loading resistors may be required at the device outputs. The blocking capacitors can also introduce low-frequency roll-off.

The impedance of the switching element varies from a few ohms to more than 100 ohms. Check the switch data sheet to determine the limits for how much power and current the switches can handle. Switch arrays, because of the physical size of the array, can have significant coupling or *crosstalk* between signal paths. Use caution when using analog switches for high-frequency signals as coupling generally increases with frequency and may compromise the isolation required for high selectivity in receivers and other RF signal processing equipment.

3.8.5 Audio Output Amplifiers

While it is possible to use op amps as low power audio output drivers for headphones, they generally have output impedances that are too high for most audio transducers such as speakers and headphones. The LM380 series of audio driver ICs has been used in radio circuits for many years and a simple schematic for a speaker driver is shown in **Fig 3.91**.

The popularity of personal music players has resulted in the creation of many new and inexpensive audio driver ICs, such as the National Semiconductor LM4800- and LM4900-series. Drivers that operate from voltages as low as 1.5 V for battery-powered devices and up to 18 V for use in vehicles are now available.

When choosing an audio driver IC for communications audio, the most important parameters to evaluate are its power requirements and power output capabilities. An overloaded or underpowered driver will result in distortion. Driver ICs intended for music players have frequency responses well in excess of the 3000 Hz required for communications. This can lead to annoying and fatiguing hiss unless steps are taken to reduce the circuit's frequency response.

Audio power amplifiers should also be carefully decoupled from the power supply and the manufacturer may recommend specific circuit layouts to prevent oscillation or feedback. Check the device's data sheet for this information.

3.8.6 Temperature Sensors

Active temperature sensors use the temperature-dependent properties of semiconductor devices to create voltages that correspond to absolute temperature in degrees Fahrenheit (LM34) or degrees Celsius (LM35). These sensors (of which many others are available than the two examples given here) are available in small plastic packages, both leaded and surface-mount, that respond quickly to temperature changes. They are available with 1% and better accuracy, requiring only a source of voltage at very low current and ground. Complete application information is available in the manufacturer data sheets. Thermistors, a type of passive temperature sensor, are discussed in the **Electrical Fundamentals** chapter. Temperature sensors are used in radio mostly in cooling and thermal management systems.

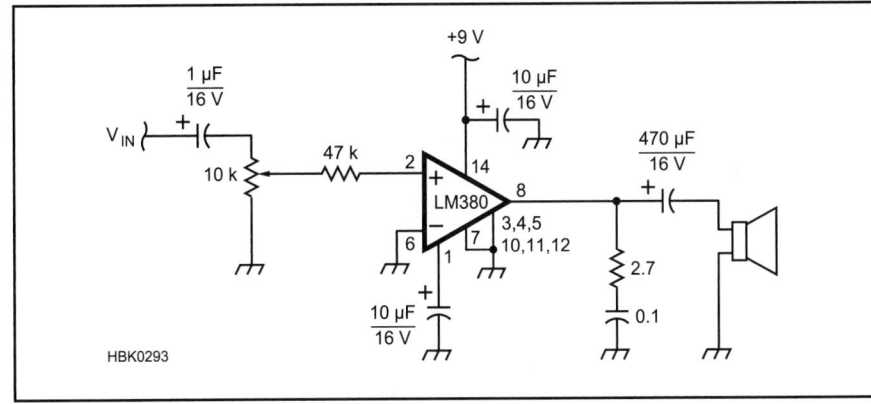

Fig 3.91 — Speaker driver. The LM380-series of audio output drivers are well-suited for low-power audio outputs, such as for headphones and small speakers. When using IC audio output drivers, be sure to refer to the manufacturer's data sheet for layout and power supply guidelines.

3.8.7 Electronic Subsystems

As a particular technology becomes popular, a wave of integrated circuitry is developed to service that technology and reduce its cost of production and service. A good example is the wireless telephony industry. IC manufacturers have developed a large number of devices targeting this industry; receivers, transmitters, couplers, mixers, attenuators, oscillators and so forth. In addition, other wireless technologies such as data transmission provide opportunities for manufacturers to create integrated circuits that implement radio-related functions at low cost. Analog ICs used in the construction of various radio systems and supporting equipment are discussed in the appropriate chapters of this book.

3.9 Analog Glossary

AC ground — A circuit connection point that presents a very low impedance to ac signals.

Accuracy — The ability of an analog-to-digital conversion to assign the correct code to an analog value or create the true analog value from a specific code.

Active — A device that requires power to operate.

Active region — The region in the characteristic curve of an analog device in which it is capable of processing the signal linearly.

Amplification — The process by which amplitude of a signal is increased. Gain is the amount by which the signal is amplified.

Analog signal — A signal that can have any amplitude (voltage or current) value and exists at any point in time.

Analog-to-digital converter (ADC) — Circuit (usually an IC) that generates a digital representation of an analog signal.

Anode — The element of an analog device that accepts electrons or toward which electrons flow.

Attenuation — The process of reducing the amplitude of a signal.

Avalanche breakdown — Current flow through a semiconductor device in response to an applied voltage beyond the device's ability to control or block current flow.

Base — The terminal of a bipolar transistor in which control current flows.

Beta (β) — The dc current gain of a bipolar transistor, also designated h_{FE}.

Biasing — The addition of a dc voltage or current to a signal at the input of an analog device, changing or controlling the position of the device's operating point on the characteristic curve.

Bipolar transistor — An analog device made by sandwiching a layer of doped semiconductor between two layers of the opposite type: PNP or NPN.

Black box — Circuit or equipment that is analyzed only with regards to its external behavior.

Bode plot — Graphs showing amplitude response in dB and phase response in degrees versus frequency on a logarithmic scale.

Buffer — An analog stage that prevents loading of one analog stage by another.

Carrier — (1) Free electrons and holes in semiconductor material. (2) An unmodulated component of a modulated signal.

Cascade — Placing one analog stage after another to combine their effects on the signal.

Cathode — The element of an analog device that emits electrons or from which electrons are emitted or repelled.

Characteristic curve — A plot of the relative responses of two or three analog-device parameters, usually of an output with respect to an input. (Also called *I-V* or *V-I* curve.)

Class — For analog amplifiers (Class A, B, AB, C), a categorization of the fraction of the input signal cycle during which the amplifying device is active. For digital or switching amplifiers (Class D and above), a categorization of the method by which the signal is amplified.

Clipping — A nonlinearity in amplification in which the signal's amplitude can no longer be increased, usually resulting in distortion of the waveform. (Also called *clamping* or *limiting*.)

Closed-loop gain — Amplifier gain with an external feedback circuit connected.

Collector — The terminal of a bipolar transistor from which electrons are removed.

Code — One possible digital value representing an analog quantity. (Also called *quantization code*.)

Common — A terminal shared by more than one port of a circuit or network.

Common mode — Signals that appear equally on all terminals of a signal port.

Comparator — A circuit, usually an amplifier, whose output indicates the relative amplitude of two input signals.

Compensation — The process of counteracting the effects of signals that are inadvertently fed back from the output to the input of an analog system. Compensation increases stability and prevents oscillation.

Compression — Reducing the dynamic range of a signal in order to increase the average power of the signal or prevent excessive signal levels.

Conversion efficiency — The amount of light energy converted to electrical energy by a photoelectric device, expressed in percent.

Conversion rate — The amount of time in which an analog-digital conversion can take place. *Conversion speed* is the reciprocal of conversion rate.

Coupling (ac or dc) — The type of connection between two circuits. DC coupling allows dc current to flow through the connection. AC coupling blocks dc current while allowing ac current to flow.

Cutoff frequency — Frequency at which a circuit's amplitude response is reduced to one-half its mid-band value (also called *half-power* or *corner* frequency).

Cutoff (region) — The region in the characteristic curve of an analog device in which there is no current through the device. Also called the OFF region.

Degeneration (emitter or source) — Negative feedback from the voltage drop across an emitter or source resistor in order to stabilize a circuit's bias and operating point.

Depletion mode — An FET with a channel that conducts current with zero gate-to-source voltage and whose conductivity is progressively reduced as reverse bias is applied.

Depletion region — The narrow region at a PN junction in which majority carriers have been removed. (Also called *space-charge* or *transition* region.)

Digital-to-analog converter (DAC) — Circuit (usually an IC) that creates an analog signal from a digital representation.

Diode — A two-element semiconductor with a cathode and an anode that conducts current in only one direction.

Drain — The connection at one end of a field-effect-transistor channel from

which electrons are removed.

Dynamic range — The range of signal levels over which a circuit operates properly. Usually refers to the range over which signals are processed linearly.

Emitter — The terminal of a bipolar transistor into which electrons are injected.

Enhancement mode — An FET with a channel that does not conduct with zero gate-to-source voltage and whose conductivity is progressively increased as forward bias is applied.

Feedback — Routing a portion of an output signal back to the input of a circuit. Positive feedback causes the input signal to be reinforced. Negative feedback results in partial cancellation of the input signal.

Field-effect transistor (FET) — An analog device with a semiconductor channel whose width can be modified by an electric field. (Also called *unipolar transistor.*)

Forward bias — Voltage applied across a PN junction in the direction to cause current flow.

Forward voltage — The voltage required to cause forward current to flow through a PN junction.

Free electron — An electron in a semiconductor crystal lattice that is not bound to any atom.

Frequency response — A description of a circuit's gain (or other behavior) with frequency.

Gain — see **Amplification**.

Gain-bandwidth product — The relationship between amplification and frequency that defines the limits of the ability of a device to act as a linear amplifier. In many amplifiers, gain times bandwidth is approximately constant.

Gate — The control electrode of a field-effect transistor.

High-side — A switch or controlling device connecting between a power source and load.

Hole — A positively charged carrier that results when an electron is removed from an atom in a semiconductor crystal structure.

Hysteresis — In a comparator circuit, the practice of using positive feedback to shift the input setpoint in such a way as to minimize output changes when the input signal(s) are near the setpoint.

Integrated circuit (IC) — A semiconductor device in which many components, such as diodes, bipolar transistors, field-effect transistors, resistors and capacitors are fabricated to make an entire circuit.

Isolation — Eliminating or reducing electrical contact between one portion of a circuit and another or between pieces of equipment.

Junction FET (JFET) — A field-effect transistor whose gate electrode forms a PN junction with the channel.

Linearity — Processing and combining of analog signals independently of amplitude.

Load line — A line drawn through a family of characteristic curves that shows the operating points of an analog device for a given load or circuit component values.

Loading — The condition that occurs when the output behavior of a circuit is affected by the connection of another circuit to that output.

Low-side — A switch or controlling device connected between a load and ground.

Metal-oxide semiconductor (MOSFET) — A field-effect transistor whose gate is insulated from the channel by an oxide layer. (Also called *insulated gate FET* or *IGFET*)

Multivibrator — A circuit that oscillates between two states.

NMOS — N-channel MOSFET.

N-type impurity — A doping atom with an excess of valence electrons that is added to semiconductor material to act as a source of free electrons.

Network — General name for any type of circuit.

Noise — Any unwanted signal, usually random in nature.

Noise figure (NF) — A measure of the noise added to a signal by an analog processing stage, given in dB. (Also called *noise factor.*)

Open-loop gain — Gain of an amplifier with no feedback connection.

Operating point — Values of a set of circuit parameters that specify a device's operation at a particular time.

Operational amplifier (op amp) — An integrated circuit amplifier with high open-loop gain, high input impedance, and low output impedance.

Optoisolator — A device in which current in a light-emitting diode controls the operation of a phototransistor without a direct electrical connection between them.

Oscillator — A circuit whose output varies continuously and repeatedly, usually at a single frequency.

P-type impurity — A doping atom with a shortage of valence electrons that is added to semiconductor material to create an excess of holes.

Passive — A device that does not require power to operate.

Peak inverse voltage (PIV) — The highest voltage that can be tolerated by a reverse biased PN junction before current is conducted. (See also **avalanche breakdown**.)

Photoconductivity — Phenomenon in

which light affects the conductivity of semiconductor material.

Photoelectricity — Phenomenon in which light causes current to flow in semiconductor material.

PMOS — P-channel MOSFET.

PN junction — The structure that forms when P-type semiconductor material is placed in contact with N-type semiconductor material.

Pole — Frequency at which a circuit's transfer function becomes infinite.

Port — A pair of terminals through which a signal is applied to or output from a circuit.

Quiescent (Q-) point — Circuit or device's operating point with no input signal applied. (Also called *bias point*.)

Pinch-off — The condition in an FET in which the channel conductivity has been reduced to zero.

Products — Signals produced as the result of a signal processing function.

Rail — Power supply voltage(s) for a circuit.

Range — The total span of analog values that can be processed by an analog-to-digital conversion.

Recombination — The process by which free electrons and holes are combined to produce current flow across a PN junction.

Recovery time — The amount of time required for carriers to be removed from a PN junction device's depletion region, halting current flow.

Rectify — Convert ac to pulsating dc.

Resolution — Smallest change in an analog value that can be represented in a conversion between analog and digital quantities. (Also called *step size*.)

Reverse bias — Voltage applied across a PN junction in the direction that does not cause current flow.

Reverse breakdown — The condition in which reverse bias across a PN junction exceeds the ability of the depletion region to block current flow. (See also *avalanche breakdown*.)

Roll-off — Change in a circuit's amplitude response per octave or decade of frequency.

Safe operating area (SOA) — The region of a device's characteristic curve in which it can operate without damage.

Sample — A code that represents the value of an analog quantity at a specific time.

Saturation (Region) — The region in the characteristic curve of an analog device in which the output signal can no longer be increased by the input signal. See *Clamping*.

Schottky barrier — A metal-to-semiconductor junction at which a depletion region is formed, similarly to a PN junction.

Semiconductor — (1) An element such as silicon with bulk conductivity between that of an insulator and a metal. (2) An electronic device whose function is created by a structure of chemically-modified semiconductor materials.

Signal-to-noise ratio (SNR) — The ratio of the strength of the desired signal to that of the unwanted signal (noise), usually expressed in dB.

Slew rate — The maximum rate at which a device can change the amplitude of its output.

Small-signal — Conditions under which the variations in circuit parameters due to the input signal are small compared to the quiescent operating point and the

device is operating in its active region.

Source — The connection at one end of the channel of a field-effect transistor into which electrons are injected.

Stage — One of a series of sequential signal processing circuits or devices.

Substrate — Base layer of material on which the structure of a semiconductor device is constructed.

Superposition — Process in which two or more signals are added together linearly.

Total harmonic distortion (THD) — A measure of how much noise and distortion are introduced by a signal processing function.

Thermal runaway — The condition in which increasing device temperature increases device current in a positive feedback cycle.

Transconductance — Ratio of output current to input voltage, with units of Siemens (S).

Transfer characteristics — A set of parameters that describe how a circuit or network behaves at and between its signal interfaces.

Transfer function — A mathematical expression of how a circuit modifies an input signal.

Unipolar transistor — see *Field-effect transistor (FET)*.

Virtual ground — Point in a circuit maintained at ground potential by the circuit without it actually being connected to ground.

Zener diode — A heavily-doped PN-junction diode with a controlled reverse breakdown voltage, used as a voltage reference or regulator.

Zero — Frequency at which a circuit's transfer function becomes zero.

3.10 References and Bibliography

REFERENCES

1. Ebers, J., and Moll, J., "Large-Signal Behavior of Junction Transistors," *Proceedings of the IRE*, 42, Dec 1954, pp 1761-1772.
2. Getreu, I., *Modeling the Bipolar Transistor* (Elsevier, New York, 1979). Also available from Tektronix, Inc, Beaverton, Oregon, in paperback form. Must be ordered as Part Number 062-2841-00.

FURTHER READING

Alexander and Sadiku, *Fundamentals of Electric Circuits* (McGraw-Hill)

Hayward, W., *Introduction to Radio Frequency Design* (ARRL, 2004)

Hayward, Campbell and Larkin, *Experimental Methods in RF Design* (ARRL, 2009)

Millman and Grabel, *Microelectronics: Digital and Analog Circuits and Systems* (McGraw-Hill, 1988)

Mims, F., *Timer, Op Amp & Optoelectronic Circuits & Projects* (Master Publishing, 2004)

Jung, W. *IC Op Amp Cookbook* (Prentice-Hall, 1986)

Hill and Horowitz, *The Art of Electronics* (Cambridge University Press, 1989)

Analog-Digital Conversion Handbook (by the staff of Analog Devices, published by Prentice-Hall)

Safe-Operating Area for Power Semiconductors (ON Semi), **www. onsemi.com/pub_link/Collateral/ AN875-D.PDF**

"Selecting the Right CMOS Analog Switch", Maxim Semiconductor, **www.maxim-ic.com/appnotes.cfm/ an_pk/638**

Hyperphysics Op-Amp Circuit Tutorials, **hyperphysics.phy-astr.gsu.edu/Hbase/ Electronic/opampvar.html#c2**

Contents

**Chapter 4 —
CD-ROM Content**

Supplemental Articles

- "Interfacing to the Parallel Port" and supporting files by Paul Danzer, N1II
- "Learning to PIC with a PIC-EL" (Parts 1 and 2) by Craig Johnson, AA0ZZ
- "Pickle with USB I/O" by Craig Johnson, AA0ZZ

Digital Basics

Radio amateurs have been involved with digital technology since the first spark transmitters, a form of pulse-coded transmission, were connected to an "aerial." Modern digital technology use by radio amateurs probably arrived first in CW keyers, where hams learned about flip-flops and gates to replace their semi-automatic mechanical "bug" keys.

Amateur use of digital technology echoed public use of these new abilities, starting with using the first home computers for calculations and later digital communications terminals. Today's Amateur Radio digital applications range from simple shack accessories like keyers and timers, to computer based digital modes such as PSK31 and Hellschreiber, to computer networking using TCP/IP over AX.25, computer controlled rigs, digital signal processing and software defined radios.

This chapter was written by Dale Botkin, NØXAS, building on material in previous editions by Christine Montgomery, KGØGN and Paul Danzer, N1II. It presents digital theory fundamentals and some applications of that theory in Amateur Radio. The fundamentals introduce digital mathematics, including number systems, logic devices and simple digital circuits. Next, the implementation of these simple circuits is explored in integrated circuits, their families and interfacing. Finally, some Amateur Radio applications are discussed involving digital logic, embedded microcontrollers and interfacing to personal computers.

4.1 Digital vs Analog

An analog signal can represent an infinitely variable indication of voltage, current, frequency, the position of a dial, or some other condition or value. As an example, using a potentiometer as a volume control will give you infinitely variable control over the volume of a signal. In theory, there is no limit to the difference in volume that can be produced. Though the control may be marked from 1 to 10, the actual value would have to be represented by a real number somewhere between 0 and 10. There are an infinite number of settings in between.

In its simplest form, a digital signal simply indicates the *on* or *off* state of some value or input signal. For example, the straight key you may use to key your CW transmitter (or the PTT switch of your voice transmitter) produces an on or off binary signal. In one state the transmitter produces an output signal of some sort; in the other state it does not. Another example is a simple light switch. The light is either on, or it's off. We represent these two states using 0 for off and 1 for on.

Digital electronics gets more interesting when we combine several or many simple on/off digital states to perform more complex tasks. For example, a relatively simple digital circuit can connect the antenna to either the transmitter or the receiver depending on a PTT or other keying signal. It can turn a preamp on or off depending on the state of the transmitter, mute the speaker while transmitting, and even select an antenna based on the selected frequency band. No special digital integrated circuits (chips) are needed to do any of these tasks; we can simply use bipolar transistors or MOSFETs, driven to saturation, as on/off switches. Simple circuits like this can often even be implemented with relays or diodes. The important fact is that the system is digital. There is no "almost transmitting" or "PTT switch partially pressed" state — it's either on, or it's off.

A very useful aspect of digital electronics is our ability to construct simple circuits that can maintain their on/off state indefinitely, until some event causes them to change. These *flip-flop* circuits can be used in various combinations to form *registers* that store information for later use or *counters* that count events and can be read or reset when needed. All of these circuits can be combined in ever larger groups until we finally arrive at the modern microprocessor. A microprocessor can accept input signals from many sources, follow a stored program to perform complex data storage and mathematical calculations, and produce output that we can use to do things that would be far more difficult with analog circuits.

So let's revisit our volume control example from the earlier paragraph. Let's assume we have a volume control, but it is used as an input to a digital system that will produce output at the desired level. This is quite

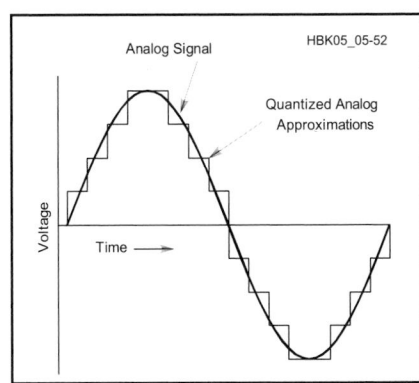

Fig 4.1 — An analog signal and its analog approximation. Note that the analog waveform has continuously varying voltage while the approximated waveform is composed of discrete steps.

common in modern equipment, whether it is amateur or consumer gear. Since the control is now digital, we know we can't have an infinite number of values. However, a simple on/off volume control would not be very useful. Using digital electronics, we can break the range between "off" and "fully on" into as many discrete steps as we need. With enough steps, we can give the user of the equipment an approximation of the original analog control while keeping the actual control digital.

By using coding, as discussed in the following pages, the two binary values (off and on, or 0 and 1) can represent any number of real values. **Fig 4.1** illustrates the contrast of an analog signal (in this case a sine wave) and its digital approximation. Four positive and four negative values are shown as an approximation to the sine wave, but any number of coded value steps can be used as an approximation.

The more values are used to approximate the wave, the closer you can come to the actual wave form.

While the focus in this chapter will be on digital theory, many circuits and systems involve *both* digital and analog components. Often, a designer may choose between using digital technology, analog technology or a combination.

4.2 Number Systems

If you have been around computer hobbyists, some of whom are also hams, you may have seen a T-shirt or bumper sticker that reads, "There are 10 kinds of people in the world: those who understand binary, and those who don't." If this has puzzled you in the past, after reading this chapter you will be able to laugh with the rest of us.

In order to understand digital electronics, you must first understand the binary numbering system. Any number system has two distinct characteristics: a set of *symbols* (digits or numerals) and a *base* or radix. A *number* is a collection of these digits, where the leftmost digit is the *most significant digit (MSD)* and the right-most digit is the *least significant digit (LSD)*. The value of this number is a weighted sum of its digits. The *weights* are determined by the system's base and the digit's position relative to the decimal point.

While these definitions may seem strange with all the technical terms, they will be more familiar when seen in a decimal system example. See **Table 4.1**. This is the "traditional" number system with which we are all familiar. In the *base-10* or decimal numbering system we use every day, the digits used are 0 through 9. The weights are powers of ten: 10^0 or 1 for the right-most column, 10^1 or 10 for the next column, 10^2 or 100 for the next and so on. Thus the number 548 represents five hundreds, four tens and eight ones. In this case, 5 is the MSD, and 8 is the LSD. Once you understand this concept, it can be applied to numbering systems using bases other than 10 such as base-2, base-8, or even base-16.

4.2.1 Binary

Binary is a *base-2* number system and therefore limited to two symbols: {0, 1}. The weight factors are now powers of 2, like 2^0, 2^1 and 2^2. For example, the decimal number, 163

and its equivalent binary number, 10100011, are shown in **Table 4.2**.

The digits of a binary number are now *bits* (short for binary digit). The MSD is the *most significant bit (MSB)* and the LSD is the *least significant bit (LSB)*. Four bits make a *nibble* (which you will occasionally see spelled *nybble*) and two nibbles, or eight bits, make a *byte*. The length of a *word* is dependent upon the hardware; it generally can consist of two or four or more bytes, but occasionally will be some other number of bits. These groupings are useful when converting to hexadecimal notation, which is explained later. It is important to remember that while everyone agrees on the meaning of a bit, a nibble (regardless of spelling) and a byte, the meaning of *word* can vary.

Counting in binary follows the same pattern we would use for decimal or any other number system. Consider the three digit binary number XXX. First fill up the right-hand column.

Binary Number	Decimal Equivalent
0000	0
0001	1

The column has been filled, and much quicker than with decimal, since there are only two values instead of 10. But just as we would with a decimal number, we now reset the right-hand column to 0, increase the next column by 1, and continue.

Table 4.1
Decimal Numbers

Example: 548

Digit = 5; Weight = 10; Position = 2

548	=	5(10^2)	+	4(10^1)	+	8(10^0)
	=	5(100)	+	4(10)	+	8(1)
	=	500	+	40	+	8
	=	5		4		8
		MSD				LSD

Table 4.2
Decimal and Binary Number Equivalents

163	=	128	+	0	+	32	+	0	+	0	+0	+	2+	1	decimal
	=	1(128)	+	0(64)	+	1(32)	+	0(16)	+	0(8)	+0(4)	+	1(2)	+1(1)	
	=	1(2^7)	+	0(2^6)	+	1(2^5)	+	0(2^4)	+	0(2^3)	+0(2^2)	+	1(2^1)	+1(2^0)	

10100011	=	1	0	1	0	0	0	1	1	binary
		MSB							LSB	

Nibble		Nibble

Byte = 8 digits

0010	2
0011	3

Now the first two columns are full, so reset both back to 0 and increase the next column by 1 and continue:

0100	4
0101	5
0110	6
0111	7
1000	8
…	
1111	15

And so on.

4.2.2 Hexadecimal

The *hexadecimal*, or hex, *base-16* number system is widely used in computer systems for its ease in conversion to and from binary numbers and the fact that it is somewhat more human-friendly than long strings of 1s and 0s. A base-16 number requires 16 symbols. Since our normal mathematical number, as set up in the decimal system, has only 10 digits (0 through 9), a set of additional new symbols is required. Hex uses both numbers and characters in its set of sixteen symbols: {0, 1, 2, 3, 4, 5,6, 7, 8, 9, A, B, C, D, E, F}. Here, the letters A to F have the decimal equivalents of 10 to 15 respectively: A=10, B=11, C=12, D=13, E=14 and F=15. Again, the weights are powers of the base, such as 16^0, 16^1 and 16^2.

The four-bit binary listing in the previous paragraph shows that the individual 16 hex digits can be represented by a four-bit binary number. Since a byte is equal to eight binary digits, two hex digits provide a byte — the equivalent of 8 binary digits. Conversion from binary to hex is therefore simplified. Take a binary number, divide it into groups of four binary digits starting from the right, and convert each of the four binary digits to an individual value.

Conversion from hex to binary is equally convenient; simply replace each hex digit with its four-bit binary equivalent. As an example, the decimal number 163 is shown in Table 4.2 as binary 10100011. Divide the binary number in groups of four, so 1010 is equivalent to decimal 10 or "A" hex, and 0011 is equivalent to decimal 3, thus decimal number 163 is equivalent to hex A3.

4.2.3 Binary Coded Decimal (BCD)

The binary number system representation is the most appropriate form for fast internal computations since there is a direct mathematical relationship for every bit in the number. To interface with a human user — who usually wants to see inputs and outputs in terms of decimal numbers — other codes are more useful. The *Binary Coded Decimal*

(BCD) system is a simple method for converting binary values to and from decimal for inputs and outputs for user-oriented digital systems. Back in the days when the most common method of presenting output to a user was via seven-segment LED displays, BCD was widely used. Since we now mostly use powerful microprocessors that can easily present information in decimal form, BCD is not nearly as common as it once was. You may, however, run into BCD when using or repairing older digital gear. It is also used in some chips intended for use in digital voltmeters.

In the BCD system, each decimal digit is expressed as a corresponding 4-bit binary number. In other words, the decimal digits 0 to 9 are encoded as the bit strings 0000 to 1001. To make the number easier to read, a space is left between each 4-bit group. For example, the decimal number 163 is equivalent to the BCD number 0001 0110 0011, as shown in **Table 4.3**.

The important difference between BCD and the previous number systems is that, starting with decimal 10, BCD loses the standard mathematical relationship of a weighted sum. BCD is simply a cut-off hexadecimal. Instead of using the 4-bit code strings 1010 to 1111 for decimal 10 to 15, BCD uses 0001 0000 to 0001 0101. This is one of the reasons that we have moved away from BCD.

4.2.4 Conversion Techniques

An easy way to convert a number from decimal to another number system is to do repeated division, recording the remainders

in a tower just to the right. The converted number, then, is the remainders, reading up the tower. This technique is illustrated in **Table 4.4** for hexadecimal and binary conversions of the decimal number 163.

For example, to convert decimal 163 to hex, repeated divisions by 16 are performed. The first division gives 163/16 = 10 remainder 3. The remainder 3 is written in a column to the right. The second division gives 10/16 = 0 remainder 10. Since 10 decimal = A hex, A is written in the remainder column to the right. This division gave a divisor of 0 so the process is complete. Reading up the remainders column, the result is A3. The most common mistake in this technique is to forget that the Most Significant Digit ends up at the bottom.

Another technique that should be briefly mentioned can be even easier: use a calculator with a binary and/or hex mode option. Many inexpensive and readily available calculators intended for scientific and programming use will convert between number systems quite easily. In addition, calculator programs are available for all types of personal computers regardless of the operating system used.

One warning for this technique: this chapter doesn't discuss negative binary numbers. If your calculator does not give you the answer you expected, it may have interpreted the number as negative. This would happen when the number's binary form has a 1 in its MSB, such as the highest (leftmost) bit for the binary mode's default size. To avoid learning about negative binary numbers the hard way, always use a leading 0 when you enter a number in binary or hex into your calculator.

Table 4.3
Binary Coded Decimal Number Conversion

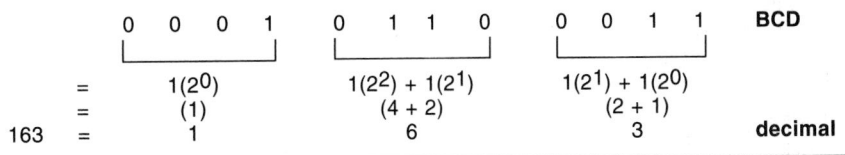

	0 0 0 1	0 1 1 0	0 0 1 1	**BCD**
=	$1(2^0)$	$1(2^2) + 1(2^1)$	$1(2^1) + 1(2^0)$	
=	(1)	(4 + 2)	(2 + 1)	
163 =	1	6	3	**decimal**

Table 4.4
Number System Conversions

Hex		Remainder		Binary		Remainder	
16	|163			2	|163		
	|10	3	LSB		|81	1	LSB
	|0	A	MSB		|40	1	
					|20	0	
					|10	0	
					|5	0	
					|2	1	
					|1	0	
					|0	1	MSB
A3 hex				1010 0011 binary			

4.3 Physical Representation of Binary States

4.3.1 State Levels

Most digital systems use the binary number system because many simple physical systems are most easily described by two state levels (0 and 1). For example, the two states may represent "on" and "off" or a "mark" and "space" in a communications transmission. In electronic systems, state levels are physically represented by voltages. A typical choice is

state 0 = 0 V
state 1 = 5 V

Since it is unrealistic to obtain these exact voltage values, a more practical choice is a range of values, such as

state 0 = 0.0 to 0.4 V
state 1 = 2.4 to 5.0 V

Fig 4.2 illustrates this representation of states by voltage levels. The undefined region between the two binary states is also known as the *transition region* or *noise margin*.

4.3.2 Transition Time

The gap in Fig 4.2, between binary 0 and binary 1, shows that a change in state does not occur instantly. There is a *transition time* between states. This transition time is a result of the time it takes to charge or discharge the stray capacitance in wires and other components because voltage cannot change instantaneously across a capacitor. (Stray inductance in the wires also has an effect because the current through an inductor can't change instantaneously.) The transition from a 0 to a 1 state is called the *rise time*, and is usually specified as the time for the pulse to rise from 10% of its final value to 90% of its final value. Similarly, the transition from a 1 to a 0 state is called the *fall time,* with a similar 10% to 90% definition. Note that these times need not be the same. **Fig 4.3A** shows an ideal signal, or *pulse*, with zero-time switching. Fig 4.3B shows a typical pulse, as it changes between states in a smooth curve.

Rise and fall times vary with the logic family used and the location in a circuit. Typical values of transition time are in the microsecond to nanosecond range. In a circuit, distributed inductances and capacitances in wires or

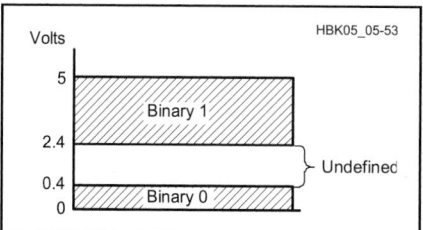

Fig 4.2 — Representation of binary states 1 and 0 by a selected range of voltage levels.

PC-board traces may cause rise and fall times to increase as the pulse moves away from the source. One reason rise and fall times may be of interest to the radio designer is because of the possibility of generating RF noise in a digital circuit.

4.3.3 Propagation Delay

Rise and fall times only describe a relationship within a pulse. For a circuit, a pulse input into the circuit must propagate through the circuit; in other words it must pass through each component in the circuit until eventually it arrives at the circuit output. The time delay between providing an input to a circuit and seeing a response at the output is the *propagation delay* and is illustrated by **Fig 4.4**.

For modern switching logic, typical propagation delay values are in the 1 to 15 nanosecond range. (It is useful to remember that the propagation delay along a wire or printed-circuit-board trace is about 1.0 to 1.5 ns per inch.) Propagation delay is the result of cumulative transition times as well as transistor switching delays, reactive element charging times and the time for signals to travel through wires. In complex circuits, different propagation delays through different paths can cause problems when pulses must arrive somewhere at exactly the same time.

The effect of these delays on digital devices can be seen by looking at the speed of the digital pulses. Most digital devices use *clock*

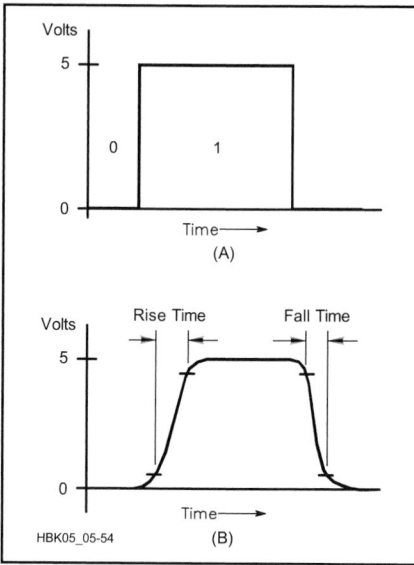

Fig 4.3 — (A) An ideal digital pulse and (B) a typical actual pulse, showing the gradual transition between states.

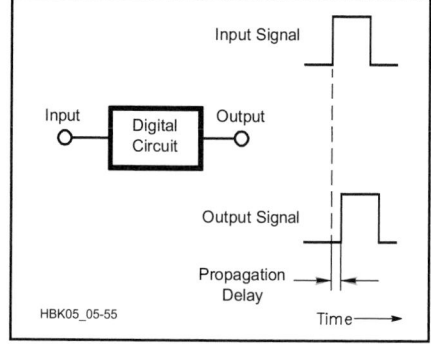

Fig 4.4 — Propagation delay in a digital circuit.

pulses. If two pulses are supposed to arrive at a logic circuit at the same time, or very close to the same time, the path length for the two signals cannot be any different than two to three inches. This can be a very significant design problem for high-speed logic designs.

4.4 Combinational Logic

Having defined a way to use voltage levels to physically represent digital numbers, we can apply digital signal theory to design useful circuits. Digital circuits combine binary inputs to produce a desired binary output or combination of outputs. This simple combination of 0s and 1s can become very powerful, implementing everything from simple switches to powerful computers.

A digital circuit falls into one of two types: combinational logic or sequential logic. In a *combinational logic* circuit, the output depends only on the *present inputs* (if we ignore propagation delay). In contrast, in a *sequential logic* circuit, the output depends on the present inputs, the *previous sequence of inputs* and often a clock signal. Later sections of this chapter will examine some circuits

built using the basics established here.

4.4.1 Boolean Algebra and the Basic Logical Operators

Combinational circuits are composed of logic gates, which perform binary operations. Logic gates manipulate binary numbers, so you need an understanding of the algebra of binary numbers to understand how logic gates operate. *Boolean algebra* is the mathematical system used to describe and design binary digital circuits. It is named after George Boole, the mathematician who developed the system. Standard algebra has a set of basic operations: addition, subtraction, multiplication and division. Similarly, Boolean algebra has a set of basic operations, called *logical operations*: NOT, AND and OR.

The function of these operators can be described by either a Boolean equation or a truth table. A Boolean *equation* describes an operator's function by representing the inputs and the operations performed on them. An equation is of the form "B = A," while an *expression* is of the form "A." In an assignment equation, the inputs and operations appear on the right and the result, or output, is assigned to the variable on the left.

A *truth table* describes an operator's function by listing all possible inputs and the corresponding outputs. Truth tables are sometimes written with Ts and Fs (for true and false) or with their respective equivalents, 1s and 0s. In company databooks (catalogs of logic devices a company manufactures), truth tables are usually written with Hs and Ls (for high and low). In the figures, 1 will mean high and 0 will mean low. This representation is called positive logic. The meaning of different logic types and why they are useful is discussed in a later section.

Each Boolean operator also has two circuit symbols associated with it. The traditional symbol — used by ARRL and other US publications — appears on top in each of the figures; for example, the triangle and bubble for the NOT function in Fig 4.7. In the traditional symbols, a small circle, or *bubble*, always represents "NOT." (This *bubble* is called a state indicator.)

Appearing just below the traditional symbol is the newer ANSI/IEEE Standard symbol. This symbol is always a square box with notations inside it. In these newer symbols, a small triangular flag represents "NOT." The new notation is an attempt to replace the detailed logic drawing of a complex function with a simpler block symbol. Adoption of the newer symbols has been spotty, and you are therefore still more likely to see the traditional symbols for basic logic functions than the ANSI/IEEE symbols.

4.4.2 Common Gates

Figs 4.5, 4.6 and 4.7 show the truth tables, Boolean algebra equations and circuit symbols for the three basic Boolean operations: AND, OR and NOT, respectively. All combinational logic functions, no matter how complex, can be described in terms of these three operators. Each truth table can be converted into words. The truth table for the two-input AND gate can be expressed as "the output C

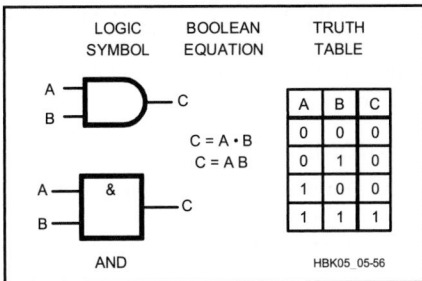
Fig 4.5 — Two-input AND gate.

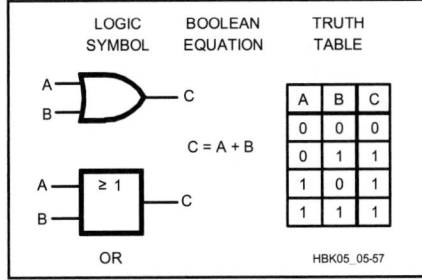
Fig 4.6 — Two-input OR gate.

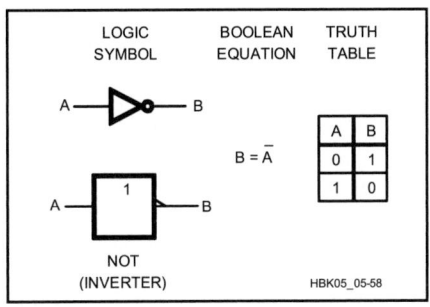
Fig 4.7 — Inverter.

is a 1 only when the inputs are both 1s." This can be seen by examining the output column C — it remains at a 0 and becomes a 1 only when the input column A and the input column B are both 1s — the last line of the table.

The NOT operation is also called *inversion, negation* or *complement*. The circuit that implements this function is called an *inverter* or *inverting buffer*. The most common notation for NOT is a bar over a variable or expression. For example, NOT A is denoted $\overline{A}$. This is read as either "Not A" or as "A bar." A less common notation is to denote Not A by A', which is read as "A prime." You will also see various other notations in schematic diagrams and component data sheets, such as a leading exclamation point or has symbol — !A or #A indicating "Not A."

While the inverting buffer and the noninverting buffer covered later have only one input and output, many combinational logic elements can have multiple inputs. When a combinational logic element has two or more inputs and one output, it is called a *gate*. (The term "gate" has a number of different but

specific technical uses. For a clarification of the many definitions of gate, see the section on **Synchronicity and Control Signals**, later in this chapter.) For simplicity, the figures and truth tables for multiple-input elements will show the operations for only two inputs, the minimum number. Remember, though, that it is quite common to have gates with more than two inputs. A three-, four-, or eight-input gate works in the exact same manner as a two-input gate.

The output of an AND function is 1 only if *all* of the inputs are 1. Therefore, if *any* of the inputs are 0, then the output is 0. The notation for an AND is either a dot (•) between the inputs, as in C = A•B, or nothing between the inputs, as in C = AB. Read these equations as "C equals A AND B."

The OR gate detects if one or more inputs are 1. In other words, if *any* of the inputs are 1, then the output of the OR gate is 1. Since this includes the case where more than one input may be 1, the OR operation is also known as an INCLUSIVE OR. The OR operation detects if *at least one* input is 1. Only if all the inputs are 0, then the output is 0. The notation for an OR is a plus sign (+) between the inputs, as in C = A + B. Read this equation as "C equals A OR B."

4.4.3 Additional Gates

More complex logical functions are derived from combinations of the basic logical operators. These operations — NAND, NOR, XOR and the noninverter or buffer — are illustrated in **Figs 4.8** through **4.11**, respectively. As before, each is described by a truth table, Boolean algebra equation and circuit symbols. Also as before, except for the noninverter, each could have more inputs than the two illustrated.

The NAND gate (short for NOT AND) is equivalent to an AND gate followed by a NOT gate. Thus, its output is the complement of the AND output: The output is a 0 only if all the inputs are 1. If any of the inputs is 0, then the output is a 1.

The NOR gate (short for NOT OR) is equivalent to an OR gate followed by a NOT gate. Thus, its output is the complement of the OR output: If any of the inputs are 1, then the output is a 0. Only if all the inputs are 0, then the output is a 1.

The operations so far enable a designer to determine two general cases: (1) if *all* inputs have a desired state or (2) if *at least one* input has a desired state. The XOR and XNOR gates enable a designer to determine if *one and only one* input of a desired state is present.

The XOR gate (read as EXCLUSIVE OR) is a combination of an OR and a NAND gate. It has an output of 1 if one and only one of the inputs is a 1 state. The output is 0 otherwise. The symbol for XOR is ⊕. This is easy to remember if you think of the "+" OR symbol

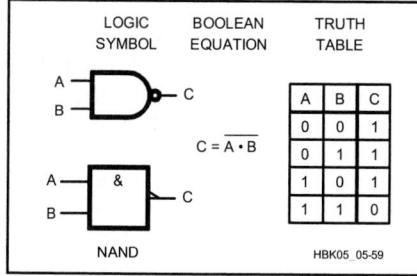

Fig 4.8 — Two-input NAND gate.

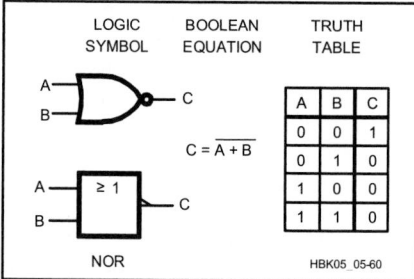

Fig 4.9 — Two-input NOR gate.

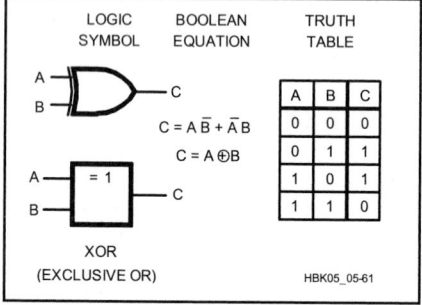

Fig 4.10 — Two-input XOR gate.

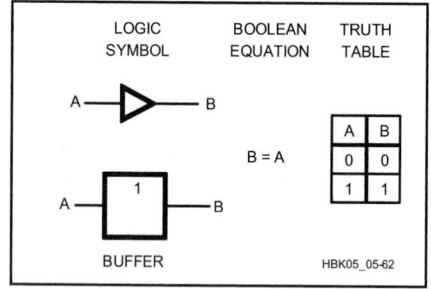

Fig 4.11 — Noninverting buffer.

enclosed in an "O" for *only one*.

The XOR gate is also known as a "half adder," because in binary arithmetic it does everything but the "carry" operation. The following examples show the possible binary additions for a two-input XOR.

A	0	0	1	1
B	0	1	0	1
Sum	0	1	1	0

The XNOR gate (read as EXCLUSIVE

NOR) is the complement of the XOR gate. The output is 0 if one and only one of the inputs is a 1. The output is 1 either if all inputs are 0 or more than one input is 1.

NONINVERTERS (BUFFERS)

A *noninverter*, also known as a *buffer*, *amplifier* or *driver*, at first glance does not seem to do anything. It simply receives an input and produces the same output. In reality, it is changing other properties of the signal in a useful fashion, such as amplifying the current level. While not useful for logical operations, applications of a noninverter include providing sufficient current to drive a number of gates or some other circuit such as a relay; interfacing between two logic families; obtaining a desired pulse rise time; and providing a slight delay to make pulses arrive at the proper time.

TRI-STATE GATES

Under normal circumstances, a logic element can drive or feed several other logic elements. A typical AND gate might be able to drive or feed 10 other gates. This is known as *fan-out*. However, with certain exceptions only one gate output can be connected to a single wire. If you have two possible driving sources to feed one particular wire, some logic network that probably includes a number OR gates must be used.

In many applications, including computers, data is routed internally on a set of wires called *buses*. The data on the bus can come from many circuits or drivers, and many other devices may be *listening* on the bus. To eliminate the need for the network of OR gates to drive each bus wire, a set of gates known as *tri-state* gates are used.

The symbol and truth table for a tri-state gate are shown in **Fig 4.12**. A tri-state gate can be any of the common gates previously described, but with one additional control lead. When this lead is enabled (it can be designed to allow either a 0 or a 1 to enable it) the gate operates normally, according to the truth table for that type of gate. However, when the gate is not enabled, the output goes to a high impedance (Hi-Z), and so far as the output wire is concerned, the gate does not exist.

Each device that has to send data down a bus wire is connected to the bus wire through a tri-state gate. However, as long as only one device, through its tri-state gate, is enabled, it is as though all the other connected tri-state gates do not exist.

4.4.4 Boolean Theorems

The analysis of a circuit starts with a logic diagram and then derives a circuit description. In digital circuits, this description is in the form of a truth table or logical equation. The

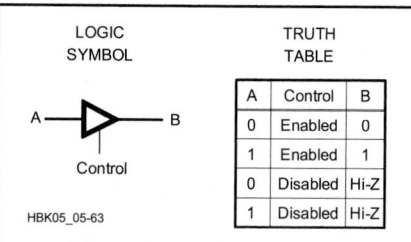

Fig 4.12 — Tri-State gate.

synthesis, or design, of a circuit goes in the reverse: starting with an informal description, determining an equation or truth table and then expanding the truth table to components that will implement the desired response. In both of these processes, we need to either simplify or expand a complex logical equation.

To manipulate an equation, we use mathematical *theorems*. Theorems are statements that have been proven to be true. The theorems of Boolean algebra are very similar to those of standard algebra, such as commutivity and associativity. Proofs of the Boolean algebra theorems can be found in an introductory digital design textbook.

BASIC THEOREMS

Table 4.5 lists the theorems for a single variable and **Table 4.6** lists the theorems for two or more variables. These tables illustrate the *principle of duality* exhibited by the Boolean theorems: Each theorem has a dual in which, after swapping all ANDs with ORs and all 1s with 0s, the statement is still true.

The tables also illustrate the *precedence* of the Boolean operations: the order in which operations are performed when not specified by parenthesis. From highest to lowest, the precedence is NOT, AND then OR. For example, the distributive law includes the expression "A + B•C." This is equivalent to

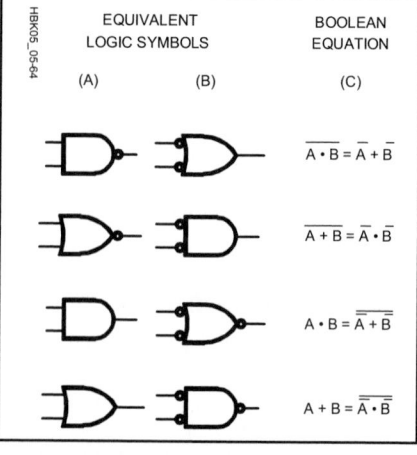

Fig 4.13 — Equivalent gates from DeMorgan's Theorem: Each gate in column A is equivalent to the opposite gate in column B. The Boolean equations in column C formally state the equivalences.

"A + (B•C)." The parenthesis around (B•C) can be left out since an AND operation has higher priority than an OR operation. Precedence for Boolean algebra is similar to the convention of standard algebra: raising to a power, then multiplication, then addition.

DeMORGAN'S THEOREM

One of the most useful theorems in Boolean algebra is DeMorgan's Theorem:

$\overline{A \bullet B} = \overline{A} + \overline{B}$

and its dual

$\overline{A + B} = \overline{A} \bullet \overline{B}$.

Table 4.5
Boolean Algebra Single Variable Theorems

Identities:	$A \bullet 1 = A$	$A + 0 = A$
Null elements:	$A \bullet 0 = 0$	$A + 1 = 1$
Idempotence:	$A \bullet A = A$	$A + A = A$
Complements:	$A \bullet \overline{A} = 0$	$A + \overline{A} = 1$
Involution:	$\overline{(A)} = A$	

The truth table in **Table 4.7** proves these statements. DeMorgan's Theorem provides a way to simplify the complement of a large expression. It also enables a designer to interchange a number of equivalent gates, as shown by **Fig 4.13**.

The equivalent gates show that the duality principle works with symbols the same as it does for Boolean equations: just swap ANDs with ORs and switch the bubbles. For example, the NAND gate — an AND gate followed by an inverter bubble — becomes an OR gate preceded by two inverter bubbles. DeMorgan's Theorem is important because it means any logical function can be implemented using either inverters and AND gates or inverters and OR gates. Also, the ability to change placement of the bubbles using DeMorgan's Theorem is useful in dealing with mixed logic, to be discussed next.

POSITIVE AND NEGATIVE LOGIC

The truth tables shown in the figures in this chapter are drawn for positive logic. In *positive logic*, or *high true*, a higher voltage means true (logic 1) while a lower voltage means false (logic 0). This is also referred to as *active high*: a signal performs a named action or denotes a condition when it is "high" or 1. In *negative logic*, or *low true*, a lower voltage means true (1) and a higher voltage means false (0). An *active low* signal performs an action or denotes a condition when it is "low" or 0.

In both logic types, true = 1 and false = 0; but whether true means high or low differs. Company databooks are drawn for general truth tables: an H for high and an L for low. (Some tables also have an X for a "don't care" state. "Don't care" means that the output does not depend on the state of that variable.) The function of the table can differ depending on whether it is interpreted for positive logic or negative logic.

Device data sheets often show positive logic convention, or positive logic is assumed. However, a signal into an IC is represented with a bar above it, indicating that the "enable" on that wire is active low — it does *not* mean negative logic (0 V = a logical 1) is used! Similarly a bubble on the input of a logic element also usually means active low. These can be sources of confusion.

Fig 4.14 shows how a general truth table differs when interpreted for different logic types. The same truth table gives two equivalent gates: positive logic gives the function of

Table 4.6
Boolean Algebra Multivariable Theorems

Commutativity:	$A \bullet B = B \bullet A$ $A + B = B + A$
Associativity:	$(A \bullet B) \bullet C = A \bullet (B \bullet C)$ $(A + B) + C = A + (B + C)$
Distributivity:	$(A + B) \bullet (A + C) = A + B \bullet C$ $A \bullet B + A \bullet C = A \bullet (B + C)$
Covering:	$A \bullet (A + B) = A$ $A + A \bullet B = A$
Combining:	$(A + B) \bullet (A + \overline{B}) = A$ $A \bullet B + A \bullet B = A$
Consensus:	$A \bullet B + \overline{A} \bullet C + B \bullet C = A \bullet B + \overline{A} \bullet C$ $(A + B) \bullet (\overline{A} + C) \bullet (B + C) = (A + B) \bullet (\overline{A} + C)$ $A + \overline{A}B = A + B$

A	B	C
L	L	H
L	H	H
H	L	H
H	H	L

A	B	C
0	0	1
0	1	1
1	0	1
1	1	0

A	B	C
1	1	0
1	0	0
0	1	0
0	0	1

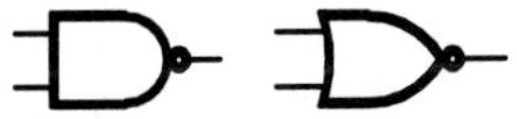

HBK0667

(A) (B) True = 1 = High (C) True = 1 = Low

Fig 4.14 — (A) A general truth table, (B) a truth table and NAND symbol for positive logic and (C) a truth table and NOR symbol for negative logic.

Table 4.7
DeMorgan's Theorem

(A) $\overline{A \bullet B} = \overline{A} + \overline{B}$
(B) $\overline{A + B} = \overline{A} \bullet \overline{B}$
(C)

(1)	(2)	(3)	(4)	(5)	(6)	(7)	(8)	(9)	(10)
A	B	$\overline{A}$	$\overline{B}$	$A \bullet B$	$\overline{A \bullet B}$	$A + B$	$\overline{A + B}$	$\overline{A} \bullet \overline{B}$	$\overline{A} + \overline{B}$
0	0	1	1	0	1	0	1	1	1
0	1	1	0	0	1	1	0	0	1
1	0	0	1	0	1	1	0	0	1
1	1	0	0	1	0	1	0	0	0

(A) and (B) are statements of DeMorgan's Theorem. The truth table at (C) is proof of these statements: (A) is proven by the equivalence of columns 6 and 10 and (B) by columns 8 and 9.

a NAND gate while negative logic gives the function of a NOR gate.

Note that these gates correspond to the equivalent gates from DeMorgan's Theorem. A bubble on an input or output terminal indicates an active low device. The absence of bubbles indicates an active high device.

Like the bubbles, signal names can be used to indicate logic states. These names can aid the understanding of a circuit by indicating control of an action (GO, /ENABLE) or detection of a condition (READY, /ERROR). The action or condition occurs when the signal is in its active state. When a signal is in its active state, it is called *asserted*; a signal not in its active state is called *negated* or *deasserted*.

A prefix can easily indicate a signal's active state. Active low signals are preceded by a symbol such as /, |, ! or # (for example /READY or !READY). Active low signals are also denoted by an overscore, such as $\overline{CL}$. Active high signals have no prefix or overscore. As an example, see the truth table for a flip-flop later in this chapter. Standard practice is that the signal name and input pin match (have the same active level). For example, an input with a bubble (active low) may be called /READY, while an input with no bubble (active high) is called READY. Output signal names should always match the device output pin.

In this chapter, positive logic is used unless indicated otherwise. Although using mixed logic can be confusing, it does have some advantages. Mixed logic combined with DeMorgan's Theorem can promote more effective use of available gates. Also, well-chosen signal names and placement of bubbles can promote more understandable logic diagrams.

4.5 Sequential Logic

The previous section discussed combinational logic, whose outputs depend only on the present inputs. In contrast, in *sequential logic* circuits, the new output depends not only on the present inputs but also on the present outputs. The present outputs depended on the previous inputs and outputs and those earlier outputs depended on even earlier inputs and outputs and so on. Thus, the present outputs depend on the previous *sequence of inputs* and the system has *memory*. Having the outputs become part of the new inputs is known as *feedback*.

4.5.1 Synchronicity and Control Signals

When a combinational circuit is given a set of inputs, the outputs take on the expected values after a propagation delay during which the inputs travel through the circuit to the output. In a sequential circuit, however, the travel through the circuit is more complicated. After application of the first inputs and one propagation delay, the outputs take on the resulting state; but then the outputs start trickling back through and, after a second propagation delay, new outputs appear. The same happens after a third propagation delay. With propagation delays in the nanosecond range, this cycle around the circuit is rapidly and continually generating new outputs. A user needs to know when the outputs are valid.

There are two types of sequential circuits: synchronous circuits and asynchronous circuits, which are analyzed differently for valid outputs. In *asynchronous* operation, the outputs respond to the inputs immediately after the propagation delay. To work properly, this type of circuit must eventually reach a *stable* state: the inputs and the fed back outputs result in the new outputs staying the same. When the nonfeedback inputs are changed, the feedback cycle needs to eventually reach a new stable state. Generally, the output of this type of logic is not valid until the last input has changed, and enough time has elapsed for all propagation delays to have occurred.

In *synchronous* operation, the outputs change state only at specific times. These times are determined by the presence of a particular input signal: a clock, toggle, latch or enable. Synchronicity is important because it ensures proper timing: all the inputs are present where needed when the control signal causes a change of state.

CONTROL SIGNALS

Some authors vary the meanings slightly for the different control signals. The following is a brief illustration of common uses, as well as showing uses for noun, verb and adjective. *Enabling* a circuit generally means the control signal goes to its asserted level, allowing the circuit to change state. *Latch* implies memory: a latch circuit can store a bit of information. A latch signal can cause a circuit to keep its present state indefinitely. *Gate* can have several meanings, some unrelated to synchronous control. For example, a gate can be a signal used to trigger the passage of other signals through a circuit. A gate can also be a logic circuit with two or more inputs and one output, as used earlier in this chapter. Of course, "gate" can also be one of the electrodes of an FET as described in another chapter. To *toggle* means a signal changes state, from 1 to 0 or vice versa. A *clock* signal is one that toggles at a regular rate.

Clock control is the most common method of synchronizing logic circuits, so it has some additional terms as illustrated by **Fig 4.15**. The *clock period* is the time between successive transitions in the same direction; the *clock frequency* is the reciprocal of the period. A *pulse* or *clock tick* is the first edge in a clock period, or sometimes the period itself or the first half of the period. The *duty cycle* is the percentage of time that the clock signal is at its asserted level. A common application of the use of clock pulses is to limit the input to a logic circuit such that the circuit is only enabled on one clock phase; that is the inputs occur before the clock changes to a logic 1.

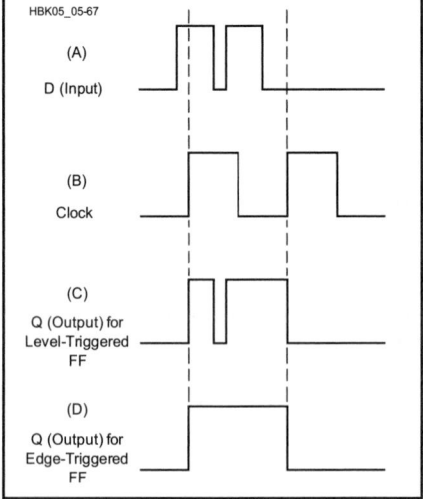

Fig 4.16 — Level-triggered vs edge-triggered for a D flip-flop: (A) D input, (B) clock input, (C) output Q for level-triggered: circuit responds whenever clock is 1. (D) output Q for edge-triggered: circuit responds only at rising edge of clock. Notice that the short negative pulse on the D input is not reproduced by the edge-triggered flip-flop.

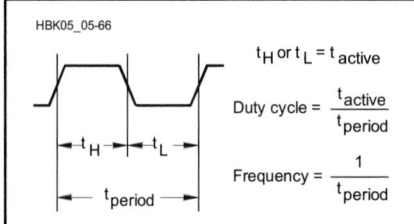

Fig 4.15 — Clock signal terms. The duty cycle would be t_H / t_{PERIOD} for an active high signal and t_L / t_{PERIOD} for an active low signal.

The outputs are sampled only after this point; perhaps when the clock next changes back to a logic 0.

The reaction of a synchronous circuit to its control signal is *static* or *dynamic*. Static, *gated* or *level-triggered* control allows the circuit to change state whenever the control signal is at its active or asserted level. Dynamic, or *edge-triggered*, control allows the circuit to change state only when the control signal *changes* from unasserted to asserted. By convention, a control signal is active high if state changes occur when the signal is high or at the rising edge and active low in the opposite case. Thus, for positive logic, the convention is enable = 1 or enable goes from 0 to 1. This transition from 0 to 1 is called *positive edge-triggered* and is indicated by a small triangle inside the circuit box. A circuit responding to the opposite transition, from 1 to 0, is called *negative edge-triggered*, indicated by a bubble with the triangle.

Whether a circuit is level-triggered or edge-triggered can affect its output, as shown by **Fig 4.16**. The D input includes a very brief pulse, called a *glitch*, which may be caused by noise. The differing results at the output illustrate how noise can cause errors. We have both edge and level triggered circuits available so that we can meet the requirements of our particular design.

4.5.2 Flip-Flops

Flip-flops are the basic building blocks of sequential circuits. A *flip-flop* is a device with two stable states: the *set* state (1) and the *reset* or *cleared* state (0). The flip-flop can be placed in one or the other of the two states by applying the appropriate input. Since a common use of flip-flops is to store one bit of information, some use the term *latch* interchangeably with flip-flop. A set of latches, or flip-flops holding an n-bit number is called a *register*. While gates have special symbols, the schematic symbol for most sequential logic components is a rectangular box with the circuit name or abbreviation, the signal names and assertion bubbles. For flip-flops, the circuit name is usually omitted since the signal names are enough to indicate a flip-flop and its type. The four basic types of flip-flops are the S-R, D, T and J-K. The most common flip-flops available to Amateurs today are the J-K and D- flip-flops; the others can be synthesized if needed by utilizing these two varieties.

TRIGGERING A FLIP-FLOP

Although the S-R (Set-reset) flip flop is no longer generally available or used, it does provide insight in basic flip-flop operations and triggering. It is also not uncommon to build S-R flip-flops out of gates for jobs such as switch contact debouncing. In **Fig 4.17** the

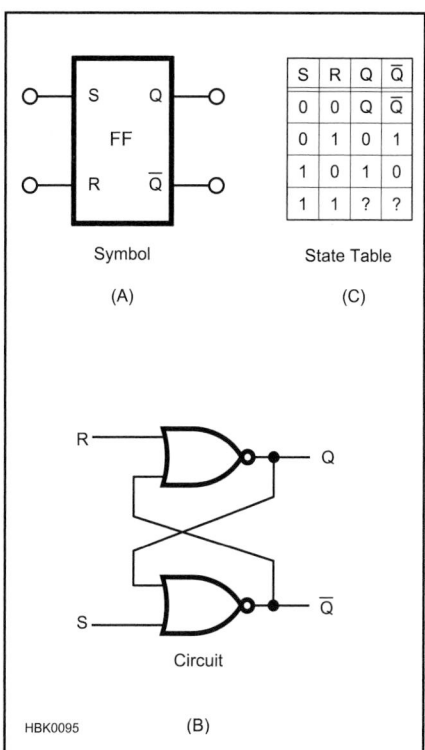

Fig 4.17 — Unclocked S-R Flip-Flop. (A) schematic symbol. (B) circuit diagram. (C) state table or truth table.

symbol for an S-R flip flop and its truth table are accompanied by a logic implementation, using NAND gates. As the truth table shows, this basic implementation requires a positive or logic 1 input on the set input to put the flip-flop in the Q or set state. Remove the input, and the flip flop stays in the Q state, which is what is expected of a flip-flop. Not until the S input receives a logic 1 input does the flip-flop change state and go to the reset or Q=0 state.

Note that the input can be a short pulse or a level; as long as it is there for some minimum duration (established by the propagation delay of the gates used), the flip-flop will respond. By contrast the clocked S-R flip-flop in **Fig 4.18** requires both a positive level to be present at either the S or R inputs and a positive clock pulse. The clock pulse is ANDed with the S or R input to trigger the flip–flop. In this case the flip-flop shown is implemented with a set of NOR gates.

A final triggering method is edge triggering. Here, instead of using the clock pulse as shown in the timing diagram of Fig 4.18, just the edge of the clock pulse is used. The edge-triggered flip-flop helps solves a problem with noise. Edge-triggering minimizes the time during which a circuit responds to its inputs: the chance of a glitch occurring during

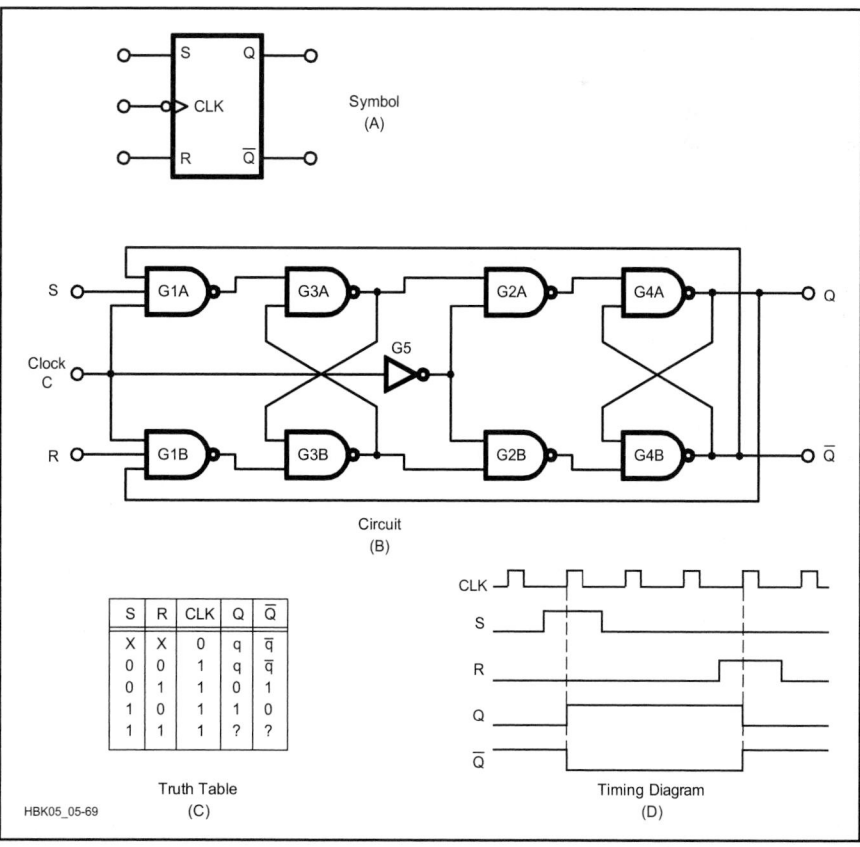

Fig 4.18 — Master-Slave Flip-Flop. (A) logic symbol. (B) NAND gate implementation. (C) truth table. (D) timing diagram.

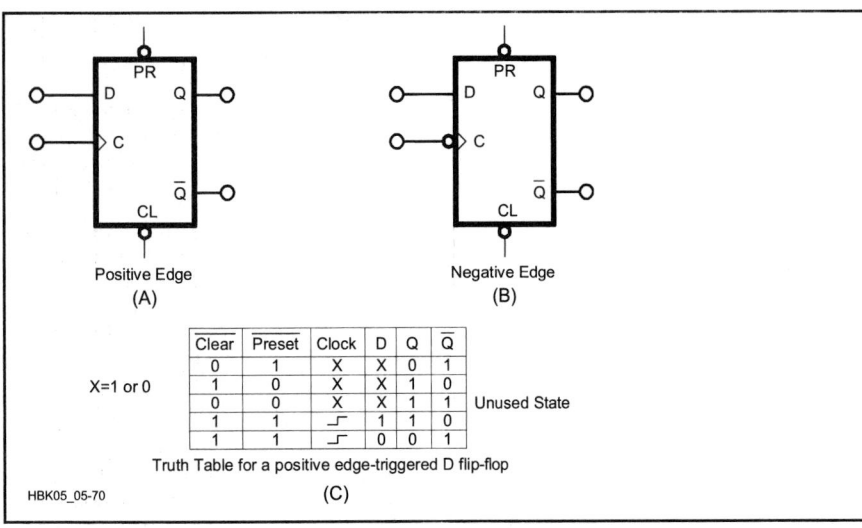

Clear	Preset	Clock	D	Q	$\overline{Q}$	
0	1	X	X	0	1	
1	0	X	X	1	0	
0	0	X	X	1	1	Unused State
1	1	⌐	1	1	0	
1	1	⌐	0	0	1	

X=1 or 0

Truth Table for a positive edge-triggered D flip-flop

HBK05_05-70

(C)

Fig 4.19 — (A & B) The D flip-flop. (C) A truth table for the positive edge-triggered D flip-flop.

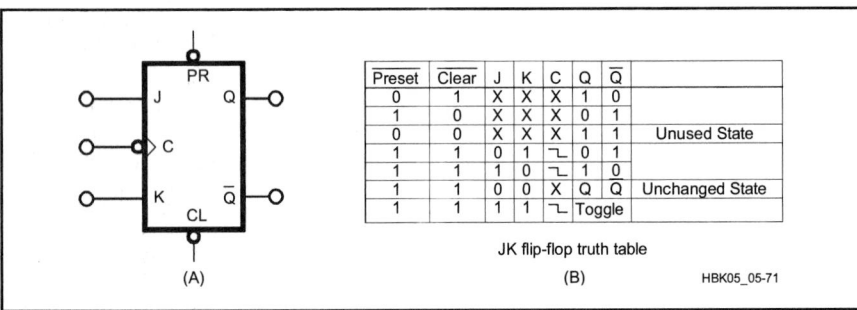

Preset	Clear	J	K	C	Q	$\overline{Q}$	
0	1	X	X	X	1	0	
1	0	X	X	X	0	1	
0	0	X	X	X	1	1	Unused State
1	1	0	1	⌐	0	1	
1	1	1	0	⌐	1	0	
1	1	0	0	X	Q	Q	Unchanged State
1	1	1	1	⌐	Toggle		

JK flip-flop truth table

(A)   (B)   HBK05_05-71

Fig 4.20 — (A) JK flip-flop. (B) JK flip-flop truth table.

the nanosecond transition of a clock pulse is remote. A side benefit of edge-triggering is that only one new output is produced per clock period. Edge-triggering is denoted by a small rising-edge or falling-edge symbol in the clock column of the flip-flop's truth table. It can also appear, instead of the clock triangle, inside the schematic symbol.

MASTER/SLAVE FLIP-FLOP

One major problem with the simple flip-flop shown up to now is the question of when is there a valid output. Suppose a flip-flop receives input that causes it to change state; at the same time the output of this flip-flop is being sampled to control some other logic element. There is a real risk here that the output will be sampled just as it is changing and thus the validity of the output is questionable.

A solution to this problem is a circuit that samples and stores its inputs before changing its outputs. Such a circuit is built by placing two flip-flops in series; both flip-flops are triggered by a common clock but an inverter on the second flip-flop's clock input causes it to be asserted only when the first flip-flop is not asserted. The action for a given clock pulse is as follows: The first, or master, flip-flop can

change only when the clock is high, sampling and storing the inputs. The second, or slave, flip-flop gets its input from the master and changes when the clock is low. Hence, when the clock is 1, the input is sampled; then when the clock becomes 0, the output is generated. Note that a bubble may appear on the schematic symbol's clock input, reminding us that the output appears when the clock is asserted low. This is conventional for TTL-style J-K flip-flops, but it can be different for CMOS devices.

The master/slave method isolates output changes from input changes, eliminating the problem of series-fed circuits. It also ensures only one new output per clock period, since the slave flip-flop responds to only the single sampled input. A problem can still occur, however, because the master flip-flop can change more than once while it is asserted; thus, there is the potential for the master to sample at the wrong time. There is also the potential that either flip-flop can be affected by noise.

A master-slave, S-R clocked input flip-flop synthesized from NAND gates, Fig 4.18B, is accompanied by its logic symbol, Fig 4.18A. From the logic symbols you can tell that the

output changes on a negative-going clock edge.

G3A and G3B form the master set-rest flip-flop, and G4A and G4B the slave flip flop. The input signals S and R are controlled by the positive going edge of the clock through gates G1A and G1B. G2A and G2B control the inputs into the slave flip-flop; these inputs are the outputs of the master flip-flop. Note G5 inverts the clock; thus while the positive-going edge places new data into the master flip-flop, the other edge of the clock transfers the output of the master into the slave on the following negative clock edge.

D FLIP-FLOP

In a D (data) flip-flop, the *data* input is transferred to the outputs when the flip-flop is enabled. The logic level at the D input is transferred to Q when the clock is positive; the Q output retains this logic level until the next positive clock pulse (see **Fig 4.19**). The truth table summarizes this operation. If D = 1 the next clock pulse makes Q = 1. If D = 0, the next clock pulse makes Q = 0. A D flip-flop is useful to store one bit of information. A collection of D flip-flops forms a register.

J-K FLIP-FLOP

The J-K flip-flop, shown schematically in **Fig 4.20A**, has five inputs. The unit shown uses both positive active inputs (the J and K inputs) and negative active inputs (note the bubbles on the C or clock, PR or preset and CL or clear inputs). With these inputs almost any other type of flip-flop may be synthesized.

The truth table of Fig 4.20B provides an explanation. Lines (rows) 1 and 2 show the preset and clear inputs and their use. These are active low, meaning that when one (and only one) of them goes to a logic 0, the flip-flop responds, just as if it was a S-R or set-reset flip-flop. Make PR a logic 0, and leave CL a logic 1, and the flip-flop goes into the Q = 1 state (line 1). Do the reverse (line 2) – PR = 1, CL = 0 and the flip-flop goes into a Q' = 1 state. When these two inputs are used, J, K and C are marked as X or don't care, because the PR and CL inputs override them. Line 3 corresponds to the unused state of the R-S flip-flop.

Line 5 shows that if J = 1 and K = 0, the next clock transition from high to low sets Q = 1 and Q' = 0. Alternately, line 4 shows J = 0 and K = 1 sets Q = 0 and Q' = 1. Therefore if a signal is applied to J, and the inverted signal sent to K, the J-K flip-flop will mimic a D flip-flop, echoing its input.

The most unique feature of the J-K flip-flop is line 7. If both J and K are connected to a 1, then each clock 1 to 0 transition will flip or toggle the flop-flop. Thus the J-K flip-flop can be used as a T flip-flop, as in a ripple counter (see the following **Counters** section.)

Table 4.8 summarizes D and J-K flip-flops.

Table 4.8

Summary of Standard Flip-Flops

q = current state   Q = next state   X = don't care

Flip-Flop Type	Symbol	Truth Table	Characteristic Equation	Excitation Table
D		D CLK | Q X ⌐ | q 0 ⌐ | 0 1 ⌐ | 1	$Q = D \cdot CLK$	q Q | D CLK 0 0 | X ⌐ 0 1 | 1 ⌐ 1 0 | 0 ⌐ 1 1 | X ⌐
J K		J K CLK | Q X X ⌐ | q 0 0 ⌐ | q 0 1 ⌐ | 0 1 0 ⌐ | 1 1 1 ⌐ | t	$Q = (J \cdot \bar{q} + \bar{K} \cdot q) \cdot CLK$ Positive Edge Clock	q Q | J K CLK 0 0 | 0 X ⌐ 0 1 | 1 X ⌐ 1 0 | X 1 ⌐ 1 1 | X 0 ⌐
J K		J K CLK | Q 0 0 ⌐ | 0 0 1 ⌐ | 0 1 0 ⌐ | 1 1 1 ⌐ | t	$Q = (J \cdot \bar{q} + \bar{K} \cdot q) \cdot CLK$ Negative Edge Clock	q Q | J K CLK 0 0 | 0 X ⌐ 0 1 | 1 X ⌐ 1 0 | X 1 ⌐ 1 1 | X 0 ⌐

HBK0668   t: If J = K, the clock toggles the flip-flop

4.5.3 Groups of Flip-Flops

COUNTERS

Groups of flip-flops can be combined to make counters. Intuitively, a counter is a circuit that starts at state 0 and sequences up through states 1, 2, 3, to m, where m is the maximum number of states available. From state m, the next state will return the counter to 0. This describes the most common counter: the *n-bit binary counter*, with n outputs corresponding to 2^n = m states. Such a counter can be made from n flip-flops, as shown in **Fig 4.21**. This figure shows implementations for each of the types of synchronicity. Both circuits pass the data count from stage to stage. In the asynchronous counter, Fig 4.21A, the clock is also passed from stage to stage and the circuit is called *ripple* or *ripple-carry*.

The J-K flip-flop truth table shows that with PR (Preset) and CL (Clear) both positive, and therefore not effecting the operation, the flip-flop will toggle if J and K are tied to a logic 1. In Fig 4.21A the first stage has its J and K inputs permanently tied to a logic 1, and each succeeding stage has its J and K inputs tied to Q of the proceeding stage. This provides a direct ripple counter implementation.

Design of a synchronous counter is bit more involved. It consists of determining, for a particular count, the conditions that will make the next stage change at the same clock edge when all the stages are changing.

To illustrate this, notice the binary counting table of Fig 4.21. The right-hand column represents the lowest stage of the counter. It alternates between 1 and 0 on every line. Thus, for the first stage the J and K inputs are tied to logic 1. This provides the alternation required by the counting table.

The middle column or second stage of the counter changes state right after the lower stage is a 1 (lines C, E and G). Thus if the Q output of the lowest stage is tied to the J and K inputs of the second stage, each time the output of the lowest stage is a 1 the second stage toggles on the next clock pulse.

Finally, the third column (third stage) toggles when both the first stage and the second stage are both 1s (line D). Thus by ANDing the Q outputs of the first two stages, and then connecting them to the J and K inputs of the third stage, the third stage will toggle whenever the first two stages are 1s.

There are formal methods for determining the wiring of synchronous counters. The illustration above is one manual method that may be used to design a counter of this type.

The advantage of the synchronous counter is that at any instant, except during clock pulse transition, all counter stage outputs *are correct* and delay due to propagation through the flip flops is not a problem.

In the synchronous counter, Fig 4.21B, each stage is controlled by a common clock signal.

There are numerous variations on this first example of a counter. Most counters have the ability to *clear* the count to 0. Some counters can also *preset* to a desired count. The clear and preset control inputs are often asynchronous — they change the output state without being clocked. Counters may either count up (increment) or down (decrement). *Up/down* counters can be controlled to count in either direction. Counters can have sequences other than the standard numbers, for example a BCD counter.

Counters are also not restricted to changing state on every clock cycle. An n-bit counter that changes state only after m clock pulses is called a *divider* or *divide-by-m* counter. There are still 2^n = m states; however, the output after p clock pulses is now p / m. Combining different divide-by-m counters can result in almost any desired count. For example, a base 12 counter can be made from a divide-by-2 and a divide-by-6 counter; a base 10 (decade) counter consists of a divide-by-2 and a BCD divide-by-5 counter.

The outputs of these counters are binary. To produce output in decimal form, the output of a counter would be provided to a binary-to-decimal decoder chip and/or an LED display.

REGISTERS

Groups of flip-flops can be combined to make registers, usually implemented with D flip-flops. A *register* stores n bits of information, delivering that information in response to a clock pulse. Registers usually have asynchronous *set* to 1 and *clear* to 0 capabilities.

Storage Register

A storage register simply stores temporary information, for example, incoming information or intermediate results. The size is related to the basic size of information handled by a computer: 8 flip-flops for an 8-bit or *byte register* or 16 bits for a *word register*. **Fig 4.22** shows a typical circuit and schematic symbols for an 8-bit storage register.

Shift Register

Shift registers also store information and provide it in response to a clock signal, but they handle their information differently: When a clock pulse occurs, instead of each flip-flop passing its result to the output, the flip-flops pass their data to each other, up and down the row. For example, in up mode, each flip-flop receives the output of the preceding flip-flop. A data bit starting in flip-flop D0 in a left

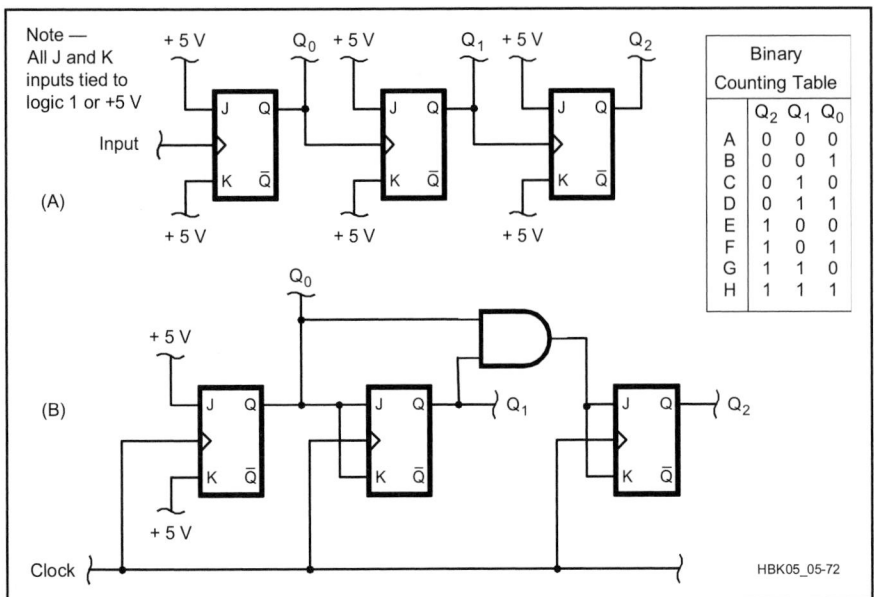

	Binary Counting Table		
	Q_2	Q_1	Q_0
A	0	0	0
B	0	0	1
C	0	1	0
D	0	1	1
E	1	0	0
F	1	0	1
G	1	1	0
H	1	1	1

Fig 4.21 — Three-bit binary counter using J-K flip-flops: (A) asynchronous or ripple, (B) synchronous.

shifter would move to D1, then D2 and so on until it is shifted out of the register. If a 0 was input to the least significant bit, D0, on each clock pulse then, when the last data bit has been shifted out, the register contains all 0s.

Shift registers can be left shifters, right shifters or controlled to shift in either direction. The most general form, a *universal shift register*, has two control inputs for four states: Hold, Shift right, Shift left and Load. Most also have asynchronous inputs for preset, clear and parallel load. The primary use of shift registers is to convert parallel information to serial or vice versa. Additional uses for a shift register are to delay or synchronize data, and to multiply or divide a number by a factor 2^n. Data can be delayed simply by

taking advantage of the Hold feature of the register control inputs. Multiplication and division with shift registers is best explained by example: Suppose a 4-bit shift register currently has the value $1000 = 8$. A right shift results in the new parallel output $0100 = 4 = 8 / 2$. A second right shift results in $0010 = 2 = (8 / 2) / 2$. Together the 2 right shifts performed a division by 2^2. In general, shifting right n times is equivalent to dividing by 2^n. Similarly, shifting left multiplies by 2^n. This can be useful to compiler writers to make a computer program run faster.

4.5.4 Multivibrators

Multivibrators are a general type of circuit

with three varieties: bistable, monostable and astable. The only truly digital multivibrator is bistable, having two stable states. The flip-flop is a *bistable multivibrator*: both of its two states are stable; it can be triggered from one stable state to the other by an external signal. The other two varieties of multivibrators are partly analog circuits and partly digital. While their output is one or more pulses, the internal operation is strictly analog.

MONOSTABLE MULTIVIBRATOR

A *monostable* or *one-shot* multivibrator has one energy-storing element in its feedback paths, resulting in one stable and one quasi-stable state. It can be switched, or *triggered*, to its quasi-stable state; then returns to the stable state after a time delay. Thus, when triggered, the one-shot multivibrator puts out a pulse of some duration, T.

A very common integrated circuit used for non-precision generation of a signal pulse is the 555 timer IC. **Fig 4.23** shows a 555 connected as a one-shot multivibrator. The one-shot is activated by a negative-going pulse between the trigger input and ground. The trigger pulse causes the output (Q) to go positive and capacitor C to charge through resistor R. When the voltage across C reaches two-thirds of V_{CC}, the capacitor is quickly discharged to ground and the output returns to 0. The output remains at logic 1 for a time determined by

$$T = 1.1\, RC$$

where:
R = resistance in ohms, and
C = capacitance in farads.

A very common, but again, non-precision application of this circuit is the generation of a delayed pulse. If there is a requirement

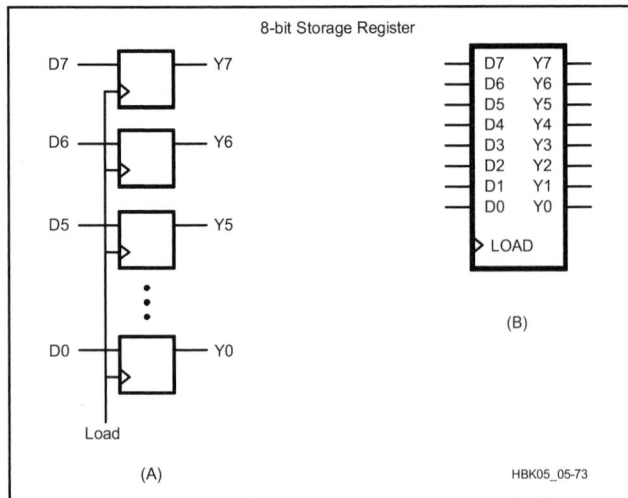

Fig 4.22 — An eight-bit storage register: (A) circuit and (B) schematic symbol.

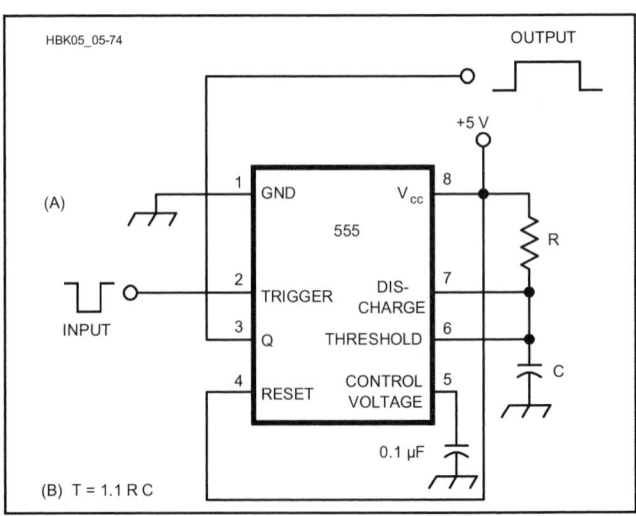

Fig 4.23 — (A) A 555 timer connected as a monostable multivibrator. (B) The equation to calculate values where T is the pulse duration in seconds, R is in ohms and C is in farads.

to generate a 50 μs pulse, but delayed from a trigger by 20 ms, two 555s might be used. The first 555, configured as an astable multivibrator, generates the 20-ms pulse, and the trailing edge of the 20-ms pulse is used to trigger a second 555 that in turn generates the 10 μs pulse. See the **Analog Basics** chapter for more information on the 555 timer and related circuits.

ASTABLE MULTIVIBRATOR

An *astable* or *free-running* multivibrator has two energy-storing elements in its feedback paths, resulting in two quasi-stable states. It continuously switches between these two states without external excitation. Thus, the astable multivibrator puts out a sequence of pulses. By properly selecting circuit components, these pulses can be of a desired frequency and width.

Fig 4.24 shows a 555 timer IC connected as an astable multivibrator. The capacitor C charges to two-thirds V_{CC} through R1 and R2 and discharges to one-third V_{CC} through R2. The ratio R1 : R2 sets the asserted high duty cycle of the pulse: t_{HIGH} / t_{PERIOD}. The output frequency is determined by:

$$f = \frac{1.46}{(R1 + 2\,R2)\,C}$$

where:
R1 and R2 are in ohms,
C is in farads and
f is in hertz.

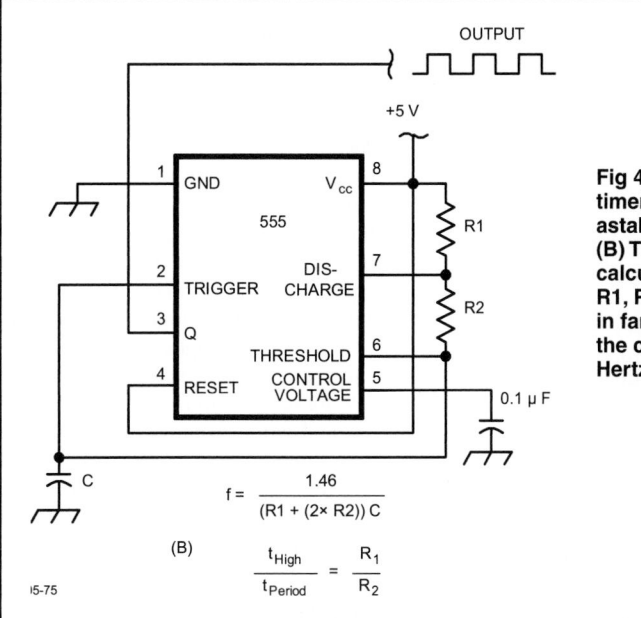

Fig 4.24 — (A) A 555 timer connected as an astable multivibrator. (B) The equations to calculate values for R1, R2 in ohms and C in farads, where f is the clock frequency in Hertz.

It may be difficult to produce a 50% duty cycle due to manufacturing tolerance for the resistors R1 and R2. One way to ensure a 50% duty cycle is to run the astable multivibrator at 2f and then divide by 2 with a toggle flip-flop.

Astable multivibrators, and the 555 integrated circuit in particular, are very often used to generate clock pulses. Although this is a very inexpensive and minimum hardware approach, the penalty is stability with temperature. Since the frequency and the pulse dimensions are set by resistors and capacitors, drift with temperature and to some extent aging of components will result in changes with time. This is no different than the problem faced by designers of L-C controlled VFOs.

4.6 Digital Integrated Circuits

Integrated circuits (ICs) are the cornerstone of digital logic devices. Modern technology has enabled electronics to become smaller and smaller in size and less and less expensive. Much of today's complex digital equipment would be impossible to build with discrete transistors and discrete components.

An IC is a miniature electronic module of components and conductors manufactured as a single unit. All you see is a ceramic or black plastic package and the silver-colored pins sticking out. Inside the package is a piece of material, usually silicon, created (fabricated) in such a way that it conducts an electric current to perform logic functions, such as a gate, flip-flop or decoder.

As each generation of ICs surpassed the previous one, they became classified according to the number of gates on a single chip. These classifications are roughly defined as:

Small-scale integration (SSI):
10 or fewer gates on a chip.

Medium-scale integration (MSI):
10-100 gates.

Large-scale integration (LSI):
100-1000 gates.

Very-large-scale integration (VLSI):
1000 or more gates.

Though SSI and MSI logic chips are still useful for building circuits to handle very simple tasks, it is more common to see them either used along with or completely replaced by programmable logic arrays and microcontrollers. In many cases you will see the smaller logic circuits referred to as "glue logic."

4.6.1 Comparing Logic Families

When selecting devices for a circuit, a designer is faced with choosing between many families and subfamilies of logic ICs. The determination of which logic subfamily is right for a specific application is based upon several desirable characteristics: logic speed, power consumption, fan-out, noise immunity and cost. From a practical viewpoint, the primary IC families available and in common use today are CMOS, with TTL a distant second place. Within these families, there are tradeoffs that can be made with respect to individual circuit capabilities, especially in the areas of speed and power consumption. Except under the most de-- manding circumstances, normal commercial

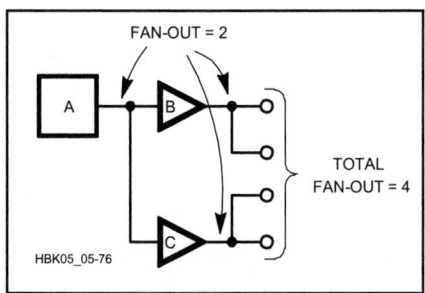

Fig 4.25 — Nonverting buffers used to increase fan-out: Gate A (fan-out = 2) is connected to two buffers, B and C, each with a fan-out of 2. Result is a total fan-out of 4.

grade temperature rating will do for amateur service.

FAN-OUT

Fan-out is a term with which you will need to become familiar when working with TTL logic families such as 7400, 74LS or 74S. A gate output can supply only a limited amount of current. Therefore, a single output can only drive a limited number of inputs. The measure of driving ability is called fan-out, expressed as the number of inputs (of the same subfamily) that can be driven by a single output. If a logic family that is otherwise desirable does not have sufficient fan-out, consider using noninverting buffers to increase fan-out, as shown by **Fig 4.25**.

Another approach is to use a CMOS logic family. These families typically have output drivers capable of sourcing or sinking 20 to 25 mA, and input current leakage in the microampere range. Thus, fan-out is seldom a problem when using these devices.

NOISE IMMUNITY

The noise margin was illustrated in Fig 4.2. The choice of voltage levels for the binary states determines the noise margin. If the gap is too small, a spurious signal can too easily produce the wrong state. Too large a gap, however, produces longer, slower transitions and thus decreased switching speeds.

Circuit impedance also plays a part in noise immunity, particularly if the noise is from external sources such as radio transmitters. At low impedances, more energy is needed to change a given voltage level than at higher impedances.

4.6.2 Bipolar Logic Families

Two broad categories of digital logic ICs are *bipolar* and *metal-oxide semiconductor* (MOS). Numerous manufacturing techniques have been developed to fabricate each type. Each surviving, commercially available family has its particular advantages and disadvantages and has found its own special niche in the market. The designer is cautioned, how-

ever, that sometimes this niche is simply the ongoing maintenance of old products. There are still very old logic families available for reasonable prices that would be considered quite obsolete and generally not suitable for new designs.

Bipolar semiconductor ICs usually employ NPN junction transistors. (Bipolar ICs can be manufactured using PNP transistors, but NPN transistors make faster circuits.) While early bipolar logic was faster and had higher power consumption than MOS logic, the speed difference has largely disappeared as manufacturing technology has developed.

There are several families of bipolar logic devices, and within some of these families there are subfamilies. The most-used bipolar logic family is transistor-transistor logic (TTL). Another bipolar logic family, Emitter Coupled Logic (ECL), has exceptionally high speed but high power consumption.

TRANSISTOR-TRANSISTOR LOGIC (TTL)

The TTL family saw widespread acceptance through the 1960s, 1970s and 1980s because it was fast compared to early MOS and CMOS logic, and has good noise immunity. It was by far the most commonly used logic family for a couple of decades. Though TTL logic is not in widespread use today for new designs, the device numbering system devised for TTL chips survives to this day for newer technologies. You will also often see TTL, especially the later low power, higher speed TTL subfamilies, in various equipment you may use and repair.

TTL Subfamilies

The original standard TTL used bipolar transistors and "totem-pole" outputs (see **Fig 4.26A** and B), which were a great improvement over the earlier diode-transistor logic (DTL) and resistor-transistor logic (RTL). Still, TTL logic consumed quite a bit of power

even at idle, and there were limits on how many inputs could be driven by a single output. Later versions used Schottky diodes to greatly improve switching speed, and reduced power requirements were introduced.

TTL IC identification numbers begin with either 54 or 74. The 54 prefix denotes an extended military temperature range of –55 to 125 °C, while 74 indicates a commercial temperature range of 0 to 70 °C. The next letters, in the middle of the TTL device number, indicate the TTL subfamily. Following the subfamily designation is a 2, 3 or 4-digit device-identification number. For example, a 7400 is a standard TTL NAND gate and a 74LS00 is a low-power Schottky NAND gate (The NAND gate is the workhorse TTL chip). A partial list of TTL subfamilies includes:

	74xx	standard TTL
	74xx	standard TTL
H	74Hxx	High-speed
L	74Lxx	Low-power
S	74Sxx	Schottky
F	74Fxx	Fairchild Advanced Schottky
LS	74LSxx	Low-power Schottky
AS	74ASxx	Advanced Schottky
ALS	74ALSxx	Advanced Low-power Schottky

Each subfamily is a compromise between speed and power consumption. **Table 5.9** shows some of these characteristics. Because the speed-power product is approximately constant, less power consumption generally results in lower speed and vice versa. The advanced low power Schottky devices (ALS, F) offer both increased speed and reduced power consumption. Historically, an additional consideration to the speed-versus-power trade-off has been the cost trade-off. For the amateur, this is not nearly the factor it once was as component costs are relatively low for the newer, faster, lower powered parts.

When a TTL gate changes state, the amount of current that it draws changes rapidly. These

Table 4.9
TTL and CMOS Subfamily Performance Characteristics

TTL Family	Propagation Delay (ns)	Per Gate Power Consumption (mW)	Speed Power Product (pico-joules)
Standard	9	10	90
L	33	1	33
H	6	22	132
S	3	20	60
F	3	8.5	25.5
LS	9	2	18
AS	1.6	20	32
ALS	5	1.3	6.5

CMOS Family Operating with							
$4.5 < V_{CC} < 5.5$ V	f=100 kHz	f=1 MHz	f=10 MHz	f=100 kHz	f=1 MHz	f=10 MHz
HC	18	0.0625	0.6025	6.0025	1.1	10.8	108
HCT	18	0.0625	0.6025	6.0025	1.1	10.8	108
AC	5.25	0.080	0.755	7.505	0.4	3.9	39
ACT	4.75	0.080	0.755	7.505	0.4	3.6	36

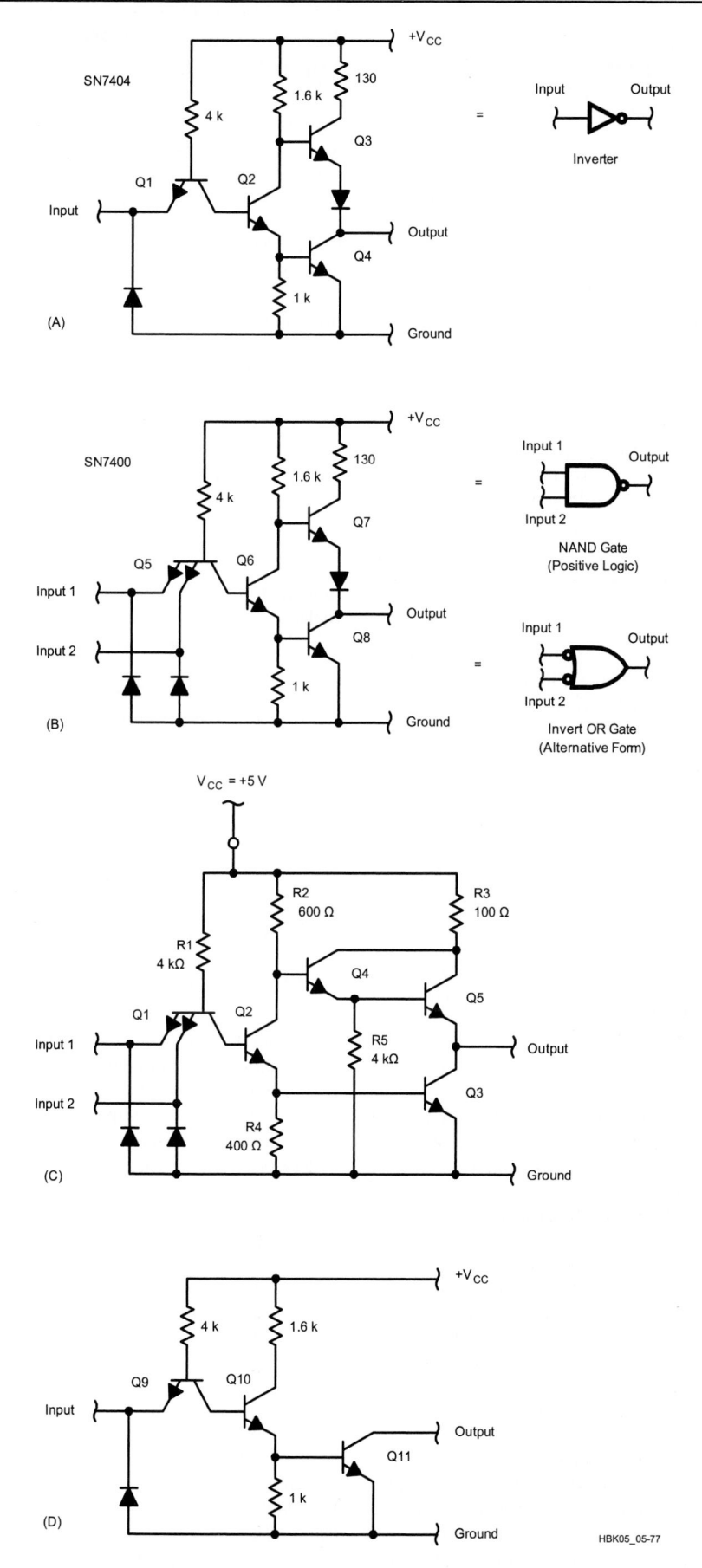

changes in current, called switching transients, appear on the power supply line and can cause false triggering of other devices. For this reason, the power bus should be adequately decoupled. For proper decoupling of TTL circuits, connect a 0.01 to 0.1 μF capacitor from V_{CC} to ground near each device to minimize the transient currents caused by device switching and magnetic coupling. These capacitors must be low-inductance, high-frequency RF capacitors (ceramic capacitors are preferred). In addition, a large-value (50 to 100 μF) capacitor should be connected from V_{CC} to ground somewhere on the board to accommodate the continually changing I_{CC} requirements of the total V_{CC} bus line. These are generally low-inductance tantalum capacitors.

Darlington and Open-Collector Outputs

Fig 4.26C and D show variations from the totem-pole configuration. They are the Darlington transistor pair and the open-collector configuration respectively.

The Darlington pair configuration replaces the single transistor Q4 with two transistors, Q4 and Q5. The effect is to provide more current-sourcing capability in the high state. This has two benefits: (1) the rise time is decreased and (2) the fan-out is increased.

Transistor(s) on the output in both the totem-pole and Darlington configurations provide active pull-up. Omitting the transistor(s) and providing an external resistor for passive pull-up gives the open-collector configuration. This configuration, unfortunately, results in slower rise time, since a relatively large external resistor must be used. The technique has some very useful applications, however: driving other devices, performing wired logic, busing and interfacing between logic devices.

Devices that need other than a 5-V supply can be driven with the open-collector output by substituting the device for the external resistor. Example devices include LEDs, relays and solenoids. Inductive devices like relay coils and solenoids need a protection diode across the coil. You must pay attention to the current ratings of open-collector outputs in such applications. You may need a switching transistor to drive some relays or other high-current loads.

Fig 4.26 — Example TTL circuits and their equivalent logic symbols: (A) an inverter and (B) a NAND gate, both with totem-pole outputs. (C) A NAND gate with a Darlington output. (D) A NAND gate with an open-collector output. (Indicated resistor values are typical. Identification of transistors is for text reference only. These are not discrete components but parts of the silicon die.)

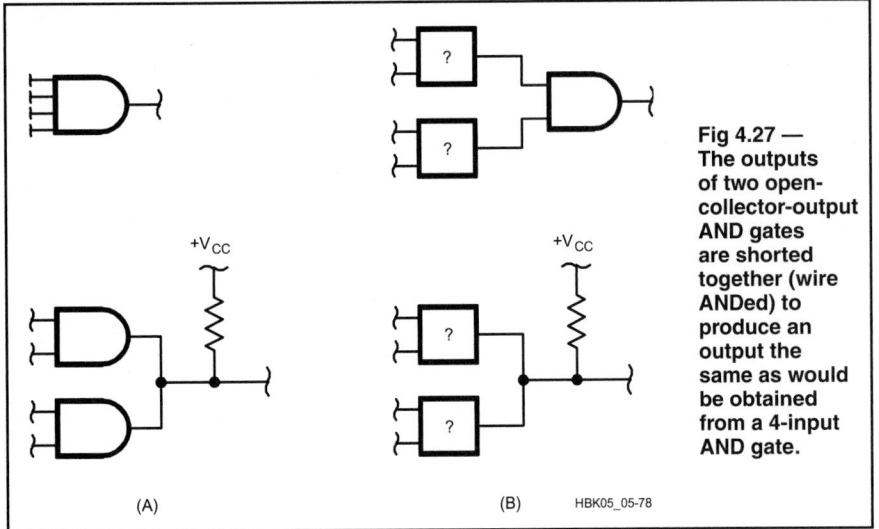

Fig 4.27 — The outputs of two open-collector-output AND gates are shorted together (wire ANDed) to produce an output the same as would be obtained from a 4-input AND gate.

HBK05_05-78

Open-collector outputs can perform wired logic, rather than gated IC logic, by wire-ANDing the outputs. This can save the designer an AND gate, potentially simplifying the design. Wire-ANDed outputs are several open-collector outputs connected to a single external pull-up resistor. The overall output, then, will only be high when all pull-down transistors are OFF (all connected outputs are high), effectively performing an AND of the connected outputs. If any of the connected outputs are low, the output after the external resistor will be low. **Fig 4.27** illustrates the wire-ANDing of open-collector outputs.

The wire-ANDed concept can be applied to several devices sharing a common bus. At any time, all but one device has a high-impedance (off) output. The remaining device, enabled with control circuitry, drives the bus output.

Open-collector outputs are also useful for interfacing TTL gates to gates from other logic families. TTL outputs have a minimum high level of 2.4 V and a maximum low level of 0.4 V. When driving non-TTL circuits, a pull-up resistor (typically 2.2 kΩ) connected to the positive supply can raise the high level to 5 V. If a higher output voltage is needed, a pull-up resistor on an open-collector output can be connected to a positive supply greater than 5 V, so long as the chip output voltage and current maximums are not exceeded.

Three-State Outputs

While open-collector outputs can perform bus sharing, a more popular method is three-state output, or tristate, devices. The three states are low, high and high impedance, also called Hi-Z or *floating*. An output in the high-impedance state behaves as if it is disconnected from the circuit, except for possibly a small leakage current. Three-state devices have an additional disable input. When the enable input is active, the device provides high and low outputs just as it would normally; when enable is inactive, the device goes into its high-impedance state.

A bus is a common set of wires, usually used for data transfer. A three-state bus has several three-state outputs wired together. With control circuitry, all devices on the bus but one have outputs in the high-impedance state. The remaining device is enabled, driving the bus with high and low outputs. Care should be taken to ensure only one of the output devices can be enabled at any time, since simultaneously connected high and low outputs may result in an incorrect logic voltage. (The condition when more than one driver is enabled at the same time is called *bus contention*.) Also, the large current drain from V_{CC} to ground through the high driver to the low driver can potentially damage the circuit or produce noise pulses that can affect overall system behavior.

Unused TTL Inputs

A design may result in the need for an n-input gate when only an n + m input gate is available. In this case, the recommended solution for extraneous inputs is to give the extra inputs a constant value that won't affect the output. A low input is easily provided by connecting the input to ground. A high input can be provided with either an inverter whose input is ground or with a pull-up resistor. The pull-up resistor is preferred rather than a direct connection to power because the resistor limits the current, thus protecting the circuit from transient voltages. Usually, a 1-kΩ to 5-kΩ resistor is used; a single 1-kΩ resistor can handle up to 10 inputs.

It's important to properly handle all inputs. Design analysis would show that an unconnected, or floating TTL input is usually high but can easily be changed low by only a small amount of capacitively-coupled noise.

4.6.3 Metal Oxide Semiconductor (MOS) Logic Families

While bipolar devices use junction transistors, MOS devices use field effect transistors (FETs). MOS is characterized by simple device structure, small size (high density) and ease of fabrication. MOS circuits use the NOR gate as the workhorse chip rather than the NAND. MOS families, specifically CMOS, are used extensively in most digital devices today because of their low power consumption and high speed.

P-CHANNEL MOS (PMOS)

The first MOS devices to be fabricated were PMOS, conducting electrical current by the flow of positive charges (holes). PMOS power consumption is much lower than that of bipolar logic, but its operating speed is also lower. The only extensive use of PMOS was in calculators and watches, where low speed is acceptable and low power consumption and low cost are desirable. PMOS was replaced by NMOS, which offered substantially higher switching speeds.

N-CHANNEL MOS (NMOS)

With improved fabrication technology, NMOS became feasible and provided improved performance and TTL compatibility. The speed of NMOS is at least twice that of PMOS, since electrons rather than holes carry the current. NMOS also has greater gain than PMOS and supports greater packaging density through the use of smaller transistors. NMOS has been almost completely obsoleted by CMOS.

COMPLEMENTARY MOS (CMOS)

CMOS combines both P-channel and N-channel devices on the same substrate to achieve high noise immunity and low power consumption: less than 1 mW per gate and negligible power during standby. This accounts for the widespread use of CMOS in battery-operated equipment. The high impedance of CMOS gates makes them susceptible to electromagnetic interference, however, particularly if long traces are involved. Consider a trace ¼-wavelength long between input and output. The output is a low-impedance point, hence the trace is effectively grounded at this point. You can get high RF potentials ¼-wavelength away, which disturbs circuit operation.

A notable feature of CMOS devices is that the logic levels swing to within a few millivolts of the supply voltages. The input-switching threshold is approximately one half the supply voltage ($V_{DD} - V_{SS}$). This characteristic contributes to high noise immunity on the input signal or power supply lines. CMOS input-current drive requirements are minuscule, so the fan-out is great, at least in low-speed systems. For high-speed systems, the input capacitance increases the dynamic power dissipation and limits the fan-out.

CMOS Subfamilies

There are a large number of CMOS subfamilies available. Like TTL, the original CMOS has largely been replaced by later subfamilies using improved technologies; in turn, these will be replaced with even newer families as technology evolves. The original family, called the 4000 series, has numbers beginning with 40 or 45 followed by two or three numbers to indicate the specific device. 4000B is second generation CMOS. When introduced, this family offered low power consumption but was fairly slow and not easy to interface with TTL.

Later CMOS subfamilies offer improved performance and, in some cases, TTL compatibility. For simplicity, the later subfamilies were given numbers similar to the TTL numbering system, with the same leading numbers, 54 or 74, followed by 1 to 4 letters indicating the subfamily and as many as 5 numbers indicating the specific device. The subfamily letters usually include a "C" to distinguish them as CMOS.

Following is a description of some of the CMOS device families available. As there are a substantial number of families offered by specific manufacturer, and there are new families being introduced frequently, this information is by no means exhaustive. A check of IC suppliers' and manufacturers' Web sites will provide the designer with a complete selection of choices to meet his or her requirements.

| 4000 | 4071B | *Standard CMOS* |
| C | 74Cxx | *CMOS versions of TTL* |

Largely obsolete today, devices in the 74C subfamily are pin and functional equivalents of many of the most popular parts in the 7400 TTL family. It may be possible to replace all TTL ICs in a particular circuit with 74C-series CMOS, but this family should not be mixed with TTL in a circuit without careful design considerations. Devices in the C series are typically 50% faster than the 4000 series.

| HC | 74HCxx | *High-speed CMOS* |

Devices in the 74HC subfamily have speed and drive capabilities similar to Low-power Schottky (LS) TTL but with better noise immunity and greatly reduced power consumption. High-speed refers to faster than the previous CMOS family, the 4000-series.

| HCT | 74HCTxx | *High-Speed CMOS, TTL compatible* |

Devices in this subfamily were designed to interface TTL to CMOS systems. The HCT inputs recognize TTL levels, while the outputs are CMOS compatible. HCT chips are commonly used as lower powered, drop-in replacements for their pin compatible LS TTL counterparts.

| AC | 74ACxxxxx | *Advanced CMOS* |

Devices in this family have reduced propagation delays, increased drive capabilities and can operate at higher speeds than standard CMOS. They are comparable to Advanced Low-power Schottky (ALS) TTL devices.

| ACT | 74ACTxxxxx | *Advanced CMOS, TTL compatible* |

This subfamily combines the improved performance of the AC series with TTL-compatible inputs.

| AHCT | 74AHCTxxxx | *Advanced High Speed CMOS, TTL Compatible* |

This subfamily is substantially faster than HC and HCT, but only slightly faster than ACT in switching speed. It has lower output drive capacity than AC/ACT.

New CMOS subfamilies are being introduced regularly. The current move is to lower voltage operation; where 5 V was the most common supply voltage until a few years ago, 3.3 V and lower supplies are becoming more common. Many newer microprocessors, microcontrollers and intelligent peripheral chips require 3.3 V logic.

| LVC | 74LVCxxxx | *Low Voltage CMOS* |

This subfamily uses a 3.3 V supply rather than 5 V. It offers low propagation delay (under 5 ns typical), extremely low supply current, robust output drive capabilities, and 5-V TTL compatible inputs.

| ALVC | 74ALVCxxxx | *Advanced Low Voltage CMOS* |

Even faster than LVC, this family offers propagation delays if under 3ns. Current demands are slightly higher, but still lower than 5V CMOS.

| VCX | 74VCXxxxx | *Low voltage CMOS* |

This family operates at supply voltages of 1.8 or 3.6 V. Offering high switching speeds and low propagation delays, it does not have TTL-compatible buffered inputs.

As with TTL, each CMOS subfamily has characteristics that may make it suitable or unsuitable for a particular design. You should consult the manufacturer's data books for complete information on each subfamily being considered.

CMOS Circuits

A simplified diagram of a CMOS logic inverter is shown in **Fig 4.28**. When the input is low, the resistance of Q2 is low so a high current flows from V_{CC}. Since Q1's resistance is high, the high current flows to the output. When the input is high, the opposite occurs: Q2's resistance is low, Q1's is high and the output is low. The diodes are to protect the circuit against static charges.

Special Considerations

Some of the diodes in the input- and output-protection circuits are an inherent part of the manufacturing process. Even with the protection circuits, however, CMOS ICs are susceptible to damage from static charges. To protect against damage from static, the pins should not be inserted in Styrofoam as is sometimes done with other components. Instead, a spongy conductive foam, usually black or pink in color, is available for this purpose. Before removing a CMOS IC from its protective material, make certain that your body is grounded. Touching nearly any large metal object before handling the ICs is probably adequate to drain any static charge off your body. Some people prefer to touch a grounded metal object or to use a conductive bracelet connected to the ground terminal of a three-wire ac outlet through a 10-MΩ resistor. Since wall outlets aren't always wired properly, you should measure the voltage between the ground terminal and any metal objects you might touch. Connecting yourself to ground through a 1-MΩ to 10-MΩ resistor will limit any current that might flow through your body.

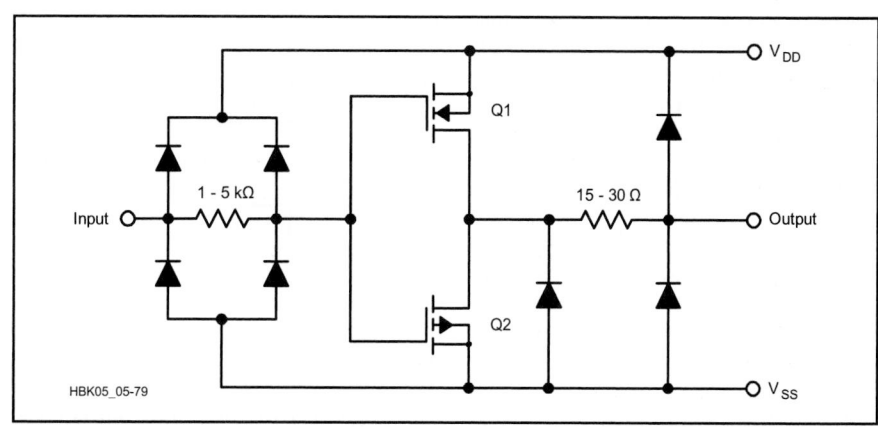

Fig 4.28 — Internal structure of a CMOS inverter.

All CMOS inputs should be tied to an input signal. A positive supply voltage or ground is suitable if a constant input is desired. Undetermined CMOS inputs, even on unused gates, may cause gate outputs to oscillate. Oscillating gates draw high current, and may overheat and self destruct.

The low power consumption of CMOS ICs made them attractive for satellite applications, but standard CMOS devices proved to be sensitive to low levels of radiation — cosmic rays, gamma rays and X rays. Later, radiation-hardened CMOS ICs, able to tolerate 10^6 rads, made them suitable for space applications. (A rad is a unit of measurement for absorbed doses of ionizing radiation, equivalent to 10^{-2} joules per kilogram.)

SUMMARY

There are many types of logic ICs, each with its own advantages and disadvantages. Regardless of the application, consult up-to-date product specification sheets and manufacturer literature when designing logic circuits. IC data sheets, application notes, databooks and more are available from IC manufacturers via their Web sites. By using a search engine and entering a few key word specifications, you will locate application notes, tutorials and a host of other information.

4.6.4 Interfacing Logic Families

Each semiconductor logic family has its own advantages in particular applications. When a design mixes ICs from different logic families, the designer must account for the differing voltage and current requirements each logic family recognizes. The designer must ensure the appropriate interface exists between the point at which one logic family ends and another begins. Knowledge of the specific input/output (I/O) characteristics of each device is necessary, and knowledge of the general internal structure is desirable to ensure reliable digital interfaces. Typical internal structures have been illustrated for each common logic family. **Fig 4.29** illustrates the logic level changes for different TTL and CMOS families. Data sheets should be consulted for manufacturer's specifications.

Often more than one conversion scheme is possible, depending on whether the designer wishes to optimize power consumption or speed. Usually one quality must be traded off for the other. The following section discusses some specific logic conversions. Where an electrical connection between two logic systems isn't possible, an optoisolator can sometimes be used.

TTL DRIVING CMOS

TTL and low-power TTL can drive 74C

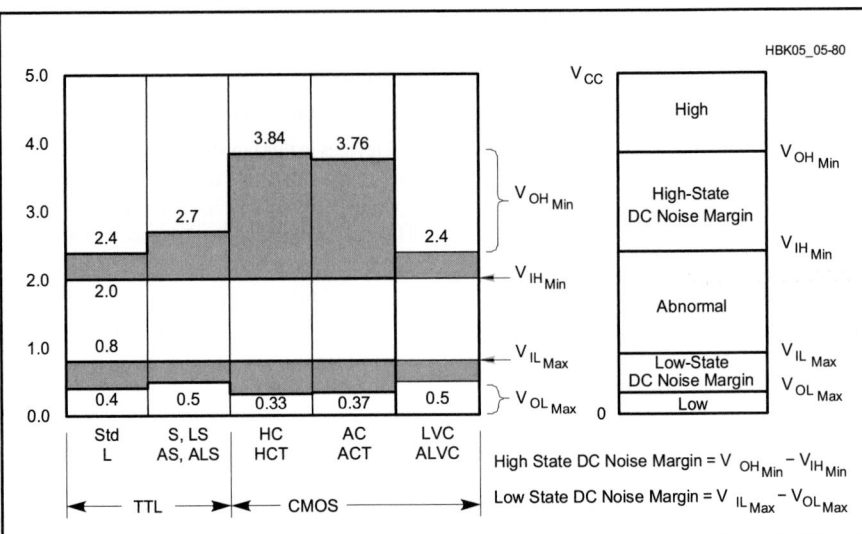

Fig 4.29 — Differences in logic levels for some TTL and CMOS families.

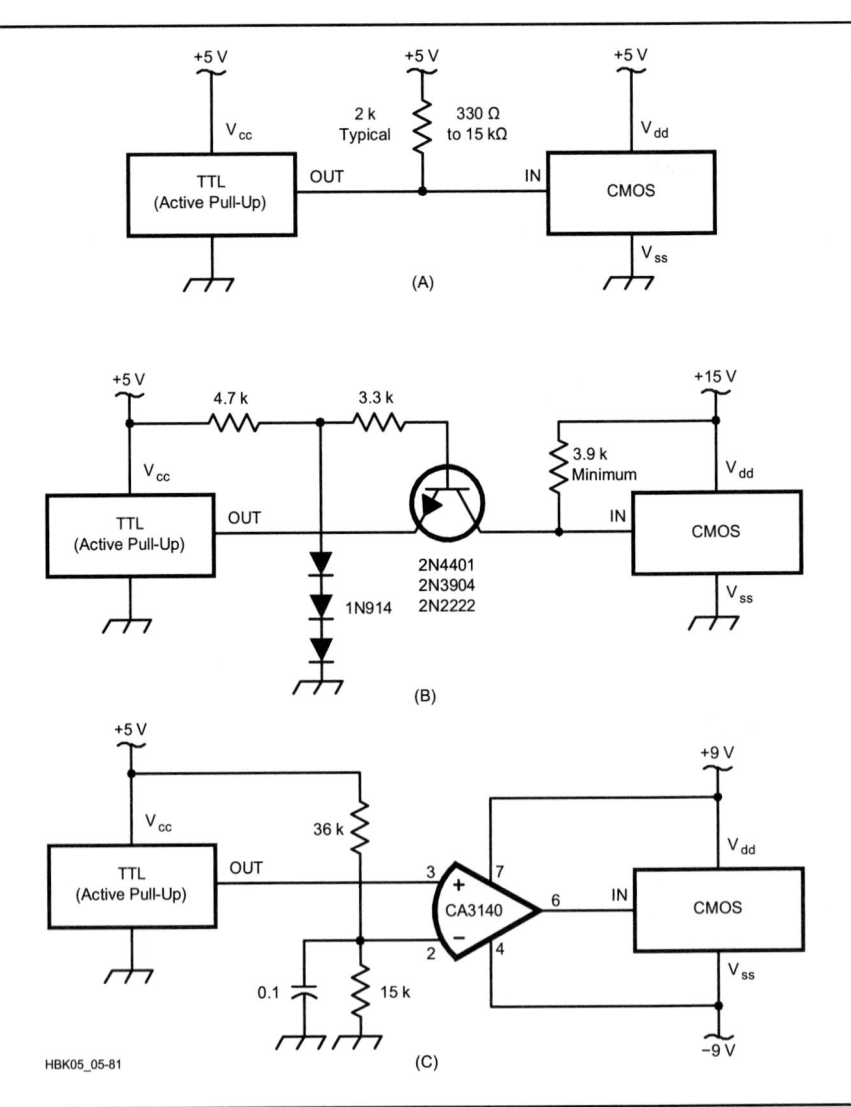

Fig 4.30 — TTL to CMOS interface circuits: (A) pull-up resistor, (B) common-base level shifter and (C) op amp configured as a comparator.

series CMOS directly over the commercial temperature range without an external pull-up resistor. However, they cannot drive 4000-series CMOS directly, and for HC-series devices, a pull-up resistor is recommended. The pull-up resistor, connected between the output of the TTL gate and V_{CC} as shown in **Fig 4.30A**, ensures proper operation and enough noise margin by making the high output equal to V_{DD}. Since the low output voltage will also be affected, the resistor value must be chosen with both desired high and low voltage ranges in mind. Resistor values in the range 1.5 kΩ to 4.7 kΩ should be suitable for all TTL families under worst conditions. A larger resistance reduces the maximum possible speed of the CMOS gate; a lower resistance generates a more favorable RC product but at the expense of increased power dissipation.

HCT-series and ACT-series CMOS devices were specifically designed to interface non-CMOS devices to a CMOS system. An HCT device acts as a simple buffer between the non-CMOS (usually TTL) and CMOS device and may be combined with a logic function if a suitable HCT device is available.

When the CMOS device is operating from a power supply other than +5 V, the TTL interface is more complex. One fairly simple technique uses a TTL open-collector output connected to the CMOS input, with a pull-up resistor from the CMOS input to the CMOS power supply. Another method, shown in Fig 4.30B, is a common-base level shifter. The level shifter translates a TTL output signal to a +15 V CMOS signal while preserving the full noise immunity of both gates. An excellent converter from TTL to CMOS using dual power supplies is to configure an operational amplifier as a comparator, as shown in Fig 4.30C. An FET op amp is shown because its output voltage can usually swing closer to the rails (+ and – supply voltages) than a bipolar device.

CMOS DRIVING TTL

Certain CMOS devices (including most modern 5 V powered CMOS logic families) can drive TTL loads directly. The output voltages of CMOS are compatible with the input requirements of TTL, but the input-current requirement of TTL limits the number of TTL loads that a CMOS device can drive from a single output (the fan-out).

Interfacing CMOS to TTL is a bit more complicated when the CMOS is operating at a voltage other than +5 V. One technique is shown in **Fig 4.31A**. The diode blocks the high voltage from the CMOS gate when it is in the high output state. A germanium diode is preferred because its lower forward-voltage drop provides higher noise immunity for the TTL device in the low state. The 68-kΩ resistor pulls the input high when the diode is reverse biased.

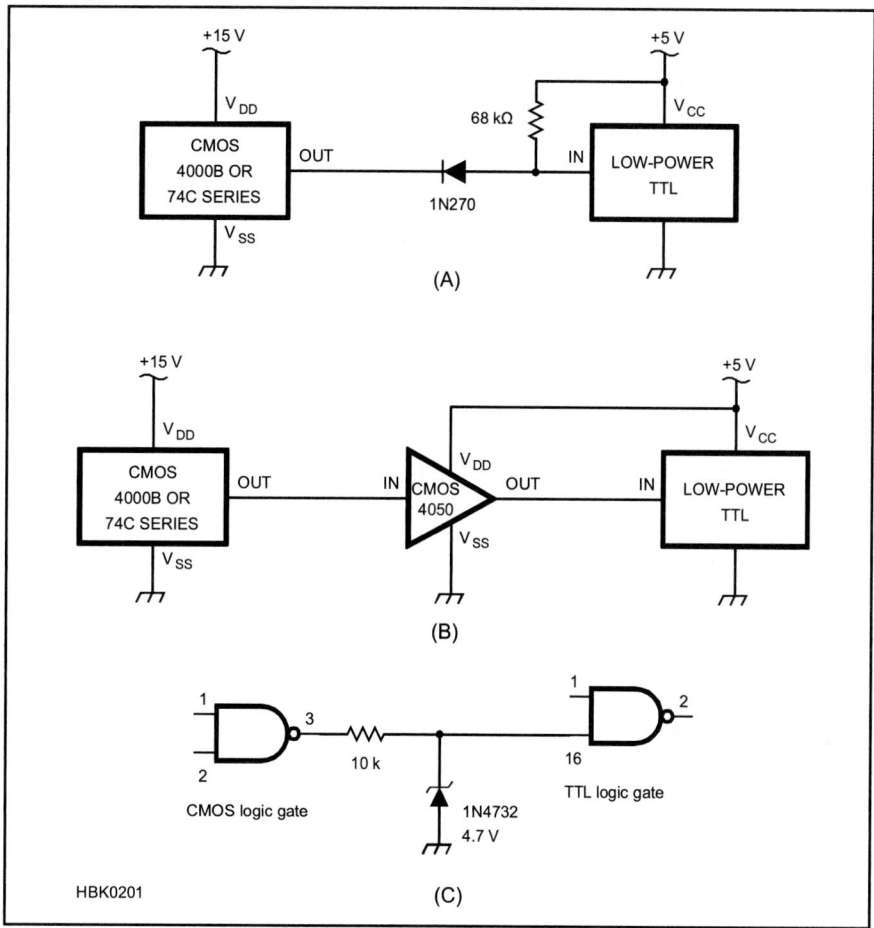

Fig 4.31 — CMOS to TTL interface circuits. (A) blocking diode chosen when different supply voltages are used. The diode is not necessary if both devices operate with a +5 V supply. (B) CMOS noninverting buffer IC. (C) Resistor/Zener circuit clamps the voltage to the TTL input at 4.7 V.

A simple resistor/Zener circuit, shown in Fig. 4.31C, can also be used. This clamps the voltage to the TTL input at 4.7 V.

There are two CMOS devices specifically designed to interface CMOS to TTL when TTL is using a lower supply voltage. The CD4050 is a noninverting buffer that allows its input high voltage to exceed the supply voltage. This capability allows the CD4050 to be connected directly between the CMOS and TTL devices, as shown in Fig 4.31B. The CD4049 is an inverting buffer that has the same capabilities as the CD4050.

5 V DRIVING 3.3 V LOGIC

Low voltage logic operating with 3.3 to 3.6 V supplies is becoming more and more common. Some of these logic families have 5 V tolerant inputs and can be driven directly by TTL levels; others require buffering to keep inputs at or below the supply voltage. Output levels may or may not be sufficient to drive TTL level inputs reliably under all conditions. The use of TTL compatible buffers (74LVC, 74ALVC) on input signals is a safe way to drive these devices. Since the low logic volt-

ages are compatible, a simple Zener clamp arrangement may also be sufficient.

3.3 V DRIVING 5 V LOGIC

Output voltages of most 3.3 V logic families are sufficient to drive the inputs of TTL-level devices. In some cases, the high level output voltage of a 3.3 V powered device may approach the lower end of the TTL level device's input range. In these cases, a pull-up resistor to 3.3 V will usually be sufficient. It is also possible to use an IC or transistor buffer.

4.6.5 Programmable Logic

As digital logic designs became more and more complex, the size and power consumption of their implementations grew. PLDs (*Programmable Logic Devices*) were introduced to allow the designer to produce application-specific logic without the expense and delay of fabricating custom chips. As explained earlier, the design of a logic circuit begins with a description; this description is used to determine an equation; then the equation is expanded to components. With programmable logic we can shortcut this process by implementing the equations themselves, programming them into a generic array of gates.

Several families of programmable logic devices have been introduced over the years. The earliest were Programmable Array Logic (PAL) chips, consisting of an array of gates whose inputs, interconnections and outputs could be configured by blowing internal fuses according to a specific design. PALs were followed by Generic Array Logic (GALs), which have more gates and more flexible I/O pin logic. GALs are also erasable and reprogrammable.

Later developments led to the Field Programmable Gate Array (FPGA) and other programmable logic arrays. These devices can contain anywhere from several hundred to millions of logic gates, and up to more than a thousand I/O pins. The connections between gates, and thus the device's function, is determined by complex program code. PLDs can be made to replace large numbers of individual SSI, MSI and even LSI chips, up to and including entire microprocessor cores.

4.7 Analog-Digital Interfacing

Quite often, logic circuits must either drive or be driven from non-logic sources. A very common requirement is sensing the presence or absence of a high (as compared to +5 V) voltage or perhaps turning on or off a 120 V ac device or moving the motor in an antenna rotator. A similar problem occurs when two different units in the shack must be interfaced because induced ac voltages or ground loops can cause problems with the desired signals.

A slow speed but safe way to interface such circuits is to use a relay. This provides absolute isolation between the logic circuits and the load. **Fig 4.32A** shows the correct way to provide this connection. The relay coil is selected to draw less than the available current from the driving logic circuit. The diode, most often a 1N914 or equivalent switching diode, prevents the inductive load from back-biasing the logic circuit and possibly destroying it.

It is often not possible to find a relay that meets the load requirements *and* has a coil that can be driven directly from the logic output. Fig 4.32B shows two methods of using transistors to allow the use of higher power relays with logic gates.

Electro-optical couplers such as opto-isolators and solid state relays can also be used for this circuit interfacing. Fig 4.32C uses an optoisolator to interface two sets of logic circuits that must be kept electrically isolated, and Fig 4.32D uses a solid state relay to control an ac line supply to a high current load. Note that this example uses a solid state relay with internal current limiting on the input side; the LED input has an impedance of approximately 300 Ω. Some devices may need a series resistor to set the LED current; always consult the device data sheet to avoid exceeding device limits of the relay or the processor's I/O pin. See the **Analog Basics** chapter for more information on optiosolators.

For safely using signals with voltages higher than logic levels as inputs, the same simple resistor and Zener diode circuit similar to that shown in Fig 4.31C can be used to clamp the input voltage to an acceptable level. Care

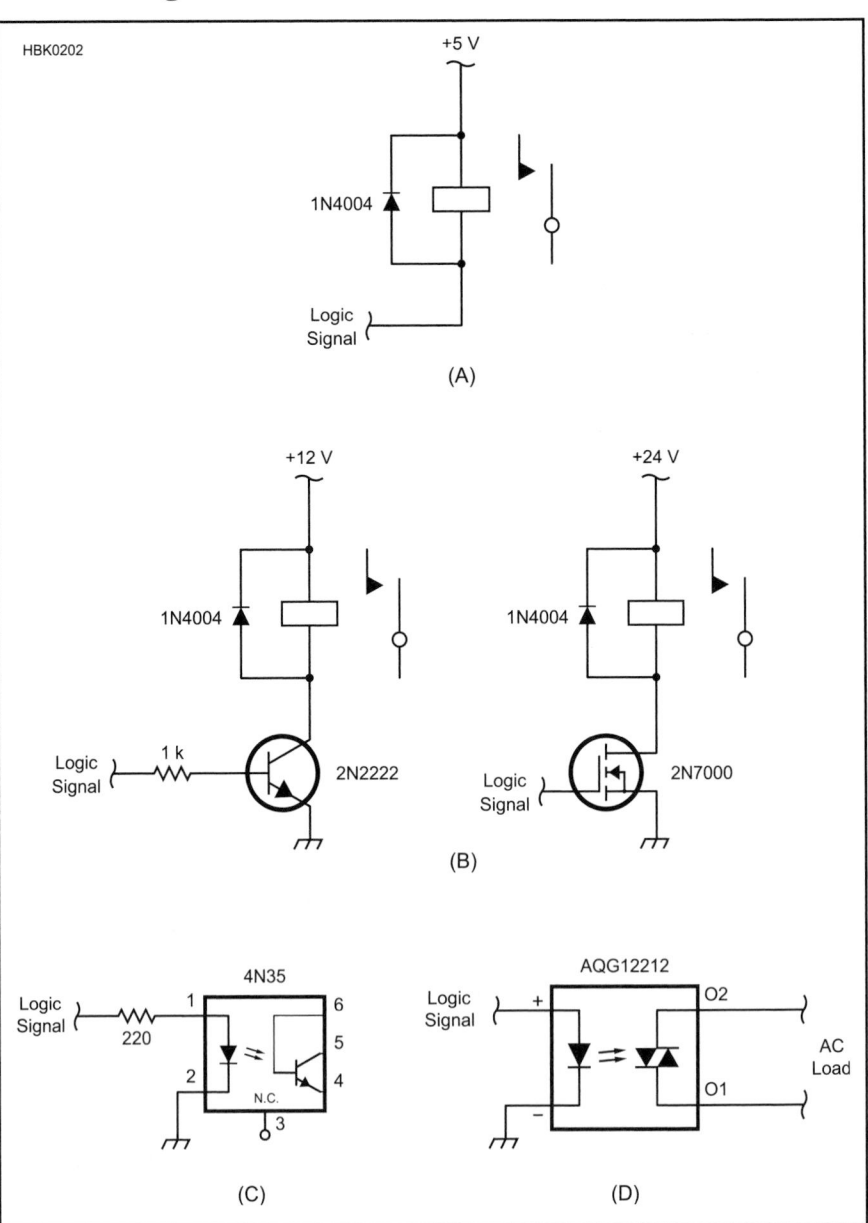

Fig 4.32— Interface circuits for logic driving real-world loads. (A) driving a relay from a logic output; (B) using a bipolar transistor or MOSFET to boost current capacity; (C) using an optoisolator for electrical isolation; (D) using a solid-state relay for switching ac loads.

must be used to choose a resistor value that will not load the input signal unacceptably.

4.7.1 Analog-Digital Converters

Of course not all signals in the real world are digital. It is often desirable to know the exact voltage of an analog sensor, for example, or to measure the level of something — light, water, current or some other input. Similarly, an analog output can be quite useful for controlling or indicating things that are better left in the analog domain. For this reason we have two types of converters. The analog-to-digital converter (ADC) and digital-to-analog converter (DAC) handle these tasks for us, and they are often fairly easy to use. Additional information on ADCs and DACs may be found in the **Analog Basics** chapter.

ADC AND DAC DIGITAL INTERFACES

Interfaces for ADC and DAC chips are generally classed as serial or parallel. Serial interfaces can vary in speed, complexity and the number of wires required for operation. A serial interface has the advantage of requiring a small number of processor I/O pins to accommodate data of any length. Whether your ADC is providing 8, 10 or 12 bits, the same small number of I/O signals are used. Some of the most common serial interfaces used for ADC and DAC chips are as follows:

Serial Peripheral Interconnect (SPI). This four-wire, synchronous bus and protocol use a common CLK signal, plus data lines for master-to-slave (Master Out-Slave In, or MOSI) and slave-to-master (Master In-Slave Out, or MISO). Multiple devices can be used by providing each with its own Slave Select (SS) signal. Speeds can range up to 70 Mbit/s, depending on the capabilities of the master and slave devices.

Microwire. This earlier predecessor to SPI implements a half-duplex subset of SPI using the same signals.

Inter-IC Communication (I2C) bus. Originally developed by Philips, this synchronous two-wire bus and protocol use a pair of open-drain lines, Serial Clock (SCL) and Serial Data (SDA). Communication is controlled by a master node, though there may be more than one master attached to the bus. Many peripheral devices can be attached to an I2C bus; a device address is sent by the master to initiate communication with a slave. Speeds range from 10 kbit/s (low speed) to 100 kbit/s (standard) to 400 kbit/s (high speed) and higher.

Parallel interfaces generally have eight or more data lines, plus a chip select, read/write and interrupt controls. The read and write signals may be separate, or may be a single read/write signal — low for WR, and high for RD, for example, or vice versa. The interrupt signal can be used to indicate the end of a conversion cycle to the CPU. In this way the processor can tell the converter to start a conversion, then continue processing until the conversion is complete. This can allow more processing to be done without waiting idly for the ADC or DAC to complete its conversion.

PROGRAMMING AND COMMUNICATION

Many ADC chips have multiplexed inputs that allow you to use one chip to sample and digitize more than one analog input. In these devices, there is a single converter but several inputs that can be internally switched. Additionally, it is common to be able to program the inputs as single-ended or differential. In the case of the popular ADC0832, for example, you can use its two analog input pins as two separate inputs, or as a single differential input. The ADC0838 expands this to eight single-ended or four differential inputs. In the case of the ADC0832, modes can be mixed. This allows the use of various combinations of single-ended and differential inputs, as needed.

Programming and selecting the inputs is done by sending a series of bits from the processor to the ADC at the beginning of the conversion cycle. **Fig 4.33** shows a diagram of the signal timing used with the ADC0834 ADC. In the case of the ADC0834, a series of four bits are sent by the CPU to the ADC to select single ended or differential mode, the polarity of the input, and which of the four inputs (or two input pairs) is to be selected. Immediately following the fourth bit, the ADC starts its conversion and starts sending the resulting data to the CPU. Other chips in this family and many from different manufacturers use a similar scheme; the number of program bits sent by the CPU depends on the number of inputs present.

Some chips are also configurable for data format as well. For example, a 12 bit ADC

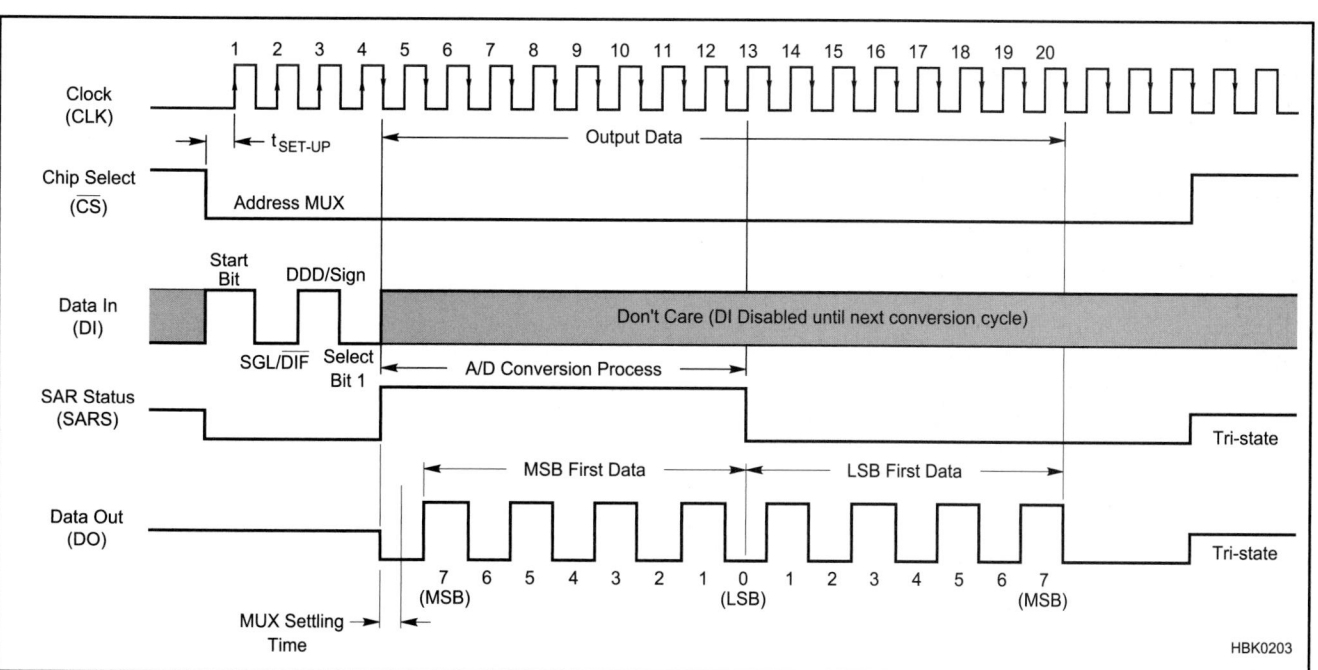

Fig 4.33 — ADC0834 timing diagram. Much more information about this device is available from the National Semiconductor datasheet.

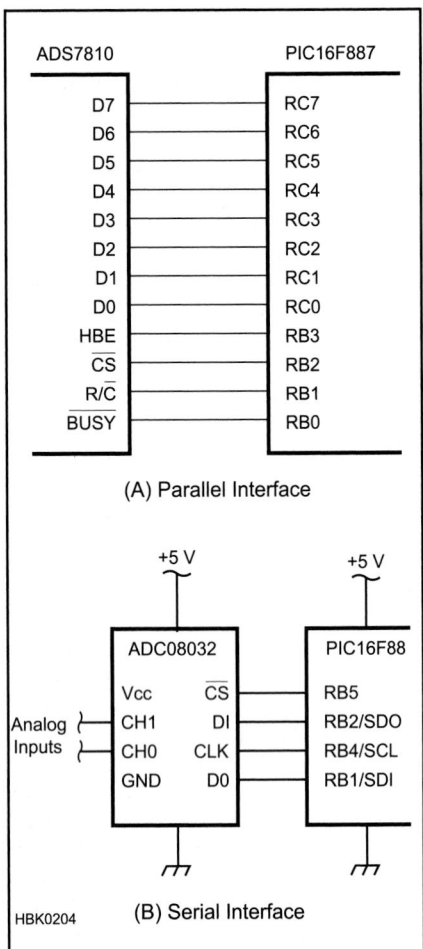

Fig 4.34 — A/D converter to microcontroller interfacing: (A) parallel; (B) serial.

may be programmable to send a sign bit for a total of 13 data bits rather than 12. Data may also be sent LSB first or MSB first. Some chips are configurable for either parallel or serial data transfer; this selection is generally made by pulling a pin or pins high or low to select the mode. In all cases, a careful reading of the device data sheet is your best bet for successful integration of the converter into your project.

INTERFACING TO YOUR MICROCONTROLLER

Probably the simplest way to interface an external ADC to your microcontroller is by using a parallel interface. In this case, one simply manipulates the I/O pins from the CPU to start the conversion, then reads the resulting data from the data lines. This method is often less desirable than a serial connection, though, due to the limited number of I/O pins usually available. **Fig 4.34A** shows an example of a parallel interface between a 12 bit ADC (the ADS7810) and a PIC microcontroller (the PIC16F887). In this example, only eight data lines are used for twelve data bits. The HBE signal is used to gate the four high order bits onto the data lines after the low order bits have been read. This saves four I/O pins on the microcontroller.

Serial connections save I/O pins, allow the use of physically smaller IC packages, and can be quite easy to implement. Fig 4.34B shows an example of a serial interface. If your CPU has a hardware SPI/Microwire interface built in, communication with the ADC can be quite simple from a program standpoint. Simply send the program word, if required, and read the resulting data — either one or two bytes, depending on the ADC resolution. If you do not have built-in hardware SPI, the program code to "bit-bang" the interface is fairly simple to implement. (See **www.dlpdesign.com/images/bit-bang-usb.pdf** for more information on bit-bang.)

Note that Fig 4.34B shows separate grounds for the processor and ADC. This is required to keep switching noise and transients from affecting the accuracy of readings. A separate analog supply is also recommended. The analog supply should be well filtered using a combination of smaller ceramic or monolithic, and bulk tantalum or other low-ESR capacitors. This should be done for the analog supply and reference voltage supplies. The more noise you can eliminate from the analog supply, the more reliable your readings will be.

MICROCONTROLLERS WITH BUILT-IN CONVERTERS

It is not at all unusual to find microcontrollers with built-in analog-to-digital converters, complete with multiplexers, analog supply pins and V_{REF} inputs. For example, many Microchip PIC processors are equipped with 10 bit ADCs, and have 5, 8 or more pins that can be configured as multiplexed analog inputs. A few also have digital-to-analog, though this is not nearly as common. While these devices can be extremely useful, there are some limitations. Reference voltages are generally limited to the device's supply voltage. Internal CPU-generated noise may also somewhat limit the usefulness of the converter. With careful attention to device specifications and limitations, though, a built-in ADC may meet the needs of a wide range of uses.

4.8 Microcontroller Overview

This section surveys microcontroller characteristics and applications as they pertain to typical Amateur Radio projects and equipment. The reader interested in using microcontrollers is directed to the many books and Internet resources that exist for learning how to apply microcontrollers. Manufacturer websites such as Microchip (**www.microchip.com**), Freescale (**www. freescale.com**) and Atmel (**www.atmel.com**) are a good place to start. There are also a number of message boards, forums and email lists devoted to specific processor families, such as the PICList (**www.piclist.com**) or the Arduino User Forum (**http://arduino. cc/forum**). Local users groups are often active in large cities. Online communities of microprocessor users are very fluid — use Internet search engines to find current resources.

4.8.1 Selecting a Microcontroller

There are a number of microcontroller families available from several manufactures, each with its own advantages and disadvantages. At the time of this writing, the more popular options for amateur use come from Microchip (PIC product line), Atmel (AVR, ATMega, 8051 and other product lines), Freescale (68HC11, HC08, HC16 and other product lines), and many manufacturers who are building chips based on the venerable Intel 8051 architecture. In addition to these you will see and hear of the Texas Instruments MSP430, ARM, and many others.

Designing projects using microcontrollers can range from relatively simple to quite complex. On the simpler end of the scale are several products intended to bring microcontroller based design ability to people who would otherwise not be able to use these devices. Various breadboard kits and development systems are available to simplify hardware development and program downloading. The following three systems are particular popular:

• Arduino (**www.arduino.cc**) — based on an AVR microcontroller, Arduino boards are available with a variety of capabilities and interfaces. Add-on modules called "shields" are added to interfaces provided on the Arduino board. Software is written in C or C++ and downloaded to the microcontroller. The manufacturer provides free development software.

• Parallax BASIC Stamp (**www.parallax. com**) — a single-board system containing a PIC microcontroller preloaded with PBASIC. Software is written in an editor program on the PC and then downloaded.

• PICAXE — a variety of PIC microcontroller with a special form of BASIC. Various sizes of microcontrollers are available with different numbers of I/O pins and clock speeds. The user is responsible for building the circuit board that supports the microcontroller.

In each of these cases, the user does not necessarily need to have a deep understanding of the details of the microcontroller or its operation. The circuit can be designed with attention paid to keeping signal levels within the limits of the device. Then, one simply writes a program to perform the functions required. To be sure, there are trade-offs. The built-in firmware required to allow the user to write a program may be in a simple language such as BASIC that takes up space and processor cycles, so execution speed is not as fast and programs are somewhat limited.

The BASIC Stamp and PICAXE use specific variants of the PIC microcontroller, so if you need features that are not present in those chips you will need to take a different approach. Still, these can be extremely valuable tools that the average amateur can use with very little time and money invested.

There is a very broad range of microcontrollers available to meet nearly any need. Available devices include 8 bit, 16 bit and 32 bit processors with internal program memory ranging from 512 words up to 256 kbytes and more. At the low end are very low cost, physically small devices (down to 8 pin ICs) that can perform relatively simple tasks with ease. Larger, faster processors may include more memory, more I/O pins, and specialized peripherals. Some of the more common features are ADCs, DACs, pulse width modulation (PWM) outputs, USB interface hardware or digital signal processing.

I/O REQUIREMENTS

Selecting a microcontroller for a new project involves evaluating your requirements and prior experience. First, evaluate the number of inputs and outputs you are likely to need. If, for example, you need to sample an analog voltage and two switch closures and generate a serial data stream, you will need one analog input and three or more digital I/O pins. Depending on memory requirements, even some of the smallest devices will fit the bill. For a more elaborate design, you may need to scan a 4×4 key switch matrix, drive an LCD module with a parallel interface, control a number of LEDs and interface to a PC through a USB port. In this case, the number of I/O pins needed will eliminate a large number of devices from consideration.

CPU SIZE, PERFORMANCE AND MEMORY

If you are a computer user, you are used to seeing 32 bit and 64 bit microprocessors, many with more than one CPU core on the same chip. In the world of embedded microcontrollers, things are a little different. Instead of handling an operating system and a large number of different programs, the microcontroller only has to execute one program and is completely dedicated to that task. Unless you require very high speed I/O, need very intensive processing or will be using DSP, a simple 8 bit processor may be more than sufficient. There are dozens of ham related products that use embedded controllers — iambic keyers, APRS transponders, repeater controllers, antenna rotator controls and many other accessories that all use simple 8 bit processors with surprisingly small amounts of program memory. On the other hand, if you are designing a new rig or antenna tuner that will require DSP, chances are you may need a more robust processor.

If you are unsure of what your project will require, you may want to pick a processor family that has scalable, pin-compatible members. For example, Microchip offers processors ranging from 8 bit and 5 million instructions per second (MIPS) with 7 kbytes of program memory, to 16 bit with 48 kbytes of program memory at 30 MIPS. All are pin-compatible, so if you find you need more speed, more memory or more built-in peripherals, you can drop in a different processor without changing the hardware.

Also take into account the physical packaging. Some products are available in through-hole, dual inline (DIP) packages; others are available only in surface mount versions. This can make a difference if you lack the equipment, expertise or desire to work with surface mount components.

SPECIALIZED PERIPHERALS

Will you need a USB interface? Analog inputs? DSP? Precision timekeeping? It may be easier to select a processor that has the features you need built in. On the other hand, it is relatively easy to add an external real-time clock, ADC or DAC, and it takes only a few I/O pins. There are several USB interface chips available that will handle communication with your PC, and present an asynchronous serial data interface to the microcontroller. This can save a lot of firmware development time and frustration.

HARDWARE COST AND DEVELOPMENT TOOLS

The cost of the actual microcontroller may be a factor, or it may not. If you are building a one-off project for your own use, there is not much difference between using a $3 chip and a $30 chip with a built-in BASIC interpreter. The difference in time and effort to develop the firmware may make it well worth the extra money for the easier-to-use solution. If you plan to manufacture your device for a larger audience, the cost difference may lead you to make a different choice.

Take into account the cost of the software and hardware tools needed for development. Does the manufacturer offer a free set of firmware development tools, or will you have to spend extra money for an assembler or compiler? Is there a low cost development kit available with the PC interface you need? Does the device require a simple PC interface to program, or a more expensive dedicated programmer? Do you have any previous experience with the processor family, or will it be a completely new effort for you?

4.9 Personal Computer Interfaces

This section provides an overview of the personal computer (PC) digital interfaces used in the amateur station. Detailed information on using these interfaces is available online, especially in users groups and specialized forums for specific software and hardware applications. This book's chapters on **DSP and Software Radio Design, Digital Modes,** and the CD-ROM chapter **Digital Communications** provide more information on specific applications and techniques. The **Station Accessories** chapter includes PC interface construction projects, as well.

4.9.1 Parallel vs Serial Signaling

To communicate a word to someone across the room, you could hold up flash cards displaying the letters of the word. If you hold up four flash cards, each with a letter on it, all at once, then you are transmitting in parallel. If, instead, you hold up each of the flashcards one at a time, then you are transmitting in serial. *Parallel* means all the bits in a group are handled exactly at the same time. *Serial* means each of the bits is sent in turn over a single channel or wire, according to an agreed sequence. **Fig 4.35** illustrates parallel and serial signaling.

Both parallel and serial signaling are appropriate for certain circumstances. Parallel signaling is faster, since all bits are transmitted simultaneously, but each bit needs its own conductor, which can be expensive. Parallel signaling is more likely to be used for internal communications. For spanning longer distances, such as to an external device, serial signaling is more appropriate. Each bit is sent in turn, so communication is slower, but it also is less expensive, since fewer channels are needed between the devices.

Most amateur digital communications use serial transmission to minimize cost and complexity. The number of channels needed for signaling also depends on the operational mode. One channel is required per bit for simplex (one-way, from sender to receiver only) and for half-duplex (two-way communication, but only one person can talk at a time), but two channels per bit

are needed for full-duplex (simultaneous communications in both directions).

PARALLEL I/O INTERFACING

Fig 4.36 shows an example of a parallel input/output interface, similar to the parallel printer ports once common in PCs. Typically, they have eight data lines and one or more handshaking lines. *Handshaking* involves a number of functions to coordinate the data

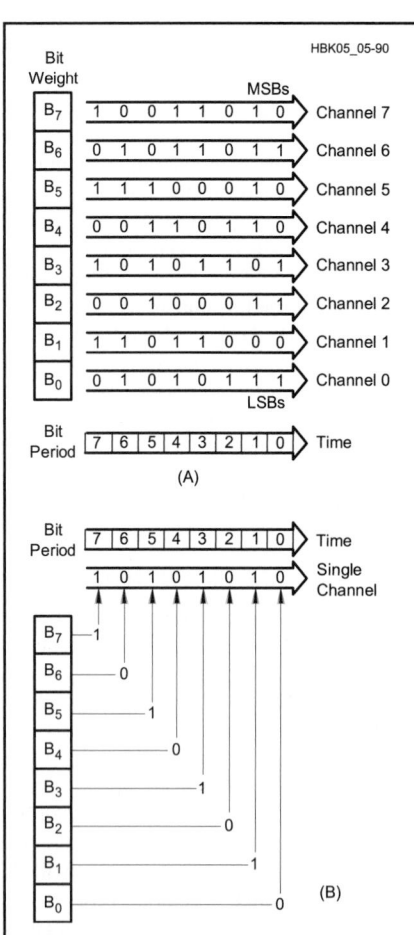

Fig 4.35 — Parallel (A) and serial (B) signaling. Parallel signaling in this example uses 8 channels and is capable of transferring 8 bits per bit period. Serial transfer only uses 1 channel and can send only 1 bit per bit period.

transfer. For example, the READY line indicates that data are available on all eight data lines. If only the READY line is used, however, the receiver may not be able to keep up with the data. Thus, the STROBE line is added so the receiver can determine when the transmitter is ready for the next character.

Interfacing to the parallel port is very simple and can be done in many languages and under many operating systems. It is an easy way to get a PC to act as a controller in the shack and around the home, with from 8 to 12 independent input or output wires. You can get an old computer either virtually free or for just a few dollars and use it exclusively as a controller. (New computers usually do not have a parallel port.)

By searching the Internet for a combination of "parallel port interface" and the language of your choice (for example, Visual BASIC or C), you can find detailed interfacing and programming instructions. It is a good and relatively simple place to "roll your own," compared to the protocol requirements of interfacing with the serial port or USB.

For more information, including practical circuits, see "Interfacing to the Parallel Port" by Paul Danzer, N1II, on the CD-ROM included with this book.

SERIAL I/O INTERFACING

Serial input/output interfacing is more complex than parallel, since the data must be transmitted based on an agreed sequence.

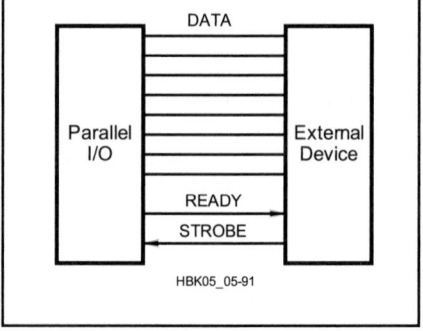

Fig 4.36 — Parallel interface with READY and STROBE handshaking lines.

For example, transmitting the 8 bits (b7, b6, . . . b0) of a word includes specifying whether the least significant bit, b0, or the most significant bit, b7, is sent first. Fortunately, a number of standards have been developed to define the agreed sequence, or encoding scheme.

In order to use a serial interface between a computer and a serial device such as a modem, printer or a new transceiver, several things must be set the same on both ends of the connection. The first is the data rate, the choices for which are usually limited to specific values between 300 and 115200 bits per second for historical reasons. The next is the data format — 7 or 8 bits, and with even, odd or no parity. Finally, both devices must be using the same handshaking method. The usual choices are RTS/CTS, DSR/DTR, or XON/XOFF. The first two use extra signals (and therefore extra wires) to indicate when the transmitting device has data to send, and when the receiving device is able to accept data. In the third method, XON/XOFF, the device receiving data will send an XOFF control character to the transmitting device to indicate that it can no longer accept data. When the transmitting device sees the XOFF it stops sending until it receives an XON from the other end. Each method has its advantages and disadvantages. For example, XON/XOFF needs no extra wires, but response is not instantaneous. Both ends must have some data buffer space and logic to support the handshake.

4.9.2 Standard Data Interfaces

This section covers a set of common data interfaces used in the amateur station. See the **Component Data and References** chapter for details on computer connector pinouts.

SIGNALING LEVELS

Inside equipment and for short runs of wire between equipment, the normal practice is to use neutral keying; that is, simply to key a voltage such as +5 V on and off. In neutral keying, the off condition is considered to be 0 V. Over longer runs of wire, the line is viewed as a transmission line, with distributed inductance and capacitance. It takes longer to make the transition from 0 to 1 or vice versa because of the additional inductance and capacitance. This decreases the maximum speed at which data can be transferred on the wire and also may cause the 1s and 0s to be different lengths, called bias distortion. Also, longer lines are more likely to pick up noise, which can make it difficult for the receiver to decide exactly when the transition takes place. Because of these problems, bipolar keying is used on longer lines. Bipolar keying uses one polarity (for example +) for a logical 1 and the other (– in this example) for a 0. This means

that the decision threshold at the receiver is 0 V. Any positive voltage is taken as a 1 and any negative voltage as a 0.

EIA-RS-232

The serial bus protocol EIA-RS-232 addresses this issue (however, a Mark "1" is a negative voltage and a Space "0" is positive). Generally called RS-232, this protocol defines connectors and voltages between data terminal equipment (DTE) such as a PC, and data communications equipment (DCE), such as a modem or TNC.

The connector is the DB-25 (25 pin), or DB-9 (9 pin) version — though RS-232 interfaces have also been implemented using nonstandard connectors such as RJ45, 3.5 mm audio plugs and many others. Signaling voltages are defined between +3 V and +25 V for logic "0" and between –3 V and –25 V for logic "1." Although the top data rate addressed in the specification is only 20 kbit/s, speeds of up to 115 kbit/s are commonly used. Communications distances of hundreds of meters are possible at reasonable data rates.

Since neutral keying is usually used inside equipment and bipolar keying for lines leaving equipment, signals must be converted between bipolar and neutral. Discrete level shifters or op amp circuits may perform this task, or low cost specialized IC line drivers and receivers are available.

UNIVERSAL SERIAL BUS (USB)

USB is a computer standard for an intelligent serial data transfer protocol, and it has become the standard for nearly all connections between a PC and its external peripherals. It has largely replaced serial, parallel, keyboard and mouse ports as well as SCSI and numerous other buses in consumer PCs.

In addition to higher speed than RS-232 and parallel ports, USB offers reasonable power availability to its loads, or functions. Under certain circumstances, up to 127 hubs and functions may connect to a single computer. USB requires that each function have on-board intelligence and that it negotiate with the host for power and bandwidth allocation. USB also has the major advantage of hot-pluggablity — the PC need not reboot when new functions are added.

The USB connectors use four-conductor cable, with two bidirectional, differential data lines, power, and ground. Approximately 5 V at 100 mA is allowed per function, with up to 500 mA available if the host system has the capability. This means that relatively sophisticated devices, such as modems, small video cameras, or hand-held scanners may operate from the bus without additional power supplies. It has also become common to use USB for connecting and charging a wide range

of devices such as cellular phones, cameras and GPS units. When a USB connecting cable is used, proper connections and proper flow are ensured by using a rectangular (USB "A") connector on the host and a different connector (USB "B", Mini-B or Micro-B) on the attached function.

There are currently two USB standards in general use. USB 1.1, somewhat obsolete but common in PCs just a few years old, is capable of 12 megabits per second (12 Mbit/s). USB 2.0 is the later standard and is rated up to 480 Mbit/s. Most USB 2.0 ports will allow the use of older USB 1.1 devices — that is, they are backward compatible. However maximum cable lengths and available power to devices may be affected. USB 3.0 uses additional wires and some new connectors to enable communication at up to 5 Gbit/s in its SuperSpeed configuration.

Older PCs that have just one or two USB ports can have additional ports by adding an inexpensive USB port expander, commonly called a USB hub. Some units obtain their power from the host computer's USB port, and distribute only the USB signals along with the remaining power available from the host. Others come with a small power supply that provides the normal power to each new USB port.

If the PC does not have any USB ports, an expansion card can be used to add USB ports. It is quite common to see new personal computers without any serial or parallel ports, with USB replacing these functions. If you need to use your older serial or parallel peripheral devices with these "legacy-free" PC systems, adapter cables are available to connect them to USB ports.

ETHERNET

For wired networks, 100BaseT or "fast Ethernet" has been the standard for a number of years, with low cost hubs and switches commonly available. The 100BaseT systems are rated to 100 Mbit/s and use unshielded, twisted-pair Category 5 network cable. Each computer on the network connects to a central hub or switch.

More recently, 1000BaseT (gigabit Ethernet, or GigE) and is quickly replacing 100BaseT even at the consumer level. Gigabit Ethernet also uses Cat 5 cable, though Cat 5e or Cat 6 is often recommended. Explanation of the MBaseN and Category N terminology can be found in many available networking books.

Mutual interference with ham station operation is not uncommon. A 10 Mbit/s signal, if all bit positions are filled, means that unshielded wire is carrying 10 MHz and various harmonics of 10 MHz around the shack. When digital words are going through the network, any number of frequencies may be present on the wires and the wires may be very susceptible to pick-up from HF, VHF and UHF signals. In addition, cable and DSL modems may not only transmit various frequencies (especially on VHF) but lose data when a few hundred watts on a ham band is present. Some RF in the shack that can normally be ignored can easily bring down a network. Unfortunately, it is also common for this equipment to have "noisy" switching power supplies that can wreak havoc on a ham's sensitive receiver. See the **RF Interference** chapter for more information on dealing with RFI from networking equipment.

WIRELESS NETWORKS — WIFI

The other technology commonly found in home and office networks is IEEE 802.11 wireless. These networks use low-power, spread spectrum transceivers operating in the 2.4 and 5.2-5.8 GHz range to transfer data at nominal rates up to 54 Mbit/s and higher. Older equipment used the 801.11b protocol at 11 Mbit/s; newer gear commonly found at low prices uses 802.11g at 54 Mbit/s maximum data rate. The 802.11n protocol supports data rates up to several hundred megabits per second in the 2.4 and 5 GHz ranges.

These data rates are the maximum under ideal conditions, and it is not uncommon to see links running at significantly lower speeds as the *wireless access point* (WAP) and the wireless adapter dynamically adapt to the conditions. Still, performance is adequate for most uses, and the lack of network cabling makes wireless networking attractive in many situations.

Wireless networks actually appear to be less susceptible to mutual interference, since they are not connected to long runs of unshielded wire that easily act as both transmitting and receiving antennas. The frequency bands used for wireless networks are reasonably far removed from normal ham operations at frequencies below 2.4 GHz, but a high power VHF or UHF station may interfere with such a network. Also, some of the channels used for 802.11b and 802.11g network equipment fall within the amateur spectrum allocation in the 2.4 GHz band.

CABLE REPLACEMENT — BLUETOOTH AND ZIGBEE

A number of low power, short range data links have been created to replace cable based interfaces such as USB and RS-232. These are digital protocols that use unlicensed spectrum in the ISM bands. Two of the most popular as of early 2013 are Bluetooth (**www.bluetooth. com**) and Zigbee (**www.zigbee.org**). Most technologies of this type are either initially or permanently developed as implementations of an IEEE 802.x standard.

Bluetooth data links operate in the 2.4 GHz band and are primarily used for streaming audio and high speed digital links up to several megabits per second. Bluetooth links use frequency-hopping spread-spectrum technology. It can also support personal area networks (piconets or PANs) at various levels of security. Up to seven devices can share a Bluetooth interface with a host. A process called synchronization is used to associate the various devices.

Zigbee is designed for ad-hoc mesh networking applications and operates in the 902 MHz ISM band in North America. Zigbee's maximum data rate is 250 kbit/sec. Because of the lower data rate, most Zigbee applications are designed for control and monitoring applications.

REPLACING COM AND LPT PC INTERFACES

USB-to-serial and USB-to-parallel adapters are relatively inexpensive and easy to find. Note, however, that not all of them support all of the handshaking signals we may require, and some may have delays or inadequate drivers. Be prepared to try several different brands to find one that works with your equipment. Users groups for specific equipment or software are good resources to find out what brands and models of port replacement accessories are compatible and operate properly. While they are becoming less common, there is also still a market for add-in PCI bus cards to add serial, parallel and other ports to desktop computers.

4.10 Glossary of Digital Electronics Terms

AND gate — A logic circuit whose output is 1 only when all of its inputs are 1.

Astable (free-running) multivibrator — A circuit that alternates between two unstable states. This circuit could be considered as an oscillator that produces square waves.

Asynchronous flip-flop — A circuit, also called a *latch*, that changes output state depending on the data inputs, without requiring a clock signal.

Binary — A base-2 number system used in digital electronics that uses the symbols 0 and 1.

Binary coded decimal (BCD) — A simple method for converting binary values to and from decimal for inputs and outputs for user-oriented digital systems. BCD was widely used in the days of 7-segment LED displays but is not common today.

Bistable multivibrator — Another name for a flip-flop circuit that has two stable output states.

Boolean algebra — The mathematical system used to describe and design binary used digital circuits, named after George Boole.

Bus —A set of wires through which data is routed internally within computers and other digital devices.

Clock — A signal that toggles at a regular rate. Clock control is the most common method of synchronizing logic circuits.

Combinational logic — A type of circuit element in which the output depends on the present inputs. (Also see **Sequential logic**.)

Complementary metal-oxide semiconductor (CMOS) — A type of construction used to make digital integrated circuits. CMOS is composed of both N-channel and P-channel MOS devices on the same chip.

Counter (divider, divide-by-n counter) — A circuit that is able to change from one state to the next each time it receives an input signal. A counter produces an output signal every time a predetermined number of input signals have been received.

Decimal — The base-10 number system we use every day that uses the symbols 0 through 9.

DeMorgan's Theorem — In Boolean algebra, a way to simplify the complement of a large expression or to enable a designer to interchange a number of equivalent gates.

Digital IC — An integrated circuit whose output is either on (1) or off (0).

Dynamic (edge-triggered) input — A control signal that allows a circuit to change state only when the control signal changes from unasserted to asserted.

Exclusive OR gate — A logic circuit whose output is 1 when either of two inputs is 1 and whose output is 0 when neither input is 1 or when both inputs are 1.

Fan-out — The ability of a logic element

to drive or feed several other logic elements.

Field Programmable Gate Array (FPGA) — A type of programmable logic array that can contain several hundred to millions of logic gates and up to 1000 or more I/O pins.

FireWire (IEEE-1394) — A very high speed serial protocol capable of up to 400 Mbit/s of sustained transfer.

Flip-flop (bistable multivibrator) — A circuit that has two stable output states, and which can change from one state to the other when the proper input signals are detected.

Gate — A combinational logic element with two or more inputs and one output. The output state depends upon the state of the inputs.

Handshaking — Functions to coordinate data transfer.

Hexadecimal — A base-16 number system widely used in computer systems that uses the 0,1,2,3,4,5,6,7,8,9,A,B,C,D,E,F.

Integrated circuit — A device composed of many bipolar or field-effect transistors manufactured on the same chip, or wafer, of silicon.

Inverter — A logic circuit with one input and one output. The output is 1 when the input is 0, and the output is 0 when the input is 1.

Latch — Another name for a bistable multivibrator (flip-flop) circuit. The term latch reminds us that this circuit serves as a memory unit, storing a bit of information.

Linear IC — An integrated circuit whose output voltage is a linear (straight line) representation of its input voltage.

Logic probe — A simple piece of test equipment used to indicate high or low

logic states (voltage levels) in digital-electronic circuits.

Microcontroller — A "computer on a chip" that usually consists of a relatively small microprocessor along with some amount of program memory, data memory, input/output ports and often some specialized peripheral devices.

Monostable multivibrator (one shot) — A circuit that has one stable state. It can be forced into an unstable state for a time determined by external components, but it will revert to the stable state after that time.

NAND (NOT AND) gate — A logic circuit whose output is 0 only when both inputs are 1.

Noninverter — A logic circuit with one input and one output, and whose output state is the same as the input state (0 or 1). Sometimes called a *noninverting buffer*.

NOR (NOT OR) gate — A logic circuit whose output is 0 if either input is 1.

OR gate — A logic circuit whose output is 1 when either input is 1.

Parallel — A digital signaling method in which all the bits in a group are handled exactly at the same time.

Programmable logic device (PLD) — A device that includes a generic array of gates that can be controlled by program code. PLDs can be made to replace large numbers of individual ICs.

Propagation delay — The time delay between providing an input to a digital circuit and seeing a response at the output.

Register — A set of latches or flip-flops storing an n-bit number.

RS-232 — The most common serial bus protocol.

RS-422 — A serial protocol similar to RS-232, but employing fully differential data lines.

Serial — A digital signaling method in which each bit is sent in turn over a single channel or wire, according to an agreed sequence.

Sequential logic — A type of circuit element in which the output depends on the present inputs, the previous sequence of inputs and often a clock signal. (Also see **Combinational logic**.)

Square wave — A periodic waveform that alternates between two values, and spends an equal time at each level. It is made up of sine waves at a fundamental frequency and all odd harmonics.

Static (level-triggered, or gated) input — A control signal that allows the circuit to change state whenever the control signal is at its active or asserted level.

Synchronous flip-flop — A circuit whose output state depends on the data inputs, but that will change output state only when it detects the proper clock signal.

Transition region — The undefined region between the two binary states. Also known as the *noise margin*.

Transition time — The time it takes a digital circuit to change state. The transition from a 0 to a 1 state is called the *rise time*, and the transition from a 1 to a 0 state is called the *fall time*.

Tri-state gate — A gate with one additional control lead. When enabled, the gate operates normally; when not enabled, the output goes to a high impedance.

Truth table — A chart showing the outputs for all possible input combinations to a digital circuit.

Universal serial bus (USB) — A computer standard for an intelligent serial data transfer protocol.

4.11 References and Bibliography

DIGITAL ELECTRONICS

Bignell and Donovan, *Digital Electronics* (Delmar Learning, 2006)

Holdsworth, B., *Digital Logic Design* (Newnes, 2002)

Lancaster and Berlin, *CMDS Cookbook* (Newnes, 1997)

Tocci, Widmer and Moss, *Digital Systems: Principles and Applications* (Prentice Hall, 2006)

Tokheim, R., *Digital Electronics: Principles and Applications* (McGraw-Hill, 2008)

Logic simulation software: "Getting Started with Digital Works" **www.spu.edu/cs/faculty/bbrown/circuits/howto.html**

MICROPROCESSORS AND MICROCONTROLLERS

Dumas, J., *Computer Architecture: Fundamentals and Principles of Computer Design* (CRC Press, 2005)

Gilmore, C., *Microprocessors: Principles and Applications* (McGraw-Hill, 1995)

Klotz, L., WA5ZNU (Ed.), *Ham Radio for Arduino and PICAXE* (ARRL, 2013)

Korneev and Kiselev, *Modern Microprocessors* (Charles River Press, 2004)

Spencer, M., WA8SME, *ARRL's PIC Programming for Beginners* (ARRL, 2010)

Tocci and Ambrosio, *Microprocessors and Microcomputers: Hardware and Software* (Prentice-Hall, 2002)

Contents

Chapter 5

RF Techniques

This chapter is a compendium of material from ARRL publications and other sources. It assumes the reader is familiar with the concepts introduced in the **Electrical Fundamentals** and **Analog Basics** chapters. The topics and techniques discussed here are associated with the special demands of circuit design in the HF and VHF ranges. The material is collected from previous editions of this book written by Leonard Kay, K1NU; *Introduction to Radio Frequency Design* by Wes Hayward, W7ZOI; and *Experimental Methods in RF Design* by Wes Hayward, W7ZOI, Rick Campbell, KK7B, and Bob Larkin, W7PUA. Material on ferrites is drawn from publications by Jim Brown, K9YC. The section on Noise was written by Paul Wade, W1GHZ, with contributions from Joe Taylor, K1JT. The editor was Ward Silver, NØAX.

Chapter 5 — CD-ROM Content

Supplemental Articles
- "Reflections on the Smith Chart" by Wes Hayward, W7ZOI
- Tuned Networks
- "Simplified Design of Impedance-Matching Networks," Parts I through III by George Grammer, W1DF (SK)
- *LTSpice* simulation files for Section 5.3, Effects of Parasitic Characteristics

5.1 Introduction

When is an inductor not an inductor? When it's a capacitor! This statement may seem odd, but it suggests the main message of this chapter. In the earlier chapter, **Electrical Fundamentals**, the basic components of electronic circuits were introduced. As you may know from experience, those simple component pictures are ideal. That is, an ideal component (or element) by definition behaves exactly like the mathematical equations that describe it, and only in that fashion. For example, the current through an ideal capacitor is equal to the capacitance times the rate of change of the voltage across it without consideration of the materials or techniques by which a real capacitor is manufactured.

It is often said that, "Parasitics are anything you don't want," meaning that the component is exhibiting some behavior that detracts from or compromises its intended use. Real components only approximate ideal components, although sometimes quite closely. Any deviation from ideal behavior a component exhibits is called *non-ideal* or *parasitic*. The important thing to realize is that *every* component has parasitic aspects that become significant when it is used in certain ways. This chapter deals with parasitic effects that are commonly encountered at radio frequencies.

Knowing to what extent and under what conditions real components cease to behave like their ideal counterparts, and what can be done to account for these behaviors, allows the circuit designer or technician to work with circuits at radio frequencies. We will explore how and why the real components behave differently from ideal components, how we can account for those differences when analyzing circuits and how to select components to minimize, or exploit, non-ideal behaviors.

5.2 Lumped-Element versus Distributed Characteristics

Most electronic circuits that we use every day are inherently and mathematically considered to be composed of *lumped elements*. That is, we assume each component acts at a single point in space, and the wires that connect these lumped elements are assumed to be perfect conductors (with zero resistance and insignificant length). This concept is illustrated in **Fig 5.1**. These assumptions are perfectly reasonable for many applications, but they have limits. Lumped element models break down when:

- Circuit impedance is so low that the small, but non-zero, resistance in the wires is important. (A significant portion of the circuit power may be lost to heat in the conductors.)
- Operating frequency is high enough that the length of the connecting wires is a significant fraction (>0.1) of the wavelength causing the propagation delay along the conductor or radiation from it to affect the circuit in which it is used.
- Transmission lines are used as conductors. (Their characteristic impedance is usually significant, and impedances connected to them are transformed as a function of the line length. See the **Transmission Lines** chapter for more information.)

Effects such as these are called *distributed*, and we talk of *distributed elements* or effects to contrast them to lumped elements.

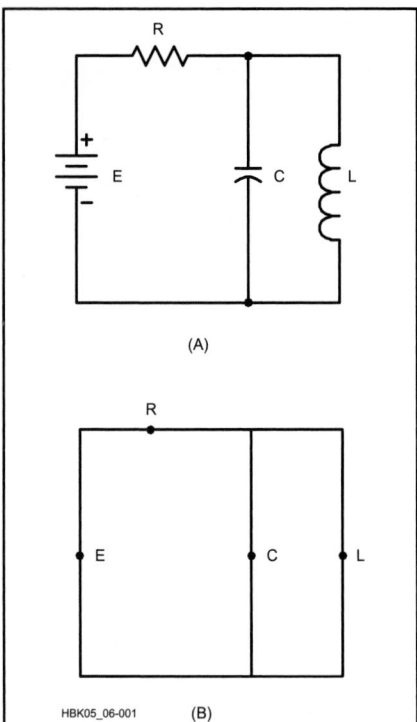

Fig 5.1 — The lumped element concept. Ideally, the circuit at A is assumed to be as shown at B, where the components are isolated points connected by perfect conductors. Many components exhibit nonideal behavior when these assumptions no longer hold.

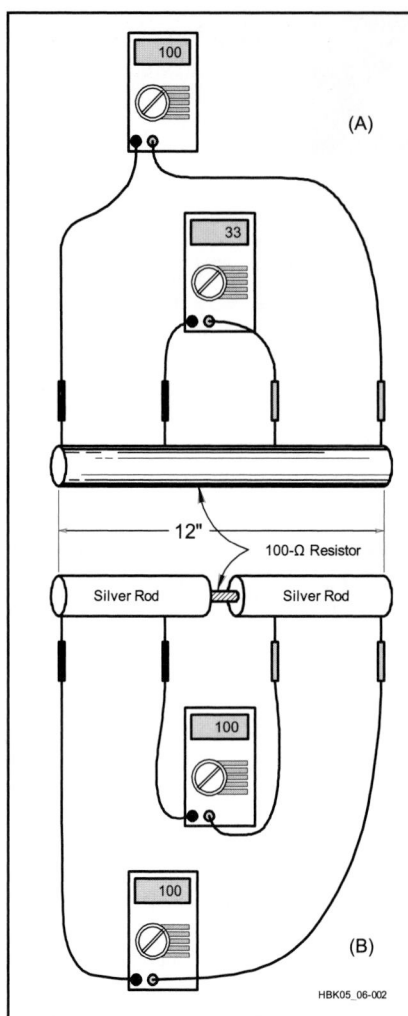

Fig 5.2 — Distributed (A) and lumped (B) resistances. See text for discussion.

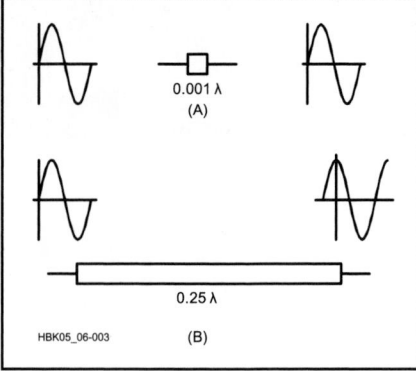

Fig 5.3 — The effects of distributed resistance on the phase of a sinusoidal current. There is no phase delay between ends of a lumped element.

To illustrate the differences between lumped and distributed elements, consider the two resistors in **Fig 5.2**, which are both 12 inches long. The resistor at A is a uniform rod of carbon. The second "resistor" B is made of two 6-inch pieces of silver rod (or other highly conductive material), with a small resistor soldered between them. Now imagine connecting the two probes of an ohmmeter to each of the two resistors, as in the figure. Starting with the probes at the far ends, as we slide the probes toward the center, the carbon rod will display a constantly decreasing resistance on the ohmmeter. This represents a distributed resistance. On the other hand, the ohmmeter connected to the other 12-inch "resistor" will display a constant resistance as long as one probe remains on each side of the small resistance and as long as we neglect the resistance of the silver rods! This represents a lumped resistance connected by perfect conductors.

Lumped elements also have the very desirable property that they introduce no phase shift resulting from propagation delay through the element. (Although combinations of lumped elements can produce phase shifts by virtue of their R, L and C properties.) Consider a lumped element that is carrying a sinusoidal current, as in **Fig 5.3A**. Since the element has negligible length, there is no phase difference in the current between the two sides of the element — *no matter how high the frequency* — precisely *because* the element length is negligible. If the physical length of the element were long, say 0.25 wavelength (0.25 λ) as shown in Fig 5.3B, the current phase would *not* be the same from end to end. In this instance, the current is delayed by 90 electrical degrees as it moves along

the element. The amount of phase difference depends on the circuit's electrical length.

Because the relationship between the physical size of a circuit and the wavelength of an ac current present in the circuit will vary as the frequency of the ac signal varies, the ideas of lumped and distributed effects actually occupy two ends of a spectrum. At HF (30 MHz and below), where λ ≥ 10 m, the lumped element concept is almost always valid. In the UHF and microwave region (300 MHz and above), where λ ≤ 1 m and physical component size can represent a significant fraction of a wavelength, nearly all components and wiring exhibits distributed effects to one degree or another. From roughly 30 to 300 MHz, whether the distributed effects are significant must be considered on a case-by-case basis.

Of course, if we could make resistors, capacitors, inductors and so on, very small, we could treat them as lumped elements at much higher frequencies. For example, surface-mount components, which are manufactured in very small, leadless packages, can be used at much higher frequencies than leaded components and with fewer non-ideal effects.

It is for these reasons that circuits and equipment are often specified to work within specific frequency ranges. Outside of these ranges the designer's assumptions about the physical characteristics of the components and the methods and materials of the circuit's assembly become increasingly invalid. At frequencies sufficiently removed from the design range, circuit behavior often changes in unpredictable ways.

5.3 Effects of Parasitic Characteristics

At HF and above (where we do much of our circuit design) several other considerations become very important, in some cases dominant, in the models we use to describe our components. To understand what happens to circuits at RF we turn to a brief discussion of some electromagnetic and microwave theory concepts.

Parasitic effects due to component leads, packaging, leakage and so on are relatively common to all components. When working at frequencies where many or all of the parasitics become important, a complex but completely general model such as that in **Fig 5.4** can be used for just about any component, with the actual component placed in the box marked *. Parasitic capacitance, C_p, and leakage conductance, G_L, appear in parallel across the device, while series resistance, R_s, and parasitic inductance, L_s, appear in series with it. Package capacitance, C_{pkg}, appears as an additional capacitance in parallel across the whole device.

These small parasitics can significantly affect frequency responses of RF circuits. Either take steps to minimize or eliminate them, or use simple circuit theory to predict and anticipate changes. This maze of effects may seem overwhelming, but remember that it is very seldom necessary to consider all parasitics at all frequencies and for all applications. The **Computer-Aided Circuit Design** chapter shows how to incorporate the effect of multiple parasitics into circuit design and performance modeling. Files for the *LTSpice* simulation package that include parasitic characteristics for a resistor, capacitor and inductor are provided on the CD-ROM that comes with this *Handbook*.

5.3.1 Parasitic Inductance

Maxwell's equations — the basic laws of electromagnetism that govern the propagation of electromagnetic waves and the operation of all electronic components — tell us that any wire carrying a current that changes with time (one example is a sine wave) develops a changing magnetic field around it. This changing magnetic field in turn induces an opposing voltage, or *back EMF*, on the wire. The back EMF is proportional to how fast the current changes (see **Fig 5.5**).

We exploit this phenomenon when we make an inductor. The reason we typically form inductors in the shape of a coil is to concentrate the magnetic field and thereby maximize the inductance for a given physical size. However, *all* wires carrying varying currents have these inductive properties. This includes the wires we use to connect our circuits, and even the *leads* of capacitors, resistors and so on. The inductance of a straight, round,

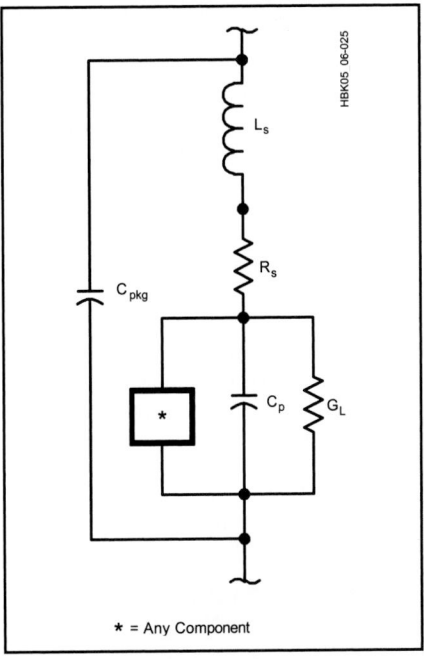

Fig 5.4 — A general model for electrical components at VHF frequencies and above. The box marked * represents the component itself. See text for discussion.

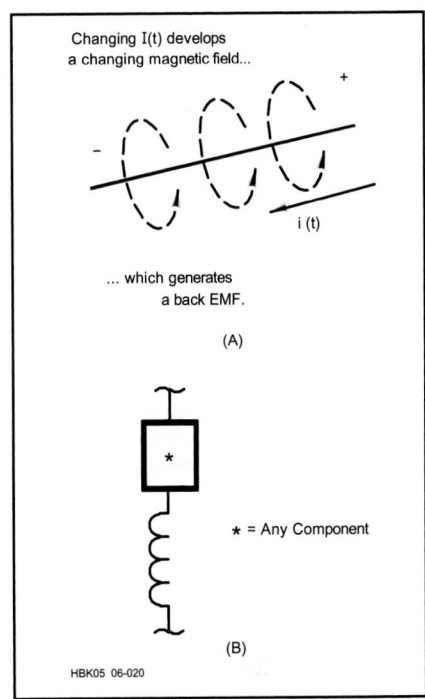

Fig 5.5 — Inductive consequences of Maxwell's equations. At A, any wire carrying a changing current develops a voltage difference along it. This can be mathematically described as an effective inductance. B adds parasitic inductance to a generic component model.

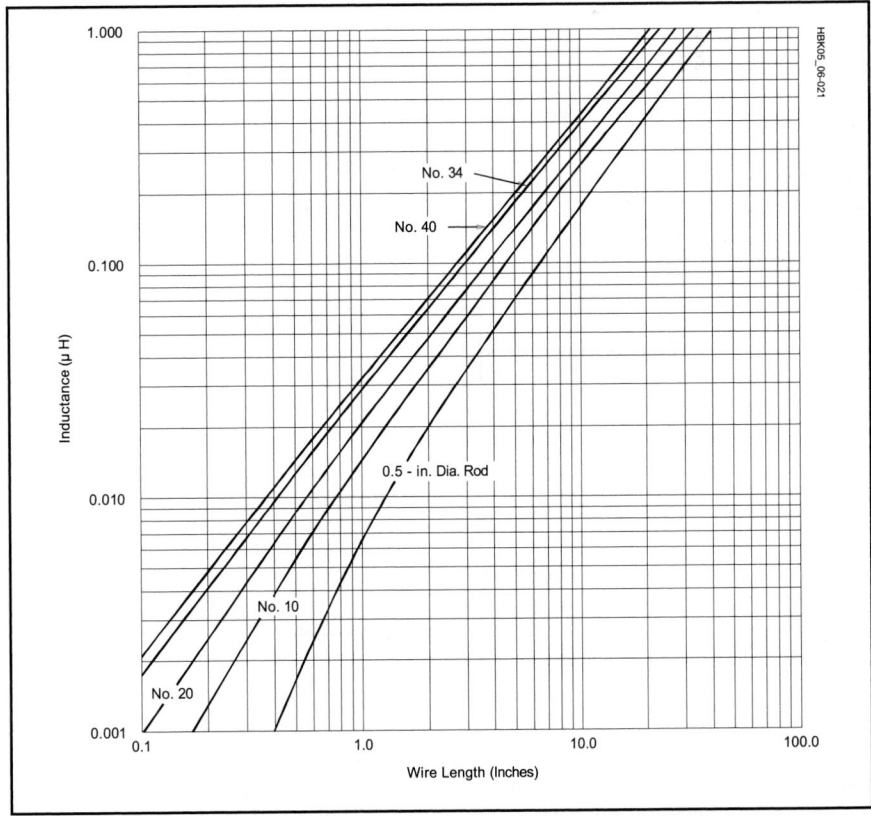

Fig 5.6 — A plot of inductance vs length for straight conductors in several wire sizes.

nonmagnetic wire in free space is given by:

$$L = 0.00508 \, b \left[\ln \left(\frac{2b}{a} \right) - 0.75 \right] \quad (1)$$

where

L = inductance, in µH
a = wire radius, in inches
b = wire length, in inches
ln = natural logarithm (2.303 × log₁₀)

Skin effect (discussed below) changes this formula slightly at VHF and above. As the frequency approaches infinity, the value 0.75 in the above equation increases to approach 1. This effect usually causes a change of no more than a few percent.

As an example, let's find the inductance of a typical #18 wire (diameter = 0.0403 inch and a = 0.0201) that is 4 inches long (b = 4):

$$L = 0.00508 \, (4) \left[\ln \left(\frac{8}{0.0201} \right) - 0.75 \right]$$

$$= 0.0203 \left[5.98 - 0.75 \right] = 0.106 \, \mu H$$

Wire of this diameter has an inductance of about 25 nH per inch of length. In circuits operating at VHF and higher frequencies, including high-speed digital circuits, the inductance of component leads can become significant. (The #24 AWG wire typically used for component leads has an inductance on the order of 20 nH per inch.) At these frequencies, lead inductance can affect circuit behavior, making the circuit hard to reproduce or repair.

Good design and construction practice is to minimize the effects of lead inductance by using surface-mount components or trimming the leads to be as short as possible.

The impact of reactance due to parasitic inductance is usually very small; at AF or LF, parasitic inductive reactance of most components is practically zero. To use this example, the reactance of a 0.106 µH inductor even at 10 MHz is only 6.6 Ω. **Fig 5.6** shows a graph of the inductance for wires of various gauges (radii) as a function of length. Whether the reactance is significant or not depends on the application and the frequency of use.

We can represent parasitic inductance in component models by adding an inductor of appropriate value in series with the component since the wire leads are in series with the element. This (among other reasons) is why minimizing lead lengths and interconnecting wires becomes very important when designing circuits for VHF and above.

PARASITIC INDUCTANCE IN RESISTORS

The basic construction of common resistor types is shown in **Fig 5.7**. The primary parasitic effect associated with resistors is parasitic inductance. (Some parasitic capacitance exists between the leads or electrodes due to packaging.) **Fig 5.8** shows some more accurate circuit models for resistors at low to medium frequencies. The type of resistor with the most parasitic inductance are wire-wound resistors, essentially inductors used as resistors. Their

use is therefore limited to dc or low-frequency ac applications where their reactance is negligible. Remember that this inductance will also affect switching transient waveforms, even at dc, because the component will act as an RL circuit. The inductive effects of wire-wound resistors begin to become significant in the audio range above a few kHz.

As an example, consider a 1-Ω wire-wound resistor formed from 300 turns of #24 wire closely-wound in a single layer 6.3 inches long on a 0.5-inch diameter form. What is its approximate inductance? From the inductance formula for air-wound coils in the **Electrical Fundamentals** chapter:

$$L = \frac{d^2 n^2}{18d + 401} = \frac{0.5^2 \times 300^2}{(18 \times 0.5) + (40 \times 6.3)} = 86 \, \mu H$$

If we want the inductive reactance to be less than 10% of the resistor value, then this resistor cannot be used above f = 0.1 / (2π × 86 µH) = 185 Hz! Real wire-wound resistors have multiple windings layered over each other to minimize both size and parasitic inductance (by winding each layer in opposite directions, much of the inductance is canceled). If we assume a five-layer winding, the length is reduced to 1.8 inches and the inductance to approximately 17 µH, so the resistor can then be used below 937 Hz. (This has the effect of increasing the resistor's parasitic capacitance, however.)

The resistance of certain types of tubular film resistors is controlled by inscribing a spiral path through the film on the inside of the tube. This creates a small inductance that may be significant at and above the higher audio frequencies.

NON-INDUCTIVE RESISTORS

The resistors with the least amount of parasitic inductance are the bulk resistors, such as carbon-composition, metal-oxide, and ceramic resistors. These resistors are made from a single linear cylinder, tube or block of resis-

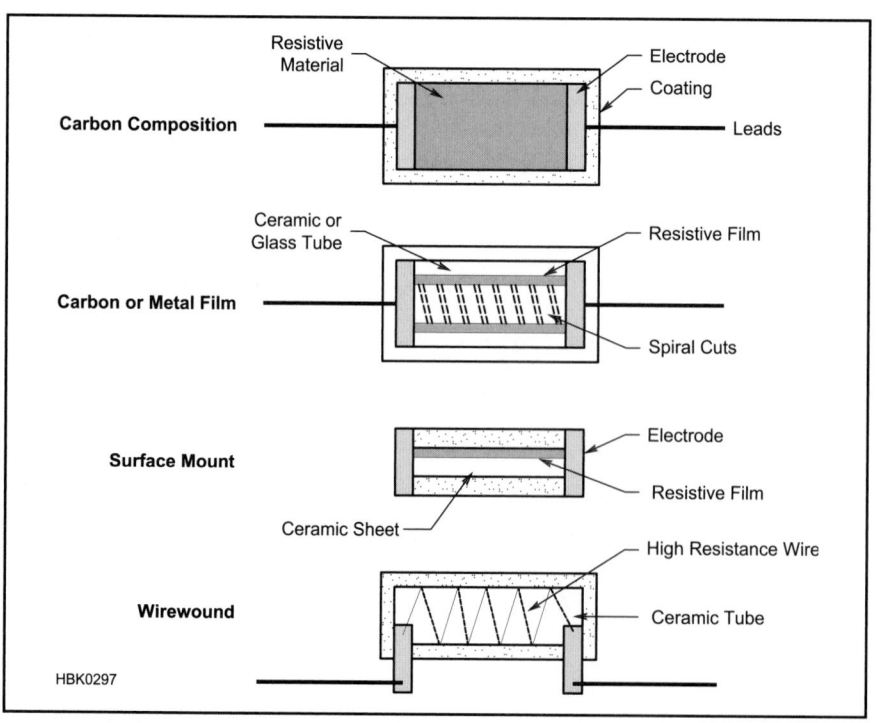

Fig 5.7 — The electrical characteristics of different resistor types are strongly affected by their construction. Reactance from parasitic inductance and capacitance strongly impacts the resistor's behavior at RF.

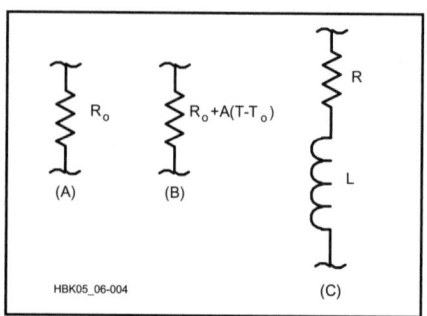

Fig 5.8 — Circuit models for resistors. The wire-wound model with associated inductance is shown at C. B includes the effect of temperature (T). For designs at VHF and higher frequencies, the model at C could be used with L representing lead inductance.

tive material so that inductance is minimized. Each type of resistor has a maximum usable frequency, above which parasitic capacitance and inductance begin to become significant. Review the manufacturer's data sheet for the component to learn about its performance at high frequencies.

Some resistors advertised as "noninductive" are actually wire-wound resistors with a special winding technique that minimizes inductance. These resistors are intended for use at audio frequencies and are not suitable for use at RF. If you are not sure, ask the vendor if the resistors are suitable for use in RF circuits.

Because resistors are manufactured with an insulating coating, it can be difficult to determine their internal structure and thus estimate their parasitic inductance. In cases where a surplus or used component is to be included, it is recommended that you test the component with an impedance meter or make some other type of reactance measurement if you are unable to access the manufacturer's specifications for the resistor.

PARASITIC INDUCTANCE IN CAPACITORS

The size and shape of a capacitor's plates and the leads used to connect them to circuits create parasitic inductance, often referred to as *equivalent series inductance* (ESL) by capacitor manufacturers. **Figs 5.9** and **5.10** show reasonable models for capacitors that are good up to VHF.

Fig 5.11 shows a roll-type capacitor made of two strips of very thin metal foil and separated by a dielectric. After leads are attached to the foil strips, the sandwich is rolled up and either placed in a metal can or coated with plastic. *Radial leads* both stick out of one end of the roll and *axial leads* from both ends along the roll's axis. Because of the rolled strips, the ESL is high. Electrolytic and many types of film capacitors are made with roll construction. As a result, they are generally not useful in RF circuits.

In the stack capacitor, thin sheets of dielectric are coated on one side with a thin metal layer. A stack of the sheets is placed under pressure and heated to make a single solid unit. Metal side caps with leads attached contact the metal layers. The ESL of stack capacitors is very low and so they are useful at high frequencies. Ceramic and mica capacitors are the most common stack-style capacitor.

Parallel-plate air and vacuum capacitors used at RF have relatively low parasitic inductance, but transmitting capacitors made to withstand high voltages and current are large enough that parasitic inductance becomes significant, limiting their use to low-VHF and lower frequencies. Adjustable capacitors (air variables and compression or piston trimmer

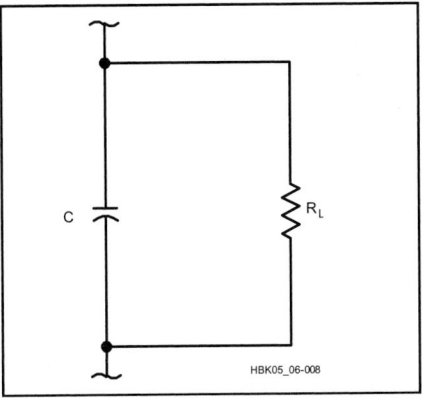

Fig 5.9 — A simple capacitor model for frequencies well below self-resonance.

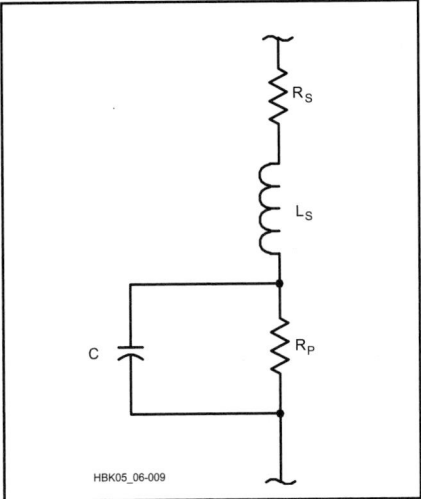

Fig 5.10 — A capacitor model for VHF and above including series resistance and distributed inductance.

capacitors) for tuning low-power circuits are much smaller and so have correspondingly lower parasitic inductance.

It is difficult for a single capacitor to work well over a very wide frequency range, so capacitors are often placed in parallel as discussed in the section below on Bypassing and Decoupling. It is often suggested that different types of capacitors be connected in parallel to avoid the effects of parasitic inductance at different frequencies, but without testing and careful modeling the results are often unpredictable or even counterproductive as the referenced discussion shows.

5.3.2 Parasitic Capacitance

Maxwell's equations also tell us that if the voltage between any two points changes with time, a displacement current is generated between these points as illustrated in **Fig 5.12**. This *displacement current* results from the propagation of the electromagnetic field between the two points and is not to be confused with *conduction current*, which is the movement of electrons. Displacement current is directly proportional to the rate at which the voltage is changing.

When a capacitor is connected to an ac voltage source, a steady ac current can flow because taken together, conduction current and displacement current "complete the loop" from the positive source terminal, across the plates of the capacitor, and back to the negative terminal.

In general, parasitic capacitance shows up *wherever* the voltage between two points is changing with time, because the laws of electromagnetics require a displacement current to flow. Since this phenomenon

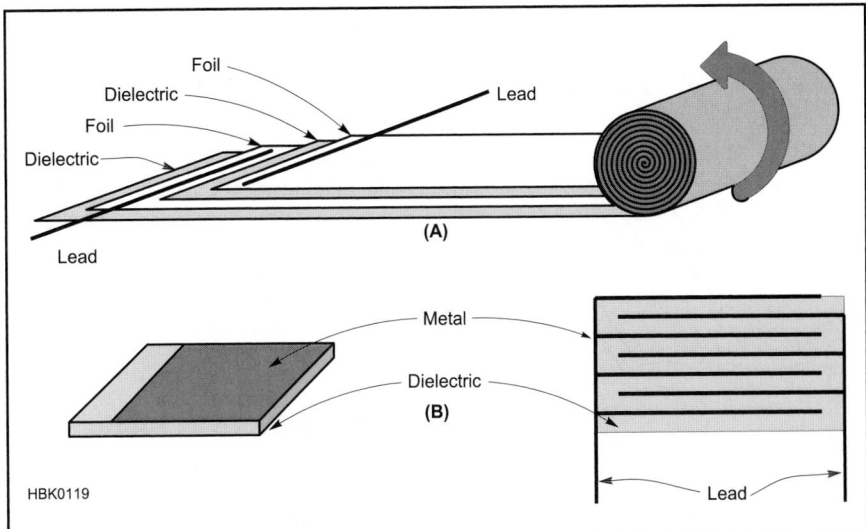

Fig 5.11 — Two common types of capacitor construction. (A) Roll construction uses two strips of foil separated by a strip of dielectric. (B) Stack construction layers dielectric material (such as ceramic or film), one side coated with metal. Leads are attached and the assembly coated with epoxy resin.

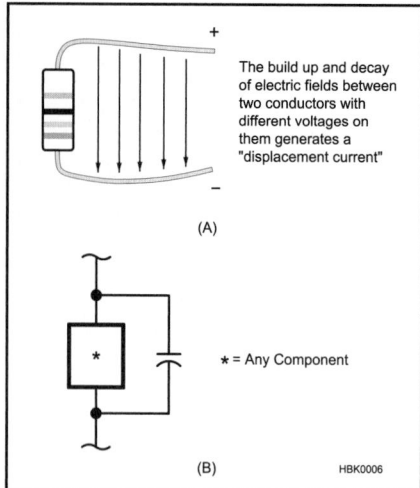

Fig 5.12 — Capacitive consequences of Maxwell's equations. A: Any changing voltage between two points, for example along a bent wire, generates a displacement current running between them. This can be mathematically described as an effective capacitance. B adds parasitic capacitance to a generic component model.

represents an *additional* current path from one point in space to another, we can add this parasitic capacitance to our component models by adding a capacitor of appropriate value in *parallel* with the component. These parasitic capacitances are typically less than 1 pF, so that below VHF they can be treated as open circuits (infinite reactances) and thus neglected.

PACKAGE CAPACITANCE

Another source of capacitance, also in the 1-pF range and therefore important only at VHF and above, is the packaging of the com-

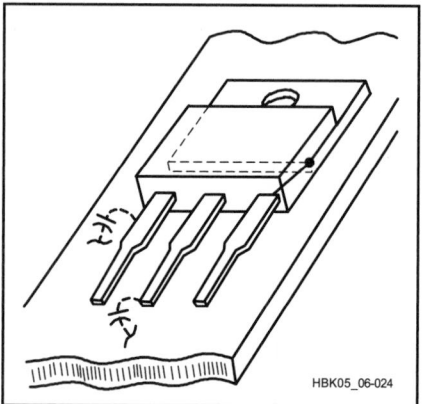

Fig 5.13 — Unexpected stray capacitance. The mounting tab of TO-220 transistors is often connected to one of the device leads. Because one lead is connected to the chassis, small capacitances from the other lead to the chassis appear as additional package capacitance at the device. Similar capacitance can appear at any device with a conductive package.

ponent itself. For example, a power transistor packaged in a TO-220 case (see **Fig 5.13**), often has either the emitter or collector connected to the metal tab itself. This introduces an extra *inter-electrode capacitance* across the junctions.

The copper traces on a PC board also create capacitance with the circuit components, to other traces, and to power and ground planes. Double-sided PC boards have a certain capacitance per square inch between the layers of copper on each side of the board. (Multi-layer PC boards have higher values of capacitance due to the smaller separation between layers.) It is possible to create capacitors by leaving unetched areas of copper on both sides of the board. The dielectric constant of inexpensive PC board materials intended for use at low-frequencies is not well-controlled, however, leading to significant variations in capacitance. For this reason, the copper on one side of a double-sided board should be completely removed under frequency-determining circuits such as VFOs.

Stray capacitance (a general term used for any "extra" capacitance that exists due to physical construction) appears in any circuit where two metal surfaces exist at different voltages. Such effects can be modeled as an extra capacitor in parallel with the given points in the circuit. A rough value can be obtained with the parallel-plate formula given in the chapter on **Electrical Fundamentals**. Similar to parasitic inductance, *any* circuit component that has wires attached to it, or is fabricated from wire, or is near or attached to metal, will have a parasitic capacitance associated with it, which again, becomes important only at RF.

Stray capacitance can be difficult to account for in circuit design because it exists *between* components and other circuit structures, depending on the physical orientation of the component. Its presence may allow signals to flow in ways that disrupt the normal operation of a circuit and may have a greater affect in a high-impedance circuit because the capacitive reactance may be a greater percentage of the circuit impedance. Also, because stray capacitance often appears in parallel with the circuit, the stray capacitor may bypass more of the desired signal at higher frequencies. Careful physical design of an RF circuit and selection of components can minimize the effects of stray capacitance.

5.3.3 Inductors at Radio Frequencies

Inductors are perhaps the component with the most significant parasitic effects. Where there are many different varieties of form for capacitors and resistors, most inductors are fundamentally similar: a coil of wire on a tubular or toroidal form. As such, they are

affected by both parasitic resistance and capacitance as shown in the simple inductor model of **Fig 5.14**.

While the leakage conductance of a capacitor is usually negligible, the series resistance of an inductor often is not. This is caused by the long lengths of thin wire needed to create typical inductance values used in RF circuits and the skin effect (discussed below). Consider a typical air-core inductor, with L = 33 μH and a minimum Q of 30 measured at 2.5 MHz. This would indicate a series resistance of $R_s = 2 \pi f L / Q = 17 \, \Omega$ that could significantly alter the bandwidth of a circuit. The skin effect also results in parasitic resistance in the upper HF ranges and above. To minimize losses due to the skin effect, inductors at these frequencies, particularly those intended for use in transmitters and amplifiers and that are expected to carry significant currents, are often made from large-diameter wire or tubing or even flat strap to maximize the surface area for current flow. For coils with many turns for which large conductors are impractical, *Litz wire* is sometimes used. Litz wire is made of many fine insulated strands woven together, each with a diameter smaller than the skin depth at the expected frequency of use, thus presenting a larger surface area than a solid wire or normal stranded wire.

If an inductor is wound on a magnetic core, the core itself can have losses that are treated as parasitic resistance. Each type of core — iron, powdered iron or ferrite — has a frequency range over which it is designed for maximum efficiency. Outside of that range, the core may have significant losses, raising the parasitic resistance of the inductor.

Parasitic capacitance is a particular concern

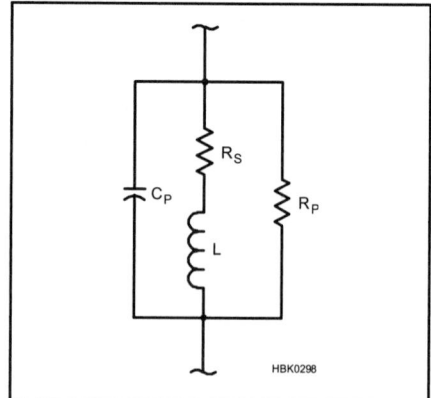

Fig 5.14 — General model for inductor with parasitic capacitance and resistance. The parasitic capacitance represents the cumulative effect of the capacitance between the different turns of wire. Parasitic resistance, R_S, depends on frequency due to the skin effect. R_P represents leakage resistance. At low frequencies, parasitic capacitance, C_P, can be neglected.

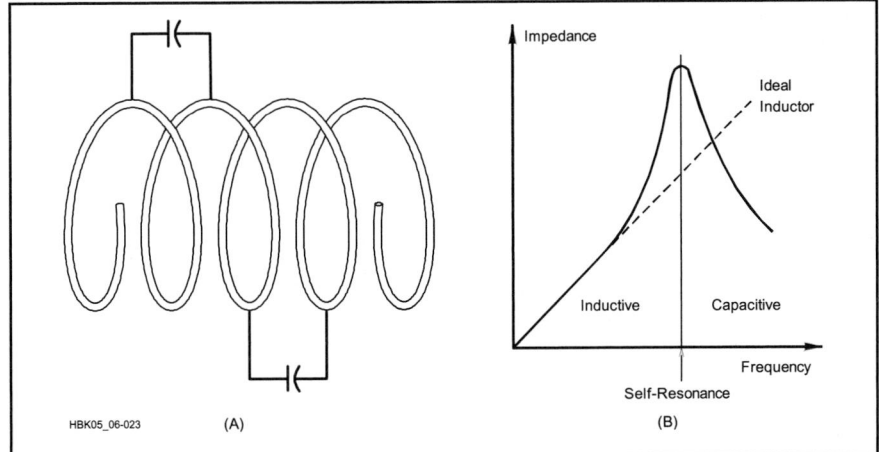

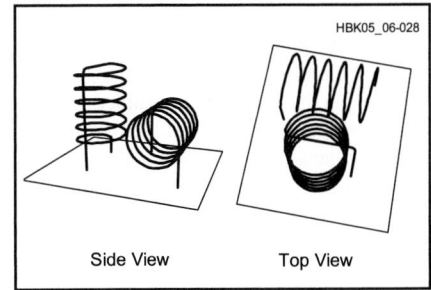

Fig 5.16 — Unshielded coils in close proximity should be mounted perpendicular to each other to minimize coupling.

Fig 5.15 — Coils exhibit distributed capacitance as explained in the text. The graph at B shows how distributed capacitance resonates with the inductance. Below resonance, the reactance is predominantly inductive which increases as frequency increases. However, above resonance, the reactance becomes predominantly capacitive which decreases as frequency increases.

for inductors because of their construction. Consider the inductor in **Fig 5.15**. If this coil has n turns, then the ac voltage between identical points of two neighboring turns is 1/n times the ac voltage across the entire coil. When this voltage changes due to an ac current passing through the coil, the effect is that of many small capacitors acting in parallel with the inductance of the coil. Thus, in addition to the capacitance resulting from the leads, inductors have higher parasitic capacitance due to their physical shape.

These various effects illustrate why inductor construction has such a large effect on performance. As an example, assume you're working on a project that requires you to wind a 5 µH inductor. Looking at the coil inductance formula in the **Electrical Fundamentals** chapter, it comes to mind that many combinations of length and diameter could yield the desired inductance. If you happen to have both 0.5 and 1-inch coil forms, why should you select one over the other? To eliminate some other variables, let's make both coils 1 inch long, close-wound, and give them 1-inch leads on each end.

Let's calculate the number of turns required for each. On a 0.5-inch-diameter form:

$$n = \frac{\sqrt{L(18d + 40l)}}{d} =$$

$$\frac{\sqrt{5[(18 \times 0.5) + (40 \times 1)]}}{0.5} = 31.3 \text{ turns}$$

This means coil 1 will be made from #20 AWG wire (29.9 turns per inch). Coil 2, on the 1-inch form, yields

$$n = \frac{\sqrt{5[(18 \times 1) + (40 \times 1)]}}{1} = 17 \text{ turns}$$

which requires #15 AWG wire in order to be close-wound.

What are the series resistances associated with each? For coil 1, the total wire length is 2 inches + (31.3 × π × 0.5) = 51 inches, which at 10.1 Ω /1000 ft gives $R_s = 0.043$ Ω at dc. Coil 2 has a total wire length of 2 inches + (17.0 × π × 1) = 55 inches, which at 3.18 Ω /1000 ft gives a dc resistance of $R_s = 0.015$ Ω, or about ⅓ that of coil 1. Furthermore, at RF, coil 1 will begin to suffer from skin effect at a frequency about 3 times lower than coil 2 because of its smaller conductor diameter. Therefore, if Q were the sole consideration, it would be better to use the larger diameter coil.

Q is not the only concern, however. Such coils are often placed in shielded enclosures. A common rule of thumb says that to prevent the enclosure from affecting the inductor, the enclosure should be at least one coil diameter from the coil on all sides. That is, 3×3×2 inches for the large coil and 1.5×1.5×1.5 for the small coil, a volume difference of over 500%.

INDUCTOR COUPLING

Mutual inductance (see the **Electrical Fundamentals** chapter) will also have an effect on the resonant frequency and Q of RF circuits. For this reason, inductors in frequency-critical circuits should always be shielded, either by constructing compartments for each circuit block or through the use of can-mounted coils. Another helpful technique to minimize coupling is to mount nearby coils with their axes perpendicular as in **Fig 5.16**.

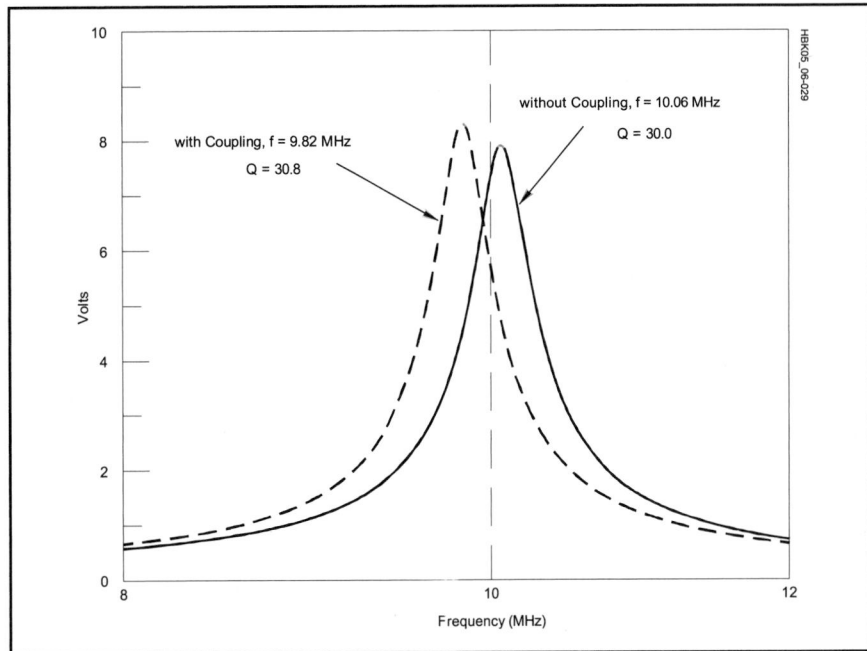

Fig 5.17 — Result of light coupling (k = 0.05) between two identical circuits of Fig 5.19A on their frequency responses.

As an example, assume we build an oscillator circuit that has both input and output resonant circuits. If we are careful to keep the two coils in these circuits uncoupled, the frequency response of either of the two circuits is that of the solid line in **Fig 5.17**.

If the two coils are coupled either through careless placement or improper shielding, the resonant frequency and Q will be affected. The dashed line in Fig 5.17 shows the frequency response that results from a coupling coefficient of k = 0.05, a reasonable value for air-wound inductors mounted perpendicular in close proximity on a circuit chassis. Note the resonant frequency shifted from 10.06 to 9.82 MHz, or 2.4%. The Q has gone up slightly from 30.0 to 30.8 as a result of the slightly higher inductive reactance at the resonant frequency.

5.3.4 Skin Effect

The resistance of a conductor to ac is different than its value for dc because of the way ac fields interact with the conductor. As a result, thick, near-perfect conductors (such as metals) conduct ac only to a certain *skin depth*, δ, inversely proportional to the square root of the frequency of the current. Called the *skin effect*, this decreases the effective cross-section of the conductor at high frequencies and thus increases its resistance.

$$\delta = \frac{1}{\sqrt{\pi f \mu \sigma}} \qquad (2)$$

where

μ is the conductor's permeability, and

σ is the conducting material's conductance.

The increase in resistance caused by the skin effect is insignificant at and below audio frequencies, but beginning around 1 MHz (depending on the size of the conductor) it is so pronounced that practically all the current flows in a very thin layer near the conductor's surface. For example, at 10 MHz, the skin depth of a copper conductor is about 0.02 mm (0.00079 inch). For this reason, at RF a hollow tube and a solid rod of the same diameter and made of the same metal will have the same resistance and the RF resistance of a conductor is often much higher than its dc resistance. A rough estimate of the frequency above which a nonmagnetic wire (one made of nonferrous metal) will begin to show appreciable skin effect can be calculated from

$$f = \frac{124}{d^2} \qquad (3)$$

where

f = frequency, in MHz

d = diameter in mils (a mil is 0.001 inch)

Above this frequency, increase the resis-

tance of the wire by 10× for every 2 decades of frequency (roughly 3.2× for every decade). For example, say we wish to find the RF resistance of a 2-inch length of #18 AWG copper wire at 100 MHz. From the wire tables in the **Component Data and References** chapter, we see that this wire has a dc resistance of 2 inches × 6.386 Ω/1000 ft = 1.06 mΩ. From the above formula, the frequency is found to be 124 / 40.3^2 = 76 kHz. Since 100 MHz is roughly three decades above this (100 kHz to 100 MHz), the RF resistance will be approximately 1.06 mΩ × 3.2^3 = 1.06 mΩ × 32.8 = 34.8 mΩ. Again, values calculated in this manner are approximate and should be used qualitatively — like when you want an answer to a question such as, "Can I neglect the RF resistance of this length of connecting wire at 100 MHz?" Several useful charts regarding skin effect are available in *Reference Data for Engineers*, listed in the References section of this chapter.

Losses associated with skin effect can be reduced by increasing the surface area of the conductor carrying the RF current. Flat, solid strap and tubing are often used for that reason. In addition, because the current-carrying layer is so thin at UHF and microwave frequencies, a thin highly conductive layer at the surface of the conductor, such as silver plating, can lower resistance. Silver plating is too thin to improve HF conductivity significantly, however.

5.3.5 RF Heating

RF current often causes component heating problems where the same level of dc or low frequency ac current may not. These losses result from both the skin effect and from dielectric losses in insulating material, such as a capacitor dielectric.

An example is the tank circuit of an RF oscillator. If several small capacitors are connected in parallel to achieve a desired capacitance, skin effect will be reduced and the total surface area available for heat dissipation will be increased, thus significantly reducing the RF heating effects as compared to a single large capacitor. This technique can be applied to any similar situation; the general idea is to divide the heating among as many components as possible.

An example is shown in the circuit block in **Fig 5.18**, which is representative of the input tank circuit used in many HF VFOs. Along with L, C_{main}, C_{trim}, C1 and C3 set the oscillator frequency. Therefore, temperature effects are critical in these components. By using several capacitors in parallel, the RF current (and resultant heating, which is proportional to the square of the current) is reduced in each component. Parallel combinations are used for the feedback capacitors for the same reason.

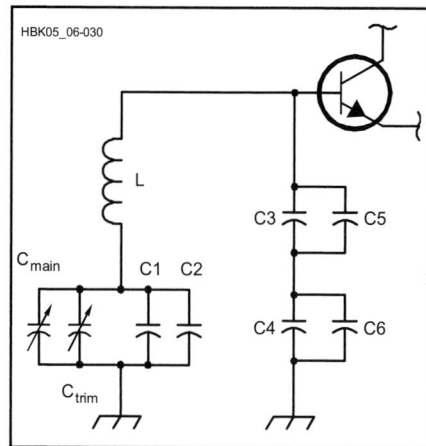

Fig 5.18 — A tank circuit of the type commonly used in VFOs. Several capacitors are used in parallel to distribute the RF current, which reduces temperature effects.

At high power levels, losses due to RF heating — either from resistive or dielectric losses — can be significant. Insulating materials may exhibit dielectric losses even though they are excellent insulators. For example, nylon plastic insulators may work very well at low frequencies, but being quite lossy at and above VHF, they are unsuitable for use in RF circuits at those frequencies. To determine whether material is suitable for use as an insulator at RF, a quick test can be made by heating the material in a microwave oven for a few seconds and measuring its temperature rise. (Take care to avoid placing parts or materials containing any metal in the microwave!) Insulating materials that heat up are lossy and thus unsuitable for use in RF circuits.

5.3.6 Effect on Q

Recall from the **Electrical Fundamentals** chapter that circuit Q, a useful figure of merit for tuned RLC circuits, can be defined in several ways:

$$Q = \frac{X_L \text{ or } X_C}{R} \qquad (4)$$

$$= \frac{\text{energy stored per cycle}}{\text{energy dissipated per cycle}}$$

Q is also related to the bandwidth of a tuned circuit's response by

$$Q = \frac{f_0}{BW_{3\,dB}} \qquad (5)$$

Parasitic inductance, capacitance and resistance can significantly alter the performance and characteristics of a tuned circuit if the design frequency is close to the self-resonant frequencies of the components.

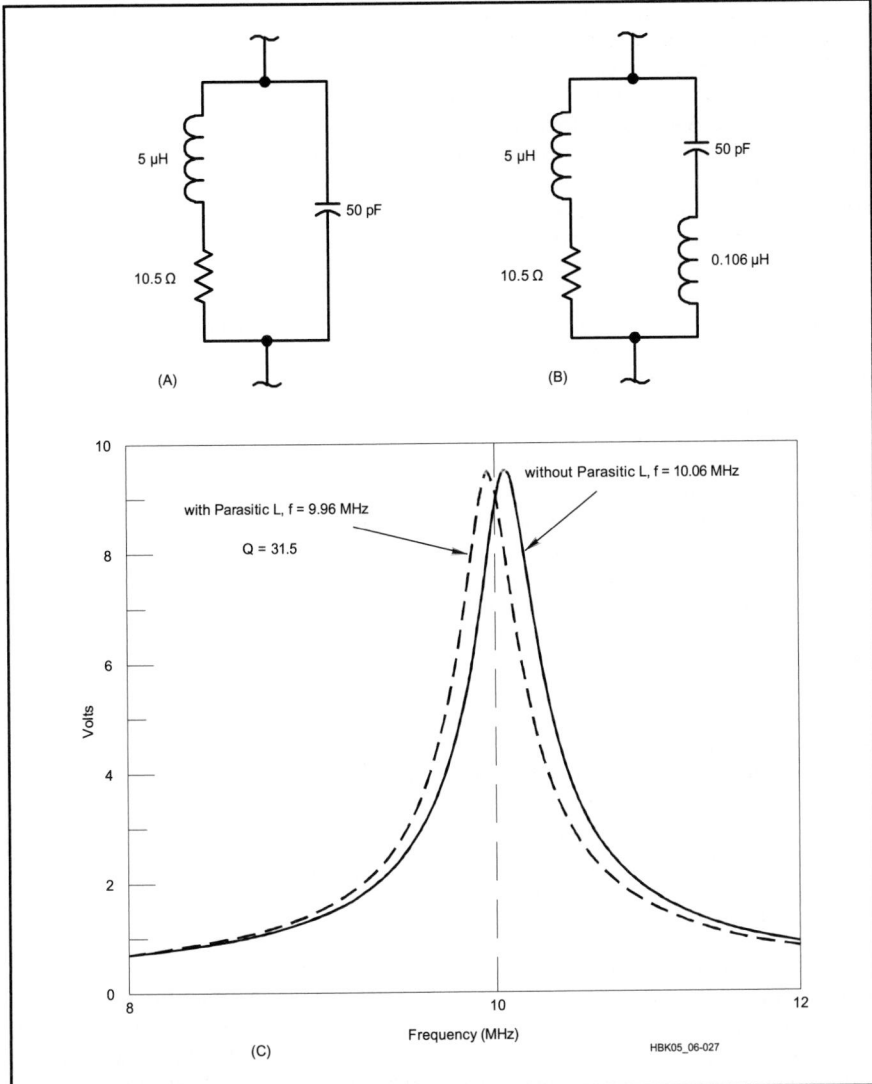

Fig 5.19 — A is a tank circuit, neglecting parasitics. B is the same circuit including L_p on capacitor. C is the frequency response curves for A and B. The solid line represents the unaltered circuit (also see Fig 5.17) while the dashed line shows the effects of adding parasitic inductance.

As an example, consider the resonant circuit of **Fig 5.19A**, which could represent the frequency-determining resonant circuit of an oscillator. Neglecting any parasitics,

$$f_0 = \frac{1}{2\pi\sqrt{LC}} = 10.07 \text{ MHz}$$

As in many practical cases, assume the resistance arises entirely from the inductor series resistance. The data sheet for the inductor specified a minimum Q of 30, so assuming Q = 30 yields an R value of

$$\frac{X_L}{Q} = \frac{2\pi\,(10.06 \text{ MHz})\,(5 \text{ }\mu\text{H})}{30} = 10.5\,\Omega$$

Next, let's include the parasitic inductance of the capacitor (Fig 5.19B). A reasonable assumption is that this capacitor has the same physical size as the example from the section on Parasitic Inductance for which we calculated $L_s = 0.106$ μH. This would give the capacitor a self-resonant frequency of 434 MHz — well above our frequency of interest. However, the added parasitic inductance does account for an extra 0.106/5.0 = 2% inductance. Since this circuit is no longer strictly series or parallel, we must convert it to an equivalent form before calculating the new f_0.

An easier and faster way is to *simulate* the altered circuit by computer. This analysis was performed on a desktop computer using *SPICE*, a standard circuit simulation program described in the **Computer-Aided Circuit Design** chapter. The voltage response of the circuit (given an input current of 1 mA) was calculated as a function of frequency for both cases, with and without parasit-

ics. The results are shown in the plot in Fig 5.19C, where we can see that the parasitic circuit has an f_0 of 9.96 MHz (a shift of 1%) and a Q (measured from the −3 dB points) of 31.5. For comparison, the simulation of the unaltered circuit does in fact show $f_0 = 10.06$ MHz and Q = 30.

5.3.7 Self-Resonance

Because of parasitic effects, a capacitor or inductor — all by itself — exhibits the properties of a resonant RLC circuit at frequencies for which the parasitic effects are significant. Figs 5.10 and 5.14 illustrate RF models for the capacitor and inductor, which are based on the general model in Fig 5.4, leaving out the packaging capacitance. Note the slight difference in configuration; the pairs C_p-R_p and L_s-R_s are in series in the capacitor but in parallel in the inductor. This is because of the different physical structure of the components.

At some sufficiently high frequency, both inductors and capacitors become *self-resonant* when the parasitic reactance cancels or equals the intended reactance, creating a series-resonant circuit. Similar to a series-resonant circuit made of discrete components, above the self-resonant frequency a capacitor will appear inductive, and an inductor will appear capacitive.

For an example, let's calculate the approximate self-resonant frequency of a 470-pF capacitor whose leads are made from #20 AWG wire (0.032-inch diameter), with a total length of 1 inch. Using equation 1, we calculate the approximate parasitic inductance

$$L\,(\mu\text{H}) = 0.00508\,(1)\left[\ln\left(\frac{2(1)}{(0.032/2)}\right) - 0.75\right]$$

$$= 0.021\,\mu\text{H}$$

and the self-resonant frequency is roughly

$$f = \frac{1}{2\pi\sqrt{LC}} = 50.7 \text{ MHz}$$

Similarly, an inductor can also have a parallel self-resonance.

The purpose of making these calculations is to provide a feel for actual component values. They could be used as a rough design guideline, but should not be used quantitatively. Other factors such as lead orientation, shielding and so on, can alter the parasitic effects to a large extent. Large-value capacitors tend to have higher parasitic inductances (and therefore a lower self-resonant frequency) than small-value capacitors.

Self-resonance becomes critically important at VHF and UHF because the self-resonant frequency of many common components is at or below the frequency where the component will be used. In this case, either special techniques can be used to construct components to operate at these frequencies, by reducing the parasitic effects, or else the idea of lumped elements must be abandoned altogether in favor of microwave techniques such as striplines and waveguides.

5.3.8 Dielectric Breakdown and Arcing

Anyone who has ever seen an arc form across a transmitting capacitor's plates, seen static discharges jump across an antenna insulator or touched a metal doorknob on a dry day has experienced the effects of dielectric breakdown.

In the ideal world, we could take any two conductors and put as large a voltage as we want across them, no matter how close together they are. In the real world, there is a voltage limit (*dielectric strength*, measured in kV/cm and determined by the insulating material between the two conductors) above which the insulator will break down.

Because they are charged particles, the electrons in the atoms of a dielectric material feel an attractive force when placed in an electric field. If the field is sufficiently strong, the force will strip the electron from the atom. This electron is available to conduct current, and furthermore, it is traveling at an extremely high velocity. It is very likely that this electron will hit another atom, and free another electron. Before long, there are many stripped electrons producing a large current, forming an *arc*. When this happens, we say the dielectric has suffered *breakdown*. Arcing in RF circuits is most common in transmitters and transmission line components where high voltages are common, but it is possible anywhere two components at significantly

different voltage levels are closely spaced.

If the dielectric is liquid or gas, it will repair itself when the applied voltage is removed and the molecules in the dielectric return to their normal state. A solid dielectric, however, cannot repair itself because its molecules are fixed in place and the low-resistance path created by the arc is permanent. A good example of this is a CMOS integrated circuit. When exposed to the very high voltages associated with static electricity, the electric field across the very thin gate oxide layer exceeds the dielectric strength of silicon dioxide, and the device is permanently damaged by the resulting hole created in the oxide layer.

The breakdown voltage of a dielectric layer depends on its composition and thickness (see **Table 5.1**). The variation with thickness is not linear; doubling the thickness does not quite double the breakdown voltage. Breakdown voltage is also a function of geometry: Because of electromagnetic considerations, the breakdown voltage between two conductors separated by a fixed distance is less if the surfaces are pointed or sharp-edged than if they are smooth or rounded. Therefore, a simple way to help prevent breakdown in many projects is to file and smooth the edges of conductors. (See the **Power Sources** and **RF Power Amplifier** chapters for additional information on high voltage applications.)

Capacitors are, by nature, the component most often associated with dielectric failure. To prevent damage, the working voltage of a capacitor — and there are separate dc and ac ratings — should ideally be two or three times the expected maximum voltage in the circuit. Capacitors that are not air-insulated or have a *self-healing dielectric* should be replaced if a dielectric breakdown occurs.

Resistors and inductors also have voltage ratings associated with breakdown of their insulating coating. High-value resistors, in particular, can be bypassed by leakage current flowing along the surface of the resistor. High-voltage resistors often have elongated bodies to create a long *leakage path* to present a high resistance to leakage current. Cleaning the bodies of resistors and inductors in high-voltage circuits helps prevent arcs from forming and minimizes leakage current.

5.3.9 Radiative Losses

Any conductor placed in an electromagnetic field will have a current induced in it. We put this principle to good use when we make an antenna. The unwelcome side of this law of nature is the phrase "any conductor"; even conductors we don't intend to act as antennas will respond this way.

Fortunately, the efficiency of such "antennas" varies with conductor length. They will be of importance only if their length is a significant fraction of a wavelength. When we

make an antenna, we usually choose a length on the order of λ/2. Therefore, when we *don't* want an antenna, we should be sure that the conductor length is *much less* than λ/2, no more than 0.1 λ. This will ensure a very low-efficiency antenna. This is why even unpaired 60-Hz power lines do not lose a significant fraction of the power they carry — at 60 Hz, 0.1 λ is about 300 miles!

In addition, we can use shielded cables. Such cables do allow some penetration of EM fields if the shield is not solid, but even 95% coverage is usually sufficient, especially if some sort of RF choke is used to reduce shield current.

Radiative losses and coupling can also be reduced by using twisted or parallel pairs of conductors — the fields tend to cancel. In some applications, such as audio cables, this may work better than shielding. Critical stages such as tuned circuits should be placed in shielded compartments where possible.

This argument also applies to large components — remember that a component or long wire can both radiate and receive RF energy. Measures that reduce radiative losses will also reduce unwanted RF pickup. See the **RF Interference** chapter for more information.

5.3.10 Bypassing and Decoupling

BYPASSING

Circuit models showing ac behavior often show ground connections that are not at dc ground. Bias voltages and currents are neglected in order to show how the circuit responds to ac signals. Rather, those points are "signal grounded" through bypass capacitors. Obtaining an effective bypass can be difficult and is often the route to design difficulty. The problem is parasitic inductance. Although we label and model parts as "capacitors," a more complete model is needed. The better model is a series RLC circuit, shown in **Fig 5.20**. Capacitance is close to the marked value while inductance is a small value that grows with component lead length. Resistance is a loss term, usually controlled by the Q of the parasitic inductor. Even a leadless SMT (surface-mount technology) component will display inductance commensurate with its dimensions. As shown in the Parasitic Inductance section, wire component leads have an inductance of about 1 nH per mm of length (20-25 nH per inch).

Bypass capacitor characteristics can be measured in the home lab with the test setup of **Fig 5.21**. **Fig 5.22** shows a test fixture with an installed 470-pF leaded capacitor. The fixture is used with a signal generator and spectrum analyzer to evaluate capacitors. Relatively long capacitor leads were required to interface to the BNC connectors,

Table 5.1
Dielectric Constants and Breakdown Voltages

Material	Dielectric Constant*	Puncture Voltage**
Aisimag 196	5.7	240
Bakelite	4.4-5.4	240
Bakelite, mica filled	4.7	325-375
Cellulose acetate	3.3-3.9	250-600
Fiber	5-7.5	150-180
Formica	4.6-4.9	450
Glass, window	7.6-8	200-250
Glass, Pyrex	4.8	335
Mica, ruby	5.4	3800-5600
Mycalex	7.4	250
Paper, Royalgrey	3.0	200
Plexiglas	2.8	990
Polyethylene	2.3	1200
Polystyrene	2.6	500-700
Porcelain	5.1-5.9	40-100
Quartz, fused	3.8	1000
Steatite, low loss	5.8	150-315
Teflon	2.1	1000-2000

*At 1 MHz
**In volts per mil (0.001 in.)

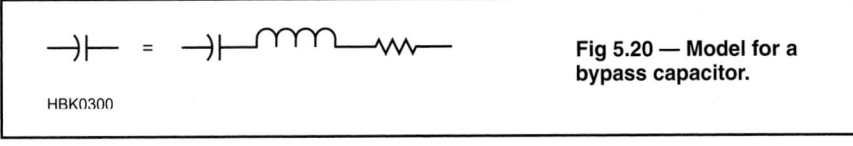

Fig 5.20 — Model for a bypass capacitor.

HBK0300

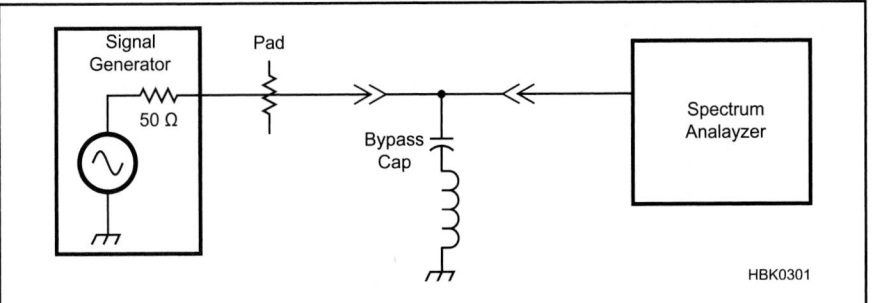

Fig 5.21 — Test set for home lab measurement of a bypass capacitor.

HBK0301

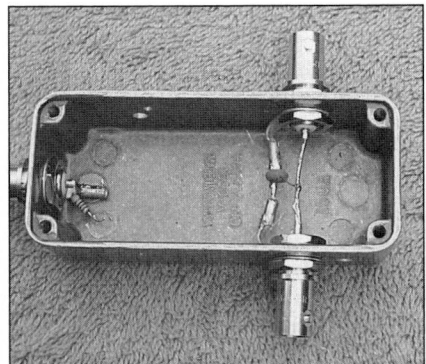

Fig 5.22 — Test fixture for measuring self-resonant frequency of capacitors.

even though the capacitor itself was small. The signal generator was tuned over its range while examining the spectrum analyzer response, showing a minimum at the series resonant frequency. Parasitic inductance is calculated from this frequency. The C value was measured with a low-frequency LC meter. Additional test instruments and techniques are discussed in the **Test Equipment and Measurement** chapter.

The measured 470-pF capacitor is modeled as 485 pF in series with an inductance of 7.7 nH. The L is larger than we would see with shorter leads. A 0.25-inch 470-pF ceramic disk capacitor with zero lead length will show a typical inductance closer to 3 nH. The measured capacitor Q was 28 at its self-resonance of 82 MHz but is higher at lower frequency. Data from a similar measurement, but with a network analyzer is shown in **Fig 5.23**. Two 470-pF capacitors are measured, one surface mounted and the other a leaded part with 0.1-inch leads.

It is often suggested that the bandwidth for bypassing can be extended by paralleling a capacitor that works well at one frequency with another to accommodate a different part of the spectrum. Hence, paralleling the 470 pF with a 0.01-µF capacitor should extend the bypassing to lower frequencies. The calculations are shown in the plots of **Fig 5.24**. The results are terrible! While the low frequency bypassing is indeed improved, a high impedance response is created at 63 MHz. This complicated behavior is again the result of inductance. Each capacitor was assumed to have a series inductance of 7 nH. A parallel resonance is approximately formed between the L of the larger capacitor and the C of the smaller. The Smith Chart plot shows us that the impedance is nearly 50 Ω at 63 MHz. Impedance would be even higher with greater capacitor Q.

Bypassing can be improved by paralleling capacitors. However, the capacitors should be approximately identical. **Fig 5.25** shows the result of paralleling two capacitors of about the same value. They differ slightly at 390 and 560 pF, creating a hint of resonance. This appears as a small perturbation in the reactance plot and a tiny loop on the Smith Chart. These

anomalies disappear as the C values become equal. Generally, paralleling is the scheme that produces the best bypassing. The ideal solution is to place an SMT capacitor on each side of a printed circuit trace or wire at a point that is to be bypassed.

Matched capacitor pairs form an effective bypass over a reasonable frequency range. Two 0.01-µF disk capacitors have a reactance magnitude less than 5 Ω from 2 to 265 MHz. A pair of the 0.1-µF capacitors was even better, producing the same bypassing impedance from 0.2 to 318 MHz. The 0.1-µF capacitors are chip-style components with attached wire leads. Even better results can be obtained with multi-layer ceramic chip capacitors. Construction with multiple layers creates an integrated paralleling.

Some applications (for example, IF amplifiers) require effective bypassing at even lower frequencies. Modern tantalum electrolytic capacitors are surprisingly effective through the RF spectrum while offering high enough C to be useful at audio. In critical applications, however, the parts should be tested to be sure of their effectiveness.

DECOUPLING

The bypass capacitor usually serves a dual role, first creating the low impedance needed to generate a "signal" ground. It also becomes part of a decoupling low-pass filter that passes dc while attenuating signals. The attenuation must function in both directions, suppressing noise in the power supply that might reach an amplifier while keeping amplifier signals from reaching the power supply. A low-pass filter is formed with alternating series and parallel component connections. A parallel bypass is followed by a series impedance, ideally a resistor.

Additional shunt elements can then be added, although this must be done with care. An inductor between shunt capacitors should have high inductance. It will resonate with the

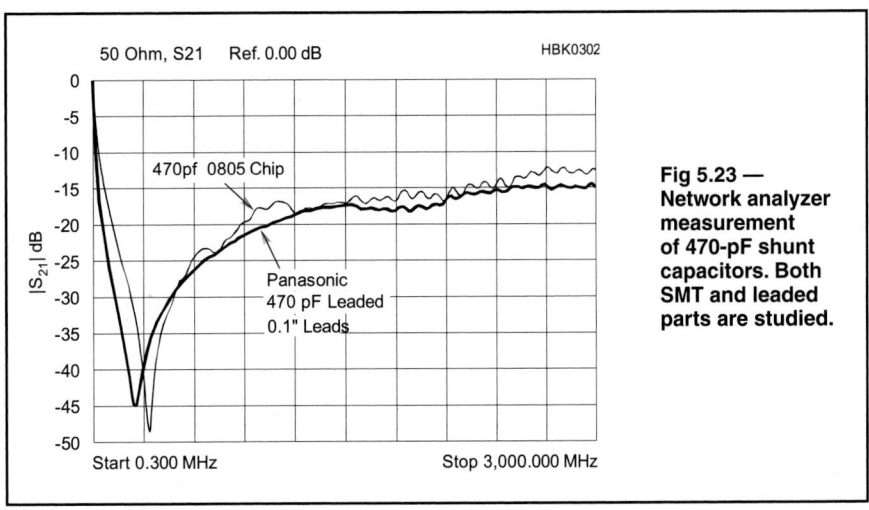

Fig 5.23 — Network analyzer measurement of 470-pF shunt capacitors. Both SMT and leaded parts are studied.

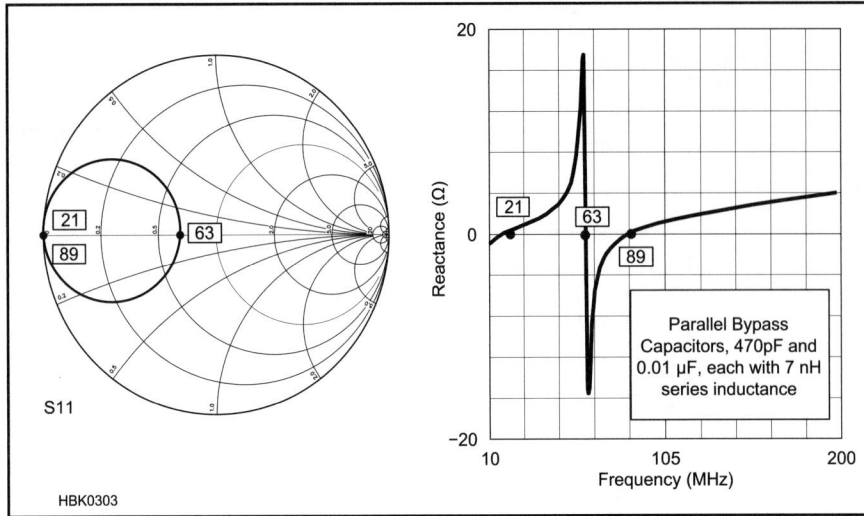

Fig 5.24 — The classic technique of paralleling bypass capacitors of two values, here 470 pF and 0.01 µF. This is a *terrible* bypass! See text.

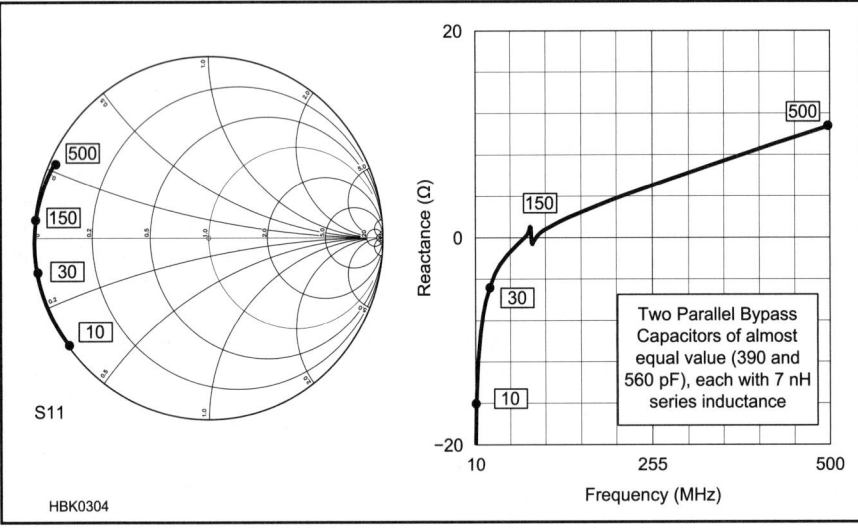

Fig 5.25 — Paralleling bypass capacitors of nearly the same value. This results in improved bypassing without complicating resonances.

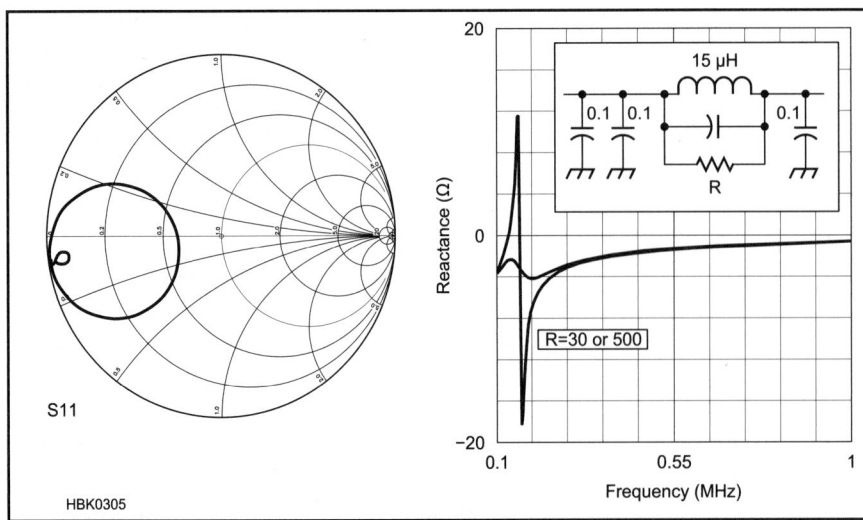

Fig 5.26 — Two different resistor values parallel a decoupling choke. The lower, 30-Ω value is more effective. See text.

shunt capacitors to create high impedances just like those that came from parasitic inductance in the bypasses. This makes it desirable to have an inductance that is high enough that any resonance is below the band of interest. But series inductors have their own problems; they have parasitic capacitance, creating their own self-resonance. As an example, a pair of typical RF chokes (RFCs) were measured (now as series elements) as described earlier. A 2.7-µH molded choke was parallel resonant at 200 MHz, indicating a parallel capacitance of 0.24 pF. The Q at 20 MHz was 52. A 15-µH molded choke was parallel resonant at 47 MHz, yielding a parallel C of 0.79 pF. This part had a Q of 44 at 8 MHz.

Large inductors can be fabricated from series connections of smaller ones. The best wideband performance will result only when all inductors in a chain have about the same value. The reasons for this (and the mathematics that describe the behavior) are identical with those for paralleling identical capacitors.

Low inductor Q is often useful in decoupling applications, encouraging us to use inductors with ferrite cores. Inductors using the Fair-Rite Type #43 material have low Q in the 4 to 10 range over the HF spectrum. One can also create low-Q circuits by paralleling a series inductor of modest Q with a resistor.

Fig 5.26 shows a decoupling network and the resulting impedance when viewed from the "bypass" end. The 15-µH RFC resonates with a 0.1-µF capacitor to destroy the bypass effect just above 0.1 MHz. A low-value parallel resistor fixes the problem.

A major reason for careful wideband bypassing and decoupling is the potential for amplifier oscillation. Instability that allows oscillations is usually suppressed by low impedance terminations. The base and collector (or gate and drain) should both "see" low impedances to ensure stability. But that must be true at all frequencies where the device can produce gain. It is never enough to merely consider the operating frequency for the amplifier. A parallel resonance in the base or collector circuits can be a disaster. When wideband bypassing is not possible, negative feedback that enhances wideband stability is often used.

Emitter bypassing is often a critical application. As demonstrated in the **Analog Basics** chapter, a few extra ohms of impedance in the emitter circuit can drastically alter amplifier performance. A parallel-resonant emitter bypass could be a profound difficulty while a series-resonant one can be especially effective.

Capacitors also appear in circuits as blocking elements. A blocking capacitor, for example, appears between stages, creating a near short circuit for ac signals while accommodating different dc voltages on the two sides. A blocking capacitor is not as critical as a bypass, for the impedances on either side will usually be higher than that of the block.

With parasitic effects having the potential to strongly affect circuit performance, the circuit designer must account for them wherever they are significant. The additional complexity of including parasitic elements can result in a design too complex to be analyzed manually. Clearly, detailed modeling is the answer to component selection and the control of parasitic effects.

5.3.11 Effects on Filter Performance

LC bandpass filters perform a critical function in determining the performance of a typical RF system such as a receiver. An input filter, usually a bandpass, restricts the frequency range that the receiver must process. Transmitters use LC filters to reduce harmonic output. Audio and LF filters may also use LC elements. The LC filters we refer to in this section are narrow with a bandwidth from 1 to 20% of the center frequency. Even narrower filters are built with resonators having higher Q; the quartz crystal filter discussed in the **RF and AF Filters** chapter is used where bandwidths of less than a part per thousand are possible. The basic concepts that we examine with LC circuits will transfer to the crystal filter.

LOSSES IN FILTERS AND Q

The key elements in narrow filters are tuned circuits made from inductor-capacitor pairs, quartz crystals, or transmission line sections. These resonators share the property that they store energy, but they have losses. A chime is an example. Striking the chime with a hammer produces the waveform of **Fig 5.27**. The rate at which the amplitude decreases with time after the hammer strike is determined by the filter's Q, which is discussed in the chapter on **Analog Basics**. The higher the Q, the longer it takes for the sound to disappear. The oscillator amplitude would not decrease if it were not for the losses that expend energy stored in the resonator. The mere act of observing the oscillation will cause some energy to be dissipated.

A chime is an acoustic resonator, but the same behavior occurs in electric resonators. A pulse input to an LC circuit causes it to ring; losses cause the amplitude to diminish. The most obvious loss in an LC circuit is conductor resistance, including that in the inductor wire. This resistance is higher than the dc value owing to the skin effect, which forces high frequency current toward the conductor surface. Other losses might result from hysteresis losses in an inductor core or dielectric losses in a capacitor.

An inductor is modeled as an ideal part with a series or a parallel resistance. The resistance will depend on the Q if the inductor was part of a resonator with that quality. The two resistances are shown in **Fig 5.28**.

$$R_{Series} = \frac{2\pi f L}{Q} \quad \text{and}$$

$$R_{Parallel} = Q \times 2\pi f L \quad (6)$$

The higher the inductor Q, the smaller the series resistance, or the larger the parallel resistance is needed to model that Q. It does not matter which configuration is used. The Q of a resonator is related to the bandwidth of the tuned circuit by

$$BW = \frac{f_C}{Q} \quad (7)$$

where f_C is the tuned circuit's center frequency. This is also the Q of an inductor in a tuned circuit if the capacitor is lossless.

The single-tuned circuit is presented in two different forms in **Fig 5.29**. In the top, a parallel tuned circuit consisting of L and C has loss modeled by three resistors. The resistor R_p is the parallel loss resistance representing the non-ideal nature of the inductor. (Another might be included to represent capacitor losses.) But the LC is here paralleled by three resistors: the source, the load and the loss element. R_p would disappear if the tuned circuit was built from perfect components. The source and load remain; they represent the source resistance that must be present if power is available and a load resistance that must be included if power is to be extracted.

Equation 6 and 7 can be applied in several ways. If the resonator is evaluated with only its intrinsic loss resistance (in either series or parallel form) the resulting Q is called the *unloaded Q*, or Q_u. If, however, the net resistance is used, which is the parallel combination of the load, the source, and the loss in the parallel tuned circuit, the resulting Q is called the *loaded Q*, Q_L. If we were working with the series tuned circuit form, the loaded Q would be related to the total series R.

Consider an example, a parallel tuned circuit (Fig 5.29A) with a 2-μH inductor tuned to 5 MHz with a 507-pF capacitor. Assume the parallel loss resistor was 12.57 kΩ. The unloaded Q calculated from equation 6 is 200. The unloaded bandwidth would be 5 MHz/200 = 25 kHz. Assume that the source and load resistors were equal, each 2 kΩ. The net resistance paralleling the LC would then be the combination of the three resistors, 926 Ω. The loaded Q becomes 14.7 with a loaded bandwidth of 339 kHz. The loaded Q is also the filter Q, for it describes the bandwidth of the single-tuned circuit, the simplest of band-pass filters.

This filter also has *insertion loss* (IL). This is illustrated in **Fig 5.30** which shows the

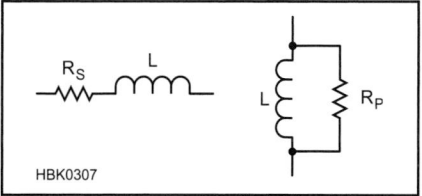

Fig 5.28 — Inductor Q may be modeled with either a series or a parallel resistance.

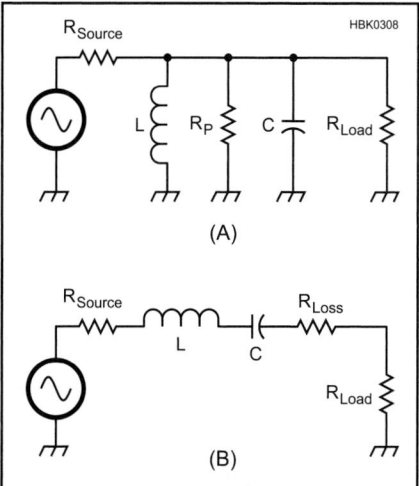

Fig 5.29 — Two simple forms of the single-tuned circuit.

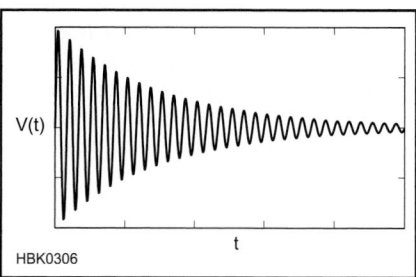

Fig 5.27 — The amplitude of a chime's ring after being struck by a hammer. Units are arbitrary.

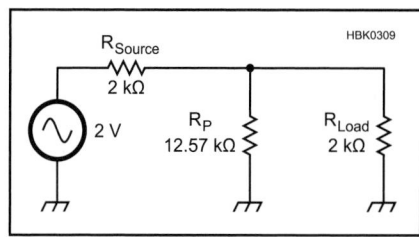

Fig 5.30 — Simplified parallel tuned circuit at resonance. The effect of loss is illustrated by removing the parallel-LC combination which at resonance has infinite impedance.

parallel-resonant LC combination removed (at resonance, the LC combination has infinite impedance), leaving only the loss resistance of 12.57 kΩ. We use an arbitrary open circuit source voltage of 2. The available power to a load is then 1 V across a resistance equaling the 2-kΩ source. If the resonator had no internal losses, this available power would be delivered to the 2-kΩ load. However, the loss R parallels the load, causing the output voltage to be 0.926 V, a bit less than the ideal 1 V. Calculation of the output power into the 2-kΩ load resistance and the available power shows that the insertion loss is 0.67 dB.

This exercise illustrates two vital points that are general for all bandpass filters. First, the bandwidth of any filter must always be larger than the unloaded bandwidth of the resonators used to build the filter. Second, any filter built from real-world components will have an insertion loss. The closer the Q of the filter approaches the unloaded resonator Q, the greater the insertion loss becomes. A parallel tuned circuit illustrated these ideas; the series tuned filter would have produced identical results. Generally, the insertion loss of a single-tuned circuit relates to loaded and unloaded Q by

$$IL(dB) = -20 \log \left(1 - \frac{Q_L}{Q_U} \right) \qquad (8)$$

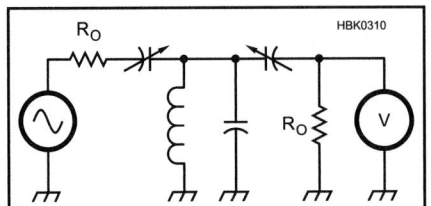

Fig 5.31 — Test setup for measuring the Q of a resonator. The source and load impedances, R_O, are assumed to be identical. The two coupling capacitors are adjusted to be equal to each other. The output signal is measured with an appropriate ac voltmeter, an oscilloscope or a spectrum analyzer.

The Q of a tuned LC circuit is easily measured with a signal generator of known output impedance (usually 50 Ω) and a sensitive detector, again with a known impedance level, often equal to that of the generator. The test configuration is shown in **Fig 5.31**. It uses equal loads and equal capacitors to couple from the terminations to the resonator. Equal capacitors, C1 and C2 guarantees that each termination contributes equally to the resonator parallel load resistance. The voltmeter across the load is calibrated in dB.

To begin measurement we remove the tuned circuit and replace it with a direct connection from generator to load. The available power delivered to the load is calculated after the voltage is measured. The resonator is then inserted between the generator and load, and the generator is tuned for a peak. The measured power is less than that available from the source, with the difference being the insertion loss for the simple filter. Capacitors C1 and C2 are adjusted until the loss is 30 dB or more. With loss this high, the intrinsic loss resistance of the resonator will dominate the loss. The generator is now tuned first to one side of the peak, and then to the other, noting the frequencies where the response is lower than the peak by 3 dB. The unloaded bandwidth, ΔF, is the difference between the two 3 dB frequencies. The unloaded Q is calculated as

$$Q_U = \frac{F}{\Delta F} \qquad (9)$$

This method for Q measurement is quite universal, being effective for audio tuned circuits, simple LC RF circuits, VHF helical resonators, or microwave resonators. The form of the variable capacitors, C1 and C2, may be different for the various parts of the spectrum, but the concepts are general. Indeed, it is not even important how the coupling occurs. The Q measurement normally determines an unloaded value, but loaded values are also of interest when testing filters.

5.4 Semiconductor Circuits at RF

The models used in the **Analog Basics** chapter are reasonably accurate at low frequencies, and they are of some use at RF, but more sophisticated models are required for consistent results at higher frequencies. This section notes several areas in which the simple models must be enhanced. Circuit design using models accurate at RF, particularly for large signals, is performed today using design software as described in the **Computer-Aided Circuit Design** chapter of this book. In-depth discussions of the elements of RF circuit design appear in Hayward's *Introduction to Radio Frequency Design*, an excellent text for the beginning RF designer (see the References section).

5.4.1 The Diode at High Frequencies

A DIODE AC MODEL

At high frequencies, the diode's behavior becomes less like a switch, especially for small signals that do not cause the diode's operating point to move by large amounts.

In this case, the diode's small-signal dynamic behavior becomes important.

Fig 5.32 shows a simple resistor-diode circuit to which is applied a dc bias voltage plus an ac signal. Assuming that the voltage drop across the diode is 0.6 V and R_f is negligible, we can calculate the bias current to be I = (5 – 0.6) / 1 kΩ = 4.4 mA. This point is marked on the diodes I-V curve in Fig 5.32. If we draw a line tangent to this point, as shown, the slope of this line represents the *dynamic resistance* of the diode, R_d, experienced by a small ac signal. At room temperature, dynamic resistance is approximately

$$R_d = \frac{25}{I} \Omega \qquad (10)$$

where I is the diode current in mA. Note that this resistance changes with bias current and should not be confused with the dc forward resistance discussed in the **Analog Basics** chapter, which has a similar value but represents a different concept. **Fig 5.33** shows a low-frequency ac model for the diode, including the dynamic resistance and junction

capacitance. This model should only be used when the diode's dc operating parameters can be neglected.

SWITCHING TIME

If you change the polarity of a signal applied to the ideal diode, current flow stops or starts instantaneously. Current in a real diode cannot do this, as a finite amount of time is required to move electrons and holes in or out of the diode as it changes states. Effectively, the diode junction capacitance, C_J, in Fig 5.33 must be charged or discharged. As a result, diodes have a maximum useful frequency when used in switching applications.

The operation of diode switching circuits can often be modeled by the circuit in **Fig 5.34**. The approximate switching time (in seconds) for this circuit is given by

$$\tau_s = \tau_p \frac{\left(- \right)}{\left(- \right)} = \tau_p - \qquad (11)$$

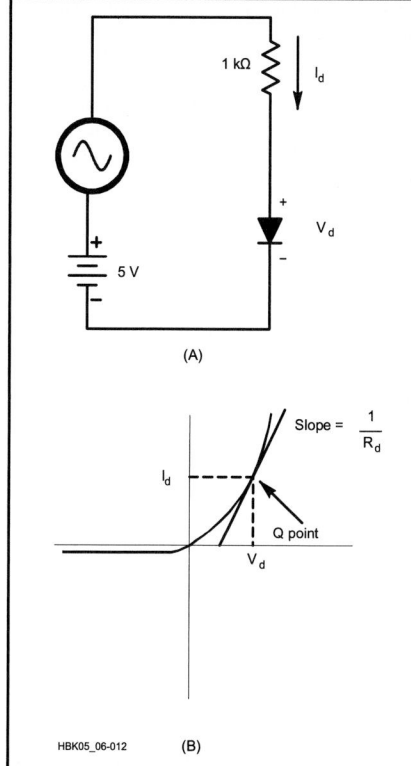

Fig 5.32 — A simple resistor-diode circuit used to illustrate dynamic resistance. The ac input voltage "sees" a diode resistance whose value is the slope of the line at the Q-point, shown in B.

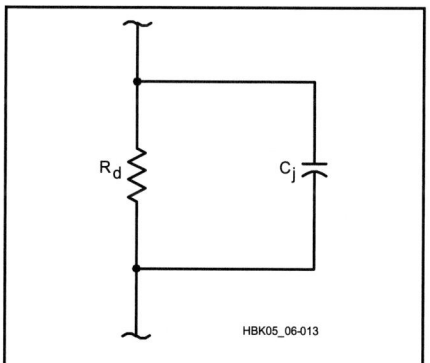

Fig 5.33 — An ac model for diodes. R_d is the dynamic resistance and C_j is the junction capacitance.

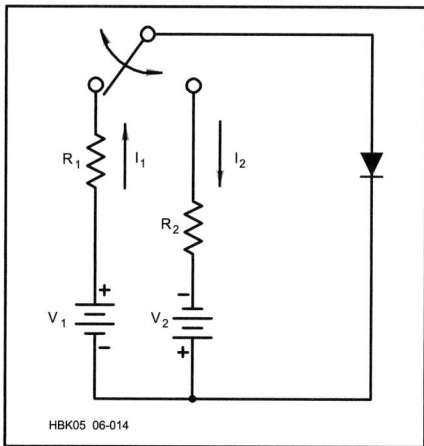

Fig 5.34 — Circuit used for computation of diode switching time.

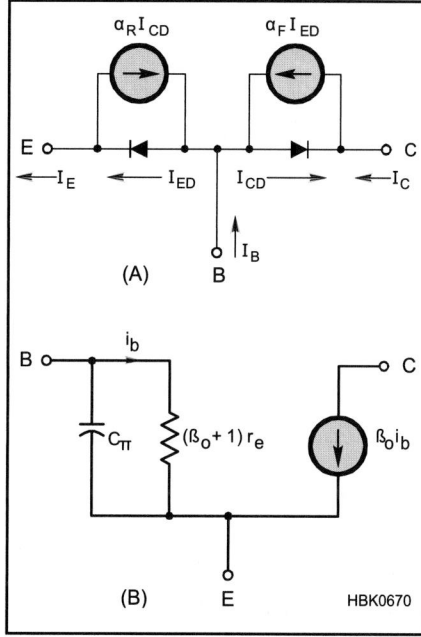

Fig 5.35 — (A) The Ebers-Moll model of the bipolar junction transistor (BJT) is used at dc and low frequencies when the transistor is in its active region. (B) The hybrid-pi model includes frequency dependence.

where τ_p is the minority carrier lifetime of the diode, a material constant determined during manufacture, on the order of 1 ms. I_1 and I_2 are currents that flow during the switching process. The minimum time in which a diode can switch from one state to the other and back again is therefore $2\,t_s$, and thus the maximum usable switching frequency is f_{sw} (Hz) $= \frac{1}{2}t_s$. It is usually a good idea to stay below this by a factor of two. Diode data sheets usually give typical switching times and show the circuit used to measure them.

Note that f_{sw} depends on the forward and reverse currents, determined by I_1 and I_2 (or equivalently V_1, V_2, R_1, and R_2). Within a reasonable range, the switching time can be reduced by manipulating these currents. Of course, the maximum power that other circuit elements can handle places an upper limit on switching currents.

5.4.2 The Transistor at High Frequencies

The development and selection of suitable models for active devices at RF is an involved and nuanced process. This section presents a few of the models used for bipolar junction transistors (BJT) and field-effect transistors (FET) as examples of the issues involved. For more information on model details, start with the *SPICE* home page at **bwrc.eecs.berkeley.edu/classes/icbook/spice** and the user's manual for the simulation software you intend to use.

GAIN VERSUS FREQUENCY

The circuit design equations in **Analog Basics** generally assumed that current gain in the bipolar transistor was independent of frequency. As signal frequency increases, however, current gain decreases. The low-frequency current gain, β_0, is constant through the audio spectrum, but it eventually decreases, and at some high frequency it will drop by a factor of 2 for each doubling of signal frequency. (The h-parameter [h_{FE}] is often substituted for β at dc and h_{FE} for ac current gain. H-parameters are discussed later in this chapter.) A transistor's frequency vs current gain relationship is specified by its *gain-bandwidth product*, or F_T, the frequency at which the current gain is 1. Common transistors for lower RF applications might have $\beta_0 = 100$ and $F_T = 500$ MHz. The frequency at which current gain equals β_0 is called F_b and is related to F_T by $F_b = F_T / \beta_0$.

The *Ebers-Moll model* for the bipolar transistor, shown in **Fig 5.35A**, is a "large signal model." It is used when the transistor is in its active mode and gives good results for dc collector and emitter currents. This would be a good model to use when designing a pass-transistor voltage regulator circuit, for example.

The frequency dependence of current gain is modeled by adding a capacitor across the base resistor of Fig 5.35B, the *hybrid-pi* model results. The capacitor's reactance should equal the low-frequency input resistance, $(\beta + 1)r_e$, at F_b. This simulates a frequency-dependent current gain.

Similar considerations affect the transconductance of FETs as the gate-to-source and gate-to-drain capacitances act to reduce high-frequency gain. Instead of F_T, most FETs designed for amplifier use specify input and output transconductance, susceptance, and power gain at different frequencies.

Recognizing that transistor gain varies with frequency, it is important to know the conditions under which the simpler model is useful. Calculations show that the simple model is valid, with $\beta = F_T/F$, for frequencies well above F_b. The approximation worsens, however, as the operating frequency (f_0) approaches F_T.

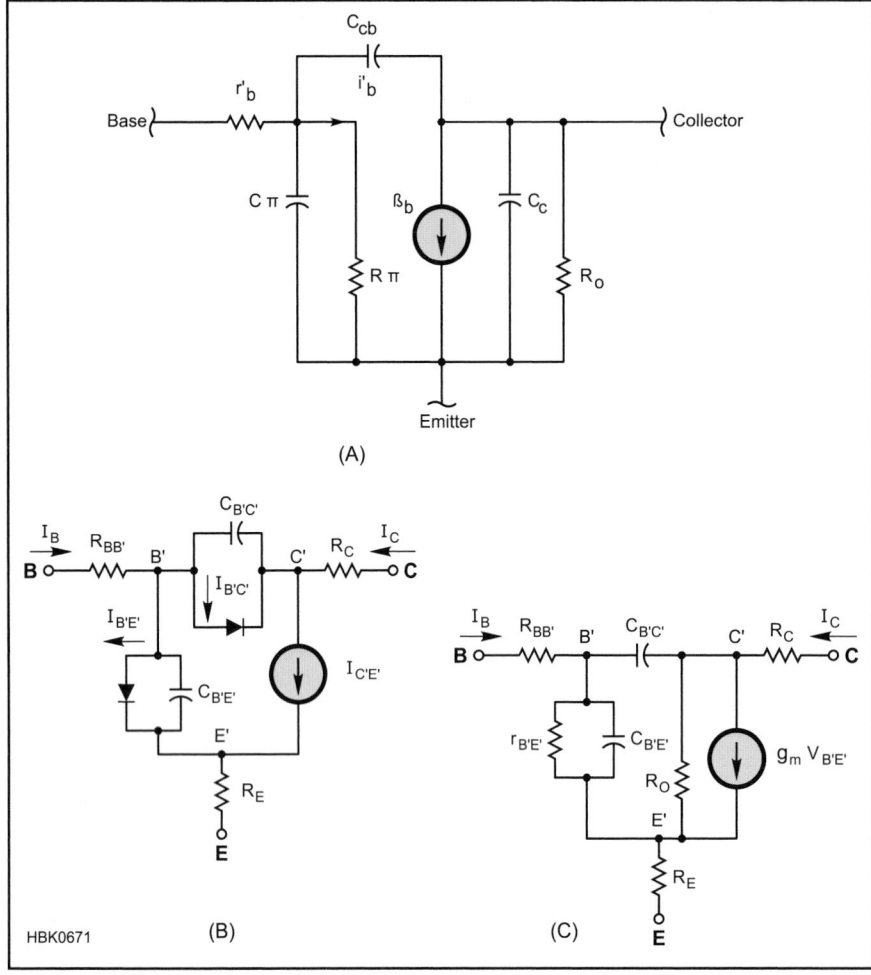

Fig 5.36 — High-frequency models for the bipolar junction transistor. (A) An improved version of the hybrid-pi model including additional device capacitances. (B) is the large-signal Gummel-Poon model used at dc and low-frequencies — the standard used for bipolar transistors by simulators based on *SPICE3* and (C) is the small-signal ac version.

SMALL-SIGNAL DESIGN AT RF

The simplified hybrid-pi model of Fig 5.35B is often suitable for non-critical designs, but **Fig 5.36A** shows a better small-signal model for RF design that expands on the hybrid-pi. Consider the physical aspects of a real transistor: There is some capacitance across each of the PN junctions (C_{cb} and C_p) and capacitance from collector to emitter (C_c). There are also capacitances between the device leads (C_e, C_b and C_c). There is a resistance in each current path, emitter to base and collector. From emitter to base, there is r_π from the hybrid-pi model and r´b, the "base spreading" resistance. From emitter to collector is R_o, the output resistance. The leads that attach the silicon die to the external circuit present three inductances.

Manual circuit analysis with this model is best tackled with the aid of a computer and specialized software. Other methods are presented in Hayward's *Introduction to Radio Frequency Design*. Surprisingly accurate results may be obtained, even at RF, from the simple models. Simple models also give a better "feel" for device characteristics that might be obscured by the mathematics of a more rigorous treatment. Use the simplest model that describes the important features of the device and circuit at hand.

The *Gummel-Poon model* shown in Figs 5.36B and 5.36C is the standard used by simulation software based on *SPICE3*. By adding additional parasitic elements, such as lead inductance and lead-to-package capacitances, the small-signal ac model can be used accurately at high frequencies. Note how the ac model becomes progressively more sophisticated from Fig 5.35A through 5.36C.

HIGH FREQUENCY FET MODELS

At low frequencies the FET can be treated as a simple current source controlled by the gate-to-source voltage as in Fig 3.49 in the **Analog Basics** chapter. As frequency increases, the inter-electrode capacitances become significant and must be included, such as in the model in **Fig 5.37A**. As with the BJT,

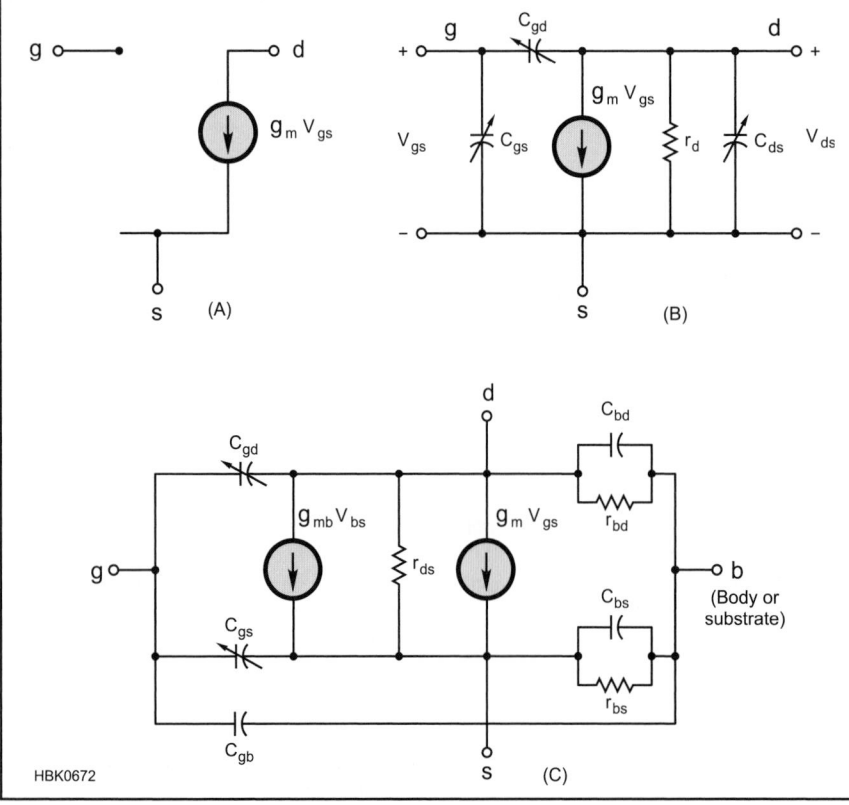

Fig 5.37 — Small-signal models for the field-effect transistor (FET). (A) is a simple low-frequency model. (B) is a common *SPICE3* large-signal model and (C) is a typical high-frequency model for the JFET.

separate models are used for small-signal and large-signal applications. Fig 5.37B shows a small-signal *SPICE3* simulation model for the JFET. All of the model's capacitances vary with temperature and device operating characteristics. Fig 5.37C shows a typical high-frequency MOSFET model used by *SPICE3* (there are several) that includes the effect of the body or substrate elements.

The advances in CMOS integrated circuit technology have resulted in transistors (and capacitors and inductors) with sufficient performance for use in RF design work. In fact, most radio transceivers for cellular telephones, wireless LANs and similar high-volume applications are implemented almost exclusively in CMOS technology due to the high integration available. MOSFET RF design principles are similar to BJT and JFET designs, and most circuit simulators provide good models for MOSFET devices. These additional techniques are enabled by the excellent device matching available in IC technology and the ability to integrate additional circuitry at low incremental cost.

PARASITIC ELEMENTS AND NOISE

For accurate RF simulation the effects of additional parasitic elements must be included. The most significant is lead inductance — generally inserted in series with the device's external connections. Lead-to-package and lead-to-lead capacitances are also included in most high-frequency simulations. Depending on the simulation package used, the user may be required to install these parasitic elements as separate circuit elements or the simulator or manufacturer may provide special models for high-frequency simulation.

Simulation of device noise is not covered here, but many simulator packages include separate noise models for various active devices. Review the simulator's documentation for information about how to include noise in your simulation.

5.4.3 Amplifier Classes

The class of operation of an amplifier is determined by the fraction of a drive cycle during which conduction occurs in the amplifying device or for switchmode devices. (Switchmode amplifiers use different criteria.) The Class A amplifier conducts for 100% of the cycle. It is characterized by constant flow of supply current, regardless of the strength of the driving signal. Most of the amplifiers we use for RF applications and many audio circuits in receivers operate in Class A.

A Class B amplifier conducts for 50% of the cycle, which is 180 degrees if we examine the circuit with regard to a driving sine wave. A Class B amplifier draws no dc current when no input signal is applied, but current begins to flow with any input, growing with the input

strength. A Class B amplifier can display good *envelope linearity*, meaning that the output amplitude at the drive frequency changes linearly with the input signal. The total absence of current flow for half of the drive cycle will create harmonics of the signal drive.

A Class C amplifier is one that conducts for less than half of a cycle. No current flows without drive. Application of a small drive produces no output and no current flow. Only after a threshold is reached does the device begin to conduct and provide output. A bipolar transistor with no source of bias for the base typically operates in Class C.

The large-signal models discussed earlier are suitable for the analysis of all amplifier classes. Small-signal models are generally reserved for Class A amplifiers. The most common power amplifier class is a cross between Class A and B — the Class AB amplifier that conducts for more than half of each cycle. A Class AB amplifier at low drive levels is indistinguishable from a Class A design. However, increasing drive produces greater collector (or drain) current and greater output.

Amplifier class letter designators for vacuum tube amplifiers were augmented with a numeric subscript. A Class AB_1 amplifier operates in AB, but with no grid current flowing. A Class AB_2 amplifier's grid is driven positive with respect to the cathode and so some grid current flows. In solid-state amplifiers, which have no grids, no numeric subscripts are used.

While wide-bandwidth Class A and Class B amplifiers are common, most circuits operating in Class C and higher are tuned at the output. The tuning accomplishes two things. First, it allows different terminations to exist for different frequencies. For example, a resistive load could be presented at the drive frequency while presenting a short circuit at some or all harmonics. The second consequence of tuning is that reactive loads can be created and presented to the amplifier collector or drain. This then provides independent control of current and voltage waveforms.

While not as common as A, B, and C, Class D and E amplifiers are of increasing interest. The Class D circuit is a balanced (two transistor) switching format where the input is driven hard enough to produce square wave collector waveforms. Class E amplifiers usually use a single switching device with output tuning that allows high current to flow in the device only when the output voltage is low. Other "switching amplifier" class designators refer to the various techniques of controlling the switched currents and voltages.

Class A and AB amplifiers are capable of good envelope linearity, so they are the most common formats used in the output of SSB amplifiers. Class B and, predominantly, Class C amplifiers are used for CW and FM

applications, but lack the envelope linearity needed for SSB.

Efficiency varies considerably between amplifier class. The Class A amplifier can reach a collector efficiency of 50%, but no higher, with much lower values being typical. Class AB amplifiers are capable of higher efficiency, although the wideband circuits popular in HF transceivers typically offer only 30% at full power. A Class C amplifier is capable of efficiencies approaching 100% as the conduction cycle becomes small, with common values of 50 to 75%. Both Class D and E are capable of 90% and higher efficiency. An engineering text treating power amplifier details is Krauss, Bostian, and Raab's *Solid State Radio Engineering*. A landmark paper by a Cal Tech group led by David Rutledge, KN6EK, targeted to the home experimenter, *High-Efficiency Class-E Power Amplifiers*, was presented in *QST* for May and June, 1997.

5.4.4 RF Amplifiers with Feedback

The feedback amplifier appears frequently in amateur RF circuits. This is a circuit with two forms of negative feedback with (usually) a single transistor to obtain wide bandwidth, well-controlled gain and well-controlled, stable input and output resistances. Several of these amplifiers can be cascaded to form a high gain circuit that is both stable and predictable.

The small-signal schematic for the feedback amplifier is shown in **Fig 5.38** without bias components or power supply details. The design begins with an NPN transistor biased for a stable dc current. Gain is reduced with emitter degeneration, increasing input resistance while decreasing gain. Additional feedback is then added with a parallel feedback resistor, R_f, between the collector and base. This is much like the resistor between an op-amp output and the inverting input which reduces gain and *decreases* input resistance.

Several additional circuits are presented

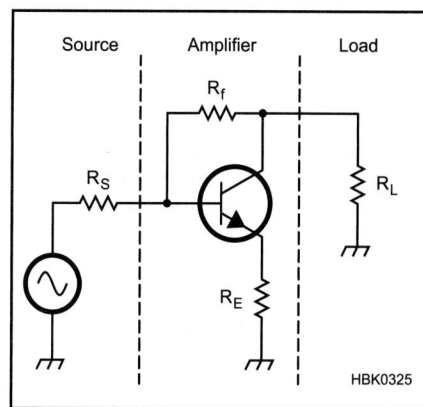

Fig 5.38 — Small-signal circuit for a feedback amplifier.

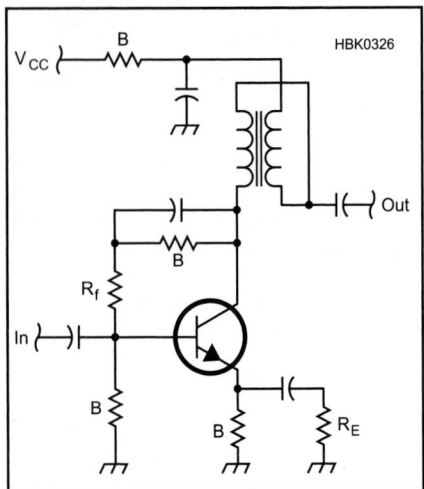

Fig 5.39 — A practical feedback amplifier. Components marked with "B" are predominantly for biasing. The 50-Ω output termination is transformed to 200 Ω at the collector. A typical RF transformer is 10 bifilar turns of #28 on a FT-37-43 ferrite toroid. The inductance of one of the two windings should have a reactance of around 250 Ω at the lowest frequency of operation.

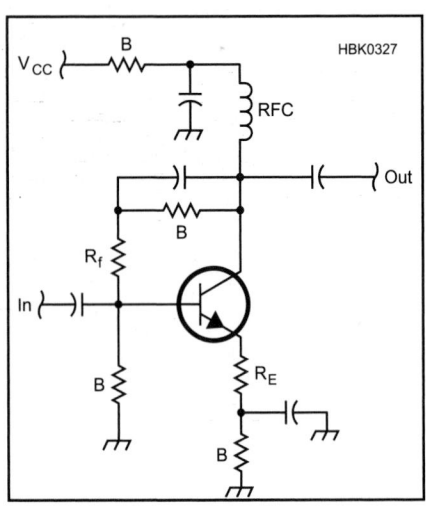

Fig 5.40 — A variation of the feedback amplifier with a 50-Ω output termination at the collector.

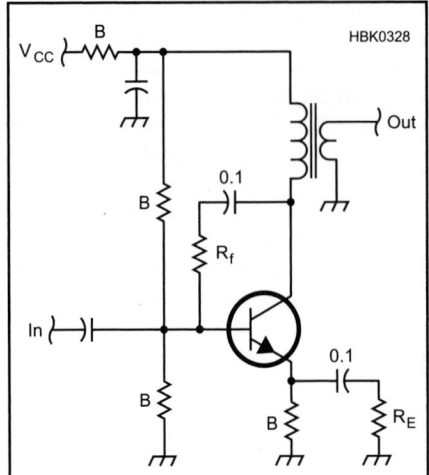

Fig 5.41 — This form of the feedback amplifier uses an arbitrary transformer. Feedback is isolated from bias components.

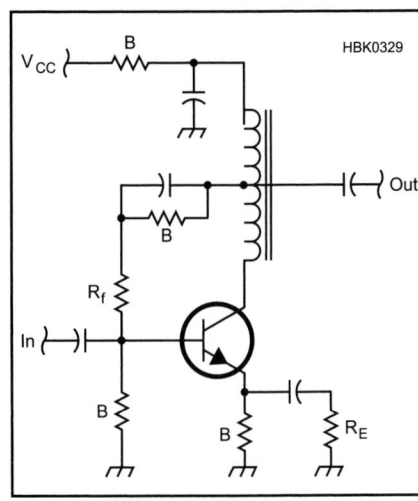

Fig 5.42 — A feedback amplifier with feedback from the output transformer tap. This is common, but can produce unstable results.

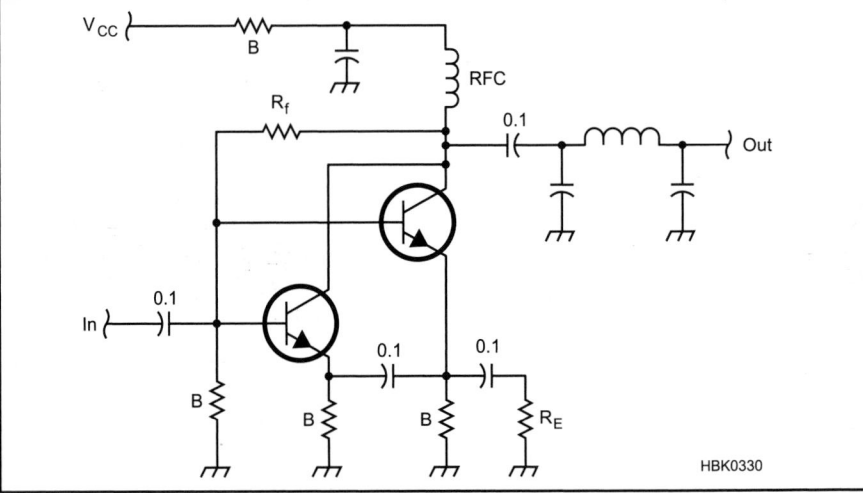

Fig 5.43 — Feedback amplifier with two parallel transistors.

showing practical forms of the feedback amplifier. **Fig 5.39** shows a complete circuit. The base is biased with a resistive divider from the collector. However, much of the resistor is bypassed, leaving only R_f active for actual signal feedback. Emitter degeneration is ac-coupled to the emitter. The resistor R_E dominates the degeneration since R_E is normally much smaller than the emitter bias resistor. Components that are predominantly used for biasing are marked with "B." This amplifier would normally be terminated in 50 Ω at both the input and output. The transformer has the effect of making the 50-Ω load "look like" a

larger load value, $R_L = 200$ Ω to the collector. This is a common and useful value for many HF applications.

Fig 5.40 differs from Fig 5.39 in two places. First, the collector is biased through an RF choke instead of a transformer. The collector circuit then "sees" 50 Ω when that load is connected. Second, the emitter degeneration is in series with the bias, instead of the earlier parallel connection. Either scheme works well, although the parallel configuration affords experimental flexibility with isolation between setting degeneration and biasing. Amplifiers without an output transformer are not constrained by degraded transformer performance and often offer constant or "flat" gain to several GHz.

The variation of **Fig 5.41** may well be the most general. It uses an arbitrary transformer to match the collector. Biasing is traditional

and does not interact with the feedback.

Feedback is obtained directly from the output tap in the circuit of **Fig 5.42**. While this scheme is common, it is less desirable than the others, for the transformer is part of the feedback loop. This could lead to instabilities. Normally, the parallel feedback tends to stabilize the amplifiers.

The circuit of **Fig 5.43** has several features. Two transistors are used, each with a separate emitter biasing resistor. However, ac coupling causes the pair to operate as a single device with degeneration set by R_E. The parallel feedback resistor, R_f, is both a signal feedback element and part of the bias divider. This constrains the values slightly. Finally, an arbitrary output load can be presented to the composite collector through a π-type matching network. This provides some low pass filtering, but constrains the amplifier bandwidth.

5.5 Ferrite Materials

Ferrites are ceramics consisting of various metal oxides formulated to have very high permeability. Iron, manganese, manganese zinc (MnZn), and nickel zinc (NiZn) are the most commonly used oxides. Ferrite cores and beads are available in many styles, as shown in **Fig 5.44**. Wires and cables are then passed through them or wound on them.

When ferrite surrounds a conductor, the high permeability of the material provides a much easier path for magnetic flux set up by current flow in the conductor than if the wire were surrounded only by air. The short length of wire passing through the ferrite will thus see its self-inductance "magnified" by the relative permeability of the ferrite. The ferrites used for suppression are soft ferrites — that is, they are not permanent magnets.

Recalling the definition in the **Electrical Fundamentals** chapter, *permeability* (μ) is the characteristic of a material that quantifies the ease with which it supports a magnetic field. Relative permeability is the ratio of the permeability of the material to the permeability of free space. The relative permeability of nonmagnetic materials like air, copper, and aluminum is 1, while magnetic materials have a relative permeability much greater than 1. Typical values (measured at power frequencies) for stainless steel, steel and mu-metal are on the order of 500, 1000 and 20,000 respectively. Various ferrites have values from the low tens to several thousand. The permeability of these materials changes with frequency and this affects their suitability for use as an inductor's magnetic core or as a means of providing EMI suppression.

Manufacturers vary the chemical composition (the *mix* or *material "type"*) and the dimensions of ferrites to achieve the desired electrical performance characteristics. It is important to select a mix with permeability and loss characteristics that are appropriate for the intended frequency range and application. (The **Component Data and References** chapter includes tables showing the data for different types of ferrite and powdered-iron cores and appropriate frequency ranges for each.)

Fair-Rite Products Corp. supplies most of the ferrite materials used by amateurs, and their website (**www.fair-rite.com**) includes extensive technical data on both the materials and the many parts made from those materials. Much can be learned from the study of this data, the most extensive of any manufacturer. The website includes two detailed application notes on the use of ferrite materials, *How to Choose Ferrite Components for EMI Suppression* and *Use of Ferrites in Broadband Transformers*.

5.5.1 Ferrite Permeability and Frequency

Product data sheets characterize ferrite materials used as chokes by graphing their series equivalent impedance, and chokes are usually analyzed as if their equivalent circuit had only a series resistance and inductance, as shown in **Figs 5.45A** and 5.45B. (Fig 5.45 presents small-signal equivalent circuits. Is-

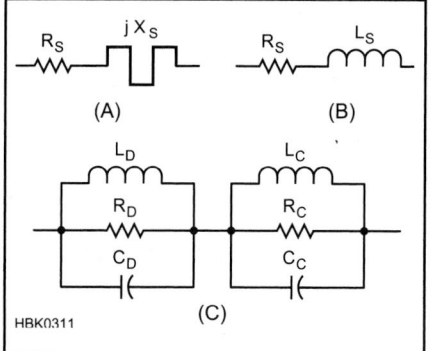

Fig 5.45 — Equivalent circuits for ferrite material. (A) shows the equivalent series circuit specified by the data sheet. (B) shows an over-simplified equivalent circuit for a ferrite choke. (C) shows a better equivalent circuit for a ferrite choke.

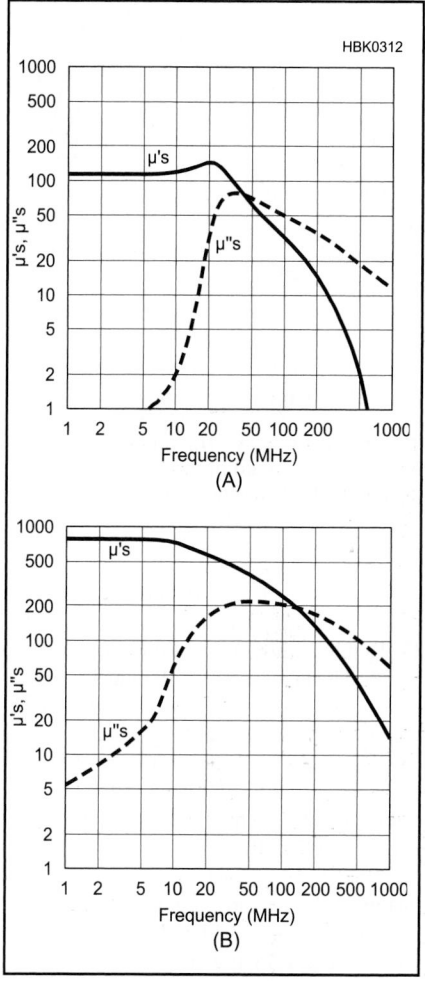

Fig 5.46 — Permeability of a typical ferrite material, Fair-Rite Type #61 (A) and of Type #43 material (B). (Based on product data published by Fair-Rite Products Corp.)

Fig 5.44 — Ferrites are made in many forms. Beads are cylinders with small center holes so that they can be slid over wires or cables. Toroidal cores are more ring-like so that the wire or cable can be passed through the center hole multiple times. Snap-on or split cores are made to be clamped over cables too large to be wound around the core or for a bead to be installed.

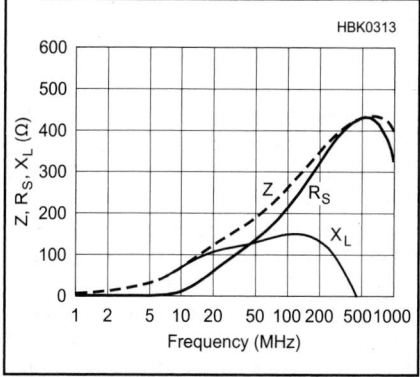

Fig 5.47 — Impedance of a bead for EMI suppression at UHF, Fair-Rite Type #61. (Based on product data published by Fair-Rite Products Corp.)

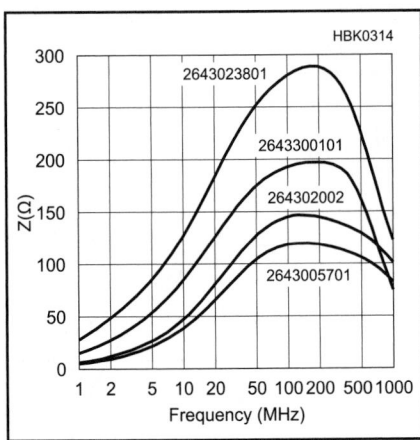

Fig 5.48 — Impedance of different length beads of Fair-Rite Type #43 material. The longer the bead, the higher the impedance. (Based on product data published by Fair-Rite Products Corp.)

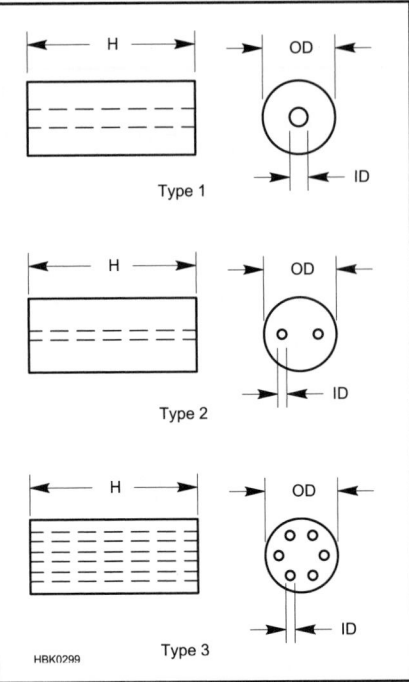

Fig 5.49 — Typical one-piece ferrite bead configurations. Component leads or cables can be inserted through beads one or more times to create inductance and loss as described in the text.

Fig 5.50 — A ferrite choke consisting of multiple turns of cable wound on a 2.4-in. OD × 1.4-in. ID × 0.5-in. toroid core.

sues of temperature and saturation at high power levels are not addressed in this section.) The actual equivalent circuit is closer to Fig 5.45C. The presence of both inductance and capacitance creates resonances visible in the graphs of impedance versus frequency as discussed below. One resonance is created by the pair $L_D C_D$ and the other by $L_C C_C$.

Fig 5.46A graphs the permeabilities μ'_S and μ''_S for Fair-Rite Type #61 ferrite material (Fair-Rite products are identified by a material "Type" and a number), one of the many mix choices available. This material is recommended for use in inductive applications below 25 MHz and for EMI suppression at frequencies above 200 MHz.

For the simple series equivalent circuit of Fig 5.45A and B, the permeability constant for ferrite is actually complex; $\mu = \mu'_S + j\,\mu''_S$. In this equation, μ'_S represents the component of permeability defining ordinary inductance, and μ''_S describes the component that affects losses in the material. You can see that up to approximately 20 MHz, μ'_S is nearly constant, meaning that an inductor wound on a core of this material will have a stable inductance. Below about 15 MHz, the chart shows that μ''_S is much smaller than μ'_S, so that losses are small, making this material good for high power inductors and transformers in this frequency range.

Above 15 MHz, μ''_S increases rapidly and so do the material's associated losses, peaking between 300 and 400 MHz. That makes the inductor very lossy at those frequencies and good for suppressing EMI by absorbing energy in the unwanted signal. **Fig 5.47** shows the manufacturer's impedance data for one "turn" of wire through a cylindrical ferrite bead (a straight wire passing through the bead) made of mix #61 with the expected peak in impedance above 200 MHz. Interestingly,

X_L goes below the graph above resonance, but it isn't zero. The negative reactance is created by the capacitors in Fig 5.45C.

The Type #43 mix (see Fig 5.46B) used for the beads of **Fig 5.48** is optimized for suppres-

sion at VHF (30-300 MHz). Type #43 material's μ'_S is much higher than type #61, meaning that fewer turns are required to achieve the needed inductance. But μ''_S for #43 remains significant even at low frequencies, limiting its usefulness for inductor and transformer cores that must handle high power. The figure shows the impedance data for several beads of different lengths. The longer the bead, the higher the impedance. The same effect can be obtained by stringing multiple beads together on the same wire or cable.

5.5.2 Resonances of Ferrite Cores

Below resonance, the impedance of a wire passing through a ferrite cylinder is proportional to the length of the cylinder. Fig 5.48

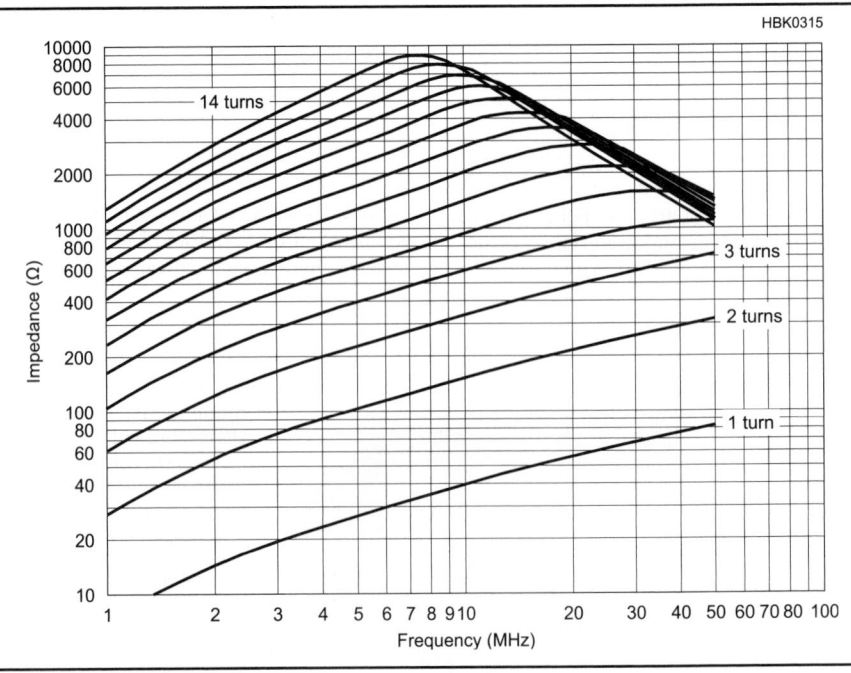

Fig 5.51 — Impedance of multi-turn choke wound on a Fair-Rite Type #43 core as shown in Fig 5.50. Type #43 material is optimized for use as a choke for the VHF range. (Measured data)

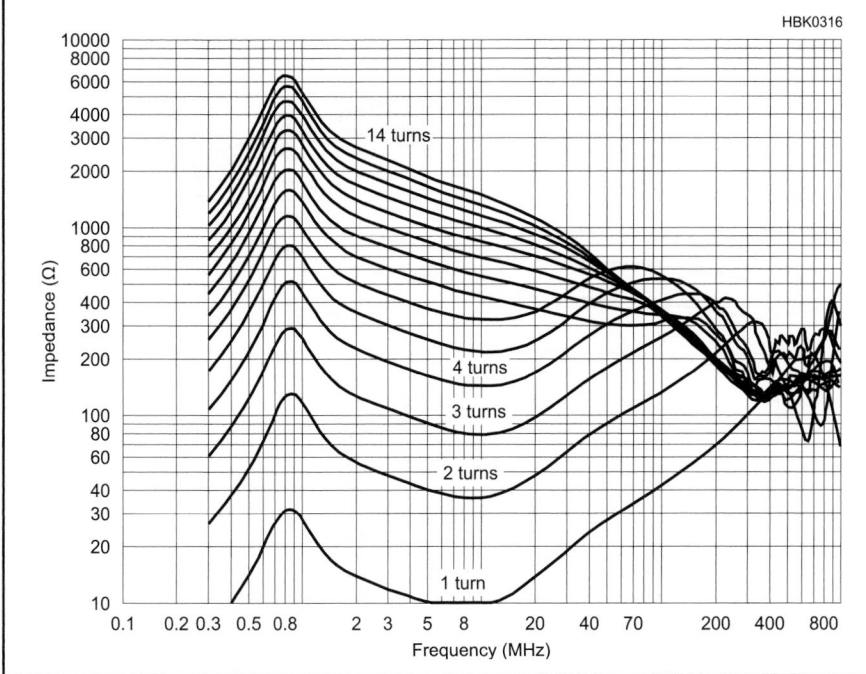

Fig 5.52 — Impedance of multi-turn choke wound on a Fair-Rite Type #78 core as shown in Fig 5.50. Type #78 material is optimized for use as a choke below 2 MHz. (Measured data)

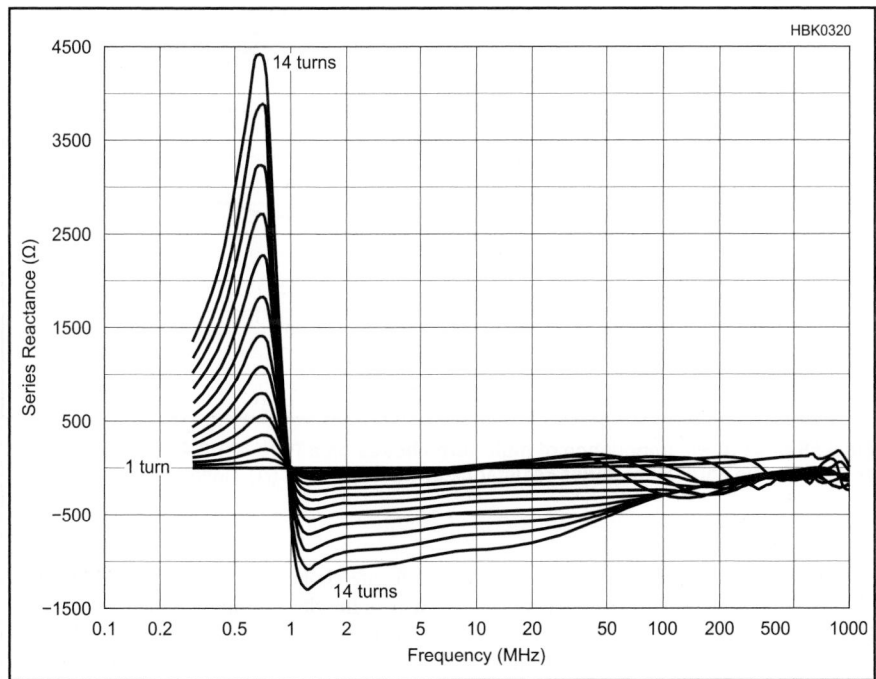

Fig 5.53 — Series reactive component of the chokes of Fig 5.52. (Measured data)

shows the impedance of a family of beads that differ primarily in their length. There are also small differences in their cross section, which is why the resonant frequency shifts slightly as discussed below. **Fig 5.49** shows some typical ferrite bead configurations.

Like all inductors, the impedance of a ferrite choke below resonance is approximately proportional to the square of the number of turns passing through the core. **Fig 5.50** shows a multi-turn choke wound on a 2.4-inch OD × 1.4-inch ID × 0.5-inch ferrite toroid core. **Fig 5.51** is measured data for the choke in Fig 5.50 made of ferrite optimized for the VHF

range (30-300 MHz). The data of **Fig 5.52** are for toroids of the same size, but wound on a material optimized for use below 2 MHz.

We'll study the $L_D C_D$ resonance first. A classic text (*Soft Ferrites, Properties and Applications*, by E. C. Snelling, published in 1969), shows that there is a dimensional resonance within the ferrite related to the velocity of propagation (V_P) within the ferrite and standing waves that are set up in the cross-sectional dimensions of the core. In general, for any given material, the smaller the core, the higher will be the frequency of this resonance, and to a first approximation, the resonant frequency will double if the core dimension is halved. In Fig 5.45C, L_D and C_D account for this dimensional resonance, and R_D for losses within the ferrite. R_D is mostly due to eddy currents (and some hysteresis) in the core.

Now it's time to account for R_C, L_C and C_C. Note that there are two sets of resonances for the chokes wound around the Type #78 material (Fig 5.52), but only one set for the choke of Fig 5.51. For both materials, the upper resonance starts just below 1 GHz for a single turn and moves down in frequency as the number of turns is increased. **Fig 5.53**, the reactance for the chokes of Fig 5.52, also shows both sets of resonances. That's why the equivalent circuit must include two parallel resonances!

The difference between these materials that accounts for this behavior is their chemical composition (the mix). Type #78 is a MnZn ferrite, while Type #43 is a NiZn ferrite. The velocity of propagation (V_P) in NiZn ferrites is roughly two orders of magnitude higher than for MnZn, and, at those higher frequencies, there is too much loss to allow the standing waves that establish dimensional resonance to exist.

To understand what's happening, we'll return to our first order equivalent circuit of a ferrite choke (Fig 5.45C). L_C, and R_C, and C_C are the inductance, resistance, (including the effect of the μ of the ferrite), and stray capacitance associated with the wire that passes through the ferrite. This resonance moves down in frequency with more turns because both L and C increase with more turns. The dimensional resonance does not move, since it depends only on the dimensions V_P of the ferrite.

What is the source of C_C if there's no "coil," only a single wire passing through a cylinder? It's the parasitic capacitance from the wire at one end of the cylinder to the wire at the other end, with the ferrite acting as the dielectric. Yes, it's a very small capacitance, but you can see the resonance it causes in the measured data and on the data sheet.

5.5.3 Ferrite Series and Parallel Equivalent Circuits

Let's talk briefly about series and parallel equivalent circuits. Many impedance analyzers express the impedance between their ter-

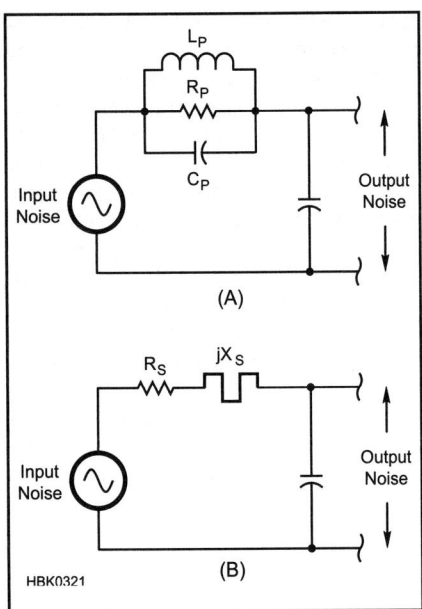

Fig 5.54 — At A, the series element of the divider is a parallel-resonant circuit. At B, the series element of the divider is the series-equivalent circuit used by ferrite data sheets.

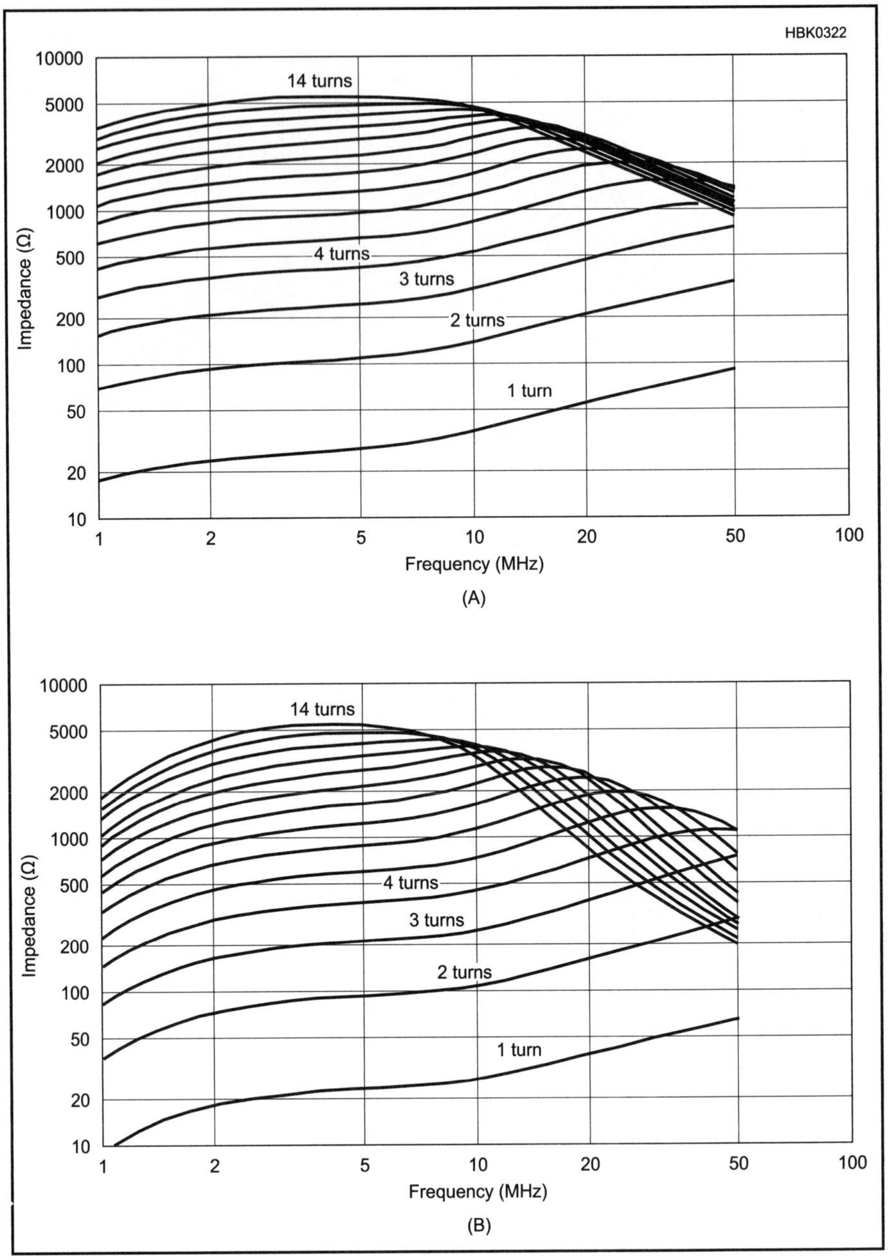

Fig 5.55 — At A, the impedance of multi-turn chokes on a Type #31 2.4-in. OD toroidal core. At B, the equivalent series resistance of the chokes of (A). (Measured data)

minals as Z with a phase angle, and the series equivalent R_S, and X_S. They could just have easily expressed that same impedance using the parallel equivalent R_P and X_P but R_P and X_P will have values that are numerically different from R_S and X_S. (See the section on series-parallel impedance transformation in the **Electrical Fundamentals** chapter.)

It is important to remember that in a series circuit, the larger value of R_S and X_S has the greatest influence, while in a parallel circuit, the smaller value of R_P and X_P is dominant. In other words, for R_P to dominate, R_P must be small.

Both expressions of the impedance (series or parallel) are correct at any given frequency, but whether the series or parallel representation is most useful will depend on the physics of the device being measured and how that device is used in a circuit. We've just seen, for example, that the parallel equivalent circuit is a more realistic representation of a ferrite choke — the values of R_P, L_P, and C_P will come much closer to remaining constant as frequency changes than if we use the series equivalent. (R_P, L_P and C_P won't be precisely constant though, because the physical properties of all ferrites — permeability, resistivity and permittivity — all vary with frequency.)

But virtually all product data for ferrite chokes is presented as the series equivalent R_S and X_S. Why? First, because it's easy to measure and understand, second, because we tend to forget there is stray capacitance, and third because ferrite beads are most often used as chokes to reduce current in a series circuit! **Fig 5.54A** and Fig 5.54B are both useful representations of the voltage divider formed by

a ferrite choke and a small bypass capacitor across the device input. Which we use will depend on what we know about our ferrite.

If we know R_P, L_P and C_P and they are constant over the frequency range of interest, Fig 5.54A may be more useful, because we can insert values in a circuit model and perhaps tweak the circuit. But if we have a graph of R_S and X_S vs frequency, Fig 5.54B

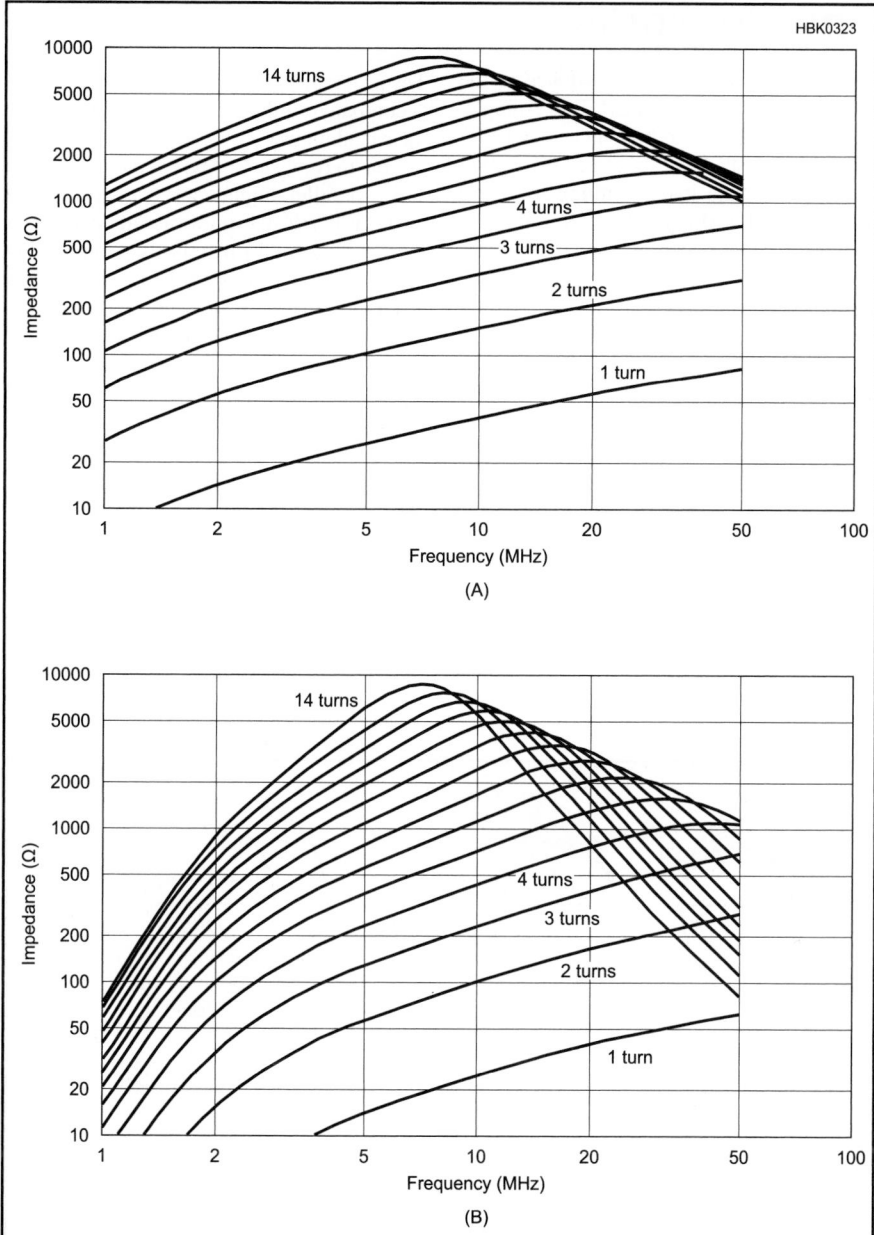

Fig 5.56 — At A, the impedance of multi-turn chokes on a Type #43 2.4-in. OD toroidal core. At B, the equivalent series resistance of the chokes of A. (Measured data).

Because the C_P of many practical chokes is quite small and their impedance at resonance is often rather high, they are quite difficult to measure accurately, especially with reflection-based measurement systems. See the **Test Equipment and Measurements** chapter for a simple measurement method that can provide good results. More information on the use of ferrite cores and beads for EMI suppression is available in the chapter on **RF Interference** and in *The ARRL RFI Book*. A thorough treatment of the use of these ferrite materials for EMI suppression is continued in the online publication *A Ham's Guide to RFI, Ferrites, Baluns, and Audio Interfacing* at **www.audio-systemsgroup.com/publish**. The use of ferrite beads and cores for transmitting chokes is presented in the **Transmission Lines** chapter.

5.5.4 Type 31 Material

The relatively new Type #31 material made by Fair-Rite Products is extremely useful as a choke core, especially if some component of your problem occurs below 5 MHz. Measured data for the new material is displayed in **Figs 5.55A** and 5.55B. Compare it with **Figs 5.56A** and 5.56B, which are corresponding plots for the older Type #43 material. By comparison, Type #31 provides nearly 7 dB greater choking impedance at 2 MHz, and at least 3 dB more on 80 meters. At 10 MHz and above, the two materials are nearly equivalent, with Type #43 being about 1 dB better. If your goal is EMI suppression or a feed line choke (a so-called "current balun"), the Type #31 material is the best all round performer to cover all HF bands, and is clearly the material of choice at 5 MHz and below. Between 5 MHz and 20 MHz, Type #43 has a slight edge (about 1 dB), and above 20 MHz they're equivalent. (Baluns are discussed more in the section below on Transmission Line Transformers and in the chapters on **Transmission Lines** and **Antennas**.)

The new Type #31 material is useful because it exhibits both of the resonances in our equivalent circuit — that is, the dimensional resonance of the core and the resonance of the choke with the lossy permeability of the core material. Below 10 MHz, these two resonances combine (in much the manner of a stagger-tuned IF) to provide significantly greater suppression bandwidth (roughly one octave, or one additional harmonically related ham band). The result is that a single choke on a Type #31 core can be made to provide very good suppression over about 8:1 frequency span, as compared to 4:1 for Type #43. Type #31 also has somewhat better temperature characteristics at HF.

will give us a good answer faster. Because we will most often be dealing with R_S and X_S data, the series circuit equivalents are used most often. Another reason for using R_S and X_S is that the impedance of two or more ferrite chokes in series can be computed simply by adding their R_S and X_S components, just as with any other series impedances! When you look at the data sheet plots of R_S, X_L and Z for a standard ferrite part, you are looking at the series equivalent parameters of their dominant resonance. For most MnZn materials, it is dimensional resonance, while, for most NiZn materials, it is the circuit resonance.

Values for R_P, L_P and C_P for nearly any ferrite choke can be obtained by curve-fitting the data for its parallel-resonance curve. (Self-resonance is discussed earlier in this chapter.)

5.6 Impedance Matching Networks

An impedance transforming or matching network is one that accepts power from a generator with one characteristic impedance, the source, and delivers virtually all of that power to a load at a different impedance. The simple designs in this section provide matching at only one frequency. More refined methods discussed in the reference texts can encompass a wide band of frequencies.

Both source and load impedances are likely to be complex with both real and imaginary (reactive) parts. The procedures given in this section allow the reader to design the common impedance-matching networks based on the simplification that both the source and load impedance are resistive (with no reactive component). To use these procedures with a complex load or source impedance, the usual method is to place a reactance in series with the impedance that has an equal and opposite reactance to cancel the reactive component of the impedance to be matched. Then treat the resulting purely-resistive impedance as the resistance to be transformed. For example, if a load impedance to be matched is 120 + j 40 Ω, add capacitive reactance of $-j$ 40 Ω in series with the load, resulting in a load impedance of 120 + j 0 Ω so that the following procedures can be used. The series reactance-canceling component may then be combined with the matching network's output component in some configurations.

A set of 14 simple resonant networks, and their equations, is presented in **Fig 5.58**. Note that in these diagrams R_S is the low impedance side and R_L is the high impedance side and that the X values are calculated in the top-down order given.

The formulas for the various networks use and produce values of reactance. To convert the reactances to L and C values, use the formulas

$$C = \frac{1}{2\pi f X_c} \text{ and } L = \frac{X_L}{2\pi f}$$

The program *MATCH.EXE* on the CD-ROM accompanying this *Handbook* can perform the calculations.

You may wish to review the section on series-parallel impedance transformations in the **Electrical Fundamentals** chapter as those techniques form the basis of impedance-matching network design. There is additional discussion of L networks in the chapters on **Transmission Lines** and of Pi networks in the **RF Power Amplifiers** chapter. These discussions apply to more specific applications of the networks.

5.6.1 L Networks

Perhaps the most common LC impedance transforming network is the L network, so named because it uses two elements — one series element and one parallel — resembling the capital L on its side. There are eight different types of L networks as shown in **Fig 5.57**.

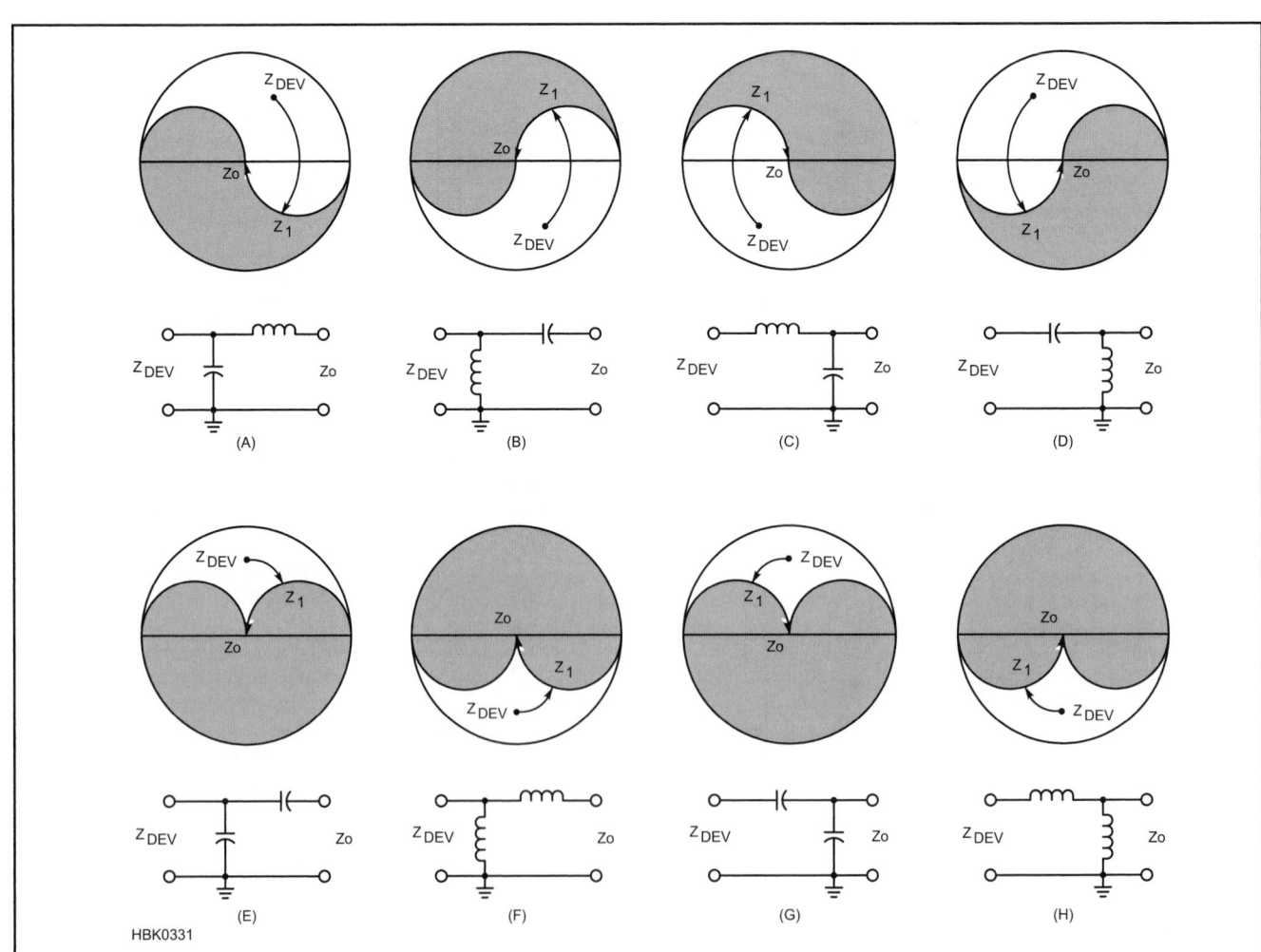

Fig 5.57 — L-networks which will match a complex impedance (shown here as Z_{DEV}, the output impedance of a device) to Z_0, a resistive source or load. Impedances within the shaded portion of the simplified Smith Chart cannot be matched by the network. Z_1 represents the impedance that is transformed from Z_{DEV} by the series element.

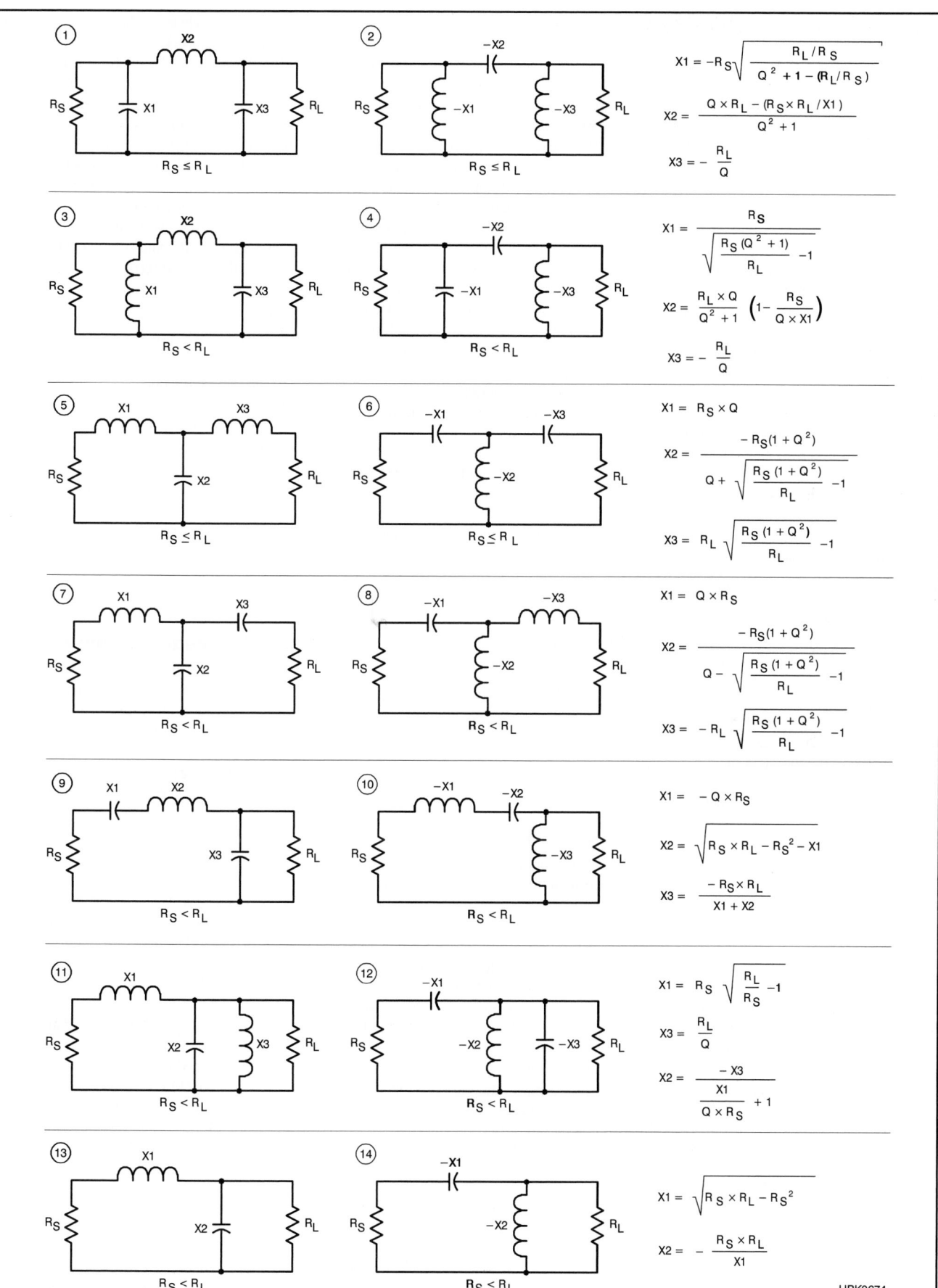

Equations for networks 1 and 2:

$$X1 = -R_S\sqrt{\frac{R_L/R_S}{Q^2 + 1 - (R_L/R_S)}}$$

$$X2 = \frac{Q \times R_L - (R_S \times R_L/X1)}{Q^2 + 1}$$

$$X3 = -\frac{R_L}{Q}$$

Equations for networks 3 and 4:

$$X1 = \frac{R_S}{\sqrt{\frac{R_S(Q^2 + 1)}{R_L} - 1}}$$

$$X2 = \frac{R_L \times Q}{Q^2 + 1}\left(1 - \frac{R_S}{Q \times X1}\right)$$

$$X3 = -\frac{R_L}{Q}$$

Equations for networks 5 and 6:

$$X1 = R_S \times Q$$

$$X2 = \frac{-R_S(1 + Q^2)}{Q + \sqrt{\frac{R_S(1 + Q^2)}{R_L} - 1}}$$

$$X3 = R_L\sqrt{\frac{R_S(1 + Q^2)}{R_L} - 1}$$

Equations for networks 7 and 8:

$$X1 = Q \times R_S$$

$$X2 = \frac{-R_S(1 + Q^2)}{Q - \sqrt{\frac{R_S(1 + Q^2)}{R_L} - 1}}$$

$$X3 = -R_L\sqrt{\frac{R_S(1 + Q^2)}{R_L} - 1}$$

Equations for networks 9 and 10:

$$X1 = -Q \times R_S$$

$$X2 = \sqrt{R_S \times R_L - R_S^2} - X1$$

$$X3 = \frac{-R_S \times R_L}{X1 + X2}$$

Equations for networks 11 and 12:

$$X1 = R_S\sqrt{\frac{R_L}{R_S} - 1}$$

$$X3 = \frac{R_L}{Q}$$

$$X2 = \frac{-X3}{\frac{X1}{Q \times R_S} + 1}$$

Equations for networks 13 and 14:

$$X1 = \sqrt{R_S \times R_L - R_S^2}$$

$$X2 = -\frac{R_S \times R_L}{X1}$$

HBK0674

Fig 5.58 — Fourteen impedance transforming networks with their design equations (for lossless components).

(Types G and H are not as widely used and not covered here.) The simplified Smith Chart sketches show what range of impedance values can be transformed to the required impedance Z_0 and the path on the chart by which the transformation is achieved by the two reactances. (An introductory discussion of the Smith Chart is provided on this book's CD-ROM and a detailed treatment is available in *The ARRL Antenna Book*.) Z_{DEV} represents the output impedance of a device, such as a transistor amplifier or any piece of equipment.

The process of transformation works the same in both "directions." That is, a network designed to transform Z_{DEV} to Z_0 will also transform Z_0 to Z_{DEV} if reversed. The L network (and the Pi and T networks discussed below) is *bilateral,* as are all lossless networks.

Purely from the standpoint of impedance matching, the L-network can be constructed with inductive or capacitive reactance in the series arm and the performance will be exactly the same in either case. However, the usual case in amateur circuits is to place the inductive reactance in the series arm to act as a low pass filter. Your particular circumstances may dictate otherwise, however. For example, you may find it useful that the series-C/parallel-L network places a dc short circuit across one side of the network while blocking dc current through it.

The design procedure for L network configurations A through F in Fig 5.57 is as follows:

Given the two resistance values to be matched, connect the series arm of the circuit to the smaller of the two (R_S) and the parallel arm to the larger (R_P).

Find the ratio R_P/R_S and the L network's Q:

$$Q = \sqrt{\frac{R_P}{R_S} - 1}$$

Calculate the series reactance $X_S = QR_S$
Calculate the parallel reactance $X_P = R_P/Q$.

5.6.2 Pi Networks

Another popular impedance matching network is the Pi network shown in circuits 1 and 2 of Fig 5.58. The Pi network can be thought of as two L networks "back to back." For example, the Pi network in circuit 1 can be split into the L network of circuit 13 on the right and its mirror image on the

left. The two inductors in series are combined into the single inductance of the Pi network. There are other forms of the Pi network with different configurations of inductance and capacitance, but the version shown is by far the most common in amateur circuits.

The use of two transformations allows the designer to choose Q for the Pi network, unlike the L network for which Q is determined by the ratio of the impedances to be matched. This is particularly useful for matching amplifier outputs as discussed in the **RF Power Amplifier** chapter because it allows more control of the network's frequency response and of component values. Q must be high enough that $(Q^2 + 1) > (R_1 / R_2)$. If these two quantities are equal, X_{C2} becomes infinite, meaning zero capacitance, and the Pi network reduces to the L network in Fig 5.57A. The design procedure for the Pi network in Fig 5.58 is as follows:

Determine the two resistance values to be matched, $R_1 > R_2$, and select a value for Q. Follow the calculation sequence for circuit 1 or 2 in Fig 5.58.

Calculate the value of the parallel reactance $X_{C1} = R_1 / Q$.

Calculate the value of the parallel reactance X_{C2}

$$X_{C2} = R_2 \sqrt{\frac{R_1 / R_2}{Q^2 + 1 - R_1 / R_2}}$$

Calculate the value of the series reactance X_L

$$X_L = \frac{QR_1 + R_1R_2 / X_{C2}}{Q^2 + 1}$$

5.6.3 T Networks

Many amateur transmission line impedance matching units ("antenna tuners") use the version of the T network in circuit 6 of Fig 5.58. The circuits are constructed using variable capacitors and a tapped inductor. This is easier and less expensive to construct than a fully-adjustable version of circuit 5 in which two variable inductors are required. Circuit 5 is especially useful in matching relatively low input impedances from solid-state amplifier outputs to 50-Ω loads with low Q and good harmonic suppression due to the series inductances. Circuit 7 is also useful in solid-state amplifier design as described in the

reference texts listed at the end of this chapter.

The T network shown in circuit 5 of Fig 5.58 is especially useful for matching to relatively low impedance from 50-Ω sources with practical components and low Q. Like the Pi network, Q must be high enough that $(Q^2+1) > (R_1/R_2)$. Designing the component values for this network requires the calculation of a pair of intermediate values, A and B, to make the equations more manageable.

Determine the two resistance values to be matched, $R_1 > R_2$, and select a value for Q.

Calculate the intermediate variables A and B

$$A = R_1(Q^2 + 1) \text{ and } B = \sqrt{\left(\frac{A}{R_2} - 1\right)}$$

Calculate the value of the input series reactance $X_{L1} = R_1 Q$.

Calculate the value of the output series reactance $X_{L2} = R_2 B$.

Calculate the value of the parallel reactance $X_C = A / (Q + B)$

Convert the reactances to component values:

$$C = \frac{1}{2\pi f X_c} \text{ and } L = \frac{X_L}{2\pi f}$$

5.6.4 Impedance Inversion

Symmetrical Pi and T networks have the useful property of *impedance inversion* when the reactances of all elements are the same at the design frequency: $X_C = X_L = |X|$, resulting in a Q of 1. For either type of network, the impedance looking into the network, Z_{IN}, will be the load impedance, Z_{OUT}, inverted about X:

$$Z_{IN} = \frac{X^2}{Z_{OUT}}$$

This is the same effect as if the network were replaced with a ¼-wavelength transmission line with $Z_0 = |X|$. Since the network is symmetrical, the inversion occurs in either direction through the network. The result is true only at the frequency for which all reactances are equal.

For example, to invert all impedances about 50 Ω, set X = 50 Ω at the design frequency. The input impedance will then be $50^2/Z_{OUT}$. If Z = 10 + j10 Ω is connected to one end of the network, the impedance looking into the other end of the network will be 2500 / (10 + j10) = 125 − j125 Ω.

5.7 RF Transformers

5.7.1 Air-Core Nonresonant RF Transformers

Air-core transformers often function as mutually coupled inductors for RF applications. They consist of a primary winding and a secondary winding in close proximity. Leakage reactances are ordinarily high, however, and the coefficient of coupling between the primary and secondary windings is low. Consequently, unlike transformers having a magnetic core, the turns ratio does not have as much significance. Instead, the voltage induced in the secondary depends on the mutual inductance.

In a very basic transformer circuit operating at radio frequencies, such as in **Fig 5.59**, the source voltage is applied to L1. R_S is the series resistance inherent in the source. By virtue of the mutual inductance, M, a voltage is induced in L2. A current flows in the secondary circuit through the reactance of L2 and the load resistance of R_L. Let X_{L2} be the reactance of L2 independent of L1, that is, independent of the effects of mutual inductance. The impedance of the secondary circuit is then:

$$Z_S = \sqrt{R_L{}^2 + X_{L2}{}^2} \qquad (12)$$

where
  Z_S = the impedance of the secondary circuit in ohms,
  R_L = the load resistance in ohms, and
  X_{L2} = the reactance of the secondary inductance in ohms.

The effect of Z_S upon the primary circuit is the same as a coupled impedance in series with L1. Fig 5.60 displays the coupled impedance (Z_P) in a dashed enclosure to indicate that it is not a new physical component. It has the same absolute value of phase angle as in the secondary impedance, but the sign of the reactance is reversed; it appears as a capacitive reactance. The value of Z_P is:

$$Z_P = \frac{(2 \pi f M)^2}{Z_S} \qquad (13)$$

where
  Z_P = the impedance introduced into the primary,
  Z_S = the impedance of the secondary circuit in ohms, and
  $2 \pi f M$ = the mutual reactance between the reactances of the primary and secondary coils (also designated as X_M).

5.7.2 Air-Core Resonant RF Transformers

The use of at least one resonant circuit in place of a pair of simple reactances elimi-

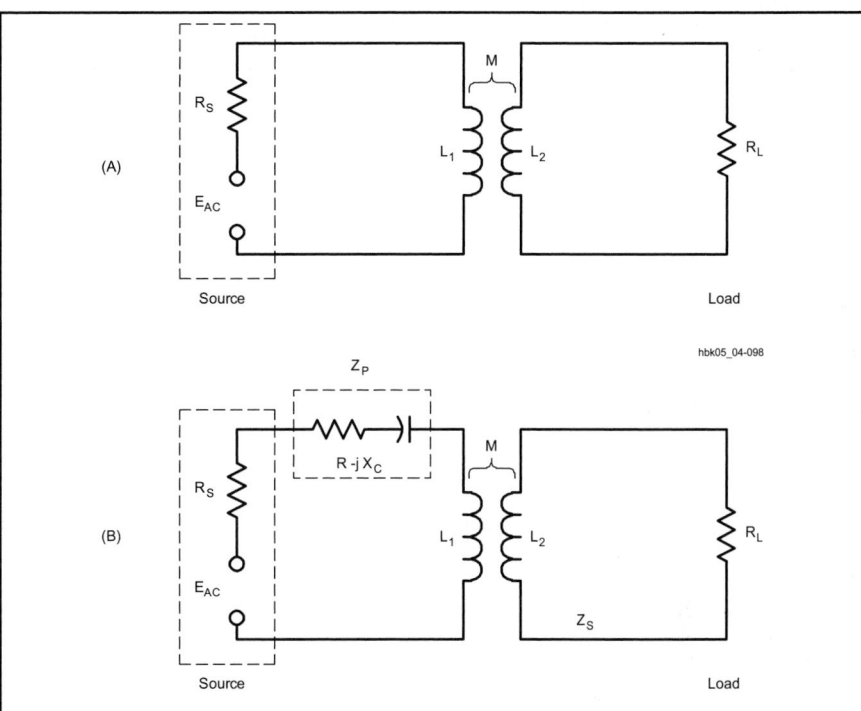

Fig 5.59 — The coupling of a complex impedance back into the primary circuit of a transformer composed of nonresonant air-core inductors.

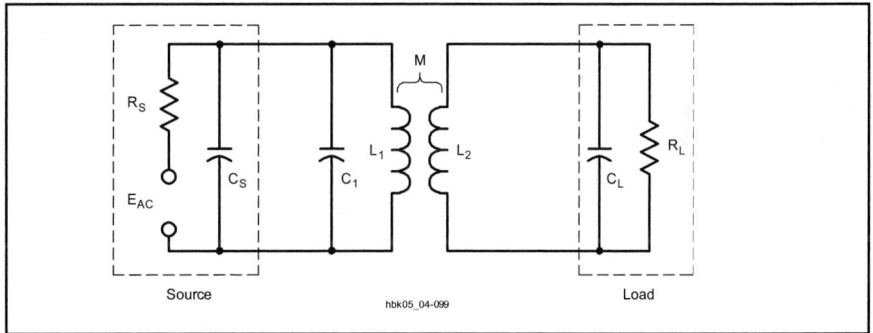

Fig 5.60 — An air-core transformer circuit consisting of a resonant primary circuit and an untuned secondary. R_S and C_S are functions of the source, while R_L and C_L are functions of the load circuit.

nates the reactance from the transformed impedance in the primary. For loaded or operating Q of at least 10, the resistances of individual components is negligible. **Fig 5.60** represents just one of many configurations in which at least one of the inductors is in a resonant circuit. The reactance coupled into the primary circuit is cancelled if the circuit is tuned to resonance while the load is connected. If the reactance of the load capacitance, C_L is at least 10 times any stray capacitance in the circuit, as is the case for low impedance loads, the value of resistance coupled to the primary is

$$R1 = \frac{X_M{}^2 R_L}{X_2{}^2 + R_L{}^2} \qquad (14)$$

where:
  R1 = series resistance coupled into the primary circuit,
  X_M = mutual reactance,
  R_L = load resistance, and
  X_2 = reactance of the secondary inductance.

The parallel impedance of the resonant circuit is just R1 transformed from a series to a parallel value by the usual formula, $R_P = X_2 / R1$.

The higher the loaded or operating Q of the circuit, the smaller the mutual inductance required for the same power transfer. If both the primary and secondary circuits consist of resonant circuits, they can be more loosely coupled than with a single tuned circuit for the same power transfer. At the usual loaded Q of 10 or greater, these circuits are quite selective, and consequently narrowband.

Although coupling networks have to a large measure replaced RF transformer coupling that uses air-core transformers, these circuits are still useful in antenna tuning units and other circuits. For RF work, powdered-iron toroidal cores have generally replaced air-core inductors for almost all applications except where the circuit handles very high power or the coil must be very temperature stable. Slug-tuned solenoid coils for low-power circuits offer the ability to tune the circuit precisely to resonance. For either type of core, reasonably accurate calculation of impedance transformation is possible. It is often easier to experiment to find the correct values for maximum power transfer, however.

5.7.3 Broadband Ferrite RF Transformers

The design concepts and general theory of ideal transformers presented in the **Electrical Fundamentals** chapter apply also to transformers wound on ferromagnetic-core materials (ferrite and powdered iron). As is the case with stacked cores made of laminations in the classic I and E shapes, the core material has a specific permeability factor that determines the inductance of the windings versus the number of wire turns used. (See the earlier discussion on Ferrite Materials in this chapter.)

Toroidal cores are useful from a few hundred hertz well into the UHF spectrum. The principal advantage of this type of core is the self-shielding characteristic. Another feature is the compactness of a transformer or inductor. Therefore, toroidal-core transformers are excellent for use not only in dc-to-dc converters, where tape-wound steel cores are employed, but at frequencies up to at least 1000 MHz with the selection of the proper core material for the range of operating frequencies. Toroidal cores are available from micro-miniature sizes up to several inches in diameter that can handle multi-kW military and commercial powers.

One of the most common ferromagnetic transformers used in amateur circuits is the conventional broadband transformer. Broadband transformers with losses of less than 1 dB are employed in circuits that must have a uniform response over a substantial frequency range, such as a 2- to 30-MHz broadband amplifier. In applications of this sort, the reactance of the windings should be at least

four times the impedance that the winding is designed to look into at the lowest design frequency.

Example: What should be the winding reactances of a transformer that has a 300-Ω primary and a 50-Ω secondary load? Relative to the 50-Ω secondary load:

$$X_S = 4 Z_S = 4 \times 50 \ \Omega = 200 \ \Omega$$

and the primary winding reactance (X_P) is:

$$X_P = 4 Z_P = 4 \times 300 \ \Omega = 1200 \ \Omega$$

The core-material permeability plays a vital role in designing a good broadband transformer. The effective permeability of the core must be high enough to provide ample winding reactance at the low end of the operating range. As the operating frequency is increased, the effects of the core tend to disappear until there are scarcely any core effects at the upper limit of the operating range. The limiting factors for high frequency response are distributed capacity and leakage inductance due to uncoupled flux. A high-permeability core minimizes the number of turns needed for a given reactance and therefore also minimizes the distributed capacitance at high frequencies.

Ferrite cores with a permeability of 850 are common choices for transformers used between 2 and 30 MHz. Lower frequency ranges, for example, 1 kHz to 1 MHz, may require cores with permeabilities up to 2000. Permeabilities from 40 to 125 are useful for VHF transformers. Conventional broadband transformers require resistive loads. Loads with reactive components should use appropriate networks to cancel the reactance.

The equivalent circuit in Fig 5.33 applies to any coil wound on a ferrite core, including transformer windings. (See the section on Ferrite Materials.) However, in the series-equivalent circuit, μ'S is not constant with frequency as shown in Fig 5.34A and 5.34B. Using the low-frequency value of μ'S is a useful approximation, but the effects of the parallel R and C should be included. In high-power transmitting and amplifier applications, the resistance R may dissipate some heat, leading to temperature rise in the core. Regarding C, there are at least two forms of stray capacitance between windings of a transformer; from wire-to-wire through air and from wire-to-wire through the ferrite, which acts as a dielectric material. (Ferrites with low iron content have a relative dielectric constant of approximately 10-12.)

Conventional transformers are wound in the same manner as a power transformer. Each winding is made from a separate length of wire, with one winding placed over the previous one with suitable insulation between. Unlike some transmission-line transformer

designs, conventional broadband transformers provide dc isolation between the primary and secondary circuits. The high voltages encountered in high-impedance-ratio step-up transformers may require that the core be wrapped with glass electrical tape before adding the windings (as an additional protection from arcing and voltage breakdown), especially with ferrite cores that tend to have rougher edges. In addition, high voltage applications should also use wire with high-voltage insulation and a high temperature rating.

Fig 5.61 illustrates one method of transformer construction using a single toroid as the core. The primary of a step-down impedance transformer is wound to occupy the entire core, with the secondary wound over the primary. The first step in planning the winding is to select a core of the desired permeability. Convert the required reactances determined earlier into inductance values for the lowest frequency of use. To find the number of turns for each winding, use the A_L value for the selected core and the equation for determining the number of turns:

$$L = \frac{A_L \times N^2}{1000000} \qquad (15)$$

where
  L = the inductance in mH
  A_L = the inductance index in mH per 1000 turns, and
  N = the number of turns.

Be certain the core can handle the power by calculating the maximum flux and comparing the result with the manufacturer's guidelines.

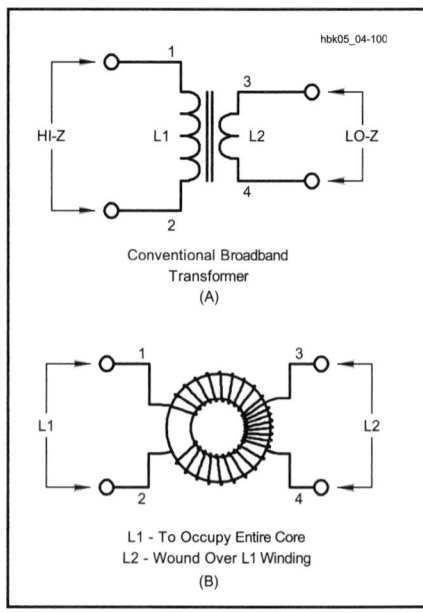

Fig 5.61 — Schematic and pictorial representation of a conventional broadband transformer wound on a toroid core. The secondary winding (L2) is wound over the primary winding (L1).

$$B_{max} = \frac{E_{RMS} \times 10^8}{4.44 \times A_e \times N \times f} \qquad (16)$$

where

B_{max} = the maximum flux density in gauss
E_{RMS} = the voltage across the inductor
A_e = the cross-sectional area of the core in square centimeters
N = the number of turns in the inductor, and
f = the operating frequency in Hz.

(Both equations are from the section on ferrite toroidal inductors in the **Electrical Fundamentals** chapter and are repeated here for convenience.)

Example: Design a small broadband transformer having an impedance ratio of 16:1 for a frequency range of 2.0 to 20.0 MHz to match the output of a small-signal stage (impedance ≈ 500 Ω) to the input (impedance ≈ 32 Ω) of an amplifier.

Since the impedance of the smaller winding should be at least 4 times the lower impedance to be matched at the lowest frequency,

$$X_S = 4 \times 32 \, \Omega = 128 \, \Omega$$

The inductance of the secondary winding should be

$$L_S = \frac{X_S}{2 \pi f} = \frac{128}{6.2832 \times 2.0 \times 10^6 \text{ Hz}}$$

$$= 0.0101 \text{ mH}$$

Select a suitable core. For this low-power application, a ³⁄₈ inch. ferrite core with permeability of 850 is suitable. The core has an A_L value of 420. Calculate the number of turns for the secondary.

$$N_S = 1000 \sqrt{\frac{L}{A_L}} = 1000 \sqrt{\frac{0.010}{420}}$$

$$= 4.88 \text{ turns}$$

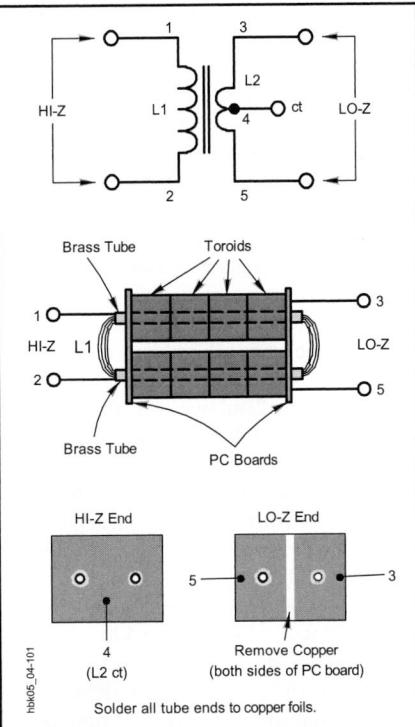

Fig 5.62 — Schematic and pictorial representation of a "binocular" style of conventional broadband transformer. This style is used frequently at the input and output ports of transistor RF amplifiers. It consists of two rows of high-permeability toroidal cores, with the winding passed through the center holes of the resulting stacks.

A 5-turn secondary winding should suffice. The primary winding derives from the impedance ratio:

$$N_P = N_S \sqrt{\frac{Z_P}{Z_S}} = 5 \sqrt{\frac{16}{1}}$$

$$= 5 \times 4 = 20 \text{ turns}$$

This low-power application will not approach the maximum flux density limits for the core, and #28 AWG enamel wire should both fit the core and handle the currents involved.

A second style of broadband transformer construction appears in **Fig 5.62**. The key elements in this transformer are the stacks of ferrite cores aligned with tubes soldered to pc-board end plates. This style of transformer is suited to high power applications, for example, at the input and output ports of transistor RF power amplifiers. Low-power versions of this transformer can be wound on "binocular" cores having pairs of parallel holes through them.

For further information on conventional transformer matching using ferromagnetic materials, see the **RF Power Amplifiers** chapter. Refer to the **Component Data and References** chapter for more detailed information on available ferrite cores. A standard reference on conventional broadband transformers using ferromagnetic materials is *Ferromagnetic Core Design and Applications Handbook* by Doug DeMaw, W1FB, published by MFJ Enterprises.

NOTES ON TOROID WINDINGS

Toroids do have a small amount of leakage flux. Toroid coils are wound in the form of a helix (screw thread) around the circular length of the core. This means that there is a small component of the flux from each turn that is perpendicular to the circle of the toroid (parallel to the axis through the hole) and is therefore not adequately linked to all the other turns. This effect is responsible for a small leakage flux and the effect is called the "one-turn" effect, since the result is equivalent to one turn that is wound around the outer edge of the core and not through the hole. Also, the inductance of a toroid can be adjusted. If the turns can be pressed closer together or separated a little, inductance variations of a few percent are possible. A grounded aluminum shield between adjacent toroidal coils can eliminate any significant capacitive or inductive (at high frequencies) coupling.

5.8 Noise

The following material was contributed by Paul Wade, W1GHZ. The section on background noise by Joe Taylor, K1JT, is reproduced from his discussion of Earth-Moon-Earth (EME) communications on the CD-ROM accompanying this book. Additional discussion of noise measurement is available in the **Test Instruments and Measurements** *chapter and in the Noise Instrumentation document provided on the CD-ROM.*

As anyone who has listened to a receiver suspects, everything in the universe generates noise. In communications, the goal is to maximize the desired signal in relation to the undesired noise we hear. In order to accomplish this goal, it would be helpful to understand where noise originates, how much our own receiver adds to the noise we hear, and how to minimize it.

It is difficult to improve something unless we are able to measure it. Measurement of noise in receivers does not seem to be clearly understood by many amateurs, so this section attempts to explain the concepts and clarify the techniques, and to describe the standard "measure of merit" for receiver noise performance: "noise figure." In addition, the Noise Instrumentation document on the CD-ROM describes how to build your own noise generator for noise figure measurements.

A number of equations are included, but only a few need be used to perform noise figure measurements. The rest are included to as an aid to understanding supported by explanatory text.

5.8.1. Noise Power

The most pervasive source of noise is *thermal noise* (also called *Johnson* or *Johnson-Nyquist noise*), due to the motion of thermally agitated free electrons in a conductor. Since everything in the universe is at some temperature above absolute zero, every conductor must generate noise.

Every resistor (and all conductors have resistance) generates an RMS noise voltage:

$$e = \sqrt{4kTRB}$$

where R is the resistance, T is the absolute temperature in kelvins (K), B is the bandwidth in hertz, and k is Stefan-Boltzmann's constant, 1.38×10^{-23} joules /K (or J K^{-1}).

Converting to power, e^2/R, and adjusting for the Gaussian distribution of noise voltage, the noise power generated by the resistor is:

$$P_n = kTB \text{ (watts)}$$

which is independent of the resistance. Thus, all resistors at the same temperature generate

the same noise power.

Thermal noise is *white noise*, meaning that the power density does not vary with frequency, but always has a *power density* or *spectral density* of kT watts/Hz. (The corresponding noise voltage distribution is a *spectral voltage density*, measured in volts / $\sqrt{Hz}$, spoken as "volts per root hertz".) More important is that the noise power is directly proportional to absolute temperature T, since k is a constant. At the nominal ambient temperature of 290 K**,** we can calculate this power; converted to dBm, we get the familiar –174 dBm/Hz. Multiply by the bandwidth in hertz to get the available noise power at ambient temperature. The choice of 290 K for ambient might seem a bit cool, since the equivalent 17° C or 62° F would be a rather cool room temperature, but the value 290 makes for an easier-to-remember numeric calculation of $P_n = (1.38 \times 10^{-23} \times 290) B = 400 \times 10^{-23} B$.

The *instantaneous* noise voltage has a *Gaussian distribution* around the RMS value. The Gaussian distribution has no limit on the peak amplitude so at any instant the noise voltage may have any value from –infinity to +infinity. For design purposes we can use a value that will not be exceeded more than 0.01% of the time. This voltage is 4 times the RMS value, or 12 dB higher, so our system must be able to handle peak powers 12 dB higher than the average noise power if we are to measure noise without errors. (See Pettai in the Reference section.)

5.8.2. Signal to Noise Ratio

Now that we know the noise power in a given bandwidth, we can easily calculate how much signal is required to achieve a desired *signal to noise ratio, S/N* or *SNR*. For SSB, perhaps 10 dB SNR is required for good communications; since ambient thermal noise in a 2.5 kHz bandwidth is –140 dBm, calculated as follows:

$$P_n = kTB = 400 \times 10^{-23} \times 2500$$
$$= 1.0 \times 10^{-17} \text{ W}$$

$$dBm = 10 \log (P_n \times 1000) \text{ [multiplying}$$
$$\text{watts by 1000 converts to milliwatts]}$$

The signal power must be 10 dB larger, so minimum signal level of –130 dBm is required for a 10 dB S/N. This represents the noise and signal power levels at the antenna. We are then faced with the task of amplifying the signal without degrading SNR.

5.8.3. Noise Temperature

There are many types of noise, but most have similar characteristics to thermal noise

and are often added together, creating a single equivalent noise source whose output power per unit of bandwidth is P_N. The *noise temperature* of the source is defined as the temperature T = P_N / k at which a resistor would generate the same noise power per unit of bandwidth as the source. This is a useful way to characterize the various sources of noise in a communications system.

All amplifiers add additional noise to the noise present at their input. The input noise per unit of bandwidth is $N_i = kT_g$, where T_g is the noise temperature at the amplifier's input. Amplified by power gain G, the output noise is kT_gG. The additional noise contributed by the amplifier can also be represented as a noise temperature, T_n. The noise power added by the amplifier, kT_n, is then added to the amplified input noise to produce a total output noise:

$$N_o = kT_gG + kT_n$$

We can treat the amplifier as ideal and noise-free but with an additional noise-generating resistor of temperature $T_e = T_n$ / G at the input so that all sources of noise can be treated as inputs to the amplifier as illustrated by **Fig 5.63**. The output noise is then:

$$N_o = kG (T_g + T_e)$$

The noise added by an amplifier can then be represented as kGT_e, which is amplifier's noise temperature amplified by the amplifier gain. T_e is sometimes referred to as *excess temperature*.

Note that while the noise temperature of a resistor is the same as its physical temperature, the noise temperature of a device such as a diode or transistor can be many times the physical temperature.

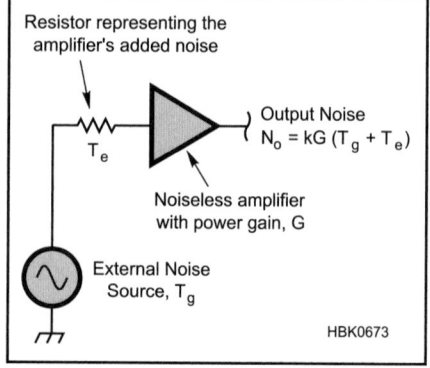

Fig 5.63 — The noise generated by an amplifier can be represented as an external resistor with a noise temperature of T_e connected at the input of a noiseless amplifier.

5.8.4. Noise Factor and Noise Figure

The *noise factor*, F, of an amplifier is the ratio of the total noise output of an amplifier with an input T_g of 290 K to the noise output of an equivalent noise-free amplifier. A more useful definition is to calculate it from the excess temperature T_e:

$$F = 1 + T_e / T_g, \text{ where } T_g = 290 \text{ K}$$

It is often more convenient to work with *noise figure, NF*, the logarithm of noise factor expressed in dB:

$$NF = 10 \log (1 + T_e / T_g) = 10 \log F$$

$$F = \log^{-1} (NF/10)$$

Expressed in terms of signal, S, and noise power, N, at the input and output of a device:

$$F = (S_{in}/N_{in})/(S_{out}/N_{out}) \text{ and}$$

$$F = G_{noise}/G_{signal}$$

where G_{signal} is the device's power gain and G_{noise} is the device's *noise gain*. If SNR in dB is known at the input and output:

$$NF = SNR_{in} - SNR_{out}$$

If NF or F is known, then T_e may be calculated as:

$$T_e = (F - 1) T_g$$

Typically, T_e is specified for very low noise amplifiers where the NF would be fraction of a dB. NF is used when it seems a more manageable number than thousands of K.

Noise figure is sometimes stated as *input noise figure* to emphasize that all noise sources and noise contributions are converted to an equivalent set of noise sources at the input of a noiseless device. In this way, noise performance can be compared on equal terms across a wide variety of devices.

Noise figure is particularly important at VHF and UHF where atmospheric and other artificial noise is quite low. Typical noise figures of amateur amplifiers range from 1 to 10 dB. Mixers are generally toward the high end of that range. Modern GaAsFET and HEMT preamplifiers are capable of attaining an NF of 0.1 to 0.2 dB at UHF with NF under 1 dB even at 10 GHz.

5.8.5. Losses

We know that any loss or attenuation in a system reduces the signal level. If attenuation also reduced the noise level then we could suppress thermal noise by adding attenua-

tion. We know intuitively that this can't be true — the attenuator or any lossy element has a noise temperature, T_x, which contributes noise to the system while the input noise is being attenuated.

The output noise after a loss L (expressed as ratio) expressed as an equivalent input noise temperature is:

$$T_g' = T_g /L + [(L - 1)/L] \, T_x$$

If the original source temperature, T_g, is higher than the attenuator temperature, T_x, then the noise contribution is found by adding the loss in dB to the NF. However, for low source temperatures the degradation can be much more dramatic. If we do a calculation for the effect of 1 dB of loss (L = 1.26) on a T_g of 25 K:

$$T_g' = 25/1.26 + (0.26/1.26) \times 290 = 80 \text{ K}$$

The resulting T_g' is 80 K, a 5 dB increase in noise power (or 5 dB degradation of signal to noise ratio). Since noise power = kT and k is a constant, the increase is the ratio of the two temperatures, 80/25, or in dB, 10 log (80/25) = 5 dB.

It is also useful to note that for linear, passive devices, such as resistors or resistive attenuators, noise figure is the same as loss in dB. A resistive attenuator with 6 dB loss has a noise figure of 6 dB which is equal to a noise factor of 4.

5.8.6. Cascaded Amplifiers

If several amplifiers are cascaded, the output noise N_o of each becomes the input noise Tg to the next stage. We can create a single equation for the total system of amplifiers. After removing the original input noise term, we are left with the added noise:

$$N_{added} = (k \, T_{e1}G_1G_2... \, G_N) + (kT_{e2}G_2... \, G_N) + ... + (kT_{eN}G_N)$$

where N is the number of stages cascaded.

Substituting in the total gain $G_T = (G_1G_2... G_N)$ results in the total excess noise:

$$T_{eT} = T_{e1} + \frac{T_{e2}}{G_1} + \frac{T_{e2}}{G_1G_2} + ... + \frac{T_{eN}}{G_1G_2...G_{N-1}}$$

with the relative noise contribution of each succeeding stage reduced by the gain of all preceding stages.

The *Friis formula for noise (a.k.a. the Friis equation)* expresses this in terms of noise factor:

$$F = F_1 + \frac{F_2 - 1}{G_1} + \frac{F_3 - 1}{G_1G_2} + ... + \frac{F_N - 1}{G_1G_2...G_{N-1}}$$

Clearly, if the gain of the first stage, G_1, is large, then the noise contributions of the succeeding stages become too small to be significant. In addition, the noise temperature of the first stage is the largest contributor to the overall system noise because it is amplified by all remaining stages. The effect on overall noise figure of adding a low-noise preamplifier ahead of a noisy receiver are illustrated in **Fig 5.64**, in which the system's noise figure changes from 20 dB for the receiver alone to 7.1 dB with the preamplifier added.

Any lossy component of an antenna system, such as the feed line, increases the noise figure at its input by an amount equal to the loss. As a result, it is important to concentrate noise-reduction efforts on the first amplifier or preamplifier in a system. Because noise performance is so important in early stages of cascaded systems such as receivers, low-noise VHF+ preamplifiers are usually mounted at the antenna so that their gain occurs ahead of the feed line loss. **Fig 5.65** compares the results of adding a preamplifier before and after 1.5 dB of feed line loss. Moving the preamplifier to the antenna improves the system's noise figure from 2.57 to 1.13 dB.

5.8.7. Antenna Temperature

Antenna temperature, T_A, is a way of describing how much noise an antenna produces. It is not the physical temperature of the an-

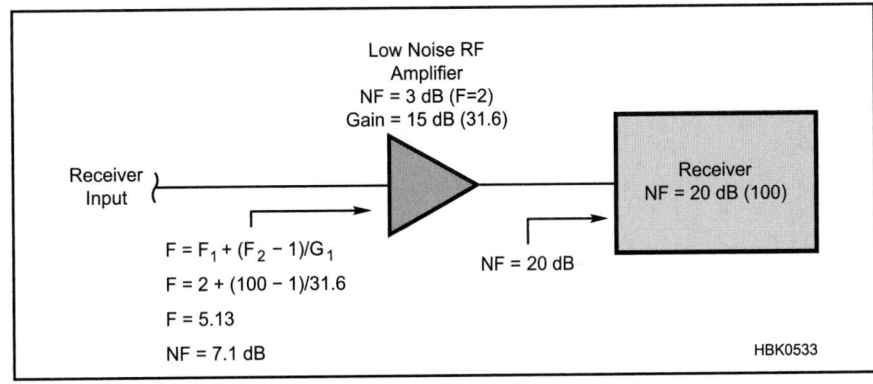

Fig 5.64 — The effect of adding a low-noise preamplifier in front of a noisy receiver system.

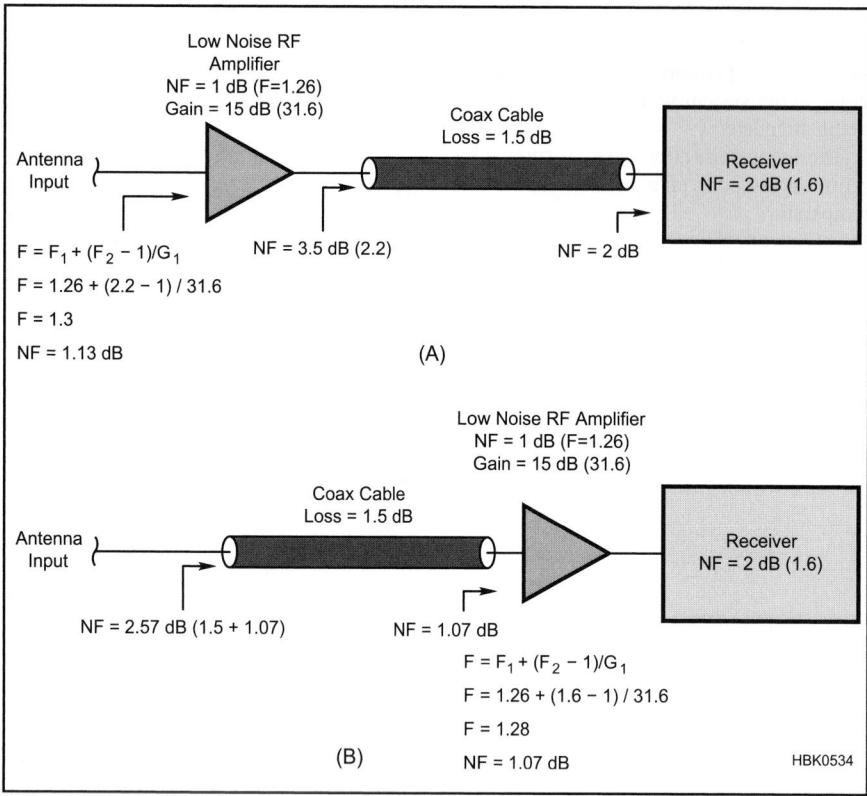

Fig 5.65 — The effect of adding a low-noise preamplifier at the antenna (A) compared to adding it at the receiver input (B).

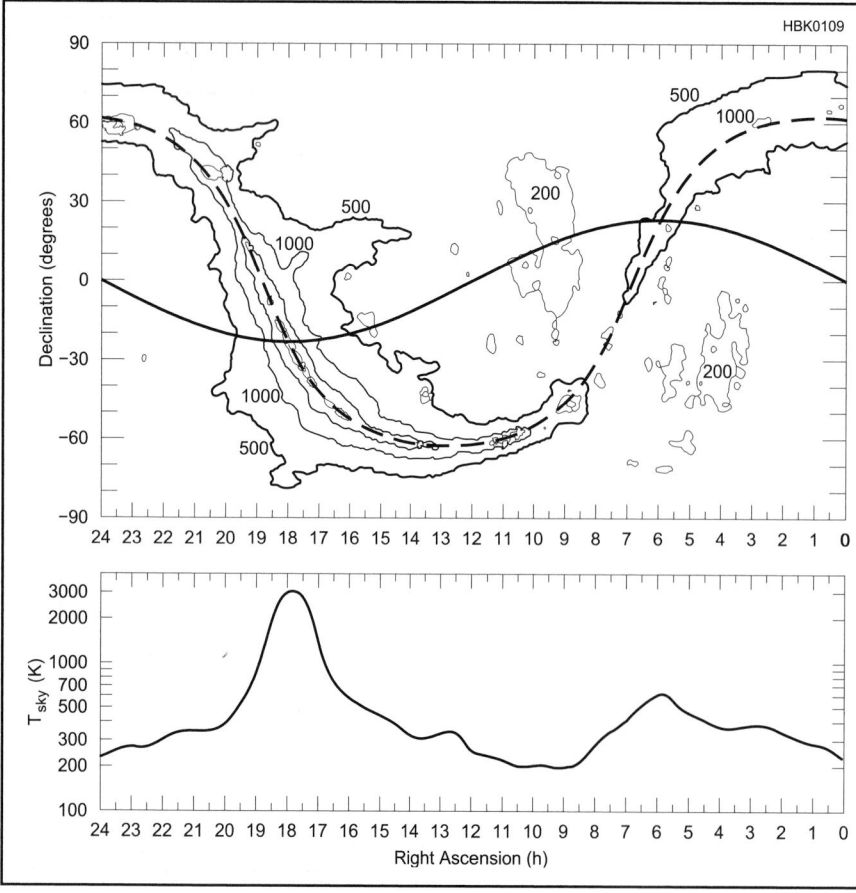

tenna because the antenna gathers noise from the environment according to its radiation pattern. If the antenna is directional and looks at a warm environment, T_A will be higher than if the antenna is looking at something cooler.

For example, if a lossless dish antenna is receiving signals from space rather than the warm Earth then the background noise is much lower than the warmer ambient temperature of 290 K or so. The background temperature of the universe has been measured as about 3.2 K. An empirical temperature for a 10 GHz antenna pointing into clear sky is about 6 K, since the antenna must always look through attenuation and temperature of the atmosphere. (See Graves in the Reference section.)

If the antenna's radiation pattern has any *spillover* (reception of signals from directions away from where the dish is pointed), that must be accounted for in the total noise received by the antenna. This raises the noise temperature. If a warm body, such as the sun, moves into the antenna's view, the additional *sun noise* will raise T_A as well. If the antenna is looking directly at the Earth, T_A will be close to ambient temperature. As an example, T_A will vary with frequency, but a good EME antenna might have a T_A of around 20 K at UHF and higher frequencies.

5.8.8. Image Response

Most receiving systems use at least one frequency converting mixer which has two responses: the desired frequency and an image frequency above and below the frequency of the local oscillator. If the image response is not filtered out, it will add additional noise to the mixer output. Since most preamps are sufficiently broadband to have significant gain (and thus, noise output) at both the desired frequency, $G_{desired}$, and at the image frequency, G_{image}, an image filter must be placed between the preamplifier and the mixer. The total NF including image response is:

$$NF = 10 \log \left[\left(\frac{1 + T_e}{T_0} \right) \left(1 + \frac{G_{image}}{G_{desired}} \right) \right]$$

Fig 5.66 —Top: All-sky contour map of sky background temperature at 144 MHz. The dashed curve indicates the plane of our Galaxy, the Milky Way; the solid sinusoidal curve is the plane of the ecliptic. The sun follows a path along the ecliptic in one year; the moon moves approximately along the ecliptic (± 5°) each month. Map contours are at noise temperatures 200, 500, 1000, 2000 and 5000 K. Bottom: One-dimensional plot of sky background temperature at 144 MHz along the ecliptic, smoothed to an effective beamwidth of 15°.

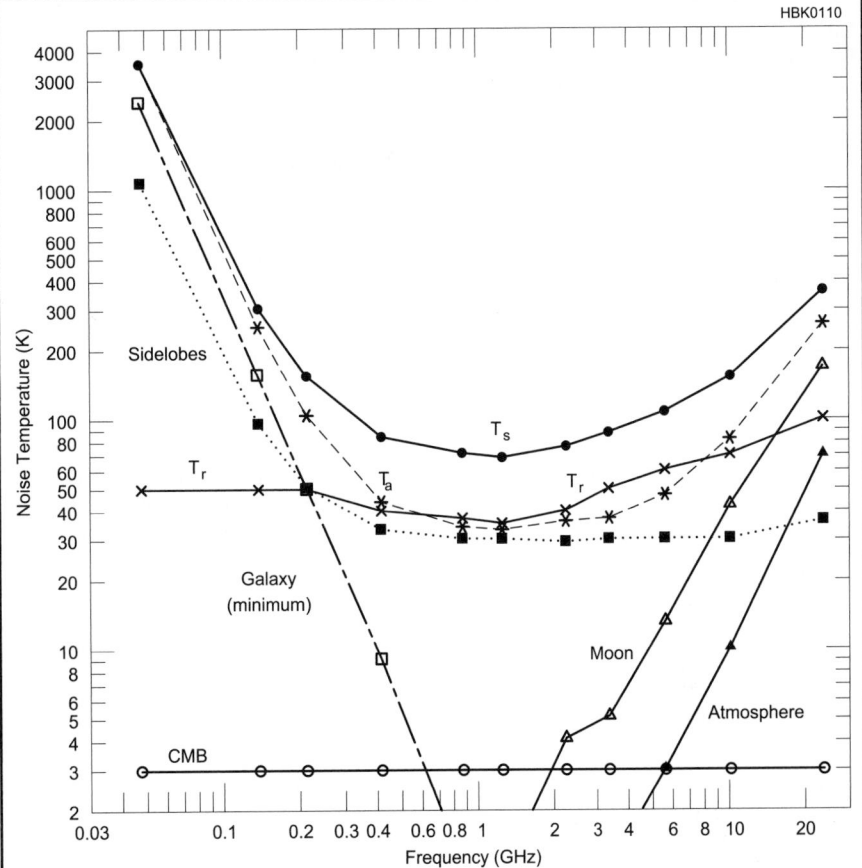

Fig 5.67 — Typical contributions to system noise temperature T_s as function of frequency. See text for definitions and descriptions of the various sources of noise.

assuming equal noise bandwidth for the desired and image responses.

Without any filtering, $G_{image} = G_{desired}$ so $G_{image}/G_{desired} = 1$, doubling the noise figure, which is the same as adding 3 dB. Thus, without any image rejection, the overall noise figure is at least 3 dB regardless of the NF of the preamplifier. For the image to add less than 0.1 dB to the overall NF, gain at the image frequency must be at least 16 dB lower than at the operating frequency.

5.8.9. Background Noise

A received signal at VHF and higher frequencies necessarily competes with noise generated in the receiver as well as that picked up by the antenna, including contributions from the warm Earth, the atmosphere, the lunar surface, the diffuse galactic and cosmic background and possibly the sun and other sources, filling the entire sky. If P_n is the total noise power collected from all such noise sources expressed in dBW (dB with respect to 1 W), we can write the expected signal-to-noise ratio of the EME link as:

$$SNR = P_r - P_n = P_t + G_t + L + G_r - P_n$$

where P_r is received power, P_n is noise power, P_t is transmitted power, G_t is gain of the transmitting antennas in dBi, L is *isotropic path loss*, and G_r is gain of the receiving antennas. All powers are expressed in dBW and gain in dBi. (Isotropic path loss is explored further in the material on Earth-Moon-Earth (EME) communications on the CD-ROM accompanying this book.)

Since isotropic path loss, L, is essentially fixed by choice of a frequency band, optimizing the signal-to-noise ratio generally involves trade-offs designed to maximize P_r and minimize P_n — subject, of course, to such practical considerations as cost, size, maintainability and licensing constraints.

It is convenient to express P_n (in dBW) in terms of an equivalent system noise temperature, T_s, in kelvins (K), the receiver bandwidth, B, in Hz and Stefan-Boltzmann's constant $k = 1.38 \times 10^{-23}$ J K^{-1}:

$$P_n = 10 \log (kT_s B)$$

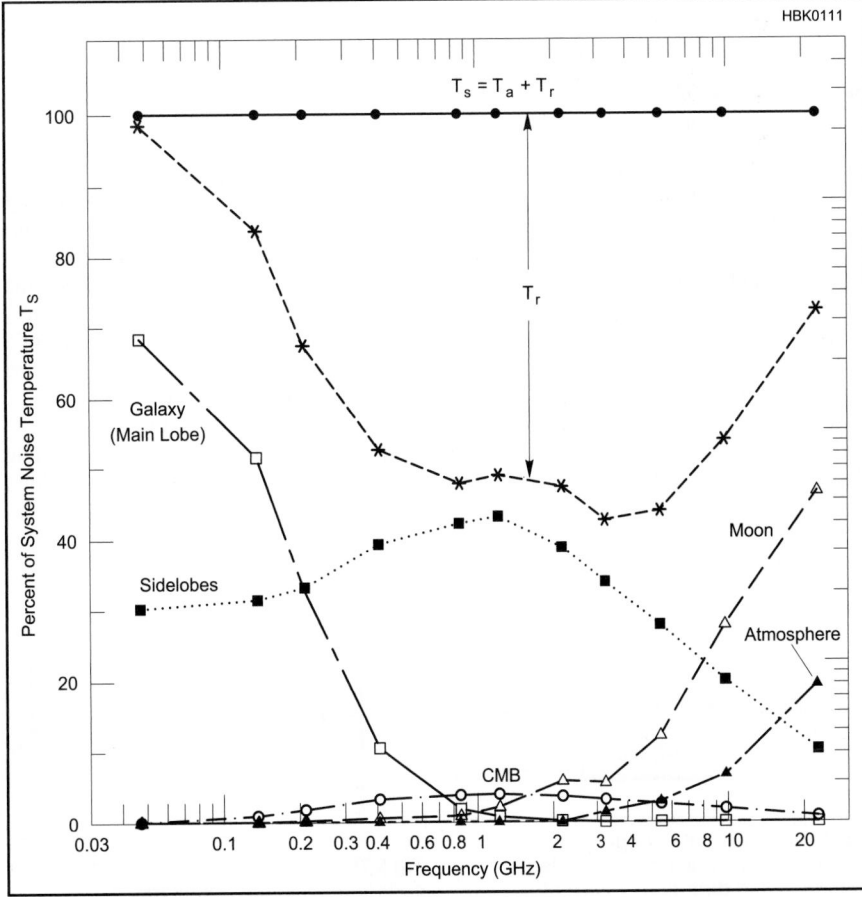

Fig 5.68 — Percentage contributions to system noise temperature as a function of frequency.

Table 5.2
Typical Contributions to System Noise Temperature

Freq (MHz)	CMB (K)	Atm (K)	Moon (K)	Gal (K)	Side (K)	T_a (K)	T_r (K)	T_s (K)
50	3	0	0	2400	1100	3500	50	3500
144	3	0	0	160	100	260	50	310
222	3	0	0	50	50	100	50	150
432	3	0	0	9	33	45	40	85
902	3	0	1	1	30	35	35	70
1296	3	0	2	0	30	35	35	70
2304	3	0	4	0	30	37	40	77
3456	3	1	5	0	30	40	50	90
5760	3	3	13	0	30	50	60	110
10368	3	10	42	0	30	85	75	160
24048	3	70	170	0	36	260	100	360

The system noise temperature may in turn be written as

$$T_s = T_r + T_a$$

Here, T_r is receiver noise temperature, related to the commonly quoted noise figure (NF) in dB by:

$$T_r = 290 \, (10^{0.1 \, NF} - 1)$$

Antenna temperature, T_a, includes contributions from all noise sources in the field of view, weighted by the antenna pattern. Side lobes are important, even if many dB down from the main beam, because their total solid angle is large and therefore they are capable of collecting significant unwanted noise power.

At VHF the most important noise source is diffuse background radiation from our gal-

axy, the Milky Way. An all-sky map of noise temperature at 144 MHz is presented in the top panel of **Fig 5.66**. This noise is strongest along the plane of the galaxy and toward the galactic center. Galactic noise scales as frequency to the –2.6 power, so at 50 MHz the temperatures in Fig 5.66 should be multiplied by about 15, and at 432 divided by 17. At 1296 MHz and above galactic noise is negligible in most directions. (See the previously mentioned CD-ROM EME material for the effects of lunar noise.)

By definition the sun also appears to an observer on Earth to move along the ecliptic and during the day solar noise can add significantly to P_n if the moon is close to the sun or the antenna has pronounced side lobes. At frequencies greater than about 5 GHz the Earth's atmosphere also contributes significantly. An ultimate noise floor of 3 K, independent of frequency, is set by cosmic background radiation that fills all space. A practical summary of significant contributions to system noise temperature for the amateur bands 50 MHz through 24 GHz is presented in **Table 5.2** and **Fig 5.67** and **Fig 5.68**.

5.9 Two-Port Networks

A *two-port network* is one with four terminals. The terminals are arranged into pairs, each being called a *port*. The general network schematic is shown in **Fig 5.69**. The input port is characterized by input voltage and current, V_1 and I_1, and the output is described by V_2 and I_2. By convention, currents into the network are usually considered positive.

Many devices of interest have three terminals rather than four. Two-port methods are used with these by choosing one terminal to be common to both input and output ports. The two-port representations of the common emitter, common base and common collector connections of the bipolar transistor are shown in **Fig 5.70**. Similar configurations

may be used with FETs, vacuum tubes, ICs or passive networks.

The general concepts of two-port theory are applicable to devices with a larger number of terminals. The theory is expandable to any number of ports. Alternatively, the bias on some terminals can be established with attention fixed only upon two ports of a multi-element device. An example would be a dual-gate MOSFET in a common-source configuration as shown in **Fig 5.71**. The input port contain the source and gate 1 while the output port contains the source and drain leads. The fourth device terminal, gate 2, has

a fixed bias potential and is treated as an ac ground. Signal currents at this terminal are ignored in the analysis.

5.9.1 Two-Port Parameters

There are four variables associated with any two-port network; two voltages and two currents. These are signal components. Any two variables may be picked as independent. The remaining variables are then dependent variables. These are expressed as an algebraic linear combination of the two independent quantities. The following overview are in-

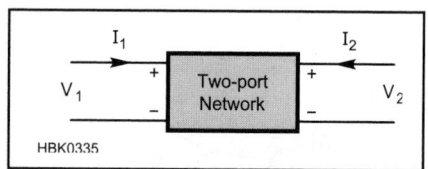

Fig 5.69 — General configuration of a two-port network. Note the voltage polarities and direction of currents.

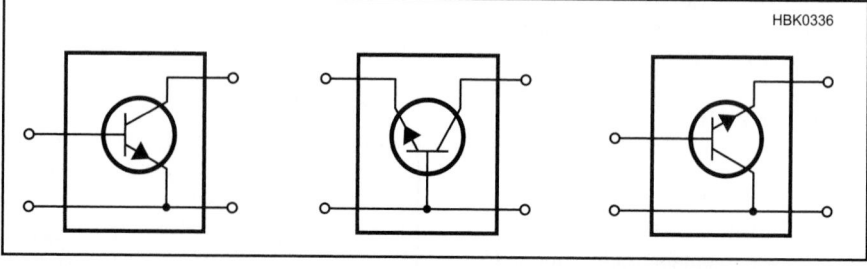

Fig 5.70 — Two-port representations of the common emitter, common base and common collector amplifiers.

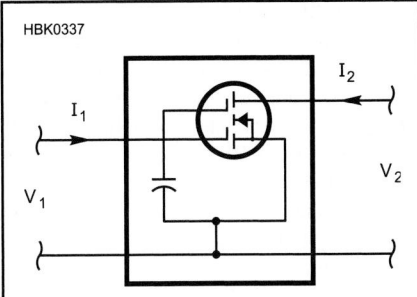

Fig 5.71 — A dual-gate MOSFET treated as a three-terminal device in a two-port network.

tended for definition purposes. A complete discussion of the use of two-port parameters can be found in the reference texts at the end of this chapter and examples of their use in RF circuit design in Hayward's *Introduction to Radio Frequency Design.*

Y AND Z PARAMETERS

Assume that the two voltages are chosen as independent variables. The two currents are then expressed as linear combinations of the voltages, $I_1 = K_a V_1 + K_b V_2$ and $I_2 = K_c V_1 + K_d V_2$. The constants of proportionality, K_a through K_d, have the dimensions of admittance. The usual representation is

$$I_1 = y_{11}V_1 + y_{12}V_2$$
$$I_2 = y_{21}V_1 + y_{22}V_2 \qquad (22)$$

The independent and dependent variable sets are column vectors, leading to the equivalent matrix representation

$$\begin{pmatrix} I_1 \\ I_2 \end{pmatrix} = \begin{pmatrix} y_{11} & y_{12} \\ y_{21} & y_{22} \end{pmatrix} \begin{pmatrix} V_1 \\ V_2 \end{pmatrix} \qquad (23)$$

The y matrix for a two-port network uniquely describes that network. The set of y_{11} through y_{22} are called the two-port network's *Y parameters* or *admittance parameters*. Consider the y parameters from an experimental viewpoint. The first y parameter, y_{11}, is the input admittance of the network with V_2 set to zero. Hence, it is termed the *short-circuit input admittance*. y_{21} is the *short-circuit forward transadmittance*, the reciprocal of transconductance. Similarly, if V_1 is set to zero, realized by short circuiting the input, y_{22} is the *short-circuit output admittance* and y_{12} is the *short-circuit reverse transadmittance*.

The matrix subscripts are sometimes replaced by letters. The set of y parameters can be replaced by y_i, y_r, y_f, and y_o where the subscripts indicate respectively input, reverse, forward, and output. The subscripts are sometimes modified further to indicate the connection of the device. For example, the short circuit forward transfer admittance

of a common emitter amplifier would be y_{21e} or y_{fe}.

The y parameters are only one set of two-port parameters. The open-circuited *Z parameters* or *impedance parameters* result if the two currents are treated as independent variables

$$\begin{pmatrix} V_1 \\ V_2 \end{pmatrix} = \begin{pmatrix} z_{11} & z_{12} \\ z_{21} & z_{22} \end{pmatrix} \begin{pmatrix} I_1 \\ I_2 \end{pmatrix} \qquad (24)$$

The parameter sets describe the same device; hence, they are related to each other. If equation 22 is multiplied by y_{12} and the resulting equations are subtracted, the result is the input voltage as a function of the currents

$$V_1 = \frac{I_1 y_{22} - y_{12} I_2}{y_{11} y_{22} - y_{12} y_{21}}$$

A similar procedure is used to find the output voltage as a function of the currents, leading to the general relationships

$$z_{11} = \frac{y_{22}}{\Delta y} \quad z_{12} = \frac{-y_{12}}{\Delta y}$$
$$z_{21} = \frac{-y_{21}}{\Delta y} \quad z_{22} = \frac{y_{11}}{\Delta y} \qquad (25)$$

where Δy is the determinant of the y matrix, $y_{11}y_{22} - y_{12}y_{21}$. The inverse transformations, yielding the y parameters when z parameters are known, are exactly the same as those in equation 25, except that the y_{jk} and z_{jk} values are interchanged. The similarity is useful when writing transformation programs for a programmable calculator or computer.

H PARAMETERS

The *H parameters* or *hybrid parameters* are defined if the input current and output voltage are selected as independent variables

$$\begin{pmatrix} V_1 \\ I_2 \end{pmatrix} = \begin{pmatrix} h_{11} & h_{12} \\ h_{21} & h_{22} \end{pmatrix} \begin{pmatrix} I_1 \\ V_2 \end{pmatrix} \qquad (26)$$

The input term, h_{11}, is an impedance, while h_{22} represents an output admittance. The forward term, h_{21}, is the ratio of the output to the input current, beta for a bipolar transistor. The reverse parameter, h_{12}, is a voltage ratio. The mixture of dimensions accounts for the "hybrid" name of the set.

SCATTERING (S) PARAMETERS

The two-port parameters presented above deal with four simple variables; input and output voltage and current at the ports. The variables are interrelated by appropriate matrices. The choice of which matrix is used depends upon which of the four variables are chosen to be independent.

There is no reason to limit the variables to simple ones. Linear combinations of the simple variables are just as valid. The more complicated variables chosen should be lin-

early independent and, ideally, should have some physical significance.

A transformation to other variables is certainly not new. For example, logarithmic transformations such as the dB or dBm are so common that we used them interchangeably with the fundamental quantities without even mentioning that a transformation has occurred. Such a new viewpoint can be of great utility in working with transmission line when an impedance is replaced by a reflection coefficient, $\Gamma = (Z - Z_0) / (Z + Z_0)$.

Scattering parameters or *S parameters* are nothing more than a repeat of this viewpoint. Instead of considering voltages and currents to be the fundamental variables, we use four "voltage waves." They are interrelated through an appropriate matrix of s parameters.

Fig 5.72 shows the traditional two-port network and an alternate one with voltage waves incident on and reflected from the ports. The voltage waves are defined with the letters a_1, b_1, a_2, and b_2. The a waves are considered to be incident waves on the parts and are the independent variables. The b waves are the result of reflection or "scattering" and are the dependent variables. The waves are related to voltages and currents and defined with respect to a characteristic impedance, Z_0.

The scattered waves are related to the incident ones with a set of linear equations just as the port currents were related to the port voltages with y parameters. The relating equations are

$$b_1 = S_{11}a_1 + S_{12}a_2$$
$$b_2 = S_{21}a_1 + S_{22}a_2 \qquad (27)$$

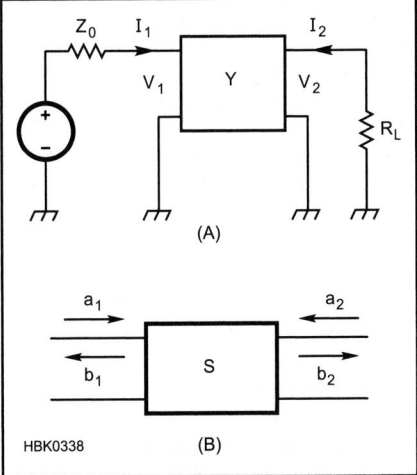

Fig 5.72 — A two-port network viewed as being driven by voltages and currents (A) or voltage waves (B). The voltages and currents are related by Y parameters while the voltage waves are related by scattering or S parameters.

or, in matrix form

$$\begin{pmatrix} b_1 \\ b_2 \end{pmatrix} = \begin{pmatrix} S_{11} & S_{12} \\ S_{21} & S_{22} \end{pmatrix} \begin{pmatrix} a_1 \\ a_2 \end{pmatrix} \qquad (28)$$

Consider the meaning of S_{11}. If the incident wave at the output, a_2, is set to zero, the set of equations 27 reduce to $b_1 = S_{11}a_1$ and $b_2 = S_{21}a_1$. S_{11} is the ratio of the input port reflected wave to the incident one. This reduces, using the defining equations for a_1 and b_1 to

$$S_{11} = \frac{Z - Z_0}{Z + Z_0} \qquad (29)$$

This is the *input port reflection coefficient*. Similarly, S_{21} is the voltage wave emanating from the output as the result of an incident wave at the network input. In other words, S_{21} represents a forward gain. The other two S parameters have similar significance. S_{22} is the *output reflection coefficient* when looking back into the output port of the network with the input terminated in Z_0. S_{12} is the reverse gain if the output is driven and the signal at the input port detected.

The reflection coefficient nature of S parameters makes them especially convenient for use in design and specification and even more so when displayed on a Smith Chart.

5.9.2 Return Loss

Although SWR as described in the **Trans-** mission Lines chapter is usually used by amateurs to describe the relationship between a transmission line's characteristic impedance and a terminating impedance, the engineering community generally finds it more convenient to use *return loss, RL,* instead.

Return loss and SWR measure the same thing — how much of the incident power, P_{INC}, in the transmission line is transferred to the load and how much is reflected by it, P_{REFL} — but state the result differently.

$$\text{Return Loss(dB)} = -10 \log\left(\frac{\text{REFL}}{\text{INC}}\right) \quad (30)$$

Because P_{REFL} is never greater than P_{FWD}, RL is always positive. The more positive RL, the less the amount of power reflected from the load compared to forward power. If all the power is transferred to the load because $Z_L = Z_0$, RL $= \infty$ dB. If none of the power is transferred to the load, such as at an open- or short-circuit, RL $= 0$ dB. (You may encounter negative values for RL in literature or data sheets. Use the absolute magnitude of these values — the negative value does not indicate power gain.)

RL can also be calculated directly from power ratios, such as dBm (decibels with respect to 1 mW) or dBW (decibels with respect to 1 watt). In this case, RL $= P_{INC} - P_{REFL}$ because the logarithm has already been taken in the conversion to dBm or dBW. (Ratios in dB are computed by subtraction, not division.) For example, if $P_{INC} = 10$ dBm and $P_{REFL} =$ 0.5 dBm, RL $= 10 - 0.5 = 9.5$ dB. Both power measurements must have the same units (dBm, dBW, and so on) for the subtraction to yield the correct results — for example, dBW can't be subtracted from dBm directly.

Since SWR and RL measure the same thing — reflected power as a fraction of forward power — they can be converted from one to the other. Start by converting RL back to a power ratio:

$$\frac{P_{REFL}}{P_{INC}} = \log^{-1}(-0.1 \times RL) \qquad (31)$$

Now use the equation for computing SWR from forward and reflected power (see the **Transmission Lines** chapter):

$$SWR = \frac{1 + \sqrt{\dfrac{P_{REFL}}{P_{INC}}}}{1 - \sqrt{\dfrac{P_{REFL}}{P_{INC}}}} \qquad (32)$$

SWR can also be converted to RL by using the equation for power ratio in terms of SWR:

$$\frac{P_{REFL}}{P_{INC}} = \left[\frac{SWR - 1}{SWR + 1}\right]^2 \qquad (33)$$

Then convert to RL using equation 30.

5.10 RF Techniques Glossary

Arc — Current flow through an insulator due to breakdown from excessive voltage.

Balun — A device that transfers power between *bal*anced and *un*balanced systems, sometimes transforming the impedance level as well (see also **unun**).

Bead — Hollow cylinder of magnetic material through which a wire is threaded to form an inductor.

Bilateral — A network that operates or responds in the same manner regardless of the direction of current flow in the network.

Choke balun — see **current balun.**

Core — Magnetic material around which wire is wound or through which it is threaded to form an inductor.

Current balun — A balun that transfers power from an unbalanced to a balanced system by forcing current flow in the balanced system to be balanced as well (also called a **choke balun**).

Dielectric strength — The rated ability of an insulator to withstand voltage.

Distributed element — Electronic component whose effects are spread out over a significant distance, area or volume.

Dynamic resistance — The change in current in response to a small change in voltage.

Equivalent Series Inductance (ESL) — A capacitor's parasitic inductance.

Ferrite — A ferromagnetic ceramic.

Gain-bandwidth product — The frequency at which a device's gain drops unity. Below that frequency the product of the device's gain and frequency tends to be constant.

Hybrid-pi — High-frequency model for a bipolar transistor.

Impedance inversion — Dividing a characteristic impedance by the ratio of the impedance to be inverted to the characteristic impedance. For example, 25 Ω inverted about 50 Ω is 100 Ω and 200 Ω inverted about 50 Ω is 12.5 Ω.

Insertion loss (IL) — The loss inherent in a circuit due to parasitic resistance.

Inter-electrode capacitance — Capacitance between the internal elements of a semiconductor or vacuum tube.

Lumped element — Electronic component that exists at a single point.

Mix — The chemical composition of a ferrite or powdered-iron material (also called **type**).

Noise — Any unwanted signal, usually refers to signals of natural origins or random effects resulting from interfering signals.

Noise factor (F) — The amount by which noise at the output of a device is greater than that at the input multiplied by the gain of the device. A measure of how much noise is generated by a device.

Noise figure (NF) — 10 log (noise factor).

Noise gain — Circuit output noise power divided by the available input noise power. This is not always equal to signal gain, depending on the source of the noise and the location of the noise source in the circuit.

Nonideal — Behavior that deviates from that of an ideal component (see also **parasitic**).

Nonlinear — A component that acts on a signal differently depending on the signal's amplitude.

Parasitic — Unintended characteristic related to the physical structure of a component.

Permeability — The ability of a material to support a magnetic field.

Return loss (RL) — The difference in dB between forward and reflected power at a network port.

Self-resonant — Resonance of a component due to parasitic characteristics.

Simulate — Model using numerical methods, usually on a computer.

Skin effect — The property of a conductor that restricts high-frequency ac current flow to a thin layer on its surface.

Skin depth — The depth of the layer at the surface of a conductor to which ac current flow is restricted (see **skin effect**).

Spectral Power Density — The amount of power per unit of bandwidth, usually "root-Hz" or √Hz, the square root of the measurement bandwidth.

Stray — see **parasitic**.

Toroid (toroidal) — A ring-shaped continuous core.

Two-port network — A network with four terminals organized in two pairs, each pair called a port.

Two-port parameters — A set of four parameters that describe the relationship between signals at the network's two ports.

Unun — A device that transfers power between two unbalanced systems, usually performing an impedance transformation (see also **balun**).

5.11 References and Bibliography

Alley, C. and Atwood, K., *Electronic Engineering* (John Wiley & Sons, New York, 1973)

Brown, J., K9YC, "Measured Data For HF Ferrite Chokes," **www.audiosystemsgroup.com/K9YC/K9YC.htm**

Brown, J., K9YC, "A Ham's Guide to RFI, Ferrites, Baluns, and Audio Interfacing," **www.audiosystemsgroup.com/K9YC/K9YC.htm**

Carr, J., *Secrets of RF Circuit Design* (McGraw-Hill/TAB Electronics, 2000)

Counselman, C., W1HIS, "Common-Mode Chokes," **www.yccc.org/Articles/W1HIS/CommonModeChokesW1HIS2006Apr06.pdf**

DeMaw, D., *Practical RF Design Manual* (MFJ Enterprises, 1997)

Dorf, R., Ed., *The Electrical Engineering Handbook* (CRC Press, 2006)

Grammer, G., W1DF, "Simplified Design of Impedance-Matching Networks," *QST*, Part I, Mar 1957, pp 38-42; Part II, Apr 1957, pp 32-35; and Part III, May 1957, pp 32-35.

Graves, M.B., WRØI, "Computerized Radio Star Calibration Program," Proceedings of the 27th Conference of the Central States VHF Society, ARRL, 1993, pp. 19-25.

Hayward and DeMaw, *Solid-State Design for the Radio Amateur* (ARRL, out of print)

Hayward, Campbell, and Larkin, *Experimental Methods in Radio Frequency Design* (ARRL, 2009)

Hayward, W., *Introduction to RF Design* (ARRL, 1994)

Kaiser, C., *The Resistor Handbook* (CJ Publishing, 1994)

Kaiser, C., *The Capacitor Handbook* (CJ Publishing, 1995)

Kaiser, C., *The Inductor Handbook* (CJ Publishing, 1996)

Maxwell, Walt W2DU, "Reflections III," *CQ Communications*, 2010

Pettai, R., *Noise in Receiving Systems*, (Wiley, 1984)

Pozar, D., *Microwave Engineering*, Fourth Edition (John Wiley & Sons, 2012)

Terman, F., *Electronic and Radio Engineering* (McGraw-Hill, New York, 1955)

Van Valkenburg, M., *Reference Data for Engineers* (Newnes, 2001)

Contents

Chapter 6 — CD-ROM Content

- "The Dangers of Simple Usage of Microwave Software" by Ulrich Rohde, N1UL and Hans Hartnagel
- "Using Simulation at RF" by Ulrich Rohde, N1UL
- "Mathematical Stability Problems in Modern Non-Linear Simulation Programs" by Ulrich Rohde, N1UL and Rucha Lakhe
- Examples of Circuit Simulation by David Newkirk, W9VES

Computer-Aided Circuit Design

This chapter provides an overview of computer-aided design (CAD) for electronic design and PCB layout. These tools enable the hobbyist to harness some of the circuit simulation power employed by professional electronic and RF engineers in the product and system design cycle.

Material originally contributed by David Newkirk, W9VES, addresses generic circuit simulation tools. Dr Ulrich Rohde, N1UL, surveys issues associated with linear and nonlinear RF simulation and contributes three extensive papers on the accompanying CD-ROM. Dale Grover, KD8KYZ, presents a comprehensive introduction to the use of PCB design and layout software.

The purpose of this chapter is not to provide detailed instructions for using any particular software package, but to explain the basic operations, limitations and vocabulary for CAD packages commonly used by amateurs. This chapter covers circuit simulators and PCB design tools. Software to aid design and analysis in specialized areas, for example filter design, switchmode power supplies, transmission lines, and RF power amplifiers, is covered in other chapters. Schematic capture is addressed as an element of both simulation and PCB design.

6.1 Circuit Simulation Overview

Mathematics can predict and analyze the action of electromagnetic signals and the radio-electronic circuitry we build to produce and process them. Program an electronic computer — which, at base, is a generic math machine — to do the radio/electronics math in practically applicable ways, and you're ready to do computer-aided design (CAD) of radio and electronic circuits. (Program a computer to do radio-electronics math *in real time*, and you're ready to replace radio-electronics hardware with software, as this book's **DSP and Software Radio Design** chapter describes.)

6.1.1 Hobby versus Professional Circuit Simulation Tools

Professional grade circuit simulation software exists to facilitate the construction of tightly packaged, highly integrated, no-tweaking-required modern electronics/RF products. These products work predictably well even when reproduced by automated processes in large quantities — quantities that may, with sufficient marketing success and buyer uptake, exceed millions of units.

Manufacturers of specialized *electronic design automation* (*EDA*) software serve the engineering needs of this industry. Through comprehensive CAD suites, one may proceed from graphical component level circuit and/or IC design (*schematic capture*), through simulation of circuit and IC behavior (often using a variant of the simulator called *SPICE*, but increasingly with non-*SPICE* simulators more fluent in issues of RF and electromagnetic design), through design of PC board and IC masks suitable for driving validation, testing and production. Comprehensive EDA CAD reduces costs and speeds time to market with the help of features that can automatically modify circuits to achieve specific performance goals (*optimization*); predict effects of component tolerances and temperature on circuit behavior across large populations of copies (*Monte Carlo analysis*); and generate bills of materials (BOMs) suitable for driving purchasing and procurement at every step of the way.

Demonstration or student versions are available for some EDA CAD products at no or low cost (see **Table 6.1**), and a subset of these are especially useful for hobby purposes. Although these *demoware* tools come to us with a large-scale-production pedigree, they are greatly (and strategically) feature-limited. Only a relatively few components, often representing only a subset of available component models, may be used per simulation. Monte Carlo analysis, optimization, BOM generation and similar enhancements are usually unavailable. The licenses for these packages often limits the use of the software to noncommercial applications. Demoware is intended to drive software purchasing decisions and serve as college level learning aids — learning aids in college study toward becoming electronics/RF professionals who will each day work with the unlimited, full versions of the demoware. *Freeware* versions of simulation and layout software with considerable power are also available and may also have some restrictions. In either case, read the licensing agreement to become aware of any obligations on your part.

The radio hobbyist's circuit simulation needs are much simpler than the professional. Most of us will build only one copy of a given design — a copy that may be lovingly tweaked and refined to our hearts' content far beyond "good enough." Many of us may build as much with the intent of learning about and exploring the behavior of circuits as achieving practical results

Table 6.1
Some Sources of Freeware/Demoware Electronic CAD Software

Source	Address	Resource
Ansoft (now Ansys)	www.ansys.com	Ansoft Designer SV 2 (schematic, linear RF simulator, planar electromagnetic simulator, layout [PCB] design), more. No longer available but older copies of the program may be available
Cadence Design Systems	www.cadence.com	OrCAD 16 (schematic, SPICE simulator, layout [PCB] design)
CadSoft	www.cadsoft.de www.cadsoftusa.com	EAGLE schematic and layout design
gEDA	www.gpleda.org	GPLed suite of electronic design automation tools
Kicad	http://iut-tice.ujf-grenoble.fr/kicad	GPLed full-function schematic and layout design
Linear Technology Corp	www.linear.com	LTSpice (schematic, SPICE simulator enhanced for power-system design)

with them. A demoware circuit simulator can accelerate such self-driven exploration and education in electronics and RF.

6.1.2 The Design Cycle

The components we use to build real circuits always operate to the full extent of their actual properties, regardless of our relative ignorance of what those properties may be and how and why they operate the way they do. *No real-world component operates ideally.* So it is that we may set out to design, build and publish an amplifier circuit only to discover at power-up that we have instead built a persistent oscillator. Or if the prototype is an oscillator, that for 3 out of every 100 subsequent reader-builders the circuit does not oscillate at all!

The electrical and electronic components available in circuit simulators are only mathematical *models* of real world components — because every modeled behavior of a component is only an approximation relative to the behaviors of its real world counterpart. Simulated component characteristics and behaviors approach those of real world components only as closely as science may allow and only as closely as the model's description of the real world behavior.

Were this chapter a textbook, or part of a textbook, on computer-aided circuit design, we might begin exploring simulation by reviewing the basics of what electronic circuits are and do, following this with a discussion of what computerized circuit simulation is and how it works. An excursion into the arcane world of active device modeling — the construction and workings of mathematical electrical equivalents to the transistors, diodes and integrated circuits that await us at our favorite electronics suppliers and in our junk boxes — might follow. Finally, we might systematically proceed through a series of simulation examples from the basic to the more complex, progressively building our store of understood, trustworthy and applicable-to-future-work concepts as we go.

But this is a chapter in a handbook, not a textbook, and following a sequence of abstract basics to concrete practice very likely does not reflect the process most of us have followed, and follow yet, in learning and using what we know about electronics and radio. More realistically, our approach is more like this: We find ourselves in need of a solution to a problem, identify one, and attempt to apply it. If it works, we move on, likely having learned little if we have not had to troubleshoot. If the solution does *not* work, we may merely abandon it and seek another, or — better, if we are open to learning — we may instead seek to understand why, with the happily revised aim of understanding what we need to understand to make the solution work. Even if we must ultimately abandon the solution as unworkable in favor of another, we do not consider our time wasted because we have further accelerated our deepening intuition by taking the initiative to understand why.

There is no smell of burning resistor or overheated transistor in a simulation. The placement of components on the screen has no effect on the behavior of the circuit, so a high gain stage whose input is too close to its output will never break into oscillation. The dc power sources are free of ripple and noise. These effects and many more can only be experienced (and remedies learned) by building real circuits.

Fig 6.1 shows the process by which you really learn circuit design from concept to finished project. The first step is to select a type of circuit and describe what it is supposed to do — these are the *performance requirements.* For example, an amplifier will need to achieve some level of gain over some frequency range. You may need a certain input impedance and output impedance. Armed with that information, choose a circuit and come up with a preliminary set of component values by using pencil and paper or a computer design tool. This is your *design.*

Next, *simulate* the circuit's performance. If the result satisfies your performance requirements, you can move to the next step. If not, change the circuit in some way (or change your requirements) until you are satisfied.

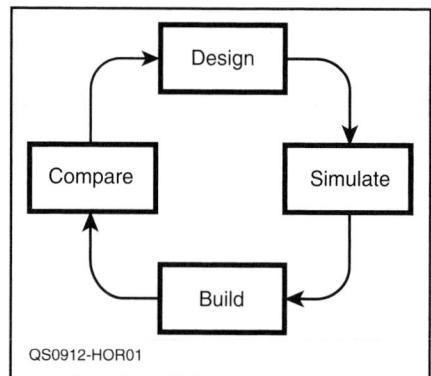

QS0912-HOR01

Fig 6.1 — Getting the most out of circuit simulation requires that you compare what the simulator predicts with how the actual circuit behaves.

Now *build* your design as a real world collection of components and verify that the circuit works. This is where the real fun begins as the effects of construction and actual component variation take effect. Are you done? Not yet!

To soak up every bit of design experience and know-how, go back and *compare* your actual measured performance to what the simulator predicted, particularly near the limits of the circuit's function. Look for *design sensitivities* by substituting different parts or values. If the circuit's behavior diverges from the simulator's predictions, now is the time to take a closer look. You may not be able to say exactly why differences are present, but you'll be aware they exist.

This continual *design cycle* simultaneously builds your knowledge of how real circuits and components behave, how your simulation tools work, and — most importantly — the circumstances for which the two are likely to be different.

The CD-ROM that comes with this book includes a number of design examples developed by the original author of this section, David Newkirk, W9VES. In these examples, he creates real world circuits, simulates their behavior, compares it to actual performance, and explains why there are differences. Later in this chapter, a section by Ulrich Rohde, N1UL, discusses the differences among the different types of simulation tools used at RF — there are even differences among the different tools!

6.1.3 Schematic Capture and Simulation Tools

Almost any circuit-simulation program or electronic design automation (EDA) suite that uses schematic circuit capture can serve as a first rate schematic editor. (See Table 6.1) Although demoware component library limitations usually restrict the types of components you can use — in CAD-speak, *place* — in a design, *part count* limitations usually operate only at simulation time. Restrictions in physical size of the output drawing and layer count will likely apply to whatever layout design facilities may be available. After all, the main purpose of demoware is to let students and potential buyers taste the candy without giving away the store.

Excellent simulation-free schematic capture and layout design products exist, of course. The schematic style long standard in ARRL publications comes from the use of Autodesk *AutoCAD*, a fully professional product with a fully professional price. Long popular with radio amateurs and professionals alike is CadSoft *EAGLE*, a schematic capture and layout design product available in freeware and affordable full version forms. You can even export *EAGLE* schematics to a *SPICE* simulator and back with Beige Bag Software's *B2 Spice* (**www.beigebag.com**). The full function

freeware schematic and PCB layout application *Kicad* and the EDA suite *gEDA* come to us from the open source community.

The basic drawing utility included with your computer's operating system, such as *Paint* which comes with *Windows*, can also serve as a limited do-it-yourself schematic capture tool. The ARRL also provides a limited set of schematic symbols that can

be used with *PowerPoint* at the Hands-On Radio web page, **www.arrl.org/hands-on-radio**. Cutting, copying, moving and pasting components snipped from favorite graphical schematic files and adding new connections as graphical lines is enough to create a picture of the schematic but without any of the underlying tools or facilities of a true schematic capture tool.

CAD Software and Your Computer's Operating System

The *OrCAD 16.0* and *Ansoft Designer SV 2* demoware packages used in this chapter, and most other CAD products you're likely to use, are compiled to run under Microsoft *Windows*. So what if you want to run RF and electronics CAD software under *Linux* or on the Mac?

You're in luck. *EAGLE* schematic and layout software is available in native versions for *Windows*, *Linux*, and the Macintosh. The GPLed EDA application suite, *gEDA*, are primarily developed on *Linux*, but are intended to run under, or at least be portable to, *POSIX*-compliant systems in general. The GPLed schematic and layout editor *Kicad* runs natively under *Windows* and *Linux*, and has been tested under *FreeBSD* and *Solaris*. Further, the great strides made in the *Wine* translation layer (**www.winehq.com/**) allow many applications written for *Windows* to run well under the operating systems supported by *Wine*, including *Linux* and the Macintosh.

MicroSim *DesignLab 8* (a widely distributed precursor to *OrCAD 16* that can run all of the *SPICE* examples described in this chapter) and Ansoft *Serenade SV 8.5* (a precursor to *Ansoft Designer SV 2*) can be run in *Wine* under *Linux* with few artifacts and their expected schematic capture and simulation capabilities intact. Cursor handling in *OrCAD 16* installs under *Wine* readily enough, but cursor handling artifacts in its schematic editor, at least in the computers tried, seems to preclude its use under *Wine* for now. *Ansoft Designer SV 2* installs but does not properly start.

All things considered, however, especially as *Wine* and CAD applications continue to strengthen and mature, running your favorite *Windows* based applications under *Wine* is well worth a try. You may also consider purchasing an inexpensive used computer that runs one of the later versions of *Windows*, such as XP, and dedicating it to running the simulation

---

6.2 Simulation Basics

This section is a collection of notes and illustrations that address various important circuit simulation concepts. In this section, conventional *SPICE* notation and vocabulary are used unless specifically noted differently. Not all simulation tools use exactly the same words and phrases to label and explain their features. When in doubt, refer to the software's user manual or HELP system. There are a number of excellent textbooks about using *SPICE*-based simulators Widely used simulation tools almost always have online communities of users, all of whom were beginners once, too. Joining one of these groups is highly recommended.

Simulation tool users groups frequently develop and maintain a library of tutorials, Frequently Asked Questions (FAQ), accessory programs and utilities, even models and complex circuit models. Before you ask questions, consult the available resources, such as searchable message archives, to see if your

question has been answered before — it usually has! The other users will appreciate your diligence before asking the entire group.

6.2.1 *SPICE* — History

SPICE — *Simulation Program with Integrated-Circuit Emphasis* — originates from the Electrical Engineering and Computer Sciences Department of the University of California at Berkeley and first appeared under its current name as *SPICE1* in 1972. "*SPICE*," write the maintainers of the official *SPICE* homepage at **http://bwrcs.eecs.berkeley.edu/Classes/IcBook/SPICE** "is a general-purpose circuit simulation program for nonlinear dc, nonlinear transient, and linear ac analyses. Circuits may contain resistors, capacitors, inductors, mutual inductors, independent voltage and current sources, four types of dependent sources, lossless and lossy transmission lines (two separate implementa-

tions), switches, uniform distributed RC lines, and the five most common semiconductor devices: diodes, BJTs, JFETs, MESFETs, and MOSFETs."

That *SPICE* is "general-purpose" does not mean that its usefulness is unfocused, but rather that it is well established as a circuit simulation mainstay of comprehensive power. A wide, deep *SPICE* community exists as a result of decades of its daily use, maintenance and enhancement by industrial, academic and hobby users. Many excellent commercial versions of *SPICE* exist — versions that may be improved for workhorse use in particular sub-disciplines of electronics, power and RF design.

6.2.2 Conventions
SCALE FACTORS

SPICE's use of unit suffixes — *scale factors* in *SPICE*-speak — differs from what we are generally accustomed to seeing in electrical schematics, and that we have multiple options for specifying values numerically using integer and decimal floating point numbers. The scale factors available in *SPICE* include:

- F (femto) — 1E–15
- G (giga) — 1E9
- K (kilo) — 1E3
- M (milli) — 1E–3
- MEG (mega) — 1E6
- MIL (0.001 inch) — 25.4E–6 meter
- N (nano) — 1E–9
- P (pico) — 1E–12
- T (tera) — 1E12
- U (micro) — 1E–6

Specifying the value of a resistor as 1M would cause *SPICE* to assign it the value of 1 milliohm (0.001 Ω). This would probably create a wildly different result than expected! Specifying the value of R1 as 1MEG or 1000K would be correct alternatives. *SPICE* scale factors are case insensitive. Until you are thoroughly familiar with using circuit simulation, take the time to double-check component and parameter values. It will save you a lot of time tracking down errors caused by mistaken component values.

Notice that *SPICE* assumes unit *dimensions* — ohms, farads, henrys and so on — from component name context; in specifying resistance, we need not specify ohms. In parsing numbers for scale factors, *SPICE* detects only scale factors it knows and, having found one, ignores any additional letters that follow. This lets us make our schematics more readable by appending additional characters to values — as long as we don't confuse *SPICE* by running afoul of existing scale factors. We may therefore specify "100pF" or "2.2uF" for a capacitance rather than just "100p" or "2.2u" — a plus for schematic readability. (On the reduced readability side, however, *SPICE* requires that there be *no space* between a value and its scale factor — a limitation that stems from programming expediency and is present in many circuit-simulation programs.)

SOURCES

There are two basic types of sources used in simulation models — voltage sources and current sources — as discussed in the **Electrical Fundamentals** and **Analog Basics** chapters. Sources can be *independent* with an assigned value or characteristic that does not change, or *dependent* with a value or characteristic that depends on some other circuit value. An example of an independent source would be a fixed voltage power supply (dc) or a sinusoidal signal source (ac). An example of a dependent source is a bipolar transistor model's collector current source that has a value of βI_B.

COMPONENT MODELS

All *real* inductors, capacitors, and resistors — all real components of any type — are non-ideal in many ways. For starters, as **Fig 6.2** models for a capacitor, every real *L* also exhibits some *C* and some *R*; every real *C*, some *L* and *R*; every real *R*, some *L* and *C*. These unwanted qualities may be termed *parasitic*, like the parasitic oscillations that sometimes occur in circuits that we want to act only as amplifiers. The **RF Techniques** chapter discusses parasitics for various components.

For simulating many ham-buildable circuits that operate below 30 MHz, the effects of component parasitic R, L and C can usually be ignored unless guidance or experience suggests otherwise. In oscillator and filter circuits and modeled active devices, however, and as a circuit's frequency of operation generally increases, neglecting to account for parasitic L, C and R can result in surprising performance shortfalls in real world *and* simulated performance. In active device modeling realistic enough to accurately simulate oscillator phase noise and amplifier phase shift and their effects on modern, phase-error-sensitive data communication modes, device-equivalent models must even include *nonlinear* parasitic inductances and capacitances — Ls and Cs that vary as their associated voltages and currents change.

Figuring out what the circuit of an appropriate model should be is one thing; measuring and/or realistically calculating real world values for R_S, L_S and R_P for application in a circuit simulator is a significant challenge. How and to what degree these parasitic character-

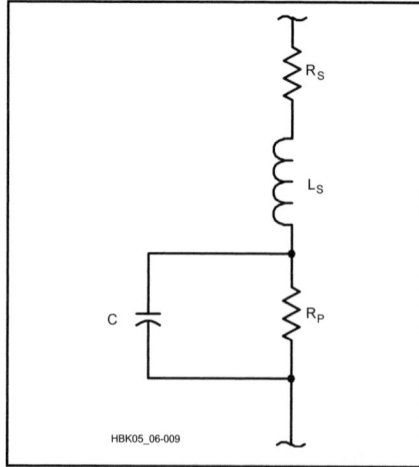

Fig 6.2 — A capacitor model that aims for improved realism at VHF and above. R_S models the net series resistance of the capacitor package; L_S, the net equivalent inductance of the structure. R_P, in parallel with the capacitance, models the effect of leakage that results in self-discharge.

istics may cause the electrical behavior of a component to differ from the ideal depends on its role in the circuit that includes it and the frequency at which the circuit operates.

Designers aiming for realism in simulating power circuits that include magnetic core inductors face the additional challenge that all real magnetic cores are nonlinear. Their magnetization versus magnetic field strength (*B-H*) characteristics exhibit hysteresis. They can and will saturate (that is, fail to increase their magnetic field strength commensurately with increasing magnetization) when overdriven. Short of saturation, the permeability of magnetic cores varies, hence changing the inductance of coils that include them, with the flow of dc through their windings. These effects can often be considered negligible in modeling ham-buildable low power circuits.

As active device operation moves from small signal — in which the signals handled by a circuit do not significantly shift the dc bias points of its active devices — to large signal — in which applied signals significantly shift active device dc bias and gain — the reality of *device self-heating* must be included in the device model. Examples: When amplitude stabilization occurs in an oscillator or gain reduction occurs in an amplifier as a result of voltage or current limiting or saturation.

While manufacturers often provide detailed models for their devices, avoid the temptation to assume that models will behave in a simulation just as an actual component will behave in a real circuit. *Absolutely every desired behavior exhibited by a simulated device must be explicitly built into the model*, mathematical element by mathematical element. A simulated device can reliably simulate real world behavior only to the extent that it has been programmed and configured to do so. A 1N4148 diode from your junk box "knows" exactly what to do when ac is applied to it, regardless of the polarity and level of the signal. Mathematically *modeling* the for-ward and reverse biased behavior of the real diode is almost like modeling two different devices. Realistically modeling the smooth transition between those modes, especially with increasing frequency, is yet another challenge.

Mathematical transistor modeling approaches the amazingly complex, especially for devices that must handle significant power at increasingly high frequencies, and especially as such devices are used in digital communication applications where phase relationships among components of the applied signal must be maintained to keep bit error rates low. The effect of nonlinear reactances — for instance, device capacitances that vary with applied signal level — must be taken into account if circuit simulation is to accurately predict oscillator phase noise and effects of the large signal phenomenon known as *AM-to-PM conversion*, in which changes in signal amplitude cause shifts in signal phase. In effect, different aspects of device behavior require greatly different models — for instance, a dc model, a small signal ac model, and a large signal ac model. Of *SPICE*'s bipolar-junction-transistor (BJT) model, we learn from the *SPICE* web pages that "The bipolar junction transistor model in *SPICE* is an adaptation of the integral charge control model of Gummel and Poon introduced in the **RF Techniques** chapter. This modified Gummel-Poon model extends the original model to include several effects at high bias levels. The model automatically simplifies to the simpler Ebers-Moll model when certain parameters are not specified."

As an illustration of device model complexity, **Fig 6.3** shows as a schematic the BIP linear bipolar junction transistor model from *ARRL Radio Designer*, a linear circuit simulator published by ARRL in the late 1990s. This model worked well at audio and relatively low radio frequencies. The model must be even more complex for accurate results in the upper HF and higher frequency ranges.

Especially in the area of MOSFET and MESFET device modeling, and large signal device modeling in general (of critical importance to designers of RF integrated circuits [RFICs] for use at microwave frequencies), *SPICE* and RF-fluent non-*SPICE* simulators include active device models home experimenters are unlikely to use. Help in deciding on what version of a model is appropriate for your application is another reason to join your simulation tool's users group.

Most of us will go (and need go) no further into the arcanities of device modeling than using *SPICE*'s JFET model for FETs such as the 2N3819, J310, and MPF102, and *SPICE*'s BJT model for bipolar transistors such as the 2N3904. *OrCAD 16* includes preconfig-

ured 1N914, 1N4148, 2N2222, 2N2907A, 2N3819, 2N3904 and 2N3906 devices, among many others, and these will be sufficient for many a ham radio simulation session. Getting the hang of the limitations and quirks of these models may well provide challenge enough for years of modeling exploration.

To get parameters for other devices, especially RF devices and specialty components like transformers, we must search the Internet in general and device manufacturer websites

in particular to find the data we need. The manufacturer sites listed in **Table 6.2** will get you started. Because the use of simulation and *SPICE* models is so widespread, manufacturers routinely make those models available at no cost. The usual place to find them is via the device's online data sheet through a hyperlink or in a special model library on the manufacturer's website — enter "models" into the site's search function if you can't find the models.

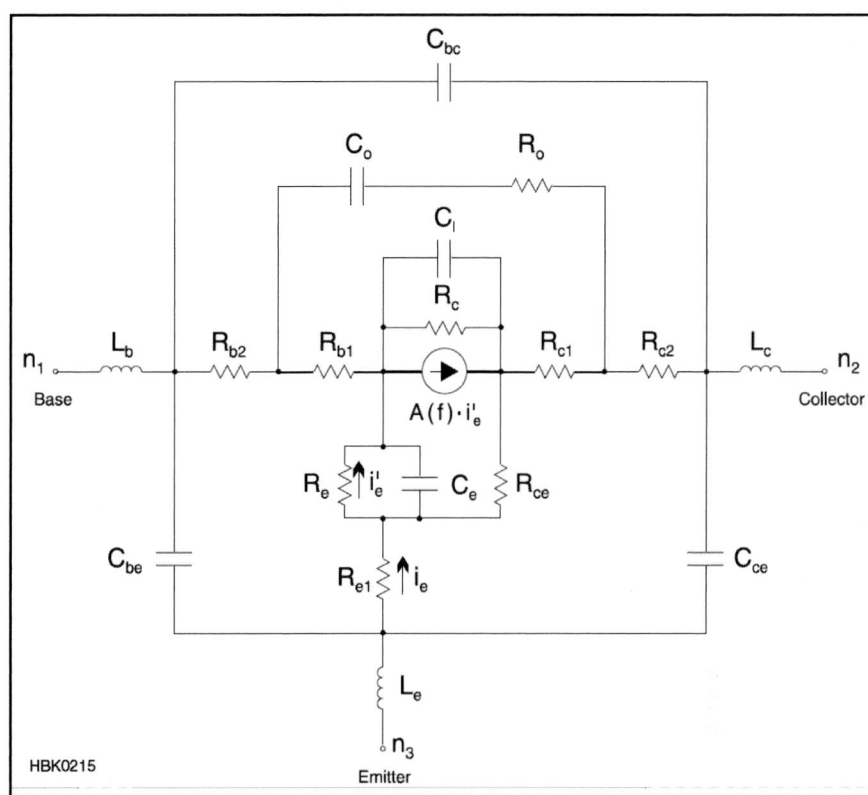

Fig 6.3 — The linear BJT model, BIP, from *ARRL Radio Designer*, a now-discontinued circuit simulation product published by ARRL in the late 1990s.

Table 6.2
Device Parameter Sources

Source	Address	Resource
Cadence Design Systems	www.cadence.com/products/orcad/pages/downloads.aspx#cd	*OrCAD*-ready libraries of manufacturer-supplied device models
California Eastern Laboratories	www.cel.com	Data and models NEC RF transistors
Duncan's Amp Pages	www.duncanamps.com/spice.html	*SPICE* models for vacuum tubes
Infineon	www.infineon.com	Data and models for Siemens devices
Fairchild Semiconductor	www.fairchildsemi.com	Device data and *SPICE* models
National Semiconductor and Texas Instruments	www.ti.com	Data and *SPICE* models for National and Texas Instruments op amps
NXP	www.nxp.com/models/index.html	Data and models for Philips devices
Freescale	www.freescale.com	Data and models for Freescale (previously Motorola) devices
On Semiconductor	www.onsemi.com	Data and models for On Semiconductor (previously Motorola) devices

Models for any particular device are generally available through a link on the online version of the device's data sheet. Internet searches for the device's part number and "model" usually works, as well.

```
                    2N2222
                    NPN
          IS    14.340000E-15
          BF   255.9
          NF     1
          VAF   74.03
          IKF     .2847
          ISE   14.340000E-15
          NE     1.307
          BR     6.092
          NR     1
          RB    10
          RC     1
          CJE   22.010000E-12
          MJE     .377
          CJC    7.306000E-12
          MJC     .3416
           TF  411.100000E-12
          XTF     3
          VTF     1.7
          ITF     .6
           TR   46.910000E-09
          XTB     1.5
          CN     2.42
           D      .87

.model Q2N2222 npn (IS=2.48E-13 VAF=73.9 BF=400 IKF=0.1962 NE=1.2069
+ ISE=3.696E-14 IKR=0.02 ISC=5.00E-09 NC=2 NR=1 BR=5 RC=0.3
CJC=7.00E-12
+ FC=0.5 MJC=0.5 VJC=0.5 CJE=1.80E-11 MJE=0.5 VJE=1 TF=4.00E-10
+ ITF=2 VTF=10 XTF=10 RE=0.4 TR=4.00E-08)
```

Fig 6.4 — A list of parameters specified for the Gummel-Poon model of a 2N2222A transistor and the condensed form usually seen in available model files.

A typical *SPICE* model is shown in **Fig 6.4**. This particular model is for the familiar 2N2222A NPN bipolar transistor. Each parameter in the vertical list (IS, BF, NF and so forth) is one element of the Gummel-Poon model for the generic transistor used by *SPICE* simulation programs. Typically, the model is provided as plain ASCII text in the compressed form at the bottom of the figure with several parameters per line. Every parameter between the opening and closing parentheses defines some element of the underlying mode. (See the *SPICE* references listed at the end of this chapter for detailed information about the syntax of *SPICE* models.)

NETLISTS

A *netlist* is a specialized table that names a circuit's components, specifies their electrical characteristics, and maps in text form the electrical interconnections among them. Uniquely numbered nodes or *nets* — in effect, specifications for each of the connections in the simulated circuit — serve as interconnects between components, with each component defined by a statement comprising one or more netlist lines.

The netlist served as the original means of circuit capture for all simulators known to the writer, including *SPICE*; *schematic capture* as we imagine it today, using graphic symbols, came later. Further reflecting *SPICE*'s pregraphical heritage is the fact that, to this day a *SPICE* netlist may be referred to by long-time *SPICE* users as a *SPICE deck*, as in "deck of Hollerith punch cards." In *SPICE*'s early days, circuit definitions and simulation instructions (netlist statements that begin with a period [.]) were commonly conveyed to the simulation engine in punched-paper-card form. Netlists are still the means of conveying circuit topology and simulation instructions to most simulators, although they are text files today.

In the models and netlists that you will encounter, statements that must span multiple lines include *continuation characters* (+) to tell the netlist parser to join them at line breaks. Asterisk (*) or other non-alphanumeric characters denote comments — informational-to-human lines to be ignored by the simulator.

SUBCIRCUITS

Fig 6.5 illustrates the level of detail involved in more accurate device modeling typical for devices used at VHF and UHF. The device is a California Eastern Labs NE46134, a surface mount BJT intended to serve as a broadband linear amplifier at collector currents up to 100 mA and collector voltages up to 12.5 V.

This manufacturer supplied model for the NE46134 embeds an unpackaged device chip (NE46100, shown as Q1 in the figure) within a *subcircuit*. (Note the .SUBCKT label on the first non-comment line of the model which indicates that the following information defines a subcircuit.) The subcircuit includes Q1 plus the parasitic reactances contributed by the transistor package, such as CCBpkg — a collector-to-base package capacitance. The chip leads themselves are modeled as transmission lines, TB and TE.

Following the lines that define of all of the subcircuit's component values that represent the parasitic reactances, the NE46100 transistor model is provided, beginning with the line .MODEL NE46100 NPN. That information will be used to define Q1 when referenced in the subcircuit model above.

The overall model then appears to the external circuit like a regular three terminal transistor with a base, emitter and collector. In this fashion nested subcircuits can be used to create arbitrarily complex models that can be used as components by the designer.

Figs 6.6A and 6.6B show a schematic of a typical RF circuit — a 7 MHz double-tuned filter. Fig 6.6A is the usual depiction you would see in a magazine or book article about the circuit. It has the usual symbols and variable capacitors of a tunable filter. Fig 6.6B shows the circuit after schematic capture by the *OrCAD Capture CIS* tool. An ac voltage source, V1, is placed at the input with a 50 Ω series resistor, R1, to create a 50 Ω signal source impedance. The transformers TX1 and TX2 actually contain subcircuits consisting of primary and secondary inductances and a coupling element with 0.22 Ω resistors that simulate loss resistance. The paralleled fixed and variable capacitors are combined into single fixed value capacitors. A 50 Ω load, R2, has been attached to the output. Ground symbols now have a 0 nearby to indicate zero voltage.

TIME STEP

Simulators are *discrete time* devices — calculations are performed throughout the circuit and results obtained, then time is changed by a fixed amount (the *time step*) and the calculations run again. The size of the time step can be automatically chosen by the simulator (a usually reasonable value based on a default setting or some evaluation of the circuit component values) or set to specific values by the user.

While using very small time steps makes for a smooth looking output and wide frequency ranges, it makes the simulation run slower (less of an issue as computers get faster) and makes the output data files quite a bit bigger (less of an issue as hard drives get bigger). While today's computers make short

```
* FILENAME:        NE46134.CIR
* from http://www.cel.com/pdf/models/ne46134.cir
* NEC PART NUMBER: NE46134
* LAST MODIFIED:   11/97
* BIAS CONDITIONS: Vce=5V, to 12.5V,   Ic=50mA to 100mA
* FREQ RANGE:      0.1GHz TO 2.5GHz
*
*                               CCBpkg
*                       .-------||--------.
*                       |        CCB       |
*                       |     .--||---.    |
*                       |     |       |    |         COLLECTOR
* BASE                  |     |       |    |
*          __Tb_        |     |  /----o-o-------o----o
*  o---------|____|-o-LB---o--|Q1    |        |
*                       |     B  \  --         |
*                       --     |   -- CCE      --
*                   -- CBEpkg E o----'      --CCEpkg
*                       |        |          |
*                       |        LE         |
*                       `---------o---------'
*                                |
*                                -
*                               | |  Te
*                               | |
*                                -
*                                |
*                                o EMITTER
*
*
*
*
*        CCB    = 0.03 pF      LB  = 1.2nH       Tb/Te:
*        CCE    = 0.5 pF       LE  = 1.2nH       z=60 ohms
*        CCBpkg= 0.18pF                          l=50 mils
*        CCEpkg= 0.18pF                          a=0.0001
*        CBEpkg= 0.01pF                          f=0.9GHz
*
*                     c b e
.SUBCKT NE46134/CEL 2 1 3
Q1 2 6 7 NE46100
CCB 6 2  0.03E-12
CCE 2 7  0.5E-12
LB  4 6  1.2E-9
LE  7 5  1.2E-9
TB 1 0 4 0 Z0=60 TD=9.63E-12
TE 5 0 3 0 Z0=60 TD=9.63E-12
CCBPKG 4 2 0.18E-12
CCEPKG 2 5 0.18E-12
CBEPKG 4 5 0.01E-12
.MODEL NE46100 NPN
+( IS=8.7e-16    BF=185.0      NF=0.959    VAF=30.0     IKF=0.20
+ ISE=5.70e-13   NE=1.80       BR=5.0      NR=1.0       VAR=12.4
+ IKR=0.018      ISC=1.0e-14   NC=1.95     RE=0.630     RB=6.0
+ RBM=4.0        IRB=0.004     RC=3.0      CJE=4.9e-12  VJE=0.60
+ MJE=0.450      CJC=2.50e-12  VJC=0.830   MJC=0.330    XCJC=0.20
+ CJS=0.0        VJS=0.750     MJS=0.0     FC=0.50      TF=12.9e-12
+ XTF=1.60       VTF=19.9      ITF=0.40    PTF=0.0      TR=1.70e-8
+ EG=1.11        XTB=0.0       XTI=3.0     KF=0.0       AF=1.0 )
.ENDS
*$
```

Fig 6.5 — California Eastern Laboratories model of the NE46134 linear broadband transistor. The transistor model is constructed as a subcircuit consisting of several parasitic components connected to the NE46100 transistor (Q1).

work of simulations that would have brought the previous era's mainframes to a halt, there is no need to generate vastly more data than you need for the job at hand.

If you set the time step to at least 10 times smaller than the reciprocal of the highest frequency in which you're interested, that will strike a nice balance. For example, when simulating an audio circuit, you might be interested in signals out to 100 kHz so a good time step size would be $\frac{1}{10}$ of $\frac{1}{100}$ kHz or about 1 μs. Similarly, if the circuit was supposed to operate up to 10 MHz, the time step should

be no smaller than 10 ns. Your simulation software manual will have guidelines for that particular package as well.

6.2.3 Types of Simulations

Once the circuit has been entered into the simulator via schematic capture and all error checks have been completed, simulation can begin in earnest. There are several ways in which the circuit can be simulated, corresponding to the various tests that you would run on a real circuit on the workbench. The

following paragraphs are only definitions — for a complete description of these simulations and how to perform them, the reader is directed to the References section of this chapter and reminded that the software provider and associated users groups will also have a lot of information and background material.

DC OPERATING POINT OR BIAS POINT

The first step in simulation is usually to run a *DC Operating Point* or *DC Bias Point*

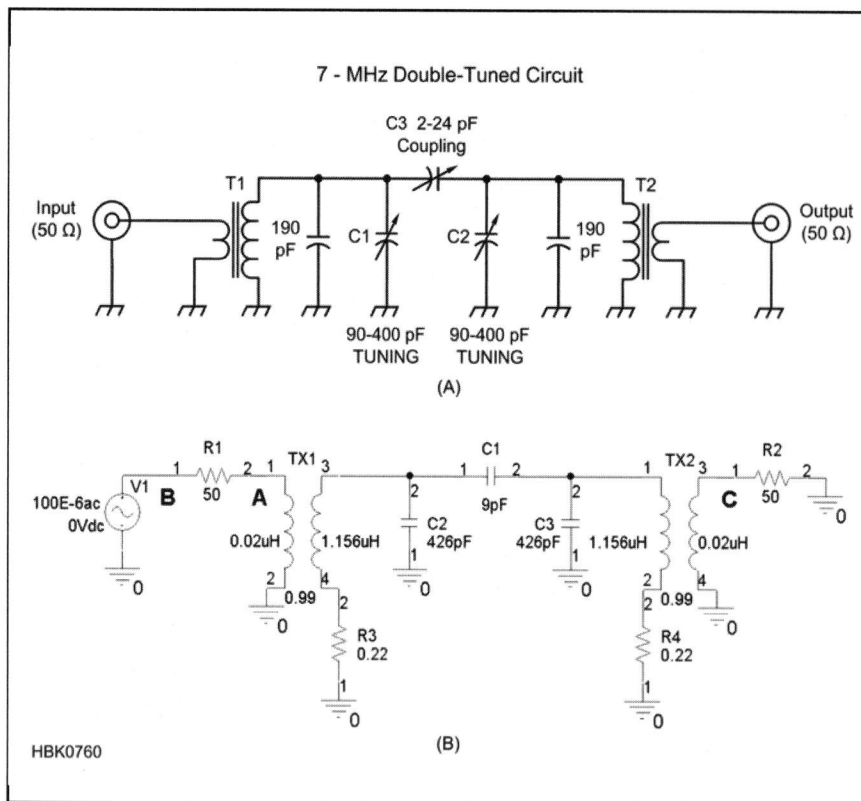

Fig 6.6 — A is a 7 MHz double-tuned filter as drawn for real world builders. B shows the filter schematic as it is typically drawn for construction and service.

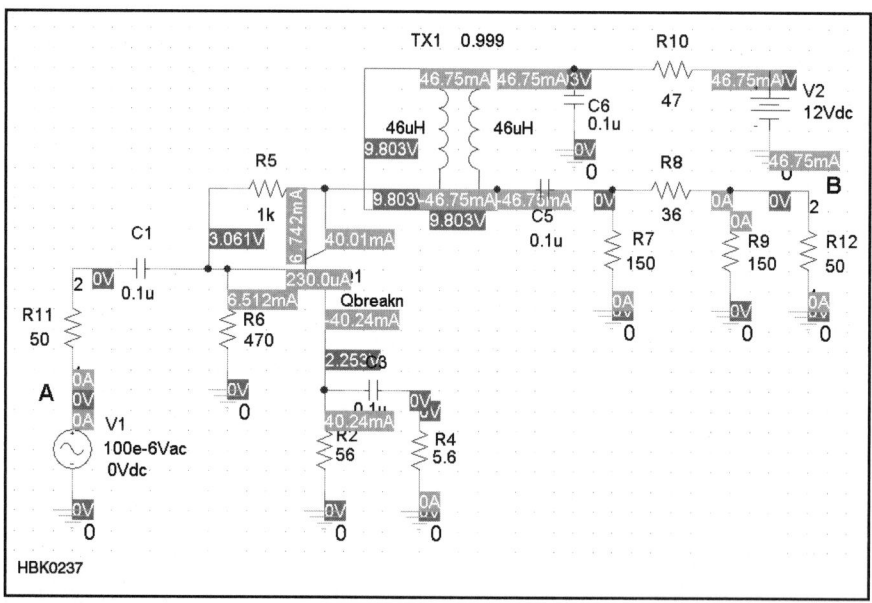

Fig 6.7 — A dc operating point simulation allows the designer to verify that the circuit biasing is realistic for all devices and that all voltages and currents are as expected.

analysis. (Remember that terminology usually changes from program to program.) In this analysis, the simulator "applies power" to the circuit and calculates the dc voltages and currents for all circuit components.

Fig 6.7 shows the results of a typical DC

Operating Point analysis for an amplifier circuit. The labels obscure the circuit elements a bit in this print image but you can see a transistor (Qbreakn) in the middle, an input voltage source at the lower left (V1), the power supply voltage source at the upper right

(V2), and a variety of components making up the circuit. The darker labels show voltages such as 3.061 V (transistor base voltage) and 46.75 mA (the power supply output current). The information is generally available as a text file, as well. Most simulators also allow you to flag certain points in the circuit for which voltage and current (ac and/or dc) are displayed on screen at all times.

It is good practice, particularly for beginning and occasional modelers, to verify that the dc operating point is as expected for all circuit components before moving ahead to the more interesting ac waveform and spectrum displays. Many hours have been wasted troubleshooting a circuit's simulated ac performance when the problem is really that the dc operating point isn't right!

TIME DOMAIN VS FREQUENCY DOMAIN

Simulators can provide two primary types of ac simulation results. The time domain output displays one or more voltages or currents on an X-Y display as traces with amplitude on the vertical axis and time on the horizontal axis. In other words, it simulates an oscilloscope type display. **Fig 6.8** shows an example time domain display of an R-C oscillator starting to oscillate. The single trace is of the oscillator's ac output voltage, showing that oscillation begins about 30 ms after the simulation begins and quickly stabilizes to a steady sine wave output.

Fig 6.9 shows a frequency domain output that looks very much like a spectrum analyzer display. This particular simulation is of the same amplifier shown in Fig 6.7 with two input signals (the large components at 7 and 10 MHz) and uses a logarithmic vertical scale to show the harmonics and intermodulation products. *SPICE*-type simulators first perform time domain ac analysis to generate a waveform then use Fast Fourier Transform techniques to calculate a frequency domain output. This technique is often insufficient to obtain highly accurate frequency domain simulations, such as noise and intermodulation performance, leading to alternatives such as harmonic balance simulations that are called *RF-fluent* since they give more accurate results at high frequencies. (See the sections on RF-Fluent Simulators and Limitations of Simulation at RF in this chapter.)

TRANSIENT

Transient simulation involves putting the circuit in a stable, known state and then applying a controlled disturbance of some sort. The response to the disturbance is then calculated and displayed over some time period in either time domain or frequency domain form. Typical inputs to transient analysis are steps (a parameter changes very quickly from one value to another), delta pulses (infinitely

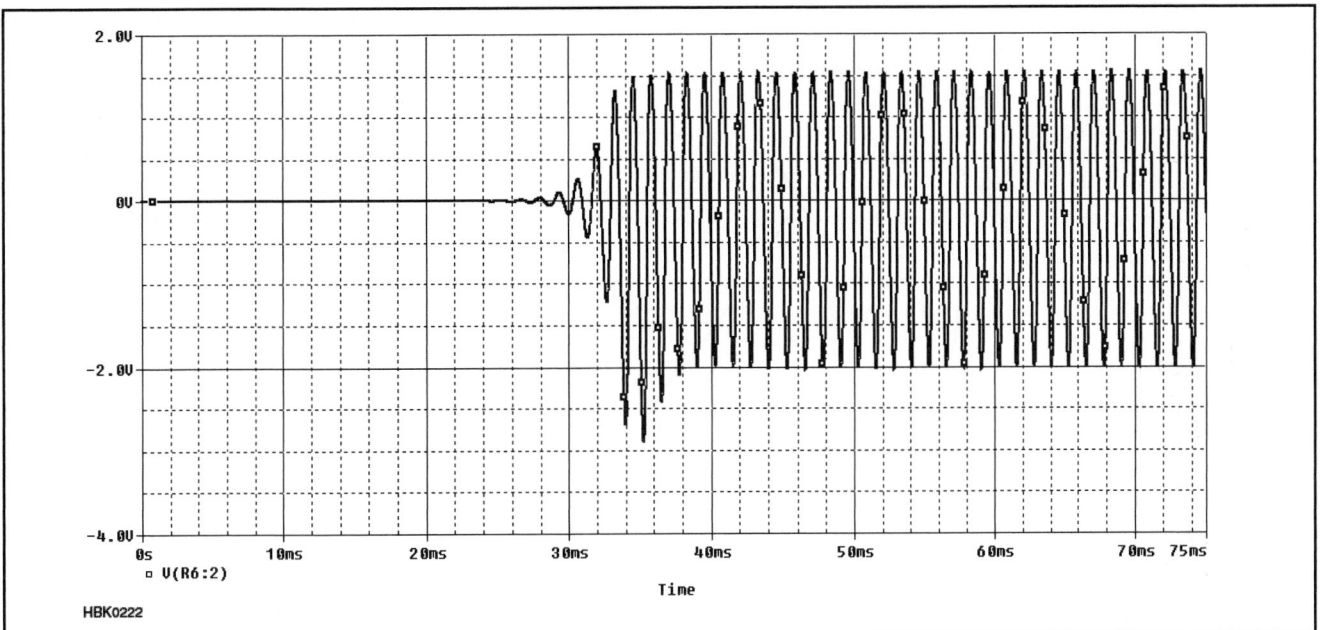

Fig 6.8 — Time domain display of an R-C oscillator startup showing voltage on the vertical axis and time on the horizontal. The display is very much like that of an oscilloscope.

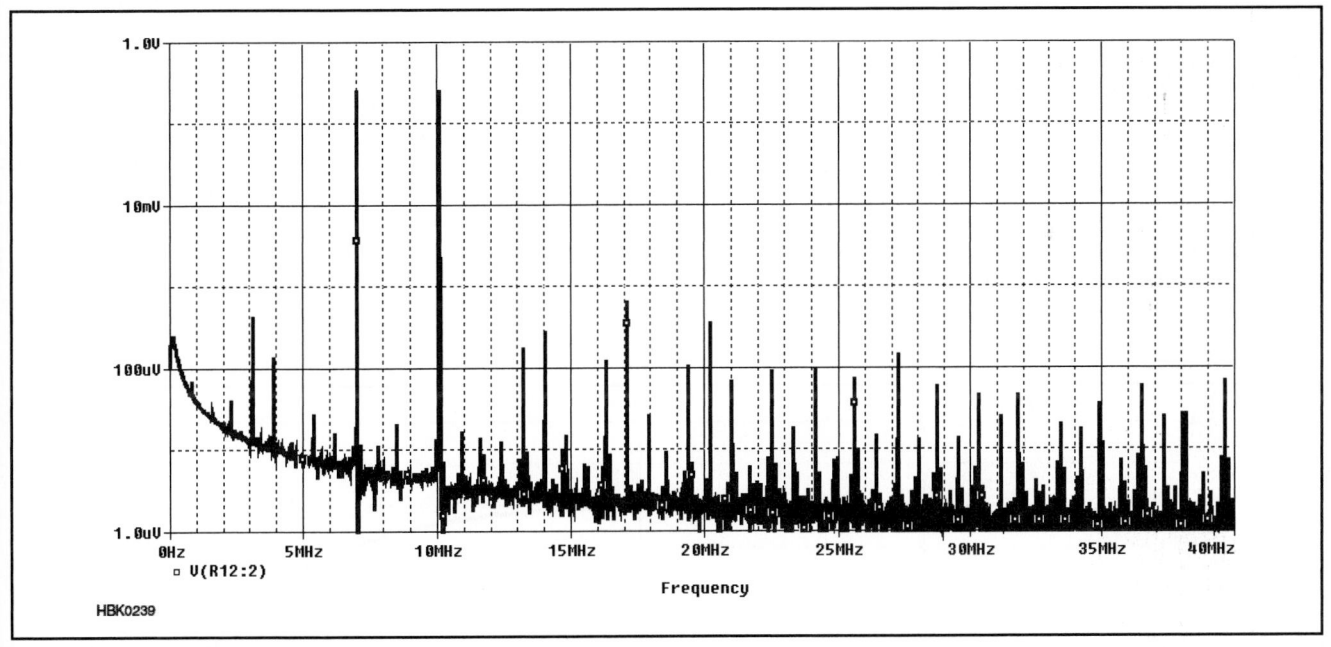

Fig 6.9 — Frequency domain display of an amplifier undergoing a two-tone test. Amplitude in dB is shown on the vertical axis and frequency on the horizontal axis. This display is very much like that of a spectrum analyzer.

narrow pulses with some finite energy), rectangular pulses, ramps, and various other selectable waveforms.

DC AND AC SWEEPS

Once a circuit is working, it is useful to characterize its performance over a range of conditions, called a *sweep*: power supply voltage, input voltage, input frequency and so on. Even temperature can be swept to see how the circuit behaves in different environments. Note that this requires the behavior of components with changing temperature to be accounted for in the component models.

One of the most common swept simulations for ham radio circuits is a frequency sweep to determine the frequency response of an amplifier or filter. **Fig 6.10** shows the insertion loss and return loss (see the **Analog Basics** chapter) for the 7 MHz filter in Fig 6.6 across a range of 6.8 to 7.4 MHz. **Fig 6.11** shows the gain of the amplifier in Fig 6.7 with the input frequency swept across a range of 1 to 100 MHz.

6.2.4 RF-Fluent Simulators

SPICE-based simulators can do wonders in many classes of circuit simulation. For RF use, however, *SPICE* has significant draw-

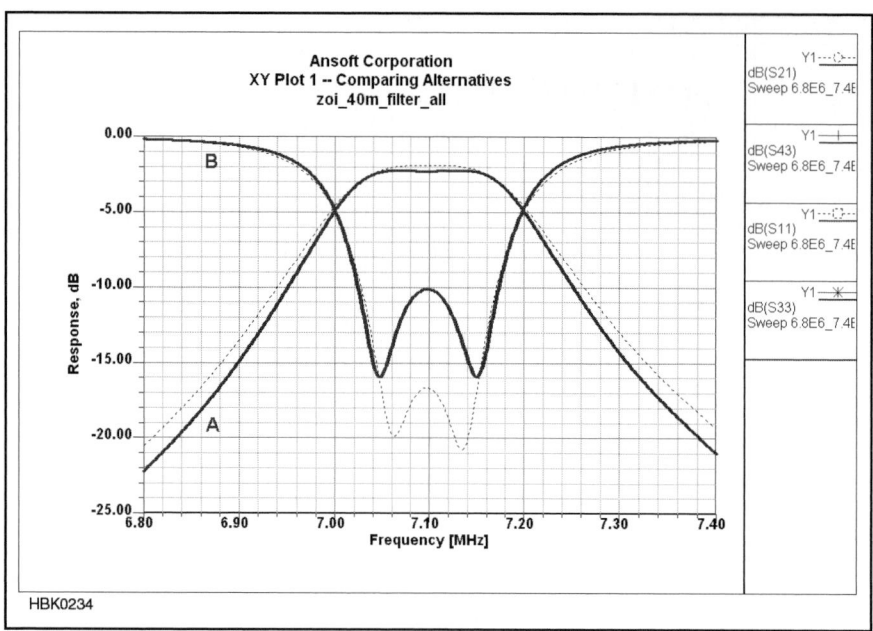

Fig 6.10 — Insertion loss (A) and return loss (B) of the 7 MHz double-tuned filter in Fig 6.6.

backs. For starters, *SPICE* is not RF-fluent in that it does not realistically model *physical* distributed circuit elements — microstrip, stripline and other distributed circuit elements based on transmission lines. It cannot directly work with network parameters (*S*, *Y*, *Z* and more — see the **Analog Basics** chapter), stability factor and group delay. It cannot simulate component *Q* attributable to skin effect. It cannot simulate noise in nonlinear circuits, including oscillator phase noise. It cannot realistically simulate intermodulation

and distortion in high-dynamic-range circuits intended to operate linearly. This also means that it cannot simulate RF mixing and inter-modulation with critical accuracy.

The feature-unlimited version of *Ansoft Designer* and competing RF-fluent simula-tion products can do these things and more excellently — but many of these features, es-pecially those related to nonlinear simulation, are unavailable in the student/demoware ver-sions of these packages if such versions exist.

From 2000 to 2005, the free demoware

precursor of *Ansoft Designer SV 2*, Ansoft *Serenade SV 8.5*, brought limited use of non-linear simulation tools to students and experi-menters. With *Serenade SV 8.5*, you could simulate mixers, including conversion gain and noise figure of a mixer. Amplifier two-tone IMD could be simulated as in Fig 6.9. Optimization was enabled. Realistic nonlinear libraries were included for several Siemens — now Infineon — parts. You could accurately predict whether or not a circuit you hoped would oscillate would *actually* oscillate, and assuming that it would, you could accurately predict its output power and frequency. **Figs 6.12**, **6.13** and **6.14** give some examples of the power of RF-fluent simulation software, in this case, Ansoft (now ANSYS) *Serenade Designer SV 8.5*.

The harmonic balance techniques used by Ansoft's nonlinear solver — and by the nonlinear solvers at the core of competing RF-fluent CAD products, such as Agilent *Advanced Design System* (ADS) — allowed you to simulate crystal oscillators as rapidly as you can simulate lower-*Q* oscillators based on LC circuits. (In *SPICE*, getting a crystal oscillator to start may be impossible without presetting current and/or voltages in key com-ponents to nonzero values.)

Although these features are no longer available as student/demoware/freeware, if you're serious about pushing into RF CAD beyond what *SPICE* can do, see if you can find a used copy of *Serenade SV 8.5* or find a friend working with the professional tools who will let you use it occasionally for hobby applications.

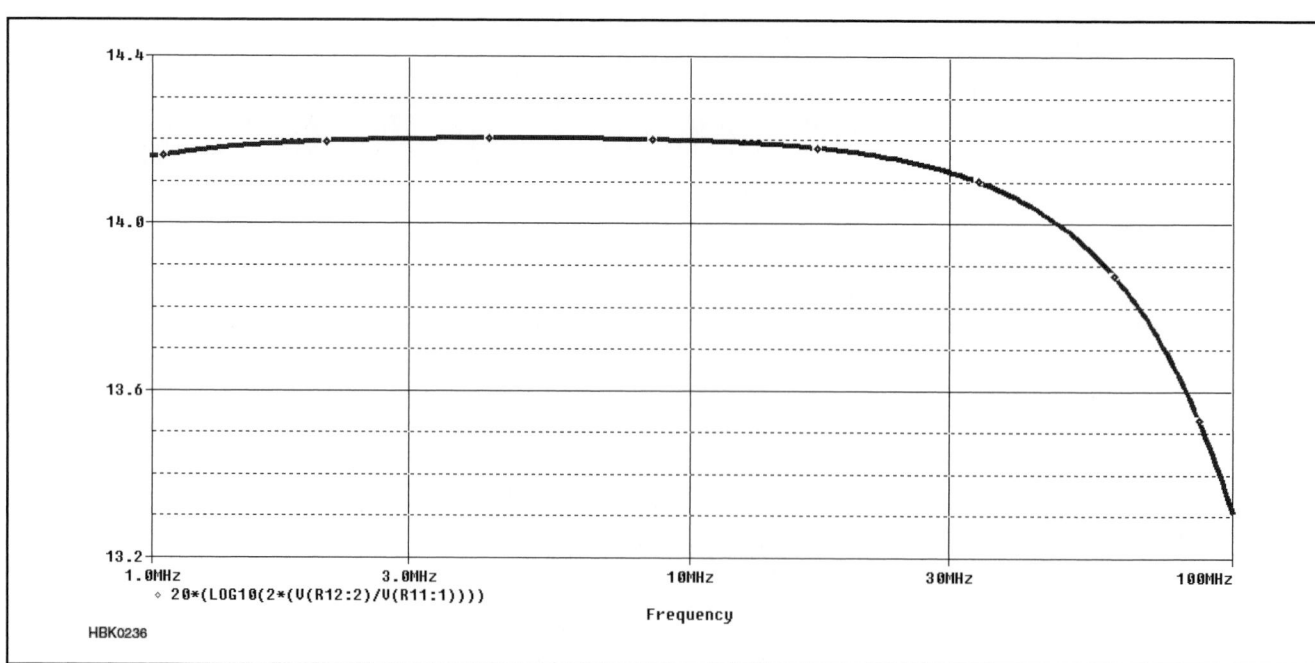

Fig 6.11 — The gain in dB of the amplifier in Fig 6.6 over a range of 1 to 100 MHz.

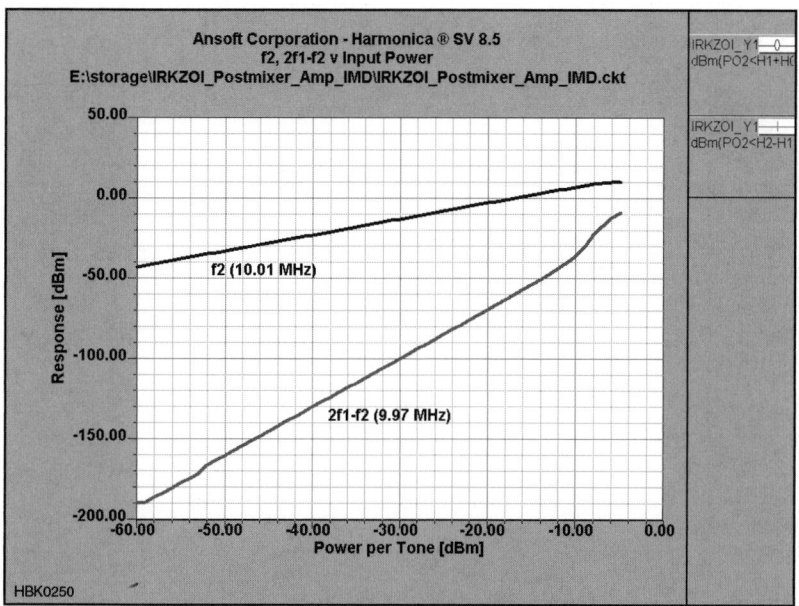

Fig 6.12 — A two-tone nonlinear simulation of third-order IMD performed by Ansoft *Serenade Designer SV 8.5.*

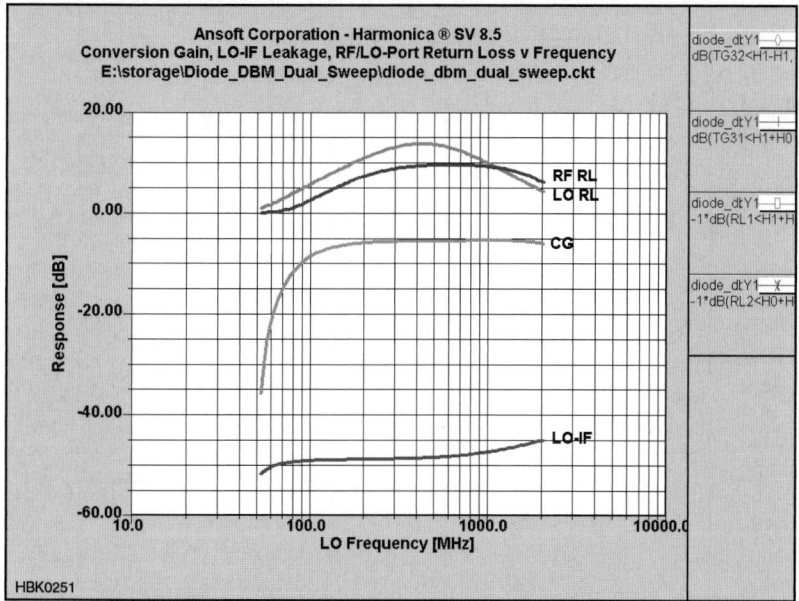

Fig 6.13 — Professional RF circuit simulators can also simulate mixing and the small signal characteristics of mixers, such as port return loss, conversion gain, and port-to-port isolation. (*Serenade SV*

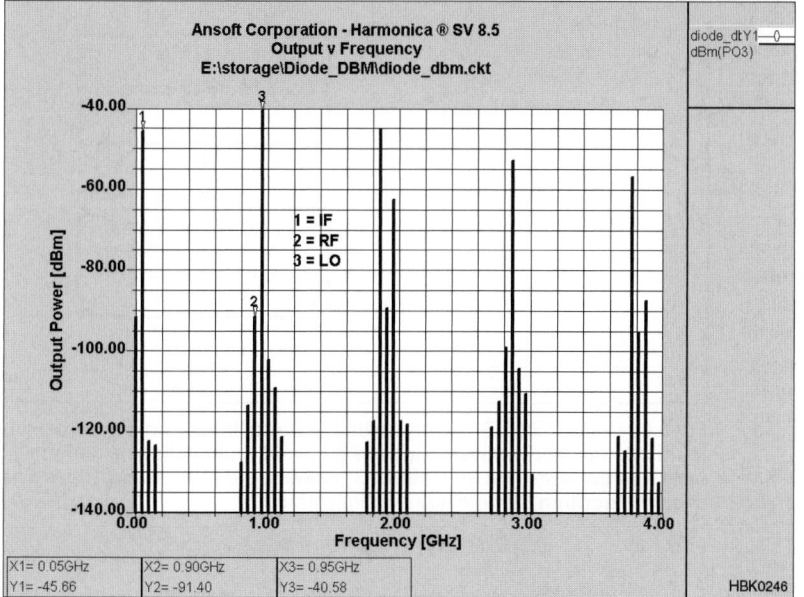

Fig 6.14 — Output spectrum of a diode-ring doubly balanced mixer as simulated by *Serenade SV 8.5.* Note the dynamic range implicit in this graph: In a simulation that includes a local oscillator (LO) signal at 7 dBm, accurate values are calculated for IMD products nearly *140 dB weaker* without encountering mathematical noise — an achievement unapproachable with *SPICE*-based simulators.

6.3 Limitations of Simulation at RF

[Experienced users of circuit simulation software are wary of using any software near or outside the boundaries of circuits and parameters for which it was intended and tested. RF simulation can present just such situations, leading to software failure and unrealistic results. Introduced and summarized in this section, several detailed papers by Dr Ulrich Rohde, N1UL, exploring simulation at RF are provided on the CD-ROM accompanying this *Handbook*. The papers are:

• "Using Simulation at RF" by Rohde, a survey of the issues of RF simulation and the techniques used in current modeling programs.

• "The Dangers of Simple Usage of Microwave Software" by Rohde and Hartnagel, a discussion of inaccuracies introduced by device parameter measurement and model characteristics.

• "Mathematical Stability Problems in Modern Non-Linear Simulations Programs" by Rohde and Lakhe, presenting various approaches to dealing with nonlinear circuit simulation.

In addition, there are many online resources to help you obtain trustworthy simulation results with a simulator designed for RF. For the interested reader with some technical background, the online paper "Introduction to RF Simulation and its Application" by Ken Kundert (**http://icslwebs.ee.ucla.edu/ dejan/researchwiki/images/3/30/Rf-sim. pdf**) provides an introduction to RF simulation methods and how they account for the characteristics of RF circuits when generating common RF measurements. The website The Designer's Guide (**www.designers-guide. org**) also provides many tutorials, technical guides, models, and other resources for analog and RF simulation users.

While the precise lower bound of "RF" is ill-defined, RF effects start already at about 100 kHz. This was first noticed as self-resonance of high-Q inductors for receivers. In response, Litz wire was invented in which braided copper wires were covered with cotton and then braided again to reduce self-resonance effects.

As frequencies get higher, passive elements will show the effects of parasitic elements such as lead inductance and stray capacitance. At very high frequencies, the physical dimensions of components and their interconnections reach an appreciable fraction of the signal wavelength and their RF performance can change drastically.

RF simulators fall in the categories of *SPICE*, harmonic balance (HB) programs

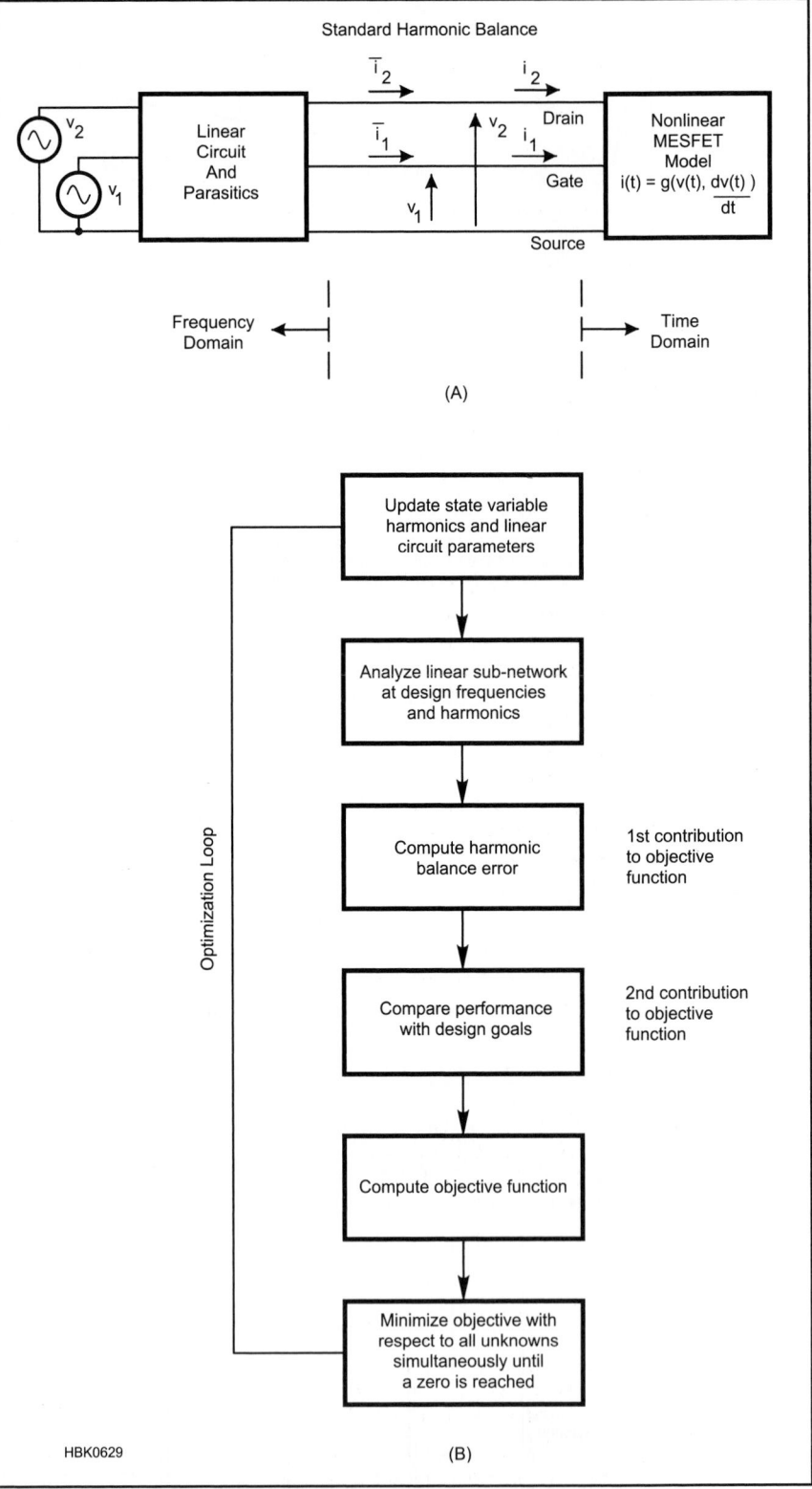

Fig 6.15 — (A) MESFET circuit partitioned into linear and nonlinear sub-circuits for harmonic balance analysis. Applied gate and drain voltages, and relevant terminal voltages and currents, are indicated. (B) Flowchart of a general purpose harmonic balance design algorithm that includes optimization.

and EM (electromagnetic) programs. The EM simulators are more exotic programs. Two types are common, the 2D (2.5) or 2-dimensional and the full 3-dimensional versions. They are used to analyze planar circuits, including vias (connections between layers) and wraparounds (top-to-ground plane connections), and solid-shapes at RF. They go far beyond the *SPICE* concept.

6.3.1 *SPICE*-based Simulators

SPICE was originally developed for low frequency and dc analysis. (Modern *SPICE* programs are based on *SPICE3* from University of California — Berkeley.) While doing dc, frequency, and time domain simulations very well, *SPICE*-based simulation has some problems. The time domain calculation uses the very complex mathematics of the Newton-Raphson solution to nonlinear equations. These methods are not always stable. All kind of adjustments to the program settings may be necessary for the calculations to converge properly. Knowledge of the specifics of different types of electronic circuits can assist the user in finding an accurate solution by specifying appropriate analysis modes, options, tolerances, and suitable model parameters. For example, oscillators require certain initializations not necessary for amplifiers and bipolar transistors may need different convergence tolerances than do MOS circuits. Generally, *SPICE* finds a solution to most circuit problems. However, because of the nonlinearity of the circuit equations and a few imperfections in the analytical device models, a solution is not always guaranteed when the circuit and its specification are otherwise correct.

The next problem at RF is that the basic *SPICE* simulator uses ideal elements and some transmission line models. As we approach higher frequencies where the lumped elements turn into distributed elements and special connecting elements become necessary, the use of the standard elements ends. To complicate matters, active elements such as diodes and transistors force the designer to more complex simulators. Adding the missing component elements leads to highly complex models and problems of convergence in which the simulator gives an error advising of a numerical problem or more likely by failing to generate a solution.

SPICE also has problems with very high-Q circuits and noise analysis. Questions of the noise figure of amplifiers or phase noise of an oscillator cannot be answered by a *SPICE*-based program accurately. Noise analysis, if not based on the noise correlation matrix approach, will not be correct if the feedback capacitance ($\text{Im}(Y_{12})$, the imaginary component of Y-parameter Y_{12}) is significant at the frequencies involved. Analysis of oscillators

in *SPICE* does not give a reliable output frequency and some of the latest *SPICE* programs resort to some approximation calculations.

6.3.2 Harmonic Balance Simulators

Harmonic balance (HB) analysis is performed using a spectrum of harmonically related frequencies, similar to what you would see by measuring signals on a spectrum analyzer. The fundamental frequencies are the frequencies whose integral combinations form the spectrum of harmonic frequency components used in the analysis. On a spectrum analyzer you may see a large number of signals, even if the input to your circuit is only one or two tones. The harmonic balance analysis must truncate the number of harmonically related signals so it can be analyzed on a computer.

The modern HB programs have found better solutions for handling very large numbers of transistors (>1 million transistors) and their math solutions are much more efficient, lead-

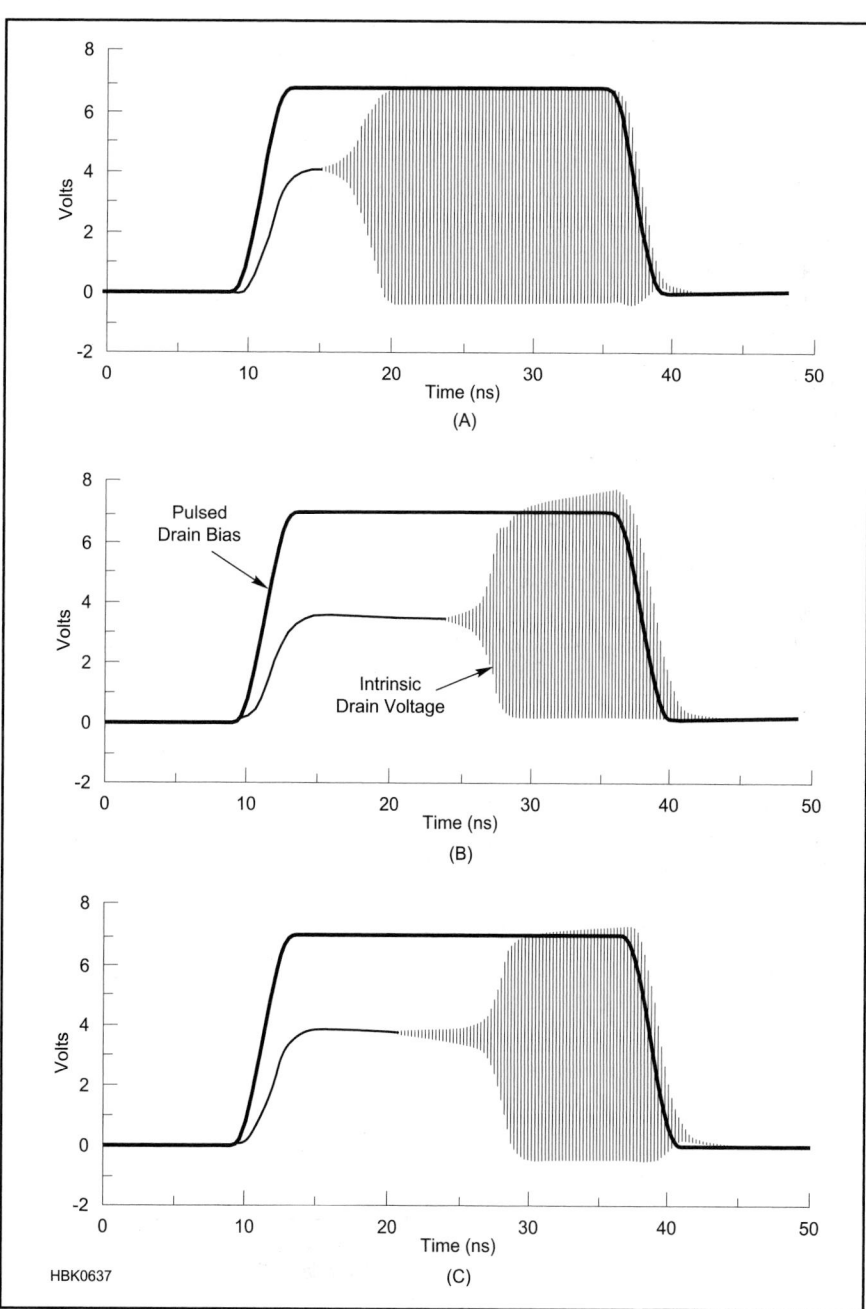

Fig 6.16 — (A) is the initial simulation of a *SPICE*-based simulator. (B) is the correct response of a pulsed microwave oscillator obtained by harmonic balance simulation using the Krylov-subspace solution. (C) is the *SPICE*-based simulation after 80 pulses of the drain voltage.

ing to major speed improvements. Memory management through the use of matrix formulations reduces the number of internal nodes and solving nonlinear equations for transient analysis are some of the key factors to this success.

HB analysis performs steady-state analysis of periodically excited circuits. The circuit to be analyzed is split into linear and nonlinear sub-circuits. The linear sub-circuit is analyzed in the frequency domain by using distributed models. In particular, this enables straightforward intermodulation calculations and mixer analysis. The nonlinear sub-circuit is calculated in the time domain by using nonlinear models derived directly from device physics. This allows a more intuitive and logical circuit representation.

Fig 6.15A diagrams the harmonic balance approach for a MESFET amplifier. Fig 6.15B charts a general purpose nonlinear design algorithm that includes optimization. Modern analysis tools that must provide accurate phase noise calculation should be based on the principle of harmonic balance.

Analysis parameters such as Number of Harmonics specify the truncation and the set of fundamental frequencies used in the analysis. The fundamental frequencies are typically not the lowest frequencies (except in the single-tone case) nor must they be the frequencies of the excitation sources. They simply define the base frequencies upon which the complete analysis spectrum is built.

6.3.3 Contrasts in Results

The following time domain analysis is a good example of differences between *SPICE* and harmonic balance simulation. A microwave oscillator is keyed on and off and a transient analysis is performed. When using the standard *SPICE* based on *SPICE3*, the initial calculation shows an incorrect response after one iteration as seen in **Fig 6.16A**. It takes about 80 pulses (80th period of the pulsed drain voltage) until the simulation attains the correct answer (Fig 6.16C) of the Krylov-subspace-

based harmonic balance in Fig 6.16B.

The frequencies involved need not be in the GHz range. Oscillators, in particular, can be very difficult to analyze at any frequency as shown by simulations of a low-MHz phase shift oscillator and a 10 MHz Colpitts oscillator in the referenced papers.

Validating the harmonic balance approach, **Fig 6.17** shows a BJT microwave oscillator entered into the schematic capture module of a commercially available HB simulator (Ansoft *Serenade 8.0*); **Fig 6.18** compares this oscillator's simulated phase noise to measured data. HB analysis gives similarly accurate results for mixers.

6.3.4 RF Simulation Tools

PSPICE: This popular version of *SPICE*, available from OrCAD (now Cadence Design Systems, **www.cadence.com**) runs under the PC and Macintosh platforms. An evaluation version, which can handle small circuits with up to 10 transistors, is freely

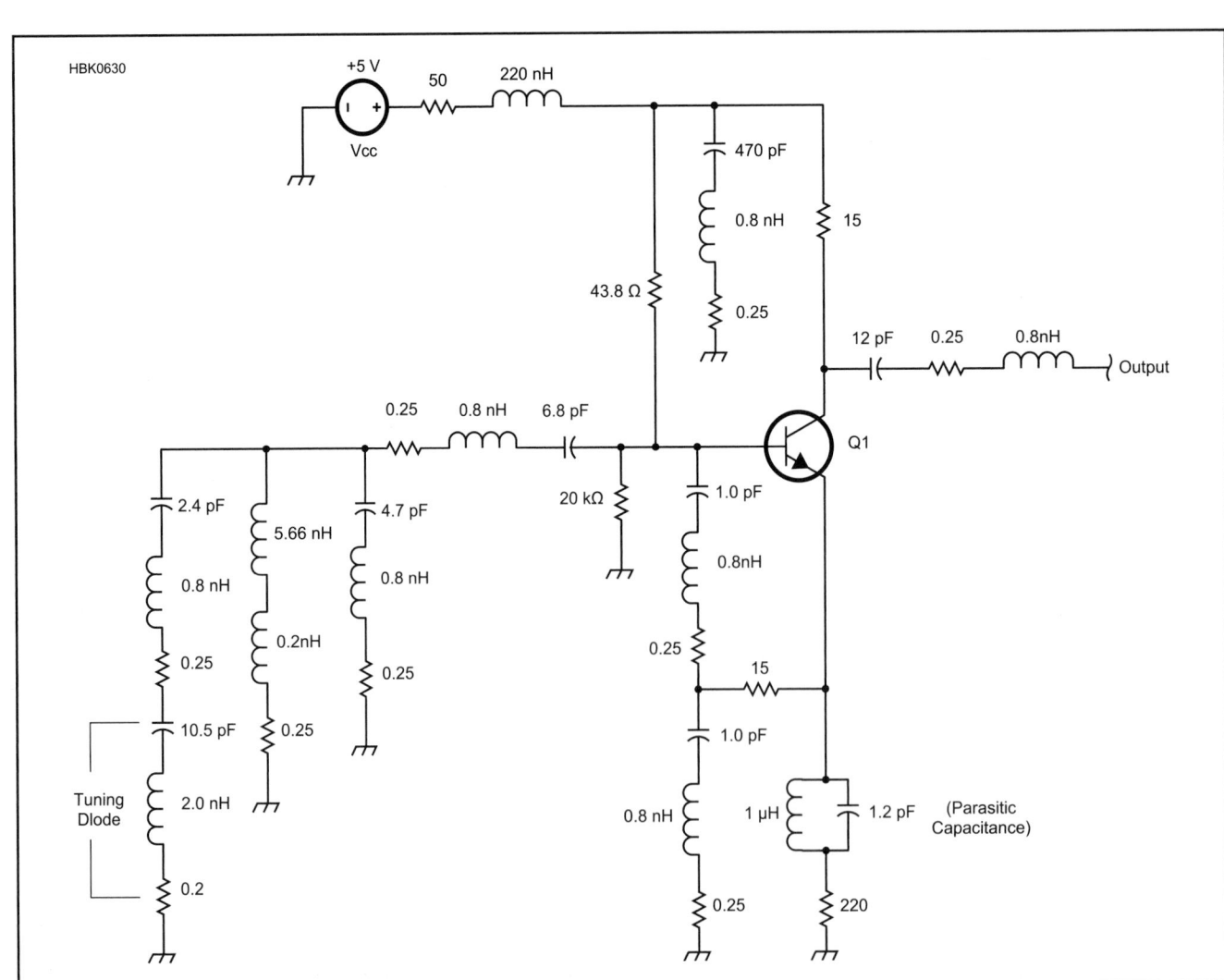

Fig 6.17 — Colpitts oscillator for 800 MHz with lumped elements modeled by their real values.

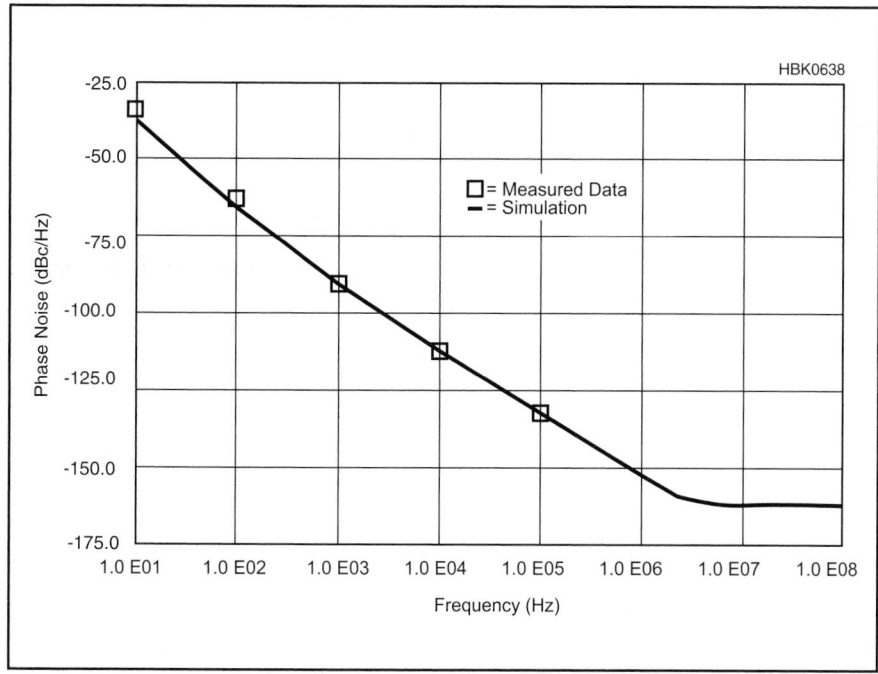

Fig 6.18 — Comparison between predicted and measured phase noise for the oscillator shown in Fig 6.17.

available, such as from **www.electronics-lab.com**. Contact Cadence for a full version or for more information. *AIM-spice* (**www.aimspice.com**) is a PC version of *SPICE* with a revised user interface, simulation control, and with extra models. A student version can be downloaded. Table 6.1 earlier in this chapter shows other free *SPICE* offerings.

There are a number of PC based *SPICE* programs in the $1000 range but they are designed more for switching power supplies and logic circuits optimization than RF. *ICAP4* (**www.intusoft.com/demos.htm**) and *MICROCAP9* (**www.spectrum-soft.com/index.shtm**) both have demonstration/evaluation versions available for download

Agilent, AWR, Ansys, and Synopsis offer very modern mixed-mode CAD tools and they combine the concept of *SPICE* with the advanced technologies. These are professional quality tools, but if one can arrange to make use of them through a friend or associate, the results are worth investing the time to learn their use.

6.4 CAD for PCB Design

[With numerous PCB design software packages available and low quantity, low cost PCB manufacturing services accepting orders electronically, the development of PCBs has never been easier for the amateur. As with any assembly or manufacturing process, it is important to understand the vocabulary and technology in order to achieve the desired result. Thus, this section provides a detailed description of the entire process of PCB design. — Ed.]

The primary goal of using software for printed circuit board (PCB) design is the production of so-called PCB artwork — the graphic design used to create the patterns of traces that establish connectivity on the PCB. Historically, PCB artwork was created by hand on clear film using black tape and special decals which were then photographically reduced. However, free and low cost programs specifically for the PCB design process are now widely available. These programs not only allow the creation of artwork efficiently and accurately, but produce the required ancillary files for commercial production, exchange information with schematic capture software, produce Bills of Materials (parts lists), and even include such features as three dimensional visualization of the finished board. While artwork files can be shared with other people for PCB production, the "source" files used by the CAD program can

typically only be used by other people who share the same program.

The decision to produce a PCB must take into account the nature of the circuit itself (for example, high frequency, low noise and high current circuits require additional care). Other considerations are time available, expense, available alternatives, quantity required, ability to share and replicate the design, and non-electrical characteristics such as thermal and mechanical, as well as desired robustness.

6.4.1 Overview of the PCB Design Process

The PCB design process begins with establishing the list of components in the circuit, the connections between the components, the physical outline/size of the board, and any other physical, thermal and electrical constraints or design goals. Much of the connectivity and component information is reflected in the schematic for a circuit, so in many cases the PCB layout process begins by entering the schematic in a schematic capture program which may be integrated with the PCB CAD program or standalone. (Schematic capture is not required for PCB layout.) Once the schematic is entered, there may be other options possible such as simulating the circuit as described in the preceding sections. A clean, well organized schematic that is easily modi-

fied is an asset regardless of the circuit production and construction methods.

With input from the schematic and other information, the board outline is created, mounting and other holes placed, the components positioned, and the pattern of traces created. Once the layout is complete, in many cases it is possible to run a design rules check — the equivalent of a "spell checker." Design rules include component connections and other information to check for problems related to connectivity and manufacturability. This step can save a great deal of time and expense by catching errors that could be fixed by hand, but would otherwise negate some of the benefits of a PCB.

The final step in the PCB layout program is to produce the collection of up to a dozen or so different files required for PCB production. In brief, the list includes the artwork for the pattern of traces, files for producing the board outline, solder masks, silk screens and holes.

The user then uploads the set of files to a PCB manufacturer. As quickly as two to three days later an envelope will be delivered with the freshly minted boards ready for assembly! Alternatively, the user may create the board "in house" using photomechanical or other processes based on the output files from the software, as is discussed in the **Construction Techniques** chapter along with PCB assembly techniques.

6.4.2 Types of PCB Design Software

PCB software varies in features, function and cost, but for the radio amateur, the most interesting software for introductory use fall into the following categories:

1) Open Source: PCB design software such as *GNU PCB* (**http://pcb.gpleda.org**, for *Linux*, Mac OS X) and *KiCad* (**http://kicad. sourceforge.net/wiki/index.php/Main_ Page**, for *Linux*, Mac OS X, and *Windows*, includes schematic capture) are free to use and have no artificial restrictions. Support is through user forums. Source code is available for the user to modify. *gschem* is a schematic capture sister program to *GNU PCB*.

2) Free, restricted use/restricted feature commercial: At least one company makes a version of their PCB and schematic software that is free to use for noncommercial purposes. Though it is restricted in number of layers (two) and maximum board size (4 × 3.2 inches), *Eagle PCB "Light Edition"* (**www.cadsoftusa.com**, for *Linux*, Mac OS X, *Windows*) is very popular among hobbyists. Files can be shared with others; the resulting industry standard files can be sent to nearly any PCB manufacturer. *Eagle* also contains a schematic entry program.

3) Free, restricted output commercial: Several PCB manufacturers offer schematic and PCB software with a proprietary output format tied to their PCB manufacturing service. *PCB123* from Sunstone Circuits (**www. sunstone.com**, for *Windows*) is one such offering, including schematic capture and layout software with up to four layers and board sizes up to 12 × 18 inches (double sided). For an additional fee (per design), industry standard files can be exported. Schematic entry is included. *Express PCB* (**www.expresspcb. com**, for *Windows*) also provides schematic capture and PCB layout capability, tied to the Express PCB board fabrication services, including the fixed size (3.8 × 2.5 inches) Miniboard service. Advanced Circuit's proprietary *PCB Artist* software (**www.4pcb. com**, for *Windows*) includes the ability to import netlists.

4. Low cost commercial: Eagle and many other companies offer PCB and schematic software at a range of prices from $50 to many thousands of dollars. Several versions are typically offered from each company, usually based on limitations on board size, schematic size/complexity and features such as autorouting. Schematic entry may be included in some packages, or be a separate purchase.

PCB design software manuals and tutorials discuss the basic operation but also special keystrokes and other shortcuts that make operations such as routing traces much more efficient.

The first time designing and ordering a PCB can be daunting, so keep the initial job simple and pay attention to details (and read the instructions). When starting to use a specific software package, join a user's support group or forum if one is available. Request sample designs from other users and experiment with them to see how they are constructed and what files are required in the output data set. Once you are comfortable with the tools, you can begin on a design of your own.

6.4.3 Schematic Capture

The first step in PCB design is to create a schematic. It is possible to design a layout directly from a paper schematic, but it is much easier if the schematic is entered (or "captured") in electronic form. Schematic capture software has two outputs — the visual schematic and the component and connectivity data for subsequent PCB layout. These two separate requirements can make some operations during schematic entry more complicated than what would seem at first glance necessary. Bear in mind however, that the user is creating not only a clear graphic representation of the circuit, but of the underlying electrical connectivity.

Schematics are generally entered on a (virtual) page usually corresponding to common paper sizes — for example, 11 × 17 inches. More complicated schematics can span multiple pages, using special labels or components to indicate both visual and electrical connectivity. Often one can group logically related elements into a module that can then be referenced as a "black box" on a higher level schematic. For complex circuits, these features are extremely useful and make the difference between a jumbled diagram that is difficult to use and an organized, compact diagram that efficiently communicates the function and operation of the circuit.

COMPONENTS

The components (resistors, capacitors, etc) on a schematic are either selected from an existing library or created by the user and stored in a custom library. It is also possible to find components and/or additional libraries on the Internet, although each program has its own specific format.

Each component includes a great deal more than shape and pin numbers. A typical component library entry includes:

Symbol — This is the graphic representation shown on the schematic. Many components may have the same symbol (eg, the op amp symbol may be shared by many different types of op amps)

Pins — For each pin or point of electrical connection, the component model may specify the pin number, label (eg, "V_{DD}"), pin type (inverting, noninverting) or pin functions (common).

PCB footprint — A given component may be available in a number of different packages (eg, DIP or surface mount). Many components may have the same physical footprint (eg, op amps, comparators and optoisolators could all map to the same eight-pin DIP footprint). Footprints include the electrical connections (pins) as well as mechanical mounting holes and pad sizes, and the component outline.

Value — Many components such as resistors and capacitors will have identical information except for a difference in value. All ¼ W resistors may be instances of the same component, differing only in value and designator.

Designator — The unique reference to the component, such as R1, C7, D3. This is assigned when the component is used (often automatically and in sequence).

Source information — Part number, vendor, cost, etc. This information is for the Bill of Materials.

Components are typically placed on the schematic by opening a library and searching for the desired component. It may be tempting for the beginner to select a component that looks "about right" when faced by a long list of components in some libraries. However, even at this early stage, the physical PCB often must be taken into account. For example, either "1/8W Resistor, Axial" or "1W Resistor, Upright" will result in the same neatly drawn resistor symbol on the schematic but in the subsequent step of using the component data to create a PCB, the footprints will be dramatically different.

It is not at all uncommon to add new components to the library in the course of creating a schematic. Since many components are closely related to existing devices, the process often consists of selecting an existing schematic symbol, editing the shape and/or component data, creating a new label, and associating the part with an existing footprint. Adding a specific type of op amp is an example. This usually only needs to be done once since symbols can be saved in a personal library (and shared with others). It is usually easier to modify a part that is close to what is desired than to "build" a new part from scratch.

Component symbols can generally be rotated and flipped when placing the component instance on the schematic. Designators (R1, T34, etc) can be assigned and modified by the user although the default designators are usually selected sequentially.

CONNECTIONS

The schematic software will have a mode for making electrical connections, called "nets." For example, one might click on the "draw net" symbol then draw a line using the mouse from one pin to another pin, using intermediate mouse clicks to route the line neatly with 90° turns on a uniform grid. Internally, the software must not only draw the visual line, but recognize what electrical connectivity that connection represents. So one must click (exactly) on a component pin to start or end a line or when making a connection between two lines that intersect, explicitly indicate a net-to-net connection (often with a special "dot" component). The connections on a schematic can often be assigned additional information, such as the desired width of the trace for this connection on the PCB or a name assigned by the user, such as "input signal."

Not all connections on a schematic are drawn. To make any schematic — electronic or hand drawn — more readable, conventions are often employed such as ground or power symbols or grouping similar connections into busses. Schematic capture software often supports these conventions. In some cases, components may be created with implicit power connections; in these cases the connections may not even be noted on the schematic but will be exported to the PCB software. However, as a general rule, software aimed at beginning PCB designers will not require the use of these advanced features.

Since it is often possible for component pins to be assigned attributes such as "power input", "output," "input," and so on, some schematic entry programs allow one to do an early design check. The program can then flag connections between two outputs, inputs that are missing connections, and so on. This is not nearly as helpful or complete as the Design Rule Check discussed below.

Free text can be placed on the schematic and there will be a text block in a corner for date, designer, version, title and the other information that identifies the schematic.

NETLISTS

Once the components are placed and connections made, the schematic may be printed and any output files for the PCB layout software produced. The connectivity and component information needed for PCB layout is captured in a *netlist* file. The flow from schematic entry to PCB may be tightly integrated, in which case the user may switch between schematic and PCB like two views of the same design (which they are). However, most schematic software will generate a separate netlist to be used by PCB layout software, whether integrated or a separate program. The netlist can also be exported to an external circuit simulation program or be used by an integrated simulator program. (See the first part of this chapter for more information on circuit simulation.)

Netlists are often human readable text files and in most cases it is possible to create a netlist file manually. In the absence of a schematic entry program, this allows the user to take a hand drawn schematic, extract the connectivity information, and create the netlist for the PCB program to perform design rule checks. However, a netlist is generally not required for the PCB layout software; the user will also have the option to create a PCB on-the-fly, adding components and connections as they wish.

ANNOTATION AND BILL OF MATERIALS

The important features of *forward* and *backward annotation* enter at the interface between schematic entry and PCB layout. It is not uncommon during the PCB layout process to either come across some design deficiency or realize that a change to the schematic could produce a design that would be easier to lay out. Likewise, a review of the schematic partway through the PCB layout process could reveal some needed design change. In the case of changes to the PCB (perhaps changing some pins on a connector to make routing easier), back annotation can propagate the changes "backward" to the schematic. The connectivity data will be updated; however the user may need to manually route the connection lines to neaten up the schematic. Likewise, changes to the schematic when the PCB is already (partially) routed are known as forward annotations and like the schematic, while the connectivity is updated the user will likely need to manually route the traces. Neither forward nor back annotation is necessary, but is useful in keeping the schematic and PCB consistent. In their absence, the user is strongly urged to keep the schematic and PCB up to date manually to avoid time consuming problems later on.

Finally, the underlying data in the schematic can be used to produce a *Bill of Materials* (BOM). A BOM lists all the components of the schematic, typically ordered by reference designator(s), and may even be exportable for online ordering.

6.4.4 PCB Characteristics
PCB CONSTRUCTION

It is useful to know a little bit about PCB construction in order to make sense of the PCB design process. **Fig 6.19** shows the basic structure of a PCB and some of its design elements (discussed in later sections). The lami-

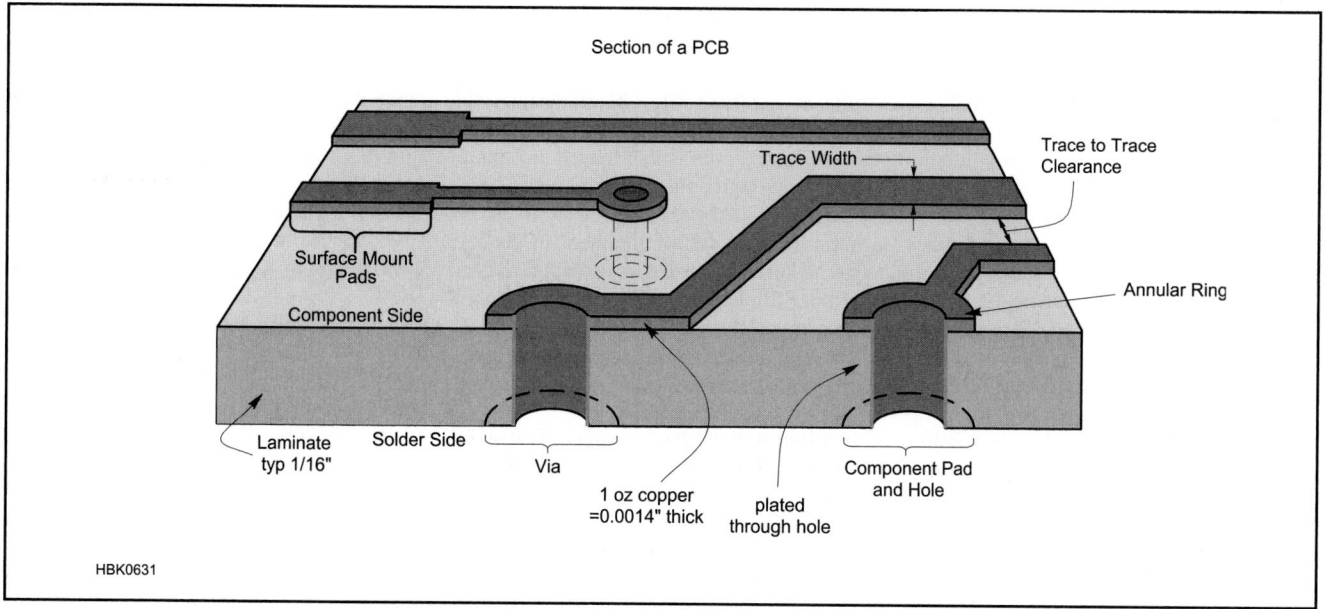

Fig 6.19 — The various elements of PCB construction and specification.

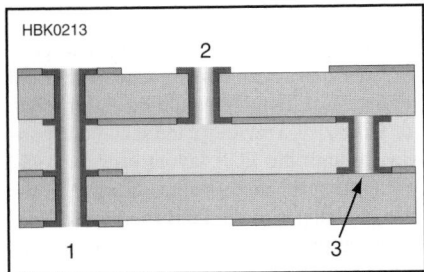

Fig 6.20 — *Via* refers to a plated-through hole that connects one board layer to another. Vias are used for signal, power, or ground connections and even for ventilation. Different via types include through-hole (1); blind (2), and buried (3).

nate material provides a stable, insulating substrate with other known characteristics (thermal, dielectric, etc). Copper is bonded to one or both sides and selectively removed (usually chemically) to leave traces and pads. The pads provide points of connection for components. Though electrical connectivity is crucial, it is important to remember that the solder and pads provide mechanical and thermal connectivity as well. Pads may be drilled for mounting *through-hole* components or left undrilled for *surface-mount* components.

A separate electrochemical process plates the inside surface of *plated-through holes* to provide connectivity between upper and lower pads. Plated-through holes whose sole purpose is to provide electrical connectivity between layers of a PCB are known as *vias*, shown in **Fig 6.20**. Since they do not need to accommodate a component lead, their hole and pad size are smaller.

While two layer boards can mount components on either side, most PCBs will have a primary side called the *component side* upon which most of the components will be placed, and a *solder side* dominated by soldered pins and traces. Where high density is required, surface mount (and sometimes through-hole) devices are mounted on both sides, but this is considerably more complex.

Multi-layer boards are essentially a stack of two or more two-layer boards, with an insulating layer between each board. Plated-through holes make connections possible on every layer, and the laminate material is proportionally thinner so the entire multi-layer board is roughly the same thickness as a regular two-layer board. Vias that join selected, adjacent copper layers without connecting the entire stack of layers are called "buried" or "blind" vias and are typically only needed for very dense designs. Multi-layer boards provide much more flexibility in routing signals and some other benefits such as dedicated layers for power distribu-

tion and grounding, but at often substantial additional cost.

PCB MANUFACTURING SPECIFICATIONS

Unless the board is manufactured by the hobbyist, the PCB files are sent out to be manufactured by a *board house*. The most important issue for the amateur may be the pricing policies of the board house. Board size, quantity, delivery time, number of layers, number and/or density of holes, presence of solder masks and silk screens, minimum trace/separation width, type of board material, and thickness of copper will all influence pricing. One cost saving option of the past, a single-layer board, may not be offered with low-cost, low-volume services — two layers may be the simplest option and it results in a more robust board. [Note that most ordering specifications use English units of inches and ounces. Offshore board houses may use both English and metric units, or be metric-only. English units are used here because they are the most common encountered by hobbyists. — *Ed.*]

The second issue to consider is manufacturing capabilities and ordering options. These will vary with pricing and delivery times, but include the following:

Board material and thickness — FR-4 is the most popular board material for low volume PCBs; it consists of flame-resistant woven fiberglass with epoxy binder. Typical thickness is 0.062 inch ($\frac{1}{16}$ inch), but thinner material is sometimes available. Flexible laminates are also available at greater cost and longer delivery time. Special board laminates for microwave use or high-temperature applications are also available.

Copper thickness — Expressed in ounces per square foot, typical values are 1-2 oz (1 oz corresponds to 0.0014 inch of thickness.) Other values may not be available inexpensively for small volumes. Inner layers on multi-layer boards may be thinner — check if this is important. Most board designs can assume at least 1 oz copper for double-sided boards; trace width is then varied to accommodate any high current requirements.

Layers — Two-layer boards are the most common. Because of the way PCBs are manufactured, the number of copper layers will be multiples of two. For quick-turn board houses, usually only two or four layer boards are available. PCBs with more than two layers will always be more expensive and often take longer to manufacture.

Minimum hole size, number, and density of holes — Minimum hole size will rarely be an issue, but unusual board designs with high hole density or many different hole sizes may incur additional costs. Be sure to include vias when specifying minimum hole size. Some board houses may have a specific list of drill sizes they support. Note that you can often

just edit the drill file to reduce the number of different drill sizes.

Minimum trace width and clearance — Often these two numbers are close in value. Most board houses are comfortable with traces at least 0.010 inch in width, but 0.008 and 0.006 inch are often available, sometimes at a higher cost.

Minimum annular ring — A minimum amount of copper is required around each plated-through hole, since the PCB manufacturing process has variations. This may be expressed as the ratio of the pad size to hole size, but more commonly as the width of the ring.

Edge clearances — Holes, pads, and traces may be too close to the edge of the board.

Board outline and copies — There may be options to route the outline of the board in other shapes than a rectangle, perhaps to accommodate a specific enclosure or optimize space. If multiple copies of a board are ordered, some board houses can *panelize* a PCB, duplicating it multiple times on a single larger PCB (with a reduction in cost per board). These copies may be cut apart at the board house or small tabs left to connect the boards so assembly of multiple boards can be done as a single unit.

Tin plating — Once the traces and pads have been etched and drilled, tin plating is usually applied to the exposed copper surfaces for good soldering.

Solder mask — This is a solder-resistant coating applied after tin plating to both sides of the board covering everything except the component mounting pads. It prevents molten solder from bridging the gaps between pads and traces. Solder mask is offered except by the quickest turn services. Green is the most common color, but other colors may be available.

Silkscreen — This is the ink layer, usually white, on top of the solder mask that lays out component shapes and designators and other symbols or text. A minimum line width may be specified — if not specified, try to avoid thin lines. All but the quickest turn services typically offer silk screening on one or both sides of the PCB.

6.4.5 PCB Design Elements

The schematic may not note the specific package of a part, nor the width or length of a connection. The PCB, being a physical object, is composed of specific instances of components (not just "a resistor," but a "$\frac{1}{4}$ W, axial-lead resistor mounted horizontally," for example) plus traces — connections between pins of components with a specific width and separation from other conductors. Before discussing the process of layout, we briefly discuss the nature of components and connections in a PCB.

COMPONENTS

A component in a PCB design is very similar to its counterpart on the schematic. **Fig 6.21** shows the PCB footprint of an opto-interrupter, including graphics and connectivity information. The footprint of a component needs to specify what the footprint is like on all applicable copper layers, any necessary holes including non-electrical mounting holes or slots, and any additional graphics such a silkscreen layer.

Take a common ¼ W axial-lead resistor as an example. This footprint will have two pins, each associated with a pad, corresponding to the resistor's two leads. This pad will appear on both the top and bottom layers of the board, but will also have a smaller pad associated with inner layers, should there be any. The hole's size will be based on the nominal lead diameter, plus some allowance (typically 0.006 inch). The pad size will be big enough to provide a reasonable annular ring, but is usually much larger so as to allow good quality soldering. The pins will be labeled in a way that corresponds to the pin numbering on the schematic symbol (even though for this component, there is no polarity). A silkscreen layer will be defined, usually a box within which the value or designator will appear. The silkscreen layer is particularly useful for indicating orientation of parts with polarity.

More complicated parts may require additional holes which will not be associated with a schematic pin (mounting hole, for example). These are usually added to the part differently than adding a hole with a pad — in this case, the hole is desired without any annular ring or plating. The silkscreen layer may be used to outline the part above and beyond what is obvious from the pads, for example, a TO-220 power transistor laying on its back, or the plastic packaging around the opto-interrupter in Fig 6.21.

As with schematic entry, it is not uncommon to have to modify or create a new PCB footprint. Good technical drawings are often available for electronic parts; when possible the user should verify these dimensions against a real part with an inexpensive dial caliper. It is also useful to print out the finished circuit board artwork at actual size and do a quick check against any new or unusual parts.

TRACES

Traces are the other main element of PCB construction — the copper pathways that connect components electrically. PCB traces are merely planar, flat wires — they have no magical properties when compared to an equivalent thickness of copper wire. At VHF/UHF/microwave frequencies and for high-speed digital signals, PCB traces act as transmission lines and these properties need to be accounted for, and can be used

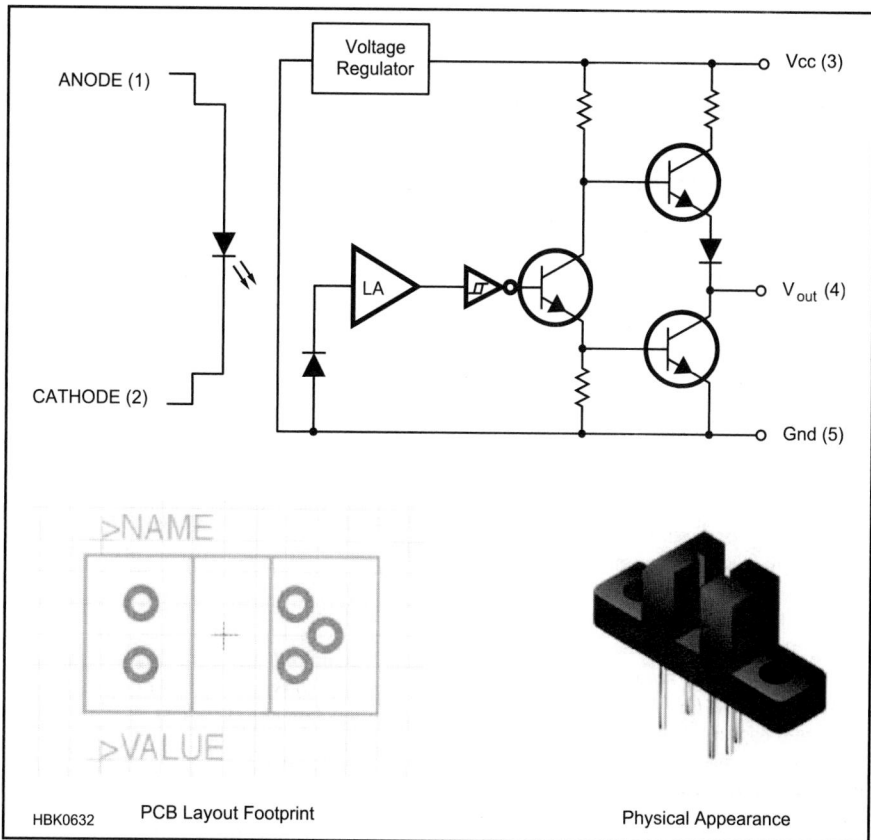

Fig 6.21 — The PCB footprint for a component, such as the opto-interrupter shown here, combines electrical connectivity as defined in the part's schematic and the part's physical attributes.

to advantage, in the design.

There are few constraints on traces apart from those such as minimum width and clearance imposed by the board house. They are created by chemical etching and can take arbitrary shapes. In fact, text and symbols may be created on copper layers which may be handy if a silkscreen is not included. Traces may be of any length, vary in width, incorporate turns or curves, and so on. However, most traces will be a uniform width their entire length (a width they will likely share with other traces carrying similar signals), make neat 45° or 90° corners, and on two-layer boards have a general preference for either horizontal travel on the component side or vertical travel on the solder side.

The same considerations when building a circuit in other methods applies to PCB design, including current capacity (width of trace, thickness of copper), voltage (clearance to other signals), noise (shielding, guarding, proximity to other signals), impedance of ground and power supplies, and so on.

6.4.6 PCB Layout

With a schematic and netlist ready and all of the PCB characteristics defined, the actual layout of the PCB can begin.

BOARD SIZE AND LAYERS

The first step in PCB layout is to create the board outline to contain not only the circuit itself and any additional features such as mounting holes. For prototype or one-off designs, the board is often best made a bit larger to allow more space between components for ease in testing and debugging. (Some low cost or freeware commercial PCB software imposes limits on board size and number of components.) The board outline may be provided in a default size that the user can modify, or the user may need to enter the outline from scratch.

As discussed above, rectangular board shapes are generally acceptable, but many board houses can accommodate more complex outlines, including curves. These outlines will be routed with reasonable accuracy and may save an assembly step if the PCB needs a cutout or odd shape to fit in a specific location.

While the software may not require deciding at the start how many layers the PCB will use, this is a decision the user should make as early as possible, since the jump from two to four or more will have a big impact on routing the traces as well as cost!

For your initial design, start with a two

layer board for a simple circuit that you have already built and tested. This will reduce the number of decisions you have to make and remove some of the unknowns from the design process.

COMPONENT PLACEMENT

Good component placement is more than half the battle of PCB layout. Poor placement will require complicated routing of traces and make assembly difficult, while good placement can lead to clean, easy-to-assemble designs.

The first elements placed should be mounting holes or other fixed location features. These are often placed using a special option selected from a palette of tools in the software rather than as parts from a library. Holes sufficient for a #4 or #6 screw are usually fine; be sure to leave room around them for the heads of the screws and nut driver or standoff below. These will be non-plated-through holes with no pad (though the board house may plate all the holes in a board, regardless).

Depending on the software and whether schematic capture was performed, the board outline may already contain the footprints of all the circuit components (sometimes stacked in a heap in one corner of the board) and the netlist will already be loaded. In this case, components may be placed by clicking and dragging the components to the desired location on the board. Most PCB programs have a "rat's nest" option that draws a straight line for each netlist connection of a component, and this is a great aid in placement as the connections between components are apparent as the components are moved around. (See **Fig 6.22**) However, connections are shown to the nearest pin sharing that electrical connection; thus, components such as decoupling capacitors (which are often meant to be near a specific component) will show rats nest connections to the nearest power and ground pins and not the pins the designer may have intended. These will have to be manually edited.

The PCB layout software may offer *autoplacement* in which the components are initially arranged automatically. The beginner should certainly feel free to experiment and see how well this tool performs, but it is likely not useful for the majority of designs.

PCBs need not be arranged to precisely mimic the schematic, but it is appropriate to place components in a logical flow when possible so as to minimize the length of traces in the signal path. Sensitive components may need to be isolated or shielded from other components, and grounding and decoupling attended to, just as one would do with a point-to-point soldered version.

If the PCB is being designed "on-the-fly" or using an imported netlist, components may need to be selected and placed on the

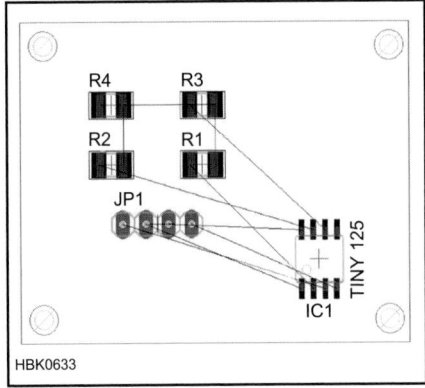

Fig 6.22 — The rat's nest view during PCB layout shows the direct connections between component pins. This helps the designer with component placement and orientation for the most convenient routing.

board manually using the libraries of parts in the PCB software. Not all design software makes this task simple or fast — in particular, the description of component footprints may be confusing. The use of highly condensed industry standard or non-uniform naming conventions often means the user needs to browse through the component library to see the different types of components. Resistors, diodes and capacitors seem particularly prone to a propagation of perplexing options. One solution is to open an example PCB layout and see what library element that designer used for resistors, LEDs and so on. Here, the PCB layout software directed at hobbyists may be superior in that there are fewer options than in professional programs.

During placement the user will find that different orientations of components simplify routing (for example, minimizing the number of traces that have to cross over each other or reducing the trace lengths). Components are generally rotated in increments of 90°, although free rotation may be an option. The user is strongly urged to maintain the same orientation on like devices as much as possible. Mixing the position of pin 1 on IC packages, or placing capacitors, diodes, and LEDs with random orientation invites time consuming problems during assembly and testing that can be minimized by consistent, logical layout.

Placement and orientation of components can also affect how easily the final PCB can be assembled. Allow plenty of space for sockets, for example, and for ICs to be inserted and removed. Components with a mechanical interface such as potentiometers and switches should be positioned to allow access for adjustment. Any components such as connectors, switches, or indicators (eg, LEDs) should be positioned carefully, especially if they are

to protrude through a panel. Often this will involve having the component overhang the edge of the PCB. (Beware of the required clearance between copper traces and pads to the edge of the PCB.)

Components should include a silkscreen outline that shows the size of the whole component — for example, a transistor in a TO-220 package mounted flat against the PCB should have an outline that shows the mounting hole and the extent of the mounting tab. The user should also consider the clearance required by any additional hardware for mounting a component, such as nuts and bolts or heatsinks — including clearance for nut drivers or other assembly tools.

Take care to minimize the mechanical stress on the PCB, since this can result in cracked traces, separated pads, or other problems. Utilize mounting holes or tabs when possible for components such as connectors, switches, pots. Use two-layer boards with plated-through holes even if the design can be single-layer. Component leads soldered to plated-through holes produce much stronger mechanical connections than single-layer boards in which the soldered pad is held only by the bond between copper and laminate and is easily lifted if too much heat or stress is imposed.

When prototyping a new design, add a few unconnected pads on the circuit board for extra components (eg, a 16 pin DIP, 0.4 inch spaced pads for resistors and other discrete components). Include test points and ground connections. These can be simply pads to which cut off leads can be soldered to provide convenient test points for ground clips or to monitor signals.

Wires or cables can be directly soldered to the PCB, but this is inconvenient when swapping out boards, and is not very robust. Connectors are much preferred when possible

PCB Design and EMC

While amateur projects are rarely subject to electromagnetic compatibility (EMC) standards, using good engineering practices when designing the board still reduces unwanted RF emissions and susceptibility to RF interference. For example, proper layout of a microprocessor circuit's power and ground traces can reduce RF emissions substantially. Proper application of ground planes, bypass capacitors and especially shield connections can have a dramatic effect on RFI performance. (See the **RF Interference** chapter for more on RFI.) A good reference on RFI and PCB design is *Electromagnetic Compatibility Engineering*, by Henry Ott, WA2IRQ.

and often provide strain relief for the wire or cable. However, if a wire is directly soldered to the PCB, the user should consider adding an unplated hole nearby just large enough to pass the wire including insulation. The wire can then be passed from the solder side through the unplated hole, then soldered into the regular plated-through hole. This provides some measure of strain relief which can be augmented with a dollop of glue if desired.

ROUTING TRACES

After placing components, mounting holes and other fixed location features that limit component or trace placement, traces can be *routed*. That is, to complete all the connections between pins without producing short circuits.

Most PCB design programs allow components and traces to be placed on a regular grid, similarly to drawing programs. There may be two grids — a visible coarse pitch grid, and a "snap" fine pitch grid, to which components and other objects will be aligned when placed. It is good practice to use a 0.1 or 0.050 inch grid for component placement and to route traces on a 0.025 inch grid. While the "snap to grid" feature can usually be turned off to allow fine adjustment of placement, a board routed on a grid is likely to look cleaner and be easier to route.

The trace starts at a component pin and wends its way to any other pins to which it should be connected. Traces should start and end at the center of pads, not at the edge of a pad, so that the connection is properly recorded in the program's database. If a netlist has been loaded, most PCB software will display a rat's nest line showing a direct connection between pads. Once the route is completed, the rat's nest line for that connection disappears. The rat's nest line is rarely the desired path for the trace and often not the correct destination. For example, when routing power traces, the user should use good design sense rather than blindly constructing a Byzantine route linking pins together in random order. For this reason, routing the power and ground early is a good practice.

High speed, high frequency, and low noise circuits will require additional care in routing. In general, traces connecting digital circuits such as microprocessors and memories should not cross or be in close proximity to traces carrying analog or RF signals. Please refer to the **RF Techniques** and **Construction Techniques** chapters of this *Handbook*, and the references listed at the end of this section.

Manual routing is a core skill of PCB design, whether or not auto-routing is used. The process is generally made as simple as possible in the software, since routing will take up most of the PCB design time. A trace will be routed on the copper layer currently selected. For a single-sided board, there is only

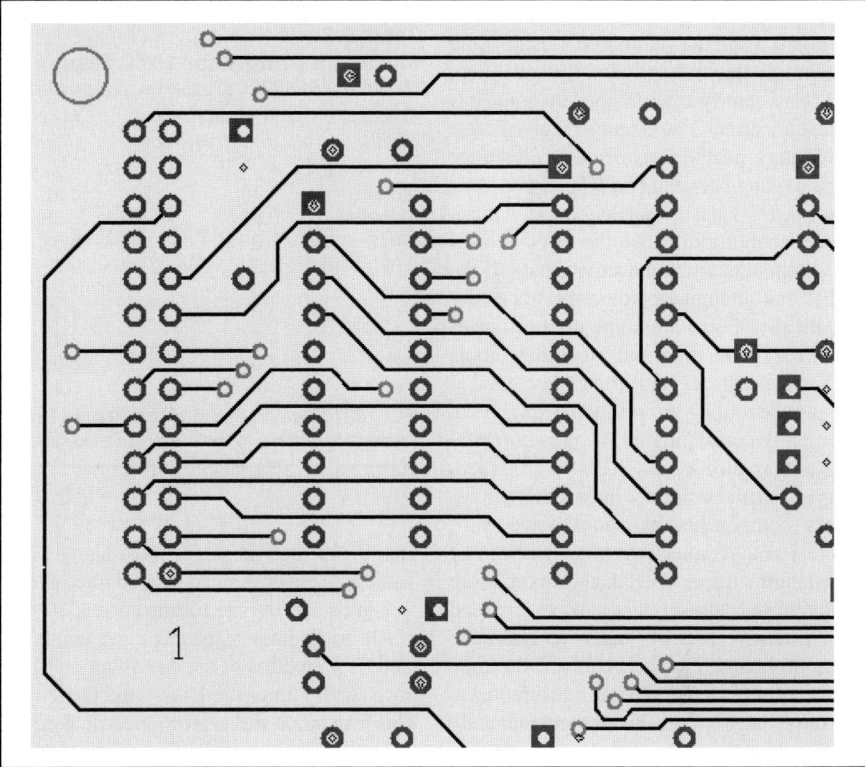

Fig 6.23 — This example shows traces on the side of the PCB for horizontal routing. Traces are routed between pins of ICs. The smaller pads are for vias to a different layer of the PCB.

one layer for routing; for a two-layer board the component side and solder side can have traces; and for multi-layer boards additional inner layers can have traces. Often a single keystroke can change the active copper layer (sometimes automatically inserting a via if a route is in progress). The trace is drawn in straight segments and ends at the destination pin. When routing, 90° corners are normally avoided — a pair of 45° angles is the norm. **Fig 6.23** shows some sample traces.

It is good practice on a double-sided board to have one side of the board laid out with mostly horizontal traces, and the other side laid out with mostly vertical traces. A trace that needs to travel primarily vertically can do so on the side with vertical traces and use a via to move to the other side to complete the horizontal part of the route.

It is easiest during testing and debugging to route most traces on the bottom (solder) side of the board — traces on the component side often run under ICs or other components, making them hard to access or follow. It is often much clearer to connect adjacent IC or connector pins by routing a trace that leaves one pad, moves away from the IC or connector, then heads back in to the adjacent pad to connect. This makes it clear the connection on the assembled board is not a solder bridge, which a direct connection between the two pads would resemble.

It may be the case that no amount of vias or wending paths can complete a route. The one remaining tool for the PCB designer is a jumper — a wire added as a component during assembly just for the purpose of making a connection between two points on the board. Jumpers are most often required for single-sided boards; when the "jump" is rather small, uninsulated wire can be used. Jumpers are usually straight lines, and can be horizontal or vertical. Professional production PCBs use machine-insertable zero-ohm resistors as jumpers. Jumpers on double-sided boards are usually not viewed very favorably, but this is an aesthetic and efficiency issue, not a functional one.

Multi-layer boards clearly offer additional routing options, but again having some dominant routing direction (vertical or horizontal) on each layer is recommended, since mixing directions tends to cause routing problems. However, it is not uncommon to devote one or two inner layers to power and ground, rather than merely be additional layers for routing signals. This allows power and ground to be routed with minimal resistance and exposes the traces carrying interesting signals on the component and solder sides where they are available to be probed or modified. It is very difficult to modify traces on inner layers, needless to say!

Before routing too many traces, it is helpful to run the *Design Rule Check (DRC)* on the board. (See the section on Design Rule Checking below.) Applied early and often, DRC can identify areas of concern when it is easiest to correct. For example, a given trace width may provide insufficient clearance when passing between two IC pads.

Some PCB design packages offer *auto-router* capability in which the software uses the component and connectivity data of the netlist and attempts to route the traces automatically. There are some circumstances when they save time, but view these tools with some caution. Auto-routers are good at solving the routing puzzle for a given board, but merely connecting all the pins correctly does not produce a good PCB design. Traces carrying critical signals may take "noisy" routes; components that should have short, low resistance connections to each other may have lengthy traces instead, and so on. More sophisticated auto-routers can be provided with extensive lists of "hints" to minimize these problems. For the beginner, the time spent conveying this design information to the auto-router is likely better spent manually routing the traces.

If an auto-router is used, at a minimum, critical connections should be first routed manually. These include sensitive signals, connections whose length should be minimized, and often power and ground (for both RFI and trace width reasons). Better still is to develop a sense of what a good layout looks like (which will come with practice and analyzing well designed boards), and learn at what stage the auto-router can be "turned loose" to finish the routing puzzle.

TRACE WIDTH AND SPACING

All traces will have some width — the width may be the default width, the last width selected, or a width provided from data in the netlist. It may be tempting to route all but the power traces using the smallest trace width available from the board house (0.008 inch or smaller), since this allows the highest density of traces and eases routing. A better design practice is to use wider traces to avoid hard-to-detect trace cracking and improve board reliability. The more common traces 0.012 inch wide can be run in parallel on a 0.025 inch grid and can pass between many pads on 0.1 inch centers. Even wider traces will make the board easier to produce "in house," though the exact process used (CNC routing, chemical etching, etc) will limit the resolution. Note that it is possible to "neck down" traces where they pass between IC or connector pads — that is, the regular, thick trace is run up close to the narrow gap between the pads, passes between the pads with a narrow width, then expands back to the original width. There is little reason to use traces wider

Table 6.3
Maximum Current for 10 °C Rise, 1 oz/ft² Copper
Based on IPC-2221 standards (not an official IPC table)

Trace Width (inches)	Max. Current (External Trace) (A)	Max. Current (Internal Trace) (A)	Resistance (ohms/inch)
0.004	0.46	0.23	0.13
0.008	0.75	0.38	0.063
0.012	1.0	0.51	0.042
0.020	1.5	0.73	0.025
0.040	2.4	1.2	0.013
0.050	2.8	1.4	0.010
0.100	4.7	2.4	0.0051
0.200	7.8	3.9	0.0025
0.400	13	6.4	0.0013

IPC-2221 Generic Standard on Printed Circuit Design, Institute for Interconnecting and Packaging Electronic Circuits, **www.ipc.org**

than 0.030 inch or so for most signals (see **Table 6.3**) but power and ground trace widths should be appropriate for the current.

All traces have resistance, and this resistance is a function of the cross section of the trace (width times thickness) and the length. This resistance will convert electrical power to heat. If the heat exceeds a relatively high threshold, the trace becomes a fairly expensive and difficult-to-replace fuse. The trace width should be selected such that for the worst case expected current, heat rise is limited to some threshold, often 10 °C. In practice, power traces (especially grounds) are often made as wide as practical to reduce resistance, and they greatly exceed the width required by heat rise limits alone.

Table 6.3 summarizes maximum currents for external (component and solder side) and internal traces for some common trace widths. Internal traces (on inner layers of multi-layer boards) can carry only about half the current of external traces for the same width since the internal layers do not dissipate heat to the ambient air like external traces can. (Note that trace widths are also sometimes expressed in "mils." 1 mil = 0.001 inch; it is shorthand for "milli-inch", not millimeter!)

There is no upper bound on the effective trace width. It is common to have large areas of the board left as solid copper. These *copper fill* areas can serve as grounds, heat sinks, or may just simplify board production (especially homemade boards). It is not a good idea to place a component hole in the middle of a copper fill — the copper is a very efficient heat sink when soldering. Instead, a "wagon wheel" pattern known as a thermal relief is placed (sometimes automatically) around the solder pad, providing good electrical connectivity but reducing the heat sinking. Often, copper fill areas can be specified using a polygon and the fill will automatically flow around pads and traces in that area, but can lead to isolated pads of copper.

In practice, most boards will have only two or perhaps three different trace widths; narrow widths for signals, and a thicker width for power (usually with a healthy margin).

One final note on trace width — vias are typically one size (ie, small), but multiple vias can be used to create low resistance connections between layers. Spacing the vias so their pads do not touch works well; the pads are then shorted on both top and bottom layers.

Voltage also figures into the routing equation, but instead of trace width, higher voltages should be met with an increased clearance between the trace and other copper. The IPC-2221 standard calls for a clearance of 0.024 inch for traces carrying 31-150 V (peak) and 0.050 inch for traces carrying 151-300 V (peak); these are external traces with no coating. (With the appropriate polymer solder mask coating, the clearances are 0.006 inch for 31-100 V and 0.016 inch for 101-300 V. Internal traces also have reduced clearance requirements.) Fully addressing the safety (and regulatory) issues around high voltage wiring is outside the scope of this brief review, however, and the reader is urged to consult UL or IPC standards.

SILKSCREEN AND SOLDER MASK

The silkscreen (or "silk") layer contains the text and graphics that will be silkscreened on the top of the board, shown in **Fig 6.24**. Components will generally have elements on the silkscreen layer that will automatically appear, such as designators and values, but other elements must be created and placed manually. Common silkscreen elements include: Circuit name, date, version, designer (and call sign), company name, power requirements (voltage, current, and fusing), labels for connections (eg, "Mic input"), warnings and cautions, labels for adjustments and switches. A solid white rectangle on the silkscreen layer can provide a good space to write a serial number or test information.

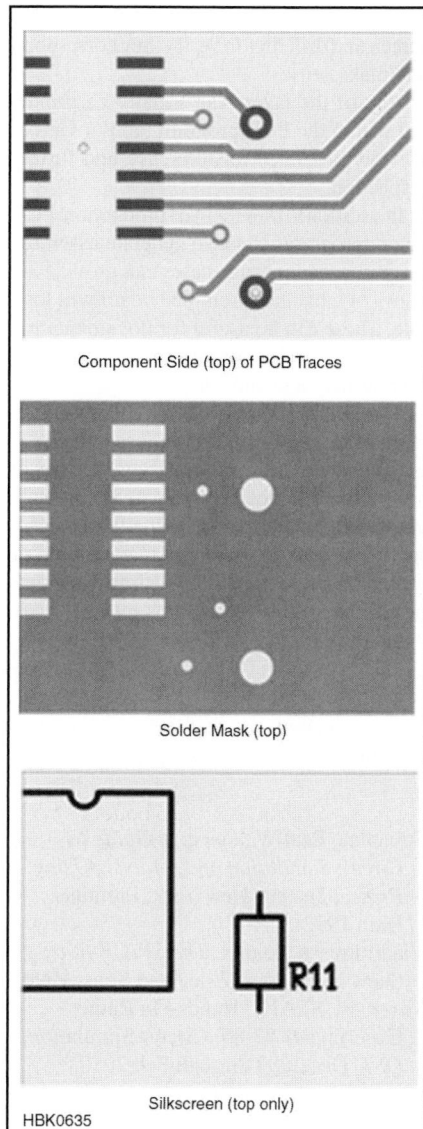

Component Side (top) of PCB Traces

Solder Mask (top)

Silkscreen (top only)

HBK0635

Fig 6.24 — The relationship between the layout's top copper layer with traces and pads, the solder mask that covers the copper (a separate solder mask is required for the top and bottom layers of the PCB) and the silkscreen information that shows component outlines and designators.

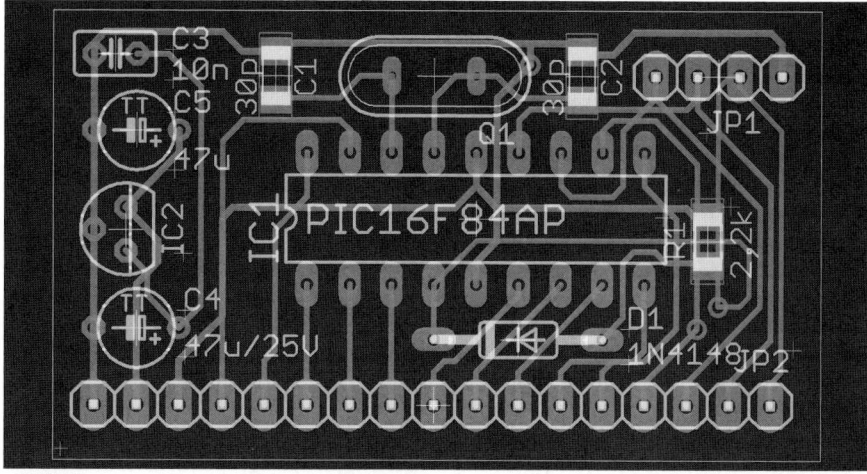

HBK0243

Fig 6.25 — A completed microprocessor board as it is seen in a typical PCB layout editor (*Eagle*). Solder mask layers are omitted for visibility. Traces that appear to cross each other are on different sides of the board and are in different colors in the layout software. The silkscreen layer is shown in white.

The board house will specify the minimum width for silkscreen lines, including the width of text. Text and graphics can be placed anywhere on the solder mask, but not on solder pads and holes.

Many quick-turn board houses omit the silkscreen for prototype boards. As noted earlier, many of the text elements above can be placed on the external copper layers. Component outlines are not possible since the resulting copper would short out traces, but component polarization can be noted with symbols such as a hand-made "+" made from two short traces, or a "1" from a single short trace. (Note that some component footprints follow a practice of marking the pad for pin 1 with a square pad while others are round or oval.)

The solder mask is a polymer coating that is screened onto the board before the silkscreen graphics. As shown in Fig 6.24, it covers the entire surface of the board except for pads and vias. Solder masking prevents solder bridges between pads and from pads to traces during assembly and is particularly important for production processes that use wave soldering or reflow soldering. There is one solder mask layer for the top layer and another for the bottom layer. Internal layers do not need a solder mask. Solder masking may be omitted for a prototype board, but care must be taken to keep solder from creating unwanted bridges or short circuits.

During the PCB layout process, solder mask layers are generally not shown because they do not affect connectivity. **Fig 6.25** shows a typical PCB as it appears when the PCB layout process is complete.

DESIGN RULE CHECK

If a netlist has been provided from the schematic capture program, a *design rule check* (*DRC*) can be made of the board's layout. The PCB software will apply a list of rules to the PCB, verifying that all the connections in the netlist are made, that there is sufficient clearance between all the traces, and so on. These rules can be modified based on the specific board house requirements. As stated above, it is useful to run the DRC even before all the traces have been routed — this can identify clearance or other issues that might require substantial re-routing or a different approach.

If the user has waited until all the routing is done before running the DRC, the list of violations can be daunting. However, it is often the case that many if not all of the violations represent issues that may prevent the board from operating as wished. Whenever possible, all DRC violations should be rectified before fabrication.

6.4.7 Preparation for Fabrication
LAYOUT REVIEW

Once the board has passed DRC, the electrical connectivity and basic requirements for manufacturability have likely been satisfied. However, the design may benefit from an additional review pass. Turn off all the layers but one copper layer and examine the traces — often simplifications in routing will be apparent without the distractions of the other copper layers. For example, a trace can be moved to avoid going between two closely spaced pins. Densely spaced traces could be spaced farther apart. There may be opportunities to reduce vias by routing traces primarily on one layer even if that now means both vertical and horizontal travel. Repeat the exercise for all the copper layers.

Review the mechanical aspects of the board as well, including the proximity of traces to hardware. If your prototype PCB does not have a solder mask, traces that run underneath components such as crystals in a conductive case or too close to mounting hardware can form a short circuit. An insulator must be provided or the trace can be re-routed.

GENERATING OUTPUT FILES

Once the PCB design is complete, the complete set of design description files can be generated for producing the PCB. These are:

Copper layers — One file per copper layer. These are known as *Gerber files* and were text files of commands originally intended to drive a *photoplotter*. Gerber was the primary manufacturer of photoplotters, machines that moved a light source of variable width (apertures) from one location to another to draw patterns on photographic film. While photoplotters have been replaced by digital technology, the format used by Gerber has been standardized as RS-274X and is universally used except by PCB software tied to a specific manufacturer. RS-274X is related to RS-274D ("G-Code") used by machinists to program CNC machinery but is an *additive* description (essentially saying "put copper here"), rather than describing the movements of a tool to remove material. A program is thus required to translate between Gerber and G-Code if a CNC machine is used to make a PCB by mechanically removing copper.

Drill file — The file containing the coordinates and drill sizes for all the holes, plated or not. Also called the *NC* or *Excellon* file, some board houses may require a specific format for the coordinates, but these are usually available to be set as options in the PCB program. There is only one drill file for a PCB, since the holes are drilled from one side. (Exotic options such as buried vias will require more information.) Like RS-274X apertures, the drill file will generally contain a drill table.

Silkscreen — Also in RS-274 format. Some board houses can provide silkscreen on both sides of the board, which will require two files.

Solder mask — The solder mask file is used by the board house to create the solder mask. One file per side is required.

A Gerber preview program such as *Gerbv* (**gerbv.gpleda.org**, open source, *Linux*, Mac OS X) or *GC-Prevue* (**www.graphicode.com**, *Windows*) can be used to review the trace layout Gerber files. This is a good test — the board house will make the boards from the Gerber files, not the PCB design file. Gerber previewers can import the copper layers, silkscreen and drill files to verify they correspond and make sense.

Any of the layers can usually be printed out within the PCB program (and/or Gerber preview program) for reference and further inspection.

In addition to files for PCB production, PCB layout programs can also generate assembly diagrams, and in some cases can provide 3D views of what the assembled board will look like. These can be useful for documentation as well as verification of mechanical issues such as height clearance.

Sending the files to the PCB manufacturer or board house and ordering PCBs is explained on the manufacturing website or a customer service representative can walk you through the process. Some firms accept sets of files on CD-ROM and may also offer a design review service for first-time customers or on a fee basis.

6.5 References and Bibliography

SIMULATION REFERENCES

Allen, J. Wayde "Gain Characterization of the RF Measurement Path," *NTIA Report TR-04-410* (Washington: US Department of Commerce: 2004). Available as **www.its.bldrdoc.gov/pub/ntia-rpt/04-410/04-410.pdf**.

Hayward, W. W7ZOI; R. Campbell, KK7B; and B. Larkin, W7PUA, *Experimental Methods in RF Design*, 2nd ed (ARRL: Newington, 2009).

P. Horowitz and W. Hill, *The Art of Electronics*, 2nd ed (Cambridge: Cambridge University Press, 1989).

Kundert, K., "Introduction to RF Simulation and its Application," **http://icslwebs.ee.ucla.edu/dejan/researchwiki/images/3/30/Rf-sim.pdf**

NXP Semiconductor: **www.nxp.com/models/bi_models/mextram/**

Newkirk, D. WJ1Z, "Math in a Box, Transistor Modeling, and a New Meeting Place," Exploring RF, *QST*, May 1995, pp 90-92.

Newkirk, D. WJ1Z, "Transistor Modeling with ARRL Radio Designer, Part 2: Optimization Produces Realistic Transistor Simulations," Exploring RF, *QST*, Jul 1995, pp 79-81.

Newkirk, D. WJ1Z, "ARRL Radio Designer as a Learning (and Just Plain Snooping) Tool," Exploring RF, *QST*, Nov 1995, pp 89-91.

Newkirk, D. WJ1Z, "An ARRL Radio Designer Voltage Probe Mystery: The 3-dB Pad that Loses 9 dB," Exploring RF, *QST*, Jan 1996, pp 79-80.

Newkirk, D. WJ1Z, "ARRL Radio Designer versus Oscillators, Part 1," Exploring RF, *QST*, Jul 1996, pp 68-69.

Newkirk, D. WJ1Z, "ARRL Radio Designer Versus Oscillators, Part 2," Exploring RF, *QST*, Sep 1996, pp 79-80.

Newkirk, D. W9VES, "Simulating Circuits and Systems with *Serenade SV*," *QST*, Jan 2001, pp 37-43.

"The Spice Page," (**http://bwrcs.eecs.berkeley.edu/Classes/IcBook/SPICE/**). The home page for *SPICE*.

Tuinenga, Paul W., *Spice: A Guide to Circuit Simulation and Analysis Using Pspice*, 2nd ed (New York: Prentice-Hall, 1992).

Vladimirescu, Andrei, *The SPICE Book* (New York: John Wiley and Sons, 1994).

Silver, W. NØAX, "Hands-On Radio Experiments 83-85: Circuit Simulation," *QST*, Dec 2009 through Feb 2010.

PCB CAD REFERENCES

Analog Devices, *High Speed Design Techniques*, Analog Devices, 1996.

Johnson, Howard, and Graham, Martin, *High Speed Digital Design: A Handbook of Black Magic*, Prentice Hall, 1993.

Ott, Henry, *Electromagnetic Compatibility Engineering*, Wiley Press, 2009.

Pease, Robert A, *Troubleshooting Analog Circuits*, Butterworth-Heinemann, 1991.

Silver, W. NØAX, "Hands-On Radio Experiments 107-110: PCB Layout," *QST*, Dec 2011 through Mar 2010.

Contents

Power Sources

Our transceivers, amplifiers, accessories, computers and test equipment all require power to operate. This chapter illustrates the various techniques, components and systems used to provide power at the voltage and current levels our equipment needs. Topics range from basic transformers, rectifiers and filters to linear voltage regulation, switchmode power conversion, high voltage techniques and batteries. Material on switchmode conversion was contributed by Rudy Severns, N6LF and Chuck Mullett, KR6R. A new section on batteries was contributed by Isidor Buchmann from his book *Batteries in a Portable World*. Alan Applegate, KØBG contributed the section on selecting batteries for mobile use.

Acknowledging the changing nature of amateur power requirements, the title of this chapter has updated from the traditional *Power Supplies* to *Power Sources*. More mobile and portable operation relies on power from batteries, for example. Hybrids of ac and dc power sources are becoming more common, blurring what has traditionally been known as a "power supply." In response, the scope of this chapter now includes more of the changing power environment in the amateur station. (Generators are covered in the Portable Installations section of the chapter on **Building a Station**.)

7.1 Power Processing

Fig 7.1 illustrates the concept of a power processing unit inserted between the energy source and the electronic equipment or load. The *power processor* is often referred to as the *power supply*. That's a bit misleading in that the energy "supply" actually comes from some external source (battery, utility power and so forth), which is then converted to useful forms by the power processor. Be that as it may, in practice the terms "power supply" and "power processor" are used interchangeably.

The real world is even more arbitrary. Power processors are frequently referred to as *power converters* or simply as *converters*, and we will see other terms used later in this chapter. It is usually obvious from the context of the discussion what is meant and the

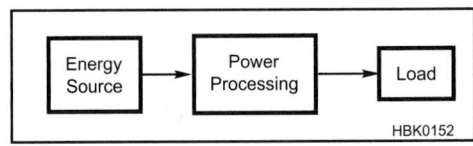

Fig 7.1 — Basic concept of power processing.

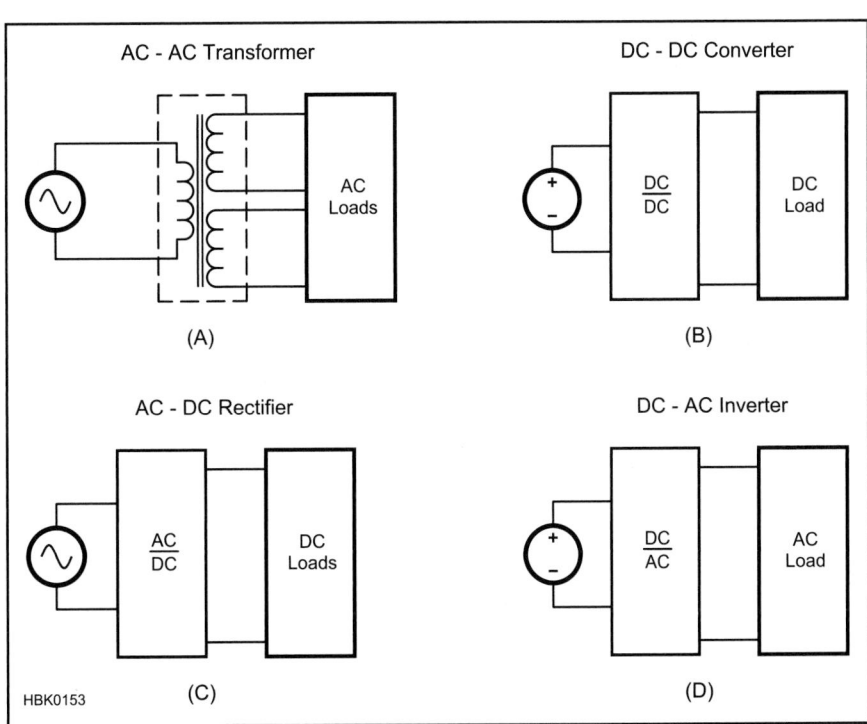

Fig 7.2 — Four power processing schemes: ac-ac, dc-dc, ac-dc and dc-ac.

glossary at the end of this chapter gives some additional information.

Power conversion schemes can take the form of: ac-to-ac (usually written ac-ac), ac-dc, dc-ac and dc-dc. Examples of these schemes are given in **Fig 7.2**. Specific names may be given to each scheme: ac-dc => rectifier, dc-dc => converter and dc-ac => inverter.

These are the generally recognized terms but you will see exceptions.

Power conversion normally includes voltage and current regulation functions. For example, the voltage of a vehicle battery may vary from 14 V when being charged down to 10 V or less when discharged. A converter and regulator are required to maintain adequate

voltage to mobile equipment at both over- and under-voltage conditions. Commercial utility power may vary from 90 to 270 V ac depending on where you are in the world. AC power converters are frequently required to handle that entire voltage range while still providing tightly regulated dc power.

7.2 AC-AC Power Conversion

In most US residences, three wires are brought in from the outside electrical-service mains to the house distribution panel. In this three-wire system, one wire is neutral and should be at earth ground potential. (See the **Safety** chapter for information on electrical safety.) The neutral connection to a ground rod or electrode is usually made at the distribution panel. The voltage between the other two wires is 60-Hz ac with a potential difference of approximately 240 V RMS. Half of this voltage appears between each of these wires and the neutral, as indicated in **Fig 7.3A**. In systems of this type, the 120 V household loads are divided at the breaker panel as evenly as possible between the two sides of the power mains. Heavy appliances such as electric stoves, water heaters, central air conditioners and so forth, are designed for 240 V operation and are connected across the two ungrounded wires.

Both hot wires for 240 V circuits and the single hot wire for 120 V circuits should be protected by either a fuse or breaker. A fuse or breaker or any kind of switch should *never* be used in the neutral wire. Opening the neutral wire does not disconnect the equipment from an active or "hot" line, possibly creating a potential shock hazard between that line and earth ground.

Another word of caution should be given at this point. Since one side of the ac line is grounded (through the green or bare wire — the standard household wiring color code) to earth, all communications equipment should be reliably connected to the ac-line ground through a heavy ground braid or bus wire of #14 or heavier-gauge wire. This wire must be a separate conductor. You must not use the power-wiring neutral conductor for this safety ground. (A properly-wired 120 V outlet with a ground terminal uses one wire for the ac hot connection, one wire for the ac neutral connection and a third wire for the safety ground connection.) This provides a measure of safety for the operator in the event of accidental short or leakage of one side of the ac line to the chassis.

Remember that the antenna system is frequently bypassed to the chassis via an RF choke or tuned circuit, which could make the antenna electrically "live" with respect

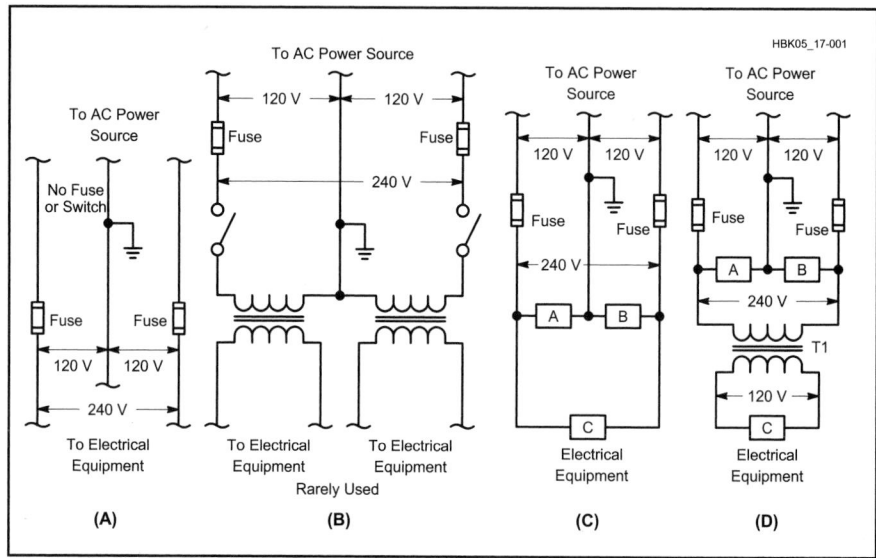

Fig 7.3 — Three-wire power-line circuits. At A, normal three-wire-line termination. No fuse should be used in the grounded (neutral) line. The ground symbol is the power company's ground, not yours! Do not connect anything other than power return wiring, including the equipment chassis, to the power neutral wire. At B, the "hot" lines each have a switch, but a switch in the neutral line would not remove voltage from either side of the line and should never be used. At C, connections for both 120 and 240 V transformers. At D, operating a 120 V plate transformer from the 240 V line to avoid light blinking. T1 is a 2:1 step-down transformer.

to the earth ground and create a potentially lethal shock hazard. A *ground fault circuit interrupter* (GFCI or GFI) is also desirable for safety reasons, and should be a part of the shack's electrical power wiring.

7.2.1 Fuses and Circuit Breakers

All transformer primary circuits should be fused properly and multiple secondary outputs should also be individually fused. To determine the approximate current rating of the fuse or circuit breaker on the line side of a power supply it is necessary to determine the total load power. This can be done by multiplying each current (in amperes) being drawn by the load or appliance, by the voltage at which the current is being drawn. In the case of linear regulated power supplies, this voltage has to be the voltage appearing at the output of the rectifiers before being applied to the regulator stage. Include the current drawn by bleeder

resistors and voltage dividers. Also include filament power if the transformer is supplying vacuum tube filaments. The National Electrical Code (NEC) also specifies maximum fuse ratings based on the wire sizes used in the transformer and connections.

After multiplying the various voltages and currents, add the individual products. This is the total power drawn from the line by the supply. Then divide this power by the line voltage and add 10 to 30% to account for the inefficiency of the power supply itself. Use a fuse or circuit breaker with the nearest larger current rating. Remember that the charging of filter capacitors can create large surges of current when the supply is turned on. If fuse blowing or breaker tripping at turn on is a problem, use slow-blow fuses, which allow for high initial surge currents.

For low-power semiconductor circuits, use fast-blow fuses. As the name implies, such fuses open very quickly once the current exceeds the fuse rating by more than 10%.

7.3 Power Transformers

Numerous factors are considered to match a transformer to its intended use. Some of these parameters are:

1. Output voltage and current (volt-ampere rating).
2. Power source voltage and frequency.
3. Ambient temperature.
4. Duty cycle and temperature rise of the transformer at rated load.
5. Mechanical considerations like weight, shape and mounting.

7.3.1 Volt-Ampere Rating

In alternating-current equipment, the term *volt-ampere* (VA) is often used rather than the term watt. This is because ac components must handle reactive power as well as real power. If this is confusing, consider a capacitor connected directly across the secondary of a transformer. The capacitor appears as a reactance that permits current to flow, just as if the load were a resistor. The current is at a 90° phase angle, however. If we assume a perfect capacitor, there will be no heating of the capacitor, so no real power (watts) will be delivered by the transformer. The transformer must still be capable of supplying the voltage, and be able to handle the current required by the reactive load. The current in the transformer windings will heat the windings as a result of the I^2R losses in the winding resistances. The product of the voltage and current in the winding is referred to as "volt-amperes," since "watts" is reserved for the real, or dissipated, power in the load. The volt-ampere rating will always be equal to, or greater than, the power actually being drawn by the load.

The number of volt-amperes delivered by a transformer depends not only upon the dc load requirements, but also upon the type of dc output filter used (capacitor or choke input), and the type of rectifier used (full-wave center tap or full-wave bridge). With a capacitive-input filter, the heating effect in the secondary is higher because of the high peak-to-average current ratio. The volt-amperes handled by the transformer may be several times the power delivered to the load. The primary winding volt-amperes will be somewhat higher because of transformer losses. This point is treated in more detail in the section on ac-dc conversion. (See the **Electrical Fundamentals** chapter for more information on transformers and reactive power.)

7.3.2 Source Voltage and Frequency

A transformer operates by producing a magnetic field in its core and windings. The intensity of this field varies directly with the instantaneous voltage applied to the transformer primary winding. These variations, coupled to the secondary windings, produce the desired output voltage. Since the transformer appears to the source as an inductance in parallel with the (equivalent) load, the primary will appear as a short circuit if dc is applied to it. The unloaded inductance of the primary (also known as the *magnetizing inductance*) must be high enough so as not to draw an excess amount of input current at the design line frequency (normally 60 Hz in the US). This is achieved by providing a combination of sufficient turns on the primary and enough magnetic core material so that the core does not saturate during each half-cycle.

The voltage across a winding is directly related to the time rate of change of magnetic flux in the core. This relationship is expressed mathematically by $V = N \, d\Phi/dt$ as described in the section on Inductance in the **Electrical Fundamentals** chapter. The total flux in turn is expressed by $\Phi = A_e B$, where A_e is the cross-sectional area of the core and B is the flux density.

The maximum value for *flux density* (the magnetic field strength produced in the core) is limited to some percentage (< 80% for example) of the maximum flux density that the core material can stand without saturating, since in saturation the core becomes ineffective and causes the inductance of the primary to plummet to a very low level and input current to rise rapidly. Saturation causes high primary currents and extreme heating in the primary windings.

At a given voltage, 50 Hz ac creates more flux in an inductor or transformer core because the longer time period per half-cycle results in more flux and higher magnetizing current than the same transformer when excited by same 60-Hz voltage. For this reason, transformers and other electromagnetic equipment designed for 60-Hz systems must not be used on 50-Hz power systems unless specifically designed to handle the lower line frequency.

7.3.3 How to Evaluate an Unmarked Power Transformer

Hams who regularly visit hamfests frequently develop a junk box filled with used and unmarked transformers. Over time, transformer labels or markings on the coil wrappings may come off or be obscured. There is a good possibility that the transformer is still useable, but the problem is to determine what voltages and currents the transformer can supply. First consider the possibility that you may have an audio transformer or other impedance-matching device rather than a power transformer. If you aren't sure, don't connect it to ac power!

If the transformer has color-coded leads, you are in luck. There is a standard for transformer lead color-coding, as is given in the **Component Data and References** chapter. Where two colors are listed, the first one is the main color of the insulation; the second is the color of the stripe.

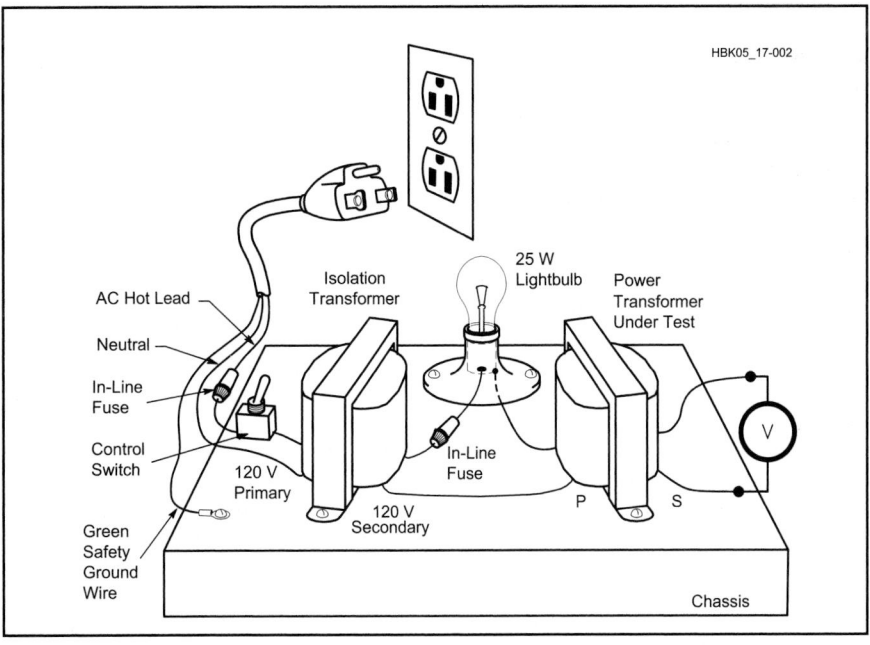

Fig 7.4 — Use a test fixture like this to test unknown transformers. Don't omit the isolation transformer, and be sure to insulate all connections before you plug into the ac mains.

Check the transformer windings with an ohmmeter to determine that there are no shorted (or open) windings. In particular, check for continuity between any winding and the core. If you find that a winding has been shorted to the core, do not use the transformer! The primary winding usually has a resistance higher than a filament winding and lower than a high-voltage winding.

Fig 7.4 shows that a convenient way to test the transformer is to rig a pair of test leads to an electrical plug with a 25 W household light bulb in series to limit current to safe (for the transformer) levels. For safety reasons use an isolation transformer and be sure to insulate all connections before you plug into the ac mains. Switch off the power while making or changing any connections. You can be electrocuted if the voltmeter leads or meter insulation are not rated for the transformer output voltage! If in doubt, connect the meter with the circuit turned off, then apply power while you are not in contact with the circuit. *Be careful! You are dealing with hazardous voltages!*

Connect the test leads to each winding separately. The filament/heater windings will cause the bulb to light to full brilliance because a filament winding has a very low impedance and almost all the input voltage will be across the series bulb. The high-voltage winding will cause the bulb to be extremely dim or to show no light at all because it will have a very high impedance, and the primary winding will probably cause a small glow. The bulb glows even with the secondary windings open-circuited because of the small magnetizing current in the transformer primary.

When the isolation transformer output is connected to what you think is the primary winding, measure the voltages at the low-voltage windings with an ac voltmeter. If you find voltages close to 6 V ac or 5 V ac, you know that you have identified the primary and the filament windings. Label the primary and low voltage windings.

Even with the light bulb, a transformer can be damaged by connecting ac mains power to a low-voltage or filament winding. In such a case the insulation could break down in a primary or high-voltage winding because of the high turns ratio stepping up the voltage well beyond the transformer ratings.

Connect the voltmeter to the high-voltage windings. Remember that the old TV transformers will typically supply as much as 800 V_{pk} or so across the winding, so make sure that your meter can withstand these potentials without damage and that you use the voltmeter safely.

Divide 6.3 (or 5) by the voltage you measured across the 6.3 V (or 5 V) winding in this test setup. This gives a multiplier that you can use to determine the actual no-load voltage rating of the high-voltage secondary. Simply multiply the ac voltage measured across the high-voltage winding by the multiplier.

The current rating of the windings can be determined by loading each winding with the primary connected directly (no bulb) to the ac line. Using power resistors, increase loading on each winding until its voltage drops by about 10% from the no-load figure. The current drawn by the resistors is the approximate winding load-current rating.

7.4 AC-DC Power Conversion

One of the most common power supply functions is the conversion of ac power to dc, or *rectification*. The output from the rectifier will be a combination of dc, which is the desired component, and ac *ripple* superimposed on the dc. This is an undesired but inescapable component. Since most loads cannot tolerate more than a small amount of ripple on the dc voltage, some form of filter is required. The result is that ac-dc power conversion is performed with a rectifier-filter combination as shown in **Fig 7.5**.

As we will see in the rectifier circuit examples given in the next sections, sometimes the rectifier and filter functions will be separated into two distinct parts but very often the two will be integrated. This is particularly true for voltage and current multipliers as described in the sections on multipliers later in the chapter. Even when it appears that the rectifier and filter are separate elements, there will still be a strong interaction where the design and behavior of each part depends heavily on the other. For example the current waveforms in the rectifiers and the input source are functions of the load and filter characteristics. In turn the voltage waveform applied to the filter depends on the rectifier circuit and the input source voltages. To simplify the discussion

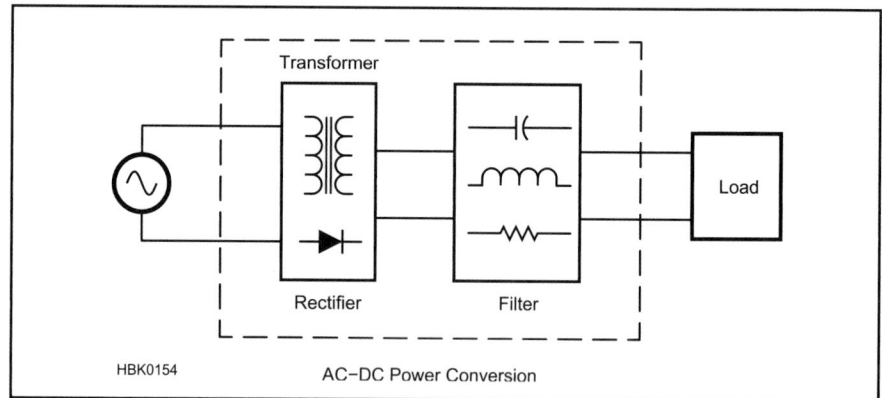

Fig 7.5 — **Ac-dc power conversion with a rectifier and a filter.**

we will treat the rectifier connections and the filters separately but always keeping in mind their interdependence.

The following rectifier-filter examples assume a conventional 60 Hz ac sine wave source, but these circuits are frequently used in switching converters at much higher frequencies and with square wave or quasi-square wave voltage and current waveforms. The component values may be different but

the basic behavior will be very similar.

There are many different rectifier circuits or "connections" that may be used depending on the application. The following discussion provides an overview of some of the more common ones. The circuit diagrams use the symbol for a semiconductor diode, but the same circuits can be used with the older types of rectifiers that may be encountered in older equipment.

For each circuit we will show the voltage and current waveforms in the circuit for resistive, capacitive and inductive loads. The inductive and capacitive loads represent commonly used filters. We will be interested in the peak and average voltages as well as the RMS currents.

7.4.1 Half-Wave Rectifier

Fig 7.6 shows several examples of the half-wave rectifier circuit. It begins with a simple transformer with a resistive load (Fig 7.6A)

and goes on to show how the output voltage and transformer current varies when a diode and filter elements are added.

Without the diode (Fig 7.6A) the output voltage (V_R) and current are just sine waves, and the RMS current in the transformer windings will be the same as the load (R) current.

Next, add a rectifier diode in series with the load (Fig 7.6B). During one half of the ac cycle, the rectifier conducts and there is current through the rectifier to the load. During the other half cycle, the rectifier is reverse-biased and there is no current (indicated by the

broken line in Fig 7.6B) in R. The output voltage is pulsating dc, which is a combination of two components: an average dc value of $0.45\,E_{RMS}$ (the voltage read by a dc voltmeter) and line-frequency ac ripple. The transformer secondary winding current is also pulsating dc. The power delivered to R is now ½ that for Fig 7.6A but the secondary RMS winding current in Fig 7.6B is still 0.707 times what it was in Fig 7.6A. For the same winding resistance, the winding loss, in proportion to the output power, is twice what it was in Fig 7.6A. This is an intrinsic limitation of the half-wave rectifier circuit — the RMS winding current is larger in proportion to the load power. In addition, the dc component of the secondary winding current may bias the transformer core toward saturation and increased core loss.

A filter can be used to smooth out these variations and provide a higher average dc voltage from the circuit. Because the frequency of the pulses (the ripple frequency) is low (one pulse per cycle), considerable filtering is required to provide adequately smooth dc output. For this reason the circuit is usually limited to applications where the required current is small. Parts C, D and E in Fig 7.6 show some possible capacitive and inductive filters.

As shown in Fig 7.6C and D, when a capacitor is used for filtering the output dc voltage will approach

$$V_{pk} = \sqrt{2} \times E_{RMS} \qquad (1)$$

and the larger we make the filter capacitance, the smaller the ripple will be.

Unfortunately, as we make the filter capacitance larger, the diode, capacitor and transformer winding currents all become high-amplitude narrow pulses which will have a very high RMS value in proportion to the power level. These current pulses are also transmitted to the input line and inject currents at harmonics of the line frequency into the power source, which may result in interference to other equipment. Narrow high-amplitude current pulses are characteristic of capacitive-input filters in all rectifier connections when driven from voltage sources.

As shown in Fig 7.6E, it is possible to use an inductive filter instead, but a second diode (D2, sometimes called a *free-wheeling diode*) should be used. Without D2 the output voltage will get smaller as we increase the size of L to get better filtering, and the output voltage will vary greatly with load. By adding D2 we are free to make L large for small output ripple but still have reasonable voltage regulation. Currents in D1 and the winding will be approximately square waves, as indicated. This will reduce the line harmonic currents injected into the source but there will still be some.

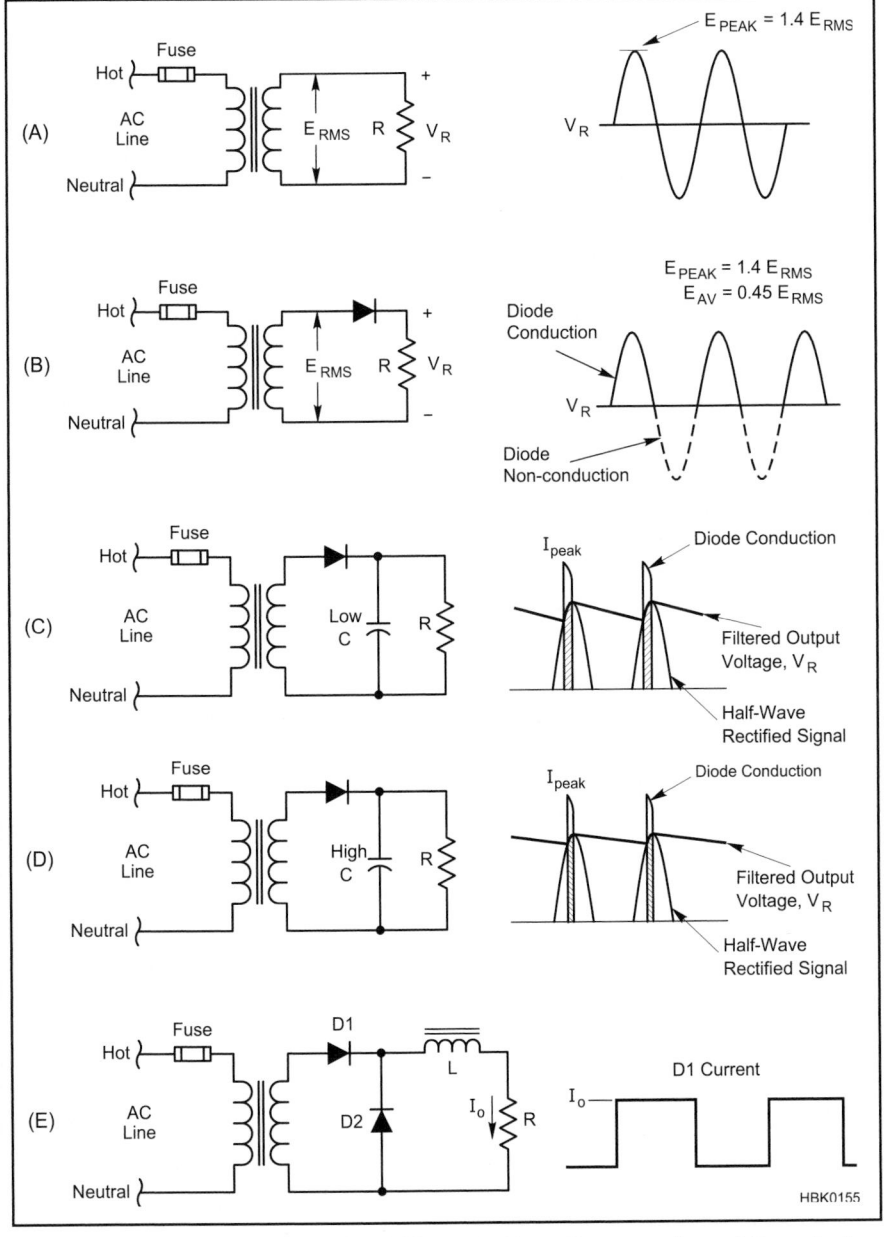

Fig 7.6 — Half-wave rectifier circuits. A illustrates the voltage waveform at the output without a rectifier. B represents the basic half-wave rectifier and the output waveform. C and D illustrate the impact of small and large filter capacitors on the output voltage and input current waveforms. E shows the effect of using an inductor filter with the half-wave rectifier. Note the addition of the shunt diode (D2) when using inductive filters with this rectifier connection.

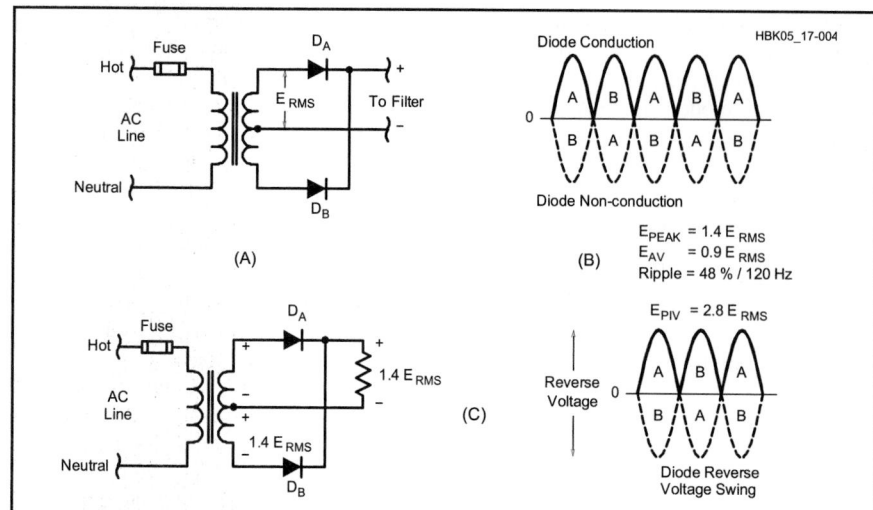

Fig 7.7 — Full-wave center-tap rectifier circuits. A illustrates the basic circuit. Diode conduction is shown at B with diodes A and B alternately conducting. The peak inverse voltage for each diode is 2.8 E$_{RMS}$ as depicted at C.

Peak inverse voltage (PIV) is the maximum voltage the rectifier must withstand when it isn't conducting. This varies with the load and rectifier connection. In the half-wave rectifier, with a resistive load the PIV is the peak ac voltage (1.4 × E$_{RMS}$); with a capacitor filter and a load drawing little or no current, the PIV can rise to 2.8 × E$_{RMS}$.

7.4.2 Full-Wave Center-Tapped Rectifier

The full-wave center-tapped rectifier circuit is shown in **Fig 7.7**. It is essentially an arrangement where the outputs of two half-wave rectifiers are combined so that both halves of the ac cycle are used to deliver power to the output. A transformer with a center-tapped secondary is required.

The average output voltage of this circuit is 0.9 × E$_{RMS}$ of half the transformer secondary (the center-tap to one side); this is the maximum that can be obtained with a suitable choke-input filter. The peak output voltage is 1.4 × E$_{RMS}$ of half the transformer secondary; this is the maximum voltage that can be obtained from a capacitor-input filter.

As can be seen in Fig 7.7C, the PIV impressed on each diode is independent of the type of load at the output. This is because the peak inverse voltage condition occurs when diode D$_A$ conducts and diode D$_B$ is not conducting. The positive and negative voltage peaks occur at precisely the same time, a condition different from that in the half-wave circuit. As the cathodes of diodes D$_A$ and D$_B$ reach a positive peak (1.4 E$_{RMS}$), the anode of diode D$_B$ is at a negative peak, also 1.4 E$_{RMS}$, but in the opposite direction. The total peak inverse voltage is therefore 2.8 E$_{RMS}$.

Fig 7.7C shows that the ripple frequency is twice that of the half-wave rectifier (two times the line frequency). Substantially less filtering is required because of the higher ripple frequency. Since the rectifiers work alternately, each handles half of the load current. The current rating of each rectifier need be only half the total current drawn from the supply.

The problem with dc bias in the transformer core associated with the half-wave connection is largely eliminated with this circuit and the RMS current in the primary winding will also be reduced.

7.4.3 Full-Wave Bridge Rectifier

Another commonly used rectifier circuit that does not require a center-tapped transformer is illustrated in **Fig 7.8**. In this arrangement, two rectifiers operate in series on each half of the cycle, one rectifier being in the lead supplying current to the load, the other being the current return lead. As shown in Figures 7.8A and B, when the top lead of the transformer secondary is positive with respect to the bottom lead, diodes D$_A$ and D$_C$ will conduct while diodes D$_B$ and D$_D$ are reverse-biased. On the next half cycle, when the top lead of the transformer is negative with respect to the bottom, diodes D$_B$ and D$_D$ will conduct while diodes D$_A$ and D$_C$ are reverse-biased.

The output voltage wave shape and ripple frequency are the same as for the full-wave center-tapped circuit. The average dc output voltage into a resistive load or choke-input filter is 0.9 times E$_{RMS}$ delivered by the transformer secondary; with a capacitor filter and a light load, the maximum output voltage is 1.4 times the secondary E$_{RMS}$ voltage.

Fig 7.8C shows the PIV to be 1.4 E$_{RMS}$ for each diode which is half that of the full-wave center-tapped circuit for the same output voltage. When an alternate pair of diodes (such as D$_A$ and D$_C$) is conducting, the other diodes are essentially connected in parallel

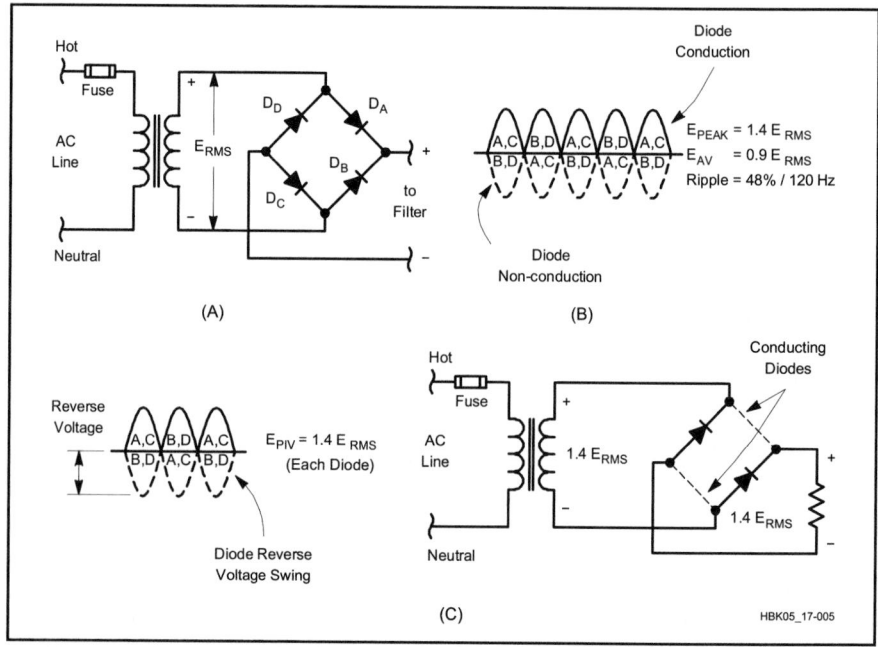

Fig 7.8 — Full-wave bridge rectifier circuits. The basic circuit is illustrated at A. Diode conduction and nonconduction times are shown at B. Diodes A and C conduct on one half of the input cycle, while diodes B and D conduct on the other. C displays the peak inverse voltage for one half cycle. Since this circuit reverse-biases two diodes essentially in parallel, 1.4 E$_{RMS}$ is applied across each diode.

(the conducting diodes are essentially short circuits) in a reverse-biased direction. The reverse stress is then 1.4 E_{RMS}. Each pair of diodes conducts on alternate half cycles, with the full load current through each diode during its conducting half cycle. Since each diode is not conducting during the other half cycle the average diode current is one-half the total load current drawn from the supply.

Compared to the half-wave and full-wave center-tapped circuit, the full-wave bridge circuit further reduces the transformer RMS winding currents. In the case of a resistive load the winding currents are the same as when the resistive load is connected directly across the secondary. The RMS winding currents will still be higher when inductive and especially capacitive filters are used because of the pulsating nature of the diode and winding currents.

7.4.4 Comparison of Rectifier Circuits

Comparing the full-wave center-tapped and the full-wave bridge circuits, we can see that the center-tapped circuit has half the number of rectifiers as the bridge but these rectifiers have twice the PIV rating requirement of the bridge diodes. The diode current ratings are identical for the two circuits. The bridge makes better use of the transformer's secondary than the center-tapped rectifier, since the transformer's full winding supplies power during both half cycles, while each half of the center-tapped circuit's secondary provides power only during its positive half-cycle.

The full-wave center-tapped rectifier is typically used in high-current, low-voltage applications because only one diode conducts at a time. This reduces the loss associated with diode conduction. In the full-wave bridge circuit there are two diodes in series in conduction simultaneously, which leads to higher loss. The full-wave bridge circuit is typically used for higher output voltages where this is not a serious concern. The lower diode PIV and better utilization of the transformer windings makes this circuit very attractive for higher output voltages and higher powers typical of high voltage amplifier supplies.

Because of the disadvantages pointed out earlier, the half-wave circuit is rarely used in 60-Hz rectification except for bias supplies or other small loads. It does see considerable use, however, in high-frequency switchmode power supplies.

7.5 Voltage Multipliers

Other rectification circuits are sometimes useful, including *voltage multipliers*. These circuits function by the process of charging one or more capacitors in parallel on one half cycle of the ac waveform, and then connecting that capacitor or capacitors in series with the opposite polarity of the ac waveform on the alternate half cycle. In full-wave multipliers, this charging occurs during both half-cycles.

Voltage multipliers, particularly *voltage doublers*, find considerable use in high-voltage supplies. When a doubler is employed, the secondary winding of the power transformer need have only half the voltage that would be required for a bridge rectifier. This reduces voltage stress in the windings and decreases the transformer insulation requirements. This is not without cost, however, because the transformer-secondary *current* rating has to be correspondingly doubled for a given load current and charging of the capacitors leads to narrow high-RMS current waveforms in the transformer windings and the capacitors.

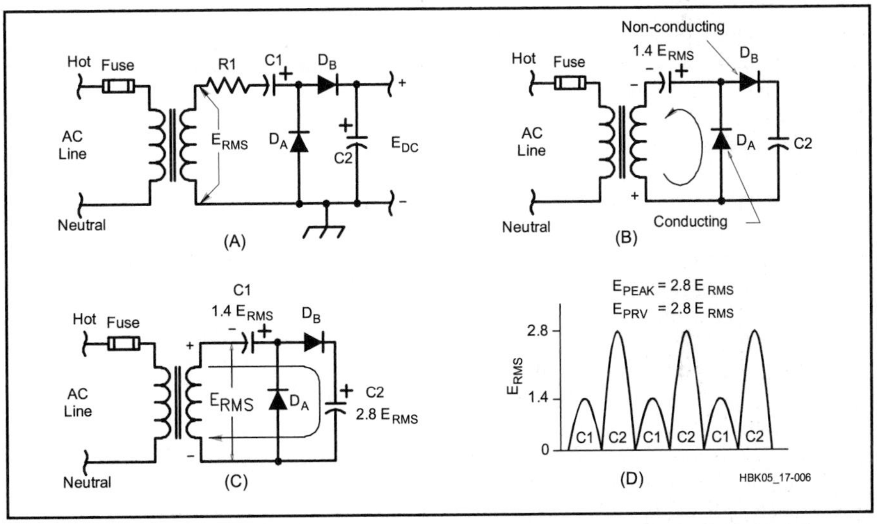

Fig 7.9 — Part A shows a half-wave voltage-doubler circuit. B displays how the first half cycle of input voltage charges C1. During the next half cycle (shown at C), capacitor C2 charges with the transformer secondary voltage plus that voltage stored in C1 from the previous half cycle. The arrows in parts B and C indicate the conventional current. D illustrates the levels to which each capacitor charges over several cycles.

7.5.1 Half-Wave Voltage Doubler

Fig 7.9 shows the circuit of a half-wave voltage doubler and illustrates the circuit operation. For clarity, assume the transformer voltage polarity at the moment the circuit is activated is that shown at Fig 7.9B. During the first negative half cycle, D_A conducts (D_B is in a nonconductive state), charging C1 to the peak rectified voltage (1.4 E_{RMS}). C1 is charged with the polarity shown in Fig 7.9B. During the positive half cycle of the secondary voltage, D_A is cut off and D_B conducts, charging capacitor C2. The amount

of voltage delivered to C2 is the sum of the transformer peak secondary voltage plus the voltage stored in C1 (1.4 E_{RMS}). On the next negative half cycle, D_B is non-conducting and C2 will discharge into the load. If no load is connected across C2, the capacitors will remain charged — C1 to 1.4 E_{RMS} and C2 to 2.8 E_{RMS}. When a load is connected to the circuit output, the voltage across C2 drops during the negative half cycle and is recharged up to 2.8 E_{RMS} during the positive half cycle.

The output waveform across C2 resembles that of a half-wave rectifier circuit because C2 is pulsed once every cycle. Fig 7.9D illustrates the levels to which the two capacitors

are charged throughout the cycle. In actual operation the capacitors will usually be large enough that they will discharge only partially, not all the way to zero as shown.

7.5.2 Full-Wave Voltage Doubler

Fig 7.10 shows the circuit of a full-wave voltage doubler and illustrates the circuit operation. During the positive half cycle of the transformer secondary voltage, as shown in Fig 7.10B, D_A conducts charging capacitor C1 to 1.4 E_{RMS}. D_B is not conducting at this time.

During the negative half cycle, as shown in Fig 7.10C, D_B conducts, charging capacitor C2 to 1.4 E_{RMS}, while D_A is non-conducting. The output voltage is the sum of the two capacitor voltages, which will be 2.8 E_{RMS} under no-load conditions. Fig 7.10D illustrates that each capacitor alternately receives a charge once per cycle. The effective filter capacitance is that of C1 and C2 in series, which is less than the capacitance of either C1 or C2 alone.

Resistors R1 and R2 in Fig 7.10A are used to limit the surge current through the rectifiers. Their values are based on the transformer voltage and the rectifier surge-current rating, since at the instant the power supply is turned on, the filter capacitors look like a short-circuited load. Provided the limiting resistors can withstand the surge current, their current-handling capacity is based on the maximum load current from the supply. Output voltages approaching twice the peak voltage of the transformer can be obtained with the voltage doubling circuit shown in Fig 7.10.

Fig 7.11 shows how the voltage depends upon the ratio of the series resistance to the load resistance, and the load resistance times the filter capacitance. The peak inverse voltage across each diode is 2.8 E_{RMS}. As indicated by the curves in Fig 7.11, the output voltage regulation of this doubler connection is not very good and it is not attractive for providing high voltages at high power levels.

There are better doubler connections for higher power applications, and two possibilities are shown in **Fig 7.12**. The connection in Fig 7.12A uses two bridge rectifiers in series with capacitive coupling between the ac terminals of the bridges. At the expense of more diodes, this connection will have much better output voltage regulation at higher power levels. Even better regulation can be achieved by using the connection shown in Fig 7.12B. In this example, two windings on the transformer are used. It is not essential that both windings have the same voltage, but both must be capable of providing the desired output current. In addition, the insulation of the upper winding must be adequate to accommodate the additional dc bias applied to it from the lower winding.

7.5.3 Voltage Tripler and Quadrupler

Fig 7.13A shows a voltage-tripling circuit. On one half of the ac cycle, C1 and C3 are charged to the source voltage through D1, D2 and D3. On the opposite half of the cycle, D2 conducts and C2 is charged to twice the source voltage, because it sees the transformer plus the charge in C1 as its source (D1 is cut off during this half cycle). At the same time, D3 conducts, and with the transformer and the

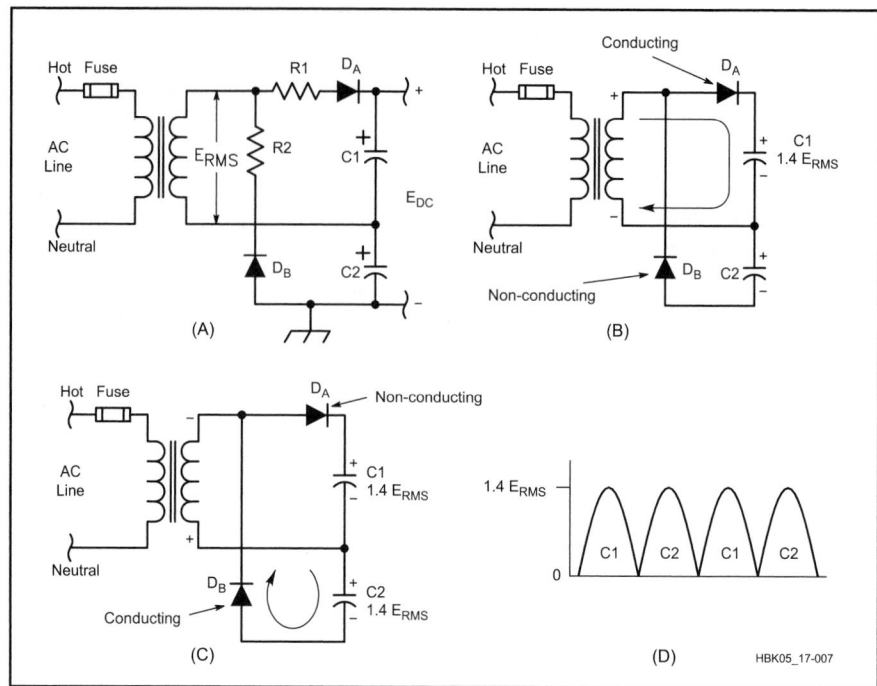

Fig 7.10 — Part A shows a full-wave voltage-doubler circuit. One-half cycle is shown at B and the next half cycle is shown at C. Each capacitor receives a charge during every input-voltage cycle. D illustrates how each capacitor is charged alternately.

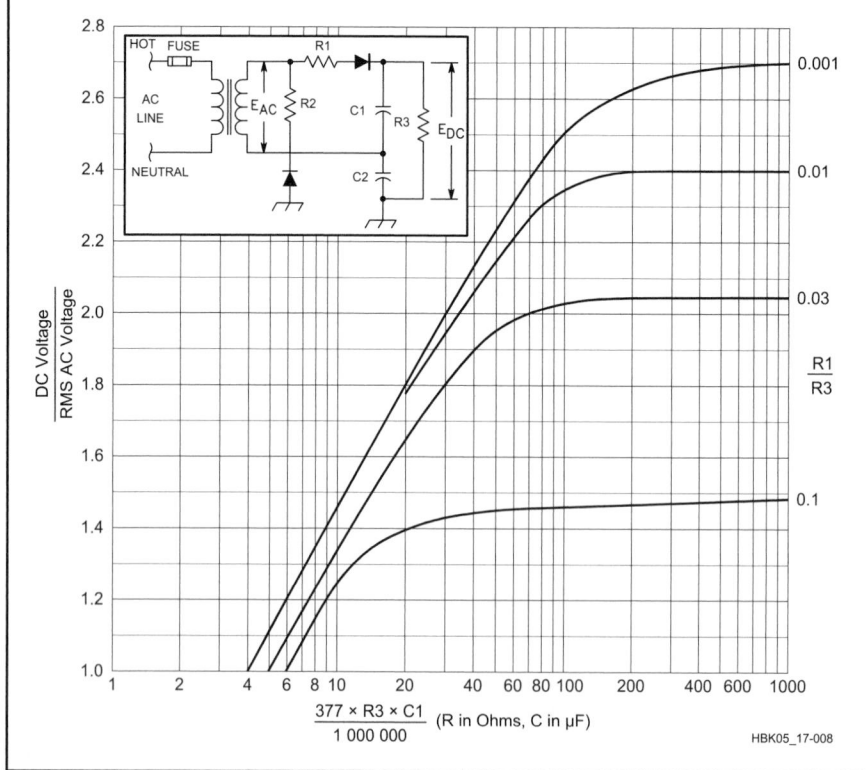

Fig 7.11 — DC output voltages from a full-wave voltage-doubler circuit as a function of the filter capacitances and load resistance. For the ratio R1 / R3 and for the R3 × C1 product, resistance is in ohms and capacitance is in microfarads. Equal resistance values for R1 and R2, and equal capacitance values for C1 and C2 are assumed. These curves are adapted from those published by Otto H. Schade in "Analysis of Rectifier Operation," *Proceedings of the I. R. E.*, July 1943.

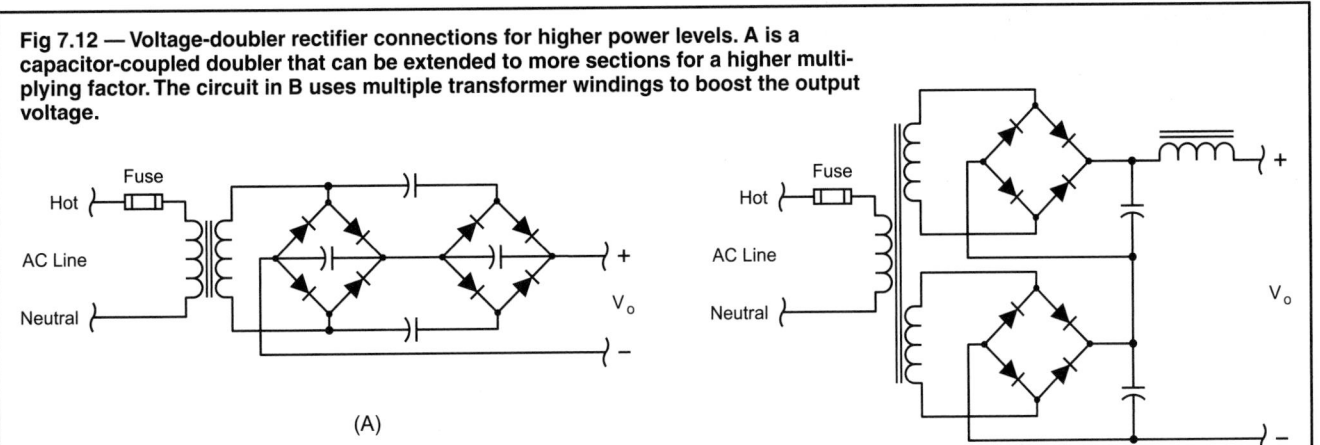

Fig 7.12 — Voltage-doubler rectifier connections for higher power levels. A is a capacitor-coupled doubler that can be extended to more sections for a higher multiplying factor. The circuit in B uses multiple transformer windings to boost the output voltage.

HBK0156

(A)

(B)

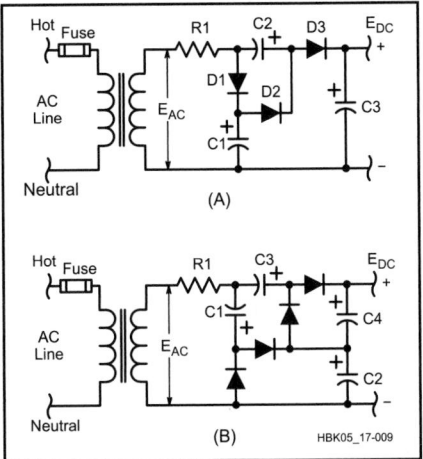

(A)

(B)

HBK05_17-009

Fig 7.13 — Voltage-multiplying circuits with one side of the transformer secondary used as a common connection. A shows a voltage tripler and B shows a voltage quadrupler. Capacitances are typically 20 to 50 μF, depending on the output current demand. Capacitor dc ratings are related to E_{PEAK} (1.4 E_{RMS}):
C1 — Greater than E_{PEAK}
C2 — Greater than 2 E_{PEAK}
C3 — Greater than 3 E_{PEAK}
C4 — Greater than 2 E_{PEAK}

charge in C2 as the source, C3 is charged to three times the transformer voltage.

The voltage-quadrupling circuit of Fig 7.13B works in similar fashion. In either of the circuits of Fig 7.13, the output voltage will approach an exact multiple of the peak ac voltage when the output current drain is low and the capacitance values are large.

7.6 Current Multipliers

Just as there are voltage multiplier connections for high-voltage, low-current loads, there are current multiplier connections for low-voltage, high-current loads. An example of a current-doubler is given in **Fig 7.14A.**

To make the circuit operation easier to visualize, we can represent L1 and L2 as current sources (Fig 7.14B) which is a good approximation for steady-state operation. When terminal 1 of the secondary winding is positive with respect to terminal 2, diode D_A will be reverse-biased and therefore nonconducting. The current flows within the circuit are shown in Fig 7.14B. Note that all of the output current (I_o) flows through D_B but only half of I_o flows through the winding. At the cathode of D_B the current divides with half going to L2 and the other half to the transformer secondary. The output voltage will be one-half the voltage of the average winding voltage (0.45 E_{RMS}). This rectifier connection divides the voltage and multiplies the current! Because of the need for two inductors, this circuit is seldom used in line-frequency applications but it is very useful in high-frequency switchmode regulators with very low output voltages (<10 V) because it makes the secondary winding design easier and can improve circuit efficiency. At high frequencies, the inductors can be quite small.

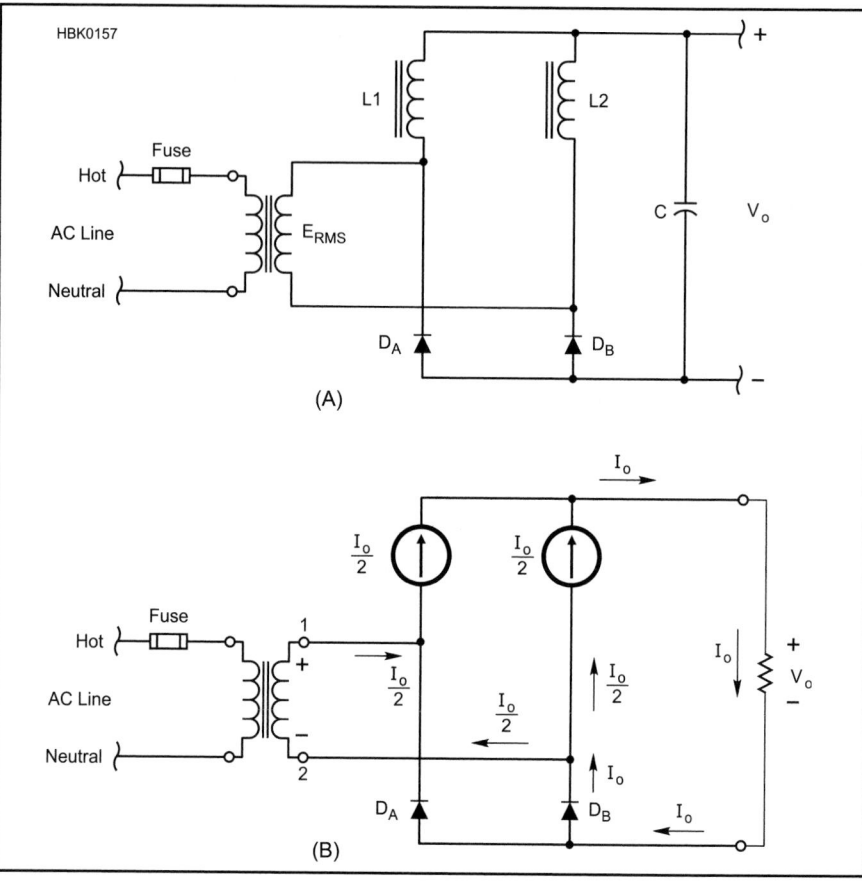

Fig 7.14 — A current doubler rectifier connection. A is the basic circuit. B illustrates current flow within the circuit.

7.7 Rectifier Types

Rectifiers have a long history beginning with mechanical rectifiers in the 1800s to today's abundant variety of semiconductor devices. While many different devices have been created for this purpose, they all have the characteristic that they block current flow in the reverse direction, withstanding substantial reverse voltage and allowing current flow in the forward direction with minimum voltage drop. The simplest rectifiers are diodes, but it is also possible to have three-terminal devices (such as a thyristor) that can be controlled to regulate the output dc in addition to providing rectification. It is also possible to use devices like MOSFETs as synchronous rectifiers with very low forward drop during conduction. This is typically done to improve efficiency for very low voltage outputs. The following is a brief description of several of the more common examples. The sections on vacuum tube and other obsolete types of rectifiers from previous editions are available as a PDF article on the CD-ROM accompanying this book.

7.7.1 Semiconductor Diodes

Rectifier diodes can be made from a number of different semiconductor materials such as germanium, silicon, silicon-carbide or gallium-arsenide, and no doubt other materials will appear in the future. The choice will depend on the application and as always cost is a factor.

Germanium diodes were the first of the solid-state semiconductor rectifiers. They have an extremely low forward voltage drop but are relatively temperature sensitive, having high reverse leakage currents at higher temperatures. They can be easily destroyed by overheating during soldering as well. Germanium diodes are no longer used as power rectifiers.

Today, silicon diodes are the primary choice for virtually all power rectifier applications. They are characterized by extremely high reverse resistance (low reverse leakage), forward drops of a volt or less and operation at junction temperatures up to 125 °C. Some

multi-junction HV diodes will have forward drops of several volts, but that is still low compared to the voltage at which they are being used.

Many different types of silicon diodes are available for different applications. Silicon rectifiers fall into to two general categories: PN-junction diodes and Schottky barrier diodes (see the **Analog Basics** chapter). Schottky diodes are the usual choice for low output voltages (<20 V) where their low forward conduction drop is critical for efficiency. For higher voltages however, the high reverse leakage of Schottky diodes is not acceptable and PN-junction diodes are normally chosen.

For 50/60 Hz applications, diodes with reverse recovery times of a microsecond or even more are suitable and very economical. For switchmode converters and inverters that regularly operate at 25 kHz and higher frequencies, fast-recovery diodes are needed. These converters typically have waveform transitions of less than 1 µs within the circuit.

MOSFET power transistors often have transitions of less than 100 ns.

During the switching transitions, previously conducting diodes see a reversal of current direction. This change tends to reverse-bias those diodes, and thereby put them into an open-circuit condition. Unfortunately, as explained in the **Analog Basics** chapter, solid-state rectifiers cannot be made to cease conduction instantaneously. As a result, when the opposing diodes in a bridge rectifier or full-wave rectifier become conductive at the time the converter switches states, the diodes being turned off will actually conduct in the reverse direction for a brief time. That effectively short circuits the converter for a period of time depending on the reverse recovery characteristics of the rectifiers. This characteristic can create high current transients that stress the switching transistors and lead to increased loss and electromagnetic interference. As the switching frequency increases, more of these transitions happen each second, and more power is lost because of diode cross-conduction.

These current transients and associated losses are reduced by using *fast recovery* diodes, which are specially doped diodes designed to minimize storage time. Diodes with recovery times of 50 ns or less are available.

7.7.2 Rectifier Strings or Stacks

DIODES IN SERIES

When the PIV rating of a single diode is not sufficient for the application, similar diodes may be used in series. (Two 500 PIV diodes in series will withstand 1000 PIV and so on.) There used to be a general recommendation to place a resistor across each diode in the string to equalize the PIV drops. With modern diodes, this practice is no longer necessary.

Modern silicon rectifier diodes are constructed to have an avalanche characteristic. Simply put, this means that the diffusion process is controlled so the diode will exhibit a Zener characteristic in the reverse-biased direction before destructive breakdown of the junction can occur. This provides a measure of safety for diodes in series. A diode will go into Zener conduction before it self-destructs. If other diodes in the chain have not reached their avalanche voltages, the current through the avalanched diode will be limited to the leakage current in the other diodes. This should normally be very low. For this reason, shunting resistors are generally not needed across diodes in series rectifier strings. In fact, shunt resistors can actually create problems because they can produce a low-impedance source of damaging current to any diode that may have reached avalanche potential.

DIODES IN PARALLEL

Diodes can be placed in parallel to increase current-handling capability. Equalizing resistors should be added as shown in **Fig 7.15**. Without the resistors, one diode may take most of the current. The resistors should be selected to have a drop of several tenths of a volt at the expected peak current. A disadvantage of this form of forced current sharing will be the increase in power loss because of the added resistors.

7.7.3 Rectifier Ratings versus Operating Stress

Power supplies designed for amateur equipment use silicon rectifiers almost exclusively. These rectifiers are available in a wide range of voltage and current ratings: PIV ratings of 600 V or more and current ratings as high as 400 A are available. At 1000 PIV, the current ratings may be several amperes. It is possible to stack several units in series for higher voltages. Stacks are available commercially that will handle peak inverse voltages up to 10 kV at a load current of 1 A or more.

7.7.4 Rectifier Protection

The discussion of rectifier circuits included the peak reverse voltage seen by the rectifiers in each circuit. You will need this information to select the voltage rating of the diodes in given application. It is normal good practice to not expose the diodes to more than 75% of their rated voltage for the worst case reverse voltage. This will probably be when operating at the highest input voltage but should also take into account transients that may occur.

The important specifications of a silicon diode are:

1. PIV — the peak inverse voltage.
2. I_0 — the average dc current rating.
3. I_{REP} — the peak repetitive forward current.
4. I_{SURGE} — a non-repetitive peak half-sine wave of 8.3 ms duration (one-half cycle of 60-Hz line frequency).
5. Switching speed or reverse recovery time.
6. Power dissipation and thermal resistance.

The first two specifications appear in most catalogs. I_{REP} and I_{SURGE} are not often specified in catalogs, but they are very important. Except in some switching regulator and capacitive filter circuits, rectifier current typically flows half the time — when it does conduct, the rectifier has to pass at least twice the average direct current. With a capacitor-input filter, the rectifier conducts much less than half the time. In this case, when it does conduct, it may pass as much as 10 to 20 times the average dc current, under certain conditions.

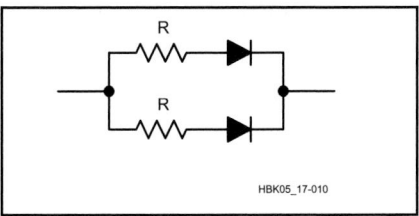

Fig 7.15 — Diodes can be connected in parallel to increase the current-handling capability of the circuit. Each diode should have a series current-equalizing resistor, with a value selected to have a drop of several tenths of a volt at the expected current.

CURRENT INRUSH

When the supply is first turned on, the filter capacitors are discharged and act like a dead short. The result can be a very heavy current surge through the diode for at least one half-cycle and sometimes more. This current transient is called I_{SURGE}. The maximum surge current rating for a diode is usually specified for a duration of one-half cycle (at 60 Hz), or about 8.3 ms. Some form of surge protection is usually necessary to protect the diodes until the filter capacitors are nearly charged, unless the diodes used have a very high surge-current rating (several hundred amperes). If a manufacturer's data sheet is not available, an educated guess about a diode's capability can be made by using these rules of thumb for silicon diodes commonly used in Amateur Radio power supplies:

Rule 1. The maximum I_{REP} rating can be assumed to be approximately four times the maximum I_0 rating, where I_0 is the average dc current rating.

Rule 2. The maximum I_{SURGE} rating can be assumed to be approximately 12 times the maximum I_0 rating. This figure should provide a reasonable safety factor. Silicon rectifiers with 750 mA dc ratings, for example, seldom have 1-cycle surge ratings of less than 15 A; some are rated up to 35 A or more. From this you can see that the rectifier should be selected on the basis of I_{SURGE} and not on I_0 ratings.

Although you can sometimes rely on the resistance of the transformer windings to provide surge-current limiting, this is seldom adequate in high-voltage power supplies. Series resistors are often installed between the secondary and the rectifier strings or in the transformer's primary circuit, but these can be a deterrent to good voltage regulation.

One way to have good surge current limiting at turn-on without affecting voltage regulation during normal operation is to have a resistor in series with the input, along with a relay across the resistor that shorts it out after 50 ms or so. This kind of arrangement is particularly important in HV supplies.

VOLTAGE TRANSIENTS

Vacuum-tube rectifiers had little problem with voltage transients on the incoming power lines. The possibility of an internal arc was of little consequence, since the heat produced was of very short duration and had little effect on the massive plate and cathode structures.

Unfortunately, such is not the case with silicon diodes. Because of their low forward voltage drop, silicon diodes create very little heat with high forward current and therefore have tiny junction areas. However, conduction in the reverse direction beyond the normal reverse recovery time (reverse avalanching) can cause junction temperatures to rise extremely rapidly with the resulting destruction of the semiconductor junction.

To protect semiconductor rectifiers from voltage transients, special surge-absorption devices are available for connection across the incoming ac bus or transformer secondary. These devices operate in a fashion similar to a Zener diode; they conduct heavily when a specific voltage level is reached. Unlike Zener diodes, however, they have the ability to absorb very high transient energy levels without damage. With the clamping level set well above the normal operating voltage range for the rectifiers, these devices normally appear as open circuits and have no effect on the power-supply circuits. When a voltage transient occurs, however, these protection devices clamp the transient and thereby prevent destruction of the rectifiers.

Transient protectors are available in three basic varieties:

1. *Silicon Zener diodes* — large junction Zeners specifically made for this purpose and available as single junction for dc (unipolar) and back-to-back junctions for ac (bipolar). These silicon protectors are available under the trade name of TransZorb from General Semiconductor Corporation and are also made by other manufacturers. They have the best transient-suppressing characteristics of the three varieties mentioned here, but are expensive and have the least energy absorbing capability per dollar of the group.

2. *Varistors* — made of a composition metal-oxide material that breaks down at a certain voltage. Metal-oxide varistors, also known as MOVs, are cheap and easily obtained, but have a higher internal resistance, which allows a greater increase in clamped voltage than the Zener variety. Varistors can also degrade with successive transients within their rated power handling limits (this is not usually a problem in the ham shack where transients are few and replacement of the varistor is easily accomplished).

Varistors usually become short-circuited when they fail. Large energy dissipation can result in device explosion. Therefore, it is a good idea to include a fuse that limits the short-circuit current through the varistor, and to protect people and circuitry from debris.

3. *Gas tube* — similar in construction to the familiar neon bulb, but designed to limit conducting voltage rise under high transient currents. Gas tubes can usually withstand the highest transient energy levels of the group. Gas tubes suffer from an ionization time problem, however. A high voltage across the tube will not immediately cause conduction. The time required for the gas to ionize and clamp the transient is inversely proportional to the level of applied voltage in excess of the device ionization voltage. As a result, the gas tube will let a little of the transient through to the equipment before it activates.

In installations where reliable equipment operation is critical, the local power is poor and transients are a major problem, the usual practice is to use a combination of protectors. Such systems consist of a varistor or Zener protector, combined with a gas-tube device. Often there is an indicator light to warn when a surge has blown out the varistor. Operationally, the solid-state device clamps the surge immediately, with the beefy gas tube firing shortly thereafter to take most of the surge from the solid-state device.

HEAT

The junction of a diode is quite small, so it must operate at a high current density. The heat-handling capability is, therefore, quite small. Normally, this is not a prime consideration in high-voltage, low-current supplies. Use of high-current rectifiers at or near their maximum ratings (usually 2 A or larger, stud-mount rectifiers) requires some form of heat sinking. Frequently, mounting the rectifier on the main chassis — directly or with thin thermal insulating washers — will suffice.

When a rectifier is directly mounted on the heatsink it is good practice to use a thin layer of thermal grease between the diode and the heat sink to assure good heat conduction. Most modern insulating thermal washers do not require the use of grease, but the older mica and other washers may benefit from a *very thin layer* of grease. Thermal grease and heat conducting insulating washers and pads are standard products available from mail-order component sellers.

Large, high-current rectifiers often require special heat sinks to maintain a safe operating temperature. Forced-air cooling from a fan is sometimes used as a further aid. Safe case temperatures are usually given in the manufacturer's data sheets and should be observed if the maximum capabilities of the diode are to be realized. See the thermal design section in the chapter on **Electrical Fundamentals** for more information.

7.8 Power Filtering

Most loads will not tolerate the ripple (an ac component) of the pulsating dc from the rectifiers. Filters are required between the rectifier and the load to reduce the ripple to a low level. As pointed out earlier, some capacitances or inductances may be inherent in the rectifier connection, reducing the ripple amplitude. In most cases, however, additional filtering is required. The design of the filter depends to a large extent on the dc voltage output, the desired voltage regulation of the power supply and the maximum load current. Power-supply filters are low-pass devices using series inductors and/or shunt capacitors.

7.8.1 Load Resistance

In discussing the performance of power-supply filters, it is sometimes convenient to characterize the load connected to the output as a resistance. This *load resistance* is equal to the output voltage divided by the total load current, including the current drawn by the bleeder resistor.

7.8.2 Voltage Regulation

In an unregulated supply, the output voltage usually decreases as more current is drawn. This happens not only because of increased voltage drops in the transformer and filter chokes, but also because the output voltage at light loads tends to soar to the peak value of the transformer voltage as a result of charging the first capacitor. Proper filter design can reduce this effect. The change in output voltage with load is called the *voltage regulation* and is expressed as a percentage.

$$\text{Percent Regulation} = \frac{(E1 - E2)}{E2} \times 100\%$$

$$(2)$$

where

E1 = the no-load voltage
E2 = the full-load voltage.

A steady load, such as that represented

by a receiver, speech amplifier or unkeyed stages of a transmitter, does not require good (low) regulation as long as the proper voltage is obtained under load conditions. The filter capacitors must have a voltage rating safe for the highest value to which the voltage will rise when the external load is removed.

Typically the output voltage will display a larger change with long-duration changes in load resistance than with short transient changes. The reason for this is that transient load currents are supplied from energy stored in the output capacitance. The regulation with long-term changes is often called the *static regulation*, to distinguish it from the *dynamic regulation* (transient load changes). A load that varies at a syllabic or keyed rate, as represented by some audio and RF amplifiers, usually requires good dynamic regulation (<15%) if distortion products are to be held to a low level. The dynamic regulation of a power supply can be improved by increasing the output capacitance.

When essentially constant voltage is required, regardless of current variation (for stabilizing an oscillator, for example), special voltage regulating circuits described later in this chapter are used.

7.8.3 Bleeder Resistors

A *bleeder resistor* is a resistance (R) connected across the output terminals of the power supply as shown in **Fig 7.16A**. Its functions are to discharge the filter capacitors as a safety measure when the power is turned off and to improve voltage regulation by providing a minimum load resistance. When voltage regulation is not of importance, the resistance may be as high as 100 Ω per volt of output voltage. The resistance value to be used for voltage-regulating purposes is discussed in later sections. From the consideration of safety, the power rating of bleeder resistors should be as conservative as possible — having a burned-out bleeder resistor is dangerous!

7.8.4 Ripple Frequency and Voltage

The ripple at the output of the rectifier is an alternating current superimposed on a steady direct current. From this viewpoint, the filter may be considered to consist of: 1) shunt capacitors that short circuit the ac component while not interfering with the flow of the dc component; and/or 2) series chokes that readily pass dc but will impede the ac component.

The effectiveness of the filter can be expressed in terms of percent ripple, which is the ratio of the RMS value of the ripple to the dc value in terms of percentage.

$$\text{Percent Ripple (RMS)} = \frac{E1}{E2} \times 100\% \quad (3)$$

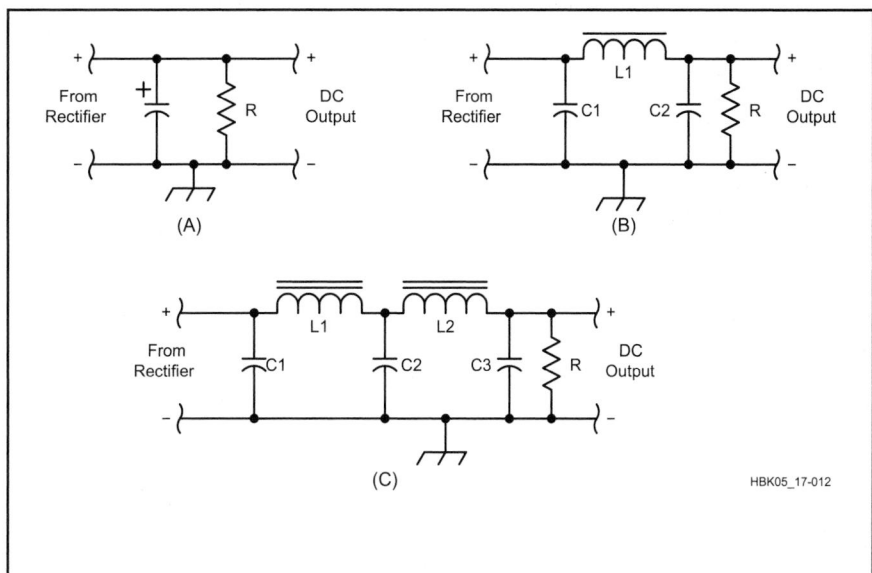

Fig 7.16 — Capacitor-input filter circuits. At A is a simple capacitor filter. B and C are single- and double-section filters, respectively.

where
    E1 = the RMS value of ripple voltage
    E2 = the steady dc voltage.

Any frequency multiplier or amplifier supply in a CW transmitter should have less than 5% ripple. A linear amplifier can tolerate about 3% ripple on the plate voltage. Bias supplies for linear amplifiers should have less than 1% ripple. VFOs, speech amplifiers and receivers may require no greater than 0.01% ripple.

Ripple frequency refers to the frequency of the pulsations in the rectifier output waveform — the number of pulsations per second. The ripple frequency of half-wave rectifiers is the same as the line-supply frequency — 60 Hz with a 60-Hz supply. Since the output pulses are doubled with a full-wave rectifier, the ripple frequency is doubled — to 120 Hz with a 60-Hz supply.

The amount of filtering (values of inductance and capacitance) required to give adequate smoothing depends on the ripple frequency. More filtering is required as the ripple frequency is reduced. This is why the filters used for line-frequency rectification are much larger than those used in switchmode converters where the ripple frequency is often in the hundreds of kHz.

7.8.5 Capacitor-Input Filters

Typical capacitor-input filter systems are shown in Fig 7.16. The ripple can be reduced by making C1 larger, but that can lead to very large capacitances and high inrush currents at turn-on. Better ripple reduction will be obtained when moderate values for C1

are employed and LC sections are added as shown in Fig 7.16B and C.

INPUT VERSUS OUTPUT VOLTAGE

The average output voltage of a capacitor-input filter is generally poorly regulated with load-current variations. As shown earlier (Fig 7.6) the rectifier diodes conduct for only a small portion of the ac cycle to charge the filter capacitor to the peak value of the ac waveform. When the instantaneous voltage of the ac passes its peak, the diode ceases to conduct. This forces the capacitor to support the load current until the ac voltage on the opposing diode in the bridge or full wave rectifier is high enough to pick up the load and recharge the capacitor. For this reason, the peak diode currents are usually quite high.

Since the cyclic peak voltage of the capacitor-filter output is determined by the peak of the input ac waveform, the minimum voltage and, therefore, the ripple amplitude, is determined by the amount of voltage discharge, or "droop," occurring in the capacitor while it is discharging and supporting the load. Obviously, the higher the load current, the proportionately greater the discharge, and therefore the lower the average output.

There is an easy way to approximate the peak-to-peak ripple for a certain capacitor and load by assuming a constant load current. We can calculate the droop in the capacitor by using the relationship:

$$C \times E = I \times t \quad (4)$$

where
    C = the capacitance in microfarads

E = the voltage droop, or peak-to-peak ripple voltage

I = the load current in milliamperes

t = the length of time in ms per cycle during which the rectifiers are not conducting, during which the filter capacitor must support the load current. For 60-Hz, full-wave rectifiers, t is about 7.5 ms.

As an example, let's assume that we need to determine the peak-to-peak ripple voltage at the dc output of a full-wave rectifier/ filter combination that produces 13.8 V dc and supplies a transceiver drawing 2.0 A. The filter capacitor in the power supply is 5000 μF. Using the above relationship:

$$C \times E = I \times t \qquad (5)$$

$$5000 \; \mu F \times E = 2000 \; mA \times 7.5 \; ms$$

$$E = \frac{2000 \; mA \; \times \; 7.5 \; ms}{5000 \; \mu F} = 3 \; V \; P\text{-}P$$

Obviously, this is too much ripple. A capacitor value of about 20,000 μF would be better suited for this application. If a linear regulator is used after this rectifier/filter combination, then it is possible to trade off higher ripple voltage against high power dissipation in the regulator. A properly designed linear regulator can reduce the ripple amplitude to a very small value.

7.8.6 Choke-Input Filters

Choke-input filters provide the benefits of greatly improved output voltage stability over varying loads and low peak-current surges in the rectifiers. On the negative side, the output voltage will be lower than that for a capacitor-input filter.

In line-frequency power supplies, choke-input filters are less popular than they once were. This change came about in part because

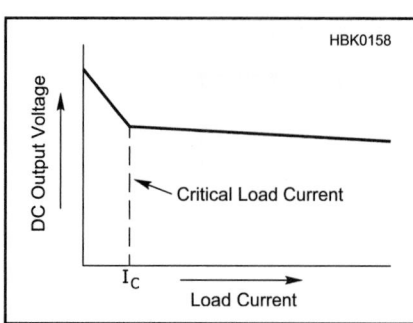

Fig 7.17 — Inductive filter output voltage regulation as a function of load current. The transition from capacitive peak charging to inductive averaging occurs at the critical load current, I_C.

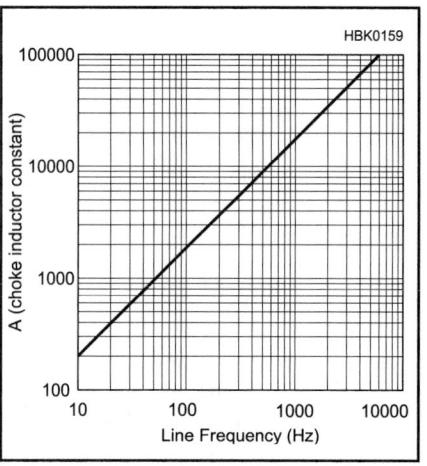

Fig 7.18 — The choke inductor constant, A, is used to solve equation 6.

of the high surge current capability of silicon rectifiers, but more importantly because size, weight and cost are reduced when large filter chokes are eliminated. However, choke input filters are frequently used in high-frequency switchmode converters where the chokes will be much smaller.

As long as the inductance of the choke is large enough to maintain a continuous current over the complete cycle of the input ac waveform, the filter output voltage will be the average value of the rectified output. The average dc value of a full-wave rectified sine wave is 0.637 times its peak voltage. Since the RMS value is 0.707 times the peak, the output of the choke input filter will be (0.637 / 0.707), or 0.9 times the RMS ac voltage.

As shown in **Fig 7.17**, there is a minimum or "critical" load current below which the choke does not provide the necessary filtering. For light loads, there may not be enough energy stored in the choke during the input waveform crest to allow continuous current over the full cycle. When this happens, the filter output voltage will rise as the filter assumes more and more of the characteristics of a capacitor-input filter. One purpose of the bleeder resistor is to keep the minimum load current above the critical value.

The value for the critical (or minimum) inductance for a given maximum value of load resistance in a single phase, full-wave rectifier with a sine wave source voltage can be approximated from:

$$L_c = R/A \qquad (6)$$

where

R = the maximum load resistance

A = a constant obtained from Fig 7.18, derived from the frequency of the input current (see Reference 1).

Low values for minimum load current (high minimum load R) can lead to large values for L_C which may not be practical. Standard filter inductors typically have a relatively constant value for L as the dc current is varied but it is possible to use a swinging choke instead. This is an inductor which has a high inductance at low currents and much lower inductance at high currents. Using a swinging choke will usually result in a much smaller filter choke and/or better output regulation.

7.9 Power Supply Regulation

The output of a rectifier/filter system may be usable for some electronic equipment, but for today's transceivers and accessories, further measures may be necessary to provide power sufficiently clean and stable for their needs. Voltage regulators are often used to provide this additional level of conditioning.

Rectifier/filter circuits by themselves are unable to protect the equipment from the problems associated with input-power-line fluctuations, load-current variations and residual ripple voltages. Regulators can eliminate these problems, but not without costs in circuit complexity and power-conversion efficiency.

7.9.1 Zener Diodes

A Zener diode (developed by Dr Clarence Zener) can be used to maintain the voltage applied to a circuit at a practically constant value, regardless of the voltage regulation of the power supply or variations in load current. The typical circuit is shown in **Fig 7.19**. Note that the cathode side of the diode is connected to the positive side of the supply.

Zener diodes are available in a wide variety of voltages and power ratings. The voltages range from less than 2 V to a few hundred volts, while the power ratings (power the diode can dissipate) run from less than 0.25 W to 50 W. The ability of the Zener diode to stabilize a voltage depends on the diode's conducting impedance. This can be as low as 1 Ω or less in a low-voltage, high-power diode or as high as 1000 Ω in a high-voltage, low-power diode.

The circuit in Fig 7.19 is a *shunt* regulator in that it "shunts" current through a controlling device (the Zener diode) to maintain a constant output voltage. To design a Zener shunt regulator, you must know the minimum and maximum input voltage (E_{DC}); the

$$V_{REG} = \frac{R1 + R2}{R2} V_{REF}$$

(A)

$$V_{REG} = \frac{R1 + R2}{R2} V_{REF}$$

(B)

(C)

HBK05_17-014

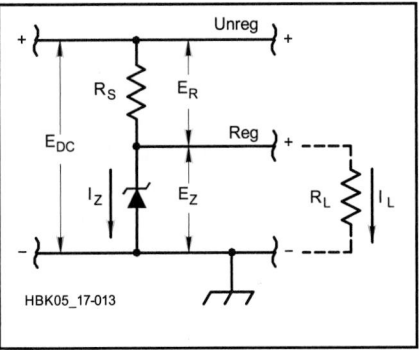

Fig 7.19 — Zener-diode voltage regulation. The voltage from a negative supply may be regulated by reversing the power-supply connections and the diode polarity.

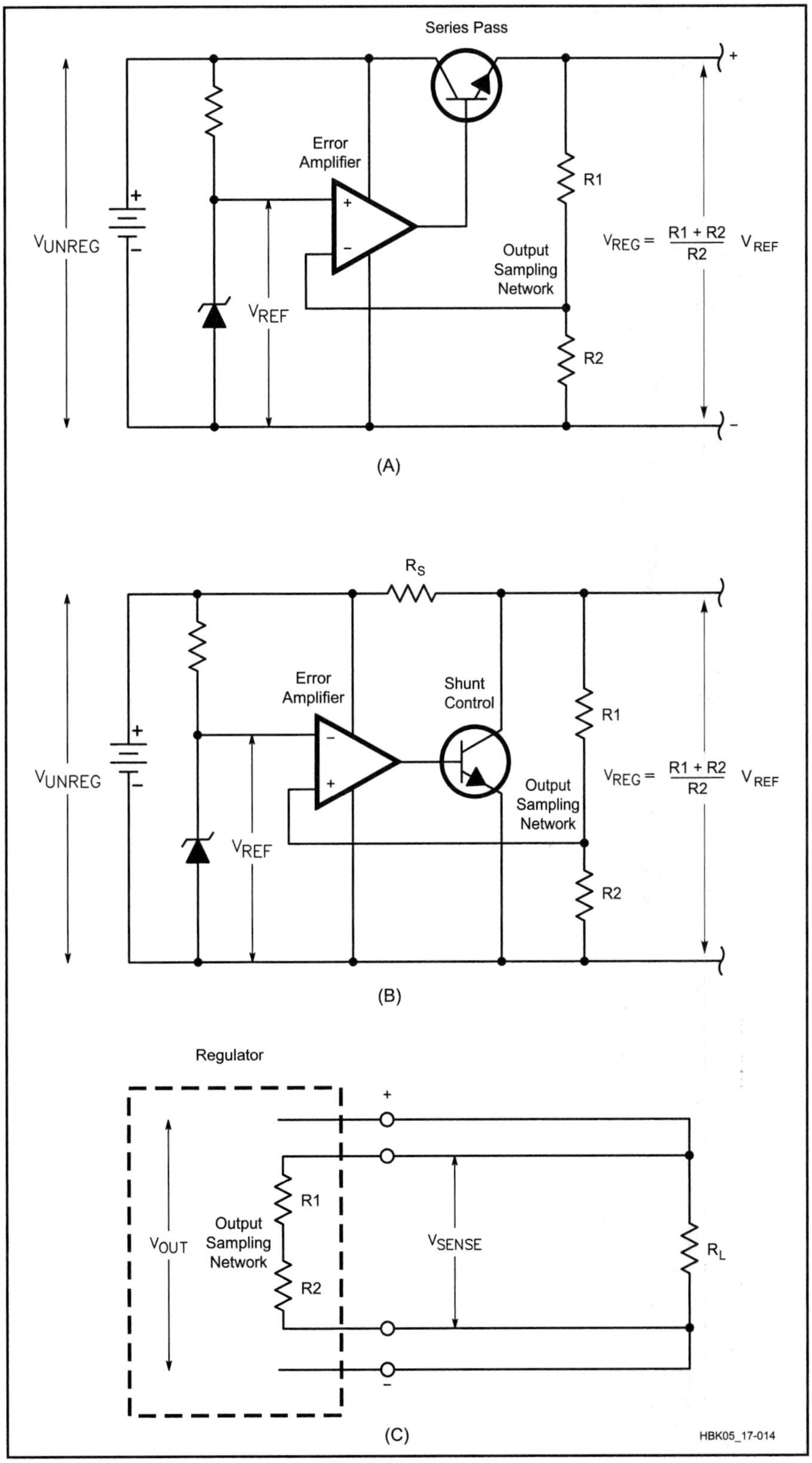

Fig 7.20 — Linear electronic voltage regulator circuits. In these diagrams, batteries represent the unregulated input-voltage source. A transformer, rectifier and filter would serve this function in most applications. Part A shows a series regulator and Part B shows a shunt regulator. Part C shows how remote sensing overcomes poor load regulation caused by the IR drop in the connecting wires by bringing them inside the feedback loop. The use of extra connections to sense voltage is called a "four wire Kelvin connection."

output voltage, which is equal to the Zener diode voltage (E_Z); and the minimum and maximum load current (I_L) through R_L. If the input voltage is variable, you must specify the maximum and minimum values for E_{DC}. As a rule of thumb, the current through the Zener should be 10% of the maximum load current for good regulation and must be greater than the $I_{Z(min)}$ at which the Zener diode maintains its constant voltage drop. Once these quantities are known the series resistance, R_S, can be determined:

$$R_S = \frac{E_{DC(min)} - E_Z}{1.1\, I_{L(max)}} \qquad (7)$$

The power dissipation of the Zener diode, P_D, is

$$P_{DZ} = \left[\frac{E_{DC(max)} - E_Z}{R_S} - I_{L(min)} \right] E_Z \quad (8)$$

and of the series resistor, R_S,

$$P_{DR} = \frac{\left(E_{DC(max)} - E_Z \right)^2}{R_S} \qquad (9)$$

It is good practice to provide a five times rated power dissipation safety margin for both the series resistor and the Zener diode. This avoids heating in the Zener and the resulting drift in voltage. High-power Zener diodes (10 W dissipation or more) will require heat-sinking as discussed in the section on Managing Heat in the **Electrical Fundamentals** chapter.

7.9.2 Linear Regulators

Linear regulators come in two varieties, *series* and *shunt*, as shown in **Fig 7.20**. The shunt regulator is simply an electronic (also called "active") version of the Zener diode. For the most part, the active shunt regulator (Fig 7.20B) is rarely used since the series regulator is a superior choice for most applications.

The series regulator (Fig 7.20A) consists of a stable voltage reference, which is usually established by a Zener diode, a transistor in series between the power source and the load (called a *series pass transistor*), and an error amplifier. In critical applications a temperature-compensated reference diode would be used instead of the Zener diode.

The output voltage is sampled by the error amplifier, which compares the output (usually scaled down by a voltage divider) to the reference. If the scaled-down output voltage becomes higher than the reference voltage, the error amplifier reduces the drive current to the pass transistor, thereby allowing the output voltage to drop slightly. Conversely, if the load pulls the output voltage below the desired value, the amplifier drives the pass transistor into increased conduction.

The "stiffness" or tightness of regulation

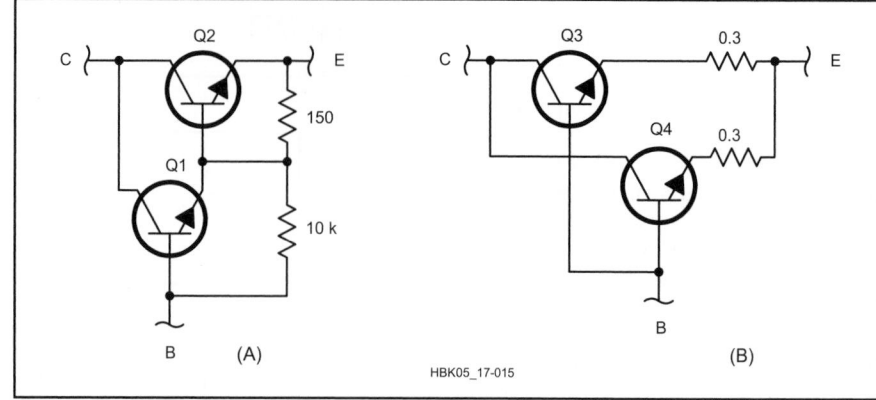

Fig 7.21 — At A, a Darlington-connected transistor pair for use as the pass element in a series-regulating circuit. At B, the method of connecting two or more transistors in parallel for high-current output. Resistances are in ohms. The circuit at A may be used for load currents from 100 mA to 5 A, and the one at B may be used for currents from 6 A to 10 A.

Q1 — NPN transistor, MJE340 or equivalent

Q2-Q4 — Power transistor such as 2N3055 or 2N3772

of a linear regulator depends on the gain of the error amplifier and the ratio of the output scaling resistors. In any regulator, the output is cleanest and regulation stiffest at the point where the sampling network or error amplifier is connected. If heavy load current is drawn through long leads, the voltage drop can degrade the regulation at the load. To combat this effect, the feedback connection to the error amplifier can be made directly to the load. This technique, called *remote sensing*, or a *four wire Kelvin connection*, moves the point of best regulation to the load by bringing the connecting loads inside the feedback loop. This is shown in Fig 7.20C.

INPUT VERSUS OUTPUT VOLTAGE

In a series regulator, the pass-transistor power dissipation is directly proportional to the load current and input/output voltage differential. The series pass element can be located in either leg of the supply. Either NPN or PNP devices can be used, depending on the ground polarity of the unregulated input.

The differential between the input and output voltages is a design tradeoff. If the input voltage from the rectifiers and filter is only slightly higher than the required output voltage, there will be minimal voltage drop across the series pass transistor. A small drop results in minimal thermal dissipation and high power-supply efficiency. The supply will have less capability to provide regulated power in the event of power line brownout and other reduced line voltage conditions, however. Conversely, a higher input voltage will provide operation over a wider range of input voltage, but at the expense of increased heat dissipation.

7.9.3 Linear Regulator Pass Transistors

DARLINGTON PAIRS

A simple Zener-diode reference or IC op-amp error amplifier may not be able to source enough current to a pass transistor that must conduct heavy load current. The Darlington configuration of **Fig 7.21A** multiplies the pass-transistor beta, thereby extending the control range of the error amplifier. If the Darlington arrangement is implemented with discrete transistors, resistors across the base-emitter junctions may be necessary to prevent collector-to-base leakage currents in Q1 from being amplified and turning on the transistor pair. These resistors are contained in the envelope of a monolithic Darlington device.

When a single pass transistor is not available to handle the current required from a regulator, the current-handling capability may be increased by connecting two or more pass transistors in parallel. The circuit of Fig 7.21B shows the method of connecting these pass transistors. The resistances in the emitter leads of each transistor are necessary to equalize the currents.

TRANSISTOR RATINGS

When bipolar (NPN, PNP) power transistors are used in applications in which they are called upon to handle power on a continuous basis, rather than switching, there are four parameters that must be examined to see if any maximum limits are being exceeded. Operation of the transistor outside these limits can easily result in device failure, and these parameters must be considered during the design process.

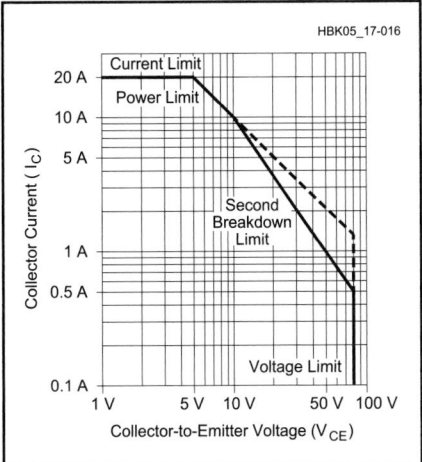

Fig 7.22 — Typical graph of the safe operating area (SOAR) of a transistor. See text for details. Safe operating conditions for specific devices may be quite different from those shown here.

The four limits are maximum collector current (I_C), maximum collector-emitter voltage (V_{CEO}), maximum power and *second breakdown* (I_{SB}). All four of these parameters are graphically shown on the transistor's data sheet on what is known as a *safe operating area (SOA)* graph. (see **Fig 7.22**) The first three of these limits are usually also listed prominently with the other device information, but it is often the fourth parameter — secondary breakdown — that is responsible for the "sudden death" of the power transistor after an extended operating period.

The maximum current limit of the transistor ($I_{C\ MAX}$) is usually the current limit for fusing of the bond wire connected to the emitter, rather than anything pertaining to the transistor chip itself. When this limit is exceeded, the bond wire can melt and open circuit the emitter. On the operating curve, this limit is shown as a horizontal line extending out from the Y-axis and ending at the voltage point where the constant power limit begins.

The maximum collector-emitter voltage limit of the transistor ($V_{CE\ MAX}$) is the point at which the transistor junctions can no longer withstand the voltage between collector and emitter.

With increasing collector-emitter voltage drop at maximum collector current, a point is reached where the power in the transistor will cause the junction temperature to rise to a level where the device leakage current rapidly increases and begins to dominate. In this region, the product of the voltage drop and the current would be constant and represent the maximum power (P_t) rating for the transistor; that is, as the voltage drop continues to increase, the collector current must decrease to maintain the power dissipation at a constant value.

With most transistors rated for higher voltages, a point is reached on the constant power portion of the curve whereby, with further increased voltage drop, the maximum power rating is *not* constant, but decreases as the collector to emitter voltage increases. This decrease in power handling capability continues until the maximum voltage limit is reached.

This special region is known as the *forward bias second breakdown (FBSB)* area. Reduction in the transistor's power handling capability is caused by localized heating in certain small areas of the transistor junction ("hot spots"), rather than a uniform distribution of power dissipation over the entire surface of the device.

The region of operating conditions contained within these curves is called the safe operating area, or SOA. If the transistor is always operated within these limits, it should provide reliable and continuous service for a long time.

MOSFET TRANSISTORS

The bipolar junction transistor (BJT) is rapidly being replaced by the MOSFET in new power supply designs because MOSFETs are easier to drive. The N-channel MOSFET (equivalent to the NPN bipolar) is more popular than the P-channel for pass transistor applications.

There are some considerations that should be observed when using a MOSFET as a linear regulator series pass transistor. Several volts of gate drive are needed to start conduction of the device, as opposed to less than 1 V for the BJT. MOSFETs are inherently very-high-frequency devices and will readily oscillate with stray-circuit capacitances. To prevent oscillation in the transistor and surrounding circuits, it is common practice to insert a small resistor of about 100 Ω directly in series with the gate of the series-pass transistor to reduce the gate circuit Q.

OVERCURRENT PROTECTION

Damage to a pass transistor can occur when the load current exceeds the safe amount. **Fig 7.23A** illustrates a simple current-limiter circuit that will protect Q1. All of the load current is routed through R1. A voltage difference will exist across R1; the value will depend on the exact load current at a given time. When the load current exceeds a predetermined safe value, the voltage drop across R1 will forward-bias Q2 and cause it to conduct. Because Q2 is a silicon transistor, the voltage drop across R1 must exceed 0.6 V to

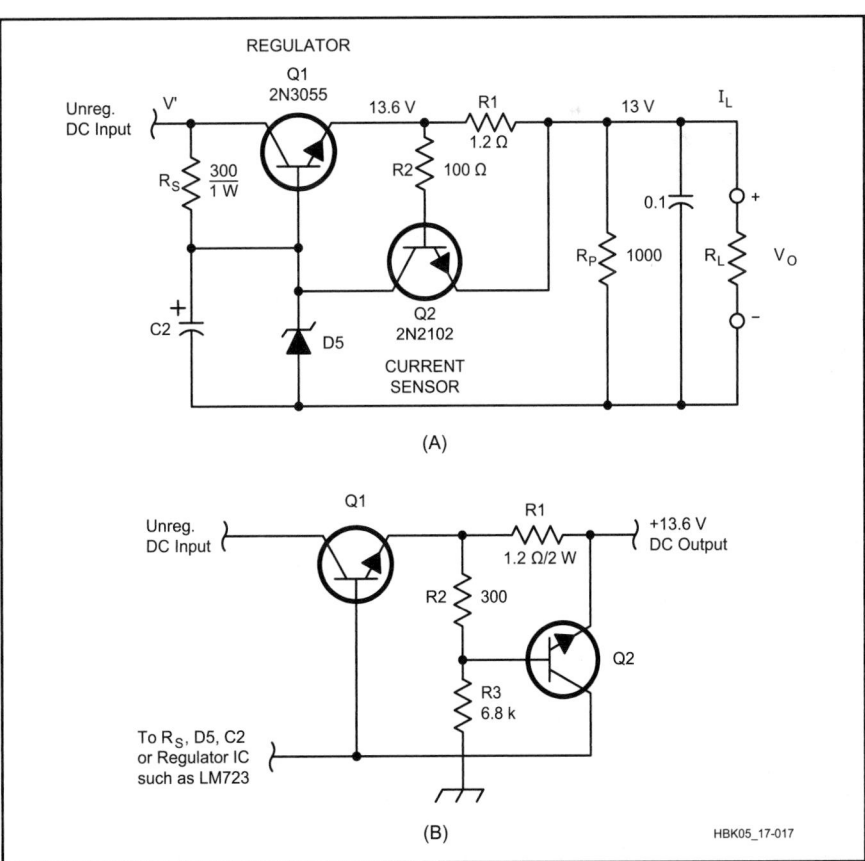

Fig 7.23 — Overload protection for a regulated supply can be implemented by addition of a current-overload-protective circuit, as shown at A. At B, the circuit has been modified to employ current-foldback limiting.

turn Q2 on. This being the case, R1 is chosen for a value that provides a drop of 0.6 V when the maximum safe load current is drawn. In this instance, the drop will be 0.6 V when I_L reaches 0.5 A. R2 protects the base-emitter junction of Q2 from current transients, or from destruction in the event Q1 fails under short-circuit conditions.

When Q2 turns on, some of the current through R_S flows through Q2, thereby depriving Q1 of some of its base current. This action, depending upon the amount of Q1 base current at a precise moment, cuts off Q1 conduction to some degree, thus limiting current through it.

FOLDBACK CURRENT LIMITING

Under short-circuit conditions, a constant-current type current limiter must still withstand the full source voltage and limited short-circuit current simultaneously, which can impose a very high power dissipation or second breakdown stress on the series pass transistor. For example, a 12 V regulator with current limiting set for 10 A and having a source of 16 V will have a dissipation of 40 W [(16 V – 12 V) × 10 A] at the point of current limiting (knee). But its dissipation will rise to 160 W under short-circuit conditions (16 V × 10 A).

A modification of the limiter circuit can cause the regulated output current to decrease with decreasing load resistance beyond the over-current knee. With the output shorted, the output current is only a fraction of the knee current value, which protects the series-pass transistor from excessive dissipation and possible failure. Using the previous example of the 12 V, 10 A regulator, if the short-circuit current is designed to be 3 A (the knee is still 10 A), the transistor dissipation with a short circuit will be only 16 V × 3 A = 48 W.

Fig 7.23B shows how the current-limiter example given in the previous section would be modified to incorporate foldback limiting. The divider string formed by R2 and R3 provides a negative bias to the base of Q2, which prevents Q2 from turning on until this bias is overcome by the drop in R1 caused by load current. Since this hold-off bias decreases as the output voltage drops, Q2 becomes more sensitive to current through R1 with decreasing output voltage. See **Fig 7.24**.

The circuit is designed by first calculating the value of R1 for short-circuit current. For example, if 0.5 A is chosen, the value for R1 is simply 0.6 V / 0.5 A = 1.2 Ω (with the output shorted, the amount of hold-off bias supplied by R2 and R3 is very small and can be neglected). The knee current is then chosen. For this example, the selected value will be 1.0 A. The divider string is then proportioned to provide a base voltage at the knee that is just sufficient to turn on Q2 (a value of 13.6 V for 13.0 V output). With 1.0 A flowing through R1, the voltage across the divider will be 14.2 V. The voltage dropped by R2 must then be 14.2 V –13.6 V, or 0.6 V. Choosing a divider current of 2 mA, the value of R2 is then 0.6 V / 0.002 A = 300 Ω. R3 is calculated to be 13.6 V / 0.002 A = 6800 Ω.

7.9.4 Three-Terminal Voltage Regulators

The modern trend in regulators is toward the use of three-terminal devices commonly referred to as *three-terminal regulators*. Inside each regulator is a voltage reference, a high-gain error amplifier, temperature-compensated voltage sensing resistors and a pass element. Many currently available units have thermal shut-down, overvoltage protection and current foldback, making them virtually destruction-proof. It is easy to see why regulators of this sort are so popular when you consider the low price and the number of individual components they can replace.

Three-terminal regulators have connections for unregulated dc input, regulated dc output and ground, and they are available in a wide range of voltage and current ratings. Fixed-voltage regulators are available with output ratings in most common values between 5 and 28 V. Other families include devices that can be adjusted from 1.25 to 50 V.

The regulators are available in several different package styles, depending on current ratings. Low-current (100 mA) devices frequently use the plastic TO-92 and DIP-style cases. TO-220 packages are popular in the 1.5 A range, and TO-3 cases house the larger 3 A and 5 A devices. They are available in surface mount packages too.

Three-terminal regulators are available as positive or negative types. In most cases, a positive regulator is used to regulate a positive voltage and a negative regulator a negative voltage. Depending on the system ground requirements, however, each regulator type may be used to regulate the "opposite" voltage.

Fig 7.25A and B illustrate how the regulators are used in the conventional mode. Several regulators can be used with a common-input supply to deliver several voltages with a common ground. Negative regulators may be used in the same manner. If no other common supplies operate from the input supply to the regulator, the circuits of Fig 7.25C and D may be used to regulate positive voltages with a negative regulator and vice versa. In these configurations the input supply is floated; neither side of the input is tied to the system ground.

Manufacturers have adopted a system of

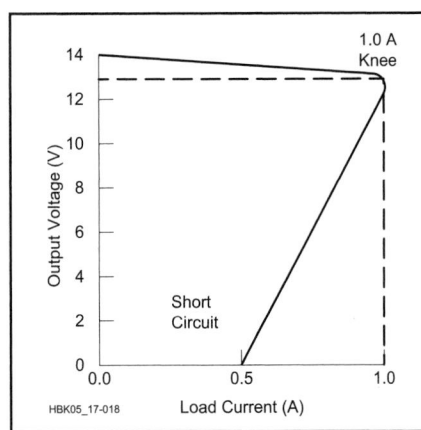

Fig 7.24 — The 1 A regulator shown in Fig 7.23B will fold back to 0.5 A under short-circuit conditions. See text.

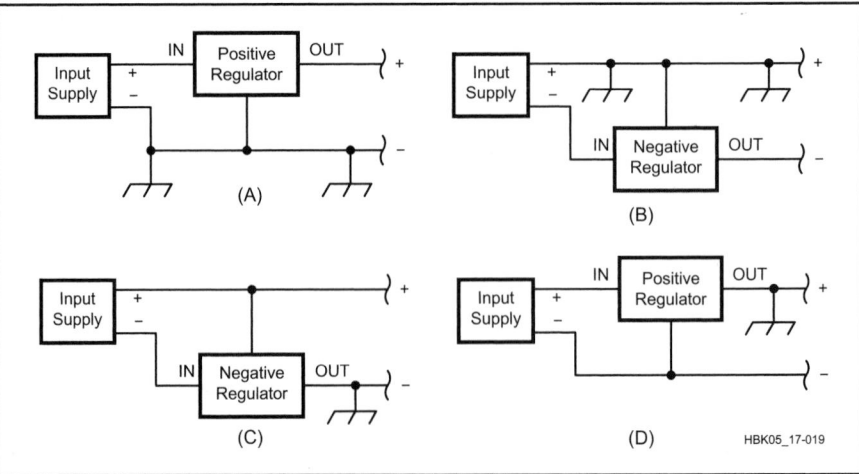

Fig 7.25 — Parts A and B illustrate the conventional manner in which three-terminal regulators are used. Parts C and D show how one polarity regulator can be used to regulate the opposite-polarity voltage.

family numbers to classify three-terminal regulators in terms of supply polarity, output current and regulated voltage. For example, 7805 describes a positive 5 V, 1.5 A regulator and 7905 a negative 5 V, 1.5 A unit. Depending on the manufacturer, the full part number might have a prefix such as LM, UA or MC, along with various suffixes (for example, LM7805CT or MC7805CTG). There are many such families with widely varied ratings available from manufacturers. More information may be found in the **Component Data and References** chapter.

SPECIFYING A REGULATOR

When choosing a three-terminal regulator for a given application, the most important specifications to consider are device output voltage, output current, minimum and maximum input-to-output differential voltages, line regulation, load regulation and power dissipation. Output voltage and current requirements are determined by the load with which the supply will ultimately be used.

Input-to-output differential voltage is one of the most important three-terminal regulator specifications to consider when designing a supply. The differential value (the difference between the voltage applied to the input terminal and the voltage on the output terminal) must be within a specified range. The minimum differential value, usually about 2.5 V, is called the *dropout voltage*. If the differential value is less than the dropout voltage, no regulation will take place. Special *low dropout regulators* with lower minimum differential values are available as well. At the other end of the scale, maximum input-output differential voltage is generally about 40 V. If this differential value is exceeded, device failure may occur.

Increases in either output current or differential voltage produce proportional increases in device power consumption. By employing current foldback, as described above, some manufacturers ensure that maximum dissipation will never be exceeded in normal operation. **Fig 7.26** shows the relationship between output current, input-output differential and current limiting for a three-terminal regulator nominally rated for 1.5 A output current. Maximum output current is available with differential voltages ranging from about 2.5 V (dropout voltage) to 12 V. Above 12 V, the output current decreases, limiting the device dissipation to a safe value. If the output terminals are accidentally short circuited, the input-output differential will rise, causing current foldback, and thus preventing the power-supply components from being over stressed. This protective feature makes three-terminal regulators particularly attractive in simple power supplies.

When designing a power supply around

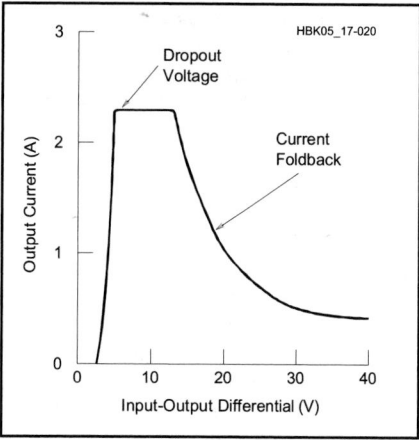

Fig 7.26 — Effects of input-output differential voltage on three-terminal regulator current.

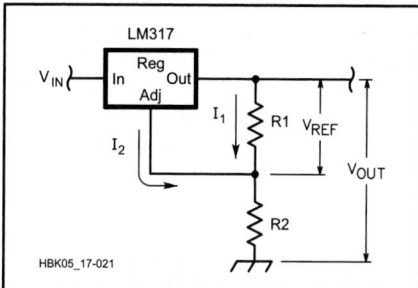

Fig 7.27 — By varying the ratio of R2 to R1 in this simple LM317 schematic diagram, a wide range of output voltages is possible. See text for details.

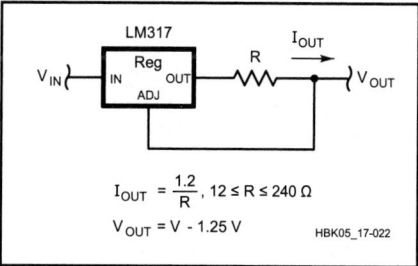

Fig 7.28 — The basic LM317 voltage regulator is converted into a constant-current source by adding only one resistor.

a particular three-terminal regulator, input-output voltage characteristics of the regulator should play a major role in selecting the transformer-secondary and filter-capacitor component values. The unregulated voltage applied to the input of the three-terminal device should be higher than the dropout voltage, yet low enough that the regulator does not go into current limiting caused by an excessive differential voltage. If, for

example, the regulated output voltage of the device shown in Fig 7.26 were 12 V, then unregulated input voltages of between 14.5 and 24 V would be acceptable if maximum output current is desired.

In use, all but the lowest-current regulators generally require an adequate external heat sink because they may be called on to dissipate a fair amount of power. Also, because the regulator chip contains a high-gain error amplifier, bypassing of the input and output leads is essential for stable operation.

Most manufacturers recommend bypassing the input and output directly at the leads where they protrude through the heat sink. Solid tantalum capacitors are usually recommended because of their good high-frequency capabilities.

External capacitors used with IC regulators may discharge through the IC junctions under certain circuit conditions, and high-current discharges can harm ICs. Look at the regulator data sheet to see whether protection diodes are needed, what diodes to use and how to place them in any particular application.

Adjustable Regulators

In addition to fixed-output-voltage ICs, high-current, adjustable voltage regulators are available. These ICs require little more than an external potentiometer for an adjustable output range from 5 to 24 V at up to 5 A. The unit price on these items is only a few dollars, making them ideal for test-bench power supplies. A very popular low current, adjustable output voltage three terminal regulator, the LM317, is shown in **Fig 7.27**. It develops a steady 1.25 V reference, V_{REF}, between the output and adjustment terminals. By installing R1 between these terminals, a constant current, I1, is developed, governed by the equation:

$$I1 \quad \frac{REF}{R1} \qquad (10)$$

Both I1 and a 100 µA error current, I2, flow through R2, resulting in output voltage V_O. V_O can be calculated using the equation:

$$V_O = V_{REF}\left(1 + \frac{R2}{R1}\right) + I2 \times R2 \qquad (11)$$

Any voltage between 1.2 and 37 V may be obtained with a 40 V input by changing the ratio of R2 to R1. At lower output voltages, however, the available current will be limited by the power dissipation of the regulator.

Fig 7.28 shows one of many flexible applications for the LM317. By adding only one resistor with the regulator, the voltage regulator can be changed into a constant-current source capable of charging NiCd batteries, for example. Design equations are given in the figure. The same precautions should be taken with adjustable regulators as with the fixed-voltage units. Proper heat sinking and lead bypassing are essential for proper circuit operation.

INCREASING REGULATOR OUTPUT CURRENT

When the maximum output current from an IC voltage regulator is insufficient to operate the load, discrete power transistors may be connected to increase the current capability. **Fig 7.29** shows two methods for boosting the output current of a positive regulator, although the same techniques can be applied to negative regulators.

In A, an NPN transistor is connected as an emitter follower, multiplying the output current capacity by the transistor beta. The shortcoming of this approach is that the base-emitter junction is not inside the feedback loop. The result is that the output voltage is reduced by the base-emitter drop, and the load regulation is degraded by variations in this drop.

The circuit at B has a PNP transistor

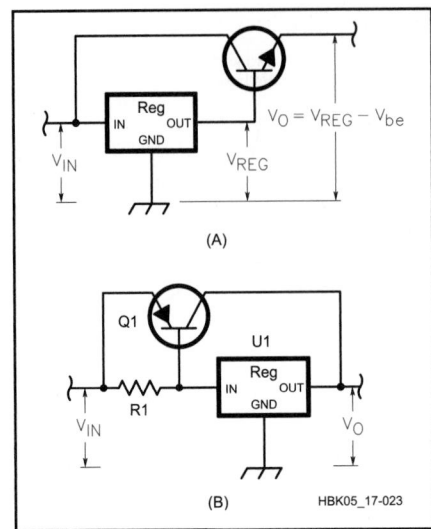

Fig 7.29 — Two methods for boosting the output-current capacity of an IC voltage regulator. Part A shows an NPN emitter follower and B shows a PNP "wrap-around" configuration. Operation of these circuits is explained in the text.

"wrapped around" the regulator. The regulator draws current through the base-emitter junction, causing the transistor to conduct. R1 provides bias voltage for turning on Q1

so that U1 doesn't see the excess current. For example, a 6 Ω resistor will limit the current U1 sees to 100 mA. The IC output voltage is unchanged by the transistor because the collector is connected directly to the IC output (sense point). Any increase in output voltage is detected by the IC regulator, which shuts off its internal-pass transistor, and this stops the boost-transistor base current.

7.10 "Crowbar" Protective Circuits

Electronic components *do* fail from time to time. In a regulated power supply, the only component standing between an elevated dc source voltage and your transceiver is one transistor, or a group of transistors wired in parallel. If the transistor, or one of the transistors in the group, happens to short internally, your equipment could suffer lots of damage.

To safeguard the load equipment against possible overvoltage, some power-supply manufacturers include a circuit known as a *crowbar*. This circuit usually consists of a

silicon-controlled rectifier (SCR) or thyristor connected directly across the output of the power supply, with an over-voltage-sensing trigger circuit tied to its gate. The SCR is large enough to take the full short-circuit output current of the supply, as if a crowbar were placed across the output terminals, thus the name.

In the event the output voltage exceeds the trigger set point, the SCR will fire, and the output is short circuited. The resulting high current in the power supply (shorted

output in series with a series pass transistor failed short) will blow the power supply's line fuses. This is a protection for the supply as well as an indicator that something has malfunctioned internally. For these reasons, never replace blown fuses with ones that have a higher current rating.

An example of a crowbar overvoltage protection circuit can be found as a project at the end of this chapter. It provides basic design equations that can be adapted to a wide range of power supply applications.

7.11 DC-DC Switchmode Power Conversion

Very often the power source is dc, such as a battery or solar cell, or the output of an unregulated rectifier connected to an ac source. In most applications high conversion efficiency is desired, both to conserve energy from the source and to reduce heat dissipation in the converter. When high efficiency is needed, some form of switching circuit will be employed for dc-dc power conversion. Besides being more efficient, switching circuits are usually much smaller and lighter than conventional 60 Hz, transformer-rectifier circuits because they operate at much higher frequencies — from 25 to 400 kHz or even higher. Switching circuits go by many names; *switching regulators* and *switchmode converters* are just two of the more common names.

The possibility of achieving high conversion efficiency stems directly from the use of switches for the power conversion process, along with low-loss inductive and capacitive elements. An active switch is a device that is either ON or OFF and the state of the switch (ON or OFF) can be controlled with an external signal. The loss in the switch is always the product of the voltage across the switch and the current flowing through it ($P = E \times I$).

In the ON (conducting) state the voltage drop across the switch is small, and in the OFF state the current through the switch is small. In both cases the losses can be small relative to the power level of the converter. During transitions between ON and OFF states, however, there will simultaneously be both substantial voltage across the switch and significant current flowing through it. This results in power dissipation in the switch, called *switching loss*. This loss is minimized by making the switching transitions as short as possible. In this way even though the instantaneous power dissipation may be high, the average loss is low because of the small duty cycle of the transitions. Of course the more frequently the switch operates (higher switching frequency), the higher the average loss will be and this eventually limits the maximum operating frequency. A limitation on the lower end of the switching frequency range is that it needs to be above audible frequencies (>25 kHz).

This is quite different from the linear regulators discussed earlier in which there is always some voltage drop across the pass transistor (which is acting as a controlled variable resistor) while current is flowing through it. As a result the efficiency of a linear regulator can be very low, often 65% or less. Switchmode converters on the other hand will typically have efficiencies in the range of 85 to 95%.

Switchmode circuits can also generate radio-frequency interference (RFI) through VHF because of switching frequency harmonics and ringing induced by the rapid rise and fall times of voltage and current. In attempting to minimize the ON-OFF transition time, significant amounts of RF energy can be generated. To prevent RFI to sensitive receivers, careful bypassing, shielding and filtering of both input and output circuits is required. (RFI from switchmode or "switching" supplies is also discussed in the chapter on **RF Interference**.)

There are literally hundreds of different switchmode circuits or "topologies" (see Reference 2) but we will only look at a few of those most commonly used by amateurs. Fortunately, the characteristics of the simpler circuits are to a large extent replicated in more complex circuits so that an understanding of the basic circuits provides an entry point to many other circuits.

The following discussion is only an introduction to the basics of switchmode converters. To really get a handle on designing these circuits the reader will have to do some additional reading. Fortunately a very large amount of useful information is freely available on-line. Many useful application notes are available from semiconductor manufacturers, and additional information can be found on the websites of filter capacitor and ferrite core manufacturers (see References 3 and 4). There are also numerous books on the subject, often available in libraries and used book stores.

Switchmode circuits can and have been implemented with many different kinds of switches, from mechanical vibrators to vacuum and gas tubes to semiconductors. Today however, semiconductors are the universal choice. For converters in the power range typical of amateur applications (a few watts through 2-3 kW) the most common choices of semiconductor switches would be either bipolar junction transistors (BJTs) or power MOSFETs. BJTs have a long history of use in switchmode applications, but to employ BJTs in a reliable and trouble-free circuit requires a relatively sophisticated understanding of them. While the MOSFET must also be used carefully, it is generally easier for a newcomer. The following circuit diagrams will show MOSFETs or generic switch symbols for the switches but keep in mind that all of the circuits can be implemented with other types of switches.

7.11.1 The Buck Converter

A schematic for a *buck converter* is shown in **Fig 7.30A**. This circuit is called a "buck" converter because the output voltage is always less than or equal to the input voltage. Power is supplied from the dc source (V_S) through

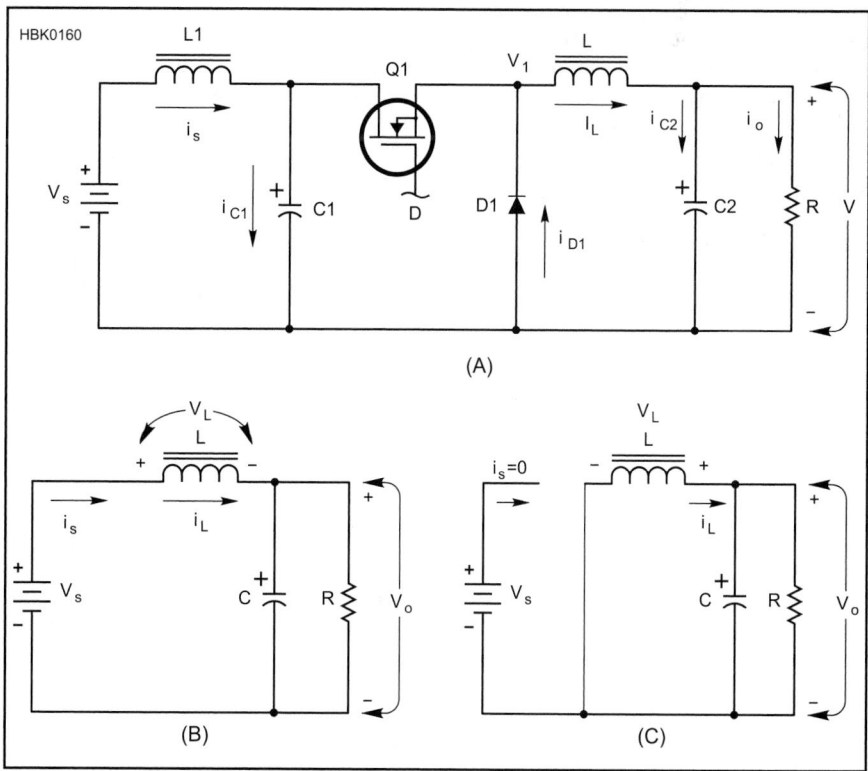

(A)

(B)

(C)

Fig 7.30 — Typical buck converter.

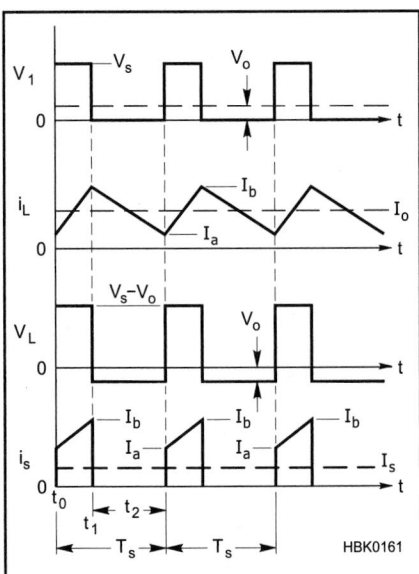

Fig 7.31 — Waveforms in a buck converter.

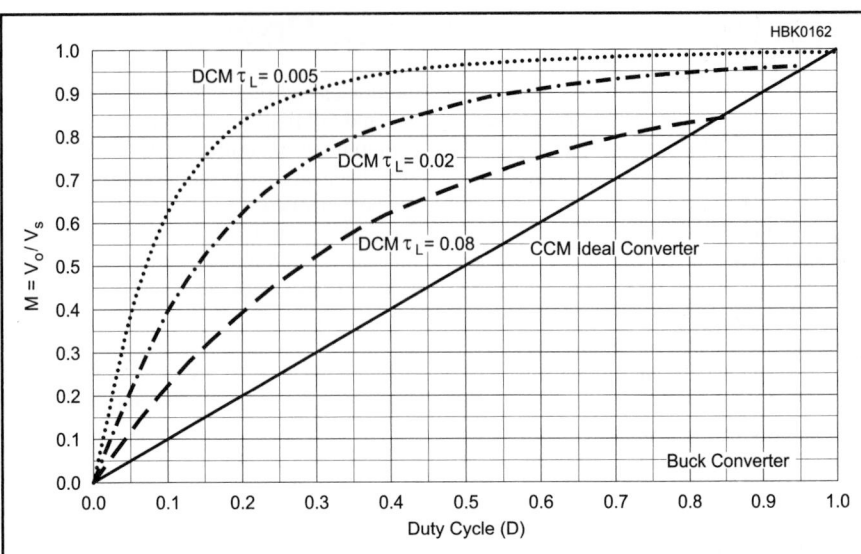

Fig 7.32 — Change in output voltage to input voltage ratio (M=VO/VS) as a function of switch duty cycle (D).

the input filter (L1, C1) to the drain of the switch (Q1). The load (R) is connected across the output filter (L, C2).

The equivalent circuit when Q1 is ON is shown in Fig 7.30B (where the input filter components are assumed to be part of V_S). V_S is connected to one end of the output inductor (L) and, because $V_O < V_S$, current flows from the source to the output delivering energy from the source to the output and also storing energy in L. At some point Q1 is turned OFF and the current flowing in L commutates (switches) to D1, as shown in Fig 7.30C. (The current in an inductor cannot change instantaneously, so when Q1 turns OFF, the collapsing magnetic field in L pulls current through shunt diode D1, called a *free-wheeling diode*.) The energy in L is now being discharged into the output. This cycle is repeated at the switching frequency (f_s). The ratio of the switch ON-time to the total switching period ($T_s = t_{on} + t_{off} = 1/f_s$) is called the *duty cycle (D)*.

Typical voltage and current waveforms for a buck converter are shown in **Fig 7.31** where V_1 is the voltage at the junction of Q1, D1 and L (see Fig 7.30A).

The interval 0-t_1 corresponds to the ON time of Q1 and $T_s - t_1$ corresponds to the OFF time ($T_s = t_1 + t_2$). From the waveforms it can be seen that the current in the inductor rises while Q1 is ON and falls while Q1 is OFF. The energy in the inductor is proportional to $LI^2/2$.

The input current (I_s) is pulsating at the switching frequency. This pulsation in the input current is the reason for the input filter (L1 and C1). We need to keep this high frequency noise (the ac component of the pulsation) out of the source. All switchmode converters require some form of filter on the input to

keep switching noise out of the source. Some form of filter is also required on the output to keep the switching noise out of the load. In Fig 7.30, L and C2 form the output filter. For applications requiring low ripple it would not be unusual to see an additional stage of L-C filtering added to the output.

The voltage waveform (V_1) at the input to the output filter (L and C2) is pulse-width modulated (PWM) and V_O (the dotted line in the waveform for V_1 in Fig 7.31A) is the time average of V_1. V_O is controlled or regulated by adjusting the duty cycle of Q1 in response to input voltage or output load changes. If we increase D we increase V_O up to the point where D = 1 (the switch is ON all the time) and $V_O \approx V_S$. If we never turn Q1 on (D = 0) then $V_O = 0$. Normal operation will be somewhere between these extremes.

The inductor current waveform in Fig 7.31 does not go to zero while Q1 is OFF. In other words, not all of the energy stored in L is discharged into the output by the end of each switching cycle. This is a common mode of operation for heavier loads. It is referred to as the *continuous conduction mode* or CCM, where the conduction referred to is the current in L. There is another possibility: during the time Q1 is off all of the energy in L may be discharged and for some period of time the inductor current is zero until Q1 is turned ON again. This is referred to as the *discontinuous conduction mode* or DCM. A typical converter will operate in CCM for heavy loads but as the load is reduced, at some point the circuit operation will change to DCM. CCM is often referred to as the "heavy load" condition and DCM as the "light load" condition.

This distinction turns out to be very important because the behavior of the circuit, for

both small signal and large signal, is radically different between the two modes. **Fig 7.32** shows the relationship between the duty cycle (D) of Q1 and the ratio of the output voltage to the input voltage ($M = V_O/V_S$) as a function of τ_L in Fig 7.32:

$$\tau_L = \left(\frac{L}{R}\right) f_S \qquad (12)$$

τ_L is just a convenient way to make Fig 7.32 more general by tying together the variables L, R and f_s. Smaller values for τ_L correspond to lighter loads (larger values for R). As can been seen in Fig 7.32, for very light loads the higher values for D have little effect on the output/input voltage ratio. This is basically charging of the output capacitor (C2) to the peak value of $V_1 \approx V_S$.

For CCM operation, M is a linear function of D. As we vary the load, V_O will remain relatively constant. But when we go into DCM, the relation between D and M is no longer linear and in addition is heavily dependent on the load: V_O will vary as the load is changed unless the duty cycle is varied to compensate. This kind of behavior is typical of all switching regulators, even those not directly related to the buck converter. In fact, we have seen this behavior already in the section on choke input filters in Fig 7.17 where the output voltage is close to the peak input voltage for light loads and decreases as the load increases until a point is reached (I_C) where V_O stabilizes.

In a buck converter the value for the critical inductance (L_C) can be found from:

$$L_c = R\left(\frac{1-M}{2 f_S}\right) \qquad (13)$$

This looks just like equation 6 (in the earlier section on choke-input filters) with

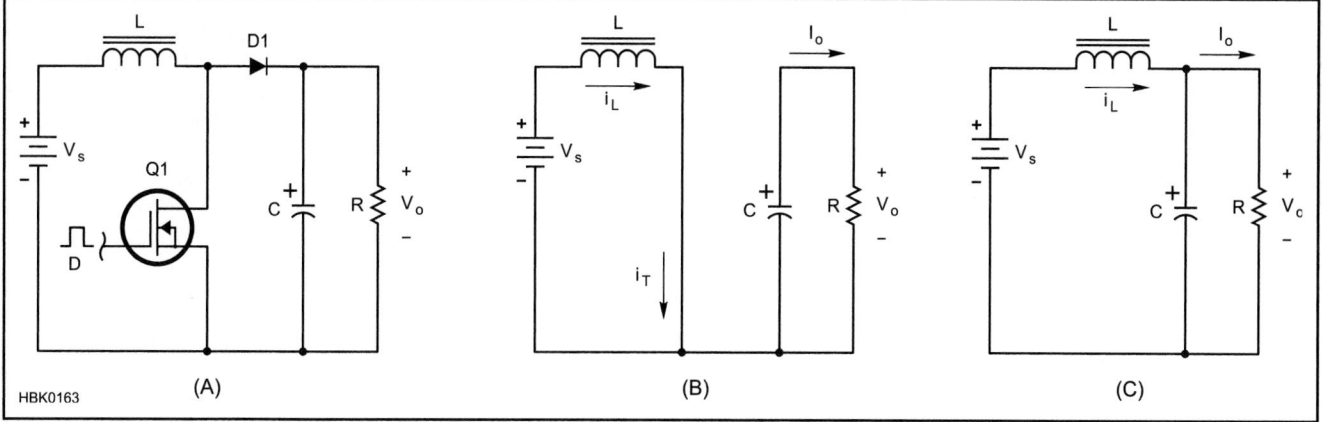

Fig 7.33 — Typical boost converter.

$A = 2fs / (1 – M)$! The two phenomena are the same: peak charging versus averaging of V1.

7.11.2 The Boost Converter

A *boost converter* circuit is shown in **Fig 7.33A**. This circuit is called a "boost" converter because the output voltage is always greater than or equal to the input voltage.

When Q1 is ON (Fig 7.33B), L is connected in parallel with the source (V_S) and energy is stored in it. During this time interval load energy is supplied from C. When Q1 turns OFF (Fig 7.33C), L is connected via D1 to the output and the energy in L, plus additional energy from the source, is discharged into the output. The value for L, V_S and length of time it is charged determines the energy stored in L. Again we have the possibilities that either some (CCM) or all (DCM) the energy in L is discharged during the OFF-time of Q1. The variation in the ratio of the output-to-input voltage (M) with duty cycle is shown in **Fig 7.34**.

As we saw in the buck converter, the conduction mode of L is important. In an ideal converter operating in CCM, the output voltage is substantially independent of load and $M \approx 1 / (1 – D)$. In realistic boost converters there is an important limit on the CCM value for M. In an ideal boost converter you could make the boost ratio (M) as large as you wish but in real converters the parasitic resistances associated with the components will limit the maximum value of M as indicated by the dashed line for CCM operation. The exact shape of this part of the control function will depend on the ratio of the parasitic resistance within the converter to the load resistance (see Reference 2).

There is also another very important practical effect of this limitation on the peak value for M. When the duty cycle is increased beyond the point of maximum M (this occurs at D=0.9 in Fig 7.34), the sign of the slope of the control function changes so that the control loop will change from negative feedback to

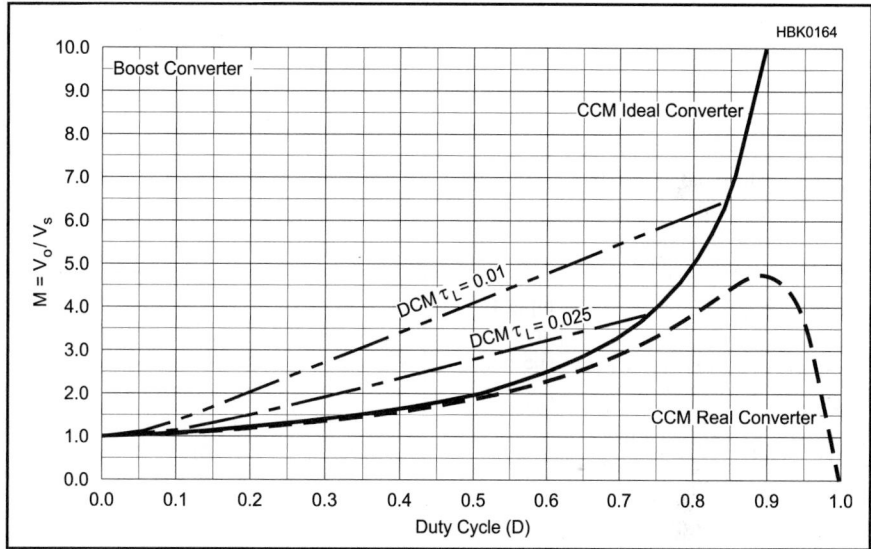

Fig 7.34 — DC control characteristic for a boost converter. M = V$_O$/V$_S$, D = duty cycle of Q1 and $\tau_L = f_s$ (L/R).

positive feedback. This can cause the converter to latch up under some load conditions if the control circuit allows D to exceed the value at maximum M. In boost converters the design of the control circuit must limit the maximum value for D so that latch-up is not possible although this may be difficult for overload conditions.

As in the case of the buck converter, in DCM the control function for M is very different from what it is in CCM and is strongly dependent on the load R. One advantage of the DCM operation is that the limitation on the maximum value for M because of parasitic circuit resistances is not nearly so pronounced. By operating in DCM it is possible to have M > 5.

An important limitation of the boost converter is that with Q1 turned OFF, you have no control over V_O. V_O will simply equal V_S. In addition, when V_S is turned on there will be an inrush current through D1, into C which

cannot be controlled by Q1. In the case of the buck converter, if Q1 is kept OFF when V_S is turned on, there will be no inrush current charging the output filter capacitor (C2). In the buck converter C2 can be charged slowly by ramping up the duty cycle of Q1 during turn-on but this is not possible in the boost converter.

7.11.3 Buck-Boost and Flyback Converters

Fig 7.35 shows an example of a *buck-boost converter*. The name "buck-boost" comes from the fact that the magnitude of the output voltage can be either greater or less than the input voltage.

When S1 is ON, V_S is applied across L and energy is stored in it. During the ON time of Q1, D1 is reverse-biased and non-conducting. The output voltage (V_O) across the load (R) is supported solely from the energy

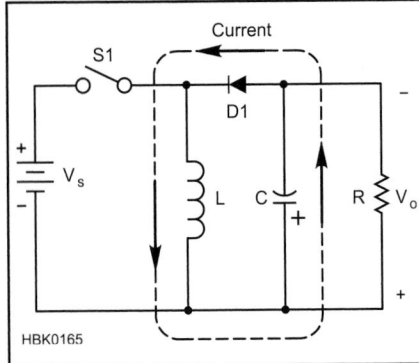

Fig 7.35 — An example of a buck-boost converter.

stored in C. When S1 is OFF, the energy in L is discharged into C through D1. Note that the polarity of V_O is inverted from V_S: *positive V_S means negative V_O*. The relationship between V_O and V_S as a function of duty cycle is shown in **Fig 7.36**.

This graph closely resembles that for the boost converter (Fig 7.34) with one important exception: |M| begins at zero. This allows the magnitude of the output voltage to be either below or above the source voltage, hence the name "buck-boost."

FLYBACK CONVERTERS

Simple buck-boost converters are occasionally used, but much more often it is the transformer-coupled version of this converter, referred to as a *flyback converter*, that is employed. The relationship between the flyback and buck-boost converters is illustrated in

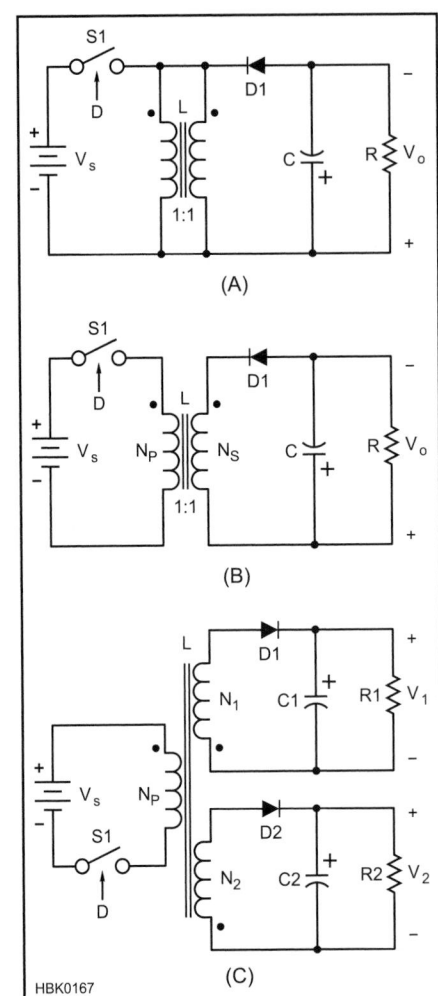

Fig 7.37 — The relationship between buck-boost and flyback converters.

Fig 7.37. In Fig 7.37A we have a standard buck-boost converter, the only change from Fig 7.35 being two parallel and equal windings on the inductor. In Fig 7.37B we remove the links at the top and bottom of the two parallel windings, converting the inductor into a transformer with primary and secondary windings. The only change is that when S1 is ON, current flows in the primary of the transformer and when S1 is OFF, the current flows in the secondary winding delivering the stored energy to the output. At this point the circuit operation is the same except that we have introduced primary-to-secondary galvanic isolation.

We are now free to change the turns ratio from 1:1 to anything we wish. We can also change the polarity of the output voltages and/or add more windings with other voltages and additional loads as shown in Fig 7.37C, a typical example of a flyback converter. These are most often used in the power range of a few watts to perhaps 200 W. For higher power levels other circuits are generally more useful.

The advantages of the flyback converter lie in its simplicity. It requires only one power switch and one diode on each of the output windings. The inductor is also the isolation transformer so you have only one magnetic component. The disadvantage of this circuit is that both in the input and output current waveforms are pulsating. The result is that more filtering is required and the filter capacitors are exposed to high RMS currents relative to the power level.

7.11.4 The Forward Converter

The buck converter has many useful properties but it lacks input-to-output galvanic isolation, the ability to produce output voltages higher than the input voltage, and/or multiple isolated output voltages for multiple loads. We can overcome these drawbacks by inserting a transformer between the switch (S1) and the shunt diode (D1). To make this simple idea work however, we also have to add a diode in series with the output of the transformer (D2), and a third winding (N_R) with another diode (D3) to the transformer. This is done to provide a means for resetting the core (returning the magnetic flux to zero) by the end of each switching cycle. The result is the *forward converter* shown in **Fig 7.38**.

The circuit in A is the one just described. The variation in B uses two switches (S1 and S2, which switch ON and OFF *simultaneously*) instead of one but eliminates the need for a reset winding on the transformer. For the circuit in A and a given input voltage, the voltage across the switch (in the OFF state) will be equal to V_S plus the reset voltage during the OFF-time. Typically the peak switch voltage will be about $2V_S$. In the circuit shown in B the switch voltage is limited to V_S which is

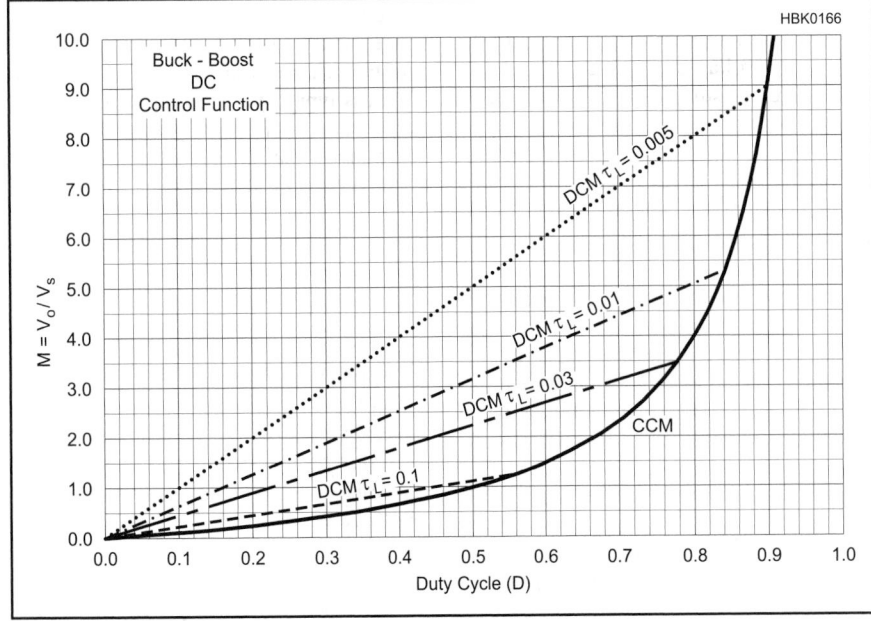

Fig 7.36 — DC control function for an ideal buck-boost converter. M = V_O/V_S, D = duty cycle of Q1 and $\tau_L = f_s$ (L/R).

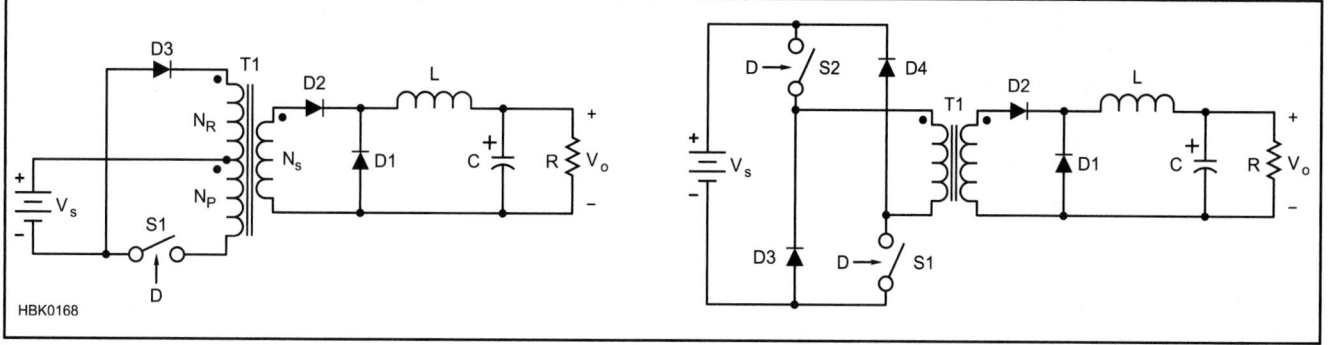

Fig 7.38 — Example of a forward converter.

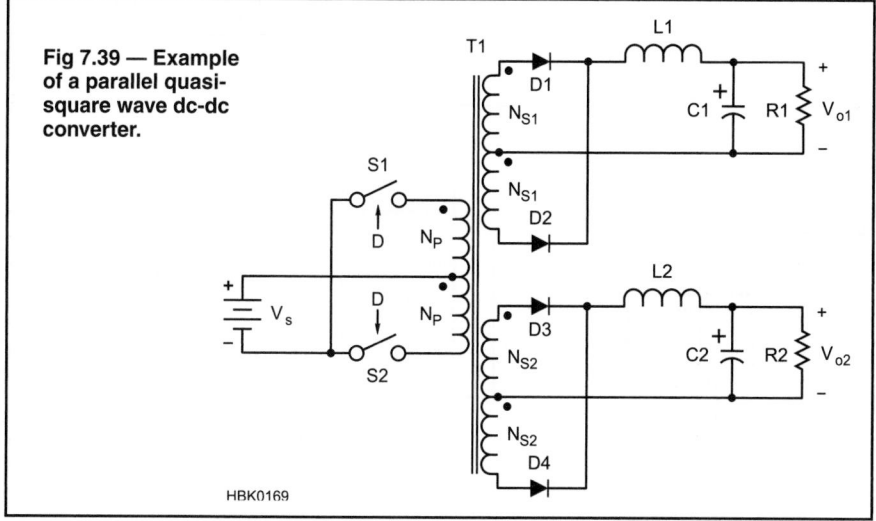

Fig 7.39 — Example of a parallel quasi-square wave dc-dc converter.

very helpful at higher line voltages. These circuits behave just like a buck converter except you can now have multiple isolated outputs with arbitrary voltages and polarities. These circuits are typically used in converters with power levels of 100 to 500 W.

7.11.5 Parallel, Half and Full-Bridge Converters

As the power level increases it becomes advantageous to use more switches in a somewhat more complex circuit. For applications where the input voltage is low (<100V) and the current high, the push-pull *quasi-square wave* circuit shown in **Fig 7.39** is often used.

S1 and S2 are switched on alternately with duty cycles <0.5. The output voltages are controlled by the duty cycle, D. The peak switch currents are equal to the input current but the peak switch voltages will be $2V_S$.

This circuit is still just a buck converter, but with an isolation transformer added that allows multiple outputs with different voltages above or below the input voltage. It would be very common to have a 5 V, high-current

output and ±12 V, lower-current outputs, for example. This converter is typically used for operation from vehicular power with loads up to several hundred watts.

Operating directly from rectified ac utility power usually means that V_S will be 200 V or more. For these applications, **Fig 7.40** shows how the switches are configured in either a half- (Fig 7.40A) or full-bridge (Fig 7.40B) circuit.

In A, S1 and S2 switch alternately and are pulse-width modulated (PWM) to control the output. CA and CB are large capacitors that form a voltage divider with $V_S/2$ across each capacitor. The peak switch voltages will be equal to V_S but the switch currents will be $2I_S$. The peak voltage across the primary winding will be $V_S/2$. This circuit would typically be used for off-line applications with output powers of 500 W or so.

In B, S1 and S4 switch simultaneously alternating with S2 and S3 which also switch simultaneously. The output is controlled by PWM. The peak switch voltage will be equal to V_S and the peak switch current close to I_S (the peak value is a little higher due to ripple on the inductor current which is reflected back into the primary winding). The full-bridge circuit is typically used for power levels of 500 W to several kW. C_S is present in the full-bridge circuit to prevent core saturation due to any asymmetry in the primary PWM voltage waveforms.

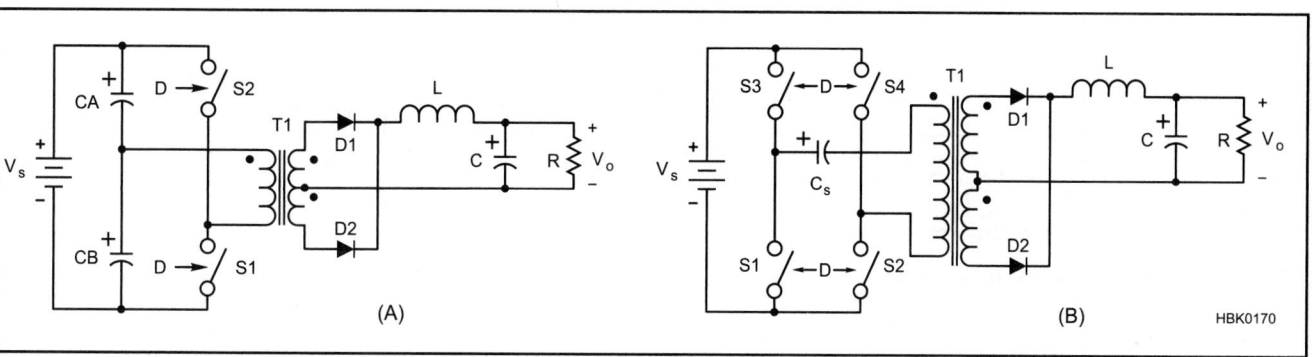

Fig 7.40 — Examples of half and full-bridge quasi-square wave dc-dc converters.

7.11.6 Building Switchmode Power Supplies

Selecting a switching circuit or topology is just the first step in building a practical switch-mode power supply. In the actual circuit you will have to sample the output voltage, provide a voltage reference against which to compare the output, include a modulator that will convert the error voltages to a variable-duty cycle signal and finally provide correct drive to the power switches. Fortunately, all of these functions can be provided with readily available integrated circuits and a few external components. These ICs typically come with extensive applications information.

Particularly for low power applications (<100 W) there are ICs that provide all the

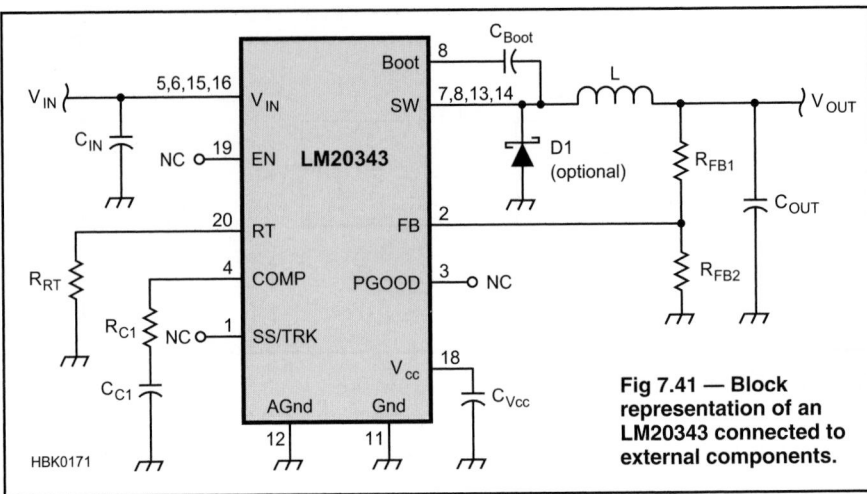

Fig 7.41 — Block representation of an LM20343 connected to external components.

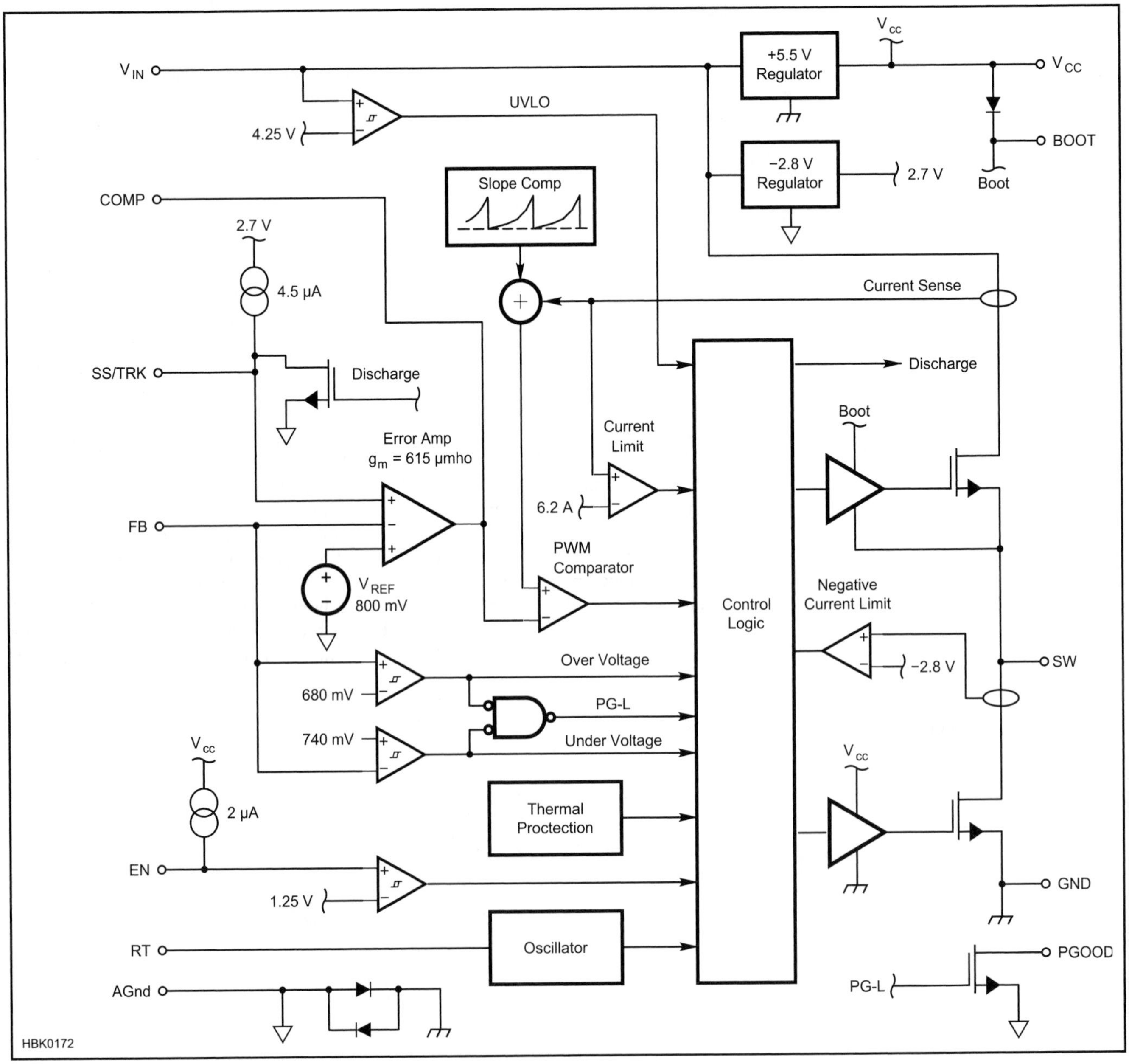

Fig 7.42 — Internal block diagram for the LM20343.

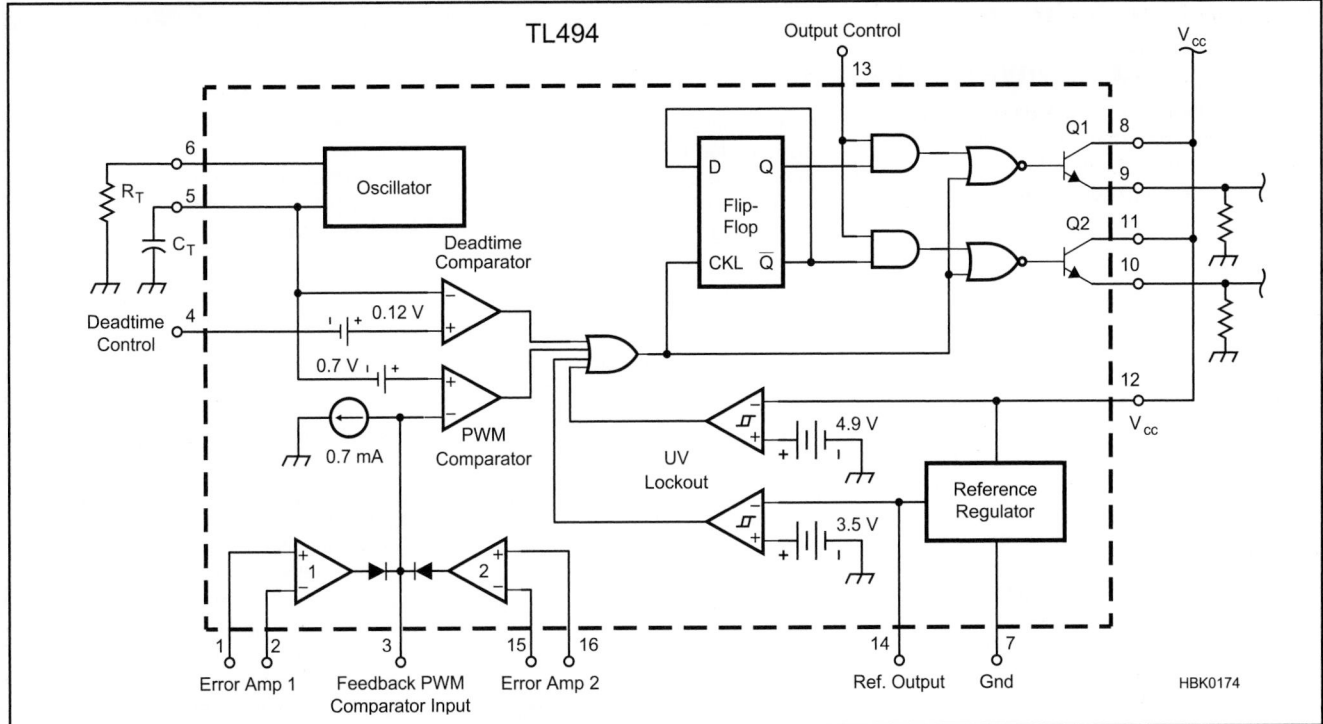

Fig 7.43 — Internal block diagram representation for the TL494.

control functions *and* the power switch. In some cases a power diode is also included. **Fig 7.41** gives an example of one of these ICs. Similar ICs are made by many different manufacturers.

The IC provides most of the components for a buck converter but some external components are still needed: filter capacitors (C_{IN}, C_{OUT}, C_{VCC}), the output inductor (L), output voltage sensing (R_{FB1} and R_{FB2}) and components to set the switching frequency (R_{C1} and C_{C1}). The shunt diode (D1) is marked as optional because there is an internal switch that can perform this function for lower powers.

Fig 7.42 and **Fig 7.43** show the internal block-diagram of typical voltage regulator control ICs. (See the manufacturer data sheets for application information and reference designs.) In addition to the control functions there are also some protective functions such as input under- and overvoltage protection, over-temperature (thermal) protection, over-current limiting and slow-starting (slowly ramping up D) to limit inrush current at turn-on. ICs similar to this one are available for boost, buck-boost and flyback converters, as

well as many other topologies. There are also ICs available that implement more complex control schemes such as current mode control and feed-forward compensation.

7.11.7 Switchmode Control Loop Issues

We do not have the space in this book to explain all the control issues associated with switchmode converters, but it is important to recognize that designing stable, high performance control loops for switchmode converters is a much more complex task than it would be for a series pass-regulator or an audio amplifier. All we can do here is to alert the reader to some of the issues and suggest consulting Reference 2 with detailed explanations and examples. In addition, there are numerous books and applications notes on switchmode converter design that go into the necessary detail.

The complexities arise from the inherent behavior of switchmode circuits. This behavior is often non-intuitive and sometimes even bizarre. As we've seen already, the dc beha-

vior differs dramatically between CCM and DCM modes of operation. This difference is even more pronounced in the small-signal control-to-output characteristics. It is very common to see poles and zeros in the control system's response (see the **Analog Basics** chapter) that move with input voltage, duty cycle and/or load. Fixed double-poles can change to moving single-poles. Moving right-half-plane zeros (this is a zero with decreasing phase-shift instead of increasing phase-shift with frequency) are inherent in CCM operation of boost and buck-boost circuits and their derivatives. There can also be large signal instabilities such as oscillations at sub-harmonics of the switching frequency and occasionally chaotic response to large load current or line voltage changes.

This is not intended to discourage readers from working with switchmode converters, but to simply alert them in advance to some of the difficulties. Once a basic understanding is achieved, the design of switchmode converters can be very interesting and rewarding.

Switchmode Converter Design Aids

Today, there are hundreds of ICs on the market, from dozens of manufacturers, aimed at controlling the power conversion process. These devices have grown from their simplest forms when introduced over 20 years ago to include a long list of ancillary functions aimed at reducing the supporting circuitry and enhancing the quality and efficiency of the power conversion process. As a result, the task of designing these ICs into the final power conversion circuit has actually become more complex. To the circuit designer who is below the expert level, the task can seem daunting, indeed.

In answer to this problem, many IC manufacturers provide computer-aided design tools that greatly simplify the task of using their products. These tools take several forms, from the most rudimentary cookbook-style guides, to full-fledged circuit simulation tools like *SPICE*. The purpose of these tools is to help the designer pick appropriate resistors, capacitors and magnetic components for the design, and also to estimate the stresses on the power-handling components.

These design tools fall generally into two types: equation-based design tools and iterative circuit simulators. The differences are that the equation-based tools simply automate the basic design equations of the circuit, providing recommended choices for the components and then solving equations to compute the voltage, current and power stresses on the components. In some cases, thermal analysis is also included.

The simulators, on the other hand, use detailed mathematical models for the components and provide both dc and ac simulation of the circuits. This allows the user to see the dynamic behavior of the circuit. The simulation includes details during the switching intervals and shows rise times, parasitic effects caused by component capacitance, internal resistance and other characteristics.

In modern power converter design, the magnetic components — transformers, power inductors and filter inductors — can be a challenge to the designer not versed in magnetic design. As a result, the vendors of these parts usually furnish design tools created specifically to the design of these components and/or the proper choice of off-the-shelf versions.

The reader is encouraged to explore design aids available from the websites of manufacturers (**Table 7.A1**). The following is far from complete, but should give the reader valuable insight into the vast array of available tools. Remember also that manufacturers of passive components such as capacitors, resistors, heat sinks and thermal hardware may also have helpful and informative design aids on their websites. — *Chuck Mullett, KR6R*

Table 7.A1

Switchmode Converter Design Aids

Vendor	Main Web Site	Design Tools
Fairchild Semiconductor	**www.fairchildsemi.com**	From the home page, select "Design Center"
Infineon	**www.infineon.com**	From the home page Search window, select "Website" and "Search Technical Documents", then select "Power Management IC's" from the pull-down menu. From the subsequent list, select "Application Notes"
International Rectifier	**www.irf.com/indexsw.html**	**www.irf.com/design-center/mypower/**
Intersil	**www.intersil.com**	**www.intersil.com/design/**
National Semiconductor	**www.national.com/analog**	**www.national.com/analog/power/simple_switcher**
ON Semiconductor	**www.onsemi.com**	**onsemi.com/PowerSolutions/supportDoc.do?type=tools**
Texas Instruments	**www.power.ti.com**	From the Power Management page, click "Tools and Software"
Linear Technology Corp	**www.linear.com**	Search for "LTSpice" in the home page window to access the simulation tools page.

7.12 High-Voltage Techniques

The construction of high-voltage supplies requires special considerations in addition to the normal design and construction practices used for lower-voltage supplies. In general, the builder needs to remember that physical spacing between leads, connections, parts and the chassis must be sufficient to prevent arcing. Also, the series connection of components such as capacitor and resistor strings needs to be done with consideration for the distribution of voltage stresses across the components. High-voltages can constitute a safety hazard and great care must be taken to limit physical access to components and wiring while high potentials are present.

7.12.1 High-Voltage Capacitors

For reasons of economy and availability, electrolytic capacitors are frequently used for output filter capacitors. Because these capacitors have relatively low voltage ratings (<600 V), in HV applications it will usually be necessary to connect them in series strings to form an equivalent capacitor with the capability to withstand the higher applied voltage. Electrolytic capacitors have relatively high leakage currents (low leakage resistance) especially at higher temperatures.

To keep the voltages across the capacitors in the series string relatively constant, equal-value bypassing resistors are connected across each capacitor. These *equalizing resistors* should have a value low enough to equalize differences in capacitor leakage resistance between the capacitors, while high enough not to dissipate excessive power. Also, capacitor bodies need to be insulated from the chassis and from each other by mounting them on insulating panels, thereby preventing arcing to the chassis or other capacitors in the string. The insulated mounting for the capacitors is often plastic or other insulating material in board form. A typical example from a commercial amplifier that implements some the guidelines given in this section is shown in **Fig 7.44**.

Equalizing resistors are needed because of differences in dc leakage current between different capacitors in the series string. The data sheet for an electrolytic capacitor will usually give the dc leakage current at 20 °C in the form of an equation. For example:

$$I_{leakage} = 3\sqrt{CV} \qquad (14)$$

where
C = the value in µF
V = the working voltage
$I_{leakage}$ = leakage current in µA.

Keep in mind that the leakage current will increase as the capacitor temperature rises above 20 °C. A value of 3 to 4 times the 20 °C value would not be unusual because of normal heating.

We can use an approximation which includes allowance for heating to determine the maximum value of the divider resistors:

$$R \leq \frac{V_r - V_m}{I_{leakage}} \qquad (15)$$

where
V_r = the voltage rating of the capacitor
V_m = the voltage across the capacitor during normal operation.

A typical 3000 V power supply might use eight 330 µF, 450 V capacitors in series. In that case, V_r = 450 V, V_m = 375 V and $I_{leakage}$ = 1.2 mA. This would make R = 63 kΩ or a total of 504 kΩ for the resistor string. With 3000 V across the resistor string, the total power dissipation would be 18 W or about 2.25 W per resistor. To be conservative you should use resistors rated for at least 4 W each.

The equalizing resistors may dissipate significant power and become quite warm. It is important that these resistors do not heat the capacitors they are associated with, as this will increase the capacitor leakage current. You also have to be careful that the heat from the resistors is not trapped under the plastic support panels. The best practice is to place the resistors above the capacitors and their mounting structure allowing the heat to rise unobstructed as shown in Fig 7.44.

OIL-FILLED CAPACITORS

For high voltages, oil-filled paper-dielectric capacitors are superior to electrolytics because they have lower internal impedance at high frequencies, higher leakage resistance and are available with much higher working voltages. These capacitors are available with values of several microfarads and have working voltage ratings of thousands of volts. On the other hand, they can be expensive, heavy and bulky.

Oil-filled capacitors are frequently offered for sale at flea markets at attractive prices. One caution: It is best to avoid older oil-filled capacitors because they may contain polychlorinated biphenyls (PCBs), a known cancer-causing agent. Newer capacitors have eliminated PCBs and have a notice on the case to that effect. Older oil-filled capacitors should be examined carefully for any

Fig 7.44 — Example of the filter capacitor/bleeder resistor installation in a high voltage power supply.

signs of leakage. Contact with leaking oil should be avoided, with careful washing of the hands after handling. Do not dispose of any oil-filled capacitors with household trash, particularly older units. Contact your local recycling agency for information about how to dispose of them properly.

7.12.2 High-Voltage Bleeder Resistors

Bleeder resistors across the output are used to discharge the stored energy in the filter capacitors when the power supply is turned off and should be given careful consideration. These resistors provide protection against shock when the power supply is turned off and dangerous wiring is exposed. A general rule is that the bleeder should be designed to reduce the output voltage to 30 V or less within 30 seconds of turning off the power supply.

Take care to ensure that the maximum voltage rating of the resistor is not exceeded. In a typical divider string, the resistor values are high enough that the voltage across the resistor is not dissipation limited. The voltage limit is typically related to the insulation intrinsic to the resistor. Resistor maximum voltage ratings are usually given in the manufacturer's data sheet and can be found online. Two major resistor manufacturers are Ohmite Electronics (**www.ohmite.com**) and Stackpole Electronics (**www.seielect.com**).

A 2 W carbon composition resistor will have a maximum voltage rating of 500 V. As a rough estimate, larger wire-wound power resistors are typically rated at 500 V_{RMS} per inch of length — but check with the manufacturer to be sure.

The bleeder will consist of several resistors in series. Typically wire-wound power resistors are used for this application. One additional recommendation is that two separate (redundant) bleeder strings be used, to provide safety in the event one of the strings fails. When electrolytic capacitors are used, the equalizing resistors can also serve as the bleeder resistor but they should be redundant. Again, give careful attention to keeping the heat from the resistors away from the capacitors as shown in Fig 7.44.

In the example given above for calculating the equalizing resistor value, eight 330 μF capacitors in series created the equivalent of a 41 μF capacitor with a 3000 V rating. The total resistance across the capacitors was $8 \times 63\,k\Omega$ = 504 kΩ. This gives us a time constant (R × C) of about $R \times C \approx 21$ seconds. To discharge the capacitors to 30 V from 3000 V would take about four time constants (about 84 seconds) which is well over a minute. To get the discharge time down to 30 seconds would require reducing the equalizing resistors to 25 kΩ each. The bleeder power dissipation

would then be:

$$P = \frac{V^2}{R} = \frac{3000^2}{200,000} = 45\ W \qquad (16)$$

For a 2 or 3 kV power supply, this is a reasonable value but it is still a significant amount of power and you need to make sure the resulting heat is properly managed.

7.12.3 High-Voltage Metering Techniques

Special considerations should be observed for metering of high-voltage supplies, such as the plate supplies for linear amplifiers. This is to provide safety to both personnel and to the meters themselves.

To monitor the current, it is customary to place the ammeter in the supply return (ground) line. This ensures that both meter terminals are close to ground potential. Placing the meter in the positive output line creates a hazard because the voltage on each meter terminal would be near the full high-voltage potential. Also, there is the strong possibility that an arc could occur between the wiring and coils inside the meter and the chassis of the amplifier or power supply itself. This hazardous potential cannot exist with the meter in the negative leg.

Another good safety practice is to place a low-voltage Zener diode across the terminals of the ammeter. This will bypass the meter in the event of an internal open circuit in the meter. A 1 W Zener diode will suffice since the current in the metering circuit is low.

The chapter on **RF Power Amplifiers** contains examples of how to perform current and voltage metering in high voltage supplies and amplifiers. In the past, amplifier articles have shown the meters mounted on plastic boards with stand-off insulators behind a plastic window in the amplifier or power supply front panel. While this can work it is not considered good practice today. It is usually possible to arrange the metering so that the meters are close to ground potential and may be safely mounted on the front panel of either the amplifier or the power supply.

For metering of high voltage, the builder should remember that resistors to be used in multiplier strings will often have voltage-breakdown ratings well below the total voltage being sampled. Usually, several identical series resistors will be used to reduce voltage stress across individual resistors. A basic rule of thumb is that these resistors should be limited to a maximum of 200 V, unless rated otherwise. For example, in a 2000 V power supply, the voltmeter multiplier should have a string of 10 resistors connected in series to distribute the voltage equally. The comments on resistor voltage rating in the sections on capacitor equalization and bleeder resistors apply to this case, as well.

7.12.4 High-Voltage Transformers and Inductors

Usable transformers and filter inductors can often be found at amateur flea markets but frequently these are very old units, often dating from WWII. The hermetically-sealed military components are likely to still be good, especially if they have not been used. Open-frame units, with insulation that has been directly exposed to the atmosphere and moisture for long periods of time, should be considered suspect. These units can be checked by running what is called a "hipot test" (high potential test). This involves a low current, variable output voltage power supply with a high-value resistor in series with the output. Unfortunately this equipment is seldom available to amateurs. Some motor repair shops will have an insulation tester called a "Megger" and may be willing to perform an insulation test on a transformer or inductor for you. This is not a completely definitive test but will certainly detect any gross problems with the insulation.

An alternative would be to perform the transformer tests discussed earlier with a variable autotransformer on the ac input. Slowly increase the input voltage until full line voltage is reached and let the transformer run for an hour or two while watching to see if there are any signs of failure. Doing this test in a dark room makes it easier to see any visible corona discharge, another sign of insulation problems. Because the transformer terminals will be at full voltage great care should be taken to avoid contact with the HV terminals. Some form of transparent insulating shield for the test setup is necessary.

7.12.5 Construction Techniques for High-Voltage Supplies

Layout and component arrangement in HV supplies requires some additional care beyond those for lower voltage projects. The photographs in the **RF Power Amplifiers** chapter are a good start, but there are some points which may not be obvious from the pictures.

SHARP EDGES

Sharp points, board edges and/or hardware with ragged edges can lead to localized intensification of the electric field's strength, resulting in a possible breakdown. One common offender is the component leads on the soldered side of a printed-circuit board. These are usually soldered and then cut off leaving a small but often very sharp point. The best way to handle this is to cut the lead as close to the board as practical and form a small solder ball or mound around the cutoff lead end. An example of these suggestions is given in **Fig 7.45**. The protruding component wire

Fig 7.45 — Example of a high voltage board with a solder ball on a protruding lead and the use of a cap nut to terminate the end of a screw thread.

near the top would have a high field gradient. Below that we see a wire cut short with a rounded mound of solder covering the end of the wire.

The ends of bolts with sharp threads often protrude beyond the associated nut. One way to eliminate this is to use the dome-style nuts (cap nuts) that capture the end of the bolt or screw, forming a nicely rounded surface. An example of a cap nut is shown in Fig 7.45.

Sheet metal screws, with their needle-sharp ends, should be avoided if possible. Sheet metal screws used to close metal housings at high potential should have the screw tips inside the enclosure, which would be normal in most cases. You must also be careful of the tips of sheet metal screws that protrude on the inside of an outer enclosure. If these are in proximity to circuits at high potential, they can also lead to arcing. Keeping sheet metal screw tips well away from high voltage circuitry is the best defense.

Small pieces of sheet metal that may be part of a structure at high potential should have their edges rounded off with a file. Copper traces near the edges of circuit boards do not benefit from rounding, however. In fact, filing may make the edge sharper, ragged and more prone to breakdown. In critical areas, a small solid round wire can be soldered to the edges of a copper trace to form a rounded edge as illustrated at the right side of Fig 7.45.

INSULATORS

Some portions of the circuit may be mounted on insulators or plastic sheets. A new, clean insulator should easily withstand 10 kV per inch without creepage or breakdown across the insulating surface, but over time that surface will accumulate dirt and dust. Reducing the high-voltage stress across insulators to 5 kV per inch would be more conservative.

In theory the spacing between two smooth surfaces across an air gap can be much smaller, perhaps 20 to 30 kV per inch or even higher. But given the uncertainties of layout and voltage gradients around hardware, wiring, components and board edges, sticking with 5 kV per inch is good idea even for air gaps.

This separation will seldom cause construction layout problems unless you are trying to build a very compact unit.

FUSES

Sometimes a fuse will be placed in series with a high voltage output to provide protection in case of load arcing. These fuses pose special problems. When a fuse blows, the fuse element will at least melt and perhaps even vaporize. There may be an interval when most, if not all, the output voltage appears across the fuse but the fuse has not stopped conducting. That is, there can be a sustained arc in the fuse.

Fuses have voltage ratings that are the maximum voltages across the fuse for which the arc can be expected to quench quickly. The standard 0.25 × 1.25 inch fuses used in the input ac line are typically rated for 250 V. While no doubt these ratings are conservative, this type of fuse cannot be expected to reliably clear an arc with 2 to 3 kV across the fuse and should not be used in series with a high-voltage output.

Fuses with very high voltage ratings do exist however. Unfortunately, for the most part these HV fuses are for utility applications. They are physically large and not available in low current ratings. Fuses specifically rated for low currents (<2A) and several kV exist but may be very hard to locate. Even if you have a suitable fuse, care must be taken that the fuse holder can withstand the voltage. Normally high voltage fuses are mounted with end clips on an insulated board.

7.12.6 High-Voltage Safety Considerations

The voltages present in HV power supplies are potentially lethal. Every effort must be made to restrict physical access to any high potentials. A number of steps can be taken:

1) Build the power supply within a closed box, preferably a metal one.

2) Install interlocks on removable panels used for access to the interior. An interlock is usually a normally-open microswitch in series with the input power line. The microswitch is

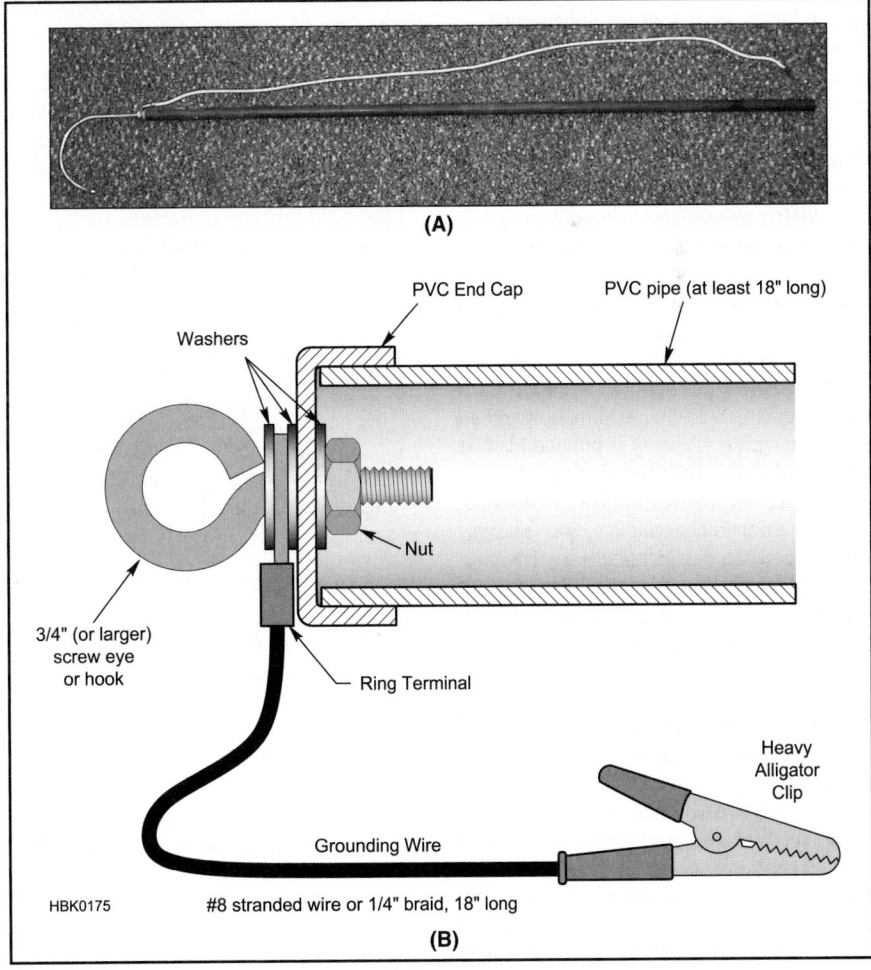

Fig 7.46 — Example of a grounding stick or hook to discharge capacitor energy safely. The end of the wire is connected to ground and the hook is touched to the capacitor terminals to discharge them. A diagram showing how to construct a grounding stick is shown at B.

positioned so that it can be closed only when the overlying panel has been secured in place. When the panel is removed the switch moves to the open position, removing power from the supply.

3) As noted previously, the use of a bleeder resistor to rapidly discharge the capacitors is mandatory.

4) To further protect the operator when accessing the inside of a high-voltage power supply, a grounding hook like that shown in **Fig 7.46A** should be used to positively discharge each capacitor by touching the hook to each capacitor terminal. The hook is connected to the chassis via the wire shown and held by the insulated handle. It is normal practice to leave the hook in place across the capacitors while the supply is being worked on, just in case primary power is inadvertently applied. When the work is finished, remove the hook and replace the covers on the power supply.

Fig 7.46B shows how to construct a grounding stick of your own. A large screw eye or hook is substituted for the hook. It is important that the grounding wire be left *outside* the handle so that any hazardous voltages or currents will not be present near your hand or body.

7.13 Batteries

A battery is a group of individual *cells*, usually series-connected to give some desired multiple of the *cell voltage*. The cell voltage is usually in the range of 1 to 4 V. Each chemical system used in the cell gives a particular nominal voltage that will vary with temperature and state of charge. This must be taken into account to make up a particular battery voltage. For example, four 1.5 V carbon-zinc cells make a 6 V battery and six 2 V lead-acid cells make a 12 V battery. **Table 7.1** lists the dimensions of commonly used batteries.

The following sections on battery types and usage consist primarily of excerpts from *Batteries in a Portable World* by Isidor Buchmann, CEO of Cadex Electronics (**www.cadex.com**), a leading manufacturer of battery charging and related equipment. The ARRL appreciates having been given permission to use this material. The book discusses the issues summarized here (and many others) in detail. The reader is directed to the original text for more complete information. This summary is intended to present and compare the various options commonly used by amateurs. Additional information and an extensive battery glossary is provided online at **www.batteryuniversity.com**.

Batteries are divided into two categories: *primary* (non-rechargeable) and *secondary* (rechargeable). Secondary batteries are expected to account for more than 80% of global battery use by 2015. The most common types of battery chemistry are lithium, lead and nickel-based systems.

Batteries store energy well and for a considerable length of time. Primary batteries hold more energy than secondary, and their *self-discharge* is lower. Alkaline cells are good for 10 years with minimal losses. Lead, nickel and lithium-based batteries need periodic recharges to compensate for lost power. **Fig 7.47** illustrates the energy and power densities of lead acid, nickel-cadmium (NiCd), nickel-metal-hydride (NiMH), and the lithium-ion family (Li-ion).

Rather than giving batteries unique names by type, they are broadly distinguished by the following characteristics:

Battery Safety

In addition to the precautions given for each type of battery, the following precautions apply to all battery types. Always follow the manufacturer's advice! Extensive application information can be found on manufacturer's websites.

Hydrogen gas escaping from storage batteries can be explosive. Keep flames or any kind of burning material away, including cigarettes, cigars, pipes and so on. Use and charge batteries in well-ventilated areas to prevent hydrogen gas from building up.

No battery should be subjected to unnecessary heat, vibration or physical shock. The battery should be kept clean. Frequent inspection for leaks is a good idea. Electrolyte that has leaked or sprayed from the battery should be cleaned from all surfaces. The electrolyte is chemically active and electrically conductive, and may ruin electrical equipment. Acid may be neutralized with sodium bicarbonate (baking soda), and alkalis (found in NiCd batteries) may be neutralized with a weak acid such as vinegar. Both neutralizers will dissolve in water and should be quickly washed off. Do not let any of the neutralizer enter the battery.

Keep a record of the battery use, and include the last output voltage and, for lead-acid storage batteries, the hydrometer reading. This allows prediction of useful charge remaining, and the recharging or procuring of extra batteries, thus minimizing failure of battery power during an excursion or emergency.

Batteries can contain a number of hazardous materials such as lead, cadmium, mercury or acid, and some thought is needed for their disposal at the end of useful life. Municipal and county waste disposal sites and recycling centers will usually accept lead acid batteries because they can readily be recycled. Other types of batteries are typically not recycled and should be treated as hazardous waste. Most disposal sites and recycling centers will have occasional special programs for accepting household hazardous waste. Hardware and electronic stores may have battery recycling programs, as well. Take advantage of them.

Table 7.1

Dimensions of Common Standard Cells

Size	Dimensions	Notes
D	34 × 61 mm	
C	25.5 × 50 mm	
A	17 × 50 mm	Only available for NiCd; also available in half-length size
AA	14.5 × 50 mm	
AAA	10.5 × 44.5 mm	
AAAA	8.3 × 42.5 mm	Typical size of cell making up 9 V batteries
N	12 × 32 mm	
9 V	48.5 × 26.5 × 17.5 mm	Contains six AAAA cells in series, snap terminals
18650	18 × 65 mm	Commonly used in lithium-ion battery packs
Lantern	115 × 68.2 × 68.2	Spring terminals
CR2016	20 × 1.6 mm	Coin cell
CR2025	20 × 2.5 mm	Coin cell
CR2032	20 × 3.2 mm	Coin cell

Information courtesy of Cadex and from **http://en.wikipedia.org/wiki/List_of_battery_sizes**

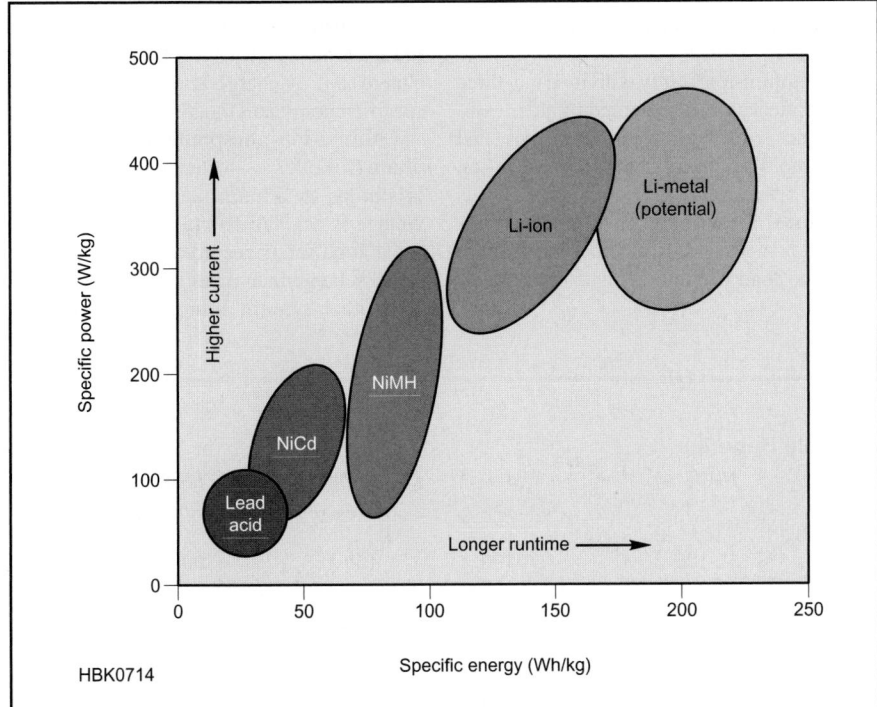

Fig 7.47 — Specific energy and specific power of rechargeable batteries. (Courtesy of Cadex)

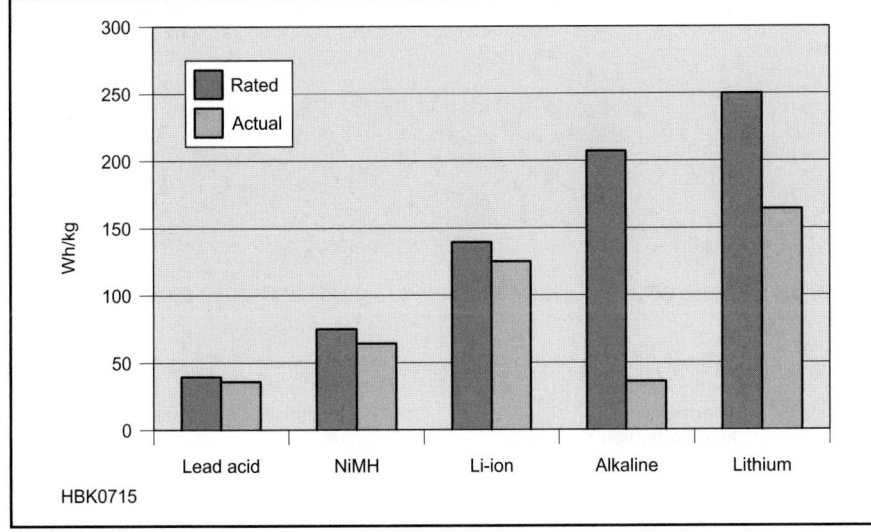

Fig 7.48 — Energy comparison of different battery types at a load of 1 C. (Courtesy of Cadex)

Table 7.2

Alkaline Specifications of Standard Batteries at Low Load

Battery Type	Nominal Voltage (V)	Rated Capacity (mAh)	Voltage Cutoff (V)	Rated Load (Ω)	Discharge C-rate
9 V	9	570	4.8	620	0.025
AAA	1.5	1150	0.8	75	0.017
AA	1.5	2870	0.8	75	0.007
C	1.5	7800	0.8	39	0.005
D	1.5	17,000	0.8	39	0.0022

(Table courtesy of Cadex)

Chemistry — the families of chemicals used to make the battery. The most common chemistries are lead, nickel and lithium.

Voltage — the nominal open circuit voltage (OCV), which varies with chemistry and the number of cells connected in series.

Size — standard sizes of batteries or cells, such as AAA, AA, C and D.

Capacity (C) — the specific energy in ampere-hours (Ah).

Cold Cranking Amps (CCA) — a starter battery's ability to supply high load current at –18° C (0° F) as specified by the vehicle standard SAE J357.

Specific energy — battery capacity in watt-hours per kilogram (Wh/kg)

Specific power — indicates the loading capability or the amount of current the battery can provide in watts/kilogram (W/kg)

C-rate — charge and discharge rate specified in units of C (capacity). At 1 C, the battery charges or discharges at a current numerically equal to its Ah rating. For example, a 2000 mAh battery discharging at 1 C is supplying a current of 2 A. At 0.5 C, the current would be 1 A, and so forth.

Load — whatever draws energy from the battery. Internal battery resistance and depleting the battery's state of charge cause the voltage to drop.

Primary batteries have one of the highest energy densities. One of the most common primary batteries is the *alkaline-manganese*, or alkaline battery. The *carbon-zinc* or *Leclanché battery* is less expensive but holds less energy than the alkaline battery. Although secondary batteries have improved, a regular household alkaline cell provides 50% more energy than lithium-ion.

Primary batteries tend to have high internal resistance, which limits the discharge to light loads. This reduces the battery's specific power, even though its specific energy may be quite high. Non-rechargeable lithium metal and alkaline batteries are commonly used in low power applications such as clocks, meter LCDs, keyers and so forth.

Manufacturers of primary batteries specify specific energy at a small fraction of C and the batteries are allowed to discharge to a very low voltage of 0.8 volts per cell. **Fig 7.48** compares primary and secondary batteries discharged at a rate of 1 C. The primary (alkaline and lithium) batteries are unable to deliver their rated specific energy at heavy loads. **Table 7.2** gives typical specifications for alkaline batteries at light loads, such as operating a handheld radio during receive.

If recharging is available, using primary batteries can be expensive — about thirty-fold higher than for secondary batteries. Amateurs using batteries in the field, particularly for emergency communications, may want to maintain the ability to use primary batteries for times when recharging for secondary batteries is not available.

7.13.1 Choices of Secondary Batteries

The following discussion examines today's most popular secondary battery systems according to specific energy, years of service life, load characteristics, safety, price, self-discharge, environmental issues, maintenance and disposal. **Table 7.3** compares the characteristics of four commonly used rechargeable battery systems showing average performance rating at the time of publication in 2013.

The lithium-ion family is divided into three major battery types, so named by their cathode oxides, which are cobalt, manganese and phosphate. The characteristics of these Li-ion systems are as follows:

Lithium-ion-cobalt or **lithium-cobalt** ($LiCoO_2$) — Has high specific energy with moderate load capabilities and modest service life.

Lithium-ion-manganese or **lithium-manganese** ($LiMn_2O_4$) — Is capable of high charge and discharge currents but has low specific energy and modest service life.

Lithium-ion-phosphate or **lithium-phosphate** ($LiFePO_4$) — often abbreviated LiPo. Is similar to lithium-manganese; nominal voltage is 3.3 V/cell; offers long cycle life, has a good safety record but exhibits higher self-discharge than other Li-ion systems.

Another type of lithium-ion cell is the

Table 7.3
Characteristics of Commonly Used Rechargeable Batteries

Specification	Lead Acid	NiCd	NiMH	Li-ion Cobalt	Li-ion Manganese	Li-ion Phosphate
Specific energy (Wh/kg)	30-50	45-80	60-120	150-190	100-135	90-120
Internal resistance[1] (mΩ)	<100 12 V pack	100-200 6 V pack	200-300 6 V pack	150-300 7.2 V	25-75[2] per cell	25-50[2] per cell
Cycle life[4] (80% DoD)	200-300	1000[3]	300-500[3]	500-1000	500-1000	1000-2000
Fast-charge time	8-16 h	1 h typical	2-4 h	2-4 h	1 h or less	1 h or less
Overcharge tolerance	High	Moderate	Low	Low. Cannot tolerate trickle charge	Low. Cannot tolerate trickle charge.	Low. Cannot tolerate trickle charge.
Self-discharge/month (room temp)	5%	20%[5]	30%[5]	<10%[6]	<10%[6]	<10%[6]
Cell voltage (nominal)	2 V	1.2 V[7]	1.2 V[7]	3.6 V[8]	3.8 V[8]	3.3 V
Peak load current (best result)	5C[9] (0.2 C)	20 C (1 C)	5 C (0.5 C)	>3 C (<1 C)	>30 C (<10 C)	>30 C (<10 C)
Operating temp.[10] (discharge only)	−20 to 60° C	−40 to 60° C	−20 to 60° C	−20 to 60° C	−20 to 60° C	−20 to 60° C
Maintenance requirement	3-6 months[11]	30-60 days	60-90 days	Not required	Not required	Not required
Safety requirements	Thermally stable	Thermally stable, fuse protection common	Thermally stable, fuse protection common	Protection circuit mandatory	Protection circuit mandatory	Protection circuit mandatory
In use since	Late 1800s	1950	1990	1991	1996	2006
Toxicity	Very high	Very high	Low	Low	Low	Low

Table Notes:
[1] Internal resistance of a battery pack varies with milliampere-hour (mAh) rating, wiring and number of cells. Protection circuit of lithium-ion adds about 100 mΩ.
[2] Based on 18650 cell size. Cell size and design determines internal resistance.
[3] Cycle life is based on battery receiving regular maintenance.
[4] Cycle life is based on the depth of discharge (DoD). Shallow DoD improves cycle life.
[5] Self-discharge is highest immediately after charge. NiCd loses 10% in the first 24 hours, then declines to 10% every 30 days. High temperature increases self-discharge.
[6] Internal protection circuits typically consume 3% of the stored energy per month
[7] The traditional voltage is 1.25 V; 1.2 V is more commonly used.
[8] Low internal resistance reduces the voltage drop under load and Li-ion is often rated higher than 3.6 V/cell. Cells marked 3.7 V and 3.8 V are fully compatible with 3.6 V.
[9] Capable of high current pulses; needs time to recuperate.
[10] Applies to discharge only; charge temperature is more confined.
[11] Maintenance may be in the form of equalizing or topping charge to prevent sulfation
(Table courtesy of Cadex)

popular *lithium-ion-polymer* or *Li-polymer*. While Li-ion systems get their name from their unique cathode materials, Li-polymer differs by having a distinct architecture in which a gelled electrolyte replaces the usual porous separator between the anode and cathode. The gelled electrolyte acts as a catalyst to enhance the electrical activity of the other battery materials.

7.13.2 Lead Acid Batteries

Most lead acid batteries used today are *maintenance-free* types in which the liquid electrolyte is sealed inside the battery in liquid or gel form. Two similar types are used: the *sealed lead acid* (SLA) and the *valve-regulated lead acid* (VRLA). SLA batteries have capacities up to about 30 Ah. VRLA batteries are larger and have capacities from 30 Ah to several thousand Ah. **Table 7.4** summarizes the advantages and limitations of lead acid battery systems.

Applying the right voltage limit when charging lead acid systems is critical and will be a compromise between capacity when recharged and maintaining the battery's internal materials. A low charge limit voltage may shelter the battery but this causes poor performance and a buildup of *sulfation* on the negative plate. A high voltage limit improves performance but it promotes irreversible grid corrosion on the positive plate. Temperature also changes the voltage threshold.

Lead acid does not lend itself to fast charging and a fully saturated charge requires 14 to 16 hours. The battery must always be stored at full state-of-charge to avoid sulfation. Lead acid is not subject to memory, but correct charge and float voltages are important to get a long life (see the section on charging below). While NiCd loses approximately 40% of its stored energy in three months, lead acid self-discharges the same amount in one year.

Lead acid batteries are inexpensive on cost-per-watt basis but are less suitable for repeated deep cycling. A full discharge causes strain and each discharge/charge cycle permanently robs the battery of a small amount of capacity. The fading becomes more acute as the battery falls below 80% of its nominal capacity. Depending on the depth of discharge and operating temperature, lead acid for deep-cycle applications provides 200 to 300 discharge/charge cycles.

High temperature reduces the number of available cycles. As a guideline, every 8° C (13° F) rise above the optimum temperature of 25° C (77° F) cuts battery life in half.

Lead acid batteries are rated at a 5-hour (0.2 C) and 20-hour (0.05 C) discharge, and the battery performs best when discharged slowly. Lead acid can deliver high pulse currents of several C if done for only a few seconds, making lead acid well suited as a *starter battery*.

STARTER AND DEEP-CYCLE BATTERIES

The *starter* battery is designed to crank an engine with a momentary high power burst; the *deep-cycle* battery, on the other hand, is built to provide continuous power. The starter battery is made for high peak power and does not tolerate deep cycling well. The deep-cycle battery has a moderate power output but permits cycling.

Starter batteries have a CCA rating in amperes and a very low internal resistance. Deep-cycle batteries are rated in Ah or minutes of runtime. A starter battery cannot be swapped with a deep-cycle battery and vice versa because of their different internal construction. **Table 7.5** compares the typical life of starter and deep-cycle batteries when deep-cycled.

ABSORBENT GLASS MAT (AGM)

AGM is an improved lead acid battery in which the electrolyte is absorbed in a mat of fine glass fibers. This makes the battery spill-proof. AGM has very low internal resistance, is capable of delivering high currents and offers long service even if *occasionally* deep-cycled. It also stands up well to high and low temperatures and has a low self-discharge. AGM has a higher specific power rating for high load currents and allows faster charge times (up to five times faster) than conventional lead acid. AGM has a slightly lower specific energy and higher manufacturing cost.

AGM batteries are sensitive to overcharging. They can be charged to 2.40 V/cell without problems but the float charge should be reduced to 2.25 to 2.30 V/cell and summer temperatures may require lower voltages. Automotive charging systems designed for flooded lead acid often have a fixed float voltage setting of 14.40 V (2.40 V/cell) and a direct replacement with an AGM battery could result in overcharge on a long drive.

Heat can be a problem for AGM and other gelled electrolyte batteries. Manufacturers recommend halting charge if the battery core reaches 49° C (120° F). However, AGM batteries can sit in storage for long periods before a recharge to prevent sulfation becomes necessary.

7.13.3 Nickel-based Batteries
NICKEL-CADMIUM (NiCd)

The standard NiCd remains one of the most rugged and forgiving batteries but needs proper care to attain longevity. Nickel-based batteries also have a flat discharge curve that ranges from 1.25 to 1.0 V/cell. **Table 7.6** summarizes the advantages and limitations of NiCd battery systems.

NICKEL-METAL-HYDRIDE (NiMH)

NiMH provides 40% higher specific energy than standard NiCd, but the decisive advantage is the absence of toxic metals. NiMH also has two major advantages over Li-ion: price and safety. NiMH is offered in AA and AAA sizes.

NiMH is not without drawbacks. It has a

Table 7.4

Advantages and Limitations of Lead Acid Batteries

AdvantagesInexpensive and simple to manufacture; lowest cost per watt-hour
Mature and well-understood technology; provides dependable service
Low self-discharge; lowest among rechargeable batteries
High specific power, capable of high discharge currents
LimitationsLow specific energy; poor weight-to-energy ratio
Slow charge; fully saturated charge takes 14 hours
Must always be stored in charged condition
Limited cycle life; repeated deep-cycling reduces battery life
Flooded version requires watering
Not environmentally friendly
Transportation restrictions on the flooded type

(Table courtesy of Cadex)

Table 7.5

Cycle Performance of Starter and Deep-Cycle Batteries

Depth of Discharge	Starter Battery	Deep-cycle Battery
100%	12-15 cycles	150-200 cycles
50%	100-120 cycles	400-500 cycles
30%	130-150 cycles	1000 and more cycles

(Table courtesy of Cadex)

Table 7.6

Advantages and Limitations of NiCd Batteries

Advantages	Fast and simple charging even after prolonged storage
	High number of charge/discharge cycles; provides
	over 1000 charge/discharge cycles with proper maintenance
	Good load performance; rugged and forgiving if abused
	Long shelf life; can be stored in a discharged state
	Simple storage and transportation; not subject to regulatory control
	Good low-temperature performance
	Economically priced; NiCd is the lowest in terms of cost per cycle
	Available in a wide range of sizes and performance options
Limitations	Relatively low specific energy compared with newer systems
	Memory effect; needs periodic full discharges
	Environmentally unfriendly; cadmium is a toxic metal and cannot be
	disposed of in landfills
	High self-discharge; needs recharging after storage

(Table courtesy of Cadex)

Table 7.7

Advantages and Limitation of NiMH Batteries

Advantages	30-40% higher capacity than a standard NiCd
	Less prone to memory than NiCd
	Simple storage and transportation; not subject to regulatory control
	Environmentally friendly; contains only mild toxins
	Nickel content makes recycling profitable
Limitations	Limited service life; deep discharge reduces service life
	Requires complex charge algorithm
	Does no absorb overcharge well; trickle charge must be kept low
	Generates heat during fast-charge and high-load discharge
	High self-discharge; chemical additives reduce self-discharge at the expense
	of capacity
	Performance degrades if stored at elevated temperatures; should be stored
	in a cool place at about 40% state-of-charge

(Table courtesy of Cadex)

lower specific energy than Li-ion and also has high self-discharge, losing about 20% of its capacity within the first 24 hours, and 10% per month thereafter. New types of NiMH such as the Eneloop from Sanyo, ReCyko by GP and others have reduced the self-discharge by a factor of six, increasing storage life by the same amount at the sacrifice of some capacity. **Table 7.7** summarizes the advantages and limitations of NiMH battery systems.

7.13.4 Lithium-based Batteries

The specific energy of Li-ion is twice that of NiCd and the high nominal cell voltage of 3.60 V as compared to 1.20 V for nickel systems contributes to this gain. The load characteristics are good, and the flat discharge curve offers effective utilization of the stored energy in a desirable voltage spectrum of 3.70 to 2.80 V/cell. Li-ion batteries vary in performance according to the choice of cathode materials. **Table 7.8** presents the characteristics of common Li-ion battery chemistries and **Table 7.9** summarizes the advantages and limitations of lithium-ion battery systems.

Lithium-polymer employs a gelled electrolyte in a micro-porous separator between the anode and cathode. Li-polymer is a construction technique and not a type of battery chemistry, so it can be applied to any of the lithium battery chemistries. Most Li-polymer battery packs are based on Li-cobalt. Charge and discharge characteristics of Li-polymer are identical to other Li-ion systems and this chemistry does not require a special charger.

Fig 7.49 compares the specific energy of lead, nickel and lithium-based systems. While Li-cobalt is the clear winner in terms of higher

Table 7.8

Characteristics of the Four Most Commonly Used Lithium-ion Batteries

Specifications	Li-cobalt $LiCoO_2$ (LCO)	Li-manganese $LiMn_2O_4$ (LMO)	Li-phosphate $LiFePO_4$ (LFP)	NMC[1] $LiNiMnCoO_2$
Voltage (V)	3.60	3.80	3.30	3.60/3.70
Charge limit (V)	4.20	4.20	3.60	4.20
Cycle life[2]	500-1000	500-1000	1000-2000	1000-2000
Operating temperature	Average	Average	Good	Good
Specific energy (Wh/kg)	150-190	100-135	90-120	140-180
Specific Power (C)	1	10, 40 pulse	35 continuous	10
Safety	Average[3]	Average[3]	Very safe[4]	Safer than Li-cobalt[4]
Thermal runaway[5] (°C/°F)	150/302	250/482	270/518	210/410
Cost	Raw material high	30% less than cobalt	High	High
In use since	1994	1996	1999	2003

Table Notes

[1] NMC (nickel-manganese-cobalt), NCM, CMN, CNM, MNC and MCN are basically the same. The order of Ni, Mn and Co does not matter much

[2] Application and environment govern cycle life; the numbers do not always apply correctly

[3] Requires protection circuit and cell balancing of multi cell pack. Requirements for small formats with 1 or 2 cells can be relaxed

[4] Needs cell balancing and voltage protection

[5] A fully charged battery raises the thermal runaway temperature, a partial charge lowers it

(Table courtesy of Cadex)

Table 7.9

Advantages and Limitations of Li-ion Batteries

Advantages	High specific energy
	Relatively low self-discharge; less than half that of NiCd and NiMH
	Low maintenance. No periodic discharge is needed; no memory
Limitations	Requires protection circuit to limit voltage and current
	Subject to aging, even if not in use (life span is similar to other chemistries)
	Transportation regulations apply when shipping in larger quantities

(Table courtesy of Cadex)

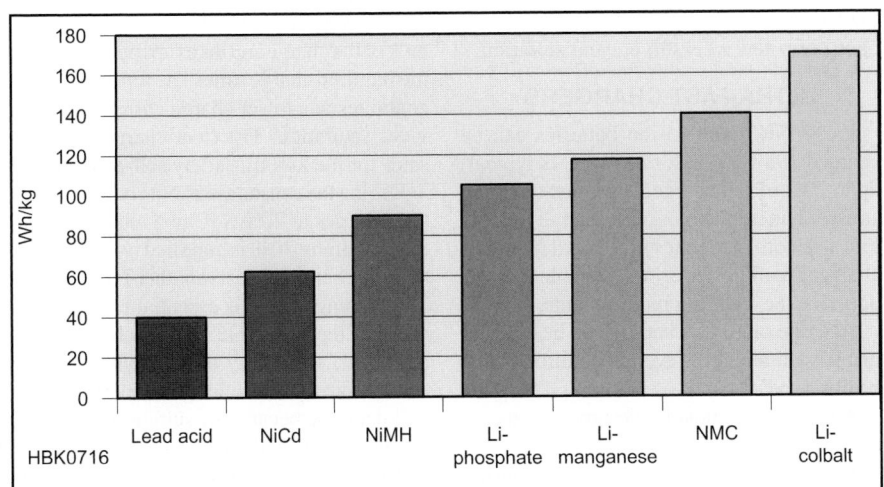

Fig 7.49 — Typical energy densities of lead, nickel- and lithium-based batteries

capacity than other systems, this only applies to specific energy. In terms of specific power and thermal stability, Li-manganese and Li-phosphate are superior.

7.13.5 Charging Methods

The performance and longevity of rechargeable batteries are to a large extent governed by the quality of the charger. Choosing a quality charger is important considering the costs of battery replacement and poor performance.

CHARGERS

This discussion focuses on third-party chargers designed to charge and maintain individual cells or batteries. *Fleet chargers* designed for maintaining many batteries of a common type in a company or agency environment are often provided by the equipment OEM and usually have special features for battery conditioning.

"Smart" chargers include valuable additional features beyond simply applying current to a battery. Temperature protection is particularly important to slow or prevent charging below freezing or above recommended thresholds when the battery is hot. Advanced lead acid chargers adjust the float and trickle charge thresholds based on temperature and battery age. **Table 7.10** summarizes the types of third-party chargers.

There are two common charge methods: voltage limiting (VL) and current limited (CL). Lead and lithium-based chargers cap the voltage at a fixed threshold. When reaching the cutoff voltage, the battery begins to saturate and the current drops while receiving the remaining charge on its own timetable. Full charge detection occurs when the current drops to a designated level.

Nickel-based batteries charge with a controlled current and the voltage is allowed to fluctuate freely. A slight voltage drop after a steady rise indicates a fully charged battery. The voltage drop method works well in terminating a fast charge, but the charger should detect and protect against shorted or mismatched cells. Most chargers include temperature sensors to end the charge if the temperature exceeds a safe level. (Some batteries also have internal temperature sensors.)

A temperature rise is normal as nickel-based batteries approach full charge. When in "ready" mode, the battery must cool down to room temperature. Heat causes stress and prolonged exposure to elevated temperature shortens battery life. Extended trickle charge also inflicts damage on nickel-based batteries, which should not be left in the charger continuously or beyond a few days.

A lithium-based battery should not get warm in a charger, and if this happens either the battery or charger may be faulty. Li-ion chargers do not apply a trickle charge and disconnect the battery electrically when fully charged. If the battery is left in the charger, a recharge may occur when its open circuit voltage drops below a set threshold. It is not necessary to remove Li-ion batteries from a charger.

SIMPLE GUIDELINES WHEN BUYING A CHARGER

- Use the correct charger for battery chemistry. Most chargers service one chemistry only.
- The battery voltage must agree with the charger. Do not charge if different.
- The Ah rating of a battery can be somewhat higher or lower than specified on the charger. A larger battery will take longer to charge than a smaller one and vice versa.
- The higher the amperage of the charger, the shorter the charge time will be, subject to limits on how fast the battery can be charged.
- Accurate charge termination and correct trickle charge prolong battery life.
- When fully saturated, a lead acid charger should switch to a lower voltage; a nickel-based charger should have a trickle charge;

Table 7.10

Charger Characteristics

Type	Chemistry	C-rate (C)	Time	Charge Termination
Slow	NiCd, Lead acid	0.1	14 h	Continuous low charge or fixed timer. Subject to overcharge. Remove battery when charged.
Rapid	NiCd, NiMH, Li-ion	0.3-0.5	3-6 h	Senses battery by voltage, current, temperature, and time-out timer
Fast	NiCd, NiMH, Li-ion	1	1+ h	Same as rapid charger with faster service
Ultra-Fast	Li-ion, NiCd, NiMH	1-10	10-60 min	Applies ultra-fast charge to 70% SoC; limited to specialty batteries

All values specified over 0°C to 45°F (32°F to 113°F) range

(Table courtesy of Cadex)

a Li-ion charger provides no trickle charge.

• Chargers should have a temperature override to end charge on a malfunctioning battery.

• Observe the temperature of the charger and battery. Lead acid batteries stay cool during charge; nickel-based batteries elevate the temperature toward the end of charge and should cool down after charge; Li-ion batteries should stay cool throughout charge.

SLOW CHARGERS

This type of charger applies a fixed current of about 0.1 C (one-tenth of the rated capacity) as long as the battery is connected. Slow chargers have no full-charge detection, charge current is always applied, and the charge time for a fully discharged battery is 14 to 16 hours. Most slow chargers have no "ready" indicator.

When fully charged, a slow charger keeps NiCd batteries lukewarm to the touch. Some overcharge is acceptable and the battery does not need to be removed immediately when ready. Leaving the battery in the charger can cause internal crystal growth that leads to "memory effects" in NiCd batteries.

Charging a battery with a lower Ah rating than specified for the charger will cause the battery to heat up as it approaches full charge due to the higher charging rate. Because slow chargers have no provision to lower the current or terminate the charge, the excessive heat will shorten the life of the battery.

The opposite can occur when a slow charger is charging a larger battery than it is rated for. In this case, the battery may never reach full charge and remains cold. Battery performance will be poor because of insufficient charge. Repeated partial charging can also cause crystal growth and memory effects.

RAPID CHARGERS

Falling between a slow and fast charger, the rapid charger is designed for nickel and lithium-based batteries. Unless specially designed, the rapid charger cannot service both types of batteries.

Rapid chargers are designed to charge fully discharged batteries and battery packs in 3 to 6 hours. When full charge is reached, the charger switches to a "ready" state. Most rapid chargers include temperature protection.

FAST CHARGERS

The fast charger typically applies charge at a 1 C rate so that a fully discharged battery is recharged in a little over 1 hour. As the battery approaches full charge, the charger may reduce the charge current (particularly for NiCd), and when the battery is fully charged, the charger switches to a trickle or maintenance charge mode.

Most nickel-based fast chargers accommodate NiCd and NiMH batteries and apply the same charging algorithm, but cannot charge Li-ion batteries. To service a Li-ion

pack, specialty dual-mode chargers can read a security code on the battery to switch to the right charger setting.

Lead acid batteries cannot be fast-charged and the term "fast charge" is a misnomer for lead acid chargers. Most lead acid chargers charge the battery in 14 hours; anything slower may be a compromise. Lead acid can be charged relatively quickly to 70% of full charge with the important saturation charge consuming the remaining time. A partial charge at high rate is acceptable provided the battery receives a fully saturated charge once every few weeks to prevent sulfation.

ULTRA-FAST CHARGERS

Large NiCd and Li-ion batteries can be charged at a very high rate (10 C is typical) up to 70% of full charge. Ultra-fast charging stresses batteries. If possible, charge the battery at a more moderate current. An ultra-fast charger should offer user-selectable rates to optimize the charging requirements.

At a rate of 10 C, a battery can be charged in a few minutes but several conditions must be observed:

• The battery must be designed to accept an ultra-fast charge.

• Ultra-fast charging only applies during the first charge phase and charge current must be lowered once the 70% state-of-charge (SoC) threshold is reached.

• All cells in a pack must be balanced and in good condition. Older batteries with high internal resistance will heat up and are no longer suitable for ultra-fast charging.

• Ultra-fast charging can only be done at moderate temperatures. Low temperatures slow the chemical reaction and the unabsorbed energy results in gassing and heat buildup.

• The charge must include temperature compensation and other safety provisions to end the charge if the battery is overly stressed.

CHARGING FROM A USB PORT

The universal serial bus (USB) interface has become ubiquitous on computers and consumer electronics. It is increasingly used on radio equipment. A drawing and pin connections for the USB interface are available in the **Component Data and References** chapter.

USB hubs are specified to provide 5 V and 500 mA of current. (Current can only flow out of a USB interface.) While this would be enough to charge a small Li-ion battery, it could overload the hub if other devices are attached. Many hubs limit the current and will shut down if overloaded, so charging capacity is quite limited.

The most common USB chargers are designed for single-cell Li-ion batteries. The charge begins with a constant current charge to 4.20 V/cell, at which point the voltage levels off and current begins to decrease. Due to

voltage drops in the USB cable and charger circuit, the hub may not be able to fully charge the battery. This will not damage a Li-ion battery but it will deliver shorter than expected runtimes.

CHARGING LEAD ACID

Lead acid batteries should be charged in three stages as shown in **Fig 7.50**:

1 — constant-current charge
2 — topping charge
3 — float charge

The constant-current charge applies the bulk of the charge and takes up roughly half of the required charge time. The topping charge continues at a lower charge current and provides saturation. The float charge compensates for the loss caused by self-discharge.

During the constant-current charge the battery charges to 70% SoC in 5 to 8 hours and the remaining 30% is supplied by the slower topping charge that lasts another 7 to 10 hours. The topping charge is essential for the well-being of the battery. If topping charge is not performed, the battery will eventually lose the ability to accept a full charge and performance will decrease because of sulfation. The float charge maintains the battery at full charge.

The switch to topping charge happens when the battery reaches the set voltage limit. Current begins to drop as the battery starts to saturate and full charge is reached when the current decreases to 3% of the rated current. A battery with high leakage may never reach this level and a timer is required to start charging termination.

The correct setting of the charge voltage is critical, ranging from 2.3 to 2.45 V per cell. The threshold is selected as a compromise between charging to maximum capacity and creating internal corrosion and gassing. The battery voltage also shifts with temperature, with warmer temperatures requiring slightly lower voltage thresholds. If variable voltage thresholds are not available in the charger, it is better to use a lower voltage threshold for safety.

Once fully charged through saturation, the battery should not dwell at the *topping voltage* for more than 48 hours and must be reduced to the *float voltage* level. This is especially critical for sealed systems because they are less able to tolerate overcharge than the flooded type due to heating and gas (hydrogen) generation. The recommended float voltage of most lead acid batteries is 2.25 to 2.27 V/cell. Manufacturers recommend lowering the float charge at ambient temperatures above 29° C (85° F). Not all chargers feature float charge. If the charger remains at the topping voltage and does not drop below 2.30 V/cell, remove the charger after a maximum of 48 hours of charge.

Aging batteries develop imbalances between cells that can result in overcharge and

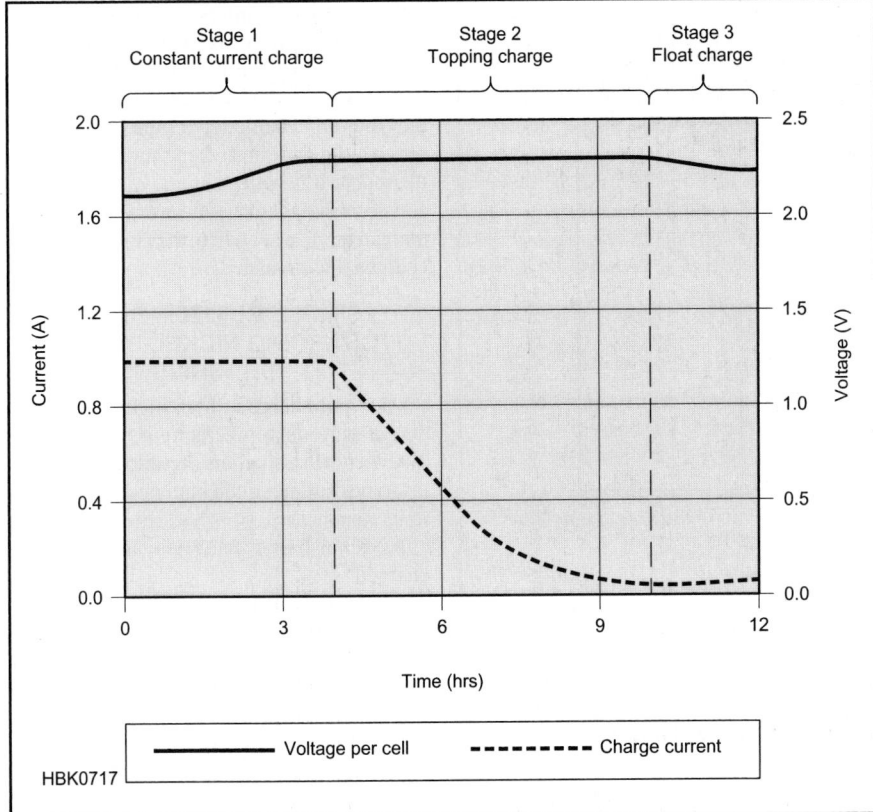

Fig 7.50 — Charge stages of a lead acid battery. The battery is fully charged when the current drops to a pre-determined level or levels out in stage 2. The float voltage must be reduced at full charge. (Courtesy of Cadex)

gassing from weak cells. This can also cause a strong cell to be undercharged and develop sulfation. Some battery manufacturers have developed cell-balancing devices that compensate for cell imbalance.

Lead acid batteries must always be stored in a charged state. A topping charge should be applied every six months to prevent the voltage from dropping below 2.10 V/cell. With AGM, these requirements can be somewhat relaxed.

Measuring the open circuit voltage (OCV) while in storage provides a reliable indication of the battery's state-of-charge (SoC). A voltage of 2.10 V at room temperature indicates a charge of about 90%. (That is equivalent to 12.6 V for a typical six-cell "12 V" lead acid battery.) Such a battery is in good condition and needs only a brief full charge prior to use. If the OCV is lower, the battery must be charged to prevent sulfation. Cool temperatures increase OCV slightly and warm temperatures lower it. Use OCV as an SoC indicator after the battery has rested for a few hours to allow the effects of charging to dissipate.

SIMPLE GUIDELINES FOR CHARGING LEAD ACID BATTERIES

• Charge in a well-ventilated area. Hydrogen gas generated during charging is explosive.

• Choose the appropriate charge program for flooded, gel and AGM batteries. Follow the manufacturer's specifications of voltage thresholds.

• Charge lead acid batteries after each use to prevent sulfation. Do not store on low charge.

• The plates of flooded batteries must always be fully submerged in electrolyte. Fill battery with distilled or de-ionized water to cover the plates if low. Never add electrolyte.

• Fill water level to designed level *after* charging. Overfilling a discharged battery can result in overflow and acid spillage.

• Formation of gas bubbles in a flooded lead acid battery is an indication of approaching full charge.

• Reduce float charge if the ambient temperature is higher than 29° C (85° F).

• Do not allow a lead acid battery to freeze and never charge a frozen battery.

• Do not charge at temperatures above 49° C (120° F).

CHARGING NICKEL-CADMIUM

Battery manufacturers recommend that new NiCd batteries be slow-charged for 16 to 24 hours before use. A slow charge brings all cells in a battery pack to an equal charge level. This is important because each cell within the

NiCd battery may have self-discharged at its own rate. Furthermore, during long storage the electrolyte tends to gravitate to the bottom of the cell and the initial trickle charge helps redistribute the electrolyte to eliminate dry spots on the separator. The cells will reach optimal performance after several charge/discharge cycles.

Full-charge Detection by Temperature

Full-charge detection of sealed nickel-based batteries is more complex than for lead acid and lithium-ion systems. Low-cost chargers often use temperature sensing to end the fast-charge, but this can be inaccurate due to internal and external temperature differences. Charger manufacturers use 50° C (122° F) as the temperature cutoff. Although any prolonged temperature above 45° C (113° F) is harmful, brief overshoot is acceptable if temperature will drop quickly when the charger changes to the "ready" state.

Some microprocessor-controlled chargers sense the rate of temperature increase with time, using the rapid temperature rise toward the end of charge to trigger the "ready" state. This is referred to as *delta temperature over delta time* or dV/dt. A rate of 1° C (1.8° F) per minute terminates charging. This keeps the battery cooler, but the cells need to charge reasonably fast for temperature to rise at the required rate. An absolute temperature of 60° C (140° F) terminates charging under any circumstances.

Chargers relying on temperature inflict harmful overcharges when fully charged batteries are inserted into the charger, such as if a hand-held radio is left in the charger between each use. This is not the case with Li-ion batteries where the charger uses voltage as the SoC indicator.

Full-charge Detection by Voltage Signature

Advanced chargers terminate charging when a defined voltage signature or profile with time occurs, referred to as *negative delta V* or NDV. This provides more precise full-charge detection for nickel-based batteries than temperature-based methods. Charging is terminated when the battery voltage drops as full charge is reached. NDV is the recommended method for NiCd cells that do not include an internal thermistor for temperature control and avoids overcharging of fully-charged batteries. NDV requires a charge rate of at least 0.5 C to generate a reliably measurable change in voltage and works best with fast charging. At a charge rate of 1 C, a fully discharged battery is recharged in about an hour.

Fig 7.51 illustrates the relationship of cell voltage, pressure, and temperature of a charging NiCd battery. Up to about 70% SoC, the

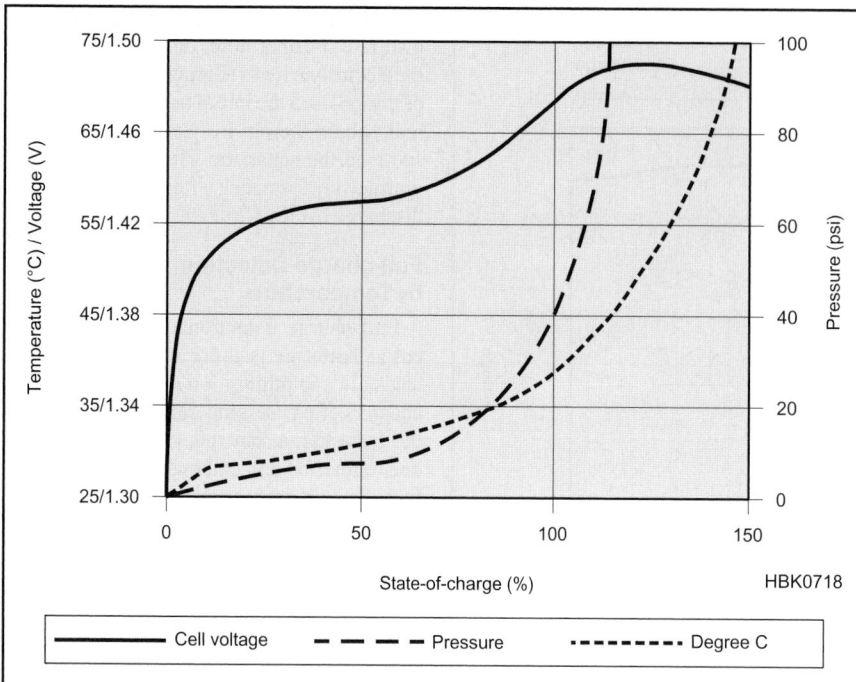

Fig 7.51 — Charge characteristics of a NiCd cell. Above 70% state-of-charge, temperature and cell pressure rise quickly. NiMH has similar charge characteristics. (Courtesy of Cadex)

battery accepts almost all of the energy supplied (called *charge efficiency*). Above 70%, the battery loses ability to accept charge, begins to generate gases so that pressure rises, and temperature increases rapidly.

Ultra-high-capacity NiCd batteries tend to heat up more than standard batteries when charging at 1 C and higher rates due to their higher internal resistance. Applying a high current during initial charge and tapering to a lower rate achieves good results with all nickel-based batteries and moderates temperature rise.

Some chargers can "burp" a charging battery by applying a load to generate a discharge pulse to cause gases to recombine and lower internal pressure. The result is a cooler and more effective charge than with conventional dc charging. Pulse charging does not apply to lead and lithium-based systems.

After full charge, the NiCd battery receives a trickle charge of between 0.05 C and 0.1 C to compensate for self-discharge. To avoid possible overcharge, trickle charging should be done at the lowest possible rate and the batteries should be removed from the charger after more than a few days.

CHARGING NICKEL-METAL-HYDRIDE

The charging algorithm for NiMH is similar to NiCd with the exception that NiMH is more complex. The NDV method of measuring full charge has difficulty because the voltage drop at full charge is very small — about

5 mV/cell. As a result, modern chargers combine the various methods of measuring voltage and temperature into a composite algorithm that reacts depending on battery condition.

Some advanced chargers apply an initial fast charge at 1 C. After reaching a certain voltage threshold, a few minutes rest is taken to allow the battery to cool. Charging then resumes at lower currents as the charge progresses to full charge. This is known as the "step-differential charge" and works well for all nickel-based batteries, achieving an extra capacity of about 6% above basic chargers. A drawback of this method is that the fast-charge stress on the battery will shorten overall battery life by 10 to 20%.

NiMH cannot absorb overcharge well and the trickle charge current must be limited to around 0.05 C. In comparison, a basic NiCd charger trickle charges at 0.1 C. This higher trickle charge and the need for sensitive full-charge detection render the basic NiCd chargers unsuitable for NiMH batteries. On the other hand, NiCd cells can be charged in a NiMH charger at the lower trickle charge rate.

Slow charging should not be used for NiMH batteries. At the charging rate of 0.1 to 0.3 C, the voltage and temperature profiles make it very difficult to measure full charge accurately so the charger must depend on a timer. Harmful overcharge will occur if a fixed timer is used, particularly when charging partially or fully charged batteries. The same is true for charging old batteries with reduced capacity.

Inexpensive chargers are prone to incorrect charging because of the difficulty in correctly sensing full charge. Remove the batteries from the charger when you think they are fully charged. For high charge rates, remove the batteries when they are warm to the touch. It is better to remove the batteries and recharge them before use than to leave them in the charger where they might be overcharged and damaged.

SIMPLE GUIDELINES ON CHARGING NICKEL-BASED BATTERIES

- Do not charge at high or freezing temperatures; room temperature is best.
- Do not use chargers that allow the batteries to heat; remove the batteries when warm to the touch.
- Nickel-based batteries are best fast charged.
- NiMH chargers can charge NiCd batteries but not vice versa.
- High charge current or overcharging on an aging battery may cause heat build-up.
- Do not leave a nickel-based battery in the charger for long periods, even with correct trickle charge. Remove and apply a brief charge before use.
- Nickel- and lithium-based batteries require different charge algorithms and cannot share the same charger unless it can switch between the different chemistries.

CHARGING LITHIUM-ION

The Li-ion charger is a voltage-limiting device that is similar to the lead acid system. The difference lies in a higher voltage per cell, tighter voltage tolerance and the absence of trickle or float charge at full charge. Li-ion cannot accept overcharge — any extra charging causes stress.

Most Li-ion cells charge to 4.20 V/cell with a tolerance of ±50 mV/cell. **Fig 7.52** shows the voltage and current signature as lithium-ion passes through the stages for constant current and topping charge. The charge rate of a typical consumer Li-ion battery is between 0.5 and 1 C in Stage 1 and the charge time is about three hours. Manufacturers recommend charging the 18650 cell at 0.8 C or less. The cell should remain cool during the charging process although there may be a slight temperature rise of a few degrees when reaching full charge. Full charge occurs when the battery reaches the voltage threshold and the current drops to 3% of the rated current, or if the charging current reaches a constant value and does not decrease further. The latter may be due to elevated self-discharge.

Li-ion does not need to be fully charged, as is the case with lead acid, nor is it desirable to do so. In fact, it is better not to fully charge so as not to stress the battery. Choosing a lower voltage threshold or eliminating saturation

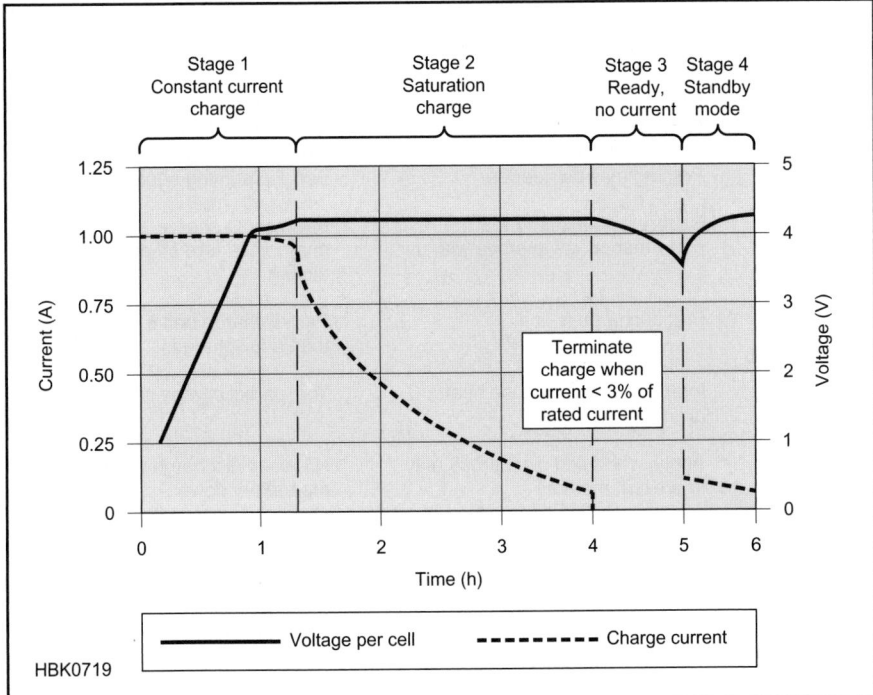

Fig 7.52 — Charge stages of lithium ion. Li-ion is fully charged when the current drops to a predetermined level or levels out at the end of Stage 2. In lieu of trickle charge, some chargers apply a topping charge when the voltage drops to 4.05 V/cell (Stage 4). (Courtesy of Cadex)

charge prolongs battery life at the cost of reduced runtime. Without a saturation stage, the battery is usually charged to around 85% of capacity.

Once charging is terminated, the battery voltage begins to drop and this eases the voltage stress. Over time, the open circuit voltage (OCV) will settle to between 3.60 and 3.90 V/cell. A battery receiving a fully saturated charge will keep the higher voltage longer than a battery that was fast charged and terminated without a saturation charge.

If a lithium-ion battery must be left in the charger for operational readiness, some chargers apply a brief topping charge to compensate for the small self-discharge of the battery and its protective circuit. It is common to let the battery voltage drop to 4.00 V/cell and then recharge to 4.05 V/cell to reduce voltage-related stress and prolong battery life. Battery-powered devices should be turned off when charging their battery. Otherwise, the *parasitic load* of the device can confuse the charger, distorting the charge cycle and stressing the battery.

Overcharging Lithium-ion

Lithium-ion systems operate safely within the designated operating voltages; however, the battery becomes unstable if inadvertently charged to a voltage higher than specified. Prolonged charging above 4.30 V/cell forms plating of metallic lithium on the anode, while

the cathode material becomes an oxidizing agent, loses stability and produces CO_2. Cell pressure rises until the internal *current interrupt device* (CID) disconnects the current at 1380 kPa (200 psi).

Should the pressure rise further, a safety membrane bursts open at 3450 kPa (500 psi) and the cell might eventually vent with flame. The thermal runaway moves lower when the battery is fully charged; for Li-cobalt this threshold is between 130-150° C (266-302° F), for nickel-manganese-cobalt (NMC) between 170-180° C (338-356° F), and manganese is 250° C (482° F). Li-phosphate enjoys similar and better temperature stabilities than manganese.

Lithium-ion is not the only battery that is a safety hazard if overcharged. Lead and nickel-based batteries are also known to melt down and cause fires if improperly handled. Properly designed charging equipment is paramount for all battery systems.

SIMPLE GUIDELINES FOR CHARGING LITHIUM-BASED BATTERIES

• A battery-powered device should be turned off while charging.

• Charge at a moderate temperature. Do not charge below freezing.

• Lithium-ion does not need to be fully charged; a partial charge is better.

• Chargers use different methods for

"ready" indication and may not always indicate a full charge.

• Discontinue using a charger and/or battery if the battery gets excessively warm.

• Before prolonger storage, apply some charge to bring a pack to about half charge.

SUMMARY OF CHARGING

Batteries have unique needs and **Table 7.11** explains how to satisfy these needs with correct handling. Because of similarities within the battery families, only lead, nickel and lithium systems are covered. Along with these guidelines, you can prolong battery life by following three simple rules: keep the battery at moderate temperatures, control the level and rate of discharge, and avoid abusing the battery.

7.13.6 Discharge Methods
C-RATE

According to the definition of *coulomb,* a current of 1 ampere is a flow of 1 coulomb (C) of charge per second. Today, the battery industry uses *C-rate* to scale the charge and discharge current of a battery.

Most portable batteries are rated at 1 C, meaning that a 1000 mAh battery that is discharged at 1 C should under ideal conditions provide a current of 1000 mA for 1 hr. The same battery discharging at 0.5 C would provide 500 mA for 2 hours, and at 2 C, the 1000 mAh battery would deliver 2000 mA for 30 minutes. 1 C is also known as a one-hour discharge; a 0.5 C is a two-hour discharge, a 2 C is a half-hour discharge, and so on.

A battery's capacity—the amount of energy a battery can hold — can be measured with a *battery analyzer*. The analyzer discharges the battery at a calibrated current while measuring the time it takes to reach its specified end-of-discharge voltage. If a 1000 mAh battery could provide 1000 mA for 1 hour, 100% of the battery's nominal energy rating would be reached.

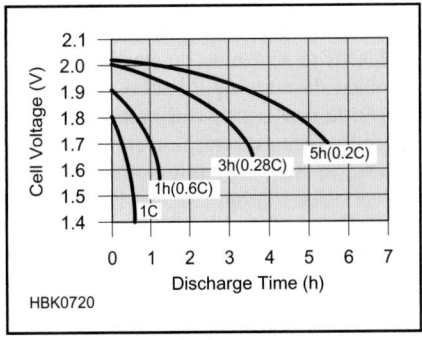

Fig 7.53 — Typical discharge curves of lead acid as a function of C-rate. Smaller batteries are rated at a 1 C discharge rate. Due to sluggish behavior, lead acid is rated at 0.2 C (5 hours) and 0.05 C (20 hours). (Courtesy of Cadex)

Table 7.11

Best Charging Methods

Frequently Asked Question	Lead Acid (Sealed, flooded)	Nickel-Based (NiCd and NiMH)	Lithium-ion (Li-ion, Polymer)
How should I prepare? a new battery	Battery comes fully charged. Apply topping charge	Charge 14-16 h. Priming may be needed.	Apply a topping charge before use. No priming needed
Can I damage a battery with incorrect use?	Yes, do not store partially charged, keep fully charged	Battery is robust and the performance will improve with use.	Keep some charge. Low charge can turn off protection circuit.
Do I need to apply a full charge?	Yes, partial charge causes sulfation.	Partial charge is fine.	Partial charge better than a full charge.
Can I disrupt a charge cycle?	Yes, partial charge causes no harm.	Interruptions can cause heat buildup.	Yes, partial charge causes no harm.
Should I use up all battery energy before charging?	No, deep discharge wears the battery down. Charge more often.	Apply scheduled discharges only to prevent memory.	No, deep discharge wears the battery down.
Do I have to worry about "memory"?	No memory	Discharge NiCd every 1-3 months.	No memory
How do I calibrate a "smart" battery?	Not applicable	Apply discharge/charge when the fuel gauge gets inaccurate. Repeat every 1-3 months.	Apply discharge/charge when the fuel gauge gets inaccurate. Repeat every 1-3 months.
Must I remove the battery when fully charged?	Depends on charger; needs correct float voltage	Remove after a few days in charger.	Not necessary; charger turns off
How do I store my battery?	Keep cells above 2.10 V, charge every 6 months	Store in cool place; a total discharge causes no harm.	Store in cool place partially charged, do not fully drain
Is the battery allowed to heat up during charge?	Battery may get lukewarm toward the end of charge.	Battery gets warm but must cool down on ready.	Battery may get lukewarm toward the end of charge.
How do I charge when cold?	Slow charge (0.1C): 0°-45°C (32°-113° F) Fast charge (0.5-1C): 5°-45° C (41°-113° F)	Slow charge (0.1C): 0°-45° C (32°-113° F) Fast charge (0.5-1C): 5°-45° C (41°-113° F)	Do not charge above 50° C (122° F) Do not charge above 50° C (122° F)
Can I charge at hot temperatures?	Above 25° C, lower threshold by 3 mV/°C.	Battery will not fully charge when hot	Do not charge above 50° C (122° F)
What should I know about chargers?	Charger should float at 2.25-2.30 V/cell when ready.	Battery should not get too hot; should include temp sensor.	Battery must stay cool; no trickle charge when ready.

(Table courtesy of Cadex)

If the discharge only lasted 30 minutes before reaching the specified voltage, the battery has a capacity of 50% of its nominal rating.

When discharging a battery at different rates a higher C-rate will produce a lower capacity reading due to the internal resistance turning some of the energy into heat instead of delivering it as current to a load. Lower C-rate discharges will produce a higher capacity.

For example, to obtain a reasonably good capacity rating, manufacturers commonly rate lead acid batteries at 0.05 C, or a 20-hour discharge. **Fig 7.53** illustrates the discharge times of a lead acid battery at various loads expressed in C-rate.

DEPTH OF DISCHARGE

The end-of-discharge voltage for lead acid is 1.75 V/cell; for nickel-based systems it is 1.00 V/cell; and for most Li-ion it is 3.00 V/cell. At this level, roughly 95% of the battery's stored energy has been spent and voltage would drop rapidly if discharge were to continue. Most devices prevent operation beyond the specified end-of-discharge voltage. When removing the load after discharge, the voltage of a healthy battery gradually recovers toward the nominal voltage.

Because of internal resistance, wiring, protection circuits and contact resistance, a high load current lowers the battery voltage and the end-of-discharge voltage threshold should be lowered accordingly. The cutoff voltage should also be lowered when discharging at very cold temperatures. **Table 7.12** shows typical end-of-discharge voltages of various

battery chemistries. The lower end-of-discharge voltage for higher loads compensates for the losses from internal battery resistance.

Since the cells in a battery pack can never be perfectly matched, a negative voltage potential can occur across a weaker cell on a multi-cell pack if the discharge is allowed to continue beyond a safe cutoff point. Known as *cell reversal*, the weak cell suffers damage to the point of developing a permanent electrical short circuit. The larger the number of cells in the pack, the greater the likelihood that a cell might reverse under heavy load. Over-discharge, particularly at low temperatures, is a large contributor to battery failure of cordless power tools, especially for nickel-based packs. Li-ion packs have protection circuits and the failure rate is lower.

Table 7.12
Recommended End-of-Discharge Voltages in V/Cell

End-of-discharge	Li-manganese	Li-phosphate	Lead-acid	NiCd/NiMH
Normal load	3.00	2.70	1.75	1.00
Heavy load	2.70	2.45	1.40	0.90

(Table courtesy of Cadex)

DISCHARGING AT HIGH AND LOW TEMPERATURES

Batteries achieve optimum service life if used at 20° C (68° F) or slightly below, and nickel-based chemistries degrade rapidly when cycled at high ambient temperatures. Higher temperature operation lowers internal resistance and speeds up the chemical reactions but shortens service life if prolonged.

The performance of all battery chemistries drops drastically at low temperatures. At –20° C (–4° F) most nickel, lead, and lithium-based batteries stop functioning. Although NiCd can be used down to –40° C (–40° F), the permissible discharge is only 0.2 C (5-hour rate). Lead acid also has the problem of the electrolyte freezing which can crack the enclosure. Lead acid electrolyte also freezes more easily at a low charge.

SIMPLE GUIDELINES FOR DISCHARGING BATTERIES

• Battery performance decreases with cold temperature and increases with heat.

• Heat increases battery performance but shortens cycle life by a factor of two for every 10° C (18° F) above 25-30° C (18° F above 77-86° F).

• Only charge at moderate temperatures. Check the manufacturer's specifications for charging below freezing.

• Use heating blankets if batteries need rapid charging at cold temperatures.

• Prevent over-discharging. Cell reversal can cause an electrical short circuit.

• Use a larger battery if repetitive deep discharge cycles cause stress.

• A moderate dc discharge is better for a battery than pulsed loads.

• Lead acid systems are sluggish and require a few seconds of recovery between heavy loads.

7.13.7 Battery Handling

This section touches on the most important aspects of handling batteries when they are new, during their service life, and how to store and dispose of them.

FORMATTING AND PRIMING BATTERIES

Rechargeable batteries may not deliver their full rated capacity when new and will require *formatting* — a process that essentially completes the manufacturing process. Li-ion systems require less care in this regard, but cycling these batteries after long storage has been reported to improve performance. *Priming* is a conditioning cycle that is applied to improve battery performance during usage or after prolonged storage. Priming applies mainly to nickel-based batteries.

Formatting of lead acid batteries occurs by applying a charge, followed by a discharge and recharge as part of regular use. Gradually increase the load on a new battery, allowing it to reach full capacity after 50 to 100 cycles.

Manufacturers advise to trickle charge a nickel-based battery pack for 16 to 24 hours when new and after a long storage. This allows the individual cells to reach an equal charge level. A slow charge also helps to redistribute the electrolyte to eliminate dry spots on the separator that may have formed due to gravity. Applying several charge/discharge cycles through normal use or with a battery analyzer completes the formatting process. This can require from five to seven cycles or as many as 50 cycles depending on battery quality.

Cycling also restores lost capacity when a nickel-based battery has been stored for six months or longer. Storage time, state-of-charge, and storage temperature all affect battery recovery. The longer the storage and the higher the temperature, the more cycles are required to regain full capacity.

Lithium-ion does not need formatting when new, nor does it require the level of maintenance that nickel-based batteries do. Maximum capacity is available immediately. A discharge/charge cycle may be beneficial for calibrating a "smart" battery but this does not improve the internal chemistry.

STORING BATTERIES

The recommended storage temperature for most batteries is 15° C (59° F) and the extreme allowable temperature is –40° C to 50° C (–40° F to 122° F) for most chemistries. While lead acid must be kept at full charge during storage, nickel and lithium-based chemistries should be stored at around 40% state-of-charge.

Storage will always cause batteries to age. **Table 7.13** illustrates the *recoverable capacity* of lithium and nickel-based batteries at various temperatures and charge levels over one year. Recoverable capacity is the available battery capacity after storage with a full charge.

A sealed lead acid battery can be stored up to two years. It is important to apply a charge when the battery falls to 70% SoC, typically 2.07 V/cell or 12.42 V for a 12 V pack.

Nickel-metal-hydride can be stored for about three years. The capacity drop that occurs during storage can partially be reversed with priming.

Primary alkaline and lithium batteries can be stored for up to 10 years with minimum capacity loss.

SIMPLE GUIDELINES FOR STORING BATTERIES

• Remove batteries from equipment and store in a dry and cool place.

• Avoid freezing. Batteries freeze more easily if in a discharged state.

• Charge lead acid before storing and monitor the voltage frequently; apply a charge if below 2.10 V/cell.

• Nickel-based batteries can be stored for five years and longer, prime before use.

Testing and Monitoring Batteries

Several sections of the "Testing and Monitoring" chapter from *Batteries in a Portable World* are provided on the CD-ROM accompanying this book. The information covers measuring internal resistance and state-of-charge, measuring capacity, and special techniques for measuring nickel- and lithium-based batteries.

Table 7.13
Estimated Recoverable Capacity After Storage For 1 Year

Temp (°C)	Lead acid at full charge	Nickel-based at any charge	Lithium-ion (Li-cobalt)	
			40% charge	100% charge
0	97%	99%	98%	94%
25	90%	97%	96%	80%
40	62%	95%	85%	65%
60	38% (after 6 months)	70%	75%	60% (after 3 months)

(Table courtesy of Cadex)

- Lithium-ion must be stored in a charged state, ideally 40%.
- Discard Li-ion if the voltage has stayed below 2.00 V/cell for more than a week.

RECYCLING BATTERIES

The main objective for recycling batteries is to prevent hazardous materials from entering landfills. Lead acid and NiCd batteries are of special concern.

Under no circumstances should batteries be incinerated, as fire can cause an explosion. Wear approved gloves when touching electrolyte. On exposure to skin, flush with water immediately. If eye exposure occurs, flush with water for 15 minutes and consult a physician immediately.

Automotive and larger lead acid batteries can be recycled through auto parts stores and battery dealers. A recycling fee is usually charged when the battery is purchased.

Smaller batteries, including smaller SLA, can be recycled at many electronics and hardware stores, or your local municipal recycling center. Perform an Internet search for battery recyclers in your area.

It is helpful to create a specific location or designate a container for spent batteries at your home, office or workbench. This makes it easy to recycle the batteries by keeping them together in one spot.

7.13.8 DC-AC Inverters

For battery-powered operation of ac-powered equipment, dc-ac inverters are used. An inverter is a dc-to-ac converter that provides 120 V ac. Inverters come with varying degrees of sophistication. The simplest type of inverter switches directly at 60 Hz to produce a square-wave output. This is no problem for lighting and other loads that don't care about the input waveform. However, some equipment will work poorly or not at all when supplied with square wave power because of the high harmonic content of the waveform.

The harmonic content of the inverter output waveform can be reduced by the simple expedient of reducing the waveform duty cycle from 50% (for the square wave) to about 40%. For many loads, such as computers and other electronic devices, this may still not be adequate, and so many inverters use waveform shaping to approximate a sine-wave output. The simplest of these methods is a resonant inductor-capacitor filter. This adds significant weight and size to the inverter. Most modern inverters use high-frequency pulse-width modulation (PWM) techniques to synthesize the 60 Hz sinusoidal output waveform, much like a switching power supply. See **Fig 7.54**.

Inverters are usually rated in terms of their VA or "volt-ampere product" capability although sometimes they will be rated in watts. Care is required in interpreting inverter

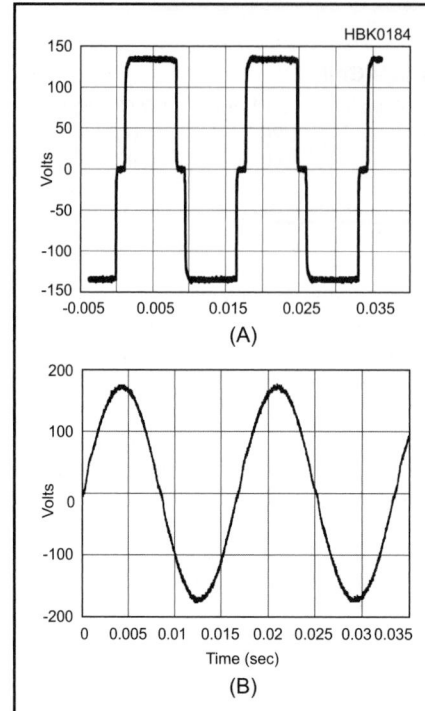

Fig 7.54 — Output waveforms of typical dc-ac inverters. At A, the output of a modified sine wave inverter. Note the stepped square waves. At B, the output of a "pure sine wave" inverter. Note the close approximation of a commercial ac sine wave.

ratings. A purely resistive load operating from a sinusoidal voltage source will have a sinusoidal current flowing in phase with the voltage. In this case, VA, the product of the voltage (V) and the current (A), will equal the actual power in watts delivered to the load so that the VA and the watt ratings are the same. Some loads, such as motors and many rectifier power supplies, will shift the phase of load current away from the source voltage; or the load current will flow in short pulses as shown earlier for capacitor input filters. In these cases (which are very common), the VA product for that type of load can be much larger than the delivered power in watts. In the absence of detailed knowledge of the load characteristics it is prudent to select an inverter with a VA or wattage capability of 25% or more above the expected load.

7.13.9 Selecting a Battery for Mobile Operation

There are two basic modes of mobile operation: *in-motion*, and *stationary*. Each mode has unique power requirements and thus different battery requirements. To satisfy these needs, there are three basic battery types. It is important to understand the differences between the types and to apply them properly.

Standard vehicle batteries are referred to as *SLI* (starting, lights, ignition) or just *starter* batteries. Their primary function is to start the engine and then act as a power filter for the alternator which is the actual long-term power source. The most important rating of an SLI battery is the *cold cranking amps* (CCA) rating — the number of amps that the battery can produce at 32 °F (0 °C) for 30 seconds.

Batteries designed for repeated cycles of charging and discharging use are often called *deep-cycle* although the term is widely overused. A true deep-cycle battery is designed to be repeatedly discharged to 20% remaining capacity. The term "deep cycle" is a misnomer, as all lead-acid batteries are considered discharged when their output voltage drops below 10.5 V at some specified current draw as outlined by the Battery Council Institute (BCI — **www.batterycouncil.org**). At 10.5 V, a six-cell lead-acid battery is considered discharged.

Marine batteries are designed to be stored without charging for up to two years, yet still maintain enough power to start a marine engine. Contrary to common practice, they're not really designed for extended low-current power delivery. Marine batteries often have hybrid characteristics between SLI and deep-cycle batteries.

To differentiate true deep-cycle batteries from SLI and marine batteries, examine a battery's *reserve capacity* (RC). A battery's RC rating is the number of minutes that the battery can deliver 25 A while maintaining an output voltage above 10.5 V. RC batteries typically have RC ratings 20% or more higher than SLI batteries and perhaps 50% higher under ICAS (intermittent, commercial, and amateur service) conditions. A deep-cycle battery has a lower CCA rating than an SLI battery due to its internal construction that favors long-term power delivery over high-current starting loads.

With these facts in mind, we can now select the correct battery for our style of mobile operating. For in-motion operation with power outputs up to 200 W, a second trunk-mounted battery is seldom needed if the power cabling wire size is chosen correctly. (See the **Assembling a Station** chapter for information on wire sizes in mobile applications.)

When an amplifier is added for higher output power levels, it is often less expensive to add a second trunk-mounted battery than to install larger cables to the main battery. In these cases, the second battery can be of almost any type, as long as it is lead-acid. The second battery should be connected in parallel to the vehicle's main SLI battery. The battery's ampere-hour rating should be close to that of the vehicle's main SLI battery.

All secondary wiring should be properly fused, as outlined in the **Assembling a**

Station chapter's section on mobile installations. The use of relays and circuit breakers should be avoided. Remember, should a short circuit occur, good-quality lead-acid batteries can deliver upwards of 3000 A which exceeds the break circuit ratings of most relays and circuit breakers. A better solution is a FET switch such as those made by Perfect Switch (**www.perfectswitch.com**).

Assuming the second battery is mounted inside the vehicle's passenger compartment or in the trunk, it should be an AGM type. AGM (Absorption Glass Mat) batteries do not outgas explosive hydrogen gas under normal operating conditions. Flooded (liquid electrolyte) batteries should *never* be used in an enclosed environment.

For stationary operation, select a battery with a large RC rating because it will not be continuously charged. There are two main considerations; the ampere-hour rating (Ah) and the reserve capacity rating, typically listed as C/8, C/10, or C/20, with units of hours. (C is the battery's capacity in Ah.) Dividing the Ah rating by the load amperage (8, 10 or 20 A) will give you the reserve capacity in hours, but the actual ampere-hours any given battery can deliver before the voltage reaches 10.5 V (nominal discharge level) will vary with the load, both average and peak. Heavier loads will reduce the actual ampere-hours available.

Automotive batteries are arranged in BCI group sizes (**www.rtpnet.org/teaa/bci-group.html**), from 21 through 98. Generally speaking, the larger the group size the larger the battery and the higher the Ah rating. For example, size 24 (small car) has an average rating of 40 Ah, and size 34 (large car) has an average rating of 55 Ah. Exact ratings, including their reserve capacity, are available from the manufacturers' websites listed below. A good rule of thumb is to select a battery as physically large as you have room for, consistent with the highest RC rating for any given Ah rating.

Batteries are heavy, and need to be properly secured inside a battery box or by using factory-supplied brackets. For example, a BIC group 34 (average SLI size) battery weighs about 55 pounds. Some battery models (such as the Optima) come supplied with mounting brackets and terminal protection covers. Even though battery boxes aren't always needed, they should be used as a safety precaution to prevent accidental contact with the terminals and can protect the battery from external items. Battery restraints should be adequate to provide 6 Gs of lateral and 4 Gs of vertical retention, ruling out sheet metal screws and most webbing material. Use the proper brackets!

There are three other considerations: isolating the battery electrically, recharging the battery, and output voltage regulation. Diode-based battery isolators are not all equal. Models with FET bypass switches are the preferred type because of the low voltage-drop across the FET.

If you have wired the battery in parallel with the vehicle's main SLI battery, recharging is taken care of whenever the vehicle is running. If you plan on operating in stationary mode, you'll need a separate recharging system. Most vehicle factory-installed trailer wiring systems also include a circuit for charging RV or boat "house" batteries. Check with your dealer's service personnel about these options.

Voltage regulators, commonly called "battery boosters," are almost a necessity for stationary operation. A model with a low-voltage cutoff should be used to avoid discharging the battery below 10.5 V, as discharging a lead-acid battery beyond this point drastically reduces its charge-cycle life — the number of full-charge/full-discharge cycles. (See the November 2008 *QST* Product Review column.)

For additional information on battery ratings, sizes, and configuration, visit these websites:
optimabatteries.com, www.exide.com
www.interstatebatteries.com
www.lifelinebatteries.com

7.14 Glossary of Power Source Terms

Bleeder — A resistive load across the output or filter of a power supply, intended to quickly discharge stored energy once the supply is turned off.

Boost converter — A switchmode converter in which the output voltage is always greater than or equal to the input voltage.

Buck converter — A switchmode converter in which the output voltage is always less than or equal to the input voltage.

Buck-boost converter — A switchmode converter in which the magnitude of the output voltage can be either greater or less than the input voltage.

C-rate — The charging rate for a battery, expressed as a ratio of the battery's ampere-hour rating.

CCA (cold cranking amps) — A measure of a battery's ability to deliver high current to a starter motor.

Circular mils — A convenient way of expressing the cross-sectional area of a round conductor. The area of the conductor in circular mils is found by squaring its diameter in mils (thousandths of an inch), rather than squaring its radius and multiplying by pi. For example, the diameter of 10-gauge wire is 101.9 mils (0.1019 inch). Its cross-sectional area is 10380 CM, or 0.008155 square inches.

Core saturation (magnetic) — That condition whereby the magnetic flux in a transformer or inductor core is more than the core can handle. If the flux is forced beyond this point, the permeability of the core will decrease, and it will approach the permeability of air.

Crowbar — A last-ditch protection circuit included in many power supplies to protect the load equipment against failure of the regulator in the supply. The crowbar senses an overvoltage condition on the supply's output and fires a shorting device (usually an SCR) to directly short-circuit the supply's output and protect the load. This causes very high currents in the power supply, which blow the supply's input-line fuse.

Darlington transistor — A package of two transistors in one case, with the collectors tied together, and the emitter of one transistor connected to the base of the other. The effective current gain of the pair is approximately the product of the individual gains of the two devices.

DC-DC converter — A circuit for changing the voltage of a dc source to ac, transforming it to another level, and then rectifying the output to produce direct current.

Deep-cycle — A battery designed for repeated charge-discharge cycles to 20% of remaining capacity.

Equalizing resistors — Equal-value bypassing resistors placed across capacitors connected in series for use in a high-voltage power supply to keep the voltages across the capacitors in the string relatively constant.

Fast recovery rectifier — A specially doped rectifier diode designed to minimize the time necessary to halt conduction when the diode is switched from a forward-biased state to a reverse-biased state.

Flyback converter — A transformer-coupled version of the **buck-boost converter**.

Forward converter — A **buck converter** with multiple isolated outputs at different voltage levels and polarities.

Foldback current limiting — A special type of current limiting used in linear power supplies, which reduces the current through the supply's regulator to a low value under short circuited load conditions in order to protect the series pass transistor from excessive power dissipation and possible destruction.

Ground fault (circuit) interrupter (GFI or GFCI) — A safety device installed between the household power mains and equipment where there is a danger of personnel touching an earth ground while operating the equipment. The GFI senses any current flowing directly to ground and immediately switches off all power to the equipment to minimize electrical shock. GFCIs are now standard equipment in bathroom and outdoor receptacles.

Input-output differential — The voltage drop appearing across the series pass transistor in a linear voltage regulator. This term is usually stated as a minimum value, which is that voltage necessary to allow the regulator to function and conduct current. A typical figure for this drop in most three-terminal regulator ICs is about 2.5 V. In other words, a regulator that is to provide 12.5 V dc will need a source voltage of at least 15.0 V at all times to maintain regulation.

Inverter — A circuit for producing ac power from a dc source.

Li-ion — Lithium-ion, a type of rechargeable battery that is about ⅓ the weight and ½ the volume of a **NiCd** battery of the same capacity.

Low dropout regulator — A three-terminal regulator designed to work with a low minimum input-output differential value.

Marine — A battery designed to retain significant energy over long periods of time without being continuously charged.

NiCd — Nickel cadmium, a type of rechargeable battery.

NiMH — Nickel metal hydride, a type of rechargeable battery that does not contain toxic substances.

Peak inverse voltage (PIV) — The maximum reverse-biased voltage that a semiconductor is rated to handle safely. Exceeding the peak inverse rating can result in junction breakdown and device destruction.

Power converter — Another term for a power supply.

Power processor — Another term for a power supply.

Primary battery — A battery intended for one-time use and then discarded.

RC (reserve capacity) — A measure of a battery's ability to deliver current over long periods.

Regulator — A device (such as a Zener diode) or circuitry in a power supply for maintaining a constant output voltage over a range of load currents and input voltages.

Resonant converter — A form of dc-dc converter characterized by the series pass switch turning on into an effective series-resonant load. This allows a zero current condition at turn-on and turn-off. The resonant converter normally operates at frequencies between 100 kHz and 500 kHz and is very compact in size for its power handling ability.

Ripple — The residual ac left after rectification, filtration and regulation of the input power.

RMS — *R*oot *M*ean *S*quare. Refers to the effective value of an alternating voltage or current, corresponding to the dc voltage or current that would cause the same heating effect.

Secondary battery — A battery that may be recharged many times. Also called a *storage battery*.

Secondary breakdown — A runaway failure condition in a transistor, occurring at higher collector-emitter voltages, where hot spots occur due to (and promoting) localization of the collector current at that region of the chip.

Series pass transistor, or pass transistor — The transistor(s) that control(s) the passage of power between the unregulated dc source and the load in a regulator. In a linear regulator, the series pass transistor acts as a controlled resistor to drop the voltage to that needed by the load. In a switch-mode regulator, the series pass transistor switches between its ON and OFF states.

SLI (starter, lights, ignition) — An automotive battery designed to start the vehicle and provide power to the lighting and ignition systems.

SOA (Safe Operating Area) — The range of permissible collector current and collector-emitter voltage combinations where a transistor may be safely operated without danger of device failure.

Surge — A moderate-duration perturbation on a power line, usually lasting for hundreds of milliseconds to several seconds.

Switching regulator — Another name for a switchmode converter.

Switchmode converter — A high-efficiency switching circuit used for dc-dc power conversion. Switching circuits are usually much smaller and lighter than conventional 60 Hz, transformer-rectifier circuits because they operate at much higher frequencies — from 25 to 400 kHz or even higher.

Three-terminal regulator — A device used for voltage regulation that has three leads (terminals) and includes a voltage reference, a high-gain error amplifier, temperature-compensated voltage sensing resistors and a pass element.

Transient — A short perturbation or "spike" on a power line, usually lasting for microseconds to tens of milliseconds.

Varistor — A surge suppression device used to absorb transients and spikes occurring on the power lines, thereby protecting electronic equipment plugged into that line. Frequently, the term MOV (*M*etal *O*xide *V*aristor) is used instead.

Volt-Amperes (VA) — The product obtained by multiplying the current times the voltage in an ac circuit without regard for the phase angle between the two. This is also known as the apparent power delivered to the load as opposed to the actual or real power absorbed by the load, expressed in watts.

Voltage multiplier — A type of rectifier circuit that is arranged so as to charge a capacitor or capacitors on one half-cycle of the ac input voltage waveform, and then to connect these capacitors in series with the rectified line or other charged capacitors on the alternate half-cycle. The voltage doubler and tripler are commonly used forms of the voltage multiplier.

Voltage regulation — The change in power supply output voltage with load, expressed as a percentage.

7.15 References and Bibliography

REFERENCES

1) Landee, Davis and Albrecht, *Electronic Designer Handbook*, 2nd edition, (McGraw-Hill, 1977), page 12-9. This book can frequently be found in technical libraries and used book stores. The power supply section is well worth reading.

2) Severns and Bloom, *Modern DC-to-DC Switchmode Power Converter Circuits,* (Van Nostrand Reinhold, 1984, ISBN: 0-442-21396-4). A reprint of this book is currently available at the Power Sources Manufacturers Association website, **www.psma.com**.

3) Fair-Rite website, **www.fair-rite.com**

4) **www.mag-inc.com**

OTHER RESOURCES

M. Brown, *Power Supply Cookbook,* (Butterworth-Heinemann, 2001).

I. Buchmann, *Batteries In a Portable World,* (Cadex Electronics, 2011).

J. Fielding, ZS5JF, *Power Supply Handbook,* (RSGB, 2006).

K. Jeffrey, *Independent Energy Guide,* (Orwell Cove Press, 1995). Includes information on batteries and inverters.

7.16 Power Source Projects

Construction of a power supply or accessory — basic to all of the radio equipment we operate and enjoy — can be one of the most rewarding projects undertaken by a radio amateur. Final testing and adjustment of most power-supply projects requires only a voltmeter, and perhaps an oscilloscope — tools commonly available to most amateurs.

General construction techniques that may be helpful in building the projects in this chapter are outlined in the **Construction Techniques** chapter. Other chapters in the *Handbook* contain basic information about the components that make up power supplies.

Safety must always be carefully considered during design and construction of any power supply. Power supplies contain potentially lethal voltages, and care must be taken to guard against accidental exposure. For example, electrical tape, insulated tubing ("spaghetti") or heat-shrink tubing is recommended for covering exposed wires, components leads, component solder terminals and tie-down points. Whenever possible, connectors used to mate the power supply to the outside world should be of an insulated type designed to prevent accidental contact.

Connectors and wire should be checked for voltage and current ratings. Always use wire with an insulation rating higher than the working voltages in the power supply. For supply voltages above 300 V, use wire with insulation rated accordingly. The **Component Data and References** chapter contains a table showing the current-carrying capability of various wire sizes. Scrimping on wire and connectors to save money could result in flashover, meltdown or fire.

All fuses and switches should be placed in the hot circuit(s) only. The neutral circuit should not be interrupted. Use of a three-wire (grounded) power connection will greatly reduce the chance of accidental shock. The proper wiring color code for 120 V circuits is: black — hot; white — neutral; and green

— ground. For 240 V circuits, the second hot circuit generally uses a red wire.

POWER SUPPLY PRIMARY-CIRCUIT CONNECTOR STANDARD

The International Commission on Rules for the Approval of Electrical Equipment (CEE) standard for power-supply primary-circuit connectors for use with detachable cable assemblies is the CEE-22. The CEE-22 has been recognized by the ARRL and standards agencies of many countries. Rated for up to 250 V, 6 A at 65 °C, the CEE-22 is the most commonly used three-wire (grounded), chassis-mount primary circuit connector for electronic equipment in North America and Europe. It is often used in Japan and Australia as well.

When building a power supply requiring 6 A or less for the primary supply, a builder would do well to consider using a CEE-22 connector and an appropriate cable assembly, rather than a permanently installed line cord. Use of a detachable line cord makes replacement easy in case of damage. CEE-22 compatible cable assemblies are available with a wide variety of power plugs including most types used overseas.

Some manufacturers even supply the CEE-22 connector with a built-in line filter. These connector/filter combinations are especially useful in supplies that are operated in RF fields. They are also useful in digital equipment to minimize conducted interference to the power lines.

CEE-22 connectors are available in many styles for chassis or PC-board mounting. Some have screw terminals; others have solder terminals. Some styles even contain built-in fuse holders.

7.16.1 Four-Output Switching Bench Supply

This project by Larry Cicchinelli, K3PTO,

describes the four-output bench power supply shown in **Fig 7.55** with three positive outputs and one negative output. The three positive outputs use identical switching regulator circuits that can be set independently to any voltage between 3.3 V and 20 V at up to 1 A. The fourth output is a negative regulator capable of about 250 mA. As built, the supply has two fixed outputs and two variable outputs, but any module can be built with variable output. (Construction diagrams and instructions, a complete parts list, and additional design details are included on this *Handbook's* CD-ROM.)

The only dependency among the outputs is that they are all driven by a single transformer. The transformer used is rated at 25 V and 2 A — good for 50 W. Assuming that the regulator IC being used has a 75% efficiency, a total of about 37 W is available from the power supply outputs.

One of the features of a switching regulator is that you can draw more current from the outputs than what the transformer is supplying — at a lower voltage, of course — as long as you stay within the 37 W limit and maximum current for the regulator. Most of the discussion in this article will be about the

Fig 7.55 — The front panel of the four-output switching supply.

positive regulators as the negative regulator was an add-on after the original system was built.

POSITIVE REGULATOR

Fig 7.56 is the circuit for the positive regulator modules — a buck-type regulator. There are several variations of the circuit, any of which you can implement.

• L2 and C4 are optional. These two components implement a low-pass filter that will decrease high frequency noise that might otherwise appear at the output.

• The pads for R1 will accommodate a small, multi-turn potentiometer. You can insert one here or you can use the pads to connect a panel-mounted potentiometer.

• If you want a fixed output you can simply short out R1 and use R2 by itself.

• You can also insert a fixed resistor in the R1 position in the case where the calculated value is non-standard and you want to use two fixed resistors.

The formula for setting output voltage using the 3.3 V version of the regulator is based on knowing the current (in mA) through the regulator's internal voltage divider = 3.3 V / 2.7 kΩ = 1.22 mA. The sum of R1 and R2 must cause the voltage at the regulator FB pin to equal 3.3 V. Thus, R1 + R2 in kΩ = (V_{out} – 3.3) / 1.22 and V_{out} = 1.22 (R1 + R2) + 3.3. If R1 = R2 = 0, a direct connection from the output voltage to the FB pin, the calculation results in an output of 3.3 V. The leakage current of the Error Amplifier in the regulator is somewhat less than 25 nA so it can be ignored. The values for R1 and R2 are shown in the caption for Fig 7.56.

The only critical parts are R1 and R2 which form the voltage dividers for the regulator module. Even their values can be changed, within reason, as long as the ratios are maintained. If you want to have an accurate, fixed output voltage, select a value for R2 that is lower than the calculated value and use a

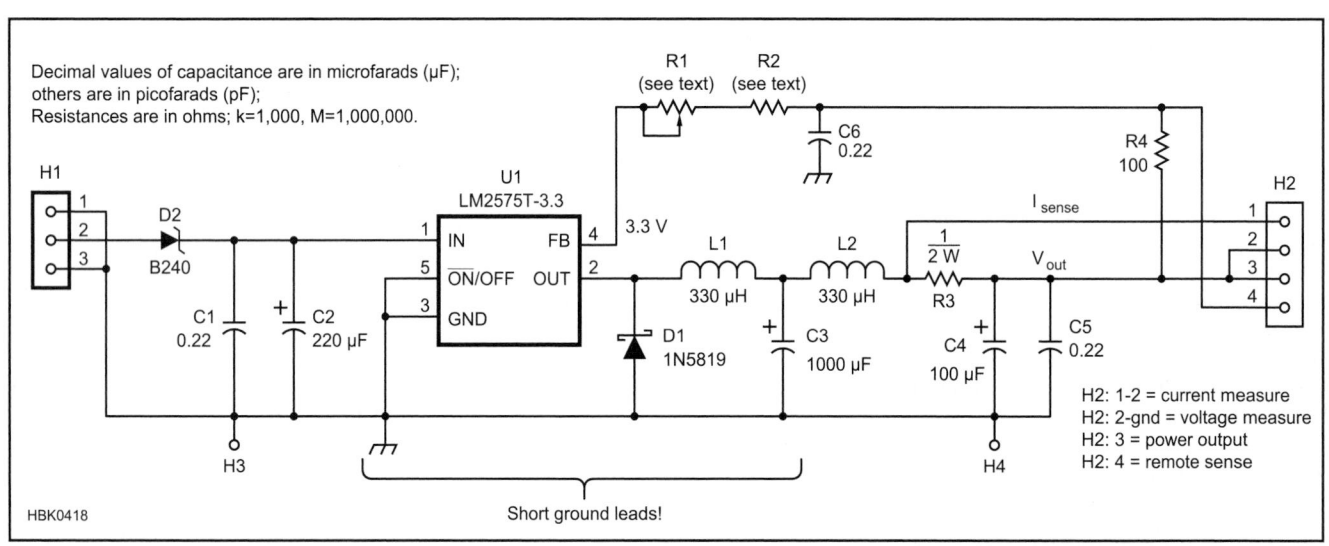

Fig 7.56 — The positive buck-type switchmode regulator uses the LM2575-3.3, a fixed-voltage regulator, with an external voltage-set resistor (R1 + R2). See text for details of the calculations needed to determine the value of R1 and R2. As noted in the text, these values are the total resistance for both parts, and can be made from one fixed resistor, one variable resistor or a combination. Some common values (R1 + R2 total) are: For a 12 V fixed supply, 7.1 kΩ; for 5 V, 1.4 kΩ; for a 3.3 to 20 V variable supply, 0-13.7 kΩ (use a 15 kΩ pot); for 5 to 15 V, 1.4-9.6 kΩ (use 1 kΩ fixed-value resistor and a 1 kΩ pot). A full parts list is included on the *Handbook* CD-ROM.

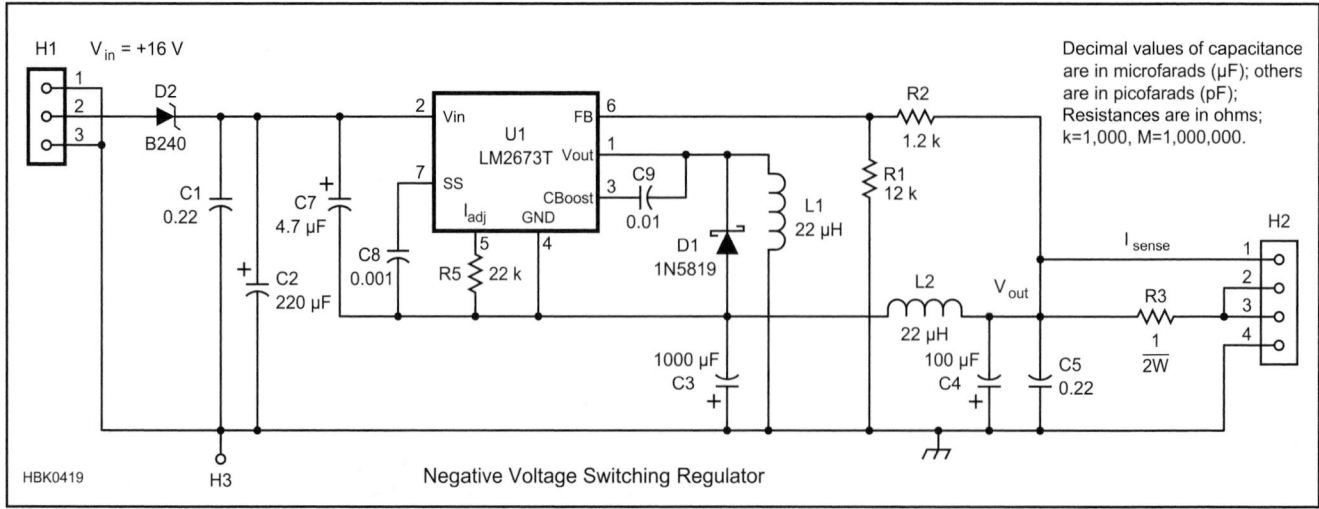

Fig 7.57 — The negative regulator uses the LM2673T in a buck-boost circuit. This circuit inverts the output voltage from the input voltage. A full parts list is included on the *Handbook* CD-ROM.

potentiometer for R1 to set the voltage exactly. The value of C3 is not especially critical; however, it should be a low-ESR (equivalent series resistance) type that is intended for use in switchmode circuits.

NEGATIVE REGULATOR

The negative regulator is a buck-boost configuration — it converts a positive voltage into a negative one — see **Fig 7.57**. This design uses many of the same component values as the positive regulators except the regulator IC is an LM2673 to improve circuit stability. The author was unable to implement the current measuring circuit within a feedback loop. Several configurations introduced a significant low frequency noise component to the output voltage. There was also some 50 kHz noise present on the output, but an additional low-pass filter on the output reduced it considerably.

REMOTE SENSING

Many power supplies use remote sensing to compensate electronically for the voltage drop in the wires carrying current to the load. Even with relatively short wires, there can be significant voltage drop between the regulator and its load. There is provision for remote sensing in this circuit described in the support information for this project on the book's CD-ROM.

If you are not going to use remote sensing then you should insert a jumper in place of R4 in Fig 7.56. R4 (100 Ω) is there for protection just in case the remote sense connection is missing. If you do not want to use remote sensing you can simplify the digital panel meter (DPM) switch wiring to use a two-pole switch instead of the three-pole model listed. In this case, do not use S2.2 and connect S1.2 to the common of S2.3 instead of S2.2.

SUB-CIRCUIT INTERCONNECTION

Fig 7.58 shows the connections among the parts of the system: regulator boards, the digital panel meter (DPM), and rectifier circuit. The components used for the main rectifier circuit can be mounted on a terminal strip and do not need to be on a printed-circuit board.

THE DIGITAL PANEL METER

Another feature of the unit is the DPM which can be switched to measure the output voltage (H2 pin 1 to ground) as well as the current draw (voltage between H2 pins 1 and 2) for each of the positive supplies. Fig 7.58 shows the 3-pole, 4-position rotary switch (S2) that selects which power supply to monitor and a 3PDT toggle switch (S1) that selects between measuring voltage and current.

In order to measure the voltage drop across the 1 Ω current sense resistors, the DPM needs either an isolated power supply or some more circuitry. This system uses an isolated power supply. A series regulator is used simply because they are somewhat easier to implement and the DPM has a very low current requirement. All components except the transformer are mounted on a piece of perforated board. Since the 1.2 mA current for the feedback circuit flows through the current sense resistor it will be included in the value displayed by the DPM when current is selected.

The DPM also has a set of jumpers that allow you to set the decimal point location. As can be seen in Fig 7.58, one pole of the toggle switch selects its location.

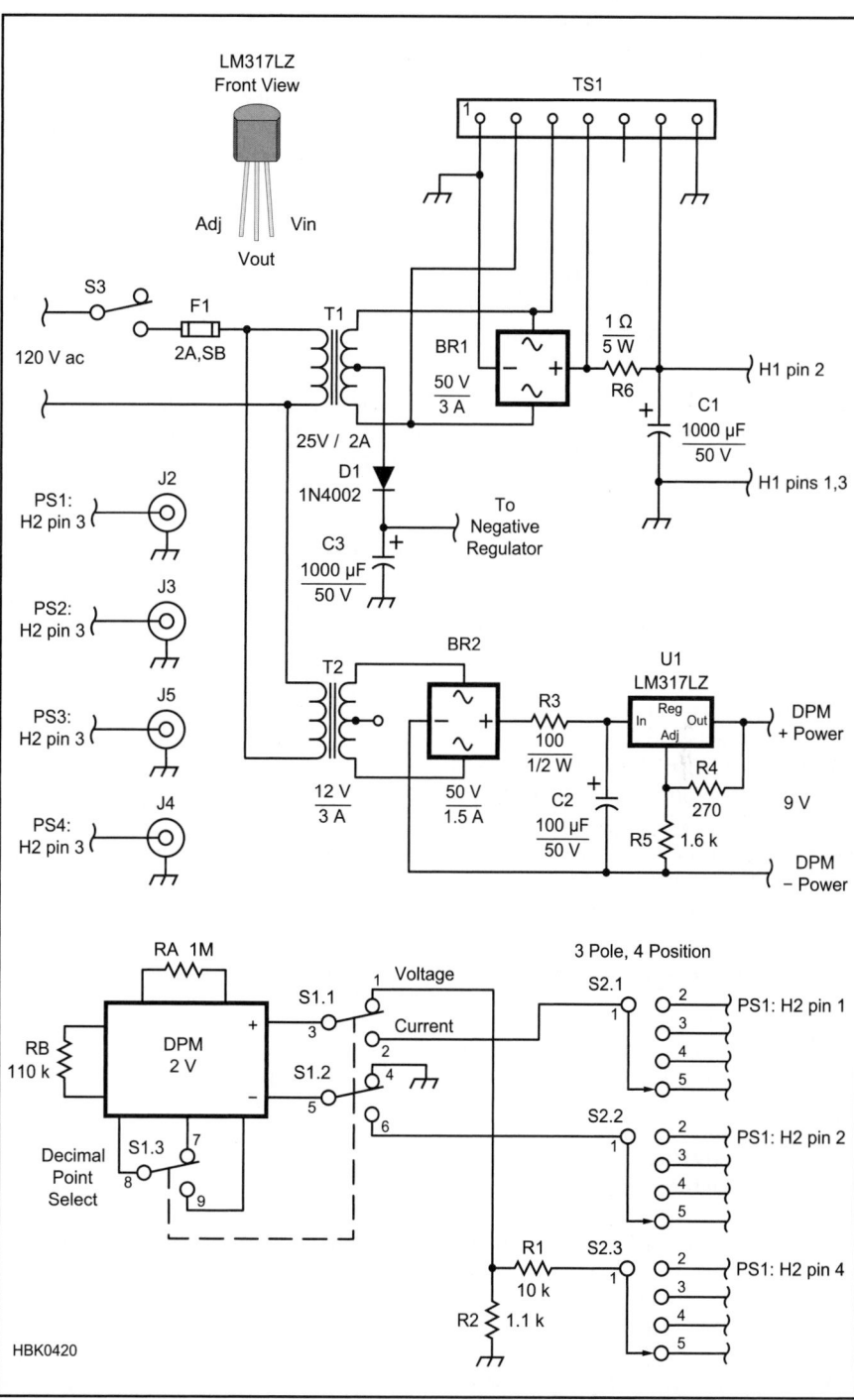

Fig 7.58 — Rectifier and metering schematic. The panel meter is switched between the four modules with a rotary switch (S2) and between voltage and current with a 3PDT toggle (S1). A separate rectifier provides power for the negative supply and a separate three-terminal regulator circuit provides power to the DPM. A full parts list is included on the *Handbook* CD-ROM.

CONSTRUCTION DETAILS

Both the DPM and the regulators use pin headers for all of the connections that come off the boards (see the parts list for details). This allows assembly of the subsystems without having to consider any attached wires. Wire lengths can be determined later, then install the mating connectors on the wires and simply push them onto the pins.

PC boards are available from FAR Circuits (**www.farcircuits.net**), a company that provides a lot of boards for ham-related projects. A caution regarding the circuit boards is in order — the boards do not have plated through holes so you will have to be sure that you solder the through-hole components on *both* sides of the board.

Artwork for the PC board layout, Gerber files, and a drill file are available on the CD-ROM included with this book. The schematic capture software *DipTrace* was used in the development of this project. Source files for the schematic and PCB files are also available on the CD-ROM.

7.16.2 12 V, 15 A Linear Power Supply

This power supply is a linear 12 V, 15 A design by Ed Oscarson, WA1TWX. It is suitable for typical mobile radios and offers adjustable output voltage and current limiting. Supply regulation is excellent, typically exhibiting a change of less than 20 mV from no load to 15 A. This basic design, with heftier components and additional pass transistors, can deliver over 30 A — enough to supply a 100 W class transceiver. (All numbered notes, additional circuit design information, a discussion of how to change the supply voltage and/or current ratings, construction and testing notes, a PCB template and a complete parts list are available on the CD-ROM included with this *Handbook*.)

CIRCUIT DESCRIPTION

Fig 7.59 is the supply's schematic. The ac line input is fused by F1, switched on and off by S1 and filtered by FL1. F1 and S1 are rated at about ¼ of the output current requirement (for 15 A output, use a 4 or 5 A slow-blow fuse or a similarly rated circuit breaker). FL1 prevents any RF from the secondary or load from coupling into the power line and prevents RF on the power line from disturbing supply operation. If your ac power line is clean, and you experience no RF problems, you can eliminate FL1, but it's inexpensive insurance.

When discharged, filter capacitor C1 looks like a short circuit across the output of rectifier U2 when ac power is applied. That usually subjects the rectifier and capacitor to a large inrush current, which can damage them. Fortunately a simple and inexpensive

means of inrush-current limiting is available. Keystone Carbon Company (and others) produce a line of inrush-current limiters (thermistors) for this purpose. The device (RT1) is placed in series with one of the transformer primary leads. RT1 has a current rating of 6 A[1], and a cold resistance of 5 Ω. When it's hot, RT1's resistance drops to 0.11 Ω. Such a low resistance has a negligible effect on supply operation. Thermistors run *hot* so they must be mounted in free air, and away from anything that can be damaged by heat.[2,3]

The largest and most important part in the power supply is the transformer (T1). If purchased new, it can also be the most costly. Fortunately, a number of surplus dealers offer power transformers that can be used in this supply.

T1 produces 17 V ac RMS at 20 A; the center tap is not used. Bridge rectifier U2 provides full-wave rectification. Full-wave rectification reduces the ripple component of current that flows in the filter capacitor, resulting in less power dissipation in the capacitor's internal resistance. U2's voltage rating should be at least 50 V, and its current rating about 25% higher than the normal load requirement; a 2 A bridge rectifier will do. U2 is secured to the chassis or a heat sink because it dissipates heat.

C1 is a computer-grade electrolytic. Any capacitor value from 15,000 to 30,000 μF will suffice. This version uses a 19,000 μF, 40 V capacitor. The capacitor's voltage rating should be at least 50% higher than the expected no-load rectified de voltage. In this supply, that voltage is 25 V, and a 40 V capacitor provides enough margin,

R5, a 75 Ω, 20 W bleeder resistor, is connected across C1's terminals to discharge the supply when no load is attached or one is removed. Any resistance value from 50 Ω to 200 Ω is fine; adjust the resistor's wattage rating appropriately.

At the terminals of C1, we have a dc voltage, but it varies widely with the load applied. When keying a CW transmitter or switching a rig from receive to full output, 5 V swings can result. The dc voltage also has an ac ripple component of up to 1.5 V under full load. Adding a solid-state regulator (U1) provides a stable output voltage even with a varying input and load.

VOLTAGE REGULATOR IC AND PASS TRANSISTORS

The LM723 used at U1 has a built-in voltage reference and sense amplifier, and a 150 mA drive output for a pass-transistor array. U1's voltage reference provides a stable point of comparison for the internal regulator circuitry. In this supply, it's connected to the

non-inverting input of the voltage-sense op amp. The reference is set internally to 7.15 V, but the absolute value is not critical because an output-voltage adjustment (R12) is provided. What is important is that the voltage is stable, with a specified variation of 0.05% per 1000 hours of operation. This is more than adequate for the supply.

For the regulator to work properly, its ground reference must be at the same point as the output ground terminal. The best way to ensure this is to use the output GROUND terminal (J4) as a single-point ground for all of the supply grounds. Run wires to J4 from each component requiring a ground connection. Fig 7.59 attempts to show this graphically through the use of parallel connections to a single circuit node.

The output pass transistor array consists of a TIP112 Darlington-pair transistor (Q5) driving three 2N3055 power transistors (Q1-Q3). This two-stage design is less efficient than connecting the power transistors directly to the LM723, but Q5 can provide considerably more base current to the 2N3055s than the 150 mA maximum rating of the LM723. You can place additional 2N3055s in parallel to increase the output current capacity of the supply.

This design is not fussy about the pass transistors or the Darlington transistor used. Just ensure all of these devices have voltage ratings of at least 40 V. Q5 must have a 5 A (or greater) collector-current rating and a beta of over 100. The pass transistors should be rated for collector currents of 10 A or more, and have a beta of at least 10.[5]

Resistors R17, R18 and R19 prevent leakage current through the collector-base junction from turning on the transistor by diverting it around the base-emitter junction. When the pass transistors are hot, at the V_{CE} encountered in this design, the leakage current can be as high as 3 mA. The resulting drop across the 33 Ω resistors is 0.1 V — safely below the turn-on value for V_{BE}.

When unmatched transistors are simply connected in parallel they usually don't equally share the current.[6] By placing a low-value resistor in each transistor's emitter lead (emitter-ballasting resistors, R1-R3), equal current sharing is ensured. When a transistor with a lower voltage drop tries to pass more current, the emitter resistor's voltage drop increases, allowing the other transistors to provide more current. Because the voltage-sense point is on the load side of the resistors, the transistors are forced to dynamically share the load current.

With a 5 A emitter current, 0.25 V develops across each 0.05 Ω resistor, producing 1.25 W of heat. Ideally, a resistor's power rating should be at least twice the power it's called upon to dissipate. To help the resistors dissipate the heat, mount them on a heat sink,

---

[1]See the full article on the *Handbook* CD-ROM for a list of numbered notes.

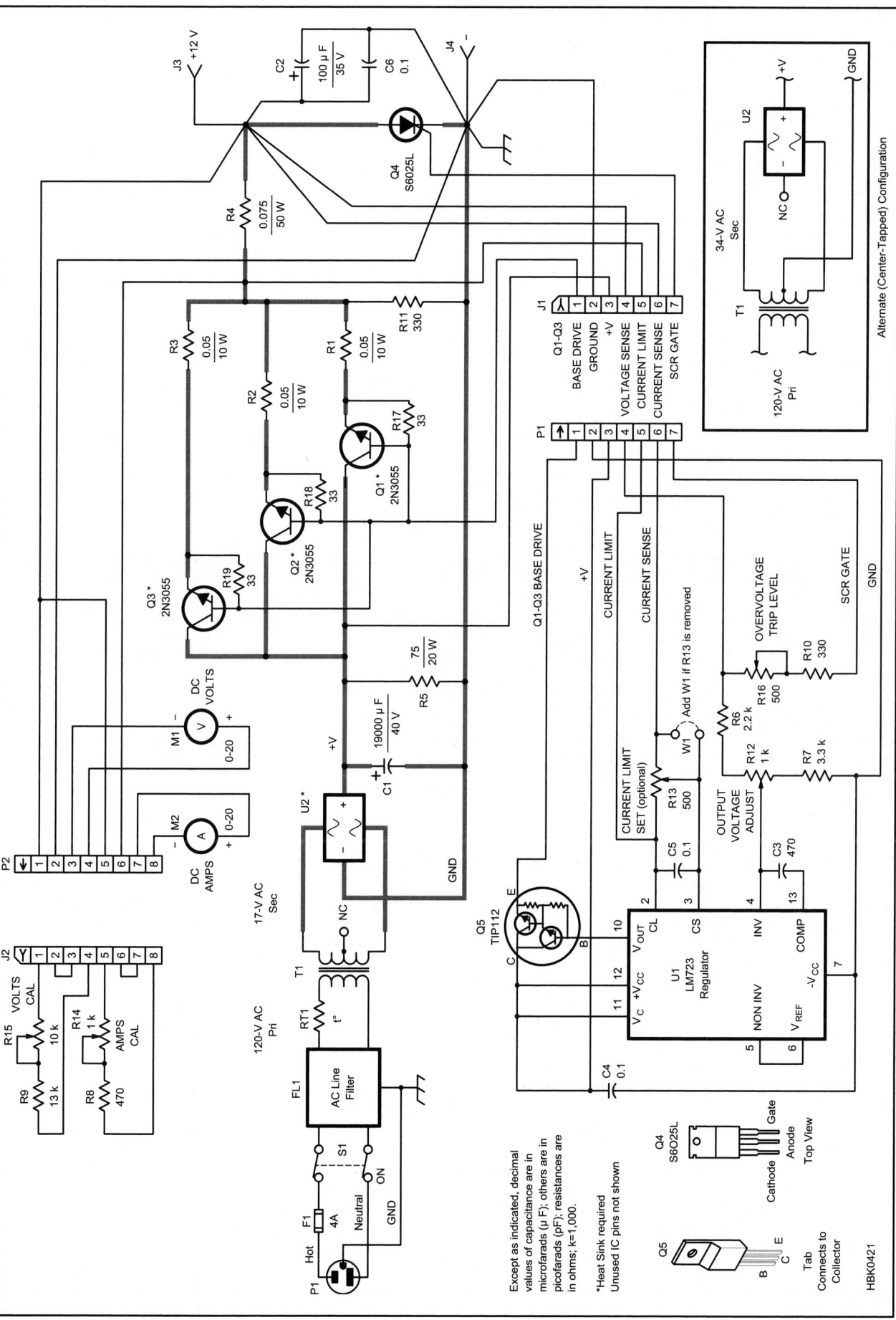

Fig 7.59 — Schematic of the 12 V, 15 A power supply. A parts list may be found on the *Handbook* CD-ROM. Equivalent parts can be substituted. The bold lines indicate high-current paths that should use heavy-gauge (#10 or #12 AWG) wire. This schematic graphically shows wiring to a single-point ground; see text. The majority of the parts used in this supply are available as surplus components.

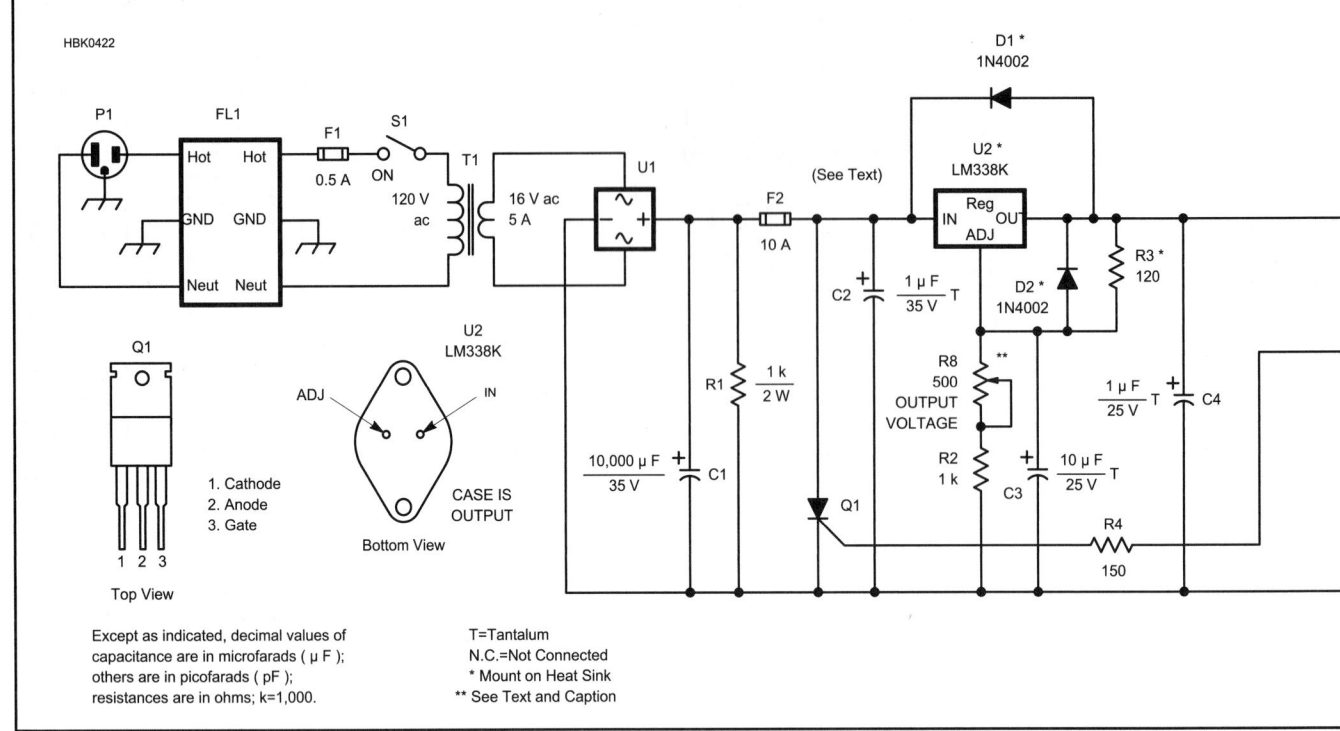

Fig 7.60 — Schematic for the 13.8 V, 5 A power supply. Unless otherwise specified, resistors are ¼ W, 5% tolerance. A PC board and U3 are available from FAR Circuits (www.farcircuits.net). The PC board has mounting holes and pads to allow for handling different trimmer-potentiometer footprints. A PC board template is available on the CD-ROM for this book.

C1 — 10,000 µF, 35 V electrolytic.
C2, C4 — 1 µF, 35 V tantalum.
C3 — 10 µF, 35 V tantalum.
C5 — 0.1 µF, 25 V ceramic disc.
D1-D3 — 1N4002.
DS1 — Red LED.
F2 — Fast-acting 10 A fuses; three required (see text).
F1 — Fast-acting 0.5 A fuse.
FL1— Ac-line filter.

Q1 — BT152 400 V, 25 A SCR in TO-220A package (NTE5554).
R8 — 500 Ω, single-turn trimmer potentiometer.
R9 — 500 Ω or 1 kΩ, single-turn trimmer potentiometer.
S1 — SPST panel-mount switch.
T1 — 120 V primary, 16- to 20 V, 5 A secondary.
U1 — 100-PIV, 6 A bridge rectifier.

U2 — LM338K 5 A adjustable power regulator in a TO-3 package.
U3 — MC3423P1 overvoltage protection IC.
Misc: two panel-mount fuse holders; line cord; heat sinks for TO-3 case transistors; TO-3 mounting kit and heat-sink grease; black and red binding posts; chassis or cabinet; PC board; hardware, rubber hoods, heat-shrink tubing or electrical tape for F1 and FL1, hook-up wire.

or secure them to a metal chassis. You can use any resistor with a value between 0.065 and 0.1 Ω, but remember that the power dissipated is higher with higher-value resistors (10 W resistors are used here).

At the high output currents provided by this supply, the pass transistors dissipate considerable power. With a current of 5 A through each transistor-and assuming a 9 V drop across the transistor — each device dissipates 45 W. Because the 2N3055's rating is 115 W when used with a properly sized heat sink, this dissipation level shouldn't present a problem. If the supply is to be used for continuous-duty operation, increase the size of the heat sink and mount it with the fins oriented vertically to assist in air circulation.

The output-voltage sense is connected through a resistive divider to the negative input of U1. U1 uses the difference between its negative and positive inputs to control the pass transistors that in turn provide the output current. C3, a compensation capacitor, is

connected between this input and a dedicated compensation pin to prevent oscillation. The output voltage is adjusted by potentiometer R12 and two fixed-value resistors, R6 and R7.[7] The voltage-sense input is connected to the supply's positive output terminal, J3.

Current sensing is done through R4, a 0.075 Ω, 50 W resistor connected between the emitter-ballasting resistors and J3. R4's power dissipation is much higher than that of R1, R2 or R3 because it sees the total output current. At 15 A, R4 dissipates 17 W. At 20 A, the dissipated power increases to 30 W.

U1 provides current limiting via two sense inputs connected across R4. Limiting takes place when the voltage across the sense inputs is greater than 0.65 V.[8] For a 15 A maximum output-current limit, this requires a 0.043 Ω resistor. By using a larger-value sense resistor and a potentiometer, you can vary the current limit. Connecting potentiometer R13 across R4 provides a current-limiting range from full limit voltage (8.7 A limit) to no limit voltage.

This allows the current limit to be fine-tuned, if needed, and also permits readily available resistor values (such as the author's 0.075 Ω resistors) to be used. A current limit of 20 A is at the top end of the ammeter scale.

R20 maintains a small load of 35-40 mA depending on power supply output voltage. This reduces the effect of leakage current in the pass transistors and keeps the regulator's feedback action active, even if no external load is connected.

METERING

Voltmeter M1 is a surplus meter. R8 and potentiometer R15 provide for voltmeter calibration. If the correct fixed-value resistor is available, R15 can be omitted. The combined value of the resistor and potentiometer is determined by the full-scale current requirement of the meter used.[9]

Ammeter M2 is actually a voltmeter (also surplus) that measures the potential across R4. The positive side of M2 connects to the high

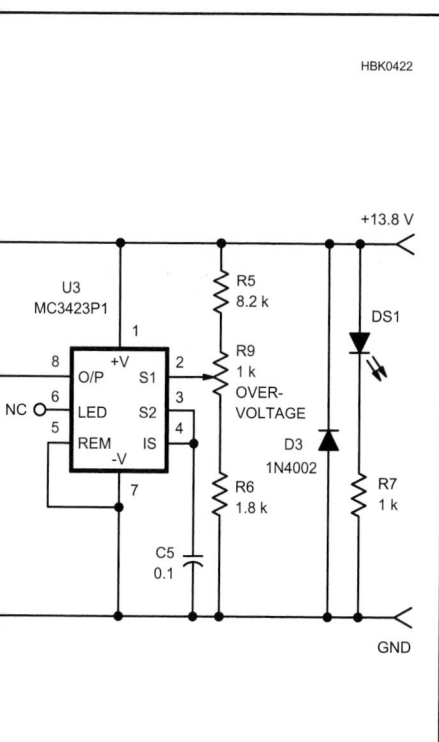

HBK0422

side of R4. R8 and potentiometer R14 connect between the positive output terminal (J3) and the negative side of M2 to provide calibration adjustment. The values of R8 and R14 are determined by the coil-current requirements of the meter used. (Digital panel meters or a dedicated DMM can be used instead of separate analog meters.)

OUTPUT WIRING AND CROWBAR CIRCUIT

The supply output is connected to the outside world by two heavy-duty banana jacks, J3 and J4. C2, a 100 µF capacitor, is soldered directly across the terminals to prevent low-frequency oscillation. C6, a 0.1 µF capacitor, is included to shunt RF energy to ground. Heavy-gauge wire must be used for the connections between the pass transistors and J3 and between chassis ground and J4. The voltage-sense wire must connect directly to J3 and U2's ground pin must connect directly to J4 (see Fig 7.59). This provides the best output voltage regulation.

An over-voltage crowbar circuit prevents the output voltage from exceeding a preset limit. If that limit is exceeded, the output is shunted to ground until power is removed. If the current-limiting circuitry in the supply is working properly, the supply current-limits to the preset value. If the current limiting is not functioning, the crowbar causes the ac-line fuse to blow. Therefore, it's important to use the correct fuse size: 4 to 5 A for a 15 A supply.

The crowbar circuit is a simple design based

on an SCR's ability to latch and conduct until the voltage source is removed. The SCR (Q4) is connected across output terminals J3 and J4. (The SCR can also be connected directly across the filter capacitor, C1, for additional protection.) R10 and potentiometer R16 in series with the Q4's gate provide a means of adjusting the trip voltage. The prototype crowbar is set to conduct at 15 V. The S6025L SCR is rated at 25 A and should be mounted on a metal chassis or heat sink. (Note: Some SCRs are isolated from their mounting tabs, others are not. The S6025L and the 65-ampere S4065J are isolated types. If the SCR you use is not isolated, use a mica washer or thermal pad to insulate it from the chassis or heat sink.)

The bold lines in Fig 7.59 indicate high-current paths that should use heavy-gauge (#10 or #12 AWG) wire. Traces that are connected to the output terminals in the schematic by individual lines should be connected directly to the terminals by individual wires. This establishes a 4-wire measurement, where the heavy wires carry the current (and have voltage drops) and the sense wires carry almost no current and therefore voltage errors are not caused by voltage drops in the wiring. If desired, the sense wire can be carried out to the load, but that may introduce noise into the sense feedback circuit, so use caution if that is done.

7.16.3 13.8 V, 5 A Linear Power Supply

This power supply was designed by Ben Spencer, G4YNM, provides 13.8 V dc at 5 A, suitable for many low-power transceivers and accessories. It features time-dependent current limiting and short-circuit protection, thermal overload protection within the safe operating area of the regulator IC, and overvoltage protection for the equipment it powers. The prototype supply powers a 25 W transmitter that continually draws 4.5 A.

Construction, testing and calibration are straightforward, requiring no special skills or equipment. Many of the components can be found in junk boxes, or purchased at hamfests or from mail-order suppliers.

CIRCUIT DESCRIPTION

Fig 7.60 is the power supply schematic. Incoming ac line current is filtered by a chassis-mounted line filter (FL1) and, after passing through the fuse (F1), is routed S1 to T1.

U1 rectifies, and C1 filters, the ac output of T1. U2 is an LM338K voltage regulator. This IC features a continuous output of 5 A, with a guaranteed peak output of 7 A, on-chip thermal and safe-operating-area protection for itself, and current limiting. U2's output voltage is set by two resistors (R2 and R3) and a trimmer potentiometer (R8), which allows for adjustment over a small range. U2's input and output are bypassed by C2, C3, and C4.

D1 and D2 protect U2 against these capacitors discharging through it.

Overvoltage protection is provided by an SCR, Q1, across the regulator input. Normally, Q1 presents an open circuit, but under fault conditions, it's triggered and short-circuits the unregulated dc input to ground. This discharges C1 and blows F2, reducing the possibility of damage to any connected equipment.

U3, an overvoltage-protection IC, continuously monitors the output voltage. When the output voltage rises above a predetermined level, U3 starts charging C5. If the overvoltage duration is sufficiently long, U3 triggers Q1. This built-in delay (about 1 ms) allows short transient noise spikes on the output voltage to be safely ignored while still triggering the SCR if a true fault occurs. The monitored voltage is set by R5 and R6 and trimmer potentiometer R9, which allows for adjustment over a limited range.

D3 protects the supply from reverse-polarity discharge from connected equipment. The presence of output voltage is indicated by an LED, DS1. R7 is a current limiting resistor for DS1.

CONSTRUCTION

How you construct your supply depends on the size of the components and enclosure you use. General physical layout is not important, although there are a couple of areas that require some attention. In the unit shown in **Fig 7.61**, FL1, the fuse holders, S1, the heat sink. DS1 and the binding posts are mounted on the front and rear enclosure panels. T1 and the PC board are secured to the enclosure's bottom plate. C1's mounting clamp is attached to the rear panel. Bleeder resistor R1 is connected directly across C1's terminals. D3 is soldered directly across the output binding posts.

U2, D1, D2 and R3 are all mounted directly on the heat sink with the transistor pins and solder lugs acting as a terminal strip. It's important to keep R3 attached as closely as possible to U2's terminals to prevent instability. Use a TO-3 mounting kit and heat-conductive grease or thermal pad to electrically isolate U2 from the heat sink.

Mount U2, C1, and the PC board close to each other and keep the wire runs between these components as short as possible. Excessively long wire runs may lead to unpredictable behavior.

Cover all ac-input wiring (use insulated wire and heat-shrink tubing) to prevent electrical shock and route the ac wiring away from the dc wiring. Mount the heat sink on the enclosure with fins oriented vertically. Louvers or ventilation holes in the cabinet will help cool internal components.

TEST AND CALIBRATION

An accurate multimeter covering ranges of

30 V dc and 10 A dc is required. A variable resistive load with a power rating of 100 W is also needed; this can he made using a heat-sink-mounted 2N3055 power transistor and a couple of components as shown in the next project.

First, set R8 (OUTPUT VOLTAGE) fully clockwise and R9 (OVERVOLTAGE) fully counterclockwise. Insert a fuse in the dc line at F2. Connect the ac line, turn on S1 and check that DS1 lights. Measure the output voltage: it should be about 12 V. Adjust R8 counterclockwise until you obtain 14.2 V output; this sets the trip voltage.

While monitoring the output voltage, gradually adjust R9 clockwise until the voltage suddenly falls to zero. This indicates that the SCR has triggered and blown F2. Disconnect the ac line cord from the wall socket. *Don't make any adjustment to R9!* Instead, adjust R8 fully clockwise.

With the ac line cord removed, check that F2 is open. Replace F2 with a new fuse (now you know why two of the three fuses are called for). Reconnect the ac line cord, and while continually monitoring the output voltage, gradually adjust R8 until Q1 again triggers at 14.2 V, blowing F2. If you find the adjustment of R9 to be too sensitive, use a 500 Ω potentiometer in its place and reduce the value of R5 (if necessary) to provide the required adjustment range.

Again disconnect the line cord from the wall socket, set R8 fully clockwise, replace F2 (there's the third fuse!) and reset R8 for 13.8 V. This completes the voltage calibration and overvoltage protection tests. The power supply is now set to 13.8 V output, with the overvoltage protection set for 14.2 V.

Adjust the variable resistive load to maximum resistance and connect it to the power supply output in series with the ammeter. Turn on the power supply and gradually adjust R10 until a current of 5 A flows. Decrease the resistance further and check that the current limits between 5.5 A and 0.5 A.

Finally, be thoroughly unpleasant and apply a short circuit via the ammeter. Check that the current-limiting feature operates correctly. The prototype limited at approximately 3.5 A. Disconnect the ac line cord and test equipment, dose up the case and your power supply is ready for service.

7.16.4 Adjustable Resistive Load

Fig 7.62 shows the schematic of an adjustable resistive load that can be used to test and adjust power supplies at currents up to 10 A if the 2N3055 transistor is mounted on an adequate heat sink. R1 and R2 vary the base bias to control the collector current of Q1. For extended use, be sure to use a large heat sink with adequate ventilation. Use heavy wire

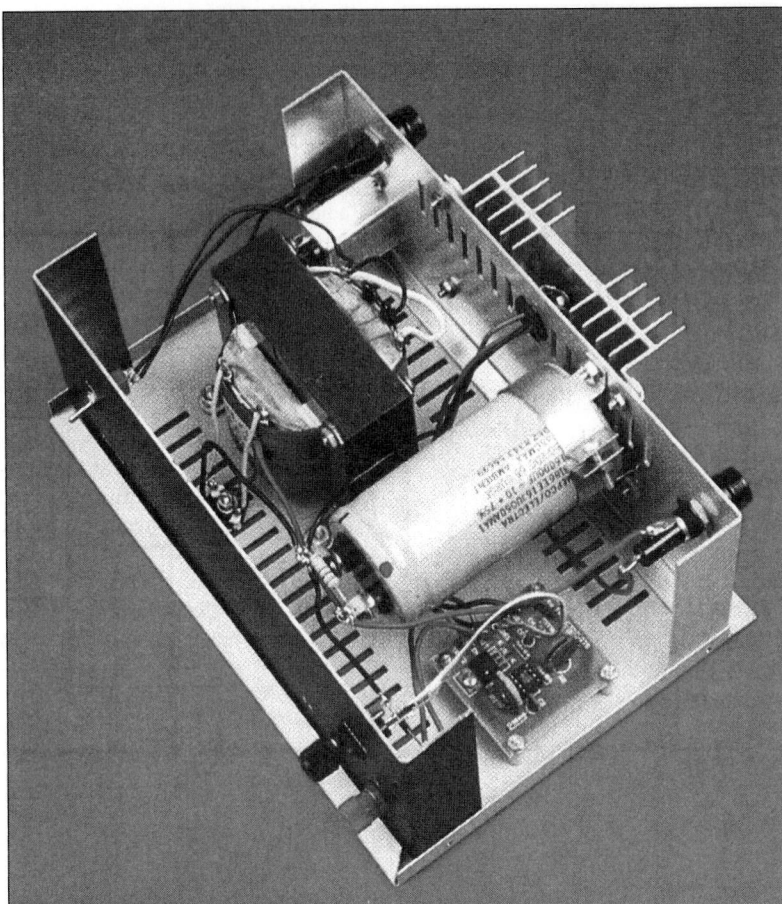

Fig 7.61 — Physical layout of the 13.8 V, 5 A power supply. On the rear panel, left, are the ac-line filter and F1. The regulator's heat sink is at the middle of the panel and F2 is to the right. At the bottom of the enclosure, in front of T1, is the diode bridge rectifier. Because C1 is too tall to mount vertically within the Hammond #14260 cabinet, its mounting clamp is secured to the inside rear panel. Immediately to the right of C1 is the PC board. On the front panel are the on/off switch, LED power-on indicator and output-voltage binding posts.

through the current meter to the collector and from the emitter of Q1.

7.16.5 Inverting DC-DC Converter

It's often the case that you need +V and –V when all you have is +V. For example, you need +12 V and –12 V, but all that's available is +12 V. It would be really handy to have a "black box" that would give you –V out when you put +V in, and work over a range of voltages without adjustment. This project by Jim Stewart, which originally appeared in the January 2013 issue of *Nuts and Volts Magazine* (**www.nutsvolts.com**), fills that need by using a switchmode voltage mirror to supply more than 100 mA without a significant drop in voltage.

The following text summarizes how the circuit works. A PDF version of the complete article is included on the CD-ROM that comes with this book. It contains more details about the circuit's design, plus construction and testing information. A PCB and parts kit is available from the *Nuts and Volts* online store, as well. An ExpressPCB file (PV2NV. PCB) is available for download at **www. nutsvolts.com/index.php?/magazine/ article/january2013_Stewart**.

CIRCUIT DESIGN

The circuit in **Fig 7.63** is a buck-boost dc-dc converter as described earlier in this chapter. This particular design has five important parameters:

- Input voltage: V_{IN}
- Output voltage: V_{OUT}
- Load resistance: R_{LOAD}
- Oscillation frequency: f
- Inductance value: L

The parameters are related to each other by $f \times L < (V_{IN} / V_{OUT})^2 \times (R_{LOAD} / 8)$. For $V_{IN} = V_{OUT}$, the equation simplifies to

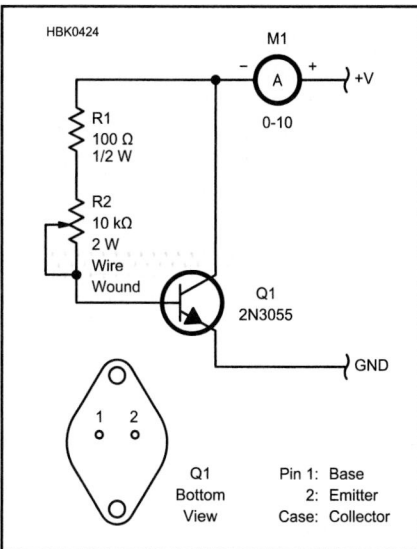

Fig 7.62 — An active resistive load to use in testing power supplies up to 10 A with adequate heat sinking. Adjustment of R2 is sensitive.

M1 — Multimeter or ammeter capable of measuring 10 A.

Q1 — 2N3055; mount on heat sink.

$f \times L < (R_{LOAD} / 8)$. (See the complete article on the CD-ROM for the derivation of these equations.)

Fig 7.63 shows the schematic of the circuit. A square wave oscillator (U2) drives a MOSFET (Q2) that, in turn, drives a PNP switching transistor (Q3). The oscillator is enabled/disabled by the output of the comparator (U1) with inputs that are a set point voltage (V_S on the comparator's + input) and a feedback voltage (V_F at the comparator's – input). (Comparators are described in the **Analog Fundamentals** chapter.) The frequency of the oscillator is $f = 1/2.2 (R7 \times C3)$.

When the output voltage, V_{OUT}, is the correct value, V_S exceeds V_F by a few µV and the oscillator is disabled by Q1 so that the PNP switch is off. When V_S is less than V_F, the oscillator is enabled and the PNP transistor switches on and off. V_S is half the input voltage $V_S = V_{IN}/2$ as set by the R3-R4 voltage divider.

V_F is set by the R1-R2 divider to equal $V_{IN}/2$ when the output voltage $V_{OUT} = -V_{IN}$. D2 is a commutating diode that blocks +V from the output while charging the inductor. It then provides a current path for the discharging inductor to transfer charge to the output capacitor.

The switching action continues until the voltage on the capacitor equals $-V_{IN}$. At that point, the comparator disables the oscillator and Q3 stays OFF. When voltage across

the capacitor drops, the comparator enables the oscillator and the inductor is pumped up again.

Since there is a single input voltage, each op-amp uses a resistor divider to "split the rail" to create a signal ground. That allows the negative input to sense positive and negative voltages with respect to the positive input. C2 and C4 bypass the signal grounds to the supply voltage ground.

C1 and C5 are low-ESR (equivalent series resistance — see the **Electrical Fundamentals** chapter) aluminum electrolytic capacitors. Tantalum capacitors might be a bit better, but are more expensive. ESR determines how much power the capacitor can safely dissipate. $P_{DISS} = I^2_{RC} \times ESR$, where I_{RC} is the ac ripple current in the capacitor. The capacitors chosen are rated for a maximum ripple current of 840 mA.

D1 and D2 are Schottky diodes that can go from conducting to non-conducting very quickly, allowing a high switching frequency. They also have a low voltage drop when conducting to reduce power loss. D1 limits the output voltage in case the feedback fails.

Q3 is a ZTX550 PNP transistor, chosen because its specifications suit this application well:

• Maximum power dissipation, $P_{MAX} = 1$ W @ 25 °C

• Maximum continuous collector current, $I_C = 1$ A

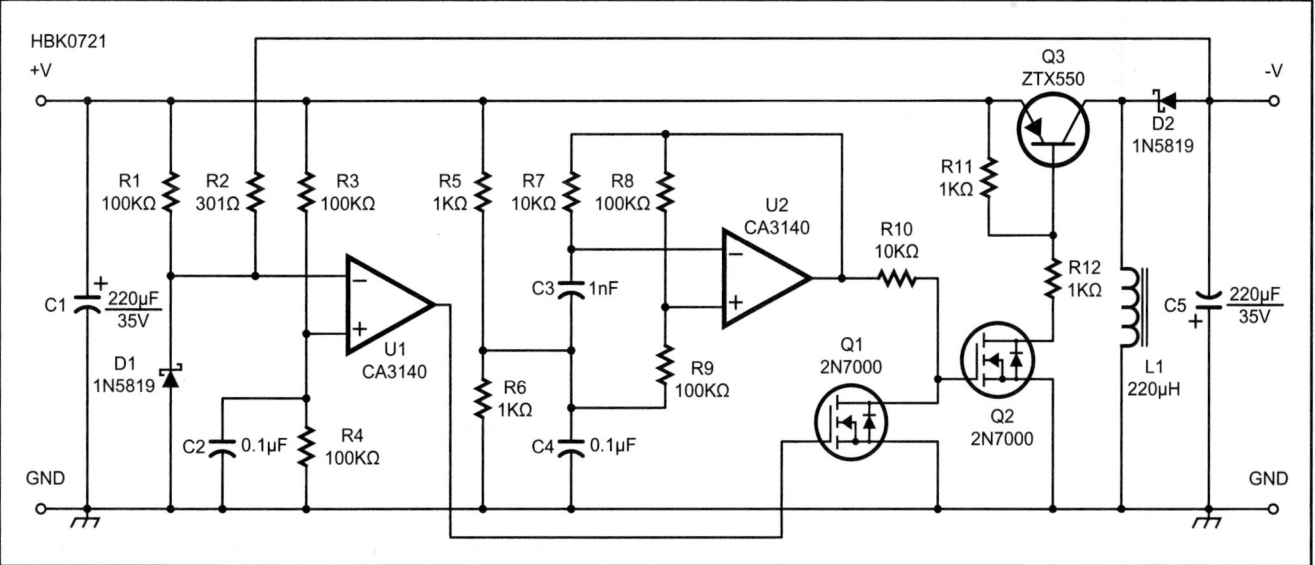

Figure 7.63 — A voltage mirror dc-dc power converter based on a buck-boost converter. The negative output —V is equal in magnitude to the positive input voltage +V at an output current up to approximately 100 mA

C1 — 220 µF, 35 V (Digi-Key part #493-1578-ND or equiv).
C2 — 0.1 µF, 50 V, ceramic.
C3 — 1 nF film, 5% (Digi-Key part #399-5871-ND or equiv).
C4 — 0.1 µF, 50 V, ceramic.
C5 — 220 µF, 35 V (Digi-Key part #493-1578-ND or equiv).

D1, D2 — Schottky, 1 A, 1N5819 or equiv.
L1 — 220 µH (Digi-Key part #811-1316-ND or equiv).
Q1, Q2 — 2N7000.
Q3 — ZTX550 PNP.
R1, R3, R4, R8, R9 — 100 kΩ, ¼ W, 1%.
R2 — 301 kΩ, ¼ W, 1%.

R5, R6, R11, R12 — 1 kΩ, ¼ W, 1%.
R7, R10 — 10 kΩ, ¼ W, 1%.
U1, U2 — CA3140.
Terminal Blocks (optional) — Two-position, 5 mm spacing (Jameco part #2094485 or equiv).

- Maximum collector-base voltage, V_{CBO} = 60 V
- Maximum saturation voltage, V_{CE} = 0.25 V @ I_C = 150 mA
- Transition frequency, f_T = 150 MHz minimum
- Package: E-Line (slightly smaller than TO-92)

For this design, the chosen values are L = 220 µH, f = 45 kHz, and R_{LOAD} = 100 Ω. Verifying that f × L is less than R_{LOAD} / 8, f × L = $(45 \times 10^3) \times (220 \times 10^{-6})$ = 9900 × 10^{-3} = 9.9 Ω and R_{LOAD} / 8 = 100/8 = 12.5 Ω which is less than R_{LOAD} / 8 = 12.5. Higher values of R_{LOAD} (lower output current) also satisfy the equation. Increasing the value of L or f will require that the circuit be tested to verify that it works.

7.16.6 High-Voltage Power Supply

This two-level, high-voltage power supply was designed and built by Dana G. Reed, W1LC. It was designed primarily for use with an RF power amplifier. The supply is rated at a continuous output current of 1.5 A, and will easily handle intermittent peak currents of 2 A. The 12 V control circuitry and the low-tap setting of the plate transformer secondary make it straightforward to adapt the design to homemade tube amplifiers.

The step-start circuit is straightforward and ensures that the rectifier diodes are current-limited when the power supply is first turned on. A 6 kV meter is used to monitor high-voltage output.

Fig 7.64 is a schematic diagram of the bi-level supply. An ideal power supply for a high-power linear amplifier should operate from a 240 V circuit, for best line regulation. A special, hydraulic/magnetic circuit breaker also serves as a disconnect for the plate transformer primary. Don't substitute a standard circuit breaker, switch or fuses for this breaker; fuses won't operate quickly enough to protect the amplifier or power supply in case of an operating abnormality. The 100 kΩ, 3 W bleeder resistors are of stable metal-oxide film design. These resistors are wired across each of the 14 capacitors to equalize voltage drops in the series-connected bank. This choice of bleeder resistor value provides a lighter load (less than 25 W total under high-tap output) and benefits mainly the capacitor-bank filter by yielding much less heat as a result. A reasonable, but longer bleed-down time to fully discharge the capacitors results — about nine minutes after power is removed. A small fan is included to remove any excess heat from the power supply cabinet during operation.

POWER SUPPLY CONSTRUCTION

The power supply can be built in a 23½ × 10¾ × 16-inch cabinet. The plate transformer is quite heavy, so use ⅛-inch aluminum for the cabinet bottom and reinforce it with aluminum angle for extra strength and stability. The capacitor bank will be sized for the specific capacitors used. This project employed ⅜-inch thick polycarbonate for reasonable mechanical stability and excellent high-voltage isolation. The full-wave bridge consists of four commercial diode block assemblies.

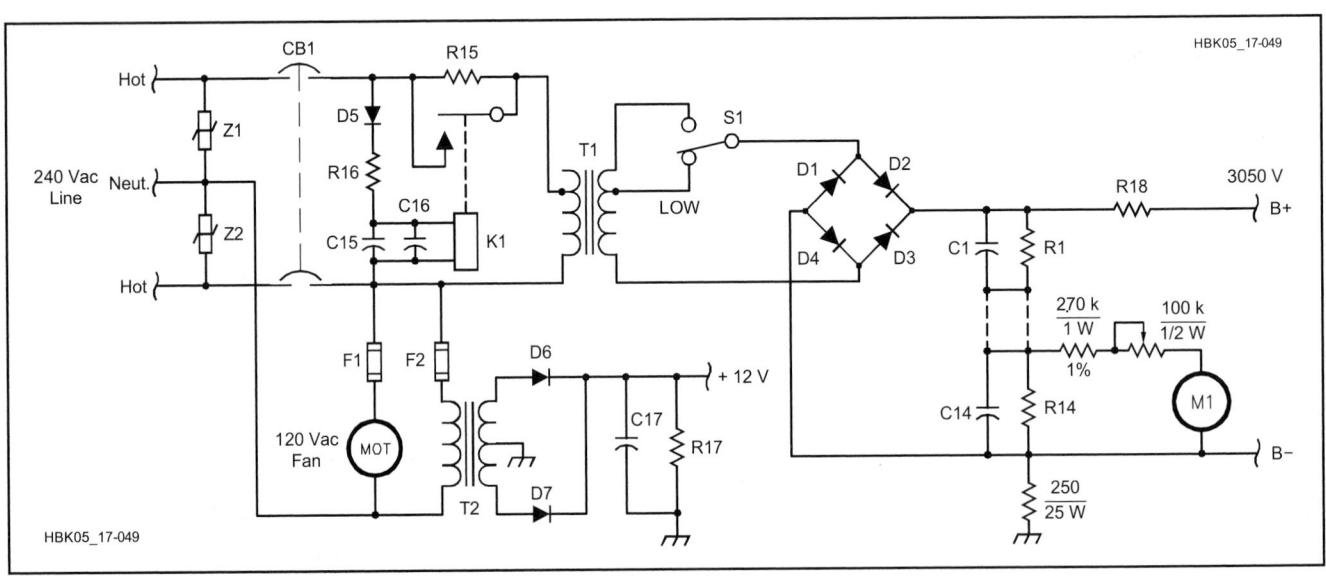

Fig 7.64 — Schematic diagram of the 3050 V/5400 V high-voltage power supply.

C1-C14 — 800 µF, 450 V electrolytic.
C15, C16 — 4700 µF, 50 V electrolytic.
C17 — 1000 µF, 50 V electrolytic.
CB1 — 20 A hydraulic/magnetic circuit breaker (TE Connectivity/Potter & Brumfield W68-X2Q12-20 or equiv). 40 A version required for commercial applications/ service (TE Connectivity/Potter & Brumfield W92-X112-40).
D1-D4 — String of 1000 PIV, 6 A diodes (6A10 or equiv).
D5 — 1000 PIV, 3 A, 1N5408 or equiv.
D6, D7 — 200 PIV, 3 A, 1N5402 or equiv.
F1, F2 — 0.5 A, 250 V (Littelfuse 313 Series, 3AG glass body or equiv).
K1 — DPDT power relay, 24 V dc coil; both poles of 240 V ac/25 A contacts in parallel (TE Connectivity/Potter & Brumfield PRD-11DYO-24 or equiv).

M1 — High-voltage meter, 6 kV dc full scale. (Important: Use a 1 mA or smaller meter movement to minimize parallel-resistive loading at R14. Also, select series meter-resistor and adjustment-potentiometer values to calibrate your specific meter. Values shown are for a 1 mA meter movement.)
MOT1 — Cooling fan, 119 mm, 120 V ac, 30-60 CFM, (EBM Pabst 4800Z or equiv).
R1-R14 — Bleeder resistor, 100 kΩ, 3 W, metal oxide film.
R15 — 50 Ω, 100 W.
R16 — 3.9 kΩ, 25 W.
R17 — 30 Ω, 25 W.
R18 — 20 Ω, 50 W.
S1 — Ceramic rotary, 2 position tap-select switch (optional). Voltage rating between tap positions should be at least

2.5 kV. Mount switch on insulated or un-grounded material such as a metal plate on standoff insulators, or an insulating plate, and use only a *nonconductive* or otherwise *electrically-isolated* shaft through the front panel for safety.
T1 — High-voltage plate transformer, 220/240 V primary, 2000/3500 V, 1.5 A CCS JK secondary, Hypersil C-core (Hammond Engineering, www.hammfg.com). Primary 220 V tap fed with nominal 240 V ac line voltage to obtain modest increase in specified secondary voltage levels.
T2 — 120 V primary, 18 V CT, 2 A secondary (Mouser 41FJ020).
Z1-Z2 — 130 V MOV.

POWER SUPPLY OPERATION

When the front-panel breaker is turned on, a single 50 Ω, 100 W power resistor limits primary inrush current to a conservative value as the capacitor bank charges. After approximately two seconds, step-start relay K1 actuates, shorting the 50 Ω resistor and allowing full line voltage to be applied to the plate transformer. No-load output voltages under low- and high-tap settings as configured and shown in Fig 7.64 are 3050 V and 5400 V, respectively. Full-load levels are somewhat lower, approximately 2800 V and 4900 V. If a tap-select switch is used as described in the schematic parts list, it should only be switched when the supply is off.

7.16.7 Reverse-Polarity Protection Circuits

The following material was collected from various public-domain sources by Terry Fletcher, WA0ITP(www.wa0itp.com/revpro. html) and published in the QRP Quarterly, *Spring 2012 issue. (www.qrparci.org)*

DC power is the standard for most amateur radios and accessories, usually a nominal 12 V (10.5 to 13.8 V). There are many different types of connectors used for dc power — from screw terminals to custom-molded multi-pin designs. This makes it easy to accidentally apply power with reversed polarity and damage equipment. Even a few milliseconds of reversed power can be sufficient to destroy a semiconductor or burn out a narrow PCB trace. This can be a particular problem when using 9 V batteries as it is easy to reverse the snap-on connector when changing or installing a battery.

The following collection of circuits illustrates ways to protect equipment from reverse-polarity dc power. The suitability of the circuits depends on the equipment and power source. All dc power sources, particularly batteries, should be fused or current-limited to mitigate fire hazards and other damage from overheating wires and other conductors.

Fig 7.65 shows several passive circuits that dissipate some power due to the series forward voltage drop of the diodes. At currents above 1 A, the power dissipated can easily exceed 1 W and the maximum junction temperature of the diode can be exceeded without some sort of thermal protection or heat sinking. (The shunt diode circuit in Fig 7.65D does not dissipate power.) **Fig 7.66** shows two methods of using an electromechanical relay that avoid the forward voltage drop of diode-based protection circuits. Which circuit you choose depends on the type of equipment and constraints on power dissipation and voltage drop.

PASSIVE CIRCUITS

Blocking diode (Fig 7.65A) — A series

diode is very simple and inexpensive. Its PIV rating should be at least twice the expected applied voltage — 50 V PIV is a good minimum value for automotive and 12 V dc use. Its maximum average forward current rating should be several times the expected maximum steady-state current draw.

Remember that a silicon junction diode's forward voltage drop, V_f, is at least 0.6 V and can approach 1.0 V at high forward current. This can result in significant power dissipation ($P = V_f \times I_f$) and cause the diode's maximum rated junction temperature to be exceeded unless some means of cooling the diode is provided.

The forward voltage drop of the diode will also reduce the voltage available to the equipment being powered. This will raise the minimum allowable power supply voltage for the equipment to operate properly. For example, if a piece of equipment is rated to operate properly at or above 11 V, a series diode with $V_f = 0.6$ V raises the minimum allowable power supply voltage to 11 + 0.6 = 11.6 V. This may be significant in battery-powered installations.

The Schottky barrier diode shown as an alternate may be a better choice due to its forward voltage drop being lower by several tenths of a volt, reducing power dissipation. A Schottky diode may be used in place of any of the diodes in Figures 7.65 and 7.66. Be sure the reverse current leakage of the Schottky diode is acceptable.

Full-wave bridge rectifier (Fig 7.65B) — The full-wave circuit has the advantage of always supplying voltage with the proper polarity to the equipment being powered.

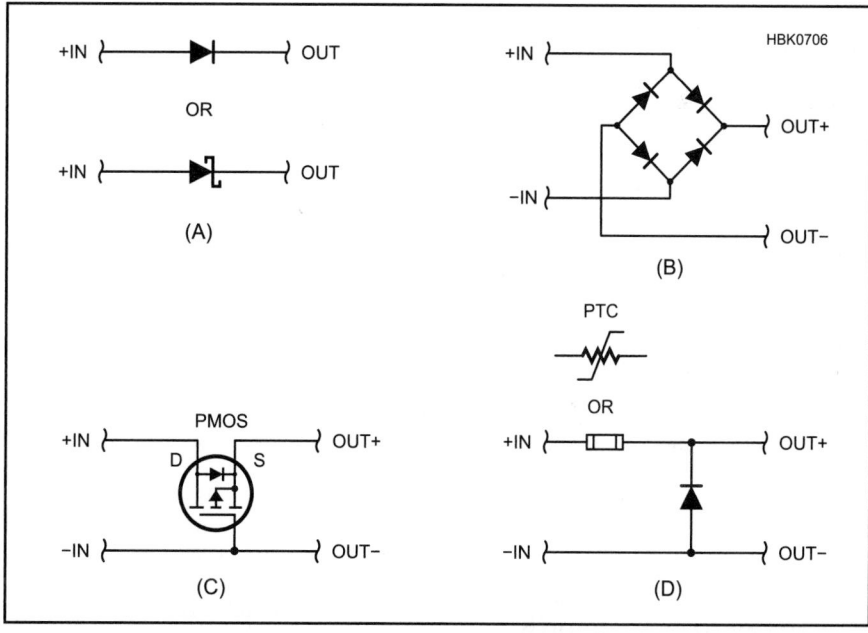

Fig 7.65 — Passive circuits for reverse-polarity protection. Series diode (A). Full-wave rectifier (B). PMOS MOSFET with integral body diode shown as separate component (C). Shunt diode (D). Schottky barrier diodes may be used in all circuits as a substitute for silicon junction rectifiers. See text for circuit comparison.

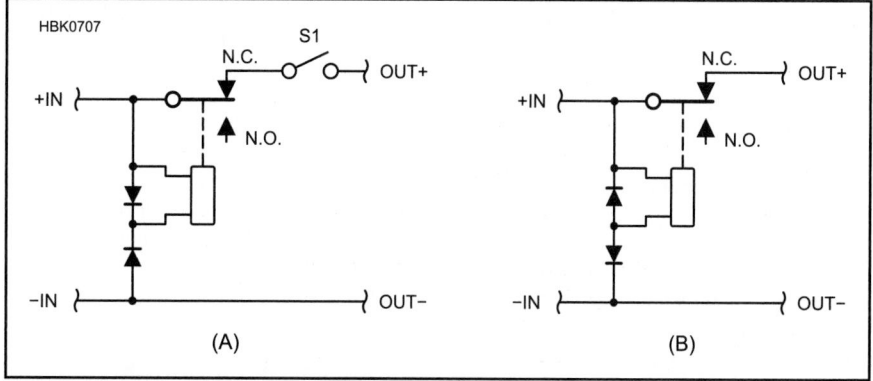

Fig 7.66 — Relay-based circuits for reverse-polarity protection. Normally-closed contacts (A) and normally-open contacts (B). See text for circuit comparison.

Full-wave rectifiers are also available as integrated packages, making them easy to install. Remember that the forward voltage drop and power dissipation of this circuit will be twice that of the single series diode because two diodes are always in series with supply current.

PMOS P-channel MOSFET (Fig 7.65C) — This circuit uses a P-channel enhancement-mode (PMOS) MOSFET that conducts current with the gate connected as shown. PMOS devices have low on-resistance ($R_{ds(on)}$) and high maximum current ratings. Devices with on-resistance of 0.050 Ω and lower are commonly available. For more information about using PMOS and NMOS devices for polarity protection see Maxim Electronics Application Note 636, "Reverse-Current Circuitry Protection" (**www.maxim-ic.com/app-notes/index.mvp/id/636**). N-channel devices may also be used in the current return or ground lead, but opening the return connection can create other problems inside the equipment and with other devices on the same power circuits.

Shunt diode with fuse (Fig 7.65D) — These configurations use a single diode that acts to blow a fuse (either a fusible-link or positive temperature coefficient PTC resettable device) if reverse polarity voltage is applied. The advantage of this circuit is that no power is dissipated by the diode during normal operation. The diode must be sufficiently rated to handle the high surge current from shorting the power source and have current ratings significantly higher than the fuse current rating.

If the diode fails shorted or in a low-resistance state, it will continue to blow the fuse until replaced. If the diode fails open or high-resistance, it will no longer protect the circuit. If shunt diode protection is used and the fuse opens, check the diode to be sure it has not failed as well.

RELAY-BASED CIRCUITS

Relay-based circuits have the advantage of little to no voltage drop, even at high currents as long as the contact ratings are sufficient. The circuits are more complex than the diode-based circuits in the preceding section but can generally handle more current and are not damaged by reverse polarity voltages. The circuits reset themselves automatically.

Relay with normally-closed contacts (Fig 7.66A) — There is no current drain through the relay coil until reverse-polarity is applied. However, there will be a few milliseconds during which reverse polarity voltage is applied if no power switch (S1) is used or the power switch is closed. This is generally enough time for damage to occur so this circuit is only recommended if a power switch is used to turn the equipment ON and OFF.

Relay with normally-open contacts (Fig 7.66B) — The relay contacts close and supply power to the equipment only when applied voltage has the proper polarity. The relay coil draws current continuously during normal operation. This may be unacceptable for low-power and battery-powered equipment.

7.16.8 Automatic Sealed Lead-Acid Battery Charger

After experiencing premature failure of the battery in his Elecraft K2 transceiver, Bob Lewis, AA4PB, began searching for an automatic battery charger. Although this charger was designed specifically for use with the Power-Sonic PS-1229A sealed lead-acid (SLA) battery used in the Elecraft K2 transceiver, its design concepts have wide ranging applications for battery operated QRP rigs of all types. Comments pertaining to the SLA batteries and chargers apply across the board and the charger described here can be used with any similar battery. (See the Batteries section of this chapter for more information on SLA batteries.)

USING A THREE-MODE CHARGER

The author first attempted to use a commercial automatic three-mode charger. However, most three-mode chargers work by sensing current and are never intended to charge a battery under load.

Three-mode chargers begin the battery charging process by applying a voltage to the battery through a 500 mA current limiter. This stage is known as *bulk-mode* charging. As the battery charges, its voltage begins to climb. When the battery voltage reaches 14.6 V the charger maintains the voltage at that level and monitors the battery charging current. This is known as the *absorption mode*, sometimes called the *overcharge mode*. By this time, the battery has achieved 85% to 95% of its full charge. As the battery continues to charge — with the voltage held constant at 14.6 V — the charging current begins to drop.

When the charging current falls to 30 mA, the three-mode charger switches to *float mode* and lowers the applied voltage to 13.8 V. At 13.8 V, the battery becomes self-limiting, drawing only enough current to offset its normal self-discharge rate. This works great — until you attach a light load to the battery, such as a receiver. The K2 receiver normally draws about 220 mA. When the charger detects a load current above 30 mA, it's fooled into thinking that the battery needs charging, so it reverts to the absorption mode, applying 14.6 V to the battery. If left in this condition, the battery is overcharged, shortening its service life.

UC3906 IC CHARGERS

Chargers using the UC3906 SLA charge-controller IC work just like the three-mode charger described earlier except that their return from float mode to absorption mode is based on voltage rather than current. Typically, once the charger is in float mode it won't return to absorption mode until the battery voltage drops to 90% of the float-mode voltage (or about 12.4 V). Although this is an improvement over the three-mode charger, it still has the potential for overcharging a battery to which a light load is attached.

First, let's look at the situation where a UC3906-controlled charger is in absorption mode and you turn on the K2 receiver, applying a load. The battery is fully charged, but because the load is drawing 220 mA, the charging current never drops to 30 mA and the charger remains in absorption mode, thinking that it is the battery that is asking for the current. As with the three-mode charger, the battery is subject to being overcharged.

If we remove the load by turning off the K2, the current demand drops below 30 mA and the charger switches to float mode (13.8 V). When the K2 is turned on again, because the charger is able to supply the 220 mA for the receiver, the battery voltage doesn't drop, so the charger stays in float mode and all is well. However, if the transmitter is keyed (increasing the current demand), the charger can't supply the required current, so it's taken from the battery and the battery voltage begins to drop. If we un-key the transmitter before the battery voltage reaches 12.4 V, the charger stays in float mode. Now it takes much longer for the charger to supply the battery with the power used during transmit than it would have if the charger had switched to absorption mode.

Let's key the transmitter again, but this time keep it keyed until the battery voltage drops below 12.4 V. At this point, the charger switches to the absorption mode. When we un-key the transmitter, we're back to the situation where the charger is locked in absorption mode until we turn off the receiver.

WHY WORRY?

So why this concern about overcharging an SLA battery? At 13.8 V, the battery self-limits, drawing only enough current to offset its self-discharge rate (typically about 0.001 times the battery capacity, or 2.9 mA for a 2.9 Ah battery). An SLA battery can be left in this float-charge condition indefinitely without overcharging it. At 14.6 V, the battery takes more current than it needs to offset the self-discharge. Under this condition, oxygen and hydrogen are generated faster than they can be recombined, so pressure inside the battery increases. Plastic-cased SLA batteries such as the PS-1229A have a one-way vent that opens at a couple of pounds per square inch pressure (PSI) and releases the gases into the atmosphere. This results in drying the gelled electrolyte and shortening the battery's service life. Both undercharging and overcharging need to be avoided to get maximum service life from the battery.

Continuing to apply 14.6 V to a 12 V SLA battery represents a relatively minor amount of overcharge and results in a gradual deterioration of the battery. Applying a potential of 16 V or excessive bulk-charging current to a small SLA battery from an uncontrolled solar panel can result in serious overcharging. Under these conditions, the overcharging can cause the battery to overheat, which causes it to draw more current and result in *thermal runaway*, a condition that can warp electrodes and render a battery useless in a few hours. To prevent thermal runaway, the maximum current and the maximum voltage need to be limited to the battery manufacturer's specifications.

DESIGN DECISION

To avoid the potential of overcharging a battery with an automatic charger locked up by the load, the author decided to design a charger that senses battery voltage rather than current in order to select the proper charging rate. A 500 mA current limiter sets the maximum bulk rate charge to protect the battery and the charger's internal power supply.

Like the three-mode chargers, when a battery with a low terminal voltage is first connected to the charger, a constant current of 500 mA flows to the battery. As the battery charges, its voltage begins to climb. When the battery voltage reaches 14.5 V, the charger switches off. With no charge current flowing to the battery, its voltage now begins to drop. When the current has been off for four seconds, the charger reads the battery voltage. If the potential is 13.8 V or less, the charger switches back on. If the voltage is still above 13.8 V, the charger waits until it drops to 13.8 V before turning on. The result is a series of 500 mA current pulses varying in width and duty cycle to provide an average current just high enough to maintain the battery in a fully charged condition. Because the repetition rate is very low (a maximum of one current pulse every four seconds) no RFI is generated that could be picked up by the K2 receiver. Because the K2's critical circuits are all well-regulated, slowly cycling the battery voltage between 13.8 V and 14.5 V has no ill effects on the transmitted or received signals.

As the battery continues to charge, the pulses get narrower and the time between pulses increases (a lower duty cycle). Now when the K2 receiver is turned on and begins drawing 220 mA from the battery, the battery voltage drops more quickly so the pulses widen (the duty cycle increases) to supply a higher average current to the battery and make up for that taken by the receiver. When the

K2 transmitter is keyed, it draws about 2 to 3 A from the battery. Because the charger is current limited to 500 mA, it is not able to keep up with the transmitter demands. The battery voltage drops and the charger supplies a constant 500 mA. The battery voltage continues to drop as it supplies the required transmit current. When the transmitter is unkeyed, the battery voltage again begins to rise as the charger replenishes the energy used during transmit. After a short time, (depending on how long the transmitter was keyed) the battery voltage reaches 14.5 V and the pulsing begins again. The charger is now fully automatic, maintaining the battery in a charged condition and adjusting to varying load conditions.

The great thing about this charging system is that during transmit the majority of the required 2 to 3 A is taken from the battery. When you switch back to receive, the charger is able to supply the 220 mA needed to run the receiver and deliver up to 280 mA to the battery to replenish what was used during transmit. This means that the power source need only supply the average energy used over time, rather than being required to supply the peak energy needed by the transmitter. (You don't need to carry a heavy 3 A regulated power supply with your K2.) As long as you don't transmit more than about 9% of the time, this system should be able to power a K2 indefinitely.

Have you ever noticed that sometimes when your handheld radio has a low battery and you drop it into its charger you hear hum on the received signals? This charger's power supply is well filtered to ensure that there is no ripple or ac hum to get into the K2 under low battery voltage conditions.

CIRCUIT DESCRIPTION

The charger schematic is shown in **Fig 7.67**; the unit is dubbed the PCR12-500A, short for Pulsed-Charge Regulator for 12 V SLA batteries with maximum bulk charge rates of 500 mA. U1, an LM317 three-terminal voltage regulator, is used as a current limiter, voltage regulator and charge-control switch. A 15 V Zener diode (D2) sets U1 to deliver a no-load output of 16.2 V. R3 sets U1 to limit the charging current to 500 mA. When Q1 is turned on by the LM555 timer (U2), the ADJ pin of U1 is pulled to ground, lowering its output voltage to 1.2 V. D4 effectively disconnects the battery by preventing battery current from flowing back into U1. A Schottky diode is used at D4 because of its low voltage drop (0.4 V).

An LM358 (U4A) operates as a voltage comparator. U5, an LM336, provides a 2.5 V reference to the positive input (pin 3) of U4. R11, R12 and R13 function as a voltage divider to supply a portion of the battery voltage

to pin 2 of U4A. R13 is adjusted so that when the battery terminal voltage reaches 14.5 V, the negative input of U4A rises slightly above the 2.5 V reference and its output switches from +12 V to 0 V. When this happens, the 1 MΩ resistor (R7) causes the reference voltage to drop a little and provide some hysteresis. The battery voltage must now drop to approximately 13.8 V before U4A turns back on.

U4B is a voltage follower. It pulls the trigger input (pin 2) of U2 to 0 V, causing its output to go to 12 V. U4B's output remains at 12 V until C5 has charged through R6 (approximately four seconds) and the trigger has been released by U4A sensing the battery dropping to 13.8 V or less. While the output of U2 is at 12 V, emitter/base current for Q1 flows via R5 and Q1's collector pulls U1's ADJ pin to ground, turning off the charging current.

The output of U2 also provides either +12 V or 0 V to the bicolor LED, D3. R14 and R15 form a voltage divider to provide a reference voltage to D3 such that D3 glows red when U2's output is +12 V and green when U2's output is at 0 V. When ac power is applied but U1 is switched off and not supplying current to the battery, D3 glows red. When U1 is on and supplying current to the battery, D3 is green. As the battery reaches full charge, D3 blinks green at about a four-second rate. As the battery charge increases, the on time of the green LED decreases and the off time increases. A fully charged battery may show green pulses as short as a half-second and the time between pulses may be 60 seconds or more.

T1, D1, C1 and C2 form a standard full-wave-bridge power supply providing an unregulated 20 V dc at 500 mA. U3, an LM78L12 three-terminal regulator, provides a regulated 12 V source for the control circuits.

Note that the mounting tab on U1 is not at ground potential. U1 should be mounted to a heat sink with suitable electrically insulated but thermally conductive mounting hardware to avoid short circuits. Suitable mounting hardware is included with the PC board.

OTHER BULK-CHARGE RATES

The maximum bulk-charge rate is set by the value of R3 in the series regulator circuit. The formula used to determine the value of this resistor is R (Ω) = 1200 / 1 (mA). T1 must be capable of supplying the bulk charge current and U1 must be rated to handle this current. The LM317T used here is rated for a maximum current of 1.5 A, provided it has a heat sink sufficiently large enough to dissipate the generated heat. If you increase the bulk-charge rate, you'll definitely need to increase the size of the on-board heat sink. Mounting U1 directly to the housing (be sure to use an insulator) may be a good option.

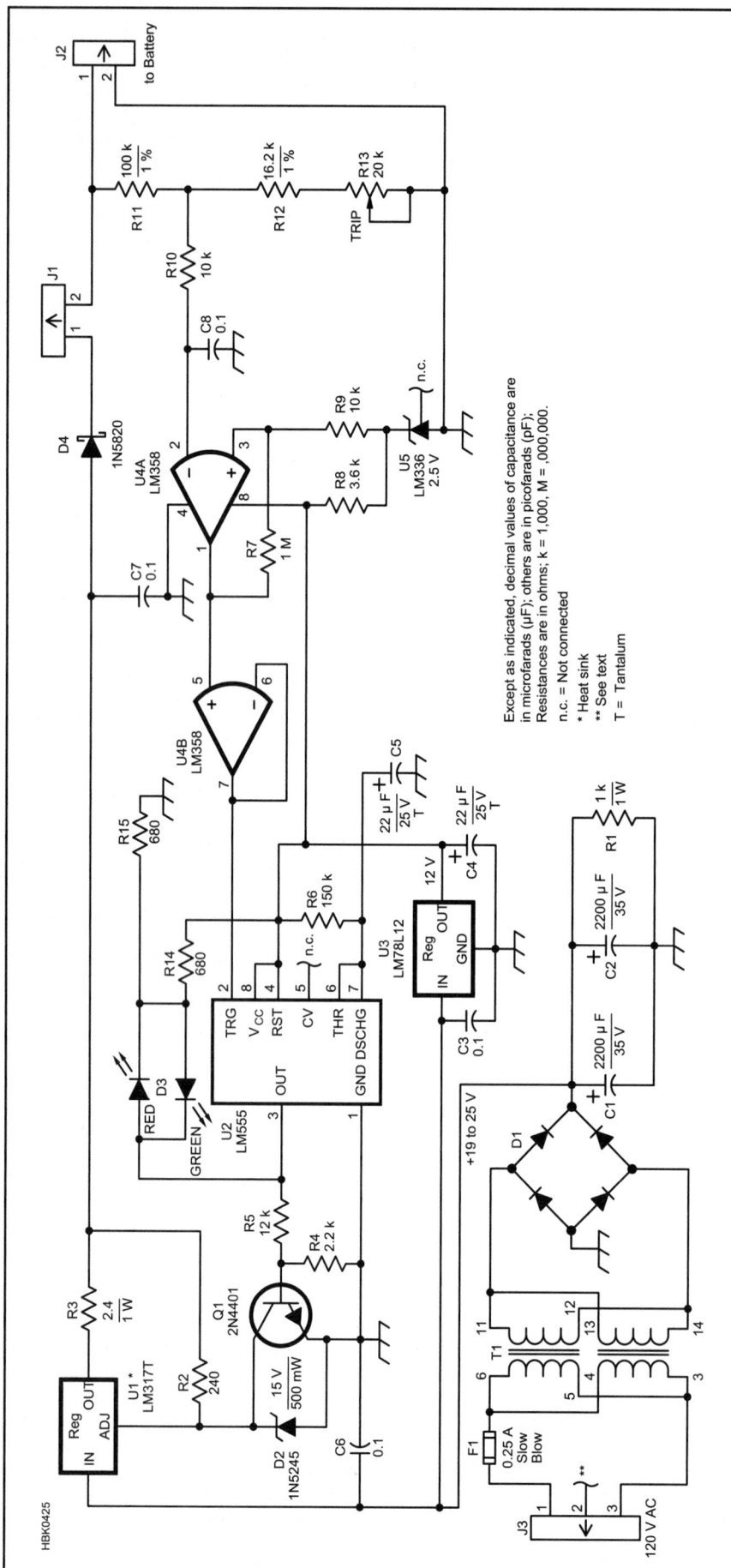

Fig 7.67 — Schematic of the SLA charger. Unless otherwise specified, resistors are ¼ W, 5%-tolerance. Equivalent parts can be substituted; n.c. indicates no connection. A PC board is available from FAR Circuits (www.farcircuits.net).

C1, C2 — 2200 µF, 35 V electrolytic.
C3, C6, C7, C8 — 0.1 µF, 50 V metalized-film.
C4, C5 — 22 µF, 25 V tantalum.
D1 — 400 V, 4 A bridge rectifier.
D2 — 1N5245 Zener diode, 15 V, 500 mW.
D3 — Bicolor LED, red-green.
D4 — 1N5820 Schottky diode.
F1 — 0.25 A slow-blow fuse.
J1 — 2-pin header, PC mount.
J2 — 2-pin connector, PC mount.
J3 — 3-pin connector, PC mount.
Q1 — 2N4401 NPN transistor.
R13 — 20 kΩ multi-turn potentiometer.
T1 — PC board mount transformer, 15 V ac, 666 mA (Digi-Key TE70043-ND). See text.
U1 — LM317T voltage regulator, TO-220 case.
U2 — LM555 timer.
U3 — LM78L12 voltage regulator, TO-92 case.
U4 — LM358 dual op amp.
U5 — LM336, 2.5 V voltage reference, TO-92 case.
Misc: PC board;TO-220 heat sink; five ¼-inch, #4-40 stand-offs; two fuse-holder clips, PC mount; two-pin shunt; two-pin connector housing; three-pin connector housing; four housing pins; enclosure

TRANSFORMER SUBSTITUTION

The transformer used for T1 is small in size and offers PC-board mounting. You can substitute any transformer rated at 15 or 16 V ac (RMS) at 500 mA or more. You may find common frame transformers to be more readily available. You can mount such a transformer to an enclosure wall and route the transformer leads to the appropriate PC-board holes.

CONSTRUCTION

There is nothing critical about building this charger. You can assemble it on a prototyping board, but a PC board and heat sink are available (**www.farcircuits.net**; see **Fig 7.68**) The specially ordered heat sink supplied with the PC board is ¼-inch higher than the one identified in the parts list and results in slightly cooler operation of U1. The remaining parts are available from Digi-Key.

Be sure to space R1 and R3 away from the board by ¼ inch or so to provide proper cooling. R13 can be a single-turn or a multi-turn pot. You'll probably find a multi-turn pot makes it easier to set the cutoff voltage to exactly 14.5 V.

R13 Adjustment

To check for proper operation and to set the trip point to 14.5 V dc, we need a test-voltage source variable from 12 to 15 V dc. A convenient means of obtaining this test voltage is to connect two 9 V transistor-radio batteries in series to supply 18 V as shown in **Fig 7.69**. Connect a 1 kΩ resistor (R2) in series with a 1 kΩ potentiometer (R1) and connect this series load across the series batteries with the fixed-value resistor to the negative lead. The voltage at the pot arm should now be adjustable from 9 to 18 V. During the following procedure, be sure to adjust the voltage with the test supply connected to the charger at 12 V because the charger loads the test-voltage supply and causes the voltage to drop a little when it's connected.

Remove the jumper at J1 and apply ac line voltage to the unit at J3. Turn R13 fully counterclockwise. D3 should glow green. Connect the test voltage to J2 and adjust R1 of Fig 7.68 for an output of 14.5 V. Slowly adjust R13 clockwise until D3 glows red. To test the circuit, wait at least four seconds, then gradually reduce the test voltage until D3 turns green. At that point, the test voltage should be approximately 13.8 V. Slowly increase the test voltage again until D3 turns red. The test voltage should now read 14.5 V. If it is not exactly 14.5 V, make a minor adjustment to R13 and try again. The aim of this adjustment is to have D3 glow red just as the test voltage reaches 14.5 V.

To test the timer functioning, remove the test voltage from J2 and set it for about 15 V. Momentarily apply the test voltage to J2. D3 should turn red for approximately four seconds, then turn green. The regulator is now calibrated and ready for operation. Remove the test voltage and ac power and install the jumper at J1.

The prototype used an 8 × 3 × 2.75-inch LMB Perf137 box (Digi-Key L171-ND) to house the charger. An alternative enclosure is the Bud CU482A Convertabox, which measures 8 × 4 × 2 inches (available from Mouser). If you use the Convertabox, be sure to add some ventilation holes directly above the board-mounted heat sink. The LMB Perf box comes with a ventilated cover. If you are inclined to do some metal work, you could build your own enclosure using aluminum angle stock and sheet and probably reduce the size to perhaps 8 × 3 × 2 inches. If you use a PC-board-mounted power transformer, watch out for potential shorts between the transformer pins (especially the 120 V ac-line pins) and the case. If you use a metal enclosure, connect the safety ground (green) wire of the ac-line cord directly to the case.

OPERATION

It is very important that this charger be connected directly to the SLA battery with no diodes, resistors or other electronics in between the two. The charger works by reading the battery voltage, so any voltage drop across

Fig 7.68 — Complete SLA charger PC board.

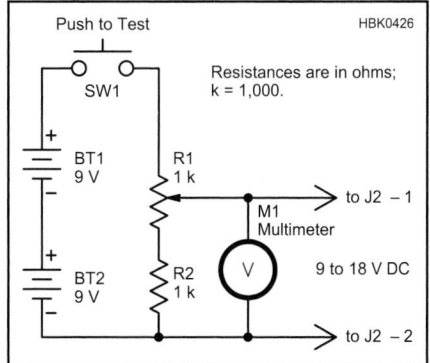

Fig 7.69 — Test voltage source for the battery charger.

an external series component results in an incorrect reading and improper charging. For example, the Elecraft K2 has internal diodes in the power-input circuit, so it's necessary to add a charging jack to the transceiver that provides a direct connection to the battery. Now the K2 can be left connected to the charger at all times and be assured that its internal battery is fully charged and ready to go at a moment's notice.

7.16.9 Overvoltage Protection for AC Generators

When using portable generators, there is always a possibility of damage to expensive equipment as a result of generator failure, especially from overvoltage. If the generator supplying power to this equipment puts out too much voltage, you run the risk of burning up power supplies or other electronic components. This project, by Jerry Paquette, WB8IOW, addresses the problem of increased voltage (not lower voltage) or surges and spikes lasting for a few microseconds.

Using a portable generator overvoltage protection circuit and ground-fault circuit interrupter (GFCI) as shown in **Fig 7.70** is good insurance. This overvoltage protection device must be used in conjunction with a

GFCI at each station! (More information on GFCIs may be found in the **Safety** chapter.)

CIRCUIT DESCRIPTION

Refer to Fig 7.70 for this description. R1 places an intentional fault on the load side of the GFCI. With the value resistor used, the fault is limited to 10 mA. (The normal tripping threshold of a GFCI is 5 mA. This current forces the GFCI to trip in just a few milliseconds. This circuit will not function at all without the use of a GFCI. A GFCI must be used at each station. If a single GFCI were used at the generator, rather than one at each location, premature tripping could occur. Several hundred feet of extension cords could have enough leakage to trip the GFCI.

You can see that the GFCI has separate lines (inputs) and loads (outputs). GFCI input terminals must be connected to the generator output. The GFCI ground must be tied to the ground of the generator. The load (computers, radios, etc) will plug into the GFCI or are wired to the load side of the GFCI. The primary of T1 is wired to the load side of the GFCI. The 12 kΩ/2 W resistor, however, must be wired to neutral on the *line* side of the GFCI in order for it to trip when used with generators that have windings isolated from ground. For safety, construct the entire unit in a single enclosure including the GFCI and its wiring. The generator connection can be made through a wired plug or using a male receptacle mounted on the enclosure.

T1 can be any 120 V to 12.6 V transformer capable of delivering 100 mA or more. Mounting of this transformer varies depending on the type used. All remaining components mount on a circuit board. D1 rectifies the ac from T1 and the 100 µF capacitor filters the dc. This voltage provides the power to the 723 voltage regulator.

Two fixed resistors and a potentiometer form the voltage-divider network supplying voltage to the LM723 input, pin 5. R1, the board-mounted potentiometer, has only three leads, but there are four pads on the circuit board, to accommodate different styles of pots. The 2.2 µF capacitor provides a slight

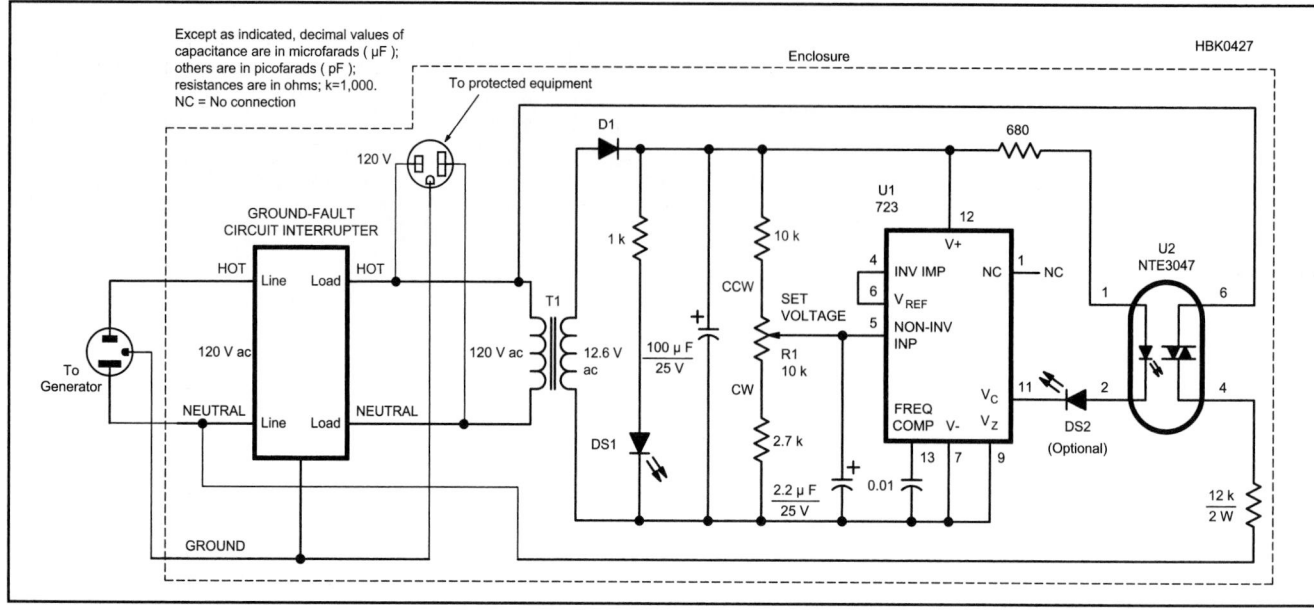

Fig 7.70 — Schematic of the Field Day equipment overvoltage-protection circuit. This circuit must be used in conjunction with a ground fault circuit interrupter (GFCI). A separate GFCI must be installed at each station. Unless otherwise specified, resistors are ¼ W, 5% tolerance. A PC board is available from FAR Circuits (www.farcircuits.net).

D1 — 200 PIV, 1 A diode; 1N4003 or equiv.
DS1, DS2 — Small LEDs.
R1 — 10 kΩ board-mounted, multi-turn
 potentiometer

T1 — 12.6 V ac transformer (see text).
U1 — 723 adjustable voltage regulator IC.

U2 — Optoisolator with TRIAC output;
 NTE3047 or equiv.

delay, to prevent false tripping when the circuit is powered up. The 0.01 µF capacitor from pin 13 of the 723 to the negative supply bus should always be used. When the voltage at pin 5 goes higher than the reference voltage at pins 4 and 6, pin 11 goes low, turning on the trip indicator LED DS2 and the optical coupler LED. LED current is limited by the 1 kΩ resistor. The optical coupler turns on the TRIAC, which creates a 10 mA fault current between the hot wire and ground of the GFCI. DS2 will remain lit as an indicator until the 100 µF capacitor is discharged.

ADJUSTMENT

Adjustment is simple. You'll need a variable ac transformer (Powerstat or Variac). Turn R1 fully clockwise and use the variable transformer to adjust input to 130 V ac. Turn the pot counterclockwise until the GFCI trips.

7.16.10 Overvoltage Crowbar Circuit

If the regulator circuitry of a power supply should fail — for example, if a pass transistor should short — the unregulated supply voltage could appear at the output terminals. This could cause a failure in the equipment connected to the supply. The overvoltage "crowbar" circuit shown here has been shown to be effective as a last line of defense against

power supply overvoltage failures. When an overvoltage condition is detected, the heavy-duty SCR is fired, becoming a short circuit — as if a crowbar (representing any over-sized conductor capable of handling the supply's short-circuit output current) had been connected across the supply, thus the name.

This circuit shown in **Fig 7.71** was originally designed for the 28 V power supply project available on the CD-ROM for this book. The use of the MC3423 overvoltage protection IC provides quicker triggering and more reliable gate drive to the SCR than comparable Zener-based circuits. Power supply builders can incorporate the overvoltage protection circuit into any dc power supply with the required component value adjustments.

The crowbar circuit is usually connected with the power input and the sense input connected together at the supply positive output and COMMON to the supply negative output. Another option is to connect power input and common across the rectifier filter capacitor and the sense input to the power supply output. In either case, the power input and common wires should be adequately sized to handle the full short-circuit current and are shown as heavy lines on the schematic.

The crowbar circuit functions as follows: the 4.7 kΩ resistor and Zener diode D1 create a supply voltage for the MC3423. U1 will function properly with a supply voltage of

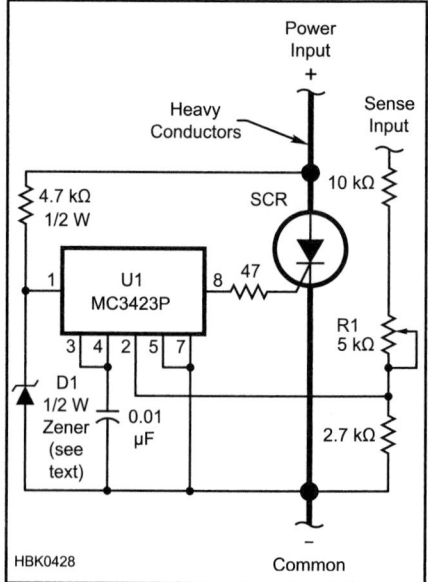

Fig 7.71 — Schematic of the overvoltage crowbar circuit. Unless otherwise specified, resistors are ¼ W, 5% tolerance. A PC board is available from FAR Circuits (www.farcircuits.net).
SCR — C38M stud-mount (TO-65 package).
U1 — MC3423P or NTE7172 voltage protection circuit, 8-pin DIP.
D1 —Zener diode, ½ W (see text).
R1 — 5 kΩ, PC board mount trimmer potentiometer.

4.5-40 V. Use a Zener diode with a voltage rating a few volts below that of the crowbar circuit positive-to-negative power input voltage. For example, if the crowbar is connected to a 12 V supply output, D1 voltage should be 6 to 9 V. The exact value is not critical.

U1 contains a 2.5 V reference and two comparators. When the voltage at pin 2 (sense terminal) reaches 2.5 V, the output voltage (pin 8) changes from the negative input voltage to the positive input voltage. This drives the gate of the SCR through the 47 Ω resistor. The trip voltage is set by the resistive divider across the + and − inputs:

$$V_{trip} = 2.5 \left(1 + \frac{10 \text{ k}\Omega + R}{2.7 \text{ k}\Omega} \right)$$

The application notes for the MC3423 recommend that the resistance from the sense input to the negative input be less than 10 kΩ for minimum drift, suggesting the value of 2.7 kΩ. The value of 10 kΩ for the fixed portion of the adjustable resistance is selected for V_{trip} = 15 V at the midpoint of the 5 kΩ potentiometer (R1) travel. For other trip voltages, the fixed resistor value should be the closest standard value to

$$R_{fixed} = 2.7 \text{ k}\Omega \left(\frac{V_{trip}}{2.5} - 1 \right) - 2.5 \text{ k}\Omega$$

assuming the potentiometer value remains at 5 kΩ.

When the SCR turns on, it short-circuits the inputs, causing any protective fuses or circuit breakers to open. The SCR will stay on until the current through drops below the "keep alive" threshold, at which point the SCR turns off. The SCR will stay on even if the input voltage to U1 drops below 4.5 V.

If the crowbar circuit fires due to RFI, an additional 0.01 μF capacitor should be connected from pin 2 of U1 to common and another across the D1.

More information may be found in the MC3423 datasheet and "Semiconductor Consideration for DC Power Supply Voltage Protector Circuits," ON Semi AN004E/D, both available from ON Semiconductor.

Additional Projects and Information

Additional power source projects and supporting files are included on the *Handbook* CD-ROM.

Contents

Modulation

8.1 Introduction

Radio amateurs use a wide variety of modulation types to convey information. This chapter, written by Alan Bloom, N1AL, explores analog, digital and image modulation techniques, as well as impairments that affect our ability to communicate. A glossary of terms and suggestions for additional reading round out the chapter.

The purpose of Amateur Radio transmissions is to send information via radio. The one possible exception to that is a beacon station that transmits an unmodulated carrier for propagation testing. In that case the only information being sent is, "Am I transmitting or not?" One can think of it as a single data bit with two states, *on* or *off*. However, in reality even a beacon station or test transmission must periodically identify with the station call sign, so we can say that modulation is an essential ingredient of Amateur Radio.

To represent the information being sent, the radio signal must periodically change its state in some way that can be detected by the receiver. In the early days of radio, the only way to do that was on-off keying using Morse code. By alternating the on and off states with the proper sequence and rhythm, a pair of highly-skilled operators can exchange textual data at rates up to perhaps 60 WPM. Later, engineers figured out how to amplify the signal from a microphone and use it to vary the power of the radio signal continuously. Thus was born amplitude modulation, which allowed transmitting voices at full speaking speeds, that is, up to about 200 WPM. That led to analog modes such as television with even faster information rates. Today's digital radio systems are capable of transferring tens of megabits of information per second, equivalent to tens of millions of words per minute.

Modulation is but one component of any transmission mode. For example AM voice and NTSC television (the analog US system) both use amplitude modulation, but the transmission protocol and type of information sent are very different. Digital modes also are generally defined not only by the modulation but also by multiple layers of protocol and data coding. Of the three components of a mode (modulation, protocol and information), this chapter will concentrate on the first. The **Digital Modes** chapter covers digital transmission protocols and the **Digital Communications** supplement on the *Handbook* CD covers the practical aspects of operating the various digital modes.

EMISSION DESIGNATORS

We tend to think of a radio signal as being "on" a particular frequency. In reality, any modulated signal occupies a band of frequencies. The bandwidth depends on the type of modulation and the data rate. A Morse code signal can be sent within a bandwidth of a couple hundred Hz at 60 WPM, or less at lower speeds. An AM voice signal requires about 6 kHz. For high-fidelity music, more bandwidth is needed; in the United States, an FM broadcast signal occupies a 200-kHz channel. Television signals need about 6 MHz, while 802.11n, the latest generation of "WiFi" wireless LAN, uses up to 40 MHz for data transmission.

The International Telecommunication Union (ITU) has specified a system for designating radio emissions based on the bandwidth, modulation type and information to be transmitted. The emission designator begins with the bandwidth, expressed as a maximum of five numerals and one letter. The letter occupies the position of the decimal point and represents the unit of bandwidth, as follows: H = hertz, K = kilohertz, M = megahertz and G = gigahertz. The bandwidth is followed by three to five emission classification symbols, as defined in the sidebar, "Emission Classifications." The first three symbols are mandatory; the fourth and fifth symbols are supplemental. These designators are found in Appendix 1 of the ITU Radio Regulations, ITU-R Recommendation SM.1138 and in the FCC rules §2.201.

For example, the designator for a CW signal might be 150H0A1A, which means 150 Hz

Emission Classifications

Emissions are designated according to their classification and their necessary bandwidth. A minimum of three symbols is used to describe the basic characteristics of radio waves. Emissions are classified and symbolized according to the following characteristics:

I. First symbol—Type of modulation of the main carrier
II. Second symbol—Nature of signal(s) modulating the main carrier
III. Third symbol—Type of information to be transmitted
Note: A fourth and fifth symbol are provided for in the ITU Radio Regulations. Use of the fourth and fifth symbol is optional.
IV. Details of signal(s)
V. Nature of multiplexing

First symbol—type of modulation of the main carrier
  (1) Emission of an unmodulated carrier .... N
  (2) Emission in which the main carrier is amplitude-modulated (including cases where subcarriers are angle-modulated):
    - Double sideband .... A
    - Single sideband, full carrier............................. H
    - Single sideband, reduced or variable level carrier... R
    - Single sideband, suppressed carrier .... J
    - Independent sidebands .... B
    - Vestigial sideband C
  (3) Emission in which the main carrier is angle-modulated:
    - Frequency modulation .... F
    - Phase modulation .... G
    Note: Whenever frequency modulation (F) is indicated, phase modulation (G) is also acceptable.
  (4) Emission in which the main carrier is amplitude and angle-modulated either simultaneously or in a pre-established sequence.................................... D
  (5) Emission of pulses[1]
    - Sequence of unmodulated pulses .... P
    - A sequence of pulses:
    - Modulated in amplitude .... K
    - Modulated in width/duration L
    - Modulated in position/phase .... M
    - In which the carrier is angle-modulated during the period of the pulse .... Q
    - Which is a combination of the foregoing or is produced by other means V
  (6) Cases not covered above, in which an emission consists of the main carrier modulated, either simultaneously or in a pre-established sequence in a combination of two or more of the following modes: amplitude, angle, pulse .... W
  (7) Cases not otherwise covered X

Second symbol—nature of signal(s) modulating the main carrier
  (1) No modulating signal 0
  (2) A single channel containing quantized or digital information without the use of a modulating subcarrier, excluding time-division multiplex 1
  (3) A single channel containing quantized or digital information with the use of a modulating subcarrier, excluding time-division multiplex 2
  (4) A single channel containing analog information .. 3
  (5) Two or more channels containing quantized or digital information 7
  (6) Two or more channels containing analog information 8
  (7) Composite system with one or more channels containing quantized or digital information, together with one or more channels containing analog information 9
  (8) Cases not otherwise covered X

Third symbol—type of information to be transmitted[2]
  (1) No information transmitted .... N
  (2) Telegraphy, for aural reception .... A
  (3) Telegraphy, for automatic reception .... B
  (4) Facsimile C
  (5) Data transmission, telemetry, telecommand .... D
  (6) Telephony (including sound broadcasting) .... E
  (7) Television (video) .... F
  (8) Combination of the above .... W
  (9) Cases not otherwise covered X

Where the fourth or fifth symbol is used it shall be used as indicated below. Where the fourth or the fifth symbol is not used this should be indicated by a dash where each symbol would otherwise appear.

Fourth symbol—Details of signal(s)
  (1) Two-condition code with elements of differing numbers and/or durations .... A
  (2) Two-condition code with elements of the same number and duration without error-correction.... B
  (3) Two-condition code with elements of the same number and duration with error-correction C
  (4) Four-condition code in which each condition represents a signal element (of one or more bits) .... D
  (5) Multi-condition code in which each condition represents a signal element (of one or more bits) .... E
  (6) Multi-condition code in which each condition or combination of conditions represents a character .... F
  (7) Sound of broadcasting quality (monophonic) ..... G
  (8) Sound of broadcasting quality (stereophonic or quadraphonic) .... H
  (9) Sound of commercial quality (excluding categories given in (10) and (11) below .... J
  (10) Sound of commercial quality with frequency inversion or band-splitting................................. K
  (11) Sound of commercial quality with separate frequency-modulated signals to control the level of demodulated signal L
  (12) Monochrome .... M
  (13) Color .... N
  (14) Combination of the above .... W
  (15) Cases not otherwise covered X

Fifth symbol—Nature of multiplexing
  (1) None .... N
  (2) Code-division multiplex[3] C
  (3) Frequency-division multiplex .... F
  (4) Time-division multiplex .... T
  (5) Combination of frequency-division and time-division multiplex .... W
  (6) Other types of multiplexing X

---

[1] Emissions where the main carrier is directly modulated by a signal which has been coded into quantized form (eg, pulse code modulation) should be designated under (2) or (3).

[2] In this context the word "information" does not include information of a constant unvarying nature such as is provided by standard frequency emissions, continuous wave and pulse radars, etc.

[3] This includes bandwidth expansion techniques.

bandwidth, double sideband, digital information without subcarrier, and telegraphy for aural reception. SSB would be 2K5J3E, or 2.5 kHz bandwidth, single sideband with suppressed carrier, analog information, and telephony. The designator for a PSK31 digital signal is 60H0J2B, which means 60 Hz bandwidth, single sideband with suppressed carrier, digital information using a modulating subcarrier, and telegraphy for automatic reception.

Authorized modulation modes for Amateur Radio operators depend on frequency, license class, and geographical location, as specified in the FCC regulations §97.305. Technical standards for amateur emissions are specified in §97.307. Among other things, they require that no amateur station transmission shall occupy more bandwidth than necessary for the information rate and emission type being transmitted, in accordance with good amateur practice. We will discuss the necessary bandwidth for each type of modulation as it is covered in the following sections.

8.2 Analog Modulation

While the newer digital modes are all the rage today, traditional analog modulation is still quite popular on the amateur bands. There are many reasons for that, some technical and some not. Analog equipment tends to be simpler; no computer is required. For example, a low-power CW transceiver is easy to construct and simple to understand. CW and SSB are both narrowband modes with spectral efficiency that is quite competitive with their digital equivalents. While heated arguments have been made about the relative power efficiency of analog vs digital modes, it is clear that in the hands of skilled operators CW and SSB are at least in the same ballpark with, if perhaps not better than, many digital modes.

When copying under difficult conditions, the human brain is capable of higher-level signal processing that is difficult to duplicate on a computer. For example, in a DX pileup, an operator may realize that the signal he wants is the weak one and mentally filter out all the stronger stations based on signal strength alone. Other aspects of the signal may be used for filtering as well, such as the operator's accent (on voice) or "fist" (on CW), propagation effects, the operator's speed and operating pattern, CW hum and chirp, and other factors. Different portions of the other station's call sign may be copied at different times and stitched together in the operator's brain. Missing portions of a transmission can be filled in based on context. Many operators enjoy the challenge of using their ears and brain to decipher a weak signal, an aspect of operating that is lost with the digital modes.

8.2.1 Amplitude-Modulated Analog Signals

Of the various properties of a signal that can be modulated to transmit voice information, amplitude was the first to be used. Not only are modulation and demodulation of AM signals simple in concept, but they are simple to implement as well. Since the RF output voltage of a class-C amplifier is approximately proportional to the power supply voltage, an AM modulator needs only to vary the power supply voltage in step with the audio signal, as shown in **Fig 8.1**.

To prevent overmodulation and the resulting distortion, the modulating signal should not try to reduce the power supply voltage to less than zero volts or more than twice the average voltage. The *modulation index* of an AM signal is the ratio of the peak modulation to those limits. The resulting RF signal might look like the waveform in **Fig 8.2C**.

Note that the shape of the envelope of the modulated RF signal matches the shape of the modulating signal. That suggests a possible demodulation method. A diode detector puts out a signal proportional to the envelope of the RF signal, recovering the original modulation. See **Fig 8.3**. The capacitor should be large enough that it filters out most of the RF ripple but not so large that it attenuates the higher audio frequencies.

So far we have been discussing signals as a function of time — all the graphs have time as the horizontal axis. It is equally valid

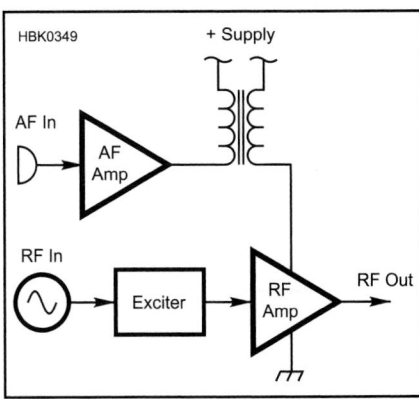

Fig 8.1 — An AM modulator. With a vacuum-tube RF power amplifier this method is known as *plate modulation* since the modulated power supply voltage is applied to the vacuum tube plate.

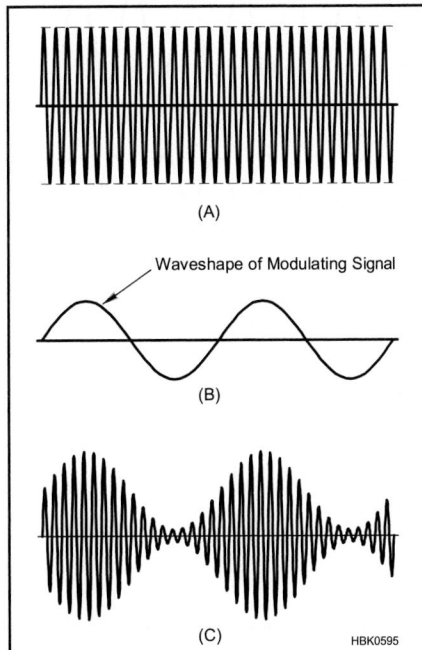

Fig 8.2 — Graphical representation of amplitude modulation. In the unmodulated carrier (A) each RF cycle has the same amplitude. When the modulating signal (B) is applied, the RF amplitude is increased or decreased according to the amplitude of the modulating signal (C). A modulation index of approximately 75% is shown. With 100% modulation the RF power would just reach zero on negative peaks of the modulating signal.

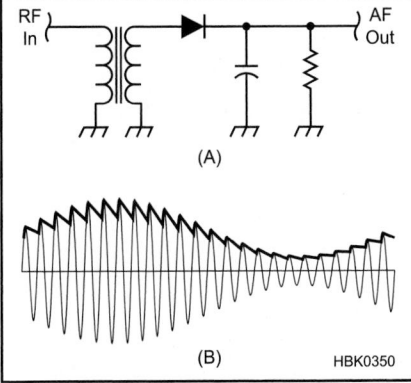

Fig 8.3 — A simple diode-type AM detector, also known as an *envelope detector* (A). The demodulated output waveform has the same shape as the envelope of the RF signal (B). In (B) the thin line is the RF signal modulated with a sine wave and the darker line is the demodulated audio frequency with some residual RF ripple.

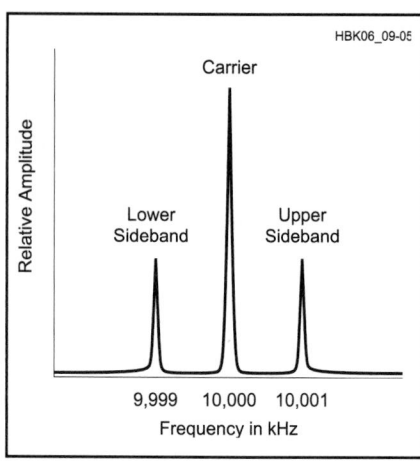

Fig 8.4 — A 10-MHz carrier that is AM-modulated by a 1-kHz sine wave.

to look in the frequency domain. It turns out that the modulation spectrum consists of a pair of sidebands above and below the RF carrier. **Fig 8.4** shows a 10-MHz signal modulated with a 1-kHz sine wave. The upper sideband is a single frequency at 10 MHz + 1 kHz = 10.001 MHz. The spectrum of the lower sideband is inverted, so it is at

10 MHz – 1 kHz = 9.999 MHz.

To see why that happens, refer to **Fig 8.5**. Simply adding the first three sine waves shown, representing a carrier and upper and lower sidebands, results in the 100%-modulated AM waveform at the bottom. Note that each sideband has half the amplitude of the carrier, which corresponds to one-quarter the power or –6 dB. Each sideband in an AM signal that is 100% modulated with a sine wave is therefore –6 dB with respect to the carrier.

If a more complex modulating signal is used, each frequency component of the modulation f_m appears at $f_c + f_m$ and $f_c - f_m$, where f_c is the carrier frequency. An AM modulator is really nothing more than a type of mixer that generates new output frequencies at the sum and difference of the carrier and modulation frequencies. The upper sideband is simply a frequency-shifted version of the modulation spectrum and the lower sideband is both frequency-shifted and inverted. For communications purposes, the audio frequency response typically extends up to about 2.5-3 kHz, resulting in a 5-6 kHz RF bandwidth.

One problem with AM is that if the amplitude of the carrier becomes attenuated for

any reason, then the modulation is distorted, especially the negative-going portion near the 100%-modulation (zero power) point. This can happen due to a propagation phenomenon called *selective fading*. It occurs when the signal arrives simultaneously at the receive antenna via two or more paths, such as ground wave and sky wave. If the difference in the distance of the two paths is an odd number of half-wavelengths, then the two signals are out of phase. If the amplitudes are nearly the same, they cancel and a deep fade results. Since wavelength depends on frequency, the fading is frequency-selective. On the lower-frequency amateur bands it is possible for an AM carrier to be faded while the two sidebands are still audible.

A solution is to regenerate the carrier in the receiver. Since the carrier itself carries no information about the modulation, it is not necessary for demodulation. The transmitted signal may be a standard full-carrier AM signal or the carrier may be suppressed, resulting in *double sideband, suppressed carrier* (DSBSC). An AM detector that regenerates the carrier from the signal is known as a *synchronous detector*. Often the regenerated carrier oscillator is part of a phase-locked loop (PLL) that locks onto the incoming carrier. See **Fig 8.6**. Synchronous detectors not only reduce the effects of selective fading but also are usually more linear than diode detectors so they have less distortion. Some commercial short-wave broadcasts include a reduced, but not suppressed, carrier (DSBRC) to allow operation with PLL-type synchronous detectors.

Since the advent of single sideband in the 1960s, full-carrier double-sideband AM has become something of a niche activity in Amateur Radio. It does retain several advantages however. We have already mentioned the simplicity of the circuitry. Another advantage is that, because of the presence of the

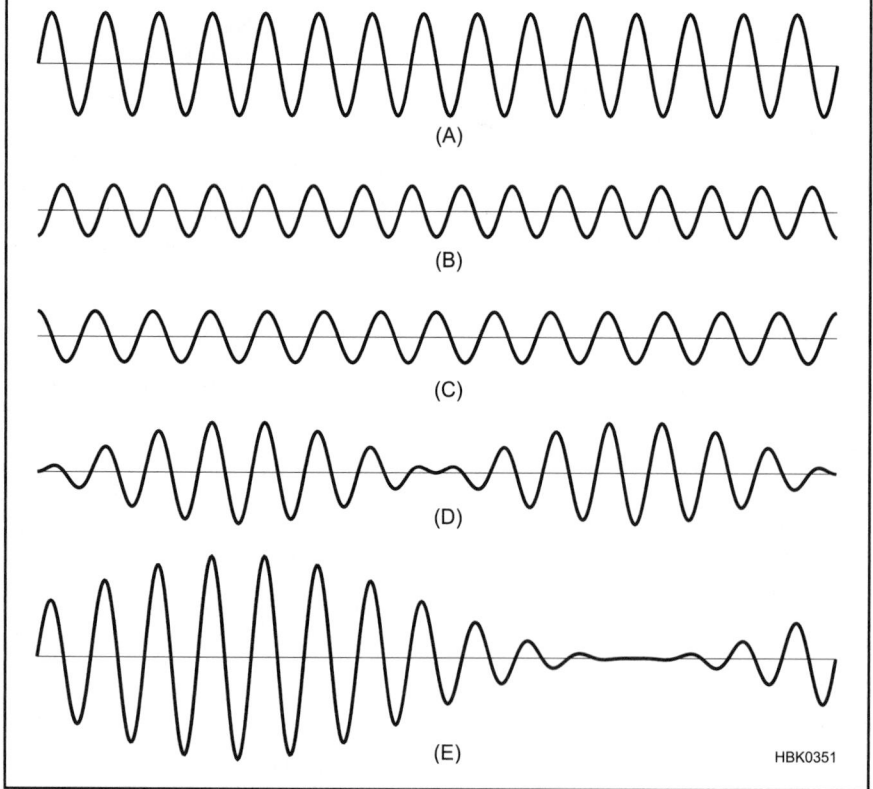

Fig 8.5 — At A is an unmodulated carrier. If the upper (B) and lower (C) sidebands are added together a double-sideband suppressed carrier (DSBSC) signal results (D). If each sideband has half the amplitude of the carrier, then the combination of the carrier with the two sidebands results in a 100%-modulated AM signal (E). Whenever the two sidebands are out of phase with the carrier, the three signals sum to zero. Whenever the two sidebands are in phase with the carrier, the resulting signal has twice the amplitude of the unmodulated carrier.

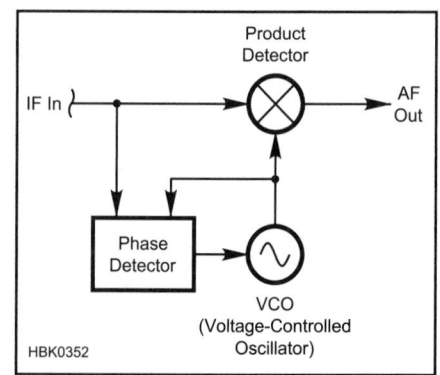

Fig 8.6 — Block diagram of a synchronous detector. The voltage-controlled oscillator (VCO) is part of a phase-locked loop that locks the oscillator to the carrier frequency of the incoming AM or DSBRC signal.

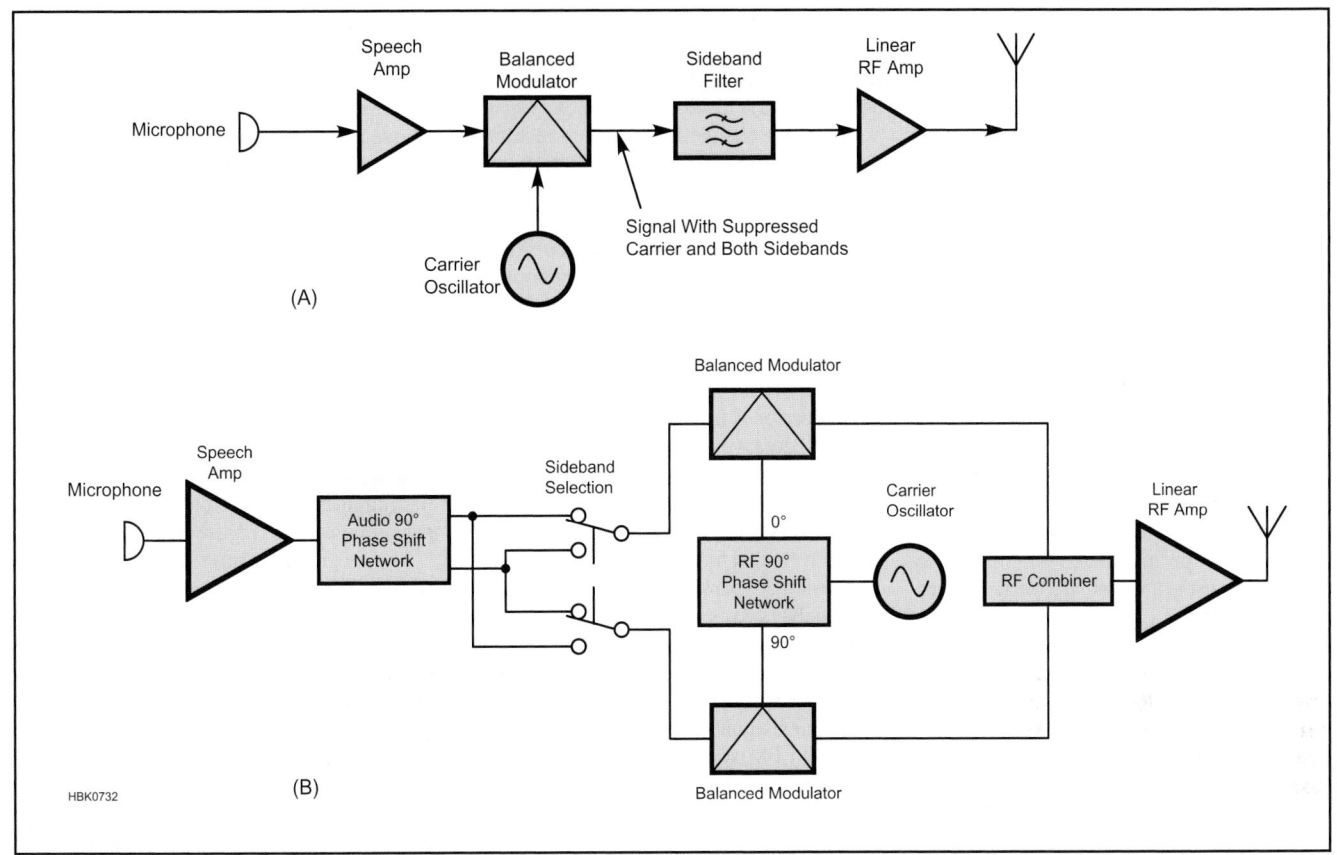

Fig 8.7 — Block diagram of a filter-type (A) and a phasing-type (B) SSB generator.

carrier, the automatic gain control system in the receiver remains engaged at all times, ensuring a constant audio level. Unlike with SSB, there is no rush of noise during every pause in speech. Also tuning is less critical than with SSB. There is no "Donald Duck" sound if the receiver is slightly mistuned. Finally, the audio quality of an AM signal is usually better than that of SSB because of the lack of a crystal filter in the transmit path and the wider filter in the receiver.

SINGLE SIDEBAND

As we have seen, since the carrier itself contains no modulation, it does not need to be transmitted, which saves at least 67% of the transmitted power. In addition, since the two sidebands carry identical information, one of them may be eliminated as well, saving half the bandwidth. The result is *single sideband, suppressed carrier* (SSBSC), which is commonly referred to simply as "SSB." If the lower sideband is eliminated, the result is called *upper sideband* (USB). If the upper sideband is eliminated, you're left with *lower sideband* (LSB). The bandwidth is a little less than half that of an equivalent AM signal. A typical SSB transmitter might have an overall audio response from 300 Hz to 2.8 kHz, resulting in 2.5 kHz bandwidth.

As with an AM synchronous detector, the carrier is regenerated in the receiver but since only one sideband is present, the synchronous detector's phase-locked loop is not possible. Instead, the detector uses a free-running *beat-frequency oscillator* (BFO). The detector itself is called a *product detector* because its output is the mathematical product of the BFO and the SSB signal. The BFO must be tuned to the same frequency as the suppressed carrier to prevent distortion of the recovered audio. In a superheterodyne receiver, that is done by carefully tuning the local oscillator (the main tuning dial) such that after conversion to the intermediate frequency, the suppressed carrier aligns with the BFO frequency.

In the transmitter, the SSB modulator must suppress both the carrier and the unwanted sideband. Carrier suppression is normally accomplished with a *balanced modulator*, a type of mixer whose output contains the sum and difference frequencies of the two input signals (the modulating signal and the carrier) but not the input signals themselves. There are several ways to eliminate the unwanted sideband, but the most common, shown in **Fig 8.7A**, is to pass the output of the balanced mixer through a crystal filter that passes the wanted sideband while filtering out the unwanted one. This is convenient in a transceiver since the same filter can be used in the receiver

by means of a transmit-receive switch.

Another method to generate single sideband is called the *phasing method*. See Fig 8.7B. Using trigonometry, it can be shown mathematically that the sum of the signals from two balanced modulators, each fed with audio signals and RF carriers that are 90° out of phase, consists of one sideband only. The other sideband is suppressed. The output can be switched between LSB and USB simply by reversing the polarity of one of the inputs, which changes the phase by 180°. The phasing method eliminates the need for an expensive crystal filter following the modulator or, in the case of a transceiver, the need to switch the crystal filter between the transmitter and receiver sections. In addition, the audio quality is generally better because it eliminates the poor phase dispersion that is characteristic of most crystal filters. In the old days, designing and building an audio phase-shift network that accurately maintained a 90° differential over a decade-wide frequency band (300 Hz to 3000 Hz) was rather complicated. With modern DSP techniques, the task is much easier.

Compared to most other analog modulation modes, SSB has excellent power efficiency because the transmitted power is proportional to the modulating signal and no power at all is transmitted during pauses in speech. Another advantage is that the lack of a carrier results in less interference to other stations. Back in the days when AM ruled the HF amateur voice bands, operators had to put up with a cacophony of heterodynes emanating from their headphones. SSB transceivers also are well-suited for the narrowband digital modes. Because an SSB transceiver simply frequency-translates the baseband audio signal to RF in the transmitter and back to audio again in the receiver, digital modulation may be generated and detected at audio frequencies using the sound card in a personal computer.

8.2.2 Angle-Modulated Analog Signals

Angle modulation varies the phase angle of an RF signal in response to the modulating signal. In this context, "phase" means the phase of the modulated RF sine wave with respect to the unmodulated carrier. Phase modulation is one type of angle modulation, but so is frequency modulation because any change in frequency results in a change in phase. For example, the way to smoothly ramp the phase from one value to another is to change the frequency and wait. If the frequency is changed by +1 Hz, then after 1 second the phase will have changed by +360°. After 2 seconds, the phase will have changed by +720°, and so on. Change the frequency

in the other direction and the phase moves in the opposite direction as well. With sine-wave modulation, the frequency and phase both vary in a sinusoidal fashion. See **Fig 8.8**.

Since any change in frequency results in a change in phase and vice versa, *frequency modulation* (FM) and *phase modulation* (PM) are fundamentally the same. The only difference is the audio frequency response. An FM transmitter with 6 dB/octave pre-emphasis of the modulating signal is indistinguishable from a PM transmitter. A PM transmitter with 6 dB/octave de-emphasis is indistinguishable from an FM transmitter. The reverse happens at the receiver. A frequency detector followed by a 6 dB/octave de-emphasis network acts like a phase detector. It is interesting to note that most VHF and UHF amateur "FM" transceivers should really be called "PM" transceivers due to the pre-emphasis and de-emphasis networks used in the transmitters and receivers respectively.

The term *frequency deviation* means the amount the RF frequency deviates from the center (carrier) frequency with a given modulating signal. The instantaneous deviation of an FM signal is proportional to the instantaneous amplitude of the modulating signal. The instantaneous deviation of a PM signal is proportional to the instantaneous *rate of change* of the modulating signal.

Since the rate of change of a sine wave is proportional to its frequency as well as its amplitude, the deviation of a PM signal is proportional to both the amplitude and the frequency of the modulating signal.

Most FM and PM transmitters include some kind of audio compressor before the modulator to limit the maximum deviation. Common usage of the term *deviation* is that it refers to the maximum peak deviation allowed by the audio compressor. If the frequency swings a maximum of 5 kHz above the center frequency and 5 kHz below the center frequency, we say the deviation is 5 kHz. The *modulation index* is the ratio of the peak deviation to the highest audio frequency. The *deviation ratio* is the ratio of maximum permitted peak deviation to the maximum permitted modulating frequency.

FCC regulations limit the modulation index at the highest modulating frequency to a maximum of 1.0 for frequencies below 29 MHz. If the audio is low-pass filtered to 3 kHz, then the deviation may be no more than 3 kHz. For that reason, FM transmitters for frequencies below 29 MHz are usually set for 3 kHz deviation while FM transmitters at higher frequencies are typically set for about 5 kHz deviation. The term *narrowband FM* generally refers to deviation of no more than 3 kHz and *wideband FM*

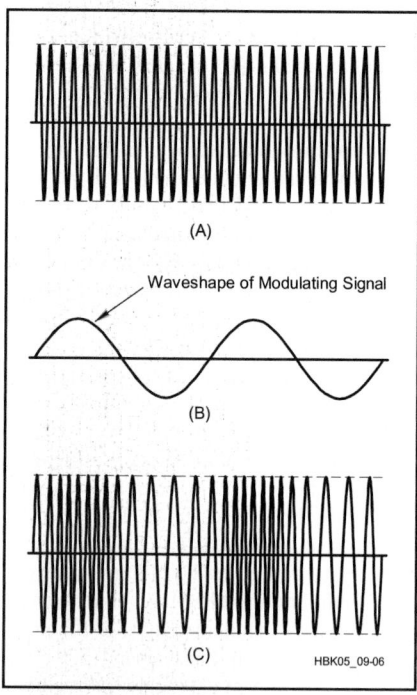

Fig 8.8 — Graphical representation of frequency modulation. In the unmodulated carrier (A) each RF cycle occupies the same amount of time. When the modulating signal (B) is applied, the radio frequency is increased or decreased according to the amplitude and polarity of the modulating signal (C).

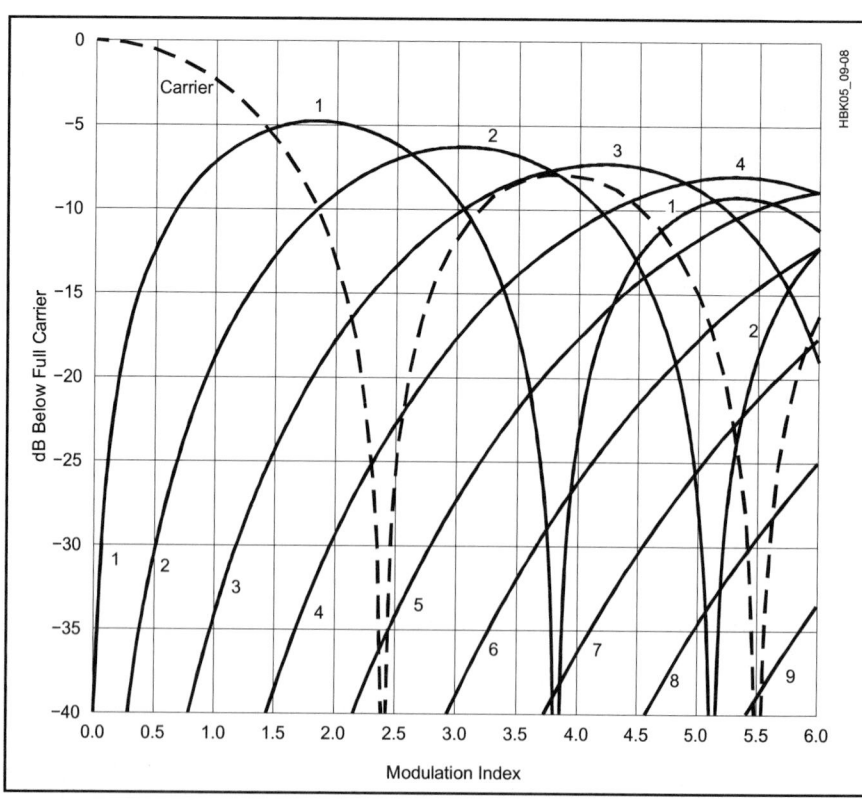

Fig 8.9 — Amplitude of the FM carrier and the first nine sidebands versus modulation index. This is a graphical representation of mathematical functions developed by F. W. Bessel. Note that the carrier completely disappears at modulation indices of 2.405 and 5.52.

refers to deviation greater than 3 kHz.

Unlike amplitude modulation, angle modulation is nonlinear. Recall that with amplitude modulation, the shape of the RF spectrum is the same as that of the modulation spectrum, single-sided with SSB and double-sided with DSB. That is not true with angle modulation. A double-sideband AM signal with audio band-limited to 3 kHz has 6 kHz RF bandwidth. It is easy to see that an FM transmitter with the same audio characteristics but with, say, 5 kHz deviation must have a bandwidth of at least 10 kHz. And it is even worse than that. While you might think that the bandwidth equals twice the deviation, in reality the transmitted spectrum theoretically extends to infinity, although it does become vanishingly small beyond a certain point. As a rule of thumb, approximately 98% of the spectral energy is contained within the bandwidth defined by *Carson's rule*:

$$BW = 2(f_d + f_m) \qquad (1)$$

where

f_d = the peak deviation, and
f_m = the highest modulating frequency.

For example, if the deviation is 5 kHz and the audio is limited to 3 kHz, the bandwidth is approximately 2 (5 + 3) = 16 kHz.

With sine-wave modulation, the RF frequency spectrum from an angle-modulated transmitter consists of the carrier and a series of sideband pairs, each spaced by the frequency of the modulation. For example, with 3 kHz sine-wave modulation, the spectrum includes tones at ±3 kHz, ±6 kHz, ±9 kHz and so on with respect to the carrier. When the modulation index is much less than 1.0, only the first sideband pair is significant, and the spectrum looks similar to that of an AM signal (although the phases of the sidebands are different). As the modulation index is increased, more and more sidebands appear within the bandwidth predicted by Carson's rule. Unlike with AM, the carrier amplitude also changes with deviation and

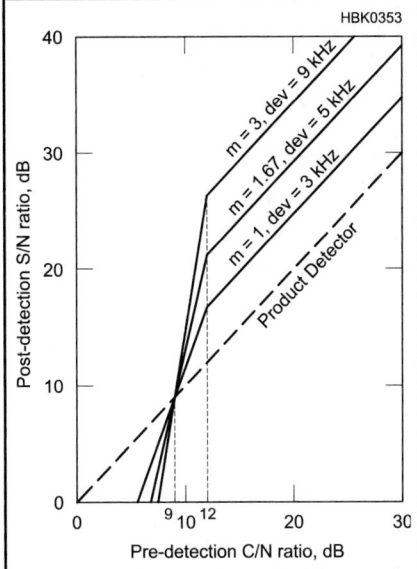

Fig 8.10 — A plot of post-detection signal-to-noise ratio (S/N) versus input carrier-to-noise ratio (C/N) for an FM detector at various modulation indices, m. For each modulation index the deviation is also noted assuming a maximum modulating frequency of 3 kHz. For comparison, the performance of an SSB product detector is also shown.

actually disappears altogether for certain modulation indices. Because the amplitude of an angle-modulated signal does not vary with modulation, the total power of the carrier and all sidebands is constant.

The amplitudes of the carrier and the various sidebands are described by a series of mathematical equations called Bessel functions, which are illustrated graphically in **Fig 8.9**. Note that the carrier disappears when the modulation index equals 2.405 and 5.52. That fact can be used to set the deviation of an FM transmitter. For example, to set 5 kHz deviation, connect the microphone input to a sine-wave generator set for a frequency of 5 / 2.405 = 2.079 kHz, listen to the carrier on

a narrowband receiver, and adjust the deviation until the carrier disappears.

Since the amplitude carries no information, FM receivers are designed to be as insensitive to amplitude variations as possible. Because noise tends to be mostly AM in nature, that results in a quieter demodulated signal. Typically the receiver includes a *limiter*, which is a very high-gain amplifier that causes the signal to clip, removing any amplitude variations, before being applied to the detector. Unlike with AM, as an FM signal gets stronger the volume of the demodulated audio stays the same, but the noise is reduced. Receiver sensitivity is often specified by how much the noise is suppressed for a certain input signal level. For example, if a 0.25 µV signal causes the noise to be reduced by 20 dB, then we say the receiver sensitivity is "0.25 µV for 20 dB of quieting."

The limiter also causes a phenomenon known as *capture effect*. If more than one signal is present at the same time, the limiter tends to reduce the weaker signal relative to the stronger one. We say that the stronger signal "captures" the receiver. The effect is very useful in reducing on-channel interference.

The suppression of both noise and interference is greater the wider the deviation. FM signals with wider deviation do take up more bandwidth and actually have a poorer signal-to-noise (S/N) ratio at the detector output for weak signals but have better S/N ratio and interference rejection for signal levels above a certain threshold. See **Fig 8.10**.

In addition to the noise and interference-reduction advantage, angle-modulated signals share with full-carrier AM the advantages of non-critical frequency accuracy and the continuous presence of a signal, which eases the task of the automatic gain control system in the receiver. In addition, since the signal is constant-amplitude, the transmitter does not need a linear amplifier. Class-C amplifiers may be used, which have greater power efficiency.

8.3 Digital Modulation

The first digital modulation used in Amateur Radio was on-off keying using Morse code. However, the term "digital" is commonly understood to refer to modes where the sending and receiving is done by machine. By that definition, the first popular amateur digital mode (called radioteleprinter, or RTTY) used frequency-shift keying with

all the coding and decoding done by mechanical teleprinters. You still hear RTTY on the bands today, although most operators are using computers these days rather than Teletype machines. More-modern digital modes use more efficient modulation formats as well as error detection and correction to ensure reliable communications. Data can be sent

much more rapidly than with Morse code or voice, with less effort, and with fewer errors. In some systems, stations operate unattended 24 hours per day, transferring data with little or no operator intervention. For point-to-point transmission of large quantities of text or other data, these modern digital modes are the obvious choice.

8.3.1 Amplitude-Modulated Digital Signals

We start this section by discussing *on-off keying* (OOK), which is a special case of *amplitude-shift keying* (ASK) and is normally sent using Morse code. For historical reasons dating from the days of spark transmitters, amateurs often refer to this as *continuous wave* (CW) even though the signal is actually keyed on and off. While most amateurs think of CW as being an analog mode, we include it here because it has technical characteristics that are more similar to other digital modes than to analog.

You can think of OOK as being the same as analog AM that is modulated with a two-level signal that switches between full power and zero power. For example, imagine that the modulation is a 10 Hz square wave, equivalent to sending a series of dits at 24 WPM. A square wave may be decomposed into a sine wave of the same frequency and a theoretically infinite succession of odd harmonics. Recall that the lower and upper sidebands of an AM signal are simply the inverted and non-inverted spectra of the baseband modulation. The RF spectrum therefore contains sidebands at the carrier frequency ±10 Hz, ±30 Hz, ±50 Hz, and so on to plus and minus infinity. This phenomenon is called *key clicks*. Stations listening on nearby frequencies hear a click upon every key closure and opening.

To prevent interference to other stations, the modulation must be low-pass filtered, which slows down the transition times between the on and off states. See **Fig 8.11**. If the transitions are too fast, then excessive key clicks occur. If the transitions are too slow, then at high speeds the previous transition may not have finished before the next one starts, which makes the signal sound mushy and hard to read. Traditionally, filtering was done with a simple resistor-capacitor low-pass filter on the keying line, but using a transition with a raised-cosine shape allows faster transition times without excessive key clicks. Some modern transceivers use DSP techniques to generate such a controlled transition shape.

The optimum transition time, and thus bandwidth, depends on keying speed. It also depends on propagation conditions. When the signal is fading, the transitions must be sharper to allow good copy. **Fig 8.12** gives recommended keying characteristics based on sending speed and propagation. As a compromise, many transmitters use a 5 ms rise and fall time. That limits the bandwidth to approximately 150 Hz while allowing good copy up to 60 WPM on non-fading channels and 35 WPM on fading channels, which covers most requirements.

With any digital system, the number of

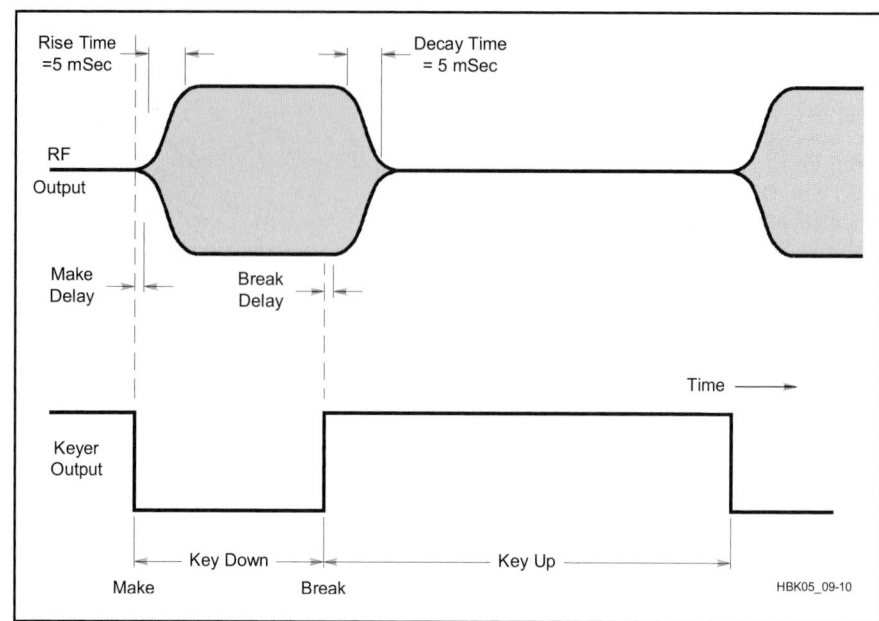

Fig 8.11 — A filtered CW keying waveform. The on-off transitions of the RF envelope should be as smooth as possible while transitioning as quickly as possible. The shape of the RF Output waveform is nearly optimum in that respect.

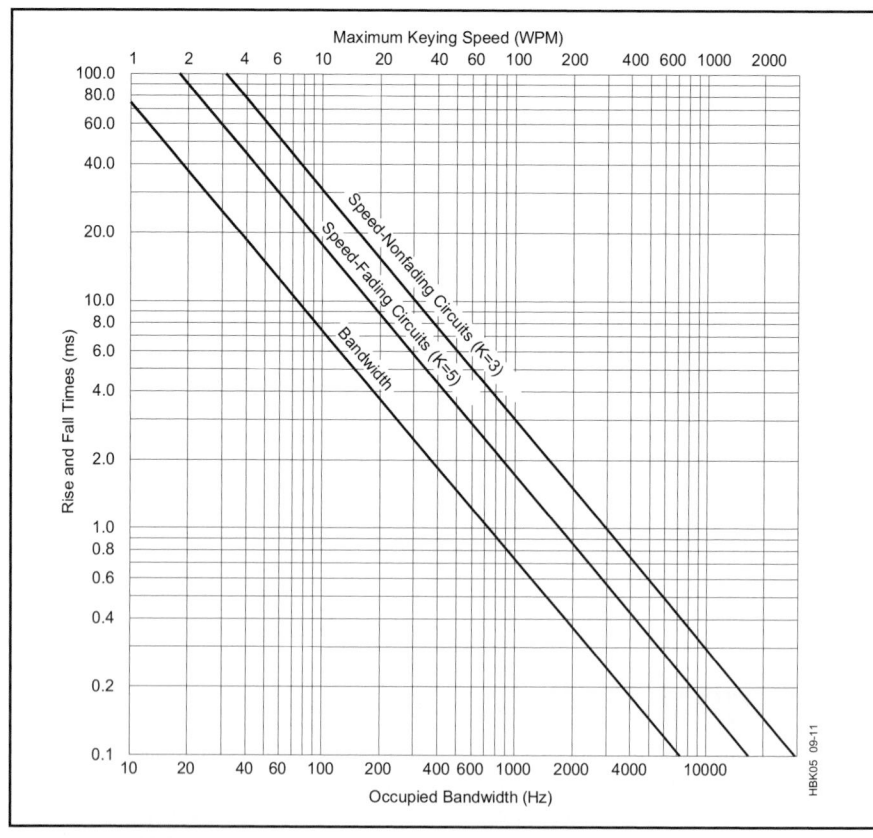

Fig 8.12 — Keying speed vs rise and fall times vs bandwidth for fading and non-fading communications circuits. For example, to optimize transmitter timing for 20 WPM on a non-fading circuit, draw a vertical line from the WPM axis to the K = 3 line. From there draw a horizontal line to the rise/fall axis (approximately 15 ms). Draw a vertical line from where the horizontal line crosses the bandwidth line and see that the bandwidth is about 50 Hz.

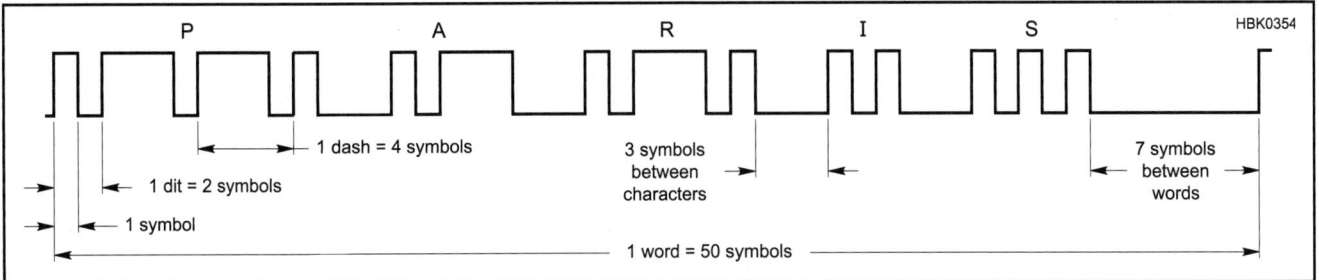

Fig 8.13 — CW timing of the word PARIS, which happens to have a length equal to the standard 50 symbols per word. By programming it multiple times into a memory keyer, the speed may be calibrated simply by counting the number of times the word is completed in one minute.

changes of state per second is called the *baud rate* or the *symbol rate*, measured in *bauds* or *symbols per second*. Sending a single Morse code dit requires two equal-length states, or symbols: *on* for the length of the dit and *off* for the space between dits. Thus, a string of dits sent at a rate of 10 dits per second has a symbol rate of 20 bauds.

Refer to **Fig 8.13**. A Morse code dash is on for three times the length of a dit. Including one symbol for the off time, the total time to send a dash is four symbols, twice the time to send a dit. For example, at a baud rate of 20 bauds, there are 10 dits per second or 5 dashes per second. The spacing between characters within a word is three symbols, two more than the normal space between dits and dashes. The spacing between words is seven symbols.

For purposes of computing sending speed, a standard word is considered to have five characters, plus the inter-word spacing. On average, that results in 50 symbols per word. From that, the speed in words per minute may be computed from the baud rate:

$$WPM = \frac{60 \,(\text{sec} \,/\, \text{min})}{50 \,(\text{symbols} \,/\, \text{word})} \times \text{bauds}$$

$$(2)$$

$$= 1.2 \times \text{bauds} = 2.4 \times \text{dits} \,/\, \text{sec}$$

Characters in Morse code do not all have the same length. Longer codes are used for characters that are used less frequently while the shortest codes are reserved for the most common characters. For example, the most common letter in the English language, E, is sent as a single dit. In that way, the average character length is reduced, resulting in a faster sending speed for a given baud rate. Such a variable-length code is known as *varicode*, a technique that has been copied in some modern digital modes.

PULSE MODULATION

Another type of digital amplitude modulation comes under the general category of *pulse modulation*. The RF signal is broken up into a series of pulses, which are usually equally-spaced in time and separated by peri-

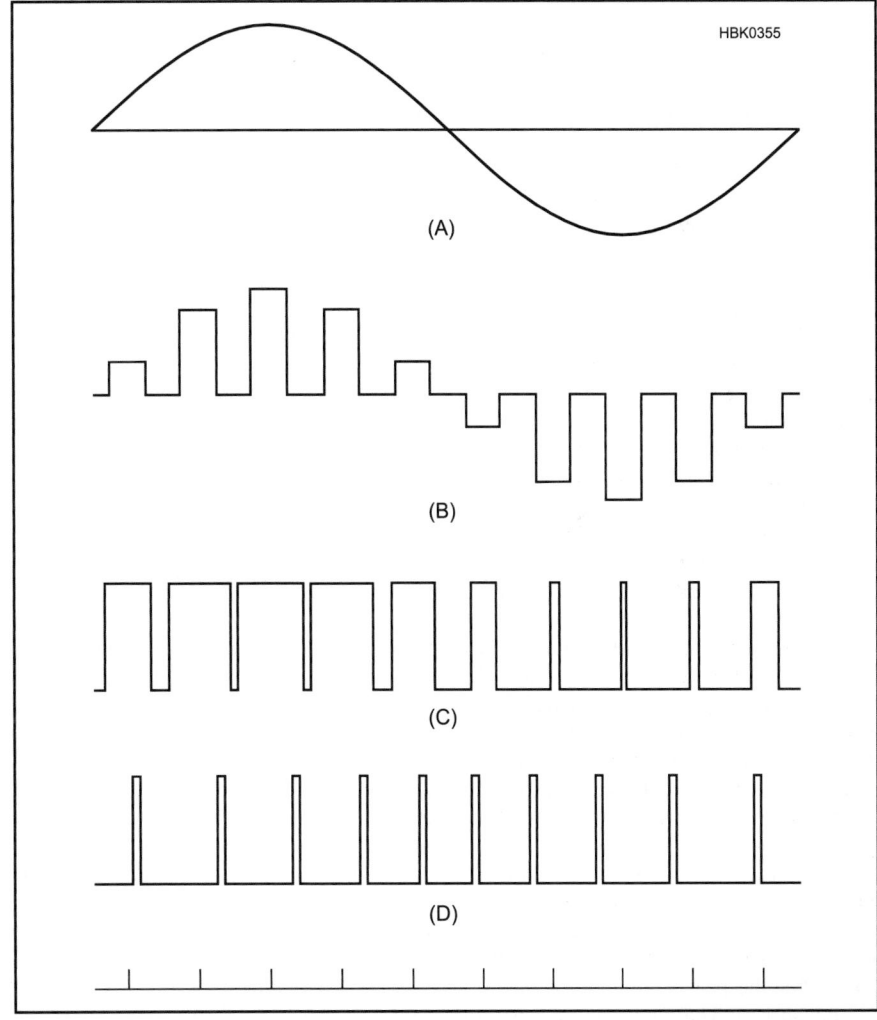

Fig 8.14 — Three types of pulse-code modulation. A sine-wave modulating signal (A) is shown at the top. Pulse-amplitude modulation (B) varies the amplitude of the pulses, pulse-width modulation (C) varies the pulse width, and pulse-position modulation (D) varies the pulse position, proportional to the modulating signal. The tic marks show the nominal pulse times. The RF signal is created by an AM modulator using (B), (C) or (D) as the modulating signal.

ods of no signal. We will discuss three types of pulse modulation, PAM, PWM and PPM.

Pulse-amplitude modulation (PAM) consists of a series of pulses of varying amplitude that correspond directly to the amplitude of the modulating signal. See **Fig 8.14B**. The

pulses can be positive or negative, depending on the polarity of the signal. The modulating signal can be recovered simply by low-pass filtering the pulse train. The effect of the negative pulses is to reverse the polarity of the RF signal. The result is very similar to a double-sideband, suppressed-carrier signal that is rapidly turned on and off at the pulse rate of the PAM. In other words, the DSBSC signal is periodically sampled at a certain pulse repetition rate.

For the signal to be properly represented, its highest modulating frequency must be less than the *Nyquist frequency*, which is one-half the sample rate. That condition, known as the *Nyquist criterion*, applies not only to PAM but to any digital modulation technique.

There are two variations of PAM that should be mentioned. *Natural sampling* does not hold the amplitude of each pulse constant throughout the pulse as shown in Fig 8.14B, but rather follows the shape of the analog modulation. The *single-polarity method* adds a fixed offset, or pedestal, to the modulating signal before the PAM modulator. As long as the pedestal is greater than or equal to the peak negative modulation, then the RF phase never changes and the signal is equivalent to sampled full-carrier AM rather than DSBSC.

PAM is rarely used on the amateur bands because it increases the transmitted bandwidth and adds circuit complexity with no improvement in signal-to-noise ratio for most types of noise and interference. The concept is useful, however, because PAM is similar to the signal generated by the sample-and-hold circuit that is used at the input to an analog-to-digital converter (ADC). An ADC is used in virtually every digital transmitter to convert the analog voice signal to a digital signal suitable for digital signal processing.

Pulse-width modulation (PWM) is a series of pulses whose width varies in proportion to the amplitude of the modulating signal. Fig 8.14C shows the pulses centered on the sample times, but in some systems the sample times may correspond to the leading or trailing pulse edges. With either method, the modulating signal can be recovered by low-pass filtering the pulse train and passing it through a coupling capacitor to remove the DC component.

Pulse-position modulation (PPM), Fig 8.14D, varies the position, or phase, of the pulses in proportion to the amplitude of the modulating signal. With both PWM and PPM, the peak amplitude of the signal is constant. That allows the receiver to be designed to be insensitive to amplitude variations, which can result in a better post-detection signal-to-noise ratio, in a manner similar to analog angle modulation.

8.3.2 Angle-Modulated Digital Signals

Just as with analog modulation, angle-modulated digital signals have constant amplitude and vary the phase and frequency to represent the modulating signal. They share some of the same advantages as their ana-log counterparts. The receiver can be made relatively insensitive to amplitude noise and on-channel interference. The constant-amplitude signal simplifies the job of the receiver's automatic-gain-control system. The transmitter does not require a linear amplifier, so a more-efficient class-C amplifier may be used. However digital angle modulation also shares analog's nonlinearity so that the bandwidth is not as well-constrained as it is with linear modulation systems.

Frequency-shift keying (FSK) was the first digital angle-modulated format to come into common use. FSK can be thought of as being the same as analog FM that is modulated with a binary (two-level) signal that causes the RF signal to switch between two frequencies. It can also be thought of as equivalent to two on-off-keyed (OOK) signals on two nearby frequencies that are keyed in such a way that whenever one is on the other is off. Just as with OOK, if the transitions between states are instantaneous, then excessive bandwidth occurs — causing interference on nearby channels similar to key clicks. For that reason, the modulating signal must be low-pass filtered to slow down the speed at which the RF signal moves from one frequency to the other.

Although FSK is normally transmitted as a true constant-amplitude signal with only the frequency changing between symbols, it does not have to be received that way. The receiver can treat the signal as two OOK signals and demodulate each one separately. This is an advantage when HF propagation conditions exhibit selective fading — even if one frequency fades out completely the receiver can continue to copy the other, a form of *frequency diversity*. To take advantage of it, wide shift (850 Hz) must normally be used. With narrow-shift FSK (170 Hz), the two tones are generally too close to exhibit selective fading.

As previously discussed, selective fading is caused by the signal arriving at the receiver antenna by two or more paths simultaneously. The same phenomenon can cause another signal impairment known as *inter-symbol interference*. See **Fig 8.15.** If the difference in the two path lengths is great enough, then the signal from one path may arrive delayed by one entire symbol time with respect to the other. The receiver sees two copies of the signal that are time-shifted by one symbol. In effect, the signal interferes with itself. One solution is to slow down the baud rate so that the symbols do not overlap. It is for this reason that symbol rates employed on HF are usually no more than 50 to 100 bauds.

Multi-level FSK (MFSK) is one method to reduce the symbol rate. Unlike with conventional binary FSK, more than two shift frequencies are allowed. For example, with eight frequencies, each symbol can have eight possible states. Since three bits are required to represent eight states, three bits are transmitted per symbol. That means you get three times the data rate without increasing the baud rate. The disadvantage is a reduced signal to noise and signal to interference ratio. If the maximum deviation is the same, then the frequencies are seven times closer with 8FSK than with binary FSK and the receiver is theoretically seven times (16.9 dB) more susceptible to noise and interference. However the IF limiter stage removes most amplitude variations before the signal arrives at the FSK detector, so the actual increase in susceptibility is less for signal levels above a certain threshold.

With any type of FSK, you can theoretically make the shift as narrow as you like. The main disadvantage is that the receiver becomes more susceptible to noise and interference, as explained above. In addition, the bandwidth is not reduced as much as you might expect. Just as with analog FM, you still get sidebands whose extent depends on the symbol rate, no matter how small the deviation.

Minimum-shift keying (MSK) is FSK with a deviation that is at the minimum practical level, taking bandwidth and signal-to-noise ratio into account. That turns out to be a frequency shift from the center frequency of 0.25 times the baud rate or, using the common definition of frequency shift, a difference between the two tones of 0.5 times the baud rate. If, on a given symbol, the frequency is shifted to the upper tone, then the phase of the RF signal will change by 0.25 cycles, or 90°, during the symbol period. If the lower tone is selected, the phase shift is –90°. Thus, MSK may be regarded as either FSK with a frequency shift of 0.5 times the symbol rate, or as *differential phase-shift keying* (DPSK) with a phase shift of ±90°. The binary data cause the phase to change by either +90 or –90° from one symbol to the next.

Gaussian minimum-shift keying (GMSK) refers to MSK where the modulating signal has been filtered with a low-pass filter that has a Gaussian frequency response. As mentioned before, with any type of angle modulation the spectrum of the modulation is not duplicated at RF, but spreads out into an increased band-

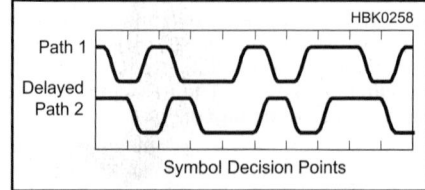

Fig 8.15 — Multipath propagation can cause inter-symbol interference (ISI). At the symbol decision points, which is where the receiver decoder samples the signal, the path 2 data is often opposing the data on path 1.

width. For that reason, there is no point in using a modulation filter with a sharp cutoff since the RF spectrum will be wider anyway. It turns out that a Gaussian filter, with its gradual transition from passband to stopband, has the optimum shape for an angle-modulated digital system. However, a Gaussian filter also has a gradual transition from symbol-to-symbol in the time domain as well. The transition is not totally completed by the time of the next symbol, which means that there is some inter-symbol interference in the transmitted signal, even in the absence of propagation impairments. With the proper choice of filter bandwidth, however, the ISI is small enough not to seriously affect performance.

Binary phase-shift keying (BPSK), often referred to simply as phase-shift keying (PSK), is included in this section because the name suggests that it is true constant-amplitude angle modulation. It is possible to implement it that way. An example is MSK which, as described above, can be considered to be differential BPSK with a ±90° phase shift. However the term BPSK is normally understood to refer to phase-shift keying with a 180° phase difference between symbols. To transition from one state to the other, the modulation filter smoothly reduces the amplitude to zero where the polarity reverses (phase changes 180°) before smoothly ramping up to full amplitude again. For that reason,

BPSK as usually implemented really should not be considered to be PSK, but rather a form of amplitude-shift keying (ASK) with two modulation amplitudes, +1 and −1. Unlike true angle modulation, it is linear so that the spectrum of the modulation filter is duplicated at RF. The transmitter must use a linear amplifier to prevent distortion and excessive bandwidth similar to the splatter that results from an over-driven SSB transmitter.

8.3.3 Quadrature Modulation

Quadrature modulation encodes digital signals using a combination of amplitude and phase modulation. With two types of modulation to work with, it is possible to cram more data bits into each modulation symbol, which allows more throughput for a given bandwidth. Modulation formats in use today have up to 8 or more bits per symbol which would be impractical for ASK, FSK or PSK alone.

Since quadrature-modulated symbols are defined by both amplitude and phase, the most common way to represent symbol states is with a polar plot, called a *constellation diagram*. See **Fig 8.16**, which shows a *four-level quadrature amplitude modulation* (4-QAM) signal, often referred to simply as QAM. The distance of each state from the origin represents the amplitude. The phase angle with

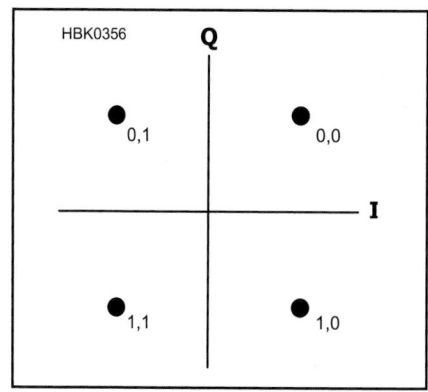

Fig 8.16 — Constellation diagram of a 4-QAM signal, also known as QPSK. The four symbol locations all have the same amplitude and have phase angles of 45°, 135°, −135° and −45°. 4-QAM is a two-bits-per-symbol format. The four symbols are selected by the four possible states of the two data bits, which can be assigned in any order. With the assignment shown, the Q value depends only on the first bit and the I value depends only on the second, an arrangement that can simplify symbol encoding.

respect to the +I axis represents the phase. The four states shown have phase angles of +45°, +135°, −45° and −135°. Normally, the receiver has no absolute phase information and can only detect phase differences between

Digital TV Modulation

DTV operates on the same 6 MHz-wide channels used for analog TV. However, instead of each channel slot carrying an analog NTSC television signal (visual carrier, color subcarrier and aural carrier), the channel slot carries an 8-VSB digitally modulated signal. The 6 MHz-wide over-the-air channel slots themselves didn't change, just the signals carried in them! So, channel 2 is still 54-60 MHz, channel 3 is 60-66 MHz, and so forth, up to channel 51. Channels 2-6 have largely been abandoned in the US as stations moved to broadcast digital TV on UHF channels. UHF channels 52-69 will be reallocated to other uses.

The designation "8-VSB" refers to 8-level vestigial sideband modulation. This is similar to 256-QAM, which means 256-state *quadrature amplitude modulation* — the 256 "states" are 256 combinations of signal phase and amplitude values that represent the 256 different transmitted symbols — a digital format used in cable TV networks (64-QAM is also used by cable companies). In the case of 8-VSB, the "8" refers to the eight-level baseband DTV signal that amplitude modulates an IF signal. For more info about 8-VSB modulation, see the online article, "What Exactly Is 8-VSB Anyway?" by David Sparano at **www.arrl.org/files/file/Technology/TV_Channels/8_Bit_VSB. pdf** and other references at **www.arrl.org/tv-channels-air-catv**.

Digitally modulated signals used to transport video — whether 8-VSB for over-the-air or 64-QAM or 256-QAM in cable networks — are noise-like signals over their 6 MHz bandwidth. If a digital TV signal interferes with analog radio communications such as FM or SSB, the effect is similar to degraded signal-to-noise ratio. Indeed, a digital TV signal can be thought of as a 6 MHz wide "pile of noise." On a spectrum analyzer, it looks like a "haystack" as shown in **Fig 8.A** Of course, the IF bandwidth

of the amateur receiver is quite narrow relative to the 6 MHz-wide digital TV channel, so the actual noise level seen by the receiver (assuming no overload or IMD problems) will in part be determined by the receiver's IF bandwidth. Still, the effect on the receiver is what amounts to an elevated noise floor, similar to the effects of wide-spectrum broadband noise.

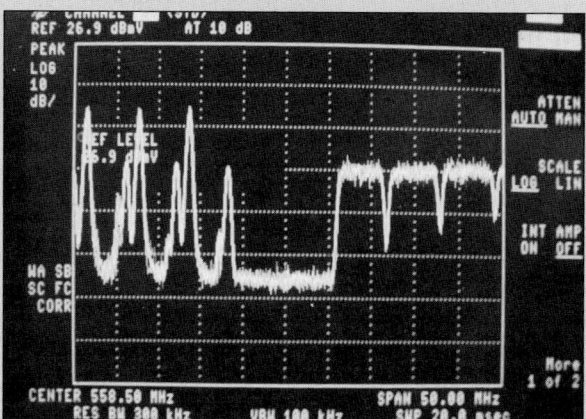

Fig 8.A — The spectrum of three analog NTSC TV channels (left of center) and three QAM signals (right of center). The digital channel power of the QAM "haystacks" is about 6 dB below the analog visual carrier PEP.

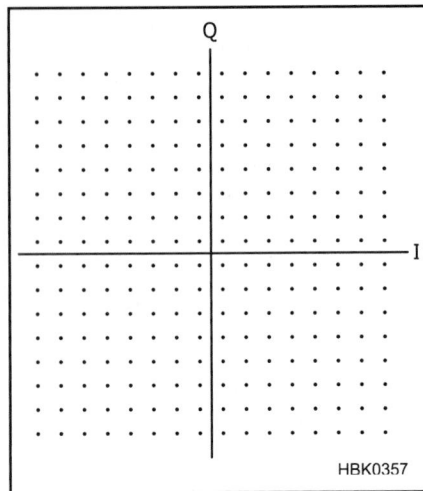

Fig 8.17 — Constellation diagram of a 256-QAM signal. Since an 8-bit number has 256 possible states, each symbol represents 8 bits of data.

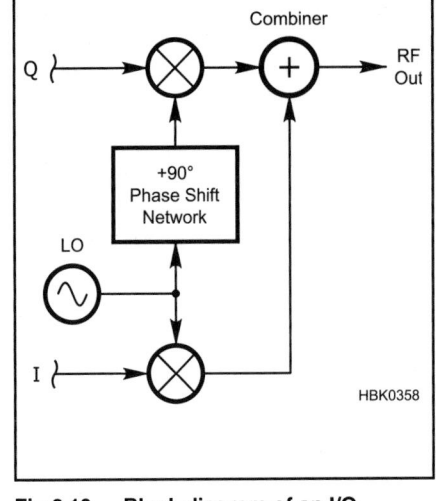

Fig 8.18 — Block diagram of an I/Q modulator. Note the similarity to Fig 8.7. By connecting an audio 90° phase-shift network to the I and Q inputs, an I/Q modulator can generate an analog SSB signal by the phasing method.

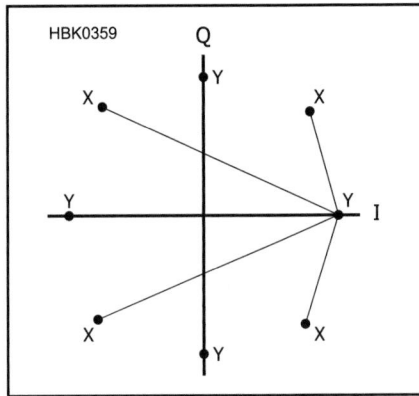

Fig 8.19 — Constellation diagram of a π/4 DQPSK signal. There are eight possible symbol locations. If the current symbol location is at one of the four positions labeled "X" then the next symbol will be at one of the four locations labeled "Y" and vice versa. That guarantees that no possible symbol trajectories (the lines between symbol locations) can ever pass through the origin.

the states. For that reason, we could just as easily have drawn the four states directly on the I and Q axes, at 0°, +90°, 180° and –90°. Note that each of the four states has the same amplitude, differing only in phase. For that reason 4-QAM is often referred to as *quadrature phase-shift keying* (QPSK), in the same manner that two-level ASK is normally referred to as BPSK.

Since 4-QAM has four possible states, there are two data bits per symbol. However QAM is not limited to four states. **Fig 8.17** illustrates a 256-QAM signal. It takes 8 bits to represent 256 states, so 256-QAM packs 8 data bits into each symbol. The disadvantage of using lots of states is that the effect of noise and interference is worse at the receiver. Since, for the same peak power, the states of a 256-QAM signal are 15 times closer together than with 4-QAM, the receiver's decoder has to determine each symbol's location to 15 times greater accuracy. That means the ratio of peak power to noise and interference must be 20 log(15) = 23.5 dB greater for accurate decoding. That is why QAM with a large number of states is normally not used on the HF bands where fading, noise and interference are common. A more common application is digital cable television where the coaxial-cable transmission channel is much cleaner. For example, the European DVB-C standard provides for 16-QAM through 256-QAM, depending on bandwidth and data rate.

There are at least two ways to generate a QAM signal. One is to combine a phase modulator and an amplitude modulator. The phase modulator places the symbol location at the correct phase angle and the amplitude modulator adjusts the amplitude so the symbol is the correct distance from the origin. That method is seldom used, both because

of the circuit complexity and because the nonlinear phase modulator makes the signal difficult to filter so as to limit the bandwidth.

A much more common method of generating QAM is with an *I/Q modulator*. See **Fig 8.18**. By using two modulators (mixers) fed with RF sine waves in quadrature (90° out of phase with each other), any amplitude and phase may be obtained by varying the amplitudes of the two modulation inputs. The input labeled I (for in-phase) moves the symbol location horizontally in the constellation diagram and the one labeled Q (for quadrature) moves it vertically. For example, to obtain an amplitude of 1 and a phase angle of –45°, set I to +0.707 and set Q to –0.707.

It is possible to generate virtually any type of modulation using an I/Q modulator. For example, to generate BPSK or on-off keying, simply disconnect the Q input and apply the modulation to I. For angle modulation, such as FM or PM, a waveform generator applies a varying signal to I and Q in such a manner to cause the symbol to rotate at constant amplitude with the correct phase and frequency (rotation rate). Even a multi-carrier signal may be generated with a single I/Q modulator by applying the sum of a number of signals, each representing one carrier, to the I and Q inputs. The phase of each signal rotates at a rate equal to the frequency offset of its carrier from the center.

One problem with QAM is that whenever the signal trajectory between two symbol states passes through the origin, the signal amplitude momentarily goes to zero. That imposes stringent linearity requirements on the RF power amplifier, since many amplifiers exhibit their worst linearity near zero

power. One solution is *offset QPSK* (OQPSK) modulation. In this case, the symbol transitions of the I and Q channels are offset by half a symbol. That is, for each symbol, the I channel changes state first then the Q channel changes half a symbol time later. That allows the symbol trajectory to sidestep around the origin, allowing use of a higher-efficiency or lower-cost power amplifier that has worse linearity.

Another solution to the zero-crossing problem is called *PI over 4 differential QPSK* (π/4 DQPSK). See **Fig 8.19**. This is actually a form of 8-PSK, where the eight symbol locations are located every 45° around a constant-amplitude circle. On any given symbol, however, only four of the symbol locations are used. The symbol location always changes by an odd multiple of 45° (π/4 radians). If the current symbol is located on the I or Q axis, then on the next symbol only the four non-axis locations are available and vice versa. As with OQPSK, that avoids transitions that pass through the origin.

Another advantage of π/4 DQPSK which is shared with other types of differential modulation is that absolute phase doesn't matter. The information is encoded only in the *difference* in the phase of successive symbols. That greatly simplifies the job of the receiver's demodulator.

The block diagram of an I/Q demodulator looks like an I/Q modulator drawn backward. See **Fig 8.20**. If the local oscillator is not tuned to exactly the same frequency as the one in the transmitter (after downconversion to the IF, assuming a superhet receiver) then the demodulated signal will rotate in the I/Q plane at a rate equal to the frequency error.

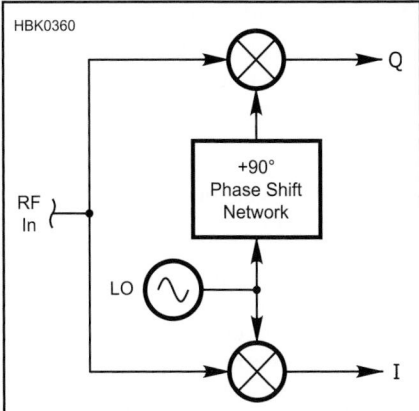

Fig 8.20 — Block diagram of an I/Q demodulator.

Most receivers include a *carrier-recovery* circuit, which phase-locks the local oscillator to the average frequency of the incoming signal to obtain stable demodulation. While that corrects the frequency error, it does not correct the phase, which must be accounted for in some other manner such as by using differential modulation or some kind of symbol-recovery mechanism.

8.3.4 Multi-carrier Modulation

An effective method to fit more data bits into each symbol is to use more than one separately-modulated signal at a time, each on its own carrier frequency spaced an appropriate distance from the frequencies of the other carriers. An example is multi-carrier FSK. This is not to be confused with MFSK which also uses multiple frequencies, but only one at a time. With multi-carrier FSK, each carrier is present continuously and is frequency-shifted in response to a separate data stream. The total data rate equals the data rate of one carrier times the number of carriers. A disadvantage of multi-carrier FSK is that the resulting signal is no longer constant-amplitude—a linear amplifier must be used. In general, multi-carrier signals using any modulation type on each carrier tend to have high peak-to-average power ratios.

In the presence of selective fading, one or more of the carriers may disappear while the others are still present. An advantage of multi-carrier modulation is that error-correcting coding can use the unaffected carriers to reconstruct the missing data. Also, since each carrier signal is relatively narrowband, propagation conditions are essentially constant within that bandwidth. That makes it easier for the receiver to correct for other frequency-selective propagation impairments such as phase distortion. If a single-carrier signal of the same total bandwidth had been used instead, the receiver would need an adaptive equalizer to correct for the amplitude and phase variations across the transmission channel.

By using multiple carriers each with multiple-bit-per-symbol modulation it is possible to obtain quite high data rates while maintaining the low symbol rates that are required to combat the effects of multi-path

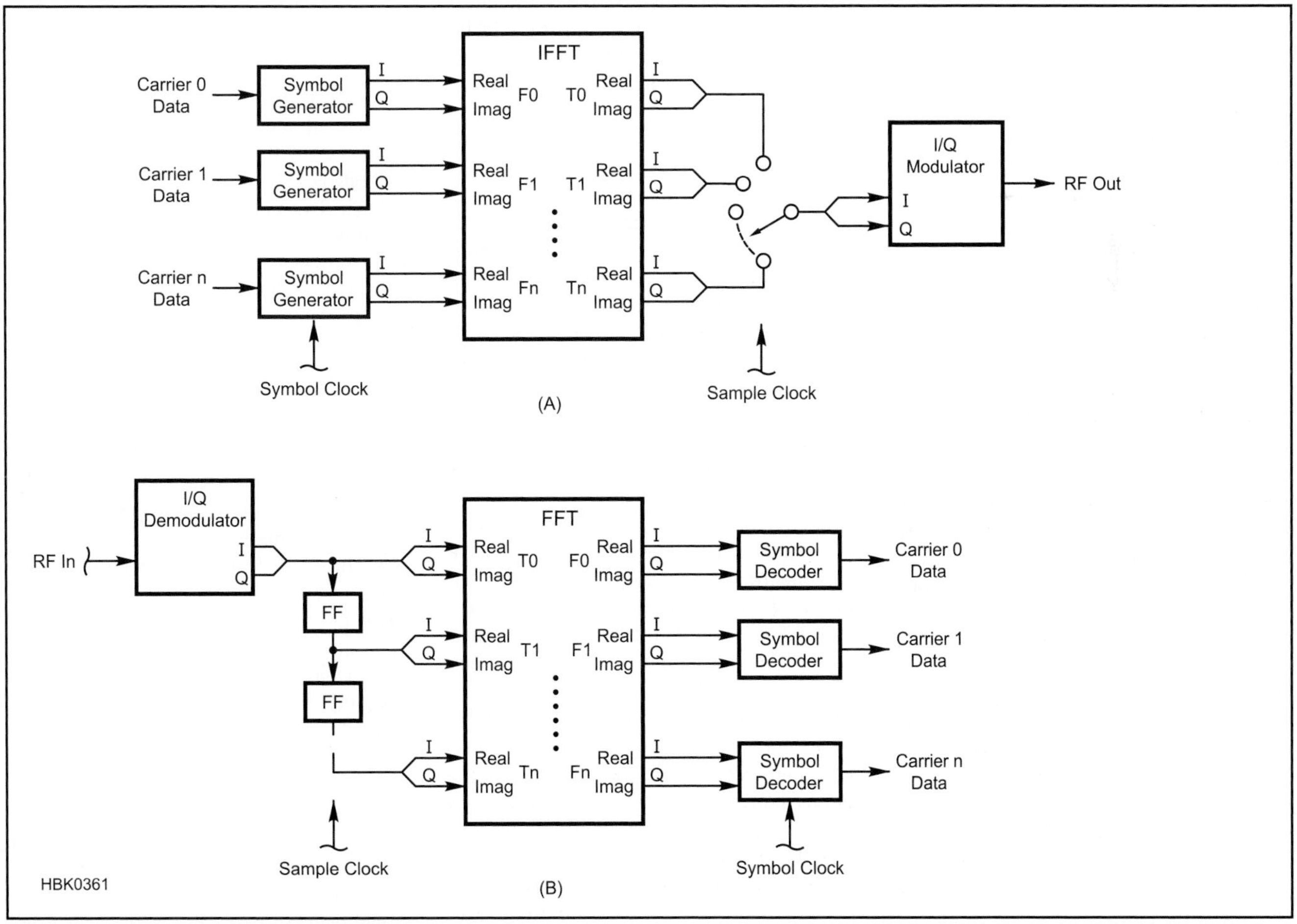

Fig 8.21 — Block diagram of an OFDM modulator (A) and demodulator (B) using the FFT/IFFT technique. The number of carriers is n, and the sample rate is n times the symbol rate. Once per symbol, all the symbol generators in the modulator are loaded with new data and the inverse fast-Fourier transform (IFFT) generates n output samples, which are selected in succession by the switch. In the demodulator, n samples are stored in a shift register (string of flip-flops) for each symbol, then the FFT generates one "frequency" output for each carrier frequency. From the amplitude and phase of each frequency, the symbol decoders can determine the symbol locations and thus the data.

propagation on the HF bands. For example, PACTOR-III achieves a raw data rate of 3600 bits per second with a 100-baud symbol rate using 18 carriers of DQPSK. (100 bauds × 2 bits/symbol × 18 carriers = 3600 bits per second.) Similarly, Clover-2000 modulation gets 3000 bits per second with a 62.5-baud symbol rate using eight carriers of 16-DPSK combined with 4-level DASK. (62.5 bauds x (4+2) bits/symbol × 8 carriers = 3000 bits per second.) Decoding is rather fragile using these complex modulation techniques, so PACTOR and Clover include means to automatically switch to simpler, more-robust modulation types as propagation conditions deteriorate.

What is the minimum carrier spacing that can be used without excessive interference between signals on adjacent frequencies? The answer depends on the symbol rate and the filtering. It turns out that it is easy to design the filtering to be insensitive to interference on frequencies that are spaced at integer multiples of the symbol rate. (See the following section on filtering and bandwidth.) For that reason, it is common to use a carrier spacing equal to the symbol rate. The carriers are said to be *orthogonal* to each other since each theoretically has zero correlation with the others.

Orthogonal frequency-division multiplexing (OFDM), sometimes called *coded OFDM* (COFDM), refers to the multiplexing of multiple data streams onto a series of such orthogonal carriers. The term usually implies a system with a large number of carriers. In that case, an efficient decoding method is to use a DSP algorithm called the *fast Fourier transform* (FFT). The FFT is the software equivalent of a hardware spectrum analyzer. It gathers a series of samples of a signal taken at regular time intervals and outputs another series of samples representing the frequency spectrum of the signal. See **Fig 8.21B**. If the length of the series of input samples equals one symbol time and if the sample rate is selected properly, then each frequency sample of the FFT output corresponds to one carrier. Each frequency sample is a complex number (containing a "real" and "imaginary" part) that represents the amplitude and phase of one of the carriers during that symbol period. Knowing the amplitude and phase of a carrier is all the information required to determine the symbol location in the I/Q diagram and thus decode the data. At the transmitter end of the circuit, an *inverse FFT* (IFFT) can be used to encode the data, that is, to convert the amplitudes and phases of each of the carriers into a series of I and Q time samples to send to the I/Q modulator. See Fig 8.21A.

One advantage of OFDM is high spectral efficiency. The carriers are spaced as closely as theoretically possible and, because of the narrow bandwidth of each carrier, the overall spectrum is very square in shape with a sharp drop-off at the passband edges. One disadvantage is that the receiver must be tuned very accurately to the transmitter's frequency to avoid loss of orthogonality, which causes cross-talk between the carriers.

8.3.5 AFSK and Other Audio-Based Modulation

Audio frequency-shift keying (AFSK) is the generation of radio-frequency FSK using an audio-frequency FSK signal fed into the microphone input of an SSB transmitter. Assume the SSB transmitter is tuned to 14.000 MHz, USB. If the audio signal consists of a sine wave that shifts between 2125 Hz and 2295 Hz (170-Hz frequency shift), then the RF signal is a sine wave shifting between 14.002125 and 14.002295 MHz. The frequency shift and spectral characteristics are theoretically unchanged, other than being translated 14 MHz upward in frequency. The RF signal should be indistinguishable from one generated by varying the frequency of an RF oscillator directly.

This technique works not only for FSK but also for nearly any modulation type with a bandwidth narrow enough to fit within the passband of an SSB transmitter and receiver. The most common non-voice analog modulation type to use this technique on the amateur bands is *slow-scan television* (SSTV), which uses frequency modulation for the video signal. In addition, nearly all narrowband digital signals today are generated in this manner. The audio is generated and received either from a dedicated hardware modulator/demodulator (modem) or using the sound card that is found in virtually all personal computers.

Whatever the method, it is important to ensure that unwanted interference is not caused by audio distortion or by insufficient suppression of the carrier and unwanted sideband. For example, with the AFSK tone frequencies mentioned above, 2125 and 2295 Hz, the tone harmonics cause no trouble because they fall outside of the transmitter's passband. However, some AFSK modems use 1275 and 1445 Hz (to accommodate 850-Hz shift without changing the 1275-Hz mark frequency). In that case, the second harmonics at 2550 and 2890 Hz must be suppressed since those frequencies are not well-attenuated by the transmitter. With non-constant-envelope modulation types such as QPSK or the various multi-carrier modes, it is important to set the amplitude of the audio input to the SSB transmitter below the level that activates the transmitter's automatic level control (ALC). That is because the ALC circuit itself generates distortion of signals within the bandwidth of its feedback loop.

8.3.6 Spread Spectrum

A *spread-spectrum* (SS) system is one that intentionally increases the bandwidth of a digital signal beyond that normally required by means of a special spreading code that is independent of the data sequence. There are several reasons for spreading the spectrum in that way.

Spread spectrum was first used in military systems, where the purpose was to encrypt the transmissions to make it harder for the enemy to intercept or jam them. Amateurs are not allowed to encrypt transmissions for the purpose of concealing the information, but reducing interference, intentional or otherwise, is an obvious benefit. The signal is normally spread in such a fashion that it appears like random noise to a receiver not designed to receive it, so other users of the band may not even be aware that an SS signal is present.

Another advantage to spreading the spectrum is that it can make frequency accuracy less critical. In addition, the wide bandwidth means that expensive narrow-bandwidth filters are not required in the receiver. It also provides a measure of frequency diversity. If certain frequencies are unusable because of interference or selective fading, the signal can often be reconstructed using information in the rest of the bandwidth.

There are several ways to spread the spectrum — we will cover the two most common methods below — but they all share certain characteristics. Imagine that the unspread signal occupies a 10 kHz bandwidth and it is spread by a factor of 100. The resulting SS signal is 1000 kHz (1 MHz) wide. Each 10-kHz channel contains 1/100 of the total signal power, or −20 dB. That means that any narrowband stations using one of those 10-kHz channels experience a 20 dB reduction in interference, but also are more likely to be interfered with because of the 100-times greater bandwidth of the SS station's emissions.

How is the spread-spectrum station affected by interference from narrowband stations? In effect, the SS receiver attenuates the signal received on each 10 kHz channel by 20 dB in order to obtain a full-power signal when all 100 channels are added together. That means that the interference from a narrowband station is reduced by 20 dB but, again, the interference is more likely to occur because of the 100-times greater bandwidth of the SS station's receiver.

How is the spread-spectrum station affected by interference from another SS station on the same frequency? It turns out that if the other station is using a different orthogonal spreading code then, once again, the interference reduction is 20 dB for 100-times spreading. That means that many SS stations can share the same channel without interference as long as they are all received at roughly the same signal level. Commercial mobile-telephone SS networks use an elaborate system of power control with real-time feedback to ensure that the signals from all the mobile stations arrive at the base station at approximately the same level.

That scheme works well in a one-to-many (base station to mobile stations) system architecture but would be much more difficult to implement in a typical amateur many-to-many arrangement because of the different distances and thus path losses between each pair of stations in the network. On the HF bands it is not uncommon to see differences in signal levels of 80 to 90 dB or more. (For example, the difference between S1 and 40 dB over S9 is 88 dB, assuming 6 dB per S-unit.) A spread spectrum signal at S9 + 40 dB with a spreading ratio of 100 times would interfere with any other signals below about S9 + 20 dB. It works the same in the other direction as well. The SS signal would experience interference from any other stations that are more than 20 dB louder than the desired signal.

Normally, increasing the bandwidth of a transmission degrades the signal-to-noise (S/N) ratio at the receiver. A 100-times greater bandwidth contains 100 times as much noise, which causes a 20 dB reduction in S/N ratio. However SS receivers benefit from a phenomenon known as *processing gain*. Just as the receiver is insensitive to other SS signals with different orthogonal spreading codes, so it is insensitive to random noise. The improvement in S/N ratio due to processing gain is:

$$\text{Processing Gain} = 10 \times \log\left(\frac{\text{spread bandwidth}}{\text{unspread bandwidth}}\right) \text{ dB} \quad (3)$$

That is exactly equal to the reduction in S/N ratio due to the increased bandwidth. The net result is that an SS signal has neither an advantage nor a disadvantage in signal-to-noise ratio compared to the unspread version of the same signal. When someone states that, because of processing gain, an SS receiver can receive signals that are below the noise level (signals that have a negative S/N ratio), that is a true statement. However, it does not imply better S/N performance than could be obtained if the signal were not spread.

FREQUENCY HOPPING SPREAD SPECTRUM

One simple way to spread the spectrum of a narrowband signal is to repetitively sweep it across the frequency range of the wider spectrum, either continuously or in a series of steps at discrete frequencies. That technique, called *chirp modulation*, can be considered a special case of *frequency-hopping spread spectrum* (FHSS), in which the narrowband signal covers the expanded spectrum by rapidly hopping back and forth from frequency to frequency in a pseudo-random manner. On average, each frequency is used the same percentage of time so that the average spectrum is flat across the bandwidth of the FHSS signal. See **Fig 8.22**.

If the receiver hops in step with the transmitter, using the same pseudo-random sequence synchronized to the one in the transmitter, then the transmitter and receiver are always tuned to the same frequency and the receiver's detector sees a continuous, non-hopped narrowband signal which can be demodulated in the normal way. We say that the signal has been *de-spread*, that is, returned to its normal narrowband form. Synchronization between the receiver and transmitter is one of the challenges in an FHSS system. If the timing of the two sequences differs by even one hop, then the receiver is always tuned to the wrong frequency, unless the same frequency happens to occur twice in succession in the pseudo-random sequence. Any signal with an unsynchronized sequence, or with a different sequence, is reduced in amplitude by the processing gain.

There are two types of FHSS based on the rate at which the frequency hops take place. *Slow-frequency hopping* refers to a hop rate slower than the baud rate. Several symbols are sent per hop. With *fast-frequency hopping*, the hop rate is faster than the baud rate. Several hops occur during each symbol. The term *chip* refers to the shortest-duration modulation state in the system. For slow-frequency hopping, that is the baud rate. For fast-frequency hopping it is the hop rate. Fast-frequency hopping can be useful in reducing the effects of multi-path propagation. If the hop period is less then the typical time delay of secondary propagation paths, then those signals are uncorrelated to the main path and are attenuated by the processing gain.

A diagram of a frequency-hopping spread spectrum system is shown in **Fig 8.23**. In

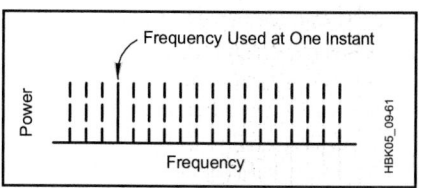

Fig 8.22 — Power versus frequency for a frequency-hopping spread spectrum signal. Emissions jump around to discrete frequencies in pseudo-random fashion. Normally the spacing of the frequencies is approximately equal to the bandwidth of the unspread signal so that the average spectrum is approximately flat.

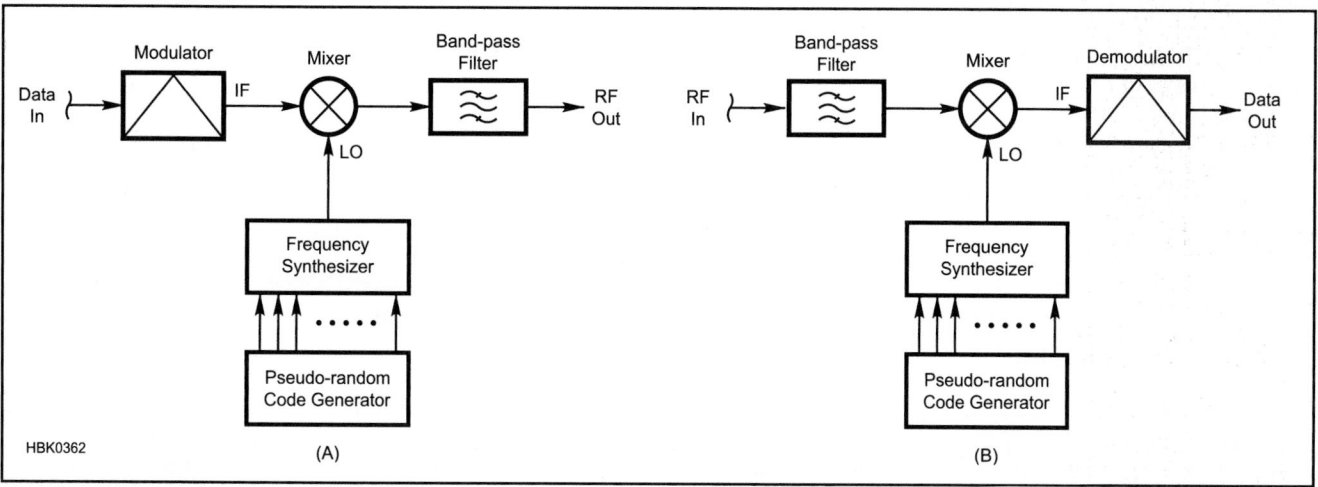

Fig 8.23 — Block diagram of an FHSS transmitter (A) and receiver (B). The receiver may be thought of as a conventional superhet with a local oscillator (LO) that is continually hopping its frequency in response to a pseudo-random code generator. The transmitter has a similar architecture to up-convert a conventionally-modulated intermediate frequency (IF) to a frequency-hopped radio frequency (RF).

both the transmitter and the receiver, a pseudo-random code generator controls a frequency synthesizer to hop between frequencies in the correct order. In this way, the narrowband signal is first spread by the transmitter, then sent over the radio channel, and finally de-spread at the receiver to obtain the original narrowband signal again. One issue with FHSS is that many synthesizers do not maintain phase coherence over successive frequency hops. That means the basic (non-spread) modulation must be a type that does not depend on phase information. That rules out PSK, QPSK and QAM. That is the reason that modulation types that do not depend on phase, such as FSK and MFSK with non-coherent detection, are frequently used as the base modulation type in FHSS systems.

DIRECT SEQUENCE SPREAD SPECTRUM

Whereas an FHSS system hops through a pseudo-random sequence of frequencies to spread the signal, a *direct-sequence spread spectrum* (DSSS) system applies the pseudo-random sequence directly to the data in order to spread the signal. See **Fig 8.24**. The binary data is considered to be in *polar* form, that is, the two possible states of each bit are represented by −1 and +1. The data bits are multiplied by a higher-bit-rate pseudo-random sequence, also in polar form, which chops the data into smaller time increments called *chips*. The ratio of the chip rate to the data's bit rate equals the ratio of the spread bandwidth to the unspread bandwidth, which is just the processing gain:

$$\text{Processing Gain} = 10 \times \log\left(\frac{\text{chip rate}}{\text{bit rate}}\right) \text{ dB} \quad (4)$$

Although it doesn't have to, DSSS nor-

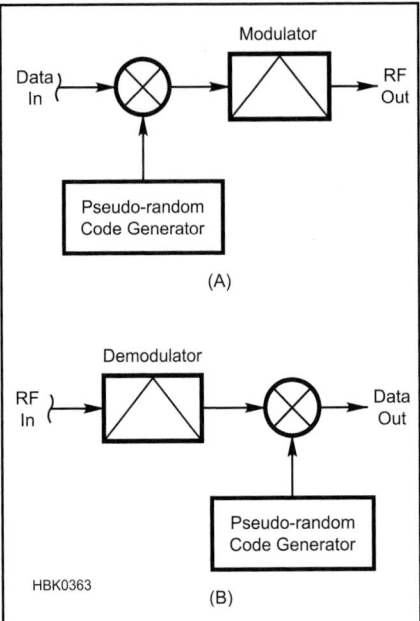

(A)

(B)

HBK0363

Fig 8.24 — Block diagram of a DSSS transmitter (A) and receiver (B). For BPSK modulation, the modulation has the same format as the bipolar data (±1), so the modulator and demodulator could be moved to the other side of the multiplier if desired.

mally uses a one-bit-per-symbol modulation type such as BPSK. In that case, the modulator and demodulator in Fig 8.24 would consist simply of a mixer, which multiplies the RF local oscillator by the bipolar DSSS modulating signal. Since unfiltered BPSK is constant-envelope, a nonlinear class-C power amplifier may be used for high efficiency. The demodulator in the receiver would also be a mixer, which multiplies the RF signal by a local oscillator to regenerate the DSSS modulating signal. Not shown are additional

mixers and filters that would be used in a superheterodyne receiver to convert the received signal to an intermediate frequency before demodulation and decoding.

The unfiltered spectrum has the form of a sinc function. That shows up clearly in the spectrum of an actual DSSS signal in **Fig 8.25A**, which is plotted on a logarithmic scale calibrated in decibels. The humps in the response to the left and right (and additional ones not shown off scale) are not needed for communications and should be filtered out to avoid excessive occupied bandwidth. Fig 8.25B shows the DSSS signal in the presence of noise at the input to the receiver, and Fig 8.25C illustrates the improvement in signal-to-noise ratio of the de-spread narrowband signal.

CODE-DIVISION MULTIPLE ACCESS (CDMA)

As mentioned before, to receive an SS signal, the de-spreading sequence in the receiver must match the spreading sequence in the transmitter. The term *orthogonal* refers to two sequences that are coded in such a way that they are completely uncorrelated. The receiver's response to an orthogonal code is the same as to random noise, that is, it is suppressed by a factor equal to the processing gain. One can take advantage of this property to allow multiple SS stations to access the same frequency simultaneously, a technique known as *code-division multiple access* (CDMA). Each transmitter is assigned a different orthogonal code. A receiver can "tune in" any transmitter's signal by selecting the correct code for de-spreading.

If multiple stations want to be able to transmit simultaneously without using spread spectrum, they must resort to either *frequency-division multiple access* (FDMA), where each station transmits on a different frequency channel, or *time-division multiple*

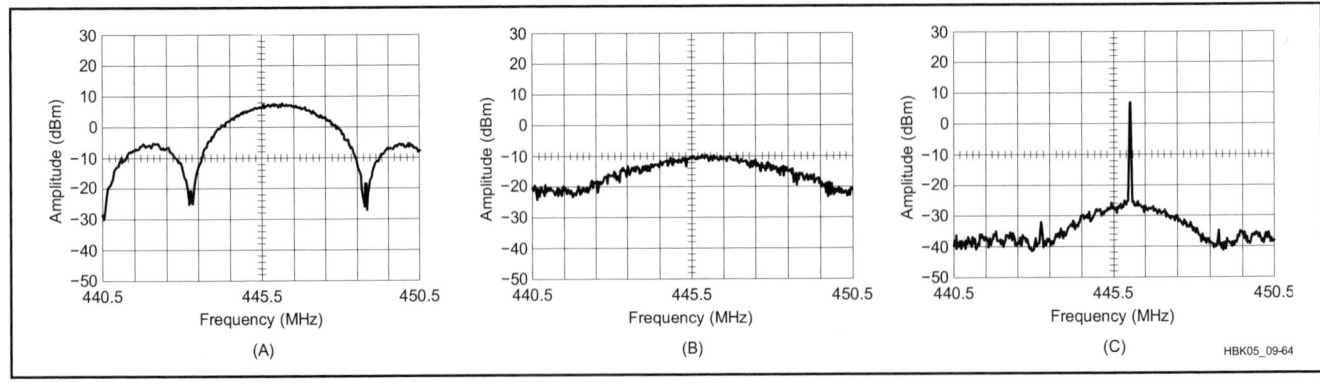

Fig 8.25 — (A) The frequency spectrum of an actual unfiltered biphase-modulated spread spectrum signal as viewed on a spectrum analyzer. In this practical system, band-pass filtering is used to confine the spread spectrum signal to the amateur band. (B) At the receiver end of the line, the filtered spread spectrum signal is apparent only as a 10-dB hump in the noise floor. (C) The signal at the output of the receiver de-spreader. The original carrier — and any modulation components that accompany it — has been recovered. The peak carrier is about 45 dB above the noise floor — more than 30 dB above the hump shown at B. (These spectrograms were made at a sweep rate of 0.1 s/division and an analyzer bandwidth of 30 kHz; the horizontal scale is 1 MHz/division.)

access (TDMA), where each transmission is broken up into short time slots which are interleaved with the time slots of the other stations. Compared to TDMA, CDMA has the advantage that it does not require an external synchronization network to make sure that different stations' time slots do not overlap. Compared to both TDMA and FDMA, CDMA has the further advantage that it experiences a gradual degradation in performance as the number of stations on the channel increases. It is relatively easy to add new users to the system. Also CDMA has inherent resistance to interference due to multi-path propagation or intentional jamming. The primary disadvantages of CDMA are the relative complexity and the necessity for accurate power-level control to make sure that unwanted signals do not exceed the level that can be rejected through processing gain.

8.3.7 Filtering and Bandwidth of Digital Signals

We have already touched on this topic in previous sections, but let us now cover it a little more systematically. The bandwidth required by a digital signal depends on the filtering of the modulation, the symbol rate and the type of modulation. For linear modulation types such as OOK, BPSK, QPSK and QAM, the bandwidth depends only on the symbol rate and the modulation filter.

As an example, an unfiltered BPSK modulating signal with alternating ones and zeroes for data (10101010...) is a square wave at one-half the symbol rate. Like any square wave, its spectrum can be broken down into a series of sine waves at the fundamental frequency (symbol rate / 2) and all the odd harmonics. If the data consists of alternating pairs of ones and pairs of zeroes (11001100...) then we have a square wave at one-fourth the symbol rate and the spectrum is a series of sine waves at one-fourth the symbol rate and all its odd harmonics. Random data contains energy at all frequencies from zero to half the symbol rate and all the odd harmonics of those frequencies. The harmonics are not needed for proper demodulation of the signal, so they can be filtered out with a low-pass filter with a cutoff frequency of one-half the symbol rate.

With random data, the shape of the unfiltered spectrum is a sinc function,

$$\text{sinc}(f / f_s) = \frac{\sin(\pi f / f_s)}{\pi f / f_s} \qquad (5)$$

where f_s is the symbol rate. See **Fig 8.26**. Note that the response is zero (minus infinity dB) whenever f is an integer multiple of the symbol rate. That is why multi-carrier modulation generally uses a carrier spacing equal to the symbol rate.

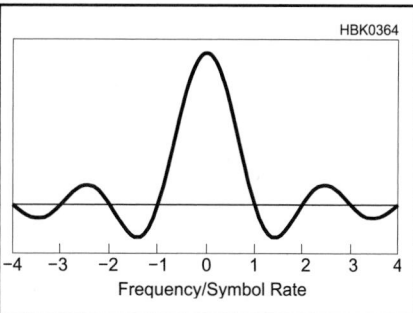

Fig 8.26 — The sinc function, which is the spectrum of an unfiltered BPSK modulating signal with random data, plotted with a linear vertical scale. The center (zero) point corresponds to the RF carrier frequency, To see what the double-sided RF spectrum looks like on a logarithmic (dB) scale, see Fig 8.25A.

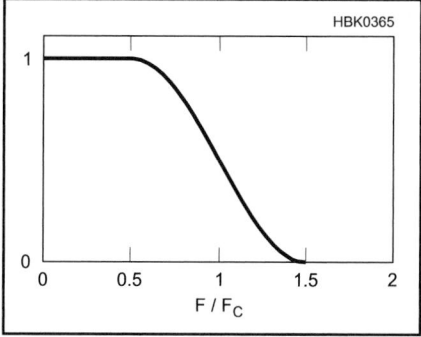

Fig 8.27 — Amplitude versus frequency for a 0.5-alpha, raised-cosine filter. The vertical scale is linear, not logarithmic as would be seen on a spectrum analyzer. The amplitude is 0.5 (−6 dB) at the cutoff frequency, Fc. The amplitude is 1.0 for frequencies less than Fc × (1 − alpha) and is 0.0 for frequencies above Fc × (1 + alpha).

The previous discussion applies to the baseband signal, before it modulates the RF carrier. BPSK is a double-sideband-type modulation so the baseband spectrum appears above and below the RF carrier frequency, doubling the bandwidth to 2 × (symbol rate / 2) = symbol rate. In reality, since no practical filter has an infinitely-sharp cutoff, the occupied bandwidth of a BPSK signal must be somewhat greater than the symbol rate. That is also true for all other double-sideband linear modulation types, such as OOK and the various forms of QPSK and QAM.

If the low-pass filtering is not done properly, it may slow down the transition between symbols to the point where one symbol starts to run into another, causing *inter-symbol interference* (ISI). A type of filter that avoids that problem is called a *Nyquist filter*. It ensures that each symbol's contribution to the

modulating signal passes through zero at the center of all other symbols, so that no ISI occurs. The most common type of Nyquist filter is the *raised-cosine* filter, so-called because the frequency response (plotted on a linear scale) in the passband-to-stopband transition region follows a raised-cosine curve. See **Fig 8.27**. The sharpness of the frequency cutoff is specified by a parameter called *alpha*. If alpha is 1.0, then the transition from passband to stopband is very gradual — it starts to roll off right at zero hertz and finally reaches zero response at two times the nominal cutoff frequency. An alpha of 0.0 specifies an ideal "brick-wall" filter that transitions instantaneously from full response to zero right at the cutoff frequency. Values in the range of 0.3 to 0.5 are common in communications systems.

Unfortunately, if any additional filter is placed before or after the Nyquist filter it destroys the anti-ISI property. In order to allow filtering in both the transmitter and the receiver many systems effectively place half the Nyquist filter in each place. Because the frequency response of each filter is the square root of the response of a Nyquist filter, they are called *root-Nyquist* filters. The *root-raised-cosine* filter is an example. While a Nyquist or root-Nyquist response theoretically could be approximated with an analog filter, they are almost always implemented as digital filters. More information on digital filters appears in the **DSP and Software Radio Design** chapter.

As mentioned before, filtering is more difficult with angle modulation because it is nonlinear. The RF spectrum is not a linear transposition of the baseband spectrum as it is with linear modes and Nyquist filtering doesn't work. Old-fashioned RTTY transmitters traditionally just used an R-C low-pass filter to slow down the transitions between mark and space. While that does not limit the bandwidth to the minimum value possible, the baud rate is low enough that the resulting bandwidth is acceptable anyway. In more modern systems, there is a tradeoff between making the filter bandwidth as narrow as possible for interference reduction and widening the bandwidth to reduce inter-symbol interference. For example, the GSM (Global System for Mobile communica-tions) standard, used for some cellular telephone systems, uses minimum-shift keying and a Gaussian filter with a BT (bandwidth symbol-time product) of 0.3. A 0.3 Gaussian filter has a 0.3 ratio of 3-dB bandwidth to baud rate, which results in a small but acceptable amount of ISI and a moderate amount of adjacent-channel interference.

CHANNEL CAPACITY

It is possible to increase the quantity of error-free data that can be transmitted over a communications channel by using an error-correcting code. That involves adding additional error-correction bits to the transmitted data. The more bits that are added, the greater the errors that can be corrected. However, the extra bits increase the data rate, which requires additional bandwidth, which increases the amount of noise. For that reason, as you add more and more error-correction bits, requiring more and more bandwidth, you eventually reach a point of limited additional return. In the 1940s, Claude Shannon worked out a formula (called the *Shannon-Hartley theorem*) for the maximum capacity possible over a communications channel, assuming a theoretically-perfect error-correction code:

$$C = B \log_2 \left(1 + \frac{S}{N}\right) \text{ bits/s} \qquad (6)$$

where

C = the net channel capacity, not including error-correcting bits,
B = the bandwidth in Hz, and
S/N = the signal-to-noise ratio, expressed as a power ratio.

Note that as B increases, N increases in the same proportion. **Fig 8.28** is a plot of channel capacity versus bandwidth based on the formula. As bandwidth is increased, channel capacity increases rapidly until the point where the S/N ratio drops to unity (labeled Bandwidth = 1 in the graph) after which channel capacity increases much more slowly.

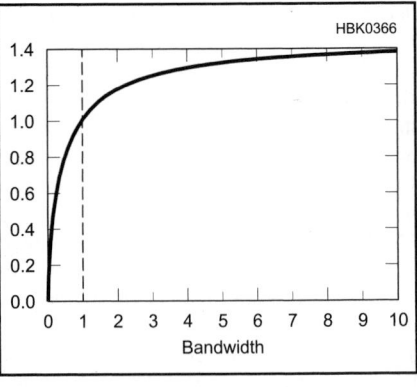

Fig 8.28 — Plot of channel capacity versus bandwidth, calculated by the Shannon-Hartley theorem. The S/N ratio has been selected to be unity at Bandwidth = 1.

8.4 Image Modulation

The following section covers the modulation aspects of amateur television and facsimile communications. More detailed information on protocols and operating standards can be found in the **Image Communications** supplement on the *Handbook* CD.

8.4.1 Fast-Scan Television

Amateur fast-scan television (ATV) is a wideband mode used in the amateur bands above 420 MHz. It is called "fast scan" to differentiate it from slow-scan TV.

The most popular ATV mode is based on NTSC, the same technical standard used by commercial analog television stations in the United States before most switched to digital TV in 2009. **Fig 8.29** shows the spectrum of an NTSC analog TV channel. It's basically a full-carrier, double-sideband AM signal, with filtering to partially remove the lower sideband. The partially-filtered *vestigial sideband* (VSB) extends 1.25 MHz below the carrier frequency. The channel is 6 MHz wide to accommodate the composite video and two subcarriers, one at 3.58 MHz for the color burst and the other an FM-modulated signal at 4.5 MHz for the sound. For simplicity, amateur stations often transmit unfiltered full-DSB AM, but the normally-removed portion of the lower sideband is unused by the TV receiver. Since less than 5% of the video energy appears more than 1 MHz below the carrier, little of the transmitter power is wasted. If needed, to reduce interference to other band users, a VSB filter in the antenna line can attenuate the lower sideband color and sound subcarrier frequencies by 20-30 dB.

The video signal includes pulses to synchronize the vertical and horizontal scan oscillators in the receiver. See **Fig 8.30**. The

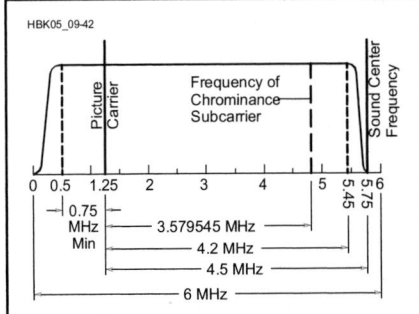

Fig 8.29 — An analog NTSC 6-MHz video channel with the video carrier 1.25 MHz up from the lower edge. The color subcarrier is at 3.58 MHz and the sound subcarrier at 4.5 MHz above the video subcarrier.

sync pulses and the "front porch" and "back porch" areas that bracket them are "blacker than black" so that the signal is blanked during retrace. The video-to-sync ratio must remain constant throughout all of the linear amplifiers in the transmit chain as the video level from the camera changes. To maintain the sync tips at 100% of peak power, the modulator usually contains a clamp circuit that also acts as a sync stretcher to compensate for amplifier gain compression.

Given NTSC's 525 horizontal lines and its 30 frames per second scan rate, the resulting horizontal resolution bandwidth is 80 lines per MHz. Therefore, with the typical TV set's 3-dB rolloff at 3 MHz (primarily in the IF filter), up to 240 vertical black lines can be

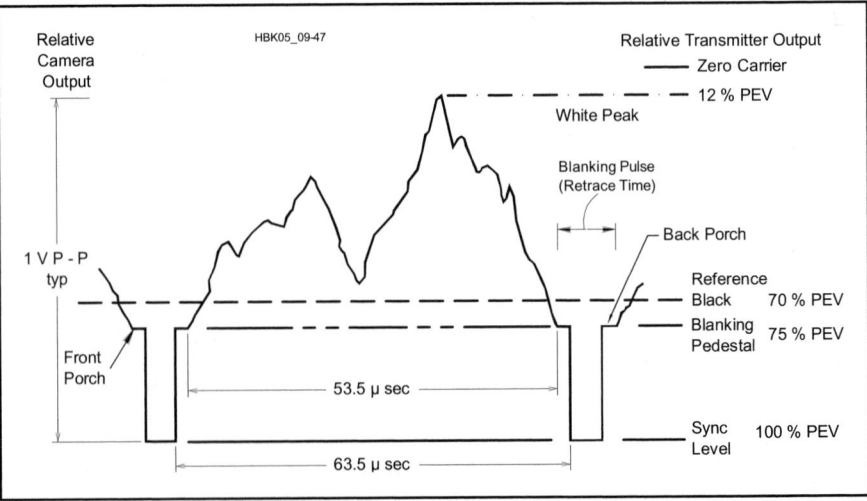

Fig 8.30 — An ATV waveform, showing the relative camera output as well as the transmitter output RF power during one horizontal line scan for black-and-white TV. (A color camera would generate a "burst" of 8 cycles at 3.58 MHz on the back porch of the blanking pedestal.) Note that "black" corresponds to a higher transmitter power than "white."

seen, corresponding to 480 pixels per line. Color bandwidth in a TV set is less than that, resulting in up to 100 color lines. (Lines of resolution should not be confused with the number of horizontal scan lines per frame.)

The PAL analog TV system in Europe used frequency modulation rather than AM. As described previously, FM achieves superior noise and interference suppression for signal levels above a certain threshold, although AM seems to work better for receiver signal levels below about 5 µV. FM ATV in the United States typically uses 4 MHz deviation with NTSC video and a 5.5-MHz sound subcarrier set to 15 dB below the video level. Using Carson's rule, the occupied bandwidth comes out to just under 20 MHz. Most available FM ATV equipment is made for the 1.2, 2.4 and 10.25-GHz bands.

Digital fast-scan TV has been explored by amateurs in Europe and the US using the commercial DVB-S satellite digital video broadcasting standard. It uses MPEG2 audio and video data compression and QPSK modulation with symbol rates up to 20 Mbauds. Much of work has been on the 23 cm band since inexpensive set-top boxes are available that cover that frequency range.

In the United States and some other countries, commercial over-the-air digital television uses a modulation type called *8-level vestigial sideband* (8-VSB), which is basically 8-level amplitude-shift keying with the lower sideband filtered out. The eight levels result in three bits per symbol, resulting in approximately 32 Mbits/second raw data rate with a 10.76 Mbaud symbol rate. The net data rate with error correction and other overhead is 19.39 Mbits/second. The DTV signal fits into the same 6 MHz channel as analog TV.

8.4.2 Facsimile

Amateur *facsimile* (fax) is a method for sending high-resolution still pictures by radio. Standard 120-line-per-minute fax uses a frequency-modulated tone between 1500 Hz (black) and 2300 Hz (white), which may be sent and received as an audio signal via any voice transmitter and receiver. Commercially, fax has been used for many years to send newspaper photos over the landline and to receive weather images from satellites.

The resolution of typical fax images greatly exceeds what can be obtained using SSTV or even conventional television. Typical images are made up of 480 to 1600 scanning lines, compared to the 240 lines in a typical SSTV image. This high resolution is achieved by slowing down the rate at which the lines are transmitted, resulting in image transmission times typically in the range of 4 to 13 minutes. For color, transmission times are longer since the image is sent three times to include the red, green and blue components.

Besides the increased resolution, the main difference with SSTV is that many fax systems use no sync pulses. Instead, they rely on the extreme accuracy of external time bases at both the transmitting and receiving ends of the circuit to maintain precise synchronization.

Modern personal computers have virtually eliminated bulky mechanical fax machines from most amateur installations. Now the incoming image can be stored in computer memory and viewed on a high-resolution computer graphics display. The use of a color display system makes it entirely practical to transmit color fax images when band conditions permit the increased transmission time.

8.4.3 Slow-Scan Television

Despite its name, so-called *slow-scan television* (SSTV) is actually a method for sending still images, like facsimile. The original monochrome analog SSTV format illustrated in **Fig 8.31** takes approximately 8 seconds to send one complete frame. The 1500 to 2300-Hz frequency-modulated audio tone resembles that of a fax signal, but is sent at a faster rate and includes pulses at 1200 Hz for synchronization.

Color may be sent using any of several methods. The first to be used was the *frame-sequential* method, in which each of the three primary colors (red, green and blue) is sent sequentially, as a complete frame. That has the disadvantage that you have to wait for the third frame to begin before colors start to become correct, and any noise or interference is three times more likely to corrupt the image and risks ruining the image registration (the overlay of the frames) and thus spoil the picture.

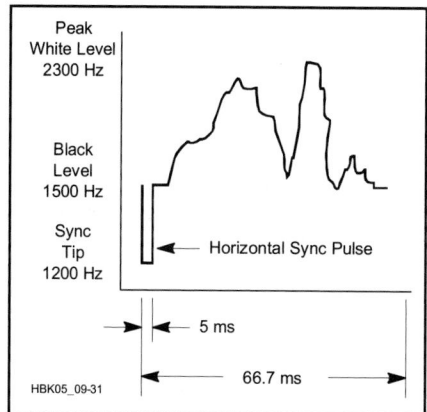

Fig 8.31 — The basic 8-second black and white transmission format developed by early SSTV experimenters. The sync pulses are "blacker than black" to blank the signal during retrace. A complete frame has 120 lines (8 seconds at 15 lines per second). Horizontal sync pulses occur at the beginning of every line; a 30-ms vertical sync pulse precedes each frame.

In the *line-sequential* method, each line is electronically scanned three times: once each for red, green and blue. Pictures scan down the screen in full color as they are received and registration problems are reduced. The Wraase SC-1 modes are examples of early line-sequential color transmission. They have a horizontal sync pulse for each of the color component scans. The major weakness of this method is that if the receiving system gets out of step, it doesn't know which scan represents which color.

Rather than sending color images with the usual RGB (red, green, blue) components, Robot Research decided to use luminance and chrominance signals for their 1200C modes. The first half or two thirds of each scan line contains the luminance information, which is a weighted average of the R, G and B components. The remainder of each line contains the chrominance signal with the color information. Existing black-and-white equipment could display the B&W-compatible image on the first part of each scan line and the rest would go off the edge of the screen. That compatibility was very beneficial when most people still had only B&W equipment.

The luminance-chrominance encoding makes more efficient use of the transmission time. A 120-line color image can be sent in 12 seconds, rather than the usual 24. Our eyes are more sensitive to details in changes of brightness than color, so the time is used more efficiently by devoting more time to luminance than chrominance. The NTSC and PAL broadcast standards also take advantage of this vision characteristic and use less bandwidth for the color part of the signal. For SSTV, luminance-chrominance encoding offers some benefits, but image quality suffers. It is acceptable for most natural images but looks bad for sharp, high-contrast edges, which are more and more common as images are altered via computer graphics. As a result, all newer modes have returned to RGB encoding.

The 1200C introduced another innovation, called *vertical interval signaling* (VIS). It encodes the transmission mode in the vertical sync interval. By using narrow FSK encoding around the sync frequency, compatibility is maintained. This new signal just looks like an extra-long vertical sync to older equipment.

The Martin and Scottie modes are essentially the same except for the timings. They have a single horizontal sync pulse for each set of RGB scans. Therefore, the receiving end can easily get back in step if synchronization is temporarily lost. Although they have horizontal sync, some implementations ignore them on receive. Instead, they rely on very accurate time bases at the transmitting and receiving stations to keep in step. The advantage of this "synchronous" strategy is that missing or corrupted sync pulses won't disturb the received image. The disadvantage is that even slight timing inaccuracies produce slanted pictures.

In the late 1980s, yet another incompatible mode was introduced. The AVT mode is different from all the rest in that it has *no horizontal sync*. It relies on very accurate oscillators at the sending and receiving stations to maintain synchronization. If the beginning-of-frame sync is missed, it's all over. There is no way to determine where a scan line begins. However, it's much harder to miss the 5-s header than the 300-ms VIS code. Redundant information is encoded 32 times and a more powerful error-detection scheme

is used. It's only necessary to receive a small part of the AVT header in order to achieve synchronization. After this, noise can wipe out parts of the image, but image alignment and colors remain correct.

Digital images may be sent over Amateur Radio using any of the standard digital modulation formats that support binary file transfer. *Digital SSTV* (DSSTV) is one method of transmitting computer image files, such as JPEG or GIF, as described in an article by Ralph Taggart, WB8DQT, in the Feb 2004 issue of *QST*. This format phase-modulates a total of eight subcarriers (ranging from 590 to 2200 Hz) at intervals of 230 Hz. Each subcarrier has nine possible modulation states. This

signal modulation format is known as *redundant digital file transfer* (RDFT) developed by Barry Sanderson, KB9VAK.

Most digital SSTV transmission has switched to using Digital Radio Mondiale (DRM), derived from a system developed for shortwave digital voice broadcasting. The DRM digital SSTV signal occupies the bandwidth between 350 and 2750 Hz. As many as 57 subcarriers may be sent simultaneously, all at the same level. Three pilot carriers are sent at twice the level of the other subcarriers. The subcarriers are modulated using OFDM and QAM, which were described earlier in this chapter. DRM SSTV includes several methods of error correction.

8.5 Modulation Impairments

Most of the previous discussion of the various modulation types has assumed the modulation is perfect. With analog modulation, that means the audio or video modulating signal is perfectly reproduced in the RF waveform without distortion, spurious frequencies or other unwanted artifacts. With digital modulation, the symbol timing and the locations and trajectories in the I/Q constellation are perfectly accurate. In all cases, the RF power amplifier is perfectly linear, if so required by the modulation type, and it introduces no noise or other spurious signals close to the carrier frequency.

In the real world, of course, such perfection can never be achieved. Some modulation impairments are caused by the transmitting system, some by the transmission medium through which the signal propagates, and some in the receiving system. This section will concentrate on impairments caused by the circuitry in the transmitter and, to some extent, in the receiver. Signal impairments due to propagation are covered in detail in the **Propagation of Radio Signals** chapter.

8.5.1 Intermodulation Distortion

To minimize distortion of the modulation, each stage in the signal chain must be linear, from the microphone or modem, through all the intermediate amplifiers and processors, to the modulator itself. For the linear modulation types, all the amplifiers and other stages between the modulator and the antenna must be linear as well.

If the modulation consists of a single sine wave, then nonlinearity causes only harmonic distortion, which produces new frequencies at integer multiples of the sine-wave frequency. If multiple frequencies are present in the modulation, however, then *intermodulation*

distortion (IMD) products are produced. IMD occurs when a nonlinear amplifier or other device acts as a mixer, producing sum and difference frequencies of all the pairs of frequencies and their harmonics. For example if two frequencies, F1 and F2 > F1, are present, then IMD will cause spurious frequencies to appear at F1 + F2, F2 – F1, 2F1, 2F2, 2F1 – F2, 2F2 – F1, 2F2 – 2F1, 3F1, 3F2, and so on. *Odd-order* products are those that include the original frequencies an odd number of times, such as 3F1, 2F1 + F2, 3F1 – 2F2, and so on. *Even-order* products contain an even number of the original frequencies, such as 2F2, F1 + F2, 3F1 + F2, and so on. If more than two frequencies are present in the undistorted modulation, then the number of unwanted frequencies increases exponentially.

Although intermodulation distortion that occurs before modulation is fundamentally the same as IMD that occurs after the modulator (at the intermediate or radio frequency), the effects are quite different. Consider two frequency components of a modulating signal at, for example, 1000 Hz and 1200 Hz that modulate an SSB transmitter tuned to 14.000 MHz, USB. The desired RF signal has components at 14.001 and 14.0012 MHz. If the IMD occurs before modulation, then the F1 + F2 distortion product occurs at 1000 + 1200 = 2200 Hz. Since that is well within the audio passband of the SSB transmitter, there is no way to filter it out so it shows up at 14.0022 MHz at the RF output. However, if the distortion had occurred after the modulator, the F1 + F2 product would be at 14.001 + 14.0012 = 28.0022 MHz which is easily filtered out by the transmitter's low-pass filter.

That explains why RF speech processors work better than audio processors. A speech processor clips or limits the peak amplitude of the modulating signal to prevent over-modulation in the transmitter. However, the limit-

ing process typically produces considerable intermodulation distortion. If the limiter is located after the modulator rather than before it, then it is an RF signal being clipped, rather than audio. If the RF limiter is followed by a band-pass filter then many of the distortion products are removed, resulting in a less-distorted signal.

If the distortion is perfectly symmetrical (for example equal clipping of positive and negative peaks) then only odd-order products are produced. Most distortion is not symmetrical so that both even and odd-order products appear. However, the odd-order products are of particular interest when measuring the linearity of an RF power amplifier. The reason is that even-order products that occur after the modulator occur only near harmonics of the RF frequency where they are easy to filter out.

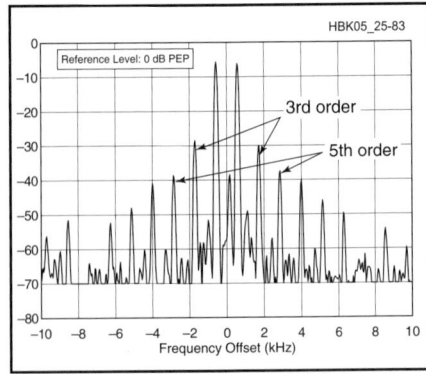

Fig 8.32 — Intermodulation products from an SSB transmitter. The two signals at the center are the desired frequencies. The frequencies of the third-order products are separated from the frequencies of the desired tones by the tone spacing. The fifth-order frequencies are separated by two times the tone spacing, and so on for higher-order products.

Odd-order products can fall within the desired channel, where they cause distortion of the modulation, or at nearby frequencies, where they cause interference to other stations.

The informal term for such IMD that interferes with other stations outside the desired channel is *splatter*. It becomes severe if the linear amplifier is over-driven, which causes clipping of the modulation envelope with the resulting odd-order IMD products.

The method commonly used to test the linearity of an SSB transmitter or RF power amplifier is the *two-tone test*. Two equal-amplitude audio-frequency tones are fed into the microphone input, the transmitter and/or amplifier is adjusted for the desired power level, and the output signal is observed on a spectrum analyzer. See **Fig 8.32**. The third-order products that occur near the desired signal are at $2F1 - F2$ and $2F2 - F1$. The fifth-order products occur at $3F1 - 2F2$ and $3F2 - 2F1$. If, for example, the two tones are spaced 1 kHz apart, then the two third-order products show up at 1 kHz above the high tone and 1 kHz below the low tone. The fifth-order products are at 2 kHz above and below the high and low tones respectively.

8.5.2 Transmitted Bandwidth

We have already discussed the necessary bandwidth for each of the various modulation types. The previous section explained how intermodulation distortion is one phenomenon that can cause unwanted emissions outside of the desired communications channel. Another is failing to properly low-pass filter the modulating signal to the minimum necessary bandwidth. That is especially a concern for linear modulation modes such as SSB, AM, OOK (CW), BPSK and QAM. For angle-modulated modes like FM and FSK, excessive bandwidth can result from simply setting the deviation too high.

There are other modulation impairments that cause emissions outside the desired bandwidth. For example, in an SSB transmitter, if the unwanted sideband is not sufficiently suppressed, the occupied bandwidth is up to twice as large as it should be. Also, an excessively strong suppressed carrier causes particularly-annoying heterodyne interference to stations tuned near that frequency. In some SSB modulators, there is an adjustment provided to optimize carrier suppression.

The carrier suppression may be degraded by a modulating signal that is too low in amplitude. For example, if the signal from the microphone to an SSB transmitter is one-tenth (−20 dB) of the proper amplitude, and if the gain of the RF amplifier stages is increased to compensate, then the carrier suppression is degraded by 20 dB.

The term *adjacent-channel power* (ACP) refers to the amount of transmitted power

that falls into an adjacent communications channel above or below the desired channel. Normally the unwanted out-of channel power is worse for the immediately-adjacent channels than for those that are two channels away, the so-called *alternate* channels. ACP is normally specified as a power ratio in dB. It is measured with a spectrum analyzer that can measure the total power within the desired channel and the total power in an adjacent channel, so that the dB difference can be calculated.

The *occupied bandwidth* is the bandwidth within which a specified percentage of the total power occurs. A common percentage used is 99%. For a properly-adjusted, low-distortion transmitting system, the occupied bandwidth is determined mainly by the modulation type and filtering and, in the case of digital modulation, the symbol rate. For example, the IS-54 TDMA format that has been used in some US digital cellular networks has about 30 kHz occupied bandwidth using 24.3-kilosymbols/sec, π/4-DQPSK modulation with a 0.35-alpha root-raised-cosine filter. The GSM cellular standard requires about a 350-kHz occupied bandwidth for its 270.833-kilosymbols/sec, 0.3 Gaussian-filtered MSK signal.

8.5.3 Modulation Accuracy

For analog modes, modulation accuracy is mainly a question of maintaining the proper frequency response across the desired bandwidth with minimal distortion and unwanted signal artifacts. In-band artifacts like noise and spurious signals should not be a problem with any reasonably-well-designed system. Maintaining modulation peaks near 100% for AM signals or the proper deviation for FM signals is facilitated by an audio compressor. It can be either the type that uses a detector and an automatic-gain-control feedback loop to vary the gain in the modulation path or a clipper-type compressor that limits the peak amplitude and then filters the clipped signal to remove the harmonics and intermodulation products that result. SSB transmitters can also use audio speech compression to maintain the proper peak power level although, as explained previously, clipping of the signal before it reaches the modulator can cause unacceptable distortion unless special techniques are used.

For digital signals, there are a number of other possible sources of modulation inaccuracy. For modes that use Nyquist filtering, the cutoff frequency and filter shape must be accurate to ensure no inter-symbol interference (ISI). Fortunately, that is easy to do with digital filters. However any additional filtering in the signal path can degrade the ISI. For example, most HF digital modes use an analog SSB transceiver to up-convert the

signal from audio to RF for transmission and to down-convert from RF to audio again at the receiver end. The crystal filters used in the transmitter and receiver can significantly degrade group delay flatness, especially near the edges of the filter passband. That is why most HF digital modes use a bandwidth substantially less than a typical transceiver's passband and attempt to center the signal near the crystal filters' center frequency.

Distortion can also impair the proper decoding of digital signals, especially formats with closely-spaced symbols such as 256-QAM. Any "flat-topping" in the final amplifier causes the symbols at the outermost corners of the constellation to be closer together than they should be. The accuracy of the symbol clock is critical in formats like JT65 that integrate the detection process over a large number of symbols. Perhaps surprisingly, the clock rate on some inexpensive computer sound cards can be off on the order of a percent, which results in a similar error in symbol rate.

The modulation accuracy of an FSK signal is normally characterized by the frequency shift at the center of each symbol time, the point at which the receiver usually makes the decision of which symbol is being received. For other digital formats, the modulation accuracy is normally characterized by the amplitude and phase at the symbol decision points. Amplitude error is typically measured as a percentage of the largest symbol amplitude. The phase error is quoted in degrees. In both cases one can specify either the average RMS value or the peak error that was detected over the measurement period. Amplitude error is the most important consideration for modulation types such as BPSK where the

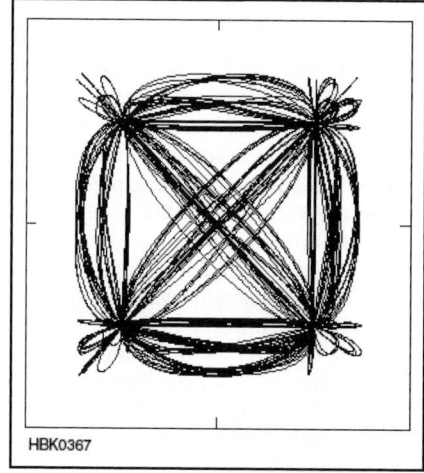

HBK0367

Fig 8.33 — Simulated I/Q constellation display of the trajectory of a QPSK signal over an extended period. A 0.5-alpha raised-cosine filter was used. Because it is a Nyquist filter, the trajectories pass exactly through each symbol location.

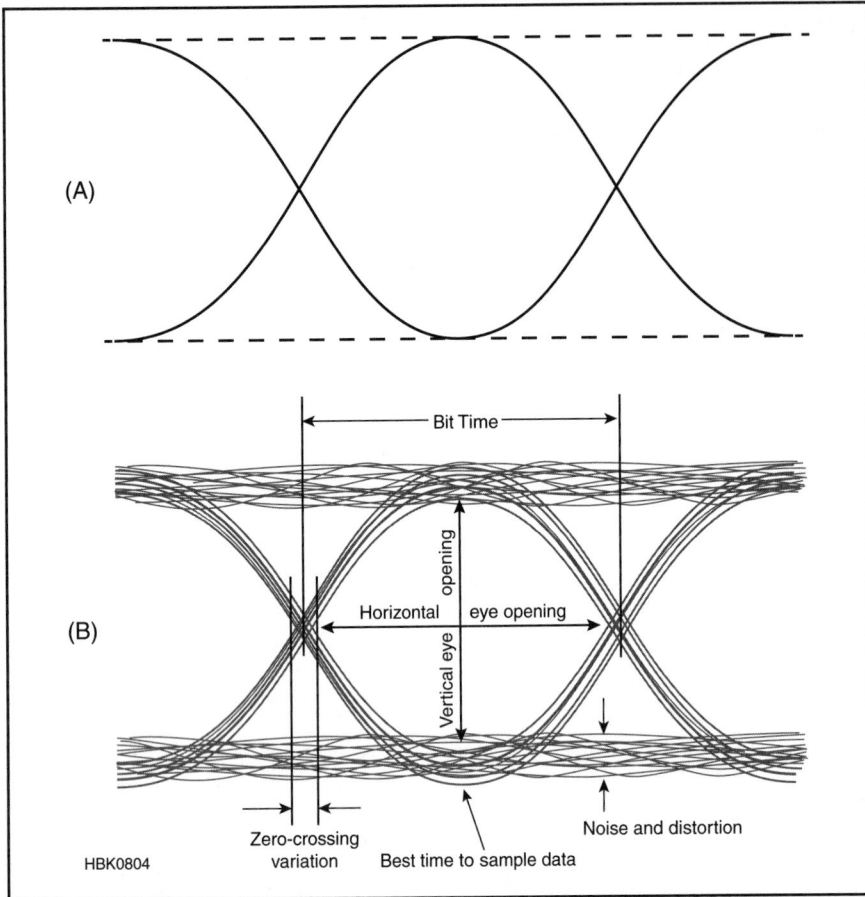

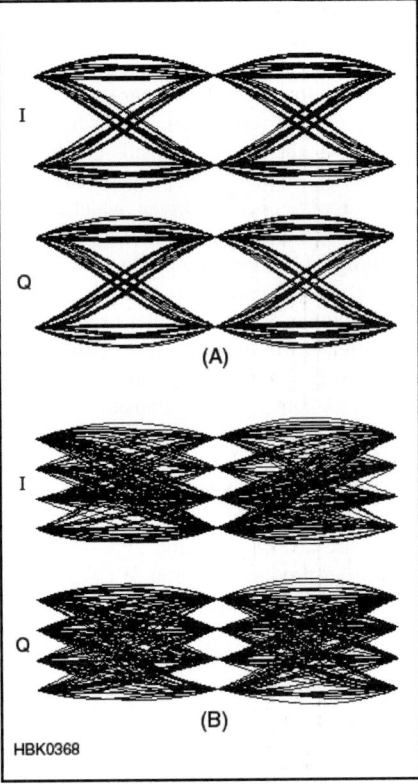

Fig 8.35 — Eye diagrams of the I and Q outputs of a QPSK generator (A) and a 16-QAM generator (B). The "eye" is the empty area between the symbol decision points, visible as the points where all the symbol trajectories come together. The bigger the "eye" the easier the signal is to decode.

Fig 8.34 — The eye diagram is an oscilloscope-style diagram showing repeated samples of the digital signal as it transitions between points in the constellation. The eye diagram at A is for a perfect, undistorted signal. The eye diagram in B shows a signal with noise, distortion, and timing variations.

information is encoded as an amplitude. For a constant-envelope format like MSK, phase error is a more important metric.

For formats like QPSK and QAM, both amplitude and phase are important in determining symbol location. A measurement that includes both is called *error vector magnitude* (EVM). It is the RMS average distance between the ideal symbol location in the constellation diagram and the actual signal value at the symbol decision point, expressed as a percentage of either the RMS signal power or the maximum symbol amplitude. This is a measurement that requires specialized test equipment not generally available to home experimenters. However it is possible to estimate the effect on EVM of various design choices using computer simulations.

Previously, we have plotted constellation diagrams with the transitions between symbol locations indicated by straight lines. In an actual system, both the I and Q signals are low-pass filtered which makes the symbol transitions smoother, with no abrupt changes of direction at the symbol locations. **Fig 8.33** shows the actual symbol trajectories for a

QPSK signal using a 0.5-alpha raised-cosine filter with random data. Since a raised-cosine is a type of Nyquist filter, the trajectories pass exactly through each symbol location. This is what you would see if you connected the I and Q outputs from a QPSK baseband generator to the X (horizontal) and Y (vertical) inputs of an oscilloscope. It is also what you would see in a receiver at the input to the symbol detector after the carrier and clock-recovery circuits have stabilized the signal. If the symbol trajectories were not accurate, then the dark areas at the symbol points would be less distinct and more spread out. Such a constellation diagram is also a good troubleshooting tool for other purposes, for example to check for amplitude distortion or to see if a faulty symbol encoder is missing some symbols or placing them at the wrong location.

In a system where a root-Nyquist filter is used in both the transmitter and receiver to obtain a net Nyquist response, the transmitter output will not display sharply defined symbol locations as in the figure. In that case, the measuring instrument must supply the missing root-Nyquist filter that is normally

in the receiver, in order to obtain a clean display. Professional modulation analyzers normally include a means of selecting the proper matching filter.

Another way to view modulation accuracy is with an *eye diagram*. See **Figs 8.34** and **8.35**. In this case the oscilloscope's horizontal axis is driven by its internal sweep generator, triggered by the symbol clock. The two vertical channels of the oscilloscope are connected to the I and Q outputs of the baseband generator. The "eye" is the empty area at the center of the I and Q traces in between the symbol decision points, where all the traces come together. The eye should be as "open" as possible to make the job of the receiver's symbol decoder as easy as possible. The oscilloscope would typically be set for infinite persistence, so that the worst-case excursions from the ideal symbol trajectory are recorded. QPSK has four symbol locations, one in each quadrant, such that I and Q each only have two possible values and there is only one "eye" for each. 16-QAM has 16 symbol locations with four possible locations for I and Q, which forms three "eyes."

8.6 Modulation Glossary

GENERAL

AM — Amplitude modulation.

Angle modulation — Modulation of an RF carrier by varying the phase or frequency.

Emission designator — Official ITU code to specify the bandwidth and modulation type of a radio transmission.

Frequency diversity — The use of a wideband signal to compensate for selective fading. While one band of frequencies is faded, the data can be reconstructed from other frequencies that are not faded.

FM — Frequency modulation.

IMD — Intermodulation distortion. Unwanted frequencies that occur at the sum and difference of the desired frequencies and their harmonics.

ITU — International Telecommunications Union. An agency of the United Nations that coordinates and recommends technical standards for electronic communications.

Modulation — The periodic alteration of some parameter of a carrier wave in order to transmit information.

PM — Phase modulation.

Protocol — A formal set of rules and procedures for the exchange of information within a communications network.

Selective fading — A propagation phenomenon in which closely-spaced frequencies experience markedly different fading at the same time.

Telemetry — The use of telecommunication for automatically indicating or recording measurements at a distance from the measuring instrument.

Telephony — A form of telecommunication primarily intended for the exchange of information in the form of speech.

Telegraphy — A form of telecommunication in which the transmitted information is intended to be recorded on arrival as a graphic document; the transmitted information may sometimes be presented in an alternative form or may be stored for subsequent use.

Television — A form of telecommunication for the transmission of transient images of fixed or moving objects.

ANALOG MODULATION (INCLUDING IMAGE)

ATV — Amateur fast-scan television.

BFO — Beat frequency oscillator. In an SSB or DSBSC receiver, the intermediate-frequency oscillator in the receiver that re-inserts the suppressed carrier.

Carson's rule — A rule of thumb to calculate FM bandwidth that says that 98% of the energy is typically contained within a bandwidth equal to two times the sum of the peak frequency deviation and the highest modulating frequency.

Capture effect — The tendency of the strongest signal to suppress other signals in an FM detector, which improves the signal-to-noise and signal-to-interference ratio.

Deviation ratio — The ratio of the maximum permitted peak deviation of an angle-modulated signal to the maximum permitted modulating frequency.

DSBSC — Double sideband, suppressed carrier. An AM signal in which the carrier has been removed but both sidebands remain.

Fax — Facsimile. The sending of still images by wire or by radio.

Frame — In television, one complete scanned image. On systems with interlaced scanning, there are two vertical scans per frame.

Frequency deviation — The amount the RF frequency of an angle-modulated signal deviates from the center (carrier) frequency in response to the modulating signal. The term is often understood to mean the maximum deviation available in a given system.

Limiter — A high-gain amplifier in an FM receiver that limits the peak amplitude of the signal in order to eliminate any AM component.

LSB — Lower sideband. An SSB signal with the upper sideband removed.

Martin — A series of analog SSTV formats, especially popular in Europe.

Modulation index (AM) — The ratio of the peak value of the modulation of an AM signal to the value that just causes the modulation envelope of the RF signal to reach zero on negative peaks and twice the average value on positive peaks.

Modulation index (FM) — The ratio of the peak deviation of an angle-modulated signal to the highest modulating frequency.

Narrowband FM — An FM signal with a modulation index less than or equal to 1.0.

NTSC — National Television System Committee. The analog television standard used in the US, Japan and several other countries.

PAL — Phase alteration line. The analog television standard used in many parts of Europe.

Phasing method — A method of generating an SSBSC signal that does not require a filter to remove the unwanted sideband.

Pixel — Picture element. The dots that make up images on a computer's monitor.

Product detector — A detector that multiplies a BFO signal with the received signal, typically SSB or DSBSC-modulated.

RGB — Red, green and blue. The three primary colors required to transmit a full-color image in many television and facsimile systems.

Scottie — A series of analog SSTV formats, especially popular in the US.

SSB — Single sideband. An AM signal in which one sideband has been removed. The term is usually understood to mean SSBSC.

SSBSC — Single sideband, suppressed carrier. An AM signal in which one sideband and the carrier have been removed.

SSTV — Slow scan TV, a system for sending and receiving still images.

Sync — Modulation pulses used in ATV and SSTV to synchronize the horizontal and/or vertical scanning.

Synchronous detector — A type of AM detector in which the carrier is regenerated in the receiver.

Two-tone test — A procedure for testing the IMD of an SSB transmitter. Two equal-amplitude tones are fed to the microphone input and the transmitter RF output is examined with a spectrum analyzer to determine the amplitudes of the IMD products.

USB — Upper sideband. An SSB signal with the lower sideband removed.

Vestigial sideband — The filtering of all but the bottom MHz or so of the lower sideband of an ATV or NTSC television signal.

VIS — Vertical interval signaling. Digital encoding of the transmission mode during the vertical sync interval of an SSTV frame.

Wideband FM — An FM signal with a modulation index greater than 1.0.

DIGITAL MODULATION

AFSK — Audio FSK. The use of an SSB transceiver to transmit and receive FSK using an audio-frequency modem.

ASK — Amplitude-shift keying. Digital amplitude modulation in which the amplitude depends on the modulating code.

Baud rate — The rate at which a digital signal transitions between symbol states. Symbol rate.

Bauds — Symbols per second. Unit of baud rate or symbol rate.

Bit rate — The total number of physically transferred bits per second over a communication link. Bit rate can be used interchangeably with **baud rate** only when each modulation transition carries exactly one bit of data.

BPSK — Binary PSK. PSK with only two possible states. The term is usually understood to mean a non-constant-envelope signal in which the two states differ by 180°.

Chirp — Incidental frequency modulation of a carrier as a result of oscillator instability during keying.

CW — Continuous wave. The term used for on-off keying using Morse code.

Constellation diagram — A diagram showing the constellation of possible symbol locations on a polar plot of modulation amplitude and phase.

DBPSK — Differential BPSK.

Decision point — The point, typically in the center of a symbol time, at which the receiver decides which symbol is being sent.

Differential modulation — A modulation technique that encodes the information in the difference between subsequent symbols, rather than in the symbols themselves.

Equalization — Correction for variations in amplitude and/or phase versus frequency across a communications channel.

DQPSK — Differential QPSK.

Eye diagram — An oscilloscope measurement of a digital modulating signal with the horizontal sweep synchronized to the symbol times. With Nyquist filtering, there should be a clear separation, or "eye", in the trajectories at the symbol decision points.

EVM — Error vector magnitude. A measure of the RMS error in the symbol locations at the symbol decision points in the constellation plot of a digital signal.

FFT — Fast Fourier transform. The Fourier transform is a mathematical function that calculates the frequency spectrum of a signal. The FFT is a software algorithm that does the calculations very efficiently.

FSK — Frequency-shift keying. A form of digital frequency modulation in which the frequency deviation depends on the modulating data.

GMSK — Gaussian MSK. MSK with a Gaussian modulation filter.

I/Q modulation — Quadrature modulation implemented with an I/Q modulator, one that uses in-phase (I) and quadrature (Q) modulating signals to generate the zero-degree and 90° components of the RF signal.

ISI — Inter-symbol interference. Interference of a signal with itself, caused when energy from one symbol is delayed long enough to interfere with a subsequent symbol.

MFSK — Multi-level FSK. FSK with more than two states represented by different frequency deviations.

Modem — Modulator/demodulator. A device that generates and demodulates digital modulation signals, usually at audio frequencies. It connects between the data terminal (usually a computer) and the radio.

MSK — Minimum-shift keying. A form of FSK with a frequency shift equal to one-half the symbol rate.

Nyquist criterion — A principle that states that the sampling frequency must be greater than twice the highest frequency in the sampled signal.

Nyquist frequency — One-half the sampling frequency.

OFDM — Orthogonal frequency-division multiplexing. A transmission mode that uses multiple carriers, spaced such that modulation on each carrier is orthogonal with the others.

OOK — On-off keying. A type of ASK with only two states, on and off.

OQPSK — Offset QPSK. By offsetting in time the symbol transitions of the I and Q channels, symbol trajectories through the origin are eliminated.

Orthogonal — Refers to streams of data that are uncorrelated with each other such that there is no mutual interference.

PAM — Pulse amplitude modulation. A type of pulse modulation where the modulating signal is encoded in the pulse amplitude.

π/4 DQPSK — PI-over-four differential QPSK. A form of differential 8PSK in which the only allowed transitions are ±45 and ±135°, resulting in four allowed states per symbol.

PPM — Pulse position modulation. A type of pulse modulation where the modulating signal is encoded in the pulse position.

PSK — Phase-shift keying. A form of digital phase modulation in which the phase of the RF signal depends on the modulating code. The term often is understood to refer to BPSK.

Pulse modulation — The modulating signal is sampled at regular intervals to generate a series of modulation symbols in the form of pulses.

PWM — Pulse width modulation. A type of pulse modulation where the modulating signal is encoded in the pulse width.

QAM — Quadrature amplitude modulation. A digital modulation type in which both amplitude and phase are varied. The number that precedes it, for example 64QAM, is the number of different possible states of the amplitude and phase.

QPSK — PSK with four possible states. The term is usually understood to be equivalent to four-level QAM, in which the four states have the same amplitude and differ in phase by 90°.

Quadrature modulation — Refers to modulation using two RF carriers in phase quadrature, that is, 90° out of phase.

Symbol rate — The rate at which a digital signal transitions between different states. Baud rate.

Varicode — A coding method in which the length of each character code depends on its frequency of occurrence. It is used to optimize the ratio of characters per second to baud rate.

SPREAD SPECTRUM

CDMA — Code-division multiple access. A method of allowing several stations to use the same frequency band simultaneously by assigning each station a different orthogonal spreading code.

Chip — The shortest-duration modulation state in a SS system.

De-spreading — Conversion of an SS signal back to its narrowband equivalent by convolving it with the spreading code.

DSSS — Direct-sequence SS. Spreading of a signal by multiplying the data stream by a higher-rate pseudo-random digital sequence.

FHSS — Frequency-hopping SS. Spreading of a signal by means of pseudo-random frequency-hopping of the unspread signal.

Processing gain — The increase in signal-to-noise ratio that occurs when an SS signal is de-spread.

SS — Spread spectrum. A system that intentionally increases the bandwidth of a digital signal by means of a special spreading code.

FILTERING AND BANDWIDTH

ACP — Adjacent channel power. The amount of transmitted power that falls into a communications channel immediately adjacent to the desired one.

Alpha — A design parameter for a Nyquist or root Nyquist filter. The smaller the alpha, the sharper the passband-to-stopband transition at the cutoff frequency.

Cutoff frequency — The frequency at which a filter response changes from the passband to the stopband.

Gaussian filter — A modulation filter with a Gaussian frequency response. The transition between passband and stopband is more gradual than with most Nyquist filters.

Key clicks — Out-of-channel interference from an OOK signal caused by too-sharp transitions between the on and off states.

Nyquist filter — A filter that causes no inter-symbol interference (ISI). It is so-called because the cutoff frequency is at one-half the symbol rate, the Nyquist frequency.

Nyquist criterion — The rule that states that the sample rate must be greater than twice the highest frequency to be sampled.

Occupied bandwidth — The bandwidth within which a specified percentage of the total power occurs, typically 99%.

Raised-cosine filter — A type of Nyquist filter whose passband-to-stopband transition region has the shape of the first half-cycle of a cosine raised so that the negative peak is at zero.

Root Nyquist filter — A filter that, when cascaded with another identical filter, forms a Nyquist filter. The frequency response is the square root of that of the Nyquist filter.

Root raised-cosine filter — A root Nyquist filter with a frequency response that is the square root of a raised-cosine filter.

Shannon-Hartley theorem — A formula that predicts the maximum channel capacity that is theoretically possible over a channel of given bandwidth and signal-to-noise ratio.

Sinc function — The spectrum of a pulse or random series of pulses, equal to $\sin(x)/x$.

Splatter — Out-of-channel interference from an amplitude-modulated signal such as SSB or QPSK caused by distortion, typically in the power amplifier stages.

8.7 References and Further Reading

Agilent Technologies, "Digital Modulation in Communications Systems — An Introduction" Application Note 1298. **http://cp.literature.agilent.com/litweb/pdf/5965-7160E.pdf**

Costas, J. P., "Poisson, Shannon, and the Radio Amateur," *Proceedings of the IRE*, vol 47, pp 2058-2068, Dec 1959.

Ford, S., WB8IMY, *ARRL's HF Digital Handbook* (ARRL, 2008)

Ford, S., WB8IMY, *ARRL's VHF Digital Handbook* (ARRL, 2008)

Haykin, S., *Digital Communications* (Wiley, 1988)

Langner, J. WB2OSZ, "SSTV Transmission Modes," **www.comunicacio.net/digigrup/ccdd/sstv.htm**

McConnell, Bodson, and Urban, *FAX: Facsimile Technology and Systems* (Artech House 1999)

Nagle, J., K4KJ, "Diversity Reception: an Answer to High Frequency Signal Fading," *Ham Radio*, Nov 1979, pp 48-55.

Sabin, W., et al, *Single-Sideband Systems and Circuits* (McGraw Hill, 1987)

Shavkoplyas, A., VE3NEA, "CW Shaping in DSP Software," *QEX*, May/Jun 2006, pp 3-7.

Seiler, T., HB9JNX/AE4WA, et al, "Digital Amateur TeleVision (D-ATV)," proc. 2001 ARRL/TAPR Digital Communications Conference, **www.baycom.org/~tom/ham/dcc2001/datv.pdf**

Silver, H. W., NØAX, "About FM," *QST*, Jul 2004, pp 38-42.

Smith, D., KF6DX, "Distortion and Noise in OFDM Systems," *QEX*, Mar/Apr 2005, pp 57-59.

Smith, D., KF6DX, "Digital Voice: The Next New Mode?," *QST*, Jan 2002, pp 28-32.

Stanley, J., K4ERO, "Observing Selective Fading in Real Time with Dream Software," *QEX*, Jan/Feb 2007, pp 18-22.

Taggart, R., WB8DQT, "An Introduction to Amateur Television," *QST*, Apr, May and Jun 1993.

Taggart, R., WB8DQT, *Image Communications Handbook* (ARRL, 2002)

Taggart, R., WB8DQT, "Digital Slow-Scan Television," *QST*, Feb 2004, pp 47-51.

Van Valkenburg, M., et al, *Reference Data for Engineers: Radio, Electronics, Computer and Communications*, chapters 23, 24 and 25, 9th Ed. (Newnes, 2001)

Contents

**Chapter 9 —
CD-ROM Content**

Supplemental Files

- Measuring Receiver Phase Noise
- "Oscillator Design Using *LTSpice*" by David Stockton,
  GM4ZNX (includes *LTSpice* simulation files in
  SwissRoll folder)
- Using the MC1648 in Oscillators
- "Novel Grounded Base Oscillator Design for VHF-UHF"
  by Dr Ulrich Rohde, N1UL
- "Optimized Oscillator Design" by Dr Ulrich Rohde, N1UL
- "Oscillator Phase Noise" by Dr Ulrich Rohde, N1UL
- "Some Thoughts On Crystal Oscillator Design"
  by Dr Ulrich Rohde, N1UL

Oscillators and Synthesizers

RF signal paths all need to start somewhere. These starting places are oscillators — circuits that generate a periodic output signal without any input signal. In general, we use sinusoidal waveforms for communication signals, and square-wave outputs for digital signals (clocks). Ham operators care greatly about oscillator design. In a crowded band, a good oscillator is necessary for the receiver to resolve each of the signals present. Similarly, a good oscillator is needed for all transmitters so that our signals use only the amount of spectrum necessary for communication.

This chapter has been updated and rewritten for the 90th edition by Earl McCune, WA6SUH, from material originally written by David Stockton, GM4ZNX, and Frederick J. Telewski, WA7TZY. The sidebar on Phase-Locked Loops was contributed by Jerry DeHaven, WAØACF. Dr Ulrich Rohde, N1UL, contributed material on oscillator designs along with articles on phase noise and oscillator design which may be found on the *Handbook* CD-ROM.

The sheer number of different oscillator circuits seen in the literature can be intimidating, but their great diversity is an illusion that evaporates once their underlying pattern is seen. Despite the number of combinations that are possible, a manageably small number of oscillator types will cover all but very special requirements. Look at an oscillator circuit and "read" it: What form of filter — resonator — does it use? What form of amplifier? How have the amplifier's input and output been coupled into the filter? How is the filter tuned? These are simple, easily answered questions that put oscillator types into appropriate categories and make them understandable. The questions themselves make more sense when we understand the mechanics of oscillation, in which resonance plays a major role.

9.1 How Oscillators Work

Any oscillator is a fundamentally nonlinear device. Among other reasons, this circuit has an RF output even without any RF input. Any linear circuit can change the magnitude and phase shift on the signal only at its input.

9.1.1 Resonance

We are all familiar with pendulums. A weight (a mass in a gravitational field) hanging from some kind of string will swing back and forth with a very regular period. It has been known for millennia that this period does not change as long as the length of the string does not change, regardless of the amount of weight or how far it swings. Only a few centuries ago did Isaac Newton invent enough calculus to show why this is true. Yet even before Newton's seminal work, people were comfortable enough with this very predictable swinging period that pendulums were adopted as the basis of clocks and timekeeping.

Newton showed that the pendulum works by continuously exchanging its energy between potential (height) and kinetic (moving) forms. The mass moves fastest when it has minimum height (all energy is kinetic), and stops moving (all energy is potential) when it has maximum height. Further, when the energy is all in one form, the ingredients are in place to assure that conversion to the other form will happen. At the bottom, when the mass is moving fastest, its momentum assures that it will keep moving. The string constrains that movement and forces it to rise. When the mass stops rising, the string pulls on it in a different direction from gravity, which is pulling it down. This assures that the mass will fall and pick up speed — and the cycle repeats.

An electrical equivalent is created by the combination of an inductor and a capacitance. The inductor stores energy in a magnetic field when there is a current (motion of charge) flowing through it. The capacitor stores energy in an electric field when there is a voltage (presence of charge) in it. Connect the two together and the charge in the capacitor (electric potential energy) forces a current to begin flowing through the inductor. When the charge in the capacitor is zero, the current in the inductor is maximum (magnetic energy) and the magnetic field forces the current to continue, which recharges the capacitor with the opposite polarity. Just like the pendulum, the combination of inductor and capacitor trade energy back and forth between two types. Because energy is being stored in this circuit, we call this a *tank circuit*.

The common features between a pendulum and an electrical tank circuit are shown in **Fig 9.1**. For the pendulum, the speed of the mass and the displacement (height) are both sinusoidal with time and offset in phase by 90 degrees. Similarly, the voltage and current in

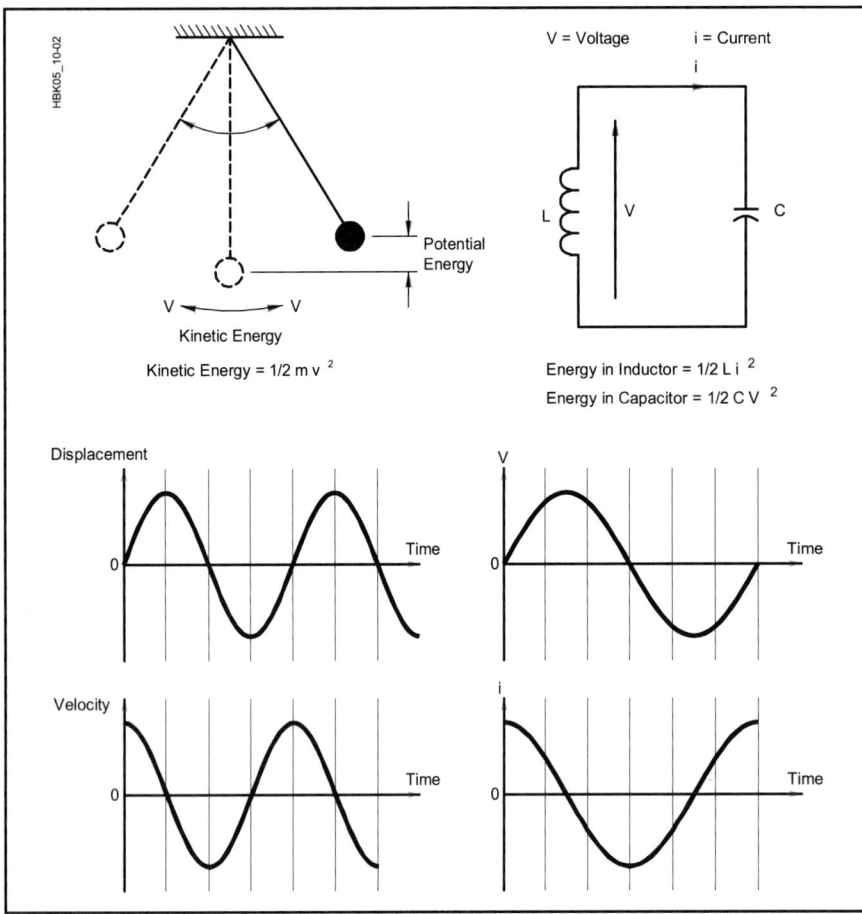

Fig 9.1 — A resonator lies at the heart of every oscillatory mechanical and electrical system. A mechanical resonator (here, a pendulum) and an electrical resonator (here, a tuned circuit consisting of L and C in parallel) share the same mechanism: the regular movement of energy between two forms — potential and kinetic in the pendulum, electric and magnetic in the tuned circuit. Both of these resonators share another trait: Any oscillations induced in them eventually die out because of losses — in the pendulum, due to drag and friction; in the tuned circuit, due to the resistance, coupling to other circuits, and radiation. Note that the curves corresponding to the pendulum's displacement vs velocity and the tuned circuit's voltage vs current differ by one-quarter of a cycle, or 90°.

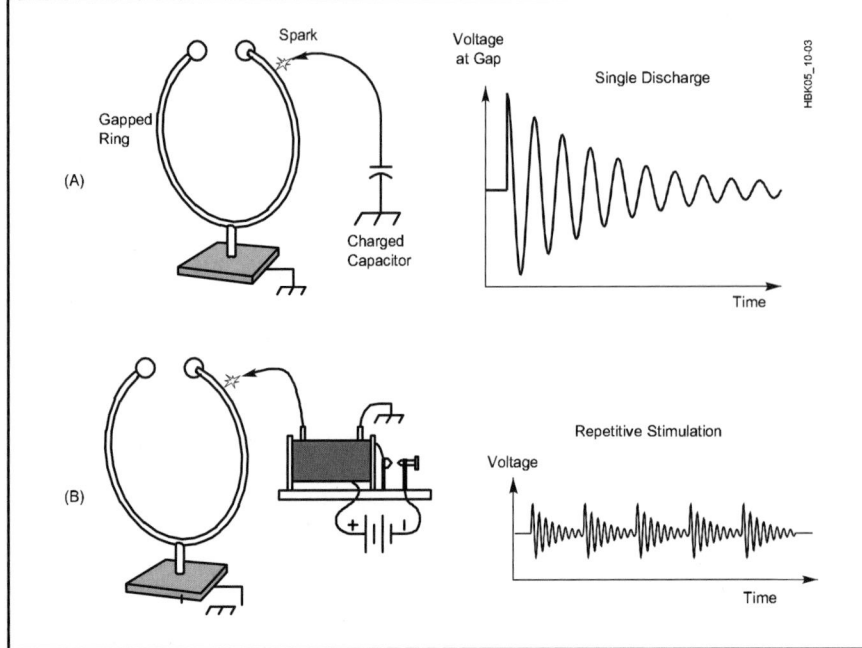

the LC tank circuit are sinusoidal and offset by 90 degrees.

Unfortunately, these oscillations do not continue forever. When you pull a pendulum aside and let it go, it will swing for a while and eventually stop. This is due to losses that take a little bit of energy away from the pendulum each time it swings. For a pendulum, one major loss mechanism is air drag: As the mass moves through the air, some energy is spent to move the air out of its way. There are losses from the string as well: Every time the string is bent, it warms up just a little bit. This also takes energy away from the pendulum. With these losses, the energy conversion is not complete between kinetic and potential forms, and each cycle of the oscillation is slightly smaller than the one before. We say that the oscillation is *damped*.

The electrical tank circuit works similarly. When an oscillation is started, say with a spark charging the capacitance as shown in **Fig 9.2A**, we see that this oscillation is also damped. The primary losses include resistance in the inductor and interactions of the electric and magnetic fields with other conductors. There is also direct radiation of electromagnetic fields — exactly what we want to happen for radio communication, but counts as an energy loss from the resonator.

It was not hard to figure out that a damped oscillation is not very useful for radio communication. We need our signals to last for a longer time so they can carry useful information. The first attempt to solve this problem was to repeatedly apply sparks to the tank circuit in a manner like that of Fig 9.2B. This still is not a continuous sine wave and it is very noisy because, among other things, the timing of the sparks is extremely difficult to hold in exact synchronization with the tank oscillation. There has to be a better way…

9.1.2 Maintained Resonance (CW: Continuous Waves)

What we need is a way to make up for the resonator losses, but only just enough to maintain the oscillation indefinitely. Invention of the escapement achieved this for the pendulum, and that's what provides the distinctive "tick-tock" of a pendulum clock. Electrically we use an amplifier to do the same thing.

Fig 9.2 — Stimulating a resonance, 1880s style. Shock-exciting a gapped ring by a charged capacitor causes the ring to oscillate at its resonant frequency. The result is a damped wave, each successive alternation of which is weaker than its predecessor because of resonator losses. Repetitively stimulating the ring produces trains of damped waves, but oscillation is not continuous.

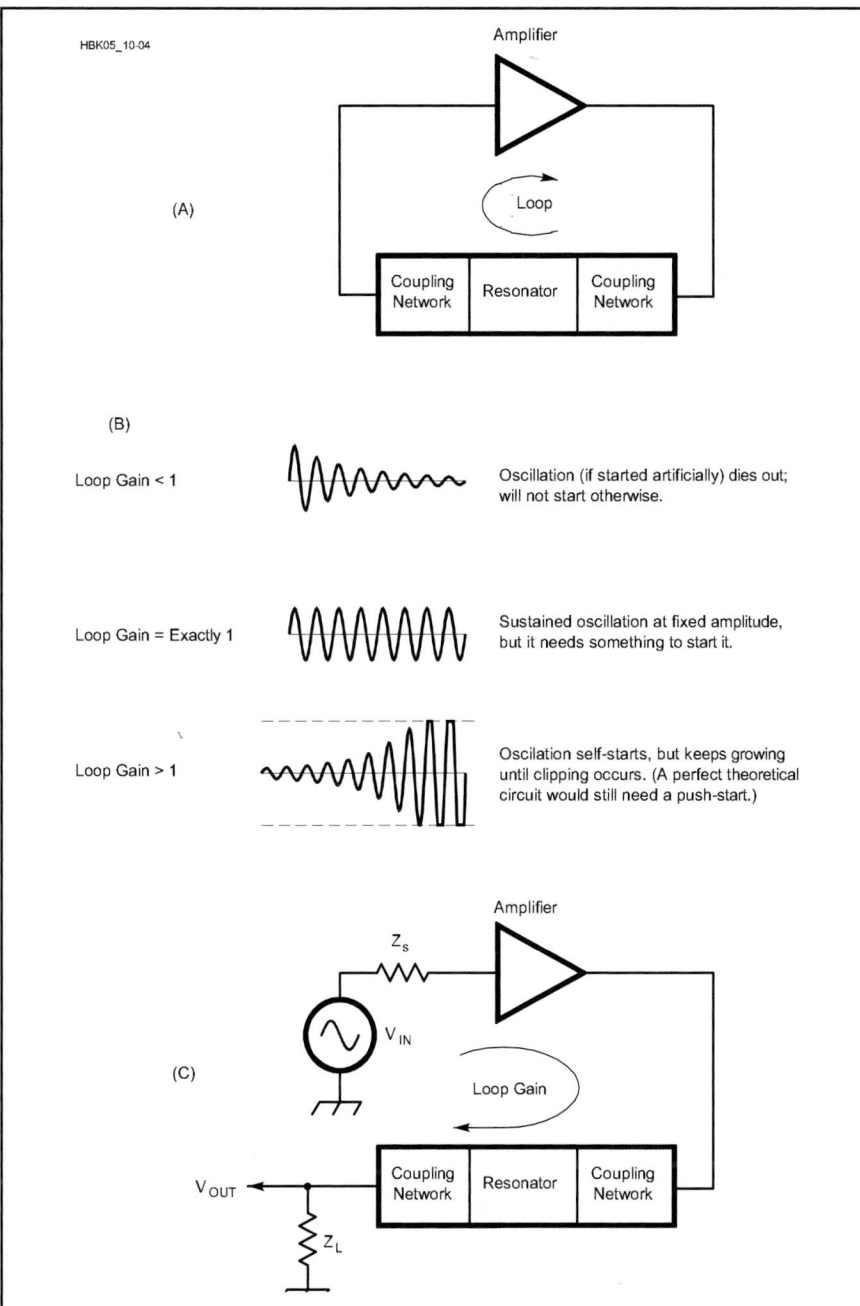

(A) Amplifier — Loop — Coupling Network — Resonator — Coupling Network

(B)

Loop Gain < 1 — Oscillation (if started artificially) dies out; will not start otherwise.

Loop Gain = Exactly 1 — Sustained oscillation at fixed amplitude, but it needs something to start it.

Loop Gain > 1 — Oscilation self-starts, but keeps growing until clipping occurs. (A perfect theoretical circuit would still need a push-start.)

(C) Amplifier — Z_s — V_{IN} — Loop Gain — V_{OUT} — Coupling Network — Resonator — Coupling Network — Z_L

Fig 9.3 — A practical oscillator requires networks to couple power in and out of the resonator (A). At (B), we see the amount of loop gain determines whether the oscillations will die out (loop gain < 1), grow until the waveform is no longer a sine wave (>1), or sustain at a steady amplitude (=1). Breaking the loop, inserting a test signal and measuring the loop's overall gain allows us to determine whether the system can oscillate, sustain oscillation or clip (C).

By adding an amplifier to the tank circuit or *resonator*, we can take a small amount of the energy from the resonator, amplify it, and inject a part of the amplified tank signal back into the resonator. This process is shown in **Fig 9.3A**. Unlike in general amplifier design, here we are not interested in maximizing power transfer. Instead, we want to couple just enough energy into the resonator to overcome its losses, and to take only the minimum amount of energy out of the resonator

needed to generate the restoring energy. Thus Fig 9.3A shows coupling networks instead of matching networks. We end up with a loop: The amplifier input comes from the output of the resonator, and the amplifier output goes to the input of the resonator.

The trick to oscillator design is to get the coupling networks and the amplifier gain working together just right. What this means is shown in Fig 9.3B. If we want to get a sine wave output, we need to get the loop gain —

gain computed as the signal passes through the combination of amplifier, resonator, and both coupling networks — exactly equal to one, also called *unity*. If the loop gain is even slightly less than one, then not enough energy is supplied to overcome resonator losses and the damping still happens. If the loop gain is greater than one, then the magnitude of the sine wave will grow until the waveform is no longer sinusoidal.

There is more to making an oscillator work than getting the loop gain right. We also need to be sure that the energy from the amplifier output is applied to the resonator having the correct phase alignment with the oscillating signal in the resonator. Taken together these requirements are known as the *Barkhausen criteria*: to achieve oscillation the loop must 1) have a net zero phase shift (or some integer multiple of 360°) and 2) have a loop gain equal to one. Both of these are critically important. If either of these criteria is not met, the circuit will not oscillate.

How do we find out if we are meeting the Barkhausen criteria? We need to break the loop, usually at the input of the amplifier as shown in Fig 9.3C. We apply a signal slightly below the desired oscillation frequency, then sweep the frequency through and slightly past the desired oscillation frequency. During the sweep we monitor both the magnitude and phase response at the output of the resonator output coupling network. If everything is right, then at the desired frequency of oscillation the input and output signals will look exactly the same on an oscilloscope. If that doesn't happen, we have work to do.

Getting the phase response of the loop correct is usually the harder problem. Much of this difficulty comes from the many types of resonators that can be used, including:

• LC tank
• Quartz crystal (and other piezoelectric materials)
• Transmission line (stripline, microstrip, open-wire, coax, and so on)
• Microwave cavities, YIG spheres, dielectric resonators
• Surface-acoustic-wave (SAW) devices

Each of these also can be called a filter. It is true that any filter can also be used as a resonator in an oscillator. The more complicated phase responses of filters makes meeting both of the Barkhausen criteria more challenging, but certainly possible.

The phase response of the resonator is related to its Q (quality factor — see the **Electrical Fundamentals** chapter). If we have a parallel LC tank, then at frequencies well below resonance the tank looks inductive ($X_L \ll X_C$) and the current lags the voltage. At frequencies well above resonance the tank looks capacitive ($X_C \ll X_L$) so the current leads the voltage. In between, the phase shift of the current relative to the voltage depends on the actual frequency. The rapidity of the

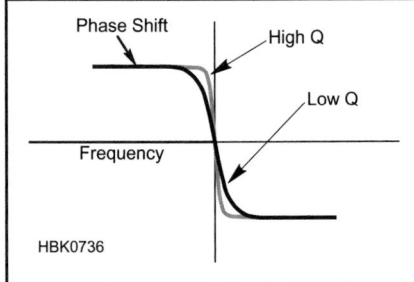

Fig 9.4 — Phase shift through the resonator depends on resonator Q. The higher resonator Q, the more abruptly phase changes through the resonator as frequency changes.

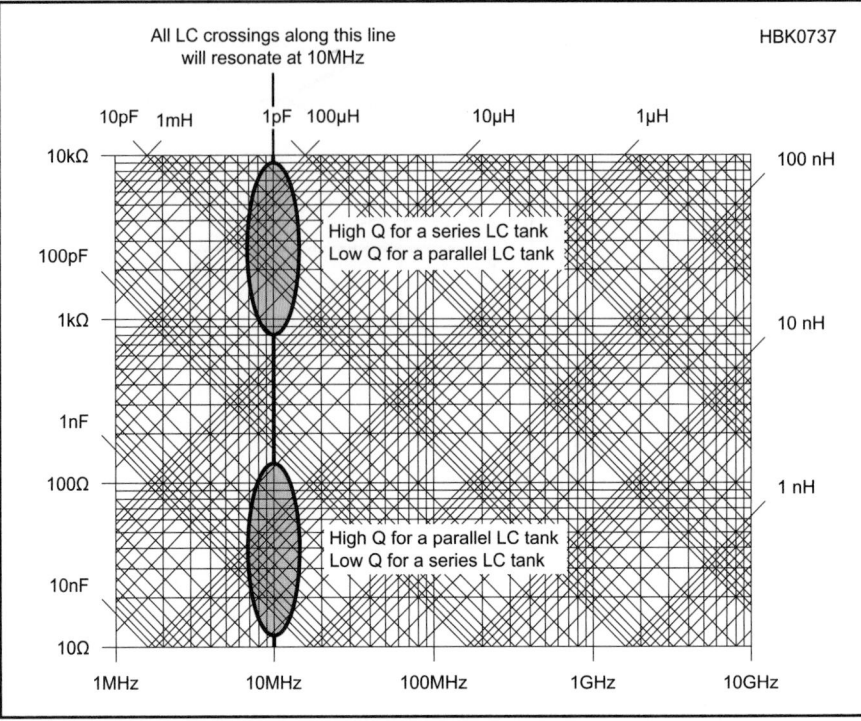

Fig 9.5 — There are an infinite number of combinations of L and C for a given resonant frequency. For a series-LC tank circuit, larger ratios of L to C result in higher values of Q and vice versa for parallel-LC tank circuits.

phase shift with changing frequency is governed by resonator Q: the phase shift happens very rapidly when Q is high. This is shown in **Fig 9.4**. When the resonator Q is low then the phase changes much more slowly at different frequencies.

If we have a series resonator then the phase change profile is reversed (capacitive to inductive) but otherwise the shape is the same. There are an infinite number of combinations of L and C values that provide the same resonant frequency. If we are interested in high Q, for a series tank we want to select large inductor values and small capacitor values. For a high-Q parallel tank we want the opposite — large capacitor values and small inductor values. **Fig 9.5**, a modified version of the reactance vs frequency chart in the **Electrical Fundamentals** chapter, shows these regions for a resonant frequency of 10 MHz.

9.1.3 Oscillator Start-Up

Looking at Fig 9.3B we see that the oscillation will build up in magnitude only when the loop gain is greater than unity. Thus this is an important additional criterion for oscillator design. The loop gain must be slightly greater than unity for it to start oscillating. Otherwise we have the undesired condition in which there is no signal in the resonator, so we sample nothing from it and put nothing back into it.

Still we need one more thing — noise. Some kind of signal needs to be injected into the resonator for the oscillation to start building up. We are fortunate that all amplifiers have output noise when power is applied, even if there is no input signal. It is this noise that allows any oscillator to start. Amplifier noise initiates a resonant signal in the tank, and if loop gain is slightly greater than unity, more signal is fed back into the resonator than the resonator loses. The output signal builds up until something stops it.

Here we take advantage of an amplifier characteristic where gain is reduced as the output reaches some predefined value. This

is referred to as *compression*: when the output signal gets large enough, the amplifier gain is reduced a little bit. When compression balances steady output amplitude with a loop gain of exactly unity, the oscillator has reached steady state operation and we are ready to go!

Why don't we have noise at the oscillator output instead of a sine wave? Because of the filtering action of the resonator, the filtered noise waveform appears mostly sinusoidal. The narrower the bandwidth of the resonator (the higher its Q), the more sinusoidal the waveform, but it actually consists of very well-filtered noise. It is impossible to get only a pure sine wave. Noise is inevitable, and tremendous efforts are spent in reducing this noise. This is discussed in section 9.2.

9.1.4 Negative Resistance Oscillators

It is possible to build an oscillator without using an amplifier. Recognizing that losses in the tank are represented by a resistor, if the loss resistor is canceled with a negative resistance then there will be no operating losses and the tank oscillation will continue without damping. This structure is shown in **Fig 9.6**.

Negative resistance devices definitely exist. Anything that draws lower current as the applied voltage increases exhibits negative resistance. Gunn diodes are classic examples and are widely used in microwave oscillators. Lower frequency devices include

tunnel diodes and lambda diodes. With this type of oscillator, the equivalent of having the loop gain slightly greater than unity at startup is for the negative resistance to be slightly more negative than the tank loss resistance is positive.

Amplitude limiting for this type of oscillator occurs when the signal amplitude increases enough to drive the negative resistance device to a less-negative value. All negative resistance devices listed here only exhibit their negative resistance over a finite range, so any signal approaching or exceeding this range will effectively reduce the negative resistance, as desired.

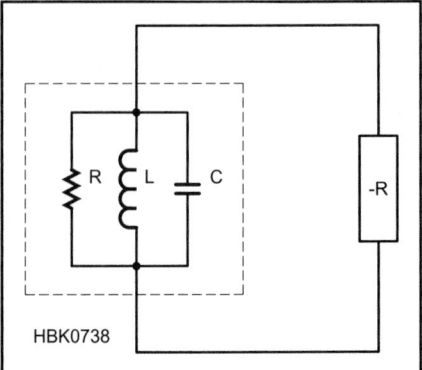

Fig 9.6 — The amplifier in an oscillator can be replaced by a device with negative resistance, such as a Gunn diode, and oscillation will still result.

9.2 Phase Noise

No oscillator output signal is perfect. Viewing an oscillator as a filtered-noise generator, as in the previous sections, is relatively modern. The older approach is to think of an oscillator making a pure sine wave with an added, unwanted noise signal. These are just different ways of visualizing the same thing. They are equally valid views which are used interchangeably, depending on which best makes some point clear.

For example, it is instructive to use the pure-sine-wave-plus-noise view to see relationships between AM noise and PM (phase) noise processes, shown in **Fig 9.7**. Adding Gaussian noise to a pure sine wave is usefully modeled using phasors as shown in Fig 9.7B. This generates both AM due to noise, and PM due to noise as shown in Fig 9.7C. Passing this signal through a limiter process, such as an amplifier in compression or through a switching mixer, leaves only the phase noise depicted in Fig 9.7D. Because it is so easy to remove AM noise but not phase noise, there is essentially no discussion about oscillator AM noise. Phase noise is the critical problem.

Phase modulation and frequency modulation are closely related. (See the **Modulation** chapter for a detailed description of each.) Phase is the integral of frequency, so phase modulation resembles frequency modulation, where the frequency deviation increases with increasing modulating frequency. Thus, there is no need to talk of "frequency noise" because phase noise already covers it.

A thorough analysis of oscillator phase noise is beyond the scope of this section. However, a detailed, state-of-the-art treatment by Dr Ulrich Rohde, N1UL, of free-running oscillators using nonlinear harmonic-balance techniques is presented as a supplement on the CD-ROM accompanying this book.

Because of the dynamic range required to measure phase noise, it is one of the most difficult measurements in all of electrical engineering. The sidebar "Transmitter Phase-Noise Measurement in the ARRL Lab" illustrates the lengths to which one must go to obtain repeatable, reliable measurements. An additional article on measuring receiver phase noise is included on the *Handbook* CD-ROM.

Fig 9.7 — At A, a vector (left) and phasor (right) diagram of an ideal oscillator with no noise. Added noise creates a region of uncertainty in the phasor's length and position (B). AM noise varies the phasor's length; PM noise varies the phasor's relative angular position (C). Limiting a signal that contains both AM and PM noise strips off the AM and leaves the PM (D).

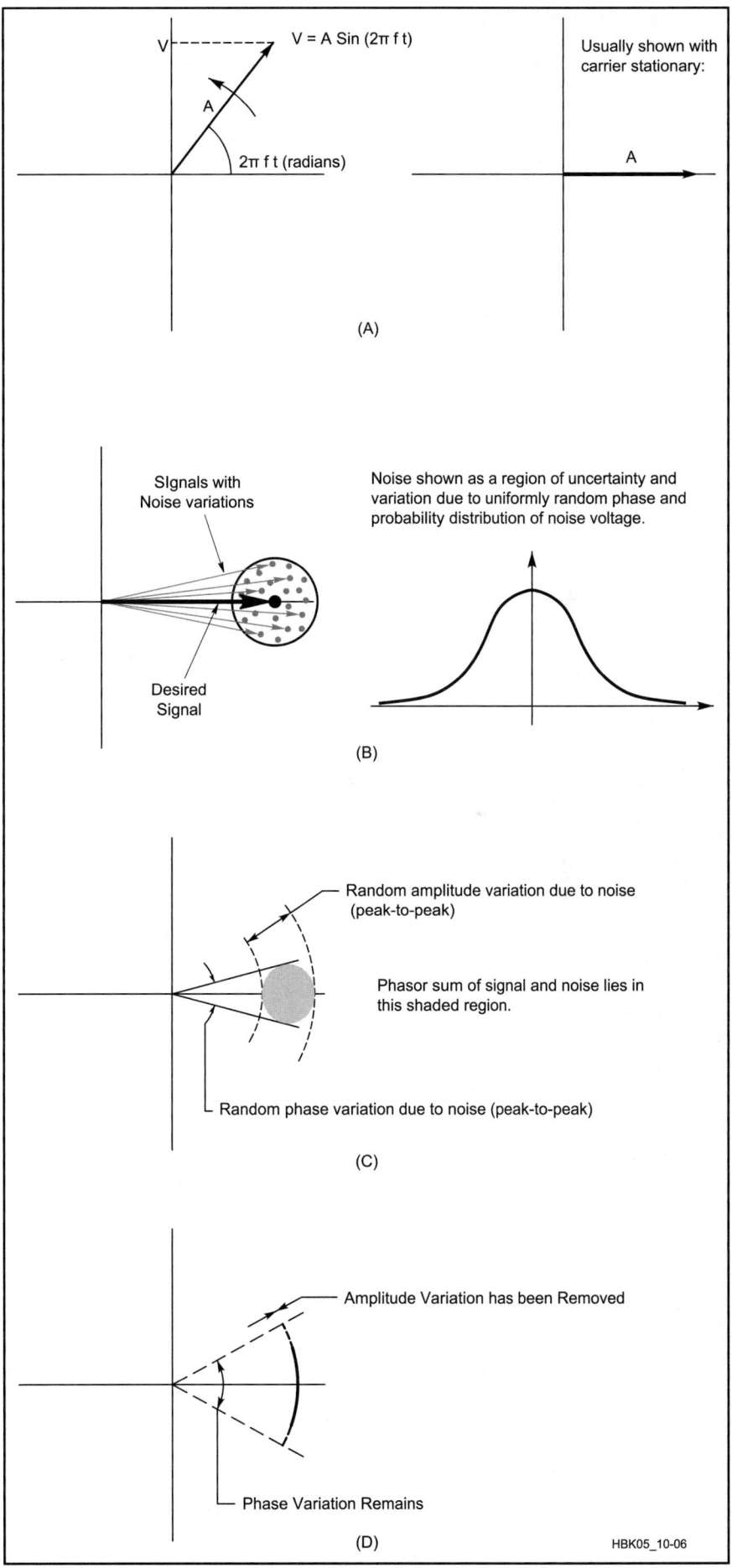

Transmitter Phase-Noise Measurement in the ARRL Lab

Here is a brief description of the technique used in the ARRL Lab to measure transmitter phase noise. The system consists of a phase noise test set with low-noise variable crystal oscillators for reference signals. The test set zero-beats the oscillator to the transmitter signal using phase detectors on both signals. As shown in **Fig 9.A1**, we use an attenuator after the transmitter to bring the signal down to a suitable level for the phase noise test set. The crystal oscillator output is low level, so it does not require attenuation.

The phase noise test set is an automated system run by a PC that steps through the test, setting parameters in the spectrum analyzers and reading the data back from them, in addition to performing system calibration measurements at the start of the test. The computer screen displays the transmitted phase-noise spectrum, which can be printed or saved to a file on the hard drive. To test the baseline phase noise of the system, two identical crystal oscillators are tested against each other. It is quite important to be sure that the phase noise of the reference source is lower than that of the signal under test.

A sample phase-noise plot for an amateur transceiver is shown in **Fig 9.A2**. It was produced with the test setup shown in Fig 9.A1. These plots do not necessarily reflect the phase-noise characteristics of all units of a particular model.

The reference level (the top horizontal line on the scale in the plot) represents 0 dBc/Hz. Because each vertical division represents 20 dB, the plot shows the noise level between 0 dBc/Hz (the top horizontal line) and −160 dBc/Hz (the bottom horizontal line). The horizontal scale is logarithmic, with one decade per division (the first division shows noise from 100 Hz to 1000 Hz offset, whereas the last division shows noise from 100 kHz through 1 MHz offset).

What Do the Phase-Noise Plots Mean?

Although they are useful for comparing different radios, plots can also be used to calculate the amount of interference you may receive from a nearby

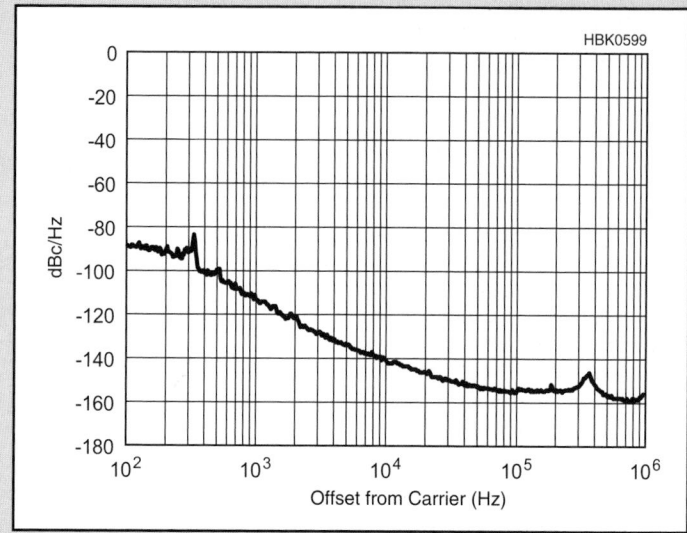

Fig 9.A2 — Sample phase noise plot for an amateur HF transceiver as published in Product Review in *QST*. This is the Elecraft K3.

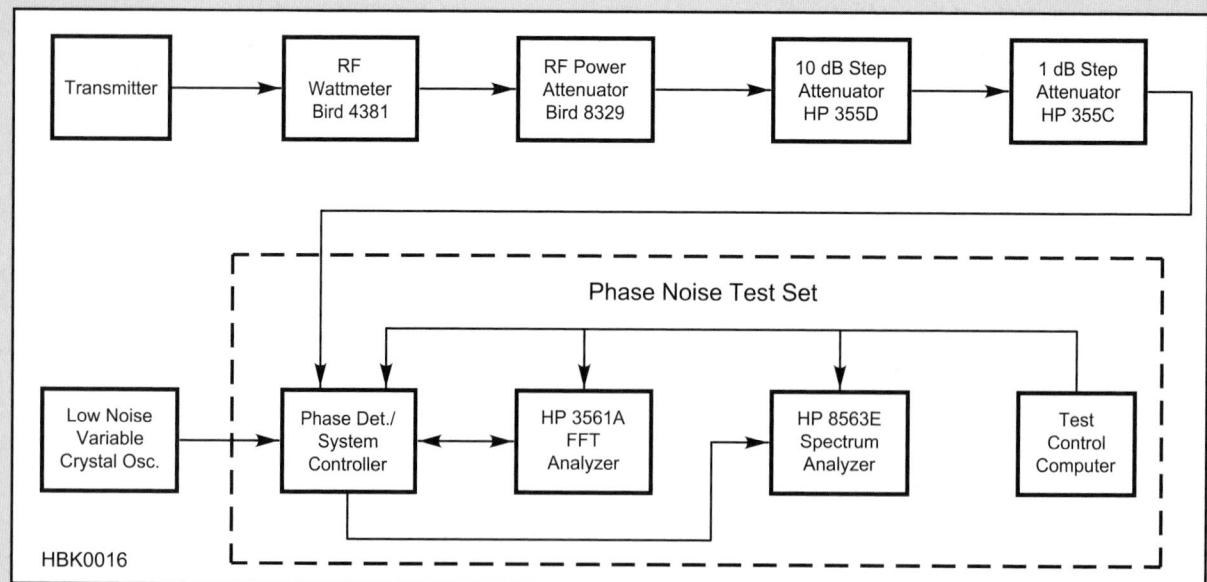

Fig 9.A1 — ARRL Lab phase-noise measurement setup.

transmitter with known phase-noise characteristics. An approximation is given by

$$A_{QRM} = NL + 10 \times \log(BW)$$

where

A_{QRM} = Interfering signal level, dBc
NL = noise level on the receive frequency, dBc
BW = receiver IF bandwidth, in Hz

For instance, if the noise level is –90 dBc/Hz and you are using a 2.5 kHz SSB filter, the approximate interfering signal will be –56 dBc. In other words, if the transmitted signal is 20 dB over S9, and each S unit is 6 dB, the interfering signal will be as strong as an S3 signal.

The measurements made in the ARRL Lab apply only to transmitted signals. It is reasonable to assume that the phase-noise characteristics of most transceivers are similar on transmit and receive because the same oscillators are generally used in local-oscillator (LO) chain.

In some cases, the receiver may have better phase noise characteristics than the transmitter. Why the possible difference? The most obvious reason is that circuits often perform less than optimally in strong RF fields, as anyone who has experienced RFI problems can tell you. A less obvious reason results from the way that many high-dynamic-range receivers work. To get good dynamic range, a sharp crystal filter called a *roofing filter* is often placed immediately after the first mixer in the receive line. This filter removes all but a small slice of spectrum for further signal processing. If the desired filtered signal is a product of mixing an incoming signal with a noisy oscillator, signals far away from the desired one can end up in this slice. Once this slice of spectrum is obtained, however, unwanted signals cannot be reintroduced, no matter how noisy the oscillators used in further signal processing. As a result, some oscillators in receivers don't affect phase noise.

The difference between this situation and that in transmitters is that crystal filters are seldom used for reduction of phase noise in transmitting because of the high cost involved. Equipment designers have enough trouble getting smooth, click-free break-in operation in transceivers without having to worry about switching crystal filters in and out of circuits at 40 WPM keying speeds! — *Zack Lau, W1VT, and Michael Tracy, KC1SX*

9.2.1 Effects of Phase Noise

Phase noise becomes a problem when it is more noticeable than other limitations. It degrades all signals, but whether it is important or not depends on the application. For voice signals it sounds like background "hiss" in headphones or speakers. It also limits the dynamic range of receivers with closely separated signals, or receiving signals with widely different input powers.

Phase noise became a significant problem for amateurs when the use of frequency synthesizers supplanted conventional LC VFOs in amateur equipment. For reasons discussed in the Synthesizers section of this chapter, it is a major task to develop a synthesizer that tunes in steps fine enough for use with SSB and CW operation while competing with the phase-noise performance of a reasonable-quality LC VFO. Many synthesizers fall far short of this target. Along with the problems with frequency synthesizers, phase noise always gets worse at higher frequencies. The trend toward general-coverage, up-converting architectures has required local oscillators to operate at higher and higher frequencies, making phase noise even more of a problem. Examples of phase noise and the problems it causes are detailed in the next two sections.

9.2.2 Reciprocal Mixing

All mixers are symmetrical, meaning that the output IF signal depends on the characteristics of both input signals: the local oscillator (LO) and the desired signal. A change to either signal shows up at the IF, where there is no way of knowing if the signal characteristics seen are from the input signal itself, or from the LO. We usually assume that the LO is very pure so only input signal modulations show up at the IF.

When there is phase noise on the LO signal, the situation changes. Noise on the LO transfers to *all* input signals at the mixer output as if it was originally present on each input signal and the LO was perfectly clean — the IF circuits can't tell the difference. This process is called *reciprocal mixing*, where noise on the LO appears as noise on the desired output signal. This is a serious limitation on a receiver's weak-signal ability.

How reciprocal mixing of LO phase noise can limit receiver dynamic range is shown in **Fig 9.8**. One possible scenario of band activity is shown as the set of input signals in Fig 9.8A. Fig 9.8B shows the phase noise profile for the LO of this receiver. Reciprocal mixing is illustrated in Fig 9.8C, showing that the LO phase noise is added to each of the signals in the band in proportion to its input power. The

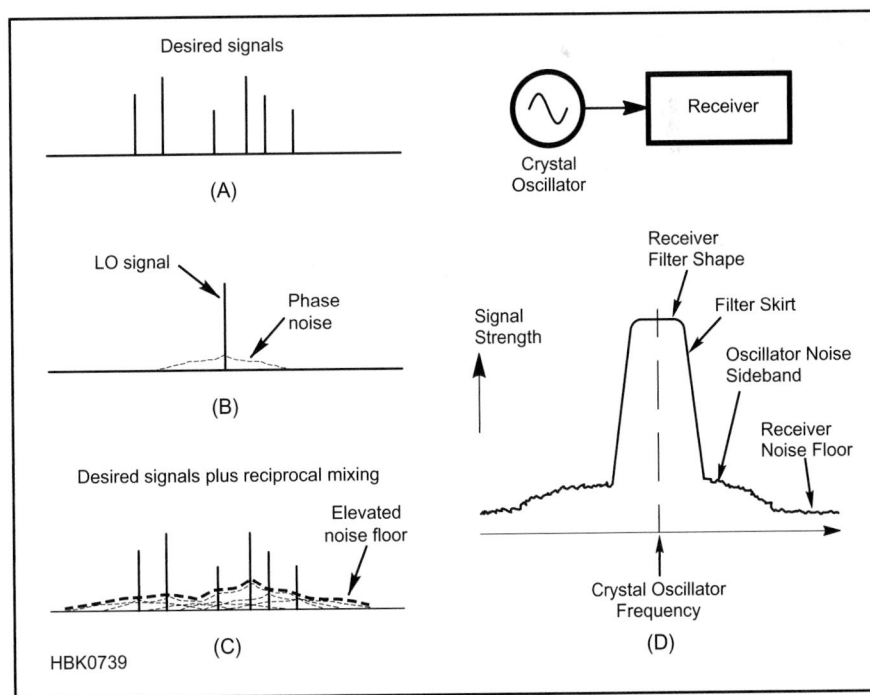

Fig 9.8 — A typical set of input signals is shown at A and an LO signal with phase noise at B. When the LO signal at B is mixed with the input signals at A, the result is a set of mixing products each having phase noise added, raising the noise floor across the band as shown at C. Phase noise in your receiver can be heard by tuning to a strong, clean crystal-oscillator signal as shown in D.

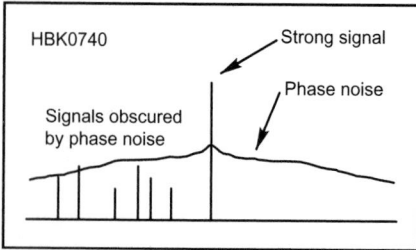

Fig 9.9 — A strong received signal can also cause such severe reciprocal mixing that desired signals are completely obscured by noise. If the signal is transmitted with phase noise from the transmitter LO, the effect is the same as noise covers the desired signals, sometimes across a very wide range of frequencies.

total noise seen by the receiver demodulator is the sum of all transferred phase noise profiles at the received frequency. This raises the apparent noise floor of the receiver.

Another receiver problem due to LO phase noise occurs if a very large signal appears in the receiver passband but at a frequency well removed from the desired signal. One example of this is shown in **Fig 9.9**. If the LO phase noise is excessive, then reciprocal mixing with this large signal can completely block the ability to demodulate the desired signals. Indeed, it is possible to make a wide swath of spectrum relatively useless. (If the strong signal is strong enough, it can result in *blocking* by causing a reduction of receiver gain. This is described in the **Receivers** and the **Test Equipment and Measurements** chapters.)

Reciprocal mixing in a receiver does not affect the operating ability of other stations. The solution is to reduce the phase noise on the receiver's LO.

9.2.3 A Phase Noise Demonstration

Healthy curiosity demands some form of demonstration so the scale of a problem can be judged "by ear" before measurements are attempted. We need to be able to measure the noise of an oscillator alone (to aid in the development of quieter ones) and we also need to be able to measure the phase noise of the oscillators in a receiver (a transmitter can be treated as an oscillator). Conveniently, a receiver contains most of the functions needed to demonstrate its own phase noise.

Because reciprocal mixing adds the LO's sidebands to clean incoming signals, in the same proportion to the incoming carrier as they exist with respect to the LO carrier, all we need do is to apply a strong, clean signal wherever we want within the receiver's tuning range. This signal's generator must have lower phase noise than the radio being evaluated. A general-purpose signal generator is unlikely to be good enough; a crystal oscillator is needed.

It's appropriate to set the oscillator's signal level into the receiver to about that of a strong broadcast carrier, say S9 + 40 dB. Set the receiver's mode to SSB or CW and tune around the test signal, looking for an increasing noise floor (higher hiss level) as you tune closer toward the signal, as shown in Fig 9.8D. Switching in a narrow CW filter allows you to hear noise closer to the carrier than is possible with an SSB filter. This is also the technique used to measure a receiver's effective selectivity, and some equipment reviewers kindly publish their plots in this format. *QST* reviews, done by the ARRL Lab, often include the results of specific phase-noise measurements.

9.2.4 Transmitted Phase Noise

Phase noise on an LO used to generate a transmitted signal will also be amplified and transmitted along with the desired output signal. This obscures reception by raising the noise floor even for receivers with low phase noise because they also receive the noise from the transmitter. In this case, the phase noise is generated externally to the receiver and must be removed at the transmitter.

If the transmitter is operating linearly, the strength of the transmitted noise is of the same proportion to transmitter output power as the phase noise is to the oscillator signal power. The noise may even extend well beyond the band in which the desired signal is transmitted unless the signal passes through narrow-band filtering that limits its bandwidth.

This transmitted noise is wasted power that is not useful for communication and unfortunately makes for a noisier band. If you are working a weak station and a nearby transmitter with a noisy oscillator comes on the air with a high power signal on a different frequency, it is possible that the output noise from this off-frequency transmitter may completely block your ability to continue working that weak station.

In bad cases, reception of nearby stations can be blocked over many tens of kilohertz above and below the frequency of the offending station. This is seen in **Fig 9.9**, where the offending signal and its associated noise are from a nearby transmitter. Frequencies close to that of the nearby transmitter are suddenly useless.

Transmitted phase noise can present serious problems for multi-station operation such as at Field Day or during emergency communications where several transmitters and receivers are in close proximity. This is a particular problem with transceivers that use early PLL-synthesized VFOs. If you are planning such an operation, be sure the level of transmitted phase noise is acceptable for the transceivers you plan on using.

At your receiver there is nothing you can do about transmitted phase noise. When there is noise power present at the same frequency as a weak (far) signal you are receiving, your receiver cannot separate them. The only solution is for the problem station to use a transmitter with a "cleaner" LO (less phase noise). This is a more serious problem than reciprocal mixing since this transmitted noise affects the ability of many other stations to use that band.

9.3 Oscillator Circuits and Construction

In this section we introduce some important oscillator circuits and provide well-tested guidelines on how to successfully build one. The following section presents a design example.

There are thousands of oscillator circuits, but just a few principal designs. One of the principal oscillator circuits is shown in **Fig 9.10**. An LC tank circuit is shown in Fig 9.10A. The usual single capacitance is replaced with two series capacitors having the same equivalent capacitance, C_{EQUIV}. The two capacitors act as an ac voltage divider, so that the voltage V1 at the midpoint is less than the total ac voltage of the tank, V_{TANK}.

What happens when we try to force V1 to be the same as V_{TANK} by adding an amplifier with unity voltage gain as shown in Fig 9.10B? The voltage division action of the capacitive divider does not change. If V1 is forced to be equal to V_{TANK} by the amplifier, then V_{TANK} will become greater than before. But then V1 will take on the new value of V_{TANK} and so forth. This is positive feedback and we have created an oscillator. Connecting an amplifier with low voltage gain to a split tank capacitance leads to the *Colpitts* group of oscillator designs.

It is certainly possible to "split" the inductor instead of the capacitor in a tank circuit and apply the same amplifier trick as before. This is shown in Fig 9.10C and describes the *Hartley* group of oscillators.

9.3.1 LC VFO Circuits

The LC oscillators used in radio equipment are usually designed to be variable frequency oscillators (VFOs). Tuning is achieved by either varying part of the capacitance of the resonator or, less commonly, by using a movable magnetic core to vary the inductance. Since the early days of radio, there has been a huge quest for the ideal, low-drift VFO. Amateurs and professionals have made immense efforts in this pursuit. A brief search of the literature reveals a large number of VFO designs, many accompanied by claims of high stability. The quest for stability has been solved by the development of low-cost frequency synthesizers, which give crystal-controlled stability. Synthesizers are generally noisier than a good VFO, so the VFO still has much to offer in terms of signal cleanliness, cost, and power consumption, making it attractive for homebrew construction. No single VFO circuit has any overwhelming advantage over any other — component quality, mechanical design and the care taken in assembly are much more important.

Fig 9.11 shows three popular oscillator circuits stripped of any non-essential components so they can be more easily compared. The original Colpitts circuit (Fig 9.11A) is now often referred to as the parallel-tuned

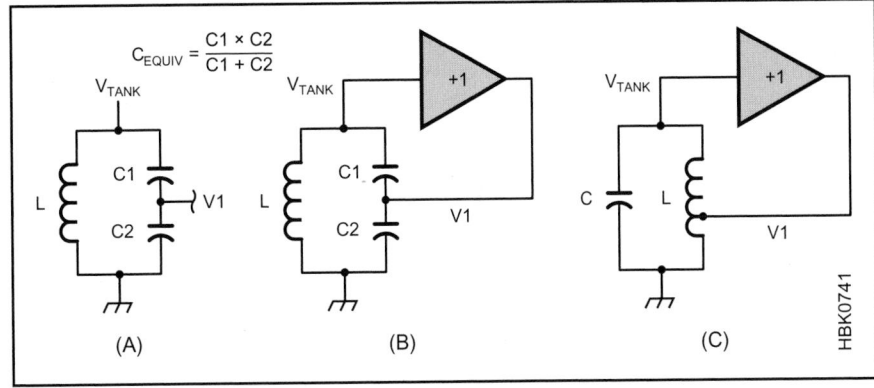

Fig 9.10 — The capacitor in an LC tank circuit (A) can be split into two capacitors with the same equivalent capacitance, C_{EQUIV}, so that the resonant frequency of the tank circuit is unchanged. (B) shows a unity-gain amplifier connected so that it forces V1 = V_{TANK}, creating positive feedback and a Colpitts oscillator results. (C) shows the same technique applied to the tank circuit inductor, creating a Hartley oscillator.

Colpitts because its series-tuned derivative (Fig 9.11B) has become the most common. All three of these circuits use an amplifier with a voltage gain less than unity, but large current gain. The N-channel JFET source follower shown appears to be the most popular current choice.

PARALLEL-TUNED COLPITTS VFO

In the parallel-tuned Colpitts, C3 and C4 are large values, perhaps 10 times larger than typical values chosen for C1 and C2 to resonate L at the desired frequency. This means only a small fraction of the total tank voltage is applied to the FET gate, and the FET can be considered to be only lightly coupled into the tank. Equally important, the values of C3 and C4 must be much larger than the FET device internal capacitances to provide good stability for the output frequency. This keeps small variations in the FET capacitances from having a significant effect on oscillator operation.

The FET is driven by the sum of the voltages across C3 and C4, while it drives the voltage across C4 alone. This means that the tank operates as a resonant, voltage-step-up transformer compensating for the less-than-unity-voltage-gain amplifier. The resonant circuit consists of L, C1, C2, C3 and C4. The resonant frequency can be calculated by using the standard formulas for capacitors in series and parallel to find the resultant capacitance effectively connected across the inductor, L, and then use the standard formula for LC resonance:

$$f = \frac{1}{2 \pi \sqrt{LC}} \quad (9.1)$$

where
  f = frequency in hertz
  L = total inductance in henries
  C = total capacitance in farads.

Equation 9.1 holds for all cases. Its accuracy is dependent on the designer's ability to account for all of the contributions to tank inductance and capacitance. There are several rules of thumb that help a VFO designer identify the various "stray" contributions to tank inductance and capacitance:

1. Wherever there is voltage, stray capacitance must be considered.

2. Wherever there is current, stray mutual inductance must be considered.

3. All currents form loops, creating inductance.

These "rules" will help point out where to look for contributions to inductance and capacitance. At frequencies below 10 MHz the actual components you see and touch tend to be all you need to consider. As the oscillator frequency increases, consideration of stray capacitance and stray inductance — reactance effects that are not associated directly with a visible component — becomes increasingly important. If the oscillator frequency is noticeably lower than what you predict from using Eq 9.1, it is almost certain that there are important stray reactances that you need to account for.

Getting back to the circuit of Fig 9.11A, for a wide tuning range C2 must be kept small to reduce the swamping effect of C3 and C4 on the variable capacitor C1. (For more information on component selection, the chapters on oscillators in *Experimental Methods for RF Design* and *Introduction to Radio Frequency Design* listed in the References provide excellent material.) "Swamping" refers to a much larger value component reducing the effect of a small component connected to it.

A parallel-tuned Colpitts oscillator is the subject of the detailed paper "A Design Example for an Oscillator for Best Phase Noise and Good Output Power" by Dr Ulrich

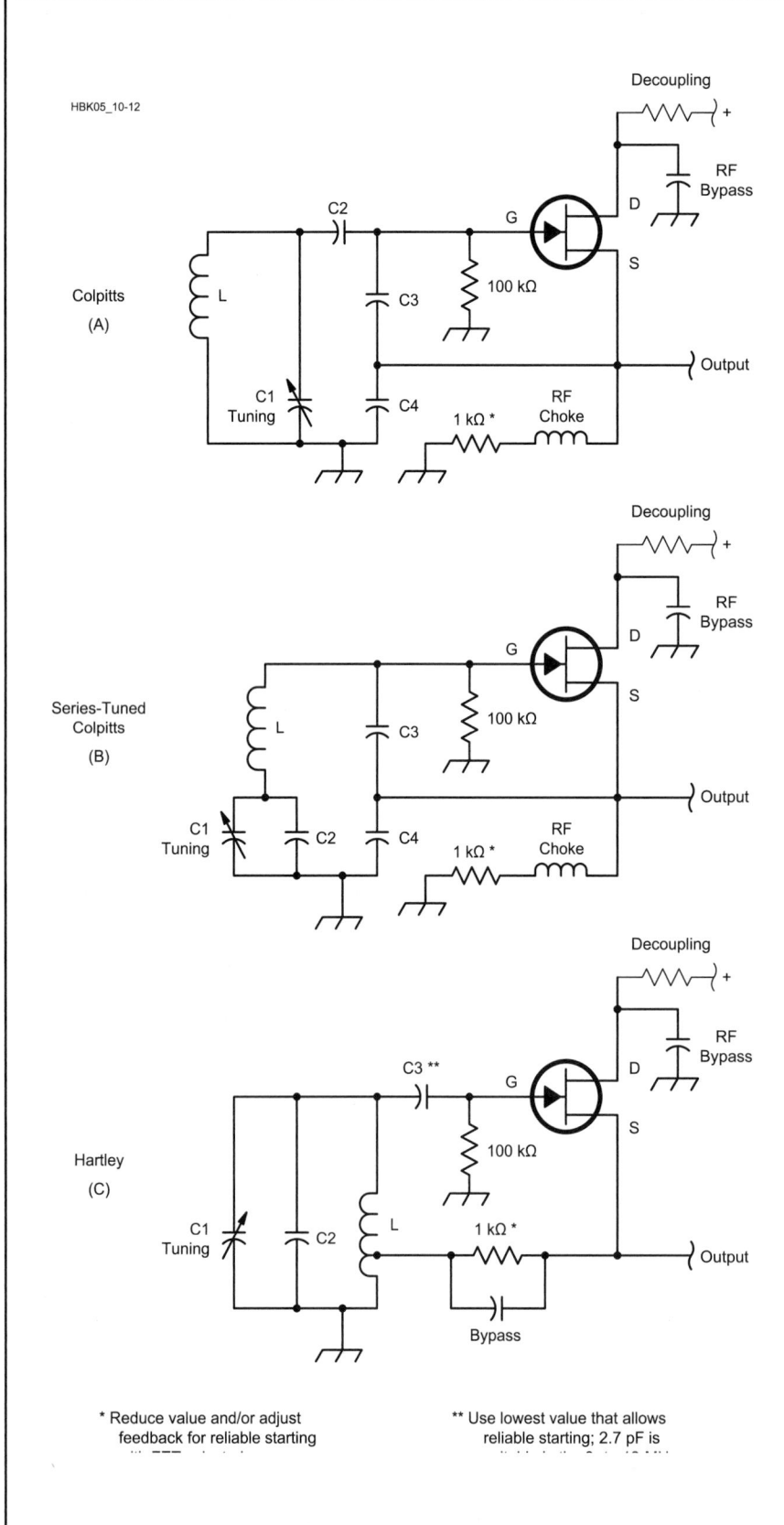

Fig 9.11 — The Colpitts (A), series-tuned Colpitts (B) and Hartley (C) oscillator circuits. Rules of thumb: C3 and C4 at A and B should be equal and valued such that their X_C = 45 Ω at the operating frequency; for C2 at A, X_C = 100 Ω. For best stability, use C0G or NP0 units for all capacitors associated with the FETs' gates and sources. Depending on the FET chosen, the 1-kΩ source-bias-resistor value shown may require adjustment for reliable starting.

* Reduce value and/or adjust feedback for reliable starting

** Use lowest value that allows reliable starting; 2.7 pF is

Rohde, N1UL, available on the CD-ROM accompanying this book. The paper shows the design process by which both noise and power can be optimized in a simple oscillator.

SERIES-TUNED COLPITTS VFO

The series-tuned Colpitts circuit works in much the same way. The difference is that the variable capacitor, C1, is positioned so that it is well-protected from being swamped by the large values of C3 and C4. In fact, small values of C3, C4 would act to limit the tuning range. Fixed capacitance, C2, is often added across C1 to allow the tuning range to be reduced to that required without interfering with C3 and C4, which set the amplifier coupling. The series-tuned Colpitts has a reputation for better stability than the parallel-tuned original. Note how C3 and C4 swamp the capacitances of the amplifier in both versions.

HARTLEY VFO

The Hartley oscillator of Fig 9.11C is similar to the parallel-tuned Colpitts, but the JFET amplifier source is connected to a tap on the tank inductance instead of the tank capacitance. A typical tap placement is 10% to 20% of the total turns up from the cold end of the inductor. (It's usual to refer to the lowest-signal-voltage end of an inductor as "cold" and the other, with the highest signal voltage as "hot".) C2 limits the tuning range as required.

C3 is reduced to the minimum value that allows reliable starting. This is necessary because the Hartley's lack of the Colpitts' capacitive divider would otherwise couple the FET's internal capacitances to the tank more strongly than in the Colpitts, potentially affecting the circuit's frequency stability.

VFO DESIGN NOTES

Generally, VFOs can be adapted to work at other frequencies (within the limits of the active device). To do so, compute an adjustment factor: f_{old} / f_{new}. Multiply the value of each frequency determining or feedback L or C by that factor. As frequency increases and amplifier gain drops, it may be required to increase feedback more than indicated by the factor.

In all three circuits, there is a 1 kΩ resistor in series with the source bias circuit. This resistor does a number of desirable things. It spoils (lowers) the Q of the inevitable low-frequency resonance of the choke with the tank tap circuit. It reduces tuning drift due to choke impedance and winding capacitance variations. It also protects against spurious oscillation at undesired frequencies due to internal choke resonances. Less obviously, it acts to stabilize the loop gain of the built-in AGC action of this oscillator. Stable operating conditions act to reduce frequency drift.

Some variations of these circuits may be found with added resistors providing a dc bias to stabilize the system quiescent current. More elaborate still are variations characterized by a constant-current source providing bias. This can be driven from a separate AGC detector system to give very tight level control. The gate-to-ground clamping diode (1N914 or similar) long used by radio amateurs as a means of avoiding gate-source conduction has been shown by Dr Ulrich Rohde, N1UL, to degrade oscillator phase-noise performance, and its use is virtually unknown in professional circles.

Fig 9.12 shows two more VFOs to illustrate the use of different devices. The vacuum-tube triode Hartley shown features permeability tuning, which has no sliding contact like that of a capacitor's rotor and can be made reasonably linear by artfully spacing the coil turns. The slow-motion drive can be done with a screw thread. The disadvantage is that special care is needed to avoid backlash and eccentric rotation of the core. If a non-rotating core is used, the slides have to be carefully designed

to prevent motion in unwanted directions. The Collins Radio Company made extensive use of tube-based permeability tuners, and a semiconductor version can still be found in a number of Ten-Tec radios.

Vacuum tubes cannot run as cool as competitive semiconductor circuits, so care is needed to keep the tank circuit away from the tube heat. In many amateur and commercial vacuum-tube oscillators, oscillation drives the tube into grid current at the positive peaks, causing rectification and producing a negative grid bias. The oscillator thus runs in Class C, in which the conduction angle reduces as the signal amplitude increases until the amplitude stabilizes. As in the FET circuits of Fig 9.11, better stability and phase-noise performance can be achieved in a vacuum-tube oscillator by moving its operating point out of true Class C — that is, by including a bypassed cathode-bias resistor (the resistance appropriate for Class A operation is a good starting value). Vacuum-tube radios are still maintained in operation and occasionally constructed for enjoyment but the

semiconductor long ago achieved dominance in VFOs.

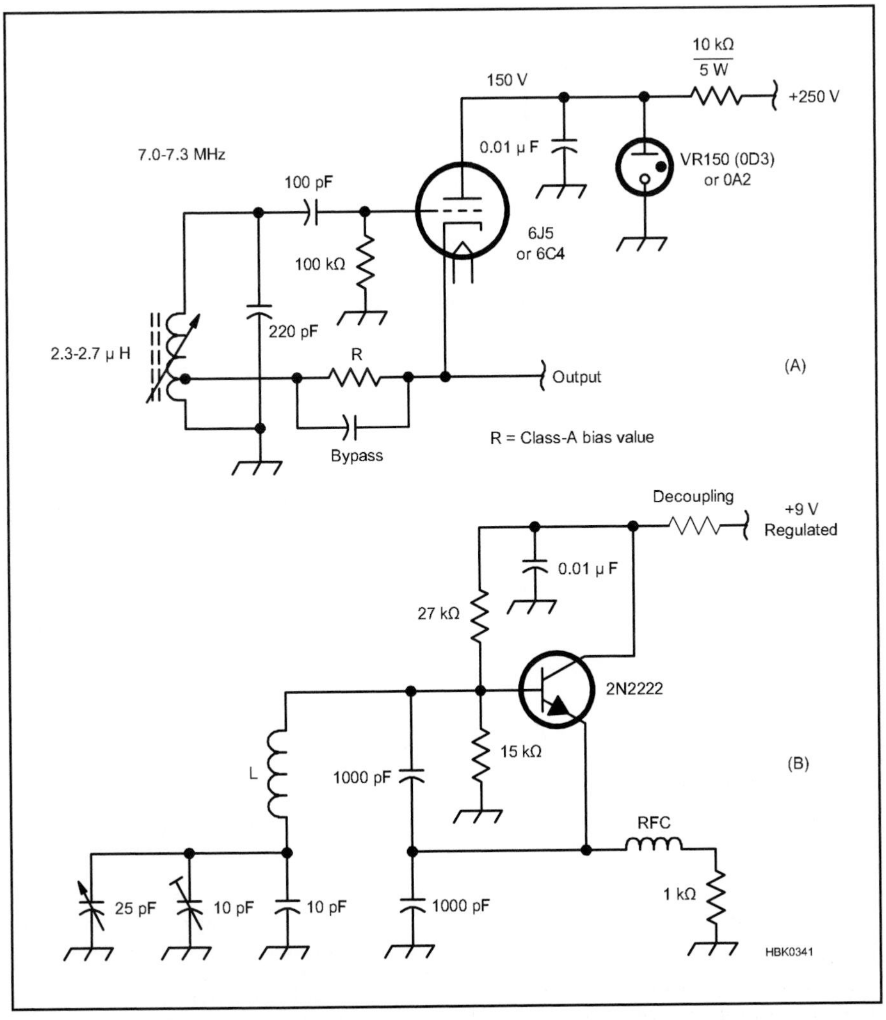

Compared to the more frequently used JFET, bipolar transistors like the one used in Fig 9.12B are relatively uncommon in oscillators because their relatively low input and output impedances are more difficult to couple into a high-Q tank without loading it excessively. Bipolar devices do tend to give better sample-to-sample amplitude uniformity for a given oscillator circuit, because many of the bipolar transistor characteristics are due less to manufacturing than physics. This is not true for JFETs of any given type; JFETs tend to vary more in their characteristics because manufacturing variations directly affect their threshold or pinch-off voltage.

SQUEGGING

The effect called *squegging* (or *squeeging*) can be loosely defined as oscillation on more than one frequency at a time, but may also manifest itself as RF oscillation interrupted at an audio frequency rate, as in a super-regenerative detector. One form of squegging occurs when an oscillator is fed from a power supply with high source impedance. The power supply charges up the decoupling capacitor until oscillation starts. The oscillator draws current and pulls down the capacitor voltage until it has starved itself of power and oscillation stops. The oscillator stops drawing current and the decoupling capacitor then recharges until oscillation restarts. The process, the low-frequency cycling of which is a form of relaxation oscillation, repeats indefinitely. The oscillator output can clearly be seen to be pulse modulated with an oscilloscope.

Squegging can be a consequence of poor design in battery-powered radios. As the cells become exhausted, their internal resistance often rises and circuits that they power can begin to misbehave. In audio stages, such misbehavior may manifest itself in the "putt-putt" sound of the slow relaxation oscillation called "motorboating."

9.3.2 RC VFO Circuits

Due to the requirements for physical length and area to create inductance, actual inductors cannot be reduced in size to the degree required for use in RF integrated circuits. Resistors and capacitors can be made very small, a fraction of a percent of the size of any inductor. Since size is money in RF integrated circuits, there is a great motivation to eliminate inductors. The problem is that any oscillator without an inductor must use resistors to develop the necessary phase shift, and resistors add loss and noise.

A 180° phase-shift network and an inverting amplifier can be used to make an oscillator with the required 360° of phase shift. A single-stage RC low-pass network cannot

Fig 9.12 — Two additional oscillator examples: at A, a triode-tube Hartley; at B, a bipolar junction transistor in a series-tuned Colpitts.

produce more than 90° of phase shift and only approaches that as the signal becomes infinitely attenuated. Two stages can only approach 180° and then also with immense attenuation. A practical RC phase-shift oscillator requires three stages to produce 180° of phase shift without destroying the loop gain. **Fig 9.13A** shows the basic form of the phase-shift oscillator; Fig 9.13B is an example of a phase-shift oscillator commonly used in commercial transceivers as an audio sidetone oscillator.

An RC oscillator can be made extremely small and to tune over a huge bandwidth (1000:1 is common). Unfortunately, Q of the phase-shift circuit is less than one, giving very poor noise performance that's unsuitable even to the least-demanding RF application.

RC oscillators are common though in digital clock generators. A common example uses a Schmitt trigger inverter as shown in Fig 9.13C. (See the discussion of comparators in the **Analog Basics** chapter.) This oscillator alternately charges and discharges the capacitor through the feedback resistor. Oscillation is assured because the inverter has hysteresis: the threshold at which output switches high is lower than the threshold at which the output switches low. The voltage across the capacitor moves between these two switching thresholds. The tuning bandwidth is the ratio of the fixed feedback resistor value to its sum with the potentiometer at full value. This ratio can be as high as 1000:1.

9.3.3 Three High-Performance HF VFOs

THE N1UL MODIFIED VACKAR VFO

The oscillator circuit of **Fig 9.14A** is contributed by Dr Ulrich Rohde, N1UL. It is a modified Vackar design (see the sidebar) in which a small coupling capacitor (8 pF) and voltage divider capacitor (18 pF) isolate the resonator circuit (10 μH and 50 pF tuning capacitor) from the oscillator transistor.

The oscillator transistor is followed by a buffer stage to isolate the oscillator from the load. Because the coupling between the transistor base and the resonator is fixed and light, stability of the oscillator is high across a wide tuning range from 5.5 to 6.6 MHz. Either the inductor or capacitor may be varied to tune the oscillator, but a variable capacitor is recommended as more practical and gives better performance.

Because of the oscillator transistor's large capacitors from base to ground (220 pF) and collector to ground (680 pF), the various parameters of the oscillator transistor have little practical influence on circuit performance. The widely available 2N3904 performs well for both the oscillator and buffer transistors.

Practical resonator coil and the tuning capacitors will have a positive temperature

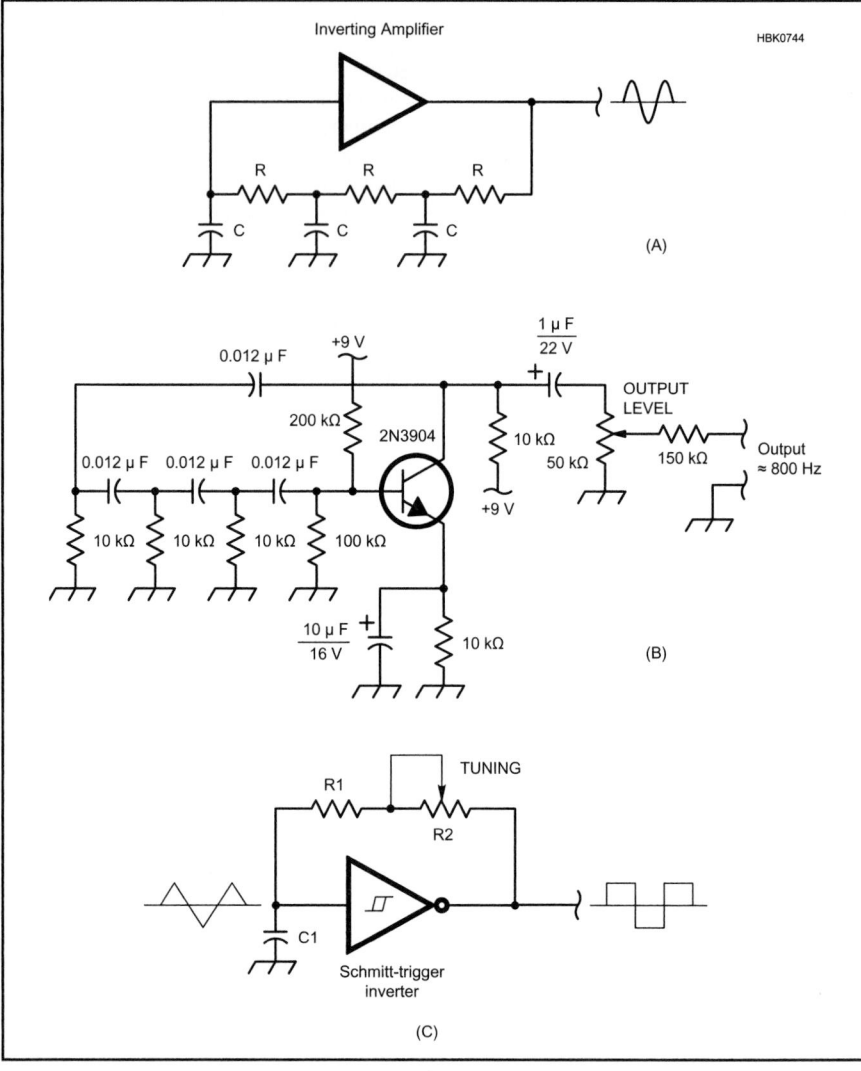

Fig 9.13 — In the basic phase-shift oscillator (A), a chain of RC networks — three or more — provide the feedback and phase shift necessary for oscillation, at the cost of low Q and considerable loop loss. Many commercial transceivers have used a phase-lead oscillator similar to that shown at B as a sidetone generator. Using a single RC circuit with a Schmitt trigger at C produces a simple oscillator with a very wide tuning range and a square wave output.

coefficient. The 8 pF and the 18 pF capacitors should have an N150 temperature coefficient to partially compensate for their drift. After 1 hour, the observed frequency drift for this circuit was less than 10 Hz / hour.

THE K7HFD LOW-NOISE DIFFERENTIAL OSCILLATOR

The other high performance oscillator example, shown in **Fig 9.15**, is designed for low-noise performance by Linley Gumm, K7HFD, and appears on page 126 of the ARRL's *Solid State Design for the Radio Amateur* (out of print, but available used and through libraries). This circuit uses no unusual components and looks simple, yet it is a subtle and sophisticated circuit.

The effects of limiting in reducing AM oscillator noise were covered previously. However, because AM noise sidebands can

The Vackar Oscillator

The original Vackar oscillator is named for Jiří Vackár, who invented the circuit in the late 1940s. The circuit's description — a refinement of the Clapp oscillator — can be found in older editions of the Radio Society of Great Britain's *Radio Communication Handbook*, with some further comments on the oscillator in RSGB's *Amateur Radio Techniques*. The circuit is also described in the Nov 1955 *QST* "Technical Correspondence" column by W9IK. The Vackar circuit optimized the Clapp oscillator for frequency stability: the oscillator transistor is isolated from the resonator, tuning does not affect the feedback coupling, and the transistor's collector output impedance is kept low so that gain is the minimum necessary to sustain oscillation.

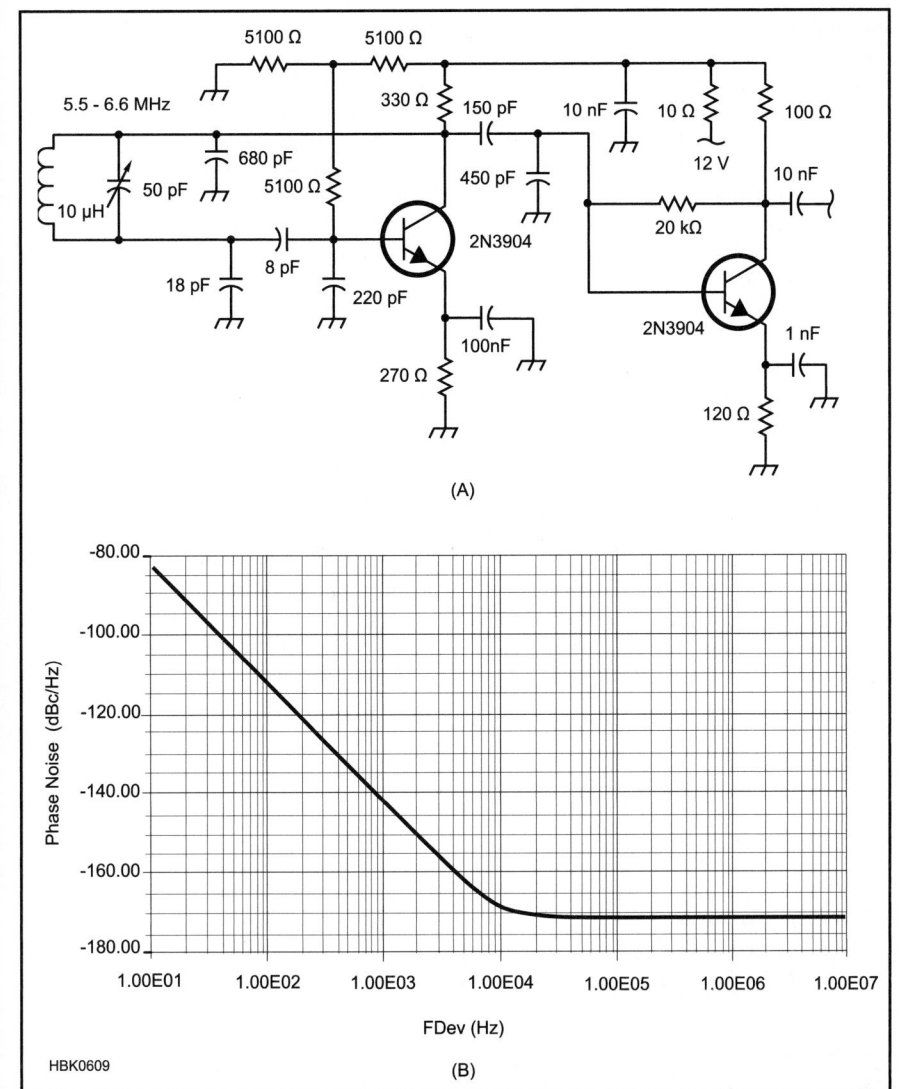

(A)

(B)

HBK0609

Fig 9.14 — At A, N1UL's Modified Vackar VFO is tuned from 5.5 to 6.6 MHz using the 50 pF capacitor. Tuning may be restricted to narrower ranges by placing a fixed capacitor in parallel with a smaller variable capacitor. The resonant frequency of the oscillator is determined by the 10 µH inductor and 50 pF tuning capacitor. B shows the excellent phase noise performance of the modified Vackar VFO in this *Harmonica* simulation. At 1 kHz from the carrier, noise is –144 dBc.

Fig 9.15 — At A, this low-noise oscillator design by K7HFD operates at an unusually high power level to achieve a high C/N (carrier-to-noise) ratio. L1 is 1.2 µH and uses 17 turns of wire on a T-68-6 toroid core. The tap is at 1 turn. Q at 10 MHz is 160. L2 is a 2 turn link over L1. At B, modeling of the differential oscillator by Dr Ulrich Rohde, N1UL, shows its excellent phase-noise performance.

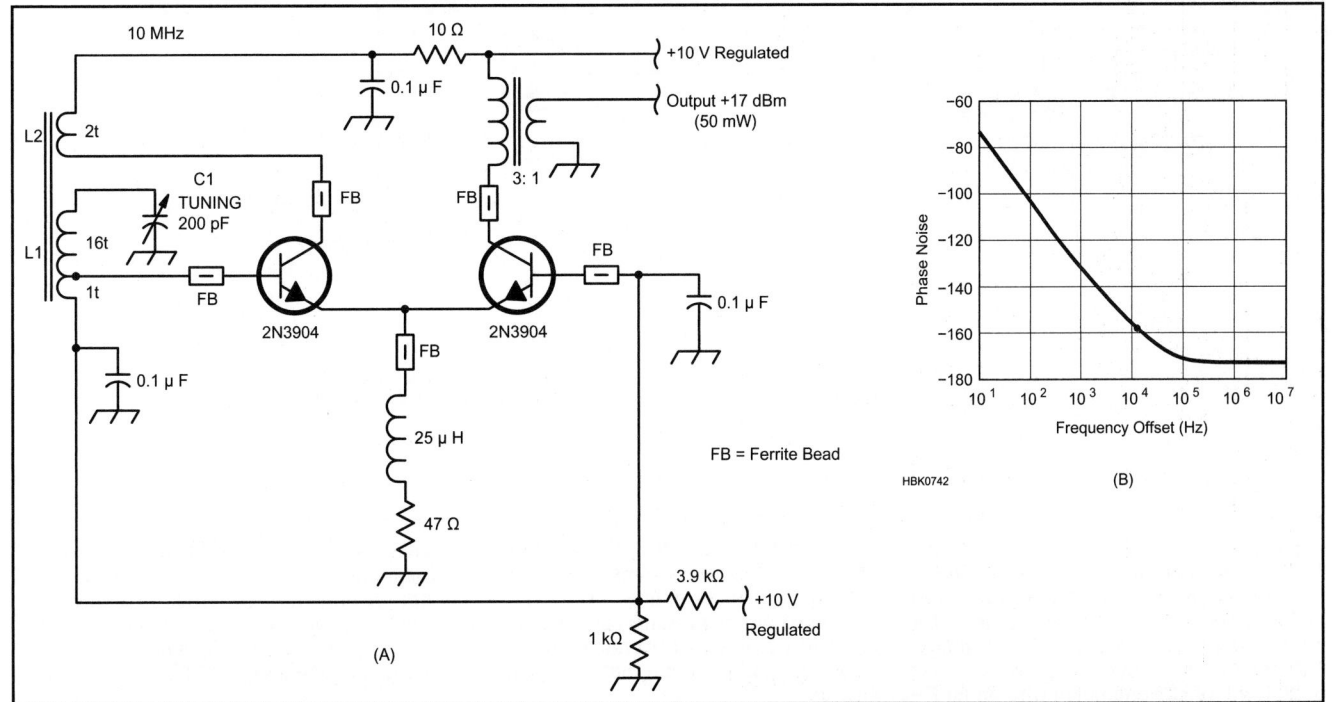

(A)

(B)

HBK0742

get translated into PM noise sidebands by imperfect limiting, there is an advantage to stripping off the AM as early as possible, in the oscillator itself. An ALC system in the oscillator will counteract and cancel only the AM components within its bandwidth, but an oscillator based on a limiter will do this over a broad bandwidth. K7HFD's oscillator uses a differential pair of bipolar transistors as a limiting amplifier. The dc bias voltage at the bases and the resistor in the common emitter path to ground establishes a controlled dc bias current, here 25 to 27 mA. The ac voltage between the bases switches this current between the two collectors. This applies a rectangular pulse of current into link winding L2, which drives the series-resonant tank L1-C1. The output impedance of the collector is high in both the current-on and current-off states. Along with the small number of turns of the link winding, this presents very high impedance to the tank circuit, which minimizes degradation of the tank Q. The input impedance of this limiter is also quite high and is applied across only a one-turn tap of L1, which similarly minimizes any impact on the tank Q. The input transistor base is driven into conduction only on one peak of the tank waveform. The output transformer has the inverse of the current pulse applied to it, so the output is not a low distortion sine wave, although the output harmonics will not be as extensive as simple theory would suggest because the circuit's high output impedance allows stray capacitances to attenuate high-frequency components. The low-frequency transistors used here also act to reduce the harmonic power.

With an output of +17 dBm, this is a power oscillator, running with nearly 300 mW of dc input power, so appreciable heating is present that can cause temperature-induced drift. The circuit's high-power operation is a deliberate ploy to create a high signal-to-noise ratio by having as high a signal power as possible. This also reduces the problem of the oscillator's broadband noise output. The limitation on the signal level in the tank is the transistors' base-emitter-junction breakdown voltage. The circuit runs with a few volts peak-to-peak across the one-turn tap, so the full tank is running at over 50 V_{P-P}.

Excessive voltage levels for the transistors can easily be generated by this circuit. The single easiest way to damage a bipolar transistor is to reverse bias the base-emitter junction until it avalanches. Most devices are rated to withstand only 5 V applied this way, the current needed to do damage is small, and very little power is needed. If the avalanche current is limited to less than that needed to perform immediate destruction of the transistor, it is likely that there will be some degradation of the device, a reduction in its bandwidth and gain along with an increase in its noise. These changes are irreversible and cumulative. Small, fast signal diodes have breakdown voltages of over 30 V and less capacitance than the transistor bases, so one possible experiment would be to try the effect of adding a fast signal diode in series with the base of each transistor and running the circuit at even higher levels.

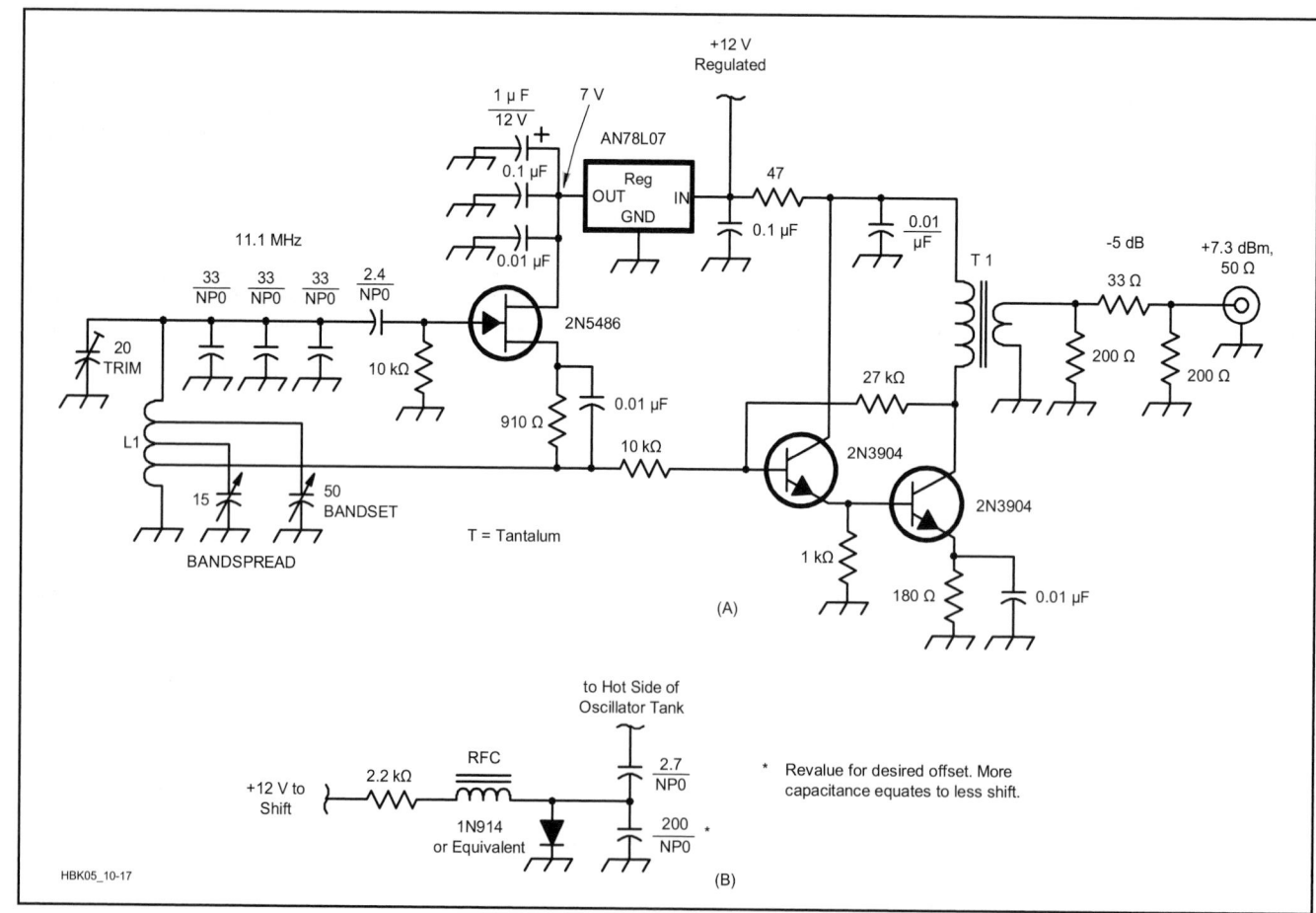

Fig 9.16 — Incorporating ideas from N1UL, KA7EXM, W7ZOI and W7EL, the oscillator at A achieves excellent stability and output at 11.1 MHz without the use of a gate-clamping diode, as well as end-running the shrinking availability of reduction drives through the use of bandset and bandspread capacitors. L1 consists of 10 turns of B & W #3041 Miniductor (#22 tinned wire, ⅝ inch in diameter, 24 turns per inch). The source tap is 2½ turns above ground; the tuning-capacitor taps are positioned as necessary for bandset and bandspread ranges required. T1's primary consists of 15 turns of #28 enameled wire on an FT-37-72 ferrite core; its secondary, 3 turns over the primary. B shows a system for adding fixed TR offset that can be applied to any LC oscillator. The RF choke consists of 20 turns of #26 enameled wire on an FT-37-43 core.

The oscillation amplitude is controlled by the drive current limit. The voltage on L2 must never allow the collector of the transistor driving it to go into saturation, for if this happens the transistor presents very low impedance to L2 and badly loads the tank, wrecking the Q and the noise performance. The circuit can be checked to verify the margin from saturation by probing the hot end of L2 and the emitter with an oscilloscope. Another, less obvious, test is to vary the power-supply voltage and monitor the output power. While the circuit is under current control, there is very little change in output power, but if the supply is low enough to allow saturation, the output power will change significantly with varying supply voltage.

The use of the 2N3904 is interesting, as it is not normally associated with RF oscillators. It is a cheap, plain, general-purpose type more often used at dc or audio frequencies. There is evidence that suggests some transistors that have good noise performance at RF have worse noise performance at low frequencies, and that the low-frequency noise they create can modulate an oscillator, creating noise sidebands. Experiments with low-noise audio transistors may be worthwhile, but many such devices have high junction capacitances.

In the description of this circuit in *Solid State Design for the Radio Amateur*, the results of a phase-noise test made using a spectrum analyzer with a crystal filter as a preselector are given. Ten kilohertz out from the carrier, in a 3 kHz measurement bandwidth, the noise was more than 120 dB below the carrier level. This translates into better than $-120 - 10 \log (3000)$, which equals -154.8 dBc/Hz, SSB, consistent with the modeled phase noise performance shown in Fig 9.15B. At this offset, -140 dBc is usually considered

VFO Construction

These suggestions by G3PDM can be applied to any analog VFO circuit. Following these guidelines minimizes the effects of temperature change and mechanical vibration (microphonics) on the VFO.

- Use silvered-mica or other highly-stable capacitors for all fixed-value capacitors in the oscillator circuit. Power-supply decoupling capacitors may be any convenient type, such as ceramic or film.
- Tuning capacitors should be a high-quality component with double ball-bearings and silver-plated surfaces and contacts.
- The resonant circuit inductor should be wound on a ceramic form and solidly mounted.
- All oscillator components should be clean and attached to a solid support to minimize thermal changes and mechanical vibration.
- The enclosure should be solid and isolated from mechanical vibration.
- The power supply should be well-regulated, liberally decoupled and free of noise.
- Keep component leads short and if point-to-point wiring is employed, use heavy wire (#16 to #18 AWG).
- Single-point grounding of the oscillator components is recommended to avoid stray inductance and to minimize noise introduced from other sources. If a PCB is used, include a ground-plane.

to be excellent. This VFO provides state-of-the art performance by today's standards — in a 1977 publication.

A JFET HARTLEY VFO

Fig 9.16 shows an 11.1 MHz version of a VFO and buffer closely patterned after that used in 7 MHz transceiver designs published by Roger Hayward, KA7EXM, and Wes Hayward, W7ZOI ("The Ugly Weekender") and Roy Lewallen, W7EL ("The Optimized QRP Transceiver"). In it, a Hartley oscillator using a 2N5486 JFET drives the two-2N3904 buffer attributed to Lewallen. This version diverges from the originals in that its JFET uses source bias (the bypassed 910 Ω resistor) instead of a gate-clamping diode and is powered from a low-current 7 V regulator

IC instead of a Zener diode and dropping resistor. The 5 dB pad sets the buffer's output to a level appropriate for "Level 7" (+7 dBm LO) diode ring mixers.

The circuit shown was originally built with a gate-clamping diode, no source bias and a 3 dB output pad. Adjusting the oscillator bias as shown increased its output by 2 dB without degrading its frequency stability (200 to 300 Hz drift at power up, stability within ±20 Hz thereafter at a constant room temperature).

In recognition that precision mechanical tuning components are hard to obtain, the resonator uses "Bandset" and "Bandspread" variable capacitors. These terms are from the early days of radio: bandset is for coarse-tuning and bandspread is for fine-tuning.

9.4 Building an Oscillator

We've covered a lot of ground about how oscillators work, their limitations and a number of interesting circuits, so the inevitable question arises of how to design one. Let's make an embarrassing confession right here: Very few oscillators you see in published circuits or commercial equipment were designed by the equipment's designer. Almost all have been adopted from other sources. While recycling in general is important for the environment, it means in this case that very few professional or amateur designers have ever designed an oscillator from scratch. We all have collections of circuits we've "harvested," and we adjust a few values or change a device type to produce something to suit a new project.

Oscillators aren't designed, they evolve. They seem to have a life of their own. The

Clarke & Hess book listed in the references contains one of the few published classical design processes. The ARRL book *Experimental Methods in RF Design* contains extensive material on oscillator circuits that is well worth reading. A demonstration by David Stockton, GM4ZNX, of oscillator design based on a simulation of how oscillators start up by the free *LTSpice* simulation program is available on the CD-ROM.

9.4.1 VFO Components and Construction

TUNING CAPACITORS AND REDUCTION DRIVES

As most commercially made radios now use frequency synthesizers, it has become

increasingly difficult to find certain key components needed to construct a good VFO. Slow-motion drives and variable capacitors are available from *QST* advertiser National RF (**www.nationalrf.com**), Dan's Small Parts and Kits (**www.danssmallpartsandkits.net**), and Antique Electronic Supply (**www.tubesandmore.com**). An alternate approach is also available: Scavenge suitable parts from old equipment; use tuning diodes instead of variable capacitors — an approach that, if uncorrected through phase locking, generally degrades stability and phase-noise performance; or use two tuning capacitors, one with a capacitance range ⅕ to ⅒ that of the other, in a bandset/bandspread approach.

Assembling a variable capacitor to a chassis and its reduction drive to a front panel can result in *backlash* — an annoying tuning effect in which rotating the capacitor shaft deforms the chassis and/or panel rather than tuning the capacitor. One way of minimizing this is to use the reduction drive to support the capacitor, and use the capacitor to support the oscillator circuit board, which is then attached to the chassis.

FIXED CAPACITORS

Traditionally, silver-mica fixed capacitors have been used extensively in oscillators, but their temperature coefficient is not as low as can be achieved with other types, and some silver micas have been known to behave erratically. Polystyrene film has become a proven alternative. One warning is worth noting: polystyrene capacitors exhibit a permanent change in value should they ever be exposed to temperatures much above 70 °C; they do not return to their old value on cooling. Particularly suitable for oscillator construction are the low-temperature-coefficient ceramic capacitors, often described as NP0 or C0G types. (NP0 and

C0G are equivalent) These names are actually temperature-coefficient codes. **Fig 9.17** contains graphs showing the behavior of three common temperature coefficients. Some ceramic capacitors are available with deliberate, controlled temperature coefficients so that they can be used to compensate for other causes of frequency drift with temperature. For example, the code N750 denotes a part with a temperature coefficient of –750 parts per million per degree Celsius. These parts are now somewhat difficult to obtain, so other methods are needed. (Values for temperature coefficients and other attributes of capacitors are presented in the **Component Data and References** chapter.)

In a Colpitts circuit, the two large-value capacitors that form the voltage divider for the active device still need careful selection. It is tempting to use any available capacitor of the right value, because the effect of these components on the tank frequency is reduced by the proportions of the capacitance values

in the circuit. This reduction is not as great as the difference between the temperature stability of an NP0 ceramic part and some of the low-cost, decoupling-quality X7R-dielectric ceramic capacitors. It's worth using low-temperature coefficient parts even in the seemingly less-critical parts of a VFO circuit — even for the bypass capacitors. Chasing the cause of temperature drift is more challenging than fun. Buy critical components like high-stability capacitors from trustworthy sources.

TUNING CAPACITOR NETWORKS

Often an available variable capacitor has greater capacitance than required for a desired frequency range. While plates can sometimes be removed, a better solution embeds the variable capacitor in a network of fixed capacitors. The evolution of this network is shown in the middle section of **Fig 9.18**. The variable capacitor, C_V, and C_2 are paralleled to form the equivalent C_{2V}. This is then placed in series with C_1 for the equivalent C_{12V}. This

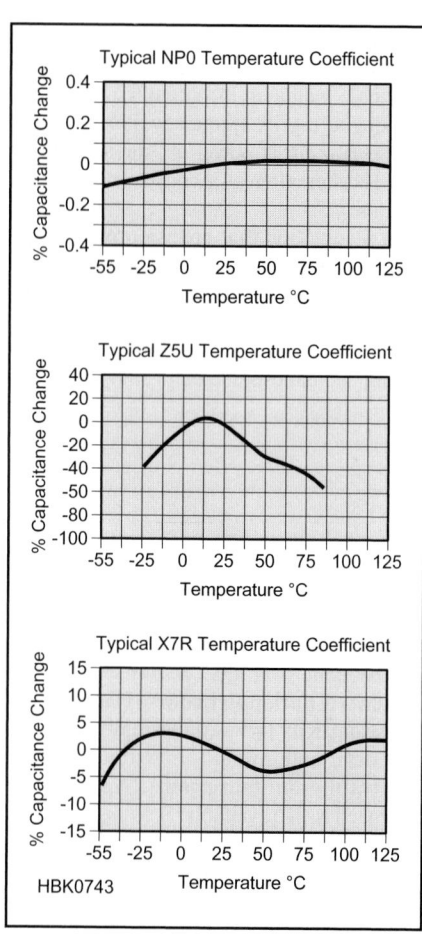

Fig 9.17 — EIA capacitor temperature coefficients specify change in capacitance with temperature. See the Component Data and References chapter for a complete table of temperature coefficient identifiers and characteristics.

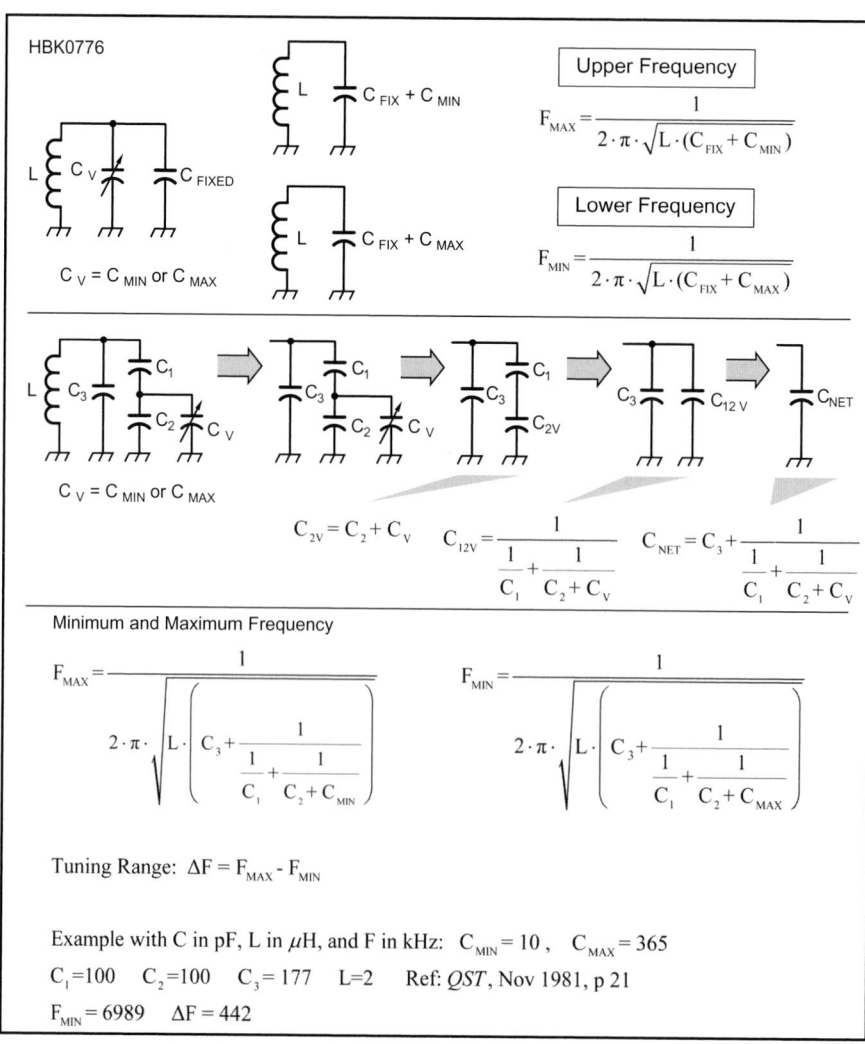

Fig 9.18 — A simple resonant circuit is tuned with parallel capacitors as shown in the top section. The tuning range is controlled by the ratio of the variable capacitance to the fixed capacitance.

is, in turn, paralleled by C_3 to form the total capacitor, C_{NET}. The overall frequency is calculated from the usual resonance relationship. The equations are shown, with capacitance in farads, inductance in henries, and frequency in Hz.

There is considerable flexibility available to the designer, afforded by picking C_1 and C_2 values. Some combinations with C_1 much smaller than the variable capacitor can produce highly nonlinear tuning.

INDUCTORS

Ceramic coil forms can give excellent results, as can self-supporting air-wound coils (Miniductor). If you use a magnetic core, support it securely and use powdered iron. Do not use ferrite because of its temperature instability. Stable VFOs have been made using toroidal cores. Micrometals mix #6 has a low temperature coefficient and works well in conjunction with NP0 ceramic capacitors. Coil forms in other materials have to be assessed on an individual basis.

A material's temperature stability will not be apparent until you try it in an oscillator, but you can apply a quick test to identify those nonmetallic materials that are lossy enough to spoil a coil's Q. Put a sample of the coil-form material into a microwave oven along with a glass of water and cook it about 10 seconds on low power. Do not include any metal fittings or ferromagnetic cores. Good materials will be completely unaffected; poor ones will heat and may even melt, smoke, or burst into flame. (This operation is a fire hazard if you try more than a tiny sample of an unknown material. Observe your experiment continuously and do not leave it unattended!)

W7ZOI suggests annealing toroidal VFO coils after winding. W7EL reports achieving success with this method by boiling his coils in water and letting them cool in air.

VOLTAGE REGULATORS

VFO circuits are often run from locally regulated power supplies, usually from resistor/Zener diode combinations. Zener diodes have some idiosyncrasies that could spoil the oscillator. They are noisy, so decoupling is needed down to audio frequencies to filter this out. Zener diodes are often run at much less than their specified optimum bias current. Although this saves power, it results in a lower output voltage than intended and the diode's impedance is much greater, increasing its sensitivity to variations in input voltage, output current and temperature. Some common Zener types may be designed to run at as much as 20 mA; check the data sheet for your diode family to find the optimum current.

True Zener diodes are low-voltage devices; above a couple of volts, so-called Zener diodes are actually avalanche types. The temperature coefficient of these diodes depends on their breakdown voltage and crosses through zero for diodes rated at about 5 V. If you intend to use nothing fancier than a common-variety Zener, designing the oscillator to run from 5 V and using a 5.1 V Zener will give you a free advantage in voltage-versus-temperature stability.

There are some diodes available with especially low temperature coefficients, usually referred to as reference or temperature-compensated diodes. These usually consist of a series pair of diodes designed to cancel each other's temperature drift. The 1N821A

diode has a temperature coefficient of ± 100 ppm/°C. Running at 7.5 mA, the 1N829A provides 6.2 V $\pm 5\%$ and a temperature coefficient of just ± 5 parts per million (ppm) maximum per degree Celsius. A change in bias current of 10% will shift the voltage less than 7.5 mV, but this increases rapidly for greater current variation. The curves in **Fig 9.19** show how the temperature coefficients of these diodes are dependent on bias current.

The LM399 is a complex IC that behaves like a superb Zener at 6.95 V, ± 0.3 ppm/°C. Precision, low-power, three-terminal regulators are also available that are designed to be used as voltage references, some of which can provide enough current to run a VFO. There are comprehensive tables of all these devices between pages 334 and 337 of Horowitz and Hill, *The Art of Electronics, 2nd ed.*

OSCILLATOR DEVICES

The 2N3819 FET, a classic from the 1960s, has proven to work well in VFOs but, like the MPF102 which is also long-popular with ham builders, it is manufactured to wide tolerances. Considering an oscillator's importance in receiver stability, you should not hesitate to spend a bit more on a better device. The 2N5484, 2N5485 and 2N5486 are worth considering; together, their transconductance ranges span that of the MPF102, but each is a better-controlled subset of that range. The 2N5245 is a more recent device with better-than-average noise performance that runs at low currents like the 2N3819. The 2N4416/A, also available as the plastic-cased PN4416, is a low-noise device, designed for VHF/UHF amplifier use, which has been featured in a

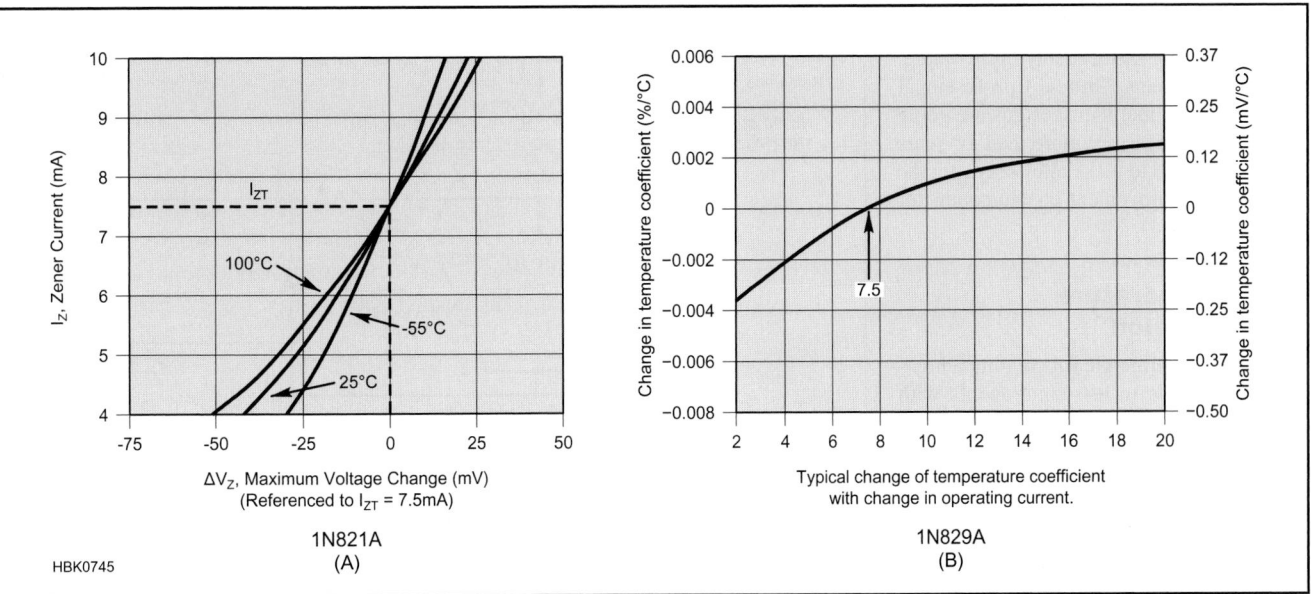

1N821A
(A)

1N829A
(B)

HBK0745

Fig 9.19 — The temperature coefficient of temperature-compensated diodes varies with bias current. To obtain the best temperature performance, use the specific bias current for the diode.

number of good oscillators up to the VHF region. Its low internal capacitance contributes to low frequency drift. The J310 (plastic; the metal-cased U310 is similar) is another popular JFET for use in oscillators.

The 2N5179 (plastic, PN5179 or MPS5179) is a bipolar transistor capable of good performance in oscillators up to the top of the VHF region. Care is needed because its absolute-maximum collector-emitter voltage is only 12 V, and its collector current must not exceed 50 mA. Although these characteristics may seem to convey fragility, the 2N5179 is sufficient for circuits powered by stabilized 6 V power supplies.

VHF-UHF capable transistors are not really necessary in LC VFOs because such circuits are rarely used above 10 MHz. (High-bandwidth transistors also increase high-frequency harmonic content in the output signal.) Absolute frequency stability is progressively harder to achieve with increasing frequency, so free-running oscillators are used only rarely to generate VHF-UHF signals for radio communication. Instead, VHF-UHF radios usually use voltage-tuned, phase-locked oscillators in some form of synthesizer. Bipolar devices like the BFR90 and MRF901, with f_T in the 5 GHz region and mounted in stripline packages, are needed for successful oscillator design at UHF.

The popular SA/NE602 mixer IC has a built-in oscillator and can be found in many published circuits. This device has separate input and output pins to the tank and has proved to be quite tame. It may not have been "improved" yet (so far, it has progressed from the SA/NE602 to the SA/NE602A, the A version affording somewhat higher dynamic range than the original SA/NE602). It might be a good idea for anyone laying out a board using one to take a little extra care to keep PCB traces short in the oscillator section to build in some safety margin so that the board can be used reliably in the future. Experienced (read: "bitten") professional designers know that their designs are going to be built for possibly more than 10 years and have learned to make allowances for the progressive improvement of semiconductor manufacture.

9.4.2 Temperature Compensation

The general principle for creating a high-stability VFO is to use components with minimal temperature coefficients in circuits that are as insensitive as possible to changes in components' secondary characteristics. Even after careful minimization of the causes of temperature sensitivity, further improvement can still be desirable. The traditional method was to split one of the capacitors in the tank so that it could be apportioned between low-temperature-coefficient parts

and parts with deliberate temperature dependency. Only a limited number of different, controlled temperature coefficients are available, so the proportioning between low coefficient and controlled coefficient parts was varied to "dilute" the temperature sensitivity of a part more sensitive than desired. This is a tedious process, involving much trial and error, an undertaking made more complicated by the difficulty of arranging means of heating and cooling the unit being compensated. (Hayward described such a means in December 1993 *QST*.) As commercial and military equipment have been based on frequency synthesizers for some time, supplies of capacitors with controlled temperature sensitivity are drying up. An alternative approach is needed.

A temperature-compensated crystal oscillator (TCXO) is an improved-stability version of a crystal oscillator that is used widely in industry. Instead of using controlled-temperature coefficient capacitors, most TCXOs use a network of thermistors and normal resistors to control the bias of a tuning diode. Manufacturers measure the temperature vs. frequency characteristic of sample oscillators, and then use a computer program to calculate the optimum normal resistor values for production. This can reliably achieve at least a tenfold improvement in stability. We here are not interested in mass manufacture, but the idea of a thermistor tuning a varactor is worth adopting. The parts involved are likely to be available for a long time.

Browsing through component suppliers' catalogs shows ready availability of 4.5 to 5 kΩ bead thermistors intended for temperature-compensation purposes, at less than a

dollar each. **Fig 9.20** shows a circuit based on this form of temperature compensation. Commonly available thermistors have negative temperature coefficients, so as temperature rises, the voltage at the counterclockwise (CCW) end of R8 increases, while that at the clockwise (CW) end drops. Somewhere near the center there is no change. Increasing the voltage on the tuning diode decreases its capacitance, so settings toward R8's CCW end simulate a negative-temperature-coefficient capacitor; toward its clockwise end, a positive-temperature-coefficient part. Choose R1 to pass 8.5 mA from whatever supply voltage is available to the 6.2 V reference diode, D1. The 1N821A/1N829A-family diode used has a very low temperature coefficient and needs 7.5 mA bias for best performance; the bridge takes the other 1 mA. R7 and R8 should be good-quality multi-turn trimmers. D2 and C1 need to be chosen to suit the oscillator circuit. Choose the least capacitance that provides enough compensation range. This reduces the noise added to the oscillator. (It is possible, though tedious, to solve for the differential varactor voltage with respect to R2 and R5, via differential calculus and circuit theory. The equations in Hayward's 1993 article can then be modified to accommodate the additional capacitors formed by D2 and C1.) Use a single ground point near D2 to reduce the influence of ground currents from other circuits. Use good-quality metal-film components for the circuit's fixed resistors.

The novelty of this circuit is that it is designed to have an easy and direct adjustment process. The circuit requires two adjustments, one at each of two different temperatures, and achieving them requires a stable frequency

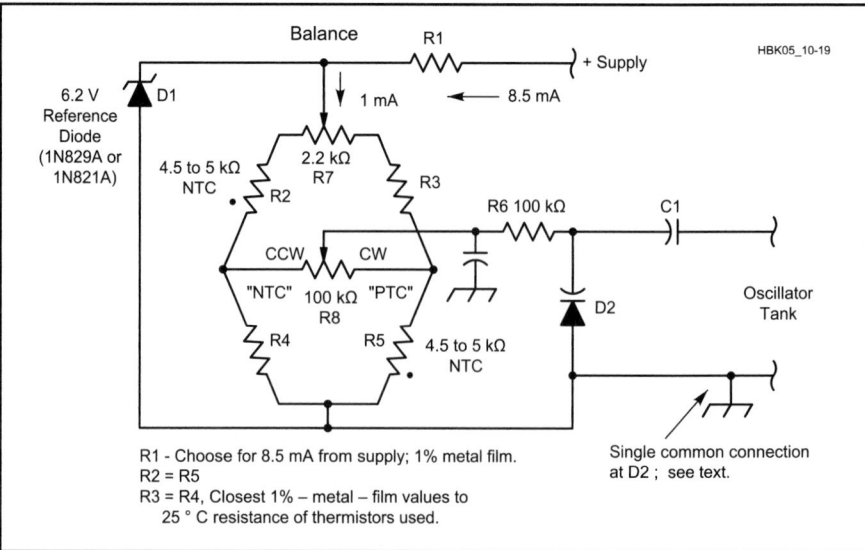

R1 - Choose for 8.5 mA from supply; 1% metal film.
R2 = R5
R3 = R4, Closest 1% – metal – film values to
 25 ° C resistance of thermistors used.

Fig 9.20 — Oscillator temperature compensation is difficult because of the scarcity of negative-temperature-coefficient capacitors. This circuit by GM4ZNX uses a bridge containing two identical thermistors to steer a tuning diode for drift correction. The 6.2 V Zener diode used (a 1N821A or 1N829A) must be a temperature-compensated part; just any 6.2 V Zener will not do.

counter that can be kept far enough from the radio so that the radio, not the counter, is subjected to the temperature extremes. (Using a receiver to listen to the oscillator under test can speed the adjustments.) After connecting the counter to the oscillator to be corrected, run the radio containing the oscillator and compensator in a room-temperature, draft-free environment until the oscillator's frequency reaches its stable operating temperature (rise over the ambient temperature). Lock its tuning, if possible. Adjust R7 to balance the bridge. This causes a drop of 0 V across R8, a condition you can reach by winding R8 back and forth across its range while slowly adjusting R7. When the bridge is balanced and 0 V appears across R8, adjusting R8 causes no frequency shift. When you've found this R7 setting, leave it there, set R8 to the exact center of its range and record the oscillator frequency.

Run the radio in a hot environment and allow its frequency to stabilize. Adjust R8 to restore the frequency to the recorded value. The sensitivity of the oscillator to temperature should now be significantly reduced between the temperatures at which you performed the adjustments. You will also have somewhat improved the oscillator's stability outside this range.

For best results with any temperature-compensation scheme, it's important to group all the oscillator and compensator components in the same enclosure, avoiding differences in airflow over components. A good oscillator should not dissipate much power, so it's feasible, even advisable, to mount all of the oscillator components in an unventilated box. In the real world, temperatures change and if the components being compensated and the components doing the compensating have different thermal time constants, a change in

temperature can cause a temporary change in frequency until the slower components have caught up. One cure for this is to build the oscillator in a thick-walled metal box that's slow to heat or cool, and so dominates and reduces the possible rate of change of temperature of the circuits inside. This is sometimes called a *cold oven*.

9.4.3 Shielding and Isolation

It is important to remember that any inductor acts as half of a transformer. Oscillators contain inductors running at moderate power levels and so can radiate strong enough signals to cause interference with other parts of a radio, or with other radios. This is the tank (or other) inductor behaving as a transformer primary. Oscillators are also sensitive to radiated signals or other nearby varying magnetic fields. This is the tank (or other) inductor also behaving as a transformer secondary. Effective shielding is therefore vital.

Any oscillator is particularly sensitive to interference on the same or very nearby frequency. If this interference is strong enough an effect called *injection locking* will occur. The oscillator effectively stops oscillating and instead directly follows the interfering signal. This situation is very bad. For example, a VFO used to directly drive a power amplifier and antenna (to form a simple CW transmitter) can prove surprisingly difficult to shield well enough because of injection locking from any leakage of the power amplifier's high-level signal back into the oscillator. Even if injection locking does not fully kick in, the wrestling inside the VFO between its own oscillation and the interference can affect its frequency, resulting in a very poor transmitted note. If the radio gear is in the station antenna's near field, there are also

strong fields that are coherent with the VFO oscillation, making sufficient shielding even more difficult.

The following rules of thumb continue to serve ham builders well:

• Use a complete metal box, with as few holes drilled in it as possible, with good contact around surface(s) where its lid(s) fit(s) on.

• Use feedthrough capacitors on power and control lines that pass in and out of the VFO enclosure and on the transmitter or transceiver enclosure as well.

• Use buffer amplifier circuitry that amplifies the signal by the desired amount and provides sufficient attenuation of signal energy flowing in the reverse direction. This is known as *reverse isolation* and is a frequently overlooked loophole in shielding. Figs 9.15 and 9.16 include buffer circuitry of proven performance. Another (and higher-cost) option is to consider using a high-speed buffer-amplifier IC (such as the LM6321N by National Semiconductor, a part that combines the high input impedance of an op amp with the ability to drive 50-Ω loads directly up into the VHF range).

• Use a mixing-based frequency-generation scheme instead of one that operates straight through or by means of multiplication. Such a system's oscillator stages can operate on frequencies with no direct frequency relationship to its output frequency. This essentially eliminates the possibility of injection locking the VFO.

• Use the time-tested technique of running your VFO at a sub-harmonic of the output signal desired — say, 3.5 MHz in a 7 MHz transmitter — and multiply its output frequency in a suitably nonlinear stage for further amplification at the desired frequency. This does reduce the tendency to injection lock.

9.5 Crystal Oscillators

Because crystals afford Q values and frequency stabilities that are orders of magnitude better than those achievable with LC circuits, fixed-frequency oscillators usually use quartz crystal resonators. Master references for frequency counters and synthesizers are always based on crystal oscillators.

So glowing is the executive summary of the crystal's reputation for stability that newcomers to radio experimentation naturally believe that the presence of a crystal in an oscillator will force oscillation at the frequency stamped on the can. This impression is usually revised after the first few experiences to the contrary! There is no sure-fire crystal oscillator circuit (although some are better than others); reading and experience soon provide a learner with plenty of anecdotes

to the effect that:

• Some circuits have a reputation of being temperamental, even to the point of not always starting.

• Crystals sometimes mysteriously oscillate on unexpected frequencies.

Even crystal manufacturers have these problems, so don't be discouraged from building crystal oscillators. The occasional uncooperative oscillator is a nuisance, not a disaster, and it just needs a little individual attention. Knowing how a crystal behaves is the key to a cure.

Dr Ulrich Rohde, N1UL, has generously contributed a pair of detailed papers that discuss the crystal oscillator along with several HF and VHF designs. Both papers, "Quartz Crystal Oscillator Design" and "A Novel

Grounded Base Oscillator Design for VHF/UHF Frequencies" are included on the CD-ROM accompanying this book.

9.5.1 Quartz and the Piezoelectric Effect

Quartz is a crystalline material with a regular atomic structure that can be distorted by the simple application of force. Remove the force, and the distorted structure springs back to its original form with very little energy loss. This property allows *acoustic waves* — sound — to propagate rapidly through quartz with very little attenuation, because the velocity of an acoustic wave depends on the elasticity and density (mass/volume) of the medium through which the wave travels.

If you heat a material, it expands. Heating may cause other characteristics of a material to change — such as elasticity, which affects the speed of sound in the material. In quartz, however, expansion and change in the speed of sound are very small and tend to cancel, which means that the transit time for sound to pass through a piece of quartz is very stable.

The third property of this wonder material is that it is *piezoelectric*. Apply an electric field to a piece of quartz and the crystal lattice distorts just as if a force had been applied. The electric field applies a force to electrical charges locked in the lattice structure. These charges are captive and cannot move around in the lattice as they can in a semiconductor, for quartz is an insulator. A capacitor's dielectric stores energy by creating physical distortion on an atomic or molecular scale. In a piezoelectric crystal's lattice, the distortion affects the entire structure. In some piezoelectric materials, this effect is sufficiently pronounced that special shapes can be made that bend *visibly* when a field is applied.

Consider a rod made of quartz. Any sound wave propagating along it eventually hits an end, where there is a large and abrupt change in acoustic impedance. Just as when an RF wave hits the end of an unterminated transmission line, a strong reflection occurs. The rod's other end similarly reflects the wave. At some frequency, the phase shift of a round trip will be such that waves from successive round trips exactly coincide in phase and reinforce each other, dramatically increasing the wave's amplitude. This is *resonance*.

The passage of waves in opposite directions forms a standing wave with antinodes at the rod ends. Here we encounter a seeming ambiguity: not just one, but a family of different frequencies, causes standing waves — a family fitting the pattern of ½, ⅔, ⁵⁄₂, ⅞ and so on, wavelengths into the length of the rod. And this is the case: A quartz rod can resonate at any and all of these frequencies.

The lowest of these frequencies, where the crystal is ½ wavelength long, is called the *fundamental* mode. The others are named the third, fifth, seventh and so on, *overtones*. There is a small phase-shift error during reflection at the ends, which causes the frequencies of the overtone modes to differ slightly from odd integer multiplies of the fundamental. Thus, a crystal's third overtone is very close to, but not exactly, three times its fundamental frequency. Many people are confused by overtones and harmonics. Harmonics are additional signals at exact integer multiples of the fundamental frequency. Overtones are not signals at all; they are additional resonances that can be exploited if a circuit is configured to excite them.

The crystals we use most often resonate in the 1 to 30 MHz region and are of the *AT-cut, thickness shear* type, although these last two characteristics are rarely mentioned.

A 15 MHz fundamental crystal of this type is about 0.15 mm thick. Because of the widespread use of reprocessed war-surplus, pressure-mounted FT-243 crystals, you may think of crystals as small rectangles on the order of a half-inch in size. The crystals we commonly use today are discs, etched and/or doped to their final dimensions, with metal electrodes deposited directly on the quartz. A crystal's diameter does not directly affect its frequency; diameters of 8 to 15 mm are typical. (Quartz crystals are also discussed in the **RF and AF Filters** chapter.)

AT-cut is one of a number of possible standard designations for the orientation at which a crystal disc is sawed from the original quartz crystal. The crystal lattice atomic structure is asymmetric, and the orientation of this with respect to the faces of the disc influences the crystal's performance. *Thickness shear* is one of a number of possible orientations of the crystal's mechanical vibration with respect to the disc. In this case, the crystal vibrates perpendicularly to its thickness. This is not easy to visualize, and diagrams don't help much, but **Fig 9.21** is an attempt at illustrating this. Place a moist bathroom sponge between the palms of your hands, move one hand up and down, and you'll see thickness shear in action.

There is a limit to how thin a disc can be made, given requirements of accuracy and price. Traditionally, fundamental-mode crystals have been made up to 20 MHz, although 30 MHz is now common at a moderately

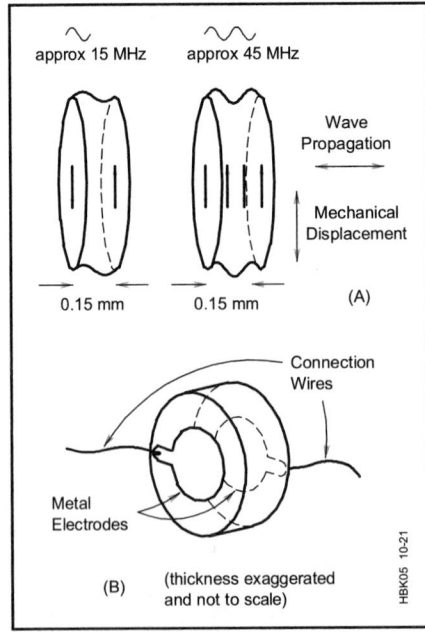

Fig 9.21 — **Thickness-shear vibration at a crystal's fundamental and third overtone (A); B shows how the modern crystals commonly used by radio amateurs consist of etched quartz discs with electrodes deposited directly on the crystal surface.**

raised price. Using techniques pioneered in the semiconductor industry, crystals have been made with a central region etched down to a thin membrane, surrounded by a thick ring for robustness. This approach can push fundamental resonances to over 100 MHz, but these are more lab curiosities than parts for everyday use. The easy solution for higher frequencies is to use a nice, manufacturably-thick crystal on an overtone mode. All crystals have multiple modes, so if you order a 28.060 MHz, third-overtone unit for a little QRP transmitter, you'll get a crystal with a fundamental resonance somewhere near 9.353333 MHz, but its manufacturer will have adjusted the thickness to plant the third overtone exactly on the ordered frequency. An accomplished manufacturer can do tricks with the flatness of the disc faces to make the wanted overtone mode a little more active and the other modes a little less active. (As some builders discover, however, this does not *guarantee* that the wanted mode is the most active!)

Quartz's piezoelectric property provides a simple way of driving the crystal electrically. Early crystals were placed between a pair of electrodes in a case. This gave amateurs the opportunity to buy surplus crystals, open them and grind them a little to reduce their thickness, thus moving them to higher frequencies. The frequency could be reduced very slightly by loading the face with extra mass, such as by blackening it with a soft pencil. Modern crystals have metal electrodes deposited directly onto their surfaces (Fig 9.21B), and such tricks no longer work.

The piezoelectric effect works both ways. Deformation of the crystal produces voltage across its electrodes, so the mechanical energy in the resonating crystal can also be extracted electrically by the same electrodes. Seen electrically, at the electrodes, the mechanical resonances look like electrical resonances. Their Q is very high. A Q of 10,000 would characterize a *poor* crystal today; 100,000 is often reached by high-quality parts. For comparison, a Q of over 200 for an LC tank is considered good.

9.5.2 Frequency Accuracy

A crystal's frequency accuracy is as outstanding as its Q. Several factors determine a crystal's frequency accuracy. First, the manufacturer makes parts with certain tolerances: ±200 ppm for a low-quality crystal for use as in a microprocessor clock oscillator, ±10 ppm for a good-quality part for professional radio use. Anything much better than this starts to get expensive! A crystal's resonant frequency is influenced by the impedance presented to its terminals, and manufacturers assume that once a crystal is brought within several parts per million of the nominal frequency, its user will perform fine adjustments electrically.

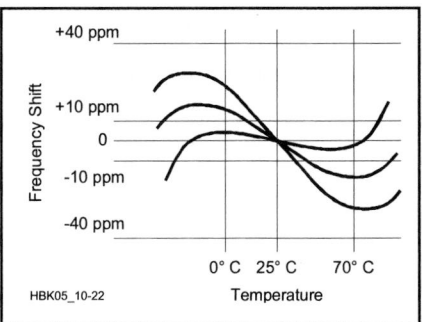

Fig 9.22 — Slight changes in a crystal cut's orientation shift its frequency-versus-temperature curve.

Second, a crystal ages after manufacture. Aging could give increasing or decreasing frequency; whichever, a given crystal usually keeps aging in the same direction. Aging is rapid at first and then slows down. Aging is influenced by the care in polishing the surface of the crystal (time and money) and by its holder style. The cheapest holder is a soldered-together, two-part metal can with glass bead insulation for the connection pins. Soldering debris lands on the crystal and affects its frequency. Alternatively, a two-part metal case can be made with flanges that are pressed together until they fuse, a process called *cold-welding*. This is much cleaner and improves aging rates roughly fivefold compared to soldered cans. An all-glass case can be made in two parts and fused together by heating in a vacuum. The vacuum raises the Q, and the cleanliness results in aging that's roughly 10 times slower than that achievable with a soldered can. The best crystal holders borrow from vacuum-tube assembly processes and have a *getter*, a highly reactive chemical substance that traps remaining gas molecules, but such crystals are used only for special purposes.

Third, temperature influences a crystal. A reasonable, professional quality part might be specified to shift not more than ±10 ppm over 0 to 70 °C. An AT-cut crystal has an *S*-shaped frequency-versus-temperature characteristic, which can be varied by slightly changing the crystal cut's orientation. **Fig 9.22** shows the general shape and the effect of changing the cut angle by only a few seconds of arc. Notice how all the curves converge at 25 °C. This is because this temperature is normally chosen as the reference for specifying a crystal. The temperature stability specification sets how accurate the manufacturer must make the cut. Better stability may be needed for a crystal used as a receiver frequency standard, frequency counter clock and so on. A crystal's temperature characteristic shows a little hysteresis. In other words, there's a bit of offset to the curve depending on whether temperature is increasing or decreasing. This is usually of no consequence except in the

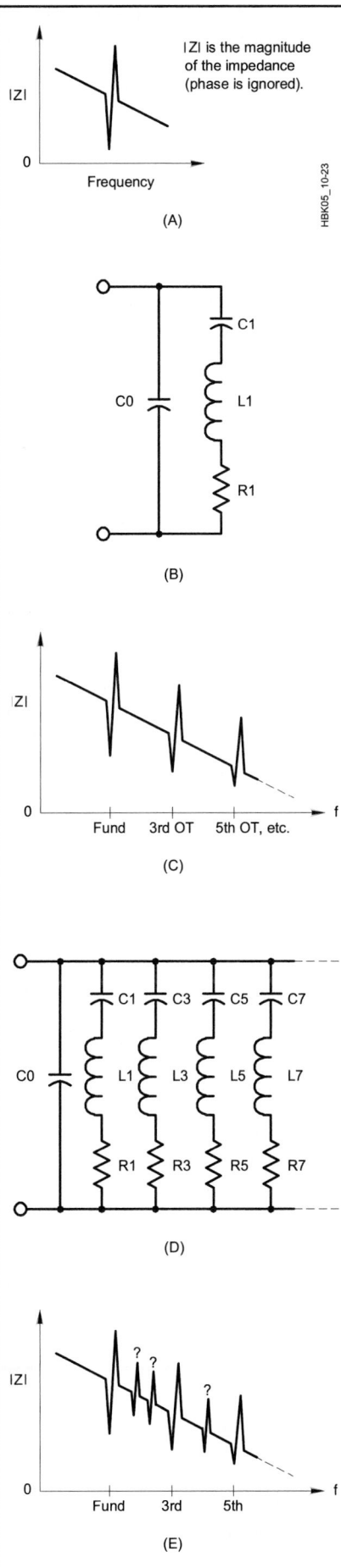

highest-precision circuits.

It is the temperature of the quartz that is important, and as the usual holders for crystals all give effective thermal insulation, only a couple of milliwatts dissipation by the crystal itself can be tolerated before self-heating becomes troublesome. Because such heating occurs in the quartz itself and does not come from the surrounding environment, it defeats the effects of temperature compensators and ovens.

The techniques shown earlier for VFO temperature compensation can also be applied to crystal oscillators. An after-compensation drift of 1 ppm is routine and 0.5 ppm is good. The result is a *temperature-compensated crystal oscillator* (*TCXO*). Recently, oscillators have appeared with built-in digital thermometers, microprocessors and ROM look-up tables customized on a unit-by-unit basis to control a tuning diode via a digital-to-analog converter (DAC) for temperature compensation. These *digitally temperature-compensated oscillators* (*DTCXOs*) can reach 0.1 ppm over the temperature range. With automated production and adjustment, they promise to become the cheapest way to achieve this level of stability.

Oscillators have long been placed in temperature-controlled *ovens*, which are typically held at 80 °C. Stability of several parts per billion can be achieved over temperature, but this is a limited benefit as aging can easily dominate the accuracy. These are usually called *oven-controlled crystal oscillators* (*OCXOs*).

Fourth, the crystal is influenced by the impedance presented to it by the circuit in which it is used. This means that care is needed to make the rest of an oscillator circuit stable, in terms of impedance and phase shift.

Gravity can slightly affect crystal resonance. Turning an oscillator upside down usually produces a small frequency shift, usually much less than 1 ppm; turning the oscillator back over reverses this. This effect is quantified for the highest-quality reference oscillators.

9.5.3 The Equivalent Circuit of a Crystal

Because a crystal is a passive, two-terminal device, its electrical appearance is that of an impedance that varies with frequency. **Fig 9.23A** shows a very simplified sketch

Fig 9.23 — Exploring a crystal's impedance (A) and equivalent circuit (B) through simplified diagrams. C and D extend the investigation to include overtones; E, to spurious responses not easily predictable by theory or controllable through manufacture. A crystal may oscillate on any of its resonances under the right conditions.

of the magnitude (phase is ignored) of the impedance of a quartz crystal. The general trend of dropping impedance with increasing frequency implies capacitance across the crystal. The sharp fall to a low value resembles a series-tuned tank, and the sharp peak resembles a parallel-tuned tank. These are referred to as series and parallel resonances. Fig 9.23B shows a simple circuit that will produce this impedance characteristic. The impedance looks purely resistive at the exact centers of both resonances, and the region between them has impedance increasing with frequency, which looks inductive.

C1 (sometimes called *motional capacitance*, C_m, to distinguish it from the lumped capacitance it approximates) and L1 (*motional inductance*, L_m) create the series resonance, and as C0 and R1 are both fairly small, the impedance at the bottom of the dip is very close to R1. At parallel resonance, L1 is resonating with C1 and C0 in series, hence the higher frequency. The impedance of the parallel tank is extremely high; the terminals are connected to a capacitive tap, which causes them to see only a small fraction of what is still a very large impedance. The overtones should not be neglected, so Figs 9.23C and 9.23D include them. Each overtone has series and parallel resonances and so appears as a series tank in the equivalent circuit. C0 again provides the shifted parallel resonance.

This is still simplified, because real-life crystals have a number of spurious, unwanted modes that add yet more resonances, as shown in Fig 9.23E. These are not well controlled and may vary a lot even between crystals made to the same specification. Crystal manufacturers work hard to suppress these spurs and have evolved a number of recipes for shaping crystals to minimize them. Just where they switch from one design to another varies from manufacturer to manufacturer.

Always remember that the equivalent circuit is just a representation of crystal behavior and does not represent circuit components actually present. Its only use is as an aid in designing and analyzing circuits using crystals. **Table 9.1** lists typical equivalent-circuit values for a variety of crystals. It is impossible to build a circuit with 0.026 to 0.0006 pF capacitors; such values would simply be swamped by strays. Similarly, the inductor must have a Q that is orders of magnitude better than is practically achievable,

and impossibly low stray C in its winding.

The values given in Table 9.1 are nothing more than rough guides. A crystal's frequency is tightly specified, but this still allows inductance to be traded for capacitance. A good manufacturer could hold these characteristics within a ±25% band or could vary them over a 5:1 range by special design. Similarly marked parts from different sources vary widely in motional inductance and capacitance.

Quartz is not the only material that behaves in this way, but it is the best. Resonators can be made out of lithium tantalate and a group of similar materials that have lower Q, allowing them to be *pulled* over a larger frequency range in VCXOs. Much more common, however, are ceramic resonators based on the technology of the well-known ceramic IF filters. These have much lower Q than quartz and much poorer frequency precision. They serve mainly as clock resonators for cheap microprocessor systems in which every last cent must be saved. A ceramic resonator could be used as the basis of a wide range, cheap VXO, but its frequency stability would not be as good as a good LC VFO.

9.5.4 Crystal Oscillator Circuits

(See also the papers "Quartz Crystal Oscillator Design" and "A Novel Grounded Base Oscillator Design for VHF/UHF Frequencies" by Dr Ulrich Rohde, N1UL, included on the CD-ROM accompanying this book.)

Crystal oscillator circuits are usually categorized as series- or parallel-mode types, depending on whether the crystal's low- or high-impedance resonance comes into play at the operating frequency. The series mode is now the most common; parallel-mode operation was more often used with vacuum tubes. **Fig 9.24** shows a basic series-mode oscillator. Some people would say that it is an overtone circuit, used to run a crystal on one of its overtones, but this is not necessarily true. The tank (L-C1-C2) tunes the collector of the common-base amplifier. C1 is larger than C2, so the tank is tapped in a way that transforms to lower impedance, decreasing signal voltage, but increasing current. The current is fed back into the emitter via the crystal. The common-base stage provides a current gain

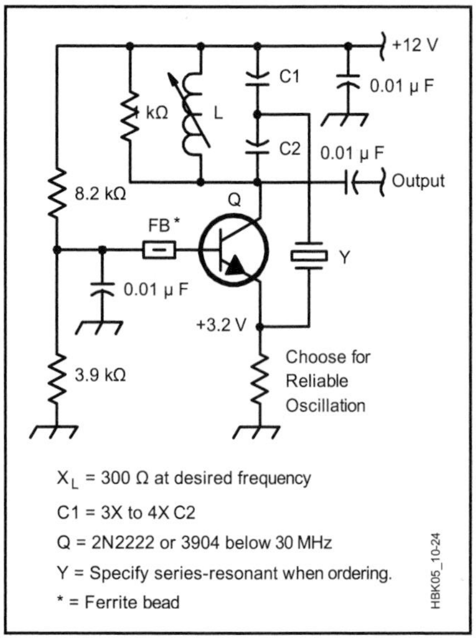

Fig 9.24 — A basic series-mode crystal oscillator. A 2N5179 can be used in this circuit if a lower supply voltage is used; see text.

of less than unity, so the transformer in the form of the tapped tank is essential to give loop gain. There are *two* tuned circuits, the obvious collector tank and the series-mode one "in" the crystal. The tank kills the amplifier's gain away from its tuned frequency, and the crystal will only pass current at the series resonant frequencies of its many modes. The tank resonance is much broader than any of the crystal's modes, so it can be thought of as the crystal setting the frequency, but the tank selecting which of the crystal's modes is active. The tank could be tuned to the crystal's fundamental, or one of its overtones.

Fundamental oscillators can be built without a tank quite successfully, but there is always the occasional one that starts up on an overtone or spurious mode. Some simple oscillators have been known to change modes while running (an effect triggered by changes in temperature or loading) or to not always start in the same mode! A series-mode oscillator should present low impedance to the crystal at the operating frequency. In Fig 9.24, the tapped collector tank presents a transformed fraction of the 1-kΩ collector load resistor to one end of the crystal, and the emitter presents a low impedance to the other. To build a practical oscillator from this circuit, choose an inductor with a reactance of about 300 Ω at the wanted frequency and calculate C1 in series with C2 to resonate with it. Choose C1 to be 3 to 4 times larger than C2. The amplifier's quiescent ("idling") current sets the gain and hence the operating level. This is not easily calculable, but can be found by experiment. Too little quiescent current and the oscillator will not start reliably; too much and the transistor can drive

Table 9.1
Typical Equivalent Circuit Values for a Variety of Crystals

Crystal Type	Series L	Series C (pF)	Series R (Ω)	Shunt C (pF)
1 MHz fundamental	3.5 H	0.007	340	3.0
10 MHz fundamental	9.8 mH	0.026	7	6.3
30 MHz third overtone	14.9 mH	0.0018	27	6.2
100 MHz fifth overtone	4.28 mH	0.0006	45	7.0

itself into saturation. If an oscilloscope is available, it can be used to check the collector waveform; otherwise, some form of RF voltmeter can be used to allow the collector voltage to be set to 2 to 3 V RMS. 3.3 kΩ would be a suitable starting point for the emitter bias resistor. The transistor type is not critical; 2N2222A or 2N3904 would be fine up to 30 MHz; a 2N5179 would allow operation as an overtone oscillator to over 100 MHz (because of the low collector voltage rating of the 2N5179, a supply voltage lower than 12 V is required). The ferrite bead on the base gives some protection against parasitic oscillation at UHF.

If the crystal is shorted, this circuit should still oscillate. This gives an easy way of adjusting the tank; it is even better to temporarily replace the crystal with a small-value (tens of ohms) resistor to simulate its *equivalent series resistance* (ESR), and adjust L until the circuit oscillates close to the wanted frequency. Then restore the crystal and set the quiescent current. If a lot of these oscillators were built, it would sometimes be necessary to adjust the current individually due to the different equivalent series resistance of individual crystals. One variant of this circuit has the emitter connected directly to the C1/C2 junction, while the crystal is a decoupler for the transistor base (the existing capacitor and ferrite bead not being used). This works, but with a greater risk of parasitic oscillation.

We commonly want to trim a crystal oscillator's frequency. While off-tuning the tank a little will pull the frequency slightly, too much detuning spoils the mode control and can stop oscillation (or worse, make the circuit unreliable). The answer to this is to add a trimmer capacitor, which will act as part of the equivalent series tuned circuit, in series with the crystal. This will shift the frequency in one way only, so the crystal frequency must be re-specified to allow the frequency to be varying around the required value. It is common to specify a crystal's frequency with a standard load (30 pF is commonly specified), so that the manufacturer grinds the crystal such that the series resonance of the specified mode is accurate when measured with a capacitor of this value in series. A 15 to 50 pF trimmer can be used in series with the crystal to give fine frequency adjustment. Too little capacitance can stop oscillation or prevent reliable starting. The Q of crystals is so high that marginal oscillators can take several seconds to start!

This circuit can be improved by driving the crystal's lower series-resonant impedance with an emitter follower as in **Fig 9.25**. This is the *Butler* oscillator. Again the tank controls the mode to either force the wanted overtone or protect the fundamental mode. The tank need not be tapped because Q2 provides current gain, although the circuit is sometimes seen with C split, driving Q2

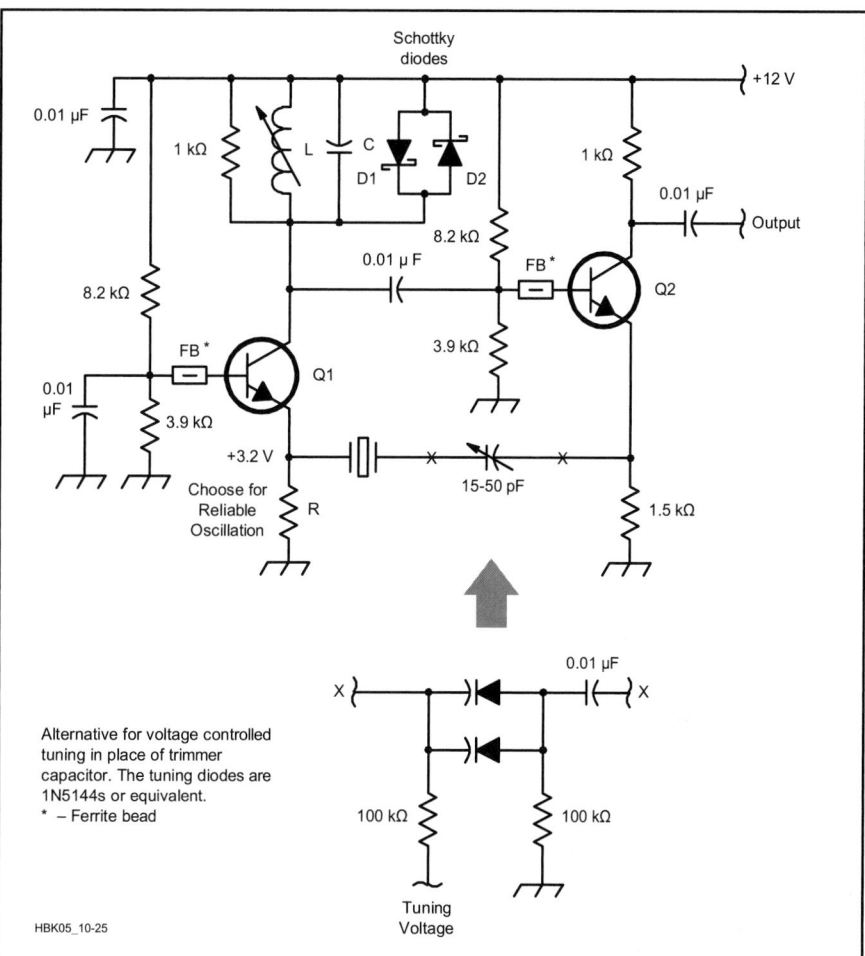

Fig 9.25 — A Butler crystal oscillator with Q2 connected as an emitter follower to drive the crystal's low series-resonant impedance.

from a tap. The position between the emitters offers a good, low-impedance environment to keep the crystal's in-circuit Q high. R, in the emitter of Q1, is again selected to give reliable oscillation. The circuit has been shown with a capacitive load for the crystal, to suit a unit specified for a 30 pF load. An alternative circuit to give electrical fine tuning is also shown. The diodes across the tank act as limiters to stabilize the operating amplitude and limit the power dissipated in the crystal by clipping the drive voltage to Q2. The tank should be adjusted to peak at the operating frequency, not used to trim the frequency. The capacitance in series with the crystal is the proper frequency trimmer.

The Butler circuit works well, and has been used in critical applications to 140 MHz (seventh-overtone crystal, 2N5179 transistor). Although the component count is high, the extra parts are cheap ones. Increasing the capacitance in series with the crystal reduces the oscillation frequency but has a progressively diminishing effect. Decreasing the capacitance pulls the frequency higher, to a point at which oscillation stops; before this

point is reached, start-up will become unreliable. The possible amount of adjustment, called *pulling range*, depends on the crystal; it can range from less than 10 to several hundred parts per million. Overtone crystals have much less pulling range than fundamental crystals on the same frequency; the reduction in pulling is roughly proportional to the square of the overtone number.

LOW-NOISE CRYSTAL OSCILLATORS

Fig 9.26A shows a crystal operating in its series mode in a series-tuned Colpitts circuit. Because it does not include an LC tank to prevent operation on unwanted modes, this circuit is intended for fundamental mode operation only and relies on that mode being the most active. If the crystal is ordered for 30 pF loading, the frequency trimming capacitor can be adjusted to compensate for the loading of the capacitive divider of the Colpitts circuit. An unloaded crystal without a trimmer would operate slightly off the exact series resonant frequency in order to create an inductive impedance to resonate

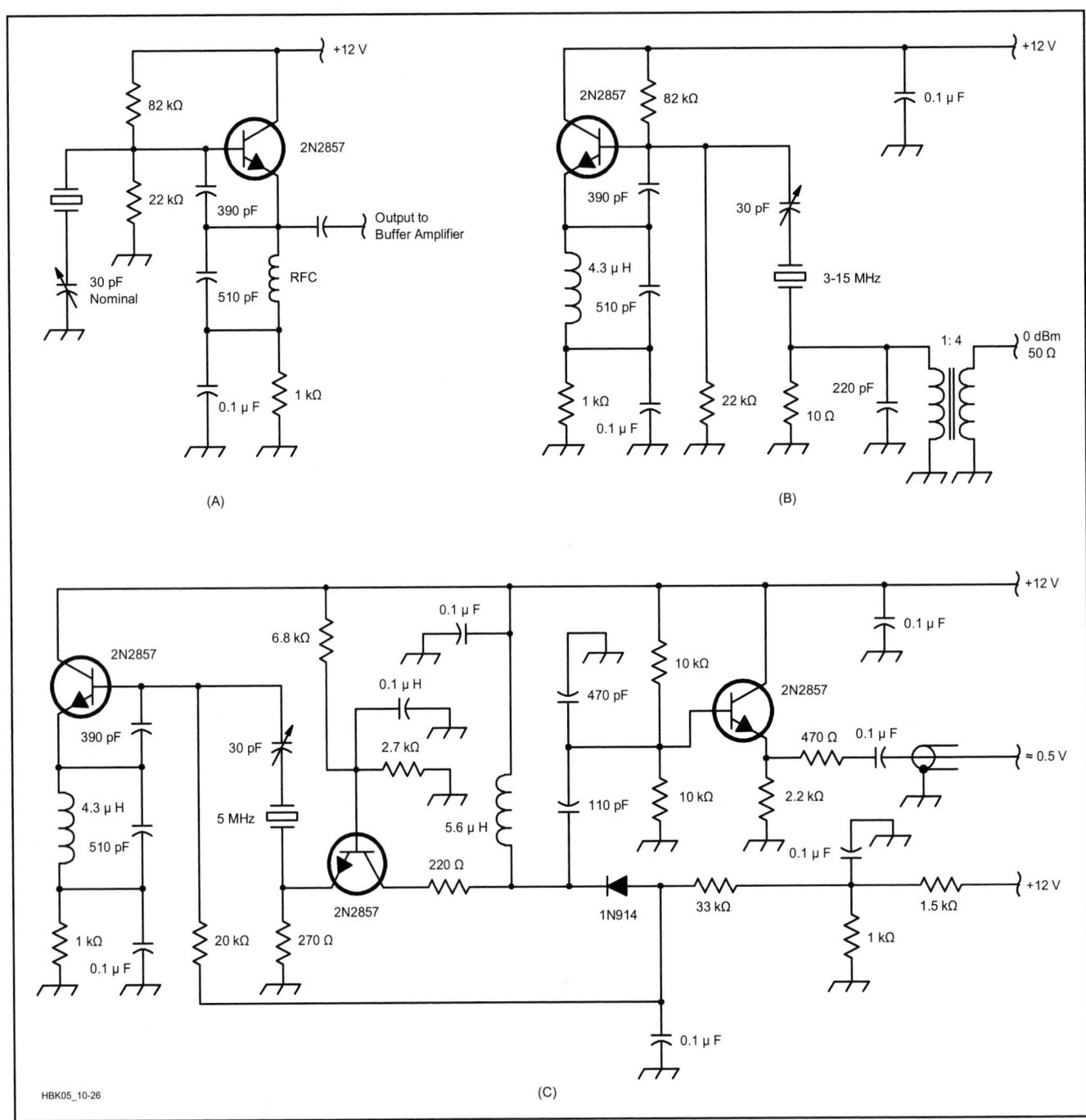

Fig 9.26 — The crystal in the series-tuned Colpitts oscillator at A operates in its series-resonant mode. B shows N1UL's low-noise version, which uses the crystal as a filter and features high harmonic suppression (from Rohde, *Microwave and Wireless Synthesizers Theory and Design*; see references). The circuit at C builds on the B version by adding a common-base output amplifier and ALC loop.

with the divider capacitors. Dr Ulrich Rohde, N1UL, in Figure 4-47 of his book *Digital PLL Frequency Synthesizers — Theory and Design*, published an elegant alternative method of extracting an output signal from this type of circuit, shown in Fig 9.26B. This taps off a signal from the current in the crystal itself. This can be thought of as using the crystal as a band-pass filter for the oscillator output. The RF choke in the emitter keeps the emitter bias resistor from loading the tank

and degrading the Q. In this case (3 MHz operation), it has been chosen to resonate close to 3 MHz with the parallel capacitor (510 pF) as a means of forcing operation on the wanted mode. The 10-Ω resistor and the transformed load impedance will reduce the in-circuit Q of the crystal, so a further development substituted a common base amplifier for the resistor and transformer. This is shown in Fig 9.26C. The common-base amplifier is run at a large quiescent current to give a very

low input impedance. Its collector is tuned to give an output with low harmonic content and an emitter follower is used to buffer this from the load. This oscillator sports a simple ALC system, in which the amplified and rectified signal is used to reduce the bias voltage on the oscillator transistor's base. This circuit is described as achieving a phase noise level of –168 dBc/Hz a few kilohertz out from the carrier. This may seem far beyond what may ever be needed, but frequency multiplication

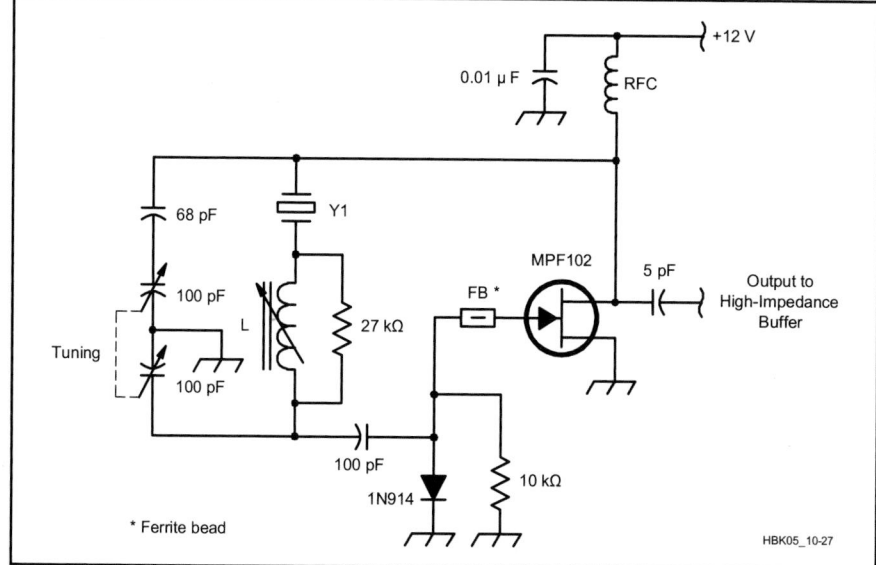

Fig 9.27 — A wide-range variable-crystal oscillator (VXO) by W7ZOI and W1FB. It was originally designed for use in low-power radios without the usual wide-range VFO.

to high frequencies, whether by classic multipliers or by frequency synthesizers, multiplies the deviation of any FM/PM sidebands as well as the carrier frequency. This means that phase noise worsens by 20 dB for each tenfold multiplication of frequency. A clean crystal oscillator and a multiplier chain is still the best way of generating clean microwave signals for use with narrow-band modulation schemes.

It has already been mentioned that overtone crystals are much harder to pull than fundamental ones. This is another way of saying that overtone crystals are less influenced by their surrounding circuit, which is helpful in a frequency-standard oscillator like this one. Even though 5 MHz is in the main range of fundamental-mode crystals and this circuit will work well with them, an overtone crystal has been used. To further help stability, the power dissipated in the crystal is kept to about 50 μW. The common-base stage is effectively driven from a higher impedance than its own input impedance, under which conditions it gives a very low noise figure.

9.5.5 VXOs

Some crystal oscillators have frequency trimmers. If the trimmer is replaced by a variable capacitor as a front-panel control, we have a *variable crystal oscillator* (*VXO*): a crystal-based VFO with a narrow tuning range, but good stability and noise performance. VXOs are often used in small, simple QRP transmitters to tune a few kilohertz around common calling frequencies. Artful constructors, using optimized circuits and components, have achieved 1000-ppm tuning ranges. Poor-quality "soft" crystals are more pull-able than high-Q ones. Overtone

crystals are not suited to VXOs. For frequencies beyond the usual limit for fundamental mode crystals, use a fundamental unit and frequency multipliers.

ICOM and Mizuho made some 2 meter SSB transceivers based on multiplied VXO local oscillators. This system is simple and can yield better performance than many expensive synthesized radios. SSB filters are available at 9 or 10.7 MHz, to yield sufficient image rejection with a single conversion. Choice of VXO frequency depends on whether the LO is to be above or below signal frequency and how much multiplication can be tolerated. Below 8 MHz, multiplier filtering is difficult. Above 15 MHz, the tuning range per crystal narrows. A 50-200 kHz range per crystal should work with a modern front-end design feeding a good 9 MHz IF, for a contest quality 2 meter SSB receiver.

The circuit in **Fig 9.27** is a JFET VXO from Wes Hayward, W7ZOI, and Doug DeMaw, W1FB, optimized for wide-range pulling. Published in *Solid State Design for the Radio Amateur*, many have been built and its ability to pull crystals as far as possible has been proven. Dr Ulrich Rohde, N1UL, has shown that the diode arrangement as used here to make signal-dependent negative bias for the gate confers a phase-noise disadvantage, but oscillators like this that pull crystals as far as possible need any available means to stabilize their amplitude and aid start-up. In this case, the noise penalty is worth paying. This circuit can achieve a 2000-ppm tuning range with amenable crystals. If you have some overtone crystals in your junk box whose fundamental frequency is close to the wanted value, they are worth trying.

This sort of circuit doesn't necessarily stop pulling at the extremes of the possible tuning

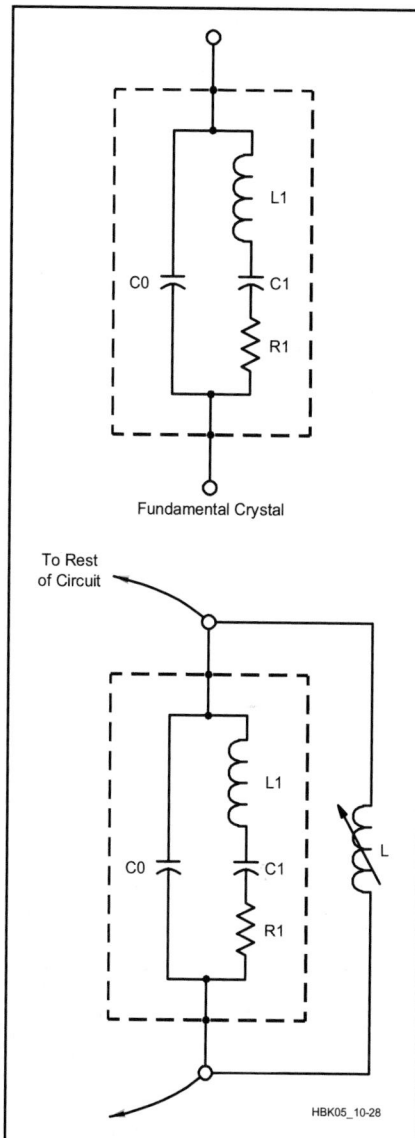

Fig 9.28 — Using an inductor to "tune out" C0 can increase a crystal oscillator's pulling range.

range; sometimes the range is set by the onset of undesirable behavior such as jumping mode or simply stopping oscillating. L was a 16 μH slug-tuned inductor for 10 MHz operation. It is important to minimize the stray and interwinding capacitance of L since this dilutes the range of impedance presented to the crystal.

One trick that can be used to aid the pulling range of oscillators is to tune out the C0 of the equivalent circuit with an added inductor. **Fig 9.28** shows how. L is chosen to resonate with C0 for the individual crystal, turning it into a high-impedance parallel-tuned circuit. The Q of this circuit is orders of magnitude less than the Q of the true series resonance of the crystal, so its tuning is much broader. The value of C0 is usually just a few picofarads, so L has to be a fairly large value considering the frequency at which it is resonated. This means that L has to have low stray capacitance or else it will self-resonate at a lower frequency. The tolerance on C0 and the variations of the stray C of the inductor means that individual adjustment is needed. This technique can also work wonders in crystal ladder filters.

9.5.6 Logic-Gate Crystal Oscillators

The frequency-determining network of an RC oscillator has a Q of less than one, which means that phase shift changes very slowly with different frequencies (see section 9.3.2 on RC Oscillators above). The Pierce crystal oscillator discussed previously is a converted phase-shift oscillator, with the crystal taking the place of one series resistor. The crystal provides a much faster phase shift than an RC stage, so the crystal "controls" the frequency of oscillation.

The actual frequency of oscillation is the frequency at which the Barkhausen criteria are met. The crystal must operate near its series resonance in order for the loop gain to be unity. Oscillation then settles at the frequency where the crystal phase shift, added to the RC phase shifts, equals 180 degrees. The crystal usually provides between 45 and 60 degrees of phase shift, so the oscillation frequency is above the series resonance and below the parallel resonance of the crystal where the crystal behaves as a large inductor. (See the figure showing crystal response in the section Quartz Crystal Filters in the **RF and AF Filters** chapter.)

The Pierce circuit is rarely seen in this full form. Instead, a cut-down version is the

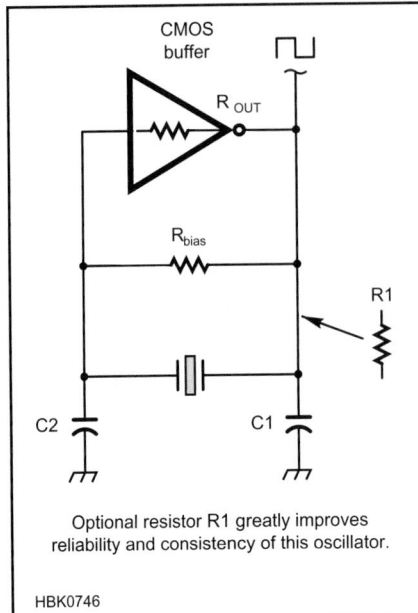

Fig 9.29 — The simplified Pierce oscillator using a logic-gate for the inverting amplifier. Adding R1 improves oscillator design and reliability.

most common circuit in many microprocessors and other digital ICs that need a crystal-controlled clock. **Fig 9.29** shows this minimalist Pierce, using a CMOS logic inverter as the amplifier. R_{bias} provides dc negative feedback to bias the gate into its linear region. This value is not critical, anything between 100 kΩ and 10 MΩ works fine. The RC phase shifts needed to make this

work come from R_{out}-C1 and $R_{crystal}$-C2. This circuit has a reputation of being temperamental, mainly because neither R_{out} nor $R_{crystal}$ are well documented or controlled in manufacturing. There is also a general belief that this circuit requires C1 = C2, which is not true.

A great improvement in performance is achieved by adding one resistor, R1, as shown in Fig 9.29. R_{bias} remains connected directly across the CMOS inverter. R1 is inserted in series from the gate output to the feedback network. The benefits are multiple:

• The edge speed of modern CMOS gates is extremely fast, so R1C1 eliminates the possibility of this fast edge exciting overtone operation.

• Phase shift into the crystal can be intentionally designed by R1 and C1 value selection.

• Drive from the inverter output is reduced into the crystal, possibly saving it from being damaged.

• Loop gain can now be controlled for best waveform and startup characteristics.

Design values are strongly dependent on the actual crystal frequency needed. C2 is selected first, to provide around 60 degrees of phase shift working against the crystal equivalent series resistance (usually a few tens of ohms, but the value needs to be verified!). The time constant R1 C1 is usefully chosen to be the reciprocal of the crystal radian frequency ($1/2\varpi f_{XTAL}$). Higher values of R1 reduce the loop gain and provide better protection for the crystal, until the loop gain gets too small and oscillation stops.

9.6 Oscillators at UHF and Above

The traditional way to make signals at higher frequencies is to make a signal at a lower frequency (where oscillators are easier) and multiply it up to the wanted range. Frequency multiplication is still one of the easiest and best ways of making a clean UHF/microwave signal. The design of a multiplier depends on whether the multiplication factor is an odd or even number. For odd multiplication, a Class-C biased amplifier can be used to create a series of harmonics; a filter selects the one wanted. For even multiplication factors, a full-wave-rectifier arrangement of distorting devices can be used to create a series of harmonics with strong even-order components, with a filter selecting the wanted component. At higher frequencies, diode-based passive circuits are commonly used. Oscillators using some of the LC circuits already described can, with care in construction, be used in the VHF range. At UHF, different approaches become necessary.

9.6.1 UHF Oscillators: Intentional and Accidental

The biggest change when working at UHF and microwave frequencies is that lumped circuit elements effectively disappear. Stray reactances and resistance dominates everything. Success at these high frequencies requires making peace with the situation and developing a very good sense of where "stray" circuit elements reside in your layout and their approximate magnitude.

Fig 9.30 shows a pair of oscillators based on a resonant length of line, which is a distributed circuit element. The first one is a return to basics: a resonator, an amplifier and

Fig 9.30 — Oscillators that use transmission-line segments as resonators. Such oscillators are more common than many of us may think, as Fig 9.31 reveals.

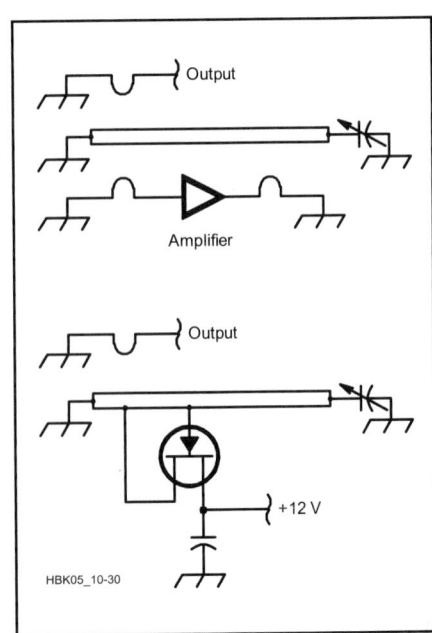

a pair of coupling loops. The amplifier can be a single bipolar or FET device or one of the monolithic microwave integrated circuit (MMIC) amplifiers. The second circuit is really a Hartley oscillator, and one is made as a test oscillator for the 70 cm band from a 10 cm length of wire suspended 10 mm over an unetched PC board as a ground plane, bent down and soldered at one end, with a trimmer at the other end. The FET is a BF981 dual-gate device used as a source follower.

No free-running oscillator is really stable enough on these bands. Oscillators in this range are almost invariably tuned with tuning diodes controlled by phase-locked-loop synthesizers, which are themselves controlled by a crystal oscillator. This transfers the same stability of the crystal oscillator to the UHF oscillator.

There is one extremely common UHF oscillator that is almost always an undesired accident, being a very common form of spurious VHF/UHF oscillation in circuitry intended to process lower-frequency signals. **Fig 9.31A** shows the circuit in its simplest

form. Analyzing this circuit using a comprehensive model of the UHF transistor reveals that the emitter presents an impedance that is small, resistive, and negative to the outside world. If this negative resistance is large enough to more than cancel the effective series resistance of a tank placed on the emitter, oscillation will occur.

At UHF your schematic diagram probably will not show a tank circuit is present. But the high frequency transistor will know it is there, since it "sees" all of the stray reactances present in the layout. Fig 9.31B shows a very basic emitter-follower circuit with some capacitance to ground on both the input and output. If the capacitor shunting the input is a distance away from the transistor, the trace to the transistor's base looks like an inductor. The trace at the emitter of the transistor also looks like an inductor, and any nearby conductor will look like a capacitor to that trace (two conductors separated by a dielectric, which here is air and the PC board material). Any intentional capacitors add to this. If the transistor has gain at any frequency where the phasing from all of these stray reactances is right for oscillation (see Barkhausen criteria in previous sections) then oscillation will happen. This circuit is effectively the same as that in Fig 9.31A. This is a good reason to use the lowest-frequency transistor that you can for any application.

Stray reactances do not always have to cause headaches. Indeed you can use them, knowing that they are there. The circuit of Fig 9.31A can be deliberately built as a useful wide-tuning oscillator covering say, 500 MHz to 1 GHz! This circuit is well-suited to construction with printed-circuit inductors. Common FR4 glass-epoxy board is lossy at these frequencies; better performance is achieved by using (the much more expensive) glass-Teflon board. If you can get surplus pieces of this type of material, it has many uses at UHF and microwave, but it is difficult to use as the adhesion between the copper and the substrate can be weak. A high-UHF transistor with a 5 GHz f_T such as the BFR90 is suitable; the base inductor can be 30 mm of 1 mm trace folded into a hairpin shape (inductance, less than 10 nH).

The upshot of this is that there is no longer any branch of electronics where RF design and layout techniques can be ignored safely. A circuit must not just be designed to do what it should do; it must also be designed so that it cannot do what it should not do.

If you have an accidental oscillator, there are three ways of taming such a circuit: adding a small resistor of perhaps 50 to 100 Ω in the collector lead close to the transistor, or adding a similar resistor in the base lead, or by fitting a ferrite bead over the base lead under the transistor. Extra resistors can disturb dc conditions, depending on the circuit and its operating currents. Ferrite beads have

the advantage that they can be easily added to existing equipment and have no effect at dc and low frequencies. Beware of some electrically conductive ferrite materials that can short transistor leads! If an HF oscillator uses beads to prevent any risk of spurs (such as shown in Fig 9.15), the beads should be anchored with a spot of adhesive to prevent movement which can cause small frequency shifts. Ferrite beads of Fair-Rite #43 material are especially suitable for this purpose; they are specified in terms of impedance, not inductance. Ferrites at frequencies above their normal usable range become very lossy and can make a lead look not inductive, but like a few tens of ohms, resistive.

9.6.2 Microwave Oscillators

Using conventional PC-board techniques with surface-mount components and extraordinarily careful layout allows the construction of circuits up to 4 GHz or so. Above this, commercial techniques and fancier materials become necessary unless we take the step to the ultimate distributed circuit — the cavity resonator.

Older than stripline techniques and far more amenable to home construction, cavity-based oscillators can give the highest possible performance at microwave frequencies. Air is a very low-loss dielectric with a dielectric constant of 1, so it gives high Q and does not force excessive miniaturization. **Fig 9.32** shows a series of structures used by G. R. Jessop, G6JP, to illustrate the evolution of a cavity from a tank made of lumped components. All cavities have a number of different modes of resonance, the orientation of the

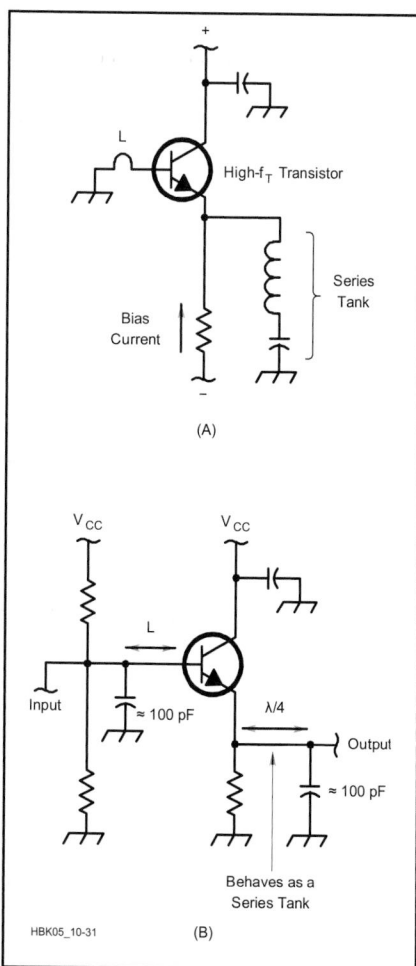

Fig 9.31 — High device gain at UHF and resonances in circuit board traces can result in spurious oscillations even in non-RF equipment.

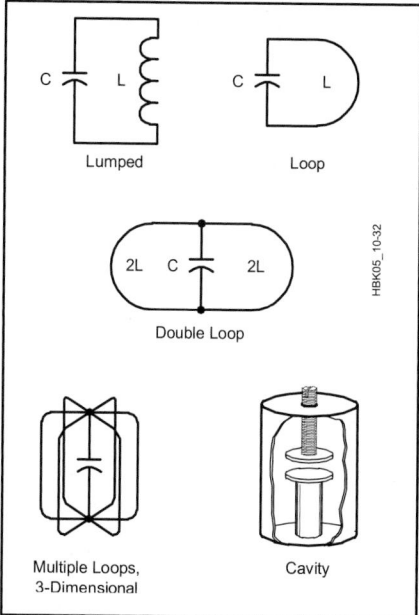

Fig 9.32 — Evolution of the cavity resonator.

currents and fields are shown in **Fig 9.33**. The cavity can take different shapes, but the one shown here has proven to suppress unwanted modes well. The gap need not be central and is often right at the top. A screw can be fitted through the top, protruding into the gap, to adjust the frequency.

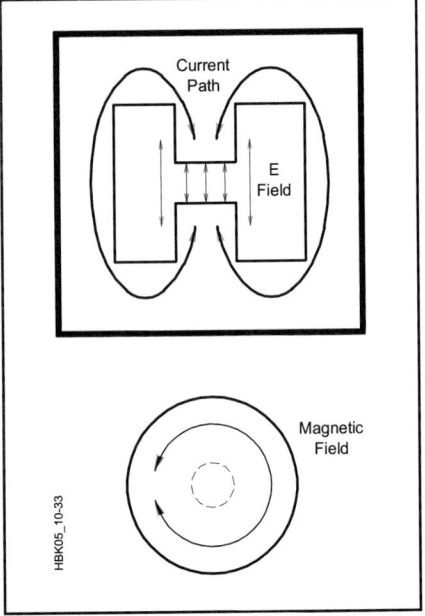

Fig 9.33 — Currents and fields in a cavity.

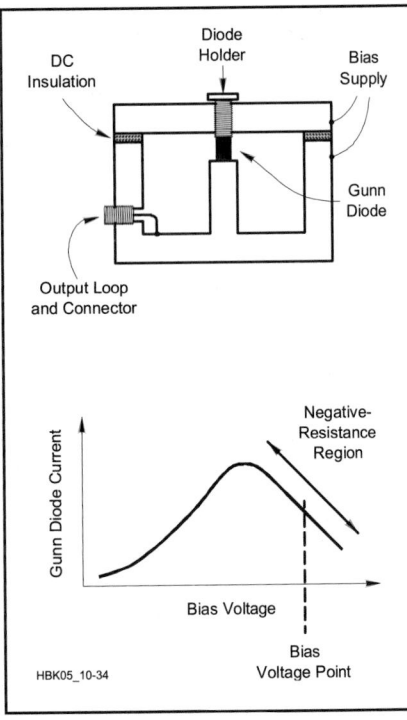

Fig 9.34 — A Gunn diode oscillator uses negative resistance and a cavity resonator to produce radio energy.

SEMICONDUCTOR CAVITY OSCILLATORS

To make an oscillator using a cavity, an amplifier is needed. Gunn and tunnel diodes have regions in their characteristics where their current falls with increasing bias voltage. This is negative resistance. If such a device is mounted in a loop in a cavity and bias is applied, the negative resistance can more than cancel the effective loss resistance of the cavity, causing oscillation. These diodes are capable of operating at extremely high frequencies and were discovered long before transistors were developed that had any gain at microwave frequencies.

A Gunn-diode cavity oscillator is the basis of many of the Doppler radar modules used to detect traffic or intruders and of the Gunnplexer 10 GHz transceiver modules used by amateurs. **Fig 9.34** shows a common configuration. The coupling loop and coax output connector could be replaced with a simple aperture to couple into waveguide or a mixer cavity. **Fig 9.35** shows a transistor cavity oscillator version using a modern microwave transistor, which can be either a FET or bipolar device. The two coupling loops are electrically completed by the capacitance of the feedthrough capacitors.

DIELECTRIC-RESONATOR OSCILLATORS (DRO)

The dielectric-resonator oscillator (DRO) is a very common microwave oscillator, as it is used in the downconverter of satellite TV receivers. The dielectric resonator itself is a ceramic cylinder, like a miniature hockey puck, several millimeters in diameter. The ceramic has a very high dielectric constant, so the surface (where ceramic meets air in an abrupt dielectric mismatch) reflects electromagnetic waves and makes the ceramic body act as a resonant cavity. It is mounted on a substrate and coupled to the active device of the oscillator by a stripline that runs past it. At 10 GHz, a FET made of gallium arsenide (GaAsFET), rather than silicon, is normally used. The dielectric resonator elements are made at frequencies appropriate to mass ap-

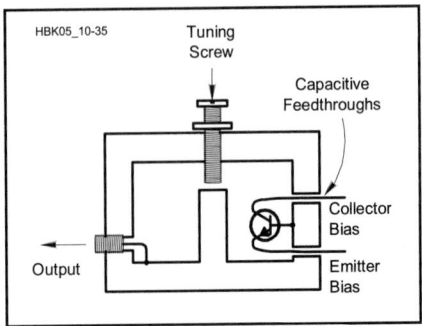

Fig 9.35 — A transistor can also directly excite a cavity resonator.

plications like satellite TV. The setup charge to manufacture small quantities at special frequencies is likely to be prohibitive for the foreseeable future.

The challenge with these devices is to devise new ways of using oscillators on industry standard frequencies. Their chief attraction is their low cost in large quantities and compatibility with microwave stripline (microstrip) techniques. Frequency stability and Q are competitive with good cavities, but are inferior to that achievable with a crystal oscillator and chain of frequency multipliers.

YIG TUNED OSCILLATORS (YTO)

The yttrium-iron garnet (YIG) oscillator is a fundamental microwave source with a very wide tuning range and a linear tuning characteristic. Many YIG oscillators can be tuned over more than an octave and some tune more than 5 octaves! They are complete units that appear as heavy blocks of metal with low-frequency connections for power supplies and tuning, and an SMA connector for the RF output. The manufacturer's label usually states the tuning range and often the power supply voltages. This is very helpful because, with new units being very expensive, it is important to be able to identify the characteristics of surplus units. The majority of YTOs operate within the 2 to 18 GHz region, although units down to 500 MHz and up to 40 GHz are occasionally found. At microwave frequencies there is no octave-tunable device that can equal their signal cleanliness and stability. A typical stability for a 3 GHz YTO is less than 1 kHz per second drift. This may seem poor — until we realize that this is 0.33 ppm per second. Still, any YIG application involving narrow-band modulation will require some form of frequency stabilizer.

Nearly all surplus RF spectrum analyzers use YTOs as their first LO. For example, a 0 to 1500 MHz analyzer usually uses a 2 to 3.5 GHz YTO with a 2.0 GHz first IF. To understand the YTO tuning circuits, should there be need for troubleshooting and repair, a basic understanding of YIG oscillators definitely helps.

Fig 9.36 shows the construction of a YIG oscillator. A YTO is based on a YIG sphere that is carefully oriented within a coupling loop. This resonator is connected to a negative-resistance device and the whole assembly is placed between the poles of an electromagnet. Negative-resistance (Gunn) diodes were originally used in these oscillators, but transistor circuits are now essentially universal and use much less power. The support for the YIG sphere often contains a controlled heater to reduce temperature variation. YIG spheres are resonant at a frequency controlled not only by their physical dimensions, but also by any magnetic field around them. Hence the electromagnet: by varying

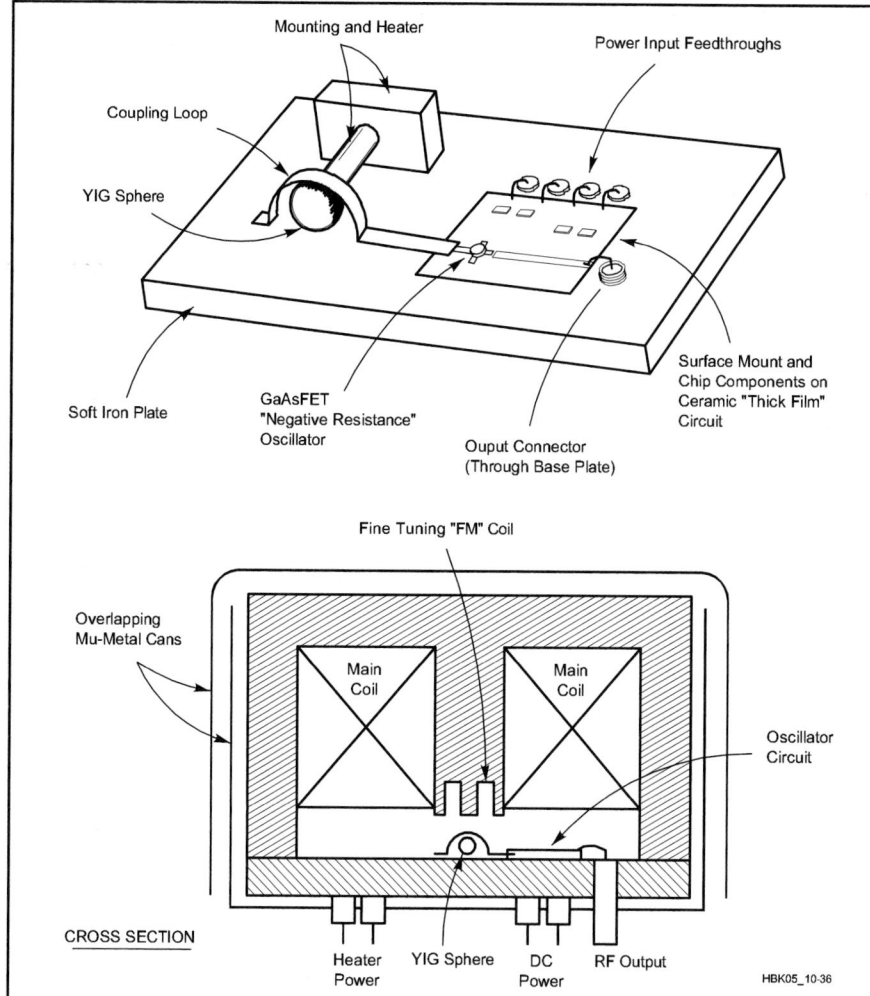

Fig 9.36 — A yttrium-iron-garnet (YIG) sphere serves as the resonator in the sweep oscillators used in many spectrum analyzers.

up to hundreds of megahertz. At frequencies higher than this, the amount of time that the electrons take to move between the cathode and the plate becomes a limiting factor. There are several special tubes designed to work at microwave frequencies, usually providing more power than can be obtained from solid-state devices.

Two of the following tubes (klystrons and traveling wave tubes) use the principle of *velocity-modulation* to extract RF energy from an electron beam. The general principles of velocity modulation and basic properties of devices using it are presented in the online tutorial **www.radartutorial.eu/druck/Book5.pdf**. Additional resources are described below.

THE KLYSTRON

The klystron tube uses the principle of velocity modulation of the electrons to avoid transit time limitations. The beam of electrons travels down a metal drift tube that has interaction gaps along its sides. RF voltages are applied to the gaps and the electric fields that they generate accelerate or decelerate the passing electrons. The relative positions of the electrons shift due to their changing velocities, causing the electron density of the beam to vary. This variation of electron beam density is used to perform amplification or oscillation.

Klystron tubes can be relatively large, and they can easily provide hundreds of watts to hundreds of kilowatts of microwave power. These power levels are useful for UHF broadcasting and particle accelerators, for example. Unfortunately klystrons have relatively narrow bandwidths, and may not be re-tunable for operation on amateur frequencies. A video titled "How a Klystron Tube Works" can be found at the YouTube online video service and a detailed history and tutorial is available at **www.slac.stanford.edu/cgi-wrap/getdoc/slac-pub-7731.pdf**.

THE MAGNETRON

The magnetron tube is an efficient oscillator for microwave frequencies. Magnetrons are most commonly found in microwave ovens and high-powered radar equipment. The anode of a magnetron is made up of a number of coupled resonant cavities that surround the cathode. The applied magnetic field causes the electrons to rotate around the cathode and the energy that they give off as they approach the anode adds to the RF electric field. The RF power is obtained from the anode through a vacuum window.

Magnetrons are self-oscillating with the frequency determined by the construction of their anodes; however, they can be tuned by coupling either inductance or capacitance to the resonant anode. The range of frequencies depends on how fast the tuning must be ac-

the current through the electromagnet's windings, a controlled variable magnetic field is applied across the YIG sphere. This tunes the oscillator across a very wide range. The frequency/current relationship can have excellent linearity, typically around 20 MHz/mA.

Magnetically tuned oscillators bring some unique problems with them. The first problem is that magnetic fields, especially at low frequencies, are extremely difficult to shield. Therefore YTO tuning is influenced by any local magnetic fields, causing frequency modulation. The YTO's magnetic core must be carefully designed to be all-enclosing in an attempt at self-shielding, and then one or more nested mu-metal cans are fitted around everything. It is still important to place the YTO away from obvious sources of magnetic fields, like power transformers. Cooling fans are also sources of fluctuating magnetic fields, with some fans generating fields 20 dB than from a well-designed 200 W 50/60-Hz transformer.

The second new problem is that the YTO's internal tuning coil needs significant current

from the power supply to create the strong tuning field. This can be eased by adding a permanent magnet as a fixed "bias" field, but the bias shifts as the magnet ages. The main tuning coil still has many turns, and therefore high inductance (often more than 0.1 henrys). Large inductances require a high supply voltage for rapid tuning, with correspondingly high power consumption. The usual compromise is to have dual coils: One with many turns for slow tuning over a wide range, and a second coil with far fewer turns for fast tuning or FM over a limited range. This "FM coil" has a sensitivity around 1% to 2% of the main coil, perhaps 500 kHz/mA.

9.6.3 Klystrons, Magnetrons, and Traveling Wave Tubes

There are a number of thermionic (vacuum-tube) devices that are widely used as amplifiers or fundamental oscillators at microwave frequencies. Standard vacuum tubes (see the **RF Amplifiers** chapter for an introduction to vacuum tubes) work well for frequencies

complished. The tube may be tuned slowly over a range of approximately 10% of the center frequency. If faster tuning is necessary, such as is required for frequency modulation, the range decreases to about 5%.

Excellent drawings showing how magnetrons work are available at **hyperphysics. phy-astr.gsu.edu/hbase/waves/magnetron. html** and a thorough introduction for the interested reader is can be downloaded from **www.cpii.com/docs/related/2/Mag%20 tech%20art.pdf**.

THE TRAVELING WAVE TUBE

A third type of tube operating in the microwave range is the traveling wave tube (TWT). For wide-band amplifiers in the microwave range this is the tube of choice. Either permanent magnets or electromagnets are used to focus the beam of electrons that travels through the TWT. This electron beam passes through a helical slow-wave structure, in which electrons are accelerated or decelerated, providing density modulation due to the applied RF signal, similar to that in the klystron. The modulated electron beam induces voltages in the helix that provides an amplified output signal whose gain is proportional to the length of the slow-wave structure. After the RF energy is extracted from the electron beam by the helix, the electrons are collected and recycled to the cathode. Traveling wave tubes can often be operated outside their designed frequencies by carefully optimizing the beam voltage.

9.7 Frequency Synthesizers

Like many of our modern technologies, the origins of frequency synthesis can be traced back to WW II. The driving force was the desire for stable, rapidly switchable and accurate frequency control technology to meet the demands of narrow-band, frequency-agile HF communications systems without resorting to large banks of switched crystals. Early synthesizers were cumbersome and expensive, and therefore their use was limited to only the most sophisticated communications systems. With the help of the same technologies that have taken computers from room-sized to now fitting into the palms of our hands, frequency synthesis techniques have become one of the most enabling technologies in modern communications equipment.

Every communications device manufactured today, be it a handheld transceiver, cell phone, pager, AM/FM entertainment radio, scanner, television, HF communications equipment, or test equipment, contains a frequency synthesizer. Synthesis is the technology that allows an easy interface with both computers and microprocessor controllers. It provides amateurs with many desirable features, such as the feel of an analog knob with precision 10-Hz frequency increments, wideband accuracy and stability determined by a single precision crystal oscillator, frequency memories, and continuously variable precision frequency splits. Now reduced in size to small integrated circuits, frequency synthesizers have long replaced the cumbersome chains of frequency multipliers and filters in VHF, UHF and microwave equipment, giving rise to the highly portable communications devices we use today. Frequency synthesis has also had a major impact in lowering the cost of modern equipment, particularly by reducing manufacturing complexity.

Frequency synthesizers are categorized in two general types: *direct synthesizers*, where the output signal is the result of operations directly performed on the input signal; and *indirect synthesizers*, where selected characteristics of the input signal are transferred onto a separately generated output signal.

One defining feature of any direct synthesizer is that no feedback is used, so there is never any dynamic stability problem. Indirect synthesizers always include feedback control loops, so dynamic stability is a major design concern.

Direct synthesizers include the major techniques of *direct analog synthesis* (DAS) and *direct digital synthesis* (DDS). Direct analog synthesizers consist primarily of frequency multipliers, frequency dividers, mixers, and filters. DAS is very useful for generating small numbers of signals, with widely spaced frequencies, from a reference oscillator. DAS is particularly useful when more than one output signal is required at the same time. Direct digital synthesizers are essentially dedicated microprocessors that have one program, to create an output signal waveform given a desired frequency (in digital form) using the applied reference signal. Unlike DAS, the output frequencies from a DDS can easily be separated by millihertz (0.001 Hz) while keeping all of the stability of the reference oscillator. Both being direct techniques, neither DAS nor DDS use feedback control loops, so switching from one frequency to another happens in nanoseconds.

Indirect synthesizers include the major techniques of frequency-locked loops (FLL) and phase-locked loops (PLL). The whole idea behind any indirect synthesizer is to transfer characteristics of one signal onto another separate signal. An FLL transfers only frequency characteristics of the input signal onto the output signal. Major FLL applications used by radio amateurs include *automatic frequency control* (AFC) loops and tone decoders. PLLs are more precise because not only frequency characteristics of the reference oscillator, but also its phase characteristics, are transferred to the oscillator generating the output signal. There are two main application classes in which PLLs find wide use. The first is as a *frequency generator*, where we call it a *frequency synthesizer*. In transmitters this synthesizer may also include modulation, particularly for FM and FSK

signals. The second PLL application class is as an angle (frequency or phase) demodulator, where we call it a *tracking demodulator* or *synchronizer*. Tracking demodulators used in deep space communication and clock recovery used in digital communication are major applications today.

It is curious to note that using modern digital circuitry, it actually is easier to make a PLL than an FLL. This turns out to be fortuitous! This section will focus on the PLL used as a frequency synthesizer. For readers interested in PLL use as a demodulator or as a synchronizer, there are many textbooks available that discuss these applications in great detail.

9.7.1 Phase-Locked Loops

As mentioned above, the PLL is a key component of any indirect synthesizer. The sidebar "An Introduction to Phase-Locked Loops" provides an overview of this technology. This section presents a detailed discussion of design and performance topics of PLL synthesizers and the circuits used to construct them. The digital logic circuits and terms included in this discussion are covered in the **Digital Basics** chapter.

ARCHITECTURE

The principle of the PLL synthesizer is straightforward. A tunable oscillator is first built to cover the required output frequency range; then the PLL system is constructed around this oscillator to keep it tuned to the desired frequency. This is done by continuously comparing the phase of the oscillator to a stable reference, such as a crystal oscillator, and using the result to steer tuning of the oscillator. If the oscillator frequency is too high, the phase of its output will start to lead that of the reference by an increasing amount. This is detected and is used to decrease the frequency of the oscillator. Too low a frequency causes an increasing phase lag, which is detected and used to increase the frequency of the oscillator. In this way,

An Introduction to Phase-Locked Loops

By Jerry DeHaven, WA0ACF

Phase locked loops (PLLs) are used in many applications from tone decoders, to stabilizing the pictures in your television set, to demodulating your local FM station or 2 meter repeater. They are used for recovering weak signals from deep space satellites and digging out noisy instrumentation signals. Perhaps the most common usage for PLLs in Amateur Radio is to combine the variability of an LC oscillator with the long term stability and accuracy of a crystal oscillator in PLL frequency synthesizers. Strictly speaking a PLL is not necessarily a frequency synthesizer and a frequency synthesizer is not necessarily a PLL, although the terms are used interchangeably.

PLL Block Diagram

The block diagram in **Fig 9.A3** shows a basic *control loop*. An example of a control loop is the furnace or air conditioning system in your house or the cruise control in your car. You input a desired output and the control loop causes the system output to change to and remain at that desired output (called a *setpoint*) until you change the setpoint. Control loops are characterized by an input, an output and a feedback mechanism as shown in the simple control loop diagram in Fig 9.A3. The setpoint in this general feedback loop is called the *reference*.

In the case of a heating system in your house, the reference would be the temperature that you set at the thermostat. The feedback element would be the temperature sensor inside the thermostat. The thermostat performs the comparison function as well. The generator would be the furnace which is turned on or off depending on the output of the comparison stage. In general terms, you can see that the reference input is controlling the output of the generator.

Like other control loops, the design of a PLL is based on feedback, comparison and correction as shown in Fig 9.A3. In this section we will focus on the concepts of using a phase locked loop to generate (synthesize) one or more frequencies based on a single reference frequency.

In a typical PLL frequency synthesizer there are six basic functional blocks as shown in **Fig 9.A4**. The PLL frequency synthesizer block diagram is only slightly more complicated than the simple control loop diagram, so the similarity should be evident. After a brief description of the function of each block we will describe the operational behavior of a PLL frequency synthesizer.

In Fig 9.A4, the general control loop is implemented with a bit more detail.

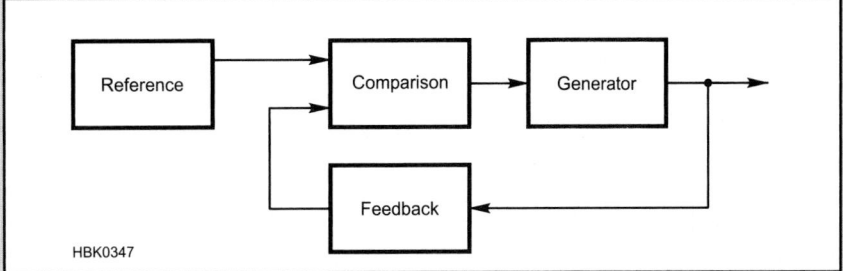

Fig 9.A3 — Simple control loop. The Comparison block creates a control signal for the Generator by comparing the output of the Reference and Feedback blocks.

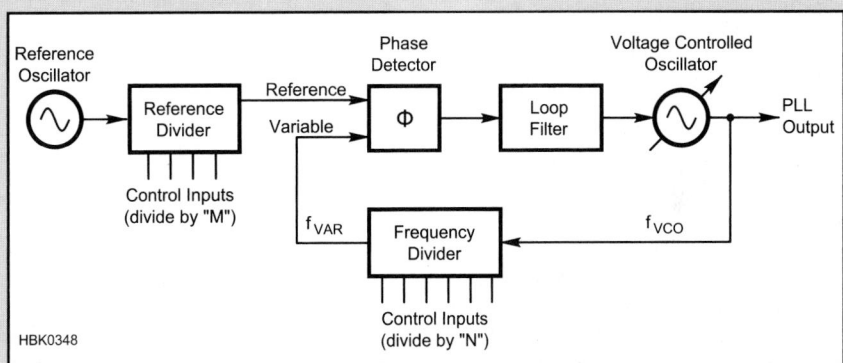

Fig 9.A4 — PLL frequency synthesizer block diagram. The PLL is locked when the frequency PLL Output / N is the same as that of the Reference Oscillator / M.

The Reference block is composed of the Reference Oscillator and the Reference Divider. The Comparison block consists of the Phase Detector and Loop Filter. The Generator is replaced by the Voltage-Controlled Oscillator (or VCO) and the Feedback block is replaced by the Frequency Divider. The two signals being compared are both frequencies; the reference frequency and variable frequency signals. The output of the Loop Filter is made up of dc and low frequency ac components that act to change the VCO frequency. The output of the PLL is an integer multiple of the reference frequency (the output of the Reference Divider). It is probably not obvious why, but read on to find out!

The Reference Oscillator — The reference oscillator is usually a crystal oscillator with special attention paid to thermal stability and low mechanical and electronic noise. The main function of the reference oscillator is to generate a stable frequency for the PLL. Let us make a distinction between the reference oscillator frequency and the reference frequency. The output of the reference oscillator is at the crystal frequency, say 5.000 MHz. The choice of the actual crystal frequency depends on the PLL design, the availability of another oscillator in the radio, the

avoidance of spurious responses in the receiver, and so on.

The *reference frequency* is the output of the reference (crystal) oscillator divided by the integer M to a relatively low frequency, say 10 kHz. In this case, 10 kHz is the reference frequency. The reference frequency equals the step size between the PLL output frequencies. In this simple example the crystal frequency can be any frequency that is an integer multiple of 10 kHz and within the operating range of the Reference Divider.

The Phase Detector — There are many types of phase detectors, but for now let us consider just a basic mixer, or a multiplier. The phase detector has two inputs and one output. In its simplest form, the output of the phase detector is a dc voltage proportional to the phase difference between the two inputs. In practice, a phase detector can be built using a diode double-balanced mixer or an active multiplier.

The output of the mixer consists of products at many frequencies both from multiplication and from nonlinearities. In this application, the desired output is the low frequency and dc terms so all of the RF products are

[continued on next 2 pages]

terminated in a load and the low frequency and dc output is passed to the PLL loop filter.

When the PLL is locked — meaning that the output frequency is the desired multiple of the reference frequency — the phase detector output is a steady dc voltage somewhere between ground and the PLL power supply voltage. When the PLL is commanded to change to another frequency, the phase detector output will be a complex, time-varying signal that gradually settles in on its final value.

The Loop Filter —The loop filter is a low-pass filter that filters the output of the phase detector. The loop filter can be a simple resistor-capacitor (RC) low-pass filter or an active circuit built with bipolar transistors or operational amplifiers. The cutoff frequency of the low pass filter is on the order of 100 Hz to 10 kHz. Although the frequencies involved are dc and low audio frequencies the design of the loop filter is critical for good reference suppression and low noise performance of the PLL frequency synthesizer.

The output of the loop filter is a dc voltage that "steers" the following stage, the Voltage Controlled Oscillator. The dc value can vary over a substantial portion of the PLL power supply. For example, if the loop filter runs off a 9 V power supply you may expect to see the output voltage range from about 2 V to 7 V depending on the VCO frequency range.

The Voltage Controlled Oscillator (*VCO*)—The VCO is probably the most critical stage in the PLL frequency synthesizer. This stage is the subject of many conflicting design goals. The VCO must be electrically tunable over the desired frequency range with low noise and very good mechanical stability. Ideally, the output frequency of the VCO will be directly proportional to the input control voltage. A key characteristic is the *VCO constant* or *tuning gain*, usually expressed in MHz per volt (MHz/V). The VCO has one input, the dc voltage from the loop filter, and usually two outputs; one for the PLL output, the other driving the feedback stage — the frequency divider.

The Frequency Divider — The VCO frequency divider that acts as the feedback block is a programmable frequency divider or counter. The division ratio N is set either by thumbwheel switches, diode arrays or a microprocessor. The input to the divider is the VCO output frequency. The output of the divider is the VCO output frequency divided by N.

When the PLL is unlocked, such as when the PLL is first turned on, or commanded to change to another frequency, the frequency divider's output frequency will vary in a nonlinear manner until the PLL locks. Under locked conditions, the divider's

output frequency (VCO Output / N) is the same as the reference frequency (Reference Oscillator / M).

The function of the frequency divider as the feedback element is easier to understand with an example. Let's assume that the desired PLL output frequency is 14.000 MHz to 14.300 MHz and it should change in steps of 10 kHz (0.010 MHz). A step size of 10 kHz means that the reference frequency must be 10 kHz. There are 31 output frequencies available — one at a time, not simultaneously. (Don't forget to count 14.000 MHz.) Let us arbitrarily choose the reference oscillator as a 5.000 MHz crystal oscillator. To generate the 10 kHz reference frequency, the reference divider must divide by M = 500 = 5.000 MHz / 0.010 MHz. (Use the same units, don't divide MHz by kHz.)

To generate a matching 10 kHz signal from the VCO output frequency, the Frequency Divider (the feedback stage) must be programmable to divide by N = 1400 to 1430 = 14.000 to 14.300 MHz / 0.010 MHz. The reference divisor (M = 500) is fixed, but the Frequency Divider stage needs to be programmable to divide by N = 1400 to 1430 in steps of 1. In this way, the output frequency of the PLL is compared and locked to the stable, crystal-generated value of the reference frequency.

PLL Start-up Operation

Let's visualize the PLL start-up from power on to steady state, visualizing in sequence what each part of the PLL is doing. Some portions of the loop will act quickly, in microseconds; other parts will react more slowly, in tens or hundreds of milliseconds.

The reference oscillator and its dividers will probably start oscillating and stabilize within tens of microseconds. The reference oscillator divider provides the reference frequency to one input of the phase detector.

The VCO will take a bit longer to come up to operation because of extensive power supply filtering with high-value capacitors. The VCO and the programmable dividers will probably reach full output within a few hundred microseconds, although the VCO will not yet be oscillating at the correct frequency.

At this point the phase detector has two inputs but they are at different frequencies. The mixer action of the phase detector produces a beat note at its output. The beat note, which could be tens or hundreds of kHz, is low-pass filtered by the PLL loop filter. That low-pass filter action averages the beat note and applies a complicated time-varying ac/dc voltage to the VCO input.

Now the VCO can begin responding to the control voltage applied to its input. An important design consideration is that the filtered VCO control signal must steer the VCO in the

correct direction. If the polarity of the control signal is incorrect, it will steer the VCO *away* from the right frequency and the loop will never lock. Assuming correct design, the control voltage will begin steering the VCO in such a direction that the VCO divider frequency will now get closer to the reference frequency.

With the phase detector inputs a little bit closer in frequency, the beat note will be lower in frequency and the low pass filtered average of the phase detector output will change to a new level. That new control voltage will continue to steer the VCO in the right direction even closer to the "right" frequency. With each successive imaginary trip around the loop you can visualize that eventually a certain value of VCO control voltage will be reached at which the VCO output frequency, when divided by the frequency divider, will produce zero frequency difference at the inputs to the phase detector. In this state, the loop is *locked*. Once running, the range over which a PLL can detect and lock on to a signal is its *capture range*.

PLL Steady-state Operation

When the divided-down VCO frequency matches the reference frequency, the PLL will be in its steady-state condition. Since the comparison stage is a phase detector, not just a frequency mixer, the control signal from the loop filter to the VCO will act in such a way that maintains a constant relationship between the *phase* of the reference frequency and the frequency divider output. For example, the loop might stabilize with a 90° difference between the inputs to the phase detector. As long as the phase difference is constant, the reference frequency and the output of the frequency divider (and by extension, the VCO output) must also be the same.

As the divisor of the frequency divider is changed, the loop's control action will keep the divided-down VCO output phase locked to the reference frequency, but with a phase difference that gets closer to 0 or 180°, depending on which direction the input frequency changes. The range over which the phase difference at the phase detector's input varies between 0° and 180° is the widest range over which the PLL can keep the input and VCO signals locked together. This is called the loop's *lock range*.

If the divisor of the frequency divider is changed even further, the output of the loop filter will actually start to move in the opposite direction and the loop will no longer be able to keep the VCO output locked to the reference frequency and the loop is said to be *out of lock* or *lost lock*.

Loop Bandwidth

Whether you picture a control loop

such as the thermostat in your house, the speed control in your car or a PLL frequency synthesizer, they all have "bandwidth," that is, they can only respond at a certain speed. Your house will take several minutes to heat up or cool down, your car will need a few seconds to respond to a hill if you are using the speed control. Likewise, the PLL will need a finite time to stabilize, several milliseconds or more depending on the design.

This brings up the concept of *loop bandwidth*. The basic idea is that disturbances outside the loop have a low-pass characteristic up to the loop bandwidth, and that disturbances inside the loop have a high-pass characteristic up to the loop bandwidth. The loop bandwidth is determined by the phase detector gain, the loop filter gain and cutoff frequency, the VCO gain and the divider ratio. Loop bandwidth and the phase response of the PLL determine the stability and transient response of the PLL.

PLLs and Noise

Whether the PLL frequency synthesizer output is being used to control a receiver or a transmitter frequency, the spectral cleanliness of the output is important. Since the loop filter output voltage goes directly to the VCO, any noise or spurious frequencies on the loop filter output will modulate the VCO causing FM sidebands. On your transmitted signal, the sidebands are heard as noise by adjacent stations. On receive, they can mix with other signals, resulting in a higher noise floor in the receiver.

A common path for noise to contaminate the loop filter output is via the power supply. This type of noise tends to be audible in your receiver or transmitter audio. Use a well-regulated supply to minimize 120-Hz power supply ripple. Use a dedicated low-power regulator separate from the receiver audio stages so you do not couple audio variations from the speaker amplifier into the loop filter output. If you have a computer / sound card interface, be careful not to share power supplies or route digital signals near the loop filter or VCO.

A more complicated spectral concern is the amount of reference frequency feed-through to the VCO. This type of spurious output is above the audio range (typically a few kHz) so it may not be heard but it will be noticed, for example, in degraded receiver adjacent channel rejection. In the example above, Fig 9.A4, the reference frequency is 10 kHz. High amounts of reference feed-through on the VCO control input will result in FM sidebands on the VCO output.

Loop filter design and optimization is a complicated subject with many trade-offs. This subject is covered in detail in

the PLL references, but the basic trade-off is between loop bandwidth (which affects the speed at which a PLL can change frequency and lock) and reference rejection. Improved phase detectors can be used which will minimize the amount of reference feed-through for the loop filter to deal with. Another cause of high reference sidebands is inadequate buffering between the VCO output and the frequency divider stage.

One indication of PLL design difficulty is the spread of the VCO range expressed either in percent or in octaves. It is fairly difficult to make a VCO that can tune over one octave, a two-to-one frequency spread. Typically, the greater the spread of the VCO tuning range, the worse the noise performance gets. The conflicting design requirements can sometimes be conquered by trading off complexity, cost, size, and power consumption.

A common source of noise in the VCO is microphonics. Very small variations in capacitance can frequency modulate the VCO and impose audible modulation on the VCO output. Those microphonic products will be heard on your transmit audio or they will be superimposed on your receiver audio output. Under vibration, an air-wound inductor used in a VCO can move slightly relative to a metal shield for example. A common technique to minimize microphonics is to cover the sensitive parts with wax, Q-Dope, epoxy or a silicone sealer. As with the loop filter use a well-filtered, dedicated regulator for the VCO to minimize modulation by conducted electrical noise.

Types of PLLs

One of the disadvantages of a single loop PLL frequency synthesizer is that the reference frequency determines the step size of the output frequency. For two-way radio and other systems with fixed channel spacing that is generally not a problem. For applications like amateur radio CW and SSB operation, tuning steps smaller than 10 Hz is almost a necessity. Like so many design requirements, there are trade-offs with making the reference frequency smaller and smaller. Most evident is that the loop filter bandwidth would have to keep getting lower and lower to preserve spectral cleanliness and loop stability. Because low frequency filters take longer to settle, a single-loop PLL with a 10 Hz loop bandwidth would take an unacceptably long time to settle to the next frequency.

Creative circuit architectures involving two or more PLLs operating together achieve finer tuning increments with only moderate complexity. Once you understand the basics of a single loop PLL you will be able to learn and understand the operation of a multi-loop PLL.

the oscillator is locked into a fixed-phase relationship with the reference, which means that their frequencies are held exactly equal.

A representative block diagram of a PLL is shown in **Fig 9.37**. Measurement of any error in the output signal phase is made by the *phase detector* (PD). The measured error is supplied to the *loop filter* and *loop amplifier* which work together to tune the *voltage controlled oscillator* (VCO) just enough to make the error go to zero. The phase detector determines the actual frequency and phase of the VCO through the *feedback divider* (N). The phase detector operates at a lower frequency which is a sub-multiple of both the crystal oscillator frequency reference (f_{XO}) and the output frequency (f_{OUT}). This lower frequency is correctly called the PLL *reference frequency* (f_{REF}) because it sets the operating conditions of the PLL.

There is unfortunate history about the term reference frequency. It is often used to refer to both the operating frequency of the phase detector and to the frequency reference applied to the PLL. If the *reference divider* (R) is not present then of course these two frequencies are the same. But in general, the reference divider is present and these frequencies are different. Ambiguity here is widespread and a big problem, even today many decades into the widespread use of PLLs. We must be very careful in the words we choose to use!

FREQUENCY RESOLUTION (STEP SIZE)

Frequency of the output signal can be easily changed by changing the divide ratio of the feedback divider. For example if the reference frequency (at the phase detector!) is 100 kHz and the output frequency is 147.5 MHz, the feedback divider number must be N = 1475. If we changed the value of N to 1474, the frequency at the output of the feedback divider becomes 100.068 kHz. The phase detector notices that the frequency of the VCO is too high, and "tells" the loop filter to reduce the frequency of the VCO until the frequency out of the feedback divider becomes exactly 100.000 kHz. This will happen when the VCO is retuned to a frequency of 147.4 MHz.

By changing the value of N by 1, we have just tuned the frequency synthesizer by its smallest available step. This is called the *frequency resolution* of the synthesizer, or equivalently the synthesizer *step size*. Here we note that this step size is (147.5 – 147.4) = 0.1 MHz. It is no accident that this frequency is exactly equal to the PLL reference frequency f_{REF}. This result is an important reason to avoid all ambiguity in the term "reference frequency." f_{REF} is equal to the synthesizer output frequency step size.

There certainly are applications where we want the step size to be a small number, say 10 Hz. You might guess — correctly — that there are problems in building a PLL synthe-

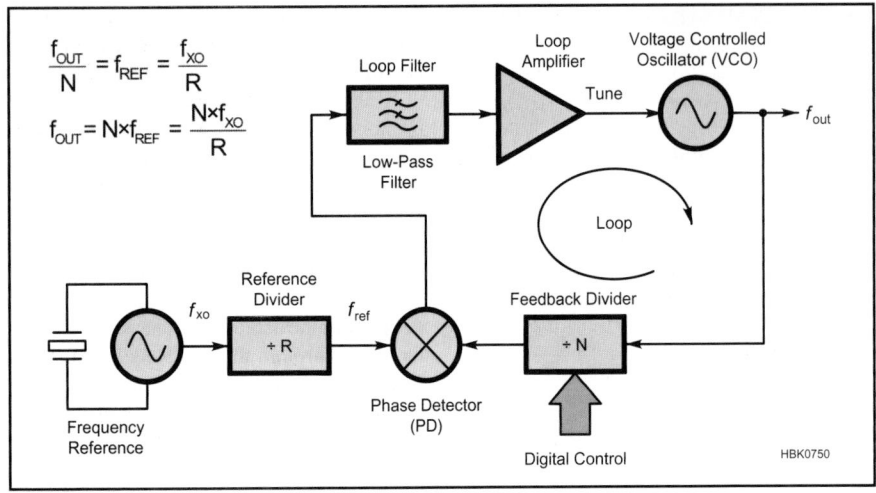

$$\frac{f_{OUT}}{N} = f_{REF} = \frac{f_{XO}}{R}$$

$$f_{OUT} = N \times f_{REF} = \frac{N \times f_{XO}}{R}$$

Loop Filter

Low-Pass Filter

Loop Amplifier

Tune

Voltage Controlled Oscillator (VCO)

f_{out}

Loop

Reference Divider

f_{xo} ÷ R f_{ref}

Feedback Divider

÷ N

Phase Detector (PD)

Frequency Reference

Digital Control

HBK0750

Fig 9.37 — A basic phase-locked-loop (PLL) synthesizer acts to keep the divided-down signal from its voltage-controlled oscillator (VCO) phase-locked to the divided-down signal from its reference oscillator. Fine tuning steps are therefore possible without the complication of direct synthesis.

sizer for a very low f_{REF}. The usual solution is to combine multiple PLLs with some DAS techniques to get around these problems.

DYNAMIC STABILITY AND LOOP BANDWIDTH

Because feedback control is required to transfer the frequency and phase characteristics from the crystal oscillator onto the VCO, all of the stability problems inherent in feedback control apply to PLL design. If the gain and phase responses are not well designed the PLL can oscillate. In this case instead of getting the stability transfer we want, the output is essentially a frequency modulated signal spread across the entire tuning bandwidth of the synthesizer — clearly a very bad thing. How to perform this design is discussed in section 9.7.2.

For the moment let us assume that our PLL dynamics are properly designed. One other characteristic of feedback control is now important: how fast can the output frequency be changed from one value to another? This is controlled by the *loop bandwidth* of the PLL. For practical reasons the loop bandwidth should be less than 5% of f_{REF}. Using the example above in which $f_{REF} = 100$ kHz, the loop bandwidth we design for cannot exceed 5 kHz. To answer our question, we can use the rule of thumb that in a well-designed loop, the settling time should be between 1 and 2 times the reciprocal of the loop bandwidth. For a 5 kHz loop bandwidth the settling time should not exceed 2/5000 = 400 µs. We can begin to see one problem with PLL design if f_{REF} is very small: To get a 1 kHz step size from a PLL synthesizer the maximum loop bandwidth available is 50 Hz. This is impractically small.

This loop bandwidth also influences how we can modulate our PLL synthesizer in an FM transmitter. Imagine now that we apply a very small amount of FM to the reference oscillator f_{XO}. The amount of deviation will be amplified by N/R — but this is true only for low modulating frequencies. If the modulating frequency is increased, the VCO has to change frequency at a higher rate. As the modulating frequency continues to increase, eventually the VCO has to move faster than

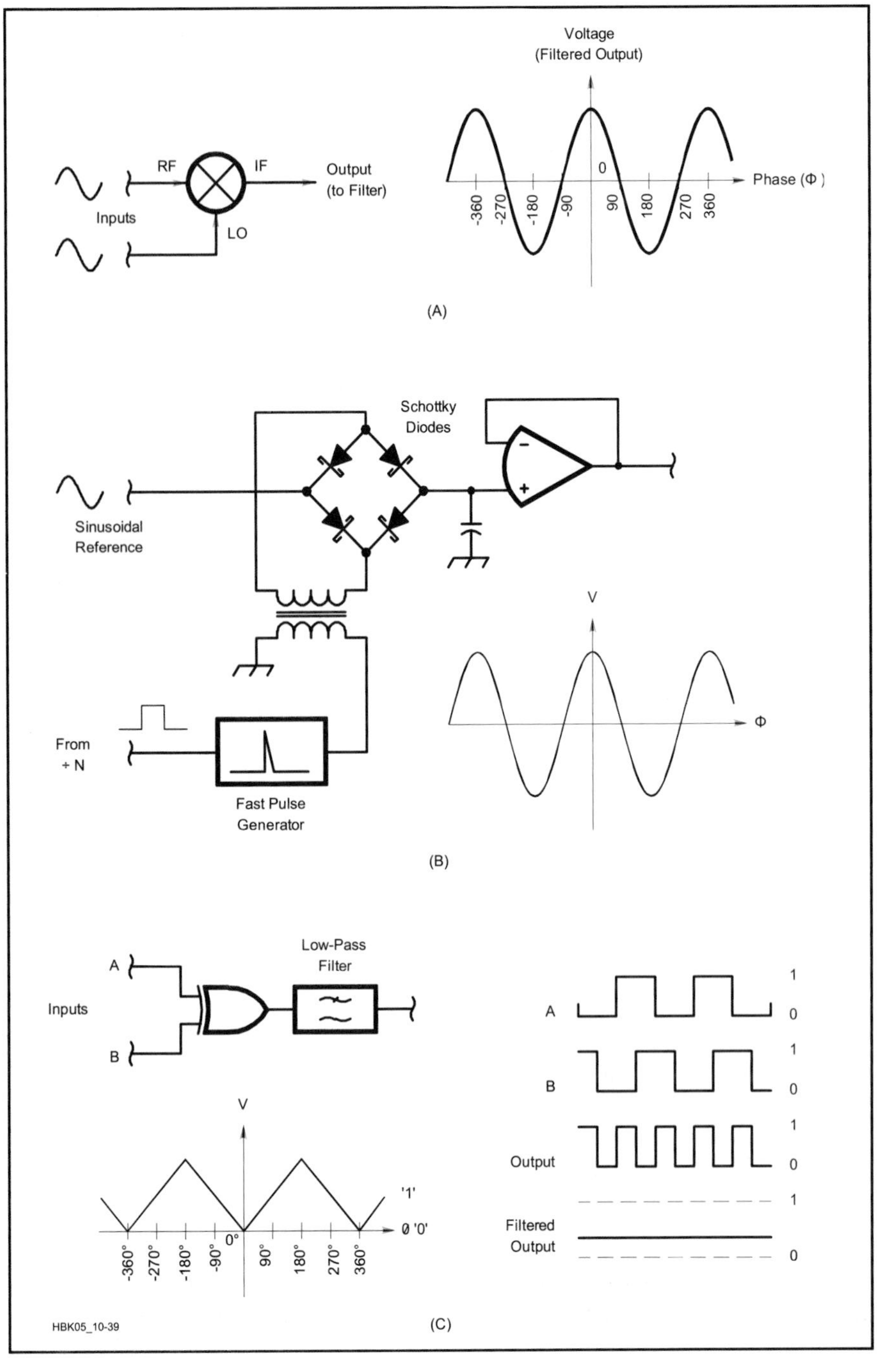

Voltage (Filtered Output)

RF IF

Inputs

LO

Output (to Filter)

0

Phase (Φ)

-360 -270 -180 -90 90 180 270 360

(A)

Schottky Diodes

Sinusoidal Reference

From ÷ N

Fast Pulse Generator

V

Φ

(B)

Low-Pass Filter

Inputs

A

B

V

-360° -270° -180° -90° 0° 90° 180° 270° 360°

'1'

Ø '0'

A

B

Output

Filtered Output

1 0

1 0

1 0

1 0

HBK05_10-39

(C)

Fig 9.38 — Simple phase detectors: a mixer (A), a sampler (B) and an exclusive-OR gate (C).

the available *settling time*. The PLL now acts as a low-pass filter, reducing the available deviation from higher modulating frequencies. The cut-off frequency of this PLL low-pass action is equal to this same loop bandwidth.

PLL COMPONENTS

Let us continue our discussion of phase-locked loop synthesizers by examining the role of each of the component pieces of the system. They are the phase detector, the VCO, the dividers (with possible prescalers), the loop filter, and of course the reference oscillator.

Phase Detectors

Phase detection is the key operation in any PLL. Remembering that we are making a phase locked loop, what we really need is not *absolute* phase measurement but a *relative* phase measurement. We need to know only how the phase of the VCO output signal f_{OUT} is changing with respect to the phase of the crystal reference f_{XO}.

As usual, there are both analog and digital circuit techniques available to perform phase detection. For frequency synthesizer design the analog techniques are seldom used. Analog phase detectors are used only in receiver and demodulator applications. There are two that sometimes still appear in ham equipment, the multiplier/mixer of **Fig 9.38A** and the sampling phase detector shown in Fig 9.38B.

The mixer is a very low noise device when used as a phase detector, which explains why it is not yet completely gone from synthesizer design. It remains only in PLL designs where f_{REF} is very high, greater than 10-100 MHz. The main reason for this is that a PLL using a mixer for the PD is subject to a phenomenon called *false lock*, where the PLL may act like it is locked even though it really isn't. Detecting and eliminating a false lock condition is much easier to do at higher frequencies. The sampling phase detector is now used only in synthesizers with output frequencies in the high microwave region, typically well above 10 GHz. The gain of this sampling circuit is very low, so it is used only if there is no viable alternative.

The most widely used digital phase detector is the exclusive-OR (XOR) gate shown in Fig 9.38C. (See the **Digital Basics** chapter for XOR gate operation.) If inputs A and B in Fig 9.38C are almost in phase, the output will be low most of the time, and its average filtered value will be close to the logic 0 level. If inputs A and B are almost in phase opposition, the output will be high most of the time, and its average voltage will be close to logic 1. The average voltage is near its mid-value when the inputs are shifted in phase by 90 degrees. In this respect the XOR gate is much like the analog mixer. To achieve this circuit's full output-voltage range, it's important that the reference and VCO signals at its input have a 50% duty cycle.

Phase-Frequency Detector

One problem common to *all* phase detectors is the possibility of false lock. This happens because the output of any phase detector can have an average value of zero even when the frequencies at its inputs are not equal. This is a very serious problem that requires any PLL using them to have additional circuitry to serve as an acquisition aid. If the PLL

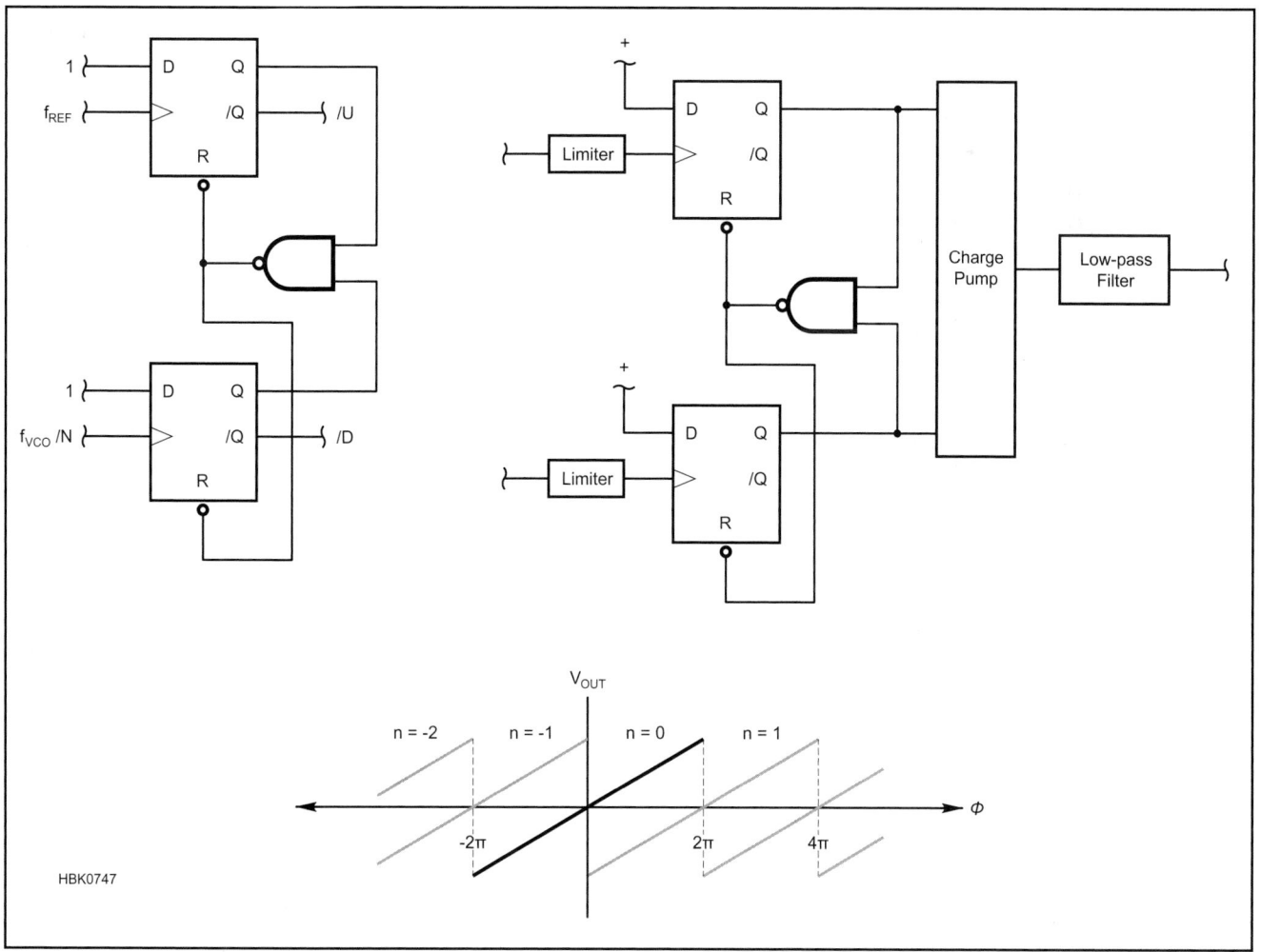

Fig 9.39 — The phase frequency detector (PFD) makes a measurement of the time difference between two rising edges of the input signals.

tries to settle when the PD inputs are not at the same frequency, this acquisition aid must prod the PLL to keep moving toward the true lock frequency. Such acquisition aids are not needed when a *phase-frequency detector* (PFD) is used.

The benefit of the PFD is that if the input signals are at different frequencies, even if they are at very different frequencies, the output is never zero. The PFD output always produces a dc shift in the direction the PLL needs to move until the lock condition is achieved. No acquisition aids are ever needed. This alone is enough to explain its nearly universal adoption today.

As seen in **Fig 9.39A**, the PFD is a simple digital circuit. This also helps make it attractive to use in this era of CMOS integrated circuits. Other designs for the PFD exist, but they all follow the logic shown in Fig 9.39A. Of particular importance is that the PFD has two outputs.

The PFD operates by measuring the time difference between the rising edges of the reference signal and the divided VCO. The first signal to arrive sets its flip-flop. The second edge to arrive also sets its flip-flop, which immediately causes both flip-flops to be reset, ready for the next set of signal edges. By operating at high speed, the PFD is very sensitive to phase differences between the input signals. The PFD does require an equal number of edges for each signal, so it is not useful when the signals have varying numbers of edges, such in a receiver's demodulator or a synchronizer.

If the VCO frequency is too low only the /U output is active, causing the loop filter to raise the VCO frequency. On the other hand, when the VCO frequency is high only the /D output is active, causing the loop filter to lower the VCO frequency. It is interesting to note that unlike the mixer and XOR phase detectors, which require the inputs to be in phase quadrature, the PFD locks when the input signals are in exact phase alignment.

Being an edge triggered circuit, PFD operation is essentially independent of input signal duty cycle. This eliminates the need for input waveform processing. If the input frequencies are different, the time differences measured by the PFD will be either constantly increasing or decreasing. As a result only one of the PFD output signals will be active, informing the PLL unambiguously which way to correct the controlled source to achieve phase lock.

Charge Pumps

Because the PFD has two outputs and most loop filters have only one input, something is needed to bridge the PFD and the loop filter. The usual technique is the *charge pump*, shown in Fig 9.39B. The charge pump gets its name by allowing current to flow only for the brief interval, here when the /D or

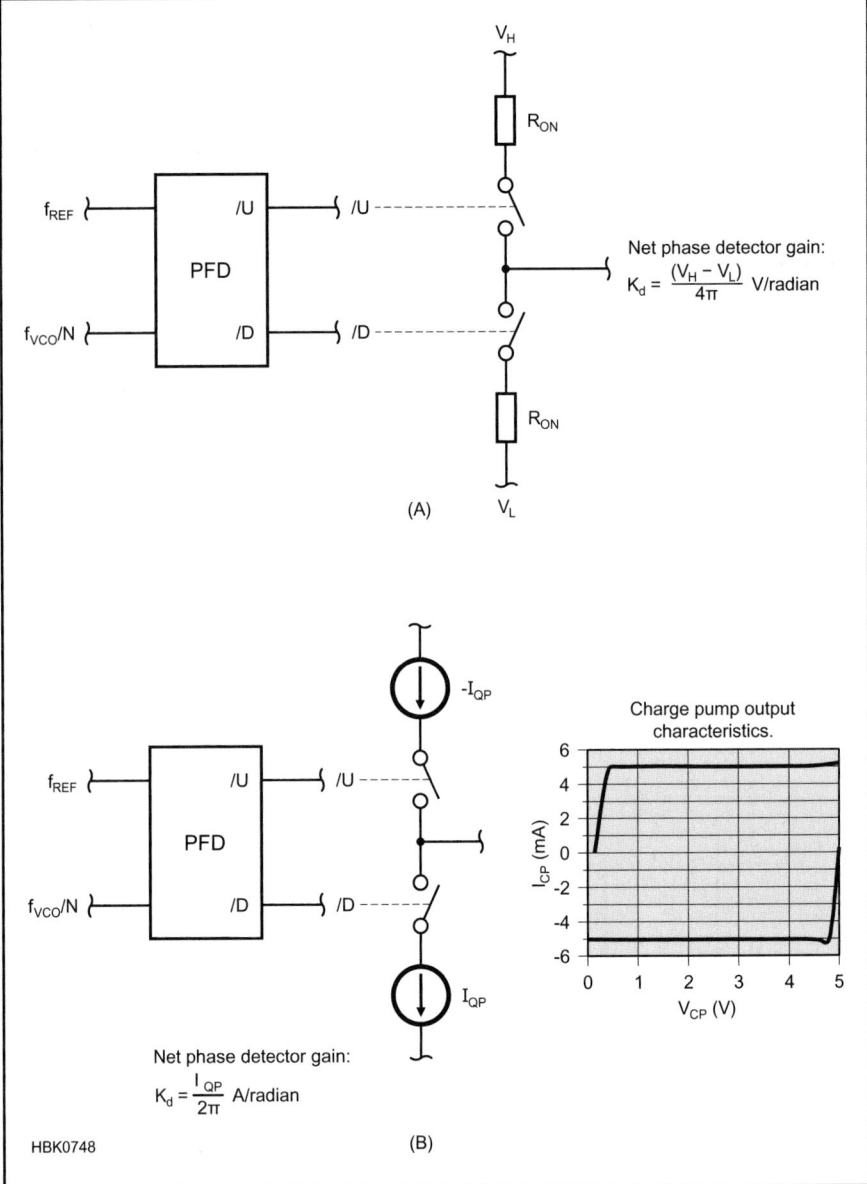

Fig 9.40 — The charge pump delivers short bursts of current to the loop filter under the control of the phase frequency detector (PFD) outputs. A voltage mode charge pump is shown at A and a current mode charge pump at B.

/U outputs are active. Current flow in short bursts is equivalent to a finite charge transfer.

Charge pumps come in two configurations, voltage mode and current mode, as shown in **Fig 9.40**. Voltage mode (Fig 9.40A) was the original version, though now current mode (Fig 9.40B) is most widely used with integrated circuit PLLs.

The voltage mode charge pump directly connects voltage sources to the loop filter when the PFD output signals are active. When the PFD output signals are at rest (both high) both voltage sources are disconnected from the loop filter. These voltage spikes transfer charge to the integrator in accordance with the PLL loop filter time constant. One voltage source inputs current to the loop filter and the other removes it.

The current mode charge pump is architecturally similar to the voltage mode charge pump, with voltage sources replaced by current sources. Like the voltage mode charge pump, this circuit disconnects both sources when the PFD output signals are at rest. In this case, current spikes transfer charge to and from the loop filter integrator. This transfer is independent of the loop filter series resistances.

The current mode charge pump has some advantages to the frequency synthesizer designer. Three are particularly significant:

• Pump current flows at a constant value independent of the loop filter voltage. Unless V_H and V_L are both very large compared with any possible loop filter voltage, the amount of charge transfer will vary with different loop filter voltage values.

• During the PFD reset time, both output sources are active. The voltage mode charge pump will attempt to connect both voltage sources to the loop filter during this brief period, which can generate some undesirably large power supply current spikes. During PFD reset, the current mode charge pump loads the power supply with a single current value of I_{QP}. If both current sources are well matched, their currents completely cancel at the loop filter input, effectively disconnecting them that much earlier.

• At low offset frequencies, the loop gain flattens out with a voltage mode charge pump. Loop gain continues to increase with a current mode charge pump, effectively matching active filter performance.

VCO

The design and characteristics of oscillators, including tunable oscillators, is the major topic of this chapter so it will not be repeated here. Of particular importance to PLL design are the tuning characteristic of the VCO and the VCO tuning bandwidth.

The tuning characteristic of an oscillator shows the output frequency versus the tuning voltage, such as the example of **Fig 9.41A**. The slope of this curve is called the VCO's *modulation gain* (K_0) and it is this modulation gain that is most important to PLL design. For wideband VCOs the variation of K_0 can exceed 5:1 from minimum to maximum frequency. In section 9.7.2 the design method shows how a wide gain variation such as this is handled with a single loop filter design.

It is important to realize that the VCO has no idea that it is controlled by a PLL. While it is oscillating it continues to drift and jitter with time and temperature like any other oscillator. You must characterize these variations to be assured that the tuning capability has sufficient range for the PLL to "find" a tuning voltage value that will retune the VCO to the required frequency, under all circumstances.

The primary VCO characteristic of interest to the synthesizer designer is the tuning gain "constant," K_0. Almost never a constant, K_0 is a local measure of how much the VCO output frequency changes with a change in tuning voltage.

VCO designers go to great lengths to make K_0 an actual constant. This additional effort often increases the VCO design complexity and usually results in a more expensive design. If no linearization effort is undertaken, K_0 can easily vary from 3:1 to 10:1 over the tuning band. This also complicates the frequency synthesizer design. Specific applications will dictate where effort needs to be expended: in the VCO design or the PLL design.

All PLL design algorithms assume that there is no delay between when the loop filter

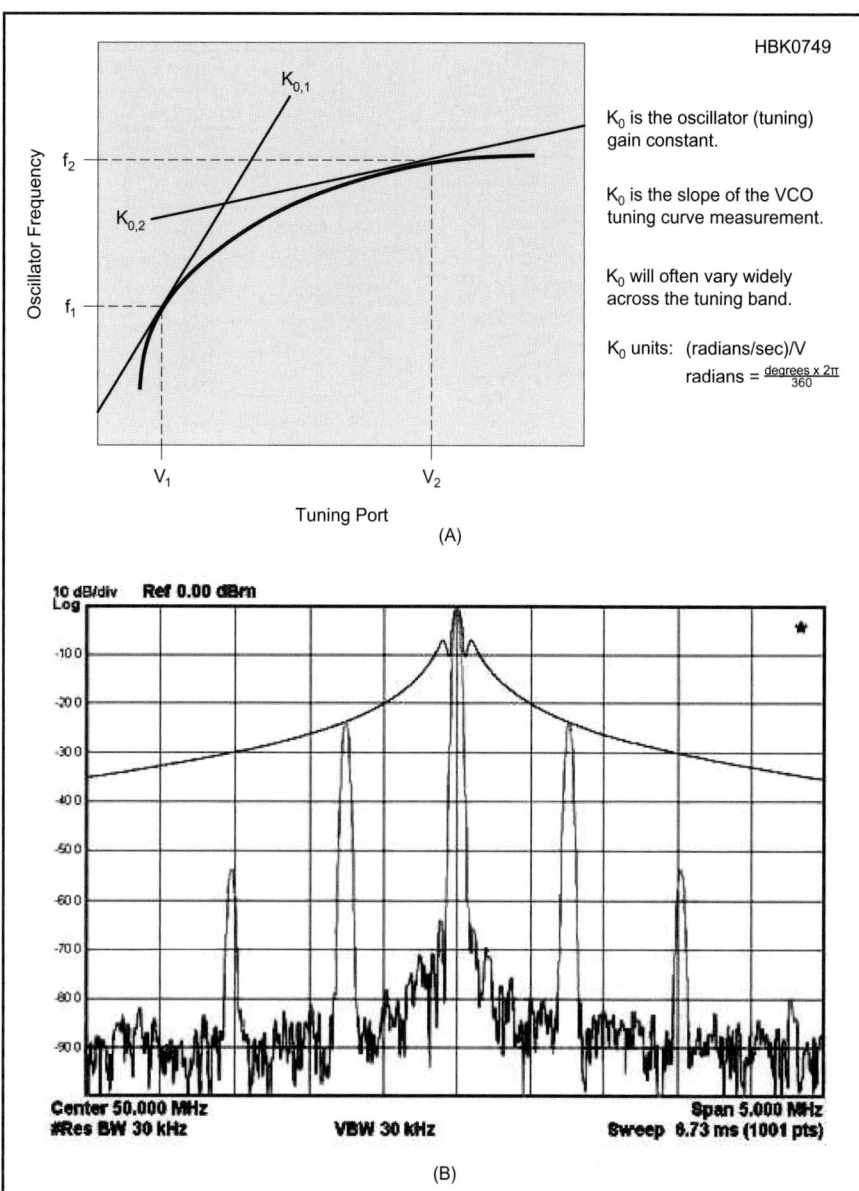

HBK0749

K_0 is the oscillator (tuning) gain constant.

K_0 is the slope of the VCO tuning curve measurement.

K_0 will often vary widely across the tuning band.

K_0 units: (radians/sec)/V
radians = $\frac{degrees \times 2\pi}{360}$

Fig 9.41 — (A) Measurement of the VCO tuning characteristic provides a curve describing frequency versus tuning voltage. The VCO tuning gain at a particular frequency (or tuning voltage) is the slope of this curve at that point. (B) A spectrum analyzer is used to measure frequency deviation of the VCO and the modulating frequency varies. FM sideband peak amplitudes follow the top profile when deviation is constant. VCO tuning bandwidth is determined by increasing the modulating frequency until the first sidebands are no longer 10 dB greater than all other sidebands.

presents a tuning change command to the VCO and when the VCO frequency actually changes. This situation is true when the VCO tuning bandwidth is much greater than the PLL loop bandwidth. If the VCO tuning bandwidth is not much greater than the PLL loop bandwidth, there are additional phase shifts that will cause problems with PLL stability. It is usually much easier to narrow the loop bandwidth than to redesign the VCO. If redesigning the VCO is a possibility, reduce the capacitance seen at the tuning input to increase the tuning bandwidth.

The VCO tuning bandwidth is important

mainly for PLL designs that have wide loop bandwidths. Measuring the frequency deviation as the modulating frequency varies is best done by looking at the FM sidebands on a spectrum analyzer, as shown in Fig 9.41B. For constant deviation, FM sidebands follow the top profile. When the modulating frequency doubles, the sidebands drop below the profile by 6 dB, and if it triples, by 10 dB. The first FM sideband must exceed all other sidebands by 10 dB for proper VCO *characterization* — the process of determining the VCO's characteristic parameters such as the tuning sensitivity, noise spectrum, dynamic

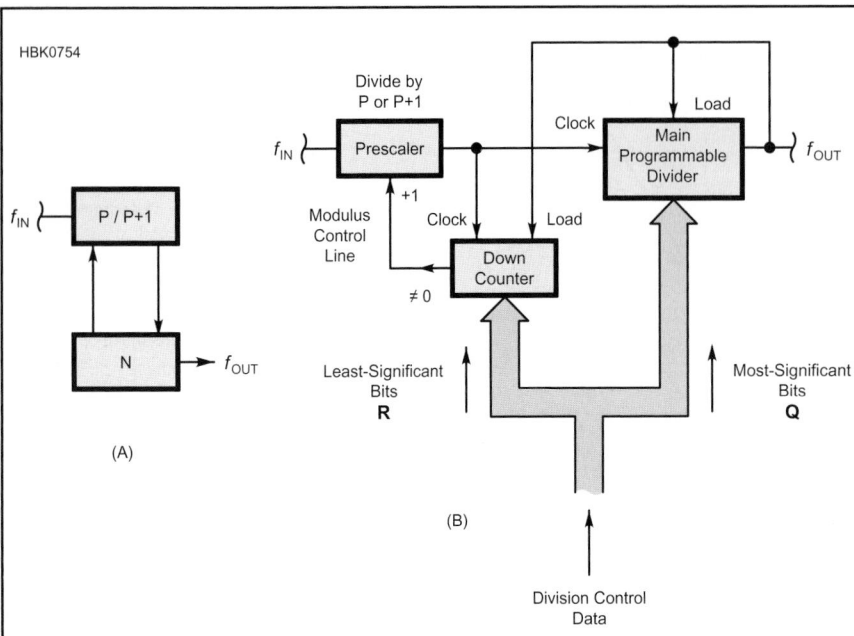

Fig 9.42 — A typical implementation of a dual-modulus prescaler.

response to changes in the tuning input signal, and so on. (See the Agilent application note "Boosting PLL Design Efficiency" in the References section.) VCO tuning bandwidth is determined by the point at which increasing the modulating frequency deviates from (drops below) the profile shown in Fig 9.41B.

Reference Divider

The reference divider is one of the easiest parts of a PLL to design, though technically is it outside the PLL. The input frequency is well known — it is the crystal reference. The output frequency is also well known, being the phase detector frequency f_{REF}. The ratio of these two frequencies is the required counter divide-by ratio or *modulus*.

When designing PLL synthesizers, f_{REF} is usually specified but f_{XO} is not. This provides some flexibility in design to use easy counter implementations if the resulting f_{XO} is also easy to get. The easiest digital counters to use for any divider are those operating with binary numbers (divide by 2, 4, 8, 16, ..., 128, ...). For example, if $f_{REF} = 20$ kHz and you choose R = 512 then $f_{XO} = 10.24$ MHz, which is a readily available crystal frequency.

Feedback Divider and Prescalers

The feedback divider is more of a challenge. First and foremost, this divider must work properly with whatever possible frequency it may ever see at its input. Whatever the highest and lowest frequencies the VCO may be, the divider must handle them all.

The feedback divider is almost always programmable so that the output frequency can be changed. Programmable counters

are always slower than non-programmable designs, so some speed limitation occurs. Fortunately in this era of tiny CMOS integrated circuits the problems of past years in getting programmable counters to operate at hundreds of MHz are nearly over. Older equipment still has feedback divider designs that are carefully crafted to handle the frequencies at which they must operate.

Fixed Prescaler

When the programmable counter does not have sufficient speed to handle the frequencies required, then the VCO frequency must be divided down ahead of the programmable counter to assure that everything works reliably. This additional divider usually has a fixed binary value so it will go fast enough without using too much power. Called a *prescaler*, as long as the output frequency from the prescaler is always within the operating range of the programmable divider the PLL will be reliable.

Of course, there are consequences: when a prescaler is used, the PLL step size is multiplied by the prescaler divider value. For our $f_{REF} = 100$ kHz, if we adopt a divide-by-16 prescaler then the step size increases by a factor of 16 to 1.6 MHz. The direct way to correct for this is to reduce f_{REF} by the same factor of 16, which unfortunately means that the PLL loop bandwidth is also reduced by a factor of 16. Is there a solution with fewer compromises?

Dual-Modulus Prescaling

Yes, there is such a solution. If the prescaler is designed to divide by two values separated

by 1, say 16 and 17, or 8 and 9, then we call this a *dual-modulus P/P+1 prescaler*. Using dual-modulus prescaling we can leave f_{REF} unchanged, keeping our step size and loop bandwidth while gaining much higher operating frequency for the feedback divider.

Dual modulus prescaling is a cooperative effort between a high speed prescaler and a much lower speed programmable counter. The result of this cooperation is a programmable counter that operates at a very high input frequency while maintaining unit division resolution. Dual-modulus prescaling works by allocating quotient and remainder values from the division N/P as shown in **Fig 9.42A**.

The improvement comes from viewing the operation of division in a slightly different way. The division ratio (N) of any two integers will always provide a quotient (Q) and a remainder (R). If Q counts of prescaled f_{in}/P are followed by R counts of f_{in} directly, then N = QP + R. One way to build this counter would be to first count a quotient's worth of prescaled input, and then count a remainder's amount of the input frequency directly. This last step is hard because of the high frequency.

A second approach is to add the remainder counts to the prescaled output as follows: If Q – R counts of prescaled f_{in}/P are followed by R counts of prescaled $f_{in}/(P + 1)$, then N = (Q – R) P + R (P + 1). This is the same as N = QP – RP + RP + R = QP + R.

If the remainder counts are distributed among the quotient counts by increasing the prescaler modulus by one from P to P + 1 for the remainder's amount of cycles, then returning the prescaler to its nominal division ratio while the rest of the quotient counts are made, the same result is achieved. This design always operates the programmable counter at a much lower frequency than the input frequency, which is a much more robust design.

The catch, however, is that there have to be enough quotient counts over which to distribute any amount of remainder counts. If you run out of quotient counts before remainder counts the technique falls apart and that particular loop divisor will not be *realizable*. For real dividers, all of the terms — Q, P, and R — must be positive integers, so Q – R must be greater than zero for N = QP+R to be realizable.

The maximum remainder, $R_{max} = P - 1$. The minimum quotient is equal to R_{max}. So the minimum remainder, $R_{min} = 0$. That means:

$$N_{min} = Q_{min} P + R_{min} = (P - 1) P + 0$$
$$= P^2 - P$$

If the PLL design always requires N to be greater than $P^2 - P$ then there are no problems. If N does need to go below $P^2 - P$, then the only solution is to choose a dual modulus prescaler with a smaller value of P. Various values of P and N_{min} are:

P	N_{min}
8	56
16	240
64	4032
128	16256

Dual-modulus prescaling therefore implies a minimum divisor value before continuous divider value coverage is realized. The value of this minimum divisor depends on the base modulus of the prescaler, P, and increases quadratically. Proper choice of prescaler modulus is one of the important decisions the frequency synthesizer designer has to make.

Fig 9.42B shows how a typical dual-modulus prescaler is implemented. Each cycle of the system begins with the last output pulse having loaded the frequency control word, shown as "Division Control Data," into both the main divider and the prescaler controller. If the division control data's least significant bits (which make up R) loaded into the prescaler controller are not zero, the prescaler is set to divide by P + 1, 1 greater than its normal ratio, P. The main divider counts Q pulses from the prescaler.

Each cycle out of the prescaler then clocks the down counter. Eventually, the down counter reaches zero, and two things happen: The counter is designed to hold (stop counting) at zero (and it will remain held until it is next reloaded) and the prescaler is switched back to its normal ratio, P, until the next reload.

Because the technique is widely used, dual-modulus prescaler ICs are widely used and widely available. Devices for use to a few hundred megahertz are cheap, and devices in the 2.5 GHz region are commonly available. Common prescaler IC division ratio pairs are: 8-9, 10-11, 16-17, 32-33, 64-65 and so on. Many ICs containing programmable dividers are available in versions with and without built-in prescaler controllers.

Loop Filter

When designing PLL synthesizers we usually just have to measure and accept the VCO characteristics of K_0 and the PFD output characteristics. Output frequency range and step size are also not flexible. We pull all of these together into a working and stable synthesizer by proper design of the loop filter.

Loop filter design is usually shrouded in mystery, which has given PLL synthesizer design an aura of a "black art." This is not at all warranted because very reliable design algorithms have been around for decades. Unfortunately these have usually been published in obscure places so they are not well known. The best algorithms known to this author are presented in section 9.7.2. These methods have served extremely well for over 30 years and should provide you with fast

design times and very stable PLL synthesizers. No iteration should be needed.

Loop filters come in active and passive structures. In general the easiest designs use passive (RC) loop filters if the output voltage range from the phase detector (including the PD or PFD and its charge pump) is sufficient to tune the VCO over its required frequency range. If the VCO needs a larger tuning voltage range then an active loop filter structure is necessary.

Passive loop filter circuits for current mode charge pump outputs are shown in **Fig 9.43**. The loop filter time constant is set by the charge pump current and C1, independent of R and of the loop filter output voltage. R provides a phase leading zero for PLL stability. The first-order filter in Fig 9.43A has a minimum parts count and results in loop dynamics that are consistent with the output frequency.

A second-order filter has the undesirable characteristic of allowing step and impulse changes in the input to appear at the output. This results in significant reference sidebands — a bad thing. Fig 9.43B shows a third-order filter in which C2 changes step changes at the input to slower ramps by rolling off the high frequency response of the loop filter. This results in a significant reduction in reference sidebands.

Active loop filters offer more structures that work well and are easier to design, though more complicated to build. When a voltage mode charge pump is being used, an active loop filter is strongly desired since it removes the PLL response variations with different loop filter voltage values.

In **Fig 9.44**, the gain of the op-amp isolates the actual loop filter output voltage from the charge pump output voltage, controlling the loop so that the charge pump output voltage is V_{REF}. Each passive component has the same basic function here as in the passive filter.

The third-order active filter in Fig 9.44 adds an additional pole within the feedback loop and reference frequency sidebands are significantly reduced. Charge pump current impulses now cause ramps instead of steps in the output voltage, but they still cause problems at the amplifier inputs. Good design algorithms include this extra pole as part of the fundamental dynamics — not as a disturbance of conventional second-order dynamics.

Though the design in **Fig 9.45** adds another capacitor by splitting the input resistor, the added complexity is usually well worth it. Charge pumps put out very narrow, spiky signals that are usually of sufficient amplitude and duration to temporarily drive the op-amp input into saturation. If this happens the PLL response becomes non-linear and problems often arise.

The improvement of this design is to low-pass filter the charge pump pulses be-

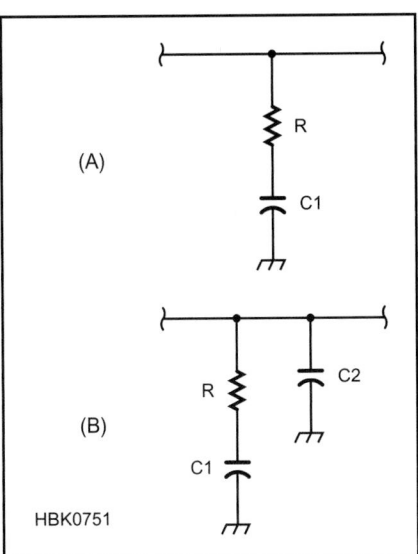

Fig 9.43 — Passive loop filters for current mode charge pump PFD outputs.

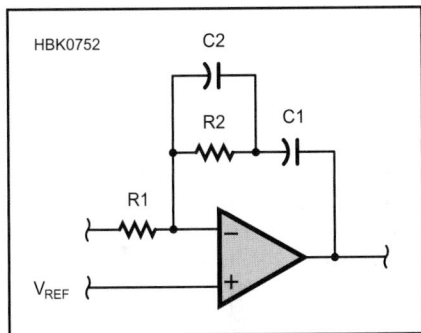

Fig 9.44 — Third-order active loop filter isolates the loop filter output voltage from the charge pump outputs.

fore being applied to the op-amp input. The op-amp far prefers the resulting ramps, remaining linear and therefore predictable in behavior. This results in further reduction in reference frequency sidebands and is easily adapted to difference-output digital PDs, including the PFD.

Design equations for these loop filters will be presented after the remaining PLL component blocks are discussed.

Reference Oscillator

Any PLL is simply a stability transfer mechanism, so the behavior of the reference oscillator will not be improved on. The fractional frequency error of the overall frequency synthesizer will match that of the reference oscillator. For example if the output of the PLL synthesizer changes 1 kHz at an output frequency of 1 GHz [1000/1,000,000,000 = 1:1 million, or 1 part per million (ppm)] then the reference has changed that same fractional amount. If the crystal reference

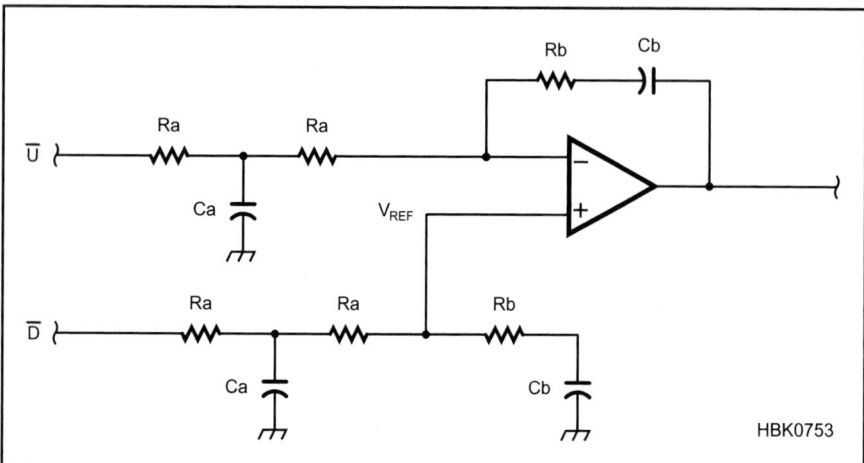

Fig 9.45 — Adding additional low-pass filtering by splitting the input resistor and adding C_A improves loop stability and reduces reference frequency sidebands. This design is well-suited for differential-output PFDs.

is 10 MHz (a very common frequency for crystal references) its frequency drift was (1 ppm) × (10 MHz) = 10 Hz. This is not much frequency drift, but it is significant to the ultimate output frequency.

9.7.2 PLL Loop Filter Design

When it is time to design a PLL synthesizer, it is best to do it in steps. Begin with knowing how the hardware you are using behaves:
• VCO tuning characteristics, K_0
• PFD characteristic (with charge pump), K_d
• Crystal reference frequency, f_{XO}
  Next, list the requirements of your application
• Output frequency range, f_{OUT}
• Output frequency step size, f_{STEP}
Now you are ready to choose the two main design parameters:
• Loop bandwidth — this should not exceed 5% of $f_{REF} = f_{STEP}$.
• Phase Margin — this sets the overall stability of the PLL and can be any value between 50 and 60 degrees. A good value is 54 degrees.

With all of this information in place it is time to calculate the values for the two frequency dividers and the loop filter components. Begin with the divider values:

$R = f_{XO}/f_{REF}$

$N = f_{OUT}/f_{REF}$

$N_{MIN} = f_{OUT,MIN}/f_{REF}$

$N_{MAX} = f_{OUT,MAX}/f_{REF}$

The best design algorithm for stable PLL design calls for using the geometric mean of the feedback divider value range in the loop

filter design. This same idea is used to manage the range of VCO gain values for K_0.

$$N_{DESIGN} = \sqrt{N_{MAX} \times N_{MIN}}$$

K_0 design target value =

$$\sqrt{K_{0,MAX} \times K_{0,MIN}}$$

PFD gain depends on whether the charge pump is voltage mode or current mode:

(voltage mode) $K_d = (V_H - V_L)/4\pi$

(current mode) $K_d = I_{QP}/2\pi$

For all of the active loop filter structures

use the voltage mode value.

Everything is now ready. Place these design values into the appropriate design equations for the selected loop filter structure. For a passive loop filter use **Fig 9.46**. If you are using an active loop filter the appropriate equations are in **Fig 9.47** or **Fig 9.48**. The resistor and capacitor values from these calculations are not critical. Choose values within 30% of what you calculate and the design will be fine. Truly!

Following any loop filter design, it is always a good idea to check the results by directly calculating the filter time constants from the components chosen. Compare these time constants with those derived theoretically. If they match within 10-20%, there should be no problem with loop stability.

The passive third order loop filter design equations of Fig 9.46 look very similar to those of its active filter counterpart. The major difference is that this procedure yields two time constants instead of three, and a term equaling the sum of both filter capacitances.

Unlike the active filter form, there are no arbitrary component choices with the passive filter. From the time constants and the total capacitance, all three loop filter components are uniquely determined.

Since the concepts of natural frequency and damping no longer apply (by definition) in a third-order filter, the more general stability concepts of gain crossover and phase margin are used. A value of 50-60 degrees is usually used for the design phase margin.

The equations for Fig 9.47 and Fig 9.48 include the rolloff pole within the loop dynamics, rather than treating it as a loop perturbation. This allows a greater amount of high frequency rolloff to be achieved as it "kicks in" at a much lower frequency than perturbation techniques would allow.

A usual design strategy is to choose the

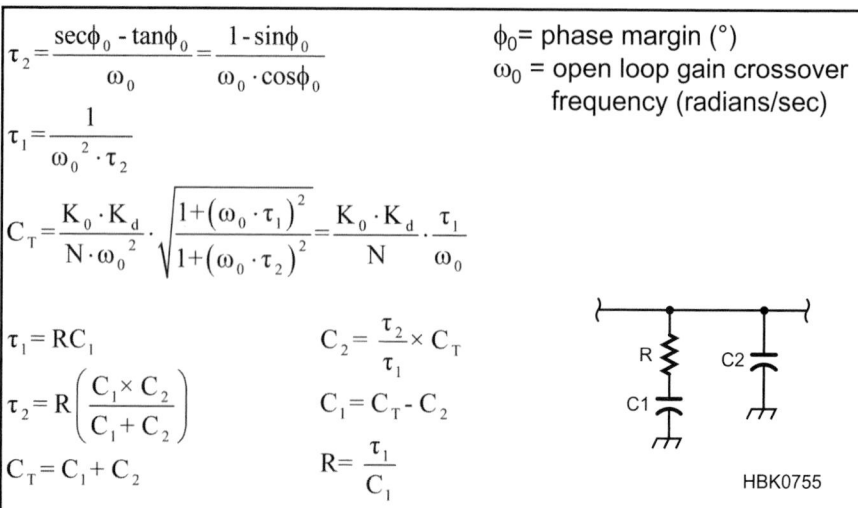

Fig 9.46 — Design equations for passive loop filters.

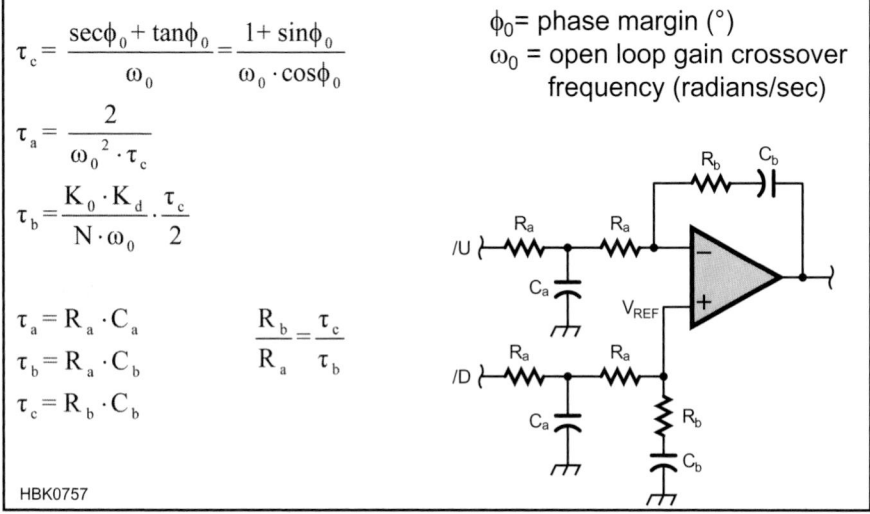

$$\tau_3 = \frac{\sec\phi_0 - \tan\phi_0}{\omega_0} = \frac{1 - \sin\phi_0}{\omega_0 \cdot \cos\phi_0}$$

$$\tau_2 = \frac{1}{\omega_0^2 \cdot \tau_3}$$

$$\tau_1 = \frac{K_0 \cdot K_d}{N \cdot \omega_0} \cdot \left| \frac{-j\omega_0\tau_2 - 1}{j\omega_0\tau_3 + 1} \right| = \frac{K_0 \cdot K_d}{N} \cdot \frac{\tau_2}{\omega_0}$$

$$\tau_1 = R_1 C_1$$

$$\tau_2 = R_2(C_1 + C_2) \qquad \frac{R_2}{R_1} = \frac{\tau_2 - \tau_3}{\tau_1}$$

$$\tau_3 = R_2 C_2$$

ϕ_0 = phase margin (°)
ω_0 = open loop gain crossover frequency (radians/sec)

HBK0756

Fig 9.47 — Design equations for third-order active loop filters

$$\tau_c = \frac{\sec\phi_0 + \tan\phi_0}{\omega_0} = \frac{1 + \sin\phi_0}{\omega_0 \cdot \cos\phi_0}$$

$$\tau_a = \frac{2}{\omega_0^2 \cdot \tau_c}$$

$$\tau_b = \frac{K_0 \cdot K_d}{N \cdot \omega_0} \cdot \frac{\tau_c}{2}$$

$$\tau_a = R_a \cdot C_a \qquad \frac{R_b}{R_a} = \frac{\tau_c}{\tau_b}$$

$$\tau_b = R_a \cdot C_b$$

$$\tau_c = R_b \cdot C_b$$

ϕ_0 = phase margin (°)
ω_0 = open loop gain crossover frequency (radians/sec)

HBK0757

Fig 9.48 — Design equations for differential-input active loop filters

resistors first to achieve the design ratio. This may provide several possible resistor pairs. Calculate the resulting capacitor values for each resistor pair possibility, and choose the set that provides the closest realizable values.

9.7.3 Fractional-N Synthesis

There is one technique that allows the relationship f_{REF} = step size to be broken. Until now the PLL synthesizer designs have all considered the feedback divider to always operate at a single programmed value. This is called an *integer-N PLL*. In a more complicated system we can change the value programmed into the feedback divider according to some pattern. For example we can use the 2 meter synthesizer from the section Frequency Resolution (Step Size) in the section on Phase-Locked Loops and divide by 1474 eight times and then divide by 1475 twice. Repeating this pattern gives an average effective divider value, averaged over the pattern, of $(8 \times 1474 + 2 \times 1475)/10 = 1474.2$.

A PLL that operates with this kind of varied programming value for the feedback divider is called a *fractional-N PLL*.

By using fractional-N techniques the step size gets smaller while f_{REF} does not change. The price for this is increased spurious sidebands and a higher noise level at the PLL output. This technique is very complicated to implement well. Fortunately several suppliers of PLL integrated circuits have done all the homework for us. It is far easier to use any of these PLL parts than to do a fractional-N design from scratch.

9.7.4 PLL Synthesizer Phase Noise

Differences in resonator Q usually make the phase-noise sidebands of a loop's reference oscillator much smaller than those of the VCO. Within its loop bandwidth, a PLL acts toward matching the phase-noise components of its VCO to those of the reference. There are several processes which get in the

way of this actually happening, which are outlined here.

Dividing the reference oscillator signal f_{XO} to produce the signal f_{REF} which is applied to the phase detector also divides the deviation of the reference oscillator's phase-noise sidebands, translating to a 20 dB reduction in phase noise per decade of division. This models as a factor of $-20 \log (R)$ dB, where R is the reference divisor value, since f_{REF} is less than f_{XO}. We also know that within its loop bandwidth the PLL acts as a frequency multiplier and this multiplies the deviation of the phase noise sidebands present at the PFD, again by 20 dB per decade or equivalently by a factor of $20\log(N)$ dB, where N is the loop divider's value. Between the reference oscillator and the PLL output, the sidebands on the reference signal are increased by 20 log (N/R) dB.

COUNTER AND PFD NOISE

Logic circuits have a noise characteristic that is extremely important to PLL design. Any circuit, when you really get down to it, is always an analog circuit. In this case the switching threshold of the logic circuits has a tiny amount of noise on it. This small amount of noise manifests itself in a very slight variation in when the logic circuit actually switches, even if the input clock is perfectly clean. We can think of this small timing jitter as a small shift in the output signal "zero crossing," which is directly equivalent to a phase modulation.

The amount of this timing jitter is constant, no matter the operating frequency. Thus, the amount of phase shift this represents depends on the operating frequency: 1 nanosecond of jitter at 1 kHz is very small, but represents 36 degrees at 100 MHz. As a result we get different measures of the noise floor from digital circuits that depend on the frequency at which they are operating. This digital noise floor interferes with the PLL action to match the VCO noise to the divided reference oscillator noise as measured at the PFD inputs.

VCO NOISE

The PLL can track only slowly moving noise, well within its bandwidth. Therefore VCO noise components that change slowly enough for the PLL to measure and track will be increasingly canceled successfully. Any VCO noise component that changes at a rate faster than the PLL bandwidth will not be changed at all. At offset frequencies beyond the PLL bandwidth the VCO noise is still present as if the PLL were not even there.

Does this imply that to get a low-noise synthesizer we are encouraged to make the PLL bandwidth very wide? Well, yes, to a certain degree. When we build PLL synthesizers we soon find that the logic noise from the prior section begins to get in the way if the loop bandwidth gets too wide. When this happens the total PLL output noise actually begins to increase as the bandwidth gets wider. There is a range of loop bandwidth values where the total PLL output noise has a minimum. This range is centered at the offset frequency where the logic noise floor, multiplied by 20 log (f_{OUT}/f_{REF}), has the same value as the VCO intrinsic phase noise.

Clearly, having a VCO design with minimum phase noise is very important to a good, low-noise PLL synthesizer. Much effort is spent on this area in industry. A summary of important results from these noise reduction efforts is presented in the section Improving VCO Noise Performance below.

FRACTIONAL DIVISION NOISE

From the brief discussion on fractional-N principles, we note that this whole idea is based on changing the feedback divider value in a semi-randomized way. This inherently jitters the output from the feedback divider, which is a source of phase noise into the PFD. This noise source is algorithmic instead of physical, but the noise is a big problem nonetheless.

OTHER NOISE SOURCES

Phase noise can be introduced into a PLL by other means. Any amplifier stages between the VCO and the circuits that follow it (such as the loop divider) will contribute some noise, as will microphonic effects in loop- and reference-filter components (such as those due to the piezoelectric properties of ceramic capacitors and the crystal filters sometimes used for reference-oscillator filtering). Noise on the power supply to the system's active components can modulate the loop. The fundamental and harmonics of the system's ac line supply can be coupled into the VCO directly or by means of ground loops.

9.7.5 Improving VCO Noise Performance

It is tempting to regard VCO design as a matter of coming up with a suitable oscillator topology with a variable capacitor, simply replacing the variable capacitor with a suitable varactor diode and applying a tuning voltage to the diode. Unfortunately, things are not this simple (as if even this is simple!). There is the matter of applying the tuning voltage to the diode without significantly disturbing the oscillator performance. There is also an issue of not introducing parasitic oscillation with the varactor circuit.

Next, as mentioned earlier, one would not like the tuning voltage to drop below the voltage swing in the oscillator tank. If this is allowed to occur, the tuning diodes go into conduction and the oscillator noise gets worse. A good first approach is to use varactor diodes back to back as shown in **Fig 9.49**. This allows the tank voltage swing

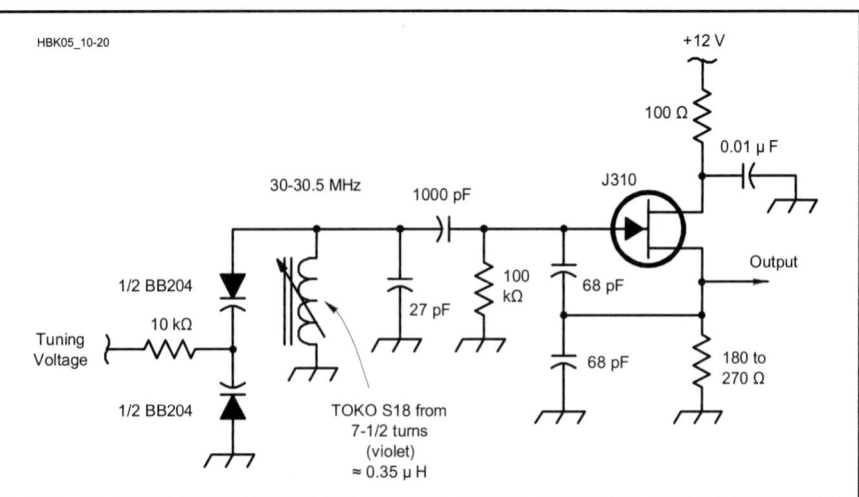

Fig 9.49 — A practical VCO. The tuning diodes are halves of a BB204 dual, common-cathode tuning diode (capacitance per section at 3 V, 39 pF) or equivalent. The ECG617, NTE617 and MV104 are suitable dual-diode substitutes, or use pairs of 1N5451s (39 pF at 4 V) or MV2109s (33 pF at 4 V).

to be developed across two diodes instead of one, as well as allow for a more balanced loading of the oscillator's tank. The semiconductor industry has realized this and there are a number of varactors available prepackaged in this configuration today.

IMPROVING VCO OPERATING Q

Often the Q of passive capacitors available for use in an oscillator tank exceeds the Q of available varactors. One way of reducing the influence of the varactor is to use only the amount of varactor capacitance required to tune the oscillator over its desired range. The balance of the capacitance is supplied to the tank in the form of a higher-Q fixed capacitor. This has the advantage of not requiring the varactor to supply all the capacitance needed to make the circuit function, and often allows for the use of a lower capacitance varactor.

Lower capacitance varactors typically exhibit higher Q values than their larger capacitance counterparts. This concept can be extended by splitting the oscillator tuning range, say 70 MHz to 98 MHz (for a typical lower-side up-converter receiver design popular for the last few decades), into multiple bands. For this example let us consider four oscillators, each with 7 MHz of tuning range. We desire the varactor Q effects to be swamped by high-Q fixed capacitors. This can be further improved through the use of a "segment-tuned VCO" discussed below. A secondary but important benefit of this is to reduce the effective tuning gain (MHz/volt, K_0) of the oscillators, making them less susceptible to other noise voltage sources in the synthesizer loop. These noise sources can come from a variety of places including, but not limited to, varactor leakage current, varactor tuning drive impedance and output noise of the driving operational amplifier or charge pump.

Segment-tuned VCOs provide the designer with additional benefits, but also with additional challenges. By segmented we mean that circuit elements, which could be both inductance and capacitance, are selected for each range that the VCO is expected to tune. **Fig 9.50** shows the frequency range-switching section of a typical VCO in which several diode switches are used to alter the total tank circuit inductance in small steps. These segments create a type of "coarse tuning" for the VCO, and the output of the PLL loop filter performs "fine tuning" with the varactor diode. Usually the component values are arranged in some sort of binary tree. In some integrated designs, 64 or even 128 sub-bands are available from the VCO.

Many times the segmented-VCO concept is applied to an oscillator design that is required to cover several octaves in frequency. For example, this would be the case with a single-IF radio with the IF at about 8 MHz. The VCO would be required to cover 8 MHz to 48 MHz for 100 kHz to 50 MHz coverage of a typical transceiver.

While segmentation allows one to build a multi-octave tunable oscillator, segmentation also imposes some additional constraints. Recalling the earlier section in this chapter on oscillators, the condition for oscillation is an oscillator loop gain of 1 at a phase angle of 0 degrees. Over several octaves, the gain of the oscillating device (typically a transistor) varies, decreasing as operating frequency increases. While conventional limiting in the oscillator circuit will deal with some of this, it is not good practice to let the normal limiting process handle the entire gain variation. This becomes the role of automatic gain control (AGC) in multi-octave oscillator designs. Application of AGC allows for the maintenance of the oscillation criteria, a uniform

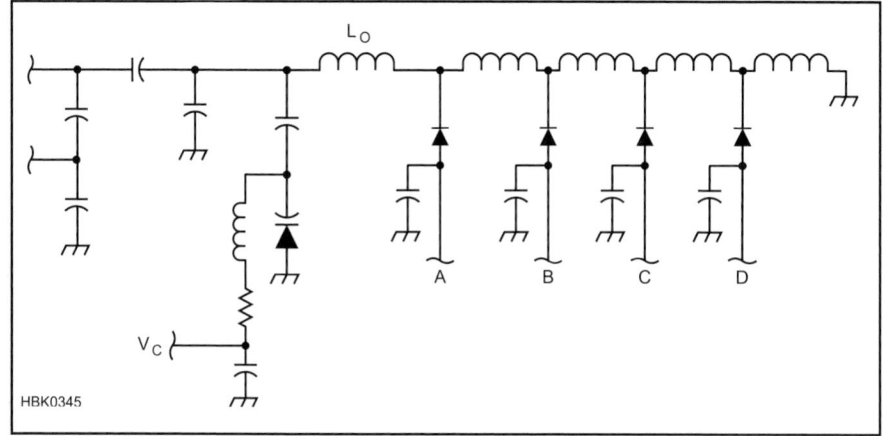

Fig 9.50 — The resonator portion of an inductor-switched segment-tuned VCO. A varactor diode provides tuning over a small range. Diode switches will alter the total inductance, changing the frequency in larger steps. Inductor-switched, capacitor-switched, and combinations are all used in VCO designs. (From Hayward, *Introduction to Radio Frequency Design*, Chapter 7. See references.)

output and good starting characteristics, and lower noise across the operating bandwidth.

Finally, a properly designed segmented VCO can improve synthesizer switching speed. The segmentation allows the designer to "pre-steer" the PLL system and reduce the time required for the loop filter to slew the VCO to the desired frequency for lock. Frequency-locked loops (FLL) are often employed for this steering operation to be sure that it happens properly.

9.7.6 A PLL Design Example

As our design example, let us consider a synthesized local oscillator chain for a 10 GHz transverter. **Fig 9.51** is a simplified block diagram of this 10 GHz converter. This example is chosen because it is a departure from the traditional multi-stage multiply-and-filter approach. It permits realization of the oscillator system with two simple loops and minimal RF hardware. It is also representative of what is achievable with current hardware, and can fit in a space of 2 to 3 square inches. This example is intended to be a vehicle to explore the loop design aspects and is not offered as a "construction project." The multiple design details required are beyond the scope of this chapter.

Two synthesized frequencies, 10 GHz and 340 MHz, are required. Since 10.368 GHz is one of the popular X-band traffic frequencies, we initially mix this with the 10 GHz LO to produce an IF of 368 MHz. The 368 MHz IF signal is subsequently mixed with the 340 MHz LO to produce a 28 MHz final IF, which can be fed into the 10 meter input of any amateur transceiver. We focus our attention on the design of the 10 GHz synthesizer only. Once this is done, the same principles are applied to the 340 MHz section. Our goal is to design a low-noise LO system (by adopt-

ing minimum division ratios) with a loop reference oscillator that is an integer multiple of 10 MHz. Using this technique allows the entire system to be locked at a later time to a 10 MHz standard for precise frequency control.

For the microwave synthesizer we consider using a line of microwave integrated circuits made by Hittite Microwave in gallium-arsenide (GaAs) material. These devices include a selection of prescalers operating to 12 GHz, a 5-bit counter that operates to 2.2 GHz and a phase/frequency detector that operates up to 1.3 GHz. If we use a crystal reference frequency f_{XO} of 100 MHz, and also apply

that as f_{REF} into the PFD (therefore R = 1), then the feedback divider number needs to be N = 10,000/100 = 100. One way to realize this in hardware is to represent 100 = 4 × 25, starting with a divide-by-4 prescaler and finishing with the 5 bit counter programmed to divide by 25. The divide-by-4 prescaler output at lock will be 2500 MHz, which is too high for the 2200 MHz limited programmable counter. We need a design change.

The obvious solution is to select a divide-by-8 prescaler, providing a 1250 MHz output from the 10 GHz input. We cannot program the counter to half of 25, so we need to leave it programmed at 25. This makes the output from the feedback divider at 10,000/(8×25) = 50 MHz. We need to set R = 2.

Before even thinking about designing the loop filter, we need to know the VCO gain, VCO noise performance, divider noise performance, phase detector gain, phase detector noise performance and finally reference noise performance. For the 10 GHz VCO, we are always looking for parts that are easily available, useable and economical, so salvaging a dielectric resonator from a Ku band LNB is promising (see "SHF Super Regenerative Reception by Andre Jamet, F9HX, in Jan-Feb 2002 *QEX*). These high-Q oscillators can be fitted with a varactor and tuned over a limited range with good results. The tuning sensitivity of our dielectric resonator VCO is about 10 MHz per volt and the phase noise at 10 kHz offset is –87 dBc/Hz, and –107 dBc/Hz at 100 kHz.

The phase detector and divider information is available from the device data sheets. The

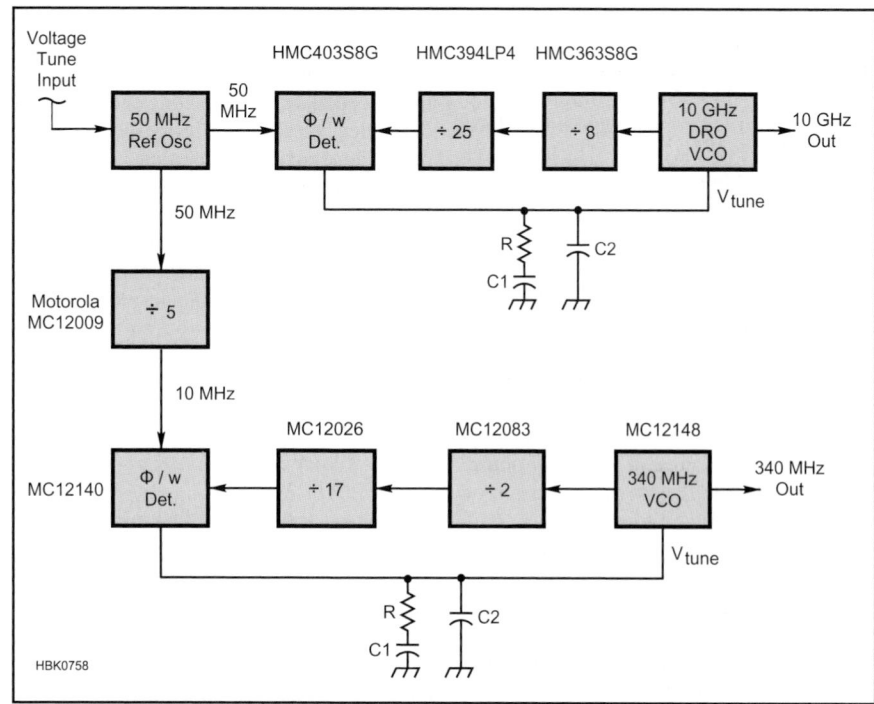

Fig 9.51 — A simplified block diagram of a PLL local oscillator for a 10 GHz converter.

HMC363 divide-by-8 operates up to 12 GHz and has a programmable charge pump that sets the phase detector gain K_d. It is usually good to keep this gain high, so we choose to use a 2 mA charge pump current, giving K_d = 0.32 mA/radian. The data sheet also says that this PFD has an output noise floor of –153 dBc/Hz measured at 100 kHz offset. The HMC394 programmable divider operated up to 2.2 GHz with the same output noise floor. The HMC984 PFD operated up to 350 MHz with a noise figure of merit (FOM) of –231 dBc/Hz/Hz-f_{REF}, so with f_{REF} = 50 MHz the PFD output noise floor is –231 + 10 log (50,000,000) = –154 dBc/Hz. Assuming the R divider is implemented in CMOS, the output noise floor from it is –163 + 10 log (50,000,000/125000) = –137 dBc/Hz. This noise is much higher than the noise from the GaAs dividers, so it is actually better here to use a GaAs divide-by-2 for the R counter instead of using CMOS.

The logic noise floor for the entire PLL is the sum of the individual noises from each divider and the PFD. The result is –147 dBc/Hz. This is determined at the PFD. To calculate how this noise will measure at the PLL output we need to add 20LogN = 46 dB. Thus the output noise floor due to PLL logic devices is –101 dBc/Hz. This is close to the VCO phase noise at an offset of 100 kHz, so minimum noise design suggests that we select 100 kHz as our loop bandwidth.

An alternative is to eliminate the R counter and directly use a 50 MHz reference. An excellent choice for a low noise reference is the one described by John Stephensen on page 13 in Nov/Dec 1999 QEX. The noise performance of this VCXO is in the order of –160 dBc/Hz at 10 kHz offset at the fundamental frequency. Translating this to the output frequency means adding 20 log N, for a result of –114 dBc/Hz. Eliminating the R counter also lowers the logic noise floor to –104 dBc/Hz at the output. The translated reference oscillator noise is 10 dB below the noise floor from the logic devices, so the logic noise floor dominates.

Using this information we can now design the loop filter. Assuming that the VCO tuning voltage is between 0.5 and 4.5V we can directly use the charge pump included in this PFD and therefore select the passive third order loop filter from Fig 9.43. We use the equations in Fig 9.46 and calculate loop filter component values of R = 6977 ohms, C1 = 724 pF, and C2 = 80 pF. We therefore select 6.8k, 820 pF, and 82 pF for these component values respectively. The Bode plot of the theoretical and actual PLL dynamics is shown in **Fig 9.52**. It is nearly impossible to tell them apart!

If the VCO requires higher tuning voltages, you can insert a noninverting op-amp gain stage between the output of this loop filter and the VCO. Choose a low input current

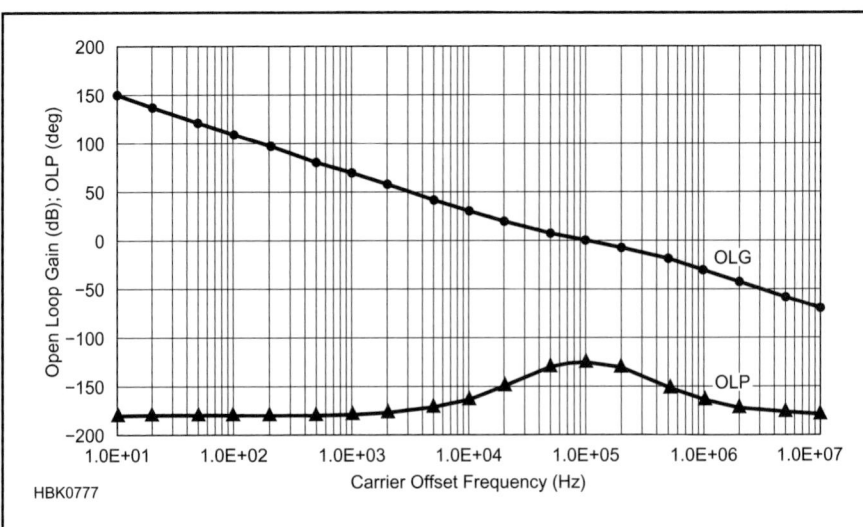

Fig 9.52 — Comparisons of the theoretical and actual PLL dynamics show extremely good correspondence.

op-amp to minimize leakage from the loop filter capacitors. Be careful — the gain of this amplifier stage changes the effective VCO gain that the loop experiences. When designing the loop filter components the "design" VCO gain must be replaced by the product of the actual VCO gain and the amplifier gain. It is also essential to assure that the 1 dB bandwidth of the voltage amplifier greatly exceeds the loop bandwidth so that it does not degrade the PLL phase response and hence loop stability.

9.7.7 PLL Measurements and Troubleshooting

VCO GAIN

One of the first things we need to measure when designing a PLL is the VCO gain. The tools needed include a voltmeter, some kind of frequency measuring device like a receiver or frequency counter and a clean source of variable dc voltage. The circuit in **Fig 9.53** containing one or more 9 V batteries and a 10-turn, 10 kΩ pot does nicely. One simply varies the voltage some amount and then records the associated frequency of the VCO.

The gain of the VCO is then the change in f divided by the change in voltage (Hz / V).

LOOP BANDWIDTH

Measuring the actual bandwidth of the PLL usually means measuring the entire gain and phase responses of the PLL. There is a much easier way that needs only a square-wave generator and an oscilloscope. This method uses the PLL testing structure of **Fig 9.54** developed by Glenn Ewart that injects a low-value square wave into the loop at a low impedance point so that impulse response can be observed directly. The test signal is also ground-referenced and is independent of the loop locking voltage.

Start by swapping the series RC components in the loop filter so the resistor is connected to ground and the capacitor attaches to the VCO tuning line. Then split the resistor into two resistances that together add up to the total resistance needed by the filter. The bottom resistor connected to ground is very small — 50 Ω is a nice choice when the

Fig 9.53 — Clean, variable dc voltage source used to measure VCO gain in a PLL design.

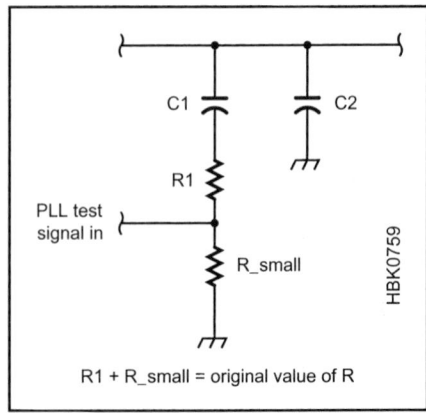

Fig 9.54 — A simple testing structure for PLL dynamics.

total resistor value is much greater than 50 Ω. The top resistor is temporarily replaced by a potentiometer that can make up the desired remaining resistance. Across the bottom resistor we connect the square-wave generator and set it to a small amplitude (100 or 200 mV is common) at a frequency a few percent of the loop bandwidth, and operate the PLL. Looking at the VCO tuning line with an oscilloscope we see the impulse response of the PLL.

Now adjust the potentiometer way off-value to the low side. The impulse response will now show ringing. The frequency of this ringing is a good approximation of the PLL loop bandwidth.

SETTLING TIME

Having measured the loop bandwidth, readjust the potentiometer back to its nominal value. The impulse response should look *much* cleaner! It is tempting to look at the settling time of the impulse response and say that this is the settling time of the PLL. This unfortunately is not quite true.

The PLL is not fully settled until capacitor C1 is fully charged. This usually takes slightly longer than what the impulse response itself shows. Move the scope probe to the "ground" side of C1 and measure when this voltage reaches zero. This means there is no more current flowing through the filter resistors and the capacitor is fully charged.

TROUBLESHOOTING PLLS

Here are some frequently encountered problems in PLL designs:

• The outputs of the phase detector are inverted. This results in the loop slewing to one or the other power supply rails. The loop cannot possibly lock in this condition. Solution: Swap the phase detector outputs.

• The loop cannot comply with the tuning voltage requirements of the VCO. If the loop runs out of tuning voltage before the required voltage for a lock is reached, the locked condition is not possible. Solution: Re-center the VCO at a lower tuning voltage or increase the rail voltages on the op amp.

The loop is very noisy and the tuning voltage is very low. The tuning voltage on the varactor diodes in the VCO should not drop below the RF voltage swing in the oscillator tank circuit. Solution: Adjust the VCO so that the loop locks with a higher tuning voltage.

9.7.8 Commercial Synthesizer ICs

In this section, we explore using commer-cially available synthesizer chips and the role of DDS in more recent hybrid architectures. The following is not intended to be project oriented, but rather is designed to expose the reader to additional concepts that cannot be fully explored here. The reader, being made aware of these ideas, may wish to examine them in detail using references at the end of this chapter.

Many synthesizer chips have been introduced in recent years for cellular and Wi-Fi applications. One might reasonably ask if any of these devices are well suited to amateur applications. The short answer is — probably not. Applications using these chips strive for a minimum of external components, so they are usually not flexible in design. These chips conform to a strict applications profile in which the communication link is typically sending a few tens of megabits for distances of less than 3 miles (5 km). Some are even designated as "low noise" with respect to other chips of their ilk, but these noise levels are not really low with respect to most amateur requirements. The phase noise, while adequate for their intended application, is not good enough for many HF, VHF and UHF applications without additional measures being taken.

Despite all this, they do have some interesting properties and capabilities that can be exploited with additional design. Some properties that can be exploited are programmable charge pump current, a rich set of division options for creating multiple-loop synthesizers and typically low power consumption.

One of the easiest ways to improve the overall performance is to follow one of these chips with additional frequency division. Consider the following example: One of these chips could be used as a 500 to 550 MHz synthesizer, followed with two cascaded decade dividers for a total division of 100. This division would reduce the phase noise profile by 40 dB, making a reasonably quiet synthesizer for the 5 to 5.5 MHz range. The step size would also be reduced by a factor of 100, making the spacing required in the 500 to 550 MHz range equal to 1 kHz for a step size of 10 Hz at 5 MHz. This would make a good local oscillator for a simple traditional radio that covers 80 and 20 meters using both mixer products against a 9 MHz SSB generator.

There are also other techniques that can prove helpful. Decoupling the VCO from the chip will permit one to avoid much of the "on chip" noise that VCOs are susceptible to at the expense of some more complexity. If the VCO is implemented externally, there is an opportunity to design it with increased operating Q, thereby improving the overall phase noise performance. An external VCO also opens other opportunities.

As mentioned earlier, these chips are designed for operation at a V_{CC} of 3 to 5 V, typically using an internal charge pump to generate the operating voltage. This limits the tuning voltage swing to V_{CC}. One must be able to fit the entire tuning range voltage and the ac voltage excursion in the VCO tank circuit within the V_{CC} range. As mentioned at the end of the earlier section Improving VCO Noise Performance, this problem can be overcome with the addition of voltage gain in an external operational amplifier, running at a higher voltage. This allows the signal-to-noise ratio of the oscillator to be improved by increasing the tank voltage swing and still having adequate voltage range to perform the tuning function. There are certainly more examples of how the performance of these devices could be improved for amateur applications.

The use of DDS (Direct Digital Synthesis) in an amateur transceiver goes back to at least 1987 when Bob Zavrel, W7SX, and the author used a unit from Digital RF Solutions as the VFO for an HF rig. Spurious performance of this DDS was below −75 dBc and it worked very well. More modern integrated DDS devices are improving rapidly and are capable of producing a good and economical medium performance transceiver; they are not as yet capable of "competition grade" performance. Using these chips still requires some sort of analog/digital hybrid to achieve this. (DSP and DDS are discussed in the **DSP and Software Radio Design** chapter.)

One method of incorporating DDS within a hybrid synthesizer system is to have the DDS supply the least significant digits of the frequency resolution and to sum that DDS output into a conventional divide-by-N loop that supplies the most significant digits. If desired, this signal can be passed through a sufficiently narrow filter to further attenuate any DDS spurious signals. This method is employed in a number of commercial signal generators as well as some of the more modern transceivers. Another widely used approach is to take advantage of the extremely small step size available from DDS devices and use this as the reference frequency into a conventional PLL synthesizer. The loop bandwidth of the PLL acts now as a tracking band-pass filter to reduce the DDS output spurious signals. Most modern FM broadcast transmitters use the latter technique when broadcasting CD-ROM and other digitized material.

9.8 Glossary of Oscillator and Synthesizer Terms

Buffer — A circuit that amplifies the output of a circuit while isolating it from the load.

Bypass — Create a low ac impedance to ground at a point in the circuit.

Cavity — A hollow structure used as an electrical resonator.

Closed-loop — Operation under the control of a feedback loop (see also **open-loop**).

Coupling — The transfer of energy between circuits or structures.

Damping (factor) — the characteristics of the decay in a system's response to an input signal. The **damping factor,** ζ, is a numeric value specifying the degree of damping. An **underdamped** system alternately overshoots and undershoots the eventual steady-state output. An **overdamped** system approaches the steady-state output gradually, without overshoot. A **critically-damped** system approaches the steady-state output as quickly as possible without overshoot.

dBc — Decibels with respect to a carrier level.

DC-FM — control of a signal generator's output frequency by a dc voltage.

Decouple — To provide isolation between circuits, usually by means of filtering.

Direct digital synthesis (DDS) — Generation of signals by using counters and accumulators to create an output waveform.

Distributed — Circuit elements that are inherent properties of an extended structure, such as a transmission line.

ESR — Equivalent series resistance.

Free-running — Oscillating without any form of external control.

Fundamental — Lowest frequency of natural vibration or oscillation.

Integrator — A low-pass filter whose output is approximately the integral of the input signal.

Intermodulation — Generation of distortion products from two signals interacting in a nonlinear medium, device, or connection.

Isolation — Preventing signal flow between two circuits or systems. **Reverse isolation** refers to signal flow against the desired signal path.

Jitter (phase jitter) — Random variations of a signal in time, usually refers to random variations in the transition time of digital signals between states.

Linearization — Creation of a linear amplification or frequency characteristic through corrections supplied by an external system.

Loop gain — The total gain applied to a signal traveling around a feedback control loop.

Lumped (element) — Circuit elements whose electrical functions are concentrated at one point in the form of an electronic component.

Match — Equal values of impedance.

Modulus — The number of states of a digital counter or divider.

Motional capacitance (inductance) — The electrical effect of a crystal's mechanical properties, modeled as a capacitance (inductance).

Natural frequency (ω_n) — Frequency at which a system oscillates without any external control.

Noise bandwidth — The width of an ideal rectangular filter that would pass the same noise power from white noise as the filter being compared (also called **equivalent noise bandwidth**).

Open-loop — Operation without controlling feedback.

Oscillation — Repetitive mechanical motion or electrical activity created by the application of positive feedback.

Overtones — Vibration or oscillation at frequencies above the **fundamental**, usually harmonically related to the fundamental.

Permeability tuning — Varying the permeability of the core of an inductor used to control an oscillator's frequency.

Phase-lock — Maintain two signals in a fixed phase relationship by means of a control system.

Phase noise — Random variations of a signal in time, expressed as variations in phase of a sinusoidal signal.

Phasor — Representation of a sinusoidal signal as an amplitude and phase, often drawn as a vector.

Power density — Amount of power per unit of frequency, usually specified as dBc/Hz or as RMS voltage/□Hz.

Prescaler — A frequency divider used to reduce the frequency of an input signal for processing by slower circuitry.

Preselector — Filters applied at a receiver's input to reject out-of-band signals.

Pull — Change the frequency at which a crystal oscillates by changing reactance of the circuit in which it is installed.

Quadrature — A 90° phase difference maintained between two signals.

Reciprocal mixing — Noise in a mixer's output due to the LO's noise sidebands mixing with those of the desired signal.

Relaxation oscillation — Oscillation produced by a cycle of gradual accumulation of energy followed by its sudden release.

Resonator — Circuit or structure whose resonance acts as a filter.

Simulation — Calculate a circuit's behavior based on mathematical models of the components.

Spurious (spur) — A signal at an undesired frequency, usually unrelated to the frequency of a desired frequency.

Squegg (squeeg) — Chaotic or random jumps in an oscillator's amplitude and/or frequency.

Static (synthesizer) — A synthesizer designed to output a signal whose frequency does not change or that is not changed frequently.

Synthesis (frequency) — The generation of variable-frequency signals by means of nonlinear combination and filtering (direct synthesis) or by using phase-lock or phase-control techniques (indirect synthesis).

TCXO — Temperature-compensated crystal oscillator. A **digitally temperature-compensated oscillator (DTCXO)** is controlled by a microcontroller or computer to maintain a constant frequency. **Oven-controlled crystal oscillators (OCXO)** are placed in a heated enclosure to maintain a constant temperature and frequency.

Temperature coefficient (tempco) — The amount of change in a component's value per degree of change in temperature.

Temperature compensation — Causing a circuit's behavior to change with temperature in such as way as to oppose and cancel the change with temperature of some temperature-sensitive component, such as a crystal.

Varactor (Varicap) — Reverse-biased diode used as a tunable capacitor.

VCO — Voltage-controlled oscillator (also called **voltage-tuned oscillator**).

VFO — Variable-frequency oscillator.

VXO — Variable crystal oscillator, whose frequency is adjustable around that of the crystal.

9.9 References and Bibliography

Agilent application note 5989-9848EN, "Boosting PLL Design Efficiency," — describes the characterization of VCOs using the E5202B Signal Source Analyzer. **cp.literature.agilent.com/ litweb/pdf/5989-9848EN.pdf**

Clarke and Hess, *Communications Circuits; Analysis and Design* (Addison-Wesley, 1971; ISBN 0-201-01040-2). Wide coverage of transistor circuit design, including techniques suited to the design of integrated circuits. Its age shows, but it is especially valuable for its good mathematical treatment of oscillator circuits, covering both frequency- and amplitude-determining mechanisms. Look for a copy at a university library or initiate an interlibrary loan.

F. Gardner, *Phaselock Techniques*, (John Wiley and Sons, 1966, ISBN 0-471-29156-0).

J. Grebenkemper, "Phase Noise and Its Effects on Amateur Communications," *Part 1, QST.* Mar 1988, pp 14-20; *Part 2,* Apr 1988 pp 22-25. Also see Feedback, *QST,* May 1988, p 44.

W. Hayward and D. DeMaw, *Solid State Design for the Radio Amateur* (Newington, CT: ARRL, 1986). Out of print, this is dated but a good source of RF design ideas.

W. Hayward, W7ZOI, R. Campbell, KK7B, and B. Larkin, W7PUA, *Experimental Methods in RF Design* (Newington, CT: ARRL, 2003). A good source of RF design ideas, with good explanation of the reasoning behind design decisions.

W. Hayward, W7ZOI, *Introduction to Radio Frequency Design* (Newington, CT: ARRL, 1994). Out of print, good in-depth treatments of circuits and techniques used at RF.

Hewlett-Packard Application Note 150-4, "Spectrum Analysis…Random Noise Measurements."\ Source of correction factors for noise measurements using spectrum analyzers. Available through Agilent as application note 5952-1147. **cp.literature.agilent.com/litweb/ pdf/5952-1147.pdf**

I. Keyser, "An Easy to Set Up Amateur Band Synthesizer," *RADCOM* (RSGB), Dec 1993, pp 33-36.

V. Manassewitsch, *Frequency Synthesizers Theory and Design* (Wiley-Interscience, 1987, ISBN 0 471-01116-9).

Motorola Application Note AN-551, "Tuning Diode Design Techniques" No longer supplied by Freescale but available through various on-line sources. A concise explanation of varactor diodes, their characteristics and use.

E.W. Pappenfus. W. Bruene and E.O. Schoenike, *Single Sideband Circuits and Systems*, (McGraw-Hill, 1964). A book on HF SSB transmitters, receivers and accoutrements by Rockwell-Collins staff. Contains chapters on synthesizers and frequency standards. The frequency synthesizer chapter predates the rise of the DDS and other recent techniques, but good information about the effects of synthesizer performance on communications is spread throughout the book.

B. Parzen, A. Ballato. *Design of Crystal and Other Harmonic Oscillators* (Wiley-Interscience, 1983, ISBN 0-471-08819-6). This book shows a thorough treatment of modern crystal oscillators including the optimum design. It is also the only book that looks at the base and collector limiting effect. The second author is affiliated with the US Army Electronics Technology & Devices Laboratory, Fort Monmouth, New Jersey.

B. E. Pontius, "Measurement of Signal Source Phase Noise with Low-Cost Equipment," *QEX,* May-Jun 1998, pp 38-49.

U. Rohde, *Microwave and Wireless Synthesizers Theory and Design* (John Wiley and Sons, 1997, ISBN 0-471-52019-5). This book contains the textbook-standard mathematical analyses of frequency synthesizers combined with unusually good insight into what makes a better synthesizer and *a lot* of practical circuits to entertain serious constructors. A good place to look for low-noise circuits and techniques.

U. Rohde, D. Newkirk, *RF/Microwave Circuit Design for Wireless Applications* (John Wiley and Sons, 2000, ISBN 0-471-29818-2) While giving a deep insight into circuit design for wireless applications, Chapter 5 (RF/Wireless Oscillators) shows a complete treatment of both discrete and integrated circuit based oscillators. Chapter 6 (Wireless Synthesizers) gives more insight in synthesizers and large list of useful references.

U. Rohde, J. Whitaker, *Communications Receivers DSP, Software Radios, and Design*, 3rd ed., (McGraw-Hill, 2001, ISBN 0-07-136121-9). While mostly dedicated to communication receivers, this book has a useful chapter on Frequency Control and Local Oscillators, specifically on Fractional Division Synthesizers.

U. Rohde, A. Poddar, G. Boeck, *The Design of Modern Microwave Oscillators for Wireless Applications Theory and Optimization* (Wiley-Interscience, 2005, ISBN 0-471-72342-8). This is the latest and most advanced textbook on oscillator design for high frequency/microwave application. It shows in detail how to design and optimize high frequency and microwave VCOs. It also contains a thorough analysis of the phase noise as generated in oscillators. While highly mathematical, the appendix gives numerical solutions that can be easily applied to various designs. It covers both bipolar transistors and FETs.

U. Rohde, "All About Phase Noise in Oscillators," Part 1, *QEX,* Dec 1993 pp 3-6; Part 2, *QEX,* Jan 1994 pp 9-16; Part 3, *QEX,* Feb 1994, pp 15-24.

U. Rohde, "Key Components of Modern Receiver Design," Part 1, *QST,* May 1994, pp 29-32; Part 2, *QST,* June 1994, pp 27-31; Part 3, *QST,* Jul 1994, pp 42-45. Includes discussion of phase-noise reduction techniques in synthesizers and oscillators.

U. Rohde, "A High-Performance Hybrid Frequency Synthesizer," *QST,* Mar 1995, pp 30-38.

D. Stockton, "Polyphase noise-shaping fractional-N frequency synthesizer", U.S. Patent 6,509,800.

F. Telewski and E. Drucker, "Noise Reduction Method and Apparatus for Phase-locked Loops," U.S. Patent 5,216,387.

G. Vendelin, A. Pavio, U. Rohde, *Microwave Circuit Design Using Linear and Nonlinear Techniques*, 2nd ed., (Wiley-Interscience, 2005, ISBN 0-471-414479-4). This is a general handbook on microwave circuit design. It covers (in 200 pages) a large variety of oscillator design problems including microwave applications. It provides a step-by-step design guide for both linear and nonlinear oscillator designs.

INTERESTING MODERN JOURNAL PUBLICATIONS
OSCILLATOR REFERENCES

A. D. Berny, A. M. Niknejad, and R. G. Meyer, "A 1.8 GHz LC VCO with 1.3 GHz Tuning Range and Digital Amplitude Calibration," *IEEE Journal of SSC,* vol. 40. no. 4, 2005, pp 909-917.

A. Chenakin, "Phase Noise Reduction in Microwave Oscillators," *Microwave Journal,* Oct 2009, pp 124-140.

A, P. S. (Paul) Khanna, "Microwave Oscillators: The State of The Technology," *Microwave Journal,* Apr 2006, pp. 22-42.

N. Nomura, M. Itagaki, and Y. Aoyagi, "Small packaged VCSO for 10 Gbit Ethernet Application," *IEEE IUFFCS*, 2004, pp. 418-421.

A. K. Poddar, "A Novel Approach for Designing Integrated Ultra Low Noise Microwave Wideband Voltage-Controlled Oscillators," Dr.-Ing. Dissertation, TU- Berlin, Germany, 14 Dec 2004.

A. Ravi, B. R. Carlton, G. Banerjee, K. Soumyanath, "A 1.4V, 2.4/5.2 GHz, 90 nm CMOS system in a Package Transreceiver for Next Generation WLAN," *IEEE Symp. on VLSI Tech.*, 2005.

U.L. Rohde, "A New Efficient Method of Designing Low Noise Microwave Oscillators," Dr.-Ing. Dissertation, TU-Berlin, Germany, 12 Feb 2004.

Sheng Sun and Lei Zhu, "Guided-Wave Characteristics of Periodically Nonuniform Coupled Microstrip Lines-Even and Odd Modes," *IEEE Transaction on MTT*, Vol. 53, No. 4, Apr 2005, pp. 1221-1227.

C.-C. Wei, H.-C. Chiu, and W.-S. Feng, "An Ultra-Wideband CMOS VCO with 3-5 GHz Tuning Range," IEEE, *International Workshop on Radio-Frequency Integration Technology*, Nov 2005, pp 87-90.

M-S Yim, K. K. O., "Switched Resonators and Their Applications in a Dual Band Monolithic CMOS LC Tuned VCO," *IEEE MTT*, Vol. 54, Jan 2006, pp 74-81.

SYNTHESIZER REFERENCES

H. Arora, N. Klemmer, J. C. Morizio, and P. D. Wolf, "Enhanced Phase Noise Modeling of Fractional-N Frequency Synthesizer," IEEE Trans. Circuits and Systems-I: Regular Papers, vol 52, Feb 2005, pp 379-395.

R. van de Beek. D. Leenaerts, G. van der Weide, "A Fast-Hopping Single-PLL 3-band UWB Synthesizer in 0.25um SiGe BiCMOS," Proc. European Solid-State Circuit Conf. 2005, pp 173-176.

R. E. Best, *Phase-Locked Loops: Design, Simulation, and Applications*, 5th ed., McGraw-Hill, 2003.

C. Lam, B. Razavi, "A 2.6/5.2 GHz Frequency Synthesizer in 0.4um CMOS Technology," IEEE JSSC, vol 35, May 2000, pp 788-79.

J. Lee, "A 3-to-8 GHz Fast-Hopping Frequency Synthesizer in 0.18-um CMOS Technology," IEEE Journal of SSC, vol 41, no 3, Mar 2006, pp 566-573.

A. Natrajan, A. Komijani, and A. Hajimiri, "A Fully 24-GHz Phased-Array Transmitter in CMOS," IEEE Journal of SSC, vol 40, no 12, Dec 2005, pp 2502-2514.

W. Rahajandraibe, L. Zaid, V. C. de Beaupre, and G. Bas, "Frequency Synthesizer and FSK Modulator For IEEE 802.15.4 Based Applications," 2007 RFIC Symp. Digest, pp 229-232.

C. Sandner, A. Wiesbauer, "A 3 GHz to 7 GHz Fast-Hopping Frequency Synthesizer For UWB," Proc. Int. Workshop on Ultra Wideband Systems 2004, pp 405-409.

R. B. Staszewski and others, "All Digital PLL and Transmitter for Mobile Phones," IEEE J. Solid-State Circuits, vol 40, no 12, Dec 2005, pp 2469-2482.

Contents

**Chapter 10 —
CD-ROM Content**

Supplemental Articles

- "Modern Receiver Mixers for High Dynamic Range" by Doug DeMaw, W1FB (SK) and George Collins, KC1V
- "Performance Capability Of Active Mixers" by Dr Ulrich Rohde, N1UL

Mixers, Modulators and Demodulators

At base, radio communication involves translating information into radio form, letting it travel for a time as a radio signal, and translating it back again. Translating information into radio form entails the process we call modulation, and demodulation is its reverse. One way or another, every transmitter used for radio communication, from the simplest to the most complex, includes a means of modulation; one way or another, every receiver used for radio communication, from the simplest to the most complex, includes a means of demodulation.

Modulation involves varying one or both of a radio signal's basic characteristics — amplitude and frequency (or phase) — to convey information. A circuit, stage or piece of hardware that modulates is called a modulator.

Demodulation involves reconstructing the transmitted information from the changing characteristic(s) of a modulated radio wave. A circuit, stage or piece of hardware that demodulates is called a demodulator.

Many radio transmitters, receivers and transceivers also contain mixers — circuits, stages or pieces of hardware that combine two or more signals to produce additional signals at sums of and differences between the original frequencies.

This chapter, by David Newkirk, W9VES, and Rick Karlquist, N6RK, examines mixers, modulators and demodulators. Related information may be found in the **Modulation** chapter, and in the chapters on **Receivers** and **Transmitters**. Quadrature modulation implemented with an I/Q modulator, one that uses in-phase (I) and quadrature (Q) modulating signals to generate the 0° and 90° components of the RF signal, is covered in the chapters on **Modulation** and **DSP and Software Radio Design**.

Amateur Radio textbooks have traditionally handled mixers separately from modulators and demodulators, and modulators separately from demodulators. This chapter examines mixers, modulators and demodulators together because the job they do is essentially the same. Modulators and demodulators translate information into radio form and back again; mixers translate one frequency to others and back again. All of these translation processes can be thought of as forms of frequency translation or frequency shifting — the function traditionally ascribed to mixers. We'll therefore begin our investigation by examining what a mixer is (and isn't), and what a mixer does.

10.1 The Mechanism of Mixers and Mixing

10.1.1 What is a Mixer?

Mixer is a traditional radio term for a circuit that shifts one signal's frequency up or down by combining it with another signal. The word *mixer* is also used to refer to a device used to blend multiple audio inputs together for recording, broadcast or sound reinforcement. These two mixer types differ in one very important way: A radio mixer makes new frequencies out of the frequencies put into it, and an audio mixer does not.

MIXING VERSUS ADDING

Radio mixers might be more accurately called "frequency mixers" to distinguish them from devices such as "microphone mixers," which are really just signal *combiners*, *summers* or *adders*. In their most basic, ideal forms, both devices have two inputs and one output. The combiner simply *adds* the instantaneous voltages of the two signals together to produce the output at each point in time (**Fig 10.1**). The mixer, on the other hand, *multiplies* the instantaneous voltages of the two signals together to produce its output signal from instant to instant (**Fig 10.2**). Comparing the output spectra of the combiner and mixer, we see that the combiner's output contains only the frequencies of the two inputs, and nothing else, while the mixer's output contains *new* frequencies. Because it combines one energy with another, this process is sometimes called *heterodyning*, from the Greek words for *other* and *power*. The sidebar, "Mixer Math: Mixing as Multiplication," describes this process mathematically.

The key principle of a radio mixer is that in mixing multiple signal voltages together, *it adds and subtracts their frequencies to produce new frequencies.* (In the field of signal processing, this process, *multiplication in the time domain,* is recognized as equivalent to the process of *convolution in the frequency domain.* Those interested in this alternative approach to describing the generation of new frequencies through mixing can find more information about it in the many textbooks available on this subject.)

The difference between the mixer we've been describing and any mixer, modulator or demodulator that you'll ever use is that it's ideal. We put in two signals and got just two signals out. *Real* mixers, modulators and demodulators, on the other hand, also produce *distortion* products that make their output spectra "dirtier" or "less clean," as well as putting out some energy at input-signal frequencies and their harmonics. Much of the art and science of making good use of multiplication in mixing, modulation and demodulation goes

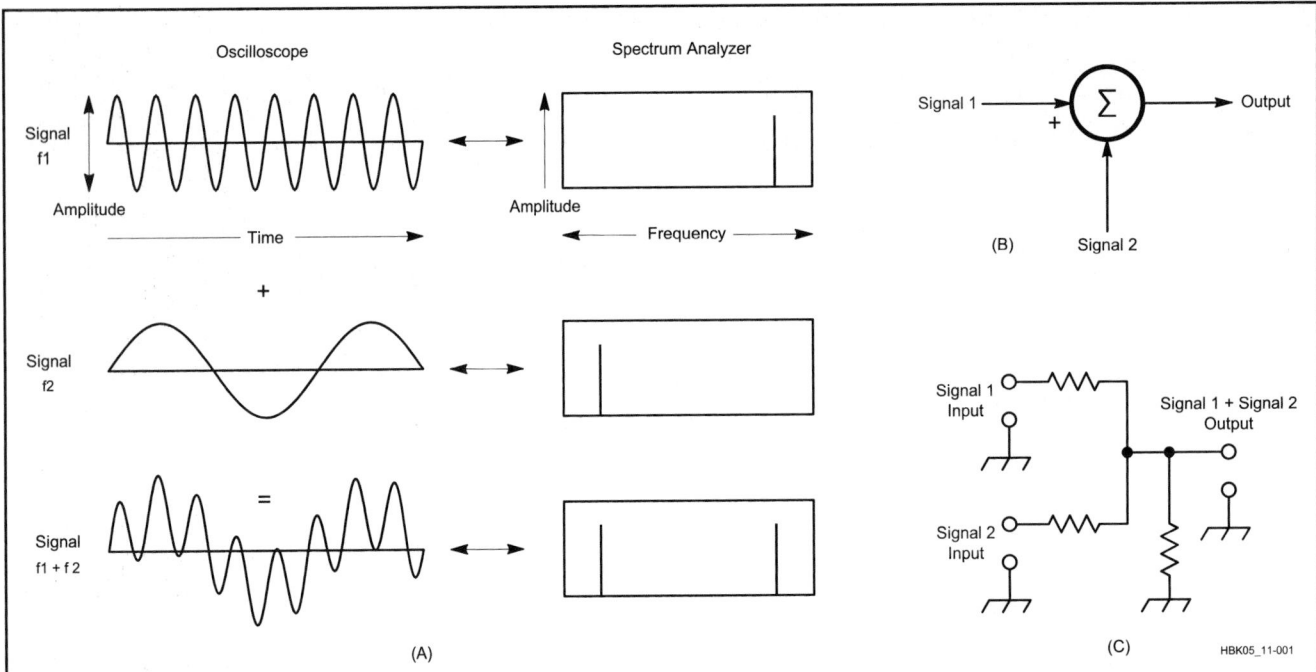

Fig 10.1 — *Adding or summing* two sine waves of different frequencies (f1 and f2) combines their amplitudes without affecting their frequencies. Viewed with an *oscilloscope* (a real-time graph of amplitude versus time), adding two signals appears as a simple superimposition of one signal on the other. Viewed with a *spectrum analyzer* (a real-time graph of signal amplitude versus frequency), adding two signals just sums their spectra. The signals merely coexist on a single cable or wire. All frequencies that go into the adder come out of the adder, and no new signals are generated. Drawing B, a block diagram of a summing circuit, emphasizes the stage's mathematical operation rather than showing circuit components. Drawing C shows a simple summing circuit, such as might be used to combine signals from two microphones. In audio work, a circuit like this is often called a mixer — but it does not perform the same function as an RF mixer.

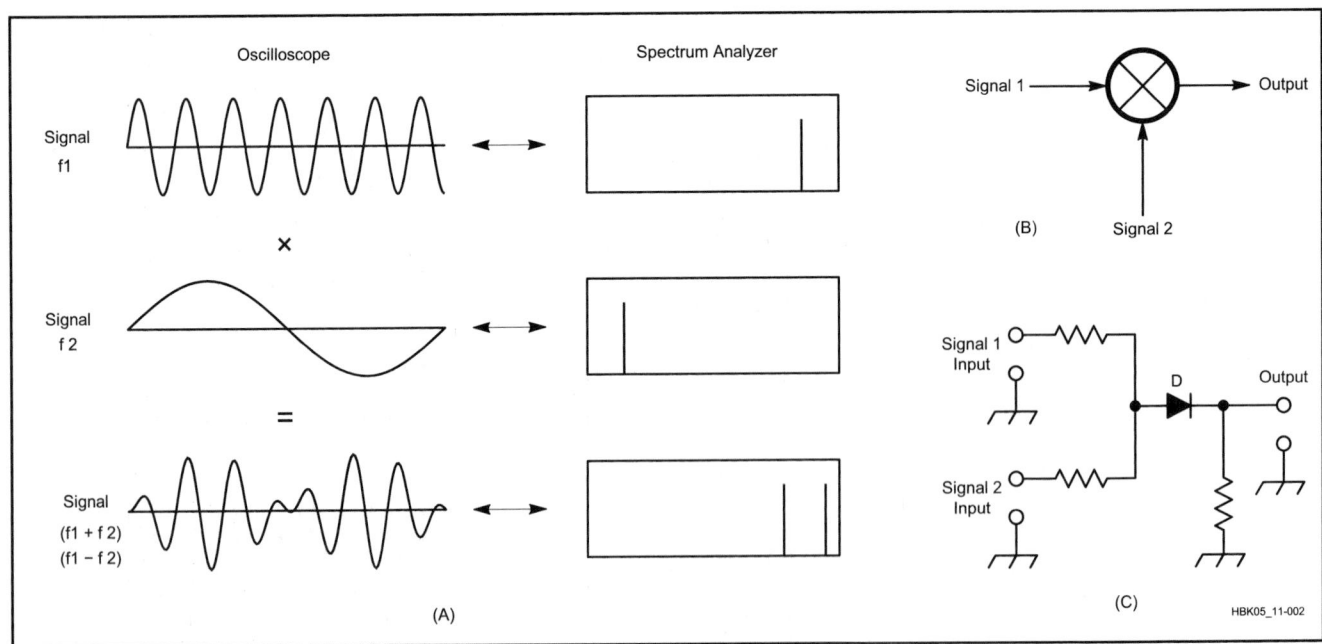

Fig 10.2 — *Multiplying* two sine waves of different frequencies produces a new output spectrum. Viewed with an oscilloscope, the result of multiplying two signals is a composite wave that seems to have little in common with its components. A spectrum-analyzer view of the same wave reveals why: The original signals disappear entirely and are replaced by two new signals — at the *sum* and *difference* of the original signals' frequencies. Drawing B diagrams a multiplier, known in radio work as a mixer. The *X* emphasizes the stage's mathematical operation. (The circled *X* is only one of several symbols you may see used to represent mixers in block diagrams, as Fig 10.3 explains.) Drawing C shows a very simple multiplier circuit. The diode, D, does the mixing. Because this circuit does other mathematical functions and adds them to the sum and difference products, its output is more complex than f1 + f2 and f1 − f2, but these can be extracted from the output by filtering.

Mixer Math: Mixing as Multiplication

Since a mixer works by means of multiplication, a bit of math can show us how they work. To begin with, we need to represent the two signals we'll mix, A and B, mathematically. Signal A's instantaneous amplitude equals

$$A_a \sin 2\pi f_a t \tag{1}$$

in which A is peak amplitude, f is frequency, and t is time. Likewise, B's instantaneous amplitude equals

$$A_b \sin 2\pi f_b t \tag{2}$$

Since our goal is to show that multiplying two signals generates sum and difference frequencies, we can simplify these signal definitions by assuming that the peak amplitude of each is 1. The equation for Signal A then becomes

$$a(t) = A \sin (2\pi f_a t) \tag{3}$$

and the equation for Signal B becomes

$$b(t) = B \sin (2\pi f_b t) \tag{4}$$

Each of these equations represents a sine wave and includes a subscript letter to help us keep track of where the signals go.

Merely combining Signal A and Signal B by letting them travel on the same wire develops nothing new:

$$a(t) + b(t) = A \sin (2\pi f_a t) + B \sin (2\pi f_b t) \tag{5}$$

As simple as equation 5 may seem, we include it to highlight the fact that multiplying two signals is a quite different story. From trigonometry, we know that multiplying the sines of two variables can be expanded according to the relationship

$$\sin x \sin y = \frac{1}{2} \left[\cos (x - y) - \cos (x + y) \right] \tag{6}$$

Conveniently, Signals A and B are both sinusoidal, so we can use equation 6 to determine what happens when we multiply Signal A by Signal B. In our case, x = $2\pi f_a t$ and $y = 2\pi f_b t$, so plugging them into equation 6 gives us

$$a(t) \times b(t) = \frac{AB}{2} \cos \left(2\pi \left[f_a - f_b \right] t \right) - \frac{AB}{2} \cos \left(2\pi \left[f_a + f_b \right] t \right) \tag{7}$$

Now we see two momentous results: a sine wave at the frequency *difference* between Signal A and Signal B $2\pi(f_a - f_b)t$, and a sine wave at the frequency *sum* of Signal A and Signal B $2\pi(f_a + f_b)t$. (The products are cosine waves, but since equivalent sine and cosine waves differ only by a phase shift of 90°, both are called *sine waves* by convention.)

This is the basic process by which we translate information into radio form and translate it back again. If we want to transmit a 1-kHz audio tone by radio, we can feed it into one of our mixer's inputs and feed an RF signal — say, 5995 kHz — into the mixer's other input. The result is two radio signals: one at 5994 kHz (5995 − 1) and another at 5996 kHz (5995 + 1). We have achieved modulation.

Converting these two radio signals back to audio is just as straightforward. All we do is feed them into one input of another mixer, and feed a 5995-kHz signal into the mixer's other input. Result: a 1-kHz tone. We have achieved demodulation; we have communicated by radio.

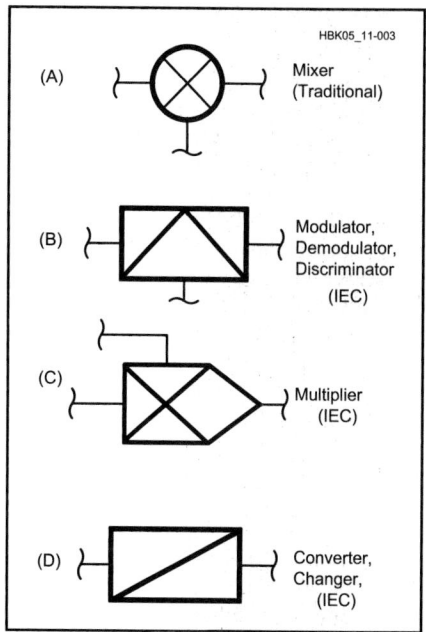

Fig 10.3 — We commonly symbolize mixers with a circled *X* (A) out of tradition, but other standards sometimes prevail (B, C and D). Although the converter/changer symbol (D) can conceivably be used to indicate frequency changing through mixing, the three-terminal symbols are arguably better for this job because they convey the idea of two signal sources resulting in a new frequency. (*IEC* stands for *International Electrotechnical Commission.*)

how to get rid of it in a *balanced modulator*?" A transmitter enthusiast may ask "Why didn't you mention *sidebands* and how we conserve spectrum space and power by getting rid of one and putting all of our power into the other?" A student of radio receivers, on the other hand, expects any discussion of the same underlying multiplication issues to touch on the topics of *LO feedthrough, mixer balance (single* or *double?), image rejection* and so on.

You likely expect this book to spend some time talking to you about these things, so it will. But *this* radio-amateur-oriented discussion of mixers, modulators and demodulators will take a look at their common underlying mechanism *before* turning you loose on practical mixer, modulator and demodulator circuits. Then you'll be able to tell the forest from the trees. **Fig 10.3** shows the block symbol for a traditional mixer along with several IEC symbols for other functions mixers may perform.

It turns out that the mechanism underlying multiplication, mixing, modulation and demodulation is a pretty straightforward thing: Any circuit structure that *nonlinearly distorts* ac waveforms acts as a multiplier to some degree.

into minimizing these unwanted multiplication products (or their effects) and making multipliers do their frequency translations as efficiently as possible.

10.1.2 Putting Multiplication to Work

Piecing together a coherent picture of how multiplication works in radio communica-

tion isn't made any easier by the fact that traditional terms applied to a given multiplication approach and its products may vary with their application. If, for instance, you're familiar with standard textbook approaches to mixers, modulators and demodulators, you may be wondering why we didn't begin by working out the math involved by examining *amplitude modulation*, also known as *AM*. "Why not tell them about the *carrier* and

NONLINEAR DISTORTION?

The phrase *nonlinear distortion* sounds redundant, but isn't. Distortion, an externally imposed change in a waveform, can be linear; that is, it can occur independently of signal amplitude. Consider a radio receiver front-end filter that passes only signals between 6 and 8 MHz. It does this by *linearly distorting* the single complex waveform corresponding to the wide RF spectrum present at the radio's antenna terminals, reducing the amplitudes of frequency components below 6 MHz and above 8 MHz relative to those between 6 and 8 MHz. (Considering multiple signals on a wire as one complex waveform is just as valid, and sometimes handier, than considering them as separate signals. In this case, it's a bit easier to think of distortion as something that happens to a waveform rather than something that happens to separate signals relative to each other. It would be just as valid — and certainly more in keeping with the consensus view — to say merely that the filter attenuates signals at frequencies below 6 MHz and above 8 MHz.) The filter's output waveform certainly differs from its input waveform; the waveform has been distorted. But because this distortion occurs independently of signal level or polarity, the distortion is linear. No new frequency components are created; only the amplitude relationships among the wave's existing frequency components are altered. This is *amplitude* or *frequency* distortion, and all filters do it or they wouldn't be filters.

Phase or *delay distortion*, also linear, causes a complex signal's various component frequencies to be delayed by different amounts of time, depending on their frequency but independently of their amplitude. No new frequency components occur, and amplitude relationships among existing frequency components are not altered. Phase distortion occurs to some degree in all real filters.

The waveform of a non-sinusoidal signal can be changed by passing it through a circuit that has only linear distortion, but only *nonlinear distortion* can change the waveform of a simple sine wave. It can also produce an output signal whose output waveform changes as a function of the input amplitude, something not possible with linear distortion. Nonlinear circuits often distort excessively with overly strong signals, but the distortion can be a complex function of the input level.

Nonlinear distortion may take the form of *harmonic distortion*, in which integer multiples of input frequencies occur, or *intermodulation distortion (IMD)*, in which different components multiply to make new ones.

Any departure from absolute linearity results in some form of nonlinear distortion, and this distortion can work for us or against us. Any amplifier, including a so-called linear amplifier, distorts nonlinearly to some degree; any device or circuit that distorts

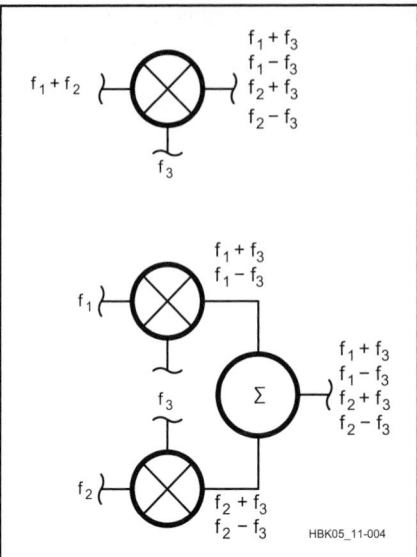

Fig 10.4 — Feeding two signals into one input of a mixer results in the same output as if f_1 and f_2 are each first mixed with f_3 in two separate mixers, and the outputs of these mixers are combined.

nonlinearly can work as a mixer, modulator, demodulator or frequency multiplier. An amplifier optimized for linear operation will nonetheless mix, but inefficiently; an amplifier biased for nonlinear amplification may be practically linear over a given tiny portion of its input-signal range. The trick is to use careful design and component selection to maximize nonlinear distortion when we want it (as in a mixer), and minimize it when we don't. Once we've decided to maximize nonlinear distortion, the trick is to minimize the distortion products we don't want, and maximize the products we want.

KEEPING UNWANTED DISTORTION PRODUCTS DOWN

Ideally, a mixer multiplies the signal at one of its inputs by the signal at its other input, but does not multiply a signal at the same input by itself, or multiple signals at the same input by themselves or by each other. (Multiplying a signal by itself — squaring it — generates harmonic distortion [specifically, *second-harmonic* distortion] by adding the signal's frequency to itself per equation 7. Simultaneously squaring two or more signals generates simultaneous harmonic and intermodulation distortion, as we'll see later when we explore how a diode demodulates AM.)

Consider what happens when a mixer must handle signals at two different frequencies (we'll call them f_1 and f_2) applied to its first input, and a signal at a third frequency (f_3) applied to its other input. Ideally, a mixer multiplies f_1 by f_3 and f_2 by f_3, but does not multiply f_1 and f_2 by each other. This produces output at the sum and difference of f_1

and f_3, and the sum and difference of f_2 and f_3, but *not* the sum and difference of f_1 and f_2. **Fig 10.4** shows that feeding two signals into one input of a mixer results in the same output as if f_1 and f_2 are each first mixed with f^3 in two separate mixers, and the outputs of these mixers are combined. This shows that a mixer, even though constructed with nonlinearly distorting components, actually behaves as a *linear frequency shifter*. Traditionally, we refer to this process as mixing and to its outputs as *mixing products*, but we may also call it frequency *conversion*, referring to a device or circuit that does it as a *converter*, and to its outputs as *conversion products*. If a mixer produces an output frequency that is higher than the input frequency, it is called an upconverter; if the output frequency is lower than the input, a downconverter.

Real mixers, however, at best act only as *reasonably* linear frequency shifters, generating some unwanted IMD products — spurious signals, or *spurs* — as they go. Receivers are especially sensitive to unwanted mixer IMD because the signal-level spread over which they must operate without generating unwanted IMD is often 90 dB or more, and includes infinitesimally weak signals in its span. In a receiver, IMD products so tiny that you'd never notice them in a transmitted signal can easily obliterate weak signals. This is why receiver designers apply so much effort to achieving "high dynamic range."

The degree to which a given mixer, modulator or demodulator circuit produces unwanted IMD is often *the* reason why we use it, or don't use it, instead of another circuit that does its wanted-IMD job as well or even better.

OTHER MIXER OUTPUTS

In addition to desired sum-and-difference products and unwanted IMD products, real mixers also put out some energy at their input frequencies. Some mixer implementations may *suppress* these outputs — that is, reduce one or both of their input signals by a factor of 100 to 1,000,000, or 20 to 60 dB. This is good because it helps keep input signals at the desired mixer-output sum or difference frequency from showing up at the IF terminal — an effect reflected in a receiver's *IF rejection* specification. Some mixer types, especially those used in the vacuum-tube era, suppress their input-signal outputs very little or not at all.

Input-signal suppression is part of an overall picture called *port-to-port isolation*. Mixer input and output connections are traditionally called *ports*. By tradition, the port to which we apply the shifting signal is the *local-oscillator (LO)* port. By convention, the signal or signals to be frequency-shifted are applied to the *RF (radio frequency)* port, and the frequency-shifted (product) signal or signals emerge at the *IF (intermediate frequency*) port. This illustrates the function of a mixer in a receiver: Since it is often impractical to achieve the desired gain and filtering at the incoming signal's frequency (at RF), a mixer is used to translate the incoming RF signal to an intermediate frequency (the IF), where gain and filtering can be applied. The IF maybe be either lower or higher than the incoming RF signal. In a transmitter, the modulated signal may be created at an IF, and then translated in frequency by a mixer to the operating frequency.

Some mixers are *bilateral*; that is, their RF and IF ports can be interchanged, depending on the application. Diode-based mixers are usually bilateral. Many mixers are not bilateral (*unilateral*); the popular SA602/612 Gilbert cell IC mixer is an example of this.

It's generally a good idea to keep a mixer's input signals from appearing at its output port because they represent energy that we'd rather not pass on to subsequent circuitry. It therefore follows that it's usually a good idea to keep a mixer's LO-port energy from appearing at its RF port, or its RF-port energy from making it through to the IF port. But there are some notable exceptions.

10.2 Mixers and Amplitude Modulation

Now that we've just discussed what a fine thing it is to have a mixer that doesn't let its input signals through to its output port, we can explore a mixing approach that outputs one of its input signals so strongly that the fed-through signal's amplitude at least equals the combined amplitudes of the system's sum and difference products! This system full-carrier, *amplitude modulation*, is the oldest means of translating information into radio form and back again. It's a frequency-shifting system in which the original unmodulated signal, traditionally called the *carrier*, emerges from the mixer along with the sum and difference products, traditionally called *sidebands*. The sidebar, "Mixer Math: Amplitude Modulation," describes this process mathematically.

10.2.1 Overmodulation

Since the information we transmit using AM shows up entirely as energy in its sidebands, it follows that the more energetic we make the sidebands, the more information energy will be available for an AM receiver to "recover" when it demodulates the signal. Even in an ideal modulator, there's a practical limit to how strong we can make an AM signal's sidebands relative to its carrier, however. Beyond that limit, we severely distort the waveform we want to translate into radio form.

We reach AM's distortion-free modulation limit when the sum of the sidebands and carrier at the modulator output *just reaches zero* at the modulating wave-form's most negative peak (**Fig 10.5**). We call this condition *100% modulation*, and it occurs when *m* in equation 8 equals 1. (We enumerate *modulation percentage* in values from 0 to 100%. The lower the number, the less information energy is in the sidebands. You may also see modulation enumerated in terms of a *modulation factor* from 0 to 1, which directly equals *m*; a modulation factor of 1 is the same as 100% modulation.) Equation 9 shows that each sideband's voltage is half that of the carrier. Power varies

as the square of voltage, so the power in each sideband of a 100%-modulated signal is therefore $(\frac{1}{2})^2$ times, or $\frac{1}{4}$, that of the carrier. A transmitter capable of 100% modulation when operating at a carrier power of 100 W therefore puts out a 150-W signal at 100% modulation, 50 W of which is attributable to the sidebands.

(The *peak* envelope power [PEP] output of a double-sideband, full-carrier AM transmitter at 100% modulation is four times its carrier PEP. This is why our solid-state, "100-W" MF/HF transceivers are usually rated for no more than about 25 W carrier output at 100% amplitude modulation.)

Mixer Math: Amplitude Modulation

We can easily make the carrier pop out of our mixer along with the sidebands merely by building enough *dc level shift* into the information we want to mix so that its waveform never goes negative. Back at equations 1 and 2, we decided to keep our mixer math relatively simple by setting the peak voltage of our mixer's input signals directly equal to their sine values. Each input signal's peak voltage therefore varies between +1 and −1, so all we need to do to keep our modulating- signal term (provided with a subscript *m* to reflect its role as the modulating or information waveform) from going negative is add 1 to it. Identifying the carrier term with a subscript c, we can write

$$\text{AM signal} = (1 + m \sin 2\pi f_m t) \sin 2\pi f_c t \tag{8}$$

Notice that the modulation ($2\pi f_m t$) term has company in the form of a coefficient, *m*. This variable expresses the modulating signal's varying amplitude — variations that ultimately result in amplitude modulation. Expanding equation 8 according to equation 6 gives us

$$\text{AM signal} = \sin 2\pi f_c t + \frac{1}{2} m \cos (2\pi f_c - 2\pi f_m) t - \frac{1}{2} m \cos (2\pi f_c + 2\pi f_m) t \tag{9}$$

The modulator's output now includes the carrier ($\sin 2\pi f_c t$) in addition to sum and difference products that vary in strength according to m. According to the conventions of talking about modulation, we call the sum product, which comes out at a frequency higher than that of the carrier, the *upper sideband (USB)*, and the difference product, which comes out a frequency lower than that of the carrier, the *lower sideband (LSB)*. We have achieved amplitude modulation.

Why We Call It Amplitude Modulation

We call the modulation process described in equation 8 *amplitude modulation* because the complex waveform consisting of the sum of the sidebands and carrier varies with the information signal's magnitude (m). Concepts long used to illustrate AM's mechanism may mislead us into thinking that the *carrier* varies in strength with modulation, but careful study of equation 9 shows that this doesn't happen. The carrier, sin $2\pi f_c t$, goes into the modulator — we're in the modulation business now, so it's fitting to use the term *modulator* instead of *mixer* — as a sinusoid with an unvarying maximum value of |1|. The modulator multiplies the carrier by the dc level (+1) that we added to the information signal (m sin $2\pi f_m t$). Multiplying sin $2\pi f_c t$ by 1 merely returns sin $2\pi f_c t$. We have proven that the carrier's amplitude does not vary as a result of amplitude modulation — a fact that makes sense when we realize that in simple AM receivers the carrier serves as an AGC control signal and (during "diode" detection) provides, at the correct power level and frequency, the "LO" signal necessary to heterodyne the sidebands back to baseband.

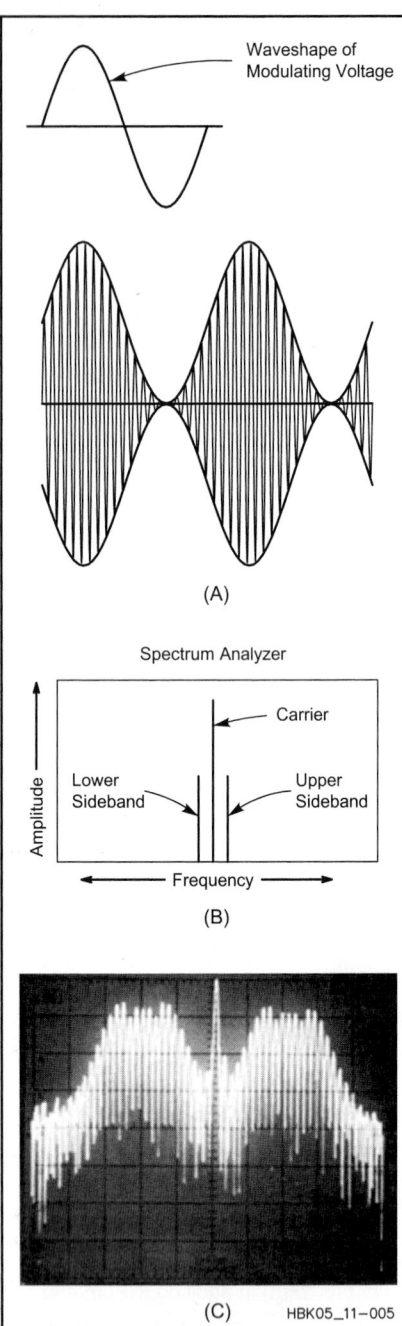

Fig 10.5 — Graphed in terms of amplitude versus time (A), the *envelope* of a properly modulated AM signal exactly mirrors the shape of its modulating waveform, which is a sine wave in this example. This AM signal is modulated as fully as it can be — 100% — because its envelope *just* hits zero on the modulating wave's negative peaks. Graphing the same AM signal in terms of amplitude versus frequency (B) reveals its three spectral components: Carrier, upper sideband and lower sideband. B shows sidebands as single-frequency components because the modulating waveform is a sine wave. With a complex modulating waveform, the modulator's sum and difference products really do show up as *bands* on either side of the carrier (C).

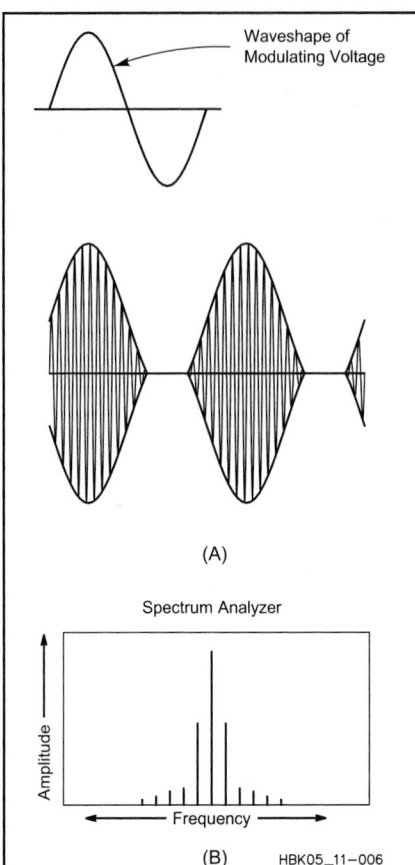

Fig 10.6 — Negative-going overmodulation of an AM transmitter results in a modulation envelope (A) that doesn't faithfully mirror the modulating waveform. This distortion creates additional sideband components that broaden the transmitted signal (B). Positive-going modulation beyond 100% is used by some AM broadcasters in conjunction with negative-peak limiting to increase "talk power" without causing negative overmodulation.

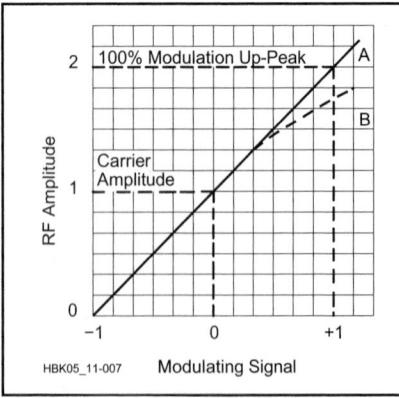

Fig 10.7 — An ideal AM transmitter exhibits a straight-line relationship (A) between its instantaneous envelope amplitude and the instantaneous amplitude of its modulating signal. Distortion, and thus an unnecessarily wide signal, results if the transmitter cannot respond linearly across the modulating signal's full amplitude range.

One-hundred-percent negative modulation is a brick-wall limit because an amplitude modulator can't reduce its output to less than zero. Trying to increase negative modulation beyond the 100% point results in *over-modulation* (**Fig 10.6**), in which the modulation envelope no longer mirrors the shape of the modulating wave (Fig 10.6A). A negatively overmodulated wave contains more energy than it did at 100% modulation, but some of the added energy now exists as *harmonics of the modulating waveform* (Fig 10.6B). This distortion makes the modulated signal take up more spectrum space than it needs. In voice operation, overmodulation commonly happens only on syllabic peaks, making the distortion products sound like transient noise we refer to as *splatter*.

MODULATION LINEARITY

If we increase an amplitude modulator's modulating-signal input by a given percentage, we expect a proportional modulation increase in the modulated signal. We expect good *modulation linearity*. Suboptimal amplitude modulator design may not allow this, however. Above some modulation percentage, a modulator may fail to increase modulation in proportion to an increase in its input signal (**Fig 10.7**). Distortion, and thus an unnecessarily wide signal, results.

10.2.2 Using AM to Send Morse Code

Fig 10.8A closely resembles what we see when a properly adjusted CW transmitter sends a string of dots. Keying a carrier on and off produces a wave that varies in amplitude and has double (upper and lower) sidebands that vary in spectral composition according to the duration and envelope shape of the on-off transitions. The emission mode we call *CW* is therefore a form of AM. The concepts of modulation percentage and overmodulation are usually not applied to generating an on-off-keyed Morse signal, however. This is related to how we copy CW by ear, and the fact that, in CW radio communication, we usually don't translate the received signal all the way back into its original pre-modulator (*baseband*) form, as a closer look at the process reveals.

In CW transmission, we usually open and close a keying line to make dc transitions that turn the transmitted carrier on and off. See Fig 10.8B. CW reception usually does not entirely reverse this process, however. Instead of demodulating a CW signal all the way back to its baseband self — a shifting dc level — we want the presences and absences of its carrier to create long and short audio tones. Because the carrier is RF and not AF, we must mix it with a locally generated RF signal — from a *beat-frequency oscillator*

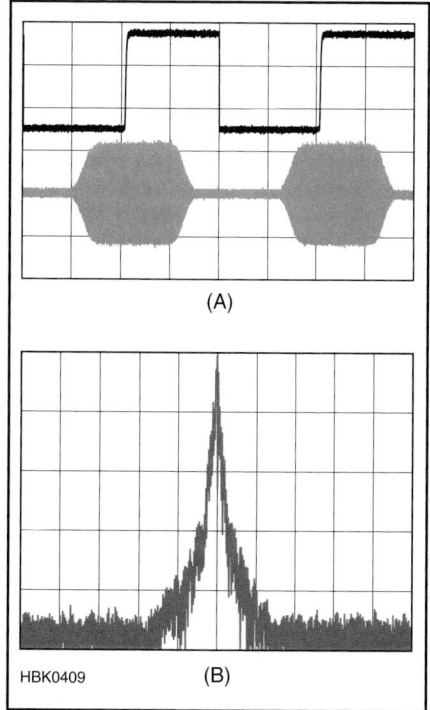

Fig 10.8 — Telegraphy by on-off-keying a carrier is also a form of AM, called *CW* (short for *continuous wave*) for reasons of tradition. *Waveshaping* in a CW transmitter often causes a CW signal's RF envelope (lower trace in the amplitude-versus-time display at A) to contain less harmonic energy than the abrupt transitions of its key closure waveform (upper trace in A) suggest should be the case. B, an amplitude-versus-frequency display, shows that even a properly shaped CW signal has many sideband components.

(BFO) — that's close enough in frequency to produce a difference signal at AF (this BFO can, of course, also be inserted at an IF stage). What goes into our transmitter as shifting dc comes out of our receiver as thump-delimited tone bursts of dot and dash duration. We have achieved CW communication.

It so happens that we always need to hear one or more harmonics of the fundamental keying waveform for the code to sound sufficiently crisp. If the transmitted signal will be subject to fading caused by varying propagation — a safe assumption for any long-distance radio communication — we can *harden* our keying by making the transmitter's output rise and fall more quickly. This puts more energy into keying sidebands and makes the signal more copiable in the presence of fading — in particular, *selective fading*, which linearly distorts a modulated signal's complex waveform and randomly changes the sidebands' strength and phase relative to the carrier and each other. The appropriate keying hardness also depends on the keying speed. The faster the keying

in WPM, the faster the on-off times — the harder the keying — must be for the signal to remain ear- and machine-readable through noise and fading.

Instead of thinking of this process in terms of modulation percentage, we just ensure that a CW transmitter produces sufficient keying-sideband energy for solid reception. Practical CW transmitters ususally do not do their keying with a modulator stage as such. Instead, one or more stages are turned on and off to modulate the carrier with Morse, with rise and fall times set by R and C values associated with the stages' keying and/or power supply lines. A transmitter's CW *waveshaping* is therefore usually hardwired to values appropriate for reasonably high-speed sending (35 to 55 WPM or so) in the presence of fading. However, some transceivers allow the user to vary keying hardness at will as a menu option. Rise and fall times of 1 to 5 ms are common; 5-ms rise and fall times equate to a keying speed of 36 WPM in the presence of fading and 60 WPM if fading is absent.

The faster a CW transmitter's output changes between zero and maximum, the more bandwidth its carrier and sidebands occupy. See Fig 10.8B. Making a CW signal's keying too hard is therefore spectrum-wasteful and unneighborly because it makes the signal wider than it needs to be. Keying sidebands that are stronger and wider than necessary are traditionally called *clicks* because of what they sound like on the air. A more detailed discussion of keying waveforms appears in the **Transmitters** chapter of this *Handbook*.

10.2.3 The Many Faces of Amplitude Modulation

We've so far examined mixers, multipliers and modulators that produce complex output signals of two types. One, the action of which equation 7 expresses, produces only the frequency sum and frequency difference between its input signals. The other, the amplitude modulator characterized by equations 8 and 9, produces carrier output in addition to the frequency sum of and frequency difference between its input signals. Exploring the AM process led us to a discussion of on-off-keyed CW, which is also a form of AM.

Amplitude modulation is nothing more and nothing less than varying an output signal's amplitude according to a varying voltage or current. All of the output signal types mentioned above are forms of amplitude modulation, and there are others. Their names and applications depend on whether the resulting signal contains a carrier or not, and both sidebands or not. Here's a brief overview of AM-signal types, what they're called, and some of the jobs you may find them doing:

• *Double-sideband (DSB), full-carrier AM* is often called just *AM*, and often what's

meant when radio folk talk about just *AM*. (When the subject is broadcasting, *AM* can also refer to broadcasters operating in the 535- to 1705-kHz region, generically called *the AM band* or *the broadcast band* or *medium wave*. These broadcasters used only double-sideband, full-carrier AM for many years, but many now use combinations of amplitude modulation and *angle modulation*, explored later in this chapter, to transmit stereophonic sound in analog and digital form.) Equations 8 and 9 express what goes on in generating this signal type. What we call *CW* — Morse code done by turning a carrier on and off — is a form of DSB, full-carrier AM.

• *Double-sideband, suppressed-carrier AM* is what comes out of a circuit that does what equation 7 expresses — a sum (upper sideband), a difference (lower sideband) and no carrier. We didn't call its sum and difference outputs upper and lower sidebands earlier in equation 7's neighborhood, but we'd do so in a transmitting application. In a transmitter, we call a circuit that suppresses the carrier while generating upper and lower sidebands a *balanced modulator*, and we quantify its *carrier suppression*, which is always less than infinite. In a receiver, we call such a circuit a *balanced mixer*, which may be *single-balanced* (if it lets either its RF signal or its LO [carrier] signal through to its output) or *double-balanced* (if it suppresses both its input signal and LO/carrier in its output), and we quantify its *LO suppression* and *port-to-port isolation*, which are always less than infinite. (Mixers [and amplifiers] that afford no balance whatsoever are sometimes said to be *single-ended*.) Sometimes, DSB suppressed-carrier AM is called just *DSB*.

• *Vestigial sideband (VSB), full-carrier AM* is like the DSB variety with one sideband partially filtered away for bandwidth reduction. Some amateur television stations use VSB AM, and it was used by commercial television systems that transmitted analog AM video in the pre-digital TV days.

• *Single-sideband, suppressed-carrier AM* is what you get when you generate a DSB, suppressed carrier AM signal and suppress one sideband with filtering or phasing. We usually call this signal type just *single sideband (SSB)* or, as appropriate, *upper sideband (USB)* or *lower sideband (LSB)*. In a modulator or demodulator system, the *unwanted sideband* — that is, the sum or difference signal we don't want — may be called just that, or it may be called the *opposite sideband*, and we refer to a system's *sideband rejection* as a measure of how well the opposite sideband is suppressed. In receiver mixers not used for demodulation and transmitter mixers not used for modulation, the unwanted sum or difference signal, or the input signal that produces the unwanted sum or difference, is the *image*, and we refer to a system's *image rejection*.

A pair of mixers specially configured to suppress either the sum or the difference output is an *image-reject mixer (IRM)*. In receiver demodulators, the unwanted sum or difference signal may just be called the opposite sideband, or it may be called the *audio image*. A receiver capable of rejecting the opposite sideband or audio image is said to be capable of *single-signal* reception.

• *Single-sideband, full-carrier AM* is akin to full-carrier DSB with one sideband missing. Commercial and military communicators may call it *AM equivalent (AME)* or *compatible AM (CAM)* — *compatible* because it can be usefully demodulated in AM and SSB receivers and because it occupies about the same amount of spectrum space as SSB.)

• *Independent sideband (ISB) AM* consists of an upper sideband and a lower sideband containing different information (a carrier of some level may also be present). Radio amateurs sometimes use ISB to transmit simultaneous slow-scan-television and voice information; international broadcasters sometimes use it for point-to-point audio feeds as a backup to satellite links.

10.2.4 Mixers and AM Demodulation

Translating information from radio form back into its original form — demodulation — is also traditionally called *detection*. If the information signal we want to detect consists merely of a baseband signal frequency-shifted into the radio realm, almost any low-distortion frequency-shifter that works according to equation 7 can do the job acceptably well.

Sometimes we recover a radio signal's information by shifting the signal right back to its original form with no intermediate frequency shifts. This process is called *direct conversion*. More commonly, we first convert a received signal to an *intermediate frequency* so we can amplify, filter and level-control it prior to detection. This is *superheterodyne* reception, and most modern radio receivers work in this way. Whatever the receiver type, however, the received signal ultimately makes its way to one last mixer or demodulator that completes the final translation of information back into audio, video, or into a signal form suitable for device control or computer processing. In this last translation, the incoming signal is converted back to recovered-information form by mixing it with one last RF signal. In heterodyne or *product* detection, that final frequency-shifting signal comes from a BFO. The incoming-signal energy goes into one mixer input port, BFO energy goes into the other, and audio (or whatever form the desired information takes) results.

If the incoming signal is full-carrier AM and we don't need to hear the carrier as a tone,

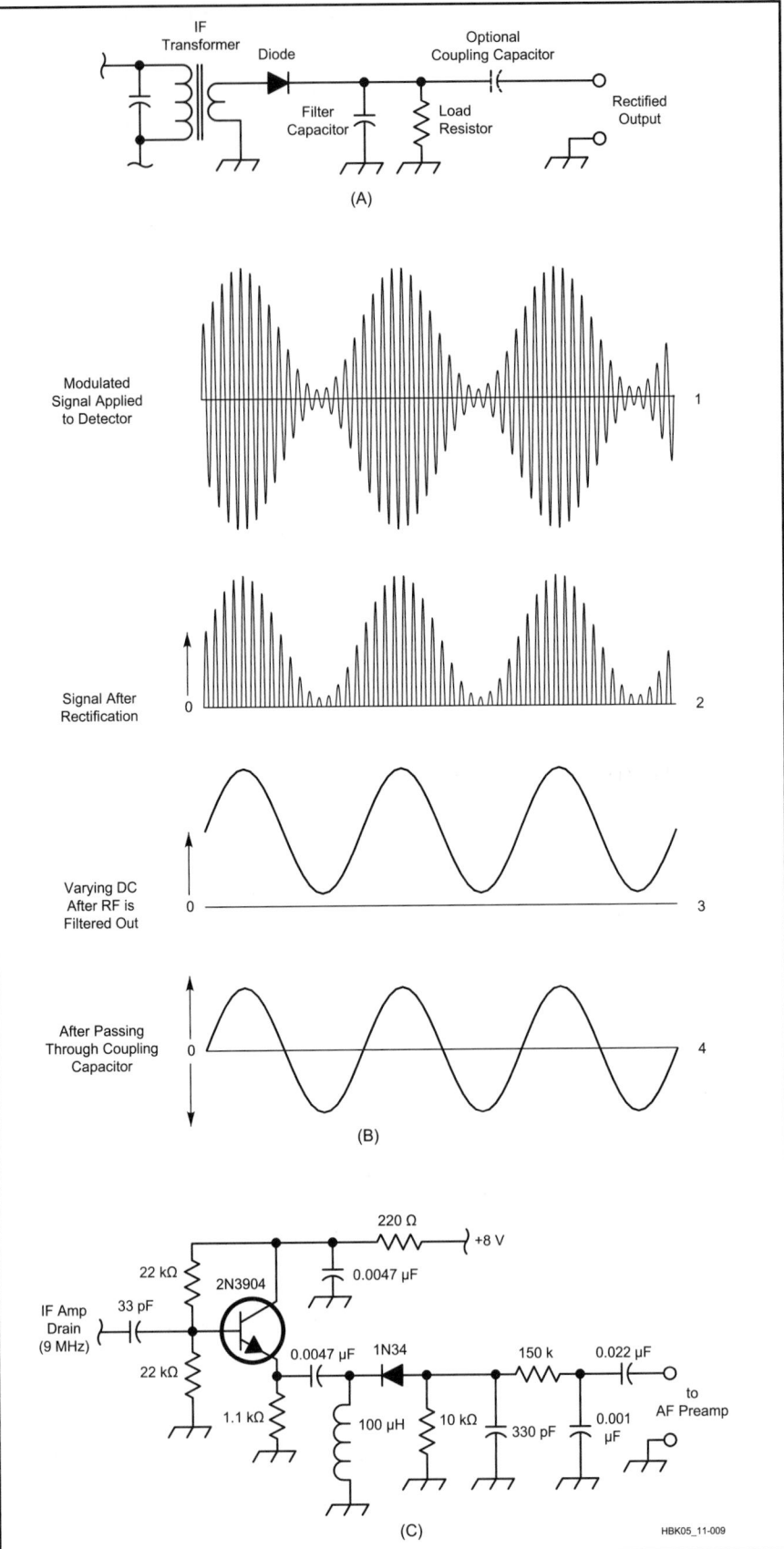

Fig 10.9 — Radio's simplest demodulator, the diode rectifier (A), demodulates an AM signal by acting as a switch that multiplies the carrier and sidebands to produce frequency sums and differences, two of which sum into a replica of the original modulation (B). Modern receivers often use an emitter follower to provide low-impedance drive for their diode detectors (C).

we can modify this process somewhat, if we want. We can use the carrier itself to provide the heterodyning energy in a process called *envelope detection*.

ENVELOPE DETECTION AND FULL-CARRIER AM

Fig 10.5 graphically represents how a full-carrier AM signal's *modulation envelope* corresponds to the shape of the modulating wave. If we can derive from the modulated signal a voltage that varies according to the modulation envelope, we will have successfully recovered the information present in the sidebands. This process is called envelope detection, and we can achieve it by doing nothing more complicated than half-wave-rectifying the modulated signal with a diode (**Fig 10.9**).

That a diode demodulates an AM signal by allowing its carrier to multiply with its sidebands may jar those long accustomed to seeing diode detection ascribed merely to "rectification." But a diode is certainly nonlinear. It passes current in only one direction, and its output voltage is (within limits) proportional to the square of its input voltage. These nonlinearities allow it to multiply.

Exploring this mathematically is tedious with full-carrier AM because the process squares three summed components (carrier, lower sideband and upper sideband). Rather than fill the better part of a page with algebra, we'll instead characterize the outcome verbally: In "just rectifying" a DSB, full-carrier AM signal, a diode detector produces

• Direct current (the result of rectifying the carrier);
• A second harmonic of the carrier;
• A second harmonic of the lower sideband;

• A second harmonic of the upper sideband;
• Two difference-frequency outputs (upper sideband minus carrier, carrier minus lower sideband), each of which is equivalent to the modulating wave-form's frequency, and both of which sum to produce the recovered information signal; and
• A second harmonic of the modulating waveform (the frequency difference between the two sidebands).

Three of these products are RF. Low-pass filtering, sometimes little more than a simple RC network, can remove the RF products from the detector output. A capacitor in series with the detector output line can block the carrier-derived dc component. That done, only two signals remain: the recovered modulation and, at a lower level, its second harmonic — in other words, second-harmonic distortion of the desired information signal.

10.3 Mixers and Angle Modulation

Amplitude modulation served as our first means of translating information into radio form because it could be implemented as simply as turning an electric noise generator on and off. (A spark transmitter consisted of little more than this.) By the 1930s, we had begun experimenting with translating information into radio form and back again by modulating a radio wave's angular velocity (frequency or phase) instead of its overall amplitude. The result of this process is *frequency modulation (FM)* or *phase modulation (PM)*, both of which are often grouped under the name *angle modulation* because of their underlying principle.

A change in a carrier's frequency or phase for the purpose of modulation is called *deviation*. An FM signal deviates according to the amplitude of its modulating waveform, independently of the modulating waveform's frequency; the higher the modulating wave's amplitude, the greater the deviation. A PM signal deviates according to the amplitude *and frequency* of its modulating waveform; the higher the modulating wave's amplitude *and/or frequency*, the greater the deviation. See the sidebar, "Mixer Math: Angle Modulation" for a numerical description of these processes.

10.3.1 Angle Modulation Sidebands

Although angle modulation produces un-

Mixer Math: Angle Modulation

An angle-modulated signal can be mathematically represented as

$$f_c(t) = \cos(2\pi f_c t + m \sin(2\pi f_m t))$$

$$= \cos(2\pi f_c t) \cos(m \sin(2\pi f_m t)) - \sin(2\pi f_c t) \sin(m \sin(2\pi f_m t)) \qquad (10)$$

In equation 10, we see the carrier frequency ($2\pi f_c t$) and modulating signal (sin $2\pi f_m t$) as in the equation for AM (equation 8). We again see the modulating signal associated with a coefficient, m, which relates to degree of modulation. (In the AM equation, *m* is the modulation factor; in the angle-modulation equation, *m* is the *modulation index* and, for FM, equals the deviation divided by the modulating frequency.) We see that angle-modulation occurs as the cosine of the sum of the carrier frequency ($2\pi f_c t$) and the modulating signal (sin $2\pi f_m t$) times the modulation index (*m*). In its expanded form, we see the appearance of sidebands above and below the carrier frequency.

Angle modulation is a multiplicative process, so, like AM, it creates sidebands on both sides of the carrier. Unlike AM, however, angle modulation creates an *infinite* number of sidebands on either side of the carrier! This occurs as a direct result of modulating the carrier's angular velocity, to which its frequency and phase directly relate. If we continuously vary a wave's angular velocity according to another periodic wave's cyclical amplitude variations, the rate at which the modulated wave repeats *its* cycle — its frequency — passes through an infinite number of values. (How many individual amplitude points are there in one cycle of the modulating wave? An infinite number. How many corresponding discrete frequency or phase values does the corresponding angle-modulated wave pass through as the modulating signal completes a cycle? An infinite number!) In AM, the carrier frequency stays at one value, so AM produces two sidebands — the sum of its carrier's unchanging frequency value and the modulating frequency, and the difference between the carrier's unchanging frequency value and the modulating frequency. In angle modulation, the modulating wave shifts the frequency or phase of the carrier through an infinite number of different frequency or phase values, resulting in an infinite number of sum and difference products.

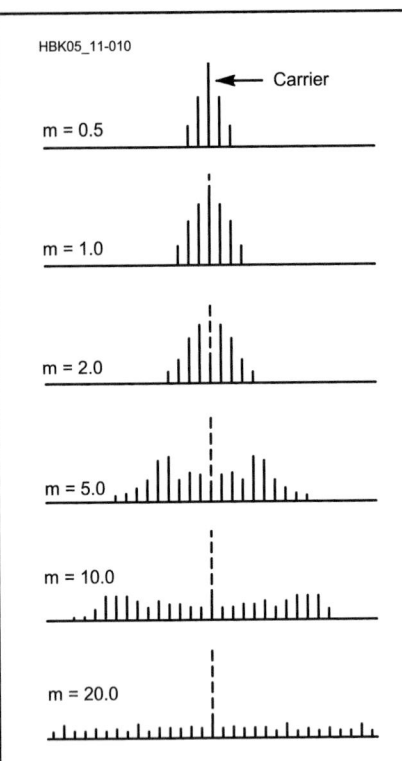

Fig 10.10 — Angle-modulation produces a carrier and an infinite number of upper and lower sidebands spaced from the average ("resting," unmodulated) carrier frequency by integer multiples of the modulating frequency. (This drawing is a simplification because it only shows relatively strong, close-in sideband pairs; space constraints prevent us from extending it to infinity.) The relative amplitudes of the sideband pairs and carrier vary with modulation index, *m*.

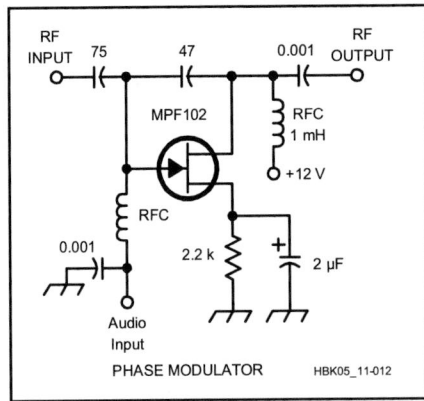

Fig 10.12 — A series reactance modulator acts as a variable shunt around a reactance — in this case, a 47-pF capacitor — through which the carrier passes.

countable sum and difference products, most of them are vanishingly weak in practical systems. They emerge from the modulator spaced from the average ("resting," unmodulated) carrier frequency by integer multiples of the modulating frequency (**Fig 10.10**). The strength of the sidebands relative to the carrier, and the strength and phase of the carrier itself, vary with the degree of modulation — the modulation index. (The *overall* amplitude of an angle-modulated signal does not change with modulation, however; when energy goes out of the carrier, it shows up in the sidebands, and vice versa.) In practice, we operate angle-modulated transmitters at modulation indexes that make all but a few of their infinite sidebands small in amplitude.

(A mathematical tool called *Bessel functions* helps determine the relative strength of the carrier and sidebands according to modulation index. The **Modulation** chapter includes a graph to illustrate this relationship.) Selectivity in transmitter and receiver circuitry further modify this relationship, especially for sidebands far away from the carrier.

10.3.2 Angle Modulators

If you vary a reactance in or associated with an oscillator's frequency-determining element(s), you vary the oscillator's frequency. If you vary the tuning of a tuned circuit through which a signal passes, you vary the signal's phase. A circuit that does this is called a *reactance modulator*, and can be little more than a tuning diode or two connected to a tuned circuit in an oscillator or amplifier (**Fig 10.11**). Varying a reactance through which the signal passes (**Fig 10.12**) is another way of doing the same thing.

The difference between FM and PM depends solely on how, and not how much, deviation occurs. A modulator that causes deviation in proportion to the modulating wave's amplitude and frequency is a phase modulator. A modulator that causes deviation only in proportion to the modulating signal's amplitude is a frequency modulator.

INCREASING DEVIATION BY FREQUENCY MULTIPLICATION

Maintaining modulation linearity is just as important in angle modulation as it is in AM, because unwanted distortion is always our enemy. A given angle-modulator circuit can frequency- or phase-shift a carrier only so much before the shift stops occurring in strict proportion to the amplitude (or, in PM, the amplitude and frequency) of the modulating signal.

If we want more deviation than an angle modulator can linearly achieve, we can operate the modulator at a suitable sub-harmonic — submultiple — of the desired frequency, and process the modulated signal through a series of *frequency multipliers* to bring it up to the desired frequency. The deviation also increases by the overall multiplication factor, relieving the modulator of having to do it all directly. A given FM or PM radio design may achieve its final output frequency through a combination of mixing (frequency shift, no deviation change) and frequency multiplication (frequency shift *and* deviation change).

THE TRUTH ABOUT "TRUE FM"

Something we covered a bit earlier bears closer study:

"An FM signal deviates according to the amplitude of its modulating waveform, independently of the modulating wave-form's frequency; the higher the modulating wave's

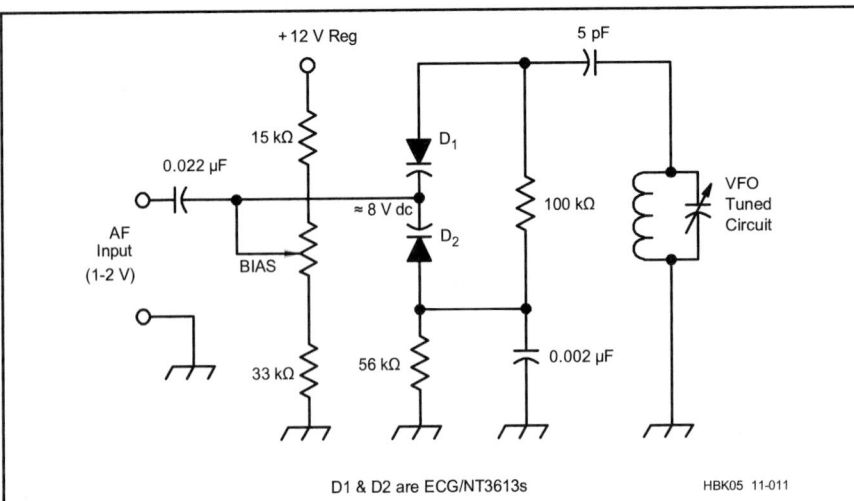

Fig 10.11 — One or more tuning diodes can serve as the variable reactance in a reactance modulator. This HF reactance modulator circuit uses two diodes in series to ensure that the tuned circuit's RF-voltage swing cannot bias the diodes into conduction. D1 and D2 are "30-volt" tuning diodes that exhibit a capacitance of 22 pF at a bias voltage of 4. The BIAS control sets the point on the diode's voltage-versus-capacitance characteristic around which the modulating waveform swings.

amplitude, the greater the deviation. A PM signal deviates according to the amplitude *and frequency* of its modulating waveform; the higher the modulating wave's amplitude *and/or frequency*, the greater the deviation."

The practical upshot of this excerpt is that we can use a phase modulator to generate FM. All we need to do is run a PM transmitter's modulating signal through a low-pass filter that (ideally) halves the signal's amplitude for each doubling of frequency (a reduction of "6 dB per octave," as we sometimes see such responses characterized) to compensate for its phase modulator's "more deviation with higher frequencies" characteristic. The result is an FM, not PM, signal. FM achieved with a phase modulator is sometimes called *indirect FM* as opposed to the *direct FM* we get from a frequency modulator.

We sometimes see claims that one piece of gear is better than another solely because it generates "true FM" as opposed to indirect FM. We can debunk such claims by keeping in mind that direct and indirect FM *sound exactly alike in a receiver* when done correctly.

CONVEYING DC LEVELS WITH ANGLE MODULATION

Depending on the nature of the modulation source, there *is* a practical difference between a frequency modulator and a phase modulator. Answering two questions can tell us whether this difference matters: Does our modulating signal contain a dc level or not? If so, do we need to accurately preserve that dc level through our radio communication link for successful communication? If both answers are *yes*, we must choose our hardware and/or information-encoding approach carefully, because a frequency modulator can convey dc-level shifts in its modulating waveform, while a phase modulator, which responds only to instantaneous changes in frequency and phase, cannot.

Consider what happens when we want to frequency-modulate a phase-locked-loop-synthesized transmitted signal. **Fig 10.13** shows the block diagram of a PLL frequency modulator. Normally, we modulate a PLL's VCO because it's the easy thing to do. As long as our modulating frequency results in frequency excursions too fast for the PLL to follow and correct — that is, as long as our modulating frequency is outside the PLL's *loop bandwidth* — we achieve the FM we seek. Trying to modulate a dc level by pushing the VCO to a particular frequency and holding it there fails, however, because a PLL's loop response includes dc. The loop, therefore, detects the modulation's dc component as a correctable error and "fixes" it. FMing a PLL's VCO therefore can't buy us the dc response "true FM" is supposed to allow.

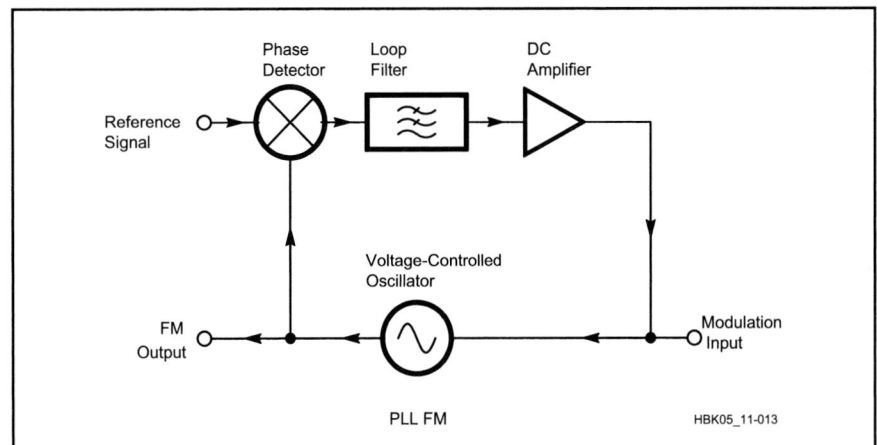

Fig 10.13 — Frequency modulation, PLL-style.

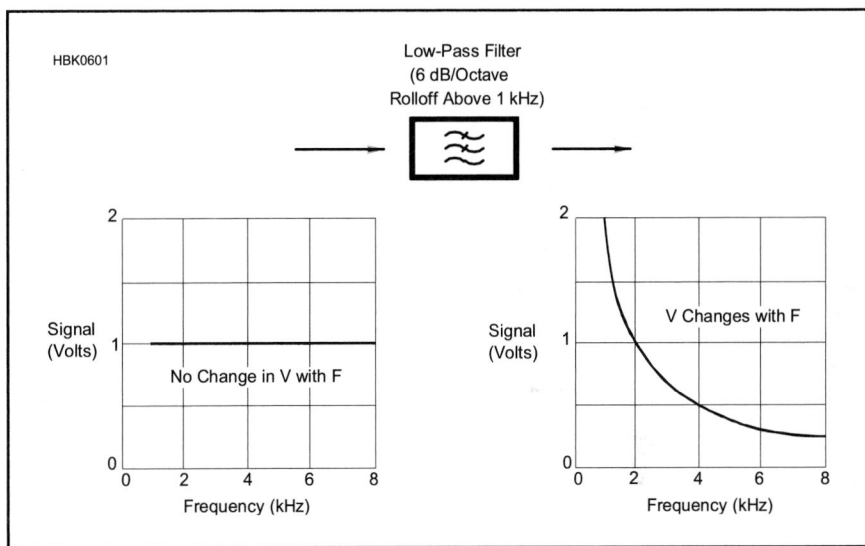

Fig 10.14 — Frequency-sweeping a constant-amplitude signal and passing it through a low-pass filter results in an output signal that varies in amplitude with frequency. This is the principle behind the angle-demodulation process called *frequency discrimination*.

We *can* dc-modulate a PLL modulator, but we must do so by modulating the frequency of the loop *reference*. The PLL then adjusts the VCO to adapt to the changed reference, and our dc level gets through. In this case, the modulating frequency must be *within* the loop bandwidth — which dc certainly is — or the VCO won't be corrected to track the shift.

10.3.3 Mixers and Angle Demodulation

With the awesome prospect of generating an infinite number of sidebands still fresh in our minds, we may be a bit disappointed to learn that we commonly demodulate angle modulation by doing little more than turning it into AM and then envelope- or product-

detecting it! But this is what happens in many of our FM receivers and transceivers, and we can get a handle on this process by realizing that a form of angle-modulation-to-AM conversion begins quite early in an angle-modulated signal's life because of linear distortion of the modulation by amplitude-linear circuitry — something that happens to angle-modulated signals, it turns out, in any linear circuit that doesn't have an amplitude-versus-frequency response that's utterly flat out to infinity.

Think of what happens, for example, when we sweep a constant-amplitude signal up in frequency — say, from 1 kHz to 8 kHz — and pass it through a 6-dB-per-octave filter (**Fig 10.14**). The filter's rolloff causes the output signal's amplitude to decrease as frequency increases. Now imagine that we linearly sweep our constant-amplitude signal *back and forth* between 1 kHz and 8 kHz at a constant rate of 3 kHz per second. The filter's output *amplitude* now varies cyclically over time as the input signal's *frequency* varies cyclically over time. Right before our eyes, a frequency change turns into an amplitude change. The process of converting angle modulation to amplitude modulation has begun.

This is what happens whenever an angle-modulated signal passes through circuitry with an amplitude-versus-frequency response that isn't flat out to infinity. As the signal deviates across the frequency- response curves of whatever circuitry passes it, its angle modulation is, to some degree, converted to AM — a form of crosstalk between the two modulation types, if we wish to look at it that way. (Variations in system phase linearity also cause distortion and FM-to-AM conversion, because the sidebands do not have the proper phase relationship with respect to each other and with respect to the carrier.)

All we need to do to put this effect to practical use is develop a circuit that does this frequency-to-amplitude conversion linearly across the frequency span of the modulated signal's deviation. Then we envelope-demodulate the resulting AM, and we have achieved angle demodulation.

Fig 10.15 shows such a circuit — a *discriminator* — and the sort of amplitude-versus-frequency response we expect from it. It's actually possible to use an AM receiver to recover understandable audio from a narrow angle-modulated signal by "off-tuning" the signal so its deviation rides up and down on one side of the receiver's IF selectivity curve. This *slope detection* process served as an early, suboptimal form of frequency discrimination in receivers not designed for FM It is always worth trying as a last-resort-class means of receiving narrowband FM with an AM receiver.

QUADRATURE DETECTION

It's also possible to demodulate an angle-modulated signal merely by multiplying it with a time-delayed copy of itself in a double-balanced mixer as shown in **Fig 10.16**; the sidebar, "Mixer Math: Quadrature Demodulation," explains the process numerically.

An ideal quadrature detector puts out 0 V dc when no modulation is present (with the carrier at f_c). The output of a real quadrature detector may include a small *dc offset* that requires compensation. If we need the detector's response all the way down to dc, we've got it; if not, we can put a suitable

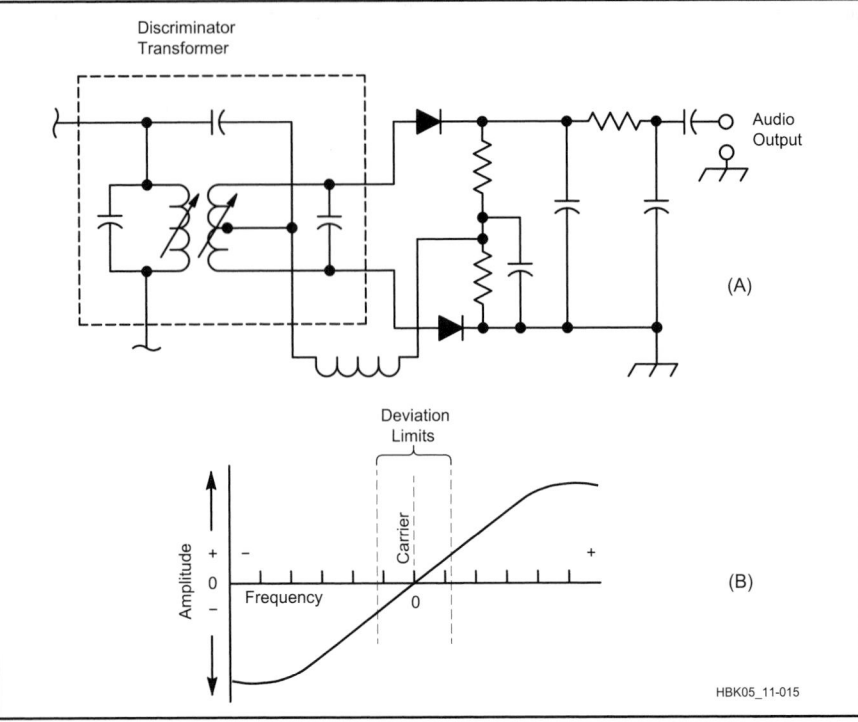

Fig 10.15 — A *discriminator* (A) converts an angle-modulated signal's deviation into an amplitude variation (B) and envelope-detects the resulting AM signal. For undistorted demodulation, the discriminator's amplitude-versus-frequency characteristic must be linear across the input signal's deviation. A *crystal discriminator* uses a crystal as part of its frequency-selective circuitry.

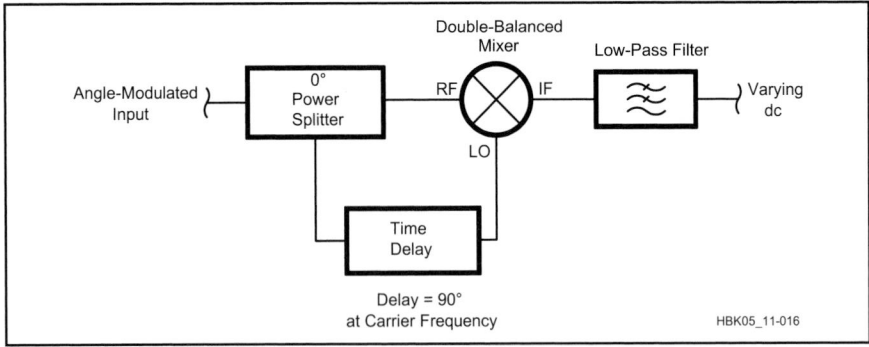

Fig 10.16 — In *quadrature detection*, an angle-modulated signal multiplies with a time-delayed copy of itself to produce a dc voltage that varies with the amplitude and polarity of its phase or frequency excursions away from the carrier frequency. A practical quadrature detector can be as simple as a 0° power splitter (that is, a power splitter with in-phase outputs), a diode double-balanced mixer, and a length of coaxial cable ¼-λ (electrical) long at the carrier frequency, and a bit of low-pass filtering to remove the detector output's RF components. IC quadrature detectors achieve their time delay with one or more resistor-loaded tuned circuits (Fig 10.17).

blocking capacitor in the output line for ac-only coupling.

Quadrature detection is more common than frequency discrimination nowadays because it doesn't require a special discrimi-

nator transformer or resonator, and because the necessary balanced-detector circuitry can easily be implemented in IC structures along with limiters and other receiver circuitry. The NXP Semiconductor SA604A FM IF IC is

Mixer Math: Quadrature Demodulation

Demodulating an angle-modulated signal merely by multiplying it with a time-delayed copy of itself in a double-balanced mixer results in quadrature demodulation (Fig 10.16). To illustrate this mathematically, for simplicity's sake, we'll represent the mixer's RF input signal as just a sine wave with an amplitude, A:

$$A \sin(2\pi ft) \qquad (11)$$

and its time-delayed twin, fed to the mixer's LO input, as a sine wave with an amplitude, A, and a time delay of d:

$$A \sin[2\pi f(t+d)] \qquad (12)$$

Setting this special mixing arrangement into motion, we see

$$A \sin(2\pi ft) \times A \sin(2\pi ft + d)$$

$$= \frac{A^2}{2}\cos(2\pi fd) - \frac{A^2}{2}\cos(2\pi fd)\cos(2 \times 2\pi ft) + \frac{A^2}{2}\sin(2\pi fd)\sin(2 \times 2\pi ft) \qquad (13)$$

Two of the three outputs — the second and third terms — emerge at twice the input frequency; in practice, we're not interested in these, and filter them out. The remaining term — the one we're after — varies in amplitude and sign according to how far and in what direction the carrier shifts away from its resting or center frequency (at which the time delay, d, causes the mixer's RF and LO inputs to be exactly 90° out of phase — in *quadrature* — with each other). We can examine this effect by replacing f in equations 11 and 12 with the sum term $f_c + f_s$, where f_c is the center frequency and f_s is the frequency shift. A 90° time delay is the same as a quarter cycle of f_c, so we can restate d as

$$d = \frac{1}{4f_c} \qquad (14)$$

The first term of the detector's output then becomes

$$\frac{A^2}{2}\cos\left(2\pi(f_c + f_s)d\right)$$

$$= \frac{A^2}{2}\cos\left(2\pi(f_c + f_s)\frac{1}{4f_c}\right)$$

$$= \frac{A^2}{2}\cos\left(\frac{\pi}{2} + \frac{\pi f_s}{2f_c}\right) \qquad (15)$$

When f_s is zero (that is, when the carrier is at its center frequency), this reduces to

$$\frac{A^2}{2}\cos\left(\frac{\pi}{2}\right) = 0 \qquad (16)$$

As the input signal shifts higher in frequency than f_c, the detector puts out a positive dc voltage that increases with the shift. When the input signal shifts lower in frequency than f_c, the detector puts out a negative dc voltage that increases with the shift. The detector therefore recovers the input signal's frequency or phase modulation as an amplitude-varying dc voltage that shifts in sign as f_s varies around f_c — in other words, as ac. We have demodulated FM by means of quadrature detection.

one example of this; **Fig 10.17** shows another, the Freescale Semiconductor (formerly Motorola) MC3359 (equivalent, NTE860).

PLL ANGLE DEMODULATION

Back at Fig 10.14, we saw how a phase-locked loop can be used as an angle modulator. A PLL also makes a fine angle *de*modulator. Applying an angle-modulated signal to a PLL keeps its phase detector and VCO hustling to maintain loop lock through the input signal's angle variations. The loop's error voltage therefore tracks the input signal's modulation, and its variations mirror the modulation signal. Turning the loop's varying dc error voltage into audio is just a blocking capacitor away.

Although we can't convey a dc level by directly modulating the VCO in a PLL angle modulator, a PLL demodulator can respond down to dc quite nicely. A constant frequency offset from f_c (a dc component) simply causes a PLL demodulator to swing its VCO over to the new input frequency, resulting in a proportional dc offset on the VCO control-voltage line. Another way of looking at the difference between a PLL angle modulator and a PLL angle demodulator is that a PLL demodulator works with a varying reference signal (the input signal), while a PLL angle modulator generally doesn't.

AMPLITUDE LIMITING REQUIRED

By now, it's almost household knowledge that FM radio communication systems are superior to AM in their ability to suppress and ignore static, manmade electrical noise and (through a characteristic called *capture effect*) co-channel signals sufficiently weaker than the desired signal. AM-noise immunity and capture effect are not intrinsic to angle modulation, however; they must be designed into the angle-modulation receiver in the form of signal amplitude *limiting*.

If we note the progress of A from the left to the right side of the equal sign in equation 13, we realize that the amplitude of a quadrature detector's input signal affects the amplitude of a quadrature detector's three output signals. A quadrature detector therefore responds to AM, and so does a frequency discriminator. To achieve FM's storied noise immunity, then, these angle demodulators must be preceded by *limiter* circuitry that removes all amplitude variations from the incoming signal.

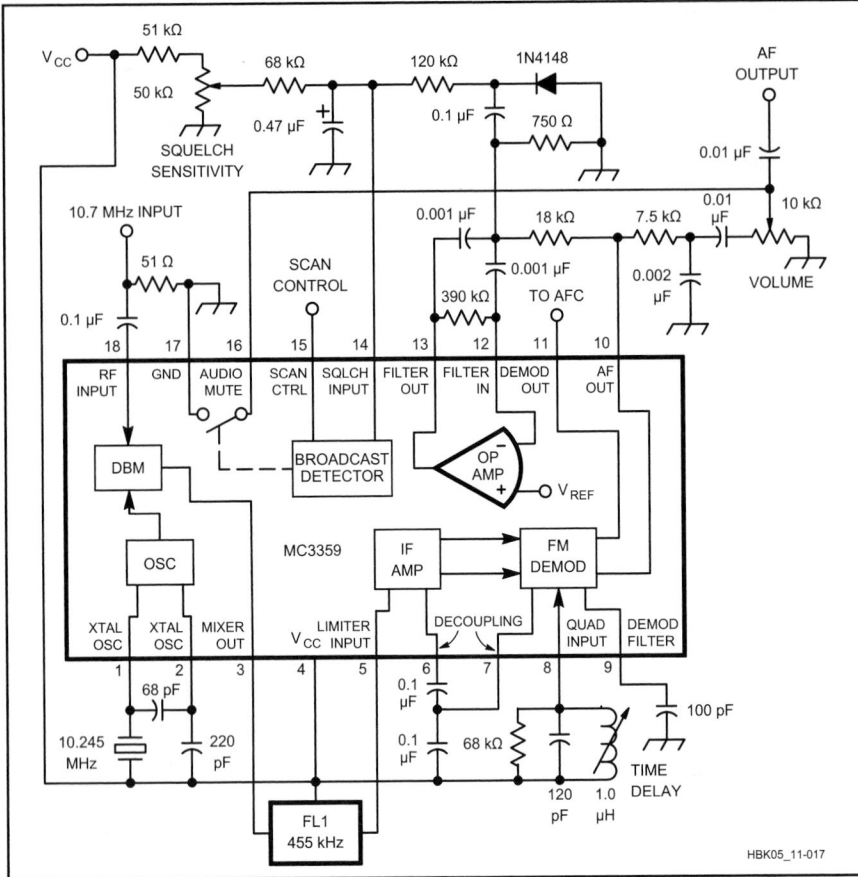

Fig 10.17 — The Freescale MC3359/NTE680 is one of many FM subsystem ICs that include limiter and quadrature-detection circuitry. The TIME DELAY coil is adjusted for minimum recovered-audio distortion.

10.4 Putting Mixers, Modulators and Demodulators to Work

Variations on relatively few mixer types provide mixing, modulation, and demodulation functions in the majority of Amateur Radio receivers, transmitters, and transceivers. Comparing their behaviors and specifications begins with a grounding in key mixer characteristics related to *dynamic range* — the span of signal strengths a circuit can handle without generating false signals that can interfere with communication.

10.4.1 Dynamic Range: Compression, Intermodulation and More

The output of a linear stage — including a mixer, which we want to act as a linear frequency shifter — tracks its input signal decibel by decibel, every 1-dB change in its input signal(s) corresponding to an identical 1-dB output change. This is the stage's *first-order* response.

Because no device is perfectly linear, however, two or more signals applied to it generate sum and difference frequencies. These IMD products occur at frequencies and amplitudes that depend on the order of the IMD response as follows:

• *Second-order* IMD products change 2 dB for every decibel of input-signal change (this figure assumes that the IMD comes from equal-level input signals), and appear at frequencies that result from the simple addition and subtraction of input-signal frequencies. For example, assuming that its input bandwidth is sufficient to pass them, an amplifier subjected to signals at 6 and 8 MHz will produce second-order IMD products at 2 MHz

Putting Mixer Math to Work through DSP

In describing the idealized behavior of mixing, modulation and demodulation circuits and processes through mathematics, this chapter's Mixer Math sidebars reflect a profound underlying reality: If what a mixer, modulator or demodulator does can be described with math, then *all a mixer, modulator or demodulator does is math*. This is true not only of mixers, modulators and demodulators, but of *all* electrical and electronic circuitry, analog and digital. An electrical or electronic circuit is merely a math machine — a machine that, for much of the history of human use of radio and electronics, has necessarily been hardwired to be capable of solving only one equation or set or class of equations. When you build an electronic circuit, interconnecting its resistors, capacitors, inductors, active devices and other components, you build out the signal-processing equation it will solve, atom by mathematical atom. Adjusting the controls of such a purpose-built math machine, every switch you throw or control modifies the equation it solves, and therefore the work it performs.

No matter how painstakingly and elegantly engineered and constructed such a practical realization of math may be, however, it is only so operationally flexible. If you want to significantly alter the equation(s) it solves, you must change its circuitry by replacing and/or reconnecting components and/or by adjusting whatever controls you may have included to tweak its operation on the fly.

The **Computer-Aided Circuit Design** chapter of this *Handbook* also characterizes electrical and electronic circuits as math machines. It does so for the purpose of exploring the use of personal computers — which are generic, arbitrarily and readily programmable and reprogrammable math machines — as engineering tools capable of simulating, through mathematical modeling, the electrical and electronic behavior of real-world circuits. The next logical expansion of the use of electronic computers in radio is to move beyond just *simulating* the mathematical signal-processing functions of circuits to actually *performing those functions in real time through mathematics applied in real time* —as physical circuitry has done for decades and still does, but better, and more flexibly, repeatably, and modifiably. Such real-time applied-math techniques, well-established and increasingly widely applied as *digital signal processing* (DSP) and fundamental to the design, evolution, and growing power of *software-defined radios* (SDRs), are likely already in use, or will soon be, in a radio you own. You can learn more about DSP and its techniques, applications and limitations in this *Handbook*'s **DSP and Software Radio Design** chapter.

(8–6) and 14 MHz (8+6).

• *Third-order* IMD products change 3 dB for every decibel of input-signal change (this also assumes equal-level input signals), and appear at frequencies corresponding to the sums and differences of twice one signal's frequency plus or minus the frequency of another. Assuming that its input bandwidth is sufficient to pass them, an amplifier subjected to signals at 14.02 MHz (f_1) and 14.04 MHz (f_2) produces third-order IMD products at 14.00 ($2f_1 - f_2$), 14.06 ($2f_2 - f_1$), 42.08 ($2f_1 + f_2$) and 42.10 ($2f_2 + f_1$) MHz. The subtractive products (the 14.02 and 14.04-MHz products in this example) are close to the desired signal and can cause significant interference. Thus, third-order-IMD performance is of great importance in receiver mixers and RF amplifiers.

It can be seen that the IMD order determines how rapidly IMD products change level per unit change of input level. Nth-order IMD products therefore change by n dB for every decibel of input-level change.

IMD products at orders higher than three can and do occur in communication systems, but their magnitudes are usually much smaller than the lower-order products. The second- and third-order products are most important in receiver front ends. In transmitters, third- and higher-odd-order products are important because they widen the transmitted signal.

10.4.2 Intercept Point

The second type of dynamic range concerns the receiver's *intercept point*, sometimes simply referred to as *intercept*. Intercept point is typically measured (or calculated by ex-

trapolation of the test results) by applying two or three signals to the antenna input, tuning the receiver to count the number of resulting spurious responses, and measuring their level relative to the input signal.

Because a device's IMD products increase more rapidly than its desired output as the input level rises, it might seem that steadily increasing the level of multiple signals applied to an amplifier would eventually result in equal desired-signal and IMD levels at the amplifier output. Real devices are incapable of doing this, however. At some point, every device *overloads*, and changes in its output level no longer equally track changes at its input. The device is then said to be operating in *compression*; the point at which its first-order response deviates from linearity by 1 dB is its *1-dB compression point*. Pushing the process to its limit ultimately leads to *saturation*, at which point input-signal increases no longer increase the output level.

The power level at which a device's second-order IMD products would equal its first-order output (a point that must be extrapolated because most likely the device is in compression by this point) is its *second-order intercept point*. Likewise, its *third-order intercept point* is the power level at which third-order responses would equal the desired signal. **Fig 10.18** graphs these relationships.

Input filtering can improve second-order intercept point; device nonlinearities determine the third, fifth and higher-odd-number intercept points. In preamplifiers and active mixers, third-order intercept point is directly related to dc input power; in passive switching mixers, to the local-oscillator power applied.

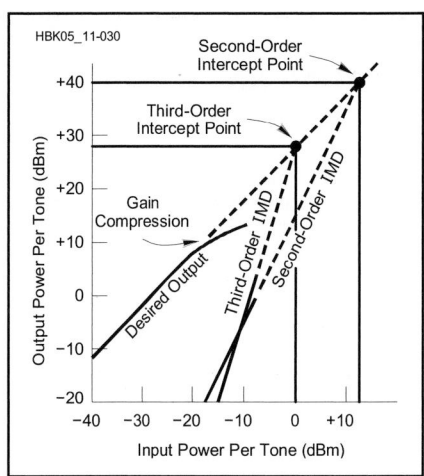

HBK05_11-030

Fig 10.18 — A linear stage's output tracks its input decibel by decibel on a 1:1 slope — its *first-order* response. *Second-order intermodulation distortion (IMD) products produced by two equal-level input signals ("tones") rise on a 2:1 slope — 2 dB for every 1 dB of input increase. Third-order IMD products* likewise increase 3 dB for every 1 dB of increase in two equal tones. For each IMD order *n*, there is a corresponding *intercept point* IP$_n$ at which the stage's first-order and *n*th order products are equal in amplitude. The first-order output of real amplifiers and mixers falls off (the device overloads and goes into *compression*) before IMD products can intercept it, but intercept point is nonetheless a useful, valid concept for comparing radio system performance. The higher an amplifier or mixer's intercept point, the stronger the input signals it can handle without overloading. The input and output powers shown are for purposes of example; every device exhibits its own particular IMD profile. (After W. Hayward, *Introduction to Radio Frequency Design*, Fig 6.17)

Testing and Calculating Intermodulation Distortion in Receivers

Second and third-order IMD can be measured using the setup of **Fig 10.A1**. The outputs of two signal generators are combined in a 3-dB hybrid coupler. Such couplers are available from various companies, and can be homemade. The 3-dB coupler should have low loss and should itself produce negligible IMD. The signal generators are adjusted to provide a known signal level at the output of the 3-dB coupler, say, −20 dBm for each of the two signals. This combined signal is then fed through a calibrated variable attenuator to the device under test. The shielding of the cables used in this system is important: At least 90 dB of isolation should exist between the high-level signal at the input of the attenuator and the low-level signal delivered to the receiver.

The measurement procedure is simple: adjust the variable attenuator to produce a signal of known level at the frequency of the expected IMD product ($f_1 \pm f_2$ for second-order, $2f_1 - f_2$ or $2f_2 - f_1$ for third-order IMD).

To do this, of course, you have to figure out what equivalent input signal level at the receiver's operating frequency corresponds to the level of the IMD product you are seeing. There are several ways of doing this. One way — the way used by the ARRL Lab in their receiver tests — uses the minimum discernible signal. This is defined as the signal level that produces a 3-dB increase in the receiver audio output power. That is, you measure the receiver output level with no input signal, then insert a signal at the operating frequency and adjust the level of this input signal until the output power is 3 dB greater than the no-signal power. Then, when doing the IMD measurement, you adjust the attenuator of Fig 10.A1 to cause a 3-dB increase in receiver output. The level of the IMD product is then the same as the MDS level you measured.

There are several things I dislike about doing the measurement this way. The problem is that you have to measure noise power. This can be difficult. First, you need an RMS voltmeter or audio power meter to do it at all. Second, the measurement varies with time (it's noise!), making it difficult to nail down a

number. And third, there is the question of the audio response of the receiver; its noise output may not be flat across the output spectrum. So I prefer to measure, instead of MDS, a higher reference level. I use the receiver's S meter as a reference. I first determine the input signal level it takes to get an S1 reading. Then, in the IMD measurement, I adjust the attenuator to again give an S1 reading. The level of the IMD product signal is now equal to the level I measured at S1. Note that this technique gives a different IMD level value than the MDS technique. That's OK, though. What we are trying to determine is the *difference* between the level of the signals applied to the receiver input and the level of the IMD product. Our calculations will give the same result whether we measure the IMD product at the MDS level, the S1 level or some other level.

An easy way to make the reference measurement is with the setup of Fig 10.A1. You'll have to switch in a lot of attenuation (make sure you have an attenuator with enough range), but doing it this way keeps all of the possible variations in the measurement fairly constant. And this way, the difference between the reference level and the input level needed to produce the desired IMD product signal level is simply the difference in attenuator settings between the reference and IMD measurements.

Calculating Intercept Points

Once we know the levels of the signals applied to the receiver input and the level of the IMD product, we can easily calculate the intercept point using the following equation:

$$IP_n = \frac{n \times P_A - P_{IM_n}}{n - 1} \tag{A}$$

Here, n is the order, P_A is the receiver input power (of one of the input signals), P_{IMn} is the power of the IMD product signal,

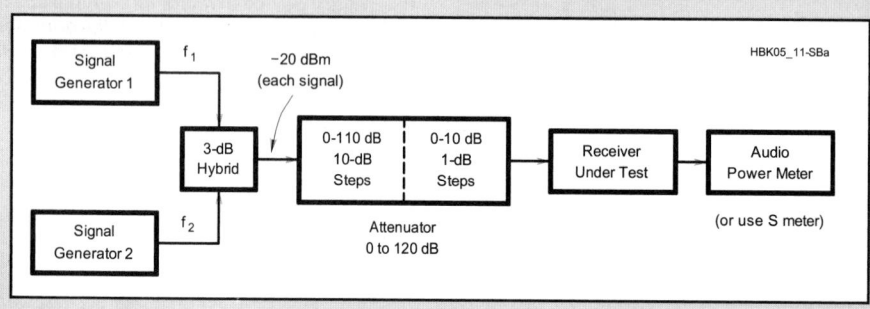

Fig 10.A1 — Test setup for measurement of IMD performance. Both signal generators should be types such as HP 608, HP 8640, or Rohde & Schwarz SMDU, with phase-noise performance of −140 dBc/Hz or better at 20 kHz from the signal frequency.

10.5 A Survey of Common Mixer Types

Depending on the application, frequency-mixer implementations may vary from the extremely simple to the complex. For example, a simple half-wave rectifier (a signal diode, such as a 1N34 [germanium] or a 1N914 [silicon]) can do the job (as we saw at Fig 10.9 in our discussion of full-carrier AM demodulation); this is an example of a *switching mixer*, in which mixing occurs as

one signal — in this case, the carrier, which in effect turns the diode on and off as its polarity reverses — interrupts the transmission of another (in demodulation of full-carrier AM, the sideband[s]).

A switch can be thought of as an amplifier toggled between two gain states, off and on, by a control signal. It turns out that a binary amplifier is not necessary; *any* device that

can be gain-varied in accordance with the amplitude of a control signal can serve as a frequency mixer.

10.5.1 Gain-Controlled Analog Amplifiers As Mixers

Fig 10.19 shows two analog-amplifier mixing arrangements commonly encoun-

and IP$_n$ is the nth-order intercept point. All powers should be in dBm. For second and third-order IMD, equation A results in the equations:

$$IP_2 = \frac{2 \times P_A - P_{IM_2}}{2-1}$$ (B)

$$IP_3 = \frac{3 \times P_A - P_{IM_3}}{3-1}$$ (C)

You can measure higher-order intercept points, too.

Example Measurements

To get a feel for this process, it's useful to consider some actual measured values.

The first example is a Rohde & Schwarz model EK085 receiver with digital preselection. For measuring second-order IMD, signals at 6.00 and 8.01 MHz, at –20 dBm each, were applied at the input of the attenuator. The difference in attenuator settings between the reference measurement and the level needed to produce the desired IMD product signal level was found to be 125 dB. The calculation of the second-order IP is then:

$$IP_2 = \frac{2(-20\ dBm) - (-20\ dBm - 125\ dB)}{2-1}$$

$$= -40\ dBm + 20\ dBm + 125\ dB = +105\ dB$$

For IP$_3$, we set the signal generators for 0 dBm at the attenuator input, using frequencies of 14.00 and 14.01 MHz. The difference in attenuator settings between the reference and IMD measurements was 80 dB, so:

$$IP_3 = \frac{3(0\ dBm) - (0\ dBm - 80\ dB)}{3-1}$$

$$= \frac{0\ dBm + 80\ dB}{2} = +40\ dBm$$

We also measured the IP$_3$ of a Yaesu FT-1000D at the same frequencies, using attenuator-input levels of –10 dBm. A difference in attenuator readings of 80 dB resulted in the calculation:

$$IP_3 = \frac{3(-10\ dBm) - (-10\ dBm - 80\ dB)}{3-1}$$

$$= \frac{-30\ dBm + 10\ dBm + 80\ dB}{2}$$

$$= \frac{-20\ dBm + 80\ dB}{2}$$

$$= +30\ dBm$$

Synthesizer Requirements

To be able to make use of high third-order intercept points at these close-in spacings requires a low-noise LO synthesizer. You can estimate the required noise performance of the synthesizer for a given IP$_3$ value. First, calculate the value of receiver input power that would cause the IMD product to just come out of the noise floor, by solving equation A for P$_A$, then take the difference between the calculated value of P$_A$ and the noise floor to find the dynamic range. Doing so gives the equation:

$$ID_3 = \frac{2}{3}(IP_3 + P_{min})$$ (D)

Where ID$_3$ is the third-order IMD dynamic range in dB and P$_{min}$ is the noise floor in dBm. Knowing the receiver bandwidth, BW (2400 Hz in this case) and noise figure, NF (8 dB) allows us to calculate the noise floor, P$_{min}$:

$$P_{min} = -174\ dBm + 10\ \log(BW) + NF$$

$$= -174\ dBm + 10\ \log(2400) + 8$$

$$= -132\ dBm$$

The synthesizer noise should not exceed the noise floor when an input signal is present that just causes an IMD product signal at the noise floor level. This will be accomplished if the synthesizer noise is less than:

$$ID + 10\ \log(BW) = 114.7\ dB + 10\ \log(2400)$$

$$= 148.5\ dBc/Hz$$

in the passband of the receiver. Such synthesizers hardly exist.
— *Dr Ulrich L. Rohde, N1UL*

---

tered in vacuum-tube-based vintage radio gear. In one (Fig 10.19A), the RF and LO signals are applied to the control grid of a pentode amplifier; in the second, the RF and LO signals are applied to different grids of a tube specifically designed to act as a mixer. RF, LO, and IF (RF and LO sum and difference frequencies) energy is present at the output of both. In both circuits, a double-tuned transformer selects the desired output frequency and provides some reduction of unwanted components. The pentagrid mixer circuit (Fig 10.19A) provides better isolation between its RF and LO inputs because of its separate RF and LO-injection grids.

Fig 10.20 shows a simple JFET front end that consists of an AGCed amplifier (2N4393) and 2N3819 or 2N5454 mixer with source LO injection. As in the tube circuits of Fig 10.19, RF, LO, and IF spectral components are all present in the output of this simple circuit.

Mixers based on FET cascode amplifiers (**Fig 10.21**) can provide excellent performance, and were common (in dual-gate MOSFET form) across at least two generations of Amateur Radio gear. As with the circuits presented in Figs 10.19 and 10.20,

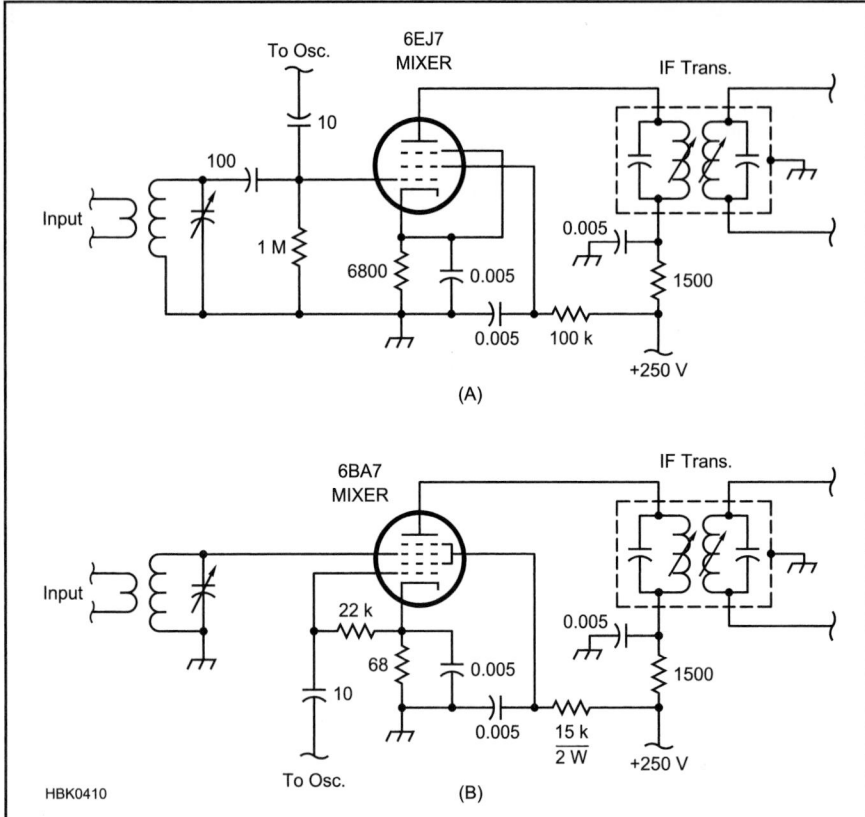

Fig 10.19 — Vacuum-tube amplifiers as frequency mixers. At A, RF and LO signals are applied to the control grid of a 6EJ7 pentode; at B, a RF and LO signals are applied to grids 1 and 3, respectively, of a 6BA7 pentagrid mixer. The 6BA7 and its relatives and similars (6A8, 6K8, 6SA7, 6SB7, and 6BE6, among others), designed specifically for frequency-conversion use, can be configured to generate their own LO signal; such a self-oscillating tube mixer is commonly called a *converter*.

RF, LO, and IF spectral components are present in the output of these mixers, and filtering is necessary to select the desired component and reduce unwanted ones.

10.5.2 Switching Mixers

Most modern radio mixers act more like fast analog switches than analog multipliers. In using a mixer as a fast switching device, we apply a square wave to its LO input with a square wave rather than a sine wave, and feed sine waves, audio, or other complex signals to the mixer's RF input. The RF port serves as the mixer's "linear" input, and therefore must preferably exhibit low intermodulation and harmonic distortion. Feeding a ±1-V square wave into the LO input alternately multiplies the linear input by +1 or –1. Multiplying the RF-port signal by +1 just transfers it to the output with no change. Multiplying the RF-port signal by –1 does the same thing, except that the signal inverts (flips 180° in phase). The LO port need not exhibit low intermodulation and harmonic distortion; all it has to do is preserve the fast rise and fall times of the switching signal.

REVERSING-SWITCH MIXERS

We can multiply a signal by a square wave without using an analog multiplier at all. All we need is a pair of balun transformers and four diodes (**Fig 10.22A**).

With no LO energy applied to the circuit, none of its diodes conduct. RF-port energy (1)

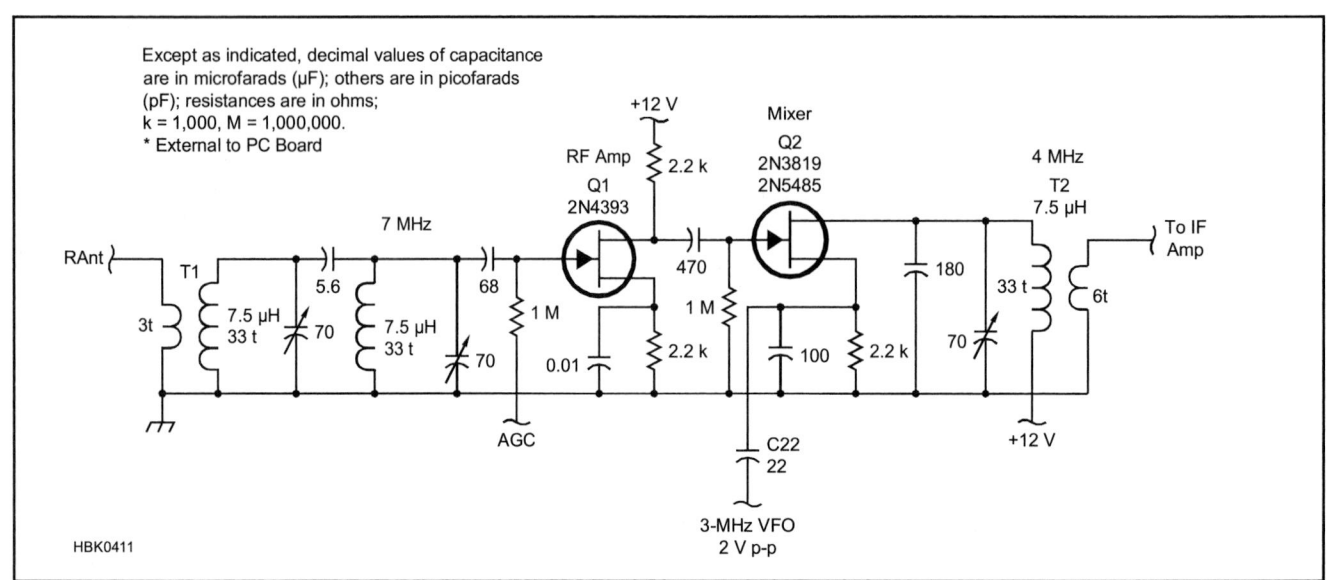

Fig 10.20 — This simple JFET receiver front end for 7 MHz features an AGCed RF amplifier (Q1, a 2N4393) and gain-controlled-amplifier mixer (Q2, a 2N3819 or 2N5485). LO voltage applied to the source of Q2 varies the stage's gain to produce sum and difference outputs. T2 selects the product (4 MHz). Any JFET with a pinchoff voltage of –1 V is suitable for Q1; the point labeled AGC can be grounded to operate the RF stage at full gain. The inductors consist of no. 22 enameled wire wound on plastic bobbins used to hold the lower thread in "New Home" brand sewing machines; if toroidal equivalents are substituted, the turns ratios in T1 and T2 should be preserved. This circuit was described in "Build the No-Excuses QRP Transceiver," December 2000 *QST*.

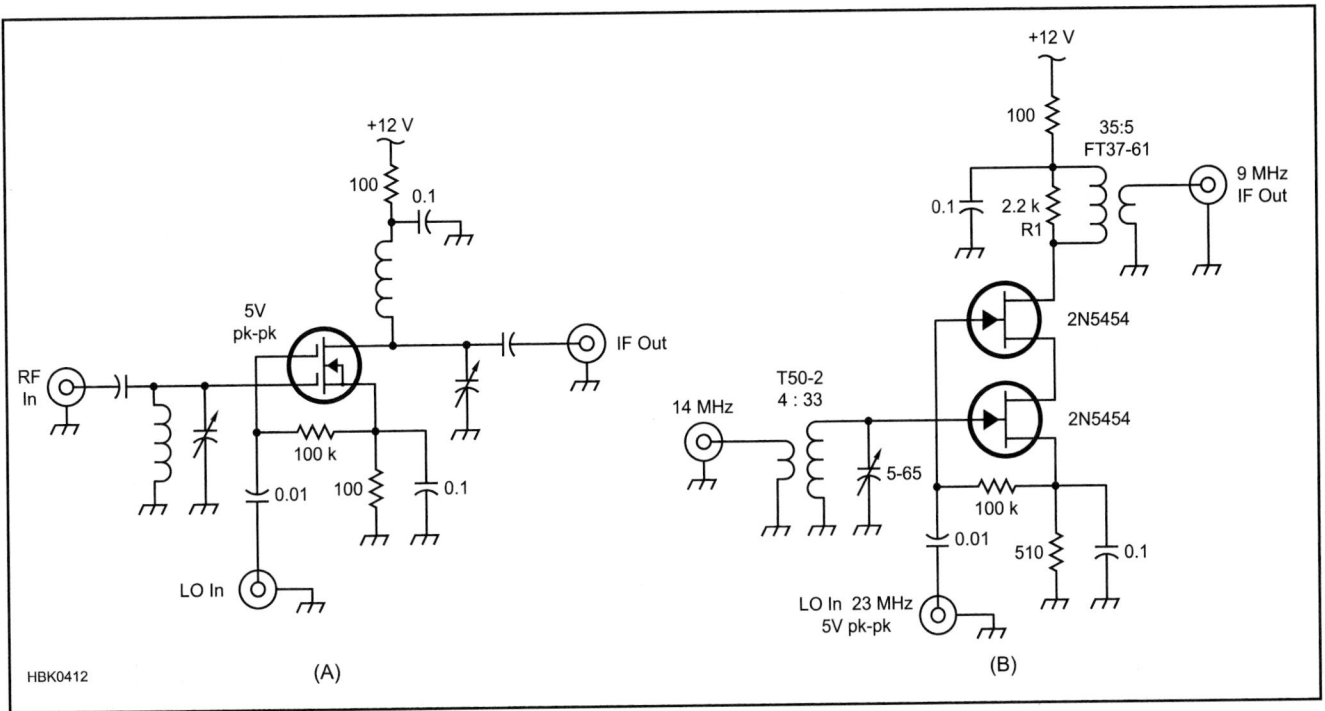

Fig 10.21 — Cascode FET amplifiers used as mixers, as presented by Hayward, Campbell and Larkin in _Experimental Methods in RF Design_. RF is applied to the gate of one device, and LO and a small positive bias are applied to the gate of the other; commercial applications of the MOSFET version commonly include positive bias on the RF-port gate. A dual-gate MOSFET is shown at A; suitable devices include the 3N187, 3N211, 40673 and BF998. (Internally, a dual-gate MOSFET consists of two single-gate MOSFETs in cascode.) Applying AGC rather than this circuit's combination of LO and source-derived bias to gate 2 of the MOSFET would turn this arrangement into one of the AGCed IF amplifiers used in commercial ham gear for over 40 years. Two JFETs in cascode are shown at B; suitable devices include the 2N4416, 2N3819, 2N5400-series parts, J310/U310 and MPF102, among others. R1 loads the circuit to eliminate oscillation at the RF-port resonance; it may be increased in value or eliminated altogether if the circuit is stable without it. Ratios accompanying the transformers at B convey the number of turns. Per _EMRFD_, the JFET circuit shown, biased for 3.4 mA at 12 V, operates with a conversion gain of 8 dB; a noise figure of 10 dB; and a third-order intercept point of +5 dBm.

can't make it to the LO port because there's no direct connection between the secondaries of T1 and T2, and (2) doesn't produce IF output because T2's secondary balance results in energy cancellation at its center tap, and because no complete IF-energy circuit exists through T2's secondary with both of its ends disconnected from ground.

Applying a square wave to the LO port biases the diodes so that, 50% of the time, D1 and D2 are on and D3 and D4 are reverse-biased off. This unbalances T2's secondary by leaving its upper wire floating and connecting its lower wire to ground through T1's secondary and center tap. With T2's secondary unbalanced, RF-port energy emerges from the IF port.

The other 50% of the time, D3 and D4 are on and D1 and D2 are reverse-biased off. This unbalances T2's secondary by leaving its lower wire floating, and connects its upper wire to ground through T1's secondary and center tap. With T2's secondary unbalanced, RF-port energy again emerges from the IF port — shifted 180° relative to the first case because T2's active secondary wires are now, in effect, transposed relative to its primary.

A reversing switch mixer's output spectrum is the same as the output spectrum of a multiplier fed with a square wave. This can be analyzed by thinking of the square wave in terms of its Fourier series equivalent, which consists of the sum of sine waves at the square wave frequency and all of its odd harmonics. The amplitude of the equivalent series' fundamental sine wave is $4/\pi$ times (2.1 dB greater than) the amplitude of the square wave. The amplitude of each harmonic is inversely proportional to its harmonic number, so the third harmonic is only ⅓ as strong as the fundamental (9.5 dB below the fundamental), the 5th harmonic is only ⅕ as strong (14 dB below the fundamental) and so on. The input signal mixes with each harmonic separately from the others, as if each harmonic were driving its own separate mixer, just as we illustrated with two sine waves in Fig 10.4. Normally, the harmonic outputs are so widely removed from the desired output frequency that they are easily filtered out, so a reversing-switch mixer is just as good as a sine-wave-driven analog multiplier for most practical purposes, and usually better — for radio purposes — in terms of dynamic range and noise.

An additional difference between multiplier and switching mixers is that the signal flow in a switching mixer is reversible (that is, bilateral). It really only has one dedicated input (the LO input). The other terminals can be thought of as I/O (input/output) ports, since either one can be the input as long as the other is the output.

CONVERSION LOSS IN SWITCHING MIXERS

Fig 10.22B shows a perfect _multiplier_ mixer. That is, the output is the product of the input signal and the LO. The LO is a perfect square wave. Its peak amplitude is ±1.0 V and its frequency is 8 MHz. Fig 10.22C shows the output waveform (the product of two inputs) for an input signal whose value is 0 dBm and whose frequency is 2 MHz. Notice that for each transition of the square-wave LO, the sine-wave output waveform polarity reverses. There are 16 transitions during the interval shown, at each zero-crossing point of the output waveform. Fig 10.22D shows the mixer output spectrum. The principle components are at 6 MHz and 10 MHz, which are the sum and difference of the signal and LO frequencies. The amplitude of each of these is –3.9 dBm. Numerous other pairs of output frequencies occur that are also spaced 4 MHz

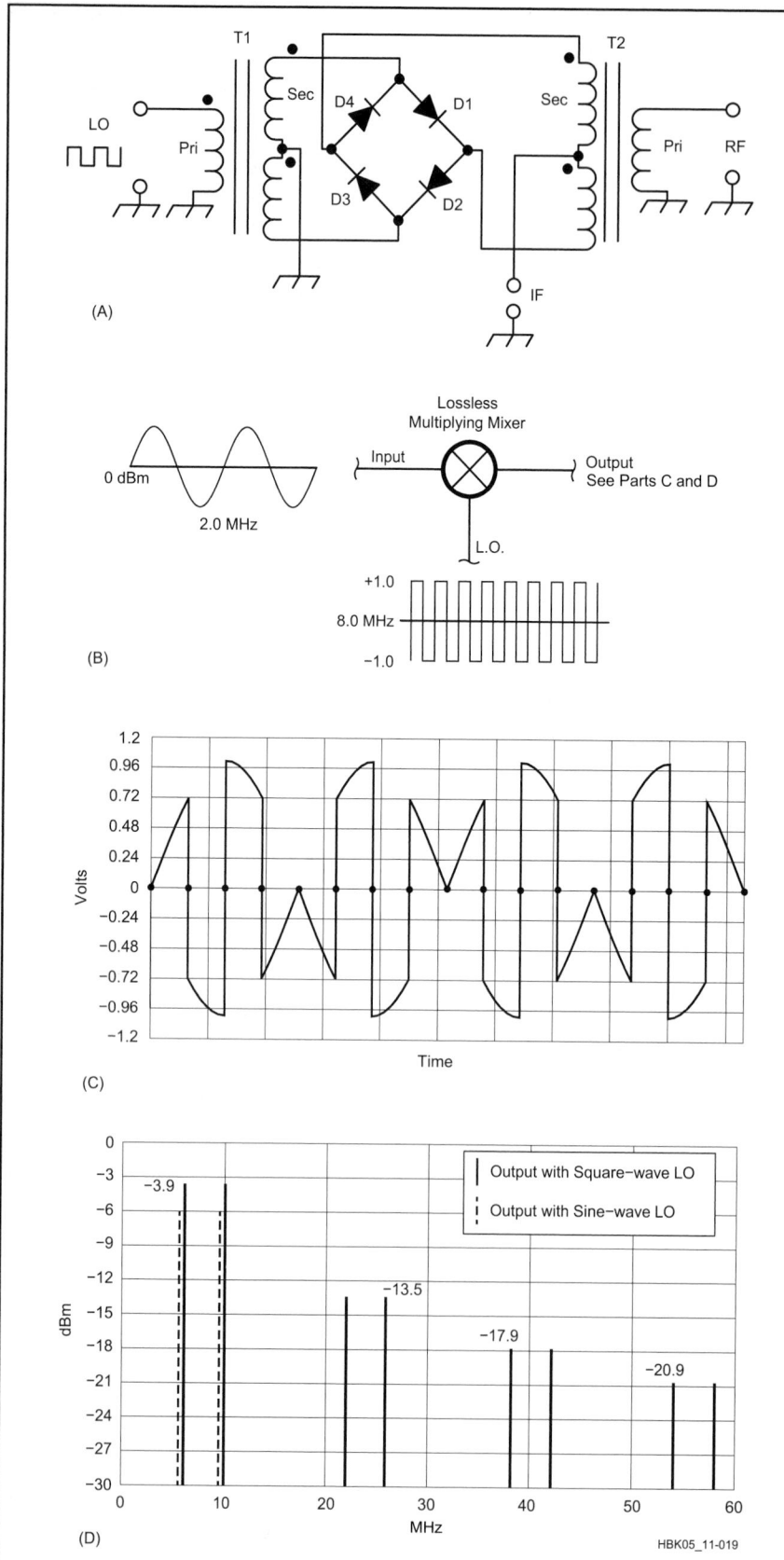

(A)

(B)

(C)

(D)

HBK05_11-019

apart and centered at 24 MHz, 40 MHz and 56 MHz and higher odd harmonics of 8 MHz. The ones shown are at –13.5 dBm, –17.9 dBm and –20.9 dBm. Because the mixer is lossless, the sum of all of the outputs must be exactly equal to the value of the input signal. As explained previously, this output spectrum can also be understood in terms of each of the odd-harmonic components of the square-wave LO operating independently.

If the mixer switched losslessly, such as in Fig 10.22A, with diodes that are perfect switches, the results would be mathematically identical to the above example. The diodes would commutate the input signal exactly as shown in Fig 10.22C.

Now consider the perfect multiplier mixer of Fig 10.22B with an LO that is a perfect sine wave with a peak amplitude of ±1.0 V. In this case the dashed lines of Fig 10.22D show that only two output frequencies are present, at 6 MHz and 10 MHz (see also Fig 10.2). Each component now has a –6 dBm level. The product of the 0 dBm sine-wave input at one frequency and the ±1.0V sine-wave LO at another frequency (see equation 6 in this chapter) is the –3 dBm total output.

These examples illustrate the difference between the square-wave LO and the sine-wave LO, for a perfect multiplier. For the same peak value of both LO waves, the square-wave LO delivers 2.1 dB more output at 6 MHz and 10 MHz than the sine-wave LO. An actual diode mixer such as Fig 10.22A behaves more like a switching mixer. Its sine-wave LO waveform is considerably flattened by interaction between the diodes and the LO generator, so that it looks somewhat like a square wave. The diodes have nonlinearities, junction voltages, capacitances, resistances and imperfect parameter matching. (See the **RF Techniques** chapter.) Also, "re-mixing" of a diode mixer's output with the LO and the input is a complicated possibility. The practical end result is that diode double-balanced mixers have a conversion loss, from input to each of the two major output frequencies, in the neighborhood of 5 to 6 dB. (Conversion loss is discussed in a later section.)

10.5.3 The Diode Double-Balanced Mixer: A Basic Building Block

The diode *double-balanced mixer* (DBM) is standard in many commercial, military and amateur applications because of its excellent balance and high dynamic range. DBMs can serve as mixers (including image-reject types), modulators (including single- and double-sideband, phase, biphase, and quadrature-phase types) and demodulators, limiters, attenuators, switches, phase detectors and frequency doublers. In some of these applica-

Fig 10.22 — Part A shows a general-purpose diode *reversing-switch* mixer. This mixer uses a square-wave LO and a sine-wave input signal. The text describes its action. Part B is an ideal *multiplier* mixer. The square-wave LO and a sine-wave input signal produce the output waveform shown in part C. The solid lines of part D show the output spectrum with the square-wave LO. The dashed lines show the output spectrum with a sine-wave LO.

tions, they work in conjunction with power dividers, combiners and hybrids.

THE BASIC DBM CIRCUIT

We have already seen the basic diode DBM circuit (Fig 10.22A). In its simplest form, a DBM contains two or more unbalanced-to-balanced transformers and a Schottky-diode ring consisting of 4 × n diodes, where n is the number of diodes in each leg of the ring. Each leg commonly consists of up to four diodes.

As we've seen, the degree to which a mixer is *balanced* depends on whether either, neither or both of its input signals (RF and LO) emerge from the IF port along with mixing products. An unbalanced mixer suppresses neither its RF nor its LO; both are present at its IF port. A single-balanced mixer suppresses its RF or LO, but not both. A double-balanced mixer suppresses its RF *and* LO inputs. Diode and transformer uniformity in the Fig 10.22 circuit results in equal LO potentials at the center taps of T1 and T2. The LO potential at T1's secondary center tap is zero (ground); therefore, the LO potential at the IF port is zero.

Balance in T2's secondary likewise results in an RF null at the IF port. The RF potential between the IF port and ground is therefore zero — except when the DBM's switching diodes operate, of course!

The Fig 10.22 circuit normally also affords high RF-IF isolation because its balanced diode switching precludes direct connections between T1 and T2. A diode DBM can be used as a current-controlled switch or attenuator by applying dc to its IF port, albeit with some distortion. This causes opposing diodes (D2 and D4, for instance) to conduct to a degree that depends on the current magnitude, connecting T1 to T2.

TRIPLE-BALANCED MIXERS

The triple-balanced mixer shown in **Fig 10.23** (sometimes called a "double double-balanced mixer") is an extension of the single-diode-ring mixer. The diode rings are fed by power-splitting baluns at the RF and LO ports. An additional balun is added at the IF output. The circuit's primary advantage is that the IF output signal is balanced and isolated from the RF and LO ports over a large bandwidth — commercial mixer IF ranges of 0.5 to 10 GHz are typical.

It has higher signal-handling capability and dynamic range (a 1-dB compression point within 3 to 4 dB below LO signal levels) and lower intermodulation levels (by 10 dB or more) than a single-ring mixer. The triple-balanced mixer is used when a very wide IF range is required.

Adding the balancing transformer in the IF output path increases IF-to-LO and IF-to-RF isolation. This makes the conversion process much less sensitive to IF impedance mis-

Testing Mixer Performance

In order to make proper tests on mixers using signal generators, a hybrid coupler with at least 40 dB of isolation between the two input ports and an attenuator are required. The test set-up provided by DeMaw in *QST*, Jan 1981, shown in **Fig 10.B1** is ideal for this. Two signal generators operating near 14 MHz are combined in the hybrid coupler, then isolated from the mixer under test (MUT) by a variable attenuator. The LO is supplied by a VFO covering 5.0-5.5 MHz and applied to the MUT through another variable attenuator. The output is isolated with another attenuator, amplified and applied to a spectrum analyzer for analysis.

Attenuation should be sufficient to provide isolation (minimum of 6 to 10 dB required) and to result in signal levels to the mixer under test (MUT) appropriate for the required testing and as suitable for the particular mixer device.

The 2N5109 amplifier shown may not be sufficient for extremely high intercept point tests as this stage may no longer be transparent (operate linearly) at high signal levels. For stability tests, it is recommended to have a reactive network at the output of the mixer for the sole purpose of checking whether the mixer can become unstable.

The two 14 MHz oscillators must have extremely low harmonic content and very low noise sidebands. A convenient oscillator circuit is provided in **Fig 10.B2**, based on a 1975 *Electronic Design* article. — Dr Ulrich L. Rohde, N1UL

DeMaw, W1FB, and Collins, ADØW, "Modern Receiver Mixers for High Dynamic Range," *QST*, Jan 1981, p 19.

U. Rohde, "Crystal Oscillator Provides Low Noise," *Electronic Design*, Oct 11, 1975.

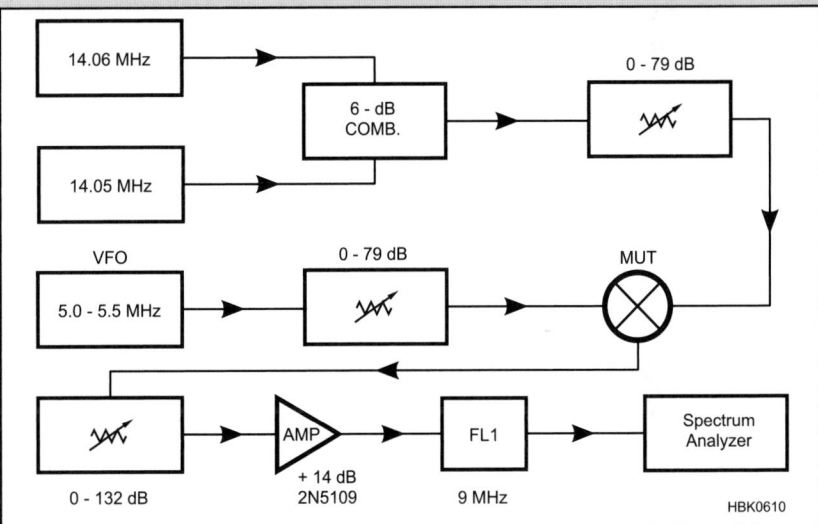

Fig 10.B1 — The equipment setup for measuring mixer performance at HF.

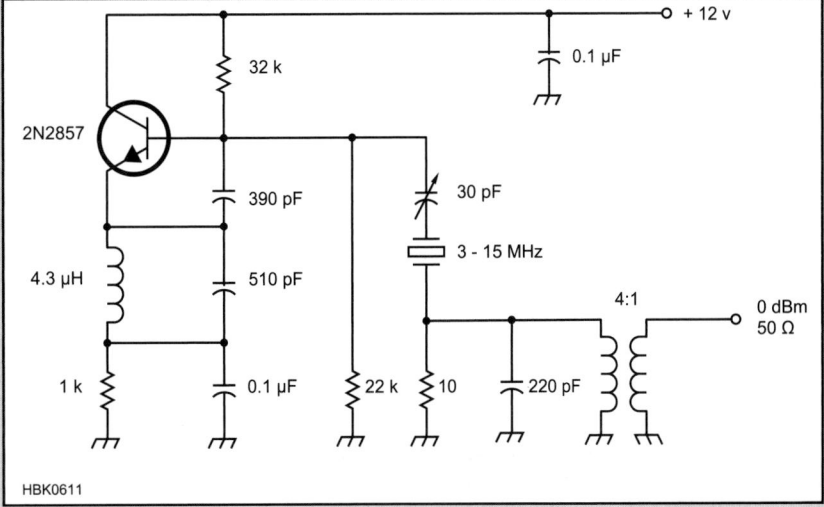

Fig 10.B2 — A low-noise VXO circuit for driving the LO port of the mixer under test in Figure 10.B1.

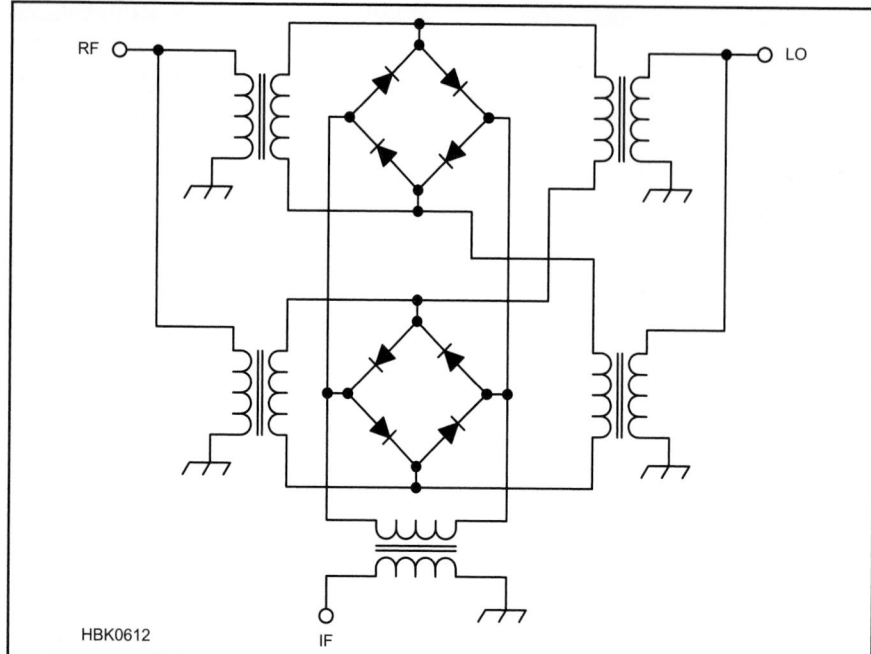

Fig 10.23 — The triple-balanced mixer uses a pair of diode rings and adds an additional balancing transformer to the IF port.

matches. Since the IF port is isolated from the RF and LO ports, the three frequency ranges (RF, LO and IF) can overlap. A disadvantage of IF transformer coupling is that a dc (or low-frequency) IF output is not available, so the triple-balanced mixer cannot be used for direct-conversion receivers.

DIODE DBM COMPONENTS

Commercially manufactured diode DBMs generally consist of a supporting base, a diode ring, two or more ferrite-core transformers commonly wound with two or three twisted-pair wires, encapsulating material, an enclosure.

Diodes

Hot-carrier (Schottky) diodes are the devices of choice for diode-DBM rings because of their low ON resistance, although ham-built DBMs for non-critical MF/HF use commonly use switching diodes like the 1N914 or 1N4148. The forward voltage drop, V_f, across each diode in the ring determines the mixer's optimum local-oscillator drive level. Depending on the forward voltage drop of each of its diodes and the number of diodes in each ring leg, a diode DBM will often be specified by the optimum LO drive level in dBm (typical values are 0, 3, 7, 10, 13, 17, 23 or 27). As a rule of thumb, the LO signal must be 20 dB stronger than the RF and IF signals for proper operation. This ensures that the LO signal, rather than the RF or IF signals, switches the mixer's diodes on and off — a critical factor in minimizing IMD and maximizing dynamic range.

Transformers

From the DBM schematic shown in Fig 10.22, it's clear that the LO and RF transformers are unbalanced on the input side and balanced on the diode side. The diode ends of the balanced ports are 180° out of phase throughout the frequency range of interest. This property causes signal cancellations that result in higher port-to-port isolation. **Fig 10.24A** plots LO-RF and LO-IF isolation versus frequency for Synergy Microwave's CLP-403 DBM, which is specified for +7 dBm LO drive level. Isolations on the order of 70 dB occur at the lower end of the band as a direct result of the balance among the four diode-ring legs and the RF phasing of the balanced ports.

As we learned in our discussion of generic switching mixers, transformer efficiency plays an important role in determining a mixer's conversion loss and drive-level requirement. Core loss, copper loss and impedance mismatch all contribute to transformer losses. Ferrite in toroidal, bead, balun (multi-hole) or rod form can serve as DBM transformer cores. Radio amateurs commonly use Fair-Rite Mix 43 ferrite ($\mu = 950$).

RF transformers combine lumped and distributed capacitance and inductance. The interwinding capacitance and characteristic impedance of a transformer's twisted wires sets the transformer's high-frequency response. The core's μ and size, and the number of winding turns, determine the transformer's lower frequency limit. Covering a specific frequency range requires a compromise in the number of turns used with a given core. Increasing a transformer's core size and number of turns improves its low-frequency response. Cores may be stacked to meet low-frequency performance specs.

Inexpensive mixers operating up to 2 GHz most commonly use twisted trifilar (three-wire) windings made of a wire size between #36 and #32. The number of twists per unit length of wire determines a winding's characteristic impedance. Twisted wires are analogous to transmission lines. The transmission-line effect predominates at the higher end of a transformer's frequency range.

DIODE DBMS IN PRACTICE

Important DBM specifications include conversion loss and amplitude flatness across the required IF bandwidth; variation of conversion loss with input frequency; variation of conversion loss with LO drive, 1-dB compression point; LO-RF, LO-IF and RF-IF isolation; intermodulation products; noise figure (usually within 1 dB of conversion loss); port SWR; and dc offset, which is directly related to isolation among the RF, LO and IF ports. Most of these parameters also apply to other mixer types.

Conversion Loss

Fig 10.24B shows conversion loss versus intermediate frequency in a typical DBM. The curves show conversion loss for two fixed RF-port signals, one at 100 kHz and

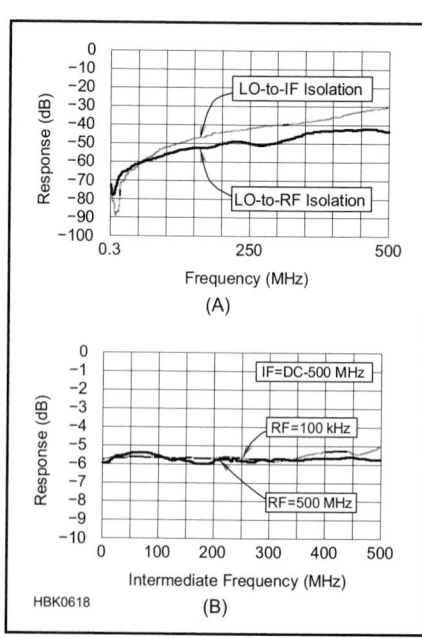

Fig 10.24 — The port-to-port isolation of a diode DBM depends on how well its diodes match and how well its transformers are balanced. (A) shows LO-IF and LO-RF isolation versus frequency and (B) shows conversion loss for a typical diode DBM, the Synergy Microwave CLP-403 mixer. In (B), LO driver level is +7 dBm.

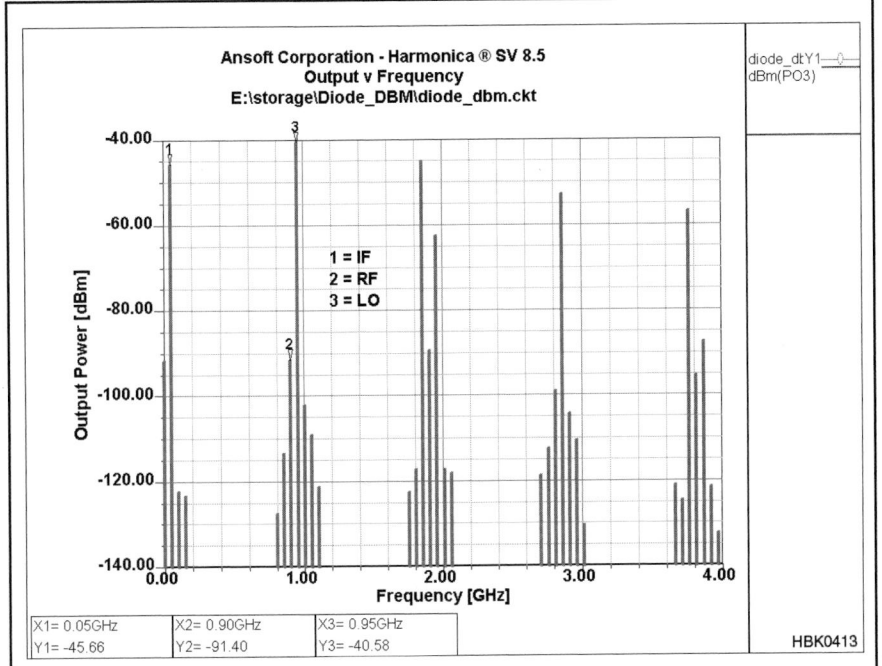

Fig 10.25— Simulated diode-DBM output spectrum with four LO harmonics evaluated. Note that the desired output products (the highest two products, RF − LO and RF + LO) emerge at a level 5 to 6 dB below the mixer's RF input (−40 dBm). This indicates a mixer conversion loss of 5 to 6 dB. (*Serenade SV8.5* simulation.)

the another at 500 MHz, while varying the LO frequency from 100 kHz to 500 MHz.

Fig 10.25 graphs a diode DBM's simulated output spectrum. Note that the RF input (900 MHz) is −40 dBm and the desired IF output (51 MHz, the frequency difference between the RF and LO signals) is −46 dBm, implying a conversion loss of 6 dB. Very nearly the same value (5 dB) applies to the sum of both signals (RF + LO). We minimize a diode DBM's conversion loss, noise figure and intermodulation by keeping its LO drive high enough to switch its diodes on fully and rapidly. Increasing a mixer's LO level beyond that sufficient to turn its switching devices all the way on merely makes them dissipate more LO power without further improving performance.

Insufficient LO drive results in increased noise figure and conversion loss. IMD also increases because RF-port signals have a greater chance to control the mixer diodes when the LO level is too low.

APPLYING DIODE DBMS

At first glance, applying a diode DBM is easy: We feed the signal(s) we want to frequency-shift (at or below the maximum level called for in the mixer's specifications, such as −10 dBm for the Mini-Circuits SBL-1 and TUF-3, and Synergy Microwave S-1, popular 7 dBm LO power parts) to the DBM's RF port, feed the frequency-shifting signal (at the proper level) to the LO port, and extract

the sum and difference products from the mixer's IF port.

There's more to it than that, however, because diode DBMs (along with most other modern mixer types) are *termination-sensitive*. That is, their ports — particularly their IF (output) ports — must be resistively terminated with the proper impedance (commonly 50 Ω, resistive). A wideband, resistive output termination is particularly critical if a mixer is to achieve its maximum dynamic range in receiving applications. Such a load can be achieved by:

• Terminating the mixer in a 50-Ω resistor or attenuator pad (a technique usually avoided in receiving applications because it directly degrades system noise figure);

• Terminating the mixer with a low-noise, high-dynamic-range *post-mixer amplifier* designed to exhibit a wideband resistive input impedance; or

• Terminating the mixer in a *diplexer*, a frequency-sensitive signal splitter that appears as a two-terminal resistive load at its input while resistively dissipating unwanted outputs and passing desired outputs through to subsequent circuitry.

Termination-insensitive mixers are available, but this label can be misleading. Some termination-insensitive mixers are nothing more than a termination-sensitive mixer packaged with an integral post-mixer amplifier. True termination-insensitive mixers are less common and considerably more

elaborate. Amateur builders will more likely use one of the many excellent termination-sensitive mixers available in connection with a diplexer, post-mixer amplifier or both.

Fig 10.26 shows one diplexer implementation. In this approach, L1 and C1 form a series-tuned circuit, resonant at the desired IF, that presents low impedance between the diplexer's input and output terminals at the IF. The high-impedance parallel-tuned circuit formed by L2 and C2 also resonates at the desired IF, keeping desired energy out of the diplexer's 50-Ω load resistor, R1.

The preceding example is called a *band-pass diplexer*. **Fig 10.27** shows another type: a *high-pass/low-pass diplexer* in which each inductor and capacitor has a reactance of 70.7 Ω at the 3-dB cutoff frequency. It can be used after a "difference" mixer (a mixer in which the IF is the difference between the signal frequency and LO) if the desired IF and its image frequency are far enough apart so that the image power is "dumped" into the network's 51-Ω resistor. (For a "summing" mixer — a mixer in which the IF is the sum of the desired signal and LO — interchange the 50-Ω idler load resistor and the diplexer's "50-Ω Amplifier" connection.) Richard Weinreich, KØUVU, and R. W. Carroll described this circuit in November 1968 *QST* as one of several absorptive TVI filters.

Fig 10.28 shows a BJT post-mixer amplifier design made popular by Wes Hayward, W7ZOI, and John Lawson, K5IRK. RF feedback (via the 1-kΩ resistor) and emitter degeneration (the ac-coupled 5.6-Ω emitter resistor) work together to keep the stage's input impedance near 50 Ω and uniformly resistive across a wide bandwidth. Performance comparable to the Fig 10.28 circuit can be obtained at MF and HF by using paralleled 2N3904s as shown in **Fig 10.29**.

AMPLITUDE MODULATION WITH A DBM

We can generate DSB, suppressed-carrier AM with a DBM by feeding the carrier to its RF port and the modulating signal to the IF port. This is a classical *balanced modulator*, and the result — sidebands at radio frequencies corresponding to the carrier signal plus audio and the RF signal minus audio — emerges from the DBM's LO port. If we also want to transmit some carrier along with the sidebands, we can dc-bias the IF port (with a current of 10 to 20 mA) to upset the mixer's balance and keep its diodes from turning all the way off. (This technique is sometimes used for generating CW with a balanced modulator otherwise intended to generate DSB as part of an SSB-generation process.) **Fig 10.30** shows a more elegant approach to generating full-carrier AM with a DBM.

As we saw earlier when considering the many faces of AM, two DBMs, used in con-

Fig 10.26 — A diplexer resistively terminates energy at unwanted frequencies while passing energy at desired frequencies. This band-pass diplexer (A) uses a series-tuned circuit as a selective pass element, while a high-C parallel-tuned circuit keeps the network's terminating resistor R1 from dissipating desired-frequency energy. Computer simulation of the diplexer's response with *ARRL Radio Designer 1.0* characterizes the diplexer's insertion loss and good input match from 8.8 to 9.2 MHz (B) and from 1 to 100 MHz (C); and the real and imaginary components of the diplexer's input impedance from 8.8 to 9.2 MHz with a 50-Ω load at the diplexer's output terminal (D). The high-C, low-L nature of the L2-C2 circuit requires that C2 be minimally inductive; a 10,000-pF chip capacitor is recommended. This diplexer was described by Ulrich L. Rohde and T. T. N. Bucher in *Communications Receivers: Principles and Design*.

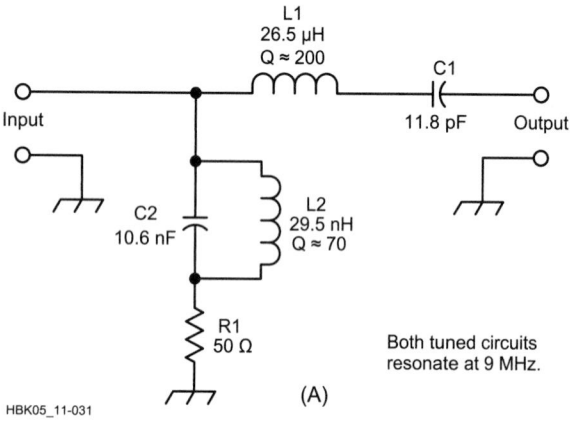

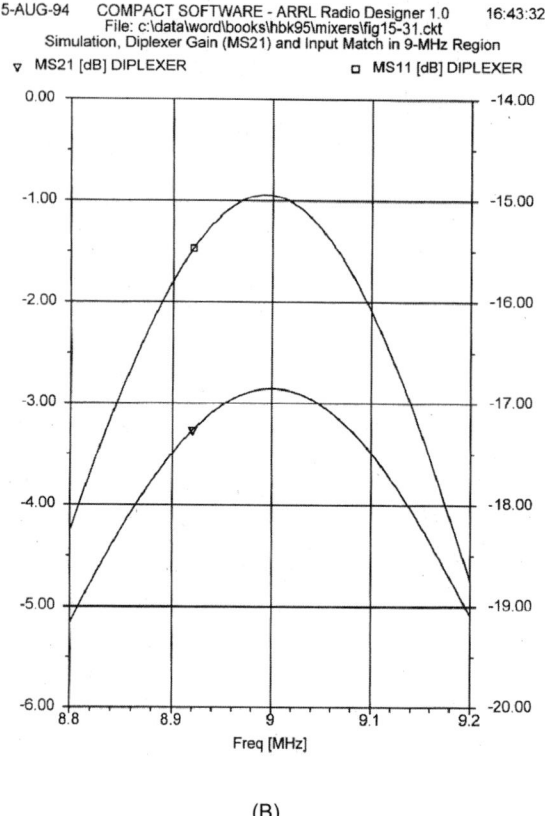

(B)

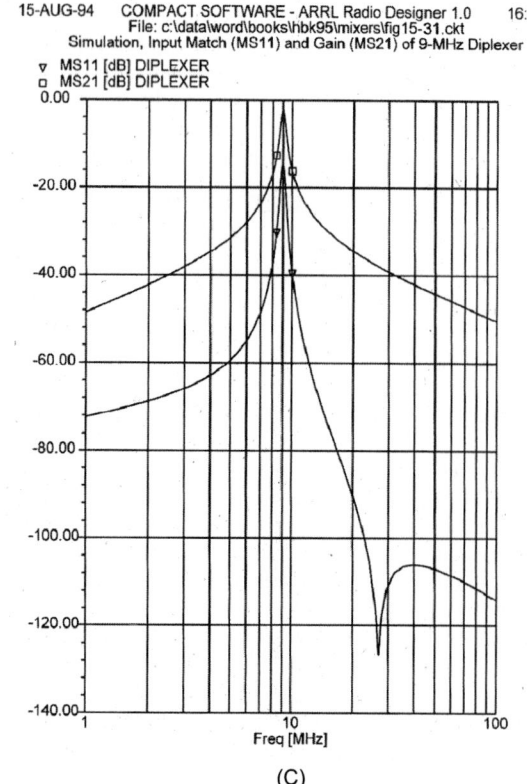

(A)

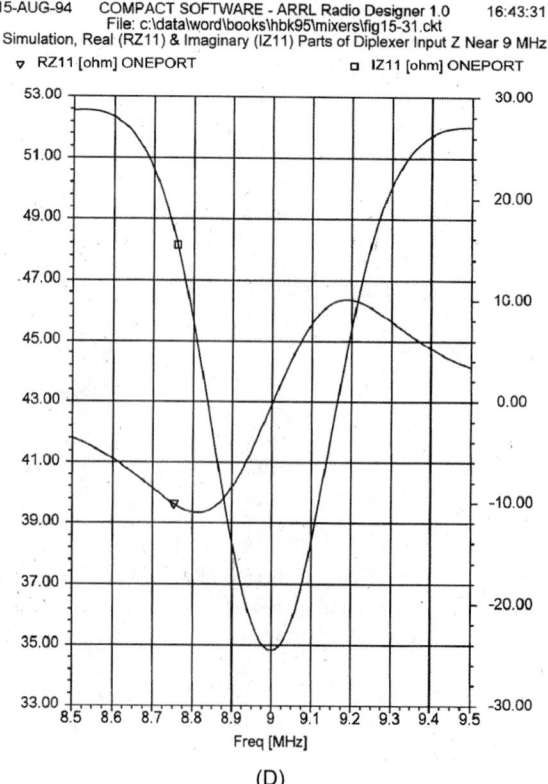

(C)

(D)

junction with carrier and audio phasing, can be used to generate SSB, suppressed-carrier AM. Likewise, two DBMs can be used with RF and LO phasing as an image-reject mixer.

PHASE DETECTION WITH A DBM

As we saw in our exploration of quadrature detection, applying two signals of equal frequency to a DBM's LO and RF ports produces an IF-port dc output proportional to the cosine of the signals' phase difference (**Fig 10.31**). This assumes that the DBM has a dc-coupled IF port, of course. If it doesn't — and some DBMs don't — phase-detector operation is out. Any dc output offset introduces error

into this process, so critical phase-detection applications use low-offset DBMs optimized for this service.

BIPHASE-SHIFT KEYING MODULATION WITH A DBM

Back in our discussion of square-wave mixing, we saw how multiplying a switching mixer's linear input with a square wave causes a 180° phase shift during the negative part of the square wave's cycle. As **Fig 10.32** shows, we can use this effect to produce *biphase-shift keying (BPSK)*, a digital system that conveys data by means of carrier phase reversals. A related system, *quadrature*

phase-shift keying (QPSK) uses two DBMs and phasing to convey data by phase-shifting a carrier in 90° increments.

10.5.4 Active Mixers — Transistors as Switching Elements

We've covered diode DBMs in depth because their home-buildability, high performance and suitability for direct connection into 50-Ω systems makes them attractive to Amateur Radio builders. The abundant availability of high-quality manufactured diode mixers at reasonable prices makes them

Fig 10.27 — All of the inductors and capacitors in this high-pass/low-pass diplexer (A) exhibit a reactance of 70.7 Ω at its tuned circuits' 3-dB cutoff frequency (the geometric mean of the IF and IF image). B and C show *ARRL Radio Designer* simulations of this circuit configured for use in a receiver that converts 7 MHz to 3.984 MHz using a 10.984-MHz LO. The IF image is at 17.984 MHz, giving a 3-dB cutoff frequency of 8.465 MHz. The inductor values used in the simulation were therefore 1.33 μH (Q = 200 at 25.2 MHz); the capacitors, 265 pF (Q = 1000). This drawing shows idler load and "50-Ω Amplifier" connections suitable for a receiver in which the IF image falls at a frequency *above* the desired IF. For applications in which the IF image falls below the desired IF, interchange the 50-Ω idler load resistor and the diplexer's "50-Ω Amplifier" connection so the idler load terminates the diplexer low-pass filter and the 50-Ω amplifier terminates the high-pass filter.

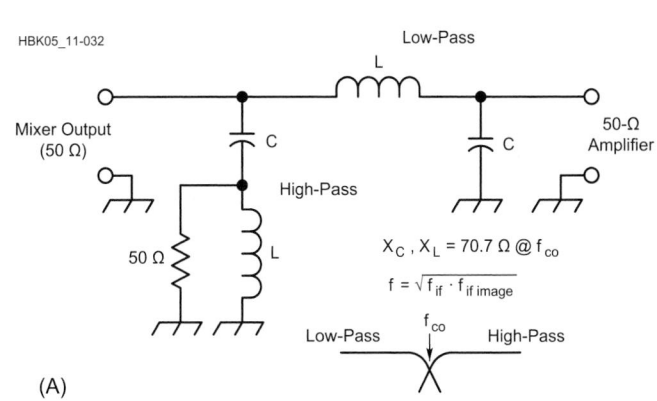

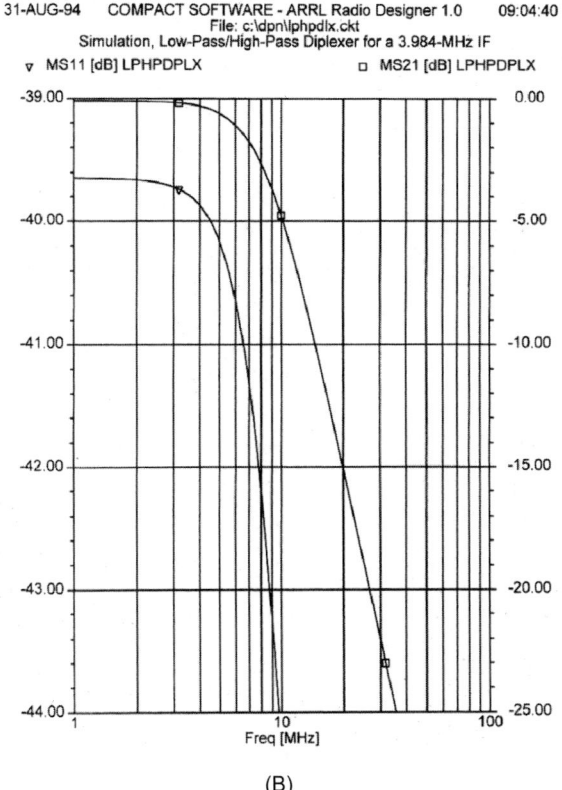

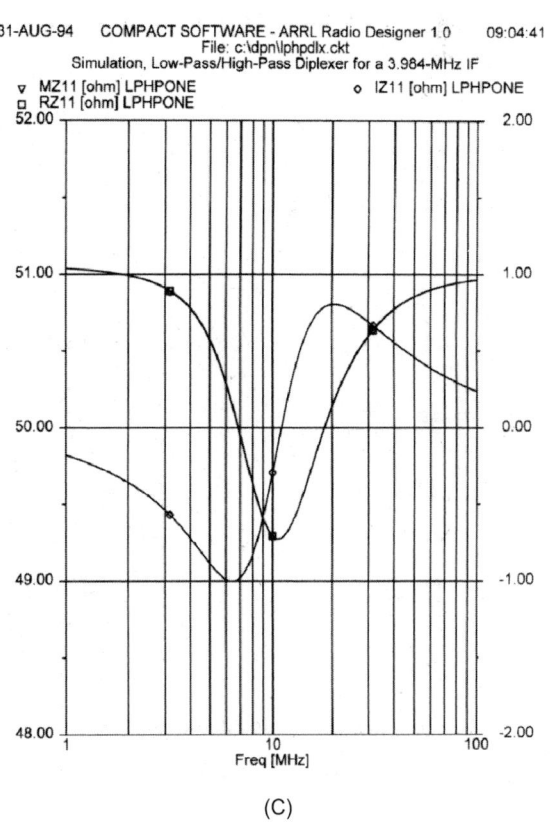

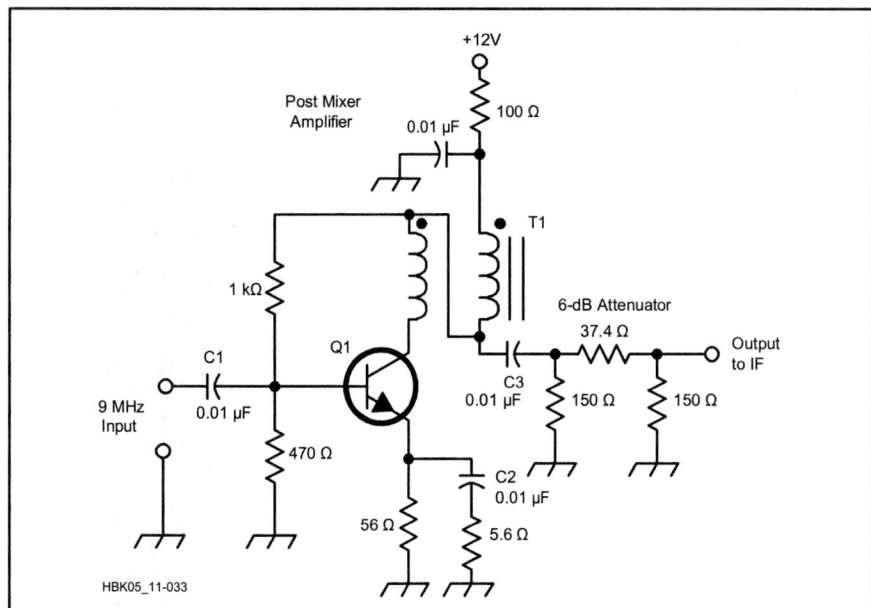

Fig 10.28 — The post-mixer amplifier from Hayward and Lawson's Progressive Communications Receiver (November 1981 *QST*). This amplifier's gain, including the 6-dB loss of the attenuator pad, is about 16 dB; its noise figure, 4 to 5 dB; its output intercept, 30 dBm. The 6-dB attenuator is essential if a crystal filter follows the amplifier; the pad isolates the amplifier from the filter's highly reactive input impedance. This circuit's input match to 50 Ω below 4 MHz can be improved by replacing 0.01-µF capacitors C1, C2 and C3 with low-inductance 0.1-µF units (chip capacitors are preferable). Q1 — TO-39 CATV-type bipolar transistor, f_T = 1 GHz or greater. 2N3866, 2N5109, 2SC1252, 2SC1365 or MRF586 suitable. Use a small heat sink on this transistor. T1 — Broadband ferrite transformer, ≈42 µH per winding: 10 bifilar turns of #28 enameled wire on an FT 37-43 core.

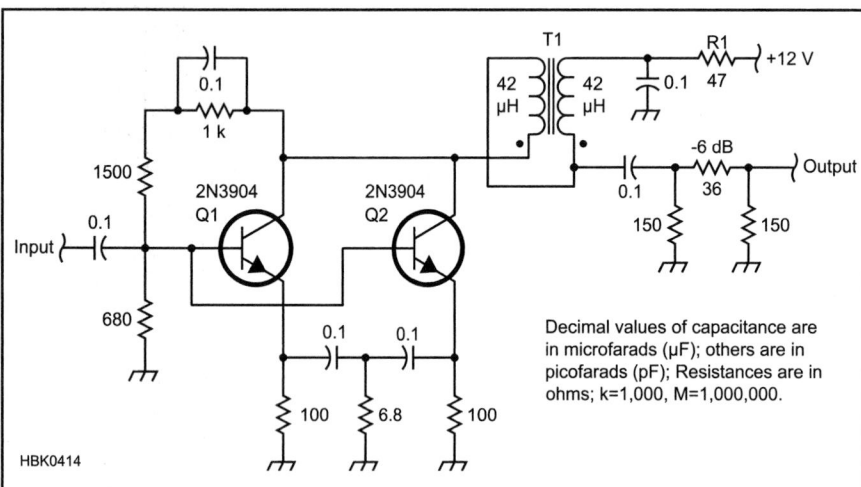

Fig 10.29 — At MF and HF, paralleled 2N3904 BJTs can provide performance comparable to that of the Fig 10.28 circuit with sufficient attention paid to device standing current, here set at ≈30 mA for the pair. The value of decoupling resistor R1 is critical in that small changes in its value cause a relatively large change in the 2N3904s' bias point. This circuit is part of the EZ-90 Receiver, described by Hayward, Campbell and Larkin in *Experimental Methods in RF Design*.

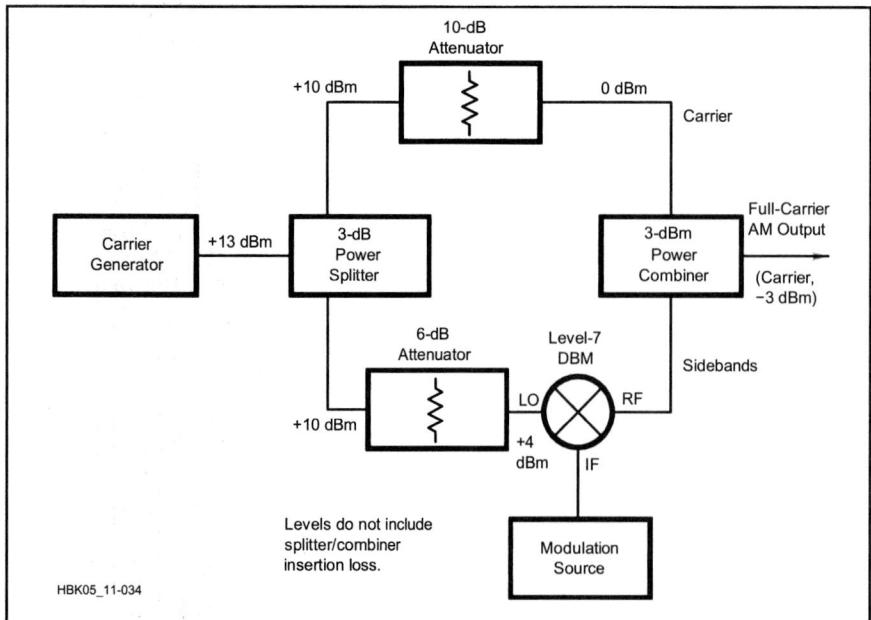

Fig 10.30 — Generating full-carrier AM with a diode DBM. A practical modulator using this technique is described in *Experimental Methods in RF Design*.

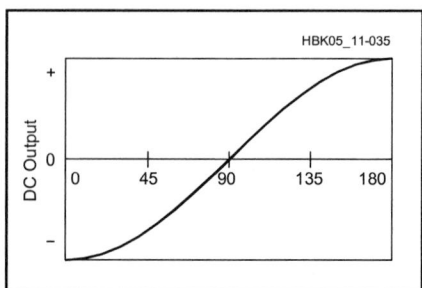

Fig 10.31 — A phase detector's dc output is the cosine of the phase difference between its input and reference signals.

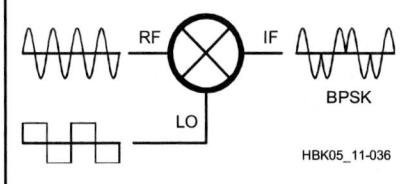

Fig 10.32 — Mixing a carrier with a square wave generates biphase-shift keying (BPSK), in which the carrier phase is shifted 180° for data transmission. In practice, as in this drawing, the carrier and data signals are phase-coherent so the mixer switches only at carrier zero crossings.

excellent candidates for home construction projects. Although diode DBMs are common in telecommunications as a whole, their conversion loss and relatively high LO power requirement have usually driven the manufacturers of high-performance MF/HF Amateur Radio receivers and transceivers to other solutions. Those solutions have generally involved single- or double-balanced FET mixers — MOSFETs in the late 1970s and early 1980s, JFETs from the early 1980s to date. A comprehensive paper that explores the differences between various forms of active mixers, "Performance Capabilities of Active Mixers," by Ulrich Rohde, N1UL, is included on the CD-ROM accompanying this *Handbook*.

Many of the JFET designs are variations of a single-balanced mixer circuit introduced to *QST* readers in 1970! **Fig 10.33** shows the circuit as it was presented by William Sabin in "The Solid-State Receiver," *QST*, July 1970. Two 2N4416 JFETs operate in a common-source configuration, with push-pull RF input and parallel LO drive. **Fig 10.34** shows a similar circuit as implemented in the ICOM IC-765 transceiver. In this version, the JFETs (2SK125s) operate in common-gate, with the LO applied across a 220-Ω resistor between the gates and ground.

Current state of the art for active mixers in the HF through GHz range replaces discrete device designs with integrated designs such as the Analog Devices AD8342 (**www.analog.com**). Using an IC greatly improves matching of the active devices, improving circuit balance. The AD8342 has a conversion gain of 3.7 dB, a noise figure of 12.2 dB, and an input IP3 of 22.7 dBm. The device operates with a single-voltage power supply and is well-suited to interface with digital hardware, such as for SDR applications. Reference circuits for applications at HF and VHF/UHF are provided in the device's datasheet.

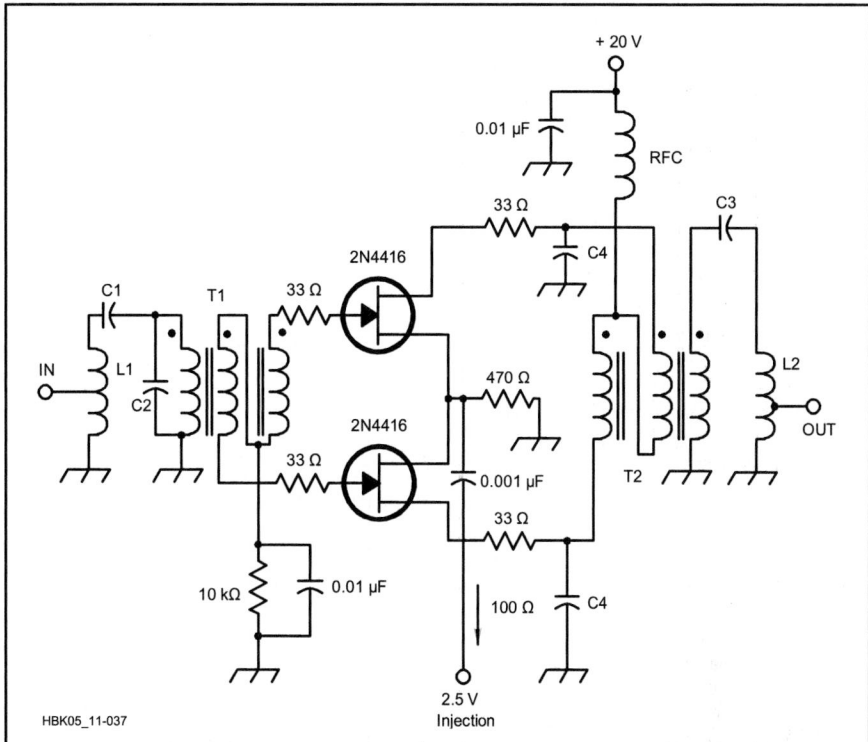

Fig 10.33 — Two 2N4416 JFETs provide high dynamic range in this mixer circuit from Sabin, *QST*, July 1970. L1, C1 and C2 form the input tuned circuit; L2, C3 and C4 tune the mixer output to the IF. The trifilar input and output transformers are broadband transmission-line types.

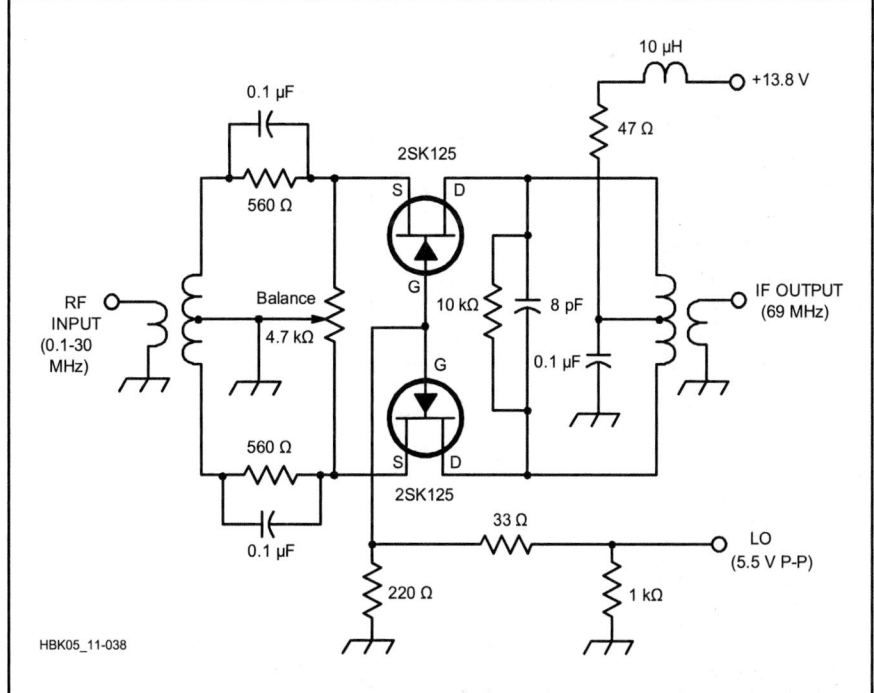

Fig 10.34 — The ICOM IC-765's single-balanced 2SK125 mixer achieves a high dynamic range (per *QST* Product Review, an IP₃ of 10.5 dBm at 14 MHz with preamp off). The first receive mixer in many commercial Amateur Radio transceiver designs of the 1980s and 1990s used a pair of 2SK125s or similar JFETs in much this way.

10.5.5 The Tayloe Mixer

[The following description of the Tayloe Product Detector (a.k.a. — the Tayloe Mixer) is adapted from the July 2002 QEX article, "Software-Defined Radio For the Masses, Part 1" by Gerald Youngblood AC5OG. — Ed.]

The beauty of the Tayloe detector (see references by Tayloe) is found in both its design elegance and its exceptional performance. In its simplest form, you can build a complete quadrature downconverter with only three or four ICs (less the local oscillator) at a cost of less than $10.

Fig 10.35 illustrates a single-balanced version of the Tayloe detector. It can be visualized as a four-position rotary switch revolving at a rate equal to the carrier frequency. The 50-Ω antenna impedance is connected to the rotor and each of the four switch positions is connected to a *sampling capacitor*. Since the switch rotor is turning at exactly the RF carrier frequency, each capacitor will track the carrier's amplitude for exactly one-quarter of the cycle and will then hold its value for the remainder of the cycle. The rotating switch will therefore sample the signal at 0°, 90°, 180° and 270°, respectively.

As shown in **Fig 10.36**, the 50-Ω impedance of the antenna and the sampling capacitors form an R-C low-pass filter during the period when each respective switch is turned on. Therefore, each sample represents the integral or average voltage of the signal during its respective one-quarter cycle. When the switch is off, each sampling capacitor will hold its value until the next revolution. If the RF carrier and the rotating frequency were exactly in phase, the output of each capacitor will be a dc level equal to the average value of the sample.

If we differentially sum outputs of the 0° and 180° sampling capacitors with an op amp (see Fig 10.35), the output would be a dc voltage equal to two times the value of the individually sampled values when the switch rotation frequency equals the carrier frequency. Imagine, 6 dB of noise-free gain! The same would be true for the 90° and 270° capacitors as well. The 0°/180° summation forms the *I* channel and the 90°/270° summation forms the *Q* channel of a quadrature downconversion. (See the **Modulation** chapter for more information on I/Q modulation.)

As we shift the frequency of the carrier away from the sampling frequency, the values of the inverting phases will no longer be dc levels. The output frequency will vary according to the "beat" or difference frequency between the carrier and the switch-rotation frequency to provide an accurate representation of all the signal components converted to baseband.

Fig 10.37 is the schematic for a simple, single-balanced Tayloe detector. It consists of a PI5V331, 1:4 FET demultiplexer (an analog switch) that switches the signal to each of the four sampling capacitors. The 74AC74 dual flip-flop is connected as a divide-by-four Johnson counter to provide the two-phase clock to the demultiplexer chip. The outputs of the sampling capacitors are differentially summed through the two LT1115 ultra-low-noise op amps to form the *I* and *Q* outputs, respectively.

Note that the impedance of the antenna forms the input resistance for the op-amp gain as shown in Equation 17. This impedance may vary significantly with the actual antenna. In a practical receiver, a buffer amplifier should be used to stabilize and control the impedance presented to the mixer.

Since the duty cycle of each switch is 25%, the effective resistance in the RC network is the antenna impedance multiplied by four in the op-amp gain formula:

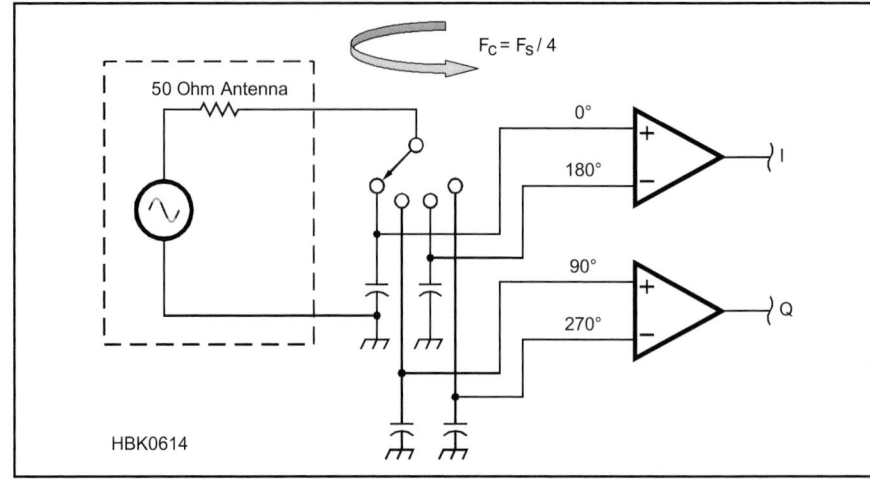

Fig 10.35 — Tayloe detector: The switch rotates at the carrier frequency so that each capacitor samples the signal once each revolution. The 0° and 180° capacitors differentially sum to provide the in-phase (I) signal and the 90° and 270° capacitors sum to provide the quadrature (Q) signal.

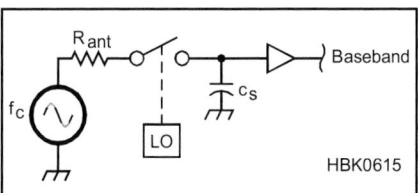

Fig 10.36 — Track-and-hold sampling circuit: Each of the four sampling capacitors in the Tayloe detector form an RC track-and-hold circuit. When the switch is on, the capacitor will charge to the average value of the carrier during its respective one-quarter cycle. During the remaining three-quarters cycle, it will hold its charge. The local-oscillator frequency is equal to the carrier frequency so that the output will be at baseband.

$$G = \frac{R_f}{4R_{ant}} \tag{17}$$

For example, with a feedback resistance, R_f, of 3.3 kΩ and antenna impedance, R_{ant}, of 50 Ω, the resulting gain of the input stage is:

$$G = \frac{3300}{4 \times 50} = 16.5$$

The Tayloe detector may also be analyzed as a *digital commutating filter* (see reference by Kossor). This means that it operates as a very-high-Q tracking filter, where Equation 18 determines the bandwidth and n is the number of sampling capacitors, R_{ant} is the antenna impedance and C_s is the value of the individual sampling capacitors. Equation 19 determines the Q_{det} of the filter, where f_c is the center frequency and BW_{det} is the bandwidth of the filter.

$$BW_{det} = \frac{1}{\pi \, n \, R_{ant} \, C_s} \tag{18}$$

$$Q_{det} = \frac{f_c}{BW_{det}} \tag{19}$$

By example, if we assume the sampling capacitor to be 0.27 μF and the antenna impedance to be 50 Ω, then BW and Q at an operating frequency of 14.001 MHz are computed as follows:

$$BW_{det} = \frac{1}{\pi \times 4 \times 50 \times (2.7 \times 10^{-7})} = 5895 \text{ Hz}$$

$$Q_{det} = \frac{14.001 \times 10^{-6}}{5895} = 2375 \tag{20}$$

The real payoff in the Tayloe detector is its performance. It has been stated that the

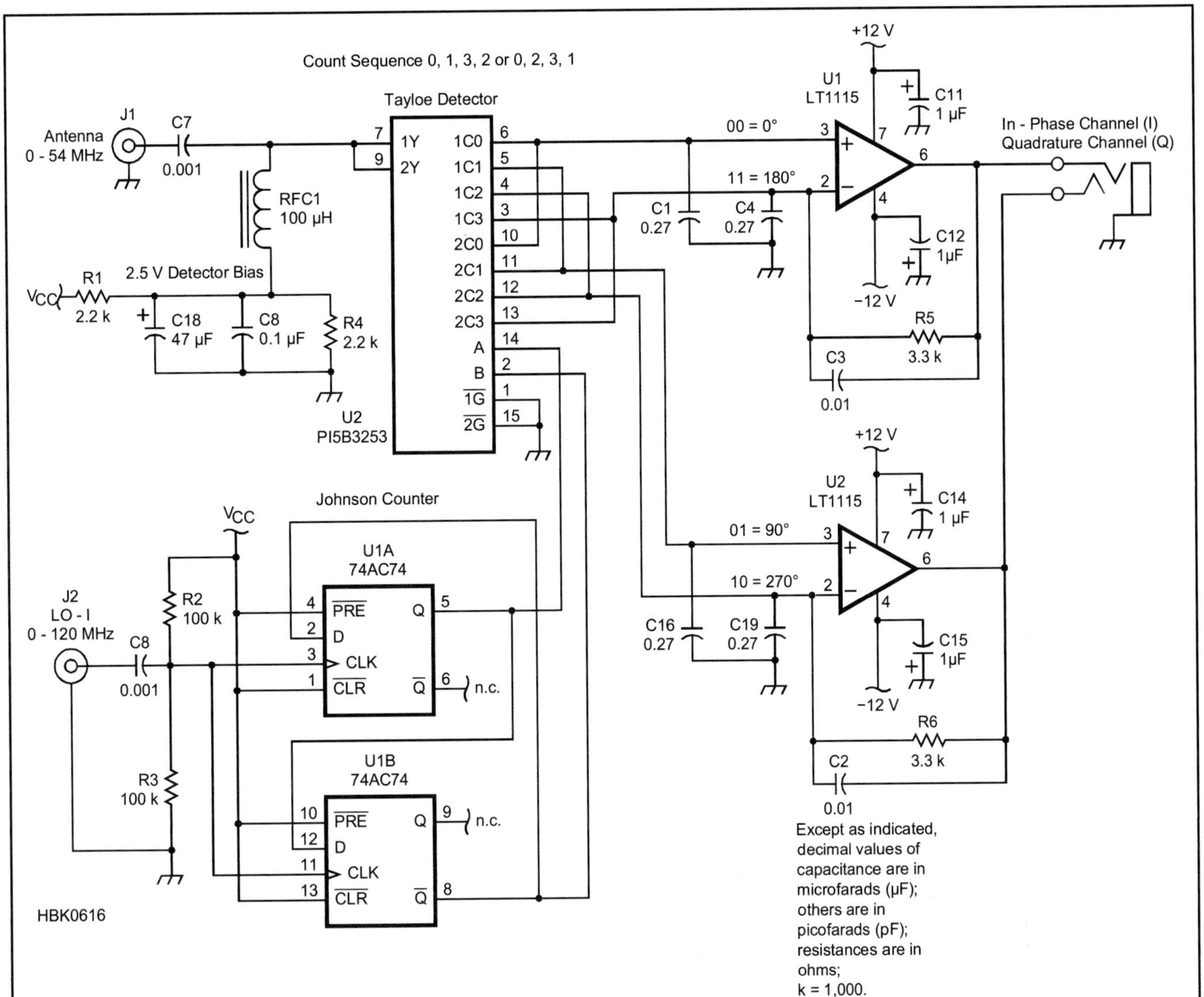

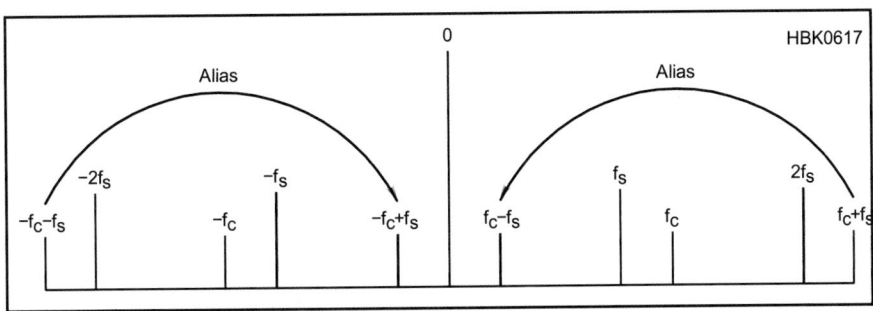

Fig 10.37 — Single balanced Tayloe detector.

Count Sequence 0, 1, 3, 2 or 0, 2, 3, 1

Tayloe Detector

Antenna 0 - 54 MHz

J1
C7 0.001

RFC1 100 µH

R1 2.2 k
2.5 V Detector Bias
V$_{CC}$
C18 47 µF
C8 0.1 µF
R4 2.2 k

U2 PI5B3253

1Y 7
2Y 9
1C0 6
1C1 5
1C2 4
1C3 3
2C0 10
2C1 11
2C2 12
2C3 13
A 14
B 2
1G 1
2G 15

Johnson Counter

V$_{CC}$

U1A 74AC74
PRE 4
D 2
CLK 3
CLR 1
Q 5
Q̄ 6) n.c.

J2 LO - I 0 - 120 MHz
C8 0.001
R2 100 k

R3 100 k

U1B 74AC74
PRE 10
D 12
CLK 11
CLR 13
Q 9) n.c.
Q̄ 8

HBK0616

+12 V
U1 LT1115
C11 1 µF
00 = 0°
C1 0.27
C4 0.27
11 = 180°
3 +
2 −
7
6
4
−12 V
C12 1µF
In - Phase Channel (I)
Quadrature Channel (Q)
R5 3.3 k
C3 0.01

+12 V
U2 LT1115
C14 1 µF
01 = 90°
C16 0.27
C19 0.27
10 = 270°
3 +
2 −
7
6
4
−12 V
C15 1µF
R6 3.3 k
C2 0.01

Except as indicated, decimal values of capacitance are in microfarads (µF); others are in picofarads (pF); resistances are in ohms; k = 1,000. n.c. = No connection

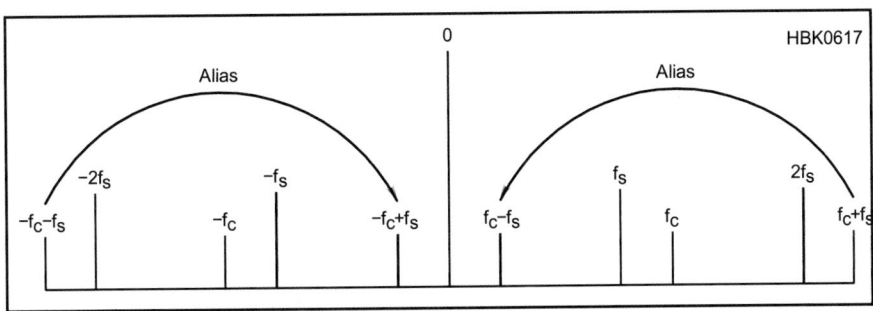

HBK0617

Alias
−2f$_s$
−f$_s$
0
Alias
f$_s$
2f$_s$
−f$_c$−f$_s$
−f$_c$
−f$_c$+f$_s$
f$_c$−f$_s$
f$_c$
f$_c$+f$_s$

Fig 10.38 — Alias summing on Tayloe detector output: Since the Tayloe detector samples the signal, the sum frequency $(f_c + f_s)$ and its image $(-f_c - f_s)$ are located at the first alias frequency. The alias signals sum with the baseband signals to eliminate the mixing product loss associated with traditional mixers. In a typical mixer, the sum frequency energy is lost through filtering thereby increasing the noise figure of the device.

ideal commutating mixer has a minimum conversion loss (which equates to noise figure — see the **RF Techniques** chapter) of 3.9 dB. Typical high-level diode mixers have a conversion loss of 6-7 dB and noise figures 1 dB higher than the loss. The Tayloe detector has less than 1 dB of conversion loss, remarkably. How can this be? The reason is that it is not really a mixer but a sampling detector in the form of a quadrature track-and-hold

circuit. This means that the design adheres to discrete-time sampling theory, which, while similar to mixing, has its own unique characteristics. Because a track and hold actually holds the signal value between samples, the signal output never goes to zero. (See the **DSP and Software Radio Design** chapter for more on sampling theory.)

This is where aliasing can actually be used to our benefit. Since each switch and capaci-

tor in the Tayloe detector actually samples the RF signal once each cycle, it will respond to alias frequencies as well as those within the Nyquist frequency range. In a traditional direct-conversion receiver, the local-oscillator frequency is set to the carrier frequency so that the difference frequency, or IF, is at 0 Hz and the sum frequency is at two times the carrier frequency. We normally remove the sum frequency through low-pass filter-

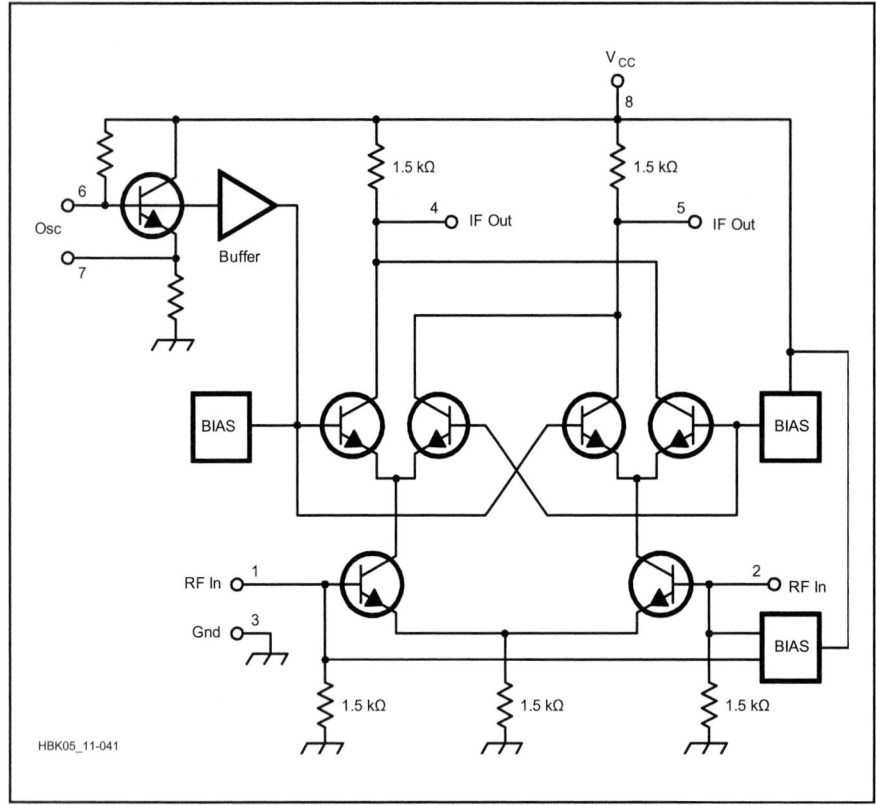

Fig 10.39 — The SA602/612's equivalent circuit reveals its Gilbert-cell heritage.

ing, resulting in conversion loss and a corresponding increase in noise figure. In the Tayloe detector, the sum frequency resides at the first alias frequency as shown in **Fig 10.38**. Remember that an alias is a real signal and will appear in the output as if it were a baseband signal. Therefore, the alias adds to the baseband signal for a theoretically lossless detector. In real life, there is a slight loss, usually less than 1 dB, due to the resistance of the switch and aperture loss due to imperfect switching times.

10.5.6 The NE602/SA602/ SA612: A Popular Gilbert Cell Mixer

Introduced as the Philips NE602 in the mid-1980s, the NXP SA602/SA612 mixer-oscillator IC has become greatly popular with amateur experimenters for receive mixers, transmit mixers and balanced modulators. The SA602/612's mixer is a *Gilbert cell* multiplier. **Fig 10.39** shows its equivalent circuit. A Gilbert cell consists of two differential transistor pairs whose bias current is controlled by one of the input signals. The other signal drives the differential pairs' bases, but only after being "predistorted" in a diode circuit. (This circuit distorts the signal equally and oppositely

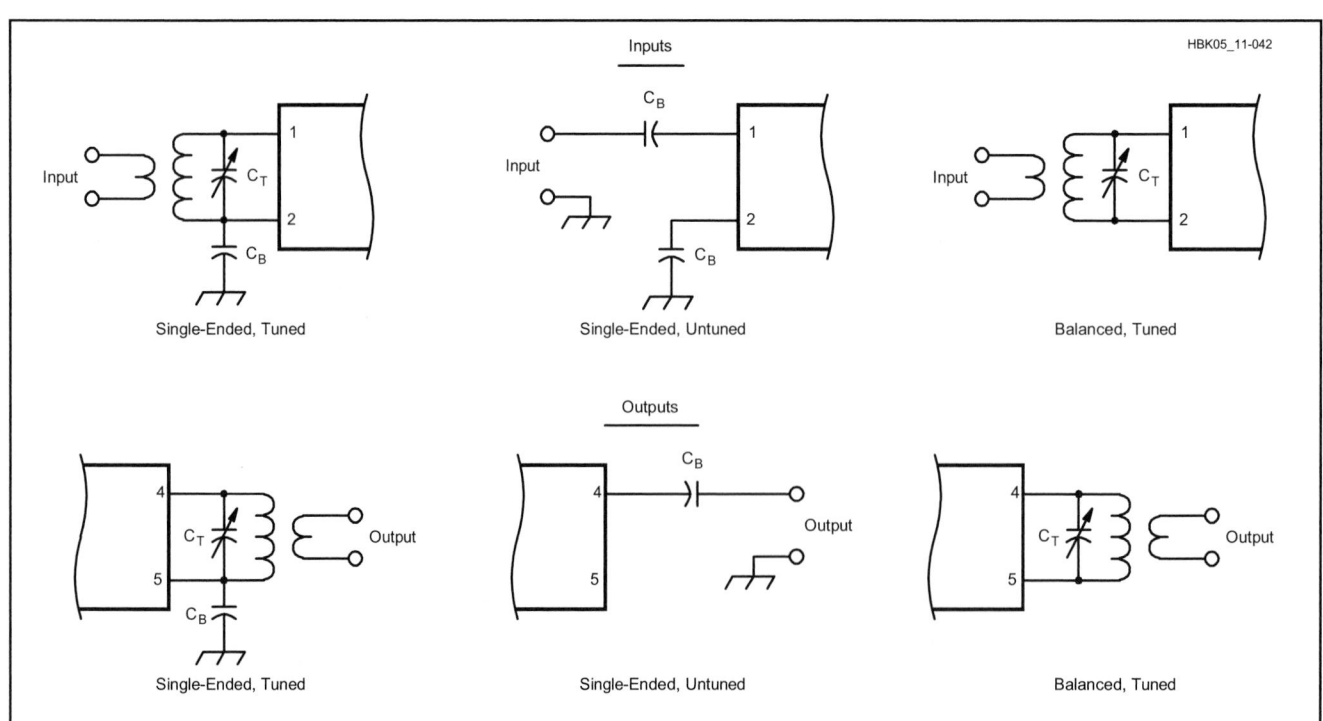

Fig 10.40 — The SA602/612's inputs and outputs can be single- or double-ended (balanced). The balanced configurations minimize second-order IMD and harmonic distortion, and unwanted envelope detection in direct-conversion service. C_T tunes its inductor to resonance; C_B is a bypass or dc-blocking capacitor. The arrangements pictured don't show all the possible input/output configurations; for instance, a center-tapped broadband transformer can be used to achieve a balanced, untuned input or output.

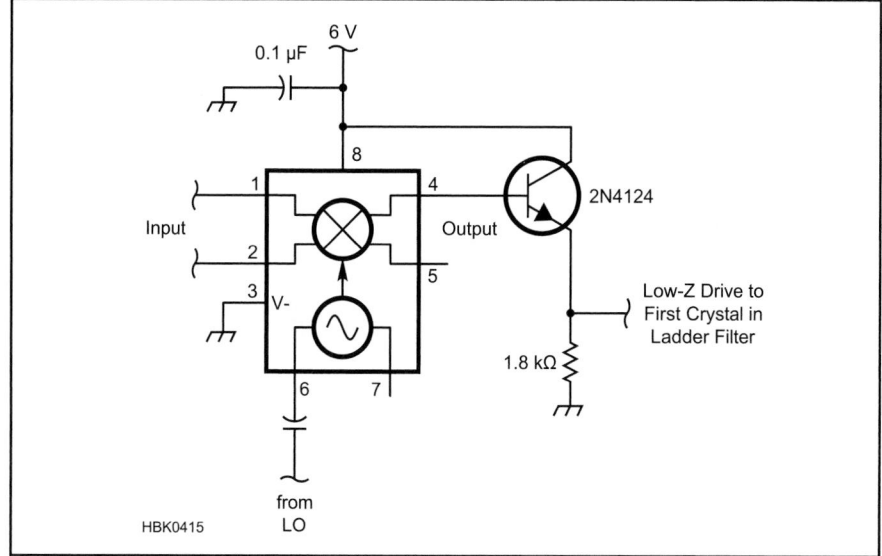

Fig 10.41 — An NPN transistor at the output of an SA602/612 mixer provides power gain and low-impedance drive for a 4.914-MHz crystal filter. A low-reactance coupling capacitor can be added between the emitter and the circuitry it drives if dc blocking is necessary. (Circuit from the Elecraft KX1 transceiver courtesy of Wayne Burdick, N6KR)

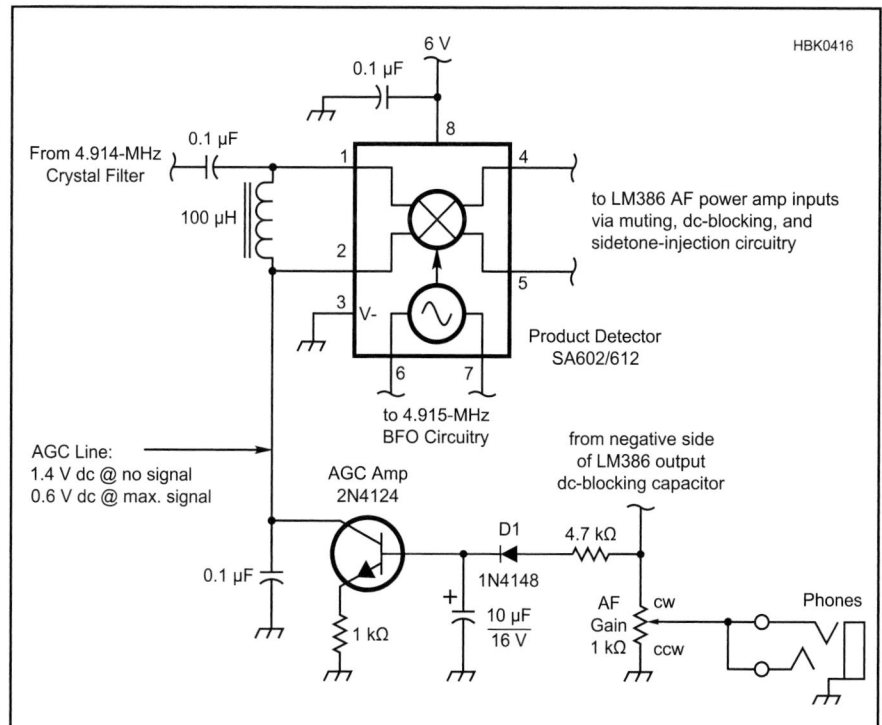

Fig 10.42 — SA602/612 product detector AGC from the Elecraft KX1 transceiver. Designed by Wayne Burdick, N6KR, this circuit first appeared in the Wilderness Radio SST transceiver with an LED used at D1 for simultaneous signal indication and rectification. The selectivity provided by the crystal filter preceding the detector works to mitigate the effects of increasing detector distortion with gain reduction, (Circuit courtesy of Wayne Burdick, N6KR, and Bob Dyer, K6KK)

to the inherent distortion of the differential pair.) The resulting output signal is an accurate multiplication of the input voltages.

SA602/612A VARIANTS

The SA602/612 began life as the NE602/SA602. SA-prefixed 602/612 parts are specified for use over a wider temperature range than their NE-prefixed equivalents. Parts without the A suffix have a slightly lower IP$_3$ specification than their A counterparts. The pinout-identical NE612A and SA612A cost less than their 602 equivalents as a result of wider tolerances. All variants of this popular part should work satisfactorily in most "NE602" experimenter projects. The same mixer/oscillator topology, modified for slightly higher dynamic range at the expense of somewhat less mixer gain, is also available in the mixer/oscillator/FM IF chips NE/SA605 (input IP$_3$ typically –10 dBm) and NE/SA615 (input IP$_3$ typically –13 dBm).

SA602/612 USAGE NOTES

The SA602/612's typical current drain is 2.4 mA; its supply voltage range is 4.5 to 8.0 V. Its inputs (RF) and outputs (IF) can be single- or double-ended (balanced) according to design requirements (**Fig 10.40**). The equivalent ac impedance of each input is approximately 1.5 kΩ in parallel with 3 pF; each output's resistance is 1.5 kΩ. **Fig 10.41** shows the use of an NPN transistor at the SA602/612 output to obtain low-impedance drive for a crystal filter; **Fig 10.42** shows how AGC can be applied to an SA602/612.

The SA602/612 mixer can typically handle signals up to 500 MHz. At 45 MHz, its noise figure is typically 5.0 dB; its typical conversion gain, 18 dB. Note that in contrast to the diode-based mixers described earlier, which have conversion *loss*, most Gilbert-cell mixers have conversion *gain*. Considering the SA602/612's low current drain, its input IP$_3$ (measured at 45 MHz with 60-kHz spacing) is usefully good at –15 dBm. Factoring in the mixer's conversion gain results in an equivalent output IP$_3$ of about 3 dBm.

The SA602/612's on-board oscillator can operate up to 200 MHz in LC and crystal-controlled configurations (**Fig 10.43** shows three possibilities). Alternatively, energy from an external LO can be applied to the chip's pin 6 via a dc blocking capacitor. At least 200 mV P-P of external LO drive is required for proper mixer operation.

The SA602/612 was intended to be used as the second mixer in double-conversion FM cellular radios, in which the first IF is typically 45 MHz, and the second IF is typically 455 kHz. Such a receiver's second mixer can be relatively weak in terms of dynamic range because of the adjacent-signal protection afforded by the high selectivity of the first-IF filter preceding it. When

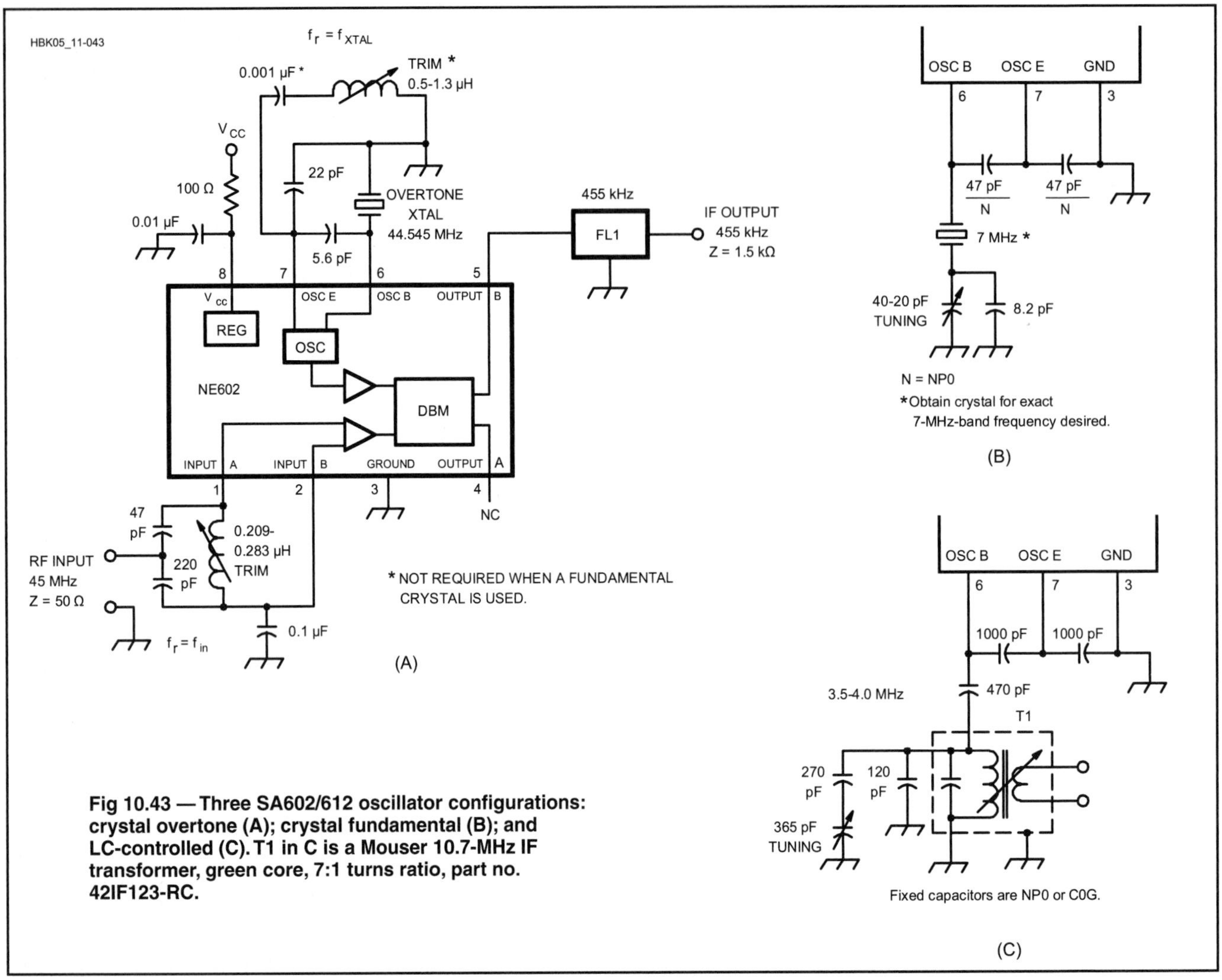

Fig 10.43 — Three SA602/612 oscillator configurations: crystal overtone (A); crystal fundamental (B); and LC-controlled (C). T1 in C is a Mouser 10.7-MHz IF transformer, green core, 7:1 turns ratio, part no. 42IF123-RC.

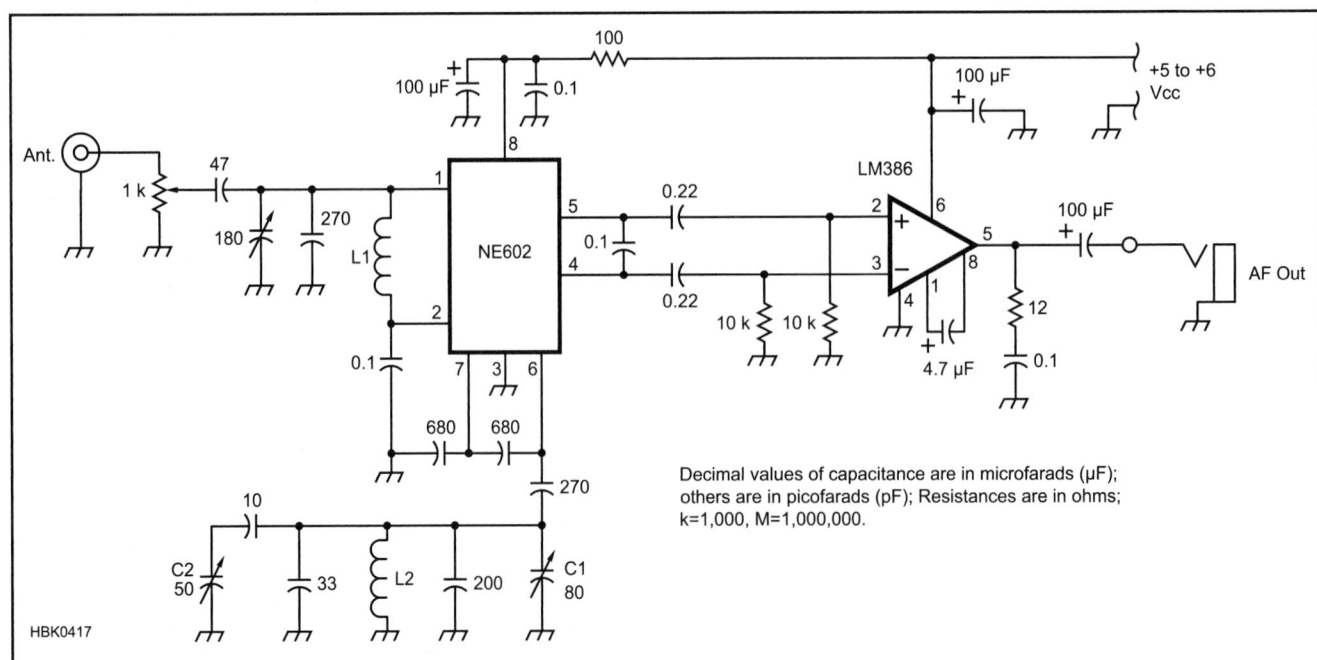

Fig 10.44 — A 7-MHz direct-conversion receiver based on the NE602/SA602/612. Equipped with a stage or two of audio filtering and a means of muting during transmit periods, such a receiver is entirely sufficient for basic Amateur Radio communication at MF and HF. L1 and L2 are 1.2 µH. This receiver is described in greater detail in *Experimental Methods in RF Design*.

HBK0417

Decimal values of capacitance are in microfarads (µF); others are in picofarads (pF); Resistances are in ohms; k=1,000, M=1,000,000.

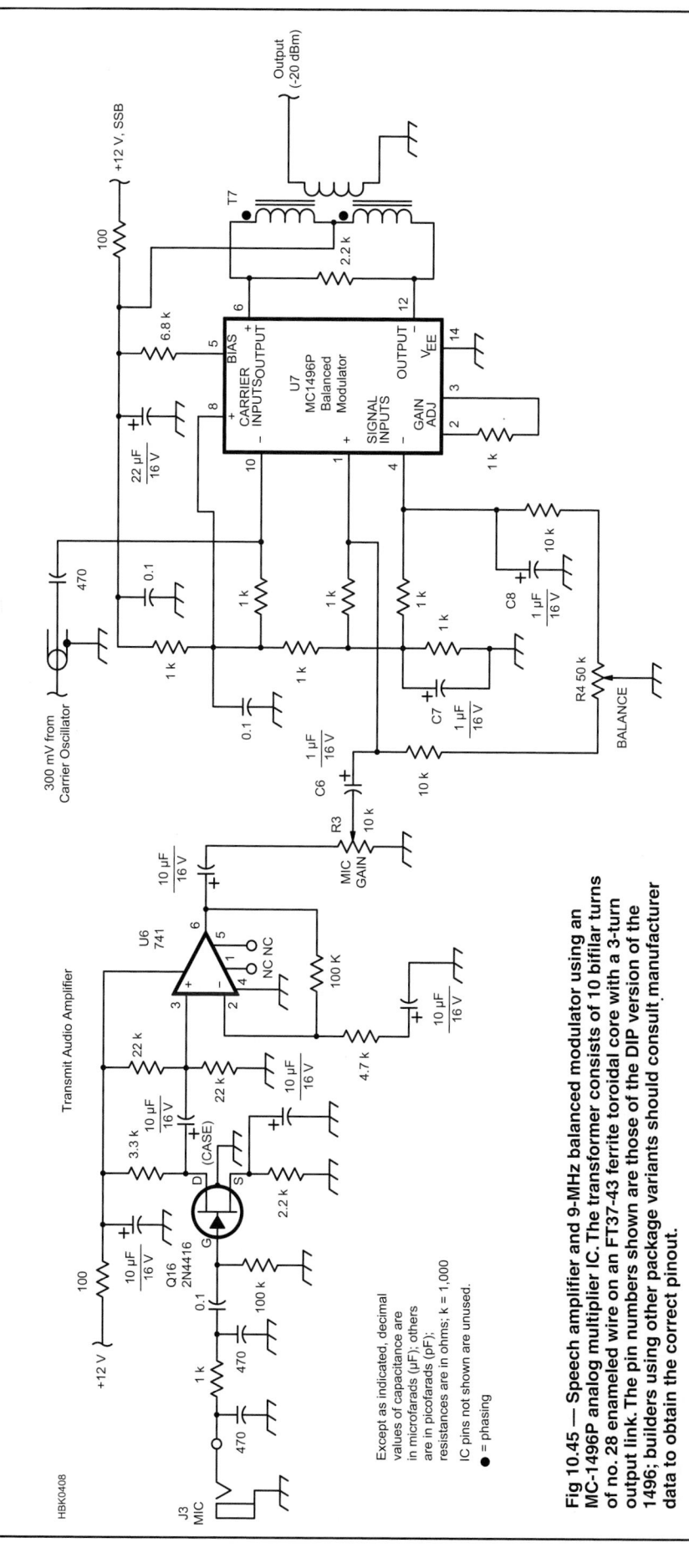

Fig 10.45 — Speech amplifier and 9-MHz balanced modulator using an MC-1496P analog multiplier IC. The transformer consists of 10 bifilar turns of no. 28 enameled wire on an FT37-43 ferrite toroidal core with a 3-turn output link. The pin numbers shown are those of the DIP version of the 1496; builders using other package variants should consult manufacturer data to obtain the correct pinout.

used as a first mixer, the SA602/612 can provide a two-tone third-order dynamic range between 80 and 90 dB, but this figure is greatly diminished if a preamplifier is used ahead of the SA602/612 to improve the system's noise figure.

When the SA602/612 is used as a second mixer, the sum of the gains preceding it should not exceed about 10 dB. An SA602/612 can serve as low-distortion (THD <1%) product detector if overload is avoided through the use of AGC and appropriate attenuation between the '602/612 and the IF strip that drives it.

The SA602/612 is generally not a good choice for VHF and higher-frequency mixers because of its input noise and diminishing IMD performance at high frequencies. There are applications, however, where 6-dB noise figure and 60- to 70-dB dynamic range performance is adequate. If your target specifications exceed these numbers, you should consider other mixers at VHF and up.

Fig 10.44 shows the schematic of a complete 7-MHz direct-conversion receiver based on the SA602/612 and the widely used LM386 AF power amplifier IC. Such simple product-detector-based receivers sometimes suffer from incidental envelope detection, which causes audio from strong, full-carrier-AM shortwave or mediumwave broadcast stations to be audible regardless of where the receiver LO is tuned. RF attenuation and/or band-limiting the receiver input with a double- or triple-tuned-circuit filter can usually reduce this effect to inaudibility.

10.5.7 An MC1496P Balanced Modulator

Although it predates the SA602/612, Freescale's MC1496 Gilbert cell multiplier remains a viable option for product detection and balanced modulator service. **Fig 10.45** shows an MC1496-based balanced modulator that is capable of a carrier suppression greater than 50 dB. Per its description in Hayward, Campbell, and Larkin's *Experimental Methods in RF Design*, its output with audio drive should be kept to about –20 dBm with this circuit. LO drive should be 200 to 500 mV P-P.

10.5.8 An Experimental High-Performance Mixer

Examining the state of the art we find that the best receive-mixer dynamic ranges are achieved with quads of RF MOSFETs operating as passive switches, with no drain voltage applied. The best of these techniques involves following a receiver's first mixer with a diplexer and low-loss roofing crystal filter, rather than terminating the mixer in a strong wideband amplifier.

Colin Horrabin's (G3SBI) experimenta-

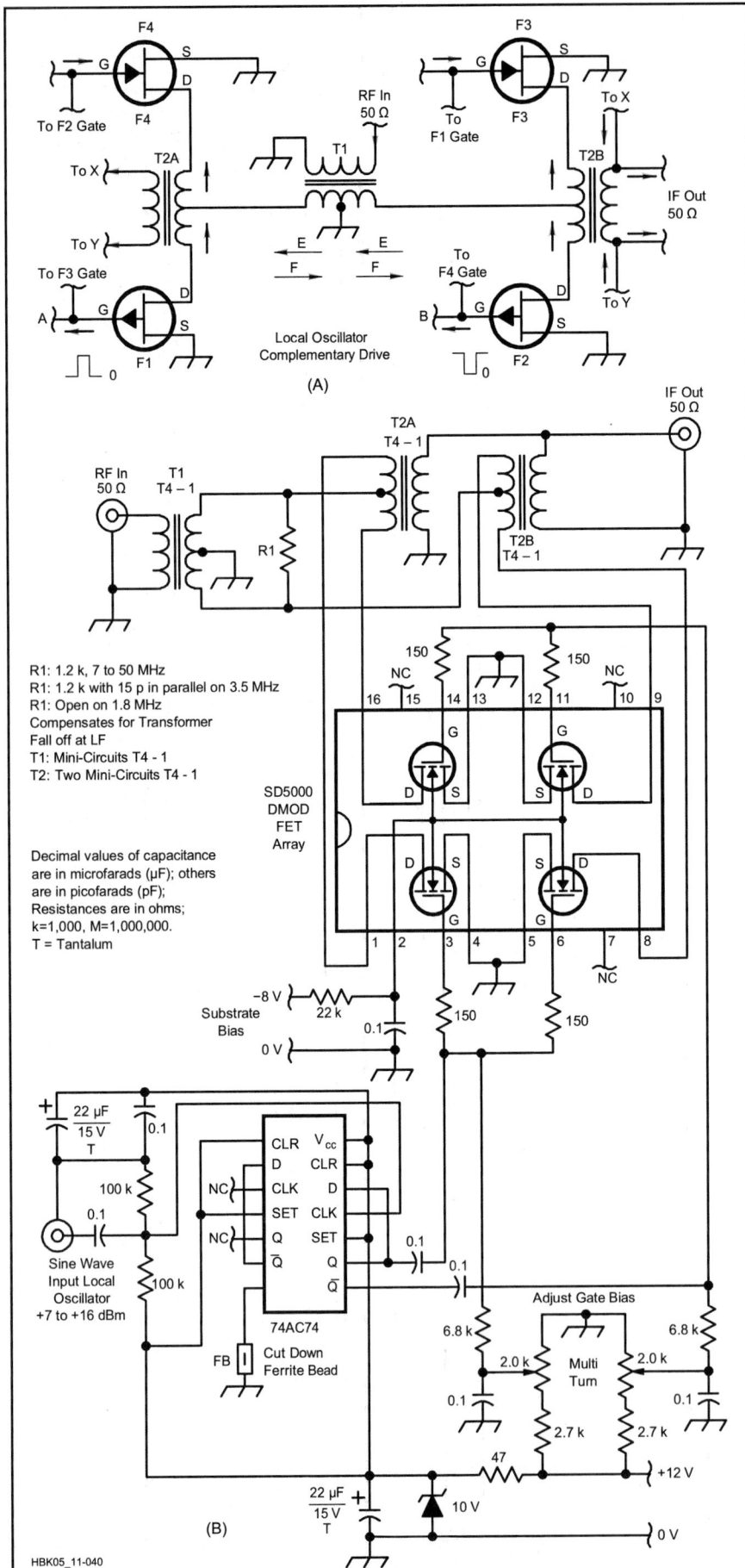

Fig 10.46 — A shows the operation of the H-mode switching mixer developed by Colin Horrabin, G3SBI. B shows the actual mixer circuit implemented with the Linear Integrated Systems SD5000 DMOS FET quad switch IC and a 74AC74 flip-flop.

R1: 1.2 k, 7 to 50 MHz
R1: 1.2 k with 15 p in parallel on 3.5 MHz
R1: Open on 1.8 MHz
Compensates for Transformer
Fall off at LF
T1: Mini-Circuits T4 - 1
T2: Two Mini-Circuits T4 - 1

Decimal values of capacitance
are in microfarads (µF); others
are in picofarads (pF);
Resistances are in ohms;
k=1,000, M=1,000,000.
T = Tantalum

tions with variations of an original high-performance mixer circuit by Jacob Mahkinson, N6NWP, led to the development of a new mixer configuration, called an H-mode mixer. This name comes from the signal path through the circuit. (See **Fig 10.46A**.) Horrabin is a professional scientist/engineer at the Science and Engineering Research Council's Darebury Laboratory, which has supported his investigative work on the H-mode switched-FET mixer, and consequently holds intellectual title to the new mixer. This does not prevent readers from taking the development further or using the information presented here.

Inputs A and B are complementary square-wave inputs derived from the sine-wave local oscillator at twice the required square-wave frequency. If A is ON, then FETs F1 and F3 are ON and F2 and F4 are OFF. The direction of the RF signal across T1 is given by the E arrows. When B is ON, FETs F2 and F4 are ON and F1 and F3 are OFF. The direction of the RF signal across T1 reverses, as shown by the F arrows.

This is still the action of a switching mixer, but now the source terminal of each FET switch is grounded, so that the RF signal switched by the FET cannot modulate the gate voltage. In this configuration the transformers are important: T1 is a Mini-Circuits type T4-1 and T2 is a pair of these same transformers with their primaries connected in parallel.

10.6 References and Bibliography

The following mixer references include books, journals and other publications as well as websites. They are arranged by publication date.

BOOKS

E. E. Bucher, *The Wireless Experimenter's Manual* (New York: Wireless Press, Inc., 1920).

F. E. Terman, *Radio Engineers' Handbook* (New York: McGraw-Hill, 1943).

R. W. Landee, D. C. Davis and A. P. Albrecht, *Electronic Designer's Handbook*, 2nd ed, Section 22 (New York: McGraw-Hill, 1957).

A. A. M. Saleh, *Theory of Resistive Mixers* (Cambridge, MA: MIT Press, 1971).

L.G. Giacoletto, *Electronics Designers' Handbook*, 2nd ed (New York: McGraw-Hill 1977).

P. Horowitz, W. Hill, *The Art of Electronics,* 2nd ed (New York: Cambridge University Press, 1989).

RF/IF Designer's Handbook (Brooklyn, NY: Scientific Components, 1992).

RF Communications Handbook (Philips Components-Signetics, 1992).

W. Hayward, *Introduction to Radio Frequency Design* (Newington, CT: ARRL, 1994).

L. E. Larson, *RF and Microwave Circuit Design for Wireless Communication*, (Artech House, 1996).

S. A. Maas, The RF and Microwave Circuit Design Cookbook (Artech House, 1998).

U. L Rohde, D. P. Newkirk, *RF/Microwave Circuit Design for Wireless Applications* (New York: John Wiley & Sons, 2000). (Large number of useful references in the mixer chapter.)

U. L. Rohde, J. Whitaker, *Communications Receivers: DSP, Software Radios, and Design*, 3rd ed (New York: McGraw-Hill, 2001. (Large number of useful references in the mixer chapter.)

H. I. Fujishiro, Y. Ogawa, T. Hamada and T. Kimura, *Electronic Letters*, Vol 37, Iss 7, Mar 2001, pp 435-436. See IET Digital Library (**www.ietdl.org**).

S. A. Maas, *Nonlinear Microwave and RF Circuits*, 2nd ed (Artech House, 2003).

S. A. Maas, Noise in Linear and Nonlinear Circuits (Artech House, 2005).

G. D. Vendelin, A. M. Pavio and U. L. Rohde, *Microwave Circuit Design Using Linear and Nonlinear Techniques* (New York: John Wiley & Sons, 2005). (Large number of useful references in the mixer chapter.)

Synergy Microwave Corporation Product Handbook (Paterson, NJ: Synergy Microwave Corporation, 2007).

JOURNALS AND OTHER PUBLICATIONS

R. Weinreich and R. W. Carroll, "Absorptive Filter for TV Harmonics," *QST*, Nov 1968, pp 20-25.

B. Gilbert, "A Precise Four-Quadrant Multiplier with Sub Nanosecond Response," *IEEE Journ Solid-State Circuits*, Vol 3, 1968, pp 364-373.

W. Sabin, "The Solid-State Receiver," *QST,* Jul 1970, pp 35-43.

G. Begemann, A. Hecht, "The Conversion Gain and Stability of MESFET Gate Mixers" 9th EMC, 1979, pp 316-319.

P. Will, "Broadband Double Balanced Mixer Having Improved Termination Insensitivity Characteristics", Adams-Russell Co, Waltham, MA, US Patent No 4,224,572, Sep 23, 1980.

C. Tsironis, R. Meierer, R. Stahlmann, "Dual-Gate MESFET Mixers," *IEEE Trans MTT*, Vol 32, No 3, 04-1984.

B. Henderson and J. Cook, "Image-Reject and Single-Sideband Mixers", Watkins-Johnson Tech Note, Vol 12 No 3, 1985.

H. Statz, P. Newman, I. Smith, R. Pucel and H. Haus, *IEEE Trans Electron Devices* 34 (1987) 160.

J. Dillon, "The Neophyte Receiver," *QST,* Feb 1988, pp 14-18.

B. Henderson, "Mixers in Microwave Systems (Part 1)", Watkins-Johnson Tech Note, Vol 17 No 1, 1990, Watkins-Johnson.

B. Henderson, "Mixers in Microwave Systems (Part 2)", Watkins-Johnson Tech Note, Vol 17 No 2, 1990, Watkins-Johnson.

R. Zavrel, "Using the NE602," Technical Correspondence, *QST*, May 1990, pp 38-39.

D. Kazdan "What's a Mixer?" *QST,* Aug 1992, pp 39-42.

L. Richey, "W1AW at the Flick of a Switch," *QST*, Feb 1993, pp 56-57.

J. Makhinson, "High Dynamic Range MF/HF Receiver Front End," *QST,* Feb 1993, pp 23-28. Also see Feedback, *QST,* Jun 1993, p 73.

J. Vermusvuori, "A Synchronous Detector for AM Transmisions," *QST*, pp 28-33. Uses SA602/612 mixers as phase-locked and quasi-synchronous product detectors.

P. Hawker, Ed., "Super-Linear HF Receiver Front Ends," Technical Topics, *Radio Communication,* Sep 1993, pp 54-56.

S. Joshi, "Taking the Mystery Out of Double-Balanced Mixers," *QST,* Dec 1993, pp 32-36.

U. L. Rohde, "Key Components of Modern Receiver Design," Part 1, *QST,* May 1994, pp 29-32; Part 2, *QST,* Jun 1994, pp 27-31; Part 3, *QST,* Jul 1994, 42-45.

U. L. Rohde, "Testing and Calculating Intermodulation Distortion in Receivers," *QEX*, Jul 1994, pp 3-4.

B. Gilbert, "Demystifying the Mixer," self-published monograph, 1994.

S. Raman, F. Rucky, and G. M. Rebeiz, "A High Performance W-band Uniplanar Subharmonic Mixer," *IEEE Trans MTT*, Vol 45, pp 955-962, Jun 1997.

M. J. Roberts, S. Iezekiel and C. M. Snowdan, "A Compact Subharmonically Pumped MMIC Self Oscillating Mixer for 77 GHz Applications," *IEEE MTT-S Digest*, 1998, pp 1435-1438.

M. Kossor, WA2EBY, "A Digital Commutating Filter," *QEX*, May/Jun 1999, pp 3-8.

K. S. Ang, A. H. Baree, S. Nam and I. D. Robertson, "A Millimeter-Wave Monolithic Sub-Harmonically Pumped Resistive Mixer," *Proc APMC*, Vol 2, 1999, pp 222-225.

C. H. Lee, S. Han, J. Laskar, "GaAs MESFET Dual-Gate Mixer with Active Filter Design for Ku-Band Application," *IEEE MTT-S Digest*, 1999, pp 841-844.

H. Darabi and A. A. Abidi, "Noise in RF-CMOS Mixers: A Simple Physical Model," *IEEE JSSC*, Vol 35, Jan 2000, pp 15-25.

P. S. Tsenes, G. E. Stratakos, N. K. Uzunoglu, M. Lagadas, G. Deligeorgis, "An X-band MMIC Down-Converter," WOCSDICE 2000, pp X.7-X.8

M. T. Terrovitis and R. G. Meyer, "Intermodulation Distortion in Current-Commutating CMOS Mixers," *IEEE Journ of SSC*, Vol 35, No 10, October 2000.

M. W. Chapman and S. Raman, "A 60 GHz Uniplanar MMIC 4X Subharmonic Mixer," *IEEE MTT-S Int Microwave Symp Dig*, 2001, pp 95–98, 2001.

M. Sironen, Y. Qian and T. Itoh, "A Subharmonic Self-Oscillating Mixer with Integrated Antenna for 60 GHz Wireless Applications," *IEEE Trans MTT*, 2001, pp 442–450

D. Tayloe, N7VE, "Letters to the Editor, Notes on 'Ideal' Commutating Mixers (Nov/Dec 1999)," *QEX*, March/April 2001, p 61.

K. Kanaya, K. Kawakami, T. Hisaka, T. Ishikawa and S. Sakmoto, "A 94 GHz High Performance Quadruple Subharmonic Mixer MMIC," *IEEE MTT-S Int Microwave Symp Dig*, 2002, pp 1249–1252.

J. Kim, Y. Kwon, "Intermodulation Analysis of Dual-Gate FET Mixers," *IEEE Trans MTT*, Vol 50, No 6, Jun 2002.

B. Matinpour, N. Lal, J. Laskar, R. E. Leoni and C. S. Whelan, "A Ka Band Subharmonic Down-Converter in a GaAs Metamorphic HEMT Process," *IEEE MTT-S Int Microwave Symp Dig*, 2001, pp 1337–1339.

T. S. Kang, S. D. Lee, B. H. Lee, S. D. Kim, H. C. Park, H. M. Park and J. K. Rhee, *Journ Korean Phys Soc 41*, 2002, 533.

D. An, B. H. Lee, Y. S. Chae, H. C. Park, H. M. Park and J. K. Rhee, *Journ Korean Phys Soc 41*, 2002, 1013.

D. Metzger, "Build the 'No Excuses' Transceiver," *QST*, Dec 2002, pp 28–34.

W. S. Sul, S. D. Kim, H. M. Park and J. K. Rhee, *Jpn Journ Appl Phys*, Vol 42, 2003, pp 7189-7193 (see **jjap.ipap.jp/link?JJAP/42/7189/**).

H. Morkner, S. Kumar and M. Vice, "An 18-45 GHz Double-Balanced Mixer with Integrated LO Amplifier and Unique Suspended Broadside-Coupled Balun," *2003 Gallium Arsenide Integrated Circuit Symp*, Nov 2003, pp 267-270.

D. Tayloe, N7VE, "A Low-noise, High-performance Zero IF Quadrature Detector/Preamplifier", March 2003, *RF Design*.

D. Tayloe, N7VE, US Patent 6,230,000, "A Product Detector and Method Therefore".

M.-F, Lei, P.-S, Wu, T.-W. Huang, H. Wang, "Design and Analysis of a Miniature W-band MMIC Sub Harmonically Pumped Resistive Mixer," *2004 IEEE MTT-S Int Microwave Symp*, Vol 1, Jun 2004, pp 235-238.

Ping-Chun Yeh, Wei-Cheng Liu and Hwann-Kaeo Chiou, "Compact 28-GHz Subharmonically Pumped Resistive Mixer MMIC Using a Lumped-Element High-Pass/Band-Pass Balun," *IEEE Microwave and Wireless Component Letters*, Vol 15, No 2, Feb 2005.

J. Browne, "Wideband Mixers Hit High Intercept Points," *Microwave & RF Journal*, Sep 2005, pp 98-104.

Chien-Chang H. and Wei-Ting C., "Multi-Band/Multi-Mode Current Folded Up-Convert Mixer Design," *Proc APMC 2006*, Japan.

X. Wang, M. Chen and O. B. Lubecke, "0.25 um CMOS Resistive Sub threshold Mixer," *Proc APMC 2006*, Japan.

Ro-Min Weng, Jing-Chyi Wang and Hung-Che Wei, "A 1V 2.4 GHz Down Conversion Folded Mixer", *IEEE APCCAS*, Dec 4-7, 2006, Singapore, pp 1476-1478.

B. M. Motlagh, S. E. Gunnarsson, M. Ferndahl and H. Zirath, "Fully Integrated 60-GHz Single-Ended Resistive Mixer in 90-nm CMOS Technology," *IEEE Microwave and Wireless Component Letters*, Vol 16, No 1, Jan 2006.

F. Ellinger, "26.5-30-GHz Resistive Mixer in 90-nm VLSI SOI CMOS Technology With High Linearity for WLAN" *IEEE Trans MTT*, Vol 53, No 8, Aug 2005.

U. L. Rohde and A. K. Poddar, "High Intercept Point Broadband, Cost Effective and Power Efficient Passive Reflection FET DBM," *EuMIC Symposium*, 10-15 Sep 2006, UK.

U. L. Rohde and A. K. Poddar, "A Unified Method of Designing Ultra-Wideband, Power-Efficient, and High IP3 Reconfigurable Passive FET Mixers," *IEEE/ICUWB*, Sep 24-27, 2006, MA, USA.

U. L. Rohde and A. K. Poddar, "Low Cost, Power-Efficient Reconfigurable Passive FET Mixers", 20th *IEEE CCECE* 2007, 22-26 Apr 2007, British Columbia, Canada.

INTERESTING WEB SITES FOR MIXERS

www.analog.com/en/content/0,2886,770%255F848%255F67465,00.html
www.maximic.com/solutions/base_stations/index.mvp?CMP=3758&gclid=CIzn1J_lm40CFSgRGgodD2Go6g
www.pulsarmicrowave.com/products/mixers/mixers.htm
www.synergymwave.com/products/mixers/?LocTitle=Mixers
news.thomasnet.com/fullstory/507272
www.wj.com/products/HighInterceptFETMixers.aspx

Contents

RF and AF Filters

This chapter discusses the most common types of filters used by radio amateurs. The sections describing basic concepts, lumped element filters and some design examples were initially prepared by Jim Tonne, W4ENE, and updated by Ward Silver, NØAX. The CD-ROM's design example for active filters was provided by Dan Tayloe, N7VE, and the section on crystal filters was developed by Dave Gordon-Smith, G3UUR.

11.1 Introduction

Electrical filters are circuits used to process signals based on their frequency. For example, most filters are used to pass signals of certain frequencies and reject others. The electronics industry has advanced to its current level in large part because of the successful use of filters. Filters are used in receivers so that the listener can hear only the desired signal; other signals are rejected. Filters are used in transmitters to pass only one signal and reject those that might interfere with other spectrum users. **Table 11.1** shows the usual signal bandwidths for several signal types.

The simplified SSB receiver shown in **Fig 11.1** illustrates the use of several common types of filters. A *preselector* filter is placed between the antenna and the first mixer to pass all frequencies within a given amateur band with low loss. Strong out-of-band signals from broadcast, commercial or military stations are rejected to prevent them from overloading the first mixer. The preselector filter is almost always built with *lumped-element* or "LC" technology.

An intermediate frequency (IF) filter is placed between the first and second mixers to pass only the desired signal. Finally, an audio filter is placed somewhere between the detector and the speaker rejects unwanted products of detection, power supply hum and noise. The audio

Chapter 11 — CD-ROM Content

Supplemental Articles

- "Using Active Filter Design Tools" by Dan Tayloe, N7VE

Projects

- "An Easy-to-Build, High-Performance Passive CW Filter" by Ed Wetherhold, W3NQN
- "A High Performance, Low Cost 1.8 to 54 MHz Low Pass Filter" by Bill Jones, K8CU

Software

The *Handbook CD* includes these useful *Windows* programs by Jim Tonne, W4ENE:
- *DiplexerDesigner* for design and analysis of diplexer filters
- *Elsie* for design and analysis of lumped-element LC filters
- *Helical* for design and analysis of helical-resonator bandpass filters
- *SVC Filter Designer* for design and analysis of lumped-element high-pass and low-pass filters
- *QuadNet* for design and analysis of active all-pass networks for SSB operation

Table 11.1
Typical Filter Bandwidths for Typical Signals

Source	Required Bandwidth
Fast-scan analog television (ATV)	4.5 MHz
Broadcast-quality speech and music	15 kHz (from 20 Hz to 15 kHz)
Communications-quality speech	3 kHz (from 300 Hz to 3 kHz)
Slow-scan television (SSTV)	3 kHz (from 300 Hz to 3 kHz)
HF RTTY	250 to 1000 Hz (depends on frequency shift)
Radiotelegraphy (Morse code, CW)	200 to 500 Hz
PSK31 digital modulation	100 Hz

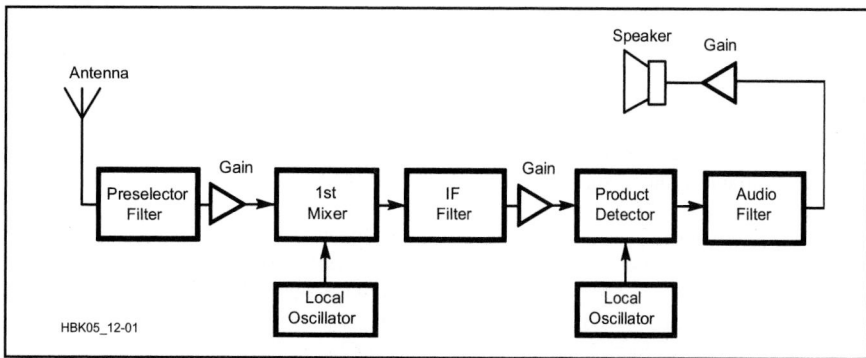

HBK05_12-01

Fig 11.1 — Single-band SSB receiver. At least three filters are used between the antenna and speaker.

filter is often implemented with *active filter* technology.

The complementary SSB transmitter block diagram is shown in **Fig 11.2**. A similar array of filters appears in reverse order. First, an audio filter between the microphone and the balanced mixer rejects out-of-band audio signals such as 60 Hz hum. Since the balanced mixer generates both lower and upper sidebands, an IF filter is placed at the mixer output to pass only the desired lower (or upper) sideband. Finally, a filter at the output of the transmit mixer passes only signals within the amateur band in use rejects unwanted frequencies generated by the mixer to prevent them from being amplified and transmitted. Filters at the transmitter output attenuate the harmonics of the transmitted signals.

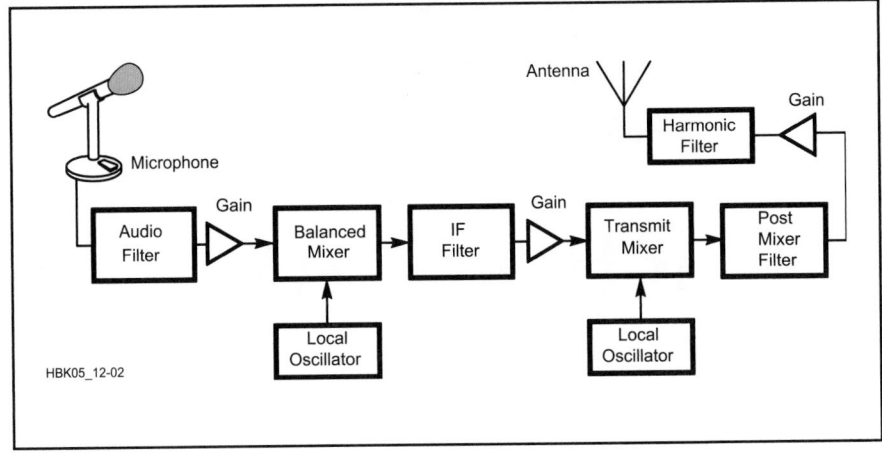

Fig 11.2 — Single-band SSB transmitter. At least three filters are needed to ensure a clean transmitted signal.

11.2 Filter Basics

A common type of filter, the *band-pass*, passes signals in a range of frequencies — the *passband* — while rejecting signals outside that range — the *stop band*. To pass signals from dc up to some *cutoff frequency* at which output power is halved (reduced by 3 dB) we would use a *low-pass* filter. To pass signals above a cutoff frequency (also called the "3-dB frequency") we would use a *high-pass* filter. Similarly, to pass signals within a range of frequencies we would use a *band-pass* filter. To pass signals at all frequencies *except* those within a specified range requires a *band-stop* or *notch filter*. The region between the passband and the stop band is logically called the *transition region*. The attenuation for signal frequencies in the stop band far from the cutoff frequency is called the *ultimate attenuation*.

Many of the basic terms used to describe filters are defined in the chapters on **Electrical Fundamentals** and **Analog Basics**. In addition, a glossary is provided at the end of this chapter.

11.2.1 Filter Magnitude Responses

Fig 11.3 illustrates the *magnitude response* of a low-pass filter. Signals lower than the cutoff frequency (3 MHz in this case) are passed with some small amount of attenuation while signals higher than that frequency are attenuated. The degree of attenuation is dependent on several variables, filter complexity being a major factor.

Of the graphs in this chapter that show a filter's magnitude response, the vertical axes are labeled "Transmission (dB)" with 0 dB at the top of the axis and negative values increasing towards the x-axis. Increasingly negative values of transmission in dB are the same as increasingly positive values of attenuation in

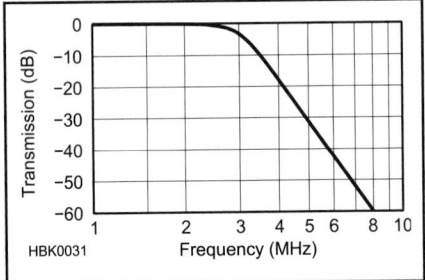

Fig 11.3 — Example of a low-pass magnitude response.

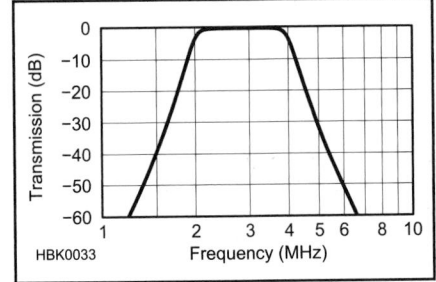

Fig 11.5 — Example of a band-pass magnitude response.

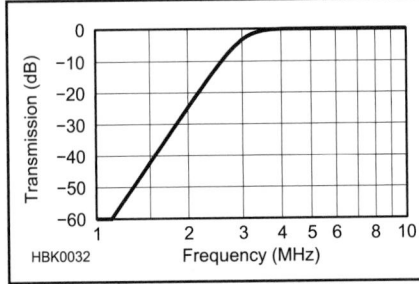

Fig 11.4 — Example of a high-pass magnitude response.

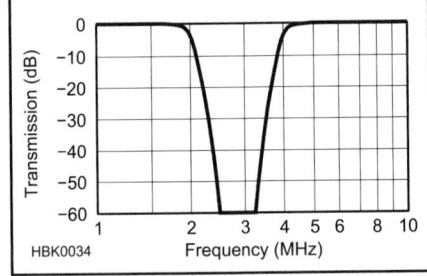

Fig 11.6 — Example of a band-stop magnitude response.

dB. For example, –40 dB transmission is the same as 40 dB of attenuation.

Fig 11.4 illustrates the magnitude response of a high-pass filter. Signals above the 3 MHz cutoff frequency are passed with minimum attenuation while signals below that frequency are attenuated. Again, the degree of attenuation is dependent on several variables.

Fig 11.5 illustrates the magnitude response of a band-pass filter. Signals within the band-pass range (between the lower and upper cutoff frequencies) are passed with minimum attenuation while signals outside that range

are attenuated. In this example the filter was designed with cutoff frequencies of 2 MHz and 4 MHz, for a passband width of 2 MHz.

Fig 11.6 illustrates the magnitude response of a band-stop filter. Signals within the band-stop range are attenuated while all other signals are passed with minimum attenuation. A notch filter is a type of band-stop filter with a narrow stop-band in which the attenuation is a maximum at a single frequency.

An *ideal filter* — a low-pass filter, for example — would pass all frequencies up to some point with no attenuation at all and

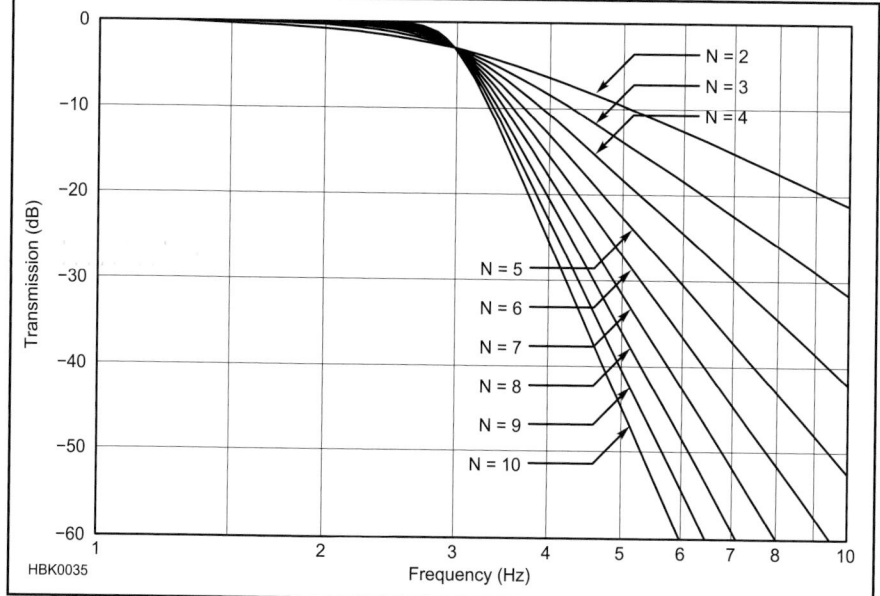

Fig 11.7 — Magnitude responses for low-pass filters with orders ranging from very simple (N=2) to very complex (N=10).

totally reject everything beyond that point. This is known as a *brick wall* response because the filter's passband and stop band are *flat*, meaning no attenuation, and the rolloff in the transition region is infinitely steep. The magnitude response of such a filter would be drawn as a flat line representing 0 dB attenuation up to the cutoff frequency that then abruptly changes at the cutoff frequency to a flat line at infinite attenuation throughout the stop band to infinite frequency.

11.2.2 Filter Order

The steepness of the descent from the passband to the region of attenuation — the stop band — is dependent on the complexity of the filter, properly called the *order*. In a lumped element filter made from inductors and capacitors, the order is determined by the number of separate energy-storing elements (either L or C) in the filter. For example, an RC filter with one resistor and one capacitor has an order of 1. A notch filter made of a single series-LC circuit has an order of 2. The definition of order for active and digital signal processing filters can be more complex, but the general understanding remains valid that the order of a filter determines how rapidly its response can change with frequency.

For example, **Fig 11.7** shows the magnitude response of one type of low-pass filter with orders varying from very simple ("N=2") to the more complex ("N=10"). In each case on this plot the 3-dB point (the filter's cutoff or corner frequency) remained the same, at 3 Hz.

As frequency increases through the transition region from the passband into the stop band, the attenuation increases. *Rolloff*, the

rate of change of the attenuation for frequencies above cutoff, is expressed in terms of dB of change per octave of frequency. Rolloff for this type of filter design is always 6.02 times the order, regardless of the circuit used to fabricate the circuit. As the order goes up, the rate of change also goes up. A filter's response with large values of rolloff is called "sharp" or "steep" and "soft" or "shallow" if rolloff is low.

11.2.3 Filter Families

There is no single "best" way to develop the parts values for a filter. Instead we have to decide on some traits and then choose the most appropriate *family* for our design. Different families have different traits, and filter families are commonly named after the mathematician or engineer responsible for defining their behavior mathematically.

Real filters respond more gradually and with greater variation than ideal filters. Two primary traits of the most importance to amateurs are used to describe the behavior of filter families: *ripple* (variations in the magnitude response within the passband and stop band) and rolloff. Different families of filters have different degrees of ripple and rolloff (and other characteristics, such as phase response).

With a flat magnitude response (and so no ripples in the passband) we have what is known as a *Butterworth* family design. This family also goes by the name of *Maximally Flat Magnitude* or *Maximally Flat Gain*. An example of the magnitude response of a filter from the Butterworth family is shown in both **Fig 11.7** and **Fig 11.8**. (Most filter discussions

are based on low-pass filters because the same concepts are easily extended to other types of filters and much of the mathematics behind filters is equivalent for the various types of frequency responses.)

By allowing magnitude response ripples in the passband, we can get a somewhat steeper rolloff from the passband into the stop band, particularly just beyond the cutoff frequency. A family that does this is the *Chebyshev* family. **Fig 11.9** illustrates how allowing magnitude ripple in the passband provides a sharper filter rolloff. This plot compares the 1-dB ripple Chebyshev with the no-ripple Butterworth filter down to 12 dB of attenuation.

Fig 11.9 also illustrates the usual definition of bandwidth for a low-pass Butterworth filter — the filter's 3-dB or cutoff frequency. For a low-pass Chebyshev filter, *ripple bandwidth* is used — the frequency range over which the filter's passband ripple is no greater than the specified limit. For example, the ripple bandwidth of a low-pass Chebyshev filter designed to have 1 dB of passband ripple is the highest frequency at which attenuation is 1 dB or less. The 3-dB bandwidth of the frequency of the filter will be somewhat greater.

A Butterworth filter is defined by specifying the order and bandwidth. The Chebyshev filter is defined by specifying the order, the ripple bandwidth, and the amount of passband ripple. In Fig 11.9, the Chebyshev filter has 1 dB of ripple; its ripple bandwidth is 1000 Hz.

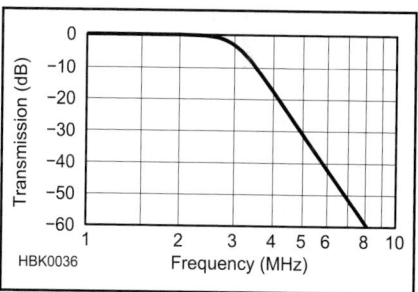

Fig 11.8 — Filters from the Butterworth family exhibit flat magnitude response in the passband.

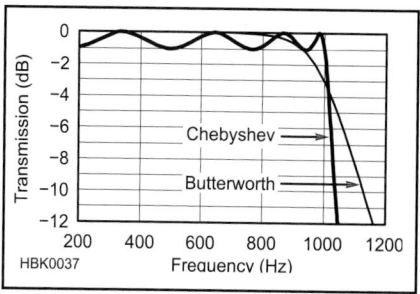

Fig 11.9 — This plot compares the response of a Chebyshev filter with a 1-dB ripple bandwidth of 1000 Hz and a Butterworth filter 3-dB bandwidth of 1000 Hz.

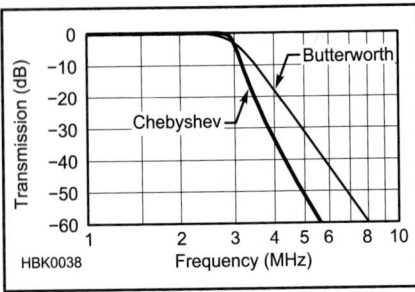

Fig 11.10 — A Chebyshev filter (0.2 dB passband ripple) allows a sharper cutoff than a Butterworth design with no passband ripple.

The Butterworth filter has a 3-dB bandwidth also of 1000 Hz. Some filter textbooks use the 3-dB point to define Chebyshev filters; most use the ripple bandwidth as illustrated here. The schematics (if you ignore parts values) of those two families are identical.

Even small amounts of ripple can be beneficial in terms of increasing a filter's rolloff. **Fig 11.10** compares a Butterworth filter (with the narrow line plot, no ripple in the passband) with a Chebyshev filter (wide line plot, 0.2 dB of ripple in the passband) down to 60 dB of attenuation. (For this comparison the cutoff frequencies at 3 dB of attenuation are the same for each filter.) Even that small amount of ripple in the Chebyshev filter passband allows a noticeably steeper rolloff between the passband and the stop band. As the passband ripple specification is increased, the steepness of the transition from passband to stop band increases, compared to the Butterworth family

although the rolloff of the two families eventually becomes equal.

For Chebyshev filters, when the value of the passband ripple is changed, the magnitude response in the stop band region also changes. **Fig 11.11** compares the stop band response of Chebyshev filters with passband ripple ranging from 0.01 to 1 dB.

It is possible to obtain even steeper rolloff into the stop band by adding "traps" whose frequencies are carefully calculated. The resonant frequencies of those traps are in the stop band region and are set to yield best performance. When this is done, we have a *Cauer* family design (also called the *elliptic-function* design). The rolloff of the Cauer filter from the passband into the stop band is the steepest of all analog filter types provided that the behavior in the passband is uniform (either no ripple or a uniform amount of ripple). **Fig 11.12** shows the response of the Chebyshev and Cauer designs for comparison.

The Cauer filter is defined by specifying the order, the ripple bandwidth, and the passband ripple, just as for the Chebyshev. Again, an alternative bandwidth definition is to use the 3-dB frequency instead of the ripple bandwidth. The Cauer family requires one more specification: the *stop band frequency* and/or the *stop band depth*. The stop band frequency is the lowest frequency of a null or notch in the stop band. The stop band depth is the minimum amount of attenuation allowed in the stop band. In Fig 11.12, the stop band frequency is about 3.5 MHz and the stop band depth is 50 dB. For this comparison both designs have the same passband ripple value of 0.2 dB.

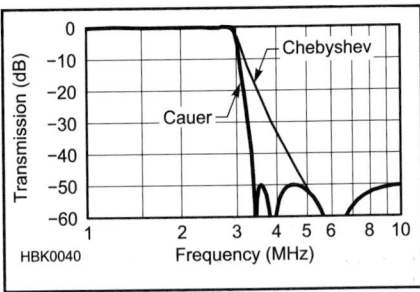

Fig 11.12 — The Cauer family has an even steeper rolloff from the passband into the stop band than the Chebyshev family. Note that the ultimate attenuation in the stop band is a design parameter rather than ever-increasing as is the case with other families.

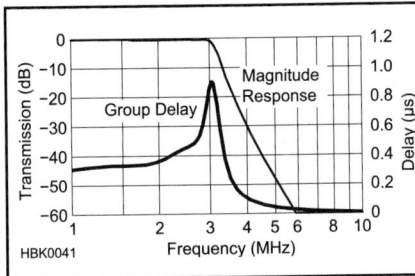

Fig 11.13 — Magnitude response and group delay of a Chebyshev low-pass filter.

A downside to the Cauer filter is that the ultimate attenuation is some chosen value rather than ever increasing, as is the case with the other families. In the far stop band region (where frequency is much greater than the cutoff frequency) the rolloff of Cauer filters ultimately reaches 6 or 12 dB per octave, depending on the order. Odd-ordered Cauer filters have an ultimate rolloff rate of about 6 dB per octave while the even-ordered versions have an ultimate rolloff rate of about 12 dB per octave.

11.2.4 Group Delay

Another trait that can influence the choice of which filter family to use is the way the transit time of signals through the filter varies with frequency. This is known as *group delay*. ("Group" refers to a group of waves of similar frequency and phase moving through a media, in this case, the filter.) The wide line (lower plot) in **Fig 11.13** illustrates the group delay characteristics of a Chebyshev low-pass filter, while the upper plot (narrow line) shows the magnitude response. As shown in Fig 11.13, the group delay of components near the cutoff frequency becomes quite large when compared to that of components at lower and higher frequencies. This is a result of the phase shift of the filter's transmission being nonlinear with frequency; it is usually greater

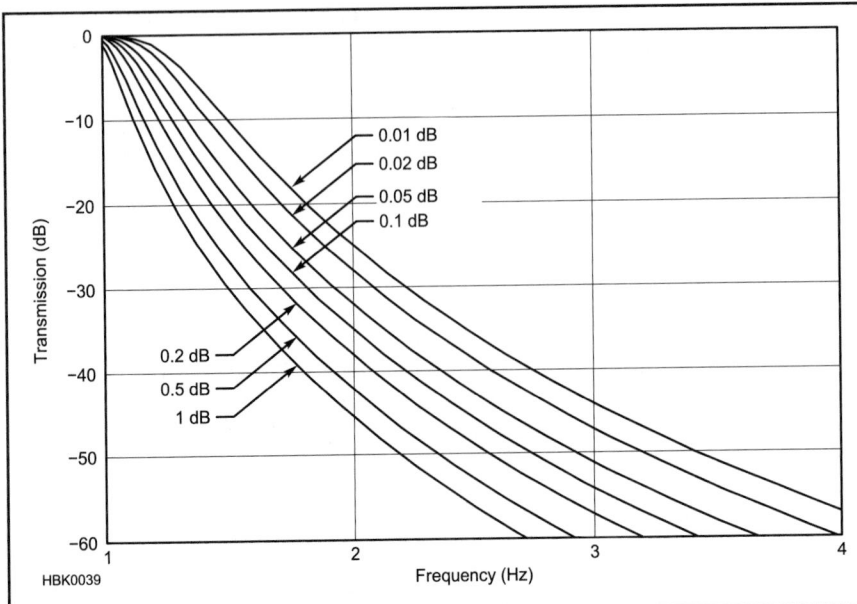

Fig 11.11 — These stop band response plots illustrate the Chebyshev family with various values of passband ripple. These plots are for a seventh-order low-pass design with ripple values from 0.01 to 1 dB. Ignoring the effects in the passband of high ripple values, increasing the ripple will allow somewhat steeper rolloff into the stop band area, and better ultimate attenuation in the stop band.

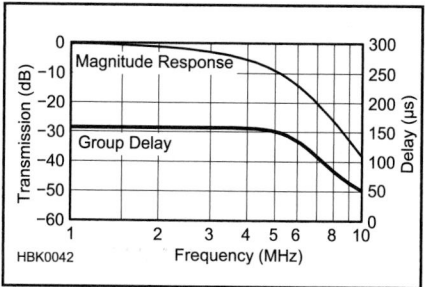

Fig 11.14 — Magnitude response and group delay of a Bessel low-pass filter.

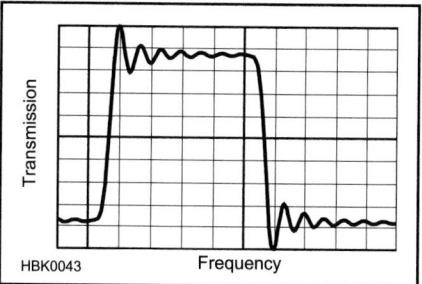

Fig 11.15 — Transient response of a Chebyshev low-pass filter.

near the cutoff frequency. By delaying signals at different frequencies different amounts, the signal components are "smeared" in time. This causes distortion of the signal and can seriously disrupt high-speed data signals.

If a uniform group delay for signals throughout the passband is needed, then the *Bessel* filter family should be selected. The Bessel filter can be used as a delay-line or time-delay element although the gentle rolloff of the magnitude response may need to be taken into account. The magnitude and delay characteristics for the Bessel family are shown in **Fig 11.14**. Comparing the Bessel filter's response to that of the Butterworth in

Fig 11.8 shows the difference in roll off.

A downside of Bessel family filters is that the rolloff characteristic is quite poor; they are not very good as a magnitude-response-shaping filter. The Bessel family is characterized largely by its constant group delay in the passband (for a low- or high-pass filter) shown as the bottom plot in Fig 11.14. The *constant-delay* characteristic of the Bessel extends into the stop band. The Bessel filter bandwidth is commonly defined by its 3-dB point (as with the Butterworth).

Because the Bessel filter is used when phase response is important, it is often characterized by the frequency at which a specific amount of phase shift occurs, usually one radian. (One radian is equal to $360 / 2\pi = 57.3°$.) Since the delay causes the output signal to lag behind the input signal, the filter is specified by its *one-radian lag frequency*.

11.2.5 Transient Response

Some applications require that a signal with sharp rising and falling edges (such as a digital data waveform) applied to the input of a low-pass design have a minimum *overshoot* or *ringing* as seen at the filter's output. Overshoot (and undershoot) occurs when a signal exceeds (falls below) the final amplitude temporarily before settling at its final value. Ringing is a repeated sequence of overshoot and undershoot. If a sharp-cutoff filter is used in such an application, overshoot or ringing will occur. The appearance of a signal such as a square wave with sharp rising and falling edges as it exits from the filter may be as shown in **Fig 11.15**. The scales for both the X- and the Y-axes would depend on the frequency and magnitude of the waveform.

The square wave used in this discussion as a test waveform is composed of a fundamental and an endless series of odd harmonics. If harmonics only up to a certain order are used to create the square wave — that is, if the square wave is passed through a sharp-cutoff low-pass

filter that attenuates higher frequencies — then that waveform will have the overshoot or ringing as shown as it exits from the filter.

All of the traits mentioned so far in this chapter apply to a filter regardless of how it is implemented, whether it is fabricated using *passive* lumped-element inductors and capacitors or using op amps with resistors and capacitors (an *active* filter). The traits are general descriptions of filter behavior and can be applied to any type of filter technology.

11.2.6 Filter Family Summary

Selecting a filter family is one of the first steps in filter design. To make that choice easier, the following list of filter family attributes is provided:

• Butterworth — No ripple in passband, smooth transition region, shallow rolloff for a given filter order, high ultimate attenuation, smooth group delay change across transition region

• Chebyshev — Some passband ripple, abrupt transition region, steep rolloff for a given filter order, peak in group delay near cutoff frequency, high ultimate attenuation

• Cauer (or Elliptical-function) — Some passband ripple, abrupt transition region, steepest roll off, ripple in stop band due to traps, ultimate attenuation smaller than Butterworth and Chebyshev, group delay peaks near cutoff frequency and in stop band

• Bessel — No ripple in passband, smooth transition region, constant group delay in passband, shallow rolloff for a given filter order, smooth changes in group delay, high ultimate attenuation

All characterizations such as "steep" and "abrupt" are relative with respect to filter designs from other families with similar orders. Other factors, such as number of components, sensitivity to component value and so on may need to be considered when selecting a filter family for a specific application.

11.3 Lumped-Element Filters

This part of the chapter deals with passive LC filters fabricated using discrete inductors and capacitors (which gives rise to their name, *lumped-element*). We will begin with a discussion of basic low-pass filters and then generalize to other types of filters.

11.3.1 Low-Pass Filters

A very basic LC filter built using inductors and capacitors is shown in **Fig 11.16**. In the first case (Fig 11.16A), less power is delivered to the load at higher frequencies because the reactance of the inductor in series with the

load increases as the test frequency increases. The voltage appearing at the load goes down as the frequency increases. This configuration would pass direct current (dc) and reject higher frequencies, and so it would be a low-pass filter.

With the second case (Fig 11.16B), less power is delivered to the load at higher frequencies because the reactance of the capacitor in parallel with the load decreases as the test frequency increases. Again, the voltage appearing at the load goes down as the frequency increases and so this, too, would be called a low-pass filter. In the real world, combinations of both series and parallel com-

ponents are used to form a low-pass filter.

A high-pass filter can be made using the opposite configuration — series capacitors and shunt inductors. And a band-pass (or band-stop) filter can be made using pairs of series and parallel tuned circuits. These filters, made from alternating LC elements or LC tuned circuits, are called *ladder filters*.

We spoke of filter order, or complexity, earlier in this chapter. **Fig 11.17** illustrates *capacitor-input* low-pass filters with orders of 3, 4 and 5. Remember that the order corresponds to the number of energy-storing elements. For example, the third-order filter

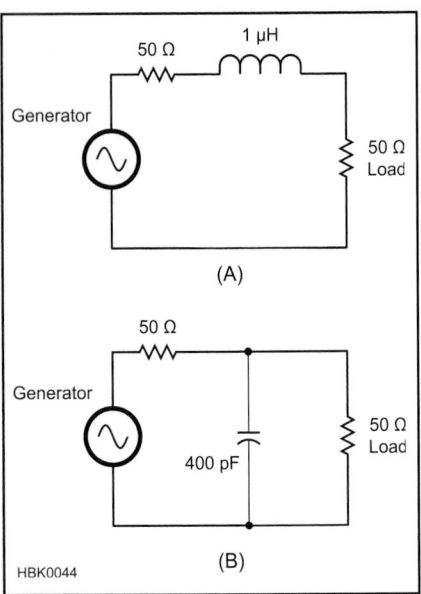

Fig 11.16 — A basic low-pass filter can be formed using a series inductor (A) or a shunt capacitor (B).

in Fig 11.17A has three energy-storing elements (two capacitors and one inductor), while the fifth-order design in Fig 11.17C has five elements total (three capacitors, two inductors). For comparison, a third-order filter is illustrated in **Fig 11.18**. The filters in Fig 11.17 are *capacitor-input* filters because a capacitor is connected directly across the input source. The filter in Fig 11.18 is an *inductor-input* filter.

As mentioned previously, the Cauer family has traps (series or parallel tuned circuits) carefully added to produce dips or notches (prop-

erly called *zeros*) in the stop band. Schematics for the Cauer versions of capacitor-input low-pass filters with orders of 3, 4 and 5 are shown in **Fig 11.19**. The capacitors in parallel with the series inductors create the notches at calculated frequencies to allow the Cauer filter to be implemented.

The capacitor-input and the inductor-input versions of a given low-pass design have identical characteristics for their magnitude, phase and time responses, but they differ in the impedance seen looking into the filter. The capacitor-input filter has low impedance in the stop band while the inductor-input filter has high impedance in the stop band.

11.3.2 High-Pass Filters

A high-pass filter passes signals above its cutoff frequency and attenuates those below. Simple high-pass equivalents of the filters in Fig 11.16 are shown in **Fig 11.20**.

The reactance of the series capacitor in Fig 11.20A increases as the test frequency is lowered and so at lower frequencies there will be less power delivered to the load. Similarly, the reactance of the shunt inductor in Fig 11.20B decreases at lower frequency, with the same effect. Similarly to the low-pass filter designs presented in Fig 11.17, a high-pass filter in a real world design would typically use both series and shunt components but with the positions of inductors and capacitors exchanged.

An example of a high-pass filter application would be a broadcast-reject filter designed to pass amateur-band signals in the range of 3.5 MHz and above while rejecting broadcast signals at 1.7 MHz and below. A high-pass

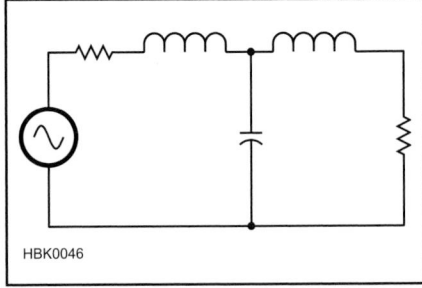

Fig 11.18 — A third-order inductor-input low-pass filter.

filter with a design cutoff of 2 MHz is illustrated in **Fig 11.21**. It can be implemented as a capacitor-input (Fig 11.21A) or inductor-input (Fig 11.21B) design. In each case, signals above the cutoff are passed with minimum attenuation while signals below the cutoff are attenuated, in a manner similar to the action of a low-pass filter.

11.3.3 Low-Pass to Band-Pass Transformation

A band-pass filter is defined in part by a *bandwidth* and a *center frequency*. (An alternative method is to specify a lower and an upper cutoff frequency.) A low-pass design such as the one shown in Fig 11.17A can be converted to a band-pass filter by resonating each of the elements at the center frequency. **Fig 11.22** shows a third-order low-pass filter with a design bandwidth of 2 MHz for use in a 50-Ω system. It should be mentioned that the next several designs and the various manipulations were done using a computer; other design methods will be shown later in the chapter.

If the shunt elements are now resonated with a parallel component, and if the series elements are resonated with a series component, the result is a band-pass filter as shown in **Fig 11.23**. The series inductor value and the shunt capacitor values are the same as those for the original low-pass design. Those components have been resonated at the center frequency for the filter (2.828 MHz in this case).

Now we can compare the magnitude response of the original low-pass filter with the band-pass design; the two responses are shown in **Fig 11.24**. The magnitude response of the low-pass version of this filter is down 3 dB at 2 MHz. The band-pass response is down 3 dB at 2 MHz and 4 MHz (the difference from the high side to the low side is 2 MHz). The response of the

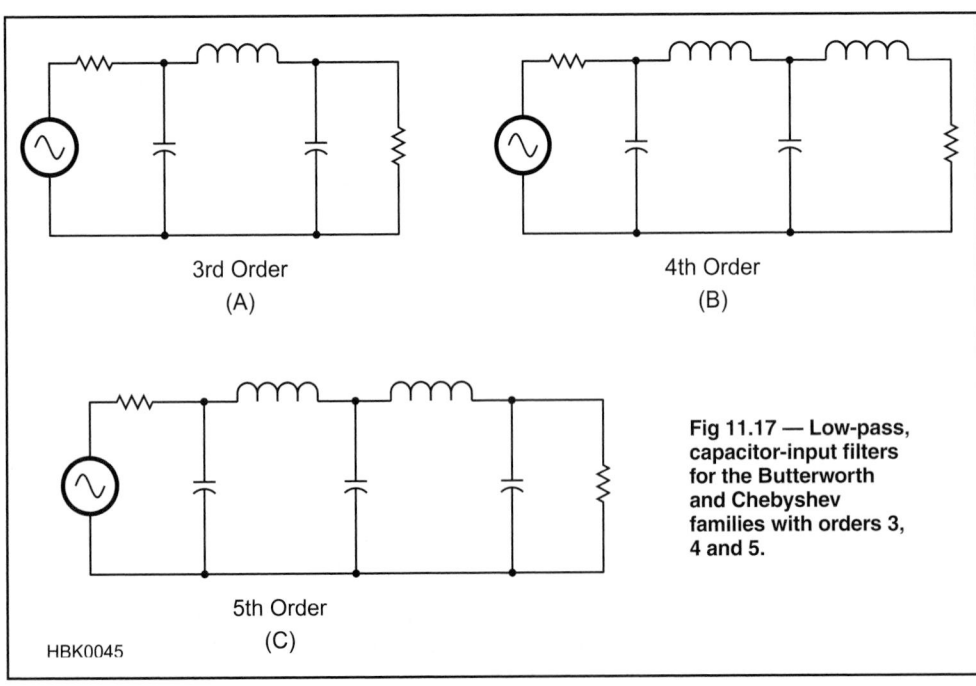

3rd Order
(A)

4th Order
(B)

5th Order
(C)

Fig 11.17 — Low-pass, capacitor-input filters for the Butterworth and Chebyshev families with orders 3, 4 and 5.

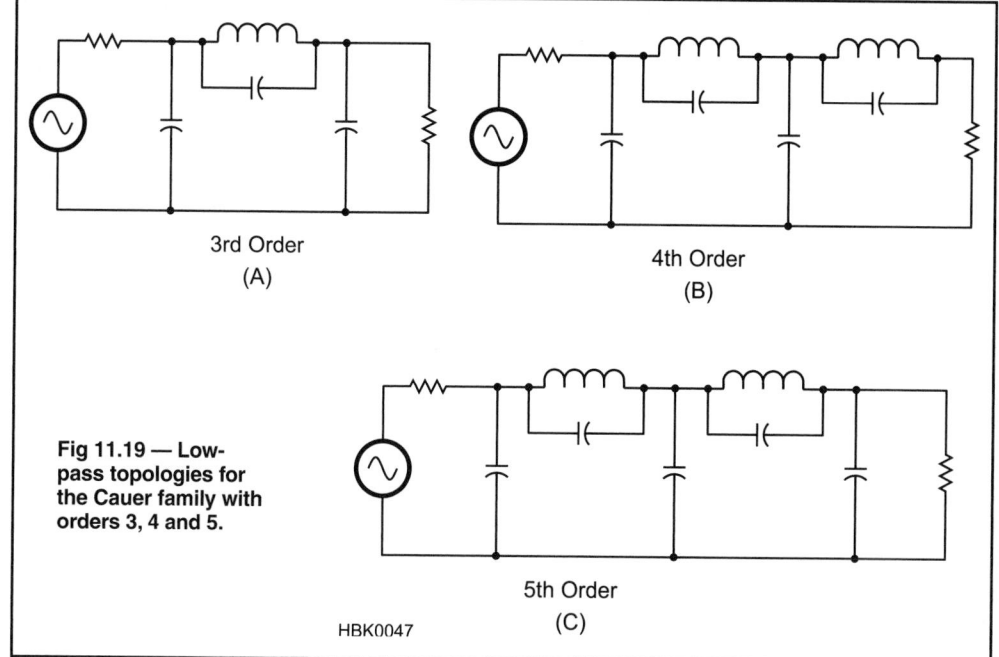

Fig 11.19 — Low-pass topologies for the Cauer family with orders 3, 4 and 5.

3rd Order (A)

4th Order (B)

5th Order (C)

HBK0047

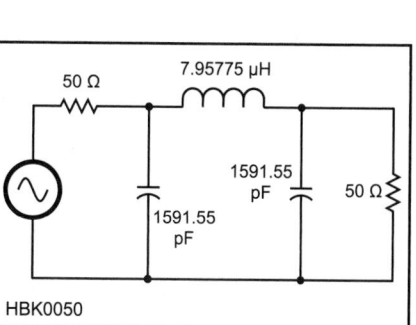

Fig 11.20 — A basic high-pass filter can be formed using a series capacitor (A) or a shunt inductor (B).

HBK0048

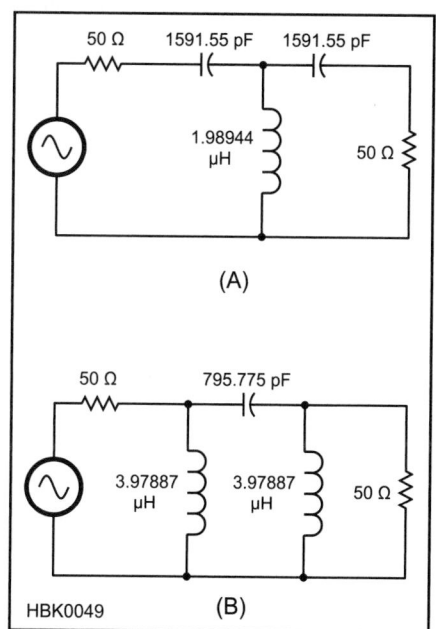

Fig 11.21 — Capacitor-input (A) and inductor-input (B) high-pass filters. Both designs have a 2 MHz cutoff.

HBK0049

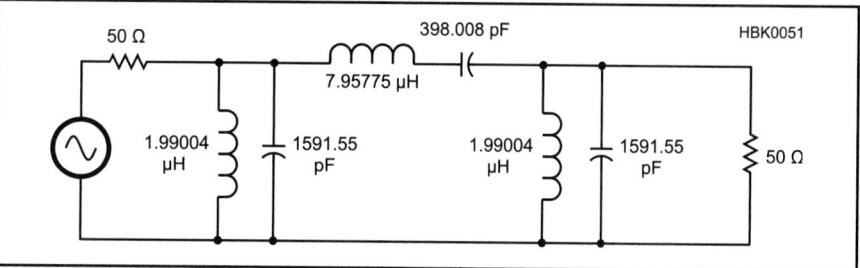

HBK0050

Fig 11.22 — Third-order low-pass filter with a bandwidth of 2 MHz in a 50 Ω system.

Fig 11.23 — The low-pass filter of Fig 11.22 can be transformed to a band-pass filter by resonating the shunt capacitors with a parallel inductor and resonating the series inductor with a series capacitor. Bandwidth is 2 MHz and the center frequency is 2.828 MHz.

low-pass design is down 33 dB at 7 MHz while the response of the band-pass version is down 33 dB at 1 and 8 MHz (the difference from high side to low side is 7 MHz).

11.3.4 High-Pass to Band-Stop Transformation

Just as a low-pass filter can be transformed to a band-pass type, a high-pass filter can be transformed into a band-stop (also called a *band-reject*) filter. The procedures for doing this are similar in nature to those of the transformation from low-pass to band-pass. As with the band-pass filter example, to transform a high-pass to a band-stop we need to specify a center frequency. The bandwidth of a band-stop filter is measured between the frequencies at which the magnitude response drops 3 dB in the transition region into the stop band.

Fig 11.25 shows how to convert the 2 MHz capacitor-input high-pass filter of Fig 11.21A to a band-stop filter centered at 2.828 MHz. The original high-pass components are resonated at the chosen center frequency to form a band-stop filter. Either a capacitor-input high-pass or an inductor-input high-pass may be transformed in this manner. In this case the series capacitor values and the shunt inductor values for the band-stop are the same as those for the high-pass. The series elements are resonated with an element in parallel with them. Similarly, the shunt elements are also resonated with an element in series with them. In each case the pair resonate at the center frequency of the band-stop. **Fig 11.26** shows the responses of both the original high-pass and the resulting band-stop filter.

11.3.5 Refinements in Band-Pass Design

The method of transforming a low-pass filter to a band-pass filter described previously is

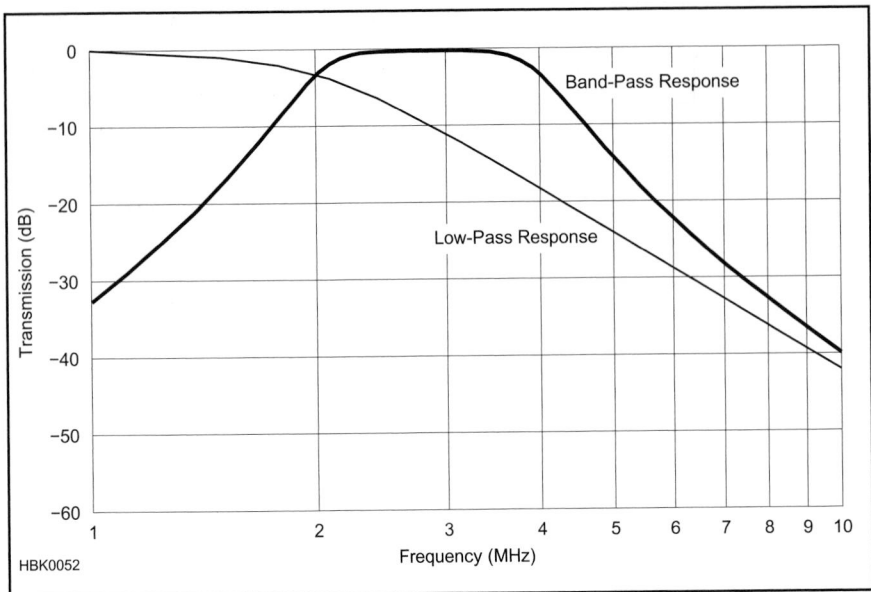

Fig 11.24 — Comparison of the response of the low-pass filter of Fig 11.22 with the band-pass design of Fig 11.23.

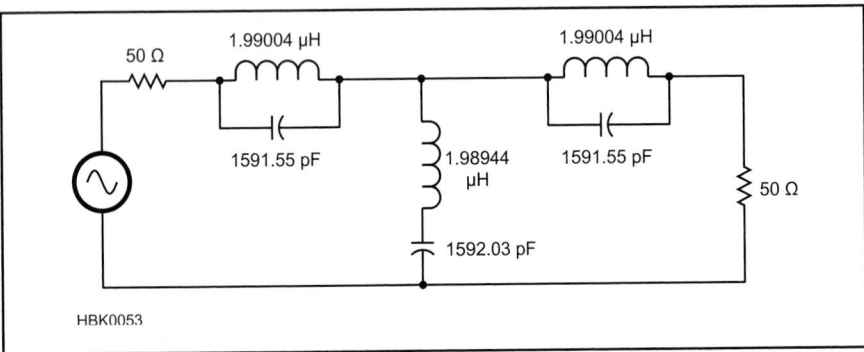

Fig 11.25 — The high-pass filter of Fig 11.21A can be converted to a band-stop filter with a bandwidth of 2 MHz, centered at 2.828 MHz.

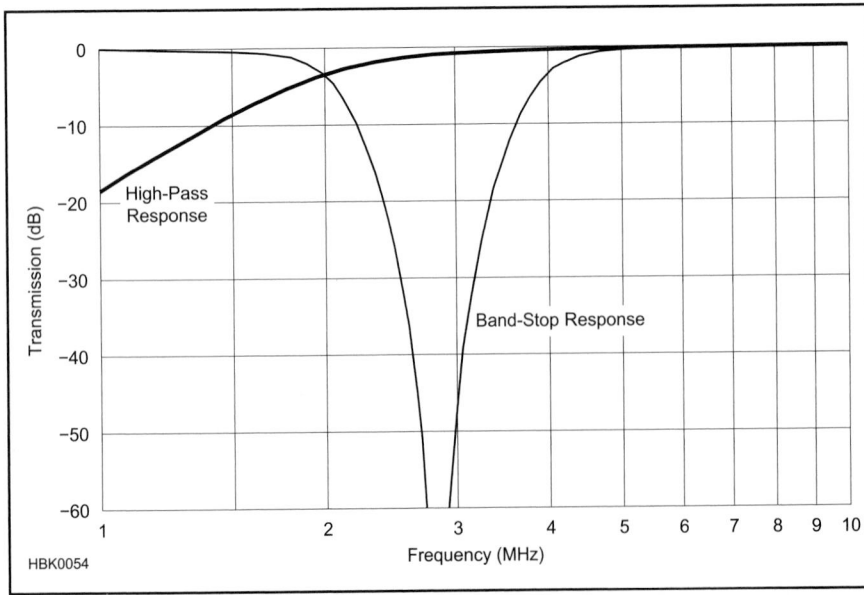

Fig 11.26 — Comparison of the response of the high-pass filter of Fig 11.21A with the band-stop design of Fig 11.25.

simple, but it has a major drawback. The resulting component values are often awkward for building a narrowband band-pass filter or inappropriate for the frequencies involved. This can be addressed through additional transformation techniques or by changing the filter to a different *topology* (the general organization of the filter circuit elements and how they are connected).

The following discussion shows an example of band-pass filter design, starting with a low-pass filter and then applying various techniques to create a filter with appropriate component values and improved response characteristics. These transformations are tedious and are best done using filter design software. Some manual design methods are presented later in this chapter.

Fig 11.27 shows a band-pass filter centered at 3 MHz with a width of 100 kHz that was designed by resonating the elements of a 3-MHz low-pass ladder-type filter as previously described. This filter is impractical for several reasons. The shunt capacitors (0.0318 µF) will probably have poor characteristics at the center frequency of 3 MHz because of the series inductance likely to be present in a practical capacitor of that value. Similarly, because of parasitic capacitance, the series inductor (159 µH) will certainly have a parallel self-resonance that is quite likely to alter the filter's response. For a narrowband band-pass filter, this method of transformation from low-pass to band-pass is not practical. (See the **RF Techniques** chapter for a discussion of parasitic effects.)

A better approach to narrowband band-pass filter design may be found by changing the filter topology. There are many different filter topologies, such as the LC "ladder" filters discussed in the preceding sections. One such topology is the *nodal-capacitor-coupled* design and **Fig 11.28** shows how the filter in Fig 11.27 can be redesigned using such an approach.

Some component values are improved, but the shunt capacitor values are still quite large for the center frequency of 3 MHz. In this case, the technique of *impedance scaling* offers some improvement. The filter was originally designed for a 50-Ω system. By scaling the filter impedance upwards from 50 to 500 Ω, the reactances of all elements are multiplied by a factor of 500 / 50 = 10. (Capacitors will get smaller and inductors larger.) The 0.03-µF capacitors will decrease in value to only 3000 pF, a much better value for a center frequency of 3 MHz. The shunt inductor values and the nodal coupling capacitor values (about 75 pF in this case) are also realistic. The 500 Ω version of the filter in Fig 11.28 is shown in **Fig 11.29**.

From a component-value viewpoint this is a better design, but it must be terminated in 500 Ω at each end. To use this filter in a 50-Ω system, impedance matching components

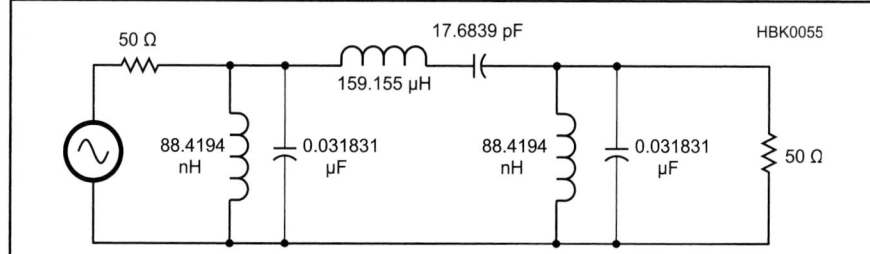

Fig 11.27 — This band-pass filter, centered at 3 MHz with a bandwidth of 100 kHz, was designed using a simple transformation. The resulting component values make the design impractical, as discussed in the text.

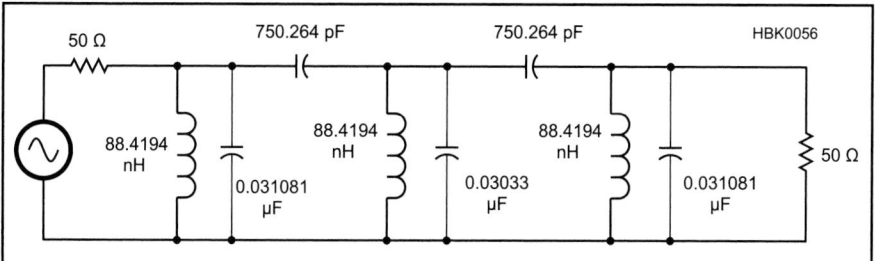

Fig 11.28 — A narrowband band-pass filter using a nodal-capacitor-coupled design improves component values somewhat.

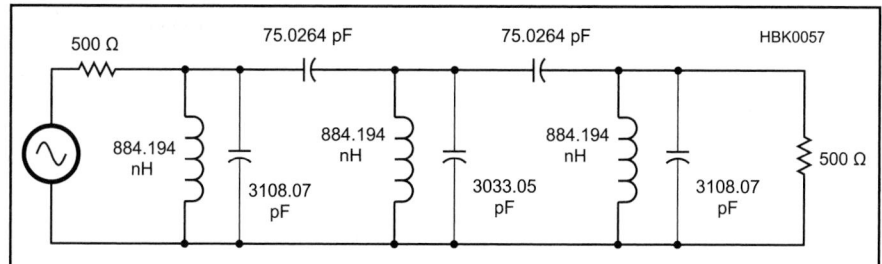

Fig 11.29 — The filter in Fig 11.28 scaled to an impedance of 500 Ω.

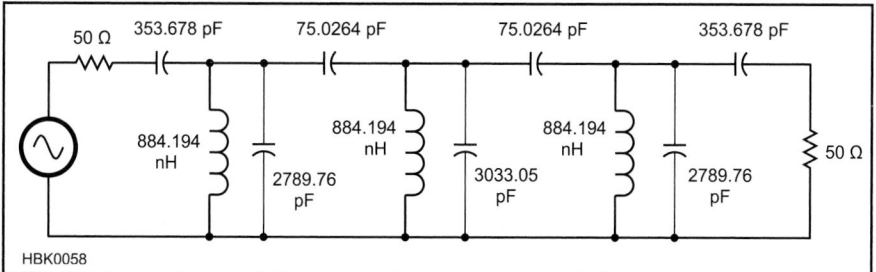

Fig 11.30 — Final design of the narrowband band-pass filter with 50 Ω terminations.

11.31C. The last two designs have identical magnitude responses.

Two other topologies are shown for comparison. The design in Fig 11.31B has only three capacitors (a minimum-capacitor design) while the design in Fig 11.31C has only three inductors (a minimum-inductor design).

The magnitude response of the capacitor-coupled design in Fig 11.31A is shown in **Fig 11.32** as the solid line. The designs in Fig 11.31B and C have magnitude responses as shown in Fig 11.32 as a dashed line. Note that the rolloff on the high-frequency side is steeper for the designs in Fig 11.31B and C. This will be the case for those filters with relatively wide (percentage-wise) designs. Such a design would be useful where harmonics of a signal are to be especially attenuated. It might also improve image rejection in a receiver RF stage where high-side injection of the local oscillator is used.

11.3.6 Effect of Component Q

When components with less-than-ideal characteristics are used to fabricate a filter, the performance will also be less than ideal. One such item to be concerned about is component "Q." As described in the **Electrical Fundamentals** chapter, Q is a measure of the loss in an inductor or capacitor, as determined by its resistive component. Q is the ratio of component's reactance to the loss resistance and is specified at a given test frequency. The loss resistance referred to here includes not only the value as measured by an ohmmeter but includes all sources of loss, such as skin effect, dielectric heating and so on.

The Q values for capacitors are usually greater than 500 and may reach a few thousand. Q values for inductors seldom reach 500 and may be as low as 20 or even worse for miniaturized parts. A good toroidal inductor can have a Q value in the vicinity of 250 to 400.

Q values can affect both the *insertion loss* of signals passing through the filter and

must be added as shown in **Fig 11.30**. This topology is typical of a radio receiver front-end preselector or anywhere else that a narrow — in terms of percentage of the center frequency — filter is needed.

When the nodal capacitor-coupled band-pass topology is used for a filter whose bandwidth is wide (generally, a bandwidth 20% or more of the center frequency, f_0, resulting in a filter Q = f_0 / BW less than 5), the attenuation at frequencies above the center frequency will

be less than below the center frequency. This characteristic should be taken into account when attenuation of harmonics of signals in the passband is of concern. The filter shown in **Fig 11.31A** is designed to pass the amateur 75 meter band and suffers from this defect. By going to the *nodal inductor-coupled* topology we can correct this problem. The resulting design is shown in Fig 11.31B. Another way to accomplish the task is by going to a *mesh capacitor-coupled* design as shown in Fig

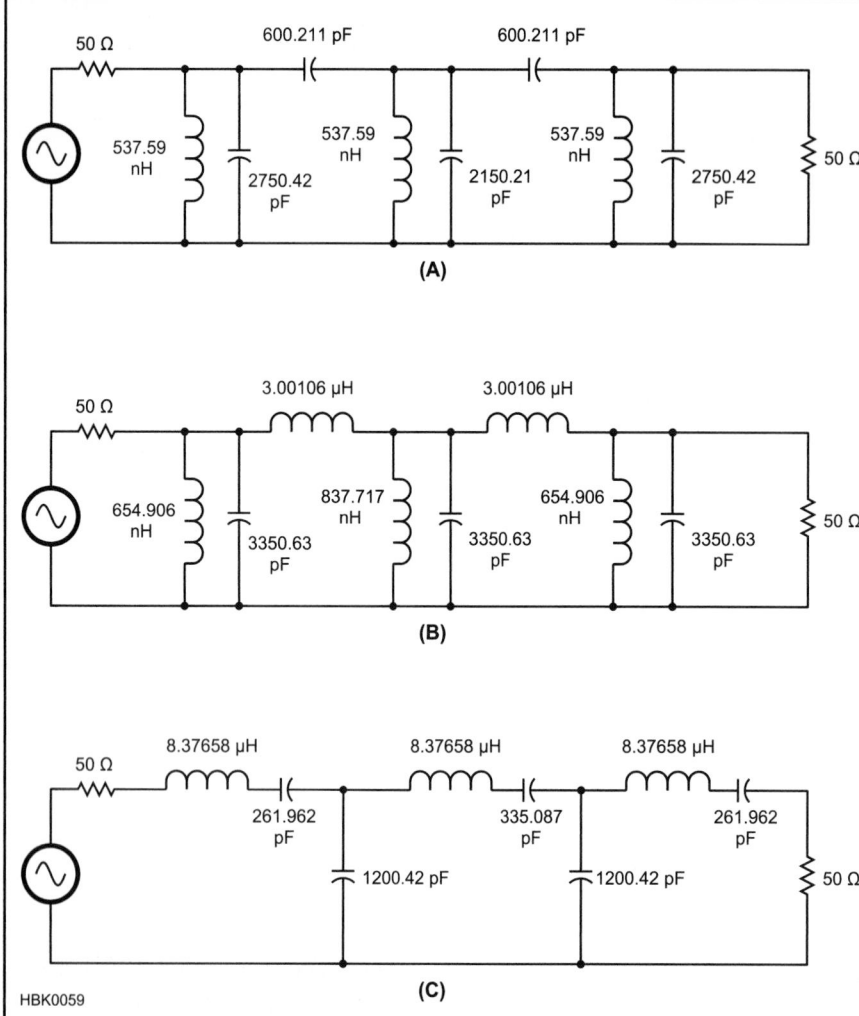

Fig 11.31 — Three band-pass filters designed to pass the 75 meter amateur band. The nodal-capacitor-coupled topology is shown at A, nodal-inductor-coupled at B, and mesh capacitor-coupled at C.

the steepness of the filter's rolloff. Band-pass filters (and especially narrowband band-pass filters) are more vulnerable to this problem than low-pass and high-pass filters. **Fig 11.33** shows the effect of finite values of inductor Q values on the response of a low-pass filter. The Q values for each plot are as shown.

Inadequate component Q values introduce loss and more importantly they compromise the filter's response at cutoff, especially problematic in the case of narrowband band-pass filters. **Fig 11.34** illustrates the effect of finite inductor Q values on a narrowband band-pass filter. In the case of a band-pass filter, the Q values required to support a given response shape are much higher than those required for the low-pass or high-pass filter (by the ratio of center to width). Capacitor Q values are generally much high than inductor Q values and so contribute far less to this effect.

In general, component-value adjustment will not be able to fully compensate for inadequate component Q values. However, if the filter is deliberately mismatched (by changing the input and/or output terminations) then a limited amount of *response-shape* correction can sometimes be achieved by network component value optimization ("tweaking"). The loss caused by Q problems (at dc in the case of a low-pass or at the center frequency in the case of a band-pass) may increase if such correction is attempted.

11.3.7 Side Effects of Passband Ripple

Especially in RF applications it is desirable to design a filter such that the impedance seen looking into the input side remains fairly constant over the passband. Increasing the value of passband ripple increases the rate of descent from the passband into the stop band, giving a

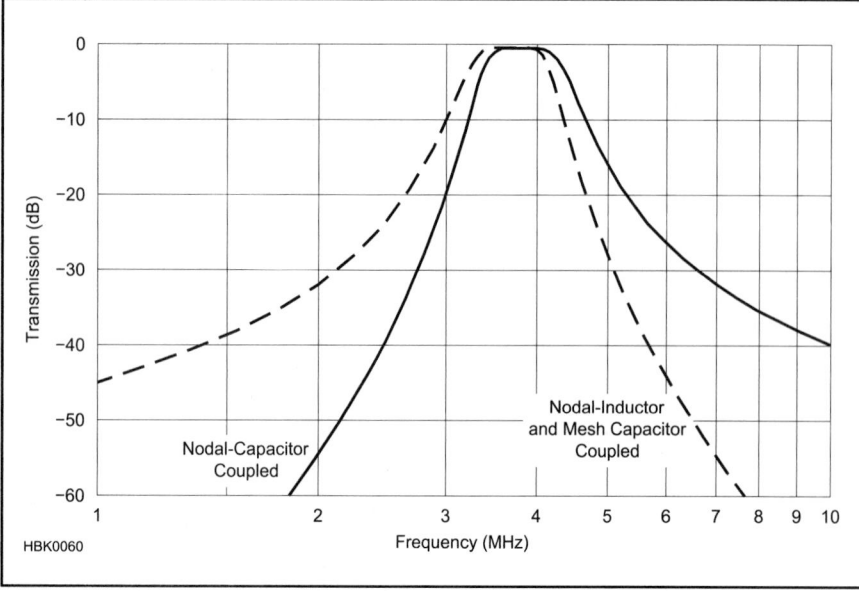

Fig 11.32 — Responses of the filters in Fig 11.31.

Table 11.2
Passband Ripple, VSWR and Return Loss

Passband Ripple (dB)	VSWR	Return Loss (dB)
0.0005	1.022	39.38
0.001	1.031	36.37
0.002	1.044	33.36
0.005	1.07	29.39
0.01	1.101	26.38
0.02	1.145	23.37
0.05	1.24	19.41
0.1	1.355	16.42
0.2	1.539	13.46
0.5	1.984	9.636
1	2.66	6.868

Note: As the passband ripple specification is changed so do the other items. Conversely, to get a particular value of VSWR or return loss, use this table to find the passband ripple that should be used to design the filter.

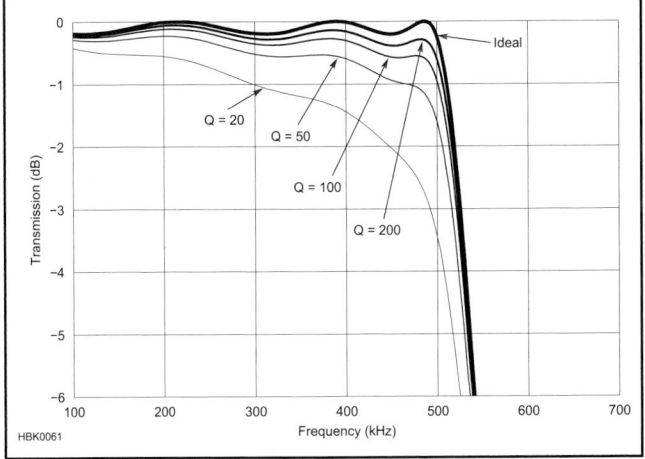

Fig 11.33 — Effect of inductor Q values on a low-pass filter.

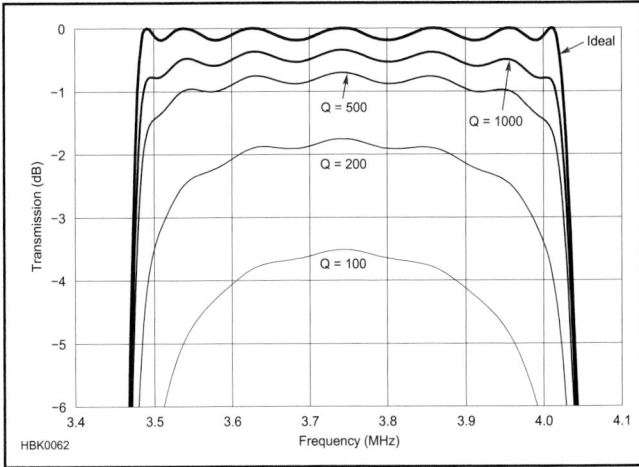

Fig 11.34 — Effect of inductor Q values on a narrowband band-pass filter.

sharper cutoff. But it also degrades the uniformity of the impedance across the passband as seen looking into the input of the filter. This may be shown in terms of VSWR or return loss; those are simply different ways of stating the same effect. (*Return loss* is explained in the **RF Techniques** chapter.)

Designs using a low value of passband ripple are preferred for RF work. Audio-frequency applications are generally not as critical, and so higher ripple values (up to about 0.2 dB) may be used in audio work.

Table 11.2 shows the values of VSWR and return loss for various values of passband ripple. Note that lower values for passband ripple yield better values for VSWR and return loss. Specifying a filter to have a passband ripple value of 0.01 dB will result in that filter's VSWR figure to be about 1.1:1. Or restated, the return loss will be about 26 dB. These values will be a function of frequency and at some test frequencies may be much better.

11.4 Filter Design Examples

Each topology of filter has its own mathematical techniques to determine the filter components, given the frequencies and attenuations that define the filter behavior. These techniques are very precise, involving the use of tables and scaling techniques as discussed in the following sections. With the availability of filter design software, such as *Elsie* and *SVC Filter Designer* (both available on the CD-ROM included with this book) filters using different topologies can be designed easily, however. (Additional programs and design resources are listed in the References section at the end of this chapter.)

Whether designing a filter manually or through the use of software, the design process begins with developing a set of filter *performance requirements* for:

- Type of response — low-pass, high-pass, band-pass, band-stop, or all-pass.
- Passband ripple — none or some maximum amount.
- Cutoff frequency (or frequencies) or bandwidth.
- Steepness of rolloff in transition region.
- Minimum stop-band attenuation and stop-band depth (optional).
- Group delay (optional).

The requirements shown as optional may not be important for simple filters. The next step is to identify the filter families that can meet those requirements. For example, if no ripple is allowed in the passband, the Butterworth or Bessel filter families would be suitable. If a very steep rolloff is required, try the Chebyshev or Cauer family.

After entering the basic requirements into the filter design software and observing the results of the calculation, the design is likely to require some adjustment. Component values can be unrealistic or the required order may be too high for practical implementation. At this point, your experience with filter design helps guide the choices of what changes to make — a different filter family, raising or lowering some of your requirements and so forth. The key is to experiment and observe to build understanding of filters.

Filter design and analysis programs show the responses of the resulting network in graphic form. This makes it easy to compare your filter to your requirements, making design adjustments quickly. Many design packages also allow the selection of nearest-5% values, tuning and other useful features. For example, the program *SVC Filter Designer* (SVC is *standard value component*) automatically selects the nearest 5% capacitor values. It selects the nearest 5% inductor values as an option. It also shows the resulting response degradation (which may be minor).

Although filter-design software has greatly

simplified the design of lumped-element filters, going through the manual design process, which is based on the use of tables of component values, will give some insight into how to make better use of the software to develop and refine a filter.

11.4.1 Normalized Values

The equations used to design filters are quite complex, so to ease the design process, tables of component values have been developed for different families of filters. Developing a set of

Table 11.3
Butterworth Normalized Values

Order	G(1)	G(2)	G(3)	G(4)	G(5)	G(6)	G(7)	G(8)	G(9)	G(10)	G(11)	R_{load}
3	1	2	1									1
4	0.7654	1.848	1.848	0.7654								1
5	0.618	1.618	2	1.618	0.618							1
6	0.5176	1.414	1.932	1.932	1.414	0.5176						1
7	0.445	1.247	1.802	2	1.802	1.247	0.445					1
8	0.3902	1.111	1.663	1.962	1.962	1.663	1.111	0.3902				1
9	0.3473	1 1.532	1.879	2	1.879	1.532	1	0.3473				1
10	0.3129	0.908	1.414	1.782	1.975	1.975	1.782	1.414	0.908	0.3129		1
11	0.2846	0.8308	1.31	1.683	1.919	2	1.919	1.683	1.31	0.8308	0.2846	1

Table 11.4
Bessel Normalized Values

Order	G(1)	G(2)	G(3)	G(4)	G(5)	G(6)	G(7)	G(8)	G(9)	G(10)	R_{load}
3	2.203	0.9705	0.3374								1
4	2.24	1.082	0.6725	0.2334							1
5	2.258	1.111	0.804	0.5072	0.1743						1
6	2.265	1.113	0.8538	0.6392	0.4002	0.1365					1
7	2.266	1.105	0.869	0.702	0.5249	0.3259	0.1106				1
8	2.266	1.096	0.8695	0.7303	0.5936	0.4409	0.2719	0.0919			1
9	2.265	1.086	0.8639	0.7407	0.6306	0.5108	0.377	0.2313	0.078		1
10	2.264	1.078	0.8561	0.742	0.6493	0.5528	0.4454	0.327	0.1998	0.0672	1

Table 11.5
Chebyshev Normalized Values
Passband ripple 0.01 dB

Order	G(1)	G(2)	G(3)	G(4)	G(5)	G(6)	G(7)	G(8)	G(9)	G(10)	G(11)	R_{load}
3	0.6292	0.9703	0.6292									1
4	0.7129	1.2	1.321	0.6476								0.9085
5	0.7563	1.305	1.577	1.305	0.7563							1
6	0.7814	1.36	1.69	1.535	1.497	0.7098						0.9085
7	0.7969	1.392	1.748	1.633	1.748	1.392	0.7969					1
8	0.8073	1.413	1.782	1.683	1.853	1.619	1.555	0.7334				0.9085
9	0.8145	1.427	1.804	1.713	1.906	1.713	1.804	1.427	0.8145			1
10	0.8196	1.437	1.819	1.731	1.936	1.759	1.906	1.653	1.582	0.7446		0.9085
11	0.8235	1.444	1.83	1.744	1.955	1.786	1.955	1.744	1.83	1.444	0.8235	1

Table 11.6
Chebyshev Normalized Values
Passband ripple 0.044 dB

Order	G(1)	G(2)	G(3)	G(4)	G(5)	G(6)	G(7)	G(8)	G(9)	G(10)	G(11)	R_{load}
3	0.855	1.104	0.855									1
4	0.9347	1.293	1.581	0.7641								0.8175
5	0.9747	1.372	1.805	1.372	0.9747							1
6	0.9972	1.413	1.896	1.55	1.729	0.8153						0.8175
7	1.011	1.437	1.943	1.621	1.943	1.437	1.011					1
8	1.02	1.452	1.969	1.657	2.026	1.61	1.776	0.8341				0.8175
9	1.027	1.462	1.986	1.677	2.067	1.677	1.986	1.462	1.027			1
10	1.031	1.469	1.998	1.69	2.091	1.709	2.067	1.633	1.797	0.8431		0.8175
11	1.035	1.474	2.006	1.698	2.105	1.727	2.105	1.698	2.006	1.474	1.035	1

Table 11.7
Chebyshev Normalized Values
Passband ripple 0.2 dB

Order	G(1)	G(2)	G(3)	G(4)	G(5)	G(6)	G(7)	G(8)	G(9)	G(10)	G(11)	R_{load}
3	1.228	1.153	1.228									1
4	1.303	1.284	1.976	0.8468								0.65
5	1.339	1.337	2.166	1.337	1.339							1
6	1.36	1.363	2.239	1.456	2.097	0.8838						0.65
7	1.372	1.378	2.276	1.5	2.276	1.378	1.372					1
8	1.38	1.388	2.296	1.522	2.341	1.493	2.135	0.8972				0.65
9	1.386	1.394	2.309	1.534	2.373	1.534	2.309	1.394	1.386			1
10	1.39	1.398	2.318	1.542	2.39	1.554	2.372	1.507	2.151	0.9035		0.65
11	1.393	1.402	2.324	1.547	2.401	1.565	2.401	1.547	2.324	1.402	1.393	1

Table 11.8
Cauer Normalized Values
Passband ripple 0.01 dB, Stop band depth 30 dB

Order	G(1)	G(2)	H(2)	G(3)	G(4)	H(4)	G(5)	G(6)	H(6)	G(7)	R_{load}	F_{stop}
3	0.5835	0.8848	0.06895	0.5835							1	3.524
4	0.4844	0.954	0.1799	1.109	0.6392	0					1	2.215
5	0.6173	1.106	0.1913	1.264	0.7084	0.6533	0.3403				1	1.418
6	0.4158	0.936	0.3958	1.074	0.7642	0.7979	0.9396	0.7292	0		1	1.252
7	0.6095	1.117	0.2497	1.027	0.5503	1.459	0.8809	0.5513	1.175	0.1594	1	1.105

Table 11.9
Cauer Normalized Values
Passband ripple 0.01 dB, Stop band depth 40 dB

Order	G(1)	G(2)	H(2)	G(3)	G(4)	H(4)	G(5)	G(6)	H(6)	G(7)	R_{load}	F_{stop}
3	0.6074	0.9299	0.03092	0.6074							1	5.12
4	0.5463	1.057	0.09503	1.144	0.634	0					1	2.885
5	0.6678	1.179	0.1179	1.359	0.9065	0.3574	0.4904				1	1.687
6	0.5089	1.063	0.2549	1.208	0.992	0.4761	1.066	0.7275	0		1	1.416
7	0.6636	1.197	0.173	1.186	0.7766	0.8845	1.04	0.7415	0.7093	0.3321	1	1.191

Table 11.10
Cauer Normalized Values
Passband ripple 0.01 dB, Stop band depth 50 dB

Order	G(1)	G(2)	H(2)	G(3)	G(4)	H(4)	G(5)	G(6)	H(6)	G(7)	R_{load}	F_{stop}
3	0.6194	0.9518	0.01415	0.6194							1	7.47
4	0.5813	1.116	0.05176	1.165	0.6309	0					1	3.797
5	0.7007	1.225	0.07357	1.432	1.045	0.2096	0.5872				1	2.045
6	0.5739	1.156	0.168	1.317	1.167	0.3023	1.157	0.7255	0		1	1.634
7	0.7019	1.252	0.122	1.319	0.9711	0.5823	1.192	0.899	0.4614	0.4575	1	1.309

Table 11.11
Cauer Normalized Values
Passband ripple 0.01 dB, Stop band depth 60 dB

Order	G(1)	G(2)	H(2)	G(3)	G(4)	H(4)	G(5)	G(6)	H(6)	G(7)	R_{load}	F_{stop}
3	0.6246	0.9617	0.00652	0.6246							1	10.94
4	0.6011	1.15	0.02862	1.178	0.629	0					1	5.025
5	0.7209	1.254	0.04608	1.482	1.137	0.1269	0.6486				1	2.514
6	0.6191	1.222	0.1121	1.399	1.296	0.1978	1.223	0.7238	0		1	1.913
7	0.7283	1.291	0.08673	1.425	1.131	0.3982	1.322	1.024	0.312	0.5489	1	1.463

Table 11.12
Cauer Normalized Values
Passband ripple 0.044 dB, Stop band depth 30 dB

Order	G(1)	G(2)	H(2)	G(3)	G(4)	H(4)	G(5)	G(6)	H(6)	G(7)	R_{load}	F_{stop}
3	0.7873	0.9891	0.09924	0.7873							1	2.788
4	0.6172	1.046	0.2284	1.239	0.8101	0					1	1.883
5	0.794	1.119	0.2452	1.353	0.6862	0.8268	0.4694				1	1.286
6	0.5391	0.9646	0.4632	1.064	0.7337	0.9564	0.9908	0.8958	0		1	1.171
7	0.7808	1.11	0.3034	1.058	0.4585	1.892	0.8492	0.5271	1.379	0.2962	1	1.065

Table 11.13
Cauer Normalized Values
Passband ripple 0.044 dB, Stop band depth 40 dB

Order	G(1)	G(2)	H(2)	G(3)	G(4)	H(4)	G(5)	G(6)	H(6)	G(7)	R_{load}	F_{stop}
3	0.8229	1.05	0.04455	0.8229							1	4.02
4	0.6958	1.176	0.1209	1.287	0.8064	0					1	2.429
5	0.8597	1.211	0.1509	1.491	0.9058	0.4527	0.6448				1	1.504
6	0.6471	1.112	0.299	1.227	0.9893	0.5657	1.126	0.8992	0		1	1.304
7	0.8479	1.204	0.2085	1.242	0.6828	1.118	1.045	0.7251	0.833	0.4797	1	1.132

Table 11.14
Cauer Normalized Values
Passband ripple 0.044 dB, Stop band depth 50 dB

Order	G(1)	G(2)	H(2)	G(3)	G(4)	H(4)	G(5)	G(6)	H(6)	G(7)	R_{load}	F_{stop}
3	0.8401	1.079	0.02038	0.8401							1	5.85
4	0.7411	1.253	0.06596	1.316	0.8039	0					1	3.178
5	0.9012	1.269	0.094	1.594	1.064	0.2658	0.761				1	1.803
6	0.724	1.22	0.1977	1.362	1.192	0.3584	1.227	0.9004	0		1	1.487
7	0.8928	1.27	0.1462	1.401	0.8857	0.7262	1.232	0.8927	0.5426	0.6158	1	1.229

Table 11.15
Cauer Normalized Values
Passband ripple 0.044 dB, Stop band depth 60 dB

Order	G(1)	G(2)	H(2)	G(3)	G(4)	H(4)	G(5)	G(6)	H(6)	G(7)	R_{load}	F_{stop}
3	0.8481	1.093	0.0094	0.8481							1	8.553
4	0.767	1.297	0.0365	1.333	0.8025	0					1	4.192
5	0.9275	1.307	0.05888	1.667	1.172	0.1613	0.8371				1	2.199
6	0.7782	1.298	0.1323	1.465	1.346	0.2346	1.301	0.9008	0		1	1.725
7	0.9275	1.317	0.104	1.531	1.056	0.4957	1.395	1.027	0.3689	0.7207	1	1.359

Table 11.16
Cauer Normalized Values
Passband ripple 0.2 dB, Stop band depth 30 dB

Order	G(1)	G(2)	H(2)	G(3)	G(4)	H(4)	G(5)	G(6)	H(6)	G(7)	R_{load}	F_{stop}
3	1.116	0.9996	0.1584	1.116							1	2.207
4	0.8176	1.087	0.3037	1.335	1.065	0					1	1.61
5	1.084	1.031	0.3443	1.482	0.5907	1.153	0.6819				1	1.18
6	0.7319	0.9415	0.5675	1.024	0.6677	1.186	1.002	1.149	0		1	1.106
7	1.065	1.01	0.4035	1.124	0.3345	2.756	0.8246	0.451	1.785	0.5138	1	1.036

Table 11.17
Cauer Normalized Values
Passband ripple 0.2 dB, Stop band depth 40 dB

Order	G(1)	G(2)	H(2)	G(3)	G(4)	H(4)	G(5)	G(6)	H(6)	G(7)	R_{load}	F_{stop}
3	1.174	1.08	0.07104	1.174							1	3.147
4	0.9224	1.254	0.1607	1.399	1.067	0					1	2.047
5	1.175	1.14	0.2106	1.686	0.8175	0.6264	0.8974				1	1.351
6	0.8638	1.107	0.3667	1.221	0.9522	0.6869	1.139	1.164	0		1	1.21
7	1.156	1.116	0.2739	1.35	0.5366	1.558	1.073	0.6422	1.069	0.7205	1	1.084

Table 11.18
Cauer Normalized Values
Passband ripple 0.2 dB, Stop band depth 50 dB

Order	G(1)	G(2)	H(2)	G(3)	G(4)	H(4)	G(5)	G(6)	H(6)	G(7)	R_{load}	F_{stop}
3	1.203	1.118	0.03253	1.203							1	4.554
4	0.9842	1.355	0.08773	1.438	1.067	0					1	2.654
5	1.233	1.211	0.1309	1.842	0.9885	0.3676	1.047				1	1.595
6	0.9599	1.23	0.2433	1.387	1.19	0.4317	1.246	1.173	0		1	1.36
7	1.213	1.192	0.1904	1.548	0.7322	0.9884	1.313	0.8105	0.694	0.8771	1	1.161

Table 11.19
Cauer Normalized Values
Passband ripple 0.2 dB, Stop band depth 60 dB

Order	G(1)	G(2)	H(2)	G(3)	G(4)	H(4)	G(5)	G(6)	H(6)	G(7)	R_{load}	F_{stop}
3	1.216	1.137	0.01501	1.216							1	6.64
4	1.02	1.413	0.04859	1.46	1.067	0					1	3.484
5	1.272	1.256	0.08207	1.953	1.108	0.2237	1.149				1	1.924
6	1.029	1.32	0.1634	1.517	1.375	0.2818	1.328	1.178	0		1	1.56
7	1.257	1.244	0.1348	1.716	0.9025	0.6672	1.527	0.9477	0.4718	1.001	1	1.269

tables for all possible filter requirements would be impractical, so *tables of normalized values* are used based on a low-pass filter with a bandwidth of 1 radian per second and with 1-Ω input termination impedance. Performance requirements for the desired filter are *normalized* to 1 rad/sec bandwidth and 1-Ω terminations and the tables used to determine the various component values for those frequencies and impedance. That filter is then transformed to the desired response if necessary — high-pass, band-pass, etc. Then the filter is converted to the desired frequency and impedance by *denormalizing* or *scaling* the component values with another set of transformations. In this way, a (relatively) small set of tables can be used to design filters of any type as you'll see in the following examples.

Summarizing, the table-based design process consists of:
- Determining performance requirements.
- Normalizing the requirements to a 1-Hz, 1-Ω, low-pass response.
- Selecting a family and order.
- Obtaining normalized component values from the tables.
- Transforming the filter to the desired response.
- Denormalizing the component values for frequency and impedance.

11.4.2 Filter Family Selection

The tables presented here are for low-pass filters from different families (Butterworth, Bessel, Chebyshev and Cauer), with various specifications for passband ripple and so on. In the normalized tables, capacitor values are in farads (F) and inductor values are in henries (H). These are converted to actual component values through the process of denormalization described below.

The Butterworth family (described by normalized component values in **Table 11.3**) is used when there should be no ripple at all in the passband, so that the response is to be as flat as possible. This trait is particularly apparent near dc for the low-pass response, at the center frequency for a band-pass response, and at infinity for the high-pass response. The resulting magnitude response will also have a relatively gentle transition from the passband into the stop band.

When a low-pass filter with constant group delay throughout the passband is desired then the Bessel family, whose normalized component values are shown in **Table 11.4**, should be used.

The Chebyshev family is used when a sharper cutoff is desired for a given number of components and where at least a small amount of ripple is allowable in the passband. The Chebyshev filter tables are broken into groups according to logically chosen values of passband ripple. The tables presented here

are for passband ripple values of 0.01 dB (**Table 11.5**, for critical RF work), 0.044 dB (**Table 11.6**, an intermediate value offering 20 dB of return loss or 1.2:1 VSWR) and 0.2 dB (**Table 11.7**, for audio and less-critical work where a steeper rolloff into the stop band is of primary concern). Some published tables show figures for passband ripple values less than 0.01 dB. These are difficult to implement because of the tight component tolerances required to achieve the expected responses.

When steepness of rolloff from passband into stop band is the item of greatest importance, then the Cauer filter family is used. Cauer filters involve a more complicated set of choices. In addition to selecting a passband ripple, the designer must also assign a stop band depth (or stop band frequency). Some of the items interact; they can't all be selected arbitrarily.

The most likely combinations of items to be chosen for Cauer filters are presented in **Tables 11.8** to **11.19**. These tables are for passband ripple values of 0.01, 0.044 and 0.2 dB (the same values that were chosen for the Chebyshev family) and stop band depths of 30, 40, 50 and 60 dB. The frequency at which the attenuation first reaches the design stop band depth value is shown in the final column, labeled F_{stop}. As with the Chebyshev filters, some published tables show ripple values of less than 0.01 dB but such designs are difficult to implement in practice because of the tight tolerances required on all of the components.

DESIGN EXAMPLE — CHEBYSHEV LOW-PASS

Here is an example of using normalized-value tables to design a low-pass Chebyshev filter with 0.01 dB of passband ripple, a bandwidth of 4.2 MHz, a rolloff requirement of 25 dB of attenuation one octave above cutoff, and an input termination of 50 Ω. **Fig 11.35** illustrates the attenuation to be expected for various orders (N) of a Chebyshev low-pass filter with 0.01 dB of passband ripple. The next

step is to determine the lowest order that can meet the requirements for roll off. Based on the rolloff requirement, a fifth-order filter is the lowest order filter for which the attenuation curve is below 25 dB at twice the normalized cutoff frequency of 1. (If 14 dB of attenuation one octave above the cutoff frequency had been required, a fourth-order filter would suffice.)

Fig 11.36 shows the schematic of a fifth-order low-pass filter that will meet the requirements. (If a higher- or lower-order filter is required, add or subtract elements, beginning with G(1) at the filter input, with capacitors as the parallel or shunt elements and inductors as the series elements.)

Next, refer to Table 11.5 to obtain the normalized component values for the 1-Ω and 1 radian/second low-pass filter. Choosing the table row with component values for Order = 5, G(1) is 0.7563, G(2) is 1.305, G(3) is 1.577, G(4) is 1.305 and G(5) is 0.7563. The last value, R_{load}, is used to calculate the output termination.

The equations to calculate the actual component values are shown in Fig 11.36A. Ω_c is the denormalizing factor used to scale the normalized filter component values to the desired frequency

$$\Omega_c = 2\pi\, F_{width} \tag{1}$$

In this case, $\Omega_c = 2 \times 3.1416 \times 4.2 \times 10^6 = 26.389 \times 10^6$. R is the filter's input termination, in this case 50 Ω. The termination on the output side of the filter will be the input termination value multiplied by R_{load}. In this case, $R_{load} = 1$, so the input and output terminations are equal. (Chebyshev *even-ordered* low-pass and high-pass filters have an output termination different from the input, as shown in the last column under R_{load}.) After denormalizing, the resulting component values are shown in Fig 11.36B. At this point in a table-based design process, the closest available standard value or adjustable components would be used to fabricate the actual circuit.

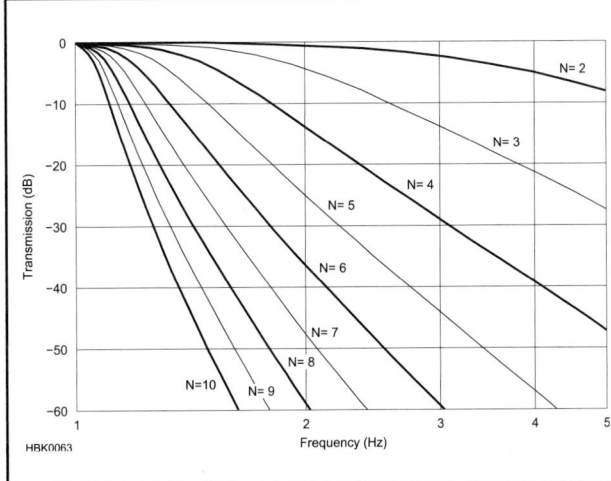

Fig 11.35 — This graph shows transmission for Chebyshev low-pass filters with 0.01 dB passband ripple filters with orders from 2 through 10. For a given order, higher passband ripple values will have higher attenuation (less transmission) at a given frequency while Butterworth filters will have less attenuation at a given frequency.

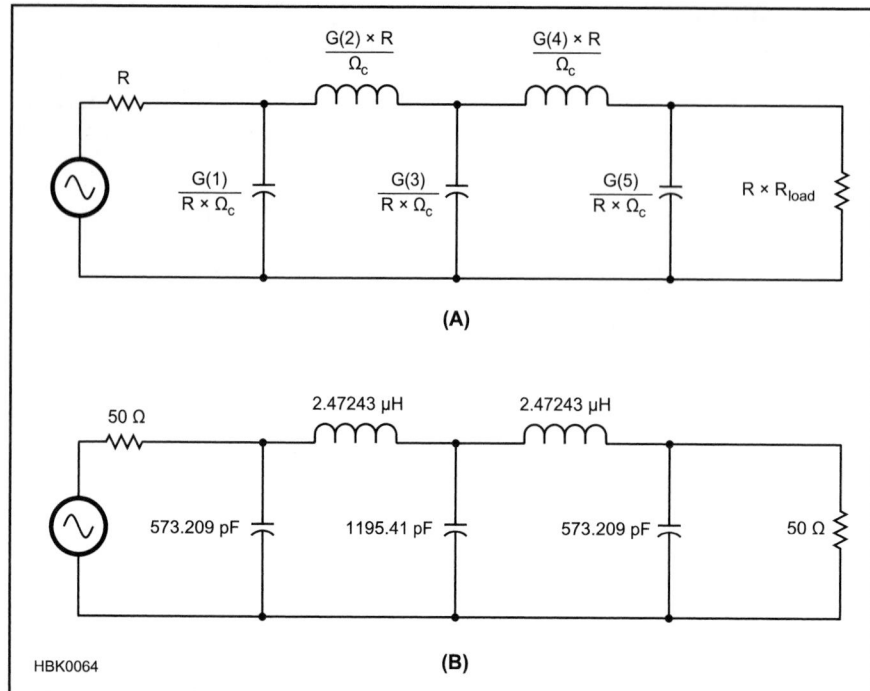

Fig 11.36 — Design example for a Chebyshev low-pass filter using the normalized filter tables. The topology and design equations are shown at A, with the resulting calculated parts values at B.

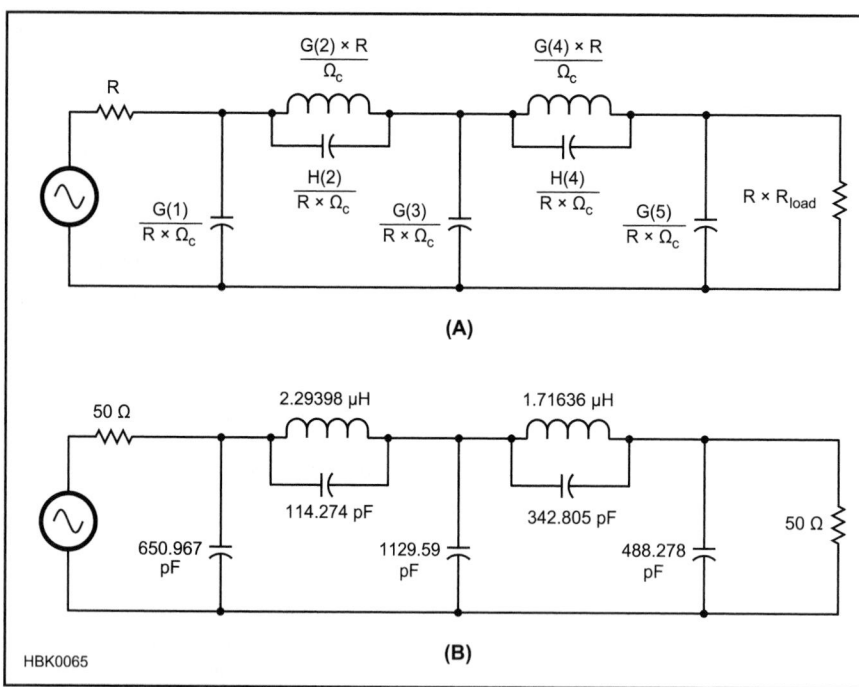

Fig 11.37 — Design example for a Cauer low-pass filter using the normalized filter tables. The topology and design equations are shown at A, with the resulting calculated parts values at B.

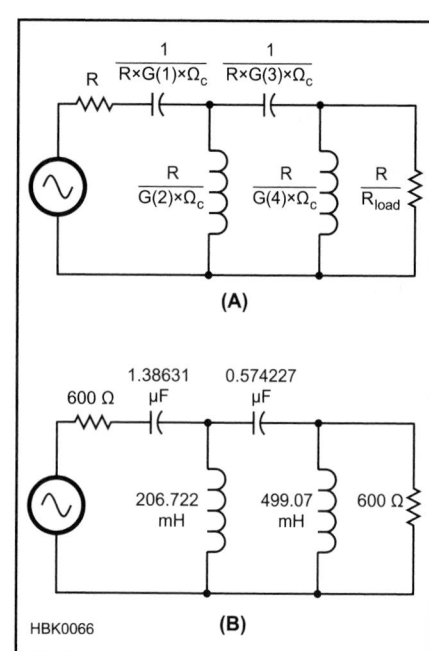

Fig 11.38 — Design example for a Butterworth high-pass filter using the normalized filter tables. The topology and design equations are shown at A, with the resulting calculated parts values at B. For a Cauer family high-pass, trap capacitors would be inserted in series with the shunt inductors. Their value is calculated by the same expression as the series capacitors, but using the $H(n)$ values from the Cauer tables.

DESIGN EXAMPLE — CAUER LOW-PASS

The design of a Cauer low-pass filter is very similar to the design of a Chebyshev, but it has added capacitors across the series inductors to form the traps that create the stop band notches. Graphs showing the stop band performance or loss of the various Cauer filters in the tables are impractical because of the large number of options to be chosen.

This example illustrates the design of a fifth-order Cauer capacitor-input low-pass filter with 0.044 dB passband ripple and 40-dB stop band depth. The input termination will be 50 Ω and the ripple bandwidth will again be 4.2 MHz.

Fig 11.37 shows the filter schematic. The first step is to obtain from Table 11.13 the normalized 1-Ω and 1 radian/second component values. Choosing the table row of component values for Order = 5, G(1) is 0.8597, G(2) is 1.211, G(3) is 1.491, G(4) is 0.9058 and G(5) is 0.6448. In addition, the "trap" capacitor values must be retrieved. The trap capacitors are shown as "H" values. H(2) is 0.1509 and H(4) is 0.4527. The output termination is 1 and so is the same as the input termination.

The filter schematic with the denormalization equations appears in Fig 11.37A. Ω_c and R have the same definitions as in the previous example. The resulting actual component values for the required frequency and impedance are shown in Fig 11.37B.

DESIGN EXAMPLE — HIGH-PASS BUTTERWORTH

The design of a high-pass filter will now be illustrated. Fig 11.35 can be used to estimate the response of a high-pass design just as it was for a low-pass. The only manipulation that needs to be done is to use the *reciprocal* of frequency when estimating the magnitude response. For example, if the attenuation requirement for the high-pass filter is for 10 dB of attenuation at 0.2 times the cutoff frequen-cy, then when using the low-pass filter response graphs, look up the attenuation at 1/0.2 = 5 times the cutoff frequency, instead.

For this example, we will design a fourth-order capacitor-input Butterworth high-pass response with the 3-dB cutoff frequency at 250 Hz and system impedance of 600 Ω. The filter schematic is shown in **Fig 11.38**. (Remember that even-order ladder filters can have either a series or parallel element at the input.)

The normalized values for the low-pass response are taken from Table 11.3, and the design equations are shown in Fig 11.38A. From the table row of component values for Order = 4, the value for G(1) is 0.7654, G(2) is 1.848, G(3) is 1.848 and G(4) is 0.7654.

Ω_c is computed the same way as in the previous examples. For the high-pass configuration, the output termination is based on the *reciprocal* of R_{load}. In this case, the table shows that the output termination is 1, so it is the same as the input. (Unlike the Chebyshev family, the output termination for the Butterworth family is always the same as the input termination.) Fig 11.38B shows the resulting denormalized component values for the actual filter.

DESIGN EXAMPLE — WIDE BAND-PASS

The design of a band-pass filter of appreciable percentage bandwidth will be looked at next. For the purposes of this discussion, "appreciable" means a bandwidth of 20% of the center frequency or higher (a filter Q of 5 or less).

Fig 11.39 shows this example, a third-order shunt-input Chebyshev type with

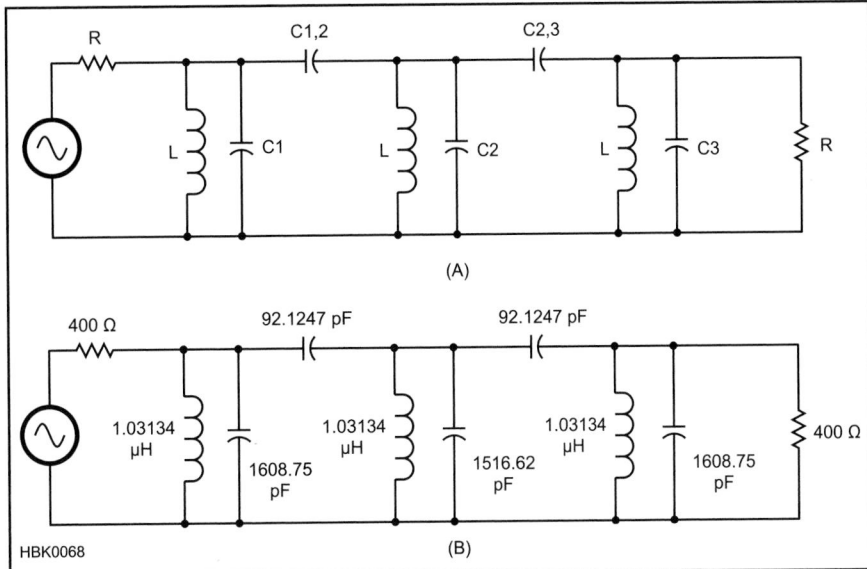

Fig 11.40 — Design example for a Chebyshev narrow band-pass filter using the normalized filter tables. The topology is shown at A and the design equations are given in the text, with the resulting calculated parts values at B.

0.2 dB of passband ripple with an input termination of 75 Ω. The center frequency is to be 4 MHz and the ripple bandwidth is to be 1 MHz. Again, Fig 11.35 can be used to determine the required filter order, bearing in mind that it is precise for passband ripple of 0.01 dB.

The normalized values for the low-pass are taken from Table 11.7. Choosing the table row of component values for Order = 3, the value for G(1) is 1.228, G(2) is 1.153 and G(3) is 1.228.

In the equations used to calculate the actual

parts values, Q_L is the loaded Q. This is the ratio of F_{center} to F_{width}. Note that in this topology both Ω_c and Ω_o are used. Ω_c is defined as in the previous examples, while

$$\Omega_o = 2\pi\, F_{center} \tag{2}$$

The filter schematic with the denormalizing equations appears in Fig 11.39A, and calculated values for the actual filter are shown in Fig 11.39B.

DESIGN EXAMPLE — NARROW BAND-PASS

Narrow band-pass filters involve more calculations than wide band-pass types. The filter topology used in this discussion is shown in **Fig 11.40** — a set of shunt-connected parallel-tuned resonators coupled by relatively small-value coupling capacitors.

This example will be for a band-pass filter with a bandwidth of 200 kHz, centered at 3.8 MHz, using the Chebyshev family with a passband ripple of 0.044 dB. The order will be three, and the normalized values are found in Table 11.6.

As in the wide band-pass example, the equations used to calculate the actual parts values use Q_L (the loaded Q), which is the ratio of F_{center} to F_{width}. Once again both Ω_c and Ω_o are used as before The same inductor, L, is used for each resonator. The value of that inductor is given by:

$$L = \frac{R}{\Omega_o \times Q_L \times G(1)} \tag{3}$$

Each resonator has a basic tuning capacitor whose value is given by:

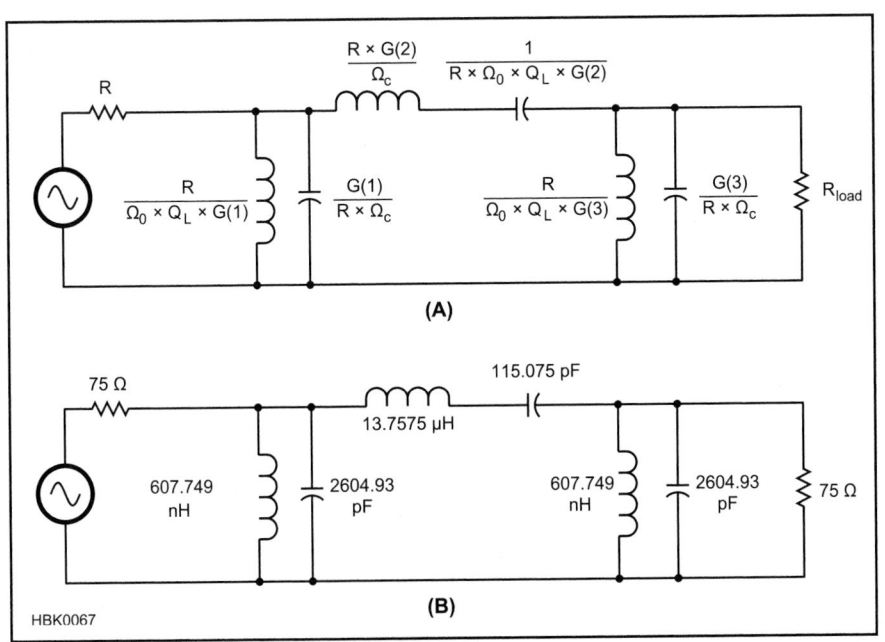

Fig 11.39 — Design example for a Chebyshev wide band-pass filter using the normalized filter tables. The topology and design equations are shown at A, with the resulting calculated parts values at B.

$$C_{basic} = \frac{G(1)}{R \times \Omega_c} \quad (4)$$

$$C2,3 = \frac{G(1)}{R \times \Omega_o}\sqrt{\frac{1}{G(2) \times G(3)}} \quad (6)$$

$$C1 = C_{basic} - C1,2$$
$$C2 = C_{basic} - C1,2 - C2,3$$
$$C3 = C_{basic} - C2,3$$

Next calculate the inter-resonator coupling capacitors:

$$C1,2 = \frac{G(1)}{R \times \Omega_o}\sqrt{\frac{1}{G(1) \times G(2)}} \quad (5)$$

The actual shunt tuning capacitors for each resonator are then the basic tuning capacitor minus the coupling capacitors on each side, as shown here:

The completed design is shown in Fig 11.40B. It should be evident how to extend this procedure to higher-order filters.

11.5 Active Audio Filters

Below RF, in what is broadly referred to as the "audio" range between a few Hz and a few hundred kHz, designers have several choices of filter technology.
- Passive LC
- Digital Signal Processing (DSP)
- Switched-Capacitor Audio Filter (SCAF)
- Active RC

LC audio filters are not used much in current designs except in high-power audio applications, such as speaker crossover networks, and are not covered here. LC filters were once popular as external audio filters for CW reception. These designs tended to be large and bulky, often using large surplus 44 or 88 mH core inductors. (A classic *ARRL Handbook* project to construct a passive LC CW filter can be found on the CD-ROM included with this book.) LC filters used in very low-level receiver applications can be very compact, but these tend to suffer from relatively high insertion loss, which reduces the receiver sensitivity (if no preamp is used ahead of the filter) or its large-signal dynamic range (if a loss-compensating preamp is used ahead of the filter). In addition, LC filters used in low-level receiver applications tend to pick up 60/120 Hz hum from stray magnetic fields from ac power transformers. Even "self shielded" inductors can produce noticeable ac hum pick up when followed by 90 to 120 dB of audio gain!

DSP filtering can yield filters that are superior to analog filters. (See the **DSP and Software Radio Design** chapter.) High signal-level DSP-based external audio filters such as the popular Timewave DSP-9 (**www.timewave.com**) and MFJ Enterprises MFJ-784B (**www.mfjenterprises.com**) have been available for some time. In addition, if the DSP is coupled to a very high-performance audio A/D converter such as an Asahi-Kasei AK5394A, DSP filtering can provide excellent filtering for even very low-level audio signals.

The main drawback to DSP processing is that of expense. A low-end DSP external audio filter costs at least $100, while a DSP with a high-end A/D converter (such as an audio sound card front-end to a PC) can run in the $150 to $400+ price range, which does not include the cost of the host PC.

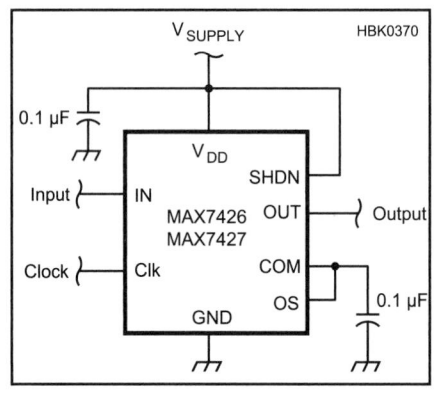

Fig 11.41 — Simple SCAF low pass filter, taken from the MAX7426 data sheet.

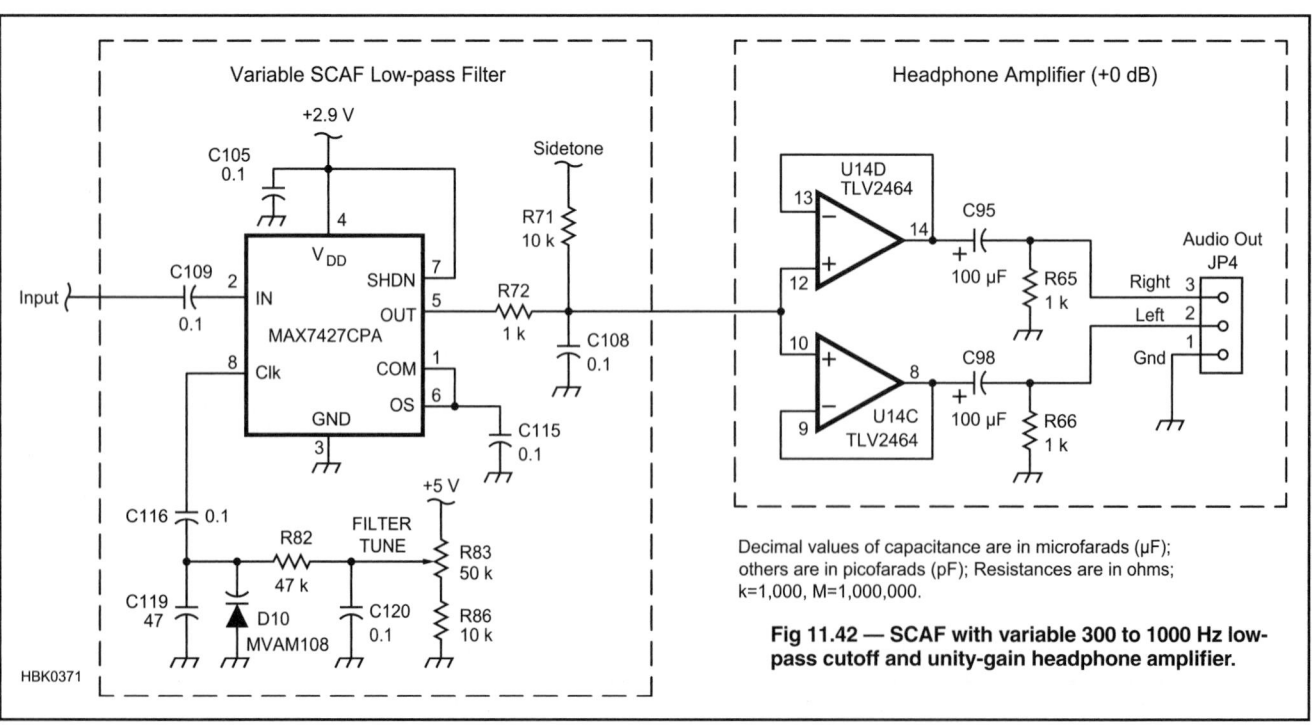

Decimal values of capacitance are in microfarads (µF); others are in picofarads (pF); Resistances are in ohms; k=1,000, M=1,000,000.

Fig 11.42 — SCAF with variable 300 to 1000 Hz low-pass cutoff and unity-gain headphone amplifier.

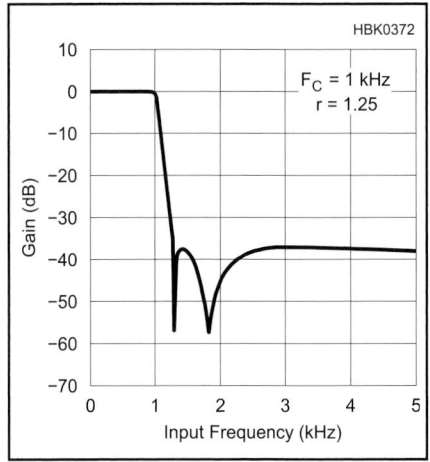

Fig 11.43 — Frequency response of a low-pass filter using the MAX7426 set to a 1 kHz cutoff frequency.

11.5.1 SCAF Filters

Simple SCAF filter designs can produce extremely effective low-power audio filters. The implementation of these filters in practical IC form usually involves small-value capacitors and high-value resistors. The use of high-value resistors tends to generate enough noise to make these filters unsuitable for very low-level audio processing such as the front end of a receiver audio chain. Thus, SCAF filters tend to be limited to filtering at the output end of the audio chain — headphone or speaker level audio applications, an area in which they very much excel. The cutoff frequency of a SCAF filter is set by an external clock signal. Thus, this class of filter naturally lends itself to use in variable frequency filters. For simplicity, excellent frequency selectivity and relatively low cost, SCAF filters are highly recommended when additional audio filtering is desired on the output of an existing receiver.

An example of a very simple, but highly effective SCAF low-pass filter IC is the Maxim MAX7426 (5 V supply) or MAX7427 (3 V supply) in **Fig 11.41.** The filter's cutoff frequency can be set by placing an appropriately sized capacitor across the clock oscillator inputs because the part can generate its own internal clock signal. For example, connecting a 180 pF capacitor across the MAX7426's clock inputs will produce a 1 kHz low-pass filter. The MAX7427 (3-V version) was used in the NC2030 QRP transceiver along with a MVAM108 varactor diode to create a low-pass filter that was tunable from 300 Hz to 1 kHz.

A portion of that schematic is shown in **Fig 11.42**. The SCAF low-pass filter is followed by a 3-V unity-gain headphone amplifier (U14C and U14D) as the filter IC itself cannot directly drive headphones. The low-pass cutoff frequency is tuned by using R83 to vary the voltage applied across the MVAM108 varactor diode D10, thus changing the capacitance across the clock input (pin 8) of the chip. C116 was used to isolate the dc voltage across the varactor diode from the bias voltage on the clock input line. The frequency response of such a filter is shown in **Fig 11.43**.

As can be seen, this is an extremely sharp filter, producing almost 40 dB of attenuation very close to the cut off frequency — not too bad for an inexpensive part. The slightly more expensive MAX7403 gives an even steeper 80 dB cutoff. Both parts are specified for an 80 dB signal-to-noise ratio based on an assumed 4-V_{P-P} signal. Sensitive modern in-ear type headphones require only 20 mV_{P-P} to produce a fairly loud signal. A 20 mV signal is 46 dB below 4 V_{P-P}, so at such headphone levels, the noise floor is actually only 34 dB below the 4 V_{P-P} signal — fairly quiet, but is a lot less noise margin than one would tend to think given the 80 dB specification.

Again, with a noise floor only 34 dB below typical headphone levels, SCAF filters such as these are fine at the end of the receiver chain, but are not useful as an audio filter at very low signal levels early on in a receiver.

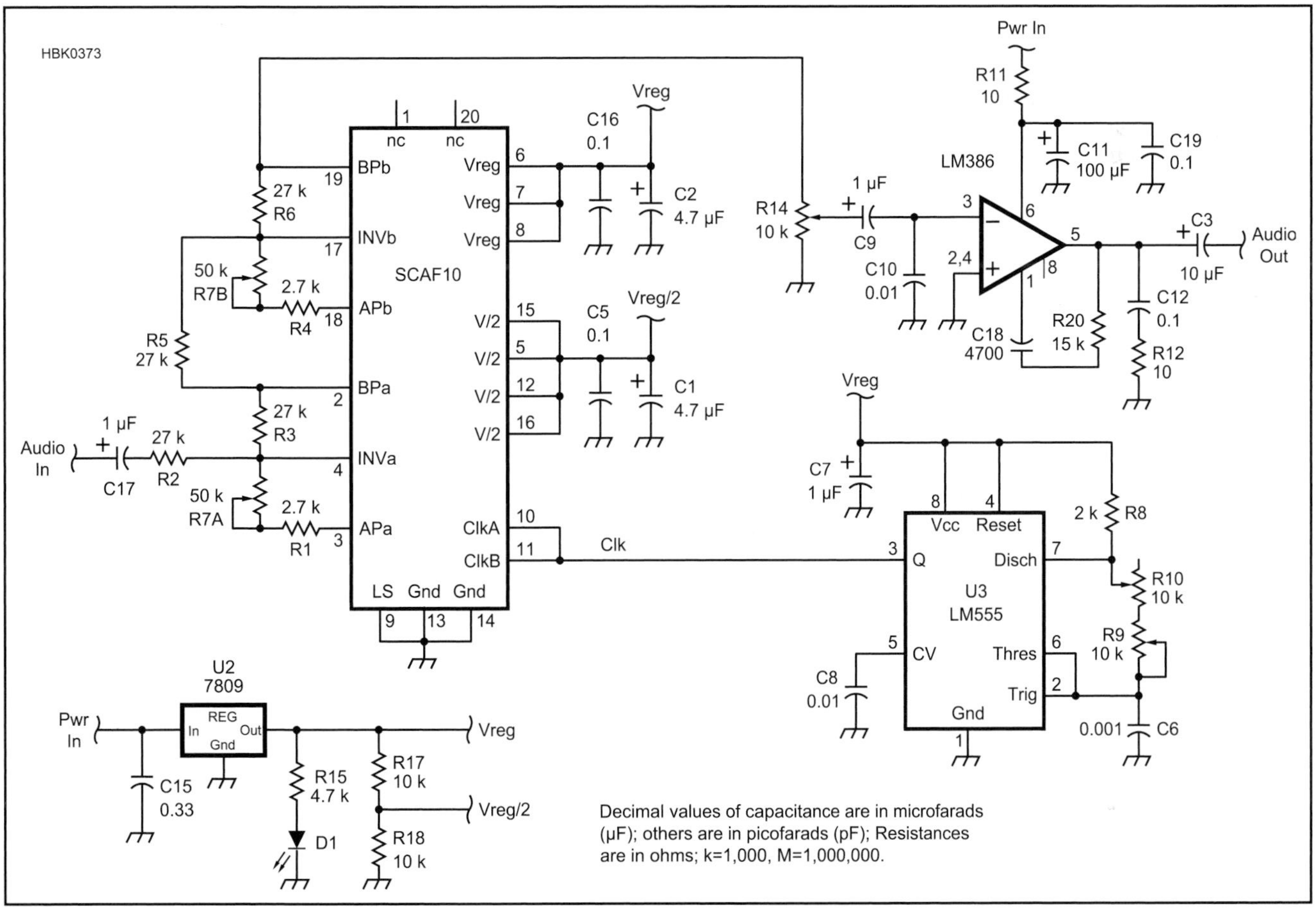

Fig 11.44 — Example of a simple SCAF band-pass filter from a New England QRP Club kit.

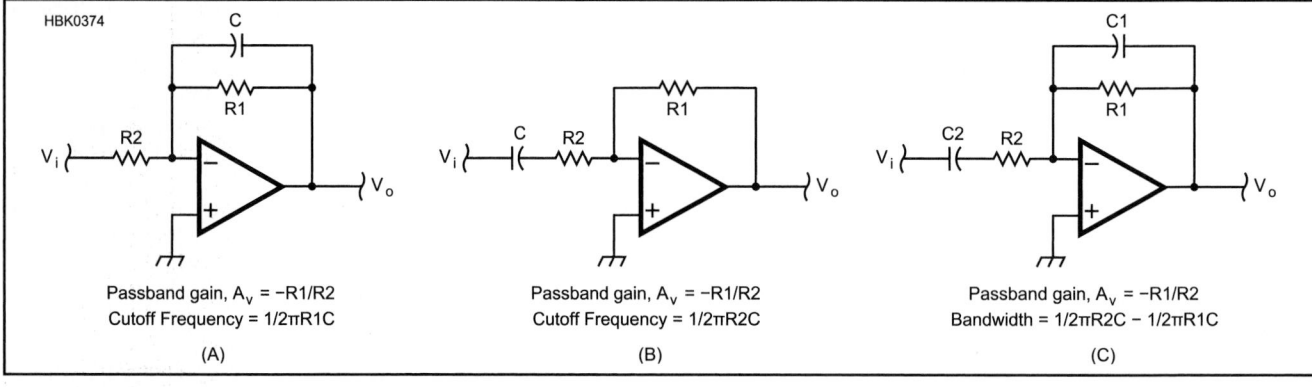

Fig 11.45 — Simple active filters. A low-pass filter is shown at A; B is a high-pass filter. The circuit at C combines the low- and high-pass filters into a wide band-pass filter.

Unless otherwise specified, values of R are in ohms, C is in farads, F in hertz and ω in radians per second. Calculations shown here were performed on a scientific calculator.

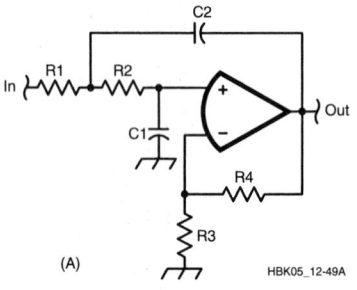

(A) HBK05_12-49A

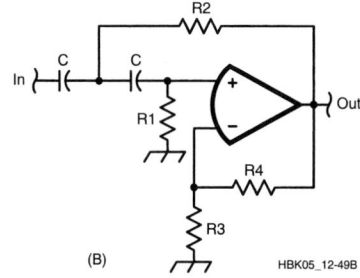

(B) HBK05_12-49B

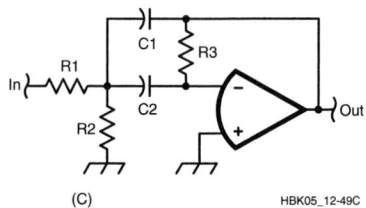

(C) HBK05_12-49C

Low-Pass Filter

$$C_1 \le \frac{\left[a^2 + 4(K-1)\right]C_2}{4}$$

$$R_1 = \frac{2}{\left[aC_2 + \sqrt{\left[a^2 + 4(K-1)\right]C_2^2 - 4C_1C_2}\right]\omega_C}$$

$$R_2 = \frac{1}{C_1C_2R_1\omega_C^2}$$

$$R_3 = \frac{K(R_1 + R_2)}{K-1} \quad (K>1)$$

$$R_4 = K(R_1 + R_2)$$

where
   K = gain
   f_c = −3 dB cutoff frequency
   $\omega_c = 2\pi f_c$
   C_2 = a standard value near $10/f_c$ (in μF)
Note: For unity gain, short R4 and omit R3.

Example:
   a = 1.414 (see table, one stage)
   K = 2
   f = 2700 Hz
   ω_c = 16,964.6 rad/sec
   C_2 = 0.0033 μF
   C1 ≤ 0.00495 μF (use 0.0050 μF)
   R1 ≤ 25,265.2 Ω (use 24 kΩ)
   R2 = 8,420.1 Ω (use 8.2 kΩ)
   R3 = 67,370.6 Ω (use 68 kΩ)
   R4 = 67,370.6 Ω (use 68 kΩ)

High-Pass Filter

$$R_1 = \frac{4}{\left[a + \sqrt{a^2 + 8(K-1)}\right]\omega_C C}$$

$$R_2 = \frac{1}{\omega_C^2 C^2 R_1}$$

$$R_3 = \frac{KR_1}{K-1} \quad (K>1)$$

$$R_4 = KR_1$$

where
   K = gain
   f_c = −3 dB cutoff frequency
   $\omega_c = 2\pi f_c$
   C = a standard value near $10/f_c$
   (in μF)

Note: For unity gain, short R4 and omit R3.

Example:
   a = 0.765 (see table, first of two stages)
   K = 4
   f = 250 Hz
   ω_c = 1570.8 rad/sec
   C = 0.04 μF (use 0.039 μF)
   R1 = 11,123.2 Ω (use 11 kΩ)
   R2 = 22,722 Ω (use 22 kΩ)
   R3 = 14,830.9 Ω (use 15 kΩ)
   R4 = 44,492.8 Ω (use 47 kΩ)

Band-Pass Filter

Pick K, Q, $\omega_o = 2\pi f_c$
where f_c = center freq.
Choose C
Then

$$R1 = \frac{Q}{K_0\omega_0 C}$$

$$R2 = \frac{Q}{\left(2Q^2 - K_0\right)\omega_0 C}$$

$$R3 = \frac{2Q}{\omega_0 C}$$

Example:
   K = 2, f_o = 800 Hz, Q = 5 and C = 0.022 μF
   R1 = 22.6 kΩ (use 22 kΩ)
   R2 = 942 Ω (use 910 Ω)
   R3 = 90.4 kΩ (use 91 kΩ)

Fig 11.46 — Equations for designing a low-pass RC active audio filter are given at A. B, C and D show design information for high-pass, band-pass and band-reject filters, respectively. All of these filters will exhibit a Butterworth response. Values of K and Q should be less than 10. See Table 11.20 for values of "a".

Fig 11.44 is an example of a slightly more complex SCAF band-pass filter. This filter features both a variable center frequency (450 Hz to 1000 Hz) and a variable bandwidth (90 Hz to 1500 Hz). This very popular filter was designed by the New England QRP Club (**www.newenglandqrp.org**) and is currently offered in kit form.

This filter features both sections of the SCAF10 IC configured as identical band-pass filters. The bandwidth (Q) is adjustable via R7A and R7B, while the center frequency is adjustable by R10 and the trimmer R9 as these resistors set the frequency of the LM555 clock generator.

As in the previous SCAF filter example, this filter also includes an audio amplifier (LM386) for driving either an external speaker or headphones as none of the SCAF filter ICs are capable of driving headphones directly.

11.5.2 Active RC Filters

Active RC filters based on op amp circuits can be used in either high-level audio output or very low-level direct-conversion receiver front-end filtering applications and thus are extremely flexible. Unlike LC filters, active RC filters can provide both filtering and gain at the same time, eliminating the relatively high insertion loss of a physically small, sharp LC filter. In addition, an active RC filter is not susceptible to the same ac hum pickup in low signal level applications as LC filters.

Active filters can be designed for gain and they offer excellent stage-to-stage isolation. The circuits require only resistors and capacitors, avoiding the limitations associated with inductors. By using gain and feedback, filter Q is controllable to a degree unavailable to passive LC filters. Despite the advantages, there are also some limitations. They require power, and performance may be limited by the op amp's finite input and output levels, gain and bandwidth. Active filters that drive speakers or other heavy loads usually employ an audio output amplifier circuit to boost the output power after a low-power filter stage.

A particular advantage of active RC filters for use in receivers is that they can be capable of extremely low noise operation, allowing the filtering of extremely small signals. At the same time they are capable of handling extremely large signals. Op amps are often capable of using ±18-V dual supply voltages or 36-V single supply voltages allowing the construction of a very high performance audio filter/amplifier chain that can handle signals up to 33 V_{P-P}. The ability to provide gain while also providing filtering provides a lot of flexibility in managing the sensitivity of a receiver audio chain.

The main disadvantage of active RC filters is their relatively high parts count compared to other filter types. In addition, active RC filters tend to be fixed-frequency designs unlike SCAF filters whose frequency can be moved simply by changing the clock frequency that drives the SCAF IC.

11.5.3 Active Filter Responses

Active filters can implement any of the passive LC filter responses described in the preceding section: low-pass, high-pass, band-pass, band-stop and all-pass. Filter family responses such as Butterworth, Chebyshev, Bessel and Cauer (elliptic) can be realized. All of the same family characteristics apply equally to passive and active filters and will not be repeated here. (Op amps are discussed in the **Analog Basics** chapter.)

Fig 11.45 presents circuits for first-order low-pass (Fig 11.45A) and high-pass (Fig 11.45B) active filters. The frequency response of these two circuits is the same as a parallel and series RC circuit, respectively, except that these two circuits can have a passband gain greater than unity. Rolloff is shallow at 6 dB/octave. The two responses can be combined to form a simple band-pass filter (Fig 11.45C). This combination cannot produce sharp band-pass filters because of the shallow rolloff.

To achieve high-order responses with steeper rolloff and narrower bandwidths, more complex circuits are required in which combinations of capacitors and resistors create *poles* and *zeroes* in the frequency response. (Poles and zeros are described in the **Analog Basics** chapter.) The various filter response families are created by different combinations of additional poles and zeroes. There are a variety of circuits that can be configured to implement the equations that describe the various families of filter responses. The most common circuits are *Sallen-Key* and *multiple-feedback*, but there are numerous other choices.

There are many types of active filters — this section presents some commonly used circuits as examples. A book on filter design (see the References section) will present more choices and how to develop designs based on the different circuit and filter family types. In addition, op amp manufacturer's publish numerous application notes and tutorials on active filter design. Several are listed in the References section of this chapter.

The set of tutorials published by Analog Devices is particularly good.

SECOND-ORDER ACTIVE FILTERS

Fig 11.46 shows circuits for four second-order filters: low-pass (Fig 11.46A), high-pass (Fig 11.46B), band-pass (Fig 11.46C) and band-reject or notch (Fig 11.46D). Sequences of these filters are used to create higher-order circuits by connecting them in series. Two second-order filter stages create a fourth-order filter, and so forth.

The low-pass and high-pass filters use the Sallen-Key circuit. Note that the high-pass circuit is just the low-pass circuit with the positions of R1-R2 and C1-C2 exchanged. R3 and R4 are used to control gain in the low- and high-pass configurations. The band-pass filter is a multiple-feedback design. The notch filter is based on the twin-T circuit. All of the filter design equations and tables will result in a Butterworth family response. (For the Chebyshev and other filter responses, consult the references listed at the end of the chapter.)

The circuits in Fig 11.46 assume the op amp is operating from a balanced, bipolar power supply, such as ±12 V. If a single supply is used (such as +12 V and ground), the circuit

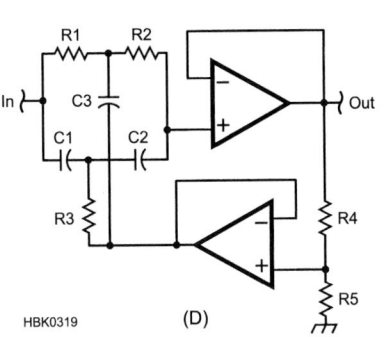

HBK0319   (D)

Band-Reject Filter

$$F_0 = \frac{1}{2\pi R1\,C1}$$

$$K = 1 - \frac{1}{4Q}$$

$$R \gg (1-K)R1$$

where

$$C1 = C2 = \frac{C3}{2} = \frac{10\,\mu F}{f_0}$$

$R1 = R2 = 2R3$
$R4 = (1 - K)R$
$R5 = K \times R$

Example:
f_0 = 500 Hz, Q=10
K = 0.975
C1 = C2 = 0.02 μF (or use 0.022 μF)
C3 = 0.04 μF (or use 0.044 μF)
R1 = R2 = 15.92 kΩ (use 15 kΩ)
R3 = 7.96 kΩ (use 8.2 kΩ)
R ≫ 1 kΩ
R4 = 25 Ω (use 24 Ω)
R5 = 975 Ω (use 910 Ω)

Table 11.20
Factor "a" for Low- and High-Pass Filters in Fig 11.46

No. of Stages	Stage 1	Stage 2	Stage 3	Stage 4
1	1.414	–	–	–
2	0.765	1.848	–	–
3	0.518	1.414	1.932	–
4	0.390	1.111	1.663	1.962

These values are truncated from those of Appendix C of Williams, *Electronic Filter Design Handbook*, for even-order Butterworth filters

must have a dc offset added and blocking capacitors between filter sections to prevent the dc offset from causing the op amp to saturate. The references listed at the end of the chapter provide more detailed information on single-supply circuit design.

Avoid electrolytic or tantalum capacitors as frequency-determining components in active filter design. These capacitors are best used for bypassing and power filtering as their tolerance is generally quite low, they have significant parasitic effects, and are usually polarized. Very small values of capacitance (less than 100 pF) can be affected by stray capacitance to other circuit components and wiring. High-order and high-Q filters require close attention to component tolerance and temperature coefficients, as well.

SECOND-ORDER ACTIVE FILTER DESIGN PROCEDURES

The following simple procedures are used to design filters based on the schematics in Fig 11.46. Equations and a design example are provided in the figure.

Low- and High-Pass Filter Design

To design a low- or high-pass filter using the circuits in Fig 11.46, start by determining your performance requirements for filter order (2, 4, 6, or 8), gain (K) and cutoff frequency (f_C). Calculate $\omega_C = 2\pi f_C$ and C_2 or C as required.

Table 11.20 provides design coefficients to create the Butterworth response from successive Sallen-Key low- and high-pass stages. A different coefficient is used for each stage. Obtain design coefficient "a" from Table 11.20. Calculate the remaining component values from the equations provided.

Band-Pass Filter Design

To design a band-pass filter as shown in Fig 11.46C, begin by determining the filter's required Q, gain, and center frequency, f_0. Choose a value for C and solve for the resistor values. Very high or very low values of resistance (above 1 MΩ or lower than 10 Ω) should be avoided. Change the value of C until suitable values are obtained. High gain and Q may be difficult to obtain in the same stage with reasonable component values. Consider a separate stage for additional gain or to narrow the filter bandwidth.

Band-Reject (Notch) Filter Design

Band-reject filter design begins with the selection of center frequency, f_0, and Q. Calculate the value of K. Choose a value for C1 that is approximately 10 μF / f_0. This determines the values of C2 and C3 as shown. Calculate the value for R1 that results in the desired value for f_0. This determines the values of R2 and R3 as shown. Select a convenient value for R4 + R5 that does not load the output amplifier. Calculate R4 and R5 from the value of K.

The depth of the notch depends on how closely the values of the components match the design values. The use of 1%-tolerance resistors is recommended and, if possible, matched values of capacitance. If all identical components are used, two capacitors can be paralleled to create C3 and two resistors in parallel create R3. This helps to minimize thermal drift.

11.5.4 Active Filter Design Tools

While the simple active filter examples presented above can be designed manually, more sophisticated circuits are more easily designed using filter-design software. Follow the same general approach to determining the filter's performance requirements and then the filter family as was presented in the section on LC filters. You can then enter the values or make the necessary selections for the design software. Once a basic design has been calculated, you can then "tweak" the design performance, use standard value components and make other adjustments. The design example presented below shows how a real analog design is assembled by understanding the performance requirements and then using design software to experiment for a "best" configuration.

Op amp manufacturers such as Texas Instruments and National Semiconductor (originally separate companies but now merged) have made available sophisticated "freeware" filter design software. These packages are extremely useful in designing active RC filters. They begin by collecting specifications from the user and then creating a basic circuit. Once the initial circuit has been designed, the user can adjust specifications, component values, and op amp types until satisfied with the final design.

Free filter design software can be obtained from the following sources:

• Analog Devices — *Analog Filter Wizard 2.0* (**www.analog.com/designtools/en/filterwizard**)

• National Semiconductor — *Webench* (**http://webench.national.com**)

• Texas Instruments — *FilterPro* (**www.ti.com**, search for "FilterPro")

Even if a manual design process is followed, using a software tool to double-check the results is a good way to verify the design before building the circuit.

An extended design example using the Texas Instruments *FilterPro* software is included on this book's CD-ROM. In the example, Dan Tayloe, N7VE, explains the process of designing a high-performance 750 Hz low-pass filter, illustrating the power of using sophisticated interactive tools that enable design changes on-the-fly. The reader is encouraged to follow along and experiment with *FilterPro* as a means of becoming familiar with the software so that it can be used for other filter design tasks. Similar processes apply to other filter design software tools.

11.6 Quartz Crystal Filters

Inductor Q values effectively limit the minimum bandwidth that can be achieved with LC band-pass filters. Higher-Q circuit elements, such as quartz crystal, PLZT ceramic and constant-modulus metal alloy resonators, are required to extend these limits. Quartz crystals offer the highest Q and best stability with time and temperature of all available resonators. They are manufactured for a wide range of frequencies from audio to VHF, using cuts (crystal orientations) that suit the frequency and application of the resonator. The AT cut is favored for HF fundamental and VHF overtone use, whereas other cuts (DT, SL and E) are more convenient for use at lower frequencies.

Each crystal plate has several modes of mechanical vibration. These can be excited electrically thru the piezoelectric effect, but generally resonators are designed so as to maximize their response on a particular operating frequency using a crystal cut that provides low loss and a favorable temperature coefficient. Consequently, for filter design, quartz crystal resonators are modeled using the simplified equivalent circuit shown in **Fig 11.47**. Here L_m, C_m and r_m represent the *motional* parameters of the resonator at the main operating frequency — r_m being the loss resistance, which is also known as the *equivalent series resistance*, or ESR. C_o is a combination of the static capacitance formed between the two metal electrodes with the quartz as dielectric ($\varepsilon_r = 4.54$ for AT-cut crystals) and some additional capacitance introduced by the metal case, base and mount. There is a physical relationship between C_m and the static capacitance formed by the resonator electrodes, but, unfortunately, the added holder capacitance masks this direct relationship causing C_o/C_m to vary from 200 to over 500. However, for modern fundamental AT-cut crystals between 1 and 30 MHz, their values of C_m are typically between 0.003 and 0.03 pF. Theoretically, the motional inductance of a quartz crystal should the same whether it is operated on the fundamental or one of its overtones, making the motional capacitance at the nth overtone $1/n^2$ of the value at the fundamental. However, this relationship is modified by the effect of the metal electrodes deposited on either side of the crystal plate and in practice the motional inductance increases with the frequency of the overtone. This makes the motional capacitance at the overtone substantially less than C_m/n^2.

An important parameter for crystal filter design is the unloaded Q at f_s, the resonant frequency of the series arm. This is usually denoted by Q_U.

$$Q_u = 2 \pi f_s L_m/r_m \qquad (7)$$

Q_U is very high, often exceeding 100,000 in the

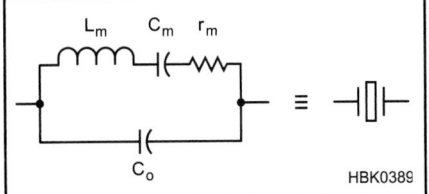

Fig 11.47 — Equivalent circuit of a quartz crystal and its circuit symbol.

lower HF region. Even VHF overtone crystals can have Q_U values over 20,000, making it possible to design quartz crystal band-pass filters with a tremendously wide range of bandwidths and center frequencies.

The basic filtering action of a crystal can be seen from **Fig 11.48**, which shows a plot of attenuation vs frequency for the test circuit shown in the inset. The series arm of the crystal equivalent circuit forms a series-tuned circuit, which passes signals with little attenuation at its resonant frequency, f_s, but appears inductive above this frequency and parallel resonates with C_o at f_p to produce a deep notch in transmission. The difference between f_s and f_p is known as the pole-zero separation, or PZ spacing, and is dependent on C_m/C_o as well as f_s. Further information on quartz crystal theory

and operation can be found in the **Oscillators and Synthesizers** chapter and Ref 1.

A simple crystal filter developed in the 1930s is shown in **Fig 11.49**. The voltage-reversing transformer T1 was usually an IF transformer, but nowadays could be a bifilar winding on a ferrite core. Voltages V_a and V_b have equal magnitude but 180° phase difference. When $C_1 = C_o$, the effect of C_o will disappear and a well-behaved single resonance will occur as indicated by the solid line in Fig 11.49B. However, if C_1 is adjusted to unbalance the circuit, a transmission zero (notch) is produced well away from the pass band and by increasing the amount of imbalance this can be brought back towards the edge of the pass band to attenuate close-by interfering CW signals. If C_1 is reduced in value from the balanced setting, the notch comes in from the high side and if C_1 is increased, it comes in from the low side. The dotted curve in Fig 11.49B illustrates how the notch can be set with C_1 less than C_o to suppress adjacent signals just above the pass band. In practice a notch depth of up to 60 dB can be achieved. This form of "crystal gate" filter, operating at 455 kHz, was present in many high-quality amateur communications receivers from the 1930s through the 1960s. When the filter was switched into the receiver IF amplifier the bandwidth was reduced to a few hundred Hz for CW reception. The close-in

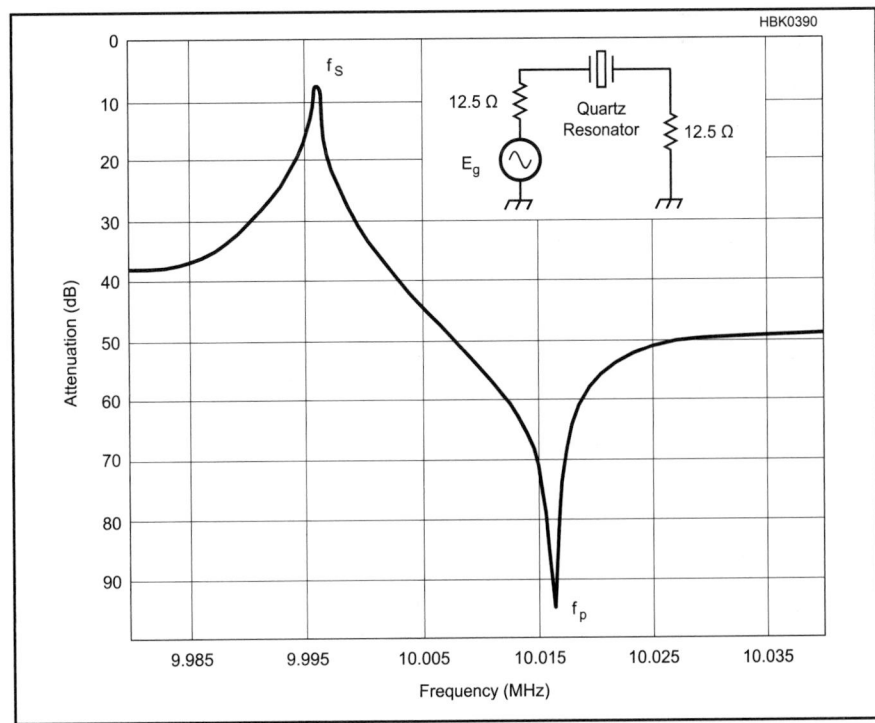

Fig 11.48 — Response of 10 MHz crystal (Cm = 0.0134 pF, Lm = 18.92 mH, ESR = 34 Ω) in a series test circuit (see inset) showing peak of transmission (lowest attenuation) at the series resonant frequency, f_s, and a null (maximum attenuation) caused by the parallel resonance due to C_o at f_p.

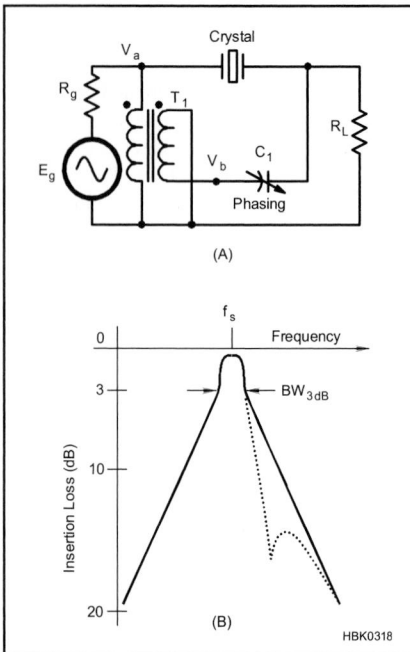

Fig 11.49 — Classic single-crystal filter in A has the response shown in B. The phasing capacitor can be adjusted to balance out C_o (solid line), or set to a lesser or greater value to create a movable null to one side, or other, of the passband (dotted line).

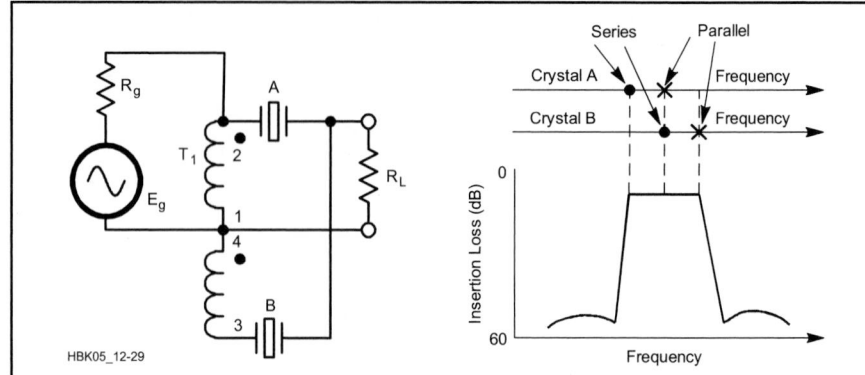

Fig 11.50 — A half-lattice crystal filter. The C_o of one crystal can be made to balance the C_o of the other, or C_o across the higher crystal can be deliberately increased to create nulls on either side of the passband.

range of the notch was sometimes improved by making C_1 part of a differential capacitor that could add extra capacitance to either the C_1 or C_o side of the IF transformer. This design could also be used to good effect at frequencies up to 1.7 MHz with an increased minimum bandwidth. However, any crystal gate requires considerable additional IF filtering to achieve a reasonable ultimate attenuation figure, so it should not be the only form of selectivity used in an IF amplifier.

The half-lattice filter shown in **Fig 11.50** offers an improvement in performance over a single-crystal filter. The quartz crystal static capacitors, C_o, cancel each other. The remaining series-resonant arms, if offset in frequency and terminated properly, will produce an approximate 2-pole Butterworth or Chebyshev response. The crystal spacing for simple Chebyshev designs is usually around two-thirds of the bandwidth. Half-lattice filter sections can be cascaded to produce composite filters with multiple poles. Many of the older commercial filters are coupled half-lattice types using 4, 6 or 8 crystals, and this is still the favored technology for some crystal filters at lower frequencies. Very often extra capacitance was added across one crystal in each half lattice to unbalance C_o and provide deep transmission zeroes on either side of the pass band to sharpen up the close-in response at the expense of the attenuation further out. Ref 11

discusses the computer design of half-lattice filters.

Most commercial HF and VHF crystal filters produced today use dual monolithic structures as their resonant elements. These are a single quartz plate onto which two sets of metal electrodes have been deposited, physically separated to control the acoustic (mechanical) coupling between them. An example of a dual monolithic filter (2-pole) is shown in **Fig 11.51**. These are available with center frequencies from 9 MHz to well over 120 MHz. In effect, the dual monolithic structure behaves just like a pair of coupled crystal resonators. There is a subtle difference, however, because the static capacitance in duals appears across the input and output terminations and doesn't produce a null above the pass band, as it would for two electrically coupled crystal resonators. Multi-pole filters can be built by coupling duals together using external capacitors, the C_o from each dual input and output being absorbed in the coupling capacitors or terminations, or

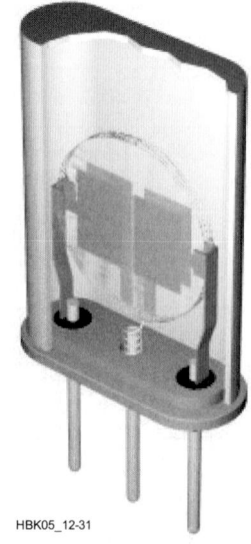

Fig 11.51 — A dual monolithic crystal filter has two sets of electrodes acoustically coupled to provide a two-pole response in a single crystal unit. The center lead is connected to the case and grounded in normal operation.

by including more acoustically coupled resonators in the electrode structure on a single quartz plate. Though not common, 3-pole and 4-pole monolithic filters housed in standard, single crystal holders are available from some manufacturers.

11.6.1 Filter Parameters

An ideal band-pass filter would pass the desired signal with no loss and completely attenuate everything else. Practically, it is not possible to achieve such a response with a finite number of resonators, and approximations to this ideal have to be accepted. Greater stop-band attenuation and steeper sides can be achieved if more and more crystals are used, and the response gets nearer to the ideal "brick-wall" one. This feature of a filter is expressed as a *shape factor*, which specifies the ratio of the bandwidth at an attenuation of 60 dB to the bandwidth at 6 dB — both these levels being taken relative to the actual pass band peak to eliminate insertion loss from the calculation. An ideal brickwall filter would have a shape factor of 1, and practical filters have shape factors that depend on the number of crystals used in the design and the type of response chosen for the pass band. A 1 dB Chebyshev design, for example, can typically produce shape factors that vary from about 4.1 for a 4-pole to 1.5 for a 10-pole, but the actual figures obtained in practice are very much dependent on Q_u and how much greater it is than the filter Q, defined by f_o/BW, where f_o is the center frequency and BW the 3 dB bandwidth. The ratio of Q_u to f_o/BW is often quoted as Q_o, or q_o, and this, along with the order and type of response, determines the insertion loss of a crystal filter. Q_o also determines how closely the pass band follows the design response, and how much the passband ripple is smoothed out and the edges of the response rounded off by crystal loss.

Commercial filter manufacturers usually choose the Chebyshev equiripple design for SSB, AM and FM bandwidths because it gives the best compromise between passband response and the steepness of the sides, and 1 dB-ripple Chebyshev designs are pretty standard for speech bandwidths. Tolerances in component values and crystal frequencies can cause the ripple in the pass band to exceed 1 dB, so often the maximum ripple is specified as 2 dB even though the target ripple is lower. Insertion loss, the signal loss going thru the filter, also varies with the type of response and increases as the order of the filter increases. The insertion loss for a given order and bandwidth is higher for high-ripple Chebyshev designs than it is for low-ripple ones, and Butterworth designs have lower insertion loss than any Chebyshev type for a given bandwidth. Pass-band amplitude response and shape factor are important parameters for assessing the performance of filters used for speech communications.

However, group delay is also important for data and narrow bandwidth CW reception. Differential group delay can cause signal distortion on data signals if the variations are greater than the automatic equalizer can handle. Ringing can be an annoying problem when using very narrow CW filters, and the group delay differential across the pass band must be minimized to reduce this effect. When narrow bandwidth filters are being considered, shape factor has to be sacrificed to reduce differential group delay and its associated ringing problems. It's no good having a narrow Chebyshev design with a shape factor of 2 if the filter produces unacceptable ringing and is intolerable to use in practice.

Both Bessel and linear phase (equiripple 0.05°) responses have practically constant, low group delay across the entire pass band and well beyond on either side, making either a good choice for narrow CW or specialized data use. They also have the great advantage of offering the lowest possible insertion loss of all the types of response currently in use, which is important when Q_o is low, as it often is for very narrow bandwidth filters. The insertion loss of the Bessel design is marginally lower than that of the linear phase, but the latter has a superior shape factor giving it the best balance of low group delay and good selectivity. A 6-pole linear phase (equiripple 0.05°) design has a shape factor of 3.39, whereas a 6-pole Bessel has 3.96. The Gaussian-to-12 dB response has a better shape factor than that of either the linear phase or Bessel designs but the group delay across the pass band is not as flat and pronounced peaks (ears) are beginning to appear at the band edges with a 6-pole design. The Gaussian-to-6 dB group delay is reasonably flat for 3- and 4-pole filters, but significant ears appear towards the passband edges in designs with 5 poles or more.

11.6.2 Crystal Filter Design

A wide variety of crystals are produced for use with microprocessors and other digital integrated circuits. They are offered in several case styles, but the most common are HC-49/U and HC-49/US. Crystal resonators in the larger HC-49/U style case are fabricated on 8 mm diameter quartz discs, whereas those in the squat HC-49/US cases are fabricated on 8 mm by 2 mm strips of quartz. At any given frequency, C_m will be lower for HC-49/US crystals because the active area is smaller than it is in the larger HC-49/U crystals. Both types are cheap and have relatively small frequency spreads, making them ideal for use in the LSB ladder configuration suggested by Dishal (Ref 2) — see **Fig 11.52**. This arrangement requires the motional inductances of all the crystals to be identical, and each loop in isolation (crystal and coupling capacitors either side) to be resonant at the same frequency. Series capacitors to trim individual crystals are needed to achieve this in some of the more advanced designs, where production frequency spreads are not sufficient to satisfy the latter requirement. Refs 3, 4, 5, 6, 7, 8 and 11 contain design information on Dishal LSB crystal ladder filters. Design software associated with Ref

11 (**11x09_Steder-Hardcastle.zip**) can also be downloaded from **www.arrl.org/qexfiles**.

The *min-loss* form of Cohn ladder filter, where $C_{12} = C_{23} = C_{34}$, has become very popular in recent years because it's so simple to design and build. However, it suffers from the drawback that the ripple in its passband response increases dramatically with increasing order, and ringing can be a real problem at bandwidths below 500Hz for Cohn *min-loss* filters of 6th-order, or more. The ripple may not be a problem in most narrow filters because it's smoothed out almost completely by loss, but the ringing can be tiring. For wider bandwidths, where the ratio of Q_u to f_o/BW is much greater, the ripple is not smoothed out, and is very evident. One way round this problem, without sacrificing simplicity, is to use the arrangement shown in **Fig 11.53**, which was originally designed for variable bandwidth applications. It was devised to make the mesh frequencies track together as variable coupling changed the bandwidth. However, it has the great advantage that it can be optimized for almost equal ripple at maximum bandwidth, making it an ideal al-

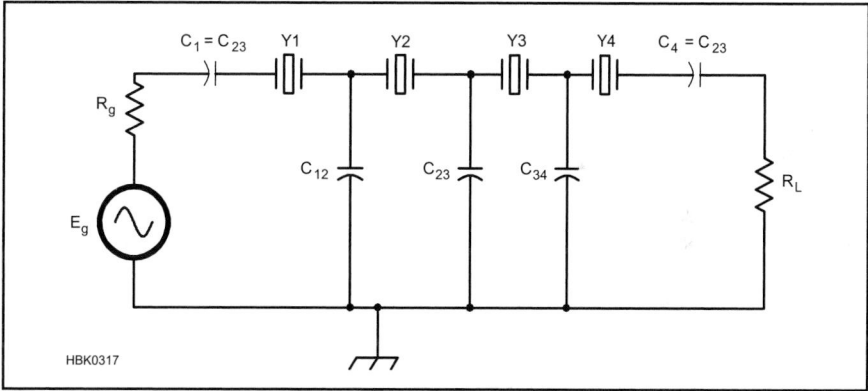

Fig 11.52 — Dishal LSB crystal-ladder filter configuration. Crystals must have identical motional inductances, and the coupling capacitors and termination resistors are selected according to the bandwidth and type of passband response required.

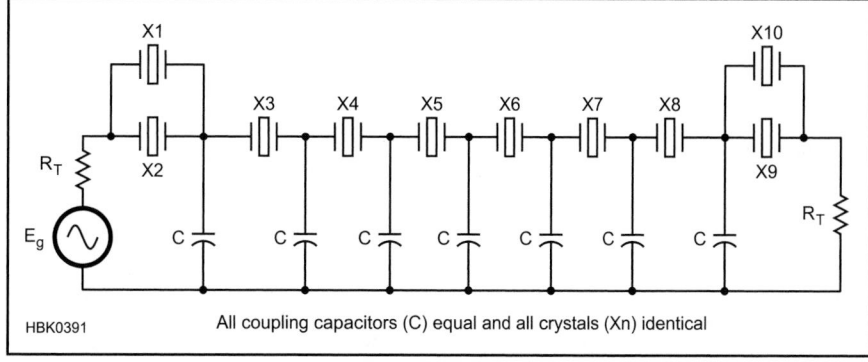

Fig 11.53 — Configuration of improved crystal ladder filter using identical crystals and equal coupling capacitor values. The parallel resonator end-sections (PRES) can provide excellent passband responses, giving either quasi-equiripple (QER) or minimum-loss (PRESML) responses with just a change in termination resistance.

Table 11.21

3 dB-down k & q Values for Quasi-Equiripple (QER) Ladder Filters

Order	q	k_{12}	k_{23}	Shape Factor	Max Ripple (dB)
4	0.9942	0.7660	0.5417	4.56	0.002
5	1.0316	0.7625	0.5391	3.02	0.018
6	1.0808	0.7560	0.5346	2.31	0.09
7	1.1876	0.7459	0.5275	1.90	0.16
8	1.2532	0.7394	0.5228	1.66	0.31
9	1.3439	0.7335	0.5187	1.50	0.42
10	1.4115	0.7294	0.5158	1.40	0.60
11	1.4955	0.7261	0.5134	1.33	0.72
12	1.5506	0.7235	0.5116	1.28	0.90

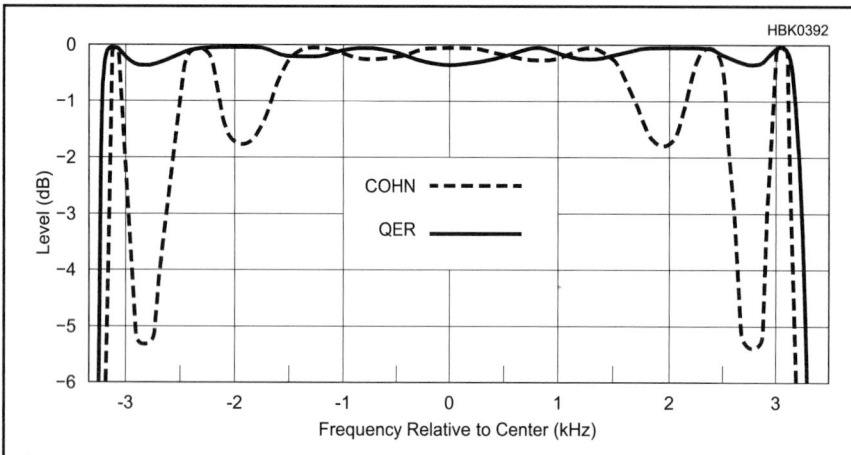

Fig 11.54 — Comparison of 8-pole Cohn min-loss passband response with that of the quasi-equiripple (QER) type. Note the almost equal ripple in the passband of the QER response.

ternative for fixed bandwidth speech applications where the Cohn *min-loss* pass band is poor. Two crystals are used in parallel to halve the motional inductance and double the motional capacitance of the resonators in the end sections, and although the two additional crystals do not increase the order of the filter by two, they do reduce the passband ripple substantially while maintaining the simplicity of design and construction offered by the Cohn *min-loss* filter. In addition, the group delay of the parallel-resonator-end-section (PRES) configuration is less than that of the Cohn *min-loss*. All the coupling capacitors are equal and the filter can be terminated to achieve a quasi-equiripple response (QER), so that its pass band resembles that of a Chebyshev design, or minimum loss (PRESML) with a response like that of the Cohn *min-loss*. **Fig 11.54** shows the Cohn *min-loss* and QER passband responses with infinite crystal Q_u for comparison. Values of k and q for QER filters from 4 to 12 poles are given in **Table 11.21,** along with the maximum ripple and shape factor for each order. The coupling capacitor value for any bandwidth can be determined from k_{23} using Eq 8. The end-section resonators formed by the two parallel crystals have twice the effective

motional capacitance of the inner resonators, and since k_{12} is always 1.414 times k_{23} for the QER design the value of C_{12} will be the same as that calculated for C_{23} and all the other coupling capacitors, C, in the design.

$$C = f_o \, C_m / (BW \, k_{23}) \qquad (8)$$

The termination resistance, R_T, must be calculated using half the motional inductance of a single crystal, as illustrated in Eq 9, where L_m is the motional inductance of one of the parallel crystals used in the end-sections.

$$R_T = 2 \pi \, BW \, L_m / 2q = \pi \, BW \, L_m / q \qquad (9)$$

An alternative way of producing a QER filter, which has the added benefit of improving the symmetry of the ladder filter response, is to add transformers with bifilar windings to each end of the filter so that the end sections are effectively like crystal gate filters with C_1 approximately equal to $2C_o$. Normally, Dishal LSB ladder filters have an asymmetrical response because each crystal produces a transmission zero at f_p, on the high side of the filter pass band This asymmetry can be counteracted to a certain extent if the value of C_1 in each end

section is adjusted to over-compensate for C_o and produce nulls on the low side of the pass band to even up the overall response. Whereas the presence of C_o causes the motional inductance of crystals to increase and their motional capacitance to decrease throughout the pass band, over compensating the end crystals to produce nulls on the low side has the opposite effect and the end-section crystals appear as if their motional capacitance is higher and their motional inductance lower, like two crystals in parallel. Therefore, they can be made to produce the QER type of response with suitable coupling and terminations. Ref 12 provides more information on this topic.

11.6.3 Crystal Characterization

Simple crystal filters can be constructed using cut-and-try methods, but sometimes the results are very disappointing. The only sure way to guarantee good results is to fully characterize the crystals beforehand, so that only the most suitable ones are used in a design appropriate for the crystal motional parameters and the application. When crystal parameters are known, computer modeling can be used to assess the effect of Q_u and C_o on bandwidth, before proceeding to the construction phase. In addition to the *Elsie* design program on the CD-ROM, AADE Filter Design and Analysis (**www.aade.com**) provides a free filter modeling program.

Crystal characterization can be done with very limited or very advanced test equipment, the main difference being the accuracy of the results. The *phase-zero* method for measuring C_m used in industry can be implemented by amateurs if a dual-beam oscilloscope is available to substitute as a phase detector — Ref 7 gives details of this method. However, many successful crystal filter constructors have achieved excellent results with a very limited amount of home-built test equipment. Ref 6 describes a simple switched-capacitor test oscillator for measuring C_m that was developed by G3UUR. His technique requires a frequency counter, a small 12 V power supply and very little construction effort. Using care and a more exact expression for C_m than the one given in Ref 6, this oscillator method can achieve results that are comparable with professional techniques. The circuit can also be modified to include relative Q or ESR measurement if a multimeter is available.

Values of Q_u for crystals can vary considerably, even within the same batch, and the ratio of the best to the worst, excluding dead ones, can be as high as 6 for cheap mass-produced crystals. This ratio can still be more than 2 for batches of high quality crystals. The relative activity of each crystal needs to be established to weed out the poor ones, and an estimate for Q_u is required to more accurately model

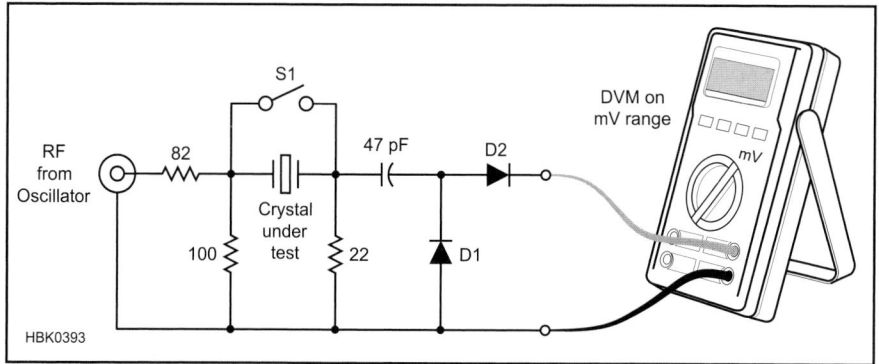

Fig 11.55 — A simple jig for measuring crystal ESR. The jig may be driven by a signal generator, or a modified crystal test oscillator, and the signal through the crystal detected on a DMM using the mV dc range. Resistors are ⅛ or ¼ W, 5%. D1 and D2 are small signal germanium or Schottky barrier diodes. S1 is a miniature pushbutton switch.

the filter performance prior to construction. A modified version of the switched-capacitor crystal test oscillator is shown in **Fig 11.55**. The RF detector circuit at the output of the oscillator provides a means of assessing the relative activity of each crystal being tested. A DMM on a suitable dc voltage range attached across points M and G (ground) will give a digital readout roughly equal to the peak-to-peak RF voltage produced by each crystal. Crystals with higher Q values will produce higher output voltages, so each crystal can be ranked according to its output reading relative to others in the batch. For convenience, a socket should be used for the crystal under test. If crystals with wire leads are to be measured, this can be fashioned from a dual-in-line IC or transistor socket. There are also small PCB connectors that might make suitable sockets.

The formula for C_m presented in Ref 6 is much simplified and less accurate than the exact derivation for C_m. Better accuracy can be achieved with Eq 10, where F_1 is the fre-

quency obtained with S1 closed and F_2 is the one with S1 open.

$$C_m = 2 (F_2 - F_1) [C_S + C_O + C_R]/F_1 \qquad (10)$$

All capacitances in this equation are in pF and the frequencies in Hz. $C_R = 4C_S \, C_O/C_F$ and C_o is the total parallel capacitance of the crystal, including the contribution from the metal case — it's assumed that the metal case is floating during these measurements and is not grounded, or being held. The feedback capacitors, C_F, are 390 pF in **Fig 11.56** and the series capacitor, C_S, is 68 pF. C_o typically varies from 2.5 to 5.5 pF for HC49/U crystals in the 4 to 12 MHz range and about 1.5 to 3.5 pF for the smaller HC49/US crystals. LC meters that can measure down to 0.01 pF and 1 nH can be constructed using PIC technology. Commercial LC meters with amazingly good specifications are also available at moderate prices if a PIC LC meter seems too ambitious a project for home construction at this stage.

Obviously, great care must be exercised to avoid stray capacitance and errors in setting zero when measuring such small values of capacitance.

When the value of C_o cannot be determined easily, or the utmost accuracy is not required, a simplified version of Eq 10 may be used. An average value for the type of crystals being tested can be assigned to C_o at the expense of a few percentage points loss in accuracy. For HC49/U crystals, a reasonable average for C_o is 3.75 pF and using the values for C_F and C_S shown in Fig 11.56, the more exact expression simplifies to Eq 11.

$$C_m = 148 (F_2 - F_1)/F_1 \qquad (11)$$

Again, F_1 is the frequency registered on the counter when S1 is closed and F_2 when it's open. Both frequencies are in Hz and C_m comes out in pF. If more accuracy is required, the value of C_m obtained with Eq 11 can be used to estimate C_o from Eq 12 and that value used in Eq 10 to achieve a closer estimate for C_m.

$$C_o = 175 \, C_m + 0.95 \qquad (12)$$

Also, the series-resonant frequency f_S for each crystal can be estimated by calculating the amount by which F_1 is higher than f_s and then subtracting that from F_1. This frequency difference, ΔF_1, is given by Eq 13 for the value of C_F used in Fig 11.56.

$$\Delta F_1 = C_m \, F_1/400 \qquad (13)$$

Ranking crystals according to their oscillator output is sufficient to be able to select the best (highest Q) crystals for a filter, but if you want to use computer modeling to correct for the influence of loss on bandwidth, then an estimate of ESR or Q_u is necessary. This can be done quite simply with the test oscillator shown in Fig 11.56, but requires just a little more time and effort than just ranking relative Q by RF output. There will be a spread of output levels corresponding to the range of Q values. Pair up two crystals from the low end of the spread with similar output figures (within 5%) and reasonably closely matched frequencies (within 50 Hz). Pair up another couple of crystals from the high end. Connect each pair of similar crystals in parallel and place them in the test oscillator a pair at a time with S_1 and S_2 closed. The output level should be much higher using a parallel pair than for each crystal alone. Now open S_2 and adjust the variable resistor VR1 until an output level is reached that is equal to the average of the levels previously obtained with the two crystals individually. After removing each pair check the resistance of VR1 using your DMM. Test points TP1 and TP2 are provided for such measurements and should

Fig 11.56 — Circuit of a modified switched-capacitor test oscillator which can be used to measure crystal motional parameters C_m and r_m.

be feed-through types mounted on the front panel or side of the oscillator box. The ESR value of the two similar crystals in the pair is approximately twice the value of the resistance measured across the test points. This method may be very crude, but it will give you an ESR value that is certainly better than 20%, and probably within 10% of the value measured by more accurate means. Once ESR values for the crystals at the extremes of the spread are established, one can be done in the middle of the range for good measure and values roughly assigned to those in between. Then, an average of the motional parameters and Q values of the set of crystals chosen for a particular filter can be used for modeling purposes.

Should a more accurate means of measuring crystal ESR be required, the phase-zero method or a VNA should be considered. Whatever the means of crystals characterization, a spreadsheet to record the data should

be prepared beforehand and some means of marking the crystals with a number or letter organized. Sticky white dots are probably the most convenient way to identify each crystal. They adhere well to metal surfaces and can be written on with a ball-point pen. Alternatively, a permanent marker pen directly on the metal case could be tried, but is sometimes not all that permanent.

11.6.4 Crystal Filter Evaluation

The simplest means of assessing the performance of a crystal filter is to temporarily install it in a finished transceiver, or receiver, and use a strong on-air signal, or locally generated carrier, to run thru the filter pass band and down either side to see if there are any anomalies. Provided that the filter crystals have been carefully characterized in the first instance,

and computer modeling has shown that the design is close to what's required, this may be all that's required to confirm a successful project. However, more elaborate checks on both the pass band and stop band can be made if a DDS signal generator, or vector network analyzer (VNA), is available. A test set-up for evaluating the response of a filter using a DDS generator requires an oscilloscope to display the response, whereas a VNA controlled via the USB port of a PC can display the response on the PC screen with suitable software — see the N2PK website **www.n2pk.com**. In addition, it can measure the phase and work out the group delay of the filter. Construction of a VNA is a very worthwhile project, and in addition to its many other applications it can be used to characterize the crystals prior to making the filter as well as evaluating filter performance after completion.

11.7 SAW Filters

The resonators in a monolithic crystal filter are coupled together by bulk acoustic waves. These acoustic waves are generated and propagated in the interior of a quartz plate. It is also possible to launch, by an appropriate transducer, acoustic waves that propagate only along the surface of the quartz plate. These are called "surface-acoustic-waves" because they do not appreciably penetrate the interior of the plate.

A *surface-acoustic-wave* (SAW) filter consists of thin aluminum electrodes, or fingers, deposited on the surface of a piezoelectric substrate as shown in **Fig 11.57**. Lithium Niobate (LiNbO₃) is usually favored over quartz because it yields less insertion loss. The electrodes make up the filter's trans-

ducers. RF voltage is applied to the input transducer and generates electric fields between the fingers. The piezoelectric material vibrates in response, launching an acoustic wave along the surface. When the wave reaches the output transducer it produces an electric field between the fingers. This field generates a voltage across the load resistor.

Since both input and output transducers are not entirely unidirectional, some acoustic power is lost in the acoustic absorbers located behind each transducer. This lost acoustic power produces a mid-band electrical insertion loss typically greater than 10 dB. The SAW filter frequency response is determined by the choice of substrate material and finger pattern.

The finger spacing, (usually one-quarter wavelength) determines the filter center frequency. Center frequencies are available from 20 to 1000 MHz. The number and length of fingers determines the filter loaded Q and shape factor.

Loaded Qs are available from 2 to 100, with a shape factor of 1.5 (equivalent to a dozen poles). Thus the SAW filter can be made broadband much like the LC filters that it replaces. The advantage is substantially reduced volume and possibly lower cost. SAW filter research was driven by military needs for exotic amplitude-response and time-delay requirements. Low-cost SAW filters are presently found in television IF amplifiers where high mid-band loss can be tolerated.

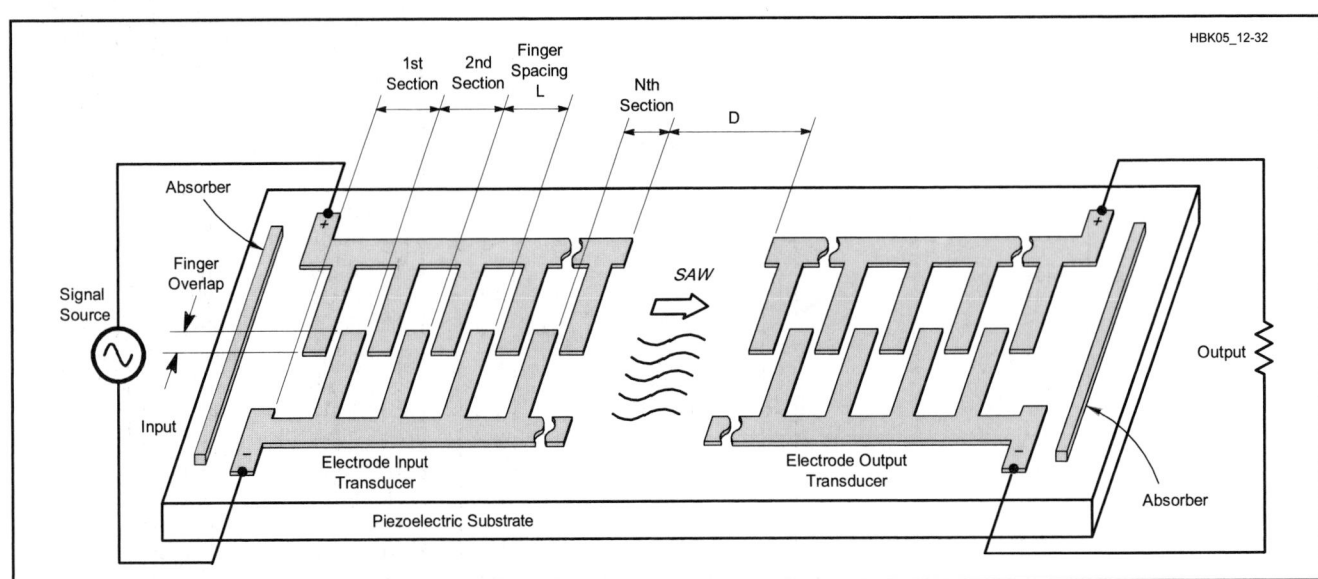

Fig 11.57 — The *interdigitated* transducer, on the left, launches SAW energy to a similar transducer on the right (see text).

11.8 Transmission Line Filters

LC filter calculations are based on the assumption that the reactances are *lumped*—that the physical dimensions of the components are considerably less than the operating wavelength. In such cases the unavoidable inter-turn parasitic capacitance associated with inductors and the unavoidable series parasitic inductance associated with capacitors are neglected as being secondary effects. If careful attention is paid to circuit layout and miniature components are used, lumped LC filter technology can be used up to perhaps 1 GHz.

Replacing lumped reactances with selected short sections of *Transverse Electromagnetic Mode* (TEM) transmission lines results in transmission line filters. (In TEM the electric and magnetic fields associated with a transmission line are at right angles (transverse) to the direction of wave propagation.) Coaxial cable, stripline and microstrip are examples of TEM components. Waveguides and waveguide resonators are not TEM components.

Coaxial cable transmission line filters are often used at HF and VHF frequencies. Stripline and microstrip transmission-line filters predominate from 500 MHz to 10 GHz. In addition they are often used down to 50 MHz when narrowband ($Q_L > 10$) band-pass filtering is required. In this application they exhibit considerably lower loss than their LC counterparts and are useful at frequencies where coaxial transmission lines are too lossy. A detailed treatment of the use of coaxial cable to form transmission line filters is presented in the **Transmission Lines** chapter. This section focuses on stripline and microstrip filters used at VHF and above.

11.8.1 Stripline and Microstrip Filters

Fig 11.58 shows three popular transmission lines used in transmission line filters. The circular coaxial transmission line (*coax*) shown in Fig 11.58A consists of two concentric metal cylinders separated by dielectric (insulating) material. The first transmission-line filters were built from sections of coaxial line. Their mechanical fabrication is expensive and it is difficult to provide electrical coupling between line sections.

Fabrication difficulties are reduced by the use of shielded strip transmission line (*stripline*) shown in Fig 11.58B. The outer conductor of stripline consists of two flat parallel metal plates (ground planes) and the inner conductor is a thin metal strip. Sometimes the inner conductor is a round metal rod. The dielectric between ground planes and strip can be air or a low-loss plastic such as polyethylene. The outer conductors (ground planes or shields)

are separated from each other by distance *b*.

Striplines can be easily coupled together by locating the strips near each other as shown in Fig 11.58B. Stripline Z_0 vs width (*w*)

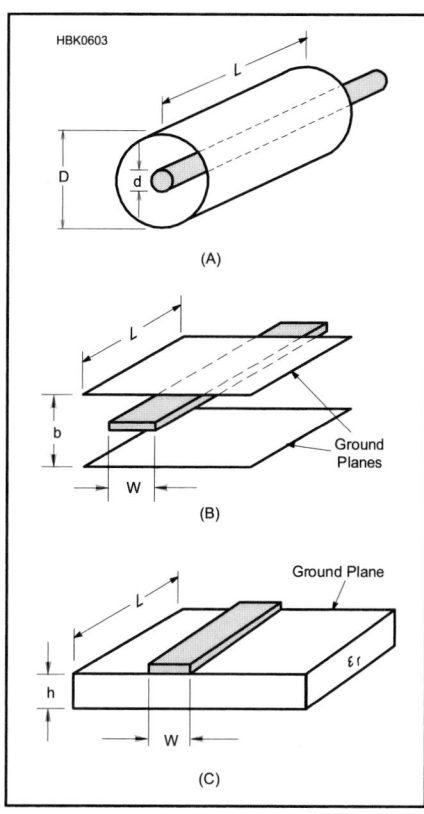

Fig 11.58 — Transmission lines. A: Coaxial line. B: Coupled stripline, which has two ground planes. C: Microstrip has only one ground plane.

is plotted in **Fig 11.59**. Air-dielectric stripline technology is best for low bandwidth ($Q_L > 20$) band-pass filters.

The most popular transmission line at UHF and microwave is *microstrip* (unshielded stripline), shown in Fig 11.58C. It can be fabricated with standard printed-circuit processes and is the least expensive configuration. In microstrip the outer conductor is a single flat metal ground-plane. The inner conductor is a thin metal strip separated from the ground-plane by a solid dielectric substrate. Typical substrates are 0.062 inch G-10 fiberglass ($\varepsilon = 4.5$) for the 50 MHz to 1 GHz frequency range and 0.031 inch Teflon ($\varepsilon = 2.3$) for frequencies above 1 GHz. Unfortunately, microstrip has the most loss of the three types of transmission line; therefore it is not suitable for narrow, high-Q, band-pass filters.

Conductor separation must be minimized or radiation from the line and unwanted coupling to adjacent circuits may become problems. Microstrip characteristic impedance and the effective dielectric constant (ε) are shown in **Fig 11.60**. Unlike coax and stripline, the effective dielectric constant is less than that of the substrate since a portion of the electromagnetic wave propagating along the microstrip travels in the air above the substrate.

The characteristic impedance for stripline and microstrip-lines that results in the lowest loss is not 75 Ω as it is for coax. Loss decreases as line width increases, which leads to clumsy, large structures. Therefore, to conserve space, filter sections are often constructed from 50 Ω stripline or microstrip stubs even though the loss at that characteristic impedance is not a minimum for that type of transmission line.

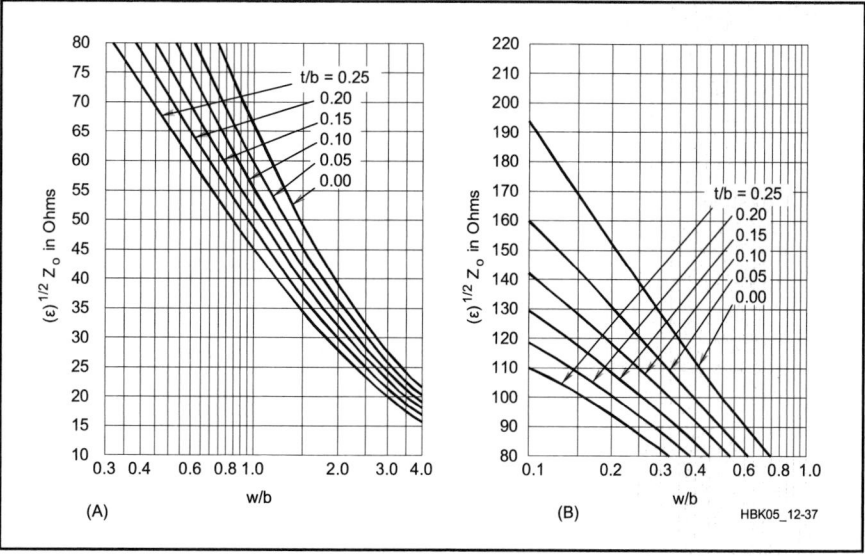

Fig 11.59 — The Z_0 of stripline varies with *w*, *b* and *t* (conductor thickness). See Fig 11.58B. The conductor thickness is *t* and the plots are normalized in terms of *t/b*.

Z_0 Ω	ε =1 (AIR) W/h	ε =2.3 (RT/Duroid) W/h	$\sqrt{\varepsilon_e}$	ε =4.5 (G–10) W/h	$\sqrt{\varepsilon_e}$
25	12.5	7.6	1.4	4.9	2.0
50	5.0	3.1	1.36	1.8	1.85
75	2.7	1.6	1.35	0.78	1.8
100	1.7	0.84	1.35	0.39	1.75
	$\sqrt{\varepsilon}=1$				

HBK05_12-38

Fig 11.60 — Microstrip parameters (after H. Wheeler, *IEEE Transactions on MTT*, March 1965, p 132). ε_e is the effective ε.

11.8.2 Transmission Line Band-Pass Filters

Band-pass filters can also be constructed from transmission line stubs. (See the **Transmission Lines** chapter for information on stub behavior and their use as filters at HF and VHF.) At VHF the stubs can be considerably shorter than a quarter-wavelength ($\frac{1}{4}\lambda$), yielding a compact filter structure with less mid-band loss than its LC counterpart. The single-stage 146 MHz stripline band-pass filter shown in **Fig 11.61** is an example. This filter consists of a single inductive 50 Ω strip-line stub mounted into a 2 × 5 × 7 inch aluminum box. The stub is resonated at 146 MHz with the "APC" variable capacitor, C1. Coupling to the 50 Ω generator and load is provided by the coupling capacitors C_c. The measured performance of this filter is: f_o = 146 MHz, BW = 2.3 MHz (Q_L = 63) and mid-band loss = 1 dB.

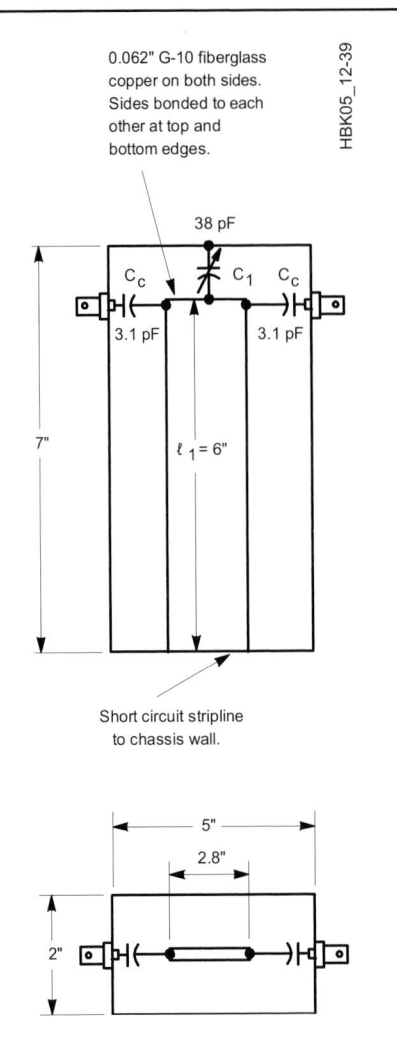

Fig 11.61 — This 146 MHz stripline band-pass filter has been measured to have a Q_L of 63 and a loss of approximately 1 dB.

Single-stage stripline filters can be coupled together to yield multistage filters. One method uses the capacitor coupled band-pass filter synthesis technique to design a 3-pole filter. Another method allows closely spaced inductive stubs to magnetically couple to each other. When the coupled stubs are grounded on the same side of the filter housing, the structure is called a "combline filter." Three examples of combline band-pass filters are shown in **Fig 11.62**. These filters are constructed in 2 × 7 × 9 inch aluminum boxes.

11.8.3 Quarter-Wave Transmission Line Filters

The reactance of a $\frac{1}{4}\lambda$ shorted-stub is infinite, as discussed in the **Transmission Lines** chapter. Thus, a $\frac{1}{4}\lambda$ shorted stub behaves like a parallel-resonant LC circuit. Proper input and output coupling to a $\frac{1}{4}\lambda$ resonator yields a practical band-pass filter. Closely spaced $\frac{1}{4}\lambda$ resonators will couple together to form a multistage band-pass filter. When the resonators are grounded on opposite walls of the filter housing, the structure is called an *interdigital filter* because the resonators look like interlaced fingers. Two examples of 3-pole UHF interdigital filters are shown in **Fig 11.63**. Design graphs for round-rod interdigital filters are given in Ref 9. The $\frac{1}{4}\lambda$ resonators may be tuned by physically changing their lengths or by tuning the screw opposite each rod.

If the short-circuited ends of two $\frac{1}{4}\lambda$ resonators are connected to each other, the resulting $\frac{1}{2}\lambda$ stub will remain in resonance, even when the connection to the ground-plane is removed. Such a floating $\frac{1}{2}\lambda$ microstrip line, when bent into a U-shape, is called a *hairpin* resonator. Closely coupled hairpin resonators can be arranged to form multistage band-pass

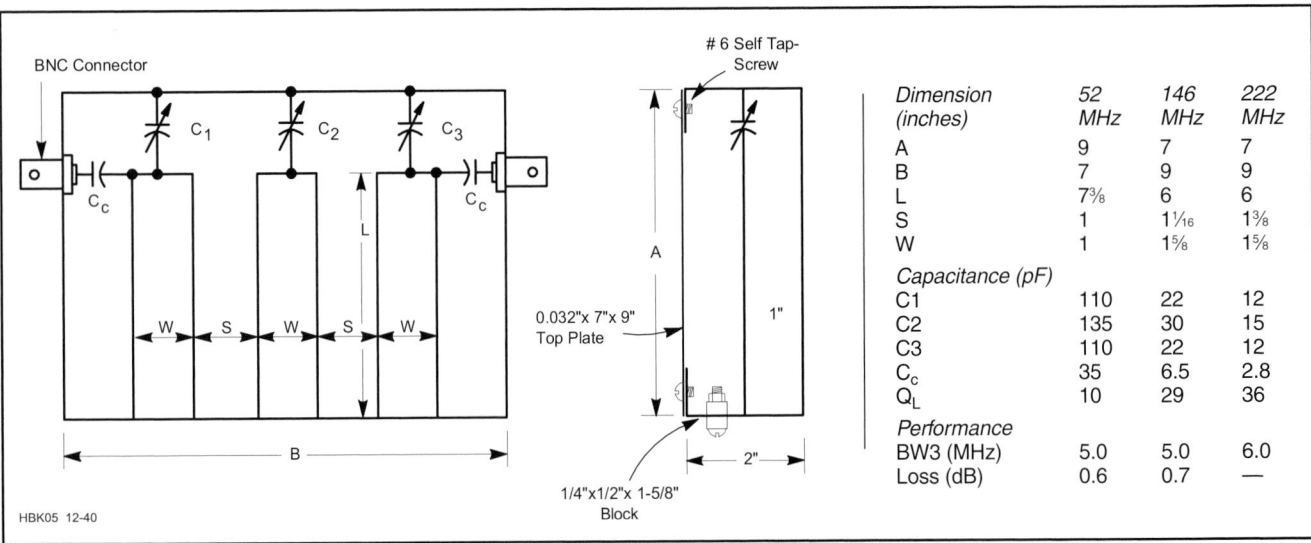

Dimension (inches)	52 MHz	146 MHz	222 MHz
A	9	7	7
B	7	9	9
L	7⅜	6	6
S	1	1¹⁄₁₆	1⅜
W	1	1⅝	1⅝
Capacitance (pF)			
C1	110	22	12
C2	135	30	15
C3	110	22	12
C_c	35	6.5	2.8
Q_L	10	29	36
Performance			
BW3 (MHz)	5.0	5.0	6.0
Loss (dB)	0.6	0.7	—

Fig 11.62 — This Butterworth filter is constructed in combline. It was originally discussed by R. Fisher in December 1968 *QST*.

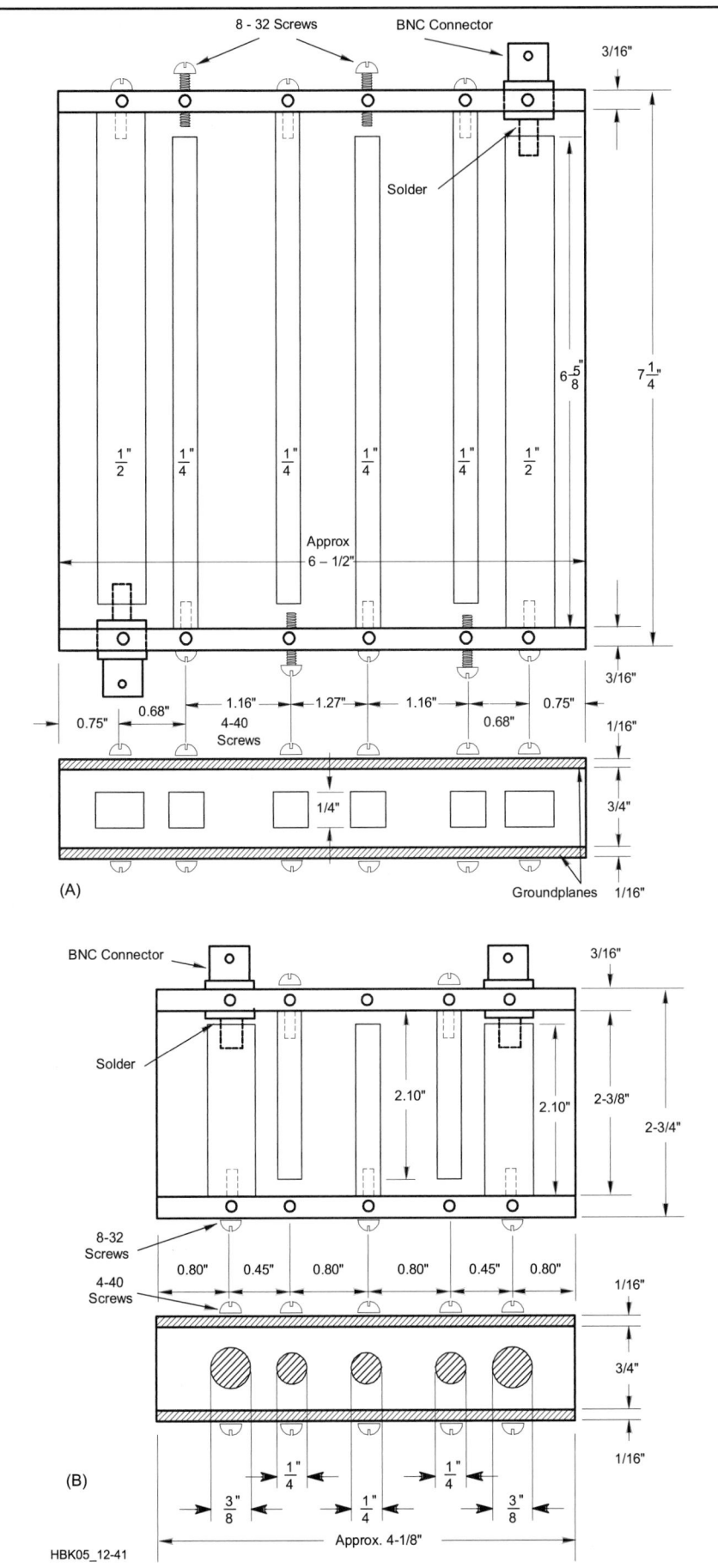

(A)

(B)

HBK05_12-41

Fig 11.63 — These 3-pole Butterworth filters for 432 MHz (shown at A, 8.6 MHz bandwidth, 1.4 dB passband loss) and 1296 MHz (shown at B, 110 MHz bandwidth, 0.4 dB passband loss) are constructed as interdigitated filters. The material is from R. E. Fisher, March 1968 *QST*.

filters. Microstrip hairpin band-pass filters are popular above 1 GHz because they can be easily fabricated using photo-etching techniques. No connection to the ground-plane is required.

11.8.4 Emulating LC Filters with Transmission Line Filters

Low-pass and high-pass transmission-line filters are usually built from short sections of transmission lines (stubs) that emulate lumped LC reactances. Sometimes low-loss lumped capacitors are mixed with transmission line inductors to form a hybrid component filter. For example, consider the 720 MHz, 3-pole microstrip low-pass filter shown in **Fig 11.64A** (on page 32) that emulates the LC filter shown in Fig 11.64B. C1 and C3 are replaced with 50 Ω open-circuit shunt stubs ℓ_C long. L2 is replaced with a short section of 100-Ω line ℓ_L long. The LC filter, Fig 11.64B, was designed for $f_c = 720$ MHz. Such a filter could be connected between a 432 MHz transmitter and antenna to reduce harmonic and spurious emissions. A reactance chart shows that X_C is 50 Ω, and the inductor reactance is 100 Ω at f_c. The microstrip version is constructed on G-10 fiberglass 0.062 inch thick, with $\varepsilon = 4.5$. Then, from Fig 11.60, w is 0.11 inch and $\ell_C = 0.125 \lambda_g$ for the 50 Ω capacitive stubs. Also, from Fig 11.60, w is 0.024 inch and ℓ_L is 0.125 ℓ_g for the 100-Ω inductive line. The inductive line length is approximate because the far end is not a short circuit. ℓ_g is $300/(720 \times 1.75) = 0.238$ m, or 9.37 inches Thus ℓ_C is 1.1 inch and ℓ_L is 1.1 inches.

This microstrip filter exhibits about 20 dB of attenuation at 1296 MHz. Its response rises again, however, around 3 GHz. This is because the fixed-length transmission line stubs change in terms of wavelength as the frequency rises. This particular filter was designed to eliminate third-harmonic energy near 1296 MHz from a 432 MHz transmitter and does a better job in this application han the Butterworth filter in Fig 11.63 which has spurious responses in the 1296 MHz band.

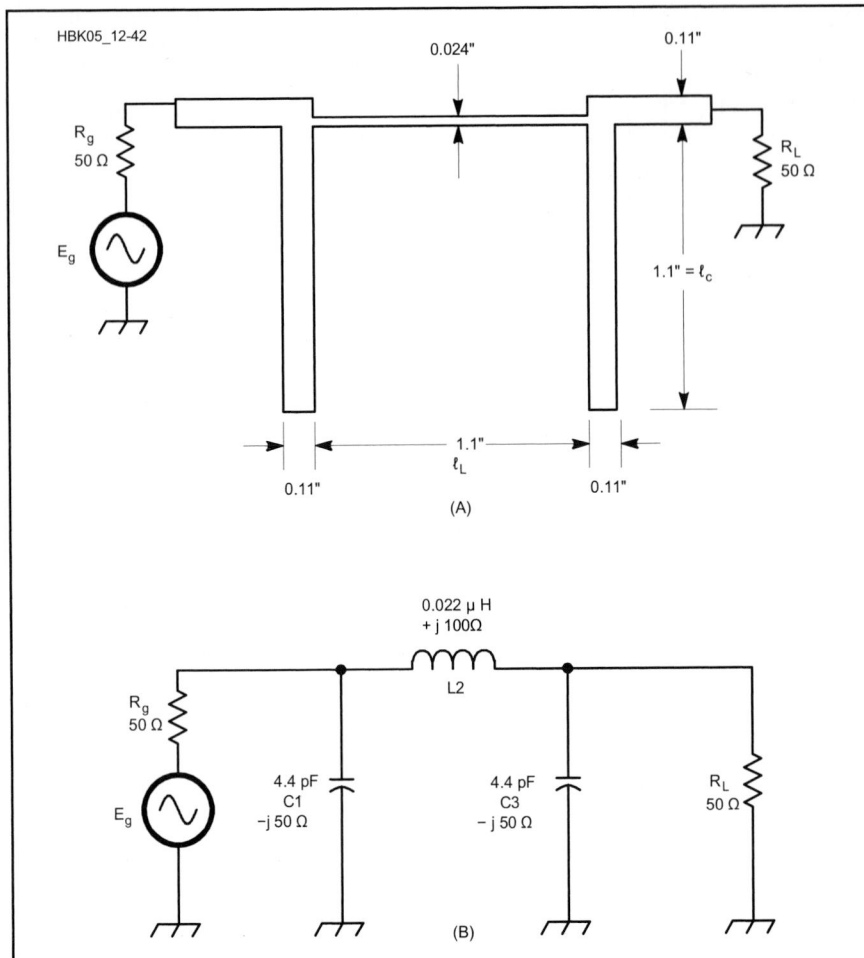

0.024"

0.11"

R_g
50 Ω

E_g

R_L
50 Ω

1.1" = ℓ_c

1.1"
ℓ_L

0.11"

0.11"

(A)

0.022 μ H
+ j 100Ω

L2

R_g
50 Ω

4.4 pF
C1
– j 50 Ω

4.4 pF
C3
– j 50 Ω

R_L
50 Ω

E_g

(B)

Fig 11.64 — A microstrip 3-pole emulated-Butterworth low-pass filter with a cutoff frequency of 720 MHz. A: Microstrip version built with G-10 fiberglass board (ε = 4.5, h = 0.062 inches). B: Lumped LC version of the same filter. To construct this filter with lumped elements very small values of L and C must be used and stray capacitance and inductance would have to be reduced to a tiny fraction of the component values.

11.9 Helical Resonators

Ever-increasing occupancy of the radio spectrum brings with it a parade of receiver overload and spurious responses. Overload problems can be minimized by using high-dynamic-range receiving techniques, but spurious responses (such as the image frequency) must be filtered out before mixing occurs. Conventional tuned circuits cannot provide the selectivity necessary to eliminate the plethora of signals found in most urban and many suburban environments. Other filtering techniques must be used.

Helical resonators are usually a better choice than ¼ λ cavities on 50, 144 and 222 MHz to eliminate these unwanted inputs because they are smaller and easier to build than coaxial cavity resonators, although their Q is not as high as that of cavities. In the frequency range from 30 to 100 MHz it is difficult to build high-Q inductors, and coaxial cavities are very large. In this frequency range the helical resonator is an excellent choice.

At 50 MHz for example, a capacitively tuned, ¼ λ coaxial cavity with an unloaded Q of 3000 would be about 4 inches in diameter and nearly 5 ft long. On the other hand, a helical resonator with the same unloaded Q is about 8.5 inches in diameter and 11.3 inches long. Even at 432 MHz, where coaxial cavities are common, the use of helical resonators results in substantial size reductions.

The helical resonator was described by the late Jim Fisk, W1HR, in a June 1976 *QST* article. [Ref 10] The resonator is described as a coil surrounded by a shield, but it is actually a shielded, resonant section of helically wound transmission line with relatively high characteristic impedance and low propagation velocity along the axis of the helix. The electrical length is about 94% of an axial ¼ λ or 84.6°. One lead of the helical winding is connected directly to the shield and the other end is open circuited as shown in **Fig 11.65**. Although the shield may be any

shape, only round and square shields will be considered here.

11.9.1 Helical Resonator Design

The unloaded Q of a helical resonator is determined primarily by the size of the shield. For a round resonator with a copper coil on a low-loss form, mounted in a copper shield, the unloaded Q is given by

$$Q_U = 50 \, D \sqrt{f_0} \qquad (14)$$

where
D = inside diameter of the shield, in inches
f_0 = frequency, in MHz.

D is assumed to be 1.2 times the width of one side for square shield cans. This formula includes the effects of losses and imperfec-

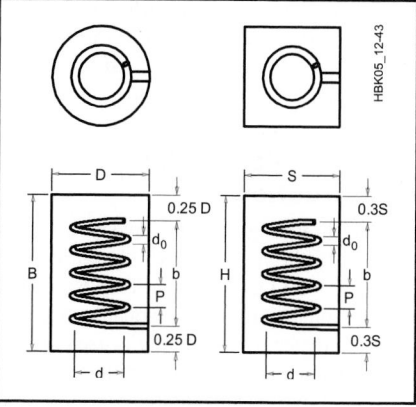

Fig 11.65 — Dimensions of round and square helical resonators. The diameter, D (or side, S) is determined by the desired unloaded Q. Other dimensions are expressed in terms of D or S (see text).

tions in practical materials. It yields values of unloaded Q that are easily attained in practice. Silver plating the shield and coil increases the unloaded Q by about 3% over that predicted by the equation. At VHF and UHF, however, it is more practical to increase the shield size slightly (that is, increase the selected QU by about 3% before making the calculation). The fringing capacitance at the open-circuit end of the helix is about 0.15D pF (that is, approximately 0.3 pF for a shield 2 inches in diameter). Once the required shield size has been determined, the total number of turns, N, winding pitch, P and characteristic impedance, Z_0, for round and square helical resonators with air dielectric between the helix and shield, are given by:

$$N = \frac{1908}{f_0 D} \quad (15A)$$

$$P = \frac{f_0 D^2}{2312} \quad (15B)$$

$$Z_0 = \frac{99{,}000}{f_0 D} \quad (15C)$$

$$N = \frac{1590}{f_0 S} \quad (15D)$$

$$P = \frac{f_0 S^2}{1606} \quad (15E)$$

$$Z_0 = \frac{82{,}500}{f_0 S} \quad (15F)$$

In these equations, dimensions D and S are in inches and f_0 is in megahertz. The design nomograph for round helical resonators in **Fig 11.66** is based on these formulas.

Although there are many variables to consider when designing helical resonators, certain ratios of shield size to length and coil diameter to length, provide optimum results. For helix diameter, d = 0.55 D or d = 0.66 S. For helix length, b = 0.825D or b = 0.99S. For shield length, B = 1.325 D and H = 1.60 S.

Design of filter dimensions can be done using the nomographs in this section or with computer software. The program *Helical* for designing and analyzing helical filters is included on this book's accompanying CD-ROM. Use of the nomographs is described in the following paragraphs.

Fig 11.67 simplifies calculation of these dimensions. Note that these ratios result in a helix with a length 1.5 times its diameter,

Helical Filter Design Software

The *ARRL Handbook CD* at the back of this book includes *Helical*, a *Windows* program by Jim Tonne, W4ENE, for design and analysis of helical-resonator bandpass filters usually used in the VHF and UHF frequency ranges.

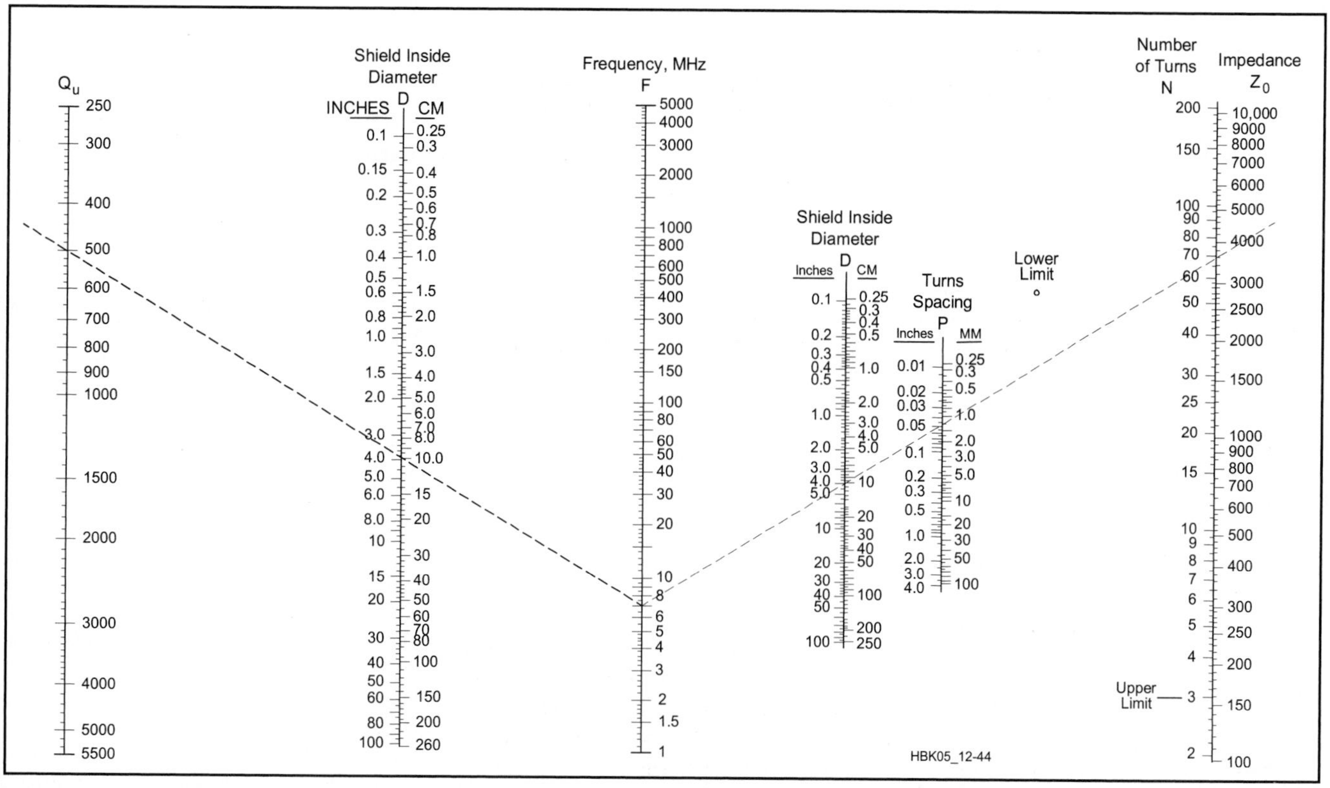

Fig 11.66 — The design nomograph for round helical resonators starts by selecting Q_U and the required shield diameter. A line is drawn connecting these two values and extended to the frequency scale (example here is for a shield of about 3.8 inches and Q_U of 500 at 7 MHz). Finally the number of turns, N, winding pitch, P, and characteristic impedance, Z_0, are determined by drawing a line from the frequency scale through selected shield diameter (but this time to the scale on the right-hand side. For the example shown, the dashed line shows P ≈ 0.047 inch, N = 70 turns, and Z_n = 3600 Ω).

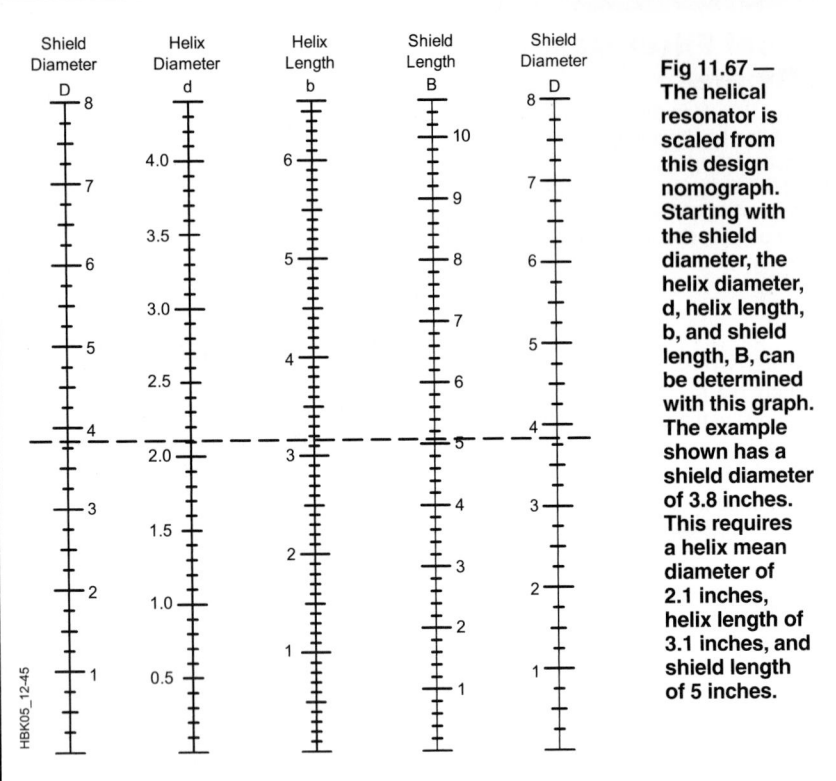

Fig 11.67 — The helical resonator is scaled from this design nomograph. Starting with the shield diameter, the helix diameter, d, helix length, b, and shield length, B, can be determined with this graph. The example shown has a shield diameter of 3.8 inches. This requires a helix mean diameter of 2.1 inches, helix length of 3.1 inches, and shield length of 5 inches.

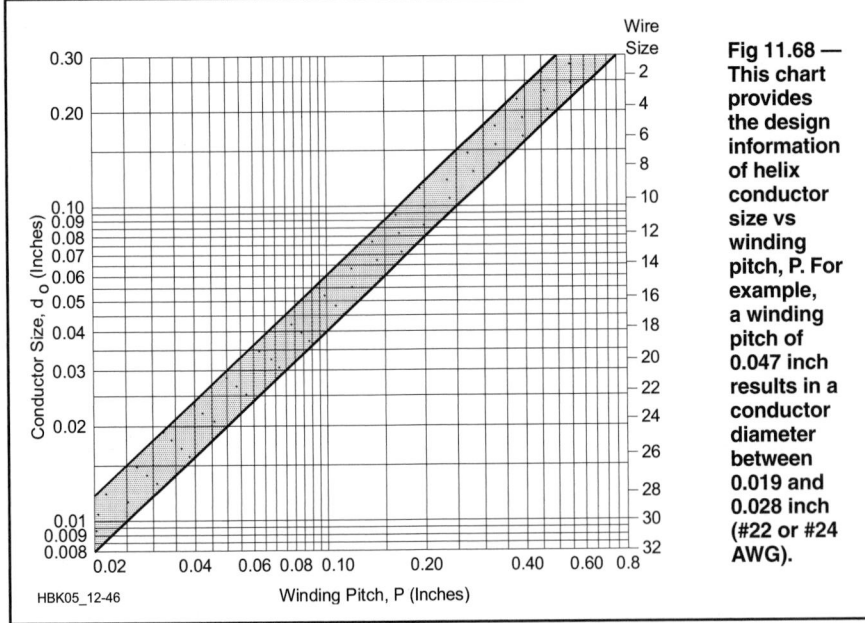

Fig 11.68 — This chart provides the design information of helix conductor size vs winding pitch, P. For example, a winding pitch of 0.047 inch results in a conductor diameter between 0.019 and 0.028 inch (#22 or #24 AWG).

the condition for maximum Q. The shield is about 60% longer than the helix — although it can be made longer — to completely contain the electric field at the top of the helix and the magnetic field at the bottom.

The winding pitch, P, is used primarily to determine the required conductor size. Adjust the length of the coil to that given by the equations during construction. Conductor size ranges from 0.4 P to 0.6 P for both round and square resonators and are plotted graphically in **Fig 11.68**.

Obviously, an area exists (in terms of frequency and unloaded Q) where the designer must make a choice between a conventional cavity (or lumped LC circuit) and a helical resonator. The choice is affected by physical shape at higher frequencies. Cavities are long and relatively small in diameter, while the length of a helical resonator is not much greater than its diameter. A second consideration is that point where the winding pitch, P, is less than the radius of the helix (otherwise

the structure tends to be non-helical). This condition occurs when the helix has fewer than three turns (the "upper limit" on the design nomograph of Fig 11.66).

11.9.2 Helical Filter Construction

The shield should not have any seams parallel to the helix axis to obtain as high an unloaded Q as possible. This is usually not a problem with round resonators because large-diameter copper tubing is used for the shield, but square resonators require at least one seam and usually more. The effect on unloaded Q is minimized if the seam is silver soldered carefully from one end to the other.

Results are best when little or no dielectric is used inside the shield. This is usually no problem at VHF and UHF because the conductors are large enough that a supporting coil form is not required. The lower end of the helix should be soldered to the nearest point on the inside of the shield.

Although the external field is minimized by the use of top and bottom shield covers, the top and bottom of the shield may be left open with negligible effect on frequency or unloaded Q. Covers, if provided, should make electrical contact with the shield. In those resonators where the helix is connected to the bottom cover, that cover must be soldered solidly to the shield to minimize losses.

11.9.3 Helical Resonator Tuning

A carefully built helical resonator designed from the nomograph of Fig 11.66 will resonate very close to the design frequency. Slightly compress or expand the helix to adjust resonance over a small range. If the helix is made slightly longer than that called for in Fig 11.67, the resonator can be tuned by pruning the open end of the coil. However, neither of these methods is recommended for wide frequency excursions because any major deviation in helix length will degrade the unloaded Q of the resonator.

Most helical resonators are tuned by means of a brass tuning screw or high-quality air-variable capacitor across the open end of the helix. Piston capacitors also work well, but the Q of the tuning capacitor should ideally be several times the unloaded Q of the resonator. Varactor diodes have sometimes been used where remote tuning is required, but varactors can generate unwanted harmonics and other spurious signals if they are excited by strong, nearby signals.

When a helical resonator is to be tuned by a variable capacitor, the shield size is based on the chosen unloaded Q at the operating frequency. Then the number of turns, N and the winding pitch, P, are based on resonance at 1.5 f_0. Tune the resonator to the desired operating frequency, f_0.

11.9.4 Helical Resonator Insertion Loss

The insertion loss (dissipation loss), I_L, in decibels, of all single-resonator circuits is given by

$$I_L = 20 \log_{10} \left(\frac{1}{1 - \dfrac{Q_L}{Q_U}} \right) \qquad (16)$$

where

Q_L = loaded Q
Q_U = unloaded Q

This is plotted in **Fig 11.69**. For the most practical cases ($Q_L > 5$), this can be closely approximated by $I_L \approx 9.0\ (Q_L/Q_U)$ dB. The selection of Q_L for a tuned circuit is dictated primarily by the required selectivity of the circuit. However, to keep dissipation loss to 0.5 dB or less (as is the case for low-noise VHF receivers), the unloaded Q must be at least 18 times the Q_L.

11.9.5 Coupling Helical Resonators

Signals are coupled into and out of helical resonators with inductive loops at the bottom of the helix, direct taps on the coil or a combination of both. Although the correct tap point can be calculated easily, coupling by loops and probes must be determined experimentally.

The input and output coupling is often pro-

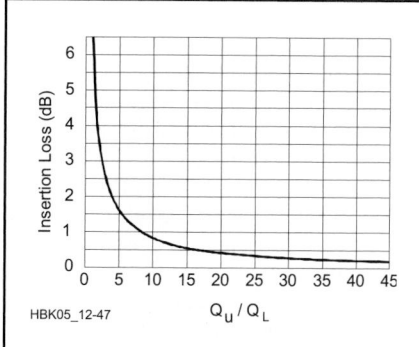

Fig 11.69 — The ratio of loaded (Q_L) to unloaded (Q_U) Q determines the insertion loss of a tuned resonant circuit.

vided by probes when only one resonator is used. The probes are positioned on opposite sides of the resonator for maximum isolation. When coupling loops are used, the plane of the loop should be perpendicular to the axis of the helix and separated a small distance from the bottom of the coil. For resonators with only a few turns, the plane of the loop can be tilted slightly so it is parallel with the slope of the adjacent conductor.

Helical resonators with inductive coupling (loops) exhibit more attenuation to signals above the resonant frequency (as compared to attenuation below resonance), whereas resonators with capacitive coupling (probes) exhibit more attenuation below the passband, as shown for a typical 432 MHz resonator in **Fig 11.70**. Consider this characteristic when choosing a coupling method. The passband can be made more symmetrical by using a combination of coupling methods (inductive input and capacitive output, for example).

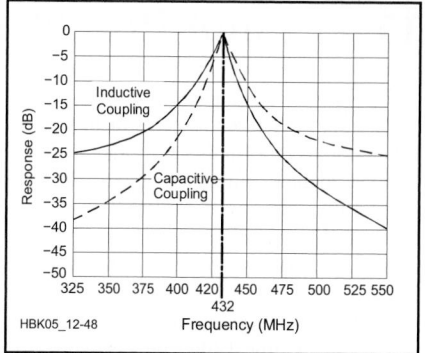

Fig 11.70 — This response curve for a single-resonator 432 MHz filter shows the effects of capacitive and inductive input/output coupling. The response curve can be made symmetrical on each side of resonance by combining the two methods (inductive input and capacitive output, or vice versa).

If more than one helical resonator is required to obtain a desired band-pass characteristic, adjacent resonators may be coupled through apertures in the shield wall between the two resonators. Unfortunately, the size and location of the aperture must be found empirically, so this method of coupling is not very practical unless you're building a large number of identical units.

Since the loaded Q of a resonator is determined by the external loading, this must be considered when selecting a tap (or position of a loop or probe). The ratio of this external loading, R_b, to the characteristic impedance, Z_0, for a $\frac{1}{4} \lambda$ resonator is calculated from:

$$K = \frac{R_b}{Z_0} = 0.785 \left(\frac{1}{Q_L} - \frac{1}{Q_U} \right) \qquad (17)$$

11.10 Use of Filters at VHF and UHF

Even when filters are designed and built properly, they may be rendered totally ineffective if not installed properly. Leakage around a filter can be quite high at VHF and UHF, where wavelengths are short. Proper attention to shielding and good grounding is mandatory for minimum leakage. Poor coaxial cable shield connection into and out of the filter is one of the greatest offenders with regard to filter leakage. Proper dc-lead bypassing throughout the receiving system is good practice, especially at VHF and above. Ferrite beads placed over the dc leads may help to reduce leakage. Proper filter termination is required to minimize loss.

Most VHF RF amplifiers optimized for noise figure do not have a 50 Ω input impedance. As a result, any filter attached to the input of an RF amplifier optimized for noise figure will not be properly terminated and filter loss may rise substantially. As this loss is directly added to the RF amplifier noise figure, carefully choose and place filters in the receiver.

11.11 Filter Projects

The filter projects to follow are by no means the only filter projects in this book. Filters for specific applications may be found in other chapters of this *Handbook*. Receiver input filters, transmitter filters, inter-stage filters and others can be extracted from the various projects and built for other applications. Since filters are a first line of defense against *electromagnetic interference* (EMI) problems, additional filter projects appear in the **RF Interference** chapter.

11.11.1 Crystal Ladder Filter for SSB

One of the great advantages of building your own crystal filter is the wide choice of crystal frequencies currently available. This allows the filter's center frequency to be chosen to fit more conveniently between adjacent amateur bands than those commercially available on 9 and 10.7 MHz do.

The 8.5 MHz series crystals used in this project were chosen in preference to ones on 9 MHz to balance the post-mixer filtering requirements on the 30 and 40 meter bands, and reduce the spurious emission caused by the second harmonic of the IF when operating on the 17 meter band. The crystals have the equivalent circuit shown in **Fig 11.71** and were obtained from Digi-Key (part number X418-ND).

The crystal-to-case capacitance becomes part of the coupling capacitance when the case is grounded, as it should be for best ultimate attenuation. The motional capacitance and inductance can vary by as much as ±5% from crystal to crystal, although the resonant frequency only varies by ±30 ppm. The ESR (equivalent series resistance) exhibits the most variation with mass-produced crystals, and in the case of these 8.5 MHz crystals can be anything from under 25 Ω to over 125 Ω.

Since the frequency variation has a spread of more than twice what can be tolerated in a 2.4 kHz SSB filter design, and the ESR is so variable, individual crystals need to be checked and selected for frequency matching and ESR. In order to allow plenty for selection, 30 crystals were purchased for the prototype. A frequency counter capable of making measurements with 10 Hz resolution, a DMM with a dc mV range and two simple self-constructed test circuits are required for crystal selection.

The project filter shown in **Fig 11.72** is based on a 7-pole QER (quasi-equiripple) design that has a shape factor of 1.96 and offers simplicity, flexibility, and a great passband. The 39 pF coupling capacitors can be silver mica or low-k disc ceramic types with a tolerance of ±2%. The termination resistance shown in the diagram has been reduced to allow for the loss in the end crystals (roughly 36 Ω for each parallel pair). The total termination resistance should be 335 Ω, theoretically, and the actual value used should be adjusted to reflect the effective ESR of your end pairs.

If more than nine suitable crystals are available from the batch tested, the order of the filter can be increased without changing the value of the coupling capacitors. The bandwidth will change very little — by less than +1% per unit increment in order. The termination resistance will need to be reduced as the order is increased, however, by the ratio of the q1 values given in Table 11.21. Increasing the order will improve the shape factor and further reduce adjacent channel interference, but also increase the passband ripple.

Effectively, each crystal in the middle section of the filter has a load of around 20.25 pF because of the 39 pF capacitors and 1.5 pF crystal-to-case capacitance on either side. The end crystal pairs are also shifted up in frequency as if they have the same load. Therefore, for best matching, all crystals should be checked in an oscillator with this load capacitance. The oscillator for this test is shown in **Fig 11.73**, and it will be seen that it shares many common parts with the VXO circuit used for ESR measurements shown in Fig 11.56. The 22 pF input capacitor and two 470 pF feedback capacitors in series present about 20 pF to the crystal under test. A transistor socket can be used as a quick means of connecting the crystals in circuit, rather than soldering and de-soldering each one in turn. If test crystals are soldered, adequate time should be allowed for them to cool so that their frequency stabilizes before

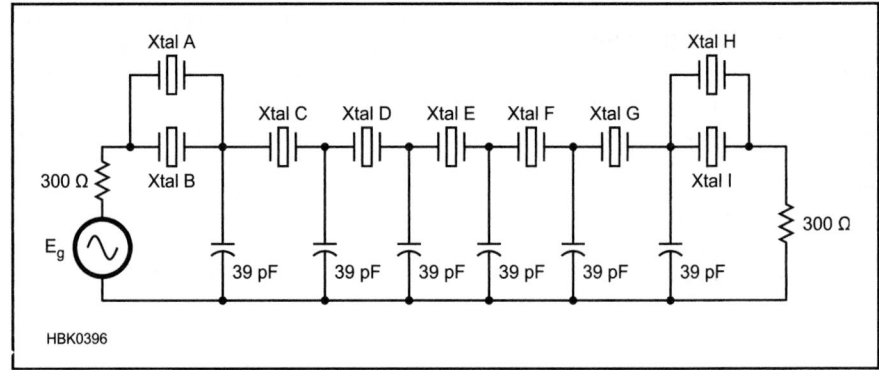

Fig 11.72 — 7-Pole 2.4 kHz QER ladder filter using ECS 8.5 MHz crystals (ECS-85-S-4) from Digi-Key (X418-ND).

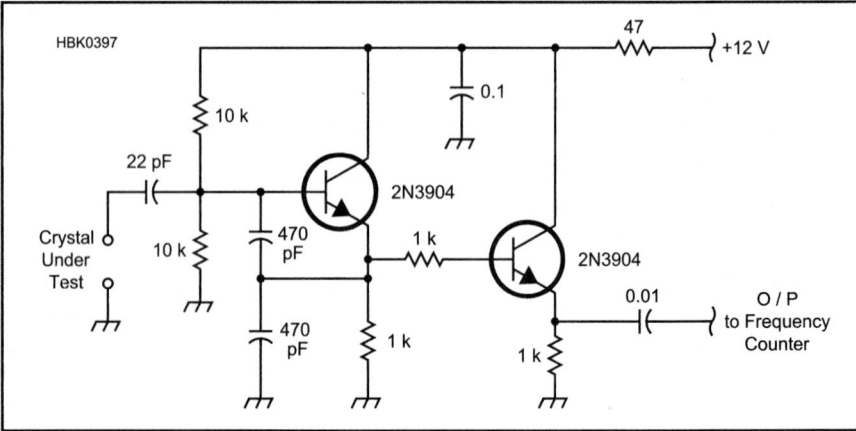

Fig 11.73 —Test oscillator with 20 pF capacitive load for matching crystal frequencies.

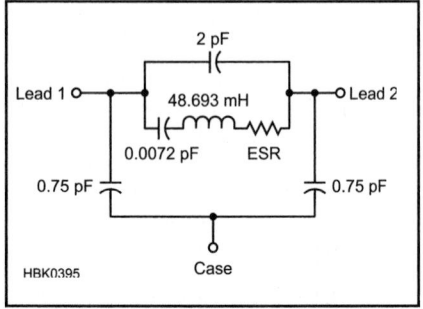

Fig 11.71 — Real electrical equivalent circuit of an ECS 8.5 MHz crystal with its metal case grounded.

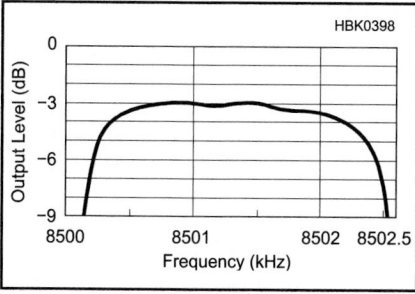

Fig 11.74 — Passband response of prototype 8.5 MHz QER crystal ladder filter with 3 dB insertion loss.

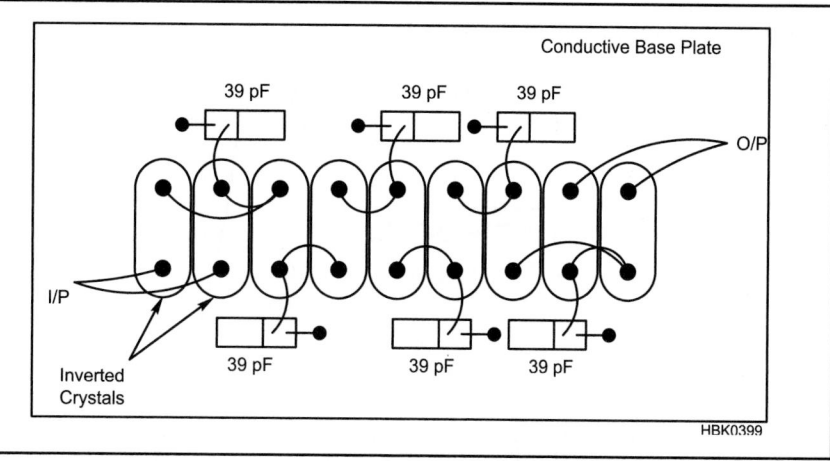

Fig 11.75 — Suggested construction arrangement using inverted crystals with their cases soldered to a conductive base plate and direct point-to-point wiring using only component leads.

Table 11.22

Measured Parameters for the 9 Crystals Selected for Use in the Prototype 7-pole 2.4 kHz SSB Filter.

Xtal	Freq (20pF)	ESR (Ω)	Q
A	8501.342 kHz	87	30k
B	8501.441 kHz	60	43k
C	8501.482 kHz	33	78k
D	8501.557 kHz	32	81k
E	8501.530 kHz	27	97k
F	8501.372 kHz	28	93k
G	8501.411 kHz	52	50k
H	8501.477 kHz	108	24k
I	8501.400 kHz	54	48k

a reading is taken.

The crystals need to be numbered in sequence with a permanent marker to identify them, and their 20 pF load frequencies recorded in a table or spreadsheet as they are measured. When this is complete, the oscillator can then be converted to a VXO, as shown in Fig 11.56, and using this and the jig in Fig 11.55 the ESR of each crystal can be assessed by comparison with metal or carbon film resistors that give the same output readings on the DMM.

The variable capacitor used in the VXO to check the crystals for the prototype filter was a polyvaricon type with maximum capacitance of 262 pF and the inductor was a 10 µH miniature RF choke. If smaller values of variable capacitor are used the inductor value will need to be increased in order to ensure the frequency swing is great enough to tune thru the peak of the lowest frequency crystal in the batch.

Once all the crystals have been checked and their 20 pF frequencies and ESR values recorded, the best selection strategy is to look over the figures for a group of nine crystals that have a spread of frequencies of less than 10% of the 2.4 kHz bandwidth. They can then be considered for position in the filter on the basis of their ESR values. The ones with the lowest values should be selected for the middle positions, with the lowest of all as the central crystal.

The crystals with the highest ESR values should be paired up for use as the end parallel crystals because their loss can be absorbed in the terminations. Try to use a pair of crystals with similar values of overall ESR if you want the two terminating resistors to have the same value after subtracting the effective loss resistance of each pair of end crystals from the required theoretical termination resistance of 335 Ω. **Table 11.22** shows how the nine crystals chosen from the batch of 30 obtained for the prototype were selected for position to obtain the passband curve shown in **Fig 11.74**. It can be seen that the spread of frequencies for a load of 20 pF in this case was 215 Hz, and the values of ESR varied from 27 Ω (best) to 108 Ω (worst). The prototype bandwidth (–6 dB) came out at 2.373 kHz with 39 pF coupling capacitors that were all approximately 1% high of their nominal value. A random selection of ±2% capacitors should produce a bandwidth of between 2.35 and 2.45 kHz. If a wider bandwidth is required, 33 pF coupling can be used instead of 39 pF, with the theoretical termination resistance increased to 393 Ω. This should provide a bandwidth of around 2.8 kHz.

The motional capacitance of the ECS 9 MHz series crystals available from Digi-Key (part number X419-ND) should be only slightly higher than that of the 8.5 MHz crystals, so they could be used in this design with a corresponding increase in bandwidth — probably around 300 Hz, making the overall filter bandwidth at 9 MHz about 2.7 kHz (±60 Hz with 2% tolerance, 39 pF capacitors).

Construction can be a matter of availability and ingenuity. Reclaimed filter cans from unwanted wide-bandwidth commercial crystal filters could be utilized if they can be picked up cheaply enough. Otherwise, inverted crystals can be soldered side by side, and in line, to a base plate made of a piece of PCB material, copper, or brass sheet as shown in **Fig 11.75**. The coupling capacitors can then be soldered

between the crystal leads and the base plate, or crystal cases, depending on which is more convenient. Excellent ultimate attenuation (>120 dB) can be achieved using direct wiring and good grounding like this, particularly if lead lengths are kept as short as possible and if additional shielding is added at each end of the filter to prevent active circuitry at either end of the filter from coupling to and leaking signal around it. A case made from pieces of the same material can be soldered around the base plate to form a fully shielded filter unit with feed-thru insulators for the input and output connections.

11.11.2 Broadcast-Band Rejection Filter

Inadequate front-end selectivity or poorly performing RF amplifier and mixer stages often result in unwanted cross-talk and overloading from adjacent commercial or amateur stations. The filter shown is inserted between the antenna and receiver. It attenuates the out-of-band signals from broadcast stations but passes signals of interest (1.8 to 30 MHz) with little or no attenuation.

The high signal strength of local broadcast stations requires that the stop-band attenuation of the high-pass filter also be high. This filter provides about 60 dB of stop-band attenuation with less than 1 dB of attenuation above 1.8 MHz. The filter input and output ports match 50 Ω with a maximum SWR of 1.353:1 (reflection coefficient = 0.15). A 10-element filter yields adequate stop-band attenuation and a reasonable rate of attenuation rise. The design uses only standard-value capacitors.

BUILDING THE FILTER

The filter parts layout, schematic diagram, response curve and component values are shown in **Fig 11.76**. The standard capa-

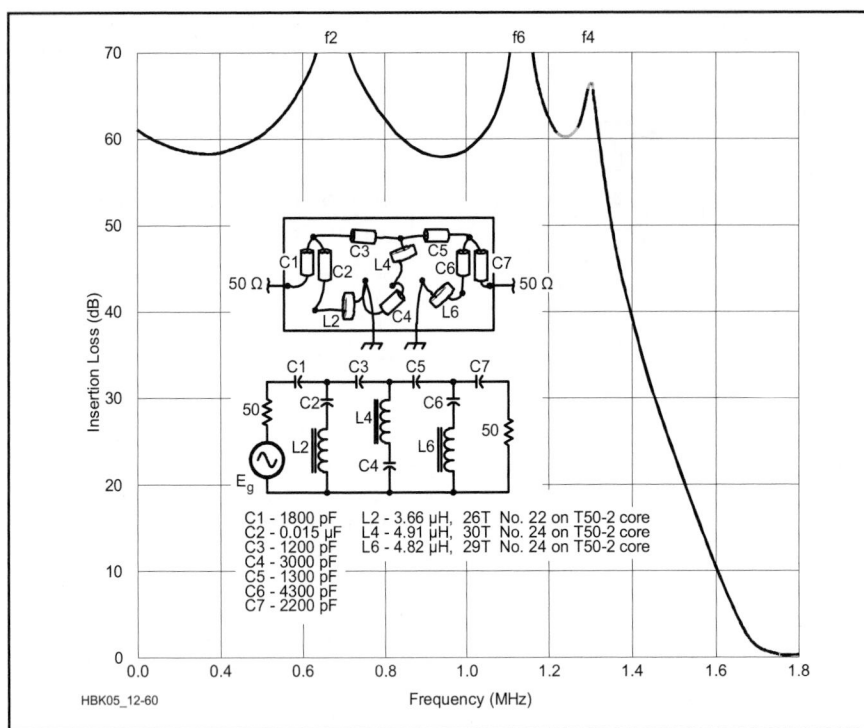

Fig 11.76 — Schematic, layout and response curve of the broadcast band rejection filter.

C1 - 1800 pF
C2 - 0.015 µF
C3 - 1200 pF
C4 - 3000 pF
C5 - 1300 pF
C6 - 4300 pF
C7 - 2200 pF

L2 - 3.66 µH, 26T No. 22 on T50-2 core
L4 - 4.91 µH, 30T No. 24 on T50-2 core
L6 - 4.82 µH, 29T No. 24 on T50-2 core

11.11.3 Wave Trap for Broadcast Stations

Nearby medium-wave broadcast stations can sometimes cause interference to HF receivers over a broad range of frequencies. A wave trap can catch the unwanted frequencies and keep them out of your receiver.

CIRCUIT DESCRIPTION

The way the circuit works is quite simple. Referring to **Fig 11.78**, you can see that it consists essentially of only two components, a coil L1 and a variable capacitor C1. This series-tuned circuit is connected in parallel with the antenna circuit of the receiver. The characteristic of a series-tuned circuit is that the coil and capacitor have a very low impedance (resistance) to frequencies very close to the frequency to which the circuit is tuned. All other frequencies are almost unaffected. If the circuit is tuned to 1530 kHz, for example, the signals from a broadcast station on that frequency will flow through the filter to ground, rather than go on into the receiver. All other frequencies will pass straight into the receiver. In this way, any interference caused in the receiver by the station on 1530 kHz is significantly reduced.

CONSTRUCTION

This is a series-tuned circuit that is adjustable from about 540 kHz to 1600 kHz. It is built into a metal box, **Fig 11.79**, to shield it from other unwanted signals and is connected

ci-tor values listed are within 2.8% of the design values. If the attenuation peaks (f2, f4 and f6) do not fall at 0.677, 1.293 and 1.111 MHz, tune the series-resonant circuits by slightly squeezing or separating the inductor windings.

Construction of the filter is shown in **Fig 11.77**. Use polypropylene film-type capacitors. These capacitors are available through Digi-Key and other suppliers. The powdered-iron T50-2 toroidal cores are available through Amidon, Palomar Engineers and others.

For a 3.4 MHz cutoff frequency, divide the L and C values by 2. (This effectively doubles the frequency-label values in Fig 11.76.) For the 80 meter version, L2 through L6 should be 20 to 25 turns each, wound on T50-6 cores. The actual turns required may vary one or two

from the calculated values. Parallel-connect capacitors as needed to achieve the nonstandard capacitor values required for this filter.

FILTER PERFORMANCE

The measured filter performance is shown in Fig 11.76. The stop-band attenuation is more than 58 dB. The measured cutoff frequency (less than 1 dB attenuation) is under 1.8 MHz. The measured passband loss is less than 0.8 dB from 1.8 to 10 MHz. Between 10 and 100 MHz, the insertion loss of the filter gradually increases to 2 dB. Input impedance was measured between 1.7 and 4.2 MHz. Over the range tested, the input impedance of the filter remained within the 37- to 67.7-Ω input-impedance window (equivalent to a maximum SWR of 1.353:1).

Fig 11.79 — The wave trap can be roughly calibrated to indicate the frequency to which it is tuned.

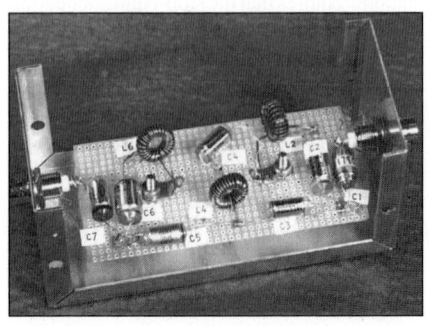

Fig 11.77 — The filter fits easily in a 2 × 2 × 5 inch enclosure. The version in the photo was built on a piece of perfboard.

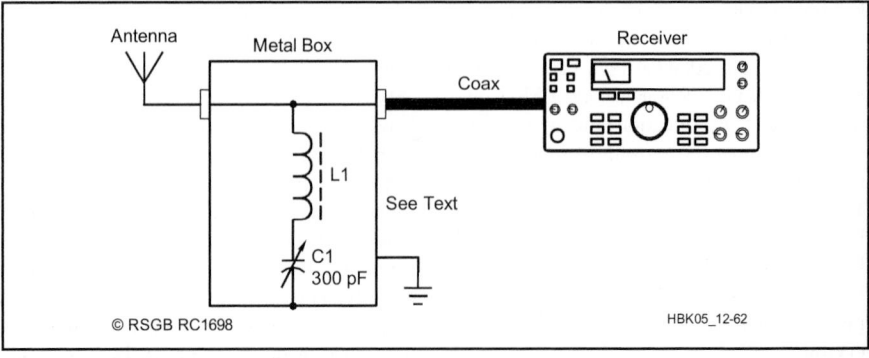

Fig 11.78 — The wave trap consists of a series tuned circuit, which 'shunts' signals on an unwanted frequency to ground.

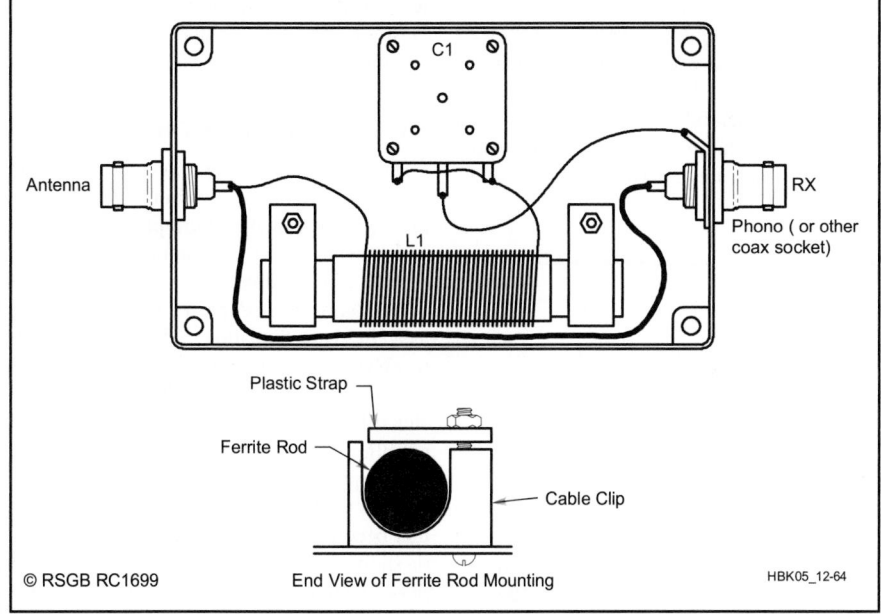

Fig 11.80 — Wiring of the wave trap. The ferrite rod is held in place with cable clips.

C1 — 300 pF polyvaricon variable.
L1 — 80 turns of 30 SWG enameled wire, wound on a ferrite rod.

Associated items: Case (die-cast box), knobs to suit, connectors to suit, nuts and bolts, plastic cable clips.

as shown in Fig 11.78. To make the inductor, first make a *former* by winding two layers of paper on the ferrite rod. Fix this in place with black electrical tape. Next, lay one end of the wire for the coil on top of the former, leaving about an inch of wire protruding beyond the end of the ferrite rod. Use several turns of electrical tape to secure the wire to the former. Now, wind the coil along the former, making sure the turns are in a single layer and close together. Leave an inch or so of wire free at the end of the coil. Once again, use a couple of turns of electrical tape to secure the wire to the former. Finally, remove half an inch of enamel from each end of the wire.

Alternatively, if you have an old AM transistor radio, a suitable coil can usually be recovered already wound on a ferrite rod. Ignore any small coupling coils. Drill the box to take the components, then fit them in and solder together as shown in **Fig 11.80**. Make sure the lid of the box is fixed securely in place, or the wave trap's performance will be adversely affected by pick-up on the components.

CONNECTION AND ADJUSTMENT

Connect the wave trap between the antenna and the receiver, then tune C1 until the interference from the offending broadcast station is a minimum. You may not be able to eliminate interference completely, but this handy little device should reduce it enough to listen to the amateur bands. Let's say you live near an AM transmitter on 1530 kHz, and the signals break through on

your 1.8 MHz receiver. By tuning the trap to 1530 kHz, the problem is greatly reduced. If you have problems from more than one broadcast station, the problem needs a more complex solution.

11.11.4 Optimized Harmonic Transmitting Filters

Low-pass filters should be placed at the output of transmitters to ensure that they meet the various regulatory agency requirements for harmonic suppression. These are commonly designed to pass a single amateur band and provide attenuation at harmonics of that band sufficient to meet the requirements. The material presented here by Jim Tonne, W4ENE, is based on material originally published in the September/October 1998 issue of *QEX*. The basic approach is to use a computer to optimize the performance in the passband (a single amateur band) while simultaneously maximizing the attenuation at the second and third harmonic of that same band. When this is done, the higher harmonics will also be well within spec.

The schematic of this filter along with parts

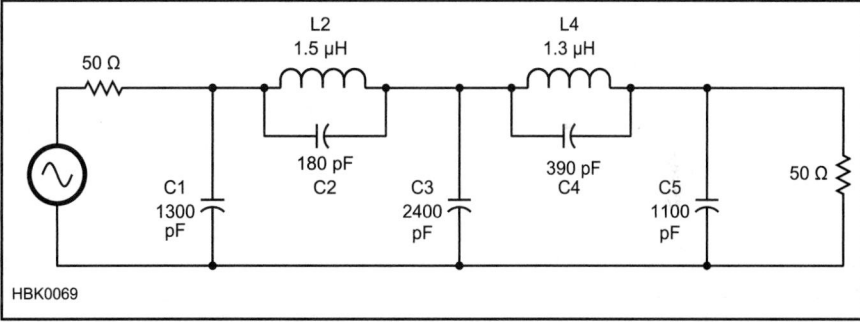

Fig 11.81 — Optimized low-pass filter. This design is for the 80 meter amateur band. It is similar to a Cauer design but the parts values have been optimized as described in the text and in the Sep/Oct 1998 issue of *QEX*.

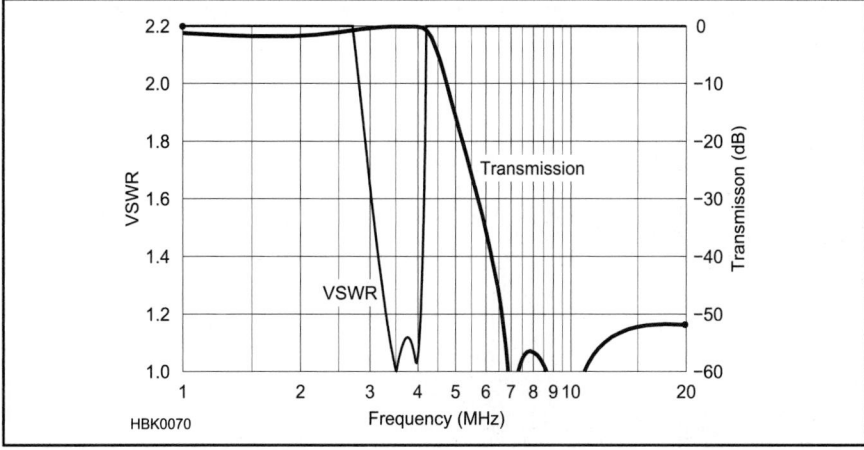

Fig 11.82 — Responses of the filter shown in Fig 11.81. Note the low values of SWR from 3.5 to 4 MHz. At the same time the harmonics are attenuated to meet regulations. Responses for the other amateur bands are very similar except for the frequency scaling.

values for the 3.5 to 4.0 MHz amateur band is shown in **Fig 11.81**. The responses of that filter are shown in **Fig 11.82**.

Component values for the 160 meter through the 6 meter amateur bands are shown in **Table 11.23**. The capacitors are shown in pF and the inductors in μH. The capacitors

are the nearest 5% values; both the nearest 5% and the exact inductor values are shown.

Using the nearest-5% inductor values will result in satisfactory operation. If the construction method is such that exact-value (adjustable) inductors can be used then the "Exact" values are preferred. These values

were obtained from the program *SVC Filter Designer* which is on the *ARRL Handbook CD* included with this book.

11.11.5 Diplexer Filter

This section, covering diplexer filters, was written by William E. Sabin, WØIYH. The diplexer is helpful in certain applications, such as frequency mixer terminations.

(The terms "diplexer" and "duplexer" are often confused. A duplexer is a device such as a circulator that allows a transmitter and a receiver to use the same antenna *without* the use of filters. Diplexers use filters so that the signal frequencies must be far apart, such as on different bands. Diplexers have a constant filter-input resistance that extends to the stop band as well as the passband. Ordinary filters that become highly reactive or have an open or short-circuit input impedance outside the passband may degrade performance of the devices to which they are attached. (For example, impedances far from 50 Ω outside the operating frequency range may cause an amplifier to develop parasitic oscillations.)

Fig 11.83 shows a *normalized* prototype 5-element, 0.1-dB Chebyshev low-pass/high-pass (LP/HP) filter. This idealized filter is driven by a voltage generator with zero internal resistance, has load resistors of 1.0 Ω and a cutoff frequency of 1.0 radian per second (0.1592 Hz). The LP prototype values are taken from standard filter tables.[1] The first element is a series inductor. The HP prototype is found by:

a) replacing the series L (LP) with a series C (HP) whose value is 1/L, and

b) replacing the shunt C (LP) with a shunt L (HP) whose value is 1/C.

For the Chebyshev filter, the return loss is improved several dB by multiplying the prototype LP values by an experimentally derived number, K, and dividing the HP values by the same K. You can calculate the LP values in henrys and farads for a 50 Ω RF application with the following formulas:

$$L_{LP} = \frac{KL_{P(LP)} R}{2\pi f_{CO}} \; ; \; C_{LP} = \frac{KC_{P(LP)}}{2\pi f_{CO} R}$$

where

$L_{P(LP)}$ and $C_{P(LP)}$ are LP prototype values
K = 1.005 (in this specific example)
R = 50 Ω
f_{CO} = the cutoff (–3 dB response) frequency in Hz.

For the HP segment:

$$L_{HP} = \frac{L_{P(HP)} R}{2\pi f_{CO} K} \; ; \; C_{HP} = \frac{C_{P(HP)}}{2\pi f_{CO} KR}$$

where $L_{P(HP)}$ and $C_{P(HP)}$ are HP prototype values.

Fig 11.84 shows the LP and HP responses of a diplexer filter for the 80 meter band. The

Table 11.23
Values for the Optimized Harmonic Filters

Band (meters)	C1 (pF)	L2, 5% (μH)	L2, Exact (μH)	C2 (pF)	C3 (pF)	L4, 5% (μH)	L4, Exact (μH)	C4 (pF)	C5 (pF)
160	2400	3.0	2.88	360	4700	2.4	2.46	820	2200
80	1300	1.5	1.437	180	2400	1.3	1.29	390	1100
60	910	1.0	1.029	120	1600	0.91	0.8897	270	750
40	680	0.75	0.7834	91	1300	0.62	0.6305	220	560
30	470	0.56	0.5626	68	910	0.47	0.4652	160	430
20	330	0.39	0.3805	47	620	0.33	0.3163	110	300
17	270	0.30	0.3063	36	510	0.27	0.2617	82	240
15	220	0.27	0.2615	30	430	0.22	0.2245	68	200
12	200	0.24	0.241	27	390	0.20	0.2042	62	180
10	180	0.20	0.2063	24	330	0.18	0.1721	56	150
6	91	0.11	0.108	13	180	0.091	0.0911	30	82

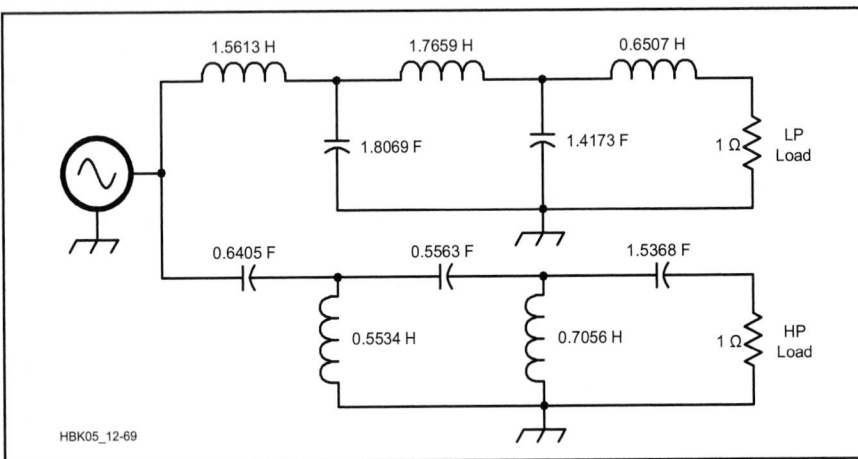

HBK05_12-69

Fig 11.83 — Low-pass and high-pass prototype diplexer filter design. The low-pass portion is at the top, and the high-pass at the bottom of the drawing. See text.

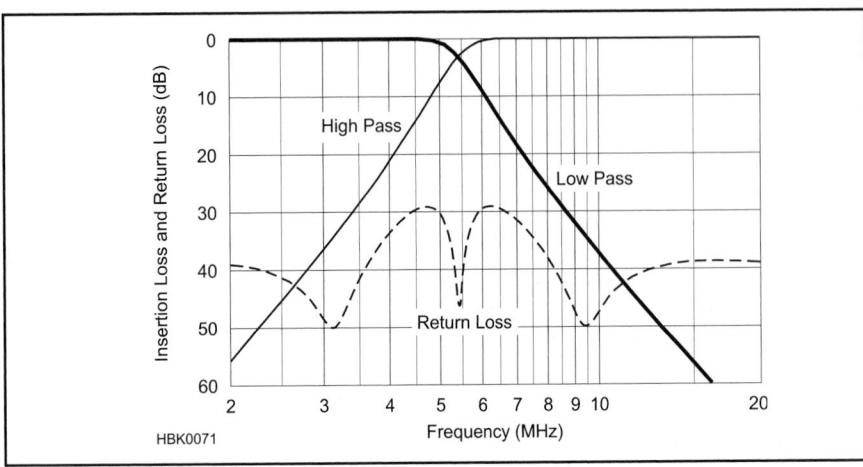

HBK0071

Fig 11.84 — Response for the low-pass and high-pass portions of the 80 meter diplexer filter. Also shown is the return loss of the filter.

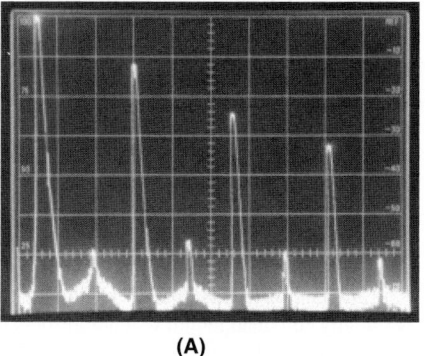
(A)

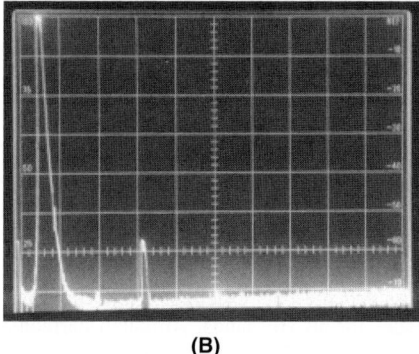
(B)

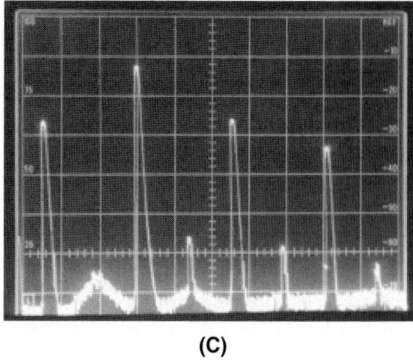
(C)

Fig 11.85 — At A, the output spectrum of a push-pull 80 meter amplifier. At B, the spectrum after passing through the low-pass filter. At C, the spectrum after passing through the high-pass filter.

following items are to be noted:

- The 3 dB responses of the LP and HP meet at 5.45 MHz.
- The input impedance is close to 50 Ω at all frequencies, as indicated by the high value of return loss (SWR <1.07:1).
- At and near 5.45 MHz, the LP input reactance and the HP input reactance are conjugates; therefore, they cancel and produce an almost perfect 50 Ω input resistance in that region.
- Because of the way the diplexer filter is derived from synthesis procedures, the transfer characteristic of the filter is mostly independent of the actual value of the amplifier dynamic output impedance.[2] This is a useful feature, since the RF power amplifier output impedance is usually not known or specified.
- The 80 meter band is well within the LP response.
- The HP response is down more than 20 dB at 4 MHz.
- The second harmonic of 3.5 MHz is down only 18 dB at 7.0 MHz. Because the second harmonic attenuation of the LP is not great, it is necessary that the amplifier itself be a well-balanced push-pull design that greatly rejects the second harmonic. In practice this is not a difficult task.
- The third harmonic of 3.5 MHz is down almost 40 dB at 10.5 MHz.

Fig 11.85A shows the unfiltered output of a solid-state push-pull power amplifier for the 80 meter band. In the figure you can see that:

- The second harmonic has been suppressed by a proper push-pull design.
- The third harmonic is typically only 15 dB or less below the fundamental.

The amplifier output goes through our diplexer filter. The desired output comes from the LP side, and is shown in Fig 11.85B. In it we see that:

- The fundamental is attenuated only about 0.2 dB.
- The LP has some harmonic content; however, the attenuation exceeds FCC requirements for a 100 W amplifier.

Fig 11.85C shows the HP output of the di-

plexer that terminates in the HP load or *dump* resistor. A small amount of the fundamental frequency (about 1%) is also lost in this resistor. Within the 3.5 to 4.0 MHz band, the filter input resistance is almost exactly the correct 50 Ω load resistance value. This is because power that would otherwise be *reflected* back to the amplifier is absorbed in the dump resistor.

Solid state power amplifiers tend to have stability problems that can be difficult to debug.[3] These problems may be evidenced by level changes in: load impedance, drive, gate or base bias, supply voltage, etc. Problems may arise from:

- The reactance of the low-pass filter outside the desired passband. This is especially true for transistors that are designed for high-frequency operation.
- Self-resonance of a series inductor at some high frequency.
- A stop band impedance that causes voltage, current and impedance reflections back to the amplifier, creating instabilities within the output transistors.

Intermodulation performance can also be degraded by these reflections. The strong third harmonic is especially bothersome for these problems.

The diplexer filter is an approach that can greatly simplify the design process, especially for the amateur with limited PA-design experience and with limited home-lab facilities. For these reasons, the amateur homebrew enthusiast may want to consider this solution, despite

its slightly greater parts count and expense.

The diplexer is a good technique for narrowband applications such as the HF amateur bands.[4] From Fig 11.84, we see that if the signal frequency is moved beyond 4.0 MHz the amount of desired signal lost in the dump resistor becomes large. For signal frequencies below 3.5 MHz the harmonic reduction may be inadequate. A single filter will not suffice for all the HF amateur bands.

This treatment provides you with the information to calculate your own filters. A *QEX* article has detailed instructions for building and testing a set of six filters for a 120 W amplifier. These filters cover all nine of the MF/HF amateur bands.[5]

You can use this technique for other filters such as Bessel, Butterworth, linear phase, Chebyshev 0.5, 1.0, etc.[6] However, the diplexer idea does *not* apply to the elliptic function types.

The diplexer approach is a resource that can be used in any application where a constant value of filter input resistance over a wide range of passband and stop band frequencies is desirable for some reason. Computer modeling is an ideal way to finalize the design before the actual construction. The coil dimensions and the dump resistor wattage need to be determined from a consideration of the power levels involved.

Another significant application of the diplexer is for elimination of EMI, RFI and TVI energy. Instead of being reflected and very possibly escaping by some other route, the unwanted energy is dissipated in the dump resistor.[7]

Diplexer Filter Design Software

The *ARRL Handbook CD* at the back of this book includes *DiplexerDesigner*, a *Windows* program for design and analysis of diplexer filters by Jim Tonne, W4ENE. The software is interactive and plots performance changes as various parameters are altered.

Notes

[1]Williams, A. and Taylor, F., *Electronic Filter Design Handbook*, any edition, McGraw-Hill.
[2]Storer, J.E., *Passive Network Synthesis*, McGraw-Hill 1957, pp 168-170. This book shows that the input resistance is ideally constant in the passband and the stop band and that the filter transfer characteristic is ideally independent of the generator impedance.
[3]Sabin, W. and Schoenike, E., *HF Radio Systems and Circuits*, Chapter 12, Noble

Publishing, 1998. Also the previous edition of this book, *Single-Sideband Systems and Circuits*, McGraw-Hill, 1987 or 1995.

[4]Dye, N. and Granberg, H., *Radio Frequency Transistors, Principles and Applications*, Butterworth-Heinemann, 1993, p 151.

[5]Sabin, W.E. WØIYH, "Diplexer Filters for the HF MOSFET Power Amplifier," *QEX*, Jul/Aug, 1999. Also check the ARRL website at **www.arrl.org/qex**.

[6]See note 1. *Electronic Filter Design Handbook* has LP prototype values for various filter types, and for complexities from 2 to 10 components.

[7]Weinrich, R. and Carroll, R.W., "Absorptive Filters for TV Harmonics," *QST*, Nov 1968, pp 10-25.

11.11.6 High-Performance, Low-Cost 1.8 to 54 MHz Low-Pass Filter

The low-pass filter shown in **Fig 11.86** offers low insertion loss, mechanical simplicity, easy construction and operation on all amateur bands from 160 through 6 meters. Originally built as an accessory filter for a 1500 W 6 meter amplifier, the filter easily handles legal limit power. No complicated test equipment is necessary for alignment. It was originally described by Bill Jones, K8CU, in November 2002 *QST*. The complete original *QST* article for this project is included on the CD-ROM included with this book. The article supplies complete assembly and alignment drawings.

Although primarily intended for coverage of the 6 meter band, this filter has low insertion loss and presents excellent SWR characteristics for all HF bands. Although harmonic attenuation at low VHF frequencies near TV channels 2, 3 and 4 does not compare to filters designed only for HF operation, the use of this filter on HF is a bonus to 6 meter operators who also use the HF bands. Six meter operators may easily tune this filter for low insertion loss and SWR in any favorite band segment, including the higher frequency FM portion of the band.

ELECTRICAL DESIGN

The software tool used to design this low-pass filter is *Elsie* by Jim Tonne, W4ENE, which may be found on the CD included with this book. The *Elsie* format data file for this filter, DC54.lct, may be downloaded for your own evaluation from the author's website at **www.realhamradio.com**.

Fig 11.87 is a schematic diagram of the filter. The use of low self-inductance capacitors with Teflon dielectric easily allows legal limit high power operation and aids in the ultimate stop band attenuation of this filter. Capacitors with essentially zero lead length will not introduce significant series inductance that upsets filter operation. This filter also uses a trap that greatly attenuates second harmonic frequencies of the 6 meter band. The parts list for the filter is given in **Table 11.24**.

PERFORMANCE DISCUSSION

Assuming the 6 meter SWR is set to a low value for a favorite part of the band, the worst case calculated forward filter loss is about 0.18 dB. The forward loss is better in the HF bands, with a calculated loss of only 0.05 dB from 1.8 through 30 MHz. The filter cutoff frequency is about 56 MHz, and the filter response drops sharply above this.

Fig 11.88 shows the calculated filter response from 1 to 1000 MHz. The impressive notch near 365 MHz is because of these inherent stray capacitances across each of the coils. Slight variations in each coil will make slightly different tuned traps. This will introduce a stagger-tuned effect that results in a broader notch.

Table 11.24
Low-Pass Filter Parts List

Qty	Description
1	Miniature brass strip, 1 × 12 in., 0.032 in. thick (C3)
1	Miniature brass strip, 2 × 12 in., 0.064 in. thick (C1, C2)
5 ft	⅛ inch diameter soft copper tubing
4	¼ × 20 × ½ in. long hex head bolt
4	Plastic spacer or washer, 0.5 in. OD, 0.25 in. ID, 0.0625 in. thick
6	¼ × 20 hex nut with integral tooth lock washer
1	¼ × 20 × 4 in. long bolt
1	¼ × 20 threaded nut insert, PEM nut, or "Nutsert"
1	1 × 0.375 in. diameter nylon spacer. ID smaller than 0.25 in. (used for C3 plunger).
4	Nylon spacer, 0.875 in. OD, 0.25 to 0.34 in. ID, approx. 0.065 in. or greater thickness (used to attach brass capacitor plates).

Aluminum diecast enclosure is available from Jameco Electronics (**www.jameco.com**) part no. 11973. The box dimensions are 7.5 × 4.3 × 2.4 in. The 0.03125 in. thick Teflon sheet is available from McMaster-Carr Supply Co (**www.mcmaster.com**), item #8545K21 is available as a 12 × 12 in. sheet.

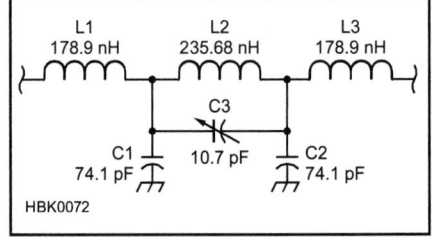

Fig 11.87 — The low-pass filter schematic. See construction details in the article on the CD-ROM.

C1, C2 — 74.1 pF. 2 × 2.65 inch brass plate sandwiched with 0.03125 inch thick Teflon sheet. The metal enclosure is the remaining grounded terminal of this capacitor.

C3 — Homemade variable using brass and Teflon.

L1, L3 — 178.9 nH. Wind with ⅛ inch OD soft copper tubing, 3.5 turns, 0.75 inch diameter form, 0.625 inch long, ¼ inch lead length for soldering to brass plate. The length of the other lead to RF connector as required.

L2 — 235.68 nH. Wind with ⅛ inch OD soft copper tubing, 5 turns, 0.75 inch diameter form, 1.75 inches long. Leave ¼ inch lead length for soldering.

Fig 11.86 — The 1.8 to 54 MHz low-pass filter is housed in a die-cast box.

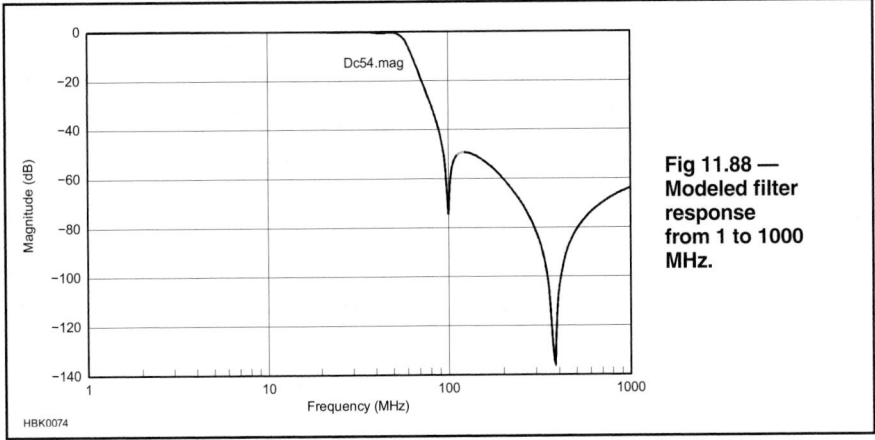

Fig 11.88 —
Modeled filter
response
from 1 to 1000
MHz.

HBK0074

(A)

(B)

(C)

HBK0708

11.11.7 Band-Pass Filter for 145 MHz

The following project is based on a design from the RSGB *Radio Communication Handbook, 11th edition.* This filter is intended to reduce harmonics and other out-of-band spurious emissions when transmitting and suppress strong out-of-band incoming signals which could overload the receiver. The filter's schematic and response curve are shown in **Fig 11.89.** This filter design is not suitable for use as a repeater duplexer for in-band signals sharing a common antenna — a high-Q cavity resonator is required.

Direct inductive coupling from the proximity of the coils is used between the first two and the last two resonant circuits. C_C performs "top coupling" between the center two sections where a shield prevents stray coupling. The input and output connections are tapped down on their respective coils to transform the 50 Ω source and load into the proper impedances for terminating the filter

At VHF, self-supporting coils and mica, ceramic or air-dielectric trimmer capacitors give adequate results for most applications. Piston ceramic trimmers can be used for receiving or for low-power signals of a few watts. At higher power, use an air-variable capacitor. The filter is assembled in a 4 × 2½ × 1½ inch die-cast aluminum box so there is room for either piston or air-variable capacitors.

The band-pass filter is made from four parallel resonant circuits formed by the inductance of the coils and their associated parasitic inter-turn capacitance described in the **RF Techniques** chapter. A 1-6 pF series capacitor (C in Fig 11.89) connected to ground adjusts the resonant frequency of each coil. A ceramic capacitor should be used for C_C which has a value of 0.5 pF.

The dimensions of the coils are important to control the self-resonant frequency. Each coil is constructed from 6½ turns of #17 AWG solid, bare wire (1.15 mm dia), ⅜ inch (9.5 mm) in diameter. Each turn is spaced 1 wire diameter apart. The original design used British #18 SWG wire which is slightly thicker (1.22 mm) so coils made with #17 AWG wire will have a slightly higher resonant frequency. The taps for the coils are made 1 turn from the grounded end of the coil as shown in the figure.

Adjustment will be required after assembly to tune out the stray capacitances and inductances. A sweep generator and oscilloscope (see the **Test Equipment and Measurements** chapter) provide the most practical adjustment method. A variable oscillator with frequency counter and a voltmeter with RF probe, plus a good deal of patience, can also do the job.

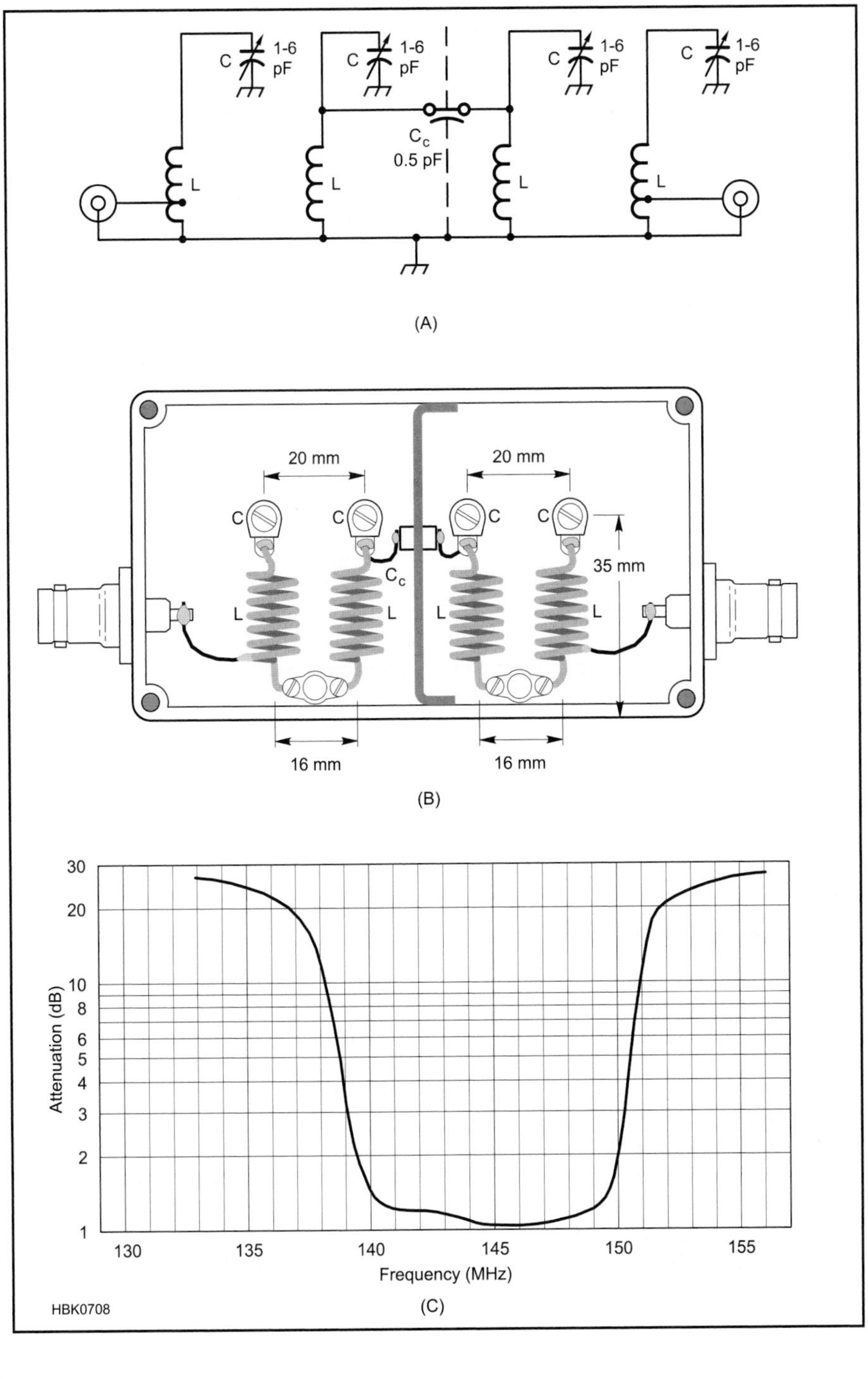

Fig 11.89 — A four-section band-pass filter for 145 MHz for attenuating strong out-of-band signals and reducing harmonics or other out-of-band spurious emissions.

11.12 Filter Glossary

Active filter — A filter that uses active (powered) devices to implement its function.

All-pass — Filter response in which the magnitude response does not change with frequency, but the phase response does change with frequency.

Amplitude response — see **magnitude response**

Band-pass — Filter response in which signals are passed in a range of frequencies and rejected outside that range.

Band-stop — Filter response in which signals are rejected in a range of frequencies and passed outside that range (also called *band-reject* or *notch* filter).

Bandwidth — Range of frequencies over which signals are passed (low-pass, high-pass, band-pass) or rejected (band-stop).

Brick wall response — An ideal filter response in which signals are either passed with no attenuation or attenuated completely.

Cutoff frequency — Frequency at which a filter's output is 3 dB below its passband output (also called *corner frequency* or *3 dB frequency*).

Decade — A ratio of 10 in frequency.

Equiripple — Equalized ripple in a filter's magnitude response across the passband, stop band, or both.

Flat — Refers to a filter's magnitude response that is constant across a range of frequencies.

Group delay — The transit time of signals through a filter.

High-pass — Filter response in which signals above the cutoff frequency are passed and rejected at lower frequencies.

Ideal filter — Filter that passes signals without loss or attenuates them completely. An ideal filter has no transition regions. (See also **brick wall response**).

Insertion loss — The loss incurred by signals in a filter's passband.

Low-pass — Filter response in which signals below the cutoff frequency are passed and rejected at higher frequencies.

Lumped elements — Discrete inductors and capacitors; a lumped-element filter made from discrete inductors and capacitors.

Magnitude response — Graph of a filter's output amplitude versus frequency.

Microstrip — A type of transmission line made from a strip of metal separated from a ground plane by a layer of insulating material, such as on a printed-circuit board..

Normalize — The technique of converting numeric values to their ratio with respect to some reference value. (To denormalize is to reverse the normalization, converting the ratios back to the original values.)

Notch filter — see **band-stop filter**.

Octave — A ratio of two in frequency (see also **decade**).

Overshoot — The condition in which the output of a circuit, in responding to a change in its input, temporarily exceeds the steady-state value that the input should cause.

Overtone — Vibration mode at a higher frequency than the fundamental mode, usually harmonically related.

Passband — The range of frequencies passed by a filter.

Passive filter — A filter that does not require power to perform its function (see also **lumped element**).

Phase response — Graph of the difference in angular units (degrees or radians) between a filter's input and output versus frequency.

Radian — Unit of angular measurement equal to $1/2\pi$ of a circle, equal to $360 / 2\pi$ degrees

Ringing — The condition in which the output of a circuit, in responding to a change in its input, exhibits a damped alternating sequence of exceeding and falling below the steady-state value that the input should cause before settling at the steady-state value.

Ripple — A regular variation with frequency in a filter's magnitude response.

Rolloff — The rate of change in a filter's magnitude response in the transition region and stop band.

Scaling — Changing a filter's impedance or frequency characteristics through multiplication or division by a constant.

Shape factor — The ratio of a filter's bandwidth between the points at which its magnitude response is 6 dB and 60 dB below the response in the filter's passband.

Stop band — The range of frequencies that are rejected by a filter.

Stripline — A transmission line consisting of a metal strip suspended between two ground planes.

TEM — Transverse electromagnetic mode in which the electric and magnetic fields of electromagnetic energy are aligned perpendicularly to the direction of motion.

Topology — The arrangement of connections of components in the filter. For example, "capacitor-input" and "inductor-input" are two different topologies.

Transition region — Range of frequencies between a filter's passband and stop band.

11.13 References and Bibliography

REFERENCES

1. Virgil E. Bottom, "Introduction to Quartz Crystal Unit Design," Van Nostrand Reinhold Company, 1982, ISBN 0-442-26201-9.
2. M. Dishal, "Modern Network Theory Design of Single Sideband Crystal Ladder Filters," *Proc IEEE*, Vol 53, No 9, Sep 1965.
3. J. A. Hardcastle, G3JIR, "Ladder Crystal Filter Design," *QST*, Nov 1980, p 20.
4. W. Hayward, W7ZOI, "A Unified Approach to the Design of Ladder Crystal Filters," *QST*, May 1982, p 21.
5. J. Makhinson, N6NWP, "Designing and Building High-Performance Crystal Ladder Filters," *QEX*, Jan 1995, pp 3-17.
6. W. Hayward, W7ZOI, "Refinements in Crystal Ladder Filter Design," *QEX*, June 1995, pp 16-21.
7. J.A. Hardcastle, G3JIR, "Computer-Aided Ladder Crystal Filter Design," *Radio Communication*, May 1983.
8. John Pivnichny, N2DCH, "Ladder Crystal Filters," MFJ Publishing Co, Inc.
9. W. S. Metcalf, "Graphs Speed Interdigitated Filter Design," *Microwaves*, Feb 1967.
10. J. Fisk W1DTY, "Helical-Resonator Design Techniques," *QST*, June 1976, pp 11-14.
11. H. Steder, DJ6EV, and J. Hardcastle, G3JIR, "Crystal Ladder Filters for All," *QEX*, Nov-Dec 2009, pp 14-18.
12. D. Gordon-Smith, G3UUR, "Extended Bandwidth Crystal Ladder Filters With Almost Symmetrical Responses," *QEX*, Jul/Aug 2011, pp 36-44.

FOR FURTHER READING

General Filter Design

Analog Devices. "Using the Analog Devices Active Filter Design Tool"

Applied Radio Labs, Design Note DN004, "Group Delay Explanations and Applications," **www.radiolab.com.au**

S. Butterworth, "On the Theory of Filter Amplifiers," *Experimental Wireless and Wireless Engineer*, Oct 1930, pp 536-541.

Laplace Transforms: P. Chirlian, *Basic Network Theory,* McGraw Hill, 1969.

S. Darlington, "Synthesis of Reactance 4-Poles Which Produce Prescribed Insertion Loss Characteristics," *Journal of Mathematics and Physics*, Sep 1939, pp 257-353.

M. Dishal, "Top Coupled Band-pass Filters," *IT&T Handbook*, 4th edition, American Book, Inc, 1956, p 216.

Fourier Transforms: *Reference Data for Engineers*, Chapter 7, 7th edition, Howard Sams, 1985.

Phillip R. Geffe, *Simplified Modern Filter Design* (New York: John F. Rider, 1963, OCLC: 2470870).

Maxim Integrated, Tutorial 733, "A Filter Primer," 2008.

Randall W. Rhea, *HF Filter Design and Computer Simulation* (New York: McGraw-Hill, ISBN 0070520550, OCLC 32013486).

Cauer Elliptic Filters: R. Saal, *The Design of Filters Using the Catalogue of Normalized Low-Pass Filters*,(German; brief introduction in English) (AEH Telefunken, 1968, OCLC 13988270). Also *Reference Data for Radio Engineers*, pp 9-5 to 9-11.

L. Weinberg, "Network Design by use of Modern Synthesis Techniques and Tables," *Proceedings of the National Electronics Conference*, vol 12, 1956.

Donald R.J. White, *A Handbook on Electrical Filters (Synthesis, Design and Applications)* (White Electromagnetics, 1980, ISBN 0932263070).

A. B. Williams, *Electronic Filter Design Handbook* (New York: McGraw-Hill, 1981).

O. Zobel, "Theory and Design of Electric Wave Filters," *Bell System Technical Journal*, Jan 1923.

Zumbahlen, H. AN-281 "Passive and Active Filtering," Analog Devices.

Zumbahlen, H., Mini-tutorial MT-202, "All-pass Filters," Analog Devices, 2012.

Zumbahlen, H., Mini-tutorial MT-204, "The Bessel Response," Analog Devices, 2012.

Zumbahlen, H., Mini-tutorial MT-205, "Bi-quadratic (Biquad) Filters," Analog Devices, 2012.

Zumbahlen, H., Mini-tutorial MT-206, "The Chebyshev Response," Analog Devices, 2012.

Zumbahlen, H., Mini-tutorial MT-210, "F_0 and Q in Filters," Analog Devices, 2012.

Zumbahlen, H., Mini-tutorial MT-215 "Low-Pass to Band-Pass Filter Transformation," Analog Devices, 2012.

Zumbahlen, H., Mini-tutorial MT-216 "Low-Pass to Band-Stop (Notch) Filter Transformation," Analog Devices, 2012.

Zumbahlen, H., Mini-tutorial MT-217 "Low-Pass to High-Pass Filter Transformation," Analog Devices, 2012.

Zumbahlen, H., Mini-tutorial MT-220, "Multiple Feedback Filters," Analog Devices, 2012.

Zumbahlen, H., Mini-tutorial MT-222, "Sallen-Key Filters," Analog Devices, 2012.

Zumbahlen, H., Mini-tutorial MT-223, "State Variable Filters," Analog Devices, 2012.

Zumbahlen, H., Mini-tutorial MT-224, "The Butterworth Response," Analog Devices, 2012.

Zumbahlen, H., Mini-tutorial MT-225, "Twin T Notch Filter," Analog Devices, 2012.

A.I. Zverev, *Handbook of Filter Synthesis* (New York: John Wiley and Sons, 1967, ISBN 0471986801, OCLC 972252).

Crystal, Helical, and Interdigital Filter Design

R. Fisher, W2CQH, "Combline VHF Bandpass Filters," *QST*, Dec 1968, p 44.

R. Fisher, W2CQH, "Interdigital Bandpass Filters for Amateur VHF/UHF Applications," *QST*, Mar 1968, p 32.

U. L. Rohde, DJ2LR, "Crystal Filter Design with Small Computers" *QST*, May 1981, p 18.

Practical Filter Designs

P. Antoniazzi, IW2ACD and M. Arecco, IK2WAQ, "Easy Microwave Filters Using Waveguides and Cavities," *QEX*, Sep/Oct 2006, pp 37-42.

B. Bartlett, VK4UW and J. Loftus, VK4EMM, "Band-Pass Filters for Contesting," *National Contest Journal*, Jan 2000, pp 11-15.

T. Cefalo Jr, WA1SPI, "Diplexers, Some Practical Applications," *Communications Quarterly*, Fall 1997, pp 19-24.

P.R. Cope, W2GOM/7, "The Twin-T Filter," *QEX*, July/Aug 1998, pp 45-49.

D. Gordon-Smith, G3UUR, "Seventh-Order Unequal-Ripple Low-pass Filter Design," *QEX*, Nov/Dec 2006, pp 31-34.

D. Gordon-Smith, G3UUR, "Fifth-Order Unequal-Ripple Low-pass Filter Design," *QEX*, Nov/Dec 2010, pp 42-47.

W. Hayward, W7ZOI, "Extending the Double-Tuned Circuit to Three Resonators," *QEX*, Mar/Apr 1998, pp 41-46.

F. Heemstra, KT3J, "A Double-Tuned Active Filter with Interactive Coupling," *QEX*, Mar/Apr 1999, pp 25-29.

D. Jansson, WD4FAB and E. Wetherhold, W3NQN, "High-Pass Filters to Combat TVI," *QEX*, Feb 1987, pp 7-8 and 13.

Z. Lau, W1VT, "A Narrow 80-Meter Band-Pass Filter," *QEX*, Sept/Oct 1998, p 57.

R. Lumachi, WB2CQM. "How to Silverplate RF Tank Circuits," *73 Amateur Radio Today*, Dec 1997, pp 18-23.

T. Moliere, DL7AV, "Band-Reject Filters for Multi-Multi Contest Operations," *CQ Contest*, Feb 1996, pp 14-22.

W. Rahe, DC8NR, "High Performance Audio Speech Low-Pass and CW Band-Pass Filters in SVL Design," *QEX*, Jul/Aug 2007, pp 31-39.

W. Sabin, WØIYH, "Diplexer Filters for the HF MOSFET Power Amplifier," *QEX*, July/Aug 1999, pp 20-26.

W. Sabin, WØIYH, "Narrow Band-Pass Filters for HF," *QEX*, Sept/Oct 2000, pp 13-17.

J. Tonne, WB6BLD, "Harmonic Filters, Improved," *QEX*, Sept/Oct 1998, pp 50-53.

E. Wetherhold, W3NQN, "Modern Design of a CW Filter Using 88 and 44-mH Surplus Inductors," *QST*, Dec 1980, pp14-19. See also Feedback in Jan 1981 *QST*, p 43.

E. Wetherhold, W3NQN, "Band-Stop Filters for Attenuating High-Level Broadcast-Band Signals," *QEX*, Nov 1995, pp 3-12.

E. Wetherhold, W3NQN, " CW and SSB Audio Filters Using 88-mH Inductors," *QEX*, Dec 1988, pp 3-10; *Radio Handbook*, 23rd edition, W. Orr, editor, p 13-4, Howard W. Sams and Co., 1987; and, "A CW Filter for the Radio Amateur Newcomer," *Radio Communication* (RSGB) Jan 1985, pp 26-31.

E. Wetherhold, W3NQN, "Clean Up Your Signals with Band-Pass Filters," Parts 1 and 2, *QST*, May and June 1998, pp 44-48 and 39-42.

E. Wetherhold, W3NQN, "Second-Harmonic-Optimized Low-Pass Filters," *QST*, Feb 1999, pp 44-46.

E. Wetherhold, W3NQN, "Receiver Band-Pass Filters Having Maximum Attenuation in Adjacent Bands," *QEX*, July/Aug 1999, pp 27-33.

Contents

**Chapter 12 —
CD-ROM Content**

Supplemental Files
- "Amateur Radio Equipment
  Development — An Historical
  Perspective" by Joel Hallas, W1ZR

Project Files
- Rock Bending Receiver PCB
  template by Randy Henderson,
  WI5W
- 10 GHz preamp PCB template by
  Zack Lau, W1VT
- Binaural Receiver project by Rick
  Campbell, KK7B

Chapter 12

Receivers

Receivers are discussed in this chapter from the standpoint of their architecture, composed of functional blocks such as mixers and oscillators and amplifiers that are discussed in the other chapters of this book. Other functions such as AGC that control the receiver from outside the primary signal path are dealt with both functionally and with examples at the circuit level. Special techniques associated with VHF, UHF and microwave receivers are reviewed.

The performance of receivers is described by standard terminology and as a set of measurements that allow a wide range of designs to be evaluated in ways that reflect their behavior in actual use.

This chapter, originally written by Joel Hallas, W1ZR, includes an update of existing material from previous editions of this book. Joel's summary of receiver and transmitter development, "Amateur Radio Equipment Development — An Historical Perspective," provides valuable insight as to why radios are constructed the way they are and is included on the CD-ROM accompanying this book.

The major subsystems of a radio receiving system are the antenna, the receiver and the information processor. The antenna's task is to provide a transition from an electromagnetic wave in space to an electrical signal that can be conducted on wires. The receiver has the job of retrieving the information content from a particular ac signal coming from the antenna and presenting it in a useful format to the processor for use.

The processor typically is an operator, but can also be an automated system. When you consider that most "processors" require signals in the range of volts (to drive an operator's speaker or headphones, or even the input of an A/D converter), and the particular signal of interest arrives from the antenna at a level of mere microvolts, the basic function of the receiver is to amplify the desired signal by a factor of a million. It must do this, while in the presence of signals many orders of magnitude greater and of completely different characteristics, without distortion of the desired signal or loss of the information it carries.

12.1 Characterizing Receivers

As we discuss receivers we will need to characterize their performance, and often their performance limitations, using certain key parameters. The most commonly encountered are as follows:

Sensitivity — This parameter is a measure of how weak a signal the receiver can extract information from. This generally is expressed at a particular signal-to-noise ratio (SNR) since noise is generally the limiting factor. A typical specification might be: "Sensitivity: 1 μV for 10 dB SNR with 3 kHz bandwidth." The bandwidth is stated because the amount (or power) of the noise, the denominator of the SNR fraction, increases directly with bandwidth. Generally the noise parameter refers to the noise generated within the receiver, often less than the noise that arrives with the signal from the antenna.

Selectivity — Selectivity is just the bandwidth discussed above. This is important because to a first order it identifies the receiver's ability to separate stations. With a perfectly sharp filter in an ideal receiver, stations within the bandwidth will be heard, while those outside it won't be detected. The selectivity thus describes how closely spaced adjacent channels can be. With a perfect 3 kHz bandwidth selectivity, and signals restricted to a 3 kHz bandwidth at the transmitter, a different station can be assigned every 3 kHz across the spectrum. In a less than ideal situation, it is usually necessary to include a *guard band* between channels.

Note that the word *channel* is used here in its generic form, meaning the amount of spectrum occupied by a signal, and not defining a fixed frequency such as an AM broadcast channel. A CW channel is about 300 Hz wide, a SSB channel about 2.5-3 kHz wide, and so forth. "Adjacent channel" refers to spectrum immediately higher or lower in frequency.

Dynamic Range — In the case of all real receivers, there is a range of signals that a receiver can respond to. This is referred to as dynamic range and, as will be discussed in more detail, can be established based on a number of different criteria. In its most basic form, dynamic range is the range in amplitude of signals that can be usefully received, typically from as low as the receiver's noise level, or *noise floor*, to a level at which stages overload in some way.

The type and severity of the overload is often part of the specification. A straightforward example might be a 130 dB dynamic range with less than 3% distortion. The nature of the distortion will determine the observed phenomenon. If the weakest and strongest signals are both on the same channel, for example, we would not expect to be able to process the weaker of the two. However, the more interesting case would be with the strong signal in an adjacent channel. In an ideal receiver, we would never notice that the adjacent signal was there. In a real receiver with a finite dynamic range or nonideal selectivity, there will be some level of adjacent channel signal or signals that will interfere with reception of the weaker on-channel signal.

The parameters described above are often the key performance parameters, but in many cases there are others that are important to specify. Examples are audio output power, power consumption, size, weight, control capabilities and so forth.

12.2 Basics of Heterodyne Receivers

The *heterodyne* receiver combines the input signal with a signal from a *local oscillator* (LO) in a nonlinear device called a *mixer* to result in the sum and difference frequencies as shown in **Fig 12.1**. The receiver may be designed so the output signal is anything from dc (a so-called *direct conversion* receiver) to any frequency above or below either of the two frequencies. The major benefit is that most of the gain, bandwidth setting and processing are performed at a single frequency.

By changing the frequency of the LO, the operator shifts the input frequency that is *translated* to the output, along with all its modulated information. In most receivers the mixer output frequency is designed to be an RF signal, either the sum or difference — the other being filtered out at this point. This output frequency is called an *intermediate frequency* or *IF*. The IF amplifier system can be designed to provide the selectivity and other desired characteristics centered at a single fixed frequency.

When more than one frequency conversion process is used, the receiver becomes a *superheterodyne* or *superhet*. A block diagram of a typical superhet receiver is shown in **Fig 12.2**. In traditional form, the RF filter is used to limit the input frequency range to those frequencies that include only the desired sum or difference but not the other — the so-called *image* frequency. The dotted line represents the fact that in receivers with a wide tuning range, such as a simple

AM broadcast receiver that tunes from 500 to 1700 kHz, a more than 3:1 range, the input RF amplifier and filter is often tracked along with the local oscillator. The IF filter is used to establish operating selectivity — that required by the information bandwidth. Circuits from the antenna input through and including the mixer (the first mixer if more than one mixing stage is used) are generally referred to as the receiver's *front-end*.

For reception of suppressed carrier single-sideband voice (SSB) or on-off or frequency-shift keyed (FSK) signals, a second *beat frequency oscillator* or *BFO* is employed to provide an audible voice, an audio tone or tones at the output for operator or FSK processing. This is the same as a heterodyne mixer with an output centered at dc, although the IF filter is usually designed to remove one of the output products.

Modern receivers using digital signal processing for operating bandwidth and information detection convert the incoming signal to digital form at one of the intermediate frequencies. Advancing speed and declining costs of analog-to-digital converters (ADC) and processors are moving the conversion closer and closer to the incoming signal's frequency, in some cases converting directly at RF, which is called *direct sampling*. See the **DSP and Software Radio Design** chapter for more information on these techniques.

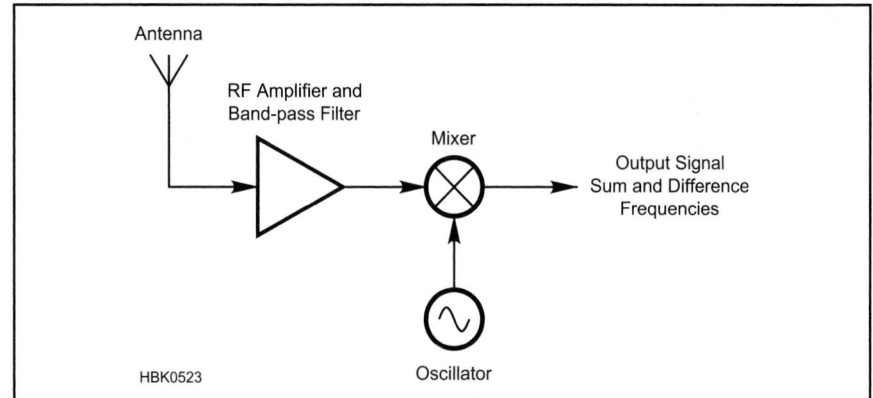

Fig 12.1 — Basic architecture of a heterodyne direct conversion receiver.

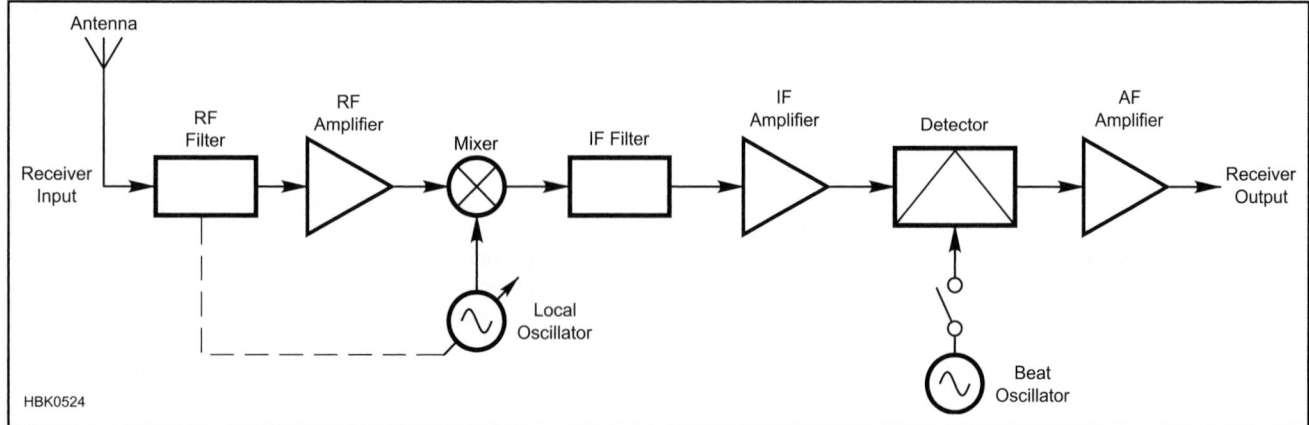

Fig 12.2 — Elements of a traditional superheterodyne radio receiver

12.2.1 The Direct Conversion Receiver

The heterodyne process can occur at a number of different points in the receiver. The simplest form of heterodyne receiver is called a *direct conversion* receiver because it performs the translation directly from the signal frequency to the audio output. It is, in effect, just the BFO and detector of the general superhet shown in Fig 12.2. In this case, the detector is often preceded by an RF amplifier with a typical complete receiver shown in **Fig 12.3**. Such a receiver can be very simple to construct, yet can be quite effective — especially for the ultra-compact low-power consumer-oriented portable station known as the mobile telephone! In fact, given the worldwide presence of the mobile telephone, the direct conversion receiver is the most widely used of all receivers.

The basic function of a mixer is to multiply two sinusoidal signals and generate two new signals with frequencies that are the sum and difference of the two original signals. This function can be performed by a linear multiplier, a switch that turns one input signal on and off at the frequency of the other input signal, or a nonlinear circuit such as a diode. (The output of a nonlinear circuit is made up of an infinite series of products, all different combinations of the two input signals, as described in the sidebar on Nonlinear Signal Combinations.) Much more information about the theory, operation and application of mixers may be found in the **Mixers, Modulators and Demodulators** chapter.

Fig 12.4 shows the progression of the spectrum of an on-off keyed CW signal through such a receiver based on the relationships described above. In 12.4C, we include an undesired image signal on the other side of the local oscillator that also shows up in the output of the receiver. Note each of the desired and undesired responses that occur as outputs of the mixer.

Some mixers are designed to be *balanced* in order to cancel one of the input signals at the output while a *double-balanced* mixer cancels both. A double-balanced mixer simplifies the output filtering job as shown in Fig 12.4D.

Products generated by nonlinearities in the mixing process (see the sidebar on Nonlinear Signal Combinations) are heard as intermodulation distortion signals that we will discuss later. Note that the nonlinearities also allow mixing with unwanted signals near multiples of local oscillator frequency. These signals, such as those from TV or FM broadcast stations, must be eliminated in the filtering before the mixer since their audio output will be right in the desired passband on the output of the mixer.

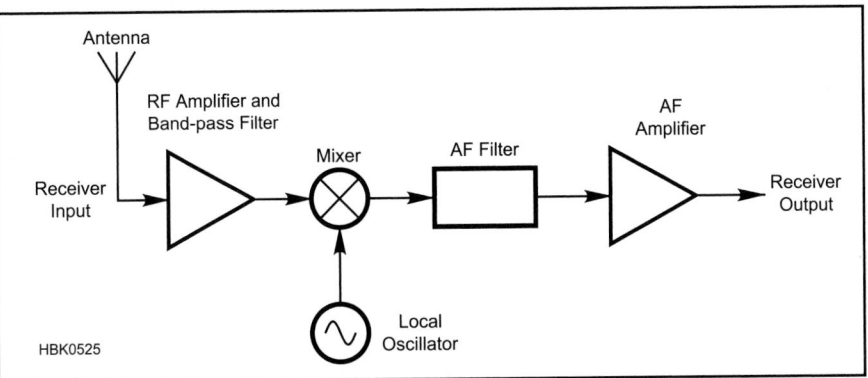

Fig 12.3 — Block diagram of a direct conversion.

Nonlinear Signal Combinations

Although a mixer is often thought of as nonlinear, it is neither necessary nor desirable for a mixer to be nonlinear. An ideal mixer is one that linearly multiplies the LO voltage by the signal voltage, creating two products at the sum and difference frequencies and only those two products. From the signal's perspective, it is a perfectly linear but time-varying device. Ideally a mixer should be as linear as possible.

If a signal is applied to a nonlinear device, however, the output will not be just a copy of the input, but can be described as the following infinite series of output signal products:

$$V_{OUT} = K_0 + K_1 \times V_{IN} + K_2 \times V_{IN}^2 + K_3 \times V_{IN}^3 + \ldots + K_N \times V_{IN}^N \qquad (A)$$

What happens if the input V_{IN} consists of two sinusoids at F_1 and F_2, or $A \times [\sin(2\pi_1) \times t]$ and $B \times [\sin(2\pi F_2) \times t]$? Begin by simplifying the notation to use angular frequency in radians/second ($2\pi F = \omega$). Thus V_{IN} becomes $A\sin\omega_1 t$ and $B\sin\omega_2 t$ and equation A becomes:

$$V_{OUT} = K_0 + K_1 \times (A\sin\omega_1 t + B\sin\omega_2 t) + K_2 \times (A\sin\omega_1 t + B\sin\omega_2 t)^2 + K_3 \times (A\sin\omega_1 t + B\sin\omega_2 t)^3 + \ldots + K_N \times (A\sin\omega_1 t + B\sin\omega_2 t)^N \qquad (B)$$

The zero-order term, K_0, represents a dc component and the first-order term, $K_1 \times (A\sin\omega_1 t + B\sin\omega_2 t)$, is just a constant times the input signals. The second-order term is the most interesting for our purposes. Performing the squaring operation, we end up with:

$$\text{Second order term} = [K_2 A^2 \sin^2\omega_1 t + 2K_2 AB(\sin\omega_1 t \times \sin\omega_2 t) + K_2 B^2 \sin^2\omega_2 t] \qquad (C)$$

Using the trigonometric identity (see Reference 1):

$$\sin á \sin â = \tfrac{1}{2} \{\cos(á - â) - \cos(á + â)\}$$

the product term becomes:

$$K_2 AB \times [\cos(\omega_1 - \omega_2)t - \cos(\omega_1 + \omega_2)t] \qquad (D)$$

Here are the products at the sum and difference frequency of the input signals! The signals, originally sinusoids are now cosinusoids, signifying a phase shift. These signals, however, are just two of the many products created by the nonlinear action of the circuit, represented by the higher-order terms in the original series.

In the output of a mixer or amplifier, those unwanted signals create noise and interference and must be minimized or filtered out. This nonlinear process is responsible for the distortion and intermodulation products generated by amplifiers operated nonlinearly in receivers and transmitters.

Project: A Rock-Bending Receiver for 7 MHz

There are many direct conversion receiver (DC) construction articles in the amateur literature. Home builders considering construction of a DC receiver should read Chapter 8 in *Experimental Methods in RF Design*, which outlines many of the pitfalls and design limitations of such receivers, as well as a providing suggestions for making a successful DC receiver.

This DC receiver design by Randy

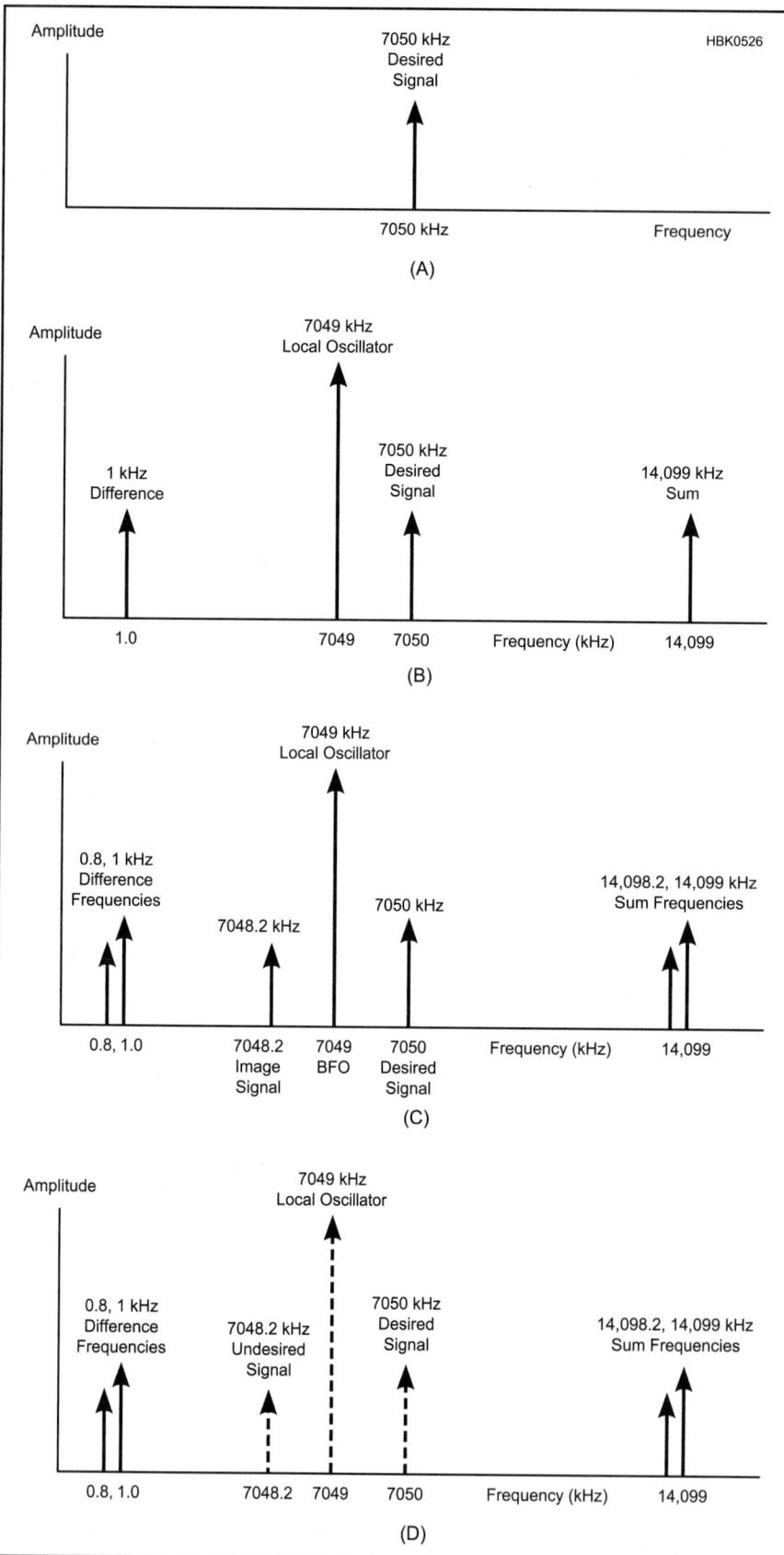

Fig 12.4 — Frequency relationships in DC receiver. At (A), desired receive signal from antenna, 7050 kHz on-off keyed carrier. At (B), internal local oscillator and receive frequency relationships. At (C), frequency relationships of mixer/detector products (not to scale).At (D), sum and difference outputs from double balanced mixer (not to scale). Note the balanced out mixer inputs that cancel at the output (dashed lines).

Henderson, WI5W, is presented here as an example. This receiver represents about as simple a receiver as can be constructed that will offer reasonable performance (see the References entry). The user of such a DC receiver will appreciate its simplicity and the nice sounding response of CW and SSB signals that it receives, but only if there aren't many signals sharing the band.

A major shortcoming of the DC receiver architecture will soon become evident — it has no rejection of *image* signals. Looking again at Fig 12.4B, while the 7049 kHz oscillator will translate the desired 7050 kHz signal to an audio tone of 1 kHz, it will equally well translate a signal at 7048 kHz to the same frequency. This is called an *image* and results in interference. The same kind of thing happens while receiving SSB; the entire channel on the other side of the BFO is received as well.

Building a stable oscillator is often the most challenging part of a simple receiver. This one uses a tunable crystal-controlled oscillator that is both stable and easy to reproduce. All of its parts are readily available from multiple sources and the fixed value capacitors and resistors are common components available from many electronics parts suppliers.

THE CIRCUIT

This receiver works by mixing two radio-frequency signals together. One of them is the signal you want to hear, and the other is generated by an oscillator circuit (Q1 and associated components) in the receiver. In **Fig 12.5**, mixer U1 puts out sums and differences of these signals and their harmonics. We don't use the sum of the original frequencies, which comes out of the mixer in the vicinity of 14 MHz. Instead, we use the frequency *difference* between the incoming signal and the receiver's oscillator — a signal in the audio range if the incoming signal and oscillator frequencies are close enough to each other. This signal is filtered in U2, and amplified in U2 and U3. An audio transducer (a speaker or headphones) converts U3's electrical output to audio.

How the Rock Bender Bends Rocks

The oscillator is a tunable crystal oscillator — a variable crystal oscillator, or *VXO*. Moving the oscillation frequency of a crystal like this is often called *pulling*. Because crystals consist of precisely sized pieces of quartz, crystals have long been called *rocks* in ham slang — and receivers, transmitters and transceivers that can't be tuned around due to crystal frequency control have been said to be *rockbound*. Widening this rockbound receiver's tuning range with crystal pulling made *rock bending* seem just as appropriate!

L2's value determines the degree of pulling available. Using FT-243 style crystals and larger L2 values, the oscillator reliably tunes

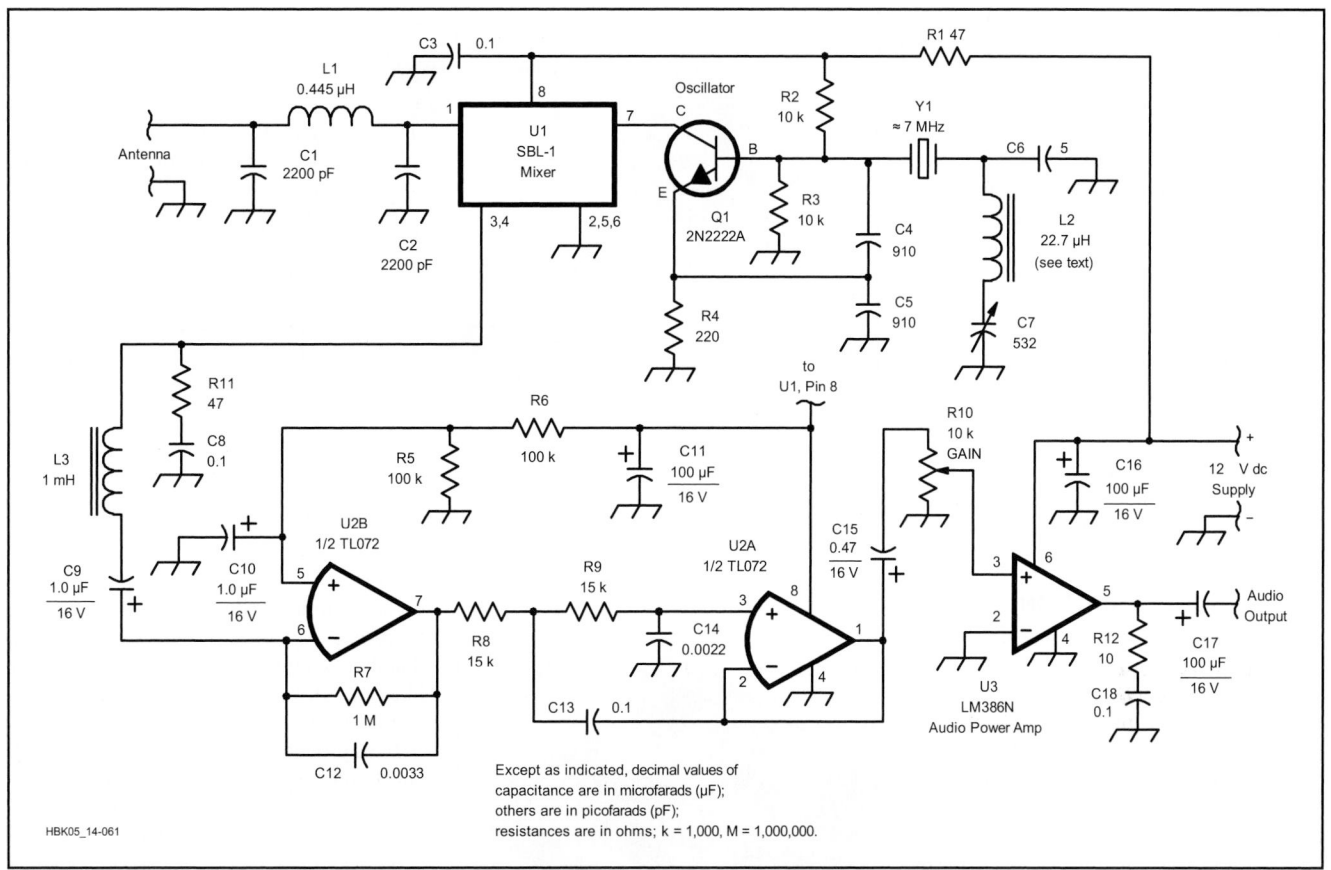

Fig 12.5 — An SBL-1 mixer (U1, which contains two small RF transformers and a Schottky-diode quad), a TL072 dual op-amp IC (U2) and an LM386 low-voltage audio power amplifier IC (U3) do much of the Rock-Bending Receiver's magic. Q1, a variable crystal oscillator (VXO), generates a low-power radio signal that shifts incoming signals down to the audio range for amplification in U2 and U3. All of the circuit's resistors are ¼ W, 5% tolerance types; the circuit's polarized capacitors are 16 V electrolytics, except C10, which can be rated as low as 10 V. The 0.1 µF capacitors are monolithic or disc ceramics rated at 16 V or higher.

C1, C2 — Ceramic or mica, 10% tolerance.
C4, C5, and C6 — Polystyrene, dipped silver mica, or C0G (formerly NP0) ceramic, 10% tolerance.
C7 — Dual gang broadcast variable capacitor (14-380 pF per section), ¼ inch dia shaft, available as #BC13380 from Ocean State Electronics. A rubber equipment foot serves as a knob. (Any variable capacitor with a maximum capacitance of 350 to 600 pF can be substituted; the wider the capacitance range, the better.)
C12, C13, C14 — 10% tolerance. For SSB, change C12, C13 and C14 to 0.001 µF.
U2 — TL072CN or TL082CN dual JFET op amp.

L1 — Four turns of AWG #18 wire on ¾ inch PVC pipe form. Actual pipe OD is 0.85 inch. The coil's length is about 0.65 inch; adjust turns spacing for maximum signal strength. Tack the turns in place with cyanoacrylic adhesive, coil dope or RTV sealant. (As a substitute, wind 8 turns of #18 wire around 75% of the circumference of a T-50-2 powdered-iron core. Once you've soldered the coil in place and have the receiver working, expand and compress the coil's turns to peak incoming signals, and then cement the winding in place.)
L2 — Approximately 22.7 µH; consists of one or more encapsulated RF chokes in series (two 10-µH chokes [Mouser #43HH105 suitable] and one 2.7-µH

choke [Mouser #43HH276 suitable] used by author). See text.
L3 — 1 mH RF choke. As a substitute, wind 34 turns of #30 enameled wire around an FT-37-72 ferrite core.
Q1 — 2N2222, PN2222 or similar small-signal, silicon NPN transistor.
R10 — 5 or 10 kΩ audio-taper control (RadioShack No. 271-215 or 271-1721 suitable).
U1 — Mini-Circuits SBL-1 mixer.
Y1 — 7 MHz fundamental-mode quartz crystal. Ocean State Electronics carries 7030, 7035, 7040, 7045, 7110 and 7125 kHz units.
PC boards for this project are available from FAR Circuits.

from the frequency marked on the holder to about 50 kHz below that point with larger L2 values. (In the author's receiver a 25 kHz tuning range was achieved.) The oscillator's frequency stability is very good.

Inductor L2 and the crystal, Y1, have more effect on the oscillator than any other components. Breaking up L2 into two or three series-connected components often works better than using one RF choke. (The author used three molded RF chokes in series — two

10 µH chokes and one 2.7 µH unit.) Making L2's value too large makes the oscillator stop.

The author tested several crystals at Y1. Those in FT-243 and HC-6-style holders seemed more than happy to react to adjustment of C7 (TUNING). Crystals in the smaller HC-18 metal holders need more inductance at L2 to obtain the same tuning range. One tiny HC-45 unit from International Crystals needed 59 µH to eke out a mere 15 kHz of tuning range.

Input Filter and Mixer

C1, L1, and C2 form the receiver's input filter. They act as a peaked *low-pass* network to keep the mixer, U1, from responding to signals higher in frequency than the 40 meter band. (This is a good idea because it keeps us from hearing video buzz from local television transmitters, and signals that might mix with harmonics of the receiver's VXO.) U1, a Mini-Circuits SBL-1, is a passive diode-ring mixer. Diode-ring mixers usually perform better if the output is terminated properly. R11 and C8 provide a resistive termination at RF without disturbing U2A's gain or noise figure.

Audio Amplifier and Filter

U2B amplifies the audio signal from U1. U2A serves as an active low-pass filter. The values of C12, C13 and C14 are appropriate for listening to CW signals. If you want SSB stations to sound better, make the changes shown in the parts list for Fig 12.5.

U3, an LM386 audio power amplifier IC, serves as the receiver's audio output stage. The audio signal at U3's output is much more powerful than a weak signal at the receiver's input, so don't run the speaker/earphone leads near the circuit board. Doing so may cause a squeal from audio oscillation at high volume settings.

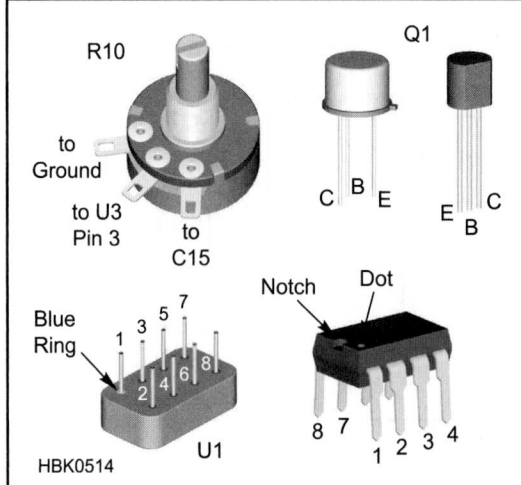

Fig 12.6 — Two Q1-case styles are shown because plastic or metal transistors will work equally well for Q1. If you build your Rock-Bending Receiver using a prefab PC board, you should mount the ICs in 8-pin mini-DIP sockets rather than just soldering the ICs to the board.

CONSTRUCTION

If you're already an accomplished builder, you know that this project can be built using a number of construction techniques, so have at it! If you're new to building, you should consider building the Rock-Bending Receiver on a printed circuit (PC) board. (The parts list tells where you can buy one ready-made.) See **Fig 12.6** for details on the physical layout of several important components used in the receiver. The receiver can be constructed either using a PC board or by using ground-plane (a.k.a "ugly") construction. **Fig 12.7** shows a photo of the receiver built on a PC board, but you can also build one on a ground-plane formed by an unetched piece of PC board material.

If you use the PC board available from FAR Circuits or a homemade double-sided circuit board based on the PC pattern on the *Handbook* CD, you'll notice that it has more holes than it needs to. The extra holes (indicated in the part-placement diagram with square pads) allow you to connect its ground plane to the ground traces on its foil side. (Doing so reduces the inductance of some of the board's ground paths.) Pass a short length of bare wire (a clipped-off component lead is fine) into each of these holes and solder on both sides. Some of the circuit's components (C1, C2 and others) have grounded leads accessible on both sides of the board. Solder these leads on both sides of the board.

Another important thing to do if you use a homemade double-sided PC board is to countersink the ground plane to clear all ungrounded holes. (Countersinking clears copper away from the holes so components won't short-circuit to the ground plane.) A ¼-inch diameter drill bit works well for this. Attach a control knob to the bit's shank and you can safely use the bit as a manual countersinking tool. If you countersink your board in a drill press, set it to about 300 rpm or less, and use

Fig 12.7 — Ground-plane construction or PC-board construction as shown here — either approach can produce the same good Rock Bending Receiver performance.

very light pressure on the feed handle.

Mounting the receiver in a metal box or cabinet is a good idea. Plastic enclosures can't shield the TUNING capacitor from the presence of your hand, which may slightly affect the receiver tuning. You don't have to completely enclose the receiver — a flat aluminum panel screwed to a wooden base is an acceptable alternative. The panel supports the tuning capacitor, GAIN control and your choice of audio connector. The base can support the circuit board and antenna connector.

CHECKOUT

Before connecting the receiver to a power source, thoroughly inspect your work to spot obvious problems like solder bridges, incorrectly inserted components or incorrectly wired connections. Using the schematic (and PC-board layout if you built your receiver on a PC board), recheck every component and connection one at a time. If you have a

digital voltmeter (DVM), use it to measure the resistance between ground and everything that should be grounded. This includes things like pin 4 of U2 and U3, pins 2, 5, 6 of U1, and the rotor of C7.

If the grounded connections seem all right, check some supply-side connections with the meter. The connection between pin 6 of U3 and the positive power-supply lead should show less than 1 Ω of resistance. The resistance between the supply lead and pin 8 of U1 should be about 47 Ω because of R1.

If everything seems okay, you can apply power to the receiver. The receiver will work with supply voltages as low as 6 V and as high as 13.5 V, but it's best to stay within the 9 to 12 V range. When first testing your receiver, use a current-limited power supply (set its limiting between 150 and 200 mA) or put a 150 mA fuse in the connection between the receiver and its power source. Once you're sure that everything is working as it should, you can remove the fuse or turn off the current limiting.

If you don't hear any signals with the antenna connected, you may have to do some troubleshooting. Don't worry; you can do it with very little equipment.

TROUBLE?

The first clue to look for is noise. With the GAIN control set to maximum, you should hear a faint rushing sound in the speaker or headphones. If not, you can use a small metallic tool and your body as a sort of test-signal generator. (If you have any doubt about the safety of your power supply, power the Rock-Bending Receiver from a battery during this test.) Turn the GAIN control to maximum. Grasp the metallic part of a screwdriver, needle or whatever in your fingers, and use the tool to touch pin 3 of U3. If you hear a loud scratchy popping sound, that stage is working. If not, then something directly related to U3 is the problem.

You can use this technique at U2 (pin 3, then pin 6) and all the way to the antenna. If you hear loud pops when touching either end of L3 but not the antenna connector, the oscillator is probably not working. You can check for oscillator activity by putting the receiver near a friend's transceiver (both must be in the same room) and listening for the VXO. Be sure to adjust the TUNING control through its range when checking the oscillator.

The dc voltage at Q1's base (measured without the RF probe) should be about half the supply voltage. If Q1's collector voltage is about equal to the supply voltage, and Q1's base voltage is about half that value, Q1 is probably okay. Reducing the value of L2 may be necessary to make some crystals oscillate.

OPERATION

Although the Rock-Bending Receiver uses only a handful of parts and its features are limited, it performs surprisingly well. Based on tests done with a Hewlett-Packard HP 606A signal generator, the receiver's minimum discernible signal (by ear) appears to be 0.3 µV. The author could easily copy 1 µV signals with his version of the Rock-Bending Receiver.

Although most HF-active hams use transceivers, there are advantages in using separate receivers and transmitters. This is especially true if you are trying to assemble a simple home-built station.

12.3 The Superheterodyne Receiver

In many instances, it is not possible to achieve all the receiver design goals with a single-conversion receiver and multiple conversion steps are used, creating the superheterodyne architecture. Traditionally, the first conversion is tasked with removing the RF image signals, while the second allows processing of the IF signal to provide the information based IF processing.

The superheterodyne receiver applies the principles of the heterodyne receiver at least twice. The concept was introduced by Major Edwin Armstrong, a US Army artillery officer, just as WW I was coming to a close. He is the same Armstrong who invented frequency modulation (FM) some years later and who held many radio patents between WW I and WW II.

In a superhet, a local oscillator and mixer are used to translate the received signal to an intermediate frequency or IF rather than directly to audio. This provides an opportunity for additional amplification and processing. Then a second mixer is used as in the DC receiver to detect the IF signal, translating it to audio. The configuration was shown previously in Fig 12.2.

An example will illustrate how this works. Let's pick a common IF frequency used in a simple AM broadcast radio, 455 kHz. If we want to listen to a 600 kHz broadcast station, the RF stage would be set to amplify the 600 kHz signal and the LO should be set to 600 + 455 kHz or 1055 kHz. The 600 kHz signal, along with any audio information it contains, is translated to the IF frequency and is amplified and then detected.

Note that we could have also set the local oscillator to 600 − 455 kHz or 145 kHz. By setting it to the sum, we reduce the relative range that the oscillator must tune. To cover the 500 to 1700 kHz with the difference, our LO would have to cover from 45 to 1245, a 28:1 range. Using the sum requires LO coverage from 955 to 2155 kHz, a range of about 2.5:1 — much easier to implement.

Note that to detect standard AM signals, the receiver's second oscillator, the beat frequency oscillator (BFO), is turned off since the AM station provides its own carrier signal over the air. Receivers designed only for standard AM reception generally don't have a BFO at all.

It's not clear yet that we've gained anything by doing this; so let's look at another example. If we decide to change from listening to the station at 600 kHz and want to listen to another station at, say, 1560 kHz, we can tune the single dial of our superhet to 1560 kHz. The RF stage is tuned to 1560 kHz, the LO is set to 1560 + 455 or 2010 kHz, and now the desired station is translated to our 455 kHz IF where the bulk of our amplification can take place. Note also that with the superheterodyne configuration, selectivity (the ability to separate stations) occurs primarily in the intermediate-frequency (IF) stages and is thus the same no matter what frequencies we choose to listen to. This simplifies the design of each stage considerably.

12.3.1 Superhet Bandwidth

Now we will discuss the bandwidth requirements of different operating modes and how that affects superhet design. One advantage of a superhet is that the operating bandwidth can be established by the IF stages, and further limited by the audio system. It is thus independent of the RF frequency to which the receiver is tuned. It should not be surprising that the detailed design of a superhet receiver is dependent on the nature of the signal being received. We will briefly discuss the most commonly received modulation types and the bandwidth implications of each below. (Each modulation type is discussed in more detail in the **Modulation** or **Digital Modes** chapters.)

AMPLITUDE MODULATION (AM)

As shown in Fig 12.4, multiplying (in other words, modulating) a carrier with a single tone results in the tone being translated to frequencies of the sum and difference of the two. Thus, if a transmitter were to multiply a 600 Hz tone and a 600 kHz carrier signal, we would generate additional new frequencies at 599.4 and 600.6 kHz. If instead we were to modulate the 600 kHz carrier signal with a band of frequencies corresponding to human speech of 300 to 3300 Hz (the usual range of communication quality voice signals), we would have a pair of bands of information carrying waveforms extending from 596.7 to 603.3 kHz, as shown in **Fig 12.8**. These bands are called *sidebands*, and some form of these is present in any AM signal that is carrying information.

Note that the total bandwidth of this AM voice signal is twice the highest frequency transmitted, or 6600 Hz. If we choose to transmit speech and limited-range music, we might allow modulating frequencies up to 5000 Hz, resulting in a bandwidth of 10,000 Hz or 10 kHz. This is the standard channel spacing that commercial AM broadcasters use in the US. (9 kHz is used in Europe) In actual use, the adjacent channels are generally geographically separated, so broadcasters can extend some energy into the next channels for improved fidelity. We would refer to this as a *narrow-bandwidth* mode.

What does this say about the bandwidth needed for our receiver? If we want to receive the full information content transmitted by a US AM broadcast station, then we need to set the bandwidth to at least 10 kHz. What if our receiver has a narrower bandwidth? Well, we will lose the higher frequency components of

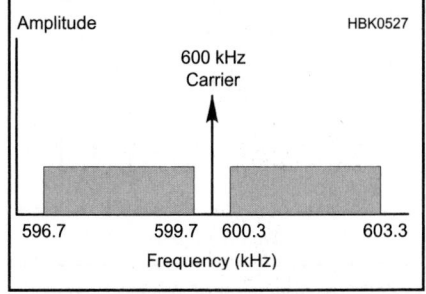

Fig 12.8 — Spectrum of sidebands of an AM voice signal sent on a 600 kHz carrier.

the transmitted signal — perhaps ending up with a radio suitable for voice but not very good at reproducing music.

On the other hand, what is the impact of having too wide a bandwidth in our receiver? In that case, we will be able to receive the full transmitted spectrum but we will also receive some of the adjacent channel information. This will sound like interference and reduce the quality of what we are receiving. If there are no adjacent channel stations, we will get any additional noise from the additional bandwidth and minimal additional information. The general rule is that the received bandwidth should be matched to the bandwidth of the signal we are trying to receive to maximize SNR and minimize interference.

As the receiver bandwidth is reduced, intelligibility suffers, although the SNR is improved. With the carrier centered in the receiver bandwidth, most voices are difficult to understand at bandwidths less than around 4 kHz. In cases of heavy interference, full carrier AM can be received as if it were SSB, as described below, with the carrier inserted at the receiver, and the receiver tuned to whichever sideband has the least interference.

SINGLE-SIDEBAND SUPPRESSED-CARRIER MODULATION (SSB)

In looking at Fig 12.8, you might have noticed that both sidebands carry the same information, and are thus redundant. In addition, the carrier itself conveys no information. It is thus possible to transmit a *single sideband* and *no carrier*, as shown in **Fig 12.9**, relying on the BFO (beat frequency oscillator) in the receiver to provide a signal with which to multiply the sideband in order to provide demodulated audio output. The implications in the receiver are that the bandwidth can be slightly less than half that required for double sideband AM (DSB). There must be an additional mechanism to carefully replace the missing carrier within the receiver. This is the function of the BFO, which must be at just exactly the right frequency. If the frequency is improperly set a baritone can come out sounding like a soprano and vice versa! The effect is quite audible even for small frequency errors of a few tens of Hz.

This results in a requirement for a much more stable receiver design with a much finer tuning system — a more expensive proposition. An alternate is to transmit a reduced level carrier and have the receiver lock on to the weak carrier, usually called a *pilot* carrier. Note that the pilot carrier need not be of sufficient amplitude to demodulate the signal, just enough to allow a BFO to lock to it. These alternatives are effective, but tend to make SSB receivers expensive, complex and most appropriate for the case in where a small number of receivers are listening to a single transmitter, as is the case of two-way amateur communication.

Note that the bandwidth required to effectively demodulate an SSB signal is actually less than half that required for the AM signal because the range centered on the AM carrier need not be received. Thus the communications-quality range of 300 to 3300 Hz can be received in a bandwidth of 3000 not 3300 Hz. Early SSB receivers typically used a bandwidth of around 3 kHz, but with the heavy interference frequently found in the amateur bands, it is more common for amateurs to use bandwidths of 1.8 to 2.4 kHz with the corresponding loss of some of the higher consonant sounds.

RADIOTELEGRAPHY (CW)

We have described radiotelegraphy as being transmitted by "on-off keying of a carrier." You might think that since a carrier takes up just a single frequency, the receive bandwidth needed should be almost zero. This is only true if the carrier is never turned on and off. In the case of CW, it will be turned on and off quite rapidly. The rise and fall of the carrier results in sidebands extending on either side of the carrier, and they must be received in order to reconstruct the signal in the receiver.

A rule of thumb is to consider the rise and fall time as about 10% of the pulse width and the bandwidth as the reciprocal of the quickest rise or fall time. This results in a bandwidth requirement of about 50 to 200 Hz for the usual CW transmission rates. Another way to visualize this is with the bandwidth being set by a high-Q tuned circuit. Such a circuit will continue to "ring" after the input pulse is gone. Thus, too narrow a bandwidth will actually "fill in" between the code elements and act like a "no bandwidth" full period carrier and this is exactly what is heard if a very narrow crystal filter is used when receiving CW.

DATA SIGNALS

(This is a short overview of amateur data signals to establish receiver requirements for the digital modes. See the **Modulation** and **Digital Modes** chapters and the **Digital Communications** supplement on the *Handbook* CD for more in-depth treatment.)

The Baudot code (used for teletype communications) and ASCII code — two popular digital communications codes used by amateurs — are constructed with sequences of elements or bits. The state of each bit — ON or OFF — is represented by a signal at one of two distinct frequencies: one designated *mark* and one designated *space*. This is referred to as *frequency shift keying* (FSK). The transmitter frequency shifts back and forth with each character's individual elements.

Amateur Radio operators typically use a 170 Hz separation between the mark and space frequencies, depending on the data rate and local convention, although 850 Hz is sometimes used. The minimum bandwidth required to recover the data is somewhat greater than twice the spacing between the tones. Note that the tones can be generated by directly shifting the carrier frequency (direct FSK), or by using a pair of 170 Hz spaced audio tones applied to the audio input of an SSB transmitter (audio FSK or AFSK). Direct FSK and AFSK sound the same to a receiver.

Note that if the standard audio tones of 2125 Hz (mark) and 2295 Hz (space) are used, they fit within the bandwidth of a voice channel and thus a voice transmitter and receiver can be employed without any additional processing needed outside the radio equipment. Alternately, the receiver can employ detectors for each frequency and provide an output directly to a computer.

If the receiver can shift its BFO frequency appropriately, the two tones can be received through a filter designed for CW reception with a bandwidth of about 300 Hz or wider. Some receivers provide such a narrow filter with the center frequency shifted midway between the tones (2210 Hz) to avoid the need for retuning. The most advanced receivers provide a separate filter for mark and space frequencies, thus maximizing interference rejection and signal-to-noise ratio (SNR). Using a pair of tones for FSK or AFSK results in a maximum data rate of about 1200 bit/s over a high-quality voice channel.

Phase shift keying (PSK) can also be used to transmit bit sequences, requiring good frequency stability to maintain the required time synchronization to detect shifts in phase. If the channel has a high SNR, as is often the case at VHF and higher, telephone network data-modem techniques can be used.

At HF, the signal is subjected to phase and amplitude distortion as it travels. Noise is also substantially higher on the HF bands. Under these conditions, modulation and demodulation techniques designed for "wire-

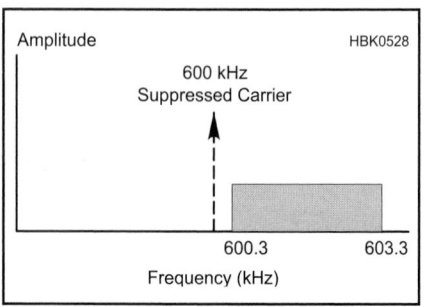

Fig 12.9 — Spectrum of single sideband AM voice signal sent adjacent to a suppressed 600 kHz carrier.

line" connections become unusable at bit rates of more than a few hundred bps. As a result, amateurs have begun adopting and developing state of the art digital modulation techniques. These include the use of multiple carriers (MFSK, Clover, PACTOR III, etc.), multiple amplitudes and phase shifts (QAM and QPSK techniques), and advanced error detection and correction methods to achieve a data throughput as high as 3600 bits per second (bps) over a voice-bandwidth channel. Newly developed coding methods for digital voice using QPSK modulation result in a signal bandwidth of less than 1200 Hz. (See **http://freedv.org** for more information.) Spread-spectrum techniques are also being adopted on the amateur UHF bands, but are beyond the scope of this discussion.

The bandwidth required for data communications can be as low as 100 Hz for PSK31 to 1 kHz or more for the faster speeds of PACTOR III and Clover. Beyond having sufficient bandwidth for the data signal, the primary requirements for receivers used for data communications are linear amplitude and phase response over the bandwidth of the data signal to avoid distorting these critical signal characteristics. The receiver must also have excellent frequency stability to avoid drift and frequency resolution to enable the receiver filters to be set on frequency.

FREQUENCY MODULATION (FM)

Another popular voice mode is frequency modulation or FM. FM can be found in a number of variations depending on purpose. In Amateur Radio and commercial mobile communication use on the shortwave bands, it is universally *narrow band FM* or *NBFM*. In NBFM, the frequency deviation is limited to around the maximum modulating frequency, typically 3 kHz. The bandwidth requirements at the receiver can be approximated by Carson's Rule of $BW = 2 \times (D + M)$, where D is the deviation and M is the maxi-

mum modulating frequency. Thus 3 kHz deviation and a maximum voice frequency of 3 kHz results in a bandwidth of 12 kHz, not far beyond the requirements for broadcast AM. (Additional signal components extend beyond this bandwidth, but are not required for voice communications.)

In contrast, broadcast or *wideband FM* or *WBFM* occupies a channel width of 150 kHz. Originally, this provided for a higher modulation index, even with 15 kHz audio that resulted in an improved SNR. However, with multiple channel stereo and sub-channels all in the same allocated bandwidth the deviation is around the maximum transmitted signal bandwidth.

In the US, FCC amateur rules limit wideband FM use to frequencies above 29 MHz. Some, but not all, HF communication receivers provide for FM reception. For proper FM reception, two changes are required in the receiver architecture as shown within the dashed line in **Fig 12.10**. The fundamental change is that the detector must recover information from the frequency variations of the input signal. The most common such detector is called a *discriminator*. The discriminator does not require a BFO, so that is turned off, or eliminated in a dedicated FM receiver. Since amplitude variations convey no information in FM, they are generally eliminated by a *limiter*. The limiter is a high-gain IF amplifier stage that clips the positive and negative peaks of signals above a certain threshold. Since most noise of natural origins is amplitude modulated, the limiting process also strips away noise from the signal.

SUMMARY OF RECEIVER BANDWIDTH REQUIREMENTS

We now have briefly discussed the typical operating modes expected to be encountered by an HF communications receiver. These are tabulated in **Table 12.1** and will be used as design requirements as we develop the various receiver architectures.

Table 12.1
Typical Communications Bandwidths for Various Operating Modes

Mode	Bandwidth (kHz)
FM Voice	15
AM Broadcast	10
AM Voice	4-6.6
SSB Voice	1.8-3
Digital Voice	1.0-1.2
RTTY (850 Hz shift)	0.3-1.0
CW	0.1-0.5

12.3.2 Selection of IF Frequency for a Superhet

Now that we have established the range of bandwidths that our receiver will need to pass, we are in a position to discuss the selection of the IF frequency at which those bandwidths will be established.

IF IMAGE RESPONSE

As noted earlier, a superhet with a single local oscillator or LO and specified IF can receive two frequencies, selected by the tuning of the RF stage. For example, using a receiver with an IF of 455 kHz to listen to a desired signal at 7000 kHz can use an LO of 7455 kHz. However, the receiver will also receive a signal at 455 kHz above the LO frequency, or 7910 kHz. This undesired signal frequency, located at twice the IF frequency from the desired signal, is called an *image*.

Images will be separated from the desired frequency by twice the IF and the filter ahead of the associated mixer must reduce the image signal by the amount of the required *image rejection*. For a given IF, this gets more difficult as the received frequency is increased. For example, with a 455 kHz single conversion system tuned to 1 MHz, the image will be at 1.91 MHz, almost a 2:1 frequency ratio and relatively easy to reject with a filter. The same receiver tuned to 30 MHz, would have an image at 30.91 MHz,

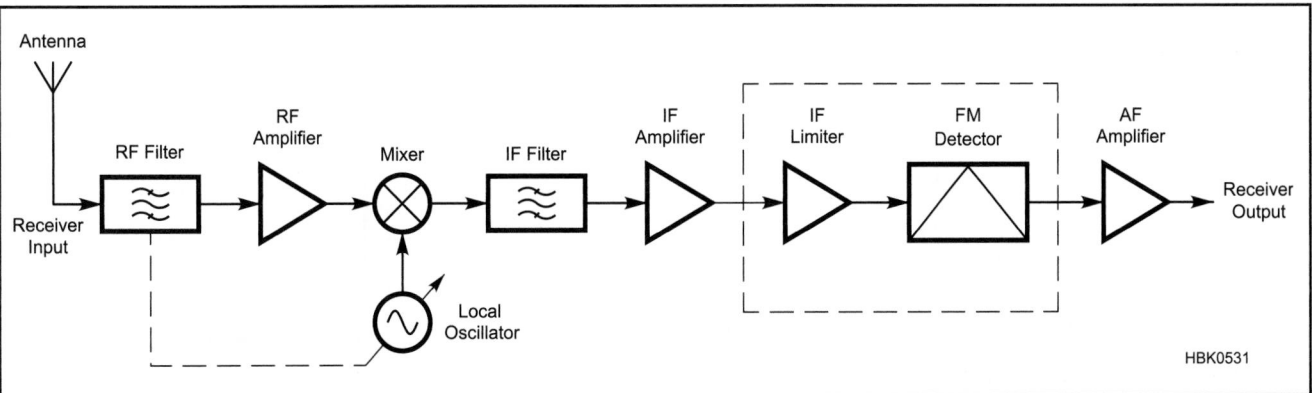

Fig 12.10 — Block diagram of an FM superhet. Changes are in dashed box.

a much more difficult filtering problem

While an image that falls on an occupied channel is obviously a problem — it's rarely desirable to receive two signals at the same time — problems occur even if the image frequency is clear of signals. This is because the atmospheric noise in the image bandwidth is added to the noise of the desired channel, as well as any internally-generated noise in the RF amplifier stage. If the image response is at the same level as the desired

signal response, there will be a 3 dB reduction in SNR.

REDUCING IMAGES WITH A HIGHER IF FREQUENCY

An obvious solution to the RF image response is to raise the IF frequency high enough so that signals at twice the IF frequency from the desired signal are sufficiently attenuated by filters ahead of the mixer. This can easily be done with IF stages operating at 5

to 10% of the highest receiving frequency (1.5 to 3 MHz for a receiver that covers the 3-30 MHz HF band). The concept is used at higher frequencies as well. The FM broadcast band (150 kHz wide channels over 87.9 to 108 MHz in the US) is generally received on superhet receivers with an IF of 10.7 MHz, which places all image frequencies outside the FM band, eliminating interference from other FM stations.

The use of higher-frequency tuned cir-

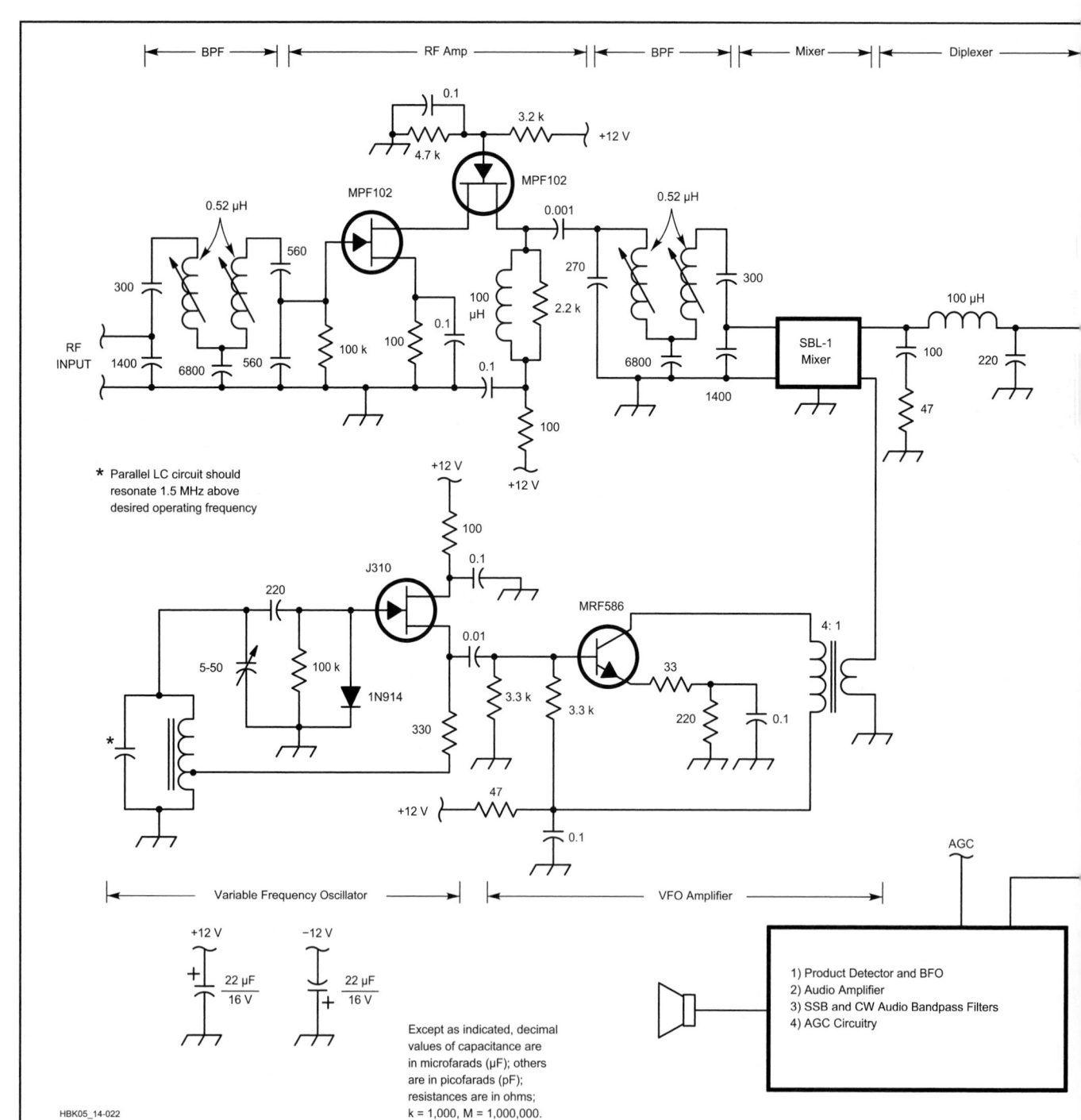

Fig 12.11 — Schematic of representative superhet using crystal lattice filters to achieve selectivity with a high IF frequency.

cuits for IF selectivity works well for the 150 kHz wide FM broadcast channels, but not so well for the relatively narrow channels encountered on HF or lower, or even for many V/UHF narrowband services. Fortunately, there are three solutions that were commonly used to resolve this problem.

The first, *double conversion*, converted the desired signal to a relatively high IF followed by a second conversion to a lower IF to set the selectivity. This was a popular technique in the 1950s. Improvements on the double conversion technique led to *triple conversion* with a very low, highly selective third IF. The *Collins system* of moving a single-range VFO to the second mixer and using switchable crystal oscillators for the first mixer also became popular. The *pre-mixed* arrangement, a third approach to double conversion was a combination of the two, used a single variable oscillator range, as with the Collins, but mixes the VFO and LO before applying them to the first mixer – outside of the signal path.

These methods are obsolete for current receivers but are commonly encountered in the vintage equipment popular with many hams. They are discussed in previous editions of the *Handbook*.

High Frequency Crystal Lattice Filters

Commercial quartz-crystal filters with bandwidths appropriate for CW and SSB

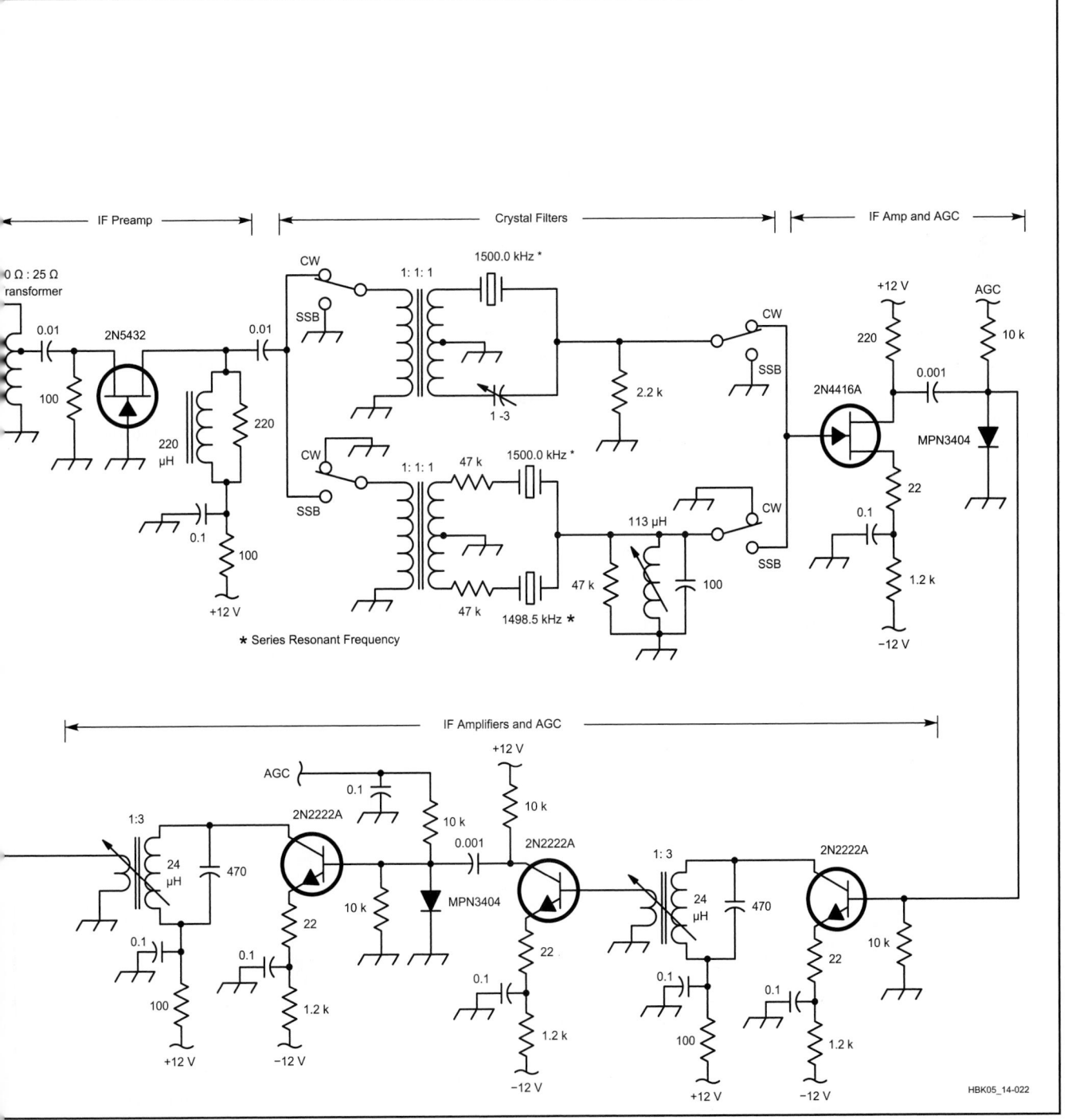

became available in the 1970s with center frequencies into the 10 MHz range. This allowed a single-conversion receiver (see Fig 12.2) with an IF in the HF range to provide both high image rejection and needed channel selectivity. This single-conversion architecture remains popular among designers of portable and low-power equipment. Crystal filter design is discussed in the **RF and AF Filters** chapter.

Fig 12.11 shows an example front-end and IF sections of a single-conversion superhet using simple filters centered 1500 kHz. While the filters shown are actually buildable by amateurs at low cost, multiple-section filters with much better performance can be purchased or constructed. Other IF frequencies can be used, depending on crystal or filter availability.

The circuit shown demonstrates the concepts involved and can be reproduced at low cost. Remaining receiver functional blocks such as the AGC circuitry, detectors and BFO, and audio amplifiers and filters can be found elsewhere in the book.

The Image-Rejecting Mixer

Another technique for reduction of image response in receivers is not as commonly encountered in HF receivers as the preceding designs, but it deserves mention because it has some very significant applications. The *image-rejecting* mixer requires phase-shift networks, as shown in **Fig 12.12**. Frequency F_1 represents the input frequency while F_2 is that of the local oscillator. Note that the two 90° phase shifts are applied at different frequencies. The phase shift network following Mixer 1 is at a fixed center frequency corresponding to the IF, while the phase shift network at F_2 must provide the required phase shift as the local oscillator tunes across the band.

If the local oscillator is required to tune over a limited fractional frequency range, this is a very feasible approach. On the other hand, maintaining a 90° phase shift over a wide range can be tricky. The good news is that this approach provides image reduction that is independent of, and in addition to, any other mechanisms such as filters that are employed toward that end.

Additionally, with the ability of DSP components to operate at higher and higher frequencies, the necessary operations seen in Fig 12.12 can be performed in software which does not depend on precision hardware design to maintain nearly exact phase relationships.

The image-rejecting mixing process has several attractive features:

• It is the only way to provide "single signal" reception with a direct conversion receiver, effectively reducing the audio image. This can make the DC receiver a very good performer, although the added complexity is not always warranted in typical amateur DC applications.

• This option is frequently found in microwave receivers in which sufficiently selective RF filtering can be difficult to obtain. Since they often operate on fixed frequencies, maintaining the required phase shift can be straightforward.

• It is found in advanced receivers that are trying to achieve optimum performance. Even with a high first IF frequency, additional image rejection can be provided.

• In transmitters, the same system is called the *phasing method* of SSB generation. The same blocks run "backwards" — one of the phase shift networks can be applied to the speech band and used to cancel one sideband. This is discussed in the **Transmitters and Transceivers** chapter.

12.3.3 Multiple-Conversion in the Digital Age

While advances in microminiaturization of all circuit elements have had a radical change in the dimensions of communication equipment, perhaps most significant technology impacts on architecture come in two areas — the application of digital signal processing and direct digital synthesis frequency generation. Both are discussed in detail in the chapter on **DSP and Software Radio Design**.

Digital signal processing provides a level of bandwidth setting filter performance not practical with other technologies. While much better than most low frequency IF LC bandwidth filters, the very good crystal or mechanical bandwidth filters in amateur gear are not very close to the rectangular shaped frequency response of an ideal filter, but rather have skirts with a 6 to 60 dB response of perhaps 1.4 to 1. That means if we select an SSB filter with a nominal (6 dB) bandwidth of 2400 Hz, the width at 60 dB down will typically be 2400 × 1.4 or 3360 Hz. Thus a signal in the next channel that is 60 dB stronger than the signal we are trying to copy (as often happens) will have energy just as strong as our desired signal.

DSP filtering approaches the ideal response. **Fig 12.13** shows the ARRL Lab measured response of a DSP bandwidth filter with a 6 dB bandwidth of 2400 Hz. Note how rapidly the skirts drop to the noise level. In addition, while analog filtering generally requires a separate filter assembly for each desired bandwidth, DSP filtering is adjustable — often in steps as narrow as 50 Hz — in both bandwidth and center frequency. In addition to bandwidth filtering, the same DSP can often provide digital noise reduction and digital notch filtering to remove interference from fixed frequency carriers.

The application of DSP filters at higher and higher frequencies as analog-to-digital conversion samples rates and processing power increase, offers greatly improved per-

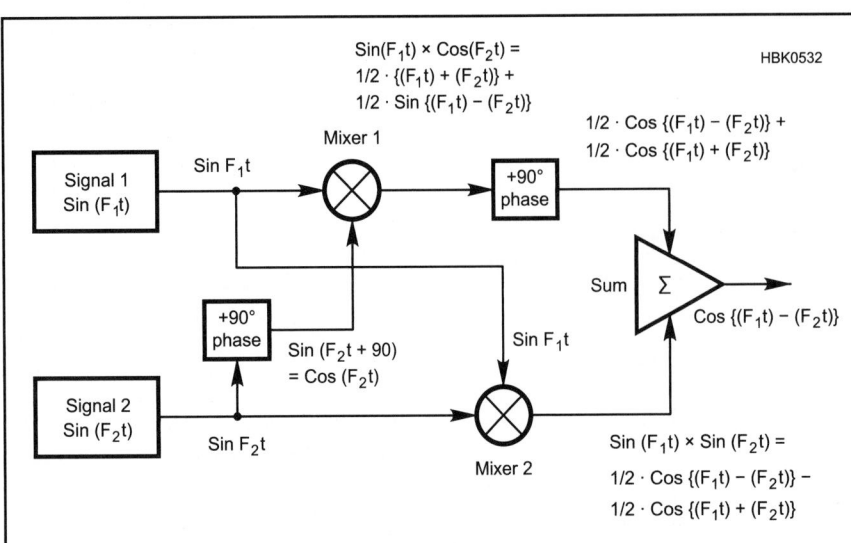

Fig 12.12 — Image rejecting mixer, block diagram and signal relationships.

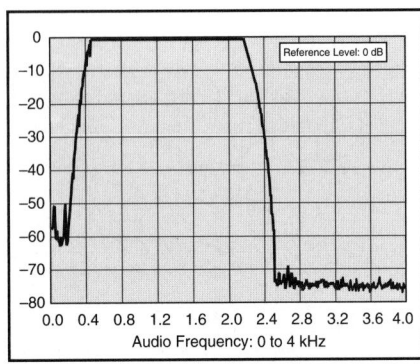

Fig 12.13 — ARRL Lab measured response of an aftermarket 2400 Hz DSP bandwidth filter.

formance for multiple-conversion receivers. The state of the art for amateur receivers is approaching direct sampling (digitization at the signal frequency) which will eliminate the need for separate IF filters, amplifiers, and detectors as all of those functions will be performed inside the DSP processor as software operations. However, it is likely that analog receivers will continue to be popular in various applications, such as portable, low-power, and home-built gear.

Project: The MicroR2 — An Easy to Build SSB or CW Receiver

The following is the description of a direct conversion receiver that uses an image-rejecting mixer as a product detector for single sideband or single-signal CW reception.

Throughout the 100-year history of Amateur Radio, the receiver has been the basic element of the amateur station. This single PC board receiver, originally described by Rick Campbell, KK7B, captures some of the elegance of the classic all-analog projects of the past while achieving the performance needed for operation in 21st-century band conditions across a selected segment of a single band.

A major advantage of a station with a separate receiver and transmitter is that each can be used to test the other. There is a particularly elegant simplicity to the combination of a crystal controlled transmitter and simple receiver. The receiver doesn't need a well-calibrated dial, because the transmitter is used to spot the frequency. The transmitter doesn't need sidetone, VFO offset or other circuitry because the crystal determines the transmitted frequency and the receiver is used to monitor the transmit note off the air. Even the receiver frequency stability may be relaxed, because there is no on-the-air penalty for touching up receiver tuning during a contact.

Let's dive into the design and construction of a simple receiver (**Fig 12.14**) that may be used as a companion for an SSB or CW transmitter on 40 meters, or as a tunable IF for higher bands. **Fig 12.15** is the block diagram of this little receiver, which can be built in a few evenings. The block diagram is similar to the simple direct conversion receivers of the early 1970s, but there the similarity ends. This radio takes advantage of 40 years of evolution of direct conversion HF receivers by some very talented designers.

"SINGLE SIGNAL" RECEPTION MAKES ALL THE DIFFERENCE

One major difference from early direct conversion receivers is the use of phasing to eliminate the opposite sideband. The receiver in Fig 12.15 is a "single-signal" receiver that only responds to signals on one side of zero beat. Basic receivers in the '50s and '60s would hear every CW signal at two places on the dial, on either side of zero beat. As you tuned through a CW signal, first you heard a high-pitched tone, and then progressively lower audio frequencies until the individual beats became audible, and then finally all the way down to zero cycles per second: zero beat. As you tuned further you heard the pitch rise back up until the high frequency CW signal became inaudible again at some high pitch.

It's not too terrible to hear the desired signal at two places on the dial — but having twice as much interference and noise is inconvenient at best. Most modern radios, even simple ones, include a simple crystal or mechanical filter for single-signal IF selectivity. Since a direct conversion receiver has no IF, selectivity is obtained by a combination of audio filtering and an image-rejecting mixer. Chapter 9 of *Experimental Methods in RF Design* (*EMRFD*) has a complete discussion of image-reject or "phasing" receivers, and an overview of both conventional superheterodyne and phasing receivers can be found elsewhere in this chapter.

BLOCK BY BLOCK DESCRIPTION

A brief description of each block in Fig 12.15 and the schematic in **Fig 12.16** follows. A parts list is given in **Table 12.2**. (Additional detail and rationale for the circuits is provided in *EMRFD*.)

The 50 Ω RF input connects to a narrow tuned-RF amplifier. The first circuit elements are a low-pass filter to prevent strong signals above the receiver tuning range — primarily local TV and FM broadcast stations — from arriving at the mixers where they would be down-converted to audio by harmonics of the local oscillator. The common-gate JFET low-noise amplifier (LNA) sets the noise figure of the receiver and provides some gain, but even more importantly, it isolates the mixers from the antenna. Isolation prevents the impedance

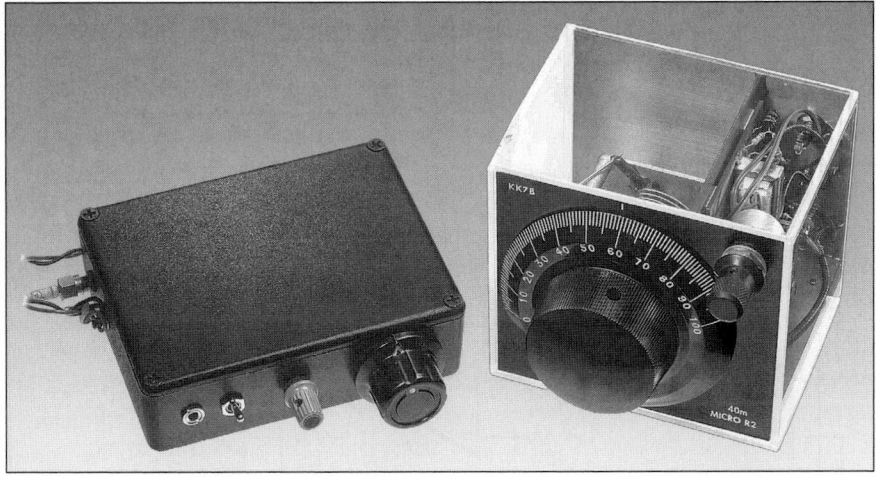

Fig 12.14 — Two versions of the MicroR2. For smooth tuning, it's hard to beat a big vernier dial!

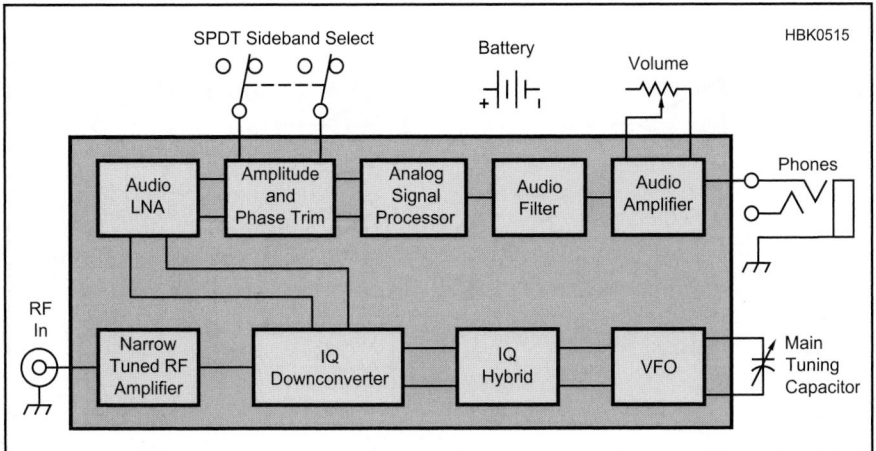

Fig 12.15 — Block diagram of the MicroR2 SSB/CW receiver.

at the RF input from affecting the opposite sideband suppression, and reverse isolation prevents local oscillator leakage from radiating out the antenna and causing interference to others as well as tunable hum.

The tuned drain circuit provides additional selectivity and greatly improves dynamic range for signals more than a few hundred kHz from the desired signal. Lifting the source of the JFET allows a convenient mute control — changing the LNA from a 10 dB amplifier to a 40 dB attenuator.

Following the LNA are the I-Q (short for "in-phase" and "quadrature," the labels we use for the two signal paths in a phasing system) down-converter and local oscillator (LO) I-Q hybrid. Many of the more subtle difficulties of phasing direct conversion reeivers are avoided by a compact, symmetrical layout and direct connection of all the mixer ports to the appropriate circuit elements without the use of transmission lines. The audio low-pass filters ensure that only audio exits the down-converter block, and set the close-in dynamic range of the receiver. The inductors in the low-pass filters will pick up magnetic hum from nearby power transformers — more on this topic later.

To the right of the I-Q hybrid is a basic JFET Hartley VFO. In this application, it is stripped of all the usual accoutrements like buffer amplifiers and receiver incremental tuning (RIT). The drain resistor and diodes in the mixers set the operating level. This simplified configuration, with a link on the VFO inductor directly driving the twisted-wire hybrid, works very well over a 5% bandwidth (350 kHz at 7 MHz) when the quadrature hybrid is optimized at the center

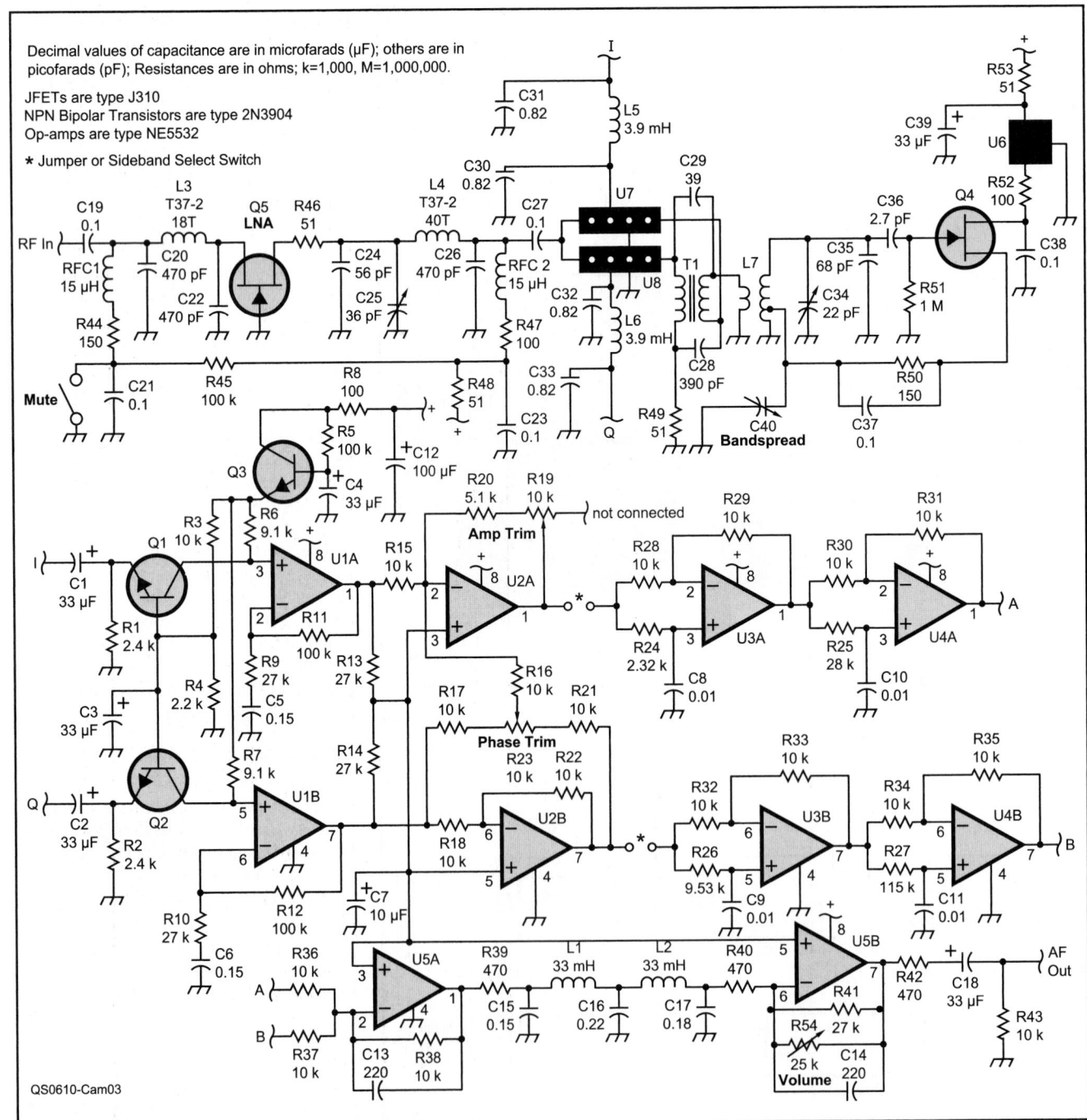

Fig 12.16 — Schematic diagram of the MicroR2 SSB/CW receiver. The parts are listed in Table 12.3.

frequency and all voltages except the LO are less than a few millivolts. (See R. Fisher, W2CQH, "Twisted-Wire Quadrature Hybrid Directional Couplers," in *QST* for Jan 1978.) The top half of the block diagram operates entirely at audio. The left block is a matched pair of audio LNAs with a noise figure of about 5 dB and near 50 Ω input impedance, to properly terminate the low-pass filters at the I-Q mixer IF ports. These are directly coupled into a circuit that performs the mathematical operations needed to remove

amplitude and phase errors from the I and Q signal paths. This is a classic use of operational amplifiers to do basic math.

The next block is a pair of second-order all-pass networks. Use of second order networks limits the opposite sideband suppression to 37 dB, which sounds exceptionally clean with multiple CW signals and static crashes in the passband.

It is straightforward to improve the opposite sideband suppression by 10 dB or so, but doing so requires much more circuitry, not

just a third-order all-pass network. Relaxing the opposite sideband suppression to a level comparable with the Drake 2B permitted cutting the parts count in half, and the construction and alignment time by a factor of 10.

The final blocks are the summer, audio low-pass filter and headphone amplifier. These are conventional, near duplicates of those used 10 years ago. When the board is finished it is an operating receiver. The completed circuitry is shown in **Fig 12.17**.

Table 12.2
MicroR2 Parts List
Parts for CW and SSB Version

C1-C4, C18, C39 — 33 µF, 16 V electrolytic.
C5, C6 — 0.15 µF, 5% polyester.
C7 — 10 µF, 16 V electrolytic.
C8-C11 — 0.01 µF, polyester matched to 1%.
C12 — 100 µF, 16 V electrolytic.
C13, C14 — 220 pF, NP0 ceramic.
C15 — 0.15 µF, 5% polyester, see Note A.
C16 — 0.22 µF, 5% polyester, see Note A.
C17 — 0.18 µF, 5% polyester, see Note A.
C19, C21, C23, C27, C37, C38 — 0.1 µF, 5% polyester.
C20, C22, C26 — 470 pF, NP0 ceramic.
C24 — 56 pF, NP0 ceramic.
C25 — 3-36 pF, poly trimmer capacitor.
C28 — 390 pF, NP0 ceramic.
C29 — 39 pF, NP0 ceramic on back of board, see Note B.
C30-C33 — 0.82 µF, 5% polyester, see Note C.

C34 — 2.2-22 pF, poly trimmer capacitor.
C35 — 68 pF, NP0 ceramic.
C36 — 2.7 pF 5% NP0 1000V ceramic.
C40 — Variable capacitor, see Note D.
L1, L2 — 33 mH inductor, see Note A.
L3 — 18t #28 T37-2, see Note E.
L4 — 40t #30 T37-2, see Note E.
L5, L6 — 3.9 mH inductor, see Note C.
L7 — 36t #28 T50-6, tap at 8 turns with a 2 turn link, see Note E.
Q1-Q3 — 2N3904.
Q4, Q5 — J310.
R1, R2 — 2.4 kΩ.
R3, R15-R18, R21, R22, R38, R43 — 10 kΩ.
R4 — 2.2 kΩ.
R5, R11, R12, R45 — 100 kΩ.
R6, R7 — 9.1 kΩ.
R8, R47, R52 — 100 Ω.
R9, R10, R13, R14, R41 — 27 kΩ.

R19, R23 — 10 kΩ trimpot.
R20 — 5.1 kΩ.
R24 — 2.32 kΩ, 1% see Note A.
R25 — 28.0 kΩ, 1% see Note A.
R26 — 9.53 kΩ, 1% see Note A.
R27 — 115 kΩ 1% see Note A.
R28-R37 — 10.0 kΩ 1%.
R39, R40, R42 — 470 Ω.
R44, R50 — 150 Ω.
R46, R48, R49, R53 — 51 Ω.
R51 — 1 MΩ.
R54 — 25 kΩ volume control.
RFC1,RFC2 — 15 µH molded RF choke.
T1 — 17 t two colors #28 bifilar T37-2, see Note E.
U1-U5 — NE5532 or equivalent dual low-noise high-output op-amp.
U6 — LM7806 or equivalent 6 V three terminal regulator.
U7, U8 — Mini-Circuits TUF-3 diode ring mixer.

Note A: Make the following parts substitutions for a CW only version:
R24 — 4.75 kΩ, 1%.
R25 — 41.2 kΩ, 1%.
R26 — 16.9 kΩ, 1%.
R27 — 147 kΩ, 1%.
C15 — 0.47 µF, 5% polyester.
C16 — 0.68 µF, 5% polyester.
C17 — 0.56 µF, 5% polyester.
L1, L2 — 100 mH inductor.

Note B: The total reactance of the parallel combination of C28 and C29 plus the capacitance between the windings of T1 is −j50 Ω at the center of the tuning range. Placing most of the capacitance at one end is a different but equivalent arrangement to the quadrature hybrid we often use with equal capacitors. C29 is only needed if there is no standard value for C28 within a few % of the required value. C29 is tack soldered to the pads provided on the back of the PC board, and may be a surface mount component if desired.

Note C: To sacrifice close-in dynamic range and selectivity for reduced 60 Hz hum pickup, make the following substitutions, as illustrated in Fig 12.18. This modification is recommended if the MicroR2 must be used near 60 Hz power transformers.
C30, C32 — 0.10 µF 5% polyester.
C31, C33 — Not used.
L5, L6 — Not used. Place wire jumper between pads.

Note D: Capacitor C40 is the tuning capacitor for the receiver. The MicroR2 was specifically designed to use an off-board tuning capacitor to provide mounting flexibility, and to encourage the substitution of whatever high-quality dual bearing capacitor and reduction drive may be in the individual builder's junk box. Any variable capacitor may be used, but values around 100 pF provide a little more than 100 kHz of tuning range on the 40 meter band, which is a practical maximum. The MicroR2 needs to be realigned (RF amplifier peaked and the amplitude and phase trimpots readjusted) when making frequency excursions of more than about ±50 kHz on the 40 meter band. To prevent tunable hum and other common ills of direct conversion receivers, the MicroR2 PC board and tuning capacitor should be in a shielded enclosure. See Chapter 8 of *EMRFD* for a complete discussion.

Note E: L3, L4, L7 and T1 are listed as number of turns on the specified core rather than a specific inductance. For those who wish to study the design with a calculator, simulator and inductance meter, L3 should be about +j100 Ω at about 8 MHz (not particularly critical). L4 and L7 should be about +j250 Ω at mid band. Each winding of T1 should be +j50 Ω at mid band.

Fig 12.17 — Photo of the inside of the receiver. The PC board and parts are available from Kanga US (www.kangaus.com).

Table 12.3
MicroR2 Receiver Performance

Parameter	Measured in the ARRL Lab
Frequency coverage:	7.0-7.1 or 7.2-7.3 MHz.
Power requirement:	50 mA, tested at 13.8 V.
Modes of operation:	SSB, CW.
SSB/CW sensitivity:	–131 dBm.
Opposite sideband rejection:	1000 Hz, 42 dB
Two-tone, 3rd-order IMD dynamic range:	20 kHz spacing, 81 dB
Third-order intercept:	20 kHz spacing, –5 dBm.
Third-order intercept points were determined using equivalent S5 reference.	

ALIGNMENT

The receiver PC board may be aligned on the bench without connecting the bandspread capacitor. There are only four adjustments, and the complete alignment takes just a few minutes with just an antenna, the crystal oscillator in the companion transmitter (or a test oscillator) and headphones. First, center the trimming potentiometers using the dots. Then slowly tune the VFO variable capacitor until the crystal oscillator is heard. The signal should be louder on one side of zero beat. Peak the LNA tuning carefully for the loudest signal. Then tune to the weaker side of zero beat and null the signal with the amplitude and phase trim pots. That's all there is.

If the adjustments don't work as expected or there is no opposite sideband suppression, look for a construction error. Don't forget to jumper the MUTE connection on the LNA — it is a "ground- to-receive" terminal just like simple receivers from the 1960s. Once the

receiver PC board is working on the bench, mount it in a box, attach the bandspread capacitor (with two wires — the ground connection needs to be soldered to the PC board too). Reset the VFO tuning capacitor for the desired tuning range, and do a final alignment in the center of the desired frequency range.

The simple on-board VFO is stable enough for casual listening and tuning around the band, but it can be greatly improved by eliminating the film trimmer capacitor that sets the VFO tuning range. It may be replaced by an air trimmer, a piston trimmer, a small compression trimmer, or a combination of small fixed NP0 ceramic capacitors experimentally selected to obtain the correct tuning range. The holes in the PC board are large enough to easily remove the film trimmer capacitor with solder wick. The VFO stability may be improved still further by following the thermal procedure described by W7ZOI in *EMRFD*.

PERFORMANCE

Is it working? I don't hear anything! Most of us are familiar with receivers with a lot of extra gain and noise, and AGC systems that turn the gain all the way up when there are no signals in the passband. Imagine instead a perfect receiver with no internal noise. With no antenna connected, you would hear absolutely nothing, just like a CD player with no disk. The MicroR2 is far from perfect — but it may be much quieter than your usual 40 meter receiver, and it probably has less gain. It is designed to drive good headphones. Weak signals will be weak and clear and strong signals strong and clear over the entire 70 dB in-channel dynamic range — just like a CD player. This receiver rewards listening skill and a good antenna — don't expect to hear much on a couple feet of wire in a noisy room. The measured performance is shown in **Table 12.3**.

Now, about that hum. One reason this receiver sounds so good is that the selectivity is established right at the mixer IF ports, before any audio gain. Then, after much of the system gain, a second low-pass filter provides additional selectivity. Distributed selectivity is a classic technique. Earlier designs included aggressive high-pass filtering to suppress frequencies below 300 Hz. That works well and has other advantages — but the audio sounds a bit thin. K7XNK suggested trying a receiver with the low frequency response wide open. It sounds great, but 60 Hz hum is audible if the receiver is within about a meter of a power transformer. Any closer and the hum gets loud. This makes it difficult to measure receiver performance in the lab, as there is a power transformer in every piece of test equipment.

Hum pickup may be eliminated by modifying the circuit as shown in **Fig 12.18**. This hurts performance in the presence of strong off-channel signals, but in some applications — an instrumentation receiver on the bench — it is a good trade-off. The receiver sounds the same either way. For battery operation in the hills, or on a picnic table in the backyard, hum pickup isn't a problem.

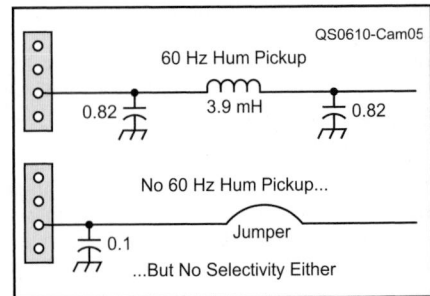

Fig 12.18 — Modified circuit to minimize hum.

12.4 Superhet Receiver Design Details

The previous sections have provided an overview of the major architectural issues involved in superheterodyne receiver design. In this section, we will cover more of the details for the superhet constructed as a hybrid of analog electronics and DSP functions applied in an IF stage, the primary receiver design architecture used in Amateur Radio today.

12.4.1 Receiver Sensitivity

The sensitivity of a receiver is a measure of the lowest power input signal that can be received with a specified signal-to-noise ratio. In the early days of radio, this was a very important parameter and designers tried to achieve the maximum practical sensitivity. In recent years, device and design technology have improved to the point that other parameters may be of higher importance, particularly in the HF region and below.

THE RELATIONSHIP OF NOISE TO SENSITIVITY

Noise level is as important as signal level in determining sensitivity. This section builds on the discussion of noise in the **RF Techniques** chapter. The most important noise parameters affecting receiver sensitivity are *noise bandwidth*, *noise figure*, *noise factor*, and *noise temperature*.

Since received noise power is directly proportional to receiver bandwidth, any specification of sensitivity must be made for a particular noise bandwidth. For DSP receivers with extremely steep filter skirts, receiver bandwidth is approximately the same as the filter or operating bandwidth. For other filter types, noise bandwidth is somewhat larger than the filter's 6 dB response bandwidth.

The relationship of noise bandwidth to noise power is one of the reasons that narrow bandwidth modes, such as CW, have a significant signal-to-noise advantage over modes with wider bandwidth, such as voice, assuming the receiver bandwidth is the minimum necessary to receive the signal. For example, compared to a 2400-Hz SSB filter bandwidth, a CW signal received in a 200 Hz bandwidth will have a 2400/200 = 12 = 10.8 dB advantage in received noise power. That is the same difference as an increase in transmitter from 100 to 1200 W.

Sources of Noise

Any electrical component will generate a certain amount of noise due to random electron motion. Any gain stages after the internal noise source will amplify the noise along with the signal. Thus a receiver with no signal input source will have a certain amount of noise generated and amplified within the receiver itself.

Table 12.4
Typical Noise Levels (Into the Receiver) and Their Source, by Frequency

Frequency range	Dominant noise sources	Typical level ($\mu v/m$)*
LF 30 to 300 kHz	atmospheric	150
MF 300 to 3000 kHz	atmospheric/man-made	70
Low HF 3 to 10 MHz	man-made/atmospheric	20
High HF 10 to 30 MHz	man-made/thermal	10
VHF 30 to 300 MHz	thermal/galactic	0.3
UHF 300 to 3000 MHz	galactic/ thermal	0.2

*The level assumes a 10 kHz bandwidth. Data from *Reference Data for Engineers*, 4th Ed, p 273, Fig 1.

Upon connecting an antenna to a receiver, there will be introduction of any noise external to the receiver that is on the received frequency. The usual sources and their properties are described below. **Table 12.5** presents typical levels of external noise in a 10 kHz bandwidth present in the environment from different sources.

Atmospheric noise. This is noise generated within our atmosphere due to natural phenomena. The principal cause is lightning which sends wideband signals great distances. All points on the Earth receive this noise, but it is much stronger in some regions than others depending on the amount of local lightning activity. This source is usually the strongest noise source in the LF range and may dominate well into the HF region, depending on the other noises in the region. The level of atmospheric noise tends to drop off by around 50 dB every time the frequency is increased by a factor of 10. This source usually drops in importance by the top of the HF range (30 MHz).

Man-made noise. This source acts in a similar manner to atmospheric noise, although it is more dependent on local activity rather than geography and weather. The sources tend to be sparks from rotating and other kinds of electrical machinery as well as gasoline engine ignition systems and some types of lighting. In recent years, noise from computing and network equipment, switchmode power supplies, and appliances has increased significantly in urban and suburban environments. All things being equal, this source, on average, drops off by about 20 dB every time the frequency is increased by a factor of ten. The slower decrease at higher frequencies is due to the sparks having faster rise times than lightning. The effect tends to be comparable to atmospheric noise in the broadcast band, less at lower frequencies and a bit more at HF.

Galactic Noise. This is noise generated by the radiation from heavenly bodies outside our atmosphere. Of course, while this is noise to *communicators*, it is the desired signal for *radio-astronomers*. This noise source is

a major factor at VHF and UHF and is quite dependent on exactly where you point an antenna (antennas for those ranges tend to be small and are often *pointable*). It also happens that the Earth turns and sometimes moves an antenna into a position where it inadvertently is aimed at a noisy area of the galaxy. If the Sun, not surprisingly the strongest signal in our solar system, appears behind a communications satellite, communications is generally disrupted until the Sun is out of the antenna's receiving pattern. Galactic noise occurs on HF, as well: Noise from the planet Jupiter can be heard on the 15 meter band under quiet conditions, for example.

Thermal Noise. Unlike the previous noise sources, this one comes from our equipment. All atomic structures have electrons that move within their structures. This motion results in very small currents that generate small amounts of wideband signals. While each particle's radiation is small, the cumulative effect of all particles becomes significant as the previous sources roll off with increasing frequency. The reason that this effect is called *thermal* noise is because the electron motion increases with the particle's temperature. In fact the noise strength is directly proportional to the temperature, if measured in terms of absolute zero (0 K). For example, if we increase the temperature from 270 to 280 K, that represents an increase in noise power of 10/270, 0.037, or about 0.16 dB. Some extremely sensitive microwave receivers use cryogenically-cooled front-end amplifiers to provide large reductions in thermal noise.

Oscillator Phase Noise. As noted in the **Oscillators and Synthesizers** chapter, real oscillators will have phase-noise sidebands that extend out on either side of the nominal carrier frequency at low amplitudes. Any such noise will be transferred to the received signal and through *reciprocal mixing* create noise products from signals on adjacent channels through the mixer. A good receiver will be designed to have phase noise that is well below the level of other internally generated noise.

Noise Power and Sensitivity

There are a number of related measures that can be used to specify the amount of noise that is generated within a receiver. If that noise approaches, or is within perhaps 10 dB of the amount of external noise received, then it must be carefully considered and becomes a major design parameter. If the internally generated noise is less than perhaps 10 dB below that expected from the environment, efforts to minimize internal noise are generally not beneficial and can, in some cases, be counterproductive.

While the total noise in a receiving system is, as discussed, proportional to bandwidth, the noise generating elements are generally not. Thus it is useful to be able to specify the internal noise of a system in a way that is independent of bandwidth. It is important to note that even though such a specification is useful, the actual noise is still directly proportional to bandwidth and any bandwidth beyond that needed to receive signal information will result in reduced SNR.

To evaluate the effect of noise power on sensitivity, we will use the following equations from the discussion of noise in the **RF Techniques** chapter:

$$F = 1 + T_E/T_0 \qquad (1)$$

where

F is the noise factor,
T_E is the equivalent noise temperature and
T_0 is the standard noise temperature, usually 290 K.

$$N_i = k \times B \times T_E \qquad (2)$$

where

N_i is the equivalent noise power in watts at the input of a perfect receiver that would result in the same noise output,
k is Boltzman's constant, 1.38×10^{-23} joules/kelvin, and
B is the system noise bandwidth in hertz.

Expressing Ni in dBm is more convenient and is calculated as:

$$dBm_i = -198.6 + (10 \times \log_{10} B) + (10 \times \log_{10} T_E) \qquad (3)$$

If N_i is greater than the noise generated internally by the receiver, the receiver's sensitivity is limited by the external noise. This is usually the case for HF receivers where atmospheric and man-made noise are much stronger than the receiver's internal noise floor. If N_i is within, perhaps, 10 dB greater than the receiver's internal noise, then the

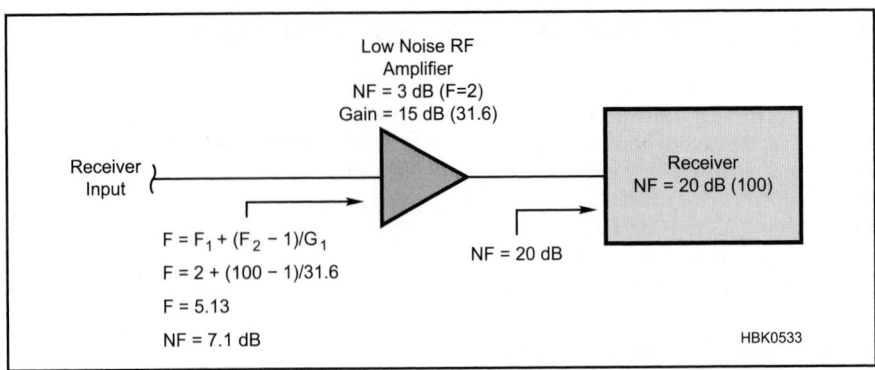

Fig 12.19 — The effect of adding a low noise preamplifier in front of a noisy receiver system.

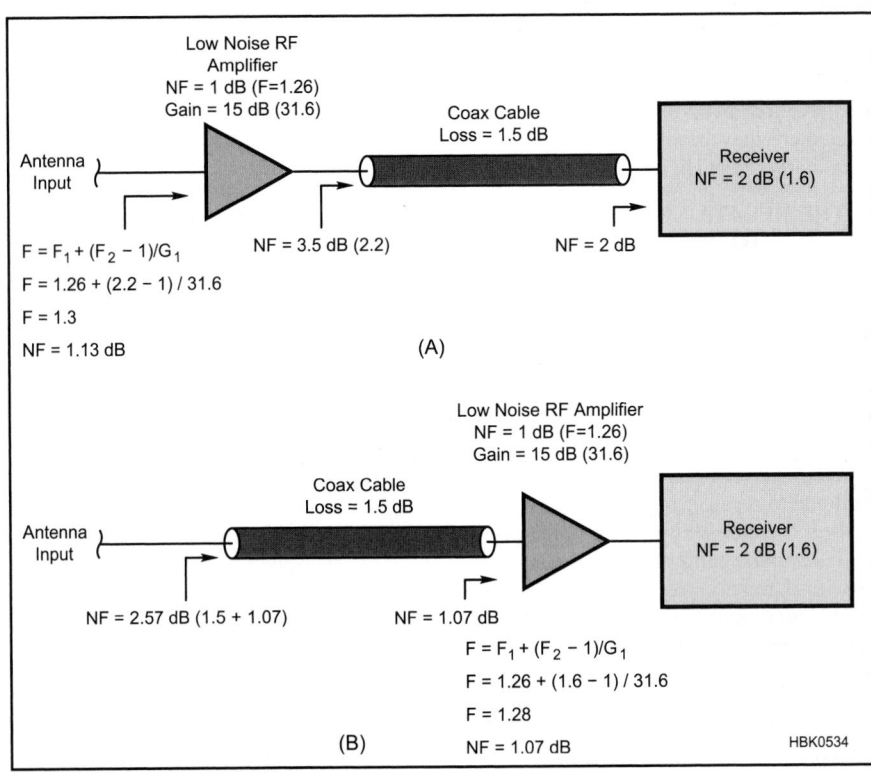

Fig 12.20 — Illustration of adding a VHF low noise preamplifier at the antenna (A) compared to one at the receiver system input (B).

effect of the receiver's internal circuits on overall system sensitivity must be taken into account.

Noise Figure of Cascaded Stages

Often we are faced with the requirement of determining the noise figure of a system of multiple stages. In general adding an amplification stage between the antenna and the rest of the system will reduce the equivalent noise figure of the system by the amount of gain of the stage but adds in the noise of the added stage directly. The formula for determining the resultant noise factor is:

$$F = F_1 + \frac{F_2 - 1}{G_1} \qquad (4)$$

where

F_1 is the noise factor of the stage closest to the input,
F_2 is the noise factor of the balance of the system, and
G_1 is the gain of the stage closest to the input.

Note that these are noise *factors* not noise *figures*. The change to noise figure is easily computed at the end, if desired.

This is commonly encountered through the addition of a *low noise preamplifier* ahead of a noisy receiver as shown in **Fig 12.19.** As shown in the figure for fairly typical values, while the addition of a low noise preamplifier does reduce the noise figure, as can be observed, the amplifier gain and noise figure of the rest of the receiver can make a big difference.

In many cases elements of a receiving system exhibit loss rather than gain. Since they reduce the desired signal, while not changing the noise of following stages, they increase the noise figure at their input by an amount equal to the loss. This is the reason that VHF low noise preamps are often antenna mounted. If the coax between the antenna and radio has a loss of 1 dB, and the preamp has a noise figure of 1 dB, the resulting noise figure with the preamp at the radio will be 2 dB, while if at the antenna, the noise added by the coax will be reduced by the gain of the preamp, resulting in a significant improvement in received SNR at VHF and above. **Fig 12.20** shows a typical example.

To determine the system noise factor of multiple cascaded stages, Equation 4 can be extended as follows:

$$F = F_1 + \frac{F_2 - 1}{G_1} + \frac{F_3 - 1}{G_1 G_2} + ... + \frac{F_N - 1}{G_1 G_2 ... G_{N-1}}$$

(5)

where F_N and G_N are the noise factor and gain of the Nth stage.

This can proceed from the antenna input to the last linear stage, stopping at the point of detection or analog-to-digital conversion, whichever occurs first.

Why Do We Care?

A receiver designer needs to know how strong the signals are to establish the range of signals the receiver will be required to handle. One may compare the equivalent noise power determined in Equation 3 with the expected external noise to determine whether the overall receiver SNR will be determined by external or internal noise. A reasonable design objective is the have the internal noise be less than perhaps 10 to 20 dB below the expected noise. As noted above, this is related closely to the frequency of signals we want to receive. Any additional sensitivity will not provide a noticeable benefit to SNR, and may result in reduced dynamic range, as will be discussed in the next section.

For frequencies at which external noise sources are strongest, the noise power (and signal power) will also be a function of the antenna design. In such cases, the signal-to-noise ratio can be improved by using an antenna that picks up more signal and less noise, such as with a directional antenna that can reject noise from directions other than that of the signal. Some antennas improve SNR simply by rejecting noise, such as the Beverage antenna. Signals from a Beverage antenna are usually much weaker than from a conventional antenna but their ability to reject noise from undesired directions cre-

ates a net improvement in the SNR at the receiver output.

12.4.2 Receiver Image Rejection

The major design choice in determining image response is selection of the first and any following IF frequencies. This is important in receiving weak signals on bands where strong signals are present. If the desired signal is near the noise, a signal at an image frequency could easily be 100 dB stronger, and thus to avoid interference, an image rejection of 110 dB would be needed. While some receivers meet that target, the receiver sections of most current amateur transceivers are in the 70 to 100 dB range.

Receiver Architectures for Image Rejection

As noted earlier, improved crystal filter technology allows *down-conversion* HF receivers to use an IF in the 4-10 MHz range. With a 10 MHz IF and an LO above the signal frequency, a 30 MHz signal would have an image at 50 MHz. This makes image-rejection filtering relatively straightforward, although many receiver IF frequencies tend to be at the lower end of the above range. Still, as will be discussed in the next section, they have other advantages.

Many current HF receivers (or receiver sections of transceivers) have elected to employ an *up-converting* architecture. They typically have an IF in the VHF range, perhaps 60 to 70 MHz, making HF image rejection easy. A 30 MHz signal with a 60 MHz IF will have an image at 150 MHz. Not only is it five times the signal frequency, but signals in this range (other than perhaps the occasional taxicab) tend to be weaker than some undesired HF signals. Receivers with this architecture have image responses at the upper end of the above range, often with the image rejected by a relatively simple low-pass filter with a cut off at the top of the receiver range.

Another advantage of this architecture is that the local oscillator can cover a wide

continuous range, making it convenient for a general coverage receiver. For example, with a 60 MHz IF, a receiver designed for LF through HF would need an LO covering 60.03 to 90 MHz, a 1.5 to 1 range, easily provided by a number of synthesizer technologies, as described in the **Oscillators and Synthesizers** chapter.

The typical upconverting receiver uses multiple conversions to move signals to frequencies at which operating bandwidth can be established. While crystal filters in the VHF range used by receivers with upconverting IFs have become available with bandwidths as narrow as around 3 kHz, they do not yet achieve the shape factor of similar bandwidth filters at MF and HF. Thus, these are commonly used as *roofing filters*, discussed in the next section, prior to a conversion to one or more lower IF frequencies at which the operating bandwidth is established. **Fig 12.21** is a block diagram of a typical upconverting receiver using DSP for setting the operating bandwidth.

12.4.3 Receiver Dynamic Range

Dynamic range can be defined as the ratio of the strongest to the weakest signal that a system, in this case our receiver, can respond to linearly. Table 12.5 gives us an idea of how small a signal we might want to receive. The designer must create a receiver that will handle signals from below the noise floor to as strong as the closest nearby transmitter can generate. Most receivers have a specified (or sometimes not) highest input power that can be tolerated, representing the other end of the spectrum. Usually the maximum power specified is the power at which the receiver will not be damaged, while a somewhat lower power level is generally the highest that the receiver can operate at without overload and the accompanying degradation of quality of reception of the desired signal.

ON-CHANNEL DYNAMIC RANGE

The signal you wish to listen to can range

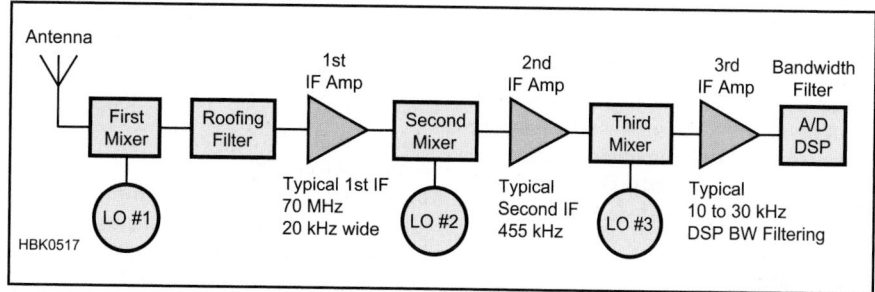

Fig 12.21 — Block diagram of a typical upconverting receiver using DSP for operating bandwidth (BW) determination. Receivers applying DSP filtering at the second IF and at higher frequencies are common.

from the strongest to the weakest, sometimes changing rapidly with conditions, or in a situation with multiple stations such as a net. While a slow change in signal level can be handled with manual gain controls, rapid changes require automatic systems to avoid overload and operator discomfort.

This is a problem that has been long solved with automatic gain control (AGC) systems. These systems are described in a further section of the chapter, but it is worth pointing out that the measurement and gain control points need to be applied carefully to the most appropriate portions of the receiver to maintain optimum performance. If all control is applied to early stages, the SNR for strong stations may suffer, while if applied in later stages, overload of early stages may occur in the presence of strong stations. Thus, gain control has to be designed into the receiver distributed from the input to the detector.

The next two sections illustrate a frequent limitation of receiver performance — dynamic range between the reception of a weak signal in the presence of one or more strong signals outside of the channel.

BLOCKING GAIN COMPRESSION

A very strong signal outside the channel bandwidth can cause a number of problems that limit receiver performance. *Blocking gain compression* (or "blocking") occurs when strong signals overload the receiver's high gain amplifiers and reduce its ability to amplify weak signals. (Note that the term "blocking" is often used outside Amateur Radio when referring to reciprocal mixing of oscillator noise with strong local signals.)

While listening to a weak signal, all stages operate at maximum gain. If the weak signal were at a level of S0, a strong signal could be at S9 + 60 dB. Using the standard of S9 representing a 50 μV input signal, and each S unit reflecting a change of 6 dB, the receiver's front-end stages would be receiving a 0.1 μV signal and a 50,000 μV into the front end at the same time. A perfectly linear receiver would amplify each signal equally until the undesired signal is eliminated at the operating bandwidth setting stage. However, in practical receivers, after a few stages of full gain amplification, the stronger signal causes amplifier clipping, which reduces the gain available to the strong signal. This is seen as a gradual reduction in gain as the input signal amplitude increases. Gain reduction also reduces the amplitude of the weaker signal which is perceived to fade as the strong signal increases in amplitude. Eventually, the weaker signal is no longer receivable and is said to have been "blocked", thus the name for the effect.

The ratio in dB between the strongest signal that a receiver can amplify linearly, with no more than 1 dB of gain reduction,

and the receiver's noise floor in a specified bandwidth is called the receiver's *blocking dynamic range* or *BDR* or the *compression-free dynamic range* or *CFDR*. In an analog superhet, BDR is established by the linear regions of the IF amplifiers and mixers. If the receiver employs DSP, the range of the analog-to-digital converter usually establishes the receiver's BDR.

A related term is "near-far interference" which is used primarily in the commercial environment to refer to a strong signal causing a receiver to reduce its gain and along with it the strength of weak received signals.

INTERMODULATION DYNAMIC RANGE

Blocking dynamic range is the straightforward response of a receiver to a single strong interfering signal outside the operating passband. In amateur operation, we often have more than one interferer. While such signals contribute to the blocking gain compression in the same manner as a single signal described above, multiple signals also result in a potentially more serious problem resulting from *intermodulation products*.

If we look again at Equation B in the earlier sidebar on Nonlinear Signal Combinations, we note that there are an infinite number of higher order terms. In general, the coefficients of these terms are progressively lower in amplitude, but they are still greater than zero. Of primary interest is third-order term, $K_3 \times V_{IN}^3$, when considering V_{IN} as the sum of two interfering signals (f_1 and f_2) near our desired signal (f_0) and within the first IF passband, but outside the operating bandwidth.

$$V_{OUT} = K_3 \times [A\sin(f_1)t + B\sin(f_2)t]^3 \qquad (6)$$

$$= K_3 \times \{A^3\sin^3(f_1)t + 3A^2B\,[\sin^2(f_1)t \times \sin(f_2)t]$$

$$+ 3AB^2\,[\sin(f_1)t \times \sin^2(f_2)t] + B^3\sin^3(f_2)t\,\}$$

The cubic terms in Equation 6 (the first and last terms) result in products at three times the frequency and can be ignored in this discussion. Using trigonometric identities to reduce the remaining $\sin^2$ terms and the subsequent $\cos(\ )\sin(\ )$ products reveal individual intermodulation (IM) products, recognizing that the signals have cross-modulated each other due to the nonlinear action of the circuit. (Math handbooks such as the *CRC Standard Mathematical Tables and Formulae* have all the necessary trigonometry information.)

IM products have frequencies that are linear combinations of the input signal frequencies, written as $n(f_1) \pm m(f_2)$, where n and m are integer values. The entire group of products that result from intermodulation are broadly referred to as *intermodulation*

distortion or IMD. The ratio in dB between the amplitude of the interfering signals, f_1 and f_2, and the resulting IM products is called the *intermodulation ratio*.

If *all* of the higher-order terms in the original equation are considered, n and m can take on any integer value. If the sum of n and m is odd, (2 and 1, or 3 and 2, or 3 and 4, etc.) the result is products that have frequencies near our desired signal, for example, $2(f_1) - 1(f_2)$. Those are called *odd-order* products. Odd-order products have frequencies close enough to those of the original signals that they can cause interference to the desired signal. If the sum of n and m is three, those are *third-order IM products* or *third-order IMD*. For fifth-order IMD, the sum of n and m is five, and so forth. The higher the order of the IM products, the smaller their amplitude, so our main concern is with third-order IMD.

If the two interfering signals have frequencies of $f_0 + \Delta$, and $f_0 + 2\Delta$, where Δ is some offset frequency, we have for the third-order term:

$$V_{OUT} = K_3 \times [A\,\sin(f_0 + \Delta)t + B\,\sin(f_0 + 2\Delta)t]^3 \qquad (7)$$

A good example would be interfering signals with offsets of 2 kHz and 2 × 2 kHz or 4 kHz from the desired frequency, a common situation on the amateur bands.

Discarding the cubic terms and applying the necessary trigonometric identities shows that a product can be produced from this combination of interfering frequencies that has a frequency of exactly f_0 — the same frequency as the desired signal! (The higher-order terms of Equation B can also produce products at f_0, but their amplitude is usually well below those of the third-order products.)

Thus we have two interfering signals that are not within our operating bandwidth so we don't hear either by themselves. Yet they combine in a nonlinear circuit and produce a signal exactly on top of our desired signal. If the interfering signals are within the passband of our first IF and are strong enough the IM product will be heard.

As the strength of the interfering signals increases, so does that of the resulting intermodulation products. For every dB of increase in the interfering signals, the third-order IM products increase by approximately 3 dB. Fifth-order IM increases by 5 dB for every dB increase in the interfering signals, and so forth. Our primary concern, however, is with the third-order products because they are the strongest and cause the most interference.

INTERCEPT POINT

Intercept point describes the IMD performance of an individual stage or a complete receiver. For example, third-order IM products increase at the rate of 3 dB for every 1-dB increase in the level of each of the interfering input signals (ideally, but not always exactly true). As the input levels increase, the distortion products seen at the output on a spectrum analyzer could catch up to, and equal, the level of the two desired signals if the receiver did not begin to exhibit blocking as discussed earlier.

The input level at which this occurs is the *input intercept point*. **Fig 12.22** shows the concept graphically, and also derives from the geometry an equation that relates signal level, distortion and intercept point. The intercept point of the most interest in receiver evaluation is that for third-order IM products and is called the *third-order intercept point* or IP_3. A similar process is used to get a second-order intercept point for second-order IMD. A higher IP_3 means that third-order IM products will be weaker for specific input signal strengths and the operator will experience less interference from IM products from strong adjacent signals.

These formulas are very useful in designing radio systems and circuits. If the input intercept point (dBm) and the gain of the stage (dB) are added the result is an output intercept point (dBm). Receivers are specified by input intercept point, referring distortion back to the receive antenna input. Intercept point is a major performance limitation of receivers used in high density contest or DX operations. Keep in mind that we have been discussing this as an effect of two signals, one that is Δ away from our operating frequency and another at twice Δ. In real life, we may be trying to copy a weak signal at f_0, and have other signals at $f_0 \pm 500$, 750, 1000, 1250…5000 Hz. There will be many combinations that produce products at or near our weak signal's frequency.

Note that the products don't need to end up exactly on top of the desired signal to cause a problem; they just need to be within the operating bandwidth. So far we have been talking about steady carriers, such as would be encountered during CW operation with interference from nearby CW stations. SSB or other wider bandwidth modes with spectrum distributed across a few kHz will have signal components that go in and out of a relationship that results in on-channel interference from IMD. This manifests itself as a time-varying synthetic noise floor, composed of all the resulting products across the channel. The difference in this low level "noise" can be dramatic between different receivers, especially when added to phase noise received from other stations and reciprocal mixing inside the receiver!

SPURIOUS-FREE DYNAMIC RANGE

IM products increase with the amplitude of the interfering signals that cause them and at some point become detectable above the receiver's noise floor. The ratio of the strength of the interfering signals to the noise floor, in dB, is the receiver's *spurious-free dynamic range* or *SFDR*. This is the range of signal strengths over which the receiver does not produce any detectable spurious products. SFDR can be specified for a specific order of IM products; for example, SFDR3 is the SFDR measured for third-order IM products only. The bandwidth for which the receiver's noise floor is measured must also be speci-fied, since smaller bandwidths will result in a lower noise floor.

RECEIVER DESIGN FOR DYNAMIC RANGE

As noted previously, there are a number of possible architectural choices for an amateur receiver. In the past, the receivers with the best close-in third order intermodulation distortion and maximum blocking dynamic range were amateur-band-only receivers, such as the primary receiver in the TEN-TEC Orion series, the receiver in the Elecraft K2 , the earlier TEN-TEC Omni VI and both receivers in the Elecraft K3. A look at a typical block diagram, as shown in **Fig 12.23**, makes it easy to see why. The problems resulting from strong unwanted signals near a desired one are minimized if the unwanted signals are kept out of the places in the receiver where they can be amplified even more and cause the nonlinear effects that we try to avoid.

Note that in Fig 12.23, the only place where the desired and undesired signals all coexist is before the first mixer. If the first mixer and any RF preamp stages have sufficient strong-signal handling capability, the undesired signals will be eliminated in the filter immediately behind the first mixer. This HF crystal filter is generally switchable to support desired bandwidths as narrow as 200 Hz. The later amplifier, mixer and DSP circuits only have to deal with the signal we want. For additional discussion of these issues, see "International Radio Roofing Filters for the Yaesu FT-1000 MP Series Transceivers," by Joel Hallas, W1ZR, in *QST* Product Review for February 2005.

Now look at a typical modern general coverage receiver as shown previously in Fig 12.21. In this arrangement, a single digital synthesizer, perhaps covering from 70 to 100 MHz, shifts the incoming signal(s) to a VHF IF, often near 70 MHz. A roofing filter at 70 MHz follows the first mixer. This arrangement offers simplified local oscillator (LO) design and the possibility of excellent image rejection. Unfortunately, crystal filter technology has only recently been able to produce narrow filters for

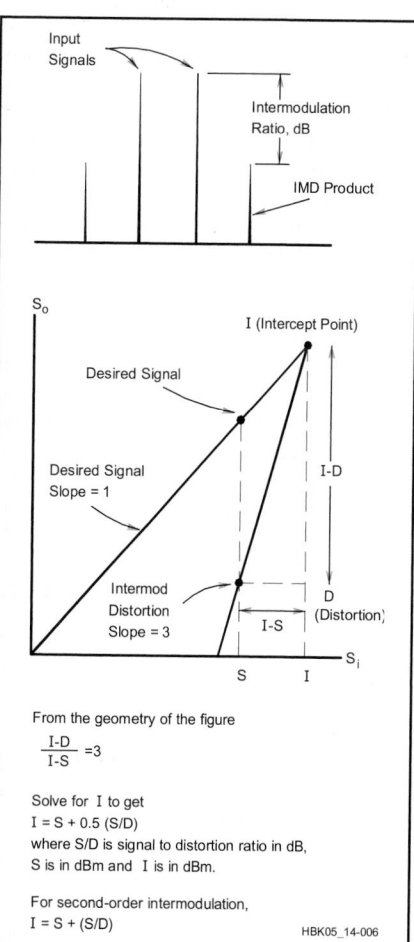

From the geometry of the figure

$$\frac{I-D}{I-S} = 3$$

Solve for I to get

$I = S + 0.5\ (S/D)$

where S/D is signal to distortion ratio in dB, S is in dBm and I is in dBm.

For second-order intermodulation,

$I = S + (S/D)$

HBK05_14-006

Fig 12.22 — Graphical representation of the third-order intercept concept.

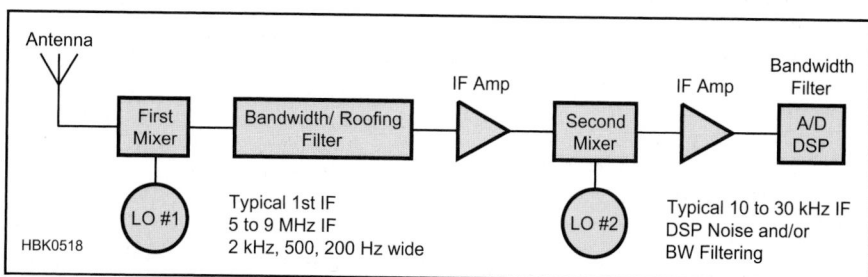

Fig 12.23 — Block diagram of a downconverting amateur band receiver with HF IF filtering to eliminate near channel interference.

70 MHz, and so far they have much wider skirts than the crystal filters used in Fig 12.23.

Many receivers and transceivers set this filter bandwidth wider than any operating bandwidth and use DSP filtering much later in the signal chain to set the final operating bandwidth. For a receiver that will receive FM and AM, as well as SSB and CW, that usually means a roofing filter with a bandwidth of around 20 kHz. With this arrangement, all signals in that 20 kHz bandwidth pass all the way through IF amplifiers and mixers and into the A/D converter before we attempt to eliminate them using DSP filters. By that time they have had an opportunity to generate intermodulation products and cause the blocking and IMD problems that we are trying to eliminate. This situation is changing with the introduction of the top of the line radios that feature both general coverage at HF and VHF roofing filters, such as the ICOM IC-7800 and Yaesu FTdx9000 transceivers. (See the Transceiver Survey by W1ZR on the *Handbook's* CD-ROM for more information on the latest models.)

A hybrid architecture has recently appeared in the TEN-TEC Omni VII transceiver that effectively combines the two technologies. The first IF has a 20 kHz wide roofing filter at 70 MHz, followed by selectable steep skirted 455 kHz Collins mechanical filters at the second IF and then DSP filters at the third IF. The topology is shown in **Fig 12.24**.

Careful attention to gain distribution among the stages between the filters maintains desired sensitivity, but not so high that the undesired products have a chance to become a serious problem. With bandwidths of 20, 6 and 2.5 kHz supplied, and 500 and 300 Hz as accessories, the undesired close-in signals are eliminated before they have an opportunity to cause serious trouble in the DSP stages that follow.

Another variation is found in the Kenwood TS-590S which switches between downconversion on the more crowded "contest bands" (160, 80, 40, 20 and 15 meters) and up-conversion on the remaining bands. In effect, trading sensitivity for dynamic range.

WHAT KIND OF PERFORMANCE CAN WE EXPECT?

As we would expect, the blocking and IMD dynamic range (IMD DR) performance of a receiver will depend on a combination of the early stage filtering, the linearity of the mixers and amplifiers, and the dynamic range of any ADC used for DSP. Product reviews of receivers and the receiver sections of transceivers published in *QST* now provide the measured dynamic range in the presence of interfering signals with spacings of 2, 5 and 20 kHz. (Details of the test procedures used are given in the **Test Equipment and Measurements** chapter.) At 20 kHz spacing, the interfering signal is usually outside of the roofing filter bandwidth of any of the above architectures. Spacing of 2 and 5 kHz represents likely conditions on a crowded band.

A look at recent receiver measurements indicates that receivers have IMD dynamic range with 2 kHz spacing results in the following ranges:

Upconverting with VHF IF (Fig 12.21): 60 to 80 dB

Downconverting with HF IF (Fig 12.23): 75 to 105 dB

Hybrid distributed architecture (Fig 12.24): Omni VII, 82 dB

Let's take an example of what this would mean. If we are listening to a signal at S3, for signals to generate a third-order IMD product at the same level in a receiver with a dynamic range of 60 dB, the $f_0 + \Delta$ and $f_0 + 2\Delta$ signals would have to have a combined power equal to S9 +27 dB, or each at S9 +24 dB. This is not unusual on today's amateur bands. On the other hand, if we had an IMD dynamic range of 102 dB, the interfering signals would have to at S9 +66 dB, much less likely. How much dynamic range you need depends in large measure on the kind of operating you do, how much gain your receiving antennas have and the closeness of the nearest station that operates on the same bands as you.

Keep in mind also that it is often difficult to tell whether or not you are suffering from IMD — it just sounds like there are many more signals than are really present. A good test to assess the source of interference is first switch off any preamplifiers and noise-blankers or noise-reduction systems that affect the receiver's linearity. Observe the level of the interference (if it's still there) and then switch in some attenuation at the front-end of the receiver. If the level of the interference goes down by *more* than the level of attenuation (estimate 6 dB per S unit), then the interference is being generated (or at least aggravated) by non-linearity inside the receiver. Continue to increase attenuation until the interference either goes away or goes down at the same rate as the attenuation is increased. You might be surprised at how much better the band "sounds" when your receiver is operating in its linear region!

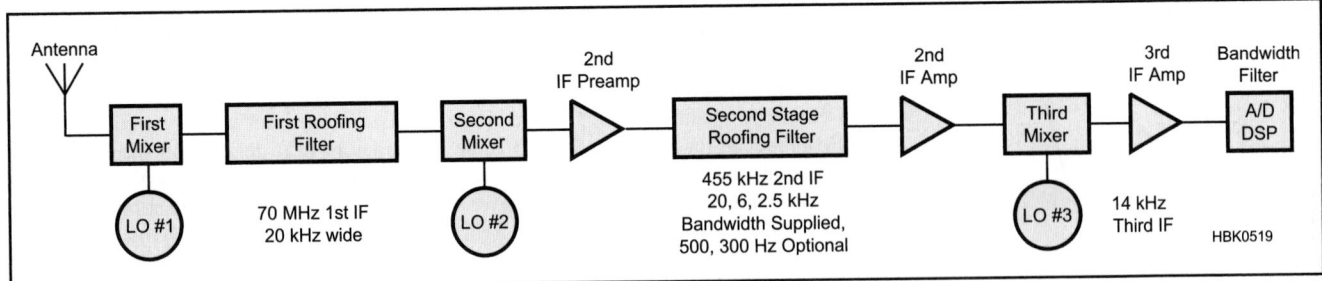

Fig 12.24 — Block diagram of a upconverting multiple conversion receiver with distributed roofing filter architecture.

12.5 Control and Processing Outside the Primary Signal Path

Discussion thus far has been about processes that occur within the primary path for signals between antenna input and transducer or system output. There are a number of control and processing subsystems that occur outside of that path that contribute to the features and performance of receiving systems.

12.5.1 Automatic Gain Control (AGC)

The amplitude of the desired signal at each point in the receiver is generally controlled by the AGC system, although manual control is usually provided as well. Each stage has a distortion versus signal level characteristic that must be known, and the stage input level must not become excessive. The signal being received has a certain signal-to-distortion ratio that must not be degraded too much by the receiver. For example, if an SSB signal has –30 dB distortion products the receiver should create additional distortion no greater than -40 dB with respect to the desired signal. The correct AGC design ensures that each stage gets the right input level. It is often necessary to redesign some stages in order to accomplish this task.

While this chapter deals exclusively with AGC in the guise of analog circuits, the same function is also implemented digitally in DSP and SDR receivers. The goal of both is the same — to maintain a signal level at all stages of the receiver that is neither too large nor too small so that the various processing systems operate properly. Whether or not the AGC offset and time constant are implemented by an analog component or by a microprocessor output is immaterial. The point is to manage the RF amplifier gain so that the overall receiver behavior is satisfactory.

The effects of an improperly operating AGC system can be quite subtle or nearly disabling to a receiver and vary with how

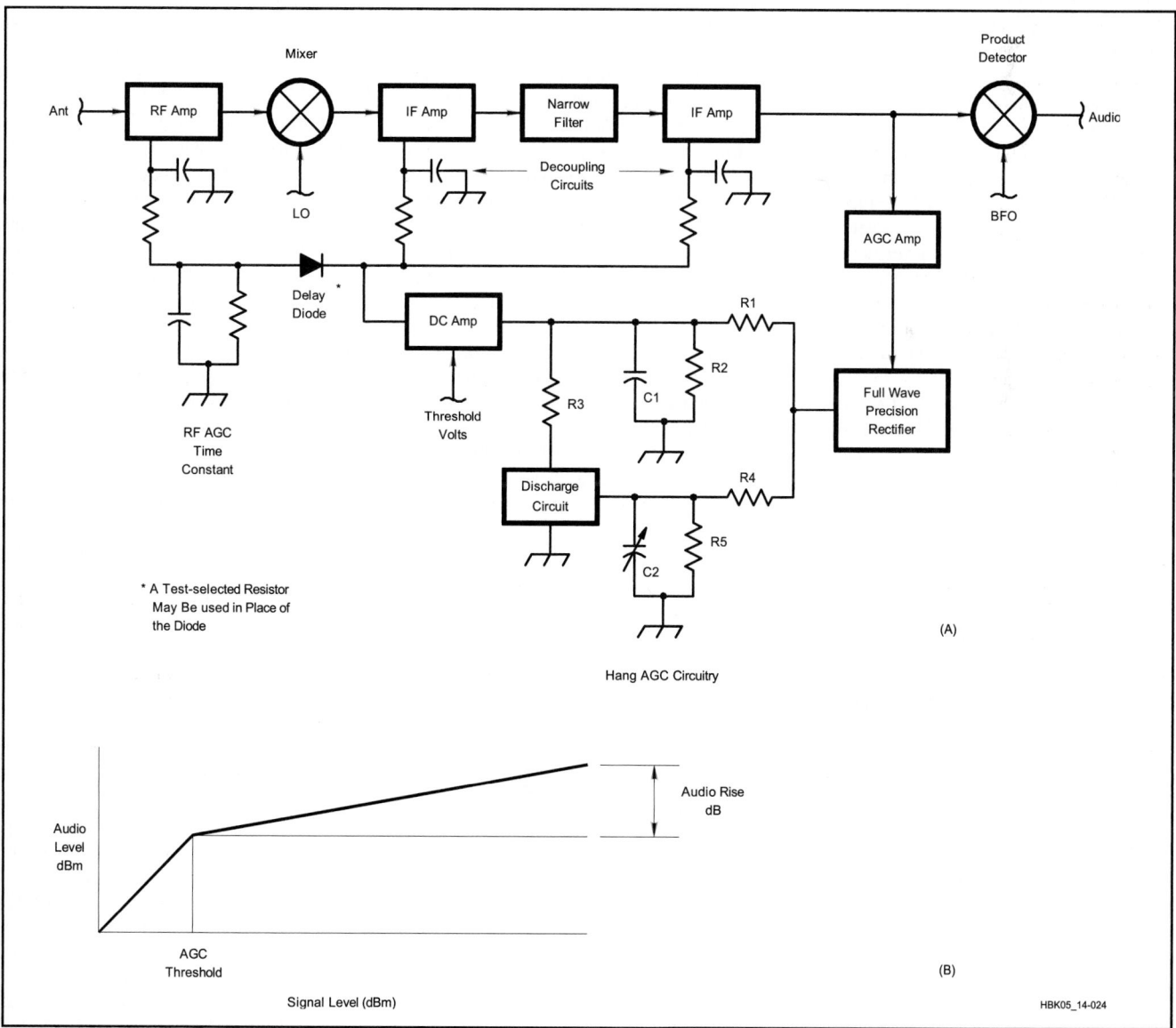

Fig 12.25 — AGC principles. At A: typical superhet receiver with AGC applied to multiple stages of RF and IF. At B: audio output as a function of antenna signal level.

the AGC system is constructed. This chapter attempts to describe the requirements for proper operation and provides some examples of implementation and common AGC failures in terms of analog circuitry which is somewhat easier to describe than software algorithms, noting that similar behaviors exist even in purely software receivers. The interested student should consider studying the AGC systems of commercial receivers to understand how professional design teams deal with the problem of managing so much gain with such stringent requirements for linearity and distortion.

THE AGC LOOP

Fig 12.25A shows a typical AGC loop that is often used in amateur superhet receivers. The AGC is applied to the stages through RF decoupling circuits that prevent the stages from interacting with each other. The AGC amplifier helps to provide enough AGC loop gain so that the gain-control characteristic of Fig 12.25B is achieved. If effect, the AGC system causes the receiver to act as a compression amplifier with lower overall gain for stronger signals.

The AGC action does not begin until a certain level, called the AGC *threshold*, is reached. The THRESHOLD VOLTS input in Fig 12.25A serves this purpose. After that level is exceeded, the audio level increases more slowly than for weaker signals. The audio rise beyond the threshold value is usually in the 5 to 10 dB range. Too much or too little audio rise are both undesirable for most operators.

As an option, the AGC signal to the RF amplifier is offset by the 0.6 V forward drop of the diode so that the RF gain does not start to decrease until larger signals appear. This prevents a premature increase of the receiver noise figure. Also, a time constant of one or two seconds after this diode helps keep the RF gain steady for the short term.

Fig 12.26 is a typical plot of the signal levels at the various stages of a certain ham band receiver using analog circuitry. Each stage has the proper level and a 115 dB change in input level produces a 10 dB change in audio level. A manual gain control could produce the same effect.

AGC TIME CONSTANTS

There are two primary AGC time constants. AGC *attack time* describes the time it takes the AGC system to respond to the presence of a signal. AGC *decay time* describes the response of the AGC system to changes in a signal that is present. The optimum time constants for the AGC system depends on the type of signal being received, the type of operation being conducted, and the operator's preference.

In Fig 12.25, following the precision rectifier, R1 and C1 set an attack time, to prevent excessively fast application of AGC. One or two milliseconds is a good value for the R1 × C1 product. If the antenna signal suddenly disappears, the AGC loop is opened because the precision rectifier stops conducting. C1 then discharges through R2 and the C1 × R2 product can be in the range of 100 to 200 ms. At some point the rectifier again becomes active, and the loop is closed again.

An optional modification of this behavior is the *hang AGC* circuit. If we make R2 × C1 much longer, say 3 seconds or more, the AGC voltage remains almost constant until the R5, C2 circuit decays with a switch selectable time constant of 100 to 1000 ms. At that time R3 quickly discharges C1 and full

receiver gain is quickly restored. This type of control is appreciated by many operators because of the lack of AGC *pumping* due to modulation, rapid fading and other sudden signal level changes.

AGC PUMPING

AGC pumping can occur in receivers in which the AGC measurement point is located ahead of the stages that determine operating bandwidth, such as when an audio filter is added to a receiver externally and outside the reach of the AGC system. If the weak signal is the only signal within the first IF passband, the AGC will cause the receiver to be at maximum gain and optimum SNR. If an interfering signal is within the first IF passband, but outside the audio DSP filter's passband, we won't hear the interfering signal, but it will enter the AGC system and reduce the gain so we might not hear our desired weak signal. AGC pumping is audible as sudden reductions in signal strength without a strong signal in the passband of the receiver

AGC LOOP RESPONSE PROBLEMS

If the various stages have the property that each 1 V change in AGC voltage changes the gain by a constant amount (in dB), the AGC loop is said to be *log-linear* and regular feedback principles can be used to analyze and design the loop. But there are some difficulties that complicate this textbook model. One has already been mentioned, that when the signal is rapidly decreasing the loop becomes open and the various capacitors discharge in an open loop manner. As the signal is increasing beyond the threshold, or if it is decreasing slowly enough, the feedback theory applies more accurately.

In SSB and CW receivers rapid changes are the rule and not the exception. It is important that the AGC loop not overshoot or ring when the signal level rises past the threshold. The idea is to design the ALC loop to be stable when the loop is closed. It obviously won't oscillate when open (during decay time). But the loop must have smooth and consistent transient response when the loop goes from open to closed state.

Another problem involves the narrow bandpass IF filter. The group delay of these filters constitutes a time lag in the loop that can make loop stabilization difficult. Moreover, these filters nearly always have much greater group delay at the edges of the passband, so that loop problems are aggravated at these frequencies. Overshoots and undershoots, called *gulping*, are very common. Compensation networks that advance the phase of the feedback help to offset these group delays. The design problem arises because some of the AGC is applied before the filter and some after the filter. It is a good idea to put as much fast AGC as possible after the filter and use a slower decaying

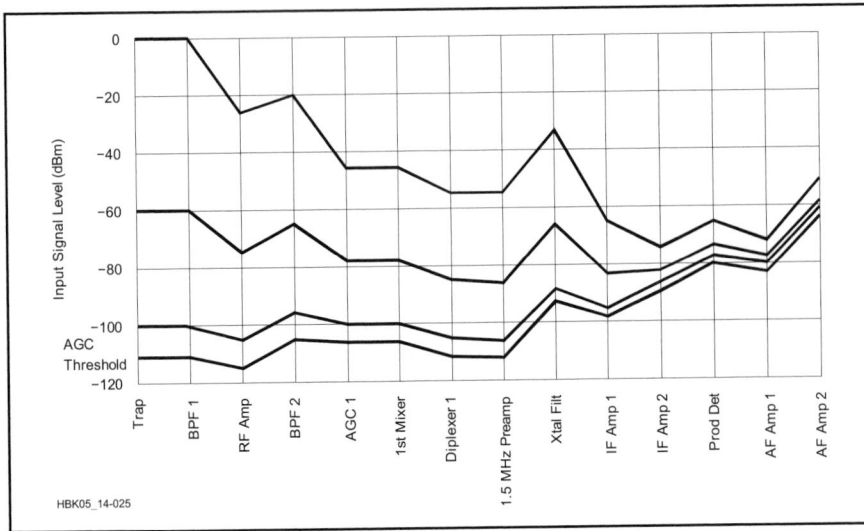

Fig 12.26 — Gain control of a ham-band receiver using AGC. A manual gain control could produce the same result.

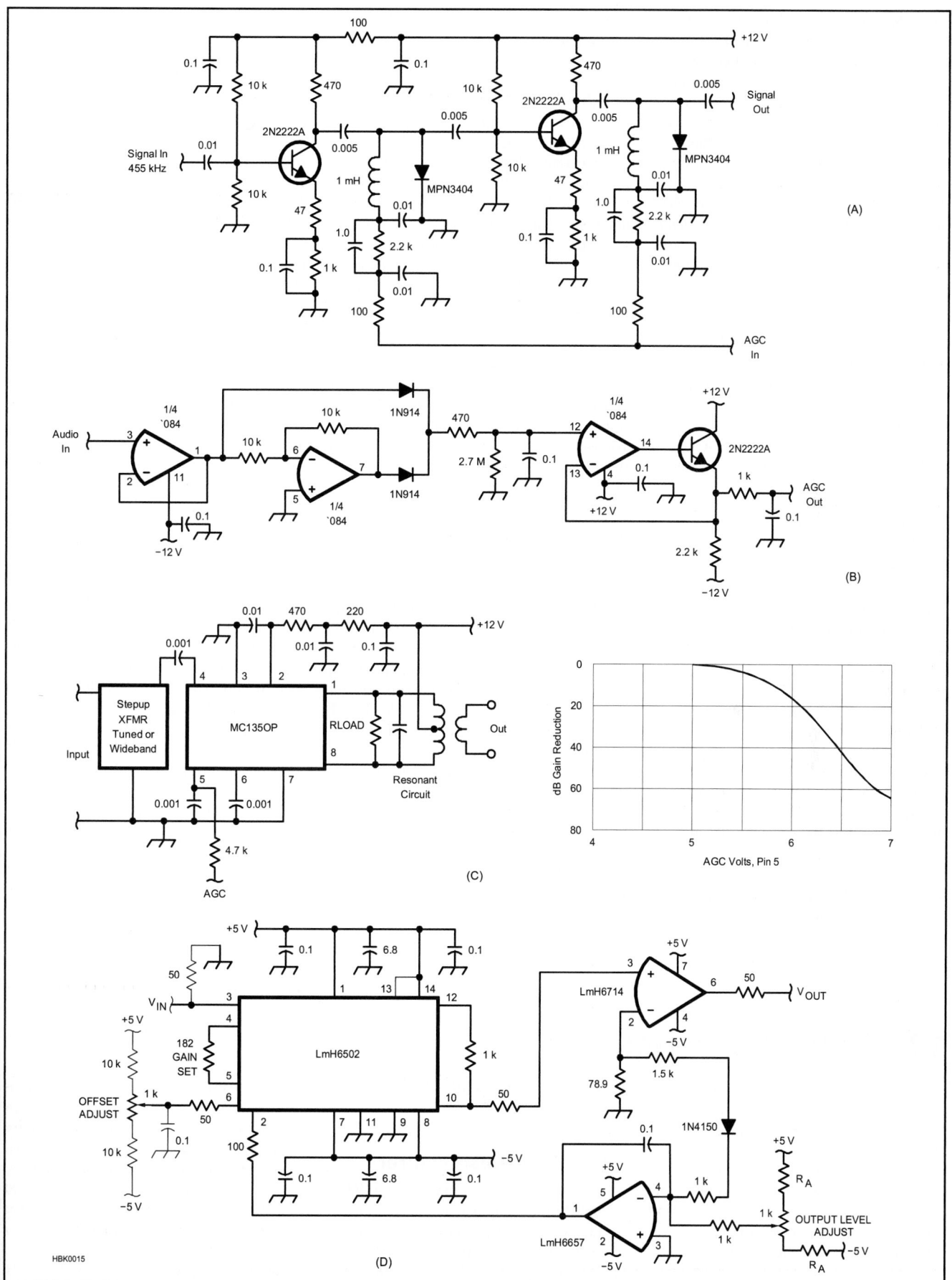

Fig 12.27 — Some gain controllable amplifiers and a rectifier suitable for audio derived AGC.

AGC ahead of the filter. The delay diode and RC in Fig 12.25A are helpful in that respect. Complex AGC designs using two or more compensated loops are also in the literature. If a second cascaded narrow filter is used in the IF it is usually a lot easier to leave the second or *downstream* filter out of the AGC loop at the risk of allowing AGC pumping as described in the preceding section.

Another problem is that the control characteristic is often not log-linear. For example, dual-gate MOSFETs tend to have much larger dB/V at large values of gain reduction. Many IC amplifiers have the same problem. The result is that large signals cause instability because of excessive loop gain. Variable gain op amps and other similar ICs are available that are intended for gain control loops.

Audio frequency components on the AGC bus can cause problems because the amplifier gains are modulated by the audio and distort the desired signal. A hang AGC circuit (essentially a low-pass filter) can reduce or eliminate this problem.

Finally, if we try to reduce the change in audio levels to a very low value, the required loop gain becomes very large, and stability problems become very difficult. It is much better to accept a 5 to 10 dB variation of audio output.

Because many parameters are involved and many of them are not strictly log-linear, it is best to achieve good AGC performance through an initial design effort and finalize the design experimentally. Use a signal generator, attenuator and a signal pulser (2 ms rise and fall times, adjustable pulse rate and duration) at the antenna and a synchronized oscilloscope to look at the IF envelope. Tweak the time constants and AGC distribution by means of resistor and capacitor decade substitution boxes. Be sure to test throughout the passband of each filter. The final result should be a smooth and pleasant sounding SSB/CW response, even with maximum RF

gain and strong signals. Patience and experience are helpful.

12.5.2 Audio-Derived AGC

Some receivers, especially direct-conversion types, use audio-derived AGC. There are problems with this approach as well. At low audio frequencies the AGC control action can be slow to develop. That is, low-frequency audio sine waves take longer to reach their peaks than the AGC time constants. During this time the RF/IF/AF stages can be over-driven. If the RF and IF gains are kept at a low level this problem can be reduced. Also, attenuating low audio frequencies prior to the first audio amplifier should help. With audio AGC, it is important to avoid so-called "charge pump" rectifiers or other slow-responding circuits that require multiple cycles of audio to pump up the AGC voltage. Instead, use a peak-detecting circuit that responds accurately on the first positive or negative half-cycle.

12.5.3 AGC Circuits

Fig 12.27 shows some gain-controllable circuits. Fig 12.27A shows a two-stage 455-kHz IF amplifier with PIN diode gain control. This circuit is a simplified adaptation from a production receiver, the Collins 651S. The IF amplifier section shown is preceded and followed by selectivity circuits and additional gain stages with AGC. The 1.0 μF capacitors aid in loop compensation. The favorable thing about this approach is that the transistors remain biased at their optimum operating point. Right at the point at which the diodes start to conduct, a small increase in IMD may be noticed, but that goes away as diode current increases slightly. Two or more diodes can be used in series, if this is a problem (it very seldom is). Another solution is to use a PIN diode that is more suitable for

such a low-frequency IF. Look for a device with τ > 10 / (2πf) where τ is the minority carrier lifetime in μs and f is the frequency in MHz.

Fig 12.27B is an audio derived AGC circuit using a full-wave rectifier that responds to positive or negative excursions of the audio signal. The RC circuit follows the audio closely.

Fig 12.27C shows a typical circuit for the MC1350P RF/IF amplifier. The graph of gain control versus AGC voltage shows the change in dB/V. If the control is limited to the first 20 dB of gain reduction this chip should be favorable for good AGC transient response and good IMD performance. Use multiple low-gain stages rather than a single high-gain stage for these reasons. The gain control within the MC1350P is accomplished by diverting signal current from the first amplifier stage into a *current sink*. This is also known as the *Gilbert cell multiplier* architecture. Another chip of this type is the NE/SA5209. This type of approach is simpler to implement than discrete circuit approaches, such as dual-gate MOSFETs that are now being replaced by IC designs.

Fig 12.27D shows the high performance National Semiconductor LMH6502MA (14-pin DIP plastic package) voltage controlled amplifier. It is specially designed for accurate log-linear AGC from 0 to 40 dB with respect to a preset maximum voltage gain from 6 to 40 dB. Its ±3 dB bandwidth is 130 MHz. It is an excellent IF amplifier for high performance receiver or transmitter projects.

Additional info on voltage-controlled amplifier ICs can be found on the Analog Devices web site (**www.analog.com**). Search the site for Tutorial MT-073, which describes the operation of various types of gain-controlled amplifiers with numerous product examples.

12.6 Pulse Noise Reduction

A major problem for those listening to receivers has historically been local impulse noise. For HF and VHF receivers it is often from the sparks of internal combustion engine spark plugs, electric fence chargers, light dimmers, faulty power-line insulators and many other similar devices that put out short duration wide band signals. In the UHF and microwave region, radar systems can cause similar problems. There have been three general methods of attempting to deal with such noise over the years, some more successful

than others. We will briefly describe the approaches.

12.6.1 The Noise Limiter

The first device used in an early (1930s) attempt to limit impulse noise was called a *noise limiter* or *clipper* circuit as originally described by H. Robinson, W3LW. (see references) This circuit would *clip* or limit noise (or signal) peaks that exceeded a preset limit. The idea was to have the limit set to about as

loud as you wanted to hear anything and nothing louder would get through. This was helpful in eliminating the loudest part of impulse noise or even nearby lightning crashes, but it had two problems. First it didn't eliminate the noise, it just reduced the peak loudness; second, it also reduced the loudness of loud non-noise signals and in the process distorted them considerably.

The second problem was fixed shortly thereafter, with the advent of the *automatic noise limiter* or ANL as described by

J. Dickert (see references). The ANL automatically set the clipping threshold to that of a loud signal. It thus would adjust itself as the loudness of signals you listened to changed with time. An ANL was fairly easy to implement and became standard equipment on amateur receivers from the late 1930s on. While ANL circuits are no longer common, simple receivers used today do sometimes incorporate passive clipping circuits to account for their limited AGC ability.

12.6.2 The Noise Blanker

It turned out that improvements in receiver selectivity over the 1950s and beyond, while improving the ability to reduce random noise, actually made receiver response to impulse noise worse. The reason for this is that a very short duration pulse will actually be lengthened while going through a narrow filter. This is due to the filter's different delay times for the pulse's wide spectrum of components, resulting in the components arriving at the filter output at different times. You can demonstrate this in your superhet receiver if it has a narrow crystal filter. Find a frequency with heavy impulse noise and switch between wide and narrow filters. If your narrow filter is 500 Hz or less, the noise pulses will likely be more prominent with the narrow filter. DSP filters with their superior group delay performance exhibit less smearing than their analog counterparts.

The noise limiters described previously were all connected at the output of the IF amplifiers and thus the noise had passed most of the selectivity before the limiter and had been widened by the receiver filters. In addition, modern receivers include *automatic gain control* (AGC), a system that reduces the receiver gain in the presence of strong signals to avoid overload of both receiver circuits and ears. In SSB receivers, since signals vary in strength as someone talks, the usual AGC responds quickly to reduce the gain of a strong signal and then slowly increases it if the signal is no longer there. This means that a strong noise pulse may reduce the receiver gain for much longer than it lasts.

The solution — a *noise blanker*. A noise blanker is almost a separate wideband receiver. It takes its input from an early stage in the receiver before much selectivity or AGC has been applied. It amplifies the wideband signal and detects the narrow noise pulses without lengthening them. The still-narrow noise pulses are used to shut off the receiver at a point ahead of the selectivity and AGC, thus keeping the noise from getting to the parts of the receiver at which the pulses would be extended. In other words, the receiver is shut

off or *gated* during the noise pulse.

A well-designed noise blanker can be very effective. Instead of just keeping the noise at the level of the signal as a noise limiter does, the noise blanker can actually *eliminate* the noise. If the pulses are narrow enough, the loss of desired signal during the time the receiver is disabled is not noticeable and the noise may seem to disappear entirely.

In addition to an ON/OFF switch, many noise blanker designs include a control labeled THRESHOLD. The THRESHOLD control adjusts the level of noise that will blank the receiver. If it is set for too low a level, it will blank on signal peaks as well as noise, resulting in distortion of the signal. The usual approach is to turn on the blanker, then adjust the THRESHOLD control until the noise is just blanked. Don't forget to turn it off when the noise goes away.

Noise blankers can also create problems. The wide-band receiver circuit that detects the noise pulses detects any signals in that bandwidth. If such a signal is strong and has sharp peaks (as voice and CW signals do), the noise blanker will treat them as noise pulses and shut down the receiver accordingly. This causes tremendous distortion and can make it sound as if the strong signal to which the noise blanker is responding is generating spurious signals that cause the distortion. Before you assume that the strong signal is causing problems, turn the noise blanker on and off to check. When the band is full of strong signals, a noise blanker may cause more problems than it solves.

12.6.3 Operating Noise Limiters and Blankers

Many current receivers include both a noise limiter and a noise blanker. If your receiver has both, they will have separate controls and it is worthwhile to try them both. There are times at which one will work better than the other, and other times when it goes the other way, depending on the characteristics of the noise. There are other times when both work better than either. In any case, they can make listening a lot more pleasant — just remember to turn them off when you don't need them since either type can cause some distortion, especially on strong signals that should otherwise be easy to listen to.

Recognizing that it is difficult for a single noise blanker to work properly with the wide variations of noise pulses, it is common for late-model receivers to have two noise blankers with different characteristics that are optimized for the different pulse types. One noise blanker is typically optimized for very short pulses and the other for longer

pulses. The operator can switch between the blankers to see which works best on the noise at hand.

OTHER TECHNIQUES

The previous techniques represent the most commonly available techniques to reduce impulse noise. There are a few other solutions as well. Note that we haven't been talking about reducing interference here. By interference, we mean another intended signal encroaching on the channel to which we want to listen. There are a number of techniques to reduce interference, and some also can help with impulse noise.

Many times impulse noise is coming from a particular direction. If so, by using a directional antenna, we can adjust the direction for minimum noise. When we think about directional antennas, the giant HF Yagi springs to mind. For receiving purposes, especially on the lower bands such as 160, 80 and 40 meters (where the impulse noise often seems the worst), a small indoor or outdoor receiving loop antenna as described in the *ARRL Antenna Book* can be very effective at eliminating either interfering stations or noise (both if they happen to be in the same direction).

Another technique that can be used to eliminate either interference or noise is to obtain a copy of the noise (or interference) that is 180° out of phase from the one you are receiving. By adjusting the amplitude to match the incoming signal, the signal can be cancelled at the input to the receiver. Several available commercial units perform this task.

Digital signal processing, described in the next section, is another multifunction system that can help with all kinds of noise.

12.6.4 DSP Noise Reduction

DSP noise reduction can actually look at the statistics of the signal and noise and figure out which is which and then reduce the noise significantly. These *adaptive filters* can't quite eliminate the noise, and need enough of the desired signal to figure out what's happening, so they won't work if the signal is far below the noise. Many DSP systems "color" the resulting audio to a degree. Nonetheless, they do improve the SNR of a signal in random or impulse noise. As with noise blankers, receivers frequently offer two or more noise reduction settings that apply different noise reduction algorithms optimized for different conditions. It's always worth experimenting with the radio's features to find out which work better. The **DSP and Software Radio Design** chapter discusses these features in more detail.

12.7 VHF and UHF Receivers

Most of the basic ideas presented in previous sections apply equally well in receivers that are intended for the VHF and UHF bands. This section will focus on the difference between VHF/UHF and HF receivers.

12.7.1 FM Receivers

Narrow-band frequency modulation (NBFM) is the most common mode used on VHF and UHF. **Fig 12.28A** is a block diagram of an FM receiver for the VHF/UHF amateur bands.

FRONT END

A low-noise front end is desirable because of the decreasing atmospheric noise level at these frequencies and also because portable gear often uses short rod antennas at ground level. Nonetheless, the possibilities for gain compression and harmonic IMD, multi-tone IMD and cross modulation are also substantial. Therefore dynamic range is an important design consideration, especially if large, high-gain antennas are used. FM limiting should not occur until after the crystal filter. Because of the high occupancy of the VHF/UHF spectrum by powerful broadcast transmitters and nearby two-way radio services, front-end preselection is desirable, so that a

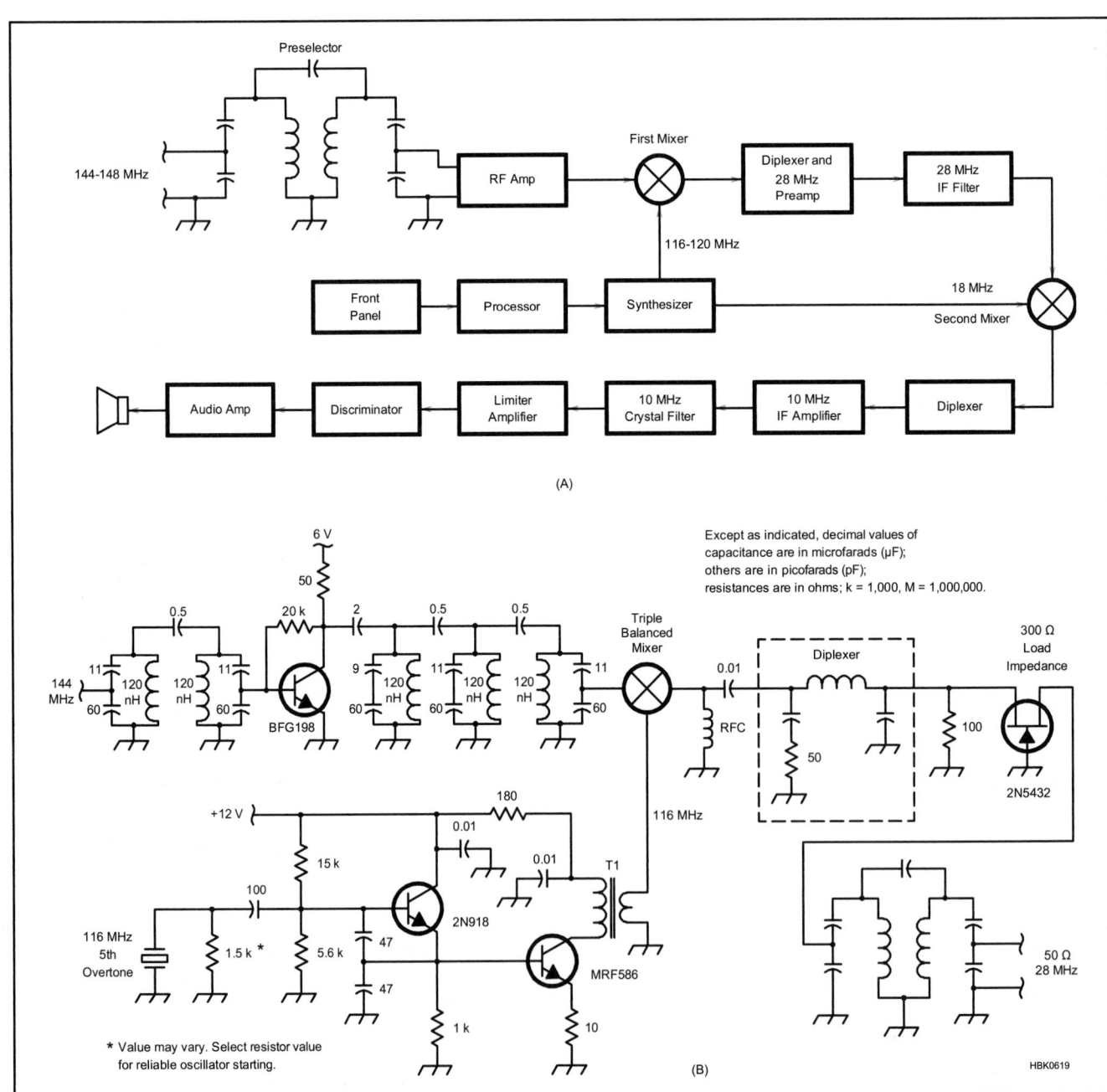

Fig 12.28 — At A, block diagram of a typical VHF FM receiver. At B, a 2 meter to 10 meter receive converter (partial schematic; some power supply connections omitted.)

low noise figure can be achieved economically within the amateur band.

DOWN-CONVERSION

Down-conversion to the final IF can occur in one or two stages. Favorite IFs are in the 5 to 10 MHz region, but at the higher frequencies rejection of the image 10 to 20 MHz away can be difficult, requiring considerable preselection. At the higher frequencies an intermediate IF in the 30 to 50 MHz region is a better choice. Fig 12.28A shows dual down-conversion.

IF FILTERS

The customary peak frequency deviation in amateur FM on frequencies above 29 MHz is about 5 kHz and the audio speech band extends to 3 kHz. This defines a maximum modulation index (defined as the deviation ratio) of 5/3 = 1.67. An inspection of the Bessel functions that describe the resulting FM signal shows that this condition confines most of the 300 to 3000 Hz speech information sidebands within a 15 kHz or so bandwidth. Using filters of this bandwidth, channel separations of 20 or 25 kHz are achievable.

Many amateur FM transceivers are channelized in steps that can vary from 1 to 25 kHz. For low distortion of the audio output (after FM detection), this filter should have good phase linearity across the bandwidth. This would seem to preclude filters with very steep descent outside the passband, which tend to have very nonlinear phase near the band edges. But since the amount of energy in the higher speech frequencies is naturally less, the actual distortion due to this effect may be acceptable for speech purposes. The normal practice is to apply pre-emphasis to the higher speech frequencies at the transmitter and de-emphasis compensates at the receiver.

LIMITING

After the filter, hard limiting of the IF is needed to remove any amplitude modulation components. In a high-quality receiver, special attention is given to any nonlinear phase shift that might result from the limiter circuit design. This is especially important in data receivers in which phase response must be controlled. In amateur receivers for speech it may be less important. Also, the *ratio detector* (see the **Mixers, Modulators and Demodulators** chapter) largely eliminates the need for a limiter stage, although the limiter approach is probably still preferred.

FM DETECTION

The discussion of this subject is deferred to the **Mixers, Modulators and Demodulators** chapter. Quadrature detection is used in some popular FM multistage ICs. An example receiver IC will be presented later.

12.7.2 FM Receiver Weak-Signal Performance

The noise bandwidth of the IF filter is not much greater than twice the audio bandwidth of the speech modulation, less than it would be in wideband FM. Therefore such things as capture effect, the threshold effect and the noise quieting effect so familiar to wideband FM are still operational, but somewhat less so, in FM. For FM receivers, sensitivity is specified in terms of a SINAD (see the **Test Equipment and Measurements** chapter) ratio of 12 dB. Typical values are −110 to −125 dBm, depending on the low-noise RF pre-amplification that often can be selected or deselected (in strong signal environments).

LO PHASE NOISE

In an FM receiver, LO phase noise superimposes phase modulation, and therefore frequency modulation, onto the desired signal. This reduces the ultimate signal-to-noise ratio within the passband. This effect is called "incidental FM (IFM)." The power density of IFM (W/Hz) is proportional to the phase noise power density (W/Hz) multiplied by the square of the modulating frequency (the familiar parabolic effect in FM). If the receiver uses high-frequency de-emphasis at the audio output (−6 dB per octave from 300 to 3000 Hz, a common practice), the IFM level at higher audio frequencies can be reduced. Ordinarily, as the signal increases the

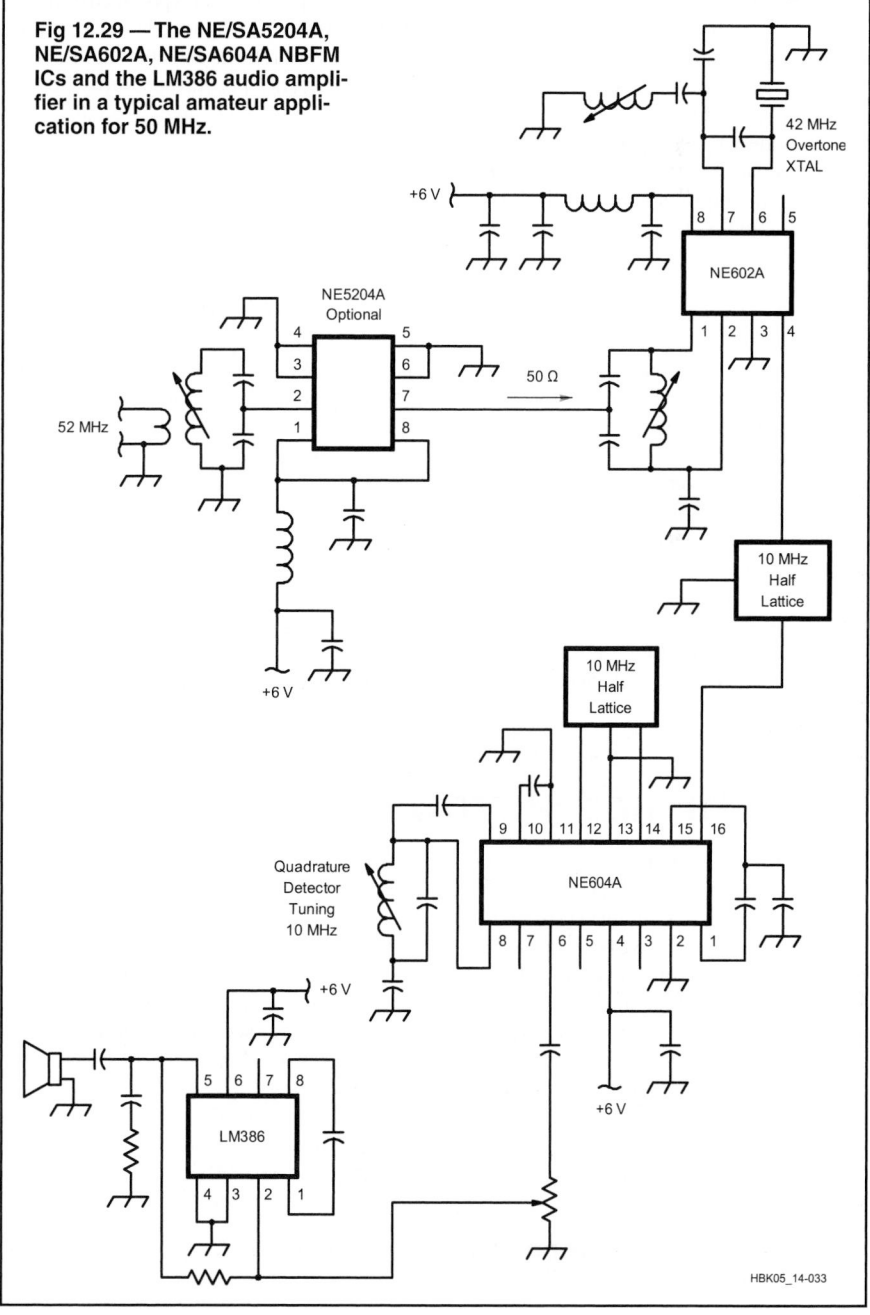

Fig 12.29 — The NE/SA5204A, NE/SA602A, NE/SA604A NBFM ICs and the LM386 audio amplifier in a typical amateur application for 50 MHz.

noise would be "quieted" (that is, "captured") in an FM receiver, but in this case the signal and the phase noise riding "piggy back" on the signal increase in the same proportion as described in the **Oscillators and Synthesizers** chapter's discuss of reciprocal mixing. IFM is not a significant problem in modern FM radios, but phase noise can become a concern for adjacent-channel interference.

As the signal becomes large the signal-to-noise ratio therefore approaches some final value. A similar ultimate SNR effect occurs in SSB receivers. On the other hand, a perfect AM receiver tends to suppress LO phase noise. (See the reference entry for Sabin.)

12.7.3 FM Receiver ICs

A wide variety of special ICs for communications-bandwidth FM receivers are available. Many of these were designed for "cordless" or mobile telephone applications and are widely used. **Fig 12.29** shows some popular versions for a 50 MHz FM receiver. One is an RF amplifier chip (NE/SA5204A) for 50 Ω input to 50 Ω output with 20 dB of gain. The second chip (NE/SA602A) is a front-end device with an RF amplifier, mixer and LO. The third is an IF amplifier, limiter and quadrature FM detector (NE/SA604A) that also has a very useful RSSI (logarithmic Received Signal Strength Indicator) output and also a "mute" function. The fourth is the LM386, a widely used audio-amplifier chip. Another FM receiver chip, complete in one package, is the MC3371P.

The NE/SA5204A plus the two tuned circuits help to improve image rejection. An alternative would be a single double-tuned filter with some loss of noise figure. The Mini-Circuits MAR/ERA series of MMIC amplifiers are excellent devices also. The

crystal filters restrict the noise bandwidth as well as the signal bandwidth. A cascade of two low-cost filters is suggested by the vendors. Half-lattice filters at 10 MHz are shown, but a wide variety of alternatives, such as ladder networks, are possible.

Another recent IC is the MC13135, which features double conversion and two IF amplifier frequencies. This allows more gain on a single chip with less of the cross coupling that can degrade stability. This desirable feature of multiple down-conversion was mentioned previously in this chapter.

The diagram in Fig 12.29 is (intentionally) only a general outline that shows how chips can be combined to build complete equipment. The design details and specific parts values can be learned from a careful study of the data sheets and application notes provided by the IC vendors. Amateur designers should learn how to use these data sheets and other information such as application notes available (usually for free) from the manufacturers or on the web.

12.7.4 VHF Receive Converters

Rather than building an entire transceiver for VHF SSB and CW, one approach is to use a receive converter. A receive converter (also called a *downconverter*) takes VHF signals and converts them to an HF band for reception using existing receiver or transceiver as a tunable IF.

Although many commercial transceivers cover the VHF bands (either multiband, multimode VHF/UHF transceivers, or HF+VHF transceivers), receive converters are sometimes preferred for demanding applications because they may be used with high-performance HF transceivers. Receive converters are often packaged with a companion transmit converter and control cir-

cuitry to make a *transverter*.

A typical 2 meter downconverter uses an IF of 28-30 MHz. Signals on 2 meters are amplified by a low-noise front-end before mixing with a 116 MHz LO. Fig 12.28B shows the schematic for a high-performance converter. The front-end design was contributed by Ulrich Rohde, N1UL, who recommends a triple-balanced mixer such as the Synergy CVP206 or SLD-K5M.

The diplexer filter at the mixer output selects the difference product: (144 to 146 MHz) – 116 = (28 to 30 MHz). A common-base buffer amplifier (the 2N5432 FET) and tuned filter form the input to the 10 meter receiver. (N1UL suggests that using an IF of 21 MHz and an LO at 165 MHz would avoid interference problems with 222 MHz band signals.) For additional oscillator designs, refer to the papers on oscillators by N1UL in the supplemental CD-ROM files for the **Mixers, Modulators and Demodulators** chapter accompanying this *Handbook*.

Based on the Philips BFG198 8 GHz transistor, the 20 dB gain front-end amplifier is optimized for noise figure (NF is approximately 2.6 dB), not for input impedance. The output circuit is optimized for best selectivity. The transistor bias is designed for dc stability at $I_C = 30$ mA and $V_C = 6$ V. Both of the transistor's emitter terminals should be grounded to prevent oscillation. NF might be improved with a higher performance transistor, such as a GaAs FET, but stability problems are often encountered with FET designs in this application.

If a mast-mounted preamplifier is used to improve the system noise figure, an attenuator should be available to prevent overload. Simulation predicts the circuit to have an IP3 figure of at least +25 dBm at 145 MHz with an I_C of 30 mA and a terminating impedance of 50 Ω.

12.8 UHF and Microwave Techniques

The ultra high frequency spectrum comprises the range from 300 MHz to 3 GHz. All of the basic principles of radio system design and circuit design that have been discussed so far apply as well in this range, but the higher frequencies require some special thinking about the methods of circuit design and the devices that are used. Additional material on construction for microwave circuits can be found in the **Construction Techniques** chapter and in the series of *QST* columns, "Microwavelengths" by Paul Wade, W1GHZ.

12.8.1 UHF Construction

Modern receiver designs make use of

highly miniaturized monolithic microwave ICs (MMICs). Among these are the Avago MODAMP and the Mini Circuits MAR and MAV/ERA lines. They come in a wide variety of gains, intercepts and noise figures for frequency ranges from dc to well into the GHz range. (See the **Component Data and References** chapter for information on available parts.)

Fig 12.30 shows the schematic diagram and the physical construction of a typical RF circuit at 430 MHz. It is a GaAsFET preamplifier intended for low noise SSB/CW, moonbounce or satellite reception. The construction uses ceramic chip capacitors, small helical inductors and a stripline surface-

mount GaAsFET, all mounted on a G10 (two layers of copper) glass-epoxy PC board. The very short length of interconnection leads is typical. The bottom of the PC board is a ground plane. At this frequency, lumped components are still feasible, while microstrip circuitry tends to be rather large.

At higher frequencies, microstrip methods become more desirable in most cases because of their smaller dimensions. However, the advent of tiny chip capacitors and chip resistors has extended the frequency range of discrete components. For example, the literature shows methods of building LC filters at as high as 2 GHz or more, using chip capacitors and tiny helical inductors. Amplifier and

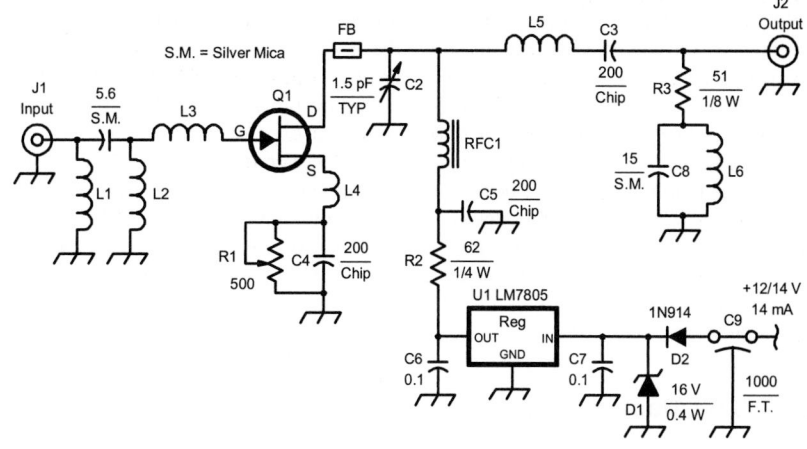

Fig 12.30 — GaAsFET preamplifier schematic and construction details for 430 MHz. Illustrates circuit, parts layout and construction techniques suitable for 430-MHz frequency range.

C1 — 5.6 pF silver-mica or same as C2.

C2 — 0.6 to 6 pF ceramic piston trimmer (Johanson 5700 series or equiv).

C3, C4, C5 — 200 pF ceramic chip.

C6, C7 — 0.1 μF disc ceramic, 50 V or greater.

C8 — 15 pF silver-mica.

C9 — 500 to 1000 pF feedthrough.

D1 — 16 to 30 V, 500 mW Zener (1N966B or equiv).

D2 — 1N914, 1N4148 or any diode with ratings of at least 25 PIV at 50 mA or greater.

J1, J2 — Female chassis-mount Type-N connectors, PTFE dielectric (UG-58 or equiv).

L1, L2 — 3t, #24 tinned wire, 0.110-inch ID spaced 1 wire dia.

L3 — 5t, #24 tinned wire, ³/₁₆-inch ID, spaced 1 wire dia. or closer. Slightly larger diameter (0.010 inch) may be required with some FETs.

L4, L6 — 1t #24 tinned wire, ⅛-inch ID.

L5 — 4t #24 tinned wire, ⅛-inch ID, spaced 1 wire dia.

Q1 — Mitsubishi MGF1402.

R1 — 200 or 500-Ω Cermet potentiometer (initially set to midrange).

R2 — 62 Ω, ¼ W.

R3 — 51 Ω, ⅛ W carbon composition resistor, 5% tolerance.

RFC1 — 5t #26 enameled wire on a ferrite bead.

U1 — 5 V, 100-mA 3 terminal regulator (LM78L05 or equiv. TO-92 package).

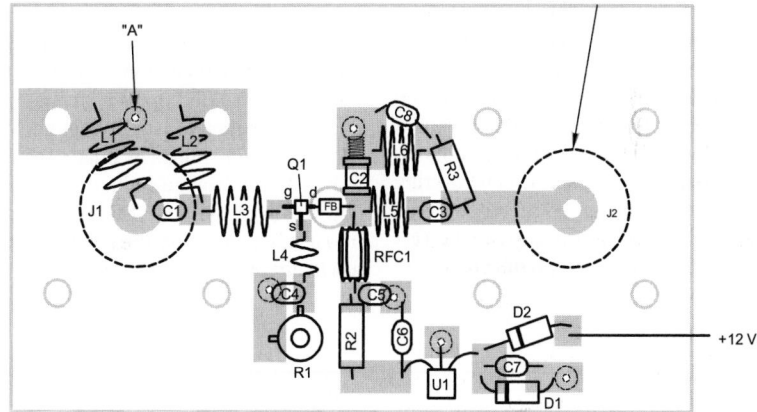

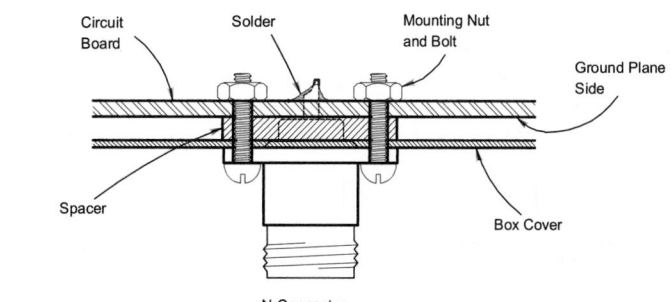

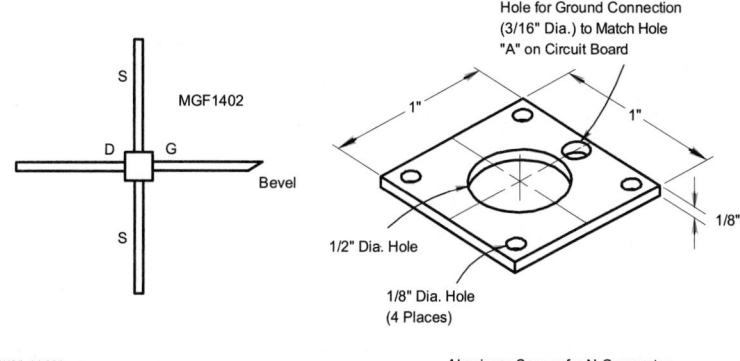

mixer circuits operate at well into the GHz range using these types of components on controlled-dielectric PC board material such as Duroid or on ceramic substrates.

Current designs emphasize simplicity of construction and adjustment, leading to "no tune" designs. The use of printed-circuit microstrip filters that require little or no adjustment, along with IC or MMIC devices, or discrete transistors, in precise PC-board layouts that have been carefully worked out, make it much easier to "get going" on the higher frequencies.

12.8.2 UHF Design Aids

Circuit design and evaluation at the higher frequencies usually require some kind of minimal lab facilities, such as a signal generator, a calibrated noise generator and, hopefully, some kind of simple (or surplus) spectrum analyzer. This is true because circuit behavior and stability depend on a number of factors that are difficult to "guess at," and intuition is often unreliable. The ideal instrument is a vector network analyzer with all of the attachments (such as an S parameter measuring setup), an instrument that has become surprisingly affordable in recent years. (See the **Test Equipment and Measurements** chapter.)

Another very desirable thing would be a circuit design and analysis program for the personal computer. Software packages created especially for UHF and microwave circuit design are available. They tend to be somewhat expensive, but worthwhile for a serious designer. Inexpensive *SPICE* programs are a good compromise but have significant limitations at VHF and above. See the chapter on **Computer-Aided Circuit Design** for information on these tools.

12.8.3 A 902 to 928 MHz (33 cm) Receiver

This 902 MHz downconverter is a fairly typical example of receiver design methods for the 500 to 3000 MHz range, in which down-conversion to an existing HF receiver (or 2 meter multimode receiver) is the most convenient and cost-effective approach for most amateurs. At higher frequencies a double down-conversion with a first IF of 200 MHz or so, to improve image rejection, might be necessary. Usually, though, the presence of strong signals at image frequencies is less likely. Image-reducing mixers plus down-conversion to 28 MHz is also coming into use, when strong interfering signals are not likely at the image frequency.

Fig 12.31A is the block diagram of the 902 MHz down-converting receiver. A cavity resonator at the antenna input provides high selectivity with low loss. The first RF amplifier is a GaAsFET. Two additional 902 MHz band-pass microstrip filters and a second RF amplifier transistor provide more gain and image rejection (at RF – 56 MHz) for the mixer. The output is at 28.0 MHz so that an HF receiver can be used as a tunable IF/demodulator stage.

CUMULATIVE NOISE FIGURE

Fig 12.31B shows the cumulative noise figure (NF) of the signal path, including the 28 MHz receiver. The 1.5 dB cumulative NF of the input cavity and first RF-amplifier combination, considered by itself, is degraded to 1.9 dB by the rest of the system following the first RF amplifier. The NF values of the various components for this example are reasonable, but may vary somewhat for actual hardware. Also, losses prior to the input such as transmission line losses (very important) are not included. They would be part of the complete receive system analysis, however. It is common practice to place a low noise preamp outdoors, right at the antenna, to overcome coax loss (and to permit use of less expensive coax).

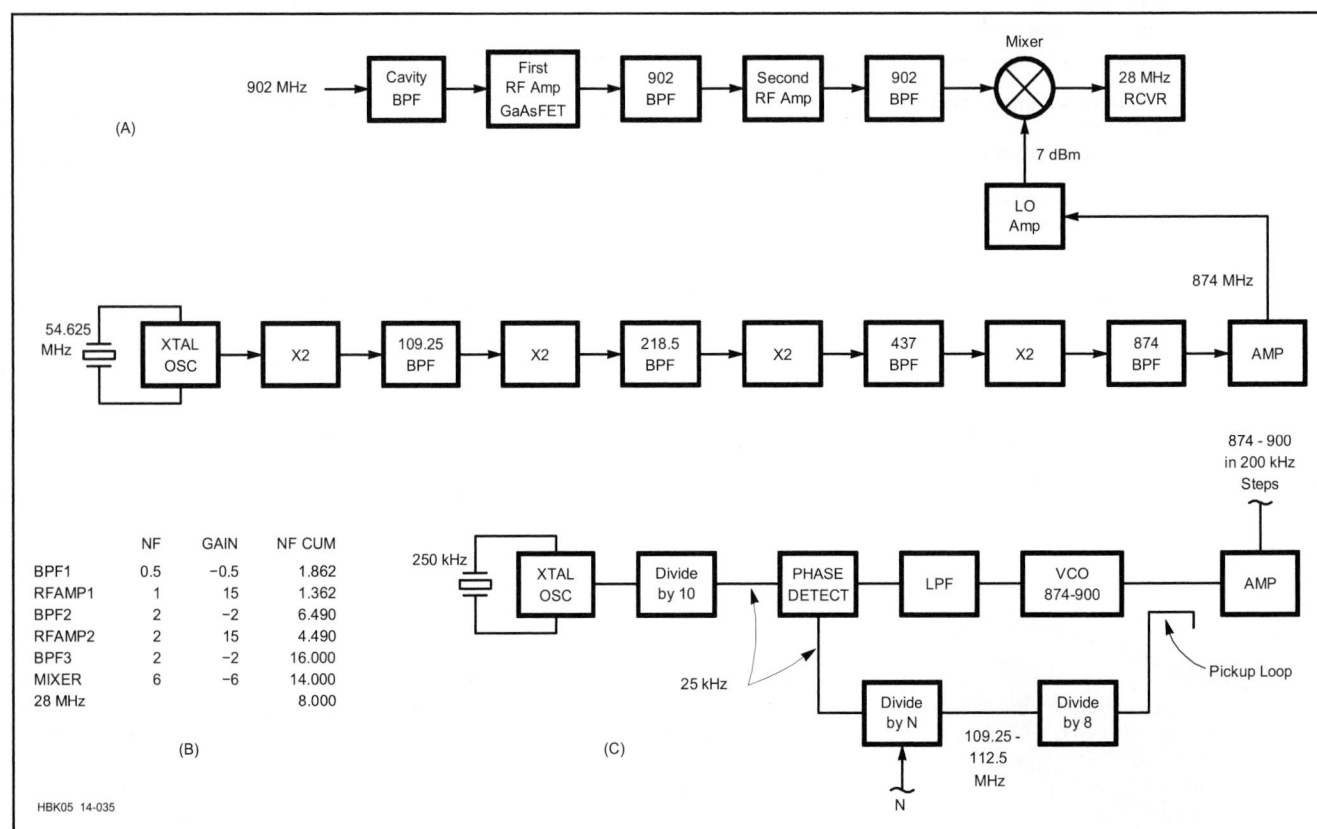

Fig 12.31 — A downconverter for the 902 to 928 MHz band. At A: block diagram; At B: cumulative noise figure of the signal path; At C: alternative LO multiplier using a phase locked loop.

LOCAL OSCILLATOR (LO) DESIGN

The +7-dBm LO at 874 to 900 MHz is derived from a set of crystal oscillators and frequency multipliers, separated by band-pass filters. These filters prevent a wide assortment of spurious frequencies from appearing at the mixer LO port. They also enhance the ability of the doubler stage to generate the second harmonic. That is, they have very low impedance at the input frequency, thereby causing a large current to flow at the fundamental frequency. This increases the nonlinearity of the circuit, which increases the second-harmonic component. The higher filter impedance at the second harmonic produces a large harmonic output.

For very narrow-bandwidth use, such as EME, the crystal oscillators are often oven controlled or otherwise temperature compensated. The entire LO chain must be of low-noise design and the mixer should have good isolation from LO port to RF port (to minimize noise transfer from LO to RF).

A phase-locked loop using GHz range prescalers (as shown in Fig 12.31C) is an alternative to the multiplier chain. The divide-by-N block is a simplification; in practice, an auxiliary dual-modulus divider (see the **Oscillators and Synthesizers** chapter) would be involved in this segment. The cascaded 902 MHz band-pass filters in the signal path should attenuate any image frequency noise (at RF–56 MHz) that might degrade the mixer noise figure.

12.8.4 Microwave Receivers

The world above 3 GHz is a vast territory with a special and complex technology well beyond the scope of this chapter. We will scratch the surface by describing a specific receiver for the 10 GHz frequency range and point out some of the important special features that are unique to this frequency range.

A 10 GHZ PREAMPLIFIER

Fig 12.32B is a schematic and parts list, Fig 12.32C is a PC board parts layout and Fig 12.32A is a photograph of a 10 GHz preamp, designed by Senior ARRL Lab Engineer Zack Lau, W1VT. With very careful design and packaging techniques a noise figure approaching the 1 to 1.5 dB range was achieved. This depends on an accurate 50-Ω generator impedance and noise matching the input using a microwave circuit-design program such as *Touchstone* or *Harmonica*. Note that microstrip capacitors, inductors and transmission-line segments are used almost exclusively. The circuit is built on a 15-mil Duroid PC board. In general, this kind of performance requires some elegant measurement equipment that few amateurs have. On the other hand, preamp noise figures in the 2 to 4-dB range are much easier to get (with simple test equipment) and are often satisfactory for amateur terrestrial communication.

Articles written by those with expertise and the necessary lab facilities almost always include PC board patterns, parts lists and detailed instructions that are easily duplicated by readers. Microwave ham clubs and their publications are a good way to get started in microwave amateur technology.

Because of the frequencies involved, dimensions of microstrip circuitry must be very accurate. Dimensional stability and dielectric constant reliability of the boards must be very good.

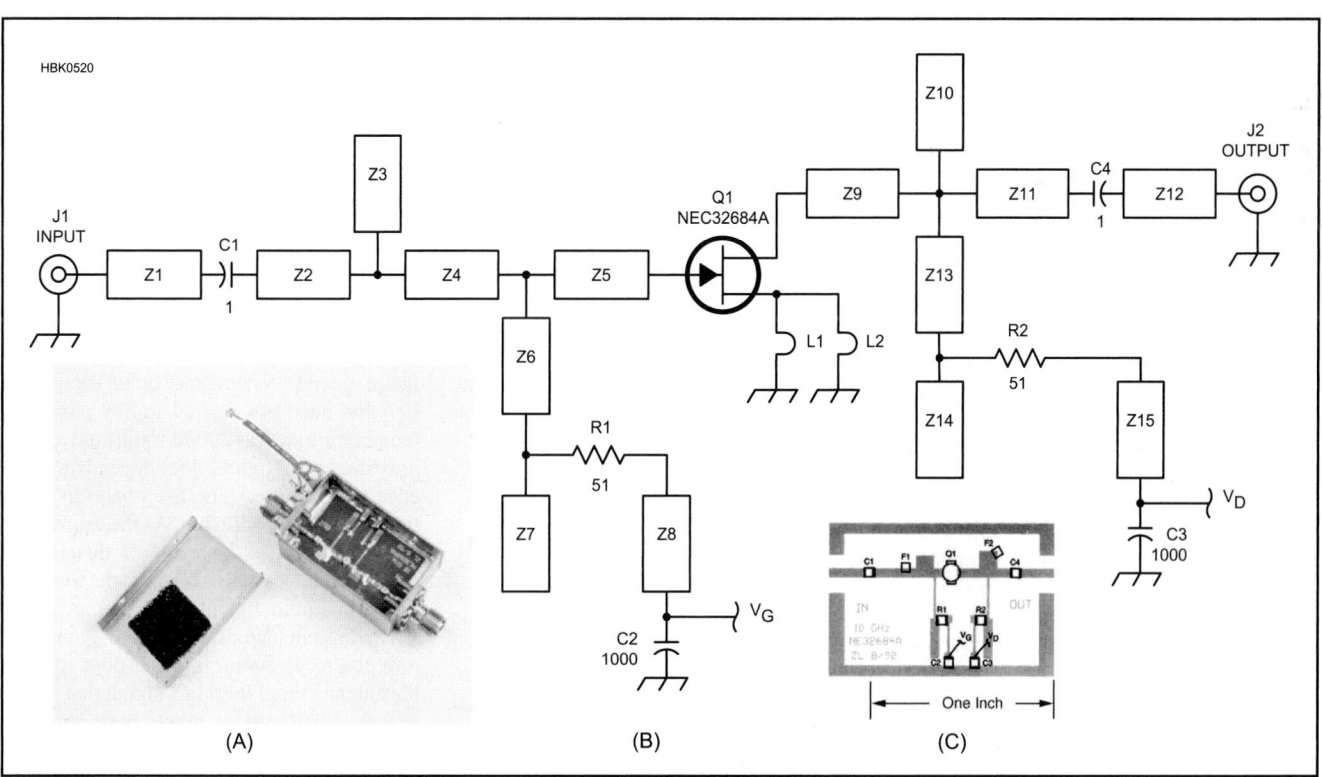

Fig 12.32 — At A, a low-noise preamplifier for 10 GHz, illustrating the methods used at microwaves. At B: schematic. At C: PC board layout. Use 15-mil 5880 Duroid, dielectric constant of 2.2 and a dissipation factor of 0.0011. A template of the PC board is available on the CD-ROM included with this book.

C1, C4 — 1 pF ATC 100 A chip capacitors. C1 must be very low loss.
C2, C3 — 1000 pF chip capacitors. (Not critical.) The ones from Mini Circuits work fine.

F1, F2 — Pieces of copper foil used to tune the preamp.
J1, J2 — SMA jacks. Ideally these should be microstrip launchers. The pin should be flush against the board.

L1, L2 — The 15 mil lead length going through the board to the ground plane.
R1, R2 — 51 Ω chip resistors.
Z1-Z15 — Microstrip lines etched on the PC board.

Analysis of a 10.368 GHz communication link with SSB modulation:

Free space path loss (FSPL) over a 50-mile line-of-sight path (S) at F = 10.368 GHz:
FSPL = 36.6 (dB) + 20 log F (MHz) + 20 log S (Mi) = 36.6 + 80.3 + 34 = 150.9 dB.

Effective isotropic radiated power (EIRP) from transmitter:
EIRP (dBm) = P_{XMIT} (dBm) + Antenna Gain (dBi)

The antenna is a 2-ft diameter (D) dish whose gain GA (dBi) is:
GA = 7.0 + 20 log D (ft) + 20 log F (GHz) = 7.0 + 6.0 + 20.32 = 33.3 dBi

Assume a transmission-line loss L_T, of 3 dB
The transmitter power P_T = 0.5 (mW PEP) = –3 (dBm PEP)
P_{XMIT} = P_T (dBm PEP) – L_T (dB) = (–3) – (3) = –6 (dBm PEP)
EIRP = P_{XMIT} + G_A = –6 + 33.3 = 27.3 (dBm PEP)
Using these numbers the received signal level is:
P_{RCVD} = EIRP (dBm) – Path loss (dB) = 27.3 (dBm PEP) – 150.9 (dB) = –123.6 (dBm PEP)
Add to this a receive antenna gain of 17 dB. The received signal is then P_{RCVD} = –123.6 +17 = –106.6 dBm

Now find the receiver's ability to receive the signal:
The antenna noise temperature T_A is 200 K. The receiver noise figure NF_R is 6 dB (FR=3.98, noise temperature T_R = 864.5 K) and its noise bandwidth (B) is 2400 Hz. The feedline loss L_L is 3 dB (F = 2.00, noise temperature T_L = 288.6 K). The system noise temperature is:
T_S =T_A + T_L + (L_L) (T_R)
T_S = 200 + 288.6 + (2.0) (864.5) = 2217.6 K
N_S = kT_SB = 1.38 × 10^{-23} × 2217.6 × 2400 = 7.34 × 10^{-17} W = –131.3 dBm
This indicates that the PEP signal is –106.6 –(–131.3) = 24.7 dB above the noise level. However, because the average power of speech, using a speech processor, is about 8 dB less than PEP, the average signal power is about 16.7 dB above the noise level.

To find the system noise factor F_S we note that the system noise is proportional to the system temperature T_S and the "generator" (antenna) noise is proportional to the antenna temperature T_A. Using the idea of a "system noise factor":

F_S = T_S / T_A = 2217.6 / 200 = 11.09 = 10.45 dB.

If the antenna temperature were 290 K the system noise figure would be 9.0 dB, which is precisely the sum of receiver and receiver coax noise figures (6.0 + 3.0).

Fig 12.33 — Example of a 10-GHz system performance calculation. Noise temperature and noise factor of the receiver are considered in detail.

System Performance

At microwaves, an estimation of system performance can often be performed using known data about the signal path terrain, atmosphere, transmitter and receivers systems. In the present context of receiver design we wish to establish an approximate goal for the receiver system, including the antenna and transmission line. **Fig 12.33** shows a simplified example of how this works.

A more detailed analysis includes terrain variations, refraction effects, the Earth's curvature, diffraction effects and interactions with the atmosphere's chemical constituents and temperature gradients.

In microwave work, where very low noise levels and low noise figures are encountered, experimenters like to use the "effective noise temperature" concept, rather than noise factor. The relationship between the two is given by

$$T_E = 290 (F – 1) \qquad (8)$$

where the noise factor F = 10$^{NF/10}$ and NF is the noise figure in dB.

T_E is a measure, in terms of temperature, of the "excess noise" of a component (such

as an amplifier). A resistor at T_E would have the same available noise power as the device (referred to the device's input) specified by T_E. For a passive device (such as a lossy transmission line or filter) that introduces no noise of its own, T_E is zero and G is a number less than one equal to the power loss of the device. The cascade of noise temperatures is similar to the formula for cascaded noise factors.

$$T_S = T_G + TE_1 + \frac{TE_1}{G_1} + \frac{TE_3}{G_1G_2} + \frac{TE_4}{G_1G_2G_3} + \dots \qquad (9)$$

where T_S is the system noise temperature (including the generator, which may be an antenna) and T_G is the noise temperature of the generator or the field of view of the antenna, usually assumed 290 °K for terrestrial communications.

The number 290 in the formulas for T_E is the standard ambient temperature (in kelvins) at which the noise factor of a two-port transducer is defined and measured, according to an IEEE recommendation. So those formulas relate a noise factor F, measured at 290 K, to the temperature T_E. In general, though, it is perfectly correct to say that the ratio $(S_I/N_I)/(S_O/N_O)$ can be thought of as

the ratio of total system output noise to that system output noise attributed to the "generator" alone, regardless of the temperature of the equipment or the nature of the generator, which may be an antenna at some arbitrary temperature, for example. This ratio is, in fact, a special "system noise factor (or figure), F_S" that need not be tied to any particular temperature such as 290 K. (Note that regular noise factor (or figure) does depend on reference temperature.) The use of the F_S notation avoids any confusion. As the example of Fig 12.33 shows, the value of this system noise factor F_S is just the ratio of the total system temperature to the antenna temperature.

Having calculated a system noise temperature, the receive system noise floor (that is, the antenna input level of a signal that would exactly equal system noise, both observed at the receiver output) associated with that temperature is:

$$N = k T_S B_N \qquad (10)$$

where
  k = 1.38 × 10^{-23} (Stefan-Boltzmann's constant) and
  B_N= noise bandwidth

The system noise figure F_S is indicated in the example also. It is higher than the sum of the receiver and coax noise figures.

The example includes a loss of 3 dB in the receiver transmission line. The formula for T_S in the example shows that this loss has a double effect on the system noise temperature, once in the second term (288.6) and again in the third term (2.0). If the receiver (or high-gain preamp with a 6 dB NF) were mounted at the antenna, the receive-system noise temperature would be reduced to 1064.5 K and a system noise figure, F_S, of 7.26 dB, a very substantial improvement. Thus, it is the common practice to mount a preamp at the antenna, although transmission line loss must still be included in system noise figure calculations.

MICROWAVE RECEIVER FOR 10 GHZ

Here is a good example of amateur techniques for the 10 GHz band. The intended use for the radio is narrowband CW and SSB work, which requires extremely good frequency stability in the LO. Here, we will discuss the receiver circuit.

Block Diagram

Fig 12.34 is a block diagram of the receiver. Here are some important facets of the design.

1) The antenna should have sufficient gain. At 10 GHz, gains of 30 dBi are not difficult to get, as the example of Fig 12.33 demonstrates. A 4-foot dish might be difficult to aim, however.

2) For best results a very low-noise pre-amp at the antenna reduces loss of system sensitivity when antenna temperature is low. For example, if the antenna temperature at a quiet direction of the sky is 50 K and the receiver noise figure is 4 dB (due in part to transmission-line loss), the system temperature is 488 K for a system noise figure of 4.3 dB. If the receiver noise figure is reduced to 1.5 dB by adding a preamp at the antenna the system temperature is reduced to 170 K for a system noise figure of 2.0 dB, which is a very big improvement.

3) After two stages of RF amplification using GaAsFETs, a probe-coupled cavity resonator attenuates noise at the mixer's image frequency, which is 10.368 − 0.288

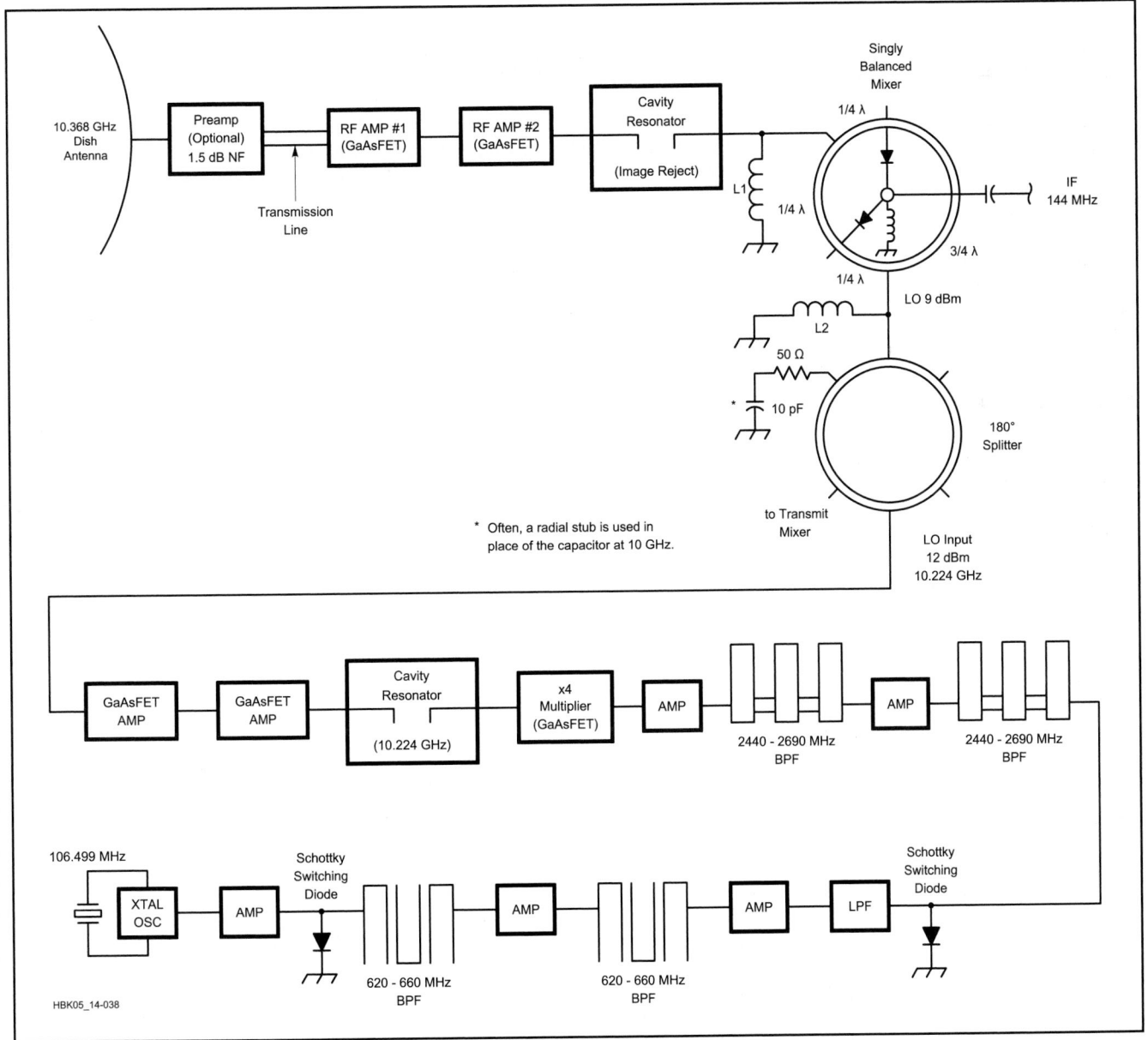

Fig 12.34 — A block diagram of the microwave receiver discussed in the text.

HBK05_14-038

= 10.080 GHz. An image reduction of 15 to 20 dB is enough to prevent image frequency noise generated by the RF amplifiers from affecting the mixer's noise figure.

4) The single-balanced diode mixer uses a "rat-race" 180° hybrid. Each terminal of the ring is ¼ wavelength (90°) from its closest neighbors. So the anodes of the two diodes are 180° (½ wavelength) apart with respect to the LO port, but in-phase with respect to the RF port. The inductors (L1, L2) connected to ground present a low impedance at the IF frequency. The mixer microstrip circuit is carefully "tweaked" to improve system performance. Use the better mixer in the transmitter.

5) The crystal oscillator is a fifth-overtone Butler circuit that is capable of high stabil-ity. The crystal frequency error and drift are multiplied 96 times (10.224/0.1065), so for narrowband SSB or CW work it may be dif-ficult to get on (and stay on) the "calling frequency" at 10.368 GHz. One acceptable (not perfect) solution might be to count the 106.5 MHz with a frequency counter whose internal clock is constantly compared with WWV. Adjust to 106.5 MHz as required. At times there may be a small Doppler shift on the WWV signal. It may be necessary to switch to a different WWV frequency, or WWV's signals may not be strong enough. Surplus frequency standards of high quality are sometimes available. Many operators just "tune" over the expected range of uncertainty.

6) The frequency multiplier chain has numerous band-pass filters to "purify" the harmonics by reducing various frequency components that might affect the signal path and cause spurious responses. The final filter is a tuned cavity resonator that reduces spurs from previous stages. Oscillator phase noise amplitude is multiplied by 96 also, so the oscillator must have very good short-term stability to prevent contamination of the de-sired signal.

7) A second hybrid splitter provides an LO output for the transmitter section of the radio. The 50-Ω resistor improves isolation between the two output ports. The original two-part *QST* article (see the references) is recommended reading for this very interest-ing project, which provides a fairly straight-forward (but not extremely simple) way to get started on 10 GHz.

12.9 References and Bibliography

J. Dickert, "A New Automatic Noise Limiter," *QST*, Nov 1938, pp 19-21.

W. Hayward, W7ZOI, R. Campbell, KK7B, and B. Larkin, W7PUA, *Experimental Methods in RF Design.* (ARRL, Newington, 2009).

R. Henderson, WI5W, "A Rock Bending Receiver for 7 MHz," *QST*, Aug 1995, pp 22-25.

McClanning and Vito, *Radio Receiver Design*, Noble Publishing Corporation, 2001.

H. Robinson, W3LW, "Audio Output Limiters for Improving the Signal-to-Noise Ratio in Reception," *QST*, Feb 1936, pp 27-29.

Rohde, Whittaker, and Bucher, *Communications Receivers*, 2nd Edition, McGraw-Hill, 1997.

Rohde and Whittaker, *Communications Receivers: DSP, Software Radios, and Design*, 3rd Edition, McGraw-Hill Professional, 2000.

W. Sabin, "Envelope Detection and AM Noise-Figure Measurement," *RF Design*, Nov 1988, p 29.

Z. Lau, W1VT, "Home-Brewing a 10-GHz SSB/CW Transverter," *QST*, May and Jun 1993.

Contents

Chapter 13 — CD-ROM Content

Supplemental Articles and Projects

- "Designing and Building Transistor Linear Power Amplifiers" Parts 1 and 2 by Rick Campbell, KK7B
- "A Fast TR Switch" by Jack Kuecken, KE2QJ
- "A Homebrew High Performance HF Transceiver — the HBR-2000" by Markus Hansen, VE7CA
- "The MicroT2 — A Compact Single-Band SSB Transmitter" by Rick Campbell, KK7B
- "The MkII — An Updated Universal QRP Transmitter" by Wes Hayward, W7ZOI
- "The Norcal Sierra: An 80-15 M CW Transceiver" by Wayne Burdick, N6KR (plus supporting files)
- "The Rockmite — A Simple Single-Band CW Transceiver" by Dave Benson, K1SWL (plus supporting files)
- "The TAK-40 SSB/CW Transceiver" by Jim Veatch, WA2EUJ
- "A Transmitter for Fox Hunting" by Mark Spencer, WA8SME
- "The Tuna Tin 2 Today" by Ed Hare, W1RFI
- "VHF Open Sources" by Rick Campbell, KK7B (plus parts placement guides)

Transmitters and Transceivers

Transmitters are the companion to the receivers discussed in the previous chapter. As with receiver design, the basic elements of transmitters such as oscillators and modulators are described in other chapters of this book. Transceivers—the combination of a transmitter and receiver in the same package — add switching and signal control circuitry.

Amplifiers for power levels above 100 W (at HF) are covered in the **RF Power Amplifiers** chapter. The **DSP and Software Radio Design** chapter has more information on digital techniques and architectures.

This chapter includes a trio of QRP transmitter projects, two transceiver projects, and supporting information and articles that can be found on the *Handbook* CD-ROM. Rick Campbell, KK7B, contributed a new section on design and construction of VHF signal sources.

Current Transceiver Overview

A supplemental article on this book's CD-ROM describes a range of commercial HF and VHF/UHF transceivers. With each subsequent edition, the overview is updated and the previous version moved to the ARRL website for future reference.

Transmitter technology has advanced in a parallel process similar to that of the technology of receivers. While transmitters are composed of many of the same named blocks as those used in receivers, it's important to keep in mind that there are significant differences. An RF amplifier in a receiver may deal with amplifying picowatts while one in a transmitter may output up to kilowatts. While the circuits may even look similar, the size of the components, especially cooling systems and power supplies, may differ significantly in scale. Still, many of the same principles apply.

Transceivers — the combination of a receiver and transmitter in a single physical piece of equipment — are the norm in Amateur Radio today. Separate receivers and transmitters are no longer offered by the major manufacturers, although many amateurs find this an easier approach when constructing homebuilt equipment. With receivers covered in their own chapter, this edition of the book combines the previously separate chapters on the closely related technology of transmitters and transceivers.

Transceivers achieve many economies by sharing receiver and transmitter elements such as high-performance components and circuits, power supplies and antenna switching circuits, as well as the physical enclosures and operating controls themselves. The sharing is facilitated by control and switching circuitry as discussed in the transceiver sections of this chapter.

Transmitters (and transceivers) may contain hazardous voltages, and at higher power levels RF exposure issues must be considered — review the **Safety** chapter for more information. Techniques for transmitter measurement are covered in the **Test Equipment and Measurements** chapter.

13.1 Transmitter Modulation Types and Methods

13.1.1 Data, Morse Code or Pulse Transmitters

The simplest transmitter consists of an oscillator generating a signal at the frequency we want to transmit. If the oscillator is connected to an antenna, the signal will propagate outward and be picked up by any receivers within range. Such a transmitter will carry little information, except perhaps for its location — it could serve as a rudimentary beacon for direction finding or radiolocation, although real beacons generally transmit identification data. It also indicates whether or not it is turned on, perhaps useful as part of an alarm system.

To actually transmit information, we must *modulate* the transmitter. The modulation process, covered in detail in the **Modulation** chapter, involves changing one or more of the signal parameters to apply the information content. This must be done in such a way that the information can be extracted at the receiver. As previously noted, the parameters available for modulation are:

Frequency — this is the number of cycles the signal makes per second.

Amplitude — although the amplitude, or strength, of a sinusoid is constantly changing with time, we can express the amplitude by the maximum value that it reaches.

Phase — the phase of a sinusoid is a measure of when a sinusoid starts compared to another sinusoid of the same frequency.

We could use any of the above parameters to modulate a simple transmitter with pulse type information, but the easiest to visualize is probably amplitude modulation. If we were to just turn the transmitter on and off, with it on for binary "one" and off for a "zero," we could surely send Morse code or other types of pulse-coded data. This type of modulation is called "On-Off-Keying" or "OOK" for short.

Some care is needed in how we implement such a function. Note that if we performed the obvious step of just removing and turning on the power supply, we might be surprised to find that it takes too much time for the voltage to rise sufficiently at the oscillator to actually turn it on at the time we make the connection. Similarly, we might be surprised to find that when we turn off the power we would still be transmitting for some time after the switch is turned. These finite intervals are referred to as *rise* and *fall times* and generally depend on the time constants of filter and switching circuits in the transmitter.

REAL WORLD CW KEYING

In the **Modulation** chapter, the importance of shaping the time envelope of the keying pulse of an on-off keyed transmitter is discussed. There are serious ramifications of not paying close attention to this design parameter. The optimum shape of a transmitter envelope should approach the form of a sinusoid raised to a power with a tradeoff between occupied bandwidth and overlap between the successive pulses. This can be accomplished either through filtering of the pulse waveform before modulation in a linear transmitter, or through direct generation of the pulse shape using DSP.

The differences between well-designed and poor pulse shaping can perhaps be best described by looking at some results. The following figures are from recent *QST* product

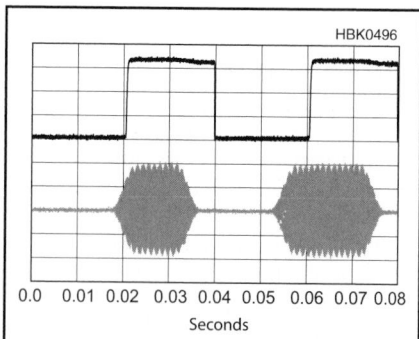

Fig 13.1 — The CW keying waveform of a transmitter with good spectrum control. The top trace is the key closure, with the start of the first contact closure on the left edge at 60 WPM using full break-in. Below that is the nicely rounded RF envelope.

reviews of commercial multimode 100 W HF transceivers. **Fig 13.1** shows the CW keying waveform of a transmitter with good spectrum control. The top trace is the key closure, with the start of the first contact closure on

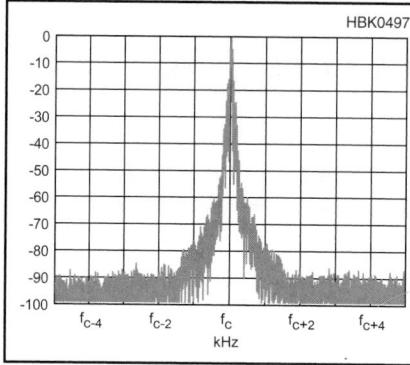

Fig 13.2 — The resultant signal spectrum from the keying shown in Fig 13.1. Note that the signal amplitude is about 80 dB down at a spacing of ±1 kHz, with a floor of –90 dB over the 10 kHz shown.

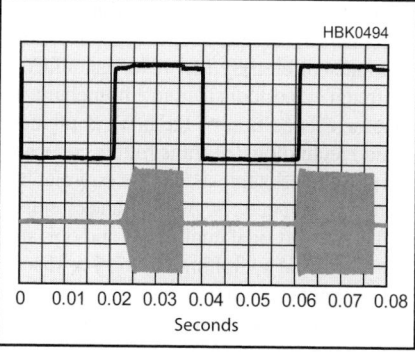

Fig 13.3 — The CW keying waveform of a transmitter with poor spectrum control. The top trace is the key closure, with the start of the first contact closure on the left edge at 60 WPM using full break-in. Note the sharp corners of the RF envelope that result in excessive bandwidth products.

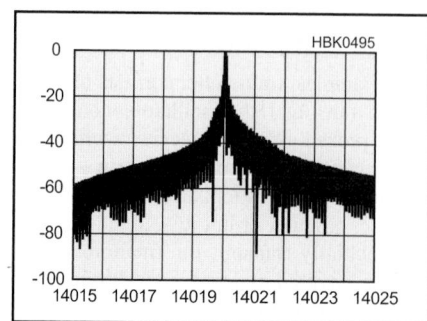

Fig 13.4 — The resultant signal spectrum from the keying shown in Fig 13.3. The resulting spectrum is not even down 40 dB at ±1 kHz and shows a floor that doesn't quite make 60 dB below the carrier across the 10 kHz.

the left edge at 60 WPM using full break-in. Below it is the nicely rounded RF envelope. **Fig 13.2** shows the resultant signal spectrum. Note that the signal amplitude is about 80 dB down at a spacing of ±1 kHz, with a floor of –90 dB over the 10 kHz shown. **Figs 13.3** and **13.4** are similar data taken from a different manufacturer's transceiver. Note the sharp corners of the RF envelope, as well as the time it takes for the first "dit" to be developed. The resulting spectrum is not even down 40 dB at ±1 kHz and shows a floor that doesn't quite make –60 dB over the 10 kHz range. It's easy to see the problems that the latter transmitter will cause to receivers trying to listen to a weak signal near its operating frequency. The unwanted components of the signal are heard on adjacent channels as sharp clicks when the signal is turned on and off, called *key clicks*. Note that even the best-shaped keying waveform in a linear transmitter will become sharp with a wide spectrum if it is used to drive a stage such as an external power amplifier beyond its linear range. This generally results in clipping or limiting with subsequent removal of the rounded corners on the envelope. Trying to get the last few dB of power out of a transmitter can often result in this sort of unintended signal impairment.

13.1.2 *Project:* Simple CW Transmitters

The schematic for a very basic low power HF CW oscillator transmitter is shown in **Fig 13.5**. Using a crystal-controlled oscillator gains frequency accuracy and stability but gives up frequency agility.

In this design, the key essentially just turns power on and off. The value of the keying line 0.1 µF bypass capacitor was chosen so it would not create an excessive rise or fall time at reasonable keying speeds. This transmitter will generate a few mW of RF power at the crystal frequency, but it has a number of limitations in common with early vacuum tube transmitters that were such an improvement over the spark transmitters that preceded them.

First, the oscillator is dependent on the environment. Changes in the antenna from wind, for example, change the load and can cause the frequency to shift. Second, the oscillator generates signals at harmonics of the fundamental frequency. Third, every time the key is closed the oscillator must start up, taking a short but audible time for the output frequency to stabilize. This creates a distinctive change in output frequency referred to as "chirp," since the signal sounds like a bird's chirp in the receiver.

By adding an amplifier stage to the oscillator transmitter, the oscillator can be isolated from changes in the environment to improve

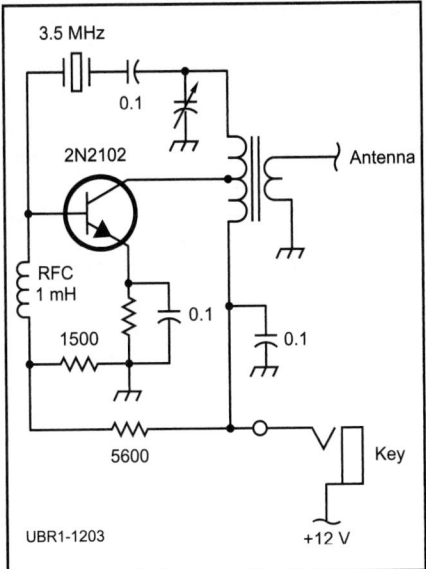

Fig 13.5 — A simple solid-state low-power HF crystal-controlled oscillator-transmitter. The unspecified tuned circuit values are resonant at the crystal frequency.

stability. Adding filtering at the output addresses the problem of harmonics.

The following project by Rev George Dobbs, G3RJV, was published in the Spring 2012 issue of *QRP Quarterly* and is reprinted courtesy of the QRP Amateur Radio Club (**www.qrparci.org**) and *Practical Wireless*

(**www.pwpublishing.ltd.uk**). An article by Ed Hare, W1RFI, describing the Tuna-Tin 2 transmitter that was included in previous editions of the *ARRL Handbook* is available on this book's CD-ROM and listed in the References section of this chapter.

Fig 13.6 is the schematic of a much better, but still simple, CW transmitter. It is an update to the "Pebble Crusher" originally designed by Doug DeMaw, W1FB (SK), and described in his *QRP Notebook* (out of print). The transmitter uses an oscillator followed by a stage of amplification to about 1/2 W of output. Note that the output circuit includes a filter to reduce any harmonic output below the levels required by FCC rules.

The circuit is a two transistor transmitter using 2N2222A devices. These are small-signal devices so the transmitter output is only about ½ W. What might surprise the reader is the number of inductors (coils) in the design, but the circuit has good performance in impedance matching and harmonic reduction. A careful design ensures a clean output waveform, low in harmonic and other spurious content.

The oscillator is a variable frequency crystal oscillator (VXO) in which the 100 pF variable capacitor between the crystal and ground enables the oscillator frequency to be shifted. Depending upon individual crystals, that movement can be on the order of 5 kHz. Increasing the variable capacitance too much may cause the transistor to cease oscillating.

The 150 pF capacitor in the emitter of the oscillator controls the oscillator feedback. This value is a compromise that appears to work well — but if the transistor fails to oscillate, try increasing this value a little.

The 4.7 µF electrolytic capacitor across the KEY connection minimizes key clicks by softening the keying waveform. Lowering this value gives a harder keying waveform and increasing it will further soften the waveform. A 10 Ω resistor is added to the base of the transistor to reduce the risk of high frequency parasitic oscillation.

A 22 µH RF choke (RFC1) provides the RF collector dc supply for the oscillator. A molded inductor was used in the original design. If this isn't available, about six turns of thin enameled wire wound through a small ferrite bead (type 43 mix) would give roughly the same inductance value.

The output from the collector goes to a single-element harmonic filter in a pi configuration. A 100 pF variable capacitor tunes the filter to resonance at the oscillator frequency. It should be adjusted for maximum output consistent with a clean CW signal.

If the intent is to use the oscillator stage as a transmitter without additional amplification, the filter is designed to have a 50 Ω output impedance for a matched 7 MHz antenna. The antenna can be connected in place of T1 in the schematic.

A 4:1 impedance ratio transformer (T1) matches the output of the oscillator to the

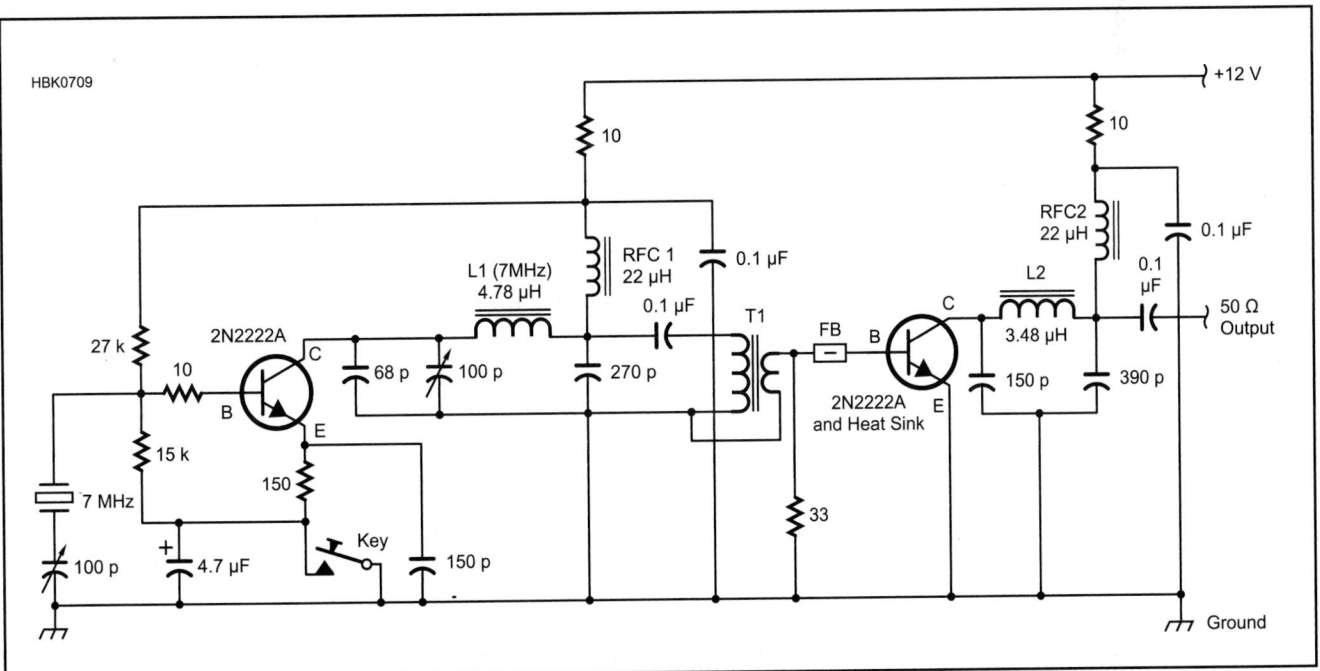

Fig 13.6 — The circuit of the Pebble Crusher transmitter. The oscillator section at left can be used as a standalone transmitter by replacing T1 with an antenna. The output of the amplifier circuit is approximately ½ W.

L1 — 32t #30 AWG on T50-6 toroid core
L2 — 28t #24 AWG on T50-6 toroid core

RFC1, RFC2 — 22 µH or 6t #30 AWG on ferrite bead (type 43 mix)
T1 — 12t #24 AWG (primary) and 6t #24

AWG (secondary) on FT37-43 toroid core (see text)
FB — ferrite bead (type 43 mix)

Fig 13.7 — G3RJV's "Anglicised" version of the Pebble Crusher is built on 0.1 inch spacing perfboard. The heat sink for the output transistor at lower right can be a commercial unit or created from metal strip or tubing (see text).

transmitter would actually be implemented. The upper portion is the RF channel, and you can think of the previously described Tuna Tin Two transmitter as a transmitter that we could use, after shifting the frequency into the voice portion of the band. The lower portion is the audio frequency or AF channel, usually called the *modulator*, and is nothing more than an audio amplifier designed to be fed from a microphone and with an output designed to match the anode or collector impedance of the final RF amplifier stage.

The output power of the modulator is applied in series with the dc supply of the output stage (only) of the RF channel of the transmitter. The level of the voice peaks needs to

core and L2 has 28 turns of #24 AWG enameled copper wire also on a T50-6 core. The transformer (T1) is wound on a ferrite FT37-43 core. Wind the primary first with 12 turns of #24 AWG spread out over three-quarters of the core circumference. Then add the secondary winding which is made up of 6 turns of #24 AWG wire between turns of the primary winding at the grounded end.

Fig 13.7 shows 2N2222A transistors in metal TO-18 cases, but the plastic TO-92 version would also work. An advantage of using the metal TO-18 is that a small heat sink can be attached to the output transistor to dissipate surplus heat. If a commercial heat sink is not available, a small piece of aluminum or copper strip can be formed into a heat sink. The strip is wrapped around a drill bit the same diameter as the transistor case, one side overlapping where the ends of the metal meet. This is then squeezed to make a tight fit on the TO-18 case. A small piece of ¼ inch OD brass tubing slightly flattened and cemented to the transistor casing with epoxy will work as well.

13.1.3 Amplitude-Modulated Full-Carrier Voice Transmission

While the telegraph key in the transmitters of the previous section can be considered a *modulator* of sorts, we usually reserve that term for a somewhat more sophisticated system that adds information to the transmitted signal. As noted earlier, there are three signal parameters that can be used to modulate a radio signal and they all can be used in various ways to add voice (or other information) to a transmitted signal.

One way to add voice to a radio signal is to first convert the analog signal to digital data and then transmit it as "ones" and "zeros." This can be done even using the simple telegraph transmitters of the last section. This is a technique frequently employed for some

applications using data applied to our "pulse" transmitter described earlier. Here we will talk about the more direct application of the analog voice signal to a radio signal.

A popular form of voice amplitude modulation is called *high-level amplitude modulation*. It is generated by mixing (or modulating) an RF carrier with an audio signal. **Fig 13.8** shows the conceptual view of this. **Fig 13.9** is a more detailed view of how such a voice

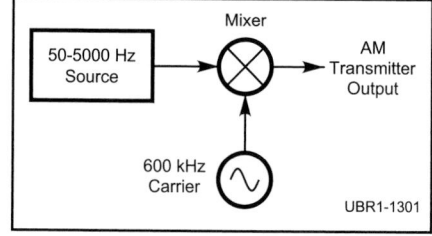

Fig 13.8 — Block diagram of a conceptual AM transmitter.

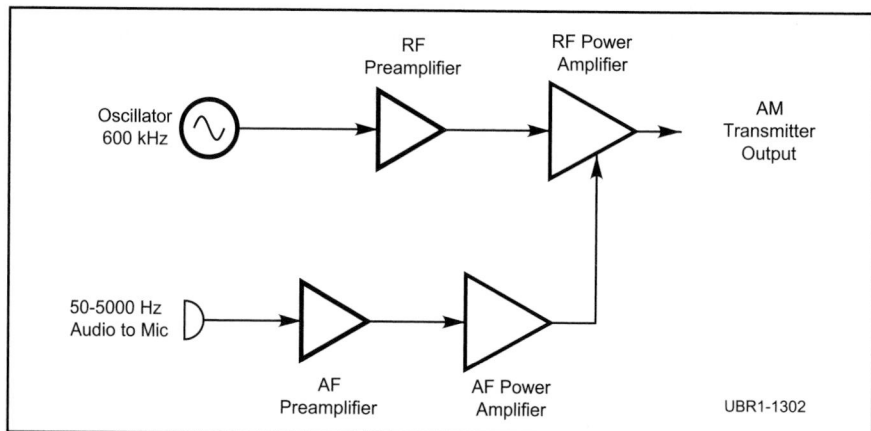

Fig 13.9 — Block diagram of a 600 kHz AM broadcast transmitter.

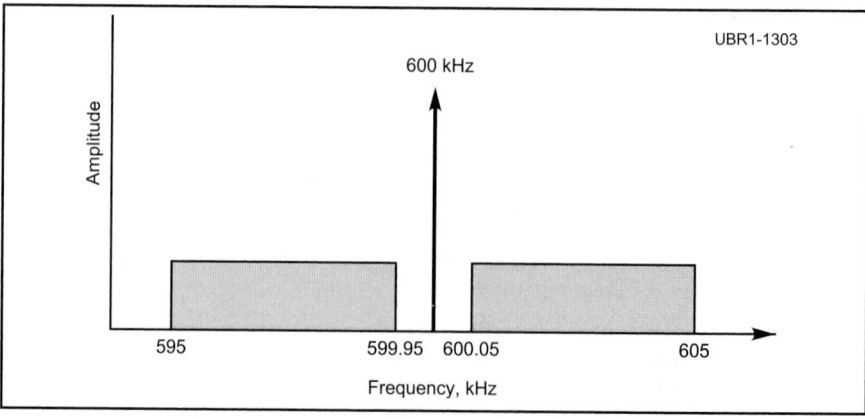

Fig 13.10 — The range of of spectrum used by a 600 kHz AM broadcast signal showing sidebands above and below a carrier at 600 kHz.

be just enough to vary the supply to the RF amplifier collector between zero volts, on negative peaks, and twice the normal supply voltage on positive voice peaks. This usually requires an AF amplifier with about half the average power output as the dc input power (product of dc collector or plate voltage times the current) of the final RF amplifier stage.

The output signal, called *full-carrier double-sideband AM*, occupies a frequency spectrum as shown in **Fig 13.10.** The spectrum shown would be that of a standard broadcast station with an audio passband from 50 Hz to 5 kHz. Note that the resulting channel width is twice the highest audio frequency transmitted. If the audio bandwidth were limited to typical "telephone quality speech" of 300 to 3300 Hz, the resulting bandwidth would be reduced to 6.6 kHz. Note also that while a perfect multiplication process would result in just two sidebands and no carrier, this implementation actually provides the sum of the carrier and the sidebands from the product terms. (See the **Receivers** chapter for the mathematical description of signal multiplication.)

Full-carrier double-sideband AM is used in fewer and fewer applications. The spectral and power efficiency are significantly lower than single sideband (SSB), and the equipment becomes quite costly as power is increased. The primary application is in broadcasting — largely because AM transmissions can be received on the simplest and least expensive of receivers. With a single transmitter and thousands of receivers, the overall system cost may be less and the audience larger than for systems that use more efficient modulation techniques. While the PEP output of an AM transmitter is four times the carrier power, none of the carrier power is necessary to carry the information, as we will discuss in the next section.

13.1.4 Single-Sideband Suppressed-Carrier Transmission

The two sidebands of a standard AM transmitter carry (reversed) copies of the same information, and the carrier carries essentially no information. We can more efficiently transmit the information with just one of the sidebands and no carrier. In so doing, we use somewhat less than half the bandwidth, a scarce resource, and also consume much less transmitter power by not transmitting the carrier and the second sideband.

SINGLE-SIDEBAND — THE FILTER METHOD

The block diagram of a simple single-sideband suppressed-carrier (SSB) transmitter is shown in **Fig 13.11.** This transmitter uses a balanced mixer as a *balanced modulator* to generate a double sideband suppressed carrier signal without a carrier. (See the **Mixers, Modulators, and Demodulators** chapter for more information on these circuits.) That signal is then sent through a filter designed to pass just one (either one, by agreement with the receiving station) of the sidebands.

Depending on whether the sideband above or below the carrier frequency is selected, the signal is called *upper sideband* (*USB*) or *lower sideband* (*LSB*), respectively. The resulting SSB signal is amplified to the desired power level and we have an SSB transmitter. Amateur practice is to use USB above 10 MHz and LSB on lower frequencies. The exception is 60 meter channels, on which amateurs are required to use USB for data and voice signals.

While a transmitter of the type in Fig 13.11 with all processing at the desired transmit frequency will work, the configuration is not often used. Instead, the carrier oscillator and sideband filter are often at an intermediate frequency that is heterodyned to the operating frequency as shown in **Fig 13.12.** The reason is that the sideband filter is a complex narrowband filter and most manufacturers would rather not have to supply a new filter design every time a transmitter is ordered for a new frequency. Many SSB transmitters can operate on different bands as well, so this avoids the cost of additional mixers, oscillators and expensive filters.

Note that the block diagram of our SSB transmitter bears a striking resemblance to the diagram of a superheterodyne receiver as shown in the **Receivers** chapter, except that the signal path is reversed to begin with information and produce an RF signal. The same kind of image rejection requirements for intermediate frequency selection that were design constraints for the superhet receiver applies here as well.

SINGLE-SIDEBAND — THE PHASING METHOD

Most current transmitters use the method

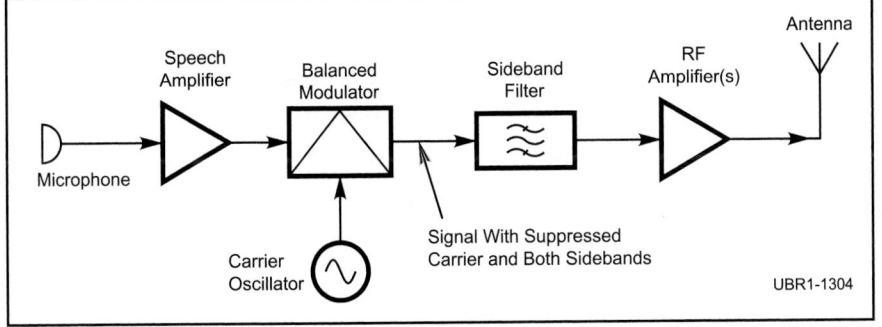

Fig 13.11 — Block diagram of a filter type single-sideband suppressed-carrier (SSB) transmitter.

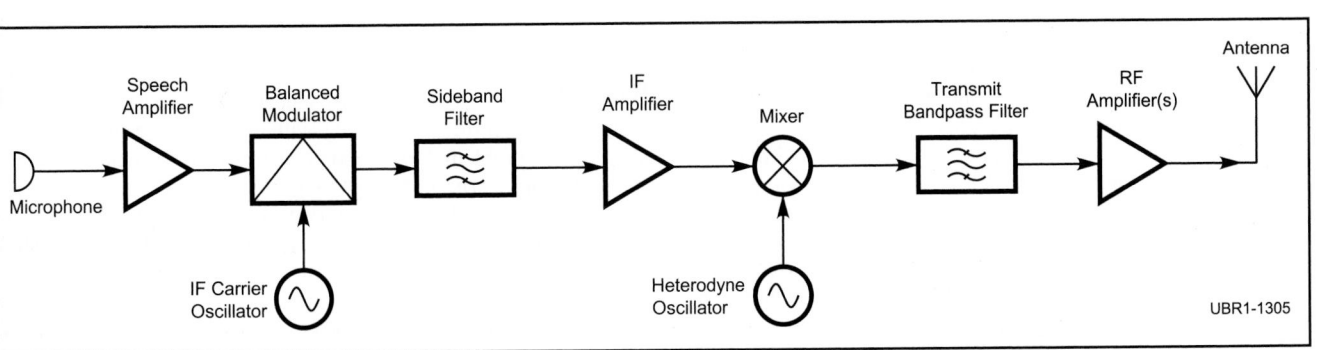

Fig 13.12 — Block diagram of a heterodyne filter-type SSB transmitter for multiple frequency operation.

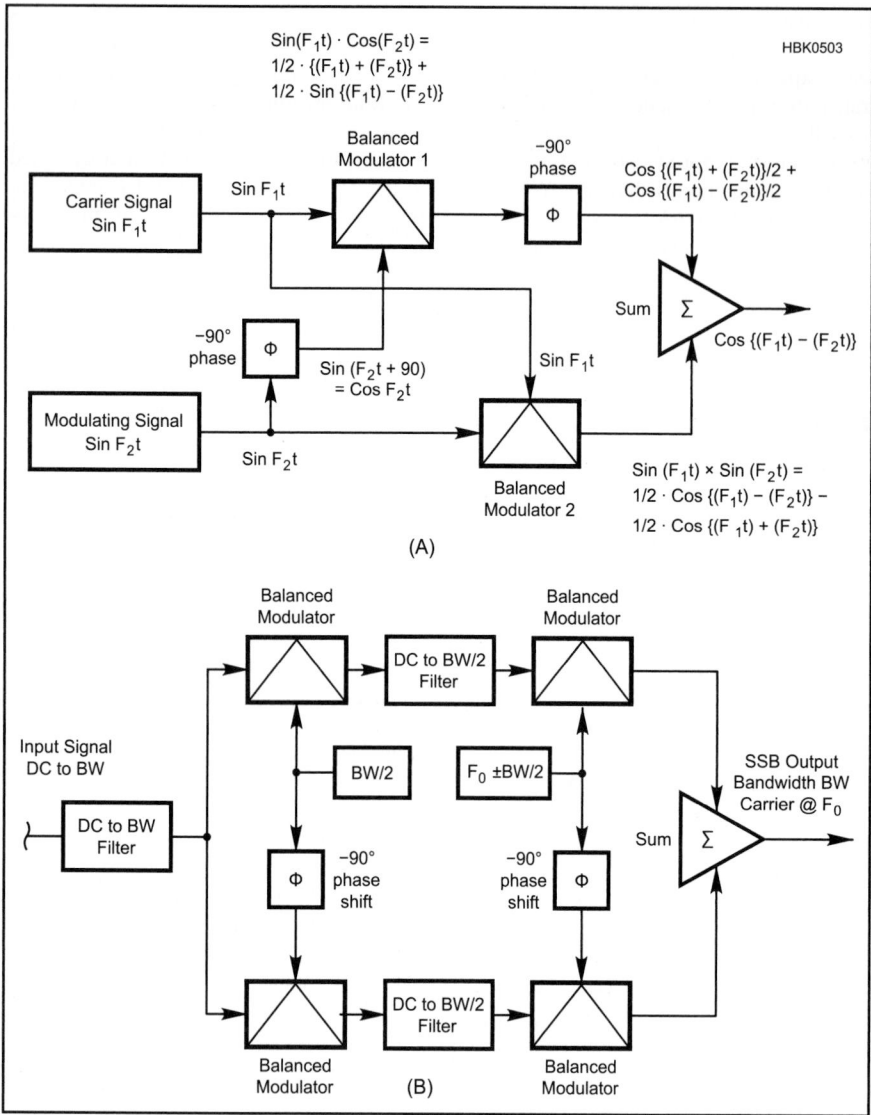

Fig 13.13 — Block diagrams of phasing type SSB transmitters for single frequency operation.

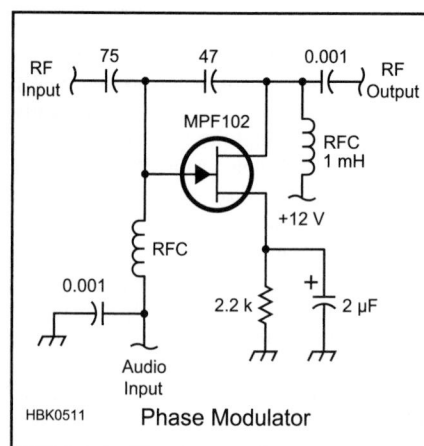

Fig 13.14 — Simple FET phase modulator circuit.

of SSB generation shown in Fig 13.11 and discussed in the previous section to generate the SSB signal. That is the *filter method*, but really occurs in two steps — first a balanced modulator is used to generate sidebands and eliminate the carrier, then a filter is used to eliminate the undesired sideband, and often to improve carrier suppression as well.

The *phasing method* of SSB generation is exactly the same as the image-rejecting mixer described in the **Receivers** chapter. This uses two balanced modulators and a phase-shift network for both the audio and RF carrier signals to produce the upper sideband signal as shown in **Fig 13.13A**. By a shift in the sign of either of the phase-shift networks, the opposite sideband can be generated. This method trades a few phase-shift networks and an extra balanced modulator for the sharp sideband filter of the filter method. While it

looks deceptively simple, a limitation is in the construction of a phase-shift network that will have a constant 90° phase shift over the whole audio range. Errors in phase shift result in less than full carrier and sideband suppression. Nonetheless, there have been some successful examples offered over the years.

SINGLE-SIDEBAND — THE WEAVER METHOD

Taking the phasing method one step further, the Weaver method solves the problem of requiring phase-shift networks that must be aligned across the entire audio range. Instead, the Weaver method, shown in Fig 13.13B, first mixes one copy of the message (shown with a bandwidth of dc to BW Hz) with an in-band signal at BW/2 Hz and another copy with a signal at BW/2 Hz that is phase-shifted by −90°. Instead of phase-shifting the mes-

sage, only the signal at BW/2 Hz must be phase-shifted — a much simpler task!

The output of each balanced modulator is filtered, leaving only components from dc to BW/2. These signals are then input to a second pair of balanced modulators with a more conventional LO signal at the carrier frequency, f_0, offset by +BW/2 for USB and −BW/2 for LSB. The output of the balanced modulators is summed to produce the final SSB signal.

The Weaver method is difficult to implement in analog circuitry, but is well-suited to digital signal processing systems. The Weaver method has become common in DSP-based equipment that generates the SSB signal digitally.

13.1.5 Angle-Modulated Transmitters

Transmitters using frequency modulation (FM) or phase modulation (PM) are generally grouped into the category of *angle modulation* since the resulting signals are often indistinguishable. An instantaneous change in either frequency or phase can create identical signals, even though the method of modulating the signal is somewhat different. To generate an FM signal, we need an oscillator whose frequency can be changed by the modulating signal.

We can make use of an oscillator whose frequency can be changed by a "tuning voltage." If we apply a voice signal to the TUNING VOLTAGE connection point, we will change the frequency with the amplitude and frequency of the applied modulating signal, resulting in an FM signal.

The phase of a signal can be varied by changing the values of an R-C phase-shift network. One way to accomplish phase modulation is to have an active element shift the phase and generate a PM signal. In **Fig 13.14**, the current through the field-effect transistor is varied with the applied modulating signal,

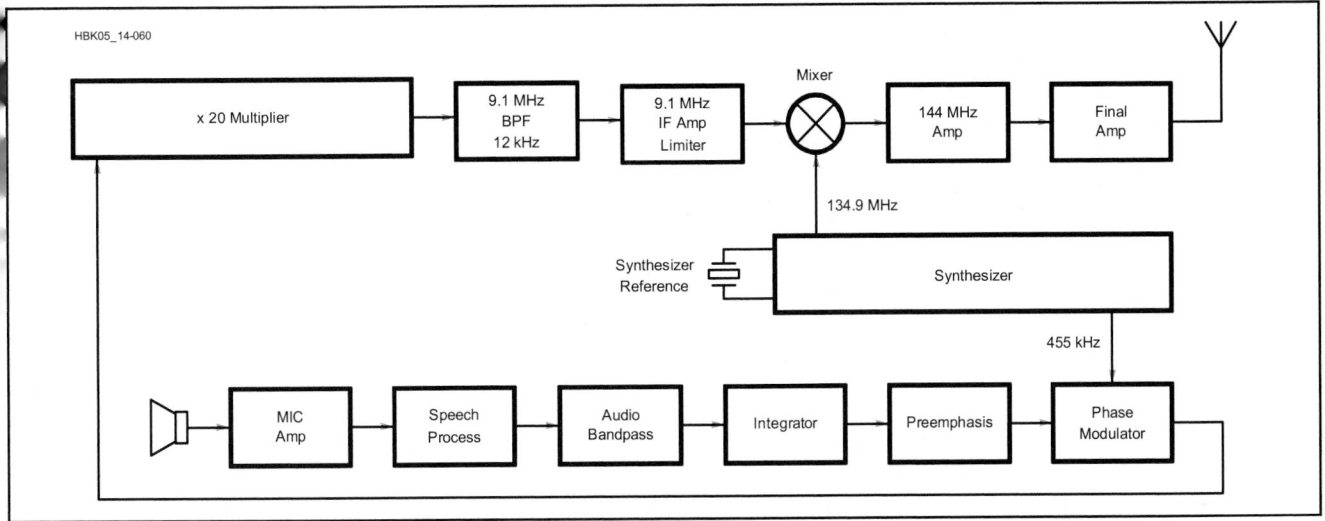

Fig 13.15 — Block diagram of a VHF/UHF NBFM transmitter using the indirect FM (phase modulation) method.

varying the phase shift at the stage's output. Because the effective load on the stage is changed, the carrier is also amplitude-modulated and must be run through an FM receiver-type limiter in order to remove the amplitude variations.

FREQUENCY MODULATION TRANSMITTER DESIGN

Frequency modulation is widely used as the voice mode on VHF for repeater and other point-to-point communications. **Fig 13.15** shows the phase-modulation method, also known as *indirect FM*, as used in many FM transmitters. It is the most widely used approach to FM. Phase modulation is performed at a low frequency, say 455 kHz. Prior to the phase modulator, speech filtering and processing perform four functions:

1. Convert phase modulation to frequency modulation (see below).

2. Apply pre-emphasis (high-pass filtering) to the speech audio higher speech frequencies for improved signal-to-noise ratio after de-emphasis (low-pass filtering) of the received audio.

3. Perform speech processing to emphasize the weaker speech components.

4. Compensate for the microphone's frequency response and possibly also the operator's voice characteristics.

Multiplier stages then move the signal to some desired higher IF and also multiply the frequency deviation to the desired final value. If the FM deviation generated in the 455 kHz modulator is 250 Hz, the deviation at 9.1 MHz is 20 × 250, or 5 kHz.

Frequency Multipliers

Frequency multipliers are frequently used in FM transmitters as a way to increase the deviation along with the carrier frequency.

They are composed of devices that exhibit high levels of harmonic distortion, usually an undesired output product. In this case the desired harmonic is selected and enhanced through filtering. The following examples show the way this can be done, both with amplifiers and with passive diode circuits.

A passive multiplier using diodes is shown in **Fig 13.16A**. The full-wave rectifier circuit can be recognized, except that the dc component is shorted to ground. If the fundamental frequency ac input is 1.0 V_{RMS}, the second harmonic is 0.42 V_{RMS} or 8 dB below the input, including some small diode losses. This value is found by calculating the Fourier series coefficients for the full-wave-rectified sine wave, as shown in many textbooks.

Transistor and vacuum-tube frequency multipliers operate on the following principle: if a sine wave input causes the plate/collector/drain current to be distorted (not a sine wave) then harmonics of the input are generated. If an output resonant circuit is tuned to a harmonic, the output at the harmonic is emphasized and other frequencies are attenuated. For a particular harmonic the current pulse should be distorted in a way that maximizes that harmonic. For example, for a doubler the current pulse should look like a half-wave rectified sine wave (180° of conduction). A transistor with Class B bias would be a good choice. For a tripler, use 120° of conduction (Class C).

An FET, biased at a certain point, is very nearly a *square-law* device as described in the **Analog Basics** chapter. That is, the drain-current change is proportional to the square of the gate-voltage change. It is then an efficient frequency doubler that also de-emphasizes the fundamental.

A push-push doubler is shown in Fig 13.16B. The FETs are biased in the square-

law region and the BALANCE potentiometer minimizes the fundamental frequency. Note that the gates are in push-pull and the drains are in parallel. This causes second harmonics to add in-phase at the output and fundamental components to cancel.

Fig 13.16C shows an example of a single-ended doubler using a bipolar transistor. The efficiency of a doubler of this type is typically 50%, that of a tripler 33%, and of a quadrupler 25%. Harmonics other than the one to which the output tank is tuned will appear in the output unless effective band-pass filtering is applied. The collector tap on L1 is placed at the point that offers the best compromise between power output and spectral purity.

A push-pull tripler is shown in Fig 13.16D. The input and output are both push-pull. The balance potentiometer minimizes even harmonics. Note that the transistors have no bias voltage in the base circuit; this places the transistors in Class C for efficient third-harmonic production. Choose an input drive level that maximizes harmonic output.

The step recovery diode (SRD) shown in **Fig 13.17A** is an excellent device for harmonic generation, especially at microwave frequencies. The basic idea of the SRD is as follows: When the diode is forward conducting, a charge is stored in the diode's diffusion capacitance, and if the diode is quickly reverse-biased, the stored charge is very suddenly released into an LC harmonic-tuned circuit. The circuit is also called a "comb generator" because of the large number of harmonics that are generated. (The spectral display looks like a comb.) A phase-locked loop (PLL) can then lock onto the desired harmonic.

A varactor diode can also be used as a multiplier. Fig 13.17B shows an example.

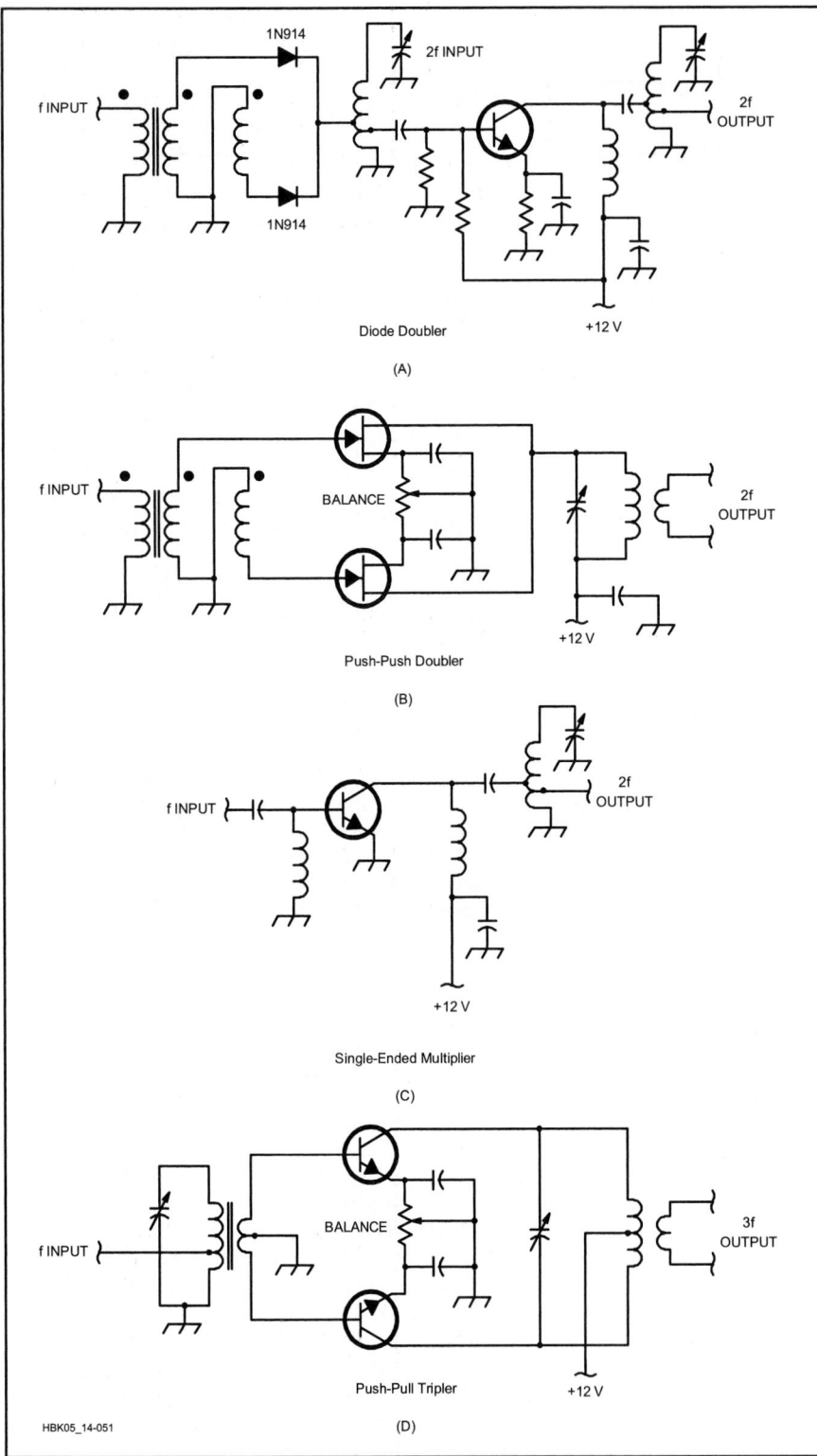

Fig 13.16 — A: diode doubler. B: push-push doubler using JFETS. C: single-ended multiplier using a BJT. D: push-pull tripler using BJTs.

This circuit depends on the fact that the capacitance of a varactor changes with the instantaneous value of the voltage across it, in this case the RF excitation voltage. This is a nonlinear process that generates harmonic currents through the diode. Power levels up to 25 W can be generated in this manner.

Following frequency multiplication, a second conversion to the final output frequency is performed. Prior to this final translation, IF band-pass filtering is performed in order to minimize adjacent channel interference

that might be caused by excessive frequency deviation. This filter needs good phase linearity to assure that the FM sidebands maintain the correct phase relationships. If this is not done, an AM component is introduced to the signal, which can cause nonlinear distortion problems in the PA stages. The final frequency translation retains a constant value of FM deviation for any value of the output signal frequency.

The IF/RF amplifiers can be nonlinear Class C amplifiers because the signal in each amplifier contains, at any one instant, only a single value of instantaneous frequency and not multiple simultaneous frequencies whose relationship must be preserved as in SSB. These amplifiers are not sources of IMD, so they need not be "linear." The sidebands that appear in the output are a result only of the FM process. (The spectrum of an FM signal is described by Bessel functions.)

In phase modulation, the frequency deviation is directly proportional to the frequency of the audio signal. (In FM, the deviation is proportional to the audio signal's amplitude.) To make deviation independent of the audio frequency, an audio-frequency response that rolls off at 6 dB per octave is needed. An op-amp low-pass circuit in the audio amplifier accomplishes this function. This process converts phase modulation to frequency modulation.

In addition, audio speech processing helps to maintain a constant value of speech amplitude, and therefore constant IF deviation, with respect to audio speech levels. Preemphasis of speech frequencies (a 6 dB per octave high-pass response from 300 to 3000 Hz) is commonly used to improve the signal-to-noise ratio at the receive end. Analysis shows that this is especially effective in FM systems when the corresponding de-emphasis (complementary low-pass response) is used at the receiver. (See reference for Schwartz.) By increasing the amplitude of the higher audio frequencies before transmission and then reducing them in the receiver, high-frequency audio noise from the demodulation process is also reduced, resulting in a "flat" audio response with lower hiss and high-frequency noise.

An IF limiter stage may be used to ensure that any amplitude changes that are created during the modulation process are removed. The indirect-FM method allows complete frequency synthesis to be used in all the transmitter local oscillators (LOs), so that the channelization of the output frequency is very accurate. The IF and RF amplifier stages are operated in a highly efficient Class-C mode, which is helpful in portable equipment operating on small internal batteries.

FM is more tolerant of frequency misalignments between the transmitter and receiver than is SSB. In commercial SSB commu-

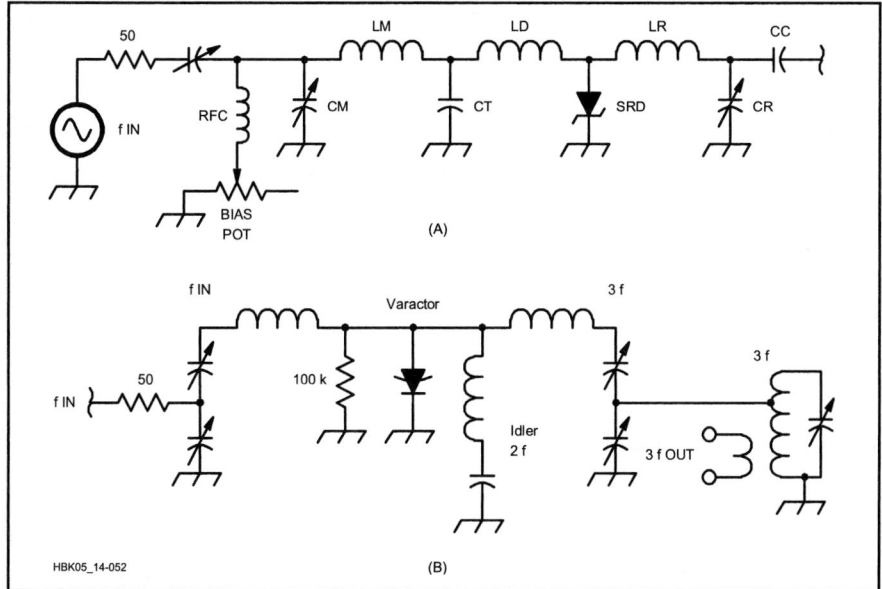

Fig 13.17 — Diode frequency multipliers. A: step-recovery diode multiplier. B: varactor diode multiplier.

nication systems, this problem is solved by transmitting a *pilot carrier* with an amplitude 10 or 12 dB below the full PEP output level. The receiver is then phase-locked to this pilot carrier. The pilot carrier is also used for squelch and AGC purposes. A short-duration "memory" feature in the receiver bridges across brief pilot-carrier dropouts, caused by multipath nulls.

In a "direct FM" transmitter, a high-fre-

quency (say, 9 MHz or so) crystal oscillator is frequency-modulated by varying the voltage on a varactor diode. The audio is pre-emphasized and processed ahead of the frequency modulator as for indirect-FM.

13.1.6 The Superhet SSB/CW Transmitter

The modern linear transmitting chain makes use of the concepts presented previously. We will now go through them in more detail. The same kind of mixing schemes, IF frequencies and IF filters that are used for superhet receivers can be, and very often are, used for a transmitter. In the following discussion, "in-band" refers to signal frequencies within the bandwidth of the desired signal. For example, for an upper sideband voice signal with a carrier frequency of 14.200 MHz, frequencies of approximately 14.2003 to 14.203 would be considered in-band. Out-of-band refers to frequencies outside this range, such as on adjacent channels.

For this example we will use the commonly-encountered dual-conversion scheme with the SSB generation at 455 kHz, a conversion to a 70 MHz IF, and then a final conversion to the HF operating frequency. **Fig 13.18** is a block diagram of the system

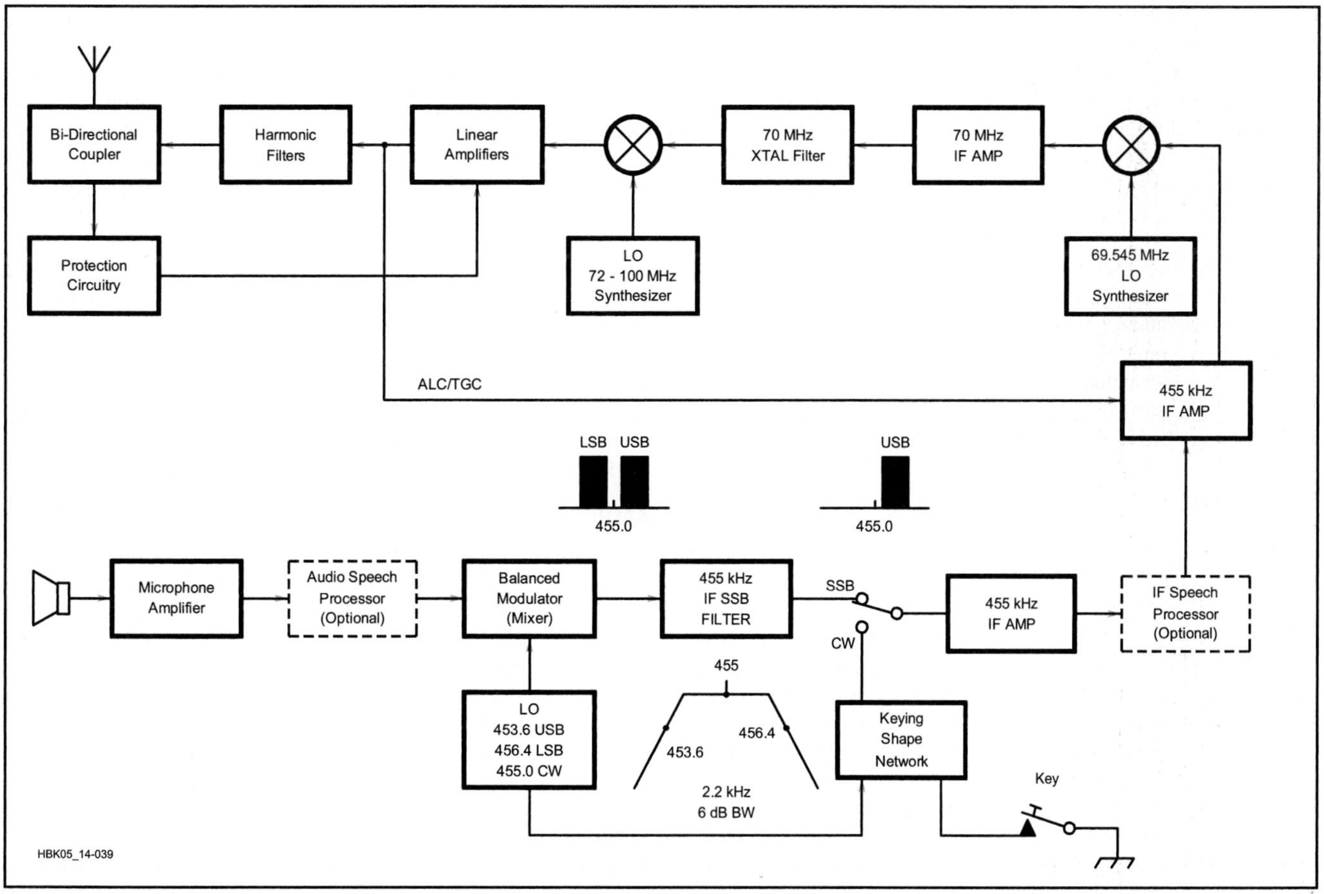

Fig 13.18 — Block diagram of an upconversion SSB/CW transmitter.

under consideration. Let's discuss the various elements in detail, starting at the microphone.

MICROPHONES

A microphone (mic) is a transducer that converts sound waves into electrical signals. For communications quality speech, its frequency response should be as flat as possible from around 200 to above 3500 Hz. Response peaks in the microphone can increase the peak to average ratio of speech, which then degrades (increases) the peak to average ratio of the transmitted signal. Some transmitters use "speech processing," which is essentially a specialized form of speech amplification either at audio or IF/RF. Since most microphones pick up a lot of background ambient noises, the output of the transmitter due to background noise pickup in the absence of speech may be as much as 20 dB greater than without speech processing. A *noise canceling* microphone is recommended to reduce this background pickup if there is much background noise. Microphone output levels vary, depending on the microphone type. Typical amateur microphones produce about 10 to 100 mV output levels.

Ceramic

Ceramic mics have high output impedances but low level outputs. They require a high-resistance load (usually about 50 kΩ) for flat frequency response and lose low-frequency response as this resistance is reduced (electrically, the mic "looks like" a small capacitor). These mics vary widely in quality, so a cheap mic is not a good bargain because of its effect on the transmitted power level and generally poor speech quality.

Dynamic

A dynamic mic resembles a small loudspeaker, with an impedance of about 600 Ω and an output of about 12 mV on voice peaks. Early dynamic mics (designed for vacuum tube transmitters) included a built-in transformer to transform the impedance to 100 kΩ suitable for high input impedance speech amplifiers. Currently available dynamic mics provide the output directly, although the transformers are available for connection outside the mic. Dynamic mics are widely used by amateurs.

Electret

Electret mics use a piece of special insulator material, such as Teflon, that contains a "trapped" polarization charge (Q) at its surfaces to create a capacitance (C). Sound waves modulate the capacitance of the material and cause a voltage change according to $\Delta V/V = -\Delta C/C$. For small changes in capacitance the change in voltage is almost linear. These mics have been greatly improved in recent years and are used in most

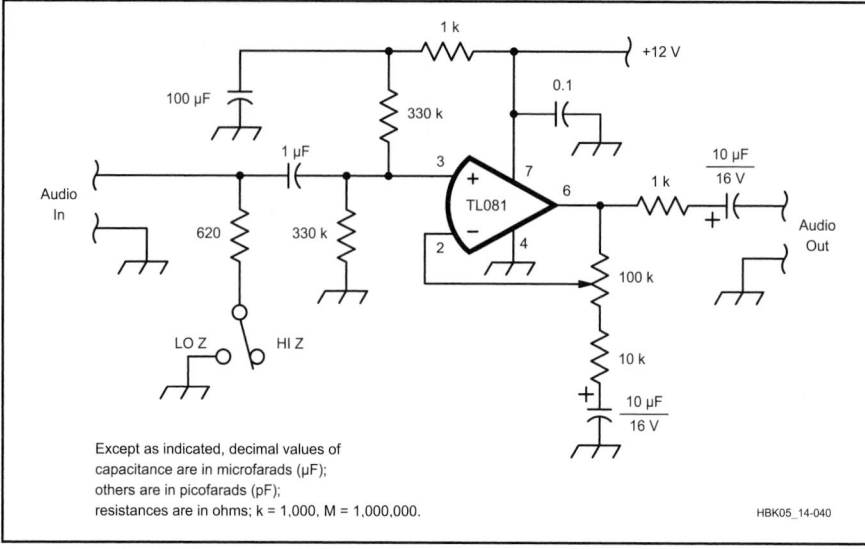

Fig 13.19 — Schematic diagram of a microphone amplifier suitable for high or low impedance microphones.

cellular handsets and computer headsets.

The output level of the electret is fairly low, and an integrated preamp is generally included in the mic cartridge. A voltage of about 4 V dc is required to power the preamp, and some commercial transceivers provide this voltage at the mic connector. The dc voltage must be blocked by a coupling capacitor if a dynamic mic element is to be used with a transceiver that supplies power to the microphone. The dynamic mic is unlikely to be damaged by the applied voltage, but the usual symptom is very low audio output with a muffled sound.

In recent years, micro electro-mechanical systems (MEMS) technology has been applied to microphones. The sound pressure is applied to a small transducer integrated on a silicon chip, which creates a mechanical motion in response to the force of the sound waves. This motion is converted into an electrical signal, usually by the varying capacitance of the transducer. Mass production of MEMS microphones is being driven by the cellular handset market.

MICROPHONE AMPLIFIERS

The balanced modulator and (or) the audio speech processor need a certain optimum level, which can be in the range of 0.3 to 0.6 V ac into perhaps 1 kΩ to 10 kΩ. Excess noise generated within the microphone amplifier should be minimized, especially if speech processing is used. The circuit in **Fig 13.19** uses a low-noise BiFET op amp. The 620 Ω resistor is selected for a low impedance microphone, and switched out of the circuit for high-impedance mics. The amplifier gain is set by the 100 kΩ potentiometer.

It is also a good idea to experiment with the low- and high-frequency responses of

the mic amplifier to compensate for the frequency response of the mic and the voice of the operator.

SPEECH PROCESSING

The output of the speech amplifier can be applied directly to the balanced modulator. The resulting signal will be reproducible with the maximum fidelity and dynamic range available for the bandwidth provided. The communications efficiency of the system will depend on the characteristics of the particular voice used. The usual peak-to-average ratio is such that the average transmitted power is on the order of 5% of the peak power.

For communications use, it is generally beneficial to increase the average-to-peak ratio by distorting the speech waveform in a measured way. If one were to merely increase the amplifier gain, a higher average power would be obtained. The problem is that clipping, heavy distortion and likely spurious signals beyond legal limits would be generated. By carefully modifying the speech waveform before application to the balanced modulator, a synthetic waveform with a higher average-to-peak waveform can be generated that will retain most of the individual voice characteristics and avoid spurious signals. We will discuss two such techniques of *speech processing* below and another as we move into the RF circuitry.

Note that speech processing should not be used with most digital mode transmissions if the digital signal is generated by audio tones applied to the transmitter's microphone input. The modulation of these signals requires linear processing and amplification to preserve the waveform shape and minimize distortion products. (The same caution applies to the ALC function as discussed in that section.)

Audio Speech Clipping

If the audio signal from the microphone amplifier is further amplified, say by as much as 12 dB, and then if the peaks are clipped (sometimes called *slicing* or *limiting*) by 12 dB by a speech clipper, the output peak value is the same as before the clipper, but the average value is increased considerably. The resulting signal contains harmonics and IMD but the speech intelligibility, especially in a white-noise background, is improved by 5 or 6 dB.

The clipped waveform frequently tends to have a square-wave appearance, especially on voice peaks. It is then band-pass filtered to remove frequencies below 300 and above 3000 Hz. The filtering of this signal can create a "repeaking" effect. That is, the peak value tends to increase noticeably above its clipped value.

An SSB generator responds poorly to a square-wave audio signal, creating significant peaks in the RF envelope. (This is described mathematically as the Hilbert Transform effect.) These peaks cause out-of-band splatter in the transmitter's linear output power amplifier unless Automatic Level Control (ALC, to be discussed later) cuts back on the RF gain. The peaks increase the peak-to-average ratio and the ALC reduces the average SSB power output, thereby reducing some of the benefit of the speech processing. The square-wave effect is also reduced by band-pass filtering (300 to 3000 Hz) the input to the clipper as well as the output.

Fig 13.20 is a circuit for a simple audio speech clipper. A CLIP LEVEL potentiometer before the clipper controls the amount of clipping and an OUTPUT LEVEL potentiometer controls the drive level to the balanced modulator. The correct adjustment of these potentiometers is done with a two-tone audio input or by talking into the microphone, rather than driving with a single tone, because single tones don't exhibit the repeaking effect.

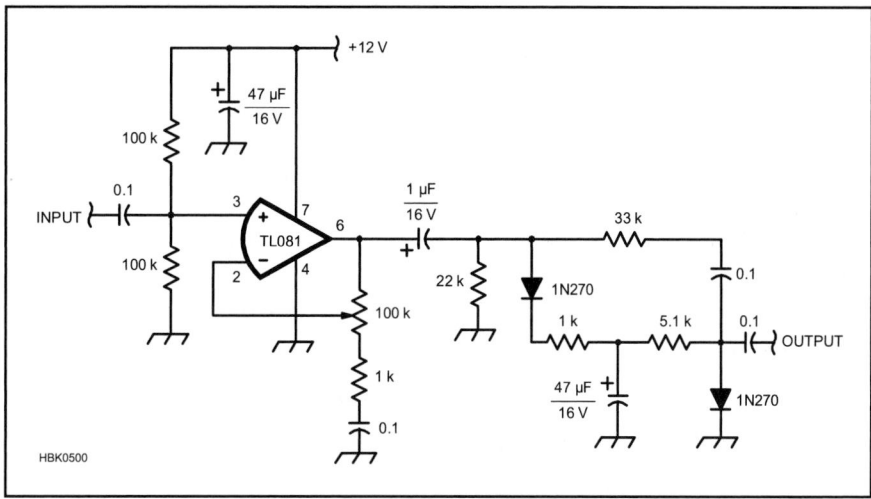

Fig 13.21 — A forward acting speech compressor circuit.

Audio Speech Compression

Although it is desirable to keep the voice level as high as possible, it is difficult to maintain constant voice intensity when speaking into the microphone. To overcome this variable output level, it is possible to use an automatic gain control that follows the average variations in speech amplitude. This can be done by rectifying and filtering some of the audio output and applying the resultant dc to a control terminal in an early stage of the amplifier.

If an audio AGC circuit derives control voltage from the output signal, the system is a closed loop. If short attack time is necessary, the rectifier-filter bandwidth must be opened up to allow syllabic modulation of the control voltage. This allows some of the voice frequency signal to enter the control terminal, causing distortion and instability. Because the syllabic frequency and speech-tone frequencies have relatively small separation, the simpler feedback AGC systems compromise fidelity for fast response.

Problems with loop dynamics in audio AGC can be side-stepped by eliminating the loop and using a forward-acting system. The control voltage is derived from the input of the amplifier, rather than from the output. Eliminating the feedback loop allows unconditional stability, but the trade-off between response time and fidelity remains. Care must be taken to avoid excessive gain between the signal input and the control voltage output. Otherwise the transfer characteristic can reverse; that is, an increase in input level can cause a decrease in output. A simple forward-acting compressor is shown in **Fig 13.21**.

BALANCED MODULATORS

A balanced modulator is a mixer. A more complete discussion of balanced modulator design was provided in the **Mixers, Modulators, and Demodulators** chapter. Briefly, the IF frequency LO (455 kHz in the example of Fig 13.18) translates the audio frequencies up to a pair of IF frequencies, the LO plus the audio frequency and the LO minus the audio frequency. The balance from the LO port to the IF output causes the LO frequency to be suppressed by 30 to 40 dB. Adjustments are provided to improve the LO null.

The filter method of SSB generation uses an IF band-pass filter to pass one of the sidebands and block the other. In Fig 13.18 the filter is centered at 455.0 kHz. The LO is offset to 453.6 kHz or 456.4 kHz so that the upper sideband or the lower sideband (respectively) can pass through the filter. This creates a problem for the other LOs in the radio, because they must now be properly offset so that the final transmit output's carrier (suppressed) frequency coincides with the frequency readout on the front panel of the radio.

Various schemes have been used to create the necessary LO offsets. One method uses two crystals for the 69.545 MHz LO that

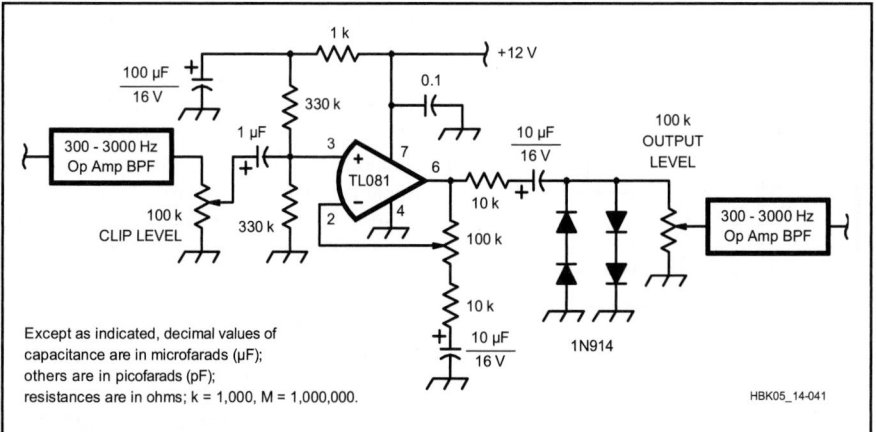

Fig 13.20 — Schematic diagram a simple audio speech clipper.

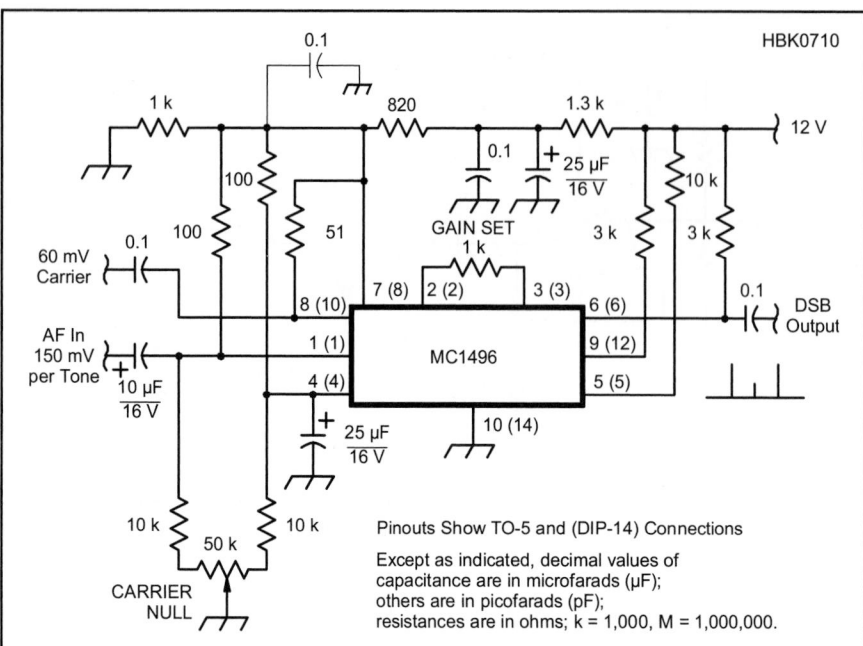

Fig 13.22 — An IC balanced modulator circuit using the MC1496 IC. The resistor between pins 2 and 3 sets the subsystem gain.

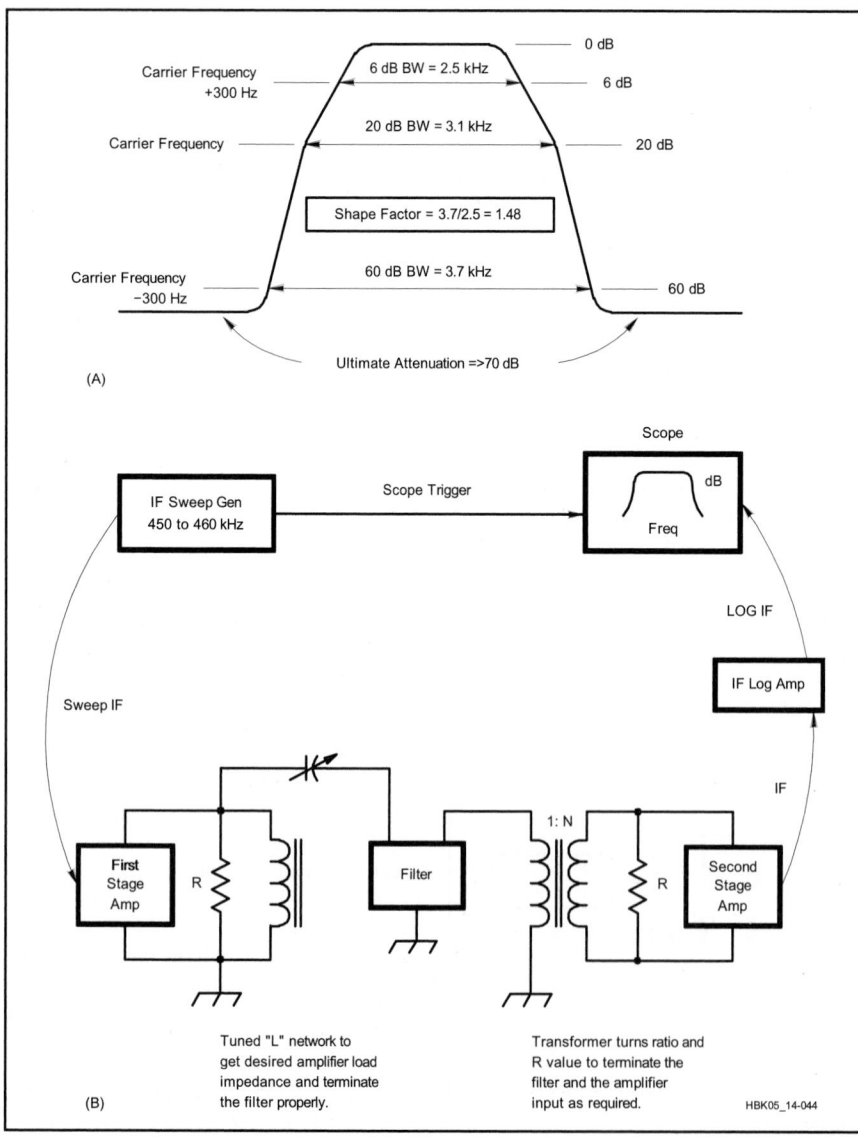

Fig 13.23 — At (A), desired response of a SSB IF filter. At (B), one method of terminating a mechanical filter that allows easy and accurate tuning adjustment and also a possible test setup for performing the adjustments.

can be selected. In synthesized radios the programming of the microprocessor controls the various LOs. Some synthesized radios use two IF filters at two different frequencies, one for USB and one for LSB, and a 455.0 kHz LO, as shown in Fig 13.18. These radios can be designed to transmit two independent sidebands (ISB) resulting in two separate channels in the spectrum space of the usual AM channel.

In times past, balanced modulators using diodes, balancing potentiometers and numerous components were used. These days it doesn't make sense to use this approach. ICs and packaged diode mixers do a much better job and are less expensive. The most widely known balanced modulator IC, the MC1496, has been around for more than 25 years and is still one of the best and least expensive. **Fig 13.22** is a typical balanced modulator circuit using the MC1496.

The data sheets for balanced modulators and mixers specify the maximum level of audio for a given LO level. Higher audio levels create excessive IMD. The IF filter following the modulator removes higher-order IMD products that are outside its passband but the in-band IMD products should be at least 40 dB below each of two equal test tones. Speech clipping (AF or IF) can degrade this to 10 dB or so, but in the absence of speech processing the signal should be clean, in-band.

IF FILTERS

The desired IF filter response is shown in **Fig 13.23A**. The reduction of the carrier frequency is augmented by the filter response. It is common to specify that the filter response be down 20 dB at the carrier frequency. Rejection of the opposite sideband should (hopefully) be 60 dB, starting at 300 Hz below the carrier frequency, which is the 300-Hz point on the opposite sideband. The ultimate attenuation should be at least 70 dB. This would represent a very good specification for a high quality transmitter. The filter passband should be as flat as possible (with passband ripple less than 1 dB or so).

Special filters, designated as USB or LSB, are designed with a steeper roll-off on the

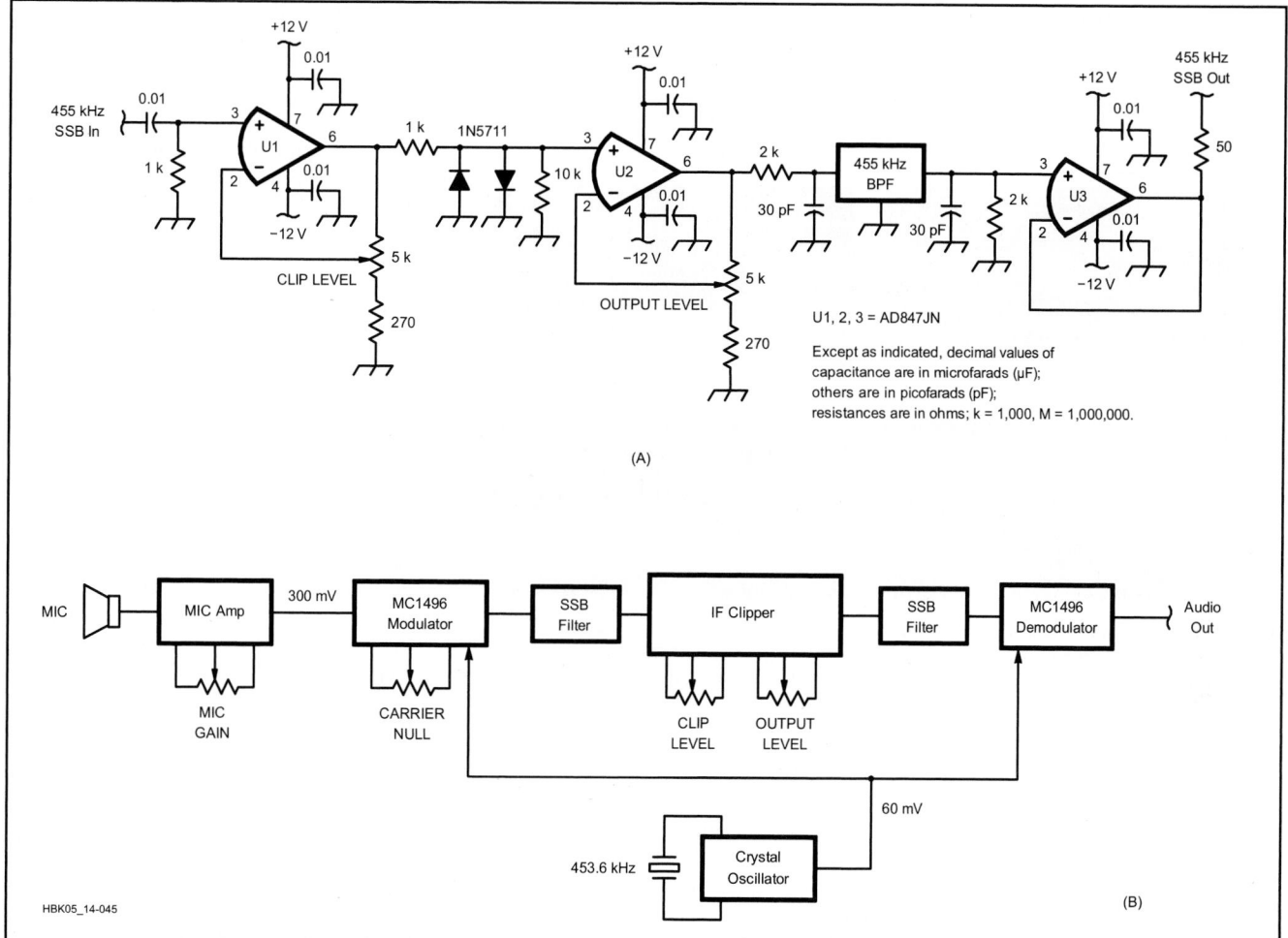

Fig 13.24 — IF speech clipping. At (A), schematic diagram of a 455 kHz IF clipper using high-frequency op amps. At (B) block diagram of an adaptation of the above system to an audio in-audio out configuration.

carrier frequency side, in order to improve rejection of the carrier and opposite sideband. Mechanical filters are available that do this. Crystal-ladder filters (see the **RF and AF Filters** chapter) are frequently called "single-sideband" filters because they also have this property. The steep skirt can be on the low side or the high side, depending on whether the crystals are across the signal path or in series with the signal path, respectively.

Filters require special attention to their terminations. The networks that interface the filter with surrounding circuits should be accurate and stable over temperature. They should be easy to adjust. One very good way to adjust them is to build a narrow-band sweep generator and look at the output IF envelope with a logarithmic amplifier, as indicated in Fig 13.23B. There are three goals:

• The driver stage must see the desired load impedance.

• The stage after the filter must see the desired source (generator) impedance.

• The filter must be properly terminated at both ends.

Lack of any of these conditions will result in loss of specified filter response. Fig 13.23B shows two typical approaches. This kind of setup is a very good way to make sure the filters and other circuitry are working properly.

Finally, overdriven filters (such as crystal or mechanical filters) can become nonlinear and generate distortion. Thus it is necessary to stay within the manufacturer's specifications. Magnetic core materials used in the tuning networks must be sufficiently linear at the signal levels encountered. They should be tested for IMD separately.

IF SPEECH CLIPPER

Audio clipper speech processors generate a considerable amount of in-band harmonics and IMD (involving different simultaneously occurring speech frequencies). The total distortion detracts somewhat from speech intelligibility. Other problems were mentioned in the earlier section on speech processing. IF clippers overcome most of these problems, especially the Hilbert Transform problem. (See Sabin and Schoenike in the References

section.)

Fig 13.24A is a schematic diagram of a 455 kHz IF clipper using high-frequency op-amps. 20 dB of gain precedes the diode clippers. A second amplifier establishes the desired output level. The clipping produces a wide band of IMD products close to the IF frequency. Harmonics of the IF frequency are easily rejected by subsequent selectivity. "Close-in" IMD distortion products are band-limited by the 2.5 kHz wide IF filter so that out-of-band splatter is eliminated. The in-band IMD products are at least 10 dB below the speech tones.

Fig 13.24B shows a block diagram of an adaptation of the above system to an audio in-audio out configuration that can be inserted into the mic input of any transmitter to provide the benefits of RF speech processing. These are sometimes offered as aftermarket accessories.

Fig 13.25 shows oscilloscope pictures of an IF clipped two-tone signal at various levels of clipping. The level of clipping in a radio can be estimated by comparing with these photos.

Listening tests verify that the IMD does not sound nearly as bad as harmonic distortion. In fact, processed speech sounds relatively clean and crisp. Tests also verify that speech intelligibility in a noise background is improved by 8 dB. (See the article on RF clippers by Sabin in the References section.)

The repeaking effect from band-pass filtering the clipped IF signal occurs, and must be accounted for when adjusting the output level. A two-tone audio test signal or a speech signal should be used. The ALC circuitry (discussed later) will reduce the IF gain to prevent splattering in the power amplifiers. If the IF filter is of high quality and if subsequent amplifiers are clean, the transmitted signal is of very high quality and is very effective in noisy situations and often also in pile-ups.

The extra IF gain implies that the IF signal entering the clipper must be free of noise, hum and spurious products. The cleanup filter also helps reduce the carrier frequency, which is outside the passband.

An electrically identical approach to the IF clipper can be achieved at audio frequencies. If the audio signal is translated to, say 455 kHz, processed as described and translated back to audio, all the desirable effects of IF clipping are retained. This output then plugs into the transmitter's microphone jack. Fig 13.24B shows the basic method. The mic amplifier and the MC1496 circuits have been previously shown and the clipper circuit can be the same as in Fig 13.24A.

The interesting operating principle in all of these examples is that the characteristics of the IF clipped (or equivalent) speech signal do not change during frequency translation, even if translated down to audio and then back up to IF in a balanced modulator.

IF Linearity and Noise

Fig 13.18 indicates that after the last SSB filter, whether it is just after the SSB modulator or after the IF clipper, subsequent BPFs are considerably wider. For example, the 70 MHz crystal filter may be 15 to 30 kHz wide. This means that there is a "gray region" in the transmitter in which out-of-band IMD that is generated in the IF amplifiers and mixers can cause adjacent-channel interference.

A possible exception, not shown in Fig 13.18, is that there may be an intermediate IF in the 10 MHz region that also contains a narrow filter.

The implication is that special attention must be paid to the linearity of these circuits. It's the designer's job to make sure that distortion in this gray area is much less than distortion generated by the PA and also less than the phase noise generated by the final mixer. Recall also that the total IMD generated in the exciter stages is the result of several amplifier and mixer stages in cascade; therefore, each element in the chain must have at least 40 to 50 dB IMD quality. The various drive levels should be chosen to guarantee this. This requirement for multistage linearity is one of the main technical and cost burdens of the SSB mode.

Of interest also in the gray region are additive white, thermal and excess noises origi-

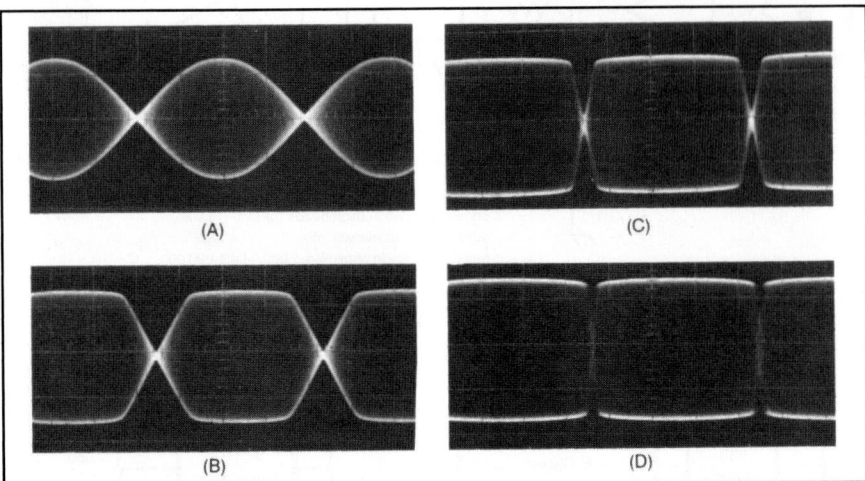

Fig 13.25 — Two-tone envelope patterns with various degrees of RF clipping. All envelope patterns are formed using tones of 600 and 1000 Hz. At A, clipping threshold; B, 5 dB of clipping; C, 10 dB of clipping; D, 15 dB of clipping.

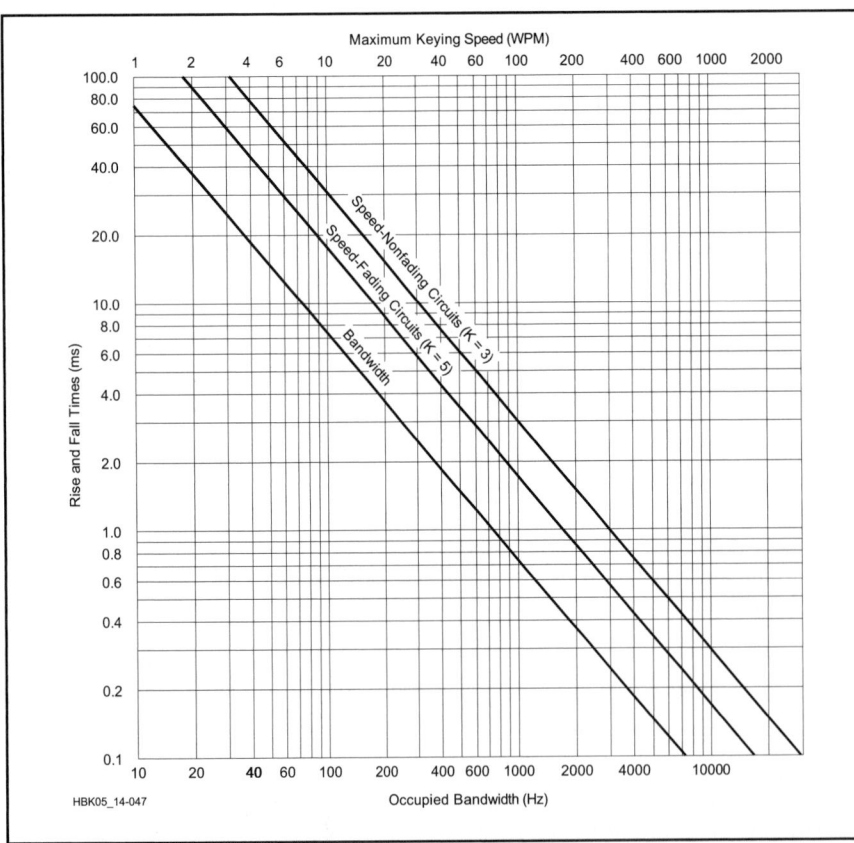

Fig 13.26 — Keying speed versus rise and fall times versus bandwidth for fading and nonfading communications circuits. For example, for transmitter output waveform rise and fall times of approximately 6 ms, draw a horizontal line from 6.0 ms on the rise and fall times scale to the bandwidth line. Then draw a vertical line to the occupied bandwidth scale at the bottom of the graph. In this case the bandwidth is about 130 Hz. Also extend the 6.0 ms horizontal line to the K = 3 line for a nonfading circuit. Finally draw a vertical line from the K = 3 line to the WPM axis. The 6 ms rise and fall time should be suitable for keying speeds up to about 50 WPM in this example.

nating in the first IF amplifier after the SSB filter and highly magnified on their way to the output. This noise can be comparable to the phase noise level if the phase noise is low, as it is in a high-quality radio. Recall also that phase noise is at its worst on modulation peaks, but additive noise may be (and often is) present even when there is no modulation. This is a frequent problem in co-located transmitting and receiving environments. Many transmitter designs do not have the benefit of the narrow filter at 70 MHz, so the amplified noise can extend over a much wider frequency range.

13.1.7 CW Operation

Radiotelegraph or CW operation can be easily obtained from the transmitter architecture design shown in Fig 13.18. For CW operation, a carrier is generated at the center of the SSB filter passband. There are two ways to make this carrier available. One way is to unbalance the balanced modulator so that the LO can pass through. Each kind of balanced modulator circuit has its own method of doing this. The approach chosen in Fig 13.18 is to

go around the modulator and the SSB filter.

A shaping network controls the envelope of the IF signal to accomplish two things: control the shape of the Morse code character in a way that limits wideband spectrum emissions that can cause interference, and make the Morse code signal easy and pleasant to copy.

RF ENVELOPE SHAPING

On-off keying (CW) is a special kind of low-level amplitude modulation (a low signal-level stage is turned on and off). It is special because the sideband power is subtracted from the carrier power, and not provided by a separate "modulator" circuit, as in high-level AM. It creates a spectrum around the carrier frequency whose amplitude and bandwidth are influenced by the rates of signal amplitude rise and fall and by the curvature of the keyed waveform. Refer to the discussion of keying speed, rise and fall times, and bandwidth in the **Modulation** chapter and the earlier discussion in this chapter for some information about this issue. For additional information see the article by Sabin on IF signal processing in the References section of this chapter.

Now look at **Fig 13.26**. The vertical axis is labeled Rise and Fall Times (ms). For a rise/fall time of 6 ms (between the 10% and 90% values) go horizontally to the line marked Bandwidth. A –20 dB bandwidth of roughly 120 Hz is indicated on the lower horizontal axis. Continuing to the K = 5 and K = 3 lines, the upper horizontal axis suggests code speeds of 30 WPM and 50 WPM respectively.

These code speeds can be accommodated by the rise and fall times displayed on the vertical axis. For code speeds greater than these the Morse code characters become "soft" sounding and difficult to copy, especially under less-than-ideal propagation conditions.

For a narrow spectrum and freedom from adjacent channel interference, a further requirement is that the spectrum must fall off very rapidly beyond the –20 dB bandwidth indicated in Fig 13.26. A sensitive narrow-band CW receiver that is tuned to an adjacent channel that is only 1 or 2 kHz away can detect keying sidebands that are 80 to 100 dB below the key-down level of a strong CW signal.

An additional consideration is that during key-up a residual signal, called "backwave,"

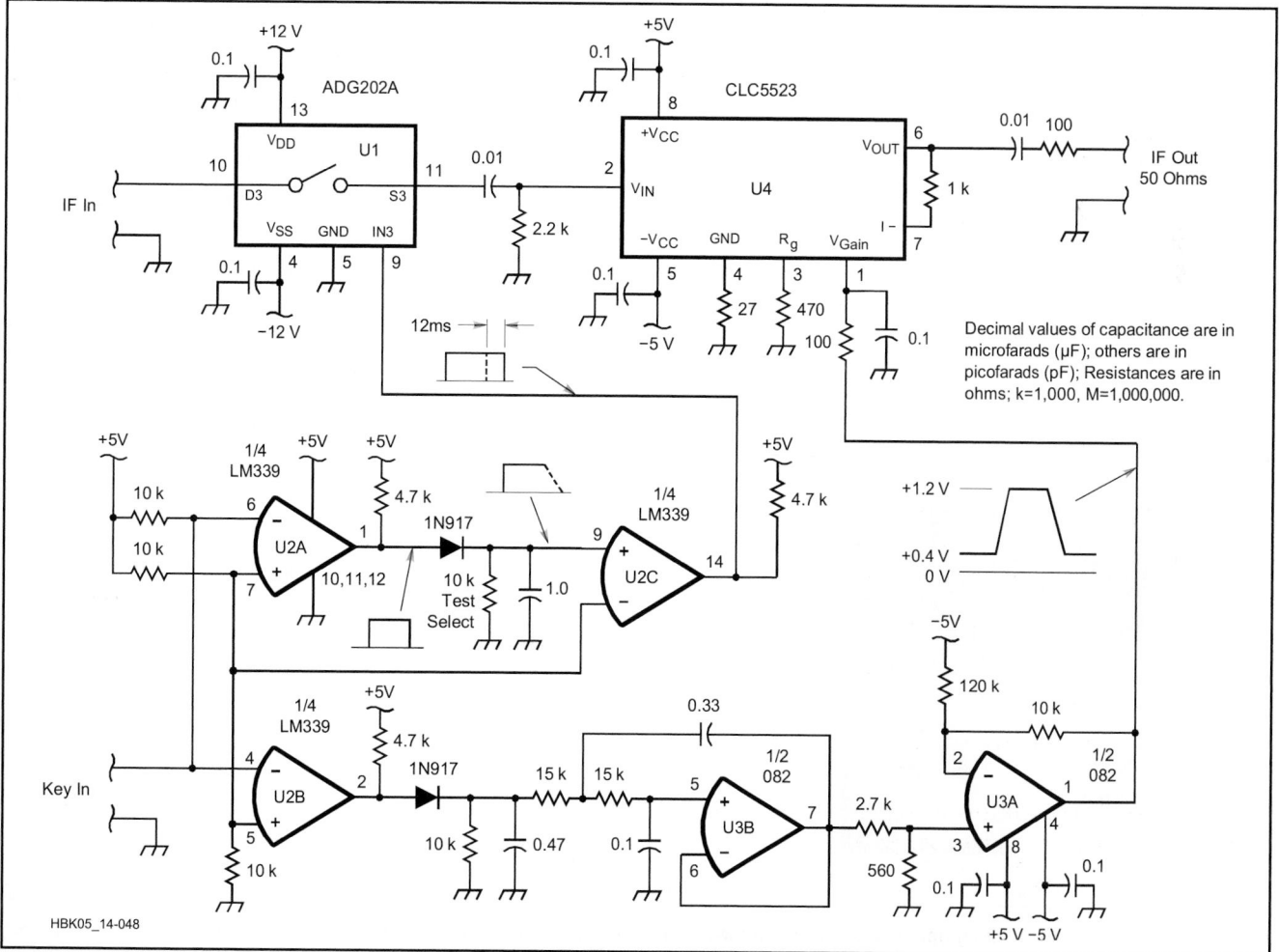

Fig 13.27 — This schematic diagram shows a CW waveshaping and keying circuit suitable for use with an SSB/CW transmitter such as is shown in Fig 13.18.

should not be noticeable in a nearby receiver. A backwave level at least 90 dB below the key-down carrier is a desirable goal.

Microprocessor-controlled transceivers manufactured today control CW keying rise- and fall-time through software. The operator generally accesses the keying shape parameter through a menu selection and adjustment process. Three to four ms is a typical value for most transceivers that balances crisp keying characteristics against excessive off-channel artifacts. See the *QST* Product Reviews for waveforms and discussions of rise- and fall-time settings.

Homebrew equipment usually relies on analog circuitry to control keying waveforms. **Fig 13.27** is the schematic of one waveshaping circuit that has been used successfully. A Sallen-Key third-order op amp low-pass filter (0.1 dB Chebyshev response) shapes the keying waveform, produces the rate of rise and fall and also softens the leading and trailing corners just the right amount. The key closure activates the CMOS switch, U1, which turns on the 455-kHz IF signal. At the key-up time, the input to the wave-shaping filter is turned off, but the IF signal switch remains closed for an additional 12 ms.

The keying waveform is applied to the gain control pin of a CLC5523 amplifier IC. This device, like nearly all gain-control amplifiers, has a *logarithmic* control of gain; therefore some experimental "tweaking" of the capacitor values was used to get the result shown in **Fig 13.28A**. The top trace shows the on/off operation of the IF switch, U1. The signal is turned on shortly before the rise of the keying pulse begins and remains on for about 12 ms after the keying pulse is turned off, so that the waveform falls smoothly to a very low value. The result is an excellent spectrum and an almost complete absence of back-

wave. Compare this to the factory transmitter waveshapes shown in Figs 13.1 and 13.3. The bottom trace shows the resulting keyed RF output waveshape. It has an excellent spectrum, as verified by critical listening tests. The thumps and clicks that are found in some CW transmitters are virtually absent. The rise and fall intervals have a waveshape that is approximately a cosine. Spread-spectrum frequency-hop waveforms have used this approach to minimize wideband interference.

Fig 13.28B is an accurate *SPICE* simulation of the wave shaping circuit output before the signal is processed by the CLC5523 amplifier. To assist in adjusting the circuit, create a steady stream of 40 ms dots that can be seen on an RF oscilloscope that is looking at the final PA output envelope. It is important to make sure that the excellent waveshape is not degraded on its way to the transmitter output. Single-sideband linear power amplifiers are well suited for a CW transmitter, but they must stay within their linear range, and the backwave problem must be resolved.

When evaluating the spectrum of an incoming CW signal during on-the-air operations, a poor receiver design can contribute problems caused by its vulnerability to a strong but clean adjacent channel signal. Clicks, thumps, front end overload, reciprocal mixing, etc can be created in the receiver. It is important to put the blame where it really belongs.

13.1.8 Wideband Noise

With receiver sensitivity, selectivity, and linearity having reached extraordinary levels of performance, a reduction in transmitted spurious emissions is clearly in the best interests of all amateurs. It does us no good to spend time and effort creating an exceptional

receiver if the channel is filled with transmitted noise and distortion products! This is of particular concern when multiple transmitters are located at the same station or facility such as at emergency communications stations, during Field Day and special events, and at multi-transmitter contest stations. (See the article by Grebenkemper in the Reference section.)

In the block diagram of Fig 13.18, the last mixer and the amplifiers after it are wideband circuits that are limited only by the harmonic filters and by any selectivity that may be in the antenna system. Wide-band phase noise transferred onto the transmitted modulation by the last LO (almost always a synthesizer of some kind) can extend over a wide frequency range, therefore LO cleanliness is always a matter of great concern.

The amplifiers after this mixer are also sources of wide-band "white" or additive noise. This noise can be transmitted even during times when there is no modulation, and it can be a source of local interference. To reduce this noise, use a high-level mixer with as much signal output as possible, and make the noise figure of the first amplifier stage after the mixer as low as possible.

Commercial and military transmitters that are used in close proximity to receivers, such as on ships and aircraft, are always designed to control wideband emissions of both additive noise and phase noise, referred to as "composite" noise.

TRANSMIT MIXER SPURIOUS SIGNALS

The last IF and the last mixer LO in Fig 13.18 are selected so that, as much as possible, harmonic IMD products are far enough away from the operating frequency that they fall outside the passband of the low-

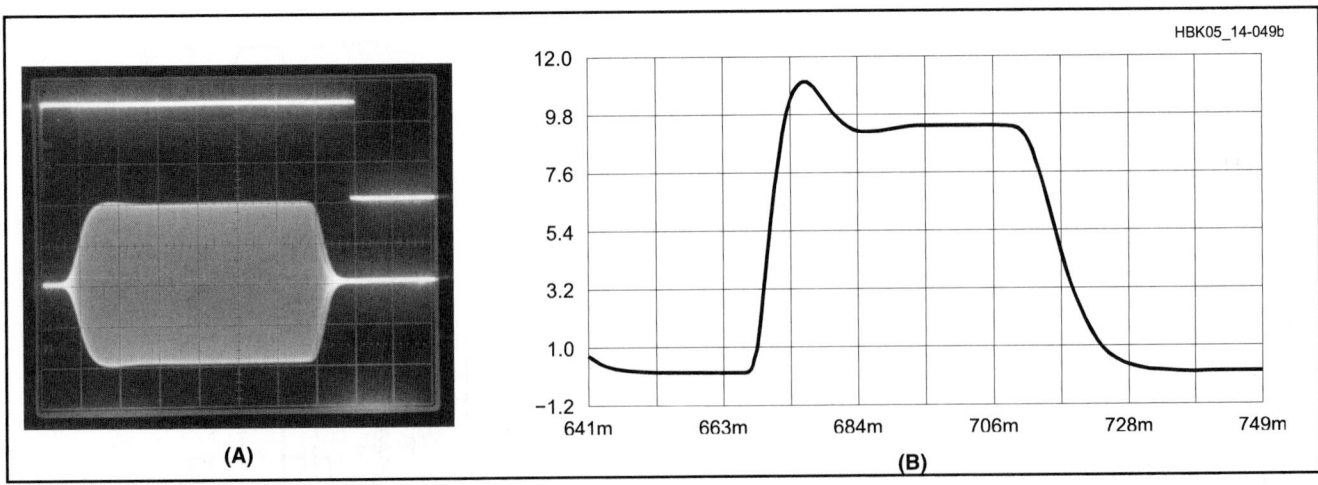

Fig 13.28 — At (A) is the oscilloscope display of the CW waveshaping and keying circuit output. The top trace is the IF keying signal applied to S1 of Fig 13.27. The bottom trace is the transmitter output RF spectrum. At (B) is a *SPICE* simulation of the waveshaping network. When this signal is applied to the logarithmic control characteristic of the CLC5523 amplifier, the RF envelope is modified slightly to the form shown in A.

pass filters and are highly attenuated. This is difficult to accomplish over the transmitter's entire frequency range. It helps to use a high-level mixer and a low enough signal level to minimize those products that are unavoidable. Low-order crossovers that cannot be sufficiently reduced are unacceptable, however; the designer must go back to the drawing board.

13.1.9 Automatic Level Control (ALC)

The purpose of ALC is to prevent the various stages in the transmitter from being overdriven. Over-drive can generate too much out-of-band distortion or cause excessive power dissipation, either in the amplifiers or in the power supply. ALC does this by sampling the peak amplitude of the modulation (the envelope variations) of the output signal and then developing a dc gain-control voltage that is applied to an early amplifier stage, as suggested in Fig 13.18.

ALC is usually derived from the last stage in a transmitter. This ensures that this last stage will be protected from overload. However, other stages prior to the last stage may not be as well protected; they may generate excessive distortion. It is possible to derive a composite ALC from more than one stage in a way that would prevent this problem. But designers usually prefer to design earlier stages conservatively enough so that, given a temperature range and component tolerances, the last stage can be the one source of ALC. The gain control is applied to an early stage so that all stages are aided by the gain reduction.

Note that ALC should be minimally active with most digital mode transmissions. The modulation of these signals requires linear amplification to preserve the waveform shape and minimize distortion products. ALC action creates distortion as it alters the power level of the signal. Adjust the radio drive levels so that the ALC is at its minimum level of activity – usually shown as the lower bar of a multi-segment LCD meter or a needle position just above zero. (The same caution applies to any form of audio or speech processing if the digital signal is generated by audio tones applied to the transmitter's microphone input.)

SPEECH PROCESSING WITH ALC

A fast response to the leading edge of the modulation is needed to prevent a transient overload. After a strong peak, the control voltage is "remembered" for some time as the voltage in a capacitor. This voltage then decays partially through a resistor between peaks. An effective practice provides two capacitors and two time constants. One capacitor decays quickly with a time constant

of, say 100 ms, the other with a time constant of several seconds. With this arrangement a small amount of speech processing, about 1 or 2 dB, can be obtained. (Explanation: The dB of improvement mentioned has to do with the improvement in speech intelligibility in a random noise background. This improvement is equivalent to what could be achieved if the transmit power were increased that same number of dB.)

The gain rises a little between peaks so that weaker speech components are enhanced. But immediately after a peak it takes a while for the enhancement to take place, so weak components right after a strong peak are not enhanced very much. **Fig 13.29A** shows a complete ALC circuit that performs speech processing.

ALC IN SOLID-STATE POWER AMPLIFIERS

Fig 13.29B shows how a dual directional

coupler can be used to provide ALC for a solid-state power amplifier (PA). The basic idea is to protect the PA transistors from excessive SWR and dissipation by monitoring both the forward power and the reflected power.

TRANSMIT GAIN CONTROL (TGC)

This is a widely used feature in commercial and military equipment. A calibrated "tune-up" test carrier of a certain known level is applied to the transmitter. The output carrier level is sampled, using a diode detector. The resulting dc voltage is used to set the gain of a low-level stage. This control voltage is digitized and stored in memory so that it is semi-permanent. A new voltage may be generated and stored after each frequency change, or the stored value may be used. A test signal is also used to do automatic antenna tuning. A dummy load is used to set the level and a low-level signal (a few mW) is used for the antenna tune-up.

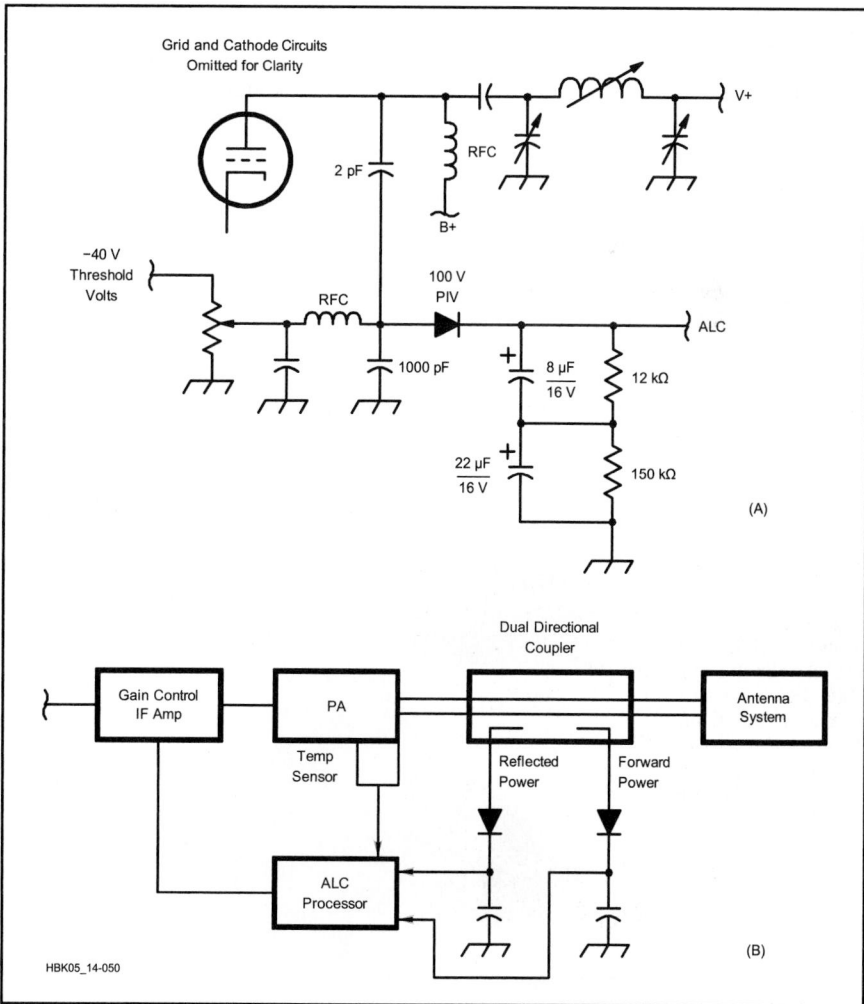

Fig 13.29 — At (A), an ALC circuit with speech processing capability. At (B), protection method for a solid-state transmitter.

13.1.10 *Project:* The MicroT2 — A Compact Single-Band SSB Transmitter

As an example of an SSB transmitter including many aspects of design covered heretofore, we present the MicroT2, shown in **Fig 13.30**, a simple SSB transmitter that generates a high-quality USB or LSB signal on any single band from 1.8 MHz to 50 MHz. Rick Campbell, KK7B, developed the MicroT2 as a companion to the MicroR2 receiver project described in the **Receivers** chapter.

While it is a bit more involved to generate an SSB signal than a CW signal, we greatly simplify the task if all the necessary circuitry is on a single PC board exciter module. Once we have a high-quality low-level SSB signal, a 5 or 500 W SSB transmitter is as easy to build as a 5 or 500 W CW transmitter. Simple transmitters are delightful, but relaxed standards are not. The MicroT2 is designed to be clean, stable and reliable, exceed FCC Part 97 requirements, and sound good, too.

A thorough description of the circuitry in this transmitter can be found in *Experimental Methods in RF Design (EMRFD)* of which the project's designer was a co-author. The complete article for this project, including schematics, parts lists, and adjustment instructions, can be found on the CD-ROM that accompanies this book.

EXCITER BLOCK DIAGRAM

RF Circuitry

Fig 13.31 is the block diagram of the circuitry on the PC board. The exciter uses the

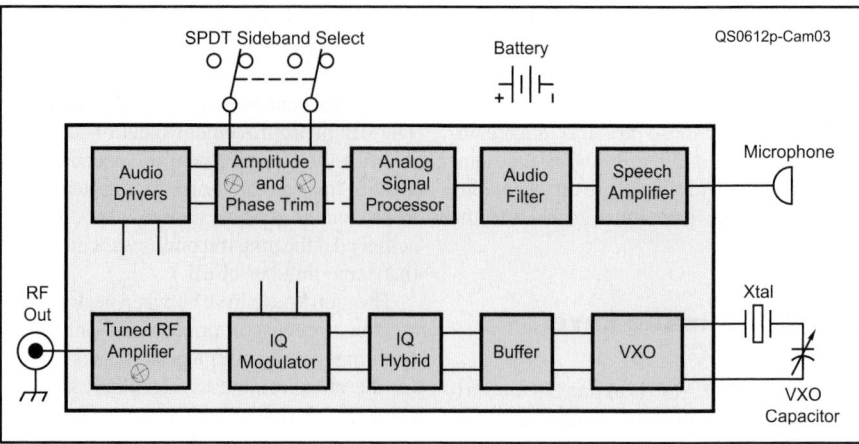

Fig 13.31 — Block diagram of the circuitry on the PC board.

phasing method of SSB generation, which makes it easy to operate on different frequencies. In a phasing SSB exciter, two identical signals with a 90° phase difference are generated and then combined so that one sideband adds and the other subtracts. The signal quality from this exciter is not merely adequate for the HF amateur bands — it is exceptional.

The frequency generator consists of a VXO, buffer amplifier and quadrature hybrid. It has a lot of parts: three transistors, a voltage regulator IC, a Zener diode, four toroids, and many resistors and capacitors. There are no adjustments. You just build it and it works. The frequency stability is better than that of most commercial radios, even when portable.

A pair of Mini-Circuits TUF-3 mixers

serves as the I and Q balanced modulators. These provide good carrier suppression without adjustment, and reasonable output at a modest distortion level. IM products in the opposite sideband are more than 30 dB down at the exciter output. The aggressive low-pass filtering right at the mixer IF ports prevents wideband noise and harmonic distortion in the audio stages from contributing to the I and Q modulation.

The exciter's output RF amplifier uses a common-gate JFET. The RF amplifier has relatively low gain, good harmonic suppression and a very clean 0 dBm SSB signal at the output. This is an appropriate level to drive a linear amplifier, balanced mixer or transverter. It is also a low enough level that it is easy to adjust the exciter with simple equipment. The exciter output signal meets FCC regulations for direct connection to an antenna for flea power experiments.

The transmitter is completed by the simple two transistor amplifier with 0.5 W PEP output. A seventh order Chebyshev low-pass filter on the output is noncritical and assures a clean signal that easily meets FCC regulations.

Audio Section

The speech amplifier drives a passive low-pass filter. The combination of this low-pass filter and the mixer IF port filters limits speech frequencies to just over 3 kHz for natural sounding speech and good spectral purity. There is no ripple in the audio passband.

The position of the sideband select switch in the signal path allows switching without readjusting the amplitude and phase trimmers. For most applications, one sideband will be used exclusively, and the sideband switch may be replaced by a pair of jumpers on the PC board. If that results in the wrong sideband, reverse the connections between the audio driver transistors and mixers.

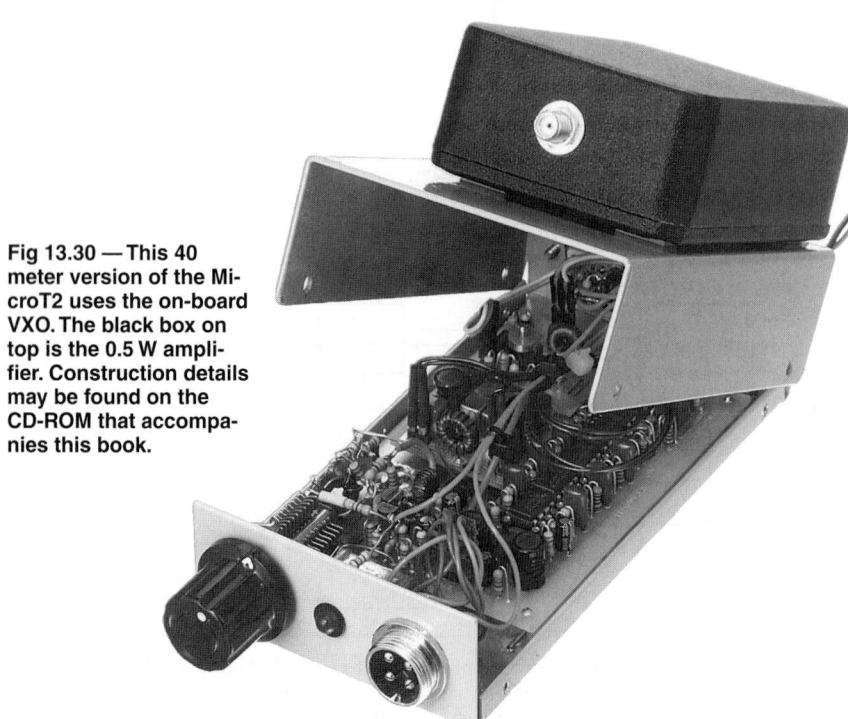

Fig 13.30 — This 40 meter version of the MicroT2 uses the on-board VXO. The black box on top is the 0.5 W amplifier. Construction details may be found on the CD-ROM that accompanies this book.

13.1.11 *Project:* The MkII — An Updated Universal QRP Transmitter

A frequently duplicated project in the now out-of-print book *Solid State Design for the Radio Amateur* (see References) was a universal QRP transmitter. This was a simple two-stage, crystal-controlled, single-band circuit with an output of about 1.5 W. The no-frills design used manual transmit-receive (TR) switching. It operated on a single frequency with no provision for frequency shift. The simplicity prompted many builders to pick this QRP rig as a first solid state project.

Wes Hayward, W7ZOI, updated the design to the three-stage MKII (**Fig 13.32** and **Fig 13.33**), develops an output of 4 W on any single band within the HF spectrum, if provided with 12 V dc. Q1 is a crystal controlled oscillator that functions with either fundamental or overtone mode crystals. It operates at relatively low power to minimize stress to some of the miniature crystals now available. The three stage design — two driver stages and a power amplifier — provides an easy way to obtain very clean keying.

The complete article for this project, including schematics, parts lists, and adjustment instructions, can be found on the CD-ROM that accompanies this book.

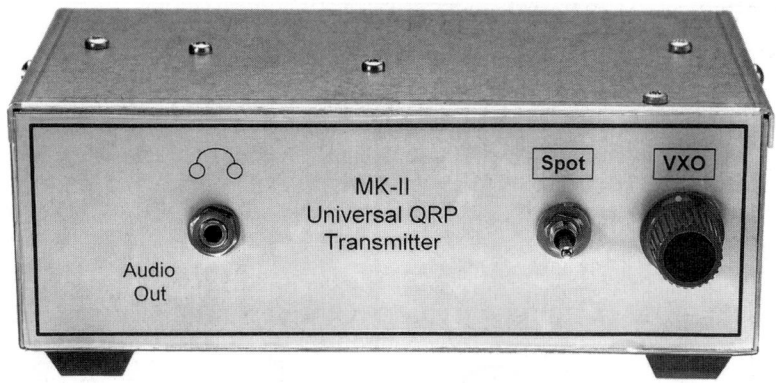

Fig 13.32 — The MKII QRP transmitter includes VXO frequency control, TR switching and a sidetone generator. Construction details may be found on the CD-ROM that accompanies this book.

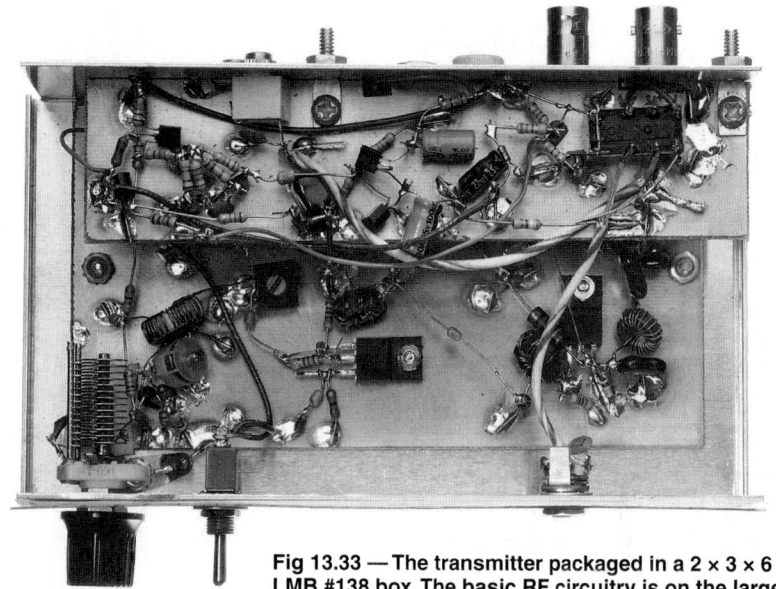

Fig 13.33 — The transmitter packaged in a 2 × 3 × 6 inch LMB #138 box. The basic RF circuitry is on the larger board. TR control is on the smaller board along the top.

13.2 VHF Signal Sources

The following section was written by Rick Campbell, KK7B, both as a technical description of circuits to produce exciter-level signals at VHF and a design tutorial for students learning about RF electronics. In both senses, the circuits are useful to amateurs who are interested in building equipment at 50 MHz and above. Amateurs may find it sufficient to use the component values published here to build either a 50 or 144 MHz exciter. The interested reader-student can adapt the circuits for any frequency between 21 and 150 MHz. An additional article with more information on the design of these sources, "VHF Open Sources" by KK7B is available on this book's CD-ROM.

Along with these general-purpose signal sources, a *QST* article by Mark Spencer, WA8SME, describing a standalone 2 meter transmitter to support direction-finding and fox-hunting activities is provided on this book's accompanying CD-ROM.

13.2.1 Overview

From 2005 through 2008, Portland State University RF Design students analyzed, simulated, built, tested, and modified beginner's transmitter and simple receiver circuits for 7 MHz. While these basic oscillator-amplifier transmitter and direct conversion receiver circuits have a number of excellent properties as both basic radios and teaching tools, they have several significant weaknesses. Many students successfully designed and built impressive small CW rigs and tested them in the lab, but never connected them to full size antennas or made contacts on the air: 40 meter antennas are too large for typical student quarters and the Technician license is not geared toward entry-level operating on 40 meter CW. Another weakness for classroom use is that many of the critically important device parasitic, lead length, and circuit transmission line challenges for 21st century engineering students may be ignored at 7 MHz.

In 2009 a new VHF signal source was developed for fundamental study that would open the 50 MHz band to student experimenters. Wire antennas for 6 meters are simple, effective, and portable, and the 6 meter band is wide open to Technician licensees. During the project development, it was determined experimentally that the same circuit board could be used on 144 MHz. Engineering students learn to design all the components for any multiple and output frequency between 21 and 150 MHz.

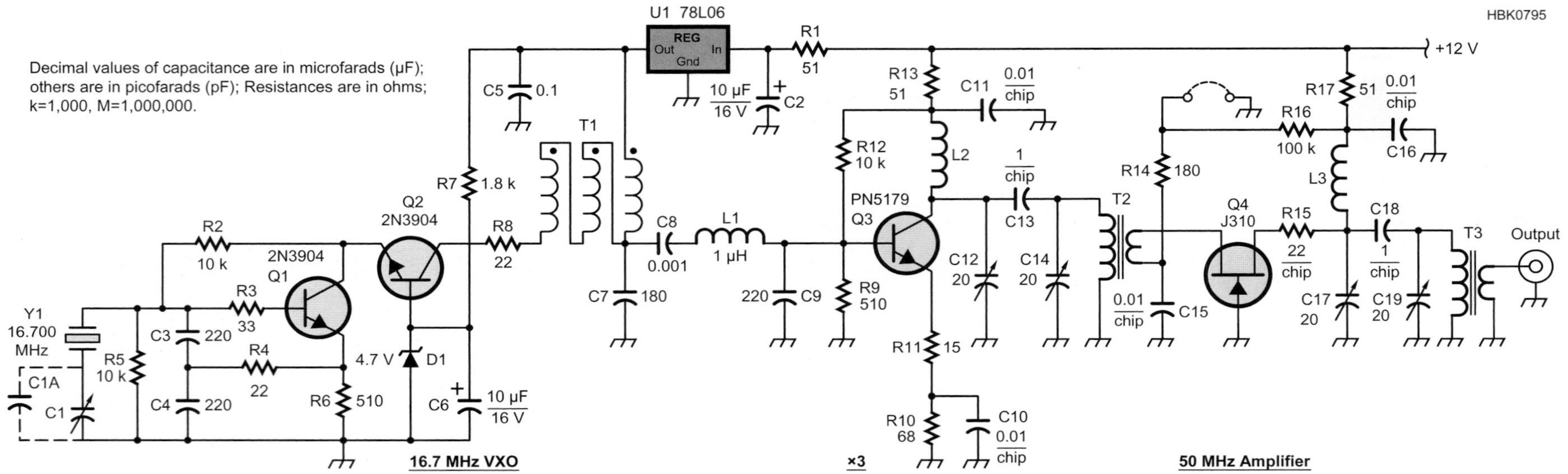

Decimal values of capacitance are in microfarads (µF); others are in picofarads (pF); Resistances are in ohms; k=1,000, M=1,000,000.

16.7 MHz VXO

50 MHz Amplifier

(A)

Fig 13.34 — (A) Schematic of the 6 meter +10 dBm signal source. Resistors are ¼ W except as noted. The output lowers gracefully to +7 dBm at 9 V with no change in frequency. Below 8 V, the 78L06 regulator drops out. (B) Photograph of the prototype 6 meter signal source from Kanga US (www.kangaus.net). A parts placement diagram for the kit PC board is available on this book's CD-ROM.

C1 — 15 pF PC mount air variable
C2,C6 — 10 µF or 6.8 µF 16 V electrolytic
C3,C4, C9 — 220 pF NP0 miniature disk ceramic
C5 — 100 nF poly
C7 — 180 pF miniature disk ceramic
C8 — 1000 pF miniature disk ceramic
C10, C11, C15, C16 — 10 nF 1206 chip ceramic
C12, C14, C17, C19 — 20 pF green film trimmer
C13, C18 — 1 pF 0806 chip ceramic
D1 — 4.7 V Zener
L1 — 20t #28 AWG on T25-6 toroid core (approx 1 µH)
L2, L3 — 14t #28 AWG on T25-6 toroid core
Q1, Q2 — 2N3904

Q3 — MPS 5179
Q4 — J310
R1, R8, R13, R15, R17 — 51 Ω
R2, R5, R12 — 10 kΩ
R3 — 33 Ω
R4 — 22 Ω
R6, R9 — 510 Ω
R7 — 1.8 kΩ
R10 — 68 Ω
R11 — 15 Ω
R14 — 180 Ω
R15 — 22 Ω 1206 chip
R16 — 100 kΩ
T1 — 6t trifilar on FT23-43 toroid core
T2, T3 — 14t primary, 2t secondary, #28 AWG on T25-6 toroid core
U1 — 78L06
Y1 — 16.700 MHz crystal

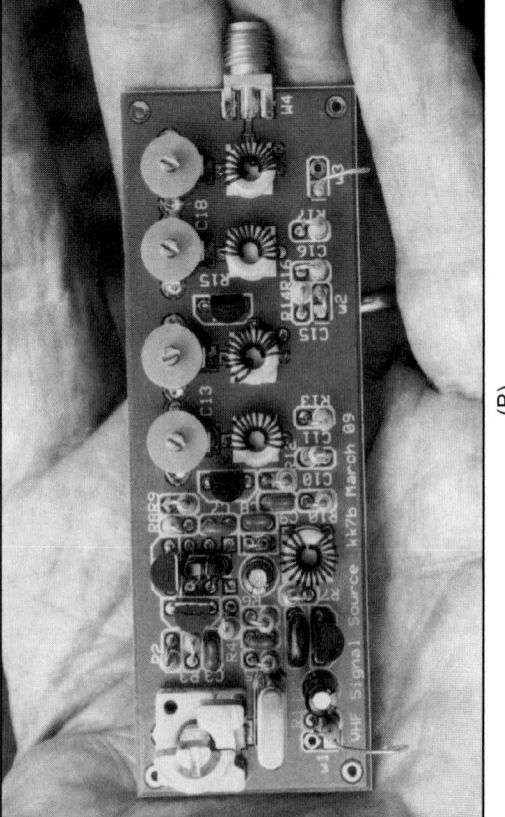

(B)

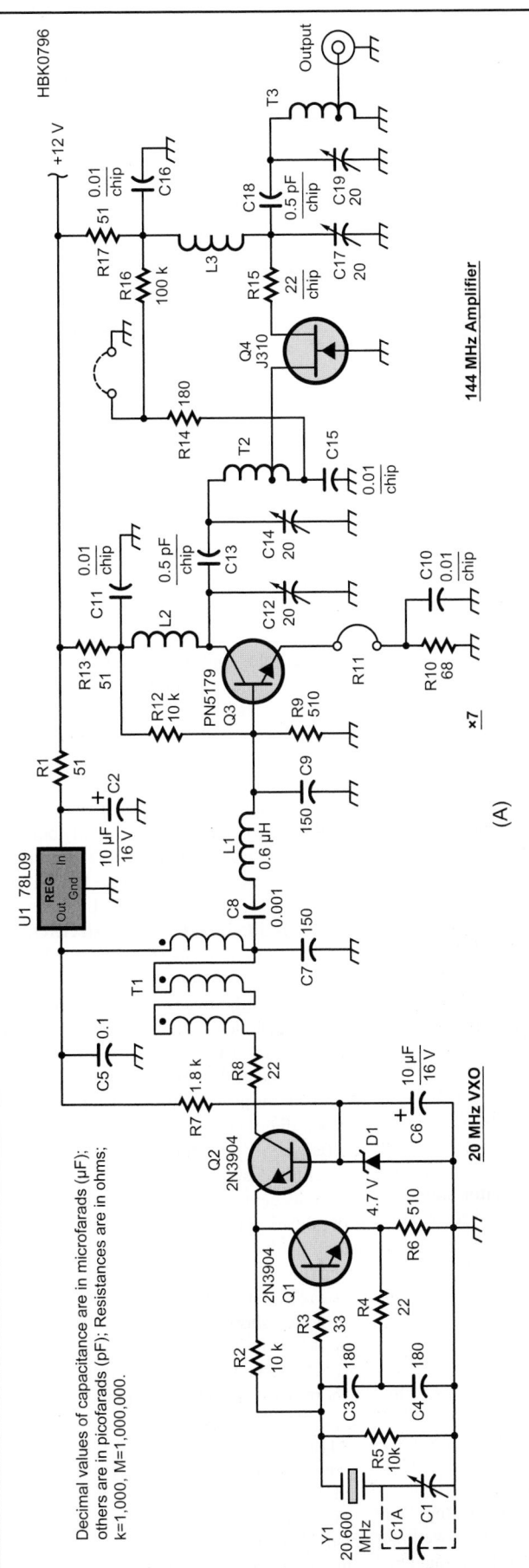

Decimal values of capacitance are in microfarads (µF); others are in picofarads (pF); Resistances are in ohms; k=1,000, M=1,000,000.

(A)

Fig 13.35 — (A) Schematic of the 2 meter +10 dBm signal source. Resistors are ¼ W except as noted. The output lowers gracefully to +8 dBm at 11 V with no change in frequency. Below 11 V, the 78L09 regulator drops out. (B) Photograph of the prototype 2 meter signal source from Kanga US (www.kangaus.net). A parts placement diagram for the kit PC board is available on this book's CD-ROM.

C1 — 15 pF PC mount air variable
C2,C6 — 10 µF, 16 V electrolytic
C3,C4 — 180 pF NP0 miniature disk ceramic
C5 — 100 nF poly
C7, C9 — 180 pF miniature disk ceramic
C8 — 1000 pF miniature disk ceramic
C10, C11, C15, C16 — 10 nF 1206 chip ceramic
C12, C14, C17, C19 — 20 pF green film trimmer
C13, C18 — 0.5 pF 0806 chip ceramic
D1 — 4.7 V Zener
L1 — 16t on T25-6 toroid core (approx 600 nH)
L2, L3 — 8t #24 AWG bare wire 1 cm long using ⅛ inch drill bit as form
Q1, Q2 — 2N3904
Q3 — MPS 5179
Q4 — J310
R1, R8, R13, R15, R17 — 51 Ω

R2, R5, R12 — 10 kΩ
R3 — 33 Ω
R4 — 22 Ω
R6, R9 — 510 Ω
R7 — 1.8 kΩ
R10 — 68 Ω
R11 — jumper
R14 — 180 Ω
R15 — 22 Ω 1206 chip
R16 — 100 kΩ
T1 — 6t trifilar on FT23-43 toroid core
T2, T3 — 8t #24 AWG bare wire 1 cm long using 1/8 inch drill bit as form; tap 1 turn from cold end
U1 — 78L09
Y1 — 20.600 MHz crystal

(B)

The nominal +10 dBm output power is ideal for driving receive mixers, modulators, power amplifiers, or antennas for low power experiments. The first prototype for the 2009 class is shown in **Fig 13.34A**, and the commercial 6 meter circuit board kit from Kanga (**www.kangaus.net**) is shown in Fig 13.34B. The 2 meter version is shown in **Fig 13.35** — note the use of small solenoid inductors wound with bare wire over a 0.1 inch form. (You may note minor component value differences between the schematics and the kit values from Kanga. This does not greatly affect performance and experimentation with the various values is encouraged. Both versions will work fine.)

The signal source may be used as a transmitter at the 10 mW level, connected directly to a dipole antenna. In 2010, RF design student teams with at least one Technician licensee built sources in pairs, along with direct conversion receivers and antennas, and carried them into the field for one-way CW communications. Best reported DX so far is 1.2 miles between dipoles, with no additional gain on receive and 10 mW output.

Any frequency between 24 and 175 MHz may be obtained by adjusting component values as noted in the schematics. For lower order N multipliers, use the 50 MHz emitter components in the frequency multiplier stage, and for N > 3 use the 2 meter components.

13.2.2 Oscillator-Buffer

The oscillator-buffer circuit shown in **Fig 13.36** is conventional, but with a number of subtle and elegant details. Q1 and associated resistors and capacitors form a Colpitts oscillator at the parallel resonant fundamental frequency of the crystal. The trimmer capacitor in series with the crystal allows the frequency to be adjusted over a few kHz. The low impedance load on the collector and the series resistors on the base and emitter suppress high frequency parasitic oscillations and reduce the impact of transistor variation on the circuit.

Buffer Q2 supplies a nearly constant 4 V at 2 mA dc to the collector of Q1, while loading the collector with a nearly constant RF impedance on the order of 10 Ω. The 4.7 V Zener on the base of Q2 sets the emitter voltage, and the capacitance of the Zener diode bypasses the base to ground at RF. Zener noise is bypassed to ground by the 10 μF electrolytic capacitor. The resistor from the base to the power supply should be adjusted for one or two mA current through the Zener. 1.5 kΩ is acceptable with a 6 volt regulated supply, and 3.3 kΩ works with the 9 V regulator used in the 144 MHz signal source.

For ultimate stability with a high quality temperature stabilized crystal, the Zener current could be experimentally optimized to the inflection point between Zener and avalanche

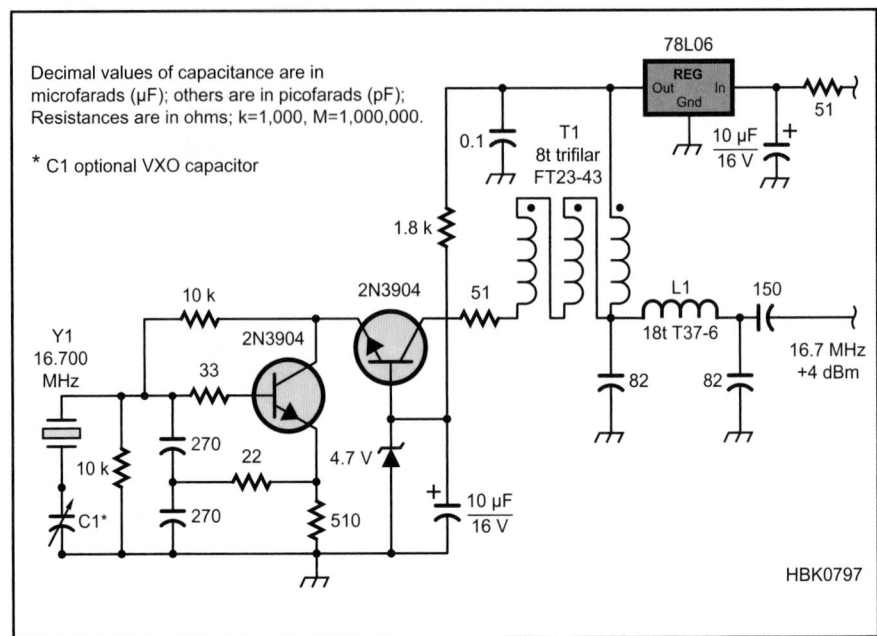

Fig 13.36 — Schematic for the oscillator-buffer circuit.

breakdown, but this is not necessary for room temperature operation of a normal crystal. The temperature of the crystal itself has a much larger impact on frequency drift and overall stability than variations in transistor parameters over temperature.

The output level of the oscillator-buffer circuit is determined by the output network and supply voltage. The peak-to-peak output signal level at the collector is somewhat less than the supply voltage, and the trifilar transformer and output pi-network step it down by a factor of 4. With a 9 volt supply, the output power is about 4 or 5 milliwatts into a 50 Ω load. The pi-network reduces harmonic content, so the output voltage appears sinusoidal on the oscilloscope.

Long- and short-term frequency stability are determined primarily by the crystal temperature and the quality of the series trimmer capacitor. For high stability applications the crystal should be wrapped in packing foam, and the trimmer capacitor should be an air variable. Placing the whole circuit in a temperature controlled environment isolated from any variable sources of heat can reduce oscillator drift to a less than a few Hz per hour.

13.2.3 Frequency Multiplier

The second block in the frequency source is a frequency multiplier. Mathematically its function is simple: for an input sine wave at frequency f the output is a sine wave at integer multiple N times f. In practice the frequency multiplier serves several other important functions, including isolation of the oscillator from the local electronic and electromagnetic environment at the output frequency.

Oscillator-multiplier frequency sources typically need less shielding than oscillators operating directly on the desired output frequency. This is of historical significance in transmitters built on wooden breadboards, and currently important in direct conversion receivers and exciters. Frequency multipliers with N less than 10 are conveniently built with single transistors and diodes. For multipliers with N greater than 10, phase locked loops (PLLs) are commonly used. PLLs are covered in the **Oscillators and Synthesizers** chapter.

Transistor frequency multipliers offer a fascinating introduction to waveform engineering. A simplified view treats the transistor as a non-linear device that generates a distorted waveform rich in harmonics, followed by a filter that selects the desired output. This simple explanation provides the right numbers for the block diagram, but doesn't explain the high efficiency or spectral purity obtained from a carefully designed frequency multiplier.

A deeper understanding considers the output tuned circuit and transistor as a coupled unit, mathematically equivalent to the piston and flywheel of an engine. The flywheel spins at a relatively constant speed, and the piston provides short power strokes over a small angle of the total rotation. The heavier the flywheel, the less the rotation speed varies as short power strokes are applied. In a single cylinder four-stroke internal combustion engine, the flywheel rotates twice for each power stroke — a frequency multiplication of 2. But the analogy goes further. When the piston is not providing power, the cylinder valves are open, and the flywheel easily drives all the mechanical parts around until it's time for the next compression and ignition stroke.

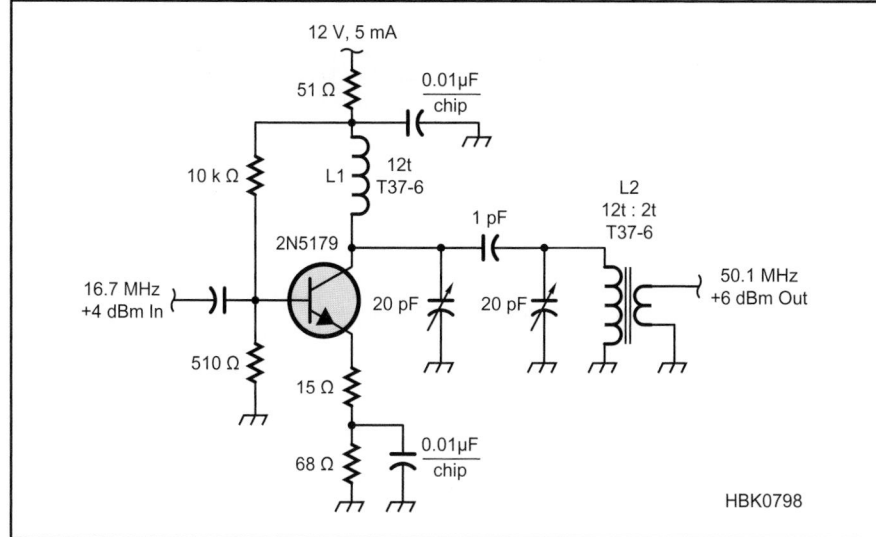

Fig 13.37 — Schematic for the 50 MHz × 3 multiplier circuit.

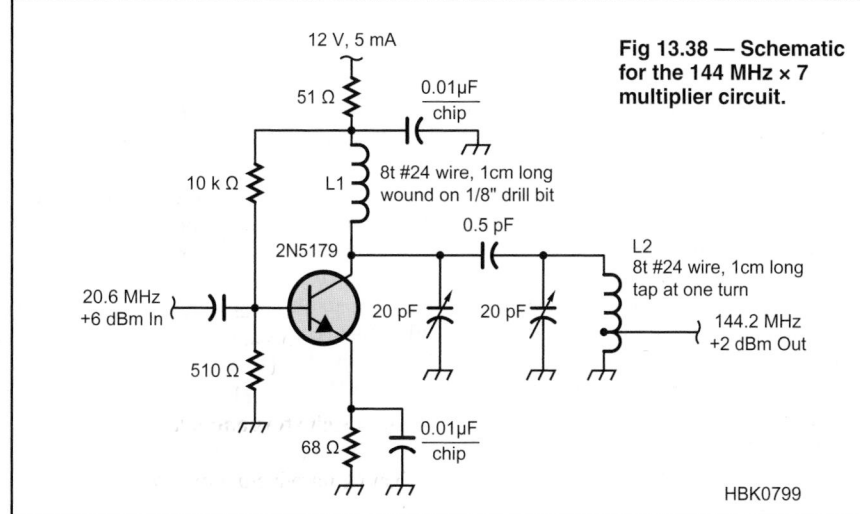

Fig 13.38 — Schematic for the 144 MHz × 7 multiplier circuit.

With a heavy flywheel and low friction, little energy is lost between power strokes.

A high-Q output circuit performs the flywheel function in a frequency multiplier. Calculating and controlling the optimum shape and timing of an engine power stroke are analogous to waveform engineering in a high efficiency Mode J or Class E power amplifier. The difference between a high efficiency power amplifier and a frequency multiplier is simply the number of power strokes per output cycle.

The frequency multiplier in **Fig 13.37** is driven by a sine wave, and the collector load is provided by the high Q tuned circuit on the output. Energy is coupled from the output tuned circuit to a second resonator through a very small capacitor. The collector current and voltage waveforms (power stroke) are determined by the collector load, drive level, and gain of the stage. Constant drive level is provided by the voltage-regulated oscillator-buffer circuit. Gain is adjusted for different harmonic numbers by changing the ac emitter resistance — note the difference between the 50 MHz times 3 and 144 MHz times 7 circuit in **Fig 13.38**. Adjusting a frequency multiplier with a tuned output requires some way of knowing that the resonant load is tuned to the correct harmonic. A spectrum analyzer is ideal, but for decades amateurs have adjusted frequency multipliers by listening on a receiver tuned to desired output frequency with a short wire antenna probe near the stage being tuned.

13.2.4 Output Bandpass Amplifier

The tuned common-gate output amplifier added to the multiplier schematic in **Fig 13.39**

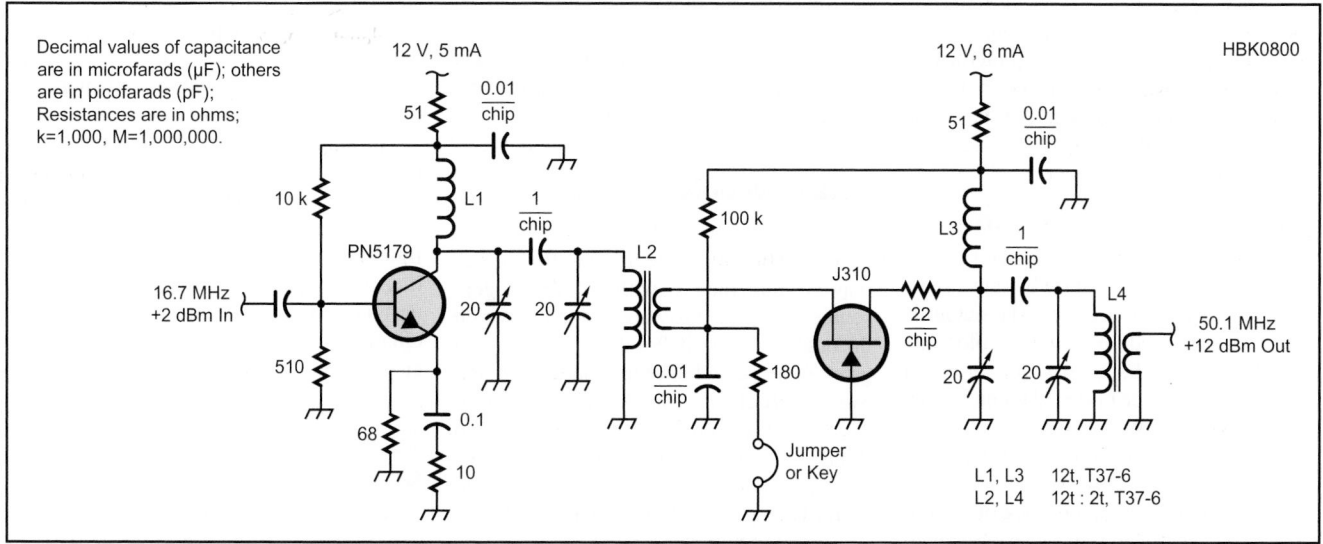

Fig 13.39 — Schematic for the tuned common-gate output amplifier. The multiplier schematic is included to illustrate the double-tuned circuits at the amplifier input and output.

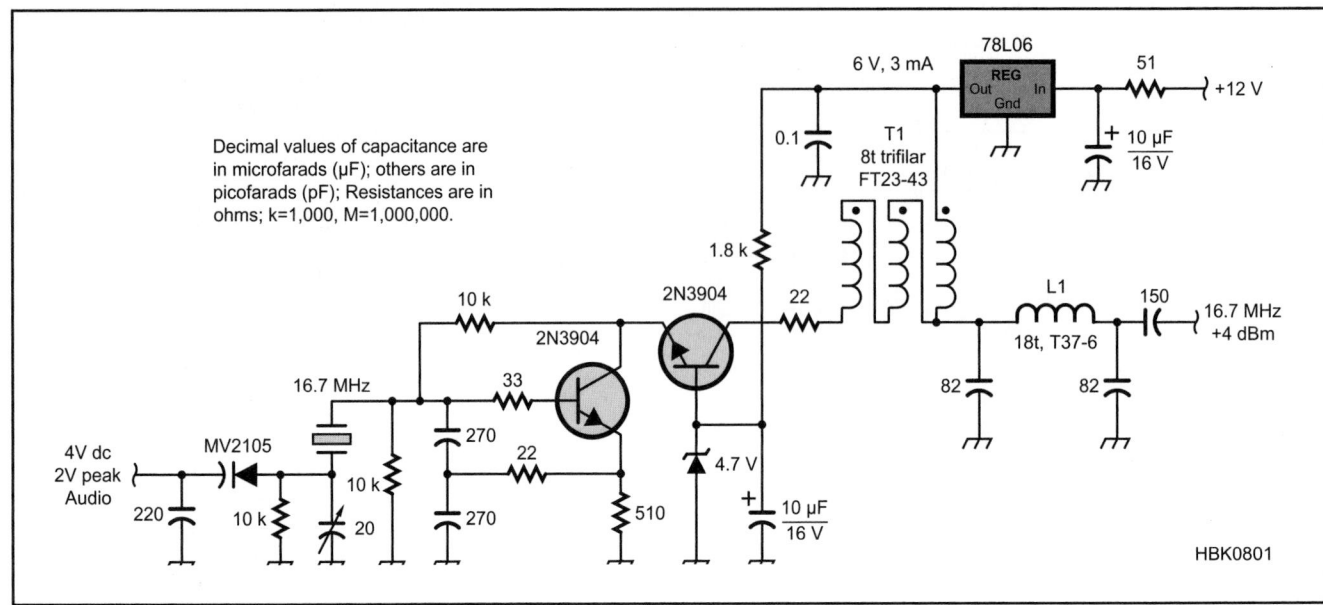

Fig 13.40 — Schematic showing how to add FM to the oscillator-multiplier boards.

provides nominal +10 dBm output power into a 50 Ω load, reduces the undesired harmonics of the fundamental crystal oscillator to below −70 dBc, and isolates the oscillator and multiplier circuit from external load variations. The jumper provides a convenient place to key the output amplifier, with key up isolation better than 40 dB.

Typical output from the 50 MHz VHF source is +12 dBm, and typical output for the 144 MHz signal source is about +8 dBm. The use of a continuous ground plane and chip capacitors for power supply bypassing reduces the need for shielding between stages on the circuit board. It is good practice to enclose any oscillator in a metal thermal and electrical shield box. The common gate configuration provides excellent isolation between the two double-tuned circuits, which means the tuning adjustments have very little interaction. Very little gain is needed, as the frequency multiplier output is a few milliwatts at the desired output frequency.

13.2.5 Tuning Up the Circuit

It is recommended that the circuit be initially built up and tested stage by stage, starting with the oscillator-buffer. The output is observed on a diode probe or oscilloscope after the pi-network, and the turns of toroid inductor L1 squeezed or spread for maximum output into 50 Ω. When experimenting with inductor values on a tight printed circuit board layout, adjustment is easier if the inductor is initially tack soldered to the back of the circuit board. Experiments are much easier with "ugly construction" (see the **Construction Techniques** chapter), and initial prototypes

of new designs are often built that way. The frequency multiplier stage components are then added, with the output connected to the link on the second tuned circuit. The correct harmonic may be identified by calculating frequency from the trace on the oscilloscope or by listening with a receiver tuned to the desired output frequency and a short wire antenna near the output. Finally, the output stage amplifier components are added and the complete circuit adjusted for maximum power output into 50 Ω at the desired output frequency.

Once a VHF signal source has been built and tested, it is easy to duplicate the circuit by copying the component placement, turns spacing on the inductors, and presetting the variable capacitors to match the working circuit. After one of the VHF signal sources is working, the rest quickly fall into place. VHF component spacing is necessarily tight and leads are short, so this is not a good first electronic construction project.

13.2.6 Modulating the VHF Sources

A number of basic modulation types may be implemented with the VHF signal sources. As previously noted, the jumper on the output stage switches the tuned common gate output stage between modest gain and over 40 dB attenuation. A telegraph key or simple push-button switch may be connected in place of the jumper for on-off keying of the VHF source output. Engineers refer to on-off keying as OOK, hams call it CW, and the general public refers to it as Morse code. In any case, it is still the most basic mode for communicating information electronically.

FM is easily implemented as shown in **Fig 13.40**. A varactor diode, 220 pF capacitor, and resistor are tacked to the back side of the circuit board. Students have experimented with Zener diodes operated below breakdown as varactor replacements with reasonable results. Several volts of audio floating at a dc offset of about 4 V is applied to the varactor. **Fig 13.41** is the microphone amplifier circuit used with 144 MHz signal sources.

The VHF signal source can produce envelope modulation — AM — using either high level modulation or by driving a diode ring modulator with a little dc offset to introduce carrier. For high level modulation, both the multiplier and output stage may be modulated by introducing an ac signal to the dc power supply. High level envelope modulation is best implemented and adjusted experimentally by watching the output envelope on an oscilloscope while trying circuit modifications and adjusting drive levels.

For single sideband or any of the digital modes, the +10 dBm output level of the VHF signal source is ideal for driving any of a number of different IQ modulator configurations. Several successful 6 meter SSB transmitters have been built by replacing the on-board HF crystal oscillator in the microT2 circuit described elsewhere in this text with an external VHF signal source.

13.2.7 Creating a Direct Conversion Receiver

As mentioned in the introduction, a simple and effective direct conversion receiver is easily built with the VHF signal source as a tunable LO. Tuning range is about 20 kHz on

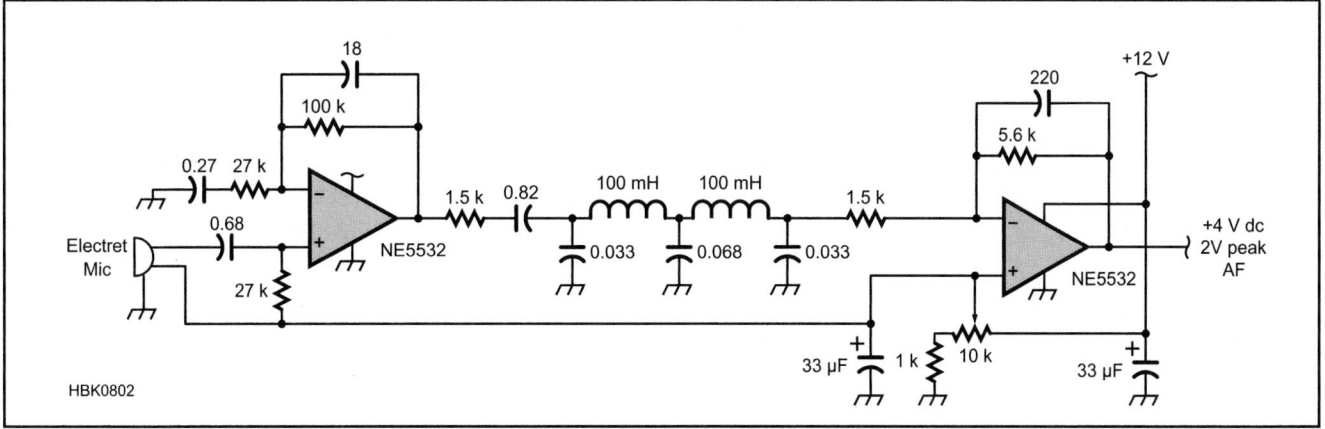

Fig 13.41 — Schematic of the microphone amplifier circuit used with the 2 meter signal sources.

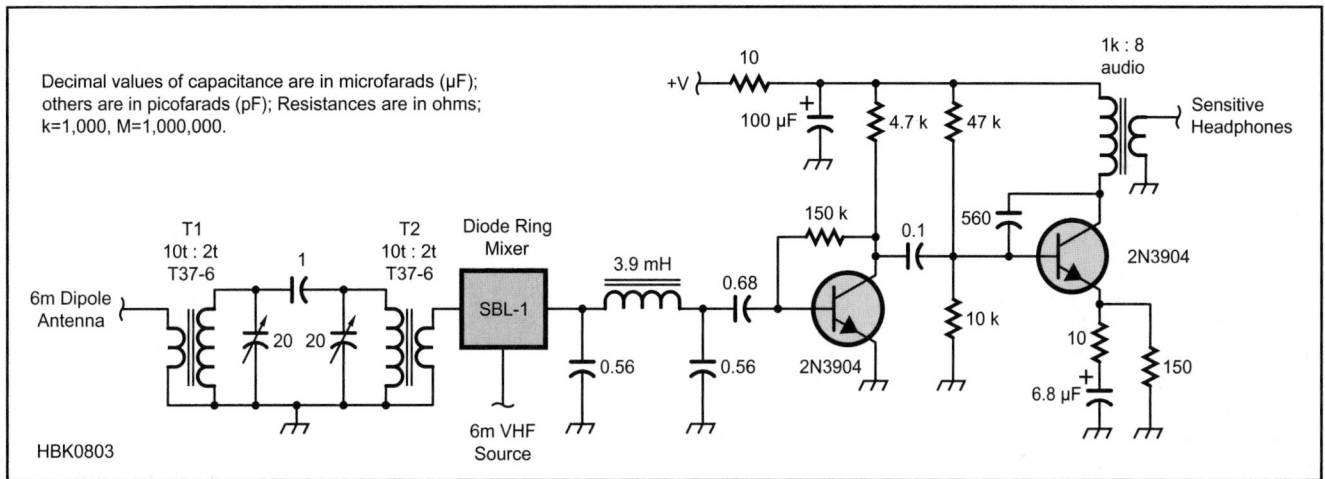

Fig 13.42 — Schematic of the simple 6 meter direct conversion receiver.

6 meters and 60 kHz on 2 meters. **Fig 13.42** is the schematic. Note how it combines and builds on previous circuits.

An LNA identical to the tuned buffer amplifier in the VHF signal source could be added after operation is confirmed, including the keying circuit as a receive mute. The diode ring mixer may be a Mini-Circuits SBL-1, TUF-1, ADE-1 or similar.

The audio amplifier with low-pass filter features a number of basic designs covered in other electronics classes, and in any of the basic electronic literature. Students are highly encouraged to build the direct conversion receiver using high performance VHF ugly construction techniques rather than laying out a circuit board. As noted by Wes Hayward in his original "Progressive Communications Receiver" article (Nov 1981 QST), when a performance difference between careful ugly construction and printed circuit board implementations of his HF receiver could be measured, the advantage went to ugly construction. For student use, ugly construction offers easy circuit modification and experimentation. Portland State design engineering students are required to modify and attempt to improve their circuits once they are operating — these are not a set of canned lab exercises!

Experiments with simple low power 50 MHz CW transmitters and direct conversion receivers using basic components rather than complex black-box integrated circuits are an excellent introduction to the fundamentals of 21st century radio. Cell phones, computers, and software applications are now taken for granted, but component-level RF electronics and electromagnetics are much closer to the underlying science. Students and practitioners of the radio frequency arts as a technical pastime can start here and follow their own experimental design paths.

13.3 Increasing Transmitter Power

The functions described so far that process input data and information and result in a signal on the desired output radio frequency generally occur at a low level. The one exception is full-carrier AM, in which the modulation is classically applied to the final amplification stage. More modern linear transmitter systems generate AM in the same way as SSB at low levels, typically between 1 mW and 1 W.

13.3.1 Types of Power Amplifiers

The **RF Power Amplifiers** chapter provides a detailed view of power amplifiers; however, we will take a quick peek here to set the stage for the following discussions. Amplifiers use dc power applied to active devices in order to increase the power or level of signals. As will all real devices, they introduce some distortion in the process, and are generally limited by the level of distortion products. Power amplifiers can be constructed using either solid-state devices or vacuum tubes as the active device. At higher powers, typically above a few hundred watts, vacuum tubes are more frequently found, although there is a clear trend toward solid state at all amateur power levels.

Independent of the device, amplifiers are divided into classes based on the fraction of the input cycle over which they conduct. A sinusoidal output signal is provided either by the *flywheel* action of a resonant circuit or by other devices contributing in turn. The usual amplifier classes are summarized in **Table 13.1**. Moving from Class A toward Class C, the amplifiers become progressively less linear but more efficient. The amplifiers with a YES in the LINEAR column thus are not all equally linear however A, AB or Class B amplifiers can be suitable for operation in a linear transmitter chain. Class C amplifiers can be used only for amplification of signals that do not have modulation information contained in the amplitude, other than on-off keyed signals. Thus class C amplifiers are useful for amplification of sinusoids, CW, FM,

or as the nonlinear stage at which high-level AM modulation is employed.

In handheld digital cellular transceivers and base stations, the power amplifier must operate in a linear fashion to meet spectral purity requirements for the complex digital modulation schemes used. Since linear amplifiers are generally not very efficient, this is a major contributor to the energy consumption. In an effort to extend battery life in portable units, and reduce wasted energy in fixed equipment, considerable research is underway to use nonlinear amplifiers to provide linear amplification. Such techniques include pre-distortion of the low-level signal, polar modulation, envelope elimination and restoration, Cartesian-loop feedback and others.

13.3.2 Linear Amplifiers

While transmitters at power levels of 1 mW to 1 W have been successfully used for communication across many portions of the spectrum, most communications systems operate with more success at higher powers. The low level stage is usually referred to as an exciter, while higher power is provided by one or more linear amplifier stages as shown in **Fig 13.43**.

The power levels shown at the various points in Fig 13.43 are fairly typical for a high powered amateur station. The 1500 W PEP output represents the legal limit for US amateurs in most bands (200 W PEP on 30 meters and 50 W ERP$_D$ on the 60 meter channels are notable exceptions). The first amplifier block may contain more than one stage, while the final output amplifier is often composed of multiple parallel active devices.

Typical power supply requirements for the amplifier stages are noted for a number of reasons. First, while power is rarely an issue at the exciter level, often it is a significant issue at the power levels shown for the ampli-

fiers. The power supplies represent a large portion of the cost and weight of the system as the power increases. Some manufacturers are beginning to use switching-type power supplies for high-power amplifiers, resulting in a major reduction in size and weight.

Note also that a gross amplifier efficiency of about 50% is assumed for the amplifiers, taking into account ancillary subsystems as well as the inefficiency of the active devices in linear mode. The 50% that doesn't result in actual RF output is radiated as heat from the amplifier and must be removed from the amplifier as it is generated to avoid component damage. This represents another cost and weight factor that increases rapidly with power level.

The voltages shown for the supplies are those typical of modern solid state amplifiers. While virtually all commercial equipment now includes solid state amplifiers at the 100 W level, vacuum tube active devices are frequently found at higher levels, although the trend is clearly moving toward solid state. Vacuum tube amplifiers typically operate at voltages in the 2 to 4 kV range, requiring stringent measures be taken to avoid arcing across components. In addition, vacuum tube amplifiers typically dissipate up to 100 W of filament power that must be added to the power supply and heat dissipation planning.

13.3.3 Nonlinear Amplifiers

Nonlinear transmitters are somewhat different in architecture than the linear systems discussed previously. The configuration of a high-level AM modulated transmitter is shown in **Fig 13.44**. Note that none of the upper RF stages (the "RF chain") need to be particularly linear. The final stage must be nonlinear to have the modulation applied. Thus the RF stages can be the more power-efficient Class C amplifiers if desired.

Table 13.1
Summary of Characteristics of Power Amplifier Classes
Values are Typical

Class	Conduction	Linear	Efficiency
A	360°	Yes	30%
AB	270°	Yes	55%
B	180°	Yes	65%
C	90°	No	74%

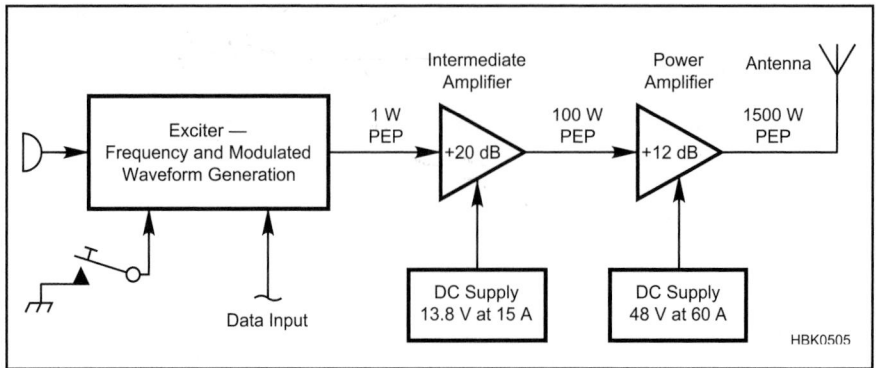

Fig 13.43 — Block diagram of a solid-state linear transmitter chain with multiple amplifier stages.

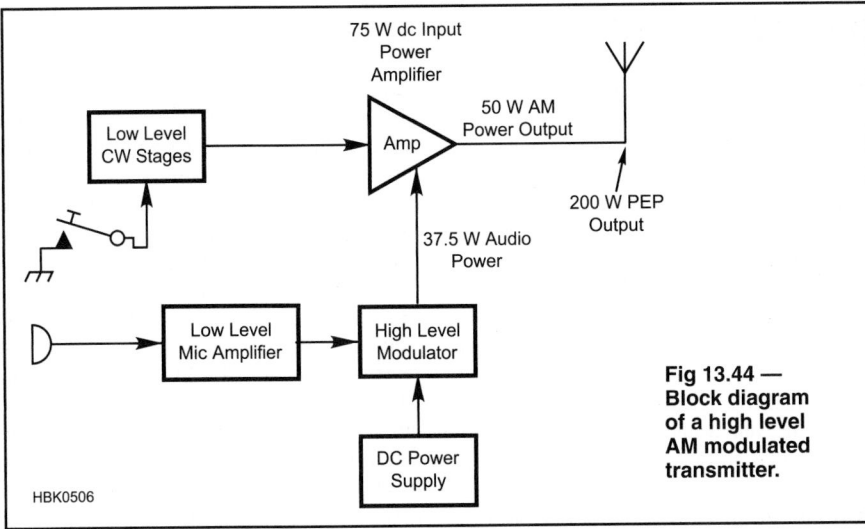

Fig 13.44 —
Block diagram
of a high level
AM modulated
transmitter.

HBK0506

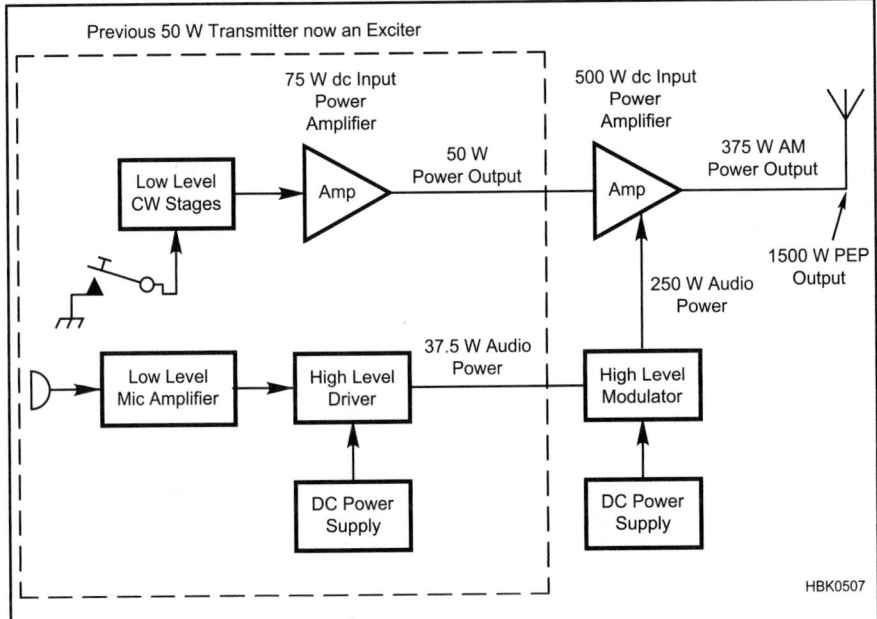

Fig 13.45 — Block diagram of a high level AM modulated transmitter with added output stage.

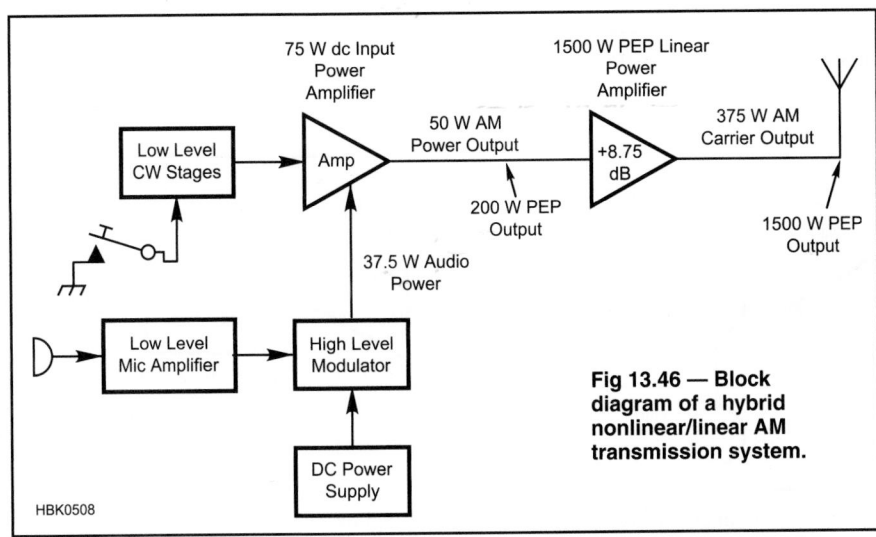

Fig 13.46 — Block diagram of a hybrid nonlinear/linear AM transmission system.

HBK0508

There are some observations to be made here. Note that the RF chain is putting out the full carrier power whenever in transmit mode, requiring a 100% duty cycle for power and amplifier components, unlike the SSB systems discussed previously. This imposes a considerable weight and cost burden on the power supply system. Note also that the PEP output of a 100% modulated AM system is equal to four times the carrier power.

The typical arrangement to increase the power of such a system is to add not only an RF amplifier stage capable of handling the desired power, but also to add additional audio power amplification to fully modulate the final RF stage. For 100% high-level plate modulation, an audio power equal to half the dc input power (plate voltage times plate current of a vacuum tube amplifier) needs to be provided. This arrangement is shown in **Fig 13.45**. In the example shown, the lower level audio stages are provided by those of the previous 50 W transmitter, now serving as an exciter for the power amplifier and as a driver for the modulating stage. This was frequently provided for in some transmitters of the AM era, notably the popular E. F. Johnson Ranger series, which provided special taps on its modulation transformer for use as a driver for higher-power systems.

It is worth mentioning that in those days the FCC US amateur power limit was expressed in terms of dc *input* to the final stage and was limited to 1000 W, rather than the 1500 W PEP *output* now specified. A fully modulated 1000 W dc input AM transmitter would likely have a carrier output of 750 W or 3000 W PEP — 3 dB above our current limit. If you end up with that classic Collins KW-1 transmitter, throttle it back to make it last and stay out of trouble!

13.3.4 Hybrid Amplifiers

Another alternative that is convenient with current equipment is to use an AM transmitter with a linear amplifier. This can be successful if the relationship that PEP = 4 × Carrier Power is maintained. **Fig 13.46** shows a 1500 W PEP output linear amplifier following a typical 50 W AM transmitter. In this example, the amplifier would be adjusted to provide a 375 W carrier output with no modulation applied to the exciter. During voice peaks the output seen on a special PEP meter, or using an oscilloscope, should be 1500 W PEP.

Note that during AM operation, the amplifier is producing a higher average power than it would without the carrier being present, as in SSB mode. The duty cycle specification of the amplifier should be checked to be sure it can handle the heavier load. If the amplifier has an RTTY rating, it should be safe to run an AM carrier at 66% of the RTTY output, following the required on and off time intervals.

13.4 Transceiver Construction and Control

With each generation of transceivers, in what has become a highly competitive marketplace, additional features were added as technology marched on. The deficiencies of early transceivers, in comparison to separate receivers and transmitters, quickly disappeared to the point that 100 W (or higher power) class transceivers exceeded the features of the best separate receivers and transmitters of the past.

The current generation of transceivers was designed using modern solid-state components that permit abundant functionality in a small enclosure. The designers and manufacturers have taken advantage of the possibilities of integrated electronics and microprocessors, incorporating many more functions into a transceiver than could have been envisioned in the early days. This is largely a function of improved technology becoming available at reduced cost.

Separate receivers and transmitters could be built with similar features and performance, but the required duplication of subsystems would make each unit cost about the same as a transceiver. Several manufacturers do make stand-alone receivers using components from transceivers, but they are generally aimed toward different markets, such as shortwave listeners or military/commercial users.

13.4.1 Upconverting Architecture

Fig 13.47 shows a traditional superheterodyne architecture for transceivers in which the sideband filter, some amplifiers, and other filters are shared between transmit and receive modes through the use of extensive switching. A limitation of that architecture is that it is not trivial to provide operation on frequencies near the first IF. The typical transceiver designer selected a first IF frequency away from the desired operating frequencies and proceeded on that basis.

New amateur bands at 30, 17, and 12 meters were approved at the 1979 ITU World Administrative Conference. The difficulties of managing image rejection on the new bands and the desire for continuous receiver coverage of LF, MF and HF bands (general-coverage receive) required a significant change in the architecture of receivers and transceivers. Thus, the upconverting architecture discussed in the **Receivers** chapter became popular in the 1980s and, with a few notable exceptions, became almost universal in commercial products over the following decade.

The solution was to move to the upconverting architecture shown in **Fig 13.48**. By selecting a first IF well above the highest receive frequency, the first local oscillator can cover the entire receive range without any gaps. With the 70 MHz IF shown, the full range from 0 to 30 MHz can be covered by an LO covering 70 to 100 MHz, less than a 1.5:1 range, making it easy to implement with modern PLL or DDS technology. Note that the high IF makes image rejection very easy and, rather than the usual tuned bandpass front end, we can use more universal low-pass filtering. The low-pass filter is generally shared with the transmit side and designed with octave cutoff frequencies to reduce transmitter harmonic content. A typical set of HF transceiver low-pass filter cut-off frequencies would be 1, 2, 4, 8, 16 and 32 MHz.

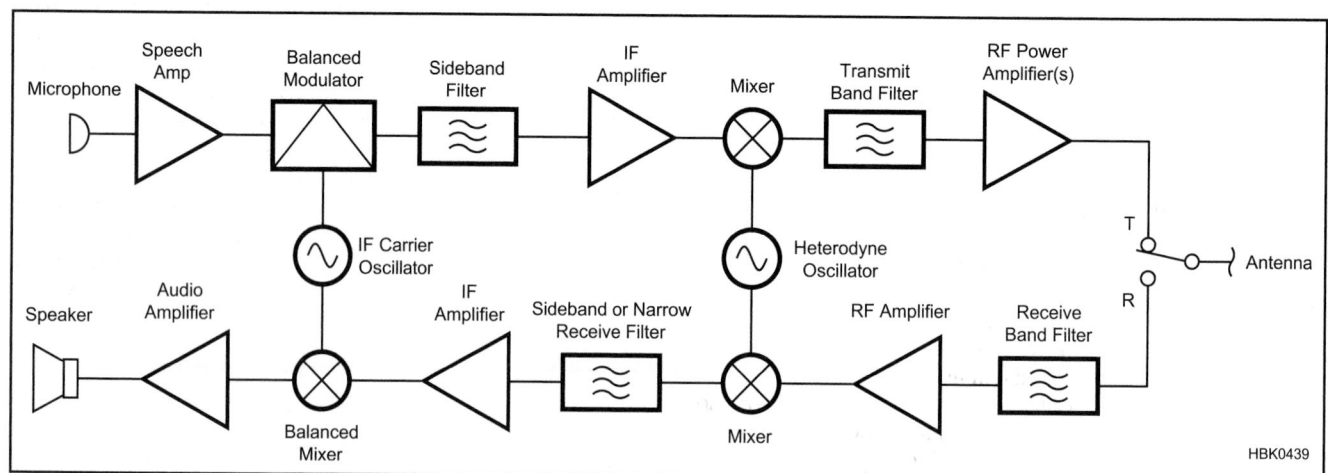

Fig 13.47 — Block diagram of a simple SSB transceiver sharing oscillator frequencies.

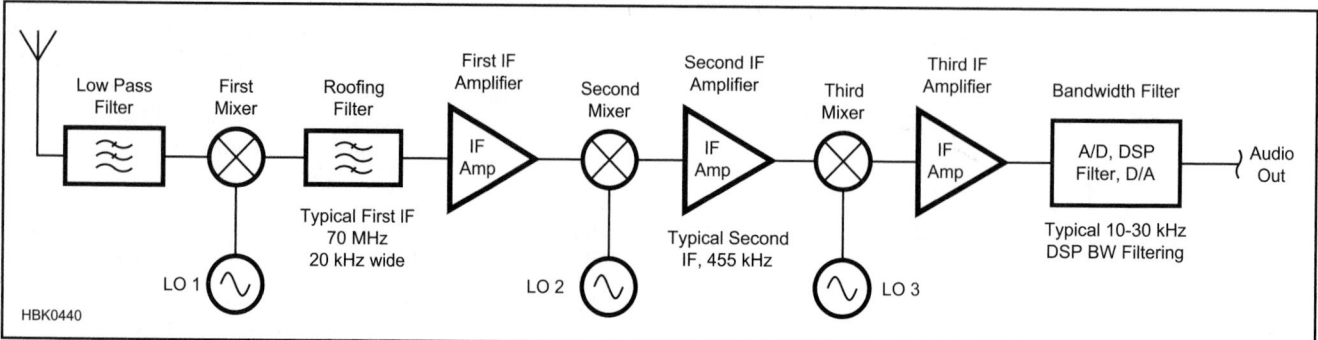

Fig 13.48 — Simplified block diagram of upconverting general coverage transceiver, receiver section shown.

This architecture offers significant benefits. By merely changing the control system programming, any frequency range or ranges can be provided with no change to the architecture or hardware implementation. Unlike the more traditional transceiver architecture (Fig 13.47), continuous receive frequency coverage over the range is actually easier to provide than to not provide, offering a marketing advantage for those who also like to do shortwave or broadcast listening.

13.4.2 Break-In CW Operation

Most current 100 W class HF transceivers use high-speed relays (with the relay actually following the CW keying) or solid-state PIN diodes to implement full break-in CW. Some RF power amplifiers use high-speed vacuum relays for the TR switching function. See the section on TR Switching later in this chapter for more information about circuits to perform this function. Two projects for adding QSK switching to linear amplifiers are included in the **Station Accessories** chapter.

The term "semi-break-in" is used to designate a CW switching system in which closing the key initiates transmission, but switching back to receive happens between words, not between individual dits. Some operators find this less distracting than full break-in, and it is easier to implement with less-expensive relays for the TR switching.

13.4.3 Push-To-Talk for Voice

Another advance in amateur station switching followed longstanding practices of aircraft and mobile voice operators who had other things to contend with besides radio switches. Microphones in those services included built-in switches to activate TR switching. Called push-to-talk (PTT), this function is perhaps the most self-explanatory description in our acronym studded environment.

Relays controlled the various switching functions when the operator pressed the PTT switch. Some top-of-the-line transmitters of the period included at least some of the relays internally and had a socket designed for PTT microphones. **Fig 13.49** is a view of the ubiquitous Astatic D-104 microphone with PTT stand, produced from the 1930s to 2004, and still popular at flea markets and auction sites. PTT operation allowed the operator to be out of reach of the radio equipment while operating, permitting "easy chair" operation for the first time.

Modern transceivers include some form of PTT (or "one switch operation"). Relays, diodes, transistors and other components seamlessly handle myriad transmit-receive changeover functions inside the transceiver. Most transceivers have additional provisions for manually activating PTT via a front-panel

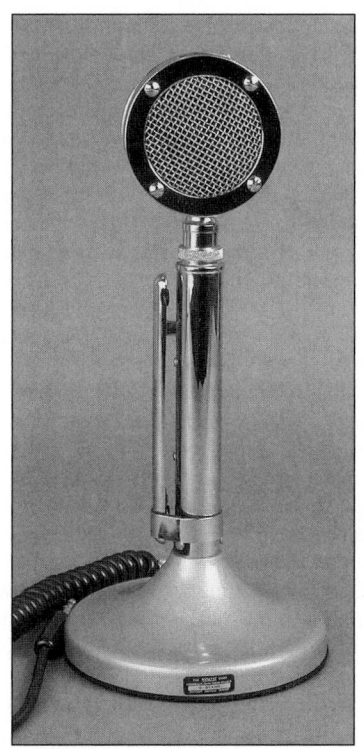

Fig 13.49 — A classic Astatic D-104 mic with PTT stand.

switch. And many have one or more jacks for external PTT control via foot switches, computer interfaces or other devices.

13.4.4 Voice-Operated Transmit-Receive Switching (VOX)

How about break-in for voice operators? SSB operation enabled the development of voice operated transmit/receive switching, or VOX. During VOX operation, speaking into the microphone causes the station to switch from receive to transmit; a pause in speaking results in switching back to receive mode. Although VOX technology can work with AM or FM, rapidly turning the carrier signal on and off to follow speech does not provide the smooth operation possible with SSB. (During SSB transmission, no carrier or signal is sent while the operator is silent.)

VOX OPERATION

VOX is built into current HF SSB transceivers. In most, but not all, cases they also provide for PTT operation, with switches or menu settings to switch among the various control methods. Some operators prefer VOX, some prefer PTT and some switch back and forth depending on the operating environment.

VOX controls are often considered to be in the "set and forget" category and thus may be controlled by a software menu or by controls on the rear panel, under the top lid or behind an access panel. The following sections discuss the operation and adjustment of radio controls associated with VOX operation. Check your transceiver's operating manual for the specifics for your radio.

Before adjusting your radio's VOX controls, it's important to understand how your particular mic operates. If it has no PTT switch, you can go on to the next section! Some mics with PTT switches turn off the audio signal if the PTT switch is released, while some just open the control contacts. If your mic does the former, you will need to lock the PTT switch closed, have a different mic for VOX, or possibly modify the internal mic connections to make it operate with the VOX. If no audio is provided to the VOX control circuit, it will never activate. If the mic came with your radio, or from its manufacturer, you can probably find out in the radio or mic manual.

VOX Gain

Fig 13.50 shows some typical transceiver VOX controls. The VOX GAIN setting determines how loud speech must be to initiate switchover, called "tripping the VOX." With a dummy load on the radio, experiment with the setting and see what happens. You should be able to advance it so far that it switches with your breathing. That is obviously too sensitive or you will have to hold your breath while receiving! If not sensitive enough, it may cause the transmitter to switch off during softly spoken syllables. Notice that the setting depends on how close you are to the microphone, as well as how loud you talk. A headset-type microphone (a "boom set") has an advantage here in that you can set the microphone distance the same every time you use it.

The optimum setting is one that switches to transmit whenever you start talking, but isn't so sensitive that it switches when the mic picks up other sounds, such as a cooling fan turning on or normal household noises.

Fig 13.50 — The function of VOX controls is described in the text. They require adjustment for different types of operating, so front-panel knobs make the most convenient control arrangement. In some radios, VOX settings are adjusted through the menu system.

VOX Delay

As soon as you stop talking, the radio can switch back to receive. Generally, if that happens too quickly, it will switch back and forth between syllables, causing a lot of extra and distracting relay clatter. The VOX DELAY control determines how long the radio stays in the transmit position once you stop talking. If set too short, it can be annoying. If set too long, you may find that you miss a response to a question because the other station started talking while you were still waiting to switch over.

You may find that different delay settings work well for different types of operation. For example, in a contest the responses come quickly and a short delay is good. For casual conversation, longer delays may be appropriate. Again, experiment with these settings with your radio connected to a dummy load.

Anti-VOX

This is a control with a name that may mystify you at first glance! While you are receiving, your loudspeaker is also talking to your mic — and tripping your VOX — even if you aren't! Early VOX users often needed to use headphones to avoid this problem. Someone finally figured out that if a sample of the speaker's audio signal were fed back to the mic input, out-of-phase and at the appropriate amplitude, the signal from the speaker could be cancelled out and would not cause the VOX circuit to activate the transmitter. The ANTI-VOX (called ANTI-TRIP in the photo) controls the amplitude of the sampled speaker audio, while the phase is set by the transceiver design.

As you tune in signals on your receiver with the audio output going to the speaker, you may find that the VOX triggers from time to time. This will depend on how far you turn up the volume, which way the speaker is pointed and how far it is from the mic. You should be able to set the ANTI-VOX so that the speaker doesn't trip the VOX during normal operation.

Generally, setting ANTI-VOX to higher values allows the speaker audio to be louder without activating the VOX circuit. Keep in mind that once you find a good setting, it may need to be changed if you relocate your mic or speaker. With most radios, you should find a spot to set the speaker, mic and ANTI-VOX so that the speaker can be used without difficulty.

13.4.5 TR Switching

As the complexity of a transceiver increases, the business of switching between receive and transmit becomes quite complex. In commercially built equipment, this function is usually controlled by a microprocessor that manages any necessary sequencing and interlock functions that would require an excessive amount of circuitry to implement with discrete components. For an example of just how complex TR switching could be in an advanced transceiver, look at the schematic for any mid-level or top-of-the-line solid-state transceiver sold in the 1980s or 1990s!

Nevertheless, the basic functions of TR switching are well within scope for the amateur building a transceiver. Understanding TR switching will also assist in troubleshooting a more complex commercial radio. Even full break-in keying is possible: Two schematics of circuits for fast TR switching from an article by Jack Kuecken, KE2QJ,

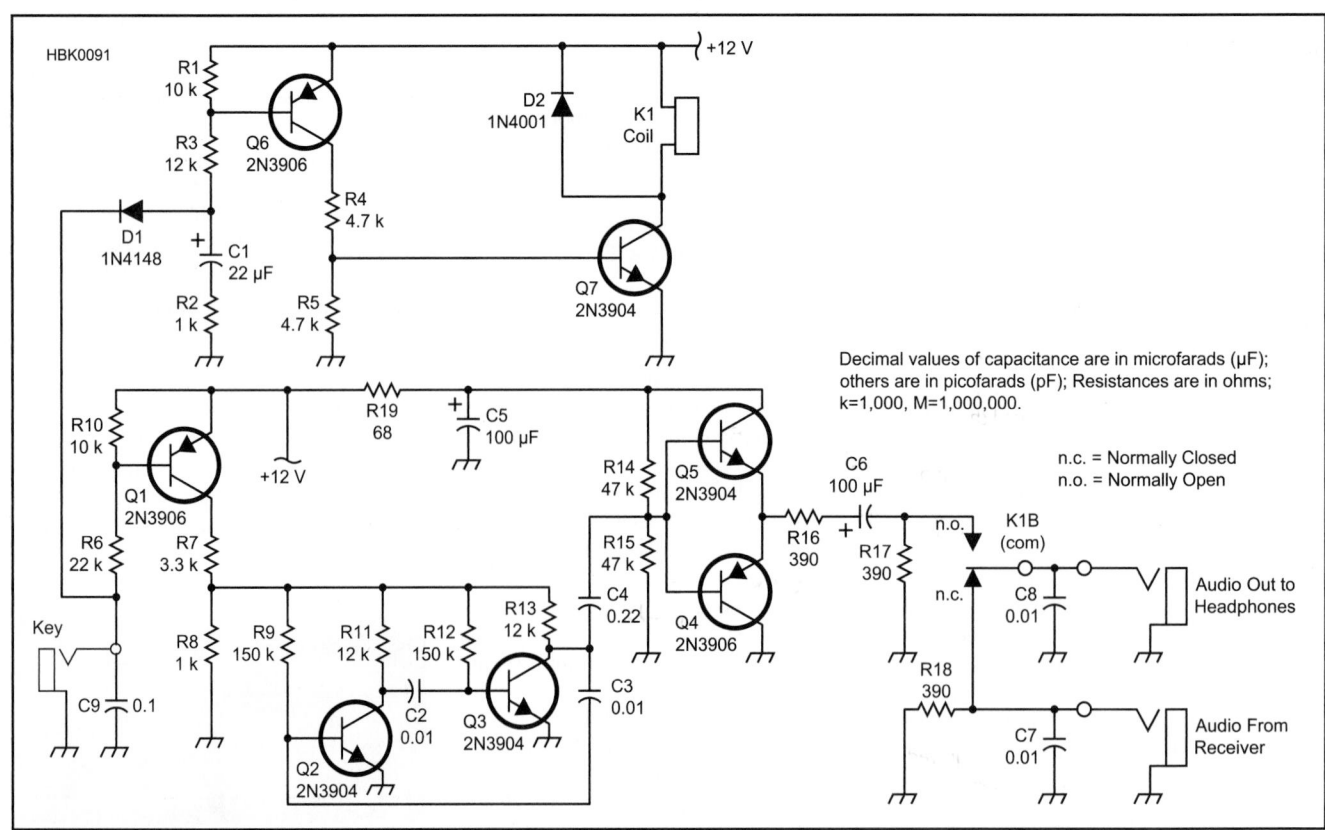

Fig 13.51 — Detailed schematic diagram and parts list for transmit-receive control section and sidetone generator of the universal QRP transmitter. Resistors are ¼ W, 5% carbon film. A kit of component parts is available from KangaUS (www.kangaus.com).

C1 — 22 µF, 25 V electrolytic
C2, C3, C7, C8 — 0.01 µF, 50 V ceramic
C4 — 0.22 µF, 50 V ceramic
C5, C6 — 100 µF, 25 V electrolytic

C9 — 0.1 µF, 50 V ceramic
K1 — DPDT 12 V coil relay. An NAIS DS2Y-S-DC12, 700 Ω, 4 ms relay was used in this example.

Q1, Q4, Q6 — 2N3906, PNP silicon small signal transistor
Q2, Q3, Q5, Q7 — 2N3904, NPN silicon small signal transistor

are included below, and the full article is included on this book's CD-ROM.

Numerous schemes are popular for switching an antenna between transmitter and receiver functions. But these schemes tend to get in the way when one is developing both simple receivers and transmitters, perhaps as separate projects. A simple relay-based TR scheme is then preferred and is presented here. In this system, used in the MkII Updated Universal QRP Transmitter by Wes Hayward, W7ZOI (see the full article on this book's CD-ROM), the TR relay not only switches the antenna from the receiver to the transmitter, but disconnects the headphones from the receiver and attaches them to a sidetone oscillator that is keyed with the transmitter.

The circuitry that does most of the switching is shown in **Fig 13.51**. A key closure discharges capacitor C1. R2, the 1 kΩ resistor in series with C1, prevents a spark at the key. Of greater import, it also does not allow us to "ask" that the capacitor be discharged instantaneously, a common request in similar published circuits. Key closure causes Q6 to saturate, causing Q7 to also saturate, turning the relay on. The relay picked for this example has a 700 Ω, 12 V coil with a measured 4 ms pull-in time.

Relay contacts B switch the audio line. R17 and R18 suppress clicks related to switching. A depressed key turns on PNP switch Q1, which then turns on the sidetone multivibrator, Q2 and Q3. The resulting audio is routed to switching amplifier Q4 and Q5. Although the common bases are biased to half of the supply voltage, emitter bias does not allow any static dc current to flow. The only current that flows is that related to the sidetone signal during key down intervals. Changing the value of R16 allows the audio volume to be adjusted, to compensate for the particular low-impedance headphones used.

Depending on the architecture of the transceiver, there will likely have to be some additional control circuitry in order to avoid annoying switching artifacts. These generally fall into the category of transients in the receiver audio and turning on the transmitter too slowly to capture initial code elements, also known as "dot shortening." A review of the transceiver design from which this circuit is taken and of other homebuilt transceiver designs will illustrate the problem and the methods used to address it.

If full break-in TR switching is required, high-speed switching components such as a reed relay or PIN diodes are required. KE2QJ provided a pair of such circuits (**Fig 13.52** and **Fig 13.53**) that can be adapted for internal use in a home-built transceiver, although their original purpose was to integrate a stand-alone receiver with a transceiver and linear amplifier. The full article is available on this book's CD-ROM.

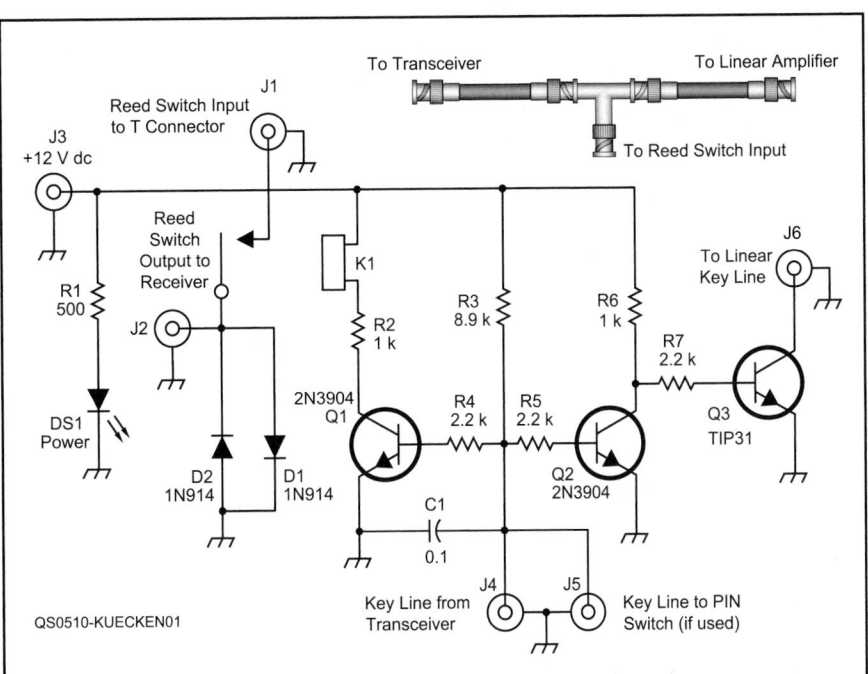

Fig 13.52 — Schematic and parts list for the reed relay TR switch. Resistors are ¼ W.

D1, D2 — High speed switching diode, 1N914 or equivalent
DS1 — LED
J1, J2 — Chassis mount BNC connectors
J3-J6 — Chassis mount phono connectors

K1 — SPST normally open reed relay with 12 V coil
Q1, Q2 — Small signal PNP transistor, 2N3904 or equivalent
Q3 — NPN power transistor, TIP31 or equivalent

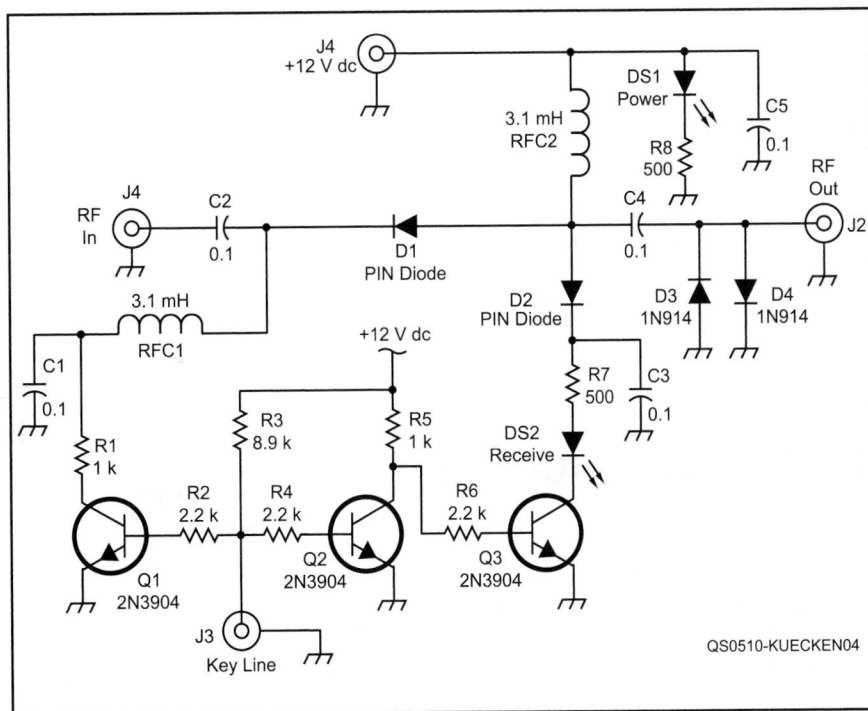

Fig 13.53 — Schematic and parts list for the PIN diode TR switch. Resistors are ¼ W.

C1-C5 — 0.1 µF ceramic capacitor
D1, D2 — PIN diode
D3, D4 — Switching diode, 1N914 or equivalent
DS1, DS2 — LED
J1, J2 — Chassis mount BNC connector
J3, J4 — Chassis mount phono connector

Q1-Q3 — Small signal PNP transistor, 2N3904 or equivalent
RFC1, RFC2 — 3.1 mH RF choke, 225 turns #30 AWG enameled wire wound on a 5/16 inch diameter, 5/8 inch long plastic tube

If you already have a receiver and transmitter and want to integrate them under the control of a separate TR switch, the K8IQY "Magic Box" (**www.4sqrp.com/MagicBox. php**) is available as a kit. This is a microprocessor-controlled design that can handle up to 10 W of transmitter power, switches at up to 50 WPM, and includes an audio sidetone output, as well. Complete documentation for the kit is available online, including schematics and design information. The kit could be extended to handle more transmit power with heavier components and the appropriate circuit changes.

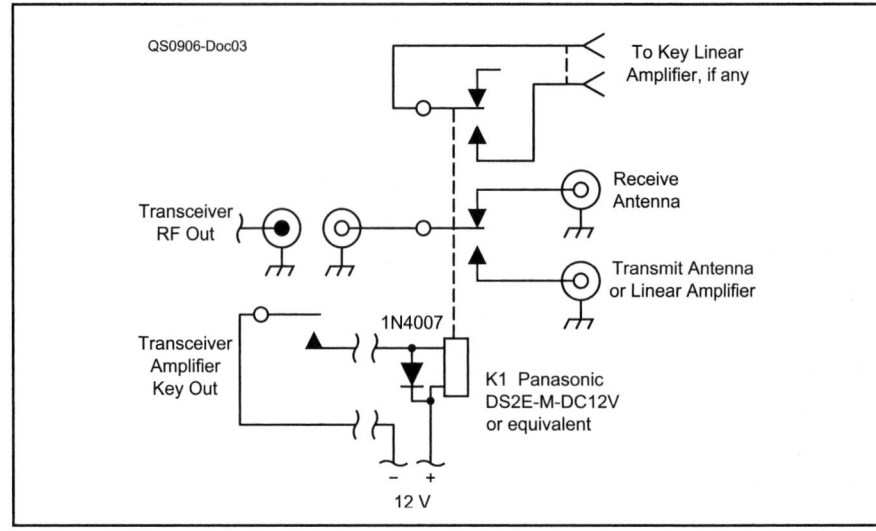

Fig 13.54 — Schematic of an external box that allows a modern transceiver to key a linear amplifier with TR switch voltage or current requirements that exceed the transceiver's ratings. As a bonus, it can also be used to allow reception from a low-noise receiving antenna.

13.4.6 TR Switching With a Linear Amplifier

Virtually every amateur HF transceiver includes a rear panel jack called something along the lines of KEY OUT, intended to provide a contact closure while in transmit mode. This jack is intended to connect to a corresponding jack on a linear amplifier called something like KEY IN. Check the transceiver and amplifier manuals to find out what they are called on your units. A diagram of the proper cabling to connect the transceiver and amplifier will be provided in the manual.

ENSURING AMPLIFIER-TRANSCEIVER COMPATIBILITY

While you're there, check the ratings to find out how much voltage and current the transceiver can safely switch. In earlier days, this switching was usually accomplished by relay contacts. More modern radios tend to use solid-state devices to perform the function. Although recent amplifiers are usually compatible with the switching capabilities of current transceivers, the voltage and/or current required to switch the relays in an older linear amplifier may exceed the ratings. Fortunately, it is pretty easy to find out what your amplifier requires, and almost as easy

to fix if it's not compatible with your radio.

If your amplifier manual doesn't say what the switching voltage is, you can find out with a multimeter or DMM. Set the meter to read voltage in a range safe for 250 V dc or higher. Connect the positive meter probe to amplifier key jack's center conductor, and connect the negative meter probe to the chassis ground (or other key jack terminal if it's not grounded). This will tell you what the open circuit voltage is on the amplifier key jack. You may need to try a lower voltage range, or an ac range, or switch the probes (if the key line is a negative dc voltage) to get a reading.

Now set the meter to read current. Start with a range that can read 1 A dc, and with the leads connected as before, you should hear the amplifier relay close and observe the current needed to operate the TR relay or circuit. Adjust the meter range, if needed,

to get an accurate reading.

These two levels, voltage and current, are what the transceiver will be asked to switch. If *either* reading is higher than the transceiver specification, do not connect the transceiver and amplifier together. Doing so will likely damage your transceiver. You will need a simple interface circuit to handle the amplifier's switching voltage and current.

The simple, low-cost relay circuit shown in **Fig 13.54** can be used to key an older amplifier with a modern transceiver. It offers an added benefit: Another potential use of the transceiver KEY OUT jack is to switch to a separate low-noise receive antenna on the lower bands. While most high-end transceivers have a separate receive-only antenna connection built in, many transceivers don't. If you don't need one of the extra functions, just leave off those wires.

13.5 Transceiver Projects

There are many transceiver designs available, aimed mainly at the advanced builder. We will provide two. One describes the construction of a 5 W single-band SSB/CW transceiver that was the winner of the ARRL's first *Homebrew Challenge*. The other is the description of the design and construction process that resulted in a 100 W transceiver for all HF bands with performance (confirmed in the ARRL Lab) that is as good as some of the best commercial products. Construction details for these projects and support files are available on the ARRL website. Additional projects may be found on the *Handbook* CD-ROM.

13.5.1 Transceiver Kits

One of the simplest kits to build is also one of the least expensive. The ARRL offers the MFJ Enterprises (**www.mfjenterprises.com**) Cub 40 meter transceiver kit (**Figs 13.55** and **13.56**) bundled with *ARRL's Low Power Communication — The Art and Science of QRP*. (See **www.arrl.org/shop**.) Other kits are available from many manufacturers.

Ten-Tec, maker of some of the best HF transceivers, offers single band CW QRP kits for your choice of 80, 40, 30 or 20 meters. Elecraft offers a range of kits from the KX1

and KX3 travel radios to the K2 and K3 high performance multiband CW and SSB transceiver semi-kit (mechanical assembly only, no soldering required). The K2 and K3 can both start out as 10 W models and be upgraded to 100 W, if desired. DZkit (**www.dzkit.com**) offers the Sienna high performance transceiver kit.

If your interest is strictly QRP transceiver kits, there are many small and specialty manufacturers. The QRP Amateur Radio Club, Inc. (QRP ARCI) maintains a manufacturer list on their website at **www.qrparci.org**, available in the "Links" section of the site.

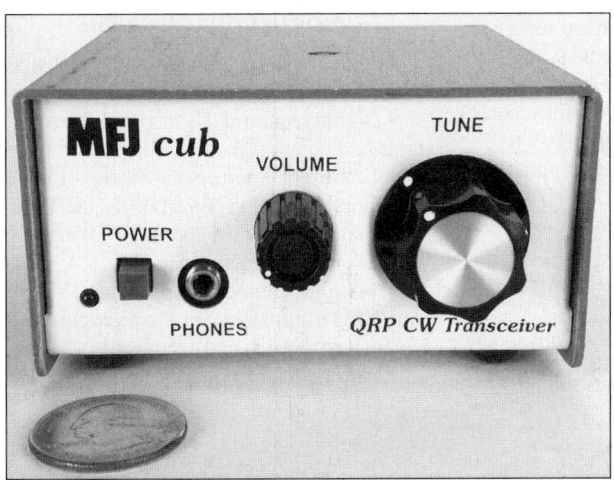

Fig 13.55 — MFJ Cub QRP transceiver built from a kit.

assembly information is available on the CD-ROM accompanying this book.

The following is a list of the criteria for the Homebrew Challenge and a brief description of how the TAK-40 meets the requirements:

• *The station must include a transmitter and receiver that can operate on the CW and voice segments of 40 meters.* The TAK-40 covers 7.0 to 7.3 MHz.

• *It must meet all FCC regulations for spectral purity.* All spurious emissions from the TAK-40 are at least 43 dB below the mean power of fundamental emission.

• *It must have a power output of at least 5 W PEP.* The TAK-40 generates at least 5 W PEP for voice and CW modes. The ALC can be set as high as 7 W if desired.

• *It can be constructed using ordinary hand tools.* Construction of the TAK-40 uses all leaded components and assembly requires only hand tools, soldering iron and an electric drill (helpful but not strictly necessary).

• *It must be capable of operation on both voice and CW.* The TAK-40 operates USB and LSB as well as CW. USB was included to allow the TAK-40 to easily operate in digital modes such as PSK31 using a PC and sound card.

• *Parts must be readily available either from local retailers or by mail order. No "flea market specials" allowed.* The TAK-40 is constructed from materials available from DigiKey, Mouser, Jameco and Amidon.

• *Any test equipment other than a multimeter or radio receiver must either be constructed as part of the project or purchased as part of the budget.* The TAK-40 requires only a multimeter for construction, and extensive built-in setup functions in the software include a frequency counter to align the oscillators and a programmable voltage source for controlling the oscillators.

• *Equipment need only operate on a single band, 40 meters. Multiband operation is acceptable and encouraged.* The TAK-40 operates on the 40 meter band.

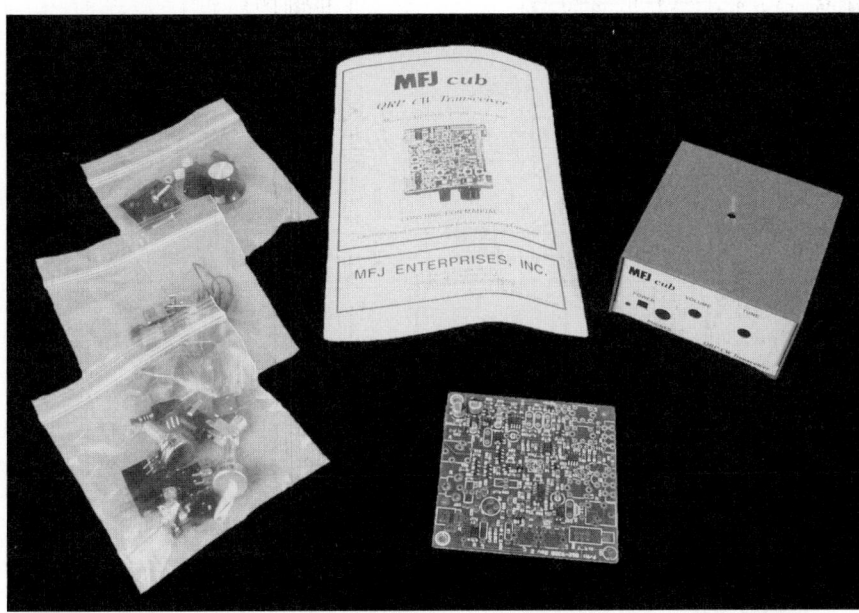

Fig 13.56 — MFJ Cub QRP transceiver kit showing parts supplied.

13.5.2 Project: The TAK-40 SSB CW Transceiver

Jim Veatch, WA2EUJ, was one of two winners of the first ARRL *Homebrew Challenge* with his TAK-40 transceiver shown in **Fig 13.57**. This challenge was to build a homebrew 5 W minimum output voice and CW transceiver using all new parts for under $50. The resulting radio had to meet FCC spurious signal requirements. The submitted radios were evaluated as to operational features and capabilities by a panel of ARRL technical staff. Jim's transceiver met the requirements for a transceiver that required a connected PC for setup — to load the PIC controller. This information was presented in May 2008 *QST* in an article provided by Jim as a challenge requirement. The information here is an overview of the project. The complete article including construction, adjustment, and final

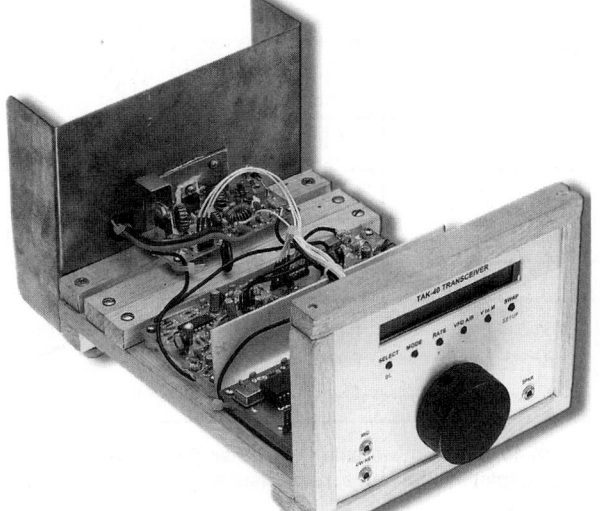

Fig 13.57 — Homebrew Challenge winner TAK-40 from Jim Veatch, WA2EUJ.

- *The total cost of all parts, except for power supply, mic, key, headphones or speaker, and usual supplies such as wire, nuts and bolts, tape, antenna, solder or glue must be less than $50.* The total cost of the parts required to build the TAK-40 was $49.50 at the time of the contest judging. It still remains an excellent price for the capabilities, even with the manufactured circuit boards as discussed below.

The TAK-40 also includes some features that make it very smooth to operate.
- Automatic Gain Control — regulates the audio output for strong and weak signals.

- S-Meter — simplifies signal reports.
- Digital frequency readout — reads the operating frequency to 100 Hz.
- Dual Tuning Rates — FAST for scanning the band and SLOW for fine tuning.
- Speech Processor — get the most from the 5 W output.
- Automatic Level Control — prevents overdriving the transmitter.
- Transmit power meter — displays approximate power output.
- Boot loader — accepts firmware updates via a computer (cable and level converter optional).

CIRCUIT DESCRIPTION

The TAK-40 transceiver is constructed as four modules:
- Digital section and front panel: The digital board contains the microprocessor, front panel controls, liquid crystal display (LCD), the digital-to-analog converter for the VFO, the beat frequency oscillator (BFO) and the oscillator switching matrix.
- Variable frequency oscillator (VFO): The VFO is implemented as a separate module for best stability and overall performance.
- Intermediate frequency (IF) board: RF

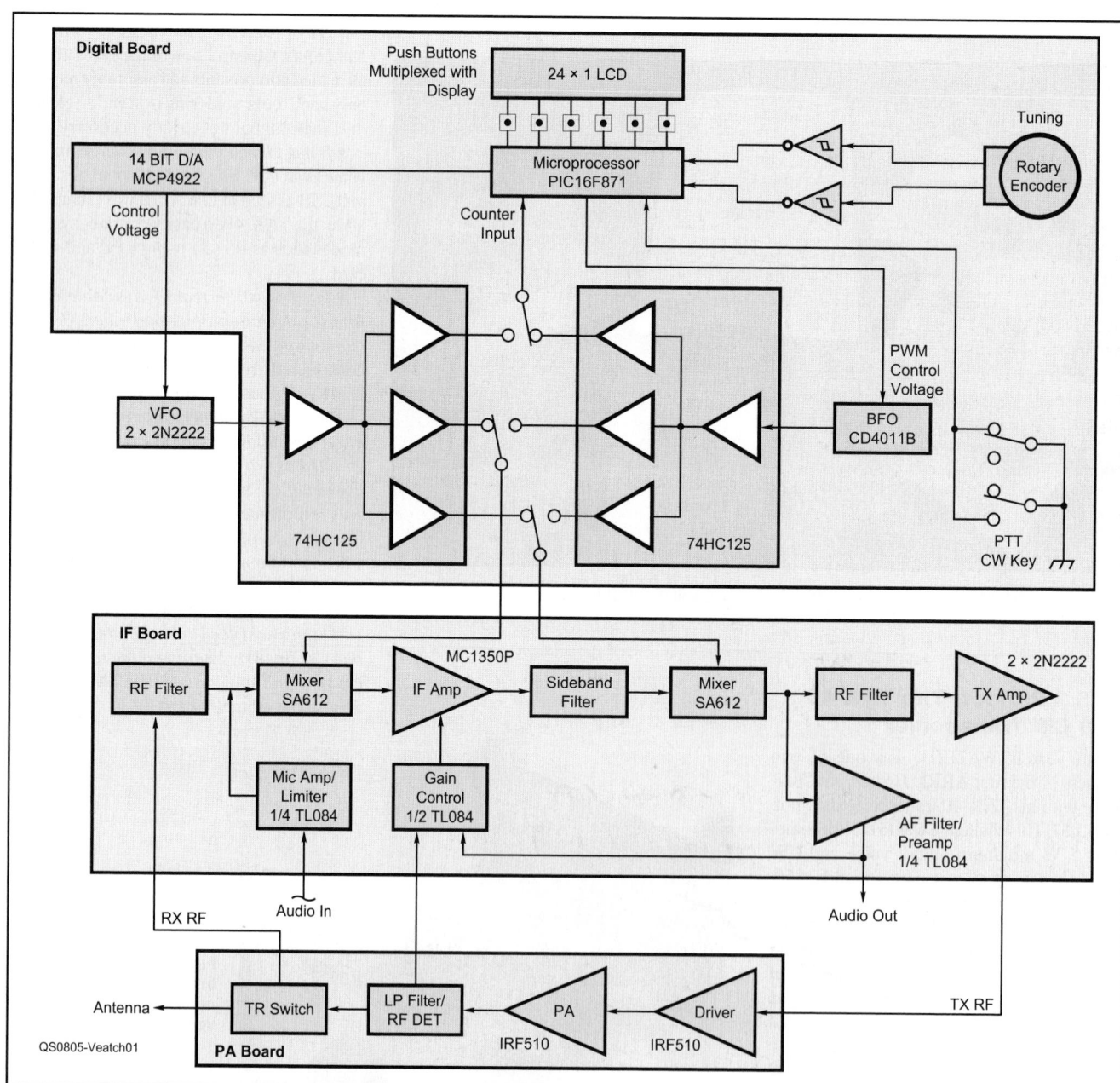

Fig 13.58 — Block diagram of TAK-40, homebrew challenge winner. This radio included many more features and capabilities than we expected to find in a $50 radio! Complete schematics, parts lists and firmware can be found at www.arrl.org/QST-in-depth. Look in the 2008 section for QS0508Veatch.zip.

filtering includes a 6-element crystal ladder filter, an integrated IF amplifier, two SA612 ICs used as a mixer and a BFO, and the audio output amplifier. The low-level RF transmit signal is also generated on this board.

- Power amplifier (PA): This module contains a driver and the final power amplifier (PA) board with filtering to meet the FCC spectral purity requirements.

The overall design is a classic superheterodyne with a 4-MHz IF and a 3- to 3.3-MHz VFO. The same IF chain is used for transmitting and receiving by switching the oscillator signals between the two mixers. **Fig 13.58** shows the block diagram of the TAK-40 transceiver. Each board is described below with a detailed schematic and parts list on the *QST* binaries website - **www.arrl. org/QST-in-depth**. Look in the 2008 section for file **QS0508Veatch.zip**.

CONSTRUCTION

The best way to build this transceiver would be to buy the printed circuit boards (PCB) but it would not have met the competition's $50 budget limit! Files included in the package on the *QST* website can be sent to **www.expresspcb.com** and they will send you two complete sets of boards for just over $100. (If you order more, the pre-set price decreases.) By ordering the PCBs you would complete the radio sooner than if you built it using any other technique and with a higher probability of success. The process by which the TAK-40 was constructed for the competition is described in the full article.

Fortunately the TAK-40 requires relatively little chassis wiring. A small harness for the LCD and push-button switches, cable for the rotary encoder audio in and out and key line wiring are all that is required for the front panel. The IF board connects to the VFO, IF and PA boards for control and metering. Two RF lines run between the IF and PA boards. The prototype is built on a wooden frame and the front panel printed on photo paper in an inkjet printer.

13.5.3 *Project:* A Homebrew High Performance HF Transceiver — the HBR-2000

Have you ever dreamed of building an Amateur Radio transceiver? Have you thought how good it would feel to say, "The rig here is homebrew"? Markus Hansen, VE7CA, shows us that it's still possible to roll your own full featured HF transceiver — and get competitive performance! The transceiver in **Fig 13.59** is a high-performance, 100 W HF transceiver that the author named the HBR-2000.

This project is a condensed version of an article that appeared in March 2006 *QST* and should provide an inspiration to all homebrewers. (The full original article is available in PDF format on the CD-ROM included with this book.) The goal of presenting the project here is not necessarily to suggest that it be replicated but to illustrate a successful approach to such a project.

The secret to being able to successfully build a major project such as this is to divide it into many small modules, as indicated in the block diagram (**Fig 13.60**). Each module represents a part of the whole, with each module built and tested before starting on the next. To choose the actual circuits that were to be built into each module, the author searched past issues of *QST, QEX, The ARRL Handbook* and publications dedicated to home brewing such as *Experimental Methods in RF Design*, published by ARRL. It's important to learn by reading, building a circuit and then taking measurements. After you build a particular circuit and measure the voltage at different points, you begin to understand how that particular circuit works.

Do not go on to the next module until finishing the previous module, including testing it to make sure it worked as expected. After building a module, follow the same process to decide on the circuit and build the next module. Then connect the two modules together and check to make sure that, when combined

together, they performed as expected. It is really that simple, one step at a time. Anyone who has had some building experience can build a receiver and transmitter using this procedure.

For a receiver, begin by building the audio amplifier and product detector module, then the BFO circuit, testing each separately and then the growing receiver, step by step. From there, build the IF/AGC module, then the VFO and the heterodyne LO system, then the receiver mixer, and on and on until reaching the antenna. At that point you have a functioning receiver. It is a thrilling day when you hook up an antenna to the receiver and tune across the Amateur Radio bands listening to signals emanating from the speaker.

After you build a particular module, you don't want RF from outside sources to get into the modules you build and you don't want RF signals produced inside the modules to travel to other parts of the receiver, other than through shielded coaxial lines. The reason that you don't want RF floating around the receiver is that stray RF can produce unwanted birdies in the receiver, adversely affect the AGC system or cause other subtle forms of mischief. To prevent this from happening, enclose each module in a separate RF tight box and use coax for all the RF lines with BNC or phono connectors on each end. All dc and control lines are connected via feed thru insulators.

Modules are enclosed in boxes made from unetched copper clad material. For the covers, cut sheet brass ½ inch wider and longer than the size of the PC box. Then lay the box on top of the brass and center the box so that there is about ¼ inch overlap around the perimeter of the box, and draw a line around the box with a felt pen. Then cut the corners out with tin snips and bend the edges of the brass cover over in a vise. By drilling small holes around the perimeter of the box, inserting wires through the holes, soldering the wires to the inside of the box and to the overlapping edges, you produce an RF-tight enclosure. See **Fig 13.61** for an example of this technique.

TYPES OF CONSTRUCTION

The audio board and IF board are etched PC boards purchased from Far Circuits. The construction method the author learned to appreciate the most is "ugly construction," discussed in the **Circuit Construction** chapter. Each module is designed for an input and output impedance of 50 Ω except for the audio output (8 Ω for speaker connection). Thus 50-Ω coax cable with BNC connectors can be employed to connect the RF paths between the different modules. The concept is shown in **Fig 13.62**.

Fig 13.59 — Head on view of the HBR-2000. This looks a lot like a commercial transceiver.

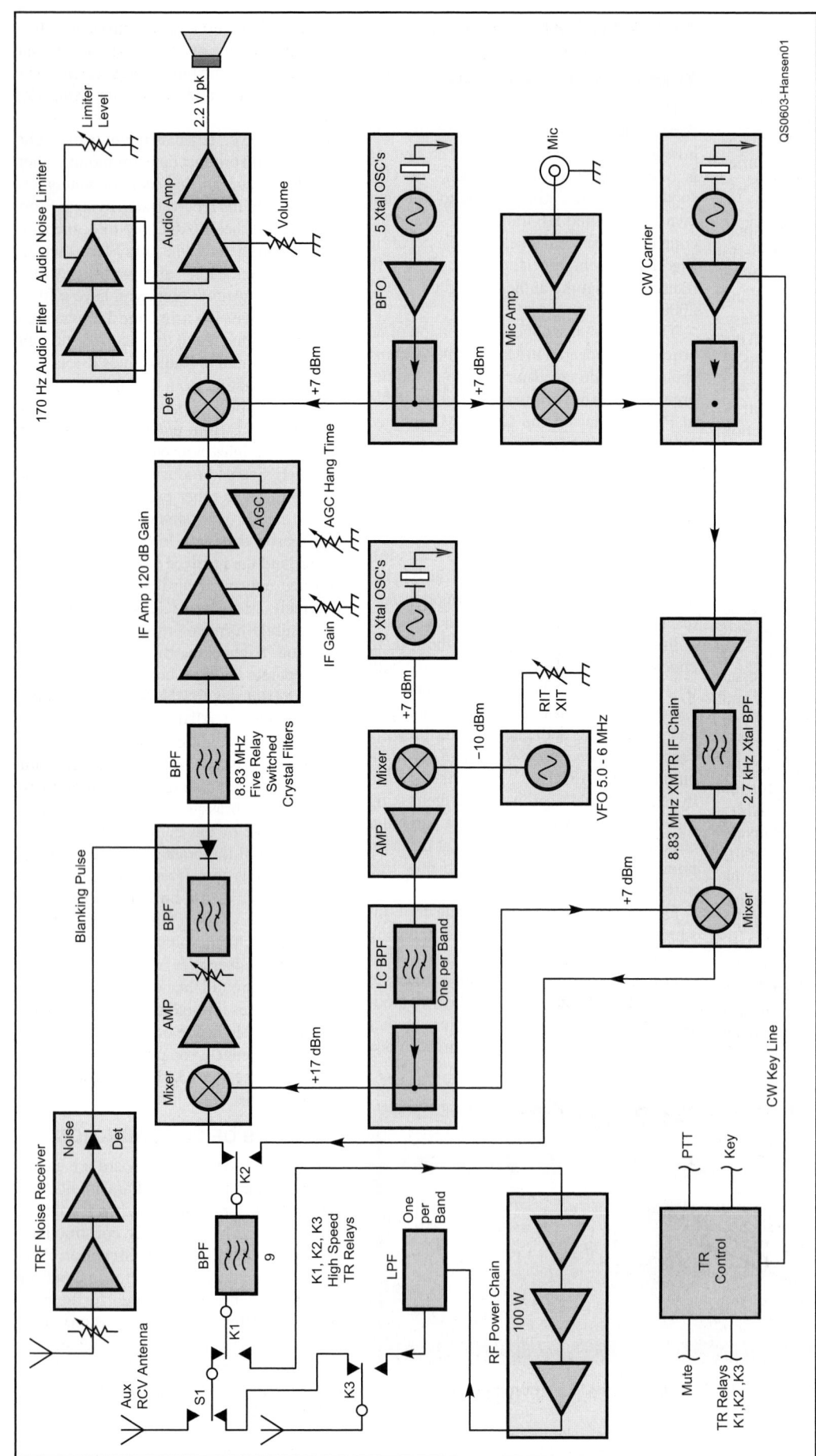

Fig 13.60 — Block diagram of the HBR-2000.

Fig 13.61 — Sample of a box made with PC board with a sheet brass cover overlapping all four sides. Note the BNC connectors and feed thru insulator.

MEASUREMENTS AND TEST EQUIPMENT

Making meaningful and accurate measurements is a major part of producing a successful project. You must make measurements as you progress or you have no way of knowing whether a module is performing according to design specifications.

One of the specifications chosen for the HBR-2000 was that it should have a very strong front end. To accomplish this, the design uses a mixer that requires that the LO port be fed with +17 dBm at 50 Ω. Having the test equipment to measure these parameters confirms the expected results.

To help in the construction of this proj-

Fig 13.62 — The inside of the HBR-2000. All the boxes are connected together with 50 Ω coax and BNC connectors for the RF runs. All dc power and control leads, in and out of the boxes, are through feed-through insulators.

ect, the author purchased surplus test instruments including an oscilloscope and signal generator. He also built test equipment such as crystal-controlled, very low power, oscillators and attenuators to measure receiver sensitivity, and high-power oscillators and a combiner to measure receiver blocking and dynamic range. He also built a spectrum analyzer, which turned out to be one of the most useful instruments because the HBR-2000 has 19 band-pass filters and 22 low-pass filters. (This instrument was described in August and September 1998 *QST* by Wes Hayward, W7ZOI, and Terry White, K7TAU.) Later he built the RF power meter featured in June 2001 *QST*, authored by W7ZOI and Bob Larkin, W7PUA. These instruments allow adjustment and measurement of the performance of each module built. A good selection of test equipment can be affordable. The author's lab is shown in **Fig 13.63**.

OTHER CIRCUITS

Here are some of the sources used for deciding on the circuits for the other modules in the HBR-2000 design. The low-distortion audio module is from the "R1 High Performance Direct Conversion Receiver" by Rick Campbell, KK7B, in August 1993 *QST*. The VFO design is from June 1991 *QST* ("Build a Universal VFO" by Doug DeMaw, W1FB).

The first mixer, post-mixer amplifier and crystal heterodyne oscillator design was taken from "A Progressive Communication Receiver" design by Wes Hayward and John Lawson, K5IRK, which appeared in November 1981 *QST* and was featured in *The ARRL Handbook* for many years. This is a classic radio article with many good circuit ideas.

The receiver input RF band-pass filter and diplexer designs along with the noise blanker and 100 W amplifier circuits were taken from the three-part series beginning in the May/June 2000 issue of *QEX* titled The "ATR-2000: A Homemade, High-Performance HF Transceiver" by John Stephenson, KD6OZH. The power amplifier and output circuit filters are in a shielded sub-enclosure as shown in **Fig 13.64** and **Fig 13.65**.

The BFO and power supply circuits were lifted right out of the *ARRL Handbook*. The transmitter portion of the transceiver consists of combinations of various circuits found in *Experimental Method in RF Design*.

RECEIVER SPECIFICATIONS

For the skeptics, the actual measured performance of the HBR-2000, as confirmed in the ARRL Lab, is shown in **Table 13.2**. Receiver sensitivity measurements on all bands are within ±0.5 dB of –130 dBm. All measurements were made with an IF filter bandwidth of 400 Hz. Test oscillators are

Fig 13.63 — The VE7CA work shop shows some of the home built and surplus test equipment used in the construction and evaluation of the HBR-2000.

Fig 13.64 — The 100 W amplifier board.

Fig 13.65 — The 100 W amplifier, 10 low-pass filters and the power supplies are located in a separate enclosure.

two separately boxed crystal oscillators, low-pass filtered and designed for a 50-Ω output impedance. MDS measurements were made with a HP-8640B signal generator and a true reading RMS voltmeter across the speaker output.

The receiver specifications were made following ARRL procedures as outlined in the ARRL "Lab Test Procedures Manual" at **www.arrl.org/product-review**. Making accurate receiver measurements is not a trivial matter and should be approached with the understanding of the limitations of the test equipment being used and thorough knowledge of the subject.

You may notice that the author made no attempt to miniaturize the HBR-2000. With large knobs and large labeling, he is able to operate the transceiver without the need for reading glasses. When you build your own equipment, you get to decide the front panel layout, what size knobs to use and where they should be located. That is a real bonus!

Why not plan to build your dream station? If you haven't built any Amateur Radio equipment, begin with a small project. As you gain experience you will eventually have the confidence to build more complex equipment. Then someday you too can say with a smile, "my rig here is homebrew."

Table 13.2
HBR-2000 Test Measurements

Spacing:	20 kHz	5 kHz	2 kHz
Two-tone blocking dynamic range:	>126 dB	124 dB	122 dB
Third-order IMD dynamic range:	103.5 dB	102.5 dB	93.0 dB
Third-order intercept:	25.5 dBm	24.0 dBm	14.5 dBm

Image rejection all bands: >135 dBm
Receive to transmit time:8 ms (incl 2 ms click filter)
CW, full QSK transmit to receive time:17 ms (30 WPM = 20 ms dot)

13.6 References and Bibliography

G. Barter, G8ATD, Editor, *International Microwave Handbook*, Second Edition (RSGB, 2008).

R. Campbell, KK7B, "High Performance Direct Conversion Receivers," *QST*, Aug 1992, pp 19-28.

D. DeMaw, W1FB, "Build a Universal VFO," *QST*, Jun 1991, pp 27-29.

J. Grebenkemper, KI6WX, "Phase Noise and its Effect on Amateur Communications," *QST*, Part 1, Mar 1988, pp 14-20; Part 2, Apr 1988, pp 22-25.

J. Hallas, W1ZR, *Basic Radio* (ARRL, 2005).

E. Hare, W1RFI, "The Tuna Tin Two Today," *QST*, Mar 2000, pp 37-40.

W. Hayward, W7ZOI, R. Campbell, KK7B, and B. Larkin, W7PUA, *Experimental Methods in RF Design* (ARRL, 2003).

W. Hayward, W7ZOI, "Crystal Oscillator Experiments," Technical Correspondence, *QST*, Jul 2006, pp 65-66.

W. Hayward, W7ZOI, and D. DeMaw, W1FB (SK), *Solid State Design for the Radio Amateur* (ARRL, 1977), pp 26-27.

W. Hayward, W7ZOI, and J. Lawson, K5IRK, "A Progressive Communication Receiver," *QST*, Nov 1981, pp 11-21. Also see Feedback in Jan 1982, p 47; Apr 1982, p 54 and Oct 1982, p 41.

J. Kuecken, KE2QJ, "A Fast T-R Switch," *QST*, Oct 2005, pp 56-59.

D. Pozar, *Microwave Engineering*, Fourth Edition (John Wiley and Sons, 2012).

W. Sabin, WØIYH, "A 455 kHz IF Signal Processor for SSB/CW," *QEX*, Mar/Apr 2002, pp 11-16.

W. Sabin and E. Schoenike, Editors, *Single-Sideband Systems and Circuits* (McGraw-Hill, 1987).

W. Sabin, "RF Clippers for SSB," *QST*, Jul 1967, pp 13-18.

J. Scarlett, KD7O, "A High-Performance Digital-Transceiver Design," Parts 1-3, *QEX*, Jul/Aug 2002, pp 35-44; Mar/Apr 2003, pp 3-12; and Nov/Dec 2003, pp 3-11.

M. Schwartz, *Information Transmission, Modulation and Noise*, Third edition (McGraw-Hill, 1980).

M. Spencer, WA8SME, "A Transmitter for Fox Hunting," *QST*, May 2011, pp 33-36.

J. Stephenson, KD6OZH, "ATR-2000: A Homemade, High-Performance HF Transceiver," *QEX*, in three parts: Mar / Apr 2000, p 3; May/Jun 2000, p 39; and Mar/Apr 2001, p 3.

Contents

Telemetry and Navigation

This brand-new chapter in the *ARRL Handbook* was created in support of a relatively new chapter in Amateur Radio — the increasing use of amateur communication as a key element of scientific experimentation. This initial foray into the use of amateur means to collect and track data focuses on high-altitude balloons which are the most popular platform used for this purpose today. Future editions will expand coverage to additional technologies, platforms and applications.

Material in this edition was contributed by Paul Verhage, KD4STH — who contributed material on remote sensing and navigation data — and Bill Brown, WB8ELK, who contributed material on payload design.

Support of scientific experimentation is hardly a new aspect of Amateur Radio — amateurs have supported science almost since the beginning when Tom Mix, 1TS, accompanied the explorer MacMillan to the Arctic aboard the *Bowdoin* in 1923. Even more directly, in 1912 the Radio Act exiled amateurs to the "useless" wavelengths above 200 meters and amateur experimenters rapidly discovered their utility for worldwide communication. The story continued with Grote Reber, W9GFZ, and radio astronomy in the 1930s, broad participation by hams during the International Geophysical Year of 1957-1958, wildlife tracking, propagation reporting, satellite construction, and numerous other instances.

In recent years we have heard more and more frequently of balloons and radio-controlled craft using Amateur Radio for their data links and control signals, even as they cross continents and entire oceans! Other platforms are land-based (such as weather stations or animal tracking) or marine (ocean or river measurement buoys). CubeSats (**www.cubesat.org**) also use Amateur Radio control and data links. Building these platforms combines several technical fields: using sensors and data acquisition systems to measure events and phenomena, mechanical and electrical engineering to construct the platform, 3-dimensional navigation, and the data link and associated radio technologies. These hybrids are attracting the experimentalists and scientists to Amateur Radio, just as they were attracted at the dawn of the wireless age.

This chapter is organized in three parts: sensors, telemetry and navigation, and platform design. In recognition of the rapid innovation in these activities, this chapter will change and expand in the coming editions. Expect to see new digital protocols and miniature telemetry and audio-image transmitters. Improvements will be forthcoming in portable-mobile power sources and antennas. The chapter will report on the engineering necessary to assemble an effective platform. There is no doubt that fulfillment of FCC Part 97.1 is thriving as hams continue to "improve the radio art" by adapting technology to new uses.

14.1 Sensors

The advent of inexpensive microcontrollers combined with Amateur Radio creates opportunities for the amateur to perform experiments in environments that are otherwise inaccessible for one reason or another. Many interesting regions of the Earth, including extremely high altitudes in the atmosphere, fall into this category. Hams can be instrumental in helping amateurs and professionals explore these environments.

The development of GPS receivers and their application in APRS has made it practical to collect data in the near-space environment. Terrestrial and marine data measurement stations and buoys, traditional weather stations, other sensors, and event logging are also being deployed. This section on sensors addresses the "front end" of remote sensing. These sources of data are analogous to the speech and video circuits of traditional amateur transmitters. The same care in their design is required if quality results are to be obtained.

These platforms are enabling *remote sensing* — observing or measuring an object or event without actually being in contact with the condition being measured. Data from the measurement is then stored on the platform for eventual collection after recovery, or transmitted to a ground station for recording and analysis (telemetry). Examples of parameters that are measured by amateur remote sensing platforms include temperature, pressure (air

Chapter 14 — CD-ROM Content

Supplemental Articles

- "A Simple Sensor Package for High Altitude Ballooning" by John Post, KA5GSQ
- "APRS Unveiled" by Bob Simmons, WB6EYV
- "APRS with a Smartphone" by Pat Cain, KØPC
- "ARRL Education and Technology Program Space/Sea Buoy" by Mark Spencer, WA8SME

and water/fluid), humidity, ozone and other gasses, acceleration and light.

The first step is to select the appropriate sensor or sensors for the parameter of interest and a means of converting sensor outputs to digital data, usually by connecting the outputs to a microcontroller. A sensor is a device that detects a specific condition of interest, such as temperature or pressure, and produces a predictable output in response. This section divides sensors into the following four types of outputs; *resistance-based, current-based, voltage-based,* and *digital.*

Four common types of outputs produced by sensors are discussed here. Not all of these outputs are directly useful to a microcontroller. However, methods exist to convert the output of these sensors into a form that can be interfaced to a microcontroller. Many microcontrollers have analog-to-digital converters built in that can digitize the data directly. (See the Analog/Digital Conversion section of the **Analog Basics** chapter.)

14.1.1 Resistance-Based Sensors

Resistance-based sensors change an internal resistance in response to the environmental variable they measure. An example includes the photocell, which is constructed of the chemical cadmium sulfide (CdS), a semiconductor that produces electrons and holes when irradiated by light. The production of free electrons and holes reduces the resistance of CdS when it is exposed to light.

In many cases, the change in resistance in response to changes in the measured condition is small. Therefore, sensor manufacturers often incorporate additional circuitry with the sensing element to convert this changing resistance into a more easily measured change in voltage. However, resistive-type sensors are still available and quite useable.

Resistance-based sensors do not create a signal that a microcontroller can measure directly. Instead, the resistance of the sensor is used to vary the voltage from a voltage source. A simple and popular circuit capable of converting a changing resistance into a changing voltage is the voltage divider as described in the **Analog Basics** chapter.

The voltage divider circuit of two resistors as shown in **Fig 14.1**. One resistor is fixed in resistance (R_F) and the other is the sensor and therefore variable in resistance (R_V). The current through the voltage divider circuit is variable. It increases as the resistance of the variable sensing element decreases and vice versa. However, the sum of the two voltage drops is always equal to the supply voltage. The preferred arrangement of the two resistors depends on the response of the sensing resistor to the condition to which it responds — temperature, humidity, illumination, and so on.

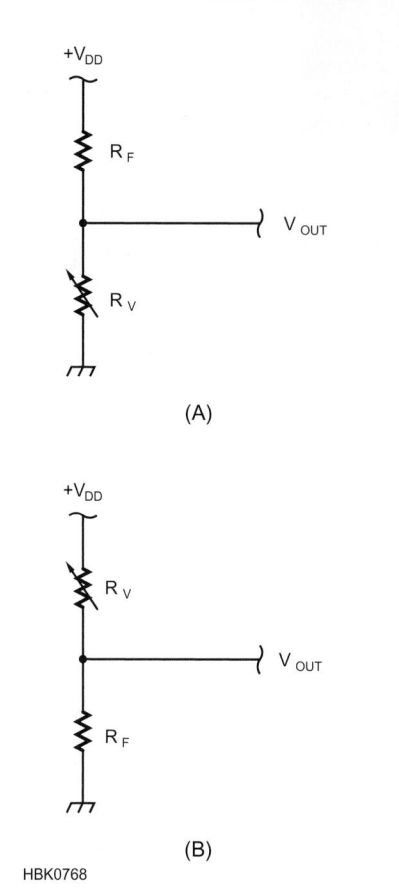

Fig 14.1 — The voltage divider circuit, a series circuit of two resistors. The orientation of resistors in A is preferable when the sensing resistor increases resistance in response to an increasing condition. Use B when the sensing resistor decreases resistance in response to an increasing condition. This results in the output voltage increasing with the increasing condition.

A microcontroller connected to the voltage divider circuit digitizes the voltage across the resistor connected to ground. This permits the design of resistance-based sensors into circuits that produce changing voltages which follow the change in the condition. For example, the resistance of photocells decreases as the light intensity increases. A microcontroller digitizing the voltage across a photocell connected as R_V in Fig 14.1A will observe V_{OUT} *increasing* as light intensity *decreases*. If however, the photocell is connected as R_V in Fig 14.1B, V_{OUT} *increases* as the light intensity *increases*. The latter case is easier to understand and work with than having output voltage and the sensed condition varying in opposite directions.

There are two equations that describe the output of the voltage divider circuit. The first describes the voltage drop across the variable resistor and the second describes the voltage drop across the fixed resistor.

For Fig 14.1A:

$$V_{OUT} = +V\ (R_V/(R_F + R_V))$$

For Fig 14.1B:

$$V_{OUT} = +V\ (R_F/(R_F + R_V))$$

OPTIMIZING R_F

The equations above show that the value selected for R_F has a large impact on the range of output voltages created by the voltage divider circuit. The precision of the sensor output is greatest when the voltage range of V_{OUT} is maximized. The value of R_F that generates the maximum range is the geometric mean of the sensor's highest expected resistance (R_H) and lowest expected resistance (R_L). The equation for calculating the best fixed resistor value (R_F) in a voltage divider circuit is shown below.

$$\text{Optimum } R_F = \sqrt{R_L \times R_H}$$

The maximum range for V_{OUT} of the voltage divider circuit is thus equal to ⅓ of the supply voltage, V_{DD}. Furthermore, the voltage range is centered at the midpoint of the supply voltage. The following three equations calculate the minimum voltage, maximum voltage, and voltage range of an optimized voltage divider.

$$V_{MIN} = V_{DD}/3$$

$$V_{MAX} = 2V_{DD}/3$$

$$\text{Range} = V_{MIN} - V_{MAX} = V_{DD}/3$$

14.1.2 Current-Based Sensors

Some types of sensors generate or change output current in response to the environmental condition they are measuring. Examples include the photodiode, solar cell and light-emitting diode (LED). All three of these devices are similar, although not used in similar ways. When a photon of light is absorbed, its energy gives an electron enough energy to jump across the PN junction inside the device. The electron creates a measurable current from the sensor.

The LED is one of the most surprising current-based sensors. While the photodiode is sensitive to a wide range of frequencies, the LED is most sensitive to light at the wavelength it emits when forward biased. This makes the LED a very inexpensive spectrally sensitive photometer.

A current-based sensor can provide useful

data when connected to a digital multimeter (DMM) set to measure milliamps of current. However, this is not a suitable configuration for a microcontroller with the capability to digitize voltage. Therefore, it is necessary to find a way to convert changing current into a changing voltage. Two popular ways to accomplish this are to use a transimpedance amplifier or by measuring the charging time of a capacitor.

THE TRANSIMPEDANCE AMPLIFIER

The transimpedance amplifier in **Fig 14.2** is a popular op amp circuit that converts input current into an output voltage.

The feedback resistor, R, sets the gain of the transimpedance amplifier. The output voltage is given by the following equation:

$$V_{OUT} = I_{IN} \times R$$

The capacitor, C, reduces gain at high frequencies above 1/RC, acting as a low-pass filter to reduce noise. A generally useful value is 220 pF with the usual values of R for LED light-sensing. You will have to take the bandwidth of your measurement into account when choosing the value of C.

It is important that the value selected for the feedback resistor does not attempt to drive the amplifier output greater than the supply voltage. It is critical that the output voltage not exceed the maximum output voltage that the amplifier is capable of producing. In those circumstances, data is lost when the sensor output is too high and the amplifier saturates.

USING CAPACITOR CHARGE TIME

A second method to digitize the current from a sensor is to measure the length of time required for a current to charge or discharge a capacitor to a certain voltage. One example

can be found in the book *Earth Measurements* by Parallax (**www.parallax.com**, manufacturer of the BASIC Stamp microcontrollers). Here, the BASIC Stamp is used to charge a capacitor. Then the Stamp measures the length of time required for the capacitor to discharge to ground through a resistor. The capacitor and resistor values are selected according to the expected current output of the sensor. The book uses the circuit in **Fig 14.3** to measure the current output of a photodiode or LED.

The program shown in **Table 14.1** (*Earth Measurements*, Program 4.2) was written to use the schematic in Fig 14.3. It assumes the circuit connects to the BASIC Stamp via I/O pin 6. Change the I/O reference to another pin as needed by your circuit.

The program reports the light intensity once per second. It begins by charging the capacitor to the same potential as the supply

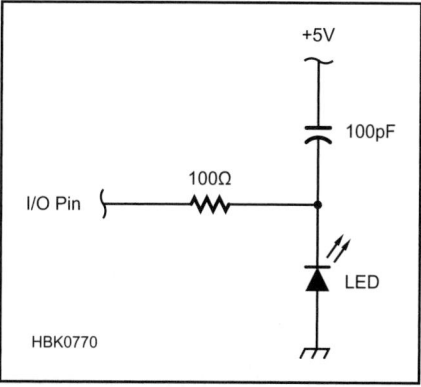

Fig 14.3 — The circuit recommended by Parallax for digitizing the current output of a photodiode or LED. See text for more information about using this circuit for current measurement.

Table 14.1
BASIC Stamp Program

This BASIC Stamp program is used with the circuit of Fig 14.3 for measuring light intensity.

```
RCT VAR Word
Light VAR Word
HIGH 6

Loop:
      RCTIME 6,1,RCT
      HIGH 6
      Light = 65535/RCT
      DEBUG "Light
Intensity: ", DEC Light
      PAUSE 1000
      GOTO Loop
```

voltage through the use of the HIGH 6 command. Afterward, the reverse current emitted by the LED, due to its exposure to light, discharges the capacitor. The changing potential of the capacitor makes the voltage drop across the LED appear to decrease from its start at +5 V. Any voltage above 1.4 V is treated as a logic high by the BASIC Stamp. Therefore, as reverse current from the LED brings the capacitor voltage lower, the voltage across the LED eventually becomes lower than 1.4 V and a logic low.

The RCTime command counts the time (in units of 2 µs) required for I/O pin 6 to change from a logic high (above 1.4 V) to a logic low (below 1.4 V). The result in units of 2 µs is stored in the variable RCT. The greater the intensity of the light shining on the LED, the faster the capacitor discharges and the smaller the value stored in the variable RCT. The value in RCT is then divided into 65535 to invert the relationship and then stored in the variable Light which then contains a value directly proportional to light intensity.

14.1.3 Voltage-Based Sensors

Some types of sensors change their voltage output in response to the condition they are measuring. Examples include the LM355 temperature sensor, Honeywell's HIH-4000 relative humidity sensor, and Silicon MicroStructures' line of pressure sensors. These devices are typically current or resistance-based sensors along with circuitry to amplify and condition the output into a useable voltage change.

The voltage-based sensors easiest to interface are those that include signal amplification and correction on the chip. The result can be a ratiometric output that is linear and proportional to the supply voltage. The conversion factor for the sensor needed to convert its voltage into the environmental condition being measured is documented in the device's datasheet.

Pressure sensors can be used as altitude sensors for an airborne platform if a digital solution, such as GPS, is not available. Absolute pressure is preferred although it must be calibrated against ground barometric pressure before launch and, if the flight covers long distance, requires additional corrections based on local pressure data.

14.1.4 Capacitance-Based Sensors

Capacitance-based sensors use changes in capacitance as their primary means of measurement. Capacitance between two electrodes of known area depends on the distance between the electrodes and the dielectric constant of the insulating material

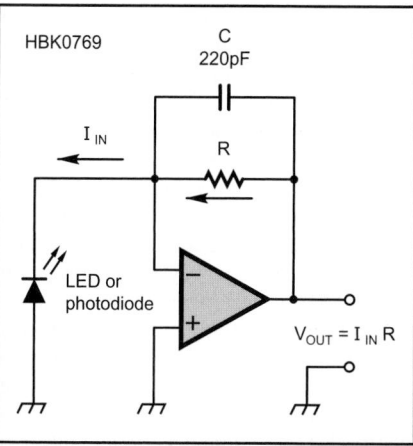

Fig 14.2 — The transimpedance amplifier converts current at its input to voltage at the output by balancing current through R with the input current.

separating them. Any process that changes either separation or dielectric constant can be then be sensed as a change in the capacitance. Parameters that are sensed in this manner include motion, moisture, fluid and material level, chemical composition and acceleration.

Sensors based on capacitance are rarely supplied without signal conditioning and linearization. Many have digital outputs that supply the measurement as a digital value. Another option is to have the sensor's capacitance control the frequency of an oscillator which can then be read by a digital circuit. For more information on capacitive sensing, the excellent overview at **www.capsense.com/capsense-wp.pdf** is recommended.

14.1.5 Sensor Calibration

Sensors come in two basic configurations: *sensing elements* and *conditioned sensors*. The voltage divider discussed earlier is an example of a sensing element. There are no electronics associated with the divider — the package contains only the two resistors and the necessary connecting wires or terminals. Conditioned sensors contain electronic circuitry that operates on the signal from the sensing element before it is made available externally. The circuits usually regulate power applied to the sensor and also *linearize* the data so that a linear range of measurements are represented by a linear change in output voltage.

All sensing elements and some conditioned sensors require a calibration equation to convert the output signal into the parameter value the sensor is measuring. In some cases, the equation is simple and linear as in the LM335 temperature sensor. In other cases, the equation may be complicated, such as for the thermistor and photocell when used in a voltage divider circuit.

It is important to understand the range over which a sensor will be measuring a condition before attempting to calibrate it. The calibration equation is usually more accurate when based on the interpolation of measurements than when based on the extrapolation of measurements. There are exceptions to this rule. For example, the calibration equations of linear sensors can be just as accurate when extrapolated, as long as the maximum operating conditions of the sensor are not exceeded. Otherwise, is it best to expose the sensor to the entire range of expected environmental conditions while collecting measurements of its output to create the calibration equation.

The ham can easily create some of these conditions, such as temperature, on the bench top. High temperatures can be created with the use of heat lamps and low temperatures created with the use of dry ice packed in

Styrofoam coolers. Other conditions might need to be simulated. For example, light intensity is easily changed by changing the distance between the sensor and a fixed light source. Recall however that light intensity decreases by a factor of $1/r^2$ when using this method to create the calibration curve of a sensor.

The spreadsheet is a powerful tool for creating calibration equations. To create the calibration equation, carefully measure the output of the sensor as the environmental condition is varied. Enter the readings and distance into a spreadsheet. In the next column, calculate the intensity of the source, based solely on its distance from the sensor. Graph the results so that the independent variable (X axis) is the distance and the dependent variable (Y axis) is the intensity. Then select the function to create a regression line from the data in the chart.

14.1.6 Digital Sensors

Some types of sensor outputs are in digital form. These sensors communicate their data as a serial protocol in which data is exchanged as a series of bits over one or more circuits. Data can be transmitted synchronized to an external timing signal (*synchronous protocol*) or synchronized to special signals embedded within the data being transmitted (*asynchronous protocol*).

Examples of synchronous serial data protocols include 1-Wire, Inter-Integrated Circuit (I2C), and Serial Peripheral Interface (SPI). Examples of asynchronous serial data transmission include USB and the RS-232 (COM) ports on PCs. These serial protocols can transfer measured data to a microprocessor without additional conditioning.

Another type of digital sensor is one in which an event's detection is signaled as a voltage pulse or as a switch closure. For example, the detection of ionizing radiation by Geiger counters is signaled by voltage pulses that occur essentially at random. These signals require additional processing, such as by a counter or register circuit that is often implemented by a microprocessor.

SYNCHRONOUS SENSOR DATA PROTOCOLS

The following protocols are by no means the only ones used by sensors, but they are the most common protocols used by sensors on amateur remote sensing platforms. The manufacturer websites mentioned below have numerous resources to support design and development with devices that support these protocols.

1-Wire

1-Wire is a communication system developed by Dallas Semiconductor (now part of Maxim Electronics — **www.maxim**

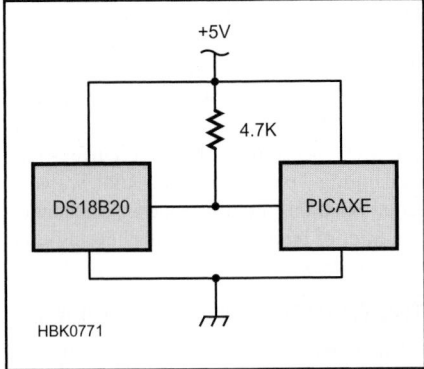

Fig 14.4 — The DS18B20 is a 1-Wire temperature sensor. In this circuit, the device does not use parasitic power and is connected to a 5 V source. A PICAXE microcontroller communicates with this device using the READTEMP or READTEMP12 command.

integrated.com) to enable communication between two or more integrated circuits. Devices on a 1-Wire network are daisy-chained together on a single-wire bus, called a *microlan*. (See **Fig 14.4**.) One device acts as the master device, and it controls communication between itself and the slave devices connected to the microlan. Some available 1-Wire devices include:

- Temperature sensor: (MAX51826)
- EEPROM memory (DS24B33)
- Low voltage sensor (DS25LV02)

Since a microlan may not include a separate power wire, many devices attached to the microlan include a small capacitor in their design. The capacitor provides *parasitic power* to the device while communications are taking place. The capacitor is necessary because communication requires the voltage on the single wire to alternate between power and ground. Without some temporary power source, devices would lose power during communications.

The master device communicates with each slave device by transmitting the slave address over the microlan prior to other commands. Because multiple devices can be connected to a microlan, each device must have a unique address to avoid confusion. Slave addresses are laser etched into 1-Wire devices. Alternatively, if a single 1-Wire device is attached to a microlan, communication on the network can ignore addressing altogether.

1-Wire is a two-way communication protocol. The master device begins communication by sending the slave device's address and then commands over the network. Only the device with the address in the message will respond to the commands. With the READTEMP command, communication between the master and slave device requires 0.75 second. The X2 series of PICAXE microcontrollers can communicate over 1-Wire using the OWIN and OWOUT commands

Fig 14.5 — A comparison between an iButton and a nickel.

native to their instruction sets.

iButtons

An iButton is a sealed 1-Wire device resembling a thick watch battery (see **Fig 14.5**). They include memory and a lithium battery. The memory contains the ID of the device and can often be used to store data. The battery permits an iButton to operate independently of a microlan. iButton devices download their stored data when connected to a microlan. The microlan connection is made by pressing the iButton device against a 1-Wire receptor. 1-Wire receptors are available for integration into microcontroller projects. Some available iButton devices include the following:

• Time and temperature loggers (DS1920 Thermochron)

• Time, temperature and humidity data loggers (DS1923 Hygrochron)

The amateur may be interested in the ongoing development of a 1-Wire weather station. Consult Maxim Integrated (**www.maxim integrated.com**) for the latest information concerning 1-Wire devices, including iButtons.

I2C

Inter-Integrated Circuit or I2C is a communication method developed by Phillips to enable communication between two or more ICs. Devices on an I2C network are daisy-chained together on a two-wire bus as in **Fig 14.6**. One device acts as the master device and it controls communications between itself and the slave devices connected to the network. The I2C network is described in detail at **www.i2c-bus.org** and in the application notes supplied by manufacturers of devices that use it.

The first connection in the I2C network is the serial data (SDA) line. This line carries slave device addresses, data, and instructions between devices. The second line is the serial clock (SCL) line. This connection provides timing pulses to synchronize the data sent from the sending IC (master) to the receiving IC (slave). In an I2C network, the SDA and SCL lines are pulled up to +5 V by pull-up resistors. A value of 4.7 kΩ also works well.

The master device communicates with each slave device by transmitting an address over the I2C bus prior to other commands. Because multiple devices can be connected to an I2C network, each device must have

a unique address to avoid confusion. Slave addresses may be designed into the IC or may be externally configured for an IC by connecting a combination of address pins to +5 V and ground.

I2C is a two-way communication protocol. The master device begins communication by sending the slave device's address and then commands over the network. Only the device receiving its address in the message will respond to the commands. Serial data can be sent in either in fast (400 kHz) or slow (100 kHz) mode. The master device sends commands and memory addresses in either one byte or one word (two bytes) long commands. Some available I2C devices include the following:

• Memory: the 24LCxxx series of I2C memory.

• Real-time clocks: DS1307

• Analog to digital converters: LTC2903 (12 bit), AD7991 (12 bit), and MCP3421 (18 bit)

SPI and Microwire

Serial Peripheral Interface or SPI is a communication method developed by Motorola (now Freescale) to enable communication between two or more ICs. Devices on a SPI network are daisy-chained together on a two- or three-wire bus. (See **Fig 14.7**.) Like I2C, one device is the master that controls communications between it and the slave devices connected to the network. The Microwire

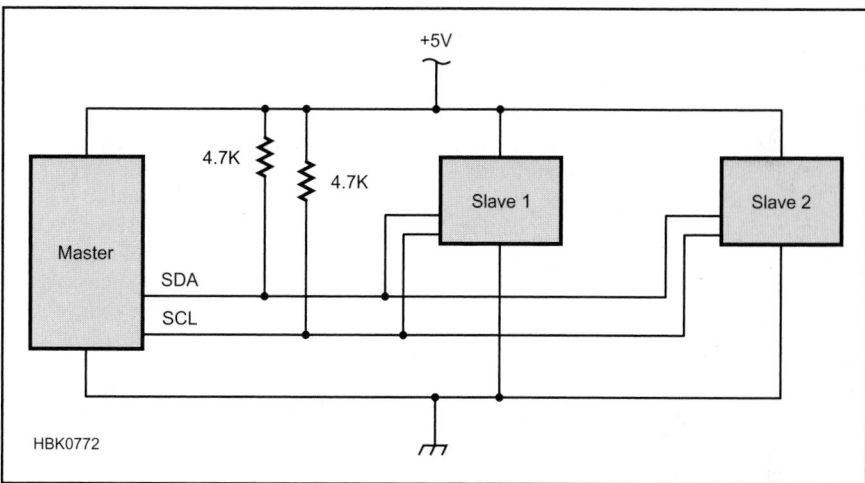

Fig 14.6 — An example of a master and two slave ICs connected via a I2C network.

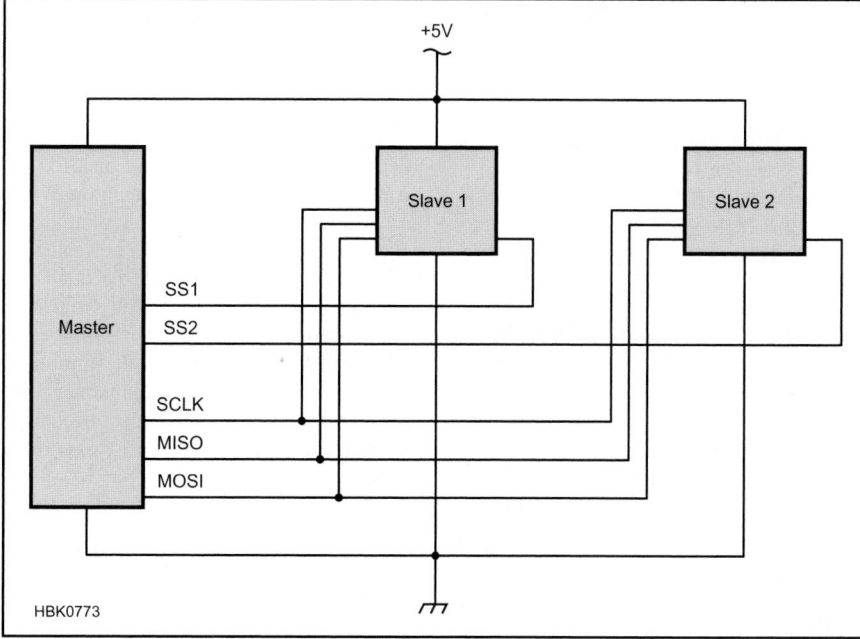

Fig 14.7 — An example of a master and two slave ICs connected via an SPI network.

network originally developed by National Semiconductor (now Texas Instruments) is essentially a subset of SPI. Microchip (manufacturer of the PIC processor family) has published an overview and tutorial about SPI at **ww1.microchip.com/downloads/en/DeviceDoc/spi.pdf**.

Two lines of the SPI bus are used to transmit data and instructions: MOSI (master out/slave in) and MISO (master in/slave out). In some cases, the MISO and MOSI lines can be combined into a single shared line. The third line of the bus is the timing clock line (SCLK) that provides timing pulses to synchronize the data sent between the master device and the slave device. None of these lines requires being pulled high by a resistor.

The master device communicates with the slave devices by activating each slave device's Slave Select (SS) line. To avoid confusion, each slave device must have a unique connection to the master device. This is a major difference between I2C and SPI. An I2C network requires only two communication lines between devices, while an SPI network requires two or three communication lines in addition to an SS line between the master and each slave. A large number of slave devices require a large number of dedicated SS lines between the master and the slave devices.

SPI is also a two-way communication protocol. The master device begins communications by activating the slave device's SS line. Only the device with the activated SS line will respond to the commands. Serial data is sent as fast as the master device pulses the SCLK line. The number of bytes in each transmission between master device and slave device is limited by the design of the slave device rather than to eight or 16 bits. Some available SPI devices include the following:

- Analog to digital converters: MAX186 (12 bit resolution)
- Temperature sensor: LM74
- Hall effect sensor: MLX90363
- Pressure sensor: MPL115A1
- Memory: AT25010B

Note that the popular Dallas Semiconductor DS1620 Temperature Sensor uses a three-wire interface similar to SPI.

ASYNCHRONOUS SENSOR DATA PROTOCOLS

Asynchronous communication is any form of communication that does not use a clock signal to maintain timing between the sender and the receiver. A message begins with a start signal that allows the receiver to synchronize with the transmitter's message. The rest of the communication follows at a predefined rate in bits of data per second or baud. (See the **Digital Modes** chapter for a discussion of data rate.) As long as the sender and receiver use equally accurate clocks, they will transmit and receive the same bits of data.

Some sensors supply their output data using RS-232 and USB ports. The data is transmitted as a stream of characters controlled by a protocol developed by the manufacturer. USB devices often conform to certain classes of data objects so that generic device drivers can be used to acquire data from the sensor. Control and configuration protocols that allow the user to interact with the sensor are usually proprietary.

Time-independent serial devices produce a change in output voltage only at the detection of an event. The primary example is the *event counter*. The simplest event counters detect the closure of a switch, which can be useful for detecting the presence of wildlife. Game cameras use switches in this way to trigger a camera to record an image of wildlife. Thermostats and thermal switches are another example.

Switches can be used to signal a microcontroller by two different methods. In the first, called *active low*, the switch connects a microcontroller I/O pin to ground at the detection of an event. When the event is not present, the I/O pin is connected to positive voltage or pulled up to a positive voltage by a pull-up resistor. In the second method, called *active high*, the switch connects a microcontroller I/O pin to positive voltage at the detection of an event. When the event is not present, the I/O pin is connected to ground. Schematics for both of these switch circuits are shown in **Fig 14.8**.

An example of a sensor that produces asynchronous output is the Geiger counter. The output of the RM-60 Geiger counter from Aware Electronics' RM-60 (**www.aw-el.com**) maintains a 5 V level until ionizing radiation is detected. Then the output voltage drops to 0 V for 20 μs. The amount of radiation detected by a RM-60 Geiger counter is measured by counting the number of pulses emitted by the sensor over a fixed period.

Other event counters can be modified for microcontroller use if they use an LED indicator or piezoelectric annunciator. When an LED is illuminated, greater than 1.4 V appears across its terminals. Wires soldered to the LED can be connected to ground and one of a microcontroller's I/O pins to permit the microcontroller to count the number of LED flashes. Care in counting the number of flashes is necessary since some inexpensive sensors may output several pulses each time the event is detected or in the case of contact bounce for a switch closure.

14.1.7 Powering Sensors

The output of sensing element sensors is typically very sensitive to power supply voltage and noise. Any changes in power supply voltage on the voltage divider also

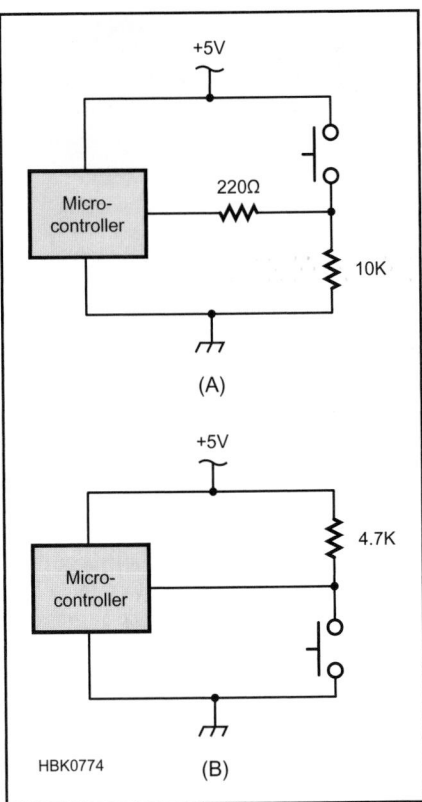

Fig 14.8 — (A) This circuit produces a logic high signal, typically 5 V, when an event is detected. (B) This circuit produces a logic low signal, typically ground or 0 V, when an event is detected.

appear, proportionally reduced, at the output of the voltage divider. This includes noise, transients, slowly dropping battery voltage — any change in the sensing element's supply voltage. The sensing element user must provide clean, filtered, regulated power to the sensor to avoid contaminating the sensor output voltage.

Loading of the sensing element is also an issue for the designer to deal with. A high-impedance sensing element will output erroneous voltages if connected to a load impedance that is too low. Be sure you know what the sensing element's ratings are!

Conditioned sensors are far less sensitive to noise and power supply variations. Some kind of voltage regulator circuit is included to make sure the electronics operate with a "clean" supply. The conditioning electronics, which often include laser-trimmed calibration circuitry, assume clean, well-regulated dc voltage from a power supply. They are much less sensitive to the effects of output loading although there are usually limits as to the amount of capacitance they can tolerate at the output, such as from a long run of wire.

In portable or mobile platforms, power is usually supplied by a battery pack. Make sure you have fresh, fully charged batteries before heading out to launch the platform for the experiment. Take into account the gradual reduction in voltage from the bat-tery pack as its charge is consumed — it's awfully hard to swap out batteries with a balloon that is in flight! In the quest to save weight in these platforms, make sure you have enough capacity in the battery pack voltage (see the **Power Sources** chapter) so that the experiment won't run out of power during its mission.

14.2 Navigation Data and Telemetry

Navigation data allows the sensor measurements to be combined with geographical data, which is important for correlating data to location (including altitude). A final step involves using Amateur Radio to either transmit the collected data to a ground station as telemetry or to track and recover a remote sensing payload for later data extraction. Since the most common use of this data is for weather balloons and other near-space missions, that context will be used.

It is important to note that transmitting data as ASCII characters (7- or 8-bit) is preferred to more compact binary formats. ASCII characters have the advantage of being human-readable so that even raw data can be inspected and used. At the low data rates of most amateur remote sensing, little overall throughput is lost by using ASCII characters. The ability to read the raw data stream directly is often invaluable during troubleshooting, as well.

14.2.1 Dead Reckoning

If a digital navigation data source such as GPS is not available, it is also possible to estimate platform position, including altitude by the process of dead reckoning. In dead reckoning, navigation (or tracking) depends on determining a known position — called a *fix* — and then calculating subsequent positions from the platform speed and direction.

Direction data can be obtained from compass sensors that output direction as an analog voltage or digitally encoded signal. Altitude can be calculated based on ground barometric pressure and absolute pressure readings from the platform.

Obtaining accurate ground speed data is difficult for mobile platforms such as balloons or water-borne instruments which move with the wind or current. If some other form of position tracking is available, it is possible to infer ground speed although rarely accurately.

14.2.2 GPS Data

As currently practiced, a GPS (Global Positioning System) module is the usual means of acquiring navigation data which is then transmitted as a telemetry stream using the Automatic Packet Reporting System (APRS). Thus the two are combined in this section.

Depending on the model, GPS receivers produce a number of navigation sentences. Most GPS receivers can form the GPGGA and GPRMC sentences described below. GPS sentences are text-based sentences. They are written in readable text that anyone can understand as long as the format of the sentences is understood. GPS sentences consist of fields separated by commas. Below is a brief description of the two more important GPS sentences (when it comes to high altitude ballooning) and their fields.

THE GPGGA SENTENCE

The GPGGA sentence is the Global Positioning System Fixed Data sentence and a typical GPGGA sentence from a balloon-based GPS looks like this.

```
$GPGGA,153919.00,4332.2076,N,
11608.6666,W,1,08,1.1,13497.
1,M,18.3,M,,*78
```

There are 13 fields in the GPGGA sentence following the sentence identifier, "$GPGGA". The fields from left to right are as follows.

1) Time in UTC (hours, minutes, seconds)
2) Latitude North (degrees and decimal minutes — note that there is no separator between degrees and minutes)
3) N (north)
4) Longitude West (degrees and decimal minutes — note that there is no separator between degrees and minutes)
5) W (west)
6) GPS Quality Indicator (0 = no GPS fix, 1 = GPS fix, and 2 = differential GPS fix)
7) Number of Satellites (number of satellites detected — not all of them may be used in determining the position)
8) Dilution of Horizontal Position (or DOHP, which is an indication of how precise the fix is and the closer to 1.0 the better)
9) Altitude (in meters)
10) M (meters)
11) Geoidal Separation (the difference in the actual height and a mathematic description of the height of an idealized Earth's surface in meters)
12) M (meters)
13) Checksum (result of exclusive ORing the sentence and used to verify that the text is not corrupted)

THE GPRMC SENTENCE

The GPRMC sentence is the Recommended Minimum Specific GPS/Transit Data sentence and a typical GPRMC sentence from a balloon-based GPS looks like this.

```
$GPRMC,153924.00,A,4332.2317,
N,11608.6330,W,24.4,46.3,2310
99,16.1,E*7E
```

There are 12 fields in the GPRMC sentence following the sentence identifier, "$GPRMC". The fields from left to right are as follows:

1) Time in UTC (hours, minutes, seconds)
2) Navigation warning (A = okay and V = warning)
3) Latitude North (degrees and decimal minutes — note that there is no separator between degrees and minutes)
4) N (north)
5) Longitude West (degrees and decimal minutes — note that there is no separator between degrees and minutes)
6) W (west)
7) Speed (in knots)
8) Heading (in degrees true north)
9) Date (day, month, and year — note that there is no separation between them)
10) Magnetic Variation (number of degrees)
11) Direction of magnetic variation (E = east and W = west)
12) Checksum (result of exclusive ORing the sentence and used to verify that the text is not corrupted)

14.2.3 Automatic Packet Reporting System (APRS)

Most balloon flights include an APRS station in order to follow the balloon's position and altitude throughout a mission to the edge of space. The APRS position reports, usually containing GPS data as described above, can be used directly to locate the position of a near-space balloon. (For more details about APRS, see the **Digital Modes** chapter.)

The usual APRS configuration is to transmit a position report once a minute with the recommended path set to WIDE2-1. A power level below 1 W is quite sufficient and many systems work quite well with just 200 mW.

There is a large network of dedicated ground

stations, digipeater and Internet gateway stations operating on the US national APRS frequency of 144.390 MHz (144.800 and other frequencies are used elsewhere in the world). Thanks to this network, the balloon's position will be plotted onto a map in near real-time. Two popular websites to view the maps are at **http://aprs.fi** and **http://findu.com**.

Chase crews collect a balloon's APRS data directly over Amateur Radio or over the Internet using a website like **aprs.fi** and **findu.com**. These sites are databases of APRS packets received and routed through APRS Internet gateways. This data is initially used to recover the platform or payload. Later the data is correlated with other sensor data and images that are stored in on-board memory.

There are a number of APRS "trackers" that combine a low-power GPS module with a VHF transmitter and microprocessor that creates the APRS message packets. For example, Byonics (**www.byonics.com**) makes a number of APRS tracking and telemetry products, including the Micro-Trak RTG FA High Altitude Combo that contains an altitude-certified GPS for balloon payloads. The RPC-Electronics (**www.rpc-electronics. com**) RTrak-HAB - High Altitude APRS Tracker Payload is specially made for high-altitude ballooning, as well.

A tracker combination built by the author is shown in **Fig 14.9**. On the right is a GPS module that creates the GPS sentences discussed previously. On the left is the MMT (Multi-Mode Transmitter) that creates and transmits the APRS packets.

APRS POSITION DATA

A simple APRS tracker can generate a stream of useful navigation data for a near-space flight. The data begins at the GPS receiver where two navigation sentences are generated. The sentences are then combined to create a position report in the required APRS format. Like GPS sentences, the raw APRS packets are also readable text that is easily interpreted.

An APRS position report uses a combination of commas and slashes as field delimiters. An example of an APRS report from a near space flight looks like this:

```
13:37:23 UTC: KD4STH-
8>APT311,WIDE1-2,qAS,KC0QBU,
133721h3836.39N/09500.
51W>160/031/A=049114
```

There are 12 fields in the APRS report. The fields from left to right are as follows:
1) Time in UTC (hours, minutes, seconds)
2) Call sign and SSID
3) Routing Information
4) GPS Time (time in UTC — note there is no separator between hours, minutes, and seconds)
5) h (hours)
6) Latitude North (degrees and decimal minutes — note that there is no separator between degrees and minutes)
7) N/ (north)
8) Longitude West (degrees and decimal minutes — note that there is no separator between degrees and minutes)
9) W> (west)
10) Heading (in degrees from true north)
11) Speed (in knots)
12) A= (altitude equals)
13) Altitude (feet)
Note that time and altitude data can be used to calculate the ascent rate of the weather balloon as a function of altitude. In addition, the same information can be used to calculate the descent rate of the parachute. Since a parachute's descent rate is a function of air drag which is controlled by air density, the parachute's descent speed during descent can be used to estimate air density as a function of altitude.

Note also that since a weather balloon is captive to the wind, measurements of altitude, speed, and heading are measurements of wind speed and direction at specific altitudes.

14.2.4 Non-Licensed Telemetry Transmissions

There is a large selection of low-power data links that use the unlicensed 915 and 2.4 GHz bands. Typically, these are intended to be used for short-range applications but with the balloon payload at great altitude, the range of these devices is much longer, particularly if a high-gain Yagi antenna is used to track the payload. (See the **Antennas** chapter for information on VHF and UHF beams.)

Many of the data link modules use a standard two-way protocol such as Zigbee and have direct analog and digital inputs and outputs. Some modules support Ethernet and Bluetooth interfaces, offering even more options for modules that can be assembled into the payload.

It is also important to note that unlicensed transmitters operating under FCC Part 15 rules are also subject to certain restrictions such as field strength. In addition, the type of antenna may be fixed and even required to be attached to the transmitter permanently. These and other restrictions are required in order to limit the range of these devices. Amateurs are used to modifying and adjusting their equipment, and this may not be allowed for some of these devices! Be sure to obtain the full documentation for any unlicensed device you plan on using and be sure you can use it in the way you expect.

14.2.5 Other Digital Modes

The usual method of communication from airborne and other remotely located platforms is via the APRS network. APRS messages are packaged in X.25 packets and usually transmitted as FSK or PSK modulation on FM transmissions. This works well and takes advantage of the extensive ground network of APRS digipeaters and servers.

Nevertheless, APRS might not be the most desirable choice in all circumstances for any number of reasons. In fact, if the payload is within line-of-sight, almost any character-based digital mode will suffice to collect data at the ground station. For example, MFSK modes such as DominoEX work just as well or better on VHF/UHF as they do at HF. Non-character modes such as Hellschreiber may also create very legible data but it will not be in numeric format.

Fig 14.9 — A two-module payload consisting of a GPS receiver module (right) and WB8ELK's MMT (Multi-Mode Transmitter) on the left.

14.3 Platform Design

This section discusses platform design in the context of high-altitude balloon-borne experiments. Similar considerations apply to terrestrial and marine experiment platforms.

14.3.1 Platform Structure

The basic structure of a remote sensing platform is shown in **Fig 14.10**. Along with the power source, there are five separate functions:

1) Sensor data or image acquisition — conversion of analog data into digital format and acquisition of still images or video

2) GPS or navigation data — acquisition of location data in digital form

3) Integration of sensor and location data — collection of all data to be stored and/or transmitted to the ground station

4) Protocol engine — packaging and encoding of data for transmission

5) Amateur transmitter — generates the digitally modulated RF signal

These functions can be implemented by separate modules or everything can be performed by a single microcontroller-based module such as one of the APRS trackers. The choice is completely up to the platform designer and varies with the requirements for the particular mission. For example, Fig 14.9 shows a two-module solution in which everything except GPS location is provided by the single MMT module. The ATV payload described later uses a separate controller to integrate the video and GPS data for transmission as part of the overall audio-video signal. The combinations are endless! The websites listed in the following section on High-Altitude Platforms (balloons) are good

places to begin looking for the right subsystems for your mission.

14.3.2 High Altitude Platform Design

It's impractical to track a weather balloon and its payload though optical means. This is why Amateur Radio is such a popular way to monitor the progress of a balloon's flight into the stratosphere. The minimum Amateur Radio system required to track a weather balloon flight consists of a GPS receiver that is certified to operate above 60,000 feet, an APRS TNC (terminal node controller), and a 2 meter FM transmitter.

If other conditions are to be measured or monitored, an additional microprocessor will be required to acquire the data and convert it to digital form. The data can then be stored in onboard memory for recovery. It can also be formatted into a digital data stream and transmitted to ground stations as a telemetry stream. The sensor data can also be integrated with the GPS data for transmission via APRS. Some APRS trackers can acquire analog and digital sensor data and integrate it into the APRS data messages.

Batteries must have sufficient capacity to operate the tracking system for its typical three-hour near-space flight plus additional time spent on the ground prior to launch and awaiting recovery. Those with experience in near-space activity encourage the use of lithium batteries since they handle the cold temperatures experienced during the flight better than alkaline or NiCds.

The maximum weight per payload is six pounds for a total of twelve pounds for the

platform. Launching additional weight requires getting special permission from the FAA. This doesn't mean that you can fly 6-pound lead weights. You have to make sure that the density of your payload will not inflict damage to others, and it also needs to protect all those expensive electronics that you have packed inside.

As long as an experiment attached to the balloon can be located and recovered, any data collected by sensors carried by the weather balloon can be analyzed and correlated to the balloon's altitude. Thus, it is extremely important that the payload be able to transmit position data during the flight while collecting data from other sensors.

Here are a few websites with a great deal of information about Amateur Radio high altitude ballooning (ARHAB):

• Amateur Radio High Altitude Ballooning (ARHAB) — **www.arhab.org**

• Edge of Space Sciences — **www.eoss. org**

• WB8ELK Balloons — **www.wb8elk. com**

• UK High Altitude Society — **www. habhub.org**

• Great Plains Super Launch — **www. superlaunch.org**

ENCLOSURE AND INSULATION

A Styrofoam box is one of the most common enclosures. The foam is very light, provides insulation against the extreme temperatures encountered during flight for the payload electronics, and helps with the impact of landing. Even on the hottest summer day on the ground, it can be approximately –60 °C at 30,000 feet above the Earth. Most battery types do not work well at these temperature extremes which are also outside the specification range of most electronic components. Fortunately, a Styrofoam box will help keep the internal temperatures well above those brutal outside conditions.

Another technique is to mount the electronics and batteries on a foam-core board and wrap everything with three layers of small-cell bubble wrap. The insulation and trapped sunlight will keep the electronics warm.

BATTERIES

Lithium batteries are recommended since they can supply power even in sub-zero conditions. Another added benefit is that they have a very high power/weight ratio. The Eveready L91 AA lithium battery is one very popular battery that is used quite often for Amateur Radio high-altitude ballooning.

ALTITUDE RATING

Many GPS receiver modules will not work

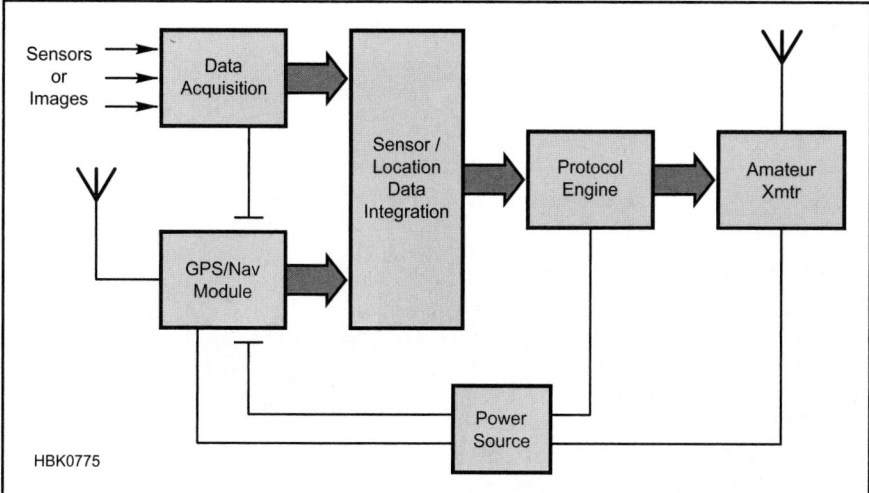

HBK0775

Fig 14.10 — The basic structure of a remote sensing platform using Amateur Radio for the telemetry link.

above 60,000 feet. When choosing a GPS receiver, make sure the datasheet specifies a maximum altitude. Some popular modules known to work at stratospheric altitudes are those made by Trimble, Garmin, u-Blox and Inventek as well as high-altitude modules offered by Byonics and Argent Data.

PREDICTING FLIGHT PATHS

When flying a high altitude balloon, it is a good idea to run a flight prediction a few days in advance as well as the night before a flight to make sure you don't land in a densely populated area or somewhere where you will have great difficulty making a successful recovery. There are two popular online prediction programs that can be accessed at the **www.arhab.org** website.

14.3.3 Types of VHF/UHF Payloads

SIMPLEX REPEATER

A high-altitude balloon at 100,000 feet has a radio line-of-sight of over 400 miles. The formula for radio signal line-of-sight in miles is

$$\text{Distance (mi)} = 1.41 \sqrt{H}$$

where H is the height in feet. Since antenna height is so important for operating on VHF and UHF, imagine having an antenna that is 19 miles high.

If you could fly a repeater to that altitude, two ground stations 800 miles apart could communicate with each other through the repeater. One simple way to do this is to fly a single handheld radio operating on 2 meters or the 70 cm band. By connecting a voice recorder and playback device to the handheld radio, you can create what is called a *simplex repeater*. One such device is offered by Argent Data and is their model ADS-SR1. A discontinued Radio Shack simplex repeater module can be sometimes found online.

It takes some practice and patience to get the hang of a simplex repeater conversation, but this provides a very simple way to make some very exciting contacts over a multi-state region using minimal equipment on the ground.

CROSSBAND REPEATER

If you use two handheld radios, one on 2 meters and one on 70 cm, you can build a crossband repeater payload. (Some handheld radios can also operate as crossband repeaters by themselves.) You'll need to build an audio level control to adjust the audio between the two radios and also provide a PTT control.

Although more complicated, heavier and more expensive than the simplex repeater,

this does provide a real-time repeater without having to worry about flying large filters to prevent desense. The input is usually on the 2 meter band with the output on the 70 cm band. Although you can set it up the other way around, the 3rd harmonic of the 2 meter transmit can cause desense issues with the 70 cm receiver.

SSTV AND STILL PHOTOGRAPHY

Many balloon enthusiasts fly either a still camera or a video camera payload. From 100,000 feet you can clearly see a spectacular view of the blackness of space and the curve of the Earth since the balloon is above 99 percent of the atmosphere. **Fig 14.11** shows a photo taken from by a balloon-launched camera. Suitable lightweight cameras are available in thumb drive (USB) formats and helmet- or bike-cams designed to be used while being worn.

A great addition to any balloon flight is the ability to actually receive live images during the flight. By using a small handheld radio on 2 meters connected to an SSTV module (for example, the Argent Data SSTVCAM) and a microcontroller, anyone who can hear the VHF signal can also view the live images. There are several programs to decode the SSTV audio signal and display the images on a computer screen. *MMSSTV*, *MixW*, *MultiPSK* and *Ham Radio Deluxe* (DM780) are a few programs that can be used. A good SSTV mode is Scottie 2, which will send down one image in 71 seconds. Although the resolution is not as clear, you can also use Robot 36 mode for quicker transmissions. (See the **Image Communications** chapter on the CD-ROM accompanying this book for more information about SSTV.)

Fig 14.11 — A balloon carrying a camera payload was launched from the Dayton Hamvention in 2010. This picture was obtained a few minutes later from an altitude of about 1000 feet.

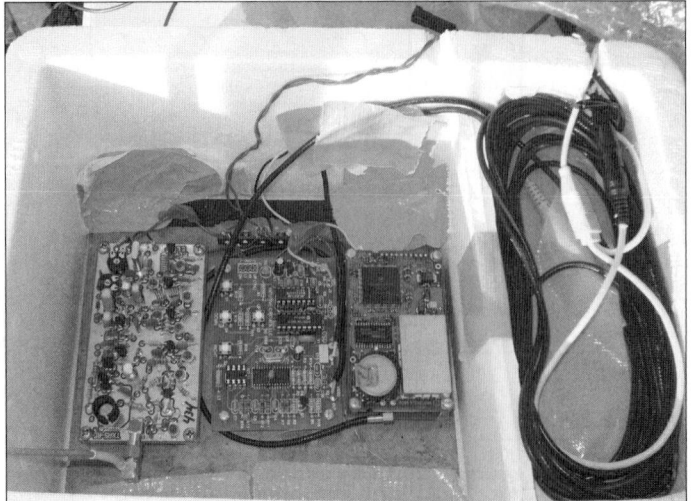

Fig 14.12 — This payload consists of a GPS receiver (right), payload controller (center), and an ATV transmitter (left). Batteries and cables are in the far-right compartment and the entire platform is contained in a Styrofoam enclosure.

ATV

There is nothing like watching a live video camera view from a flight to the edge of space. The very first few amateur high altitude balloon flights in the US carried amateur television (ATV) transmitters on them. Typically a 1 W to 3 W AM-modulated ATV transmitter on either 434 or 439.25 MHz is used for best results. You can also use FM ATV transmitters on higher frequencies, such as the 23 cm and 13 cm bands, but the path loss will be much higher on those frequencies and that will limit your maximum downrange reception distance. There are many lightweight video cameras that can be used as long as they can provide an analog video output. (See the **Image Communications** chapter on the CD-ROM accompanying this book for more information about ATV.)

Remember that the power requirements for a continually operating 1 W TV transmitter will be much higher than an APRS or audio repeater payload. You'll typically need at least 12 V with an Ah rating sufficient to allow for at least three hours of operating time. A surplus military lithium pack is a good option or you can use enough AA lithium batteries to meet the requirement.

Fig 14.12 shows a typical ATV payload in the insulating Styrofoam box enclosure. On the left is the low-power 70 cm transmitter. In the middle is the microprocessor-based controller. The GPS receiver module is on the right. The batteries and cables are placed in the separate compartment at the far right. Note that the three electronics boards are mounted over a common PCB ground-plane to provide mechanical stability and to minimize RFI from the transmitter. The antenna for the ATV link hangs below the package.

You will need a good antenna on the ground, an ATV downconverter, and an analog TV receiver. If you are flying in an area where horizontal polarization is used for local ATV activity, the "Big Wheel" antenna is a good option for the payload's ATV antenna. It provides good coverage at the horizon as well as underneath the payload. You can also use a vertical antenna, but there will be a null directly underneath a vertical radiator. PC Electronics (**www.hamtv.com/wheel. html**) carries the Olde Antenna Labs line of "wheel" antennas for various bands as well as video camera modules, ATV transmitters and receivers.

14.3.4 HF Payloads

The RF range of a high altitude balloon at peak altitude is limited to about 450 miles when using VHF and UHF. Some balloon groups have flown transmitters on the HF bands with reception reports many thousands of miles away. It's a great way to include Amateur Radio operators far outside your local region.

There are several digital modes that can be programmed into a small microcontroller without having to invoke floating point math (see **www.elktronics.com** for an example of a multi-mode HF balloon transmitter). Morse Code, RTTY, PSK31, DominoEX and Hellschreiber have all been successfully flown, as well.

Transmit power levels under 1 W will work well due to the weak signal advantage of some of these digital modes. DominoEX, Hellschreiber and PSK31 are particularly good for very weak signal reception.

It is recommended to have a way to turn the HF transmitter on or off under remote control. There are a number of inexpensive and lightweight UHF handheld radios that can be used as a control receiver along with a DTMF decoder board.

14.4 References and Bibliography

The ARRL Operating Manual, 10th edition, (ARRL, 2012).

The ARRL Antenna Book, 22nd edition, (ARRL, 2012).

Britain, K., WA5VJB, "Cheap Yagis," **www.wa5vjb.com/yagi-pdf/cheapyagi. pdf**

Buchmann, I., *Batteries in a Portable World* (Cadex, 2011).

Cain, P., KØPC, "APRS with a Smartphone," *NCJ*, Sep/Oct 2012, pp 16-17.

Horzepa, S., WA1LOU, *APRS: Moving Hams on Radio and the Internet*, (ARRL, 2004 — out of print).

Post, J., KA5GSQ, "A Simple Sensor Package for High Altitude Ballooning," *QEX*, May/Jun 2012, pp 10-19.

Simmons, B., WB6EYV, "APRS Unveiled," *QEX*, Nov/Dec 2012, pp 19-23.

Spencer, M., WA8SME, "ARRL Education and Technology Program Space/Sea Buoy," *QST*, May 2012, pp 33-35.

Williams, D., AJ5W, "APRS and High-Altitude Research Balloons," *QST*, Aug 2007, pp 48-49.

CQ VHF Magazine Articles

Benton, D., KE5URH, "Tulsa (OK) Technology Center's Innovative Approach to Training Future Engineers," Fall 2008, p 32.

Brown, B., WB8ELK, "Up in the Air: Kentucky Space Balloon-1," Spring 2009, p 57.

Brown, B., WB8ELK, "Up in the Air: HF Balloon Tracking," Winter 2011, p 63.

Dean, T., KB1JIJ, "ATV: W2KGY BalloonSat Payload," Fall 2010, p 68.

Ferguson, D., AI6RE, "Transatlantic Balloon Flight 2012, CNSP-18 K6RPT-12," Winter 2013, p 8.

Helm, M., WC5Z, "A Low Cost 70 cm Tracking Beacon for Rocket or Balloon Payloads," Fall 2010, p 31.

Whitham, R., VE3NSA and Cieszecki, J. VE4CZK, "The WinCube Project," Winter 2009, p 12.

Nuts and Volts Magazine Articles by Paul Verhage, KD4STH

"Near Space: Approaching the Final Frontier," Mar 2013, p 68.

"Near Space: Using the Nearspace Simple Flight Computer," Jan 2013, p 72.

"Near Space: Approaching the Final Frontier," Nov 2012, p 14.

"Near Space: A New BalloonSat Airframe Design," Jul 2012, p 68.

"Near Space: The NearSpace UltraLight — The Everyman's Flight Computer," Jan 2011, p 67.

"Near Space: Your Own Micro Datalogger," May 2009, p 80.

Contents

Chapter 15

DSP and Software Radio Design

In recent years, digital signal processing (DSP) technology has progressed to the point where it is an integral part of our radio equipment. DSP is rapidly replacing hardware circuits with software, offering amateurs flexibility and features only dreamed of in the past. This chapter, by Alan Bloom, N1AL, explores DSP and its use in radio design. DSP projects and additional background and support materials may be found on the *Handbook* CD.

15.1 Introduction

Digital signal processing (DSP) has been around a long time. The essential theory was developed by mathematicians such as Newton, Gauss and Fourier in the 17th, 18th and 19th centuries. It was not until the latter half of the 20th century, however, that digital computers became available that could do the calculations fast enough to process signals in real time. Today DSP is important in many fields, such as seismology, acoustics, radar, medical imaging, nuclear engineering, audio and video processing, as well as speech and data communications.

In all those systems, the idea is to process a digitized signal so as to extract information from it or to control its characteristics in some way. For example, an EKG monitor in a hospital extracts the essential characteristics of the signal from the patient's heart for display on a screen. A digital communications receiver uses DSP to filter and demodulate the received RF signal before sending it to the speaker or headphones. In some systems, the signal to be processed may have more than one dimension. An example is image data, which requires two-dimensional processing. Similarly, the controller for an electrically-steerable antenna array uses multi-dimensional DSP techniques to determine the amplitude and phase of the RF signal in each of the antenna elements. A CT scanner analyzes X-ray data in three dimensions to determine the internal structures of a human body.

SOFTWARE-DEFINED RADIO

The concept of a *software-defined radio* (SDR) became popular in the 1990s. By then, DSP technology had developed to the point that it was possible to implement almost all the signal-processing functions of a transceiver using inexpensive programmable digital hardware. The frequency, bandwidth, modulation, filtering and other characteristics can be changed under software control, rather than being fixed by the hardware design as in a conventional radio. Adding a new modulation type or a new improved filter design is a simple matter of downloading new software. In addition, with the same hardware design, a single radio can have several different modulation modes.

SDR is appealing to regulatory bodies such as the FCC because it makes possible a communications system called *cognitive radio* in which multiple radio services can share the same frequency spectrum.[1] Each node in a wireless network is programmed to dynamically change its transmission or reception characteristics to avoid interference with other users. In this way, services that in the past enjoyed fixed frequency allocations but that only use their channels a small percentage of the time can share their spectrum with other wireless users with minimal interference.

DSP ADVANTAGES

Digital signal processing has the reputation of being more complicated than the analog circuitry that it replaces. In reality, once the analog signal has been converted into the digital domain, complicated functions can be implemented in software much more simply than would be possible with analog components. For example, the traditional "phasing" method of generating an SSB signal without an expensive crystal filter requires various mixers, oscillators, filters and a wide-band audio-frequency phase-shift network built with a network of high-precision resistors and capacitors. To implement the same function in a DSP system requires adding one additional subroutine to the software program — no additional hardware is needed.

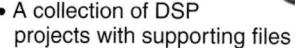

Chapter 15 — CD-ROM Content

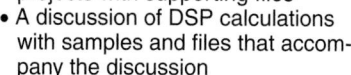

- A collection of DSP projects with supporting files
- A discussion of DSP calculations with samples and files that accompany the discussion

There are many features that are straightforward with DSP techniques but would be difficult or impractical to implement with analog circuitry. A few examples drawn just from the communications field are imageless mixing, noise reduction, OFDM modulation and adaptive channel equalization. Digital signals can have much more dynamic range than analog signals, limited only by the number of bits used to represent the signal. For example, it is easy to add an extra 20 or 30 dB of headroom to the intermediate signal processing stages to ensure that there is no measurable degradation of the signal, something that might be difficult or impossible with analog circuitry. Replacing analog circuitry with software algorithms eliminates the problems of nonlinearity and drift of component values with time and temperature. The programmable nature of most DSP systems means you can make the equivalent of circuit modifications without having to unsolder any components.

DSP LIMITATIONS

Despite its many advantages, we don't mean to imply that DSP is best in all situations. High-power and very high-frequency signals are still the domain of analog circuitry. Where simplicity and low power consumption are primary goals, a DSP solution may not be the best choice. For example, a simple CW receiver that draws a few milliamps from the power supply can be built with two or three analog ICs and a handful of discrete components. In many high-performance systems, the performance of the analog-to-digital converter (ADC or A/D converter) and digital-to-analog converter (DAC or D/A converter) are the limiting factors. That is why, even with the latest generation of affordable ADC technology, it is still possible to obtain better blocking dynamic range in an HF receiver using a hybrid analog-digital system rather than going all-digital by routing the RF input directly to an ADC.

The plan of this chapter is first to discuss the overall hardware and software structure of a DSP system, including general information on factors to be considered when designing at the system level. Then we will cover the basic theory of digital signals, with emphasis on topics relevant to radio communications. Following that is a section on digital filters and another section that describes several other miscellaneous DSP applications. The concept of analytic signals (negative frequencies and all that) is important for understanding software-defined radios, so we cover that before getting into SDRs themselves. The final two sections, on SDR hardware and software, use many of the concepts explained in previous sections to show how a radio can be built with most of the signal processing done digitally. For the fullest understanding of this chapter the reader should have a basic familiarity of the topics covered in the **Electrical Fundamentals**, **Analog Basics** and **Digital Basics** chapters as well as some high-school trigonometry.

15.2 Typical DSP System Block Diagram

A typical DSP system is conceptually very simple. It consists of only three sections, as illustrated in **Fig 15.1**. An ADC at the input converts an analog signal into a series of digital numbers that represent snapshots of the signal at a series of equally spaced sample times. The digital signal processor itself does some kind of calculations on that digital signal to generate a new stream of numbers at its output. A DAC then converts those numbers back into analog form.

Some DSP systems may not have all three components. For example, a DSP-based audio-frequency generator does not need an ADC. Similarly, there is no need for a DAC in a measurement system that monitors some sensor output, processes the signal and stores the result in a computer file or displays it on a digital readout.

The term "DSP" is normally understood to imply processing that occurs in real time, at least in some sense. For example, an RF or microwave signal analyzer might include a DSP co-processor that processes chunks of sampled data in batch mode for display a fraction of a second later. However, a computer program that analyzes historical sunspot data or stock prices normally would not be called "digital signal processing" even though the types of calculations might be very similar.

15.2.1 Data Converters

In this chapter we will discuss only briefly several aspects of ADC and DAC specifica-

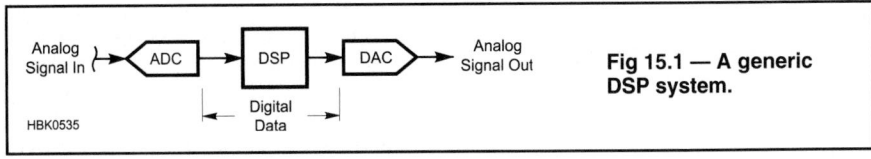

Fig 15.1 — A generic DSP system.

HBK0535

tions and performance that directly affect design decisions at the system level. The **Analog Basics** chapter has additional details that must be considered when doing an actual circuit design.

The first requirement when selecting a DAC or ADC is that it be able to handle the required *sample rate*. For communications-quality voice, a sample rate on the order of 8000 samples per second (8 ksps) should be adequate. For high-quality music, sample rates are typically an order of magnitude higher and for processing wideband RF signals, you'll need data converters with sample rates in the megasamples per second (Msps) range. In many systems the input and output sample rates are different. For example, a software-defined receiver might sample the input RF signal at 100 Msps while the output audio DAC is running at only 8 ksps.

The *resolution* of a data converter is expressed as the number of bits in the data words. For example, an 8-bit ADC can only represent the sampled analog signal as one of $2^8 = 256$ possible numbers. The smallest signal that it can resolve is therefore $\frac{1}{256}$ of full scale. Even with an ideal, error-free

ADC, the *quantization error* is up to $\pm\frac{1}{2}$ of one least-significant bit (LSB) of the digital word, or $\pm\frac{1}{512}$ of full scale with 8-bit resolution. Similarly, a DAC can only generate the analog signal to within $\pm\frac{1}{2}$ LSB of the desired value. Later in the chapter we will discuss how to determine the required sample rate and resolution for a given application.

Another important data converter specification is the *spurious-free dynamic range* (SFDR). This is the ratio, normally expressed in dB, between a (usually) full-scale sine wave and the worst-case spurious signal. While higher-resolution ADCs and DACs tend to have better SFDR, that is not guaranteed. Devices that are intended for signal-processing applications normally specify the SFDR on the data sheet.

While sample rate, resolution and SFDR are the principal selection criteria for data converters in a DSP system, other parameters such as signal-to-noise ratio, harmonic and intermodulation distortion, full-power bandwidth, and aperture delay jitter can also affect performance. Of course, basic specifications such as power requirements, interface type (serial or parallel) and cost also determine a device's suitability for a particular application. As with any electronic component, it is very important to read and fully understand the data sheet.

15.2.2 DSP

The term *digital signal processor* (DSP) is commonly understood to mean a special-purpose microprocessor with an architecture that has been optimized for signal processing. And indeed, in many systems the box labeled "DSP" in Fig 15.1 is such a device. A microprocessor has the advantage of flexibility because it can easily be re-programmed. Even with a single program, it can perform many completely-different tasks at different points in the code. On-chip hardware resources such as multipliers and other computational units are used efficiently because they are shared among various processes.

That is also the Achilles' heel of programmable DSPs. Any hardware resource that is shared among various processes can be used by only one process at a time. That can create bottlenecks that limit the maximum computation speed. Some DSP chips include multiple computational units or multiple *cores* (basically multiple copies of the entire processor) that can be used in parallel to speed up processing.

DIGITAL SIGNAL PROCESSING WITHOUT A "DSP"

Another way to speed up processing is to move all or part of the computations from the programmable DSP to an *application-specific integrated circuit* (ASIC), which has an architecture that has been optimized to perform some specific DSP function. For example, *direct digital synthesis* (DDS) frequency synthesizer chips are available that run at rates that would be impossible with a microprocessor-type DSP.

You could also design your own application-specific circuitry using a PC board full of discrete logic devices. Nowadays, however, it is more common to do that with a *programmable-logic device* (PLD). This is an IC that includes many general-purpose logic elements, but the connections between the elements are undefined when the device is manufactured. The user defines those connections by programming the device to perform whatever function is required. PLDs come in a wide variety of types, described by an alphabet soup of acronyms.

Programmable-array logic (PAL), *programmable logic array* (PLA), and *generic array logic* (GAL) devices are relatively simple arrays of AND gates, OR gates, inverters and latches. They are often used as "glue logic" to replace the miscellaneous discrete logic ICs that would otherwise be used to interface various larger digital devices on a circuit board. They are sometimes grouped under the general category of *small PLD* (SPLD). A *complex PLD* (CPLD) is similar but bigger, often consisting of an array of PALs with programmable interconnections between them.

A *field-programmable gate array* (FPGA) is bigger yet, with up to millions of gates per device. An FPGA includes an array of *complex logic blocks* (CLB), each of which includes some programmable logic, often implemented with a RAM *look-up table* (LUT), and output registers. *Input/output blocks* (IOB) also contain registers and can be configured as input, output, or bi-directional interfaces to the IC pins. The interconnections between blocks are much more flexible and complicated than in CPLDs. Some FPGAs also include higher-level circuit blocks such as general-purpose RAM, dozens or hundreds of hardware multipliers, and even entire on-chip microprocessors.

Some of the more inexpensive PLDs are *one-time programmable* (OTP), meaning you have to throw the old device away if you want to change the programming. Other devices are re-programmable or even *in-circuit programmable* (ICP) which allows changing the internal circuit configuration after the device has been soldered onto the PC board, typically under the control of an on-board microprocessor. That offers the best of both worlds, with speed nearly as fast as an ASIC but retaining many of the benefits of the reprogrammability of a microprocessor-type DSP. Most large FPGAs store their programming in *volatile memory*, which is RAM that must be re-loaded every time power is applied, typically by a ROM located on the same circuit board. Some FPGAs have programmable ROM on-chip.

Programming a PLD is quite different from programming a microprocessor. A microprocessor performs its operations sequentially — only one operation can be performed at a time. Writing a PLD program is more like designing a circuit. Different parts of the circuit can be doing different things at the same time. Special *hardware-description languages* (HDL) have been devised for programming the more complicated parts such as ASICs and FPGAs. The two most common industry-standard HDLs are Verilog and VHDL. (The arguments about which is "best" approach the religious fervor of the *Windows* vs *Linux* wars!) There is also a version of the C++ programming language called SystemC that includes a series of libraries that extend the language to include HDL functions. It is popular with some designers because it allows simulation and hardware description using the same software tool.

Despite the speed advantage of FPGAs, most amateurs use microprocessor-type devices for their DSP designs, supplemented with off-the-shelf ASICs where necessary. The primary reason is that the design process for an FPGA is quite complicated, involving obtaining and learning to use several sophisticated software tools. The steps involved in programming an FPGA are:

1. Simulate the design at a high abstraction level to prove the algorithms.
2. Generate the HDL code, either manually or using some tool.
3. Simulate and test the HDL program.
4. Synthesize the gate-level netlist.
5. Verify the netlist.
6. Perform a timing analysis.
7. Modify the design if necessary to meet timing constraints.
8. "Place and route" the chip design.
9. Program and test the part.

Table 15.1

PLD Manufacturers

Company	Devices	URL	Notes
Achronix	FPGA	www.achronix.com	High-speed FPGAs
Actel	FPGA	www.actel.com	Mixed-signal flash-based FPGAs
Altera	CPLD, FPGA, ASIC	www.altera.com	One of the two big FPGA vendors
Atmel	SPLD, CPLD, FPGA, ASIC	www.atmel.com	Fine-grain-reprogrammable FPGAs with AVR microprocessors on chip
Cypress Semiconductor	SPLD, CPLD	www.cypress.com	
Lattice Semiconductor	SPLD, CPLD, FPGA	www.latticesemi.com	Leading supplier of flash-based nonvolatile FPGAs
SiliconBlue	FPGA	www.siliconbluetech.com	Low-power FPGAs
Texas Instruments	SPLD, ASIC	www.ti.com	
Xilinx	CPLD, FPGA	www.xilinx.com	One of the two big FPGA vendors

Many of the software tools needed to perform those steps are quite expensive, although some manufacturers do offer free proprietary software for their own devices. Some principal manufacturers are listed in **Table 15.1**.

MICROPROCESSOR-TYPE DSP CHIPS

In contrast with designing an FPGA, programming a DSP chip is relatively easy. C compilers are available for most devices, so you don't have to learn assembly language. Typically you include a connector on your circuit board into which is plugged an *in-circuit programmer* (ICP), which is connected to a PC via a serial or USB cable. The software is written and compiled on the PC and then downloaded to the DSP. The same hardware often also includes an *in-circuit debugger* (ICD) so that the program can be debugged on the actual circuitry used in the design. The combination of the editor, compiler, programmer, debugger, simulator and related software is called an *integrated development environment* (IDE).

Until recently you had to use an *in-circuit emulator* (ICE), which is a device that plugs into the circuit board in place of the microprocessor. The ICE provides sophisticated debugging tools that function while the emulator runs the user's software on the target device at full speed. Nowadays, however, it is more common to use the ICD function that is built into many DSP chips and which provides most of the functions of a full-fledged ICE. It is much cheaper and does not require using a socket for the microprocessor chip.

The architecture of a digital signal processor shares some similarities to that of a general-purpose microprocessor but also differs in important respects. For example, DSPs generally don't spend much of their lives handling large computer files, so they tend to have a smaller memory address space than processors intended to be used in computers. On the other hand, the memory they do have is often built into the DSP chip itself to improve speed and to reduce pin count by eliminating the external address and data bus.

Most microprocessors use the traditional *Von Neumann architecture* in which the program and data are stored in the same memory space. However, most DSPs use a *Harvard architecture*, which means that data and program are stored in separate memories. That speeds up the processor because it can be reading the next program instruction at the same time as it is reading or writing data in response to the previous instruction. Some DSPs have two data memories so they can read and/or write two data words at the same time. Most devices actually use a modified Harvard architecture by providing some (typically slower and less convenient) method for the processor to read and write data to program memory.

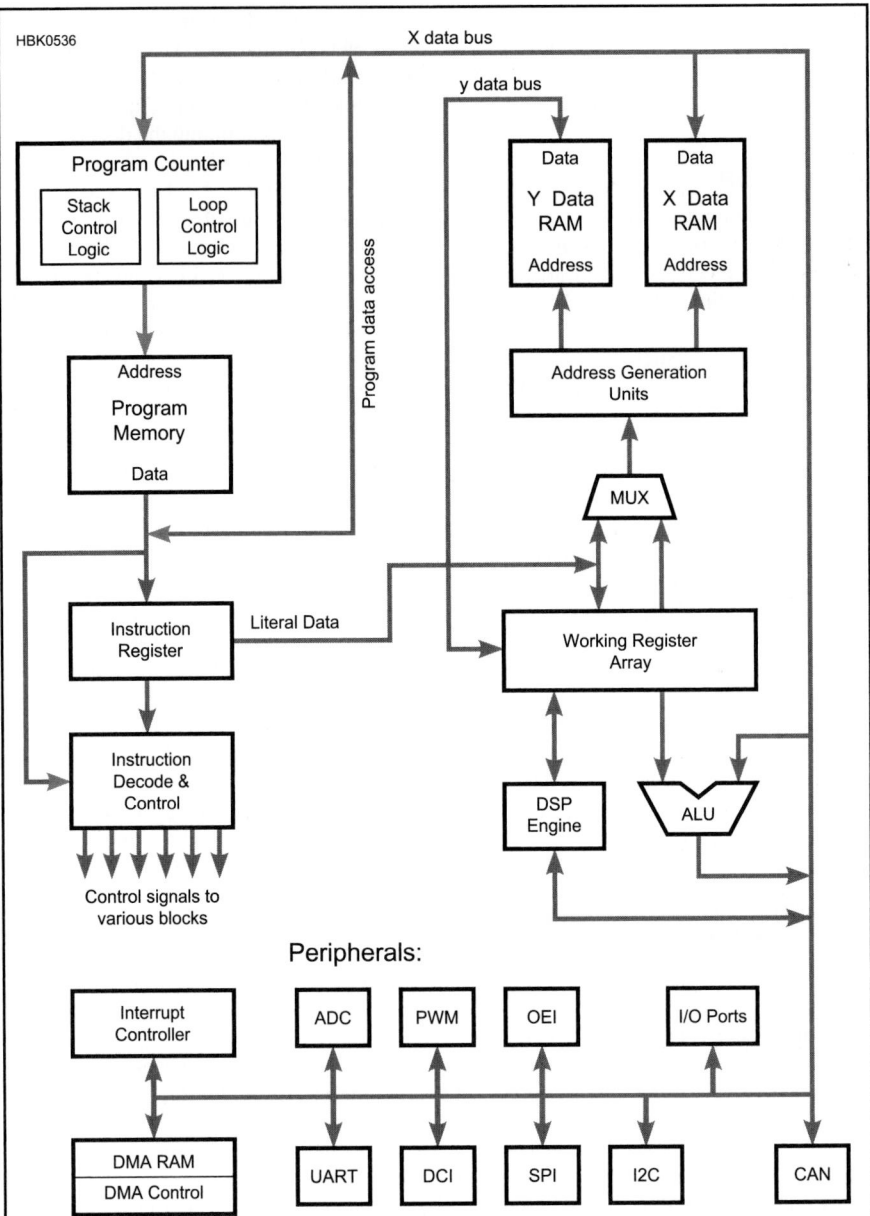

Fig 15.2 — Simplified block diagram of a dsPIC processor.

Probably the key difference between general-purpose and digital-signal processors is in the computational core, often called the *arithmetic logic unit* (ALU). The ALU in a traditional microprocessor only performs integer addition, subtraction and bitwise logic operations such as AND, OR, one-bit shifting and so on. More-complicated calculations, such as multiplication, division and operations with floating-point numbers, are done in software routines that exercise the simple resources of the ALU multiple times to generate the more-complicated results.

In contrast, a DSP has special hardware to perform many of these operations much faster. For example, the *multiplier-accumulator* (MAC) multiplies two numbers and adds (accumulates) the product with the previous results in a single step. Many common DSP algorithms involve the sum of a large number of products, so nearly all DSPs include this function. **Fig 15.2** is a simplified block diagram of the dsPIC series from Microchip. Its architecture is basically that of a general-purpose microcontroller to which has been added a DSP engine, which includes a MAC, a barrel shifter and other DSP features. It does use a modified Harvard architecture with two data memories that can be simultaneously accessed.

A *floating-point* number is the binary equivalent of scientific notation. Recall that the decimal integer 123000 is expressed as 1.23×10^5 in scientific notation. It is common practice to place the decimal point after the first non-zero digit and indicate how many digits the decimal point must be moved by the *exponent* of ten, 5 in this case. The 1.23 part is called the *mantissa*. In a computer, base-2 binary numbers are used in place of the base-10 decimal numbers used in scientific notation. The *binary point* (equivalent to the decimal point in a decimal number) is assumed to be to the left of the first non-zero bit. For example the binary number 00110100 when converted to a 16-bit floating point number would have an 11-bit mantissa of 11010000000 (with the binary point assumed to be to the left of the first "1") and a 5-bit exponent of 00110 (decimal +6).

A floating point number can represent a signal with much more dynamic range than an integer number with the same number of bits. For example, a 16-bit signed integer can vary from −32768 to +32767. The difference between the smallest (1) and largest signal that can be represented is $20 \log(65535) = 96$ dB. If the 16 bits are divided into an 11-bit mantissa and 5-bit exponent, the available range is $20 \log(2048) = 66$ dB from the mantissa and $20 \log(2^{32}) = 193$ dB from the exponent for a total of 259 dB. The disadvantage is that the mantissa has less resolution, potentially increasing noise and distortion. Normally floating-point numbers are at least 32 bits wide to mitigate that effect.

Some DSPs can process floating-point numbers directly in hardware. Fixed-point DSPs can also handle floating-point num-bers, but it must be done in software. The additional dynamic range afforded by floating-point processing is normally not needed for radio communications signals since the dynamic range of radio signals is typically less than can be handled by the 16-bit data words used by most integer DSPs. Using integer arithmetic saves the additional cost of a floating-point processor or the additional computational overhead of floating-point calculations on a fixed-point device. However, it requires careful attention to detail on the part of the programmer to make sure the signal can never exceed the maximum integer value or get so weak that the signal-to-noise ratio is degraded. If cost or computation time is not an issue, it is much easier to program in floating point since dynamic range issues can be ignored for most computations.

The term *pipeline* refers to the ability of a microprocessor to perform portions of several instructions at the same time. The sequence of operations required to perform an instruction is broken down into steps. Since each step is performed by a different chunk of hardware, different chunks can be working on different instructions at the same time. Most DSPs have at least a simple form of pipelining in which the next instruction is being fetched while the previous instruction is being executed. Some DSPs can do a multiply-accumulate while the next two multiplicands are being read from memory and the previous accumulated result is being stored so that the entire operation can occur in a single clock cycle. MACs per second is a common figure of merit for measuring DSP speed. For conventional microprocessors, a more common figure of merit is millions of instructions per second (MIPS) or floating-point operations per second (FLOPS).

Many DSPs have a sophisticated address generation unit that can automatically increment one or more data memory pointers so that repetitive calculations can step through memory without the processor having to calculate the addresses. *Zero-overhead looping* is the ability to automatically jump the address pointer back to the beginning of the array when it reaches the end. That saves several microprocessor instructions per loop that normally would be required to check the current address and jump when it reaches a predetermined value.

While most DSPs do not include a full hardware divider, some do include special instructions and hardware to speed up division calculations. A *barrel shifter* is another common DSP feature. It allows shifting a data word a specified number of bits in a single clock cycle. *Direct memory access* (DMA) refers to special hardware that can automatically transfer data between memory and various peripheral devices or ports without processor overhead.

DSP IN EMBEDDED SYSTEMS

An *embedded system* is a device that is not a computer but nevertheless has a micropro-

---

Table 15.2
Manufacturers of Embedded DSPs

Company	Family	Data Bits	Speed MMACs	Nr. of Cores	ROM (bytes)	RAM (bytes)	Notes
Analog Devices www.analog.com	ADSP-21xx	16	75-160	1	12k-144k	8k-112k	Easy assembly language
	SHARC	32/40 fp	300-900	1	2-4M	0.5-5M	Runs fixed or floating point
	Blackfin	16/32	400-2400	1-2	External	53k-328k	Many on-chip peripherals
Cirrus Logic www.cirrus.com	CS48xxxx	32	150	1		96k	Audio applications
	CS49xxxx	32	300	2	512k	296k-328k	Audio applications
Freescale www.freescale.com	DSP568xx	16	32-120	1	2k-576k	2k-128k	Also a microcontroller
	DSP563xx	24	80-275	1	External	576k	
	StarCore	16	1000-48,000	1-6	External	0-1436k	
Microchip www.microchip.com	dsPIC	16	30-70	1	6k-256k	256k-32k	Also a microcontroller Free IDE software
Texas Instruments www.ti.com	C5000	16	50-600	1	8k-256k	0-1280k	
	C6000	16/64 fp	300-24,000	1-3	0-384k	32k-3072	Fixed or floating point ver.
Zilog www.zilog.com	Z89xxx	16	20	1	4k-8k	512	

cessor or DSP chip embedded somewhere in its circuitry. Examples are microwave ovens, automobiles, mobile telephones and software-defined radios. DSPs intended for embedded systems often include a wide array of on-chip peripherals such as various kinds of timers, multiple hardware interrupts, serial ports of various types, a real-time clock, pulse-width modulators, optical encoder interfaces, A/D and D/A converters and lots of general-purpose digital I/O pins. Some DSPs not only include lots of peripherals but in addition have architectures that are well-suited for general-purpose control applications as well as digital signal processing.

Table 15.2 lists some manufacturers of DSP chips targeted to embedded systems. It should be mentioned that microprocessors intended for personal computers made by Intel and AMD also include extensive DSP capability. However, they are large, complicated, power-hungry ICs that are not often used in embedded applications.

When selecting a DSP device for a new design, often the available development environment is more important than the characteristics of the device itself. Microchip's dsPIC family of DSPs was chosen for the examples in this chapter because their integrated development environment is extensive and easy to use and the IDE software is available for free download from their Web site.[2] The processor instruction set is a superset of the PIC24 family of general-purpose microcontrollers, with which many hams are already familiar. The company offers a line of low-cost evaluation boards and starter kits as well as an inexpensive in-circuit debugger, the ICD 3. The free IDE software includes a simulator that can run dsPIC software on a PC (at a much slower rate, of course), so that you can experiment with DSP algorithms before buying any hardware.

The Microchip DSP family is limited to 70 million instructions per second. In a system with, say, a 70 kHz sample rate, 1000 instructions per sample are available which should be plenty if the calculations are not too complex. However if the sample rate is 1 MHz, then you get only 70 instructions per sample, which likely would be insufficient.

If more horsepower is required, you'll need to select a processor from a different manufacturer. Look for one with a well-integrated suite of development software that is powerful and easy to use. Also check out the cost and availability of development hardware such as evaluation kits, programmers and debuggers. Once those requirements are met, then you can move on to selecting a specific device with the performance and features required for your application. It can be helpful at the beginning of a project to first write some of the key software routines and test them on a simulator to estimate execution times, in order to determine how powerful a processor is needed.

When estimating execution time, don't forget to include the effect of interrupts. Most DSP systems require real-time response and make extensive use of interrupts to ensure that certain events happen at the correct times. Although this is hidden from the programmer's view when programming in C, the interrupt service routines contain quite a bit of overhead each time they are called (to save the processor state when responding and to recall the state just before returning from the interrupt). Sometimes an interrupt may be called more often than you expect, which can eat up processor cycles and so increase the execution time of other unrelated routines.

In the past, may embedded systems were written in assembly language so save memory and increase processing speed. Many early microprocessors and DSPs did not have enough memory to support a high-level language. Today, most processors have sufficient memory and processing speed to support a C kernel and library without difficulty. For anything but the simplest of programs, it is not only faster and easier to develop software in C but it is easier to support and maintain as well, especially if people other than the original programmer might become involved. Far more people know the C programming language than any particular processor's assembly language. It is true that the version of C used on a DSP chip is usually modified from standard ANSI C to support specific hardware features, but it would still be far easier to learn for a programmer familiar with writing C code on a PC or on a different DSP.

A common technique is first to write the entire application in C. Then, if execution time is not acceptable, analyze the system to determine in which software routines the bottlenecks are occurring. Those routines can then be re-written in assembly language. Having an already-working version written in C (even if too slow) can be helpful in testing and troubleshooting the equivalent assembly language.

15.3 Digital Signals

Digital signals differ from analog signals in two ways. One is that they are digitized in time, a process called *sampling*. The other is that they are digitized in amplitude, a process called *quantization*. Sampling and quantization affect the digitized signal in different ways so the following sections will consider their effects separately.

15.3.1 Sampling — Digitization in Time

Sampling is the process of measuring a signal at discrete points of time and recording the measured values. An example from history is recording the number of sunspots. If an observer goes out at noon every day and writes down the number of observed sunspots, then that data can be used to plot sunspot number versus time. In this case, we say the *sample rate* is one sample per day. The data can then be analyzed in various ways to determine short and long-term trends. After recording only a few months of data it will quickly become apparent that sunspot number has a marked periodicity — the numbers tend to repeat every 27 days (which happens to be the rotation rate of the sun as seen from earth).

What if, instead of taking a reading once a day, the readings were taken only once per month? With a 30-day sample period, the 27-day periodicity would likely be impossible to see. Clearly, the sample rate must be at least some minimum value to accurately represent the measured signal. Based on earlier work by Harry Nyquist, Claude Shannon proved in 1948 that in order to sample a signal without loss of information, the sample rate must be greater than the *Nyquist rate*, which is two times the bandwidth of the signal. In other words, the bandwidth must be less than the *Nyquist frequency*, which is one-half the sample rate. This is known as the *Nyquist sampling criterion*.

That simple rule has some profound implications. If all the frequency components of a signal are contained within a bandwidth of B Hz, then sampling at a rate greater than 2B

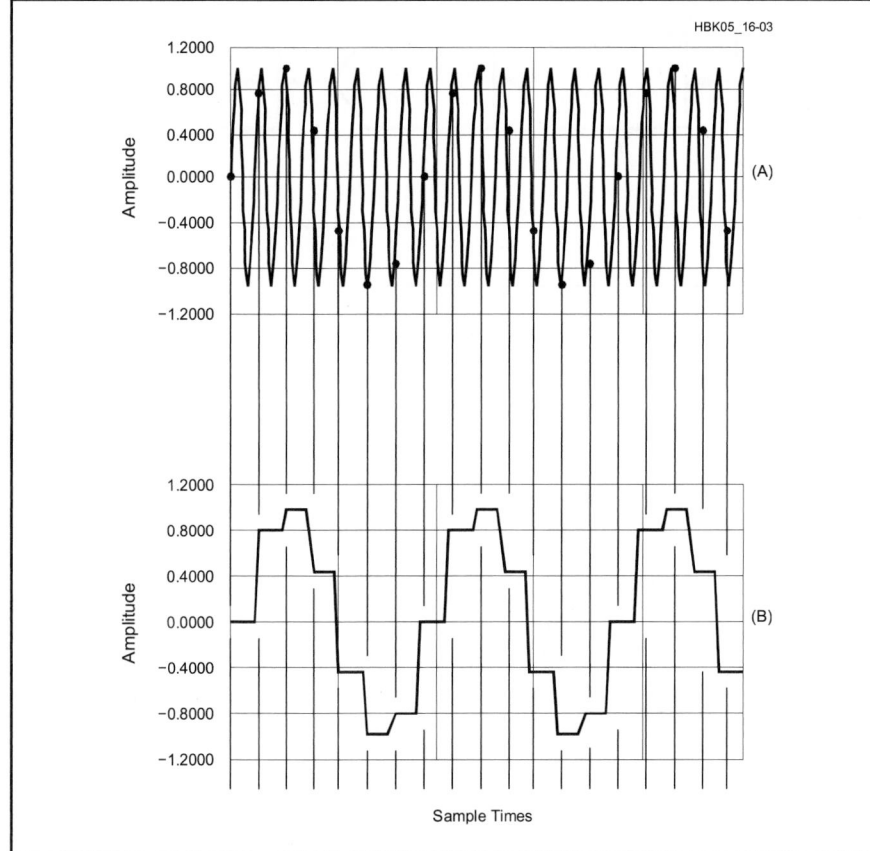

Fig 15.3 — Undersampled sine wave (A). Samples aliased to a lower frequency (B).

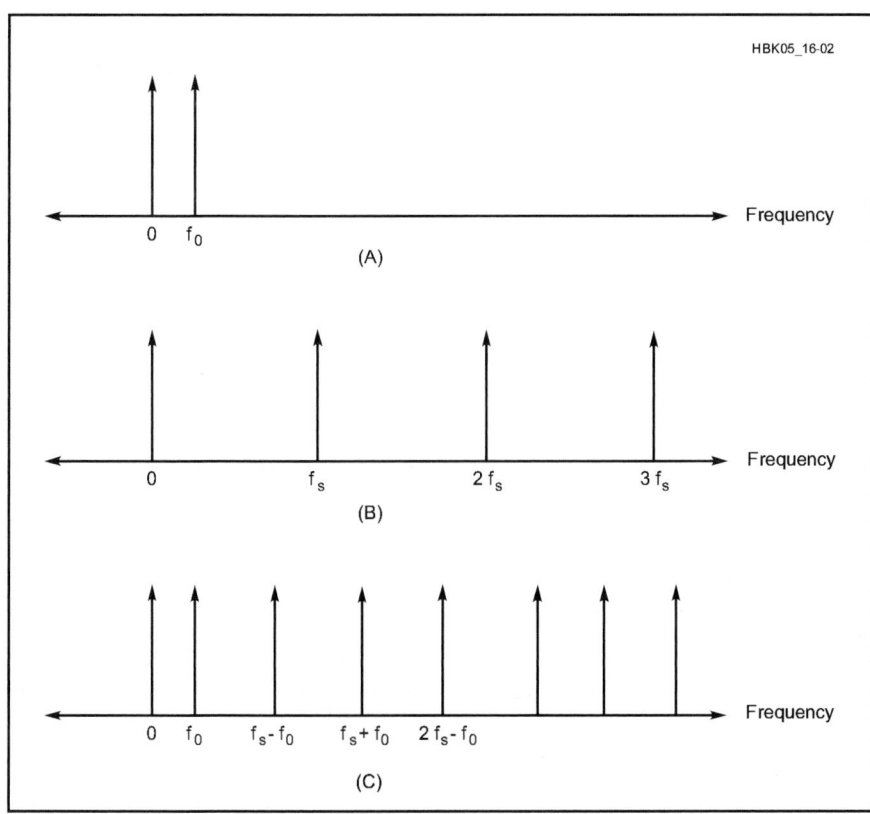

Fig 15.4 — Spectrum of an analog sine wave (A). The spectrum of the sampling function, including all harmonics (B). The spectrum of the sampled sine wave (C).

samples per second is sufficient to represent the signal with 100% accuracy and with no loss of information. It is theoretically possible to convert the samples back to an analog signal that is exactly identical to the original.

Of course, a real-world digital system measures those samples with only a finite number of bits of resolution, with consequences that we will investigate in the section on quantization that follows. In addition, sampling theory assumes that there is absolutely no signal energy outside the specified bandwidth; in other words the stopband attenuation is infinity dB. Any residual signal in the stop-band shows up as distortion or noise in the sampled signal.

To simplify the discussion, let's think about sampling a signal of a single frequency (a sine wave). **Fig 15.3** illustrates what happens if the sample rate is too low. As shown, the sample rate is approximately $7\!/\!8$ the sine-wave frequency. You can see that the sampled signal has a period about 8 times greater than the period of the sine wave, or $1\!/\!8$ the frequency. The samples are the same as if the analog signal had been a sine wave of $1\!/\!8$ the actual frequency.

That is an example of a general principle. If the sample rate is too low, the sampled signal will be *aliased* to a frequency equal to the difference between the actual frequency of the analog signal and the sample rate. In the above example, the alias frequency f_o is

$$f_o = f_{sig} - f_s = \left(1 - \frac{7}{8}\right) f_{sig} = \left(\frac{1}{8}\right) f_{sig}$$

where f_{sig} is the frequency of the signal before sampling and f_s is the sample rate.

If the analog signal's frequency is even higher, then it aliases relative to whichever harmonic of the sample rate is closest. **Fig 15.4C** shows all the signal frequencies that alias to a frequency of f_o, calculated from the equation

$$f_o = \left| f_{sig} - N f_s \right|$$

where N is the harmonic number. One way to think of it is that a sampler is a harmonic mixer. The sampled signal (equivalent to the mixer output) contains the sum and difference frequencies of the input signal and all the harmonics of the sample frequency.

To avoid aliasing, most systems use an *anti-aliasing filter* before the sampler, as shown in **Fig 15.5**. For a baseband signal (one that extends to zero Hz), the anti-aliasing filter is a low-pass filter whose stopband extends from the Nyquist frequency to infinity. Of course, practical filters do not transition instantaneously from the passband to the stopband, so the bandwidth of the passband must be somewhat less than half the sample rate.

It is actually possible to accurately sample signals above the Nyquist frequency so long

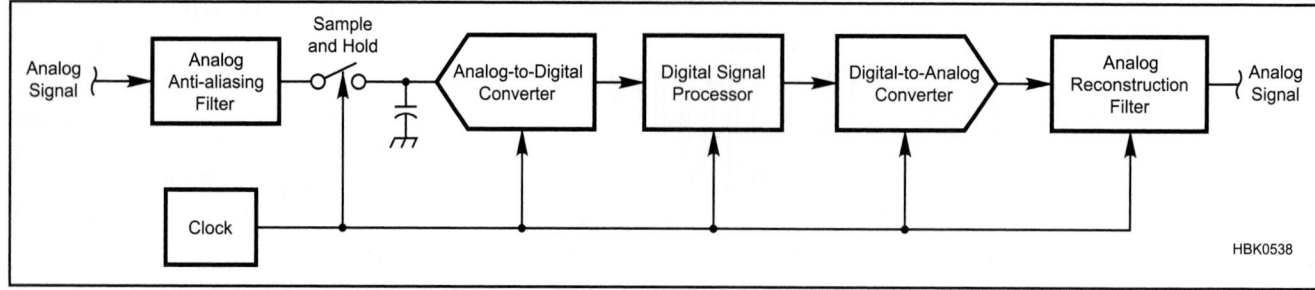

Fig 15.5 — A more complete block diagram of a DSP system.

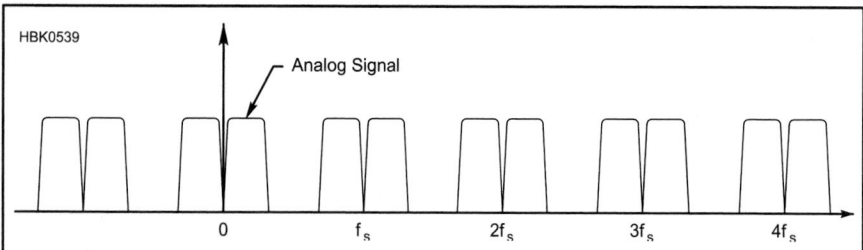

Fig 15.6 — An ideal sampled signal repeats the spectrum of the analog signal at all harmonics of the sample rate, f_s.

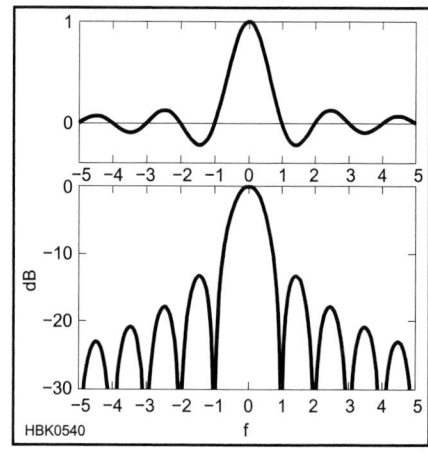

Fig 15.7 — The sinc function, where the horizontal axis is frequency normalized to the sample rate. At the bottom is the same function in decibels.

as their bandwidth does not violate the Nyquist criterion, a process called *undersampling* or *harmonic sampling*. Imagine an LSB signal at 455 kHz with a bandwidth of 3 kHz that is being sampled at a 48 kHz rate. The 455 kHz signal mixes with the ninth harmonic of the sample rate at 432 kHz, resulting in a sampled signal with its suppressed carrier at 455 – 432 = 23 kHz and extending 3 kHz below that to 20 kHz. So long as the incoming signal has no significant energy below 432 kHz or above 456 kHz [432 + (48/2)] kHz no unwanted aliasing occurs.

With harmonic sampling, the anti-alias filter must be a band-pass type. In the previous example, you'd probably need to use a crystal or mechanical filter in order to have a sufficiently sharp transition from the top edge of the passband slightly below 455 kHz to the stopband edge at 456 kHz.

Fig 15.3 shows each sample being held at a constant value for the duration of one sample period. However, sampling theory actually assumes that the sample is only valid at the instant the signal is sampled; it is zero or undefined at all other times. A series of such infinitely-narrow impulses has harmonics all the way to infinite frequency. Each harmonic has the same amplitude and is modulated by the signal being sampled. See **Fig 15.6**. When a digitized signal is converted back to analog form, unwanted harmonics must be filtered out by a *reconstruction filter* as shown in Fig 15.5. This is similar to the anti-aliasing filter used at the input in that its bandwidth should

be no greater than one-half the sample rate. It is a low-pass filter for a baseband signal and a band-pass filter for an undersampled signal.

Most DACs actually do hold each sample value for the entire sample period. This is called *zero-order hold* and results in a frequency response in the shape of a sinc function

$$\text{sinc}(f) = \frac{\sin(\pi f)}{\pi f}$$

where f is normalized to the sample rate, f = frequency / sample rate.

The graph of the sinc function in **Fig 15.7** shows both positive and negative frequencies for reasons explained in the Analytic Signals section. Note that the logarithmic frequency response has notches at the sample rate and all of its harmonics. If the signal bandwidth is much less than the Nyquist frequency, then most of the signal at the harmonics falls near the notch frequencies, easing the task of the reconstruction filter. If the signal bandwidth is small enough (sample rate is high enough), the harmonics are almost completely notched out and a reconstruction filter may not even be required.

The sin(πf)/πf frequency response also affects the passband. For example if the passband extends to sample rate / 4 (f = 0.25), then the response is

$$20 \log \frac{\sin(\pi \cdot 0.25)}{\pi \cdot 0.25} = -0.9 \text{ dB}$$

at the top edge of the passband. At the Nyquist frequency, (f = 0.5), the error is 3.9 dB. If the signal bandwidth is a large proportion of the Nyquist frequency, then some kind of digital or analog compensation filter may be required to correct for the high-frequency rolloff.

DECIMATION AND INTERPOLATION

The term *decimation* simply means reducing the sample rate. For example to decimate by two, simply eliminate every second sample. That works fine as long as the signal bandwidth satisfies the Nyquist criterion at the lower, output sample rate. If the analog anti-aliasing filter is not narrow enough, then a digital anti-aliasing filter in the DSP can be used to reduce the bandwidth to the necessary value. This must be done *before* decimation to satisfy the Nyquist criterion.

If you need to decimate by a large amount, then the digital anti-aliasing filter must have a very small bandwidth compared to the sample rate. As we will see later, a digital filter with a small bandwidth is computationally intensive. For this reason, large decimation factors are normally accomplished in multiple steps, as shown in **Fig 15.8A**. The first decimation is by a small factor, typically 2, so that the first anti-alias filter can be as simple as

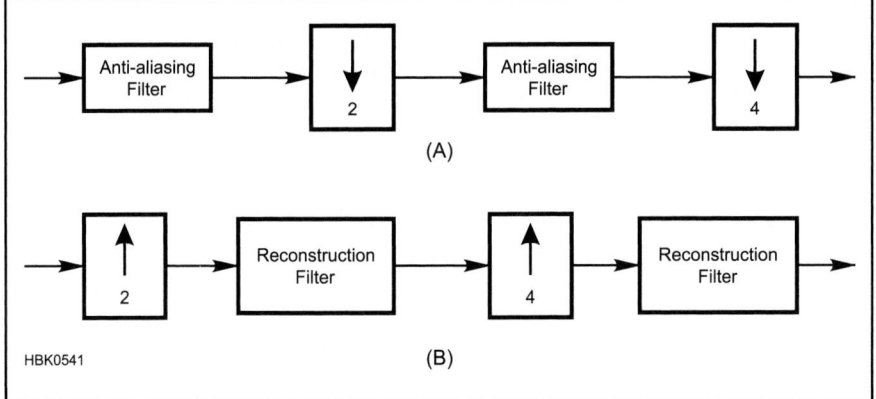

Fig 15.8 — Decimation (A) and interpolation (B). The arrow's direction indicates decimation (down) or interpolation (up) and the number is the factor.

HBK0541

(A)

(B)

possible. The second decimation stage then does not have to decimate by such a large factor, simplifying its task. In addition, since it is running at only half the input sample rate it has more time to do its calculations. Generally it is most efficient to decimate by the smallest factor in the first stage, a larger factor in the second, and the largest factors in the third and any subsequent stages. The larger the total decimation factor, the greater the number of stages is appropriate but more than three stages is uncommon.

Interpolation means increasing the sample rate. One way to do that is simply to insert additional zero-value samples, a process called *zero-stuffing*. For example, to interpolate by a factor of three, insert two zero-value samples after each input sample. That works, but may not give the results you expect. Recall that a sampled signal has additional copies of the baseband signal at all harmonics of the sample rate. All of those harmonics remain in the resampled signal, even though the sample rate is now higher. To eliminate them, the signal must be filtered after interpolation. After filtering, there is signal only at baseband and around the harmonics of the interpolated (higher-frequency) sample rate. It's as if the analog signal had been sampled at the higher rate to begin with.

Just as with decimation, interpolation by a large factor is best done in stages, as shown in Fig 15.8B. In this case, the stage running at the lowest sample rate (again the first stage) is the one with the lowest interpolation factor.

Zero-stuffing followed by filtering is not the only way to interpolate. Really what you are trying to do is to fill in between the lower-rate samples with additional samples that "connect the dots" in as smooth a manner as possible. It can be shown that that is mathematically equivalent to zero-stuffing and filtering. For example, if instead of inserting zero-value samples you instead

simply repeat the last input sample, you have a situation similar to the zero-order hold of a DAC output. It is equivalent to zero-stuffing followed by a low-pass filter with a frequency response of $\sin(\pi f)/\pi f$. If you do a straight-line interpolation between input samples (a "first-order" interpolation), it turns out that it is equivalent to a low pass filter with a frequency response of $[\sin(\pi f)/\pi f]^2$, which has a sharper cutoff and better stop-band rejection than a zero-order interpolation. Higher-order interpolations have smoother responses in the time domain which translate to better filter responses in the frequency domain.

So far we have only covered decimation and interpolation by integer factors. It is also possible to change the sample rate by a non-integer factor, which is called *resampling* or *multi-rate* conversion. For example, if you want to increase the sample rate by a factor of ⅘, simply interpolate by 4 and then decimate by 3. That method can become impractical for some resample ratios. For example, to convert an audio file recorded from a computer sound card at 48 kHz to the 44.1 kHz required by a compact disc, the resample ratio is 44,100 / 48,000 = 147 / 160. After interpolation by 147, the 44.1 kHz input file is sampled at 6.4827 MHz, which would result in excessive processing overhead.

In addition, the interpolation/decimation method only works for resample ratios that are rational numbers (the ratio of two integers). To resample by an irrational number, a different method is required. The technique is as follows. For each output sample, first determine the two nearest input samples. Calculate the coefficients of the Nth-order equation that describes the trajectory between the two input samples. Knowing the trajectory between the input samples and the output sample's relative position between them, the value of the output sample can be calculated from the equation.

15.3.2 Quantization — Digitization in Amplitude

While sampling (digitization in time) theoretically causes no loss of signal information, quantization (digitization in amplitude) always does. For example, an 8-bit signed number can represent a signal as a value from –128 to +127. For each sample, the A/D converter assigns whichever number in that range is closest to the analog signal at that instant. If a particular sample has a value of 10, there is no way to tell if the original signal was 9.5, 10.5 or somewhere in between. That information has been lost forever.

When quantizing a complex signal such as speech, this error shows up as noise, called *quantization noise*. See **Fig 15.9**. The error is random — it is equally likely to be anywhere in the range of –1/2 to +1/2 of a single step of the ADC. We say that the maximum error is one-half of one *least-significant bit* (LSB). It can be shown mathematically that a series of uniformly-distributed random numbers between +0.5 LSB and –0.5 LSB has an RMS value of

$$\frac{LSB}{\sqrt{12}}$$

which is –10.79 dB less than one LSB. Each time you add one bit to the data word, the number of LSBs in the range doubles, which means each LSB gets two times smaller reducing the noise by 6.02 dB. A full-scale sine wave has an RMS power –3.01 dB from a full-scale dc voltage. Combining that information results in the following equation for signal-to-noise ratio in decibels for a data word of width b bits:

$$SNR = 1.76 + 6.02b \;\; dB$$

For example, with 8-bit data, SNR = 49.9 dB. An ideal 16-bit ADC would achieve a 98.1 dB signal-to-noise ratio. Of course, real-world devices are never perfect so actual performance would be somewhat less.

One critical point that is sometimes overlooked is that quantization noise is spread over the entire bandwidth from zero Hz to the sample rate. If you are digitizing a 3 kHz audio channel with a 48 ksps sampler, only a fraction of the noise power is within the channel. For that reason, the effective signal-to-noise ratio depends not only on the number of bits but also the sample rate, f_s, and the signal bandwidth, B:

$$SNR_{eff} = SNR + 10\log\left(\frac{f_s}{2B}\right) \;\; dB$$

The reason for the factor of two in the denominator is that the bandwidth of a positive-frequency scalar signal should be compared to the Nyquist bandwidth, $f_s/2$. When filtering

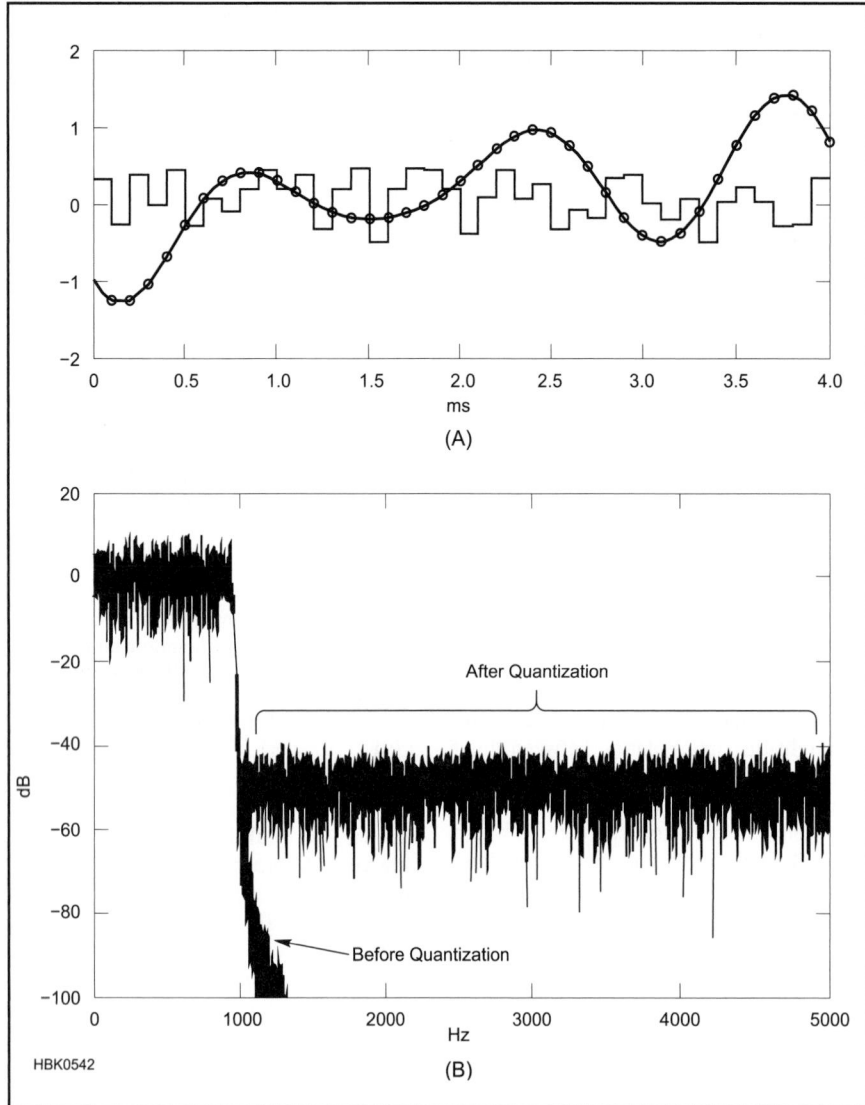

Fig 15.9 — Quantization error of a random noise signal that has been band-limited to 1 kHz to simulate an audio signal. (A) The sampler resolution is 8 bits and the sample rate is 10 kHz. Sample values are indicated by circles. Also shown is the quantization error, in units of LSB. Below is the frequency spectrum of the signal before and after quantization. (B)

a complex signal (one that contains I and Q parts), the 2B in the denominator should be replaced by B.

When choosing an A/D converter don't forget that the effective SNR depends on the sample rate. As an example, let's compare an Analog Devices AD9235 12-bit, 65 Msps ADC to an AD7653, which is a 16-bit 100 ksps ADC from the same manufacturer. Assume a 10 kHz signal bandwidth.

An ideal 12-bit ADC has a SNR of $1.76 + 6.02 \times 12 = 74.0$ dB. The AD9235's performance is not far from the ideal; its SNR is specified at 70.5 dB at its 65 Msps maximum sample rate. In a 10 kHz bandwidth, the effective SNR is $70.5 + 10 \log (65,000/20) = 105.6$ dB.

An ideal 16-bit ADC has an SNR of 98.1 dB. The AD7653 is specified at 86 dB. The effective SNR is $86 + 10\log(100/20) = 93$ dB.

So the 12-bit ADC with 70.5 dB SNR is actually 12.6 dB better than the 16-bit device with 86 dB SNR! Even an ideal 16-bit, 100 ksps ADC would only have an effective SNR of $98.1 + 10\log(100/20) = 105.1$ dB, still worse than the actual performance of the AD9235 when measured with the same bandwidth. Note that to actually realize 105.6 dB of dynamic range the signal from the ADC would need to be filtered to a 10 kHz bandwidth while increasing the bits of data resolution.

Oversampling is the name given to the

technique of using a higher-than-necessary sample rate in order to achieve an improved S/N ratio. Don't forget that when the high-sample-rate signal is decimated the data words must have enough bits to support the higher dynamic range at the lower sample rate. As a rule of thumb, the quantization noise should be at least 10 dB less than the signal noise in order not to significantly degrade the SNR. In the AD9235 example, assuming a 100 kHz output sample rate, about 18 bits would be required: $1.76 + 6.02 \times 18 + 10\log(100/20) = 117.1$ dB, which is 11.5 dB better than the 105.6 dB dynamic range of the ADC in a 10 kHz bandwidth.

Most ADCs and DACs used in high-fidelity audio systems use an extreme form of oversampling, where the internal converter may oversample by a rate of 128 or 256 times, but with very low resolution (in some cases just a 1-bit ADC!). In addition, such converters use a technique called noise shaping to push most of the quantization noise to frequencies near the sample rate, and reduce it in the audio spectrum. The noise is then removed in the decimation filter.

Although quantization error manifests itself as noise when digitizing a complex non-periodic signal, it can show up as discrete spurious frequencies when digitizing a periodic signal. **Fig 15.10** illustrates a 1 kHz sine wave sampled with 8-bit resolution at a 9.5 kHz rate. On average there are 9.5 samples per cycle of the sine wave so that the sampling error repeats every second cycle. That 500-Hz periodicity in the error signal causes a spurious signal at 500 Hz and harmonics. As the signal frequency is changed, the spurs move around in a complicated manner that depends on the ratio of sample rate to signal frequency. In real-world ADCs, nonlinearities in the transfer function can also create spurious signals that vary unpredictably as a function of the signal ampli-tude, especially at low signal levels.

In many applications, broadband noise is preferable to spurious signals on discrete frequencies. The solution is to add *dithering*. Essentially this involves adding a small amount of noise, typically on the order of an LSB or two, in order to randomize the quantization error. Some DACs have dithering capability built in to improve the SFDR, even though it does degrade the SNR slightly. Dithering is also useful in cases where the input signal to an ADC is smaller than one LSB. Even though the signal would be well above the noise level after narrow-band filtering, it cannot be detected if the ADC input is always below one LSB. In many systems there is sufficient noise at the input, both from input amplifiers as well as from the ADC itself, to cause natural dithering.

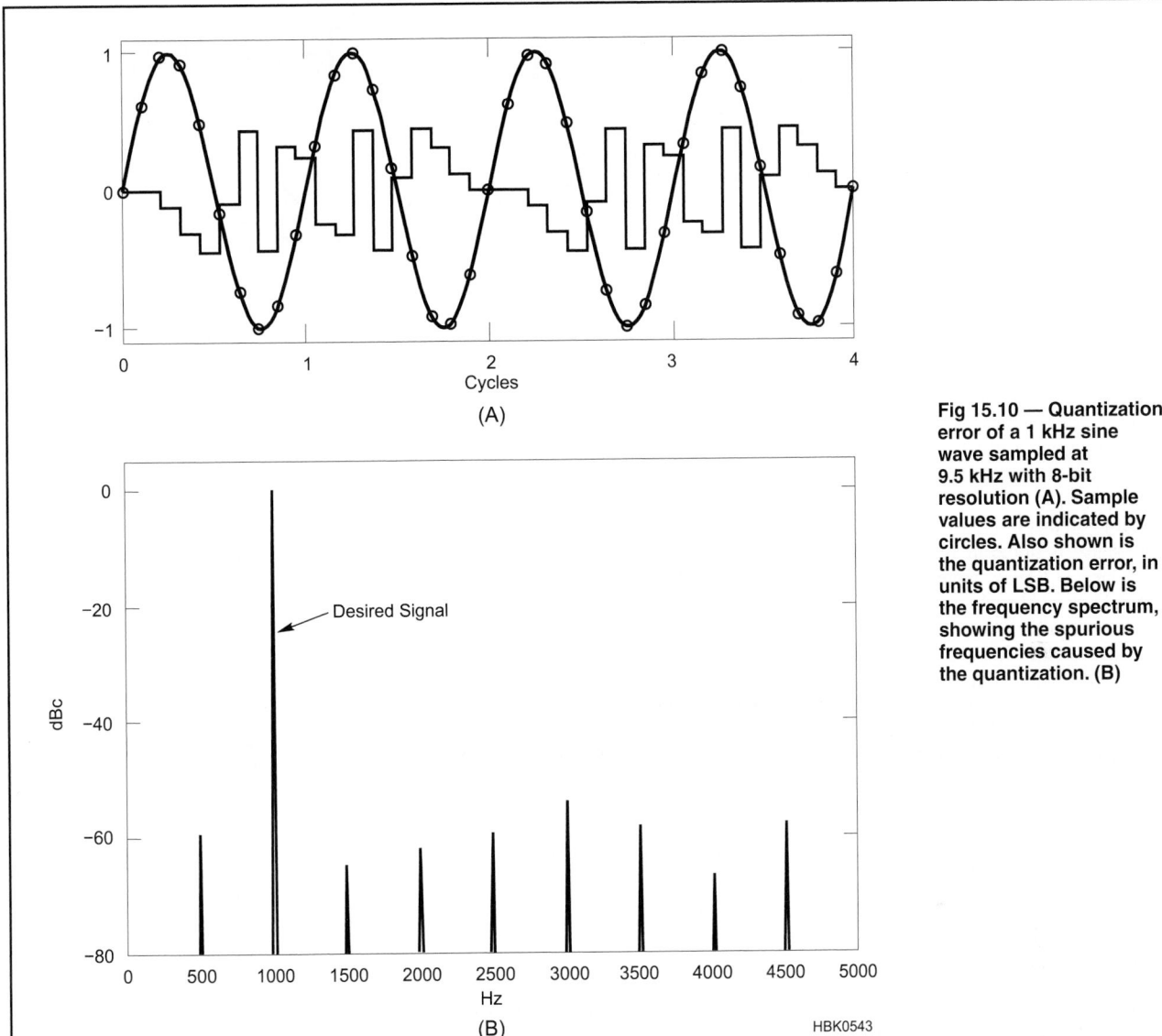

Fig 15.10 — Quantization error of a 1 kHz sine wave sampled at 9.5 kHz with 8-bit resolution (A). Sample values are indicated by circles. Also shown is the quantization error, in units of LSB. Below is the frequency spectrum, showing the spurious frequencies caused by the quantization. (B)

HBK0543

15.4 Digital Filters

As radio amateurs, most of us are well-acquainted with the concept of frequency. We know, for example, that a pure sine wave consists of a single frequency, which is inversely proportional to the wavelength. If the sine wave is distorted, additional harmonic frequencies appear at integer multiples of the fundamental. For example, a square wave consists of sine waves at the fundamental frequency and all the odd harmonics. In general, any periodic waveform can be decomposed into a combination of sine waves at various phase angles with frequencies that are integer multiples of the repetition rate of the waveform.

Even a non-periodic waveform can be decomposed into sine waves, although in this case they are not harmonically-related. For example a single pulse of width τ seconds has a frequency spectrum proportional to $\mathrm{sinc}(f\tau)$ = $\sin(\pi f\tau)/(\pi f\tau)$. You can think of this as an infinite number of sine waves spaced infinitely closely together with amplitudes that trace out that spectral shape. It is interesting to note that if τ is decreased, the value of f must increase by the same factor for any given value of $\sin(\pi f\tau)/(\pi f\tau)$. In other words, the narrower the pulse the wider the spectrum. Of course that applies to sine waves and other periodic waveforms as well — the smaller the wavelength the higher the frequency. In general, anything that makes the signal "skinnier" in the time domain makes it "fatter" in the frequency domain and vice versa.

As the pulse becomes narrower and narrower, the frequency spectrum spreads out more and more. In the limit, if the pulse is made infinitely narrow, the spectrum becomes flat from zero hertz to infinity. An infinitely-narrow pulse is called an *impulse* and is a very useful concept because of its flat frequency spectrum. If you feed an impulse into the input of a filter, the signal that comes out, the *impulse response*, has a frequency spectrum equal to the frequency response of the filter. One way to design a filter is to determine the impulse response that corresponds to the desired frequency spectrum and then design the filter to have that impulse response. That method is ideally suited for designing FIR filters.

15.4.1 FIR Filters

A *finite impulse response* (FIR) filter is a filter whose impulse response is finite, ending in some fixed time. Note that analog filters have an infinite impulse response — the output theoretically rings forever. Even a simple R-C low-pass filter's output dies exponentially toward zero but theoretically never quite reaches it. In contrast, an FIR filter's impulse response becomes exactly zero at some time after receiving the impulse and stays zero forever (or at least until another impulse comes along).

Given that you have somehow figured out the desired impulse response, how would you design a digital filter to have that response? The obvious method would be to pre-calculate a table of impulse response values, sampled at the sample rate. These are called the filter *coefficients*. When an impulse of a certain amplitude is received, you multiply that amplitude by the first entry in the coefficient table and send the result to the output. At the next sample time, multiply the impulse by the second entry, and so on until you have used up all the entries in the table.

A circuit to do that is shown in **Fig 15.11**. The input signal is stored in a shift register. Each block labeled "Delay" represents a delay of one sample time. At each sample time, the signal is shifted one register to the right. Each register feeds a multiplier and the other input to the multiplier comes from one of the coefficient table entries. All the multiplier outputs are added together. Since the input is assumed to be a single impulse, at any given time all the shift registers contain zero except one, which is multiplied by the appropriate table entry and sent to the output.

We've just designed an FIR filter! By using a shift register with a separate multiplier for each tap, the filter works for continuous signals as well as for impulses. Since this is a linear system, the continuing signal is affected by the filter the same as an individual impulse.

It should be obvious from the diagram how to implement an FIR filter in software. You set up two buffers in memory, one for the filter coefficients and one for the data. The length of each buffer is the number of filter taps. (A *tap* is the combination of one filter coefficient, one shift register and one multiplier/accumulator.) Each time a new data value is received, it is stored in the next available position in the data buffer and the accumulator is set to zero. Next, a software loop is executed a number of times equal to the number of taps. During each loop, pointers to the two buffers are incremented, the next coefficient is multiplied by the next data value and the result is added to the current accumulator value. After the last loop, the accumulator contents are the output value. Normally the buffers are implemented as *circular buffers* — when the address pointer gets to the end it is reset back to the beginning.

Now you can see why a hardware multiplier-accumulator (MAC) is such an important feature of a DSP chip. Each tap of the FIR filter involves one multiplication and one addition. With a 1000-tap FIR filter, 1000 multiplications and 1000 additions must be performed

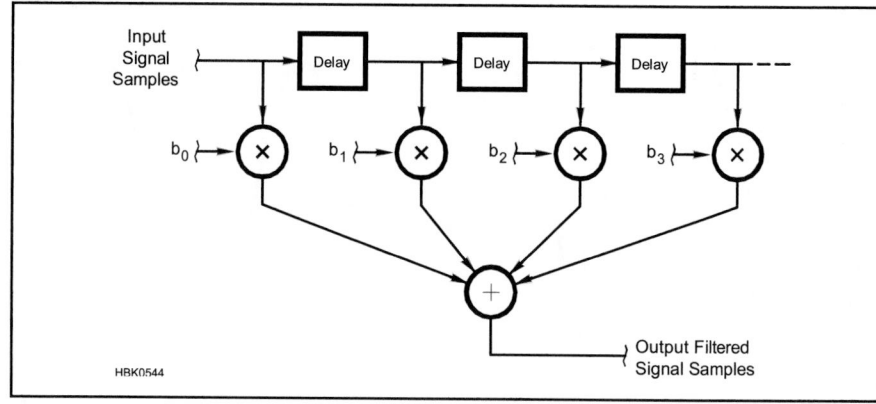

Fig 15.11 — A 4-tap FIR filter. The b_n values are the filter coefficients.

during each sample time. Being able to do each MAC in a single clock cycle saves a lot of processing time.

An FIR filter is a hardware or software implementation of the mathematical operation called *convolution*. We say that the filter *convolves* the input signal with the impulse response of the filter. It turns out that convolution in the time domain is mathematically equivalent to multiplication in the frequency domain. That means that the frequency spectrum of the output equals the frequency spectrum of the input times the frequency spectrum of the filter. Expressed in decibels, the output spectrum equals the input spectrum plus the filter frequency response, all in dB. If at some frequency the input signal is +3 dB and the filter is –10 dB compared to some reference, then the output signal will be 3 – 10 = –7 dB at that frequency.

An FIR filter whose bandwidth is very small compared to the sample rate requires a long impulse response with lot of taps. This is another consequence of the "skinny" versus "fat" relationship between the frequency and time domains. If the filter is narrow in the frequency domain, then its impulse response is wide. Actually, if you want the frequency response to go all the way to zero (minus infinity dB) throughout the stop band, then the impulse response theoretically becomes infinitely wide. Since we're designing a *finite* impulse response filter we have to truncate the impulse response at some point to get it to fit in the coefficient table. When you do that, however, you no longer have infinite attenuation in the stopband. The more heavily you truncate (the narrower the impulse response) the worse the stopband attenuation and the more ripple you get in the passband. Assuming optimum design techniques for selecting coefficients, you can estimate the minimum length L of the impulse response from the following equation:

$$L = 1 - \frac{10\log(\delta_1\delta_2) - 15}{14\left(\dfrac{f_T}{f_s}\right)} \text{ taps}$$

where

δ_1 and δ_2 = the passband and stopband ripple expressed as a fraction

f_T = the transition bandwidth (frequency difference between passband and stopband edges)

f_s = the sample rate.

For example, for a low-pass filter with a passband that extends up to 3 kHz, a stopband that starts at 4 kHz ($f_T = 4 - 3 = 1$ kHz), $f_s = 10$ kHz sample rate, ±0.1 dB passband ripple ($\delta_1 = 10^{0.1/20} - 1 = 0.0116$), and 60 dB stopband rejection ($\delta_2 = 10^{-60/20} = 0.001$), we get

$$L = 1 - \frac{10\log(0.0116 \times 0.001) - 15}{14\left(\dfrac{1}{10}\right)}$$

$$= 1 - \frac{-49.4 - 15}{1.4} = 47 \text{ taps}$$

Overflow is a potential problem when doing the calculations for an FIR filter. Multiplying two N-bit numbers results in a product with 2N bits, so space must be provided in the accumulator to accommodate that. Although the final result normally will be scaled and truncated back to N bits, it is best to carry through all the intermediate results with full resolution in order not to lose any dynamic range. In addition, the sum of all the taps can be a number with more than 2N bits. For example, if the filter width is 256 taps, then if all coefficients and data are at full scale, the final result could theoretically be 256 times larger, requiring an extra 8 bits in the accumulator. We say "theoretically" because normally most of the filter coefficients are much less than full scale and it is highly unlikely that all 256 data values would ever simultaneously be full-scale values of the correct polarity to cause overflow. The dsPIC processors use 16-bit multipliers with 32-bit results and a 40-bit accumulator, which should handle any reasonable circumstances.

After all taps have been calculated, the final

result must be retrieved from the accumulator. Since the accumulator has much more resolution than the processor's data words, normally the result is truncated and scaled to fit. It is up to the circuit designer or programmer to scale by the correct value to avoid overflow. The worst case is when each data value in the shift register is full-scale — positive when it is multiplying a positive coefficient and negative for negative coefficients. That way, all taps add to the maximum value. To calculate the worst-case accumulator amplitude, simply add the absolute values of all the coefficients. However, that normally gives an unrealistically pessimistic value because statistically it is extremely unlikely that such a high peak will ever be reached. For a low-pass filter, a better estimate is to calculate the gain for a dc signal and add a few percent safety margin. The dc gain is just the sum of all the coefficients (not the absolute values). For a band-pass filter, add the sum of all the coefficients multiplied by a sine wave at the center frequency.

CALCULATING FIR FILTER COEFFICIENTS

So far we have ignored the question of how to determine the filter coefficients. For an ideal "brick-wall" low-pass filter, the answer turns out to be pretty simple. A "brick-wall" low-pass filter is one that has a constant response from zero hertz up to the cutoff frequency and zero response above. Its impulse response is proportional to the sinc function:

$$C(n) = C_0 \; \text{sinc}(2Bn) = \frac{\sin(2\pi Bn)}{2\pi Bn}$$

where $C(n)$ are the filter coefficients, n is the sample number with $n = 0$ at the center of the impulse response, C_0 is a constant, and B is the single-sided bandwidth normalized to the sample rate, $B = $ bandwidth / sample rate.

It is interesting that this has the same form as the frequency response of a pulse, as was shown in Fig 15.7. That is because a brick-wall response in the frequency domain has the same shape as a pulse in the time domain. A pulse in one domain transforms to a sinc function in the other. This is an example of the general principle that the transformation between time and frequency domains is symmetrical. We will discuss this more later, in the section on Fourier transforms.

Normally, the filter coefficients are set up with the peak of the sinc function, sinc(0), at the center of the coefficient table so that there is an equal amount of "tail" on both sides. That points up the principle problem with this method of determining filter coefficients. Theoretically, the sinc function extends from minus infinity to plus infinity. Abruptly terminating the tails causes the frequency response to differ from an ideal brick-wall filter. There is ripple in the passband and

Table 15.3
Routine for dsPIC Processor to Calculate Filter Coefficients

```
// Calculate FIR filter coefficients
// using the windowed-sinc method
void set_coef (
 double sample_rate;
 double bandwidth;)
{
extern int c[FIR_LEN]; // Coefficient array
int i; // Coefficient index
double ph; // Phase in radians
double coef; // Filter coefficient
int coef_int; // Digitized coefficient
double bw_ratio; // Normalized bandwidth

bw_ratio = 2 * bandwidth / sample_rate;
for (i = 0; i < (FIR_LEN/2); i++) {
 // Brick-wall filter:
 ph = PI * (i + 0.5) * bw_ratio;
 coef = sin(ph) / ph;
 // Hann window:
 ph = PI * (i + 0.5) / (FIR_LEN/2);
 coef *= (1 + cos(ph)) / 2;
 // Convert from floating point to int:
 coef *= 1 << (COEF_WIDTH - 1);
 coef_int = (int)coef;
 // Symmetrical impulse response:
 c[i + FIR_LEN/2] = coef_int;
 c[FIR_LEN/2 - 1 - i] = coef_int;
 }
}
```

non-zero response in the stopband, as shown in the graph in the upper right of **Fig 15.12**. This is mainly caused by the abruptness of the truncation. In effect, all the coefficients outside the limits of the coefficient table have been set to zero. The passband and stopband response can be improved by tapering the edges of the impulse response instead of abruptly transitioning to zero.

The process of tapering the edges of the impulse response is called *windowing*. The impulse response is multiplied by a window, a series of coefficients that smoothly taper to zero at the edges. For example, a rectangular window is equivalent to no window at all. Many different window shapes have been developed over the years — at one time it seemed that every doctoral candidate in the field of signal processing did their dissertation on some new window. Each window has its advantages and disadvantages. A window that transitions slowly and smoothly to zero has excellent passband and stopband response but a wide transition band. A window that has a wider center portion and then transitions more abruptly to zero at the edges has a narrower transition band but poorer passband and stopband response. The equations for the windows in Fig 15.12 are in-

cluded in a sidebar.

The routine shown in **Table 15.3** is written for a dsPIC processor so it can be used to calculate filter coefficients "on the fly" as the operator adjusts a bandwidth control. The same code should also work using a generic C compiler on a PC so the coefficients could be downloaded into an FIR filter implemented in hardware.

The windowed-sinc method works pretty well for a simple low-pass filter, but what if some more-complicated spectral shape is desired? The same general design approach still applies. You determine the desired spectral shape, transform it to the time domain using an inverse Fourier transform, and apply a window. There is lots of (often free) software available that can calculate the fast Fourier transform (FFT) and inverse fast Fourier transform (IFFT). So the technique is to generate the desired spectral shape, transform to the time domain with the IFFT, and multiply the resulting impulse response by the desired window. Then you can transform back to the frequency domain with the FFT to see how the window affected the result. If the result is not satisfactory, you can either choose a different window or modify the original spectral shape and go through the process again.

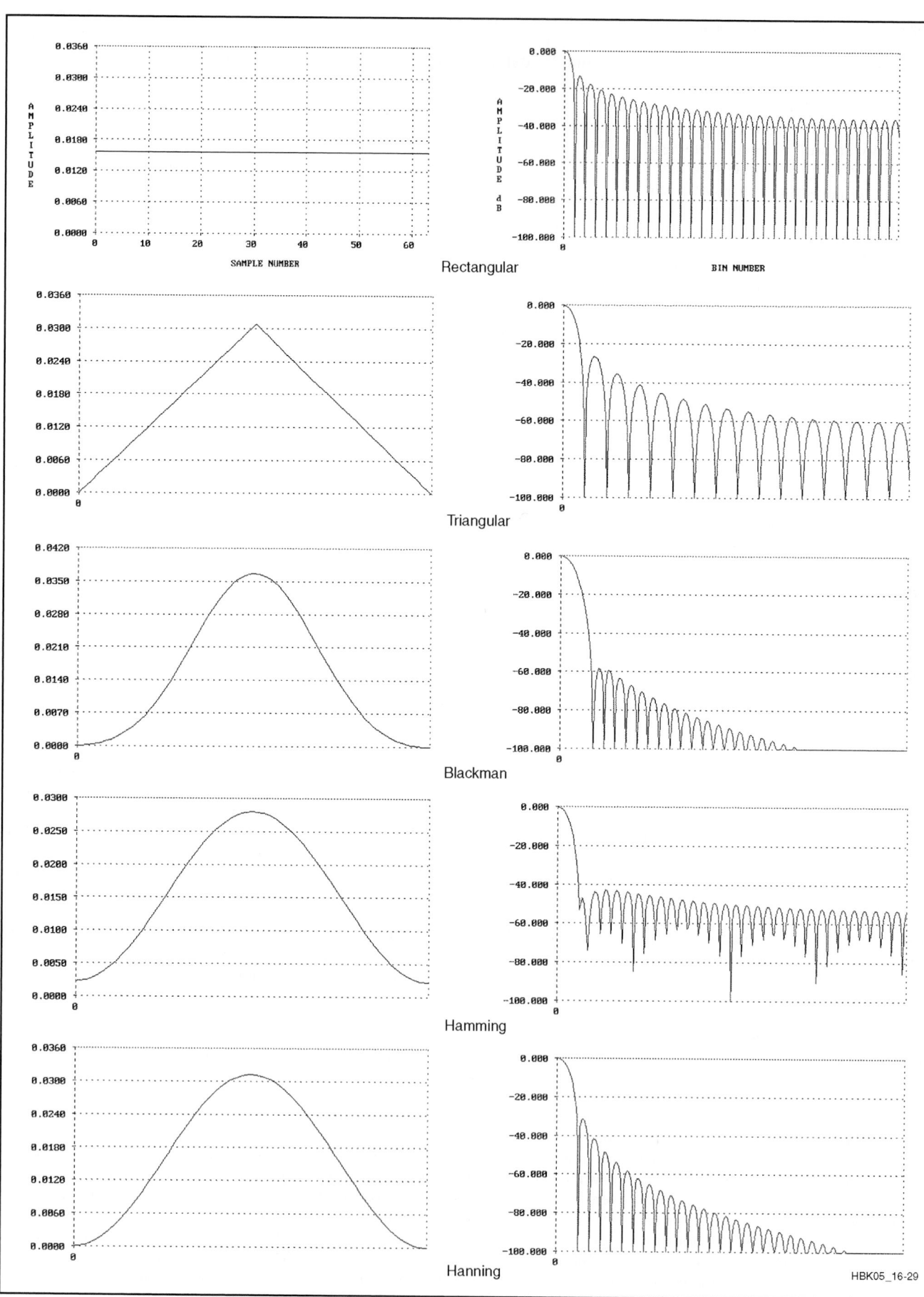

Fig 15.12 — Various window functions and their Fourier transforms.

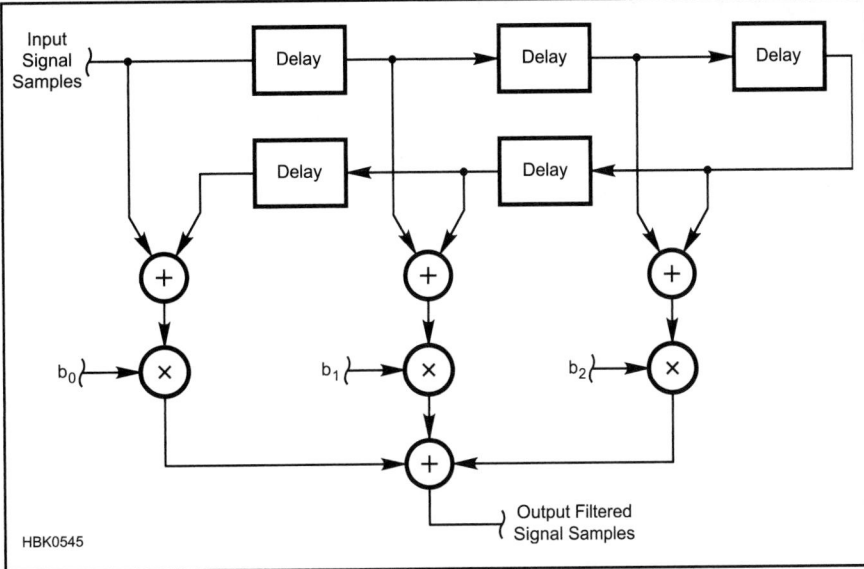

Fig 15.13 — A 6-tap FIR filter. Because the coefficients are symmetrical, the symmetrical taps may be combined before multiplication.

Equations for Window Functions

For each window function, the center of the response is considered to be at time t = 0 and the width of the impulse is L. Each window is 1.0 when t = 0 and 0.0 when the | t | > L/2.

Rectangular:

$$w(t) = 1.0$$

Triangular (Bartlett):

$$w(t) = 2\left(\frac{L/2 - |t|}{L}\right)$$

Blackman:

$$w(t) = 0.42 + 0.5\cos\left(\frac{2\pi t}{L}\right)$$

$$+0.08\cos\left(\frac{4\pi t}{L}\right)$$

Hamming:

$$w(t) = 0.54 + 0.46\cos\left(\frac{2\pi t}{L}\right)$$

Hanning (Hann):

$$w(t) = 0.5 + 0.5\cos\left(\frac{2\pi t}{L}\right)$$

Windowing methods are useful because they are simple to program and the resulting software routines execute quickly. For example, you can include a bandwidth knob on your DSP filter and calculate filter coefficients "on the fly" as the user turns the knob. However, while the filter performance that results is pretty good, it is not "optimum" in the sense that it does not have minimum passband ripple and maximum stopband attenuation for a given number of filter coefficients. For that, you need what is known as an *equal-ripple*, or *Chebyshev* filter. The calculations to determine Chebyshev filter coefficients are more complicated and time-consuming. For that reason, the coefficients are normally calculated in advance on a PC and stored in DSP program memory for retrieval as needed.

Engineers have not had much success in devising a mathematical algorithm to calculate the Chebyshev coefficients directly, but in 1972 Thomas Parks and James McClellan figured out a method to do it iteratively. The *Parks-McClellan algorithm* is supported in most modern filter-design software, includ-

ing a number of programs available for free download on the Web. Typically you enter the sample rate, the passband and stopband frequency ranges, the passband ripple and the stopband attenuation. The software then determines the required number of filter coefficients, calculates them and displays a plot of the resulting filter frequency response.

Filter design software typically presents the filter coefficients as floating-point numbers to the full accuracy of the computer. You will need to scale the values and truncate the resolution to the word size of your filter implementation. Truncation of filter coefficients affects the frequency response of the filter but does not add noise in the same manner as truncating the signal data.

As you look at impulse responses for various FIR filters calculated by various methods you soon realize that most of them are symmetrical. If the center of the impulse response is considered to be at time zero, then the value at time t equals the value at time –t for all t. If you know in advance that the filter coefficients are symmetrical, you can take advantage of that in the filter design. By re-arranging the adders and multipliers, the number of multipliers can be reduced by a factor of two, as shown in **Fig 15.13**. This trick is less useful in a software implementation of an FIR filter because the number of additions is the same and many DSPs take the same amount of time to do an addition as a multiply-accumulate.

In addition to the computational benefit, a symmetrical impulse response also has the

advantage that it is *linear phase*. The time delay through such a filter is one-half the length of the filter for all frequencies. For example, for a 1000-tap filter running at 10 kHz the delay is 500/10,000 = 0.05 second. Since the time delay is constant for all frequencies, the phase delay is directly proportional to the frequency. For example, if the phase delay at 20 Hz is one cycle (0.05 second) it is ten cycles at 200 Hz (still 0.05 second). Linear phase delay is important with digital modulation signals to avoid distortion and inter-symbol interference. It is also desirable with analog modulation where it can result in more natural-sounding audio. All analog filters are non-linear-phase; the phase distortion tends to be worse the more abrupt the transition between passband and stopband. That is why an SSB signal sounds unnatural after being filtered by a crystal filter with a small shape factor even though the passband ripple may be small and distortion minimal.

A band-pass filter can be constructed from a low-pass filter simply by multiplying the impulse response by a sine wave at the desired center frequency. This can be done before or after windowing. The linear-phase property is retained but with reference to the center frequency of the filter, that is, the phase shift is proportional to the difference in frequency from the center frequency. The frequency response is a double-sided version of the low-pass response with the zero-hertz point of the low-pass filter shifted to the frequency of the sine wave.

15.4.2 IIR Filters

An *infinite impulse response* (IIR) filter is a filter whose impulse response is infinite. After an impulse is applied to the input, theoretically the output never goes to zero and stays there. In practice, of course, the signal eventually does decay until it is below the noise level (analog filter) or less than one LSB (digital filter).

Unlike a symmetrical FIR filter, an IIR filter is not generally linear-phase. The delay through the filter is not the same for all frequencies. Also, IIR filters tend to be harder to design than FIR filters. On the other hand, many fewer adders and multipliers are typically required to achieve the same passband and stop band ripple in a given filter, so IIR filters are often used where computations must be minimized.

All analog filters have an infinite impulse response. For a digital filter to be IIR it must have feedback. That means a delayed copy of some internal computation is applied to an earlier stage in the computation. A simple but useful example of an IIR filter is the exponential decay circuit in **Fig 15.14**. In the absence of a signal at the input, the output on the next clock cycle is always $(1-\delta)$ times the current output. The time constant (the time for the output to die to $1/e = 36.8\%$ of the initial value) is very nearly

$$\tau = f_s \left(\frac{1}{\delta} - \frac{1}{2} \right)$$

where f_s is the sample rate. The circuit is the digital equivalent of a capacitor with a resistor in parallel and might be useful for example in a digital automatic gain control circuit.

One issue with IIR filters is resolution. Because of the feedback, the number of bits of resolution required for intermediate computations can be much greater than at the input or output. In the previous example, δ is very small for very long time constants. When the value in the register falls below a certain level the multiplication by $(1-\delta)$ will no longer be accurate unless the bit width is increased. In practice, the increased resolution required with IIR filters often cancels out part of the savings in the number of circuit elements.

Another issue with IIR filters is stability.

Because of the feedback it is possible for the filter to oscillate if care is not taken in the design. Stability can also be affected by non-linearity at low signal levels. A circuit that is stable with large signals may oscillate with small signals due to the round-off error in certain calculations, which causes faint tones to appear when strong signals are not present. This is known as an unstable *limit cycle*. These issues are part of the reason that IIR filters have a reputation for being hard to design.

Design techniques for IIR filters mostly involve first designing an analog filter using any of the standard techniques and then transforming the design from the analog to the digital domain. The *impulse-invariant* method attempts to duplicate the filter response directly by making the digital impulse response equal the impulse response of the equivalent analog filter. It works fairly well for low-pass filters with bandwidths much less than the sample rate. Its problem is that it tries to duplicate the frequency response all the way to infinity hertz, but that violates the Nyquist criterion resulting in a folding back of the high-frequency response down into low frequencies. It is similar to the aliasing that occurs in a DSP system when the input signal to be sampled is not band-limited below the Nyquist frequency.

The *bilinear transform* method gets around that problem by distorting the frequency axis such that infinity hertz in the analog domain becomes sample rate / 2 in the digital domain. Low frequencies are fairly accurate, but high frequencies are squeezed together more and more the closer you get to the Nyquist frequency. It avoids the aliasing problem at the expense of a change in the spectrum shape,

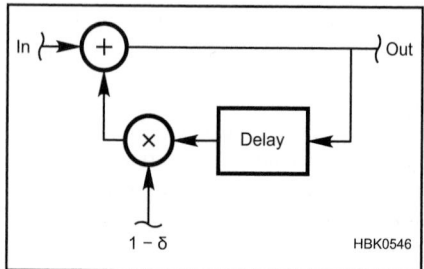

Fig 15.14 — An exponential decay circuit.

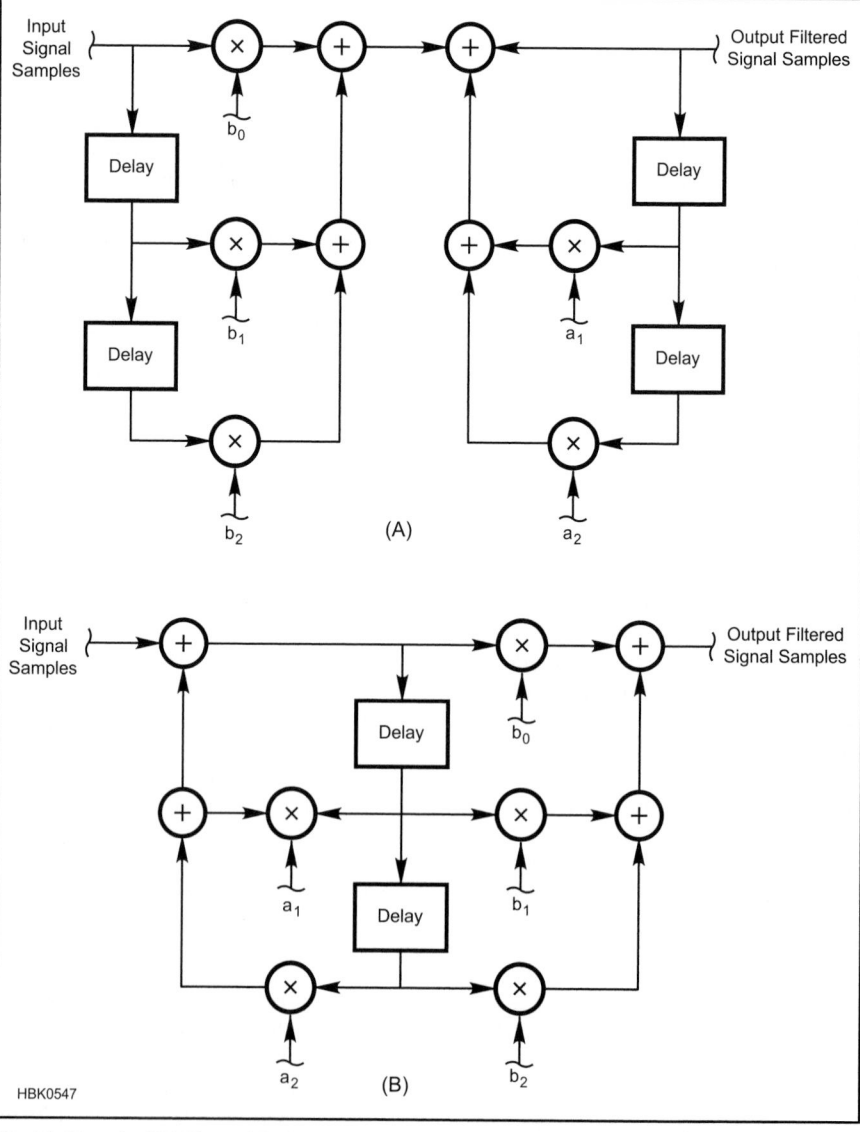

Fig 15.15 — An IIR filter with three feed-forward taps and two feed-back taps. Direct form I (A) and the equivalent direct form II (B).

especially at the high-frequency end. For example, when designing a low-pass filter it may be necessary to change the cutoff frequency to compensate. Again, the method works best for filters with passband frequencies much less than the sample rate.

In general, the output of an IIR filter is a combination of the current and previous input values (feed-forward) and previous output values (feed-back). **Fig 15.15A** shows the so-called *direct form I* of an IIR filter. The b_i coefficients represent feed-forward and the a_i coefficients feed-back. For example the previous value of the y output is multiplied by a_1, the second previous value is multiplied by a_2, and so on. Because the filter is linear, it doesn't matter whether the feed forward or feed back stage is performed first. By reversing the order, the number of shift registers is reduced as in Fig 15.15B. There are other equivalent topologies as well. The mathematics for generating the a_i and b_i coefficients for both the impulse-invariant and bilinear transform methods is fairly involved, but fortunately some filter design programs can handle IIR as well as FIR filters.

15.4.3 Adaptive Filters

An *adaptive filter* is one that automatically adjusts its filter coefficients under the control of some algorithm. This is often done in situations where the filter characteristics are not known in advance. For example, an adaptive channel equalizer corrects for the non-flatness in the amplitude and phase spectrum of a communications channel due to multipath propagation. Typically, the transmitting sta-

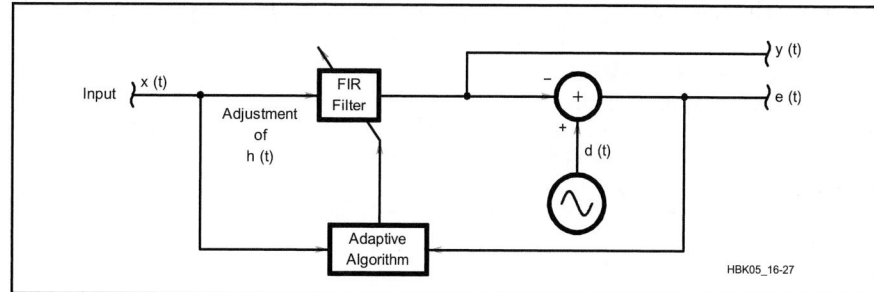

Fig 15.16 — An adaptive filter.

tion periodically sends a known sequence of data, known as a *training sequence*, which is used by the receiver to determine the channel characteristics and adjust its filter coefficients accordingly.

Another example is an automatic notch filter. An algorithm determines the frequency of an interfering tone and automatically adjusts the notch frequency to remove the tone. Noise cancellation is another application. It can be thought of as the opposite of a notch filter. In this case, all the sine-wave tones in the input signal are considered to be desired and the filter coefficients are configured to enhance them. That method works not only for CW signals but for voice as well since the human voice consists largely of discrete frequencies.

A generic block diagram of an adaptive filter is shown in **Fig 15.16**. The variable filter is typically an FIR type with coefficients calculated by the update algorithm. By some means, an estimate of the desired, unimpaired signal, d, is generated and compared to the filter output y. The difference between y and

d is the error, e, which is used by the update algorithm to modify the filter coefficients to improve the accuracy of y. The algorithm is capable of acting as a noise-reduction filter and a notch filter simultaneously. Assuming d is in the form of a pure tone (sine wave), the tone is simultaneously optimized in the y output and minimized in the e output.

A common algorithm for minimizing the error signal is called *least mean squares* (LMS). The LMS algorithm includes a performance parameter, μ, which can be adjusted between 0 and 1 to control the tradeoff between adjustment speed and accuracy. A value near 1 results in fast convergence but the convergence is not very accurate. For better accuracy at the cost of slower adjustment, lower the value of μ. Some implementations adjust μ on the fly, using a large value at first to get faster lock-in when the error is large then a smaller value after convergence to reduce the error. That works as long as the signal characteristics are not changing too rapidly.

15.5 Miscellaneous DSP Algorithms

15.5.1 Sine Wave Generation

There are a number of techniques available for building a digital sine-wave generator, either in hardware or in software. One obvious idea is to make a digital oscillator, analogous to a conventional analog oscillator. Simply design a band-pass filter at the desired frequency and include positive feedback around it with a loop gain of unity. The problem is that, because of round-off error in the digital calculations, it is difficult to get the loop gain to be *exactly* 1.0. If it is slightly less, then the oscillation will eventually die out. If it is slightly greater, then the oscillation will gradually increase in amplitude until it exceeds the maximum signal that the digital circuitry can handle. There are two techniques to handle this problem. One idea is to include an automatic gain-control circuit

to detect the amplitude, low-pass filter it and feed the result to a multiplier in the feedback path to control the gain. Another idea is to intentionally set the loop gain slightly greater than 1.0 and include a clipping stage in the feedback path that limits the peak amplitude in a controlled manner.

With both of those techniques there is a tradeoff between distortion and start-up time. It can take many oscillation cycles for the amplitude to stabilize. If the loop gain is increased or the AGC time constant is reduced to improve the start-up time, worse distortion results.

Probably the most common technique for generating sine waves is the *numerically-controlled oscillator* (NCO), or *direct digital synthesizer* (DDS) as shown in **Fig 15.17**.

At each clock cycle, the phase accumulator increments the phase by an amount equal to $360°$ times f / f_s, where f is the sine-wave frequency and f_s is the sample rate. The current phase value is used as a pointer to the proper address in a sine-wave lookup table. As the phase increases, the pointer moves through the look-up table, tracing out the sine-wave amplitude. Different frequencies can be obtained by changing the step size in the phase accumulator.

The lookup table size is a factor of two and the accumulator output is scaled such that the maximum count, corresponding to $360°$, accesses the final entry in the table. When the phase passes $360°$ it automatically jumps back to zero as required. For good frequency resolution, the word size of the phase accu-

Fourier Transform

The *Fourier transform* is the software equivalent of a hardware spectrum analyzer. It takes in a signal in the time domain and outputs a signal in the frequency domain that shows the spectral content of the input signal. The Fourier transform works on both periodic and non-periodic signals, but since the periodic case is easier to explain we will start with that.

A periodic signal is one that repeats every τ seconds, where τ is the period. That means that the signal can consist only of frequencies whose sinusoidal waveforms have an integer number of cycles in τ seconds. In other words, the signal is made up of sinusoids that are at the frequency $1/\tau$ and its harmonics. Fourier's idea was that you can determine if a frequency is present by multiplying the waveform by a sinusoid of that frequency and integrating the result. The result of the integration yields the amplitude of that harmonic. If the integration yields zero, then that frequency is not present.

To see how that works, look at **Fig 15-A1**. For the purpose of discussion, assume the signal to be tested consists of a single tone at the second harmonic as shown at (A). The first test frequency is the fundamental, shown at (B). When you multiply the two together you get the waveform at (C). Integrating that signal gives its average value, which is zero. However if you multiply the test signal by a sine wave at the second harmonic (D), the resulting waveform (E) has a large dc offset so the integration yields a large non-zero value. It turns out that all harmonics other than the second yield a zero result. That is, the second harmonic is *orthogonal* to all the others.

If the test waveform included more than one frequency, each of those frequencies would yield a non-zero result when tested with the equivalent-frequency sine wave. The presence of additional frequencies does not disturb the tests for other frequencies since they are all orthogonal with each other.

You may have noticed that this method only works if the test sine wave is in phase with the one in the signal. If they are 90° out of phase, the integration yields zero. The Fourier transform therefore multiplies the signal by both a sine wave and a cosine wave at each frequency. The results of the two tests then yield both the amplitude and phase of that frequency component of the signal using the equations

$$A = \sqrt{a^2 + b^2}$$

and

$$\varphi = \arctan\left(\frac{b}{a}\right)$$

where A is the amplitude, φ is the phase, a is the cosine amplitude and b is the sine amplitude.

If one period of the signal contains, say, 256 samples, then testing a single frequency requires multiplying the signal by the sine wave and by the cosine wave 256 times and adding the results 256 times as well, for a total of 512 multiplications and additions. There are 128 frequencies that must be tested, since the 128th harmonic is at the Nyquist frequency. The total number of calculations is therefore $512 \times 128 = 256^2$ multiplications and additions. That is a general result. For any sample size, n, calculating the digital Fourier transform requires n^2 multiply-accumulates.

The FFT

The number of calculations grows rapidly with sample size. Calculating the Fourier transform on 1024 samples requires over a million multiply-accumulates. However, you may notice that there is some redundancy in the calculations. When testing the second harmonic, for example, each of the two cycles of the test sine wave is identical. It would be possible to pre-add signal data from the first and second halves of the sequence and then just multiply once by a single cycle of the test sine wave. Also, the first quarter cycle of a sine wave is just a mirror-image of the second quarter cycle and the first half is just the negative of the second half.

In 1965, J. W. Cooley and John W. Tukey published an algorithm that takes advantage of all the symmetries inherent in the Fourier transform to speed up the calculations. The *Cooley-Tukey algorithm*, usually just called the *fast Fourier transform* (FFT), makes the number of calculations proportional to $n\log_2(n)$ instead of n^2. For a 1024-point FFT, the calculation time is proportional to $1024\log_2(1024) = 10,240$ instead of $1024^2 = 1,048,576$, more than a 100-times improvement.

You'll notice that sample sizes are usually a power of two, such as $2^7 = 128$, $2^8 = 256$ and $2^9 = 512$. That is because the FFT algorithm is most efficient with sequences of such sizes. The algorithm uses a process called

radix-2 decimation in time, that is, it first breaks the data into two chunks of equal size, then breaks each of those chunks into two still-smaller chunks of equal size, and so on. It is possible to squeeze even a little more efficiency out of the algorithm with a *radix-4* FFT which is based on decimation by four instead of by two. That is why you often see sample sizes that are powers of four, such as $4^3 = 64$, $4^4 = 256$ and $4^5 = 1024$. Other variations on the algorithm include decimation in frequency rather than time, mixed-radix FFTs that use different decimation factors at different stages in the calculation, and in-place calculation that puts the results into the same storage buffer as the input data, saving memory. The latter method causes the order of the output data to be scrambled by bit-reversing the address words. For example, address 01010000 becomes 00001010.

Non-periodic signals

So far we have assumed that the signal to be transformed is periodic, so that there is an integer number of cycles of each sine wave harmonic in the sequence. With a non-periodic signal, that is not necessarily so. The various frequencies in the signal are not exact harmonics of $1/\tau$ and are no longer

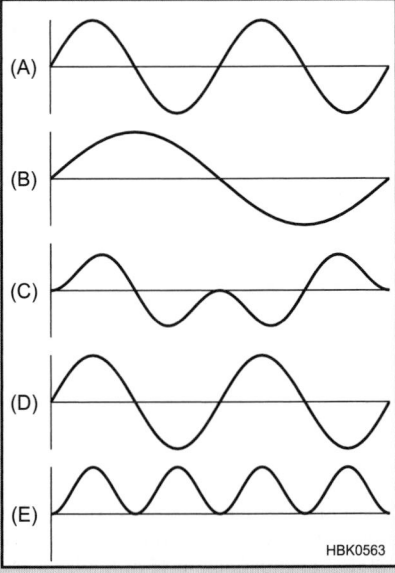

Fig 15-A1 — Signal to be tested for frequency content (A). Fundamental test frequency (B). Product of signal and fundamental (C). Second harmonic test frequency (D). Product of signal and second harmonic (E).

orthogonal to the test frequencies. The result is *spectral leakage*; a single frequency in the signal may give a non-zero result when tested at a number of different harmonic frequencies. In **Fig 15-A2**, (B) illustrates the FFT of a single sine wave at a non-harmonic frequency. You can see that the spurious response extends quite far from the actual frequency.

Those far-out spurious responses are primarily caused by the abrupt termination of the signal at the edges of the sequence. The spectrum can be cleaned up considerably by tapering the edges with a window, in a manner similar to windowing FIR filter coefficients as previously described. In fact, the same windows work for both. Fig 15-A2 part C illustrates the result of applying a Hamming window to the signal in (A) and the resulting improved spectrum is shown at (D). Just as with FIR filters, different windows excel in different areas. Windows with a gradual transition to zero at the edges do a better job of suppressing spurious responses but smear adjacent spectral lines, analogous to using a wider resolution bandwidth in an analog spectrum analyzer. Windows with a fatter center section and a more abrupt transition to zero at the edges have less smearing but worse spurious responses.

While it is interesting and instructional to write your own FFT from scratch, most programmers don't bother to try to re-invent the wheel. Many implementations have been published on the Web and in books and articles. Most of the software development systems offered by DSP vendors include an FFT library routine, which runs faster than anything you are likely to come up with on your own.

Fig 15-A2 — Illustrating the use of windowing to minimize spectral leakage, the figures show (A) a cosine waveform not at a harmonic frequency, (B) the resulting unwindowed power spectrum, (C) the same cosine waveform with a Hamming window, and (D) the much narrower power spectrum of the windowed waveform.

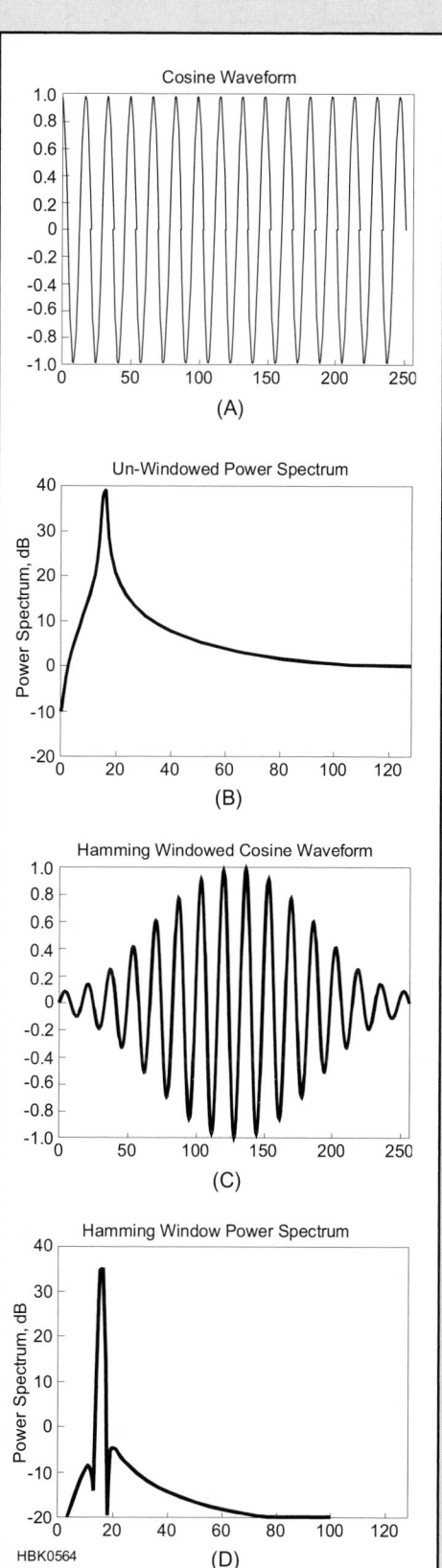

mulator, 32 in this example, is normally much greater than the address width of the look-up table, p. The less-significant accumulator bits are not used in the address. For example, with a 100 MHz clock rate and a 32-bit phase accumulator the frequency resolution is $100 \times 10^6 / 2^{32} = 0.023$ Hz.

The address width of the look-up table determines the number of table entries and thus the phase accuracy of the samples. The table size can be reduced by a factor of four by including only the first quarter-cycle of the sine wave in the table. The other three quadrants can be covered by modifying the look-up address appropriately and by negating the output when required under the control of some additional logic. For example, the method to transform quadrant three to quadrant one is illustrated in **Fig 15.18**. Another technique to improve waveform accuracy without increasing table size is to interpolate between the entries. Instead of using the look-up table output directly, the output value is calculated by interpolating between the two nearest table entries using either a straight-line interpolation as shown in **Fig 15.19** or a higher-order curve fit.

Taking that last idea to an extreme, it is possible to generate a good approximation to one quadrant of a sine wave using a fifth-order interpolation between the zero and 90° points. Assuming that the phase x has been scaled so that x = 1.0 corresponds to 90°, the sine formula is

$$\sin(x) = Ax + Bx^2 + Cx^3 + Dx^4 + Ex^5$$

where the coefficients are A = 3.140625, B = 0.02026367, C = -5.325196, D = 0.5446778 and E = 1.800293.

With those coefficients, which come from an old Analog Devices DSP manual, all the harmonics of the sine wave are more than 100 dB below the carrier.[3] With a 16-bit integer DSP the performance would be limited only by the roundoff error. The formula can be reformulated as

$$\sin(x) = (A + (B + (C + (D + Ex)x)x)x)x$$

which reduces the number of multiplications required from 15 to 5, as shown in **Fig 15.20**.

15.5.2 Tone Decoder

Tone decoders have a number of applications in Amateur Radio. A Morse code reader might use a tone decoder to determine the on and off states of the incoming CW signal. A sub-audible tone detector in a VHF FM receiver is another application. A DTMF decoder needs to detect two tones simultane-

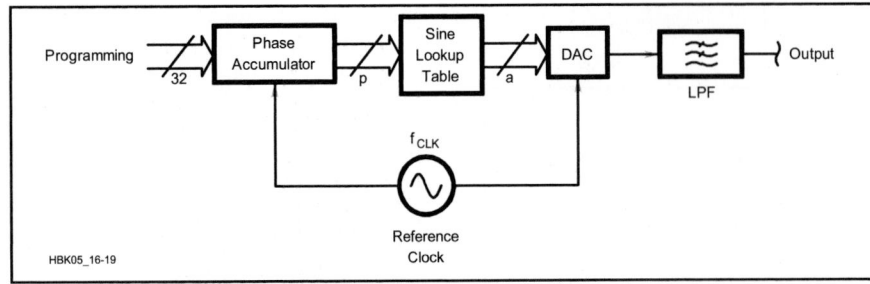

Fig 15.17 — DDS block diagram.

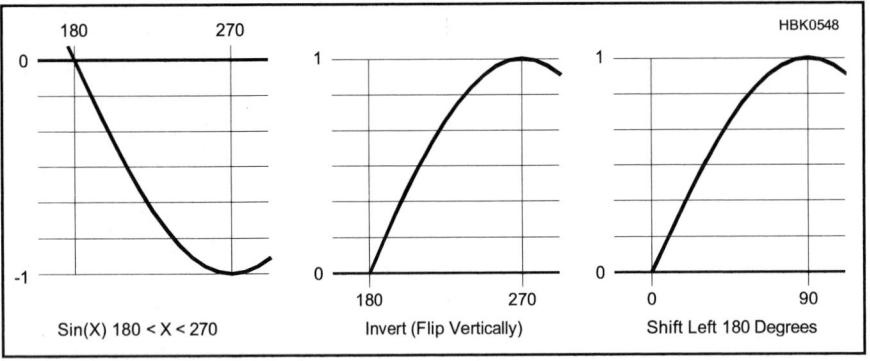

Fig 15.18 — The values of sin(x) between 180 and 270° are the same as those between 0 and 90°, after the curve has been flipped vertically and shifted 180°.

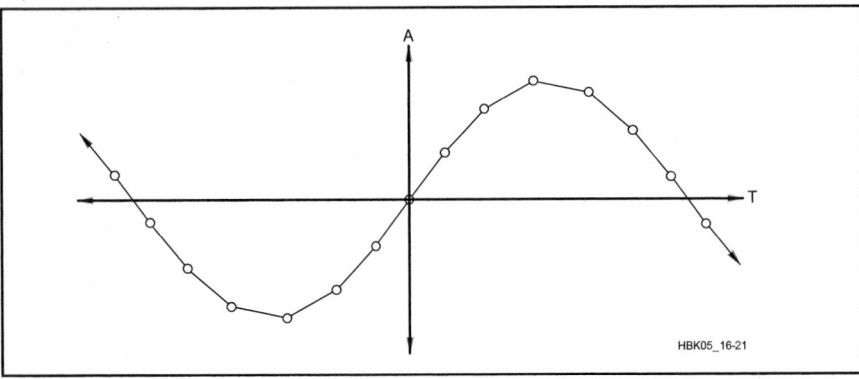

Fig 15.19 — An approximation of a sine wave using a straight-line interpolation between lookup table entries.

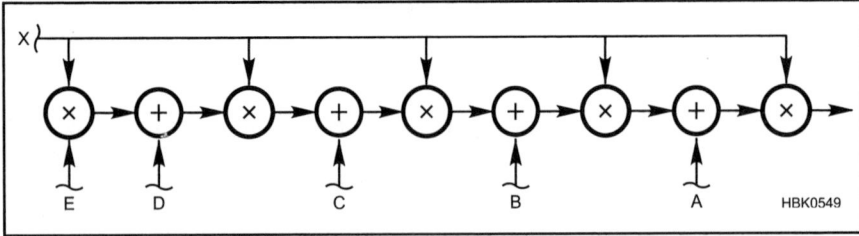

Fig 15.20 — Method to calculate a fifth-order interpolation between lookup table entries.

ously to determine which of the Touch-Tone keys has been pressed.

An FFT is one type of tone decoder. A 1024-point FFT simultaneously decodes 512 frequencies up to one-half the sample rate using a very efficient algorithm. You can control the detection bandwidth by choosing the number of points in the FFT; the spacing of the frequency samples is just the sample rate divided by the number of FFT points. However, in many applications you don't need that much resolution. For example, a DTMF signal consists of two tones, each of which can be one of four frequencies, for a total of 8 possible frequencies. Instead of performing a complete FFT, you can simply convolve sinusoidal waves of each of those 8 frequencies with the sequence of incoming samples to detect energy at those 8 frequencies. This will take less computation than a full FFT whenever the number of test frequencies is less than the base-2 logarithm of the number of points in the sequence. Since $\log_2(1024)$ is 10, decoding 8 frequencies separately would be more efficient than a 1024-point FFT. Spectral leakage is just as much of a problem with this method as with an FFT, so you still need to window the sequence of samples before performing the convolution.

If only a single frequency needs to be detected it might make more sense to mimic analog techniques and use a band-pass digital filter followed by an amplitude detector. That would have the advantage that the passband and stopband characteristics of the filter could be much more precisely controlled than with an FFT. In addition, the output is updated continuously instead of in "batch mode" after each batch of samples is collected and processed.

15.6 Analytic Signals and Modulation

In the area of modulation, the topic that seems to give people the most trouble is the concept of negative frequency. What in the world is meant by that? Consider a single-frequency signal oscillating at ω radians per second. (Recall that $\omega = 2\pi f$, where f is frequency in Hz.) Let's represent the signal by a cosine wave with a peak amplitude of 1.0, $x(t) = \cos(\omega t)$, where t is time. Changing the sign of the frequency is equivalent to running time backwards because $(-\omega)t = \omega(-t)$. By examining **Fig 15.21A** you can see that, because a cosine wave is symmetrical about the time $t = 0$ point, a negative frequency results in exactly the same signal. That is, as you may remember from high-school trigonometry, $\cos(-\omega t) = \cos(\omega t)$. If, for example, you add a positive-frequency cosine wave to its negative-frequency twin, you get the same signal with twice the amplitude.

That assumes that the phase of the signal is such that it reaches a peak at $t = 0$. What if instead we had a sine wave, which is zero at $t = 0$? From Fig 15.21B you can see that running time backwards results in a reversal of polarity, $\sin(-\omega t) = -\sin(\omega t)$. If you add positive and negative-frequency sine waves of the same frequency and amplitude, they cancel, resulting in zero net signal.

A sinusoidal wave of any arbitrary amplitude and phase may be represented by the weighted sum of a sine and cosine wave:

$$x(t) = I\cos(\omega t) + Q\sin(\omega t)$$

For computational purposes, it is convenient to consider the *in-phase* (I) and *quadrature* (Q) components separately. Since the I and Q components are 90° out of phase in the time domain, they are often plotted on a polar graph at a 90° angle from each other. See **Fig 15.22**. For example if $Q = 0$, then as time increases the signal oscillates along the I (horizontal) axis, tracing out the path back and forth between $I = +1$ and $I = -1$ in a sinusoidal fashion. Conversely, if $I = 0$, then the signal oscillates along the Q axis.

What if both I and Q are non-zero, for example $I = Q = 1$? Recall that the cosine and sine are 90° out of phase. When $t = 0$, $\cos(\omega t) = 1$ and $\sin(\omega t) = 0$. A quarter cycle later, $\cos(\omega t) = 0$ and $\sin(\omega t) = 1$. Comparing Fig 15.22 with Fig 15.21 it should not be hard to convince yourself that the signal is tracing out a circle in the counter-clockwise direction.

What about negative frequency? Again, it should not be hard to convince yourself that changing ω to $-\omega$ results in a signal that circles the origin in the clockwise direction. If you combine equal-amplitude signals of opposite frequency, the sine portions cancel out and you are left with a simple cosine wave of twice the amplitude:

$$x(t) = \left[\cos(\omega t) + \sin(\omega t)\right]$$
$$+ \left[\cos(-\omega t) + \sin(-\omega t)\right]$$
$$= 2\cos(\omega t)$$

You can see that graphically in **Fig 15.23**. Imagine the two vectors rotating in opposite directions. If you mentally add them by placing the tail of one vector on the head of the other, as shown by the dotted line, the result always lies on the I axis and oscillates between +2 and –2.

That is why we say that a single scalar sinusoidal signal, $\cos(\omega t)$, actually contains two frequencies, $+\omega$ and $-\omega$. It also offers a logical explanation of why a mixer or modulator produces the sum and difference of the frequencies of the two inputs. For example, an AM modulator produces sidebands at the carrier frequency plus and minus the modulating frequency precisely because those positive and negative frequencies are actually already present in the modulating signal.

For many purposes, it is useful to separate the portion of the signal that specifies the amplitude and phase (I and Q) from the oscillating part ($\sin(\omega t)$ and $\cos(\omega t)$). For mathematical convenience, the I/Q part is represented by a complex number, $x = I + jQ$. The oscillating part is also a complex number $e^{-j\omega t} = \cos(\omega t) - j\sin(\omega t)$. (Don't worry if you don't know where that equation comes from — concentrate on the part to the right of the equals sign.[4]) In the equations, $j = \sqrt{-1}$. Of course, –1 does not have a real square root (any real number multiplied by itself is positive) so j, or any real number multiplied by j, is called an *imaginary number*. A number with both real and imaginary parts is called a *complex number*. The total *analytic signal* is a complex number equal to

$$x(t) = xe^{-j\omega t} = (I + jQ)(\cos(\omega t) - j\sin(\omega t))$$

In the above equation, the $\cos(\omega t) - \sin(\omega t)$ portion generally represents an RF carrier, with ω being the carrier frequency (a positive or negative value). The $I + jQ$ part is the modulation. The *scalar* value of a modulated signal (what you would measure with an oscilloscope) is just the real part of the analytic signal. Using the fact that $j^2 = -1$,

$$\text{Re}[x(t)] = \text{Re}\left[(I + jQ)(\cos(\omega t) - j\sin(\omega t))\right]$$

$$\text{Re}[x(t)] = \text{Re}\begin{bmatrix}(I\cos(\omega t) + Q\sin(\omega t)) \\ +j(Q\cos(\omega t) - I\sin(\omega t))\end{bmatrix}$$

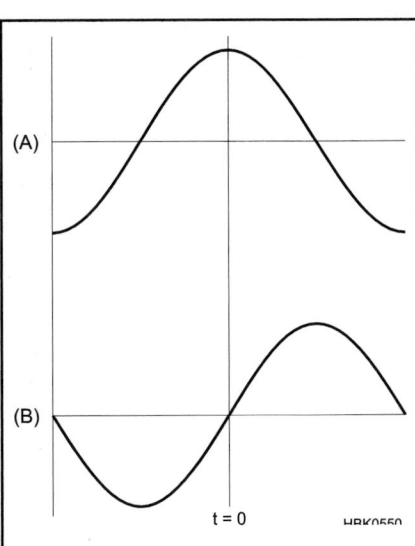

Fig 15.21 — Cosine wave (A) and sine wave (B).

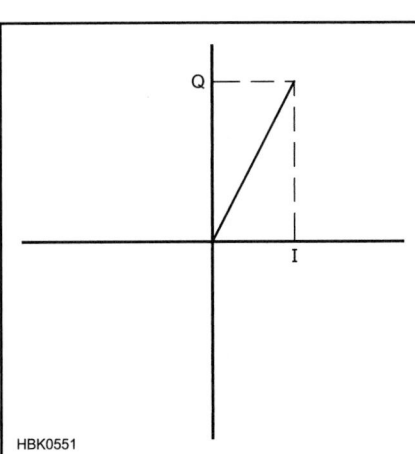

HBK0551

Fig 15.22 — In-phase (I) and quadrature (Q) portions of a signal.

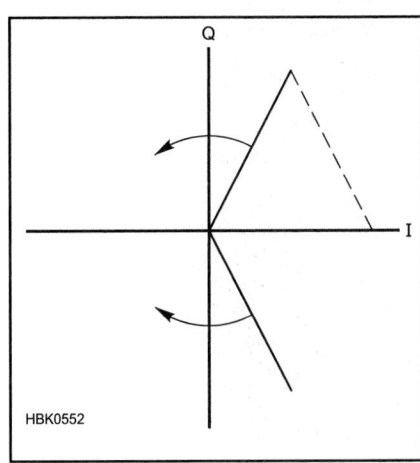

HBK0552

Fig 15.23 — A real frequency is the sum of a positive and negative analytic frequency.

$$\text{Re}[x(t)] = I\cos(\omega t) + Q\sin(\omega t)$$

Note that if the modulation (I and Q) varies with time, the above equation assumes that the modulated signal does not overlap zero Hz. That is, I and Q have no frequency components greater than ω.

Normally the I/Q diagram shows only I and Q (the modulation) and not the oscillating part. We call such a representation a *phasor diagram*. The I/Q vector represents the *difference* in phase and amplitude of the RF signal compared to the unmodulated carrier. For example, if the I/Q vector is at 90°, that means the carrier has been phase-shifted by 90° from what it otherwise would have been. If the I/Q vector is rotating counter-clockwise 10 times per second, then the carrier frequency has been increased by 10 Hz.

It is worth noting that the modulation can be specified either by the in-phase and quadrature (I and Q) values as shown or alternatively by the amplitude and phase. The amplitude is the length of the I/Q vector in the phasor diagram,

$$A = \sqrt{I^2 + Q^2}$$

The phase is the angle of the vector with respect to the +I axis,

$$\varphi = \arctan\left[\frac{Q}{I}\right]$$

An alternative expression for the modulated analytic signal using amplitude and phase is

$$x(t) = Ae^{-j(\varphi + \omega t)}$$
$$= A[\cos(\varphi + \omega t) + j\sin(\varphi + \omega t)]$$

and for the scalar signal

$$\text{Re}[x(t)] = A\cos(\varphi + \omega t)$$

One final comment. So far we have been looking at signals that consist of a single sinusoidal frequency. In any linear system, anything that is true for a single frequency is also true for a combination of many frequencies. Each frequency is affected by the system as though the others were not present. Since any complicated signal can be broken down into a (perhaps large) number of single-frequency sinusoids, all our previous conclusions apply to multi-frequency signals as well.

15.6.1 I/Q Modulation and Demodulation

An *I/Q modulator* is just a device that controls the amplitude and phase of an RF signal directly from the in-phase (I) and quadrature (Q) components. See **Fig 15.24A**. An *I/Q demodulator* is basically the same circuit in reverse. It puts out I and Q signals that represent the in-phase and quadrature components of the incoming RF signal. See Fig 15.24B. Assuming the demodulator's local oscillator is on the same frequency and is in phase with the carrier of the signal being received then the I/Q output of the receiver's demodulator is theoretically identical to the I/Q input at the transmitter end.

I/Q modulators and demodulators can be built with analog components. The LO could be a transistor oscillator and the 90° phase-shift network could be implemented with coils and capacitors. The circles with the multiplication symbol would be double-balanced mixers. Not shown in the diagram are trim adjustments to balance the amplitude between the I and Q channels and to adjust the phase shift as close as possible to 90°.

No analog circuit is perfect, however. If the 90° phase-shift network is not exactly 90° or the amplitudes of the I and Q channels are not perfectly balanced, you don't get perfect opposite-sideband rejection. The modulator output includes a little bit of signal on the unwanted sideband and the I/Q signal from the demodulator includes a small signal rotating in the wrong direction. If there is a small dc offset in the amplifiers feeding the modulator's I/Q inputs, that shows up as carrier feedthrough. On receive, a dc offset makes the demodulator think there is a small signal at a constant amplitude and phase angle that is always there even when no actual signal is being received. Nor is analog circuitry distortion-free, especially the mixers. Intermodulation distortion shows up as out-of-channel "splatter" on transmit and unwanted out-of channel responses on receive.

All those problems can be avoided by going digital. If the analog I/Q inputs to the modulator are converted to streams of digital numbers with a pair of ADCs, then the mixers, oscillator, phase-shift network and summer can all be digital. In many systems, the I and Q signals are also generated digitally, so that the digital output signal has perfect unwanted sideband rejection, no carrier feedthrough and no distortion within the dynamic range afforded by the number of bits in the data words. A similar argument holds for a digital demodulator. If the incoming RF signal is first digitized with an ADC, then the demodulation can be done digitally without any of the artifacts caused by imperfections in the analog circuitry.

You can think of an I/Q modulator as a device that converts the analytic signal I + *j*Q into a scalar signal at some RF frequency. The spectrum of the I/Q signal, both positive and negative frequencies, is translated upward in frequency so that it is centered on the carrier frequency. Thinking in terms of the phasor diagram, any components of the I/Q signal that are rotating counter-clockwise appear above the carrier frequency and clockwise components appear below.

15.6.2 SSB Using I/Q Modulators and Demodulators

As an example of how this works, let's walk through the process of generating an upper-sideband signal using an I/Q modulator. See **Fig 15.25**. We'll first describe the mathematics in the following paragraph and then give the equivalent explanation using the phasor diagram.

The modulating signal is a sine wave at a frequency of ω_m radians per second ($\omega_m / 2\pi$ cycles per second). Because ω_m is a positive frequency the signals applied to the I/Q inputs are $I(t) = \cos(\omega_m t)$ and $Q(t) = \sin(\omega_m t)$. Assume the modulating frequency ω_m is much less than the RF frequency ω. The analytic signal is

$$x(t) = [\cos(\omega_m t) + j\sin(\omega_m t)]$$
$$\times [\cos(\omega t) - j\sin(\omega t)]$$

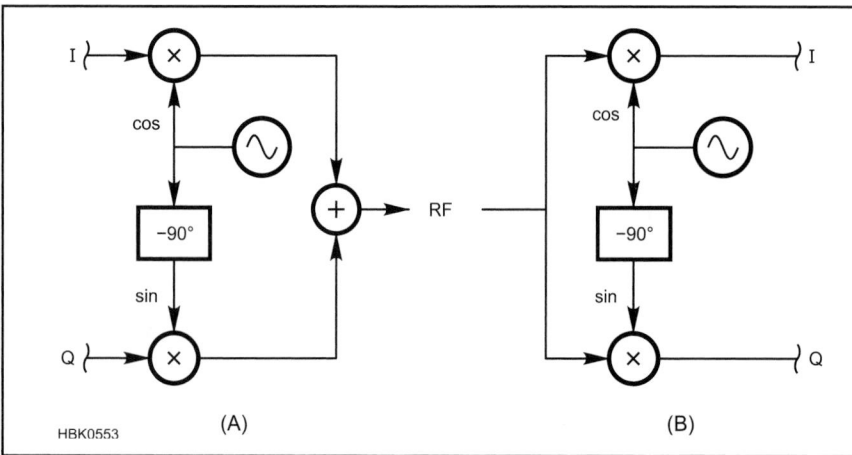

Fig 15.24 — I/Q modulator (A) and demodulator (B).

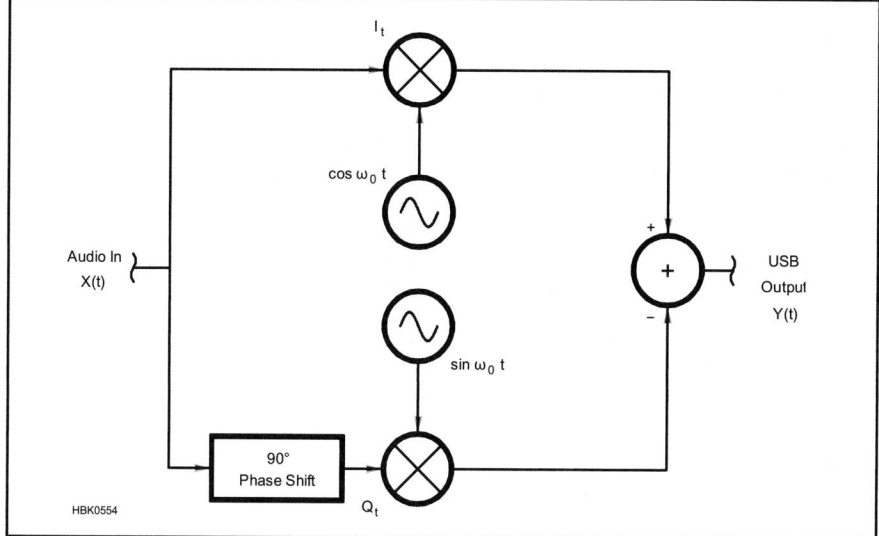

Fig 15.25 — Generating a USB signal with an I/Q modulator.

so that the real, scalar signal that appears at the modulator output is

$$\mathrm{Re}\big[x(t)\big] = \cos(\omega_m t)\cos(\omega t)$$
$$+ \sin(\omega_m t)\sin(\omega t)$$

At the moment when $t = 0$, then $\cos(\omega_m t)$ = 1 and $\sin(\omega_m t) = 0$, so the real signal is just $\cos(\omega t)$, the RF signal with zero phase. One quarter of a modulation cycle later $\omega_m t$ = $\pi/2$, so $\cos(\omega_m t) = 0$ and $\sin(\omega_m t) = 1$, and the real signal is now $\sin(\omega t)$, the RF signal with a phase of $+\pi/2$, or $+90°$. Every quarter cycle of the modulating signal, the RF phase, increases by $90°$. That means that the RF phase increases by one full cycle for every cycle of the modulation, which is another way of saying the frequency has shifted by ω_m. We have an upper sideband at a frequency of $\omega + \omega_m$.

On the phasor diagram, the I/Q signal is rotating counterclockwise at a frequency of ω_m radians per second. As it rotates it is increasing the phase of the RF signal at the same rate, which causes the frequency to increase by ω_m radians per second. To cause the phasor to rotate in the opposite direction, you could change the polarity of either I or Q or you could swap the I and Q inputs. In that case you would have a lower sideband.

For that to work, the baseband signals applied to the I and Q inputs must be $90°$ out of phase. That's not hard to do for a single sine wave, but to generate a voice SSB signal, all frequencies in the audio range must be simultaneously phase-shifted by $90°$ without changing their amplitudes. To do that with analog components requires a broadband phase-shift network consisting of an array of precision resistors and capacitors and a number of operational amplifiers.

THE HILBERT TRANSFORMER

To do that with DSP requires a *Hilbert transformer*, an FIR filter with a constant $90°$ phase shift at all frequencies. Recall that a symmetrical FIR filter has a constant *delay* at all frequencies. That means that the phase shift is not constant — it increases linearly with frequency. It turns out that with an *anti-symmetrical* filter, in which the top half of the coefficients are the negative of the mirror image of the lower half, the phase shift is $90°$ at all frequencies, which is exactly what we need to generate an SSB signal.

The Hilbert transformer is connected in series with either the I or Q input, depending on whether USB or LSB is desired. Just as with any FIR filter, a Hilbert transformer has a delay equal to half its length, so an equal delay must be included in the other I/Q channel as shown in **Fig 15.26**. It is possible to combine

the Hilbert transformer with the normal FIR filter that may be needed anyway to filter the baseband signal. The other I/Q channel then simply uses a similar filter with the same delay but without the $90°$ phase shift.

Because the RF output of the modulator is normally at a much higher frequency than the audio signal, it is customary to use a higher sample rate for the output signal than for the input. The FIR filters can still run at the lower rate to save processing time, and their output is then upsampled to a higher rate with an interpolator. It is convenient to use an output sample rate that is exactly four times the carrier frequency because each sample advances the RF phase by exactly $90°$. The sequence of values for the sine wave is 0, 1, 0 and -1. To generate the $90°$ phase shift for the cosine wave, simply start the sequence at the second sample: 1, 0, -1, 0. The complete block diagram is shown in **Fig 15.27**.

A Hilbert transformer may also be used in an SSB demodulator at the receiver end of the communications system. It is basically the same block diagram drawn backwards, as illustrated in **Fig 15.28**.

Amateurs who have been in the hobby for many years may recognize this as the "phasing method" of SSB generation. It was popular when SSB first became common on the amateur bands back in the 1950s because suitable crystal filters were expensive or difficult to obtain.[5] The phasing method had the reputation of producing signals with excellent-quality audio, no doubt due to the lack of the phase distortion caused by crystal filters.

It is important to note that an ideal Hilbert transformer is impossible to construct because it theoretically has an infinitely-long impulse response. However, with a sufficiently-long impulse response, the accuracy is much better than an analog phase-shift network. Just as with an analog network, the

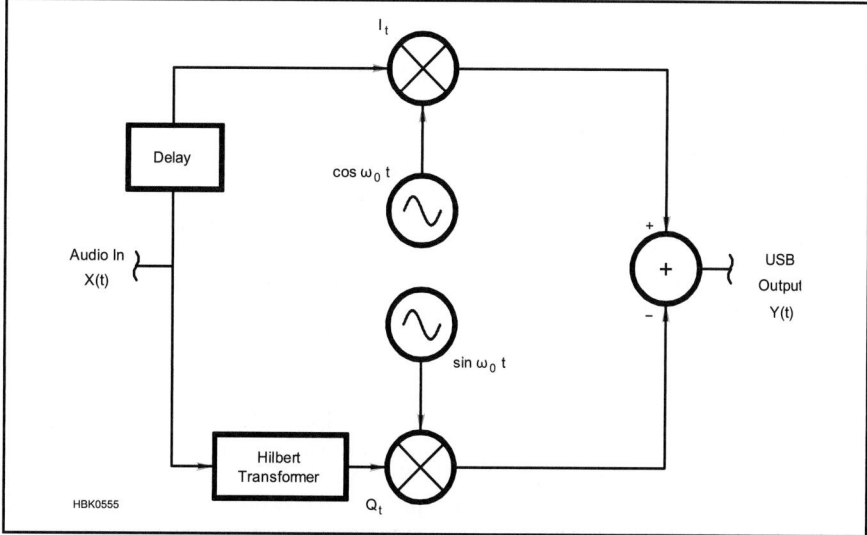

Fig 15.26 — Generating a non-sinusoidal USB signal with an I/Q modulator.

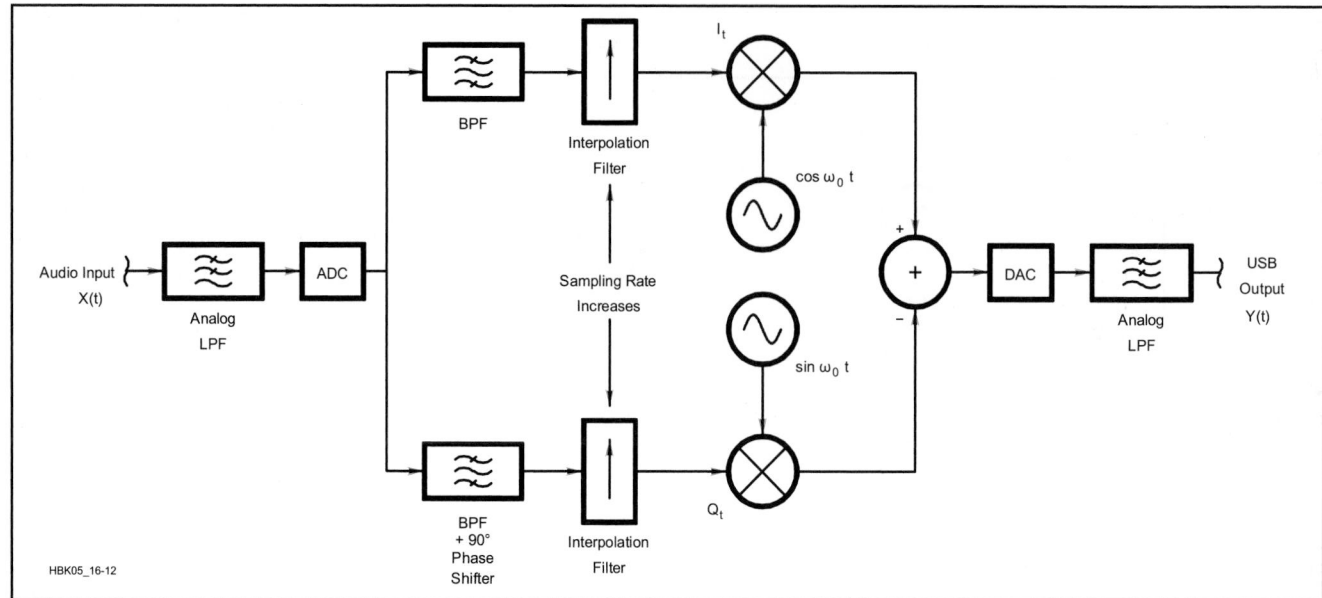

Fig 15.27 — Block diagram of a digital SSB modulator.

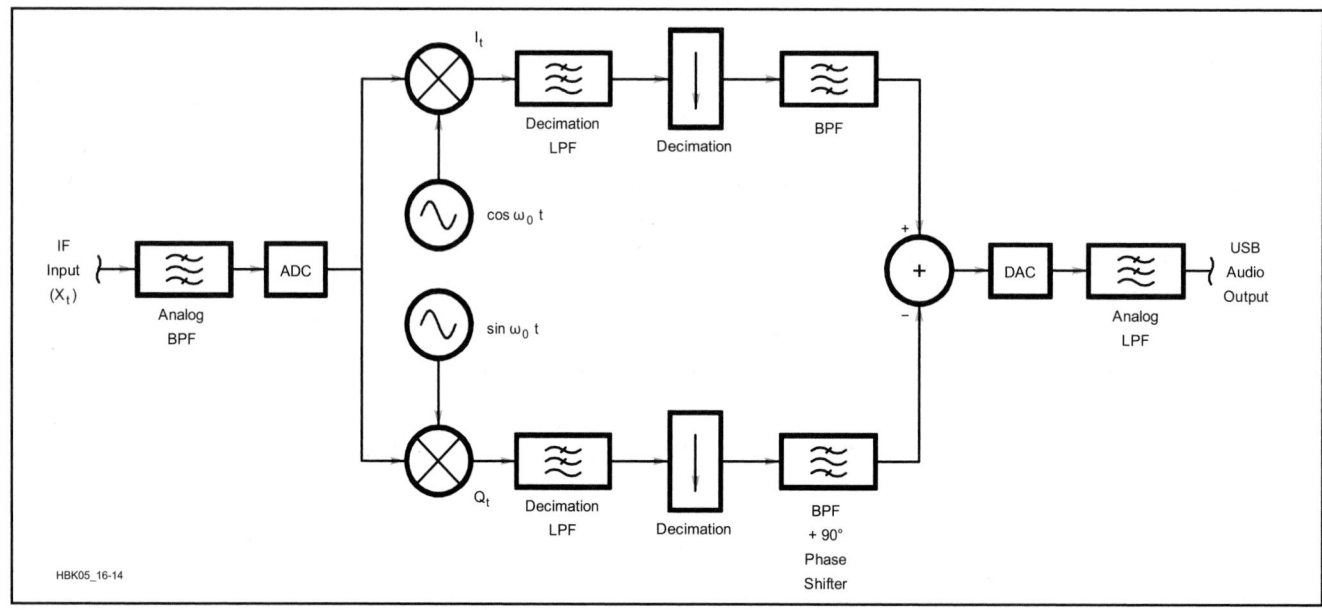

Fig 15.28 — Block diagram of a digital SSB demodulator.

frequency passband must be limited both at the low end as well as the high end. That is, the audio must be band-pass filtered before the 90° phase shift. Actually, the filtering and phase shifting can be combined into one operation using the following method.

First design a low-pass FIR filter with a bandwidth one-half the desired audio bandwidth. For example, if the desired passband is 300 to 2700 Hz, the low-pass filter bandwidth should be (2700 – 300)/2 = 1200 Hz. Then multiply the impulse response coefficients with a sine wave of a frequency equal to the center frequency of the desired passband, (2700 + 300)/2 = 1500 Hz in this case. That results in

a band-pass filter with the desired 300 – 2700 Hz response. By using sine waves 90° out of phase for the I and Q channels, you end up with two band-pass filters with the same amplitude response and delay but a 90° phase difference at all frequencies. Multiply by a cosine for zero phase and by a sine for a 90° phase shift.

Old timers may notice that this bears a striking resemblance to the Weaver method, the so-called "third method" of SSB generation, that was used back in the late 1950s to eliminate the need for a wide-band audio phase-shift network.[6,7] It is almost as if there is no such thing as truly new technology, just old ideas coming back with new terminology!

Analog modulators and demodulators using the phasing and Weaver methods are covered in the **Transmitters and Transceivers** and **Mixers, Modulators and Demodulators** chapters.

USES FOR I/Q MODULATORS AND DEMODULATORS

While I/Q modulators and demodulators can be used for analog modes such as SSB, they really shine when used with digital modulation modes. The **Modulation** chapter shows how the modulation states of the various digital formats map to positions in the phasor diagram, what is called a *constellation* *diagram.* The transmitter can generate the correct modulation states simply by placing the correct values on the I and Q inputs to the I/Q modulator. In the receiver, the filtered I and Q values are sampled at the symbol decision times to determine which modulation state they most closely match.

I/Q modulators and demodulators can also be used as so-called *imageless mixers.* A normal mixer with inputs at f_1 and f_2 produces outputs at f_1, f_2, $f_1 + f_2$, and $f_1 - f_2$. A balanced mixer eliminates the f_1 and f_2 terms but both the sum and difference terms remain, even though normally only one is desired. By feeding an RF instead of AF signal into the input of an SSB modulator, we can choose the sum or difference frequency in the same way as choosing the upper or lower sideband. If the input signal is a sine wave, the Hilbert transformer can be replaced by a simple 90° phase shifter. Similarly, a mixer with the same architecture as an SSB demodulator can be used to downconvert an RF signal to IF with zero image response. Analog imageless mixers are covered in the **Receivers** chapter. They are sometimes used in microwave receivers and transmitters where it is difficult to build filters narrow enough to reject the image response, but they typically only achieve image rejection in the 20-30 dB range. With a digital imageless mixer, the image rejection is "perfect" within the dynamic range of the bit resolution.

15.7 Software-Defined Radios (SDR)

There has been much, sometimes heated, discussion about the precise definition of a *software-defined radio* (SDR). Most feel that, at minimum, an SDR must implement in software at least some of the functions that are normally done in hardware. Others feel that a radio doesn't count as an SDR unless nearly all the signal-processing functions, from the input mixer to the audio output (for the receiver) and from the microphone ADC to the power amplifier input (for the transmitter), are done in software. Others add the requirement that the software must be reconfigurable by downloading new code, preferably open-source. For our purposes we will use a rather loose definition and consider any signal-processing function done in software to fall under the general category of SDR.

Some SDRs use a personal computer to do the computational heavy lifting and external hardware to convert the transmitted and received RF signals to lower-frequency signals that the computer's sound card can handle. Some SDRs avoid the use of the sound card by including their own audio codec and transferring the data to the PC via a USB port. Modern PCs provide a lot of computational power for the buck and are getting cheaper and more powerful all the time. They also come with a large color display, a keyboard for easy data entry, and a large memory and hard disk, which allows running logging programs and other software while simultaneously doing the signal processing required by the SDR.

Other SDRs look more like conventional analog radios with everything contained in one box, which makes for a neater, more compact installation. The signal processing is done with one or more embedded DSPs. For those who prefer a knob and button user interface, this is much preferred to having to use a mouse. Especially for contesting and competitive DXing, it is much faster to have a separate control for each critical function rather than having to select from pull-down menus. In addition SDRs of this type often have some performance advantages over PC-based SDRs, as we shall see.

Either method offers all the most-important advantages of applying DSP techniques to signal processing. The channel filter can have a much better *shape factor* (the ratio between the width of the passband and the frequency difference of the stopband edges). FIR filters are linear phase and have less ringing than analog filters of the same bandwidth and shape factor. Once the signal is in the digital domain all the fancy digital signal processing algorithms can be applied such as automatic notch filters, adaptive channel equalization, noise reduction, noise blanking, and feed-forward automatic gain control. Correcting bugs, improving performance or adding new features is as simple as downloading new software.

15.7.1 SDR Hardware

The transition between analog and digital signals can occur at any of several places in the signal chain between the antenna and the human interface. Back in 1992, Dave Hershberger W9GR designed an audio-frequency DSP filter based on the TMS320C10, one of the earliest practical DSP chips available.[8] This was an external unit that plugged into the headphone jack of a receiver and included FIR filters with various bandwidths, an automatic multi-frequency notch filter, and an adaptive noise filter. The advantage of doing the DSP at AF is that it can easily be added to an unmodified analog radio as in **Fig 15.29**. It is the technique used today to implement many digital modulation modes using the sound card of a PC connected to the audio input and output of a conventional transceiver.

A related technique is to downconvert a slice of the radio spectrum to baseband audio using a technique similar to the direct-conversion receivers popular with simple low-power CW transceivers. This idea was pioneered by Gerald Youngblood, AC5OG (now K5SDR), with the SDR-1000 transceiver, which he described in a series of *QEX* articles in 2002-2003.[9] The receiver block diagram is shown in Fig **15.30**. It uses a unique I/Q demodulator designed by Dan Tayloe, N7VE, to convert the RF frequency directly to baseband I and Q signals, which are fed to the stereo input of a PC's sound card, represented by the low-pass filters and A/D converters in the figure.[10] Software in the PC does all the signal processing and demodulation. The transmitter is the

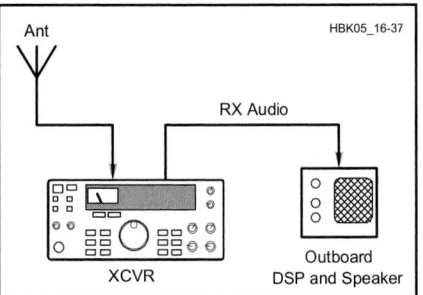

Fig 15.29 — An outboard DSP processor.

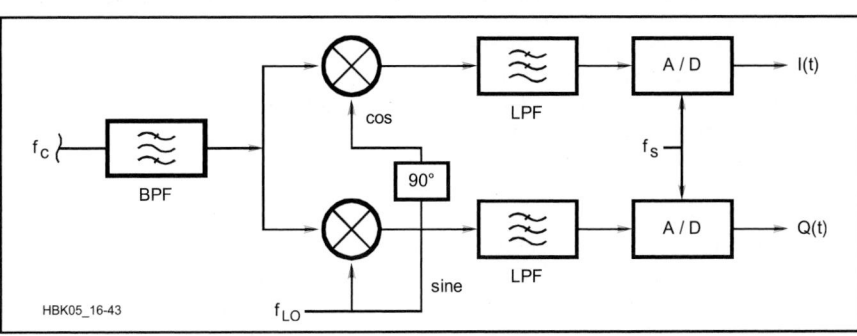

Fig 15.30 — Block diagram of K5SDR's direct-conversion software-defined receiver.

same block diagram in reverse, with an I/Q modulator converting an I/Q signal from the sound card up to the RF frequency where it is filtered and amplified to the final power level.

The sound card method manages to achieve reasonable performance with simple inexpensive hardware. Once the signal is digitized by the A/D converters in the sound card, the powerful DSP capability of the PC can do amazing things with it. The software for the SDR-1000 is open-source and available for free download on the Web.[11] In addition to implementing conventional transceiver functions such as several types of detector, variable-bandwidth filters, software AGC, an S-meter and speech compression, the software includes some extra goodies such as an automatic notch filter, noise reduction, and a panadapter spectrum display.

The simple hardware does impose some performance limitations. Because of imperfections in the analog downconverter, unwanted-sideband rejection is not perfect. This is called "image rejection" in the SDR-1000 literature. On the panadapter display, strong signals show up weakly on the opposite side of the display, equally-spaced from the center. Dc offset in the analog circuitry causes a spurious signal to appear at the center of the bandwidth. To prevent an unwanted tone from appearing in the audio output, the software demodulator is tuned slightly off frequency, but that means interference at the image frequency can cause problems because of the imperfect image rejection. The dynamic range depends on the sound card performance as well as the RF hardware. Some newer SDRs include an integrated audio codec optimized for the application so that the PC's sound card is not needed.

Among the integrated, one-box, software-defined radios, the most common place to perform the analog-digital transition is at an intermediate frequency. In the receiver, placing the ADC after a crystal IF filter improves the *blocking dynamic range* (BDR) for interfering signals that fall outside the crystal filter bandwidth. BDR is the ratio, expressed in dB, between the noise level (normally assuming a 500 Hz bandwidth) and an interfering signal strong enough to cause 1 dB gain reduction of the desired signal. With careful design, a receiver with such an architecture can achieve up to about 140 dB of BDR. The third-order dynamic range is similar to what can be achieved with a conventional analog architecture since the circuitry up to the crystal filter is the same.

Another advantage of the IF-based approach compared to sampling right at the final RF frequency is that the ADC does not have to run at such a high sample rate. In fact, because the crystal filter acts as a high-performance, narrow-bandwidth anti-aliasing filter, undersampling is possible. With bandwidths

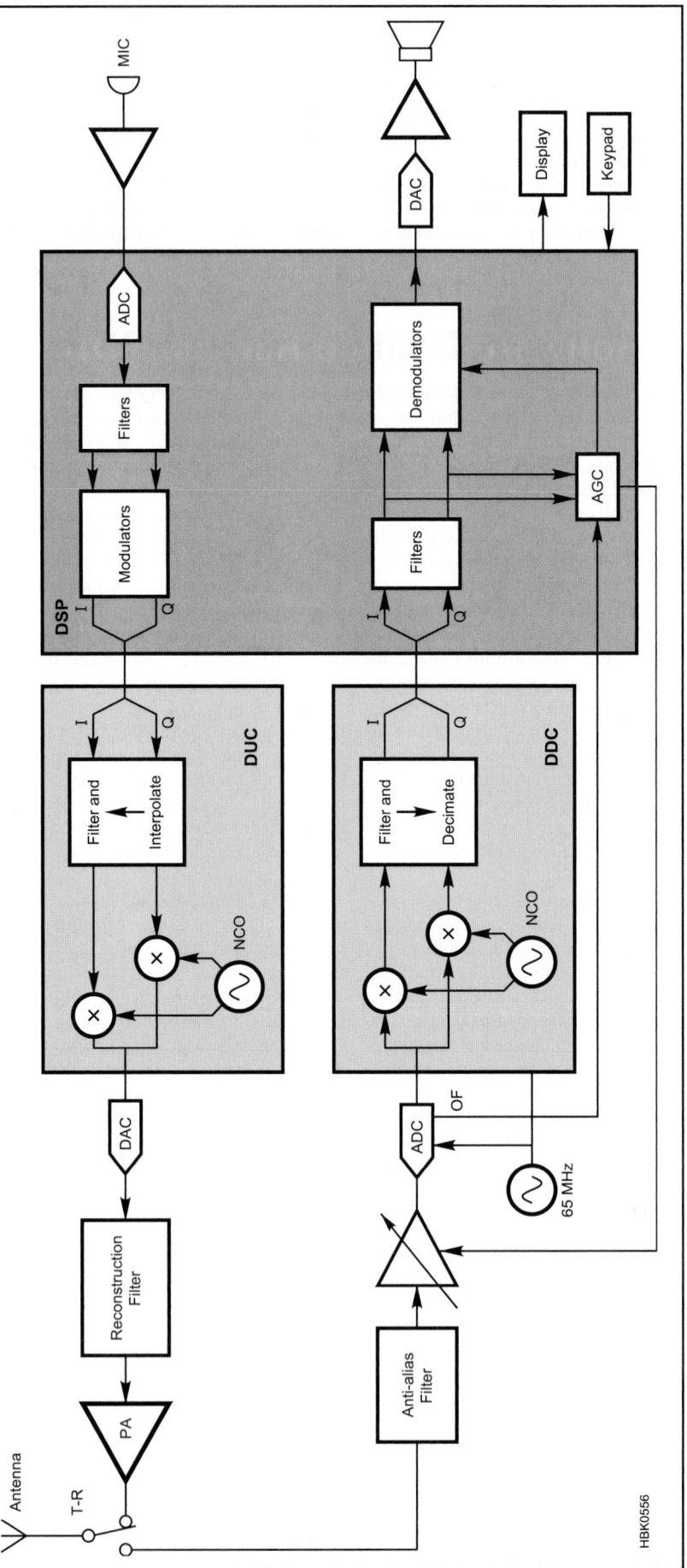

Fig 15.31 — An SDR transceiver that samples directly at the RF frequency.

of a few kHz or less, sample rates in the 10s of kHz can be used even though the center frequency of the IF signal is much higher, so long as the ADC's sample-and-hold circuit has sufficient bandwidth. Another common approach is to add a conventional analog mixer after the crystal filter to heterodyne the signal down to a second, much lower IF in the 10-20 kHz range, which is then sampled by an inexpensive, low-sample-rate ADC in the normal fashion. IF-based SDRs tend to have the highest overall performance at the expense of additional complexity.

SAMPLING AT RF

The ultimate SDR architecture is to transition between the analog and digital domains right at the frequency to be transmitted or received. In the receiver, the only remaining analog components in the signal chain are a wide-band anti-aliasing filter and an amplifier to improve the noise figure of the ADC. See **Fig 15.31**. The local oscillator, mixer, IF filters, AGC, demodulators and other circuitry are all replaced by digital hardware and software. It has only been fairly recently that low-cost high-speed ADCs have become available with specifications good enough to allow reasonable performance in a communications receiver. Today it is possible to achieve blocking dynamic range in the low 120s of dB. That is not as good as the best analog radios but is comparable to some medium-priced models currently available on the Amateur Radio market.

Third-order dynamic range is not a meaningful specification for this type of radio because it assumes that distortion products increase 3 dB for each 1 dB increase in signal level, which is not true for an ADC. The level of the distortion products in an ADC tends to be more-or-less independent of sig-

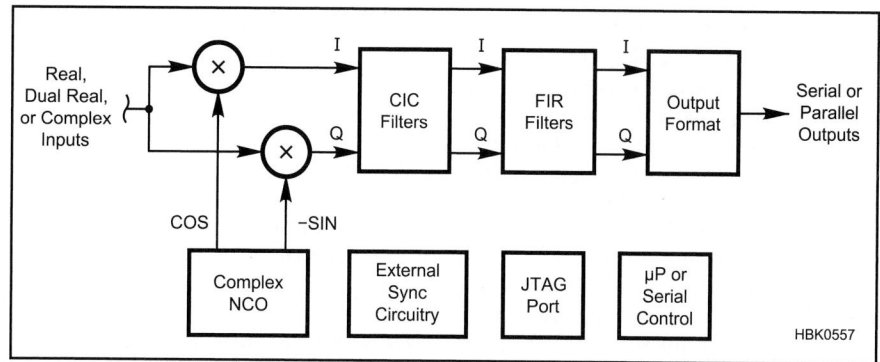

Fig 15.32 — The Analog Devices AD6620 is a digital downconverter (DDC) IC. The CIC filters and FIR filter are all decimating types.

nal level until the signal peak exceeds full scale, at which point the distortion spikes up dramatically. Compared to a conventional analog mixer, ADCs tend to give very good results with a two-tone test but don't do as well when simultaneously handling a large number of signals, which results in a high peak-to-average ratio. It is important to read the data sheet carefully and note the test conditions for the distortion measurements.

There are definite advantages to sampling at RF. For one thing, it saves a lot of analog circuitry. Even if the ADC is fairly expensive the radio may be end up being cheaper because of the reduced component count. Performance is improved in some areas. For example, image rejection is no longer a worry, as long as the anti-aliasing filter is doing its job. The dynamic range theoretically does not depend on signal spacing — close-in dynamic range is often better than with a conventional architecture that uses a wide roofing filter. With no crystal filters in the signal chain, the entire system is completely linear-phase which can improve the quality of both analog

and digital signals after demodulation.

The biggest challenge with RF sampling is what to do with the torrent of high-speed data coming out of the receiver ADC and how to generate transmit data fast enough to keep up with the DAC. To cover the 0-30 MHz HF range without aliasing requires a sample rate of at least 65 or 70 MHz. That is much faster than a typical microprocessor-type DSP can handle. The local oscillator, mixer and decimator or interpolator must be implemented in digital hardware so that the DSP can send and receive data at a more-reasonable sample rate. Analog Devices makes a series of *digital downconverters* (DDC) which perform those functions and output a lower-sample-rate digital I/Q signal to the DSP.[12] See **Fig 15.32**. It would also be possible to implement your own DDC in an FPGA. The same company also makes *digital upconverters* (DUC) that do the same conversion in reverse for the transmitter. Some of their DUCs even include the capability to encode several digital modulation formats such as GMSK, QPSK and π/4 DQPSK.

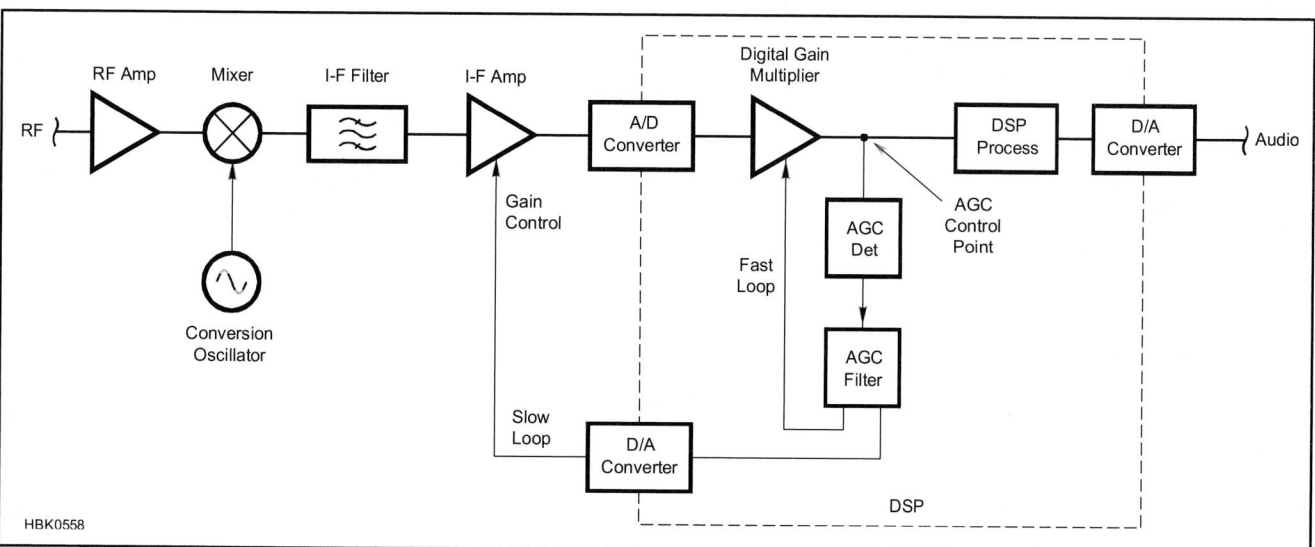

Fig 15.33 — DSP-based feedback type of AGC showing a combination of analog and digital gain-control points.

DIGITAL AGC

In the transmitter portion of a software-defined radio, dynamic range is generally not a problem because the transmitted signal always has approximately the same power level. Even if power control is implemented digitally, a 1-100 W adjustment range only adds 20 dB to the dynamic range. The receiver is another story. Assuming 6 dB per S-unit, the difference between S1 and 60 dB over S9 is 108 dB. Considering peak power rather than average, the actual range is much greater than that. While AGC implemented in software can be quite effective in regulating the signal level at the speaker output, it does nothing to prevent the ADC from being overloaded on signal peaks. For that, some kind of hardware AGC is needed in the signal path ahead of the A/D converter. This device could be a switched attenuator or variable-gain amplifier as illustrated in **Fig 15.33**. It normally runs at full gain and only attenuates the incoming signal when very strong peak signal levels are encountered. It could be totally self-contained with its own threshold detector or it can be controlled as shown by the DSP, which activates the hardware AGC whenever it detects ADC overflow.

One issue with most AGC systems is response time. The level detector is normally placed after the gain-control stage. Because of delays in the feedback loop, by the time the AGC circuit detects an over-range condition it is already too late to reduce the gain without overshoot. One of the advantages of digital AGC is that it is easy to use feed-forward rather than feed-back control. In **Fig 15.34**, the gain control stage is placed after the level detector. A small delay is included in the signal path so that the AGC circuit can reduce the gain just before the large signal arrives at the gain multiplier. With proper design, that totally eliminates overshoot and makes for a very smooth-operating AGC.

THE LOCAL OSCILLATOR

With direct-RF sampling, the digital local oscillator is normally implemented with a direct digital synthesizer, operating totally in digital hardware. DDS operation was explained previously in the sine-wave generation section. A separate DDS chip with a built-in DAC is sometimes used in IF-sampled SDRs as well as in some analog radios. One advantage that a DDS oscillator has over a phase-locked loop (PLL) synthesizer is very fast frequency changing. That can be important in transceivers that use the same local oscillator for both the receiver and transmitter. If the transmitter and receiver are tuned to different frequencies, each time the rig is keyed the LO frequency must settle at its new value before a signal is transmitted.

The phase noise of the DDS clock is just as important as the phase noise of the local oscillator in a conventional radio. Phase noise shows up as broadband noise that gradually diminishes the farther you get from the oscillator frequency. In a receiver, phase noise causes a phenomenon called *reciprocal*

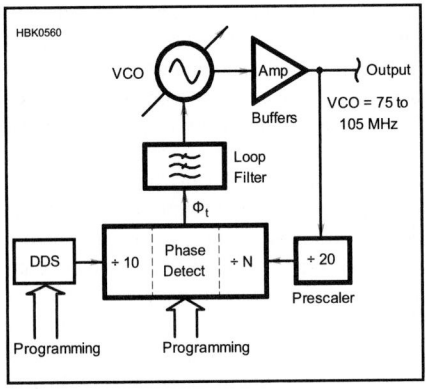

Fig 15.35 — A hybrid DDS/PLL synthesizer.

mixing, in which a strong off-channel signal mixes with off-channel phase noise to cause a noise-modulated spurious signal to appear in the receiver passband. In many receivers, dynamic range measurements are phase-noise-limited because the spurious response due to reciprocal mixing is louder than the distortion products. One way to reduce the phase noise from the DDS is to use a conventional PLL synthesizer to generate a signal with large frequency steps and combine it with a DDS synthesizer to obtain the fine-grained frequency resolution, as suggested in **Fig 15.35**. In this way, you get the phase noise of the DDS and PLL within the loop bandwidth of the PLL and the phase noise of the VCO outside that bandwidth.

One advantage a PLL has over a DDS oscillator is lower spurious signal levels. A DDS with a wideband spurious-free dynamic range (SFDR) specification of 60 dB would be better than most, but that could cause spurious responses in the receiver only 60 dB down. The hybrid PLL/DDS technique can suppress these spurs as well.

15.7.2 SDR Software

When designing DSP software, it is sometimes surprising how much of your intuition about analog circuits and systems transfers directly over to the field of digital signal processing. The main difference is that you need to forget much of what you have learned about the imperfections of analog circuitry. For example, a multiplier is the DSP equivalent of an ideal double-balanced mixer. The multiplier output contains frequencies only at the sum and difference of the two input signals. There is no intermodulation distortion to create spurious frequencies.

Multiplication by a constant is equivalent to an amplifier or attenuator, but with no dc offset and with a very precisely-set gain that

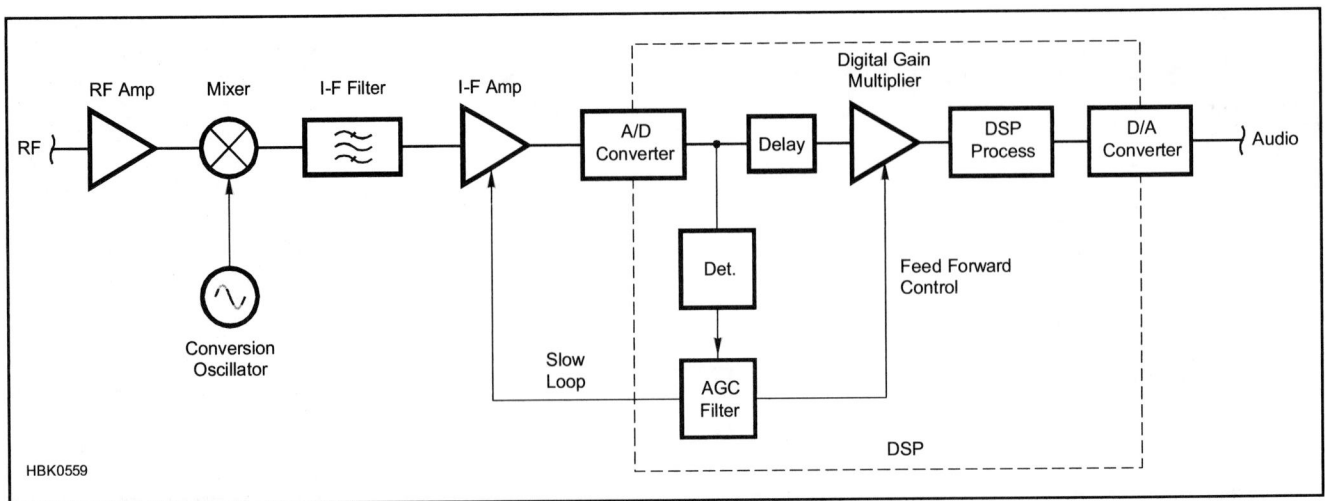

Fig 15.34 — DSP-based AGC with analog feedback and digital feed forward control.

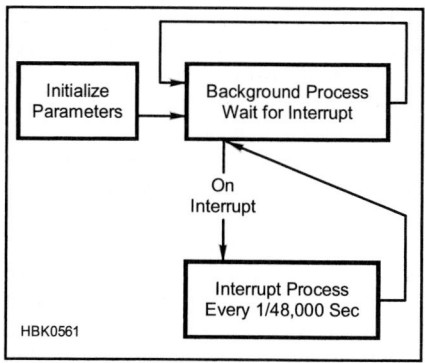

Fig 15.36 — Main program flow of a typical DSP program.

does not drift with time or temperature. A distortionless AM "diode" detector is just a software routine that forces the signal to zero whenever it is negative. To build an ideal full-wave diode detector, just take the absolute value of the signal.

When you design an analog circuit you have to take into account all the things that don't appear on the schematic. For example your crystal filter may show a beautiful frequency response in your filter design program, but in the actual circuit the passband is skewed because of the input impedance of the post-amplifier and the stopband response is degraded by signals leaking around the filter due to the PC board layout. With software, what you see in the simulation is what you get (assuming you did the calculations correctly!)

SOFTWARE ARCHITECTURE

Incoming data to the DSP from A/D converters or other devices is normally handled in an *interrupt service routine* (ISR) that is called automatically whenever new data is ready. The AM and AGC detection code in Table 15.5 could be included right in the ISR, but it is almost always better to limit the ISR function to the minimum necessary to service

the hardware that calls the interrupt and do all the signal processing elsewhere. For example, **Table 15.4** shows an ISR that inputs 16-bit I and Q data coming in on the dsPIC serial data communications interface (DCI). The first line is a secret incantation that defines this function to be the interrupt service routine for the DCI interface.

As you can see, there is minimal functionality in the ISR itself. All the heavy computational lifting is done in processes that run in the background and are interrupted periodically when new data is available. The data_flag variable is a semaphore to signal the signal-processing routine that new data is ready. **Fig 15.36** illustrates the basic program architecture of some DSP projects for the Analog Devices EZ-Kit Lite DSP development board described in the ARRL publication *Experimental Methods in RF Design*. That book, by the way, is an excellent source for practical "how-to" information on designing DSP projects.

SOFTWARE MODULATORS AND DEMODULATORS

In this chapter we've already covered many of the algorithms needed for a software-defined radio. For example, we know how to make I/Q modulators and demodulators and use them to build an SSB modulator and detector. Let's say we want our software-defined transceiver to operate on AM voice as well. How do you make an AM modulator and demodulator?

The modulator is easy. Simply add a constant value, representing the carrier, to the audio signal and multiply the result by a sine wave at the carrier frequency, as shown in **Fig 15.37**.

Demodulation is almost as easy. We could just simulate a full-wave rectifier by taking the absolute value of the signal, as mentioned previously, and low-pass filter the result to remove the RF energy. If the signal to be demodulated is complex, with I and Q com-

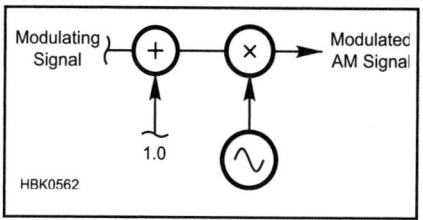

Fig 15.37 — A digital AM modulator.

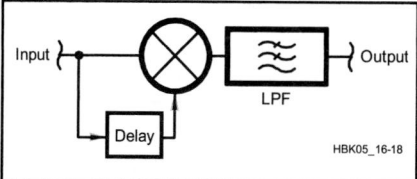

Fig 15.38 — A digital quadrature detector.

ponents, then instead of absolute value we take the magnitude

$$A = \sqrt{I^2 + Q^2}$$

The dc bias can be removed by adding a "series blocking capacitor" — a high-pass filter with a suitable cut-off frequency.

A little more elegant way to do it would be to include the AM detector as part of the AGC loop. In the C code snippet shown in **Table 15.5**, the variable "carrier" is the average AM carrier level. It is passed to another subroutine to control the gain.

Note that no "series capacitor" is needed since the audio signal is computed by subtracting the average historical value, carrier, from the magnitude of the current I/Q signal, am. A small fraction of its value is added to the historical value so that the AGC tracks the average AM carrier level. AGC speed is controlled by that fraction. Dividing by $2^{10} = 1024$ gives a time constant of about 1024 clock cycles.

Another type of detector we haven't discussed yet is for frequency modulation. For a scalar signal, the *quadrature detector* shown in **Fig 15.38** is one elegant solution. This is the same circuit whose analog equivalent is used today in millions of FM receivers around the world. In the digital implementation, the delay block is a FIFO buffer constructed from a series of shift registers. Multiplying the signal by a delayed version of itself gives an output with a cosinusoidal response versus frequency. The response crosses zero whenever the carrier

---

Table 15.4
Interrupt Service Routine

```
void __attribute__((auto_psv, interrupt)) _DCIInterrupt(void)
{
extern int i, q, data_flag;

// Clear interrupt flag:
IFS3bits.DCIIF = 0;
// Input data is 2's complement:
i = RXBUF0;
q = RXBUF1;
data_flag = 1;
return;
}
```

Table 15.5
AM Detector

```
static long int carrier;
long int am;
int i, q, signal;

/* Code that generates i and q omitted */
am = (long int)sqrt((long int)i*i + (long int)q*q);
signal = am - carrier;
// Divide signal by 2^10:
carrier += signal >> 10;
// Audio output to DAC via SPI bus:
SPI1BUF = signal;
```

frequency is $f = N/(4\tau)$, where N is an odd integer and the delay in seconds is $\tau = n/f_s$, where n is the number of samples of delay and f_s is the sample frquency. As the carrier deviates above and below the zero-crossing frequency the output varies above and below zero, just what we want for an FM detector.

For an I/Q signal, probably the most straightforward FM detector is a phase detector followed by a differentiator to remove the 6 dB per octave rolloff caused by the phase detector. The phase is just

$$\varphi = \arctan\left(\frac{Q}{I}\right)$$

You have to be a little careful since there is a 180° phase ambiguity in the arctangent function. For example,

$$\arctan\left(\frac{1}{1}\right) = \arctan\left(\frac{-1}{-1}\right)$$

Software will have to check which quadrant of the phasor diagram the I/Q signal is in and add 180° when necessary. If there is no arctan function in the library, one can be constructed using a look-up table. Frequency is the derivative of the phase. A differentiator is nothing more than a subtractor that takes the difference between successive samples.

$$f = \frac{\varphi_n - \varphi_{n-1}}{2\pi f_s}$$

where n is the sample number and f_s is the sample rate. It is important to make sure that the difference equation functions properly around 360°. If the phase variable is scaled so that 360° equals the difference between the highest and lowest representable numbers, then standard two's complement subtraction should roll over to the right value at the 360° / 0° transition. Another thing to watch out for is that the derivative of the phase may be a rather small signal, so it might be necessary to carry through all the calculations using long integers or floating point numbers.

OTHER SOFTWARE FUNCTIONS

A *carrier-locked loop* is a circuit that automatically tunes a receiver so that it is centered on the carrier of the incoming signal. One way to achieve that is to make the receiver local oscillator controllable by a frequency detector. For example if the local oscillator (LO) in the receiver were an analog voltage-controlled oscillator (VCO), the output of the FM detector described above could be applied to a DAC that generates an error voltage to pull the VCO on frequency. If the LO were an NCO or other digitally-controlled synthesizer, then the error signal could be used to control the frequency digitally. An even more elegant way to do it is to leave the LO alone and tune the frequency of the I/Q signal directly. Conceptually, you determine the amplitude

$$A = \sqrt{I^2 + Q^2}$$

and phase

$$\varphi = \arctan\left(\frac{Q}{I}\right)$$

of the signal, add or subtract the phase error from ϕ to keep its average value from changing and then convert back to $I = A\cos(\phi)$ and $Q = A\sin(\phi)$ again for further processing. This is easy to do with a signal that does not change phase, such as AM phone. For an FM or PM signal, considerable averaging must be done of the error signal so that it represents the average phase of the carrier rather than the instantaneous phase of the modulation. Speech processing is a function that lends itself well to digital signal processing. The human voice has a high peak-to-average power ratio, typically on the order 15 dB. That means, that without processing, a 100-W PEP SSB transmitter is only putting out about 3 W average! Most SSB transmitters do have an *automatic level control* (ALC) circuit that can reduce the peak-to-average ratio by 3-6 dB, but that still means your 100-W transmitter is only putting out 6-12 W on average.

The problem is that if the ALC setting is too aggressive, considerable distortion of the audio can result. A transmitter's ALC circuit operates much like the AGC in a receiver. It can do a fair job of keeping the peak power from overdriving the amplifier but it can do little to reduce the short-term power variations between speech syllables. With digital processing, it is fairly easy to use feed forward gain control rather than feed back, in a manner similar to the AGC system illustrated in Fig 15.34.

The gold standard of speech processing of SSB signals is RF clipping. By clipping at radio frequency instead of at audio, many of the distortion products fall outside the passband where they can be filtered out, by a crystal filter in an analog radio or by a digital band-pass filter in an SDR. "RF" clipping doesn't actually have to be done at a high RF frequency. An IF of a few kHz is sufficient, so long as the center frequency is greater than about twice the audio bandwidth. That can all be done in software and then the signal can be converted back to baseband audio if desired.

Developing DSP software is a wonderful homebrew activity for the Radio Amateur. As electronic devices have become smaller and smaller and more and more sophisticated it has become harder and harder to get a soldering iron on the tiny pins of surface-mount ICs. Software development allows hobbyists to experiment to their heart's content with no danger that an expensive piece of electronics will be destroyed by one moment of clumsiness. With nothing more than a PC and some free software, the enthusiast can while away hours exploring the fascinating world of digital signal processing and software radios.

15.8 Glossary

Adaptive filter — A filter whose coefficients can be changed automatically.

Analog-to-digital converter (ADC) — A device that samples an analog signal and outputs a digital number representing the amplitude of the signal.

Analytic signal — A representation of the phase and amplitude of a signal (often in the form of the in-phase and quadrature components), without explicitly including the oscillating part (the carrier).

Anti-aliasing filter — A band-limiting filter placed before a sampler to make sure the incoming signal obeys the Nyquist criterion.

Anti-symmetrical — A function that is anti-symmetrical about point x=0 has the property that $f(x) = -f(-x)$.

Application-specific integrated circuit (ASIC) — A non-programmable IC that is designed for a particular application.

Arithmetic logic unit (ALU) — The portion of a microprocessor that performs basic arithmetic and logical operations.

Automatic gain control (AGC) — The circuit in a receiver that keeps the signal level approximately constant.

Automatic level control (ALC) — The circuit in a transmitter that keeps the peak signal level approximately constant.

Barrel shifter — A circuit in a microprocessor that can bit-shift a number by multiple bits at one time.

Baseband — The low-frequency portion of a signal. This is typically the modulation.

Bilinear transform — A design technique for IIR filters in which the frequency axis is transformed to prevent the filter bandwidth from violating the Nyquist criterion.

Binary point — The symbol that separates the integer part from the fractional part of a binary number.

Blocking dynamic range (BDR) — The difference between the noise level (usually in a 500-Hz bandwidth) and the signal level that causes a 1 dB reduction in the level of a weaker signal.

Carrier-locked loop — A feedback control loop to automatically tune a receiver or demodulator to the frequency of a received carrier.

Chebyshev filter — A filter with equal ripple in the passband, stopband or both.

Circular buffers — A buffer in which the final entry is considered to be adjacent to the first.

Cognitive radio — A radio system in which a wireless node automatically changes its transmission or reception parameters to avoid interference with other nodes.

Complex number — A number that contains real and imaginary parts.

Complex PLD (CPLD) — A programmable logic device that is more complex than a small PLD, such as a PAL, but with a similar architecture.

Constellation diagram — A phasor diagram showing the locations of all the modulation states of a digital modulation type.

Convolution — A mathematical operation that modifies a sequence of numbers with another sequence of numbers so as to produce a third sequence with a different frequency spectrum or other desired characteristic. An FIR filter is a convolution engine.

Cooley-Tukey algorithm — Another name for the fast Fourier transform.

Decimation — Reduction of sample rate by an integer factor.

Decimation in time — The division of a sequence of numbers into successively smaller sub-sequences in order to facilitate calculations such as the Fourier transform.

Digital downconverter (DDC) — A device that translates a band of frequencies to baseband, typically at a lower sample rate.

Digital signal processing — The processing of sequences of digital numbers that represent signals.

Digital signal processor (DSP) — A device to do digital signal processing. The term normally is understood to refer to a microprocessor-type device with special capabilities for signal processing.

Digital-to-analog converter (DAC) — A device that converts digital numbers to an analog signal with an amplitude proportional to the digital numbers.

Digital upconverters (DUC) — A device that frequency-translates a baseband signal to a higher frequency, typically at a higher sample rate.

Direct digital synthesis (DDS) — The generation of a periodic waveform by directly calculating the values of the waveform samples.

Direct form — A circuit topology for an IIR filter that corresponds directly to the standard filter equation.

Direct memory access (DMA) — The ability of a microprocessor chip to transfer data between memory and some other device without the necessity to execute any processor instructions.

Dithering — Randomly varying the amplitude or phase of a signal in order to overcome quantization effects.

Embedded system — A system that includes a microprocessor for purposes other than general-purpose computing.

Equal-ripple filter — A filter in which the variation in passband or stopband response is constant across the band.

Exponent — The number of digits that the radix point must be moved to represent a number.

Fast Fourier transform (FFT) — An algorithm that can calculate the discrete Fourier transform with an execution time proportional to nlog(n), instead of n^2 as is required by the straight-forward application of the Fourier transform equation.

Field-programmable gate array (FPGA) — An IC that contains a large array of complex logic blocks whose function and connections can be re-programmed in the field.

Filter coefficient — One of a series of numbers that define the transfer function of a filter.

Finite impulse response (FIR) — An impulse response that is zero for all time that is greater than some finite amount from the time of the impulse.

Floating-point — Refers to a number whose value is represented by a mantissa and an exponent.

Fourier transform — A mathematical operation that derives the frequency spectrum of a time-domain signal.

Hardware-description languages (HDL) — A computer language to specify the circuitry of a digital device or system.

Harmonic sampling — The use of a sample rate that is less than twice the highest frequency of the signal to be sampled. The sample rate must be greater than two times the bandwidth of the signal.

Harvard architecture — A computer architecture in which the program and data are stored in separate memories.

Hilbert transformer — A filter that creates a constant 90° phase difference over a band of frequencies.

Imageless mixer — A mixer in which the output contains only the sum or difference of the two input frequencies, but not both.

Imaginary number — A real number multiplied by the square root of minus one.

Impulse — A pulse of finite energy with a width that approaches zero.

Impulse-invariant — A design technique for IIR filters in which the impulse response is the same as the impulse response of a certain analog filter.

Impulse response — The response versus time of a filter to an impulse.

In-circuit emulator (ICE) — A device that emulates the operation of a microprocessor while providing debugging tools to the operator. The ICE normally plugs into an IC socket that normally holds the microprocessor.

In-circuit debugger (ICD) — A device that uses debugging features built into the microprocessor so that it can be tested while in the circuit.

In-circuit programmable (ICP) — A programmable IC that can be programmed while it is connected to the application circuit.

In-circuit programmer (ICP) — A device to facilitate programming of programmable ICs while they are connected to the application circuit.

In-phase (I) — The portion of a radio signal that is in phase with a reference carrier.

Infinite impulse response (IIR) — An impulse response that theoretically never goes to zero and stays there.

Integrated development environment (IDE) — An integrated collection of software and hardware tools for developing a microprocessor project.

I/Q demodulator — A device to derive the in-phase and quadrature portions of an oscillating signal.

I/Q modulator — A device to generate an oscillating signal with specified in-phase and quadrature parts.

Interpolation — Increasing the sample rate by an integer factor.

Interrupt service routine (ISR) — A software subroutine that is called automatically when the main routine is interrupted by some event.

Least mean squares (LMS) — An algorithm for adaptive filters that minimizes the mean square error of a signal.

Least-significant bit (LSB) — When used as a measurement unit, the size of the smallest step of a digital number.

Limit cycle — A non-linear oscillation in an IIR filter.

Linear phase — Refers to a system in which the delay is constant at all frequencies, which means that the phase is linear with frequency.

Mantissa — The decimal or binary part of a logarithm or floating-point number.

Multiplier-accumulator (MAC) — A device that can multiply two numbers and add the result to a previous result all in one operation.

Multi-rate — Refers to a system with more than one sample rate.

Numerically-controlled oscillator (NCO) — An oscillator that synthesizes the output frequency from a fixed timebase. A DDS oscillator.

Nyquist criterion — The requirement that the sample rate must be at least twice the bandwidth of the signal.

Nyquist frequency — One half the sample rate.

Nyquist rate — Twice the signal bandwidth.

One-time programmable (OTP) — A programmable device that may not be re-programmed.

Orthogonal — Perpendicular. In analogy with the mathematics of perpendicular geometrical vectors, the term is used in communications to refer to two signals that produce zero when convolved.

Oversampling — Use of a sample rate higher than required by the Nyquist criterion in order to improve the signal-to-noise ratio.

Parks-McClellan algorithm — An optimized design technique for equal-ripple filters.

Phasor diagram — A polar plot of the magnitude of the in-phase and quadrature components of a signal.

Pipeline — An arrangement of computational units in a microprocessor or other digital device so that different units can be working on different instructions or signal samples at the same time.

Programmable-array logic (PAL) — A type of small PLD that consists of an array of AND gates, OR gates, inverters and latches.

Programmable-logic device (PLD) — A device with many logic elements whose connections are not defined at manufacturer but must be programmed.

Quadrature (Q) — The portion of a radio signal that is 90° out of phase with a reference carrier.

Quadrature detector — An FM detector that multiplies the signal by a delayed version of the same signal.

Quantization — The representation of a continuous analog signal by a number with a finite number of bits.

Quantization error — The difference in amplitude between an analog signal and its digital samples.

Quantization noise — Noise caused by random quantization error.

Radix – The base of a number system. Binary is radix 2 and decimal is radix 10.

Radix point — The symbol that separates the integer part from the fractional part of a number.

Reciprocal mixing — A spurious response in a receiver to an off-channel signal caused by local oscillator phase noise at the same frequency offset as the interference.

Reconstruction filter — A filter located after a digital-to-analog conversion or interpolation to filter out sampling spurs.

Resampling — Changing the sample rate by a non-integer ratio.

Resolution — The number of bits required to represent a digital number from its smallest to its largest value.

Sample rate — The rate at which samples are generated, processed or output from a system.

Sampling — The process of measuring and recording a signal at discrete points of time.

Software-defined radio (SDR) — A transmitter and/or receiver whose principal signal processing functions are defined by software.

Spectral leakage — In a Fourier transform, the indication of frequencies that are not actually present in the signal due to inadequate windowing.

Tap — One processing block, consisting of a coefficient memory, signal register, multiplier and adder, of an FIR filter.

Training sequence — A sequence of one or more known symbols transmitted for the purpose of training the adaptive filter in a receiver.

Undersampling — Harmonic sampling.

Volatile memory — A memory that requires the presence of power supply voltage for data retention.

Von Neumann architecture — A computer architecture that includes a processing unit and a single separate read/write memory to hold both program and data.

Windowing — Tapering the edges of a data sequence so that the samples do not transition abruptly to zero. This avoids passband and stopband ripple in an FIR filter and spectral leakage in a Fourier transform.

Zero-order hold — Holding of a data value for the entire sample period.

Zero-overhead looping — The ability of a microprocessor to automatically jump from the end of a memory block back to the beginning without additional instructions.

Zero-stuffing — Interpolation by inserting zero-valued samples in between the original samples.

15.9 References and Bibliography

REFERENCES

1. The FCC report and order authorizing software-defined radios in the commercial service is available at **www.fcc.gov/Bureaus/Engineering_Technology/Orders/2001/fcc01264.pdf.**
2. The Microchip IDE software can be downloaded free at **www.microchip.com**.
3. Mar, chapter 4. [See bibliography]
4. This is Euler's famous formula, $e^{jx} = \cos(x) + j\sin(x)$, named after Leonhard Euler, an 18th-century mathematician. The factor $e \approx 2.718$ is the base of the natural logarithm and $j = \sqrt{-1}$ is the imaginary unit.
5. Blanchard, R., W6UYG, "Sugar-Coated Single Sideband," *QST*, Oct 1952, p 38ff.
6. Weaver, D.K., Jr, "A Third Method of Generation and Detection of Single-Sideband Signals" *Proc. IRE*, Dec. 1956.
7. Wright, Jr., H., W1PNB, "The Third Method of S.S.B.," *QST*, Sep 1957, pp 11-15.
8. Hershberger, D., W9GR, "Low-Cost Digital Signal Processing for the Radio Amateur," *QST*, Sep 1992, pp 43-51.
9. Youngblood. (See the Articles section below.)
10. Tayloe, D., N7VE, "Notes on 'Ideal' Commutating Mixers (Nov/Dec 1999)" Letters to the Editor, *QEX*, Mar 2001, p 61.
11. *PowerSDR* software is available on the Flex Radio Web site, **www.flex-radio.com**.
12. Analog Devices has lots of good free tutorial literature available for download on their Web site, **www.analog.com**.

BIBLIOGRAPHY

(Key: **D** = disk included, **A** = disk available, **F** = filter design software)

DSP Software Tools

Alkin, O., *PC-DSP*, Prentice Hall, Englewood Cliffs, NJ, 1990 (**DF**).

Kamas, A. and Lee, E., *Digital Signal Processing Experiments*, Prentice Hall, Englewood Cliffs, NJ, 1989 (**DF**).

Momentum Data Systems, Inc., *QEDesign*, Costa Mesa, CA, 1990 (**DF**).

Stearns, S. D. and David, R. A., *Signal Processing Algorithms in FORTRAN and C*, Prentice Hall, Englewood Cliffs, NJ, 1993 (**DF**).

Textbooks

Frerking, M. E., *Digital Signal Processing in Communication Systems*, Van Nostrand Reinhold, New York, NY, 1994.

Hayward, Wes, W7ZOI et al, *Experimental Methods in RF Design*, chapter 10 "DSP Components" and chapter 11 "DSP Applications in Communications," ARRL, 2009 (**D**).

Ifeachor, E. and Jervis, B., *Digital Signal Processing: A Practical Approach*, Addison-Wesley, 1993 (**AF**).

Madisetti, V. K. and Williams, D. B., Editors, *The Digital Signal Processing Handbook*, CRC Press, Boca Raton, FL, 1998 (**D**).

Mar, Amy (Editor), *Digital Signal Processing Applications Using the ADSP-2100 Family, Volume I*, Prentice Hall 1990. While oriented to the ADSP-2100, there is much good general information on DSP algorithms. Both volume I and II are available for free download on the Analog Devices Web site, **www.analog.com**.

Oppenheim, A. V. and Schafer, R. W., *Digital Signal Processing*, Prentice Hall, Englewood Cliffs, NJ, 1975.

Parks, T. W. and Burrus, C.S., *Digital Filter Design*, John Wiley and Sons, New York, NY, 1987.

Proakis, J. G. and Manolakis, D., *Digital Signal Processing*, Macmillan, New York, NY, 1988.

Proakis, J. G., Rader, C. M., et. al., *Advanced Digital Signal Processing*, Macmillan, New York, NY, 1992.

Rabiner, L. R. and Schafer, R. W., *Digital Processing of Speech Signals*, Prentice Hall, Englewood Cliffs, NJ, 1978.

Rabiner, L. R. and Gold, B., *Theory and Application of Digital Signal Processing*, Prentice Hall, Englewood Cliffs, NJ, 1975.

Sabin, W. E. and Schoenike, E. O., Eds., *HF Radio Systems and Circuits*, rev. 2nd ed., Noble Publishing Corp, Norcross, GA, 1998.

Smith, D., *Digital Signal Processing Technology: Essentials of the Communications Revolution*, ARRL, 2001.

Widrow, B. and Stearns, S. D., *Adaptive Signal Processing*, Prentice Hall, Englewood Cliffs, NJ, 1985.

Zverev, A. I., *Handbook of Filter Synthesis*, John Wiley and Sons, New York, NY, 1967.

Articles

Albert, J. and Torgrim, W., "Developing Software for DSP," *QEX*, Mar 1994, pp 3-6.

Ahlstrom, J., N2ADR, "An All-Digital SSB Exciter for HF," *QEX*, May/Jun 2008, pp 3-10.

Ahlstrom, J., N2ADR, "An All-Digital Transceiver for HF," *QEX*, Jan/Feb 2011, pp 3-8.

Anderson, P. T., "A Simple SSB Receiver Using a Digital Down-Converter," *QEX*, Mar, 1994, pp 17-23.

Anderson, P. T., "A Faster and Better ADC for the DDC-Based Receiver," *QEX*, Sep/Oct 1998, pp 30-32.

Applebaum, S. P., "Adaptive arrays," *IEEE Transactions Antennas and Propagation*, Vol. PGAP-24, pp 585-598, September, 1976.

Åsbrink, L., "Linrad: New Possibilities for the Communications Experimenter," *QEX*, Part 1, Nov/Dec 2002; Part 2, Jan/Feb 2003; Part 3 May/Jun 2003.

Ash, J. et al., "DSP Voice Frequency Compandor for Use in RF Communications," *QEX*, Jul 1994, pp 5-10.

Bloom, J., "Measuring SINAD Using DSP," *QEX*, Jun 1993, pp 9-18.

Bloom, J., "Negative Frequencies and Complex Signals," *QEX*, Sep 1994.

Bloom, J., "Correlation of Sampled Signals," *QEX*, Feb 1996.

Brannon, B., "Basics of Digital Receiver Design," *QEX*, Sep/Oct 1999, pp 36-44.

Cahn, H., "Direct Digital Synthesis — An Intuitive Introduction," *QST*, Aug 1994, pp 30-32.

Cercas, F. A. B., Tomlinson, M. and Albuquerque, A. A., "Designing With Digital Frequency Synthesizers," *Proceedings of RF Expo East*, 1990.

de Carle, B., "A Receiver Spectral Display Using DSP," *QST* Jan 1992, pp 23-29.

Dick, R., "Tune SSB Automatically," *QEX*, Jan/Feb 1999, pp 9-18.

Dobranski, L., "The Need for Applications Programming Interfaces (APIs) in Amateur Radio," *QEX*, Jan/Feb 1999, pp 19-21.

Emerson, D., "Digital Processing of Weak Signals Buried in Noise," *QEX*, Jan 1994, pp 17-25.

Forrer, J., "Programming a DSP Sound Card for Amateur Radio," *QEX*, Aug 1994.

Forrer, J., KC7WW, "A DSP-Based Audio Signal Processor," *QEX*, Sep 1996.

Forrer, J., KC7WW, "Using the Motorola DSP56002EVM for Amateur Radio DSP Projects," *QEX*, Aug 1995.

Gradijan, S., WB5KIA, "Build a Super Transceiver — Software for Software Controllable Radios," *QEX*, Sept/Oct 2004, pp 30-34.

Green, R., "The Bedford Receiver: A New Approach," *QEX*, Sep/Oct, 1999, pp 9-23.

Hale, B., "An Introduction to Digital Signal Processing," *QST*, Sep 1992, pp 43-51.

Hershberger, D., W9GR, "DSP — An Intuitive Approach," *QST*, Feb 1996.

Hightower, M., KF6SJ, "Simple SDR Receiver," *QEX*, Mar/Apr 2012, pp 3-8.

Hill, C., W5BAA, "SDR2GO: A DSP Odyssey," *QEX*, Mar/Apr 2011, pp 36-43.

Kossor, M., "A Digital Commutating Filter," *QEX*, May/Jun 1999, pp 3-8.

Larkin, B., W7PUA, "The DSP-10: An All-Mode 2-Meter Transceiver Using a DSP IF and PC-Controlled Front Panel, Part 1," *QST*, Sep 1999, pp 33-41.

Larkin, B., W7PUA, "The DSP-10: An All-Mode 2-Meter Transceiver Using a DSP IF and PC-Controlled Front Panel, Part 2," *QST*, Oct 1999, pp 34-40.

Larkin, B., W7PUA, "The DSP-10: An All-Mode 2-Meter Transceiver Using a DSP IF and PC-Controlled Front Panel, Part 3," *QST*, Nov 1999, pp 42-45.

Mack, Ray, W5IFS, "SDR: Simplified," *QEX*, columns beginning Nov 2009.

Morrison, F., "The Magic of Digital Filters," *QEX*, Feb 1993, pp 3-8.

Nickels, R., W9RAN, "Cheap and Easy SDR," *QST*, Jan 2013, pp 30-35.

Olsen, R., "Digital Signal Processing for the Experimenter," *QST*, Nov 1994, pp 22-27.

Reyer, S. and Herschberger, D., "Using the LMS Algorithm for QRM and QRN Reduction," *QEX*, Sep 1992, pp 3-8.

Rohde, D., "A Low-Distortion Receiver Front End for Direct-Conversion and DSP Receivers," *QEX*, Mar/Apr 1999, pp 30-33.

Runge, C., *Z. Math. Physik*, Vol 48, 1903; also Vol 53, 1905.

Scarlett, J., "A High-Performance Digital Transceiver Design," *QEX*, Part 1, Jul/Aug 2002; Part 2, Mar/Apr 2003.

Smith, D., "Introduction to Adaptive Beamforming," *QEX*, Nov/Dec 2000.

Smith, D., "Signals, Samples and Stuff: A DSP Tutorial, Parts 1-4," *QEX*, Mar/Apr-Sep/Oct, 1998.

Stephensen, J., KD6OZH, Software Defined Radios for Digital Communications," *QEX*, Nov/Dec 2004, pp 23-35. [Open-source platform]

Veatch, J., WA2EUJ, "The DSP-610 Transceiver," *QST*, Aug 2012, pp 30-32.

Ward, R., "Basic Digital Filters," *QEX*, Aug 1993, pp 7-8.

Wiseman, J., KE3QG, "A Complete DSP Design Example Using FIR Filters," *QEX*, Jul 1996.

Youngblood, G., "A Software-Defined Radio for the Masses," *QEX*; Part 1, Jul/Aug 2002; Part 2, Sep/Oct 2002; Part 3, Nov/Dec 2002; Part 4, Mar/Apr 2003.

SDR and DSP Receiver Testing

Hakanson, E., "Understanding SDRs and their RF Test Requirements," Anritsu Application Note. Available online from **www.us.anritsu.com/downloads/files/11410-00403.pdf**.

Kundert, K. "Accurate and Rapid Measurement of IP2 and IP3," Designer's Guide Consulting, 2002-2006. Available online from **www.designers-guide.org/Analysis/intercept-point.pdf**.

MacLeod, J.R.; Beach, M.A.; Warr, P.A.; Nesimoglu, T., "Software Defined Radio Receiver Test-Bed," IEEE Vehicular Tech Conf, Fall 2001, VTS 54th, Vol 3, pp 565-1569. Available online from **http://ieeexplore.ieee.org/Xplore/login.jsp?url=/iel5/7588/20685/00956461.pdf**.

R. Sierra, Rhode & Schwarz, "Challenges In Testing Software Defined Radios," SDR Forum Sandiego Workshop, 2007.

Sirmans, D. and Urell, B., "Digital Receiver Test Results," Next Generation Weather Radar Program. Available online from **www.roc.noaa.gov/eng/docs/digital%20receiver%20test%20results.pdf**.

"Testing and Troubleshooting Digital RF Communications Receiver Designs," Agilent Technologies Application Note 1314. Available online from **http://cp.literature.agilent.com/litweb/pdf/5968-3579E.pdf**.

Contents

**Chapter 16 —
CD-ROM Content**

Supplemental Files

- Table of digital mode characteristics (section 16.6)
- ASCII and ITA2 code tables
- Varicode tables for PSK31, MFSK16 and DominoEX
- Tips for using *FreeDV* HF digital voice software by Mel Whitten, KØPFX

Digital Modes

The first Amateur Radio transmissions were digital using Morse code. Computers have now become so common, households often have several and there are a large number of other digital information appliances such as tablets, netbooks and smart phones. E-mail and texting are displacing traditional telephone communications and Amateur Radio operators are creating new digital modes to carry this traffic over the air or to solve a particular need. Many new modes can be entirely implemented in software for use with a computer sound card, making tools for experimentation and implementation available worldwide via the Internet.

This chapter will focus on the protocols for transferring various data types. The material was written or updated by Scott Honaker, N7SS, with support from Alan Bloom, N1AL, and Pete Loveall, AE5PL. Kok Chen W7AY contributed the sections on RTTY, PSK, and MFSK. Mel Whitten, KØPFX, and David Rowe, VK5DGR, contributed a new section on *FreeDV* digital voice. Modulation methods are covered in the **Modulation** chapter and the **Digital Communications** supplement on the *Handbook* CD discusses the practical considerations of operating using these modes. This chapter addresses the process by which data is encoded, compressed and error checked, packaged and exchanged.

There is a broad array of digital modes to service various needs with more coming. The most basic modes simply encode text data for transmission in a keyboard-to-keyboard chat-type environment. These modes may or may not include any mechanism for error detection or correction. The second class of modes are generally more robust and support more sophisticated data types by structuring the data sent and including additional error-correction information to properly reconstruct the data at the receiving end. The third class of modes discussed will be networking modes with protocols often the same or similar to versions used on the Internet and computer networks.

16.1 Digital "Modes"

The ITU uses *Emission Designators* to define a "mode" as demonstrated in the **Modulation** chapter. These designators include the bandwidth, modulation type and information being sent. This system works well to describe the physical characteristics of the modulation, but digital modes create some ambiguity because the type of information sent could be text, image or even the audio of a CW session. As an example, an FM data transmission of 20K0F3D could transmit spoken audio (like FM 20K0F3E or 2K5J3E) or a CW signal (like 150H0A1A). These designators don't help identify the type of data supported by a particular mode, the speed that data can be sent, if it's error-corrected, or how well it might perform in hostile band conditions. Digital modes have more characteristics that define them and there are often many variations on a single mode that are optimized for different conditions. We'll need to look at the specifics of these unique characteristics to be able to determine which digital modes offer the best combination of features for any given application.

16.1.1 Symbols, Baud, Bits and Bandwidth

The basic performance measure of a digital mode is the *data rate*. This can be measured a number of ways and is often confused. Each change of state on a transmission medium defines a *symbol* and the *symbol rate* is also known as *baud*. (While commonly used, "baud rate" is redundant because "baud" is already defined as the rate of symbols/second.) Modulating a carrier increases the frequency range, or bandwidth, it occupies. The FCC currently limits digital modes by symbol rate on the various bands as an indirect (but easily measurable) means of controlling bandwidth.

The *bit rate* is the product of the *symbol rate* and the number of bits encoded in each symbol. In a simple two-state system like an RS-232 interface, the bit rate will be the same as baud. More complex waveforms can represent more than two states with a single symbol so the bit rate will be higher than the baud. For each additional bit encoded in a symbol, the number of states of the carrier doubles. This makes each state less distinct from the others, which in turn makes it more difficult for the receiver to detect each symbol correctly in the presence of noise. A V.34 modem may transmit symbols at a baud rate of 3420 baud and each symbol can represent up to 10 discrete states or bits, resulting in a *gross* (or *raw*) *bit rate* of 3420 baud × 10 or 34,200 bits per second (bit/s). Framing bits and other overhead reduce the *net bit rate* to 33,800 bit/s.

Bits per second is abbreviated here as *bit/s* for clarity but is also often seen as *bps*. Bits per second is useful when looking at the protocol but is less helpful determining how long it

Table 16.1
Data Rate Symbols and Multipliers

Name	Symbol	Multiplier
kilobit per second	kbit/s or kbps	1000 or 10^3
Megabit per second	Mbit/s or Mbps	1,000,000 or 10^6
Gigabit per second	Gbit/s or Gbps	1,000,000,000 or 10^9

takes to transmit a specific size file because the number of bits consumed by overhead is often unknown. A more useful measure for calculating transmission times is *bytes per second* or *Bps* (note the capitalization). Although there are only eight bits per byte, with the addition of start and stop bits, the difference between bps and Bps is often tenfold. Since the net bit rate takes the fixed overhead into account, Bytes per second can be calculated as $bps_{net}/8$. Higher data rates can be expressed with their metric multipliers as shown in **Table 16.1**.

Digital modes constantly balance the relationship between symbol rate, bit rate, bandwidth and the effect of noise. The Shannon-Hartley theorem demonstrates the maximum channel capacity in the presence of Gaussian white noise and was discussed in the **Modulation** chapter in the Channel Capacity section. This theorem describes how an increased symbol rate will require an increase in bandwidth and how a reduced signal-to-noise ratio (SNR) will reduce the potential *throughput* of the channel.

16.1.2 Error Detection and Correction

Voice modes require the operator to manually request a repeat of any information required but not understood. Using proper phonetics makes the information more easily understood but takes longer to transmit. If 100% accuracy is required, it may be necessary for the receiver to repeat the entire message back to the sender for verification. Computers can't necessarily distinguish between valuable and unnecessary data or identify likely errors but they offer other options to detect and correct errors.

ERROR DETECTION

The first requirement of any accurate system is to be able to detect when an error has occurred. The simplest method is *parity*. With 7-bit ASCII data, it was common to transmit an additional 8th parity bit to each character. The parity bit was added to make the total number of 1 bits odd or even. The binary representation for an ASCII letter Z is 1011010. Sent as seven bits with even parity, the parity bit would be 0 because there are already an even number of 1 bits and the result would be 01011010. The limitation of parity is that it only works with an odd number of

bit inversions. If the last two bits were flipped to 01011001 (the ASCII letter Y), it would still pass the parity check because it still has an even number of bits. Parity is also rarely used on 8-bit data so it cannot be used when transferring binary data files.

Checksum is a method similar to the "check" value in an NTS message. It is generally a single byte (8-bit) value appended to the end of a packet or frame of data. It is calculated by adding all the values in the packet and taking the least significant (most unique) byte. This is a simple operation for even basic processors to perform quickly but can also be easily mislead. If two errors occur in the packet of equal amounts in the opposite direction (A becomes B and Z becomes Y), the checksum value will still be accurate and the packet will be accepted as error-free.

Cyclic redundancy check (CRC) is similar to checksum but uses a more sophisticated formula for calculating the check value of a packet. The formula most closely resembles long division, where the quotient is thrown away and the remainder is used. It is also common for CRC values to be more than a single byte, making the value more unique and likely to identify an error. Although other error detection systems are currently in use, CRC is the most common.

ERROR CORRECTION

There are two basic ways to design a protocol for an error correcting system: *automatic repeat request (ARQ)* and *forward error correction (FEC)*. With ARQ the transmitter sends the data packet with an error detection code, which the receiver uses to check for errors. The receiver will request retransmission of packets with errors or sends an acknowledgement (ACK) of correctly received data, and the transmitter re-sends anything not acknowledged within a reasonable period of time.

With forward error correction (FEC), the transmitter encodes the data with an *error-correcting code (ECC)* and sends the encoded message. The receiver is not required to send any messages back to the transmitter. The receiver decodes what it receives into the "most likely" data. The codes are designed so that it would take an "unreasonable" amount of noise to trick the receiver into misinterpreting the data. It is possible to combine the two, so that minor errors are corrected without retransmission, and major errors are detected

and a retransmission requested. The combination is called *hybrid automatic repeat request (hybrid ARQ)*.

There are many error correcting code (ECC) algorithms available. Extended Golay coding is used on blocks of ALE data, for example, as described in the section below on G-TOR. In addition to the ability to detect and correct errors in the data packets, the modulation scheme allows sending multiple data streams and interleaving the data in such a way that a noise burst will disrupt the data at different points.

16.1.3 Data Representations

When comparing digital modes, it is important to understand how the data is conveyed. There are inherent limitations in any method chosen. PSK31 might seem a good choice for sending data over HF links because it performs well, but it was only designed for text (not 8-bit data) and has no inherent error correction. It is certainly possible to use this modulation scheme to send 8-bit data and add error correction to create a new mode. This would maintain the weak signal performance but the speed will suffer from the increased overhead. Similarly, a digital photo sent via analog SSTV software may only take two minutes to send, but over VHF packet it could take 10 minutes, despite the higher speed of a packet system. This doesn't mean SSTV is more efficient. Analog SSTV systems generally transmit lower resolution images with no error correction. Over good local links, the VHF packet system will be able to deliver perfect images faster or of higher quality.

TEXT REPRESENTATIONS

Morse code is well known as an early code used to send text data over a wire, then over the air. Each letter/number or symbol is represented with a varying length code with the more common letters having shorter codes. This early *varicode* system is very efficient and minimizes the number of state changes required to send a message.

The Baudot code (pronounced "bawd-OH") was invented by Émile Baudot and is the predecessor to the character set currently known more accurately as International Telegraph Alphabet No 2 (ITA2). This code is used for radioteletype communications and contains five bits with start and stop pulses. This only allows for 2^5 or 32 possible characters to be sent, which is not enough for all 26 letters plus numbers and characters. To resolve this, ITA2 uses a LTRS code to select a table of upper case (only) letters and a FIGS code to select a table of numbers, punctuation and special symbols. The code is defined in the ITA2 codes table on the CD included with this book.

Early computers used a wide variety of alphabetic codes until the early 1960s until the advent of the American Standard Code for Information Interchange or ASCII (pronounced "ESS-key"). At that point many computers standardized on this character set to allow simple transfer of data between machines. ASCII is a 7-bit code which allows for 2^7 or 128 characters. It was designed without the *control characters* used by Baudot for more reliable transmissions and the letters appear in English alphabetical order for easy sorting. The code can be reduced to only six bits and still carry numbers and uppercase letters. Current FCC regulations provide that amateur use of ASCII conform to ASCII as defined in ANSI standard X3.4-1977. The international counterparts are ISO 646-1983 and International Alphabet No. 5 (IA5) as published in ITU-T Recommendation V.3. A table of ASCII characters is presented as "ASCII Character Set" on the CD included with this book.

ASCII has been modified and initially expanded to eight bits, allowing the addition of foreign characters or line segments. The different extended versions were often referred to as *code pages*. The IBM PC supported code page 437 which offers line segments, and English *Windows* natively supports code page 1252 with additional foreign characters and symbols. All of these extended code pages include the same first 128 ASCII characters for backward compatibility. In the early 1990s efforts were made to support more languages directly and *Unicode* was created. Unicode generally requires 16 bits per character and can represent nearly any language.

More recent schemes use varicode, where the most common characters are given shorter codes (see **http://en.wikipedia.org/wiki/Prefix_code**). Varicode is used in PSK31and MFSK to reduce the number of bits in a message. Although the PSK31 varicode contains all 128 ASCII characters, lower-case letters contain fewer bits and can be sent more quickly. Tables of PSK and MFSK varicode characters are included on the CD included with this book.

IMAGE DATA REPRESENTATIONS

Images are generally broken into two basic types, *raster* and *vector*. Raster or *bitmap* images are simply rows of colored points known as *pixels* (picture elements). Vector images are a set of drawing primitives or instructions. These instructions define shapes, placement, size and color. Similar coding is used with plotters to command the pens to create the desired image.

Bitmap images can be stored at various *color depths* or bits per pixel indicating how many colors can be represented. Common color depths are 1-bit (2 colors), 4-bit (16 colors), 8-bit (256 colors), 16-bit (65,536 colors also called *high color*) and 24-bit (16

million also called *true color*). True color is most commonly used with digital cameras and conveniently provides eight bits of resolution each for the red, green and blue colors. Newer scanners and other systems will often generate 30-bit (2^{30} colors) and 36-bit (2^{36} colors). Images with 4- and 8-bit color can store more accurate images than might be obvious because they include a *palette* where each of their 16 or 256 colors respectively are chosen from a palette of 16 million. This palette is stored in the file which works well for simple images. The GIF format only supports 256 colors (8-bit) with a palette and lossless compression.

Digital photographs are raster images at a specific resolution and color depth. A typical low resolution digital camera image would be 640×480 pixels with 24-bit color. The 24-bit color indicated for each pixel requires three bytes to store (24 bits / 8 bits per byte) and can represent one of a possible 2^{24} or 16,777,216 colors with eight bits each for red, green and blue intensities. The raw (uncompressed) storage requirement for this image would be 640×480x3 or 921,600 bytes. This relatively low resolution image would require significant time to transmit over a slow link.

Vector images are generally created with drawing or CAD packages and can offer significant detail while remaining quite small. Because vector files are simply drawn at the desired resolution, there is no degradation of the image if the size is changed and the storage requirements remain the same at any resolution. Typical drawing primitives include lines and polylines, polygons, circles and ellipses, Bézier curves and text. Even computer font technologies such as TrueType create each letter from Bézier curves allowing for flawless scaling to any size and resolution on a screen or printer.

Raster images can be resized by dropping or adding pixels or changing the color depth. The 640×480×24-bit color image mentioned above could be reduced to 320×240×16-bit color with a raw size of 153,600 bytes — a significant saving over the 921,600 byte original. If the image is intended for screen display and doesn't require significant detail, that size may be appropriate. If the image is printed full sized on a typical 300 dpi (dots per inch) printer, each pixel in the photo will explode to nearly 100 dots on the printer and appear very blocky or pixilated.

AUDIO DATA REPRESENTATIONS

Like images, audio can be stored as a sampled waveform or in some type of primitive format. Storing a sampled waveform is the most versatile but can also require substantial storage capacity. MIDI (musical instrument digital interface, pronounced "MID-ee") is a common music format that stores instrument, note, tempo and intensity information as a musical score. There are also voice coding techniques that store speech as *allophones* (basic human speech sounds).

As with images, storage of primitives can save storage space (and transmission times) but they are not as rich as a high quality sampled waveform. Unfortunately, the 44,100 Hz 16-bit sample rate of an audio compact disc (CD) requires 176,400 bytes to store each second and 10,584,000 bytes for each minute of stereo audio.

The Nyquist-Shannon sampling theorem states that perfect reconstruction of a signal is possible when the sampling frequency is at least twice the maximum frequency of the signal being sampled. With its 44,100 Hz sample rate, CD audio is limited to a maximum frequency response of 22,050 Hz. If only voice-quality is desired, the sample rate can easily be dropped to 8000 Hz providing a maximum 4000 Hz frequency response. (See the **DSP and Software Radio Design** chapter for more information on sampling.)

The *bit depth* (number of bits used to represent each sample) of an audio signal will determine the theoretical dynamic range or signal-to-noise ratio (SNR). This is expressed with the formula SNR = (1.761 + 6.0206 × bits) dB. A dynamic range of 40 dB is adequate for the perception of human speech. **Table 16.2** compares the audio quality of various formats with the storage (or transmission) requirements.

VIDEO DATA REPRESENTATIONS

The most basic video format simply stores a series of images for playback. Video can place huge demand on storage and bandwidth because 30 frames per second is a common rate for smooth-appearing video. Rates as low as 15 frames per second can still be considered "full-motion" but will appear jerky and even this rate requires substantial storage. Primitive formats are less common for video

Table 16.2
Typical Audio Formats

Audio Format	Bits per Sample	Dynamic Range	Maximum Frequency	kbytes per Minute
44.1 kHz stereo	16	98	22.050 kHz	10,584
22 kHz mono	16	98	11 kHz	2,640
8 kHz mono	8	50	4 kHz	960

but animation systems like Adobe *Flash* are often found on the web.

16.1.4 Compression Techniques

There are a large variety of compression algorithms available to reduce the size of data stored or sent over a transmission medium. Many techniques are targeted at specific types of data (text, audio, images or video). These compression techniques can be broken into two major categories: *lossless compression* and *lossy compression*. Lossless algorithms are important for compressing data that must arrive perfectly intact but offer a smaller compression ratio than lossy techniques. Programs, documents, databases, spreadsheets would all be corrupted and made worthless if a lossy compression technique were used.

Images, audio and video are good candidates for lossy compression schemes with their substantially higher compression rates. As the name implies, a lossy compression scheme deliberately omits or simplifies that data to be able to represent it efficiently. The human eye and ear can easily interpolate missing information and it simply appears to be of lower quality.

Compression of a real-time stream of data such as audio or video is performed by software or firmware *codecs* (from *co*der-*dec*oder). Codecs provide the real-time encoding or decoding of the audio or video stream. Many codecs are proprietary and have licensing requirements. Codecs can be implemented as operating system or application plug-ins or even as digital ICs, as with the P25 IMBE and D-STAR AMBE codecs.

There are also specifically designed low bit-rate codecs for voice that are more accurately called *vocoders*. Vocoders are optimized for voice characteristics and can encode it extremely efficiently. Conversely, they cannot effectively process non-voice signals. This is easily demonstrated with a mobile phone by listening to music through a voice connection, such as when a call is placed on hold. Mobile phones use highly efficient vocoders to minimize the bandwidth of each voice channel which allows more channels per tower. This also means that non-voice sounds such as music or mechanical sounds are not well reproduced.

Good vocoders are extremely valuable and vigorously guarded by their patent holders. This has made it difficult for amateurs to experiment with digital voice techniques. David Rowe, VK5DGR, worked for several years on a project called CODEC2. This is a very effective open source vocoder made available at the end of 2012. It has been incorporated into the *FDMDV* software and there has been extensive testing and tuning

made on the HF bands. More information about the project can be found at **http://codec2.org**. Digital voice is discussed elsewhere in this chapter.

LOSSLESS COMPRESSION TECHNIQUES

One of the earliest lossless compression schemes is known as *Huffman coding*. Huffman coding creates a tree of commonly used data values and gives the most common values a lower bit count. Varicode is based on this mechanism. In 1984, Terry Welch released code with improvements to a scheme from Abraham Lempel and Jacob Ziv, commonly referred to as *LZW* (Lempel-Ziv-Welch). The Lempel-Ziv algorithm and variants are the basis for most current compression programs and is used in the GIF and optionally TIFF graphic formats. *LZW* operates similarly to Huffman coding but with greater efficiency.

The actual amount of compression achieved will depend on how redundant the data is and the size of the data being compressed. Large files will achieve greater compression rates because the common data combinations will be seen more frequently. Simple text and documents can often see 25% compression rates. Spreadsheets and databases generally consist of many empty cells and can often achieve nearly 50% compression. Graphic and video compression will vary greatly depending on the complexity of the image. Simple images with solid backgrounds will compress well where complex images with little recurring data will see little benefit. Similarly, music will compress poorly but spoken audio with little background noise may see reasonable compression rates.

Run length encoding (*RLE*) is a very simple scheme supported on bitmap graphics (*Windows* BMP files). Each value in the file is a color value and "run length" specifying how many of the next pixels will be that color. It works well on simple files but can make a file larger if the image is too complex.

It is important to note that compressing a previously compressed file will often yield a larger file because it simply creates more overhead in the file. There is occasionally some minor benefit if two different compression algorithms are used. It is not possible to compress the same file repeatedly and expect any significant benefit. Modern compression software does offer the additional benefit of being able to compress groups of files or even whole directory structures into a single file for transmission.

The compression mechanisms mentioned above allow files to be compressed prior to transmission but there are also mechanisms that allow near real-time compression of the transmitted data. V.92 modems implement a LZJH adaptive compression scheme called V.44 that can average 15% better through-

put over the wire. Winlink uses a compression scheme called B2F and sees an average 44% improvement in performance since most Winlink data is uncompressed previously. Many online services offer "Web Accelerators" that also compress data going over the wire to achieve better performance. There is a slight delay (latency) as a result of this compression but over a slow link, the additional latency is minimal compared to the performance gain.

LOSSY COMPRESSION SCHEMES

Lossy compression schemes depend on the human brain to "recover" or simply ignore the missing data. Since audio, image and video data have unique characteristics as perceived by the brain, each of these data types have unique compression algorithms. New compression schemes are developed constantly to achieve high compression rates while maintaining the highest quality. Often a particular file format actually supports multiple compression schemes or is available in different versions as better methods are developed. In-depth discussion of each of these algorithms is beyond the scope of this book but there are some important issues to consider when looking at these compression methods.

Lossy Audio Compression

Audio compression is based on the psychoacoustic model that describes which parts of a digital audio signal can be removed or substantially reduced without affecting the perceived quality of the sound. Lossy audio compression schemes can typically reduce the size of the file 10- or 12-fold with little loss in quality. The most common formats currently are MP3, WMA, AAC, Ogg Vorbis and ATRAC.

Lossy Image Compression

The Joint Photographic Experts Group developed the JPEG format (pronounced "JAY-peg") in 1992 and it has become the primary format used for lossy compression of digital images. It is a scalable compression scheme allowing the user to determine the level of compression/quality of the image. Compression rates of 10-fold are common with good quality and can be over 100-fold at substantially reduced quality. The JPEG format tends to enhance edges and substantially compress fields of similar color. It is not well suited when multiple edits will be required because each copy will have generational loss and therefore reduced quality.

Lossy Video Compression

Because of the massive amount of data required for video compression it is almost always distributed with a lossy compression scheme. Lossless compression is only used when editing to eliminate generational loss. Video support on a computer is generally implemented with a codec that allows encoding and decoding the video stream. Video files are containers that can often support more than one video format and the specific format information is contained in the file. When the file is opened, this format information is read and the appropriate codec is activated and fed into the data stream to be decoded.

The most common current codecs are H.261 (video conferencing), MPEG-1 (used in video CDs), MPEG-2 (DVD Video), H.263 (video conferencing), MPEG-4, DivX, Xvid, H.264, Sorenson 3 (used by *QuickTime*), WMV (Microsoft), VC-1, RealVideo (RealNetworks) and Cinepak.

The basic mechanism of video compression is to encode a high quality "key-frame" that could be a JPEG image as a starting image. Successive video frames or "inter-

FLDIGI

By Ken Humbertson, WØKAH, and Jeff Coval, ACØSC

You may have heard of the free digital mode software *fldigi* by David H. Freese Jr., W1HKJ. It is very popular for use in emergency communications, for example, because of its multimode capabilities and its ability to work with a companion program *flmsg* that generates standard message forms to be transmitted using *fldigi*. The software supports more than 34 modes (as of early 2013) as well as variations of tones, bandwidth, baud rates, number of bits, and other variations for many modes. Versions of *fldigi* are available for *Linux/Free-BSD*, *Windows* XP/W2K/NT/Vista/Win7, and OS X from W1HKJ's download page at **www.w1hkj.com/download.html**.

If you have a computer with a sound card and microphone, you can begin using *fldigi* to receive with no additional hardware. A quick example would be to tune to 14.070 MHz or 21.070 MHz USB during the day. Set the radio speaker volume to a normal listening level. Start *fldigi* on the computer with sound card and microphone connected and you should see a waterfall display similar to **Fig 16.A** (left).

If your rig has computer control capability, *fldigi* can likely interface with it to display frequency and mode, as well as control the radio from the program. In Fig 16.A, an ICOM IC-7000 is controlled by *fldigi* and thus shows the current operating frequency and mode of 21.070 MHz USB, the QSO frequency of 21071.511 (kHz) (in the windows at the upper left corner), and the current mode of BPSK31 (lower left corner).

When you see multiple signals in the waterfall, *fldigi* will decode whichever one you choose when you place the mouse cursor over a signal of interest and line up the two vertical red lines on your screen with the sides of the signal, as shown near 1500 Hz. When you change modes or bandwidths within a mode, the spacing of the red lines will adjust to the new bandwidth. **Fig 16.B** (right) shows that by selecting RTTY-45 under the OP MODE tab, you will notice that the red line spacing increases to match the familiar RTTY tone shift of 170 Hz. Simply place the two red lines over the signal you wish to decode and the program does the rest.

Fig 16.B shows a decoded ARRL digital bulletin from January 2, 2013. While the examples show BPSK31 and RTTY, there are many operating modes to choose from, including CW.

Actions in *fldigi* are performed by action buttons that invoke *macros* — text scripts that control the program. For example, to call CQ use the mouse to click on an unoccupied spot in the waterfall display. This shifts the modulating tones of the signal to that offset within your receive bandwidth. Click the CQ action button on the *fldigi* display. The transmitted text is displayed in red above the waterfall display. When the text is finished, *fldigi* will return to receive mode. The tutorial "Beginner's Guide to Fldigi" at **www.w1hkj.com/beginners.html** is recommended and the program has an extensive Help file.

The program can be used with an internal sound card or external sound card adapter such as the Tigertronics Signalink USB (**www.tigertronics.com**) or West Mountain Radio Rigblaster series (**www.westmountainradio.com**). Setting up a sound card to use *fldigi* may require manipulation of the audio device configuration for your computer's operating system. Follow the instructions in the *fldigi* manuals and the manufacturer's manuals if you are using an external adapter. A digital data interface that allows you to connect the sound card to your rig's microphone input is recommended. (See the **Assembling a Station** chapter and the **Digital Communications** supplement on the CD-ROM.)

Fldigi is frequently updated to include new modes that are being developed. The program is a user-friendly way to become active on the digital modes, a rapidly expanding aspect of Amateur Radio.

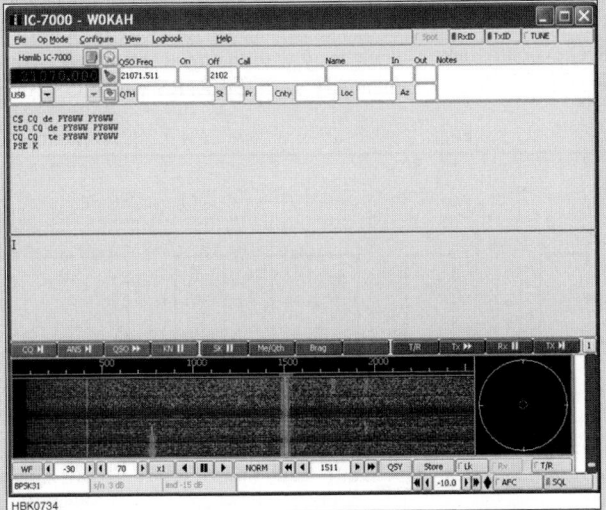

Fig 16.A

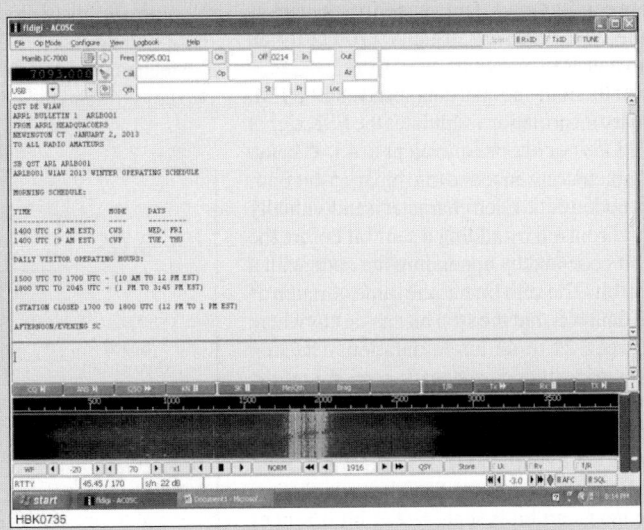

Fig 16.B

frames" contain only changes to the previous frame. After some number of frames, it is necessary to use another key-frame and start over with inter-frames. The resolution of the images, frame rate and compression quality determine the size of the video file.

BIT RATE COMPARISON

Table 16.3 provides an indication of minimum bit rates required to transmit audio and video such that the average listener would not perceive them significantly worse than the standard shown.

Table 16.3
Audio and Video Bit Rates

Audio

8 kbit/s	Telephone quality audio (using speech codecs)
32 kbit/s	AM broadcast quality (using MP3)
96 kbit/s	FM broadcast quality (using MP3)
224-320 kbit/s	Near-CD quality, indistinguishable to the average listener (using MP3)

Video

16 kbit/s	Videophone quality (minimum for "talking head")
128-384 kbit/s	Business videoconference system quality
1.25 Mbit/s	VCD (video compact disk) quality
5 Mbit/s	DVD quality
10.5 Mbit/s	Actual DVD video bit rate

16.1.5 Compression vs. Encryption

There is some confusion about compression being a form of encryption. It is true that a text file after compression can no longer be read unless uncompressed with the appropriate algorithm. In the United States the FCC defines encryption in part 97.113 as "messages encoded for the purpose of obscuring their meaning." Compressing a file with ZIP or RAR (common file compression methods) and transmitting it over the air is simply an efficient use of spectrum and time and is not intended to "obscure its meaning."

As amateur digital modes interact more with Internet-based services, the issue arises because many of these services utilize encryption of various types. Banks and other retailers may encrypt their entire transactions to insure confidentiality of personal data. Other systems as benign as e-mail may simply encrypt passwords to properly authenticate users. The FCC has offered no additional guidance on these issues.

16.2 Unstructured Digital Modes

The first group of modes we'll examine are generally considered "sound card modes" for keyboard-to-keyboard communications. Because each of these modes is optimized for a specific purpose by blending multiple features, they often defy simple categorization.

16.2.1 Radioteletype (RTTY)

Radioteletype (*RTTY*) consists of a *frequency shift keyed* (*FSK*) signal that is modulated between two carrier frequencies, called the *mark* frequency and the *space* frequency. The protocol for amateur RTTY calls for the mark carrier frequency to be the higher of the two in the RF spectrum. The difference between the mark and space frequencies is called the *FSK shift*, usually 170 Hz for an amateur RTTY signal.

At the conventional data speed of 60 WPM, binary information modulates the FSK signal at 22 ms per bit, or equivalent to 45.45 baud. Characters are encoded into binary 5-bit Baudot coded data. Each character is individually synchronized by adding a start bit before the 5-bit code and by appending the code with a stop bit. The start bit has the same duration as the data bits, but the stop bit can be anywhere between 22 to 44 ms in duration. The stop bit is transmitted as a mark carrier, and the RTTY signal "rests" at this state until a new character comes along. If the number of stop bits is set to two, the RTTY signal will send a minimum of 44 ms of mark carrier before the next start bit is sent. A start bit is sent as a space carrier. A zero in the Baudot code is sent as a space signal and a one is sent as a mark signal. **Fig 16.1** shows the character D sent by RTTY.

BAUDOT CODE

The Baudot code (see the *Handbook* CD) is a 5-bit code; thus, it is capable of encoding only 32 unique characters. Since the combination of alphabets, decimal numbers and common punctuations exceeds 32, the Baudot code is created as two sets of tables. One table is called the LTRS Shift, and consists mainly of alphabetic characters. The second table is called the FIGS Shift, and consists mainly of decimal numerals and punctuation marks. Two unique Baudot characters called LTRS and FIGS are used by the sender to command the decoder to switch between these two tables.

Mechanical teletypewriter keyboards have two keys to send the LTRS and FIGS characters. The two keys behave much like the caps lock key on a modern typewriter keyboard. Instead of locking the keyboard of a typewriter to upper case shift or lower case shift, the two teletypewriter keys lock the state of the teletypewriter into the LTRS table or the FIGS table. LTRS and FIGS, among some other characters such as the space character, appear in both LTRS and FIGS tables so that you can send LTRS, FIGS and shift no matter which table the encoder is using.

To send the letter Q, you need to first make sure that the decoder is currently using the LTRS table, and then send the Baudot codeword for Q, 17 (hexadecimal or "hex" value). If the same hex 17 is received when the decoder is in the FIGS shift, the number 1 will be decoded instead of the Q. Modern software does away with the need for the operator to manually send the LTRS and FIGS codes.

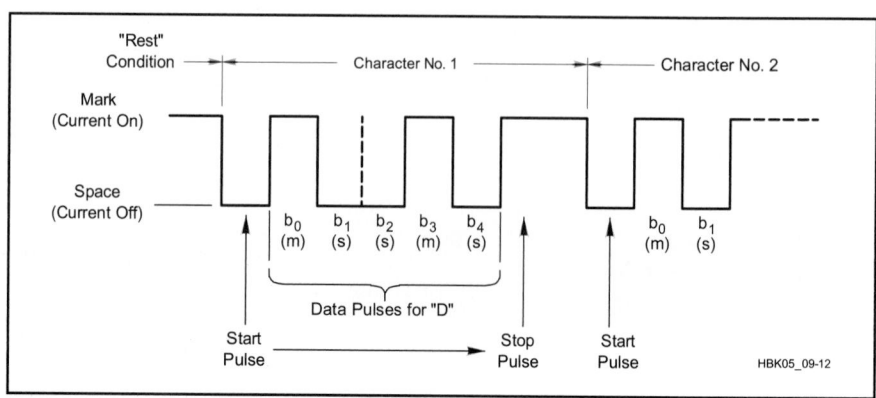

Fig 16.1 — The character "D" sent via RTTY.

When the operator sends a character from the LTRS table, the software first checks to make sure that the previous character had also used a character from the LTRS table; if not, the software will first send a LTRS character.

Noise can often cause the LTRS or FIGS character to be incorrectly received. This will cause subsequent characters to decode into the wrong Baudot character, until a correct LTRS or FIGS is received (see also USOS below). Instead of asking that the message be repeated by the sender, a trick that many RTTY operators use is to observe that on a standard QWERTY keyboard, the Q key is below the 1 key in the row above it, the W key is below the 2 key, and so on. Q and 1 happen to share the same Baudot code; W and 2 share the same Baudot code, and so forth. Given this visual aid, a printed UE can easily be interpreted by context to 73 and TOO interpreted as 599.

If the sender uses one stop bit, an RTTY character consists of a total of seven bits after adding the start and stop bits. If the sender uses 1.5 stop bits, each RTTY character has a total length of 7.5 bits. The least significant bit of the Baudot code follows the start bit of a character.

INVERTED SIGNALS

There are times when the sender does not comply with the RTTY standard and reverses the mark and space order in the spectrum. This is often called an "inverted" signal or "reverse shift." Most RTTY modulators and demodulators have a provision to reverse the shift of an inverted signal.

DEMODULATION AND DECODING

The most common way to decode an RTTY signal is to use a single sideband (SSB) receiver to first translate the two FSK carriers into two audio tones. If the carriers are 170 Hz apart, the two audio tones (called the *tone pair*) will also be 170 Hz apart. The RTTY demodulator, in the form of a terminal unit (TU) or a software modem (modulator-demodulator), then works to discriminate between the two audio tones. Some packet TNCs (terminal node controllers) can be made to function as RTTY demodulators, but often they do not work as well under poor signal-to-noise conditions because their filters are matched to packet radio FSK shifts and baud rate instead of to RTTY shift and baud rate.

As long as the tone pair separation is 170 Hz, the frequencies of the two audio tones can be quite arbitrary. Many TUs and modems are constructed to handle a range of tone pairs. If reception uses a lower sideband (LSB) receiver, the higher mark carrier will become the lower of the two audio tones. If upper sideband (USB) is used, the mark carrier will remain the higher of the tone pair. It is common to use 2125 Hz as the mark tone and 2295 Hz as a space tone. Since the mark tone (2125

Hz) is lower in frequency, the receiver will be set to LSB. In general, modem tone pairs can be "reversed" (see the Inverted Signals section), so either an LSB or a USB receiver can be used. Moreover, the tone pairs of many modems, especially software modems, can be moved to place them where narrowband filters are available on the receiver.

In the past, audio FSK demodulators were built using high-Q audio filters followed by a "slicer" to determine if the signal from the mark filter is stronger or weaker than the signal that comes out of the space filter. To counter selective fading, where ionospheric propagation can cause the mark carrier to become stronger or weaker than the space carrier, the slicer's threshold can be designed to adapt to the imbalance. Once "sliced" into a bi-level signal, the binary stream is passed to the decoder where start bit detection circuitry determines where to extract the 5-bit character data. That data is then passed to the Baudot decoder, which uses the current LTRS or FIGS state to determine the decoded character. The mark and space transmitted carriers do not overlap, although this can occur after they pass through certain HF propagation conditions. Sophisticated demodulators can account for this distortion.

A software modem performs the same functions as a hardware terminal unit, except that the software modems can apply more sophisticated mathematics that would be too expensive to implement in hardware. Modern desktop computers have more than enough processing speed to implement an RTTY demodulator. Software modems first convert the audio signal into a sequence of binary numbers using an analog-to-digital converter which is usually part of an audio chip set either on the motherboard or sound card. Everything from that point on uses numerical algorithms to implement the demodulation processes.

FSK VS AFSK MODULATION

An RTTY signal is usually generated as an F1B or F2B emission. F1B is implemented by directly shifting an RF carrier between the two (mark and space) frequencies. This method of generating FSK is often called *direct FSK*, or *true FSK*, or simply FSK. F2B is implemented by shifting between two audio tones, instead of two RF carriers. The resultant audio is then sent to an SSB transmitter to become two RF carriers. This method of first generating an audio FSK signal and then modulating an SSB transmitter to achieve the same FSK spectrum is usually called AFSK (audio frequency shift keying).

AFSK can be generated by using either an upper sideband (USB) transmitter or a lower sideband (LSB) transmitter. With a USB transmitter, the mark tone must be the *higher* of the two audio tones in the audio FSK sig-

nal. The USB modulator will then place the corresponding mark carrier at the higher of the two FSK carrier frequencies. When LSB transmission is used, the mark tone must be the *lower* of the two audio tones in the audio FSK signal. The LSB modulator will then place the corresponding mark carrier at the higher of the two FSK carrier frequencies. As when receiving, the actual audio tones are of no importance. The important part of AFSK is to have the two audio tones separated by 170 Hz, and have the pair properly flipped to match the choice of USB or LSB transmission. (See the **Modulation** chapter for more information on USB and LSB modulation and the relationship between modulating frequency and transmitted signal frequency.)

When using AFSK with older transceivers, it is wise to choose a high tone pair so that harmonic by-products fall outside the passband of the transmitter. Because of this, a popular tone pair is 2125 Hz/2295 Hz. Most transceivers will pass both tones and also have good suppression of the harmonics of the two tones. Not all transceivers that have an FSK input are FSK transmitters. Some transceivers will take the FSK keying input and modulate an internal AFSK generator, which is then used to modulate an SSB transmitter. In this case, the transmitter is really operating as an F2B emitter. This mode of operation is often called "keyed AFSK."

"SPOTTING" AN RTTY SIGNAL

By convention, RTTY signals are identified by the frequency of the mark carrier on the RF spectrum. "Spotting" the suppressed carrier frequency dial of an SSB receiver is useless for someone else unless they also know whether the spotter is using upper or lower sideband and what tone pair the spotter's demodulator is using. The mark and space carriers are the only two constants, so the amateur RTTY standard is to spot the frequency of the mark carrier.

DIDDLE CHARACTERS

In between the stop bit of a preceding character and the start bit of the next character, the RTTY signal stays at the mark frequency. When the RTTY decoder is in this "rest" state, a mark-to-space transition tells a decoder that the start of a new character has arrived. Noise that causes a start bit to be misidentified can cause the RTTY decoder to fall out of sync. After losing sync, the decoder will use subsequent data bits to help it identify the location of the next potential start bit.

Since all mark-to-space transitions are potential locations of the leading edge of a start bit, this can cause multiple characters to be incorrectly decoded until proper synchronization is again achieved. This "character slippage" can be minimized somewhat by not allowing the RTTY signal to rest for longer than its stop bit duration. An idle or *diddle character* (so called because of the sound of the demodulated audio from an idle RTTY signal sending the idle characters) is inserted immediately after a stop bit when the operator is not actively typing. The idle character is a non-printing character from the Baudot set and most often the LTRS character is used. Baudot encodes a LTRS as five bits of all ones making it particularly useful when the decoder is recovering from a misidentified start bit.

An RTTY diddle is also useful when there is selective fading. Good RTTY demodulators counter selective fading by measuring the amplitudes of the mark and space signals and automatically adjusting the decoding threshold when making the decision of whether a mark or a space is being received. If a station does not transmit diddles and has been idle for a period of time, the receiver will have no idea if selective fading has affected the space frequency. By transmitting a diddle, the RTTY demodulator is ensured of a measurement of the strength of the space carrier during each character period.

UNSHIFT-ON-SPACE (USOS)

Since the Baudot code aliases characters (for example, Q is encoded to the same 5-bit code as 1) using the LTRS and FIGS Baudot shift to steer the decoder, decoding could turn into gibberish if the Baudot shift characters are altered by noise. For this reason, many amateurs use a protocol called *unshift-on-space* (USOS). Under this protocol, both the sender and the receiver agree that the Baudot character set is always shifted to the LTRS state after a space character is received. In a stream of text that includes space characters, this provides additional, implicit, Baudot shifts.

Not everyone uses USOS. When used with messages that have mostly numbers and spaces, the use of USOS causes extra FIGS characters to be sent. A decoder that complies with USOS will not properly decode an RTTY stream that does not have USOS set. Likewise, a decoder that has USOS turned off will not properly decode an RTTY stream that has USOS turned on.

OTHER FSK SHIFTS AND RTTY BAUD RATE

The most commonly used FSK shift in amateur RTTY is 170 Hz. However, on rare occasions stations can be found using 425 and 850 Hz shifts. The wider FSK shifts are especially useful in the presence of selective fading since they provide better frequency diversity than 170 Hz.

Because HF packet radio uses 200 Hz shifts, some TNCs use 200 Hz as the FSK shift for RTTY. Although they are mostly compatible with the 170 Hz shift protocol, under poor signal to noise ratio conditions these demodulators will produce more er-

ror hits than a demodulator that is designed for 170 Hz shift. Likewise, a signal that is transmitted using 170 Hz shift will not be optimally copied by a demodulator that is designed for a 200 Hz shift.

To conserve spectrum space, amateurs have experimented with narrower FSK shifts, down to 22.5 Hz. At 22.5 Hz, optimal demodulators are designed as *minimal shift keyed* (*MSK*) instead of frequency shift keyed (FSK) demodulators.

SOME PRACTICAL CHARACTERISTICS

When the demodulator is properly implemented, RTTY can be very resilient against certain HF fading conditions, namely when selective fading causes only one of the two FSK carriers to fade while the other carrier remains strong. However, RTTY is still susceptible to "flat fading" (where the mark and space channels both fade at the same instant). There is neither an error correction scheme nor an interleaver (a method of rearranging — interleaving — the distribution of bits to make errors easier to correct) that can make an RTTY decoder print through a flat and deep fade. The lack of a data interleaver, however, also makes RTTY a very interactive mode. There is practically no latency when compared to a mode such as MFSK16, where the interleaver causes a latency of over 120 bit durations before incoming data can even be decoded. This makes RTTY attractive to operating styles that have short exchanges with rapid turnarounds, such as in contests.

Although RTTY is not as "sensitive" as PSK31 (when there is no multipath, PSK31 has a lower error rate than RTTY when the same amount of power is used) it is not affected by phase distortion that can render even a strong PSK31 signal from being copied. When HF propagation conditions deteriorate, RTTY can often function as long as sufficient power is used. Tuning is also moderately uncritical with RTTY. When the signal-to-noise ratio is good, RTTY tuning can be off by 50 Hz and still print well.

16.2.2 PSK31

PSK31 is a family of modes that uses differentially encoded varicode (see the next section), envelope-shaped phase shift keying. BPSK31, or binary PSK31, operates at 31.25 bit/s (one bit every 32 ms). QPSK31, or quadrature PSK31, operates at 31.25 baud. Each symbol consists of four possible quadrature phase change (or *dibits*) at the signaling rate of one dibit every 32 ms. QPSK31 sends a phase change symbol of 0°, 90°, 180° or 270° every 32 ms.

Characters that are typed from the keyboard are first encoded into variable-length varicode binary digits. With BPSK31, the

varicode bits directly modulate the PSK31 modulator, causing a 180° phase change if the varicode bit is a 0 and keeping a constant phase if the varicode bit is a 1. With QPSK31, the varicode bits first pass through two convolution encoders to create a sequence of bit pairs (dibits). Each dibit is then used to shift the QPSK31 modulator into one of four different phase changes.

PSK63 is a double-clock-rate version of a PSK31 signal, operating at the rate of one symbol every 16 ms (62.5 symbols per second). PSK125 is a PSK31 signal clocked at four times the rate, with one symbol every 8 ms (125 symbols per second). Although PSK63 and PSK125 are both in use, including the binary and quadrature forms, the most popular PSK31 variant remains BPSK31.

Most implementations of PSK31 first generate an audio PSK31 signal. The audio signal is then used to modulate an SSB transmitter. Since BPSK31 is based upon phase reversals, it can be used with either upper sideband (USB) or lower sideband (LSB) systems. With QPSK31 however, the 90° and 270° phase shifts have to be swapped when using LSB transmitters or receivers.

PSK31 VARICODE

Characters that are sent in PSK31 are encoded as varicode, a variable length prefix code as varicode. As described earlier, characters that occur more frequently in English text, such as spaces and the lower-case e, are encoded into a fewer number of bits than characters that are less frequent in English, such as the character q and the upper case E.

PSK31 varicode characters always end with two bits of zeros. A space character is sent as a one followed by the two zeros (100); a lower case e is sent as two ones, again terminated by the two bits of zero (1100). None of the varicode code words contain two consecutive zero bits. Because of this, the PSK31 decoder can uniquely identify the boundary between characters. A special "character" in PSK31 is the idle code, which consists of nothing but the two prefix bits. A long pause at the keyboard is encoded into a string of even numbers of zeros. The start of a new character is signaled by the first non-zero bit received after at least two consecutive zeros.

CONVOLUTION CODE

As described earlier, QPSK31 encodes the varicode stream with two convolution encoders to form the dibits that are used to modulate the QPSK31 generator. Both convolution encoders are fourth-order polynomials, shown in **Fig 16.2**.

The varicode data is first inverted before the bits are given to the convolution encoders. The first polynomial generates the most significant bit and the second polynomial generates the least significant bit of the dibit. Since we are

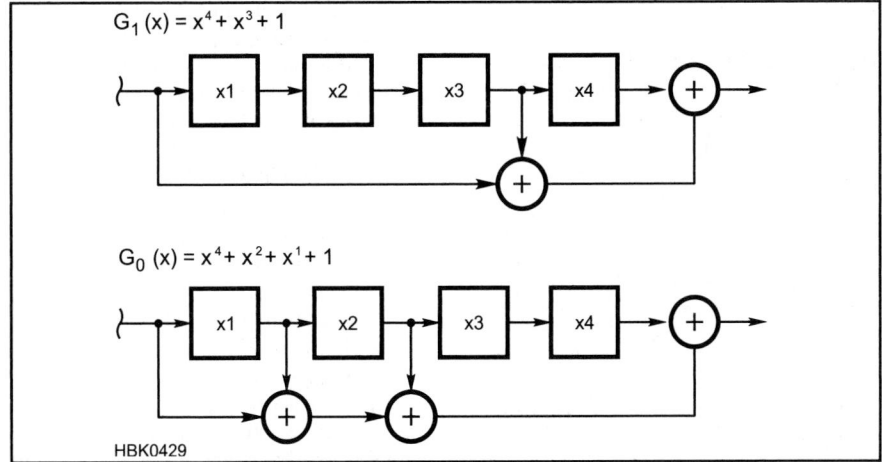

$$G_1(x) = x^4 + x^3 + 1$$

$$G_0(x) = x^4 + x^2 + x^1 + 1$$

HBK0429

Fig 16.2 — QPSK31 convolution encoders.

Table 16.4
QPSK31 Modulation

Dibit	Phase Change
00	0°
01	90°
10	180°
11	270°

working with binary numbers, the GF(2) sums shown in the above figure are exclusive-OR functions and the delay elements x1, x2, x3 and x4 are stages of binary shift registers. As each inverted varicode bit is available, it is clocked into the shift register and the two bits of the dibit are computed from the shift register taps.

MODULATION

PSK31 uses both differential phase shift modulation and envelope modulation to maintain its narrow-band characteristics. The most common way to generate an envelope-shaped PSK31 signal is to start with baseband in-phase (I) and quadrature (Q) signals. Both I and Q signals settle at either a value of +1 or a value of –1. When no phase transition is needed between symbols, the I and Q signals remain constant (at their original +1 or –1 values). See the **Modulation** chapter for information on creating I and Q signals.

To encode a 180° transition between symbols, both I and Q signals are slewed with a cosinusoidal envelope. If the in-phase signal had a value of +1 during the previous symbol, it is slewed to –1 cosinusoidally. If the in-phase signal had a value of –1 in the previous symbols, it is slewed cosinusoidally to +1. The quadrature signal behaves in a likewise manner. To encode a 90° phase shift in QPSK31, only the in-phase signal is slewed between the two symbols, the quadrature signal remains constant between the two symbols. To encode a 270° phase shift in QPSK31, only the quadrature signal is slewed between the two symbols; the in-phase signal remains constant between the two symbols.

The envelope of the real signal remains constant if there is no phase change. When the signal makes a 180° phase transition, the amplitude of a PSK31 signal will drop to zero in between the two symbols. The actual phase reversal occurs when the signal amplitude is zero, when there is no signal energy. During the 90° and 270° phase shifts between

symbols of QPSK31, the amplitude of the signal does not reach zero. It dips to only half the peak power in between the two symbols.

To provide a changing envelope when the operator is idle at the keyboard, a zero in the varicode (remember that an idle varicode consists of two 0 bits) is encoded as a 180° phase change between two BPSK31 symbols. A 1 in the varicode is encoded as no phase change from one BPSK31 symbol to the next symbol. This changing envelope allows the receiver to extract bit timing information even when the sender is idle. Bit clock recovery is implemented by using a comb filter on the envelope of the PSK31 signal.

The convolution code that is used by QPSK31 converts a constant idle stream of zeros into a stream of repeated 10 dibits. To produce the same constant phase change during idle periods as BPSK31, the QPSK31 10 dibit is chosen to represent the 180° phase shift modulating term. **Table 16.4** shows the QPSK31 modulation.

To produce bit clocks during idle, combined with the particular convolution code that was chosen for QPSK31, results in a slightly sub-optimal (non-Gray code) encoding of the four dibits. At the end of a transmission, PSK31 stops all modulation for a short period and transmits a short unmodulated carrier. The intended function is as a squelch mechanism.

DEMODULATION AND DECODING

Many techniques are available to decode differential PSK, where a reference phase is not present. Okunev has a good presentation of the methods (reference: Yuri Okunev, *Phase and Phase-difference Modulation in Digital Communications* (1997, Artech House), ISBN 0-89006-937-9).

As mentioned earlier, there is sufficient amplitude information to extract the bit clock from a PSK31 signal. The output of a differential-phase demodulator is an estimate of the phase angle difference between the centers

of one symbol and the previous one. With BPSK31, the output can be compared to a threshold to determine if a phase reversal or a non-reversal is more likely. The decoder then looks for two phase reversals in a row, followed by a non-reversal, to determine the beginning of a new character. The bits are gathered until two phase reversals are again seen and the accumulated bits are decoded into one of the characters in the varicode table.

QPSK31 decoding is more involved. As in the BPSK31 case, the phase difference demodulator estimates the phase change from one bit to another. However, one cannot simply invert the convolution function to derive the data dibits. Various techniques exist to decode the measured phase angles into dibits. The *Viterbi algorithm* is a relatively simple algorithm for the convolution polynomials used in QPSK31. The estimated phase angles can first be fixed to one of the four quadrature angles before the angles are submitted to the Viterbi algorithm. This is called a hard-decision *Viterbi decoder*.

A soft-decision Viterbi decoder (that is not that much more complex to construct) usually gives better results. The soft-decision decoder uses arbitrary phase angles and some measure of how "far" an angle is away from one of the four quadrature angles. Error correction occurs within the trellis that implements the Viterbi algorithm. References to convolution code, trellis and the Viterbi algorithm can be found at **http://en.wikipedia.org/wiki/Convolutional_code**.

16.2.3 MFSK16

MFSK16 uses M-ary FSK modulation with 16 "tones" (also known as 16-FSK modulation), where only a single tone is present at any instant. MFSK16 has a crest factor of 1, with no wave shaping performed on the data bits. The tone centers of MFSK16 are separated by 15.625 Hz. Data switches at a rate of 15.625 baud (one symbol change every 64 ms).

Characters are first encoded into the MFSK16 varicode, creating a constant bit stream whose rate is one bit every 32 ms. This results in a similar, although not identical, character rate as PSK31. The difference is due to the different varicode tables that are used by the two modes.

This bit stream is clocked into a pair of 6th-order convolution encoders. The result

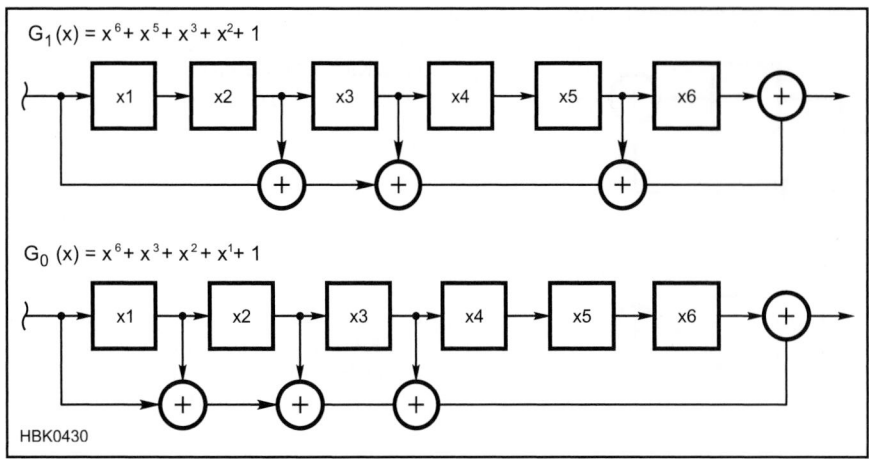

$$G_1(x) = x^6 + x^5 + x^3 + x^2 + 1$$

$$G_0(x) = x^6 + x^3 + x^2 + x^1 + 1$$

HBK0430

Fig 16.3 — MFSK16 convolution encoders.

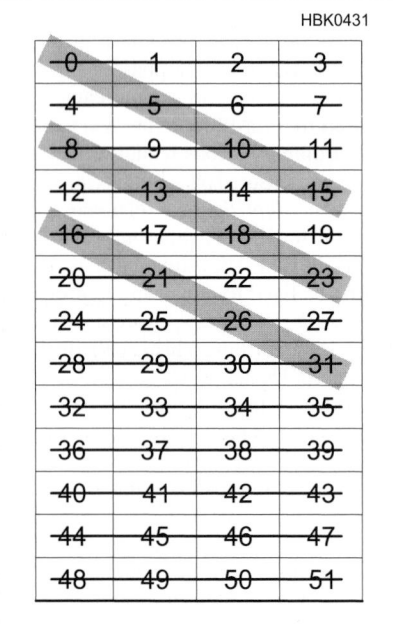

HBK0431

Fig 16.4 — IZ8BLY interleaver.

is a pair of bits (dibit) for each varicode bit, at a rate of one dibit every 32 ms. Consecutive pairs of dibits are next combined into sets of four bits (*quadbit* or *nibble*). Each nibble, at the rate of one per 64 ms, is then passed through an interleaver. The interleaver produces one nibble for each nibble that is sent to it. Each nibble from the interleaver is then Gray-coded and the resultant 4-bit Gray code is the ordered index of each 16 tones that are separated by 15.625 Hz. The result is a 16-FSK signal with a rate of one symbol per 64 ms. Since the symbol time is the reciprocal of the tone separation, a 16-point fast Fourier transform (FFT) can be conveniently used as an MFSK16 modulator.

MFSK16 VARICODE

Although the varicode table that is used by MFSK16 (see the *Handbook* CD) is not the same as the one used by PSK31, they share similar characteristics. Please refer to the PSK31 section of this chapter for a more detailed description of varicode. Unlike PSK31 varicode, MSK16 varicode encodings

can contain two or more consecutive zero bits, as long as the consecutive zeros are at the tail of a code word. Character boundaries are determined when two or more consecutive zeros are followed by a one.

CONVOLUTION CODE

As described earlier, MFSK16 encodes the varicode stream with two convolution encoders to form the dibits that are passed on to the interleaver. Both convolution encoders are sixth order polynomial, shown in **Fig 16.3**.

The first polynomial generates the most significant bit and the second polynomial generates the least significant bit of the dibit. Since we are working with binary numbers, the GF(2) sums shown in the above figure are exclusive OR functions and the delay elements x1, x2, x3, x4, x5 and x6 are stages of binary shift registers. As each varicode bit is available, it is clocked into the shift register and the two bits of the dibit are computed from the shift register taps.

INTERLEAVER

HF fading channels tend to generate "burst" errors that are clumped together. An interleaver is used to permute the errors temporally so that they appear as uncorrelated errors to the convolution decoder. MFSK16 uses 10 concatenated stages of an IZ8BLY Diagonal Interleaver. More information on the interleaver can be found at **www.qsl.net/ zl1bpu/MFSK/Interleaver.htm**. **Fig 16.4** illustrates how a single IZ8BLY interleaver spreads a sequence of bits.

The bits enter the IZ8BLY interleaver in the order 0, 1, 2, 3, 4,... and are passed to the output in the order 0, 5, 10, 15, 8, 13, ... (shown in the diagonal boxes). In MFSK16, the output of one interleaver is sent to the input of a second interleaver, for a total of 10 such stages. Each stage spreads the bits out over a longer time frame.

This concatenated 10-stage interpolator

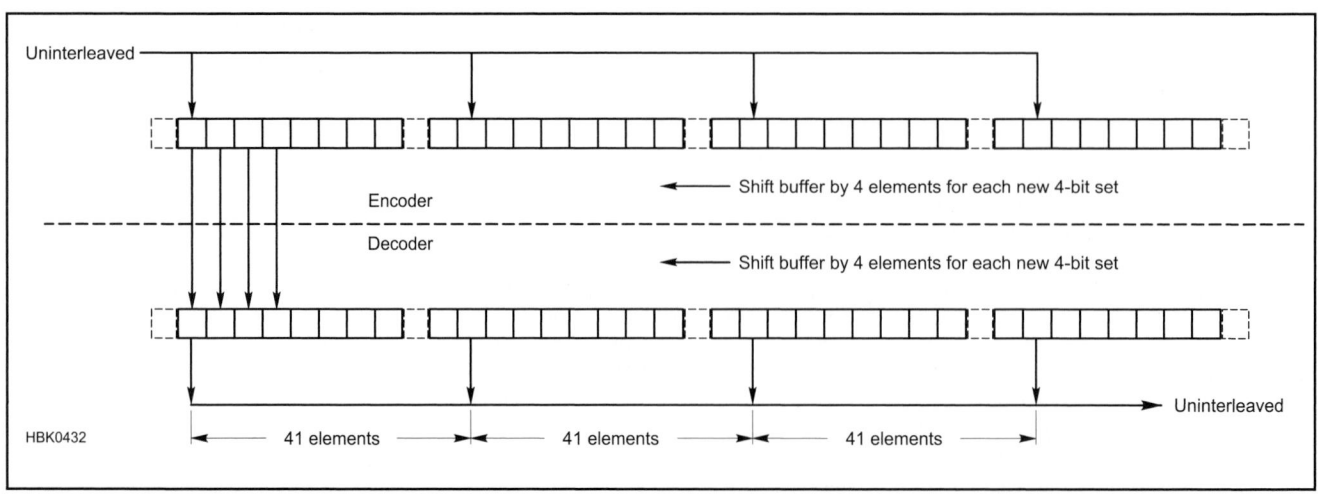

HBK0432

Fig 16.5 — Bit spreading through interpolation.

is equivalent to a single interpolator that is 123 bits long. **Fig 16.5** shows the structure of the single interpolator and demonstrates how four consecutive input bits are spread evenly over 123 time periods.

An error burst can be seen to be spread over a duration of almost two seconds. This gives MFSK16 the capability to correct errors over deep and long fades. While it is good for correcting errors, the delay through the long interleaver also causes a long decoding latency.

GRAY CODE

The Gray code creates a condition where tones that are next to one another are also different by a smaller *Hamming* distance. This optimizes the error correction process at the receiver. References to Gray code and Hamming distance can be found at **http://en.wikipedia.org/wiki/Gray_code** and **http://en.wikipedia.org/wiki/Hamming_distance** and **http://en.wikipedia.org/wiki/Hamming_distance**.

DEMODULATION AND DECODING

A 16-point FFT can be used to implement a set of matched filters for demodulating an MFSK16 signal once the input waveform is properly time aligned so that each transform is performed on an integral symbol and the signal is tuned so that the MFSK16 tones are perfectly centered in the FFT bins. A reference to a matched filter can be found at **http://en.wikipedia.org/wiki/Matched_filter**.

The 16 output bins from an FFT demodulator can first be converted to the "best" 4-bit index of an FFT frequency bin, or they can be converted to a vector of four numerical values representing four "soft" bits. The four bits are then passed through an MFSK16 de-interleaver. In the case of "soft decoding," the de-interleaver would contain numerical values rather than a 0 or 1 bit.

The output of the de-interleaver is passed into a convolution decoder. The Gray code makes sure that adjacent FFT bins also have the lowest Hamming distance; i.e., the most likely error is also associated with the closest FFT bin. The hard or the soft Viterbi Algorithm can be used to decode and correct errors.

16.2.4 DominoEX

DominoEX is a digital mode with MFSK (multi-frequency shift keying) designed for simplex chats on HF by Murray Greenman, ZL1BPU. It was designed to be easier to use and tune than other similar modes, offer low latency for contesting or other quick exchange situations, offer reliable copy down into the noise floor, and work well as an NVIS (near-vertical incidence skywave, see the **Propagation of Radio Signals** chapter) mode for emergency communications.

Generally MFSK requires a high degree of tuning accuracy and frequency stability and can be susceptible to multipath distortion. DominoEX specifically addresses these issues. To avoid tuning issues, IFK (*incremental frequency keying*) is used. With IFK, the data is represented not by the frequency of each tone, but by the frequency difference between one tone and the next. It also uses offset incremental keying to reduce inter-symbol interference caused by multipath reception. These techniques provide a tuning tolerance of 200 Hz and a drift of 200 Hz/minute. DominoEX also features an optional FEC mode that increases latency but provides communications over even more difficult channels. More information can be found online at **www.qsl.net/zl1bpu/MFSK/DEX.htm**.

DominoEX uses M-ary FSK modulation with 18 tones in which only a single tone is present at any instant. Information is sent as a sequence of separate 4-bit ("nibble") symbols. The value of each nibble is represented during transmission as the position of the single tone.

The position of the tone is computed as the difference of the current nibble value from the nibble value of the previously transmitted symbol. In addition, a constant offset of 2 is applied to this difference. Because there are 18 tones, any possible 4-bit value between 0 and 15 can be represented, including the offset of 2.

The additional offset of 2 tone positions ensures that a transmitted tone is separated from the previously transmitted tone by at least two tone positions. It is thus impossible for two sequential symbols to result in the same tone being transmitted for two sequential tone periods. This means sequential tones will always be different by at least two positions, an important consideration in maintaining sync.

This minimum separation of successive tones of incremental frequency keying (IFK) in DominoEX reduces the inter-symbol distortion that results from a pulse being temporally smeared when passing through an HF channel. The double-tone spacing of DominoEX modes (see **Table 16.5**) further reduces inter-symbol distortion caused by frequency smearing.

Incremental frequency keying allows the DominoEX nibbles to immediately decode without having to wait for the absolute tone to be identified. With MFSK16, a tone cannot be uniquely identified until the lowest and highest of the 16 tones have passed through the receiver. This contributes to the decoding latency of MFSK16. There is no such latency with IFK.

Since IFK depends upon frequency differences and not absolute tone frequencies, DominoEX tolerates tuning errors and drifting signals without requiring any additional automatic frequency tracking algorithms.

Like MFSK16, the DominoEX signal is not wave-shaped and has constant output power. The baud rates for DominoEX are shown in Table 16.5. The tone spacings for DominoEX 11, DominoEX 16 and DominoEX 16 have the same values as their baud rates. The tone spacings for DominoEX 8, DominoEX 5 and DominoEX 4 are twice the value of their baud rates.

Unlike PSK31 and MFSK16, characters in DominoEX are encoded into varicode nibbles instead of encoding into varicode bits. The DominoEX varicode table can be found in the CD-ROM included with this book.

BEACON MESSAGE

Instead of transmitting an idle varicode symbol when there is no keyboard activity, DominoEX transmits a "beacon" message from an alternate set of varicode (SECVAR columns in the DominoEX varicode table). This user-supplied repeating beacon message is displayed at the receiving station when the sending station is not actively sending the primary message. On average, the character rate of the beacon channel is about half of the character rate of the primary channel.

FORWARD ERROR CORRECTION (FEC)

When FEC is turned off, DominoEX has very low decoding latency, providing an interactive quality that approaches RTTY and PSK31. The first character is decoded virtually instantly by the receiver after it is transmitted. Because of that, FEC is not usually used even though it is available in most software that implements DominoEX.

DominoEX FEC is similar to the FEC that is used in MFSK16. When FEC is on, each 4-bit IFK symbol that is decoded by the receiver is split into two di-bits. The dibits enter an identical convolution coder to that used

Table 16.5
Comparison of DominoEX Modes

Mode	Baud/ (sec)	BW (Hz)	Tones	Speed (WPM)	FEC (WPM)	Tone Spacing
DominoEX 4	3.90625	173	18	~25	~12	Baud rate ×2
DominoEX 5	5.3833	244	18	~31	~16	Baud rate ×2
DominoEX 8	7.8125	346	18	~50	~25	Baud rate ×2
DominoEX 11	10.766	262	18	~70	~35	Baud rate ×1
DominoEX 16	15.625	355	18	~100	~50	Baud rate ×1
DominoEX 22	21.533	524	18	~140	~70	Baud rate ×1

by MFSK16. However, instead of a 10-stage IZ8BLY interleaver (see Fig 16.4), only 4 cascaded stages of the basic 4-bit interleaver are present in DominoEX. In the presence of long duration fading, the performance of the shortened interleaver is moderately poor when used with DominoEX 16 and DominoEX 22. However, the interleaver is quite efficient in countering fading when FEC is used with DominoEX 4, DominoEX 5 and DominoEX 8, with their longer symbol periods.

Since DominoEX works in dibit units rather than nibble units when FEC is turned on, it also switches to using the same binary varicode used by MFSK16 instead of using the nibble-based varicode. DominoEX does not implement Gray code as is used by MFSK16 FEC.

Even without FEC, DominoEX works well under many HF propagation conditions, including the ITU NVIS ("Mid-latitude Disturbed") propagation profile. However, there are conditions where DominoEX is not usable unless FEC is switched on, specifically the CCIR Flutter and ITU High Latitude Moderate Conditions profiles. DominoEX modes, especially those with tone spacings that are twice the baud rate, are very robust even under these extreme conditions once FEC is switched on.

DominoEX performance charts (character error rates versus signal-to-noise ratios) are included as the HTML document "DominoEX Performance" on the CD-ROM accompanying this book and online at **http://homepage.mac.com/chen/Technical/DominoEX/Measurements/index.html**.

CHIP64/128

Chip64 and Chip128 modes were released in 2004 by Antonino Porcino, IZ8BLY. The modes were tested on the air by IZ8BLY, Murray Greenman, ZL1BPU, (who also contributed in the design of the system), Chris Gerber, HB9BDM, and Manfred Salzwedel, OH/DK4ZC. According to IZ8BLY, "The design of this new digital mode served to introduce the spread spectrum technology among radio amateurs by providing a communication tool to experiment with. Its purpose was to prove that it's possible to take advantage of the spread spectrum techniques even on the HF channels, making the communication possible under conditions where traditional narrowband modes fail.

"Among the different possible implementations of spread spectrum, Chip64 uses the so-called Direct Sequence Spread Spectrum (DSSS). In a DSSS transmission, the low speed signal containing the data bits to be transmitted is mixed (multiplied) with a greatly higher speed signal called code. The result of this mixing operation, called *spreading*, is a high-speed bit stream which is then transmitted as a normal DBPSK. Indeed, a DSSS signal looks like nothing else than wideband BPSK.

"The system proved to be efficient and we found it comparable to the other modern digital modes. Being totally different in its architecture, it shows better performance during certain circumstances, while in others it shows no actual gain. In particular, it performs better under multipath where normal BPSK can't track arriving symbols, but in quiet environments it doesn't show any improvement over plain BPSK. This is expected because of the losses that occur due to the imperfect autocorrelation of the codes."

Chip64 has a total data rate of 37.5 bit/s and the more robust Chip128 is 21.09 bit/s. Both use the same varicode used by MFSK16. The software can be downloaded from **http://xoomer.virgilio.it/aporcino/Chip64/index.htm** and more information is available at **www.arrl.org/technical-characteristics**. Spread spectrum is discussed in the **Modulation** chapter, as well.

16.2.5 THROB

Throb is an experimental mode written by Lionel Sear, G3PPT, and gets the name from the "throbbing" sound it makes on the air. It uses either single tones or pairs from a possible nine tones spaced 8 or 16 Hz apart, resulting in a bandwidth of 72 or 144 Hz, respectively. It has three transmission speeds — 1, 2 and 4 throbs/s — resulting in data rates of 10, 20 and 40 WPM, respectively. The 1 and 2 throb/s speeds use a tone spacing of 8 Hz for a 72 Hz bandwidth and the 4-throb/s speed uses a spacing of 16 Hz for a 144 Hz bandwidth. It is implemented as a stand-alone application or included in a multimode package such as *MixW* (**www.mixw.net**).

16.2.6 MT63

MT63 is a mode developed by Pawel Jalocha, SP9VRC. MT63 is very complex with wide bandwidth, low speed and very high noise immunity. By using 64 different modulated tones, MT63 includes a large amount of extra data in the transmission of each character, so that the receiving equipment can work out, with no ambiguity, which character was sent, even if 25% of the character is obliterated. MT63 also features a secondary channel that operates simultaneously with the main channel that can be used for an ID or beacon.

MT63 likely has the most extensive error correction and can be quite processor intensive. It uses a Walsh function that spreads the data bits of each 7-bit ASCII character across all 64 of the tones of the signal spectrum and simultaneously repeats the information over a period of 64 symbols within any one tone. This coding takes several seconds. The combination of time domain (temporal) and frequency domain (spectral) interleaving re- sults in superb impulse noise rejection. At the same time, in the frequency domain, significant portions of the signal can be masked by unwanted noise or other transmissions without any noticeable effect on successful reception.

On each of the 64 tones, the transmission data rate is fairly slow, which suits the nature of ionospheric disturbances. Despite the low data rate, good text speed is maintained because the text is sent on many tones at once. The system runs at several different speeds, which can be chosen to suit conditions but 100 WPM is typical of the MT63-1K mode. Although the 1 kHz bandwidth mode is typical, MT63 can also run at 500 Hz and 2 kHz bandwidth where the tone spacing and baud rate are halved or doubled and the throughput is halved or doubled, respectively.

Tuning of MT63 modes is not critical. This is because the mode can use FEC techniques to examine different combinations of the 64 tones that calculate the correct location within the spectrum. As an example, MT63-1K will still work if the decoder is off-frequency by as much as 100 Hz. MT63-2K requires even less precision and can tolerate an error of 250 Hz.

The incredible noise immunity comes at a price beyond the large bandwidth required. There is a large latency caused by the error correction and interleaving process. Quick-turnaround QSOs are not possible because there is a several second delay between typing the last character and it being transmitted.

Without confirming each transmission with some type of ARQ mode, there is no more robust digital mode than MT63. The mode was evaluated and recommended for Navy MARS message handling. The evaluation is published on the Navy MARS website (**www.navymars.org**), along with other information on this mode.

16.2.7 Olivia

Olivia is an MFSK-based protocol designed to work in difficult (low signal-to-noise ratio plus multipath propagation) conditions on the HF bands. The signal can still be copied accurately at 10 dB below the noise floor. Olivia was developed in 2003 by Pawel Jalocha, SP9VRC, and performs well for digital data transfer with white noise, fading and multipath, polar path flutter and auroral conditions.

Olivia transmits a stream of 7-bit ASCII characters. The characters are sent in blocks of five with each block requiring two seconds to transmit. This results in an effective data rate of 2.5 characters/second or 150 characters/minute. A transmission bandwidth of 1000 Hz and the baud rate of 31.25 MFSK tones/second, also known as *Olivia 1000/32*, is the most common. To adapt to different propagation conditions, the number of tones and the bandwidth can be changed and the time and frequency parameters are proportionally scaled. The number of tones can be 2, 4, 8, 16, 32, 64, 128 or 256 and the bandwidth can be 125, 250, 500, 1000 or 2000 Hz.

The Olivia is constructed of two layers: the lower, modulation and FEC code layer is a classical MFSK while the higher layer is an FEC code based on Walsh functions. More detail on Walsh functions is available online at **en.wikipedia.org/wiki/Walsh_function**.

Assuming Olivia 1000/32 is being used, in the first layer the orthogonal functions are cosine functions, with 32 different tones. Since only one of those 32 tones is being sent at a time, the demodulator measures the amplitudes of all the 32 possible tones and identifies the tone with the highest amplitude. In the second layer every ASCII character is encoded as one of 64 possible Walsh functions. The receiver again measures the amplitudes for all 64 vectors and selects the greatest as the true value.

To avoid simple transmitted patterns (like a constant tone) and to minimize the chance for a false lock at the synchronizer, the characters encoded into the Walsh function pass through a scrambler and interleaver. The receiver synchronizes automatically by searching through time and frequency offsets for a matching pattern.

More information can be found online at **http://n1su.com/olivia** and Olivia is supported in a number of digital multimode packages such as *MixW*, *MultiPSK* and *Ham Radio Deluxe*.

16.3 Fuzzy Modes

There is a group of modes referred to as "fuzzy modes" because although they are machine generated and decoded, they are designed to be human-read. These include facsimile (fax), slow-scan TV (SSTV) and Hellschreiber.

16.3.1 Facsimile (fax)

Facsimile was developed as a mechanically transmitted technology where the source material was placed on a spinning drum and scanned line by line into an electrical signal which would be transmitted by wire or over the air. It is important that the receiving station have their drum spinning at the correct speed in order to correctly recreate the image. A value known as the *index of cooperation* (IOC) must also be known to decode a transmission. IOC governs the image resolution and is the product of the total line length and the number of lines per unit length divided by π. Most fax transmissions are sent with LPM (RPM) at 120 and an IOC of 576.

Facsimile is generally transmitted in single sideband with a tone of 1500 Hz representing black and 2300 Hz representing white. The *automatic picture transmission* (*APT*) format is used by most terrestrial weather facsimile stations and geostationary weather satellites. It features a start tone that triggers the receiving system, originally used to allow the receiving drum to come up to speed. It also includes a phasing signal with a periodic pulse that synchronizes the receiver so the image appears centered on the page. A stop tone, optionally followed by black, indicates the end of the transmission. The APT format is shown in **Table 16.6**.

Stations with Russian equipment sometimes use RPM 60 or 90 and sometimes an IOC of 288. Photofax transmissions such as those from North Korea use RPM 60 and an IOC 352 with gray tones, and satellite rebroadcast use also RPM 120 IOC 576, with gray tones (4 or more bit depth). For software decoding of weather fax images it is best to decode with Black and White (2-bit depth).

16.3.2 Slow-Scan TV (SSTV)

Slow-Scan TV or SSTV is similar to facsimile where a single image is converted to individual scanned lines and those lines sent as variable tones between 1500 and 2300 Hz. Modern systems use computer software and a sound card to generate and receive the required tones. (Some SSTV communication uses purely digital protocols, such as WinDRM described later, in which the picture content is sent as digital data and not directly represented in the modulation scheme.)

There are a number of different SSTV "modes" that define image resolution and color scheme. A color image takes about 2 minutes to transmit, depending on mode. Some black and white modes can transmit an image in under 10 seconds. More information about SSTV may be found in the **Image Communications** supplement on the *Handbook* CD.

16.3.3 Hellschreiber, Feld-Hell or Hell

Hellschreiber is a facsimile-based mode developed by Rudolph Hell in the 1920s. The name is German and means "bright writer" or "light writer" and is a pun on the inventor's name. In Hellschreiber, text is transmitted by dividing each column into seven pixels and transmitting them sequentially starting at the lowest pixel. Black pixels are transmitted as a signal and white as silence at 122.5 bit/s (about a 35 WPM text rate). Originally the text was printed on continuous rolls of paper so the message could be any length.

Even though each pixel is only transmitted once, they are printed twice, one below the other. This compensates for slight timing errors in the equipment that causes the text to slant. If properly in sync, the text will appear as two identical rows, one below the other or a line of text in the middle with chopped lines top and bottom. Regardless of the slant, it is always possible to read one copy of the text. Since the text is read visually, it can be sent in nearly any language and tends to look like an old dot matrix printer. More information can be found online at **www.qsl.net/zl1bpu/ FUZZY/Feld.htm** and Randy, K7AGE, has a great introduction to Hellschreiber available on YouTube at **www.youtube.com/ watch?v=yR-EmyEBVqA**.

Table 16.6
Facsimile Automatic Picture Format

Signal	Duration	IOC576	IOC288	Remarks
Start Tone	5 s	300 Hz	675 Hz	200Hz for color fax modes
Phasing Signal	30 s			White line interrupted by black pulse
Image	Variable	1200 lines	600 lines	At 120 LPM
Stop Tone	5 s	450 Hz	450 Hz	
Black	10 s			

16.4 Structured Digital Modes

This group of digital modes has more structured data. This provides more robust data connections and better weak signal performance or more sophisticated data. Each of these modes bundles data into packets or blocks that can be transmitted and error checked at the receive end.

16.4.1 FSK441

FSK441 was implemented in the *WSJT* software by Joe Taylor, K1JT, in 2001. FSK441 was designed for meteor scatter using four-tone frequency shift keying at a rate of 441 baud and was given the technical name FSK441. Meteor scatter presents an unusual problem because the signals are reflected by the ionization trails left by meteors that often last less than a second. The signals are weak, of short duration with widely varying signal strength and include some amount of Doppler shift. Since these meteors enter the atmosphere constantly, this mode can be used any time of day or year to make contacts on VHF between 500 and 1100 miles with a single large Yagi and 100 W.

FSK441 uses a 43-character alphabet compatible with the earlier PUA43 mode by Bob Larkin, W7PUA. This alphabet includes letters, numbers and six special characters and is shown in the **Table 16.7**. These characters are encoded by the four tones at 882, 1323, 1764 and 2205 Hz and are designated 0-3 in the table below. For example, the letter T has the code 210 and is transmitted by sending sequentially the tones at 1764, 1323 and 882 Hz. Since the modulation rate is specified as 441 baud, or 441 bit/s, the character transmission rate is 441/3 = 147 characters per second. At this speed, a ping lasting 0.1 s can convey 15 characters of text. Four signals (000, 111, 222 and 333) generate single solid tones that are easily decoded and are reserved for the standard meteor scatter messages — R26, R27, RRR and 73.

When receiving FSK441, the receiver time and frequency need to be synchronized precisely. The computer clock must be set as accurately as possible (ideally to WWV or other time standard) before attempting to transmit or receive. When the receiver hears a burst of signal the decoder attempts to identify the correct frequency shift and align the message. For reasons of transmission efficiency, no special synchronizing information is embedded in an FSK441 message. Instead, the proper synchronization is established from the message content itself, making use of the facts that (a) three-bit sequences starting with 3 are never used, and (b) the space character is coded as 033, as shown in FSK441 character code table. Messages sent by *WSJT* always contain at least one trailing space and the

software will insert one if necessary.

Timed message sequences are a must, and according to the procedures used by common consent in North America, the westernmost station transmits first in each minute for 30 seconds, followed by 30 seconds of receiving. *WSJT* messages are also limited to 28 characters which are plenty to send call sign and signal report information but clearly does not make *WSJT* a good ragchew mode.

The meteor scatter procedures are also well-defined and supported directly by templates in the *WSJT* software. Signal reports are conventionally sent as two-digit numbers. The first digit characterizes the lengths of pings being received, on a 1-5 scale, and the second estimates their strength on a 6-9 scale. The most common signal reports are 26 for weak pings and 27 for stronger ones, but under good conditions reports such as 38 and higher are sometimes used. Whatever signal report is sent to the QSO partner, it is important that it not be changed. You never know when pings will successfully convey fragments of your message to the other end of the path, and you want the received information to be consistent.

More information can be found at the *WSJT* website in the user manual and other articles at **http://pulsar.princeton.edu/~joe/K1JT**.

16.4.2 JT6M

JT6M is another *WSJT* mode optimized for meteor and ionospheric scatter on 6 meters. Like FSK441, JT6M uses 30 second periods for transmission and reception to look for enhancements produced by short-lived meteor trail ionization. It will also account for Doppler shift of ± 400 Hz. JT6M includes some improvements that allow working signals many dB weaker than FSK441.

JT6M uses 44-tone FSK with a synchronizing tone and 43 possible data tones — one for each character in the supported alphanumeric set, the same set used for FSK441. The sync tone is at 1076.66 Hz, and the 43 other possible tones are spaced at intervals of 11025/512 = 21.53 Hz up to 2002.59 Hz. Transmitted symbols are keyed at a rate of 21.53 baud, so each one lasts for 1/21.53 = 0.04644 s. Every 3rd symbol is the sync tone, and each sync symbol is followed by two data symbols. The transmission rate of user data is therefore (2/3)×21.53 = 14.4 characters per second. The transmitted signal sounds a bit like piccolo music.

Contacts in FSK441 and JT6M are often scheduled online on Ping Jockey at **www.pingjockey.net**.

16.4.3 JT65

JT65 is another *WSJT*-supported mode designed to optimize Earth-Moon-Earth (EME) contacts on the VHF bands, and conforms efficiently to the established standards and procedures for such QSOs. It also performs well for weak signal VHF/UHF, and for HF skywave propagation. JT65 includes error-correcting features that make it very robust, even with signals much too weak to be heard. With extended transmission duration and three decoders, JT65 can reliably decode 24 to 30 dB *below* the noise floor.

JT65 does not transmit messages character by character, as done in Morse code. Instead, whole messages are translated into unique strings of 72 bits, and from those into sequences of 63 six-bit symbols. These symbols are transmitted over a radio channel; some of them may arrive intact, while others are corrupted by noise. If enough of the symbols are correct, the full 72-bit compressed message can be recovered exactly. The decoded bits are then translated back into the human-readable message that was sent. The coding scheme and robust FEC assure that messages are never received in fragments. Message components cannot be mistaken for one another, and call signs are never displayed with a few characters missing or incorrect. There is no chance for the letter O or R in a call sign to be confused with a signal report or an acknowledgment.

JT65 uses 60-second transmit-receive sequences and carefully structured messages. Standard messages are compressed so that two call signs and a grid locator can be transmitted with just 71 bits. A 72nd bit serves as a flag to indicate that the message consists of arbitrary text (up to 13 characters) instead of

Table 16.7
FSK441 Character Codes

Character	Tones	Character	Tones
1	001	H	120
2	002	I	121
3	003	J	122
4	010	K	123
5	011	L	130
6	012	M	131
7	013	N	132
8	020	O	133
9	021	P	200
.	022	Q	201
,	023	R	202
?	030	S	203
/	031	T	210
#	032	U	211
space	033	V	212
$	100	W	213
A	101	X	220
B	102	Y	221
C	103	0	223
D	110	E	230
F	112	Z	231
G	113		

call signs and a grid locator. Special formats allow other information such as call sign prefixes (for example, ZA/PA2CHR) or numerical signal reports (in dB) to be substituted for the grid locator.

The aim of source encoding is to compress the common messages used for EME QSOs into a minimum fixed number of bits. After being compressed into 72 bits, a JT65 message is augmented with 306 uniquely defined error-correcting bits. The FEC coding rate is such that each message is transmitted with a "redundancy ratio" of 5.25. With a good error-correcting code, however, the resulting performance and sensitivity are far superior to those obtainable with simple five times message repetition. The high level of redundancy means that JT65 copes extremely well with QSB. Signals that are discernible to the software for as little as 10 to 15 seconds in a transmission can still yield perfect copy.

JT65 requires tight synchronization of time and frequency between transmitter and receiver. Each transmission is divided into 126 contiguous time intervals or symbols, each lasting 0.372 s. Within each interval the waveform is a constant-amplitude sinusoid at one of 65 pre-defined frequencies. Half of the channel symbols are devoted to a pseudo-random synchronizing vector interleaved with the encoded information symbols. The sync vector allows calibration of relative time and frequency offsets between transmitter and receiver. A transmission nominally begins at t = 1 s after the start of a UTC minute and finishes at t = 47.8 s.

WSJT does its JT65 decoding in three phases: a soft-decision *Reed-Solomon* decoder, the deep search decoder and the decoder for shorthand messages. In circumstances involving birdies, atmospherics, or other interference, operator interaction is an essential part of the decoding process. The operator can enable a "Zap" function to excise birdies, a "Clip" function to suppress broadband noise spikes and a "Freeze" feature to limit the frequency range searched for a sync tone. Using these aids and the program's graphical and numerical displays appropriately, the operator is well equipped to recognize and discard any spurious output from the decoder.

The JT65 procedures are also well-defined and supported directly by templates in the *WSJT* software. More information can be found at the *WSJT* website in the user manual and other articles at **http://pulsar.princeton.edu/~joe/K1JT**.

16.4.4 WSPR

WSPR (pronounced "whisper") implements a protocol designed for probing potential propagation paths with low-power transmissions. Each transmission carries a station's call sign, Maidenhead grid loca-

tor, and transmitter power in dBm. The program can decode signals with S/N as low as −28 dB in a 2500 Hz bandwidth. Stations with Internet access can automatically upload their reception reports to a central database called WSPRnet, which includes a mapping facility.

WSPR implements a protocol similar to JT65 called MEPT_JT (Manned Experimental Propagation Tests, by K1JT). In receive mode the program looks for all detectable MEPT_JT signals in a 200-Hz passband, decodes them, and displays the results. If nothing is decoded, nothing will be printed. In T/R mode the program alternates in a randomized way between transmit and receive sequences. Like JT65, MEPT_JT includes very efficient data compression and strong forward error correction. Received messages are nearly always exactly the same as the transmitted message, or else they are left blank.

Basic specifications of the MEPT_JT mode are as follows:
- Transmitted message: call sign + 4-character-locator + dBm Example: "K1JT FN20 30"
- Message length after lossless compression: 28 bits for call sign, 15 for locator, 7 for power level; 50 bits total.
- Forward error correction (FEC): Long-constraint convolutional code, K =32, r = 1/2.
- Number of channel symbols: nsym = $(50 + K - 1) \times 2 = 162$.
- Keying rate: 12000/8192 = 1.46 baud.
- Modulation: Continuous phase 4-FSK. Tone separation 1.46 Hz.
- Synchronization: 162-bit pseudo-random sync vector.
- Data structure: Each channel symbol conveys one sync bit and one data bit.
- Duration of transmission: $(162 \times 8192)/12000 = 110.6$ s
- Occupied bandwidth: About 6 Hz
- Minimum S/N for reception: Around −27 dB on the *WSJT* scale (2500 Hz reference bandwidth).

The current version and documentation of *WSPR* can be found at **www.physics.princeton.edu/pulsar/K1JT/wspr.html** and WSPRnet propagation data is available at **http://wsprnet.org**.

16.4.5 HF Digital Voice

AOR

In 2004, AOR Corporation introduced its HF digital voice and data modem, the AR9800. Digital voice offers a quality similar to FM with no background noise or fading as long as the signal can be properly decoded. The AR9800 can alternatively transmit binary files and images. AOR later released the AR9000 which is compatible with the AR9800 but less expensive and only sup-

ports the HF digital voice mode.

The AR9800 uses a protocol developed by Charles Brain, G4GUO. The protocol uses the AMBE (Advanced Multi-Band Excitation) codec from DVSI Inc. to carry voice. It uses 2400 bit/s for voice data with an additional 1200 bit/s for Forward Error Correction for a total 3600 bit/s data stream. The protocol is detailed below:
- Bandwidth: 300-2500 Hz, 36 carriers
- Symbol Rate: 20 ms (50 baud)
- Guard interval: 4 ms
- Tone steps: 62.5 Hz
- Modulation method: 36 carriers: DQPSK (3.6K)
- AFC: ±125 Hz
- Error correction: Voice: Golay and Hamming
- Video/Data: Convolution and Reed-Solomon
- Header: 1 s; 3 tones plus BPSK training pattern for synchronization
- Digital voice: DVSI AMBE2020 coder, decoder
- Signal detection: Automatic Digital detect, Automatic switching between analog mode and digital mode
- Video Compression: AOR original adaptive JPEG

AOR has more information online at **www.aorusa.com/ard9800.html** and Amateur Radio Video News featured a number of digital voice modes including the AOR devices available on DVD at **www.arvideonews.com/dv/index.html**.

WinDRM

Digital Radio Mondiale (DRM) — not to be confused with Digital Rights Management — is a set of commercial digital audio broadcasting technologies designed to work over the bands currently used for AM broadcasting, particularly shortwave. DRM can fit more channels than AM, at higher quality, into a given amount of bandwidth, using various MPEG-4 codecs optimized for voice or music or both. As a digital modulation, DRM supports data transmission in addition to the audio channels. The modulation used by DRM is COFDM (coded orthogonal frequency division multiplex) with each carrier coded using QAM (quadrature amplitude modulation). The use of multiple carriers in COFDM provides a reasonably robust signal on HF.

Two members of Darmstadt University of Technology in Germany, Volker Fisher and Alexander Kurpiers, developed an open-source software program to decode commercial DRM called *Dream*. Typically, the DRM commercial shortwave broadcasts use 5 kHz, 10 kHz or 20 kHz of bandwidth (monaural or stereo) and require large bandwidths in receivers.

Shortly after, Francesco Lanza, HB9TLK, modified the *Dream* software to use the DRM technology concepts in a program that only used 2.5 kHz of bandwidth for ham radio use. His software, named *WinDRM*, includes file transfer functionality and is available for download at **http://n1su.com/windrm/**.

FreeDV

FreeDV is a *Windows* and *Linux* application that allows any SSB radio to be used for low bit-rate digital voice. Speech and call sign data is compressed down to 1400 bit/s, which then modulates an 1125 Hz wide QPSK signal, which is then applied to the microphone input of an SSB radio. On receive, the signal is received as SSB, then further demodulated and decoded by *FreeDV*.

FreeDV was coded from scratch by David Witten, KDØEAG, (GUI, architecture) and David Rowe, VK5DGR, (*Codec2*, modem implementation, integration). The *FreeDV* design and user interface is based on *FDMDV*, which was developed by Francesco Lanza, HB9TLK. Francesco received advice on modem design from Peter Martinez, G3PLX. Bruce Perens, K6BP has been a thought leader on open source, patent-free voice codecs for Amateur Radio. He has inspired, promoted and encouraged the development of *Codec2* and *FreeDV*. A team of reviewers and beta testers also supported the development of *FreeDV* as credited on the **http://freedv.org** home page.

FreeDV is entirely open source — even the voice codec. This makes it unique among Amateur Radio digital voice systems that typically rely on a proprietary voice codec that is not available to ham experimentation.

FreeDV Design

- *Codec2* voice codec and *FDMDV* modem
- 14, 50 baud QPSK voice data carriers
- 1 center BPSK carrier with 2× power for fast and robust synchronization
- 1.125 kHz spectrum bandwidth (half SSB) with 75 Hz carrier spacing
- 1400 bit/s data rate with 1375 bit/s voice coding and 25 bit/s text for call sign ID
- No interleaving in time or FEC, resulting in low latency, fast synchronization and quick recovery from fades
- 44.1 or 48 kHz sample rate, sound card compatible

Key Features

- Waterfall, spectrum, scatter and audio oscilloscope displays
- Adjustable squelch

- Fast/slow SNR estimation
- Microphone and speaker signal audio equalizer
- Control of transmitter PTT via RS-232 levels
- Works with one (receive only) or two (transmit and receive) sound cards, for example a built-in sound card and USB headphones.

More details and the software can be found at **http://freedv.org** as well as written and video setup guides. A coordinating website for *FreeDV* QSOs is available at **http://qso.k7ve.org**.

16.4.6 ALE

Automatic link establishment (ALE) was created as a series of protocols for government users to simplify HF communications. The protocol provides a mechanism to analyze signal quality on various channels/bands and choose the best option. The purpose is to provide a reliable rapid method of calling and connecting during constantly changing HF ionospheric propagation, reception interference and shared spectrum use of busy or congested HF channels. It also supports text messages with a very robust protocol that can get through even if no voice-quality channel can be found.

Each radio ALE station uses a call sign or address in the ALE controller. When not actively in communication with another station, each HF SSB transceiver constantly scans through a list of frequencies, listening for its call sign. It decodes calls and soundings sent by other stations, using the bit error rate to store a quality score for that frequency and sender call sign.

To reach a specific station, the caller simply enters the call sign, just like dialing a phone number. The ALE controller selects the best available frequency and sends out brief digital selective calling signals containing the call signs. When the distant scanning station detects the first few characters of its call sign, it stops scanning and stays on that frequency. The two stations' ALE controllers automatically handshake to confirm that a link of sufficient quality is established and they are ready to communicate.

When successfully linked, the receiving station which was muted will typically emit an audible alarm and visual alert for the receiving operator of the incoming call. It also indicates the call sign of the linked station. The operators then can talk in a regular conversa-

Table 16.8
ALE Tones

Frequency	Data
750 Hz	000
1000 Hz	001
1250 Hz	011
1500 Hz	010
1750 Hz	110
2000 Hz	111
2250 Hz	101
2500 Hz	100

tion. At the conclusion of the QSO, one of the stations sends a disconnect signal to the other station, and they each return their ALE stations to the scanning mode. Some military / commercial HF transceivers are available with ALE available internally. Amateur Radio operators commonly use the PCALE soundcard software ALE controller, interfaced to a ham transceiver via rig control cable and multi-frequency antenna.

The ALE waveform is designed to be compatible with the audio passband of a standard SSB radio. It has a robust waveform for reliability during poor path conditions. It consists of 8-ary frequency-shift keying (FSK) modulation with eight orthogonal tones, a single tone for a symbol. These tones represent three bits of data, with least significant bit to the right, as shown in **Table 16.8**.

The tones are transmitted at a rate of 125 tones per second, 8 ms per tone. The resultant transmitted bit rate is 375 bit/s. The basic ALE word consists of 24 bits of information. Details can be found in Federal Standard 1045, Detailed Requirements at **www.its.bldrdoc.gov/fs-1054a/ 45-detr.htm**.

It would require a lot of time for the radio to go through the sequence of calling a station on every possible frequency to establish a link. Time can be decreased by using a "smarter" way of predictive or synchronized linking. With *Link Quality Analysis* (LQA), an ALE system uses periodic sounding and linking signals between other stations in the network to stay in touch and to predict which channel is likely to support a connection to the desired station at any given time. Various stations may be operating on different channels, and this enables the stations to find and use a common open channel.

The PCALE software developed by Charles Brain, G4GUO, is available for download at **http://hflink.com/software**. Much more ALE information and real-time data is available online at **http://hflink.com**.

16.5 Networking Modes

The modes described in this section operate using features and functions associated with computer-to-computer networking. Even though communication using these modes may not involve the creation of a network, the modes are referred to as "networking modes" because of their structure. In cases such as Winlink 2000 and D-STAR, the most common use is to implement a networked system and those features are described along with the modes and protocols used to implement communications within the network.

16.5.1 OSI Networking Model

The Open Systems Interconnection Model or *OSI Model* is an abstract description for computer network protocol design. It defines seven different *layers* or functions performed by a *protocol stack*. In the OSI model, the highest level is closest to the user and the lowest is closest to the hardware required to transport the data (network card and wire or radio). The seven layers are described in **Table 16.9**. The modes examined previously implemented the protocols as a *monolithic stack* where all the functions are performed inside a single piece of code. The modes described in this section implement networking features in a more modular fashion. This allows greater flexibility (and complexity) when mixing features.

As a data packet moves through these layers the header or preamble is removed and any required action performed before the data is passed to the next layer, much like peeling away layers of an onion until just the basic clean data is left. The OSI model does not define any interfaces between layers; it is just a conceptual model of the functions required. Real-world protocols rarely implement each layer individually and often span multiple layers.

This description is by no means exhaustive and more information can be found online and in every networking textbook. **Table 16.10** shows the placement of commonly recognized protocols within the OSI layered structure.

16.5.2 Connected and Connectionless Protocols

The protocols discussed to this point have been *connectionless* meaning they don't establish a connection with a specific machine for the purpose of transferring data. Even with packetized modes like FSK441 with a destination call sign specified, the packet is transmitted and it's up to the user to identify they are the intended recipient. In a packet-switched network, connectionless mode transmission is a transmission in which each packet is prepended with a header containing a destination address to allow delivery of the packet without the aid of additional instructions. A packet transmitted in a connectionless mode is frequently called a *datagram*.

In *connection-oriented protocols*, the stations about to exchange data need to first declare to each other they want to "establish a connection". A connection is sometimes defined as a logical relationship between the peers exchanging data. Connected protocols can use a method called *automatic repeat request* (*ARQ*) to insure accurate delivery of packets using *acknowledgements* and *timeouts*. This allows the detection and correction of corrupted packets, misdelivery, duplication, or out-of-sequence delivery of the packets.

Connectionless modes can have error correction and detection included by a higher layer of the protocol but they have no mechanism to request a correction. An advantage of connectionless mode over connection-oriented mode is that it has a low data overhead. It also allows for *multicast* and *broadcast* (net-type) operations, which may save even more network resources when the same data needs to be transmitted to several recipients. In contrast, a connected mode is always *unicast* (point-to-point).

Another drawback of the connectionless mode is that no optimizations are possible when sending several frames between the same two peers. By establishing a connection at the beginning of such a data exchange the components (routers, bridges) along the network path would be able to pre-compute (and hence cache) routing-related information, avoiding re-computation for every packet.

Many network modes incorporate both types of protocol for different purposes. In the Internet TCP/IP protocol, TCP is a connection-oriented transport protocol where UDP is connectionless.

16.5.3 The Terminal Node Controller (TNC)

While a *terminal node controller* (*TNC*) is nominally an OSI Physical layer device, the internal firmware often implements a protocol such as PACTOR that handles all the routing, and error correction through the transport layer. This greatly simplifies the coding of any protocol or application that uses these devices.

A TNC is actually a computer that contains the protocols implemented in firmware and a *modem* (modulator/demodulator). The

---

Table 16.9
OSI Seven Layer Networking Model

7 — Application Layer	End-user program or "application" that uses the network
6 — Presentation Layer	The format of data after transfer (code conversion, encryption)
5 — Session Layer	Manages the transfer process
4 — Transport Layer	Provides reliable data transfer to the upper layers
3 — Network Layer	Controls data routing
2 — Data Link Layer	Provides error detection and flow control
1 — Physical Layer	Signal used on the medium—voltage, current, frequency, etc.

Table 16.10
Networking Protocols in the OSI Model

Layer	Examples	IP Protocol Suite
7 — Application		NNTP, DNS, FTP, Gopher, HTTP, DHCP, SMTP, SNMP, TELNET
6 — Presentation	ASCII, EBCDIC, MIDI, MPEG	MIME, SSL
5 — Session	Named Pipes, NetBIOS, Half Duplex, Full Duplex, Simplex	Sockets, Session establishment in TCP
4 — Transport		TCP, UDP
3 — Network	AX.25	IP, IPsec, ICMP, IGMP
2 — Data Link	802.3 (Ethernet), 802.11a/b/g/n MAC/LLC, ATM, FDDI, Frame Relay, HDLC, Token Ring, ARP (maps layer 3 to layer 2 address)	PPP, SLIP, PPTP, L2TP
1 — Physical	RS-232, T1, 10BASE-T, 100BASE-TX, POTS, DSL, 802.11a/b/g/n, Soundcard, TNC, Radio	

TNC generally connects to a PC as a serial or USB device on one side and to the radio with appropriate audio and PTT cables on the other. Most of the newer rigs have dedicated data connections available that feature audio lines with fixed levels that are unaffected by settings in the radio. These jacks make swapping mike cables unnecessary when switching between voice and digital modes. Bypassing internal audio processing circuitry eliminates a number of issues that can cause problems with digital modes and makes the use of digital modes more reproducible/reliable by eliminating a number of variables when configuring equipment. These same data jacks are recommended when using a computer sound card.

Although many of the modes discussed can use a computer sound card to generate the required modulation and a separate mechanism to support push-to-talk (PTT), a TNC offers some advantages:

• TNC hardware can be used with any computer platform.

• A computer of nearly any vintage/performance level can be used.

• Data transmission/reception is unaffected by computer interruptions from virus checkers or other "inits."

• Initialization settings are held internal to the TNC and can easily be reset as needed — once working, they stay working.

• Virtually eliminates the computer as a problem/failure point.

• Offers features independent of the computer (digipeat, BBS, APRS beacon, telemetry, weather beacon, and so forth).

The majority of TNCs are designed for 300 or 1200-bit/s packet and implement the Bell 103 or Bell 202 modulation respectively. A *multimode communications processor (MCP)* or *multi-protocol controller (MPC)* may offer the capability to operate RTTY, CW, AMTOR, PACTOR, G-TOR, Clover, fax, SSTV and other modes in addition to packet. Some of these modes are only available in TNC hardware because the real-time operating system in the TNC provides a more reliable platform to implement the mode and it also helps protect proprietary intellectual property.

KISS-Mode TNCs have become popular. These devices simply provide the modem and filters to implement the baseband signals for a type of digital modulation. They rely on the computer software to generate the appropriate packet protocol and complete the mode. This means the software must be written specifically to support these TNCs by creating the entire AX.25 packet with the data embedded, rather than simply sending the data to the TNC expecting the TNC to frame the packet. By leaving the TNC to handle only the baseband signal generation and data recovery, much simpler, smaller and less expensive designs are possible while still retaining the platform independence and robustness of a separate TNC. The TNC-X from Coastal Chipworks at **http://tnc-x.com** is a good example of a KISS-mode TNC.

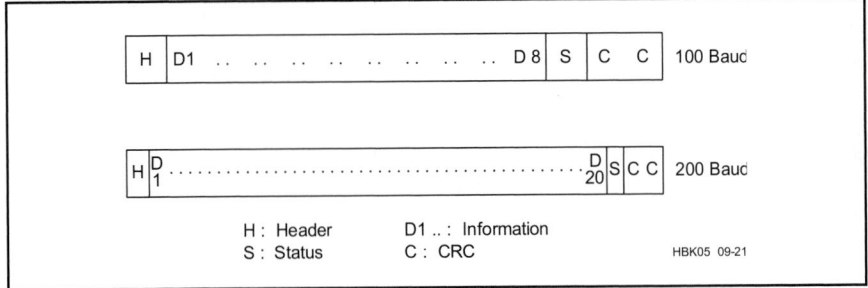

Fig 16.6 — PACTOR packet structure.

Table 16.11
PACTOR Timing

Object	Length (seconds)
Packet	0.96 (200 baud: 192 bits; 100 baud: 96 bits)
CS receive time	0.29
Control signals	0.12 (12 bits at 10 ms each)
Propagation delay	0.17
Cycle	1.25

16.5.4 PACTOR-I

PACTOR, now often referred to as PACTOR-I, is an HF radio transmission system developed by German amateurs Hans-Peter Helfert, DL6MAA, and Ulrich Strate, DF4KV. It was designed to overcome the shortcomings of AMTOR and packet radio. It performs well under both weak-signal and high-noise conditions. PACTOR-I has been overtaken by PACTOR-II and PACTOR-III but remains in use.

TRANSMISSION FORMATS

All packets have the basic structure shown in **Fig 16.6**, and their timing is as shown in **Table 16.11**.

• Header: Contains a fixed bit pattern to simplify repeat requests, synchronization and monitoring. The header is also important for the Memory ARQ function. In each packet that carries new information, the bit pattern is inverted.

• Data: Any binary information. The format is specified in the status word. Current choices are 8-bit ASCII or 7-bit ASCII (with Huffman encoding). Characters are not broken across packets. ASCII RS (hex 1E) is used as an IDLE character in both formats.

• Status word: See Table 16.11

• CRC: The CRC is calculated according to the CCITT standard, for the data, status and CRC.

The PACTOR acknowledgment signals are shown in **Table 16.12**. Each of the signals is 12-bits long. The characters differ in pairs in eight bits (Hamming offset) so that the chance of confusion is reduced. If the CS is not correctly received, the TX reacts by repeating the last packet. The request status can be uniquely recognized by the 2-bit packet number so that wasteful transmissions of pure RQ blocks are unnecessary.

The receiver pause between two blocks is 0.29 s. After deducting the CS lengths, 0.17 s remains for switching and propagation delays so that there is adequate reserve for DX operation.

CONTACT FLOW

In the listen mode, the receiver scans any received packets for a CRC match. This method uses a lot of computer processing resources, but it's flexible.

A station seeking contacts transmits CQ packets in an FEC mode, without pauses for acknowledgment between packets. The transmit time length, number of repetitions and speed are the transmit operator's choice. (This mode is also suitable for bulletins and other group traffic.) Once a listening station has copied the call, the listener assumes the TX station role and initiates a contact. Thus, the station sending CQ initially takes the RX station role. The contact begins as shown in **Table 16.13**.

With good conditions, PACTOR's normal signaling rate is 200 baud, but the system automatically changes from 200 to 100 baud and back, as conditions demand. In addition, Huffman coding can further increase the throughput by a factor of 1.7. There is no loss of synchronization speed changes; only one packet is repeated.

Table 16.12
PACTOR Control Signals

Code	Chars (hex)	Function
CS1	4D5	Normal acknowledge
CS2	AB2	Normal acknowledge
CS3	34B	Break-in (forms header of first packet from RX to TX)
CS4	D2C	Speed change request

All control signals are sent only from RX to TX

Table 16.13
PACTOR Initial Contact

Master Initiating Contact

Size (bytes)	1	8	6
Content	/Header	/SLAVECAL	/SLAVECAL/
Speed (baud)	100	100	200

Slave Response
The receiving station detects a call, determines mark/space polarity, and decodes 100 baud and 200-bd call signs. It uses the two call signs to determine if it is being called and the quality of the communication path. The possible responses are:

First call sign does not match slave's call sign (Master not calling this slave)	none
Only first call sign matches slave's call sign (Master calling this slave, poor communications)	CS1
First and second call signs both match the slaves (good circuit, request speed change to 200 baud)	CS4

When the RX receives a bad 200-baud packet, it can acknowledge with CS4. TX immediately assembles the previous packet in 100-baud format and sends it. Thus, one packet is repeated in a change from 200 to 100 baud.

The RX can acknowledge a good 100-baud packet with CS4. TX immediately switches to 200 baud and sends the next packet. There is no packet repeat in an upward speed change.

The RX station can become the TX station by sending a special change-over packet in response to a valid packet. RX sends CS3 as the first section of the changeover packet. This immediately changes the TX station to RX mode to read the data in that packet and responds with CS1 and CS3 (acknowledge) or CS2 (reject).

PACTOR provides a sure end-of-contact procedure. TX initiates the end of contact by sending a special packet with the QRT bit set in the status word and the call of the RX station in byte-reverse order at 100 baud. The RX station responds with a final CS.

16.5.5 PACTOR-II

This is a significant improvement over PACTOR-I, yet it is fully compatible with the older mode. PACTOR-II uses 16PSK to transfer up to 800 bit/s at a 100 baud rate. This keeps the bandwidth less than 500 Hz.

PACTOR-II uses digital signal processing (DSP) with Nyquist waveforms, Huffman and Markov compression and powerful Viterbi decoding to increase transfer rate and sensitivity into the noise level. The effective transfer rate of text is over 1200 bit/s. Features of PACTOR II include:

• Frequency agility — it can automatically adjust or lock two signals together over a ±100 Hz window.

• Powerful data reconstruction based upon computer power — with over 2 Mbyte of available memory.

• Cross correlation — applies analog Memory ARQ to acknowledgment frames and headers.

• Soft decision making — Uses artificial intelligence (AI), as well as digital information received to determine frame validity.

• Extended data block length — when transferring large files under good conditions, the data length is doubled to increase the transfer rate.

• Automatic recognition of PACTOR-I, PACTOR-II and so on, with automatic mode switching.

• Intermodulation products are canceled by the coding system.

• Two long-path modes extend frame timing for long-path terrestrial and satellite propagation paths.

This is a fast, robust mode that has excellent coding gain as well. PACTOR-II stations acknowledge each received transmission block. PACTOR-II employs computer logic as well

as received data to reassemble defective data blocks into good frames. This reduces the number of transmissions and increases the throughput of the data.

16.5.6 PACTOR-III

PACTOR-III is a software upgrade for existing PACTOR-II modems that provides a data transmission mode for improved speed and robustness. Both the transmitting and receiving stations must support PACTOR-III for end-to-end communications using this mode.

PACTOR-III's maximum uncompressed speed is 2722 bit/s. Using online compression, up to 5.2 kbit/s is achievable. This requires an audio passband from 400 Hz to 2600 Hz (for PACTOR-III speed level 6). On an average channel, PACTOR-III is more than three times faster than PACTOR-II. On good channels, the effective throughput ratio between PACTOR-III and PACTOR-II can exceed five. PACTOR-III is also slightly more robust than PACTOR-II at their lower SNR edges.

The ITU emission designator for PACTOR-III is 2K20J2D. Because PACTOR-III builds on PACTOR-II, most specifications like frame length and frame structure are adopted from PACTOR-II. The only significant difference is PACTOR III's multitone waveform that uses up to 18 carriers while PACTOR-II uses only two carriers. PACTOR-III's carriers are located in a 120 Hz grid and modulated with 100 symbols per second DBPSK or DQPSK. Channel coding is also adopted from PACTOR-II's Punctured Convolutional Coding.

PACTOR-III Link Establishment

The calling modem uses the PACTOR-I FSK connect frame for compatibility. When the called modem answers, the modems negotiate to the highest level of which both modems are capable. If one modem is only capable of PACTOR-II, then the 500 Hz PACTOR-II mode is used for the session. With the MYLevel (MYL) command a user may limit a modem's highest mode. For example, a user may set MYL to 1 and only a PACTOR-I connection will be made, set to 2 and PACTOR-I and II connections are available, set to 3 and PACTOR-I through III connections are enabled. The default MYL is set to 2 with the current firmware and with PACTOR-III firmware it will be set to 3. If a user is only allowed to occupy a 500 Hz channel, MYL can be set to 2 and the modem will stay in its PACTOR-II mode.

The PACTOR-III Protocol Specification is available online at **www.scs-ptc.com/pactor.html**. More information can also be found online at **www.arrl.org/technical-characteristics** or in *ARRL's HF Digital Handbook* by Steve Ford, WB8IMY.

The protocol specifications and equipment for PACTOR-IV have been released, but the mode is not yet legal for US amateurs. The symbol rate for PACTOR-IV is 1800 baud, but FCC rules limit US amateurs to 300 baud below the upper end of 10 meters. PACTOR-IV is being used outside the US by individual amateurs and by Winlink stations that are not subject to FCC rules. It is not known if or when this restriction will be lifted.

16.5.7 G-TOR

This brief description has been adapted from "A Hybrid ARQ Protocol for Narrow Bandwidth HF Data Communication" by Glenn Prescott, WBØSKX, Phil Anderson, WØXI, Mike Huslig, KBØNYK, and Karl Medcalf, WK5M (May 1994 QEX).

G-TOR is short for Golay-TOR, an innovation of Kantronics. It was inspired by HF automatic link establishment (ALE) concepts and is structured to be compatible with ALE. The purpose of the G-TOR protocol is to provide an improved digital radio communication capability for the HF bands. The key features of G-TOR are:

• Standard FSK tone pairs (mark and space)
• Link-quality-based signaling rate: 300, 200 or 100 baud
• 2.4-s transmission cycle
• Low overhead within data frames
• Huffman data compression — two types, on demand
• Embedded run-length data compression
• Golay forward-error-correction coding
• Full-frame data interleaving
• CRC error detection with hybrid ARQ
• Error-tolerant "Fuzzy" acknowledgments.

Since one of the objectives of this protocol is ease of implementation in existing TNCs, the modulation format consists of standard tone pairs (FSK), operating at 300, 200 or 100 baud, depending upon channel conditions. G-TOR initiates contacts and sends ACKs only at 100 baud. The G-TOR waveform consists of two phase-continuous tones (BFSK), spaced 200 Hz apart (mark = 1600 Hz, space = 1800 Hz); however, the system can still operate at the familiar 170 Hz shift (mark = 2125 Hz, space = 2295 Hz), or with any other convenient tone pairs. The optimum spacing for 300-baud transmission is 300 Hz, but you trade some performance for a narrower bandwidth.

Each transmission consists of a synchronous ARQ 1.92-s frame and a 0.48-s interval for propagation and ACK transmissions (2.4 s cycles). All advanced protocol features are implemented in the signal-processing software.

Data compression is used to remove re-dundancy from source data. Therefore, fewer bits are needed to convey any given message. This increases data throughput and decreases transmission time — valuable features for HF. G-TOR uses run-length encoding and two types of Huffman coding during normal text transmissions. Run-length encoding is used when more than two repetitions of an 8-bit character are sent. It provides an especially large savings in total transmission time when repeated characters are being transferred.

The Huffman code works best when the statistics of the data are known. G-TOR applies Huffman A coding with the upper- and lower-case character set, and Huffman B coding with upper-case-only text. Either type of Huffman code reduces the average number of bits sent per character. In some situations, however, there is no benefit from Huffman coding. The encoding process is then disabled. This decision is made on a frame-by-frame basis by the information sending station.

The real power of G-TOR resides in the properties of the (24, 12) extended Golay error-correcting code, which permits correction of up to three random errors in three received bytes. The (24, 12) extended Golay code is a half-rate error-correcting code: Each 12 data bits are translated into an additional 12 parity bits (24 bits total). Further, the code can be implemented to produce separate input-data and parity-bit frames.

The extended Golay code is used for G-TOR because the encoder and decoder are simple to implement in software. Also, Golay code has mathematical properties that make it an ideal choice for short-cycle synchronous communication. More information can also be found online at **www.arrl. org/technical-characteristics** or in *ARRL's HF Digital Handbook* by Steve Ford, WB8IMY.

16.5.8 CLOVER-II

The desire to send data via HF radio at high data rates and the problem encountered when using AX.25 packet radio on HF radio led Ray Petit, W7GHM, to develop a unique modulation waveform and data transfer protocol that is now called CLOVER-II. Bill Henry, K9GWT, supplied this description of the CLOVER-II system.

CLOVER modulation is characterized by the following key parameters:

• Very low base symbol rate: 31.25 symbols/second (all modes).
• Time-sequence of amplitude-shaped pulses in a very narrow frequency spectrum.
• Occupied bandwidth = 500 Hz at 50 dB below peak output level.
• Differential modulation between pulses.
• Multilevel modulation.

The low base symbol rate is very resistant to multipath distortion because the time between modulation transitions is much longer than even the worst-case time-smearing caused by summing of multipath signals. By using a time-sequence of tone pulses, Dolph-Chebychev "windowing" of the modulating signal and differential modulation, the total occupied bandwidth of a CLOVER-II signal is held to 500 Hz.

Multilevel tone, phase and amplitude modulation gives CLOVER a large selection of data modes that may be used (see **Table 16.14**). The adaptive ARQ mode of CLOVER senses current ionospheric conditions and automatically adjusts the modulation mode to produce maximum data throughput. When using the Fast bias setting, ARQ throughput automatically varies from 11.6 byte/s to 70 byte/s.

The CLOVER-II waveform uses four tone pulses that are spaced in frequency by 125 Hz. The time and frequency domain characteristics of CLOVER modulation are shown in **Figs 16.7, 16.8** and **16.9**. The time-domain shape of each tone pulse is intentionally shaped to produce a very compact frequency spectrum. The four tone pulses are spaced in time and then combined to produce the composite output shown. Unlike other modulation schemes, the CLOVER modulation spectrum is the same for all modulation modes.

Table 16.14
CLOVER-II Modulation Modes

As presently implemented, CLOVER-II supports a total of seven different modulation formats: five using PSM and two using a combination of PSM and ASM (Amplitude Shift Modulation).

Name Rate	Description	In-Block Data
16P4A	16 PSM, 4-ASM	750 bps
16PSM	16 PSM	500 bps
8P2A	8 PSM, 2-ASM	500 bps
8PSM	8 PSM	375 bps
QPSM	4 PSM	250 bps
BPSM	Binary PSM	125 bps
2DPSM	2-Channel Diversity BPSM	62.5 bps

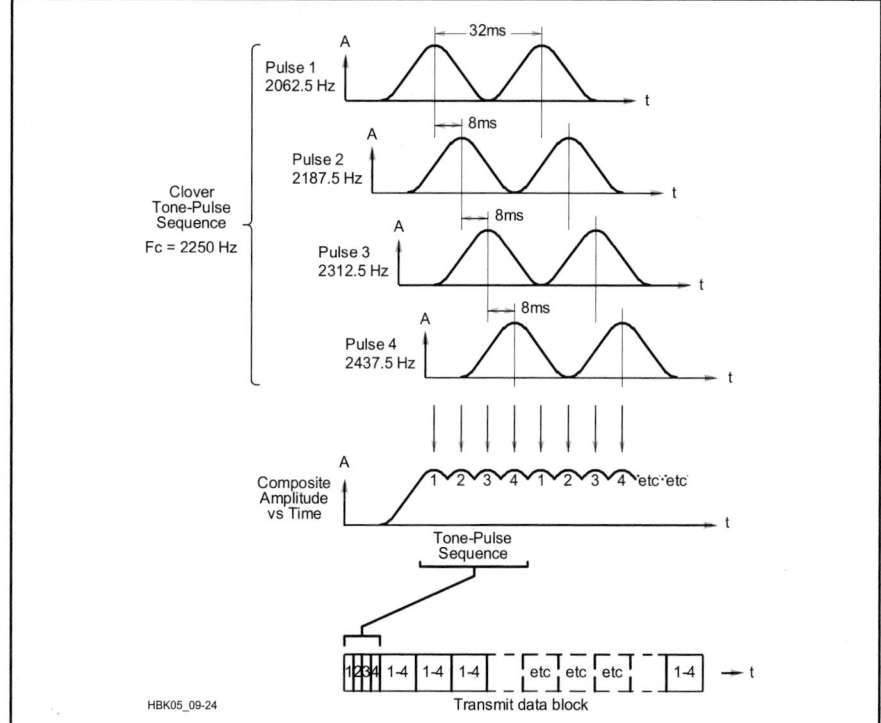

Fig 16.7 — Amplitude vs time plots for CLOVER-II's four-tone waveform.

Table 16.15
Data Bytes Transmitted Per Block

Block Size	Reed-Solomon Encoder Efficiency			
	60%	75%	90%	100%
17	8	10	12	14
51	28	36	42	48
85	48	60	74	82
255	150	188	226	252

Table 16.16
Correctable Byte Errors Per Block

Block Size	Reed-Solomon Encoder Efficiency			
	60%	75%	90%	100%
17	1	1	0	0
51	9	5	2	0
85	16	10	3	0
255	50	31	12	0

Data is modulated on a CLOVER-II signal by varying the phase and/or amplitude of the tone pulses. Further, all data modulation is differential on the same tone pulse — data is represented by the phase (or amplitude) difference from one pulse to the next. For example, when binary phase modulation is used, a data change from 0 to 1 may be represented by a change in the phase of tone pulse one by 180° between the first and second occurrence of that pulse. Further, the phase state is changed only while the pulse amplitude is zero. Therefore, the wide frequency spectra normally associated with PSK of a continuous carrier is avoided. This is true for all CLOVER-II modulation formats. The term *phase-shift modulation* (PSM) is used when describing CLOVER modes to emphasize this distinction.

CLOVER-II has four "coder efficiency" options: 60%, 75%, 90% and 100% ("effi-ciency" being the approximate ratio of real data bytes to total bytes sent). 60% efficiency corrects the most errors but has the lowest net data throughput. 100% efficiency turns the encoder off and has the highest throughput but fixes no errors. There is therefore a tradeoff between raw data throughput versus the number of errors that can be corrected without resorting to retransmission of the entire data block.

Note that while the In Block Data Rate numbers listed in the table go as high as 750 bit/s, overhead reduces the net throughput or overall efficiency of a CLOVER transmission. The FEC coder efficiency setting and protocol requirements of FEC and ARQ modes add overhead and reduce the net ef-

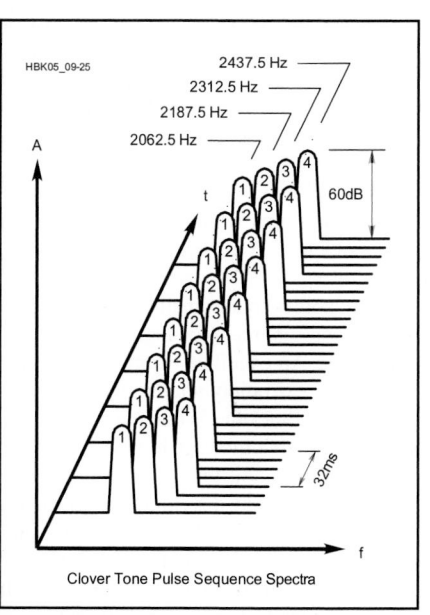

Fig 16.8 — A frequency-domain plot of a CLOVER-II waveform.

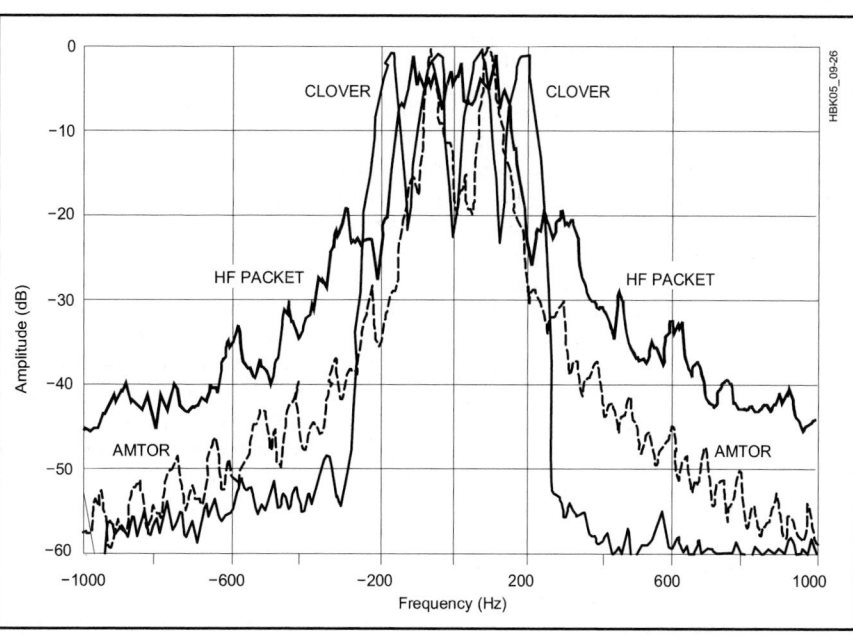

Fig 16.9 — Spectra plots of AMTOR, HF packet-radio and CLOVER-II signals.

ficiency. **Tables 16.15** and **16.16** detail the relationships between block size, coder efficiency, data bytes per block and correctable byte errors per block.

With seven different modulation formats, four data-block lengths (17, 51, 85 or 255 bytes) and four Reed-Solomon coder efficiencies (60%, 75%, 90% and 100%), there are 112 ($7 \times 4 \times 4$) different waveform modes that could be used to send data via CLOVER. Once all of the determining factors are considered, however, there are eight different waveform combinations that are actually used for FEC and/or ARQ modes.

16.5.9 CLOVER-2000

CLOVER-2000 is a faster version of CLOVER (about four times faster) that uses eight tone pulses, each of which is 250 Hz wide, spaced at 250 Hz centers, contained within the 2 kHz bandwidth between 500 and 2500 Hz. The eight tone pulses are sequential, with only one tone being present at any instant and each tone lasting 2 ms. Each frame consists of eight tone pulses lasting a total of 16 ms, so the base modulation rate of a CLOVER-2000 signal is always 62.5 symbols per second (regardless of the type of modulation being used). CLOVER-2000's maximum raw data rate is 3000 bit/s.

Allowing for overhead, CLOVER-2000 can deliver error-corrected data over a standard HF SSB radio channel at up to 1994 bit/s, or 249 characters (8-bit bytes) per second. These are the uncompressed data rates; the maximum throughput is typically doubled for plain text if compression is used. The effective data throughput rate of CLOVER-2000 can be even higher when binary file transfer mode is used with data compression.

The binary file transfer protocol used by HAL Communications operates with a terminal program explained in the HAL E2004 engineering document. Data compression algorithms tend to be context sensitive — compression that works well for one mode (say, text), may not work well for other data forms (graphics, for example). The HAL terminal program uses the PK-WARE compression algorithm, which has proved to be a good general-purpose compressor for most computer files and programs. Other algorithms may be more efficient for some data formats, particularly for compression of graphic image files and digitized voice data. The HAL Communications CLOVER-2000 modems can be operated with other data compression algorithms in the users' computers.

CLOVER-2000 is similar to the previous version of CLOVER, including the transmission protocols and Reed-Solomon error detection and correction algorithm. The original descriptions of the CLOVER Control Block (CCB) and Error Correction Block (ECB) still apply for CLOVER-2000, except for the higher data rates inherent to CLOVER-2000. Just like CLOVER, all data sent via CLOVER-2000 is encoded as 8-bit data bytes and the error-correction coding and modulation formatting processes are transparent to the data stream — every bit of source data is delivered to the receiving terminal without modification.

Control characters and special "escape sequences" are not required or used by CLOVER-2000. Compressed or encrypted data may therefore be sent without the need to insert (and filter) additional control characters and without concern for data integrity. Five different types of modulation may be used in the ARQ mode — BPSM (Binary Phase Shift Modulation), QPSM (Quadrature PSM), 8PSM (8-level PSM), 8P2A (8PSM + 2-level Amplitude-Shift Modulation) and 16P4A (16 PSM plus 4 ASM).

The same five types of modulation used in ARQ mode are also available in Broadcast (FEC) mode, with the addition of 2-Channel Diversity BPSM (2DPSM). Each CCB is sent using 2DPSM modulation, 17-byte block size and 60% bias. The maximum ARQ data throughput varies from 336 bit/s for BPSM to 1992 bit/s for 16P4A modulation. BPSM is most useful for weak and badly distorted data signals, while the highest format (16P4A) needs extremely good channels, with high SNRs and almost no multipath.

Most ARQ protocols designed for use with HF radio systems can send data in only one direction at a time. CLOVER-2000 does not need an OVER command; data may flow in either direction at any time. The CLOVER ARQ time frame automatically adjusts to match the data volume sent in either or both directions. When first linked, both sides of the ARQ link exchange information using six bytes of the CCB. When one station has a large volume of data buffered and ready to send, ARQ mode automatically shifts to an expanded time frame during which one or more 255 byte data blocks are sent.

If the second station also has a large volume of data buffered and ready to send, its half of the ARQ frame is also expanded. Either or both stations will shift back to CCB level when all buffered data has been sent. This feature provides the benefit of full-duplex data transfer but requires use of only simplex frequencies and half-duplex radio equipment. This two-way feature of CLOVER can also provide a back-channel "order-wire" capability. Communications may be maintained in this chat mode at 55 WPM, which is more than adequate for real-time keyboard-to-keyboard communications.

More information can also be found at **www.arrl.org/technical-characteristics** or in *ARRL's HF Digital Handbook* by Steve Ford, WB8IMY.

16.5.10 WINMOR

While the various PACTOR modes currently dominate and generally represent the best available performance HF ARQ protocols suitable for digital messaging, PC sound cards with appropriate DSP software can now begin to approach PACTOR performance. The WINMOR (Winlink Message Over Radio) protocol is an outgrowth of the work SCAMP (Sound Card Amateur Message Protocol) by Rick Muething, KN6KB. SCAMP put an ARQ "wrapper" around Barry Sanderson's RDFT (Redundant Digital File Transfer) then integrated SCAMP into a Client and Server for access to the Winlink message system. (More on Winlink in a later section.) SCAMP worked well on good channels but suffered from the following issues:

• The RDFT batch-oriented DLLs were slow and required frame pipelining, increasing complexity and overhead.

• RDFT only changed the RS encoding on its 8PSK multi carrier waveform to achieve a 3:1 range in speed/robustness which is not enough.

• RDFT was inefficient in Partial Frame recovery (no memory ARQ).

• RDFT was a 2.4 kHz mode and limited to narrow HF sub bands.

• SCAMP's simple multi-tone ACK/NAK did not carry session ID info, increasing chances of fatal cross session contamination.

WINMOR is an ARQ mode generated from the ground up to address the limitations of SCAMP/RDFT and leverage what was learned. Today, a viable message system (with the need for compression and binary attachments) requires true "error-free" delivery of binary data. To achieve this there must be some "back channel" or ARQ so the receiving station can notify the sender of lost or damaged data and request retransmission or repair. **Table 16.17** outlines the guidelines used in the development of WINMOR.

Perhaps the most challenging of these requirements are:

• The ability to quickly tune, lock and acquire the signal which is necessary for practical length ARQ cycles in the 2-6 s range.

• The ability to automatically adapt the modulation scheme to changing channel conditions. An excellent example of this is Pactor III's extremely wide range of speed/robustness (18:1) and is one reason it is such an effective mode in both good and poor channel conditions.

The most recent development effort has focused on 62.5 baud BPSK, QPSK and 16QAM and 31.25-baud 4FSK using 1 (200 Hz), 3 (500 Hz) and 15 (2000 Hz). With carriers spaced at twice the symbol rate. These appear to offer high throughput and

Table 16.17
WINMOR Development Guidelines

Absolute Requirements	Desirable Requirements
Work with standard HF (SSB) radios	Modest CPU & OS demands
Accommodate Automatic Connections	Bandwidth options (200, 500, 3000 Hz)
Error-free transmission and confirmation	Work with most sound cards/interfaces
Fast Lock for practical length ARQ cycles	Good bit/s/Hz performance ~ P2 goal
Auto adapt to a wide range of changing channel conditions	Efficient modulation and demodulation for acceptable ARQ latency
Must support true transparent binary to allow attachments and compression	Selective ARQ & memory ARQ to maximize throughput and robustness.
Must use loosely synchronous ARQ timing to accommodate OS and DSP demands	Near Pactor ARQ efficiency (~70% of raw theoretical throughput)

robustness especially when combined with multi-level FEC coding.

WINMOR uses several mechanisms for error recovery and redundancy.

1) FEC data encoding currently using:
- 4,8 Extended Hamming Dmin = 4 (used in ACK and Frame ID)
- 16-bit CRC for data verification

Two-level Reed-Solomon (RS) FEC for data:
- First level Weak FEC, for example RS 140,116 (corrects 12 errors)
- Second level Strong FEC, for example RS 254,116 (corrects 69 errors)

2) Selective ARQ. Each carrier's data contains a Packet Sequence Number (PSN). The ACK independently acknowledges each PSN so only carriers with failed PSNs get repeated. The software manages all the PSN accounting and re-sequencing.

3) Memory ARQ. The analog phase and amplitude of each demodulated symbol is saved for summation (phasor averaging) over multiple frames. Summation is cleared and re-started if max count reached. Reed-Solomon FEC error decoding done after summation.

4) Multiple Carrier Assignment (MCA). The same PSN can be assigned to multiple carriers (allows tradeoff of throughput for robustness). Provides an automatic mechanism for frequency redundancy and protection from interference on some carriers.

5) Dynamic threshold adjustment (used on QAM modes) helps compensate for fading

which would render QAM modes poor in fading channels.

In trying to anticipate how WINMOR might be integrated into applications they came up with a "Virtual TNC" concept. This essentially allows an application to integrate the WINMOR protocol by simply treating the WINMOR code as just another TNC and writing a driver for that TNC. Like all TNCs there are some (<10) parameters to set up: call sign, timing info, sound card, keying mechanism, etc. A sample image of the virtual TNC appears in **Fig 16.10**.

The WINMOR software DLL can even be made to appear as a physical TNC by "wrapping" the DLL with code that accesses it through a virtual serial port or a TCP/IP port. Like a physical TNC WINMOR has a "front panel" with flashing lights. But since operation is automatic with no front panel user interaction required the WINMOR TNC can be visible or hidden.

WINMOR looks promising and the testing to date confirms:
- Sound card ARQ is possible with a modern CPU and OS while making acceptable CPU processing demands. (CPU Loading of < 20% on a 1.5 GHz Celeron/Win XP)
- Throughput and robustness can be adjusted automatically to cover a wide range of bandwidth needs and channel conditions. (10:1 bandwidth range, 57:1 throughput range)
- ARQ throughput in excess of 0.5 bit/s/

Hz is possible in fair to good channels (0.68 - 0.82 bit/s/Hz measured)
- Good ARQ efficiency — 70-75%
- Throughput is currently competitive with P2 and P3 and significantly better than P1

More information about WINMOR will be available as it develops at **www.winlink.org/WINMOR**.

16.5.11 Packet Radio

Amateur packet radio began in Canada after the Canadian Department of Communications permitted amateurs to use the mode in 1978. The FCC permitted amateur packet radio in the US in 1980. In the first half of the 1980s, packet radio was the habitat of a small group of experimenters who did not mind communicating with a limited number of potential fellow packet communicators. In the second half of the decade, packet radio took off as the experimenters built a network that increased the potential number of packet stations that could intercommunicate and thus attracted tens of thousands of communicators who wanted to take advantage of this potential.

Packet radio provides error-free data transfer via the AX.25 protocol. The receiving station receives information exactly as the transmitting station sends it, so you do not waste time deciphering communication errors caused by interference or changes in propagation. This simplifies the effort required to build other protocols or applications on top of packet technology since much of the data transfer work is handled seamlessly by the TNC in the lower layers. There are a number of other advantages to the packet radio design:
- Packet uses time efficiently, since packet bulletin-board systems (PBBSs) permit packet operators to store information for later retrieval by other amateurs.
- Packet uses the radio spectrum efficiently, since one radio channel may be used for multiple communications simultaneously, or one radio channel may be used to interconnect a number of packet stations to form a "cluster" that provides for the distribution

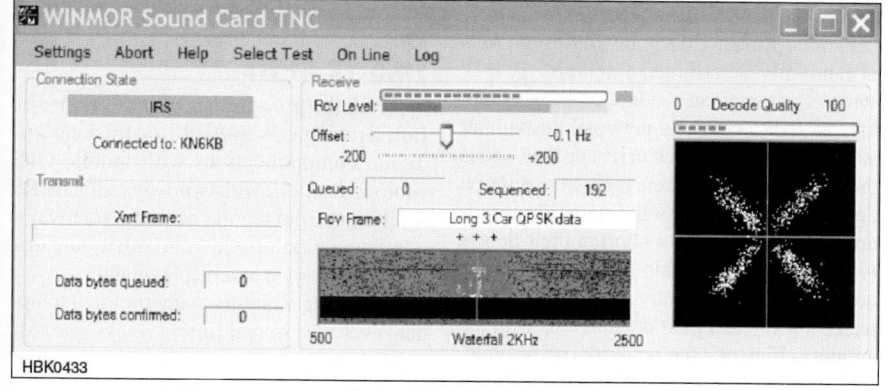

HBK0433

Fig 16.10 — WINMOR sound card TNC screen.

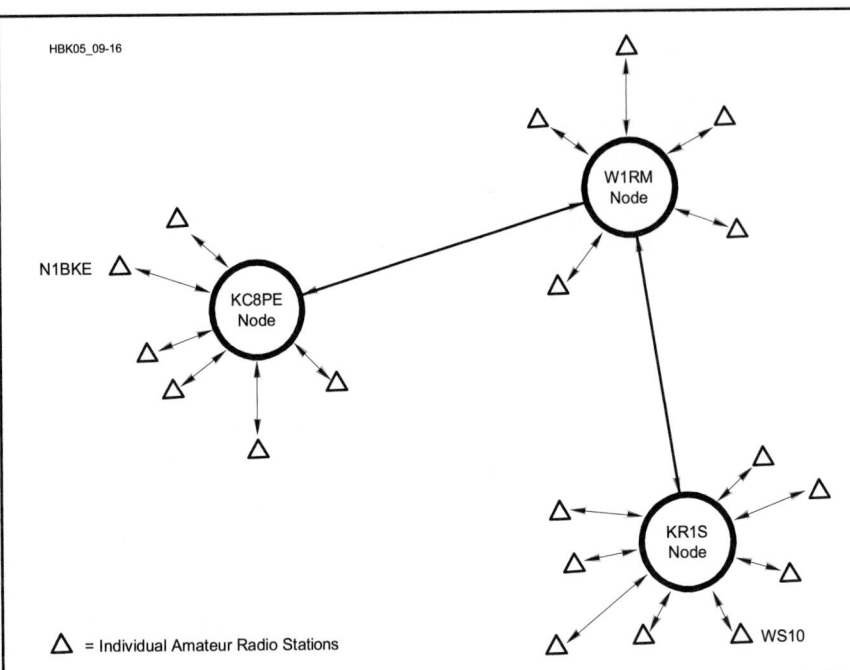

△ = Individual Amateur Radio Stations

Fig 16.11 — DX spotting clusters (based on *PacketCluster* software) are networks comprised of individual nodes and stations with an interest in DXing and contesting. In this example, N1BKE is connected to the KC8PE node. If he finds a DX station on the air, he'll post a notice — otherwise known as a spot — which the KC8PE node distributes to all its local stations. In addition, KC8PE passes the information along to the W1RM node. W1RM distributes the information and then passes it to the KR1S node, which does the same. Eventually, WS1O — who is connected to the KR1S node — sees the spot on his screen. Depending on the size of the network, WS1O will receive the information within minutes after it was posted by N1BKE. Many such networks can also be found on the web or are accessible by the use of *TELNET* software.

of information to all of the clustered stations. The DX PacketCluster nodes are typical examples (see **Fig 16.11**). Each local channel may be connected to other local channels to form a network that affords interstate and international data communications. This network can be used by interlinked packet bulletin-board systems to transfer information, messages and third party traffic via HF, VHF, UHF, satellite and the Internet.

• Packet uses other stations efficiently, since any packet-radio station can use one or more other packet-radio stations to relay data to its intended destination.

• Packet uses current station transmitting and receiving equipment efficiently, since the same equipment used for voice communications may be used for packet communications. The outlay for the additional equipment necessary to add packet capability to a voice station may be less than $100.

DIGIPEATERS

A *digipeater* is a packet-radio station capable of recognizing and selectively repeating packet frames. An equivalent term used in the network industry is *bridge*. Virtually any TNC can be used as a single-port digipeater, because the digipeater function is included

in the AX.25 Level 2 protocol firmware. The digipeater function is handy when you need a relay and no node is available, or for on-the-air testing. Digipeaters are used extensively with APRS (Automatic Packet Reporting System) covered below.

TCP/IP

Despite its name, TCP/IP (Transmission Control Protocol/Internet Protocol) is more than two protocols; it's actually a set of several protocols. TCP/IP provides a standardized set of protocols familiar to many and compatible with existing network technologies and applications. TCP/IP has a unique solution for busy networks. Rather than transmitting packets at randomly determined intervals, TCP/IP stations automatically adapt to network delays as they occur. As network throughput slows down, active TCP/IP stations sense the change and lengthen their transmission delays accordingly. As the network speeds up, the TCP/IP stations shorten their delays to match the pace. This kind of intelligent network sharing virtually guarantees that all packets will reach their destinations with the greatest efficiency the network can provide.

With TCP/IP's adaptive networking scheme, you can chat using the TELNET

protocol with a ham in a distant city and rest assured that you're not overburdening the system. Your packets simply join the constantly moving "freeway" of data. They might slow down in heavy traffic, but they will reach their destination eventually. This adaptive system is used for all TCP/IP packets, no matter what they contain.

TCP/IP excels when it comes to transferring files from one station to another. By using the TCP/IP File Transfer Protocol (FTP), you can connect to another station and transfer computer files — including software. As might be imagined, transferring large files can take time. With TCP/IP, however, you can still send and receive mail (using the SMTP protocol) or talk to another ham while the transfer is taking place.

When you attempt to contact another station using TCP/IP, all network routing is performed automatically according to the TCP/IP address of the station you're trying to reach. In fact, TCP/IP networks are transparent to the average user. To operate TCP/IP, all you need is a computer, a 2 meter FM transceiver and a TNC with KISS (keep it simple, stupid) capability. As you might guess, the heart of your TCP/IP setup is software. The TCP/IP software set was written by Phil Karn, KA9Q, and is called NOSNET or just NOS for short. There are dozens of NOS derivatives available today. All are based on the original NOSNET. NOS takes care of all TCP/IP functions, using your "KISSable" TNC to communicate with the outside world. The only other item necessary is your own IP address in Network 44, termed AMPRNet (AMateur Packet Radio Network). Individual IP Address Coordinators assign addresses to new TCP/IP users in the 44.x.x.x subnet based on physical location worldwide. Your local coordinator can be found on AMPR.org on the list at **http://noh.ucsd.edu/~brian/amprnets.txt**.

More packet information can be found in the annual ARRL/TAPR Digital Communications Conference proceedings and on the TAPR (Tucson Amateur Packet Radio) website at **www.tapr.org**. TAPR also maintains the AX.25 protocol specification at **www.tapr.org/pub_ax25.html**.

16.5.12 APRS

APRS (**http://aprs.org**) was developed by Bob Bruninga, WB4APR, for tracking and digital communications with mobile GPS equipped stations with two-way radio. APRS is different from regular packet in four ways:

• Integration of maps and other data displays to organize and display data

• By using a one-to-many protocol to update everyone in real time

• By using generic digipeating so that prior knowledge of the network is not required

• Provides worldwide transparent Internet

backbone, linking everyone worldwide

APRS turns packet radio into a real-time tactical communications and display system for emergencies, public service applications and global communications. Normal packet radio has shown usefulness in passing bulk message traffic (e-mail) from point to point. It has been difficult to apply conventional packet to real time events where information has a very short life time and needs to get to everyone.

The APRS network consists of five types of APRS stations:

- Basic — Full transmit and receive capability
- Tracker — Transmit-only device (portable or weather station)
- RELAY — Provides basic digipeating
- WIDE — Dedicated digipeaters with specific coverage areas
- IGate — Internet gateways to repeat APRS packets to servers on the Internet

APRS stations are generally set to beacon their position and other information at frequent intervals. Fixed stations don't require a GPS and should beacon infrequently to avoid crowding the channel with needless reports. Moving mobile stations can receive their coordinates from a GPS and should report more frequently. More sophisticated APRS devices support "smart beaconing" and "corner pinning" where a station will beacon more frequently when it moves faster or automatically beacon when changing direction more than a predefined angle. An APRS beacon will generally contain a call sign with *Secondary Station ID* (*SSID*), position report (lat/long), heading, speed, altitude, display icon type, status text and routing information. It may also contain antenna/power information, weather data or short message data.

Call signs generally include an SSID that historically provided information about the type of station. For example, a call sign of N7SS-9 would generally indicate a mobile station. Since there are several systems in use, the SSID information is not definitive. Recently a number of dedicated APRS radios have appeared that feature an internal TNC and APRS software. These radios include a method of displaying and entering messages without requiring the use of a computer meaning their user can directly respond to a message sent. There is value in knowing someone can respond and it is recommended they use an SSID of -7. Tactical call signs can also be used if the assigned call sign is included as part of the status text.

The APRS packet includes PATH information that determines how many times it should be digipeated. A local group should be able to offer guidance on how the PATH should be set to not overload the local network. Listening to squawking signals on the national APRS frequency of 144.390 MHz

will provide some idea how busy the channel is. In much of the country the APRS network is well developed and a beacon doesn't need more than a couple hops to cover a wide area and find an IGate.

APRS SOFTWARE

There is APRS software available for most any platform with varying levels of support. Some of the most common are DOSAPRS (DOS), WinAPRS / APRS+SA / APRSPoint / UI-View (*Windows*), MacAPRS (Macintosh), PocketAPRS (Palm devices), APRSce (*Windows* CE devices) and Xastir / X-APRS (*Linux*). Because APRS beacons find their way to the Internet via IGates, it is also possible to track station or view the network online. There is a good network view available at **http://aprs.fi** and individual station information can be found at **http://map.findu. com/call sign**. The Findu.com site also offers historical weather data, a message display and a large list of other queries.

APRS INTEGRATION WITH OTHER TECHNOLOGIES

APRS is very similar to the *AIS* (*Automatic Identification System*) used to track ships in coastal waters. The AIS transponder equipment is becoming more common in commercial ships and private vessels. The AIS information can be displayed simultaneously with APRS data on the aprs.fi website. More information about AIS can be found at **http://en.wikipedia.org/wiki/Automatic_ Identification_System**.

APRS RF networks worldwide are interconnected via APRS-IS. APRS-IS is an ad hoc network of Amateur Radio servers, gateways (IGates), and clients (display software). APRS-IS facilitates world-wide messaging without having to define special paths. Because all clients have access to all gated packets, databases such as **http://aprs. fi** and **http://findu.com** can store and parse the packets for later retrieval from browsers and other clients. APRS-IS messaging support allows many interfaces for the RF ham running an APRS client such as call sign lookups, e-mail, and calling CQ world-wide. More information on many of these features can be found at **www.aprs-is.net**.

APRS position reports are also available from GPS-equipped D-STAR radios via DPRS gateways (more on D-STAR in a later section). The DPRS specification defines how to translate the continuous stream of GPS position reports in the D-STAR DV stream to individual APRS packets. This definition restricts the number of APRS packets generated to prevent overrunning a local RF network because the D-STAR radios continuously stream new position reports while the user is talking. Most D-STAR repeaters have DPRS IGates running on the gateway provid-

ing the D-STAR radio positions to APRS-IS. Those positions can be gated to local RF by APRS IGates if the APRS IGate sysop enables this feature.

Keith Sproul, WU2Z, operates an e-mail gateway that allows APRS messages to be converted to short e-mail or text messages. When entering a message, E-MAIL is used for the destination address. The message body must then contain the actual e-mail address followed by a space and very short message. If the message is picked up by an IGate it will be sent to the WU2Z mail server and out to the recipient, with a confirmation APRS message. For dozens of other messaging options, check online at **http://aprs.org/aprs-messaging.html**.

There is connectivity between APRS and the Winlink system via APRSLink. APRSLink monitors all APRS traffic gated to the Internet, worldwide, and watches for special commands that allow APRS users to:

- Read short e-mail messages sent to their call sign@winlink.org
- Send short e-mail messages to any valid e-mail address or Winlink 2000 user
- Perform e-mail related maintenance (see commands below)
- Be notified of pending Winlink e-mail via APRS message
- Query APRSLink for information on the closest Winlink RMS packet station

Attention must be paid to the APRS traffic generated when using these features but they can be very handy. Details on the APRSLink system are available at **www.winlink.org/ aprslink**.

16.5.13 Winlink 2000

Winlink 2000 or *WL2K* is a worldwide radio messaging system that takes advantage of the Internet where possible. It does this in order to allow the end-user more radio spectrum on the crowded spectrum. The system provides radio interconnection services including: e-mail with attachments, position reporting, graphic and text weather bulletins, emergency / disaster relief communications, and message relay. The PACTOR-I, II, and III protocols are used on HF, and AX.25 Packet, D-STAR and 802.11 are used on VHF/UHF.

Winlink 2000 has been assisting the maritime and RV community around the clock for many years. More recently there has been an increasing interest in emergency communications (emcomm), and the Winlink 2000 development team has responded by adding features and functions that make the system more reliable, flexible and redundant. The role of Winlink 2000 in emergency communications is to supplement existing methodologies to add another tool in the toolkit of the volunteer services deploying emergency communications in their communities.

The Winlink 2000 system is a "star" based network containing five mirror-image, redundant Common Message Servers (CMS) located in San Diego (USA), Washington DC (USA), Vienna (Austria), Halifax (Canada) and Perth (Australia). These ensure that the system will remain in operation should any piece of the Internet become inoperative. Each Radio Message Server node (RMS) is tied together as would be the ends of a spoke on a wheel with the hub function performed by the Common Message Servers. Traffic goes in and out between the CMS and the Internet e-mail recipient, and between the end users and the Radio Message Server (RMS) gateways. Multiple radio-to-radio addresses may be mixed with radio-to-Internet e-mail addresses, allowing complete flexibility.

Because each Radio Message Server gateway is a mirror image of the next, it does not matter which station is used. Each can provide over 700 text-based or graphic weather products, and each can relay the user's position to a web-based view of reporting users and interoperates with the APRS system.

One of the most important objectives in the eyes of the Winlink development team was to reduce the use of the HF spectrum to only that required to exchange messages with a user, and to do that at full "machine" speeds. The HF spectrum is very crowded, and limiting the forwarding of messages between WL2K RMS stations to the Internet, a great deal of radio air time is eliminated, making the time and spectrum available to individual users either for message handling or for other operations. A number of the RMS PACTOR servers found on HF restrict their protocols to PACTOR II (400 to 800 bit/s) and PACTOR III (1400 to 3600 bit/s.) In doing so, these RMS stations typically have a much higher ratio of traffic minutes and message counts to connect times than do the RMS stations that also receive the slower PACTOR I (100 to 200 bps) protocol. In other words, the amount of traffic that is passed with an Express station is much greater for an equivalent amount of connect time with approximately the same number of connections. On average, this translates to a PACTOR I station downloading an 80,000 byte file in approximately 80 minutes while on PACTOR III, the same download takes approximately six minutes. Note that PACTOR IV is not yet legal for US amateurs as discussed earlier in the PACTOR section.

The RMS servers provide endpoint connectivity to users via HF or VHF and can be found worldwide. When connecting via the Internet, users can connect directly to the CMS server operational closest to them via TELNET. Formerly, the full-featured mes-

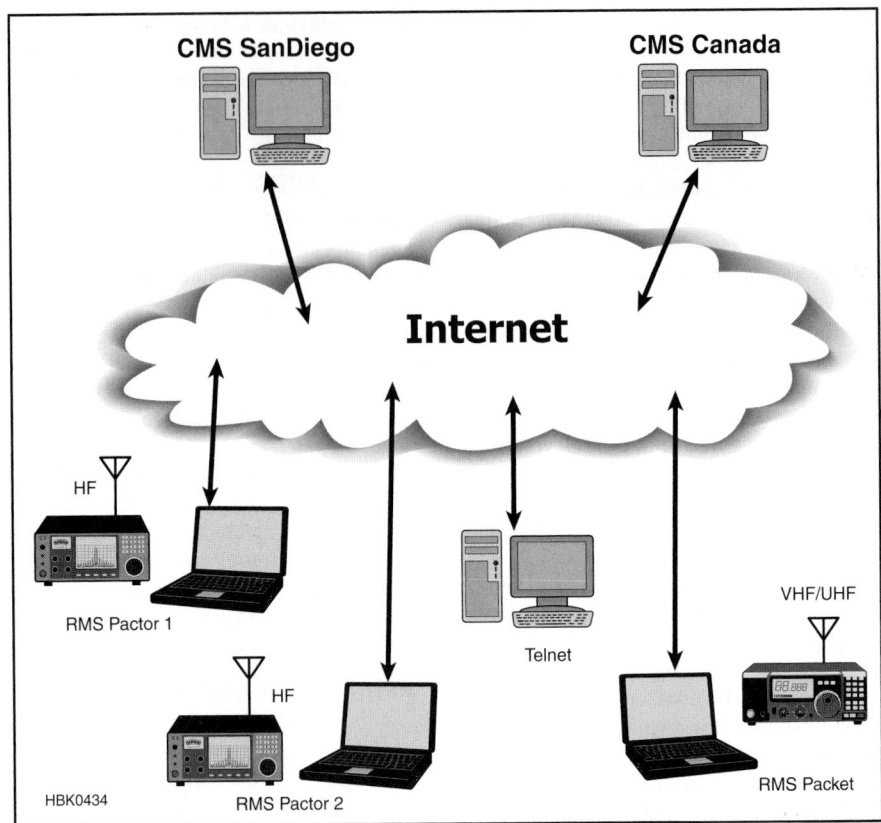

Fig 16.12 — Winlink 2000 system.

sage servers on HF were known as PMBOs (Participating Mail Box Offices) and there was a packet radio to TELNET bridge component for VHF/UHF called Telpac. These have been replaced with the newer RMS PACTOR (HF) and RMS Packet (VHF/UHF) software although the older terms are still in use. Over the air, PACTOR is used on HF, AX.25 packet on VHF/UHF and both employ the B2F compressed binary format to maximize transmission efficiency. A diagram of the Winlink 2000 architecture is shown in **Fig 16.12**.

CLIENT ACCESS TO WINLINK

The primary purpose of the Winlink 2000 network system is to assist the mobile or remotely located user and to provide emergency e-mail capabilities to community agencies. Because of this, WL2K supports a clean, simple interface to the Internet SMTP e-mail system. Any message sent or received may include multiple recipients and multiple binary attachments. The radio user's e-mail address, however, must be known to the system as a radio user or the message will be rejected. This simple Internet interface protocol has an added benefit in case of an emergency where local services are interrupted and the system must be used by non-Amateur groups as an alternative to normal SMTP e-mail.

Connecting to any one of the WL2K publicly-used RMS Packet stations via HF or the specialized non-public emcomm RMS

stations, can immediately and automatically connect a local amateur station to the Internet for emergency traffic. Using a standard SMTP e-mail client, the *Paclink* mini-e-mail server can replace a network of computers (behind a router) as a transparent substitute for normal SMTP mail. WL2K uses no external source for sending or receiving Internet e-mail. It is a stand-alone function which interacts directly with the Internet rather than through any external Internet service provider.

Airmail is independently developed, distributed and supported by Jim Corenman, KE6RK. It is the oldest and most widely used program for sending and receiving messages using the WL2K system. Airmail may be used for HF Pactor, VHF/UHF Packet, and for TELNET connections over any TCP/IP medium including the Internet and high-speed radio media, such as D-STAR and HSMM. Once connected to a WL2K station, message transfer is completely automatic. On the ham bands, Airmail can transfer messages automatically with any station supporting the BBS or F6FBB protocols, such as Winlink 2000, F6FBB, MSYS and other Airmail stations. When used with WL2K, Airmail also contains position reporting capabilities, and a propagation prediction program to determine which of the participating Winlink stations will work from anywhere on Earth. Airmail also contains a limited mail server that allows it to host e-mail independently from the Winlink network.

To obtain a copy of Airmail, including the installation and operating instructions, download the program from the Airmail web page at **www.siriuscyber.net/ham** and support is available from the Airmail user group at **http://groups.yahoo.com/group/airmail2000**.

When an RF connection is not available (or necessary) but web access is available, Winlink 2000 messages can be retrieved or sent via a web interface. The web browser access is limited to text-based messages without the use of bulletins or file attachments. Password-protected access to the Winlink mailbox is available through the Winlink page at **www.winlink.org/webmail**. There is also a terminal mode for interactive keyboard commands, allowing a terminal rather than computer-based software to connect to an RMS via RF. Because of the inefficient use of airtime, this method is discouraged but may be used for the listing and deletion of messages only.

Airmail provides a super-fast replica of WL2K radio operations while directly connected through the Internet to one of the CMS servers. This method of obtaining messages over the Internet allows multiple attachments, catalog bulletins, and all other Winlink 2000 services normally available over radio channels, but at Internet speeds. In order to use this service, a user must currently be listed as a radio user, and obtain the connection information for the closest CMS server. Both Paclink and Airmail support the TELNET client service. This operation allows regular use of the system with the same software configuration at high speed, without using RF bandwidth and provides a full-featured mechanism to access the system where no RF connection is available. An Emergency Operations Center with Internet access can use the TELNET path with RF as a fallback with just a minor change in the software. Similarly, an RV or marine user can use the software from an Internet café or home via TELNET and switch the software back to the TNC when mobile, with no change in functionality.

WINLINK FEATURES

To address the needs of mobile users for near real-time data, WL2K uses an "on-demand" bulletin distribution mechanism. (Note that such bulletins are not the same as traditional AX.25 Packet Bulletins.) Users must first select requested bulletins from an available "catalog" list managed in Airmail. When bulletin requests are received by an RMS station, a fresh locally cached copy of the requested bulletin is delivered. If no fresh locally-cached version is available, the RMS accesses the Internet and finds the bulletin which is then downloaded to the RMS and

then sent to the user. The global catalog currently includes over 700 available weather, propagation, and information bulletins, including, instructions for using the system, World news, and piracy reports. All WL2K RMS stations support a single global catalog which ensures users can access any bulletin from any RMS. Bulletins can contain basic text, graphic fax or satellite images, binary or encoded files like GRIB or WMO weather reports. Local processing is used to re-process images to sizes suitable for HF Pactor transmission. The system prevents bulletin duplication and automatically purges obsolete time-sensitive weather bulletins and replaces them with the current version.

The system also has the ability to contain bulletins with attachment information which is local to each RMS. This is especially useful for the non-public emcomm RMS which may house valuable procedural information pertinent to complex information or instruction needed by specific agencies in any community emergency.

Multiple binary or text-based file attachments of any type or number may be attached to a message by simply selecting the file to be sent from a *Windows* selection dialog in the user's HF AirMail or the Winlink 2000 VHF/UHF Paclink server with a standard SMTP e-mail client. E-mail message attachments sent through the Winlink 2000 system must be limited in size. Users are provided an option to allow this limit to be determined. When using the default B2F format, the protocol chosen by the user usually determines the file size of an attachment. A user may also turn off the ability to receive file attachments. Certain file attachment types are blocked from the system for the protection of the user from virus attacks.

The WL2K network administrator may post notices that are delivered to all individual WL2K users as a private message. This is a valuable tool for notifying users of system changes, outages, software upgrades, emergencies, etc.

The integrated nature of the system makes possible other services beyond just simple messaging. The bulletin services mentioned above is beyond normal messaging, but WL2K also provides rapid position reporting from anywhere in the world. This facility is interconnected with the APRS, ShipTrak, and YotReps networks. It supports weather reporting from cruising yachts at sea and an interconnection with the YotReps network which is used by government forecasters for weather observations in parts of the world where no others are available. It also allows the maritime user to participate in the National Weather Service's NOAA MAROB a voluntary marine observation reporting program.

To ensure equitable access to the system individual users are assigned daily time lim-

its on HF frequencies by RMS sysops. The default time per any 24-hour period is 30 minutes, however, the user may request more time from the RMS sysop should it be needed. The time limit is individual to each RMS station. Utilization of the PACTOR-II and PACTOR-III protocols are a great timesaver, allowing the user up to 18 times the volume of messages over that of PACTOR-I for the same period of time.

The system has a number of other secondary features to help keep it healthy. Extensive traffic reports are collected, the state of individual RMS stations are monitored and reported if it becomes inactive and daily backups are performed automatically at all RMS stations as well as the Common Message Servers to ensure the system integrity. Security is ensured through the vigorous updating of virus definitions and automatic virus screening for all Internet mail and files. The system has the ability to block any user by both radio (by frequency band) and Internet (by e-mail address) to prevent abuse of the system. Spam is controlled through the use of a secure "acceptance list" or "white list" methodology. More information about the Winlink 2000 system is available on the Winlink website at **www.winlink.org**.

16.5.14 D-STAR

D-STAR (Digital Smart Technologies for Amateur Radio) is a digital voice and data protocol specification developed as the result of research by the Japan Amateur Radio League (JARL) to investigate digital technologies for Amateur Radio in 2001. While there are other digital on-air technologies being used by amateurs that were developed for other services, D-STAR is one of the first on-air protocols to be widely deployed and sold by a major radio manufacturer that is designed specifically for amateur service use. D-STAR transfers both voice and data via a data stream over the 2 meter (VHF), 70 cm (UHF) and 23 cm (1.2 GHz) Amateur Radio bands either simplex or via repeater.

One of the interesting features about the D-STAR protocol is the fact the system uses Amateur Radio call signs not only as an identifier, but also for signal routing. In the most common configuration, the most vital part of a D-STAR system is the gateway server, which networks a single system into a D-STAR network via a *trust server*. The trust server provides a central, master database to look up users and their associated system. This allows Amateur Radio operators to respond to calls made to them, regardless of their location on the D-STAR network. While almost all documentation references the Internet as the connection point for a network, any IP network connectivity will work, depending on signal latency.

Additionally, D-STAR can provide a zero infrastructure support system utilizing point-to-point "backbone" 10 GHz connections. Currently, the global D-STAR trust server is maintained by a group of dedicated D-STAR enthusiasts from Dallas, Texas — the Texas Interconnect Team.

The D-STAR protocol specifies two modes; Digital Voice (DV) and Digital Data (DD). In the protocol, the DV mode provides both voice and low speed data channel on 2 meters, 70 cm and 23 cm over a 4800-bit/s data stream. In the protocol, the DV mode uses a data rate of 4800 bit/s. This data stream is broken down to three main packages: voice, forward error correction (FEC) and data. The largest portion of the data stream is the voice package, which is a total of 3600 bits/s with 1200 bit/s dedicated to forward error correction, leaving 1200 bit/s for data. This additional data contains various data flags as well as the data header, leaving about 950 bit/s available for either GPS or serial data. This portion of the data stream does not provide any type of error correction, which has been overcome by implementing error correction in the application software.

While there are various techniques of encoding and transporting a DV signal, the focus of D-STAR's design was the most efficient way to conserve RF spectrum. While D-STAR's "advertised" occupied bandwidth is 6.25 kHz, tests reveal a band plan of 10 kHz spacing is adequate to incorporate the D-STAR signal as well as provide space for channel guards.

In addition to DV mode, the D-STAR protocol outlines the high speed Digital Data (DD) mode. This higher speed data, 128 kbit/s, is available only on the 23 cm band because it requires an advertised 130 kHz bandwidth, only available at 23 cm in world-wide band plans. Unlike the DV mode repeaters, the DD mode module operates as an "access point" operating in half duplex, switching quickly on a single channel.

As with the DV mode, there is a portion of the data stream used for signal identification with the data header as well as various system flags and other D-STAR related items. Once this portion of the data stream is taken into consideration, the 128 kbit/s is reduced to approximately 100 kbit/s — still more than double a dial-up connection speed with significant range. Another consideration is the data rate specified at 128 kbit/s is the gross data rate. Therefore, the system developers are challenged by the area coverage/potential user issue. This means the higher the elevation of the system, the more potential users and the slower the system will become as all the users split the data bandwidth. Finally, there is an issue from the days of packet radio. While technically, the opportunity for "hidden transmitter" issues

Fig 16.13 — Full D-STAR system.

- D-STAR Repeater Controller
- 23 cm DD Mode Data (128 kb)
- 23 cm DV Mode Voice Repeater
- 70 cm DV Mode Voice Repeater
- 2 m DV Mode Voice Repeater
- Internet Gateway Server
- Local Application Server (Email, FTP, Chat, Web, etc.)

HBK0435

does exist and collisions do occur, the TR switching is very fast and this effect is handled by TCP/IP as it is for WiFi access points.

The simplex channel eliminates the need for duplexers at a repeater site if only the DD mode system is installed. It is still recommended to have filtering, such as a band-pass filter, in place to reduce possible interference from other digital sources close to the 23 cm band as well as to reduce RF overload from nearby RF sources. While some DD mode system owners would like more sensitivity or more output power (10 W), at the time of print, no manufacturer has developed pre-amps or RF power amplifiers with an adequate TR switching time to boost the signals.

Radios currently providing DV mode data service use a serial port for low-speed data (1200 bit/s), while the DD mode radio offers a standard Ethernet connection for high-speed (128 kbit/s) connections, to allow easy interfacing with computer equipment. The DD mode Ethernet jack allows two radios to act as an Ethernet bridge without any special software support required. This allows standard file sharing, FTP, TELNET, HTTP/web browsing, IRC chat or even remote desktop connections to function as if connected by wire.

In a Gateway configuration, *all* users must be registered in the network. This provides the DD mode sysop a layer of authorization, meaning that if someone wants to use a DD mode system, and they have not received authorization to use the gateway, their DD mode access will be denied. Any gateway registered user, on the common network, can use any DD mode system, even if the registration was not made on that system. While we are not able to use encryption in the Amateur Radio service, security can

be implemented in standard software or consumer routers and firewalls.

A D-STAR repeater system consists of at least one RF module and a controller. While any combination of RF modules can be installed, typically a full system includes the three voice modules (2 meters, 70 cm and 23 cm) and the 23 cm DD mode module shown in **Fig 16.13**. A computer with dual Ethernet ports, running the Gateway software, is required for Internet access to the global network. An additional server, as shown in the diagram, can be incorporated for local hosting of e-mail, chat, FTP, web and other services. In a D-STAR system installation, the standard repeater components (cavities, isolators, antennas and so forth) are not shown but are required as with any analog system. Some groups have removed analog gear and replaced it with D-STAR components on the same frequency with no additional work beyond connecting the power and feed lines. More information about D-STAR systems may be found in the **Repeaters** chapter and in the **Digital Communications** supplement on the *Handbook* CD.

USER-CREATED FEATURES AND TOOLS

D-STAR has already benefitted from a strong user community that has been developing applications, products and upgrades at a rapid pace. The *DPlus* gateway add-on provides a number of new features, such as an echo test, voicemail, simulcast to all modules, a linking feature, playback of voice or text messages and more. There are also a number of "D-STAR Reflectors" around the world that act like conference bridges for connecting multiple repeater systems.

One of the other projects that came from

the Open D-STAR project is the DV Dongle by Robin Cutshaw, AA4RC. The DV Dongle contains the DSVI AMBE codec chip and a USB interface to allow a computer user to talk with other D-STAR voice users much the way EchoLink users can connect to the system with a PC. The DV Tool software is written in Java to provide cross-platform support. The DV Dongle is available at ham radio dealers and the latest software and manual can be downloaded from **www.opendstar. org/tools**.

Pete Loveall, AE5PL, created a DPRS-to-APRS gateway that moves D-STAR GPS data to the APRS-IS network. Although it doesn't natively appear on the APRS RF network (unless IGated specifically), it does appear in all the online APRS tools. DPRS has been implemented in software with *DStarMonitor* running on many D-STAR gateways and *DPRS Interface* available for most computers at **www.aprs-is.net/dprs**. DPRS has been implemented in hardware with the μSmartDigi manufactured by Rich Painter, AB0VO. The μSmartDigi offers a compact, portable TNC designed to act as a gateway for position packets between a D-STAR digital network and a conventional analog APRS RF network via a D-STAR radio and a conventional FM radio. More DPRS and μSmartDigi information is available at **www.aprs-is.net/dprs** and **www.usmartdigi.com** respectively.

Satoshi Yasuda, 7M3TJZ/AD6GZ, created a D-STAR node adapter or "hot spot" that connects to a standard simplex FM radio and allows a D-STAR radio user to access the D-STAR network. The system requires the FM rig be accurately configured for a clean signal and loses the D-STAR benefit of only consuming 6.25 kHz of bandwidth. It does allow D-STAR users access to the network via RF where there is Internet access and no repeater infrastructure available. More information can be found online at **http:// d-star.dyndns.org/rig.html.en**.

Brian Roode, NJ6N, wrote one of the earliest end-user applications for D-STAR. His *d*Chat* is a *Windows*-based keyboard to keyboard communication application that uses the D-STAR DV mode data capability.

*d*Chat* is used to enable text-based communication between multiple stations on a simplex frequency or through a repeater. More information can be found online at **http://nj6n.com/dstar/dstar_chat.html**.

Dan Smith, KK7DS developed D-RATS with public service users in mind. In addition to chat, D-RATS supports file/photo downloads bulletins, forms, email and a sophisticated mapping capability. It also has a "repeater" functionality that allows sharing a D-STAR radio data stream over a LAN. The repeater feature is interesting in an environment where the radio would be physically separated from the software user, like in an emergency operations center. Dan has current developments and downloads available at **www.d-rats.com/wiki**.

D-STAR TV was developed by John Brown, GM7HHB, and provides still images over D-STAR DV mode and streaming video over DD mode. The limited bandwidth of DV mode data and the lack of error correction mean the images take a while and may have some corruption (just like SSTV). The DD mode has much more bandwidth and can support live video. More information is available online at **www.dstartv.com**.

This is just a sample of the development taking place and there will undoubtedly be more applications coming as the user base increases. The two websites with the most current D-STAR information are **www. dstarinfo.com** and **www.dstarusers.org**.

16.5.15 APCO Project 25 (P25)

Project 25 (P25) or APCO-25 is a suite of standards for digital radio communications for use by public safety agencies in North America. P25 was established by Association of Public-Safety Communications Officials (APCO) to address the need for common digital public safety radio communications standards for first responders and Homeland Security/Emergency Response professionals. In this regard, P25 fills the same role as the European Tetra protocol, although not interoperable with it.

P25 has been developed in phases with

Phase 1 completed in 1995. P25 Phase 1 radio systems operate in 12.5 kHz analog, digital or mixed mode. Phase 1 radios use Continuous 4-level FM (C4FM) modulation for digital transmissions at 4800 baud and two bits per symbol, yielding 9600 bit/s total channel throughput. In the case of data transmission, data packets basically consist of a header, containing overhead information, followed by data. In the case of digitized voice transmission, after the transmission of a header containing error protected overhead information, 2400 bit/s is devoted to periodically repeating the overhead information needed to allow for late entry (or the missed reception of the header).

After evaluating several candidates, Project 25 selected the IMBE (Improved MultiBand Excitation) vocoder from DVSI for phase 1, operating at 4400 bit/s. An additional 2800 bit/s of forward error correction is added for error correction of the digitized voice. An 88.9-bit/s low-speed data channel is provided in the digitized voice frame structure and several forms of encryption are supported.

P25 Phase 2 requires a further effective bandwidth reduction to 6.25 kHz by operating as two-slot TDMA (Time Division Multiple Access) with support for trunking systems. This provides two channels from the same 12.5 kHz spectrum when working through a repeater to provide time synchronization. The TDMA timing requirements does limit range to a published 35 km but this is plenty for public safety users. It uses the newer AMBE (Advanced MultiBand Excitation) codec from DVSI to accommodate the bit rate reduction to 4800 bit/s. Phase 2 radios will still offer backward compatibility with Phase 1. Phase 3 work has started and will support access to the new 700 MHz band in the US.

P25 radios offer a number of features of interest to public service agencies such as an emergency mode, optional text messaging, over the air programming/deactivation. More P25 information can be found at **www.apco911.org/frequency/project25/ information.html#documents** and **www. tiaonline.org/standards/technology/ project_25**.

Table 16.18
Digital Modulation Modes and Formats Used in Amateur Radio

See the text of Chapter 16 for definitions and abbreviations
Developed by Alan Bloom, N1AL and edited by Scott Honnaker, N7SS; March 2013

Mode or Format Name	Developer	Principal Freq	Principal Application	Data rate (bit/s)	Bit rate (bit/s)	Symbol rate (baud)	Modulation ("N" means multi-carrier)	Error handling
ALE	MIL-STD-188-141, FED-STD-1045	HF	Data	<375	375	125	8FSK	FEC
AMTOR-A	G3PLX	HF	Data	53	114	114	FSK	ARQ
AMTOR-B	G3PLX	HF	Data	57	114	114	FSK	FEC
AOR AMBE	AOR Corp	HF	Voice, Data	2400	3600	50	36-QPSK	FEC
APCO P25	APCO	VHF	Voice	6800	9600	4,800	4FSK/QPSK	FEC
Chip64	IZ8BLY	HF	Keyboarding	21.1/37.5	300	300	DBPSK-DSSS	FEC/FEC+ARQ
CLOVER-II	Hal Comm.	HF	Data	<37.5-750	62.5-750	31.25	4-(2-16)DPSK/(2-4)DASK	FEC/FEC+ARQ
CLOVER-2000	Hal Comm.	HF	Data	108-1994	500-3000	62.5	8-(2-16)DPSK/(2-4)DASK	FEC/FEC+ARQ
Domino	ZL2AFP	HF	Keyboarding	31/44/62?	31/44/62	7.8/11/15.6	16FSK	None
DominoEX	ZL1BPU	HF	Keyboarding	(<15.63)-86.13	15.63-86.13	3.9-21.5	18FSK	None/FEC
D-Star (DV)	JARL	VHF	Voice & data	D:960 V:2400	4800	4800	0.5 GMSK/QPSK/4PSK	FEC
D-Star (DD)	JARL	UHF	Voice & data	72k-124k	128,000	128,000	0.5 GMSK/QPSK/4PSK	FEC
Facsimile		HF	Image			120 lpm	FM, 1500-2300 Hz	None
FDMDV	G3PLX/HB9TLK	HF	Voice	1450	1450	50	15-QPSK	None
FSK441	K1JT	VHF	Meteor scatter	882	882	441	4FSK	None
G-TOR	Kantronics	HF	Data	35/75/115	80/160/240	100/200/300	FSK, 170/200 shift	FEC + ARQ
Hellschreiber (Feld)	Rudolf Hell	HF	Keyboarding	2.5 char/s	122.5	122.5	ASK	None
JT6M	K1JT	50 MHz	Meteor scatter	77.9	116.8	21.53	44FSK	FEC
JT65	K1JT	V/UHF	Moonbounce	1.54	16.1	2.7	65FSK	FEC
MFSK16	ZL1BPU/IZ8BLY	HF	Keyboarding, Data	31.25	62.5	15.625	16FSK	FEC
MT63	SP9VRC	HF	Keyboarding	35/70/140	320/640/1280	5/10/20	64-DPSK	FEC
Olivia	SP9VRC	HF	Keyboarding	8.75/17.5	78.13/156.25	15.63/31.25	32-FSK	FEC
Packet (Bell202)		VHF	Data	<1200	1200	1200	FSK	ARQ
PACTOR-I	DL6MAA/DF4KV	HF	Data, Winlink e-mail	51.2/128	100/200	100/200	FSK, 200 Hz	ARQ
PACTOR-II	Spec. Comm. Sys.	HF	Data, Winlink e-mail	100-700	200-800	100	2-DBPSK/PI-4DQPSK/8,16DPSK	FEC + ARQ
PACTOR-III	Spec. Comm. Sys.	HF	Data, Winlink e-mail	85-2722	200-3600	100	(2-18)-DBPSK/DQPSK	FEC + ARQ
PSK31	G3PLX	HF	Keyboarding	31.25	31.25	31.25	BPSK	None
QPSK31	G3PLX	HF	Keyboarding	31.25	62.5	31.25	QPSK	FEC
PSK63/125		HF	Keyboarding	62.5/125	62.5/125	62.5/125	BPSK	None
QPSK62/126		HF	Keyboarding	62.5/125	125/250	62.5/125	QPSK	FEC
Q15X25	SP9VRC	HF	Data	300/1200/2400	2,500	83.33	15-QPSK	FEC + ARQ
RTTY (Baudot)		HF	Keyboarding, contests	30.3	45.45	45.45	FSK, 170 Hz	None
SSTV (traditional)	WØORX	HF	Image	8 s/frame		120-line B/W	FM, 1200-2300 Hz	None
SSTV Martin M1	Martin Emmerson	HF	Image	114 s/frame		240-line RGB	FM, 1200-2300 Hz	None
SSTV Scottie S1	Eddie Murphy	HF	Image	110 s/frame		240-line RGB	FM, 1200-2300 Hz	None
Throb	G3PPT	HF	Keyboarding	10/20/40 wpm		1/2/4	9FSK/2-9FSK	None
WinDRM	HB9TLK	HF	Voice, Data		62.5-3750		COFDM (n-QAM)	FEC/FEC+ARQ
WINMOR	KN6KB	HF	Winlink email			31.25/62.5	(1-15)-QPSK/16QAM/4FSK	FEC + ARQ
WSPR (MEPT-JT)	K1JT	HF-VHF	Weak signal beacon	0.45	2.93	1.46	4FSK	FEC

16.6 Digital Mode Table

Table 16.18 is a summary of common digital modes used by amateurs as of early 2013 and their primary characteristics. The following text is intended to make high-level comparisons. For more information on these modes see the earlier sections of this chapter. This page is also available as a PDF file on the CD-ROM accompanying this book.

Many modes have a number of variants, with the most common shown in the table. High reliability, low data rate modes are more common on HF. Higher data rates are available on VHF/UHF, where the bands have less noise and require less effort to make them reliable. Some modes have very specific intended uses like the meteor scatter,

moonbounce and beaconing modes created by Joe Taylor, K1JT as *WSJT* (see the section Structured Digital Modes).

There can be a significant difference between data rate and bit rate in high reliability modes. The data rate is the amount of user data transmitted where the bit rate includes the packet and error correction overhead. Some of these rates are approximate and can vary based on conditions and the variant of the mode used. The symbol rate is a function of the modulation scheme and how many simultaneous carriers are used to transmit the data.

The error handling mechanism is critical to note when sending data. Some modes include

no error handling which means errors are not detected and must be addressed in a higher protocol layer (as in AX.25 packet). Forward error correction (FEC) makes a best effort to address errors in real time as part of the data sent in each packet. FEC can add substantial overhead to each packet and does not guarantee error-free delivery but does make the mode robust in high noise environments. Automatic Retry Request (ARQ) can guarantee error-free delivery of data but has no ability to actually correct received data. The combination of FEC and ARQ allows minor errors to be corrected in real time with major errors generating a retry request.

16.7 Glossary

ACK — Acknowledgment, the control signal sent to indicate the correct receipt of a transmission block.

Address — A character or group of characters that identifies a source or destination.

AFSK — Audio frequency-shift keying.

ALE — Automatic link establishment.

Algorithm — Numerical method or process.

APCO — Association of Public Safety Communications Officials.

ARQ — Automatic Repeat reQuest, an error-sending station, after transmitting a data block, awaits a reply (ACK or NAK) to determine whether to repeat the last block or proceed to the next.

ASCII — American National Standard Code for Information Interchange, a code consisting of seven information bits.

AX.25 — Amateur packet-radio link-layer protocol.

Baud — A unit of signaling speed equal to the number of discrete conditions or events per second. (If the duration of a pulse is 20 ms, the signaling rate is 50 baud or the reciprocal of 0.02, abbreviated Bd).

Baudot code — A coded character set in which five bits represent one character. Used in the US to refer to ITA2.

Bell 103 — A 300-baud full-duplex modem using 200-Hz-shift FSK of tones centered at 1170 and 2125 Hz.

Bell 202 — A 1200-baud modem standard with 1200-Hz mark, 2200-Hz space, used for VHF FM packet radio.

BER — Bit error rate.

BERT — Bit-error-rate test.

Bit stuffing — Insertion and deletion of 0s in a frame to preclude accidental occurrences of flags other than at the beginning and end of frames.

Bit — Binary digit, a single symbol, in binary terms either a one or zero.

Bit/s or *bps* — Bits per second.

Bitmap — See **raster.**

Bit rate — Rate at which bits are transmitted in bit/s or bps. **Gross** (or **raw**) **bit rate** includes all bits transmitted, regardless of purpose. **Net bit rate** (also called **throughput**) only includes bits that represent data.

BLER — Block error rate.

BLERT — Block-error-rate test.

BPSK — Binary phase-shift keying in which there are two combinations of phase used to represent data symbols.

Byte — A group of bits, usually eight.

Cache — To store data or packets in anticipation of future use, thus improving routing or delivery performance.

Channel — Medium through which data is transmitted.

Checksum — Data representing the sum of all character values in a packet or message.

CLOVER — Trade name of digital communications system developed by Hal Communications.

COFDM — Coded Orthogonal Frequency Division Multiplex, OFDM plus coding to provide error correction and noise immunity.

Code — Method of representing data.

Codec — Algorithm for compressing and decompressing data.

Collision — A condition that occurs when two or more transmissions occur at the same time and cause interference to the intended receivers.

Compression — Method of reducing the amount of data required to represent a signal or data set. **Decompression** is the method of reversing the compression process.

Constellation — A set of points that represent the various combinations of phase and amplitude in a QAM or other complex modulation scheme.

Contention — A condition on a communications channel that occurs when two or more stations try to transmit at the same time.

Control characters — Data values with special meanings in a protocol, used to cause specific functions to be performed.

Control field — An 8-bit pattern in an HDLC frame containing commands or responses, and sequence numbers.

Convolution — The process of combining or comparing signals based on their behavior over time.

Cyclic Redundancy Check (CRC) — The result of a calculation representing all character values in a packet or message. The result of the CRC is sent with a transmission block. The receiving station uses the received CRC to check transmitted data integrity.

Datagram — A data packet in a connectionless protocol.

Data stream — Flow of information, either over the air or in a network.

DBPSK — Differential binary phase-shift keying.

Dibit — A two-bit combination.

DQPSK — Differential quadrature phase-shift keying.

Domino — A conversational HF digital mode similar in some respects to MFSK16.

DRM — Digital Radio Mondiale. A consortium of broadcasters, manufacturers, research and governmental organizations which developed a system for digital sound broadcasting in bands between 100 kHz and 30 MHz.

DV —Digital voice.

Emission — A signal transmitted over the air.

Encoding — Changing data into a form represented by a particular code.

Encryption — Process of using codes and encoding in an effort to obscure the meaning of a transmitted message. **Decryption** reverses the encryption process to recover the original data.

Error Correcting Code (ECC) — Code used to repair transmission errors.

Eye pattern — An oscilloscope display in the shape of one or more eyes for observing the shape of a serial digital stream and any impairments.

Fast Fourier Transform (FFT) — An algorithm that produces the spectrum of a signal from the set of sampled value of the waveform.

FEC — Forward error correction.

Frame — Data set transmitted as one package or set.

Facsimile (fax) — A form of telegraphy for the transmission of fixed images, with or without half-tones, with a view to their reproduction in a permanent form.

FCS — Frame check sequence. (see also **CRC**)

FDM — Frequency division multiplexing.

FDMA — Frequency division multiple access.

FEC — Forward error correction, an error-control technique in which the transmitted data is sufficiently redundant to permit the receiving station to correct some errors.

FSK — Frequency-shift keying.

Gray code — A code that minimizes the number of bits that change between sequential numeric values.

G-TOR — A digital communications system developed by Kantronics.

HDLC — High-level data link control procedures as specified in ISO 3309.

Hellschreiber — A facsimile system for transmitting text.

Host — As used in packet radio, a computer with applications programs accessible by remote stations.

IA5 — International Alphabet, designating a specific set of characters as an ITU standard.

Information field — Any sequence of bits containing the intelligence to be conveyed.

Interleave — Combine more than one data stream into a single stream or alter the data stream in such a way as to optimize it for the modulation and channel characteristics being used.

IP — Internet Protocol, a network protocol used to route information between addresses on the Internet.

ISO — International Organization for Standardization.

ITU — International Telecommunication Union, a specialized agency of the United Nations. (See **www.itu.int**.)

ITU-T — Telecommunication Standardization Sector of the ITU, formerly CCITT.

Jitter — Unwanted variations in timing or phase in a digital signal.

Layer — In communications protocols, one of the strata or levels in a reference model.

Least significant bit (LSB) — The bit in a byte or word that represents the smallest value.

Level 1 — Physical layer of the OSI reference model.

Level 2 — Link layer of the OSI reference model.

Level 3 — Network layer of the OSI reference model.

Level 4 — Transport layer of the OSI reference model.

Level 5 — Session layer of the OSI reference model.

Level 6 — Presentation layer of the OSI reference model.

Level 7 — Application layer of the OSI reference model.

Lossless (compression) — Method of compression that results in an exact copy of the original data

Lossy (compression) — Method of compression in which some of the original data is lost

MFSK16 — A multi-frequency shift communications system

Modem — Modulator-demodulator, a device that connects between a data terminal and communication line (or radio). Also called **data set**.

Most significant bit (MSB) — The bit in a byte or word that represents the greatest value.

Multicast — A protocol designed to distribute packets of data to many users without communications between the user and data source.

MSK — Frequency-shift keying where the shift in Hz is equal to half the signaling rate in bit/s.

MT63 — A keyboard-to-keyboard mode similar to PSK31 and RTTY.

Nibble — A four-bit quantity. Half a byte.

Node — A point within a network, usually where two or more links come together, performing switching, routine and concentrating functions.

OFDM — Orthogonal Frequency Division Multiplex. A method of using spaced subcarriers that are phased in such a way as to reduce the interference between them.

OSI-RM — Open Systems Interconnection Reference Model specified in ISO 7498 and ITU-T Recommendation X.200.

PACTOR — Trade name of digital communications protocols offered by Special Communications Systems GmbH & Co KG (SCS).

Packet — 1) Radio: communication using the AX.25 protocol. 2) Data: transmitted data structure for a particular protocol (see also **frame**.)

Parity (parity check) — Number of bits with a particular value in a specific data element, such as a byte or word or packet. Parity can be odd or even. A **parity bit** contains the information about the element parity.

Pixel — Abbreviation for "picture element."

Primitive — An instruction for creating a signal or data set, such as in a drawing or for speech.

Project 25 — Digital voice system developed for APCO, also known as P25.

Protocol — A formal set of rules and procedures for the exchange of information.

PSK — Phase-shift keying.

PSK31 — A narrow-band digital communications system developed by Peter Martinez, G3PLX.

QAM — Quadrature Amplitude Modulation. A method of simultaneous phase and amplitude modulation. The number that precedes it, for example, 64QAM, indicates the number of discrete stages in each symbol.

QPSK — Quadrature phase-shift keying in which there are four different combinations of signal phase that represent symbols.

Raster (image) — Images represented as individual data elements, called pixels.

Router — A network packet switch. In packet radio, a network level relay station capable of routing packets.

RTTY — Radioteletype.

Sample — Convert an analog signal to a set of digital values.

Shift — 1) The difference between mark and space frequencies in an FSK or AFSK signal. 2) To change between character sets, such as between LTRS and FIGS in RTTY.

SSID — Secondary station identifier. In AX.25 link-layer protocol, a multipurpose octet to identify several packet radio stations operating under the same call sign.

State — Particular combination of signal attributes used to represent data, such as amplitude or phase.

Start (stop) bit — Symbol used to synchronize receiving equipment at the beginning (end) of a data byte.

Symbol — Specific state or change in state of a transmission representing a particular signaling event. **Symbol rate** is the number of symbols transmitted per unit of time (see also **baud**).

TAPR — Tucson Amateur Packet Radio Corporation, a nonprofit organization involved in digital mode development.

TDM — Time division multiplexing.

TDMA — Time division multiple access.

Throb — A multi-frequency shift mode like MFSK16.

TNC — Terminal node controller, a device that assembles and disassembles packets (frames).

Trellis — The set of allowed combination

of signal states that represent data. (see also **Viterbi coding** and **constellation**.)

Turnaround time — The time required to reverse the direction of a half-duplex circuit, required by propagation, modem reversal and transmit-receive switching time of transceiver.

Unicast — A protocol in which information is exchanged between a single pair of points on a network.

Varicode — A code in which the different data values are represented by codes with different lengths.

Vector — Image represented as a collection of drawing instructions.

Word — Set of bits larger than a byte.

16.8 References and Bibliography

A. Bombardiere, K2MO, "A Quick-Start Guide to ALE400 ARQ FAE," *QST*, Jun 2010, pp 34-36.

B.P. Lathi, *Modern Digital and Analog Communication Systems*, Oxford University Press, 1998.

B. Sklar, *Digital Communications, Fundamentals and Applications*, Prentice Hall, 1988.

S. Ford, WB8IMY, *HF Digital Handbook*, 4th edition, ARRL, 2007.

S. Ford, WB8IMY, *VHF Digital Handbook*, ARRL, 2008.

Contents

Chapter 17 — CD-ROM Content

Software

- *TubeCalculator* by Bentley Chan and John Stanley, K4ERO, for analysis of operation of popular high power transmitting tubes.
- *PI-EL* by Jim Tonne, W4ENE, for design and analysis of pi and pi-L networks for transmitter output.
- *SVC Filter* by Jim Tonne, W4ENE, for design and analysis of low-pass and high-pass filters using standard value components.
- *MeterBasic* by Jim Tonne, W4ENE, for design and printing of custom analog meter scales.
- *MATCH.EXE* software (for use with Tuned (Resonant) Networks discussion)

Supplemental Articles

- Tuned (Resonant) Networks (for use with MATCH.EXE)
- Design Example — RF Amplifier using 8877 Vacuum Tube by John Stanley, K4ERO
- Design Example — MOSFET Thermal Design by Dick Frey, K4XU
- Determining a Transistor's Power Rating (APT Application Note) by Dick Frey, K4XU
- *ARRL RF Amplifier Classics* Table of Contents

HF Amplifier Projects

- "The Everyham's Amp" by John Stanley, K4ERO
- Everyham's Amp files — construction notes, layouts, modifications for various tube types
- "A 3CX1500D7 RF Linear Amplifier" by Jerry Pittenger, K8RA
- 3CX1500D7 HF amplifier files — PCB layout, Pi-L values spreadsheet
- "A 250 W Broadband Solid-State Linear Amplifier" by Dick Frey, K4XU
- 250 W solid state amplifier files — PCB artwork, parts lists, photos (including updated PCB and schematic files, Mar 2013)
- "The Sunnyvale/Saint Petersburg Kilowatt-Plus," a 4CX1600B HF amplifier project by George Daughters, K6GT

VHF Amplifier Projects

- "A 6 Meter Kilowatt Amplifier" by Dick Stevens, W1QWJ (SK)
- "144 MHz Amplifier Using the 3CX1200Z7" by Russ Miller, N7ART
- "Build a Linear 2 Meter, 80 W All Mode Amplifier" by James Klitzing, W6PQL
- "Design Notes for 'A Luxury Linear' Amplifier" by Mark Mandelkern, K5AM
- "High-Performance Grounded-Grid 220-MHz Kilowatt Linear" by Robert Sutherland, W6PO (SK)

UHF/Microwave Amplifier Projects

- "432 MHz 3CX800A7 Amplifier" by Steve Powlishen, K1FO (SK)
- "A High-Power Cavity Amplifier for the New 900-MHz Band" by Robert Sutherland, W6PO (SK)
- "A Quarter-Kilowatt 23-cm Amplifier" by Chip Angle, N6CA
- "2 Watt RF Power Amplifier for 10 GHz" by Steven Lampereur, KB9MWR

RF Power Amplifiers

Amateur Radio operators typically use a very wide range of transmitted power — from milliwatts to the full legal power limit of 1.5 kW. This chapter covers RF power amplifiers beyond the 100-150 W level of the typical transceiver. The sections on tube-type amplifiers were prepared by John Stanley, K4ERO. The sections on solid state amplifiers were contributed by Richard Frey, K4XU. Roger Halstead, K8RI, contributed material on amplifier tuning and the use of surplus components in amplifier construction. This edition also includes several new amplifier projects in the print edition: The Everyham's Amplifier by K4ERO; an All-mode, 2 Meter, 80 W Linear Amp by W6PQL, and a 2 W 10 GHz Amplifier by KB9MWR. In addition, more amplifier projects and supplemental information are included on the CD-ROM.

17.1 High Power, Who Needs It?

There are certain activities where higher power levels are almost always required for the greatest success — contesting and DXing, for example. While there are some outstanding operators who enjoy being competitive in spite of that disadvantage, the high-power stations usually have the biggest scores and get through the pileups first. On the VHF and higher bands, sometimes high power is the only way to overcome path loss and poor propagation.

Another useful and important area where higher power may be needed is in net operations. The net control station needs to be heard by all potential net participants, some of whom may be hindered by a noisy location or limited antenna options or operating mobile. This can be crucial to effective emergency communications. General operation also benefits from the availability of high power when conditions demand it to maintain contact.

How do you decide that you need amplification? As a rule of thumb, if you have a good antenna and hear a lot of stations that don't seem to hear you, you probably need to run more power. If you operate on bands where noise levels are high (160 and 75/80 meters) or at times when signals are weak then you may find that running the legal limit makes operations more enjoyable. On the other hand, many stations will find that they can be heard fine with the standard 100 W transmitter or even with lower power.

Power requirements also depend on the mode being used. Some digital modes, such as PSK31, work very well with surprisingly low power. CW is more power efficient than SSB voice. Least effective is full carrier AM, which is used by vintage equipment lovers. Once you have determined that higher power will enhance your operations, you should study the material in this chapter no matter whether you plan to build or buy your amplifier.

Note that many power amplifiers are capable of exceeding the legal limit, just as most

Safety First!

YOU CAN BE KILLED by coming in contact with the high voltages inside a commercial or homebrew RF amplifier. Please don't take foolish chances. Remember that you cannot go wrong by treating each amplifier as potentially lethal! For a more thorough treatment of this all-important subject, please review the applicable sections of the **Power Sources** and the **Safety** chapters in this *Handbook*.

automobiles are capable of exceeding posted speed limits. That does not mean that every operator with an amplifier capable of more than 1.5 kW output is a scofflaw. Longer life for the amplifying devices and other amplifier components as well as a cleaner signal result from running an amplifier below its maximum rating. However, just as with automobiles, that extra capability also presents certain temptations. Remember that FCC rules require you to employ an accurate way to determine output power, especially when you are running close to the limit.

17.2 Types of Power Amplifiers

Power amplifiers are categorized by their power level, intended frequencies of operation, device type, class of operation and circuit configuration. Within each of these categories there are almost always two or more options available. Choosing the most appropriate set of options from all those available is the fundamental concept of design.

17.2.1 Why a "Linear" Amplifier?

The amplifiers commonly used by amateurs for increasing their transmitted power are often referred to as "linears" rather than amplifiers or linear amplifiers. What does this mean and why is it important?

The active device in amplifiers, either tube or transistor, is like a switch. In addition to the "on" and "off" states of a true switch, the active device has intermediate conditions where it presents a finite value of resistance, neither zero nor infinity. As discussed in more detail in the **RF Techniques** chapter, active devices may be operated in various *classes of operation*. Class A operation never turns the device fully on or off; it is always somewhere in between. Class B turns the device fully off for about half the time, but never fully on. Class C turns the device off for about 66% of the time, and almost achieves the fully on condition. Class D switches as quickly as possible between the on and off conditions. Other letters have been assigned to various rapid switching methods that try to do what Class D does, only better. Class E and beyond use special techniques to enable the device to make the switching transition as quickly as possible.

During the operating cycle, the highest efficiency is achieved when the active device spends most of its time in the on or off condition and the least in the resistive condition. For this reason, efficiency increases as we go from Class A to B to C to D.

A *linear amplifier* is one that produces an output signal that is identical to the input signal, except that it is stronger. Not all amplifiers do this. Linear amplifiers use Class A, AB or B operation. They are used for modes such as SSB where it is critical that the output be a close reproduction of the input.

The Class C amplifiers used for FM transmitters are *not* linear. A Class C amplifier, properly filtered to remove harmonics, reproduces the frequencies present in the input signal, but the *envelope* of the signal is distorted or even flattened completely. (See the **Modulation** chapter for more information on waveforms, envelopes and other signal characteristics.)

An FM signal has a constant amplitude, so it carries no information in the envelope. A CW signal does carry information in the amplitude variations. Only the on and off states must be preserved, so a Class C amplifier retains the information content of a CW signal. However, modern CW transmitters carefully shape the pulses so that key clicks are reduced to the minimum practical value. A Class C amplifier will distort the pulse shape and make the key clicks worse. Therefore, except for FM, a linear amplifier is recommended for all amateur transmission modes.

Some digital modes, such as RTTY using FSK, are a form of FM and can also use a nonlinear Class C, D or E amplifier. If these signals are not clean, however, a Class C amplifier may make them worse. Also, Class C or even D and E can be used for very slow CW, for very simple low-power CW transmitters or on uncrowded bands where slightly worse key clicks are not so serious. After all, Class C was used for many years with CW operation.

Class of operation as it relates to tube-type amplifier design is discussed in more detail in a later section of this chapter.

ACHIEVING LINEAR AMPLIFICATION

How is linear amplification achieved? Transistors and tubes are capable of being operated in a linear mode by restricting the input signal to values that fall on the linear portion of the curve that relates the input and output power of the device. Improper bias and excessive drive power are the two most common causes of distortion in linear amplifiers. All linear amplifiers can be improperly

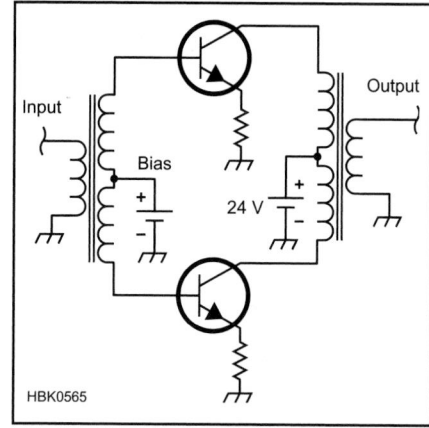

Fig 17.1 — This simple circuit can operate in a linear manner if properly biased.

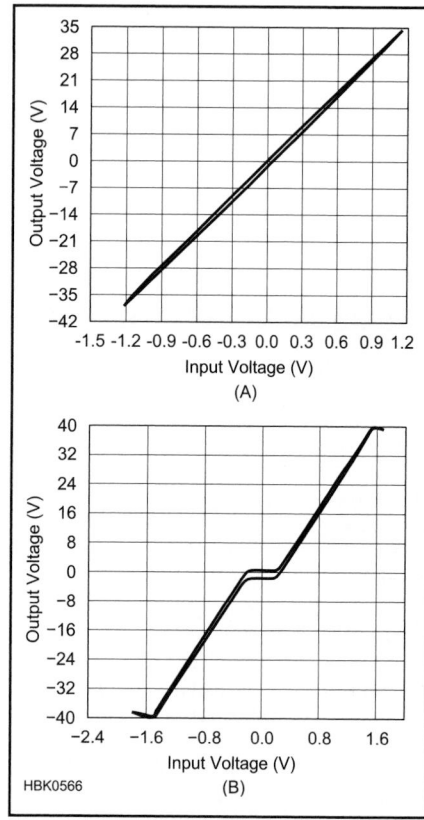

Fig 17.2 — Input versus output signals from an amplifier, as observed with the X-Y display on an oscilloscope. At A, an amplifier with proper bias and input voltage. At B, the same amplifier with improper bias and high input voltage.

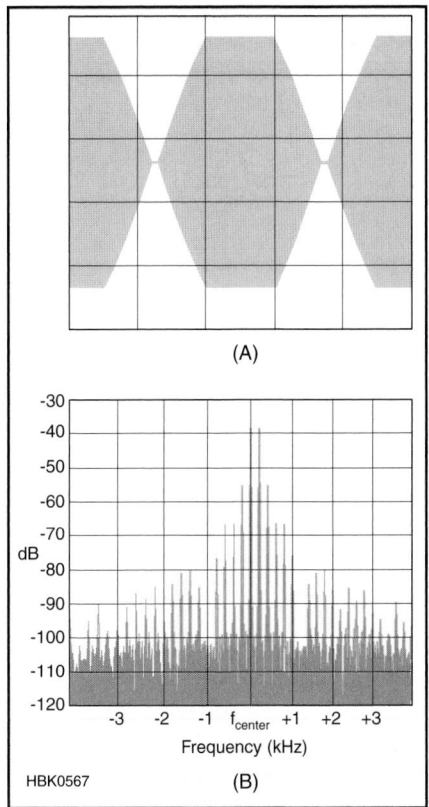

Fig 17.3 — Improper operation of a linear using a two-tone test will show peak clipping on an oscilloscope (A) and the presence of additional frequencies on a spectrum analyzer (B). When these patterns appear, your "linear" has become a "nonlinear" amplifier. Users of adjacent channels will not be happy. More information on transmitter testing may be found in the Test Equipment and Measurements chapter.

biased or overdriven, regardless of the power level or whether transistors or tubes are used.

Fig 17.1 shows a simple circuit capable of operating in a linear manner. For linear operation, the bias on the base of the transistors must be such that the circuit begins to produce an output signal even with very small input signal values. As shown in **Fig 17.2**, when properly adjusted for linear operation, the amplifier's output signal faithfully tracks the input signal. Without proper bias on the transistors, there will be no output signal until the input voltage goes above a threshold. As shown in Fig 17.2B, the improperly adjusted amplifier suddenly switches on and produces output when the input signal reaches 0.5 V.

Some tubes are designed for *zero bias* operation. This means that an optimum bias current is inherent in the design of the tube when it is operated with the correct plate voltage. Other types of tubes and all transistors must have bias applied with circuits made for that purpose.

All amplifiers have a limit to the amount of power they can produce, even if the input power is very large. When the output power runs up against this upper limit (that is, when additional drive power results in no more output power), *flat topping* occurs and the output is distorted, as shown in **Fig 17.3**. Many amplifiers use *automatic level control* (ALC) circuits to provide feedback between the amplifier and the companion transceiver or transmitter. When adjusted properly, ALC will control the transmitter output power, preventing the worst effects of overdriving the amplifier. Even with ALC, however, overdriving can occur.

17.2.2 Solid State vs Vacuum Tubes

With the exception of high-power amplifiers, nearly all items of amateur equipment manufactured commercially today use solid state (semiconductor) devices exclusively. Semiconductor diodes, transistors and integrated circuits (ICs) offer several advantages in designing and fabricating equipment. Solid state equipment is smaller, offers broadband (no-tune-up) operation, and is easily manufactured using PC boards and automated (lower cost) processes.

Based on all these facts, it might seem that there would be no place for vacuum tubes in a solid state world. Transistors and ICs do have significant limitations, however, especially in a practical sense. Individual present-day transistors cannot generally handle the combination of current and voltage needed nor can they safely dispose of the amount of heat dissipated for RF amplification to high power levels. Pairs of transistors, or even pairs of pairs, are usually employed in practical power amplifier designs at the 100 W level and beyond. Sometimes various techniques of power combination from multiple amplifiers must be used.

Tube amplifiers can be more economical to build for a given output power. Vacuum tubes operate satisfactorily at surface temperatures as high as 150-200 °C, so they may be cooled by simply blowing sufficient ambient air past or through their relative large cooling surfaces. The very small cooling surfaces of power transistors should be held to 75-100 °C to avoid drastically shortening their life expectancy. Thus, assuming worst-case 50 °C ambient air temperature, the large cooling surface of a vacuum tube can be allowed to rise 100-150 °C above ambient, while the small surface of a transistor must not be allowed to rise more than about 50 °C.

Furthermore, RF power transistors are much less tolerant of electrical abuse than are most vacuum tubes. An overvoltage spike lasting only microseconds can — and is likely to — destroy RF power transistors. A comparable spike is unlikely to have any effect on a tube. So the important message is this: designing with RF power transistors demands caution to ensure that adequate thermal and electrical protection is provided.

Even if one ignores the challenge of the RF portions of a high-power solid state amplifier, there is the dc power supply to consider. A solid state amplifier capable of delivering 1 kW of RF output might require regulated (and transient-free) 50 V at more than 40 A. Developing that much current can be challenging. A vacuum tube amplifier at the same power level might require 2000 to 3000 V, unregulated, at less than 1 A.

At the kilowatt level, the vacuum tube is still a viable option for amateur constructors because of its cost-effectiveness and ease of equipment design. Because tube amplifiers and solid state amplifiers are quite different in many ways, we shall treat them in different sections of this chapter. Also, the author of the solid state section presents a slightly different perspective on the tube-vs-solid state discussion.

17.3 Vacuum Tube Basics

The term *vacuum tube* describes the physical construction of the devices, which are usually tubular and have a vacuum inside. The British call them electron valves which describes the operation of the devices, since they control the flow of electrons, like a water valve controls the flow of water.

17.3.1 Thermionic Emission

Metals are electrical conductors because the electrons in them readily move from one atom to the next under the influence of an electrical field. It is also possible to cause the electrons to be emitted into space if enough energy is added to them. Heat is one way of adding energy to metal atoms, and the resulting flow of electrons into space is called *thermionic emission*. As each electron leaves the metal surface, it is replaced by another provided there is an electrical connection from outside the tube to the heated metal.

In a vacuum tube, the emitted electrons hover around the surface of the metal unless acted upon by an electric field. If a positively charged conductor is placed nearby, the electrons are drawn through the vacuum and arrive at that conductor, thus providing a continuous flow of current through the vacuum tube.

17.3.2 Components of a Vacuum Tube

A basic vacuum tube contains at least two parts: a *cathode* and a *plate*. The electrons are emitted from the *cathode*. The cathode can be *directly heated* by passing a large dc current through it, or it can be located adjacent to a heating element (*indirectly heated*). Although ac currents can also be used to directly heat cathodes, if any of the ac voltage mixes with the signal, ac hum will be introduced into the output. If the ac heater supply voltage can be obtained from a center tapped transformer, and the center tap is connected to the signal ground, hum can be minimized.

The difficulty of producing thermionic emission varies with the metal used for the cathode, and is called the "work function" of that metal. An ideal cathode would be made of a metal with a low work function that can sustain high temperatures without melting. Pure tungsten was used in early tubes as it could be heated to a very high temperature. Later it was learned that a very thin layer of thorium greatly increased the emission. Oxides of metals with low work functions were also developed. In modern tubes, thoriated-tungsten is used for the higher power tubes and oxide-coated metals are commonly used at lower power levels.

Filament voltage is important to the proper operation of a tube. If too low, the emission will not be sufficient. If too high, the useful life of the tube will be greatly shortened. It is important to know which type of cathode is being used. Oxide-coated cathodes can be run at 5 to 10% under their rated voltage with an increase in their life, provided performance is satisfactory. They should never be run above the nominal value. Thoriated-tungsten tubes can also gain operational life by running with reduced filament voltage, and as they approach the end of their useful life, can be run at higher than the nominal value as necessary to maintain emission. Carelessly running either type at higher than nominal voltage when new is sure to lead to shorter than normal life with no performance advantage.

Every vacuum tube needs a receptor for the emitted electrons. After moving though the vacuum, the electrons are absorbed by the *plate,* also called the *anode*. This two-element tube — anode and cathode — is called a *diode*. The diode tube is similar to a semiconductor diode: it allows current to pass in only one direction. If the plate goes negative relative to the cathode, current cannot flow because electrons are not emitted from the plate. Years ago, in the days before semiconductors, tube diodes were used as rectifiers.

TRIODES

To amplify signals, a vacuum tube must also contain a control *grid*. This name comes from its physical construction. The grid is a mesh of wires located between the cathode and the plate. Electrons from the cathode pass between the grid wires on their way to the plate. The electrical field that is set up by the voltage on these wires affects the electron flow from cathode to plate. A negative grid voltage repels electrons, blocking their flow to the plate. A positive grid voltage enhances the flow of electrons to the plate. Vacuum tubes containing a cathode, a grid and a plate are called *triode* tubes (*tri* for three components). See **Fig 17.4**.

The input impedance of a vacuum tube amplifier is directly related to the grid current. Grid current varies with grid voltage, increasing as the voltage becomes more positive. When the grid voltage is negative, no grid current flows and the input impedance of a tube is nearly infinite. When the grid is driven positive, it draws current and thus presents a lower input impedance, and requires significant drive power. The load placed across the plate of the tube strongly affects its output power and efficiency. An important part of tube design involves determining the optimum load resistance. These parameters are plotted as *characteristic curves* and are used

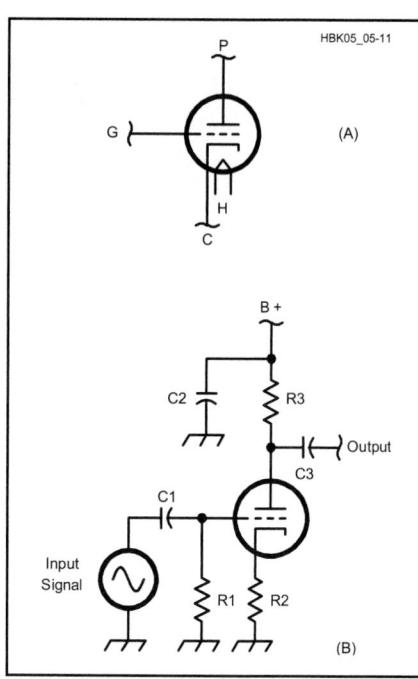

Fig 17.4 — **Vacuum tube triode. (A) Schematic symbol detailing heater (H), cathode (C), grid (G) and plate (P). (B) Audio amplifier circuit using a triode. C1 and C3 are dc blocking capacitors for the input and output signals to isolate the grid and plate bias voltages. C2 is a bypass filter capacitor to decrease noise in the plate bias voltage, B+. R1 is the grid bias resistor, R2 is the cathode bias resistor and R3 is the plate bias resistor. Note that although the cathode and grid bias voltages are positive with respect to ground, they are still negative with respect to the plate.**

to aid the design process. **Fig 17.5** shows an example.

Since the elements within the vacuum tube are conductors that are separated by an insulating vacuum, the tube is very similar to a capacitor. The capacitance between the cathode and grid, between the grid and plate, and between the cathode and plate can be large enough to affect the operation of the amplifier at high frequencies. These capacitances, which are usually on the order of a few picofarads, can limit the frequency response of an amplifier and can also provide signal feedback paths that may lead to unwanted oscillation. Neutralizing circuits are sometimes used to prevent such oscillations. Techniques for neutralization are presented later in this chapter.

TETRODES

The grid-to-plate capacitance is the chief source of unwanted signal feedback. Therefore tubes were developed with a second grid, called a *screen grid*, inserted be-

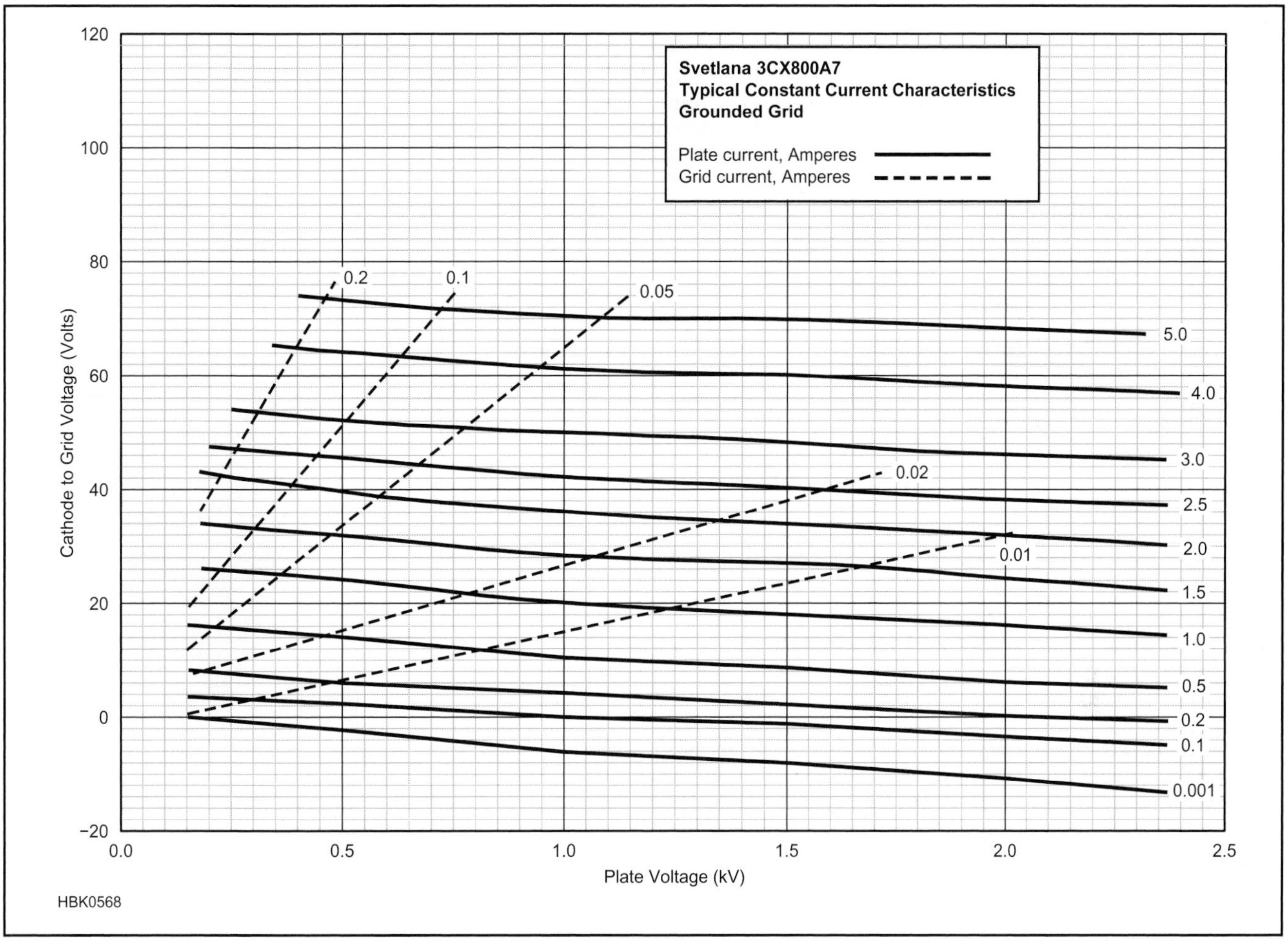

Fig 17.5 — Characteristic curves for a 3CX800A7 triode tube. Grid voltage is plotted on the left, plate voltage along the bottom. The solid lines are plate current, the dashed lines are grid current. This graph is typical of characteristic curves shown in this chapter and used with the *TubeCalculator* program described in the text and included on the *Handbook* CD.

tween the original grid (now called a *control grid*) and the plate. Such tubes are called tetrodes (having four elements). See **Fig 17.6**. This second grid is usually tied to RF ground and acts as a screen between the grid and the plate, thus preventing energy from feeding back, which could cause instability. Like the control grid, the screen grid is made of a wire mesh and electrons pass through the spaces between the wires to get to the plate.

The screen grid carries a high positive voltage with respect to the cathode, and its proximity to the control grid produces a strong electric field that enhances the attraction of electrons from the cathode. The gain of a tetrode increases sharply as the screen voltage is increased. The electrons accelerate toward the screen grid and most of them pass through the spaces and continue to accelerate until they reach the plate. In large tubes this is aided by careful alignment of the screen wires with the grid wires. The effect of the screen can also be seen in the flattening of the tube curves. Since the screen shields the

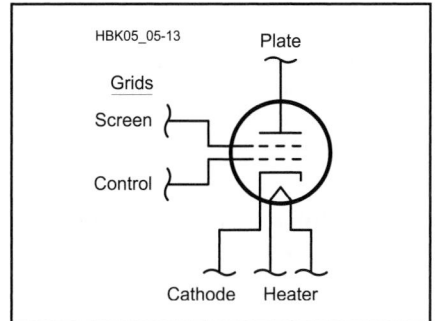

Fig 17.6 — Vacuum tube tetrode. Schematic symbol detailing heater (H), cathode (C), the two grids: control and screen, and the plate (P).

grid from the plate, the plate current vs plate voltages becomes almost flat, for a given screen and grid voltage. **Fig 17.7** shows characteristic curves for a typical tetrode, and curves for many more tube types are included on the *Handbook* CD.

A special form of tetrode concentrates the electrons flowing between the cathode and the plate into a tight beam. The decreased electron-beam area increases the efficiency of the tube. *Beam tetrodes* permit higher plate currents with lower plate voltages and large power outputs with smaller grid driving power. The 6146 is an example of a beam power tube.

PENTODES

Another unwanted effect in vacuum tubes is the so-called *secondary emission*. The electrons flowing within the tube can have so much energy that they are capable of dislodging electrons from the metal atoms in the grids and plate. Secondary emission can cause a grid, especially the screen grid, to lose more electrons than it absorbs. Thus while a screen usually draws current from its supply, it occasionally pushes current into the supply. Screen supplies must be able to absorb as well as supply current.

A third grid, called the *suppressor grid*,

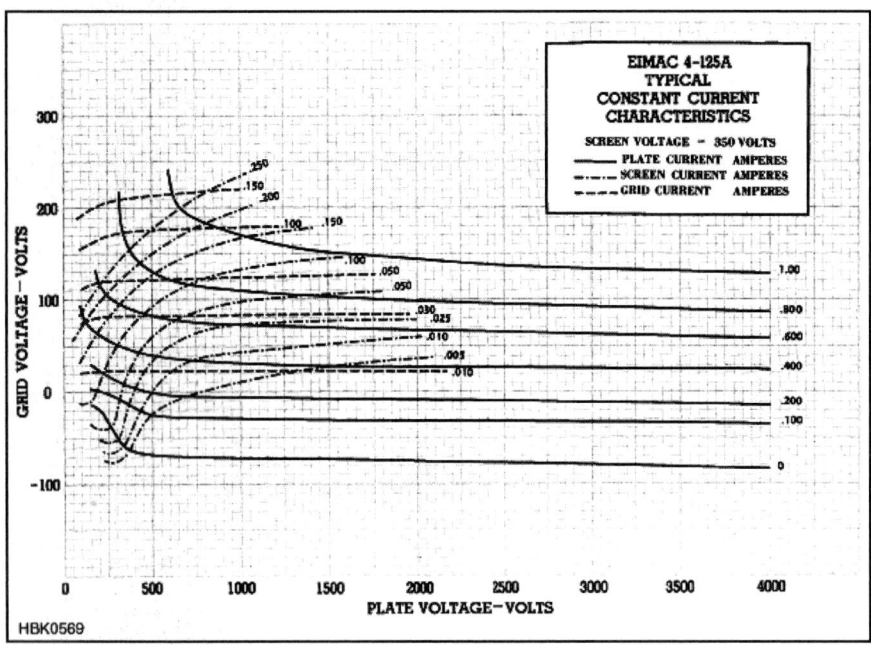

Fig 17.7 — Characteristic curves for a 4-125 tetrode tube.

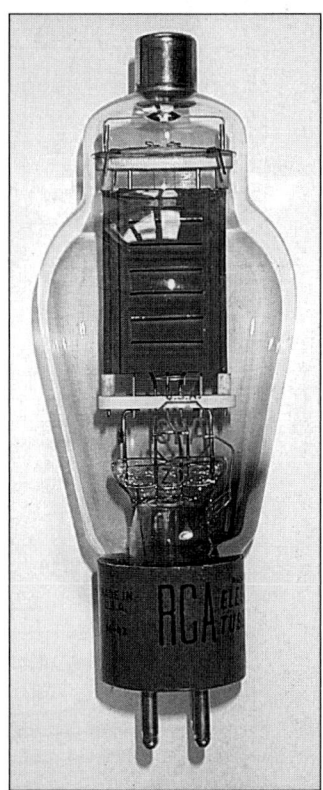

Fig 17.8 — The RCA 811A is an example of a transmitting triode with a glass envelope. [Photo courtesy the Virtual Valve Museum, www.tubecollector.org]

can be added between the screen grid and the plate. This overcomes the effects of secondary emission in tetrodes. A vacuum tube with three grids is called a *pentode* (penta- for five elements). The suppressor grid is connected to a low voltage, often to the cathode.

17.3.3 Tube Nomenclature

Vacuum tubes are constructed with their elements (cathode, grid, plate) encased in an envelope to maintain the vacuum. Tubes with glass envelopes, such as the classic transmitting tube shown in **Fig 17.8**, are most familiar. Over time, manufacturers started exploring other, more rugged, methods for making high power transmitting tubes. Modern power tubes tend to be made of metal parts separated by ceramic insulating sections (**Fig 17.9**).

Because of their long history, vacuum tube types do not all follow a single logical system of identification. Many smaller tubes types begin with an indication of the filament voltage, such as the 6AU6 or 12AT7. Other tubes such as the 811 (Fig 17.8) and 6146 were assigned numbers in a more or less chronological order, much as transistors are today. Some glass envelope power tubes follow a numbering system that indicates number of tube elements and plate dissipation — 3-500Z and 4-1000A are two common examples in amateur circles.

Some power tubes follow the 3CX and 4CX numbering system. The first number indicates a triode (3) or tetrode (4) and the C indicates ceramic/metal construction. The X indicates cooling type: X for air, W for water and V for vapor cooling. The cooling type is followed by the plate dissipation. (The tubes that amateurs use typically have three or four numbers indicating plate dissipation; those used in commercial and broadcast service can have much higher numbers.) Thus a 4CX250 is a ceramic, air cooled 250 W tetrode. A 3CX1200 is a ceramic, air cooled, 1200 W triode. Often these tubes have additional characters following the plate dissipation to indicate special features or an upgraded design. For example, a 4CX250R is a special version of the 4CX250 designed for AB1 linear operation. A 4CX1500B is an updated version of the 4CX1500A.

During the heyday of tube technology, some tube types were developed with several tubes in the same glass envelope, such as the 12AX7 (a dual triode). Except for a very few devices used in the specialty audio market, tubes of this type are no longer made.

17.3.4 Tube Mounting and Cooling Methods

Most tubes mount in some kind of socket so that they can be easily replaced when they reach the end of their useful life. Connections to the tube elements are typically made through pins on the base. The pins are arranged or keyed so that the tube can be inserted into the socket only one way and are sized and spaced to handle the operating voltages and currents involved. Tubes generally use a standard base or socket, although a great many different bases developed over the years. Tube data sheets show pinouts for the various tube elements, just like data sheets for ICs and transistors. See the **Component**

Fig 17.9 — Modern power tubes, such as this 4CX1000A tetrode, tend to use metal and ceramic construction.

Data and References chapter for base diagrams of some popular transmitting tubes.

For transmitting tubes, a common arrangement is for filament and grid connections to be made through pins in the main base, while the plate connection is made through a large pin or post at the top of the tube for easier connection to the high voltage supply and tank circuit. This construction is evident in the examples shown in Figs 17.8 and 17.9. To reduce stray reactances, in some older glass tubes the grid used a separate connection.

Heat dissipation from the plate is one of the major limiting factors for vacuum tube power amplifiers. Most early vacuum tubes were encased in glass, and heat passed through it as infrared radiation. If more cooling was needed, air was simply blown over the outside of the glass. Modern ceramic tubes suitable for powers up to 5 kW are usually cooled by forcing air directly through an external anode. These tubes require a special socket that allows free flow of air. The large external anode and cooling fins may be seen in the example in Fig 17.9. Conduction through an insulating block and water cooling are other options, though they are not often seen in amateur equipment. Practical amplifier cooling methods are discussed in detail later in this chapter.

17.3.5 Vacuum Tube Configurations

Just as the case with solid state devices described in the **Analog Basics** chapter, any of the elements of the vacuum tube can be common to both input and output. A common plate connection — called a cathode follower and similar to the emitter follower — was once used to reduce output impedance (current gain) with little loss of voltage. This application is virtually obsolete.

Most modern tube applications use either the common cathode or the common grid connection. Fig 17.4B shows the common cathode connection, which gives both current and voltage gain. The common grid (often called *grounded grid*) connection shown in **Fig 17.10** gives only voltage gain. Thus the common cathode connection is capable of 20 to 30 dB of gain in a single stage, whereas the grounded grid connection typically gives 10 to 15 dB of gain.

The input impedance of a grounded grid stage is low, typically less than a few hundred ohms. The input impedance of a grounded cathode stage is much higher.

17.3.6 Classes of Operation in Tube Amplifiers

Class of operation was discussed briefly in the previous section describing the need for linear operation of RF power amplifiers for most Amateur Radio modes except FM. The class of operation of an amplifier stage is defined by its conduction angle, the angular portion of each RF drive cycle, in degrees, during which plate current flows. The conduction angle is determined by the bias on the device, and to a lesser extent on the drive level. These, in turn, determine the amplifier's efficiency, linearity and operating impedances. Refer to **Fig 17.11** for the following discussion.

Class A is defined as operation where plate current is always flowing. For a sine wave this means during 360° of the wave. Class A has the best linearity, but poor efficiency.

Class B is when the bias is set so that the tube is cut off for negative input signals, but current flows when the input signal is positive. Thus for a sine wave, conduction occurs during 1/2 of the cycle, or 180°. Class B is linear only when two devices operate in push-pull, so as to provide the missing half of the wave, or when a tuned circuit is present to restore the missing half by "flywheel" action (discussed later in this chapter).

Class AB is defined as operation that falls between Class A and Class B. For a sine wave, the conduction angle will be more than 180°, but less than 360°. In practice Class AB amplifiers usually fall within the gray area shown in the center area of the graph. Like Class B, a push pull connection or a tuned circuit are needed for linear operation. Class AB is less efficient that class B, but better than Class A. The linearity is better than class B but worse than class A.

Class AB vacuum tube amplifiers are further defined as class AB1 or AB2. In class AB1, the grid is not driven positive, so no grid current flows. Virtually no drive power is required. In Class AB2, the grid is driven positive at times with respect to the cathode and some grid current flows. Drive power and output both increase as compared to AB1. Most linear amplifiers used in the Amateur service operate Class AB2, although for greater linearity some operate Class AB1 or even Class A.

Class C is when conduction angle is less than 180° — typically 120° to 160° for vacuum tube amplifiers or within the gray area

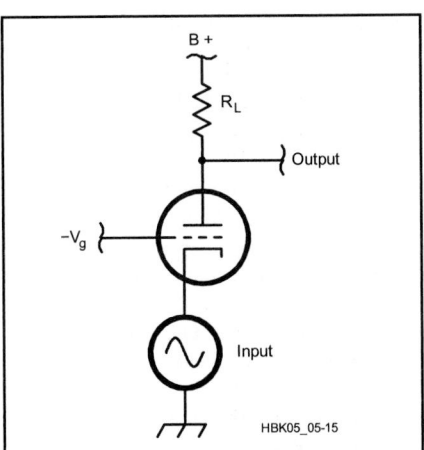

Fig 17.10 — Grounded grid amplifier schematic. The input signal is connected to the cathode, the grid is biased to the appropriate operating point by a dc bias voltage, $-V_G$, and the output voltage is obtained by the voltage drop through R_L that is developed by the plate current, I_P.

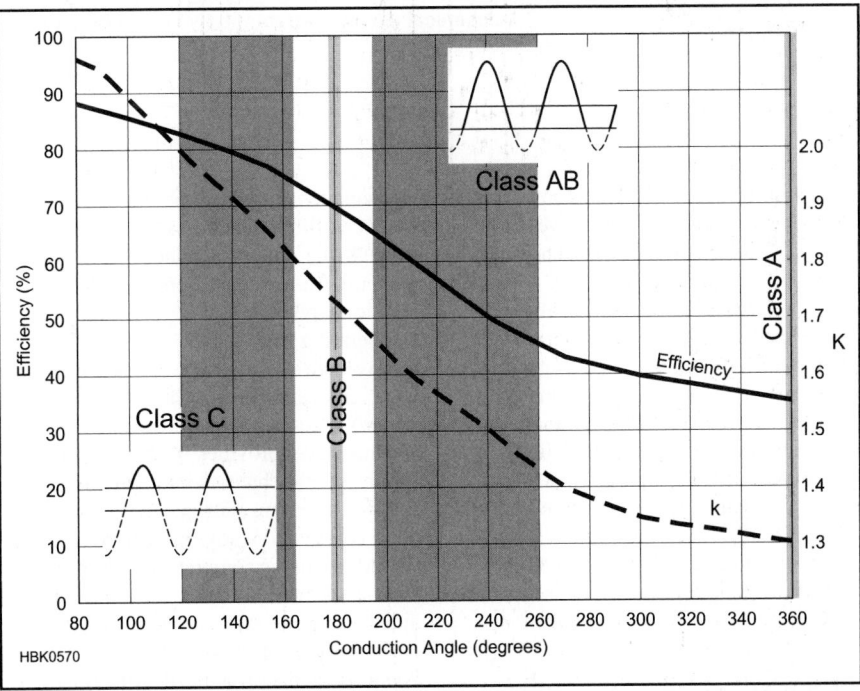

Fig 17.11 — Efficiency and K for various classes of operation. Read the solid line to determine efficiency. Read the dashed line for K, which is a constant used to calculate the plate load required, as explained in the text.

The Flywheel Effect

The operation of a resonant tank circuit (see the **Oscillators and Synthesizers** chapter for a discussion of resonant LC circuits) is sometimes referred to as the "flywheel" effect. A flywheel does illustrate certain functions of a resonant tank, but a flywheel alone is non-resonant; that is, it has no preferred frequency of operation.

A better analogy for a resonant tank — although not exact — is found in the balance wheel used in mechanical watches and clocks in which a weighted wheel rotates back and forth, being returned to its center position by a spiral spring, sometimes called a hairspring. The balance wheel stores energy as inertia and is analogous to an inductor. The hairspring also stores energy and is similar in operation to a capacitor. The spring has an adjustment for the frequency of operation or resonance (1 Hz). An escapement mechanism gives the wheel a small kick with each tick of the watch to keep it going.

A plot of the rotational (or radial) velocity of the balance wheel will give a sine function. It thus converts the pulses from the escapement into smooth sinusoidal motion. In a similar way, pulses of current in the plate of a tube are smoothed into a sine wave in the tank circuit.

The plate (or collector or drain) voltage is a sine wave, even though the plate current is made up of pulses. Just as the escarpment mechanism must apply its kicks at just the right time, the frequency at which current pulses are added to the tank must match the natural resonant frequency of the tank, which is adjusted with the "plate tune" capacitor.

Efficient operation of the tank itself occurs when the losses are small compared to the energy transferred to the output. In a similar manner, lowering the losses in a balance wheel by using jeweled bearings makes the watch more efficient. The amount of energy stored in the balance wheel should also be kept as low as practical since excess "oscillation" of the wheel wastes energy to the air and bearings. Likewise, the "circulating currents" in the tank must be limited to reduce heating from the inevitable losses in the components.

Fig 17.A1 — A balance wheel and hairspring in a mechanical clock illustrate the flywheel effect in tank circuits.

to the left in Fig 17.11. The tube is biased well beyond cutoff when no drive signal is applied. Output current flows only during positive crests in the drive cycle, so it consists of relatively narrow pulses at the drive frequency. Efficiency is high, but nonlinear operation results. Class C amplifiers always use tuned circuits at the input and output. Attempts to achieve extreme efficiency with very narrow pulses (small conduction angles) require very high drive power, so a point of diminishing returns is eventually reached.

Classes D through H use various switched mode techniques. These are used almost exclusively with solid state circuits.

17.3.7 Understanding Tube Operation

Vacuum tubes have complex current transfer characteristics, and each class of operation produces different RMS values of RF current through the load impedance. As described earlier, tube manufacturers provide characteristic curves that show how the tube behaves as operating parameters (such as plate and grid voltage and current) vary. See Figs 17.5 and 17.7 for examples of characteristic curves for two different transmitting tubes. The use of tube curves provides the best way to gain insight into the characteristics of a given tube. Before designing with a tube, get a set of these curves and study them thoroughly.

Because of the complexity and interaction among the various parameters, computer-aided design (CAD) software is useful in analyzing tube operation. One such program, *TubeCalculator*, is included on the *Handbook* CD-ROM, along with curves for many popular transmitting tubes. This software makes it easier to do analysis of a given operation

with the tube you have chosen. **Fig 17.12** shows a *TubeCalculator* screen with an example of "constant current" curves for a typical tube used in high power amplifiers and tables showing values for the various operating parameters. The curves shown have grid voltage on the vertical axis and plate voltage on the horizontal axis. Older tubes and some newer ones use a slightly different format in which plate voltage is plotted on the horizontal axis and plate current on the vertical. *TubeCalculator* allows analysis using either type.

The tube is the heart of any amplifier. Using the software to arrive at the desired operating parameters is a major step toward understanding and designing an amplifier. The second most important part of the design is the tuning components. Before they can be designed, the required plate load resistance must be determined and *TubeCalculator* will do that. In addition it will give insight into what happens when a tube is under driven or over driven, when the bias is wrong, or when the load resistance incorrect.

If a tube is to be used other than with nominal voltages and currents, analysis using the tube curves is the only solution short of trial and error. Trial and error is not a good idea because of the high voltages and currents found in high power amplifiers. It's best to conduct your analysis using curves and software, or else stick to operating the tube very close to the voltages, currents, drive levels and load values specified by the manufacturer.

ANALYZING OPERATING PARAMETERS

Characteristic curves allow a detailed look at tube operation as voltages and currents vary. For example, you can quickly see how much negative grid voltage is required to set the plate current to any desired value, depending also on the plate voltage and screen voltages. You can also see how much grid voltage is needed to drive the plate current to the maximum desired value.

With RF power amplifiers, both the grid voltage and the plate voltage will be sinusoidal and will be 180° out of phase. With constant current curves, an operating line can be drawn that will trace out every point of the operating cycle. This will be a straight line connecting two points. One point will be at the intersection of the peak plate voltage and the peak negative grid voltage. The other point will be at the intersection of the peak positive grid voltage and the minimum plate voltage.

If we plot the plate current along this line as a function of time, it will be seen that it changes in a nonlinear fashion; the exact shape of which depends on the class of operation. For class AB2 operation, which is the most commonly used in linear RF power amplifiers, it will look something like the

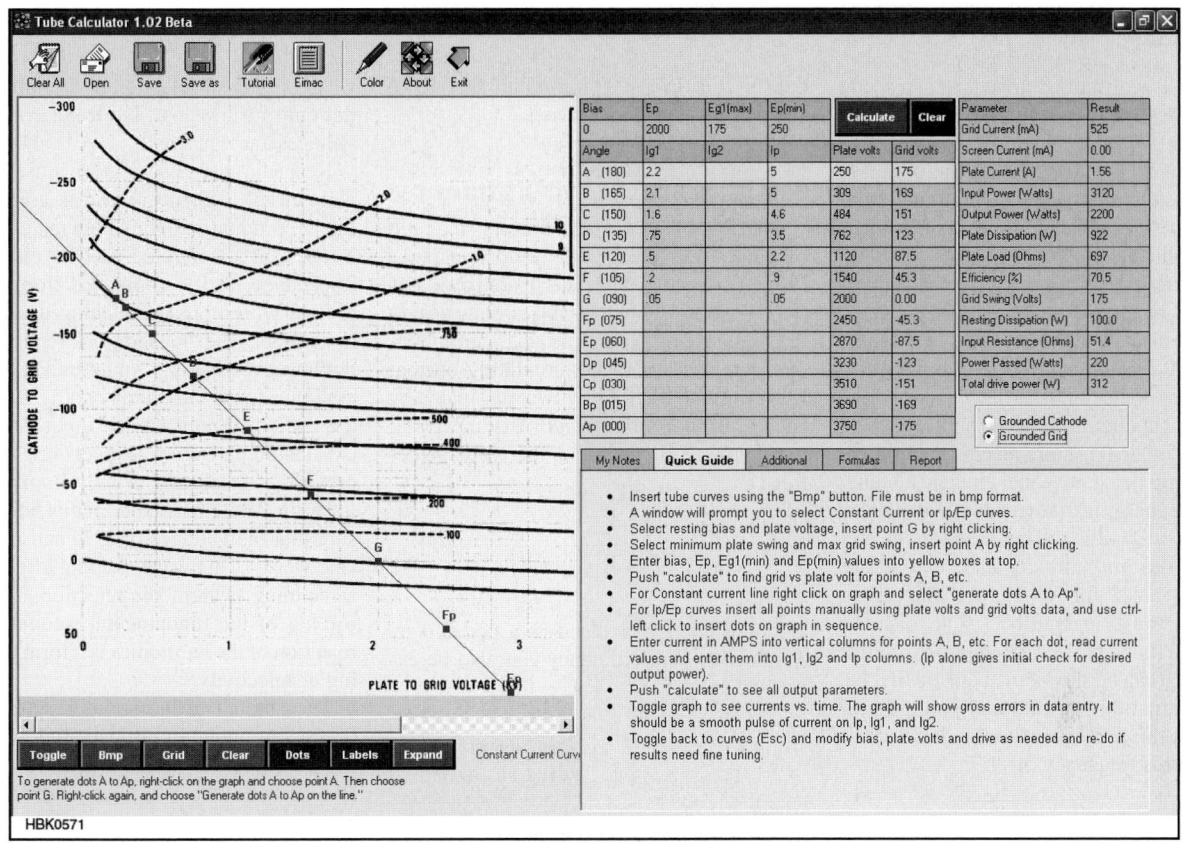

Fig 17.12 — Main screen of *TubeCalculator* program (included on the *Handbook* CD). A characteristic curve plot is loaded in the window on the left side of the screen.

plot shown in **Fig 17.13**. This rather complicated waveform is not easily evaluated using simple formulas, but it can be analyzed by taking the current values vs time and applying averaging techniques. This method was developed by Chaffee in 1936 and popularized by Eimac, the company that developed many of the power tubes used today.

With *TubeCalculator*, data can be extracted from the curves and can be converted into many useful operating parameters such as input power, output power, grid power required, grid and plate dissipation and required load resistance.

MANUAL METHODS FOR TUBE PARAMETER SELECTION

For those not wishing to use a computer for design, most tube manufacturers will supply a table of typical operating values. A summary of some of this information is available in the **Component Data and References** chapter. These values have already been determined both from the tube characteristics and actual operational tests. As long as your proposed operation is close to the typical values in terms of voltages and currents, the typical values can provide you the desired load resistance and expected output power

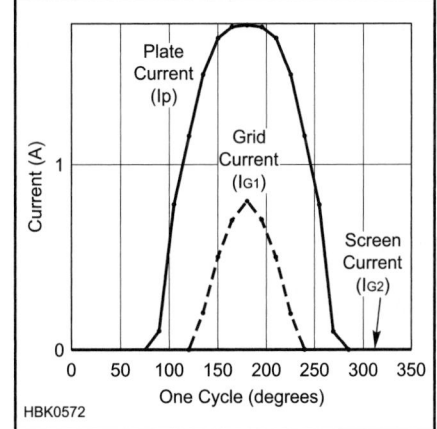

Fig 17.13 — Class AB2 plate and grid current over one cycle as plotted by the *TubeCalculator* program. This plot is for a triode, so there is no curve for screen (G2) current.

and efficiency. The typical operating parameters may also include a suggested optimum load resistance. If not, we can use well established "rules of thumb."

The optimum load resistance for vacuum-tube amplifiers can be approximated by the ratio of the dc plate voltage to the dc plate current at maximum signal, divided by a constant appropriate to each class of operation. The load resistance, in turn, determines the maximum power output and efficiency the amplifier can provide. The approximate value for tube load resistance is

$$R_L = \frac{V_p}{K \times I_p} \quad (1)$$

where

R_L = the appropriate load resistance, in ohms

V_P = the dc plate potential, in V

I_P = the dc plate current, in A

K = a constant that approximates the RMS current to dc current ratio appropriate for each class. For the different classes of operation: Class A, K ≈ 1.3; Class AB, K ≈ 1.5-1.7; Class B, K ≈ 1.57-1.8; Class C, K ≈ 2. The way in which K varies for different conduction angles is shown in Fig 17.11 (right scale).

Once we determine the optimum load resistance value for the tube(s) to be used we are ready to design the output networks for the amplifier. After tube selection, this is the most important part of the total design.

17.4 Tank Circuits

Usually we want to drive a transmission line, typically 50 Ω, with the output of our amplifier. An output network is used to transform that 50 Ω impedance to the optimum load resistance for the tube. This transformation is accomplished by resonant output networks which also serve to reduce harmonics to a suitable level. The **Electrical Fundamentals** chapter of this *Handbook* gives a detailed analysis of the operation of resonant circuits. We summarize here only the most important points.

Resonant circuits have the ability to store energy. Capacitors store electrical energy in the electric field between their plates; inductors store energy in the magnetic field induced by the coil winding. These circuits are referred to as *tank circuits* since they act as storage "tanks" for RF energy. This energy is continuously passed back and forth between the inductive storage and the capacitive storage. It can be shown mathematically that the "alternating" current and voltage produced by this process are sinusoidal in waveform with a frequency of

$$f = \frac{1}{2\pi\sqrt{LC}} \tag{2}$$

which, of course, is the resonant frequency of the tank circuit.

17.4.1 Tank Circuit Q

In order to quantify the ability of a tank circuit to store energy, a quality factor, Q, is defined. Q is the ratio of energy stored in a system during one complete RF cycle to energy lost.

$$Q = 2\pi \frac{W_S}{W_L} \tag{3}$$

where
W_S = is the energy stored
W_L = the energy lost to heat and the load

A load connected to a tank circuit has exactly the same effect on tank operation as circuit losses. Both consume energy. It just happens that energy consumed by circuit losses becomes heat rather than useful output. When energy is coupled out of the tank circuit into a load, the loaded Q (Q_L) is:

$$Q_L = \frac{X}{R_{Loss} + R_{Load}} \tag{4}$$

where R_{Load} is the load resistance. Energy dissipated in R_{Loss} is wasted as heat. And X represents the reactance of the inductor or the capacitor, assumed to be equal at resonance. Ideally, all the tank circuit energy should be

delivered to R_{Load}. This implies that R_{Loss} should be as small as possible.

17.4.2 Tank Circuit Efficiency

The efficiency of a tank circuit is the ratio of power delivered to the load resistance (R_{Load}) to the total power dissipated by losses (R_{Load} and R_{Loss}) in the tank circuit. Within the tank circuit, R_{Load} and R_{Loss} are effectively in series and the circulating current flows through both. The power dissipated by each is proportional to its resistance. The loaded tank efficiency can, therefore, be defined as

$$Tank\ Efficiency = \frac{R_{Load}}{R_{Load} + R_{Loss}} \times 100 \tag{5}$$

where efficiency is stated as a percentage. The loaded tank efficiency can also be expressed as

$$Tank\ Efficiency = \left(1 - \frac{Q_L}{Q_U}\right) \times 100 \tag{6}$$

where
Q_L = the tank circuit loaded Q, and
Q_U = the unloaded Q of the tank circuit.

For practical circuits, Q_U is very nearly the Q of the coil with switches, capacitors and parasitic suppressors making a smaller contribution. It follows, then, that tank efficiency can be maximized by keeping Q_L low which keeps the circulating current low and the I^2R losses down. Q_U should be maximized for best efficiency; this means keeping the circuit losses low. With a typical Q_L of 10, about 10% of the stored energy is transferred to the load in each cycle. This energy is replaced by energy supplied by the tube. It is interesting to contemplate that in a typical amplifier which uses a Q_L of 10 and passes 1.5 kW from the tube to the output, the plate tank is storing about 15 kW of RF energy. This is

why component selection is very important, not only for low loss, but to resist the high voltages and currents.

Resonant circuits are always used in the plate circuit. When the grid is used as the input (common cathode), both matching and a tuned circuit may be used or else a low impedance load is connected from grid to ground with a broad matching transformer or network. The "loaded grid" reduces gain, but improves stability. In grounded grid operation, a tuned circuit may not be needed in the cathode circuit, as the input Z may be close to 50 Ω, but a tuned network may improve the match and usually improves the linearity. These resonant circuits help to ensure that the voltages on grid and plate are sine waves. This wave-shaping effect is the same thing as harmonic rejection. The reinforcing of the fundamental frequency and rejection of the harmonics is a form of filtering or selectivity.

The amount of harmonic suppression is dependent upon circuit loaded Q_L, so a dilemma exists for the amplifier designer. A low Q_L is desirable for best tank efficiency, but yields poorer harmonic suppression. High Q_L keeps amplifier harmonic levels lower at the expense of some tank efficiency. At HF, a compromise value of Q_L can usually be chosen such that tank efficiency remains high and harmonic suppression is also reasonable. At higher frequencies, tank Q_L is not always readily controllable, due to unavoidable stray reactances in the circuit. Unloaded Q_U can always be maximized, however, regardless of frequency, by keeping circuit losses low.

17.4.3 Tank Output Circuits

THE PI NETWORK

The pi network with the capacitors to ground and the inductor in series is commonly used for tube type amplifier matching. This acts like a low pass filter, which is helpful for

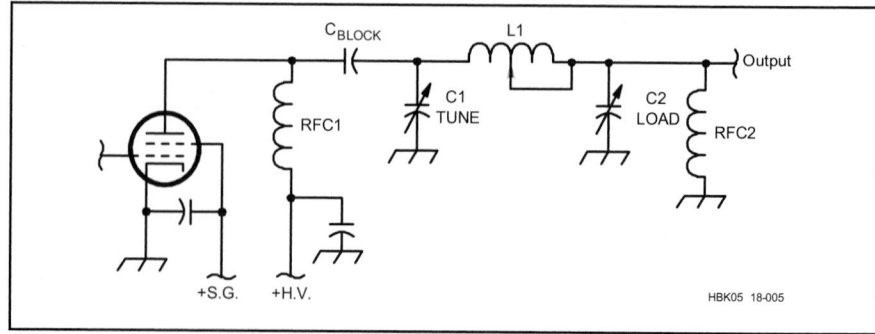

Fig 17.14 — A pi matching network used at the output of a tetrode power amplifier. RFC2 is used for protective purposes in the event C_{BLOCK} fails.

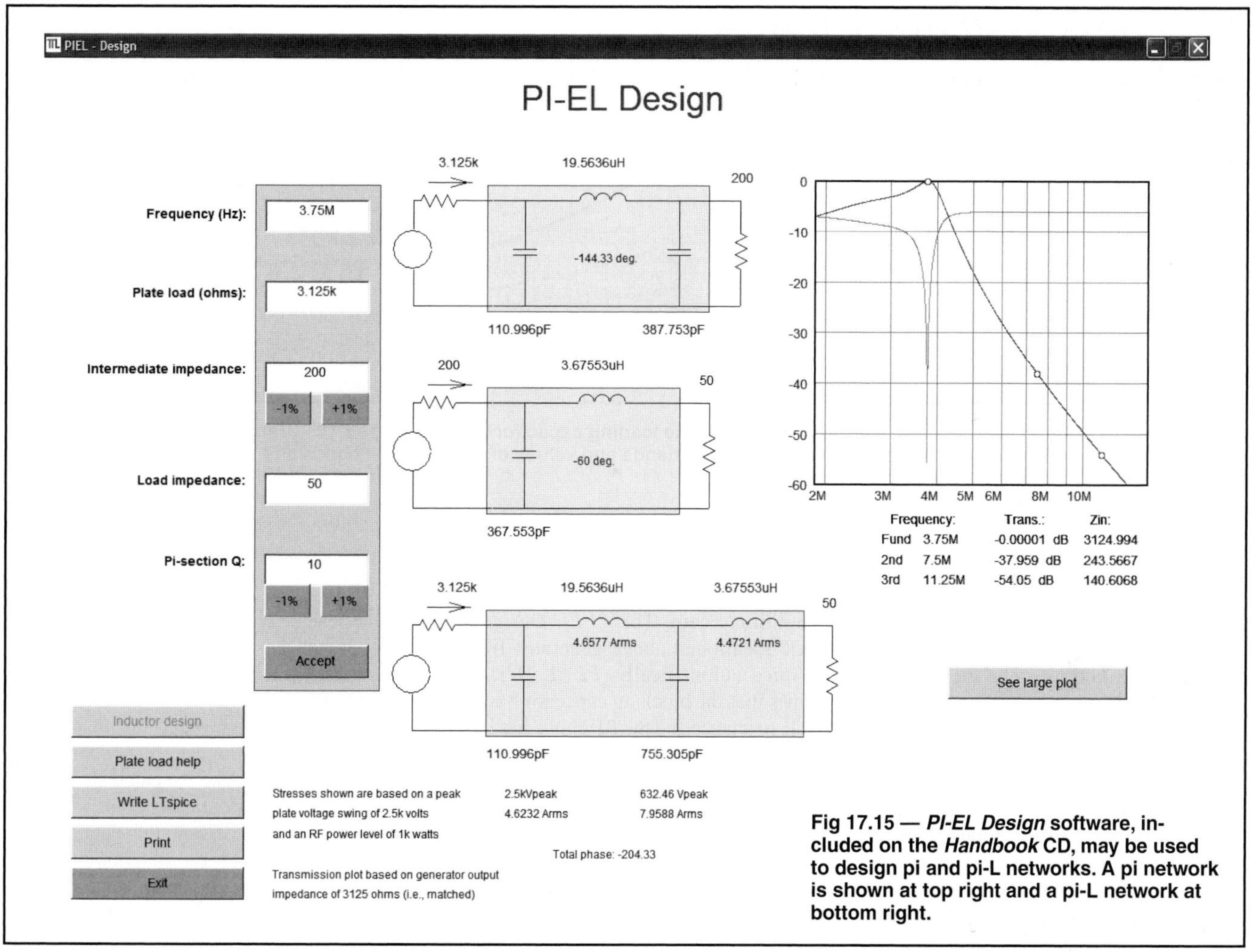

PI-EL Design

Fig 17.15 — *PI-EL Design* software, included on the *Handbook* CD, may be used to design pi and pi-L networks. A pi network is shown at top right and a pi-L network at bottom right.

getting rid of harmonics. Harmonic suppression of a pi network is a function of the impedance transformation ratio and the Q_L of the circuit. Second-harmonic attenuation is approximately 35 dB for a load impedance of 2000 Ω in a pi network with a Q_L of 10. In addition to the low pass effect of the pi network, at the tube plate the third harmonic is already typically 10 dB lower and the fourth approximately 7 dB below that. A typical pi network as used in the output circuit of a tube amplifier is shown in **Fig 17.14**. The **RF and AF Filters** chapter describes harmonic filters that can also greatly reduce harmonics. These are typically not switched but left in the circuit at all times. With such a filter, the requirements for reducing the harmonics on the higher bands with the amplifier pi network is greatly reduced.

The formulas for calculating the component values for a pi network, for those who wish to use them, are included on the CD that accompanies this *Handbook*, along with tabular data for finished designs. The input variables are desired plate load resistance, output impedance to be matched (usually 50 Ω), the desired loaded Q (typically 10 or 12) and the frequency. With these inputs one can calculate the values of the components C1, L1 and C2. These components are usually referred to as the plate tune capacitor, the tank inductor and the loading capacitor. In a multi-frequency amplifier, the coil inductance is changed with a band switch and the capacitors

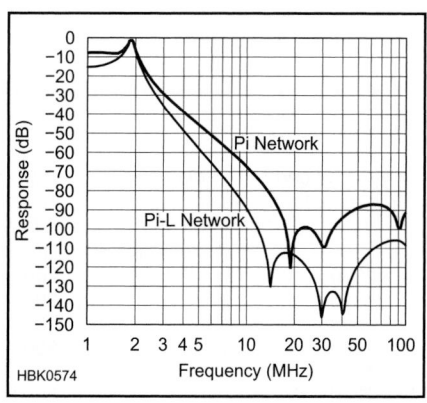

Fig 17.16 — Relative harmonic rejection of pi and pi-L circuits

are adjusted to the correct value for the band in question.

Tank circuit component values are most easily found using computer software. The program *PI-EL Design* by Jim Tonne, W4ENE, is included on the *Handbook CD* and is illustrated here. With this software, all of the components for a pi or pi-L network (described in the next section) can be quickly calculated. Since there are so many possible variables, especially with a pi-L network, it is impractical to publish graphical or tabular data to cover all cases. Therefore, the use of this software is highly recommended for those designing output networks. The software allows many "what-if" possibilities to be quickly checked and an optimum design found.

THE PI-L NETWORK

There are some advantages in using an additional inductor in the output network, effectively changing it from a pi network to a pi-L network as shown in the bottom right corner of **Fig 17.15**. The harmonic rejection is increased, as shown in **Fig 17.16**, and the

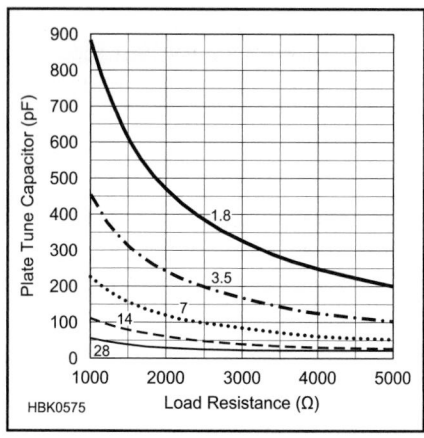

Fig 17.17 — Plate tuning capacitor values for various bands and values of load resistance. Figs 17.17 through 17.19 may be used for manual design of a tank circuit, as explained in the text.

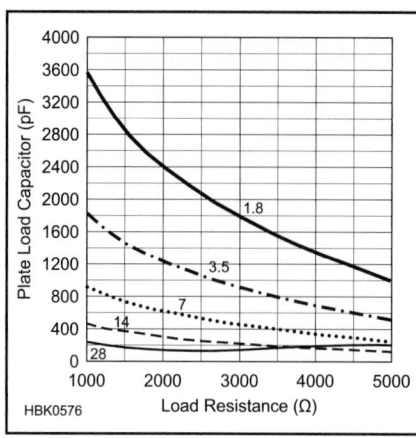

Fig 17.18 — Plate loading capacitor values for various bands and values of load resistance.

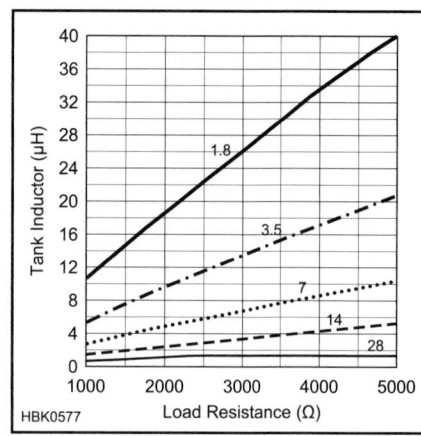

Fig 17.19 — Tank inductor values for various bands and values of load resistance.

component values may become more convenient. Alternatively, the Q can be reduced to lower losses while retaining the same harmonic rejection as the simple pi. This can reduce maximum required tuning capacitance to more easily achievable values.

With a pi-L design, there are many more options than with the simple pi network. This is because the intermediate impedance can take on any value we wish to assign between the output impedance (usually 50 Ω) and the desired plate resistance. This intermediate impedance need not be the same for each frequency, providing the possibility of further optimizing the design. For that reason, it is especially desirable that the software be used instead of using the chart values when a pi-L is contemplated. Using this software, one can quickly determine component values for the required load resistance and Q values for any frequency as well as plotting the harmonic rejection values. Even the voltage and current ratings of the components are calculated.

Further analysis can be done using various versions of *SPICE*. A popular *SPICE* version is *LTspice* available from Linear Technologies and downloadable for free on their website,

www.linear.com. The *PI-EL Design* software mentioned above generates files for *LTspice* automatically. *PI-EL Design* assumes that the blocking capacitor has negligible reactance and the RF choke has infinite reactance. There are times when these assumptions may not be valid. The effect of these components and changes to compensate for them can easily be evaluated using *LTspice*. It can also evaluate the effects of parasitic and stray effects, which all components have. The deep nulls in the response curves in Fig 17.16 are caused by the stray capacitance in the inductors. These can be used to advantage but, if not understood, can also lead to unexpected results. See the **RF Techniques** chapter for more information.

MANUAL METHODS FOR TANK DESIGN

For those who wish to try designing a pi network without a computer, pi designs in chart form are provided. These charts (**Fig 17.17** to **Fig 17.19**) give typical values for the pi network components for various bands and desired plate load resistance. For each value of load resistance the component

values can be read for each band. For bands not shown, an approximate value can be reached by interpolation between the bands shown. It's not necessary to be able to read these values to high precision. In practice, unaccounted-for stray capacitance and inductance will likely make the calculated values only an approximate starting point. For those desiring greater precision, this data is available in tabular form on the *Handbook* CD along with the formulas for calculating them.

Several things become obvious from these charts. The required capacitance is reduced and the required inductance increased when a higher load resistance is used. This means that an amplifier with higher plate voltage and lower plate current will require smaller capacitor values and larger inductors. The capacitors will, of course, also have to withstand higher voltages, so their physical size may or may not be any smaller. The inductors will have less current in them so a smaller size wire may be used. It will be obvious that for an amplifier covering many bands, the most challenging parts of the design are at the frequency extremes. The 1.8 MHz band requires the most inductance and capacitance

Table 17.1
Pi-L Values for 1.8 MHz

Plate Load Resistance (Ω)	Plate Tune Capacitor (pF)	Plate Inductor (µH)	Plate Load Capacitor (pF)	Output Inductor (µH)	Intermediate Resistance (Ω)	Loaded Q (Q_L)	Harmonic Attenuation (3rd/5th, dB)
1000	883	10.5	3550	None (pi)	None (pi)	12	30/42
1000	740	15	2370	7.7	200	12	38/54
1000	499	21.5	1810	7.7	200	8	35/51
2200	376	26.4	1938	7.7	200	12	39/55
2200	255	37.7	1498	7.7	200	8	36/52
5000	180	50.4	1558	7.7	200	12	39/55
5000	124	71	1210	7.7	200	8	36/53
5000	114	83	928	11.7	400	8	39/55

and the 28 MHz band requires the least. Often, the output capacitance of the tube plus the minimum value of the plate capacitor along with assorted stray capacitance will put a lower limit on the effective plate capacitance that can be achieved. These charts assume that value to be 25 pF and the other components are adjusted to account for that minimum value. An inevitable trade-off here is that the Q of the tank will be higher than optimum on the highest bands.

When tuning to the lower frequency bands, the maximum value of the capacitor, especially the loading capacitor, will be a limiting factor. Sometimes, fixed mica or ceramic transmitting rated capacitors will be switched in parallel with the variable capacitor to reach the total value required.

COMPONENT SELECTION FOR THE PI-L NETWORK

For those wishing to use manual methods to design a pi-L network, there are look-up tables on the CD along with the mathematical formulas from which they are derived. **Table 17.1** is an abbreviated version that shows the general trends. Values shown are for 1.8 MHz. Other bands can be approximated by dividing all component values by the frequency ratio. For example, for 18 MHz, divide all component values by 10. For 3.6 MHz, divide values by two, and so on.

For a given Q_L, the pi-L circuit has better harmonic rejection than the pi circuit. This allows the designer to use a lower Q_L, resulting in lower losses and lower capacitor values, which is an advantage on the lower frequencies. However, the inductor values will be higher, and the lower capacitor values may be unachievable at higher frequencies. With the pi-L circuit there are many more variables to work with and, thus, one can try many different possibilities to make the circuit work within the limits of the components that are available.

PROBLEMS AT VHF AND HIGHER FREQUENCIES

As the size of a circuit approaches about 5% of a wavelength, components begin to seriously depart from the pure inductance or capacitance we assume them to be. Inductors begin to act like transmission lines. Capacitors often exhibit values far different from their marked values because of stray internal reactances and lead inductance. Therefore, tuned circuits are frequently fabricated in the form of striplines or other transmission lines in order to circumvent the problem of building "pure" inductances and capacitances. The choice of components is often more significant than the type of network used.

The high impedances encountered in VHF tube-amplifier plate circuits are not easily matched with typical networks. Tube output capacitance is usually so large that most matching networks are unsuitable. The usual practice is to resonate the tube output capacitance with a low-loss inductance connected in series or parallel. The result can be a very high-Q tank circuit. Component losses must be kept to an absolute minimum in order to achieve reasonable tank efficiency. Output impedance transformation is usually performed by a link inductively coupled to the tank circuit or by a parallel transformation of the output resistance using a series capacitor.

Since high values of plate load impedance call for low values of plate tuning capacitance, one might be tempted to add additional tubes in parallel to reduce the required load impedance. This only adds to the stray plate capacitance, and the potential for parasitic oscillations is increased. For these reasons, tubes in parallel are seldom used at VHF. Push-pull circuits offer some advantages, but with modern compact ceramic tube types, most VHF amplifiers use a single tube along with distributed type tuned networks. Other approaches are discussed later in this chapter.

"COLD TUNING" AN AMPLIFIER

Because of the high voltage and current involved, as well as the danger of damaging an expensive tube or other component, it is prudent to "cold tune" an amplifier before applying power to it. This can actually be done early in the construction as soon as tank components and the tube are in place. Cold tuning requires some test equipment, but is not difficult or time consuming. Only if you have a problem in getting the tuning right will it take much time, but that is exactly the case in which you would not want to turn on the power without having discovered that there is a problem. With cold tuning, you can also add and remove additional components, such as the RF choke, and see how much it affects the tuning.

There are always stray capacitances and inductances in larger sized equipment, so there is a good chance that your carefully designed circuits may not be quite right. Even with commercial equipment, you may want to become aware of the limitations of the tuning ranges and the approximate settings for the dials for each band. The equipment manual may provide this information, but what if the amplifier you have is a bit out of calibration? In all of these cases, cold tests provide cheap insurance against damage caused by bad tuning and, at the same time, give you practice in setting up.

The first step is to ensure that the equipment is truly cold by removing the power plug from the wall. Since you will be working around the high voltage circuits, you may want to remove fuses or otherwise ensure that power cannot come on. In a well-designed amplifier there will be interlocks that prevent

turn on and, perhaps, also short out the plate voltage. If these are in a place that affects the RF circuits, they may have to be temporarily removed. Just be sure to put them back when done.

The adjustment of the plate circuit components is the most important. The easiest way to check those is to attach a resistor across the tube from plate to ground. The resistor should be the same value as the design load resistance and must be non-inductive. Several series resistors in the 500 Ω range, either carbon film or the older carbon composition type will work, but *not* wire wound. The tube must remain in its socket and all the normal connections to it should be in place. Covers should be installed, at least for the final tests, since they may affect tuning.

Connect a test instrument to the 50 Ω output. This can be an antenna bridge or other 50 Ω measuring device. Examples of suitable test equipment would be a vector network analyzer, preferably with an impedance step up transformer, a vector impedance meter or an RX meter, such as the Boonton 250A. If the wattage of the resistor across the plate can stand it, it can even be a low power transmitter.

Select the correct band and tune the plate and load capacitors so that the instrument at the output connector shows 50 Ω or a low SWR at the 50 Ω point. Since this type of circuit is bilateral, you now have settings that will be the same ones you need to transform the 50 Ω load to look like the resistive load you used at the tube plate for the test. If you have an instrument capable of measuring relatively high impedances at RF frequencies, you can terminate the output with 50 Ω and measure the impedance on the plate. It will look something like **Fig 17.20**.

A similar test will work for the input, al-

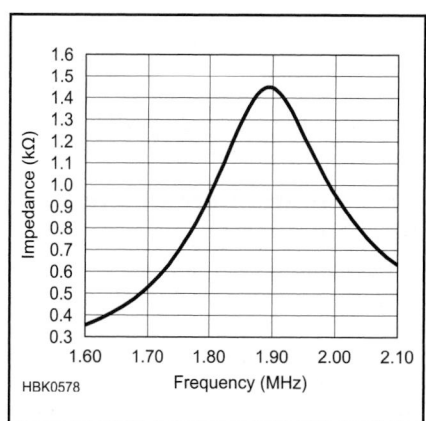

Fig 17.20 — Impedance as would be measured at the plate of an amplifier tube with the output network tuned to 1.9 MHz and the output terminated in its proper load.

Tuning Your Vacuum-Tube Amplifier

Today most hams are used to a wide variety of amplifiers using triodes that are tuned for maximum power output. Unfortunately, not all amplifiers or tubes are rugged enough for this approach. For example, amplifiers that use tetrodes employ a different tuning procedure in which tuning for maximum power may result in a destroyed control grid or screen grid! Procedures vary depending not only on the ratings of the tetrode, but the voltages as well. When tuning any amplifier, monitor grid currents closely and do not exceed the specified maximum current as those are the easiest elements of the tube to damage.

For commercial amplifiers, "Read the Manual" as the following procedures may not be exactly the same as the manufacturer's directions and can be quite different in some cases. In all cases, the last tuning adjustments should be made at full power output, not at reduced power, because the characteristics of the tube change with different power levels. Operating the amplifier at high power after tuning at low power can result in spurious emissions or over-stressing the output network components.

Begin by making sure you have all band-switching controls set properly. If the amplifier TUNE and LOAD controls (sometimes referred to as PLATE TUNE and OUTPUT, respectively) have recommended settings on a particular band, start at those settings. If your amplifier

can be set to a TUNE mode, do so. Set the initial amount of drive (input power) from the exciter — read the amplifier manual or check the tube's specifications if a manual is not available. The exciter output should be one-half or more of full power so that the exciter's ALC systems function properly.

Tuning a triode-based, grounded-grid amplifier is the simplest: Tune for maximum output power without exceeding the tube ratings, particularly grid current, or the legal output power limit. The typical procedure is, while monitoring grid current, to increase drive until the plate current equals about one-quarter to one-half of the target current (depending on the tube and grid bias) while monitoring the output on a wattmeter or the internal power meter. Adjust the TUNE control then advance the LOAD control for maximum output. Repeat the sequence of peaking TUNE then increasing LOAD until no more output power can be obtained without exceeding the ratings for the tube or the legal power limit. If necessary, increase drive and re-peak both the TUNE and LOAD controls.

Operating somewhat differently, a grid-driven tetrode (or pentode) amplifier operating near its designed output power uses the TUNE control for peaking output power and the LOAD control for increasing, but not exceeding, the maximum allowable screen current. Generally, the first part of tetrode amplifier tuning is the

same as for a triode amplifier with both the TUNE and LOAD controls adjusted for maximum power output while monitoring screen and control grid current. After the initial tuning step, the LOAD control is used to peak the screen current. Maximum power should coincide with maximum screen current. The screen current is just a better indicator. As with the triode amplifier, if drive needs to be increased, readjust the TUNE and LOAD controls.

For both triode and tetrode amplifiers, once the procedures above have been completed, try moving the TUNE control a small amount higher or lower and re-peaking the LOAD control. Also known as "rocking" the TUNE control, this small variation can find settings with a few percent more output power or better efficiency.

Once tuning has been completed, it is a good idea to mark the settings of the TUNE and LOAD controls for each band. This reduces the amount of time for on-the-air or dummy-load tuning, reducing stress to the tube and interference to other stations. Usually, a quick "fine-tune" adjustment is all that is required for maximum output. The set of markings also serves as a diagnostic tool, should the settings for maximum power suddenly shift. This indicates a change in the antenna system, such as a failing connector or antenna. — *Roger Halstead, K8RI*

though it is sometimes more difficult to know what the input impedance at the tube will be. In fact, it will change somewhat with drive level so a low power cold test may not give a full picture. However, just like the case of the plate circuit, you can put your best estimate of the input impedance at the grid using a non-inductive resistor. Then, tune the input circuits for a match.

An old timer's method for tuning these circuits, especially the plate circuit, is to use a dip meter. These are getting pretty hard to

find these days and, in any case, only show resonance. You won't know if the transformation ratio between plate and output is correct, but it is better than nothing. At least, if you can't get a dip at the proper frequency, you will know that something is definitely wrong.

17.5 Transmitting Tube Ratings

17.5.1 Plate, Screen and Grid Dissipation

The ultimate factor limiting the power-handling capability of a tube is often (but not always) its maximum plate dissipation rating. This is the measure of how many watts of heat the tube can safely dissipate, if it is cooled properly, without exceeding critical temperatures. Excessive temperature can damage or destroy internal tube components or vacuum seals — resulting in tube failure. The same tube may have different voltage, current and power ratings depending on the

conditions under which it is operated, but its safe temperature ratings must not be exceeded in any case! Important cooling considerations are discussed in more detail later in this chapter.

The efficiency of a power amplifier may range from approximately 25% to 85%, depending on its operating class, adjustment and circuit losses. The efficiency indicates how much of the dc power supplied to the stage is converted to useful RF output power; the rest is dissipated as heat, mostly by the plate. The *TubeCalculator* program will cal-

culate the dissipation of the plate as one of its outputs. Otherwise, it can be determined by multiplying the plate voltage (V) times the plate current (A) and subtracting the output power (W).

For a class AB amplifier, the resting dissipation should also be noted, since with no RF input, *all* of the dc power is dissipated in the plate. Multiply plate voltage times the resting plate current to find this resting dissipation value. Screen dissipation is simply screen voltage times screen current. Grid dissipation is a bit more complicated since some of the power into the grid goes into the bias supply, some is passed through to the output (when grounded grid is used) and some is dissipated in the grid. Some tubes have very fragile grids and cannot be run with any grid current at all.

Almost all vacuum-tube power amplifiers in amateur service today operate as linear amplifiers (Class AB or B) with efficiencies of approximately 50% to 65%. That means that a useful power output of approximately 1 to 2 times the plate dissipation generally can be achieved. This requires, of course, that the tube is cooled enough to realize its maximum plate dissipation rating and that no other tube rating, such as maximum plate current or grid dissipation, is exceeded.

Type of modulation and duty cycle also influence how much output power can be achieved for a given tube dissipation. Some types of operation are less efficient than others, meaning that the tube must dissipate more heat. Some forms of modulation, such as CW or SSB, are intermittent in nature, causing less average heating than modulation formats in which there is continuous transmission (RTTY or FM, for example).

Power-tube manufacturers use two different rating systems to allow for the variations in service. CCS (Continuous Commercial Service) is the more conservative rating and is used for specifying tubes that are in constant use at full power. The second rating system is based on intermittent, low-duty-cycle operation, and is known as ICAS (Intermittent Commercial and Amateur Service). ICAS ratings are normally used by commercial manufacturers and individual amateurs who wish to obtain maximum power output consistent with reasonable tube life in CW and SSB service. CCS ratings should be used for FM, RTTY and SSTV applications. (Plate power transformers for amateur service are also rated in CCS and ICAS terms.).

MAXIMUM RATINGS

Tube manufacturers publish sets of maximum values for the tubes they produce. No maximum rated value should ever be exceeded. As an example, a tube might have a maximum plate-voltage rating of 2500 V, a maximum plate-current rating of 500 mA, and a maximum plate dissipation rating of 350 W. Although the plate voltage and current ratings might seem to imply a safe power input of 2500 V × 500 mA = 1250 W, this is true only if the dissipation rating will not be exceeded. If the tube is used in class AB2 with an expected efficiency of 60%, the maximum safe dc power input is

$$P_{IN} = \frac{100 P_D}{100 - N_D} = \frac{100 \times 350}{100 - 60} = 875 \text{ W}$$

17.5.2 Tank Circuit Components

CAPACITOR RATINGS

The tank capacitor in a high-power amplifier should be chosen with sufficient spacing between plates to preclude high-voltage breakdown. The peak RF voltage present across a properly loaded tank circuit, without modulation, may be taken conservatively as being equal to the dc plate voltage. If the dc supply voltage also appears across the tank capacitor, this must be added to the peak RF voltage, making the total peak voltage twice the dc supply voltage. At the higher voltages, it is usually desirable to design the tank circuit so that the dc supply voltages do not appear across the tank capacitor, thereby allowing the use of a smaller capacitor with less plate spacing. Capacitor manufacturers usually rate their products in terms of the peak voltage between plates. Typical plate spacings are given in **Table 17.2**.

Output tank capacitors should be mounted as close to the tube as possible to allow short low inductance leads to the plate. Especially at the higher frequencies, where minimum circuit capacitance becomes important, the capacitor should be mounted with its stator plates well spaced from the chassis or other shielding. In circuits in which the rotor must be insulated from ground, the capacitor should be mounted on ceramic insulators of a size commensurate with the plate voltage involved and — most important of all, from the viewpoint of safety to the operator — a well-insulated coupling should be used between the capacitor shaft and the knob. The section of the shaft attached to the control knob should be well grounded. This can be done conveniently by means of a metal shaft bushing at the panel.

COIL RATINGS

Tank coils should be mounted at least half their diameter away from shielding or other large metal surfaces, such as blower housings, to prevent a marked loss in Q. Except perhaps at 24 and 28 MHz, it is not essential that the coil be mounted extremely close to the tank capacitor. Leads up to 6 or 8 inches are permissible. It is more important to keep the tank capacitor, as well as other components, out of the immediate field of the coil.

The principal practical considerations in designing a tank coil usually are to select a conductor size and coil shape that will fit into available space and handle the required power without excessive heating. Excessive power loss as such is not necessarily the worst hazard in using too-small a conductor. It is not uncommon for the heat generated to actually unsolder joints in the tank circuit and lead to physical damage or failure. For this reason it's extremely important, especially at power levels above a few hundred watts, to ensure that all electrical joints in the tank circuit are secured mechanically as well as soldered.

Table 17.3 shows recommended conductor sizes for amplifier tank coils, assuming loaded tank circuit Q of 15 or less on the 24 and 30 MHz bands and 8 to 12 on the lower frequency bands. In the case of input circuits for screen-grid tubes where driving power is quite small, loss is relatively unimportant and almost any physically convenient wire size and coil shape is adequate.

The conductor sizes in Table 17.3 are based on experience in continuous-duty amateur CW, SSB and RTTY service and assume that the coils are located in a reasonably well ventilated enclosure. If the tank area is not well ventilated and/or if significant tube heat is transferred to the coils, it is good practice to increase AWG wire sizes by two (for example, change from #12 to #10) and tubing sizes by $1/16$ inch.

Table 17.3
Copper Conductor Sizes for Transmitting Coils for Tube Transmitters

Power Output (W)	Band (MHz)	Minimum Conductor Size
1500	1.8-3.5	10
	7-14	8 or 1/8"
	18-28	6 or 3/16"
500	1.8-3.5	12
	7-14	10
	18-28	8 or 1/8"
150	1.8-3.5	16
	7-14	12
	18-28	10

*Whole numbers are AWG; fractions of inches are tubing ODs.

Table 17.2
Typical Tank-Capacitor Plate Spacings

Spacing Inches	Peak Voltage	Spacing Inches	Peak Voltage	Spacing Inches	Peak Voltage
0.015	1000	0.07	3000	0.175	7000
0.02	1200	0.08	3500	0.25	9000
0.03	1500	0.125	4500	0.35	11000
0.05	2000	0.15	6000	0.5	13000

Larger conductors than required for current handling are often used to maximize unloaded Q, particularly at higher frequencies. Where skin depth effects increase losses, the greater surface area of large diameter conductors can be beneficial. Small-diameter copper tubing, up to ⅜ inch outer diameter, can be used successfully for tank coils up through the lower VHF range. Copper tubing in sizes suitable for constructing high-power coils is generally available in 50 foot rolls from plumbing and refrigeration equipment suppliers. Silver-plating the tubing may further reduce losses. This is especially true as the tubing ages and oxidizes. Silver oxide is a much better conductor than copper oxide, so silver-plated tank coils maintain their low-loss characteristics even after years of use. (There is some debate in amateur circles about the benefits of silver plating.).

At VHF and above, tank circuit inductances do not necessarily resemble the familiar coil. The inductances required to resonate tank circuits of reasonable Q at these higher frequencies are small enough that only strip lines or sections of transmission line are practical. Since these are constructed from sheet metal or large diameter tubing, current-handling capabilities normally are not a relevant factor.

17.5.3 Other Components

RF CHOKES

The characteristics of any RF choke vary with frequency. At low frequencies the choke presents a nearly pure inductance. At some higher frequency it takes on high impedance characteristics resembling those of a parallel-resonant circuit. At a still higher frequency it goes through a series-resonant condition, where the impedance is lowest — generally much too low to perform satisfactorily as a shunt-feed plate choke. As frequency increases further, the pattern of alternating parallel and series resonances repeats. Between resonances, the choke will show widely varying amounts of inductive or capacitive reactance.

In most high-power amplifiers, the choke is directly in parallel with the tank circuit, and is subject to the full tank RF voltage. See **Fig 17.21A**. If the choke does not present a sufficiently high impedance, enough power will be absorbed by the choke to burn it out. To avoid this, the choke must have a sufficiently high reactance to be effective at the lowest frequency (at least equal to the plate load resistance) and yet have no series resonances near any of the higher frequency bands. A resonant-choke failure in a high-power amplifier can be very dramatic and damaging!

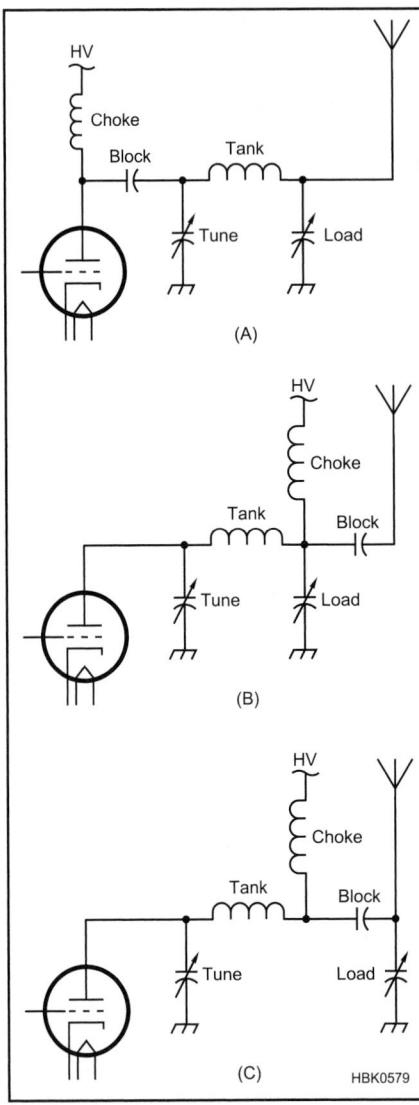

Fig 17.21 — Three ways of feeding dc to a tube via an RF choke. See text for a discussion of the tradeoffs.

Thus, any choke intended for shunt-feed use should be carefully investigated. The best way would be to measure its reactance to ground with an impedance measuring instrument. If the dip meter is used, the choke must be shorted end-to-end with a direct, heavy braid or strap. Because nearby metallic objects affect the resonances, it should be mounted in its intended position, but disconnected from the rest of the circuit. A dip meter coupled an inch or two away from one end of the choke nearly always will show a deep, sharp dip at the lowest series-resonant frequency and shallower dips at higher series resonances.

Any choke to be used in an amplifier for the 1.8 to 28 MHz bands requires careful (or at least lucky!) design to perform well on all amateur bands within that range. Most simply

put, the challenge is to achieve sufficient inductance that the choke doesn't "cancel" a large part of tuning capacitance at 1.8 MHz. At the same time, try to position all its series resonances where they can do no harm. In general, close wind enough #20 to #24 magnet wire to provide about 135 μH inductance on a ¾ to 1-inch diameter cylindrical form of ceramic, Teflon or fiberglass. This gives a reactance of 1500 Ω at 1.8 MHz and yet yields a first series resonance in the vicinity of 25 MHz. Before the advent of the 24 MHz band this worked fine. But trying to "squeeze" the resonance into the narrow gaps between the 21, 24 and/or 28 MHz bands is quite risky unless sophisticated instrumentation is available. If the number of turns on the choke is selected to place its first series resonance at 23.2 MHz, midway between 21.45 and 24.89 MHz, the choke impedance will typically be high enough for satisfactory operation on the 21, 24 and 28 MHz bands. The choke's first series resonance should be measured very carefully as described above using a dip meter and calibrated receiver or RF impedance bridge, with the choke mounted in place on the chassis.

Investigations with a vector impedance meter have shown that "trick" designs, such as using several shorter windings spaced along the form, show little if any improvement in choke resonance characteristics. Some commercial amplifiers circumvent the problem by band switching the RF choke. Using a larger diameter (1 to 1.5 inches) form does move the first series resonance somewhat higher for a given value of basic inductance. Beyond that, it is probably easiest for an all-band amplifier to add or subtract enough turns to move the first resonance to about 35 MHz and settle for a little less than optimum reactance on 1.8 MHz.

However, there are other alternatives. If one is willing to switch the choke when changing bands, it is possible to have enough inductance for 1.8 to 10 MHz, with series resonances well above 15 MHz. Then for 14 MHz and above, a smaller choke is used which has its resonances well above 30 MHz. Providing an extra pole on the band switch is, of course, the trade-off. This switch must withstand the full plate voltage. Switches suitable for changing bands for the pi network would handle this fine.

Another approach is to feed the high-voltage dc through the main tank inductor, putting the RF choke at the loading capacitor, instead of at the tube. (See Fig 17.21B) This puts a much lower RF voltage on the choke and, thus, not as much reactance is required for satisfactory rejection of the RF voltage. However, this puts both dc and RF voltages on the plate and loading capacitors which may be beyond their ratings. The blocking capacitor can be put before the loading capacitor, as in Fig 17.21C. This removes the dc from the loading capacitor, which typically has a lower voltage rating than the plate capacitor, but puts high current in the blocker.

Yet another method involves using hollow tubing for the plate tank and passing the dc lead through it. This lowers the RF voltage on the choke without putting dc voltage on the tuning components. This method works best for higher power transmitters where the tuning inductor can be made of ⅛ inch or larger copper tubing.

BLOCKING CAPACITORS

A series capacitor is usually used at the input of the amplifier output circuit. Its purpose is to block dc from appearing on matching circuit components of the antenna. As mentioned in the section on tank capacitors, output-circuit voltage requirements are considerably reduced when only RF voltage is present.

To provide a margin of safety, the voltage rating for a blocking capacitor should be at least 25% to 50% greater than the dc voltage applied. A large safety margin is desirable, since blocking capacitor failure can bring catastrophic results. The worse case is when dc is applied to the output of the transmitter and even to the antenna, with potentially fatal results. Often an RF choke is placed from the RF output jack to ground as a safety backup. A shorted blocker will blow the power supply fuse.

To avoid affecting the amplifier's tuning and matching characteristics, the blocking capacitor should have a low impedance at all operating frequencies. If it presents more than 5% of the plate load resistance, the pi components should be adjusted to compensate. Use of a *SPICE* analysis provides a useful way to see what adjustments might be required to maintain the desired match.

The capacitor also must be capable of handling, without overheating or significantly changing value, the substantial RF current that flows through it. This current usually is greatest at the highest frequency of operation where tube output capacitance constitutes a significant part of the total tank capacitance. A significant portion of circulating tank current, therefore, flows through the blocking capacitor. When using the connection of the RF choke shown in Fig 17.21C, the entire circulating current must be accommodated.

Transmitting capacitors are rated by their manufacturers in terms of their RF current-carrying capacity at various frequencies. Below a couple hundred watts at the high frequencies, ordinary disc ceramic capacitors of suitable voltage rating work well in high-impedance tube amplifier output circuits. Some larger disk capacitors rated at 5 to 8 kV also work well for higher power levels at HF. For example, two inexpensive Centralab type DD-602 discs (0.002 μF, 6 kV) in parallel have proved to be a reliable blocking capacitor for 1.5-kW amplifiers operating at plate voltages to about 2.5 kV.

At very high power and voltage levels and at VHF, ceramic "doorknob" transmitting capacitors are needed for their low losses and high current handling capabilities. When in doubt, adding additional capacitors in parallel is cheap insurance against blocking capacitor failure and also reduces the impedance. So-called "TV doorknobs" may break down at high RF current levels and should be avoided.

The very high values of Q_L found in many VHF and UHF tube-type amplifier tank circuits often require custom fabrication of the blocking capacitor. This can usually be accommodated through the use of a Teflon "sandwich" capacitor. Here, the blocking capacitor is formed from two parallel plates separated by a thin layer of Teflon. This capacitor often is part of the tank circuit itself, forming a very low-loss blocking capacitor. Teflon is rated for a minimum breakdown voltage of 2000 V per mil of thickness, so voltage breakdown should not be a factor in any practically realized circuit. The capacitance formed from such a Teflon sandwich can be calculated from the information presented elsewhere in this *Handbook* (use a dielectric constant of 2.1 for Teflon). In order to prevent any potential irregularities caused by dielectric thickness variations (including air gaps), Dow-Corning DC-4 silicone grease should be evenly applied to both sides of the Teflon dielectric. This grease has properties similar to Teflon, and will fill in any surface irregularities that might cause problems.

17.6 Sources of Operating Voltages

17.6.1 Tube Filament or Heater Voltage

A power vacuum tube can use either a directly heated filament or an indirectly heated cathode. The filament voltage for either type should be held within 5% of rated voltage. Because of internal tube heating at UHF and higher, the manufacturers' filament voltage rating often is reduced at these higher frequencies. The de-rated filament voltages should be followed carefully to maximize tube life.

Series dropping resistors may be required in the filament circuit to attain the correct voltage. Adding resistance in series will also reduce the inrush current when the tube is turned on. Cold tungsten has much lower resistance than when hot. Circuits are available that both limit the inrush current at turn on and also regulate the voltage against changes in line voltage.

The voltage should be measured with a true RMS meter at the filament pins of the

tube socket while the amplifier is running. The filament choke and interconnecting wiring all have voltage drops associated with them. The high current drawn by a power-tube heater circuit causes substantial voltage drops to occur across even small resistances. Also, make sure that the plate power drawn from the power line does not cause the filament voltage to drop below the proper value when plate power is applied.

Thoriated filaments lose emission when the tube is overloaded appreciably. If the overload has not been too prolonged, emission, sometimes, may be restored by operating the filament at rated voltage, with all other voltages removed, for a period of 30 to 60 minutes. Alternatively, you might try operating the tube at 20% above rated filament voltage for five to ten minutes.

17.6.2 Vacuum-Tube Plate Voltage

DC plate voltage for the operation of RF

amplifiers is most often obtained from a transformer-rectifier-filter system (see the **Power Sources** chapter) designed to deliver the required plate voltage at the required current. It is not unusual for a power tube to arc over internally (generally from the plate to the screen or control grid) once or twice, especially soon after it is first placed into service. The flashover by itself is not normally dangerous to the tube, provided that instantaneous maximum plate current to the tube is held to a safe value and the high-voltage plate supply is shut off very quickly.

A good protective measure against this is the inclusion of a high-wattage power resistor in series with the plate high-voltage circuit. The value of the resistor, in ohms, should be approximately 10 to 15 times the no-load plate voltage in KV. This will limit peak fault current to 67 to 100 A. The series resistor should be rated for 25 or 50 W power dissipation; vitreous enamel coated wire-wound resistors have been found to be capable of handling repeated momentary fault-current surges without damage. Aluminum-cased resistors (Dale) are not recommended for this application. Each resistor also must be large enough to safely handle the maximum value of normal plate current; the wattage rating required may be calculated from $P = I^2R$. If the total filter capacitance exceeds 25 μF, it is a good idea to use 50 W resistors in any case. Even at high plate-current levels, the addition of the resistors does little to affect the dynamic regulation of the plate supply.

Since tube (or other high-voltage circuit) arcs are not necessarily self-extinguishing, a fast-acting plate overcurrent relay or primary circuit breaker is also recommended to quickly shut off ac power to the HV supply when an arc begins. Using this protective system, a mild HV flashover may go undetected, while a more severe one will remove ac power from the HV supply. (The cooling blower should remain energized, however, since the tube may be hot when the HV is removed due to an arc.) If effective protection is not provided, however, a "normal" flashover, even in a new tube, is likely to damage or destroy the tube, and also frequently destroys the rectifiers in the power supply as well as the plate RF choke. A power tube that flashes over more than about 3 to 5 times in a period of several months likely is defective and will have to be replaced before long.

17.6.3 Grid Bias

The grid bias for a linear amplifier should be highly filtered and well regulated. Any ripple or other voltage change in the bias circuit modulates the amplifier. This causes hum and/or distortion to appear on the signal. Since most linear amplifiers draw only small amounts of grid current, these bias-supply requirements are not difficult to achieve.

Fixed bias for class AB1 tetrode and pentode amplifiers is usually obtained from a variable-voltage regulated supply. Voltage adjustment allows setting bias level to give the desired resting plate current. **Fig 17.22A** shows a simple Zener-diode-regulated bias supply. The dropping resistor is chosen to allow approximately 10 mA of Zener current. Bias is then reasonably well regulated for all drive conditions up to 2 or 3 mA of grid current. The potentiometer allows bias to be adjusted between Zener and approximately 10 V higher. This range is usually adequate to allow for variations in the characteristics of different tubes. Under standby conditions, when it is desirable to cut off the tube entirely, the Zener ground return is interrupted so the full bias supply voltage is applied to the grid.

In Fig 17.22B and C, bias is obtained from the voltage drop across a Zener diode in the cathode (or filament center-tap) lead. Operating bias is obtained by the voltage drop across D1 as a result of plate (and screen)

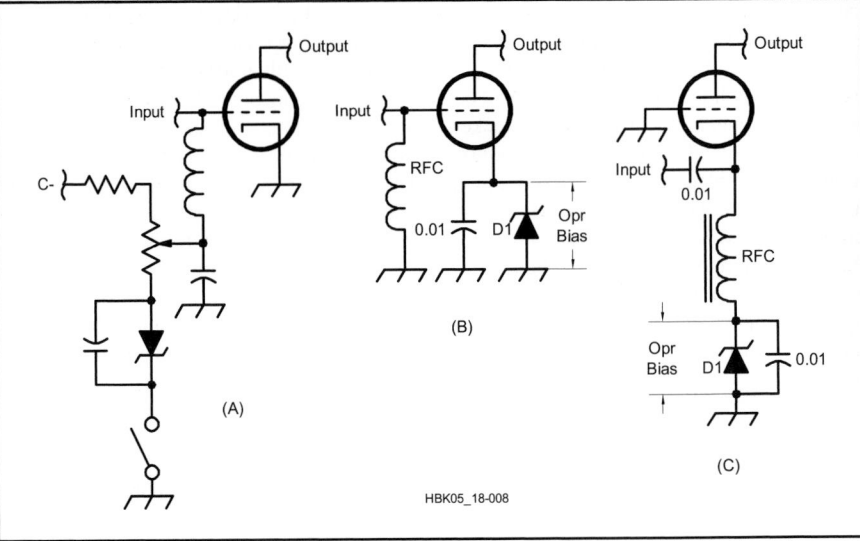

Fig 17.22 — Various techniques for providing operating bias with tube amplifiers.

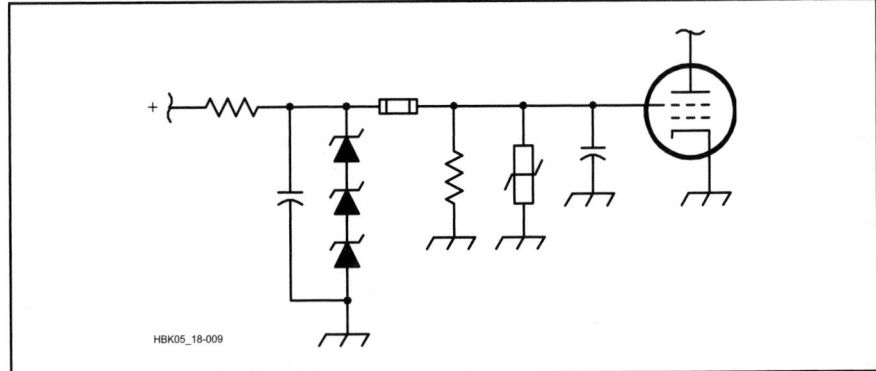

Fig 17.23 — A Zener-regulated screen supply for use with a tetrode. Protection is provided by a fuse and a varistor.

current flow. The diode voltage drop effectively raises the cathode potential relative to the grid. The grid is, therefore, negative with respect to the cathode by the Zener voltage of the diode. The Zener-diode wattage rating should be twice the product of the maximum cathode current times the rated Zener voltage. Therefore, a tube requiring 15 V of bias with a maximum cathode current of 100 mA would dissipate 1.5 W in the Zener diode. To allow a suitable safety factor, the diode rating should be 3 W or more. The circuit of Fig 17.22C illustrates how D1 would be used with a cathode driven (grounded grid) amplifier as opposed to the grid driven example at B.

In all cases, the Zener diode should be bypassed by a 0.01-μF capacitor of suitable voltage. Current flow through any type of diode generates shot noise. If not bypassed, this noise would modulate the amplified sig-

nal, causing distortion in the amplifier output.

17.6.4 Screen Voltage For Tubes

Power tetrode screen current varies widely with both excitation and loading. The current may be either positive or negative, depending on tube characteristics and amplifier operating conditions. In a linear amplifier, the screen voltage should be well regulated for all values of screen current. The power output from a tetrode is very sensitive to screen voltage, and any dynamic change in the screen potential can cause distorted output. Zener diodes are commonly used for screen regulation.

Fig 17.23 shows a typical example of a regulated screen supply for a power tetrode amplifier. The voltage from a fixed dc supply is dropped to the Zener stack voltage by the current-limiting resistor. A screen bleeder resistor is connected in parallel with the Zener stack to allow for the negative screen current developed under certain tube operating conditions. Bleeder current is chosen to be roughly 10 to 20 mA greater than the expected maximum negative screen current, so that screen voltage is regulated for all values of current between maximum negative screen current and maximum positive screen current. For external-anode tubes in the 4CX250 family, a typical screen bleeder current value would be 20 mA. For the 4CX1000 family, a screen-bleeder current of 70 mA is required.

Screen voltage should never be applied to a tetrode unless plate voltage and load also are applied; otherwise, the screen will act like an anode and will draw excessive current. Perhaps the best way to insure this is to include logic circuits that will not allow the screen supply to turn on until it senses plate voltage. Supplying the screen through a series-dropping resistor from the plate supply affords a measure of protection, since the screen voltage only appears when there is plate voltage. Alternatively, a fuse can be placed between the regulator and the bleeder resistor. The fuse should not be installed between the bleeder resistor and the tube because the tube should never be operated without a load on the screen. Without a load,

the screen potential tends to rise to the anode voltage. Any screen bypass capacitors or other associated circuits are likely be damaged by this high voltage.

In Fig 17.23, a varistor is connected from screen to ground. If, because of some circuit failure, the screen voltage should rise substantially above its nominal level, the varistor will conduct and clamp the screen voltage to a low level. If necessary to protect the varistor or screen dropping resistors, a fuse or overcurrent relay may be used to shut off the screen supply so that power is interrupted before any damage occurs. The varistor voltage should be approximately 30% to 50% higher than normal screen voltage.

17.7 Tube Amplifier Cooling

Vacuum tubes must be operated within the temperature range specified by the manufacturer if long tube life is to be achieved. Tubes having glass envelopes and rated at up to 25 W plate dissipation may be used without forced-air cooling if the design allows a reasonable amount of convection cooling. If a perforated metal enclosure is used, and a ring of ¼ to ⅜-inch-diameter holes is placed around the tube socket, normal convective airflow can be relied on to remove excess heat at room temperatures.

For tubes with greater plate dissipation ratings, and even for very small tubes operated close to maximum rated dissipation, forced-air cooling with a fan or blower is needed. Most manufacturers rate tube-cooling requirements for continuous-duty operation. Their literature will indicate the required volume of airflow, in cubic feet per minute (CFM), at some particular back pressure. Often, this data is given for several different values of plate dissipation, ambient air temperature and even altitude above sea level.

One extremely important consideration is often overlooked by power-amplifier designers and users alike: a tube's plate dissipation rating is only its maximum potential capability. The power that it can actually dissipate safely depends directly on the cooling provided. The actual power capability of virtually all tubes used in high-power amplifiers for amateur service depends on the volume of air forced through the tube's cooling structure.

17.7.1 Blower Specifications

This requirement usually is given in terms of cubic feet of air per minute (CFM), delivered into a back pressure, representing the resistance of the tube cooler to air flow, stated in inches of water. Both the CFM of airflow

Table 17.4
Specifications of Some Popular Tubes, Sockets and Chimneys

Tube	CFM	Back Pressure (inches)	Socket	Chimney
3-500Z	13	0.13	SK-400, SK-410	SK-416
3CX800A7	19	0.50	SK-1900	SK-1906
3CX1200A7	31	0.45	SK-410	SK-436
3CX1200Z7	42	0.30	SK-410	—
3CX1500/8877	35	0.41	SK-2200, SK-2210	SK-2216
4-400A/8438	14	0.25	SK-400, SK-410	SK-406
4-1000A/8166	20	0.60	SK-500, SK-510	SK-506
4CX250R/7850	6.4	0.59	SK602A, SK-610, SK-610A SK-611, SK-612, SK-620, SK-620A, SK-621, SK-630	
4CX400/8874	8.6	0.37	SK1900	SK606
4CX400A	8	0.20	SK2A	—
4CX800A	20	0.50	SK1A	—
4CX1000A/8168	25	0.20	SK-800B, SK-810B, SK-890B	SK-806
4CX1500B/8660	34	0.60	SK-800B, SK-1900	SK-806
4CX1600B	36	0.40	SK3A	CH-1600B

These values are for sea-level elevation. For locations well above sea level (5000 ft/1500 m, for example), add an additional 20% to the figure listed.

required and the pressure needed to force it through the cooling system are determined by ambient air temperature and altitude (air density), as well as by the amount of heat to be dissipated. The cooling fan or blower must be capable of delivering the specified airflow into the corresponding back pressure. As a result of basic air flow and heat transfer principles, the volume of airflow required through the tube cooler increases considerably faster than the plate dissipation, and back pressure increases even faster than airflow. In addition, blower air output decreases with increasing back pressure until, at the blower's so-called "cutoff pressure," actual air delivery is zero. Larger and/or faster-rotating blowers are required to deliver larger volumes of air at higher back pressure.

Values of CFM and back pressure required to realize maximum rated plate dissipation for some of the more popular tubes, sockets and chimneys (with 25 °C ambient air and at sea level) are given in **Table 17.4**. Back pressure is specified in inches of water and can be measured easily in an operational air system as indicated in **Figs 17.24** and **17.25**. The pressure differential between the air passage and atmospheric pressure is measured with a device called a *manometer*. A manometer is nothing more than a piece of clear tubing, open at both ends and fashioned in the shape of a "U." The manometer is temporarily connected to the chassis and is removed after the measurements are completed. As shown in the diagrams, a small amount of water is placed in the tube. At Fig 17.25A, the blower

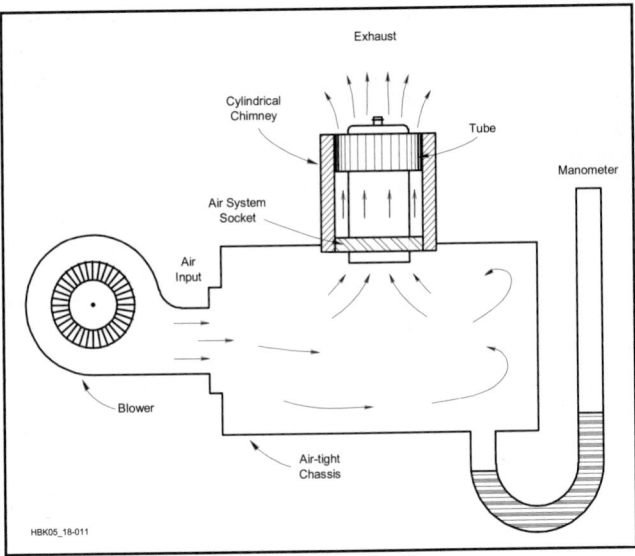

Fig 17.24 — Air is forced into the chassis by the blower and exits through the tube socket. The manometer is used to measure system back pressure, which is an important factor in determining the proper size blower.

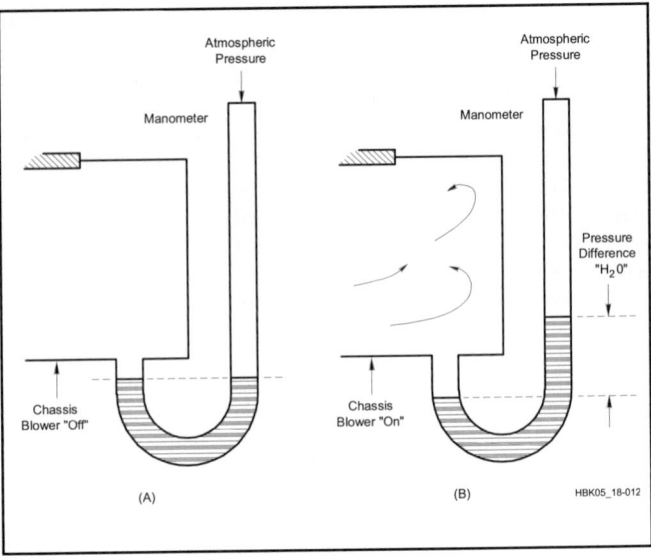

Fig 17.25 — At A the blower is "off" and the water will seek its own level in the manometer. At B the blower is "on" and the amount of back pressure in terms of inches of water can be measured as indicated.

Table 17.5
Blower Performance Specifications

Wheel Dia	Wheel Width	RPM	Free Air CFM	0.1	0.2	0.3	0.4	0.5	Stock No.
					CFM for Back Pressure (inches)				
2"	1"	3340	12	9	6	—	—	—	1TDN2
2¹⁵⁄₁₆"	1½"	3388	53	52	50	47	41	23	1TDN5
3"	1⅞"	3036	50	48	44	39	32	18	1TDN7
3"	1⅞"	3010	89	85	78	74	66	58	1TDP1
3¹⁵⁄₁₆"	1¹⁵⁄₁₆"	3016	75	71	68	66	61	56	1TDP3
3¾"	1⅞"	2860	131	127	119	118	112	105	1TDP5

Representative sample of Dayton squirrel cage blowers. More information and other models available from Grainger Industrial Supply (**www.grainger.com**).

is "off" and the water seeks its own level, because the air pressure (ordinary atmospheric pressure) is the same at both ends of the manometer tube. At B, the blower is "on" (socket, tube and chimney in place) and the pressure difference, in terms of inches of water, is measured. For most applications, a standard ruler used for measurement will yield sufficiently accurate results.

Table 17.5 gives the performance specifications for a few of the many Dayton blowers, which are available through Grainger Industrial Supply (**www.grainger.com**). Other blowers having wheel diameters, widths and rotational speeds similar to any in Table 17.5 likely will have similar flow and back pressure characteristics. If in doubt about specifications, consult the manufacturer. Tube temperature under actual operating conditions is the ultimate criterion for cooling adequacy and may be determined using special crayons or lacquers that melt and change appearance at specific temperatures. The setup of Fig 17.25, however, nearly always gives sufficiently accurate information.

17.7.2 Cooling Design Example

As an example, consider the cooling design of a linear amplifier to use one 3CX800A7 tube to operate near sea level with the air temperature not above 25 °C. The tube, running 1150 W dc input, easily delivers 750 W continuous output, resulting in 400 W plate dissipation ($P_{DIS} = P_{IN} - P_{OUT}$). According to the manufacturer's data, adequate tube cooling at 400 W P_D requires at least 6 CFM of air at 0.09 inches of water back pressure. In Table 17.5, a Dayton no. 1TDN2 will do the job with a good margin of safety.

If the same single tube were to be operated at 2.3 kW dc input to deliver 1.5 kW output (substantially exceeding its maximum electrical ratings!), P_{IN} would be about 2300 W and $P_D \approx 800$ W. The minimum cooling air required would be about 19 CFM at 0.5 inches of water pressure — doubling P_{DIS}, more than tripling the CFM of air flow required and increasing back pressure requirements on the blower by a factor of 5.5!

However, two 3CX800A7 tubes are needed to deliver 1.5 kW of continuous maximum legal output power in any case. Each tube will operate under the same conditions as in the single-tube example above, dissipating 400 W. The total cooling air requirement for the two tubes is, therefore, 12 CFM at about 0.09 inches of water, only two-thirds as much air volume and one-fifth the back pressure required by a single tube. While this may seem surprising, the reason lies in the previously mentioned fact that both the airflow required by a tube and the resultant back pressure increase much more rapidly than P_D of the tube. Blower air delivery capability, conversely, decreases as back pressure is increased. Thus, a Dayton 1TDN2 blower can cool two 3CX800A7 tubes dissipating 800 W total, but a much larger (and probably noisier) no. 1TDN7 would be required to handle the same power with a single tube.

In summary, three very important considerations to remember are these:

• A tube's actual safe plate dissipation capability is totally dependent on the amount of cooling air forced through its cooling system. Any air-cooled power tube's maximum plate dissipation rating is meaningless unless the specified amount of cooling air is supplied.

• Two tubes will always safely dissipate a given power with a significantly smaller (and quieter) blower than is required to dissipate the same power with a single tube of the same type. A corollary is that a given blower can virtually always dissipate more power when cooling two tubes than when cooling a single tube of the same type.

• Blowers vary greatly in their ability to deliver air against back pressure so blower selection should not be taken lightly.

17.7.3 Other Considerations

A common method for directing the flow of air around a tube involves the use of a pressurized chassis. This system is shown in Fig 17.24. A blower attached to the chassis forces air around the tube base, often through holes in its socket. A chimney is used to guide air leaving the base area around the tube envelope or anode cooler, preventing it from dispersing and concentrating the flow for maximum cooling.

A less conventional approach that offers a significant advantage in certain situations is shown in **Fig 17.26**. Here the anode compartment is pressurized by the blower. A special chimney is installed between the anode heat exchanger and an exhaust hole in the compartment cover. When the blower pressurizes the anode compartment, there are two parallel paths for airflow: through the anode and its chimney, and through the air system

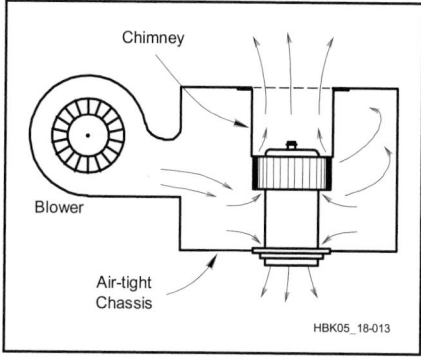

Fig 17.26 — Anode compartment pressurization may be more efficient than grid compartment pressurization. Hot air exits upwards through the tube anode and through the chimney. Cool air also goes down through the tube socket to cool tube's pins and the socket itself.

socket. Dissipation, and hence cooling air required, generally is much greater for the anode than for the tube base. Because high-volume anode airflow need not be forced through restrictive air channels in the base area, back pressure may be very significantly reduced with certain tubes and sockets. Only airflow actually needed is bled through the base area. Blower back pressure requirements may sometimes be reduced by nearly half through this approach.

Table 17.4 also contains the part numbers for air-system sockets and chimneys available for use with the tubes that are listed. The builder should investigate which of the sockets listed for the 4CX250R, 4CX300A, 4CX1000A and 4CX1600A best fit the circuit needs. Some of the sockets have certain tube elements grounded internally through the socket. Others have elements bypassed to ground through capacitors that are integral parts of the sockets.

Depending on your design philosophy and

tube sources, some compromises in the cooling system may be appropriate. For example, if glass tubes are available inexpensively as broadcast pulls, a shorter life span may be acceptable. In such a case, an increase of convenience and a reduction in cost, noise, and complexity can be had by using a pair of "muffin" fans. One fan may be used for the filament seals and one for the anode seal, dispensing with a blower and air-system socket and chimney. The airflow with this scheme is not as uniform as with the use of a chimney. The tube envelope mounted in a cross flow has flow stagnation points and low heat transfer in certain regions of the envelope. These points become hotter than the rest of the envelope. The use of multiple fans to disturb the cross airflow can significantly reduce this problem. Many amateurs have used this cooling method successfully in low-duty-cycle CW and SSB operation but it is not recommended for AM, SSTV or RTTY service.

The true test of the effectiveness of a forced air cooling system is the amount of heat carried away from the tube by the air stream. The power dissipated can be calculated from the airflow temperatures. The dissipated power is

$$P_D = 0.543 \, Q_A \, (T_2 - T_1) \qquad (7)$$

where

P_D = the dissipated power, in W
Q_A = the air flow, in CFM (cubic feet per minute).
T_1 = the inlet air temperature, °C (normally quite close to room temperature).
T_2 = the amplifier exhaust temperature, °C.

The exhaust temperature can be measured with a cooking thermometer at the air outlet. The thermometer should not be placed inside the anode compartment because of the high voltage present.

17.8 Vacuum Tube Amplifier Stabilization

Purity of emissions and the useful life (or even survival) of a tube depend heavily on stability during operation. Oscillations can occur at the operating frequency, or far from it, because of undesired positive feedback in the amplifier. Unchecked, these oscillations pollute the RF spectrum and can lead to over-dissipation and subsequent failure. Each type of oscillation has its own cause and its own cure.

17.8.1 Amplifier Neutralization

An RF amplifier, especially a linear amplifier, can easily become an oscillator at various frequencies. When the amplifier is operating, the power at the output side is large. If a fraction of that power finds its way back to the input and is in the proper phase, it can be re-amplified, repeatedly, leading to oscillation. An understanding of this process can be

had by studying the sections on feedback and oscillation in the **Analog Basics** chapter. Feedback that is self-reinforcing is called "positive" feedback, even though its effects are undesirable. Even when the positive feedback is insufficient to cause actual oscillation, its presence can lead to excessive distortion and strange effects on the tuning of the amplifier and it, therefore, should be eliminated or at least reduced. The deliberate use of "negative" feedback in amplifiers to increase linearity is discussed briefly elsewhere in this chapter.

The power at the output of an amplifier will couple back to the input of the amplifier through any path it can find. It is a good practice to isolate the input and output circuits of an amplifier in separate shielded compartments. Wires passing between the two compartments should be bypassed to ground if possible. This prevents feedback via paths external to the tube.

However, energy can also pass back through the tube itself. To prevent this, a process called neutralization can be used. Neutralization seeks to prevent or to cancel out any transfer of energy from the plate of the tube back to its input, which will be either the grid or the cathode. An effective way to neutralize a tube is to provide a grounded shield between the input and the output. In the grounded grid connection, the grid itself serves this purpose. For best neutralization, the grid should be connected through a low inductance conductor to a point that is at RF ground. Ceramic external tube types may have multiple low inductance leads to ground to enhance the shielding effect. Older glass type tubes may have significant inductance inside the tube and in the socket, and this will limit the effectiveness of the shielding effect of the grid. Thus, using a grounded grid circuit with those tube types does not rule out the need for further efforts at neutralization, especially at the higher HF frequencies.

When tetrodes are used in a grounded cathode configuration, the screen grid acts as an RF shield between the grid and plate. Special tube sockets are provided that provide a very low inductance connection to RF ground. These reduce the feed through from plate to grid to a very small amount, making the effective grid-to-plate capacitance a tiny fraction of a picofarad. If in doubt about amplifiers that will work over a large frequency range, use a network analyzer or impedance measuring instrument to verify how well grounded a "grounded" grid or screen really is. If at some frequencies the impedance is more than an ohm or two, a different grounding configuration may be needed.

For amplifiers to be used at only a single frequency, a series resonant circuit can be used at either the screen or grid to provide nearly perfect bypassing to ground. Typical values for a 50 MHz amplifier are shown in **Fig 17.27**.

For some tubes, at a certain frequency, the lead inductance to ground can just cancel the grid-to-plate capacitance. Due to this effect, some tube and socket combinations have a naturally self-neutralizing frequency based on the values of screen inductance and grid-to-plate capacitance. For example, the "self-neutralizing frequency" of a 4-1000 is about 30 MHz. This effect has been utilized in some VHF amplifiers.

BRIDGE NEUTRALIZATION

When the shielding effect of a grid or screen bypassed to ground proves insufficient, other circuits must be devised to cancel out the remaining effect of the grid-to-plate or grid-to-cathode capacitance. These, in effect, add an additional path for negative feedback that will combine with the undesired positive feedback and cancel it. The most commonly used circuit is the "bridge neutralization" circuit shown in **Fig 17.28**. This method gets its name from the fact that the four important capacitance values can be redrawn as a bridge circuit, as shown in **Fig 17.29**. Clearly when the bridge is properly balanced, there is no trans-

fer of energy from the plate to the grid tanks. Note that four different capacitors are part of the bridge. C_{gp} is characteristic of the chosen tube, somewhat affected by the screen or grid bypass mentioned earlier. The other components must be chosen properly so that bridge balance is achieved. C1 is the neutralizing capacitor. Its value should be adjustable to the point where

$$\frac{C1}{C3} = \frac{C_{gp}}{C_{IN}} \qquad (8)$$

where

C_{gp} = tube grid-plate capacitance
C_{IN} = tube input capacitance

The tube input capacitance must include all strays directly across the tube. C3 is not simply a bypass capacitor on the ground side of the grid tank, but rather a critical part of the bridge. Hence, it must provide a stable value of capacitance. Sometimes, simple bypass capacitors are of a type which change their value drastically with temperature. These are not suitable in this application.

Neutralization adjustment is accomplished

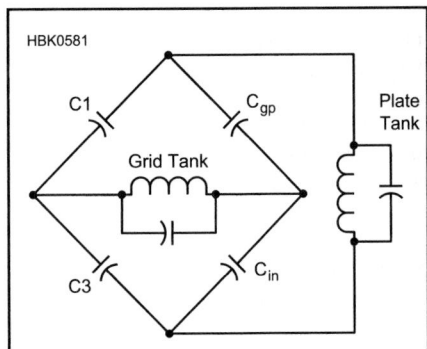

Fig 17.29 — The "bridge neutralization" circuit of Fig 17.28 redrawn to show the capacitance values.

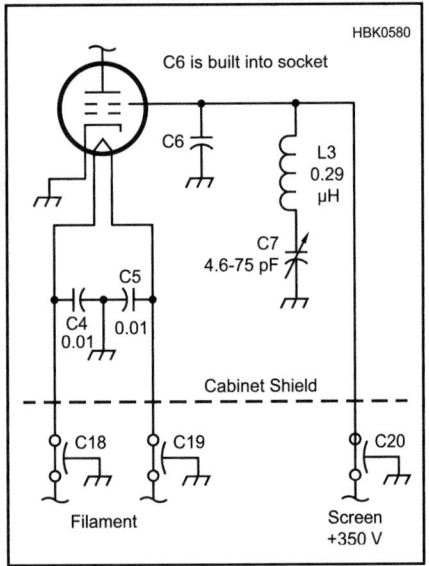

Fig 17.27 — A series-resonant circuit can be used to provide nearly perfect screen or grid bypassing to ground. This example is from a single-band 50 MHz amplifier.

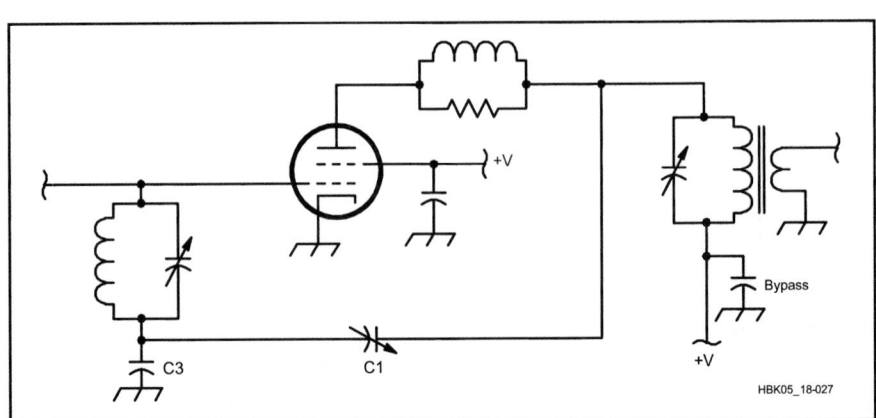

Fig 17.28 — A neutralization circuit uses C1 to cancel the effect of the tube internal capacitance.

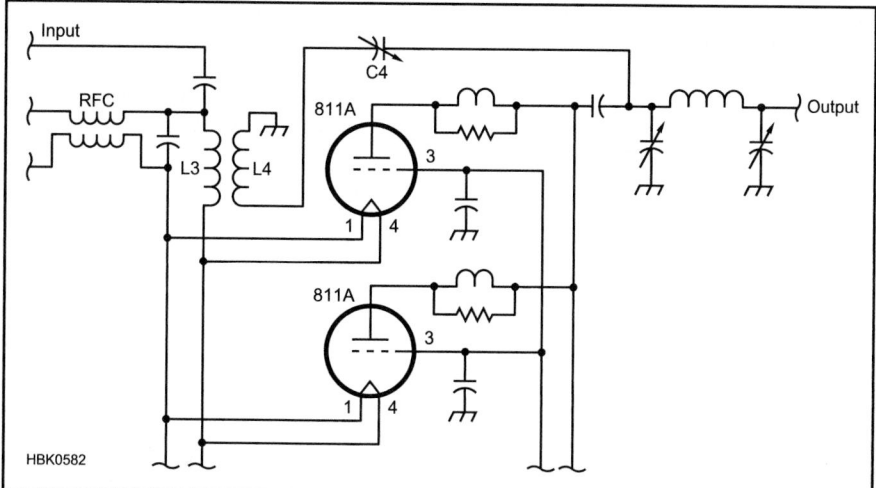

Fig 17.30 — In this neutralizing method, a broadband transformer (L3, L4) provides the needed out-of-phase signal. L3 is 6 turns of #14 wire close wound, ½ inch diameter. L4 is 5 turns of insulated wire over L3. C4 is 6 pF with 0.06-inch spacing. This circuit was originally featured in June 1961 *QST* and is still found in modern amplifiers using 811A tubes.

by applying energy to the output of the amplifier, and measuring the power fed through the input. Conversely, the power may be fed to the input and the output power measured. *with the power off*, the neutralization capacitor C1 is adjusted for minimum feed through, while keeping the output tuning circuit and the input tuning (if used) at the point of maximum response. Since the bridge neutralization circuit is essentially broad band, it will work over a range of frequencies. Usually, it is adjusted at the highest anticipated frequency of operation, where the adjustment is most critical.

BROADBAND TRANSFORMER

Another neutralizing method is shown in **Fig 17.30**, where a broadband transformer provides the needed out of phase signal. C4 is adjusted so that the proper amount of negative feedback is applied to the input to just cancel the feedback via the cathode to plate capacitance. Though many 811A amplifiers have been built without this neutralization, its use makes tuning smoother on the higher bands. This circuit was featured in June 1961 *QST* and then appeared in the *RCA*

Transmitting Tube Handbook. Amplifiers featuring this basic circuit are still being manufactured in 2009 and are a popular seller. Many thousands of hams have built such amplifiers as well.

An alternate method of achieving stable operation is to load the grid of a grounded cathode circuit with a low value of resistance.

A convenient value is 50 Ω as it provides a match for the driver. This approach reduces the power being fed back to the grid from the output to a low enough level that good stability is achieved. However, the amplifier gain will be much less than without the grid load. Also, unlike the grounded grid circuit, where much of the power applied to the input feeds through to the output, with this "loaded grid" approach, the input power is lost in the load, which must be able to dissipate such power. Distortion may be low in that the driver stage sees a very constant load. In addition, no tuning of the input is required.

17.8.2 VHF and UHF Parasitic Oscillations

RF power amplifier circuits contain parasitic reactances that have the potential to cause so-called parasitic oscillations at frequencies far above the normal operating frequency. Nearly all vacuum-tube amplifiers designed for operation in the 1.8 to 29.7 MHz frequency range exhibit tendencies to oscillate somewhere in the VHF-UHF range — generally between about 75 and 250 MHz depending on the type and size of tube. A typical parasitic resonant circuit is shown in **Fig 17.31**. Stray inductance between the tube plate and the output tuning capacitor forms a high-Q resonant circuit with the tube's C_{OUT}. C_{OUT} normally is much smaller (higher X_C) than any

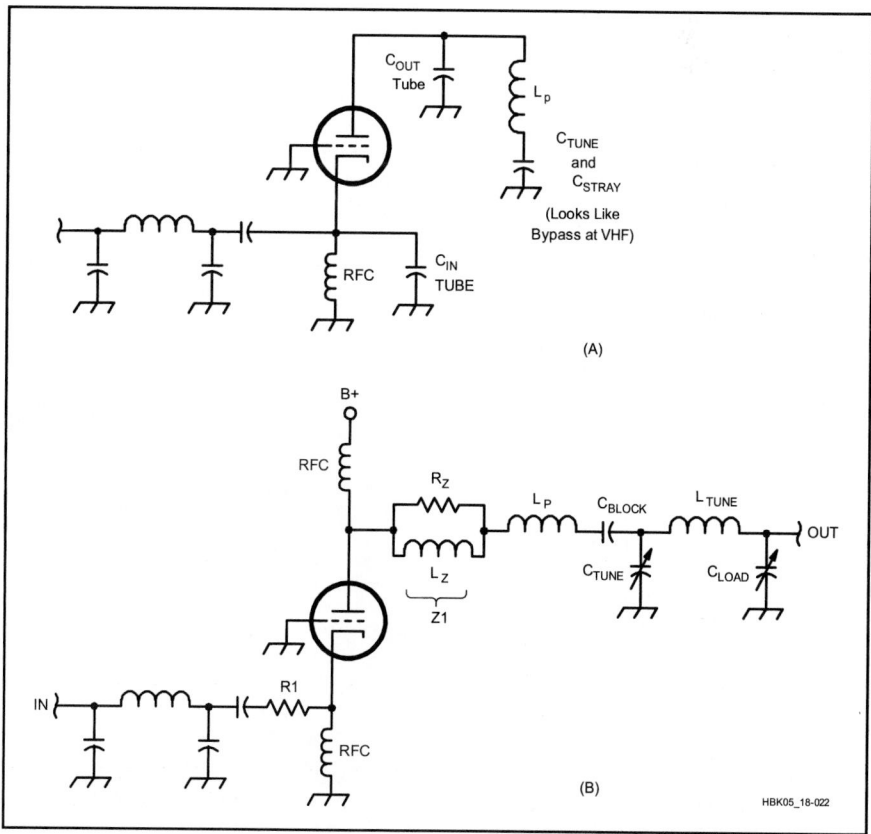

Fig 17.31 — At A, typical VHF/UHF parasitic resonance in plate circuit. The HF tuning inductor in the pi network looks like an RF choke at VHF/UHF. The tube's output capacitance and series stray inductance combine with the pi-network tuning capacitance and stray circuit capacitance to create a VHF/UHF pi network, presenting a very high impedance to the plate, increasing its gain at VHF/UHF. At B, Z1 lowers the Q and therefore gain at parasitic frequency.

of the other circuit capacitances shown. The tube's C_{IN} and the tuning capacitor C_{TUNE} essentially act as bypass capacitors, while the various chokes and tank inductances shown have high reactances at VHF. Thus, the values of these components have little influence on the parasitic resonant frequency.

Oscillation is possible because the VHF resonant circuit is an inherently high-Q parallel-resonant tank that is not coupled to the external load. The load resistance at the plate is very high and thus, the voltage gain at the parasitic frequency can be quite high, leading to oscillation. The parasitic frequency, f_r, is approximately:

$$f_r = \frac{1000}{2\pi\sqrt{L_P C_{OUT}}} \qquad (9)$$

where

f_r = parasitic resonant frequency in MHz

L_P = total stray inductance between tube plate and ground via the plate tuning capacitor (including tube internal plate lead) in μH.

C_{OUT} = tube output capacitance in pF.

In a well-designed HF amplifier, L_P might be in the area of 0.2 μH and C_{OUT} for an 8877 is about 10 pF. Using these figures, the equation above yields a potential parasitic resonant frequency of

$$f_r = \frac{1000}{2\pi\sqrt{0.2 \times 10}} = 112.5 \text{ MHz}$$

For a smaller tube, such as the 3CX800A7 with C_{OUT} of 6 pF, f_r = 145 MHz. Circuit details affect f_r somewhat, but these results do, in fact, correspond closely to actual parasitic oscillations experienced with these tube types. VHF-UHF parasitic oscillations can be prevented (*not* just minimized!) by reducing the loaded Q of the parasitic resonant circuit so that gain at its resonant frequency is insufficient to support oscillation. This is possible with any common tube, and it is especially easy with modern external-anode tubes like the 8877, 3CX800A7 and 4CX800A.

PARASITIC SUPPRESSORS

Z1 of Fig 17.31B is a parasitic suppressor. Its purpose is to add loss to the parasitic circuit and reduce its Q enough to prevent oscillation. This must be accomplished without significantly affecting normal operation. L_z should be just large enough to constitute a significant part of the total parasitic tank inductance (originally represented by L_P), and located right at the tube plate terminal(s). If L_z is made quite lossy, it will reduce the Q of the parasitic circuit as desired.

The inductance and construction of L_z depend substantially on the type of tube used.

Popular glass tubes like the 3-500Z and 4-1000A have internal plate leads made of wire. This significantly increases L_P when compared to external-anode tubes. Consequently, L_z for these large glass tubes usually must be larger in order to constitute an adequate portion of the total value of L_P. Typically a coil of 3 to 5 turns of #10 wire, 0.25 to 0.5 inches in diameter and about 0.5 to 1 inches long is sufficient. For the 8877 and similar tubes it usually is convenient to form a "horseshoe" in the strap used to make the plate connection. A "U" about 1-inch wide and 0.75 to 1 inch deep usually is sufficient. In either case, L_z carries the full operating-frequency plate current; at the higher frequencies this often includes a substantial amount of circulating tank current, and L_z must be husky enough to handle it without overheating even at 29 MHz. **Fig 17.32** shows a typical parasitic suppressor.

Regardless of the form of L_z, loss may be introduced as required by shunting L_z with one or more suitable non-inductive resistors. In high-power amplifiers, two composition or metal film resistors, each 100 Ω, 2 W, connected in parallel across L_z usually are adequate. For amplifiers up to perhaps 500 W a single 47 Ω, 2 W resistor may suffice. The resistance and power capability required to prevent VHF/UHF parasitic oscillations, while not overheating as a result of normal plate circuit current flow, depend on circuit parameters. Operating-frequency voltage drop across L_z is greatest at higher frequencies, so it is important to use the minimum necessary value of L_z in order to minimize power dissipation in R_z.

The parasitic suppressors described above very often will work without modification, but in some cases it will be necessary to experiment with both L_z and R_z to find a suitable combination. Some designers use nichrome or other resistance wire for L_z.

In exceptionally difficult cases, particularly when using glass tetrodes or pentodes, additional parasitic suppression may be attained by connecting a low value resistor (about 10 to 15 Ω) in series with the tube input, near the tube socket. This is illustrated by R1 of Fig 17.31B. If the tube has a relatively low input impedance, as is typical of grounded-grid amplifiers and some grounded-cathode tubes with large C_{IN}, R1 may dissipate a significant portion of the total drive power.

TESTING TUBE AMPLIFIERS FOR VHF-UHF PARASITIC OSCILLATIONS

Every high-power amplifier should be tested, before being placed in service, to insure that it is free of parasitic oscillations. For this test, nothing is connected to either the RF input or output terminals, and the band switch is first set to the lowest-frequency range. If the input is tuned and can be band switched separately, it should be set to the highest-frequency band. The amplifier control system should provide monitoring for both grid current and plate current, as well as a relay, circuit breaker or fast-acting fuse to quickly shut off high voltage in the event of excessive plate current. To further protect the tube grid, it is a good idea to temporarily insert in series with the grid current return line a resistor of approximately 1000 Ω to prevent grid current from soaring in the event a vigorous parasitic oscillation breaks out during initial testing.

Apply filament and bias voltages to the amplifier, leaving plate voltage off and/or cutoff bias applied until any specified tube warm-up time has elapsed. Then apply the lowest available plate voltage and switch the amplifier to transmit. Some idling plate current should flow. If it does not, it may be necessary to increase plate voltage to normal or to reduce bias so that at least 100 mA or so does flow. Grid current should be zero. Vary the plate tuning capacitor slowly from maximum capacitance to minimum, watching closely for any grid current or change in plate current, either of which would indicate a parasitic oscillation. If a tunable input net-

**Fig 17.32 —
Typical parasitic
suppressor.**

work is used, its capacitor (the one closest to the tube if a pi circuit) should be varied from one extreme to the other in small increments, tuning the output plate capacitor at each step to search for signs of oscillation. If at any time either the grid or plate current increases to a large value, shut off plate voltage immediately to avoid damage! If moderate grid current or changes in plate current are observed, the frequency of oscillation can be determined by loosely coupling an RF absorption meter or a spectrum analyzer to the plate area. It will then be necessary to experiment with parasitic suppression measures until no signs of oscillation can be detected under any conditions. This process should be repeated using each band switch position.

When no sign of oscillation can be found, increase the plate voltage to its normal operating value and calculate plate dissipation (idling plate current times plate voltage). If dissipation is at least half of, but not more than its maximum safe value, repeat the previous tests. If plate dissipation is much less than half of maximum safe value, it is desirable (but not absolutely essential) to reduce bias until it is. If no sign of oscillation is detected, the temporary grid resistor should be removed and the amplifier is ready for normal operation.

LOW-FREQUENCY PARASITIC OSCILLATIONS

The possibility of self-oscillations at frequencies lower than VHF is significantly lower than in solid state amplifiers. Tube amplifiers will usually operate stably as long as the input-to-output isolation is greater than the stage gain. Proper shielding and dc-power-lead bypassing essentially eliminate feedback paths, except for those through the tube itself.

On rare occasions, tube-type amplifiers will oscillate at frequencies in the range of about 50 to 500 kHz. This is most likely with high-gain tetrodes using shunt feed of dc voltages to both grid and plate through RF chokes. If the resonant frequency of the grid RF choke and its associated coupling capacitor occurs close to that of the plate choke and its blocking capacitor, conditions may support a tuned-plate tuned-grid oscillation. For example, using typical values of 1 mH and 1000 pF, the expected parasitic frequency would be around 160 kHz.

Make sure that there is no low-impedance, low-frequency return path to ground through inductors in the input matching networks in

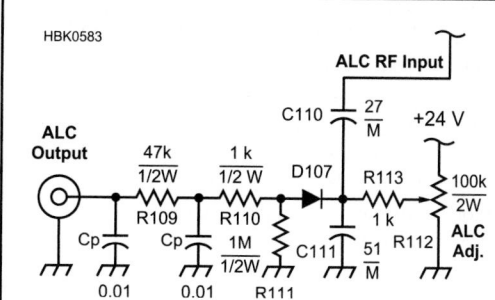

Fig 17.33 — An amplifier ALC circuit can be used to automatically limit the drive power from the transceiver to a safe level.

series with the low impedances reflected by a transceiver output transformer. Usually, oscillation can be prevented by changing choke or capacitor values to insure that the input resonant frequency is much lower than that of the output.

17.8.3 Reduction of Distortion

As mentioned previously, a common cause of distortion in amplifiers is over drive (flat topping). The use of automatic level control (ALC) is a practical way of reducing the ill effects of flat topping while still being assured of having a strong signal. This circuit detects the voltage applied to the input of the amplifier. Other circuits are based on detecting the onset of grid current flow. In either case, when the threshold is reached, the ALC circuit applies a negative voltage to the ALC input of the transceiver and forces it to cut back on the driving power, thus keeping the output power within set limits. Most transceivers also apply an ALC signal from their own output stage, so the ALC signal from the amplifier will add to or work in parallel with that. See **Fig 17.33** for a representative circuit.

Some tube types have inherently lower distortion than others. Selection of a tube specifically designed for linear amplifier service, and operating it within the recommended voltage and current ranges is a good start. In addition, the use of tuned circuits in the input circuits when running class AB2 will help by maintaining a proper load on the driver stages over the entire 360° cycle, rather than letting the load change as the tube begins to draw grid current. Another way to accomplish this is with the "loaded grid," the use of a rather low value of resistance from the grid

to cathode. Thus, when grid current flows, the change in impedance is less drastic, having been swamped by the resistive load.

For applications where the highest linearity is desired, operating class A will greatly reduce distortion but at a high cost in efficiency. Some solid state amateur transceivers have provision for such operation. The use of negative feedback is another way of greatly reducing distortion. High efficiency is maintained, but there is a loss of overall gain. Often, two stages of gain are used and the feedback applied around both stages. In this way, gain can be as high as desired, and both stages are compensated for any inherent nonlinearities. Amplifiers using RF negative feedback can achieve values of intermodulation distortion (IMD) as much as 20 dB lower than amplifiers without feedback.

It must be remembered that distortion tends to be a cumulative problem, with each nonlinear part of the transmission chain adding its part. It is not worthwhile to have a super clean transceiver if it is followed by an amplifier with poor linearity. In the same way, a very good linear will look bad if the transceiver driving it is poor. It is even possible to have a clean signal out of your amplifier, but have it spoiled by a ferrite core tuner inductor or balun that is saturated.

Distortion in a linear amplifier is usually measured with a spectrum analyzer while transmitting a two tone test. If the spectrum analyzer input is overloaded, this can also produce apparent distortion in the amplifier. Reducing the level so that the analyzer is not clipping the input signal is necessary to see the true distortion in the amplifier chain. Use of test equipment for various types of measurements is covered in the **Test Equipment and Measurements** chapter.

17.9 MOSFET Design for RF Amplifiers

There are two general classes of MOSFETs: high and low frequency designs. (See the **Analog Basics** chapter for MOSFET basics.) The low frequency types are generally optimized for high volume commercial switching applications: computer power supplies, motor controllers, inverters, and so on. They have molded plastic packages, the die are made with aluminum top side metallization, they have maximum junction temperature ratings of 150 to 175°C. Most have polysilicon gate conductors. Polysilicon is easy to manufacture consistently and it's cheap. This works very well for applications up to 200 kHz, but the gate losses start to increase dramatically when they are used at higher frequencies.

A MOSFET gate is essentially a capacitor, but its folded structure is long and skinny. The gate in a 500 W device may be more than a meter long! ("Meter" is not a misprint.) If its conductor material is a lossy material like polysilicon, the gate becomes a long distributed RC network. If an RF signal does make it all the way to the end it will be attenuated and no longer be in phase with the start. This effectively reduces the useful area of the device as frequency increases. It takes a lot of RF current to feed the gate capacitance:

$I = CVf$, where C is the gate capacitance, V is the peak gate voltage, and f is the operating frequency. If the gate capacitance is 500 pF and is being driven to 10 V at 30 MHz, the gate current is 150 mA RMS. While the current is directly proportional to the frequency, the power loss is I^2R. If the gate's top conductor is not low loss, it will fail due to I^2R losses at the gate bond pad (the metallization melts) when used at frequencies much higher than it was designed for.

There are two MOSFET manufacturers that use a metal gate instead of polysilicon for their switchmode devices, IXYS and Microsemi. While the die of these devices are quite capable of HF operation, their packaging (usually TO-247 or TO-264) does not provide an optimum HF layout. Because these are aimed at the switchmode market, their drain terminal is on the mounting surface and the source bond-wire length adds gain-killing degeneration at higher frequencies. But on the other hand, they are cheap in terms of cost per watt of dissipation and are acceptable for single-band designs through 20 meters.

When these same metal-gate MOSFET die are placed in packages that are specifically for RF use, the source is often connected to the mounting surface of the package. The source bond wires are thus short, which improves the available gain at all frequencies. This is very convenient because the source is grounded in most RF power amplifier circuits. It also eliminates the need for a mounting insulator, which in turn improves the power dissipation capability.

In MOSFETs specifically designed for RF, the main distinguishing feature is the gate structure. The channels are "shorter" (there is less distance between the gate and source) which reduces the transit time for electrons. As the active area of a device is increased by making the channel "wider," its power dissipation capability is increased. At the same time, the intrinsic (inter-electrode) capacitances also get bigger. A larger device is more difficult to use at higher frequencies because the input impedance (mostly gate capacitance) becomes ever smaller, which makes it harder to drive. In order to solve the gate loss problem mentioned earlier, the long skinny gate is folded into a comb shape. (See **Fig 17.34.**) The gate signal now only has to travel to the end of each finger. The highest frequency designs use multiple combs with shorter fingers. Several of these comb struc-

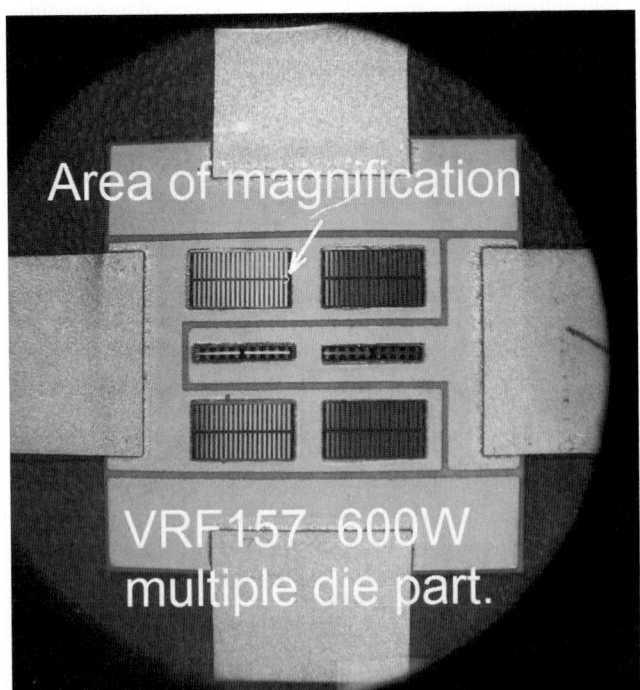

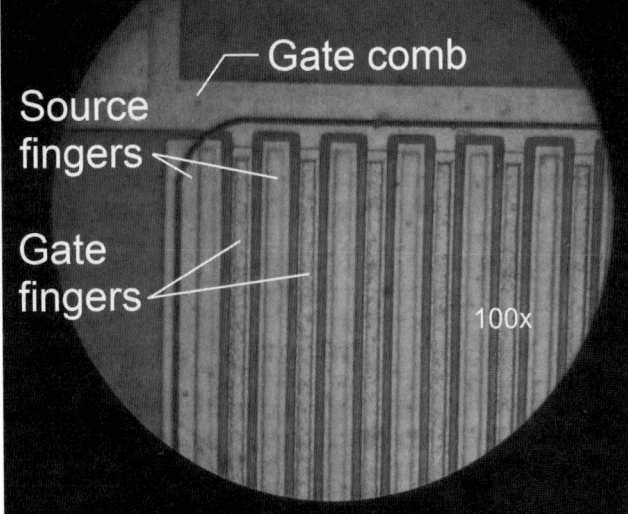

Fig 17.34 — A shows the layout of a multiple-die RF MOSFET (VRF157). B illustrates the comb structure of the gate. C is a closeup of the gate showing the interleaved source and gate finger structure. [Dick Frey, K4XU, photos]

tures are arrayed on the die and are connected in parallel when the die is wire-bonded in the package.

The top metallization for RF parts is either aluminum or gold. Aluminum is cheaper but gold is best because it has a higher operating temperature rating, up to 225 °C, and it is immune to power cycling failures due to its excellent ductility. The downside is that it is more expensive and the devices are much harder to manufacture because gold likes to dissolve into silicon. Its higher temperature rating means you can get more power from a small device, which offsets their higher cost somewhat.

17.9.1 LDMOS versus VDMOS

So far all the devices discussed are vertical MOSFETs, or VDMOS. Their gate and source electrodes are on the top surface of the die and the drain is on the bottom. For RF applications there is another type, LDMOS. This is a lateral device. Here the MOS structure is laid on edge and all the electrodes are on the top side of the die. Vertical p+ source connections are made through the die to make the bottom side of the die a source contact in order to get the optimum "common source" configuration. The channel area is low so the capacitances are smaller, especially the feedback capacitance, C_{GD}. However, the operating voltage capability is also low. There are none rated for more than 50 V operation. The gates are particularly sensitive to ESD and overdrive. However, they have spectacular high frequency capability and gain, and reasonable ruggedness. Your mobile phone would not work without LDMOS technology.

New RF amplifier designs are using LDMOS to replace bipolar transistors, which manufacturers are no longer making. The ability of LDMOS to operate at lower voltages is well suited for 12 V operation and, with its high gain and frequency response, providing 6 meter capability is simple. The downside is that these devices are not as linear as the bipolar transistor they replace.

Bipolar transistors need emitter ballasting resistors in each tiny bipolar cell so they can be paralleled in the die. The resistors also provide negative feedback, which improves linearity. MOSFETs do not require source resistors: paralleled cells will naturally share the load because their ON resistance has a positive temperature coefficient that prevents thermal runaway. As a result, LDMOS amplifiers require more negative feedback to provide comparable IMD performance, which offsets their higher gain advantage.

Regardless of the device type, the packaging is particularly important to an RF device. It must have low parasitics (see the **RF Techniques** chapter) and superior thermal qualities. The package insulator is made of ceramics, beryllia BeO (which is toxic), and/or alumina Al_2O_3, for high temperature capability. The conductors are gold-plated copper or Kovar, and the base flanges are often copper-tungsten or copper-molybdenum. The package is the major determining factor in the cost of an RF part. Parasitic inductance introduced by gate and source bond wires limits the ultimate frequency capability of a VDMOS part. LDMOS parts have gate and drain bond wires. LDMOS devices have an advantage in terms of frequency and package cost because they are free of the gain degeneration caused by source wire inductance and their die may be soldered directly to a metal mounting flange.

17.9.2 Designing Amplifiers with MOSFETs

Designing an amplifier requires a systems approach. You will need to consider how much power supply is required, as well as the cooling and control systems needed to keep it happy. If you have the transistors and want to build them into an amplifier, the design procedure is a little different. The place to start in any case is with its transistor's data sheet. This will show the voltage and power capabilities, and from these the circuit requirements can be calculated and the cooling system defined.

VOLTAGE RATINGS

Designing an amplifier with MOSFETs requires knowledge of the part being used. Generally, the cheap plastic-packaged switchmode parts will be best for single-band operation. Switchmode parts are rated by their drain breakdown voltage (BV_{DSS}), ON resistance (R_{DS}), and power dissipation. They are available in voltage ratings from as little as 5 V to over 1200 V. RF parts are sold by operating voltage, V_{DD}, power dissipation and frequency capability.

Select the part to suit your power supply requirements. A 500 V MOSFET will not work on a 12 V supply and will barely work at 50 V. This is because the MOSFET's intrinsic capacitances are higher at low voltage. Between the drain and source of every MOSFET is a parasitic "body diode" as shown in **Fig 17.35**. It's too slow to rectify RF, but like any diode, its capacitance changes with reverse bias voltage. This relationship is always given in the device's C-V curves on its data sheet. (See **Fig 17.36**.) Data sheets can be found on manufacturer or distributor websites or perform an Internet search for the part number and "data sheet."

In class AB operation, a MOSFET works best when operated at a little less than one-half of its rated breakdown voltage, BV_{DSS}. The drain voltage will swing up to 2 or even 3.562 times (for class E) the supply voltage. Under high VSWR, the drain voltage can be somewhat higher still. The RF voltage breakdown of a MOSFET is typically 20% higher than its data sheet value but is hard to specify reliability, so RF devices are rated by their

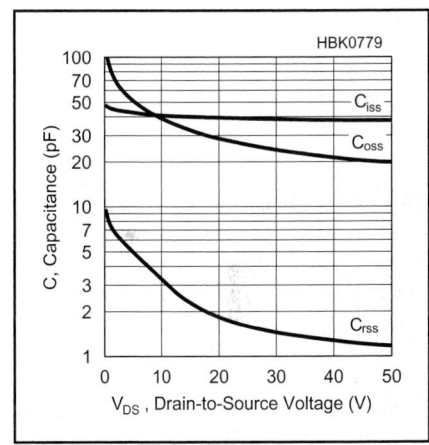

Fig 17.36 — The capacitance versus voltage (C-V) curves for the Microsemi VRF151 RF MOSFET. (Illustration courtesy Microsemi Corp.)

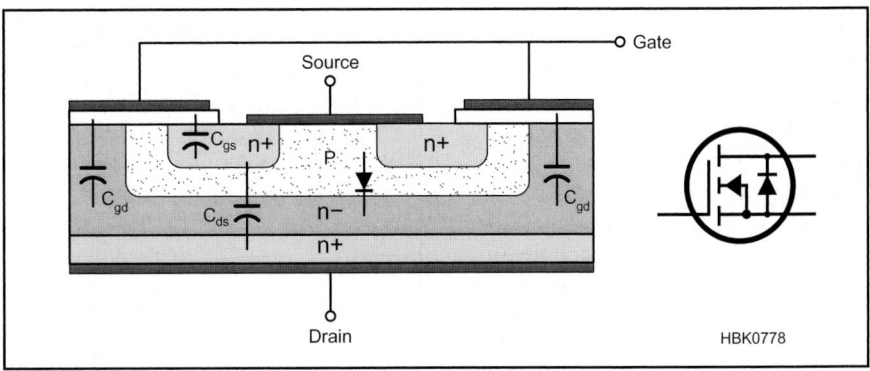

Fig 17.35 — All MOSFETs have parasitic capacitances as shown and a body diode between the drain and source in the cross-section and schematic symbol. The diode is shown for an N-channel device. It is reversed for P-channel devices.

dc operating voltages rather than BV_{DSS}. This takes into account the requirement for operating overhead.

RF parts are usually rated at a specific operating voltage such as 13.5 V, 28 V or 50 V. Originally these were common battery voltages for civilian and military vehicles and the tradition persists. The devices are optimized for their operating voltage. Choosing the operating voltage is a matter of considering many different parameters, not just the breakdown voltage.

THERMAL DESIGN

Suffice it to say that the thermal design of a high power transistor PA is often as challenging as the electrical design. It can be done "by the numbers" but the tricky part is making it fit into the available space. A thermal design example and an Advanced Power Technology application note by the author are provided as a supplemental article on the CD-ROM accompanying this book.

A word of caution is in order: tubes used in power amplifiers have a great deal of "headroom" in their specifications and are quite forgiving of momentary operator errors. RF power transistors, because they are more expensive in terms of cost per watt, are specified much closer to their limits. These limits must be observed at all times. Even though the data sheet says the device can do X watts, the designer must observe the requirements for proper cooling in order to reach this level in practice. In addition, manufacturers rate the power dissipation in theoretical terms. You will be lucky to achieve half of it.

As with tubes, there are CW ratings and SSB ratings. For transistors, the ratings are based on average power. The difference is simply the size of the heat sink required, as the peak power is the same for each. Each transistor has a thermal rating expressed as $R_{\theta JC}$, the thermal resistance from the transistor junction to the bottom of its case. Since the device has an upper junction temperature limit, somewhere between 150 and 200 °C, the power dissipation is determined by the difference between the junction and the case temperature:

$$P_d = (T_J - T_C) / R_{\theta JC}$$

where P_d is the available power dissipation, T_J is junction and T_C is case temperature. What this says is that without any cooling, the transistor's case will be almost the same as the junction temperature so its power dissipation capability is nearly zero. When placed on a heat sink, the case will be cooler and it then has power dissipation capability. It follows that the better the heat sink, the more power can be dissipated by the transistor.

There is another thermal consideration: the thermal resistance between the case of the transistor and the heat sink, $R_{\theta CS}$. Even if the base of the transistor and the heat sink are flat and smooth, microscopic air gaps still exist. These do not conduct heat and so reduce the net effectiveness of the heat sink. The solution is to use a thermal interface compound or *thermal grease*. The simplest and best is silicone oil loaded with zinc oxide powder. The oil does most of the work: the powder thickens it to a paste so it doesn't run out of the joint. It is applied as a very thin coat between the heat sink and device. When using thermal grease, always wiggle the transistor around on the sink to insure a minimum of grease between the part and the sink. Remember, thermal grease is not a good thermal conductor, it's just much better than air. Use it sparingly!

Most commercial high power broadcast amplifiers are cooled with circulated water, or a water-glycol mix if the minimum ambient temperature will be below 0 °C. While it has yet to be introduced to the amateur market, liquid cooling has great potential. The advances in plastic fittings, small pumps and heat exchangers driven by the high-performance computer market have great potential benefits to the amateur high power amplifier. Water is more than four times better than air for absorbing and moving heat. **Fig 17.37** illustrates both open- and closed-loop cooling systems. But regardless of where it is moved to, the heat must still be dissipated. It can warm the air in the shack, heat the rest of the house, or warm the septic tank.

17.9.3 The Transistor Data Sheet

Regardless of manufacturer, all data sheets contain the same basic information. The following should help make sense of what can be very confusing to a first time user.

Transistor specifications rely heavily on several ideal conditions that cannot happen in practice but since it is a common practice by all manufacturers, the numbers are very useful for comparing different devices. As long as you have the part number (and it is not a custom part), the corresponding data sheet can be found easily by searching for the part number and "data sheet."

The Microsemi VRF151 N-channel enhancement-mode VDMOS transistor will be used as an example. It is used in the 250 W broadband amplifier project presented in this chapter. The following discussion assumes the reader has obtained a data sheet for this part (see the company website at **www. microsemi.com**) and can refer to it.

MINIMUM, TYPICAL AND MAXIMUM

All of the specification parameters which the manufacturer guarantees are subject to either a minimum value or a maximum value. This is the worst case. As in an automobile, there is a maximum safe stopping distance and a minimum gas tank capacity. Most parameters also have a *typical* value that is generally representative of typical performance than the specified minimum or maximum. Some quantities have both upper and lower limits. Every parameter has specific test conditions under which it is measured.

ABSOLUTE MAXIMUM RATINGS

These are all dc ratings, easily verified with a variable power supply and a multimeter. If the manufacturer finds that any of these have been exceeded, any warranty claims are voided.

V_{DSS} is the maximum drain to source voltage rating, with the gate is shorted to the source. Think of the device as a high voltage Zener diode. As soon as it draws any current, power is dissipated, and temperature rises

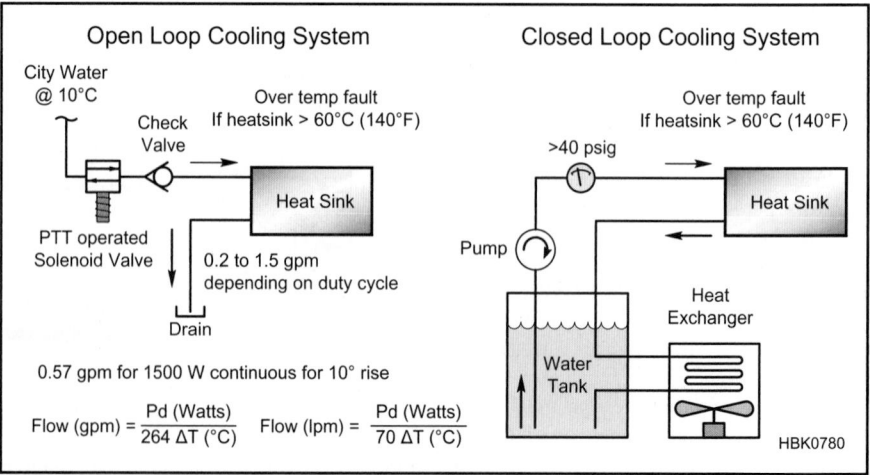

Fig 17.37 — Open- and closed-loop water cooling systems for solid-state amplifiers.

very quickly. Damage occurs either from puncturing through the junction or by arcing over the edge of the die.

Maximum drain current, I_D, is the current that will cause the device to dissipate its maximum rated power when fully turned on. Every device has an ON resistance called $R_{DS(ON)}$. The power dissipated is due to $I_D^2 \times 2.5$ $R_{DS(ON)}$. The factor of 2.5 accounts for $R_{DS(ON)}$ having a positive temperature coefficient which causes it to roughly double by the point at which the junction temperature is 200 °C.

V_{GS} is the maximum gate to source voltage. The gate is essentially a capacitor, with a SiO_2 dielectric perhaps 400 to 1000 Angstroms (10^{-10} m) thick. The limit is lower on LDMOS than VDMOS, and cannot be exceeded without destroying the device. Because LDMOS devices have much lower V_{GS} ratings and thinner dielectrics, some LDMOS manufacturers build in diode protection to make them less susceptible to electrostatic discharge, ESD.

P_D is the maximum power dissipation of the device under theoretical conditions: The bottom of the case at 25 °C and the junction at its maximum temperature, T_{jmax}. If not stated directly, the thermal resistance $R_{\theta JC}$ is equal to $P_D / (T_{jmax} - 25°C)$.

Storage temperature is straightforward. Maximum T_J is the junction temperature above which things start to come unsoldered, or the reliability seriously impaired, or smoke emitted.

STATIC ELECTRICAL CHARACTERISTICS

V_{DSS} is specified again, this time showing the measurement conditions, the guaranteed minimum, and the typical production values.

$V_{DS(ON)}$ or sometimes $R_{DS(ON)}$ is the min-imum resistance between drain and source that is obtained when the device is fully ON and carrying half the rated current. It is more commonly specified on switchmode parts, but it is of particular importance in high-efficiency saturated modes like class D and E because it limits the maximum obtainable efficiency.

I_{DSS} is the maximum drain current flowing with the gate shorted to the source. In an N-channel enhancement device one must apply a positive voltage to the gate to turn it on. The VRF151 will not conduct more than 1 mA at 100 V by itself without gate bias. This is a leakage current. In a perfect device it is zero.

Similarly, I_{GSS} is gate leakage current with the $V_{DS} = 0$. It represents a resistor in parallel with the gate capacitor. In this case, 1 μA at 20 V is 20 MΩ. This does not sound like much but if the drain has voltage on it and there is no resistor across the gate, this leakage current will cause the gate to charge from the drain, eventually turning the device fully ON with bad consequences.

Forward transconductance, g_{fs}, is the dc gain of the device expressed in terms of change in drain current per change in gate volts measured at a particular drain current. It has only a mild relationship with RF gain and too high a g_{fs} can cause bias instability and/or parasitics.

V_{TH} is the gate threshold specification. When V_{DS} is 10 V, V_{GS} of no less than 2.9 V and no more than 4.4 V will cause 100 mA of drain current to flow. A typical Class AB quiescent bias condition is 100 mA. Of all the parameters, this one has the widest window. Enough variation exists so that manufacturers sort parts into "bins" of values across the range, and assign letter codes to each which are marked on the package. This allows one to make matched pairs within a device type.

THERMAL CHARACTERISTICS

R_{JC} is equal to $P_{Dmax} / (T_{j max} - 25°C)$. Specifying the first two parameters generates the third. In this sense it is redundant but is often reiterated for those who are looking for the particular parameter. Because thermal performance is as important as RF performance, much of the information in the data sheet is concerned with it. In addition to the static thermal dissipation, $R_{\theta JC}$, there is a dynamic thermal impedance $Z_{\theta JC}$. Transistor packages have thermal mass. You cannot change its die temperature immediately For pulse operation, the effective $R_{\theta JC}$ is lower than it is for steady-state operation.

The dynamic thermal impedance is shown in **Fig 17.38** (Figure 5 of the VRF151 data sheet). It shows how the device can be used for pulse operation at higher power than it can on CW because the effective thermal impedance is lower for short pulses. This has some application to SSB, but there are other constraints on peak power that boil down to just allowing a smaller heat sink when used only for intermittent service — a bad practice for amateur amplifiers!

DYNAMIC CHARACTERISTICS

Somewhat of a misnomer, these are the parasitic capacitances between the gate, drain and gate. Except for the gate capacitance which is a fixed value based on device dimensions, the other two are a function of the drain voltage, just like varactors (voltage-variable capacitors). Because it is rather difficult to

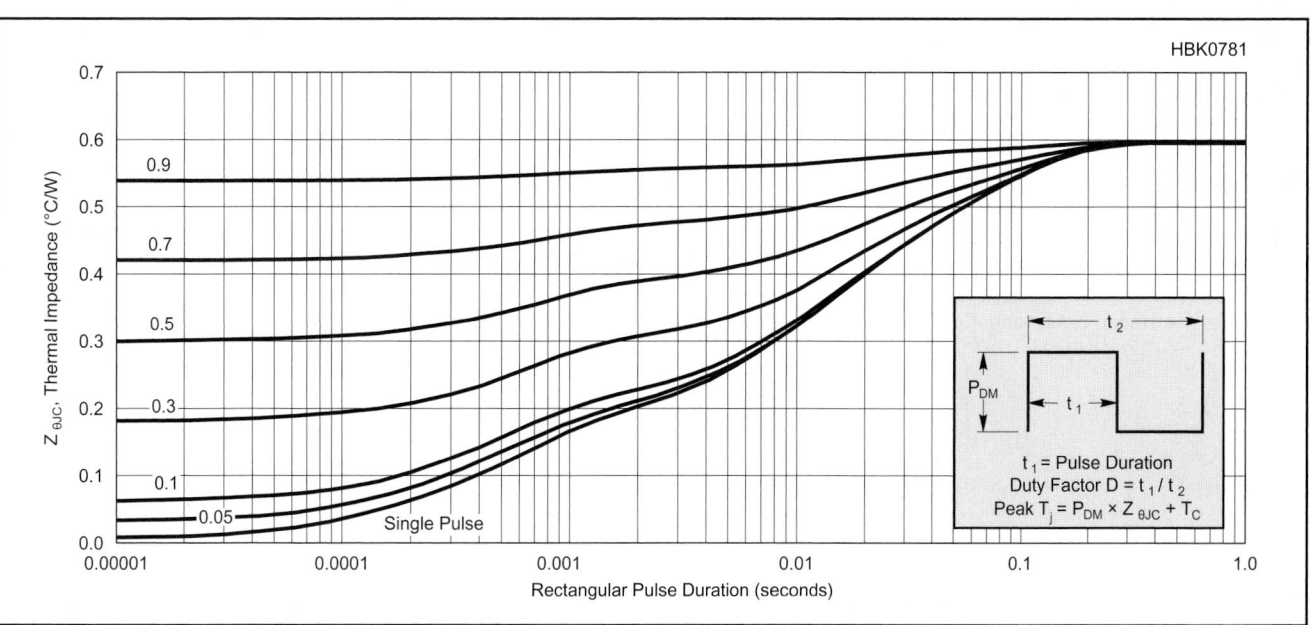

Fig 17.38 — Dynamic thermal impedance for the Microsemi VRF151 RF MOSFET. (Illustration courtesy Microsemi Corp.)

measure these parameters, the parameters are defined in a common-source configuration. C_{ISS} is the gate-to-source capacitance with the drain ac-shorted to the source. It is actually C_{GS} in parallel with C_{GD}. The three parameters are usually measured at the specified drain supply voltage with V_{GS} at zero. They are also usually displayed as in Fig 17.36 (Figure 3 of the data sheet), a graph of C vs drain voltage.

This voltage-varying capacitance is one of the causes of IMD in a transistor. Its effect is to impart some phase modulation (PM) on the signal. It can be observed as unequal IMD products on each side of the carrier pair. The PM distortion has pairs of odd-order carriers like AM distortion but the high side carriers are −180° out of phase with the AM products. This causes them to reduce the level of amplitude products on the high side of the carrier pair. (See the reference list entry for Sabin and Schoenike, *Single Sideband Systems and Circuits*.)

TRANSFER CHARACTERISTICS

As discussed above, g_{fs} describes the gain of a MOSFET as the change in drain current per change in gate voltage: $\Delta I_D / \Delta V_{GS}$. This means that g_{fs} is the slope of the transfer curve. The transfer curve for the VRF151 is shown in **Fig 17.39** (the data sheet's Figure 2). There are three curves in this graph, depicting the transfer characteristic at three different temperatures.

The three curves cross each other at the "thermal neutral point." This is where the temperature coefficient of V_{TH} changes from negative to positive and it is usually at a current much higher than the part normally operates. This explains why MOSFETs must have thermally compensated gate bias. Below the crossover point, where the part would be biased for class AB, the temperature coefficient of V_{TH} is negative. This means the gate voltage required for a given current goes down as the part heats up. Without thermal bias compensation we can have thermal runaway. The part will usually melt before it reaches the crossover point.

The transfer curve also shows that the gain of the device is not constant. It is quite low at low current and increases to a nominal value over its linear range and then the

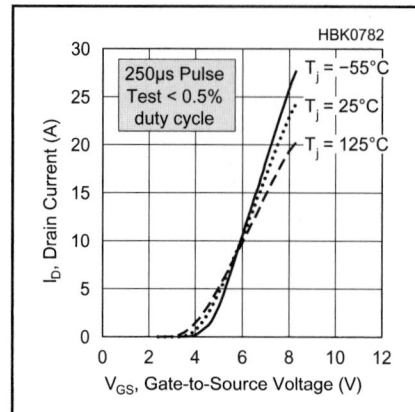

Fig 17.39 — Transfer characteristics for the Microsemi VRF151 RF MOSFET. (Illustration courtesy Microsemi Corp.)

curves flatten out at higher current as the part saturates. This demonstrates why very low distortion amplifiers are usually operated in class A.

FUNCTIONAL CHARACTERISTICS

This section is where the RF performance is specified — how much gain at what frequency, IMD performance, and ruggedness. While gain and IMD are easily understood, ruggedness is more difficult and it is very poorly defined by most manufacturers. The VRF151 has a ruggedness specification at all phase angles of a 30:1 VSWR when putting out 150 W PEP at 30 MHz. The test circuit is shown on page 4 in the data sheet. Nothing is said about how long the test takes or how well the device is cooled. If the test time and cooling conditions are not given, the specification is inadequate to guarantee that a design will remain within the device power limits. In the author's experience, 90% of all VSWR-related failures are due to over-dissipation. This says that limiting the amplifier's drain current is a simple and effective means for VSWR protection.

DATA SHEET EXTRAS

Test circuits are usually provided so customers can duplicate the test conditions for gain, IMD and output power. Sometimes circuit layouts and complete part lists are also provided. Note however, that while most high power parts are usually used in push-pull circuits, parts in data sheets are always tested one at a time in single-ended circuits. The last page of the data sheet gives the mechanical outline and sometimes mounting information. The VRF151 is "binned" (sorted into similarly performing groups) for V_{TH} as it exits the final testing and the bin letter code is marked on the package. This allows very close matching of gate threshold which is important when used in a push-pull circuit with a common bias supply for both parts. Most designers use separate bias adjustments regardless of matching, but it insures a measure of gain matching also, especially if the parts are from the same lot date code.

17.9.4 Summary Observations

MOSFETs are not perfect devices. g_{fs}, V_{TH}, BV and R_{DS} are all affected by die temperature. g_{fs} goes down with increasing die temperature. While this might cause the output power to sag a bit as the amplifier gets hotter, it is generally a benefit. MOSFETs can be paralleled without requiring source resistors because as the one carrying more current heats up, it loses gain and its resistance goes up, allowing the others to share the load. R_{DS} rises with temperature, a useful trait in hard-switching applications using paralleled devices. Breakdown voltage, BV_{DSS}, increases with temperature and so is usually not a concern in a part that heats up as it is being used. It has implications in cases of extreme cold, as in satellites.

One caveat when paralleling MOSFETs: always be mindful that they will easily oscillate at UHF. The device's inter-electrode capacitances and bonding wire inductances conspire to form a cross-coupled multivibrator. The solution is to place a small resistor in series with the gate leads, 3.3 Ω will usually suffice. This lowers the circuit Q enough to prevent oscillation from starting. [ref: Motorola *Engineering Bulletin 104*, et al]

17.10 Solid State RF Amplifiers

17.10.1 Solid State vs Vacuum Tubes

Solid state amplifiers have become the norm in transceivers, but their use in external high power amplifiers has not. The primary reason is economic. It is more expensive to generate a kilowatt or more with transistors because they are smaller and have less dissipation capability. This is changing as the broadcast and industrial RF industry converts to solid state, making RF power transistors available for amateur use at lower prices. A number of legal-limit solid state Amateur Radio amplifiers are available.

17.10.2 Classes of Operation

This topic applies to transistors as well as tubes, and it was covered earlier in this chapter and also in the **RF Techniques** chapter. In communications amplifiers, Class A is used mainly for driver stages where linearity is desired and efficiency is not a concern.

Class B is usually passed over in favor of the more linear Class AB. Class AB offers RF amplifiers increased linearity, mainly in less crossover distortion, for a very small (perhaps 1% or 2%) reduction in efficiency. It is the most commonly used class of operation for linear power amplifiers that must cover a wide range of frequencies. Broadband solid state Class AB amplifiers typically achieve 50 to 60% efficiency.

Class C is used where efficiency is important and linearity and bandwidth or harmonics are not. FM transmitters are the most common application in communications. Single-band tuned amplifiers can be as much as 80% efficient. However, in a single-ended amplifier they require a tank circuit. Class C amplifiers are *not* suitable for on-off keyed modes like CW without extensive pre-distortion of the driving signal in order to prevent key clicks.

Class D and E are most efficient, up to 95% in practical circuit applications. But both of these both require a narrow band tuned tank circuit to achieve this efficiency. They are not linear; their output is essentially either on or off. They can be used quite effectively for linear amplification by the process of EER (envelope elimination and restoration) but it is always in a single-band circuit. On-off keying can be employed if the power supply is keyed with a properly shaped envelope. EER is difficult to do well and requires very complex circuitry. Without careful system design, EER results in poor SSB performance.

17.10.3 Modeling Transistors

The design method using performance curves that was detailed earlier in this chapter is more applicable to vacuum tube amplifier design than solid state. The most common method used with solid state is electronic design analysis (EDA, also called computer-aided design, or CAD) using electronic models. *SPICE* or S-parameter models are available for some high power transistors, and simple amplifiers can be readily designed with the aid of an appropriate analysis program. (See the **Computer-Aided Circuit Design** chapter for more information on *SPICE* and related modeling techniques.)

Full-featured circuit design and analysis programs are expensive, and the resulting designs are only as good as the accuracy of the transistor models they use. A complicating factor is that any design relies heavily on models for all the passive components in the circuit. While passive part models can be obtained for some commercial components, many others — such as ferrite-loaded transformers — must also be designed before the circuit can be modeled. It is not unusual for the electronic design to take much longer and cost more than the benefits are worth.

As detailed in the **Computer-Aided Circuit Design** chapter, many of the EDA vendors offer free or inexpensive "student versions" of their products. These are fully capable up to a certain level of circuit complexity. Although they usually are not big enough to analyze a whole amplifier, student versions are still particularly useful for looking at parts of the whole.

Electronic design is very useful for getting the circuit design "in the ballpark." The design will be close enough that it will work when built, and any necessary fine tuning can be done easily once it is constructed. Computer modeling is very useful for evaluating the stresses on the circuit's passive components so they can be properly sized. Another very helpful use of CAD is in the development of the output filters. *SVC Filter Designer* by Jim Tonne, W4ENE, included with the *Handbook* CD, is exceptional in this regard.

17.10.4 Impedance Transformation — "Matching Networks"

Aside from the supply voltage, there is little difference between the operation of a tube amplifier and a transistor amplifier. Each amplifies the input signal, and each will only work into a specific load impedance. In a tube amplifier, the proper plate load impedance is provided by an adjustable pi or pi-L plate tuning network, which also transforms the impedance down to 50 Ω.

A single-transistor amplifier can be made in the same way, and in fact most single-band VHF amplifier "bricks" are. A tuned matching network provides the proper load impedance for the transistor and transforms it up to 50 Ω. The major difference is that the proper load impedance for a transistor, at any reasonable amount of power, is much *lower* than 50 Ω. For vacuum tubes it is much *higher* than 50 Ω.

BROADBAND TRANSFORMERS

Broadband transformers are often used in matching to the input impedance or optimum load impedance in a power amplifier. Unlike the tuned matching circuits, transformers can provide constant impedance transformation over a wide range of frequency without tuning. Multi-octave power amplifier performance can be achieved by appropriate application of these transformers. The input and output transformers are two of the most critical components in a broadband amplifier. Amplifier efficiency, gain flatness, input SWR, and even linearity all are affected by transformer design and application.

There are two basic RF transformer types, as described in the **RF Techniques** and **Transmission Lines** chapters: the conventional transformer and the transmission line transformer. More information on RF transformers is included on the *Handbook* CD as well.

The conventional transformer is wound much the same way as a power transformer. Primary and secondary windings are wound around a high-permeability core, usually made from a ferrite or powdered-iron material. Coupling between the secondary and primary is made as tight as possible to minimize leakage inductance. At low frequencies, the coupling between windings is predominantly magnetic. As the frequency rises, core permeability decreases and leakage inductance increases; transformer losses increase as well.

Typical examples of conventional transformers are shown in **Fig 17.40**. In Fig 17.40A, the primary winding consists of brass or copper tubes inserted into ferrite sleeves. The tubes are shorted together at one end by a piece of copper-clad circuit board material, forming a single turn loop. The secondary winding is threaded through the tubes. Since the low-impedance winding is only a single turn, the impedance transformation ratio is limited to the squares of integers — 1, 4, 9, 16 and so on.

The lowest effective transformer frequency is determined by the inductance of the one-turn winding. It should have a reactance, at the lowest frequency of intended operation, at least four times greater than the impedance

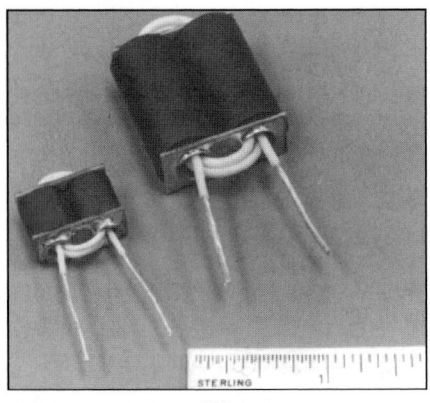

(A)

(B)

Fig 17.40 — The two methods of constructing the transformers outlined in the text. At A, the one-turn loop is made from brass tubing; at B, a piece of coaxial cable braid is used for the loop.

it is connected to. The coupling coefficient between the two windings is a function of the primary tube diameter and its length, and the diameter and insulation thickness of the wire used in the high-impedance winding. High impedance ratios, greater than 36:1, should use large-diameter secondary windings. Miniature coaxial cable (using only the braid as the conductor) works well. Another use for coaxial cable braid is illustrated in Fig 17.40B. Instead of using tubing for the primary winding, the secondary winding is threaded through copper braid. Because of the increased coupling between the primary and secondary of the transformer made with multiple pieces of coax, leakage reactance is reduced and bandwidth performance is increased.

The cores used must be large enough so the core material will not saturate at the power level applied to the transformer. Core saturation can cause permanent changes to the core permeability, as well as overheating. Transformer nonlinearity also develops at core saturation. Harmonics and other distortion products are produced — clearly an undesirable situation. Multiple cores can be used to increase the power capabilities of the transformer.

Transmission line transformers are similar to conventional transformers, but can be used over wider frequency ranges. In a conventional transformer, high-frequency performance deterioration is caused primarily by leakage inductance, the reactance of which rises with frequency. In a transmission line transformer, the windings are arranged so there is tight capacitive coupling between the two. A high coupling coefficient is maintained up to considerably higher frequencies than with conventional transformers.

The upper frequency limit of the transmission line transformer is limited by the length of the lines. As the lines approach ¼ wavelength, they start to exhibit resonant line effects and the transformer action becomes erratic.

MATCHING NETWORKS AND TRANSFORMERS

The typical tube amplifier tank circuit is an impedance transforming network in a pi or pi-L configuration. With reasonable loaded Q, it also functions as a low-pass filter to reduce the output signal harmonic levels below FCC minimums.

If a transistor amplifier uses a broadband transformer, it must be followed by a separate low-pass filter to achieve FCC harmonic suppression compliance. This is one reason broadband transistor amplifiers are operated in push-pull pairs. The balance between the circuit halves naturally discriminates against even harmonics, making the filtering job easier, especially for the second harmonic. The push-pull configuration provides double the power output when using two transistors, with very little increase in circuit complexity or component count. Push-pull pairs are also easier to match. The input and output impedance of a push-pull stage is twice that of a single-ended stage. The impedance is low, and raising it usually makes the matching task easier.

The transistor's low-impedance operation provides the opportunity to use a simple broadband transformer to provide the transformation needed from the transistor's load impedance up to 50 Ω. This low impedance also swamps out the effects of the device's output capacitance and, with some ferrite loading on the transformer, the amplifier can be made to operate over a very wide bandwidth without adjustment. This is not possible with tubes.

On the other hand, a tube amplifier with its variable output network can be adjusted for the actual output load impedance. The transistor amplifier with its fixed output network cannot be adjusted and is therefore much less forgiving of load variations away from 50 Ω. Circuits are needed to protect the transistor from damage caused by mis-

matched loads. These protection circuits generally operate "behind the scenes" without any operator intervention. They are essential for the survival of any transistor amplifier operating in the real world.

CALCULATING PROPER LOAD IMPEDANCE

The proper load impedance for a single transistor (or a tube for that matter) is defined by $R = E^2/P$. Converting this from RMS to peak voltage, the formula changes to $R = E^2/2P$. If two devices are used in push-pull (with twice the power and twice the impedance) the formula becomes $R = 2E^2/P$.

There is a constraint. Transformer impedance ratios are the square of their turns ratios. Those with single turn primaries are limited to integer values of 1, 4, 9, 16 and so on. Real-world transformers quickly lose their bandwidth at ratios larger than 25:1 due to stray capacitance and leakage inductance. A design solution must be found which uses one of these ratios. We will use the ubiquitous 100 W, 12 V transceiver power amplifier as an example. Using the push-pull formula, the required load impedance is $2 \times 12.5^2/100 = 3.125$ Ω. The required transformer impedance ratio is $50/3.125 = 16$, which is provided by a turns ratio of 4:1.

Transistor manufacturers, recognizing the broadband transformer constraint, have developed devices that operate effectively using automotive and military battery voltages and practical transformer ratios: 65 W devices for 12 V operation and 150 W devices for 48 V. Bigger 50 V devices have been designed (for example, the MRF154) that will put out 600 W, but practical transformer constraints limit 50 V push-pull output power to 900 W. There are higher voltage devices developed for the industrial markets and they are gradually finding their way into amateur designs. Being able to adjust the impedance to a convenient value for a common transformer turns ratio by adjusting the operating voltage is a powerful design option. These device and transformer limitations on output power can also be overcome by combining the outputs of several PA modules.

17.10.5 Combiners and Splitters

With some exceptions, practical solid state amplifiers have an upper power limit of about 500 W. This is a constraint imposed by the available devices and, to some extent, the ability to cool them. As devices are made more powerful by increasing the area of silicon die, the power density can become so high that only water cooling can provide adequate heat removal. Large devices also have large parasitic capacitances that make securing a broadband match over several octaves very difficult. By building a basic amplifier cell or "brick" and then combining several cells together, transmitter output powers are only limited by the complexity. Combiners and splitters have losses and add cost, so there are practical limits.

Broadband combiners usually take the form of an N-way 0° hybrid followed by an N:1 impedance transformer. The square ratio rule applies here too because the output impedance of a broadband combiner with N input ports is 50/N Ω — so combiners are usually 2, 4, 9 and 16-way. The higher the ratio, the lower the bandwidth will be.

Many types of combiners have been developed. The most common is the 4-way. It is easy to construct and has very good bandwidth. Most of the commercial "1 kW" broadband amplifiers use a 4-way combiner to sum the output of four 300 W push-pull modules operating on 48 V. Every combiner has loss. It may only be a few percent, but this represents a considerable amount of heat and loss of efficiency for a kilowatt output. This is a case where 4 × 300 does not make 1200.

The combiner approach to make a 1 kW solid state amplifier uses a large number of individual parts. A comparable 1 kW tube amplifier requires relatively few. This makes the high-powered solid state unit more expensive and potentially less reliable.

There is an alternative. If we had higher voltage transistors, we could use the same output transformer network configuration to get more output because power rises with the square of the operating voltage. There are high voltage transistors that can operate on 200 V or more. The problem is that these transistors must be capable of handling the corresponding higher power dissipation. The downside of making bigger, more powerful devices is an increase in parasitic capacitance. These bigger transistors become harder to drive and to match over broad bandwidths. However, the circuit simplicity and elimination of the combiner and its losses makes the higher voltage approach quite attractive.

TRANSMISSION LINE TRANSFORMERS AND COMBINERS

Fig 17.41 illustrates, in skeleton form, how transmission-line transformers can be used in a push-pull solid state power amplifier. The idea is to maintain highly balanced stages so that each transistor shares equally in the amplification in each stage. The balance also minimizes even-order harmonics so that low-pass filtering of the output is made much easier. In the diagram, T1 and T5 are current (choke) baluns that convert a grounded connection at one end to a balanced (floating) connection at the other end, with a high impedance to ground at both wires. T2 transforms the 50 Ω generator to the 12.5 Ω (4:1

impedance) input impedance of the first stage. T3 performs a similar step-down transformation from the collectors of the first stage to the gates of the second stage. The MOSFETs require a low impedance from gate to ground. The drains of the output stage require an impedance step up from 12.5 Ω to 50 Ω, performed by T4. Note how the choke baluns and the transformers collaborate to maintain a high degree of balance throughout the amplifier. Note also the various feedback and loading networks that help keep the amplifier frequency response flat.

Three methods are commonly used to combine modules: parallel (0°), push-pull (180°) and quadrature (90°). In RF circuit design, the combining is often done with special types of "hybrid" transformers called splitters and combiners. These are both the same type of transformer that can perform either function. The splitter is at the input, the combiner at the output.

Fig 17.42 illustrates one example of each of the three basic types of combiners. In a 0° hybrid splitter at the input the tight coupling between the two windings forces the voltages at A and B to be equal in amplitude and also equal in phase if the two modules are identical. The 2R resistor between points A and B greatly reduces the transfer of power between A and B via the transformer, but only if the generator resistance is closely equal to R. The output combiner separates the two outputs

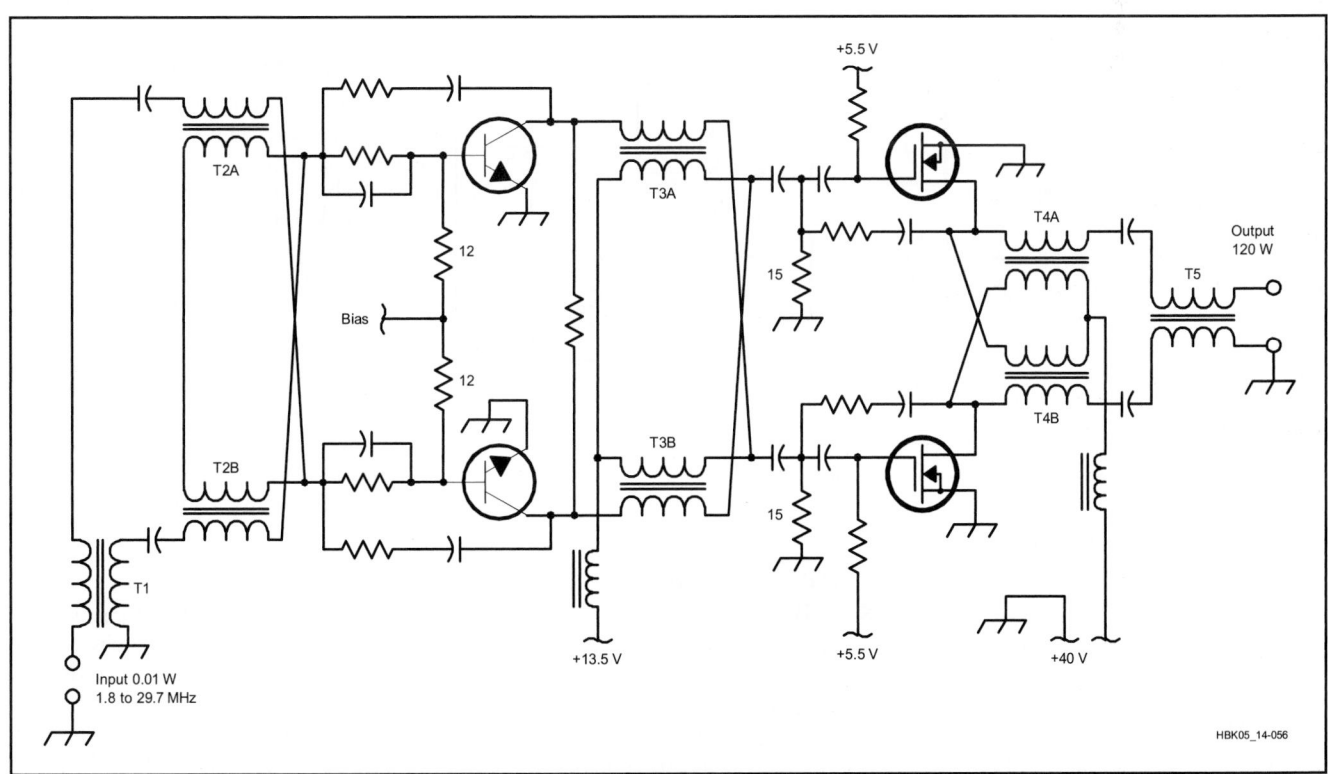

Fig 17.41 — Typical use of transmission line transformers as baluns and combiners in solid state power amplifiers.

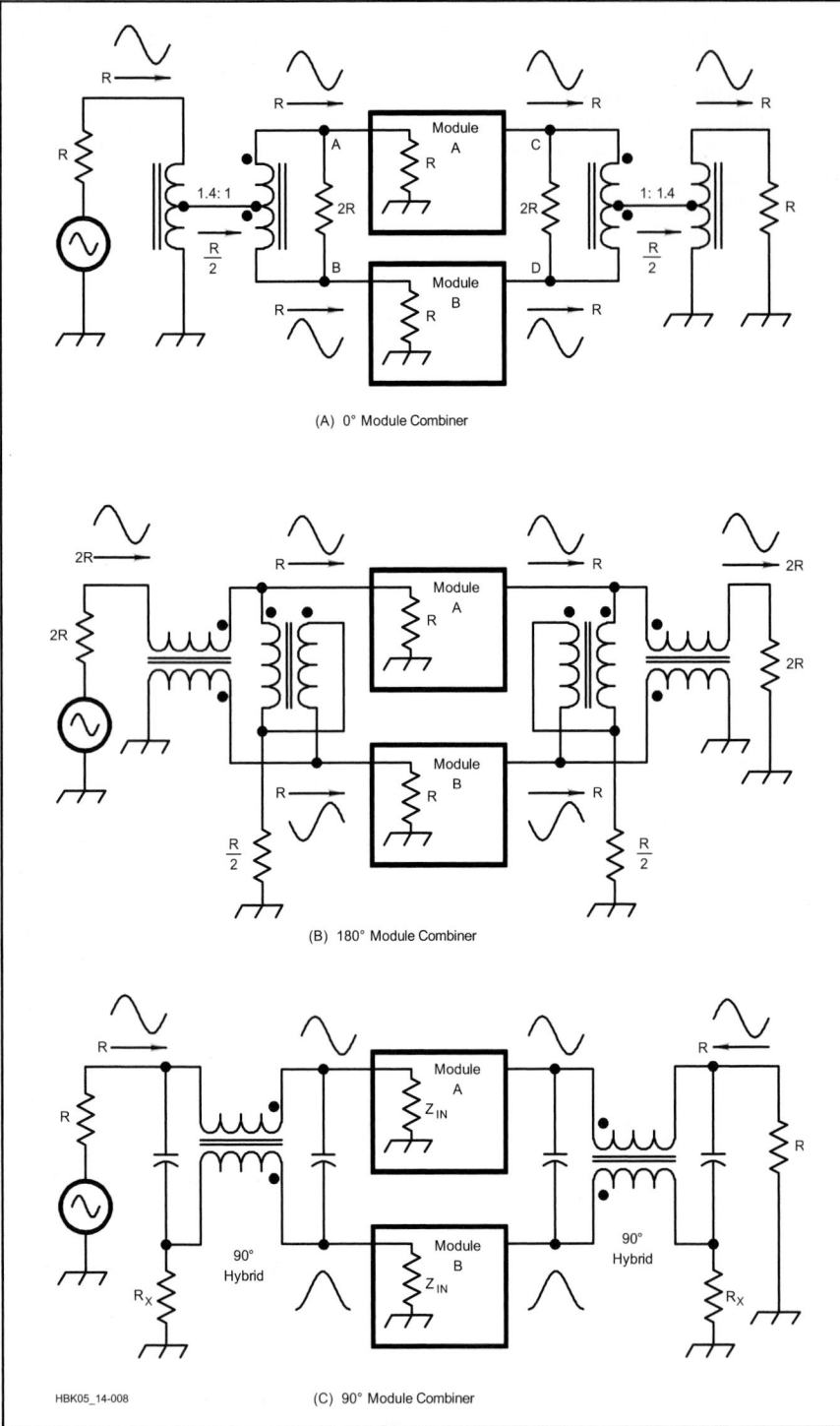

Fig 17.42 — Three basic techniques for combining modules.

C and D from each other in the same manner, if the output load is equal to R, as shown. No power is lost in the 2R resistor if the module output levels are identical. This section covers the subject of combiners very lightly. We suggest that the reader consult the considerable literature for a deeper understanding and for techniques used at different frequency ranges.

17.10.6 Amplifier Stabilization

Purity of emissions and the useful life (or even survival) of the active devices in a tube or transistor circuit depend heavily on stability during operation. Oscillations can occur at or away from the operating frequency because of undesired positive feedback in the amplifier. Unchecked, these oscillations pollute the RF spectrum and can lead to tube or transistor over-dissipation and subsequent failure. Each type of oscillation has its own cause and its own cure.

In a linear amplifier, the input and output circuits operate on the same frequency. Unless the coupling between these two circuits is kept to a small enough value, sufficient energy from the output may be coupled in phase back to the input to cause the amplifier to oscillate. Care should be used in arranging components and wiring of the two circuits so that there will be negligible opportunity for coupling external to the tube or transistor itself. A high degree of shielding between input and output circuits usually is required. All RF leads should be kept as short as possible and particular attention should be paid to the RF return paths from input and output tank circuits to emitter or cathode.

In general, the best arrangement using a tube is one in which the input and output circuits are on opposite sides of the chassis. Individual shielded compartments for the input and output circuitry add to the isolation. Transistor circuits are somewhat more forgiving, since all the impedances are relatively low. However, the high currents found on most amplifier circuit boards can easily couple into unintended circuits. Proper layout, the use of double-sided circuit boards (with one side used as a ground plane and low-inductance ground return), and heavy doses of bypassing on the dc supply lines often are sufficient to prevent many solid state amplifiers from oscillating.

PARASITIC OSCILLATIONS

In low-power solid state amplifiers, parasitic oscillations can be prevented by using a small amount of resistance in series with the base or collector lead, as shown in **Fig 17.43A**. The value of R1 or R2 typically should be between 10 and 22 Ω. The use of both resistors is seldom necessary, but an empirical determination must be made. R1 or R2 should be located as close to the transistor as practical.

At power levels in excess of approximately 0.5 W, the technique of parasitic suppression shown in Fig 17.43B is effective. The voltage drop across a resistor would be prohibitive at the higher power levels, so one or more ferrite beads placed over connecting leads can be substituted (Z1 and Z2). A bead permeability of 125 presents a high impedance at VHF and above without affecting HF performance. The beads need not be used at both circuit locations. Generally, the terminal carrying the least current is the best place for these suppression devices. This suggests that the resistor or ferrite beads should be connected in the base lead of the transistor.

C3 of Fig 17.43C can be added to some

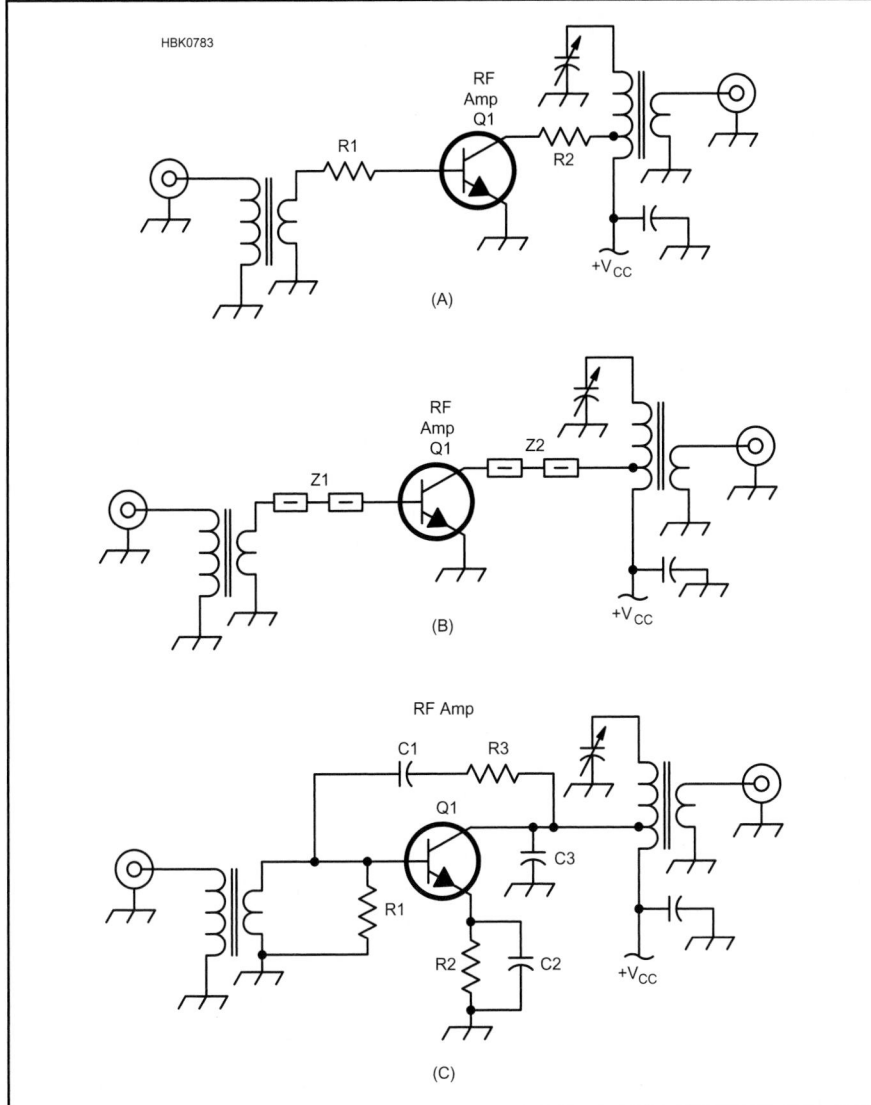

Fig 17.43 — Suppression methods for VHF and UHF parasitics in solid state amplifiers. At A, small base and collector resistors are used to reduce circuit Q. B shows the use of ferrite beads to increase circuit impedance. In circuit C, C1 and R3 make up a high-pass network to apply negative feedback.

power amplifiers to dampen VHF/UHF parasitic oscillations. The capacitor should be low in reactance at VHF and UHF, but must present a high reactance at the operating frequency. The exact value selected will depend upon the collector impedance. A reasonable estimate is to use an X_C of 10 times the collector impedance at the operating frequency. Silver-mica or ceramic chip capacitors are suggested for this application. An additional advantage is the resultant bypassing action for VHF and UHF harmonic energy in the collector circuit. C3 should be placed as close to the collector terminal as possible, using short leads.

The effects of C3 in a broadband amplifier are relatively insignificant at the operating frequency. However, when a narrow-band collector network is used, the added capacitance of C3 must be absorbed into the network design in the same manner as the C_{OUT} of the transistor.

LOW-FREQUENCY PARASITIC OSCILLATIONS

Bipolar transistors and MOSFETs exhibit a rising gain characteristic as the operating frequency is lowered. To preclude low-frequency instabilities because of the high gain, shunt and degenerative feedback are often used. In the regions where low-frequency self-oscillations are most likely to occur, the feedback increases by nature of the feedback network, reducing the amplifier gain. In the circuit of Fig 17.43C, C1 and R3 provide negative feedback, which increases progressively as the frequency is lowered. The network has a small effect at the desired operating frequency but has a pronounced effect at the lower frequencies. The values for C1 and R3 are usually chosen experimentally. C1 will usually be between 220 pF and 0.0015 µF for HF-band amplifiers while R3 may be a value from 51 to 5600 Ω.

R2 of Fig 17.43C develops emitter degeneration at low frequencies. The bypass capacitor, C2, is chosen for adequate RF bypassing at the intended operating frequency. The impedance of C2 rises progressively as the frequency is lowered, thereby increasing the degenerative feedback caused by R2. This lowers the amplifier gain. R2 in a power stage is seldom greater than 10 Ω, and may be as low as 1 Ω. It is important to consider that under some operating and layout conditions R2 can cause instability. This form of feedback should be used only in those circuits in which unconditional stability can be achieved.

R1 of Fig 17.43C is useful in swamping the input of an amplifier. This reduces the chance for low-frequency self-oscillations, but has an effect on amplifier performance in the desired operating range. Values from 3 to 27 Ω are typical. When connected in shunt with the normally low base impedance of a power amplifier, the resistors lower the effective device input impedance slightly. R1 should be located as close to the transistor base terminal as possible, and the connecting leads must be kept short to minimize stray reactances. The use of two resistors in parallel reduces the amount of inductive reactance introduced compared to a single resistor.

17.11 Solid State Amplifier Projects

This section presents three solid state amplifier designs that operate on a very wide range of frequencies. The first is an HF+6 meter design by Dick Frey, K4XU that will amplify the output of QRP radios in the 10-15 W range to 250 W. It provides an excellent design example of current solid state amplifier design practices described in the previous sections, especially the control and protection circuitry. A second amplifier by James Klitzing, W6PQL, covers 2 meters. It is a linear design suitable for both SSB/CW and FM and has an output power of around 80 W. The final design by Steve Lampereur, KB9MWR, operates on the microwave band of 10 GHz. While the 2 W output may not sound like a lot of power, at X band it is considered QRO especially when combined with a dish antenna that provides tens of dB of gain!

Complete design and construction information for all of these amplifier designs is included on the CD-ROM accompanying this book. In addition, several other amplifier designs are presented for both HF and VHF/UHF operation.

17.11.1 *Project:* A 250 W Broadband Linear Amplifier

The amplifier described here and shown in **Fig 17.44** is neither revolutionary nor daring. It uses commonly available parts — no special parts and no "flea market specials." It is based on well-proven commercial designs and "best design practices" acquired over the past 30 years as solid state technology has matured. This design project was undertaken for the 2010 edition of the *ARRL Handbook* as a detailed design example as well as a practical solid state amplifier you can build. It is a continuing project. The full description of the amplifier including construction and operating details is provided on the CD-ROM accompanying this book.

A block diagram for the project is given in **Fig 17.45**. The amplifier is built on three PC boards — a PA module, a low-pass filter assembly, and a board for control, protection and metering circuitry. The *Handbook* CD includes *ExpressPCB* files for these boards, and the artwork can be used to have boards made in small quantities (see **www.expresspcb.com** for details).

The basic PA configuration has been in the *Handbook* since the 2010 edition. It is intended to be a "ham-proof" external amplifier for QRP transceivers that put out 15 W or less. It is designed for a gain of 30×, or 15 dB. Drive power of less than 10 W will provide 250 W output from 1.8 through 51 MHz. The amplifier provides exceptionally linear performance, necessary for high quality SSB and PSK modes, and is rugged enough to withstand the most rigorous contest environment.

Amplifier design tends to focus on the RF section, but a successful stand-alone solid state amplifier is equally dependent on its control system. The control requirements for a tube amplifier are well known, while those for solid state amplifiers are not. This is mostly because the functions of a solid state amplifier's control system are generally transparent to the user. Parameters are monitored and protection is applied without any operator intervention. This must be. While tubes are fairly forgiving of abuse, semiconductors can heat so quickly that intervention *must* be automatic or they can be destroyed.

Transistors are sensitive to heat, so cooling and temperature compensation are critical to a successful design. Transistors require a heat sink. Power amplifier tubes have large surface areas and are cooled by air blown on or through them. Transistors are small. Mounting them on a heat sink increases their thermal mass and provides a much larger surface area so the heat dissipated in the devices can be removed either by convection or forced air. The thermal design of an amplifier is just as important as the electrical design. More information on thermal design may be found in the **Analog Basics** chapter.

Silicon's thermal coefficient causes the bias current to increase as the device heats up if the bias source is fixed. The increased current causes even more heating and can lead to thermal runaway. For stable Class AB linear operation, the gate bias for a MOSFET or bipolar transistor must track the temperature of the device. The control circuit typically uses another silicon device such as a diode thermally coupled to the amplifier heat sink near the transistor to sense the temperature and adjust the bias to maintain a constant bias current.

Transistor power amplifiers are designed to operate into 50 Ω. Operation into a VSWR other than 1:1 will cause an increase in device dissipation and other stress. The success of the solid state transceiver is due to its integrated PA protection system. The temperature of the heatsink, the load VSWR, the output power and the supply current are all monitored by the control system. If any of these exceed their threshold limits, the RF drive is reduced by the transceiver's ALC system.

An external solid state PA protection system must perform the same functions, but the driver's ALC circuit is not always available so other means must be used to protect the PA. This is usually accomplished simply by taking the amplifier out of the circuit. An indicator then tells the operator which condition caused the fault so appropriate action can be taken. Access to the driver's ALC system would make this protection task more automatic, smoother and less troublesome, but no two transceiver models have the same ALC characteristic. This makes the design of a universal ALC interface more difficult.

THE 1.8 TO 55 MHz PA

Fig 17.46 shows the power amplifier (PA) schematic. Two Microsemi VRF151 MOSFETs are used in this amplifier. The circuit topology is a 4:1 transmission line transformer type, rather than a "tube and sleeve" type common in many PA designs and discussed earlier. This style offers more bandwidth, necessary to provide performance on 6 meters. Typical gain is 15 dB; 10 W drive will easily provide 250 W output with a 48 V supply. There is a lot of latitude in this design. It can even be operated on an unregulated supply. As long as the maximum unloaded voltage does not exceed 65 V, the transistors will not be overstressed. Other devices such as the MRF151, SD2931 or BLF177 would probably also work but have not been tested. They will require a regulated power supply, however.

FEEDBACK — TWO KINDS

The amplifier's gain is controlled by two kinds of feedback. Shunt feedback (from drain to gate) is provided by the link on T2

Fig 17.44 — This 250 W amplifier for 160 through 6 meters provides a detailed design example as well as a practical project. Additional photos and information about the interior layout may be found on the Handbook CD-ROM.

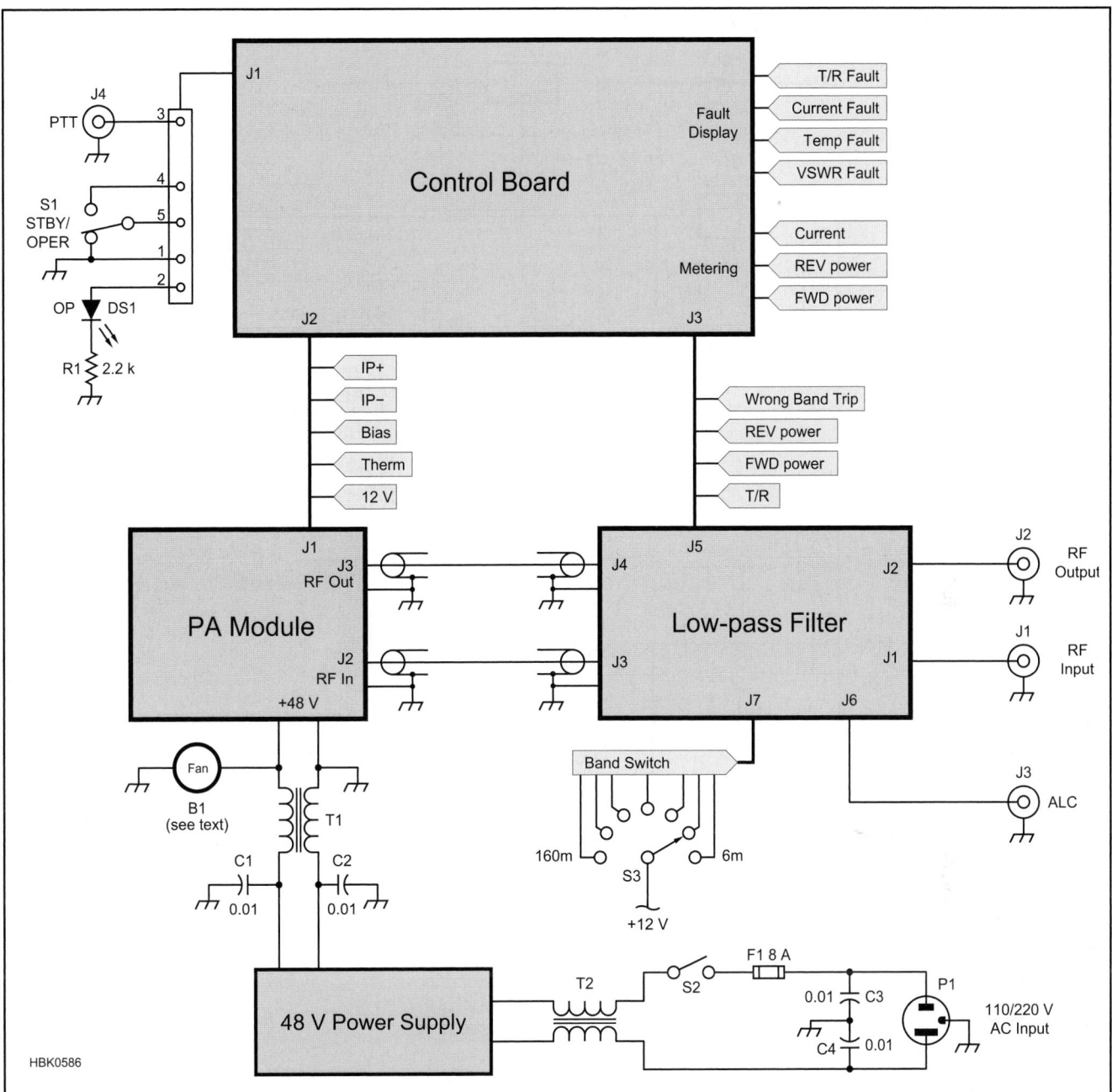

Fig 17.45 — Block diagram of the 250 W solid state amplifier. It is built on three PC boards, which are described in the text and accompanying diagrams. T1 and T2 are 2 turns through a Fair-Rite 2643540002 ferrite bead if needed to suppress RFI from the switching power supply.

through resistors R5 and R6. It tends to lower the input impedance, but it also helps to keep the gain constant over frequency and improve the linearity. Series feedback is provided by the 0.05 Ω of resistance in each source. This increases the input impedance, cuts the gain by 3 dB, and most importantly, it has a huge effect on the linearity.

Without any feedback at all, the amplifier would have more than about 30 dB of gain (×1000) at some frequencies, tending to make it unstable — prone to parasitic oscillation. And the linearity would be terrible, −25 dBc

or so IMD products. It would also be very sensitive to load changes. The input SWR is 1.2:1 on 160 meters and rises to 1.5:1 on 6 meters. The amplifier's gain is 15 dB ±0.5 dB over the same frequency range.

CONTROL AND PROTECTION

The control board appears far more complicated than the PA but in reality, it is just a few analog and logic ICs. This circuitry monitors several parameters, displays them, and if necessary, puts the amplifier into standby if one of them goes out of range. This

control system could be used on any amplifier. All solid state amplifiers need similar protection. The amplifier is protected for:

1. Over temperature, by a thermistor on the heatsink and setting a limit.

2. Over current, by measuring the PA current and setting a maximum limit.

3. High SWR, by monitoring the reflected power and setting a maximum limit.

4. Selection of a low-pass filter lower than the frequency in use.

Each of these fault trips results in forcing the amplifier into the standby position and

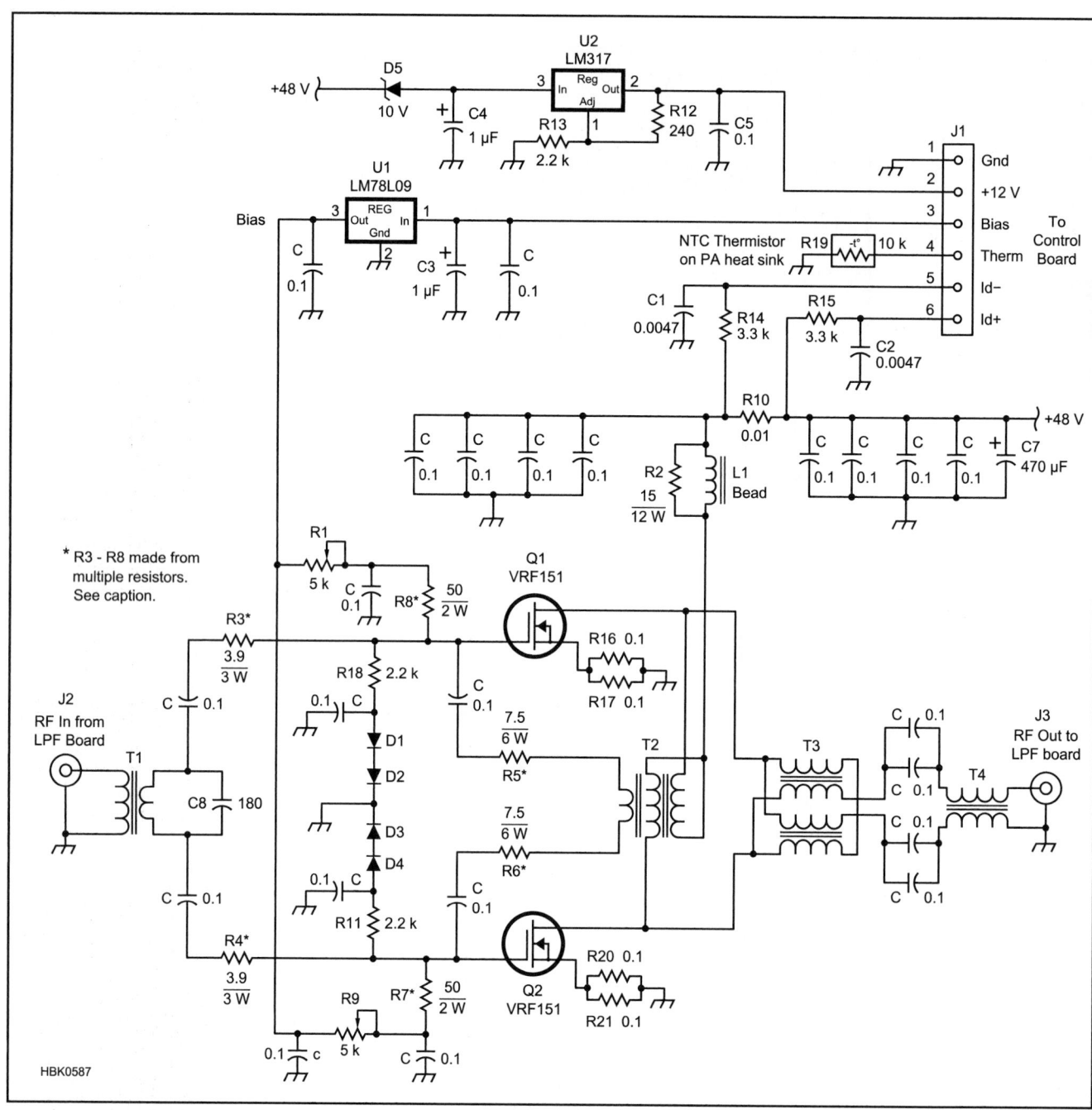

Fig 17.46 — Schematic diagram of the 250 W amplifier PA module. A complete parts list may be found on the *Handbook* CD-ROM.

out of the RF path, and lighting an error LED. There is also an ALC level detector that generates a negative-going feedback voltage for the driver when the RF drive goes above the level corresponding to maximum power. If this PA were part of a transceiver, the several faults described above would generate inputs into the ALC system and turn back the drive rather than taking it off the air. We do not always have that luxury so the best course is to take it off line until the cause can be fixed.

PERFORMANCE

The maximum for the PA design itself is

300 W. Increasing its output past 300 W to make up for filter loss quickly degrades the IMD performance. It needs some headroom. So, in very un-amateur fashion, this amplifier is conservatively rated at 250 W output. This provides a clean signal and plenty of margin for wrong antenna selection, disconnected feed lines, and all the other things that can kill amplifiers that are run too close to their limit.

The PA will provide 250 W PEP for sideband or PSK and 250 W CW. The design goal for this PA was to make it reliable and at least as good as any competitive transceiver. The

harmonics are –60 dB on HF and –70 dB on 6 meters. Transmit IMD is >38 dB down from either tone as shown in **Fig 17.47**.

Parasitics are not usually a problem in broadband amplifiers because of the feedback used. The prototype was tested into a 3:1 SWR load at all phase angles without breaking into parasitic oscillation anywhere.

17.11.2 *Project:* **All-Mode, 2 Meter, 80 W Linear Amp**

This solid state amplifier by James Klitzing, W6PQL, is designed for the many

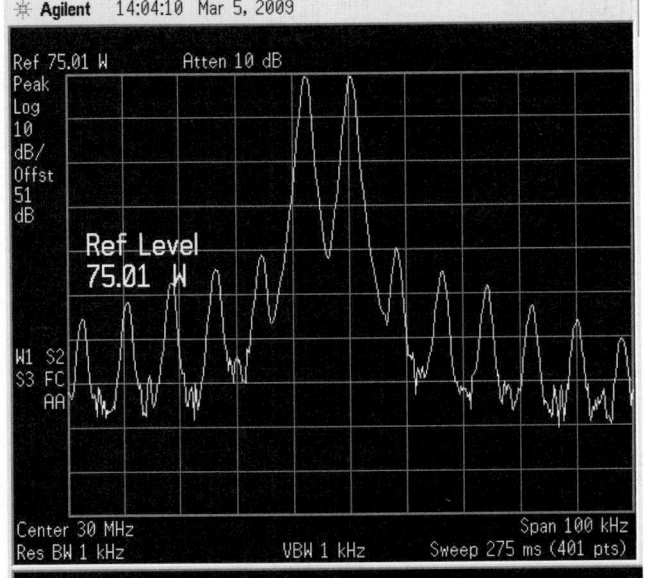

Fig 17.47 — Output of the 250-W amplifier during two-tone IMD testing. All IMD products are better than 38 dB down from either tone.

low-power 2 meter rigs, ranging from hand-held transceivers for FM to older multimode transceivers and even the newer all purpose types such as the Yaesu FT-817 or the Elecraft 2 meter transverters.

The project was featured in the May 2013 issue of *QST* and that article is available on the CD-ROM that accompanies this *Handbook*. The article explains more about the amplifier design and supplies additional construction details. Artwork for the PC board is provided on the *QST* in Depth web page (**www.arrl.org/qst-in-depth**), along with fabrication drawings for sheet metal parts.

The amplifier shown in **Fig 17.48** is low in cost and simple (no preamp or power meters), yet capable of fixed station or mobile operation in any mode and operation from the usual nominal 12 V dc power supply. The same supply that powers a 100 W HF transceiver can likely power the amplifier. The amplifier includes:

• An output low-pass filter to comply with FCC regulations for harmonic and spurious suppression.

• A low-loss antenna relay.

• An RF-sensing TR switch for remote operation, as well as a hard key option.

• TR sequencing to protect the S-AV36 module and prevent hot switching of the antenna relay.

• Indicator LEDs and control switches.

• Reverse polarity protection.

The inside construction is shown in **Fig 17.49**. The small PC board can be made at home (see the **Construction Techniques** chapter) or ordered from the author (see **www.arrl.org/qst-in-depth**). The schematic is provided in **Fig 17.50** and a complete parts list is included in the full article on the CD-ROM. The input and output power of the amplifier with the built in attenuator for a 10 W exciter is provided in **Table 17.6**. This also shows the current required at 13.8 V dc.

Fig 17.48 — The 80 W 2 meter linear amplifier is suitable for operation with any mode.

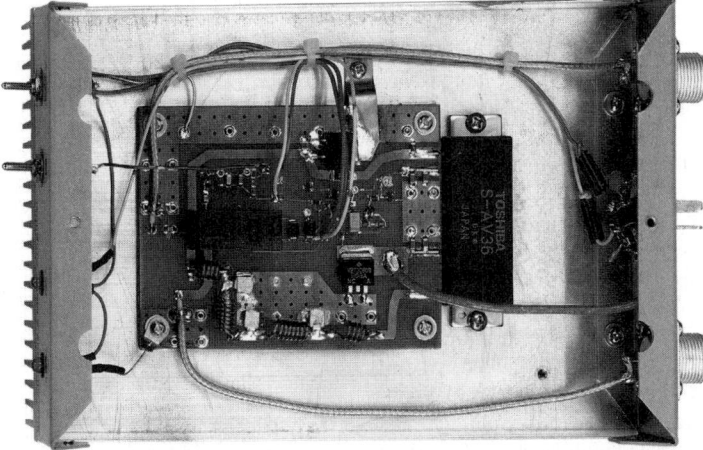

Fig 17.49 — Inside of the compact amplifier showing the simplicity of the final design. The amplifier module is the black rectangle connected to the right edge of the PC board.

Table 17.6
2 Meter Amplifier Operating Parameters
Output power and current required with resting current at 8 A

Drive Power (W)	Output Power (W)	Current (A) at 13.5 V
1	12	8.2
2	29	9.0
3	44	9.5
4	53	10.0
5	66	11.0
6	74	11.5
7	80	12.0
8	85	12.5
9	89	12.8
10	92	13.0

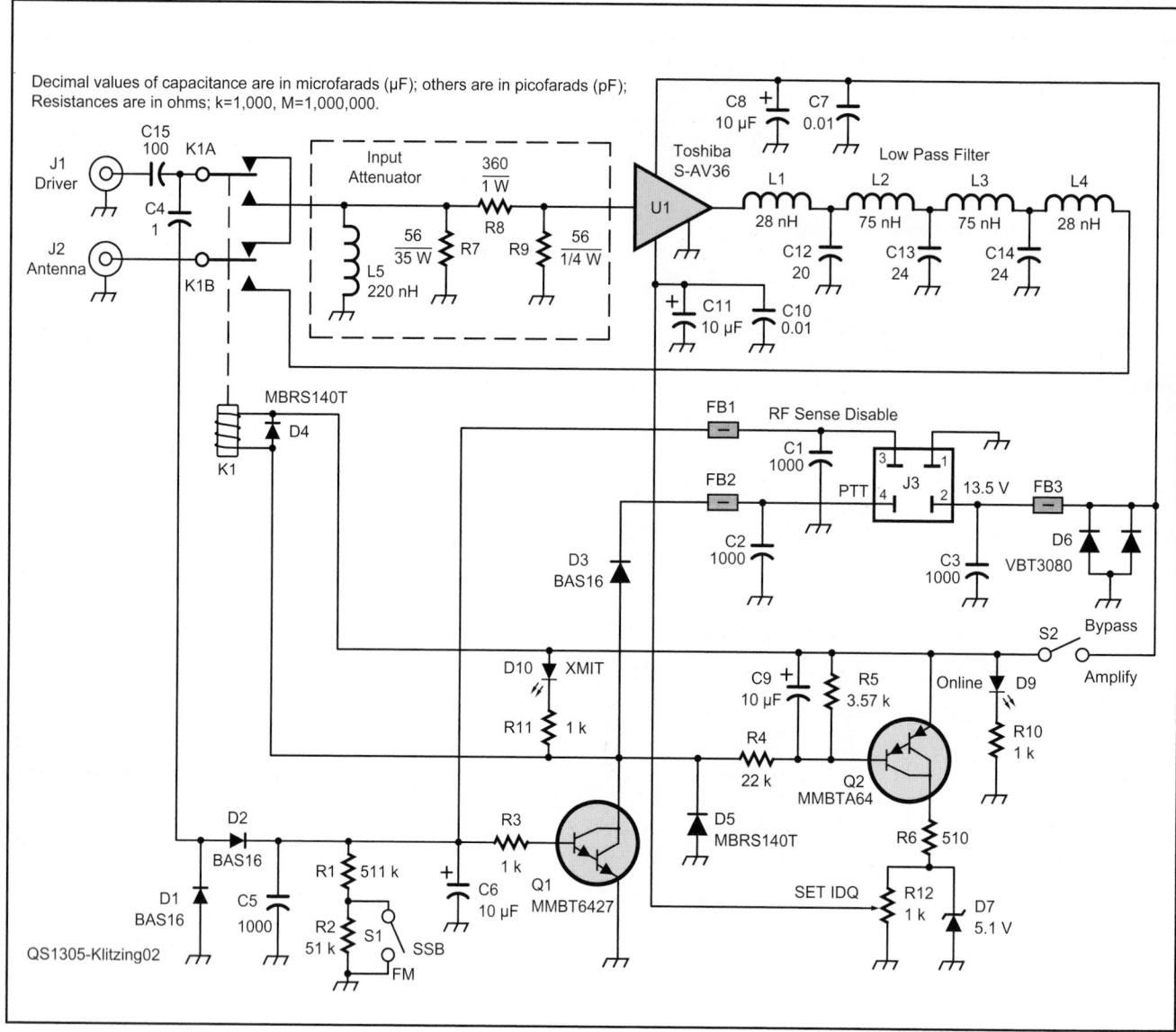

Fig 17.50 — Schematic of the 80 W, 2 meter linear amplifier. A complete parts list is included in the full article on the *Handbook* CD-ROM.

The amplifier is designed around the Toshiba S-AV36 module. The module provides 50 Ω input and output impedances and supplies enough gain that less than 50 mW can drive it to full output in any mode. This design will work with any exciter providing 1 to 10 W of drive, through the use of a built-in attenuator (R7, R8, and R9 in the schematic — see the full article for a table of resistor values that create different levels of attenuation.)

The low-pass filter (L1-L4, C12-C14) is a standard pi network, seven-pole Chebyshev filter. The design provides the required additional harmonic suppression to meet the FCC requirements with very little insertion loss at the operating frequency.

A PCB-mount DPDT general purpose relay was chosen for TR and bypass switch-

ing. The contacts are rated at 8 A. At 2 meters, a bit of reactance is introduced by this part, but compensated for by a small capacitor (C15) in series with its input.

The amplifier can be switched ON and OFF by using a control line back to the driving radio (PTT) or an RF sensing circuit is includes that samples the drive from the input connector to provide the transmit trigger.

TR switching is sequenced to prevent the module from attempting to transmit until the relay contacts have switched and settled completely. Hot-switching damages the relay contacts and stressed the amplifier module. The circuitry associated with the base of Q1 does the sampling and controls the sequence timing. Less than ½ W of drive will trigger TR operation.

17.11.3 *Project: 10 GHz 2 W Amplifier*

Generating RF power above a few milliwatts in the 10 GHz band used to be very difficult. Thankfully, Hittite Microwave Corp (**www.hittite.com**) has the HMC487 (the version currently available is the HMC487LP5), which is an easy-to-use X-band amplifier chip that requires no complicated external RF circuitry or special voltage biasing. The HMC487 costs around $60 in single quantities and the evaluation board — which is highly recommended — is a couple hundred dollars. This amplifier project by Steve Lampereur, KB9MWR, is based on the HMC487 evaluation board to help make construction of the final amplifier very easy, even

Fig 17.51 — Power supply board that generates the +7 V dc operating and -0.3 V dc bias voltages for the amplifier evaluation board.

for a beginner microwave experimenter.

The complete article is available on the CD-ROM accompanying this book, including construction details and adjustment instructions. The HMC487 data sheet is available from the manufacturer's website.

The HMC487 has around 20 dB of gain from 9 to 12 GHz and is internally matched to 50 Ω on both the RF input and output. It will easily generate 1 W (+30 dBm) of RF output with a 10 mW (+10 dBm) RF input over most of the X-band. It saturates at around 2 W (+33 dBm). The only real drawback to the HMC487 is the large amount of heat it needs to dissipate. Its RF efficiency is only around 20% and the rest of this energy will need to be dissipated as heat.

There are two components to the project — the power supply shown in **Fig 17.51** and the evaluation board. The power supply must generate +7 V at high current to supply the amplifier and –0.3 V for biasing the amplifier gate.

The evaluation board is shown in **Fig 17.52**, where it is mounted on an aluminum plate that acts as a heat sink. A 10 GHz isolator is added to the output to prevent any power reflected from the load from reaching the amplifier IC. Short cables with SMA connectors are used to connect the evaluation board to the chassis-mounted receptacles.

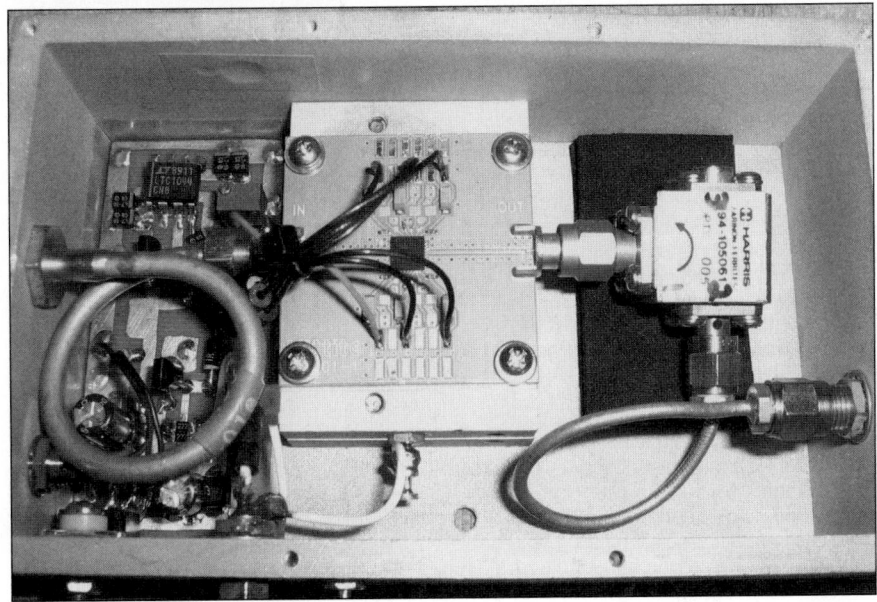

Fig 17.52 — Assembled 10 GHz amplifier. The power supply board is at the left of the enclosure. The evaluation board at center is mounted on an aluminum block to dissipate heat. An isolator at right protects the amplifier output from high SWR.

17.12 Tube Amplifier Projects

Vacuum tube amplifier projects involve several aspects that the builder, particularly beginning amplifier builders, should be aware of and respect. First and foremost are the high voltages associated with these projects — even at the low end of the output power scale. Read the section on High Voltage Techniques in the **Power Sources** chapter and the cautions on measuring high voltages in the **Test Equipment and Measurements** chapter — comply with those recommendations and cautions. Make sure your antenna system and test equipment are suitable for the power and voltages level you will encounter. Consider RF exposure, particularly at 10 meters and higher frequencies at which the Maximum

Permissible Exposure limits can be easily reached with a high-power amplifier. (See the **Safety** chapter for more information on RF exposure.) Respect the capabilities of high-power RF!

This section contains overviews of three tube amplifier projects.

• The "Everyham's Amp" is a simple, entry level amplifier design that can develop an output of several hundred watts. The amplifier can be built around several combinations of inexpensive tubes currently available new or surplus. It is intended for the first-time amp builder or for someone who needs a utility amplifier, perhaps for a single band.

• The 3CX1500D7 amplifier is a sophis-

ticated legal-limit design competitive with top-of-the-line commercial models. This model is intended for intermediate builders who have experience with basic metalworking skills and high-power amplifiers.

• A 6 meter kilowatt amplifier design is also suitable for intermediate builders and uses a single 4CX1600 ceramic tube to reach power outputs of about 1 kW, depending on drive level.

Complete design and construction information for all of these amplifier designs is included on the CD-ROM accompanying this book. In addition, several other amplifier designs are presented for both HF and VHF/UHF operation.

Using "Surplus" Parts for Your Amplifier

First-time builders of power amplifiers soon discover that buying all new electronic components from a parts list is costly. Manufacturers realize significant savings by purchasing components in bulk, but the builder of a single unit will pay top dollar for each new part thus negating any savings in the labor costs. While some impressive equipment has been built using all new parts, pride of design and workmanship are usually the goals, not cost savings.

The way to break this economic barrier is to use surplus parts. These are items left over when a project is finished, a design changes, or a war ends. Sometimes, costly parts can be recovered from relatively new but obsolete equipment. Alert dealers locate sources of surplus parts, buy them at auction, often for pennies on the dollar, and make them available in stores, on-line or at hamfests.

Parts can become available when a project is abandoned for which parts have been gathered. Many hams maintain a "junk box" of parts against future needs, only to find it has grown beyond a manageable size. This may contain a mixture of new, used and NOS parts (new old stock, meaning unused but stored for many years).

When buying electronic parts from other than a trusted supplier of new stock, some precautions are in order. Before shopping the surplus shelves or making an on-line purchase, do some research. You are leaving behind the security of buying a specific part that the designer of the amplifier has tested in actual operation. One has to be sure not only that the part is sound, but that it is suited for the intended purpose. This effort is justified by the anticipated savings of up to 80%. Using "odd" parts may make your amplifier a "one of a kind" project, since the next builder won't be able to obtain the same items. Your amplifier may not be reproducible, but it can still be an object of pride and usefulness.

Surplus vacuum tubes may be of military origin. In the 50s and 60s WWII

Fig 17.A2 — A hamfest is often a good source of used, surplus, and even new parts for building amplifiers and other high-power RF equipment, such as impedance matching units or full-power filters. Follow the cautions in this sidebar when evaluating parts and pieces.

tubes powered many ham amplifiers. Tubes made in Russia in past decades are now available.[1] Many hams have used these tubes for new designs and

even as replacements for hard-to-find or more expensive tube types in existing amplifiers. Since tubes have a limited life, one may wish to buy a spare or two

---

17.12.1 *Project:* The Everyham's Amplifier

Recent editions of the *Handbook* have featured amplifiers that are true works of art, nearly as capable and as elegant as any on the market. Amplifiers such as the HF legal limit amplifier design by Jerry Pittenger, K8RA, in this and recent editions clearly show what advanced amateurs can do.

Many hams who would like to build a basic linear amplifier for their station may not feel ready for advanced amplifier designs — either technically or financially. Contributed by John Stanley, K4ERO, the amplifier project in **Fig 17.53** is a very basic design that will satisfy the need for more power and encourage the reader to experience amplifier building, while providing only the bands and features desired. Once a simple amplifier has been constructed successfully, more advanced designs will be easier to tackle.

The following is an overview and summary of the amplifier's design. A complete description of this amplifier, including additional photographs and drawings, construction details, and additional features is available on the CD-ROM that accompanies this book. An additional three-band design by Leigh Tregellas, VK5TR, is also included on the CD-ROM.

AMPLIFIER CONFIGURATION

The design presented here is based on a modular approach. Each of the three major

before committing to using a surplus type, since future availability is always a concern. Broadcast or medical "pulls" or "pull-outs" provide a source of good used tubes. For highest equipment reliability, these tubes are replaced on a scheduled basis rather than at failure. They will have less life than a new tube, but can be so cheap that they are still a good deal.

When buying used tubes at a hamfest, a quick ohmmeter check is important. Between the filament pins there should be only a few ohms, with high resistance between all other connections. Signs of obvious overheating, such as discoloration or warping, are a concern and call at the least for a reduced price. There is no 100% guarantee that a tube is good apart from trying it in a circuit, since tube testers usually will not test large power tubes. Experience with on-line suppliers of NOS tubes indicates that a small percentage will be duds. If your supplier guarantees them you should be willing to pay a bit more. Some dealers will test or even match the tubes for parallel operation.

Finding sockets for odd tubes may be a challenge. Several enterprising hams supply sockets for popular types, such as the Russian GI7B, or you can sometimes bolt the tubes into place with straps and screws.[2]

Using vacuum capacitors really adds class and value to an amplifier. Check the vacuum by setting the capacitor to minimum capacitance and watching to see if the vacuum is sufficiently strong to pull the plates all the way in when the screw or hand pressure is released. Weak vacuum means the capacitor will not withstand its rated voltage. With the plates fully meshed, check for a short circuit, which indicates damaged plates. Be sure that the adjusting screw moves freely. Lubrication won't always free up a sticky screw. Check that the voltage and current ratings are suitable for your application. The voltage rating is usually printed on the capacitor. The current rating will be on a factory data sheet or use physical size as an indication.

Vacuum TR (transmit-receive) relays work well in an amplifier and often switch quickly enough to be used for full-QSK (full break-in) operation. If you shop carefully, a surplus relay can cost little more than a much-inferior open-frame type. Visual inspection of the contacts and an ohmmeter check on the coil will generally insure that the relay is good. If the relay's coil voltage is 24 V or some value other than what you wanted, remember that a small power supply can be added cheaply. Be aware that the amplifier keying circuits of most transceivers are not rated to switch the large coil currents of surplus relays, particularly those that operate at 24 V or higher. An outboard switching interface will be needed for these relays. Older non-vacuum relays also have long switching times that, if not accounted for in the amplifier design, may cause "hot-switching" and contact damage.

Inductors can usually be evaluated by inspection since hidden faults are rare. The current rating is determined by conductor size. With formulas found in the **Electrical Fundamentals** chapter, you can determine the inductance. Carry a hand calculator when shopping.

Choose RF switches based on physical size and inspection. Avoid badly arced-over or pitted contacts or insulators.

For power transformers, finding the exact required voltage may be difficult. You can use extra low-voltage windings to buck or boost the primary voltage. External transformers can also be used in an autotransformer connection for adjustment. Use of a variable autotransformer (Variac or similar) can adapt a transformer to your needs, but the extra winding and core losses may cause voltage sag under load. Amplifiers normally use a separate transformer for supplying the tube filaments as filaments are often turned on before the high-voltage supply. Therefore, filament windings on a plate transformer are not needed. Also, with either center-tapped, full bridge or voltage doubler circuits, you have three options for ac secondary voltage.

Transformers should be rated for 50 or 60 Hz operation, not 400 Hz as used in aircraft. Check that the current and overall VA (volt-amp) ratings are adequate. Use an ohmmeter to detect shorts or open windings, and to be sure that you understand the terminal connections. Check each winding to ground and to each other. Reject transformers with charred insulation. Use your nose to detect burned smells from internal overloads. A rusty core can be wire brushed and spray painted, but might indicate moisture has gotten to the windings. Test the primary with 120 or 240 V ac, if possible. High current or loud buzzing can indicate shorted turns. Be very careful when measuring the secondary windings as the high voltage can easily destroy your multimeter, not to mention endanger your life! (See the **Power Sources** chapter for a method of testing transformers safely.)

Blowers can be found on otherwise worthless equipment, such as a rack of old tube gear. It may be worth buying the whole rack just to get a nice blower. The pressure and volume ratings of a blower may not be known, but apart from noise, there is no disadvantage in over-sizing your blower. An over-sized blower can be slowed down by reducing the voltage or throttled back with baffles. Remember that if the voltage rating is strange, providing an odd voltage may be cheaper than buying a more expensive blower. Avoid 400 Hz aircraft blowers. For blowers needing an external capacitor, try to obtain the capacitor with the blower.

Racks and chassis components can often be refurbished for an amplifier. Standard 19-inch rack-type cabinets are widely available surplus and at hamfests. Even new, they are available at modest cost and have good fit and finish. Filling the many panel holes and painting or adding a new front panel can make them look like new, while saving a lot of expensive and time consuming metal work.[3] — *Roger Halstead, K8RI*

[1]www.nd2x.net/base-1.html
[2]http://gi7b.com
[3]W. Yoshida, KH6WZ, "Recycling Old Cabinets and Chassis Boxes," *QST*, Jul 2008, p 30.

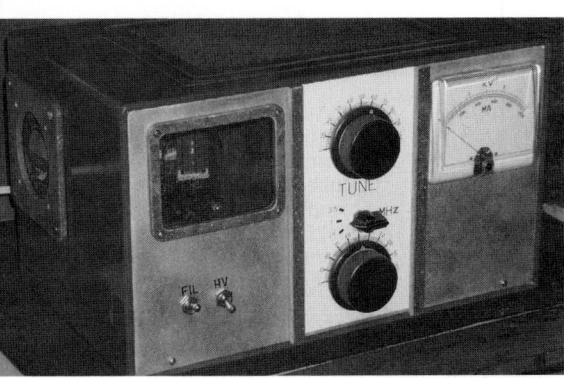

Fig 17.53 — The front panel of the Everyham's Amplifier. This version is designed to cover three bands (80, 40, and 20 meters). A window is included at left to allow viewing of the tube.

sections (tubes, tuning network, power supply) is somewhat independent and can be changed out separately. In addition, the starting design is "bare bones," containing nothing that is not absolutely essential for the amplifier to work and provide a minimum level of safety against overloads, abuse, or accidents. TR switching is included in the most basic design because it gives significant advantages at low cost. For the basic design, each part will be described and its purpose explained.

Several options for the major components are presented with the builder being able to choose what best meets his or her needs and best uses the components available in junk boxes, online, or at hamfests. Many small parts such as resistors and diodes are cheap enough that one need not buy them used. Control transformers and tubes are easy to find through online auction sites. Be sure to check shipping costs on heavy items such as transformers.

By shopping carefully, you can avoid the budget being busted by that one essential component costing lots more than expected. Don't overlook acquiring a damaged commercial amplifier or a "basket case" homebrew project — the parts and enclosure hardware available from these are often worth many times the asking price! Refer to the sidebar in this chapter, "Using 'Surplus' Parts for Your Amplifier" for more information about purchasing used and surplus amplifier parts.

Additions and modifications that will improve performance are described in the supplemental article on the CD-ROM and can be added after the initial amplifier is built and tested. Most can be installed in a day, so you need not discontinue using the amp in order to upgrade.

DESIGN OVERVIEW

The basic circuit is shown in **Fig 17.54**. The tubes used in this design are triodes, connected in a grounded-grid configuration. The amplifier can be constructed to use a single 3-500Z or a pair of 811A, 572B, GI-6B or GI-7B tubes. All of these are currently available new or surplus at reasonable cost. The input grid circuit input is driven directly. Tuned input circuits are discussed later as an option.

C3, L1 and C4 form an impedance-matching pi network that transforms the 50 Ω load at the output to the several thousand ohm load that the tube requires. In addition, this "tank" filters out the harmonics from the tube while passing the desired fundamental signal. An optional band switch, S3, is shown. In its simplest form it shorts out sections of L1 as the operating frequency increases. A single-band design doesn't need S3.

When the amplifier is OFF or the transceiver is in receive, the antenna connects directly to the transceiver antenna jack via the two relays as shown in **Fig 17.55**. (K2 and K3 in the main schematic.) When the transceiver is in transmit *and* the amplifier turned on, the transceiver output is routed through the amplifier and amplified.

As part of the TR process, we will want to bias the amplifier off during receive. This saves energy, cools the tube(s) and also removes any RF noise the tube(s) might produce that could get into the receiver. When opened during receive, K1 in Fig 17.54 lifts the center tap of the filament transformer (tube cathode) from ground. K1 can be a separate relay or an additional set of contacts on the TR relays K2 and K3 if they are internal to the amplifier. Usually a resistor (R11), perhaps 47 kΩ or so, is left between the center tap and

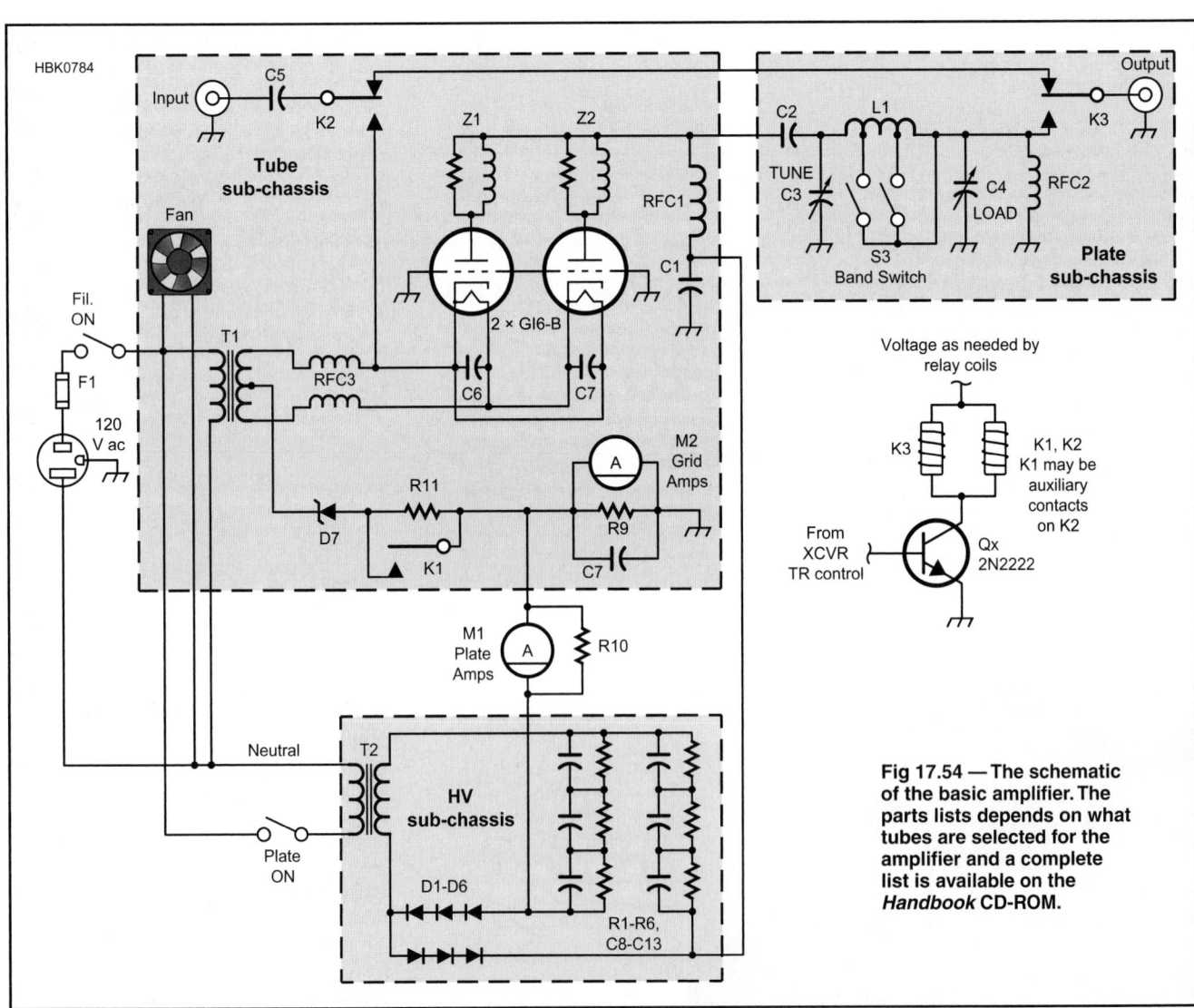

Fig 17.54 — The schematic of the basic amplifier. The parts lists depends on what tubes are selected for the amplifier and a complete list is available on the *Handbook* CD-ROM.

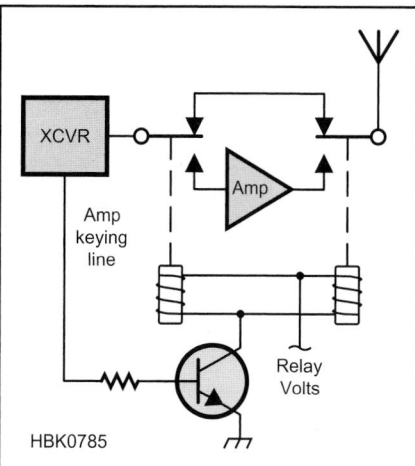

Fig 17.55 — The TR switching of transmit and receive signals for the basic amplifier is controlled by the transceiver's amplifier keying output signal, assumed here to be a positive voltage to key the amplifier with a suitable current-limiting resistor and relay drive transistor installed in the amplifier.

ground to provide a dc current path.

The power supply shown is a basic voltage doubler. (See the **Power Sources** chapter for information on rectifier and voltage multiplier circuits, as well as an alternative power supply design.) If a high-voltage secondary transformer is available, a rating of 500 VA is required for intermittent service up to 1 kW output. A 250 VA rating is sufficient for intermittent service to 500 W with a voltage quadrupler. At one time, TV transformers were commonly used for amplifiers of this size. However, they are becoming hard to find and their age leads to insulation breakdown, thus they are not recommended for this project.

For T2, the main power transformer, the circuit uses two 480/240 to 120 V "control" transformers with the two 480 V windings in series as a secondary. Control transformers are used in industry to reduce the plant's 480 or 240 V equipment wiring to 120 V for the purpose of operating instrumentation and control electronics. These typically have two 240 V windings that can be connected in series for 480 V or run in parallel for higher current. These transformers are commonly available at online auction sites.

The rectifier diodes D1-D6 are rated at 3 A and 1 kV PIV. The filter capacitors C8-C13 are 100 μF, 450 V electrolytics. Modern rectifiers do not need equalizing resistors and capacitors as seen in older articles and designs. Today, it is cheaper to overrate the diode stack instead. C8-C13 do need equalizing resistors, which also serve as bleeder resistors to discharge the capacitors when the supply is turned off.

A meter in the negative lead of the supply monitors the plate current. Several inexpensive options are included in the parts list. Some way to indicate RF output voltage or current is needed, but if you have an external SWR or forward/reflected power meter already, it need not be included in the amp.

CONSTRUCTION NOTES

The type of construction chosen will depend on the enclosure (chassis) which is required for good RF performance and for safety reasons. Having a spacious enclosure and an oversized chassis will make construction easier and upgrades simple. If your situation dictates, you can certainly build everything on a single chassis.

The power supply should be included in the enclosure and not built as a separate piece of equipment. Constructing safe high voltage connections between pieces of equipment requires experience and strict attention to safety-related details that may not be obvious to the beginning amplifier builder. For that reason, this basic design assumes an internal power supply. Rectifiers and filter caps can be mounted on an insulated board (plexiglass or other plastic), or an etched PC board. Install that board on top of the transformer for a neat and compact installation. Board layouts are included on the CD-ROM for both the doubler and quadrupler circuits.

If you find an old piece of equipment you can reclaim, that may dictate the layout. Hamfests usually provide a choice of obsolete tube-type lab or medical instrumentation with high quality cabinets at low cost. They may include desirable items such as a blower or meter in addition to the enclosure. The author chose to use a cabinet from an old piece of Heathkit gear, which had no chassis, and so built a front panel and several sub-chassis sections for the various parts of the amplifier. This allows changing out various parts of the circuit easily and even to maintain several different sub-sections to allow experiments with different tubes, frequency ranges or power supply types.

The tube circuit layout will depend on which tube type is chosen. The 811A/572B option is a great place to start. There is just no other option that can compete in price, counting the tubes and sockets. A pair is recommended for this design although up to four tubes in parallel are common in ham-built amplifiers. A single 3-500Z is another popular choice and, for higher power, a pair of them if the power supply is adequately rated. Ceramic sockets are available for the 3-500Z. Surplus Russian triodes such as the GI-6B and GI-7B are becoming popular and one or a pair is commonly used. The sockets will have to be built as described in the construction details or purchased from ham sources.

External anode tubes such as the Russian GI-6B or GI-7B will need a blower. With the 811A or 3-500Z options, a muffin fan blowing on the tubes will allow them to work much harder. This fan also keeps the other components cooler for extended operation.

Finding suitable capacitors and coils and the band switch will be a significant part of the total procurement process. This is where you can save a lot by using surplus parts. The parts list will specify certain types, mainly so you will know what to look for, but if you simply buy new parts from the list, you will spend your entire budget on these items alone. You don't need exact values for the circuit to work. Limiting the bands to be covered can greatly relax the coil, capacitor and switch requirements.

17.12.2 *Project:* 3CX1500D7 RF Linear Amplifier

This project is a 160 through 10 meter RF linear amplifier that uses the compact Eimac

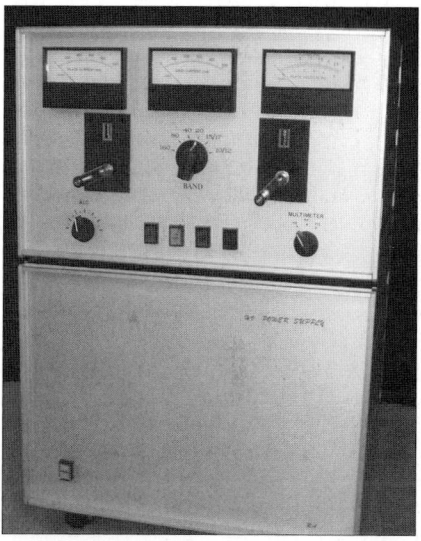

Fig 17.56 — At the top, front panel view of RF Deck and Power Supply for 3CX1500D7 amplifier. At bottom, rear view of RF Deck and Power Supply.

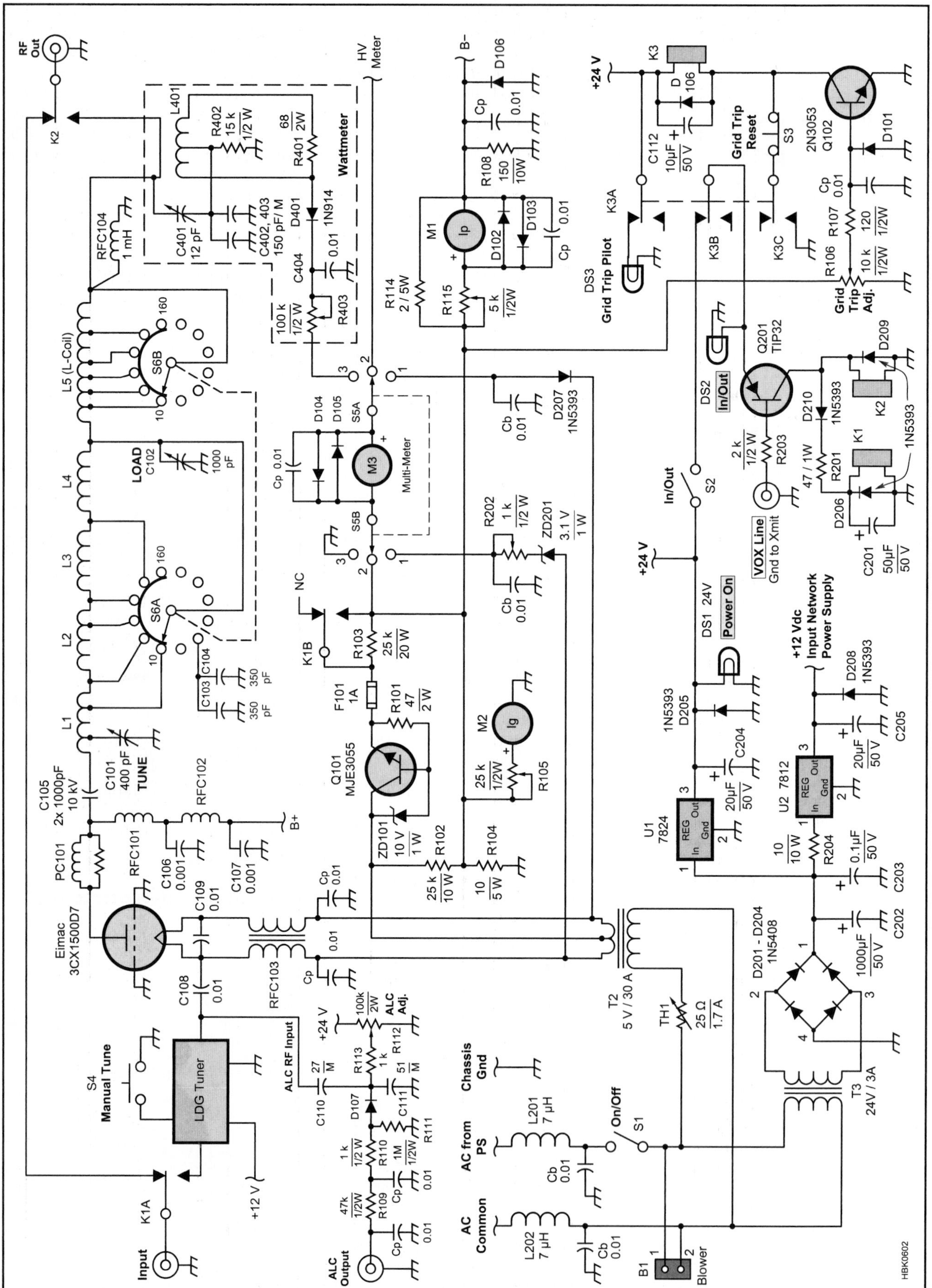

Fig 17.57 — Schematic of the RF deck and control circuitry. A complete parts list is available on the *Handbook* CD-ROM.

3CX1500D7 metal ceramic triode. It was designed and constructed by Jerry Pittenger, K8RA. The complete description of this amplifier including construction and operating details is provided on the CD-ROM accompanying this book. The 3CX1200D7 tube may be more affordable, and the design can be adapted to use it with somewhat less plate dissipation.

The amplifier features instant-on operation and provides a solid 1500 W RF output with less than 100 W drive. Specifications for this rugged tube include 1500 W anode dissipation, 50 W grid dissipation and plate voltages up to 6000 V. A matching 4000 V power supply is included. The amplifier can be easily duplicated and provides full output in key-down service with no time constraints in any mode. **Fig 17.56** shows the RF deck and power supply cabinets.

DESIGN OVERVIEW

The Eimac 3CX1500D7 was designed as a compact, but heavy duty, alternative to the popular lineup of a pair of 3-500Z tubes. It has a 5 V/30 A filament and a maximum plate dissipation of 1500 W, compared to the 1000 W dissipation for a pair of 3-500Zs. The 3CX1500D7 uses the popular Eimac SK410 socket and requires forced air through the anode for cooling. The amplifier uses a conventional grounded-grid design with an adjustable grid-trip protection circuit. See the RF Deck schematic in **Fig 17.57**.

Output impedance matching is accomplished using a pi-L tank circuit for good harmonic suppression. The 10 to 40 meter coils are hand wound from copper tubing, and they are silver plated for efficiency. Toroids are used for the 80 and 160 meter coils for compactness. The amplifier incorporates a heavy-duty shorting-type bandswitch. Vacuum variable capacitors are used for pi-L tuning and loading.

A unique feature of this amplifier is the use of a commercial computer-controlled impedance-matching module at the input. This greatly simplifies the amplifier design by eliminating the need for complex ganged switches and sometimes frustrating setup adjustments. The AT-100AMP module kit available from W4RT Electronics (**www.w4rt.com**) is an acceptable tuning unit.

An adjustable ALC circuit is also included to control excess drive power. The amplifier metering circuits allow simultaneous monitoring of plate current, grid current, and a choice of RF output, plate voltage or filament voltage.

The blower was sized to allow full 1500 W plate dissipation (65 cfm at 0.45 inches H_2O hydrostatic backpressure). The design provides for blower mounting on the rear of the RF deck or optionally in a remote location to reduce ambient blower noise in the shack. The flange on the socket for connecting an air hose was ground off for better air flow. (This is not necessary.).

The power supply is built in a separate

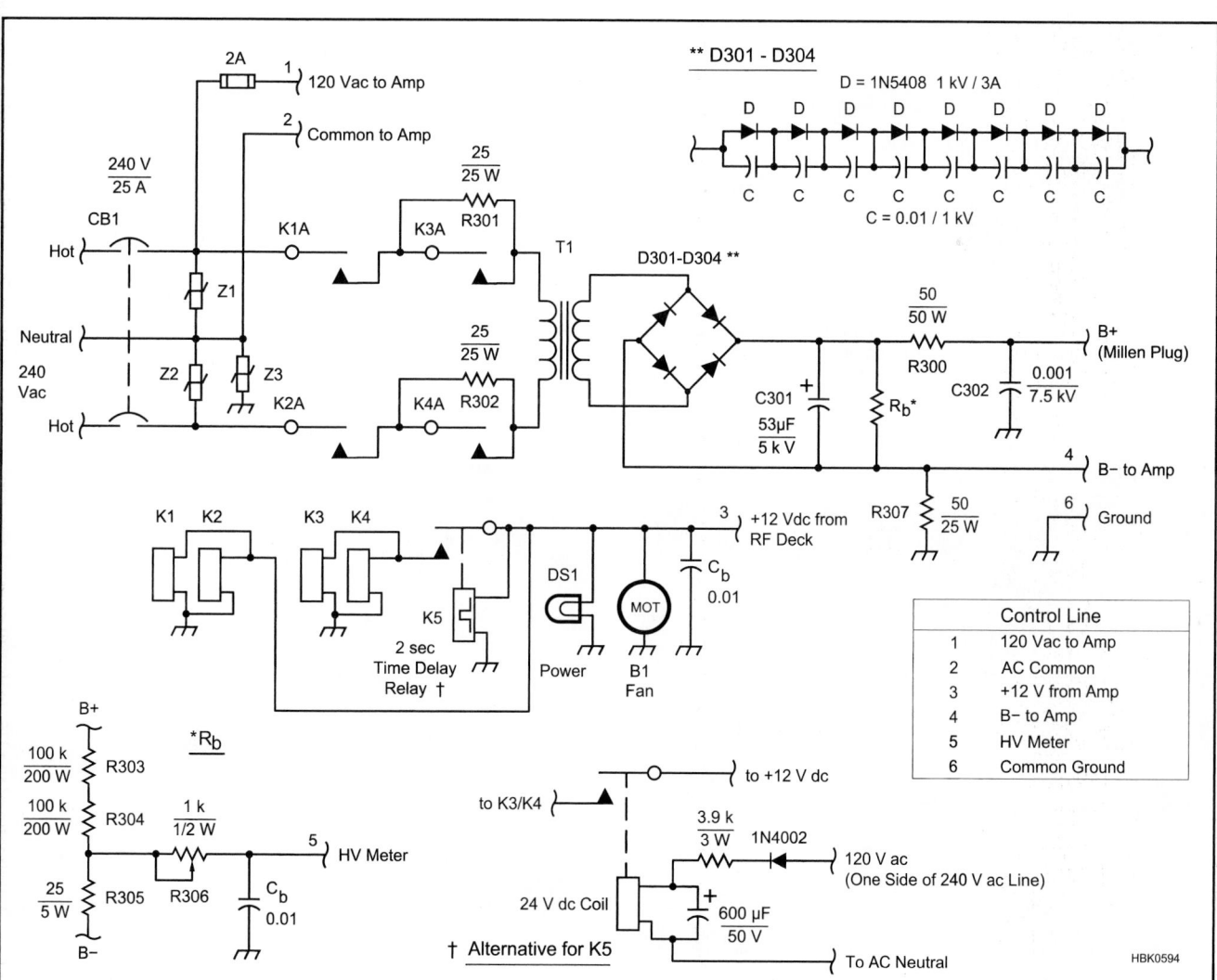

Fig 17.58 — Schematic for power supply for 3CX1500D7 amplifier. A complete parts list is available on the *Handbook* CD-ROM.

cabinet with casters and is connected to the RF deck using a 6-conductor control cable, with a separate high voltage (HV) cable. The power transformer has multiple primary taps (220/230/240 V ac) and multiple secondary taps (2300/2700/3100 V ac). No-load HV ranges can be selected from 3200 to 4600 V dc using different primary-secondary combinations. The amplifier is designed to run at 4000 V dc under load to maintain a reasonable plate resistance and component size. A step-start circuit is included to protect against current surge at turn on that can damage the diode bank. The power supply schematic is shown in **Fig 17.58** and a photo of the inside of the power supply is shown in **Fig 17.59**.

Since this project was first published, Peter W. Dahl has discontinued business and transformers built to the Dahl specifications are now available from Hammond Manufacturing Company, Inc. of Cheektowaga, NY (**www. hammfg.com**). Contact Hammond to cross-reference the Peter Dahl part numbers in the parts list for T1, T2 and RFC103 with current stock or equivalent designs.

Both +12 V and +24 V regulated power supplies are included in the power supply. The +12 V is required for the computer-controlled input network and +24 V is needed for the output vacuum relay. The input and output relays are time sequenced to avoid amplifier drive without a 50 Ω load. Relay

actuation from the exciter uses a low-voltage/ low-current circuit to accommodate the amplifier switching constraints imposed by many new solid state radios.

GENERAL CONSTRUCTION NOTES

Much thought was put into the physical appearance of the amplifier. The goal was to obtain a unit that looks commercial and that would look good sitting on the operating table. To accomplish the desired look, commercial cabinets were used. Not only does this help obtain a professional look but it eliminates a large amount of the metal work required in construction. Careful attention was taken making custom meter scales and cabinet labeling. The results are evident in the pictures provided.

The amplifier was constructed using basic shop tools and does not require access to a sophisticated metal shop or electronics test bench. Basic tools included a band saw, a jig saw capable of cutting thin aluminum sheet, a drill press and common hand tools. Some skill in using tools is needed to obtain good results and ensure safety, but most people can accomplish this project with careful planning and diligence.

The amplifier is built in modules. This breaks the project into logical steps and facilitates testing the circuits along the way. For example, modules include the HV power supply, LV power supply, input network, control circuits, tank circuit and wattmeter. Each module can be tested prior to being integrated into the amplifier.

The project also made extensive use of computer tools in the design stage. The basic layout of all major components was done using the *Visio* diagramming software pack-

Fig 17.59 — Inside view of the power supply, showing rectifier stack, control relays and HV filter capacitor with bleeder resistors. The heavy-duty high voltage transformer is at the upper left in this photo.

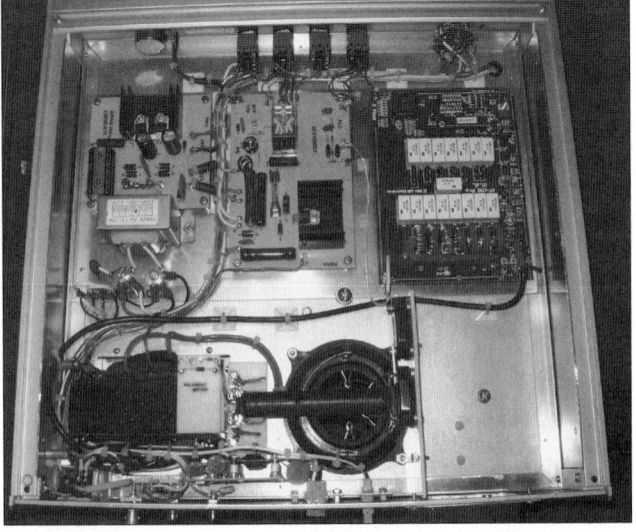

Fig 17.60 — At the left, under the chassis of the RF Deck. The autotuner used as the input network for this amplifier is at the upper right. At right, view of the Pi-L output network in the RF Deck.

age. The printed-circuit boards were designed using a free layout program called *ExpressPCB* (**www.expresspcb. com,** see also the **Computer-Aided Circuit Design** chapter). Masks were developed and the iron-on transfer technique was used to transfer the traces to copper-clad board. The boards were then etched with excellent results. The layout underneath the RF Deck is shown in **Fig 17.60A** and the top side of the RF Deck is shown in Fig 17.60B.

Meter scales were made using an excellent piece of software called *Meter* by Jim Tonne, W4ENE (**www.tonnesoftware.com**). Also, K8RA wrote an *Excel* spread-sheet to calculate the pi-L tank parameters. A copy of the spreadsheet, *Meter Basic* software and *ExpressPCB* files for the PC boards are all included on the CD-ROM that accompanies this book.

17.12.3 *Project:* A 6 Meter Kilowatt Amplifier

This amplifier is based on an original de-

sign by George Daughters, K6GT, "The Sunnyvale/Saint Petersburg Kilowatt-Plus" in 2005 and earlier editions of this book and that article is included on this book's CD-ROM for details on suitable control and power supply circuitry. This 6 meter amplifier uses the same basic design as K6GT's, except for modified input and output circuits in the RF deck. See **Fig 17.61**, a photograph of the front panel of the 6 meter amplifier built by Dick Stevens, W1QWJ (SK).

The Svetlana 4CX1600B attracted a lot of attention because of its potent capabilities and relatively low cost. (The currently available model of the tube is the 4CX1600U.) **Fig 17.62** is a schematic of the RF deck. The control and power supply circuitry are basically the same as that used in the K6GT HF amplifier, except that plate current is monitored with a meter in series with the B– lead, since the cathode in this amplifier is grounded directly

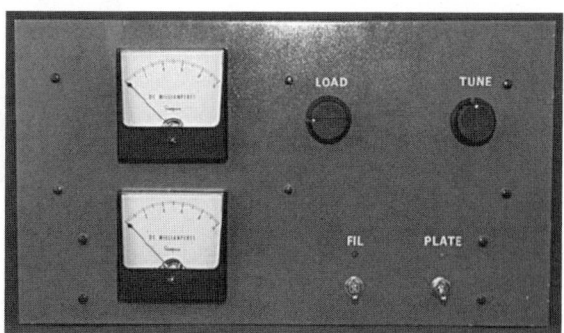

Fig 17.61 — Photo of the front panel of the 6 meter 4CX1600B amplifier.

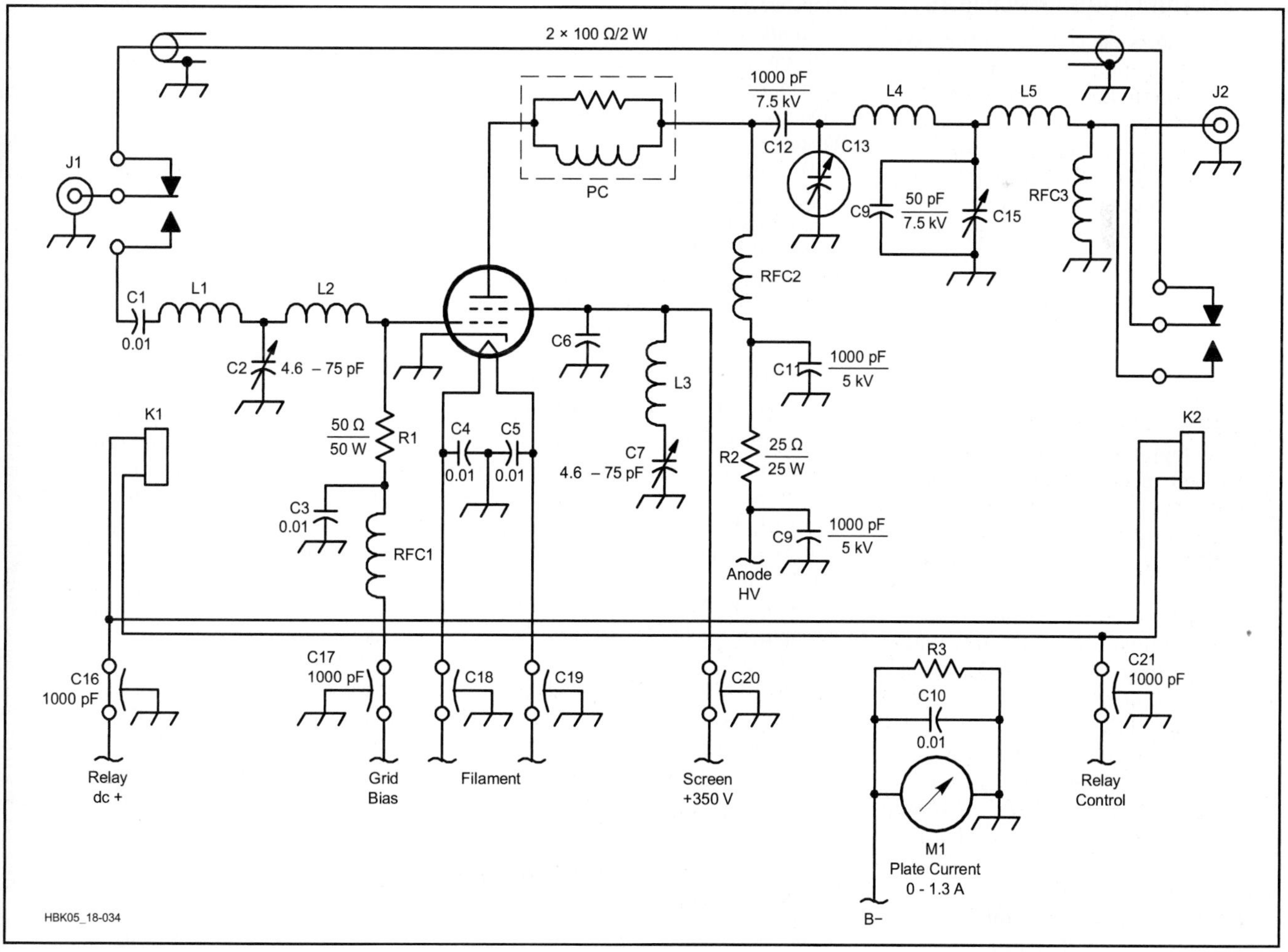

Fig 17.62 — Schematic for the RF deck for the 6 meter 4CX1600B amplifier. The power supply schematic and a complete parts list is available on the *Handbook* CD-ROM.

CONSTRUCTION

Like the K6GT amplifier, this amplifier is constructed in two parts: an RF deck and a power supply. Two aluminum chassis boxes bolted together and mounted to a front panel are used to make the RF deck. **Fig 17.63** shows the 4CX1600B tube and the 6 meter output tank circuit.

Fig 17.63 — Close-up photo of the anode tank circuit for 6 meter kW amplifier. The air-cooling chimney has been removed in this photo.

17.13 References and Bibliography

VACUUM TUBE AMPLIFIERS

ARRL members can download many *QST* articles on amplifiers from: **www.arrl. org/arrl-periodicals-archive-search**. Enter the keyword "amplifiers" in the "Title/Keywords:" search window.

Badger, G., W6TC, "The 811A: Grandfather of the Zero Bias Revolution," *QST*, Apr 1996, pp 51-53.

Care and Feeding of Power Grid Tubes, **www.cpii.com/library.cfm/9**

EIMAC tube performance computer, **www. cpii.com/library.cfm/9**

Grammer, G., "How to Run Your Linear," *QST,* Nov 1962, pp 11-14.

LTSPICE, download from Linear Technology, **www.linear.com/design-tools/software/ltspice.jsp**

Measures, R., AG6K, "Improved Parasitic Suppression for Modern Amplifier Tubes," *QST*, Oct 1988, pp 36-38, 66, 89.

Orr, W., W6SAI, *Radio Handbook*, 22nd Ed (Howard W. Sams & Co, Inc, 1981).

Pappenfus, Bruene and Schoenike, *Single Sideband Principles and Circuits* (McGraw-Hill, 1964).

Potter and Fich, *Theory of Networks and Lines* (Prentice-Hall, 1963).

RCA Transmitting Tubes Technical Manual TT-5, Radio Corporation of America, 1962.

Rauch, T., W8JI, "Neutralizing Amplifiers," **www.w8ji.com/ neutralizing_amplifier.htm**

Reference Data for Radio Engineers, ITT, Howard W. Sams Co, Inc.

Simpson, *Introductory Electronics for Scientists and Engineers* (Allyn and Bacon, 1975).

Terman, *Electronic and Radio Engineering* (McGraw-Hill).

Wilson, R., K4POZ, "All About the GI-7B," **http://gi7b.com**

Wingfield, E., "New and Improved Formulas for the Design of Pi and Pi-L Networks," *QST*, Aug 1983, pp 23-29. (Feedback, *QST*, Jan 1984, p 49.).

Wingfield, E., "A Note on Pi-L Networks," *QEX*, Dec 1983, pp 5-9.

ADDITIONAL VACUUM TUBE AMPLIFIER PROJECTS

Knadle, R., K2RIW, "A Strip-line Kilowatt Amplifier 432 MHz," *QST*, Apr 1972, pp 49-55.

Meade, E., K1AGB, "A High-Performance 50-MHz Amplifier," *QST*, Sep 1975, pp 34-38.

Meade, E., K1AGB, "A 2-KW PEP Amplifier for 144 MHz," *QST*, Dec 1973, pp 34-38.

Peck, F., K6SNO, "A Compact High-Power Linear," *QST*, Jun 1961, pp 11-14.

Additional projects are available on the *Handbook* CD-ROM.

SOLID STATE AMPLIFIERS

References for RF Power Amplifiers and Power Combining

Classic Works in RF Engineering : Combiners, Couplers, Transformers and Magnetic Materials, Edited by Myer,

Walker, Raab, Trask. (Artech House, June 2005. ISBN: 9781580530569).

Abulet, Mihai, *RF Power Amplifiers* (Noble Publishing, 2001. ISBN: 1-884932-12-6).

Blinchikoff, Zverev, *Filtering in the Time and Frequency Domains* (Noble Publishing, 2001, ISBN: 1884932177).

Bowick**,** *RF Circuit Design*, 2nd Ed. (Newnes imprint of Butterworth-Heinemann, 1982. ISBN: 0750685182).

Cripps, Steve C. *Advanced Techniques in RF Power Amplifier Design (*Artech House, 2002. ISBN: 1-58053-282-9).

Dye, Norman. and Helge Granberg, *Radio Frequency Transistors — Principles and Practical Applications* (Butterworth-Heinemann, 1993. ISBN: 0750690593; 2nd Ed, 2001. ISBN: 0750672811).

Egan, William F., *Practical RF System Design,* (Wiley, 2003. ISBN: 0471200239).

Kraus, Bostian and Raab, *Solid State Radio Engineering,* (Wiley, 1980. ISBN: 0-471-03018-x).

Ludwig and Bogdanov, *RF Circuit Design: Theory and Applications* (Prentice Hall, 2009. ISBN: 0-13-147137-6).

Oxner, Edwin, *Power FETS and Their Applications* (Prentice-Hall, 1982. ISBN: 0-13-686923-8).

Pozar, *Microwave Engineering,* 3rd Ed. (Wiley, 2004. ISBN: 0471448788).

Rutledge, David B., *The Electronics of Radio* (Cambridge University Press, August 1999. ISBN: 9780521646451).

Sabin, Schoenike, et al, *Single Sideband Systems and Circuits* (McGraw-Hill, 1993. ISBN: 0-07-912038-5).

Sevick, J., *Transmission Line Transformers,* 4th Ed., (Noble Publishing, 2001. ISBN: 1884932185).

Vizmuller, *RF Design Guide: Systems, Circuits, and Equations* (Artech House, 1995. ISBN: 0890067546).

White, Joseph F., *High Frequency Techniques: An Introduction to RF and Microwave Engineering* (Wiley, 2004. ISBN: 0471455911).

Zverev, A. I., *Handbook of Filter Synthesis* (Wiley, 1970. ISBN: 0471986801).

Articles

Davis and Rutledge, "Industrial Class-E Power Amplifiers With Low-cost Power MOSFETs and Sine-wave Drive" Conf. Papers for RF Design '97, Santa Clara, CA, Sep 1997, pp. 283-297.

Franke and Noorani, "Lumped-Constant Line Stretcher For Testing Power Amplifier Stability," *rf Design*, Mar/Apr 1983.

Frey, R., "A 50 MHz, 250 W Amplifier using Push-Pull ARF448A/B," APT Application Note APT9702.

Frey, R., "Low Cost 1000 Watt 300 V RF Power Amplifier for 27.12 MHz," APT Application Note APT9701.

Frey, R., "A push-pull 300 watt amplifier for 81.36 MHz," *Applied Microwave and Wireless,* vol 10 no 3, pp 36-45, Apr 1998.

Frey, R., "Push-Pull ARF449A/B Amplifier for 81.36 MHz," APT Application Note APT9801.

Gonzalez, Martin, Lopez "Effects of Matching on RF Power Amplifier Efficiency and Output Power," *Microwave Journal*, Apr 1998.

Granberg, H., "Broadband Transformers and Power Combining Techniques for RF," AN-749, Motorola Semiconductor Products Inc.

Hejhall, R. "Systemizing RF Power Amplifier Design," Motorola Semiconductor Products, Inc., Phoenix, AZ, Application note AN282A.

Hilbers, A.H., "Design of HF wideband power transformers," Philips Semiconductors, ECO 6907, Mar 1998.

Hilbers, A.H., "Design of HF wideband power transformers Part II," Philips Semiconductors, ECO 7213, Mar 1998.

Hilbers, A.H., "Power Amplifier Design," Philips Semiconductors, Application Note, March 1998.

Klitzing, J, W6PQL, "Build a Linear 2 Meter 80 W All Mode Amplifier," *QST*, May 2013, pp 30-34. Feedback, Jul 2013 *QST*, p 40.

Trask, C., N7ZWY, "Designing Wide-band Transformers for HF and VHF Power Amplifiers," *QEX*, Mar/Apr 2005, pp 3-15.

Microwave — Cellular Applications

Cripps, *RF Power Amplifiers for Wireless Communication* (Artech House, 1999. ISBN: 1596930187).

Groe and L. E. Larson, *CDMA Mobile Radio Design* (Artech House, 2000. ISBN: 1580530591).

Pozar, *Microwave Engineering,* 3rd Ed. (Wiley, 2004. ISBN: 0471448788).

Raghavan, Srirattana, Laskar *Modeling and Design Techniques for RF Power Amplifiers,* (Wiley-IEEE Press, 2008. ISBN: 0471717460).

van Nee and Prasad, *OFDM for Wireless Multimedia Communications*, (Artech House, 2000. ISBN: 0890065306).

York and Popovic, *Active and Quasi-Optical Arrays for Solid-State Power Combining*, (Wiley, 1997. ISBN: 0471146145).

Contents

Repeaters

For decades, FM has been a dominant mode of Amateur Radio operation. FM and repeaters fill the VHF and UHF bands, and most hams have at least one handheld or mobile FM radio. Thousands of repeaters throughout the country provide reliable communication for amateurs operating from portable, mobile and home stations. Now, Amateur Radio is beginning to see the emergence of digital voice modulation, though analog is expected to dominate for the foreseeable future. The first part of this chapter, by Paul M. Danzer, N1II, describes FM voice repeaters. Later sections, contributed by Jim McClellan, N5MIJ, and Pete Loveall, AE5PL, of the Texas Interconnect Team, cover more recent developments in D-STAR digital voice and data repeaters. Gary Pearce, KN4AQ, contributed updates to the chapter.

Chapter 18 — CD-ROM Content

Supplemental Articles
- "From Analog to D-STAR" by Gary Pearce, KN4AQ
- "Discovering D-STAR" by Larry Moxon, K1KRC
- "D-STAR Uncovered" by Pete Loveall, AE5PL
- "Operating D-STAR" by Gary Pearce, KN4AQ

18.1 A Brief History

Few hams today don't operate at least some VHF/UHF FM, and for many hams, FM *is* Amateur Radio. That wasn't always the case.

Until the late 1960s, the VHF and UHF Amateur Radio bands were home to a relatively small number of highly skilled operators who used CW and SSB for long distance communication and propagation experiments. This operation used just a small fraction of the spectrum available at 50 MHz and above. A somewhat larger number of hams enjoyed low power, local operation with AM transceivers on 6 and 2 meters. Our spectrum was underutilized, while public safety and commercial VHF/UHF two-way operation, using FM and repeaters, was expanding rapidly.

The business and public-safety bands grew so rapidly in the early 1960s that the FCC had to create new channels by cutting the existing channels in half. Almost overnight, a generation of tube-type, crystal-controlled FM equipment had to be replaced with radios that met the new channel requirements. Surplus radios fell into the hands of hams for pennies on the original dollar. This equipment was designed to operate around 150 MHz and 450 MHz, just above the 2 meter and 70 cm amateur bands. It was fairly easy to order new crystals and retune them to operate inside the ham bands. Hams who worked in the two-way radio industry led the way, retuning radios and building repeaters that extended coverage. Other hams quickly followed, attracted by the noise-free clarity of FM audio, the inexpensive equipment, and the chance to do something different.

That initial era didn't last long. The surplus commercial equipment was cheap, but it was physically large, ran hot and consumed lots of power. By the early 1970s, manufacturers recognized an untapped market and began building solid-state equipment specifically for the Amateur Radio FM market. The frequency synthesizer, perfected in the mid-1970s, eliminated the need for crystals. The stage was set for an explosion that changed the face of Amateur Radio. Manufacturers have added plenty of new features to equipment over the years, but the basic FM operating mode remains the same.

In the 1980s, hams experimenting with data communications began modulating their FM radios with tones. *Packet radio* spawned a new system of *digipeaters* (digital repeaters).

Digitized audio has been popular since audio compact discs (CDs) were introduced in the 1980s. In the 1990s, technology advanced enough to reduce the bandwidth needed for digital audio, especially voice, to be carried over the Internet and narrowband radio circuits. The first digital-voice public safety radio systems (called APCO-Project 25) appeared and a variety of Internet voice systems for conferencing and telephone-like use were developed.

Hams are using VHF/UHF digital voice technology, too. The Japan Amateur Radio League (JARL) developed a true ham radio standard called D-STAR, a networked VHF/UHF repeater system for digital voice and data that is beginning to make inroads around the world. Hams are also using surplus Project 25 (or just P25) radios. Another digital voice system, generically known as DMR (Digital Mobile Radio) and by the Motorola trade name MOTOTRBO, is being adapted from commercial service to the ham bands.

18.2 Repeater Overview

Amateurs learned long ago that they could get much better use from their mobile and portable radios by using an automated relay station called a *repeater*. Home stations benefit as well — not all hams are located near the highest point in town or have access to a tall tower. A repeater, whose basic idea is shown in **Fig 18.1**, can be more readily located where the antenna system is as high as possible and can therefore cover a much greater area.

18.2.1 Types of Repeaters

The most popular and well-known type of amateur repeater is an FM voice system on the 144 or 440 MHz bands. Amateurs operate many repeaters on the 29, 50, 222, 902 and 1240 MHz bands as well, but 2 meters and 70 cm are the most popular. Tens of thousands of hams use mobile and handheld radios for casual ragchewing, emergency communications, public service activities or just staying in touch with their regular group of friends during the daily commute.

While the digital voice modes are gaining ground, FM is still the most popular mode for voice repeaters. Operations are *channelized* — all stations operate on specific, planned frequencies, rather than the more or less random frequency selection employed in CW and SSB operation. In addition, since the repeater receives signals from mobile or fixed stations and retransmits these signals simultaneously, the transmit and receive frequencies are different, or *split*. Direct contact between two or more stations that listen and transmit on the same frequency is called operating *simplex*.

Individuals, clubs, emergency communications groups and other organizations all sponsor repeaters. Anyone with a valid amateur license for the band can establish a repeater in conformance with the FCC rules. No one owns specific repeater frequencies, but nearly all repeaters are *coordinated* to minimize repeater-to-repeater interference. Frequency coordination and interference are discussed later in this chapter. Operational aspects are covered in more detail in *The ARRL Operating Manual*.

BEYOND FM VOICE

In addition to FM voice, there are several other types of ham radio repeaters.

ATV (amateur TV) — ATV repeaters are used to relay wideband television signals in the 70 cm and higher bands. They are used to extend coverage areas, just like voice repeaters. ATV repeaters are often set up for *crossband* operation, with the input on one band (say, 23 cm) and the output on another (say, 70 cm). More information on ATV repeaters may be found in the **Image Communications** supplement on the *Handbook* CD.

Digital voice — Repeaters for D-STAR digital voice and data signals are discussed later in this chapter. There are also some P25 and DMR repeaters operating in the Amateur service (see the **Digital Communications**

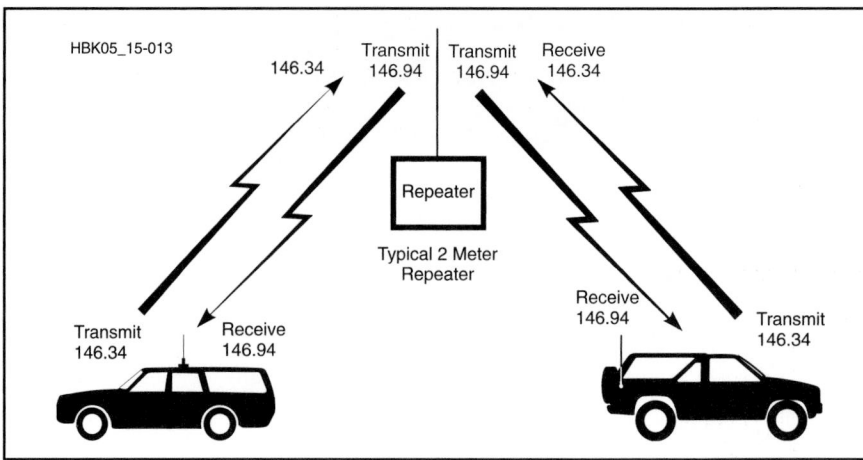

Fig 18.1 — Typical 2 meter repeater, showing mobile-to-mobile communication through a repeater station. Usually located on a hill or tall building, the repeater amplifies and retransmits the received signal on a different frequency.

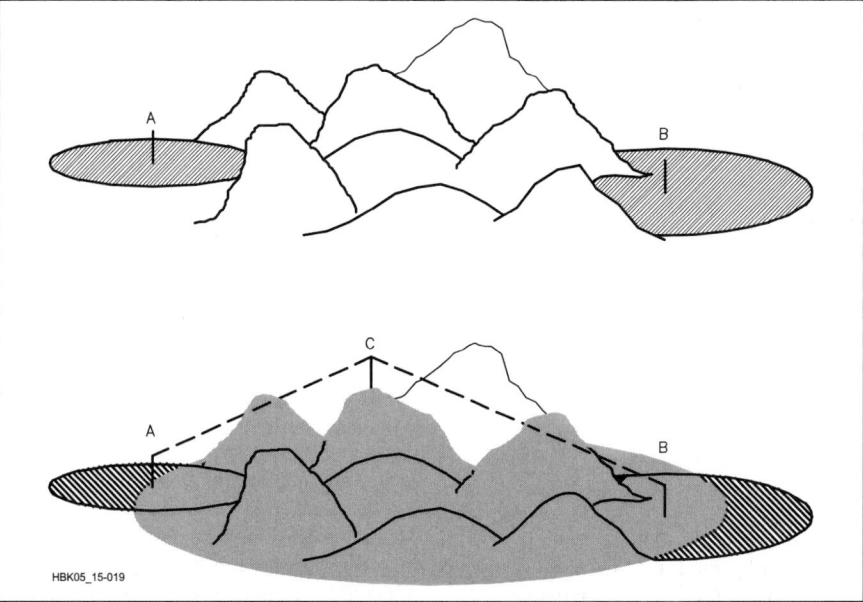

Fig 18.2 — In the upper diagram, stations A and B cannot communicate because their mutual coverage is limited by the mountains between them. In the lower diagram, stations A and B can communicate because the coverage of each station falls within the coverage of repeater C, which is on a mountaintop.

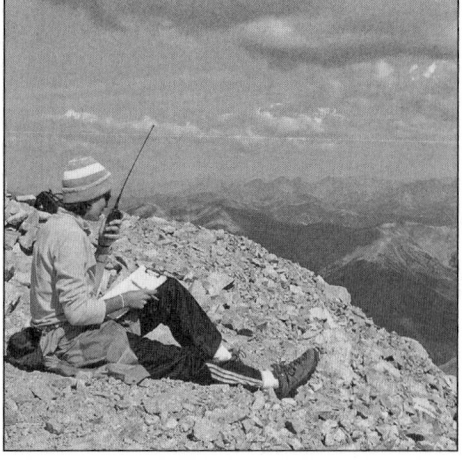

Fig 18.3 — In the Rocky Mountain west, handheld transceivers can often cover great distances, thanks to repeaters located atop high mountains. [Photo courtesy Rachel Witte, KC0ETU, and Bob Witte, K0NR]

supplement on the *Handbook* CD for additional information.)

Digipeaters — Digital repeaters are used primarily for packet communications, including APRS (the Amateur Packet/position Reporting System). They can use a single channel (single port) or several channels (multi-port) on one or more VHF and UHF bands. See the **Digital Modes** chapter and the **Digital Communications** supplement on the *Handbook* CD for details of these systems.

Multi-channel (wideband) — Amateur satellites are the best-known examples. Wide bandwidth (perhaps 50 to 200 kHz) is selected to be received and transmitted so all signals in bandwidth are heard by the satellite (repeater) and retransmitted, usually on a different VHF or UHF band. See the **Space Communications** supplement on the *Handbook* CD for details.

18.2.2 Advantages of Using a Repeater

When we use the term *repeater* we are almost always talking about transmitters and receivers on VHF or higher bands, where radio-wave propagation is normally line of sight. Don't take "line of sight" too literally. VHF/UHF radio signals do refract beyond what you can actually see on the horizon, but the phrase is a useful description. (See the **Propagation of Radio Signals** chapter for more information on these terms.)

We know that the effective range of VHF and UHF signals is related to the height of each antenna. Since repeaters can usually be located at high points, one great advantage of repeaters is the extension of coverage area from low-powered mobile and portable transceivers.

Fig 18.2 illustrates the effect of using a repeater in areas with hills or mountains. The same effect is found in metropolitan areas, where buildings provide the primary blocking structures.

Siting repeaters at high points can also have disadvantages. Since most repeaters have co-channel neighbors (other repeaters operating on the same channel) less than 150 miles away, there may be times when your transceiver can receive both. But since it operates FM, the *capture effect* usually ensures that the stronger signal will capture your receiver and the weaker signal will not be heard — at least as long as the stronger repeater is in use.

It is also simpler to provide a very sensitive receiver, a good antenna system, and a slightly higher power transmitter at just one location — the repeater — than at each mobile, portable, table or home location. A superior repeater system compensates for the low power (5 W or less), and small, inefficient antennas that many hams use to operate through them. The repeater maintains the range or coverage we

want, despite our equipment deficiencies. If both the handheld transceiver and the repeater are at high elevations, for example, communication is possible over great distances, despite the low output power and inefficient antenna of the transceiver (see **Fig 18.3**).

Repeaters also provide a convenient meeting place for hams with a common interest. It's usually geographic — your town, or your club. A few repeaters are dedicated to a particular interest such as DX or passing traffic. Operation is channelized, and usually in any area you can find out which channel — or repeater — to pick to ragchew, get highway information or whatever your need or interest is. The conventional wisdom is that you don't have to tune around and call CQ to make a contact, as on the HF bands. Simply call on a repeater frequency — if someone is there and they want to talk, they will answer you. But with a few dozen repeaters covering almost any medium size town, you probably use the scan function in your radio to seek activity.

EMERGENCY OPERATIONS

When there is a weather-related emergency or a disaster (or one is threatening), most repeaters in the affected area immediately spring to life. Emergency operation and traffic always take priority over other ham activities, and many repeaters are equipped with emergency power sources just for these occasions. See **Fig 18.4**.

Almost all Amateur Radio emergency organizations use repeaters to take advantage of

their extended range, uniformly good coverage and visibility. Most repeaters are well known — everyone active in an area with suitable equipment knows the local repeater frequencies.

18.2.3 Repeaters and the FCC

Repeaters are specifically authorized by the FCC rules. For a brief period when the repeater concept was new in the amateur service, the FCC required special repeater licenses identified by a "WR" call sign prefix and a fairly complex application process. While that complexity is gone, repeaters are still restricted to certain band segments and have lower maximum power limits. But most of the "rules" that make our repeater systems work are self-imposed and voluntary. Hams have established frequency coordination, band plans, calling frequencies, digital protocols and rules that promote efficient communication and interchange of information.

FCC rules on prohibited communication have also been somewhat relaxed, allowing hams to communicate with businesses, and allowing employees of emergency-related agencies and private companies to participate in training and drills while "on the clock." There are significant restrictions to this operation, so for the latest rules and how to interpret them, see *QST* and the ARRL website, **www. arrl.org**.

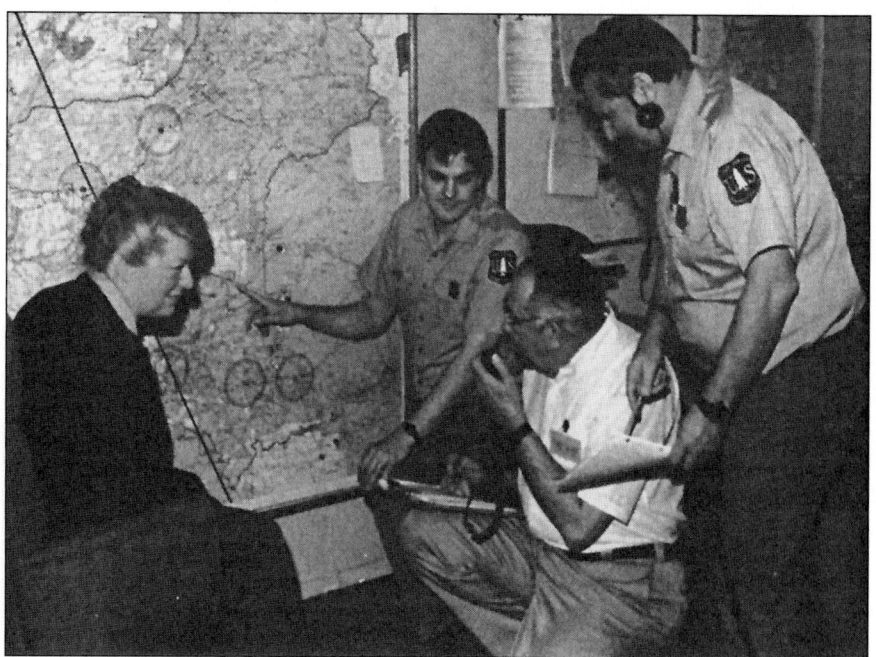

Fig 18.4 — During disasters, repeaters over a wide area are used solely for emergency-related communication until the danger to life and property is past. [Photo courtesy Eugene Kramer, WA9TZL]

18.3 FM Voice Repeaters

Repeaters normally contain at least the sections shown in **Fig 18.5**. Repeaters consist of a receiver and transmitter plus a couple more special devices. One is a *controller* that routes the audio between the receiver and transmitter, keys the transmitter and provides remote control for the repeater licensee or designated control operators.

The second device is the *duplexer* that lets the repeater transmit and receive on the same antenna. A high power transmitter and a sensitive receiver, operating in close proximity within the same band, using the same antenna, present a serious technical challenge. You might think the transmitter would just blow the receiver away. But the duplexer keeps the transmit energy out of the receiver with a series of tuned circuits. Without a duplexer, the receiver and transmitter would need separate antennas, and those antennas would need to be 100 or more feet apart on a tower. Some repeaters do just that, but most use duplexers. A 2 meter duplexer is about the size of a two-drawer filing cabinet. See **Fig 18.6**.

Receiver, transmitter, controller, and duplexer: the basic components of most repeaters. After this, the sky is the limit on imagination. As an example, a remote receiver site can be used to extend coverage (**Fig 18.7**).

The two sites can be linked either by telephone ("hard wire") or a VHF or UHF link. Once you have one remote receiver site, it is natural to consider a second site to better hear those "weak mobiles" on the other side of town (**Fig 18.8**). Some of the stations using the repeater are on 2 meters while others are on 70 cm? Just link the two repeaters! (See **Fig 18.9**).

For even greater flexibility, you can add an auxiliary receiver, perhaps for a NOAA weather channel (**Fig 18.10**).

The list goes on and on. Perhaps that is why so many hams have put up repeaters.

18.3.1 FM Repeater Operation

There are almost as many operating procedures in use on repeaters as there are repeaters. Only by listening can you determine the customary procedures on a particular machine. A number of common operating techniques are found on many repeaters, however.

One such common technique is the transmission of *courtesy tones*. Suppose several stations are talking in rotation — one following another. The repeater detects the end of a transmission of one user, waits a few seconds, and then transmits a short tone or beep. The next station in the rotation waits until the beep before transmitting, thus giving any other station wanting to join in a brief period to transmit their call sign. Thus the term *courtesy tone* — you are politely pausing to allow other stations to join in the conversation.

Another common repeater feature that encourages polite operation is the *repeater timer*. A 3-minute timer is actually designed to comply with an FCC rule for remotely controlled stations, but in practice the timer serves a more social function. Since repeater operation is channelized — allowing many stations to use the same frequency — it is

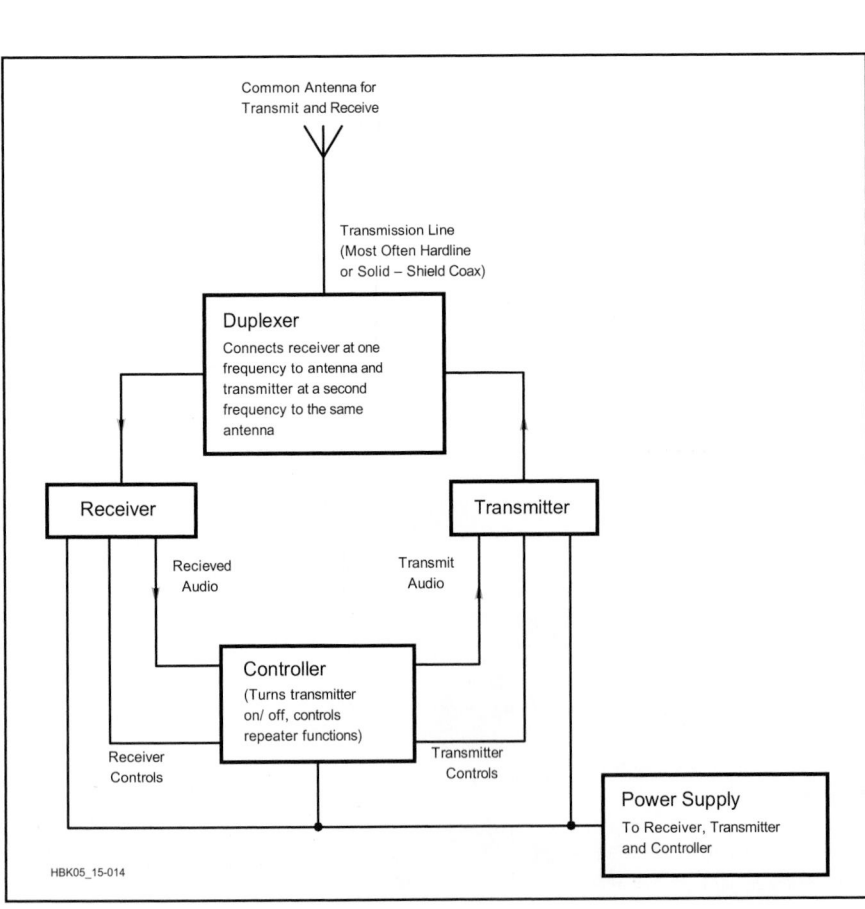

Fig 18.5 — The basic components of a repeater station. In the early days of repeaters, many were home-built. Today, most are commercial, and are far more complex than this diagram suggests.

Fig 18.6 — The W4RNC 2 meter repeater includes the repeater receiver, transmitter and controller in the rack. The large object underneath is the duplexer. [Photo courtesy Gary Pearce, KN4AQ]

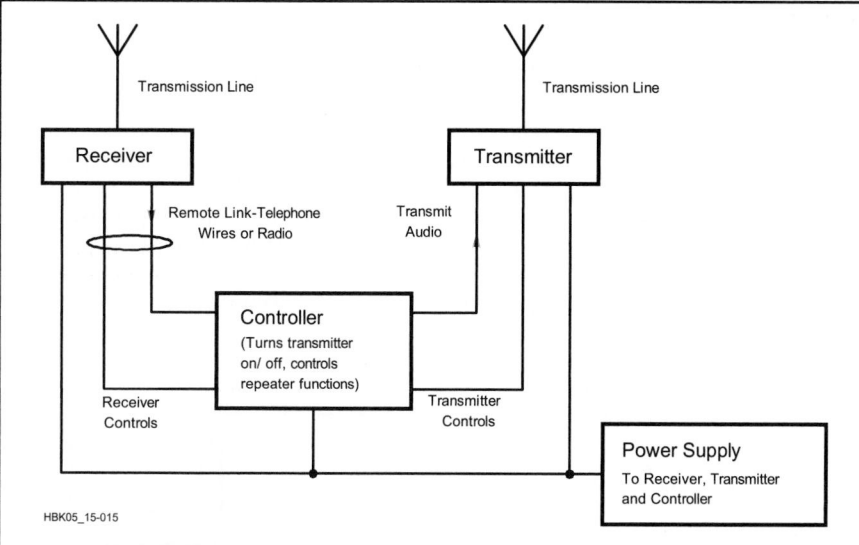

Fig 18.7 — Separating the transmitter from the receiver can extend the repeater's coverage area. The remote receiver can be located on a different building or hill, or consist of a second antenna at a different height on the tower.

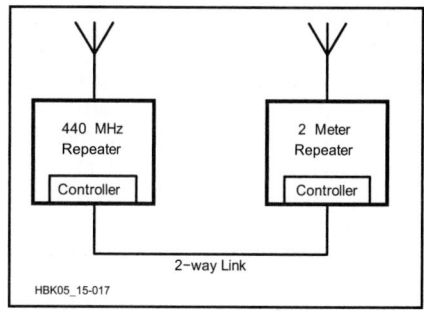

Fig 18.9 — Two repeaters using different bands can be linked for added convenience.

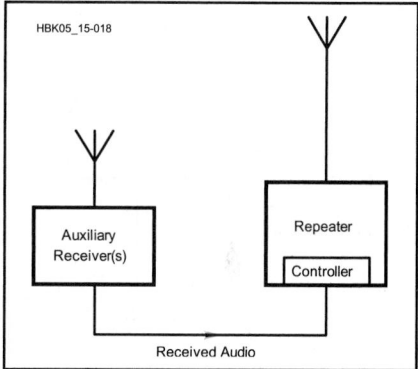

Fig 18.10 — For even greater flexibility, you can add an auxiliary receiver.

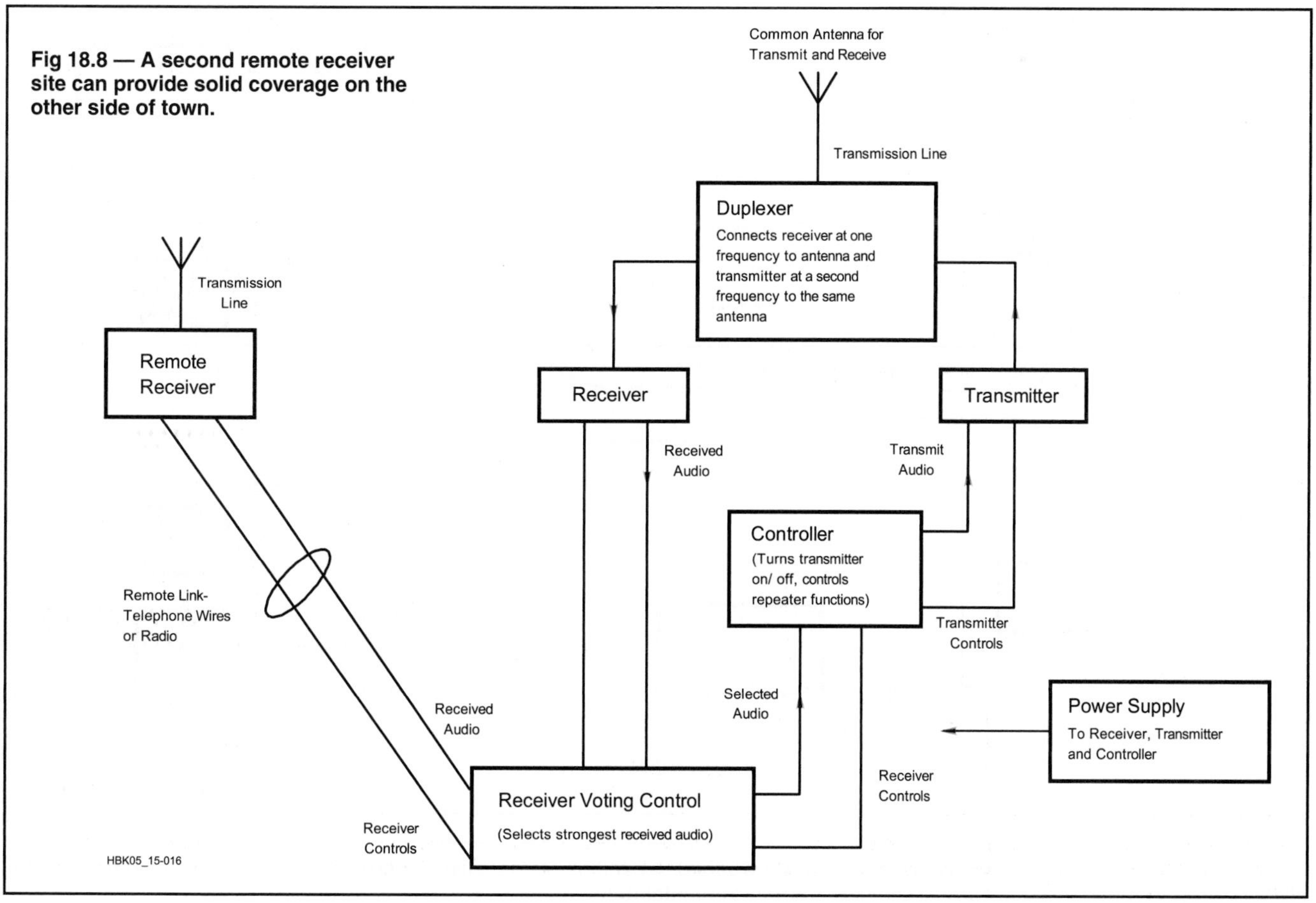

polite to keep your transmissions short. If you forget this little politeness many repeaters simply cut off your transmission after 2 or 3 minutes of continuous talking. If the repeater "times out," it remains off the air until the station on the input frequency stops transmitting.

A general rule, in fact law — both internationally and in areas regulated by the FCC — is that emergency transmissions always have priority. These are defined as relating to life, safety and property damage. Many repeaters are voluntarily set up to give mobile

Fig 18.8 — A second remote receiver site can provide solid coverage on the other side of town.

stations priority, at least in checking into the repeater. If there is going to be a problem requiring help, the request will usually come from a mobile station. This is particularly true during rush hours; some repeater owners request that fixed stations limit their use of the repeater during these hours.

Some parts of the country have one or more *closed repeaters*. These are repeaters whose owners wish, for any number of reasons, to not make them available for general use. Often they require transmission of a *subaudible* or *CTCSS* tone (discussed later). Not all repeaters requiring a CTCSS tone are closed — many open repeaters use tones to minimize interference among machines in adjacent areas using the same frequency pair. Other closed repeaters require the transmission of a special tone sequence to turn on. It is desirable that all repeaters, including generally closed repeaters, be made available at least long enough for the presence of emergency information to be made known.

Repeaters have many uses. In some areas they are commonly used for formal traffic nets, replacing or supplementing HF nets. In other areas they are used with tone alerting for severe weather nets. Even when a particular repeater is generally used for ragchewing it can be linked for a special purpose. As an example, an ARRL volunteer official may hold a periodic section meeting across her state, with linked repeaters allowing both

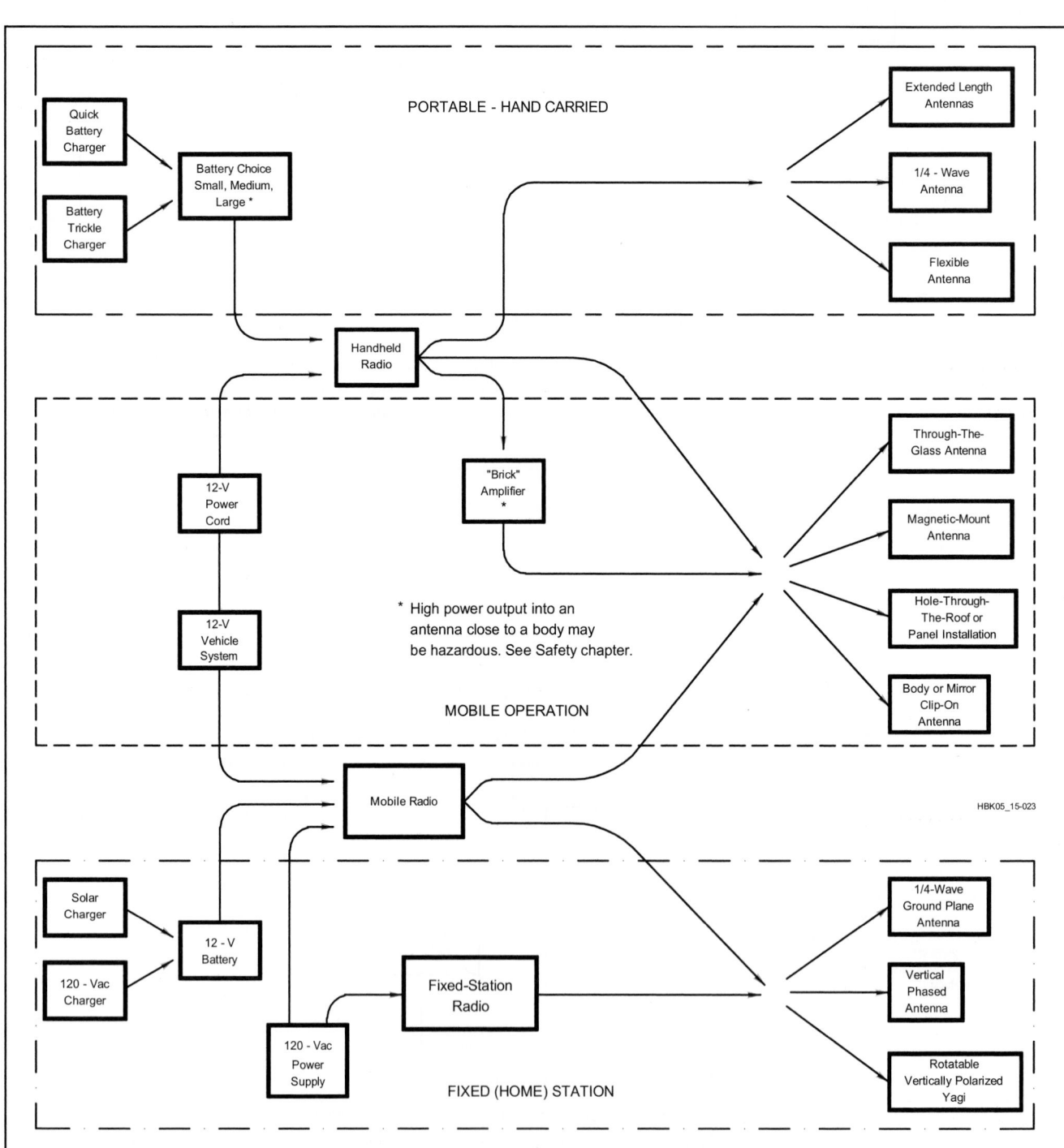

Fig 18.11 — Equipment choices for use with repeaters are varied. A handheld transceiver is perhaps the most versatile type of radio, as it can be operated from home, from a vehicle and from a mountaintop.

announcements and questions directed back to her.

One of the most common and important uses of a repeater is to aid visiting hams. Since repeaters are listed in the *ARRL Repeater Directory* and other directories, hams traveling across the country with mobile or handheld radios often check into local repeaters asking for travel route, restaurant or lodging information. Others just come on the repeater to say hello to the local group. In most areas courtesy prevails — the visitor is given priority to say hello or get the needed help.

Detailed information on repeater operating techniques is included in a full chapter of the *ARRL Operating Manual*.

18.3.2 Home, Mobile and Handheld Equipment

There are many options available in equipment used on repeaters. A number of these options are shown in **Fig 18.11**.

HANDHELD TRANSCEIVERS

A basic handheld radio with an output power of 500 mW to 5 W can be used almost anywhere — in a building, walking down a street, or in a car.

Several types of antennas can be used in the handheld mode. The smallest and most convenient is a rubber flex antenna, known as a "rubber duck," a helically wound antenna encased in a flexible tube. Unfortunately, to obtain the small size the use of a wire helix or coil produces a very low efficiency.

A quarter-wave whip, which is about 19 inches long for the 2 meter band, is a good choice for enhanced performance. The rig and your hand act as a ground plane and a reasonably efficient result is obtained. A longer antenna, consisting of several electrical quarter-wave sections in series, is also commercially available. Although this antenna usually produces extended coverage, the mechanical strain of 30 or more inches of antenna mounted on the radio's antenna connector can cause problems. After several months, the strain may require replacement of the connector.

Most newer handheld radios are supplied with lithium-ion (Li-ion) batteries. These high-capacity batteries are lightweight and allow operation for much longer periods than the classic NiCd battery pack. Charging is accomplished either with a "quick" charger in an hour or less or with a trickle charger overnight. See the **Power Sources** chapter for more information on batteries.

Power levels higher than 7 W may cause a safety problem on handheld units, since the antenna is usually close to the operator's head and eyes. See the **Safety** chapter for more information.

For mobile operation, an external antenna provides much greater range than the "rubber duck" as discussed in the following Mobile Equipment section. A power cord plugs into the vehicle cigarette lighter so that the battery remains charged, and a speaker-microphone adds convenience. In addition, commercially available "brick" amplifiers can be used to raise the output power level of the handheld radio to 50 W or more. Many hams initially go this route, but soon grow tired of frequently connecting and disconnecting all the accessories from the handheld and install a permanent mobile radio.

MOBILE EQUIPMENT

Compact mobile transceivers operate from 11-15 V dc and generally offer several transmit power levels up to about 50 W. They can operate on one or more bands. Most common are the single band and dual-band transceivers. "Dual-band" usually means 2 meters and 70 cm, but other combinations are available, as are radios that cover three or more bands.

Mobile antennas range from the quick and easy magnetic mount to "drill through the car roof" assemblies. The four general classes of mobile antennas shown in the center section of Fig 18.11 are the most popular choices. Before experimenting with antennas for your vehicle, there are some precautions to be taken.

Through-the-glass antennas: Rather than trying to get the information from your dealer or car manufacturer, test any such antenna first using masking tape or some other temporary technique to hold the antenna in place. Some windshields are metalicized for defrosting, tinting and car radio reception. Having this metal in the way of your through-the-glass antenna will seriously decrease its efficiency.

Magnet-mount or "mag-mount" antennas are convenient if your car has a metal roof or trunk. The metal also serves as the ground plane. They work well, but are not a good long-term solution. Eventually they'll scratch the car's paint, and the coax run through a door can be subject to flexing or crimping and can eventually fail.

Through-the-roof antenna mounting: Most hams are squeamish about drilling a hole in their car roof, unless they intend to keep the car for a long time. This mounting method provides the best efficiency, however, since the (metal) roof serves as a ground plane. Before you drill, carefully plan and measure how you intend to get the antenna cable down under the interior car headliner to the radio. Be especially careful of side-curtain air bags. Commercial two-way shops can install antennas and power cables, usually for a reasonable price.

Trunk lip and clip-on antennas: These antennas are good compromises. They are usually easy to mount and they perform acceptably. Cable routing must be planned. If you are going to run more than a few watts, do not mount the antenna close to one of the car windows — a significant portion of the radiated power may enter the car interior.

More information on mobile equipment may be found in the **Assembling a Station** and **Antennas** chapters.

HOME STATION EQUIPMENT

Most "base station" FM radios are actually mobile rigs, powered either from rechargeable batteries or ac-operated power supplies. Use of batteries has the advantage of providing back-up communications ability in the event of a power interruption. Some HF transceivers designed for fixed-station use also offer operation on the VHF or VHF/UHF bands, using SSB and CW in addition to FM. Using them means that you will not be able to monitor a local FM frequency while operating HF.

The general choice of fixed-location antennas is also shown in Fig 18.11. Most hams use an omnidirectional vertical, but if you are in an area between two repeaters on the same channel, a rotatable Yagi may let you pick which repeater you will use without

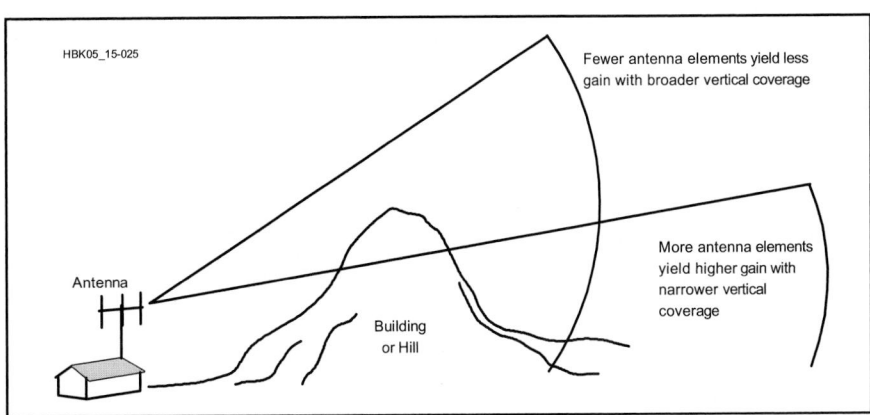

HBK05_15-025

Fewer antenna elements yield less gain with broader vertical coverage

More antenna elements yield higher gain with narrower vertical coverage

Antenna

Building or Hill

Fig 18.12 — As with all line-of-sight communications, terrain plays an important role in how your signal gets out.

interfering with the other repeater. Vertical polarization is the universal custom, since it is easiest to accomplish in a mobile installation. VHF/UHF SSB operation is customarily horizontally polarized. An operator with a radio that does both has a tough choice, as there can be a serious performance hit between stations using cross-polarized antennas.

Both commercial and homemade ¼-λ and larger antennas are popular for home use. A number of these are shown in the **Antennas** chapter. Generally speaking, ¼-λ sections may be stacked up to provide more gain on any band. As you do so, however, more and more power is concentrated toward the horizon. This may be desirable if you live in a flat area. See **Fig 18.12**.

18.3.3 Coded Squelch and Tones

Squelch is the circuit in FM radios that turns off the loud rush of noise with no signal present. Most of the time, hams use *noise squelch*, also called *carrier squelch*, a squelch circuit that lets any signal at all come through. But there are ways to be more selective about what signal gets to your speaker or keys up your repeater. That's generically known as *coded squelch*, and more than half of the repeaters on the air require you to send coded squelch to be able to use the repeater.

CTCSS

The most common form of coded squelch has the generic name *CTCSS* (Continuous Tone Coded Squelch System), but is better known by the nickname "tone." Taken from the commercial services, subaudible tones are

generally not used to keep others from using a repeater but rather are a method of minimizing interference from users of the same repeater frequency. CTCSS tones are sine-wave audio tones between 67 and 250 Hz, that are added to the transmit audio at a fairly low level. They are *sub*audible only because your receiver's audio circuit is supposed to filter them out. A receiver with CTCSS will remain silent to all traffic on a channel unless the transmitting station is sending the correct tone. Then the receiver sends the transmitted audio to its speaker.

For example, in **Fig 18.13** a mobile station on hill A is nominally within the normal coverage area of the Jonestown repeater (146.16/76). The Smithtown repeater, also on the same frequency pair, usually cannot hear stations 150 miles away. The mobile is on a hill and so he is in the coverage area of both Jonestown and Smithtown. Whenever the mobile transmits both repeaters hear him.

The common solution to this problem, assuming it happens often enough, is to equip the Smithtown repeater with a CTCSS decoder and require all users of the repeater to transmit a CTCSS tone to access the repeater. Thus, the mobile station on the hill does not come through the Smithtown repeater, since he is not transmitting the required CTCSS tone.

Table 18.1 shows the available CTCSS tones. Most radios built since the early 1980s have a CTCSS encoder built in, and most radios built since the early 1990s also have a CTCSS *decoder* built in. Newer radios have a "tone scan" feature that will hunt the tone, *if* the repeater ouput includes a tone. Most repeaters that require tone also transmit their

tone, but they don't have to. Listings in the *ARRL Repeater Directory* include the CTCSS tone required, if any.

If your local repeater sends a CTCSS tone, you can use your decoder to monitor just that repeater, and avoid hearing the co-channel neighbor, intermod or the annoying fizzes of nearby consumer electronics. Radios typically store CTCSS frequency and mode in their memory channels.

DIGITAL-CODED SQUELCH (DCS)

A newer form of coded squelch is called *DCS* (Digital-Coded Squelch). DCS appeared in commercial service because CTCSS didn't provide enough tones for the many users. DCS adds another 100 or so code options. DCS started showing up in Amateur Radio transceivers several years ago and is now a standard feature in most new radios. Open repeaters generally still use CTCSS rather than DCS, since many older radios still in use don't have DCS.

DTMF

In the days before widespread use of cell phones, one of the most attractive features of repeaters was the availability of autopatch services that allowed a mobile or portable station to use a standard telephone DTMF (dual-tone multi-frequency, or Touch-Tone) key pad to connect the repeater to the local telephone line and make outgoing calls.

Although autopatches see less use today, DTMF key pads are still used for sending

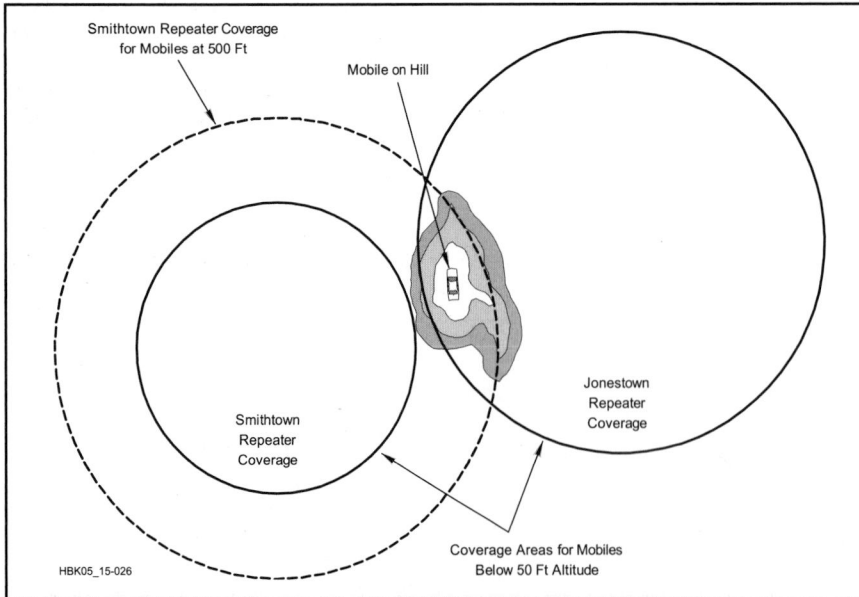

Fig 18.13 — When two repeaters operate on the same frequencies, a well-situated operator can key up both repeaters simultaneously. A directional antenna may help.

control signals. DTMF can also be used as a form of squelch, to turn a receiver on, though it's more often used to control various functions such as autopatch and talking S meters. Some repeaters that require CTCSS have a DTMF "override" that puts the repeater into carrier-squelch mode for a few minutes if you send the proper digits. Other applications for DTMF tones include controlling linked repeaters, described in a later section.

Table 18.2 shows the DTMF tones. Some keyboards provide the standard 12 sets of tones corresponding to the digits 0 through 9 and the special signs # and *. Others include the full set of 16 pairs, providing special keys A through D. The tones are arranged in two groups, usually called the low tones and high tones. Two tones, one from each group, are required to define a key or digit. For example, pressing 5 will generate a 770-Hz tone and a 1336-Hz tone simultaneously.

The standards used by the telephone company require the amplitudes of these two tones to have a certain relationship.

18.3.4 Frequency Coordination and Band Plans

Since repeater operation is channelized, with many stations sharing the same frequency pairs, the amateur community has formed coordinating groups to help minimize conflicts between repeaters and among repeaters and other modes. Over the years, the VHF and UHF bands have been divided into repeater and non-repeater sub-bands. These frequency-coordination groups maintain lists of available frequency pairs in their areas (although in most urban areas, there are no "available" pairs on 2 meters, and 70 cm pairs are becoming scarce). A complete list of frequency coordinators, band plans and repeater pairs is included in the *ARRL Repeater Directory*.

Each VHF and UHF repeater band has been subdivided into repeater and nonrepeater channels. In addition, each band has a specific *offset* — the difference between the transmit frequency and the receive frequency for the repeater. While most repeaters use these standard offsets, others use "oddball splits." These nonstandard repeaters are generally also coordinated through the local frequency coordinator. **Table 18.3** shows the standard frequency offsets for each repeater band.

FM repeater action isn't confined to the VHF and UHF bands. There are a large handful of repeaters on 10 meters around the US and the world. "Wideband" FM is permitted only above 29.0 MHz, and there are four band-plan repeater channels (outputs are 29.62, 29.64, 29.66 and 29.68 MHz), plus the simplex channel 29.60 MHz. Repeaters on 10 meters use a 100 kHz offset, so the

Table 18.2
Standard Telephone (DTMF) Tones

| | Low Tone Group | | High Tone Group | |
	1209 Hz	1336 Hz	1477 Hz	1633 Hz
697 Hz	1	2	3	A
770 Hz	4	5	6	B
852 Hz	7	8	9	C
941 Hz	*	0	#	D

Table 18.3
Standard Frequency Offsets for Repeaters

Band	Offset
29 MHz	–100 kHz
52 MHz	Varies by region
	–500 kHz, –1 MHz, –1.7 MHz
144 MHz	+ or –600 kHz
222 MHz	–1.6 MHz
440 MHz	+ or –5 MHz
902 MHz	12 MHz
1240 MHz	12 MHz

corresponding inputs are 29.52, 29.54, 29.56 and 29.58 MHz. During band openings, you can key up a repeater thousands of miles away, but that also creates the potential for interference generated when multiple repeaters are keyed up at the same time. CTCSS would help reduce the problem, but not many 10 meter repeater owners use it and too many leave their machines on "carrier access."

18.3.5 Narrowbanding

We noted in the previous section that in most urban areas, there are no "available" frequencies for new repeaters. And you might recall from our short history section at the beginning of this chapter that the Amateur Radio FM boom began when the business and public-safety services ran out of room and had to buy all new radios. Their solution to overcrowding, mandated by the FCC, was to reduce the spectrum used for each channel. In the 1960s, that meant reducing the modulation ("deviation") of FM signals from 15 kHz to 5 kHz, and splitting each channel in two. Hams inherited the 15 kHz deviation radios (called "wideband" at the time), but soon adopted the 5 kHz "narrowband" standard.

History is repeating itself. Our spectrum neighbors are again out of room, and the FCC is again requiring them to reduce deviation, from 5 kHz (now called "wideband"), to 2.5 kHz (the new "narrowband"). It's been postponed for several years, but is finally happening.

Will hams follow suit and create space for more repeaters in our own crowded bands? So far, the answer is "no." While most Amateur

Radio FM equipment built in the past decade has a "narrow" option that reduces the deviation and incorporates tighter receive filters, we still have a lot of legacy equipment in the field, and no corporate or municipal budget to draw on to replace it. Most of our repeaters are still made of old hardware converted from commercial service. Few repeater councils have seriously considered adjusting frequency coordination to accommodate narrowbanding. No one expects the FCC to require hams to adopt narrowbanding.

What *is* happening is the placement of D-STAR digital repeaters, which are especially "narrow" already, in between the channels occupied by analog FM repeaters. Since there is still some spectrum overlap, the D-STAR repeaters must also be a good distance away from their new adjacent-channel neighbors — 30 to 50 miles — to reduce the field strength of all the signals involved.

To help you understand how this all works, we'll explain that the terminology "5 kHz" and "2.5 kHz" deviation for analog FM signals is misleading. It refers to the peak frequency shift a modulated signal takes *in one direction* from the center frequency. But the FM signal moves both up and down from that center, and has some sidebands as well. The actual spectrum used by a "5 kHz" FM signal peaks at 16 kHz and the "narrow" 2.5 kHz signal hits 11 kHz on peaks — not much of a savings. The digital D-STAR signal is about 6.25 kHz wide. The digital signals fill their spectrum completely, all the time, and don't vary with modulation.

Narrowbanding may become a voluntary practice in Amateur Radio, though *your* use would be "mandated" by remaining compatible with narrowbanded repeaters. It's not on the horizon as of this edition of the *Handbook*.

18.3.6 Linked Repeaters

Most repeaters are standalone devices, providing their individual pool of coverage and that's it. But a significant number of repeaters are linked — connected to one or more other repeaters. Those other repeaters can be on other bands at the same location, or they can be in other locations, or both. Linked

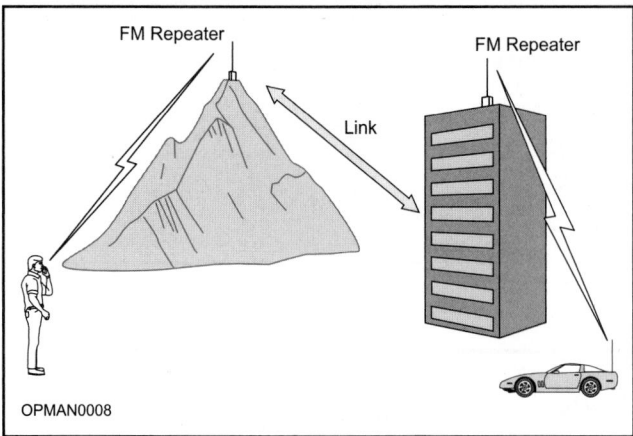

Fig 18.14 — Repeater linking can greatly expand VHF/UHF communication distances. Repeater links are commonly made via dedicated radio hardware or via the Internet.

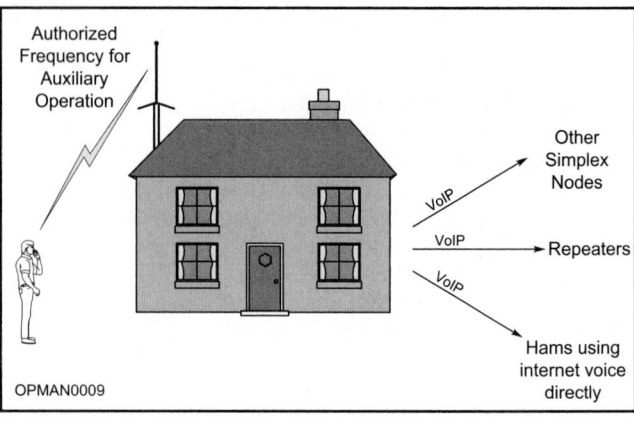

Fig 18.15 — A diagram of a VoIP simplex node. If a control operator is not physically present at the station location and the node is functioning with wireless remote control, the control link must follow the rules for *auxiliary* operation.

repeaters let users communicate between different bands and across wider geographic areas than they can on a single repeater. **Fig 18.14** shows an example.

There are many ways to link repeaters. Repeaters on the same tower can just be wired together, or they may even share the same controller. Repeaters within a hundred miles or so of each other can use a radio link — separate link transmitters and receivers at each repeater, with antennas pointed at each other. Multiple repeaters, each still about 100 miles apart, can "daisy-chain" their links to cover even wider territory. There are a few linked repeater systems in the country that cover several states with dozens of repeaters, but most radio-linked repeater systems have more modest ambitions, covering just part of one or two states.

Repeater linking via the Internet has created the ability to tie repeaters together around the world and in nearly unlimited number. We'll talk more about Internet linking in the next section.

There are several ways linked repeaters can be operated, coming under the categories of *full-time* and *on demand*. Full-time linked repeaters operate just as the name implies — all the repeaters in a linked network are connected all the time. If you key up one of them, you're heard on all of them, and you can talk to anyone on any of the other repeaters on the network at any time. You don't have to do anything special to activate the network, since it's always there.

In an on-demand system, the linked repeaters remain isolated unless you take some action, usually by sending a code by DTMF digits, to connect them. Your DTMF sequence may activate all the repeaters on the network, or the system may let you address

just one specific repeater, somewhat like dialing a telephone. When you're finished, another DTMF code drops the link, or a timer may handle that chore when the repeaters are no longer in use.

INTERNET LINKING

The Internet and *Voice over Internet Protocol* (VoIP) has expanded repeater linking exponentially, making worldwide communication through a local repeater commonplace. Two Internet linking systems, IRLP and EchoLink, have reached critical mass in the US and are available almost everywhere. The D-STAR and DMR digital systems also have a significant Internet linking component. A brief overview is included here; more information may be found in the **Digital Communications** supplement on the *Handbook* CD.

IRLP

The Internet Radio Linking Project (IRLP) is the most "radio" based linking system. User access is only via radio, using either simplex or repeater stations, while linking is done using VoIP on the Internet. An IRLP system operator establishes a *node* by interfacing his radio equipment to a computer with an Internet connection, and then running IRLP software. Once that's set up, repeater users send DTMF tones to make connections, either directly to other individual repeater or simplex nodes (**Fig 18.15**), or to *reflectors* — servers that tie multiple nodes together as one big party line.

The direct connections work like on-demand linked repeaters. You dial in the node number you want to connect to and access code (if required), and you are connected to the distant repeater. Once connected,

everyone on both ends can communicate. When finished, another DTMF sequence takes the link down. Someone from a distant repeater can make a connection to you as well.

Reflectors work like a hybrid between on-demand and full-time linked repeaters. You can connect your local repeater to a reflector and leave it there all day, or you can connect for a special purpose (like a net), and drop it when the event is over.

EchoLink

EchoLink requires a PC with sound card and appropriate software. It allows repeater connections like IRLP, and it has Conference Servers, similar to IRLP reflectors that permit multiple connections. The big difference is that EchoLink allows individuals to connect to the network from their computers, without using a radio.

The EchoLink conference servers all have more or less specific functions. Some are just regional gathering places, while some are region, topic or activity based (SKYWARN and National Hurricane Center Nets, Jamboree on the Air, and so on).

You can connect your EchoLink-enabled computer to your base station radio fairly easily through a sound card, and create an on-air node. Don't pipe EchoLink to a local repeater without permission from the repeater owner, though.

If you decide to create a full-time link from a computer to a repeater, consider using a dedicated UHF link frequency rather than just a base station on the repeater input. This applies to IRLP connections as well. Of course, the Internet is infrastructure dependent, and both power and Internet access can be interrupted during storms or other disasters.

D-STAR Network Overview

The D-STAR specification defines the repeater controller/gateway communications and defines the general D-STAR network architecture. The diagram shown here as Fig 18.A1 is taken from the English translation of the D-STAR specification:

The Comp. IP and Own IP are shown for reference if this were a DD communications. As they do not change and are not passed as part of the D-STAR protocol, they can safely be ignored for the purposes of the following explanation.

Headers 1 through 4 are W$1QQQ calling W$1WWW. Headers 5 through 8 are W$1WWW calling W$1QQQ. Note that "Own Callsign" and "Companion Callsign" are never altered in either sequence. The "Destination Repeater Callsign" and the "Departure Repeater Callsign" are changed between the gateways. This is so the receiving gateway and repeater controller know which repeater to send the bit stream to. It also makes it easy to create a "One Touch" response as ICOM has done by simply placing the received "Own Callsign" in the transmitted "Companion Callsign", the received "Destination Repeater Callsign" in the transmitted "Departure Repeater Callsign", and the received "Departure Repeater Callsign" in the transmitted "Destination Repeater Callsign".

Use of the "special" character "/" at the beginning of a call sign indicates that the transmission is to be routed to the repeater specified immediately following the slash. For instance, entering "/K5TIT B" in the "Companion Callsign" would cause the transmission to be routed to the "K5TIT B" repeater for broadcast. Using the above example, W$1QQQ would put "/W$1SSS" in the "Companion Callsign" for the same sequence 1 through 4 to occur. At the W$1VVV gateway, however, the "/W$1SSS" in the "Companion Callsign" would be changed to "CQCQCQ". All stations within range of W$1SSS would see the transmission as originating from W$1QQQ and going to CQCQCQ just like that station was local (but the "Departure Repeater Callsign" would be "W$1VVV G" and the "Destination Repeater Callsign" would be "W$1SSS"). Replying would still be done the same way as before since the received "Companion Callsign" is ignored when programming the radio to reply.

Every "terminal" (station) has an IP address assigned to it for DD purposes. The address is assigned from the 10.0.0.0/8 address range. The D-STAR gateway is always 10.0.0.2. The router to the Internet is always 10.0.0.1. The addresses 10.0.0.3-31 are reserved for local-to-the-gateway (not routable) use. What this makes possible is the ability to send Ethernet packets to another station by only knowing that terminal's IP address and the remote station can directly respond based solely on IP address. This is because the gateway software can correlate IP address with call sign and ID. This makes it possible to route DD Ethernet packets based on the "Companion Callsign" or based on IP address with "Companion Callsign" set to "CQCQCQ". — *Pete Loveall, AE5PL*

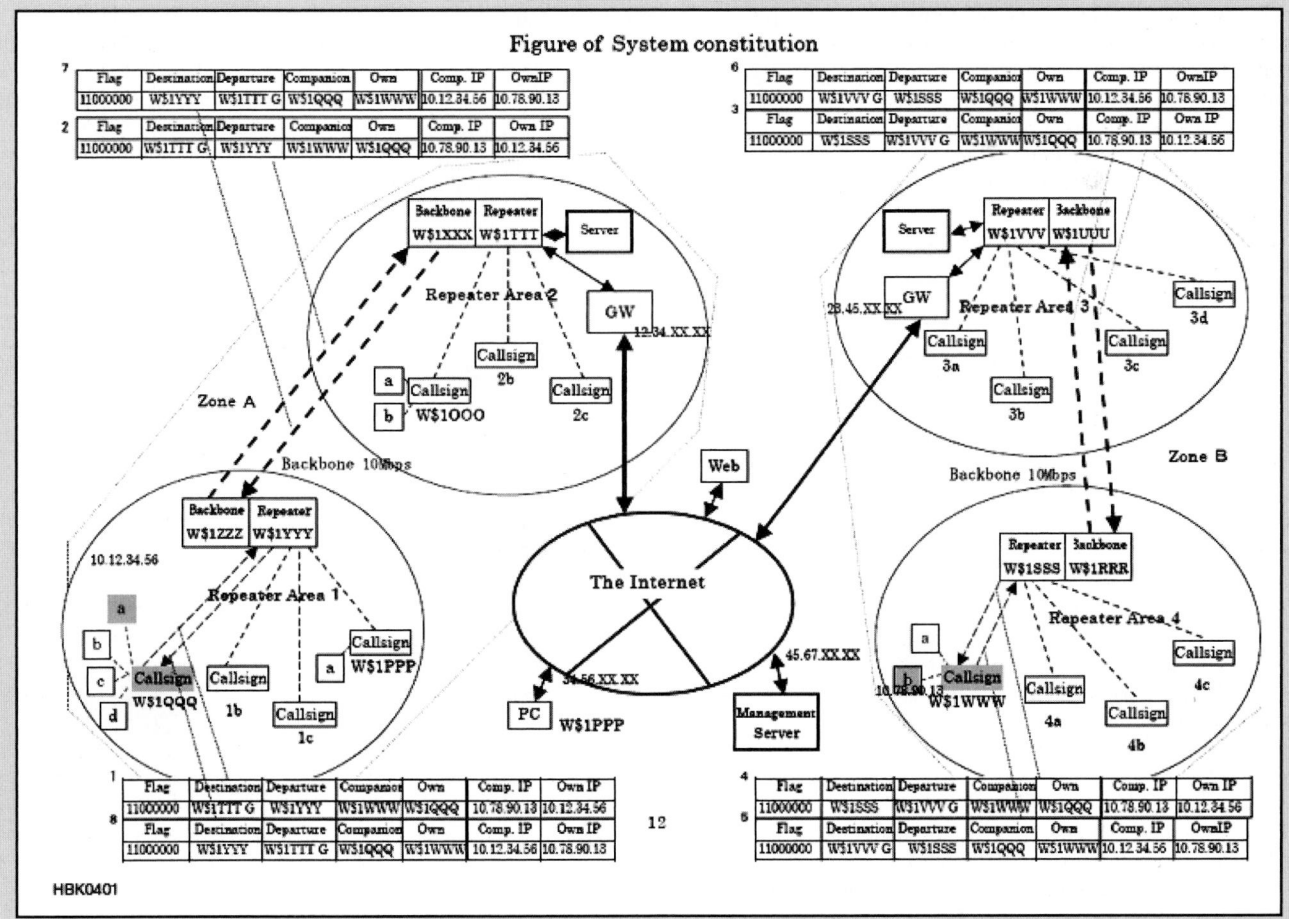

Fig 18.A1 — A D-STAR system overview.

18.4 D-STAR Repeater Systems

D-STAR is a digital protocol developed by the Japan Amateur Radio League (JARL) that takes Amateur Radio into the 21st century. D-STAR is a bit-streaming protocol able to encapsulate voice and low speed data (DV) or higher speed Ethernet data (DD). Because the protocol is entirely digital (GMSK modulation is used for amateur station transceivers and the respective repeaters), signaling is carried entirely out-of-band (ie, control codes are a separate part of the data stream from the voice or Ethernet information). When looking at the D-STAR repeater systems, it is important to keep that aspect in mind. Amateurs are more familiar with in-band signaling (DTMF tones, for example) in the analog world. Additional information on D-STAR may be found in the **Digital Communications** supplement on the *Handbook* CD.

18.4.1 D-STAR Station IDs

The D-STAR specification defines a protocol that can be used for simplex communication or repeater operations. When used simplex, the D-STAR radios function similarly to analog radios with the additional capability to enable selective listening based on station ID (call sign and a character or space). To operate through a repeater, you must know the repeater's station ID as well as the frequency the repeater is on. In this case, consider the repeater's station ID equivalent to a unique CTCSS tone in the analog world.

The D-STAR specification defines all station IDs as seven upper-case alphanumeric characters (space padded) and one upper-case alphanumeric station identifier (which may also be a space). In other words, W1AW can operate multiple D-STAR stations where the first seven characters of the station ID are "W1AW<space><space><space>" (three space characters follow W1AW) and the eighth character may be a space or upper-case alphanumeral. The G2 gateway software restricts the eighth character to a space or upper-case alphanumeral. For instance, two stations could operate at the same time using "W1AW<space><space><space> " and "W1AW<space><space><space>P".

To talk to someone via D-STAR, you need to set your radio with four station IDs:
- your station ID (MYCALL)
- their station ID (URCALL)
- your local repeater station ID (RPT1)
- your local gateway station ID (RPT2).

MYCALL is always set to your call sign, followed by a character or space. RPT1 is set to your local repeater's station ID (the repeater call sign, followed by a character or space).

If you want to talk locally through the repeater:
- Set URCALL to CQCQCQ.
- Set RPT2 to your local gateway station ID (this isn't needed for local communication, but allows some new network functions to operate, and has become the default recommendation).

The power of the D-STAR protocol becomes evident when D-STAR *gateways* are implemented, providing interconnectivity between repeater systems. This interconnectivity is the same whether you are using voice or Ethernet data.

To talk to someone elsewhere in the D-STAR world, beyond your local area:
- Set URCALL to the other station's ID.
- Set RPT2 to your local gateway station ID.

The RPT2 setting is very important. You don't need to know where the other station is. You simply tell your local repeater to send your bit stream (everything is digital) to the local gateway so that gateway can determine where to send it next. The local gateway looks at URCALL (remember, this is part of the bit stream) and determines where that station was heard last. It then sends the bit stream on to that remote gateway, which looks at URCALL again to determine which repeater at the far end should transmit the bit stream.

Sounds complex? Yes, the implementation can be complex but the user is shielded from all of this by simply setting the four station IDs. **Fig 18.16** shows an example.

18.4.2 Station Routing

There are several different ways to set the destination station ID using URCALL:
- If URCALL is set to CQCQCQ, this means "don't go any farther."
- If URCALL is set to the remote station's ID and RPT2 is set to the local gateway, this means "gateway, send my bit stream to be transmitted out the repeater that the remote station was last heard on."
- If URCALL is set to/followed by a remote *repeater* station ID, and RPT2 is set to the local gateway, this means "gateway, send my bit stream to be transmitted out the repeater designated."

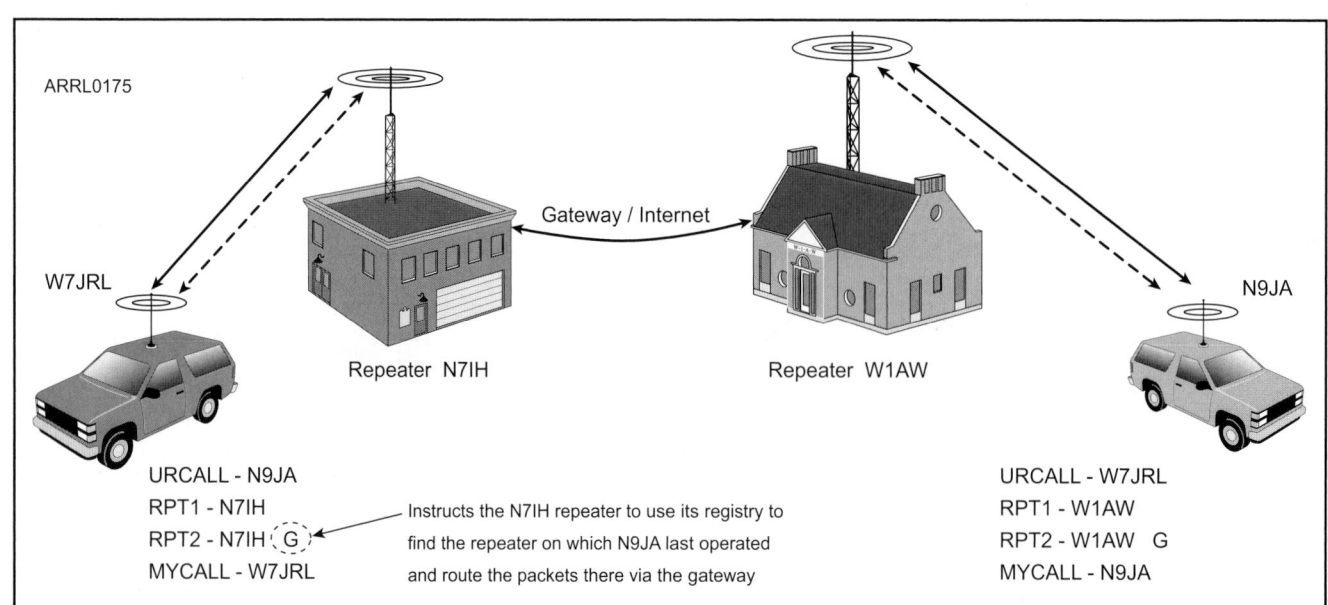

ARRL0175

Gateway / Internet

W7JRL

Repeater N7IH

Repeater W1AW

N9JA

URCALL - N9JA
RPT1 - N7IH
RPT2 - N7IH (G)
MYCALL - W7JRL

Instructs the N7IH repeater to use its registry to find the repeater on which N9JA last operated and route the packets there via the gateway

URCALL - W7JRL
RPT1 - W1AW
RPT2 - W1AW G
MYCALL - N9JA

Fig 18.16 — The necessary call sign set to make a call on a remote D-STAR repeater by using a gateway.

Some people mistakenly call this "call sign routing." In fact, this is "station routing" because you are specifying the station you want to hear your bit stream (voice or Ethernet data). This is not source routing because the source station is only defining what station they want to talk to; it is up to the gateways to determine routing (very similar to Internet routers).

You can specify a destination but you have no guarantee that:

1) The designated station is on the air to receive your transmission.

2) The repeater that the designated station is using is not busy.

3) Other factors will not prevent your bit stream from reaching the designated station.

Station routing is similar to your address book. In your address book you may have a work number, home number, cell number, fax number and email address for a person. How do you know which one to contact them at? Maybe the email address is best if you want to send them data (equivalent to the station ID for their Ethernet data radio). Maybe the cell phone number if that is what they normally have with them (equivalent to their hand-held transceiver station ID).

How do I know if they heard me? When they talk (or send Ethernet data) back to you. D-STAR is connectionless. Therefore, there is

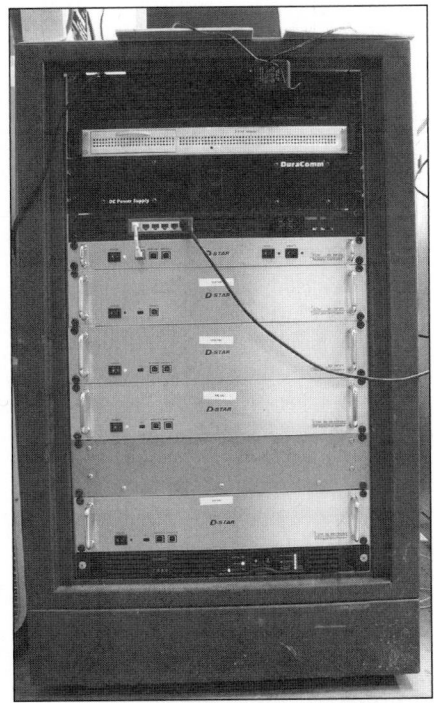

Fig 18.17 — A full rack of D-STAR equipment on the bench of Jim McClellan, N5MIJ. Top to bottom: ICOM IP-RPC2 controller, ID-RP2V 1.2-GHz voice repeater, ID-RP4000 440 MHz voice and data repeater, a blank panel and an ID-RP2000 146 MHz voice and data repeater.

no equivalence in D-STAR to repeater linking as we think of it in the analog world. However, there have been independent implementations of linking of repeaters similar to the linking we see with IRLP (see the section on *DPlus*).

In all cases of D-STAR repeater use, all digital voice (DV) signals with a proper RPT1 are always repeated so everyone hears your transmission through the repeater, regardless of the other settings. Ethernet data (DD) signals are not repeated on frequency because the "repeater" is actually a half-duplex Ethernet bridge operating on a single frequency.

For more details, see the sidebar, "D-STAR Network Overview."

18.4.3 Enhancing D-STAR Operation with *DPlus*

Because D-STAR is a true digital protocol, repeaters have no need for decoding the audio transmitted as bits from each radio. As mentioned previously, this requires all signaling to be out-of-band (with regard to the voice or Ethernet data).

Applications can be built to work with this out-of-band signaling to implement enhancements to the base product without modifying those products. One of these applications is *DPlus*, software that runs on the D-STAR Gateway computer at the repeater site. It provides many capabilities and more are being added as the software develops.

A key concept is the capability to link repeaters either *directly* (everything heard on one repeater is heard on another) or *indirectly* through a *reflector* (everything sent to a reflector is reflected back out to all linked repeaters). A reflector is a special version of *DPlus* that runs on a standalone computer that is not part of a repeater system. Linking and unlinking a repeater is done by altering the contents of URCALL according to the *DPlus* documentation. (These features continue to evolve, so specific operating commands are not covered here.)

There is no way to directly link two DD "repeaters." Because DD is Ethernet bridging, however, full TCP support is available, allowing each individual station to make connections as needed to fit their requirements.

For a station to make use of a linked repeater, the station must have URCALL set to CQCQCQ and RPT2 set to the local gateway. If RPT2 is not set to the local gateway, *DPlus* running on the local gateway computer will never see the bit stream and therefore not be able to forward the bit stream to the linked repeater or reflector. This is why setting RPT2 for your local gateway should be your default setting.

If your radio is set for automatic low-speed data transmission, remember that low-speed data is carried as part of the digital voice bit stream and is not multiplexed. Therefore, any

low-speed data transmission will block the frequency for the time the transmission is occurring. Caution: if you are using a linked repeater or if you have set URCALL to something other than CQCQCQ, your bit stream will be seen by all stations that are on the other end of that transmission. A reflector could have over 100 other repeaters listening to your transmission.

18.4.4 D-STAR Repeater Hardware

D-STAR repeaters are a bit different from the FM repeaters with which we're all familiar. A quick comparison will help to illustrate.

A complete FM repeater consists of at least three identifiable components. A receiver receives the original signal and demodulates it. The demodulated audio is routed to a controller, where it is mixed with other audio. The resulting combined audio signal is routed back to one or more transmitters. At least one additional source of audio is present in the controller, as it is a legal requirement to ID the repeater transmitter correctly. A well-constructed system includes validation that the levels and frequency response of the processed audio are consistent and true to the originally transmitted signals.

A D-STAR repeater's block diagram looks very similar, but functions very differently. A receiver receives the original signal and demodulates it. That signal is passed to a controller, which then drives one or more transmitters. The most significant difference is that there is no audio involved! D-STAR is a digital protocol. All required manipulation of information is performed in the digital domain, including the necessary ID functions. Most existing D-STAR repeaters do not contain the vocoders necessary to recover audio information, so there exists no local speaker or microphone. There is also no level-setting to consider with D-STAR.

D-STAR REPEATER OPTIONS

There is much discussion about "homebrew" D-STAR repeaters. The two most common approaches are to modify an existing FM repeater to pass the digital signal, or to wire two radios back-to-back. Both approaches provide the desired extended RF coverage, but fail to accurately process the digital signal. Thus both approaches fall short in either functional capability or in legality of the transmitted signal.

It is relatively simple to modify an existing FM repeater to pass the digital signal. The limitations of this approach are that there can be no additional capability added (for example, a D-STAR Gateway), since the digital signal is never decoded. Additionally, this method lacks the ability to encapsulate

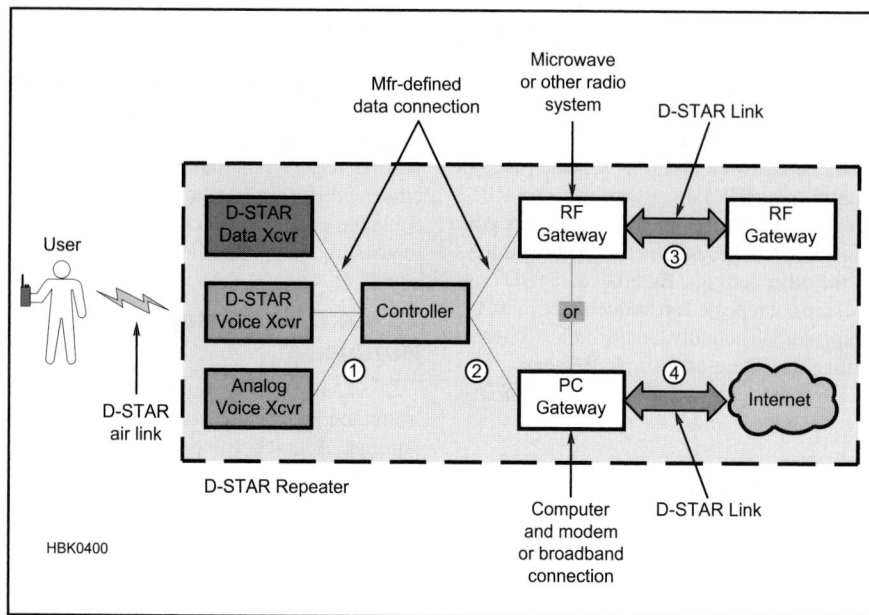

Fig 18.18 — Internal and external connections of a D-STAR repeater stack.

the ID for the repeater transmitter into the transmitted data.

It is also very simple to wire two D-STAR radios back-to-back, such that the incoming signals are retransmitted. Again, the functionality is limited by the inability to process the entire data stream. This approach presents an

additional consideration, as the radio used as a transmitter does not process the data stream as is done in the D-STAR repeater system, and the ID of the originating transmitter is lost, replaced by the ID of the "repeater" transmitter.

Current commercially produced D-STAR repeaters are designed to be used across a

broad frequency range. They do not have some of the tight front-end filtering provided by our familiar FM repeaters, so repeater builders must provide that front-end filtering externally. Installing additional band-pass filters between the antenna and the repeater will significantly improve the performance of the system. This is true for both digital voice (DV) repeaters and digital data (DD) access points.

Following good engineering practices will ensure good performance of the system. A properly constructed and installed D-STAR repeater can exhibit performance improvements of 15% or more in range, as compared to a comparably constructed FM repeater. This performance gain comes from the combination of the forward error correction (FEC) contained in the transmitted signal, and the fact that the radiated power is contained within a narrower bandwidth.

D-STAR is an exciting new system for the amateur community, providing significant opportunities for us to develop applications and implementations using capabilities we've never had before. We truly are limited only by our imaginations. The growth of the world-wide D-STAR network illustrates the level of interest in the technology by both amateurs and our served agencies. Amateurs are once again developing at the leading edge of technology. What we do with the new tools is up to us. How will you use D-STAR?

18.5 P25, DMR and Digital Voice

D-STAR is not the only digital voice mode in use by hams. P25 (or APCO Project 25) is a digital voice system designed for public safety (police, fire, EMS, etc.). It was developed in the 1990s to update the FM infrastructure. After about 10 years, the first P25 radios were retired and acquired by hams, who built P25 repeaters around the country. So far, though, they have not developed a digital P25 network. See **www.p25.com** and **http://en.wikipedia.org/wiki/Project_25** for more information on P25.

DMR — Digital Mobile Radio — is a commercial system also adapted by hams that is networked worldwide. (DMR should not be confused with DRM which is used on HF — see the **Digital Modes** chapter.) There are several DMR radio manufacturers, but Motorola's MOTOTRBO brand is practically synonymous with the mode, at least among hams in the US. In addition to Internet networking, DMR's claim to fame is that it uses two-slot TDMA (time-division multiple-access). That scheme lets two conversations occupy the same RF channel by carefully (and rapidly) turning each user's

transmitter on and off so that each is on for half the time. The repeater reconstructs the audio streams. The result is spectrum efficiency (at least when both streams are in use), and the ability to get double-duty out of one repeater/feed line/antenna system. A typical ham system will see one time slot used for a nationwide or worldwide network, and one for local or regional use. For more information on DMR, see **www.etsi.org/technologies-clusters/technologies/digital-mobile-radio** and **http://en.wikipedia.org/wiki/Digital_mobile_radio**.

No Amateur Radio manufacturers are building either P25 or DMR equipment for the ham market (Yaesu is introducing a "DMR" radio, but it's not compatible with the MOTOTRBO flavor of DMR). As a result hams must use commercial equipment. Most of the radios will cover the ham bands without retuning, but they must be programmed with proprietary software and cables. Most of the radios can't be programmed manually in the field. The radios are often more rugged and can be more immune to interference than amateur equipment.

D-STAR's advantage is that it is designed for hams, runs on flexible ham equipment and can be controlled to a large extent by the user. P25 and DMR adherents claim that their systems are more robust, especially at weak-signal levels, and don't degrade into the digital garble that D-STAR can exhibit.

Some hams, though, criticize *all* of the digital systems for relying on proprietary *vocoders* (voice encoders), the chips that do the analog to digital conversion and encoding. They believe that an amateur digital system should be open-source. The problem has been that until now there were no open-source vocoders that could generate a good-quality voice signal at the low bit-rates required for a narrowband radio channel. David Rowe, VK5DGR, has been working on *CODEC2*, an open-source vocoder designed to do just that. *CODEC2* hasn't been implemented in any commercially produced radios. But one day it may allow a manufacturer — or a ham in his workshop — to develop a digital voice plus data system to compete with D-STAR, P25 and DMR. (See the **Digital Modes** chapter for more on HF Digital Voice and *CODEC2*.)

18.6 Glossary of FM and Repeater Terminology

Access code — One or more numbers and/or symbols that are keyed into the repeater with a DTMF tone pad to activate a repeater function, such as an autopatch.

Autopatch — A device that interfaces a repeater to the telephone system to permit repeater users to make telephone calls. Often just called a "patch."

Carrier-operated relay (COR) — A device that causes the repeater to transmit in response to a received signal. Solid state versions may be called COS (carrier-operated switch).

Channel — The pair of frequencies (input and output) used by a repeater, or a single frequency used for simplex.

Channel step — The difference (in kHz) between FM channels. The common steps are 15 and 20 kHz for 2 meter repeaters, 20 kHz for 222 MHz repeaters, and 25 kHz for 440 MHz repeaters. Closer spacing is beginning to be used in some congested areas.

Closed repeater — A repeater whose access is limited to a select group (see *open repeater*).

Control operator — The Amateur Radio operator who is designated to "control" the operation of the repeater, as required by FCC regulations.

Courtesy tone — An audible indication that a repeater user may go ahead and transmit.

Coverage — The geographic area within which the repeater provides communications.

Crossband — A repeater with its input on one band and output on another.

CTCSS — Abbreviation for continuous tone-controlled squelch system, a subaudible tone sent with an FM voice transmission to access a repeater.

DCS — Digital Coded Squelch. A newer version of CTCSS that uses a subaudible digital code instead of an analog tone to selectively open a receiver's squelch.

Digipeater — A packet radio (digital) repeater, usually using store-and-forward on a single frequency.

DTMF — Abbreviation for *dual-tone multifrequency,* commonly called Touch Tone, the series of tones generated from a keypad on a ham radio transceiver (or a regular telephone).

Duplex or full duplex — A mode of communication in which a user transmits on one frequency and receives on another frequency simultaneously (see *half duplex*).

Duplexer — A device that allows the repeater transmitter and receiver to use the same antenna simultaneously.

Frequency coordinator — An individual or group responsible for assigning frequencies to new repeaters without causing interference to existing repeaters.

Full quieting — A received signal that contains no noise.

Half duplex — A mode of communication in which a user transmits at one time and receives at another time.

Handheld — A small, lightweight portable transceiver small enough to be carried easily.

Hang time — A few seconds of repeater carrier following a user transmission that allows others who want to access the repeater a chance to do so; the *courtesy beep* sounds during the hang time.

Input frequency — The frequency of the repeater's receiver (and your transceiver's transmitter).

Intermod — Intermodulation distortion (IMD), the unwanted mixing of two strong RF signals that causes a signal to be received on an unintended frequency.

Key up — To turn on a repeater by transmitting on its input frequency.

Li-ion — Lithium-ion battery. Longer life, smaller and lighter than NiCd, Li-ion batteries are becoming more popular for use with handheld radios.

Machine — A repeater system.

Mag mount — Magnetic mount, an antenna with a magnetic base that permits quick installation and removal from a motor vehicle or other metal surface.

NiCd — A nickel-cadmium battery that may be recharged many times; often used to power portable transceivers. Pronounced *NYE-cad.*

NiMH — Nickel-metal-hydride battery; rechargeable, offers more capacity and lighter weight than an NiCd battery. Often used to power portable transceivers.

Offset — the spacing between a repeater's input and output frequencies.

Open repeater — a repeater whose access is not limited.

Output frequency — the frequency of the repeater's transmitter (and your transceiver's receiver).

Over — A word used to indicate the end of a voice transmission.

Repeater Directory — An annual ARRL publication that lists repeaters in the US, Canada and other areas.

Separation — The difference (in kHz) between a repeater's transmitter and receiver frequencies, also called the *offset,* or *split.* Repeaters that use unusual separations, such as 1 MHz on 2 meters, are sometimes said to have "oddball splits."

Simplex — A mode of communication in which users transmit and receive directly on the same frequency.

Squelch tail — The noise burst heard in a receiver that follows the end of an FM transmission, before the squelch circuit turns off the speaker.

Time-out — To cause the repeater or a repeater function to turn off because you have transmitted for too long.

Timer — A device that measures the length of each transmission and causes the repeater or a repeater function to turn off after a transmission has exceeded a certain length.

Tone pad — An array of 12 or 16 numbered keys that generate the standard telephone dual-tone multifrequency (*DTMF*) dialing signals. Resembles a standard telephone keypad. (see *autopatch*).

18.7 References and Bibliography

D-STAR Users website — **www.dstarusers.org**

Ford, S., WB8IMY, *ARRL's VHF Digital Handbook* (ARRL, 2008). Chapter 4 and Appendix B cover D-STAR.

Japan Amateur Radio League, *Translated D-STAR System Specification*, **www.jarl.com/d-star**

Loveall, P., AE5PL, "D-STAR Uncovered," *QEX*, Nov/Dec 2008, pp 50-56.

Moxon, L., K1KRC, "Discovering D-STAR," *QST*, Sep 2010, p 72.

Pearce, G., KN4AQ, "ICOM IC-92AD Dual-Band Hand Held Transceiver," Product Review, *QST*, Sep 2008, pp 39-43.

Pearce, G., KN4AQ, "Operating D-STAR," *QST*, Sep 2007, pp 30-33.

Pearce, G., KN4AQ, "From Analog to D-STAR," *QST*, May 2012, pp 36-39.

Texas Interconnect Team website — **www.k5tit.org**

The Repeater Builder's Technical Information Page — **www.repeaterbuilder.com**

Wilson, M., K1RO, ed., *The ARRL Operating Manual*, 10th ed. (ARRL: 2012). Chapter 2 covers FM, Repeaters, Digital Voice and Data.

Contents

Propagation of Radio Signals

Radio waves, like light waves — and all other forms of electromagnetic radiation, normally travel in straight lines. Obviously this does not happen all the time, because long-distance communication depends on radio waves traveling beyond the horizon. How radio waves propagate in other than straight-line paths is a complicated subject, but one that need not be a mystery. This chapter, by Emil Pocock, W3EP, with updates by Carl Luetzelschwab, K9LA, provides basic understanding of the principles of electromagnetic radiation, the structure of the Earth's atmosphere and solar-terrestrial interactions necessary for a working knowledge of radio propagation. The section on VHF/UHF mobile propagation was contributed by Alan Bloom, N1AL. More detailed discussions and the underlying mathematics of radio propagation physics can be found in the references listed at the end of this chapter.

Chapter 19 — CD-ROM Content

Supplemental Articles

- "The Penticton Solar Flux Receiver" by John White, VA7JW and Ken Tapping
- "Hands-On Radio: Recording Signals" by Ward Silver, NØAX

Projects

- "Build a Homebrew Radio Telescope" by Mark Spencer, WA8SME

19.1 Fundamentals of Radio Waves

Radio belongs to a family of electromagnetic radiation that includes infrared (radiation heat), visible light, ultraviolet, X-rays and the even shorter-wavelength gamma and cosmic rays. Radio has the longest wavelength and thus the lowest frequency of this group. See **Table 19.1**.

Electromagnetic waves are composed of an inter-related electric and magnetic field. The electric and magnetic components are oriented at right angles to each other and are also perpendicular to the direction of travel. The polarization of a radio wave is usually designated the same as the orientation of its electric field. This relationship can be visualized in **Fig 19.1**. Unlike sound waves or ocean waves, electromagnetic waves need no propagating medium, such as air or water. This property enables electromagnetic waves to travel through the vacuum of space.

19.1.1 Velocity

Radio waves, like all forms of electromagnetic radiation, travel nearly 300,000 km (186,400 miles) per second in a vacuum. Radio waves travel more slowly through any other medium. The decrease in speed through the atmosphere is so slight that it is usually ignored, but sometimes even this small difference is significant. The speed of a radio wave in a piece of wire, by contrast, is about 95% that in free space, and the speed can be even slower in other media.

The speed of a radio wave is always the product of wavelength and frequency, whatever the medium. That relationship can be stated simply as:

$$c = f \lambda$$

where
  c = speed in meters/second
  f = frequency in hertz
  λ = wavelength in meters

The wavelength (λ) of any radio frequency can be determined from this simple formula by rearranging the above equation to $\lambda = c/f$. For example, in free space the wavelength of a 30 MHz radio signal is thus 10 meters. A simplified equation in metric units is λ in meters = 300 divided by the frequency in MHz. Alternately in English units, λ in feet = 984 divided by

Table 19.1

The Electromagnetic Spectrum

Radiation	Frequency	Wavelength
Radio	10 kHz – 300 GHz	30 km – 1 mm
Infrared	300 GHz – 428.6 THz	1 mm – 700 nm
Visible light	428.6 THz – 750 THz	700 nm – 400 nm
Ultraviolet	750 THz – 3×10^3 THz	400 nm – 100 nm
Extreme Ultraviolet	3×10^3 THz – 3×10^4 THz	100 nm – 10 nm
"Soft" X-ray	3×10^4 THz – 3×10^5 THz	10 nm – 1 nm
"Hard" X-ray	3×10^5 THz – 3×10^6 THz	1 nm – 0.1 nm

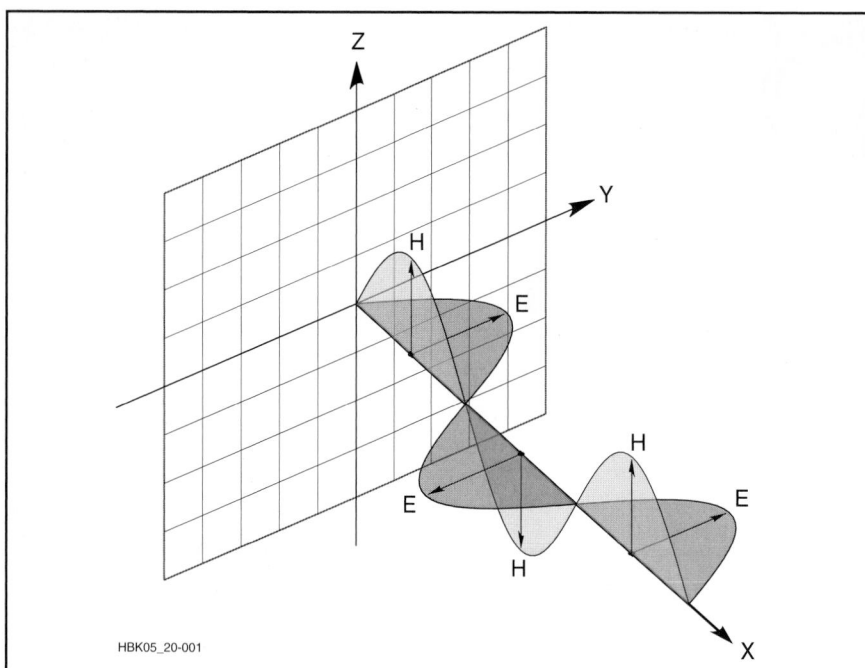

Fig 19.1 — Electric and magnetic field components of the electromagnetic wave. The polarization of a radio wave is the same direction as the plane of its electric field.

the frequency in MHz.

Wavelength decreases in other media because the propagating speed is slower. In a piece of wire, the wavelength of a 30 MHz signal shortens to about 9.5 meters. This factor must be taken into consideration in antenna designs, in transmission line designs, and in other applications.

19.1.2 Free Space Attenuation and Absorption

The intensity of a radio wave decreases as it travels. There are two mechanisms by which the intensity decreases: free space attenuation and absorption.

Free-space attenuation results from the spherical spreading of radio energy from its source. See **Fig 19.2**. Attenuation grows

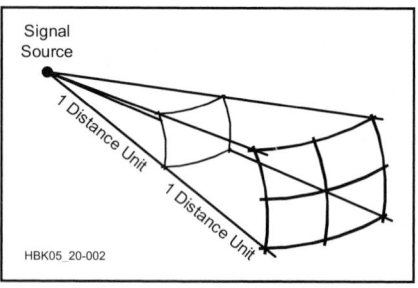

Fig 19.2 — Radio energy disperses as the square of the distance from its source. For the change of one distance unit shown the signal's power per unit of area is only one quarter as strong. Each spherical section has the same surface area.

rapidly with distance because signal strength decreases with the square of the distance from the source. (The signal's field strength in V/m decreases linearly with distance and its power density in W/m² decreases with the square of distance.) If the distance between transmitter and receiver is increased from 1 km to 10 km (0.6 to 6 miles), the signal power will be reduced by a factor of 100. Attenuation increases with frequency as well. Free space attenuation (path loss) can be expressed as

$$L_{fs} = 32.45 + 20 \log d + 20 \log f$$

where
L_{fs} = free space path loss in dB
d = distance in km
f = frequency in MHz

Free-space attenuation is a major factor governing signal strength, but radio signals undergo a variety of other losses as well. Energy is lost to absorption when radio waves travel through media other than a vacuum. Radio waves propagate through the atmosphere or solid material (like a wire) by exciting electrons, which then reradiate energy at the same frequency. This process is not perfectly efficient, so some radio energy is transformed into heat and retained by the medium. The amount of radio energy lost in this way depends on the characteristics of the medium and on the frequency. Attenuation in the atmosphere is minor from 10 MHz to 3 GHz, but at higher frequencies, absorption due to water vapor and oxygen can be high.

Radio energy is also lost during refraction,

diffraction and reflection — the very phenomena that allow long-distance propagation. Indeed, any form of useful propagation is accompanied by attenuation. This may vary from the slight losses encountered by refraction from sporadic E clouds near the maximum usable frequency, to the more considerable losses involved with tropospheric forward *scatter* (not enough ionization for refraction or reflection, but enough to send weak electromagnetic waves off into varied directions) or D Layer absorption in the lower HF bands. These topics will be covered later. In many circumstances, total losses can become so great that radio signals become too weak for communication (they are below the sensitivity of a receiver).

19.1.3 Refraction Below the Ionosphere

Electromagnetic waves travel in straight lines until they are deflected by something. Radio waves are *refracted*, or bent, slightly when traveling from one medium to another. Radio waves behave no differently from other familiar forms of electromagnetic radiation in this regard. The apparent bending of a pencil partially immersed in a glass of water demonstrates this principle quite dramatically.

Refraction is caused by a change in the velocity of a wave when it crosses the boundary between one propagating medium and another. If this transition is made at an angle, one portion of the wavefront slows down (or speeds up) before the other, thus bending the wave slightly. This is shown schematically in **Fig 19.3**.

The amount of bending increases with the ratio of the *refractive indices* of the two media. Refractive index is simply the velocity of a

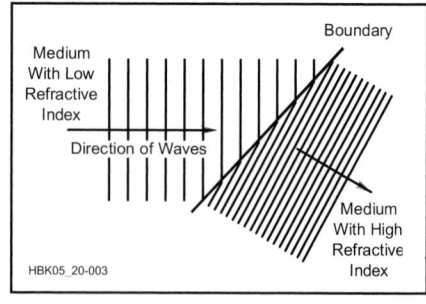

Fig 19.3 — Radio waves are refracted as they pass at an angle between dissimilar media. The lines represent the crests of a moving wave front and the distance between them is the wavelength. The direction of the wave changes because one end of the wave slows down before the other as it crosses the boundary between the two media. The wavelength is simultaneously shortened, but the wave frequency (number of crests that pass a certain point in a given unit of time) remains constant.

radio wave in free space divided by its velocity in the medium. The refractive properties of air may be calculated from temperature, moisture and atmospheric pressure. The index of refraction of air, at a very wide range of frequencies, may be calculated from:

$$N = \frac{77.6\,p}{T} + \frac{3.73 \times 10^5\,e}{T^2}$$

where

N = index of refraction, N units (number of millionths by which the index of refraction exceeds 1.0)

p = atmospheric pressure, millibars (mb)

e = partial pressure of water vapor, millibars

T = temperature, Kelvins

Note that the index of refraction below the ionosphere is greater than 1, causing more bending.

The refraction of radio signals is a function of the change in the index of refraction with altitude. N varies between 290 and 400 at the Earth's surface and normally diminishes with altitude at the rate of 40 N units per kilometer within the first few kilometers.

Radio waves are commonly refracted when they travel through different layers of the atmosphere, whether the highly charged ionospheric layers roughly 100 km (60 miles) and higher, or the weather-sensitive area near the Earth's surface. When the ratio of the refractive indices of two media is great enough, radio waves can be reflected, just like light waves striking a mirror. The Earth is a rather lossy reflector, but a metal surface works well if it is several wavelengths in diameter.

19.1.4 Scattering

The direction of radio waves can also be altered through *scattering*. The effect seen by a beam of light attempting to penetrate fog is a good example of light-wave scattering. Even on a clear night, a highly directional searchlight is visible due to a small amount of atmospheric scattering perpendicular to the beam. Radio waves are similarly scattered when they encounter randomly arranged objects of wavelength size or smaller, such as masses of electrons or water droplets. When the density of scattering objects becomes great enough, they behave more like a propagating medium with a characteristic refractive index.

If the scattering objects are arranged in some alignment or order, scattering takes place only at certain angles. A rainbow provides a good analogy for *field-aligned scattering* of light waves. The arc of a rainbow can be seen only at a precise angle away from the Sun, while the colors result from the variance in scattering across the light-wave frequency range. Ionospheric electrons can be field-aligned by magnetic forces in auroras and under other unusual circumstances. Scattering in such cases

is best perpendicular to the Earth's magnetic field lines.

19.1.5 Reflection

At amateur frequencies above 30 MHz, reflections from a variety of large objects, such as water towers, buildings, airplanes, mountains and the like, can provide a useful means of extending over-the-horizon paths several hundred km. Two stations need only beam toward a common reflector, whether stationary or moving.

Maximum range is limited by the radio line-of-sight distance of both stations to the reflector and by reflector size and shape. The reflectors must be many wavelengths in size and ideally have flat surfaces. Large airplanes make fair reflectors and may provide the best opportunity for long-distance contacts. The calculated limit for airplane reflections is 900 km (560 miles), assuming the largest jets fly no higher than 12,000 meters (40,000 ft), but actual airplane reflection contacts are likely to be considerably shorter.

19.1.6 Knife-Edge Diffraction

Radio waves can also pass behind solid objects with sharp upper edges, such as a mountain range, by *knife-edge diffraction*. This is a common natural phenomenon that affects light, sound, radio and other coherent waves,

but it is difficult to comprehend. **Fig 19.4** depicts radio signals approaching an idealized knife-edge. The portion of the radio waves that strike the base of the knife-edge is entirely blocked, while that portion passing several wavelengths above the edge travel on relatively unaffected. It might seem at first glance that a knife-edge as large as a mountain, for example, would completely prevent radio signals from appearing on the other side but that is not quite true. Something quite unexpected happens to radio signals that pass just over a knife-edge.

Normally, radio signals along a wave front interfere with each other continuously as they propagate through unobstructed space, but the overall result is a uniformly expanding wave. When a portion of the wave front is blocked by a knife-edge, the resulting interference pattern is no longer uniform. This can be understood by visualizing the radio signals right at the knife-edge as if they constituted a new and separate transmitting point, but in-phase with the source wave at that point. The signals adjacent to the knife-edge still interact with signals passing above the edge, but they cannot interact with signals that have been obstructed below the edge. The resulting *interference pattern* no longer creates a uniformly expanding wave front, but rather appears as a pattern of alternating strong and weak bands of waves that spread in a nearly 180° arc behind the knife-edge.

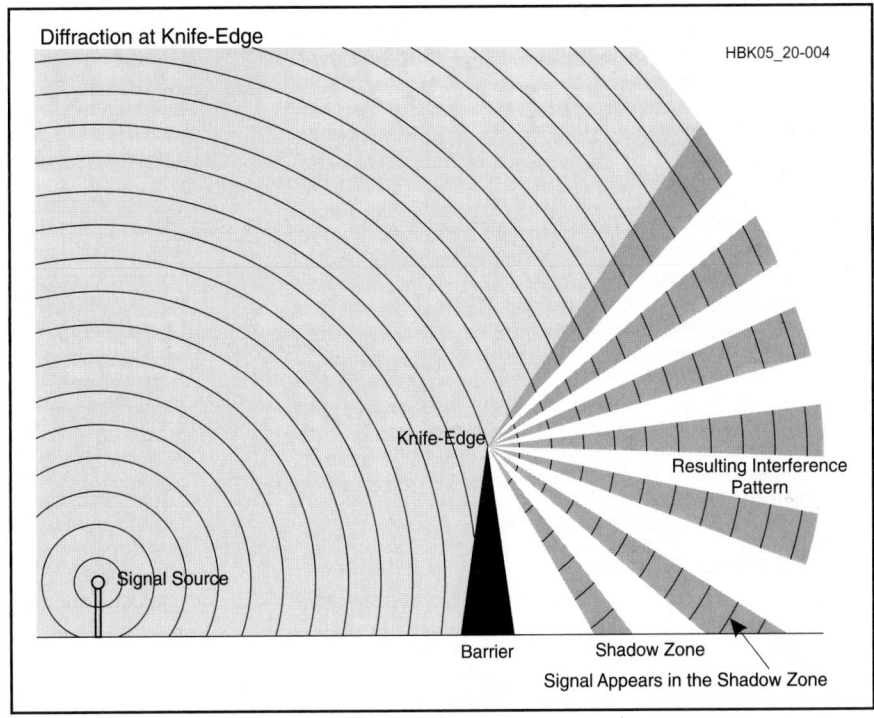

Fig 19.4 — VHF and UHF radio waves, light and other waves are diffracted around the sharp edge of a solid object that is large in terms of wavelengths. Diffraction results from interference between waves right at the knife-edge and those that are passing above it. Some signals appear behind the knife-edge as a consequence of the interference pattern. Hills or mountains can serve as natural knife-edges at radio frequencies.

Propagation Summary, by Band

OUR ONLY MEDIUM FREQUENCY (MF) BAND

1.8-2.0 MHz (160 meters)

Top band, as it is sometimes called, suffers from daytime D layer absorption. Daytime communication is limited to ground-wave coverage and a single E hop out to about 1500 km for well equipped stations (running the full legal limit, quarter-wave verticals with a good ground system, and a low noise receiving environment). At night, the D layer quickly disappears and worldwide 160 meter communication becomes possible via F_2 layer skip and ducting. Atmospheric and man-made noise limits propagation. Tropical and mid latitude thunderstorms cause high levels of static in summer, making winter evenings the best time to work DX at 1.8 MHz. A proper choice of receiving antenna (Beverage, 4-square, small loop) can often significantly reduce the amount of received noise to improve the signal-to-noise ratio.

HIGH FREQUENCY (HF) BANDS (3-30 MHz)

A wide variety of propagation modes are useful on the HF bands. The lowest two bands in this range share many daytime characteristics with 160 meters. The transition between bands primarily useful at night or during the day appears around 10 MHz. Most long-distance contacts are made via F_2 layer skip. Above 21 MHz, more exotic propagation, including TE, sporadic E, aurora and meteor scatter, begins to be practical.

3.5-4.0 MHz (80 meters for the lower end, 75 meters for the higher end)

The lowest HF band is similar to 160 meters in many respects. Daytime absorption is significant, but not quite as extreme as at 1.8 MHz. At night, signals are often propagated halfway around the world. As at 1.8 MHz, atmospheric noise is a nuisance, making winter the most attractive season for the 80/75 meter DXer.

5.3-5.4 MHz (60 meters)

The distance covered during daytime propagation will fall in between that achievable on the 80 meter and 40 meter bands. At night, worldwide propagation is possible in spite of the relatively low power limit. Signal strengths will typically be higher than on 80 meters but not as high as on 40 meters.

7.0-7.3 MHz (40 meters)

The popular 40 meter band has a clearly defined skip zone during the day due to insufficient ionization to refract high angles. D layer absorption is not as severe as on the lower bands, so short-distance skip via the E and F layers is possible. During the day, a typical station can cover a radius of approximately 800 km (500 miles). At night, reliable worldwide communication via F_2 is common on the 40 meter band.

Atmospheric noise is much less troublesome than on 160 and 80 meters, and 40 meter DX signals are often of sufficient strength to override even high-level summer static. For these reasons, 40 meters is the lowest-frequency amateur band considered reliable for DX communication in all seasons. Even during the lowest point in the solar cycle, 40 meters may be open for worldwide DX throughout the night.

10.1-10.15 MHz (30 meters)

The 30 meter band is unique because it shares characteristics of both daytime and nighttime bands. D layer absorption is not a significant factor. Communication up to 3000 km (1900 miles) is typical during the daytime, and this extends halfway around the world via all-darkness paths. The band is generally open via F_2 on a 24-hour basis, but during a solar minimum, the MUF on some DX paths may drop below 10 MHz at night. Under these conditions, 30 meters adopts the characteristics of the daytime bands at 14 MHz and higher. The 30 meter band shows the least variation in conditions over the 11-year solar cycle, thus making it generally useful for long-distance communication anytime.

14.0-14.35 MHz (20 meters)

The 20 meter band is traditionally regarded as the amateurs' primary long-haul DX favorite. Regardless of the 11-year solar cycle, 20 meters can be depended on for at least a few hours of worldwide F_2 propagation during the day. During solar-maximum periods, 20 meters will often stay open to distant locations throughout the night. Skip distance is usually appreciable and is always present to some degree. Daytime E layer propagation may be detected along very short paths. Atmospheric noise is not a serious consideration, even in the summer. Because of its popularity, 20 meters tends to be very congested during the daylight hours.

18.068-18.168 MHz (17 meters)

The 17 meter band is similar to the 20 meter band in many respects, but the effects of fluctuating solar activity on F_2 propagation are more pronounced. During the years of high solar activity, 17 meters is reliable for daytime and early-evening long-range communication, often lasting well after sunset. During moderate years, the band may open only during sunlight hours and close shortly after sunset. At solar minimum, 17 meters will open to middle and equatorial latitudes, but only for short periods during midday on north-south paths.

21.0-21.45 MHz (15 meters)

The 15 meter band has long been considered a prime DX band during solar cycle maxima, but it is sensitive to changing solar activity. During peak years, 15 meters is reliable for daytime F_2 layer DXing and will often stay open well into the night. During periods of moderate solar activity, 15 meters is basically a daytime-only band, closing shortly after sunset. During solar minimum periods, 15 meters may not open at all except for infrequent north-south transequatorial circuits. Sporadic E is observed occasionally in early summer and mid-winter, although this is not common and the effects are not as pronounced as on the higher frequencies.

24.89-24.99 MHz (12 meters)

This band offers propagation that combines the best of the 10 and 15 meter bands. Although 12 meters is primarily a daytime band during low and moderate sunspot years, it may stay open well after sunset during the solar maximum. During years of moderate solar activity, 12 meters opens to the low and middle latitudes during the daytime hours, but it seldom remains open after sunset. Periods of low solar activity seldom cause this band to go completely dead, except at higher latitudes. Occasional daytime openings, especially in the lower latitudes, are likely over north-south paths. The main sporadic E season on 24 MHz lasts from late spring through summer and short openings may be observed in mid-winter.

28.0-29.7 MHz (10 meters)

The 10 meter band is well known for extreme variations in characteristics and a variety of propagation modes. During solar maxima, long-distance F_2 propagation is so efficient that very low power can produce strong signals halfway around the globe. DX is abundant with modest equipment. Under these conditions, the band is usually open from sunrise to a few hours past sunset. During periods of moderate solar activity, 10 meters usually opens only to low and transequatorial latitudes around noon. During the solar minimum, there may be no F_2 propagation at any time during the day or night.

Sporadic E is fairly common on 10 m, especially May through August, although it may appear at any time. Short skip, as sporadic E is sometimes called on the HF bands, has little relation to the solar cycle and occurs regardless of F layer conditions. It provides single-hop communication from 300 to 2300 km (190 to 1400 miles) and multiple-hop opportunities of 4500 km (2800 miles) and farther.

Ten meters is a transitional band in that it also shares some of the propagation modes more characteristic of VHF. Meteor scatter, aurora, auroral E and transequatorial propagation provide the means of making contacts out to 2300 km (1400 miles) and farther, but these modes often go unnoticed at 28 MHz. Techniques similar to those used at VHF can be very effective on 10 meters, as signals are usually stronger and more persistent. These exotic modes can be more fully exploited, especially during the solar minimum when F_2 DXing has waned.

VERY HIGH FREQUENCY (VHF) BANDS (30-300 MHz)

A wide variety of propagation modes are useful in the VHF range. F layer skip appears on 50 MHz during solar cycle peaks. Sporadic E and several other E layer phenomena are most effective in the VHF range. Still other forms of VHF ionospheric propagation, such as field-aligned irregularities (FAI) and transequatorial propagation (TE), are rarely observed at VHF. Tropospheric propagation, which is not a factor at HF, becomes increasingly important above 50 MHz.

50-54 MHz (6 meters)

The lowest amateur VHF band shares many of the characteristics of both lower and higher frequencies. In the absence of any favorable ionospheric propagation conditions, well-equipped 50 MHz stations work regularly over a radius of 300 km (190 miles) via tropospheric scatter, depending on terrain, power, receiver capabilities and antenna. Weak-signal troposcatter allows the best stations to make 500 km (310 mile) contacts nearly any time. Weather effects may extend the normal range by a few hundred km, especially during the summer months, but true tropospheric ducting is rare.

During the peak of the 11-year sunspot cycle (especially during the winter months), worldwide 50 MHz DX is possible via the F_2 layer during daylight hours. F_2 backscatter provides an additional propagation mode for contacts as far as 4000 km (2500 miles) when the MUF is just below 50 MHz. TE paths as long as 8000 km (5000 miles) across the magnetic equator are common around the spring and fall equinoxes of peak solar cycle years.

Sporadic E is probably the most common and certainly the most popular form of propagation on the 6 meter band. Single-hop E-skip openings may last many hours for contacts from 600 to 2300 km (370 to 1400 miles), primarily during the spring and early summer. Multiple-hop E_s provides transcontinental contacts several times a year, and contacts between the US and South America, Europe and Japan via multiple-hop E-skip occur nearly every summer.

Other types of E layer ionospheric propagation make 6 meters an exciting band. Maximum distances of about 2300 km (1400 miles) are typical for all types of E layer modes. Propagation via FAI often provides additional hours of contacts immediately following sporadic E events. Auroral propagation often makes its appearance in late afternoon when the geomagnetic field is disturbed. Closely related auroral E propagation may extend the 6 meter range to 4000 km (2500 miles) and sometimes farther across the northern states and Canada, usually after midnight. Meteor scatter provides brief contacts during the early morning hours, especially during one of the dozen or so prominent annual meteor showers.

144-148 MHz (2 meters)

Ionospheric effects are significantly reduced at 144 MHz, but they are far from absent. F layer propagation is unknown except for TE, which is responsible for the current 144 MHz terrestrial DX record of nearly 8000 km (5000 miles). Sporadic E occurs as high as 144 MHz less than a tenth as often as at 50 MHz, but the usual maximum single-hop distance is the same, about 2300 km (1400 miles). Multiple-hop sporadic E contacts greater than 3000 km (1900 miles) have occurred from time to time across the continental US, as well as across Southern Europe.

Auroral propagation is quite similar to that found at 50 MHz, except that signals are weaker and more Doppler-distorted. Auroral E contacts are rare. Meteor-scatter contacts are limited primarily to the periods of the great annual meteor showers and require much patience and operating skill. Contacts have been made via FAI on 144 MHz, but its potential has not been fully explored.

Tropospheric effects improve with increasing frequency, and 144 MHz is the lowest VHF band at which terrestrial weather plays an important propagation role. Weather-induced enhancements may extend the normal 300 to 600 km (190 to 370 mile) range of well-equipped stations to 800 km (500 miles) and more, especially during the summer and early fall. Tropospheric ducting extends this range to 2000 km (1200 miles) and farther over the continent and at least to 4000 km (2500 miles) over some well-known all-water paths, such as that between California and Hawaii.

222-225 MHz (135 cm)

The 135 cm band shares many characteristics with the 2 meter band. The normal working range of 222 MHz stations is nearly as far as comparably equipped 144 MHz stations. The 135 cm band is slightly more sensitive to tropospheric effects, but ionospheric modes are more difficult to use. Auroral and meteor-scatter signals are somewhat weaker than at 144 MHz, and sporadic E contacts on 222 MHz are extremely rare. FAI and TE may also be well within the possibilities of 222 MHz, but reports of these modes on the 135 cm band are uncommon. Increased activity on 222 MHz will eventually reveal the extent of the propagation modes on the highest of the amateur VHF bands.

ULTRA-HIGH FREQUENCY (UHF) BANDS (300-3000 MHz) AND HIGHER

Tropospheric propagation dominates the bands at UHF and higher, although some forms of E layer propagation are still useful at 432 MHz. Above 10 GHz, atmospheric attenuation increasingly becomes the limiting factor over long-distance paths. Reflections from airplanes, mountains and other stationary objects may be useful adjuncts to propagation at 432 MHz and higher.

420-450 MHz (70 cm)

The lowest amateur UHF band marks the highest frequency on which ionospheric propagation is commonly observed. Auroral signals are weaker and more Doppler distorted; the range is usually less than at 144 or 222 MHz. Meteor scatter is much more difficult than on the lower bands, because bursts are significantly weaker and of much shorter duration. Although sporadic E and FAI are unknown as high as 432 MHz and probably impossible, TE may be possible.

Well-equipped 432 MHz stations can expect to work over a radius of at least 300 km (190 miles) in the absence of any propagation enhancement. Tropospheric refraction is more pronounced at 432 MHz and provides the most frequent and useful means of extended-range contacts. Tropospheric ducting supports contacts of 1500 km (930 miles) and farther over land. The current 432 MHz terrestrial DX record of more than 4000 km (2500 miles) was accomplished by ducting over water.

The crest of a range of hills or mountains 50 to 100 wavelengths long can produce knife-edge diffraction at UHF and microwave frequencies. Hillcrests that are clearly defined and free of trees, buildings and other clutter make the best knife-edges, but even rounded hills may serve as a diffracting edge. Alternating bands of strong and weak signals, corresponding to the interference pattern, will appear on the surface of the Earth behind the mountain, known as the *shadow zone*. The phenomenon is generally reciprocal, so that two-way communication can be established under optimal conditions. Knife-edge diffraction can make it possible to complete paths of 100 km or more that might otherwise be entirely obstructed by mountains or seemingly impossible terrain.

19.1.7 Ground Wave

A *ground wave* is the result of a special form of diffraction that primarily affects longer-wavelength vertically polarized radio waves. It is most apparent in the 80 and 160 meter amateur bands, where practical ground-wave distances may extend beyond 200 km (120 miles). It is also the primary mechanism used by AM broadcast stations in the medium-wave bands. The term ground wave is often mistakenly applied to any short-distance communication, but the actual mechanism is unique to the longer-wave bands.

Radio waves are bent slightly as they pass over a sharp edge, but the effect extends to edges that are considerably rounded. At medium and long wavelengths, the curvature of the Earth looks like a rounded edge. Bending results when the lower part of the wave front loses energy due to currents induced in the ground. This slows down the lower part of the wave, causing the entire wave to tilt forward slightly. This tilting follows the curvature of the Earth, thus allowing low- and medium-wave radio signals to propagate over distances well beyond line of sight.

Ground wave is most useful during the day at 1.8 and 3.5 MHz, when D layer absorption makes skywave propagation more difficult. Vertically polarized antennas with excellent ground systems provide the best results. Ground-wave losses are reduced considerably over saltwater and are worst over dry and rocky land.

19.2 Sky-Wave Propagation and the Sun

The Earth's atmosphere is composed primarily of nitrogen (78%), oxygen (21%) and argon (1%), with smaller amounts of a dozen other gases. Water vapor can account for as much as 5% of the atmosphere under certain conditions. This ratio of gases is maintained until an altitude of about 80 km (50 miles), when the mix begins to change. At the highest levels, helium and hydrogen predominate.

Solar radiation acts directly or indirectly on all levels of the atmosphere. Adjacent to the surface of the Earth, solar warming controls all aspects of the weather, powering wind, rain and other familiar phenomena. *Solar ultraviolet (UV) radiation* creates small concentrations of ozone (O_3) molecules between 10 and 50 km (6 and 30 miles). Most UV radiation is absorbed by this process and never reaches the Earth.

At even higher altitudes, UV and X-ray radiation partially ionize atmospheric gases. Electrons freed from gas atoms eventually recombine with positive ions to recreate neutral gas atoms, but this takes some time. In the low-pressure environment at the highest altitudes, atoms are spaced far apart and the gases may remain ionized for many hours. At lower altitudes, recombination happens rather quickly, and only constant radiation can keep any appreciable portion of the gas ionized.

19.2.1 Structure of the Earth's Atmosphere

The atmosphere, which reaches to more than 600 km (370 miles) altitude, is usually divided into a number of regions based on a transitioning characteristic of the atmosphere — like temperature. For propagation purposes, the important regions are shown in **Fig 19.5**. The weather-producing *troposphere* lies between the surface and an average altitude of 10 km (6 miles). Between 10 and 50 km (6 and 30 miles) are the *stratosphere* and the embedded *ozonosphere*, where ultraviolet-absorbing ozone reaches its highest concentrations. About 99% of atmospheric gases are contained within these two lowest regions.

Above 50 km to about 600 km (370 miles) is the *ionosphere*, notable for its effects on radio propagation. At these altitudes, atomic oxygen, molecular oxygen, molecular nitrogen, and nitric oxide predominate under very low pressure and are the important species to consider for propagation. High-energy solar UV and X-ray radiation ionize these constituents, creating a broad region where ions are created in relative abundance. The ionosphere is subdivided into distinctive D, E and F regions.

The *magnetosphere* begins around 600 km (370 miles) and extends as far as 160,000 km (100,000 miles) into space. The predominant component of atmospheric gases gradually shifts from atomic oxygen, to helium and finally to hydrogen at the highest levels. The lighter gases may reach escape velocity or be swept off the atmosphere by the *solar wind* (electrically charged particles emitted by the Sun and traveling through space). At about 3200 and 16,000 km (2000 and 9900 miles, respectively), the Earth's magnetic field traps energetic electrons and protons in two bands, known as the *Van Allen belts*.

These have only a minor effect on terrestrial radio propagation.

19.2.2 The Ionosphere

The ionosphere plays a basic role in long-distance communications in all the amateur bands from 1.8 MHz to 30 MHz. The effects of the ionosphere are less apparent at the very high frequencies (30-300 MHz), but they persist at least through 432 MHz. As early as 1902, Oliver Heaviside and Arthur E. Kennelly independently suggested the existence of a layer in the upper atmosphere that could account for the long-distance radio transmissions made the previous year by Guglielmo Marconi and others. Edward Appleton confirmed the existence of the Kennelly-Heaviside layer in publications beginning in 1925 and used the letter E on his diagrams to designate the strength of the electric field of the waves that were apparently reflected from the layer he measured.

In late 1927 Appleton reported the existence of an additional layer in the "ionosphere." (Robert Watson-Watt coined the term "ionosphere", but it wasn't commonly used until 1932.) The additional higher-altitude layer was named "F" and subsequently another was termed "D." For a time during the 1930s, a "C" layer was proposed, and later discarded. Appleton was reluctant to alter this arbitrary nomenclature for fear of discovering yet other lower layers, so it has stuck to the present day. The basic physics of ionospheric propagation was largely worked out by the 1930s, yet both amateur and professional experimenters made further discoveries during the 1930s, 1940s and 1950s. Sporadic E, aurora, meteor scatter and several types of field-aligned scattering were among additional ionospheric phenomena that required explanation.

Although the term "layer" is used in this chapter, this could lead to the erroneous assumption that the ionosphere consists of distinct thin sheets separated by emptiness in between. This is not so, and as we'll see later in the chapter the ionosphere is a continuous electronic density versus altitude, with definite peaks and inflections points that define the D, E, and F regions. Studies have shown that the height of the various layers may also vary by latitude. Research into ionospheric physics is ongoing in an attempt to better understand the variability of the ionosphere.

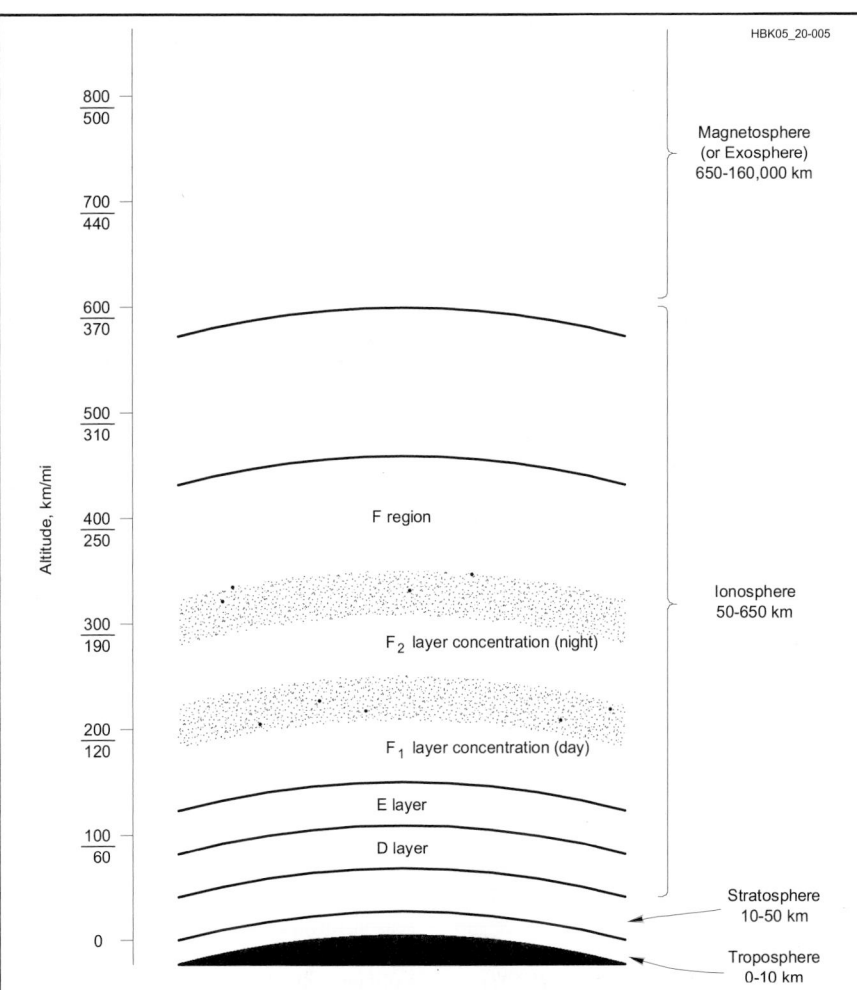

Fig 19.5 — Regions of the lower atmosphere and the ionosphere.

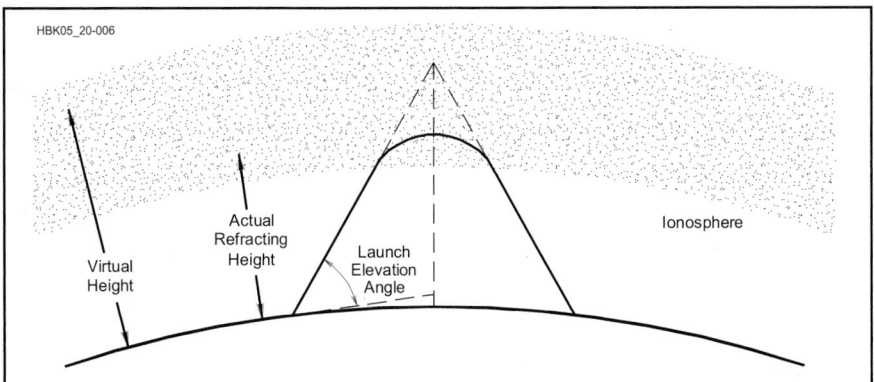

Fig 19.6 — Gradual refraction in the ionosphere allows radio signals to be propagated long distances. It is often convenient to imagine the process as a reflection with an imaginary reflection point at some virtual height above the actual refracting region. The other figures in this chapter show ray paths as equivalent reflections, but you should keep in mind that the actual process is a gradual refraction.

19.2.3 Ionospheric Refraction

The refractive index of an ionospheric layer decreases from a value of 1.00 as the density of free-moving electrons increases (this is opposite from the refractive index in the troposphere since the ionosphere is a dispersive medium). In the densest regions of the F layer, that density can reach a trillion electrons per cubic meter (10^{12} e/m^3). Even at this high level, radio waves are refracted gradually over a considerable vertical distance, usually amounting to tens of km. Radio waves become useful for terrestrial propagation only when they are refracted enough to bring them back to Earth. See **Fig 19.6**.

Although refraction is the primary mechanism of ionospheric propagation, it is usually more convenient to think of the process as a reflection. The *virtual height* of an ionospheric layer is the equivalent altitude of a reflection that would produce the same effect as the actual refraction. The virtual height of

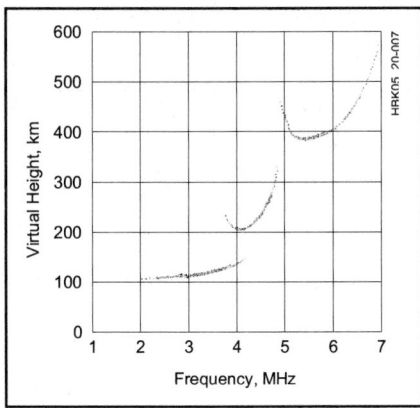

Fig 19.7 — Simplified vertical incidence ionogram showing echoes returned from the E, F₁ and F₂ layers. The critical frequencies of each layer (4.1, 4.8 and 6.8 MHz) can be read directly from the ionogram scale.

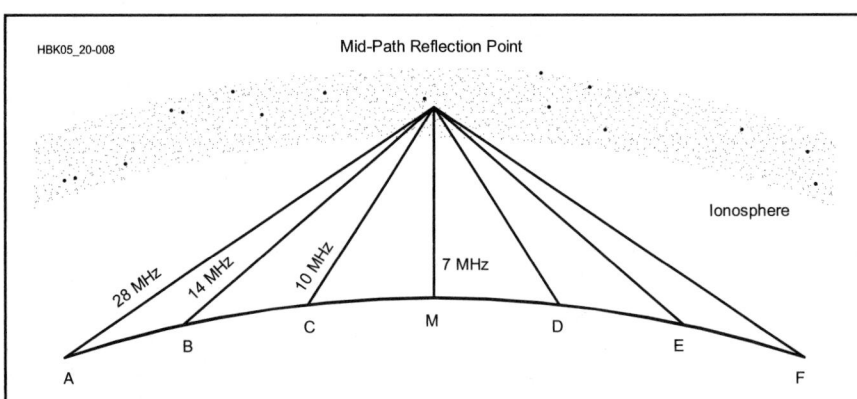

Fig 19.8 — The relationships between critical frequency, maximum usable frequency (MUF) and skip zone can be visualized in this simplified, hypothetical case. The critical frequency is 7 MHz, allowing frequencies below this to be used for short-distance ionospheric communication by stations in the vicinity of point M. These stations cannot communicate by the ionosphere at 14 MHz. Stations at points B and E (and beyond) can communicate because signals at this frequency are refracted back to Earth because they encounter the ionosphere at an oblique angle of incidence. At greater distances, higher frequencies can be used because the MUF is higher at the larger angles of incidence (low launch angles). In this figure, the MUF for the path between points A and F, with a small launch angle, is shown to be 28 MHz. Each pair of stations can communicate at frequencies at or below the MUF of the path between them, but not below the LUF — see text.

any ionospheric layer can be determined using an ionospheric sounder, or *ionosonde*, a sort of vertically oriented radar. The ionosonde sends pulses that sweep over a wide frequency range, generally from 2 MHz to 20 MHz or higher, straight up into the ionosphere. The frequencies of any echoes are recorded against time and then plotted as distance on an *ionogram*. **Fig 19.7** depicts a simple ionogram. Real-time ionograms can be found online at the Digital Ionogram Database, sponsored by the University of Massachusetts Lowell (**http://umlcar.uml.edu/DIDBase/**). For an extensive discussion of ionogram interpretation, download *UAG-23A: URSI Handbook of Ionogram Interpretation and Reduction* from the Australian IPS (Ionospheric Prediction Service) website at **www.ips.gov.au/IPSHosted/INAG/uag_23a/uag_23a.html**.

The highest frequency that returns echoes from the E and F regions at vertical incidence is known as the *vertical incidence* or *critical frequency*. (There is a D region critical frequency, but its value is well below 1 MHz and thus has minimal impact on our Amateur Radio bands.) The critical frequency is a function of ion density. The higher the ionization at a particular altitude, the higher becomes the critical frequency. Strictly speaking, the critical frequency is the term applicable to the peak electron density of a region. Physicists call any electron density in any part of the ionosphere a *plasma frequency*, because technically gases in the ionosphere are in a plasma, or partially ionized state. F layer critical frequencies commonly range from about 1 MHz to as high as 15 MHz.

19.2.4 Maximum and Lowest Usable Frequencies

When the frequency of a vertically incident signal is raised above the critical frequency of an ionospheric layer, that portion of the ionosphere is unable to refract the signal back to Earth. However, a signal above the critical frequency may be returned to Earth if it enters the layer at an *oblique angle*, rather than at vertical incidence. This is fortunate because it permits two widely separated stations to communicate on significantly higher frequencies than the critical frequency. See **Fig 19.8**.

The highest frequency supported by the ionosphere between two stations is the *maximum usable frequency* (MUF) for that path. If the separation between the stations is increased, a still higher frequency can be supported at lower launch angles. The MUF for this longer path is higher than the MUF for the shorter path because more refraction can occur

for electromagnetic waves that encounter the ionosphere at more oblique angles. When the distance is increased to the maximum one-hop distance, the launch angle of the signals between the two stations is zero (that is, the ray path is tangential to the Earth at the two stations) and the MUF for this path is the highest that can be supported by that layer of the ionosphere at that location. This maximum distance is about 4000 km (2500 miles) for the F₂ layer and about 2000 km (1250 miles) for the E layer. See **Fig 19.9**.

The MUF is a function of path, time of day, season, location, solar UV and X-ray radiation levels and ionospheric disturbances. For vertically incident waves, the MUF is the same as the critical frequency. For path lengths at the

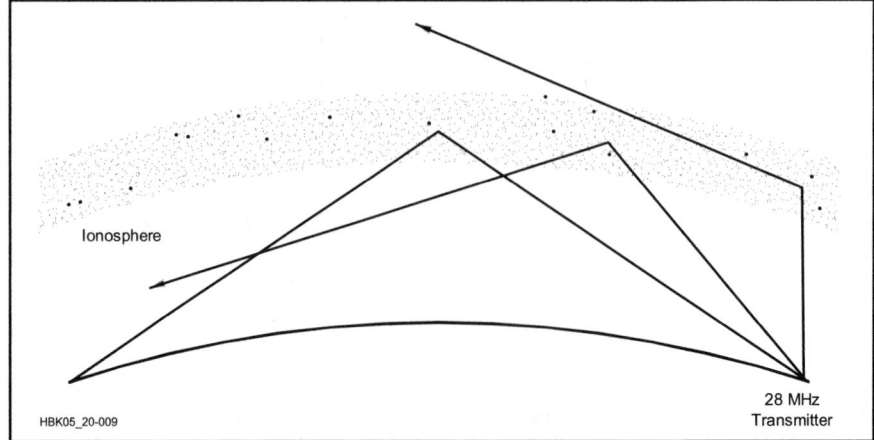

Fig 19.9 — Signals at the MUF propagated at a low angle to the horizon provide the longest possible one-hop distances. In this example, 28 MHz signals entering the ionosphere at higher angles are not refracted enough to bring them back to Earth.

limit of one-hop propagation, the MUF can be several times the critical frequency. The ratio between the MUF and the critical frequency is known as the *M-factor*.

The M-factor can be estimated using simple geometry in a spherical model of the Earth-ionosphere system. The angle of incidence on the ionosphere of an electromagnetic wave launched from the ground depends on the launch angle and the height of the ionospheric region. Due to the spherical geometry of the Earth-ionosphere system, the angle of incidence on the ionosphere does not approach zero as the launch angle approaches zero — it is limited to approximately 19° and 11° for the F_2 and E regions, respectively, which then limits the M-factor to approximately 3 and 5 for these two regions (from the equation MUF = 1 / the sine of the angle of incidence on the ionosphere). See **Table 19.2** for typical M-factors of the various regions.

The term *skip zone* is closely related to MUF. When two stations are unable to communicate with each other on a particular frequency because the ionosphere is unable to refract the signal enough from one to the other through the required angle — that is, the operating frequency is above the MUF — the stations are said to be in the skip zone for that frequency. Stations within the skip zone may be able to work each other on a lower frequency, or by ground wave or other mechanisms if they are close enough. There is no skip zone at frequencies below the MUF.

The MUF at any time on a particular path is just that — the *maximum* usable frequency. Frequencies below the MUF will also propagate along the path, but ionospheric absorption and noise at the receiving location (due to man-made noise and/or noise from local or distant thunderstorms) may make the received signal-to-noise ratio too low to be usable. In this case, the frequency is said to be below the *lowest usable frequency* (LUF). This occurs most frequently below 10 MHz, where atmospheric and man-made noises are most troublesome.

The LUF can be lowered somewhat by the use of high power and directive antennas, or through the use of communication modes that permit reduced receiver bandwidth or are less demanding of SNR — CW or PSK31 instead of SSB, for example. This is not true of the MUF, which is limited by the physics of ionospheric refraction, no matter how high your transmitter power or how narrow your receiver bandwidth. The LUF can be higher than the MUF. This is a common occurrence on 160 meters during the day due to too much absorption; another scenario would be too much noise on the higher bands due to thunderstorm activity or man-made noise. When the LUF is higher than the MUF, there is no frequency that supports communication on the particular path at that time.

19.2.5 NVIS Propagation

In the previous section, the statement was made that *stations within the skip zone may be able to work each other at a lower frequency, or by ground wave if they are close enough.* This statement summarizes the purpose of *Near Vertical Incidence Skywave (NVIS)* propagation — to bridge the gap between where ground wave is too weak and where the skip zone ends. By going to lower frequencies, communications can be maintained over these relatively short distances.

Propagation over short distances means high elevation angles — this is not DXing in which lower angles in general are most effective. For example, a path from San Francisco to Sacramento is 75 miles, and requires an average elevation angle of 78 degrees. To radiate maximum energy at these higher angles, relatively low height antennas need to be used.

In the December 2005 issue of *QST*, Dean Straw N6BV used the *VOACAP* propagation prediction program to analyze a variety of NVIS paths centered on San Francisco. His analysis showed area coverage maps (signal strength contours versus distance from the transmitter) and elevation patterns of antennas at various height. His analysis allowed him to formulate a very nice summary: *As a rule-of-thumb, for ham band NVIS, I would recommend that 40 meters be used during the day; 80 meters during the night.*

19.2.6 Ionospheric Fading

HF signal strengths typically rise and fall over periods of a few seconds to several minutes, and rarely hold at a constant level for very long. Fading is generally caused by the interaction of several radio waves from the same source arriving along different propagation paths. Waves that arrive in-phase combine to produce a stronger signal, while those out-of-phase cause destructive interference and lower net signal strength. Short-term variations in ionospheric conditions may change individual path lengths or signal strengths enough to cause fading. Even signals that arrive primarily over a single path may vary as the propagating medium changes. Fading may be most notable at sunrise and sunset, especially near the MUF, when the ionosphere undergoes dramatic transformations. Other ionospheric traumas, such as auroras and geomagnetic storms, also produce severe forms of HF fading.

19.2.7 Polarization at HF

Although the ionosphere varies on a short-term basis and results in somewhat random polarization, there is more order to polarization than realized. The ionosphere is immersed in the Earth's magnetic field, and the result of an electromagnetic wave propagating through an ionized medium (called a plasma, which is what our ionosphere is) immersed in a magnetic field is the propagation of two characteristic waves: the ordinary wave (O-wave) and the extraordinary wave (X-wave). Ordinary and extraordinary are terms borrowed from the science of optics. These two waves are generally elliptically polarized (the tip of the polarization vector traces out an ellipse), and are orthogonal to each other.

Upon entering the ionosphere, and depending on the location and the heading, a linearly polarized wave (from our horizontal or vertical antenna) will either couple all of its energy into the O-wave, all of its energy into the X-wave, or divide its energy into the O-wave and X-wave. Our horizontal antenna will couple best into one characteristic wave, and our vertical antenna will couple best into the other characteristic wave. The same coupling issue is present when the characteristic waves exit the ionosphere. The bottom line is that on the HF bands of 80 meters and higher, both characteristic waves generally propagate with similar absorption, and thus the use of a vertical or horizontal antenna is not too critical with respect to polarization. One of the characteristic waves will couple into whichever antenna is being used.

On 160 meters, though, the X-wave is heavily absorbed, leaving the O-wave. In general, vertical polarization is best for those of us in North America and at high northern latitudes. However, 160 meter operators have observed that under some conditions,

---

Table 19.2

Maximum Usable Frequency Factors (M-factors) for 2000 km E Hops and 3000 km F Hops

Layer	Maximum Critical Frequency (MHz)	M-factor	Useful Operating Frequencies (MHz)
F_2	15.0	3.3-4.0	1-60
F_1*	5.5	4.0	10-20
E*	4.0	4.8	5-20
Es	30.0	5.3	20-160

*Daylight only

horizontally polarized antennas outperform vertically polarized antennas.

19.2.8 The 11-Year Solar Cycle

The density of ionospheric layers depends on the amount of solar radiation reaching the Earth, but solar radiation is not constant. Variations result from daily and seasonal motions of the Earth, the Sun's own 27-day rotation and the 11-year cycle of solar activity. One visual indicator of both the Sun's rotation and the solar cycle is the periodic appearance of dark spots on the Sun, which have been observed continuously since the mid-18th century. On average, the number of *sunspots* reaches a maximum every 10.7 years, but the period has varied between 7 and 17 years. Cycle 19 peaked in 1958, with a smoothed sunspot number of 201, the highest recorded to date. **Fig 19.10** shows monthly mean sunspot numbers for the past five cycles (it also includes the smoothed values).

Sunspots are cooler areas on the Sun's surface associated with high magnetic activity. Active regions adjacent to sunspot groups, called *plages*, are capable of producing great flares and sustained bursts of radiation in the radio through X-ray spectrum. During the peak of the 11-year solar cycle, average solar radiation increases along with the number of flares and sunspots. The ionosphere becomes more intensely ionized as a consequence, resulting in higher critical frequencies, particularly in the F_2 layer. The possibilities for long-distance communications are considerably improved during solar maxima, especially in the higher-frequency bands.

One key to forecasting F layer critical frequencies, and thus long-distance propagation, is the intensity of ionizing UV and X-ray radiation. Until the advent of satellites, UV and X-ray radiation could not be measured directly, because they were almost entirely absorbed in the upper atmosphere during the process of ionization. The sunspot number provided the most convenient approximation of general solar activity. The sunspot number is not a simple count of the number of visual spots, but rather the result of a complicated formula that takes into consideration size, number and grouping. The smoothed sunspot number (equivalent to a running average of monthly mean sunspot numbers from 6 months before to 6 months after the desired month) varies from near zero during the solar cycle minimum to over 200 during an extremely active solar cycle.

Another method of gauging solar activity is the *solar flux*, which is a measure of the intensity of 2800 MHz (10.7 cm) radio noise coming from the Sun (10^{-22} W m^{-2} Hz^{-1}). The smoothed 2800 MHz radio flux is an indication of the intensity of ionizing UV and X-ray radiation and provides a convenient alternative to sunspot numbers. Solar flux values commonly vary on a scale of 60-300 and can be related to sunspot numbers, as shown in **Fig 19.11** (note that this is only valid for converting between smoothed values). The Dominion Radio Astrophysical Observatory, Penticton, British Columbia, measures the 2800 MHz solar flux three times a day (centered on local noon). (A supplemental article, "The Penticton Solar Flux Receiver" from February 2013 *QST* is on the CD-ROM that comes with this book.) Radio station WWV broadcasts the latest solar-flux index at 18 minutes after each hour; WWVH does the same at 45 minutes after the hour. Solar flux and other useful data is also available online at **www.swpc.noaa.gov**.

The Penticton solar flux is employed in a wide variety of other applications. Daily, weekly, monthly and even smoothed solar flux readings are commonly used in propagation predictions (but be aware that our HF propagation predictions are based on the correlation between smoothed solar flux or smoothed sunspot number and monthly median ionospheric parameters — more on this later in the chapter).

High flux values generally result in higher MUFs, but the actual procedures for predicting the MUF at any given hour and path is quite complicated. Solar flux is not the sole determinant, as the angle of the Sun to the Earth, season, time of day, exact location of the radio path and other factors must all be taken into account. MUF forecasting a few days or months ahead involves additional variables and even more uncertainties.

TRENDS IN SOLAR CYCLES

If one looks at all 23 recorded solar cycles, three characteristics stand out. First, we see a cyclic nature to the maximum smoothed sunspot numbers. Second, we've been through three high cycle periods of 50 years or so (consisting of several solar cycles) and two low cycle periods of 50 years or so (again, consisting of several solar cycles), and we appear to be headed into a third low cycle period. Third, we've lived through the highest period of high cycle activity (but this is currently being challenged by solar scientists through an

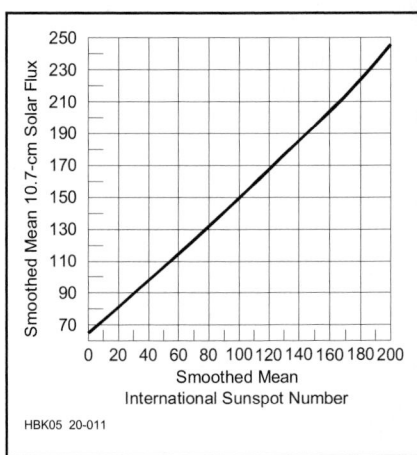

Fig 19.11 — **Approximate conversion between solar flux and sunspot number. Note that this is for smoothed values of Solar Flux and Sunspot Numbers.**

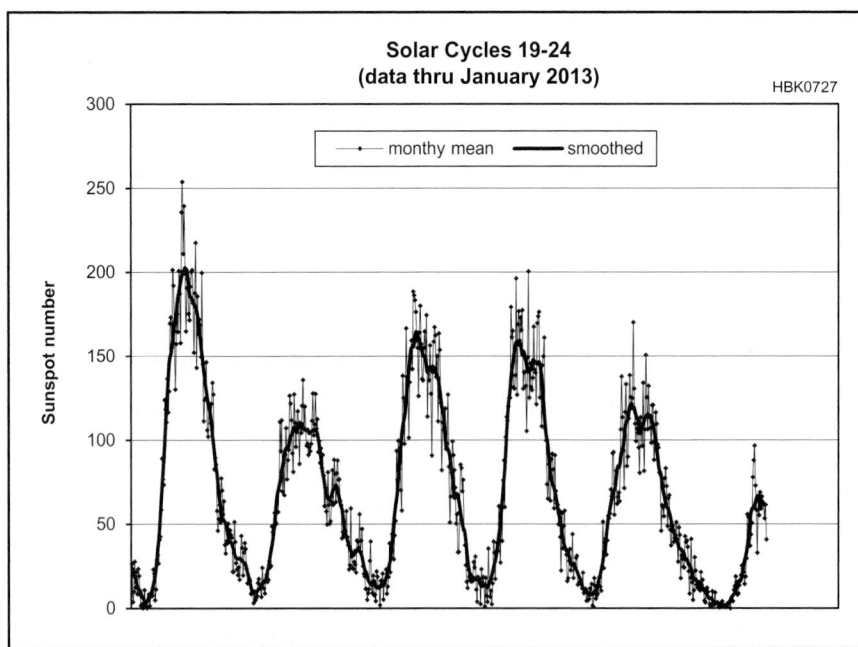

Fig 19.10 — **Monthly mean sunspot numbers for Solar Cycles 19 to 24 (through February 2013). This data also includes the smoothed values.**

Status of Solar Cycle 24

(This information was prepared in early February 2013.)

Fig 19A1 updates the monthly mean 10.7 cm solar flux data through February 2013 and the smoothed 10.7 cm solar flux through July 2012. The most obvious observation of this data is the peak in the smoothed solar flux in early 2012 (as a reminder, the smoothed data is how solar cycles are officially measured). This peak is earlier than the two most notable predictions from solar scientists (at ISES, the International Space Environment Service, and at MSFC, the Marshall Space Flight Center). Both of these predictions peg the maximum this year — not last year.

The January 2013 monthly mean data certainly showed a nice increase in solar activity. But if it doesn't continue in the months to come, then indeed early 2012 may be the peak of Cycle 24.

I believe the best way to approach this is to assume that the peak of Cycle 24 occurred, and take advantage of the higher bands (15 meters, 12 meters and especially 10 meters) *right now*. They should be good through the fall of 2013, but after that consistent worldwide propagation will begin waning. And looking at the peaks of all 23 solar cycles suggests we're headed into several cycles of low activity. See **Fig 19A2**. So taking the attitude that "I'll wait for Cycle 25" may not be a viable solution. — *Carl Luetzelschwab, K9LA*

Fig 19A2 — Maximum smoothed sunspot number of all 23 solar cycles, showing three periods of high activity (large solar cycles) and two periods of low activity (small solar cycles). We appear to be headed for a period of low activity (small cycles).

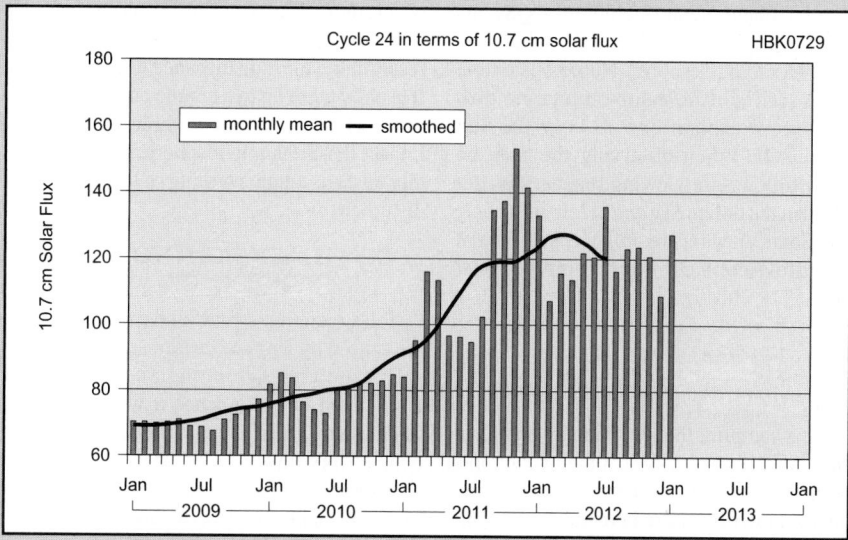

Fig 19A1 — Monthly mean and smoothed 10.7 cm solar flux data through February 2013 and July 2012, respectively.

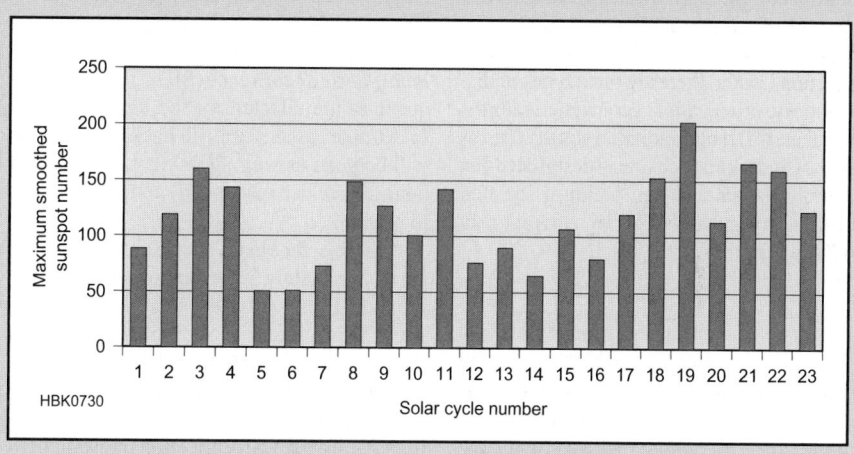

---

in-depth review of old sunspot records), which has allowed excellent worldwide propagation on the higher frequency bands and provided great enjoyment for radio amateurs.

If we look at solar cycles prior to recorded history through various proxies for solar activity (for example, carbon-14 in trees and beryllium-10 in ice cores), we'll see extended periods of very low solar activity referred to as Grand Minima (for example, the Maunder Minimum between 1640 and 1710 AD). It's likely that we'll again enter one of these extended periods, but trying to predict when this will happen is, at best, a wild guess.

With respect to solar minimum periods between solar cycles, historical data shows a great variation. The average length of solar minimum, for example defined as the number of months in which the smoothed sunspot number is below 20, is around 37 months. Using this definition, the shortest minimum was 17 months (between Cycles 1 and 2) and the longest minimum was 96 months (between Cycles 5 and 6). The minimum period between Cycle 23 and Cycle 24 turned out to be 56 months. Interestingly, up until the minimum between Cycle 23 and Cycle 24, in our lifetimes we have experienced minimum periods of approximately 24 months — much shorter than the average. This leads us to believe this recent solar minimum period was unusual. But historical data, with its great amount of variation, says otherwise.

19.2.9 The Sun's 27-Day Rotation

Sunspot observations also reveal that the Sun rotates on its own axis. The Sun is composed of extremely hot gases and does not turn uniformly (it is essentially a fluid). At the equator, the period is just over 25 days, but it approaches 35 days at the poles. Sunspots that affect the Earth's ionosphere, which appear almost entirely within 35° of the Sun's equator, take about 26 days for one rotation. After taking into account the Earth's movement around the Sun, the apparent period of solar rotation is about 27 days.

Active regions must face the Earth in the proper orientation to have an impact on the ionosphere. They may face the Earth only once before rotating out of view, but they often persist for several solar rotations. The net effect is that solar activity often appears in 27-day cycles corresponding to the Sun's rotation, even though the active regions themselves may last for several solar rotations.

The CD-ROM that accompanies this book includes supplemental articles about observing solar phenomena, including a *QST* article by Mark Spencer, WA8SME on building your own radio telescope. Numerous online sources such as **http://solar-center.stanford. edu/observe** explain how to view the Sun for yourself. When observing the Sun, be sure to follow safe viewing practices as described in the online *Sky and Telescope* article "Safe Solar Viewing" by Jeff Medkeff (**www. skyandtelescope.com/observing/objects/ sun/3309106.html**).

19.2.10 Disturbances to Propagation

Like a campfire that occasionally spits out a flaming ember, our Sun sometimes erupts spasmodically — but on a much grander scale than a summer campfire here on Earth. After all, any event that violently releases as much as 10 billion tons of solar material traveling up to four and a half million miles per hour or releases large amounts of electromagnetic radiation at extremely short wavelengths has to be considered pretty impressive!

Following the lead of the Space Weather Prediction Center, there are three types of disturbances to propagation: geomagnetic storms (designated G), solar radiation storms (designated S) and radio blackouts (designated R). For more details and the scaling associated with these designators, visit **www.swpc.noaa. gov/NOAAscales/**.

GEOMAGNETIC STORMS

Geomagnetic storms are generally caused by *coronal mass ejections* and high speed wind streams from *coronal holes*. A *coronal mass ejection* (CME) originates in the Sun's outer atmosphere — its corona. With several sophisticated satellites launched in the mid 1990s, we have gained powerful new tools to monitor the intricacies of solar activity. Using the latest satellite technology (and also some re-engineered earthbound instruments), scientists have observed many CMEs, greatly expanding our knowledge about them. Previously, the only direct observations we had of coronal activity were during solar eclipses — and eclipses don't occur very often. CMEs are observed with an instrument called a *coronagraph*, which has an occulting disk to block out the main portion of the Sun. In essence a coronagraph creates an artificial eclipse.

A *coronal hole* is a region on the Sun where the magnetic field is open to the interplanetary magnetic field (IMF) and ionized particles can escape into the solar wind. Normally the solar wind blows at approximately 400 km per second. During CMEs and coronal holes, solar wind speeds can increase to 2000 km per second.

Coronal holes and coronal mass ejections that are Earth-directed (these are also called full halo CMEs, as the explosion surrounds the occulting disk of a coronagraph) concurrent with the IMF oriented in a southerly direction result in the most disturbance to propagation. It usually takes up to a couple days for a CME or the effects of a coronal hole to reach and impact the Earth's ionosphere, so this generally gives us ample warning of the impending disturbance.

SOLAR RADIATION STORMS AND RADIO BLACKOUTS

Solar radiation storms and radio blackouts are caused by large *solar flares*. When a large solar flare erupts from the Sun's surface, it can launch out into space a wide spectrum of electromagnetic energy. Since electromagnetic energy travels at the speed of light, the first indication of a solar flare reaches the Earth in about eight minutes. A large flare shows up as an increase in visible brightness near a sunspot group, accompanied by increases in UV and X-ray radiation and high levels of noise in the VHF radio bands. It is the X-ray radiation that results in radio blackouts on the daytime side of the Earth due to increased D region absorption, and this is called a *sudden ionospheric disturbance* (SID). The lower frequencies are affected for the longest period. In extreme cases, nearly all background noise will be gone as well. SIDs may last up to an hour, after which ionospheric conditions return to normal.

A large solar flare can also release matter into space, mainly in the form of very energetic protons. These cause solar radiation storms, whereby increased absorption in the polar cap (that area inside the auroral oval) degrades over-the-pole paths. This is called a *polar cap absorption* (PCA) event. A PCA event may last for days, dramatically affecting transpolar HF propagation. An interesting fact with respect to PCAs is that they do not necessarily affect the northern and southern polar regions similarly. Thus if the short path between two points is degraded over one pole, the long path may still be available over the other pole.

At one time, scientists believed that solar flares and CMEs were causally related, but now they recognize that many CMEs occur without an accompanying flare. And while many flares do result in an ejection of some solar material, many do not. It now seems clear that flares don't cause CMEs and vice versa.

Since geomagnetic storms have the most impact to propagation, it is instructive to understand their occurrence statistics. The number of geomagnetic storms varies considerably from year to year, but peak geomagnetic activity follows the peak of solar activity. See **Fig 19.12**. Also, geomagnetic activity affects the ionosphere mostly in the equinox months (March and September).

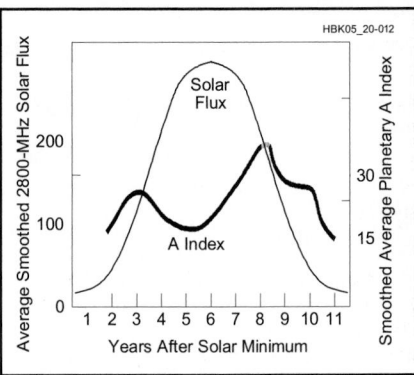

Fig 19.12 — Geomagnetic activity (measured as the A index) also follows an 11-year cycle. Average values over the past few cycles show that geomagnetic activity peaks before and after the peak of solar flux.

MONITORING GEOMAGNETIC ACTIVITY

Geomagnetic activity is monitored by devices known as *magnetometers*. These may be as simple as a magnetic compass rigged to record its movements. Small variations in the geomagnetic field are scaled to two measures known as the K and A indices. The *K* index provides an indication of magnetic activity on a finite logarithmic scale of 0-9, and it is updated every three hours. Very quiet conditions are reported as 0 or 1, while geomagnetic storm levels begin at 4. See **Table 19.3** for the latest NOAA descriptions of geomagnetic storms.

A worldwide network of magnetometers constantly monitors the Earth's magnetic field, because the Earth's magnetic field varies with location. K indices that indicate average planetary conditions are indicated as K_p. Daily geomagnetic conditions are also summarized by the open-ended linear *A index*, which corresponds roughly to the cumulative K index values (it's the daily average of the eight K indices after converting the K indices to a linear scale). The A index commonly varies between 0 and 30 during quiet to active conditions, and up to 100 or higher during geomagnetic storms.

At 18 minutes past the hour, radio station WWV broadcasts the solar flux measured at 2000 UTC, the prior day's planetary A Index and the latest mid latitude K Index. In addition,

Table 19.3

Geomagnetic Storms

Typical Kp	Description	Days per Solar Cycle
9	Extreme	4
8	Severe	60
7	Strong	130
6	Moderate	360
5	Minor	900

they broadcast a descriptive account of the condition of the geomagnetic field and a forecast for the next three hours. For more details about WWV broadcasts, visit **www.nist.gov/pml/div688/grp40/wwv_format.cfm**. You should keep in mind that the A Index is a description of what happened yesterday. Strictly speaking, the K Index is valid only for mid latitudes. However, the trend of the K Index is very important for propagation analysis and forecasting. A rising K foretells worsening HF propagation conditions, particularly for transpolar paths. At the same time, a rising K alerts VHF operators to the possibility of enhanced auroral activity, particularly when the K Index rises above 3. Another source of useful information about solar disturbances is **www.spaceweather.com**.

19.2.11 D Layer Propagation

The *D layer* is the lowest region of the ionosphere, situated between 55 and 90 km (30 and 60 miles). See **Fig 19.13**. It is ionized primarily by the strong emission of solar hydrogen at 121.5 nanometers and short-wavelength X-rays (so-called hard X-rays), both of which penetrate through the upper atmosphere to ionize nitric oxide (radiation at 121.5 nm) and all other constituents (hard X-rays). The D layer exists only during daylight, because constant radiation is needed to replenish ions that quickly recombine into neutral molecules. The D layer abruptly disappears at night so far as amateur MF and HF signals are concerned. D layer ionization varies a small amount over the solar cycle. It is unsuitable as a refracting medium for MF and HF radio signals, but it is very important for VLF signals.

During daylight hours, radio energy as high as 5 MHz is effectively absorbed by the D

layer, severely limiting the range of daytime 1.8- and 3.5 MHz signals. Signals at 7 MHz and 10 MHz pass through the D layer and on to the E and F layers only at relatively high angles. Low-angle waves, which must travel a much longer distance through the D layer, are subject to greater absorption. As the frequency increases above 10 MHz, radio waves pass through the D layer with increasing ease (less absorption).

D layer ionization falls 100-fold as soon as the Sun sets and the source of ionizing radiation is removed. Low-band HF signals are then free to pass through the E layer (also greatly diminished at night) and on to the F layer, where the MUF is almost always high enough to propagate 1.8 and 3.5 MHz signals half way around the world. Long-distance propagation at 7 and 10 MHz generally improves at night as well, because absorption is less and low-angle waves are able to reach the F layer.

19.2.12 E Layer Propagation

The *E layer* lies between 90 and 150 km (60 and 90 miles) altitude, but a narrower region centered at 95 to 120 km (60 to 70 miles) is more important for radio propagation. In the E layer nitric oxide and molecular oxygen are ionized by short-wavelength UV and long-wavelength X-ray radiation (so-called soft X-rays). The normal E layer exists primarily during daylight hours, because like the D layer, it requires a constant source of ionizing radiation. Recombination is not as fast as in the denser D layer and absorption is much less. The E layer has a daytime critical frequency that varies between 3 and 4 MHz with the solar cycle. At night, the normal E layer decays to a minimum critical frequency of 0.3 – 0.5 MHz, which is still enough to refract low elevation angle 1.8 MHz signals.

DAYTIME E LAYER

The E layer plays a small role in propagating HF signals but can be a major factor limiting propagation during daytime hours. Its usual critical frequency of 3 to 4 MHz, with an M-factor of about 5, suggests that single-hop E layer skip might be useful between 5 and 20 MHz at distances up to 2300 km (1400 miles). In practice this is not the case, because the potential for E layer skip is severely limited by D layer absorption. Signals radiated at low angles at 7 and 10 MHz, which might be useful for the longest-distance contacts, are largely absorbed by the D layer. Only high-angle signals pass through the D layer at these frequencies, but high-angle E layer skip is typically limited to 1200 km (750 miles) or so. Signals at 14 MHz penetrate the D layer at lower angles at the cost of some absorption, but the casual operator may not be able to distinguish between signals propagated by the E layer or higher-angle F layer propagation.

An astonishing variety of other propagation modes find their home in the E layer, and this perhaps more than makes up for its ordinary limitations. Each of these other modes — sporadic E, field-aligned irregularities, aurora, auroral E and meteor scatter — are related forms of E layer propagation with unique characteristics. They are primarily useful only on the highest HF and lower VHF bands.

SPORADIC E

Short skip, long familiar on the 10 meter band during the summer months, affects the VHF bands as high as 222 MHz. *Sporadic E* (E_s), as this phenomenon is properly called, commonly propagates 28, 50 and 144 MHz radio signals between 500 and 2300 km (300 and 1400 miles). Signals are apt to be exceedingly strong, allowing even modest stations to make E_s contacts. At 21 MHz, the skip distance may only be a few hundred km. During the most intense E_s events, skip may shorten to less than 200 km (120 miles) on the 10 meter band and disappear entirely on 15 meters. Multiple-hop E_s has supported contacts up to 10,000 km (6200 miles) on 28 and 50 MHz and more than 3000 km (1900 miles) on 144 MHz. The first confirmed 220 MHz E_s contact was made in June 1987, but such contacts are likely to remain very rare.

Sporadic E at mid latitudes (roughly 15° to 45°) may occur at any time, but it is most common in the Northern Hemisphere during May, June and July, with a less-intense season at the end of December and early January. Its appearance is independent of the solar cycle. Sporadic E propagation is most likely to occur from 9 AM to noon local time and again early in the evening between 5 PM and 8 PM. Mid latitude E_s events may last only a few minutes or can persist for many hours. In contrast, sporadic E is an almost constant feature of the polar regions at night and the equatorial belt during the day.

Efforts to predict mid latitude E_s have not been successful, probably because its causes are complex and not well understood. Studies have demonstrated that thin and unusually dense patches of ionization in the E layer, between 100 and 110 km (60 and 70 miles) altitude and 10 to 100 km (6 to 60 miles) in extent, are responsible for most E_s reflections. Sporadic E clouds may form suddenly, move quickly from their birthplace, and dissipate within a few hours. Professional studies have recently focused on the role of heavy metal ions, probably of meteoric origin, and wind shears as two key factors in creating the dense patchy regions of E layer ionization.

Sporadic E clouds exhibit an MUF that can rise from 28 MHz through the 50 MHz band and higher in just a few minutes. When the skip distance on 28 MHz is as short as 400 or 500 km (250 or 310 miles), it is an indication

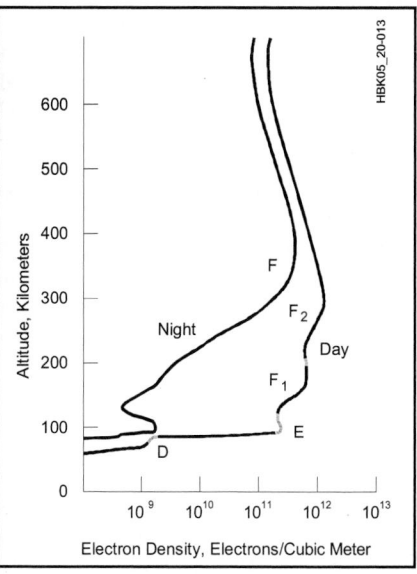

Fig 19.13 — Typical electron densities for the various ionospheric regions.

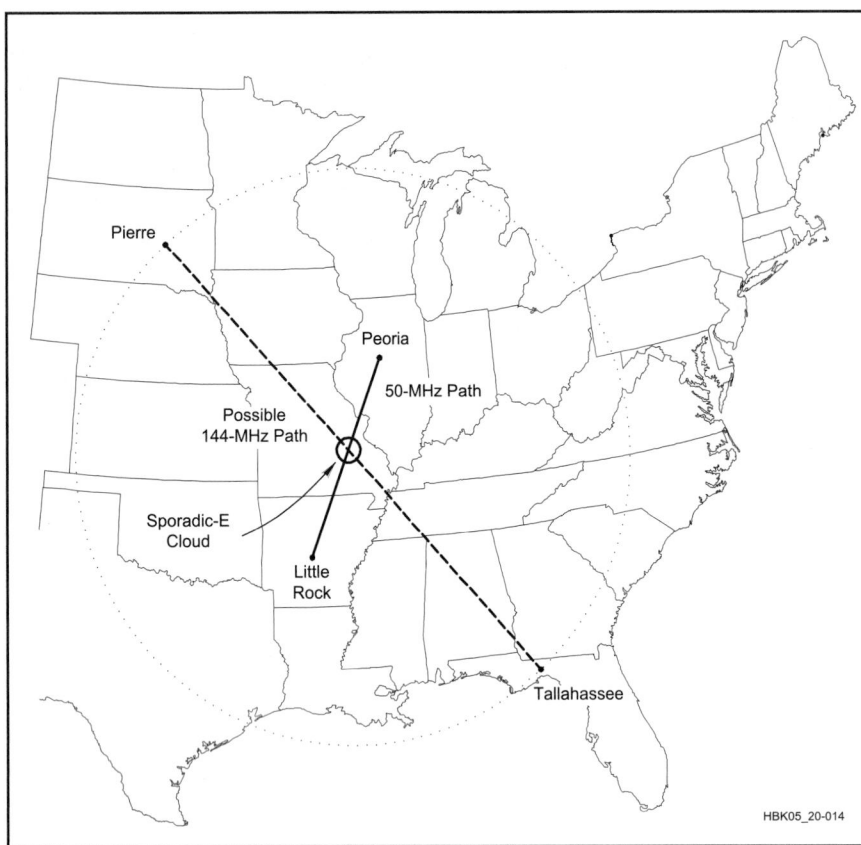

Fig 19.14 — 50 MHz sporadic E contacts of 700 km (435 miles) or shorter (such as between Peoria and Little Rock) indicate that the MUF on longer paths is above 144 MHz. Using the same sporadic E region reflecting point, 144 MHz contacts of 2200 km (1400 miles), such as between Pierre and Tallahassee, should be possible.

that the MUF has reached 50 MHz for longer paths at low launch angles. Contacts at the maximum one-hop sporadic E distance, about 2300 km (1400 miles), should then be possible at 50 MHz. *E-skip* (yet another term for sporadic E) contacts as short as 700 km (435 miles) on 50 MHz, in turn, may indicate that 144 MHz contacts in the 2300 km (1400 mile) range can be completed. See **Fig 19.14**. Sporadic E openings occur about a tenth as often at 144 MHz as they do at 50 MHz and for much shorter periods.

Sporadic E can also have a detrimental effect on HF propagation by masking the F_2 layer from below. HF signals may be prevented from reaching the higher levels of the ionosphere and the possibilities of long F_2 skip. Reflections from the tops of sporadic E clouds can also have a masking effect, but they may also lengthen the F_2 propagation path with a top-side intermediate hop that never reaches the Earth.

E LAYER FIELD-ALIGNED IRREGULARITIES

Amateurs have experimented with a little-known scattering mode known as *field-aligned irregularities* (FAI) at 50 and 144 MHz since 1978. FAI commonly appears directly after

sporadic E events and may persist for several hours. Oblique-angle scattering becomes possible when electrons are compressed together due to the action of high-velocity ionospheric acoustic (sound) waves. The resulting irregularities in the distribution of free electrons are aligned parallel to the Earth's magnetic field, in something like moving vertical rods. A similar process of electron field-alignment takes place during radio aurora, making the two phenomena quite similar.

Most reports suggest that 8 PM to midnight may be the most productive time for FAI. Stations attempting FAI contacts point their antennas toward a common scattering region that corresponds to an active or recent E_s reflection point. The best direction must be probed experimentally, for the result is rarely along the great-circle path. Stations in south Florida, for example, have completed 144 MHz FAI contacts with north Texas when participating stations were beamed toward a common scattering region over northern Alabama.

FAI-propagated signals are weak and fluttery, reminiscent of aurora signals. Doppler shifts of as much as 3 kHz have been observed in some tests. Stations running as little as 100 W and a single Yagi should be able to complete FAI contacts during the most

favorable times, but higher power and larger antennas may yield better results. Contacts have been made on 50 and 144 MHz, and 222 MHz FAI seems probable as well. Expected maximum distances should be similar to other forms of E layer propagation, or about 2300 km (1400 miles).

AURORA

Radar signals as high as 3000 MHz have been scattered by the *aurora borealis* or northern lights (*aurora australis* in the Southern Hemisphere), but amateur aurora contacts are common only from 28 through 432 MHz. By pointing directional antennas generally north toward the center of aurora activity, oblique paths between stations up to 2300 km (1400 miles) apart can be completed. See **Fig 19.15**. High power and large antennas are not necessary. Stations with small Yagis and as little as 10 W output have used auroras on frequencies as high as 432 MHz, but contacts at 902 MHz and higher are exceedingly rare. Auroral propagation works just as well in the Southern Hemisphere, in which case antennas must be pointed south.

The appearance of auroras is closely linked to solar activity. During massive geomagnetic storms, low energy particles that were trapped in the magnetosphere flow into the atmosphere near the polar regions, where they ionize the gases of the E layer and higher. This unusual ionization produces spectacular visual auroral displays, which often spread southward into the mid latitudes. Higher energy electrons get down to lower altitudes to cause auroral ionization (via collisions with neutral particles) in the E layer, and this scatters radio signals in the VHF and UHF ranges.

In addition to scattering radio signals, auroras have other effects on worldwide radio propagation. Communication below 20 MHz is disrupted in high latitudes, primarily by absorption, and is especially noticeable over polar and near-polar paths. Signals on the AM broadcast band through the 40 meter band late in the afternoon may become weak and watery. The 20 meter band may close down altogether. Satellite operators have also noticed that 144 MHz downlink signals are often weak and distorted when satellites pass near the polar regions. At the same time, the MUF in equatorial regions may temporarily rise dramatically, including an enhancement of transequatorial paths at frequencies as high as 50 MHz.

Auroras occur most often around the spring and fall equinoxes (March-April and September-October), but auroras may appear in any month. Aurora activity generally peaks about two years before and after solar cycle maximum (the "before" generally due to CMEs and the "after" generally due to high-speed wind streams from coronal holes). Radio aurora activity is usually heard first in late

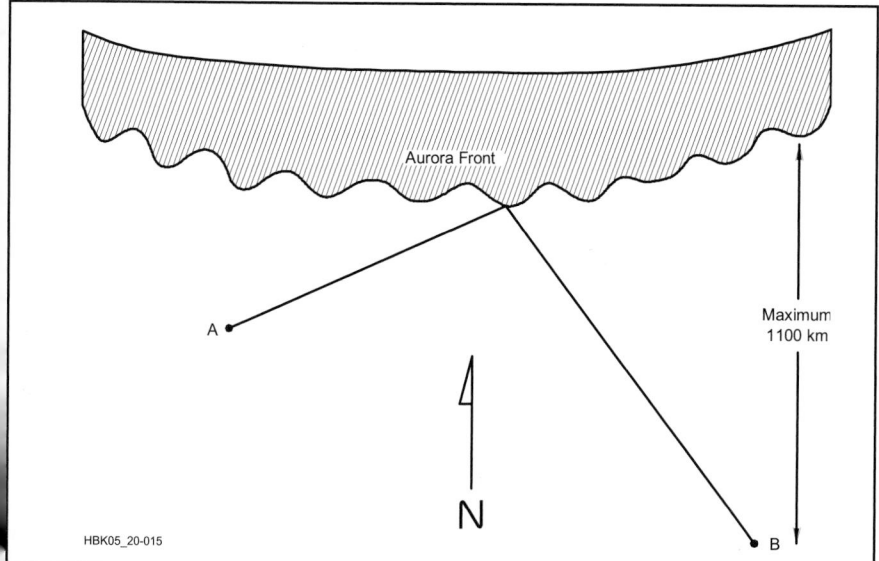

Fig 19.15 — Point antennas generally north to make oblique long-distance contacts on 28 through 432 MHz via aurora scattering. Optimal antenna headings may shift considerably to the east or west depending on the location of the aurora. This necessitates moving your antenna azimuth direction for best propagation, and continuing to do so as the aurora progresses.

afternoon and may reappear later in the evening. Auroras may be anticipated by following the A and K indices reports on WWV. A K index of five or greater and an A index of at least 30 are indications that a geomagnetic storm is in progress and an aurora likely. The probability, intensity and southerly extent of auroras increase as the two index numbers rise. Stations north of 42° latitude in North America experience many auroral openings each year, while those in the Gulf Coast states may hear auroral signals no more than once a year, if that often.

Aurora-scattered signals are easy to identify. On 28- and 50 MHz SSB, signals sound very distorted and somewhat wider than normal; at 144 MHz and above, the distortion may be so severe that only CW is useful. Auroral CW signals have a distinctive note variously described as a buzz, hiss or mushy sound. This characteristic auroral signal is due to Doppler broadening, caused by the movement of electrons within the aurora. An additional Doppler shift of 1 kHz or more may be evident at 144 MHz and several kilohertz at 432 MHz. This second Doppler shift is the result of massive electrical currents that sweep electrons toward the Sun side of the Earth during magnetic storms. Doppler shift and distortion increase with higher frequencies, while signal strength dramatically decreases.

It is not necessary to see an aurora to make auroral contacts. Useful auroras may be 500-1000 km (310-620 miles) away and below the visual horizon. Antennas should be pointed generally north and then probed east and west to peak signals, because auroral ionization is field aligned. This means that for any pair of stations, there is an optimal direction for aurora scatter. Offsets from north are usually greatest when the aurora is closest and often provide the longest contacts. There may be some advantage to antennas that can be elevated, especially when auroras are high in the sky.

AURORAL E

Radio auroras may evolve into a propagation mode known as *auroral E* at 21, 28, 50 and rarely 144 MHz. Auroral E is essentially sporadic E in the auroral zone. Doppler distortion disappears and signals take on the characteristics of sporadic E. The most effective antenna headings shift dramatically away from oblique aurora paths to direct great-circle bearings. The usual maximum distance is 2300 km (1400 miles), typical for E layer modes, but 28 and 50 MHz auroral E contacts of 5000 km (3100 miles) are sometimes made across Canada and the northern US, apparently using two hops. Contacts at 50 MHz between Alaska and the east coasts of Canada and the northern US have been completed this way. Transatlantic 50 MHz auroral E paths are also likely.

Typically, 28 and 50 MHz auroral E appears across the northern third of the US and southern Canada when aurora activity is diminishing. This usually happens after midnight on the eastern end of the path. Auroral E signals sometimes have a slightly hollow sound to them and build slowly in strength over an hour or two, but otherwise they are indistinguishable from sporadic E. Auroral E paths are almost always east-west oriented, perhaps because there are few stations at very northern latitudes to take advantage of this propagation. A common auroral E path on 28 MHz and 21 MHz is from North America in the late afternoon hours in the autumn months to the Scandinavian countries. This path is open when normal F layer propagation is not possible — even to more southern locations in Europe.

Auroral E may also appear while especially intense auroras are still in progress, as happened during the great aurora of March 1989. On that occasion, 50 MHz propagation shifted from Doppler-distorted aurora paths to clear-sounding auroral E over a period of a few minutes. Many 6 meter operators as far south as Florida and Southern California made single- and double-hop auroral E contacts across the country. At about the same time, the MUF reached 144 MHz for stations west of the Great Lakes to the Northeast, the first time auroral E had been reported so high in frequency.

METEOR SCATTER

Contacts between 800 and 2300 km (500 and 1400 miles) can be made at 28 through 432 MHz via reflections from the ionized trails left by meteors as they travel through the atmosphere. The kinetic energy of meteors no larger than grains of rice are sufficient to ionize a column of air 20 km (12 miles) long in the E layer. The particle itself evaporates and never reaches the ground, but the ionized column may persist for a few seconds to a minute or more before it dissipates. This is enough time to make very brief contacts by reflections from the ionized trails. Millions of meteors enter the Earth's atmosphere every day, but few have the required size, speed and orientation to the Earth to make them useful for meteor-scatter propagation.

Radio signals in the 30 to 100 MHz range are reflected best by meteor trails, making the 50 MHz band prime for meteor-scatter work. The early morning hours around dawn are usually the most productive, because the morning side of the Earth faces in the direction of the planet's orbit around the Sun. The relative velocity of meteors that head toward the Earth's morning side are thus increased by up to 30 km/sec, which is the average rotational speed of the Earth in orbit. See **Fig 19.16**. The maximum velocity of meteors in orbit around the Sun is 42 km/sec. Thus when the relative velocity of the Earth is considered, most meteors must enter the Earth's atmosphere somewhere between 12 and 72 km/sec.

Meteor contacts ranging from a second or two to more than a minute can be made nearly any morning at 28 or 50 MHz. Meteor-scatter contacts at 144 MHz and higher are more difficult because reflected signal strength and duration drop sharply with increasing frequency. A meteor trail that provides 30 seconds of communication at 50 MHz will last only a few seconds at 144 MHz, and less than a second at 432 MHz.

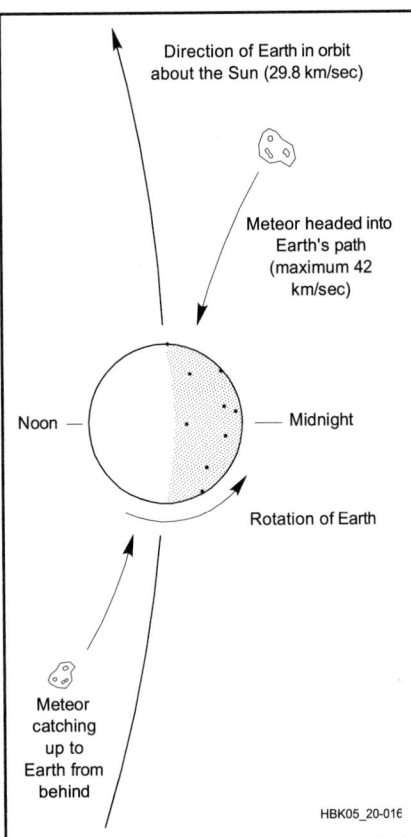

Fig 19.16 — The relative velocity of meteors that meet the Earth head-on is increased by the rotational velocity of the Earth in orbit. Fast meteors strike the morning side of the Earth because their velocity adds to the Earth's rotational velocity, while the relative velocity of meteors that "catch up from behind" is reduced.

Meteor scatter opportunities are somewhat better during July and August for North America because the average number of meteors entering the Earth's atmosphere peaks during those months. The best times are during one of the great annual *meteor showers*, when the number of useful meteors may increase ten-fold over the normal rate of five to ten per hour. See **Table 19.4**. A meteor shower occurs when the Earth passes through a relatively dense stream of particles, thought to be the remnants of a comet in orbit around the Sun. The most-productive

Table 19.4

Major Annual Meteor Showers

Name	Peak Dates	Approximate Rate (meteors/hour)
Quadrantids	Jan 3	50
Arietids	Jun 7-8	60
Perseids	Aug 11-13	80
Orionids	Oct 20-22	20
Geminids	Dec 12-13	60

showers are relatively consistent from year to year, although several can occasionally produce great storms.

Because meteors provide only fleeting moments of communication even during one of the great meteor showers, special operating techniques are often used to increase the chances of completing a contact. Prearranged schedules between two stations establish times, frequencies and precise operating standards. Usually, each station transmits on alternate 15-second periods until enough information is pieced together a bit at a time to confirm contact. High-speed Morse code of several hundred words per minute, generated and slowed down by special computer programs, can make effective use of very short meteor bursts. Non-scheduled random meteor contacts are common on 50 MHz and 144 MHz, but short transmissions and alert operating habits are required.

It is helpful to run several hundred watts to a single Yagi, but meteor-scatter can be used by modest stations under optimal conditions. During the best showers, a few watts and a small directional antenna are sufficient at 28 or 50 MHz. At 144 MHz, at least 100 W output and a long Yagi are needed for consistent results. Proportionately higher power is required for 222 and 432 MHz even under the best conditions.

A technique allowing even smaller VHF/UHF stations to take advantage of meteor scatter and other modes (ionospheric scatter and even moonbounce) is available in software called *WSJT*. This software uses the sound card in a PC and a variety of modulation and coding techniques to allow reception of signals up to 10 dB below your noise level. For more information, visit **www.physics.princeton.edu/pulsar/K1JT/.**

19.2.13 F Layer Propagation

The region of the *F layers*, from 150 km (90 miles) to over 400 km (250 miles) altitude, is by far the most important for long-distance HF communications. F-region oxygen atoms are ionized primarily by ultraviolet radiation. During the day, ionization reaches maxima in two distinct layers. The F_1 layer forms between 150 and 250 km (90 and 160 miles), is most prevalent in the summer months, and disappears at night. The F_2 layer extends above 250 km (160 miles), with a peak of ionization around 300 km (190 miles). At night, F-region ionization collapses into one broad layer at 300-400 km (190-250 miles) altitude. Ions recombine very slowly at these altitudes, because atmospheric density is relatively low. Maximum ionization levels change significantly with time of day, season and year of the solar cycle.

F_1 LAYER

The daytime F_1 layer is not very important to HF communication. In reality it is an inflection point in the electron density profile, not a peak. It exists only during daylight hours and is largely absent in winter. Radio signals below 10 MHz are not likely to reach the F_1 layer, because they are either absorbed by the D layer or refracted by the E layer. Signals higher than 20 MHz that pass through both of the lower ionospheric regions are likely to pass through the F_1 layer as well, because the F_1 MUF rarely rises above 20 MHz. Absorption diminishes the strength of any signals that continue through to the F_2 layer during the day. Some useful F_1 layer refraction may take place between 10 and 20 MHz during summer days, yielding paths as long as 3000 km (1900 miles), but these would be practically indistinguishable from F_2 skip.

F_2 AND NIGHTTIME F LAYERS

The F_2 layer forms between 250 and 400 km (160 and 250 miles) during the daytime and persists throughout the night as a single consolidated F region 50 km (30 miles) higher in altitude. Typical ion densities are the highest of any ionospheric layer, with the possible exception of some unusual E layer phenomenon. In contrast to the other ionospheric layers, F_2 ionization varies considerably with time of day, season and position in the solar cycle, but it is never altogether absent. These two characteristics make the F_2 layer the most important for long-distance HF communications.

The F_2 layer MUF is nearly a direct function of ultraviolet (UV) solar radiation, which in turn closely follows the solar cycle. During the lowest years of the cycle, the daytime MUF may climb above 14 MHz for only a few hours a day. In contrast, the MUF may rise beyond 50 MHz during peak years and stay above 14 MHz throughout the night. The virtual height of the F_2 region averages 330 km (210 miles), but varies between 200 and 400 km (120 and 250 miles). Maximum one-hop distance is about 4000 km (2500 miles). Near-vertical incidence skywave propagation just below the critical frequency provides reliable coverage out to 200-300 km (120-190 miles) with no skip zone. It is most often observed on 7 MHz during the day.

The extremely high-angle *Pedersen ray* can create effective single-hop paths of 5000 to 12,000 km under certain conditions, but most operators will not be able to distinguish Pedersen Ray paths from normal F layer propagation. A Pederson ray is a path that follows the contour of the Earth near the height of the maximum F_2 region electron density, and requires a fairly stable ionosphere. Pedersen ray paths are most evident over high-latitude east-west paths at frequencies near the MUF.

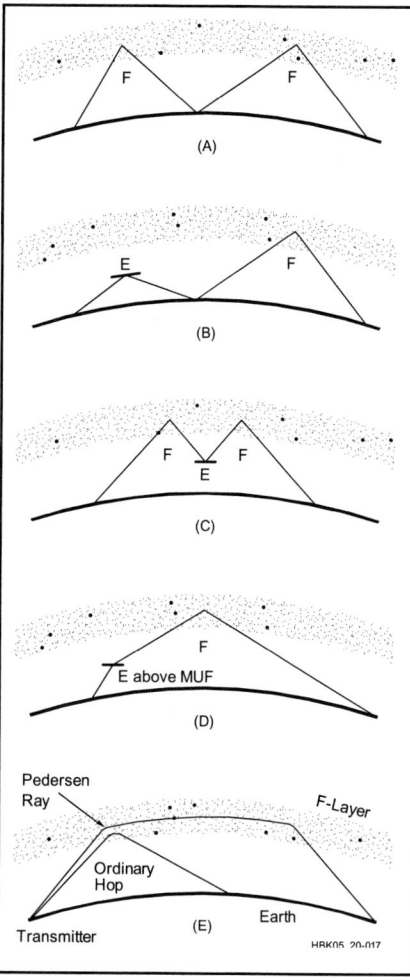

Fig 19.17 — Multihop paths can take many different configurations, including a mixture of E and F layer hops. (A) Two F layer hops. Five or more consecutive F layer hops are possible. (B) An E layer hookup to the F layer. (C) A top-side E layer reflection can shorten the distance of two F layer hops. (D) Refraction in the E layer above the MUF is insufficient to return the signal to Earth, but it can go on to be refracted in the F layer. (E) The Pedersen Ray, which originates from a signal launched at a relatively high angle above the horizon into the E or F region, may result in a single-hop path, 5000 km (3100 miles) or more. This is considerably further than the normal 4000 km (2500 mile) maximum F-region single-hop distance, where the signal is launched at a very low takeoff angle. The Pedersen Ray can easily be disrupted by any sort of ionospheric gradient or irregularity. Not shown in this figure is a chordal hop — since chordal hops are most prevalent in the equatorial ionosphere, refer to Fig 19.21 and Fig 19.22.

They appear most often about noon local time at mid-path when the geomagnetic field is very quiet. Pedersen ray propagation may be responsible for 50 MHz paths between the US Northeast and Western Europe, for example, when ordinary MUF analysis could not explain the 5000 km contacts. See part E in **Fig 19.17**.

At any given location on Earth, in general both F_2 layer ionization and MUF at that point build rapidly at sunrise, usually reaching a maximum after local noon, and then decrease to a minimum at night prior to the next sunrise. Depending on the season, the MUF is generally highest within 20° of the equator and lower toward the poles. For this reason, transequatorial paths may be open at a particular frequency when all other paths are closed

In contrast to all the other ionospheric layers, daytime ionization in the winter F_2 layer averages four times the level of the summer at the same period in the solar cycle, doubling the MUF. This so-called *winter anomaly* is caused by a seasonal increase in the ratio of atoms to molecules at F_2 layer heights (atoms are instrumental in the production of electrons, whereas molecules are instrumental in the loss

of electrons). Winter daytime F_2 conditions are much superior to those in summer, because the MUF is much higher.

MULTIHOP F LAYER PROPAGATION

Most HF communication beyond 4000 km (2500 miles) takes place via multiple ionospheric hops. Radio signals are reflected from the Earth back toward space for additional ionospheric refractions. A series of ionospheric refractions and terrestrial reflections commonly create paths half-way around the Earth. Each hop involves additional attenuation and absorption, so the longest-distance signals tend to be the weakest. Even so, it is possible for signals to be propagated completely around the world and arrive back at their originating point. Multiple reflections within the F layer may at times bypass ground reflections altogether, creating what are known as *chordal hops*, with lower total attenuation. It takes a radio signal about 0.14 seconds to make an around-the-world trip.

Multihop paths can take on many different configurations, as shown in the examples of Fig 19.17. E layer (especially sporadic E) and F layer hops may be mixed. In practice, multihop signals arrive via many different paths, which often increases the problems of fading. Analyzing multi-hop paths is complicated by the effects of D and E layer absorption, possible reflections from the tops of sporadic E layers, disruptions in the auroral zone and other phenomena.

In general, when a band is opening and closing as the MUF changes, extremely low elevation angles are dictated (less than 5°). During the main part of the opening, the elevation angle is higher — generally in the range of 5° to 20°. This is why stations with extremely

high antennas (which have higher gain at lower takeoff angles) perform better at band openings and closings.

F LAYER LONG PATH

Most HF communication takes place along the shortest great-circle path between two stations. Short-path propagation is always less than 20,000 km (12,000 miles) — halfway around the Earth. Nevertheless, it may be possible at times to make the same contact in exactly the opposite direction via the *long path*. The long-path distance will be 40,000 km (25,000 miles) minus the short-path length. Signal strength via the long path is usually considerably less than the more direct short-path. When both paths are open simultaneously, there may be a distinctive sort of echo on received signals. The time interval of the echo represents the difference between the short-path and long-path distances.

Sometimes there is a great advantage to using the long path when it is open, because signals can be stronger and fading less troublesome, or because fewer interfering signals lie along the path between the stations. There are times when the short path may be closed or disrupted by E layer *blanketing* (when the E-region ionization is high enough to keep waves from penetrating, regardless of elevation angle), D layer absorption or F layer gaps, especially when operating just below the MUF. Long paths that predominantly cross the night side of the Earth, for example, are sometimes useful because they generally avoid blanketing and absorption problems. Daylight-side long paths may take advantage of higher F layer MUFs that occur over the sunlit portions of the Earth.

F LAYER GRAY-LINE

Gray-line paths can be considered a special form of long-path propagation that takes into account the unusual ionospheric configuration along the twilight region between night and day. The gray line, as the twilight region is sometimes called, extends completely around the world. It is not precisely a line, for the distinction between daylight and darkness is a gradual transition due to atmospheric scattering. On one side, the gray line heralds sunrise and the beginning of a new day; on the opposite side, it marks sunset and the end of the day.

The ionosphere undergoes a significant transformation between night and day. As day begins, the highly absorbent D and E layers are recreated, while the F layer MUF rises from its pre-dawn minimum. At the end of the day, the D and E layers quickly disappear, while the F layer MUF continues its slow decline from late afternoon. For a brief period just along the gray-line transition, the D and E layers are not well formed, yet the F_2 MUF usually remains higher than 5 MHz. This provides a special opportunity for stations at 1.8 and 3.5 MHz.

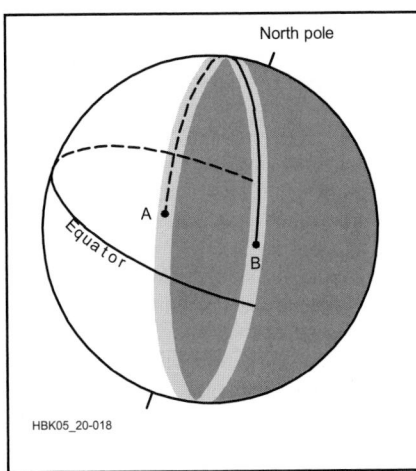

Fig 19.18 — The gray line encircles the Earth, but the tilt at the equator to the poles varies over 46° with the seasons. Long-distance contacts can often be made halfway around the Earth along the gray line, even as low as 1.8 and 3.5 MHz. The strength of the signals, characteristic of gray-line propagation, indicates that multiple Earth-ionosphere hops are not the only mode of propagation, since losses in many such hops would be very great. Chordal hops, where the signals are confined to the ionosphere for at least part of the journey, may be involved.

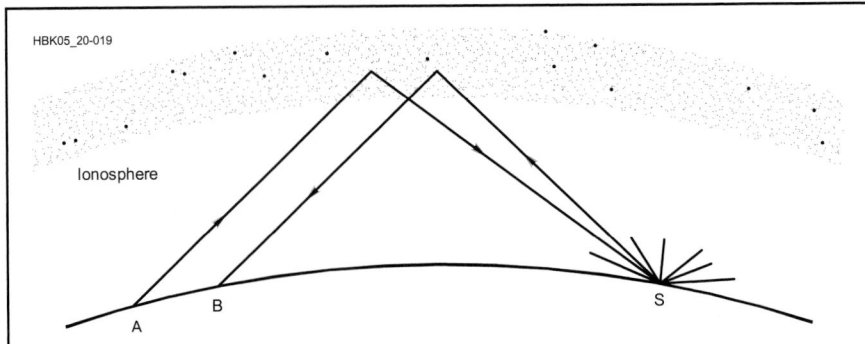

Fig 19.19 — Schematic of a simple backscatter path. Stations A and B are too close to make contact via normal F layer ionospheric refraction. Signals scattered back from a distant point on the Earth's surface (S), often the ocean, may be accessible to both and create a backscatter circuit.

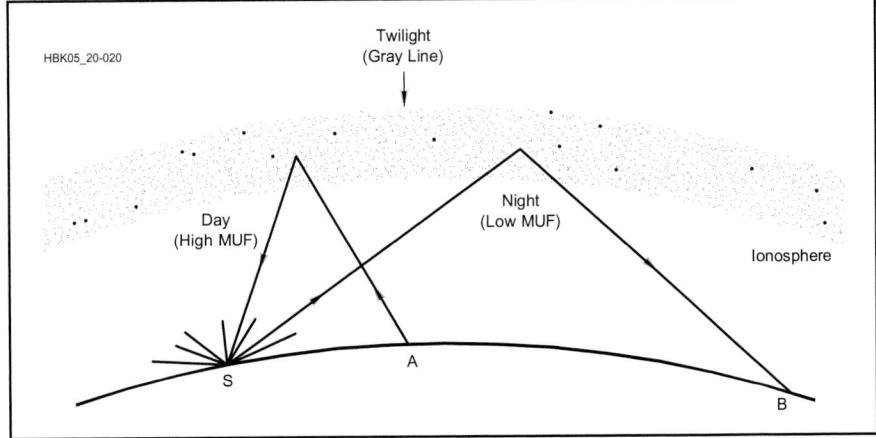

Fig 19.20 — Backscatter path across the gray line. Stations A and B are too close to make contact via normal ionospheric refraction, but may hear each other's signals scattered from point S. Station A makes use of a high-angle refraction on the day side of the gray line, where the MUF is high. Station B makes use of a night-time refraction, with a lower MUF and lower angle of propagation. Note that station A points away from B to complete the circuit.

Normally, long-distance communication on the lowest two amateur bands can take place only via all-darkness paths because of daytime D layer absorption. The gray-line propagation path, in contrast, extends completely around the world. See **Fig 19.18**. This unusual situation lasts less than an hour at sunrise and sunset when the D layer is largely absent, and may support contacts that are difficult or impossible at other times.

The gray line generally runs north-south, but it varies by 23° either side of true north as measured at the equator over the course of the year. This variation is caused by the tilt in the Earth's axis. The gray line is exactly north-south through the poles at the equinoxes (March 21 and September 21) and is at its 23° extremes on June 21 and December 20. Over a one-year period, the gray line crosses a 46° sector of the Earth north and south of the equator, providing optimum paths to slightly different parts of the world each day. Many commonly available computer programs plot the gray line on a flat map or globe.

A web based application to calculate worldwide sunrise and sunset times is available at **http://aa.usno.navy.mil/data/docs/RS_OneDay.html**. An alternative method of determining sunrise and sunset times, along with seeing great circle paths on the same plot, is W6EL's *W6ELProp* software, available free at **www.qsl.net/w6elprop**. For an online gray line map, visit **http://dx.qsl.net/propagation/greyline.html**.

F LAYER BACKSCATTER AND SIDESCATTER

Special forms of F layer scattering can create unusual paths within the skip zone. *Backscatter* and *sidescatter* signals are usually observed just below the MUF for the direct path and allow communications not normally possible by other means. Stations using backscatter point their antennas toward a common scattering region at the one-hop distance, rather than toward each other. Backscattered signals are generally weak and have a characteristic hollow sound. Useful communication distances range from 100 km (60 miles) to the normal one-hop distance of 4000 km (2500 miles).

Backscatter and sidescatter are closely related and the terminology does not precisely distinguish between the two. Backscatter usually refers to single-hop signals that have been scattered by the Earth or the ocean at some distant point back toward the transmitting station. Two stations spaced a few hundred km apart can often communicate via a backscatter path near the MUF. See **Fig 19.19**.

Sidescatter usually refers to a circuit that is oblique to the normal great-circle path and is sometimes referred to as *skew path*. Two stations can make use of a common side-scattering region well off the direct path, often toward the south. European and North American stations sometimes complete 28 MHz contacts via a scattering region over Africa. U.S. and Finnish 50 MHz operators observed a similar effect early one morning in November 1989 when they made contact by beaming off the coast of West Africa.

When backscattered signals cross an area where there is a sharp gradient in ionospheric

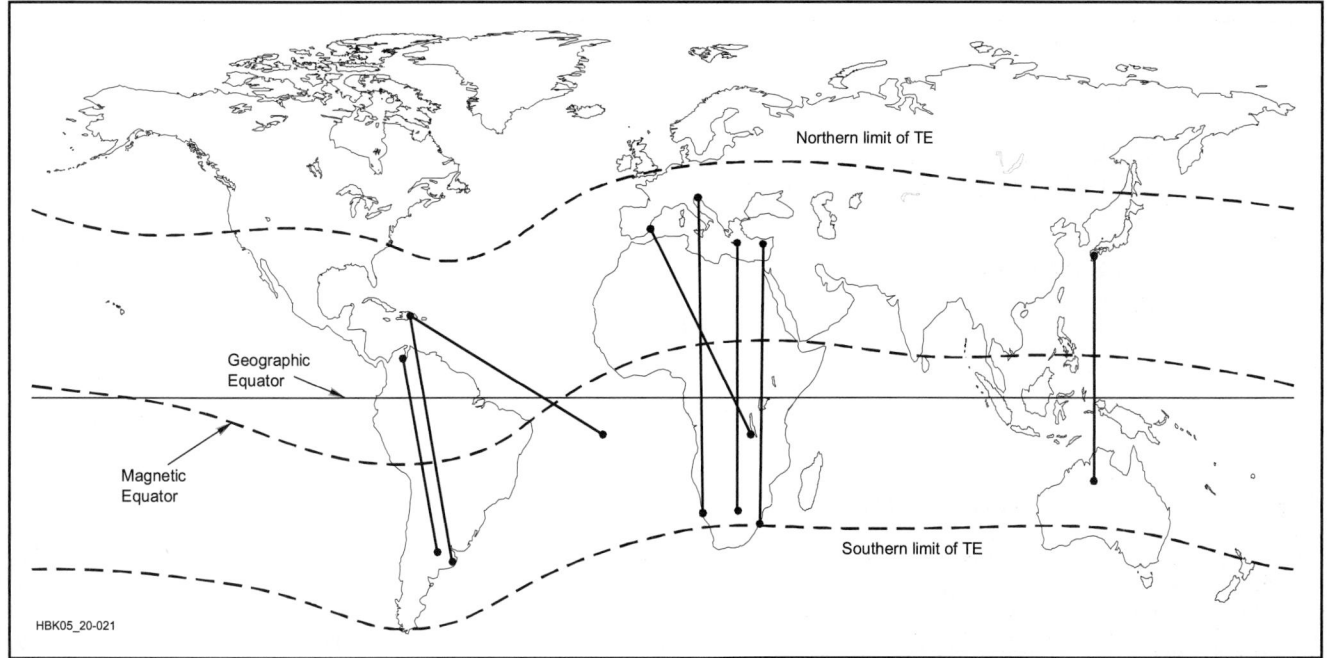

Fig 19.21 — Transequatorial propagation takes place between stations equidistant across the geomagnetic equator. Distances up to 8000 km (5000 miles) are possible on 28 through 432 MHz. Note the geomagnetic equator is considerably south of the geographic equator in the Western Hemisphere.

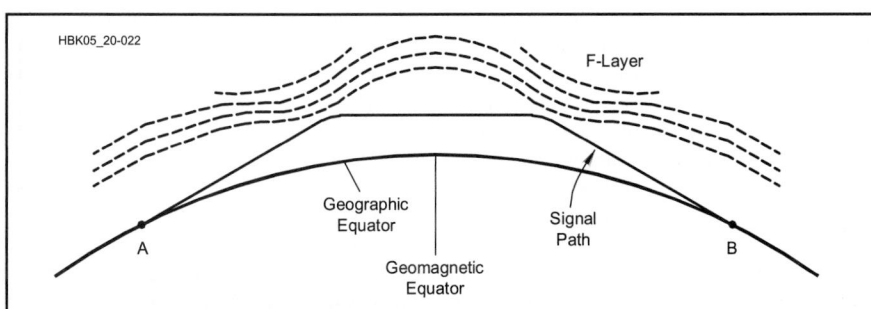

Fig 19.22 — Cross-section of a transequatorial signal path, showing the effects of ionospheric bulging and a double refraction above the normal MUF.

density, such as between night and day, the path may take on a different geometry, as shown in **Fig 19.20**. In this case, stations can communicate because backscattered signals return via the day side ionosphere on a shorter hop than the night side. This is possible because the dayside MUF is higher and thus the skip distance shorter. The net effect is to create a backscatter path between two stations within the normal skip zone.

TRANSEQUATORIAL PROPAGATION

Discovered by Amateur Radio operators in 1947, *transequatorial propagation* (commonly abbreviated TE or TEP) supports propagation between 5000 and 8000 km (3100 and 5000 miles) across the magnetic equator from 28 MHz to as high as 432 MHz. Stations attempting TE contacts must be nearly equidistant from the geomagnetic equator. Many contacts have been made at 50 and 144 MHz

between Europe and South Africa, Japan and Australia and the Caribbean region and South America. Fewer contacts have been made on the 222 MHz band, and TE signals have been heard at 432 MHz.

Unfortunately for most continental US stations, the *geomagnetic equator* dips south of the geographic equator in the Western Hemisphere, as shown in **Fig 19.21**, making only the most southerly portions of Florida and Texas within TE range. TE contacts from the southeastern part of the country may be possible with Argentina, Chile and even South Africa.

Transequatorial propagation peaks between 5 PM and 10 PM during the spring and fall equinoxes, especially during the peak years of the solar cycle. The lowest probability is during the summer. Quiet geomagnetic conditions are required for TE to form. High power and large antennas are not required to

work TE, as VHF stations with 100 W and single long Yagis have been successful.

TE propagation depends on bulges of intense F_2 layer ionization on both sides of the geomagnetic equator. This field-aligned ionization forms shortly after sunset via a process known as the *fountain effect* in an area 100-200 km (60-120 miles) north and south of the geomagnetic equator and 500-3000 km (310-1900 miles) wide. It moves west with the setting Sun. The MUF may increase to twice its normal level 15° either side of the geomagnetic equator. See **Fig 19.22**.

19.2.14 Emerging Theories of HF and VHF Propagation

Although much is known about the ionosphere and propagation, there are still many instances of unusual propagation that cannot be easily explained with our textbook knowledge. It is encouraging to note that Amateur Radio operators continue to be at the forefront of trying to understand many of these unusual modes. Several recent studies have included:

1) The tie between polar mesospheric summer echoes and 6 meter propagation in "Polar Mesospheric Summer Echoes," *Worldradio On-Line*, Feb 2009, **www.cq-amateur-radio. com**. These summer echoes are also called SSSP (summer solstice short path), a mode that Japanese amateurs have noted.

2) Theta aurora and 6 meter propagation across the polar cap in "More Alaska to EU on 6 m," at **http://myplace.frontier.com/~k9la**.

3) The impact of galactic cosmic rays

on 160 meter *ducting* in "A Theory on the Role of Galactic Cosmic Rays in 160-Meter Propagation," *CQ*, Nov 2008. *Ducting* refers to an electromagnetic wave successively refracting between a lower boundary and an upper boundary (for example, between the top of the E region and the bottom of the F region for 160 meters at night)

4) Discovery of single-day enhancements in the F_2 region that do not be appear to be tied to geomagnetic activity. This came from an interesting paper that was published in the *Journal*

of *Geophysical Research* in late 2010. The paper was by Michael David and Jan J. Sojka, who were with the Center for Atmospheric and Space Sciences at Utah State University. The title of the paper is "Single-day dayside density enhancements over Europe: A survey of a half-century of ionosonde data."

5) Although not a propagation issue, there is evidence that we may have missed a small solar cycle way back in the 1790 time frame. See the *WorldRadio Online* Propagation column by Carl Luetzelschwab K9LA, November 2011.

6) Another topic not tied directly to propagation is the on-going research into better models of the ionosphere. We already pretty much understand solar radiation and we've recently developed a much better model of how elevated geomagnetic indices affect the ionosphere (it's not simply applying a single K index). Research is on-going to better understand the physics behind events in the lower atmosphere coupling up to the ionosphere. Understanding these three processes is leading the way towards a true daily model of the ionosphere.

19.3 MUF Predictions

F layer MUF prediction is key to forecasting HF communications paths at particular frequencies, dates and times, but forecasting is complicated by several variables. Solar radiation varies over the course of the day, season, year and solar cycle. Additionally, the ionization at any given point in the world depends on geomagnetic field activity and events in the lower atmosphere coupling up to the ionosphere. These regular intervals provide the main basis for prediction, yet recurrence is far from reliable. In addition, forecasts are predicated on a quiet geomagnetic field, but the condition of the Earth's magnetic field is most difficult to predict weeks or months ahead. For professional users of HF communications, uncertainty is a nuisance for maintaining reliable communications paths, while for many amateurs it provides an aura of mystery and chance that adds to the fun of DXing. Nevertheless, many amateurs want to know what to expect on the HF bands to make best use of available on-the-air time, plan contest strategy, ensure successful net operations or engage in other activities.

19.3.1 MUF Forecasts

Long-range forecasts several months ahead provide only the most general form of prediction. A series of 48 charts on the members-only ARRL website (**www.arrl.org/propagation**), similar to **Fig 19.23**, forecast average propagation for a one-month period over specific paths. The charts assume a single average solar flux value for the entire month and they assume that the geomagnetic field is undisturbed.

The uppermost curve in Fig 19.23 shows the highest frequency that will be propagated on at least 10% of the days in the month. The given values might be exceeded considerably on a few rare days. On at least half the days, propagation should be possible on frequencies as high as the middle curve. Propagation will exceed the lowest curve on at least 90% of the days. The exact MUF on any particular

day cannot be determined from these statistical charts, but you can determine when you should start monitoring a band to see if propagation actually does occur that day — particularly at frequencies above 14 MHz.

Short-range forecasts of a few days ahead are marginally more reliable than long-range forecasts, because underlying solar indices and geomagnetic conditions can be anticipated with greater confidence. The tendency for solar disturbances to recur at 27-day intervals may enhance short-term forecasts. Daily forecasts may not be any better, as the ionosphere does not instantly react to small changes in solar flux (or sunspot number) and geomagnetic field indices. Regardless of these limitations, it is always good to know the current solar and

geophysical data, as well as understanding warnings provided by observations of the Sun in the visual to X-ray range.

The CD-ROM bundled with *The ARRL Antenna Book* contains even more detailed propagation-prediction tables from more than 150 locations around the world for six levels of solar activity, for the 12 months of the year. Again, keep in mind that these long-range forecasts assume quiet geomagnetic conditions. Real-time MUF forecasts are also available in a variety of text and graphical forms on the web. Forecasts can also be made by individuals using one of several popular programs for personal computers, including *ASAPS*, *VOACAP*, and *W6ELProp*. (See the sidebar, "MUF Prediction on the Home Computer.")

19.3.2 Statistical Nature of Propagation Predictions

The model of the ionosphere in propagation prediction software is a monthly median model — thus the results (usually MUF and signal strength) are statistical over a month's time frame. The median aspect means that the predicted value will actually occur on at least half the days of the month — but unfortunately it's tough to tell which are the "better" days and which are the "worse" days.

This monthly median model came about due to the efforts of scientists to correlate what the ionosphere was doing to what the Sun was doing. The best correlation was between a smoothed solar index (smoothed sunspot number or smoothed 10.7 cm solar flux) and monthly median ionospheric parameters. Thus our propagation prediction software was never intended to give daily predictions. One good way to achieve better short-term predictions is to use the concept of an effective sunspot number — SSNe.

EFFECTIVE SUNSPOT NUMBER (SSNe)

The concept of SSNe is quite simple. The

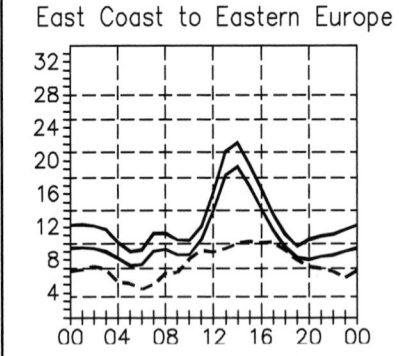

Fig 19.23 — Propagation prediction chart for East Coast to Eastern Europe from the ARRL web members-only site for April 2001. An average 2800 MHz (10.7 cm) solar flux of 159 was assumed for the month. On 10% of the days, the highest frequency propagated is predicted to be at least as high as the uppermost curve (the Highest Possible Frequency, or HPF, approximately 33 MHz), and for 50% of the days as high as the middle curve, the MUF. The lowest curve shows the Lowest Usable Frequency (LUF) for a 1500-W CW transmitter.

F_2 layer critical frequency data from worldwide ionosondes is monitored, and then the sunspot number in an F_2 layer model of the ionosphere is varied until it provides a best fit to the monitored data. The current SSNe comes from Northwest Research Associates in Tucson (AZ), and can be found at **www.nwra.com/spawx/ssne.html**. By plugging the current SSNe into your favorite propagation prediction program, you will have a better (but not perfect) picture of what the ionosphere is doing "now."

19.3.3 Direct Observation

Propagation conditions can be determined directly by listening to the HF bands. The simplest method is to tune higher in frequency until no more long-distance stations are heard. This point is roughly just above the MUF to anywhere in the world at that moment. The highest usable amateur band would be the next lowest one. If HF stations seem to disappear around 23 MHz, for example, the 15 meter band at 21 MHz might make a good choice for DXing. By carefully noting station locations as well, the MUF in various directions can also be determined quickly.

The shortwave broadcast bands (see **Table 19.5**) are most convenient for MUF browsing, because there are many high-powered stations on regular schedules. Take care to ensure that programming is actually transmitted from the originating country. A Radio Moscow or BBC program, for example, may be relayed to a transmitter outside Russia or England for retransmission. An excellent guide to shortwave broadcast stations is the *World Radio TV Handbook*, available through the ARRL.

19.3.4 WWV and WWVH

The standard time stations WWV (Fort Collins, Colorado) and WWVH (Kauai, Hawaii), which transmit on 2.5, 5, 10, 15 and 20 MHz, are also popular for propagation monitoring. They transmit 24 hours a day. Daily monitoring of these stations for signal strength and quality can quickly provide a good basic indication of propagation conditions. In addition, each hour they broadcast the geomagnetic A and K indices, the 2800 MHz (10.7 cm) solar flux, and a short forecast of conditions for the next day. These are heard on WWV at 18 minutes past each hour and on WWVH at 45 minutes after the hour. The same information, along with a lot more space weather information, is available at various websites, such as **www.swpc.noaa.gov** or **www.spaceweather.com**. The K index is updated every three hours, while the A index and solar flux are updated after 2100 UTC. These data are useful for making predictions on home computers, especially when averaged over several days of solar flux observations.

19.3.5 Beacons

Automated *beacons* in the higher amateur bands can also be useful adjuncts to propagation watching. Beacons are ideal for this purpose because most are designed to transmit 24 hours a day.

One of the best organized beacon systems is the International Beacon Project, sponsored by the Northern California DX Foundation (NCDXF) and International Amateur Radio Union (IARU). The beacons operate 24 hours a day at 14.100, 18.110, 21.150, 24.930 and 28.200 MHz. Eighteen beacons on five continents transmit in successive 10-second intervals (each beacon transmits once every 3 minutes). More on this system can be found at the Northern California DX Foundation website **www.ncdxf.org** (click on the *Beacons* link on the left).

A network of automated SDR receivers called the Reverse Beacon Network (**www.reversebeacon.net**) has been created to monitor the HF CW subbands for signals that are automatically decoded by *CW Skimmer* software by VE3NEA (**www.dxatlas.com/cwskimmer**). The network logs signal strength and location for the received signals, providing real-time information on HF propagation.

A list of many 28 MHz beacons can be found on the 10-10 International website, **www.ten-ten.org** (look for *Beacons* under the Resources link). Beacons often include location as part of their automated message, and many can be located from their call sign. Thus, even casual scanning of beacon subbands can be useful. **Table 19.6** provides the frequencies

where beacons useful to HF propagation are most commonly placed.

There are also many beacons on VHF and higher bands. "G3USF's Worldwide List of 50 MHz Beacons" may be found at **www.keele.ac.uk/depts/por/50.htm**. Information on North American beacons on 144 MHz and up is maintained by Ron Klimas, WZ1V, at **www.newsvhf.com/beacons2.html**.

19.3.6 Space Weather Information

Living in the space age results in a wealth of information about the Sun-Earth environment. Not only do we have measurements available, but we also have extensive tutorials available on what is being measured, how it is measured, and its general impact to propagation.

SPACE WEATHER SATELLITES

Three of the most beneficial satellites for propagation information are the Advanced Composition Explorer (ACE), the Solar and Heliospheric Observatory (SOHO), and the STEREO mission.

ACE is on a line from the Earth to the Sun, and is about 1 million miles from the Earth. Its primary purpose is to measure solar wind characteristics. For example, it measures the direction of the IMF (interplanetary magnetic field), the solar wind speed, and the dynamic solar wind pressure. More detailed information about ACE can be found at **www.srl.caltech.edu/ACE/**.

SOHO moves around the Sun in step with Earth — it always faces the Sun. Its primary purpose is to study the Sun. For example, it views coronal mass ejections, and in conjunction with the Michelson Doppler Imager (MDI), allows scientists to "see" sunspots on the far side of the Sun. More detailed information about SOHO and MDI can be found at **http://sohowww.nascom.nasa.gov** and **http://soi.stanford.edu**, respectively.

The STEREO mission consists of two spacecraft that are in heliocentric orbits leading and lagging the Earth. Each spacecraft provides a unique observing vantage point, and taken together, they enable a stereoscopic view of the Sun, solar activity, and the solar environment between the Sun and Earth. More information can be found at **www.swpc.noaa.gov/stereo**.

More satellite data can be found at **www.swpc.noaa.gov** and at **www.spaceweather.com**.

Table 19.5
Shortwave Broadcasting Bands

Frequency (MHz)	Band (meters)
2.300-2.495	120
3.200-3.400	90
3.900-4.000	75
4.750-5.060	60
5.959-6.200	49
7.100-7.300	41
9.500-9.900	31
11.650-12.050	25
13.600-13.800	22
15.100-15.600	19
17.550-17.900	16
21.450-21.850	13
25.600-26.100	11

Table 19.6
Popular Beacon Frequencies

(MHz)	Comments
14.100, 18.110, 21.150, 24.930, 28.200	Northern California DX Foundation beacons
28.2-28.3	Several dozen beacons worldwide
50.0-50.1	Most US beacons are within 50.06-50.08 MHz
70.03-70.13	Beacons in England, Ireland, Gibraltar and Cyprus

MUF Prediction on the Home Computer

Like predicting the weather, predicting HF propagation — even with the best computer software available — is not an exact science. The processes occurring as a signal is propagated from one point on the Earth to another are enormously complicated and subject to an incredible number of variables. Experience and knowledge of propagation conditions (as related to solar activity, especially unusual solar activity, such as flares or coronal mass ejections) are needed when you actually get on the air to check out the bands. Keep in mind, too, that ordinary computer programs are written mainly to calculate propagation for great-circle paths via the F layer. Scatter, skew-path, auroral and other such propagation modes may provide contacts when computer predictions indicate no contacts are possible.

What follows is some brief information about commercially available propagation-prediction programs for the IBM PC and compatible computers. See **Table 19.A.** These programs generally allow predictions from 3 to 30 MHz. Unfortunately prediction programs are not available on 160 meters (because of an incomplete understanding of the lower ionosphere and ducting mechanisms that contribute to propagation on that band) and on 6 meters (because of an incomplete understanding of openings when the MUF is not predicted to be high enough). As a general guideline, you should look for 160 meter openings when the path to your target station is in darkness and around sunrise/sunset, and you should look for 6 meter F_2 openings in the daytime during winter months near solar maximum.

You may find an article that appeared in the Summer 2008 issue of *CQ VHF* to be helpful with your 6m endeavors: "Predicting 6-meter F_2 Propagation," by Carl Luetzelschwab, K9LA.

ASAPS Version 6.1

An agency of the Australian government has developed the *ASAPS* program, which stands for Advanced Stand-Alone Prediction System. It rivals *IONCAP* (see below) in its analysis capability and in its prediction accuracy. It is a *Windows* program that interacts reasonably well with the user, once you become accustomed to the acronyms used. If you change transmit power levels, antennas and other parameters, you can see the new results almost instantly without further menu entries. Available from IPS Radio and Space Services. See **www.ips.gov.au/Products_and_Services/1/1**.

IONCAP and VOACAP

IONCAP, short for Ionospheric Communications Analysis and Prediction, was written by an agency of the US government and has been under development for about 30 years in one form or another. The *IONCAP* program has a well-deserved reputation for being difficult to use, since it came from the world of Fortran punch cards and mainframe computers.

VOACAP is a version of *IONCAP* adapted to Voice of America predictions, but this one includes a sophisticated Windows interface. The Voice of America (VOA) started work on *VOACAP* in the early 1990s and continued for several years before funding ran out. The program was maintained by a single, dedicated computer scientist, Greg Hand, at NTIA/ITS (Institute for Telecommunication Sciences), an agency of the US Department of Commerce in Boulder, Colorado. Greg Hand is now retired, but considerable documentation can be found at the website referenced below, along with links to other VOACAP-related websites. Although *VOACAP* is not specifically designed for amateurs (and thus doesn't include some features that amateurs are fond of, such as entry of locations by ham radio call signs and multiple receiving antennas), it is available for free by downloading from **http://elbert.its.bldrdoc.gov/hf.html**.

W6ELProp, Version 2.70

In 2001, W6EL ported his well known *MINIPROP PLUS* program into the *Windows* world. It uses the same Fricker-based F_2 region computation engine as its predecessor. (This method was developed by Raymond Fricker of the BBC.) *W6ELProp* has a highly intuitive, ham-friendly user interface. It produces detailed output tables, along with a number of useful charts and maps, including the unique and useful "frequency map," which shows the global MUFs from a given transmitting location for a particular month/day/time and solar-activity level. *W6ELProp* is available for free by downloading from **www.qsl.net/w6elprop**.

PropLab Pro, Version 3.0

PropLab Pro by Solar Terrestrial Dispatch represents the high end of propagation-prediction programs. It is the only

19.4 Propagation in the Troposphere

All radio communication involves propagation through the troposphere for at least part of the signal path. Radio waves traveling through the lowest part of the atmosphere are subject to refraction, scattering and other phenomena, much like ionospheric effects. Tropospheric conditions are rarely significant below 30 MHz, but they are very important at 50 MHz and higher. Much of the long-distance work on the VHF, UHF and microwave bands depends on some form of tropospheric propagation. Instead of watching solar activity and geomagnetic indices, those who use tropospheric propagation are much more concerned about terrestrial weather as opposed to space weather.

19.4.1 Line of Sight

At one time it was thought that communications in the VHF range and higher would be restricted to line-of-sight paths. Although this has not proven to be the case even in the microwave region, the concept of line of sight is still useful in understanding tropospheric propagation. In the vacuum of space or in a completely homogeneous medium, radio waves do travel essentially in straight lines, but these conditions are almost never met in terrestrial propagation.

Radio waves traveling through the troposphere are ordinarily refracted slightly earthward. The normal drop in temperature, pressure and water-vapor content with increasing altitude change the index of refraction of the atmosphere enough to cause refraction. Under average conditions, radio waves are refracted toward Earth enough to make the horizon appear 1.15 times farther away than the visual horizon. Under unusual conditions, tropospheric refraction may extend this range significantly.

commercial program presently available that can do complete 3D ray tracing through the ionosphere, even taking complex geomagnetic effects and electron-neutral collisions (what causes absorption) into account. The number of computations is huge, especially in the full-blown 3D mode and operation can be slow and tedious. The *Windows* based user interface is easier to use than previous versions, which were complex and difficult to learn. It is fascinating to see exactly how a signal can bend off-azimuth or how it can split into the ordinary and extraordinary waves. It is best considered a propagation analysis tool, as opposed to a propagation prediction program. See **www.spacew.com/proplab**.

Table 19.A
Features and Attributes of Several Currently Available Propagation Prediction Programs

	ASAPS V. 6.1	VOACAP Windows	W6ELProp V. 2.70	PropLab Pro 3.0
User Friendliness	Good	Good	Good	Good
Operating System	Windows	Windows	Windows	Windows
Uses K index	No	No	Yes	Yes
User library of QTHs	Yes/Map	Yes	Yes	No
Bearings, distances	Yes	Yes	Yes	Yes
MUF calculation	Yes	Yes	Yes	Yes
LUF calculation	Yes	Yes	No	Yes
Wave angle calculation	Yes	Yes	Yes	Yes
Vary minimum wave angle	Yes	Yes	Yes	Yes
Path regions and hops	Yes	Yes	Yes	Yes
Multipath effects	Yes	Yes	No	Yes
Path probability	Yes	Yes	Yes	Yes
Signal strengths	Yes	Yes	Yes	Yes
S/N ratios	Yes	Yes	Yes	Yes
Long path calculation	Yes	Yes	Yes	No
Antenna selection	Yes	Yes	Indirectly	Yes
Vary antenna height	Yes	Yes	Indirectly	Yes
Vary ground characteristics	Yes	Yes	No	No
Vary transmit power	Yes	Yes	Indirectly	Yes
Graphic displays	Yes	Yes	Yes	2D/3D
UT-day graphs	Yes	Yes	Yes	Yes
Area Mapping	Yes	Yes	Yes	Yes
Documentation	Yes	Online	Yes	Yes
Price class	$AUD350[1]	free[2]	free[3]	$240[4]

Prices are for early 2013 and are subject to change.

[1]See **www.ips.gov.au/Products_and_Services/1/1**. Be advised that the price will depend on the exchange rate.
[2]*VOACAP* available at **elbert.its.bldrdoc.gov/hf.html**
[3]*W6ELProp*, see **www.qsl.net/w6elprop**
[4]*PropLab Pro V3*, see **www.spacew.com/proplab/index.html**

A simple formula can be used to estimate the distance to the radio horizon under average conditions:

$$d = \sqrt{2h}$$

where

d = distance to the radio horizon, miles
h = height above average terrain, ft.

and

$$d = \sqrt{17h}$$

where

d = distance to the radio horizon, km
h = height above average terrain, m.

The distance to the radio horizon for an antenna 30 meters (98 ft) above average terrain is thus 22.6 km (14 miles). A station on top of a 1000 meter (3280 ft) mountain has a radio horizon of 130 km (80 miles).

ATMOSPHERIC ABSORPTION

Atmospheric gases, most notably oxygen and water vapor, absorb radio signals, but neither is a significant factor below 10 GHz. See

Fig 19.24. Attenuation from rain becomes important at 3.3 GHz, where signals passing through 20 km (12 miles) of heavy showers incur an additional 0.2 dB loss. That same rain would impose 12 dB additional losses at 10 GHz and losses continue to increase with frequency. Heavy fog is similarly a problem only at 5.6 GHz and above.

19.4.2 Tropospheric Scatter

Contacts beyond the radio horizon out to a working distance of 100 to 500 km (60-310 miles), depending on frequency, equip-

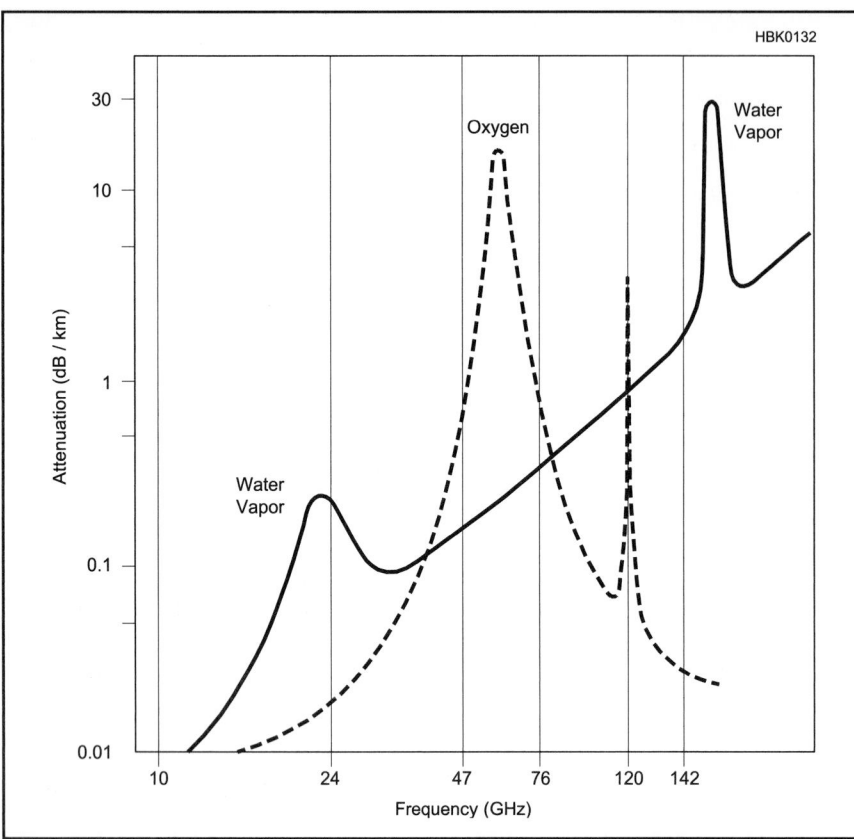

Fig 19.24 — Attenuation caused by oxygen and water vapor at 10 grams per cubic meter (equivalent to 40% humidity at 25 °C). There is little attenuation caused by atmospheric gases at 10 GHz and lower. Note that the 24 GHz band lies near the center of a peak of water vapor absorption and the 120 GHz band is near a peak of oxygen absorption.

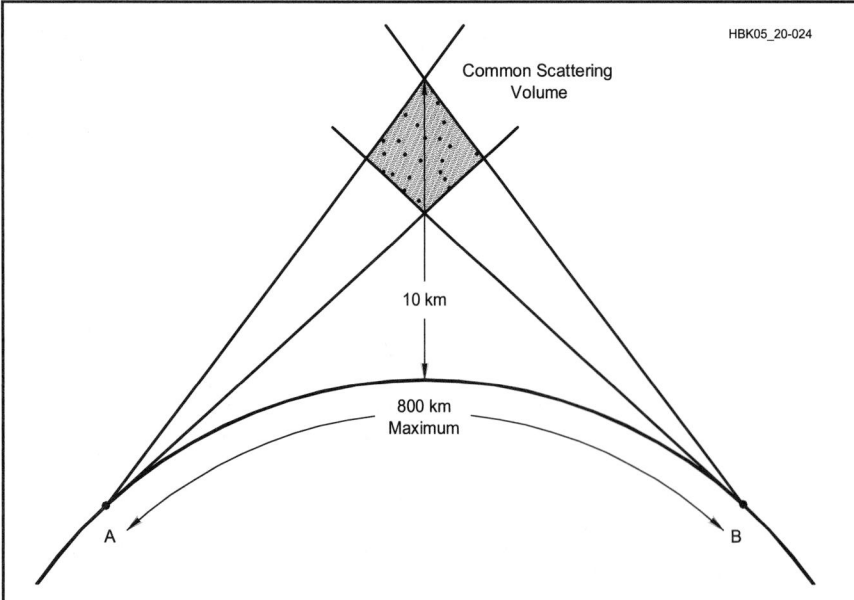

Fig 19.25 — Tropospheric-scatter path geometry. The lower boundary of the common scattering volume is limited by the take-off angle of both stations. The upper boundary of 10 km (6 miles) altitude is the limit of efficient scattering in the troposphere. Signal strength increases with the scattering volume.

ment and local geography, are made every day without the aid of obvious propagation enhancement. At 1.8 and 3.5 MHz, local communication is due mostly to ground wave. At higher frequencies, especially in the VHF range and above, the primary mechanism is scattering in the troposphere, or *troposcatter*.

Most amateurs are unaware that they use troposcatter even though it plays an essential role in most local communication. Radio signals through the VHF range are scattered primarily by wavelength sized gradients in the index of refraction of the lower atmosphere due to turbulence, along with changes in temperature. Radio signals in the microwave region can also be scattered by rain, snow, fog, clouds and dust. That tiny part that is scattered forward and toward the Earth creates the over-the-horizon paths. Troposcatter path losses are considerable and increase with frequency.

The maximum distance that can be linked via troposcatter is limited by the height of a scattering volume common to two stations, shown schematically in **Fig 19.25**. The highest altitude for which scattering is efficient at amateur power levels is about 10 km (6 miles). An application of the distance-to-the-horizon formula yields 800 km (500 miles) as the limit for troposcatter paths, but typical maxima are about half that. Tropospheric scatter varies little with season or time of day, but it is difficult to assess the effect of weather on troposcatter alone. Variations in tropospheric refraction, which is very sensitive to the weather, probably account for most of the observed day-to-day differences in troposcatter signal strength.

Troposcatter does not require special operating techniques or equipment, as it is used unwittingly all the time. In the absence of all other forms of propagation, especially at VHF and above, the usual working range is essentially the maximum troposcatter distance. Ordinary working range increases most dramatically with antenna height, because that lowers the take-off angle to the horizon. Working range increases less quickly with antenna gain and transmitter power. For this reason, a mountaintop is the choice location for extending ordinary troposcatter working distances.

RAIN SCATTER IN THE TROPOSPHERE

Scatter from raindrops is a special case of troposcatter practical in the 3.3 to 24 GHz range. Stations simply point their antennas toward a common area of rain. A certain portion of radio energy is scattered by the raindrops, making possible over-the-horizon or obstructed-path contacts, even with low power. The theoretical range for rain scatter is as great as 600 km (370 miles), but the experience of amateurs in the microwave bands suggests that expected distances are less than 200 km (120 miles). Snow and hail make less efficient scattering media unless the ice particles are partially melted. Smoke and dust particles are too small for extraordinary scattering, even in the microwave bands.

19.4.3 Refraction and Ducting in the Troposphere

Radio waves are refracted by natural gradients in the index of refraction of air with altitude, due to changes in temperature, humidity and pressure. Refraction under standard atmospheric conditions extends the radio horizon somewhat beyond the visual line of sight. Favorable weather conditions further enhance normal tropospheric refraction, lengthening the useful VHF and UHF range by several hundred kilometers and increasing signal strength. Higher frequencies are more sensitive to refraction, so its effects may be observed in the microwave bands before they are apparent at lower frequencies.

Ducting takes place when refraction is so great that radio waves are bent back to the surface of the Earth. When tropospheric ducting conditions exist over a wide geographic area, signals may remain very strong over distances of 1500 km (930 miles) or more. Ducting results from the gradient created by a sharp increase in temperature with altitude, quite the opposite of normal atmospheric conditions. A simultaneous drop in humidity contributes to increased refractivity.

Normally the temperature steadily decreases with altitude, but at times there is a small portion of the troposphere in which the temperature increases and then again begins decreasing normally. This is called an *inversion*. Useful temperature inversions form between 250 and 2000 meters (800-6500 ft) above ground. The elevated inversion and the Earth's surface act something like the boundaries of a natural open-ended waveguide. Radio waves of the right frequency range caught inside the duct will be propagated for long distances with relatively low losses. Several common weather conditions can create temperature inversions.

RADIATION INVERSIONS IN THE TROPOSPHERE

Radiation inversions are probably the most common and widespread of the various weather conditions that affect propagation. Radiation inversions form only over land after sunset as a result of progressive cooling of the air near the Earth's surface. As the Earth cools by radiating heat into space, the air just above the ground is cooled in turn. At higher altitudes, the air remains relatively warmer, thus creating the inversion. A typical radiation-inversion temperature profile is shown in **Fig 19.26A**.

The cooling process may continue through the evening and predawn hours, creating inversions that extend as high as 500 meters (1500 ft). Deep radiation inversions are most common during clear, calm, summer evenings.

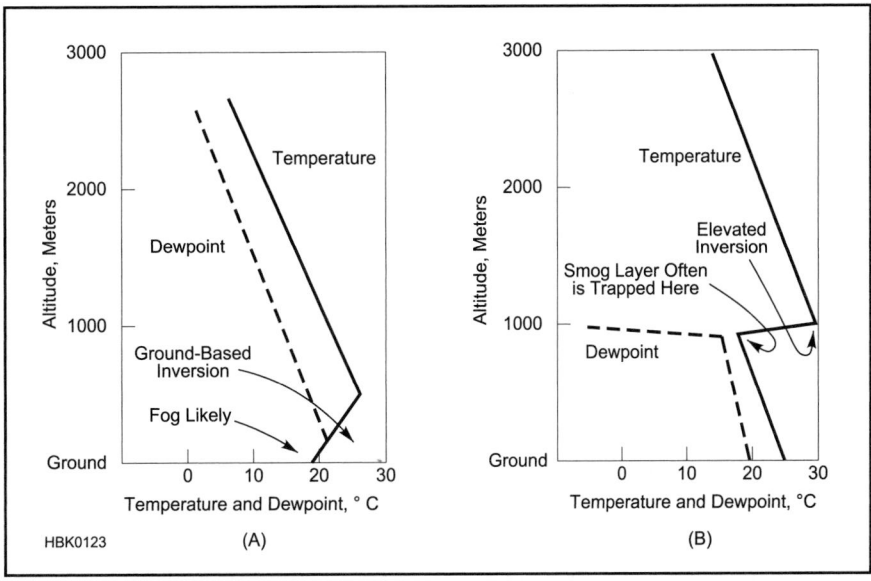

Fig 19.26 — Temperature and dew point profile of an early-morning radiation inversion is shown at A. Fog may form near the ground. The midday surface temperature would be at least 30 °C. At B, temperature and humidity profile across an elevated duct at 1000 meters altitude. Such inversions typically form in summertime high-pressure systems. Note the air is very dry in the inversion.

They are more distinct in dry climates, in valleys and over open ground. Their formation is inhibited by wind, wet ground and cloud cover. Although radiation inversions are common and widespread, they are rarely strong enough to cause true ducting. The enhanced conditions so often observed after sunset during the summer are usually a result of this mild kind of inversion.

HIGH-PRESSURE WEATHER SYSTEMS

Large, sluggish, high-pressure systems (or *anticyclones*) create the most dramatic and widespread tropospheric ducts due to *subsidence*. Subsidence inversions in high-pressure systems are created by air that is sinking. As air descends, it is compressed and heated. Layers of warmer air — temperature inversions — often form between 500 and 3000 meters (1500-10,000 ft) altitude, as shown in Fig 19.26B. Ducts usually intensify during the evening and early morning hours, when surface temperatures drop and suppress the tendency for daytime ground-warmed air to rise. In the Northern Hemisphere, the longest and strongest radio paths usually lie to the south of high-pressure centers. See **Fig 19.27**.

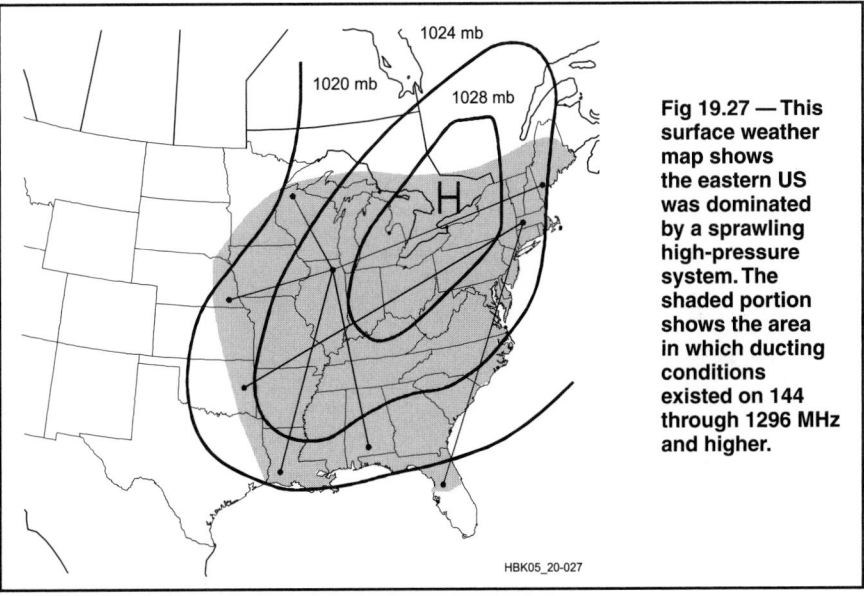

Fig 19.27 — This surface weather map shows the eastern US was dominated by a sprawling high-pressure system. The shaded portion shows the area in which ducting conditions existed on 144 through 1296 MHz and higher.

Sluggish high-pressure systems likely to contain strong temperature inversions are common in late summer over the eastern half of the US. They generally move southeastward out of Canada and linger for days over the Midwest, providing many hours of extended propagation. The southeastern part of the country and the lower Midwest experience the most high-pressure openings; the upper Midwest and East Coast somewhat less frequently; the western mountain regions rarely.

Semi-permanent high-pressure systems, which are nearly constant climatic features in certain parts of the world, sustain the longest and most exciting ducting paths. The Eastern Pacific High, which migrates northward off the coast of California during the summer, has been responsible for the longest ducting paths reported to date. Countless contacts in the 4000 km (2500 mile) range have been made from 144 MHz through 5.6 GHz between California and Hawaii. The *Bermuda High* is a nearly permanent feature of the Caribbean area, but during the summer it moves north and often covers the southeastern US. It has supported contacts in excess of 2800 km (1700 miles) from Florida and the Carolinas to the West Indies, but its full potential has not been exploited. Other semi-permanent highs lie in the Indian Ocean, the western Pacific and off the coast of western Africa.

WAVE CYCLONE

The *wave cyclone* is a more dynamic weather system that usually appears during the spring over the middle part of the American continent. The wave begins as a disturbance along a boundary between cooler northern and warmer southern air masses. Southwest of the disturbance, a cold front forms and moves rapidly eastward, while a warm front moves slowly northward on the eastward side. When the wave is in its open position, as shown in **Fig 19.28**, north-south radio paths 1500 km (930 miles) and longer may be possible in the area to the east of the cold front and south of the warm front, known as the warm sector. East-west paths nearly as long may also open in the southerly parts of the warm sector.

Wave cyclones are rarely productive for more than a day in any given place, because the eastward-moving cold front eventually closes off the warm sector. Wave-cyclone temperature inversions are created by a southwesterly flow of warm, dry air above 1000 meters (3200 ft) that covers relatively cooler and moister Gulf air flowing northward near the Earth's surface. Successive waves spaced two or three days apart may form along the same frontal boundary.

WARM FRONTS AND COLD FRONTS

Warm fronts and cold fronts sometimes bring enhanced tropospheric conditions, but rarely true ducting. A warm front marks the surface boundary between a mass of warm air flowing over an area of relatively cooler and more stationary air. Inversion conditions may be stable enough several hundred kilometers ahead of the warm front to create extraordinary paths.

A cold front marks the surface boundary between a mass of cool air that is wedging itself under more stationary warm air. The warmer air is pushed aloft in a narrow band behind the cold front, creating a strong but highly unstable temperature inversion. The best chance for enhancement occurs parallel to and behind the passing cold front.

OTHER CONDITIONS ASSOCIATED WITH DUCTS

Certain kinds of wind may also create useful inversions. The *Chinook* wind that blows off the eastern slopes of the Rockies can flood the Great Plains with warm and very dry air, primarily in the springtime. If the ground is cool or snow-covered, a strong inversion can extend as far as Canada to Texas and east to the Mississippi River. Similar kinds of *foehn* winds, as these mountain breezes are called, can be found in the Alps, Caucasus Mountains and other places.

The *land breeze* is a light, steady, cool wind that commonly blows up to 50 km (30 miles) inland from the oceans, although the distance may be greater in some circumstances. Land breezes develop after sunset on clear summer evenings. The land cools more quickly than the adjacent ocean. Air cooled over the land flows near the surface of the Earth toward the ocean to displace relatively warmer air that is rising. See **Fig 19.29**. The warmer ocean air, in turn, travels at 200-300 meters (600-1000 ft) altitude to replace the cool surface air. The land-sea circulation of cool air near the ground and warm air aloft creates a mild inversion that may remain for hours. Land-breeze inversions often bring enhanced conditions and occasionally allow contacts in excess of 800 km (500 miles) along coastal areas.

In southern Europe, a hot, dry wind known as the *sirocco* sometimes blows northward from the Sahara Desert over relatively cooler and moister Mediterranean air. Sirocco inversions can be very strong and extend from Israel and Lebanon westward past the Straits of Gibraltar. Sirocco-type inversions are probably responsible for record-breaking microwave contacts in excess of 1500 km (930 miles) across the Mediterranean.

MARINE BOUNDARY LAYER EFFECTS

Over warm water, such as the Caribbean and other tropical seas, *evaporation inversions* may create ducts that are useful in the microwave region between 3.3 and 24 GHz. This inversion depends on a sharp drop in water-vapor content rather than on an increase in temperature to create ducting conditions. Air just above the surface of water at least 30°C is saturated because of evaporation. Humidity drops significantly within 3 to 10 meters (10 to 30 ft) altitude, creating a very shallow but stable duct. Losses due to water vapor absorption may be intolerable at the highest ducting frequencies, but breezes may raise the effective height of the inversion and open the duct to longer wavelengths. Stations must be set up right on the beaches to ensure being inside an evaporation inversion.

TROPOSPHERIC DUCTING FORECASTS

To help decide if weather conditions may support tropospheric ducting, visit "William Hepburn's Worldwide Tropospheric Ducting Forecasts" at **www.dxinfocentre.com/tropo.html**. The principal content on this

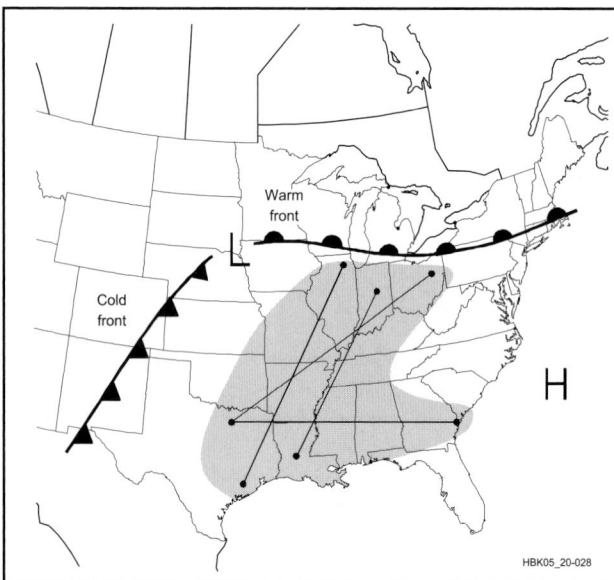

Fig 19.28 — This surface weather map shows a typical spring wave cyclone over the southeastern quarter of the US. The shaded portion shows where ducting conditions existed.

HBK05_20-028

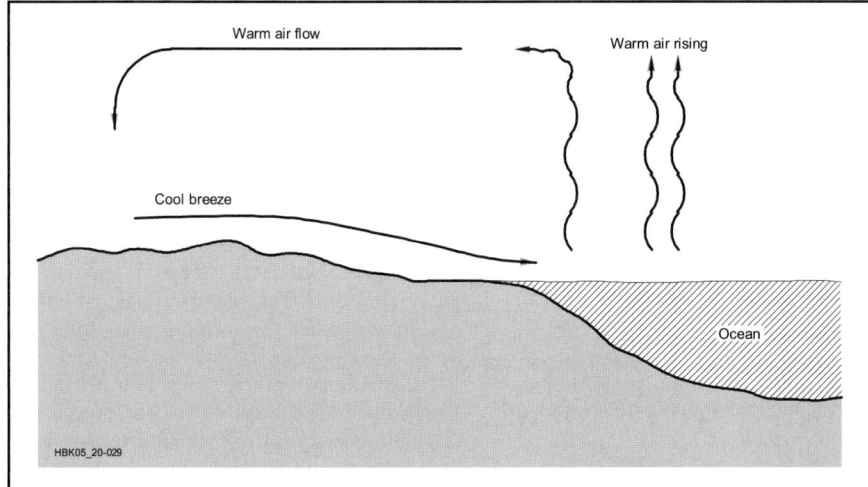

Fig 19.29 — Land-breeze convection along a coast after sunset creates a temperature inversion over the land.

could potentially disrupt signal paths and cause unusual and sometimes rapid variations in signal strengths

19.4.4 Tropospheric Fading

Because the atmosphere is not homogeneous, radio signals are refracted slightly as they encounter variations in density and humidity. A signal may arrive at the receiver by two or more different paths simultaneously, causing addition or partial cancellation depending on the relative phases and amplitudes of the paths. As the wind blows, the paths change, which causes the net amplitude of the received signal to vary slowly, a process known as *scintillation fading*. This is the same process that causes the stars to twinkle in the visible portion of the electromagnetic spectrum. The higher the frequency and the longer the path length, the more pronounced the effect. While scintillation is generally not significant for short-range VHF/UHF repeater work, it can be the main limitation on power budgets for microwave point-to-point links.

Fast-flutter fading at 28 MHz and above is often the result of an airplane that temporarily creates a second propagation path. Flutter results as the phase relationship between the direct signal and that reflected by the airplane changes with the airplane's movement.

website is the forecast maps — 6-day preview as well as 42-hour preview (in 6-hour increments) maps for virtually every region of the world.

The purpose of these maps is to display potential duct paths for VHF, UHF and microwave signals, indicated by the color shading on the maps using the Hepburn Tropo Index.

This index indicates the degree of tropospheric bending forecast to occur over a particular area. It is an indication of the overall tropospheric radio signal strength on a linear scale from 0 to 10 (with 10 representing an "extremely intense opening" for propagation by ducting).

Also shown on the maps are predicted "unstable signal areas" where weather conditions

19.5 VHF/UHF Mobile Propagation

Most amateurs are aware that radio signals in free space obey the inverse-square law: the received signal power is inversely proportional to the square of the distance between the transmitting and receiving antennas. That law applies *only* if the transmitting and receiving antennas have an obstruction-free radio path between them.

Imagine two operators using hand-held 144 MHz radios, each standing on a mountaintop so that they have a direct line of sight. How far apart can they be and still maintain reliable communications? Assume 5 W transmitter power, 0 dBi antenna gain (2.15 dB worse than a dipole), 5 dB receiver noise figure, 10 dB S/N ratio and 12 kHz receiver bandwidth.

Ask experienced VHF mobile operators that question and you'll generally get guesses in the range of 20 to 30 miles because their experience tells them that's about the most you can expect when not communicating through a repeater. The correct answer, however, is over 9500 km (5938 miles) based on the parameters given in the previous paragraph! This also explains how some operators have been able to work the Amateur Radio station on the International Space Station using

only a handheld transceiver.

The discrepancy is explained by the fact that the line-of-sight scenario is not realistic for a mobile station located close to ground level. At distances greater than a few miles, there usually is no line of sight — the signal is reflected at least once on its journey from transmitter to receiver. As a result, path loss is typically proportional to distance to the third or fourth power, not the second power as the inverse-square law implies.

19.5.1 Rayleigh Fading

Not only is the signal reflected, but it usually arrives at the receiver by several different paths simultaneously. See **Fig 19.30**. Because the length of each path is different, the signals are not in phase. If the signals on two paths happen to be 180° out of phase and at the same amplitude, they will cancel. If they are in phase then their amplitudes will add. As the mobile station moves about, the phases of the various paths vary in a random fashion. However they tend to be uncorrelated over distances greater than λ/4 or so, which is about 20 inches on the 2 meter band. That is why if the repeater you are listening to drops out when you are

stopped at a traffic light, you can often get it back again by creeping forward a few inches.

Because there are typically dozens of paths, it is rare for their amplitudes and phases to be such that they all cancel perfectly. Fades of 20 to 30 dB or more are common. The range of signal strengths has a *Rayleigh* distribution,

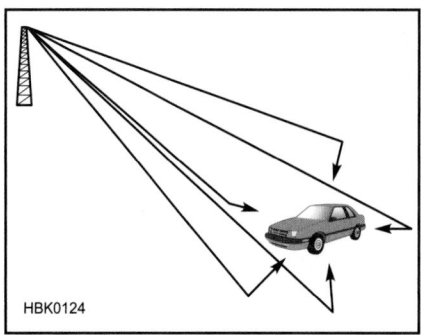

Fig 19.30 — At distances greater than a few miles, there normally is no line-of-sight path to the mobile station. Each path experiences at least one reflection. (Because the radio paths are reciprocal, propagation is the same for both receiving and transmitting.)

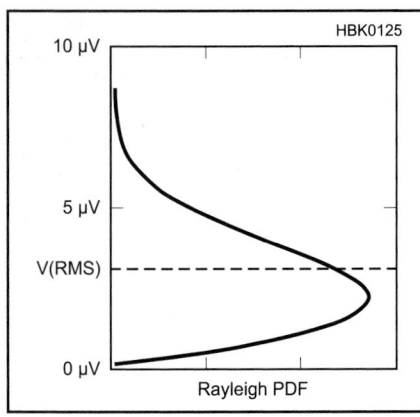

Fig 19.31 — The horizontal axis is the Rayleigh probability density function (PDF), which is the relative probability of the different signal levels shown on the vertical axis. For this graph, an RMS value of 3.16 µV has been selected to represent a typical average signal level in a marginal-coverage area.

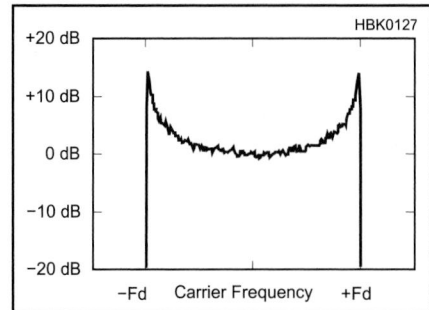

Fig 19.33 — Rayleigh fading spectrum of a CW (unmodulated) signal. The maximum deviation (F_d) from the center frequency is typically less than 10-15 Hz on the 144 MHz band.

named after the physicist/mathematician who first derived the mathematical formula. That is why the phenomenon is called *Rayleigh fading*. **Fig 19.31** shows the relative probability of various signal strengths. **Fig 19.32** is the same graph plotted on a logarithmic (dB) scale, along with a typical plot of signal strength versus time as the mobile station moves down the road.

The closer a reflecting object is to the antenna the smaller is the path loss. A reflector that is close to the transmitter or receiver antenna gives a much stronger signal than one located halfway between. Even a weak reflector, such as a tree branch or telephone pole, is significant if it is close to the mobile station. Because there are many such close-in reflectors, many rays arrive from all directions.

Rays arriving from in front of a forward-moving vehicle experience a positive Doppler frequency shift and rays from the rear have a negative Doppler shift. Those from the sides are somewhere in-between, proportional to the cosine of the angle of arrival. The received signal is the sum of all those rays, which results in *Doppler spreading* of the signal as illustrated in **Fig 19.33**. At normal vehicle speeds on the 2 meter band, the Doppler spread is only plus and minus 10 or 15 Hz, calculated from

Doppler frequency = F_c v/c

where

F_c = the carrier frequency
v = the vehicle speed
c = the speed of light

Use the same units for v and c. On an FM voice signal the only effect is a slight distortion of the audio, but Doppler spreading can severely affect digital signals, as will be discussed later.

19.5.2 Multipath Propagation

In addition to scattering by local reflectors, it is not uncommon also to have more than one main radio path caused by strong reflectors, such as large metal buildings located some distance away. See **Fig 19.34**. Each main path typically does not reach the mobile station directly, but is separately Rayleigh-faded by the local reflectors.

As the mobile station moves around, the shadowing of various paths by intervening hills, buildings and other objects causes the average signal level to fade in and out, but at a much slower rate than Rayleigh fading. This is called *shadowing* or *slow fading*. It is also called *log-normal fading* because the distribution of average signal levels tends to follow a log-normal curve. That means that the *logarithm* of the signal level (on a dB scale, if you will) has a *normal* distribution (the famous bell-shaped curve). This effect typically causes the average signal level on each path to vary plus and minus 10-20 dB (at two standard deviations) from the mean value. This is in addition to the signal variation due to Rayleigh fading.

19.5.3 Effect on the Receiver

Radio direction finding (RDF) enthusiasts have noticed that their RDF receivers usually do not give stable or accurate indications unless located on a hilltop or other location with a clear line of sight to the transmitter because of fading and reflections. The best technique is to record the bearing to the hidden transmitter from a high location clear of nearby reflecting objects, then drive in that direction and take the next reading from another (hopefully closer) hilltop. Only when close to the transmitter will the RDF equipment typically give good readings while the vehicle is in motion.

Under weak-signal conditions Rayleigh fading causes the signal to drop out periodically even though the average signal level would be high enough to maintain reliable communications if there were no fading. *Picket fencing*, as such rapid periodic dropout is called, is a common occurrence when traveling in a weak-signal area at highway speeds. With analog modulation, normally the only solution is to stop at a location where the signal is strong. Moving the vehicle forward or backward a few inches is often enough to change an unreadable signal to solid copy.

Another possible solution is to employ *diversity reception*. If two (or more) mobile receiver antennas are spaced a half-wavelength or more apart, their Rayleigh fading will be almost entirely uncorrelated. That means it is relatively rare for both to experience a deep fade at the same time. The receiving system must have circuitry to determine which of the antennas has a stronger signal at any given time and to automatically combine the signals using some scheme that minimizes the probability of signal dropout. (One engineering text

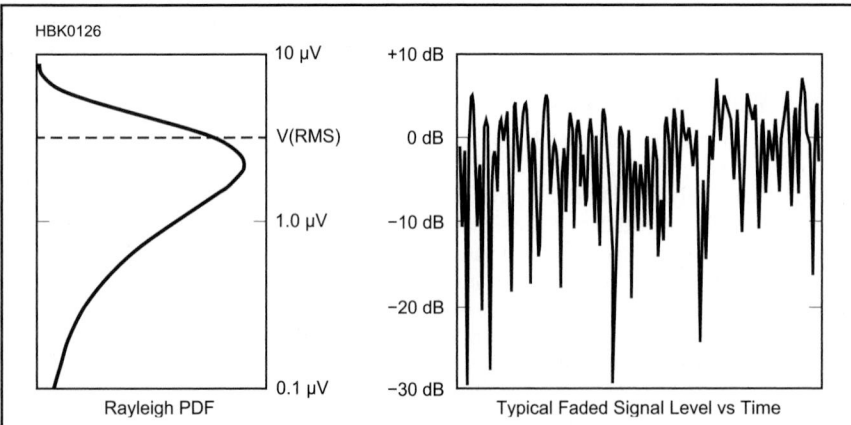

Fig 19.32 — The graph at the left is the same as Fig 19.31 except that the vertical axis is logarithmic (proportional to dB). At the right is a typical Rayleigh-faded signal of the same RMS voltage level plotted using the same vertical scale.

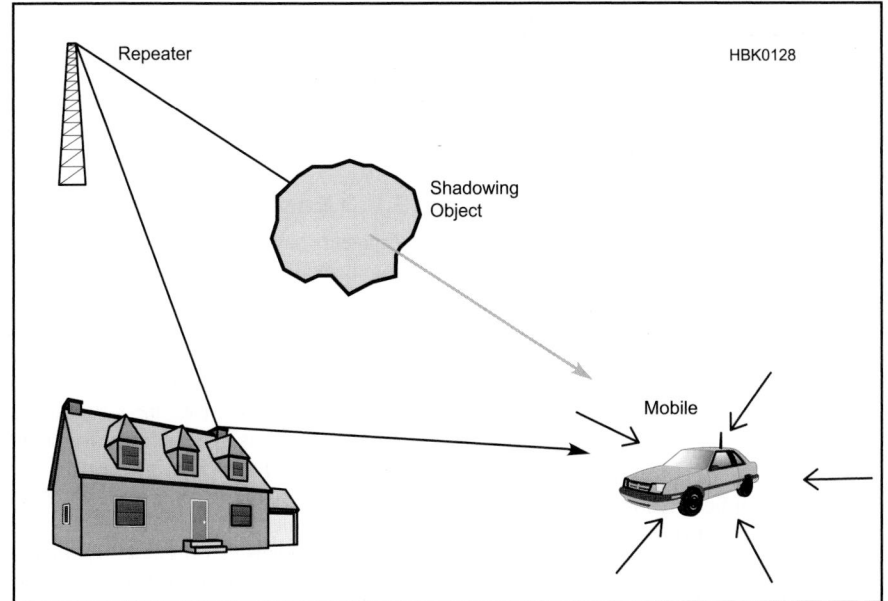

Fig 19.34 — When multipath propagation occurs, each main path is typically Rayleigh-faded by multiple reflectors close to the mobile station.

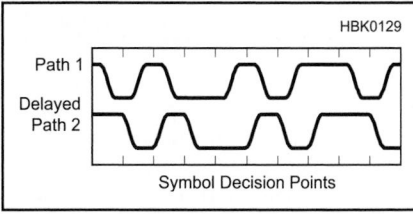

Fig 19.35 — Multipath propagation can cause inter-symbol interference (ISI). At the symbol decision points, where the receiver decoder samples the signal, the path 2 data is often opposing the data on path 1.

that has a fairly readable discussion of fading is *Cellular Radio Performance Engineering* by Asha Mehrotra; see the "References and Bibliography" section at the end of this chapter.)

One low-tech scheme is to use stereo headphones with each channel connected to a separate receiver and antenna. (The TEN-TEC Orion operating manual, downloadable from **www.tentec.com**, has a nice discussion of diversity reception.) That method works better with linear modulation (AM, SSB, CW) than with FM because of the noise burst that occurs when the FM signal drops out.

Diversity antennas can also be used at a repeater site. The conditions are different because most repeater antennas are located in the clear with few local reflectors. The diversity antennas must be located much farther apart, typically on the order of 10 to 20 wavelengths, for the fading to be uncorrelated.

FADING AND DIGITAL SIGNALS

Digitally-modulated signals, such as those used in the cellular telephone industry, use several techniques to combat Rayleigh fading. One is *error-correcting coding*, which adds redundant bits to the transmitted signal in a special way such that the receiver can "fill in" missing bits to obtain error-free reception. That technique does not work if the signal drops out completely for longer than a few

consecutive bits. The solution often employed is called *interleaving*. This takes data bits that represent points close together in time in the original voice signal and shuffles them into several different time slots for transmission. That way, a single brief dropout does not affect all the bits for that time period. Instead, the lost data are scattered over several different time periods. Since only a few bits are missing at any one point, the error-correcting decoder in the receiver has enough information to reconstruct the missing information.

As the signal gradually gets weaker, the error correction in a digital receiver produces nearly-perfect voice quality until the signal gets too weak for the decoder to handle. At that point reception stops abruptly and the call is dropped. With an analog signal, reception gets scratchier as the signal gets weaker; this gives advance warning before the signal drops out completely.

The main paths of a multipath-faded signal can differ in length by several miles. A 10 km (6 mile) difference in path length results in a difference in propagation delay of over 30 μs. While that is not noticeable on an FM voice signal, it can wreak havoc with digital signals by causing *intersymbol interference (ISI)*. See **Fig 19.35**. If the delay difference is on the order of one symbol, the receiver sees adjacent symbols superimposed. In effect, the signal interferes with itself.

One solution is to use an *equalizer*. This is a special type of digital filter that filters the demodulated signal to remove the delay differences so that multiple paths become time-aligned. Since the path characteristics change as the mobile station moves around, a special *training sequence* is sent periodically so that the receiver can re-optimize the filter coefficients based on the known symbols in the training sequence.

ISI is an even bigger problem at HF than at VHF/UHF because the path lengths are much greater. It is not unusual to have path length differences up to 3000 km (1800 miles), which correspond to propagation delay differences of up to 10 ms. That is why digital modes on HF with symbol periods of less than about 10 to 20 ms (that is, with symbol rates greater than 50 to 100 baud) are not very practical without some method of equalization.

There is much more detailed information available about mobile radio propagation, but unfortunately most of it is written at an engineering level. For example, Agilent Technologies manufactures a fading simulator that operates in conjunction with their E4438C ESG signal generator. The simulator software may be downloaded for free from **www.agilent.com** — search for the product number N5115A. Although a license must be purchased to make it actually work with a signal generator, it is instructive to play with the user interface, and there is quite a bit of tutorial material on fading in the help file.

For those who would like to explore the subject further, a *Mathcad* file with equations and explanatory text related to Rayleigh fading is available at **www.arrl.org/qst-in-depth**. Look in the 2006 section for Bloom0806.zip. The file was used to generate some of the graphics in this section. For those without access to *Mathcad*, a read-only PDF version of the file is available in the same directory.

19.6 Propagation for Space Communications

Communication of all sorts into space has become increasingly important. Amateurs confront extraterrestrial propagation when accessing satellite repeaters or using the moon as a reflector. (More information on these modes may be found in the **Space Communications** supplement on the *Handbook* CD.) Special propagation problems arise from signals that travel from the Earth through the ionosphere (or a substantial portion of it) and back again. Tropospheric and ionospheric phenomena, so useful for terrestrial paths, are unwanted and serve only as a nuisance for space communication. A phenomenon known as *Faraday rotation* may change the polarization of radio waves traveling through the ionosphere, presenting special problems to receiving weak signals. Cosmic noise also becomes an important factor when antennas are intentionally pointed into space.

19.6.1 Faraday Rotation

Magnetic and electrical forces rotate the polarization of radio waves passing through the ionosphere. For example, signals that leave the Earth as horizontally polarized and return after a reflection from the moon may not arrive with the same polarization. Additional path loss can occur when polarization is shifted by 90°, resulting in fading of the received signal.

Faraday rotation is difficult to predict and its effects change over time and with operating frequency. At 144 MHz, the polarization of space waves may shift back into alignment with the antenna within a few minutes, so often just waiting can solve the Faraday problem. At 432 MHz, it may take half an hour or longer for the polarization to become realigned. Use of circular polarization completely eliminates this problem, but creates a new problem for EME paths. The sense of circularly polarized signals is reversed with reflection, so two complete antenna systems are normally required, one with left-hand polarization and one with right-hand polarization.

19.6.2 Scintillation

Extraterrestrial signals experience scintillation fading as they travel through the lower atmosphere, as described previously in the *Tropospheric Fading* section. However, they also experience scintillation fading as they traverse the ionosphere. The rule of thumb is that the ionosphere dominates below about 2 GHz and the atmosphere is usually more significant above 2 GHz. Scintillation in the ionosphere is more complex than in the troposphere because, unlike the troposphere, the ionosphere is highly anisotropic and irregularities tend to be aligned with the Earth's magnetic field lines.

Ionospheric scintillation varies with geographical location, time of day, the 11-year sunspot cycle, and the presence of geomagnetic storms. **Fig 19.36** shows the worldwide areas where scintillation occurs. The most intense and frequent disturbances are located within about 20 degrees of the Earth's magnetic equator. Fading depth in this region can be as much as 20 to 30 dB. Scintillation fading also is common within about 30 degrees of the magnetic poles, although fade depths are generally no more than 10 dB or so. At mid latitudes ionospheric scintillation is rarely an issue. Longitude is important as well. In South America, Africa and India, the frequency of occurrence is significantly lower in the southern hemisphere winter (May-August) than during the summer, while the pattern is exactly opposite in the central Pacific.

In the polar regions disturbances occur at any time of day, but in the equatorial region they tend to start an hour or so after sunset and last a few hours, from perhaps 1900-2400 local time at the equinox. Fading events are sometimes intermittent, starting abruptly and ending abruptly some minutes later. The signal paths to slow-moving sources, such as the moon or geostationary satellites, tend to fade about once or twice a second. While not much work has been done on characterizing ionospheric scintillation fading on low Earth orbit (LEO) satellites, the fade rate should be more than an order of magnitude faster since the orbital period is typically on the order of 1.5 hours, much less than the Earth's 24-hour rotational period.

As might be expected, sunspots also are important. Ionospheric scintillation is much more severe and frequent near the peak of

the 11-year solar cycle. Geomagnetic storms also have a major effect, principally in the polar regions.

19.6.3 Earth-Moon-Earth

Amateurs have used the moon as a reflector on the VHF and UHF bands since 1960. Maximum allowable power and large antennas, along with the best receivers, are normally required to overcome the extreme free-space and reflection losses involved in Earth-Moon-Earth (EME) paths. More modest stations make EME contacts by scheduling operating times when the moon is at perigee (on the horizon). The moon, which presents a target only one-half degree wide, reflects only 7% of the radio signals that reach it. Techniques have to be designed to cope with Faraday rotation, cosmic noise, Doppler shift (due to the moon's movements) and other difficulties. In spite of the problems involved, hundreds (and possibly thousands) of amateur stations have made contacts via the moon on all bands from 50 MHz to 10 GHz. The techniques of EME communication are discussed in the **Space Communications** supplement on the *Handbook* CD.

19.6.4 Satellites

Accessing amateur satellites generally does not involve huge investments in antennas and equipment, yet station design does have to take into account special challenges of space propagation. Free-space loss is a primary consideration, but it is manageable when satellites are only a few hundred kilometers distant. Free-space path losses to satellites in high-Earth orbits are considerably

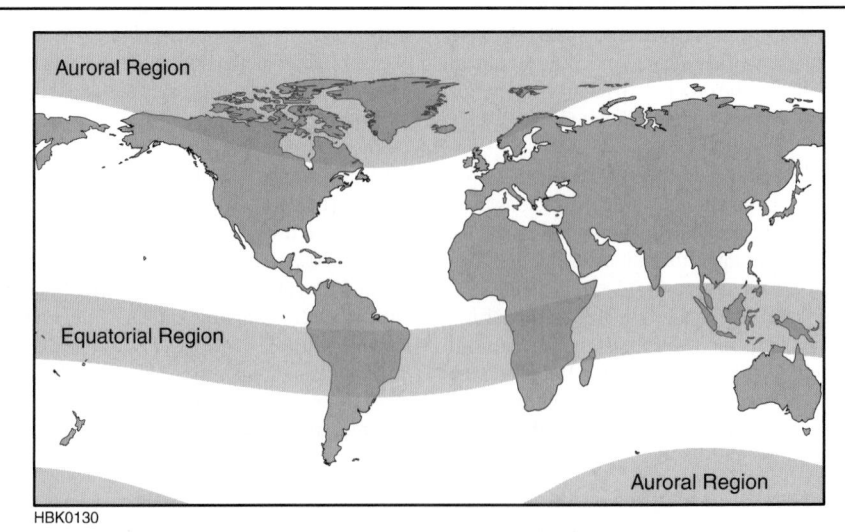

Fig 19.36 — Regions of the world (shaded) where scintillation fading effects can be especially severe.

greater, and appropriately larger antennas and higher powers are needed.

Satellite frequencies below 30 MHz can be troublesome. Ionospheric absorption and refraction may prevent signals from reaching space, especially to satellites at very low elevations. In addition, man-made and natural sources of noise are high. VHF and especially UHF are largely immune from these effects, but free-space path losses are greater. Problems related to polarization, including Faraday rotation, intentional or accidental satellite tumbling and the orientation of a satellite's antenna in relation to terrestrial antennas, are largely overcome by using circularly polarized antennas. More on using satellites can be found in the **Space Communications** supplement on the *Handbook* CD.

19.7 Noise and Propagation

Noise simply consists of unwanted electromagnetic radiation that interferes with desired communications. In some instances, noise imposes the practical limit on the lowest usable frequencies. Noise may be classified by its sources: man-made, terrestrial (atmospheric) and cosmic. Interference from other transmitting stations on adjacent frequencies is not usually considered noise and may be controlled, to some degree anyway, by careful station design or by simply moving to a frequency further away from the interference. (Additional information on noise can be found in the **RF Techniques** and **Receivers** chapters.)

19.7.1 Man-Made Noise

Many unintentional radio emissions result from man-made sources. Broadband radio signals are produced whenever there is a spark, such as in contact switches, electric motors, gasoline engine spark plugs and faulty electrical connections. Household appliances, such as fluorescent lamps, microwave ovens, lamp dimmers and anything containing an electric motor may all produce undesirable broadband radio energy. Devices of all sorts, especially computers and anything controlled by microprocessors, television receivers and many other electronics also emit radio signals that may be perceived as noise well into the UHF range. In many cases, these sources are local and can be controlled with proper measures. See the **RF Interference** chapter.

High-voltage transmission lines and associated equipment, including transformers, switches and lightning arresters, can generate high-level radio signals over a wide area, especially if they are corroded or improperly maintained. Transmission lines may act as efficient antennas at some frequencies, adding to the noise problem. Certain kinds of street lighting, neon signs and industrial equipment also contribute their share of noise.

Fig 19.37 shows typical noise levels versus frequency, in terms of power, for various noise environments. Note that man-made noise prevails below roughly 14 MHz, and it, rather than the sensitivity of your receiver (here depicted as the MDS — the minimum discernible signal), determines how weak a signal you can hear.

19.7.2 Lightning

Static is a common term given to the ear-splitting crashes of noise commonly heard on nearly all radio frequencies, although it is most severe on the lowest frequency bands. Atmospheric static is primarily caused by lightning and other natural electrical discharges. Static may result from close-by thunderstorms, but most static originates with tropical storms. Like any radio signals, lightning-produced static may be propagated over long distances by the ionosphere. Thus static is generally higher during the summer, when there are more nearby thunderstorms, and at night, when radio propagation generally improves. Static is often the limiting factor on 1.8 and 3.5 MHz, making winter a more favorable time for using these frequencies. (Note that the quiet "winter" months in the Southern Hemisphere are June through August.)

19.7.3 Precipitation Static and Corona Discharge

Precipitation static is an almost continuous hash-type noise that often accompanies various kinds of precipitation, including snowfall. Precipitation static is caused by rain drops, snowflakes or even wind-blown dust, transferring a small electrical charge on contact with an antenna. Electrical fields under thunderstorms are sufficient to place many objects such as trees, hair and antennas, into corona discharge. *Corona noise* may sound like a harsh crackling in the radio — building in intensity, abruptly ending, and then building again, in cycles of a few seconds to as long as a minute. A corona charge on an antenna may build to some critical level and then discharge in the atmosphere with an audible pop before recharging. Precipitation static and corona discharge can be a nuisance from LF to well into the VHF range.

19.7.4 Cosmic Sources

The Sun, distant stars, galaxies and other cosmic features all contribute radio noise well into the gigahertz range. These *cosmic sources* are perceived primarily as a more-or-less constant background noise at HF. In the VHF range and higher, specific sources of cosmic noise can be identified and may be a limiting factor in terrestrial and space communications. The Sun is by far the greatest source of radio noise, but its effects are largely absent at night. The center of our own galaxy is nearly as noisy as the Sun. Galactic noise is especially noticeable when high-gain VHF and UHF antennas, such as may be used for satellite or EME communications, are pointed toward the center of the Milky Way. Other star clusters and galaxies are also radio hot-spots in the sky. Finally, there is a much lower cosmic background noise that seems to cover the entire sky.

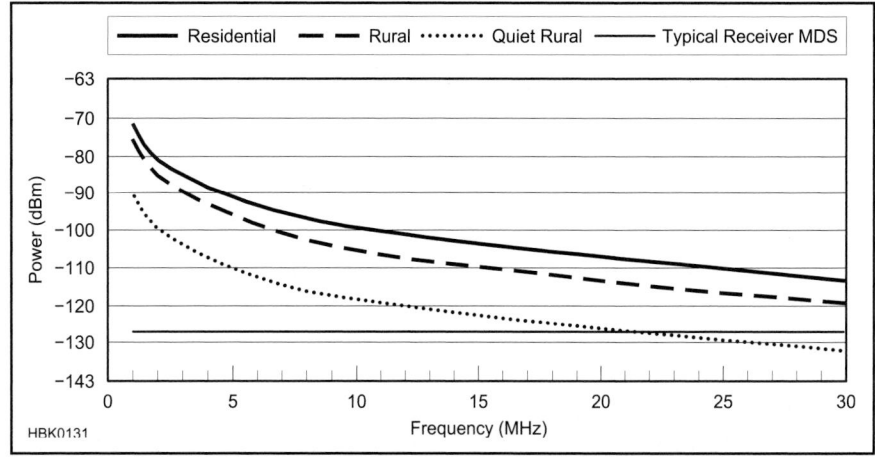

Fig 19.37 — Typical noise levels versus frequency for various environments. (Man-made noise in a 500-Hz bandwidth, from Rec. ITU-R P.372.7, *Radio Noise*)

19.8 Propagation Below the AM Broadcast Band

In his column "It Seems to Us — Two New Bands?" in February 2013 *QST*, ARRL CEO David Sumner, K1ZZ reviewed the status of Amateur Radio frequency allocations below the AM broadcast band. Two bands at these low frequencies are under consideration: 135.7-137.8 kHz and 472-479 kHz. If and when we are allowed access to these bands is anyone's guess, but it would be beneficial to review some fundamental issues with respect to propagation in these bands.

Three important parameters are involved in determining the likelihood of a signal getting from Point A to Point B. These parameters are refraction, absorption and polarization. Although propagation is certainly different on 160 meters and 6 meters, an electromagnetic wave on both of these bands follows the same laws of physics. Thus understanding how these three parameters change versus frequency will give us insight into propagation on our potentially new low frequency bands.

Fig 19.38 is a ray trace at night (0500 UTC) in mid January at moderate solar activity for frequencies from 10.65 MHz down to 150 kHz. All rays are launched at a low elevation angle of 5 degrees. (This figure was made using *Proplab Pro V3* — see the section MUF Prediction on the Home Computer for more details on this software package.)

The MUF is not high enough for 10.65 MHz under these conditions, and it goes through the ionosphere. But frequencies of 8.9 MHz and lower are successively refracted at lower heights, which results in shorter hops. So RF on 135.7-137.8 kHz and 472-479 kHz will take short hops.

Table 19.7 summarizes absorption (data from the ray trace) on 5.4 MHz, 3.65 MHz, 1.9 MHz and 150 kHz for a 1500 km hop under the same conditions as the ray traces in Fig 19.38. Note that as the frequency decreases from 5.4 MHz to 1.9 MHz, the loss due to ionospheric absorption increases as expected. But at 150 kHz the absorption is significantly reduced. This is due to the wave not getting very high into the ionosphere — in

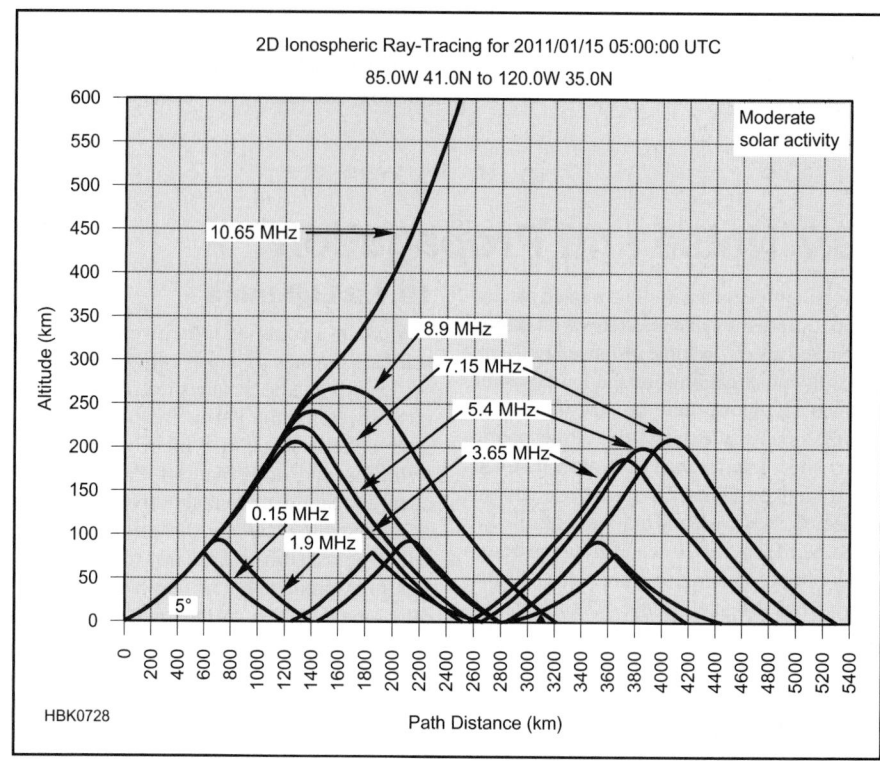

Fig 19.38 — Low frequency ray traces showing more bending as the frequency is lowered.

Table 19.7
Absorption vs Frequency

Frequency (MHz)	Absorption per Hop (dB)
5.4	0.8
3.65	2.3
1.9	17.8
0.150	4.0

fact, the residual nighttime D region ionization is sufficient to refract this low frequency with minimal absorption. So RF on 135.7-137.8 kHz and 472-479 kHz will not suffer as much absorption as on 160 meters.

In theory, highly elliptical polarization will prevail for these low frequencies. It is likely that most antennas for these bands will be vertically polarized, which should be a good match to couple the most energy into the ordinary wave.

In summary, skywave propagation on 135.7-137.8 kHz and 472-479 kHz will be via short hops between the ionosphere and the D region, but with lower absorption than on 160 meters. Tempering this lower absorption are the lower antenna efficiencies and higher received noise levels. Thus these bands will be tough, but the Amateur Radio spirit will make the most of them.

19.9 Glossary of Radio Propagation Terms

A index — An open-ended linear index that corresponds roughly to the cumulative **K index** values (it's the daily average of the eight K indices after converting the K indices to a linear scale). The A index commonly varies between 0 and 30 during quiet to active conditions, and up to 100 and higher during geomagnetic storms.

Absorption — The dissipation of the energy of a radio wave as it travels through a medium such as the ionosphere.

Antipode — Locations directly opposite each other on a globe.

Atmosphere — The mass of air surrounding the Earth. Radio signals travel through the atmosphere and different conditions in the atmosphere (in conjunction with the signal's frequency) affect how those signals travel or propagate.

Aurora — A disturbance of the atmosphere at high latitudes resulting from an interaction between electrically charged particles from the magnetosphere and the magnetic field of the Earth. *Auroral propagation* occurs when HF through UHF signals are reflected from the aurora to reach another station.

Auroral E — Sporadic E in the auroral zone.

Backscatter — Single-hop signals that have been scattered by the Earth or the ocean at some distant point back toward the transmitting station.

Beacon station — A station that transmits continuously, allowing other stations to assess propagation to and from the location of the beacon station.

Coronal hole — A region on the Sun where the magnetic field is open to the interplanetary magnetic field (IMF) and ionized particles can escape into the solar wind.

Critical angle — The largest angle at which a radio wave of a specified frequency can be returned to Earth by the ionosphere.

Critical frequency — The highest frequency that returns echoes from the E and F regions at vertical incidence.

D region — The lowest region of the ionosphere. The D region (or layer) contributes very little to short-wave radio propagation. It absorbs energy from radio waves as they pass through it. This absorption has a significant effect on signals below about 7.5 MHz during daylight.

Diffraction — Bending of waves by an edge or corner.

E region — The second lowest ionospheric region, the E region (or layer) has its highest electron density during the day and falls to a much lower electron density during the night. Under certain conditions, it may refract radio waves enough to return them to Earth.

Earth-Moon-Earth (EME) or *Moonbounce* — A method of communicating with other stations by reflecting radio signals off the Moon's surface.

Electromagnetic wave — A wave of energy composed of an electric and magnetic field.

Equinoxes — One of two points in the orbit of the Earth around the Sun at which the Earth crosses a horizontal plane extending through the equator of the Sun. The vernal equinox marks the beginning of spring and the autumnal equinox marks the beginning of autumn.

Faraday rotation — A rotation of the polarization of radio waves when the waves travel through an ionized medium that is immersed in a magnetic field (for example, the Earth's ionosphere).

F region — A combination of the two highest ionospheric regions (or layers), the F_1 and F_2 regions. The F region refracts radio waves and returns them to Earth. Its height varies greatly depending on the time of day, season of the year and amount of sunspot activity.

Field-aligned irregularities (FAI) — A propagation mechanism observed at 50 and 144 MHz that occurs when irregularities in the distribution of free electrons in the ionosphere are aligned parallel to the Earth's magnetic field.

Free-space attenuation — The dissipation of the energy of a radio wave that results from the spherical spreading of radio energy from its source.

Gray-line — A special form of **long-path** propagation that takes into account the unusual ionospheric configuration along the twilight region between night and day.

Ground-wave propagation — The method by which radio waves travel along the Earth's surface.

High frequency (HF) — The term used for the frequency range between 3 MHz and 30 MHz. The amateur HF bands are where you are most likely to make long-distance (worldwide) contacts.

Ionosphere — A region of electrically charged (ionized) gases high in the atmosphere. The ionosphere bends radio waves as they travel through it, returning them to Earth. Also see **sky-wave propagation**.

K index — A geomagnetic-field measurement that is updated every three hours at various observatories around the world.

Changes in the K index can be used to indicate HF propagation conditions. Rising values generally indicate disturbed conditions while falling values indicate improving conditions.

Line-of-sight propagation — The term used to describe VHF and UHF propagation in a straight line directly from one station to another.

Long-path propagation — Propagation between two points on the Earth's surface that follows a path along the great circle between them, but in a direction opposite from the shortest distance between them.

Lowest usable frequency (LUF) — The frequency below the **maximum usable frequency** (MUF) at which ionospheric absorption and noise at the receiving location make the received signal-to-noise ratio too low to be usable.

M-factor — The ratio between the **maximum usable frequency** (MUF) and the critical frequency.

Maximum usable frequency (MUF) — The highest-frequency radio signal that will reach a particular destination using **sky-wave propagation**, or *skip*. The MUF may vary for radio signals sent to different destinations.

Meteor-scatter communication — A method of radio communication that uses the ionized trail of a meteor that burned up in the Earth's atmosphere to reflect, refract, or scatter radio signals back to Earth.

Microwave — Radio waves or signals with frequencies greater than 1000 MHz (1 GHz). This is not a strict definition, just a conventional way of referring to those frequencies.

Moonbounce — A common name for EME communication in which signals are bounced off the Moon before being received.

Multihop propagation — Long-distance radio propagation using several skips or hops between the Earth and the ionosphere.

Multipath — A fading effect caused by the transmitted signal traveling to the receiving station over more than one path.

Near Vertical Incidence Skywave (NVIS) propagation — A propagation mechanism that allows stations located within the **skip zone,** but too far apart for ground wave propagation, to maintain communications by going to a lower frequency.

Path loss — The total signal loss between transmitting and receiving stations relative to the total radiated signal energy.

Pedersen ray — A high-angle radio wave that penetrates deeper into the F region of the ionosphere, so the wave is bent less than a lower-angle wave, and thus for some distance parallels the Earth's surface in the F region, returning to Earth at a distance farther than normally expected for single-hop propagation.

Polarization — The orientation of the electrical-field of a radio wave. An antenna that is parallel to the surface of the Earth, such as a dipole, produces horizontally polarized waves. One that is perpendicular to the Earth's surface, such as a quarter-wave vertical, produces vertically polarized waves. An antenna that has both horizontal and vertical polarization is said to be circularly polarized.

Propagation — The process by which radio waves travel.

Radio frequency (RF) signals — RF signals are generally considered to be any electrical signals with a frequency higher than 20,000 Hz, up to 300 GHz.

Radio horizon — The position at which a direct wave radiated from an antenna becomes tangent to the surface of the Earth. Note that as the wave continues past the horizon, the wave gets higher and higher above the surface.

Radiation inversion — A weather condition that affects propagation at VHF and above. Radiation inversions form only over land after sunset as a result of progressive cooling of the air near the Earth's surface.

Rain scatter — A special case of tropospheric scatter practical in the 3.3 to 24 GHz range that is caused by scatter from raindrops.

Reflection — Signals that travel by **line-of-sight propagation** are reflected by large objects like buildings.

Refraction — Bending waves by changing the velocity of propagation. Radio waves refract as they travel through the ionosphere. If the radio waves refract enough they will return to Earth. This is the basis for long-distance communication on the HF bands.

Scattering — Radio wave propagation by means of multiple reflections in the layers of the atmosphere or from an obstruction. Scatter propagation also occurs in the ionosphere when there is not enough ionization for refraction or reflection, but enough to send weak electromagnetic waves off into varied directions.

Scintillation fading — Fading that occurs when a signal arrives at the receiver by two or more different paths simultaneously, causing addition or partial cancellation depending on the relative phases and amplitudes of the paths.

Selective fading — A variation of radio-wave intensity that changes over small frequency changes. It may be caused by changes in the medium through which the wave is traveling or changes in transmission path, among other things.

Short path — The shorter of the two great circle paths between two stations.

Skip — Propagation by means of ionospheric reflection. Traversing the distance to the ionosphere and back to the ground is called a *hop*.

Skip zone — A ring-shaped area of poor radio communication, too distant for ground waves and too close for sky waves.

Sky-wave propagation — The method by which radio waves travel through the ionosphere and back to Earth. Sometimes called *skip*, sky-wave propagation has a far greater range than **line-of-sight** and **ground-wave propagation**.

Solar cycle — The approximate 11-year period of variation in solar activity.

Solar flare — An eruption on the surface of the Sun that launches a wide spectrum of electromagnetic energy into space, disrupting communications on Earth. A large flare can also release relativistic protons that cause additional absorption in the polar cap.

Solar wind — Electrically charged particles emitted by the Sun and traveling through space. Variations in the solar wind may have a sudden impact on radio communications when they arrive at the atmosphere of the Earth.

Sporadic E — A form of enhanced radio-wave propagation that occurs when radio signals are reflected from small, dense ionization patches in the E region of the ionosphere. Sporadic E is observed on the 15, 10, 6 and 2 meter bands, and occasionally on the 1.25 meter band.

Sunspot cycle — The number of **sunspots** increases and decreases in a somewhat predictable cycle that lasts about 11 years.

Sunspots — Dark spots on the surface of the Sun. When there are few sunspots, long-distance radio propagation is poor on the higher-frequency bands. When there are many sunspots, long-distance HF propagation improves.

Temperature inversion — A condition in the atmosphere in which a region of cool air is trapped beneath warmer air.

Transequatorial propagation — A form of F layer ionospheric propagation, in which signals of higher frequency than the expected MUF are propagated across the Earth's magnetic equator.

Troposphere — The region in Earth's atmosphere just above the Earth's surface and below the ionosphere.

Tropospheric bending — When radio waves are bent in the troposphere, they return to Earth farther away than the visible horizon.

Tropospheric ducting — A type of VHF propagation that can occur when warm air overruns cold air (a temperature inversion).

Tropospheric scatter — A method of radio communication at VHF and above that takes advantage of scattering in the **troposphere** to allow contacts beyond the radio horizon out to a working distance of 100 to 500 km (60 to 310 miles), depending on frequency.

Ultra high frequency (UHF) — The term used for the frequency range between 300 MHz and 3000 MHz (3 GHz). Technician licensees have full privileges on all Amateur UHF bands.

Very high frequency (VHF) — The term used for the frequency range between 30 MHz and 300 MHz. Technician licensees have full privileges on all Amateur VHF bands.

Visible horizon — The most distant point one can see by line of sight.

WWV/WWVH — Radio stations run by the US NIST (National Institute of Standards and Technology) to provide accurate time and frequencies.

19.10 References and Bibliography

A. Barter, G8ATD, ed., *International Microwave Handbook*, 2nd edition (Potters Bar: RSGB, 2008). Includes a chapter on microwave propagation.

A. Barter, G8ATD, ed., *VHF/UHF Handbook*, 2nd edition (Potters Bar: RSGB, 2008). Includes a chapter on VHF/UHF propagation.

B. R. Bean and E. J. Dutton, *Radio Meteorology* (New York: Dover, 1968).

K. Davies, *Ionospheric Radio* (London: Peter Peregrinus, 1989). Excellent though highly technical text on propagation.

R. D. Hunsucker, J.K. Hargreaves, *The High Latitude Ionosphere and Its Effect on Radio Propagation* (Cambridge University Press, 2003). Highly technical, but with an excellent chapter on fundamental physics of the ionosphere.

G. Jacobs, T. Cohen, R. Rose, *The New Shortwave Propagation Handbook*, *CQ* Communications, Inc. (Hicksville, NY: CQ Communications, 1995).

C. Luetzelschwab, K9LA, **http://myplace. frontiercom/~k9la**. Many articles about the ionosphere and propagation as they relate to 160 meters, HF, VHF, contesting and more.

L. F. McNamara, *Radio Amateur's Guide to the Ionosphere* (Malabar, Florida: Krieger Publishing Company, 1994). Excellent, quite-readable text on HF propagation.

A. Mehrotra, *Cellular Radio Performance Engineering* (Artech House, 1994). See Chapter 4 "Propagation" and the introduction to Chapter 3 "Cellular Environment." Diversity reception techniques are covered in Chapter 8.

D. Straw, N6BV, "What's the Deal About 'NVIS'?," *QST*, Dec 2005.

J. A. White, VA7JW, and K. Tapping, "The Penticton Solar Flux receiver," *QST*, Feb 2013, pp 39-45.

Contents

Chapter 20 — CD-ROM Content

Supplemental Articles

- "Multiband Operation with Open-wire Line" by George Cutsogeorge, W2VJN
- Smith Chart Supplement
- "Measuring Receiver Isolation" by George Cutsogeorge, W2VJN
- "A Commercial Triplexer Design" by George Cutsogeorge, W2VJN
- "HF Yagi Triplexer Especially for ARRL Field Day" by Gary Gordon, K6KV
- "Using *TLW* to Design Impedance Matching Networks" by George Cutsogeorge, W2VJN
- "Measuring Ferrite Chokes" by Jim Brown, K9YC
- "Microwave Plumbing" by Paul Wade, W1GHZ
- Transmission Lines in Digital Circuits
- Matching Network Material and MATCH.EXE by Bill Sabin, WØIYH

Software

- *jjSmith* from Tonne Software

Transmission Lines

RF power is rarely generated right where it will be used. A transmitter and the antenna it feeds are a good example. The most effective antenna installation is outdoors and clear of ground and energy-absorbing structures. The transmitter, however, is most conveniently installed indoors, where it is out of the weather and is readily accessible. A *transmission line* is used to convey RF energy from the transmitter to the antenna. A transmission line should transport the RF from the source to its destination with as little loss as possible. This chapter, written by Dean Straw, N6BV, and updated by George Cutsogeorge, W2VJN, explores transmission line theory and applications. Jim Brown, K9YC, contributed updated material on transmitting choke baluns.

20.1 Transmission Line Basics

There are three main types of transmission lines used by radio amateurs: coaxial, open-wire and waveguide. The most common type is the *coaxial* line, usually called *coax*, shown in various forms in **Fig 20.1**. Coax is made up of a center conductor, which may be either stranded or solid wire, surrounded by a concentric outer conductor with a *dielectric* center insulator between the conductors. The outer conductor may be braided shield wire or a metallic sheath. A flexible aluminum foil or a second braided shield is employed in some coax to improve shielding over that obtainable from a standard woven shield braid. If the outer conductor is made of solid aluminum or copper, the coax is referred to as *hardline*.

The second type of transmission line uses parallel conductors, side by side, rather than the concentric ones used in coax. Typical examples of such *open-wire* lines are 300 Ω TV ribbon line or *twin-lead* and 450 Ω ladder line (sometimes called *window line*), also shown in Fig 20.1. Although open-wire lines are enjoying a sort of renaissance in recent years because of their inherently lower losses in simple multiband antenna systems, coaxial cables are far more prevalent because they are much more convenient to use.

The third major type of transmission line is the *waveguide*. While open-wire and coaxial lines are used from power-line frequencies to well into the microwave region, waveguides are used at microwave frequencies only. Waveguides will be covered at the end of this chapter.

20.1.1 Fundamentals

In either coaxial or open-wire line, currents flowing in the two conductors travel in opposite directions as shown in Figs 20.1E and 20.1I. If the physical spacing between the two parallel conductors in an open-wire line, S, is small in terms of wavelength, the phase difference between the currents will be very close to 180°. If the two currents also have equal amplitudes, the field generated by each conductor will cancel that generated by the other, and the line will not radiate energy, even if it is many wavelengths long.

The equality of amplitude and 180° phase difference of the currents in each conductor in an open-wire line determine the degree of radiation cancellation. If the currents are for some reason unequal, or if the phase difference is not 180°, the line will radiate energy. How such imbalances occur and to what degree they can cause problems will be covered in more detail later.

In contrast to an open-wire line, the outer conductor in a coaxial line acts as a shield, confining RF energy within the line as shown in Fig 20.1E. Because of *skin effect* (see the **RF Techniques** chapter), current flowing in the outer conductor of a coax does so on the inner surface of the outer conductor. The fields generated by the currents flowing on the outer surface of the inner conductor and on the inner surface of the outer conductor cancel each other out, just as they do in open-wire line.

VELOCITY FACTOR

In free space, electrical waves travel at the speed of light, or 299,792,458 meters per second. Converting to feet per second yields 983,569,082. The length of a wave in space may be related to frequency as wavelength = λ = velocity/frequency. Thus, the wavelength of a

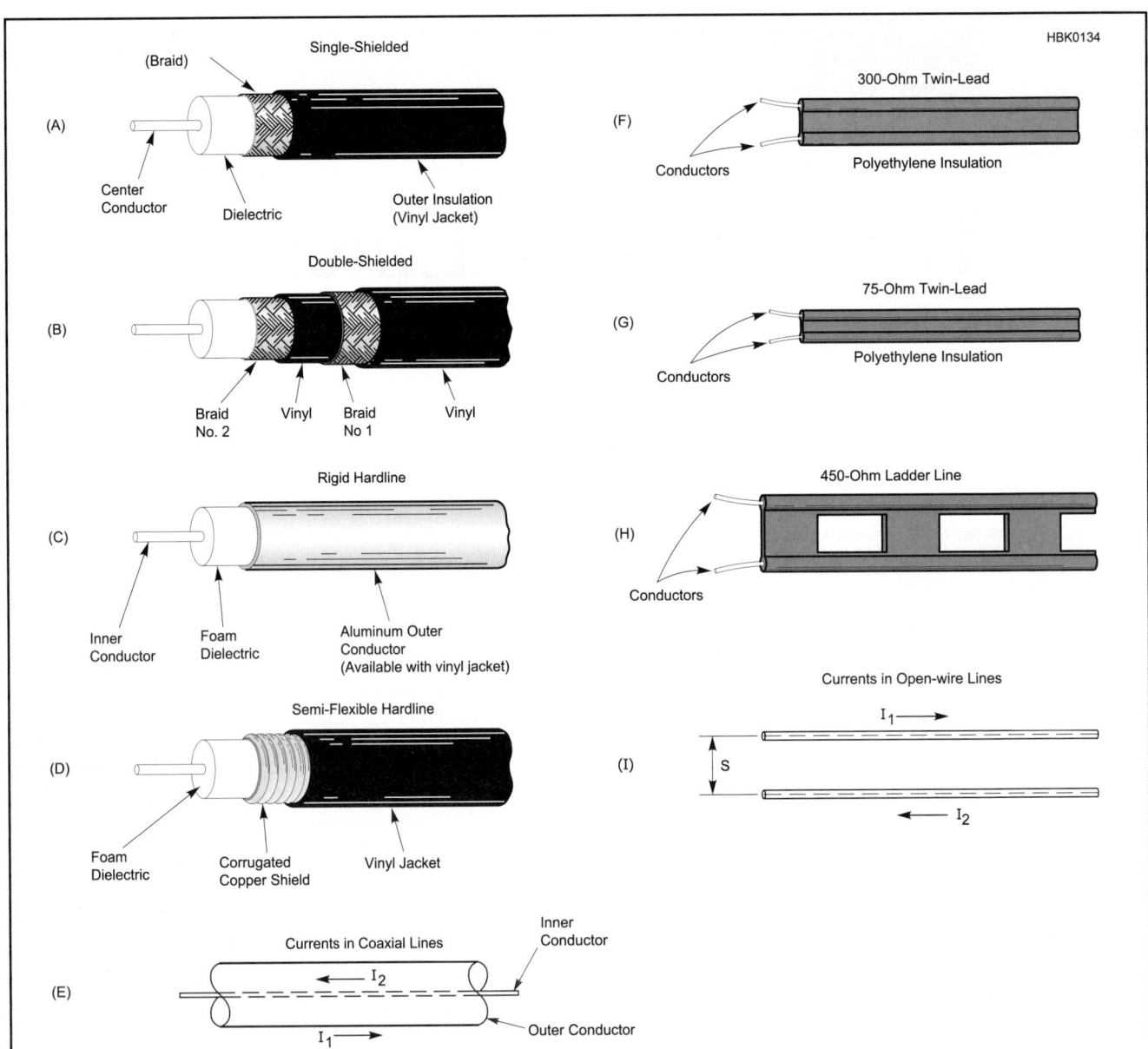

Fig 20.1 — Common types of transmission lines used by amateurs. Coaxial cable, or "coax," has a center conductor surrounded by insulation. The second conductor, called the shield, cover the insulation and is, in turn, covered by the plastic outer jacket. Various types are shown at A, B, C and D. The currents in coaxial cable flow on the outside of the center conductor and the inside of the outer shield (E). Open-wire line (F, G and H) has two parallel conductors separated by insulation. In open-wire line, the current flows in opposite directions on each wire (I).

1 Hz signal is 983,569,082 ft. Changing to a more useful expression gives:

$$\lambda = \frac{983.6}{f} \qquad (1)$$

where
　　λ = wavelength, in feet
　　f = frequency in MHz.

Thus, at 14 MHz the wavelength is 70.25 ft.

Wavelength (λ) may also be expressed in electrical degrees. A full wavelength is 360°, ½ λ is 180°, ¼ λ is 90°, and so forth.

Waves travel slower than the speed of light in any medium denser than a vacuum or free space. A transmission line may have an insulator which slows the wave travel down. The actual velocity of the wave is a function of the dielectric characteristic of that insulator. We can express the variation of velocity as the *velocity factor* for that particular type of dielectric — the fraction of the wave's velocity of propagation in the transmission line compared to that in free space. The velocity factor is related to the dielectric constant of the material in use.

$$VF = \frac{1}{\sqrt{\varepsilon}} \qquad (2)$$

where
　　VF = velocity factor
　　ε = dielectric constant.

So the wavelength in a real transmission line becomes:

$$\lambda = \frac{983.6}{f} VF \qquad (3)$$

As an example, many coax cables use polyethylene dielectric over the center conductor as the insulation. The dielectric constant for polyethylene is 2.3, so the VF is 0.66. Thus, wavelength in the cable is about two-thirds as long as a free-space wavelength.

The VF and other characteristics of many types of lines, both coax and twin lead, are shown in the table "Nominal Characteristics of Commonly used Transmission Lines" in the **Component Data and References** chapter.

There are differences in VF from batch to batch of transmission line because there are some variations in dielectric constant during the manufacturing processes. When high accuracy is required, it is best to actually measure VF by using an antenna analyzer to measure the resonant frequency of a length of cable. (The antenna analyzer's user manual will describe the procedure.)

CHARACTERISTIC IMPEDANCE

A perfectly lossless transmission line may be represented by a whole series of small inductors and capacitors connected in an infinitely long line, as shown in **Fig 20.2**. (We first consider this special case because we need not consider how the line is terminated at its end, since there is no end.)

Each inductor in Fig 20.2 represents the inductance of a very short section of one wire and each capacitor represents the capacitance between two such short sections. The inductance and capacitance values per unit of line depend on the size of the conductors and the spacing between them. Each series inductor acts to limit the rate at which current can charge the following shunt capacitor, and in so doing establishes a very important property of a transmission line: its *surge impedance*, more commonly known as its *characteristic impedance*. This is usually abbreviated as Z_0,

$$Z_0 \approx \sqrt{\frac{L}{C}}$$

where L and C are the inductance and capacitance per unit length of line.

The characteristic impedance of an air-insulated parallel-conductor line, neglecting the effect of the insulating spacers, is given by

$$Z_0 = \frac{120}{\sqrt{\varepsilon}} \cosh^{-1} \frac{S}{d} \qquad (4)$$

where

Z_0 = characteristic impedance
S = center to center distance between the conductors
d = diameter of conductors in the same units as S

When S >> d, the approximation $Z_0 = 276 \log_{10} (2S/d)$ may be used but for S < 2d gives values that are significantly higher than the correct value, such as is often the case when wires are twisted together to form a transmission line for impedance transformers.

The characteristic impedance of an air-insulated coaxial line is given by

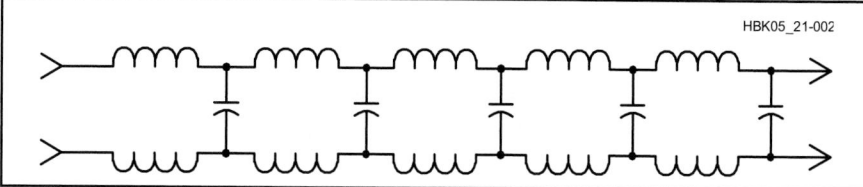

Fig 20.2 — Equivalent of an infinitely long lossless transmission line using lumped circuit constants.

$$Z_0 = 138 \log_{10} \left(\frac{b}{a} \right) \qquad (5)$$

where

Z_0 = characteristic impedance
b = inside diameter of outer conductors
a = outside diameter of inner conductor (in same units as b).

It does not matter what units are used for S, d, a or b, as long as they are the *same* units. A line with closely spaced, large conductors will have a low characteristic impedance, while one with widely spaced, small conductors will have a relatively high characteristic impedance. Practical open-wire lines exhibit characteristic impedances ranging from about 200 to 800 Ω, while coax cables have Z_0 values between 25 to 100 Ω. Except in special instances, coax used in Amateur Radio has an impedance of 50 or 75 Ω.

All practical transmission lines exhibit some power loss. These losses occur in the resistance that is inherent in the conductors that make up the line, and from leakage currents flowing in the dielectric material between the conductors. We'll next consider what happens when a real transmission line, which is not infinitely long, is terminated in an actual load impedance, such as an antenna.

20.1.2 Matched and Mismatched Lines

Real transmission lines have a definite length and are connected to, or *terminate* in, a load (such as an antenna), as illustrated in **Fig 20.3A**. If the load is a pure resistance whose value equals the characteristic impedance of the line, the line is said to be *matched*. To current traveling along the line, such a load at the end of the line appears as though it were still more transmission line of the same characteristic impedance. In a matched transmission line, energy travels along the line from the source until it reaches the load, where it is completely absorbed (or radiated if the load is an antenna).

MISMATCHED LINES

Assume now that the line in Fig 20.3B is terminated in an impedance Z_a which is not equal to Z_0 of the transmission line. The line is now a *mismatched* line. Energy reaching

Fig 20.3 — At A the coaxial transmission line is terminated with resistance equal to its Z_0. All power is absorbed in the load. At B, coaxial line is shown terminated in an impedance consisting of a resistance and a capacitive reactance. This is a mismatched line, and a reflected wave will be returned back down the line toward the generator. The reflected wave adds to the forward wave, producing a standing wave on the line. The amount of reflection depends on the difference between the load impedance and the characteristic impedance of the transmission line.

the end of a mismatched line will not be fully absorbed by the load impedance. Instead, part of the energy will be reflected back toward the source. The amount of reflected versus absorbed energy depends on the degree of mismatch between the characteristic impedance of the line and the load impedance connected to its end.

The reason why energy is reflected at a discontinuity of impedance on a transmission line can best be understood by examining some limiting cases. First, consider the rather extreme case where the line is shorted at its end. Energy flowing to the load will encounter the short at the end, and the voltage at that point will go to zero, while the current will rise to a maximum. Since the short circuit does not dissipate any power, the energy will all be *reflected* back toward the source generator.

If the short at the end of the line is replaced with an open circuit, the opposite will happen. Here the voltage will rise to maximum, and the current will by definition go to zero. The phase will reverse, and again all energy will be reflected back towards the source. By the way, if this sounds to you like what happens at the end of a half-wavelength dipole antenna, you are quite correct. However, in the case of an antenna, energy traveling along the antenna is lost by radiation on purpose, whereas a good transmission line will lose little energy to radiation because the fields from the conductors cancel outside the line.

For load impedances falling between the extremes of short and open circuit, the phase and amplitude of the reflected wave will vary. The amount of energy reflected and the amount of energy absorbed in the load will depend on the difference between the characteristic impedance of the line and the impedance of the load at its end.

What actually happens to the energy reflected back down the line? This energy will encounter another impedance discontinuity, this time at the source. Reflected energy flows back and forth between the mismatches at the source and load. After a few such journeys, the reflected wave diminishes to nothing, partly as a result of finite losses in the line, but mainly because of partial absorption at the load each time it reaches the load. In fact, if the load is an antenna, such absorption at the load is desirable, since the energy is actually radiated by the antenna.

If a continuous RF voltage is applied to the terminals of a transmission line, the voltage at any point along the line will consist of a vector sum of voltages, the composite of waves traveling toward the load and waves traveling back toward the source generator. The sum of the waves traveling toward the load is called the *forward* or *incident* wave, while the sum of the waves traveling toward the generator is called the *reflected wave*.

20.1.3 Reflection Coefficient and SWR

In a mismatched transmission line, the ratio of the voltage in the reflected wave at any one point on the line to the voltage in the forward wave at that same point is defined as the *voltage reflection coefficient*. This has the same value as the current reflection coefficient. The reflection coefficient is a complex quantity (that is, having both amplitude and phase) and is generally designated by the Greek letter ρ (rho), or sometimes in the professional literature as Γ (Gamma). The relationship between R_1 (the load resistance), X_1 (the load reactance), Z_0 (the line characteristic impedance, whose real part is R_0 and whose reactive part is X_0) and the complex reflection coefficient ρ is

$$\rho = \frac{Z_1 - Z_0}{Z_1 + Z_0} = \frac{(R_1 \pm jX_1) - (R_0 \pm jX_0)}{(R_1 \pm jX_1) + (R_0 \pm jX_0)}$$

$$(6)$$

For most transmission lines the characteristic impedance Z_0 is almost completely resistive, meaning that $Z_0 = R_0$ and $X_0 \cong 0$. The magnitude of the complex reflection coefficient in equation 6 then simplifies to:

$$|\rho| = \sqrt{\frac{(R_1 - R_0)^2 + X_1^2}{(R_1 + R_0)^2 + X_1^2}} \qquad (7)$$

For example, if the characteristic impedance of a coaxial line is 50 Ω and the load impedance is 120 Ω in series with a capacitive reactance of –90 Ω, the magnitude of the reflection coefficient is

$$|\rho| = \sqrt{\frac{(120 - 50)^2 + (-90)^2}{(120 + 50)^2 + (-90)^2}} = 0.593$$

Note that if R_1 in equation 6 is equal to R_0 and X_1 is 0, the reflection coefficient, ρ, is 0. This represents a matched condition, where all the energy in the incident wave is transferred to the load. On the other hand, if R_1 is 0, meaning that the load is a short circuit and has no real resistive part, the reflection coefficient is 1.0, regardless of the value of R_0. This means that all the forward power is reflected since the load is completely reactive.

The concept of reflection is often shown in terms of the *return loss* (RL), which is given in dB and is equal to 20 times the log of the reciprocal of the reflection coefficient.

$$RL\,(dB) = -10\log\left(\frac{P_r}{P_f}\right) = -20\log(\rho) \quad (8)$$

In the example above, the return loss is 20 log (1/0.593) = 4.5 dB. (See Table 22.65 in the **Component Data and References** chapter.)

If there are no reflections from the load, the voltage distribution along the line is constant or *flat*. A line operating under these conditions is called either a *matched* or a *flat* line. If reflections do exist, a voltage *standing-wave* pattern will result from the interaction of the forward and reflected waves along the line. For a lossless transmission line at least ¼ λ long, the ratio of the maximum peak voltage anywhere on the line to the minimum value anywhere along the line is defined as the *voltage standing-wave ratio*, or VSWR. (The line must be ¼ λ or longer for the true maximum and minimum to be created.) Reflections from the load also produce a standing-wave pattern of currents flowing in the line. The ratio of maximum to minimum current, or ISWR, is identical to the VSWR in a given line.

In amateur literature, the abbreviation SWR is commonly used for standing-wave ratio, as the results are identical when taken from proper measurements of either current or voltage. Since SWR is a ratio of maximum to minimum, it can never be less than one-to-one. In other words, a perfectly flat line has

an SWR of 1:1. The SWR is related to the magnitude of the complex reflection coefficient and vice versa by

$$SWR = \frac{1 + |\rho|}{1 - |\rho|} \qquad (9A)$$

and

$$|\rho| = \frac{SWR - 1}{SWR + 1} \qquad (9B)$$

The definitions in equations 8 and 9 are valid for any line length and for lines that are lossy, not just lossless lines longer than ¼ λ at the frequency in use. Very often the load impedance is not exactly known, since an antenna usually terminates a transmission line. The antenna impedance may be influenced by a host of factors, including its height above ground, end effects from insulators, and the effects of nearby conductors. We may also express the reflection coefficient in terms of forward and reflected power, quantities that can be easily measured using a directional RF wattmeter. The reflection coefficient and SWR may be computed as

$$|\rho| = \sqrt{\frac{P_r}{P_f}} \qquad (10A)$$

and

$$SWR = \frac{1 + \sqrt{\dfrac{P_r}{P_f}}}{1 - \sqrt{\dfrac{P_r}{P_f}}} \qquad (10B)$$

where

P_r = power in the reflected wave
P_f = power in the forward wave.

If a line is not matched (SWR > 1:1) the difference between the forward and reflected powers measured at any point on the line is the net power going toward the load from that point. The forward power measured with a directional wattmeter (often referred to as a reflected power meter or reflectometer) or a mismatched line will thus always appear greater than the forward power measured on a flat line with a 1:1 SWR.

The software program *TLW*, written by Dean Straw, N6BV, and included on the *ARRL Antenna Book* CD solves these complex equations. The characteristics of many common types of transmission lines are included in the software so that real antenna matching problems may be easily solved. Detailed instructions on using the program are included with it. *TLW* was used for the example calculations in this chapter.

20.1.4 Losses in Transmission Lines

A transmission line exhibits a certain amount of loss, caused by the resistance of the

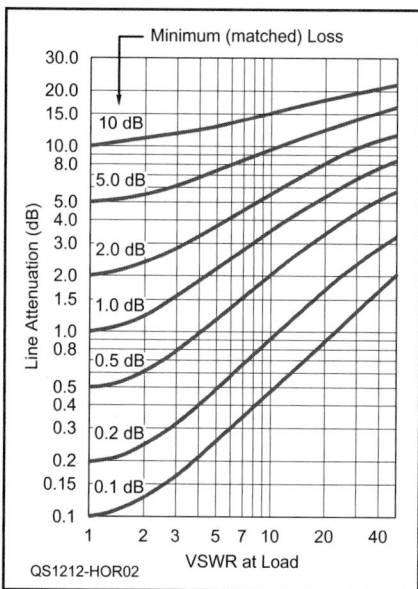

Fig 20.4 —Total insertion loss in a transmission line terminated in a mismatch. To use the chart, start with the SWR at the load. Find that value on the horizontal axis. From the SWR value on the horizontal axis, proceed vertically to the curve representing the feed line's matched loss (loss with SWR of 1:1). At the intersection, the total loss can be read on the vertical axis. (Courtesy of Joe Reisert, W1JR; see reference)

conductors used in the line and by dielectric losses in the line's insulators. The *matched-line loss* for a particular type and length of transmission line, operated at a particular frequency, is the loss when the line is terminated in a resistance equal to its characteristic impedance. The loss in a line is lowest when it is operated as a matched line.

Line losses increase when SWR is greater than 1:1. Each time energy flows from the generator toward the load, or is reflected at the load and travels back toward the generator, a certain amount will be lost along the line. The net effect of standing waves on a transmission line is to increase the average value of current and voltage, compared to the matched-line case. An increase in current raises I^2R (ohmic) losses in the conductors, and an increase in RF voltage increases E^2/R losses in the dielectric. Line loss rises with frequency, since the conductor resistance is related to skin effect, and also because dielectric losses rise with frequency.

Matched-line loss (ML) is stated in decibels per hundred feet at a particular frequency. The matched-line loss per hundred feet versus frequency for a number of common types of lines, both coaxial and open-wire balanced types, is shown graphically and as a table in the **Component Data and References** chapter. For example, RG-213 coax cable

has a matched-line loss of 2.5 dB/100 ft at 100 MHz. Thus, 45 ft of this cable feeding a 50 Ω load at 100 MHz would have a loss of

$$\text{Matched line loss} = \frac{2.5 \text{ dB}}{100 \text{ ft}} \times 45 \text{ ft}$$

$$= 1.13 \text{ dB}$$

If a line is not matched, standing waves will cause additional loss beyond the inherent matched-line loss for that line.

Total Mismatched Line Loss (dB)

$$= 10 \log \left[\frac{a^2 - |\rho^2|}{a \left(1 - |\rho^2| \right)} \right] \quad (11)$$

where
   $a = 10^{ML/10}$
   ML = the line's matched loss in dB.

The effect of SWR on line loss is shown graphically in **Fig 20.4**. The horizontal axis is the SWR at the load end of the line. The family of curves is the matched loss of the length of transmission line in use. From the SWR value on the horizontal axis, proceed vertically to the curve representing the feed line's matched loss (loss with SWR of 1:1). At the intersection, the total loss can be read on the vertical axis.

20.2 Choosing a Transmission Line

Making the best choice for a particular installation requires balancing the properties of the three common types of feed lines used by amateurs — coaxial, parallel-conductor or open-wire, and hardline — along with cost. The primary electrical considerations for feed line are characteristic impedance and loss. Mechanical concerns include weight, suitability for exposure to weather or soil, and interaction with other cables and conductors. When evaluating cost, be sure to include the cost of connectors and any auxiliary costs such as baluns and waterproofing materials.

The entire antenna system, composed of the feed line, tuners or matching networks and the antenna itself, must be included when evaluating what type of line to use. Along with loss, the effects of SWR on maximum voltage in the system must be considered if high power will be used, especially if high SWR is anticipated.

Multiband antenna systems, such as nonresonant wire antennas, can present quite a challenge because of the range of SWR values and the wide range of frequencies of use. As an example of the considerations involved, the article "Multiband Operation with Open-wire Line" is included on this book's CD-ROM. By following the general process of modeling

Transmission Lines for Microwave Frequencies

While low-loss waveguide is generally used to carry microwave frequency signals for long distances, *semi-rigid* coaxial cable — essentially miniature hardline — is used for connections inside and between pieces of equipment in the shack. Working with this type of cable requires special techniques, addressed in the supplemental article "Microwave Plumbing" on this book's CD-ROM. More information on microwave construction techniques is available in the **Construction Techniques** chapter and in the "Microwavelengths" columns by Paul Wade, W1GHZ in *QST*, available on-line to ARRL members.

or calculation and evaluation with different types of feed line, reasonable choices can be made that result in satisfactory performance.

20.2.1 Effect of Loss

For most types of line and for modest values

of SWR, the additional line loss due to SWR is of little concern. As the line's loss increases at higher frequencies, the total line loss (the sum of matched-line loss and additional loss due to SWR) can be surprisingly large at high values of SWR.

Because of losses in a transmission line, the measured SWR at the input of the line is lower than the SWR measured at the load end of the line. This does *not* mean that the load is absorbing any more power. Line loss absorbs power as it travels to the load and again on its way back to the generator, so the difference between the generator output power and the power returning from the load is higher than for a lossless line. Thus, P_r/P_f is smaller than at the load and so is the measured SWR.

For example, RG-213 solid-dielectric coax cable exhibits a matched-line loss at 28 MHz of 1.14 dB per 100 ft. A 250 ft length of this cable has a matched-line loss of 1.14 × 250/100 = 2.86 dB. Assume that we measure the SWR at the load as 6:1, the total mismatched line loss from equation 11 is 5.32 dB.

The additional loss due to the 6:1 SWR at 28 MHz is 5.32 – 2.86 = 2.46 dB. The SWR at the input of the 250 ft line is only 2.2:1, because line loss has masked the true magnitude of SWR (6:1) at the load end of the line.

The losses increase if coax with a larger matched-line loss is used under the same conditions. For example, RG-58A coaxial cable is about one-half the diameter of RG-213, and it has a matched-line loss of 2.81 dB/ 100 ft at 28 MHz. A 250 ft length of RG-58A has a total matched-line loss of 7.0 dB. With a 6:1 SWR at the load, the additional loss due to SWR is 3.0 dB, for a total loss of 10.0 dB. The additional cable loss due to the mismatch reduces the SWR measured at the input of the line to 1.33:1. An unsuspecting operator measuring the SWR at his transmitter might well believe that everything is just fine, when in truth only about 10% of the transmitter power is getting to the antenna! Be suspicious of very low SWR readings for an antenna fed with a long length of coaxial cable, especially if the SWR remains low across a wide frequency range. Most antennas have narrow SWR bandwidths, and the SWR *should* change across a band.

On the other hand, if expensive ⅞ inch diameter 50 Ω hardline cable is used at 28 MHz, the matched-line loss is only 0.19 dB/ 100 ft. For 250 ft of this hardline, the matched-line loss is 0.475 dB, and the additional loss due to a 6:1 SWR is 0.793 dB. Thus, the total loss is 1.27 dB.

At the upper end of the HF spectrum, when the transmitter and antenna are separated by a long transmission line, the use of bargain coax may prove to be a very poor cost-saving strategy. Adding a 1500 W linear amplifier (providing 8.7 dB of gain over a 200 W transmitter), to offset the loss in RG-58A compared to hardline, would cost a great deal more than higher-quality coax. Furthermore, no *transmitting* amplifier can boost *receiver* sensitivity — loss in the line has the same effect as putting an attenuator in front of the receiver.

At the lower end of the HF spectrum, say 3.5 MHz, the amount of loss in common coax lines is less of a problem for the range of SWR values typical on this band. For example, consider an 80 meter dipole cut for the middle of the band at 3.75 MHz. It exhibits an SWR of about 6:1 at the 3.5 and 4.0 MHz ends of the band. At 3.5 MHz, 250 ft of RG-58A

small-diameter coax has an additional loss of 2.1 dB for this SWR, giving a total line loss of 4.0 dB. If larger-diameter RG-213 coax is used instead, the additional loss due to SWR is 1.3 dB, for a total loss of 2.2 dB. This is an acceptable level of loss for most 80 meter operators.

The loss situation gets dramatically worse as the frequency increases into the VHF and UHF regions. At 146 MHz, the total loss in 250 ft of RG-58A with a 6:1 SWR at the load is 21.4 dB, 10.1 dB for RG-213A, and 2.7 dB for ⅞ inch, 50 Ω hardline. At VHF and UHF, a low SWR is essential to keep line losses low, even for the best coaxial cable. The length of transmission line must be kept as short as practical at these frequencies.

Table 20.1 lists some commonly used coax cables showing feet per dB of loss vs frequency. The table can help with the selection of coax cable by comparing lengths that result in 1 dB of loss for different types of cable. The larger the value in the table, the less loss in the cable per unit length. (See also Table 23-3 and 23-4 in the *ARRL Antenna Book* for information on more types of cable.) Determine the length of line your installation requires and the maximum acceptable amount of line loss in dB. Divide the total line length by the *maximum* acceptable loss to calculate the *minimum* acceptable length of line with 1 dB of loss. From the table, select a cable type that has a length per 1 dB of loss that is greater than the minimum acceptable length.

Example — An installation requires 400 feet of feed line at 14 MHz with a maximum acceptable value of 3 dB of loss. This requires cable with a minimum of 400/3 = 133 feet per dB of loss. Find a cable in Table 20.1 that shows more than 133 feet for 1 dB of loss at 14.2 MHz. RG-213 has the highest acceptable loss at this frequency: 137 ft / dB of loss.

Don't forget that you can combine types of cable and accessories to lower the total system cost while still meeting performance requirements. For example, it is common to use a single low-loss cable from the shack to a distant tower with multiple antennas. At the

tower, an antenna switch then selects short runs of less-expensive cable to the various antennas.

20.2.2 Mechanical Considerations

Coaxial cable is mechanically much easier to use than open-wire line. Because of the excellent shielding afforded by its outer shield, coax can be installed along a metal tower leg or taped together with numerous other cables, with virtually no interaction or crosstalk. Coax can be used with a rotatable antenna without worrying about shorting or twisting the conductors, which might happen with an open-wire line.

Class 2 PVC (P2) noncontaminating outer jackets are designed for long-life outdoor installations. Class 1 PVC (P1) outer jackets are not recommended for outdoor installations. (See the table of coaxial cables in the **Component Data and References** chapter.) Coax and hardline can be buried underground, especially if run in plastic piping (with suitable drain holes) so that ground water and soil chemicals cannot easily deteriorate the cable. A cable with an outer jacket of polyethylene (PE) rather than polyvinyl chloride (PVC) is recommended for direct-bury installations.

Hardline has low loss compared to flexible coax and with the proper connectors and waterproofing will last for many years. Although it is expensive if purchased new, surplus hardline is commonly available from cable TV and radio or cellular telephone system installers. 75 Ω cable introduces a 1.5:1 mismatch in 50 Ω systems that may be tolerable. If not, impedance matching transformers are available.

Hardline can be difficult to install and replace and is heavier than flexible cable, requiring stronger mechanical supports. Particularly in the larger sizes, connectors can be expensive although they are often available as surplus. There are a number of innovative adaptations of plumbing and similar fittings to allow the use of standard connectors for flexible cable. Take extra precautions with unorthodox connectors to ensure water cannot enter the feed line, or the extra expense and effort will be wasted.

Open-wire line generally has lower loss per unit of length at HF and low VHF frequencies than does coaxial or hardline, but the difference is not great for relatively short lengths of line. It is also lighter and usually lower-cost than coax and hardline of equivalent power handling capability.

The high characteristic impedance of open-wire line requires impedance matching for use in 50 Ω systems. Open-wire line generally requires balanced loads and radio equipment which requires baluns in balanced systems.

Open-wire line must carefully be spaced away from nearby conductors, by at least sev-

Table 20.1
Length in Feet for 1 dB of Matched Loss

MHz	1.8	3.6	7.1	14.2	21.2	28.4	50.1	144	440	1296
RG-58	179	122	83	59	50	42	30	18	9	5
RG-8X	257	181	128	90	74	63	47	27	14	8
RG-213	397	279	197	137	111	95	69	38	19	9
LMR-400	613	436	310	219	179	154	115	67	38	21
9913	625	435	320	220	190	155	110	62	37	20
EC4-50*				290		202		87		26
EC5-50*				787		548		239		75

Eupen manufactures EC4-50 and EC5-50.

eral times the spacing between its conductors, to minimize possible electrical imbalances between the two parallel conductors. Such imbalances lead to line radiation and extra losses.

One popular type of open-wire line is called ladder line because the insulators used to separate the two parallel, uninsulated conductors are individual rods or bars that resemble the steps of a ladder. Long lengths of ladder

line can twist together in the wind and short together if not properly supported. Window line that uses solid plastic insulation with rectangular openings or twin-lead avoids the shorting problem.

20.3 The Transmission Line as Impedance Transformer

If the complex mechanics of reflections, SWR and line losses are put aside momentarily, a transmission line can very simply be considered as an impedance transformer. A certain value of load impedance, consisting of a resistance and reactance, at the end of the line is transformed into another value of impedance at the input of the line. The amount of transformation is determined by the electrical length of the line, its characteristic impedance, and by the losses inherent in the line. The input impedance of a real, lossy transmission line is computed using the following equation

$$Z_{in} = Z_0 \times \frac{Z_L \cosh(\eta\ell) + Z_0 \sinh(\eta\ell)}{Z_L \sinh(\eta\ell) + Z_0 \cosh(\eta\ell)}$$

$$(12)$$

where

Z_{in} = complex impedance at input of line = $R_{in} \pm j\, X_{in}$

Z_L = complex load impedance at end of line = $R_l \pm j\, X_l$

Z_0 = characteristic impedance of line = $R_0 \pm j\, X_0$

η = complex loss coefficient = $\alpha + j\,\beta$

α = matched line loss attenuation constant, in nepers/unit length (1 neper = 8.688 dB, so multiply line loss in dB per unit length by 8.688)

β = phase constant of line in radians/unit length (multiply electrical length in degrees by 2π radians/360 degrees)

ℓ = electrical length of line in same units of length as used for α.

This and other complex equations describing the electrical behavior of transmission lines were traditionally solved through the use of Smith charts. (Smith charts are discussed in the article "Smith Chart Supplement" on this book's CD-ROM. Many references to Smith charts and their use may be found on-line.) While the Smith chart is an extremely effective way of visualizing and understanding the transmission line, using it directly in design has been replaced by software. Programs such as those mentioned in the sidebar "Smith Chart Software" can perform the numerical calculations and display the results on a Smith chart. The software *TLW* provided with the *ARRL Antenna Book* also solves problems of this nature, although without Smith chart graphics.

20.3.1 Transmission Line Stubs

The impedance-transformation properties of a transmission line are useful in a number of applications. If the terminating resistance is zero (that is, a short) at the end of a low-loss transmission line which is less than ¼ λ long, the input impedance consists of a reactance, which is given by a simplification of equation 12.

$$X_{in} \cong Z_0 \tan \ell \qquad (13)$$

If the line termination is an open circuit, the input reactance is given by

$$X_{in} \cong Z_0 \cot \ell \qquad (14)$$

The input of a short (less than ¼ λ) length of line with a short circuit as a terminating load appears as an inductance, while an open-circuited line appears as a capacitance. This is a useful property of a transmission line, since it can be used as a low-loss inductor or capacitor in matching networks. Such lines are often referred to as stubs.

A line that is electrically ¼ λ long is a special kind of a stub. When a ¼ λ line is short circuited at its load end, it presents an open circuit at its input. Conversely, a ¼ λ line with an open circuit at its load end presents a short circuit at its input. Such a line inverts the impedance of a short or an open circuit

at the frequency for which the line is ¼ λ long. This is also true for frequencies that are odd multiples of the ¼ λ frequency. However, for frequencies where the length of the line is ½ λ, or integer multiples thereof, the line will duplicate the termination at its end.

20.3.2 Transmission Line Stubs as Filters

The impedance transformation properties of stubs can be put to use as filters. For example, if a shorted line is cut to be ¼ λ long at 7.1 MHz, the impedance looking into the input of the cable will be an open circuit. The line will have no effect if placed in parallel with a transmitter's output terminals. However, at twice the *fundamental* frequency, 14.2 MHz, that same line is now ½ λ, and the line looks like a short circuit. The line, often dubbed a *quarter-wave stub* in this application, will act as a trap for not only the second harmonic, but also for higher even-order harmonics, such as the fourth or sixth harmonics.

This filtering action is extremely useful in multitransmitter situations, such as Field Day, emergency operations centers, portable communications facilities and multioperator contest stations. Transmission line stubs can operate at high power where lumped-constant filters would be expensive. Using stub filters reduces noise, harmonics and strong fundamental signals from the closely spaced antennas that cause overload and interference to receivers. (For information on determining isolation between radios and filter requirements, see the supplemental article "Measuring Receiver Isolation" by W2VJN on this book's CD-ROM.)

Quarter-wave stubs made of good-quality coax, such as RG-213, offer a convenient way to lower transmitter harmonic levels. Despite the fact that the exact amount of harmonic attenuation depends on the impedance (often unknown) into which they are working at the harmonic frequency, a quarter-wave stub will typically yield 20 to 25 dB of attenuation of the second harmonic when placed directly at the output of a transmitter feeding common amateur antennas.

This attenuation may be a bit higher when the output impedance of the plate tuning network in an amplifier increases with frequency, such as for a pi-L network. Attenuation may be a bit lower if the tuning network's output impedance decreases with frequency, such as for a pi network.

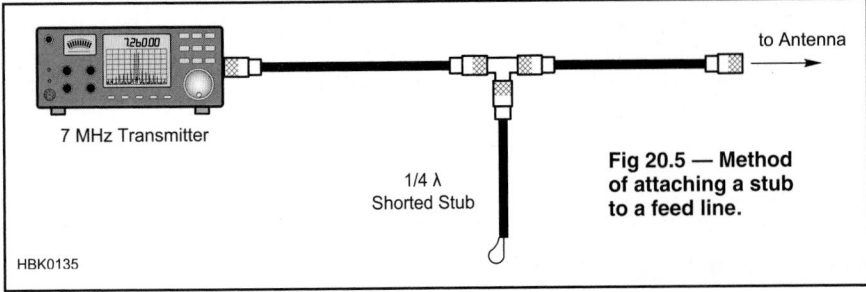

7 MHz Transmitter

1/4 λ Shorted Stub

to Antenna

Fig 20.5 — Method of attaching a stub to a feed line.

HBK0135

Because different manufacturing runs of coax will have slightly different velocity factors, a quarter-wave stub is usually cut a little longer than calculated, and then carefully pruned by snipping off short pieces, while using an antenna analyzer to monitor the response at the fundamental frequency. Because the end of the coax is an open circuit while pieces are being snipped away, the input of a ¼ λ line will show a short circuit exactly at the fundamental frequency. Once the coax has been pruned to frequency, a short jumper is soldered across the end, and the response at the second harmonic frequency is measured. **Fig 20.5** shows how to connect a shorted stub to a transmission line and **Fig 20.6** shows a typical frequency response.

The shorted quarter-wave stub shows low loss at 7 MHz and at 21 MHz where it is ¾-λ long. It nulls 14 and 28 MHz. This is useful for reducing the even harmonics of a 7 MHz transmitter. It can be used for a 21 MHz transmitter as well, and will reduce any spurious emissions such as phase noise and wideband noise which might cause interference to receivers operating on 14 or 28 MHz.

The open-circuited quarter-wave stub has a low impedance at the fundamental frequency, so it must be used at two times the frequency for which it is cut. For example, a quarter-wave open stub cut for 3.5 MHz will present a high impedance at 7 MHz where it is ½ λ long. It will present a high impedance at those frequencies where it is a multiple of ½ λ, or 7, 14 and 28 MHz. It would be connected in the same manner as Fig 20.5 shows, and the frequency plot is shown in **Fig 20.7**.

This open stub can protect a receiver operating on 7, 14, 21 or 28 MHz from interference by a 3.5 MHz transmitter. It also has nulls at 10.5, 17.5 and 24.5 MHz — the 3rd, 5th and 7th harmonics. The length of a quarter-wave stub may be calculated as follows:

$$L_e = \frac{VF \times 983.6}{4f} \qquad (15)$$

where

L_e = length in ft
VF = propagation constant for the coax in use
f = frequency in MHz.

For the special case of RG-213 (and any

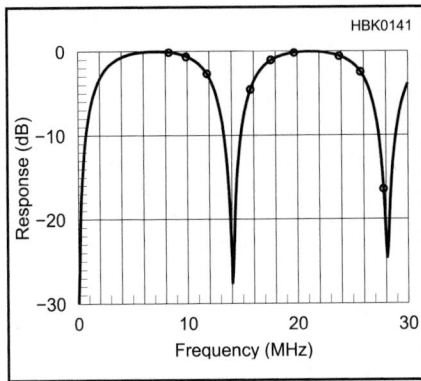

Fig 20.6 — Frequency response with a shorted stub.

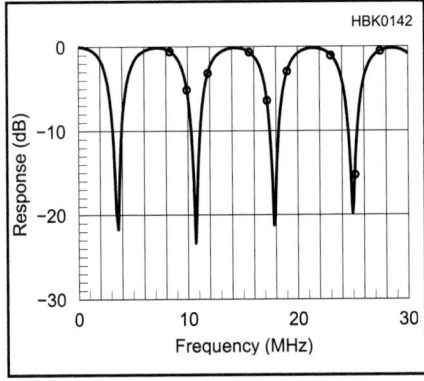

Fig 20.7 — Frequency response with an open stub.

Table 20.2

Quarter-Wave Stub Lengths for the HF Contesting Bands

Freq (MHz)	Length (L_e)*	Cut off per 100 kHz
1.8	90 ft, 10 in	57⅜ in
3.5	46 ft, 9 in	15½ in
7.0	23 ft, 4 in	4 in
14.0	11 ft, 8 in	1 in
21.0	7 ft, 9 in	⁷⁄₁₆ in
28.0	5 ft, 10 in	¼ in

*Lengths shown are for RG-213 and any similar cable, assuming a 0.66 velocity factor (L_e = 163.5/f). See text for other cables.

Commercial Triplexer Design Example

Provided on the CD-ROM accompanying this book, the article "Commercial Triplexer Design" by George Cutsogeorge, W2VJN, discusses the issues encountered in adapting a popular *QST* article on a Field Day-style triplexer to become a commercial product. The referenced *QST* article by Gary Gordon, K6KV, is provided as well.

similar cable with VF = 0.66), equation 15 can be simplified to:

$$L_e = \frac{163.5}{f} \qquad (16)$$

where

L_e = length in ft
f = frequency in MHz.

Table 20.2 solves this equation for the major contesting bands where stubs are often used. The third column shows how much of the stub to cut off if the desired frequency is 100 kHz higher in frequency. For example: To cut a stub for 14.250 MHz, reduce the overall length shown by 2.5 × 1 inches, or 2.5 inches. There is some variation in dielectric constant of coaxial cable from batch to batch or manufacturer to manufacturer, so it is always best to measure the stub's fundamental resonance before proceeding.

CONNECTING STUBS

Stubs are usually connected in the antenna feed line close to the transmitter. They may also be connected on the antenna side of a switch used to select different antennas. Some small differences in the null depth may occur for different positions.

To connect a stub to the transmission line it is necessary to insert a coaxial T (as shown in Fig 20.5). If a female-male-female T is used, the male can connect directly to the transmitter while the antenna line and the stub connect to the two females. It should be noted that the T inserts a small additional length in series with the stub that lowers the resonant frequency. The additional length for an Amphenol UHF T is about ⅜ inch. This length is negligible at 1.8 and 3.5 MHz, but on the higher bands it should not be ignored.

MEASURING STUBS WITH ONE-PORT METERS

Many of the common measuring instruments used by amateurs are *one-port devices*, meaning they have one connector at which the measurement — typically VSWR — is made. Probably the most popular instrument for this type of work is the antenna analyzer, available from a number of manufacturers.

To test a stub using an antenna analyzer, connect the stub to the meter by itself and tune the meter for a minimum impedance value, ignoring the VSWR setting. It is almost impossible to get an accurate reading on the higher HF bands, particularly with open stubs. For example, when a quarter-wave open stub cut for 20 meters was nulled on an MFJ-259 SWR analyzer, the frequency measured 14.650 MHz, with a very broad null. A recheck with a professional-quality network analyzer measured 14.018 MHz. (Resolution on the network analyzer is about ±5 kHz.) Running the same test on a quarter-wave shorted stub gave a measurement of 28.320 MHz on the MFJ-259 and 28.398 MHz on the network analyzer. (These inaccuracies are typical of amateur instrumentation and are meant to illustrate the difficulties of using inexpensive instruments for precise measurements.)

Other one-port instruments that measure phase can be used to get a more accurate reading. The additional length added by the required T adapter must be accounted for. If the measurement is made without the T and then with the T, the average value will be close to correct.

MEASURING STUBS WITH TWO-PORT INSTRUMENTS

A two-port measurement is made with a signal generator and a separate detector. A T connector is attached to the generator with the stub connected to one side. The other side is connected to a cable of any length that goes to the detector. The detector should present a 50 Ω load to the cable. This is how a network analyzer is configured, and it is similar to how the stub is connected in actual use. If the generator is calibrated accurately, the measurement can be very good. There are a number of ways to do this without buying an expensive piece of lab equipment.

An antenna analyzer can be used as the signal generator. Measurements will be quite accurate if the detector has 30 to 40 dB dynamic range. Two setups were tested by the author for accuracy. The first used a digital voltmeter (DVM) with a diode detector. (A germanium diode must be used for the best dynamic range.) Tests on open and shorted stubs at 14 MHz returned readings within 20 kHz of the network analyzer. Another test was run using an oscilloscope as the detector with a 50 Ω load on the input. This test produced results that were essentially the same as the network analyzer.

A noise generator can be used in combination with a receiver as the detector. (An inexpensive noise generator kit is available from Elecraft, **www.elecraft.com**.) Set the receiver for 2-3 kHz bandwidth and turn off the AGC. An ac voltmeter connected to the audio output of the receiver will serve as a null detector. The noise level into the receiver without the stub connected should be just at

or below the limiting level. With the stub connected, the noise level in the null should drop by 25 or 30 dB. Connect the UHF T to the noise generator using any necessary adapters. Connect the stub to one side of the T and connect the receiver to the other side with a short cable. Tune the receiver around the expected null frequency. After locating the null, snip off pieces of cable until the null moves to the desired frequency. Accuracy with this method is within 20 or 30 kHz of the network analyzer readings on 14 MHz stubs.

A *vector network analyzer* or VNA is another type of *two-port device*. VNAs can perform both reflection measurements as a one-port device and transmission measurements through a device or system at two locations. Once only available as expensive lab instruments, a number of inexpensive VNA designs that use a PC to perform display and calculations are now available to amateurs. The use of VNAs is discussed in the **Test Equipment and Measurements** chapter of this book.

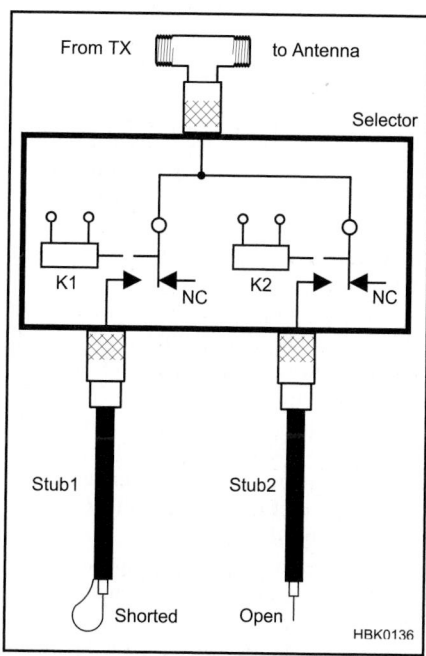

Fig 20.8 — Schematic of the Field Day stub switching relay control box. Table 20.3 shows which relays should be closed for the desired operating band.

STUB COMBINATIONS

A single stub will give 20 to 30 dB attenuation in the null. If more attenuation is needed, two or more similar stubs can be combined. Best results will be obtained if a short coupling cable is used to connect the two stubs rather than connecting them directly in parallel. The stubs may be cut to the same frequency for maximum attenuation, or to two slightly different frequencies such as the CW and SSB frequencies in one band. Open and shorted stubs can be combined together to attenuate higher harmonics as well as lower frequency bands.

An interesting combination is the parallel connection of two $\frac{1}{8} \lambda$ stubs, one open and the other shorted. The shorted stub will act as an inductor and the open stub as a capacitor. Their reactance will be equal and opposite, forming a resonant circuit. The null depth with this arrangement will be a bit better than a single quarter-wave shorted stub. This presents some possibilities when combinations of stubs are used in a band switching system.

20.3.3 Project: A Field Day Stub Assembly

Fig 20.8 shows a simple stub arrangement that can be useful in a two-transmitter Field Day station. The stubs reduce out-of-band noise produced by the transmitters that would cause interference to the other stations — a common Field Day problem where the stations are quite close together. This noise cannot be filtered out at the receiver and must be removed at the transmitter. One stub assembly would be connected to each transmitter output and manually switched for the appropriate band.

Two stubs are connected as shown. The two-relay selector box can be switched in four ways. Stub 1 is a shorted quarter-wave 40 meter stub. Stub 2 is an open quarter-wave 40 meter stub. Operation is as shown in **Table 20.3**.

The stubs must be cut and tuned while connected to the selector relays. RG-213 may be used for any amateur power level and will provide 25 to 30 dB reduction in the nulls. For power levels under 500 W or so, RG-8X may be used. It will provide a few dB less reduction in the nulls because of its slightly higher loss than RG-213.

Table 20.3

Stub Selector Operation

See Fig 20.8 for circuit details.

Relay K1 Position	Relay K2 Position	Bands Passed (meters)	Bands Nulled (meters)
Open	Open	All	None
Energized	Energized	80	40, 20, 15, 10
Energized	Open	40, 15	20, 10
Open	Energized	20, 10	40, 15

20.4 Matching Impedances in the Antenna System

Only in a few special cases is the antenna impedance the exact value needed to match a practical transmission line. In all other cases, it is necessary either to operate with a mismatch and accept the SWR that results, or else to bring about a match between the line and the antenna.

When transmission lines are used with a transmitter, the most common load is an antenna. When a transmission line is connected between an antenna and a receiver, the receiver input circuit is the load, not the antenna, because the power taken from a passing wave is delivered to the receiver.

Whatever the application, the conditions existing at the load, and *only* the load, determine the reflection coefficient, and hence the SWR, on the line. If the load is purely resistive and equal to the characteristic impedance of the line, there will be no standing waves. If the load is not purely resistive, or is not equal to the line Z_0, there will be standing waves. No adjustments can be made at the input end of the line to change the SWR at the load. Neither is the SWR affected by changing the line length, except as previously described when the SWR at the input of a lossy line is masked by the attenuation of the line.

20.4.1 Conjugate Matching

Technical literature sometimes uses the term *conjugate match* to describe the condition where the impedance seen looking toward the load from any point on the line is the complex conjugate of the impedance seen looking toward the source. This means that the resistive and reactive magnitudes of the impedances are the same, but that the reactances have opposite signs. For example, the complex conjugate of $20 + j\,75$ is $20 - j\,75$. The complex conjugate of a purely resistive impedance, such as $50 + j\,0\;\Omega$, is the same impedance, $50 + j\,0\;\Omega$. A conjugate match is necessary to achieve the maximum power gain possible from most signal sources.

For example, if 50 feet of RG-213 is terminated in a $72 - j\,34\;\Omega$ antenna impedance, the transmission line transforms the impedance to $35.9 - j\,21.9\;\Omega$ at its input. (The *TLW* program is used to calculate the impedance at the line input.) To create a conjugate match at the line input, a matching network would have to present an impedance of $35.9 + j\,21.9\;\Omega$. The system would then become resonant, since the $\pm\,j\,21.9\;\Omega$ reactances would cancel, leaving $35.9 + j\,0\;\Omega$. A conjugate match is not the same as transforming one impedance to another, such as from $35.9 - j\,0\;\Omega$ to $50 + j\,0\;\Omega$. An additional impedance transformation network would be required for that step.

Conjugate matching is often used for small-signal amplifiers, such as preamps at VHF and above, to obtain the best power gain. The situation with high-power amplifiers is complex and there is considerable discussion as to whether conjugate matching delivers the highest efficiency, gain and power output. Nevertheless, conjugate matching is the model most often applied to impedance matching in antenna systems.

20.4.2 Impedance Matching Networks

When all of the components of an antenna system — the transmitter, feed line, and antenna — have the same impedance, all of the power generated by the transmitter is transferred to the antenna and SWR is 1:1. This is rarely the case, however, as antenna feed-point impedances vary widely with frequency and design. This requires some method of *impedance matching* between the various antenna system components.

Many amateurs use an *impedance-matching unit or "antenna tuner"* between their transmitter and the transmission line feeding the antenna. (This is described in a following section.) The antenna tuner's function is to transform the impedance, whatever it is, at the transmitter end of the transmission line into the 50 Ω required by the transmitter. Remember that the use of an antenna tuner at the transmitter does *not tune the antenna*, reduce SWR on the feed line or reduce feed line losses!

Some matching networks are built directly into the antenna (for example, the gamma and beta matches) and these are discussed in the chapter on **Antennas** and in *The ARRL Antenna Book*. Impedance matching networks made of fixed or adjustable components can also be used at the antenna and are particularly useful for antennas that operate on a single band.

Remember, however, that impedance can be transformed anywhere in the antenna system to match any other desired impedance. A variety of techniques can be used as described in the following sections, depending on the circumstances.

An electronic circuit designed to convert impedance values is called an *impedance matching network*. The most common impedance matching network circuits for use in systems that use coax cable are:

1) The low-pass L network.
2) The high-pass L network.
3) The low-pass pi network.
4) The high-pass T network.

Basic schematics for each of the circuits are shown in **Fig 20.9**. Properties of the circuits are shown in **Table 20.4**. As shown in Table 20.4, the L networks can be reversed if matching does not occur in one direction. L networks are the most common for single-band antenna matching. The component in parallel is the *shunt* component, so the L networks with the shunt capacitor or inductor at the input (Figs 20.9A and 20.9C) are *shunt-input* networks and the others are *series-input* networks.

Impedance matching circuits can use fixed-value components for just one band when a particular antenna has an impedance that

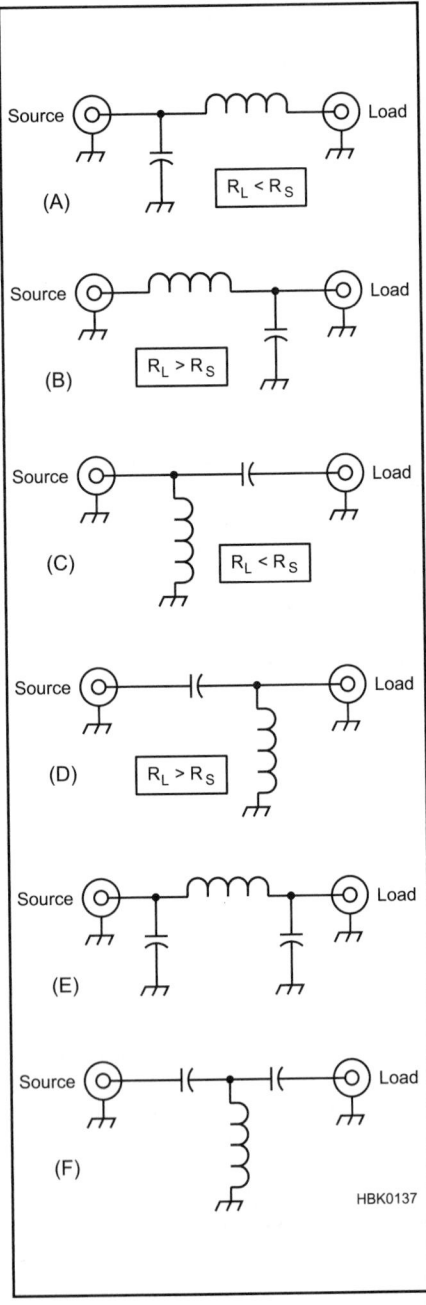

Fig 20.9 — Matching network variations. A through D show L networks. E is a pi network, equivalent to a pair of L networks sharing a common series inductor. F is a T network, equivalent to a pair of L networks sharing a common parallel inductor.

Table 20.4
Network Performance

Fig 20.9 Section	Circuit Type	Match Higher or Lower?	Harmonic Attenuation?
(A)	Low-pass L network	Lower	Fair to good
(B)	Reverse Low-pass L network	Higher	Fair to good
(C)	High-pass L network	Lower	No
(D)	Reverse high-pass L network	Higher	No
(E)	Low-pass Pi network	Lower and Higher	Good
(F)	High-pass T network	Lower and higher	No

is too high or low, or they can be made to be adjustable when matching is needed on several bands, such as for matching a dipole antenna fed with open-wire line.

Additional material by Bill Sabin, WØIYH, on matching networks can be found on the CD-ROM accompanying this book along with his program *MATCH*. The supplemental article "Using *TLW* to Design Impedance Matching Networks" by W2VJN is also on the CD-ROM.

DESIGNING AN L NETWORK

The L network, shown in Fig 20.9A through 20.9D, only requires two components and is a particularly good choice of matching network for single-band antennas. The L network is easy to construct so that it can be mounted at or near the feed point of the antenna, resulting in 1:1 SWR on the transmission line to the shack. (Note that L networks as well as pi and T networks can easily be designed with the *TLW* software.)

To design an L network, both the source and load impedances must be known. Let us assume that the source impedance, R_S, will be 50 Ω, representing the transmission line to the transmitter, and that the load is an arbitrary value, R_L.

First, determine the circuit Q.

$$Q^2 + 1 = \frac{R_L}{50} \tag{17A}$$

or

$$Q = \sqrt{\frac{R_L}{50} - 1} \tag{17B}$$

Next, select the type of L network you want from Fig 20.9. Note that the parallel component is always connected to the higher of the two impedances, source or load. Your choice should take into account whether either the source or load require a dc ground (parallel or shunt-L) and whether it is necessary to have a dc path through the network, such as to power a remote antenna switch or other such device (parallel- or shunt-C). Once you have selected a network, calculate the values of X_L and X_C:

$$X_L = Q\,R_S \tag{18}$$

and

$$X_C = \frac{R_L}{Q} \tag{19}$$

As an example, we will design an L network to match a 300 Ω antenna (R_L) to a 50 Ω transmission line (R_S). $R_L > R_S$ so we can select either Fig 20.9B or Fig 20.9D. The network in B is a low-pass network and will attenuate harmonics, so that is the usual choice.

$$Q = \sqrt{\frac{300}{50} - 1} = 2.236$$

$$X_L = 50 \times 2.236 = 112\ \Omega$$

$$X_C = \frac{300}{2.236} = 134\ \Omega$$

If the network is being designed to operate at 7 MHz, the actual component values are:

$$L = \frac{X_L}{2\pi f} = 2.54\ \mu H$$

$$C = \frac{1}{2\pi f X_C} = 170\ pF$$

The components could be fixed-value or adjustable. By running the *TLW* software, additional information is obtained: At 1500 W there will be 942 V peak across the capacitor and 5.5 A flowing in the inductor.

The larger the ratio of the impedances to be transformed, the higher Q becomes. High values of Q (10 or more) may result in impractically high or low component values. In this case, it may be easier to design the matching network as a pair of L networks back-to-back that accomplish the match in two steps. Select an intermediate value of impedance, R_{INT}, the geometric mean between R_L and the source impedance:

$$R_{INT} = \sqrt{R_L R_S}$$

Construct one L network that transforms R_L to R_{INT} and a second L network that transforms R_{INT} to R_S.

20.4.3 Matching Antenna Impedance at the Antenna

This section describes methods by which a network can be installed at or near the antenna feed point to provide matching to a transmission line. Having the matching system at the antenna rather than down in the shack at the end of a long transmission line does seem intuitively desirable, but it is not always very practical, especially in multiband antennas as discussed below.

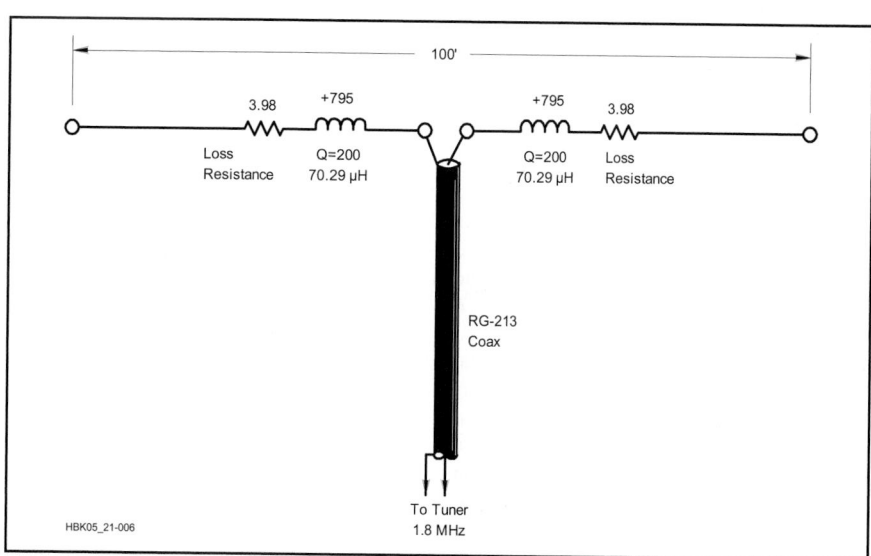

Fig 20.10 — The efficiency of a short dipole can be improved at 1.8 MHz with a pair of inductors inserted symmetrically at the feed point. Each inductor is assumed to have a Q of 200. By resonating the dipole in this fashion the system efficiency, when fed with RG-213 coax, is about 21 dB better than using this same antenna without the resonator. The disadvantage is that the formerly multiband antenna can only be used on a single band.

RESONATING THE ANTENNA

If a highly reactive antenna can be tuned to resonance, even without special efforts to make the resistive portion equal to the line's characteristic impedance, the resulting SWR is often low enough to minimize additional line loss due to SWR. For example, the multiband 100 ft long flat-top antenna in **Fig 20.10** has a feed point impedance of $4.18 - j\,1590\;\Omega$ at 1.8 MHz. Assume that the antenna reactance is tuned out with a network consisting of two symmetrical inductors whose reactance is $+j\,1590/2 = j\,795\;\Omega$ each, with a Q of 200. These inductors are 70.29 µH coils in series with inherent loss resistances of $795/200 = 3.98\;\Omega$. The total series resistance is thus $4.18 + 2 \times (3.98) = 12.1\;\Omega$. If placed in series with each leg of the antenna at the feed point as in the figure, the antenna reactance and inductor reactance cancel out, leaving a purely resistive impedance at the antenna feed point.

If this tuned system is fed with 50 Ω coaxial cable, the SWR is $50/12.1 = 4.13{:}1$, and the loss in 100 ft of RG-213 cable would be 0.48 dB. The antenna's radiation efficiency is the ratio of the antenna's radiation resistance (4.18 Ω) to the total feed point resistance including the matching coils (12.1 Ω), so efficiency is $4.18/12.1 = 34.6\%$ which is equivalent to 4.6 dB of loss compared to a 100% efficient antenna. Adding the 0.48 dB of loss in the line yields an overall system loss of 5.1 dB. Compare this to the loss of 26 dB if the RG-213 coax is used to feed the antenna directly, without any matching at the antenna. The use of moderately high-Q resonating inductors has yielded almost 21 dB of "gain" (that is, less loss) compared to the case without the inductors. The drawback of course is that the antenna is now resonant on only one frequency, but it certainly is a lot more efficient on that one frequency!

THE QUARTER-WAVE TRANSFORMER OR "Q" SECTION

The range of impedances presented to the transmission line is usually relatively small on a typical amateur antenna, such as a dipole or a Yagi when it is operated close to resonance. In such antenna systems, the impedance-transforming properties of a ¼ λ section of transmission line are often utilized to match the transmission line at the antenna.

If the antenna impedance and the characteristic impedance of a feed line to be matched are known, the characteristic impedance needed for a quarter-wave matching section of low-loss cable is expressed by another simplification of equation 12.

$$Z = \sqrt{Z_1 Z_0} \qquad (20)$$

where
  Z = characteristic impedance needed for matching section

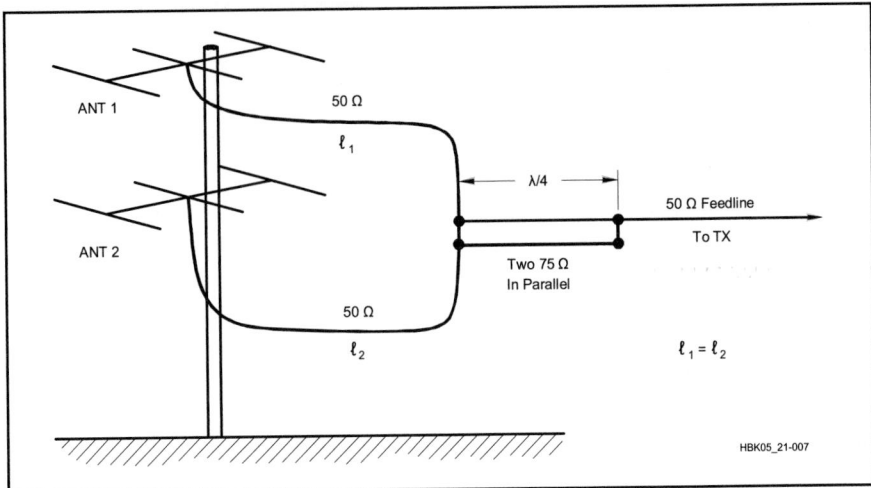

Fig 20.11 — Array of two stacked Yagis, illustrating use of ¼ λ matching sections. At the junction of the two equal lengths of 50 Ω feed line the impedance is 25 Ω. This is transformed back to 50 Ω by the two paralleled 75 Ω, ¼ λ lines, which together make a net characteristic impedance of 37.5 Ω. This is close to the 35.4 Ω value computed by the formula.

Z_1 = antenna impedance
Z_0 = characteristic impedance of the line to which it is to be matched.

Such a matching section is called a *synchronous quarter-wave transformer* or a *quarter-wave transformer*. The term synchronous is used because the match is achieved only at the length at which the matching section is exactly ¼ λ long. This limits the use of ¼ λ transformers to a single band. Different ¼ λ lengths of line are required on different bands.

Fig 20.11 shows one example of this technique to feed an array of stacked Yagis on a single tower. Each antenna is resonant and is fed in parallel with the other Yagis, using equal lengths of coax to each antenna called *phasing lines*. A stacked array is used to produce not only gain, but also a wide vertical elevation pattern, suitable for coverage of a broad geographic area. (See *The ARRL Antenna Book* for details about Yagi stacking.)

The feed point impedance of two 50 Ω Yagis fed with equal lengths of feed line connected in parallel is 25 Ω (50/2 Ω); three in parallel yield 16.7 Ω; four in parallel yield 12.5 Ω. The nominal SWR for a stack of four Yagis is 4:1 (50/12.5). This level of SWR does not cause excessive line loss, provided that low-loss coax feed line is used. However, many station designers want to be able to select, using relays, any individual antenna in the array, without having the load seen by the transmitter change. (Perhaps they might wish to turn one antenna in the stack in a different direction and use it by itself.) If the load changes, the amplifier must be retuned, an inconvenience at best.

Example: To match a 50 Ω line to a Yagi

stack consisting of two antennas fed in parallel to produce a 25 Ω load, the quarter-wave matching section would require a characteristic impedance of

$$Z = \sqrt{50 \times 25} = 35.4\;\Omega$$

A transmission line with a characteristic impedance of 35 Ω could be closely approximated by connecting two equal ¼ λ sections of 75 Ω cable (such as RG-11A) in parallel to yield the equivalent of a 37.5 Ω cable. Three Yagis fed in parallel would require a ¼ λ transformer made using a cable having a characteristic impedance of

$$Z = \sqrt{16.7 \times 25} = 28.9\;\Omega$$

This is approximated by using a ¼ λ section of 50 Ω cable in parallel with a ¼ λ section of 75 Ω cable, yielding a net impedance of 30 Ω, quite close enough to the desired 28.9 Ω. Four Yagis fed in parallel would require a ¼ λ transformer made up using cable with a characteristic impedance of

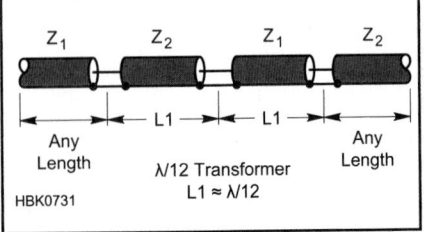

Fig 20.12 — A two-section twelfth-wave transformer can be used to match two different feed line impedances. This is a common technique for using surplus 75 Ω hardline in 50 Ω systems.

25 Ω, easily created by using two 50 Ω cables in parallel.

The 100 foot flat-top example in the previous section with the two resonating coils has an impedance of 12 Ω at the feed point. Two RG-58A cables, each ¼ λ long at 1.8 MHz (90 ft) can be connected in parallel to feed this antenna. An additional 10 ft length of RG-213 can make up the required 100 ft. The match will be almost perfect. The disadvantage of this system is that it limits the operation to one band, but the overall efficiency will be quite good.

Another use of ¼ λ transformers is in matching the impedance of full-wave loop antennas to 50 Ω coax. For example, the driven element of a quad antenna or a full-wave 40 meter loop has an impedance of 100 to 150 Ω. Using a ¼ λ transformer made from 62, 75 or 93 Ω coaxial cable would lower the line SWR to a level where losses were insignificant.

TWELFTH-WAVE TRANSFORMERS

The Q-section is really a special case of series-section impedance matching. (See the *ARRL Antenna Book* for an extended discussion of the technique.) A particularly handy two-section variation shown in **Fig 20.12** can be used to match two different impedances of transmission line, such as 50 Ω coax and 75 Ω hardline. Sections with special impedances are not required, only sections of the two feed lines to be matched.

This configuration is referred to as a *twelfth-wave transformer* because when the ratio of impedances to be matched is 1.5:1 (as is the case with 50 and 75 Ω cables) the electrical length of the two matching sections between the lines to be matched is 0.0815 λ (29.3°), quite close to λ/12 (0.0833 λ or 30°). As with all synchronous transformers, the matching function works on only one band.

MATCHING TRANSFORMERS

Another matching technique uses wideband toroidal transformers. Transformers can be made that operate over very wide frequency ranges and that will match various impedances.

A very simple matching transformer consists of three windings connected in series as shown in **Fig 20.13A**. The physical arrangement of the three windings is shown in Fig 20.13B. This arrangement gives the best bandwidth. **Fig 20.14** shows a picture of this type of transformer. An IN/OUT relay is included with the transformer. One relay pole switches the 50 Ω input port while two poles in parallel switch the 22 Ω port. Three 14 inch lengths of #14 AWG wire are taped together so they lie flat on the core. A #61 mix toroid core 2.4 inch in diameter will handle full legal power.

The impedance ratio of this design is 1:2.25

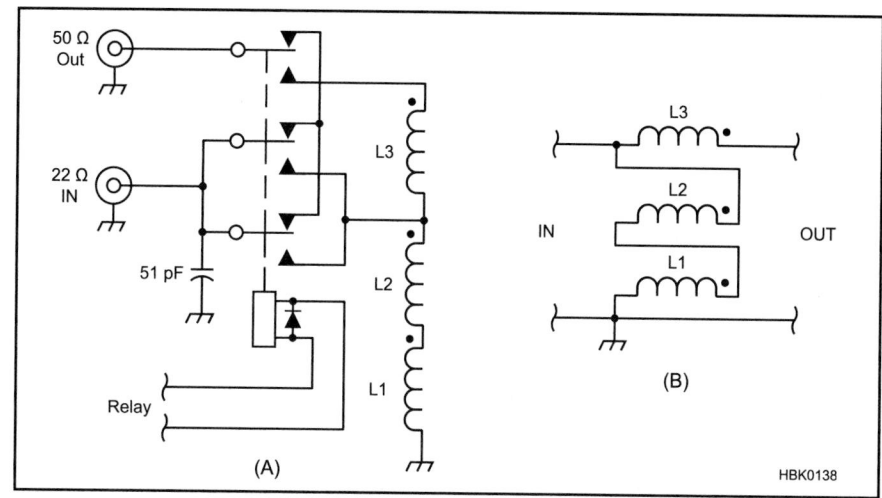

Fig 20.13 — Schematic for the impedance matching transformer described in the text. The complete schematic is shown at A. The physical positioning of the windings is shown at B.

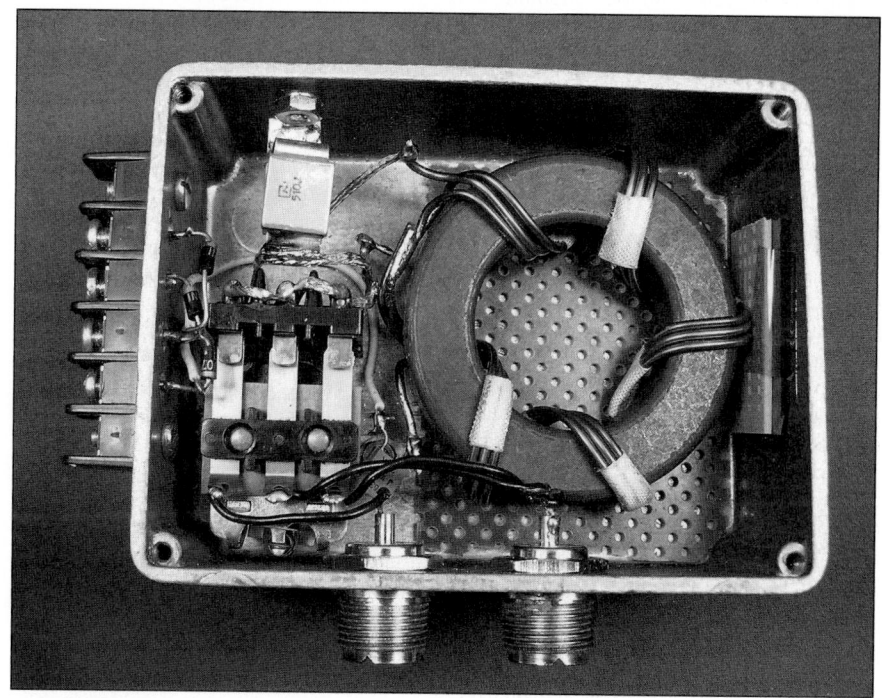

Fig 20.14 — The completed impedance matching transformer assembly for the schematic shown in Fig 20.13.

or 22.22-to-50 Ω. This ratio turns out to work well for two or three 50 Ω antennas in parallel. Two in parallel will give an SWR of 25/22.22 or 1.125:1. Three in parallel give an SWR of 22.22/16.67 or 1.33. The unit shown in Fig 20.14 has an SWR bandwidth of 1.5 MHz to more than 30 MHz. The 51 pF capacitor is connected at the low impedance side to ground and tunes out some inductive reactance.

This is a good way to stack two or three triband antennas. If they have the same length feed lines and they all point the same way, their patterns will add and some gain will result. However, they don't even need to be on the same tower or pointed in the same direction or fed with the same length lines to have some benefit. Even dissimilar antennas can sometimes show a benefit when connected together in this fashion.

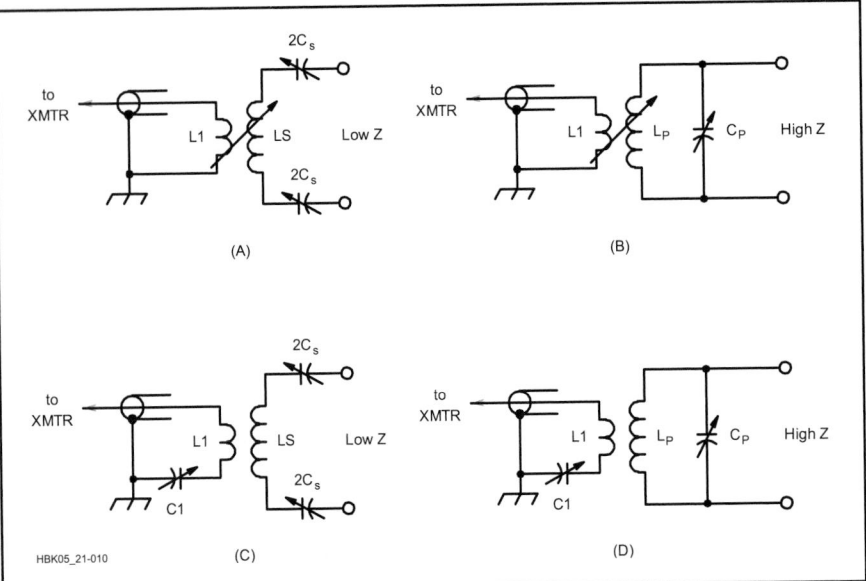

Fig 20.15 — Simple antenna tuners for coupling a transmitter to a balanced line presenting a load different from the transmitter's design load impedance, usually 50 Ω. A and B, respectively, are series and parallel tuned circuits using variable inductive coupling between coils. C and D are similar but use fixed inductive coupling and a variable series capacitor, C1. A series tuned circuit works well with a low-impedance load; the parallel circuit is better with high impedance loads (several hundred ohms or more).

20.4.4 Matching the Line to the Transmitter

So far we have been concerned mainly with the measures needed to achieve acceptable amounts of loss and a low SWR for feed lines connected to antennas. Not only is feed line loss minimized when the SWR is kept within reasonable bounds, but also the transmitter is able to perform as specified when connected to the load resistance it was designed to drive.

Most modern amateur transmitters have solid-state final amplifiers designed to work into a 50 Ω load with broadband, non-adjustable impedance matching networks. While the adjustable pi networks used in vacuum-tube amplifiers can accommodate a wide range of impedances, typical solid-state transmitters utilize built-in protection circuitry that auto-matically reduces output power if the SWR rises to more than about 2:1.

Besides the rather limited option of using only inherently low-SWR antennas to ensure that the transmitter sees the load for which it was designed, an *impedance-matching unit* or *antenna tuner* ("tuner" for short) can be used. The function of an antenna tuner is to transform the impedance at the input end of the transmission line, whatever that may be, to the 50 Ω value required by the transmitter for best performance.

Some solid-state transmitters incorporate (usually at extra cost) automatic antenna tuners so that they too can cope with practical antennas and transmission lines. The range of impedances that can be matched by the built-in tuners is typically rather limited, however, especially at lower frequencies. Most built-in tuners specify a maximum SWR of 3:1 that

can be transformed to 1:1.

Do not forget that a tuner does *not* alter the SWR on the transmission line between the tuner and the antenna; it only adjusts the impedance at the transmitter end of the feed line to the value for which the transmitter was designed. Other names for antenna tuners include *transmatch, impedance matcher, matchbox* or *antenna coupler*. Since the SWR on the transmission line between the antenna and the output of the antenna tuner is rarely 1:1, some loss in the feed line due to the mismatch is unavoidable, even though the SWR on the short length of line between the tuner and the transmitter is 1:1.

If separate feed lines are used for different bands, the tuner can be inserted in one feed line, tuned for best VSWR, and left at that setting. If a particular antenna has a minimum VSWR in the CW portion of a band and operation in the SSB end is desired, the tuner can be used for matching and switched out when not needed. Multiband operation generally requires retuning for each band in use.

Antenna tuners for use with balanced or open-wire feed lines include a balun or link-coupling circuit as seen in **Fig 20.15**. This allows a transmitter's unbalanced coaxial output to be connected to the balanced feed line. A fully-balanced tuner has a symmetrical internal circuit with a tuner circuit for each side of the feed line and the balun at the input to the tuner where the impedance is close to 50 Ω. Most antenna tuners are unbalanced, however, with a balun located at the output of the impedance matching network, connected directly to the balanced feed line. At very high or very low impedances, the balun's power rating may be exceeded at high transmitted power levels.

Automatic antenna tuners use a microprocessor to adjust the value of the internal components. Some models sense high values of SWR and retune automatically, while others require the operator to initiate a tuning operation. Automatic tuners are available for low- and high-power operation and generally handle the same values of impedance as their manually-adjusted counterparts.

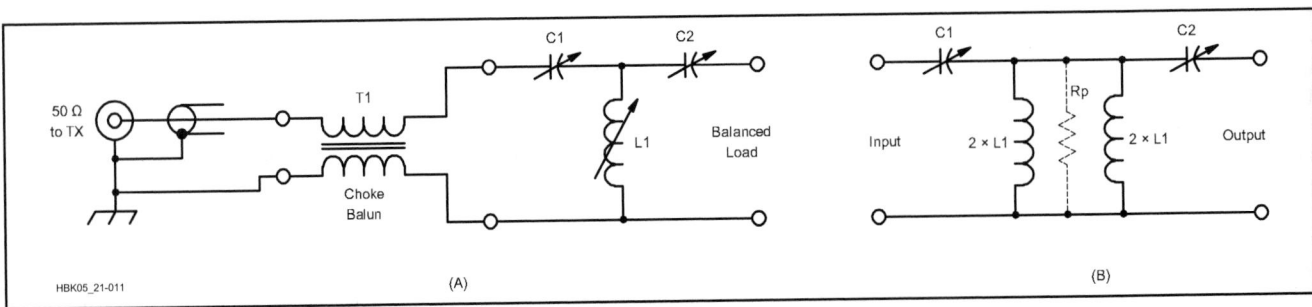

Fig 20.16 — Antenna tuner network in T configuration. This network has become popular because it has the capability of matching a wide range of impedances. At A, the balun transformer at the input of the antenna tuner preserves balance when feeding a balanced transmission line. The balun can be placed at the input or output of the tuner. At B, the T configuration is shown as two L networks back to back. (in the L network version, the two ½-value inductors replacing L1 are assumed to be adjustable with identical values).

THE T NETWORK

Over the years, radio amateurs have derived a number of circuits for use as tuners. The most common form of antenna tuner in recent years is some variation of a T network, as shown in **Fig 20.16A**. Note that the choke or current balun can be used at the input or output of the tuner to match balanced feed lines.

The T network can be visualized as being two L networks back to front, where the common element has been conceptually broken down into two inductors in parallel (see Fig 20.16B). The L network connected to the load transforms the output impedance $R_a \pm j\,X_a$ into its parallel equivalent by means of the series output capacitor C2. The first L network then transforms the parallel equivalent back into the series equivalent and resonates the reactance with the input series capacitor C1.

Note that the equivalent parallel resistance R_p across the shunt inductor can be a very large value for highly reactive loads, meaning that the voltage developed at this point can be very high. For example, assume that the load impedance at 3.8 MHz presented to the antenna tuner is $Z_a = 20 - j\,1000$. If C2 is 300 pF, then the equivalent parallel resistance across L1 is 66,326 Ω. If 1500 W appears across this parallel resistance, a peak voltage of 14,106 V is produced, a very substantial level indeed. Highly reactive loads can produce very high voltages across components in a tuner.

The ARRL computer program *TLW* calculates and shows graphically the antenna-tuner values for operator-selected antenna impedances transformed through lengths of various types of practical transmission lines. (See the supplemental article "Using *TLW* to Design Impedance Matching Networks" on this book's CD-ROM.) The **Station Accessories** chapter includes antenna tuner projects, and *The ARRL Antenna Book* contains detailed information on tuner design and construction.

ANTENNA TUNER LOCATION

The tuner is usually located near the transmitter in order to adjust it for different bands or antennas. If a tuner is in use for one particular band and does not need to be adjusted once set up for minimum VSWR, it can be placed in a weatherproof enclosure near the antenna. Some automatic tuners are designed to be installed at the antenna, for example. For some situations, placing the tuner at the base of a tower can be particularly effective and eliminates having to climb the tower to perform maintenance on the tuner.

It is useful to consider the performance of the entire antenna system when deciding where to install the antenna tuner and what types of feed line to use in order to minimize system losses. Here is an example, using the program *TLW*. Let's once again assume a flat-top antenna 50 ft high and 100 ft long and not resonant on any amateur band. As extreme

examples, we will use 3.8 and 28.4 MHz with 200 ft of transmission line. There are many ways to configure this system, but three examples are shown in **Fig 20.17**.

Example 1 in Fig 20.17A shows a 200 ft

run of RG-213 going to a 1:1 balun that feeds the antenna. A tuner in the shack reduces the VSWR for proper matching in the transmitter. Example 2 shows a similar arrangement using 300 Ω transmitting twin lead. Exam-

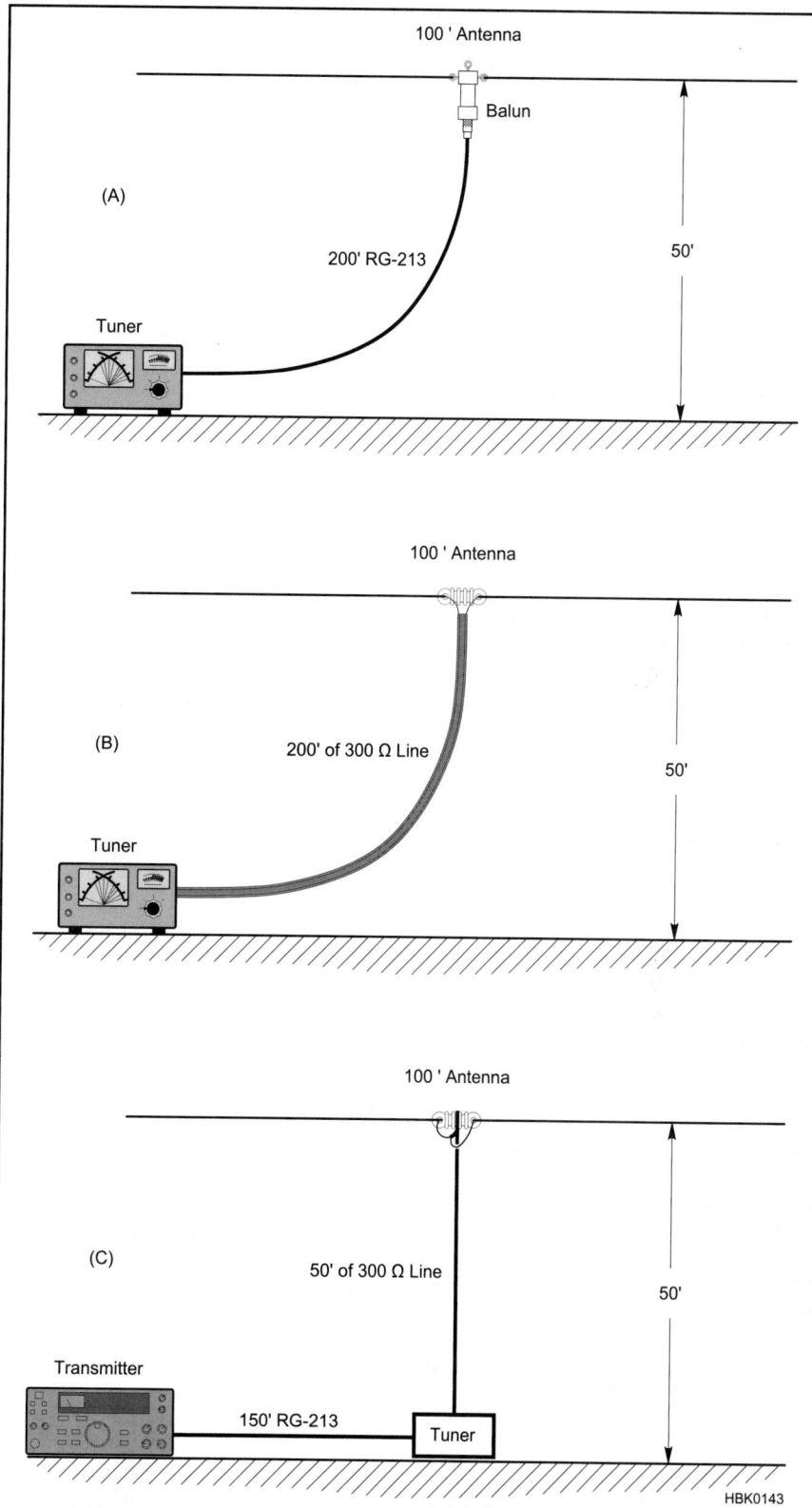

Fig 20.17 — Variations of an antenna system with different losses. The examples are discussed in the text.

Table 20.5

Tuner Settings and Performance

Example (Fig 20.17)	Frequency (MHz)	Tuner Type	L (µH)	C (pF)	Total Loss (dB)
A	3.8	Rev L	1.46	2308	8.53
	28.4	Rev L	0.13	180.9	12.3
B	3.8	L	14.7	46	2.74
	28.4	L	0.36	15.6	3.52
C	3.8	L	11.37	332	1.81
	28.4	L	0.54	94.0	2.95

ple 3 shows a 50 ft run of 300 Ω line dropping straight down to a tuner near the ground and 150 ft of RG-213 going to the shack. **Table 20.5** summarizes the losses and the tuner values required.

Some interesting conclusions can be drawn. First, direct feeding this antenna with coax through a balun is very lossy — a poor solution. If the flat-top were ½ λ long — a resonant half-wavelength dipole — direct coax feed would be a good method. In the second example, direct feed with 300 Ω low-loss line does not always give the lowest loss. The combination method in Example 3 provides the best solution.

There are other considerations as well. Supporting a balun at the antenna feed point adds stress to the wires or requires an additional support. Example 3 has some additional advantages. It feeds the antenna in a symmetrical arrangement which is best to reduce noise pickup on the shield of the feed line. The shorter feed line will not weigh down the antenna as much. The coax back to the shack can be buried or laid on the ground and it is perfectly matched. Burial of the cable will also prevent any currents from being induced on the coax shield. Once in the shack, the tuner is adjusted for minimum SWR per the manufacturer's instructions.

20.4.5 Adjusting Antenna Tuners

The process of adjusting an antenna tuner is described here and results in minimum SWR to the transmitter and also minimizes power losses in the tuner circuitry. If you have a commercial tuner and have the user's manual, the manufacturer will likely provide a method of adjustment that you should follow, including initial settings.

If you do not have a user's manual, first open the tuner and determine the circuit for the tuner. The most common circuit for commercial tuners is the high-pass T network shown in Fig 20.9F. To adjust this type of tuner:

1) Set the series capacitors to maximum value. This may not correspond to the highest number on the control scale — verify that the capacitor's plates are fully meshed.

2) Set the inductor to maximum value. This corresponds to placing a switch tap or roller inductor contact so that it is electrically closest to circuit ground.

3) If you have an antenna analyzer, connect it to the TRANSMITTER connector of the tuner. Otherwise, connect the transceiver and tune it to the desired frequency, but do not transmit.

In the following step, it is important to verify that you hear a peak in received noise before transmitting significant power through the tuner. Tuners can sometimes be adjusted to present a low SWR to the transmitter while coupling little energy to the output. Transmitting into a tuner in this configuration can damage the tuner's components.

4) Adjust the inductor throughout its range, watching the antenna analyzer for a dip in the SWR or listen for a peak in the received noise. Return the inductor to the setting for lowest SWR or highest received noise.

4a) If no SWR minimum or noise peak is detected, reduce the value of the capacitor closest to the transmitter in steps of about 20% and repeat.

4b) If still no SWR minimum or noise peak is detected, return the input capacitor to maximum value and reduce the output capacitor value in steps of about 20%.

4c) If still no SWR minimum or noise peak is detected, return the output capacitor to maximum value and reduce both input and output capacitors in 20% steps.

5) Once a combination of settings is found with a definite SWR minimum or noise peak:

5a) If you are using an antenna analyzer, make small adjustments to find the combination of settings that produce minimum SWR with the maximum value of input and output capacitance.

5b) If you do not have an antenna analyzer, set the transmitter output power to about 10 W, ensure that you won't cause interference, identify with your call sign, and transmit a steady carrier by making the same adjustments as in step 5a.

5c) For certain impedances, the tuner may not be able to reduce the SWR to an acceptable value. In this case, try adding feed line at the output of the tuner from ⅛ to ½ λ electrical wavelengths long. This

will not change the feed line SWR, but it may transform the impedance to a value more suitable for the tuner components.

In general, for any type of tuner, begin with the maximum reactance to ground (maximum inductance or minimum capacitance) and the minimum series reactance between the source and load (minimum inductance or maximum capacitance). The configuration that produces the minimum SWR with maximum reactance to ground and minimum series reactance will generally have the highest efficiency and broadest tuning bandwidth.

To reduce on-the-air tune-up time, record the settings of the tuner for each antenna and band of operation. If the tuner requires readjustment across the band, record the settings of the tuner at several frequencies across the band. Print out the results and keep it near the tuner — this will allow you to adjust the tuner quickly with only a short transmission to check or fine tune the settings. This also serves as a diagnostic, since changes in the setting indicate a change in the antenna system.

20.4.6 Myths About SWR

This is a good point to stop and mention that there are some enduring and quite misleading myths in Amateur Radio concerning SWR.

• Despite some claims to the contrary, a high SWR does not by itself cause RFI, or TVI or telephone interference. While it is true that an antenna located close to such devices can cause overload and interference, the SWR on the feed line to that antenna has nothing to do with it, providing of course that the tuner, feed line or connectors are not arcing. The antenna is merely doing its job, which is to radiate. The transmission line is doing its job, which is to convey power from the transmitter to the radiator.

• A second myth, often stated in the same breath as the first one above, is that a high SWR will cause excessive radiation from a transmission line. SWR has nothing to do with excessive radiation from a line. Imbalances in feed lines cause radiation, but such imbalances are not related to SWR. An asymmetric arrangement of a transmission line and antenna can result in current being induced on the transmission line — on the shield of coax or as an imbalance of currents in an open-wire line. This current will radiate just as if it was on an antenna. A choke balun is used on coaxial feed lines to reduce these currents as described in the section on baluns later in this chapter.

• A third and perhaps even more prevalent myth is that you can't "get out" if the SWR on your transmission line is higher than 1.5:1, or 2:1 or some other such arbitrary figure. On the HF bands, if you use reasonable lengths of good coaxial cable (or even better yet, open-wire line), the truth is that you need not be overly concerned if the SWR at the load is kept below about 6:1. This sounds pretty radical to some amateurs who have heard horror story after horror story about SWR. The fact is that if you can load up your transmitter without any arcing inside, or if you use a tuner to make sure your transmitter is operating into its rated load resistance, you can enjoy a very effective station, using antennas with feed lines having high values of SWR on them. For example, a 450 Ω open-wire line connected to the multiband dipole in the previous sections would have a 19:1 SWR on it at 3.8 MHz. Yet time and again this antenna has proven to be a great performer at many installations.

Fortunately or unfortunately, SWR is one of the few antenna and transmission-line parameters easily measured by the average radio amateur. Ease of measurement does not mean that a low SWR should become an end in itself! The hours spent pruning an antenna so that the SWR is reduced from 1.5:1 down to 1.3:1 could be used in far more rewarding ways — making contacts, for example, or studying transmission line theory.

20.5 Baluns and Transmission Line Transformers

Center-fed dipoles and loops are *balanced*, meaning that they are electrically and physically symmetrical with respect to the feed point. A balanced antenna may be fed by a balanced feeder system to preserve this symmetry, thereby avoiding difficulties with unbalanced currents on the line and undesirable radiation from the transmission line itself. Line radiation can be prevented by a number of devices which detune or *decouple* the line, greatly reducing currents induced onto the feed line from the signal radiated by the antenna.

Many amateurs use center-fed dipoles or Yagis, fed with unbalanced coaxial line. Some method should be used for connecting the line to the antenna without upsetting the symmetry of the antenna itself. This requires a circuit that will isolate the balanced load from the unbalanced line, while still providing efficient power transfer. Devices for doing this are called *baluns* (a contraction for "balanced to unbalanced"). A balanced antenna fed with balanced line, such as two-wire ladder line, will maintain its inherent balance, so long as external causes of unbalance are avoided. However, even they will require some sort of balun at the transmitter, since modern transmitters have unbalanced (coaxial) outputs.

If a balanced antenna is fed at the center by a coaxial feed line without a balun, as indicated in **Fig 20.18A**, the inherent symmetry and balance is upset because one side of the ½ λ radiator is connected to the shield while the other is connected to the inner conductor. On the side connected to the shield, current can be diverted from flowing into the antenna, and instead can flow away from the antenna on the outside of the coaxial shield. The field thus set up cannot be canceled by the field from the inner conductor because the fields inside the cable cannot escape through the shielding of the outer conductor. Hence currents flowing on the outside of the line will be responsible for some radiation from the line, just as if they were flowing on an antenna.

This is a good point at which to say that striving for perfect balance in a line and antenna system is not always absolutely mandatory. For example, if a nonresonant, center-fed dipole is fed with open-wire line and a tuner for multiband operation, the most desirable radiation pattern for general-purpose communication is actually an omnidirectional pattern. A certain amount of feed-line radiation might actually help fill in otherwise undesirable nulls in the azimuthal pattern of the antenna itself. Furthermore, the radiation pattern of a coaxial-fed dipole that is only a few tenths of a wavelength off the ground (50 feet high on the 80 meter band, for example) is not very directional anyway, because of its severe interaction with the ground.

Purists may cry out in dismay, but there are many thousands of coaxial-fed dipoles in daily use worldwide that perform very effectively without the benefit of a balun. See Fig 20.19A for a worst-case comparison between a dipole with and without a balun at its feed point. This is with a 1 λ feed line slanted downward 45° under one side of the antenna. *Common-mode currents* are conducted and induced onto the outside of the shield of the feed line, which in turn radiates. The amount of pattern distortion is not particularly severe for a dipole. It is debatable whether the bother and expense of installing a balun for such an antenna is worthwhile.

Some form of balun should be used to preserve the pattern of an antenna that is purposely designed to be highly directional, such as a Yagi or a quad. Fig 20.19B shows

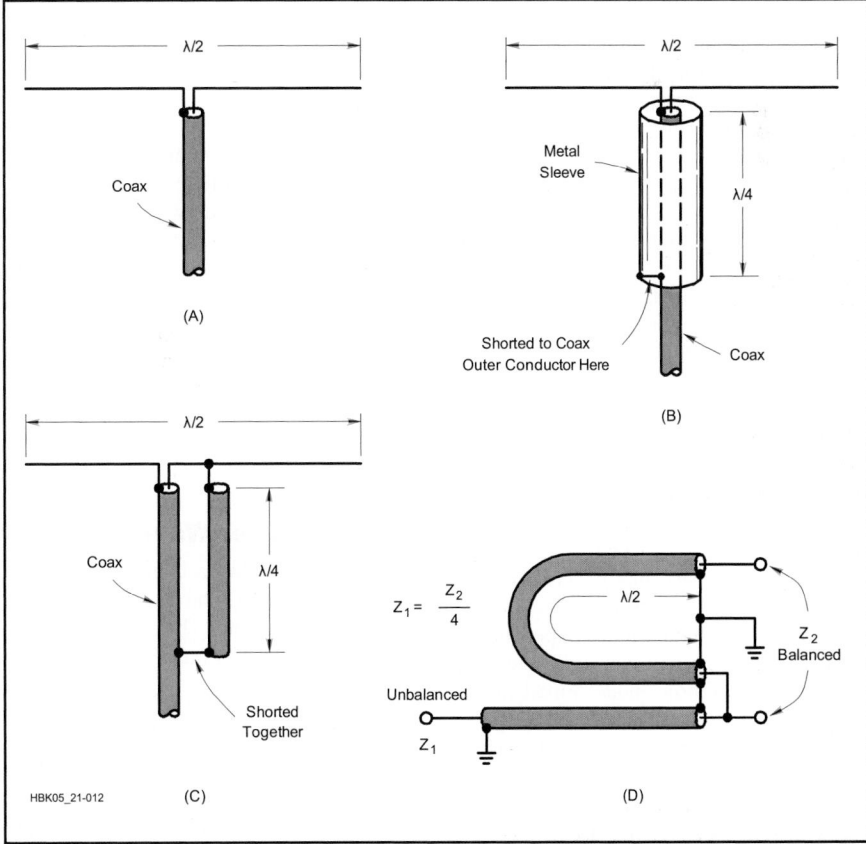

Fig 20.18 — Quarter-wavelength baluns. Radiator with coaxial feed (A) and methods of preventing unbalanced currents from flowing on the outside of the transmission line (B and C). The ½ λ phasing section shown at D is used for coupling to an unbalanced circuit when a 4:1 impedance ratio is desired or can be accepted.

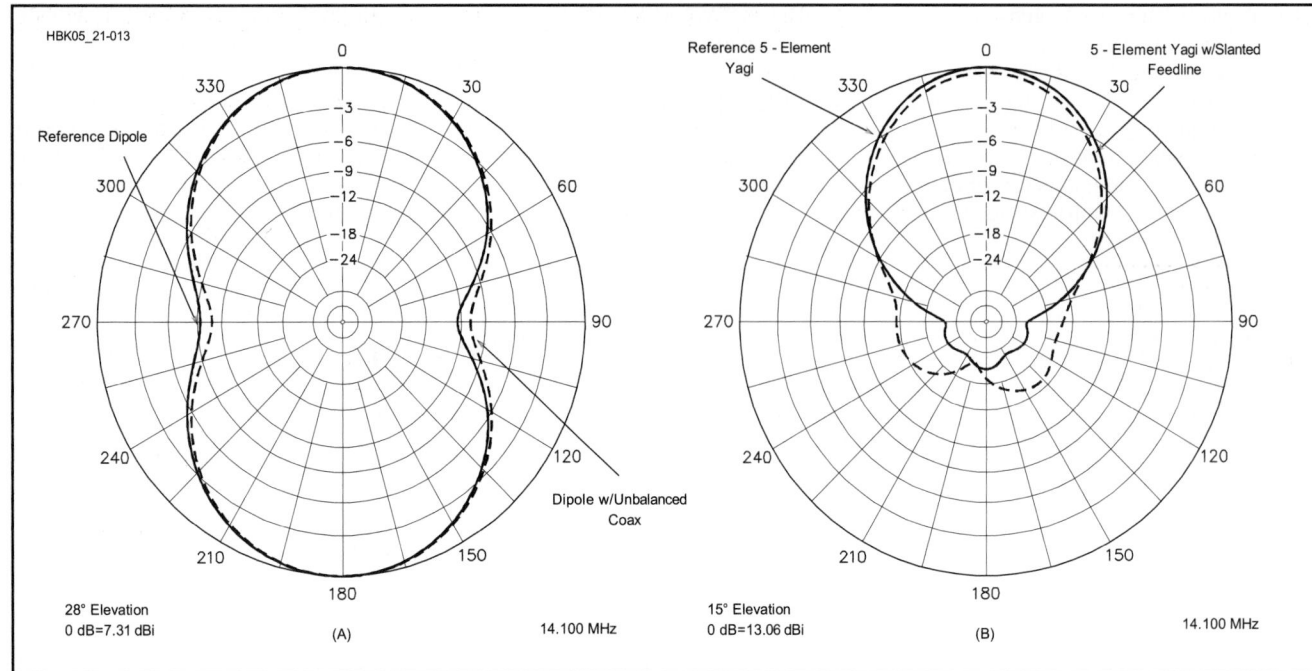

Fig 20.19 — At A, computer-generated azimuthal responses for two λ/2 dipoles placed 0.71 λ high over typical ground. The solid line is for a dipole with no feed line. The dashed line is for an antenna with its feed line slanted 45° down to ground. Current induced on the outer braid of the 1 λ long coax by its asymmetry with respect to the antenna causes the pattern distortion. At B, azimuthal response for two 5 element 20 meter Yagis placed 0.71 λ over average ground. Again, the solid line is for a Yagi without a feed line and the dashed line is for an antenna with a 45° slanted, 1 λ long feed line. The distortion in the radiated pattern is now clearly more serious than for a simple dipole. A balun is needed at the feed point, and most likely point, preferably ¼ λ from the feed point, to suppress the common-mode currents and restore the pattern.

the distortion that can result from common-mode currents conducted and radiated back onto the feed line for a 5 element Yagi. This antenna has purposely been designed for an excellent pattern but the common-mode currents seriously distort the rearward pattern and reduce the forward gain as well. A balun is highly desirable in this case.

Choke or current baluns force equal and opposite currents to flow in the load (the antenna) by creating a high common-mode impedance to currents that are equal in both conductors or that flow on the outside of coaxial cable shields, such as those induced by the antenna's radiated field. The result of using a current balun is that currents coupled back onto the transmission line from the antenna are effectively reduced, or "choked off," even if the antenna is not perfectly balanced. Choke baluns are particularly useful for feeding asymmetrical antennas with unbalanced coax line. The common-mode impedance of the choke balun varies with frequency, but the line's differential-mode impedance is unaffected.

Reducing common-mode current on a feed line also reduces:

• Radiation from the feed line that can distort an antenna's radiation pattern

• Radiation from the feed line that can cause RFI to nearby devices

• RF current in the shack and on power-line wiring

• Coupling of noise currents on the feed line to receivers and receiving antennas

• Coupling between different antennas via their feed lines

A single choke balun at the antenna feed point may not be sufficient to reduce common-mode current everywhere along a long feed line. If common-mode current on the line far from the antenna feed point is a problem, additional choke baluns can be placed at approximately ¼ λ intervals along the line. This breaks up the line electrically into segments too short to act as effective antennas. The chokes in this case function similarly to insulators used to divide tower guy wires into non-resonant lengths.

20.5.1 Quarter-Wave Baluns

Fig 20.18B shows a balun arrangement known as a *bazooka*, which uses a sleeve over the transmission line. The sleeve, together with the outside portion of the outer coax conductor, forms a shorted ¼ λ section of transmission line. The impedance looking into the open end of such a section is very high, so the end of the outer conductor of the coaxial line is effectively isolated from the part of the line below the sleeve. The length is an electrical ¼ λ, and because of the velocity factor may be physically shorter if the insulation between the sleeve and the line is not air. The bazooka has no effect on an-

tenna impedance at the frequency where the ¼ λ sleeve is resonant. However, the sleeve adds inductive shunt reactance at frequencies lower, and capacitive shunt reactance at frequencies higher than the ¼ λ resonant frequency. The bazooka is mostly used at VHF, where its physical size does not present a major problem.

Another method that gives an equivalent effect is shown at Fig 20.18C. Since the voltages at the antenna terminals are equal and opposite (with reference to ground), equal and opposite currents flow on the surfaces of the line and second conductor. Beyond the shorting point, in the direction of the transmitter, these currents combine to cancel out each other. The balancing section acts like an open circuit to the antenna, since it is a ¼ λ parallel-conductor line shorted at the far end, and thus has no effect on normal antenna operation. This is not essential to the line-balancing function of the device, however, and baluns of this type are sometimes made shorter than ¼ λ to provide a shunt inductive reactance required in certain matching systems (such as the hairpin match described in the **Antennas** chapter).

Fig 20.18D shows a balun in which equal and opposite voltages, balanced to ground, are taken from the inner conductors of the main transmission line and a ½ λ phasing section. Since the voltages at the balanced end are in

Baluns, Chokes, and Transformers

The term "balun" applies to any device that transfers differential-mode signals between a balanced (*bal-*) system and an unbalanced (*un-*) system while maintaining symmetrical energy distribution at the terminals of the balanced system. The term only applies to the function of energy transfer, not to how the device is constructed. It doesn't matter whether the balanced-unbalanced transition is made through transmission line structures, flux-coupled transformers, or simply by blocking unbalanced current flow. A common-mode *choke balun*, for example, performs the balun function by putting impedance in the path of common-mode currents and is therefore a balun.

A *current balun* forces symmetrical current at the balanced terminals. This is of particular importance in feeding antennas, since antenna currents determine the antenna's radiation pattern. A *voltage balun* forces symmetrical voltages at the balanced terminals. Voltage baluns are less effective in causing equal currents at their balanced terminals, such as at an antenna's feed point.

An *impedance transformer* may or may not perform the balun function. Impedance transformation (changing the ratio of voltage and current) is not required of a balun nor is it prohibited. There are balanced-to-balanced impedance transformers (transformers with isolated primary and secondary windings, for example) just as there are unbalanced-to-unbalanced impedance transformers (autotransformer and transmission-line designs). A *transmission-line transformer* is a device that performs the function of power transfer (with or without impedance transformation) by utilizing the characteristics of transmission lines.

Multiple devices are often combined in a single package called a "balun." For example, a "4:1 current balun" is a 1:1 current balun in series with a 4:1 impedance transformer or voltage balun. Other names for baluns are common, such as "line isolator" for a choke balun. Baluns are often referred to by their construction — "bead balun," "coiled-coax balun," "sleeve balun," and so forth. What is important is to separate the function (power transfer between balanced and unbalanced systems) from the construction.

---

series while the voltages at the unbalanced end are in parallel, there is a 4:1 step-down in impedance from the balanced to the unbalanced side. This arrangement is useful for coupling between a 300 Ω balanced line and a 75 Ω unbalanced coaxial line.

Fig 20.20 shows a variation on the quarterwave balun called the Quarter/Three-Quarter-Wave, or Q3Q, balun. It is a 1:1 decoupling balun made from two pieces of coaxial cable. One leg is ¼ λ long and the other is ¾ λ long. The two cables and the feed line are joined together with a T connector. At the antenna, the shields of the cables are connected together and the center conductors connected to the terminals of the antenna feed point. The balun has very little loss and is reported to have a bandwidth of more than 10%.

The balun works because of the current-forcing function of a transmission line is an odd number of λ/4 long. The current at the output of such a transmission line is V_{IN}/Z_0 regardless of the load impedance (less line losses — see the references for *QST* articles about this balun), similar to the behavior of a current source. Because both lines are fed with the same voltage, being connected in parallel, the output currents will also be equal.

The current out of the ¾ λ line is delayed by 180° from the current out of the ¼ λ line and so is out of phase. The result is that equal and opposite currents are forced into the terminals of the load.

20.5.2 Transmission Line Transformers

The basic transmission line transformer, from which other transformers are derived, is the 1:1 *choke balun* or *current balun*, shown in **Fig 20.21A**. To construct this type of balun, a length of coaxial cable or a pair of close-spaced, parallel wires forming a transmission line are wrapped around a ferrite rod or toroid or inserted through a number of beads. (The coiled feed line choke balun is discussed in the next section.) For the HF bands, use type 75 or type 31 material. Type 43 is used on the VHF bands. The Z_0 of the line should equal the load resistance, R.

Because of the ferrite, a high impedance exists between points A and C and a virtually identical impedance between B and D. This is true for parallel wire lines and it is also true for coax. The ferrite affects the A to C impedance of the coax inner conductor and the B to D impedance of the outer braid equally.

The conductors (two wires or coax braid and center-wire) are tightly coupled by electromagnetic fields and therefore constitute a good conventional transformer with a turns ratio of 1:1. The voltage from A to C is equal to and in-phase with that from B to D. These

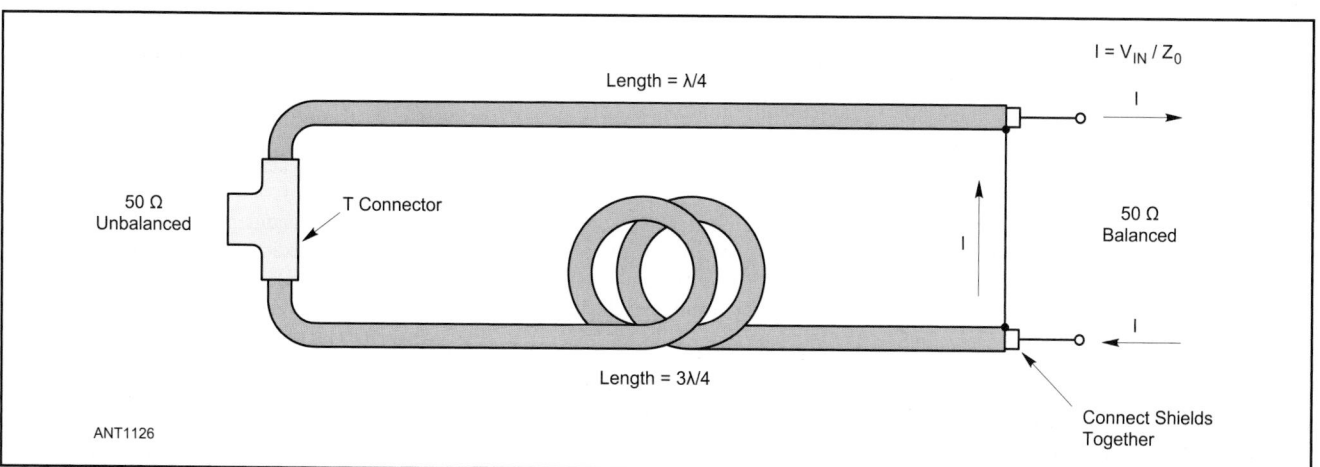

Fig 20.20 — The Quarter/Three-Quarter Wavelength (Q3Q) balun uses the current-forcing function of feed lines odd numbers of λ/4 long and the λ/2 delay of the longer line to cause equal and opposite currents to flow in the antenna terminals.

are called the *common-mode voltages* (*CM*).

A common-mode (CM) current is one that has the same value and direction in both wires (or on the shield and center conductor). Because of the ferrite, the CM current encounters a high impedance that acts to reduce (choke) the current. The normal *differential-mode (DM)* signal does not encounter this CM impedance because the electromagnetic fields due to equal and opposite currents in the two conductors cancel each other at the ferrite, so the magnetic flux in the ferrite is virtually zero. (See the section on Transmitting Ferrite Choke Baluns.)

The main idea of the transmission line transformer is that although the CM impedance may be very large, the DM signal is virtually unopposed, especially if the line length is a small fraction of a wavelength. But it is very important to keep in mind that the common-mode voltage across the ferrite winding that is due to this current is efficiently coupled to the center wire by conventional transformer action, as mentioned before and easily verified. This equality of CM voltages, and also CM impedances, reduces the *conversion* of a CM signal to an *undesired* DM signal that can interfere with the *desired* DM signal in both transmitters and receivers.

The CM current, multiplied by the CM impedance due to the ferrite, produces a CM voltage. The CM impedance has L and C reactance and also R. So L, C and R cause a broad parallel self-resonance at some frequency. The R component also produces some dissipation (heat) in the ferrite. This dissipation is an excellent way to dispose of a small amount of unwanted CM power.

Because of the high CM impedance, the two output wires of the balun in Fig 20.21A have a high impedance with respect to, and are therefore "isolated" from, the generator. This feature is very useful because now any point of R at the output can be grounded. In a well-designed balun circuit almost all of the current in one conductor returns to the generator through the other conductor, despite this ground connection. Note also that the ground connection introduces some CM voltage across the balun cores and this has to be taken into account. This CM voltage is a maximum if point C is grounded. If point D is grounded and if all "ground" connections are at the same potential, which they often are not, the CM voltage is zero and the balun may no longer be needed. In a coax balun the return current flows on the inside surface of the braid.

We now look briefly at a transmission line transformer that is based on the choke balun. Fig 20.21B shows two identical choke baluns whose inputs are in parallel and whose outputs are in series. The output voltage amplitude of each balun is identical to the common input, so the two outputs add in-phase (equal time delay) to produce twice the input voltage. It is

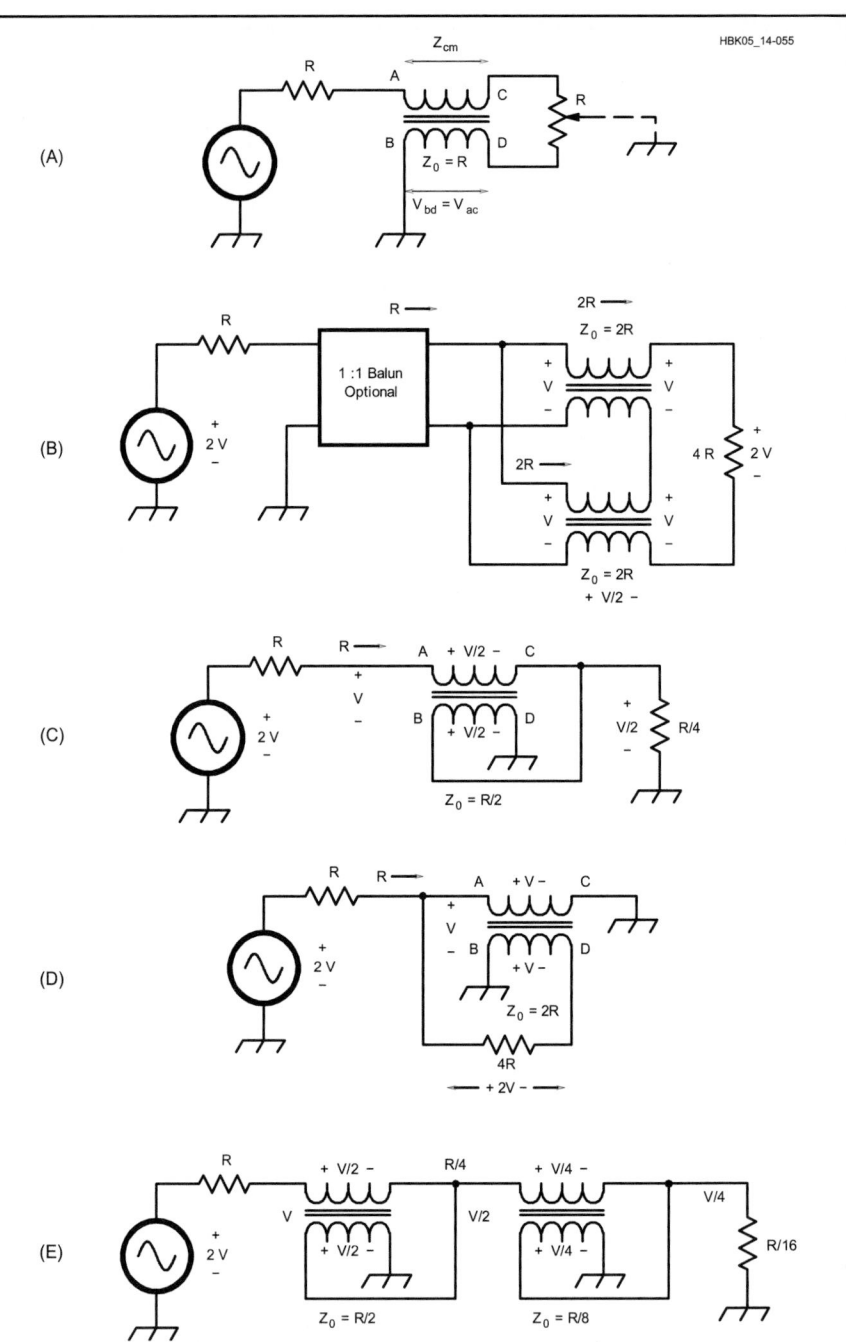

Fig 20.21 — (A) Basic current or choke balun. (B) Guanella 1:4 transformer. (C) Ruthroff 4:1 unbalanced transformer. (D) Ruthroff 1:4 balanced transformer. (E) Ruthroff 16:1 unbalanced transformer.

the high CM impedance that makes this voltage addition possible. If the power remains constant the load current must be one-half the generator current, and the load resistor is $2V/0.5I = 4V/I = 4R$.

THE GUANELLA TRANSFORMER

The CM voltage in each balun is V/2, so there is some flux in the cores. The right side floats. This is named the *Guanella* transformer. If Z_0 of the lines equals 2R and if the load is a pure resistance of 4R then the input resistance R is independent of line length. If the lines are exactly one-quarter wavelength, then $Z_{IN} = (2R)^2 / Z_L$, an impedance inverter, where Z_{IN} and Z_L are complex. The quality of balance can often be improved by inserting a 1:1 balun (Fig 20.21A) at the left end so that both ends of the 1:4 transformer are floating and a ground is at the far left side as shown. The Guanella transformer can also be operated from a grounded right end to a floating left end. The 1:1 balun at the left then allows a grounded far left end.

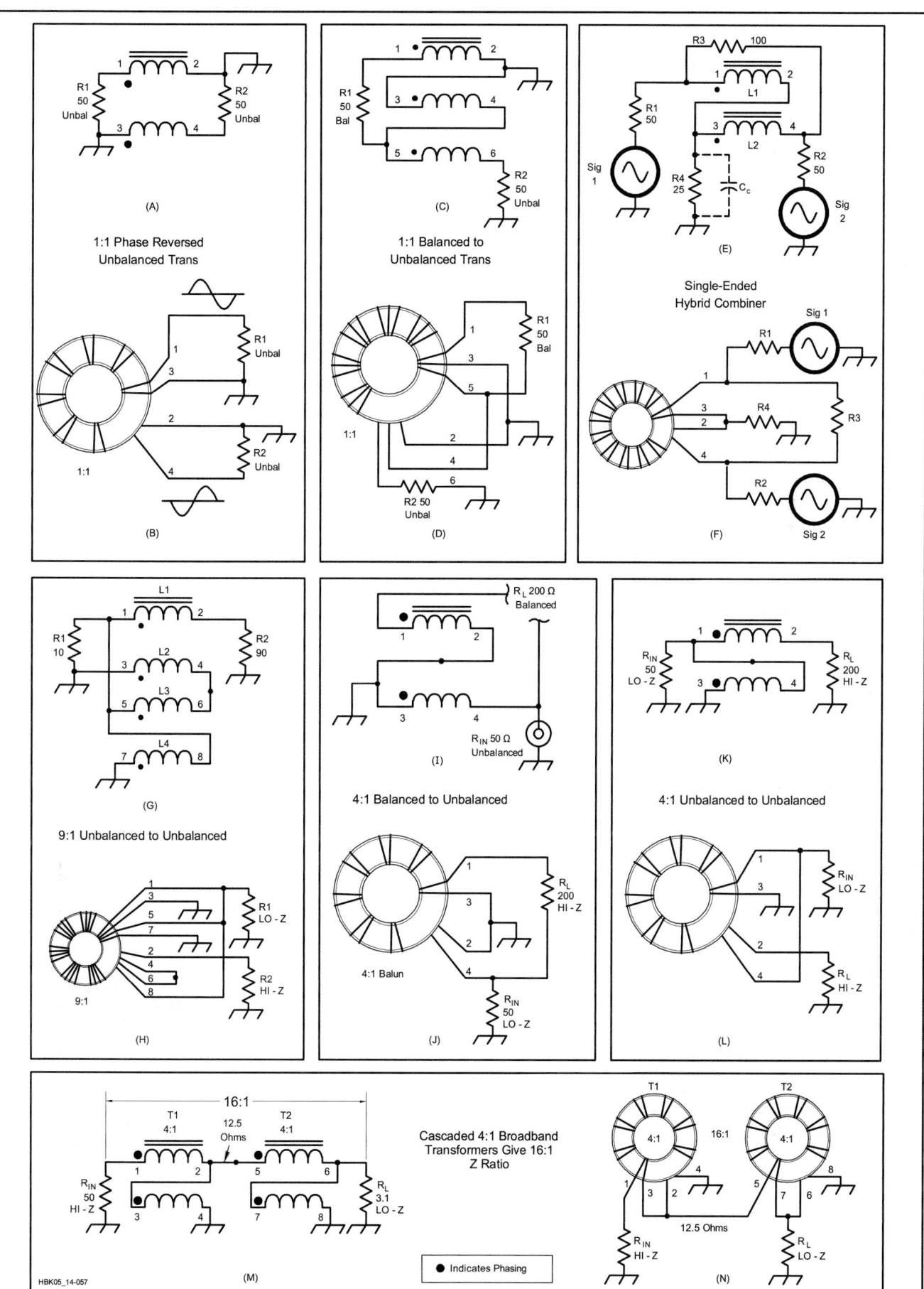

Fig 20.22 — Assembly instructions for some transmission-line transformers. See text for ferrite material type.

THE RUTHROFF TRANSFORMER

Fig 20.21C is the *Ruthroff* transformer in which the input voltage V is divided into two equal in-phase voltages AC and BD (they are tightly coupled), so the output is V/2. And because power is constant, $I_{OUT} = 2I_{IN}$ and the load is R/4. There is a CM voltage V/2 between A and C and between B and D, so in normal operation the core is not free of magnetic flux. The input and output both return to ground so it can also be operated from right to left for a 1:4 impedance step-up.

The Ruthroff transformer is often used as an amplifier interstage transformer, for example between 200 Ω and 50 Ω. To maintain low attenuation the line length should be much less than one-fourth wavelength at the highest frequency of operation, and its Z_0 should be R/2. A balanced version is shown in Fig 20.21D, where the CM voltage is V, not V/2, and transmission is from left-to-right only. Because of the greater flux in the cores, no different than a conventional transformer, this is not a preferred approach, although it could be used with air wound coils (for example in antenna tuner circuits) to couple 75 Ω unbalanced to 300 Ω balanced. The tuner circuit could then transform 75 Ω to 50 Ω.

APPLICATIONS OF TRANSMISSION-LINE TRANSFORMERS

There are many transformer schemes that use the basic ideas of Fig 20.21. Several of them, with their toroid winding instructions, are shown in **Fig 20.22**. Two of the most commonly used devices are the 1:1 current balun and 4:1 impedance transformer wound on toroid cores as shown in **Fig 20.23**.

Because of space limitations, for a comprehensive treatment we suggest *Transmission Line Transformers* and *Building and Using Baluns and Ununs* by Jerry Sevick W2FMI (SK). For applications in solid-state RF power amplifiers, see Sabin and Schoenike, *HF Radio Systems and Circuits*, Chapter 12.

20.5.3 Coiled-Coaxial Choke Baluns

The simplest construction method for a 1:1 choke balun made from coaxial feed line is simply to wind a portion of the cable into a coil (see **Fig 20.24**), creating an inductor from the shield's outer surface. This type of choke balun is simple, cheap and reduces common-mode current. Currents on the outside of the shield encounter the coil's impedance, while currents on the inside are unaffected.

A scramble-wound flat coil (like a coil of rope) shows a broad resonance that easily covers three octaves, making it reasonably effective over the entire HF range. If particular problems are encountered on a single band, a coil that is resonant on that band may be added. The choke baluns described in **Table 20.6** were constructed to have a

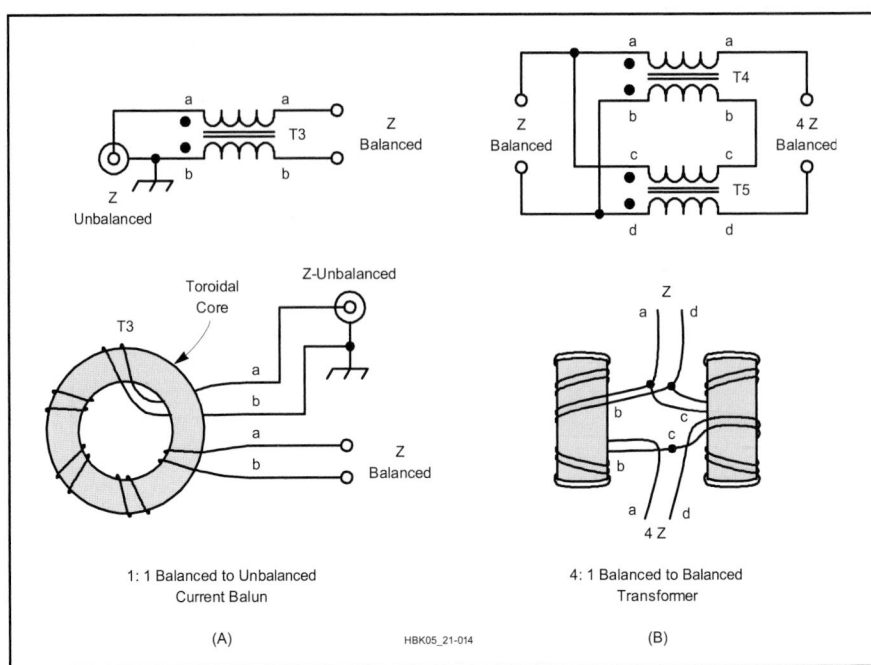

Fig 20.23 — Broadband baluns. (A) 1:1 current balun and (B) Guanella 4:1 impedance transformer wound on two cores, which are separated. Use 12 bifilar turns of #14 AWG enameled wire, wound on 2.4 inch OD cores for A and B. Distribute bifilar turns evenly around core. See text for ferrite material type.

high impedance at the indicated frequencies as measured with an impedance meter. This construction technique is not effective with open-wire or twin-lead line because of coupling between adjacent turns.

The inductor formed by the coaxial cable's shield is self-resonant due to the distributed capacitance between the turns of the coil. The self-resonant frequency can be found by using a dip meter. Leave the ends of the choke open,

couple the coil to the dip meter, and tune for a dip. This is the parallel resonant frequency and the impedance will be very high.

The distributed capacitance of a flat-coil choke balun can be reduced (or at least controlled) by winding the cable as a single-layer solenoid around a section of plastic pipe, an empty bottle or other suitable cylinder. **Fig 20.25** shows how to make this type of choke balun. A coil diameter of about 5 inches is reasonable. This type of construction reduces the stray capacitance between the ends of the coil.

For both types of coiled-coaxial chokes, use cable with solid insulation, not foamed, to minimize migration of the center conduc-

Table 20.6
Coiled-Coax Choke Baluns

Wind the indicated length of coaxial feed line into a coil (like a coil of rope) and secure with electrical tape. (Diameter 6-8 inches.)

The balun is most effective when the coil is near the antenna.

Lengths and diameter are not critical.

Single Band (Very Effective)

Freq (MHz)	RG-213, RG-8	RG-58
3.5	22 ft, 8 turns	20 ft, 6-8 turns
7	22 ft, 10 turns	15 ft, 6 turns
10	12 ft, 10 turns	10 ft, 7 turns
14	10 ft, 4 turns	8 ft, 8 turns
21	8 ft, 6-8 turns	6 ft, 8 turns
28	6 ft, 6-8 turns	4 ft, 6-8 turns

Multiple Band

Freq (MHz)	RG-8, 58, 59, 8X, 213
3.5-30	10 ft, 7 turns
3.5-10	18 ft, 9-10 turns
1.8-3.5	40 ft, 20 turns
14-30	8 ft, 6-7 turns

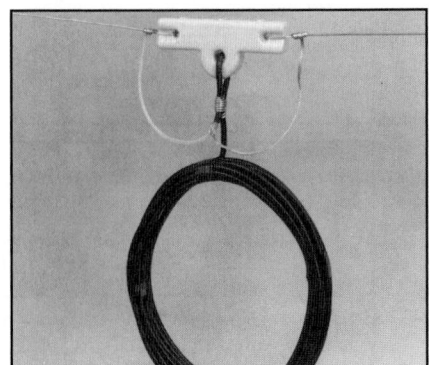

Fig 20.24 — RF choke formed by coiling the feed line at the point of connection to the antenna. The inductance of the choke isolates the antenna from the outer surface of the feed line.

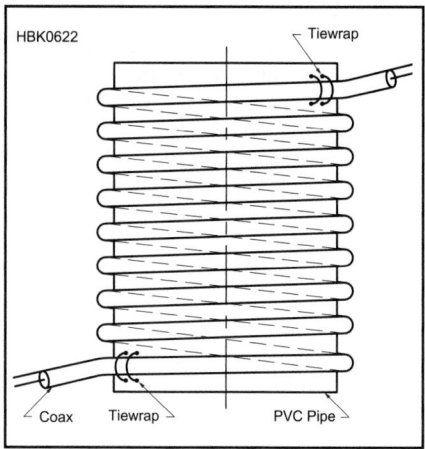

Fig 20.25 — Winding a coaxial choke balun as a single-layer solenoid typically increases impedance and self-resonant frequency compared to a flat-coil choke.

tor through the insulation toward the shield. The diameter of the coil should be at least 10 times the cable diameter to avoid mechanically stressing the cable.

20.5.4 Transmitting Ferrite Choke Baluns

A ferrite choke is simply a very low-Q parallel-resonant circuit tuned to the frequency where the choke should be effective. Passing a conductor through most ferrite cores (that is, one turn) produces a resonance around 150 MHz. By choosing a suitable core material, size and shape, and by adding multiple turns and varying their spacing, the choke can be "tuned" (optimized) for the required frequency range. The supplemental article "Measuring Ferrite Choke Impedance" by Jim Brown, K9YC is included on this book's CD-ROM.

Transmitting chokes differ from other common-mode chokes because they must be designed to work well when the line they are choking carries high power. They must also be physically larger so that the bend radius of the coax is large enough that the line is not deformed. Excellent common-mode chokes having very high power handling capability can be formed simply by winding multiple turns of coax through a sufficiently large ferrite core or multiple cores. (Chokes made by winding coaxial cable on ferrite cores will be referred to as "wound-coax chokes" to distinguish them from the coiled-coax chokes of the preceding section.)

CHOKES ON TRANSMISSION LINES

A transmission line can be wound around a ferrite core to form a common-mode choke. If the line is coax, all of the magnetic flux associated with differential mode current is confined to the dielectric (the insulating material between the center conductor and the shield). The external ferrite core carries only flux associated with common-mode current.

If the line is made up of parallel wires (a bifilar winding), a significant fraction of the flux associated with differential current will leak outside the line to the ferrite core. Leakage flux can exceed 30% of the total flux for even the most tightly-spaced bifilar winding. In addition to this leakage flux, the core will also carry the flux associated with common-mode current.

When a transformer (as opposed to a choke) is wound on a magnetic core, all of the field associated with current in the windings is carried by the core. Similarly, all forms of voltage baluns require all of the transmitted power to couple to the ferrite core. Depending on the characteristics of the core, this can result in considerable heating and power loss. Only a few ferrite core materials have loss characteristics suitable for use as the cores of high power RF transformers. Type 61 material has reasonably low dissipation below about 10 MHz, but its loss tangent rises rapidly above that frequency. The loss tangent of type 67 material makes it useful in high power transformers to around 30 MHz.

Leakage flux, corresponding to 30 to 40% of the transmitter power, causes heating in the ferrite core and attenuates the transmitted signal by a dB or so. At high power levels, temperature rise in the core also changes its magnetic properties, and in the extreme case, can result in the core temporarily losing its magnetic properties. A flux level high enough to make the core hot is also likely to saturate the core, producing distortion (harmonics, splatter, clicks).

Flux produced by common-mode current can also heat the core — if there is enough common-mode current. Dissipated power is equal to I^2R, so it can be made very small by making the common-mode impedance so large that the common-mode current is very small.

It can be shown mathematically, and experience confirms, that wound-coax chokes having a resistive impedance at the transmit frequency of at least 5000 Ω and wound with RG-8 or RG-11 size cable on five toroids are conservatively rated for 1500 W under high duty-cycle conditions, such as contesting or digital mode operation. While chokes wound with smaller coax (RG-6, RG-8X, RG-59, RG-58 size) are conservatively rated for dissipation in the ferrite core, the voltage and current ratings of those smaller cables suggests a somewhat lower limit on their power handling. Since the chokes see only the common-mode voltage, the only effect of high SWR on power handling of wound-coax chokes is the peaks of differential current and voltage along the line established by the mismatch.

Experience shows that 5000 Ω is also a good design goal to prevent RFI, noise coupling and pattern distortion. While 500-1000 Ω has long been accepted as sufficient to prevent pattern distortion, W1HIS has correctly observed that radiation and noise coupling from the feed line should be viewed as a form of pattern distortion that fills in the nulls of a directional antenna, reducing its ability to reject noise and interference.

Chokes used to break up a feed line into segments too short to interact with another antenna should have a choking impedance on the order of 1000 Ω to prevent interaction with simple antennas. A value closer to 5000 Ω may be needed if the effects of common-mode current on the feed line are filling the null of directional antennas.

BUILDING WOUND-COAX FERRITE CHOKES

Coaxial chokes should be wound with a bend radius sufficiently large that the coax is not deformed. When a line is deformed, the spacing between the center conductor and the shield varies, so voltage breakdown and heat-

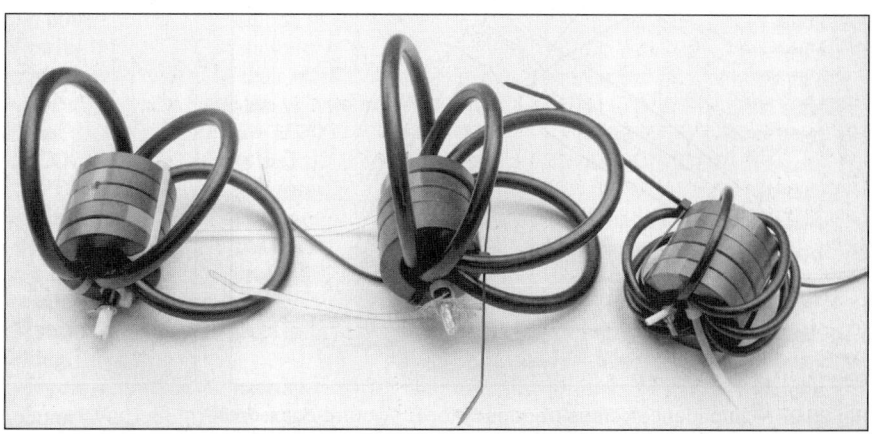

Fig 20.26 — Typical transmitting wound-coax common-mode chokes suitable for use on the HF ham bands.

ing are more likely to occur. Deformation also causes a discontinuity in the impedance; the resulting reflections may cause some waveform distortion and increased loss at VHF and UHF. (Coaxial cable has a specified "minimum bend radius".)

Chokes wound with any large diameter cable have more stray capacitance than those wound with small diameter wire. There are two sources of stray capacitance in a ferrite choke: the capacitance from end-to-end and from turn-to-turn via the core; and the capacitance from turn-to-turn via the air dielectric. Both sources of capacitance are increased by increased conductor size, so stray capacitance will be greater with larger coax. Turn-to-turn capacitance is also increased by larger diameter turns.

At low frequencies, most of the inductance in a ferrite choke results from coupling to the core, but some is the result of flux outside the core. At higher frequencies, the core has less permeability, and the flux outside the core makes a greater contribution.

The most useful cores for wound-coax chokes are the 2.4 inch OD, 1.4 inch ID toroid of type 31 or 43 material, and the 1 inch ID × 1.125 inch long clamp-on of type 31 material. Seven turns of RG-8 or RG-11 size cable easily fit through these toroids with no connector attached, and four turns fit with a PL-259 attached. Four turns of most RG-8 or RG-11 size cable fit within the 1 inch ID clamp-on. The toroids will accept at least 14 turns of most RG-6, RG-8X or RG-59 size cables.

PRACTICAL CHOKES

Fig 20.26 shows typical wound-coax chokes suitable for use on the HF ham bands. **Fig 20.27**, **Fig 20.28**, and **Fig 20.29** are graphs of the magnitude of the impedance

for HF transmitting chokes of various sizes. Fourteen close-spaced, 3 inch diameter turns of RG-58 size cable on a #31 toroid is a very effective 300 W choke for the 160 and 80 meter bands.

Table 20.7 summarizes designs that meet the 5000 Ω criteria for the 160 through 6 meter ham bands and several practical transmitting choke designs that are "tuned" or optimized for ranges of frequencies. The table entries refer to the specific cores in the preceding paragraph. If you construct the chokes using toroids, remember to make the diameter of the turns large enough to avoid deformation of the coaxial cable. Space turns evenly around the toroid to minimize inter-turn capacitance.

USING FERRITE BEADS

The early "current baluns" developed by

Walt Maxwell, W2DU, formed by stringing multiple beads in series on a length of coax to obtain the desired choking impedance, are really common-mode chokes. Maxwell's designs utilized 50 very small beads of type 73 material as shown in **Fig 20.30**. Product data sheets show that a single type 73 bead has a very low-Q resonance around 20 MHz, and has a predominantly resistive impedance of 10-20 Ω on all HF ham bands. Stringing 50 beads in series simply multiplies the impedance of one bead by 50, so the W2DU "current balun" has a choking impedance of 500-1000 Ω, and because it is strongly resistive, any resonance with the feed line is minimal.

This is a fairly good design for moderate power levels, but suitable beads are too small to fit most coax. A specialty coaxial cable such as RG-303 must be used for high-power

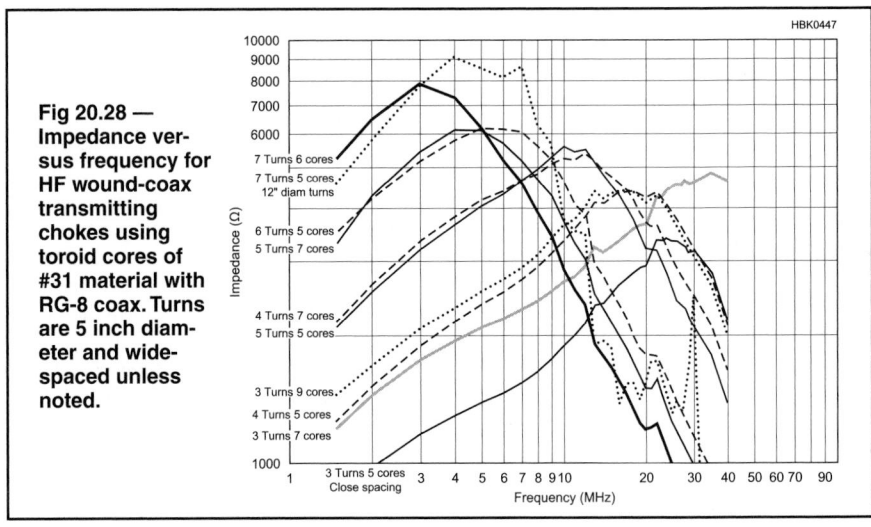

Fig 20.28 — Impedance versus frequency for HF wound-coax transmitting chokes using toroid cores of #31 material with RG-8 coax. Turns are 5 inch diameter and wide-spaced unless noted.

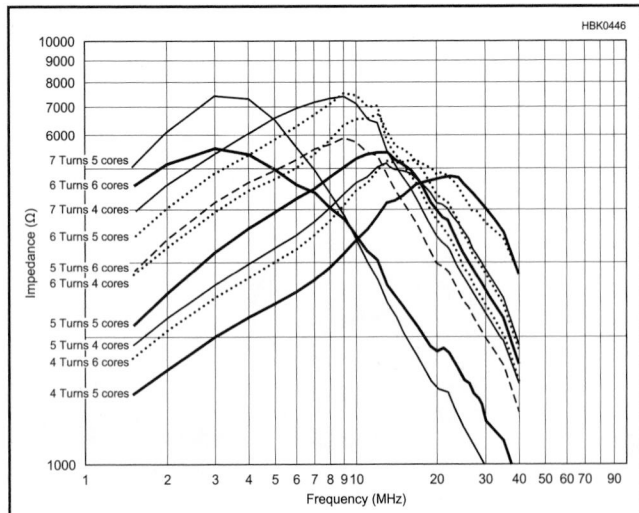

Fig 20.27 — Impedance versus frequency for HF wound-coax transmitting chokes using 2.4 inch toroid cores of #31 material with RG-8X coax.

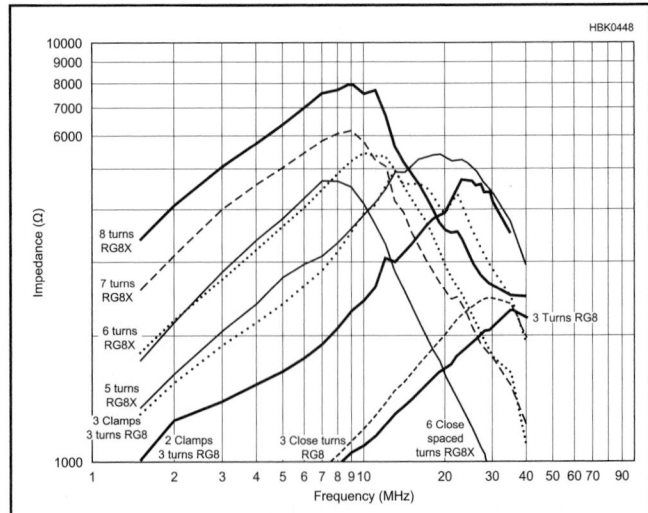

Fig 20.29 — Impedance versus frequency for HF wound-coax transmitting chokes wound on big clamp-on cores of #31 material with RG-8X or RG-8 coax. Turns are 6 inch diameter, wide-spaced except as noted.

Table 20.7
Transmitting Choke Designs

Freq Band(s) (MHz)	Mix	RG-8, RG-11 Turns	Cores	RG-6, RG-8X, RG-58, RG-59 Turns	Cores
1.8, 3.8	#31	7	5 toroids	7	5 toroids
				8	Big clamp-on
3.5-7		6	5 toroids	7	4 toroids
				8	Big clamp-on
10.1	#31 or #43	5	5 toroids	8	Big clamp-on
				6	4 toroids
7-14		5	5 toroids	8	Big clamp-on
14		5	4 toroids	8	2 toroids
		4	6 toroids	5-6	Big clamp-on
21		4	5 toroids	4	5 toroids
		4	6 toroids	5	Big clamp-on
28		4	5 toroids	4	5 toroids
				5	Big clamp-on
7-28 10.1-28 or 14-28	#31 or #43	Use two chokes in series: #1 — 4 turns on 5 toroids #2 — 3 turns on 5 toroids		Use two chokes in series: #1 — 6 turns on a big clamp-on #2 — 5 turns on a big clamp-on	
14-28		Two 4-turn chokes, each w/one big clamp-on		4 turns on 6 toroids, or 5 turns on a big clamp-on	
50		Two 3-turn chokes, each w/one big clamp-on			

Notes: Chokes for 1.8, 3.5 and 7 MHz should have closely spaced turns.
Chokes for 14-28 MHz should have widely spaced turns.
Turn diameter is not critical, but 6 inches is good.

Table 20.8
Combination Ferrite and Coaxial Coil

Freq (MHz)	-------Measured Impedance------- 7 ft, 4 turns of RG-8X	1 Core	2 Cores
1.8	—	—	520 Ω
3.5	—	660	1.4 kΩ
7	—	1.6 kΩ	3.2 kΩ
14	560 Ω	1.1 kΩ	1.4 kΩ
21	42 kΩ	500 Ω	670 Ω
28	470 Ω	—	—

material. **Fig 20.31** shows that these high-Q chokes are quite effective in the narrow frequency range near their resonance. However, the resonance is quite difficult to measure and it is so narrow that it typically covers only one or two ham bands. Away from resonance, the choke becomes far less effective, as choking impedance falls rapidly and its reactive component resonates with the line.

Air-wound coaxial chokes are less effective than bead baluns. Their equivalent circuit is also a simple high-Q parallel resonance and they must be used below resonance. They are simple, inexpensive and unlikely to overheat. Choking impedance is purely inductive and not very great, reducing their effectiveness. Effectiveness is further reduced when the inductance resonates with the line at frequencies where the line impedance is capacitive and there is almost no resistance to damp the resonance.

Adding ferrite cores to a coiled-coax balun is a way to increase their effectiveness. The resistive component of the ferrite impedance damps the resonance of the coil and increases

applications. Even with high-power coax, the choking impedance is often insufficient to limit current to a low enough value to prevent overheating. Equally important — the lower choking impedance is much less effective at rejecting noise and preventing the filling of nulls in a radiation pattern.

Newer "bead balun" designs use type 31 and 43 beads, which are resonant around 150 MHz, are inductive below resonance, and have only a few tens of ohms of strongly inductive impedance on the HF bands. Even with 20 of the type 31 or 43 beads in the string, the choke is still resonant around 150 MHz, is much less effective than a wound coaxial ferrite choke, and is still inductive on the HF bands (so it will be ineffective at frequencies

where it resonates with the line).

Joe Reisert, W1JR, introduced the first coaxial chokes wound on ferrite toroids. He used low-loss cores, typically type 61 or 67

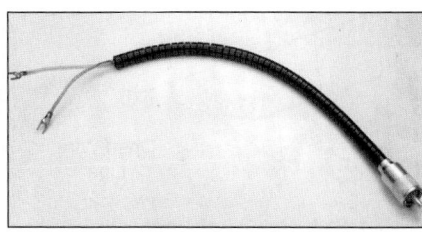

Fig 20.30 — W2DU bead balun consisting of 50 FB-73-2401 ferrite beads over a length of RG-303 coax. See text for details.

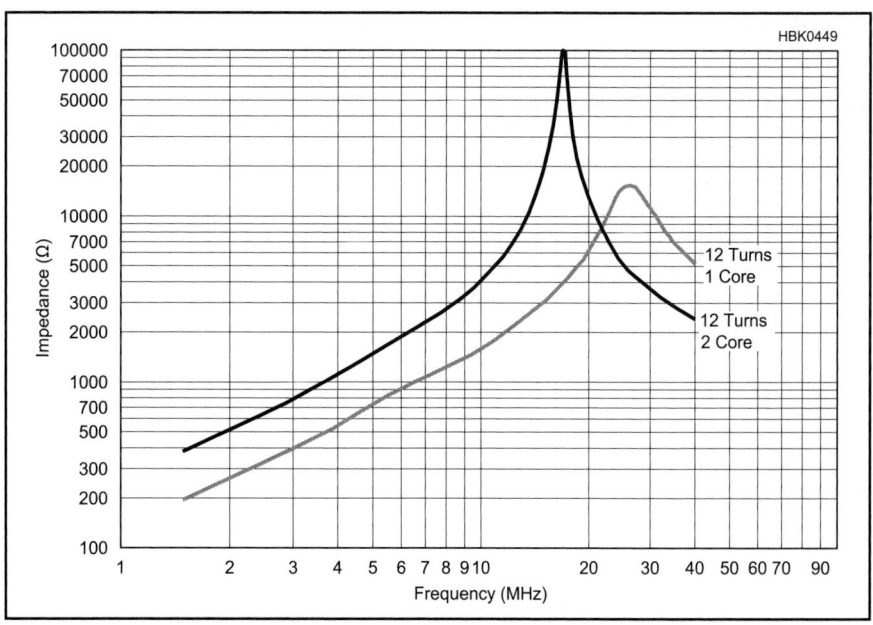

Fig 20.31 — Impedance versus frequency for HF wound-coax transmitting chokes wound with RG-142 coax on toroid cores of #61 material. For the 1 core choke: R = 15.6 kΩ, L = 25 µH, C = 1.4 pF, Q = 3.7. For the 2 core choke: R = 101 kΩ, L = 47 µH, C = 1.9 pF, Q = 20.

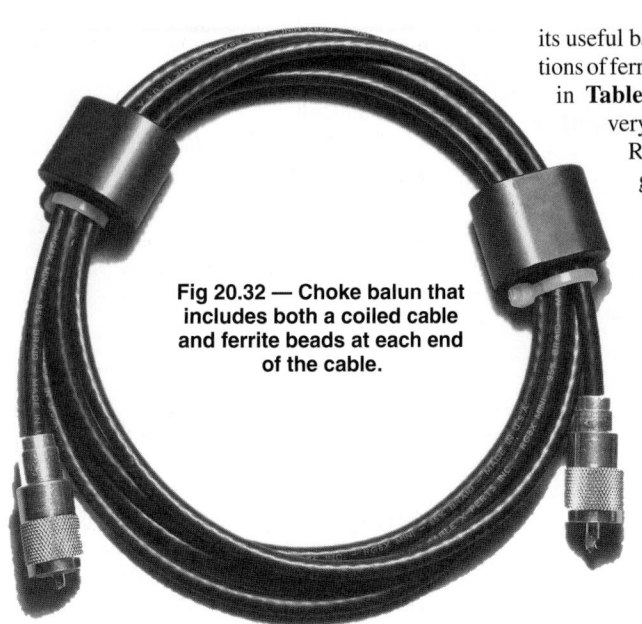

Fig 20.32 — Choke balun that includes both a coiled cable and ferrite beads at each end of the cable.

its useful bandwidth. The combinations of ferrite and coil baluns shown in **Table 20.8** demonstrate this very effectively. Eight feet of RG-8X in a 5 turn coil is a great balun for 21 MHz, but it is not particularly effective on other bands. If one type 43 core (Fair-Rite 2643167851) is inserted in the same coil of coax, the balun can be used from 3.5 to 21 MHz. If two of these cores are spaced a few inches apart on the coil as in **Fig 20.32**, the balun is more effective from 1.8 to 7 MHz and usable to 21 MHz. If type 31 material was used (the Fair-Rite 2631101902 is a similar core), low-frequency performance would be even better. The 20-turn, multiple-band, 1.8-3.5 MHz coiled-coax balun in Table 20.6 weighs 1 pound, 7 ounces. The single ferrite core combination balun weighs 6.5 ounces and the two-core version weighs 9.5 ounces.

20.6 Waveguides

A waveguide is a hollow conducting tube through which microwave energy is transmitted in the form of electromagnetic waves. The tube does not carry a current in the same sense that the wires of a two-conductor line do. Instead, it is a boundary that confines the waves to the enclosed space. Skin effect on the inside walls of the waveguide confines electromagnetic energy inside the guide in much the same manner that the shield of a coaxial cable confines energy within the coax. Microwave energy is injected at one end (either through capacitive or inductive coupling or by radiation) and is received at the other end. The waveguide merely confines the energy of the fields, which are propagated through it to the receiving end by means of reflections off its inner walls.

the plate is large enough, it will prevent the magnetic lines of force from encircling the RF current.

An infinite number of these quarter-wave stubs may be connected in parallel without affecting the standing waves of voltage and current. The transmission line may be supported from the top as well as the bottom, and when infinitely many supports are added, they form the walls of a waveguide at its *cutoff frequency*. **Fig 20.33** illustrates how a rectangular waveguide evolves from a two-wire parallel transmission line. This simplified analysis also shows why the cutoff dimension is ½ λ.

While the operation of waveguides is usually described in terms of fields, current does

flow on the inside walls, just as fields exist between the current-carrying conductors of a two-wire transmission line. At the waveguide cutoff frequency, the current is concentrated in the center of the walls, and disperses toward the floor and ceiling as the frequency increases.

Analysis of waveguide operation is based on the assumption that the guide material is a perfect conductor of electricity. Typical distributions of electric and magnetic fields in a rectangular guide are shown in **Fig 20.34**. The intensity of the electric field is greatest (as indicated by closer spacing of the lines of force in Fig 20.34B) at the center along the X dimension and diminishes to zero at the end walls. Zero field intensity is a necessary

20.6.1 Evolution of a Waveguide

Suppose an open-wire line is used to carry UHF energy from a generator to a load. If the line has any appreciable length, it must be well-insulated from the supports in order to avoid high losses. Since high-quality insulators are difficult to make for microwave frequencies, it is logical to support the transmission line with quarter-wave stubs, shorted at the far end. The open end of such a stub presents an infinite impedance to the transmission line, provided that the shorted stub is non-reactive. However, the shorting link has finite length and, therefore, some inductance. This inductance can be nullified by making the RF current flow on the surface of a plate rather than through a thin wire. If

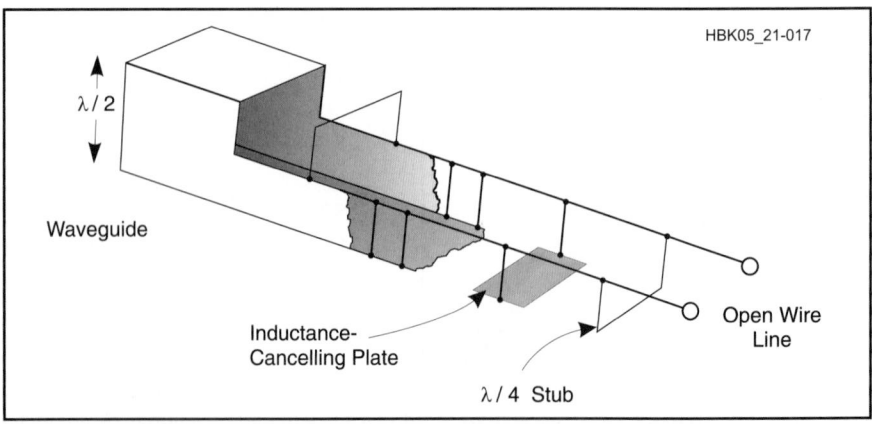

Fig 20.33 — At its cutoff frequency a rectangular waveguide can be analyzed as a parallel two-conductor transmission line supported from top and bottom by an infinite number of quarter-wave stubs.

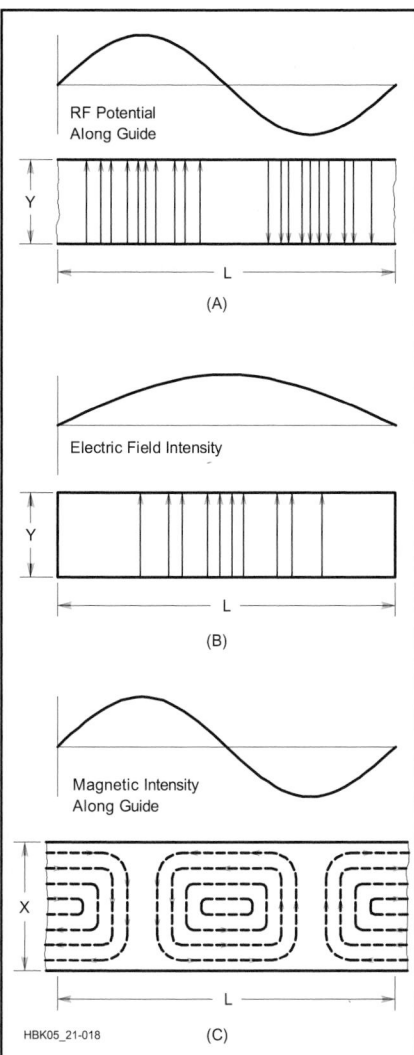

RF Potential
Along Guide

Y

L

(A)

Electric Field Intensity

Y

L

(B)

Magnetic Intensity
Along Guide

X

L

HBK05_21-018   (C)

Fig 20.34 — Field distribution in a rectangular waveguide. The TE$_{1,0}$ mode of propagation is depicted.

condition at the end walls, since the existence of any electric field parallel to any wall at the surface would cause an infinite current to flow in a perfect conductor, an impossible situation.

20.6.2 Modes of Waveguide Propagation

Fig 20.34 represents a relatively simple distribution of the electric and magnetic fields. An infinite number of ways exist in which the fields can arrange themselves in a guide, as long as there is no upper limit to the frequency to be transmitted. Each field configuration is called a *mode*. All modes may be separated into two general groups. One group, designated *TM* (transverse magnetic), has the magnetic field entirely crosswise to the direction of propagation, but has a component of electric field in the propagation direction. The other type, designated *TE* (transverse electric)

has the electric field entirely crosswise to the direction of propagation, but has a component of magnetic field in the direction of propagation. TM waves are sometimes called *E-waves*, and TE waves are sometimes called *H-waves*. The TM and TE designations are preferred, however.

The particular mode of transmission is identified by the group letters followed by subscript numbers; for example TE$_{1,1}$, TM$_{1,1}$ and so on. The number of possible modes increases with frequency for a given size of guide. There is only one possible mode (called the *dominant mode*) for the lowest frequency that can be transmitted. The dominant mode is the one normally used in practical applications.

20.6.3 Waveguide Dimensions

In rectangular guides the critical dimension (shown as X in Fig 20.34C) must be more than one-half wavelength at the lowest frequency to be transmitted. In practice, the Y dimension is usually made about equal to 1/2 X to avoid the possibility of operation at other than the dominant mode. Cross-sectional shapes other than rectangles can be used; the most important of those is the circular pipe.

Table 20.9 gives dominant-mode wavelength formulas for rectangular and circular guides. X is the width of a rectangular guide, and R is the radius of a circular guide. **Table 20.10** lists the standard waveguide designators and sizes used on the amateur microwave bands from 1-100 GHz. For additional information on waveguide sizes, see **www.rf-cafe.com/references/electrical/waveguide-chart.htm** or **www.radio-electronics.com/info/antennas/waveguide/rf-waveguide-dimensions-sizes.php**.

20.6.4 Coupling to Waveguides

Energy may be introduced into or extracted from a waveguide or resonator by means of either the electric or magnetic field. The energy transfer frequently takes place through

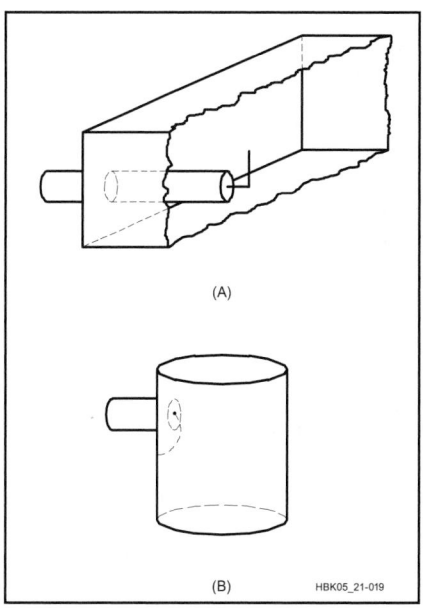

(A)

(B)   HBK05_21-019

Fig 20.35 — Coupling to waveguide and resonators. The probe at A is an extension of the inner conductor of coax line. At B an extension of the coax inner conductor is grounded to the waveguide to form a coupling loop.

Table 20.9
Wavelength Formulas for Waveguide

	Rectangular	Circular
Cut-off wavelength	2X	3.41R
Longest wavelength transmitted with little attenuation	1.6X	3.2R
Shortest wavelength before next mode becomes possible	1.1X	2.8R

Table 20.10
Standard Waveguide Designators and Sizes

Band (GHz)	WR Designation	WG Designation	Frequency Range (GHz)	Inside Dimensions (in.)
1.3	WR770	WG5	0.96 - 1.45	7.700 × 3.850
1.3	WR650	WG6	1.12 - 1.70	6.500 × 3.250
2.4	WR340	WG9A	2.20 - 3.30	3.400 × 1.700
3.4	WR229	WG11A	3.30 - 4.90	2.290 × 1.150
5.8	WR159	WG13	4.90 - 7.05	1.590 × 0.795
10	WR90	WG16	8.20 - 12.4	0.900 × 0.400
10	WR75	WG17	10.0 - 15.0	0.750 × 0.375
24	WR42	WG20	18.0 - 26.5	0.420 × 0.170
47	WR22	WG23	33.0 - 50.0	0.224 × 0.112
77	WR12	WG26	60.0 - 90.0	0.122 × 0.061

a coaxial line. Two methods for coupling are shown in **Fig 20.35**. The probe at A is simply a short extension of the inner conductor of the feed coaxial line, oriented so that it is parallel to the electric lines of force. The loop shown at B is arranged to enclose some of the magnetic lines of force. The point at which maximum coupling will be obtained depends on the particular mode of propagation in the guide or cavity; the coupling will be maximum when the coupling device is in the most intense field.

Coupling can be varied by rotating the probe or loop through 90°. When the probe is perpendicular to the electric lines the coupling will be minimum; similarly, when the plane of the loop is parallel to the magnetic lines, the coupling will be minimum. See *The ARRL Antenna Book* for more information on waveguides.

20.7 Glossary of Transmission Line Terms

Antenna tuner — A device that matches the antenna system input impedance to the transmitter, receiver or transceiver output impedance. Also called an *antenna-matching network, impedance matcher, transmatch, ATU, matchbox*.

Balanced line — A symmetrical two-conductor feed line that has uniform voltage and current distribution along its length.

Balun — Contraction of "balanced to unbalanced." A device to couple a balanced load to an unbalanced feed line or device, or vice versa. May be in the form of a choke balun, or a transformer that provides a specific impedance transformation (including 1:1). Often used in antenna systems to interface a coaxial transmission line to the feed point of a balanced antenna, such as a dipole.

Characteristic impedance — The ratio of voltage to current in a matched feed line, it is determined by the physical geometry and materials used to construct the feed line. Also known as *surge impedance* since it represents the impedance electromagnetic energy encounters when entering a feed line.

Choke balun — A balun that prevents current from flowing on the outside of a coaxial cable shield when connected to a balanced load, such as an antenna.

Coax — See **coaxial cable**.

Coaxial cable — Transmission lines that have the outer shield (solid or braided) concentric with the same axis as the inner or center conductor. The insulating material can be a gas (air or nitrogen) or a solid or foam insulating material.

Common-mode current — Current that flows equally and in phase on all conductors of a feed line or multiconductor cable.

Conductor — A metal body such as tubing, rod or wire that permits current to travel continuously along its length.

Conjugate match — Creating a purely resistive impedance by connecting an impedance with an equal-and-opposite reactive component.

Current balun — see **Choke balun**.

Decibel — A logarithmic power ratio, abbreviated dB. May also represent a voltage or current ratio if the voltages or currents are measured across (or through) identical impedances. Suffixes to the abbreviation indicate references: dBi, isotropic radiator; dBm, milliwatt; dBW, watt.

Dielectrics — Various insulating materials used in antenna systems, such as found in insulators and transmission lines.

Dielectric constant (k) — Relative figure of merit for an insulating material used as a dielectric. This property determines how much electric energy can be stored in a unit volume of the material per volt of applied potential.

Electric field — An electric field exists in a region of space if an electrically charged object placed in the region is subjected to an electrical force.

Electromagnetic wave — A wave of energy composed of an electric and magnetic field.

Feed line — See **transmission line**.

Feed point — The point at which a feed line is electrically connected to an antenna.

Feed point impedance — The ratio of RF voltage to current at the feed point of an antenna.

Ferrite — A ceramic material with magnetic properties.

Hardline — Coaxial cable with a solid metal outer conductor to reduce losses compared to flexible cables. Hardline may or may not be flexible.

Impedance match — To adjust impedances to be equal or the case in which two impedances are equal. Usually refers to the point at which a feed line is connected to an antenna or to transmitting equipment. If the impedances are different, that is a **mismatch**.

Impedance matcher — See **Antenna tuner**.

Impedance matching (circuit) — A circuit that transforms impedance from one value to another. Adjustable impedance matching circuits are used at the output of transmitters and amplifiers to allow maximum power output over a wide range of load impedances.

Impedance transformer — A transformer designed specifically for transforming impedances in RF equipment.

L network — A combination of two reactive components used to transform or match impedances. One component is connected in series between the source and load and the other shunted across either the source or the load. Most L networks have one inductor and one capacitor, but two-inductor and two-capacitor configurations are also used.

Ladder line — see **Open-wire line**.

Lambda (λ) — Greek symbol used to represent wavelength.

Line loss — The power dissipated by a transmission line as heat, usually expressed in decibels.

Load — (noun) The component, antenna, or circuit to which power is delivered; (verb) To apply a load to a circuit or a transmission line.

Loading — The process of a transferring power from its source to a load. The effect a load has on a power source.

Magnetic field — A region through which a magnetic force will act on a magnetic object.

Matched-line loss — The line loss in a feed line terminated by a load equal to its characteristic impedance.

Matching — The process of effecting an impedance match between two electrical circuits of unlike impedance. One example is matching a transmission line to the feed point of an antenna. Maximum power transfer to the load (antenna system) will occur when a matched condition exists.

Microstrip — A transmission line made from a strip of printed-circuit board conductor above a ground plane, used primarily at UHF and microwave frequencies.

Open-wire line — Parallel-conductor feed line with parallel insulators at regular intervals to maintain the line spacing. The dielectric is principally air, making it a low-loss type of line. Also known as *ladder line* or *window line*.

Output impedance — The equivalent impedance of a signal source.

Parallel-conductor line — A type of transmission line that uses two parallel wires spaced from each other by insulating material. Also known as *open-wire, ladder* or *window line*.

Phasing lines — Sections of transmission line that are used to ensure the correct phase relationship between the elements of a driven array, or between bays of an array of antennas. Also used to effect impedance transformations while maintaining the desired phase.

Q section — Term used in reference to transmission-line matching transformers and phasing lines.

Reflection coefficient (ρ) — The ratio of the reflected voltage at a given point on a transmission line to the incident voltage at the same point. The reflection coefficient is also equal to the ratio of reflected and incident currents. The Greek letter rho (ρ) is used to represent reflection coefficient.

Reflectometer — see **SWR bridge**

Resonance — (1) The condition in which a system's natural response and the frequency of an applied or emitted signal are the same. (2) The frequency at which a circuit's capacitive and inductive reactances are equal and cancel.

Resonant frequency — The frequency at which the maximum response of a circuit occurs. In an antenna, the resonant frequency is one at which the feed point impedance is purely resistive.

Return loss — The absolute value of the ratio in dB of the power reflected from a load to the power delivered to the load.

Rise time — The time it takes for a waveform to reach a maximum value.

Series-input network — A network such as a filter or impedance matching circuit in which the input current flows through a component in series with the input.

Shunt-input network — A network such as a filter or impedance matching circuit with a component connected directly across the input.

Skin effect — The phenomenon in which ac current at high frequencies flows in a thin layer near the surface of a conductor.

Smith chart — A coordinate system developed by Phillip Smith to represent complex impedances graphically. This chart makes it easy to perform calculations involving antenna and transmission-line impedances and SWR.

Standing-wave ratio (SWR) — Sometimes called voltage standing-wave ratio (VSWR). A measure of the impedance match between a feed line's characteristic impedance and the attached load (usually an antenna). VSWR is the ratio of maximum voltage to minimum voltage along the feed line, or of antenna impedance to feed line impedance.

Stacking — The technique of placing similar directive antennas atop or beside one another, forming a "stacked array." Stacking provides more gain or directivity than a single antenna.

Stub — A section of transmission line used to perform impedance matching or filtering.

Surge impedance — see **Characteristic impedance**.

SWR — see **Standing-wave ratio**.

SWR bridge — Device for measuring SWR in a transmission line. Also known as an SWR meter or reflectometer.

TE mode — Transverse electric field mode. Condition in a waveguide in which the E-field component of the traveling electromagnetic energy is ori-

ented perpendicular to (transverse) the direction the energy is traveling in the waveguide.

TM mode — Transverse magnetic field mode. Condition in a waveguide in which the H-field (magnetic field) component of the traveling electromagnetic energy is oriented perpendicular to (transverse) the direction the energy is traveling in the waveguide.

Transmatch — See **Antenna tuner.**

Transmission line — The wires or cable used to connect a transmitter or receiver to an antenna. Also called **feed line**.

Twin-lead — Parallel-conductor transmission line in which both conductors are completely embedded in continuous strip of insulating material.

Unbalanced line — Feed line with one conductor at dc ground potential, such as coaxial cable.

Universal stub system — A matching network consisting of a pair of transmission line stubs that can transform any impedance to any other impedance.

Velocity factor (velocity of propagation) — The speed at which an electromagnetic wave will travel through a material or feed line stated as a fraction of the speed of the wave in free space (where the wave would have its maximum velocity).

VSWR — Voltage standing-wave ratio. See **SWR**.

Waveguide — A hollow conductor through which electromagnetic energy flows. Usually used at UHF and microwave frequencies instead of coaxial cable.

Window line — see **Open-wire line**.

20.8 References and Bibliography

J. Brown, K9YC, "A Ham's Guide to RFI, Ferrites, Baluns, and Audio Interfacing," **audiosystemsgroup.com/RFI-Ham.pdf**

C. Counselman, W1HIS, "Common Mode Chokes," **www.yccc.org/articles/w1his/mmonModeChokesW1HIS2006Apr06.pdf**

G. Cutsogeorge, W2VJN, *Managing Interstation Interference,* Second Edition (International Radio, 2009).

H.W. Johnson and M. Graham, *High Speed Digital Design* (Prentice Hall, 1993).

R. Lewallan, W7EL, "Baluns: What They Do and How They Do It," 1985, **www.eznec.com/Amateur/Articles/Baluns.pdf**

J. Reisert, W1JR, "VHF/UHF World," *Ham Radio,* Oct 1987, pp. 27-38.

F. Regier, "Series-Section Transmission Line Impedance Matching," *QST,* Jul 1978, pp 14-16.

Sabin and Schoenike, *HF Radio Systems and Circuits,* (SciTech Publishing, 1998).

J. Sevick, W2FMI, *Transmission Line Transformers,* 4th Edition (Noble Publishing, 2001).

J. Sevick, W2FMI, *Building and Using Baluns and Ununs* (CQ Communications, 2003).

H.W. Silver, Ed., *The ARRL Antenna Book,* 22nd Edition (Newington: ARRL, 2011). Chapters 23 through 27 include material on transmission lines and related topics.

P. Smith, *Electronic Applications of the Smith Chart* (Noble Publishing, 1995).

P. Wade, W1GHZ, "Microwavelengths," *QST* column.

F. Witt, "Baluns in the Real (and Complex) World," *The ARRL Antenna Compendium Vol 5* (Newington: ARRL, 1996).

The ARRL UHF/Microwave Experimenter's Manual (Newington: ARRL, 2000). Chapters 5 and 6 address transmission lines and impedance matching. (This book is out of print.)

ONLINE RESOURCES

A collection of Smith chart references — **http://sss-mag.com/smith.html**

Ferrite and powdered iron cores are available from **www.fair-rite.com**, **www.amidoncorp.com** and **www.cwsbytemark.com**

Cores, wire and more — **http://kitsandparts.com/toroids.php**

Microwave oriented downloads — **www.microwaves101.com/content/ downloads.cfm**

Table of transmission line properties — **hf-antenna.com/trans/**

Times Microwave catalog — **www. timesmicrowave.com**

Transmission line loss factors — **www. microwaves101.com/encyclopedia/ transmission_loss.cfm**

Transmission line matching with the

Smith chart — **www.odyseus.nildram. co.uk/RFMicrowave_Theory_Files/ SmithChartPart2.pdf**

Transmission line transformer theory — **www.bytemark.com/products/ tlttheory.htm**

Contents

Chapter 21

Antennas

In the world of radio, the antenna is "where the rubber meets the road!" With antennas so fundamental to communication, it is important that the amateur have a basic understanding of their function. That understanding enables effective selection and application of basic designs to whatever communications task is at hand. In addition, the amateur is then equipped to engage in one of the most active areas of amateur experimentation, antenna design. The goal of this chapter is to define and illustrate the fundamentals of antennas and provide a selection of basic designs; simple verticals and dipoles, quads and Yagi beams, and other antennas. The reader will find additional in-depth coverage of these and other topics in the *ARRL Antenna Book* and other references provided. This chapter was originally written by Chuck Hutchinson, K8CH, and has been updated by Ward Silver, NØAX. Alan Applegate, KØBG, updated the material on mobile antennas. The section on Radio Direction Finding Antennas was written by Joe Moell, KØOV.

Chapter 21 — CD-ROM Content

Supplemental Articles

• "Direction Finding Techniques" by Joe Moell, KØOV"

Projects

• "Rotatable Dipole Inverted-U Antenna" by L.B. Cebik, W4RNL
• Construction details for "Top-Loaded Low-Band Antenna" by Dick Stroud, W9SR
• "Five-Band Two-Element Quad" by Al Doig, W6NBH, and William Stein, KC6T
• "Medium-Gain 2 Meter Yagi" by L.B. Cebik, W4RNL
• "K8SYL's 75 and 10-Meter Dipole" by Sylvia Hutchinson, K8SYL
• "A True Plumber's Delight for 2 Meters — An All-Copper J-Pole" by Michael Hood, KD8JB
• "Cheap Antennas for the AMSAT LEOs" by Kent Britain, WA5VJB
• "Wire Quad for 40 Meters" by Dean Straw, N6BV
• "Vertical Loop Antenna for 28 MHz"
• "Dual-Band Antenna for 146/446 MHz" by Wayde Bartholomew, K3MF

21.1 Antenna Basics

This section covers a range of topics that are fundamental to understanding how antennas work and defines several key terms. (A glossary is included at the end of the chapter.) While the discussion in this section uses the dipole as the primary example, the concepts apply to all antennas.

21.1.1 Directivity and Gain

All antennas, even the simplest types, exhibit directive effects in that the intensity of radiation is not the same in all directions from the antenna. This property of radiating more strongly in some directions than in others is called the *directivity* of the antenna. Directivity is the same for receiving as transmitting.

The directive pattern of an antenna at a given frequency is determined by the size and shape of the antenna, and on its position and orientation relative to the Earth and any other reflecting or absorbing surfaces.

The more an antenna's directivity is enhanced in a particular direction, the greater the *gain* of the antenna. This is a result of the radiated energy being concentrated in some directions at the expense of others. Similarly, gain describes the ability of the antenna to receive signals preferentially from certain directions. Gain does not create additional power beyond that delivered by the feed line — it only focuses that energy.

Gain is usually expressed in decibels, and is always stated with reference to a *standard* antenna — usually a dipole or an *isotropic radiator*. An isotropic radiator is a theoretical antenna that would, if placed in the center of an imaginary sphere, evenly illuminate that sphere with radiation. The isotropic radiator is an unambiguous standard, and for that reason frequently used as the comparison for gain measurements.

When the reference for gain is the isotropic radiator in *free space*, gain is expressed in dBi. When the standard is a dipole, also located in free space, gain is expressed in dBd. Because the dipole has some gain (2.15 dB) in its favorite direction with respect to the isotropic antenna (see the next section on the dipole antenna), the dipole's gain can be expressed as 2.15 dBi. Gain in dBi can be converted to dBd by subtracting 2.15 dB and from dBd to dBi by adding 2.15 dB.

Gain also takes losses in the antenna or surrounding environment into account. For example, if a practical dipole antenna's wire element dissipated 0.5 dB of the transmitter power as heat, that specific dipole's gain with respect to an isotropic antenna would be $2.15 - 0.5 = 1.65$ dBi.

21.1.2 Antenna Polarization

An electromagnetic wave has two components: an electric field and a magnetic field at right angles to each other. For most antennas, the field of primary interest is the electric, or *E-field*. The magnetic field is called the *H-field*. (The abbreviations E- and H- come from Maxwell's equations that describe electromagnetic waves.) By convention, the orientation of the E-field is the reference for determining the electromagnetic wave's *polarization*. The E-field of an electromagnetic wave can be oriented in any direction, so orientation with respect to the Earth's surface is the usual frame of reference. The wave's polarization can be vertical, horizontal, some intermediate angle, or even circular.

Antennas are considered to have polarization, too, determined by the orientation of the E-field of the electromagnetic field radiated by the antenna. Because the E-field of the radiated wave is parallel to the direction of current flow in the antenna's elements, the polarization of the wave and the orientation of the antenna elements is usually the same. For example, the E-field radiated by an antenna with linear elements is parallel to those elements, so that the polarization of the radiated wave is the same as the orientation of the elements. (This is somewhat over-simplified and additional considerations apply for elements that are not linear.) Thus a radiator that is parallel to the earth radiates a horizontally polarized wave, while a vertical antenna radiates a vertically polarized wave. If a wire antenna is slanted, it radiates waves with an E-field that has both vertical and horizontal components.

Antennas function symmetrically — a received signal will create the strongest antenna current when the antenna's elements are parallel to the E-field of the incoming wave just as the radiated wave's E-field will be strongest parallel to current in the antenna's radiating elements. This also means that for the strongest received signal, the antenna elements should have the same polarization as that of the incoming wave. Misalignment of the receiving antenna's elements with the passing wave's E-field reduces the amount of signal received. This is called *cross-polarization*. When the polarizations of antenna and wave are at right angles, very little antenna current is created by the incoming signal.

For best results in line-of-sight communications, antennas at both ends of the circuit should have the same polarization. However, it is not essential for both stations to use the same antenna polarity for ionospheric propagation or sky wave (see the **Propagation** chapter). This is because the radiated wave is bent and rotated considerably during its travel through the ionosphere. At the far end of the communications path the wave may be horizontal, vertical or somewhere in between at any given instant. For that reason, the main consideration for a good DX antenna is a low angle of radiation rather than the polarization.

Most HF-band antennas are either vertically or horizontally polarized. Although circular polarization is possible, just as it is at VHF and UHF, it is seldom used at HF. While most amateur antenna installations use the Earth's surface as their frame of reference, in cases such as satellite communication or EME the terms "vertical" and "horizontal" have no meaning with respect to polarization.

21.1.3 Current and Voltage Distribution

When power is fed to an antenna, the current and voltage vary along its length. The current is minimum at the ends, regardless of the antenna's length. The current does not actually reach zero at the current minima, because of capacitance at the antenna ends. Insulators, loops at the antenna ends, and support wires all contribute to this capacitance, which is also called the *end effect*. The opposite is true of the RF voltage. That is, there is a voltage maximum at each end.

In the case of a half-wave antenna there is a current maximum at the center and a voltage minimum at the center as illustrated in **Fig 21.1**. The voltage and current in this case are 90° out of phase. The pattern of alternating current and voltage maxima a quarter-wavelength apart repeats every half-wavelength along a resonant linear antenna as shown in Fig 21.1B. The phase of the current and voltage are inverted in each successive half-wavelength section.

Power is dissipated as heat or as signals by the resistance of the antenna, which consists of both the RF resistance of the wire (ohmic loss resistance) and the *radiation resistance*. The radiation resistance is the equivalent resistance that would dissipate the power the antenna radiates, with a current flowing in it equal to the antenna current at a current maximum. Radiation resistance represents the work done by the electrons in the antenna in transferring the energy from the signal source to the radiated electromagnetic wave. The loss resistance of a half-wave antenna is ordinarily small, compared with the radiation resistance, and can usually be neglected for practical purposes except in electrically small antennas, such as mobile HF antennas.

21.1.4 Impedance

The *impedance* at a given point in the antenna is determined by the ratio of the voltage to the current at that point. For example, if there were 100 V and 1.4 A of RF current at a specified point in an antenna and if they were in phase, the impedance would be approximately 71 Ω. The antenna's *feed point impedance* is the impedance at the point where the feed line is attached. If the feed point location changes, so does the feed point impedance.

Antenna impedance may be either resistive or complex (that is, containing resistance and reactance). The impedance of a *resonant* antenna is purely resistive anywhere on the

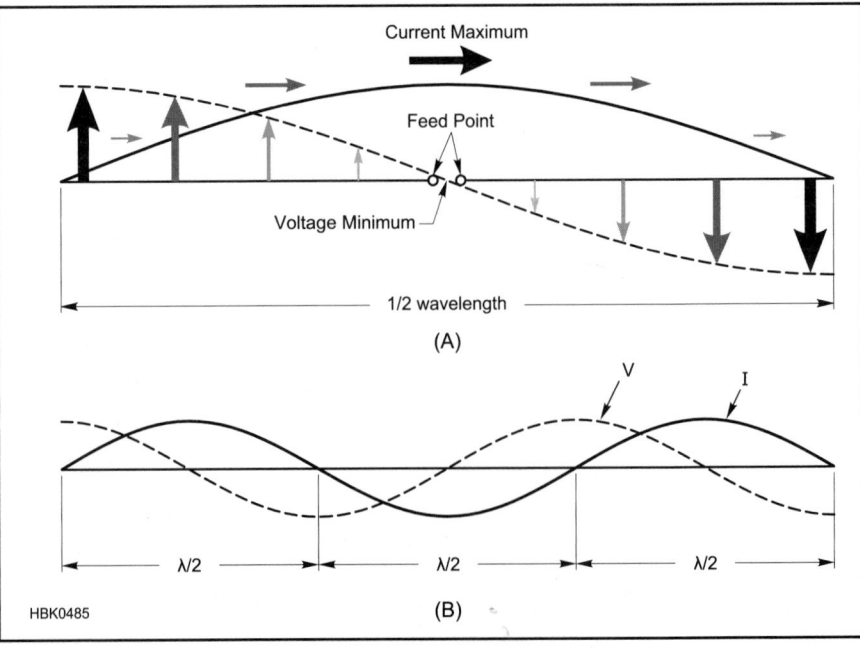

Fig 21.1 — The current and voltage distribution along a half-wave dipole (A) and for an antenna made from a series of half-wave dipoles (B).

antenna, no matter what value that impedance may be. For example, the impedance of a resonant half-wave dipole may be low at the center of the antenna and high at the ends, but it is purely resistive in all cases, even though its magnitude changes.

The feed point impedance is important in determining the appropriate method of matching the impedance of the antenna and the transmission line. The effects of mismatched antenna and feed line impedances are described in detail in the **Transmission Lines** chapter of this book. Some mistakenly believe that a mismatch, however small, is a serious matter. This is not true. The significance of a perfect match becomes more pronounced only at VHF and higher, where feed line losses are a major factor. Minor mismatches at HF are rarely significant.

21.1.5 Impedance and Height Above Ground

The feed point impedance of an antenna varies with height above ground because of the effects of energy reflected from and absorbed by the ground. For example, a ½ λ (or half-wave) center-fed dipole will have a feed point impedance of approximately 75 Ω in free space far from ground, but **Fig 21.2** shows that only at certain electrical heights above ground will the feed point impedance be 75 Ω. The feed point impedance will vary from very low when the antenna is close to the ground to a maximum of nearly 100 Ω at 0.34 λ above ground, varying between ±5 Ω as the antenna is raised farther. The 75 Ω feed point impedance is most likely to be realized in a practical installation when the horizontal dipole is approximately ½, ¾ or 1 wavelength above ground. This is why

few amateur λ/2 dipoles exhibit a center-fed feed point impedance of 75 Ω, even though they may be resonant.

Fig 21.2 compares the effects of perfect ground and typical soil at low antenna heights. The effect of height on the radiation resistance of a horizontal half-wave antenna is not drastic so long as the height of the antenna is greater than 0.2 λ. Below this height, while decreasing rapidly to zero over perfectly conducting ground, the resistance decreases less rapidly with height over actual lossy ground. At lower heights the resistance stops decreasing at around 0.15 λ, and thereafter increases as height decreases further. The reason for the increasing resistance is that more and more energy from the antenna is absorbed by the earth as the height drops below ¼ λ, seen as an increase in feed point impedance.

21.1.6 Antenna Bandwidth

The *bandwidth* of an antenna refers generally to the range of frequencies over which the antenna can be used to obtain a specified level of performance. The bandwidth can be specified in units of frequency (MHz or kHz) or as a percentage of the antenna's design frequency.

Popular amateur usage of the term antenna bandwidth most often refers to the 2:1 SWR bandwidth, such as, "The 2:1 *SWR bandwidth* is 3.5 to 3.8 MHz" or "The antenna has a 10% SWR bandwidth" or "On 20 meters, the antenna has an SWR bandwidth of 200 kHz." Other specific bandwidth terms are also used, such as the *gain bandwidth* (the bandwidth over which gain is greater than a specified level) and the *front-to-back ratio bandwidth* (the bandwidth over which front-to-back ratio is greater than a specified level).

As operating frequency is lowered, an equivalent bandwidth in percentage becomes narrower in terms of frequency range in kHz or MHz. For example, a 5% bandwidth at 21 MHz is 1.05 MHz (more than wide enough to cover the whole band) but at 3.75 MHz only 187.5 kHz! Because of the wide percentage bandwidth of the lower frequency bands 160 meters is 10.5% wide, 80 meters is 3.4% wide) it is difficult to design an antenna with a bandwidth sufficient to include the whole band.

It is important to recognize that SWR bandwidth does not always relate directly to gain bandwidth. Depending on the amount of feed line loss, an 80 meter dipole with a relatively narrow 2:1 SWR bandwidth can still radiate a good signal at each end of the band, provided that an antenna tuner is used to allow the transmitter to load properly. Broadbanding techniques, such as fanning the far ends of a dipole to simulate a conical type of dipole, can help broaden the SWR bandwidth.

21.1.7 Effects of Conductor Diameter

The impedance and resonant frequency of an antenna also depend on the diameter of the conductors that make up its elements in relation to the wavelength. As diameter of a conductor increases, its capacitance per unit length increases and inductance per unit length decreases. This has the net effect of lowering the frequency at which the antenna element is resonant, as illustrated in **Fig 21.3**. The larger the conductor diameter in terms of wavelength, the smaller its *length-to-diameter ratio (l/d)* and the lower the frequency at which a specific length of that conductor is ½ wavelength long electrically, in free space.

$$l/d = \frac{\lambda/2}{d} = \frac{300}{2f \times d} \qquad (1)$$

where f is in MHz and d is in meters. For example, a ½ wavelength dipole for 7.2 MHz made from #12 AWG wire (0.081 inch dia) has an l/d ratio of

$$l/d = \frac{300}{2f \times d} =$$

$$\frac{300}{2 \times 7.2 \times \dfrac{0.081\,in}{39.37\,in/m}} = 10,126$$

The effect of l/d is accounted for by the factor K which is based on l/d. From Fig 21.3 an l/d ratio of 10,126 corresponds to K ≈ 0.975, so the resonant length of that ½ wave dipole would be 0.975 × (300 / 2f) = 20.31 meters instead of the free-space 20.83 meters.

Most wire antennas at HF have l/d ratios in the range of 2500 to 25,000 with K = 0.97 to 0.98. The value of K is taken into account in the classic formula for ½ wave dipole length, 468/f (in MHz). If K = 1, the formula would be 492/f (in MHz). (This is discussed further in the following section on Dipoles and the Half-Wave Antenna.)

For single-wire HF antennas, the effects

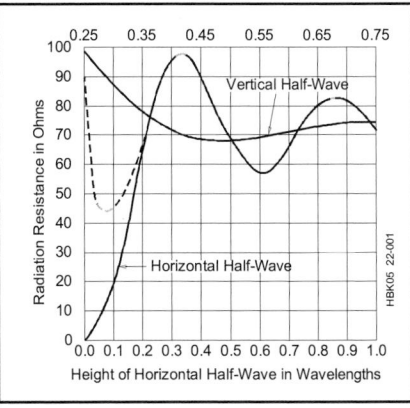

Fig 21.2 — Curves showing the radiation resistance of vertical and horizontal half-wavelength dipoles at various heights above ground. The broken-line portion of the curve for a horizontal dipole shows the resistance over *average* real earth, the solid line for perfectly conducting ground.

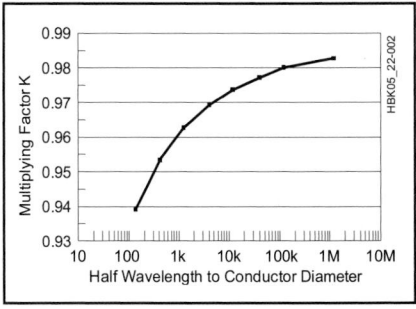

Fig 21.3 — Effect of antenna diameter on length for half-wavelength resonance, shown as a multiplying factor, K, to be applied to the free-space, half-wavelength equation.

of ground and antenna construction make a precise accounting for K unnecessary in practice. At and above VHF, the effects of l/d ratio can be of some importance, since the wavelength is small.

Since the radiation resistance is affected relatively little by l/d ratio, the decreased L/C ratio causes the Q of the antenna to decrease. This means that the change in antenna impedance with frequency will be less, increasing the antenna's SWR bandwidth. This is often used to advantage on the lower HF bands by using multiple conductors in a cage or fan to decrease the l/d ratio.

21.1.8 Radiation Patterns

Radiation patterns are graphic representations of an antenna's directivity. Two examples are given in **Figs 21.4** and **21.5**. Shown in polar coordinates (see the math references in the **Electrical Fundamentals** chapter for information about polar coordinates), the angular scale shows direction and the scale from the center of the plot to the outer ring, calibrated in dB, shows the relative strength of the antenna's radiated signal (gain) at each angle. A line is plotted showing the antenna's relative gain (transmitting and receiving) at each angle. The antenna is located at the exact center of the plot with its orientation specified separately.

The pattern is composed of *nulls* (angles at which a gain minimum occurs) and *lobes* (a range of angles in which a gain maximum occurs). The *main lobe* is the lobe with the highest amplitude unless noted otherwise and unless several plots are being compared, the peak amplitude of the main lobe is placed at the outer ring as a 0 dB reference point. The peak of the main lobe can be located at any angle. All other lobes are *side lobes* which can be at any angle, including to the rear of the antenna.

Fig 21.4 is an *azimuthal* or *azimuth pattern* that shows the antenna's gain in all horizontal directions (azimuths) around the antenna. As with a map, 0° is at the top and bearing angle increases clockwise. (This is different from polar plots generated for mathematical functions in which 0° is at the right and angle increases counter-clockwise.)

Fig 21.5 is an *elevation pattern* that shows the antenna's gain at all vertical angles. In this case, the horizon at 0° is located to both sides of the antenna and the zenith (directly overhead) at 90°. The plot shown in Fig 21.5 assumes a ground plane (drawn from 0° to 0°) but in free-space, the plot would include the missing semicircle with –90° at the bottom. Without the ground reference, the term "elevation" has little meaning, however.

You'll also encounter E-plane and H-plane radiation patterns. These show the antenna's radiation pattern in the plane parallel to the E-field or H-field of the antenna. It's important to remember that the E-plane and H-plane do not have a fixed relationship to the Earth's surface. For example, the E-plane pattern from a horizontal dipole is an azimuthal pattern, but if the same dipole is oriented vertically, the E-plane pattern becomes an elevation pattern.

Antenna radiation patterns can also be plotted on rectangular coordinates with gain on the vertical axis in dB and angle on the horizontal axis as shown in **Fig 21.6**. This is particularly useful when several antennas are being compared. Multiple patterns in polar coordinates can be difficult to read, particularly close to the center of the plot.

The amplitude scale of antenna patterns is almost always in dB. The scale rings can be calibrated in several ways. The most common is for the outer ring to represent the peak amplitude of the antenna's strongest lobe as 0 dB. All other points on the pattern represent *relative gain* to the peak gain. The antenna's *absolute gain* with respect to an isotropic (dBi) antenna or dipole (dBd) is printed as a label somewhere near the pattern. If several antenna radiation patterns are shown on the same plot for comparison, the pattern with the largest gain value is usually assigned the role of 0 dB reference.

The gain amplitude scale is usually divided in one of two ways. One common division is to have rings at 0, –3, –6, –12, –18, and –24 dB. This makes it easy to see where the gain has fallen to one-half of the reference or peak value (–3 dB), one-quarter (–6 dB), one-sixteenth (–12 dB), and so on. Another popular division of the amplitude scale is 0, –10, –20, –30, and –40 dB with intermediate rings or tick marks to show the –2, –4, –6, and –8 dB levels. You will encounter a number of variations on these basic scales.

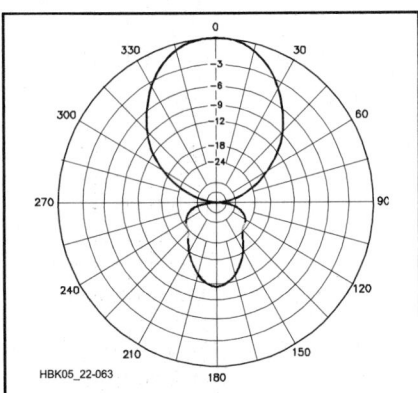

Fig 21.4 — Azimuthal pattern of a typical three-element Yagi beam antenna in free space. The Yagi's boom is along the 0° to 180° axis.

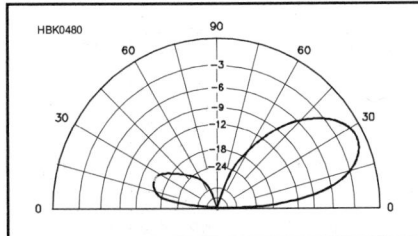

Fig 21.5 — Elevation pattern of a 3 element Yagi beam antenna placed ½ λ above perfect ground.

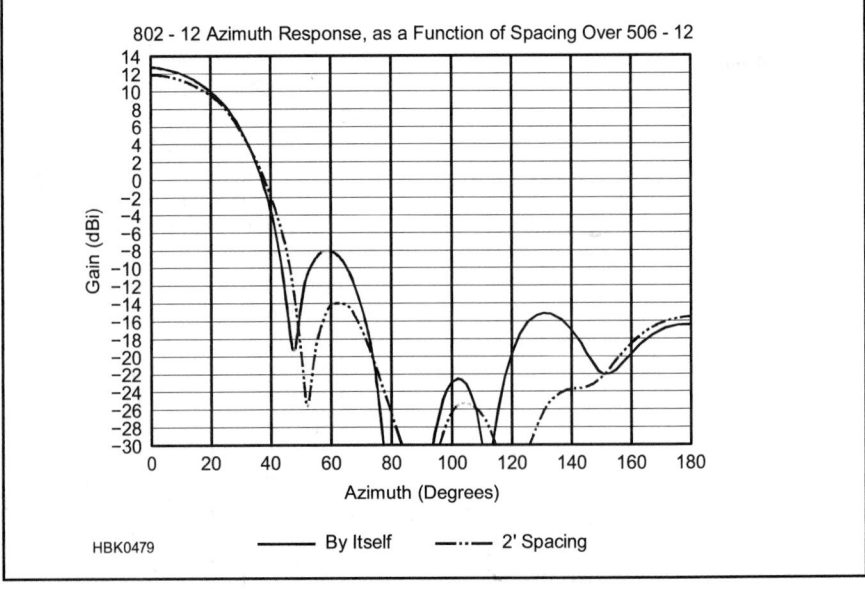

Fig 21.6 — Rectangular azimuthal pattern of an 8 element 2 meter Yagi beam antenna by itself and with another identical antenna stacked two feet above it. This example shows how a rectangular plot allows easier comparison of antenna patterns away from the main lobe.

RADIATION PATTERN MEASUREMENTS

Given the basic radiation pattern and scales, it becomes easy to define several useful measurements or metrics by which antennas are compared, using their azimuthal patterns. Next to gain, the most commonly-used metric for directional antennas is the *front-to-back ratio* (*F/B*) or just "front-to-back." This is the difference in dB between the antenna's gain in the specified "forward" direction and in the opposite or "back" direction. The front-to-back ratio of the antenna in Fig 21.4 is about 11 dB. *Front-to-side ratio* is also used and is the difference between the antenna's "forward" gain and gain at right angles to the forward direction. This assumes the radiation pattern is symmetric and is of most use to antennas such as Yagis and quads that have elements arranged in parallel planes. The front-to-side ratio of the antenna in Fig 21.4 is more than 30 dB. Because the antenna's rear-ward pattern can have large amplitude variations, the *front-to-rear ratio* is sometimes used. Front-to-rear uses the average of rear-ward gain over a specified angle, usually the 180° semicircle opposite the direction of the antenna's maximum gain, instead of a single gain figure at precisely 180° from the forward direction.

The antenna's *beamwidth* is the angle over which the antenna's main lobe gain is within 3 dB of the peak gain. Stated another way, the beamwidth is the angle between the directions at which the antenna's gain is –3 dB. In Fig 21.4, the antenna's main lobe beamwidth is about 54°, since the pattern crosses the –3 dB gain scale approximately 27° to either side of the peak direction. Antenna patterns with comparatively small beamwidths are referred to as "sharp" or "narrow."

An antenna with an azimuthal pattern that shows equal gain in all directions is called *omnidirectional*. This is not the same as an isotropic antenna that has equal gain in all directions, both vertical both horizontal.

21.1.9 Elevation Angle

For long-distance HF communication, the (vertical) *elevation angle* of maximum radiation, or *radiation angle*, is of considerable importance. You will want to erect your antenna so that its strongest radiation occurs at vertical angles resulting in the best performance at the distances over which you want to communicate. In general, the greater the height of a horizontally polarized antenna, the stronger its gain will be at lower vertical angles. **Fig 21.7** shows this effect at work in horizontal dipole antennas. (See the **Propagation**

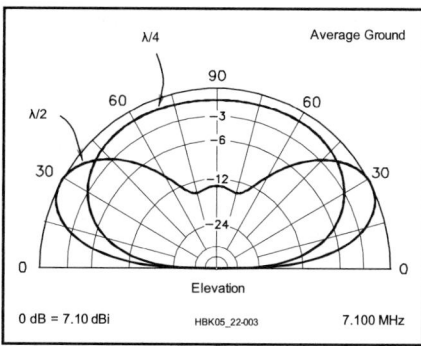

Fig 21.7 — Elevation patterns for two 40 meter dipoles over average ground (conductivity of 5 mS/m and dielectric constant of 13) at ¼ λ (33 foot) and ½ λ (66 foot) heights. The higher dipole has a peak gain of 7.1 dBi at an elevation angle of about 26°, while the lower dipole has more response at high elevation angles.

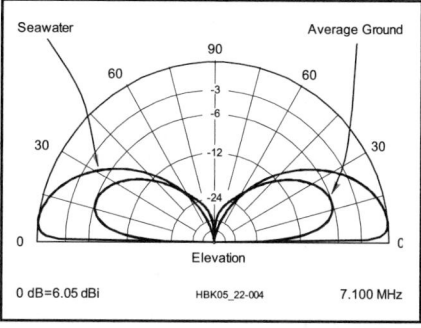

Fig 21.8 — Elevation patterns for a vertical dipole over sea water compared to average ground. In each case the center of the dipole is just over ¼ λ high. The low-angle response is greatly degraded over average ground compared to sea water, which is virtually a perfect ground.

chapter and the *ARRL Antenna Book* for more information about how to determine the best elevation angles for communication.)

Since low radiation angles usually are most effective for long distance communications, this generally means that horizontal antennas should be high — higher is usually better. (The optimum angle for intercontinental contacts on the HF bands is generally 15° or lower.) Experience shows that satisfactory results can be attained on the bands above 14 MHz with antenna heights between 40 and 70 feet.

Higher vertical angles can be useful for medium to short-range communications. For example, elevation angles between 20° and 65° are useful on the 40 and 80 meter bands over the roughly 550 mile path between Cleveland

and Boston. Even higher angles may be useful on shorter paths when using these lower HF frequencies. A 75 meter dipole between 30 and 70 feet high works well for ranges out to several hundred miles.

For even shorter-range communications centered on your location, such as for emergency communications and regional nets, a very low antenna is used, generating its strongest radiation straight up. This is referred to as *Near-Vertical Incidence Skywave* (*NVIS*) communication. The antenna should be less than ¼ λ above ground and the frequency used should be below the ionosphere's critical frequency so that the signal is completely reflected back toward the ground over a wide area.

Azimuthal patterns must also specify at what elevation angle the antenna gain is measured or calculated. While an azimuthal pattern may be in the plane of the antenna (an elevation angle of 0°), for antennas located above ground, the gain will vary strongly with elevation angle.

21.1.10 Imperfect Ground

Earth conducts, but is far from being a perfect conductor. This influences the radiation pattern of the antennas that we use. The effect is most pronounced at high vertical angles (the ones most important for short-range communications and least important for long-distance communications) for horizontal antennas. The consequences for vertical antennas are greatest at low angles, and are quite dramatic as can be clearly seen in **Fig 21.8**, where the elevation pattern for a 40 meter vertical half-wave dipole located over average ground is compared to one located over saltwater. At 10° elevation, the saltwater antenna has about 7 dB more gain than its landlocked counterpart.

An HF vertical antenna may work very well for a ham living in an area with rich soil. Ground of this type has very good conductivity. By contrast, a ham living where the soil is rocky or in a desert area may not be satisfied with the performance of a vertical HF antenna over such poorly conducting ground.

When evaluating or comparing antennas, it is also important to include the effects of ground on antenna gain. Depending on height above ground and the qualities of the ground, reflections can increase antenna gain by up to 6 dB. Because the actual installation of the antenna is unlikely to duplicate the environment in which the gain with reflections is claimed or measured, it is preferable to rely on free-space gain measurements that are independent of reflecting surfaces.

21.2 Dipoles and the Half-Wave Antenna

A fundamental form of antenna is a wire whose length is half the transmitting wavelength. It is the unit from which many more complex forms of antennas are constructed and is known as a *dipole antenna*. (The name di-meaning *two* and -pole meaning *electrical terminal* comes from the antenna having two distinct regions of electrical polarity as shown in Fig 21.1.) A dipole is resonant when it is electrically ½ λ long so that the current and voltage in the antenna are exactly 90° out of phase as shown in Fig 21.1.

The actual length of a resonant ½ λ antenna will not be exactly equal to the half wavelength of the radio wave of that frequency in free space, but depends on the thickness of the conductor in relation to the wavelength as shown in Fig 21.3. An additional shortening effect occurs with wire antennas supported by insulators at the ends because of current flow through the capacitance at the wire ends due to the end effect. Interaction with the ground and any nearby conductors also affects the resonant length of the physical antenna.

The following formula is sufficiently accurate for dipoles below 10 MHz at heights of ⅛ to ¼ λ and made of common wire sizes. To calculate the length of a half-wave antenna in feet,

$$\text{Length (ft)} = \frac{492 \times 0.95}{f\,(\text{MHz})} = \frac{468}{f\,(\text{MHz})} \tag{2}$$

Example: A half-wave antenna for 7150 kHz (7.15 MHz) is 468/7.15 = 65.5 feet, or 65 feet 6 inches.

For antennas at higher frequencies and/or higher above ground, a denominator value of 485 to 490 is more useful. In any case, be sure to include additional wire for attaching to insulators and be prepared to adjust the length of the antenna once installed in its intended position.

Above 30 MHz use the following formulas, particularly for antennas constructed from rod or tubing. K is taken from Fig 21.3.

$$\text{Length (ft)} = \frac{492 \times K}{f\,(\text{MHz})} \tag{3}$$

$$\text{Length (in)} = \frac{5904 \times K}{f\,(\text{MHz})} \tag{4}$$

Example: Find the length of a half-wave antenna at 50.1 MHz, if the antenna is made of ½ inch-diameter tubing. At 50.1 MHz, a half wavelength in space is

$$\frac{492}{50.1} = 9.82 \text{ ft}$$

The ratio of half wavelength to conductor diameter (changing wavelength to inches) is

$$\frac{(9.82 \text{ ft} \times 12 \text{ in./ft})}{0.5 \text{ in.}} = 235.7$$

From Fig 21.3, K = 0.945 for this ratio. The length of the antenna, from equation 3 is

$$\frac{492 \times 0.945}{50.1} = 9.28 \text{ ft}$$

or 9 feet 3⅜ inches. The answer is obtained directly in inches by substitution in equation 4

$$\frac{5904 \times 0.945}{50.1} = 111.4 \text{ in}$$

Regardless of the formula used to calculate the length of the half-wave antenna, the effects of ground and conductive objects within a wavelength or so of the antenna usually make it necessary to adjust the installed length in order to obtain the lowest SWR at the desired frequency. Use of antenna modeling software may provide a more accurate initial length than a single formula.

The value of the SWR indicates the quality of the match between the impedance of the antenna and the feed line. If the lowest SWR obtainable is too high, an impedance-matching network may be used, as described in the **Transmission Lines** chapter. (High SWR may cause modern transmitters with solid-state power amplifiers to reduce power output as a protective measure for the output transistors.)

21.2.1 Radiation Characteristics

The radiation pattern of a dipole antenna in free space is strongest at right angles to the wire as shown in **Fig 21.9**, a free-space radiation pattern. In an actual installation, the figure-8 pattern is less directive due to reflections from ground and other conducting surfaces. As the dipole is raised to ½ λ or greater above ground, nulls off the ends of

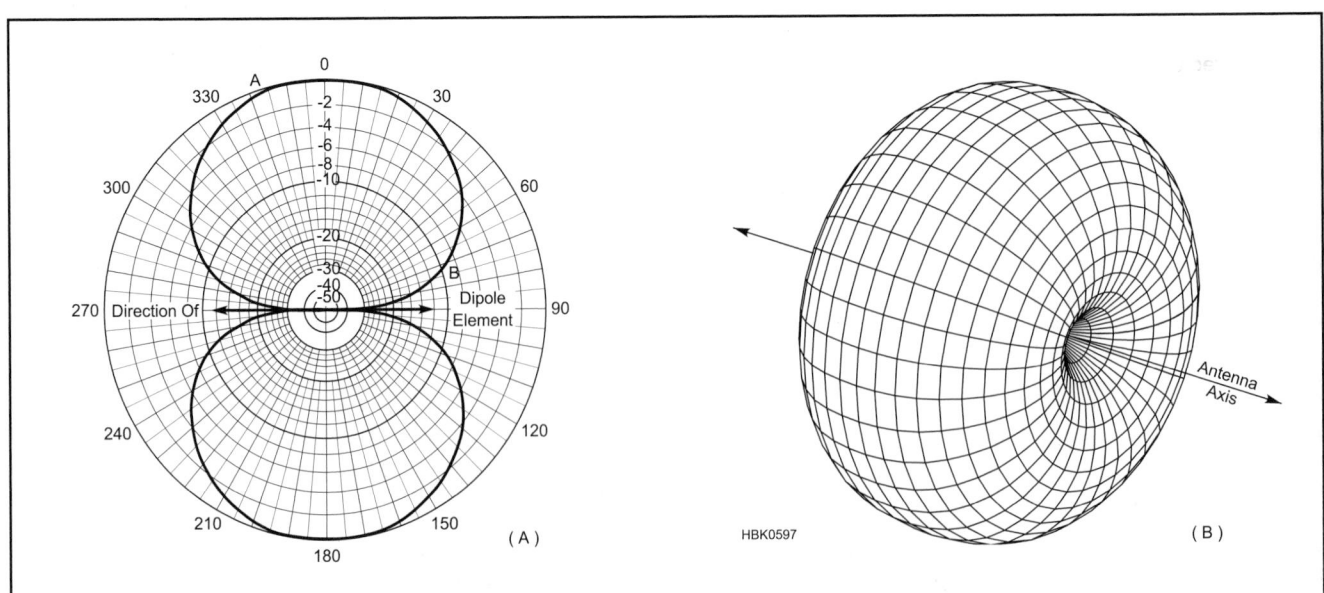

Fig 21.9 — Response of a dipole antenna in free space in the plane of the antenna with the antenna oriented along the 90° to 270° axis (A). The full three-dimensional pattern of the dipole is shown at (B). The pattern at A is a cross-section of the three-dimensional pattern taken perpendicularly to the axis of the antenna.

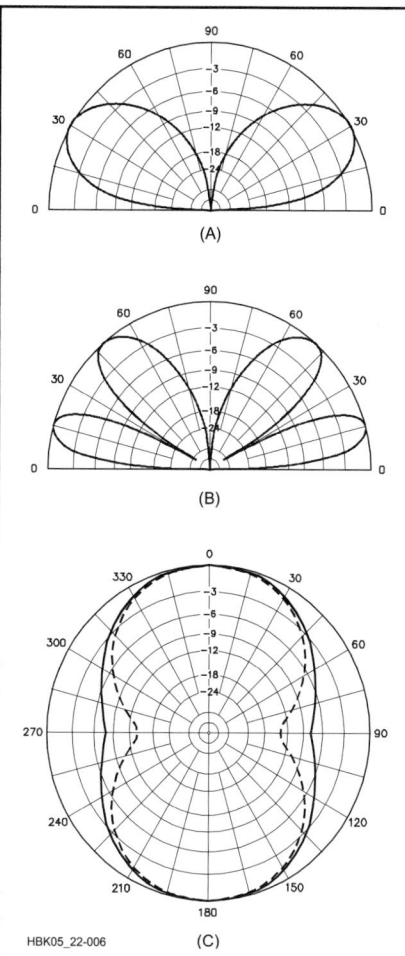

Fig 21.10 — At A, the elevation response pattern of a dipole antenna placed ½ λ above a perfectly conducting ground. At B, the pattern for the same antenna when raised to 1 λ. For both A and B, the conductor is coming out of the paper at a right angle. C shows the azimuth patterns of the dipole for the two heights at the most-favored elevation angle, the solid-line plot for the ½ λ height at an elevation angle of 30°, and the broken-line plot for the 1 λ height at an elevation angle of 15°. The conductor in C lies along the 90° to 270° axis.

the dipole become more pronounced. Sloping the antenna above ground and coupling to the feed line tend to distort the pattern slightly.

As a horizontal antenna is brought closer to ground, the elevation pattern peaks at a higher elevation angle as shown in Fig 21.7. **Fig 21.10** illustrates what happens to the directional pattern as antenna height changes. Fig 21.10C shows that there is significant radiation off the ends of a low horizontal dipole. For the ½ λ height (solid line), the radiation off the ends is only 7.6 dB lower than that in the broadside direction.

Fig 21.10 also shows that for short-range communication that depends on high vertical angles of radiation (NVIS communica-

tions), a dipole can be too high. For these applications, the dipole should be installed at or below ¼ λ so that the antenna radiates strongly at high vertical angles and with little horizontal directivity.

A classic dipole antenna is ½ λ long and fed at the center. The low feed point impedance at the dipole's resonant frequency, f_0, and its odd harmonics results in a low SWR when fed with coaxial cable feed lines. The feed point impedance and resulting SWR with coaxial cable will be high at even harmonics and other frequencies.

When fed with ladder line and a wide-range impedance-matching unit, such an antenna can be used on nearly any frequency, including non-resonant frequencies. (An example of such an antenna system is presented as a project farther along in this section.)

21.2.2 Feed Methods

The feed line is attached directly to the dipole, generally at the center, where an insulator separates the antenna's conductor into two sections. This is the antenna's *feed point*. One conductor of the feed line is attached to each section. **Figs 21.11** and **21.12** show how the two types of feed lines are attached. There are numerous variations, of course. You can make your own insulators from plastic or ceramic and there are many commercial insulators available, including some with built-in coax connectors for the feed point.

A dipole can be fed (feed line attached) anywhere along its length, although the impedance of the antenna will vary as discussed earlier. One common variation is the *off-center-fed (OCF)* dipole where the feed point is offset from center by some amount and an impedance transformer used to match the resulting moderately-high impedance to that of coaxial cable. Another variation, shown in Fig 21.12B, is the *end-fed Zepp*, named for its original application as an antenna deployed from Zeppelin airships. The feed point impedance of a "Zepp" is quite high, requiring open-wire feed line and impedance matching techniques to deliver power effectively.

Either *coaxial cable* ("coax") or *open-wire* transmission line or feed line is used to connect the transmitter and antenna. There are pro's and con's for each type of feed line. Coax is the common choice because it is readily available, its characteristic impedance is close to that of a center-fed dipole, and it may be easily routed through or along walls and among other cables. Where a very long feed line is required or the antenna is to be used at frequencies for which the feed point impedance is high, coax's increased RF loss and low working voltage (compared to that of open-wire line) make it a poor choice. Refer to the **Transmission Lines** and **Component Data and References** chapters for informa-

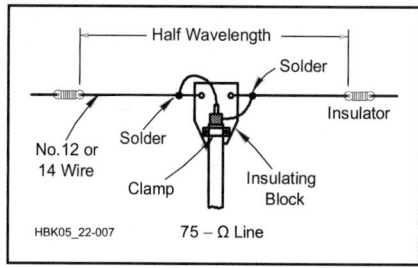

Fig 21.11 — Method of attaching feed line to the center of a dipole antenna. A plastic block is used as a center insulator. The coax is held in place by a clamp. A balun is often used to feed dipoles or other balanced antennas to minimize current on the outside of the feed line and reduce radiation pattern distortion. See text for explanation.

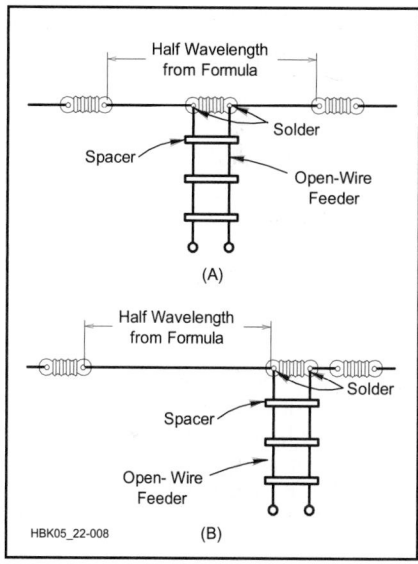

Fig 21.12 — Center-fed multiband Zepp antenna at A and an end-fed Zepp at B. See also Fig 21.15 for connection details.

tion that will help you evaluate the RF loss of coaxial cable at different lengths and SWR.

Respect coax's power-handling ratings. Cables such as RG-58 and RG-59 are suitable for power levels up to 300 W with low SWR. RG-8X cable can handle higher power and there are a number of variations of this type of cable. For legal-limit power or moderate SWR, use the larger diameter cable types, such as RG-8 or RG-213, that are 0.4 inches in diameter or larger. Subminiature cables, such as RG-174, are useful for very short lengths at low power levels, but the high RF losses associated with these cables make them unsuitable for most uses as antenna feed lines.

The most common open-wire transmission lines are *ladder line* (also known as *window line*) and *twin-lead*. Since the conductors are not shielded, two-wire lines are affected by their environment. Use standoffs and insulators to keep the line several inches from structures or other conductors. Ladder line has very low loss (twin-lead has a little more), and it can stand very high voltages (created by high SWR) as long as the insulators are clean. Twin-lead can be used at power levels up to 300 W and ladder line to the full legal power limit.

The characteristic impedance of open-wire line varies from 300 Ω for twin-lead to 450 to 600 Ω for most ladder and window line. When used with ½ λ dipoles, the resulting moderate to high SWR requires an impedance-matching unit at the transmitter. The low RF losses of open-wire lines make this an acceptable situation on the HF bands.

21.2.3 Baluns

Open-wire transmission lines and center-fed dipole antennas are *balanced*, that is, each conductor or section has the same impedance to earth ground. This is different from *unbalanced* coaxial cable, in which the shield is generally connected to an earth ground at some point, generally at the transmitter. To use balanced open-wire transmission lines with unbalanced equipment — most amateur equipment is unbalanced — a *balun* is required to make the transition between the balanced and unbalanced parts of the antenna system. "Balun" is an abbreviation of "balanced-to-unbalanced," the function of the device — it allows power to be transferred between the balanced and unbalanced portions of an antenna system in either direction. The most common application of baluns is to connect an unbalanced feed line to a balanced antenna.

Because dipoles are balanced, a balun is often used at the feed point when a dipole is fed with coax. Due to the skin effect discussed in the **RF Techniques** chapter, the inside and outside of the coaxial cable shield are separate conductors at RF. This "third conductor" of a coaxial cable unbalances the symmetry of the dipole antenna when the coax is connected directly to the dipole as shown in Fig 21.11. As a result, RF current can flow on the outside of the cable shield to the enclosures of station equipment connected to the cable.

Shield currents can impair the function of instruments connected to the line (such as SWR meters and SWR-protection circuits in the transmitter). The shield current also produces some feed line radiation, which changes the antenna radiation pattern, and allows objects near the cable to affect the antenna-system performance.

The consequences may be negligible: A slight skewing of the antenna pattern usually goes unnoticed. Or, they may be significant: False SWR readings may cause the transmitter to reduce power unnecessarily; radiating coax near a TV feed line

may cause strong local interference from overload. Therefore, it is better to eliminate feed line radiation whenever possible, and a balun should be used at any transition between balanced and unbalanced systems. Even so, balanced or unbalanced systems without a balun often operate with no apparent problems. For temporary or emergency stations, do not let the lack of a balun deter you from operating.

A balun can be constructed in a number of ways: the simplest being to coil several turns of coaxial cable at the antenna feed point. This creates inductance on the outer surface of the cable (the inner surface of the shield and the center conductor are unaffected) and the resulting reactance opposes RF current flow. There are other methods, such as the use of ferrite beads and cores, that are discussed in the **Transmission Lines** chapter.

21.2.4 Building Dipoles and Other Wire Antennas

The purpose of this section is to offer information on the actual physical construction of wire antennas. Because the dipole, in one of its configurations, is probably the most common amateur wire antenna, it is used in the following examples. The techniques described here, however, enhance the reliability and safety of all wire antennas.

WIRE

Choosing the right type of wire for the project at hand is the key to a successful antenna — the kind that works well and stays up through a winter ice storm or a gusty spring wind storm. What gauge of wire to use is the first question to settle; the answer depends on strength, ease of handling, cost, availability and visibility. Generally, antennas that are expected to support their own weight, plus the weight of the feed line should be made from #12 AWG wire. Horizontal dipoles, Zepps, some long wires and the like fall into this category. Antennas supported in the center, such as inverted-V dipoles and delta loops, may be made from lighter material, such as #14 AWG wire — the minimum size called for in the National Electrical Code.

The type of wire to be used is the next important decision. The wire specifications table in the **Component Data and References** chapter shows popular wire styles and sizes. The strongest wire suitable for antenna service is *copper-clad steel*, also known as *Copperweld*. The copper coating is necessary for RF service because steel is a relatively poor conductor. Practically all of the RF current is confined to the copper coating because of skin effect. Copper-clad steel is outstanding for permanent installations, but it can be difficult to work with because of the stiffness of the steel core.

Solid-copper wire, either hard-drawn or soft-drawn, is another popular material. Easier to handle than copper-clad steel, solid copper is available in a wide range of sizes. It is generally more expensive however, because it is all copper. Soft-drawn tends to stretch under tension, so periodic pruning of the antenna may be necessary in some cases. Enamel-coated *magnet-wire* is a suitable choice for experimental antennas because it is easy to manage, and the coating protects the wire from the weather. Although it stretches under tension, the wire may be pre-stretched before final installation and adjustment. A local electric motor rebuilder might be a good source for magnet wire.

Hook-up wire, speaker wire or even ac lamp cord are suitable for temporary installations. Almost any copper wire may be used, as long as it is strong enough for the demands of the installation.

Aluminum wire can be used for antennas, but is not as strong as copper or steel for the same diameter and soldering it to feed lines requires special techniques. Galvanized and steel wire, such as that used for electric fences, is inexpensive, but it is a much poorer conductor at RF than copper and should be avoided.

Kinking, which severely weakens wire, is a potential problem when handling any solid conductor. When uncoiling solid wire of any type — copper, steel, or aluminum — take care to unroll the wire or untangle it without pulling on a kink to straighten it. A kink is actually a very sharp twist in the wire and the wire will break at such a twist when flexed, such as from vibration in the wind.

Solid wire also tends to fail at connection or attachment points at which part of the wire is rigidly clamped. The repeated flexing from wind and other vibrations eventually causes metal fatigue and the wire breaks. Stranded wire is preferred for antennas that will be subjected to a lot of vibration and flexing. If stranded wire is not suitable, use a heavier gauge of solid wire to compensate.

Insulated vs Bare Wire

Losses are the same (in the HF region at least) whether the antenna wire is insulated or bare. If insulated wire is used, a 3 to 5% shortening from the length calculated for a bare wire is required to obtain resonance at the desired frequency. This is caused by the increased distributed capacitance resulting from the dielectric constant of the plastic insulating material. The actual length for resonance must be determined experimentally by pruning and measuring because the dielectric constant of the insulating material varies from wire to wire. Wires that might come into contact with humans or animals should be insulated to reduce the chance of shock or burns.

INSULATORS

Wire antennas must be insulated at the ends. Commercially available insulators are made from ceramic, glass or plastic. Insulators are available from many Amateur Radio dealers. RadioShack and local hardware stores are other possible sources.

Acceptable homemade insulators may be fashioned from a variety of material including (but not limited to) acrylic sheet or rod, PVC tubing, wood, fiberglass rod or even stiff plastic from a discarded container. **Fig 21.13** shows some homemade insulators. Ceramic or glass insulators will usually outlast the wire, so they are highly recommended for a safe, reliable, permanent installation. Other materials may tear under stress or break down in the presence of sunlight. Many types of plastic do not weather well.

Most wire antennas require an insulator at the feed point. Although there are many ways to connect the feed line, there are a few things to keep in mind. If you feed your antenna with coaxial cable, you have two choices. You can install an SO-239 connector on the center insulator and use a PL-259 on the end of your coax, or you can separate the center conductor from the braid and connect the feed line directly to the antenna wire. Although it costs less to connect direct, the use of connectors offers several advantages.

Coaxial cable braid acts as a wick to soak up water. If you do not adequately seal the antenna end of the feed line, water will find its way into the braid. Water in the feed line will lead to contamination, rendering the coax useless long before its normal lifetime is up. It is not uncommon for water to drip from the end of the coax inside the shack after a year or so of service if the antenna connection is not properly waterproofed. Use of a PL-259/SO-239 combination (or other connector of your choice) makes the task of waterproofing connections much easier. Another advantage to using the PL-259/SO-239 combination is that feed line replacement is much easier, should that become necessary or desirable.

Whether you use coaxial cable, ladder line, or twin lead to feed your antenna, an often-overlooked consideration is the mechanical strength of the connection. Wire antennas and feed lines tend to move a lot in the breeze, and unless the feed line is attached securely, the connection will weaken with time. The resulting failure can range from a frustrating intermittent electrical connection to a complete separation of feed line and antenna. **Fig 21.14** illustrates several different ways of attaching the feed line to the antenna. An idea for supporting ladder line is shown in **Fig 21.15**.

PUTTING IT TOGETHER

Fig 21.16 shows details of antenna construction. Although a dipole is used for

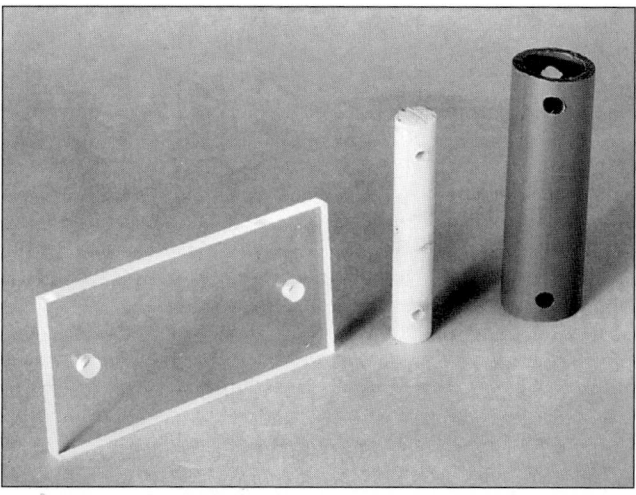

Fig 21.13 — Some ideas for homemade antenna insulators.

Fig 21.14 — Some homemade dipole center insulators. The one in the center includes a built-in SO-239 connector. Others are designed for direct connection to the feed line.

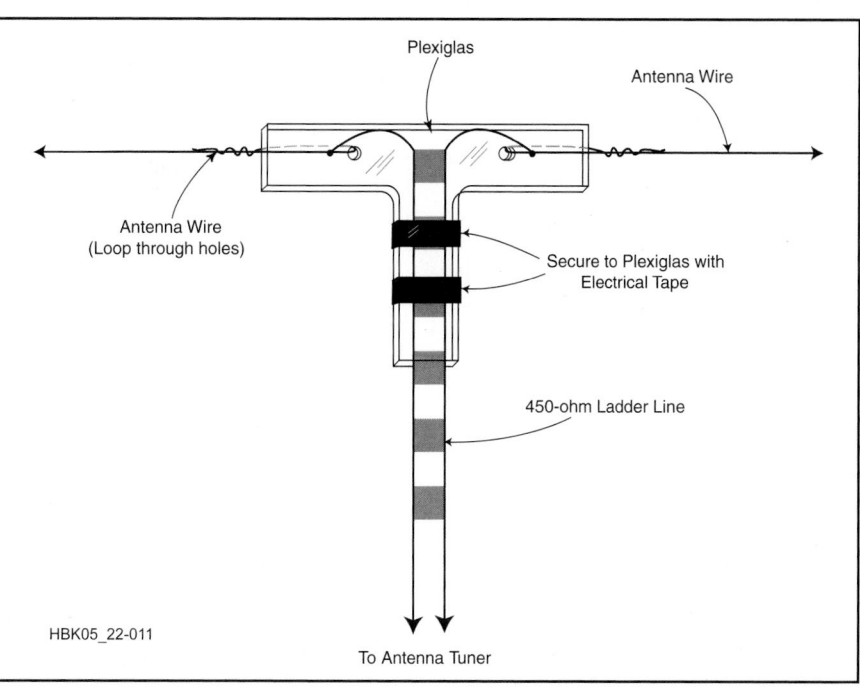

Fig 21.15 — A piece of cut Plexiglas can be used as a center insulator and to support a ladder-line feeder. The Plexiglas acts to reduce the flexing of the wires where they connect to the antenna.

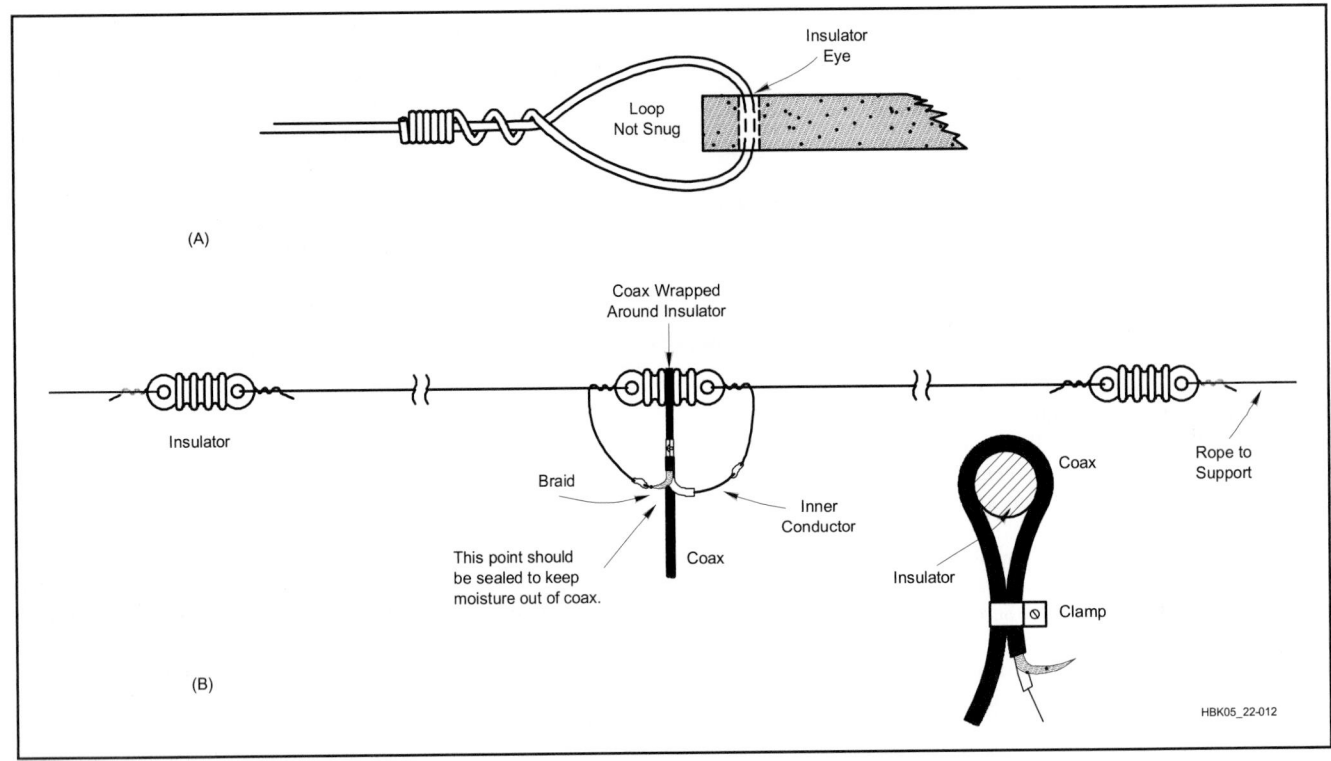

Fig 21.16 — Details of dipole antenna construction. The end insulator connection is shown at A, while B illustrates the completed antenna. This is a balanced antenna and is often fed with a balun at the feed point as described in the text.

the examples, the techniques illustrated here apply to any type of wire antenna. **Table 21.1** shows dipole lengths for the amateur HF bands. These lengths do not include the extra wire required to attach the wire to the insulator as shown in Fig 21.16. Determine the extra amount of wire required by experimenting with the insulator you intend to use. Add twice this amount of wire to the leg lengths in Table 21.1, one extra length for each insulator.

Most dipoles require a little pruning to reach the desired resonant frequency due to the effects of ground and nearby conducting objects and surfaces. So, cut the wire to result in a constructed length 2 to 3% longer than the calculated or table length and record the con-structed length with all insulators attached. (The constructed length is measured between the ends of the loops at each end of the wire.) Next, raise the dipole to the working height and find the frequency at which minimum SWR occurs. Multiply the frequency of the SWR minimum by the antenna length and divide the result by the desired f_0. The result is the finished length; trim both ends equally to reach that length and you're done. For example, if you want the SWR minimum to occur at 14.1 MHz and the first attempt with a constructed length of 33.8 feet results in an SWR minimum at 13.9 MHz, the final length for the antenna is

$$\frac{13.9}{14.1} \times 33.8 = 33.3 \text{ ft}$$

In determining how well your antenna will work over the long term, how well you put the pieces together is second only to the ultimate strength of the materials used. Even the smallest details, such as how you connect the wire to the insulators (Fig 21.16A), contribute significantly to antenna longevity. By using plenty of wire at the insulator and wrapping it tightly, you will decrease the possibility of the wire pulling loose in the wind. There is no need to solder the wire once it is wrapped. There is no electrical connection here, only mechanical. The high heat needed for soldering can anneal the wire, significantly weakening it at the solder point.

Similarly, the feed line connection at the center insulator should be made to the antenna wires after they have been secured to the insulator (Fig 21.16B). This way, you will be assured of a good electrical connection between the antenna and feed line without compromising the mechanical strength. Do a good job of soldering the antenna and feed line connections. Use a heavy iron or a torch, and be sure to clean the materials thoroughly before starting the job. If possible, solder the connections at a workbench, where the best possible joints may be made. Poorly soldered or unsoldered connections will become headaches as the wire oxidizes and the electrical integrity degrades with time. Besides degrading your antenna performance, poorly made joints can even be a cause of TVI because of rectification. Spray the connections with a UV-resistant acrylic coating for waterproofing.

So that the antenna stays up after installation, keep it away from tree branches and other objects that might rub or fall on the antenna. If the supports for the antenna move in the wind, such as trees, leave enough slack in the antenna that it is not pulled overly tight in normal winds. Other options are to use pulleys and counterweights to allow the antenna supports to flex without pulling on the antenna. (This and other installation topics are covered in the *ARRL Antenna Book*.)

If made from the right materials and installed in the clear, the dipole should give

Table 21.1
Dipole Dimensions for Amateur Bands

Freq	Overall Length		Leg Length	
(MHz)	ft	in	ft	in
28.4	16	6	8	3
24.9	18	9½	9	4¾
21.1	22	2	11	1
18.1	25	10	12	11
14.1	33	2	16	7
10.1	46	4	23	2
7.1	65	10	32	11
5.37	87	2	43	7
3.6	130	0	65	0

years of maintenance-free service. As you build your antenna, keep in mind that if you get it right the first time, you won't have to do it again for a long time.

21.2.5 Dipole Orientation

Dipole antennas need not be installed horizontally and in a straight line. They are generally tolerant of bending, sloping or drooping. Bent dipoles may be used where antenna space is at a premium. **Fig 21.17** shows a couple of possibilities; there are many more. Bending distorts the radiation pattern somewhat and may affect the impedance as well, but compromises may be acceptable when the situation demands them. When an antenna bends back on itself (as in Fig 21.17B) some of the signal is canceled; avoid this if possible. Remember that dipole antennas are RF conductors. For safety's sake, mount all antennas away from conductors (especially power lines), combustibles and well beyond the reach of passersby.

21.2.6 Inverted-V Dipole

An *inverted-V* dipole is supported at the center with a single support, such as a tree or mast. While *V* describes the shape of this antenna, this antenna should not be confused with long-wire horizontal-V antennas, which are highly directive.

The inverted-V's radiation pattern and feed point impedance depend on the *apex angle* between the legs: As the apex angle decreases, so does feed point impedance, and the radiation pattern becomes less directive. At apex angles below 90°, the antenna efficiency begins to decrease, as well.

The proximity of ground to the antenna ends will lower the resonant frequency of the antenna so that a dipole may have to be shortened in the inverted-V configuration. Losses in the ground increase when the antenna ends are close to the ground. Keeping the ends eight feet or higher above ground reduces ground loss and also prevents humans and animals from coming in contact with the antenna.

Remember that antenna current produces the radiated signal, and current is maximum at the dipole center. Therefore, performance is best when the central area of the antenna is high and clear of nearby objects.

21.2.7 Sloping Dipole

A sloping dipole is shown in **Fig 21.18**. This antenna is often used to favor one direction (the *forward direction* in the figure). With a non-conducting support and poor ground, signals off the back are weaker than those off the front. With a non-conducting mast and good ground, the response is omnidirectional.

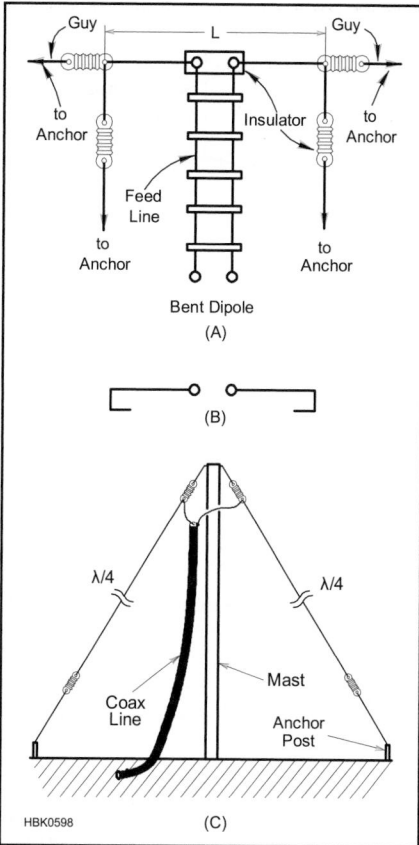

Fig 21.17 — When limited space is available for a dipole antenna, the ends can be bent downward as shown at A, or back on the radiator as shown at B. The inverted-V at C can be erected with the ends bent parallel with the ground when the available supporting structure is not high enough.

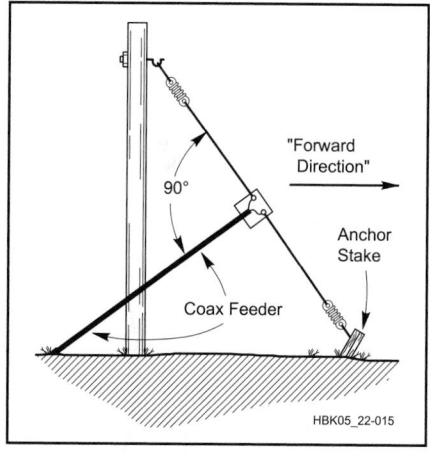

Fig 21.18 — Example of a sloping 1/2 λ dipole, or *full sloper*. On the lower HF bands, maximum radiation over poor to average earth is off the sides and in the *forward direction* as indicated, if a non-conductive support is used. A metal support will alter this pattern by acting as a parasitic element. How it alters the pattern is a complex issue depending on the electrical height of the mast, what other antennas are located on the mast and on the configuration of any guy wires.

There is no gain in any direction with a non-conducting mast.

A conductive support such as a tower acts as a parasitic element. (So does the coax shield, unless it is routed at 90° from the antenna.) The parasitic effects vary with ground quality, support height and other conductors on the support (such as a beam at the top or other wire antennas). With such variables, performance is very difficult to predict.

Losses increase as the antenna ends approach the support or the ground, so the same cautions about the height of the antenna ends applies as for the inverted-V antenna. To prevent feed line radiation, route the coax away from the feed point at 90° from the antenna as far as possible.

21.2.8 Shortened Dipoles

Inductive loading increases the electrical length of a conductor without increasing its physical length. Therefore, we can build physically-short dipole antennas by placing inductors in the antenna. These are called *loaded dipoles*, and *The ARRL Antenna Book* shows how to design them. There are some trade-offs involved: Inductively loaded antennas are less efficient and have narrower bandwidths than full-size antennas. Generally they should not be shortened more than 50%.

21.2.9 Half-Wave Vertical Dipole (HVD)

Unlike its horizontal counterpart, which has a figure-8 pattern, the azimuthal pattern of a vertical dipole is omnidirectional. In other words, it looks like a circle. Look again at Figs 21.7 and 21.8 and note the comparison between horizontal and vertical dipole elevation patterns. These two figures illustrate the fact that performance of a horizontal dipole depends to a great extent on its height above ground. By contrast, *half-wave vertical dipole (HVD)* performance is highly dependent on ground conductivity and dielectric constant.

After looking at these figures, you might easily conclude that there is no advantage to an HVD. Is that really the case? Experiments at K8CH between 2001 and 2003 showed that the HVD mounted above average ground works well for long-distance (DX) contacts. Two antennas were used in the trials. The first was a 15 meter HVD with its base 14 feet above ground (feed point at 25 feet). The second (reference) antenna was a 40 meter inverted-V modified to operate with low SWR on 15 meters. The reference antenna's feed point was at 29 feet and the ends drooped slightly for an apex angle of 160°. Signals from outside North America were usually stronger on the HVD. Antenna modeling revealed the reasons for this behavior.

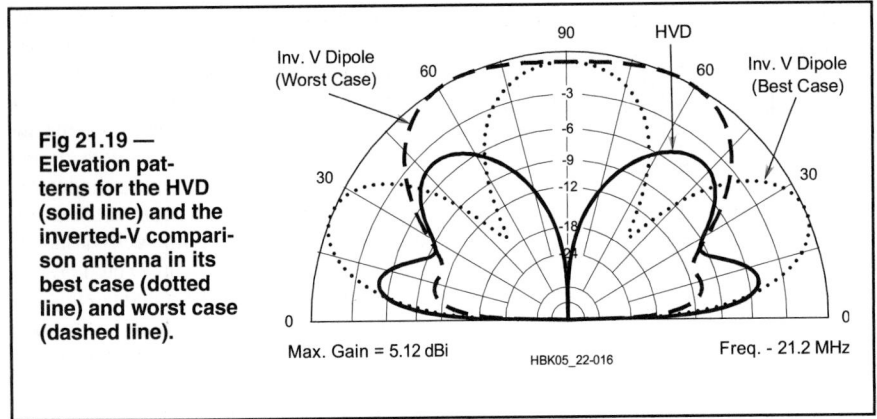

Fig 21.19 —
Elevation patterns for the HVD (solid line) and the inverted-V comparison antenna in its best case (dotted line) and worst case (dashed line).

Max. Gain = 5.12 dBi
HBK05_22-016
Freq. - 21.2 MHz

radiation resistance *and* shorten the antenna. It is possible to make a loaded vertical dipole that is half the height of an HVD and that has a good SWR when fed with 50 Ω coax.

21.2.10 Folded Dipoles

Fig 21.22 shows a *folded dipole* constructed from open-wire transmission line. The dipole is made from a ½ λ section of open-wire line with the two conductors connected together at each end of the antenna. The top conductor of the open-wire length is continuous from end to end. The lower conductor, however, is cut in the middle and the feed line attached at that point. Open-wire transmission line is

Fig 21.19 shows the elevation patterns for the vertical dipole and for the reference dipole at a pattern peak and at a null. The vertical dipole does not look impressive, does it? The large lobe in the HVD pattern at 48° is caused by the antenna being elevated 14 feet above ground. This lobe will shrink at lower heights.

Recall that the optimum elevation angles for long-distance contacts fall below 15°. Azimuthal patterns for the HVD at an elevation angle of 10° are shown in **Fig 21.20** and for 3° in **Fig 21.21**. You can clearly see in the patterns the DX potential of an HVD.

Another advantage of the HVD is its radiation resistance at low heights. Look back at Fig 21.2 at the curve for the vertical half-wave antenna. With its base just above ground, the HVD will have a radiation resistance of over 90 Ω. That can easily be turned to an advantage. Capacitive loading will lower the

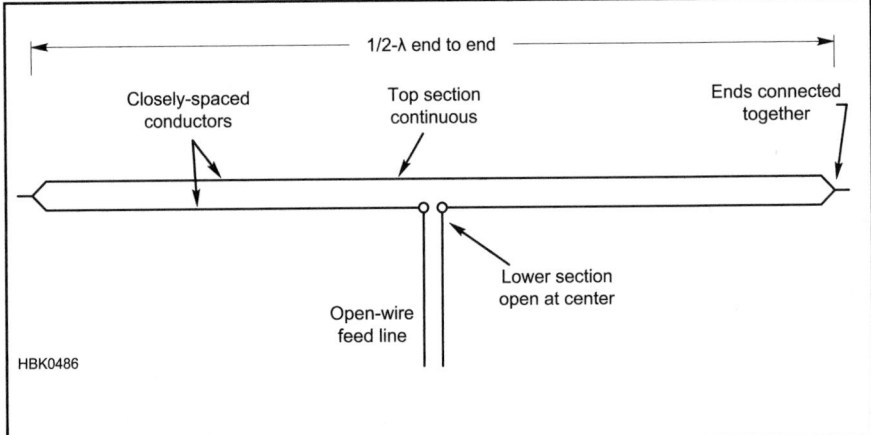

Fig 21.22 — The folded dipole is constructed from open-wire transmission line with the ends connected together. The close proximity of the two conductors and the resulting coupling act as an impedance transformer to raise the feed point impedance over that of a single-wire dipole by the square of the number of conductors used.

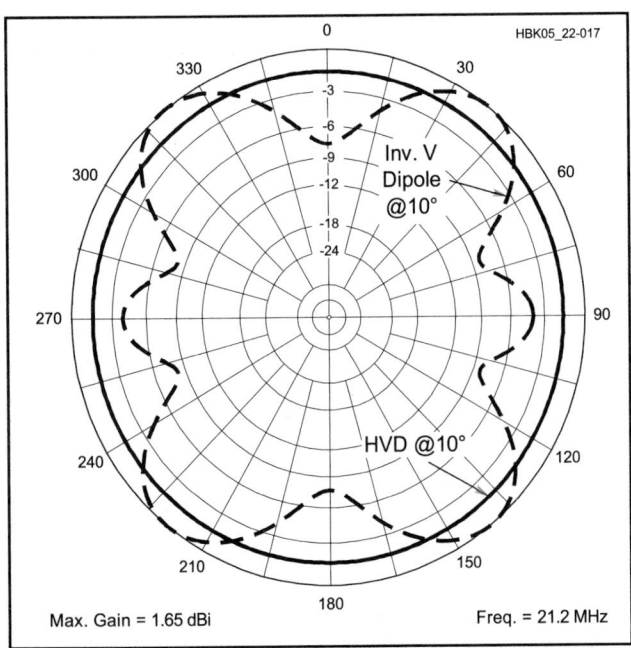

Max. Gain = 1.65 dBi
Freq. = 21.2 MHz

Fig 21.20 — Azimuth patterns at 10° elevation for the HVD (solid line) and inverted-V (dashed line).

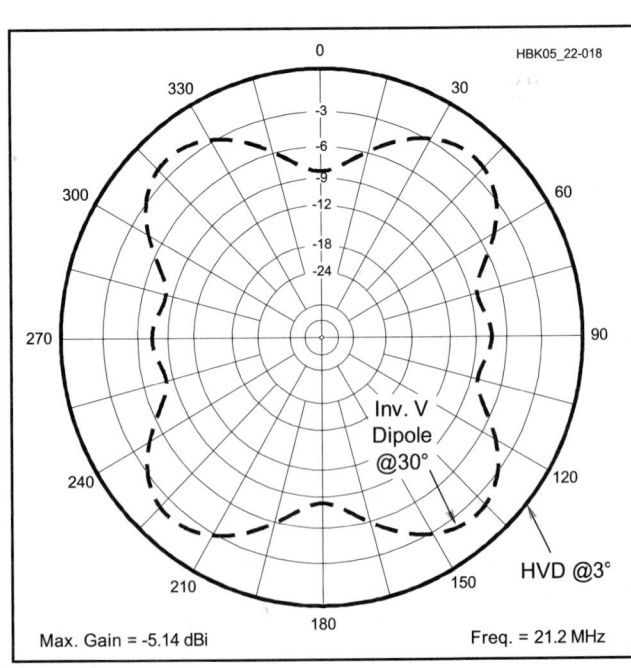

Max. Gain = -5.14 dBi
Freq. = 21.2 MHz

Fig 21.21 — Azimuth patterns at 3° elevation for the HVD (solid line) and the inverted-V (dashed line).

Turn a Horizontal Antenna Vertical

An option for adding at least one more band to a flat-top or inverted-V dipole is to turn it into a flat-top T vertical antenna. (See the T and Inverted-L Antennas section.) To do this, disconnect the feed line at the ground level, short the feed line conductors together, and connect them to the transmitter as a single wire. The remaining connection to the transmitter should be connected to a ground rod (as shown), counterpoise, or system of ground radials. (The antenna system's safety ground connection is still required.) A coaxial feed line is shown but the same technique works just as well for open-wire feed lines. An antenna tuner is required for either type of feed line.

This technique often allows a dipole to be used effectively at frequencies below those at which its horizontal section is resonant. For example, a 40 meter dipole can be used this way on 160 and 80 meters.

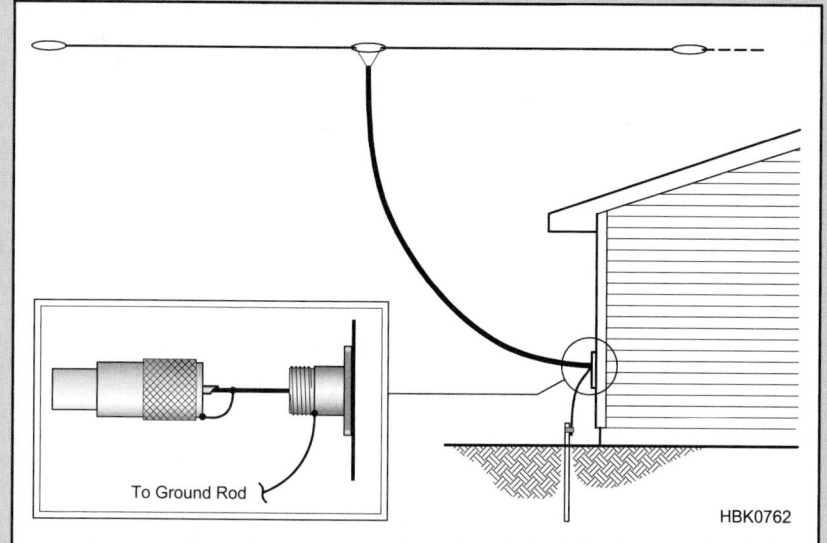

HBK0762

Fig 21.A1 – A dipole can be fed as a flat-top T vertical antenna by reconfiguring the feed line connections and exciting the antenna against ground.

A variation of the folded dipole called the *twin-folded terminated dipole* (*TFTD*) adds a resistor in the top conductor. Values of 300 to 600 Ω are used. The function of the resistor is to act as a *swamping* load, reducing the higher feed point impedances over a wide frequency range. A TFTD ½ λ long at 80 meters can be constructed to cover the entire 2 to 30 MHz range with SWR of 3:1 or less. The resistor dissipates some of the transmitter power (more than 50% at some frequencies!), but the improvement in SWR allows a coaxial feed line to be used without an impedance-matching unit. The increased convenience and installation outweigh the reduction in radiated signal. TFTD antennas are popular for emergency communications and where only a single HF antenna can be installed and high performance is not required.

21.2.11 Multiband Dipole Systems

There are several ways to construct coax-fed multiband dipole systems. These techniques apply to dipoles of all orientations. Each method requires a little more work than a single dipole, but the materials don't cost much.

PARALLEL DIPOLES

Parallel dipoles as shown in **Fig 21.23** are a simple and convenient answer. Center-fed dipoles have low feed point impedances near f_0 and its odd harmonics, and high impedances at other frequencies. This lets us construct

then used to connect the transmitter.

A folded dipole has exactly the same gain and radiation pattern as a single-wire dipole. However, because of the mutual coupling between the upper and lower conductors, the feed point impedance of a single-wire dipole is multiplied by the square of the number of conductors in the antenna. In this case, there are two conductors in the antenna, so the feed point impedance is $2^2 = 4$ times that of a single-wire dipole. (A three-wire folded dipole would have a nine times higher feed point impedance and so forth.)

A common use of the folded dipole is to raise the feed point impedance of the antenna to present a better impedance match to high impedance feed line. For example, if a very long feed line to a dipole is required, open-wire feed line would be used. By raising the dipole's feed point impedance, the SWR on the open-wire line is reduced from that of a single-wire dipole fed with open-wire feed line.

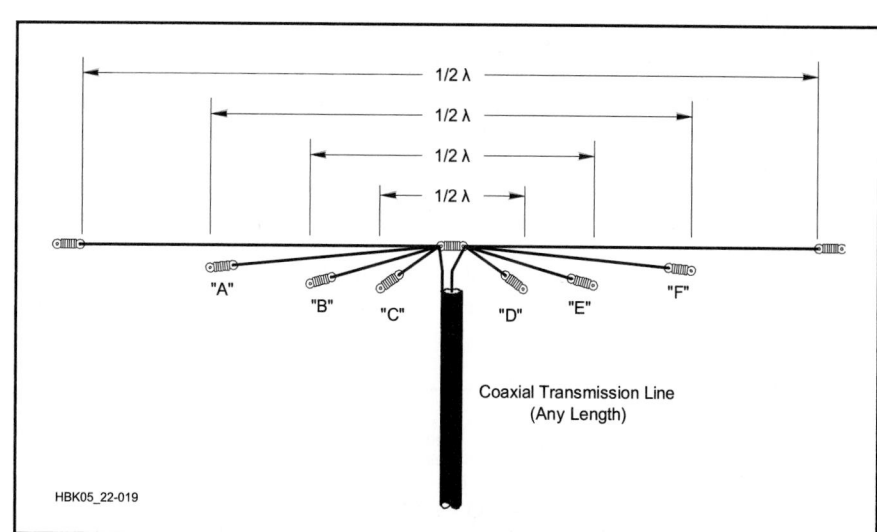

HBK05_22-019

Fig 21.23 — Multiband antenna using paralleled dipoles, all connected to a common 50 or 75 Ω coax line. The ½ λ dimensions may be either for the centers of the various bands or selected for favorite frequencies in each band. Be prepared to adjust the length of the various elements — both longer and shorter — because of interaction among them. See text.

simple multiband systems that automatically select the appropriate antenna. Consider a 50 Ω resistor connected in parallel with a 5 kΩ resistor. A generator connected across the two resistors will see 49.5 Ω, and 99% of the current will flow through the 50 Ω resistor. When resonant and non-resonant antennas are connected in parallel, the same result occurs: The non-resonant antenna has a high impedance, so little current flows in it and it has little effect on the total feed point impedance. Thus, we can connect several dipoles together at the feed point, and power naturally flows to the resonant antenna.

There are some limits, however. Wires in close proximity tend to couple due to mutual inductance. In parallel dipoles, this means that the resonant length of the shorter dipoles lengthens a few percent. Shorter antennas don't affect longer ones much, so adjust for resonance in order from longest to shortest. Mutual inductance also reduces the bandwidth of shorter dipoles, so an impedance-matching unit may be needed to achieve an acceptable SWR across all bands covered. These effects can be reduced by spreading the ends of the dipoles apart.

Also, the power-distribution mechanism requires that only one of the parallel dipoles is near resonance on any amateur band. Separate dipoles for 80 and 30 meters should not be connected in parallel because the higher band is near an odd harmonic of the lower band (80/3 ≈ 30) and center-fed dipoles have low impedance near odd harmonics. (The 40 and 15 meter bands have a similar relationship.) This means that you must either accept the performance of the low-band antenna operating on a harmonic or erect a separate antenna for those odd-harmonic bands. For example, four parallel-connected dipoles cut for 80, 40, 20 and 10 meters (fed by a single impedance-matching unit and coaxial cable) work reasonably on all HF bands from 80 through 10 meters.

TRAP DIPOLES

Trap dipoles (also called "trapped dipoles") provide multiband operation from a coax-fed single-wire dipole. **Fig 21.24** shows a two-band trap antenna. A trap consists of inductance and capacitance in parallel with a resonant frequency on the higher of the two bands of operation. The high impedance of the trap at its resonant frequency effectively disconnects the wire beyond the trap. Thus, on the higher of the two bands of operation at which traps are resonant, only the portion of the antenna between the traps is active.

Above resonance, the trap presents a capacitive reactance. Below resonance, the trap is inductive. On the lower of the two bands of operation, then, the inductive reactance of the trap acts as a loading coil to create a shortened or loaded dipole with the wire beyond the trap.

Traps may be constructed from coiled sections of coax or from discrete inductors and capacitors. (Traps are also available commercially.) Choose capacitors (C1 in the figure) that are rated for high current and voltage. Mica transmitting capacitors are good. Ceramic transmitting capacitors may work, but their values may change with temperature. Use large wire for the inductors to reduce loss. Any reactance (X_L and X_C) above 100 Ω (at f_0) will work, but bandwidth increases with reactance (up to several thousand ohms). Check trap resonance before installation. This can be done with a grid-dip meter and a receiver or with an SWR analyzer or impedance bridge.

To construct a trap antenna, build a dipole for the higher band of operation and connect the pre-tuned traps to its ends. It is fairly complicated to calculate the additional wire needed for each band, so just add enough wire to make the antenna ½ λ long on the lower band of operation, pruning it as necessary. Because the inductance in each trap reduces the physical length needed for resonance, the finished antenna will be shorter than a simple ½ λ dipole on the lower band.

21.2.12 NVIS Antennas

The use of very low dipole antennas that radiate at very high elevation angles has become popular in emergency communications ("emcomm") systems. This works at low frequencies (7 MHz and below) that are lower than the ionosphere's critical frequency — the highest frequency for which a signal traveling vertically will be reflected. (See the Propagation chapter.) The most common band for NVIS communication is 75 meters because the critical frequency is almost always above 4 MHz.

No special antenna construction techniques are required for NVIS antennas — just build a ½ λ dipole and install it at a height of 0.25-λ or below. 0.1 λ is often cited as the optimum height for NVIS antennas, but it is not critical.

At these low heights, the dipole's resonant frequency will be reduced because of the effects of ground. Shortening the antenna will restore the desired resonant frequency, although feed point impedance will drop.

Project: Multiband Center-Fed Dipole

An 80 meter dipole fed with ladder line is a versatile antenna. If you add a wide-range matching network, you have a low-cost antenna system that works well across the entire HF spectrum, and even 6 meters. Countless hams have used one of these in single-antenna stations and for Field Day operations.

Fig 21.25A shows a typical installation

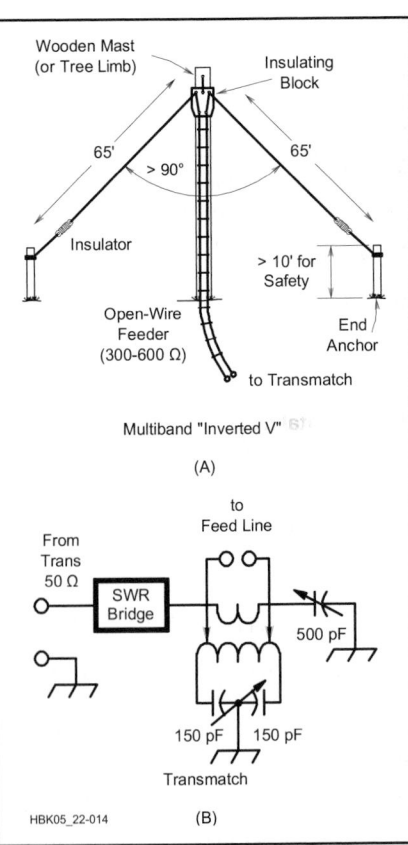

Fig 21.25 — At A, details for an inverted-V fed with open-wire line for multiband HF operation. An impedance-matching unit is shown at B, suitable for matching the antenna to the transmitter over a wide frequency range. The included angle between the two legs should be greater than 90° for best performance.

Fig 21.24 — Example of a trap dipole antenna. L1 and C1 can be tuned to the desired frequency by means of a grid-dip meter or SWR analyzer *before* they are installed in the antenna.

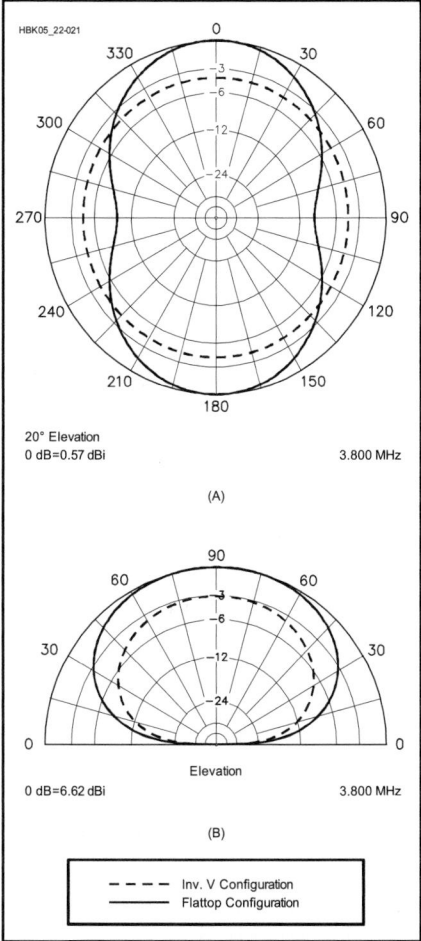

Fig 21.26 — Patterns on 80 meters for 135 foot, center-fed dipole erected as a horizontal dipole at 50 feet, and as an inverted-V with the center at 50 feet and the ends at 10 feet. The azimuth pattern is shown at A, where the conductor lies in the 90° to 270° plane. The elevation pattern is shown at B, where the conductor comes out of paper at a right angle. At the fundamental frequency the patterns are not markedly different.

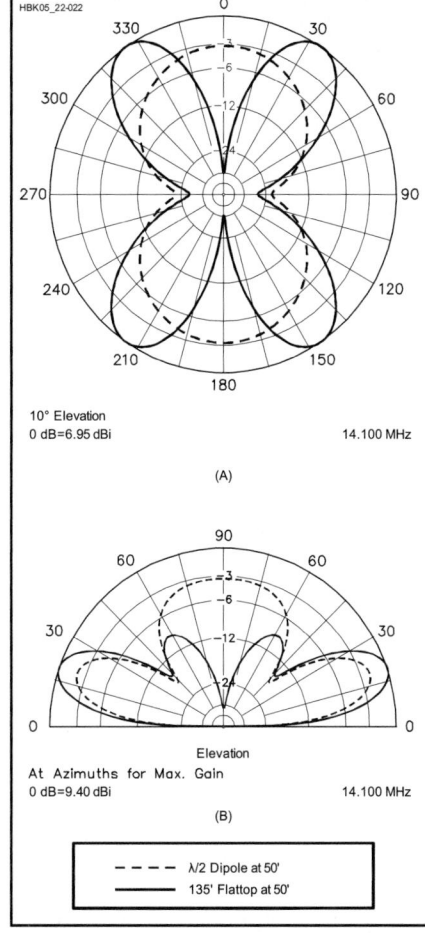

Fig 21.27 — Patterns on 20 meters comparing a standard ½ λ dipole and a multi-band 135 foot dipole. Both are mounted horizontally at 50 feet. The azimuth pattern is shown at A, where conductors lie in the 90° to 270° plane. The elevation pattern is shown at B. The longer antenna has four azimuthal lobes, centered at 35°, 145°, 215°, and 325°. Each is about 2 dB stronger than the main lobes of the ½ λ dipole. The elevation pattern of the 135 foot dipole is for one of the four maximum-gain azimuth lobes, while the elevation pattern for the ½ λ dipole is for the 0° azimuthal point.

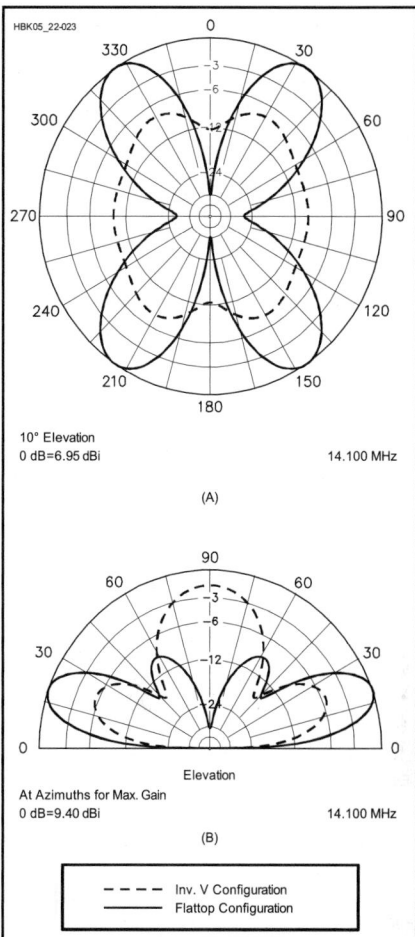

Fig 21.28 — Patterns on 20 meters for two 135 foot dipoles. One is mounted horizontally as a flat-top and the other as an inverted-V with 120° included angle between the two legs. The azimuth pattern is shown at A, and the elevation pattern is shown at B. The inverted-V has about 6 dB less gain at the peak azimuths, but has a more uniform, almost omnidirectional, azimuthal pattern. In the elevation plane, the inverted-V has a fat lobe overhead, making it a somewhat better antenna for local communication, but not quite so good for DX contacts at low elevation angles.

for such an antenna. The inverted-V configuration is shown, lowering the total antenna length to 130 feet from the 135 feet used if the entire antenna is horizontal. Either configuration will work well. Fig 21.25B shows the schematic of an impedance-matching unit or "antenna tuner" that you can build yourself. You can also use a balanced impedance-matching unit with a balun between it and the transmitter. Many amateurs are successful in using unbalanced impedance-matching units with a balun at either the output or the input of the tuner. Don't be afraid to experiment!

This configuration is popular with other lengths for the antenna:

- 105 feet — 80 through 10 meters

- 88 feet — 80 through 10 meters
- The next lower band may also be covered if the impedance-matching unit has sufficient range, although the adjustment will be fairly sharp. Six meter coverage is possible, but depends on the station layout, length of feed line, and impedance-matching unit abilities. Again, don't be afraid to experiment!

For best results place the antenna as high as you can, and keep the antenna and ladder line clear of metal and other conductive objects. Despite significant SWR on some bands, the open-wire feed line keeps system losses low as described in the **Transmission Lines** chapter.

ARRL staff analyzed a 135 foot dipole at

50 feet above typical ground and compared that to an inverted-V with the center at 50 feet, and the ends at 10 feet. The results show that on the 80 meter band, it won't make much difference which configuration you choose. (See **Fig 21.26**.) The inverted-V exhibits additional losses because of its proximity to ground.

Fig 21.27 shows a comparison between a 20 meter flat-top dipole and the 135 foot flat-top dipole when both are placed at 50 feet above ground. At a 10° elevation angle, the 135 foot dipole has a gain advantage. This advantage comes at the cost of two deep, but narrow, nulls that are broadside to the wire.

Fig 21.28 compares the 135 foot dipole to the inverted-V configuration of the same an-

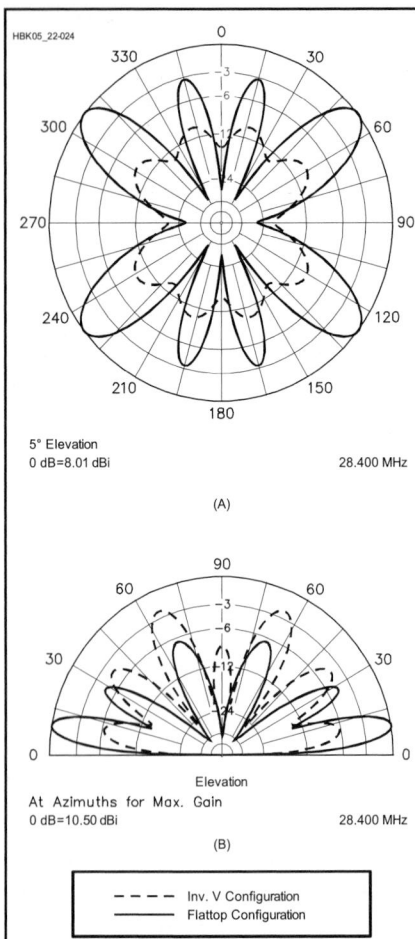

Fig 21.29 — Patterns on 10 meters for 135 foot dipole mounted horizontally and as an inverted-V, as in Fig 21.28. The azimuth pattern is shown at A, and the elevation pattern is shown at B. Once again, the inverted-V configuration yields a more omnidirectional pattern, but at the expense of almost 8 dB less gain than the flat-top configuration at its strongest lobes.

tenna on 14.1 MHz. Notice that the inverted-V pattern is essentially omnidirectional. That comes at the cost of gain, which is less than that for a horizontal flat-top dipole.

As expected, patterns become more complicated at 28.4 MHz. As you can see in **Fig 21.29**, the inverted-V has the advantage of a pattern with slight nulls, but with reduced gain compared to the flat-top configuration.

Installed horizontally, or as an inverted-V, the 135 foot center-fed dipole is a simple antenna that works well from 3.5 to 30 MHz (and on 1.8 MHz if the impedance-matching unit has sufficient range). If extremely high SWR or evidence of RF on objects at the operating position ("RF in the shack") is encountered, change the feed line length by adding or subtracting ⅛ λ at the problem frequency. A few such adjustments should yield a workable solution.

Project: 40 to 15 Meter Dual-Band Dipole

As mentioned earlier, dipoles have harmonic resonances at odd multiples of their fundamental resonances. Because 21 MHz is the third harmonic of 7 MHz, 7 MHz dipoles are harmonically resonant in the popular ham band at 21 MHz. This is attractive because it allows you to install a 40 meter dipole, feed it with coax, and use it without an antenna tuner on both 40 and 15 meters.

But there's a catch: The third harmonic resonance is actually higher than three times the fundamental resonant frequency. This is because there is no end effect in the center portion of the antenna where there are no insulators.

An easy fix for this, as shown in **Fig 21.30**, is to add capacitive loading to the antenna about ¼ λ wavelength (at 21.2 MHz) away from the feed point in both halves of the dipole. Known as *capacitance hats*, the simple loading wires shown lower the antenna's resonant frequency on 15 meters without substantially affecting resonance on 40 meters. This scheme can also be used to build a dipole that can be used on 80 and 30 meters and on 75 and 10 meters. (A project for a 75 and 10 meter dipole is included on the CD-ROM included with this book.)

Measure, cut and adjust the dipole to resonance at the desired 40 meter frequency. Then, cut two 2-foot-long pieces of stiff wire (such as #12 or #14 AWG house wire) and solder the ends of each one together to form two loops. Twist the loops in the middle to form figure-8s, and strip and solder the wires where they cross. Install these capacitance hats on the dipole by stripping the antenna wire (if necessary) and soldering the hats to the dipole about a third of the way out from the feed point (placement isn't critical) on each wire. To resonate the antenna on 15 meters, adjust the loop shapes (*not while you're transmitting!*) until the SWR is acceptable in the desired segment of the 15 meter band. Conversely, you can move the hats back and forth along the antenna until the desired SWR is achieved and then solder the hats to the antenna.

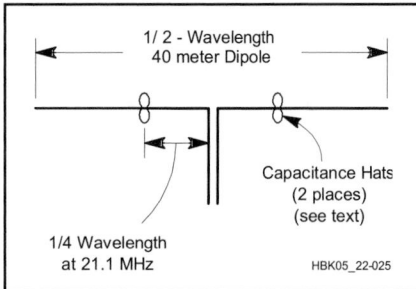

Fig 21.30 — Figure-8-shaped capacitance hats made and placed as described in the text, can make a 40 meter dipole resonate anywhere in the 15 meter band.

Project: W4RNL Rotatable Dipole Inverted-U Antenna

This simple rotatable dipole was designed and built by L.B. Cebik, W4RNL (SK), for use during the ARRL Field Day. For this and other portable operations we look for three antenna characteristics: simplicity, small size and light weight. Today, a number of lightweight collapsible masts are available. When properly guyed, some will support antennas in the 5 to 10 pound range. Most are suitable for 10 meter tubular dipoles and allow the user to hand-rotate the antenna. Extend the range of the antenna to cover 20 through 10 meters, and you put these 20 to 30 foot masts to even better use. The inverted-U meets this need. **Fig 21.31** shows the basic kit for the antenna. Complete construction details and more information about antenna performance are available on the CD-ROM that accompanies this book.

A dipole's highest current occurs within the first half of the distance from the feed point to the outer tips. Therefore, very little performance is lost if the outer end sections are bent. The W4RNL inverted-U starts with a 10 meter tubular dipole. You add wire extensions for 12, 15, 17 or 20 meters to cover those bands.

You only need enough space to erect a 10 meter rotatable dipole. The extensions hang down. **Fig 21.32** shows the relative proportions of the antenna on all bands from 10

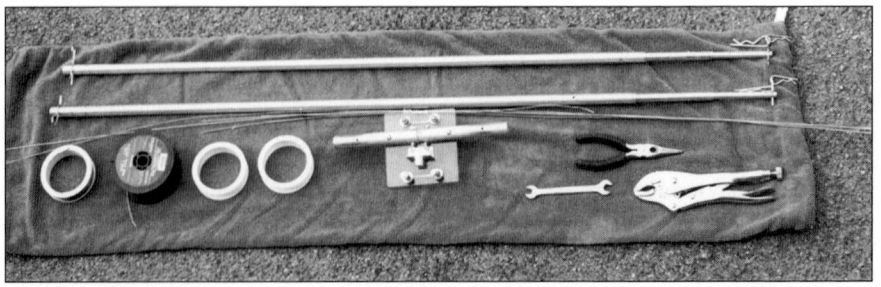

Fig 21.31 — The entire inverted-U antenna parts collection in semi-nested form, with its carrying bag. The tools stored with the antenna include a wrench to tighten the U-bolts for the mast-to-plate mount and a pair of pliers to help remove end wires from the tubing.

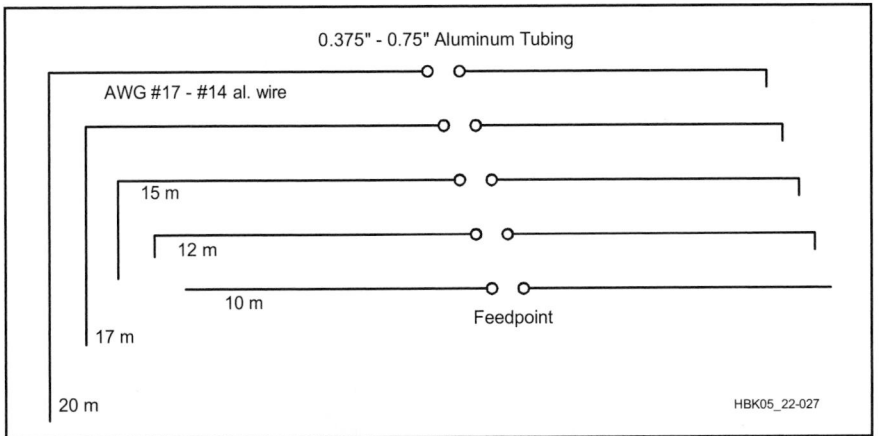

Fig 21.32 — The general outline of the inverted-U field dipole for 20 through 10 meters. Note that the vertical end extension wires apply to both ends of the main 10 meter dipole, which is constant for all bands.

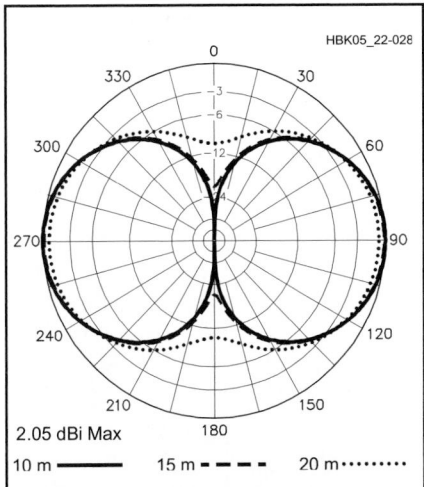

Fig 21.33 — Free-space E-plane (azimuth) patterns of the inverted-U for 10, 15, and 20 meters, showing the pattern changes with increasingly longer vertical end sections.

to 20 meters. The 20 meter extensions are the length of half the 10 meter dipole.

Not much signal strength is lost by drooping up to half the overall element length straight down. What is lost in bidirectional gain shows up in decreased side nulls. **Fig 21.33** shows the free-space E-plane (azimuth) patterns of the inverted-U with a 10 meter horizontal section. There is an undetectable decrease in gain between the 10 meter and 15 meter versions. The 20 meter version shows a little over ½ dB gain decrease and a signal increase off the antenna ends.

The feed point impedance of the inverted-U remains well within acceptable limits for virtually all equipment, even at 20 feet above ground. Also, the SWR curves are very broad, so it's not as critical to find exact dimensions,

even for special field conditions.

Changing bands is a simple matter. Remove the extensions for the band you are using and install the ends for the new band. An SWR check and possibly one more adjustment of the end lengths will put you back on the air.

With a dipole having drooping ends, safety is very important. At any power level, the ends of a dipole have high RF voltages, and we must keep them out of contact with human body parts. Do not use the antenna unless the wire ends for 20 meters are higher than any person can touch when the antenna is in use. Even with QRP power levels, the RF voltage on the wire ends can be dangerous. With the antenna at 20 feet at its center, the ends should be at least 10 feet above ground.

Project: Two W8NX Multiband, Coax-Trap Dipoles

Over the last 60 or 70 years, amateurs have used many kinds of multiband antennas to cover the traditional HF bands. The availability of the 30, 17 and 12 meter bands has expanded our need for multiband antenna coverage.

Two different antennas are described here. The first covers the traditional 80, 40, 20, 15 and 10 meter bands, and the second covers 80, 40, 17 and 12 meters. Each uses the

same type of W8NX trap — connected for different modes of operation — and a pair of short capacitive stubs to enhance coverage. The W8NX coaxial-cable traps have two different modes: a high- and a low-impedance mode. The inner-conductor windings and shield windings of the traps are connected in series for both modes. However, either the low- or high-impedance point can be used as the trap's output terminal. For low-impedance trap operation, only the center conductor turns of the trap windings are used. For high-impedance operation, all turns are used, in the conventional manner for a trap. The short stubs on each antenna are strategically sized and located to permit more flexibility in adjusting the resonant frequencies of the antenna.

80, 40, 20, 15 AND 10 M DIPOLE

Fig 21.34 shows the configuration of the 80, 40, 20, 15 and 10 meter antenna. The radiating elements are made of #14 AWG stranded copper wire. The element lengths are the wire span lengths in feet. These lengths do not include the lengths of the pigtails at the balun, traps and insulators. The 32.3-foot-long inner 40 meter segments are measured from the eyelet of the input balun to the tension-relief hole in the trap coil form. The 4.9 foot segment length is measured from the tension-relief hole in the trap to the 6 foot stub. The 16.1 foot outer-segment span is measured from the stub to the eyelet of the end insulator.

The coaxial-cable traps are wound on PVC pipe coil forms and use the low-impedance output connection. The stubs are 6 foot lengths of ⅛ inch stiffened aluminum or copper rod hanging perpendicular to the radiating elements. The first inch of their length is bent 90° to permit attachment to the radiating elements by large-diameter copper crimp connectors. Ordinary #14 AWG wire may be used for the stubs, but it has a tendency to curl up and may tangle unless weighed down at the end. You should feed the antenna with 75 Ω coax cable using a good 1:1 balun.

This antenna may be thought of as a modified W3DZZ antenna (see the References) due to the addition of the capacitive stubs. The length and location of the stub give the

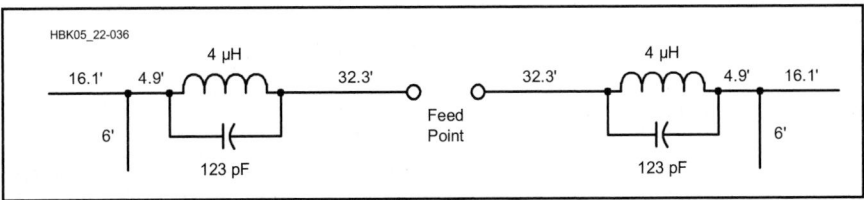

Fig 21.34 — A W8NX multiband dipole for 80, 40, 20, 15 and 10 meters. The values shown (123 pF and 4 µH) for the coaxial-cable traps are for parallel resonance at 7.15 MHz. The low-impedance output of each trap is used for this antenna.

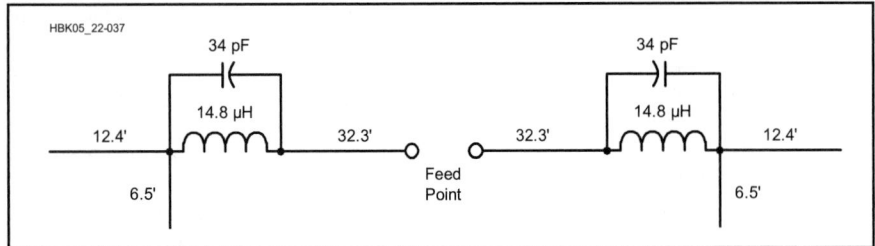

Fig 21.35 — A W8NX multiband dipole for 80, 40, 17 and 12 meters. For this antenna, the high-impedance output is used on each trap. The resonant frequency of the traps is 7.15 MHz.

antenna designer two extra degrees of freedom to place the resonant frequencies within the amateur bands. This additional flexibility is particularly helpful to bring the 15 and 10 meter resonant frequencies to more desirable locations in these bands. The actual 10 meter resonant frequency of the original W3DZZ antenna is somewhat above 30 MHz, pretty remote from the more desirable low frequency end of 10 meters.

80, 40, 17 AND 12 METER DIPOLE

Fig 21.35 shows the configuration of the 80, 40, 17 and 12 meter antenna. Notice that the capacitive stubs are attached immediately outboard after the traps and are 6.5 feet long, ½ foot longer than those used in the other antenna. The traps are the same as those of the other antenna, but are connected for the high-impedance parallel-resonant output mode. Since only four bands are covered by this antenna, it is easier to fine tune it to precisely the desired frequency on all bands. The 12.4 foot tips can be pruned to a particular 17 meter frequency with little effect on the 12 meter frequency. The stub lengths can be pruned to a particular 12 meter frequency with little effect on the 17 meter frequency. Both such pruning adjustments slightly alter the 80 meter resonant frequency. However, the bandwidths of the antennas are so broad on 17 and 12 meters that little need for such pruning exists. The 40 meter frequency is nearly independent of adjustments to the

capacitive stubs and outer radiating tip elements. Like the first antennas, this dipole is fed with a 75 Ω balun and feed line.

Fig 21.36 shows the schematic diagram of the traps. It illustrates the difference between the low and high-impedance modes of the traps. Notice that the high-impedance terminal is the output configuration used in most conventional trap applications. The low-impedance connection is made across only the inner conductor turns, corresponding to one-half of the total turns of the trap. This mode steps the trap's impedance down to approximately one-fourth of that of the high-impedance level. This is what allows a single trap design to be used for two different multiband antennas.

Fig 21.37 is a drawing of a cross-section of the coax trap shown through the long axis of the trap. Notice that the traps are conventional coaxial-cable traps, except for the added low-impedance output terminal. The traps are 8¾ close-spaced turns of RG-59 (Belden 8241) on a 2⅜ inch OD PVC pipe (schedule 40 pipe with a 2 inch ID) coil form. The forms are 4⅛ inches long. Trap resonant frequency is very sensitive to the outer diameter of the coil form, so check it carefully. Unfortunately, not all PVC pipe is made with the same wall

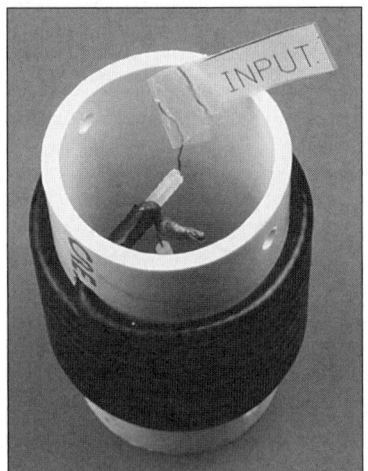

Fig 21.38 — Other views of a W8NX coax-cable trap.

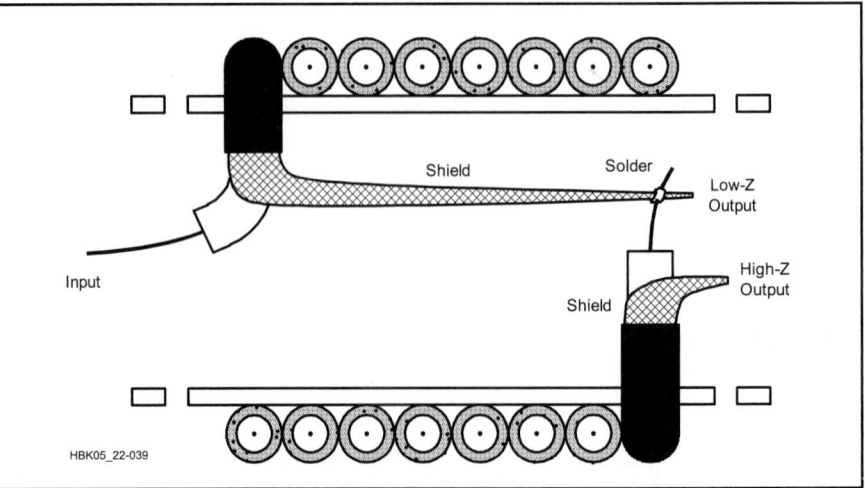

Fig 21.36 — Schematic for the W8NX coaxial-cable trap. RG-59 is wound on a 2⅜ inch OD PVC pipe.

Fig 21.37 — Construction details of the W8NX coaxial-cable trap.

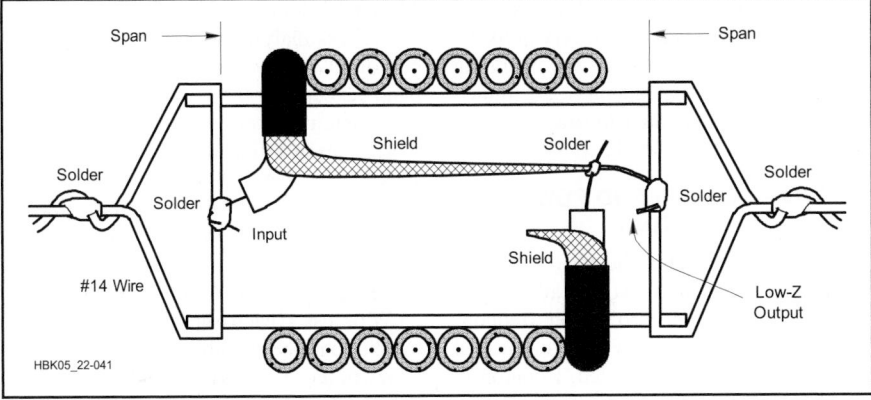

Fig 21.39 — Additional construction details for the W8NX coaxial-cable trap.

thickness. The trap frequencies should be checked with a dip meter and general-coverage receiver and adjusted to within 50 kHz of the 7150 kHz resonant frequency before installation. One inch is left over at each end of the coil forms to allow for the coax feed-through holes and holes for tension-relief attachment of the antenna radiating elements to the traps. Be sure to seal the ends of the trap coax cable with RTV sealant to prevent moisture from entering the coaxial cable.

Also, be sure that you connect the 32.3 foot wire element at the start of the inner conductor winding of the trap. This avoids detuning the antenna by the stray capacitance of the coaxial-cable shield. The trap output terminal (which has the shield stray capacitance) should be at the outboard side of the trap. Reversing the input and output terminals of the trap will lower the 40 meter frequency by approximately 50 kHz, but there will be

negligible effect on the other bands.

Fig 21.38 shows a coaxial-cable trap. Further details of the trap installation are shown in **Fig 21.39**. This drawing applies specifically to the 80, 40, 20, 15 and 10 meter antenna, which uses the low-impedance trap connections. Notice the lengths of the trap pigtails: three to four inches at each terminal of the trap. If you use a different arrangement, you must modify the span lengths accordingly. All connections can be made using crimp connectors rather than by soldering. Using a crimping tool instead of a soldering iron allows easier access to the trap's interior.

PERFORMANCE

The performance of both antennas has been very satisfactory. W8NX uses the 80, 40, 17 and 12 meter version because it covers 17 and 12 meters. (He has a tri-band Yagi for 20, 15 and 10 meters.) The radiation pattern on

17 meters is that of ³⁄₂ wave dipole. On 12 meters, the pattern is that of a ⁵⁄₂ wave dipole. At his location in Akron, Ohio, the antenna runs essentially east and west. It is installed as an inverted-V, 40 feet high at the center, with a 120° included angle between the legs. Since the stubs are very short, they radiate little power and make only minor contributions to the radiation patterns. In theory, the pattern has four major lobes on 17 meters, with maxima to the northeast, southeast, southwest and northwest. These provide low-angle radiation into Europe, Africa, South Pacific, Japan and Alaska. A narrow pair of minor broadside lobes provides north and south coverage into Central America, South America and the polar regions.

There are four major lobes on 12 meters, giving nearly end-fire radiation and good low-angle east and west coverage. There are also three pairs of very narrow, nearly broadside, minor lobes on 12 meters, down about 6 dB from the major end-fire lobes. On 80 and 40 meters, the antenna has the usual figure-8 patterns of a half-wave-length dipole.

Both antennas function as electrical half-wave dipoles on 80 and 40 meters with a low SWR. They both function as odd-harmonic current-fed dipoles on their other operating frequencies, with higher, but still acceptable, SWR. The presence of the stubs can either raise or lower the input impedance of the antenna from those of the usual third and fifth harmonic dipoles. Again W8NX recommends that 75 Ω, rather than 50 Ω, feed line be used because of the generally higher input impedances at the harmonic operating frequencies of the antennas.

The SWR curves of both antennas were

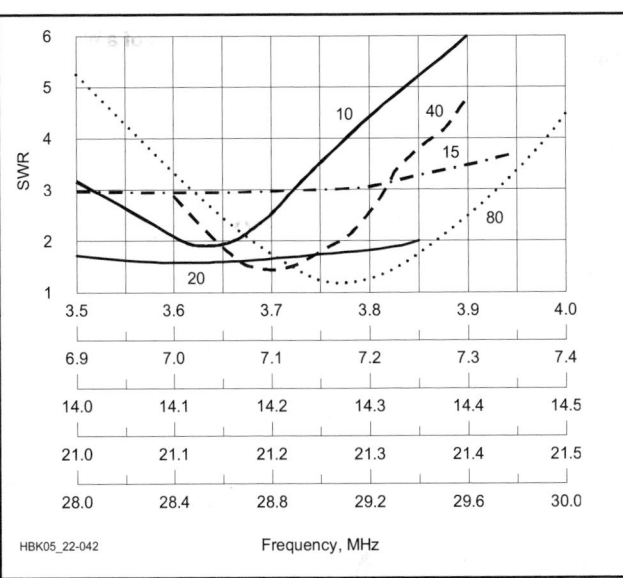

Fig 21.40 — Measured SWR curves for an 80, 40, 20, 15 and 10 meter antenna, installed as an inverted-V with 40 foot apex and 120° included angle between legs.

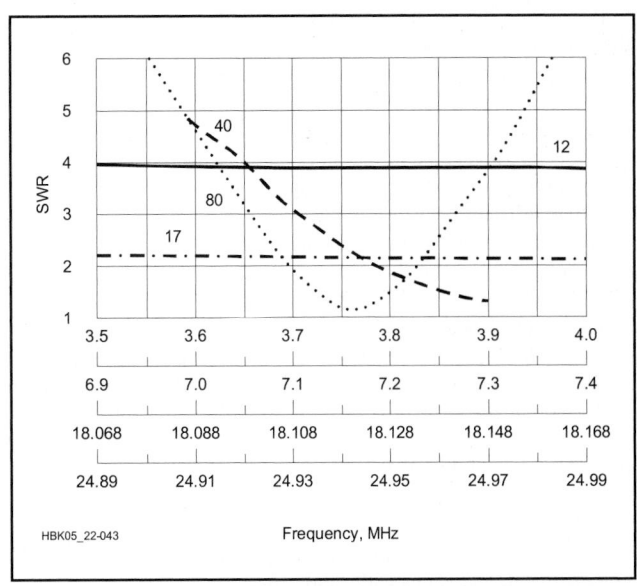

Fig 21.41 — Measured SWR curves for an 80, 40, 17 and 12 meter antenna, installed as an inverted-V with 40 foot apex and 120° included angle between legs.

carefully measured using a 75 to 50 Ω transformer from Palomar Engineers inserted at the junction of the 75 Ω coax feed line and a 50 Ω SWR bridge. The transformer is required for accurate SWR measurement if a 50 Ω SWR bridge is used with a 75 Ω line. Most 50 Ω rigs operate satisfactorily with a 75 Ω line, although this requires different tuning and load settings in the final output stage of the rig or antenna tuner. The author uses the 75 to 50 Ω transformer only when making SWR measurements and at low power levels. The transformer is rated for 100 W, and when he runs his 1 kW PEP linear amplifier the transformer is taken out of the line.

Fig 21.40 gives the SWR curves of the 80, 40, 20, 15 and 10 meter antenna. Minimum SWR is nearly 1:1 on 80, 1.5:1 on 40, 1.6:1 on 20, and 1.5:1 on 10 meters. The minimum SWR is slightly below 3:1 on 15 meters. On 15 meters, the stub capacitive reactance combines with the inductive reactance of the outer segment of the antenna to produce a resonant rise that raises the antenna input resistance to about 220 Ω, higher than that of the usual ½ wavelength dipole. An antenna tuner may be required on this band to keep a solid-state final output stage happy under these load conditions.

Fig 21.41 shows the SWR curves of the 80, 40, 17 and 12 meter antenna. Notice the excellent 80 meter performance with a nearly unity minimum SWR in the middle of the band. The performance approaches that of a full-size 80 meter wire dipole. The short stubs and the low-inductance traps shorten the antenna somewhat on 80 meters. Also observe the good 17 meter performance, with the SWR being only a little above 2:1 across the band.

But notice the 12 meter SWR curve of this antenna, which shows 4:1 SWR across the band. The antenna input resistance approaches 300 Ω on this band because the capacitive reactance of the stubs combines with the inductive reactance of the outer antenna segments to give resonant rises in impedance. These are reflected back to the input terminals. These stub-induced resonant impedance rises are similar to those on the other antenna on 15 meters, but are even more pronounced.

Too much concern must not be given to SWR on the feed line. Even if the SWR is as high as 9:1 *no destructively high voltages will exist on the transmission line.* Recall that transmission-line voltages increase as the square root of the SWR in the line. Thus, 1 kW of RF power in 75 Ω line corresponds to 274 V line voltage for a 1:1 SWR. Raising the SWR to 9:1 merely triples the maximum voltage that the line must withstand to 822 V. This voltage is well below the 3700 V rating of RG-11, or the 1700 V rating of RG-59, the

two most popular 75 Ω coax lines. Voltage breakdown in the traps is also very unlikely. As will be pointed out later, the operating power levels of these antennas are limited by RF power dissipation in the traps, not trap voltage breakdown or feed line SWR.

TRAP LOSSES AND POWER RATING

Table 21.2 presents the results of trap Q measurements and extrapolation by a two-frequency method to higher frequencies above resonance. W8NX employed an old, but recently calibrated, Boonton Q meter for the measurements. Extrapolation to higher-frequency bands assumes that trap resistance losses rise with skin effect according to the square root of frequency, and that trap dielectric losses rise directly with frequency. Systematic measurement errors are not increased by frequency extrapolation. However, random measurement errors increase in magnitude with upward frequency extrapolation. Results are believed to be accurate within 4% on 80 and 40 meters, but only within 10 to 15% at 10 meters. Trap Q is shown at both the high- and low-impedance trap terminals. The Q at the low-impedance output terminals is 15 to 20% lower than the Q at the high-impedance output terminals.

W8NX computer analyzed trap losses for both antennas in free space. Antenna-input resistances at resonance were first calculated, assuming lossless, infinite-Q traps. They were again calculated using the Q values in Table 21.2. The radiation efficiencies were also converted into equivalent trap losses in decibels. **Table 21.3** summarizes the trap-loss analysis for the 80, 40, 20, 15 and 10 meter antenna and **Table 21.4** for the 80, 40, 17 and

12 meter antenna.

The loss analysis shows radiation efficiencies of 90% or more for both antennas on all bands except for the 80, 40, 20, 15 and 10 meter antenna when used on 40 meters. Here, the radiation efficiency falls to 70.8%. A 1 kW power level at 90% radiation efficiency corresponds to 50 W dissipation per trap. In W8NX's experience, this is the trap's survival limit for extended key-down operation. SSB power levels of 1 kW PEP would dissipate 25 W or less in each trap. This is well within the dissipation capability of the traps.

When the 80, 40, 20, 15 and 10 meter antenna is operated on 40 meters, the radiation efficiency of 70.8% corresponds to a dissipation of 146 W in each trap when 1 kW is delivered to the antenna. This is sure to burn out the traps — even if sustained for only a short time. Thus, the power should be limited to less than 300 W when this antenna is operated on 40 meters under prolonged key-down conditions. A 50% CW duty cycle would correspond to a 600 W power limit for normal 40 meter CW operation. Likewise, a 50% duty cycle for 40 meter SSB corresponds to a 600 W PEP power limit for the antenna.

The author knows of no analysis where the burnout wattage rating of traps has been rigorously determined. Operating experience seems to be the best way to determine trap burn-out ratings. In his own experience with these antennas, he's had no traps burn out, even though he operated the 80, 40, 20, 15 and 10 meter antenna on the critical 40 meter band using his AL-80A linear amplifier at 600 W PEP output. He did not make continuous, key-down, CW operating tests at full power purposely trying to destroy the traps!

Some hams may suggest using a different

Table 21.2
Trap Q

Frequency (MHz)	3.8	7.15	14.18	18.1	21.3	24.9	28.6
High Z out (Ω)	101	124	139	165	73	179	186
Low Z out (Ω)	83	103	125	137	44	149	*155*

Table 21.3
Trap Loss Analysis: 80, 40, 20, 15, 10 Meter Antenna

Frequency (MHz)	3.8	7.15	14.18	21.3	28.6
Radiation Efficiency (%)	96.4	70.8	99.4	99.9	100.0
Trap Losses (dB)	0.16	1.5	0.02	0.01	0.003

Table 21.4
Trap Loss Analysis: 80, 40, 17, 12 Meter Antenna

Frequency (MHz)	3.8	7.15	18.1	24.9
Radiation Efficiency (%)	89.5	90.5	99.3	99.8
Trap Losses (dB)	0.5	0.4	0.03	0.006

Antenna Modeling by Computer

Modern computer programs have made it a *lot* easier for a ham to evaluate antenna performance. The elevation plots for the 135 foot long center-fed dipole were generated using a sophisticated computer program known as *NEC*, short for "Numerical Electromagnetics Code." *NEC* is a general-purpose antenna modeling program, capable of modeling almost any antenna type, from the simplest dipole to extremely complex antenna designs. Various mainframe versions of *NEC* have been under continuous development by US government researchers for several decades.

But because it is a general-purpose program, *NEC* can be very slow when modeling some antennas — such as long-boom, multi-element Yagis. There are other, specialized programs that work on Yagis much faster than *NEC*. Indeed, *NEC* has developed a reputation for being accurate (if properly applied!), but decidedly difficult to learn and use. A number of commercial software developers have risen to the challenge and created more *user-friendly* versions such as *EZNEC* and *4nec2* which are advertised in *QST*.

NEC uses a *Method of Moments* algorithm. The mathematics behind this algorithm are pretty formidable to most hams, but the basic principle is simple. An antenna is broken down into a set of straight-line wire *segments*. The fields resulting from the current in each segment and from the mutual interaction between segments are vector-summed in the far field to create azimuth and elevation-plane patterns.

The most difficult part of using a *NEC*-type of modeling program is setting up the antenna's geometry — you must condition yourself to think in three-dimensional coordinates. Each end point of a wire is represented by three numbers: an x, y and z coordinate. An example should help sort things out. See **Fig 21.A2**, showing a *model* for a 135 foot center-fed dipole, made of #14 wire placed 50 feet above flat ground. This antenna is modeled as a single, straight wire.

For convenience, ground is located at the *origin* of the coordinate system, at (0, 0, 0) feet, directly under the center of the dipole. The dipole runs parallel to, and above, the y-axis. Above the origin, at a height of 50 feet, is the dipole's feed point. The *wingspread* of the dipole goes toward the left (that is, in the *negative y* direction) one-half the overall length, or −67.5 feet. Toward the right, it goes +67.5 feet. The x dimension of our dipole is zero. The dipole's ends are thus represented by two points, whose coordinates are: (0, −67.5, 50) and (0, 67.5, 50) feet. The thickness of the antenna is the diameter of the wire, #14 gauge.

To run the program you must specify the number of segments into which the dipole is divided for the method-of-moments analysis. The guideline for setting the number of segments is to use at least 10 segments per half-wavelength. In Fig 21.A1, our dipole has been divided into 11 segments for 80 meter operation. The use of 11 segments, an odd rather than an even number such as 10, places the dipole's feed point (the *source* in *NEC*-parlance) right at the antenna's center and at the center of segment number six.

Since we intend to use our 135 foot long dipole on all HF amateur bands, the number of segments used actually should vary with frequency. The penalty for using more segments in a program like *NEC* is that the program slows down roughly as the square of the segments — double the number and the speed drops to a fourth. However, using too few segments will introduce inaccuracies, particularly in computing the feed point impedance. The commercial versions of *NEC* handle such nitty-gritty details automatically.

Let's get a little more complicated and specify the 135 foot dipole, configured as an inverted-V. Here, as shown in **Fig 21.A3**, you must specify *two* wires. The two wires join at the top, (0, 0, 50) feet. Now the specification of the source becomes more complicated. The easiest way is to specify two sources, one on each end segment at the junction of the two wires. If you are using the *native* version of *NEC*, you may have to go back to your high-school trigonometry book to figure out how to specify the end points of our droopy dipole, with its 120° included angle. Fig 21.A3 shows the details, along with the trig equations needed.

So, you see that antenna modeling isn't entirely a cut-and-dried procedure. The commercial programs do their best to hide some of the more unwieldy parts of *NEC*, but there's still some art mixed in with the science. And as always, there are trade-offs to be made — segments versus speed, for example.

However, once you do figure out exactly how to use them, computer models are wonderful tools. They can help you while away a dreary winter's day, designing antennas on-screen — without having to risk life and limb climbing an ice-covered tower. And in a relatively short time a computer model can run hundreds, or even thousands, of simulations as you seek to optimize an antenna for a particular parameter. Doesn't that sound better than trying to optimally tweak an antenna by means of a thousand cut-and-try measurements, all the while hanging precariously from your tower by a climbing harness?! — *R. Dean Straw, N6BV*

Fig 21.A2 — Model of a 135 foot center-fed dipole.

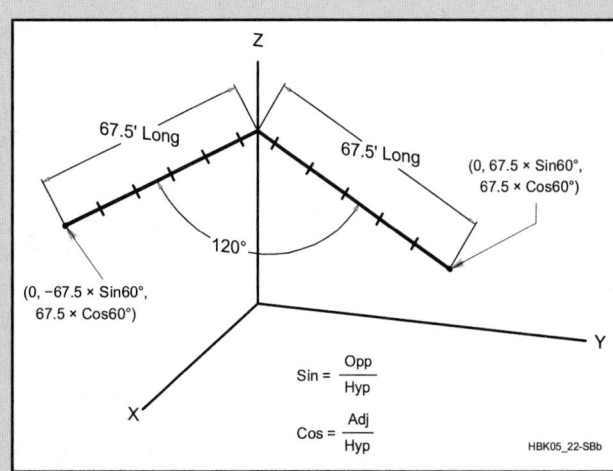

Fig 21.A3 — Model of the 135 foot dipole configured as an inverted V.

type of coaxial cable for the traps. The dc resistance of 40.7 Ω per 1000 feet of RG-59 coax seems rather high. However, W8NX has found no coax other than RG-59 that has the necessary inductance-to-capacitance ratio to create the trap characteristic reactance required for the 80, 40, 20, 15 and 10 meter antenna. Conventional traps with wide-spaced, open-air inductors and appropriate fixed-value capacitors could be substituted for the coax traps, but the convenience, weatherproof configuration and ease of fabrication of coaxial-cable traps is hard to beat.

Project: Extended Double-Zepp for 17 Meters

Although the Extended Double-Zepp (EDZ) antenna shown in **Fig 21.42** has several attractive features, it is rarely used by hams, perhaps out of concern over the Zepp's high feed point impedance. The antenna's overall length is 1.28 λ and its pattern is bidirectional broadside to the antenna. The SWR of the antenna is low enough near the design frequency that it can be fed with coax and an impedance-matching unit or open-wire line can be used for wider range and multiband use. This project describes an EDZ for 17 meters.

The Zepp antenna (a half-wave dipole, fed at one end) was introduced earlier in this section. The Zepp can be modified in two ways. The first is to double the length of the antenna and feed it in the middle, making a *double-Zepp*. This creates a one-wavelength dipole, with the expected high feed point impedance and about 1.6 dBd gain. A ½ λ center-fed

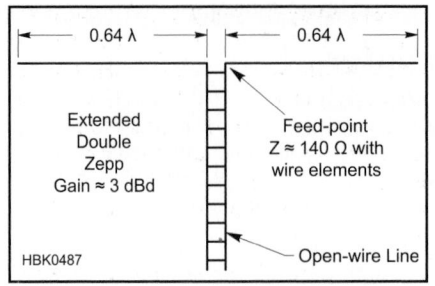

Fig 21.42 — The Extended Double-Zepp antenna consists of two 0.64 λ sections placed end to end and fed in the middle. The high-impedance points of the antenna have been moved away from the feed point, lowering feed point impedance. The antenna has gain of approximately 3 dBd broadside to the antenna.

dipole operated on its second harmonic is effectively a double-Zepp. The second modification is to extend the double-Zepp to be 0.64 λ (close to ⅝ λ) long on each side of the feed point. The feed point is then no longer at a high-impedance point on the antenna. This creates the extended, double-Zepp. (The EDZ is described in more detail in the *ARRL Antenna Book*.)

The overall length of the EDZ is calculated as follows:

$$984/f(MHz) \times 1.28 = \text{length in feet} \qquad (5)$$

Using this formula, an 18.1 MHz EDZ is 69.6 feet (69 feet, 7 inches.) long. The EDZ has 3 dBd of gain in a figure-8 pattern of two major lobes broadside to the antenna and four minor lobes at smaller angles to the axis of

the antenna. The feed point impedance is approximately 140 Ω.

The EDZ is useful at lower frequencies, as well. On 20 meters, the 17 meter EDZ is just slightly longer than the double-Zepp, with 1.6 to 2 dBd of gain and a rather high feed point impedance of several hundred ohms. On 40 meters, the antenna is a slightly long ½ λ dipole. If your antenna tuner has sufficient range, the antenna can also serve as a shortened dipole for 75/80 meters. At these lower frequencies, the antenna's radiation pattern is a single lobe, broadside to the antenna.

On higher frequencies, the pattern continues to split into more lobes. For example, on 15 meters, there are four lobes at approximately 45° from the antenna axis. On 10 meters, where the antenna is approximately two full-wavelengths long, the pattern is similar, with the lobes a bit closer to the antenna axis and smaller lobes beginning to appear.

Some hams use a 4:1 impedance transformer to reduce the feed point impedance and improve SWR as the operating frequency moves away from the design frequency. This works best if the antenna is to be used on a single band. However, if the antenna is to be used on multiple bands, a better solution is to use open-wire feed line and an antenna tuner. If you wish to operate on a frequency at which the feed point impedance is high, use a feed line length near an odd multiple of a quarter-wavelength long, presenting a lower impedance to your antenna tuner that may be easier to match.

(This project is based on a "Hints and Kinks" item by Bob Baird, W7CSD, from the January 1992 issue of *QST*.)

21.3 Vertical (Ground-Plane) Antennas

One of the more popular amateur antennas is the *vertical*. It usually refers to a single radiating element erected vertically over the ground. A typical vertical is an electrical ¼ λ long and is constructed of wire or tubing. The vertical antenna is more accurately named the *ground plane* because it uses a conductive surface (the ground plane) to create a path for return currents, effectively creating the "missing half" of a ½ λ antenna. Another name for this type of antenna is the *monopole* (sometimes *unipole*).

The ground plane can be a solid, conducting surface, such as a vehicle body for a VHF/UHF mobile antenna. At HF, this is impractical and systems of *ground radials* are used; wires laid out on the ground radially from the base of the antenna. One conductor of the feed line is attached to the vertical radiating element of the antenna and the remaining

conductor is attached to the ground plane.

Single vertical antennas are omnidirectional radiators. This can be beneficial or detrimental, depending on the situation. On transmission there are no nulls in any direction, unlike most horizontal antennas. However, QRM on receive can't be nulled out from the directions that are not of interest unless multiple verticals are used in an array.

Ground-plane antennas need not be mounted vertically. A ground-plane antenna can operate in any orientation as long as the ground plane is perpendicular to the radiating element. Other considerations, such as minimizing cross-polarization between stations, may require a specific mounting orientation though. In addition, due to the size of HF antennas, mounting them vertically is usually the most practical solution.

A vertical antenna can be mounted at the

Earth's surface, in which case it is a *ground-mounted vertical*. The ground plane is then constructed on the surface of the ground. A vertical antenna and the associated ground plane can also be installed above the ground. This often reduces ground losses, but it is more difficult to install the necessary number of radials. *Ground-independent* verticals are often mounted well above the ground because their operation does not rely on a ground plane.

21.3.1 Ground Systems

When compared to horizontal antennas, verticals also suffer more acutely from two main types of losses — *ground return losses* for currents in the near field, and *far-field ground losses*. Ground losses in the near field can be minimized by using many ground radials. This is covered in the sidebar, "Optimum Ground Systems for Vertical Antennas."

Far-field losses are highly dependent on the conductivity and dielectric constant of the earth around the antenna, extending out as far as 100 wavelengths from the base of the antenna. There is very little that someone can do to change the character of the ground that far away — other than moving to a small island surrounded by saltwater! Far-field losses greatly affect low-angle radiation, causing the radiation patterns of practical vertical antennas to fall far short of theoretical patterns over *perfect ground*, often seen in classical texts.

Fig 21.43 shows the elevation pattern response for two different 40 meter quarter-wave verticals. One is placed over a theoretical infinitely large, infinitely conducting ground. The second is placed over an extensive radial system over average soil, having a conductivity of 5 mS/m and a dielectric constant of 13. This sort of soil is typical of heavy clay found in pastoral regions of the US mid-Atlantic states. At a 10° elevation angle, the real antenna losses are almost 6 dB compared to the theoretical one; at 20° the difference is about 3 dB. See *The ARRL Antenna Book* chapter on the effects of ground for further details.

While real verticals over real ground are not a magic method to achieve low-angle radiation, cost versus performance and ease of installation are incentives that inspire many antenna builders. For use on the lower frequency amateur bands — notably 160 and 80 meters — it is not always practical to erect a full-size vertical. At 1.8 MHz, a full-sized quarter-wave vertical is 130 feet high. In such instances it is often necessary to accept a shorter radiating element and use some form of loading.

Fig 21.44 provides curves for the physi-

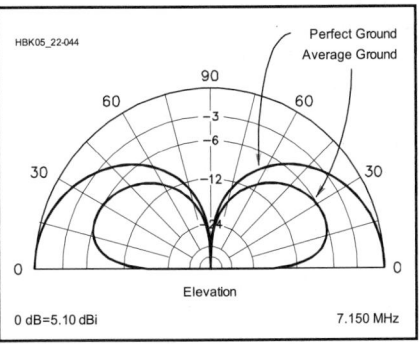

Fig 21.43 — Elevation patterns for two quarter-wave vertical antennas over different ground. One vertical is placed over perfect ground, and the other is placed over average ground. The far-field response at low elevation angles is greatly affected by the quality of the ground — as far as 100 λ away from the vertical antenna.

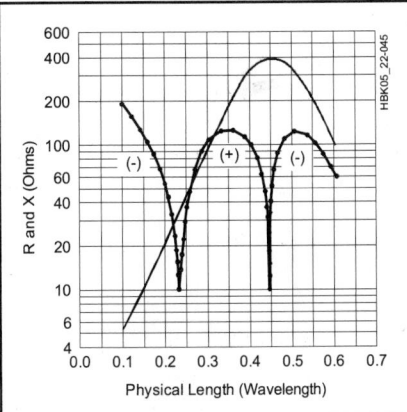

Fig 21.44 — Radiation resistance (solid curve) and reactance (dotted curve) of vertical antennas as a function of their electrical height.

Optimum Ground Systems for Vertical Antennas

A frequent question brought up by old-timers and newcomers alike is: "So, how many ground radials do I *really* need for my vertical antenna?" Most hams have heard the old standby tales about radials, such as "if a few are good, more must be better" or "lots of short radials are better than a few long ones."

John Stanley, K4ERO, eloquently summarized a study he did of the professional literature on this subject in his article "Optimum Ground Systems for Vertical Antennas" in December 1976 *QST*. His approach was to present the data in a sort of "cost-benefit" style in **Table 21.A**, reproduced here. John somewhat wryly created a new figure of merit — the total amount of wire needed for various radial configurations. This is expressed in terms of wavelengths of total radial wire.

Table 21.A
Optimum Ground-System Configurations

Configuration Designation	A	B	C	D	E	F
Number of radials	16	24	36	60	90	120
Length of each radial in wavelengths	0.1	0.125	0.15	0.2	0.25	0.4
Spacing of radials in degrees	22.5	15	10	6	4	3
Total length of radial wire installed, in wavelengths	1.6	3	5.4	12	22.5	48
Power loss in dB at low angles with a quarter-wave radiating element	3	2	1.5	1	0.5	0*
Feed point impedance in ohms with a quarter-wave radiating element	52	46	43	40	37	35

Note: Configuration designations are indicated only for text reference.
*Reference. The loss of this configuration is negligible compared to a perfectly conducting ground.

The results almost jumping out of this table are:
- If you can only install 16 radials (Case A), they needn't be very long — 0.1 λ is sufficient. You'll use 1.6 λ of radial wire in total, which is about 450 feet at 3.5 MHz.
- If you have the luxury of laying down 120 radials (Case F), they should be 0.4 λ long, and you'll gain about 3 dB over the 16 radial case. You'll also use 48 λ of total wire — For 80 meters, that would be about 13,500 feet!
- If you can't put out 120 radials, but can install 36 radials that are 0.15 λ long (Case C), you'll lose only 1.5 dB compared to the optimal Case F. You'll also use 5.4 λ of total wire, or 1,500 feet at 3.5 MHz.
- A 50 Ω SWR of 1:1 isn't necessarily a good thing — the worst-case ground system in Case A has the lowest SWR.

Table 21.A represents the case for "Average" quality soil, and it is valid for radial wires either laid on the ground or buried several inches in the ground. Note that such ground-mounted radials are detuned because of their proximity to that ground and hence don't have to be the classical quarter-wavelength that they need to be in "free space."

In his article John also made the point that ground-radial losses would only be significant on transmit, since the atmospheric noise on the amateur bands below 30 MHz is attenuated by ground losses, just as actual signals would be. This limits the ultimate signal-to-noise ratio in receiving.

So, there you have the tradeoffs — the loss in transmitted signal compared to the cost (and effort) needed to install more radial wires. You take your pick.

cal height of verticals in wavelength versus radiation resistance and reactance. Although the plots are based on perfectly conducting ground, they show general trends for installations where many radials have been laid out to make a ground screen. As the radiator is made shorter, the radiation resistance decreases — with 6 Ω being typical for a 0.1 λ high antenna. The lower the radiation resistance, the more the antenna efficiency depends on ground conductivity and the effectiveness of the ground screen. Also, the bandwidth decreases markedly as the length is reduced toward the left of the scale in Fig 21.44. It can be difficult to develop suitable matching networks when radiation resistance is very low.

Generally a large number of shorter radials results in a better ground system than a few longer ones. For example, eight ¼ λ radials are preferred over four ¼ λ radials. Optimum radial lengths are described in the sidebar.

The conductor size of the radials is not especially significant. Wire gauges from #4 to #20 AWG have been used successfully by amateurs. Copper wire is preferred, but where soil is low in acid (or alkali), aluminum wire can be used. The wires may be bare or insulated, and they can be laid on the earth's surface or buried a few inches below ground. Insulated wires will have greater longevity by virtue of reduced corrosion and dissolution from soil chemicals.

When property dimensions do not allow a classic installation of equally spaced radial wires, they can be placed on the ground as space permits. They may run away from the antenna in only one or two compass directions. They may be bent to fit on your property. Hardware cloth and chicken wire are also quite effective, although the galvanizing must be of high-quality to prevent rapid rusting.

A single ground rod, or group of them bonded together, is seldom as effective as a collection of random-length radial wires.

All radial wires should be connected together at the base of the vertical antenna. The electrical bond needs to be of low resistance. Best results will be obtained when the wires are soldered together at the junction point. When a grounded vertical is used, the ground wires should be affixed securely to the base of the driven element.

Ground return losses are lower when vertical antennas and their radials are elevated above ground, a point that is well-known by those using ground plane antennas on their roofs. Even on 160 or 80 meters, effective vertical antenna systems can be made with as few as four ¼ λ long radials elevated 10 to 20 feet off the ground.

21.3.2 Full-Size Vertical Antennas

When it is practical to erect a full-size ¼ λ vertical antenna, the forms shown in **Fig 21.45** are worthy of consideration. The example at A is the well-known *vertical ground plane*. The ground system consists of four above-ground radial wires. The length of the driven element and ¼ λ radials is derived from the standard equation

$$L(ft) = 234/f(MHz) \qquad (6)$$

With four equidistant radial wires drooped at approximately 30° (Fig 21.45A), the feed point impedance is roughly 50 Ω. When the radials are at right angles to the radiator (Fig 21.45B) the impedance approaches 36 Ω.

Besides minimizing ground return losses, another major advantage in this type of vertical antenna over a ground-mounted type is that the system can be elevated well above nearby conductive objects (power lines, trees, buildings and so on). When drooping radials are used, they can also serve as guy wires for the mast that supports the antenna. The coax shield braid is connected to the radials, and the center conductor to the driven element.

The *Marconi* vertical antenna shown in

Fig 21.45C is the classic form taken by a ground-mounted vertical. It can be grounded at the base and *shunt fed*, or it can be isolated from ground, as shown, and *series fed*. As always, this vertical antenna depends on an effective ground system for efficient performance. If a perfect ground were located below the antenna, the feed impedance would be near 36 Ω. In a practical case, owing to imperfect ground, the impedance is more likely to be in the vicinity of 50 Ω.

Vertical antennas can be longer than ¼ λ, too; ⅜ λ, ½ λ, and ⅝ λ verticals can all be used with good results, although none will present a 50 Ω feed point impedance at the base. Non-resonant lengths have become popular for the same reasons as non-resonant horizontal antennas; when fed with low-loss feed line and a wide-range impedance matching unit, the antenna can be used on multiple bands. Various matching networks, described in the **Transmission Lines** chapter, can be employed. Antenna lengths above ½ λ are not recommended because the radiation pattern begins to break up into more than one lobe, developing a null at the horizon at 1 λ.

A gamma-match feed system for a grounded ¼ λ vertical is presented in Fig 21.45D. (The gamma match is also discussed in the

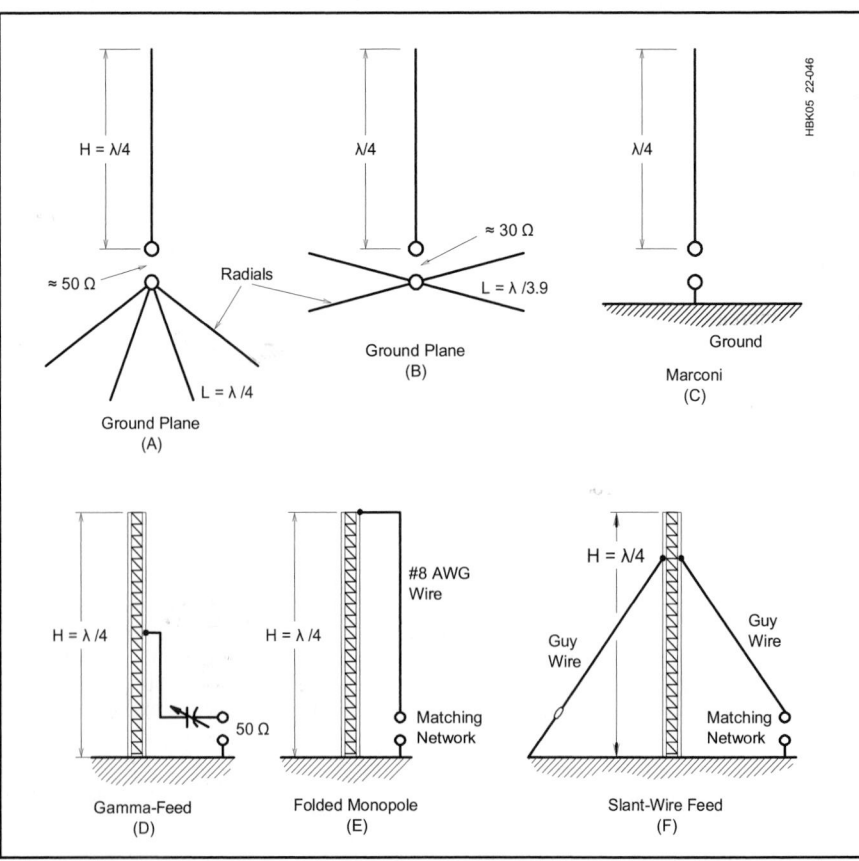

Fig 21.45 — Various types of vertical antennas.

Transmission Lines chapter.) Some rules of thumb for arriving at workable gamma-arm and capacitor dimensions are to make the rod length 0.04 to 0.05 λ, its diameter ⅓ to ½ that of the driven element and the center-to-center spacing between the gamma arm and the driven element roughly 0.007 λ. The capacitance of C1 at a 50 Ω matched condition will be about 7 pF per meter of wavelength. The absolute value of C1 will depend on whether the vertical is resonant and on the precise value of the radiation resistance. For best results, make the radiator approximately 3% shorter than the resonant length.

Amateur antenna towers lend themselves to use as shunt-fed verticals, even though an HF-band beam antenna is usually mounted on the tower. The overall system should be close to resonance at the desired operating frequency if a gamma feed is used. The HF-band beam will contribute somewhat to *top loading* of the tower. The natural resonance of such a system can be checked by dropping a #12 or #14 AWG wire from the top of the tower (connecting it to the tower top) to form a folded monopole (Fig 21.45E). A four- or five-turn link can be inserted between the lower end of the drop wire and the ground system. A dip meter is then inserted in the link to determine the resonant frequency.

If the tower is equipped with guy wires, they should be broken up with strain insulators to prevent unwanted loading of the vertical. In such cases where the tower and beam antennas are not able to provide ¼ λ resonance, portions of the top guy wires can be used as top-loading capacitance. Experiment with the guy-wire lengths (using the dip-meter technique) while determining the proper dimensions.

A folded-monopole is depicted in Fig 21.45E. This system has the advantage of increased feed point impedance. Furthermore, an impedance-matching unit can be connected between the bottom of the drop wire and the ground system to permit operation on more than one band. For example, if the tower is resonant on 80 meters, it can be used as shown on 160 and 40 meters with reasonable results, even though it is not electrically long enough on 160 to act as a full-size antenna. The drop wire need not be a specific distance from the tower, but you might try spacings between 12 and 30 inches.

The method of feed shown at Fig 21.45F is commonly referred to as *slant-wire feed*. The guy wires and the tower combine to provide quarter-wave resonance. A matching network is placed between the lower end of one guy wire and ground and adjusted for an SWR of 1:1. It does not matter at which level on the tower the guy wires are connected, assuming that the impedance-matching unit is capable of effecting a match to 50 Ω.

21.3.3 Physically Short Verticals

A group of short vertical radiators is presented in **Fig 21.46**. Illustrations A and B are for top and center loading. A capacitance hat is shown in each example. The hat should be as large as practical to increase the radiation resistance of the antenna and improve the bandwidth. The wire in the loading coil is chosen for the largest gauge consistent with ease of winding and coil-form size. The larger wire diameters will reduce the resistive (I²R) losses in the system. The coil-form material should have a medium or high dielectric constant. Phenolic or fiberglass tubing is entirely adequate.

A base-loaded vertical is shown at C of Fig 21.46. The primary limitation is that the high current portion of the vertical exists in the coil rather than the driven element. With center loading, the portion of the antenna below the coil carries high current, and in the top-loaded version the entire vertical element carries high current. Since the high-current part of the antenna is responsible for most of the radiating, base loading is the least effective of the three methods. The radiation resistance of the coil-loaded antennas shown is usually less than 16 Ω.

A method for using guy wires to top load a short vertical is illustrated in Fig 21.46D. This system works well with gamma feed. The loading wires are trimmed to provide an electrical quarter wavelength for the overall system. This method of loading will result in a higher radiation resistance and greater bandwidth than the systems shown at A through C. If an HF or VHF array is at the top of the tower, it will simply contribute to the top loading.

A three-wire monopole is shown in Fig 21.46E. Two #8 AWG drop wires are connected to the top of the tower and brought to ground level. The wires can be spaced any convenient distance from the tower — normally 12 to 30 inches from the sides. C1 is adjusted for best SWR. This type of vertical has a fairly narrow bandwidth, but because C1 can be motor driven and controlled from the operating position, frequency changes can be accomplished easily. This technique will not be suitable for matching to 50 Ω line unless the tower is less than an electrical quarter wavelength high.

A different method for top loading is shown in Fig 21.46F. Barry Boothe, W9UCW, described this method in December 1974 *QST*.

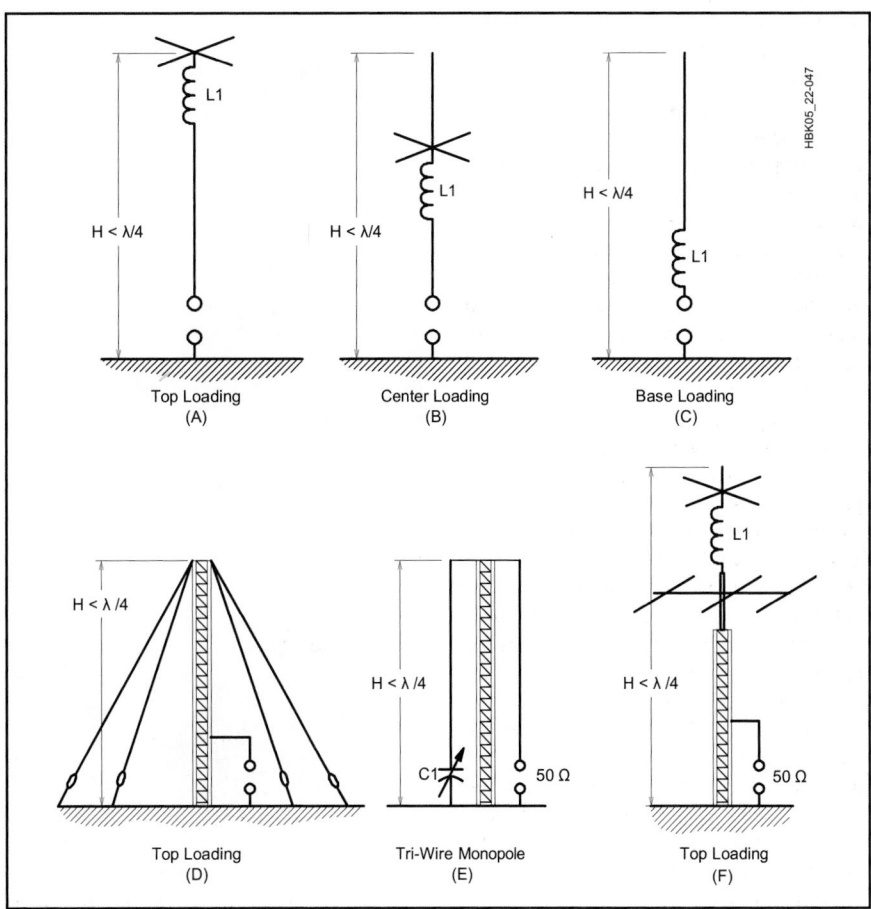

Fig 21.46 — Vertical antennas that are less than one-quarter wavelength in height.

An extension is used at the top of the tower to create an electrical quarter-wavelength vertical. L1 is a loading coil with sufficient inductance to provide antenna resonance. This type of antenna lends itself to operation on 160 meters.

Project: Top-Loaded Low-Band Antenna

The short, top-loaded vertical antenna described here by Dick Stroud, W9SR, is of interest to hams with limited space or who are portable operators. It has been used on 40, 80 and 160 meters. The antenna uses a single 10 foot TV mast section for the vertical radiator, along with a capacitance hat, loading coil and short top mast. The overall height is less than 15 feet, as seen in **Fig 21.47**. The capacitance hat and loading coil assembly can be used with longer vertical radiators with changes to the coil inductance. A drawing showing complete dimensions and construction details is available on the CD-ROM accompanying this book.

CAPACITANCE HAT

The capacitance hat consists of a hub that mounts to the top mast above the coil and six elements made from aluminum rod. The machining, drilling and tapping of the hub assembly can be done by nearly any machine shop if you don't have the facilities. Be sure to use stainless steel hardware throughout. (Thanks to Fred Gantzer, WØAWD, for building the original hub.)

The hub is made of two pieces of ½ inch thick aluminum as shown in **Fig 21.48**. The two pieces are bolted together to form the hub, which slides over the 1⅛ inch top mast. It is held in place with three 10-32 screws. The six elements of the capacitance hat are made from ³⁄₁₆ inch aluminum rod, each 4.5 feet long. These are held in place with 6-32 screws.

LOADING COIL

The coil form is made from fiberglass tubing available from Small Parts, Inc. A 6 inch length of 1.25 inch OD fiberglass tubing (part no. LFT-125/16-30) is centered over a 10 inch length of 1 inch OD tubing (part no. LFT-125/20-30). The tubes telescope tightly and it may be necessary to lightly sand the smaller tube for a smooth fit.

The loading coil is wound on the 1.25 inch OD fiberglass tube. After the coil is optimized, it is covered with a length of shrink sleeving for weather protection. Coil winding information in the drawing on this book's CD-ROM is for use with a 10 foot mast.

The bottom section of the coil assembly fits directly into the tapered upper section of the TV mast and the exposed upper fiberglass section of the coil assembly fits into a 2.5 foot long, 1.125 inch OD, aluminum upper mast. Stainless 8-32 screws join the pieces together and also provide a connection point for the loading coil wires. If you use a painted steel mast, be sure to remove paint from the connection point, and then weather-seal it after adjustment is complete.

The capacitance hat hub slides over the upper mast and is held in place with three screws. The hub is about 6 inches above the

(A)

(B)

coil, but the location can be moved to change the resonance of the antenna slightly.

The completed capacitance hat and loading coil assembly are shown in Fig 21.48. A dab of Glyptol (exterior varnish or Loctite also works) locks the screws in place once adjustments have been made and the antenna is ready for installation.

The coaxial feed line attaches to the bottom of the TV mast. Again, use an 8-32 screw for attachment and clean any paint from the metal. To match the antenna to a 50 Ω transmission line, a small parallel (shunt) inductance is needed at the base. The inductor is air-wound with #16 AWG wire (see **Table 21.5**).

MOUNTING THE ANTENNA

A glass beverage bottle serves well as the base insulator, with the neck of the bottle fitting snugly into the TV mast. To support the base insulator, drill a hole large enough to accept the base of the bottle in the center of a 2 × 6 board about 14 inches long. Nail a piece of ¼ inch plywood over the bottom to keep the bottle from slipping through. To keep the base support board from moving around, drill a couple of holes and secure it to the ground with stakes.

The antenna is top-heavy and will need to be guyed. A simple insulated guy ring can be made from a 2 inch PVC coupling and placed on the mast just below the loading coil. The PVC is locked to the mast with three ¼-20 bolts. They are 1 inch long and have nuts on the inside and outside of the PVC. Three ¼ inch holes are drilled for the guy lines, made from lengths of ³⁄₁₆ inch nonabsorbent rope.

OPERATION

On-air results with a 10 foot mast have been very good, even with low power. The ground system for the early tests was nothing more than an 8 foot ground rod hammered into Hoosier soil. With this setup, and using about 90 W, many DX contacts were made over one week's time and stateside contacts were plentiful.

There is plenty of room to experiment however. Performance could be improved by using an extended radial system or raised and insulated radials. (Expect much better performance over average ground with a

Fig 21.47 — W9SR's short vertical uses top loading and a capacitance hat with a 10 foot TV mast to make a compact antenna for 160, 80 or 40 meters.

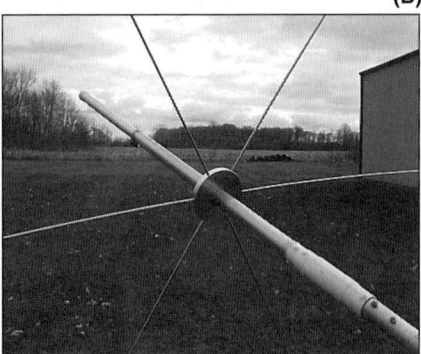

Fig 21.48 — The completed capacitance hat assembly is shown at A, and B shows a view of the upper mast assembly with the loading coil and capacitance hat.

Table 21.5
Shunt Inductor Winding Details

Frequency (MHz)	Turns
1.8	10 turns #16, spaced ⅛ inch
3.5	8.75 turns #16, spaced ⅛ inch
7	7.5 turns #16, spaced ⅛ inch

Note: All inductors are air wound, 1.75 inch ID. Dimensions shown are for use with 10 foot mast.

system of radials) Two, or even three, mast sections could be used with additional guys and proper loading coil inductance. If you use multiple masts, be sure to make a good electrical connection at the joints. The upper assembly is now permanently used to top load a 60 foot pole for transmitting on 160 meters.

21.3.4 Cables and Control Wires on Towers

Most vertical antennas of the type shown in Fig 21.45 and 21.46C-E consist of towers, usually with HF or VHF beam antennas at the top. The rotator control wires and the coaxial feeders to the top of the tower will not affect antenna performance adversely. In fact, they become a part of the composite antenna. To prevent unwanted RF currents from following the wires into the shack, simply dress them close to the tower legs and bring them to ground level. (Running the cables inside the tower works even better.) This decouples the wires at RF. The wires should then be routed along the earth surface (or buried underground) to the operating position. It is not necessary to use bypass capacitors or RF chokes in the rotator control leads if this is done, even when maximum legal power is employed.

21.3.5 Multiband Trap Verticals

The two-band trap vertical antenna of **Fig 21.49** operates in much the same manner as a trap dipole or trap Yagi. The notable difference is that the vertical is one-half of a dipole. The radial system (in-ground or above-ground) functions as a ground plane for the antenna, and provides an equivalent for the missing half of the dipole. Once again, the more effective the ground system, the better will be the antenna performance.

Trap verticals usually are designed to work as ¼ λ radiators. The portion of the antenna below the trap is adjusted as a ¼ λ radiator at the higher proposed operating frequency. That is, a 20/15 meter trap vertical would be a resonant quarter wavelength at 15 meter from the feed point to the bottom of the trap. The trap and that portion of the antenna above the trap (plus the 15 meter section below the trap) constitute the complete antenna during 20 meter operation. But because the trap is in the circuit, the overall physical length of the

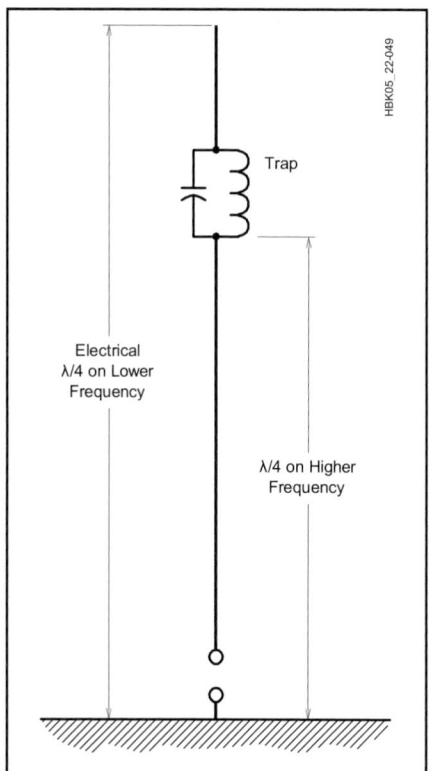

Fig 21.49 — A two-band trap vertical antenna. The trap should be resonated by itself as a parallel resonant circuit at the center of the operating range for the higher frequency band. The reactance of either the inductor or the capacitor range from 100 to 300 Ω. At the lower frequency the trap will act as a loading inductor, adding electrical length to the total antenna.

vertical antenna will be slightly less than that of a single-band, full-size 20 meter vertical.

"Ground-independent" multiband vertical antennas also have traps, but are designed to be electrically longer than ¼ λ. A common electrical length, is ⅜ λ, for example. The "traps" in these antennas are generally not parallel-LC circuits as described above. A variety of techniques are used with both parallel-LC traps and short resonant structures similar to stubs being used to change the antenna's electrical length at different frequencies.

TRAPS

The trap functions as the name implies: the high impedance of the parallel resonant circuit "traps" the 15 meter energy and con-

fines it to the part of the antenna below the trap. (See the **Electrical Fundamentals** chapter for more information on resonant LC circuits.) During 20 meter operation it allows the RF energy to reach all of the antenna. The trap in this example is tuned as a parallel resonant circuit to 21 MHz. At this frequency it electrically disconnects the top section of the vertical from the lower section because it presents a high impedance at 21 MHz, blocking 21 MHz current. Generally, the trap inductor and capacitor have a reactance of 100 to 300 Ω. Within that range it is not critical.

The trap is built and adjusted separately from the antenna. It should be resonated at the center of the portion of the band to be operated. Thus, if one's favorite part of the 15 meter band is between 21.0 and 21.1 MHz, the trap should be tuned to 21.05 MHz.

Resonance is checked by using a dip meter and detecting the dipper signal in a calibrated receiver. An SWR analyzer can also be used. Once the trap is adjusted it can be installed in the antenna, and no further adjustment will be required. It is easy, however, to be misled after the system is assembled: Attempts to check the trap resonance in the antenna will suggest that the trap has moved much lower in frequency (approximately 5 MHz lower in a 20/15 meter vertical). This is because the trap is now part of the overall antenna, and the resultant resonance is that of the total antenna. Measure the trap's resonant frequency separately from the rest of the antenna.

Multiband operation is quite practical by using the appropriate number of traps and tubing sections. The construction and adjustment procedure is the same, regardless of the number of bands covered. The highest frequency trap is always closest to the feed end of the antenna, and the lowest frequency trap is always the farthest from the feed point. As the operating frequency is progressively lowered, more traps and more tubing sections become a functional part of the antenna.

Traps should be weatherproofed to prevent moisture from detuning them. Several coatings of high dielectric compound, such as polystyrene Q Dope or Liquid Electrical Tape, are effective. Alternatively, a protective sleeve of heat-shrink tubing can be applied to the coil after completion. The coil form for the trap should be of high insulating quality and be rugged enough to sustain stress during periods of wind.

21.4 T and Inverted-L Antennas

This section covers variations on the vertical antenna. **Fig 21.50** shows a flat-top T vertical. The T is basically a shortened ¼ λ vertical with the flat-top T section acting as capacitive loading to length the antenna electrically. Dimension H should be as large as possible (up to ¼ λ) for best results. The horizontal section, L, is adjusted to a length that provides resonance. Maximum radiation is polarized vertically despite the horizontal top-loading wire because current in each horizontal half creates out-of-phase radiation that cancels.

A variation of the T antenna is depicted in **Fig 21.51**. This antenna is commonly referred to as an *inverted-L* and is basically a ⁵⁄₁₆ λ vertical bent in the middle so that the top section runs parallel to the ground. Similarly to the T antenna, the vertical section should be as long as possible. L is then added to provide an electrical ⁵⁄₁₆ λ overall.

Because the horizontal section does carry some current, there will be some horizontally-polarized radiation at high angles. This is often considered desirable because it provides local and regional coverage. The horizontal section need not be perfectly horizontal — sloping the wire at a shallow angle from horizontal does not greatly affect antenna performance. This allows the inverted-L to be constructed with a single vertical support.

A sidearm or a length of line attached to a tower can be used to support the vertical section of the T or inverted-L antenna. (Keep the vertical section of the antennas as far from the tower as is practical. Certain combinations of tower height and top loading can create a resonance that interacts severely with the antennas — a 70 foot tower and a 5-element Yagi, for example.)

Both the T and inverted-L antennas are ground-plane antennas and require a good ground system to be effective. If the T or inverted-L are used with a very good ground system, the feed-point impedance will approach 35-40 Ω so that the SWR approaches 1.4:1. The inverted-L is constructed longer than resonance as illustrated in Fig 21.51 so that the feed point resistance increases to 50 Ω plus some inductive reactance due to the extra length. A series capacitor at the feed point then cancels the reactance, leaving a 50 Ω impedance suitable for direct connection by coaxial cable.

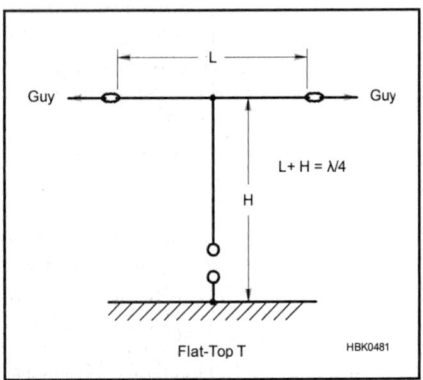

Fig 21.50 — The T antenna is basically a shortened ¼ λ vertical with the flat-top T section acting as capacitive loading to length the antenna electrically.

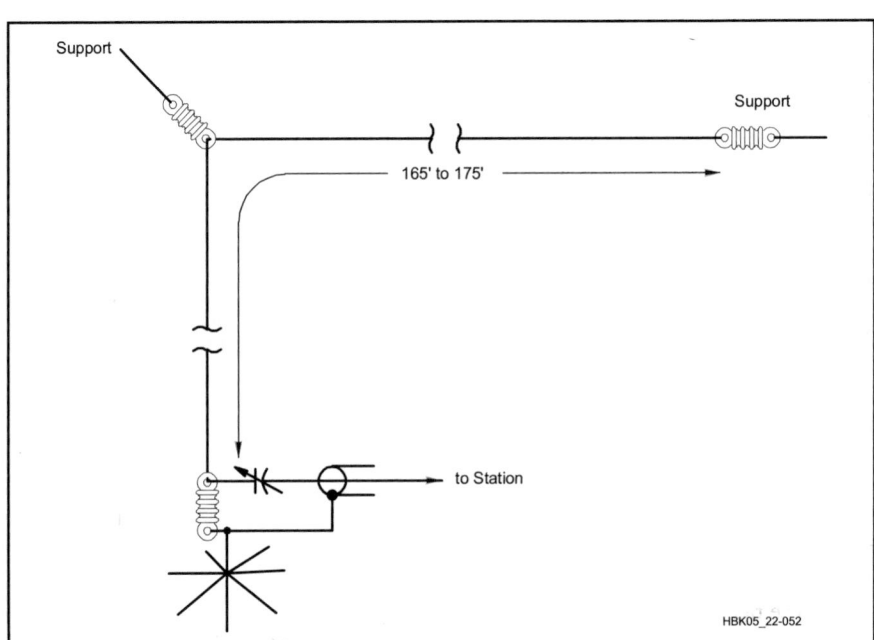

Fig 21.51 — The inverted-L antenna designed for the 1.8 MHz band. Overall wire length is 165 to 175 feet. The variable capacitor has a maximum capacitance of 500 to 800 pF.

21.5 Slopers and Vertical Dipoles

21.5.1 Half-Sloper Antenna

Many hams have had excellent results with *half-sloper* antennas, while others have not had such luck. Investigations by ARRL Technical Advisor John S. Belrose, VE2CV, have brought some insight to the situation through computer modeling and antenna-range tests. The following is taken from VE2CV's Technical Correspondence in Feb 1991 *QST*, pp 39 and 40. Essentially,

the half-sloper is a top-fed vertical antenna that uses the structure at the top of the tower plane (such as a grounded Yagi antenna) as a ground plane and the tower acts as a reflector. See **Fig 21.52**.

For half-slopers, the input impedance, the resonant length of the sloping wire and the antenna pattern all depend on the tower height, the angle (between the sloper and tower) the type of Yagi and the Yagi orientation. Here

are several configurations extracted from VE2CV's work:

At 160 meters — use a 40 meter beam on top of a 95 foot tower with a 55° sloper apex angle. The radiation pattern varies little with Yagi type. The pattern is slightly cardioid with about 8 dB front-to-back ratio at a 25° takeoff angle (see Fig 21.52B and C). Input impedance is about 50 Ω.

At 80 meters — use a 20 meter beam on

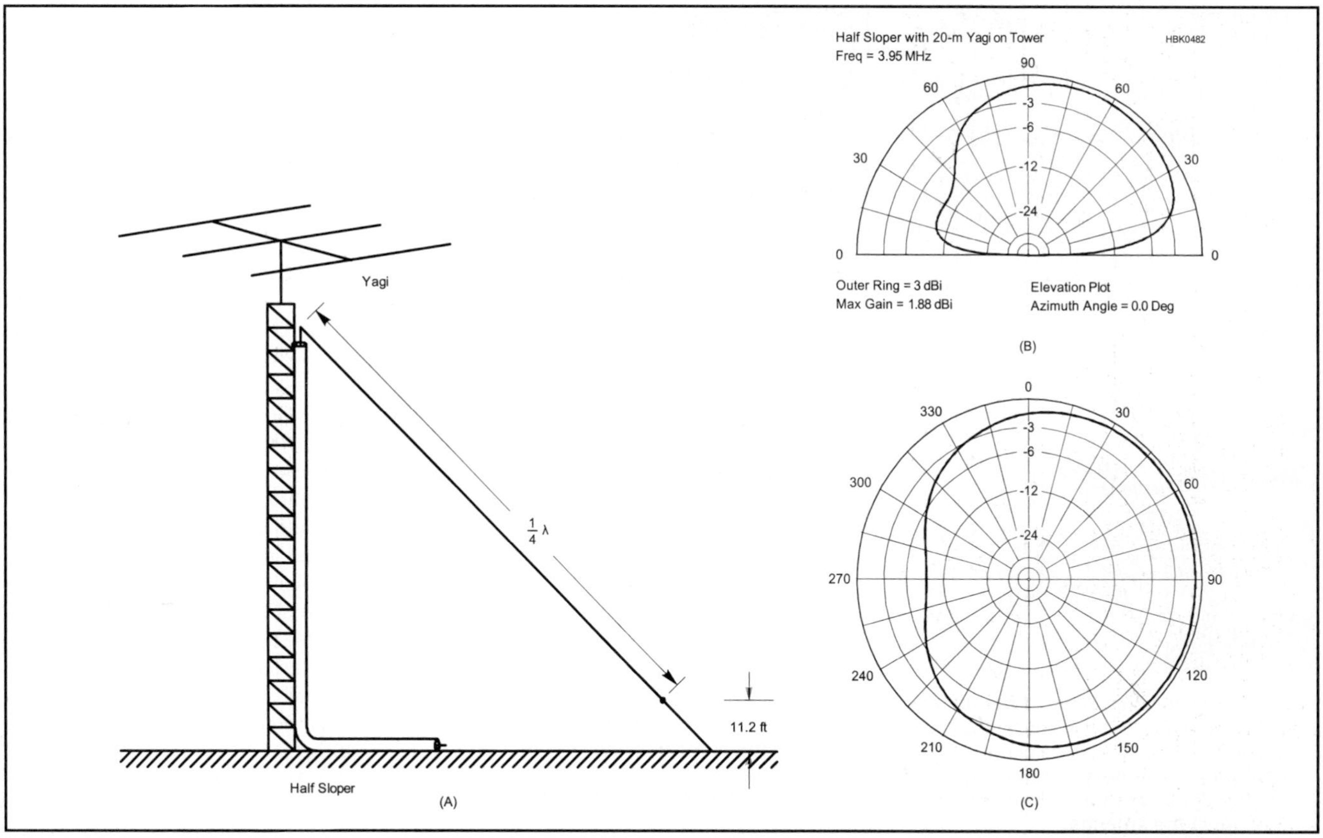

Fig 21.52 — The half-sloper antenna (A). B is the vertical radiation pattern in the plane of a half sloper, with the sloper to the right. C is the azimuthal pattern of the half sloper (90° azimuth is the direction of the sloping wire). Both patterns apply to 160 and 80 meter antennas described in the text.

top of a 50 foot tower with a 55° sloper apex angle. The radiation pattern and input impedance are similar to those of the 160 meter half-sloper.

At 40 meters — use a 20 meter beam on top of a 50 foot tower with a 55° sloper apex angle. The radiation pattern and impedance depend strongly on the azimuth orientation of the Yagi. Impedance varies from 76 to 127 Ω depending on Yagi direction.

Project: Half-Wave Vertical Dipole (HVD)

Chuck Hutchinson, K8CH, describes a 15 meter vertical dipole (HVD) that he built for the ARRL book, *Simple and Fun Antennas for Hams.* The performance of this antenna, with its base at 14 feet, compares favorably with a horizontal dipole at 30 feet when making intercontinental QSOs.

CONSTRUCTION OF A 15 METER HVD

The 15 meter HVD consists of four 6 foot lengths of 0.875 inch aluminum tube with 0.058 wall thickness. In addition there are two 1 foot lengths of 0.75 inch tubing for splices, and two 1 foot lengths of 0.75 inch

Table 21.6
HVD Dimensions
Length using 0.875-inch aluminum tubing

MHz	Feet	Inches
18.11	33	11
21.2	22	0
24.94	18	9
28.4	16	5

These lengths should divided by two to determine the length of the dipole legs

fiberglass rod for insulators. See **Table 21.6** for dimensions.

Start by cutting off 1 foot from a 6 foot length of 0.875 inch tubing. Next, insert six inches of one of the 1-foot-long 0.75 inch tubes into the machine-cut end of your tubing and fasten the tubes together. Now, slide an end of a 6 foot length of 0.875 tube over the protruding end of the 0.75 tube and fasten them together. Repeat this procedure with the remaining 0.875 inch tubing.

You should now have two 11-foot-long elements. As you can see in **Fig 21.53**, K8CH was temporarily out of aluminum pop rivets, so he used sheet metal screws. Either will

Fig 21.53 — Element splice uses a 1 foot length of 0.75 inch tubing inserted into the 0.875 inch sections to join them together. Self-tapping sheet-metal screws are used in this photo, but aluminum pop rivets or machine screws with washers and nuts can be used.

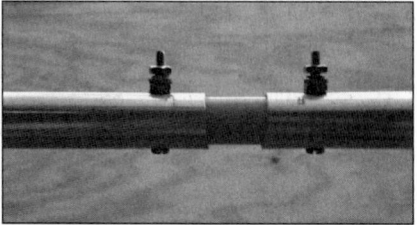

Fig 21.54 — The center insulator of the 15 meter HVD is a 1 foot length of 0.75 inch fiberglass rod. Insulator and elements have been drilled to accept #8 hardware.

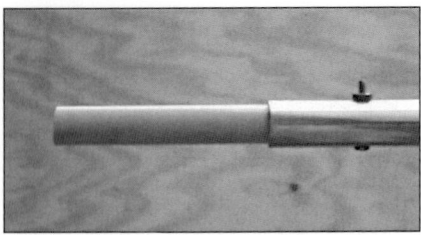

Fig 21.55 — The HVD base insulator is a 1 foot length of 0.75 inch fiberglass rod.

Fig 21.56 — Guys are made of Dacron line that is attached to the HVD by a stainless-steel worm-screw-type hose clamp. A self-tapping sheet-metal screw (not visible in the photo) prevents the clamp from sliding down the antenna.

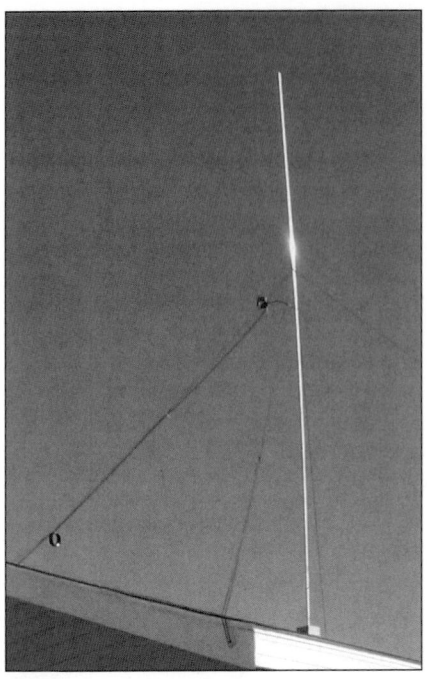

Fig 21.58 — The HVD installed at K8CH. An eye screw that is used for securing one of the guy lines is visible in the foreground. You can also see the two choke baluns that are used in the feed system (see text).

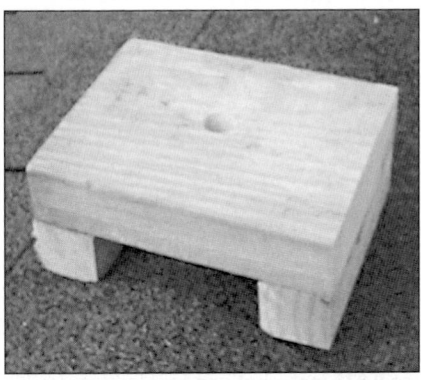

Fig 21.57 — At K8CH, the HVD base insulator sits in this saddle-shaped wooden fixture. This was photo was taken before the fixture was painted — a necessary step to protect against the weather.

work fine, but pop rivets can easily be drilled out and the antenna disassembled if you ever want to make changes.

Because hand-made cuts are not perfectly square, put those element ends at the center of the antenna. Slip these cut ends over the ends of a 1 foot length of 0.75 inch fiberglass rod. This rod serves as the center insulator. Leave about a 1 inch gap at the center. Drill aluminum and fiberglass for #8 hardware as shown in **Fig 21.54**.

Now, slip half of the remaining 1 foot length of 0.75 inch fiberglass rod into one end of the dipole. (This end will be the bottom end or base.) Drill and secure with #8 hardware. See **Fig 21.55**.

The final step is to secure the guy wires to your vertical. You can see how K8CH did that in **Fig 21.56**. Start by drilling a pilot hole and then drive a sheet metal screw into the antenna about a foot above the center. The purpose of that screw is to prevent the clamp and guys from sliding down the antenna.

The guys are clean lengths of 3/16 inch Dacron line. (The Dacron serves a dual purpose: it supports the antenna vertically, and it acts as an insulator.) Tie secure knots into the guy ends and secure these knotted ends to the antenna with a stainless-steel worm-screw-type hose clamp. Take care to not over tighten the clamps. You don't want the clamp to slip (the knots and the sheet-metal screw will help), but you especially don't want to cut your guy lines. Your antenna is ready for installation.

INSTALLATION

Installation requires two things. First, a place to sit or mount the base insulator. Second, you need anchors for the support guys.

K8CH used a piece of 2 × 6 lumber to make a socket to hold the HVD base securely in place. He drilled a ¾ inch-deep hole with a ¾ inch spade bit. A couple of pieces of 2 × 2 lumber at the ends of the base form a saddle which nicely straddles the ridge at the peak of his garage roof. You can see this in **Fig 21.57**. The dimensions are not critical. Paint your base to protect it from the weather.

BALUN

This antenna needs a common-mode choke balun to ensure that stray RF doesn't flow on the shield of the coax. (See the **Transmission Lines** chapter for more information on choke baluns.) Unlike a horizontal dipole, don't consider it an option to omit the common-mode choke when building and installing an HVD.

You can use 8 feet of the RG-213 feed line wound into 7 turns for a balun. Secure the turns together with electrical tape so that each turn lies parallel with the next turn, forming a solenoid coil. Secure the feed line and balun to one of the guy lines with UV-resistant cable ties.

Because the feed line slants away from the antenna, you'll want to do *all* that you can to eliminate common-mode currents from the feed line. For that reason, make another balun about 11.5 feet from the first one. This balun also consists of 8 feet of the RG-213 feed line wound into 7 turns. See **Fig 21.58** for a photo of the installed antenna.

Project: Compact Vertical Dipole (CVD)

An HVD for 20 meters will be about 33 feet tall, and for 30 meters, it will be around 46 feet tall. Even the 20 meter version can prove to be a mechanical challenge. The compact vertical dipole (CVD), designed by Chuck Hutchinson, K8CH, uses capacitance loading to shorten the antenna. Starting with the 15 meter HVD described in the previous project, Chuck added capacitance loading wires to lower the resonance to 30 meters. Later, he shortened the wires to move resonance to the 20 meter band. This project describes those two CVDs.

PERFORMANCE ISSUES

Shortened antennas frequently suffer reduced performance caused by the shortening. A dipole that is less than ½ λ in length is a compromise antenna. The issue becomes how much is lost in the compromise. In this case there are two areas of primary interest, radiation efficiency and SWR bandwidth.

Radiation Efficiency

Capacitance loading at the dipole ends is the most efficient method of shortening the antenna. Current distribution in the high-current center of the antenna remains virtually unchanged. Since radiation is related directly to current, this is the most desirable form of loading. Computer modeling shows that radiation from a 30 meter CVD is only 0.66 dB less than that from a full-size 30 meter HVD when both have their bases 8 feet above ground. The angle of maximum radiation shifts up a bit for the CVD. Not a bad compromise when you consider that the CVD is 22 feet long compared to the approximately 46 foot length of the HVD.

SWR and SWR Bandwidth

Shortened antennas usually have lower radiation resistance and less SWR bandwidth than the full-size versions. The amount of change in the radiation resistance is related to the amount and type of loading (shortening), being lower with shorter the antennas. This can be a benefit in the case of a shortened vertical dipole. In Fig 21.2 you can see that vertical dipoles have a fairly high radiation resistance. With the dipole's lower end $\frac{1}{8}$ λ above ground, the radiation resistance

is roughly 80 Ω. In this case, a shorter antenna can have a better SWR when fed with 50 Ω coax.

SWR bandwidth tends to be wide for vertical dipoles in general. A properly designed CVD for 7 MHz or higher should give you good SWR (1.5:1 or better) across the entire band!

As you can see, in theory the CVD provides excellent performance in a compact package. Experience confirms the theory.

CONSTRUCTION

To convert the K8CH 15 meter HVD to 20 or 30 meters, you'll need to add four loading wires at the top and four more at the bottom of the HVD. The lengths are shown in **Table 21.7**. The upper wires droop at a 45° angle and the lower wires run horizontally. The antenna is supported by four guy lines. See **Fig 21.59**. You can connect the wires to the vertical portion with #8 hardware. Crimp and solder terminals on the wire ends to make

connections easier. The technique is illustrated in **Fig 21.60**.

The upper loading wires can be extended with insulated line and used for additional guying. The lower wires are extended with insulated line and fasten to the guy lines so that the lower wires run horizontally.

Prune the lower wires for best SWR across the band of interest. The K8CH CVD has its base at 14 feet. This antenna has an SWR of less than 1.2:1 on 30 meters and less than 1.3:1 across the entire 20 meter band.

EXPERIENCE

The 30 meter CVD was compared to a ground-mounted quarter-wave vertical and a horizontal dipole at 30 feet. In tests, the CVD was always the superior antenna. Every DX station that was called by K8CH responded with one or two calls. What more could you ask for?

Later, the CVD loading wires were shortened for operation on 20 meters. Once again the results were very encouraging. Many contest QSOs were entered in the log using this antenna.

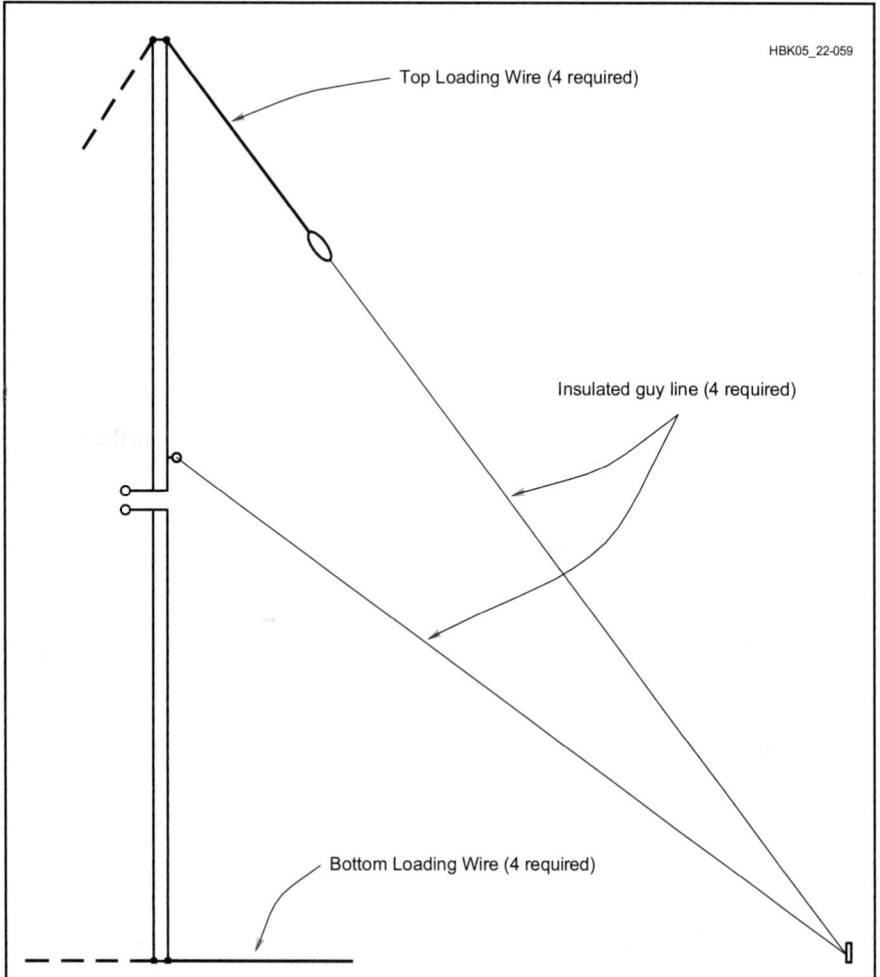

HBK05_22-059

Top Loading Wire (4 required)

Insulated guy line (4 required)

Bottom Loading Wire (4 required)

Fig 21.59 — The CVD consists of a vertical dipole and loading wires. Only one set of the four loading wires and only one guy line is shown in this drawing. See text for details.

Table 21.7
CVD Loading Wires
Length using #14 AWG insulated copper wire

MHz	Feet	Inches	
10.1	6	0	Top & Bottom
14	4	2¼	Top
14	3	½	Bottom

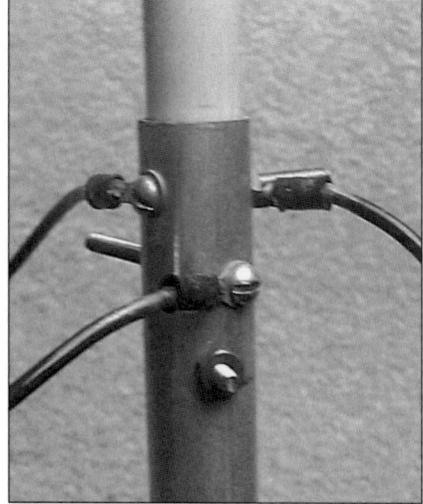

Fig 21.60 — CVD loading wires can be attached using #8 hardware. Crimp and solder terminals on the wire ends to make connections easier.

Project: All-Wire 30 Meter CVD

If you have a tree or other support that will support the upper end of a CVD at 32 feet above the ground, you might want to consider an all-wire version of the 30 meter CVD. The vertical is 24 feet long and it will have an SWR of less than 1.1:1 across the band. The four loading wires at top and bottom are each 5 feet, 2 inches long.

The configuration is shown in **Fig 21.61**. As with any vertical dipole, you'll need to use a balun between the feed line and the antenna.

Alternatively you can use two loading wires at the top and two at the bottom. In this case each of the loading wires is 8 feet, 7.5 inches long.

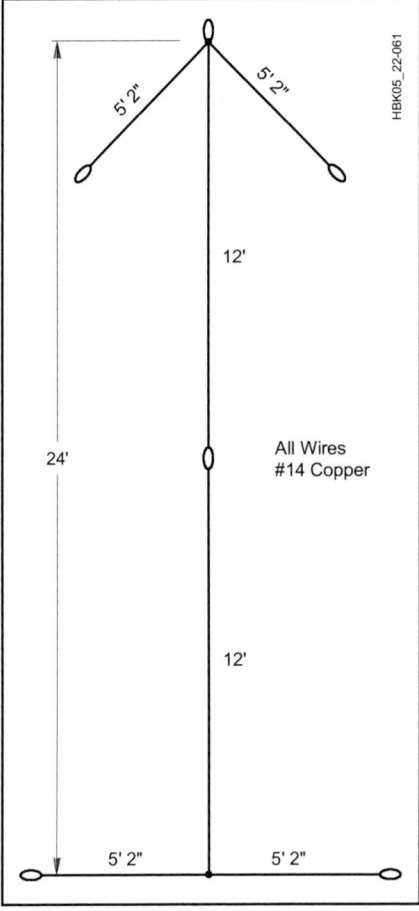

Fig 21.61 — The all-wire 30 meter CVD consists of a vertical dipole and loading wires. It can be made entirely with #14 AWG wire. Support lines have been omitted for simplicity. See text for details.

21.6 Yagi Antennas

Most antennas described earlier in this chapter have unity gain compared to a dipole, or just slightly more. For the purpose of obtaining gain and directivity it is convenient to use a Yagi-Uda *beam* antenna. The former is commonly called a *Yagi*. There are other forms of directive antennas, but the Yagi is by far the most popular used by amateurs. (For more information on phased arrays and other types of directive antennas, see the *ARRL Antenna Book*.)

Most operators prefer to erect these antennas for horizontal polarization, but they can be used as vertically polarized antennas merely by rotating the elements by 90°. In effect, the beam antenna is turned on its side for vertical polarization. The number of elements used will depend on the gain desired and the limits of the supporting structure. At HF, many amateurs obtain satisfactory results with only two elements in a beam antenna, while others have four or five elements operating on a single amateur band, called a *mono-*

band beam. On VHF and above, Yagis with many elements are common, particularly for simplex communication without repeaters. For fixed point-to-point communications, such as repeater links, Yagis with three or four elements are more common.

Regardless of the number of elements used, the height-above-ground considerations discussed earlier for dipole antennas remain valid with respect to the angle of radiation. This is demonstrated in **Fig 21.62** at A and B where a comparison of radiation characteristics is given for a 3 element Yagi at one-half and one wavelength above average ground. It can be seen that the higher antenna (Fig 21.62B) has a main lobe that is more favorable for DX work (roughly 15°) than the lobe of the lower antenna in Fig 21.62A (approximately 30°). The pattern at B shows that some useful high-angle radiation exists also, and the higher lobe is suitable for short-skip contacts when propagation conditions dictate the need.

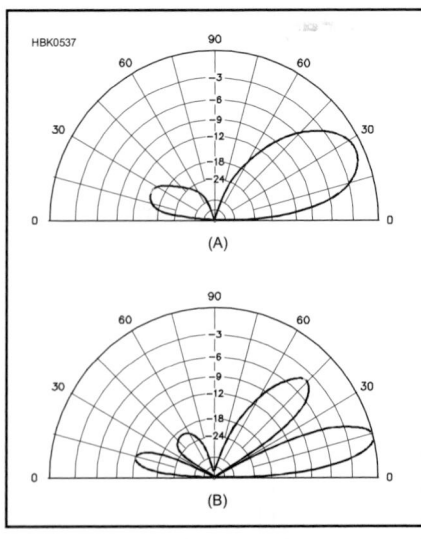

Fig 21.62 — Elevation-plane response of a 3 element Yagi placed ½ λ above perfect ground at A and the same antenna spaced 1 λ above ground at B.

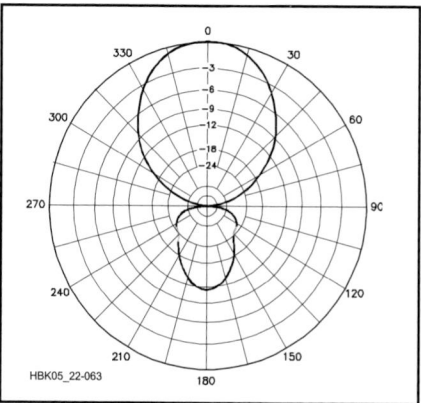

Fig 21.63 — Azimuthal pattern of a typical three-element Yagi in free space. The Yagi's boom is along the 0° to 180° axis.

The azimuth pattern for the same antenna is provided in **Fig 21.63**. (This is a free-space pattern, so the pattern is taken in the plane of the antenna. Remember that azimuth patterns taken over a reflecting surface must also specify the elevation angle at which the pattern was measured or calculated.) Most of the power is concentrated in the main lobe at 0° azimuth. The lobe directly behind the main lobe at 180° is often called the *back lobe* or *rear lobe*. The front-to-back ratio (F/B) of this antenna is just less than 12 dB — the peak power difference, in decibels, between the main lobe at 0° and the rearward lobe at 180°. It is infrequent that two 3 element Yagis with different element spacing and tuning will yield the same lobe patterns. The patterns also change with frequency of operation. The pattern of Fig 21.63 is shown only for illustrative purposes.

21.6.1 Parasitic Excitation

In a Yagi antenna only one element (the *driven element*) is connected to the feed line. The additional elements are *coupled* to the driven element because they are so close. (Element-to-element spacing in a Yagi antenna is generally on the order of $\frac{1}{10}$ to $\frac{1}{8}$ wavelength.) This *mutual coupling* results in currents being induced in the non-driven elements from the radiated field of the driven element. These elements are called *parasitic elements* and the Yagi antenna is therefore a *parasitic array*. (An antenna in which multiple elements all receive power from the transmitter is called a *driven array*.) The currents induced in the parasitic elements also result in radiated fields, just as if the current were the result of power from a feed line. This is called *re-radiation*, and it has a 180° phase shift from the current-inducing field. The combination of the field radiated by the driven element, the fields from the parasitic elements, and the physical spacing of the elements results in the

fields having the proper phase relationship so as to focus the radiated energy in the desired direction and reject it in other directions.

The parasitic element is called a *director* when it reinforces radiation along a line pointing to it from the driven element, and a *reflector* in the opposite case. Whether the parasitic element is a director or reflector depends on the parasitic element tuning, which is usually adjusted by changing its length. The structure on which the elements are mounted is called the *boom* of the antenna.

21.6.2 Yagi Gain, Front-to-Back Ratio and SWR

The gain of a Yagi antenna with parasitic elements varies with the spacing and tuning of the elements. Element tuning is a function of length, diameter and *taper schedule* (the steps in length and diameter) if the element is constructed with telescoping tubing. For any given number of elements and the spacing between them, there is a tuning condition that will result in maximum gain. However, the maximum front-to-back ratio seldom, if ever, occurs at the same condition that gives maximum forward gain. The impedance of the driven element in a parasitic array, and thus the SWR, also varies with the tuning and spacing.

It is important to remember that all these parameters change as the operating frequency is varied. For example, if you operate both the CW and phone portions of the 20 meter band with a Yagi antenna, you probably will want an antenna that *spreads out* the performance over most of the band. Such designs typically must sacrifice a little gain in order to achieve good F/B and SWR performance across the band.

Gain and F/B performance generally improve with the number of elements. In Yagi antennas with more than three elements (a driven element and one director and reflector), the additional elements are added as directors, since little additional benefit is obtained from multiple reflectors. Wider spacing also improves gain and F/B up to a certain point, depending on a number of factors, beyond which performance begins to fall. Optimizing element spacing is a complex problem and no single spacing satisfies all design requirements. For the lower HF bands, the size of the antenna quickly becomes impractical for truly *optimal* designs, and compromise is necessary.

21.6.3 Two-Element Beams

A two-element beam is useful — especially where space or other considerations prevent the use of a three-element, or larger, beam. The general practice is to tune the parasitic element as a reflector and space it about

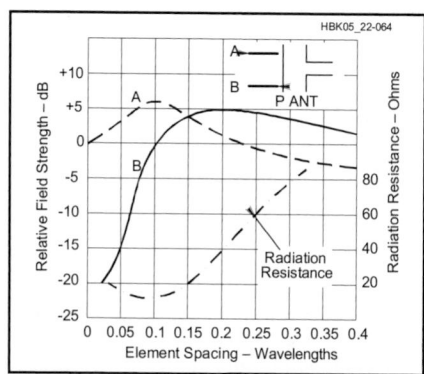

Fig 21.64 — Gain vs element spacing for a two-element Yagi, having one driven and one parasitic element. The reference point, 0 dB, is the field strength from a half-wave antenna alone.

0.15 λ from the driven element, although some successful antennas have been built with 0.1-λ spacing and director tuning.

Gain vs element spacing for a two-element antenna is given in **Fig 21.64** for the special case where the parasitic element is resonant. It is indicative of the performance to be expected under maximum-gain tuning conditions. Changing the tuning of the driven element in a Yagi or quad will not materially affect the gain or F/R. Thus, only the spacing and the tuning of the single parasitic element have any effect on the performance of a 2 element Yagi.

In Fig 21.64, the greatest gain is in the direction A (in which the parasitic element is acting as a director) at spacings of less than 0.14 λ, and in direction B (in which the parasitic element is a reflector) at greater spacings. The front-to-back ratio is the difference in decibels between curves A and B. The figure also shows variation in radiation resistance of the driven element.

These curves are for the special case of a self-resonant parasitic element, but are representative of how a two-element Yagi works. At most spacings the gain as a reflector can be increased by slight lengthening of the parasitic element; the gain as a director can be increased by shortening. This also improves the front-to-rear ratio.

Most two-element Yagi designs achieve a compromise F/R of about 10 dB, together with an acceptable SWR and gain across a frequency band with a percentage bandwidth less than about 4%.

21.6.4 Three-Element Beams

A theoretical investigation of the three-element case (director, driven element and reflector) has indicated a maximum gain of about 9.7 dBi (7.6 dBd). A number of experimental investigations have shown that the spacing between the driven element and

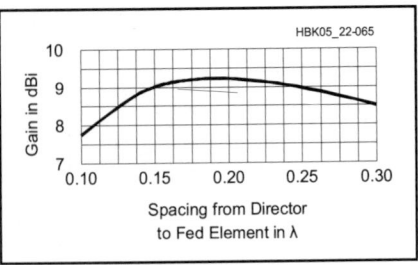

Fig 21.65 — General relationship of gain of three-element Yagi vs director spacing, the reflector being fixed at 0.2 λ. This antenna is tuned for maximum forward gain.

reflector for maximum gain is in the region of 0.15 to 0.25 λ. With 0.2 λ reflector spacing, **Fig 21.65** shows that the gain variation with director spacing is not especially critical. Also, the overall length of the array (boom length in the case of a rotatable antenna) can be anywhere between 0.35 and 0.45 λ with no appreciable difference in the maximum gain obtainable.

If maximum gain is desired, wide spacing of both elements is beneficial because adjustment of tuning or element length is less critical and the input resistance of the driven element is generally higher than with close

spacing. A higher input resistance improves the efficiency of the antenna and makes a greater bandwidth possible. However, a total antenna length, director to reflector, of more than 0.3 λ at frequencies of the order of 14 MHz introduces difficulty from a construction standpoint. Lengths of 0.25 to 0.3 λ are therefore used frequently for this band, even though they are less than optimum from the viewpoint of maximum gain.

In general, Yagi antenna gain drops off less rapidly when the reflector length is increased beyond the optimum value than it does for a corresponding decrease below the optimum

Table 21.8
Standard Sizes of Aluminum Tubing
6061-T6 (61S-T6) Round Aluminum Tube in 12-ft Lengths

OD (in)	Wall Thickness (in)	stubs ga	ID (in)	Approx Weight (lb) per ft	per length	OD (in)	Wall Thickness (in)	stubs ga	ID (in)	Approx Weight (lb) per ft	per length
3/16	0.035	no. 20	0.117	0.019	0.228	1 1/8	0.035	no. 20	1.055	0.139	1.668
	0.049	no. 18	0.089	0.025	0.330		0.058	no. 17	1.009	0.228	2.736
1/4	0.035	no. 20	0.180	0.027	0.324	1 1/4	0.035	no. 20	1.180	0.155	1.860
	0.049	no. 18	0.152	0.036	0.432		0.049	no. 18	1.152	0.210	2.520
	0.058	no. 17	0.134	0.041	0.492		0.058	no. 17	1.134	0.256	3.072
5/16	0.035	no. 20	0.242	0.036	0.432		0.065	no. 16	1.120	0.284	3.408
	0.049	no. 18	0.214	0.047	0.564		0.083	no. 14	1.084	0.357	4.284
	0.058	no. 17	0.196	0.055	0.660	1 3/8	0.035	no. 20	1.305	0.173	2.076
3/8	0.035	no. 20	0.305	0.043	0.516		0.058	no. 17	1.259	0.282	3.384
	0.049	no. 18	0.277	0.060	0.720	1 1/2	0.035	no. 20	1.430	0.180	2.160
	0.058	no. 17	0.259	0.068	0.816		0.049	no. 18	1.402	0.260	3.120
	0.065	no. 16	0.245	0.074	0.888		0.058	no. 17	1.384	0.309	3.708
7/16	0.035	no. 20	0.367	0.051	0.612		0.065	no. 16	1.370	0.344	4.128
	0.049	no. 18	0.339	0.070	0.840		0.083	no. 14	1.334	0.434	5.208
	0.065	no. 16	0.307	0.089	1.068		*0.125	1/8"	1.250	0.630	7.416
1/2	0.028	no. 22	0.444	0.049	0.588		*0.250	1/4"	1.000	1.150	14.823
	0.035	no. 20	0.430	0.059	0.708	1 5/8	0.035	no. 20	1.555	0.206	2.472
	0.049	no. 18	0.402	0.082	0.948		0.058	no. 17	1.509	0.336	4.032
	0.058	no. 17	0.384	0.095	1.040	1 3/4	0.058	no. 17	1.634	0.363	4.356
	0.065	no. 16	0.370	0.107	1.284		0.083	no. 14	1.584	0.510	6.120
5/8	0.028	no. 22	0.569	0.061	0.732	1 7/8	0.508	no. 17	1.759	0.389	4.668
	0.035	no. 20	0.555	0.075	0.900	2	0.049	no. 18	1.902	0.350	4.200
	0.049	no. 18	0.527	0.106	1.272		0.065	no. 16	1.870	0.450	5.400
	0.058	no. 17	0.509	0.121	1.452		0.083	no. 14	1.834	0.590	7.080
	0.065	no. 16	0.495	0.137	1.644		*0.125	1/8"	1.750	0.870	9.960
3/4	0.035	no. 20	0.680	0.091	1.092		*0.250	1/4"	1.500	1.620	19.920
	0.049	no. 18	0.652	0.125	1.500	2 1/4	0.049	no. 18	2.152	0.398	4.776
	0.058	no. 17	0.634	0.148	1.776		0.065	no. 16	2.120	0.520	6.240
	0.065	no. 16	0.620	0.160	1.920		0.083	no. 14	2.084	0.660	7.920
	0.083	no. 14	0.584	0.204	2.448	2 1/2	0.065	no. 16	2.370	0.587	7.044
7/8	0.035	no. 20	0.805	0.108	1.308		0.083	no. 14	2.334	0.740	8.880
	0.049	no. 18	0.777	0.151	1.810		*0.125	1/8"	2.250	1.100	12.720
	0.058	no. 17	0.759	0.175	2.100		*0.250	1/4"	2.000	2.080	25.440
	0.065	no. 16	0.745	0.199	2.399	3	0.065	no. 16	2.870	0.710	8.520
1	0.035	no. 20	0.930	0.123	1.467		*0.125	1/8"	2.700	1.330	15.600
	0.049	no. 18	0.902	0.170	2.040		*0.250	1/4"	2.500	2.540	31.200
	0.058	no. 17	0.884	0.202	2.424						
	0.065	no. 16	0.870	0.220	2.640						
	0.083	no. 14	0.834	0.281	3.372						

*These sizes are extruded; all other sizes are drawn tubes. Shown here are standard sizes of aluminum tubing that are stocked by most aluminum suppliers or distributors in the United States and Canada.

value. The opposite is true of a director. It is therefore advisable to err, if necessary, on the long side for a reflector and on the short side for a director. This also tends to make the antenna performance less dependent on the exact frequency at which it is operated. An increase above the design frequency has the same effect as increasing the length of both parasitic elements, while a decrease in frequency has the same effect as shortening both elements. By making the director slightly short and the reflector slightly long, there will be a greater spread between the upper and lower frequencies at which the gain starts to show a rapid decrease.

21.6.5 Construction of Yagi Antennas

Most beams and verticals are made from sections of aluminum tubing. Compromise beams have been fashioned from less-expensive materials such as electrical conduit (steel) or bamboo poles wrapped with conductive tape or aluminum foil. The steel conduit is heavy, is a poor conductor and is subject to rust. Similarly, bamboo with conducting material attached to it may deteriorate rapidly in the weather. Given the drawbacks of alternative materials, aluminum tubing (or rod for VHF and UHF Yagis) is far and away the best choice for antenna construction.

For reference, **Table 21.8** details the standard sizes of aluminum tubing, available in many metropolitan areas. Dealers may be found in the Yellow Pages under *Aluminum*. Tubing usually comes in 12 foot lengths, although 20 foot lengths are available in some sizes. Your aluminum dealer will probably also sell aluminum plate in various thicknesses needed for boom-to-mast and boom-to-element connections. Distributors of

antenna towers and masts often sell aluminum tubing, as well.

Aluminum is rated according to its hardness. The most common material used in antenna construction is grade 6061-T6. This material is relatively strong and has good workability. In addition, it will bend without taking a *set*, an advantage in antenna applications where the pieces are constantly flexing in the wind. The softer grades (5051, 3003 and so on) will bend much more easily, while harder grades (7075 and so on) are more brittle.

Wall thickness is of primary concern when selecting tubing. It is of utmost importance that the tubing fits snugly where the element sections join. Sloppy joints will make a mechanically unstable antenna. The magic wall thickness is 0.058 inch. For example (from Table 21.8), 1 inch outside diameter (OD) tubing with a 0.058 inch wall has an inside diameter (ID) of 0.884 inch. The next smaller size of tubing, ⅞ inch, has an OD of 0.875 inch. The 0.009 inch difference provides just the right amount of clearance for a snug fit.

Fig 21.66 shows several methods of fastening antenna element sections together. The slot and hose clamp method shown at the upper left is probably the best for joints where adjustments are needed. Generally, one adjustable joint on each side of the element is sufficient to tune the antenna — usually the tips at each end of an element are made adjustable. Stainless steel hose clamps (beware — some "stainless steel" models do not have a stainless screw and will rust) are recommended for longest antenna life.

The remaining photos show possible fastening methods for joints that are not adjustable. At the upper right, machine screws and nuts hold the elements in place. At the lower left, sheet metal screws are used. At the lower

right, rivets secure the tubing. If the antenna is to be assembled permanently, rivets are the best choice. Once in place, they are permanent. They will never work free, regardless of vibration or wind. If aluminum rivets with aluminum mandrels are employed, they will never rust. Also, being aluminum, there is no danger of corrosion from interaction between dissimilar metals. If the antenna is to be disassembled and moved periodically, either machine or sheet metal screws will work. If machine screws are used, however, take precautions to keep the nuts from vibrating free. Use of lock washers, lock nuts and flexible adhesive such as silicone bathtub sealant will keep the hardware in place. For portable or temporary use, such as Field Day, rivets may be held in place with electrical tape and removed when the operation is finished.

Use of a conductive grease at the element joints is essential for long life. Left untreated, the aluminum surfaces will oxidize in the weather, resulting in a poor connection. Some trade names for this conductive grease are Penetrox and Noalox. Many electrical supply houses carry these products.

DRIVEN ELEMENT

The ARRL recommends *plumbers delight* construction, in which all elements are mounted directly on, and grounded to, the boom. This puts the entire array at dc ground potential, affording better lightning protection. A gamma- or T-match section can be used for matching the feed line to the array.

An alternative method is to insulate the driven element from the boom, but use a *hairpin* or *beta match*, the center point of which is electrically neutral and can be attached directly to the boom, restoring the dc ground for the driven element.

Direct feed designs in which the feed point impedance of the driven element is close to 50 Ω, requiring no impedance matching structure, presents some issues. First, a current or choke balun should be used (see the **Transmission Lines** chapter) to prevent the outer surface of the feed line shield from interacting with the antenna directly or by picking up the radiated signal. Such interaction can degrade the antenna's radiation pattern, especially by compromising signal rejection to the side and rear. Second, the driven element must be insulated from the boom, requiring some additional mechanical complexity.

BOOM MATERIAL

The boom size for a rotatable Yagi or quad should be selected to provide stability to the entire system. The best diameter for the boom depends on several factors, but mostly the element weight, number of elements and overall length. Two-inch-diameter booms should not be made any longer than 24 feet

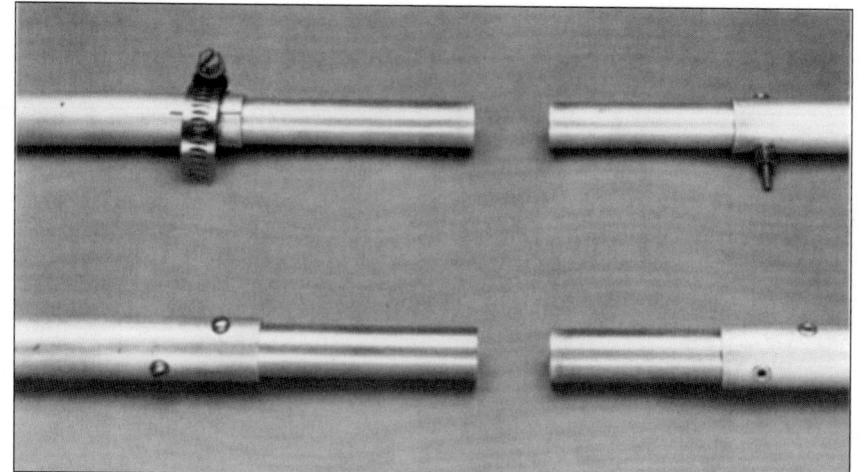

Fig 21.66 — Some methods of connecting telescoping tubing sections to build beam elements. See text for a discussion of each method.

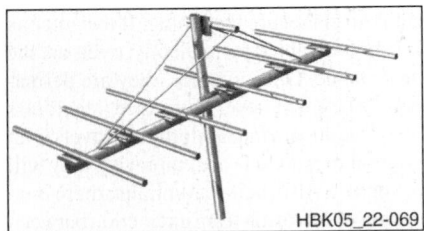

HBK05_22-069

Fig 21.67 — A long boom needs both vertical and horizontal support. The crossbar mounted above the boom can support a double truss, which will help keep the antenna in position.

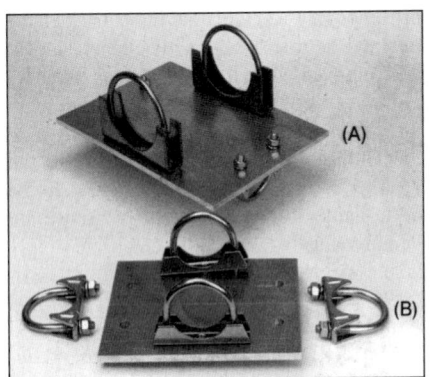

(A)

(B)

Fig 21.68 — The boom-to-element plate at A uses muffler-clamp-type U-bolts and saddles to secure the round tubing to the flat plate. The boom-to-mast plate at B is similar to the boom-to-element plate. The main difference is the size of materials used.

Table 21.9
10 Meter Optimized Yagi Designs

	Spacing Between Elements (in)	Seg 1 Length (in)	Seg 2 Length (in)	Seg 3 Length (in)	Midband Gain F/R
310-08					
Refl	0	24	18	66.750	7.2 dBi
DE	36	24	18	57.625	22.9 dB
Dir 1	54	24	18	53.125	
410-14					
Refl	0	24	18	64.875	8.4 dBi
DE	36	24	18	58.625	30.9 dB
Dir 1	36	24	18	57.000	
Dir 2	90	24	18	47.750	
510-24					
Refl	0	24	18	65.625	10.3 dBi
DE	36	24	18	58.000	25.9 dB
Dir 1	36	24	18	57.125	
Dir 2	99	24	18	55.000	
Dir 3	111	24	18	50.750	

Note: For all antennas, the tube diameters are: Seg 1=0.750 inch, Seg 2=0.625 inch, Seg 3=0.500 inch.

Table 21.10
12 Meter Optimized Yagi Designs

	Spacing Between Elements (in)	Seg 1 Length (in)	Seg 2 Length (in)	Seg 3 Length (in)	Midband Gain F/R
312-10					
Refl	0	36	18	69.000	7.5 dBi
DE	40	36	18	59.125	24.8 dB
Dir 1	74	36	18	54.000	
412-15					
Refl	0	36	18	66.875	8.5 dBi
DE	46	36	18	60.625	27.8 dB
Dir 1	46	36	18	58.625	
Dir 2	82	36	18	50.875	
512-20					
Refl	0	36	18	69.750	9.5 dBi
DE	46	36	18	61.750	24.9 dB
Dir 1	46	36	18	60.500	
Dir 2	48	36	18	55.500	
Dir 3	94	36	18	54.625	

Note: For all antennas, the tube diameters are: Seg 1 = 0.750 inch, Seg 2 = 0.625 inch, Seg 3 = 0.500 inch.

unless additional support is given to reduce both vertical and horizontal bending forces. Suitable reinforcement for a long 2 inch boom can consist of a truss or a truss and lateral support, as shown in **Fig 21.67**.

A boom length of 24 feet is about the point where a 3 inch diameter begins to be very worthwhile. This dimension provides a considerable amount of improvement in overall mechanical stability as well as increased clamping surface area for element hardware. The latter is extremely important to prevent rotation of elements around the boom if heavy icing is commonplace. Pinning an element to the boom with a large bolt helps in this regard. On smaller diameter booms, however, the elements sometimes work loose and tend to elongate the pinning holes in both the element and the boom. After some time the elements shift their positions slightly (sometimes from day to day) and give a ragged appearance to the system, even though this may not harm the electrical performance.

A 3 inch diameter boom with a wall thickness of 0.065 inch is very satisfactory for antennas up to about a five-element, 20 meter array that is spaced on a 40 foot boom. A truss is recommended for any boom longer than 24 feet. One possible source for large boom material is irrigation tubing sold at farm supply houses.

PUTTING IT TOGETHER

Once you assemble the boom and elements, the next step is to fasten the elements to the boom securely and then fasten the boom to the mast or supporting structure using mounting plates as shown in **Fig 21.68**. Be sure to leave plenty of material on either side of the U-bolt holes on the element-to-boom mounting plates. The U-bolts selected should be a snug fit for the tubing. If possible, buy muffler-clamp U-bolts that come with saddles.

The *boom-to-mast* plate shown in Fig 21.68B is similar to the *boom-to-element* plate in 21.68A. The size of the plate and number of U-bolts used will depend on the size of the antenna. Generally, antennas for the bands up through 20 meters require only two U-bolts each for the mast and boom. Longer antennas for 15 and 20 meters

Table 21.11
15 Meter Optimized Yagi Designs

	Spacing Between Elements (in)	Seg 1 Length (in)	Seg 2 Length (in)	Seg 3 Length (in)	Seg 4 Length (in)	Midband Gain F/R
315-12						
Refl	0	30	36	18	61.375	7.6 dBi
DE	48	30	36	18	49.625	25.5 dB
Dir 1	92	30	36	18	43.500	
415-18						
Refl	0	30	36	18	59.750	8.3 dBi
DE	56	30	36	18	50.875	31.2 dB
Dir 1	56	30	36	18	48.000	
Dir 2	98	30	36	18	36.625	
515-24						
Refl	0	30	36	18	62.000	9.4 dBi
DE	48	30	36	18	52.375	25.8 dB
Dir 1	48	30	36	18	47.875	
Dir 2	52	30	36	18	47.000	
Dir 3	134	30	36	18	41.000	

Note: For all antennas, the tube diameters (in inches) are: Seg 1 = 0.875, Seg 2 = 0.750, Seg 3 = 0.625, Seg 4 = 0.500.

Table 21.12
17 Meter Optimized Yagi Designs

	Spacing Between Elements (in)	Seg 1 Length (in)	Seg 2 Length (in)	Seg 3 Length (in)	Seg 4 Length (in)	Seg 5 Length (in)	Midband Gain F/R
317-14							
Refl	0	24	24	36	24	60.125	8.1 dBi
DE	65	24	24	36	24	52.625	24.3 dB
Dir 1	97	24	24	36	24	48.500	
417-20							
Refl	0	24	24	36	24	61.500	8.5 dBi
DE	48	24	24	36	24	54.250	27.7 dB
Dir 1	48	24	24	36	24	52.625	
Dir 2	138	24	24	36	24	40.500	

Note: For all antennas, tube diameters (inches) are: Seg 1=1.000, Seg 2=0.875, Seg 3=0.750, Seg 4=0.625, Seg 5=0.500.

Table 21.13
20 Meter Optimized Yagi Designs

	Spacing Between Elements (in.)	Seg 1 Length (in.)	Seg 2 Length (in.)	Seg 3 Length (in.)	Seg 4 Length (in.)	Seg 5 Length (in.)	Seg 6 Length (in.)	Midband Gain F/R
320-16								
Refl	0	48	24	20	42	20	69.625	7.3 dBi
DE	80	48	24	20	42	20	51.250	23.4 dB
Dir 1	106	48	24	20	42	20	42.625	
420-26								
Refl	0	48	24	20	42	20	65.625	8.6 dBi
DE	72	48	24	20	42	20	53.375	23.4 dB
Dir 1	60	48	24	20	42	20	51.750	
Dir 2	174	48	24	20	42	20	38.625	

Note: For all antennas, tube diameters (inches) are: Seg 1=1.000, Seg 2=0.875, Seg 3=0.750, Seg 4=0.625, Seg 5=0.500. Seg 6=0.375.

(35 foot booms and up) and most 40 meter beams should have four U-bolts each for the boom and mast because of the torque that the long booms and elements exert as the antennas move in the wind. When tightening the U-bolts, be careful not to crush the tubing. Once the wall begins to collapse, the connection begins to weaken. Many aluminum suppliers sell ¼ inch or ⅜ inch thick plates just right for this application. Often they will shear pieces to the correct size on request. As with tubing, the relatively hard 6061-T6 grade is a good choice for mounting plates.

The antenna should be put together with good-quality hardware. Stainless steel or galvanized is best for long life. Rust will attack plated steel hardware after a short while, making nuts difficult, if not impossible, to remove. If stainless or galvanized muffler clamps are not available, the next best thing is to have them plated. If you can't get them plated, then at least paint them with a good zinc-chromate primer and a finish coat or two.

Good-quality hardware is more expensive initially, but if you do it right the first time, you won't have to take the antenna down after a few years and replace the hardware. Also, when repairing or modifying an installation, nothing is more frustrating than fighting rusty hardware at the top of a tower. Stainless steel hardware can also develop surface defects called *galling* that can cause threads on nuts and bolts to seize. On hardware ¼ inch and larger, the use of an anti-seize compound is recommended.

Project: Family of Computer-Optimized HF Yagis

Yagi designers are now able to take advantage of powerful personal computers and software to optimize their designs for the parameters of gain, F/R and SWR across frequency bands. Dean Straw, N6BV, has designed a family of Yagis for HF bands. These can be found in **Tables 21.9, 21.10, 21.11, 21.12** and **21.13**, for the 10, 12, 15, 17 and 20 meter amateur bands, respectively.

For 12 through 20 meters, each design has been optimized for better than 20 dB F/R, and an SWR of less than 2:1 across the entire amateur frequency band. For the 10 meter band, the designs were optimized for the lower 800 kHz of the band, from 28.0 to 28.8 MHz. Each Yagi element is made of telescoping 6061-T6 aluminum tubing, with 0.058 inch thick walls. This type of element can be telescoped easily, using techniques shown in Fig 21.66. Measuring each element to an accuracy of ⅛ inch results in performance remarkably consistent with the computations, without any need for tweaking or fine-tuning when the Yagi is on the tower.

The dimensions shown are designed for specific telescoping aluminum elements, but the elements may be scaled to different sizes by using the information about tapering and scaling in *The ARRL Antenna Book*, although with a likelihood of deterioration in performance over the whole frequency band.

Each element is mounted above the boom with a heavy rectangular aluminum boom-to-element plate, by means of galvanized U-bolts with saddles, as shown in Fig 21.68. This method of element mounting is rugged and stable, and because the element is mounted away from the boom, the amount of element detuning due to the presence of the boom is minimal. The element dimensions given in each table already take into account any element detuning due to the mounting plate. The element mounting plate for all the 10 meter Yagis is a 0.250 inch thick flat aluminum plate, 4 inches wide by 4 inches long. For the 12 and 15 meter Yagis, a 0.375 inch thick flat aluminum plate, 5 inches wide by 6 inches long is used, and for the 17 and 20 meter Yagis, a 0.375 inch thick flat aluminum plate, 6 inches wide by 8 inches long is used. Where the plate is rectangular, the long dimension is in line with the element.

Each design table shows the dimensions for *one-half* of each element, mounted on one side of the boom. The other half of each element is the same, mounted on the other side of the boom. Use a tubing sleeve inside the center portion of the element so that the element is not crushed by the mounting U-bolts. Each telescoping section is inserted 3 inches into the next size of tubing. For example, in the 310-08 design for 10 meters (3 elements on an 8 foot boom), the reflector tip, made out of ½ inch OD tubing, sticks out 66.75 inches from the ⅝ inch OD tubing. For each 10 meter element, the overall length of each ⅝ inch OD piece of tubing is 21 inches, before insertion into the ¾ inch piece. Since the ¾ inch OD tubing is 24 inches long on each side of the boom, the center portion of each element is actually 48 inches of uncut ¾ inch OD tubing.

The boom for all these antennas should be constructed with at least 2 inch OD tubing, with 0.065 inch wall thickness. Because each boom has three inches of extra length at each end, the reflector is actually placed three inches from one end of the boom. For the 310-08 design, the driven element is placed 36 inches ahead of the reflector, and the di-

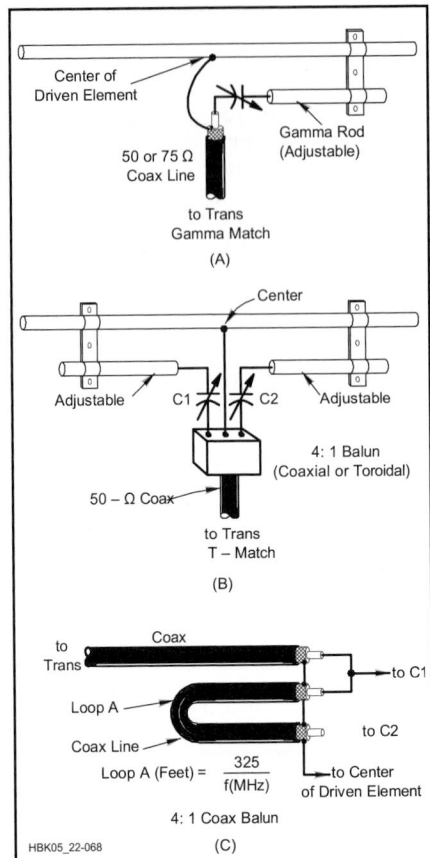

Fig 21.69 — Illustrations of gamma and T matching systems. At A, the gamma rod is adjusted along with the capacitor until the lowest SWR is obtained. A T match is shown at B. It is the same as two gamma-match rods. A 4:1 coaxial balun transformer for use with the T match is shown at C.

rector is placed 54 inches ahead of the driven element. The antenna is attached to the mast with the *boom-to-mast* mounting plate shown in Fig 21.68.

Each antenna is designed with a driven element length appropriate for a gamma or T matching network, as shown in **Fig 21.69**. The variable gamma or T capacitors can be housed in small plastic enclosures for weatherproofing; receiving-type variable capacitors with close plate spacing can be used at powers up to a few hundred watts. Maximum capacitance required is usually 140 pF at 14 MHz and proportionally less at the

higher frequencies.

The driven-element's length may require slight readjustment for best match, particularly if a different matching network is used. *Do not change either the lengths or the telescoping tubing schedule of the parasitic elements* — they have been optimized for best performance and will not be affected by tuning of the driven element.

TUNING ADJUSTMENTS

To tune the gamma match, adjust the gamma capacitor for best SWR, then adjust the position of the shorting strap or bar that connects the gamma rod to the driven element. Repeat this alternating sequence of adjustments until a satisfactory SWR is reached.

To tune the T-match, the position of the shorting straps and C1 and C2 are adjusted alternately for a best SWR. To maintain balance of the antenna, the position of the straps and capacitor settings should be the same for each side and adjusted together. A coaxial 4:1 balun transformer is shown at Fig 21.69C. A toroidal balun can be used in place of the coax model shown. The toroidal version has a broader frequency range than the coaxial one. The T match is adjusted for 200 Ω and the balun steps this balanced value down to 50 Ω, unbalanced. Or the T match can be set for 300 Ω, and the balun used to step this down to 75 Ω unbalanced.

Dimensions for the gamma and T match rods will depend on the tubing size used, and the spacing of the parasitic elements of the beam. Capacitors C1 and C2 can be 140 pF for 14 MHz beams. Somewhat less capacitance will be needed at 21 and 28 MHz.

Preliminary matching adjustments can be done on the ground. The beam should be aligned vertically so that the reflector element is closest to and a few feet off the ground, with the beam pointing upward. The matching system is then adjusted for best SWR. When the antenna is raised to its operating height, only slight touch-up of the matching network may be required.

A *choke balun* (see the **Transmission Lines** chapter) should be used to isolate the coaxial feed line shield from the antenna. Secure the feed line to the boom of the antenna between the feed point and the supporting mast.

21.7 Quad and Loop Antennas

One of the more effective DX antennas is the *quad*. It consists of two or more loops of wire, each supported by a bamboo or fiberglass cross-arm assembly. The loops are ¼ λ per side (one full wavelength overall). One loop is driven and the other serves as a parasitic element — usually a reflector. The design of the quad is similar to that of the Yagi, except that the elements are loops instead of dipoles. A two-element quad can achieve better F/R, gain and SWR across a band, at the expense of greater mechanical complexity compared to a two-element Yagi and very nearly the same performance as a three-element Yagi. A type of Yagi called the *quagi* has also been constructed with a quad element as the driven element. The larger quad driven element results in somewhat better SWR bandwidth, but gain and F/B are approximately the same as on regular Yagi antennas.

A variation of the quad is called the *delta loop*. The electrical properties of both antennas are the same. Both antennas are shown in **Fig 21.70**. They differ mainly in their physical properties, one being of plumber's delight construction, while the other uses insulating support members. One or more directors can be added to either antenna if additional gain and directivity are desired, though most operators use the two-element arrangement.

It is possible to interlace quads or deltas for two or more bands, but if this is done the lengths calculated using the formulas given in Fig 21.70 may have to be changed slightly to compensate for the proximity effect of the second antenna. Using a tuning capacitor as shown in the following project allows the antenna to be adjusted for peak performance without cumbersome adjustment of wire lengths.

If multiple arrays are used, each antenna should be tuned separately for maximum forward gain, or best front-to-rear ratio, as observed on a field-strength meter. The reflector stub on the quad should be adjusted for this condition. The resonance of the antenna can be found by checking the frequency at which the lowest SWR occurs. By lengthening or shortening it, the driven element length can be adjusted for resonance in the most-used portion of the band.

A gamma match can be used at the feed point of the driven element to match the impedance to that of coaxial cable. Because the loop's feed point impedance is *higher* than that of 50 Ω coaxial cable, a *synchronous transmission line transformer* or *Q-section* (see the **Transmission Lines** chapter) with an impedance intermediate to that of the loop and the coaxial cable can be used.

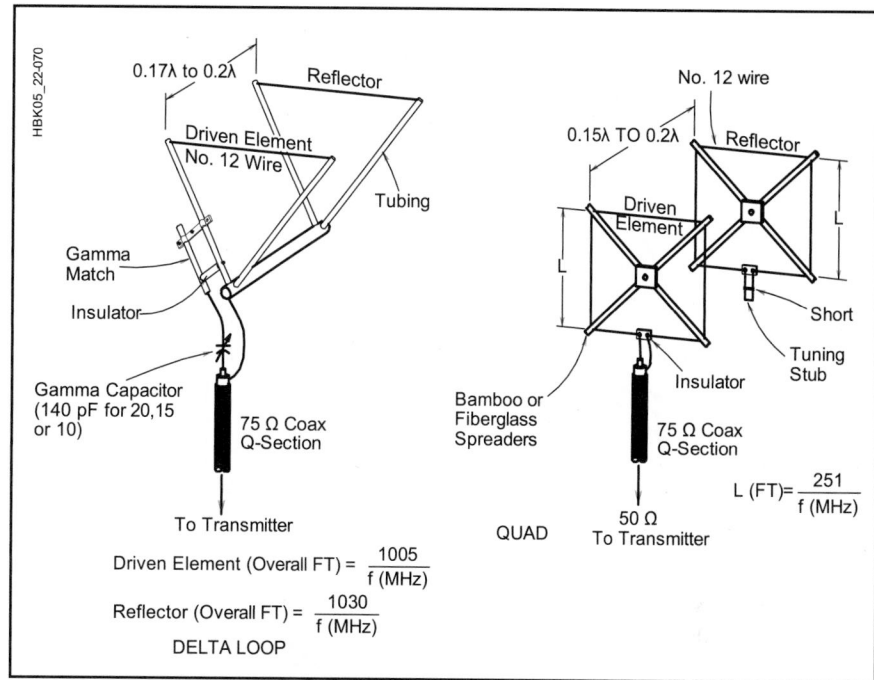

Fig 21.70 — Information on building a quad or a delta-loop antenna. The antennas are electrically similar, but the delta-loop uses plumber's delight construction. The λ/4 length of 75 Ω coax acts as a synchronous transmission-line transformer from approximate 100 Ω feed point impedance of quad to the 50 Ω feed line.

Project: Five-Band, Two-Element HF Quad

Two quad designs are described in this project, both nearly identical to the drawing in **Fig 21.71**. One was constructed by William A. Stein, KC6T, from scratch, and the other was built by Al Doig, W6NBH, using modified commercial triband quad hardware. The principles of construction and adjustment are the same for both models, and the performance results are also essentially identical. One of the main advantages of this design is the ease of (relatively) independent performance adjustments for each of the five bands. These quads were described by William A. Stein, KC6T, in *QST* for April 1992. Both models use 8-foot-long, 2 inch diameter booms, and conventional X-shaped spreaders (with two sides of each quad loop parallel to the ground).

Each driven element is fed separately, but running five separate feed lines to the shack would be unwieldy. A remote coax switch on the boom is used to select the feed line for each element. A gamma match or quarter-wave synchronous transmission line transformer is used to match the feed point impedance of the element to 50 Ω.

These designs can also be simplified to monoband quads by using the formulas in Fig 21.70 for loop dimensions and spacing. It is recommended to the antenna builder unfamiliar with quads that a monoband quad be attempted first in order to become acquainted with the techniques of building a quad. Once comfortable with constructing and erecting the quad, success with a multiband design is much easier to achieve.

Complete construction details and more information about the performance of these designs are available on the CD-ROM that accompanies this book.

21.7.1 Loop Antennas

The loop antennas described in this section are continuous loops at least one wavelength in circumference and formed into open shapes with sides that are approximately equal, such as triangles, diamonds, squares, or circles. Smaller loops used for receiving purposes are discussed in the *ARRL Antenna Book*. Loops with ratios of side lengths greater than 2 or 3:1 begin to have special characteristics beyond the scope of this chapter.

A 1 λ loop can be thought of as two ½ λ dipoles with their ends connected together and pulled apart into an open shape as described above. The feed point of one dipole is replaced with a short circuit so that there is only one feed point on the antenna. As such,

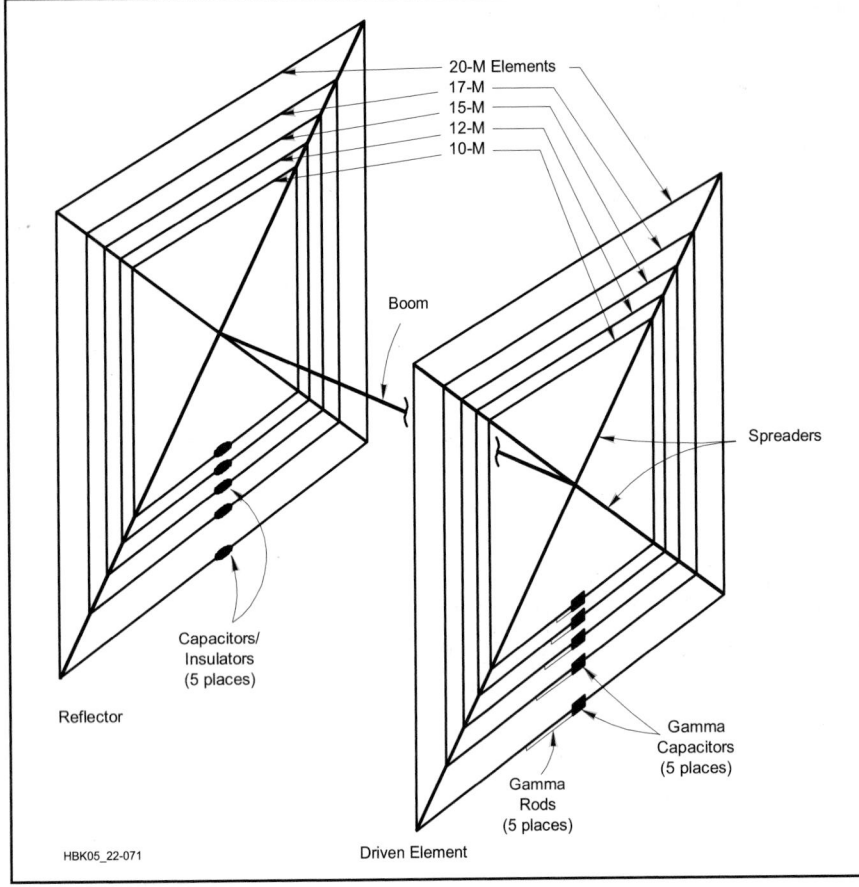

Fig 21.71 — Mechanical layout of the five-band quad. The boom is 8 feet long; see the article on the *Handbook* CD-ROM for all other dimensions.

the current and voltage distribution around the loop is an extension of Fig 21.1. Three typical loop shapes and the current distributions on them are shown in **Fig 21.72**. Note that the current flow reverses at points ¼ λ to either side of the feed point. That means the current direction opposite the feed point is the same as at the feed point.

The maximum radiation strength of a 1 λ loop is perpendicular to the plane of the loop and minimum in the plane of the loop. If the loop is horizontal, the antenna radiates best straight up and straight down and poorly to the sides. The gain of a 1 λ loop in the direction of maximum radiation is approximately 1 dBd.

If the plane of the three loops shown in Fig 21.72 is vertical, the radiation is horizontally polarized because the fields radiated by the vertical components of current are symmetrical and opposing, so they cancel, leaving only the horizontally polarized fields that reinforce each other perpendicular to the

loop plane. If the feed point of the antenna is moved to a vertical side or the antenna is rotated 90°, it is the horizontally polarized fields that will cancel, leaving a vertically polarized field, still maximum perpendicular to the plane of the loop. Feeding the loop at some other location, rotating the loop by some intermediate value, or constructing the loop in an asymmetrical shape will result in polarization somewhere between vertical and horizontal, but the maximum radiation will still occur perpendicular to the plane of the loop.

In contrast to straight-wire antennas, the electrical length of the circumference of a 1 λ loop is shorter than the actual length. For a loop made of bare #18 AWG wire and operating at a frequency of 14 MHz, so that the length-to-diameter ratio is very large, the loop will be close to resonance in free space when:

$$\text{Length (feet)} = 1032/f(\text{MHz}) \qquad (7)$$

The radiation resistance of a resonant 1 λ loop is approximately 120 Ω under these conditions. Since the loop dimensions are larger than those of a ½ λ dipole, the radiation efficiency is high and the SWR bandwidth of the antenna significantly larger than for the dipole.

The loop antenna is resonant on all frequencies at which it is an integral number of wavelengths in circumference; f_0, $2f_0$, $3f_0$, etc. That means an 80 meter 1 λ loop will also have a relatively low feed point impedance on 40, 30, 20, 15, 12, and 10 meters. As each side of the loop becomes longer electrically, the radiation pattern of the loop begins to develop nulls perpendicular to the plane of the loop and lobes that are closer to the plane of the loop. A horizontal diamond-shaped loop with legs more than a wavelength long is a *rhombic* antenna and can develop significant gain along the long axis of the antenna. (The diamond-shaped rhombic is the origin of the symbol of the ARRL and many other radio organizations.)

Project: Low-Band Quad and Delta Loops

(The following material is summarized from Chapter 10 of *ON4UN's Low-Band DXing, Fifth Edition*.) Dimensions for these designs assume an operating frequency of 3.75 MHz. The dimensions for the loops in this section may be scaled to frequencies in the 160, 60, 40 or 30 meter bands. The performance of the loops will vary with height above ground and ground conductivity.

SQUARE LOOP

Fig 21.73 shows the vertical-plane radiation patterns for a quad loop over very poor ground and over very good ground on the

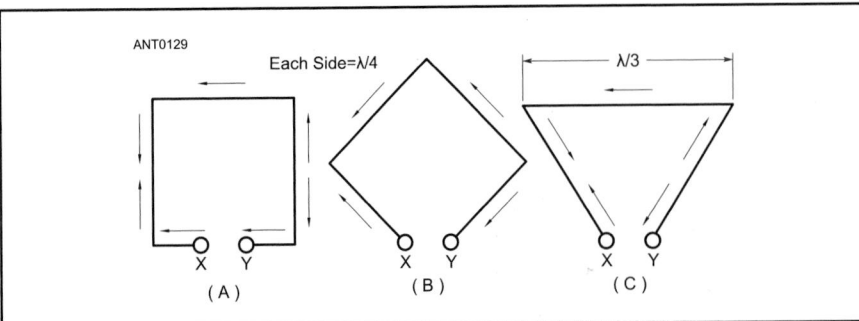

Fig 21.72 — At A and B, loops have sides ¼ λ long , and at C having sides ⅓ λ long for a total conductor length of 1 λ. The polarization depends on the orientation of the loop and on the position of the feed point (terminals X-Y) around the perimeter of the loop.

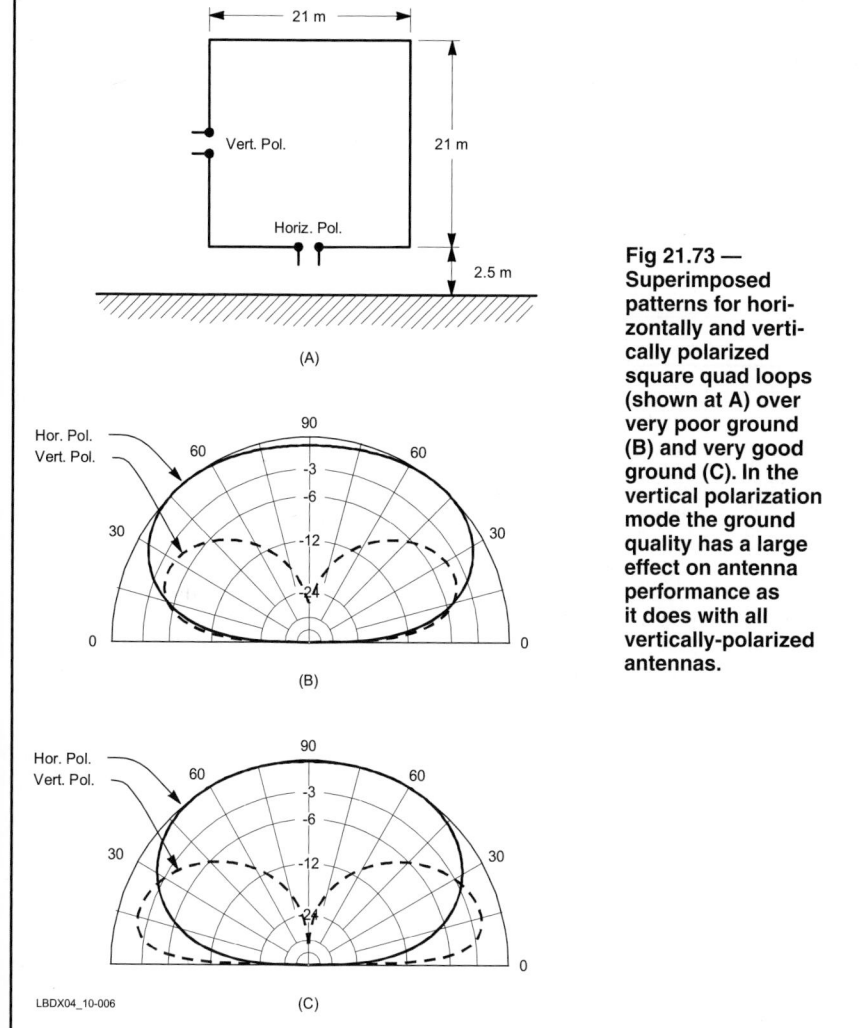

(A)

(B)

(C)

Fig 21.73 — Superimposed patterns for horizontally and vertically polarized square quad loops (shown at A) over very poor ground (B) and very good ground (C). In the vertical polarization mode the ground quality has a large effect on antenna performance as it does with all vertically-polarized antennas.

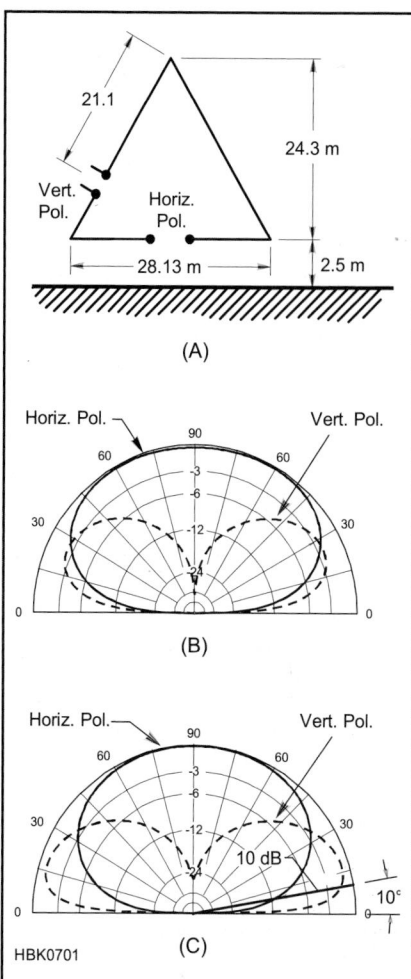

(A)

(B)

(C)

HBK0701

Fig 21.74 — Superimposed patterns for horizontally and vertically polarized delta quad loops (shown at A) over very poor ground (B) and very good ground (C). Over better ground, the vertically polarized loop performs much better at low radiation angles, while of both good and poor ground the vertically polarized loop gives good discrimination against high-angle local signals.

same dB scale for both horizontal and vertical polarization. Polarization of the loop depends on the location of the feed point as shown in the figure.

The vertically polarized quad loop can be considered as two shortened top-loaded vertical dipoles, spaced ¼ λ apart. Broadside radiation from the horizontal elements of the quad is very low because the currents in the horizontal legs are approximately equal but in opposite directions in each half of the leg. The radiation angle in the broadside direction will be essentially the same as for either of the vertical members.

The resulting radiation angle will depend on the quality of the ground up to several wavelengths away from the antenna, as is the case with all vertically polarized antennas. The quality of the ground is as important as it is for any other vertical antenna, meaning that vertically polarized loops close to the ground will not work well over poor soil. In a typical situation on 80 meters, a vertically-polarized quad loop will radiate an excellent low-angle signal (lobe peak at approximately

21°) when operated over average ground. Over poorer ground, the peak elevation angle would be closer to 30°. The horizontal directivity is rather poor and amounts to approximately 3.3 dB of side rejection at any elevation angle.

A horizontally polarized quad-loop antenna can be thought of as two stacked short dipoles with a peak elevation angle dependent on the height of the loop. The low horizontally polarized quad (top at 0.3 λ) radiates most of its energy right at or near zenith angle (straight up). At low wave angles (20° to 45°) the horizontally polarized loop shows more front-to-side ratio (5 to 10 dB) than the vertically polarized rectangular loop.

With a horizontally polarized quad loop the angle of peak radiation is very dependent on the antenna height but not so much on the quality of the ground. At very low heights, the angle of peak radiation varies between 50° and 60° (but is rather constant all the way up to 90°). This is very good for NVIS and regional communication but not very good for DX. As far as gain is concerned, there is

a 2.5-dB gain difference between very good and very poor ground, which is only half the difference found with the vertically polarized loop.

Comparing the gains of the horizontally and vertically polarized loops, Fig 21.73 shows that at very low antenna heights the gain is about 3 dB better for the horizontally polarized loop. But this gain exists at a high wave angle (50° to 90°) while the vertically polarized loop at very low heights radiates at 17° to 25°.

At heights from 3 to 6 meters for the bottom leg, feed point resistance for the horizontally polarized loop is approximately 100 to 120 Ω over average ground. For vertical polarization, feed point resistance varies from 200 to 170 Ω.

The quad loop feed point is symmetrical,

whether you feed the quad in the middle of the vertical or the horizontal wire. At the feed point, use a common-mode choke balun (see the **Transmission Lines** chapter) as current flowing on the outside of the coaxial feed line could upset the radiation pattern.

DELTA LOOP

Fig 21.74 shows the configuration as well as the superimposed elevation patterns for vertically and horizontally polarized low-height equilateral triangle delta loops over two different types of ground (same dB scale). The model was constructed for a frequency of 3.75 MHz. The base is 2.5 meters above ground, which puts the apex at 26.8 meters. Over good ground, the vertically polarized delta loop shows nearly 3 dB front-to-side ratio at the peak radiation angle of 22°. With average ground the gain is 1.3 dBi.

Over very poor ground, the horizontally polarized delta loop is better than the vertically polarized loop for all wave angles above 35°. Below 35° the vertically polarized loop takes over, but quite marginally. The maximum gain of the vertically and the horizontally polarized loops differs by only 2 dB but the big difference is that for the horizontally polarized loop, the gain occurs at almost 90°, while for the vertically polarized loop it occurs at 25°. The vertically polarized antenna also gives good high-angle rejection (rejection of local signals), while the horizontally polarized loop will not.

Over very good ground, the performance at low angles is greatly improved for both polarizations. The vertically polarized loop is still better at any elevation angle under 30° than when horizontally polarized. At a 10° radiation angle the difference is as high as 10 dB. This makes the vertically polarized delta over good ground far superior for DX operating.

Most practical delta loops show a feed point impedance between 50 and 100 Ω, depending on the exact geometry and coupling to other antennas. The antenna can be fed directly with a 50 or 70 Ω coaxial cable, or via a 70 Ω quarter-wave transformer (see the **Transmission Lines** chapter) if the feed point impedance is near 100 Ω. At the feed point, use a common-mode choke balun (see the **Transmission Lines** chapter) as current flowing on the outside of the coaxial feed line could upset the radiation pattern.

THE BOTTOM-CORNER-FED DELTA LOOP

Fig 21.75 shows the layout of the delta loop being fed at one of the two bottom corners. The antenna is slightly compressed from the previous section with a slightly lower apex and longer base than the loop described in the previous section. Because of the "incorrect" location of the feed point, cancellation of radiation from the base wire is incomplete, resulting in a significant horizontally polarized radiation component. The total field has a very uniform gain coverage (within 1 dB) from 25° to 90°. This may be a disadvantage for the rejection of high-angle signals when working DX at low angles.

Due to the "incorrect" feed-point location,

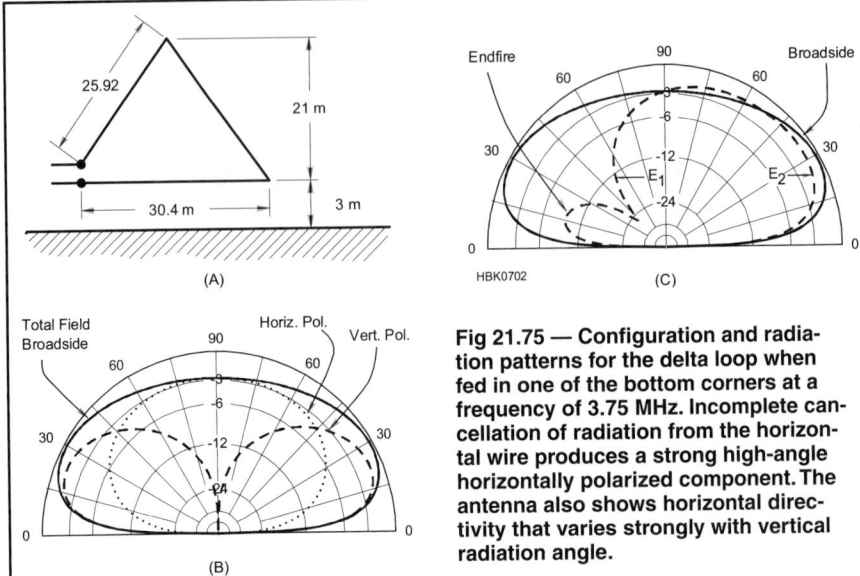

Fig 21.75 — Configuration and radiation patterns for the delta loop when fed in one of the bottom corners at a frequency of 3.75 MHz. Incomplete cancellation of radiation from the horizontal wire produces a strong high-angle horizontally polarized component. The antenna also shows horizontal directivity that varies strongly with vertical radiation angle.

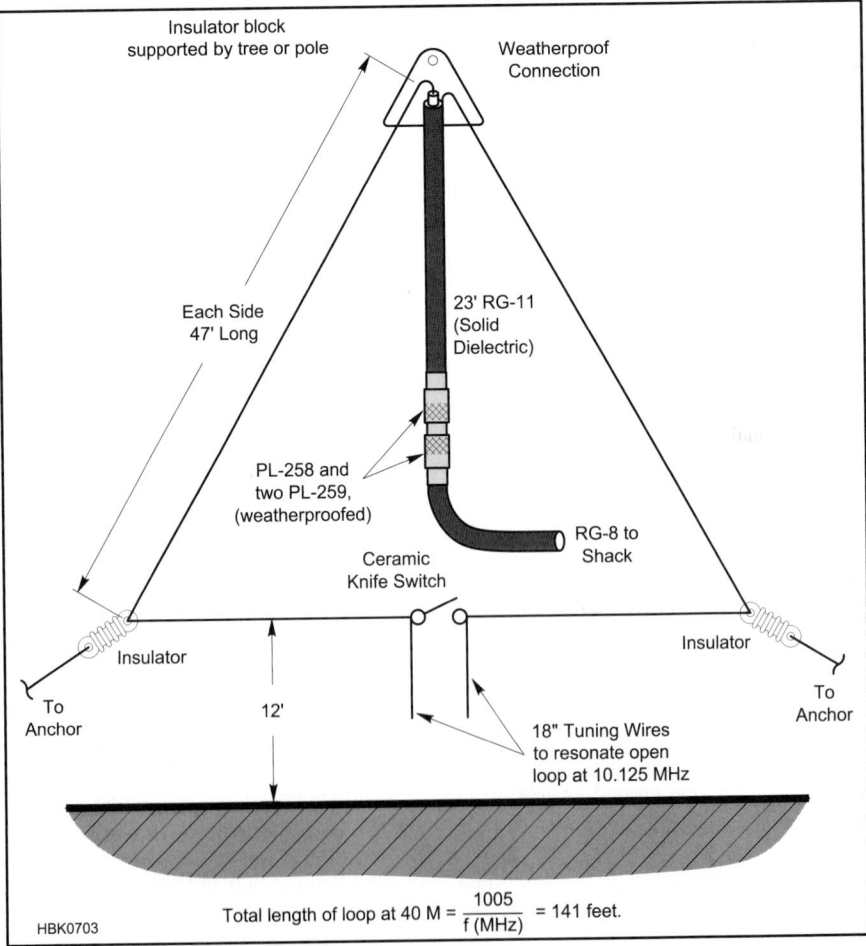

Fig 21.76 — NT4B's 30 and 40 meter loop is fed at the top via a quarter-wave 40 meter matching transformer made of 75 Ω coax. Note the 18 inch tuning wires used to lower the antenna's 30 meter resonance from 10.5 to 10.1 MHz. Adjust the length of these wires to set the 30 meter resonant frequency.

the end-fire radiation (radiation in line with the loop) has become asymmetrical with a side null of nearly 12 dB at the peak radiation angle of 29°. The loop actually radiates its maximum signal about 18° off the broadside direction. This feed point configuration greatly affects the pattern of the loop so use bottom-corner-feed with care.

Project: Two-Band Loop for 30 and 40 Meters

The following antenna design is from a *QST* Hints and Kinks entry by James Brenner, NT4B, in the May 1989 issue. The version shown in **Fig 21.76** is fed at the apex of a delta loop but can be adapted to a square or quad loop shape.

The original design was derived from "The Mini X-Q Loop" in *All About Cubical Quad Antennas* by Bill Orr, W6SAI (now out of print) which is 1½ λ in circumference, with an open circuit opposite the feed point. That antenna has approximately 1 dB of additional gain over a 1 λ loop. Since 30 and 40 meters are close to the same 1:1½ λ ratio, one loop can be converted between 1 λ on 40 meters and 1½ λ on 30 meters with a switch.

A large, ceramic SPST knife switch is installed in the center of the delta loop's bottom leg as shown in Fig 21.76. With the switch open, the loop acts a 1½ λ loop at 10.5 MHz, so 18 inch wires were added to the loop on either side of the switch to lengthen the antenna and lower the resonant frequency to 10.1 MHz. Closing the switch shorts out the wires and the loop becomes a regular 1 λ continuous loop for 40 meters.

Note that there is fairly high voltage present at the switch when transmitting on 30 meters. If a relay is used, be sure the contact spacing is sufficient to avoid arcing or use additional pairs of contacts to increase the overall spacing.

The antenna is fed through a quarter-wave transformer (see the **Transmission Lines** chapter) of 75 Ω RG-11 coax, approximately 23 feet long. According to the author, when configured for 40 meters, the loop has a satisfactory SWR of less than 2:1 on 15 meters. In addition, the 30 meter configuration can be used successfully on 80 meters with the use of an antenna tuner.

Project: Skeleton Slot for 14 to 30 MHz

(The following material is adapted from the RSGB *Radio Communications Handbook, 11th Edition*.) The Skeleton Slot is a loop antenna that was first documented in an article by G2HCG in 1953. With the dimensions shown, it will operate on the HF bands from 14 to 28 MHz using a balanced feed line and

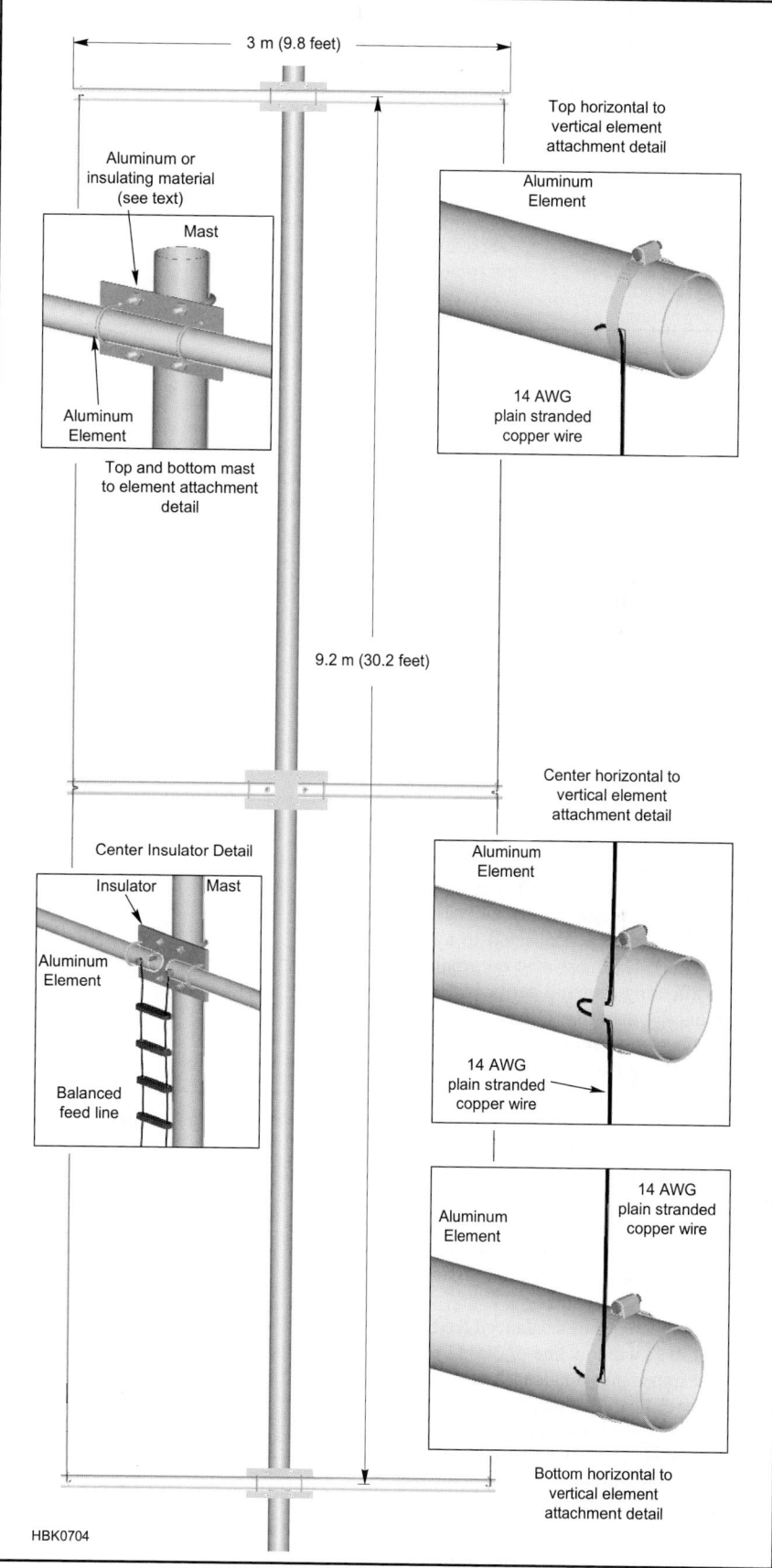

3 m (9.8 feet)

Top horizontal to vertical element attachment detail

Aluminum or insulating material (see text)

Mast

Aluminum Element

Aluminum Element

14 AWG plain stranded copper wire

Top and bottom mast to element attachment detail

9.2 m (30.2 feet)

Center horizontal to vertical element attachment detail

Center Insulator Detail

Insulator   Mast

Aluminum Element

Aluminum Element

14 AWG plain stranded copper wire

Balanced feed line

14 AWG plain stranded copper wire

Aluminum Element

Bottom horizontal to vertical element attachment detail

HBK0704

Fig 21.77 — The G3LDO multiband Skeleton Slot antenna for 14 to 28 MHz. The elements are attached to the mast and the whole mast is rotated. The wire elements are fixed to the horizontal elements with stainless steel hose clamps. The center insulator can be homebrewed from plastic sheet or a commercial unit may be used.

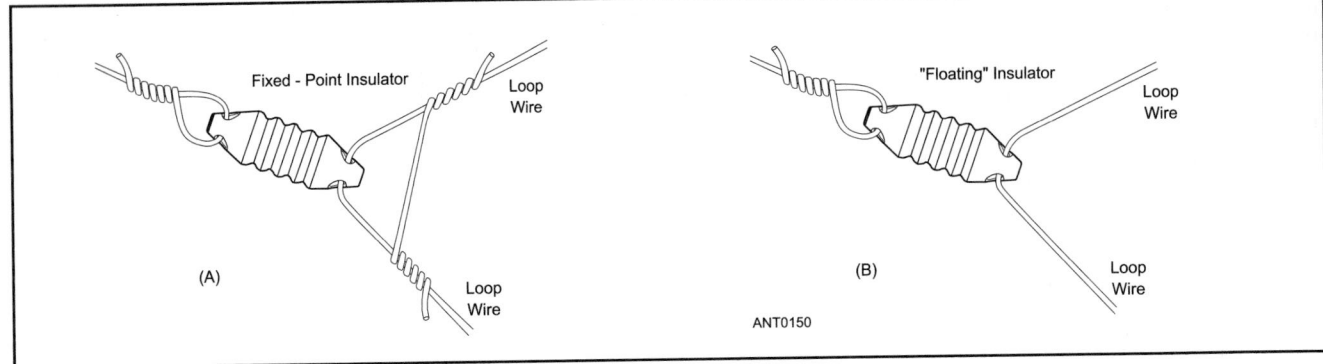

Fig 21.78 — Two methods of installing the insulators at the loop corners.

antenna tuner. It is very easy to construct and has no traps or critical adjustments. The turning radius is 1.5 meters (5 feet) even though it is 14 meters (47 feet) tall. The antenna radiation pattern is bidirectional, perpendicular to the plane of the loop, and horizontally polarized with a gain of approximately 8 dBi on 14 MHz, increasing to 11 dBi on 28 MHz. The design can be scaled to the VHF bands, as well.

This version of the antenna uses wire for the vertical elements, resulting in a simplified and rugged construction as shown in **Fig 21.77**. The structure of the antenna is provided by three aluminum tube elements attached to a central mast at 15 foot intervals with the lowest element 15 feet above the ground. The mast is electrically connected to the supporting elements, just as Yagi elements are electrically connected to a supporting boom. The mast may be grounded without affecting antenna performance. A non-conductive mast may also be used or the elements insulated from a conductive mast without affecting the antenna performance.

The center element is fed in the center with connecting wires attached from the end of the center element to the upper and lower elements as shown in Fig 21.77. Stainless steel hose clamps are used to attach the wire and aluminum. Use an anti-oxidation compound such as Penetrox or Noalox to prevent corrosion from the contact between dissimilar metals.

The antenna requires a balanced feed line such as 450 Ω window line or 300 Ω twin-lead. The impedance and length of the feed line are not critical. Use standoffs to hold the feed line away from the mast and elements until clear of the antenna. Alternatively, a 9:1 impedance transformer at the feed point may be used with coaxial cable. A choke balun to decouple the coaxial feed line's outer surface from the antenna should also be used at the feed point. (See the **Transmission Lines** chapter.)

Project: **Multiband Horizontal Loop Antenna**

Along with the multiband, non-resonant dipole, many amateurs operate on HF with great success using a horizontal loop antenna. All that is required are at least three supports able to hold the corners of the antenna 20 or more feet above the ground (and even that is negotiable) and enough room for a loop of wire one wavelength or more in circumference at the lowest frequency of operation. (Smaller loops can be used with an impedance-matching unit.)

Start by calculating the total length of wire you need using Equation 7. You'll need one insulator for each support and lengths of rope that are at least twice the height to which the insulator will be raised. You can feed the antenna at one of the corner insulators or anywhere along the wire with a separate insulator. Examples of corner insulators are shown in **Fig 21.78**. Using floating insulators allows the wire to move as the antenna flexes. One of the insulators should be of the fixed type, or the antenna can be fed at one corner with the loop wires attached to a pair of insulators sharing a common support rope. This holds the antenna feed point in place.

If the loop is only going to be used on the band for which it is resonant, coaxial cable

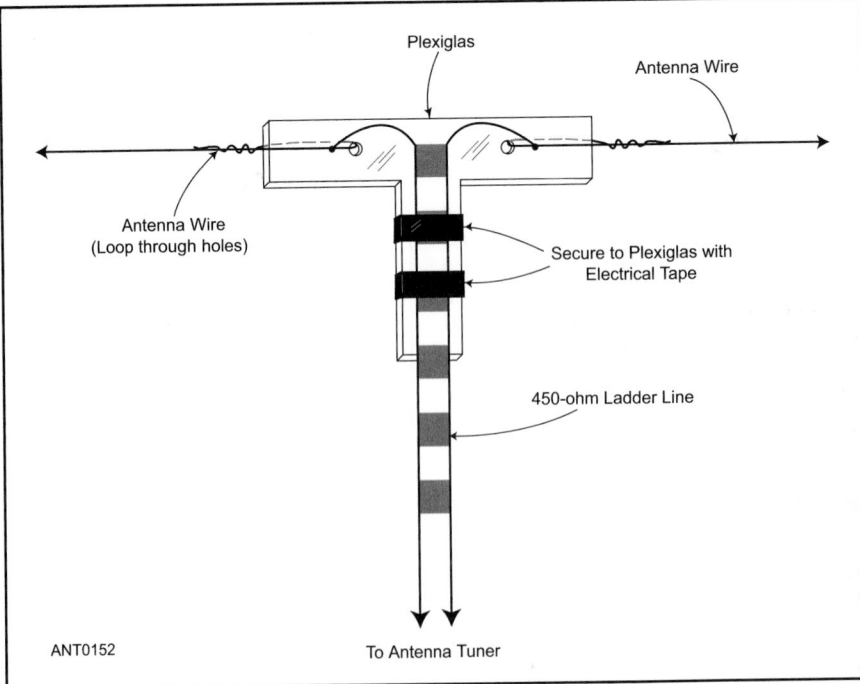

Fig 21.79 — One possible method of constructing a feed point insulator for use with open-wire line.

can be used as the feed line, since SWR will be low. A choke balun at the feed point is recommended. For multiband use, open-wire feed line should be used, with an impedance-matching unit in the shack. A feed point insulator for open-wire line is shown in **Fig 21.79**.

On its fundamental frequency, the antenna's maximum radiation will be straight up, making it most useful for regional communications at high elevation angles with the occasional DX contact. At higher frequencies, the loop will radiate more strongly at lower angles for better signal strengths at long distances.

21.8 HF Mobile Antennas

HF mobile operation has been part of Amateur Radio since the 1930s. Once primarily an HF endeavor, the FM explosion of the 1970s seemed to displace HF mobiling in favor of local repeater and "mag-mount" VHF ground planes. HF mobiling is back, though, in a big way and with the excellent equipment and antennas available, here to stay. Material in this section was contributed and updated by Alan Applegate, KØBG, whose website (**www.k0bg.com**) has many useful pages on HF mobile stations and operating.

High frequency (HF) mobile antennas come in every size and shape imaginable, from simple whips to elaborate, computer-controlled behemoths. Regardless of the type and construction, an HF mobile antenna should have a few important attributes.

• Sturdiness: It should be permanently mounted (without altering the vehicle's safety equipment) to stay upright at highway speeds with a minimum of sway.

• Mechanically stable: Sudden stops or sharp turns won't cause it to sway about, endangering others.

• Flexibly mounted: Permits bending around branches and obstacles at low speeds.

• Weatherproof: Withstands the effects of wind, rain, snow and ice.

• Tunable: If multiband operation is desired, be tunable to different HF bands without stopping the vehicle.

• Be easily removable when required.

• Be as efficient as possible.

Of all the antenna choices available, the *whip* antenna — a self-supporting rod or wire mounted at its base — has passed the test of time as providing all of these attributes in one way or another. The following sections discuss the different types of whips, how they are attached to and interact with the vehicle, and how they are connected to the transmitter.

21.8.1 Simple Whips

The simplest of antennas is a quarter-wave whip, but it's only practical on the upper HF bands because of the required length. For example, a 10 meter quarter-wavelength antenna is about 8 feet long. It doesn't require a loading coil, so its efficiency is approximately 80%. The reason efficiency isn't 100% is because of resistive losses in the whip itself, stray capacitance losses in the mounting hardware and ground losses, which we'll cover later. The end result is that the feed point impedance at the antenna's base is very close to 50 Ω and a good match to modern solid state transceivers.

As we move lower in frequency, the physical length must increase for an equivalent electrical length, but there is a limit. In most localities, the maximum height at the tip of the antenna needs to be less than 13.5 feet (4.1 meters). This generally limits whip length to 10.5 feet (3.2 meters) for an average installation on a vehicle. Creating a resonant antenna below this length on 10 or 12 meters isn't a problem. On the 15 meter and lower frequency bands, 10.5 feet is not long enough for a resonant whip antenna and additional electrical measures are required.

WHIP RADIATION RESISTANCE

The power radiated by the antenna is equal to the radiation resistance times the square of the antenna current. The radiation resistance, R_r, of an electrically small antenna is given by:

$$R_r = 395 \times \left(\frac{h}{\lambda}\right)^2 \qquad (8)$$

where

h = radiator height in meters

λ = wavelength in meters = 300 / f in MHz

Since radiation resistance of these electrically small antennas is a function of height, the antenna must be lengthened physically or electrically to increase it. Increasing radiation resistance improves efficiency as shown next.

The efficiency of the antenna, η, equals the radiation resistance, R_r, divided by the resistive component of the feed point impedance, R_{fp}, which for actual antennas includes ground losses and losses in the antenna:

$$\eta = \frac{R_r}{R_{fp}} \times 100\% \qquad (9)$$

Since an electrically short antenna has a low radiation resistance, careful attention

Table 21.14
Characteristics of an 8 foot Mobile Whip

f(MHz)	Loading L μH	R_C(Q50) Ω	R_C(Q300) Ω	R_R Ω	Feed R* Ω	Matching L μH
Base Loading						
1.8	345	77	13	0.1	23	3
3.8	77	37	6.1	0.35	16	1.2
7.2	20	18	3	1.35	15	0.6
10.1	9.5	12	2	2.8	12	0.4
14.2	4.5	7.7	1.3	5.7	12	0.28
18.1	3.0	5.0	1.0	10.0	14	0.28
21.25	1.25	3.4	0.5	14.8	16	0.28
24.9	0.9	2.6	—	20.0	22	0.25
29.0	—	—	—	—	36	0.23
Center Loading						
1.8	700	158	23	0.2	34	3.7
3.8	150	72	12	0.8	22	1.4
7.2	40	36	6	3.0	19	0.7
10.1	20	22	4.2	5.8	18	0.5
14.2	8.6	15	2.5	11.0	19	0.35
18.1	4.4	9.2	1.5	19.0	22	0.31
21.25	2.5	6.6	1.1	27.0	29	0.29

R_C = loading coil resistance; R_R = radiation resistance.
*Assuming loading coil Q = 300, and including estimated ground-loss resistance.

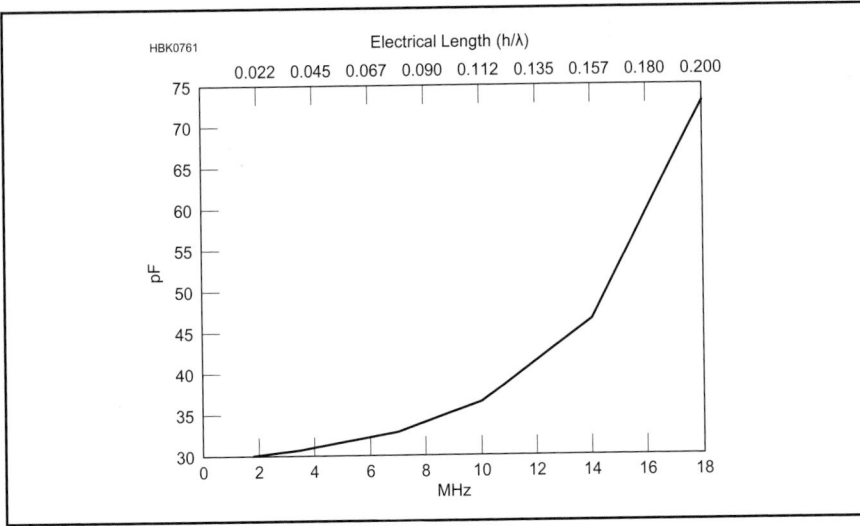

Fig 21.80 — Relationship between frequency and capacitance for a 3.2 meter vertical whip.

Table 21.15
HF Mobile Antenna Comparison

Antenna Type	Length	Frequency Coverage	Efficiency	Mounting Difficulty	Matching Required
Simple Whip	< 11 ft	15 m & up	Excellent	Easy	No
Base-Loaded	9 to 10.6 ft	160 - 6 m	Fair to good	Average	Yes
Center-Loaded	9 to 10.5 ft	160 - 6 m	Good to excellent	Average	Yes
Top-Loaded	<9 ft	160 - 6 m	Fair	Average	No
Continuous Loading	< 7 feet	80 - 6 m	Poor to fair	Easy	No
Remote-tuned Small	< 7 feet	80 - 6 m	Poor to fair	Easy	No
Remote-tuned Large	9 to 10.5 ft	160 - 6 m	Excellent	Difficult	Yes

must be paid to minimizing losses in the antenna system that can greatly reduce the antenna's effectiveness.

WHIP CAPACITANCE

As we shorten an antenna to less than ¼ λ, its radiation resistance decreases and the capacitance drops as shown in **Table 21.14**. **Fig 21.80** shows that capacitance is not very sensitive to frequency for h/λ less than 0.075.

The capacitance in pF of an electrically small antenna is given approximately by:

$$C = \frac{55.78 \times h}{[(den1) \times (den2)]} \qquad (10)$$

where
  $(den1) = (\ln (h/r) - 1)$
  $(den2) = [1 - (f \times h/75)^2]$
  ln = natural logarithm
  r = conductor radius in meters
  f = frequency in MHz

Radiation resistance rises in a nonlinear fashion and the capacitance drops just as dramatically with increase in the ratio h/λ. Fig 21.80 can be used for estimating antenna capacitance for other heights and shows that

capacitance is not very sensitive to frequency for h/λ less than 0.075 which occurs at 8 MHz in this case.

21.8.2 Coil-Loaded Whips

To bring an electrically-short whip antenna to resonance, we must add inductance in the form of a loading coil. The coil can take many forms, and it may be placed almost anywhere along the length of the radiating element. It cancels out the capacitive reactance (–j) by introducing an equal but opposite inductive reactance (+j). Some coils are mounted at the base of the mast — a base-loaded antenna — and some are mounted near the center (center-loaded) or the top (top-loaded).

As the coil is moved higher, the radiation resistance increases (a good thing), but the necessary coil reactance also increases as do resistive losses in the coil. Therefore it becomes a balancing act, which requires a thorough understanding of the parameters involved, to choose the optimal coil location.

Table 21.14 lists the characteristics of an 8 foot (2.4 meter) mobile whip in both base-loaded and center-loaded configurations. The table shows the required loading coil

inductance to bring the antenna to resonance on the different bands. The matching coil inductance is placed across the feed point to bring the impedance to 50 Ω.

Note that center-loading approximately doubles both the required inductance and the coil's resistive losses. The radiation resistance also increases, but only in that part of the antenna above the loading coil. If ground losses are included in the calculations, the coil's optimal position changes, but it is typically close to the center of the antenna.

It is important to note that the total amount of stray capacitance from the mounting hardware and proximity of the whip to the vehicle may be much higher depending on where and how the antenna is mounted. The higher the stray capacitance, the less efficient the antenna will be.

Table 21.15 compares different types of HF mobile antennas against an unloaded whip antenna. The simple whip is assumed to be a full-size ¼ wavelength. The effects of mounting are not included.

LOADING COIL Q

Antenna system Q is limited by the Q of the coil. The bandwidth between 2:1 SWR points of the system = 0.36 × f/Q. On 80 meters, the bandwidth of the 10.5 foot whip = 0.36 × 3.5/200 = 6.3 kHz. If we could double the Q of the coil, the efficiency would double and the bandwidth would be halved. The converse is also true. In the interest of efficiency, the highest possible Q should be used!

Loading coil Q is especially important on the lower HF bands where coil losses can exceed ground losses. The factors involved include wire size, wire spacing, length-to-diameter ratio and the materials used in constructing the coil. All of the factors interact with one another, making coil design a compromise — especially when wind loading and weight become major considerations.

In general — the larger the coil, the higher the Q. The more mass within the field of the coil (metal end caps for example) the lower the Q. Short, fat coils are better than long, skinny ones. However, the coil's length-to-diameter ratio (L/D) for the highest Q increases as the inductance increases. It ranges from 1:1 on the upper HF bands to as much as 4:1 on the lower bands where the inductance is large. Practical mechanical considerations for coils with large reactance values above 1000 Ω (160 meter coils for example) require the length-to-diameter ratio to increase, which lowers Q. **Table 21.16** suggests loading coil dimensions that maximize Q.

Another significant factor arises from high Q. Let's assume that we deliver 100 W on 80 meters to the 7.43 Ω at the antenna terminals. The current is 3.67 A and flows through the 1375 Ω reactance of the coil giving rise to

Table 21.16
Suggested Loading Coil Dimensions

Req'd L (µH)	Turns	Wire Size	Dia. (Inches)	Length (Inches)
700	190	22	3	10
345	135	18	3	10
150	100	16	2½	10
77	75	14	2½	10
77	29	12	5	4¼
40	28	16	2½	2
40	34	12	2½	4¼
20	17	16	2½	4¼
20	22	12	2½	2¾
8.6	16	14	2	2
8.6	15	12	2½	3
4.5	10	14	2	1¼
4.5	12	12	2½	4
2.5	8	12	2	2
2.5	8	6	2⅜	4½
1.25	6	12	1¾	2
1.25	6	6	2⅜	4½

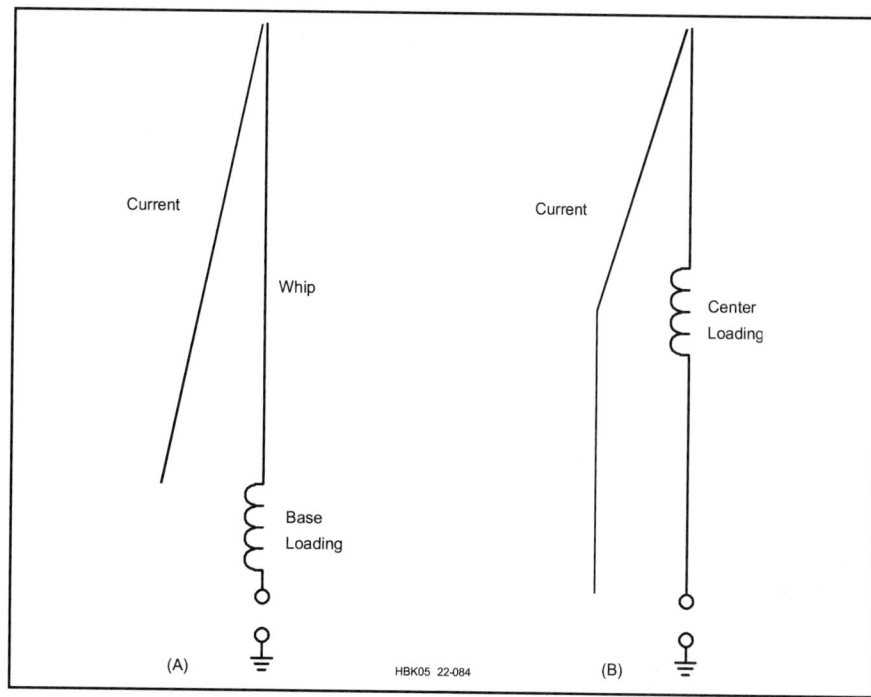

Fig 21.81 — Relative current distribution on a base-loaded antenna is shown at A and for a center-loaded antenna at B.

$1375 \times 3.67 = 5046 \, V_{RMS} \, (7137 \, V_{peak})$ across the coil! This is a significant voltage and may cause arcing if the coil is wet or dirty.

With only 30.6 pF of antenna capacitance, the presence of significant stray capacitance at the antenna base shunts currents away from the antenna. RG-58 coax presents about 21 pF/foot. A 1.5 foot length of RG-58 would halve the radiation efficiency of our example antenna. For cases like the whip at 3.5 MHz, the matching network has to be right at the antenna!

21.8.3 Base vs Center vs Continuous Loading

There are a few important aspects to be kept in mind when selecting or building an HF mobile antenna. As the antenna becomes longer, less loading inductance is required and the coil Q can become higher, improving efficiency. Also, for longer antennas, the better the mounting location has to be in order to optimize efficiency. We'll cover mounting and efficiency later.

Placing the loading coil at the base results in the current distribution shown in **Fig 21.81A**. If we move the coil to the center, the current curve looks like the one in Fig 21.81B. The location of the optimal position between the two extremes depends on the ground losses, and to a lesser degree on loading coil Q and overall length. For example, if the ground losses were zero, the best position would be at the bottom. As the ground losses increase, the optimal position gets closer to the center. If the ground losses are high enough, the optimal position is in the top one-third of the antenna's length, but efficiency is very poor.

Center-loading increases the current in the lower half of the whip as shown in Fig 21.81B.

Capacitance for the section above the coil can be calculated just as for the base-loaded antenna. This permits calculation of the loading inductance. The center-loaded antenna is often operated without any base matching in which case the resistive component can be assumed to be 50 Ω for purposes of calculating the current rating and selecting wire size for the inductor. The reduced size of the top section results in reduced capacitance which requires a much larger loading inductor.

Because of the high value of inductance

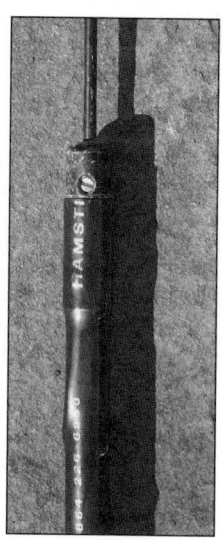

Fig 21.82 — Continuously-loaded whip antennas are short and lightweight. The base section consists of a fiberglass tube wound with wire to form the loading inductor. At the top of the base section the length of a steel whip or stinger can be adjusted to bring the antenna to resonance. [Joel Hallas, W1ZR, photo]

Designing a Base Loading System

This design procedure was contributed by Jack Kuecken, KE2QJ. To begin, estimate the capacitance, capacitive reactance and radiation resistance as shown at the beginning of this section. Then calculate the expected loss resistance of the loading coil required to resonate the antenna. There is generally additional resistance amounting to about half of the coil loss which must be added in. As a practical matter, it is usually not possible to achieve a coil Q in excess of 200 for such applications.

Using the radiation resistance plus 1.5 times the coil loss and the power rating desired for the antenna, one may select the wire size. For high efficiency coils, a current density of 1000 A/inch^2 is a good compromise. For the 3.67 A of the example we need a wire 0.068 inch diameter, which roughly corresponds to #14 AWG. Higher current densities can lead to a melted coil.

Design the coil with a pitch equal to twice the wire diameter and the coil diameter approximately equal to the coil length. These proportions lead to the highest Q in air core coils.

The circuit of **Fig 21.A4** will match essentially all practical HF antennas on a car or truck. The circuit actually matches the antenna to 12.5 Ω and the transformer boosts it up to 50 Ω. Actual losses alter the required values of both

the shunt inductor and the series capacitor. At a frequency of 3.5 MHz with an antenna impedance of 0.55 −j1375 Ω and a base capacitance of 2 pF results in the values shown in **Table 21.B.** Inductor and capacitor values are highly sensitive to coil Q.

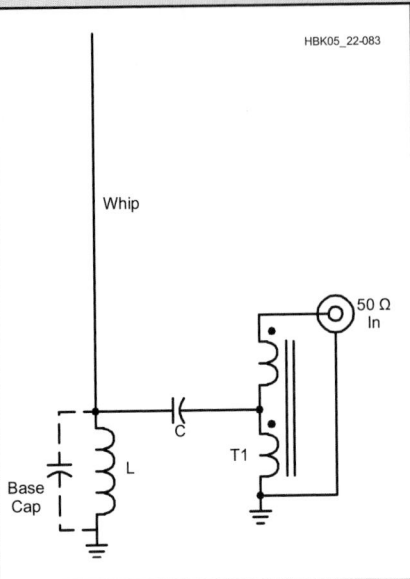

Fig 21.A4 — The base-matched mobile whip antenna

Table 21.B
Values of L and C for the Circuit of Fig 21.A4 on 3.5 MHz

Coil Q	L (µH)	C (pF)	System Efficiency (%)
300	44	11.9	8.3
200	29.14	35	3.72
100	22.2	58.1	1.4

Furthermore, the inductor values are considerably below the 62.5 µH required to resonate the antenna.

This circuit has the advantage that the tuning elements are all at the base of the antenna. The whip radiator itself has minimal mass and wind resistance. In addition, the rig is protected by the fact that there is a dc ground on the radiator so any accidental discharge or electrical contact is kept out of the cable and rig. Variable tuning elements allow the antenna to be tuned to other frequencies.

Connect the antenna, L and C. Start with less inductor than required to resonate the antenna. Tune the capacitor to minimum SWR. Increase the inductance and tune for minimum SWR. When the values of L and C are right, the SWR will be 1:1.

---

required for center-loading, high-Q coils are very large. The large wind resistance necessitates a very sturdy mount for operation at highway speed. One manufacturer of this type of coil does not recommend their use in rain or inclement weather. The higher Q of these large coils results in a lower feed point impedance, necessitating the use of a base matching element in the form of either a tapped inductor or a shunt capacitor to match to 50 Ω. (See this chapter's section on Mobile HF Antenna Matching.) Another manufacturer places the coil above the center and uses a small extendable whip or wand for tuning.

Antennas known by the trade name "Hamsticks" shown in **Fig 21.82** aren't really continuously-loaded. Instead, a small diameter enameled wire loosely wound around a fiberglass tube forms the base section of the antenna. Approximately half way up the antenna, the wire is close wound to form a lumped-element loading coil and a metal whip or stinger is attached at the top to complete the whip. A heat-shrink sleeve covers the wound section of the antenna. The stinger's length can be adjusted to tune the antenna to the desired operating frequency.

In recent years, these lightweight antennas

have become very popular. Their input impedance is near 50 Ω in part because of their low-Q base and loading coil sections. Thus they don't require matching once the length of the top whip or stinger is adjusted. They are light in weight (about a pound), short in length (typically 6 to 8 feet) and thus easy to mount. For temporary use, easy mounting may be more important than high efficiency. This type of antenna is most effective on 20 meters and higher frequency bands.

The wound base section of these antennas is a very weak radiator, acting more like a transmission line or distributed element than a linear radiating element. The long and thin coil has low Q, as well, increasing antenna loss. The close-wound section acts like a lumped element and does not radiate.

21.8.4 Top-Loaded Whips

In the interests of efficiency, electrical length matters because radiation resistance increases as the square of electrical length. Higher radiation resistance in a mobile antenna results in higher efficiency. All else being equal, a 12 foot antenna will have 4 times the radiation resistance than one 6 feet

long. As pointed out above, the maximum physical height should be less than 13.5 feet (4.1 meters). As discussed in this chapter's section on Physically Short Antennas, one way to increase the electrical length but not the physical length, is top-loading. A mobile HF antenna is top-loaded by using a *capacitance hat* or "cap hat."

As their name implies, cap hats add capacitance at the top of the antenna, above any loading coil. This increases radiation resistance by as much as four times under ideal conditions, but at the expense of increased weight, wind loading and complexity. Not all antennas, especially small screwdriver types, are sturdy enough to support a large cap hat. For those that are, or if the antenna can be guyed or stiffened, cap hats offer increased efficiency and bandwidth.

The actual placement of the cap hat is important, too. If mounted too close to the loading coil, efficiency is lower than with no cap hat at all. Thus, the best mounting location is at the very top of the antenna.

The physical design of the cap hat is limited by practical concerns. The largest are approximately 3 feet in diameter and require additional bracing or guying of the antenna.

Most are constructed of straight radial wires without an outer rim. The loop design shown in **Fig 21.83** is more efficient than straight wires, but tends to snag more on errant limbs. See the referenced *QEX* article by Griffith and the *QST* article by Clement for more on capacitance hats and top loading of mobile antennas.

21.8.5 Remotely Controlled HF Mobile Antennas

Remotely controlled (motorized) HF mobile antennas are commonly referred to as *screwdriver antennas*. Don Johnson,

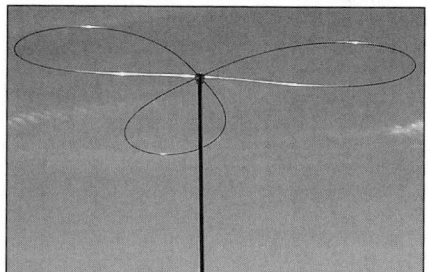

Fig 21.83 — A typical capacitance hat or "cap hat" added at the top of a mobile whip antenna. The antenna is a Scorpion (scorpionantennas.com).

W6AAQ (SK), is credited by many as the father of the screwdriver antenna. His design was not the first motorized antenna, but he certainly popularized it. There are now over 50 commercial versions available.

They're called screwdrivers because the first examples used a stripped-down rechargeable electric screwdriver assembly to adjust the resonant frequency of the antenna. The motor turns a threaded rod in and out of a nut attached to the bottom of the coil. This in turn moves the coil in and out of the lower mast section. Contacts at the top of the mast slide on the outside of the coil, thus adjusting the resonance point. Position sensors may be used to keep track of the location of the coil tap. Nowadays, calling them screwdrivers is a bit of a misnomer as the electric screwdriver motors have been replaced with much more reliable gear motors.

There are several remotely controlled HF mobile antennas that don't change length as true screwdrivers do. Both base and center loaded models are available (see **Fig 21.84**). Whether or not they're more efficient is dependent on the factors discussed in the previous section, rather than the method used to adjust the coil.

RF CHOKES

The motors and position sensors of all remotely-controlled antennas operate above

RF ground potential. The amount of RF present on the leads depends on several factors, especially where and how the antenna is mounted and its overall (electrical) length. Thus the RF current coupled onto the leads must be minimized with an RF choke before entering the vehicle. An inadequate choke may result in erratic controller operation and possible interference with transceiver operation.

The choke should have an impedance of at least two orders of magnitude greater than the impedance of the circuit. In other words, at least 5 kΩ, and perhaps two or three times that in some cases (stubby antennas and poor mounting schemes are examples). Mix 31 ferrite split beads are ideal for this application, but it takes eight turns to obtain a 5 kΩ choking impedance. Depending on the wire size and insulation, you'll need to use the ½ or ¾ inch ID cores. Snap-on ferrite beads are

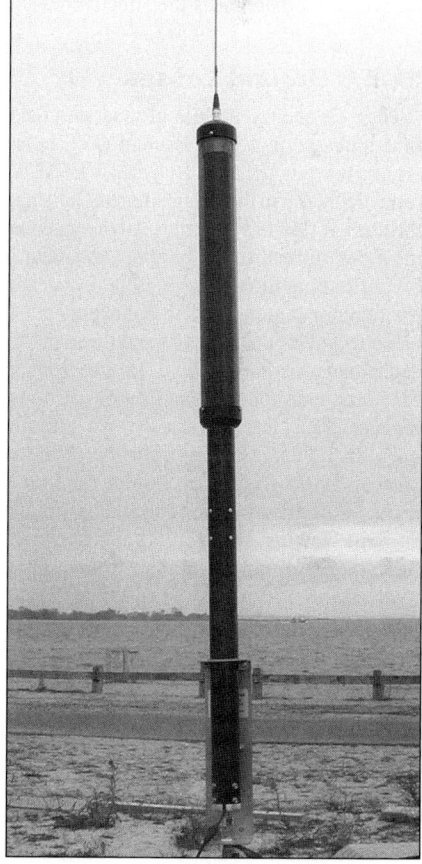

Fig 21.84 — The screwdriver-style remotely-controlled whip antenna. A small motor in the base mast moves a coil past contacts at the top of the metal base section. The top whip section is attached to the top of the coil. As the coil moves out of the mast, more inductance is connected in series between the base section and top section. Screwdriver antennas are popular because they offer multiband coverage. [Joel Hallas, W1ZR, photo]

Determining the Radiation Efficiency of a Center Loaded Mobile Whip

We can measure the radiation efficiency by measuring ground wave field strength E (dB referenced to µV/m). For the average radio amateur, a field strength meter is not a part of his ham shack gear. We can predict performance using readily available antenna modeling software (one of the many available versions of *NEC*) provided we have a measure of actual losses.

There are a number of loss parameters we do not know. We do not know the Q factor for the center loading coil (R_L), and we do not know the ground-induced loss resistance (R_g). In fact we do not know with certainty the radiation resistance (R_r), since the antenna sees an image of itself in the ground. *NEC* only gives us the sum of the various resistances.

$$R_{as} = R_r + R_C + R_g + R_L$$

R_C, the only parameter not discussed above, is the conductor loss resistance.

We need to know R_r if we are going to compute radiation efficiency, since radiation efficiency is given by:

$$\eta = \frac{R_r}{R_{as}}$$

So what do we do? We can measure R_{as} using the SWR analyzer, by adjusting the tuning so the reactance at the base of the antenna is equal to zero. We can then use *NEC* to predict the base impedance (resistive component), by changing the Q factor of the inductor so that R_{as} predicted equals R_{as} measured. We can then predict the ground field strength (dBµV/m) at say 100 m for a transmitter power of 1 kW. We then reference this predicted field strength to that for an electrically small lossless vertical antenna (129.54 dBµV/m at 100 m for 1 kW transmitter power — which corresponds to the commonly quoted value of 300 mVm at 1 km). This gives us a pretty good estimate of the radiation efficiency of our mobile whip. — *Jack Belrose, VE2CV*

Fig 21.85 — RF choke for screwdriver antenna control and power leads.

available from most Amateur Radio dealers. (More information on this type of RF choke may be found in the **RF Techniques** chapter.)

The choke shown in **Fig 21.85** consists of 13 turns of #18 AWG wire, wound on a ¾ inch ID, mix 31 split bead. It has an impedance of approximately 10 kΩ at 10 MHz. When winding the chokes, try not to overlap or twist the wires as this reduces the effectiveness.

21.8.6 Ground Losses

High frequency mobile ground loss data first appeared in a 1953 issue of *QST*, in an article written by Jack Belrose, VE3BLW (now VE2CV). In the article, Belrose said that the current flowing at the base of the antenna must be returned to the base of the antenna by currents induced in the ground beneath the radiator (antenna). These currents must be collected by the car body and through the capacitance of the car body to the ground. Since the maximum dimension of car body is considerably less than a quarter wavelength on most HF bands, only a portion of these currents will be collected by the car frame itself, and the rest will be collected by ground currents flowing through the capacitance of the car to the ground. Since the ground is not lossless, quite a large loss resistance (R_g) is found.

From that article, the accepted ground loss figure for HF mobile applications varies between 12 Ω (for 80 meters) and 2 Ω (for 10 meters). However, these figures do not include stray capacitance from the mounting location and method. Stray capacitance has the same effect as ground losses: reduced efficiency. As a result, in the real world, ground losses can be double the accepted values, reducing an otherwise efficient antenna to mediocrity.

21.8.7 Antenna Mounting
PERMANENT OR TEMPORARY

There are many reasons to install any mobile antenna permanently on a vehicle. The decision to drill holes in sheet metal to

mount antennas is hotly debated. While no-hole mounts can be used satisfactorily, it is best to look at all sides of the issue before installing any antenna.

A common concern about drilling a hole for an antenna mount is with regard to a leased vehicle. Leases don't necessarily preclude properly installed antenna mounting holes. What lease agreements are primarily concerned with is body damage such as from an accident or mistreatment. Properly installed NMO mounts, for example, are often acceptable. It's always prudent to ask before leasing the vehicle.

Drilled holes and waterproof mounts also help minimize common-mode current on the coaxial feed line. This helps reduce RFI to or from on-board computers and electrical devices. Aside from the hole itself, a permanent mount also minimizes damage to the finish.

Here is an important caveat to keep in mind: While the roof of a vehicle is a very good place to mount an antenna, more and more new vehicles are equipped with side curtain air bags. They typically are mounted along the edges of the headliner, including the rear seat area if there is one. The wiring to these devices is routed through any one (or more) of the roof pillars. Extra care is required when installing antennas in vehicles so equipped. If you are the least bit apprehensive about installing a roof-mounted antenna, seek professional help from your dealer or a qualified installer.

Mobile antenna mounting hardware runs the gamut from mundane to extravagant. Choosing the correct hardware is based on need, as well as on personal preference. There are too many variables with respect to mounting HF mobile antennas on modern vehicles to cover in a short discussion. It is easier to explain what not to do and adapt those guidelines to your own personal circumstances: the antenna mount should:

• Be permanently mounted;
• Be strong enough to support the antenna;
• Have as much metal mass under it as possible;
• Be well-grounded to the chassis or body;
• Not interfere with doors, trunks, or access panels and
• Be removable with minimal damage to the vehicle.

Aside from maximizing performance by minimizing ground and stray capacitance losses, there are also safety reasons for using a permanent mount with HF antennas. If you need to use a temporary mount, use the multiple-magnet mounts for their superior holding strength. Any mobile HF antenna attached to a vehicle traveling at highway speed by a single-magnet mount is a tenuous situation at best.

TYPE OF MOUNT

The type of mount is dictated by several conditions. These include a decision whether or not to drill holes, the size, weight and length of the antenna. If you want to operate HF mobile operation regularly, you are better off with a permanent mount. If you're not, a trunk lip, angle bracket, or license plate mount, and a lightweight, continuously-loaded antenna may meet your needs.

Ground plane losses directly affect a mobile antenna's efficiency. From this standpoint, mounts positioned high on the vehicle are preferred over a trailer hitch, bumper or other locations that place the antenna where the vehicle body will be close to the radiating element.

The ground plane of a mobile vertical antenna begins where the coax shield connects. When the feed line shield couples to the vehicle body capacitively as in a mag-mount, or when the mass of the vehicle is far below the feed point (long extension sections attached to trailer hitch mounts), ground losses escalate dramatically. Running a ground strap to the nearest connecting point to the vehicle body does not eliminate ground losses. Remember that the ground connection is part of the antenna system, and it is the efficiency of the whole system that is important.

While it is difficult to mount an HF antenna without at least part of the mast being close to the body, the coil must be kept free and clear or tuning problems and reduced efficiency will result. In some cases, vans and SUVs for example, front-mounting may become necessary to avoid excessive coil-to-body interaction.

One drawback to trunk lip and similar clip mounts is the stress imposed on them as the lids and doors are open and closed. In most cases, angle brackets that attach to the inner surfaces with screws are a better choice.

Sometimes, the only solution is a custom-made bracket like the one shown later in this chapter. Here too, the circumstances dictate the requirements. Keep in mind that removing permanently installed antenna mounts such as the ball mount shown in **Fig 21.86** will always leave some body damage, but temporary ones do, too. Only the severity is in question, and that's in the eyes of the beholder.

21.8.8 Mobile HF Antenna Matching

Modern solid-state transceivers are designed for loads close to 50 Ω impedance. Depending on the design of the antenna (primarily depending on coil position and Q), overall length and the ground losses present, the input impedance is usually closer to 25 Ω but may vary from 18 Ω to more than 50 Ω. Note that a vehicle is an inadequate ground plane for any HF mobile antenna. Typical

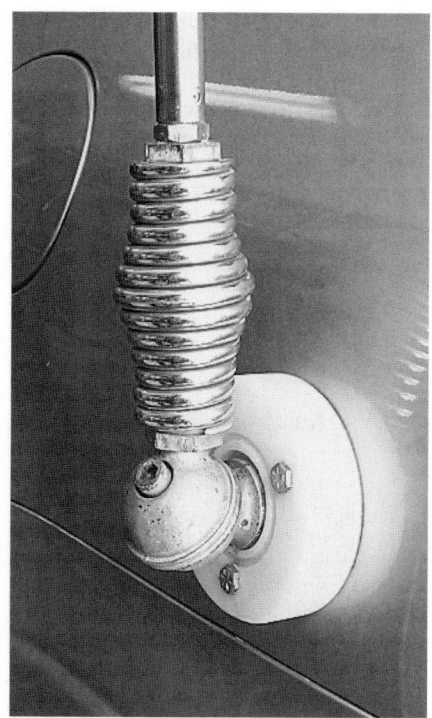

Fig 21.86 — A simple ball mount is sturdy but requires drilling holes in the vehicle body.

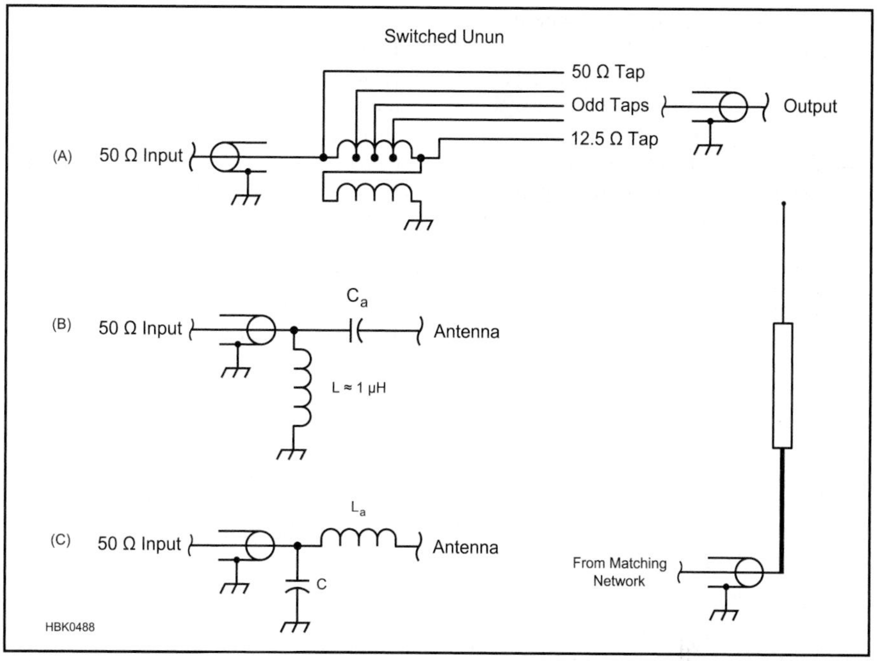

Fig 21.87 — A well-constructed and mounted HF mobile antenna will have an average input impedance around 25 Ω, requiring some matching to the feed line. The unun transmission line transformer (A) is a 4:1 configuration with taps added to match intermediate impedance values between 50 and 12.5 Ω. The high-pass L-network (B) uses some of the antenna's capacitive reactance as part of the network and the low-pass L-network (C) uses some of the antenna's inductive reactance as part of the network. Both (A) and (B) result in an antenna at dc ground, an important safety issue.

ground loss varies from 20 Ω (160 meters) to 2 Ω (10 meters). Stray capacitance losses may further increase the apparent ground losses.

It's important to remember two important facts. First, as the coil is moved past the center of the antenna toward the top, the coil's resistive losses begin to dominate, and the input impedance gets closer to 50 Ω. Second, short stubby antennas require more inductance than longer ones, which increases resistive losses in the coil (low Q). While no matching is required in either case, efficiency suffers, and may actually drop below 1% on the lower bands. Another way to look at the situation is that an antenna with no matching required implies a low efficiency.

Ground loss, coil position, coil Q, mast size, whip size and a few other factors determine the feed point impedance, which averages about 25 Ω for a typical quality antenna and mount. This represents an input SWR of 2:1, so some form of impedance matching is required for the transceiver. There are three ways to accomplish the impedance transformation: capacitive, transmission-line transformer and inductive matching as shown in **Fig 21.87**. Each has its own unique attributes and drawbacks.

Transmission line transformers, in this case an unun in Fig 21.87A, do provide a dc ground for the antenna. They can be tapped or switched to match loads as low as a few ohms. Their broadband nature makes them

ideal for HF mobile antenna matching. Since a remotely-controlled HF mobile antenna's input impedance varies over a wide range, transmission line transformers are best utilized for matching monoband antennas.

Inductive matching in Fig 21.87B borrows a little capacitance from the antenna (C_a) — the antenna is adjusted to a frequency slightly higher than the operating frequency, making the input impedance capacitive. This forms a high-pass L-network, which transforms the input impedance to the 50 Ω transmission line impedance. It is ideal for use with remotely-controlled antennas, as its reactance increases with frequency. By selecting the correct inductance, a compromise can be reached such that the impedance transformation will result in a low SWR from 160 through 10 meters. The approximate value is 1 μH, but may vary between 0.7 and 1.5 μH.

Adjusting the shunt coil can be done without transmitting by using an antenna analyzer and takes about 10 minutes. (Full instructions for properly adjusting a shunt coil may be found at **www.k0bg.com/coil.html**.) Because no further adjustment is necessary, shunt coil matching is ideal for remotely-controlled HF mobile antennas.

Capacitive matching in Fig 21.87C borrows a little inductance from the antenna (L_a) — the antenna is adjusted to a frequency slightly lower than the operating frequency, making the input impedance inductive. This

forms a low-pass L-network, which transforms the input impedance to the 50 Ω transmission line impedance. While it works quite well, it has two drawbacks. First, capacitive matching presents a dc ground for the antenna, which tends to increase the static levels on receive. Second, the capacitance changes with frequency, so changing bands also requires a change in capacitance. This can be a nuisance with a remotely controlled antenna.

An important point should be made about dc grounding in addition to the static issue. If the antenna element should come in contact with a low-hanging high tension wire, or if lightning should strike it, dc grounding offers an additional level of protection for you and your transceiver.

21.8.9 Remotely-Tuned Antenna Controllers

There are three basic types of remote controllers: manual, position sensing and SWR sensing. Manual controllers consist of a DPDT center-off switch that changes the polarity of the current to the motor. Some commercial models include an interface with the radio that causes the radio to transmit a low-power carrier for tuning. Reading the SWR is left to the user. Some manual controllers incorporate a position readout to aid the operator in correctly positioning the antenna.

Position sensing controllers incorporate a

magnet attached to the motor output shaft. The magnet opens and closes a reed switch. During set up, the antenna is set to one end of its range or the other. Then the resonant points are found (you have to do this yourself) and stored in multiple memory locations. As long as power remains applied to the controller, a simple button push will move the antenna to a specific preset point. Some controllers use band or frequency data from a port on the radio and reset the antenna to the nearest preset based on that information.

SWR-sensing controllers either read data from the radio or from a built-in SWR bridge. Depending on the make and model, a push of the radio's tuner button (or one on the controller) causes the radio to transmit at a reduced power setting. The controller then powers the antenna's tuning motor. When the preset SWR threshold is reached, the controller stops the transmission and shuts off the motor. **Fig 21.88** shows an example of a controller made to work with a specific transceiver.

Automatic controllers are far less distracting than manual ones. Most offer a parking function that collapses the coil of a screwdriver antenna into the mast (highest frequency position, lowest overall length). If you garage your vehicle, this is a welcome feature.

21.8.10 Efficiency

Length matters! All else being equal, a 9 foot antenna will be twice as efficient as a 6 foot antenna, because radiation resistance relates directly to the square of the physical length. Further, longer antennas require less reactance to resonate, hence coil Q is higher,

and resistive losses lower.

Mounting methodology matters! It is the mass under the antenna, not alongside, that counts. The higher the mounting, the less capacitive coupling there will be between the antenna and the surface of the vehicle and the lower ground losses will be.

Project: Mounts for Remotely-Tuned Antennas

Remotely tuned antennas have become very popular, but they all have one thing in common: they're difficult to mount. They require both a coaxial feed line and a dc power connection, and no one makes a universal mount for them. The short, stubby ones aren't any more difficult to mount than a small whip antenna, but the "full-sized" ones (8 feet and longer) require special consideration.

These antennas are heavy (up to 18 pounds), so the mounting medium must be extra strong, and well anchored. As a result, many hams opt for a bumper or trailer hitch mount, even though the low mounting position reduces efficiency.

For some, efficiency is paramount which dictates mounting the antenna as high as possible. Doing either low or high mounting often requires custom fabrication. The accompanying photos illustrate the two different strategies that show Amateur Radio ingenuity at its finest.

Figs 21.89 and **21.90** depict the mobile installation of Fokko Vos, PA3VOS. Except for a Hi-Q heavy-duty quick-disconnect, the complete mount was custom engineered by Fokko. The mount bolts to a frame extension, which in turn is bolted to the undercarriage using existing bolts. Note that the rear hatch

may be opened without the antenna being removed. Had the trailer hitch been used, this would not be the case.

Fig 21.91 depicts installation on a Ford F350 based motor home owned by Hal

Fig 21.89 — The PA3VOS antenna mount easily supports a large screwdriver antenna and is offset to allow the hatch to open and close without removing the antenna.

Fig 21.90 — A close-up of the PA3VOS mount.

Fig 21.88 — A screwdriver antenna controller made to work with the IC-7000 transceiver.

Fig 21.91 — The KE5DKM bracket mounts under the hood of a motor home.

Wilson, KE5DKM. Hal designed the bracket, and had a local machine shop do the hard work. It is made of ¼ inch, high-strength aluminum, and the seams are welded. A powder-coat finish tops off the fabrication.

Shown here during installation, the bracket just fits into the right-side hood seam. The piece jutting out from the mount was to be used to further brace the mount. However, after all of the bolts were installed, additional bracing became unnecessary.

Project: Retuning a CB Whip Antenna

The most efficient HF mobile antenna is a full-size quarter-wavelength whip. Wouldn't it be nice if we could use one on every band? Alas, we cannot, but we can use one easily on 10 and 12 meters since the overall length will be less than 10 feet. If we start with a standard-length 102 inch whip and its base spring, we just need to shorten it a little for 10 meters, and lengthen it a little for 12 meters. Here's how to do it.

The formula for calculating the length of a ¼ λ antenna in feet is 234/f, where f is the frequency (MHz). Since the formula is for wire antennas, and the whip is larger in

diameter, the resulting length will be slightly too long. This is a good thing because it is easier to remove a little length than it is to add some. This makes tuning easier.

Using the formula, we discover the needed length for 10 meters (28.5 MHz) is 98.5 inches. Thus, we need to remove 3.5 inches and account for the length of the base spring (about 6 inches), for a total of 9.5 inches to be removed. This is best accomplished by filing a notch on opposite sides of the tip of the whip, and snapping it in two. Protect your eyes when you do this, as shards and splinters can fly off the broken ends. If a fiberglass whip is being modified, clip the internal wire at the top of the remaining base section. Remove the plastic cap from the discarded top section and replace it over the new top of the antenna.

Depending on the mount used, the actual resonant frequency will be lower than 28.5 MHz as the mounts adds effective length. A standard CB antenna ball mount will easily support a whip. The finished antenna is shown in **Fig 21.92A**.

Once the antenna is mounted on the vehicle, simply trimming the overall length ½ inch at a time will eventually produce a low SWR at your desired frequency.

LENGTHENING FOR 12 METER OPERATION

Lengthening a CB whip for 12 meters requires a little more work. Thankfully, there's a easy solution if you have a CB radio or truck stop near you. Wilson and other vendors sell short masts designed for the CB market. They have the requisite ⅜ × 24 threads to accept a standard whip, and they come with a female-to-female coupler. The 10 inch model is ideal for our use, and costs under $10.

Using our formula 234/f, the overall length needs to be 112.6 inches for 24.93 MHz. Adding 6 inches for the spring, 10 inches for the extension mast, and 102 inches for the whip, gives us 118 inches. So we need to remove 5.4 inches from the whip. Then just trim off ½ inch at a time as above to resonate the antenna. The finished antenna is shown in Fig 21.92B.

FINISHING UP

There are two more things to consider. If a

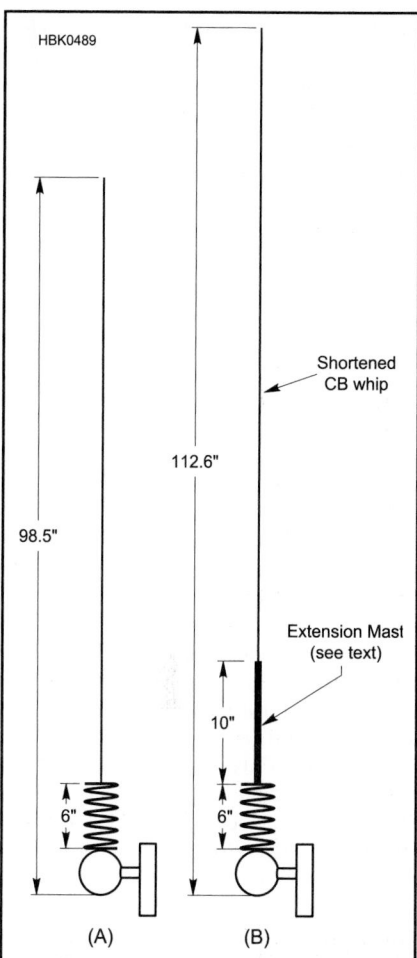

Fig 21.92 — CB whips are easily retuned for 10 meter (A) and 12 meter (B) mobile operation.

metal whip was modified, you'll need to replace the corona ball at the tip of the antenna. It helps reduce static from corona discharge. It's held on by a set screw and has little effect on tuning. Most CB shops sell them.

The other consideration is ground loss. Theoretically, a ¼ λ vertical will have an input impedance of 36 Ω. In a mobile installation, we have ground losses and stray capacitance losses in the mounting hardware. As a result, the real-world input impedance should be very close to 42 Ω, yielding a rather low SWR and good efficiency.

21.9 VHF/UHF Mobile Antennas

The simple ¼ λ, ½ λ, and ⅝ λ ground-plane whips are the most common VHF/UHF mobile antennas. Collinear antennas with higher gain are available. However, high gain and the lower radiation angle that goes with it isn't always a desirable attribute.

When using repeaters in urban areas, where higher angles of radiation are preferred, you're typically better off with a unity-gain antenna — a ¼ λ whip. It all depends on the HAAT (height above average terrain) of the repeater being used with respect to the mobile station's HAAT. In mountainous areas you're always better off with a unity gain antenna as the repeaters are much higher in elevation than the mobile station is.

If you're working simplex and living in a suburban or rural area, a high-gain antenna might have a slight edge. However, where, and how the antenna is mounted is more important than gain. A ¼ λ ground plane mounted in the center of the roof will typically out perform a gain antenna mounted on the trunk lid.

Sturdiness is also an important attribute, given the unintended abuse to which mobile antennas are subjected. The simple quarter-wave ground plane has the advantage here, as it isn't much more than a springy piece wire. If you look closely at some of the higher-gain antennas, you'll notice they have very small phasing coils, usually held together by small set screws. Hit one hard enough with a low-hanging limb and your antenna will break whereas a simple quarter-wave whip will only bend. A little straightening and you're back on the air.

21.9.1 VHF/UHF Antenna Mounts

Without doubt, the best VHF/UHF mount ever devised is the NMO (New Motorola) shown in **Fig 21.93**. When properly installed

in vehicle sheet metal, the mount will not leak even when the antenna is removed for car washing. SO-239 and threaded or snap-in mounts often leak even with the antenna attached.

Glass-mounted antennas are rather lossy, especially at lower VHF frequencies (2 meters). Many new vehicles use window glass with a metallic, anti-glare (passivated) coating that interferes with capacitive coupling through the glass. These antennas also transfer mechanical abuse to the glass, risking breaking the glass the antenna is mounted on.

MOUNTING LOCATION

As mentioned above, the center of the roof is an ideal mounting location for a VHF/UHF antenna. However, many mobile operators share a reluctance to drill the proper mounting holes fearing that doing so will depreciate the value of the vehicle. Instead, they rely on a mag-mount which has its own set of negatives including where and how to route the coax cable into the vehicle. They tend to collect metallic road debris, primarily brake pad dust, marring and scratching the painted surface under them, often causing more damage than would the hole for an NMO mount. For these reasons alone, mag-mounts should only be used for temporary installations, such as emergency communication.

Roof mounting has a few caveats. Modern vehicle roofs have strengthening supports to protect the occupants should the vehicle rollover due to a crash. Do not drill through these supports!

The various side pillars supporting the roof contain wiring for lighting, side airbags, and other accessories, which can make routing of the coax difficult. If you're at all reticent about roof mounting, see a dealer or professional installer.

As an alternative, a trunk lip mount may fit the bill. However, they too have some

drawbacks. They're stressed every time the trunk is opened or closed, and tend to work loose over time. Thus regular maintenance is required to assure a good electrical ground to the trunk lid. In any case, do not sand the paint down to bare metal as this removes the zinc undercoating which in turn promotes rust. It is also important to bond across the trunk hinges to assure a good ground.

Angle brackets work well, too. They're thin enough to fit into the seam of hoods, trunk lids, and back hatches. Coax routing under or around the weather seals can be a problem with the latter two.

ADJUSTING SWR

Setting the SWR of a VHF/UHF antenna is an important installation procedure, but not one to worry about unless the SWR suddenly makes a large change. Unlike the HF bands, most VHF antennas will cover the whole band segment without the need to retune. That is to say, the SWR will be low across the FM portion of the respective bands. An in-line SWR meter is generally not required as the vertically-polarized antenna is not intended to be used in the weak-signal portion of the bands where horizontal polarization is the norm.

21.9.2 VHF/UHF Antennas for SSB and CW

Operating SSB and CW on 6 meters, 2 meters, and 70 cm offers some exciting prospects for all license classes. While FM communications on the VHF bands are often considered line-of-site, propagation *beyond* line-of-site is common, as discussed in the

Fig 21.93 — The NMO (New Motorola) mount is widely used for VHF and UHF mobile antennas. It is available in mag-mount, trunk and lip mount and through-body mounting styles.

Fig 21.94 — The squalo (square halo) is a popular horizontally polarized VHF/UHF mobile antenna. The antenna shown here is made by M² Antenna Systems (www.m2inc.com).

Propagation of Radio Waves chapter.

Modern mobile SSB/CW transceivers usually output 100 W PEP on 6 meters and at least 50 W PEP on 2 meters and 70 cm. Under good band conditions, using horizontally polarized antennas, *beyond* line-of-sight distances can exceed 200 miles even without any sky-wave or tropospheric scatter present!

There's a catch, however. FM communications utilize vertically polarized antennas. Vertical polarization can be used for SSB and CW, but depending on the propagation path, signal strength from a vertically polarized mobile antenna can have a 20+ dB disadvantage compared to a horizontally polarized antenna due to cross-polarization.

Fortunately, horizontally-polarized antennas are of manageable size on the VHF and UHF bands, although they are not as simple to construct as vertically polarized whips. Dipoles and small beams present too much wind resistance to withstand the normal mobile environment. The usual solution is a loop antenna.

Fig 21.94 shows an M² Antenna Systems (**www.m2inc.com**) horizontally polarized 6 meter loop called a *halo* (for circular versions) or *squalo* (if square as shown). Equivalent antennas for 2 meters and 70 cm are common. Although this particular design is square, they're still called loops and have a roughly omnidirectional pattern. The "Big Wheel" design is another option. Mounting loop antennas on a vehicle can be less cumbersome than an HF antenna because they don't require a ground plane. A simple hitch-mounted mast will suffice, with no body holes needed!

21.10 VHF/UHF Antennas

Improving an antenna system is one of the most productive moves open to the VHF enthusiast. It can increase transmitting range, improve reception, reduce interference problems and bring other practical benefits. The work itself is by no means the least attractive part of the job. Even with high-gain antennas, experimentation is greatly simplified at VHF and UHF because an array is a workable size, and much can be learned about the nature and adjustment of antennas. No large investment in test equipment is necessary.

Whether we buy or build our antennas, we soon find that there is no one *best* design for all purposes. Selecting the antenna best suited to our needs involves much more than scanning gain figures and prices in a manufacturer's catalog. The first step should be to establish priorities for the antenna system as a whole. Once the objectives have been sorted out in a general way, we face decisions on specific design features, such as polarization, length and type of transmission line, matching methods, and mechanical design.

21.10.1 Gain

As has been discussed previously, shaping the pattern of an antenna to concentrate radiated energy, or received signal pickup, in some directions at the expense of others is the only possible way to develop gain. Radiation patterns can be controlled in various ways. One is to use two or more driven elements, fed in phase. Such arrays provide gain without markedly sharpening the frequency response, compared to that of a single element. More gain per element, but with some sacrifice in frequency coverage, is obtained by placing parasitic elements into a Yagi array.

21.10.2 Radiation Pattern

Antenna radiation can be made omnidirectional, bidirectional, practically unidirectional, or anything between these conditions. A VHF net operator may find an omnidirectional system almost a necessity but it may be a poor choice otherwise. Noise pickup and other interference problems tend to be greater with omnidirectional antennas. Maximum gain and low radiation angle are usually prime interests of the weak-signal DX aspirant. A clean pattern, with lowest possible pickup and radiation off the sides and back, may be important in high-activity areas, where the noise level is high, or for challenging modes like EME (Earth-Moon-Earth).

21.10.3 Height Gain

In general, the higher a VHF antenna is installed, the better will be the results. If raising the antenna clears its view over nearby obstructions, it may make dramatic improvements in coverage. Within reason, greater height is almost always worth its cost, but height gain must be balanced against increased transmission line loss. Line losses can be considerable at VHF and above, and they increase with frequency. The best available line may be none too good, if the run is long in terms of wavelength. Consider line losses in any antenna planning.

21.10.4 Physical Size

A given antenna design for 432 MHz, say a 5-element Yagi on a 1 λ boom, will have the same gain as one for 144 MHz, but being only one-third the size it will intercept only one-ninth as much energy in receiving. Thus, to be equal in communication effectiveness, the 432 MHz array should be at least equal in physical size to the 144 MHz one, requiring roughly three times the number of elements. With all the extra difficulties involved in going higher in frequency, it is better to be on the big side in building an antenna for the UHF bands.

21.10.5 Polarization

Whether to position the antenna elements vertically or horizontally has been a question since early VHF operation. Originally, VHF communication was mostly vertically polarized, but horizontal gained favor when directional arrays became widely used. Tests of signal strength and range with different polarizations show little evidence on which to set up a uniform polarization policy. On long paths there is no consistent advantage, either way. Shorter paths tend to yield higher signal levels with horizontal in some kinds of terrain. Man-made noise, especially ignition interference, tends to be lower with horizontal polarization. Vertically polarized antennas, however, are markedly simpler to use in omnidirectional systems and in mobile work, resulting in a standardization on vertical polarization for mobile and repeater operation on FM and for digital communications. Horizontal polarization is the standard for weak signal VHF and UHF operation. (Circular polarization is preferred for satellite work as described below.) A loss in signal strength of 20 dB or more can be expected with cross-polarization so it is important to use antennas with the same polarization as the stations with which you expect to communicate.

21.10.6 Circular Polarization

Polarization is described as *horizontal* or *vertical*, but these terms have no meaning once the reference of the Earth's surface is lost. Many propagation factors can cause polarization change — reflection or refraction and passage through magnetic fields (Faraday rotation), for example. Polarization of VHF waves is often random, so an antenna capable of accepting any polarization is useful. Circular polarization, generated with helical antennas or with crossed elements fed 90° out of phase, will respond to any linear polarization.

The circularly polarized wave in effect threads its way through space, and it can be left- or right-hand polarized. These polarization senses are mutually exclusive, but either will respond to any plane (horizontal or vertical) polarization. A wave generated with right-hand polarization, when reflected from the moon, comes back with left-hand polarization, a fact to be borne in mind in setting up EME circuits. Stations communicating on direct paths should have the same polarization sense.

Both senses can be generated with crossed dipoles, with the aid of a switchable phasing harness. With helical arrays, both senses are provided with two antennas wound in opposite directions.

21.10.7 Transmission Lines

The most common type of transmission line at VHF through the low microwave bands is unbalanced coaxial cable. Small coax such as RG-58 or RG-59 should never be used in VHF work if the run is more than a few feet. Half-inch lines (RG-8 or RG-11) work fairly well at 50 MHz, and runs of 50 feet or less are acceptable at 144 MHz. Lines with foam rather than solid insulation have about 30% less loss. Low-loss cable is required for all but the shortest runs above 222 MHz and *waveguide* is used on microwave frequencies. (See the **Transmission Lines** chapter for a discussion of waveguides.)

Solid aluminum-jacketed *hardline* coaxial cable with large inner conductors and foam insulation are well worth the cost. Hardline can sometimes even be obtained for free from local Cable TV operators as *end runs* — pieces at the end of a roll. The most common CATV variety is ½ inch OD 75 Ω hardline. Hardline is considered *semi-rigid* in that it can be bent, but only with a large radius to avoid kinking and repeated bending should be avoided.

Waterproof commercial connectors for hardline are fairly expensive, but enterprising amateurs have *home-brewed* low-cost connectors. If they are properly waterproofed,

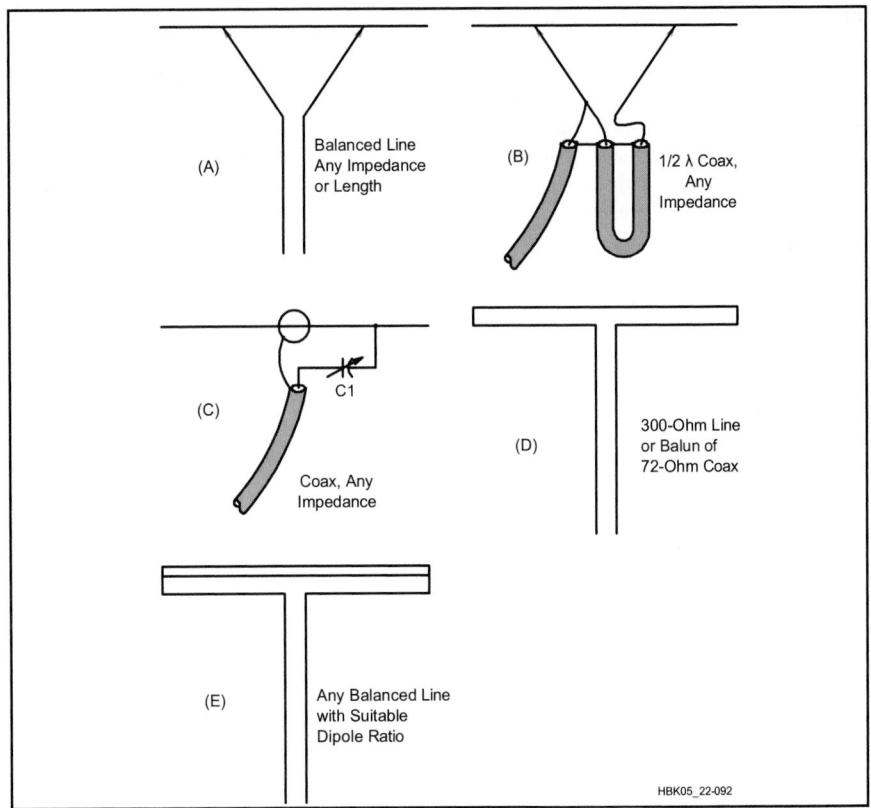

Fig 21.95 — Matching methods commonly used in VHF antennas. In the delta match, A and B, the line is fanned out to tap on the dipole at the points of best impedance match. The gamma match, C, is for direct connection of coax. C1 tunes out inductance in the arm. Folded dipole of uniform conductor size, D, steps up antenna impedance by a factor of four. Using a larger conductor in the unbroken portion of the folded dipole, E, gives higher orders of impedance transformation.

connectors and hardline can last almost indefinitely. See *The ARRL Antenna Book* for details on connectors and techniques for working with hardline.

Properly-built open-wire line can operate with very low loss in VHF and even UHF installations. A line made of #12 AWG wire, spaced ¾ inch or less with Teflon spreaders, and running essentially straight from antenna to station, can be better than anything but the most expensive hardline at a fraction of the cost. Line loss under 2 dB per 100 feet at 432 MHz is readily obtained. This assumes the use of high-quality baluns to match into and out of the balanced line, with a short length of low-loss coax for the rotating section from the top of the tower to the antenna. Such an open-wire line could have a line loss under 1 dB at 144 MHz.

Effects of weather on transmission lines should not be ignored. A well-constructed open-wire line works well in nearly any weather, and it stands up well. TV-type twin-lead is almost useless in heavy rain, wet snow or icing conditions. The best grades of coax and hardline are impervious to weather. They can be run underground, fastened to metal towers without insulation, or bent into

almost any convenient position, with no adverse effects on performance. However, beware of bargain coax. Lost transmitter power can be made up to some extent by increasing power, but once lost in the transmission line a weak signal can never be recovered in the receiver.

21.10.8 Impedance Matching

Theory and practice in impedance matching are discussed in detail in the **Transmission Lines** chapter, and in theory, at least, are the same for frequencies above 50 MHz. Practice may be similar, but physical size can be a major modifying factor in choice of methods.

DELTA MATCH

Probably the first impedance match was made when the ends of an open line were fanned out and tapped onto a half-wave antenna at the points of most efficient power transfer, as in **Fig 21.95A**. Both the side length and the points of connection either side of the center of the element must be adjusted for minimum reflected power in the line, but the impedances need not be known. The delta makes no provision for tuning out reactance, so the length of the dipole is pruned for best SWR.

Once thought to be inferior for VHF applications because of its tendency to radiate if adjusted improperly, the delta has come back to favor now that we have good methods for measuring the effects of matching. It is very handy for phasing multiple-bay arrays with low-loss open lines, and its dimensions in this use are not particularly critical.

GAMMA MATCH

The gamma match is shown in Fig 21.95C and is covered in more detail in the preceding section on HF Yagi antennas and in the **Transmission Lines** chapter. The center of a half-wave dipole being electrically neutral, the outer conductor of the coax is connected to the element at this point, which may also be the junction with a metallic or non-conductive boom. The inner conductor is connected to the element at the matching point. Inductance of the connection to the element is canceled by means of C1. Both the point of contact with the element and the setting of the capacitor are adjusted for minimum SWR using an antenna analyzer or SWR bridge.

The capacitor C1 can be a variable unit during adjustment and then replaced with a suitable fixed unit when the required capacitance value is found. Maximum capacitance should be about 100 pF for 50 MHz and 35 to 50 pF for 144 MHz. The capacitor and arm can be combined with the arm connecting to the driven element by means of a sliding clamp, and the inner end of the arm sliding inside a sleeve connected to the inner conductor of the coax. It can be constructed from concentric pieces of tubing, insulated by plastic sleeving or shrink tubing. RF voltage across the capacitor is low, once the match is adjusted properly, so with a good dielectric, insulation presents no great problem. A clean, permanent, high-conductivity bond between arm and element is important, as the RF current is high at this point.

Because it is inherently somewhat unbalanced, the gamma match can sometimes introduce pattern distortion, particularly on long-boom, highly directive Yagi arrays. The T-match, essentially two gamma matches in series creating a balanced feed system, has become popular for this reason. (See the preceding discussion on T-matches in the HF Yagi section.) A coaxial balun like that shown in Fig 21.95B is used from the balanced T-match to the unbalanced coaxial line going to the transmitter. To maintain a symmetrical pattern, the feed line should be run along the antenna boom at the centerline of the elements to the mast. A choke balun is often used to minimize currents that might be induced on the outer surface of the feed line shield.

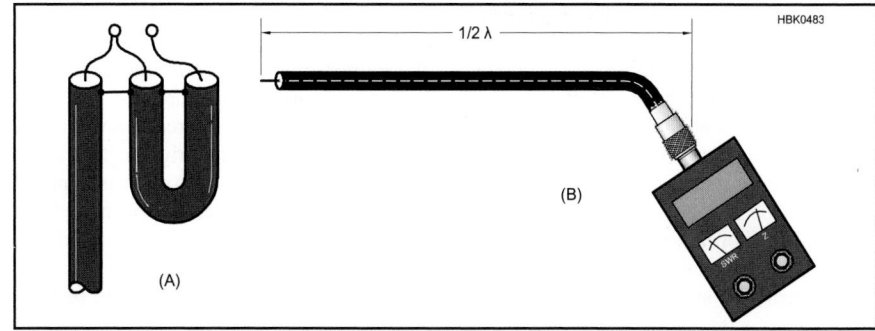

Fig 21.96 — Conversion from unbalanced coax to a balanced load can be done with a half-wave coaxial balun, A. The half-wave balun gives a 4:1 impedance step up. Electrical length of the looped section should be checked with an antenna analyzer with the far end of the line open, as in B. The lowest frequency at which the line impedance is a minimum is the frequency at which the line is ¼ λ long. Multiply that frequency by two to obtain the ½ λ frequency.

FOLDED DIPOLE

The impedance of a half-wave dipole feed point at its center is 72 Ω. If a single conductor of uniform size is folded to make a half-wave dipole, as shown in Fig 21.95D, the impedance is stepped up four times. Such a folded dipole can thus be fed directly with 300 Ω line with no appreciable mismatch. Coaxial feed line of 70 to 75 Ω impedance may then be used with a 4:1 impedance transformer. Higher impedance step-up can be obtained if the unbroken portion is made larger in cross-section than the fed portion, as in Fig 21.95E. The folded dipole is discussed further in the *ARRL Antenna Book*.

21.10.9 Baluns and Impedance Transformers

Conversion from balanced loads to unbalanced lines, or vice versa, can be performed with electrical circuits, or their equivalents made of coaxial line. A balun made from flexible coax is shown in **Fig 21.96A**. The looped portion is an electrical half-wave. This type of balun gives an impedance step-up of 4:1, 50 to 200 Ω, or 75 to 300 Ω typically. See the **RF Techniques** and **Transmission Lines** chapters for a detailed discussion of baluns and impedance transformers.

The physical length of the line section depends on the propagation factor of the line used, so it is best to check its resonant frequency, as shown at B. One end of the line is left open and an antenna analyzer used to find the lowest frequency at which the impedance at the other end of the line is a minimum, the frequency at which the section of line is ¼ λ long. Multiply the frequency by two to find the frequency at which the section is ½λ long.

Coaxial baluns giving a 1:1 impedance transfer are shown in **Fig 21.97**. The coaxial sleeve, open at the top and connected to the outer conductor of the line at the lower end (Fig 21.97A) is the preferred type. A conductor of approximately the same size as the line

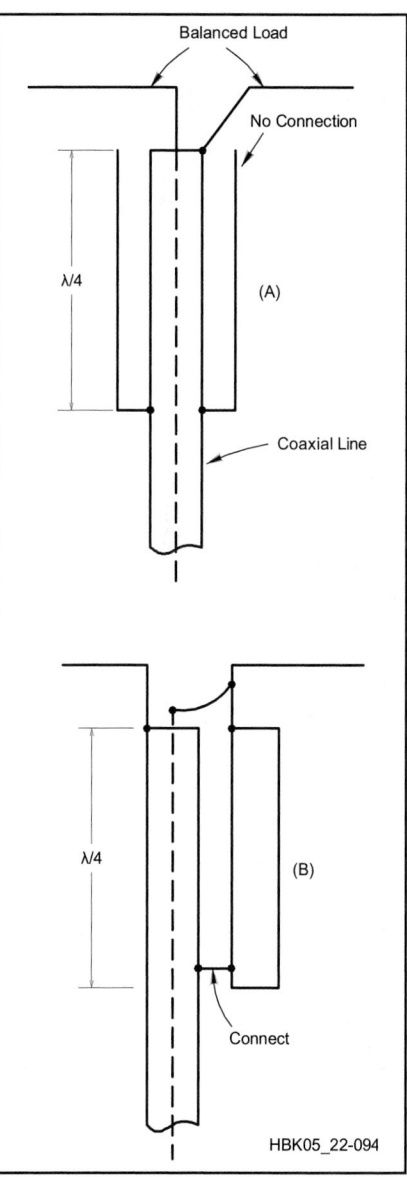

Fig 21.97 — The balun conversion function, with no impedance change, is accomplished with quarter-wave lines, open at the top and connected to the coax outer conductor at the bottom. The coaxial sleeve shown at A is preferred.

is used with the outer conductor to form a quarter-wave stub, in Fig 21.97B. Another piece of coax, using only the outer conductor, will serve this purpose. Both baluns are intended to present a high impedance to any RF current that might otherwise tend to flow on the outer conductor of the coax. Choke baluns made of ferrite beads of the proper material type or mix may also be used. See the **RF Techniques** chapter for information about ferrite use at VHF and UHF.

Project: Simple, Portable Ground-Plane Antenna

The ground-plane antenna is shown in **Fig 21.98** and uses a female chassis-mount connector to support the element and two radials. With only two radials, it is essentially two dimensional, which makes it easier to store when not in use. UHF connectors work well for 144 and 222 MHz, but you may prefer to use Type N connectors. N connectors

are recommended for 440 MHz and higher frequencies. BNC connectors can be used for the shorter antennas on 915 and 1280 MHz but are not particularly sturdy.

If the antenna is sheltered from weather, copper wire is sufficiently rigid for the radiating element and radials. Antennas exposed to the wind and weather can be made from brazing rod, which is available at welding supply stores. Alternatively, #12 or #14 AWG copper-clad steel wire could be used to construct this antenna.

To eliminate sharp ends, it's a good idea to bend the element and radial ends into a circle or to terminate them with a crimp terminal as in **Fig 21.99**. The crimp terminal approach is easier with stiff wire. Crimp and then solder the terminal to the wire. Make the overall length of the element and radials the same as shown in Fig 21.98, measuring to the outer tip of the loop or terminal.

Radials may be attached directly to the mounting holes of the coaxial connector.

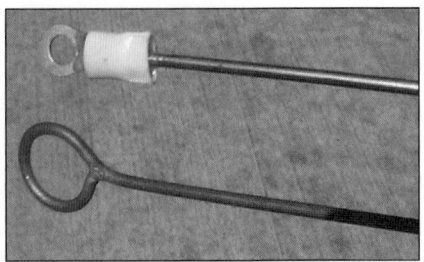

Fig 21.99 — Alternate methods for terminating element and radial tips on the simple ground-plane antenna. See text. (Photo by K8CH)

Bend a hook at one end of each radial for insertion through the connector. Solder the radials to the connector using a large soldering iron or propane torch.

Solder the element to the center pin of the connector. If the element does not fit inside the solder cup, use a short section of brass tubing as a coupler (a slotted $\frac{1}{8}$ inch ID tube will fit over an SO-239 or N receptacle center pin).

If necessary, prune the antenna to raise the frequency of minimum SWR. Then adjust the radial droop angle for minimum SWR — this should not affect the frequency at which the minimum SWR occurs.

One mounting method for fixed-station antennas appears in Fig 21.98. The feed line and connector are inside the mast, and a hose clamp squeezes the slotted mast end to tightly grip the plug body. Once the antenna is mounted and tested, thoroughly seal the open side of the coaxial connector with silicone sealant, and weatherproof the connections with rust-preventative paint.

Project: Coaxial Dipole for VHF or UHF

(The following antenna was originally described in July 2009 *QST* by John Portune, W6NBC, and was also reprinted in *The ARRL Antenna Compendium Volume 8.*)

Here is a homebrew coaxial dipole built from a small stainless whip, a length of threaded table-lamp tubing and some ¾ inch copper and PVC fittings. The one shown is for 440 MHz but it can readily be scaled for 146 or 222 MHz.

For homebrew vertical VHF antennas, coaxial dipoles often play second fiddle to J-poles. That's because the center connection to coax is often difficult to fabricate in the home workshop. Yet both antennas have the same performance. They're both full sized, half wave vertical dipoles, and the coaxial is shorter.

MAKING A COAXIAL DIPOLE

If you start with a common half wave

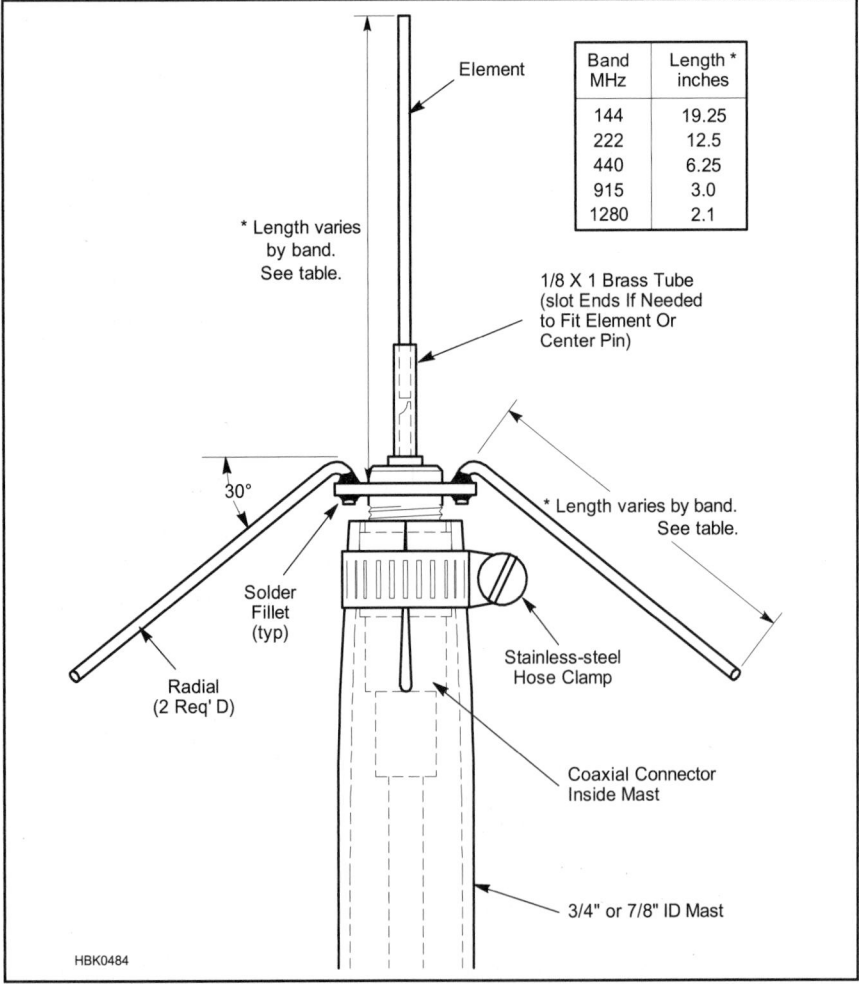

Band MHz	Length * inches
144	19.25
222	12.5
440	6.25
915	3.0
1280	2.1

Element

* Length varies by band. See table.

1/8 X 1 Brass Tube (slot Ends If Needed to Fit Element Or Center Pin)

30°

* Length varies by band. See table.

Solder Fillet (typ)

Stainless-steel Hose Clamp

Radial (2 Req' D)

Coaxial Connector Inside Mast

3/4" or 7/8" ID Mast

HBK0484

Fig 21.98 — A simple ground-plane antenna for the 144, 222 or 440 MHz bands. The feed line and connector are inside the mast, and a hose clamp squeezes the slotted mast end to tightly grip the plug body. Element and radial dimensions given in the drawing are good for the entire band.

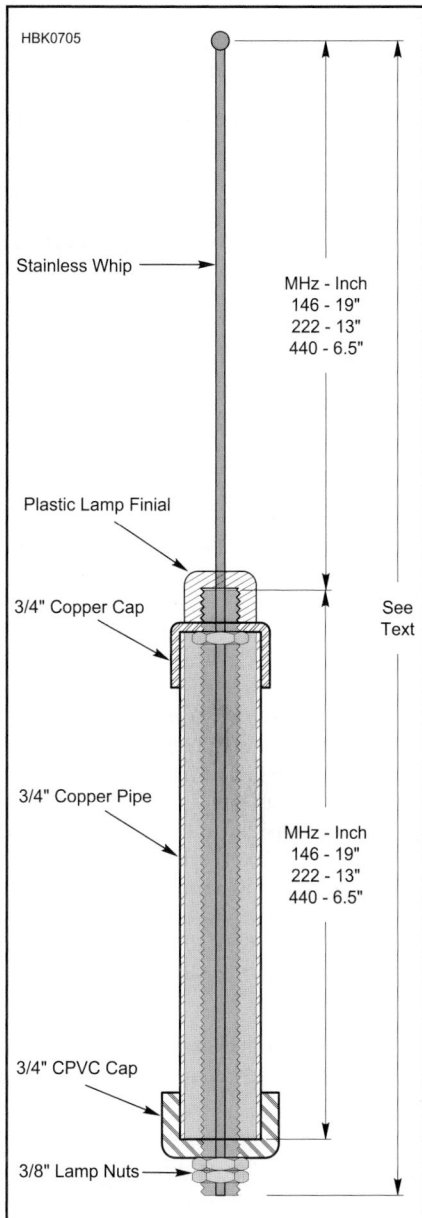

HBK0705

Stainless Whip

MHz - Inch
146 - 19"
222 - 13"
440 - 6.5"

See Text

Plastic Lamp Finial

3/4" Copper Cap

3/4" Copper Pipe

MHz - Inch
146 - 19"
222 - 13"
440 - 6.5"

3/4" CPVC Cap

3/8" Lamp Nuts

Fig 21.100 — Dimensioned drawing of coaxial dipole for three bands.

($\lambda/2$) stainless whip and extend it all the way down through a $\lambda/2$ long support tubing, here made from a threaded table lamp tube, the lower part of the whip becomes the center conductor of a short length of rigid coax feeding the center of the antenna. Now connection to normal coax is easily made below the antenna. To form the rigid coax section, you'll need to insulate the center conductor (lower part of the stainless whip) from the lamp tubing with some ¼ inch inside diameter (ID) polyethylene tubing. Hardware stores normally carry it. This short length of rigid coax formed in this way isn't precisely 50 Ω characteristic impedance, but the difference is totally insignificant. The drawing in **Fig 21.100** shows the details.

Assembly Details

The bottom half ($\lambda/4$) of the radiating dipole is a coaxial sleeve made from ¾ inch copper pipe and a pipe cap. The coax feed runs up its center to the connector at the bottom of the lamp tubing. Support and insulation of the bottom of the sleeve is provided by a ¾ inch CPVC plastic pipe cap. For those not familiar with CPVC fittings, they're made to mate with copper pipe and can handle high water temperatures. That's not true of common PVC fittings. Most hardware stores now carry CPVC. Drill a ⅜ inch hole in the center top of the copper and the CPVC caps for the lamp tubing to pass through.

The whole antenna is held together by two lamp tubing nuts and a plastic lamp finial, also readily available at hardware stores (see **Fig 21.101**). Note that a lamp tubing nut is also required inside the copper pipe cap. Drill a small hole in the middle of the lamp finial for the stainless whip. On the bottom of lamp tubing below the antenna install a 1¼ inch common PVC pipe cap, and secure it with two more lamp tubing nuts. This gives you a way to easily mount the antenna on top of any convenient length of ¼ inch PVC pipe. Run the coax feed down through the PVC pipe.

Hooking it Up

A conventional PL-259 UHF type coax connector for RG-8 coax will actually screw onto the bottom of the lamp tubing. The threads are not a perfect fit, but will tighten satisfactorily. The stainless whip runs down all the way to the very tip of the PL-259 connector. Solder it in there. Before doing so, however, install all the pieces of the antenna onto the threaded lamp tubing.

Many hams may think that stainless steel won't solder. It definitely will with a hot iron and acid flux. Scrape the end of the whip and dip it in hydrochloric swimming pool acid. With a little action from the tip of the soldering iron the whip will tin perfectly well. Before soldering, however, grind two or three small side notches in the bottom end of the whip. A Dremel tool works well for this. The notches will help the solder securely lock the whip into the tip of the PL-259 connector. Neutralize any leftover acid with baking soda solution.

Perhaps surprising to some, it really isn't necessary to solder any other parts of the antenna. There is adequate mating surface at the joints for the RF to cross over efficiently. Do, however, seal all possible water access spots with common silicone sealant and or plastic electrical tape.

MAKE IT FOR THE BAND YOU LIKE

There isn't an exact length required for the lamp tubing or the stainless whip. These

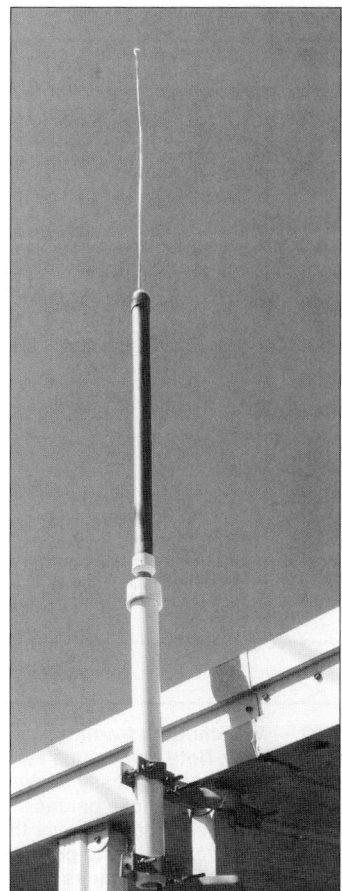

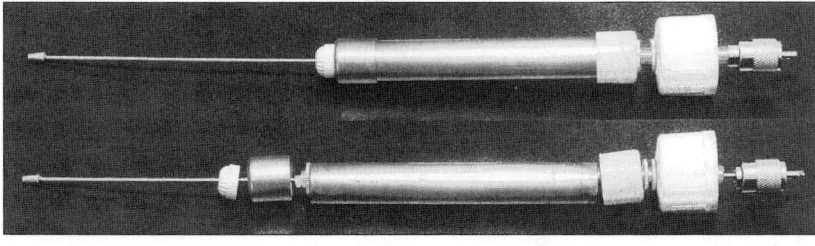

(A)

Fig 21.101 — Details of final assembly of coaxial dipole (A) and the finished product (B).

(B)

merely need to provide enough space for all the pieces of the antenna to go together. The author had a 48 inch whip on hand that he used uncut for the 146 MHz coaxial dipole and a similar 17 inch uncut whip for 440 MHz. He cut the lamp tubing to an appropriate length to fit the whips. What does matter, however, is the length of the whip above the top of the lamp tubing as well as

the length of the coaxial sleeve. These need to be close to a λ/4 — for 440 MHz, 6½ inches; for 222 MHz, 13 inches; and 19 inches for 146 MHz. These antennas are quite broad band and will cover the entire band in each case with these sizes. No cutting or pruning is necessary.

For ruggedness, or perhaps for stealth, you can install the whole antenna inside of

2 inch PVC water or ABS soil pipe and close the ends with end caps. The author lives in a mobile home park where antennas are not permitted, but the landlord thinks these coaxial dipoles (in ABS pipe) are vent pipes.

Try out one of these homebrew coaxial dipoles. You may find you prefer its smaller size, less obvious appearance and superior weatherproofing as compared to a J-pole.

21.11 VHF/UHF Yagis

Without doubt, the Yagi is king of home-station antennas these days. Today's best designs are computer optimized. For years amateurs as well as professionals designed Yagi arrays experimentally. Now we have powerful (and inexpensive) personal computers and sophisticated software for antenna modeling. These have brought us antennas with improved performance, with little or no element pruning required. A more complete discussion of Yagi design can be found earlier in this chapter and in the *ARRL Antenna Book*.

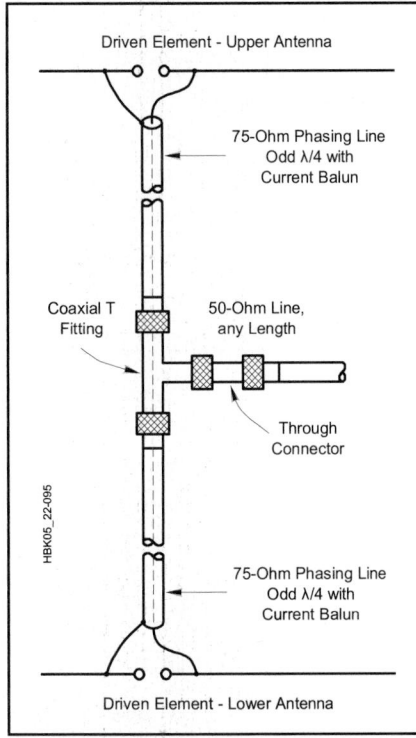

Fig 21.102 — A method for feeding a stacked Yagi array. Note that baluns at each antenna are not specifically shown. Good practice is to use choke baluns made up of ferrite beads slipped over the outside of the coax and taped to prevent movement. See the RF Techniques and Transmission Lines chapter for details.

21.11.1 Stacking Yagis

Where suitable provision can be made for supporting them, two Yagis mounted one above the other and fed in-phase may be preferable to one long Yagi having the same theoretical or measured gain. The pair will require a much smaller turning space for the same gain, and their lower radiation angle can provide interesting results. On long ionospheric paths a stacked pair occasionally may show an apparent gain much greater than the 2 to 3 dB that can be measured locally as the gain from stacking.

Optimum spacing for Yagis with booms longer than 1 λ is one wavelength, but this may be too much to handle for many builders of 50 MHz antennas. Worthwhile results are possible with separations of as little as ½ λ (10 feet), but ⅝ λ (12 feet) is markedly better. At 50 MHz, the difference between 12 and 20 foot spacing may not be worth the added structural problems.

The closer spacings give lowered measured gain, but the antenna patterns are cleaner (less power in the high-angle elevation lobes) than with 1 λ spacing. Extra gain with wider spac-

ings is usually the objective on 144 MHz and higher bands, where the structural problems are not quite as severe as on 50 MHz.

One method for feeding two 50 Ω antennas, as might be used in a stacked Yagi array, is shown in **Fig 21.102**. The transmission lines from each antenna, with a balun feeding each antenna (not shown in the drawing for simplicity), to the common feed point must be equal in length and an odd multiple of ¼ λ. This line acts as a quarter-wave (Q-section) impedance transformer, raises the feed impedance of each antenna to 100 Ω, and forces current to be equal in each driven element. When the feed lines are connected in parallel at the coaxial tee connector, the resulting impedance is close to 50 Ω.

Project: Three and Five-Element Yagis for 6 Meters

Boom length often proves to be the deciding factor when one selects a Yagi design. Dean Straw, N6BV, created the designs shown in **Table 21.17**. Straw generated the designs in the table for convenient boom lengths (6

Table 21.17
Optimized 6 Meter Yagi Designs

	Spacing From Reflector (in.)	Seg 1 Length (in.)	Seg 2 Length (in.)	Midband Gain F/R
306-06				
Refl	0	36	23.500	8.1 dBi
DE	24	36	16.000	28.3 dB
Dir 1	66	36	15.500	
506-12				
OD		0.750	0.625	
Refl	0	36	23.625	10.0 dBi
DE	24	36	17.125	26.8 dB
Dir 1	36	36	19.375	
Dir 2	80	36	18.250	
Dir 3	138	36	15.375	

Note: For all antennas, telescoping tube diameters (in inches) are: Seg1=0.750, Seg2=0.625.
Boom length should be 6 inches longer than the maximum Refl-Dir spacing to allow 3 inches on each end for element mounting hardware.

and 12 feet). The 3 element design has about 8 dBi gain, and the 5 element version has about 10 dBi gain. Both antennas exhibit better than 22 dB front-to-rear ratio, and both cover 50 to 51 MHz with better than 1.6:1 SWR.

Element lengths and spacings are given in the table. Elements can be mounted to the boom as shown in **Fig 21.103**. Two muffler clamps hold each aluminum plate to the boom, and two U bolts fasten each element to the plate, which is 0.25 inches thick and 4.4 inches square. Stainless steel is the best choice for hardware. However, galvanized hardware can be substituted. Automotive muffler clamps do not work well in this application, because they are not galvanized and quickly rust once exposed to the weather.

The driven element is mounted to the boom on a phenolic plate of similar dimension to the other mounting plates. A 12 inch piece of Plexiglas rod is inserted into the driven element halves. The Plexiglas allows the use of a single clamp on each side of the element and also seals the center of the elements against moisture. Self-tapping screws are used for electrical connection to the driven element.

Refer to **Fig 21.104** for driven element and hairpin match details. A bracket made from a piece of aluminum is used to mount the three SO-239 connectors to the driven element plate. A 4:1 transmission-line balun connects the two element halves, transforming the 200 Ω resistance at the hairpin match to 50 Ω at the center connector. Note that the electrical length of the balun is λ/2, but

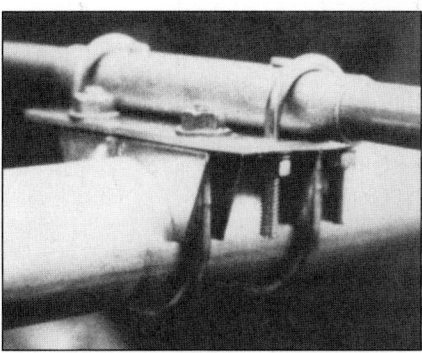

Fig 21.103 — The boom-to-element clamp. Galvanized U-bolts are used to hold the element to the plate, and 2 inch galvanized muffler clamps hold the plates to the boom.

the physical length will be shorter due to the velocity factor of the particular coaxial cable used. The hairpin is connected directly across the element halves. The exact center of the hairpin is electrically neutral and should be fastened to the boom. This has the advantage of placing the driven element at dc ground potential.

The hairpin match requires no adjustment as such. However, you may have to change the length of the driven element slightly to obtain the best match in your preferred portion of the band. Changing the driven-element length will not adversely affect antenna performance. *Do not adjust the lengths or spacings of the other elements*

— they are optimized already. If you decide to use a gamma match, add three inches to each side of the driven element lengths given in the table for both antennas.

Project: Medium-Gain 2 Meter Yagi

This project was designed and built by L. B. Cebik, W4RNL (SK). Practical Yagis for 2 meters abound. What makes this one a bit different is the selection of materials. The elements, of course, are high-grade aluminum. However, the boom is PVC and there are only two #6 nut-bolt sets and two #8 sheet metal screws in the entire antenna. The remaining fasteners are all hitch-pin clips. The result is a very durable six-element Yagi that you can disassemble with fair ease for transport. The antenna is shown in **Fig 21.105**. The complete construction details and more discussion of the antenna are included on the CD-ROM that accompanies this book.

THE BASIC ANTENNA DESIGN

The 6 element Yagi presented here is a derivative of the *optimized wide-band antenna* (OWA) designs developed for HF use by NW3Z and WA3FET. **Fig 21.106** shows the general outline. The reflector and first director largely set the impedance. The next 2 directors contribute to setting the operating bandwidth. The final director (Dir. 4) sets the gain. This account is over-simplified, since every element plays a role in every facet of

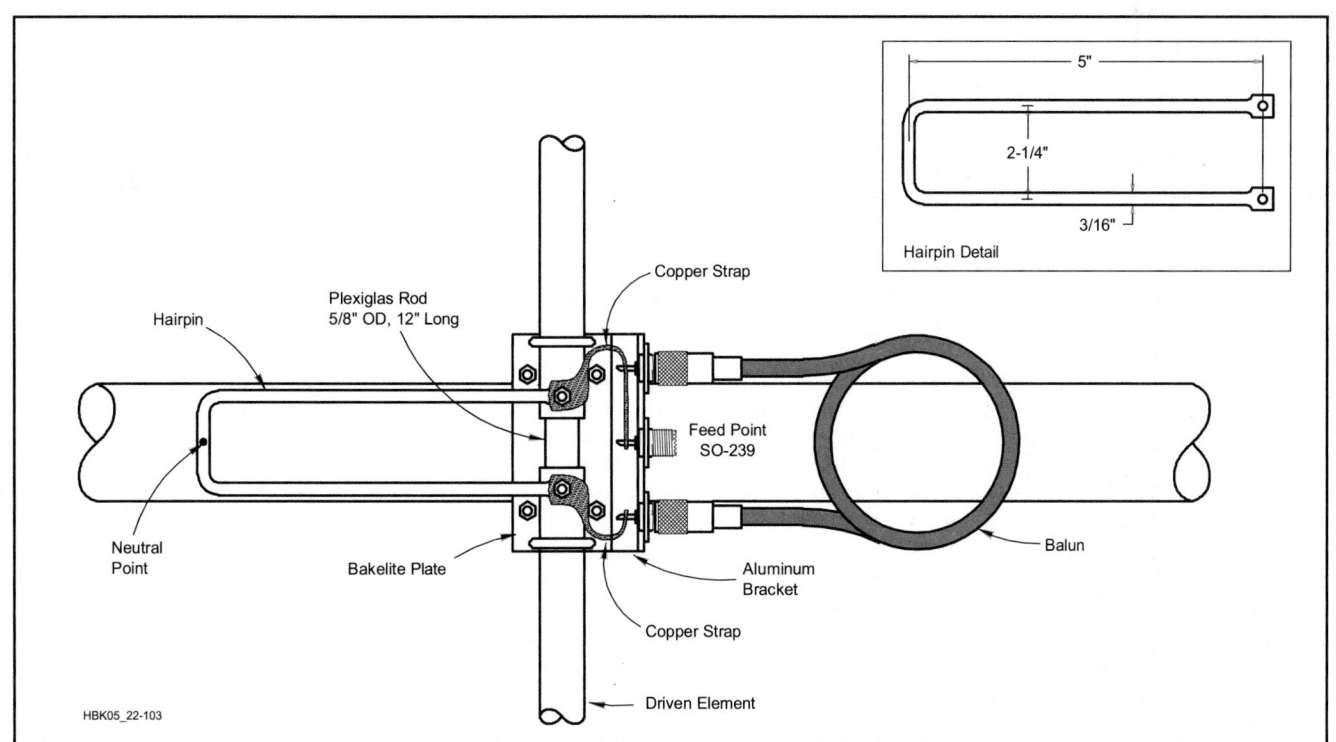

Fig 21.104 — Detailed drawing of the feed system used with the 50 MHz Yagi. Balun lengths: For cable with 0.80 velocity factor — 7 feet, 10⅜ inches. For cable with 0.66 velocity factor — 6 feet, 5¾ inches

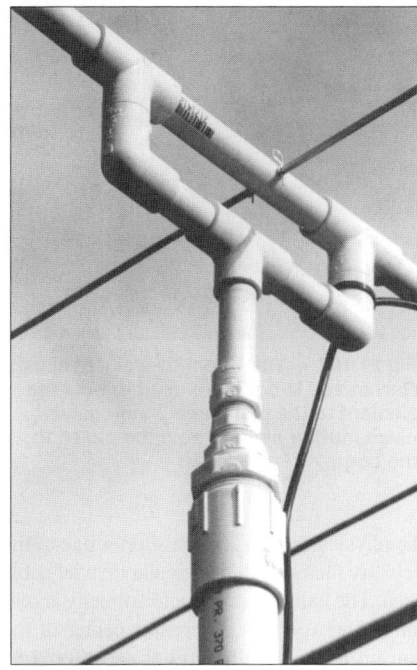

Fig 21.105 — The completed 2 meter Yagi is shown with the PVC boom and mast mount.

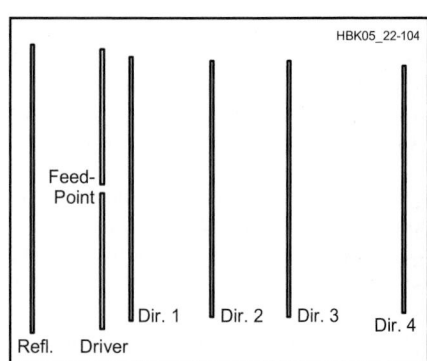

Fig 21.106 — The general outline of the 2 meter, 6 element OWA Yagi. Dimensions are given in Table 21.21.

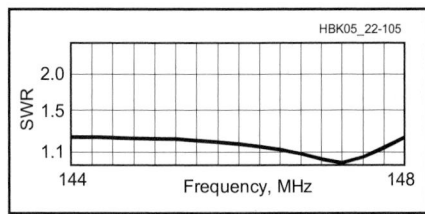

Fig 21.107 — SWR curve for the 2 meter, 6 element OWA Yagi as modeled using *NEC-4*.

Table 21.18
2 Meter OWA Yagi Dimensions

Element	Element Length (in)	Spacing from Reflector (in)	Element Diameter (in)
Version described here:			
Refl.	40.52	——	0.1875
Driver	39.70	10.13	0.5
Alt. Driver	*39.96*	*10.13*	*0.1875*
Dir. 1	37.36	14.32	0.1875
Dir. 2	36.32	25.93	0.1875
Dir. 3	36.32	37.28	0.1875
Dir. 4	34.96	54.22	0.1875
Version using ⅛-inch diameter elements throughout:			
Refl.	40.80	——	0.125
Driver	40.10	10.20	0.125
Dir. 1	37.63	14.27	0.125
Dir. 2	36.56	25.95	0.125
Dir. 3	36.56	37.39	0.125
Dir. 4	35.20	54.44	0.125

Yagi performance. However, the notes give some idea of which elements are most sensitive in adjusting the performance figures.

Designed using *NEC-4*, the antenna uses 6 elements on a 56 inch boom. **Table 21.18** gives the specific dimensions for the version described in these notes. The parasitic elements are all ³⁄₁₆ inch aluminum rods. For ease of construction, the driver is ½ inch aluminum tubing. Do not alter the element diameters without referring to a source, such as RSGB's *The VHF/UHF DX Book*, edited by Ian White, G3SEK, (Chapter 7), for information on how to recalculate element lengths.

The OWA design provides about 10.2 dBi of free-space gain with better than 20 dB

front-to-back (or front-to-rear) ratio across the entire 2 meter band. Azimuth (or E-plane) patterns show solid performance across the entire band. This applies not only to forward gain but rejection from the rear.

One significant feature of the OWA design is its direct 50 Ω feed point impedance that requires no matching network. Of course, a choke balun to suppress any currents on the feed line is desirable, and a simple ferrite bead balun (see the **Transmission Lines** and **Station Accessories** chapters) works well in this application. The SWR, shown in **Fig 21.107**, is very flat across the band and never reaches 1.3:1. The SWR and the pattern consistency together create a very useful utility

antenna for 2 meters, whether installed vertically or horizontally. The only remaining question is how to effectively build the beam in the average home shop.

The six-element OWA Yagi for 2 meters performs well. It serves as a good utility antenna with more gain and directivity than the usual three-element general-use Yagi. When vertically polarized, the added gain confirms the wisdom of using a longer boom and more elements. With a length under five feet, the antenna is still compact. The ability to disassemble the parts simplifies moving the antenna to various portable sites.

Project: Cheap Yagis by WA5VJB

If you're planning to build an EME array, don't use these antennas. But if you want to put together a VHF rover station with less than $500 in the antennas, read on as Kent Britain, WA5VJB, shows you how to put together a VHF/UHF Yagi with QRO performance at a QRP price. (This material is adapted from Kent's on-line paper "Controlled Impedance 'Cheap' Antennas" at **www.wa5vjb.com/references.html**.)

The simplified feed uses the structure of the antenna itself for impedance matching. So the design started with the feed and the elements were built around it. The antennas were designed with *YagiMax*, tweaked in *NEC*, and the driven elements experimentally determined on the antenna range.

Typically a high-gain antenna is designed in the computer, then you try to come up with a driven element matching arrangement for whatever feed point impedance the computer comes up with. In this design, compromises for the feed impedance, asymmetrical feed, simple measurements, wide bandwidth, the ability to grow with the same spacing, and trade-offs for a very clean pattern cost many dB of gain. But you can build these antennas for about $5!

Construction of the antennas is straightforward. The boom is ¾ inch square, or ½ inch by ¾ inch wood. To install an element, drill a hole through the boom and insert the element. A drop of cyanoacrylate "super glue," epoxy, or silicone adhesive is used to hold the elements in place. There is no boom-to-mast plate — drill holes in the boom and use a U-bolt to attach it to the mast!

The life of the antenna is determined by what you coat it with. The author had a 902 MHz version, varnished with polyurethane, in the air for two years with little deterioration.

The parasitic elements on prototypes have been made from silicon-bronze welding rod, aluminum rod, brass hobby tubing, and #10 or #12 AWG solid copper ground wire. So that you can solder to the driven element, use the welding rod, hobby tubing, or copper wire.

The driven element is folded at one end with its ends inserted through the boom.

Fig 21.108 shows the basic plan for the antenna and labels the dimensions that are given in the table for each band. All table dimensions are given in inches.

Fig 21.109 shows how the driven element is constructed for each antenna. Trim the free end of the driven element to tune it for minimum SWR at the desired frequency. **Fig 21.110** shows how to attach coaxial cable to the feed point. Sliding a quarter-wave

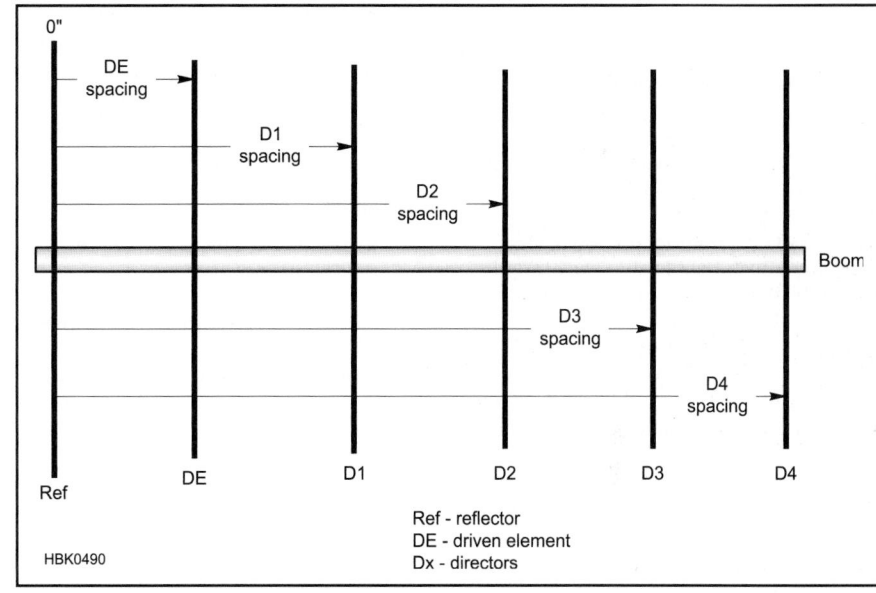

Fig 21.108 — Element spacing for the Cheap Yagis. Refer to Tables 21.19 to 21.26 for exact dimensions for the various bands.

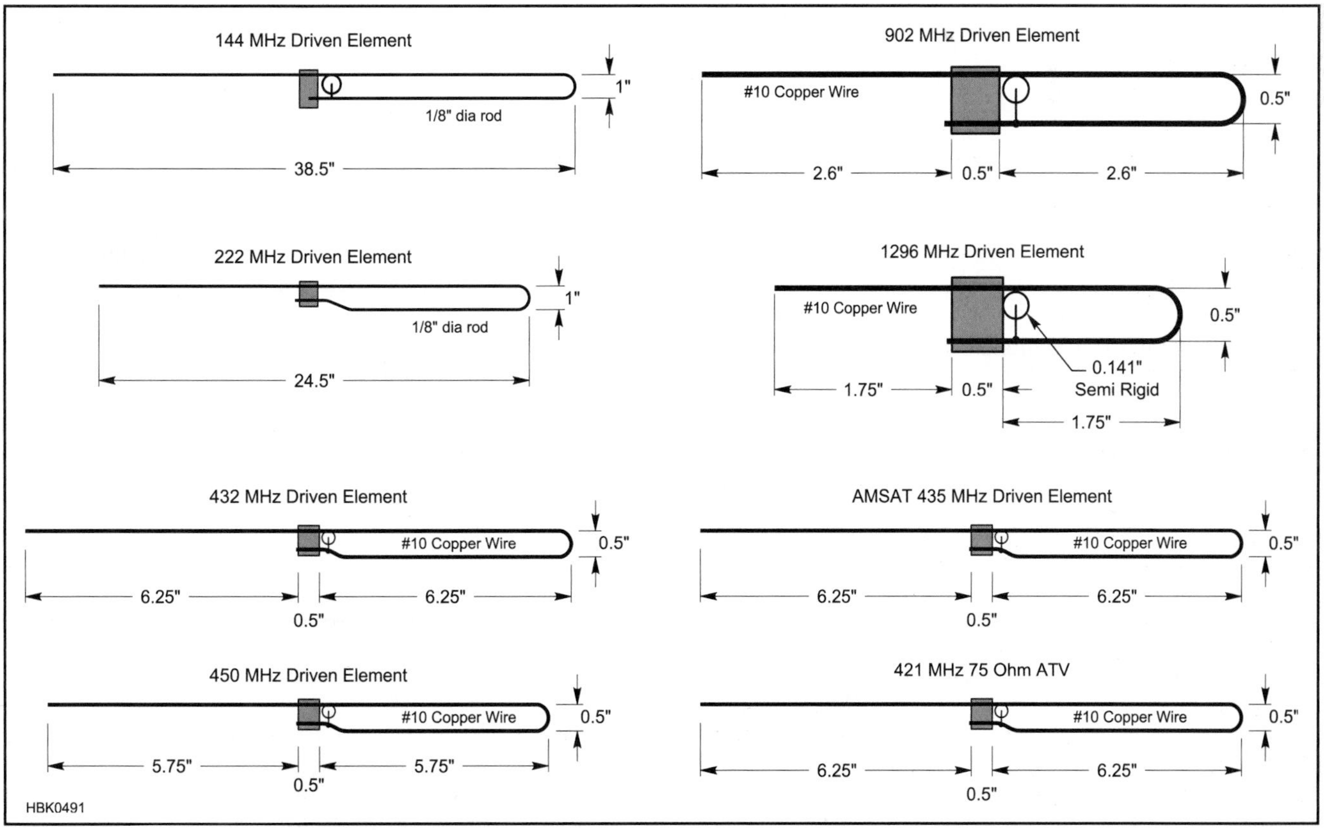

Fig 21.109 — Driven element dimensions for the Cheap Yagis. Attaching the coax shield to the center of the driven element is appropriate because that is the lowest impedance point of the element.

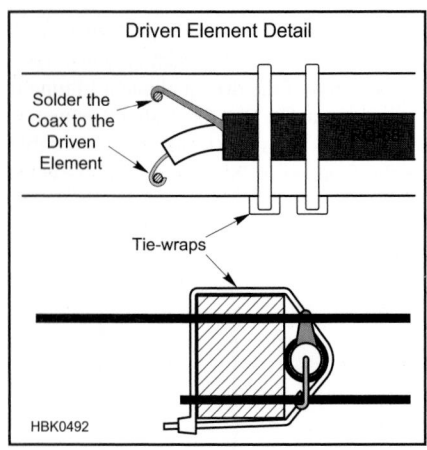

Fig 21.110 — Construction details and feed line attachment for the Cheap Yagi driven element.

sleeve along the coax had little effect, so there's not much RF on the outside of the coax. You may use a ferrite bead choke balun if you like, but these antennas are designed for minimum expense!

Finally a bit of history on the design of these antennas. In 1993 at the Oklahoma City Central States VHF Society Conference, Arnie, CO2KK spoke on the difficulties building VHF antennas in non-industrialized nations. Just run down to the store and pick up some Delrin insulators and 0.141 inch Teflon coax? Arnie's tales were the motivation to use advanced technology to come up with something simple.

144 MHz Yagi

While others have reported good luck with 16 element long-boom wood antennas, six elements was about the maximum for most rovers. The design is peaked at 144.2 MHz, but performance is still good at 146.5 MHz.

All parasitic elements are made from ³⁄₁₆ inch aluminum rod and the driven element is made from ⅛ inch rod. Lengths and spacings are given in **Table 21.19**.

222 MHz Yagi

This antenna is peaked at 222.1 MHz, but performance has barely changed at 223.5 MHz. You can drill the mounting holes to mount it with the elements horizontal or vertical. All parasitic elements are made from ³⁄₁₆ inch aluminum rod and the driven element is made from ⅛ inch rod. Lengths and spacings are given in **Table 21.20**.

432 MHz Yagi

At this band the antenna is getting very

practical and easy to build. All parasitic elements are made from ⅛ inch diameter rod and the driven element is made from #10 AWG solid copper wire. Lengths and spacings are given in **Table 21.21**.

435 MHz Yagi for AMSAT

KA9LNV provided help and motivation for these antennas. A high front-to-back ratio (F/B) was a major design consideration of all versions. The model predicts 30 dB F/B for the six-element and over 40 dB for the others. For gain, *NEC* predicts 11.2 dBi for the six-element, 12.6 dBi for the eight-element, and 13.5 dBi for the 10 element, and 13.8 dBi for the 11 element.

Using ¾ inch square wood for the boom

Table 21.19
WA5VJB 144 MHz Yagi Dimensions

		Ref	DE	D1	D2	D3	D4
3-element	Length	41.0	—	37.0			
	Spacing	0	8.5	20.0			
4-element	Length	41.0	—	37.5	33.0		
	Spacing	0	8.5	19.25	40.5		
6-element	Length	40.5	—	37.5	36.5	36.5	32.75
	Spacing	0	7.5	16.5	34.0	52.0	70.0

Dimensions in inches.

Table 21.20
WA5VJB 222 MHz Yagi Dimensions

		Ref	DE	D1	D2	D3	D4
3-element	Length	26.0	—	23.75			
	Spacing	0	5.5	13.5			
4-element	Length	26.25	—	24.1	22.0		
	Spacing	0	5.0	11.75	23.5		
6-element	Length	26.25	—	24.1	23.5	23.5	21.0
	Spacing	0	5.0	10.75	22.0	33.75	45.5

Dimensions in inches.

Table 21.21
WA5VJB 432 MHz Yagi Dimensions

		Ref	DE	D1	D2	D3	D4	D5	D6	D7	D8	D9
6-element	Length	13.5	—	12.5	12.0	12.0	11.0					
	Spacing	0	2.5	5.5	11.25	17.5	24.0					
8-element	Length	13.5	—	12.5	12.0	12.0	12.0	12.0	11.25			
	Spacing	0	2.5	5.5	11.25	17.5	24.0	30.75	38.0			
11-element	Length	13.5	—	12.5	12.0	12.0	12.0	12.0	12.0	11.75	11.75	11.0
	Spacing	0	2.5	5.5	11.25	17.5	24.0	30.75	38.0	45.5	53.0	59.5

Dimensions in inches.

Table 21.22
WA5VJB 435 MHz Yagi Dimensions

		Ref	DE	D1	D2	D3	D4	D5	D6	D7	D8	D9
6-element	Length	13.4	—	12.4	12.0	12.0	11.0					
8-element	Length	13.4	—	12.4	12.0	12.0	12.0	12.0	11.1			
10-element	Length	13.4	—	12.4	12.0	12.0	12.0	12.0	11.75	11.75	11.1	
11-element	Length	13.4	—	12.4	12.0	12.0	12.0	12.0	11.75	11.75	11.75	11.1
	Spacing	0	2.5	5.5	11.25	17.5	24.0	30.5	37.75	45.0	52.0	59.5

Dimensions in inches.

makes it easy to build two antennas on the same boom for cross-polarization. Offset the two antennas 6½ inches along the boom and feed them in-phase for circular polarization, or just use one for portable operations. All parasitic elements are made from ⅛ inch diameter rod and the driven element is made from #10 AWG solid copper wire. Lengths and spacings are given in **Table 21.22**. The same element spacing is used for all four versions of the antenna.

450 MHz Yagi

For FM, this six-element Yagi is a good, cheap antenna to get a newcomer into a repeater or make a simplex-FM QSO during a contest. RadioShack ⅛ inch diameter aluminum ground wire (catalog #15-035) was used in the prototype for all the elements except the driven element, which is made from #10 AWG solid copper wire. Other ⅛ inch diameter material could be used. Lengths and spacings are given in **Table 21.23**.

902 MHz Yagi

This was the first antenna the author built using the antenna to control the driven element impedance. The 2.5 foot length has proven very practical. All parasitic elements are made from ⅛ inch diameter rod and the driven element is made from #10 AWG solid copper wire. Lengths and spacings are given in **Table 21.24**.

1296 MHz Yagi

This antenna is the veteran of several "Grid-peditions" and has measured 13.5 dBi on the Central States VHF Society antenna range. Dimensions must be followed with great care. The driven element is small enough to allow 0.141 inch semi-rigid coax to be used. The prototype antennas use ⅛ inch silicon-bronze welding rod for the elements, but any ⅛ inch-diameter material can be used. The driven element is made from #10 AWG solid copper wire. Lengths and spacings are given in **Table 21.25**.

421.25 MHz 75 Ω Yagi for ATV

421 MHz vestigial sideband video is popular in North Texas for receiving the FM video input repeaters. These antennas are made for 421 MHz use and the driven element is designed for 75 Ω. RG-59 or an F adapter to RG-6 can be directly connected to a cable-TV converter or cable-ready TV on channel 57. All parasitic elements are made from ⅛ inch diameter rod and the driven element is made from #10 AWG solid copper wire. Lengths and spacings are given in **Table 21.26**. The same spacing is used for all versions.

Table 21.23
WA5VJB 450 MHz Yagi Dimensions

		Ref	DE	D1	D2	D3	D4
6-element	Length	13.0	—	12.1	11.75	11.75	10.75
	Spacing	0	2.5	5.5	11.0	18.0	28.5

Dimensions in inches.

Table 21.24
WA5VJB 902 MHz Yagi Dimensions

		Ref	DE	D1	D2	D3	D4	D5	D6	D7	D8
10-element	Length	6.2	—	5.6	5.5	5.5	5.4	5.3	5.2	5.1	5.1
	Spacing	0	2.4	3.9	5.8	9.0	12.4	17.4	22.4	27.6	33.0

Dimensions in inches.

Table 21.25
WA5VJB 1296 MHz Yagi Dimensions

		Ref	DE	D1	D2	D3	D4	D5	D6	D7	D8
10-element	Length	4.3	—	3.9	3.8	3.75	3.75	3.65	3.6	3.6	3.5
	Spacing	0	1.7	2.8	4.0	6.3	8.7	12.2	15.6	19.3	23.0

Dimensions in inches.

Table 21.26
WA5VJB 421.25 MHz 75-Ω Yagi Dimensions

		Ref	DE	D1	D2	D3	D4	D5	D6	D7	D8	D9
6-element	Length	14.0	—	12.5	12.25	12.25	11.0					
9-element	Length	14.0	—	12.5	12.25	12.25	12.0	12.0	11.25			
11-element	Length	14.0	—	12.5	12.25	12.25	12.0	12.0	12.0	11.75	11.75	11.5
	Spacing	0	3.0	6.5	12.25	17.75	24.5	30.5	36.0	43.0	50.25	57.25

Dimensions in inches.

21.12 Radio Direction Finding Antennas

Radio direction finding (RDF) is almost as old as radio communication. It gained prominence when the British Navy used it to track the movement of enemy ships in World War I. Since then, governments and the military have developed sophisticated and complex RDF systems. Fortunately, simple equipment, purchased or built at home, is quite effective in Amateur Radio RDF.

In European and Asian countries, direction-finding contests are foot races. The object is to be first to find four or five transmitters in a large wooded park. Young athletes have the best chance of capturing the prizes. This sport is known as *foxhunting* (after the British hill-and-dale horseback events) or *ARDF* (Amateur Radio direction finding). It is growing in popularity here in North America. Today, most competitive hunts worldwide are for 144 MHz FM signals, though other VHF bands are also used. Some international foxhunts include 3.5 MHz events.

In North America and England, most RDF contests involve mobiles — cars, trucks, and vans, even motorcycles. It may be possible to drive all the way to the transmitter, or there may be a short hike at the end, called a *sniff*. These competitions are also called foxhunting by some, while others use *bunny hunting, T-hunting* or the classic term *hidden transmitter hunting*.

Even without participating in RDF contests, you will find knowledge of the techniques useful. They simplify the search for a neighborhood source of power-line interference or TV cable leakage. RDF must be used to track down emergency radio beacons, which signal the location of pilots and boaters in distress. Amateur Radio enthusiasts skilled in transmitter hunting are in demand by agencies such as the Civil Air Patrol and the US Coast Guard Auxiliary for search and rescue support. RDF is an important part of the evidence-gathering process in interference cases.

The most basic RDF system consists of a directional antenna and a method of detecting and measuring the level of the radio signal, such as a receiver with signal strength indicator. RDF antennas range from a simple tuned loop of wire to an acre of antenna elements with an electronic beam-forming network. Other sophisticated techniques for RDF use the Doppler effect or measure the time of arrival difference of the signal at multiple antennas.

All of these methods have been used from 2 to 500 MHz and above. However, RDF practices vary greatly between the HF and VHF/UHF portions of the spectrum. For practical reasons, high gain beams, Dopplers and switched dual antennas find favor on VHF/UHF, while loops and phased arrays are the most popular choices on 6 meters and below. Signal propagation differences between HF and VHF also affect RDF practices. But many basic transmitter-hunting techniques, discussed later in this chapter, apply to all bands and all types of portable RDF equipment.

Several RDF projects may be found on the *Handbook CD* along with a thorough article on Direction-Finding Techniques and mobile RDF system installation, including some examples of mobile RDF antenna mounting.

21.12.1 RDF Antennas for HF Bands

Below 50 MHz, gain antennas such as Yagis and quads are of limited value for RDF. The typical tribander installation yields only a general direction of the incoming signal, due to ground effects and the antenna's broad forward lobe. Long monoband beams at greater heights work better, but still cannot achieve the bearing accuracy and repeatability of simpler antennas designed specifically for RDF.

RDF LOOPS

An effective directional HF antenna can be as uncomplicated as a small loop of wire or tubing, tuned to resonance with a capacitor. When immersed in an electromagnetic field, the loop acts much the same as the secondary winding of a transformer. The voltage at the output is proportional to the amount of flux passing through it and the number of turns. If the loop is oriented such that the greatest amount of area is presented to the magnetic field, the induced voltage will be the highest. If it is rotated so that little or no area is cut by the field lines, the voltage induced in the loop is zero and a null occurs.

To achieve this transformer effect, the loop must be small compared with the signal wavelength. In a single-turn loop, the conductor should be less than 0.08λ long. For example, a 28 MHz loop should be less than 34 inches in circumference, giving a diameter of approximately 10 inches. The loop may be smaller, but that will reduce its voltage output. Maximum output from a small loop antenna is in directions corresponding to the plane of the loop; these lobes are very broad. Sharp nulls, obtained at right angles to that plane, are more useful for RDF.

For a perfect bidirectional pattern, the loop must be balanced electrostatically with respect to ground. Otherwise, it will exhibit two modes of operation, the mode of a perfect loop and that of a non-directional vertical antenna of small dimensions. This dual-mode condition results in mild to severe inaccuracy, depending on the degree of imbalance, because the outputs of the two modes are not in phase.

The theoretical true loop pattern is illustrated in **Fig 21.111A**. When properly balanced, there are two nulls exactly 180° apart. When the unwanted antenna effect is appreciable and the loop is tuned to resonance, the loop may exhibit little directivity, as shown in Fig 21.111B. By detuning the loop to shift the phasing, you may obtain a useful pattern similar to Fig 21.111C. While not symmetrical, and not necessarily at right angles to the plane of the loop, this pattern does exhibit a pair of nulls.

By careful detuning and amplitude balancing, you can approach the unidirectional pattern of Fig 21.111D. Even though there

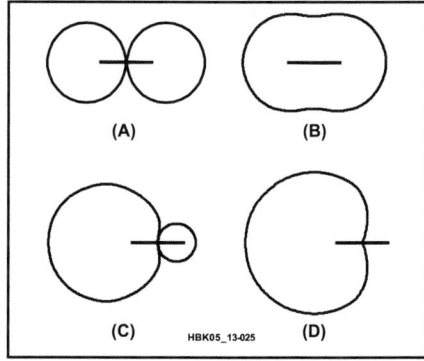

Fig 21.111 — Small loop field patterns with varying amounts of antenna effect — the undesired response of a loop acting merely as a mass of metal connected to the receiver antenna terminals. The horizontal lines show the plane of the loop turns.

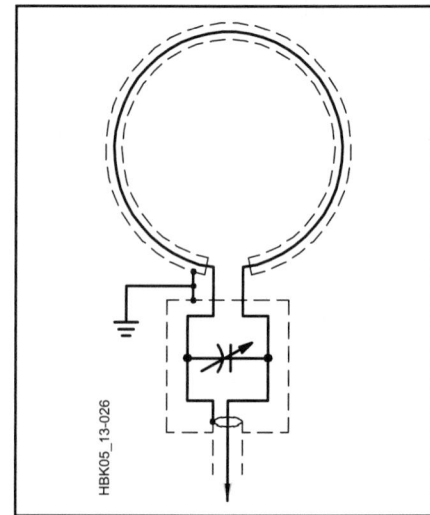

Fig 21.112 — Electrostatically-shielded loop for RDF. To prevent shielding of the loop from magnetic fields, leave the shield unconnected at one end.

may not be a complete null in the pattern, it resolves the 180° ambiguity of Fig 21.111A. Korean War era military loop antennas, sometimes available on today's surplus market, use this controlled antenna effect principle.

An easy way to achieve good electrostatic balance is to shield the loop, as shown in **Fig 21.112**. The shield, represented by the dashed lines in the drawing, eliminates the antenna effect. The response of a well-constructed shielded loop is quite close to the ideal pattern of Fig 21.111A.

For 160 through 30 meters, single-turn loops that are small enough for portability are usually unsatisfactory for RDF work. Multi-turn loops are generally used instead. They are easier to resonate with practical capacitor values and give higher output voltages. This type of loop may also be shielded. If the total conductor length remains below 0.08 λ, the directional pattern is that of Fig 21.111A.

FERRITE ROD ANTENNAS

Another way to get higher loop output is to increase the permeability of the medium in the vicinity of the loop. By winding a coil of wire around a form made of high-permeability material, such as ferrite rod, much greater flux is obtained in the coil without increasing the cross-sectional area.

Modern magnetic core materials make compact directional receiving antennas practical. Most portable AM broadcast receivers use this type of antenna, commonly called a *loopstick*. The loopstick is the most popular RDF antenna for portable/mobile work on 160 and 80 meters.

Like the shielded loop discussed earlier, the loopstick responds to the magnetic field of the incoming radio wave, and not to the electrical field. For a given size of loop, the output voltage increases with increasing flux density, which is obtained by choosing a ferrite core of high permeability and low loss at the frequency of interest. For increased output, the turns may be wound over two rods taped together. A practical loopstick antenna is described later in this chapter.

A loop on a ferrite core has maximum signal response in the plane of the turns, just as an air core loop. This means that maximum response of a loopstick is broadside to the axis of the rod, as shown in **Fig 21.113**. The loopstick may be shielded to eliminate the antenna effect; a U-shaped or C-shaped channel of aluminum or other form of "trough" is best. The shield must not be closed, and its length should equal or slightly exceed the length of the rod.

SENSE ANTENNAS

Because there are two nulls 180° apart in the directional pattern of a small loop or loopstick, there is ambiguity as to which null indicates the true direction of the target sta-

tion. For example, if the line of bearing runs east and west from your position, you have no way of knowing from this single bearing whether the transmitter is east of you or west of you.

If bearings can be taken from two or more positions at suitable direction and distance from the transmitter, the ambiguity can be resolved and distance can be estimated by triangulation, as discussed later in this chapter. However, it is almost always desirable to be able to resolve the ambiguity immediately by having a unidirectional antenna pattern available.

You can modify a loop or loopstick antenna pattern to have a single null by adding a second antenna element. This element is called a *sense antenna*, because it senses the phase of the signal wavefront for comparison with the phase of the loop output signal. The sense element must be omnidirectional, such as a short vertical. When signals from the loop and the sense antenna are combined with 90° phase shift between the two, a heart-shaped (cardioid) pattern results, as shown in **Fig 21.114A**.

Fig 21.114B shows a circuit for adding a

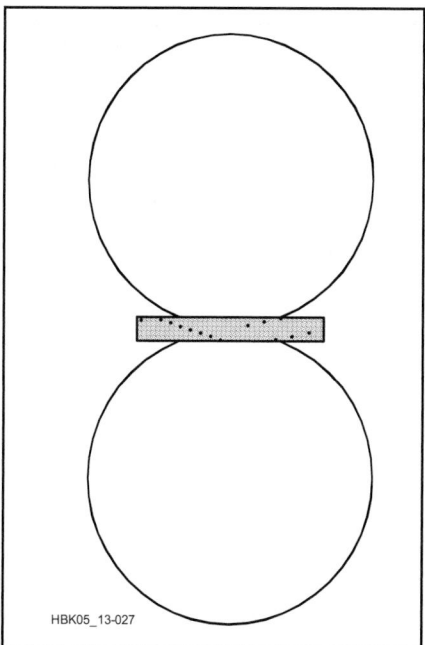

Fig 21.113 — Field pattern for a ferrite-rod antenna. The dark bar represents the rod on which the loop turns are wound.

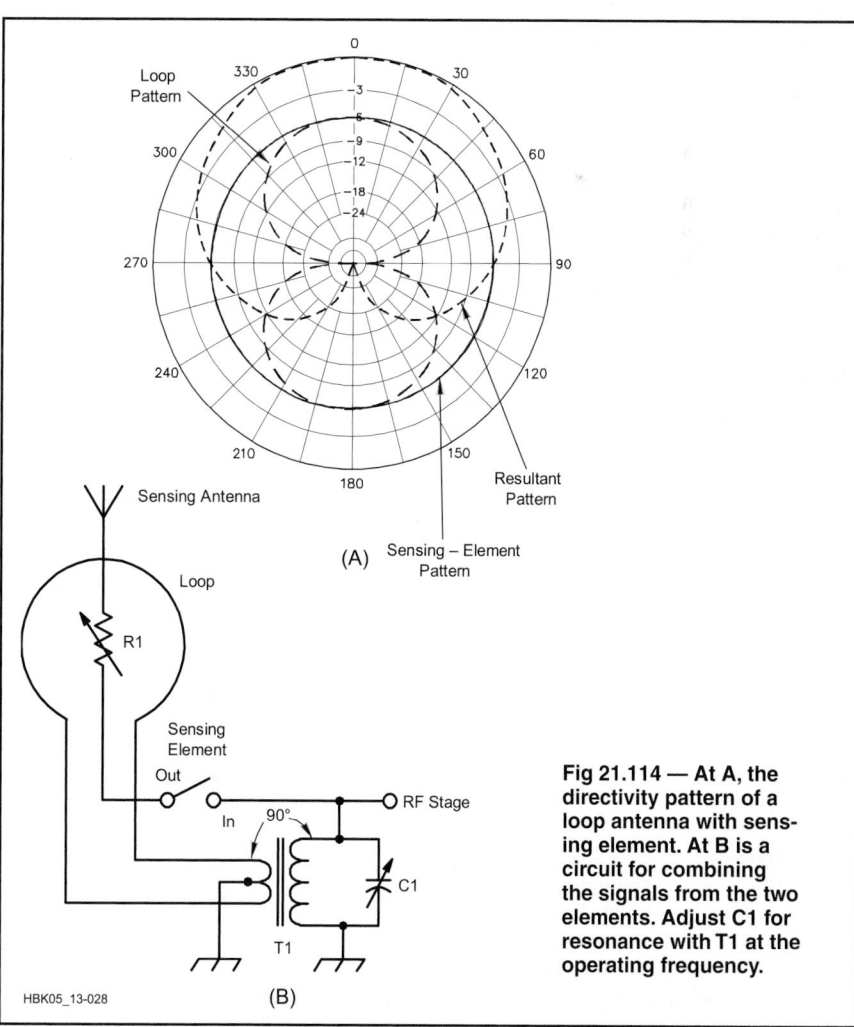

Fig 21.114 — At A, the directivity pattern of a loop antenna with sensing element. At B is a circuit for combining the signals from the two elements. Adjust C1 for resonance with T1 at the operating frequency.

sense antenna to a loop or loopstick. For the best null in the composite pattern, signals from the loop and sense antennas must be of equal amplitude. R1 adjusts the level of the signal from the sense antenna.

In a practical system, the cardioid pattern null is not as sharp as the bidirectional null of the loop alone. The usual procedure when transmitter hunting is to use the loop alone to obtain a precise line of bearing, then switch in the sense antenna and take another reading to resolve the ambiguity.

PHASED ARRAYS AND ADCOCK ANTENNAS

Two-element phased arrays are popular for amateur HF RDF base station installations. Many directional patterns are possible, depending on the spacing and phasing of the elements. A useful example is two ½ λ elements spaced ¼ λ apart and fed 90° out of phase. The resultant pattern is a cardioid, with a null off one end of the axis of the two antennas and a broad peak in the opposite direction. The directional frequency range of this antenna is limited to one band, because of the critical length of the phasing lines.

The best-known phased array for RDF is the Adcock, named after the man who invented it in 1919. It consists of two vertical elements fed 180° apart, mounted so the array may be rotated. Element spacing is not critical, and may be in the range from 0.1 to 0.75 λ. The two elements must be of identical lengths, but need not be self-resonant; shorter elements are commonly used. Because neither the element spacing nor length is critical in terms of wavelengths, an Adcock array may operate over more than one amateur band.

Fig 21.115 is a schematic of a typical Adcock configuration, called the H-Adcock because of its shape. Response to a vertically polarized wave is very similar to a conventional loop. The passing wave induces currents I1 and I2 into the vertical members. The output current in the transmission line is equal to their difference. Consequently, the directional pattern has two broad peaks and two sharp nulls, like the loop. The magnitude of the difference current is proportional to the spacing (d) and length (l) of the elements. You will get somewhat higher gain with larger dimensions. The Adcock of **Fig 21.116**, designed for 40 meters, has element lengths of 12 feet and spacing of 21 feet (approximately 0.15 λ).

Fig 21.117 shows the radiation pattern of the Adcock. The nulls are broadside to the axis of the array, becoming sharper with increased element spacing. When element spacing exceeds ¾ λ, however, the antenna begins to take on additional unwanted nulls off the ends of the array axis.

The Adcock is a vertically polarized an-

tenna. The vertical elements do not respond to horizontally polarized waves, and the currents induced in the horizontal members by a horizontally polarized wave (dotted arrows in Fig 21.115) tend to balance out regardless of the orientation of the antenna.

Since the Adcock uses a balanced feed system, a coupler is required to match the unbalanced input of the receiver. T1 is an air-wound coil with a two-turn link wrapped around the middle. The combination is reso-

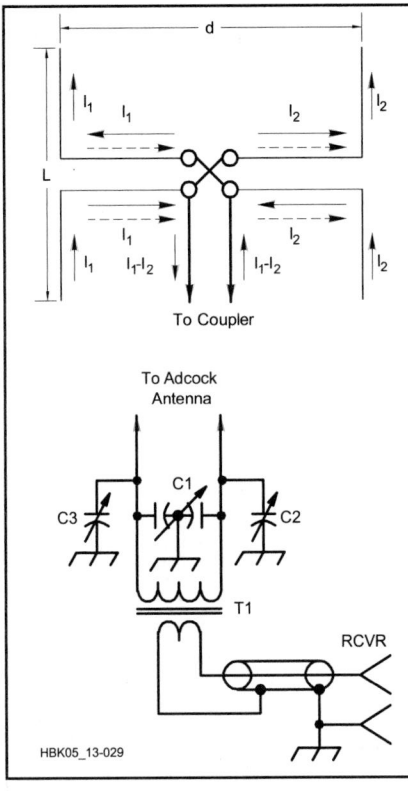

Fig 21.115 — A simple Adcock antenna and its coupler.

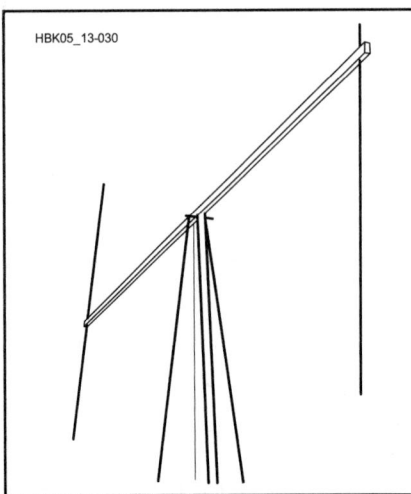

Fig 21.116 — An experimental Adcock antenna on a wooden frame.

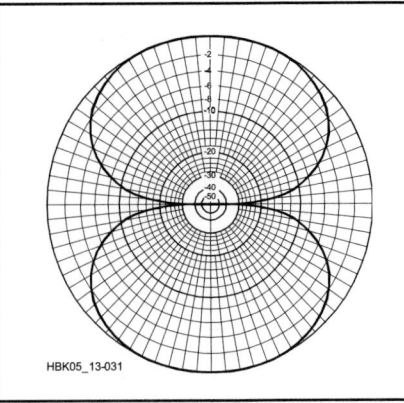

Fig 21.117 — The pattern of an Adcock array with element spacing of ½ wavelength. The elements are aligned with the vertical axis.

nated with C1 to the operating frequency. C2 and C3 are null-clearing capacitors. Adjust them by placing a low-power signal source some distance from the antenna and exactly broadside to it. Adjust C2 and C3 until the deepest null is obtained.

While you can use a metal support for the mast and boom, wood is preferable because of its non-conducting properties. Similarly, a mast of thick-wall PVC pipe gives less distortion of the antenna pattern than a metallic mast. Place the coupler on the ground below the wiring harness junction on the boom and connect it with a short length of 300 Ω twin-lead-feed line.

LOOPS VS PHASED ARRAYS

Loops are much smaller than phased arrays for the same frequency, and are thus the obvious choice for portable/mobile HF RDF. For base stations in a triangulation network, where the 180° ambiguity is not a problem, Adcocks are preferred. In general, they give sharper nulls than loops, but this is in part a function of the care used in constructing and feeding the individual antennas, as well as of the spacing of the elements. The primary construction considerations are the shielding and balancing of the feed line against unwanted signal pickup and the balancing of the antenna for a symmetrical pattern. Users report that Adcocks are somewhat less sensitive to proximity effects, probably because their larger aperture offers some space diversity.

Skywave Considerations

Until now we have considered the directional characteristics of the RDF loop only in the two-dimensional azimuthal plane. In three-dimensional space, the response of a vertically oriented small loop is doughnut-shaped. The bidirectional null (analogous to a line through the doughnut hole) is in the line of bearing in the azimuthal plane and toward the horizon in the vertical plane. Therefore, maximum null depth is achieved only on signals arriving at 0° elevation angle.

Skywave signals usually arrive at nonzero wave angles. As the elevation angle increases, the null in a vertically oriented loop pattern becomes shallower. It is possible to tilt the loop to seek the null in elevation as well as azimuth. Some amateur RDF enthusiasts report success at estimating distance to the target by measurement of the elevation angle with a tilted loop and computations based on estimated height of the propagating ionospheric layer. This method seldom provides high accuracy with simple loops, however.

Most users prefer Adcocks to loops for skywave work, because the Adcock null is present at all elevation angles. Note, however, that an Adcock has a null in all directions from signals arriving from overhead. Thus for very high angles, such as under-250-mile skip on 80 and 40 meters, neither loops nor Adcocks will perform well.

ELECTRONIC ANTENNA ROTATION

State-of-the-art fixed RDF stations for government and military work use antenna arrays of stationary elements, rather than mechanically rotatable arrays. The best-known type is the *Wullenweber antenna*. It has a large number of elements arranged in a circle, usually outside of a circular reflecting screen. Depending on the installation, the circle may be anywhere from a few hundred feet to more than a quarter of a mile in diameter. Although the Wullenweber is not practical for most amateurs, some of the techniques it uses may be applied to amateur RDF.

The device, which permits rotating the antenna beam without moving the elements, has the classic name *radio goniometer*, or simply *goniometer*. Early goniometers were RF transformers with fixed coils connected to the array elements and a moving pickup coil connected to the receiver input. Both amplitude and phase of the signal coupled into the pickup winding are altered with coil rotation in a way that corresponded to actually rotating the array itself. With sufficient elements and a goniometer, accurate RDF measurements can be taken in all compass directions.

Beam Forming Networks

By properly sampling and combining signals from individual elements in a large array, an antenna beam is electronically rotated or steered. With an appropriate number and arrangement of elements in the system, it is possible to form almost any desired antenna pattern by summing the sampled signals in appropriate amplitude and phase relationships. Delay networks and/or attenuation are added in line with selected elements before summation to create these relationships.

To understand electronic beam forming,

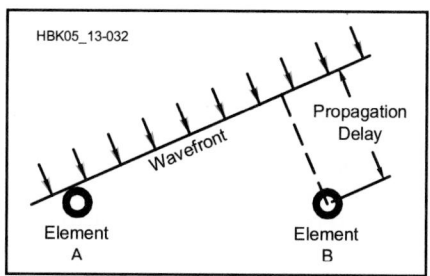

Fig 21.118 — One technique used in electronic beam forming. By delaying the signal from element A by an amount equal to the propagation delay, two signals are summed precisely in phase, even though the signal is not in the broadside direction.

first consider just two elements, shown as A and B in **Fig 21.118**. Also shown is the wavefront of a radio signal arriving from a distant transmitter. The wavefront strikes element A first, then travels somewhat farther before it strikes element B. Thus, there is an interval between the times that the wavefront reaches elements A and B.

We can measure the differences in arrival times by delaying the signal received at element A before summing it with that from element B. If two signals are combined directly, the amplitude of the sum will be maximum when the delay for element A exactly equals the propagation delay, giving an in-phase condition at the summation point. On the other hand, if one of the signals is inverted and the two are added, the signals will combine in a 180° out-of-phase relationship when the element A delay equals the propagation delay, creating a null. Either way, once the time delay is determined by the amount of delay required for a peak or null, we can convert it to distance. Then trigonometry calculations provide the direction from which the wave is arriving.

Altering the delay in small increments steers the peak (or null) of the antenna. The system is not frequency sensitive, other than the frequency range limitations of the array elements. Lumped-constant networks are suitable for delay elements if the system is used only for receiving. Delay lines at installations used for transmitting and receiving employ rolls of coaxial cable of various lengths, chosen for the time delay they provide at all frequencies, rather than as simple phasing lines designed for a single frequency.

Combining signals from additional elements narrows the broad beamwidth of the pattern from the two elements and suppress unwanted sidelobes. Electronically switching the delays and attenuations to the various elements causes the formed beam to rotate around the compass. The package of electronics that does this, including delay lines and electronically switched attenuators, is the beam-forming network.

21.12.2 Methods for VHF/UHF RDF

Three distinct methods of mobile RDF are commonly in use by amateurs on VHF/UHF bands: directional antennas, switched dual antennas and Dopplers. Each has advantages over the others in certain situations. Many RDF enthusiasts employ more than one method when transmitter hunting.

DIRECTIONAL ANTENNAS

Ordinary mobile transceivers and handhelds work well for foxhunting on the popular VHF bands. If you have a lightweight beam and your receiver has an easy-to-read S meter, you are nearly ready to start. All you need is an RF attenuator and some way to mount the setup in your vehicle.

Amateurs seldom use fractional wavelength loops for RDF above 60 MHz because they have bidirectional characteristics and low sensitivity, compared to other practical VHF antennas. Sense circuits for loops are difficult to implement at VHF, and signal reflections tend to fill in the nulls. Typically VHF loops are used only for close-in sniffing where their compactness and sharp nulls are assets, and low gain is of no consequence.

Phased Arrays

The small size and simplicity of two-element driven arrays make them a common choice of newcomers at VHF RDF. Antennas such as phased ground planes and ZL Specials have modest gain in one direction and a null in the opposite direction. The gain is helpful when the signal is weak, but the broad response peak makes it difficult to take a precise bearing.

As the signal gets stronger, it becomes possible to use the null for a sharper S meter indication. However, combinations of direct and reflected signals (called *multipath*) will distort the null or perhaps obscure it completely. For best results with this type of antenna, always find clear locations from which to take bearings.

Parasitic Arrays

Parasitic arrays are the most common RDF antennas used by transmitter hunters in high competition areas such as Southern California. Antennas with significant gain are a necessity due to the weak signals often encountered on weekend-long T-hunts, where the transmitter may be over 200 miles distant. Typical 144 MHz installations feature Yagis or quads of three to six elements, sometimes more. Quads are typically home-built, using data from *The ARRL Antenna Book* and *Transmitter Hunting* (see Bibliography).

Two types of mechanical construction are popular for mobile VHF quads. One model uses thin gauge wire (solid or stranded), suspended on wood dowel or fiberglass rod spreaders. It is lightweight and easy to turn rapidly by hand while the vehicle moves. Many hunters prefer to use larger gauge solid wire (such as #10 AWG) on a PVC plastic pipe frame. This quad is more rugged and has somewhat wider frequency range, at the expense of increased weight and wind resistance. It can get bent going under a branch, but it is easily reshaped and returned to service.

Yagis are a close second to quads in popularity. Commercial models work fine for VHF RDF, provided that the mast is attached at a good balance point. Lightweight and small-diameter elements are desirable for ease of turning at high speeds.

A well-designed mobile Yagi or quad installation includes a method of selecting wave polarization. Although vertical polarization is the norm for VHF-FM communications, horizontal polarization is allowed on many T-hunts. Results will be poor if a VHF RDF antenna is cross-polarized to the transmitting antenna, because multipath and scattered signals (which have indeterminate polarization) are enhanced, relative to the cross-polarized direct signal. The installation of **Fig 21.119** features a slip joint at the boom-to-mast junction, with an actuating cord to rotate the boom, changing the polarization. Mechanical stops limit the boom rotation to 90°.

Parasitic Array Performance for RDF

The directional gain of a mobile beam (typically 8 dB or more) makes it unexcelled for both weak signal competitive hunts and for locating interference such as TV cable leakage. With an appropriate receiver, you can get bearings on any signal mode, including FM, SSB, CW, TV, pulses and noise. Because

Fig 21.119 — The mobile RDF installation of WB6ADC features a thin wire quad that can be switched between vertical and horizontal polarization.

only the response peak is used, the null-fill problems and proximity effects of loops and phased arrays do not exist.

You can observe multiple directions of arrival while rotating the antenna, allowing you to make educated guesses as to which signal peaks are direct and which are from non-direct paths or scattering. Skilled operators can estimate distance to the transmitter from the rate of signal strength increase with distance traveled. The RDF beam is useful for transmitting, if necessary, but use care not to damage an attenuator in the coax line by transmitting through it.

The 3 dB beamwidth of typical mobile-mount VHF beams is on the order of 80°. This is a great improvement over 2 element driven arrays, but it is still not possible to get pinpoint bearing accuracy. You can achieve errors of less than 10° by carefully reading the S meter. In practice, this is not a major hindrance to successful mobile RDF. Mobile users are not as concerned with precise bearings as fixed station operators, because mobile readings are used primarily to give the general direction of travel to "home in" on the signal. Mobile bearings are continuously updated from new, closer locations.

Amplitude-based RDF may be very difficult when signal level varies rapidly. The transmitter hider may be changing power, or the target antenna may be moving or near a well-traveled road or airport. The resultant rapid S meter movement makes it hard to take accurate bearings with a quad. The process is slow because the antenna must be carefully rotated by hand to "eyeball average" the meter readings.

SWITCHED ANTENNA RDF UNITS

Three popular types of RDF systems are relatively insensitive to variations in signal level. Two of them use a pair of vertical dipole antennas, spaced ½ λ or less apart, and alternately switched at a rapid rate to the input of the receiver. In use, the indications of the two systems are similar, but the principles are different.

Switched Pattern Systems

The switched pattern RDF set (**Fig 21.120**) alternately creates two cardioid antenna patterns with lobes to the left and right. The patterns are generated in much the same way as in the phased arrays described above. PIN RF diodes select the alternating patterns. The combined antenna outputs go to a receiver with AM detection. Processing after the detector output determines the phase or amplitude difference between the patterns' responses to the signal.

Switched pattern RDF sets typically have a zero center meter as an indicator. The meter swings negative when the signal is coming from the user's left, and positive when the

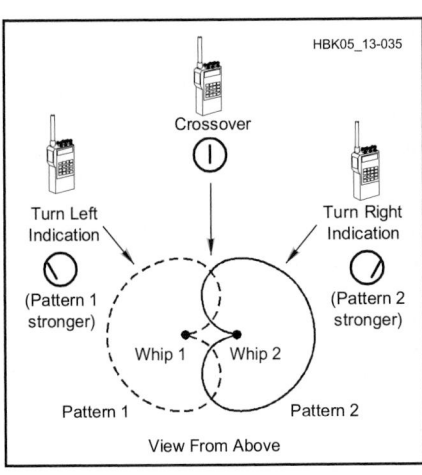

Fig 21.120 — In a switched pattern RDF set, the responses of two cardioid antenna patterns are summed to drive a zero center indicator.

signal source is on the right. When the plane of the antenna is exactly perpendicular to the direction of the signal source, the meter reads zero.

The sharpness of the zero crossing indication makes possible more precise bearings than those obtainable with a quad or Yagi. Under ideal conditions with a well-built unit, null direction accuracy is within 1°. Meter deflection tells the user which way to turn to zero the meter. For example, a negative (left) reading requires turning the antenna left. This solves the 180° ambiguity caused by the two zero crossings in each complete rotation of the antenna system.

Because it requires AM detection of the switched pattern signal, this RDF system finds its greatest use in the 120 MHz aircraft band, where AM is the standard mode. Commercial manufacturers make portable RDF sets with switched pattern antennas and built-in receivers for field portable use. These sets can usually be adapted to the amateur 144 MHz band. Other designs are adaptable to any VHF receiver that covers the frequency of interest and has an AM detector built in or added.

Switched pattern units work well for RDF from small aircraft, for which the two vertical antennas are mounted in fixed positions on the outside of the fuselage or simply taped inside the windshield. The left-right indication tells the pilot which way to turn the aircraft to home in. Since street vehicles generally travel only on roads, fixed mounting of the antennas on them is undesirable. Mounting vehicular switched-pattern arrays on a rotatable mast is best.

Time-of-Arrival Systems

Another kind of switched antenna RDF set uses the difference in arrival times of the signal wavefront at the two antennas. This narrow-aperture Time-Difference-of-Arrival (TDOA) technology is used for many sophisticated military RDF systems. The rudimentary TDOA implementation of **Fig 21.121** is quite effective for amateur use. The signal from transmitter 1 reaches antenna A before antenna B. Conversely, the signal from transmitter 3 reaches antenna B before antenna A. When the plane of the antenna is perpendicular to the signal source (as transmitter 2 is in the figure), the signal arrives at both antennas simultaneously.

If the outputs of the antennas are alternately switched at an audio rate to the receiver input, the differences in the arrival times of a continuous signal produce phase changes that are detected by an FM discriminator. The resulting short pulses sound like a tone in the receiver output. The tone disappears when the antennas are equidistant from the signal source, giving an audible null.

The polarity of the pulses at the discriminator output is a function of which antenna is closer to the source. Therefore, the pulses can be processed and used to drive a left-right zero-center meter in a manner similar to the switched pattern units described above. Left-right LED indicators may replace the meter for economy and visibility at night.

RDF operations with a TDOA dual antenna RDF are done in the same manner as with a switched antenna RDF set. The main difference is the requirement for an FM receiver in the TDOA system and an AM receiver in the switched pattern case. No RF attenuator is needed for close-in work in the TDOA case.

Popular designs for practical do-it-yourself TDOA RDF sets include the Simple Seeker (described elsewhere in this chapter) and the W9DUU design (see article by Bohrer in the Bibliography). Articles with plans for the Handy Tracker, a simple TDOA set with a delay line to resolve the dual-null ambiguity instead of LEDs or a meter, are listed in the Bibliography.

Performance Comparison

Both types of dual antenna RDFs make good on-foot "sniffing" devices and are excellent performers when there are rapid amplitude variations in the incoming signal. They are the units of choice for airborne work. Compared to Yagis and quads, they give good directional performance over a much wider frequency range. Their indications are more precise than those of beams with broad forward lobes.

Dual-antenna RDF sets frequently give inaccurate bearings in multipath situations, because they cannot resolve signals of nearly equal levels from more than one direction. Because multipath signals are a combined pattern of peaks and nulls, they appear to change in amplitude and bearing as you move the RDF antenna along the bearing path or

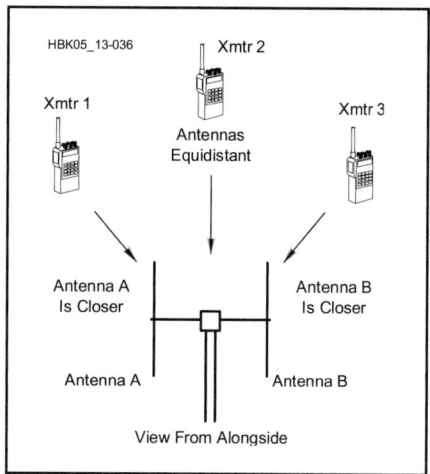

Fig 21.121 — A dual-antenna TDOA RDF system has a similar indicator to a switched pattern unit, but it obtains bearings by determining which of its antennas is closer to the transmitter.

perpendicular to it, whereas a non-multipath signal will have constant strength and bearing.

The best way to overcome this problem is to take large numbers of bearings while moving toward the transmitter. Taking bearings while in motion averages out the effects of multipath, making the direct signal more readily discernible. Some TDOA RDF sets have a slow-response mode that aids the averaging process.

Switched antenna systems generally do not perform well when the incoming signal is horizontally polarized. In such cases, the bearings may be inaccurate or unreadable. TDOA units require a carrier type signal such as FM or CW; they usually cannot yield bearings on noise or pulse signals.

Unless an additional method is employed to measure signal strength, it is easy to "overshoot" the hidden transmitter location with a TDOA set. It is not uncommon to see a TDOA foxhunter walk over the top of a concealed transmitter and walk away, following the opposite 180° null, because there is no display of signal amplitude.

DOPPLER RDF SETS

RDF sets using the Doppler principle are popular in many areas because of their ease of use. They have an indicator that instantaneously displays direction of the signal source relative to the vehicle heading, either on a circular ring of LEDs or a digital readout in degrees. A ring of four, eight or more antennas picks up the signal. Quarter-wavelength monopoles on a ground plane are popular for vehicle use, but half-wavelength vertical dipoles, where practical, perform better.

Radio signals received on a rapidly moving antenna experience a frequency shift due

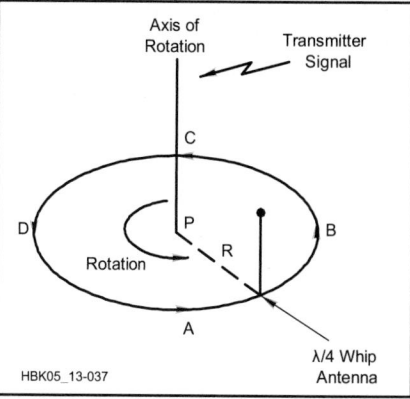

Fig 21.122 — A theoretical Doppler antenna circles around point P, continuously moving toward and away from the source at an audio rate.

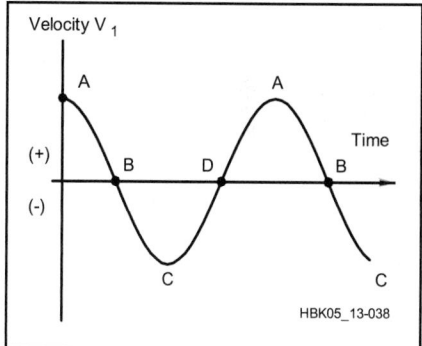

Fig 21.123 — Frequency shift versus time produced by the rotating antenna movement toward and away from the signal source.

to the Doppler effect, a phenomenon well-known to anyone who has observed a moving car with its horn sounding. The horn's pitch appears higher than normal as the car approaches, and lower as the car recedes. Similarly, the received radio frequency increases as the antenna moves toward the transmitter and vice versa. An FM receiver will detect this frequency change.

Fig 21.122 shows a ¼ λ vertical antenna being moved on a circular track around point P, with constant angular velocity. As the antenna approaches the transmitter on its track, the received frequency is shifted higher. The highest instantaneous frequency occurs when the antenna is at point A, because tangential velocity toward the transmitter is maximum at that point. Conversely, the lowest frequency occurs when the antenna reaches point C, where velocity is maximum away from the transmitter.

Fig 21.123 shows a plot of the component of the tangential velocity that is in the direction of the transmitter as the antenna moves around the circle. Comparing Figs 21.122

and 21.123, notice that at B in Fig 21.123, the tangential velocity is crossing zero from the positive to the negative and the antenna is closest to the transmitter. The Doppler shift and resulting audio output from the receiver discriminator follow the same plot, so that a negative-slope zero-crossing detector, synchronized with the antenna rotation, senses the incoming direction of the signal.

The amount of frequency shift due to the Doppler effect is proportional to the RF frequency and the tangential antenna velocity. The velocity is a function of the radius of rotation and the angular velocity (rotation rate). The radius of rotation must be less than ¼ λ to avoid errors. To get a usable amount of FM deviation (comparable to typical voice modulation) with this radius, the antenna must rotate at approximately 30,000 RPM (500 Hz). This puts the Doppler tone in the audio range for easy processing.

Mechanically rotating a whip antenna at this rate is impractical, but a ring of whips, switched to the receiver in succession with RF PIN diodes, can simulate a rapidly rotating antenna. Doppler RDF sets must be used with receivers having FM detectors. The Dopple ScAnt and Roanoke Doppler (see Bibliography) are mobile Doppler RDF sets designed for inexpensive home construction.

Doppler Advantages and Disadvantages

Ring-antenna Doppler sets are the ultimate in simplicity of operation for mobile RDF. There are no moving parts and no manual antenna pointing. Rapid direction indications are displayed on very short signal bursts.

Many units lock in the displayed direction after the signal leaves the air. Power variations in the source signal cause no difficulties, as long as the signal remains above the RDF detection threshold. A Doppler antenna goes on top of any car quickly, with no holes to drill. Many Local Interference Committee members choose Dopplers for tracking malicious interference, because they are inconspicuous (compared to beams) and effective at tracking the strong vertically polarized signals that repeater jammers usually emit.

A Doppler does not provide superior performance in all VHF RDF situations. If the signal is too weak for detection by the Doppler unit, the hunt advantage goes to teams with beams. Doppler installations are not suitable for on-foot sniffing. The limitations of other switched antenna RDFs also apply: (1) poor results with horizontally polarized signals, (2) no indication of distance, (3) carrier type signals only and (4) inadvisability of transmitting through the antenna.

Readout to the nearest degree is provided on some commercial Doppler units. This does not guarantee that level of accuracy, however. A well-designed four-monopole set is typically capable of ±5° accuracy on 2 meters, if the target signal is vertically polarized and there are no multipath effects.

The rapid antenna switching can introduce cross modulation products when the user is near strong off-channel RF sources. This self-generated interference can temporarily render the system unusable. While not a common problem with mobile Dopplers, it makes the Doppler a poor choice for use in remote RDF installations at fixed sites with high power VHF transmitters nearby.

21.13 Glossary

Antenna — An electrical conductor or array of conductors that radiates signal energy (transmitting) or collects signal energy (receiving).

Antenna tuner — A device containing variable reactances (and perhaps a balun) used to convert an antenna or feed line impedance to 50 Ω. (Also called Transmatch, impedance-matching unit).

Apex angle — The included angle between the legs of an inverted-V antenna.

Azimuth (azimuthal) pattern — A radiation pattern in a plane oriented parallel to the Earth's surface or at a specified angle to the Earth's surface.

Balanced line — A symmetrical two-conductor feed line (also called open-wire line, ladder line, window line, twin-lead).

Balun — A device that transfers energy between a balanced and unbalanced system. A balun may or may not change the impedance ratio between the systems.

Base loading — Adding a coil to the base of a ground-plane antenna to increase its electrical length.

Beamwidth — The width in degrees of the major lobe of a directive antenna between the two angles at which the relative radiated power is equal to one-half its value (–3 dB) at the peak of the lobe.

Capacitance hat — A conducting structure with a large surface area that is added to an antenna to add capacitive reactance at that point on the antenna.

Center loading — Adding a coil near the center of a ground-plane antenna to increase its electrical length.

Coaxial cable (coax) — A coaxial transmission line with a center conductor surrounded by a layer of insulation and then a tubular shield conductor and covered by an insulating jacket. (see also — unbalanced line)

Delta loop — A full-wavelength loop shaped like a triangle or delta.

Delta match — Center-feed technique used with antenna elements that are not split at the center in which the transmission is spread apart and connected to the element symmetrically, forming a triangle or delta.

Dipole — An antenna, usually one-half wavelength long, divided into two parts at a feed point.

Directivity — The property of an antenna that concentrates the radiated energy to form one or more major lobes.

Director — An antenna element in a parasitic array that causes radiated energy from the driven element to be focused along the line from the driven element to the director.

Driven array — An array of antenna elements which are all driven or excited by means of a transmission line.

Driven element — An antenna element excited by means of a transmission line.

E-plane — The plane in which the electric field of an electromagnetic wave is maximum.

Efficiency — The ratio of useful output power to input power.

Elements — The conductive parts of an antenna system that determine the antenna's characteristics.

Elevation pattern — A radiation pattern in a plane perpendicular to the Earth's surface.

End effect — The effect of capacitance at the end of an antenna element that acts to electrically lengthen the element.

Feed line — see transmission line.

Front-to-back ratio — The ratio in dB of the radiation from an antenna in a favored direction to that in the opposite direction.

Front-to-rear ratio — The ratio in dB of the radiation from an antenna is a favored direction to an average of the radiation in the opposite direction across some specified angle.

Front-to-side ratio — The ratio in dB of the radiation from an antenna in a favored direction to that at right angles to the favored direction.

Gain — The increase in radiated power in the desired direction of the major lobe.

Gamma match — A matching system used with driven antenna elements in which a conductor is placed near the element and connected to the feed line with an adjustable capacitor at the end closest to the center and connected to the element at the other.

Ground plane — A system of conductors configured to act as a reflecting surface to an antenna element and connected to one side of the transmission line.

H-plane — The plane in which the magnetic field of an electromagnetic wave is maximum.

Hairpin match — A U-shaped conductor that is connected to the two inner ends of a split antenna element for the purpose of creating a match to a feed line.

Impedance — The ratio of voltage to current in a feed line or along an antenna.

Inverted-V — A dipole antenna supported at its mid-point with halves angled down toward the ground.

Isotropic — An imaginary antenna that radiates and receives equally well in all directions.

Ladder line — See balanced line.

Line loss — The power lost in a transmission line, specified in dB per unit of length.

Load — The electrical system or component to which power is delivered.

Lobe — A region in an antenna's radiation pattern between two nulls.

Matching — The process by which power at one impedance is transferred to a system at a different impedance.

Monopole — An antenna with a single-element that functions in concert with a ground-plane.

Null — A point of minimum radiation in an antenna's radiation pattern.

Open-wire line — See balanced line.

Parasitic array — A set of elements that form a radiation pattern through coupling and re-radiation of energy from one or more driven elements.

Polarization — The orientation of the electromagnetic field, usually referring to the orientation of the E field.

Q section — A quarter-wavelength section of transmission line used for impedance-matching purposes.

Quad — A directive antenna based on the Yagi with elements that consist of one-wavelength loops.

Radiation pattern — The characteristics of an antenna's distribution of energy in a single plane. (See also elevation pattern and azimuth pattern.)

Radiation resistance — A resistance that represents the work done by the current in an antenna to radiate power.

Reflector — An antenna element in a parasitic array that causes radiated energy from the driven element to be focused along the line from the driven element away from the reflector.

Sense Antenna — An antenna added to a bidirectional array or loop that samples the incoming signal's phase for comparison to that of the main receiving antenna.

Stacking — Arranging two or more directive antennas such that their radiation pattern characteristics reinforce each other.

SWR — Standing-wave ratio. A measure of the match between a transmission line and a load such as an antenna.

T-match — A symmetrical version of the gamma match for a balanced antenna system.

Top loading — Addition of a reactance, usually capacitive, at the top of a ground-plane antenna so as to increase its electrical length.

Transmatch — See antenna tuner.

Trap — A parallel LC-circuit used as an electrical switch to isolate sections of an antenna.

Twin-lead — See balanced line.

Unipole — See monopole.

Yagi — A parasitic array consisting of a driven element and one or more director and reflectors.

Zepp — A half-wavelength antenna fed at one end by means of open-wire feed line.

21.14 References and Bibliography

J. S. Belrose, "Short Antennas for Mobile Operation," *QST*, Sep 1953, pp 30-35.

G. H. Brown, "The Phase and Magnitude of Earth Currents Near Radio Transmitting Antennas," *Proc IRE*, Feb 1935.

G. H. Brown, R. F. Lewis and J. Epstein, "Ground Systems as a Factor in Antenna Efficiency," *Proc IRE*, Jun 1937, pp 753-787.

G. H. Brown and O. M. Woodward, Jr, "Experimentally Determined Impedance Characteristics of Cylindrical Antennas," *Proc IRE*, April 1945.

C. L. Buchanen, W3DZZ, "The Multimatch Antenna System," *QST*, Mar 1955, p 22.

A. Christman, "Elevated Vertical Antenna Systems," *QST*, Aug 1988, pp 35-42.

J. Clement, VE6AB, "Gain Twist 75 Meter Mobile Monobander," *QST*, Jul 2011, pp 39-42.

J. Devoldere, ON4UN's *Low-Band DXing*, 5th ed (Newington: ARRL, 2011).

R. B. Dome, "Increased Radiating Efficiency for Short Antennas," *QST*, Sep 1934, pp 9-12.

A. C. Doty, Jr, J. A. Frey and H. J. Mills, "Characteristics of the Counterpoise and Elevated Ground Screen," Professional Program, Session 9, Southcon '83 (IEEE), Atlanta, GA, Jan 1983.

A. C. Doty, Jr, J. A. Frey and H. J. Mills, "Efficient Ground Systems for Vertical Antennas," *QST*, Feb 1983, pp 20-25.

A. C. Doty, Jr, technical paper presentation, "Capacitive Bottom Loading and Other Aspects of Vertical Antennas," Technical Symposium, Radio Club of America, New York City, Nov 20, 1987.

A. C. Doty, Jr, J. A. Frey and H. J. Mills, "Vertical Antennas: New Design and Construction Data," *The ARRL Antenna Compendium, Volume 2* (Newington: ARRL, 1989), pp 2-9.

R. Fosberg, "Some Notes on Ground Systems for 160 Meters," *QST*, Apr 1965, pp 65-67.

G. Grammer, "More on the Directivity of Horizontal Antennas; Harmonic Operation — Effects of Tilting," *QST*, Mar 1937, pp 38-40, 92, 94, 98.

H. E. Green, "Design Data for Short and Medium Length Yagi-Uda Arrays," *Trans IE Australia*, Vol EE-2, No. 1, Mar 1966.

A. Griffith, W4ULD, "Capacitance Hats for HF Mobile Antennas," *QEX*, Jul/Aug 1996, p 16.

H. J. Mills, technical paper presentation, "Impedance Transformation Provided by Folded Monopole Antennas," Technical Symposium, Radio Club of America, New York City, Nov 20, 1987.

B. Myers, "The W2PV Four-Element Yagi," *QST*, Oct 1986, pp 15-19.

L. Richard, "Parallel Dipoles of 300-Ohm Ribbon," *QST*, Mar 1957.

J. H. Richmond, "Monopole Antenna on Circular Disc," *IEEE Trans on Antennas and Propagation*, Vol. AP-32, No. 12, Dec 1984.

W. Schulz, "Designing a Vertical Antenna," *QST*, Sep 1978, pp 19- 21.

J. Sevick, "The Ground-Image Vertical Antenna," *QST*, Jul 1971, pp 16-17, 22.

J. Sevick, "The W2FMI 20-Meter Vertical Beam," *QST*, Jun 1972, pp 14-18.

J. Sevick, "The W2FMI Ground-Mounted Short Vertical," *QST*, Mar 1973, pp 13-18, 41.

J. Sevick, "A High Performance 20-, 40- and 80-Meter Vertical System," *QST*, Dec 1973.

J. Sevick, "Short Ground-Radial Systems for Short Verticals," *QST*, Apr 1978, pp 30-33.

C. E. Smith and E. M. Johnson, "Performance of Short Antennas," *Proc IRE*, Oct 1947.

J. Stanley, "Optimum Ground Systems for Vertical Antennas," *QST*, Dec 1976, pp 13-15.

R. E. Stephens, "Admittance Matching the Ground-Plane Antenna to Coaxial Transmission Line," Technical Correspondence, *QST*, Apr 1973, pp 55-57.

D. Sumner, "Cushcraft 32-19 'Boomer' and 324-QK Stacking Kit," Product Review, *QST*, Nov 1980, pp 48-49.

B Sykes, "Skeleton Slot Aerials," *RSGB* Bulletin, Jan 1953.

W. van B. Roberts, "Input Impedance of a Folded Dipole," *RCA Review*, Jun 1947.

E. M. Williams, "Radiating Characteristics of Short-Wave Loop Aerials," *Proc IRE*, Oct 1940.

TEXTBOOKS ON ANTENNAS

C. A. Balanis, *Antenna Theory, Analysis and Design* (New York: Harper & Row, 1982).

D. S. Bond, *Radio Direction Finders*, 1st ed. (New York: McGraw-Hill Book Co).

W. N. Caron, *Antenna Impedance Matching* (Newington: ARRL, 1989).

L. B. Cebik, *ARRL Antenna Modeling Course* (Newington: ARRL, 2002).

K. Davies, *Ionospheric Radio Propagation* — National Bureau of Standards Monograph 80 2(Washington, DC: U.S. Government Printing Office, Apr 1, 1965).

R. S. Elliott, *Antenna Theory and Design* (Englewood Cliffs, NJ: Prentice Hall, 1981).

A. E. Harper, *Rhombic Antenna Design* (New York: D. Van Nostrand Co, Inc, 1941).

K. Henney, *Principles of Radio* (New York: John Wiley and Sons, 1938), p 462.

C. Hutchinson and R. D. Straw, *Simple and Fun Antennas for Hams* (Newington: ARRL, 2002).

H. Jasik, *Antenna Engineering Handbook*, 1st ed. (New York: McGraw-Hill, 1961).

W. C. Johnson, *Transmission Lines and Networks*, 1st ed. (New York: McGraw-Hill Book Co, 1950).

Johnson and Jasik, *Antenna Engineering Handbook*, 2nd ed. (New York: McGraw-Hill).

R. C. Johnson, *Antenna Engineering Handbook*, 3rd ed. (New York: McGraw-Hill, 1993).

E. C. Jordan and K. G. Balmain, *Electromagnetic Waves and Radiating Systems*, 2nd ed. (Englewood Cliffs, NJ: Prentice-Hall, Inc, 1968).

R. Keen, *Wireless Direction Finding*, 3rd ed. (London: Wireless World).

R. W. P. King, *Theory of Linear Antennas* (Cambridge, MA: Harvard Univ. Press, 1956).

R. W. P. King, H. R. Mimno and A. H. Wing, *Transmission Lines, Antennas and Waveguides* (New York: Dover Publications, Inc, 1965).

King, Mack and Sandler, *Arrays of Cylindrical Dipoles* (London: Cambridge Univ Press, 1968).

M. G. Knitter, ed., *Loop Antennas — Design and Theory* (Cambridge, WI: National Radio Club, 1983).

M. G. Knitter, ed., *Beverage and Long Wire Antennas* — Design and Theory (Cambridge, WI: National Radio Club, 1983).

J. D. Kraus, *Electromagnetics* (New York: McGraw-Hill Book Co).

J. D. Kraus, *Antennas*, 2nd ed. (New York: McGraw-Hill Book Co, 1988).

E. A. Laport, *Radio Antenna Engineering* (New York: McGraw-Hill Book Co, 1952).

J. L. Lawson, *Yagi-Antenna Design*, 1st ed. (Newington: ARRL, 1986).

P. H. Lee, *The Amateur Radio Vertical Antenna Handbook*, 2nd ed. (Port Washington, NY: Cowen Publishing Co., 1984).

D. B. Leeson, *Physical Design of Yagi Antennas* (Newington: ARRL, 1992).

A. W. Lowe, *Reflector Antennas* (New York: IEEE Press, 1978).

M. W. Maxwell, *Reflections — Transmission Lines and Antennas* (Newington: ARRL, 1990).

M. W. Maxwell, *Reflections II — Transmission Lines and Antennas* (Sacramento: Worldradio Books, 2001).

G. M. Miller, *Modern Electronic Communication* (Englewood Cliffs, NJ: Prentice Hall, 1983).

V. A. Misek, *The Beverage Antenna Handbook* (Hudson, NH: V. A. Misek, 1977).

T. Moreno, *Microwave Transmission Design Data* (New York: McGraw-Hill, 1948).

L. A. Moxon, *HF Antennas for All Locations* (Potters Bar, Herts: Radio Society of Great Britain, 1982).

Ramo and Whinnery, *Fields and Waves in Modern Radio* (New York: John Wiley & Sons).

V. H. Rumsey, *Frequency Independent Antennas* (New York: Academic Press, 1966).

P. N. Saveskie, *Radio Propagation Handbook* (Blue Ridge Summit, PA: Tab Books, Inc, 1980).

S. A. Schelkunoff, *Advanced Antenna Theory* (New York: John Wiley & Sons, Inc, 1952).

S. A. Schelkunoff and H. T. Friis, *Antennas Theory and Practice* (New York: John Wiley & Sons, Inc, 1952).

J. Sevick, *Transmission Line Transformers* (Atlanta: Noble Publishing, 1996).

H. H. Skilling, *Electric Transmission Lines* (New York: McGraw-Hill Book Co, Inc, 1951).

M. Slurzburg and W. Osterheld, *Electrical Essentials of Radio* (New York: McGraw-Hill Book Co, Inc, 1944).

G. Southworth, *Principles and Applications of Waveguide Transmission* (New York: D. Van Nostrand Co, 1950).

R. D. Straw, Ed., *The ARRL Antenna Book*, 21st ed. (Newington: ARRL, 2007).

F. E. Terman, *Radio Engineers' Handbook*, 1st ed. (New York, London: McGraw-Hill Book Co, 1943).

F. E. Terman, *Radio Engineering*, 3rd ed. (New York: McGraw-Hill, 1947).

S. Uda and Y. Mushiake, *Yagi-Uda Antenna* (Sendai, Japan: Sasaki Publishing Co, 1954). [Published in English — Ed.]

P. P. Viezbicke, "*Yagi Antenna Design*," NBS Technical Note 688 (US Dept of Commerce/National Bureau of Standards, Boulder, CO), Dec 1976.

G. B. Welch, *Wave Propagation and Antennas* (New York: D. Van Nostrand Co, 1958).

The GIANT Book of Amateur Radio Antennas (Blue Ridge Summit, PA: Tab Books, 1979).

IEEE Standard Dictionary of Electrical and Electronics Terms, 3rd ed. (New York: IEEE, 1984).

Radio Broadcast Ground Systems, available from Smith Electronics, Inc, 8200 Snowville Rd, Cleveland, OH 44141.

Radio Communication Handbook, 5th ed. (London: RSGB, 1976).

RDF BIBLIOGRAPHY

Bohrer, "Foxhunt Radio Direction Finder," *73 Amateur Radio*, Jul 1990, p 9.

Bonaguide, "HF DF — A Technique for Volunteer Monitoring," *QST*, Mar 1984, p 34.

DeMaw, "Maverick Trackdown," *QST*, Jul 1980, p 22.

Dorbuck, "Radio Direction Finding Techniques," *QST*, Aug 1975, p 30.

Eenhoorn, "An Active Attenuator for Transmitter Hunting," *QST*, Nov 1992, p 28.

Flanagan and Calabrese, "An Automated Mobile Radio Direction Finding System," *QST*, Dec 1993, p 51.

Geiser, "A Simple Seeker Direction Finder," *ARRL Antenna Compendium, Volume 3*, p 126.

Gilette, "A Fox-Hunting DF Twin'Tenna," *QST*, Oct 1998, pp 41-44.

Johnson and Jasik, *Antenna Engineering Handbook*, Second Edition, New York: McGraw-Hill.

Kossor, "A Doppler Radio-Direction Finder," *QST*, Part 1: May 1999, pp 35-40; Part 2: June 1999, pp 37-40.

McCoy, "A Linear Field-Strength Meter," *QST*, Jan 1973, p 18.

Moell and Curlee, *Transmitter Hunting: Radio Direction Finding Simplified*, Blue Ridge Summit, PA: TAB/McGraw-Hill. (This book, available from ARRL, includes plans for the Roanoke Doppler RDF unit and in-line air attenuator, plus VHF quads and other RDF antennas.)

Moell, "Transmitter Hunting — Tracking Down the Fun," *QST*, Apr 1993, p 48 and May 1993, p 58.

Moell, "Build the Handy Tracker," *73 Magazine*, Sep 1989, p 58 and Nov 1989, p 52.

O'Dell, "Simple Antenna and S-Meter Modification for 2-Meter FM Direction Finding," *QST*, Mar 1981, p 43.

O'Dell, "Knock-It-Down and Lock-It-Out Boxes for DF," *QST*, Apr 1981, p 41.

Ostapchuk, "Fox Hunting is Practical and Fun!" *QST*, Oct 1998, pp 68-69.

Rickerd, "A Cheap Way to Hunt Transmitters," *QST*, Jan 1994, p 65.

The "Searcher" (SDF-1) Direction Finder, Rainbow Kits.

RDF RESOURCES

Homing In
www.homingin.com — website by KØOV on direction finding techniques and activities

Amateur Radio Direction Finding (IARU Region II)
www.ardf-r2.org/en — ARDF activities and organizations in IARU Region II

Radio Direction Finding
http://en.wikipedia.org/wiki/Direction_finding — a general site on RDF with links to related subjects

DX Zone RDF Links
www.dxzone.com/catalog/Operating_Modes/Radio_Direction_Finding — a page of links to RDF articles and websites

Contents

Component Data and References

Radio amateurs are known for electronic experimentation and homebrew building. Using the wide variety of components available, they design and build impressive radio equipment. With the industry growth of components for wireless communications and surface mount technology (SMT), the choices available seem endless and selecting the proper component can seem a daunting task.

Fortunately, most amateurs tend to use a limited number of component types that have "passed the test of time," making component selection in many cases easy and safe. Others are learning to design and build using the vast array of SMT parts.

Paul Harden, NA5N, updated this chapter for the 2010 edition and Dick Frey, K4XU, updated the power MOSFET tables for this edition.

**Chapter 22 —
CD-ROM Content**

Supplemental Files
- BNC Crimp Installation Instructions
- N Crimp Installation Instructions
- Miniature Lamp Guide
- Thermoplastics Properties
- TV Deflection Tube Guide
- Obsolete RF Power Semiconductor Tables

22.1 Component Data

This section provides reference information on the old and new components most often used by the Amateur Radio experimenter and homebrewer, and information for those wishing to learn more about component performance and selection.

22.1.1 EIA and Industry Standards

The American National Standards Institute (ANSI), the Electronic Industries Alliance (EIA), and the Electronic Components Association (ECA) establish the US standards for most electronic components, connectors, wire and cables. These standards establish component sizes, wattages, "standard values," tolerances and other performance characteristics. A branch of the EIA sets the standards for Mil-spec (standard military specification) and special electronic components used by defense and government agencies. The Joint Electron Devices Engineering Council (JEDEC), another branch of the EIA, develops the standards for the semiconductor industry. The EIA cooperates with other standards agencies such as the International Electrotechnical Commission (IEC), a worldwide standards agency. You can often find published EIA standards in the engineering library of a college or university.

And finally, the International Organization of Standardization (ISO), headquartered in Geneva, Switzerland, sets the global standards for nearly everything from paper sizes to photographic film speeds. ANSI is the US representative to the ISO.

These organizations, or their acronyms, are familiar to most of us. They are much more than a label on a component. EIA and other industry standards are what mark components for identification, establishes the "preferred standard values" and ensures their reliable performance from one unit to the next, regardless of their source. Standards require that a 1.2 kΩ 5% resistor from Ohmite Corp. has the same performance as a 1.2 kΩ 5% resistor from Vishay-Dale, or a 2N3904 to have the same performance characteristics and physical packaging whether from ON Semi or Gold Star.

Much of the component data in this chapter is devoted to presenting these component standards, physical dimensions and the various methods of component identification and marking. By selecting components manufactured under these industry standards, building a project from the *Handbook* or other source will ensure nearly identical performance to the original design.

22.1.2 Other Sources of Component Data

There are many sources you can consult for detailed component data but the best source of component information and data sheets is the Internet. Most manufacturers maintain extensive Web sites with information and data on their products. Often, the quickest route to detailed product information is to enter "data sheet" and the part number into an Internet search engine. Distributors such as Digi-Key and Mouser include links to useful information in their online catalogs as well. Some manufacturers still publish data books for the components they make, and parts catalogs themselves are often good sources of component data and application notes and bulletins.

Some of the tables printed in previous editions of this book have been moved to the accompanying CD-ROM to make room for new material. If a table or figure you need is missing, check the CD-ROM!

22.1.3 The ARRL Technical Information Service (TIS)

The ARRL Technical Information Service on the ARRL Web site (**www.arrl.org/technical-information-service**) provides technical assistance to members and nonmembers, including information about components and useful references. The TIS includes links to detailed, commonly needed information in many technical areas. Questions may also be submitted via email (**tis@arrl.org**); fax (860-594-0259); or mail (TIS, ARRL, 225 Main St, Newington, CT 06111).

22.1.4 Definitions

Electronic components such as resistors, capacitors, and inductors are manufactured with a *nominal* value — the value with which they are labeled. The component's *actual* value is what is measured with a suitable measuring instrument. If the nominal value is given as text characters, an "R" in the value (for example "4R7") stands for *radix* and is read as a decimal point, thus "4.7".

Tolerance refers to a range of acceptable values above and below the nominal compo-

nent value. For example, a 4700-Ω resistor rated for ±20% tolerance can have an actual value anywhere between 3760 Ω and 5640 Ω. You may always substitute a closer-tolerance device for one with a wider tolerance. For most Amateur Radio projects, assume a 10% tolerance if none is specified.

The *temperature coefficient* or *tempco* of a component describes its change in value with temperature. Tempco may be expressed as a change in unit value per degree (ohms per degree Celsius) or as a relative change per degree (parts per million per degree). Except for temperature sensing components that may use Fahrenheit or Kelvin, Celsius is almost always used for the temperature scale. Temperature coefficients may not be linear, such as those for capacitors, thermistors, or quartz crystals. In such cases, tempco is specified by an identifier such as Z5U or C0G and an equation or graph of the change with temperature provided by the manufacturer.

22.1.5 Surface-Mount Technology (SMT)

"SMT" is used throughout this book to refer to components, printed-circuit boards or assembly techniques that involve surface-mount technology. SMT components are often referred to by the abbreviations "SMD" and "SMC," but all three abbreviations are considered to be effectively equivalent. *Through-hole* or *leaded* components are those with wire leads intended to be inserted into holes in printed-circuit boards or used in point-to-point wiring.

Many different types of electronic components, both active and passive, are now available in surface-mount packages. Each package is identified by a code, such as 1802 or SOT. Resistors in SMT packages are referred to by package code and not by power dissipation, as through-hole resistors are. The very small size of these components leaves little space for marking with conventional codes, so brief alphanumeric codes are used to convey the most information in the smallest possible space. You will need a magnifying glass to read the markings on the bodies of SMT components.

In many cases, vendors will deliver SMT components packaged in tape from master reels and the components will not be marked. This is often the case with SMT resistors and small capacitors. However, the tape will be marked or the components are delivered in a plastic bag with a label. Take care to keep the components separated and labeled or you'll have to measure their values one by one!

22.2 Resistors

Most resistors are manufactured using EIA standards to establish common ratings for wattage, resistor values and tolerance regardless of the manufacturer. EIA marking methods for resistors utilize either an alphanumeric scheme or a color code to denote the value and tolerance.

In the earlier days of electronics, 10% and 20% tolerance resistors were the common and inexpensive varieties used by most amateurs. 1% tolerance resistors were considered the "precision resistors" and seldom used by the amateur due to their significantly higher cost.

Today, with improved manufacturing techniques, both 5% and 1% tolerance resistors are commonly available *and* inexpensive,

with precision resistors to 0.1% not uncommon.

22.2.1 Resistor Types

The major resistor types are carbon composition, carbon film, metalized film and wire-wound, as described below. (For additional discussion of the characteristics of the different types of resistors, see the **Electrical Fundamentals** chapter.)

Carbon composition resistors are made from a slurry of carbon and binder material formulated to achieve the desired resistance when compressed into a cylinder and encapsulated. This yields a resistor with tolerances in the 5% to 20% range. "Carbon comp"

resistors have a tendency to absorb moisture over time and to change value, but can withstand temporary "pulse" overloads that would damage or destroy a film-type resistor.

Carbon film resistors are made from a layer of carbon deposited on a dielectric film or substrate. The thickness of the carbon film is controlled to form the desired resistance with greater accuracy than for carbon composition. They are low cost alternatives to carbon composition resistors and are available with 1% to 5% tolerances.

Metalized film resistors replace carbon films with metal films deposited onto the dielectric using sputtering techniques to achieve very accurate resistances to 0.1% tolerances. Metal film resistors also generate less thermal noise than carbon resistors.

All three of these resistor types are normally available with power ratings from $\frac{1}{10}$ W to 2 W. **Fig 22.1** and **Tables 22.1** and **22.2** provide the body sizes and lead or pad spacing for through-hole and SMT resistors.

For new designs, carbon film and metalized film resistors should be used for their improved characteristics and lower cost compared to the older carbon composition resistors. Metalized films have lower residual inductance and often preferred at VHF. Most surface mount resistors (shown in **Fig 22.2**) are metalized films.

Wire-wound resistors, as the name implies, are made from lengths of wire wound around an insulating form to achieve the desired resistance for power ratings above 2 W. Wirewound resistors have high parasitic inductance, caused by the wire wrapped around a form similar to a coil, and thus should not be used at RF frequencies. **Fig 22.3** (A, B and D) show three types of wire-wound resistors with wattage ranges in **Table 22.3**.

An alternative to wire-wound resistors is the new generation of resistors known as *thick-film power resistors*. They are rated up to 100 W and packaged in a TO-220 or similar case which makes it easy to mount them on heat sinks and printed-circuit boards. Most varieties are non-inductive and suitable for RF use. Metal-oxide ("cement") resistors are also available in packages similar to that of Fig 22.3B. Similar to carbon composition resistors, metal-oxide resistors are non-inductive and useful at RF.

22.2.2 Resistor Identification

Resistors are identified by the EIA numerical or color code standard as shown in **Fig 22.4**. The EIA numerical code for resistor identification is widely used in industry. The nominal resistance, expressed in ohms, is identified by three digits for 2% (and greater) tolerance devices. The first two digits represent the significant figures; the last digit specifies the multiplier as the exponent of 10. (The multiplier is simply the number of zeros following the significant numerals.) For values less than 100 Ω, the letter R is substituted for one of the significant digits and represents a decimal point. An alphabetic character indicates the tolerance as shown in Table 22.2

For example, a resistor marked with "122J" would be a 1200 Ω, or a 1.2 kΩ 5% resistor. A resistor containing four digits, such as "1211," would be a 1210 Ω, or a 1.21 kΩ 1% precision resistor.

If the tolerance of the unit is narrower than ±2%, the code used is a four-digit code where the first three digits are the significant figures and the last is the multiplier. The letter R is used in the same way to represent

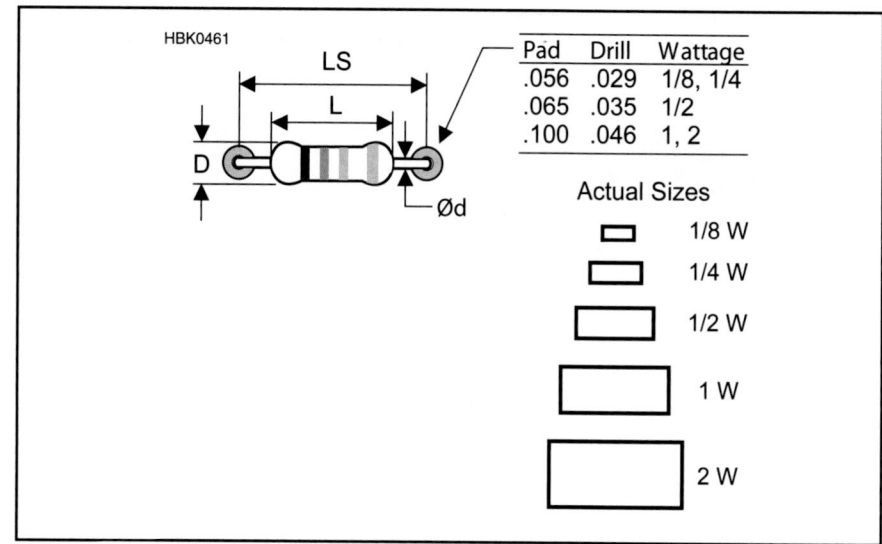

Fig 22.1 — Resistor wattages and sizes.

Table 22.1
Resistor Wattages and Sizes

Size	L	D	LS*	Ød	PCB Pad Size and Drill
⅛ W	0.165	0.079	0.25	0.020	0.056 round, 0.029 hole
¼ W	0.268	0.098	0.35	0.024	0.056 round, 0.029 hole
½ W	0.394	0.138	0.60	0.029	0.065 round, 0.035 hole
1 W	0.472	0.197	0.70	0.032	0.100 round, 0.046 hole
2 W	0.687	0.300	0.90	0.032	0.100 round, 0.046 hole

Dimensions in inches.
*LS = Recommended PCB lead bend

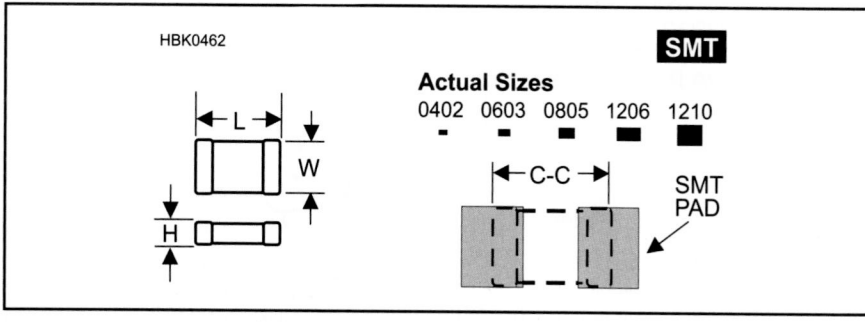

Fig 22.2 — Surface mount resistors.

Table 22.2
SMT Resistor Wattages and Sizes

Body Size	L	W	H	SMT Pad	C-C*
0402	0.039	0.020	0.014	0.025 × 0.035	0.050
0603	0.063	0.031	0.018	0.030 × 0.030	0.055
0805	0.079	0.049	0.020	0.040 × 0.050	0.075
1206	0.126	0.063	0.024	0.064 × 0.064	0.125
1210	0.126	0.102	0.024	0.070 × 0.100	0.150

Dimensions in inches.
*C-C is SMT pad center-to-center spacing

SMT Resistor Tolerance Codes

Letter	Tolerance
D	±0.5%
F	±1.0%
G	±2.0%
J	±5.0%

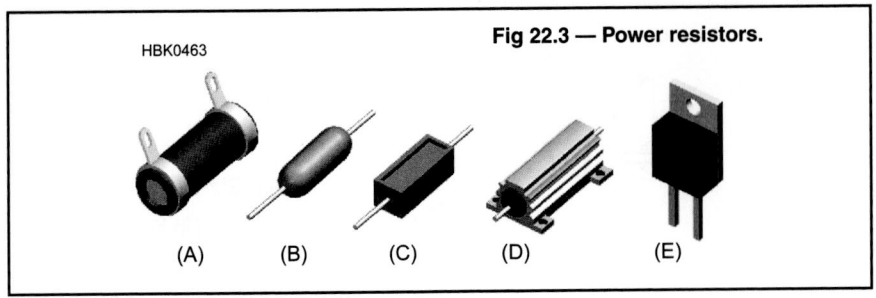

Fig 22.3 — Power resistors.

HBK0463

(A) (B) (C) (D) (E)

Table 22.3
Power Resistors

Fig 22.3	Power Resistor Type	Wattage Range
A	Wire-wound, ceramic core	10-300 W
B	Wire-wound, axial	3-10 W
C	Metal-oxide	5-25 W
D	Wire-wound, aluminum housing	3-50 W
E	Thick-film resistors*	15-100 W

*Wire-wound resistors are inductive, though seldom noted as such on the data sheets, and are not recommended for RF. Thick-film and metal-oxide power resistors are low inductance or noninductive.

Standard EIA Identification and Marking

5–20% Resistors

```
1 3 2 J
```

1st digit
2nd digit
multiplier
tolerance
J ± 5%
K ±10%
G ± 20%

Examples:
"132J"=1300=1.3KΩ 5%
"510K"=51 =51Ω 10%
"2R2G"=2.2Ω 20%

"Precision" Resistors

```
2 4 9 1 B
```

1st digit
2nd digit
3rd digit
multiplier
tolerance
B ±0.1%
D ±0.5%
F ± 1%

Examples:
"2491B"=2490=2.49K 0.1%
"5110D"=511 =511Ω 0.5%
"51R1F" =51.1=51.1Ω 1%

EIA Resistor Color Codes

Color Codes

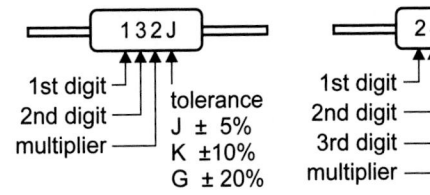

Digit	Color	
0	black	(blk)
1	brown	(brn)
2	red	(red)
3	orange	(org)
4	yellow	(ylw)
5	green	(grn)
6	blue	(blu)
7	violet	(vio)
8	gray	(gry)
9	white	(wht)

5–20% Resistors
(4-band color code)

1st digit
2nd digit
multiplier
tolerance
gold ± 5%
silver ±10%
none ±20%

"Precision" Resistors
(5-band color code)

1st digit
2nd digit
3rd digit
multiplier
tolerance
vio ±0.1%
grn ±0.5%
brn ± 1%

Examples:
brn-org-org-gold = 13KΩ 5%
grn-brn-blk-silver = 51Ω 10%
brn-org-org-brn-brn = 1.33KΩ

HBK0464

---

a decimal point. For example, 1001 indicates a 1000-Ω unit, and 22R0 indicates a 22-Ω unit.

Here are some additional examples of resistor value markings:

Code	Value
101	10 and 1 zero = 100 Ω
224	22 and 4 zeros = 220,000 Ω
1R0	1.0 and no zeros = 1 Ω
22R	22.0 and no zeros = 22 Ω
R10	0.1 and no zeros = 0.1 Ω

The resistor color code, used only with through-hole components, assigns colors to the numerals one through nine and zero, as shown in **Table 22.4**, to represent the significant numerals, the multiplier and the tolerance. The color code is often memorized with a mnemonic such as "Big boys race our young girls, but Violet generally wins" to represent the colors black (0), brown (1), red (2), orange (3), yellow (4), green (5), blue (6), violet (7), gray (8) and white (9). You will no doubt discover other versions of this memory aid made popular over the years.

For example, a resistor with color bands black (1), red (2), red (2) and gold would be a 1200 Ω, or 1.2 kΩ 5% resistor, with the gold band signifying 5% tolerance.

The resistor color code should be memorized as it is also used for identifying capacitors, and inductors. It is also handy to use when connecting multi-conductor or ribbon cables.

Resistors are also identified by an "E" series classification, such as E12 or E48. The number following the letter E signifies the number of logarithmic steps per decade. The more steps per decade, the more choices of resistor values and tighter the tolerances can be. For example, in the E12 series, there are twelve resistor values between 1 kΩ and 10 kΩ with 10% tolerance; E48 provides 48 values between 1 kΩ and 10 kΩ at 1% tolerance. This system is often used with online circuit calculators to indicate the resistor accuracy and tolerance desired. The standard resistor values of the E12 (±10%), E24 (±5%), E48 (±2%) and E96 (±1%) series are listed in **Table 22.5**.

Resistors used in military electronics (Mil-spec) use the type identifiers listed in **Table 22.6**. In addition, Mil-spec resistors with paint-stripe value bands have an extra band indicating the reliability level to which they are certified.

Fig 22.4 — Resistor value identification.

Table 22.4
Resistor Color Codes

Color	Significant Figure	Decimal Multiplier	Tolerance (%)
Black	0	1	
Brown	1	10	1
Red	2	100	2
Orange	3	1,000	
Yellow	4	10,000	
Green	5	100,000	0.5
Blue	6	1,000,000	0.25
Violet	7	10,000,000	0.1
Gray	8	100,000,000	0.05
White	9	1,000,000,000	
Gold		0.1	5
Silver		0.01	10
No color			20

Table 22.5
EIA Standard Resistor Values

±10% (E12)	±5% (E24)	±2% (E48)		±1% (E96)			
100	100	100	316	100	178	316	562
120	110	105	332	102	182	323	576
150	120	110	348	105	187	332	590
180	130	115	365	107	191	340	604
220	150	121	383	110	196	348	619
270	160	127	402	113	200	357	634
330	180	133	422	115	205	365	649
390	200	140	442	118	210	374	665
470	220	147	464	121	215	383	681
560	240	154	487	124	221	392	698
680	270	162	511	127	226	402	715
820	300	169	536	130	232	412	732
	330	178	562	133	237	422	750
	360	187	590	137	243	432	768
	390	196	619	140	249	442	787
	430	205	649	143	255	453	806
	470	215	681	147	261	464	825
	510	226	715	150	267	475	845
	560	237	750	154	274	487	866
	620	249	787	158	280	499	887
	680	261	825	162	287	511	909
	750	274	866	165	294	523	931
	820	287	909	169	301	536	953
	910	301	953	174	309	549	976

Use Table 22.5 values for each decade.
Example: 133 = 13.3 Ω, 133 Ω, 1.33 kΩ, 13.3 Ω, 133 kΩ, 1.33MΩ

Table 22.6
Mil-Spec Resistors

Wattage	Metal Film Types	Fixed Film Types	Composition Types	
1/10 W	RN50			
1/8 W	RN55	RL05	RLR05	RCR05
1/4 W	RN60	RL07	RLR07	RCR07
1/2 W	RN65	RL20	RLR20	RCR20
1 W	RN75	RL32	RLR32	RCR32
2 W	RN80	RL42	RLR62	RCR42

Tolerance Codes

B	±0.1%
C	±0.25%
D	±0.5%
F	±1%
G	±2%
J	±5%
K	±10%

Examples:
RN60D-2202F = 22 kΩ 1%
RL07S-471J = 470 Ω ±5%
RLR07C-471J = 470 Ω ±5%

Note: The RN Mil-Spec was discontinued in 1996 Still used by some manufacturers such as Vishay-Dale.

22.3 Capacitors

Capacitors exhibit the largest variety of electronic components. So many varieties and types are available that selecting the proper capacitor for a particular application can be overwhelming. Ceramic and film capacitors are the two most common types used by the amateur. (For additional information on the characteristics of the different types of capacitors, see the **Electrical Fundamentals** chapter.)

Though capacitors are classified by dozens of characteristics, the EIA has simplified the selection process by organizing ceramic capacitors into four categories called Class 1, 2, 3 and 4. Class 1 capacitors are the most stable and Class 4 the least preferred. Many catalogs now list ceramic capacitors by their class, greatly simplifying component selection.

For capacitors used in frequency-sensitive circuits, such as the frequency determining capacitors in oscillators or tuned circuits, select a Class 1 capacitor (C0G or NP0). For other applications, such as interstage coupling or bypass capacitors, components from Class 2 or Class 3 (X7R or Z5U) are usually sufficient. With modern manufacturing techniques, it is rare to find a Class 4 capacitor today.

Like resistors, capacitors are available in EIA standard series of values, E6 and E12, shown in **Table 22.7**. Most capacitors have a tolerance of 5% or greater. High-value capacitors used for filtering may have asymmetric tolerances, such as −5% and +10%, since the primary concern is for a guaranteed minimum value of capacitance.

22.3.1 Capacitor Types

Capacitor types can be grouped into ceramic dielectrics, film dielectrics and electrolytics. The type of capacitor can often be determined by their appearance, as shown in **Figs 22.5** and **22.6**. Capacitors with wire leads have two lead styles: *axial* and *radial*. Axial leads are aligned in opposite directions along a common axis, usually the largest dimension of the capacitor. Radial leads leave the capacitor body in the same direction and are usually arranged radially about the center of the capacitor body.

Disc (or disk) ceramic capacitors consist of two metal plates separated by a ceramic dielectric that establishes the desired capacitance. Due to their low cost, they are the most common capacitor type. The main disadvantage is their sensitivity to temperature changes (that is, a high temperature coefficient).

Monolithic ceramic capacitors are made by sandwiching layers of metal electrodes and ceramic layers to form the desired capacitance. "Monolithics" are physically smaller than disc ceramics for the same value of capacitance and cost, but exhibit the same high temperature coefficients.

Polyester film capacitors use layers of metal and polyester (Mylar) to make a wide

Table 22.7
EIA Standard Capacitor Values

±20% Capacitors (E6)

pF	pF	pF	µF	µF	µF	µF
1.0	10	100	0.001	0.01	0.1	1
1.5	15	150	0.0015	0.015	0.15	1.5
2.2	22	220	0.0022	0.022	0.22	2.2
3.3	33	330	0.0033	0.033	0.33	3.3
4.7	47	470	0.0047	0.047	0.47	4.7
6.8	68	680	0.0068	0.068	0.68	6.8

±10%, ±5% Capacitors (E12)

pF	pF	pF	µF	µF	µF	µF
1.0	10	100	0.001	0.01	0.1	1
1.2	12	120	0.0012	0.012	0.12	
1.5	15	150	0.0015	0.015	0.15	
1.8	18	180	0.0018	0.018	0.18	
2.2	22	220	0.0022	0.022	0.22	2.2
2.7	27	270	0.0027	0.027	0.27	
3.3	33	330	0.0033	0.033	0.33	3.3
3.9	39	390	0.0039	0.039	0.39	
4.7	47	470	0.0047	0.047	0.47	4.7
5.6	56	560	0.0056	0.056	0.56	
6.8	68	680	0.0068	0.068	0.68	
8.2	82	820	0.0082	0.082	0.82	

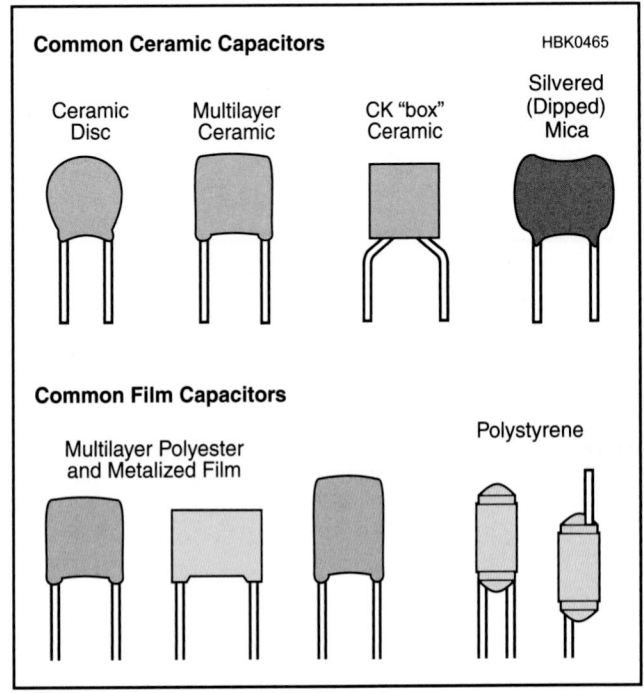

Fig 22.5 — Common capacitor types and package styles.

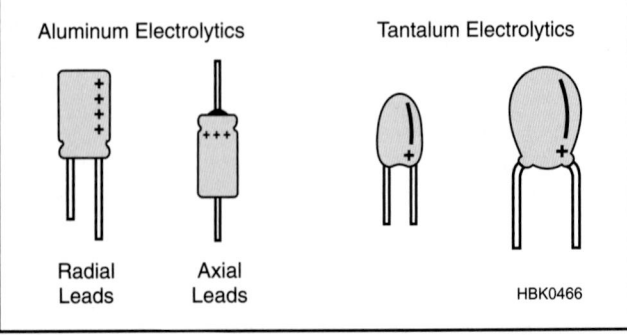

Fig 22.6 — Aluminum and tantalum electrolytic capacitors.

range of capacitances. They have poor temperature coefficients and are not recommended for high frequency use, but are suitable for low frequency and audio circuits.

To improve the performance of film capacitors, other dielectrics are used, such as polypropylene, polystyrene or polycarbonate film, or silvered-mica. These are very stable capacitors developed for RF use. Their main disadvantages are higher cost and lower working voltages than other varieties.

Capacitors are particularly sensitive to temperature changes because the physical dimensions of the capacitor determine its value. Standard temperature coefficient codes are shown in **Table 22.8**. Each code is made up of one character from each column in the table. For example, a capacitor marked Z5U is suitable for use between +10 and +85°C, with a maximum change in capacitance of –56% or +22%.

Capacitors with highly predictable temperature coefficients of capacitance are sometimes used in circuits whose performance must remain stable with temperature. If an application called for a temperature coefficient of –750 ppm/°C (N750), a capacitor marked U2J would be suitable. The older industry code for these ratings is being replaced with the EIA code shown in Table 22.8. NP0 (that is, N-P-zero) means "negative, positive, zero." It is a characteristic often specified for RF circuits requiring temperature stability, such as VFOs. A capacitor of the proper value marked C0G is a suitable replacement for an NP0 unit.

22.3.2 Electrolytic Capacitors

Aluminum electrolytic capacitors use aluminum foil "wetted" with a chemical agent and formed into layers to increase the effective area, and therefore the capacitance. Aluminum electrolytics provide high capacitance in small packages at low cost. Most varieties are polarized, that is, voltage should only be applied in one "direction." Polarized capacitors have a negative (–) and positive (+) lead. Standard dimensions of aluminum electrolytics are shown in **Fig 22.7** and **Table 22.9**. EIA standard values for aluminum electrolytics are given in **Table 22.10**.

Very old electrolytic capacitors should be used with care or, preferably, replaced. The wet dielectric agent can dry out during prolonged periods of non-use, causing the internal capacitor plates to form a short circuit when energized. Applying low voltage and gradually increasing it over a period of time may restore the capacitor to operation, but if the dielectric agent has dried out, the capacitor will have lost some or most of its value and will likely be lossy and prone to failure.

Tantalum electrolytic capacitors consist of a tantalum pentoxide powder mixed with a

wet or dry electrolyte, then formed into a pellet or slug for a large effective area. Tantalums also provide high capacitance values in very small packages. Tantalums tend to be more expensive than aluminum electrolytic capacitors. Like the aluminum electrolytic capacitor, tantalum capacitors are also polarized for which care should be exercised. Some varieties of tantalums can literally explode or burst open if voltage is applied with reverse polarity or the voltage rating is exceeded. Tantalum electrolytics are used almost exclusively as high-value SMT components due to their small sizes. Capacitance values up to 1000 µF at 4 V are available with body sizes about a quarter-inch square.

Identifying the polarity markings of aluminum and tantalum electrolytics (shown in Fig 22.6) can be confusing. Most tantalum electrolytics are marked with a solid band indicating the positive lead. Aluminum electrolytics are available with bands or symbols marking

Table 22.8
Ceramic Temperature Characteristics
Common EIA Types:

EIA Class	EIA Code	Characteristics	Temp. Range*
1	C0G	0 ± 30 ppm/°C	–55 °C to +125 °C
2	Y5P	±10%	–30 °C to + 85 °C
2	X7R	±15%	–55 °C to +125 °C
2	Y5U	±20%	–10 °C to + 85 °C
2	Z5U	±20%	+10 °C to + 85 °C
2	Z5V	+80%, –20%	–30 °C to + 85 °C
3	Y5V	+80%, –20%	–10 °C to + 85 °C

Common Industry Types:

EIA Class	EIA Code	Characteristics	Temp. Range*
1	NP0	0 ± 30 ppm/°C	–55 °C to +125 °C
2	CK05	±10%	–55 °C to +125 °C

*Temp. range for which characteristics are specified and may vary slightly between different manufacturers

Temperature Coefficient Codes

Minimum Temperature	Maximum Temperature	Maximum capacitance change over temp range
X –55 °C	2 +45°C	A ±1.0%
Y –30 °C	4 +65°C	B ±1.5%
Z +10 °C	5 +85°C	C ±2.2%
6 +105 °C		D ±3.3%
7 +125 °C		E ±4.7%
		F ±7.5%
		P ±10%
		R ±15%
		S ±22%
		T –33%, +22%
		U –56%, +22%
		V –82%, +22%

Table 22.9
Aluminum Electrolytic Capacitors Standard Sizes (Radial Leads)

H	Dia	LS	Pad Size and Drill*
0.44	0.20	0.08	0.056 round, 0.029 hole
0.44	0.25	0.10	0.056 round, 0.029 hole
0.44	0.32	0.14	0.065 round, 0.029 hole
0.52	0.40	0.20	0.080 round, 0.035 hole
0.78	0.50	0.20	0.080 round, 0.035 hole
1.00	0.63	0.30	0.100 round, 0.035 hole
1.42	0.72	0.30	0.100 round, 0.035 hole
1.60	0.88	0.40	0.100 round, 0.035 hole

Dimensions in inches.
*Customary to make "+" lead square pad on PCB

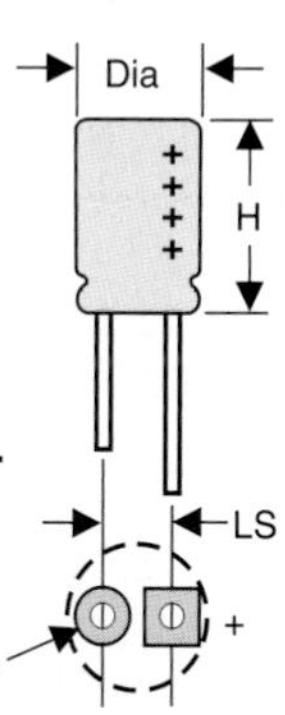

HBK0467

Fig 22.7 — Aluminum electrolytic capacitor dimensions.

Table 22.10
Aluminum Electrolytic Capacitors
EIA ±20%Standard Values

µF	µF	µF	µF	µF
0.1	1.0	10	100	1000
0.22	2.2	22	220	2200
0.33	3.3	33	330	3300
0.47	4.7	47	470	4700
0.68	6.8	68	680	6800
0.82	8.2	82	820	8200

either the negative or positive lead. The positive lead of axial-lead electrolytic capacitors is usually manufactured to be longer than the negative lead and often enters the capacitor through fiber or plastic insulating material while the negative lead is connected directly to the metallic case of the capacitor. Misidentifying the polarity of capacitors is a common error during assembly or repair.

22.3.3 Surface Mount Capacitors

SMT capacitors are generally film, ceramic or tantalum electrolytics. Body sizes are shown in **Fig 22.8** and **22.9**. Although the EIA scheme is the standard method of labeling capacitor value, you may encounter a two-character alphanumeric code (see **Table 22.11**) consisting of a letter indicating the significant digits and a number indicating the multiplier. The code represents the capacitance in picofarads. For example, a chip capacitor marked "A4" would have a capacitance of 10,000 pF, or 0.01 μF. A unit marked "N1" would be a 33-pF capacitor. If there is sufficient space on the device package, a tolerance code may be included. The standard SMT body sizes and pad spacing are provided in **Table 22.12**.

Table 22.11
SMT Capacitor Two-Character Labeling
Significant Figure Codes

Character	Significant Figures	Character	Significant Figures
A	1.0	T	5.1
B	1.1	U	5.6
C	1.2	V	6.2
D	1.3	W	6.8
E	1.5	X	7.5
F	1.6	Y	8.2
G	1.8	Z	9.1
H	2.0	a	2.5
J	2.2	b	3.5
K	2.4	d	4.0
L	2.7	e	4.5
M	3.0	f	5.0
N	3.3	m	6.0
P	3.6	n	7.0
Q	3.9	t	8.0
R	4.3	y	9.0
S	4.7		

Multiplier Codes

Numeric Character	Decimal Multiplier
0	1
1	10
2	100
3	1,000
4	10,000
5	100,000
6	1,000,000
7	10,000,000
8	100,000,000
9	0.1

Table 22.12
Surface Mount Capacitors — EIA Standard Sizes

Size	Length	Width	Height	C	SMT Pad	C-C*
0402	0.039	0.020	0.014	0.010	0.025 × 0.035	0.050
0603	0.063	0.031	0.018	0.014	0.030 × 0.030	0.055
0805	0.079	0.049	0.020	0.016	0.040 × 0.050	0.075
1206	0.126	0.063	0.024	0.020	0.064 × 0.064	0.125
1210	0.126	0.102	0.024	0.020	0.070 × 0.100	0.150

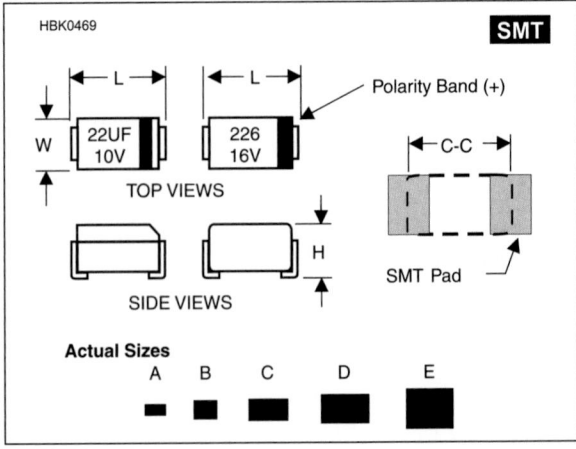

Fig 22.8 — Surface-mount capacitor packages.

Surface Mount Electrolytic Capacitors — EIA Standard Sizes

Size	Length	Width	Height	C-C*	SMT Pad
A (1206)	0.126	0.063	0.063	0.110	0.055 × 0.060
B (1411)	0.138	0.110	0.075	0.136	0.075 × 0.090
C (2412)	0.236	0.126	0.098	0.265	0.090 × 0.120
D (2916)	0.287	0.169	0.110	0.250	0.100 × 0.100
E (2924)	0.287	0.236	0.142	0.250	0.100 × 0.100

Dimensions in inches.
*C-C is SMT pad center-to-center spacing

Fig 22.9 — Surface-mount electrolytic packages.

22.3.4 Capacitor Voltage Ratings

Capacitors are also rated by their maximum operating voltage. The importance of selecting a capacitor with the proper voltage rating is often overlooked. Exceeding the voltage rating, even momentarily, can cause excessive heating, a permanent shift of the capacitance value, a short circuit, or outright destruction. As a result, the voltage rating should be at least 25% higher than the working voltage across the capacitor; many designers use 50-100%.

Following the 25% guideline, filter capacitors for a 12-V system should have at least a 15-V rating (12 V × 1.25). However, 12-V systems such as 12-V power supplies and automotive 12-V electrical systems actually operate near 13.8 V and in the case of automotive systems, as high as 15 V. In such cases, capacitors rated for 15 V would be an insufficient margin of safety; 20 to 25-V capacitors should be used in such cases.

In large signal ac circuits, the maximum voltage rating of the capacitor should be based on the peak-to-peak voltages present. For example, the output of a 5-W QRP transmitter is 16 V_{RMS}, or about 45 V_{P-P}. Capacitors exposed to the 5 W RF power, such as in the output low-pass filter, should be rated well above 50 V for the 25% rule. A 100 W transmitter produces RF voltages of about 200 V_{P-P}.

Capacitors that are to be connected to primary ac circuits (directly to the ac line) for filtering or coupling *must* be rated for ac line use. These capacitors are listed as such in catalogs and are designed to minimize fire and other hazards in case of failure. Remember, too, that ac line voltage is given as RMS, with peak-to-peak voltage 2.83 times higher: 120 V_{RMS} = 339 V_{P-P}.

Applying peak-to-peak voltages approaching the maximum voltage rating will cause excessive heating of the capacitor. This, in turn, will cause a permanent shift in the capacitance value. This could be undesirable in the output low pass filter example cited above in trying to maintain the proper impedance match between transmitter and antenna.

Exceeding the maximum voltage rating can also cause a breakdown of the dielectric material in the capacitor. The voltage can jump between the plates causing momentary or permanent electrical shorts between the capacitor plates.

In electrolytic and tantalum capacitors, exceeding the voltage rating can produce extreme heating of the oil or wetting agent used as the dielectric material. The expanding gases can cause the capacitor to burst or explode.

These over-voltage problems are easily avoided by selecting a capacitor with a voltage rating 25-50% above the normal peak-to-peak

Table 22.13
Capacitor Standard Working Voltages

Ceramic	Polyester	Electrolytic	Tantalum
		6.3 V	6.3 V
		10 V	10 V
16 V		16 V	16 V
			20 V
25 V		25 V	25 V
		35 V	35 V
50 V	50 V	50 V	50 V
		63 V	63 V
100 V	100 V	100 V	
	150 V	150 V	
200 V	200 V		
	250 V	250 V	

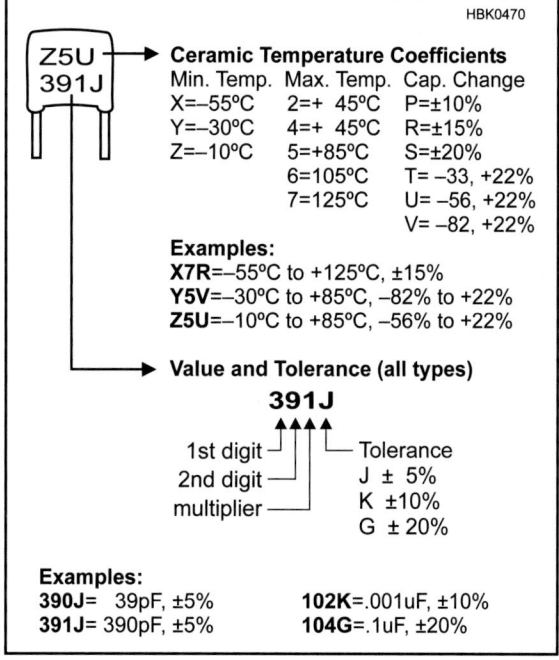

HBK0470

Ceramic Temperature Coefficients

Min. Temp.	Max. Temp.	Cap. Change
X=−55°C	2=+ 45°C	P=±10%
Y=−30°C	4=+ 45°C	R=±15%
Z=−10°C	5=+85°C	S=±20%
	6=105°C	T= −33, +22%
	7=125°C	U= −56, +22%
		V= −82, +22%

Examples:
X7R=−55°C to +125°C, ±15%
Y5V=−30°C to +85°C, −82% to +22%
Z5U=−10°C to +85°C, −56% to +22%

Value and Tolerance (all types)

391J

1st digit
2nd digit
multiplier

Tolerance
J ± 5%
K ±10%
G ± 20%

Examples:
390J= 39pF, ±5%
391J= 390pF, ±5%
102K=.001uF, ±10%
104G=.1uF, ±20%

Fig 22.10 — Abbreviated EIA capacitor identification. This method is used on SMT capacitors. An "R" in the numeric field stands for "radix" and represents a decimal point, so that "4R7" indicates "4.7" for example.

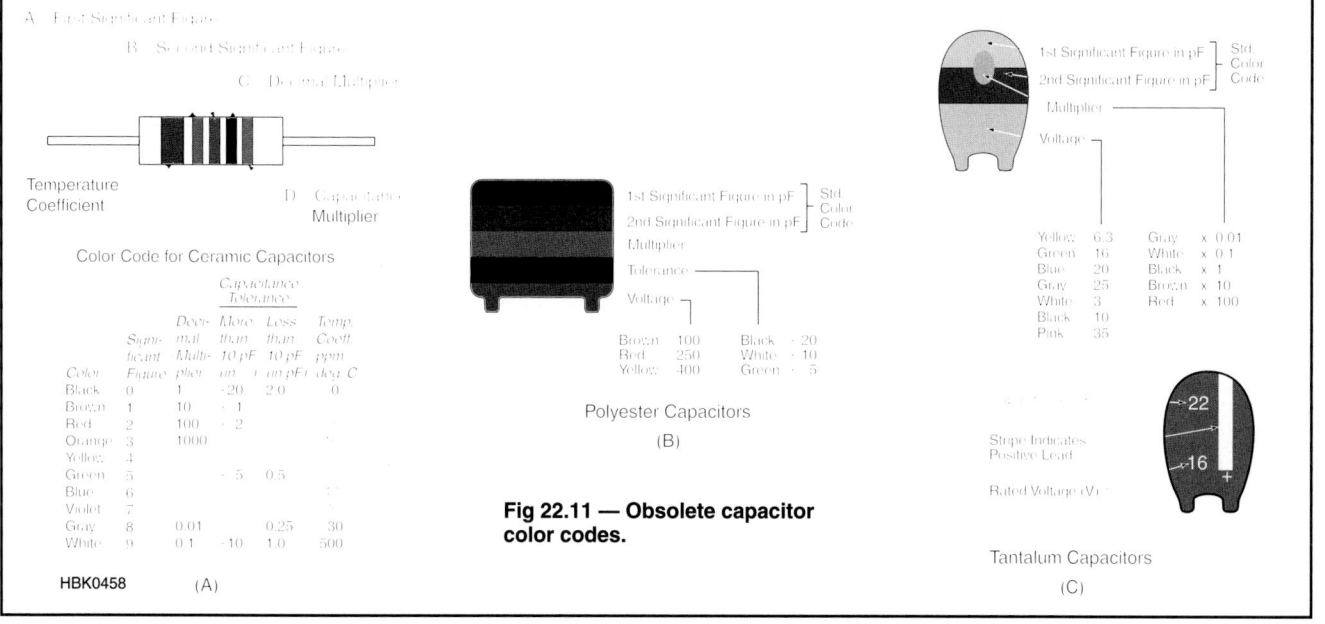

Fig 22.11 — Obsolete capacitor color codes.

operating voltage. **Table 22.13** lists standard working voltages for common capacitor types.

22.3.5 Capacitor Identification

Capacitors are identified by the EIA numerical or color code standard as shown in **Fig 22.10**. Since 2000, the EIA numerical code is the most dominant form of capacitor identification and is used on all capacitor types and body styles. Color coding schemes are becoming rare, used only by a few non-US manufacturers. Some thru-hole "gum drop" tantalum capacitors also still use the color codes of **Fig 22.11**. Electrolytic and tantalum capacitors are often labeled with capacitance and working voltage in µF and V as in Fig 22.11C.

Similar to the resistor EIA code, numerals are used to indicate the significant numerals and the multiplier, followed by an alphabetic character to indicate the tolerance. The multiplier is simply the number of zeros following the significant numerals. For example, a capacitor marked with "122K" would be a 1200 pF 10% capacitor. The use of R to denote a decimal point in a value can be confusing if pF or µF are not specified. Generally, an inspection of the capacitor will determine which is correct but a capacitance meter may be required. Additional digits and codes may be encountered as shown in **Fig 22.12**.

Military-surplus equipment using the obsolete "postage stamp" capacitors is still encountered in Amateur Radio. These capacitors used the colored dot method of value identification shown in **Fig 22.13**.

European manufacturers often use nanofarads or nF, such that 10 nF, or simply 10N, indicates 10 nanofarads. This is equivalent to 10,000 pF or 0.01 µF. This notational scheme, shown in **Table 22.14**, is more commonly found on schematic diagrams than actual part markings.

Table 22.14
European Marking Standards for Capacitors

Marking	Value
1p	1 pF
2p2	2.2 pF
10p	10 pF
100p	100 pF
1n	1 nF (= 0.001 µF)
2n2	2.2 nF (= 0.0022 µF)
10n	10 nF (= 0.01 µF)
100n	100 nF (= 0.1 µF)
1u	1 µF
5u6	5.6 µF
10u	10 µF
100u	100 µF

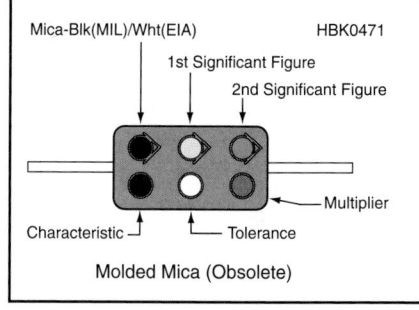

Fig 22.13 — Obsolete JAN "postage stamp" capacitor labeling.

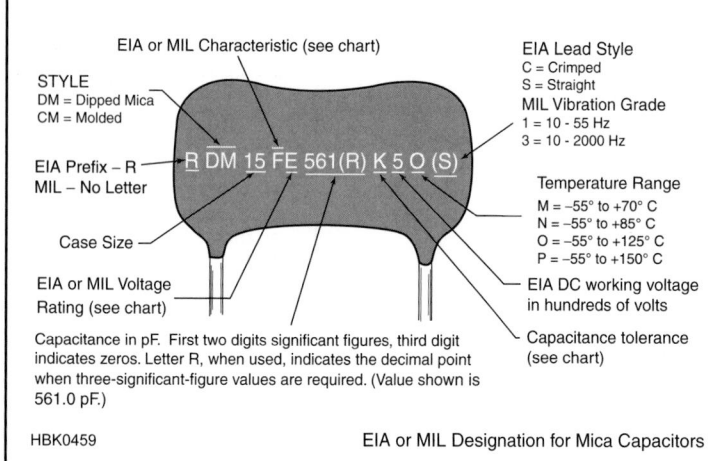

Letter Designator	"Characteristic" Max Capacitance Drift	"Characteristic" Max Range of Temp Coeff (ppm / deg. C)	MIL Voltage Rating (V)	Capacitance Tolerance (Percent)
A	-	-	100	-
B	Not Specified	Not Specified	250	-
C	±(0.5% + 0.1 pF)	±200	300	-
D	±(0.3% + 0.1 pF)	±100	500	-
E	±(0.1% + 0.1 pF)	−20 to +100	600	-
F	±(0.05% + 0.1 pF)	0 to +70	1000	±1
G	-	-	1200	±2
H	-	-	1500	-
J	-	-	2000	±5
K	-	-	2500	±10
L	-	-	3000	-
M	-	-	4000	±20

MIL voltage ratings for other letter designators: N=5000 V, P=6000 V, Q=8000 V, R=10,000 V, S=12,000 V, T=15,000 V, U=20,000 V, V=25,000 V, W=30,000 V, X=35,000 V.

Fig 22.12 — Complete EIA capacitor labeling scheme.

22.4 Inductors

Inductors, both fixed and variable are available in a wide variety of types and packages, and many offer few clues as to their values. Some coils and chokes are marked with the EIA color code shown in Table 22.4. See **Fig 22.14** for another marking system for cylindrical encapsulated RF chokes. The body of these components is often green to identify them as inductors and not resistors. Measure the resistance of the component with an ohmmeter if there is any doubt as to the identity of the component. **Table 22.15** is a list of the EIA standard inductor values.

Table 22.16 lists the properties of common powdered-iron cores. Formulas are given for calculating the number of required turns based on a given inductance and for calculating the inductance given a specific number of turns. Most powdered-iron toroid cores that amateurs use are manufactured by Micrometals (**www.micrometals.com**). Paint is used to identify the material used in the core. The Micrometals color code is part of Table 22.16. **Table 22.17** gives the physical dimensions of powdered-iron toroids.

An excellent design resource for ferrite-based components is the Fair-Rite Materials Corp on-line catalog at **www.fair-rite.com**. The Fair-Rite Web site's Technical section also has free papers on the use of ferrites for EMI suppression and broadband transformers. The

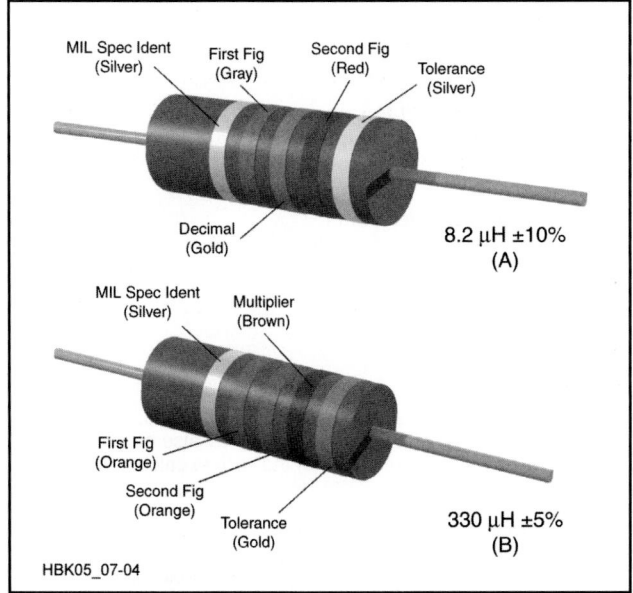

Fig 22.14 — Color coding for cylindrical encapsulated RF chokes. At A, an example of the coding for an 8.2-µH choke is given. At B, the color bands for a 330-µH inductor are illustrated. The color code is given in Table 22.4.

Table 22.15
EIA Standard Inductor Values

µH	µH	µH	mH	mH	mH
1.0	10	100	1.0	10	100
1.2	12	120	1.2	12	120
1.5	15	150	1.5	15	150
2.2	22	220	2.2	22	220
2.7	27	270	2.7	27	270
3.3	33	330	3.3	33	330
3.9	39	390	3.9	39	390
4.7	47	470	4.7	47	470
5.6	56	560	5.6	56	560
6.8	68	680	6.8	68	680
8.2	82	820	8.2	82	820

following list presents the general characteristics (material and composition and intended application) of Fair-Rite's ferrite materials:

• Type 31 (MnZn) — EMI suppression applications from 1 MHz up to 500 MHz.

• Type 43 (NiZn) — Suppression of conducted EMI, inductors and HF common-mode chokes from 20 MHz to 250 MHz.

• Type 44 (NiZn) — EMI suppression from 30 MHz to 500 MHz.

• Type 61 (NiZn) — Inductors up to 25 MHz and EMI suppression above 200 MHz.

• Type 67 (NiZn) — Broadband transformers, antennas and high-Q inductors up to 50 MHz.

• Type 73 (MnZn) — Suppression of conducted EMI below 50 MHz.

• Type 75 (MnZn) — Broadband and pulse transformers.

See **Table 22.18** for information about the magnetic properties of ferrite cores. Ferrite cores are not typically painted, so identification is often difficult. More information about the use of ferrites at RF is provided in the **RF Techniques** chapter.

Table 22.16
Powdered-Iron Toroidal Cores: Magnetic Properties

There are differing conventions for referring to the type of core material: #, mix and type are all used. For example, all of the following designate the same material: #12, Mix 12, 12-Mix, Type 12 and 12-Type.

Inductance and Turns Formula

The turns required for a given inductance or inductance for a given number of turns can be calculated from:

$$N = 100\sqrt{\frac{L}{A_L}} \qquad L = A_L\left(\frac{N^2}{10,000}\right)$$

where N = number of turns; L = desired inductance (μH); A_L = inductance index (μH per 100 turns).

The industry standard is to provide values of A_L in units of inductance per number of turns squared as in the preceding formulas. Amidon Associates gives the A_L value in units of inductance per turn. The units of inductance are generally in nH but may also be in mH. Make sure you understand which units apply and use the A_L value and formula provided by the manufacturer of the core to calculate number of turns or inductance.

Toroid diameter is indicated by the number following "T" — T-200 is 2.00 in. dia; T-68 is 0.68 in. diameter, etc.

AL Values

Size	26*	3	15	1	2	7	6	10	12	17	0
T-12	na	60	50	48	20	18	17	12	7.5	7.5	3.0
T-16	145	61	55	44	22	na	19	13	8.0	8.0	3.0
T-20	180	76	65	52	27	24	22	16	10.0	10.0	3.5
T-25	235	100	85	70	34	29	27	19	12.0	12.0	4.5
T-30	325	140	93	85	43	37	36	25	16.0	16.0	6.0
T-37	275	120	90	80	40	32	30	25	15.0	15.0	4.9
T-44	360	180	160	105	52	46	42	33	18.5	18.5	6.5
T-50	320	175	135	100	49	43	40	31	18.0	18.0	6.4
T-68	420	195	180	115	57	52	47	32	21.0	21.0	7.5
T-80	450	180	170	115	55	50	45	32	22.0	22.0	8.5
T-94	590	248	200	160	84	na	70	58	32.0	na	10.6
T-106	900	450	345	325	135	133	116	na	na	na	19.0
T-130	785	350	250	200	110	103	96	na	na	na	15.0
T-157	870	420	360	320	140	na	115	na	na	na	na
T-184	1640	720	na	500	240	na	195	na	na	na	na
T-200	895	425	na	250	120	105	100	na	na	na	na

*Mix-26 is similar to the older Mix-41, but can provide an extended frequency range.

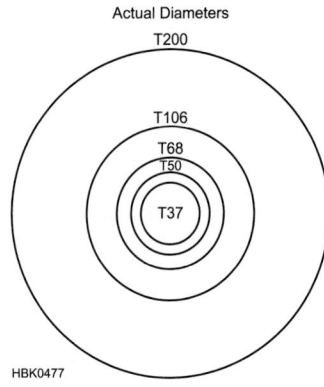

Actual Diameters

Magnetic Properties Iron Powder Cores

Mix	Color	Material	μ	Temp stability (ppm/°C)	f (MHz)	Notes
26	Yellow/white	Hydrogen reduced	75	825	dc - 1	Used for EMI filters and dc chokes
3	Gray	Carbonyl HP	35	370	0.05 - 0.50	Excellent stability, good Q for lower frequencies
15	Red/white	Carbonyl GS6	25	190	0.10 - 2	Excellent stability, good Q
1	Blue	Carbonyl C	20	280	0.50 - 5	Similar to Mix-3, but better stability
2	Red	Carbonyl E	10	95	2 - 30	High Q material
7	White	Carbonyl TH	9	30	3 - 35	Similar to Mix-2 and Mix-6, but better temperature stability
6	Yellow	Carbonyl SF	8	35	10 - 50	Very good Q and temperature stability for 20-50 MHz
10	Black	Powdered iron W	6	150	30 - 100	Good Q and stability for 40 - 100 MHz
12	Green/white	Synthetic oxide	4	170	50 - 200	Good Q, moderate temperature stability
17	Blue/yellow	Carbonyl	4	50	40 - 180	Similar to Mix-12, better temperature stability, Q drops about 10% above 50 MHz, 20% above 100 MHz
0	Tan	phenolic	1	0	100 - 300	Inductance may vary greatly with winding technique

Courtesy of Amidon Assoc and Micrometals
Note: Color codes hold only for cores manufactured by Micrometals, which makes the cores sold by most Amateur Radio distributors.

Table 22.17

Powdered-Iron Toroidal Cores: Dimensions

Toroid diameter is indicated by the number following "T" — T-200 is 2.00 in. dia; T-68 is 0.68 in. diameter, etc.

See Table 22.16 for a core sizing guide.

Red E Cores—500 kHz to 30 MHz ($\mu = 10$)

No.	OD (in)	ID (in)	H (in)
T-200-2	2.00	1.25	0.55
T-94-2	0.94	0.56	0.31
T-80-2	0.80	0.50	0.25
T-68-2	0.68	0.37	0.19
T-50-2	0.50	0.30	0.19
T-37-2	0.37	0.21	0.12
T-25-2	0.25	0.12	0.09
T-12-2	0.125	0.06	0.05

Black W Cores—30 MHz to 200 MHz ($\mu=6$)

No.	OD (In)	ID (In)	H (In)
T-50-10	0.50	0.30	0.19
T-37-10	0.37	0.21	0.12
T-25-10	0.25	0.12	0.09
T-12-10	0.125	0.06	0.05

Yellow SF Cores—10 MHz to 90 MHz ($\mu=8$)

No.	OD (In)	ID (In)	H (In)
T-94-6	0.94	0.56	0.31
T-80-6	0.80	0.50	0.25
T-68-6	0.68	0.37	0.19
T-50-6	0.50	0.30	0.19
T-26-6	0.25	0.12	0.09
T-12-6	0.125	0.06	0.05

Number of Turns vs Wire Size and Core Size

Approximate maximum number of turns—single layer wound—enameled wire.

Wire Size	T-200	T-130	T-106	T-94	T-80	T-68	T-50	T-37	T-25	T-12
10	33	20	12	12	10	6	4	1		
12	43	25	16	16	14	9	6	3		
14	54	32	21	21	18	13	8	5	1	
16	69	41	28	28	24	17	13	7	2	
18	88	53	37	37	32	23	18	10	4	1
20	111	67	47	47	41	29	23	14	6	1
22	140	86	60	60	53	38	30	19	9	2
24	177	109	77	77	67	49	39	25	13	4
26	223	137	97	97	85	63	50	33	17	7
28	281	173	123	123	108	80	64	42	23	9
30	355	217	154	154	136	101	81	54	29	13
32	439	272	194	194	171	127	103	68	38	17
34	557	346	247	247	218	162	132	88	49	23
36	683	424	304	304	268	199	162	108	62	30
38	875	544	389	389	344	256	209	140	80	39
40	1103	687	492	492	434	324	264	178	102	51

Actual number of turns may differ from above figures according to winding techniques, especially when using the larger size wires. Chart prepared by Michel J. Gordon, Jr, WB9FHC.
Courtesy of Amidon Assoc.

Table 22.18
Ferrite Toroids: A_L Chart (mH per 1000 turns) Enameled Wire

There are differing conventions for referring to the type of ferrite material: #, mix and type are all used. For example, all of the following designate the same ferrite material: #43, Mix 43, 43-Mix, Type 43, and 43-Type.

Fair-Rite Corporation (**www.fair-rite.com**) and Amidon (**www.amidoncorp.com**) ferrite toroids can be cross-referenced as follows:

For Amidon toroids, "FT-XXX-YY" indicates a ferrite toroid, with XXX as the OD in hundredths of an inch and YY the mix. For example, an FT-23-43 core has an OD of 0.23 inch and is made of type 43 material. Additional letters (usually "C") are added to indicate special coatings or different thicknesses.

For Fair-Rite toroids, digits 1 and 2 of the part number indicate product type (59 indicates a part for inductive uses), digits 3 and 4 indicate the material type, digits 5 through 9 indicate core size, and the final digit indicates coating (1 for Paralene and 2 for thermo-set). For example, Fair-Rite part number 5943000101 is equivalent to the Amidon FT-23-43 core.

Ferrite Toroids: A_L Chart (mH per 1000 turns)
Toroid diameter is specified as the outside diameter of the core. See Table 22.16 for a core sizing guide.

Core Size (in)	63/67-Mix $\mu = 40$	61-Mix $\mu = 125$	43-Mix $\mu = 850$	77 (72)-Mix $\mu = 2000$	J (75)-Mix $\mu = 5000$
0.23	7.9	24.8	188	396	980
0.37	19.7	55.3	420	884	2196
0.50	22.0	68.0	523	1100	2715
0.82	22.4	73.3	557	1170	NA
1.14	25.4	79.3	603	1270	3170
1.40	45	140	885	2400	5500
2.40	55	170	1075	2950	6850

31-Mix is an EMI suppression material and not recommended for inductive use.

Inductance and Turns Formula
The turns required for a given inductance or inductance for a given number of turns can be calculated from:

$$N = 1000\sqrt{\frac{L}{A_L}} \quad L = A_L\left(\frac{N^2}{1,000,000}\right)$$

where N = number of turns; L = desired inductance (mH); A_L = inductance index (mH per 1000 turns).

Ferrite Magnetic Properties

Property	Unit	63/67-Mix	61-Mix	43-Mix	77 (72)-Mix	J (75)-Mix	31-Mix
Initial perm.	(μ_i)	40	125	850	2000	5000	1500
Max. perm.		125	450	3000	6000	8000	Not spec.
Saturation flux density @ 10 oe	gauss	1850	2350	2750	4600	3900	3400
Residual flux density	gauss	750	1200	1200	1150	1250	2500
Curie temp.	°C	450	350	130	200	140	>130
Vol. resistivity	ohm/cm	1×10^8	1×10^8	1×10^5	1×10^2	5×10^2	3×10^3
Resonant circuit frequency	MHz	15-25	0.2-10	0.01-1	0.001-1	0.001-1	*
Specific gravity		4.7	4.7	4.5	4.8	4.8	4.7
Loss factor	$\frac{1}{\mu_i Q}$	110×10^{-6} @25 MHz	32×10^{-6} @2.5 MHz	120×10^{-6} @1 MHz	4.5×10^{-6} @0.1 MHz	15×10^{-6} @0.1 MHz	20×10^{-6} @0.1 MHz
Coercive force	Oe	2.40	1.60	0.30	0.22	0.16	0.35
Temp. Coef. of initial perm.	%/°C (20°-70°)	0.10	0.15	1.0	0.60	0.90	1.6

*31-Mix is an EMI suppression material and not recommended for inductive uses.

Ferrite Toroids—Physical Properties
All physical dimensions in inches.

OD (in)	ID (in)	Height (in)	A_e	ℓ_e	V_e
0.230	0.120	0.060	0.00330	0.529	0.00174
0.375	0.187	0.125	0.01175	0.846	0.00994
0.500	0.281	0.188	0.02060	1.190	0.02450
0.825	0.520	0.250	0.03810	2.070	0.07890
1.142	0.750	0.295	0.05810	2.920	0.16950
1.400	0.900	0.500	0.12245	3.504	0.42700
2.400	1.400	0.500	0.24490	5.709	1.39080

Different height cores may be available for each core size.
A_e — Effective magnetic cross-sectional area (in)^2
ℓ_e — Effective magnetic path length (inches)
V_e — Effective magnetic volume (in)^3
To convert from (in)^2 to (cm)^2, divide by 0.155
To convert from (in)^3 to (cm)^3, divide by 0.0610
Courtesy of Amidon Assoc. and Fair-Rite Corp.

22.5 Transformers

Many transformers, including power transformers, IF transformers, and audio transformers, are made to be installed on PC boards, and have terminals designed for that purpose. Some transformers are manufactured with wire leads that are color-coded to identify each connection. When colored wire leads are present, the color codes in **Tables 22.19, 22.20** and **22.21** usually apply. In addition, many miniature IF transformers are tuned with slugs, color-coded to signify their application. **Table 22.22** lists application versus slug color.

Table 22.20

IF Transformer Wiring Color Codes

Plate lead:	Blue
B+ lead:	Red
Grid (or diode) lead:	Green
Grid (or diode) return:	Black

Note: If the secondary of the IF transformer is center-tapped, the second diode plate lead is green-and-black striped, and black is used for the center-tap lead.

Table 22.21

IF Transformer Slug Color Codes

Frequency	Application	Slug color
455 kHz	1st IF	Yellow
	2nd IF	White
	3rd IF	Black
	Osc tuning	Red
10.7 MHz	1st IF	Green
	2nd or 3rd IF	Orange, Brown or Black

Table 22.19

Power-Transformer Wiring Color Codes

Non-tapped primary leads:	Black
Tapped primary leads:	Common: Black
	Tap: Black/yellow striped
	Finish: Black/red striped
High-voltage plate winding:	Red
Center tap:	Red/yellow striped
Rectifier filament winding:	Yellow
Center tap:	Yellow/blue striped
Filament winding 1:	Green
Center tap:	Green/yellow striped
Filament winding 2:	Brown
Center tap:	Brown/yellow striped
Filament winding 3:	Slate
Center tap:	Slate/yellow striped

Table 22.22

Audio Transformer Wiring Color Codes

Plate lead of primary	Blue
B+ lead (plain or center-tapped)	Red
Plate (start) lead on center-tapped primaries	Brown (or blue if polarity is not important)
Grid (finish) lead to secondary	Green
Grid return (plain or center tapped)	Black
Grid (start) lead on center tapped secondaries	Yellow (or green if polarity not important)

Note: These markings also apply to line-to-grid and tube-to-line transformers.

22.6 Semiconductors

Most semiconductors are labeled with industry standard part numbers, such as 1N4148 or 2N3904, and possibly a date or batch code. You will also encounter numerous manufacturer-specific part numbers and the so-called "house numbers" (marked with codes used by an equipment manufacturer instead of the standard part numbers). In such cases, it is often possible to find the standard equivalent or a suitable replacement by using one of the semiconductor cross-reference directories available from various replacement-parts distributors. If you look up the house number and find the recommended replacement part, you can often find other standard parts that are replaced by that same part.

Information on the use of semiconductors, common design practices, and the necessary circuit design equations can be found in the chapters on **Analog Basics** and **Digital**

Basics. Manufacturer Web sites are often a rich source of information on applying semiconductors, both in general and the specific devices they offer.

22.6.1 Diodes

The diode parameters of most importance are maximum forward current or power handling capacity, reverse leakage current, maximum peak inverse voltage (PIV), maximum reverse voltage and the forward voltage. (See **Table 22.23**) For switching or high-speed rectification applications, the time response parameters are also important.

Power dissipation in a diode is equal to the diode's forward voltage drop multiplied by the average forward current. Although fixed voltages are often used for diodes in small-signal applications (0.6 V for silicon

PN-junction diodes, 0.3 V for germanium, for example), the actual forward voltage at higher currents can be significantly higher and must be taken into account for high-current applications, such as power supplies.

Most diodes are marked with a part number and some means of identifying the anode or cathode. A thick band or stripe is commonly used to identify the cathode lead or terminal. Stud-mount diodes are usually labeled with a small diode symbol to indicate anode and cathode. Diodes in axial lead packages are sometimes identified with a color scheme as shown in **Fig 22.15**. The common diode packaging standards are illustrated in **Fig 22.16** and the dimensions listed in **Table 22.24**. Many surface mount diodes are packaged in the same SMT packages as resistors.

Packages containing multiple diodes and rectifier bridge configurations are also

Table 22.23

Semiconductor Diode Specifications†

Listed numerically by device

Device	Type	Material	Peak Inverse Voltage, PIV (V)	Average Rectified Current Forward (Reverse) $I_O(A)(I_R(A))$	Peak Surge Current, I_{FSM} 1 s @ 25°C (A)	Average Forward Voltage, VF (V)
1N34	Signal	Ge	60	8.5 m (15.0 μ)		1.0
1N34A	Signal	Ge	60	5.0 m (30.0 μ)		1.0
1N67A	Signal	Ge	100	4.0 m (5.0 μ)		1.0
1N191	Signal	Ge	90	15.0 m		1.0
1N270	Signal	Ge	80	0.2 (100 μ)		1.0
1N914	Fast Switch	Si	75	75.0 m (25.0 n)	0.5	1.0
1N1183	RFR	Si	50	40 (5 m)	800	1.1
1N1184	RFR	Si	100	40 (5 m)	800	1.1
1N2071	RFR	Si	600	0.75 (10.0 μ)		0.6
1N3666	Signal	Ge	80	0.2 (25.0 μ)		1.0
1N4001	RFR	Si	50	1.0 (0.03 m)		1.1
1N4002	RFR	Si	100	1.0 (0.03 m)		1.1
1N4003	RFR	Si	200	1.0 (0.03 m)		1.1
1N4004	RFR	Si	400	1.0 (0.03 m)		1.1
1N4005	RFR	Si	600	1.0 (0.03 m)		1.1
1N4006	RFR	Si	800	1.0 (0.03 m)		1.1
1N4007	RFR	Si	1000	1.0 (0.03 m)		1.1
1N4148	Signal	Si	75	10.0 m (25.0 n)		1.0
1N4149	Signal	Si	75	10.0 m (25.0 n)		1.0
1N4152	Fast Switch	Si	40	20.0 m (0.05 μ)		0.8
1N4445	Signal	Si	100	0.1 (50.0 n)		1.0
1N5400	RFR	Si	50	3.0 (500 μ)	200	
1N5401	RFR	Si	100	3.0 (500 μ)	200	
1N5402	RFR	Si	200	3.0 (500 μ)	200	
1N5403	RFR	Si	300	3.0 (500 μ)	200	
1N5404	RFR	Si	400	3.0 (500 μ)	200	
1N5405	RFR	Si	500	3.0 (500 μ)	200	
1N5406	RFR	Si	600	3.0 (500 μ)	200	
1N5408	RFR	Si	1000	3.0 (500 μ)	200	
1N5711	Schottky	Si	70	1 m (200 n)	15 m	0.41 @ 1 mA
1N5767	Signal	Si		0.1 (1.0 μ)		1.0
1N5817	Schottky	Si	20	1.0 (1 m)	25	0.75
1N5819	Schottky	Si	40	1.0 (1 m)	25	0.9
1N5821	Schottky	Si	30	3.0		
ECG5863	RFR	Si	600	6	150	0.9
1N6263	Schottky	Si	70	15 m	50 m	0.41 @ 1 mA
5082-2835	Schottky	Si	8	1 m (100 n)	10 m	0.34 @ 1 mA

Si = Silicon; Ge = Germanium; RFR = rectifier, fast recovery.

†For package shape, size and pin-connection information see manufacturers' data sheets. Many retail suppliers offer data sheets to buyers free of charge on request. Data books are available from many manufacturers and retailers.

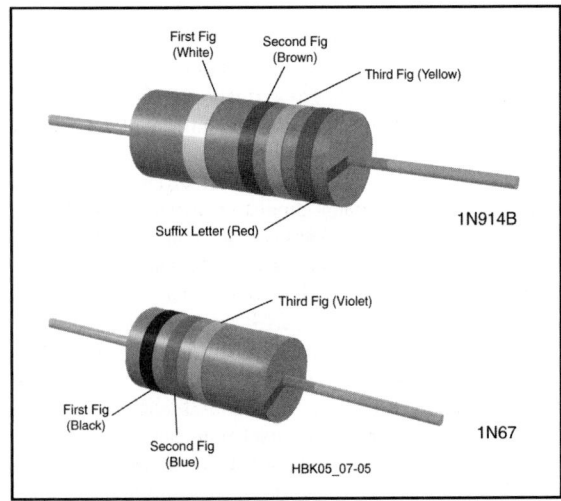

Fig 22.15 — Color-coding for semiconductor diodes. At A, the cathode is identified by the double-width first band. At B, the bands are grouped toward the cathode. Two-Fig designations are signified by a black first band. The color code is given in Table 22.4. The suffix-letter code is A-Brown, B-red, C-orange, D-yellow, E-green, F-blue. The 1N prefix is assumed.

Table 22.24
Package Dimensions for Small Signal, Rectifier and Zener diodes

Case	L	D	Ød	LS	PCB Pads*	Hole	Example
DO-35	0.166	0.080	0.020	0.30	0.056 dia	0.029	1N4148
DO-41	0.205	0.107	0.034	0.40	0.074 dia	0.040	1N4001
DO-201	0.283	0.189	0.048	0.65	0.150 dia	0.079	1N5401
DO-204	0.205	0.106	0.034	0.40	0.074 dia	0.040	1N4001

Dimensions in inches.
*Customary to make cathode lead square.

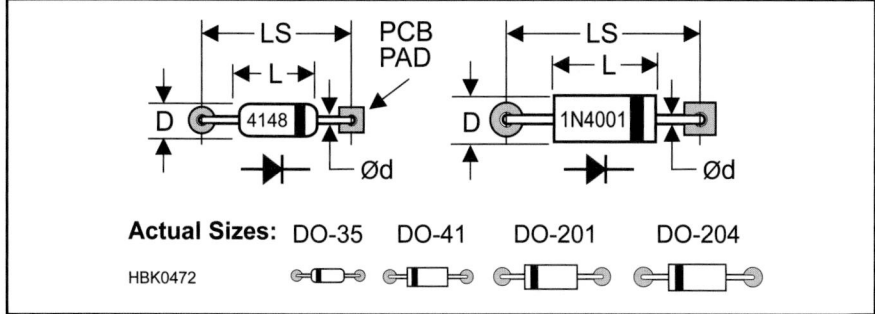

Fig 22.16 — Axial-leaded diode packages and pad dimensions.

commonly available. Full-wave bridge packages are labeled with tildes (~) for the ac inputs and + and – symbols for the rectifier outputs. High-power diodes are often packaged in TO-220 packages with two leads. The package may be labeled with a diode symbol, but if not you will have to obtain the manufacturer's data sheet to identify the anode and cathode leads.

The common 1N914, 1N4148, and 1N5767 switching diodes are suitable for most all small signal applications. The 1N4000 family is commonly used for ac voltage rectification up to 1000 volts PIV. Schottky diodes are used when low forward voltages are required, particularly at high-currents, and exhibit voltages of 0.1-0.2 V at low currents.

Zener diodes, used as voltage references, are manufactured in a wide range of voltages and power handling capacities. Power dissipation in a Zener diode is equal to the Zener voltage multiplied by the average reverse current. The Zener voltage has a significant temperature coefficient and also varies with reverse current. To avoid excessive variations in Zener voltage, limit the diode's power dissipation to no more than ½ of the rated value and for precision uses, ⅕ to ⅒ of the rated power dissipation is recommended. Common varieties of Zener diodes are listed in **Table 22.25**.

Voltage-variable capacitance diodes, also called Varicaps, varactors or tuning diodes, are used in oscillator and tuned circuits where a variable capacitor is needed. Operated with reverse bias, the depletion region forms a capacitor of variable width with a fairly linear voltage vs. capacitance function. Standard tuning diodes produce capacitances in the range of 5 to 40 pF. Hyper-abrupt tuning diodes produce variable capacitances to 100 pF or more for low frequency or wide tuning range applications. Some of the common voltage variable capacitance diodes are listed in **Table 22.26**.

Maximum capacitance occurs with minimum reverse voltage. As the reverse voltage is increased, the capacitance decreases. Tuning diodes are specified by the capacitance produced at two reverse voltages, usually 2 and 30 V. This is called the capacitance ratio and is specified in units of pF per volt. Beyond this range, capacitance change with voltage can become non-linear and may cause signal distortion.

All diodes exhibit some capacitance when reversed biased. Amateurs have learned to use reverse biased Zener and rectifier diodes to form tuning diodes with 20-30 pF maximum capacitance. These "poor man's" tuning diodes are widely used in homebrew projects. However, because the capacitance ratio varies widely from one diode to the next, requiring experimentation to find a suitable diode, they are seldom used in published construction articles.

Light emitting diodes (LED) are another common type of diode. The primary application is that of an illuminated visual indicator when forward biased. LEDs have virtually replaced miniature lamp bulbs for indicators and illumination. LED's are specified primar-

ily by their color, size, shape and output light intensity.

The "standard" size LED is the T-1¾; 5 mm or 0.20 in. diameter. The "miniature" size is the T-1; 3 mm or 0.125 in. diameter. Today, the "standard" and "miniature" size is a bit of a misnomer due to the wide variety of LED sizes and shapes, including SMT varieties. However, the T-1 and T-1¾ remain the most common for homebrew projects due to their inexpensive availability and ease of mounting with a simple panel hole. Their long leads are ideal for prototyping.

22.6.2 Transistors

The information in the tables of transistor data includes the most important parameters for typical applications of transistors of that type. The meaning of the parameters and their relationship to circuit design is covered in the **Analog Basics** chapter or in the references listed at the end of that chapter.

The tables are organized by application; small-signal, general-purpose, RF power, and so on. Some obsolete parts are listed in these tables for reference in repair and maintenance of older equipment. Before using a device in a new design, it is recommended that you check the manufacturer's Web site to be sure that the device has not been replaced by a more capable part and that it is available for future orders.

22.6.3 Voltage Regulators

For establishing a well-regulated fixed voltage reference, the linear voltage regular ICs are often preferred over the Zener diode. Three-terminal voltage regulators require no external components and most have internal current limiting and thermal shutdown circuitry, making them virtually indestructible. (Three-terminal regulators are described in the **Analog Basics** chapter.) The specifications and packages for common voltage regulators are listed in **Table 22.27**.

For fixed-voltage positive regulators, the 7800 family in the TO-220 package is the most common and reasonably priced. They are available in a variety of voltages and supply up to 1 A of current or more, depending on the input voltage. The part number identifies the voltage. For example, a 7805 is a 5-V regulator and a 7812 is a 12-V regulator. The 78L00 low-power versions in the TO-98 package or SOT-89 surface-mount package provide load currents up to 100 mA. The 317 and 340 are the most common adjustable-voltage regulators. Integrated voltage regulators can be used with a pass transistor to extend their load current capability as described in the device data sheets.

Three-terminal regulators are selected primarily for output voltage and maximum load

Table 22.25
Common Zener Diodes
Power dissipation and (package style)

Voltage (V)	¼ W (DO-35)	0.35 W (SOT-23)	½ W (SOD-123)	½ W (DO-35)	1 W (DO41)
2.7	1N4618	MMBZ5223B	MMSZ5223B	1N5223B	—
3.3	1N4620	MMBZ5226B	MMSZ5226B	1N5226B	1N4728A
3.6	1N4621	MMBZ5227B	MMSZ5227B	1N5227B	1N4729A
3.9	1N4622	MMBZ5228B	MMSZ5228B	1N5228B	1N4730A
4.3	1N4623	MMBZ5229B	MMSZ5229B	1N5229B	1N4731A
4.7	1N4624	MMBZ5230B	MMSZ5230B	1N5230B	1N4732A
5.1	1N4625	MMBZ5231B	MMSZ5231B	1N5231B	1N4733A
5.6	1N4626	MMBZ5232B	MMSZ5232B	1N5232B	1N4734A
6.0	— —	MMBZ5233B	MMSZ5233B	1N5233B	—
6.2	1N4627	MMBZ5234B	MMSZ5234B	1N5234B	1N4735A
6.8	1N4099	MMBZ5235B	MMSZ5235B	1N5235B	1N4736A
7.5	1N4100	MMBZ5236B	MMSZ5236B	1N5236B	1N4737A
8.2	1N4101	MMBZ5237B	MMSZ5237B	1N5237B	1N4738A
9.1	1N4103	MMBZ5239B	MMSZ5239B	1N5239B	1N4739A
10	1N4104	MMBZ5240B	MMSZ5240B	1N5240B	1N4740A
11	1N4105	MMBZ5241B	MMSZ5241B	1N5241B	1N4741A
12	— —	MMBZ5242B	MMSZ5242B	1N5242B	1N4742A
13	1N4107	MMBZ5243B	MMSZ5243B	1N5243B	1N4743A
15	1N4109	MMBZ5245B	MMSZ5245B	1N5245B	1N4744A
18	1N4112	MMBZ5248B	MMSZ5248B	1N5248B	1N4746A
20	1N4114	MMBZ5250B	MMSZ5250B	1N5250B	1N4747A
22	1N4115	MMBZ5251B	MMSZ5251B	1N5251B	1N4748A
24	1N4116	MMBZ5252B	MMSZ5252B	1N5252B	1N4749A
27	1N4118	MMBZ5254B	MMSZ5254B	1N5254B	1N4750A
28	1N4119	MMBZ5255B	MMSZ5255B	1N5255B	—
30	1N4120	MMBZ5256B	MMSZ5256B	1N5256B	1N4751A
33	1N4121	MMBZ5257B	MMSZ5257B	1N5257B	1N4752A
36	1N4122	MMBZ5258B	MMSZ5258B	1N5258B	1N4753A
39	1N4123			1N5259B	1N4754A
43	—			1N5260B	1N4755A
47	1N4125			1N5261B	1N4756A
51	1N4126			1N5262B	1N4757A
56				1N5263B	1N4758A
60				1N5264B	—
62				1N5265B	1N4759A
68				1N5266B	1N4760A
75				1N5267B	1N4761A
82					1N4762A
91					1N4763A
100					1N4764A

current. Dropout voltage — the minimum voltage between input and output for which regulation can be maintained — is also very important. For example, the dropout voltage for the 5-V 78L05 is 1.7 V. Therefore, the input voltage must be at least 6.7 V (5 + 1.7 V) to ensure output voltage regulation. The maximum input voltage should also not be exceeded.

Make sure to check the pin assignments for all voltage regulators. While the fixed-voltage positive regulators generally share a common orientation of input, output, and ground, negative-voltage and adjustable regulators do not. Installing a regulator with the wrong connections will usually destroy it and may allow excessive voltage to be applied to the circuit it supplies.

22.6.4 Analog and Digital Integrated Circuits

Integrated circuits (ICs) come in a variety of packages, including transistor-like metal cans, dual and single in-line packages (DIPs and SIPs), flat-packs and surface-mount packages. Most are marked with a part number and a four-digit manufacturer's date code indicating the year (first two digits) and week (last two digits) that the component was made. As mentioned in the introduction to this chapter, ICs are frequently house-marked and cross-reference directories can be helpful in identification and replacement. Another very useful reference tool for working with ICs is IC Master (**www.icmaster.com**), a master selection guide that organizes ICs by type, function and certain key parameters.

A part number index is included, along with application notes and manufacturer's information for millions of devices.

IC part numbers provide a complete description of the device's function and ratings. For example, a 4066 IC contains four independent CMOS SPST switches. The 4066 is a CMOS device available from a number of different manufacturers in different package styles and ratings. The two- or three-letter prefix of the part number is generally associated with the part manufacturer. Next, the part type (4066 in this case) shows the function and pin assignments or "pin outs." Following the part type is an alphabetic suffix that describes the version of the part, package code, temperature range, reliability rating and possibly other information. For

Table 22.26
Voltage-Variable Capacitance Diodes[†]

Listed numerically by device

Device	Nominal Capacitance pF ±10% @ V_R = 4.0 V f = 1.0 MHz	Capacitance Ratio 2-30 V Min.	Q @ 4.0 V 50 MHz Min.	Case Style	Device	Nominal Capacitance pF ±10% @ V_R = 4.0 V f = 1.0 MHz	Capacitance Ratio 2-30 V Min.	Q @ 4.0 V 50 MHz Min.	Case Style
1N5441A	6.8	2.5	450		1N5471A	39	2.9	450	
1N5442A	8.2	2.5	450		1N5472A	47	2.9	400	
1N5443A	10	2.6	400	DO-7	1N5473A	56	2.9	300	DO-7
1N5444A	12	2.6	400		1N5474A	68	2.9	250	
1N5445A	15	2.6	450		1N5475A	82	2.9	225	
1N5446A	18	2.6	350		1N5476A	100	2.9	200	
1N5447A	20	2.6	350		MV2101	6.8	2.5	450	TO-92
1N5448A	22	2.6	350	DO-7	MV2102	8.2	2.5	450	
1N5449A	27	2.6	350		MV2103	10	2.0	400	
1N5450A	33	2.6	350		MV2104	12	2.5	400	
1N5451A	39	2.6	300		MV2105	15	2.5	400	
1N5452A	47	2.6	250		MV2106	18	2.5	350	TO-92
1N5453A	56	2.6	200	DO-7	MV2107	22	2.5	350	
1N5454A	68	2.7	175		MV2108	27	2.5	300	
1N5455A	82	2.7	175		MV2109	33	2.5	200	
1N5456A	100	2.7	175		MV2110	39	2.5	150	
1N5461A	6.8	2.7	600		MV2111	47	2.5	150	TO-92
1N5462A	8.2	2.8	600		MV2112	56	2.6	150	
1N5463A	10	2.8	550	DO-7	MV2113	68	2.6	150	
1N5464A	12	2.8	550		MV2114	82	2.6	100	
1N5465A	15	2.8	550		MV2115	100	2.6	100	
1N5466A	18	2.8	500						
1N5467A	20	2.9	500						
1N5468A	22	2.9	500	DO-7					
1N5469A	27	2.9	500						
1N5470A	33	2.9	500						

[†]For package shape, size and pin-connection information, see manufacturers' data sheets.

complete information on the part — any or all of which may be significant to circuit function — use the Web sites of the various manufacturers or enter "data sheet" and the part number into an Internet search engine.

When choosing ICs that are not exact replacements, be wary of substituting "similar" devices, particularly in demanding applications, such as high-speed logic, sensitive receivers, precision instrumentation and similar devices. In particular, substitution of one type of logic family for another — even if the device functions and pin outs are the same — can cause a circuit to not function or function erratically, particularly at temperature extremes. For example, substituting LS TTL devices for HCMOS devices will result in mismatches between logic level thresholds. Substituting a lower-power IC may result in problems supplying enough output current. Even using a faster or higher clock-speed part can cause problems if signals change faster or propagate more quickly than the circuit was designed for. Problems of this sort can be extremely difficult to troubleshoot unless you are skilled in circuit design. When necessary, you can add interface circuits or buffer amplifiers that improve the input and output capabilities of replacement ICs, but auxiliary circuits cannot improve basic device ratings, such as speed or bandwidth. Whenever possible, substitute ICs that are guaranteed or "direct" replacements and that are listed as such by the manufacturer.

ICs are available in different operating temperature ranges. Three standard ranges are common:

- Commercial: 0 °C to 70 °C
- Industrial: –25 °C to 85 °C
- Automotive: –40 °C to 85 °C
- Military: –55°C to 125°C

In some cases, part numbers reflect the temperature ratings. For example, an LM301A op amp is rated for the commercial temperature range; an LM201A op amp for the industrial range and an LM101A for the military range. It is usually acceptable, all other things being equal, to substitute ICs rated for a wider temperature range, but there are often other performance differences associated with the devices meeting wider temperature specifications that should be evaluated before making the substitution.

Table 22.27

Three-Terminal Voltage Regulators

Listed numerically by device

Device	Description	Package	Voltage	Current (A)
317	Adj Pos	TO-205	+1.2 to +37	0.5
317	Adj Pos	TO-204,TO-220	+1.2 to +37	1.5
317L	Low Current Adj Pos	TO-205,TO-92	+1.2 to +37	0.1
317M	Med Current Adj Pos	TO-220	+1.2 to +37	0.5
338	Adj Pos	TO-3	+1.2 to +32	5.0
350	High Current Adj Pos	TO-204,TO-220	+1.2 to +33	3.0
337	Adj Neg	TO-205	−1.2 to −37	0.5
337	Adj Neg	TO-204,TO-220	−1.2 to −37	1.5
337M	Med Current Adj Neg	TO-220	−1.2 to −37	0.5
309		TO-205	+5	0.2
309		TO-204	+5	1.0
323		TO-204,TO-220	+5	3.0
140-XX	Fixed Pos	TO-204,TO-220	Note 1	1.0
340-XX		TO-204,TO-220		1.0
78XX		TO-204,TO-220		1.0
78LXX		TO-205,TO-92		0.1
78MXX		TO-220		0.5

Device	Description	Package	Voltage	Current (A)
78TXX		TO-204		3.0
79XX	Fixed Neg	TO-204,TO-220	Note 1	1.0
79LXX		TO-205,TO-92		0.1
79MXX		TO-220		0.5

Note 1—XX indicates the regulated voltage; this value may be anywhere from 1.2 V to 35 V. A 7815 is a positive 15-V regulator, and a 7924 is a negative 24-V regulator.

The regulator package may be denoted by an additional suffix, according to the following:

Package	Suffix
TO-204 (TO-3)	K
TO-220	T
TO-205 (TO-39)	H, G
TO-92	P, Z

For example, a 7812K is a positive 12-V regulator in a TO-204 package. An LM340T-5 is a positive 5-V regulator in a TO-220 package. In addition, different manufacturers use different prefixes. An LM7805 is equivalent to a μA7805 or MC7805.

Common Voltage Regulators — Fixed Positive Voltage

Device	Output Voltage (V)	Output Current (A)	Load Regulation (mV)	Dropout Voltage (V)	Min. Input Voltage (V)	Max. Input Voltage (V)
Surface Mount SOT-89 Case						
78L05ACPK	5.0	0.1	60	1.7	7.0	20
78L06ACPK	6.2	0.1	80	1.7	8.5	20
78L08ACPK	8.0	0.1	80	1.7	10.5	23
78L09ACPK	9.0	0.1	90	1.7	11.5	24
78L12ACPK	12	0.1	100	1.7	14.5	27
78L15ACPK	15	0.1	150	1.7	17.5	30
TO-92 Case						
78L33ACZ	3.3	0.1	60	1.7	5.0	30
78L05ACZ	5.0	0.1	60	1.7	7.0	30
78L06ACZ	6.0	0.1	60	1.7	8.5	30
78L08ACZ	8.0	0.1	80	1.7	10.5	30
78L09ACZ	9.0	0.1	80	1.7	11.5	30
78L12ACZ	12	0.1	100	1.7	14.5	35
78L15ACZ	15	0.1	150	1.7	17.5	35
TO-220 Case						
78M05CV	5.0	0.5	100	2.0	7.0	35
7805ACV	5.0	1.0	100	2.0	7.0	35
7806ACV	6.0	1.0	100	2.0	8.0	35
7808ACV	8.0	1.0	100	2.0	10.0	35
7809ACV	9.0	1.0	100	2.0	11.0	35
7812ACV	12	1.0	100	2.0	14.0	35
7815CV	15	1.0	300	2.0	17.0	35

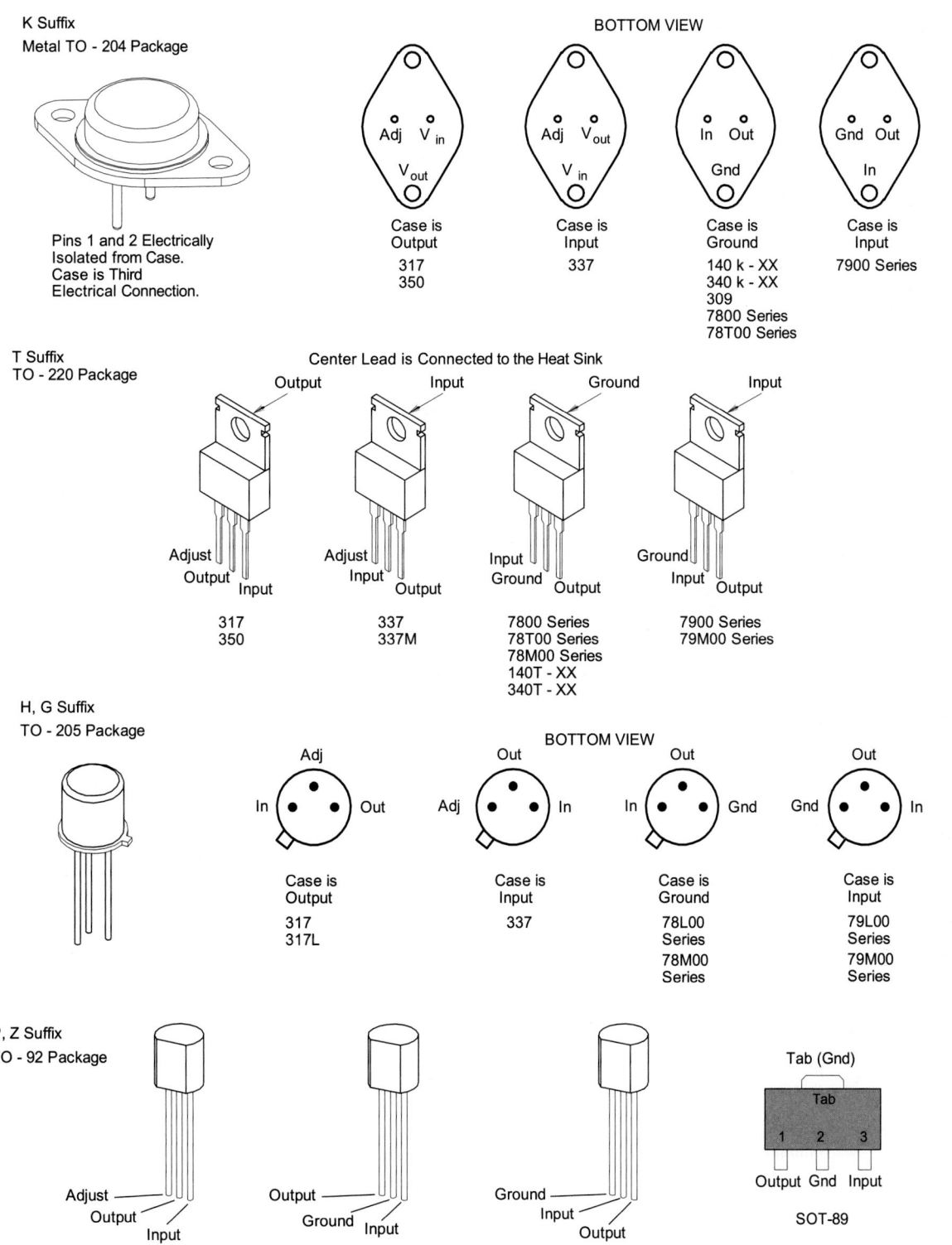

K Suffix
Metal TO - 204 Package

Pins 1 and 2 Electrically
Isolated from Case.
Case is Third
Electrical Connection.

BOTTOM VIEW

Adj   V$_{in}$

V$_{out}$

Case is
Output

317
350

Adj   V$_{out}$

V$_{in}$

Case is
Input

337

In   Out

Gnd

Case is
Ground

140 k - XX
340 k - XX
309
7800 Series
78T00 Series

Gnd   Out

In

Case is
Input

7900 Series

T Suffix
TO - 220 Package

Center Lead is Connected to the Heat Sink

Output

Adjust
Output   Input

317
350

Input

Adjust
Input   Output

337
337M

Ground

Input
Ground   Output

7800 Series
78T00 Series
78M00 Series
140T - XX
340T - XX

Input

Ground
Input   Output

7900 Series
79M00 Series

H, G Suffix
TO - 205 Package

BOTTOM VIEW

Adj

In   Out

Case is
Output

317
317L

Out

Adj   In

Case is
Input

337

Out

In   Gnd

Case is
Ground

78L00
Series
78M00
Series

Out

Gnd   In

Case is
Input

79L00
Series
79M00
Series

P, Z Suffix
TO - 92 Package

Adjust
Output
Input

317L

Output
Ground
Input

78L00 Series

Ground
Input
Output

79L00 Series

Tab (Gnd)

Tab

1   2   3

Output Gnd Input

SOT-89

HBK0478

Table 22.28
Monolithic 50-Ω Amplifiers (MMIC Gain Blocks)

Device	Case Style	Freq. Range (MHz)	Gain at 100 MHz (dB)	Gain at 1000 MHz (dB)	Output Power P1dB (dBm)	IP3 (dB)	NF (dB)	DC Conditions Vb @ Ib
Avago Technologies								
MSA-0386	B	dc-2400	12.5	11.9	+10.0	+23.0	6.0	5.0 V @ 35 mA
MSA-0486	B	dc-3200	16.4	15.9	+12.5	+25.5	6.5	5.3 V @ 50 mA
MSA-0505	A	dc-2300	7.8	7.0	+18.0	+29.0	6.5	8.4 V @ 80 mA
MSA-0611	C	dc-700	19.5	12.0	+2.0	+14.0	3.0	3.3 V @ 16 mA
MSA-0686	B	dc-800	20.0	16.0	+2.0	+14.5	3.0	3.5 V @ 16 mA
MSA-0786	B	dc-2000	13.5	12.6	+2.0	+14.5	3.0	3.5 V @ 16 mA
MSA-0886	B	dc-1000	32.5	22.4	+5.5	+19.0	5.5	4.0 V @ 22 mA
Mini-Circuits "ERA" Series								
ERA-1+	A, B	dc-8000	12.2	12.1	+11.7	+26.0	5.3	3.6 V @ 40 mA
ERA-2+	A, B	dc-6000	16.2	16.0	+12.8	+26.0	4.7	3.6 V @ 40 mA
ERA-3+	A, B	dc-3000	22.9	22.2	+12.1	+23.0	3.8	3.5 V @ 35 mA
ERA-4+	A, B	dc-4000	13.8	13.7	+17.0	+32.5	5.5	5.0 V @ 65 mA
ERA-5+	A, B	dc-4000	20.2	19.8	+18.4	+33.0	4.5	4.9 V @ 65 mA
ERA-6+	A, B	dc-4000	11.1	11.1	+18.5	+36.5	8.4	5.2 V @ 70 mA
Mini-Circuits "MAR" Series								
MAR-1SM+	A, B	dc-1000	18.5	15.5	+1.5	+14.0	5.5	5.0 V @ 17 mA
MAR-2SM+	A, B	dc-2000	12.5	12.0	+4.5	+17.0	6.5	5.0 V @ 25 mA
MAR-3SM+	A, B	dc-2000	12.5	12.0	+10.0	+23.0	6.0	5.0 V @ 35 mA
MAR-4SM+	A, B	dc-1000	8.3	8.0	+12.5	+25.5	7.0	5.3 V @ 50 mA
MAR-6SM+	A, B	dc-2000	20.0	16.0	+2.0	+14.5	3.0	3.5 V @ 16 mA
MAR-7SM+	A, B	dc-2000	13.5	12.5	+5.5	+19.0	5.0	4.0 V @ 22 mA
MAR-8SM+	A, B	dc-1000	32.5	22.5	+12.5	+27.0	3.3	7.8 V @ 36 mA
Mini-Circuits "VAM" Series								
VAM-3+	C	dc-2000	11.5	11.0	+9.0	+22.0	6.0	4.7 V @ 35 mA
VAM-6+	C	dc-2000	19.5	15.0	+2.0	+14.0	3.0	3.3 V @ 16 mA
VAM-7+	C	dc-2000	13.0	12.0	+5.5	+18.0	5.0	3.8 V @ 22 mA
Mini–Circuits "GALI" Series								
GALI-1+	D	dc-8000	12.7	12.5	+10.5	+27.0	4.5	3.4 V @ 40 mA
GALI-2+	D	dc-8000	16.2	15.8	+12.9	+27.0	4.6	3.5 V @ 40 mA
GALI-3+	D	dc-3000	22.4	21.1	+12.5	+25.0	3.5	3.3 V @ 35 mA
GALI-39+	D	dc-7000	20.8	21.1	+10.5	+22.9	2.4	3.5 V @ 35 mA
GALI-4+	D	dc-4000	14.4	14.1	+17.5	+34.0	4.0	4.6 V @ 65 mA
GALI-5+	D	dc-4000	20.6	19.4	+18.0	+35.0	3.5	4.4 V @ 65 mA
GALI-6+	D	dc-4000	12.2	12.2	+18.2	+35.5	4.5	5.0 V @ 70 mA
GALI-S66+	D	dc-3000	22.0	20.3	+2.8	+18.0	2.7	3.5 V @ 16 mA
Mini–Circuits "RAM" Series								
RAM–1+	B	dc-1000	19.0	15.5	+1.5	+14.0	5.5	5.0 V @ 17 mA
RAM–2+	B	dc-2000	12.5	11.8	+4.5	+17.0	6.5	5.0 V @ 25 mA
RAM–3+	B	dc-2000	12.5	12.0	+10.0	+23.0	6.0	5.0 V @ 35 mA
RAM–4+	B	dc-1000	8.5	8.0	+12.5	+25.5	6.5	5.3 V @ 50 mA
RAM–6+	B	dc-2000	20.0	16.0	+2.0	+14.5	2.8	3.5 V @ 16 mA
RAM–7+	B	dc-2000	13.5	12.5	+5.5	+19.0	4.5	4.0 V @ 22 mA
RAM–8+	B	dc-1000	32.5	23.0	+12.5	+27.0	3.0	7.8 V @ 36 mA

Avago — **www.avagotech.com**
Mini-Circuits Labs — **www.minicircuits.com**

22.6.5 MMIC Amplifiers

Monolithic microwave integrated circuit (MMIC) amplifiers are single-supply 50-Ω wideband gain blocks offering high dynamic range for output powers to about +15 dBm. MMIC amplifiers are becoming increasingly popular in homebrew communications circuits. With bandwidths over 1 GHz, they are well suited for HF, VHF, UHF and lower microwave frequencies.

MMIC amplifiers produce power gains from 10 dB to 30 dB. They also have a high third-order intercept point (IP3), usually in the +20 to +30 dBm range, easing the concerns about amplifier compression for most applications. They are used for RF and IF amplifiers, local oscillator amplifiers, transmitter drivers, and other medium power applications in 50-Ω systems. MMICs are especially well suited for driving 50-Ω double-balanced mixers (DBM). **Fig 22.17** shows the typical circuit arrangement for most MMIC amplifiers.

MMICs are available in a variety of packages, mostly surface mount as shown in **Fig 22.18**, requiring very few external components. Vendor data sheets and application notes, found on the manufacturer's Web sites, should be used for the proper selection of the biasing resistor, coupling capacitors, and other design criteria. Some of the popular MMIC amplifiers are listed in **Table 22.28**.

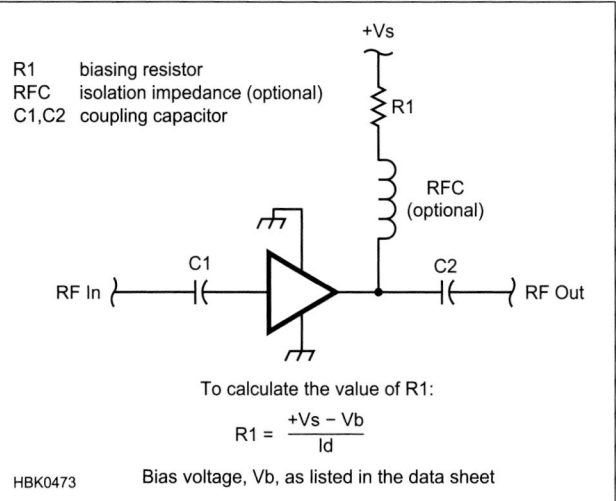

HBK0473

Fig 22.17 — MMIC application.

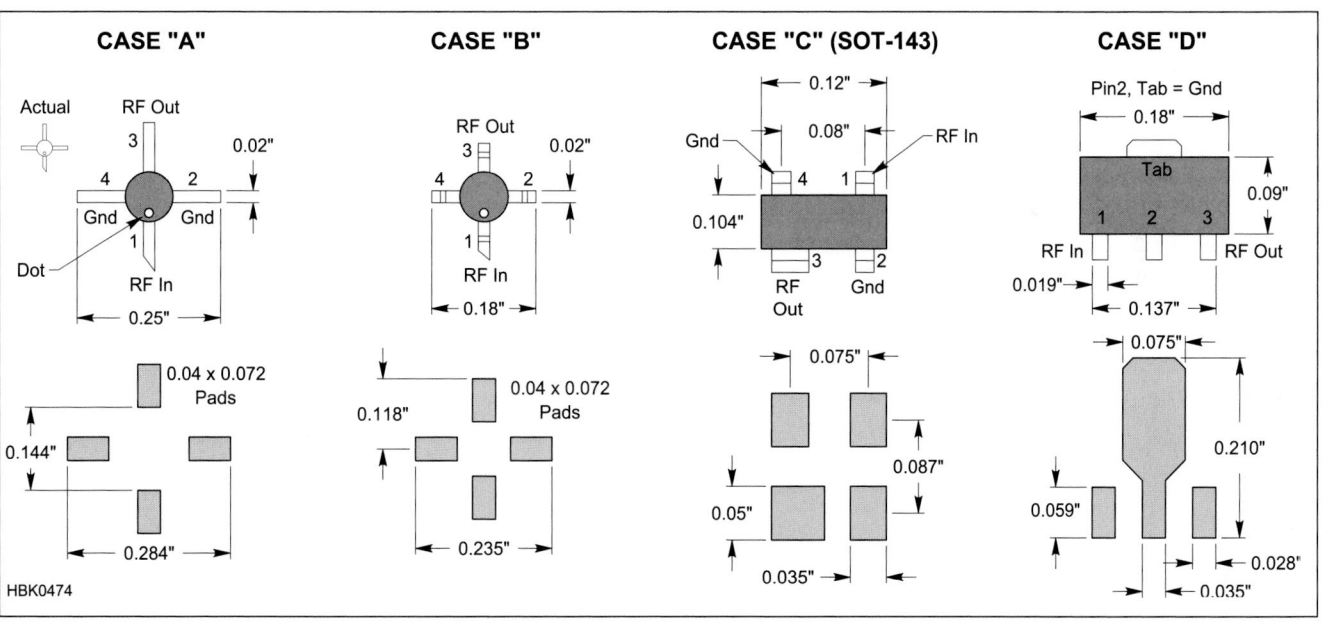

Fig 22.18 — MMIC package styles.

The main disadvantage of MMIC amplifiers is their relatively high current demands, usually in the 30 mA to 80 mA range per device, making them unsuitable for battery-powered portable equipment. On the other hand, the high current demand is what establishes their high gain and high IP3 characteristics with 50-Ω loads.

Another disadvantage is their wide gain-bandwidth. Their gain should be band-limited by input and output tuned circuits or filters to reduce the gain outside the desired ranges. For example, for an HF amplifier, 30 MHz low-pass filters can be used to reduce the gain outside the HF spectrum, or a band-pass filter used for the frequency band of interest.

Selecting the proper MMIC amplifier is fairly straightforward. First, select a device for the desired frequency bandwidth, gain, and output power. Ensure device current is compatible with the design application. Calculate the value for the bias resistor (R1 in Fig 22.17) based on the biasing voltage (V_b) listed in Table 22.28 and whatever value of supply voltage (V_s) is available.

With increasing availability and ease of use, there are many circuits where MMIC amplifiers can be used. There are many MMIC amplifiers that are relatively inexpensive for hobby use.

Table 22.29
Small-Signal FETs

Device	Type	Max Diss (mW)	Max V_{DS} (V)	$V_{GS(off)}$ (V)	Min gfs (µS)	Input C (pF)	Max ID (mA)[1]	f_{max} (MHz)	Noise Figure (typ)	Case	Base	Applications
2N4416	N-JFET	300	30	−6	4500	4	−15	450	4 dB @400 MHz	TO-72	1	VHF/UHF amp, mix, osc
2N5484	N-JFET	310	25	−3	2500	5	30	200	4 dB @200 MHz	TO-92	2	VHF/UHF amp, mix, osc
2N5485	N-JFET	310	25	−4	3500	5	30	400	4 dB @400 MHz	TO-92	2	VHF/UHF amp, mix, osc
2N5486	N-JFET	360	25	−2	5500	5	15	400	4 dB @400 MHz	TO-92	2	VHF/UHF amp. mix, osc
3N200 NTE222 SK3065	N-dual-gate MOSFET	330	20	−6	10,000	4-8.5	50	500	4.5 dB @400 MHz	TO-72	3	VHF/UHF amp, mix, osc
3N202 NTE454 SK3991	N-dual-gate MOSFET	360	25	−5	8000	6	50	200	4.5 dB @200 MHz	TO-72	3	VHF amp, mixer
MPF102 NTE451 SK9164	N-JFET	310	25	−8	2000	4.5	20	200	4 dB @400 MHz	TO-92	2	HF/VHF amp, mix, osc
MPF106 2N5484	N-JFET	310	25	−6	2500	5	30	400	4 dB @200 MHz	TO-92	2	HF/VHF/UHF amp, mix, osc
40673 NTE222 SK3050	N-dual-gate MOSFET	330	20	−4	12,000	6	50	400	6 dB @200 MHz	TO-72	3	HF/VHF/UHF amp, mix, osc
U304	P-JFET	350	−30	+10	27		−50	—	—	TO-18	4	analog switch chopper
U310	N-JFET	500 300	30 30	−6	10,000	2.5	60	450	3.2 dB @450 MHz	TO-52	5	common-gate VHF/UHF amp,
U350	N-JFET Quad	1W	25	−6	9000	5	60	100	7 dB @100 MHz	TO-99	6	matched JFET doubly bal mix
U431	N-JFET Dual	300	25	−6	10,000	5	30	100	—	TO-99	7	matched JFET cascode amp and bal mix
2N5670	N-JFET	350	25	8	3000	7	20	400	2.5 dB @100 MHz	TO-92	2	VHF/UHF osc, mix, front-end amp
2N5668	N-JFET	350	25	4	1500	7	5	400	2.5 dB @100 MHz	TO-92	2	VHF/UHF osc, mix, front-end amp
2N5669	N-JFET	350	25	6	2000	7	10	400	2.5 dB @100 MHz	TO-92	2	VHF/UHF osc, mix, front-end amp
J308	N-JFET	350	25	6.5	8000	7.5	60	1000	1.5 dB @100 MHz	TO-92	2	VHF/UHF osc, mix, front-end amp
J309	N-JFET	350	25	4	10,000	7.5	30	1000	1.5 dB @100 MHz	TO-92	2	VHF/UHF osc, mix, front-end amp
J310	N-JFET	350	25	6.5	8000	7.5	60	1000	1.5 dB @100 MHz	TO-92	2	VHF/UHF osc, mix, front-end amp
NE32684A	HJ-FET	165	2.0	−0.8	45,000	—	30	20 GHz	0.5 dB @12 GHz	84A		Low-noise amp

Notes:

[1] 25°C.

For package shape, size and pin-connection information, see manufacturers' data sheets.

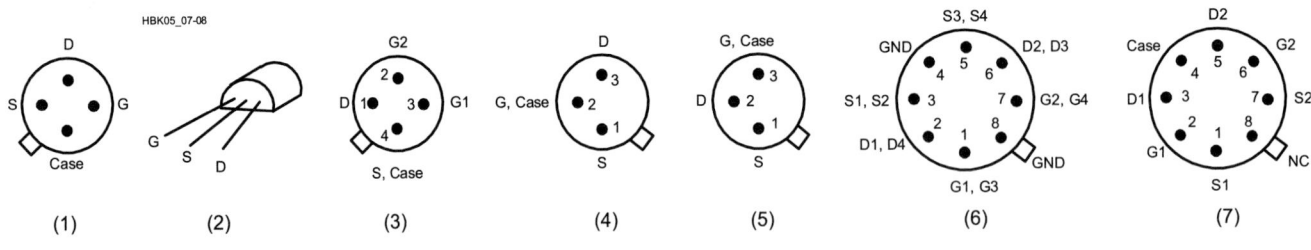

(1) (2) (3) (4) (5) (6) (7)

Table 22.30

Low-Noise Bipolar Transistors

Device	NF (dB)	F (MHz)	f_T (GHz)	I_C (mA)	Gain (dB)	F (MHz)	$V_{(BR)CEO}$ (V)	I_C (mA)	P_T (mW)	Case
MRF904	1.5	450	4	15	16	450	15	30	200	TO-206AF
MRF571	1.5	1000	8	50	12	1000	10	70	1000	Macro-X
MRF2369	1.5	1000	6	40	12	1000	15	70	750	Macro-X
MPS911	1.7	500	7	30	16.5	500	12	40	625	TO-226AA
MRF581A	1.8	500	5	75	15.5	500	15	200	2500	Macro-X
BFR91	1.9	500	5	30	16	500	12	35	180	Macro-T
BFR96	2	500	4.5	50	14.5	500	15	100	500	Macro-T
MPS571	2	500	6	50	14	500	10	80	625	TO-226AA
MRF581	2	500	5	75	15.5	500	18	200	2500	Macro-X
MRF901	2	1000	4.5	15	12	1000	15	30	375	Macro-X
MRF941	2.1	2000	8	15	12.5	2000	10	15	400	Macro-X
MRF951	2.1	2000	7.5	30	12.5	2000	10	100	1000	Macro-X
BFR90	2.4	500	5	14	18	500	15	30	180	Macro-T
MPS901	2.4	900	4.5	15	12	900	15	30	300	TO-226AA
MRF1001A	2.5	300	3	90	13.5	300	20	200	3000	TO-205AD
2N5031	2.5	450	1.6	5	14	450	10	20	200	TO-206AF
MRF4239A	2.5	500	5	90	14	500	12	400	3000	TO-205AD
BFW92A	2.7	500	4.5	10	16	500	15	35	180	Macro-T
MRF521*	2.8	1000	4.2	−50	11	1000	−10	−70	750	Macro-X
2N5109	3	200	1.5	50	11	216	20	400	2500	TO-205AD
2N4957*	3	450	1.6	−2	12	450	−30	−30	200	TO-206AF
MM4049*	3	500	5	−20	11.5	500	−10	−30	200	TO-206AF
2N5943	3.4	200	1.5	50	11.4	200	30	400	3500	TO-205AD
MRF586	4	500	1.5	90	9	500	17	200	2500	TO-205AD
2N5179	4.5	200	1.4	10	15	200	12	50	200	TO-206AF
2N2857	4.5	450	1.6	8	12.5	450	15	40	200	TO-206AF
2N6304	4.5	450	1.8	10	15	450	15	50	200	TO-206AF
MPS536*	4.5	500	5	−20	4.5	500	−10	−30	625	TO-226AA
MRF536*	4.5	1000	6	−20	10	1000	−10	−30	300	Macro-X

*denotes a PNP device

Complementary devices

NPN	PNP
2N2857	2N4957
MRF904	MM4049
MRF571	MRF521

For package shape, size and pin-connection information, see manufacturers' data sheets. Many retail suppliers and manufacturers offer data sheets on their Web sites.

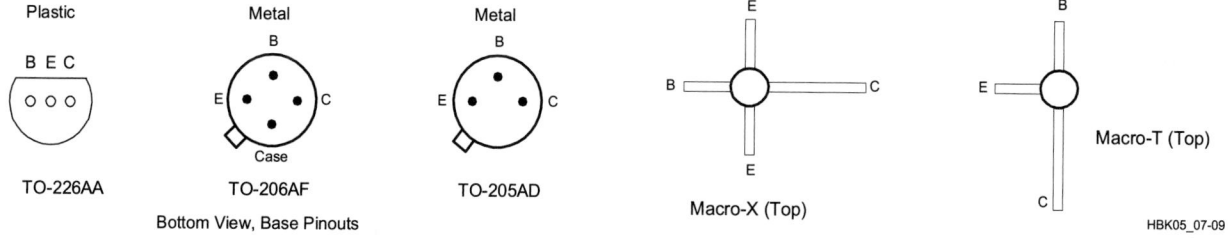

Bottom View, Base Pinouts

Device	Type	V_{CEO} Maximum Collector Emitter Voltage (V)	V_{CBO} Maximum Collector Base Voltage (V)	V_{EBO} Maximum Emitter Base Voltage (V)	I_C Maximum Collector Current (mA)	P_O Maximum Device Dissipation (W)	**Minimum DC Current Gain** $I_C = 0.1\ mA$	$I_C = 150\ mA$	Current-Gain Bandwidth Product f_T (MHz)	Noise Figure NF Maximum (dB)	Base
2N918	NPN	15	30	3.0	50	0.2	20 (3 mA)	—	600	6.0	3
2N2102	NPN	65	120	7.0	1000	1.0	20	40	60	6.0	2
2N2218	NPN	30	60	5.0	800	0.8	20	40	250		2
2N2218A	NPN	40	75	6.0	800	0.8	20	40	250		2
2N2219	NPN	30	60	5.0	800	3.0	35	100	250		2
2N2219A	NPN	40	75	6.0	800	3.0	35	100	300	4.0	2
2N2222	NPN	30	60	5.0	800	1.2	35	100	250		2
2N2222A	NPN	40	75	6.0	800	1.2	35	100	200	4.0	2
2N2905	PNP	40	60	5.0	600	0.6	35	—	200		2
2N2905A	PNP	60	60	5.0	600	0.6	75	100	200		2
2N2907	PNP	40	60	5.0	600	0.4	35	—	200		2
2N2907A	PNP	60	60	5.0	600	0.4	75	100	200		2
2N3053	NPN	40	60	5.0	700	5.0	—	50	100		2
2N3053A	NPN	60	80	5.0	700	5.0	—	50	100		2
2N3563	NPN	15	30	2.0	50	0.6	20	—	800		1
2N3904	NPN	40	60	6.0	200	0.625	40	—	300	5.0	1
2N3906	PNP	40	40	5.0	200	0.625	60	—	250	4.0	1
2N4037	PNP	40	60	7.0	1000	5.0	—	50			2
2N4123	NPN	30	40	5.0	200	0.35	—	25 (50 mA)	250	6.0	1
2N4124	NPN	25	30	5.0	200	0.35	120 (2 mA)	60 (50 mA)	300	5.0	1
2N4125	PNP	30	30	4.0	200	0.625	50 (2 mA)	25 (50 mA)	200	5.0	1
2N4126	PNP	25	25	4.0	200	0.625	120 (2 mA)	60 (50 mA)	250	4.0	1
2N4401	NPN	40	60	6.0	600	0.625	20	100	250		1
2N4403	PNP	40	40	5.0	600	0.625	30	100	200		1
2N5320	NPN	75	100	7.0	2000	10.0	—	30 (I A)			2
2N5415	PNP	200	200	4.0	1000	10.0	—	30 (50 mA)	15		2
MM4003	PNP	250	250	4.0	500	1.0	20 (10 mA)	—			2
MPSA55	PNP	60	60	4.0	500	0.625	—	50 (0.1 A)	50		1
MPS6531	NPN	40	60	5.0	600	0.625	60 (10 mA)	90 (0.1 A)			1
MPS6547	NPN	25	35	3.0	50	0.625	20 (2 mA)	—	600		1

Test conditions: $I_C = 20$ mA dc; $V_{CE} = 20$ V; f = 100 MHz

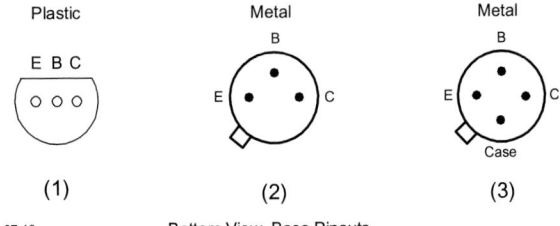

Plastic

E B C

(1)

Metal

B

E   C

(2)

Metal

B

E   C

Case

(3)

HBK05_07-10

Bottom View, Base Pinouts

Table 22.32

General Purpose Silicon Bipolar Power Transistors

TO-220 Case, Pin 1=Base, Pin 2, Case = Collector; Pin 3 = Emitter

TO-204 Case (TO-3), Pin 1=Base, Pin 2 = Emitter, Case = Collector;

NPN	PNP	I_C Max (A)	V_{CEO} Max (V)	h_{FE} Min	F_T (MHz)	Power Dissipation (W)
D44C8		4	60	100/220	50	30
	D45C8	4	60	40/120	50	30
TIP29		1	40	15/75	3	30
	TIP30	1	40	15/75	3	30
TIP29A		1	50	15/75	3	30
	TIP30A	1	60	15/75	3	30
TIP29B		1	80	15/75	3	30
TIP29C		1	100	15/75	3	30
	TIP30C	1	100	15/75	3	30
TIP47		1	250	30/150	10	40
TIP48		1	300	30/150	10	40
TIP49		1	350	30/150	10	40
TIP50		1	400	30/150	10	40
TIP110*		2	60	500	> 5	50
	TIP115*	2	60	500	> 5	50
TIP116		2	80	500	25	50
TIP31		3	40	25	3	40
	TIP32	3	40	25	3	40
TIP31A		3	60	25	3	40
	TIP32A	3	60	25	3	40
TIP31B		3	80	25	3	40
	TIP32B	3	80	25	3	40
TIP31C		3	100	25	3	40
	TIP32C	3	100	25	3	40
2N6124		4	45	25/100	2.5	40
2N6122		4	60	25/100	2.5	40
MJE1300		4	300	6/30	4	60
TIP120*		5	60	1000	> 5	65
	TIP125*	5	60	1000	> 10	65
	TIP42	6	40	15/75	3	65
TIP41A		6	60	15/75	3	65
TIP41B		6	80	15/75	3	65
2N6290		7	50	30/150	4	40
	2N6109		50	30/150	4	40
2N6292		7	70	30/150	4	40
	2N6107	7	70	30/150	4	40
MJE3055T		10	50	20/70	2	75
	MJE2955T	10	60	20/70	2	75
2N6486		15	40	20/150	5	75
2N6488		15	80	20/150	5	75
TIP140*		10	60	500	> 5	125
	TIP145*	10	60	600	> 10	125
2N3055A		15	60	20/70	0.8	115

NPN	PNP	I_C Max (A)	V_{CEO} Max (V)	h_{FE} Min	F_T (MHz)	Power Dissipation (W)
2N3055		15	60	20/70	2.5	115
	MJ2955	15	60	20/70	2.5	115
2N6545		8	400	7/35	6	125
2N5039		20	75	20/100	—	140
2N3771		30	40	15	0.2	150
2N3789		10	60	15	4	150
2N3715		10	60	30	4	150
	2N3791	10	60	30	4	150
	2N5875	10	60	20/100	4	150
	2N3790	10	80	15	4	150
2N3716		10	80	30	4	150
	2N3792	10	80	30	4	150
2N3773		16	140	15/60	4	150
2N6284		20	100	750/18K	—	160
	2N6287	20	100	750/18K	—	160
2N5881		15	60	20/100	4	160
2N5880		15	80	20/100	4	160
2N6249		15	200	10/50	2.5	175
2N6250		15	275	8/50	2.5	175
2N6546		15	300	6/30	6-28	175
2N6251		15	350	6/50	2.5	175
2N5630		16	120	20/80	1	200
2N5301		30	40	15/60	2	200
2N5303		20	80	15/60	2	200
2N5885		25	60	20/100	4	200
2N5302		30	60	15/60	2	200
	2N4399	30	60	15/60	4	200
2N5886		25	80	20/100	4	200
	2N5884	25	80	20/100	4	200
MJ802		30	100	25/100	2	200
	MJ4502	30	100	25/100	2	200
MJ15003		20	140	25/150	2	250
	MJI5004	20	140	25/150	2	250
MJ15024		25	250	15/60	4	250

▨ = Complimentary pairs

* = Darlington transistor

Useful URLs for finding transistor/IC data sheets:

1. Line of replacement transistors and ICs: **www.nteinc.com**
2. General-purpose replacements: **www.mouser.com**, **www.digikey.com**
3. NXP Semiconductors: **www.nxp.com**
4. Mitsubishi: **www.mitsubishielectric.com/semiconductors/php/eSearch.php**
5. ON Semiconductor: **www.onsemi.com**
6. M/A-COM: **www.macomtech.com**
7. Freescale: **www.freescale.com**
8. STMicroelectronics: **www.st.com**
9. Microsemi: **www.microsemi.com**

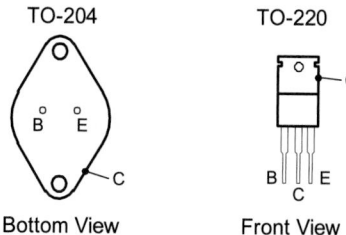

TO-204

Bottom View

TO-220

Front View

HBK05_07-11

Table 22.33
General Purpose JFETs and MOSFETs

Device	Type	VDSS min (V)	RDS(on) max (Ω)	ID max (A)	PD max (W)	Case†	Mfr
BS250P	P-channel	45	14	0.23	0.7	E-line	Z
IRFZ30	N-channel	50	0.050	30	75	TO-220	IR
IRFZ42	N-channel	50	0.035	50	150	TO-220	IR
2N7000	N-channel	60	5	0.20	0.4	E-line	Z
VN10LP	N-channel	60	7.5	0.27	0.625	E-line	Z
VN10KM	N-channel	60	5	0.3	1	TO-237	S
ZVN2106B	N-channel	60	2	1.2	5	TO-39	Z
IRF511	N-channel	60	0.6	2.5	20	TO-220AB	IR
IRF531	N-channel	60	0.180	14	75	TO-220AB	IR
IRF531	N-channel	80	0.160	14	79	TO-220	IR
ZVP3310A	P-channel	100	20	0.14	0.625	E-line	Z
ZVN2110B	N-channel	100	4	0.85	5	TO-39	Z
ZVP3310B	P-channel	100	20	0.3	5	TO-39	Z
IRF510	N-channel	100	0.6	2	20	TO-220AB	IR
IRF520	N-channel	100	0.27	5	40	TO-220AB	IR
IRF150	N-channel	100	0.055	40	150	TO-204AE	IR
IRFP150	N-channel	100	0.055	40	180	TO-247	IR
ZVP1320A	P-channel	200	80	0.02	0.625	E-line	Z
ZVN0120B	N-channel	200	16	0.42	5	TO-39	Z
ZVP1320B	P-channel	200	80	0.1	5	TO-39	Z
IRF620	N-channel	200	0.800	5	40	TO-220AB	IR
IRF220	N-channel	200	0.400	8	75	TO-220AB	IR
IRF640	N-channel	200	0.18	10	125	TO-220AB	IR

Manufacturers: IR = International Rectifier; M = Motorola; S = Siliconix; Z = Zetex.

†For package shape, size and pin-connection information, see manufacturers' data sheets. Many retail suppliers offer data sheets to buyers free of charge on request. Data books are available from many manufacturers and retailers.

Table 22.34
RF Power Transistors — By Part Number

Part Number	P_O (W)	Type	Gain (dB)	V_{DD} (V)	Package	f(MHz)	BV_{DSS}	P_D max (W)	Mfr
ARF1500	750	MOS	16	125	T1	40	500		MS
ARF1501	750	MOS	17	250	T1	40	1000		MS
ARF1505	750	MOS	17	300	T1	40	1200		MS
ARF460AG	150	MOS	15	125	TO-247s	65	500		MS
ARF461AG	150	MOS	15	250	TO-247s	65	1000		MS
ARF463AG	100	MOS	15	125	TO-247s	100	500		MS
ARF465AG	150	MOS	15	300	TO-247s	60	1200		MS
ARF466AG	300	MOS	16	200	TO-247s	45	1000		MS
ARF466FL	300	MOS	16	200	T3	45	1000		MS
ARF473	300	MOS	14	165	M244	150	500		MS
ARF475FL	450	MOS	14	165	T3	150	500		MS
ARF476FL	450	MOS	14	165	T3	150	500		MS
ARF477FL	400	MOS	16	165	T3a	100	500		MS
ARF521	150	MOS	15	165	M174	150	500		MS
BLF1043	10	MOS	16.5	26	SOT538A	1000			NXP
BLF1046	45	MOS	14	26	SOT467C	1000			NXP
BLF145	30	MOS	20	28	SOT123A	30			NXP
BLF147	150	MOS	14	28	SOT121B	175			NXP
BLF174XR	600	LDMOS	28.5	50	SOT1214A	128			NXP
BLF175	30	MOS	20	50	SOT123A	108			NXP
BLF177	150	MOS	19	50	SOT121B	108			NXP
BLF202	2	MOS	13	12.5	SOT409A	175			NXP
BLF242	5	MOS	16	28	SOT123A	200			NXP
BLF244	15	MOS	17	28	SOT123A	175			NXP
BLF245	30	MOS	15.5	28	SOT123A	175			NXP
BLF245B	30	MOS	18	28	SOT279A	175			NXP
BLF246	80	MOS	18	28	SOT121B	175			NXP
BLF246B	60	MOS	19	28	SOT161A	175			NXP
BLF278	300	MOS	16	50	SOT262A1	225			NXP
BLF369	500	LDMOS	18	32	SOT800-2	500			NXP
BLF571	20	LDMOS	27.5	50	SOT467C	500			NXP

Table 22.34 continued

Part Number	P_O (W)	Type	Gain (dB)	V_{DD} (V)	Package	f(MHz)	BV_{DSS}	P_D max (W)	Mfr
BLF573	300	LDMOS	27.2	50	SOT502A	500			NXP
BLF573S	300	LDMOS	27.2	50	SOT502B	500			NXP
BLF574	600	LDMOS	26.5	50	SOT539A	500			NXP
BLF574XR	600	LDMOS	23	50	SOT1214A	500			NXP
BLF578	300	LDMOS	26	50	SOT539A	500			NXP
BLF642	35	LDMOS	19	32	SOT467C	1400			NXP
BLF645	100	LDMOS	18	32	SOT540A	1400			NXP
BLF871	100	LDMOS	21	40	SOT467C	1000			NXP
BLF871S	100	LDMOS	21	40	SOT467B	1000			NXP
BLF881	140	LDMOS	21	50	SOT467C	1000			NXP
BLF881S	140	LDMOS	21	50	SOT467B	1000			NXP
MRF141	150	MOS	21	28	M174	175			MA
MRF141G	300	MOS	21	28	M244	175			MA
MRF148A	30	MOS	18	50	M113	175			MA
MRF150	150	MOS	20	50	M174	150			MA
MRF151	150	MOS	21	50	M177	175			MA
MRF151G	300	MOS	20	50	M244	175			MA
MRF154	600	MOS	16	50	HOG	80			MA
MRFE6VP100H	100	LDMOS	27.2	50	Flange	0 to 2000			FR
MRFE6VP5600H	600	LDMOS	24.6	50	Flange	1.8 to 600			FR
MRFE6VP61K25H	1250	LDMOS	22.9	50	Flange	1.8 to 600			FR
MRFE6VP6300H	300	LDMOS	25	50	Flange	1.8 to 600			FR
RD00HHS1	0.3	LDMOS	18.7	12.5	SOT-89	30	30	3.1	MT
RD00HVS1	0.5	LDMOS	20	12.5	SOT-89	175	30	3.1	MT
RD06HHF1	6	LDMOS	16	12.5	TO-220S	30	50	27.8	MT
RD06HVF1	6	LDMOS	16	12.5	TO-220S	175	50	27.8	MT
RD100HHF1	100	LDMOS	14	12.5	Flange large	30	50	176.5	MS
RD15HVF1	15	LDMOS	12	12.5	TO-220S	520	30	48	MT
RD16HHF1	16	LDMOS	16	12.5	TO-220S	30	50	56.8	MT
RD20HMF1	20	LDMOS	8.5	12.5	Flange small	900	30	71.4	MT
RD30HUF1	30	LDMOS	10	12.5	Flange small	520	30	75	MT
RD30HVF1	30	LDMOS	15	12.5	Flange small	175	30	75	MT
RD45HMF1	45	LDMOS	8	12.5	Flange large	900	30	125	MT
RD60HUF1	60	LDMOS	10	12.5	Flange large	520	30	150	MT
RD70HHF1	70	LDMOS	14	12.5	Flange large	30	50	150	MS
RD70HVF1	70	LDMOS	12	12.5	Flange large	175	30	150	MT
SD1274-01	30	BJT	10	13.6	M113	160			ST
SD1275-01	40	BJT	9	13.6	M113	160			ST
SD1726	150	BJT	14	50	M174	30			ST
SD1728	250	BJT	14.5	50	M177	30			ST
SD2902	15	BJT	12.5	28	M113	400			ST
SD2904	30	BJT	9.5	28	M113	400			ST
SD2918	30	MOS	18	50	M113	30			ST
SD2931-10	150	MOS	14	50	M174	175			ST
SD2932	300	MOS	15	50	M244	175			ST
SD2933	300	MOS	20	50	M177	30			ST
SD2941-10	175	MOS	15	50	M174	175			ST
SD2942	350	MOS	15	50	M244	175			ST
SD2943	350	MOS	22	50	M177	30			ST
SD3931-10	175	MOS	20	100	M174	150			ST
SD3932	350	MOS	24	100	M244	150			ST
SD3933	350	MOS	25	100	M177	30			ST
SD4931	150	MOS	14.8	50	M174	175			ST
SD4933	300	MOS	24	50	M177	30			ST
VRF141	150	MOS	13	28	M174	175	80		MS
VRF141G	300	MOS	14	28	M244	175	80		MS
VRF148A	30	MOS	16	50	M113	175	170		MS
VRF150	150	MOS	11	50	M174	150	170		MS
VRF151	150	MOS	14	50	M174	175	170		MS
VRF151E	150	MOS	14	50	M174	175	170		MS
VRF151G	300	MOS	16	50	M244	175	170		MS
VRF152	150	MOS	14	50	M174	175	170		MS
VRF154FL	600	MOS	17	50	T2	80	170		MS
VRF157FL	600	MOS	21	50	T2	80	170		MS
VRF2933	300	MOS	22	50	M177	100	170		MS

Manufacturer codes:
1. FR – Freescale: www.freescale.com
2. MA – M/A-COM: www.macomtech.com
3. MS - Microsemi: www.microsemi.com
4. MT – Mitsubishi: www.mitsubishielectric.com/semiconductors/php/eSearch.php
5. NXP – NXP Semiconductors: www.nxp.com
6. ST – STMicroelectronics: www.st.com

Table 22.35
RF Power Transistors — By Frequency and Power Output

Part Number	P_O (W)	Type	Gain (dB)	V_{DD} (V)	Package	f(MHz)	BV_{DSS}	P_D max (W)	Mfr
RD00HHS1	0.3	LDMOS	18.7	12.5	SOT-89	30	30	3.1	MT
RD06HHF1	6	LDMOS	16	12.5	TO-220S	30	50	27.8	MT
RD16HHF1	16	LDMOS	16	12.5	TO-220S	30	50	56.8	MT
BLF145	30	MOS	20	28	SOT123A	30			NXP
SD2918	30	MOS	18	50	M113	30			ST
RD70HHF1	70	LDMOS	14	12.5	Flange large	30	50	150	MS
RD100HHF1	100	LDMOS	14	12.5	Flange large	30	50	176.5	MS
SD1726	150	BJT	14	50	M174	30			ST
SD1728	250	BJT	14.5	50	M177	30			ST
SD2933	300	MOS	20	50	M177	30			ST
SD4933	300	MOS	24	50	M177	30			ST
SD2943	350	MOS	22	50	M177	30			ST
SD3933	350	MOS	25	100	M177	30			ST
ARF1500	750	MOS	16	125	T1	40	500		MS
ARF1501	750	MOS	17	250	T1	40	1000		MS
ARF1505	750	MOS	17	300	T1	40	1200		MS
ARF466AG	300	MOS	16	200	TO-247s	45	1000		MS
ARF466FL	300	MOS	16	200	T3	45	1000		MS
ARF465AG	150	MOS	15	300	TO-247s	60	1200		MS
ARF460AG	150	MOS	15	125	TO-247s	65	500		MS
ARF461AG	150	MOS	15	250	TO-247s	65	1000		MS
MRF154	600	MOS	16	50	HOG	80			MA
VRF154FL	600	MOS	17	50	T2	80	170		MS
VRF157FL	600	MOS	21	50	T2	80	170		MS
ARF463AG	100	MOS	15	125	TO-247s	100	500		MS
VRF2933	300	MOS	22	50	M177	100	170		MS
ARF477FL	400	MOS	16	165	T3a	100	500		MS
BLF175	30	MOS	20	50	SOT123A	108			NXP
BLF177	150	MOS	19	50	SOT121B	108			NXP
BLF174XR	600	LDMOS	28.5	50	SOT1214A	128			NXP
MRF150	150	MOS	20	50	M174	150			MA
ARF521	150	MOS	15	165	M174	150	500		MS
VRF150	150	MOS	11	50	M174	150	170		MS
SD3931-10	175	MOS	20	100	M174	150			ST
ARF473	300	MOS	14	165	M244	150	500		MS
SD3932	350	MOS	24	100	M244	150			ST
ARF475FL	450	MOS	14	165	T3	150	500		MS
ARF476FL	450	MOS	14	165	T3	150	500		MS
SD1274-01	30	BJT	10	13.6	M113	160			ST
SD1275-01	40	BJT	9	13.6	M113	160			ST
RD00HVS1	0.5	LDMOS	20	12.5	SOT-89	175	30	3.1	MT
BLF202	2	MOS	13	12.5	SOT409A	175			NXP
RD06HVF1	6	LDMOS	16	12.5	TO-220S	175	50	27.8	MT
BLF244	15	MOS	17	28	SOT123A	175			NXP
RD30HVF1	30	LDMOS	15	12.5	Flange small	175	30	75	MT
MRF148A	30	MOS	18	50	M113	175			MA
BLF245	30	MOS	15.5	28	SOT123A	175			NXP
BLF245B	30	MOS	18	28	SOT279A	175			NXP
VRF148A	30	MOS	16	50	M113	175	170		MS
BLF246B	60	MOS	19	28	SOT161A	175			NXP
RD70HVF1	70	LDMOS	12	12.5	Flange large	175	30	150	MT
BLF246	80	MOS	18	28	SOT121B	175			NXP
MRF141	150	MOS	21	28	M174	175			MA
MRF151	150	MOS	21	50	M177	175			MA
BLF147	150	MOS	14	28	SOT121B	175			NXP
SD2931-10	150	MOS	14	50	M174	175			ST
SD4931	150	MOS	14.8	50	M174	175			ST
VRF141	150	MOS	13	28	M174	175	80		MS
VRF151	150	MOS	14	50	M174	175	170		MS
VRF151E	150	MOS	14	50	M174	175	170		MS
VRF152	150	MOS	14	50	M174	175	170		MS
SD2941-10	175	MOS	15	50	M174	175			ST
MRF151G	300	MOS	20	50	M244	175			MA
MRF141G	300	MOS	21	28	M244	175			MA
SD2932	300	MOS	15	50	M244	175			ST
VRF141G	300	MOS	14	28	M244	175	80		MS
VRF151G	300	MOS	16	50	M244	175	170		MS
SD2942	350	MOS	15	50	M244	175			ST
BLF242	5	MOS	16	28	SOT123A	200			NXP
BLF278	300	MOS	16	50	SOT262A1	225			NXP
SD2902	15	BJT	12.5	28	M113	400			ST
SD2904	30	BJT	9.5	28	M113	400			ST
BLF571	20	LDMOS	27.5	50	SOT467C	500			NXP
BLF578	300	LDMOS	26	50	SOT539A	500			NXP
BLF573	300	LDMOS	27.2	50	SOT502A	500			NXP
BLF573S	300	LDMOS	27.2	50	SOT502B	500			NXP
BLF369	500	LDMOS	18	32	SOT800-2	500			NXP
BLF574XR	600	LDMOS	23	50	SOT1214A	500			NXP
BLF574	600	LDMOS	26.5	50	SOT539A	500			NXP
RD15HVF1	15	LDMOS	12	12.5	TO-220S	520	30	48	MT

Part Number	P_O (W)	Type	Gain (dB)	V_{DD} (V)	Package	f(MHz)	BV_{DSS}	P_D max (W)	Mfr
RD30HUF1	30	LDMOS	10	12.5	Flange small	520	30	75	MT
RD60HUF1	60	LDMOS	10	12.5	Flange large	520	30	150	MT
RD20HMF1	20	LDMOS	8.5	12.5	Flange small	900	30	71.4	MT
RD45HMF1	45	LDMOS	8	12.5	Flange large	900	30	125	MT
BLF1043	10	MOS	16.5	26	SOT538A	1000			NXP
BLF1046	45	MOS	14	26	SOT467C	1000			NXP
BLF871	100	LDMOS	21	40	SOT467C	1000			NXP
BLF871S	100	LDMOS	21	40	SOT467B	1000			NXP
BLF881	140	LDMOS	21	50	SOT467C	1000			NXP
BLF881S	140	LDMOS	21	50	SOT467B	1000			NXP
BLF642	35	LDMOS	19	32	SOT467C	1400			NXP
BLF645	100	LDMOS	18	32	SOT540A	1400			NXP
MRFE6VP100H	100	LDMOS	27.2	50	Flange	0 to 2000			FR
MRFE6VP6300H	300	LDMOS	25	50	Flange	1.8 to 600			FR
MRFE6VP5600H	600	LDMOS	24.6	50	Flange	1.8 to 600			FR
MRFE6VP61K25H	1250	LDMOS	22.9	50	Flange	1.8 to 600			FR

Manufacturer codes:
1. FR – Freescale: www.freescale.com
2. MA – M/A-COM: www.macomtech.com
3. MS - Microsemi: www.microsemi.com
4. MT – Mitsubishi: www.mitsubishielectric.com/semiconductors/php/eSearch.php
5. NXP – NXP Semiconductors: www.nxp.com
6. ST – STMicroelectronics: www.st.com

Table 22.36

RF Power Amplifier Modules

Listed by frequency

Device	Supply (V)	Frequency Range (MHz)	Output Power (W)	Power Gain (dB)	Package[†]	Mfr/ Notes
M57735	17	50-54	14	21	H3C	MI; SSB mobile
M57719N	17	142-163	14	18.4	H2	MI; FM mobile
S-AV17	16	144-148	60	21.7	5-53L	T, FM mobile
S-AV7	16	144-148	28	21.4	5-53H	T, FM mobile
MHW607-1	7.5	136-150	7	38.4	301K-02/3	FR; class C
BGY35	12.5	132-156	18	20.8	SOT132B	NXP
M67712	17	220-225	25	20	H3B	MI; SSB mobile
M57774	17	220-225	25	20	H2	MI; FM mobile
MHW720-1	12.5	400-440	20	21	700-04/1	FR; class C
MHW720-2	12.5	440-470	20	21	700-04/1	FR; class C
M57789	17	890-915	12	33.8	H3B	MI
MHW912	12.5	880-915	12	40.8	301R-01/1	FR; class AB
MHW820-3	12.5	870-950	18	17.1	301G-03/1	FR; class C
HMC487LP5/E	7	9-12 GHz	2	20	25 mm² SMT	H

Manufacturer codes: FR = Freescale; H = Hittite; MI = Mitsubishi; NXP = NXP Semiconductors; T = Toshiba.
[†]For package shape, size and pin-connection information, see manufacturers' data sheets. See Tables of RF Power Transistors for manufacturers and URL for data sheets.

Table 22.37

Digital Logic Families

Type	Propagation Delay for CL = 50 pF (ns) Typ	Max	Max Clock Frequency (MHz)	Power Dissipation (CL = 0) @ 1 MHz (mW/gate)	Output Current @ 0.5 V max (mA)	Input Current (Max mA)	Threshold Voltage (V)	Supply Voltage (V) Min	Typ	Max
CMOS										
74AC	3	5.1	125	0.5	24	0	V+/2	2	5 or 3.3	6
74ACT	3	5.1	125	0.5	24	0	1.4	4.5	5	5.5
74HC	9	18	30	0.5	8	0	V+/2	2	5	6
74HCT	9	18	30	0.5	8	0	1.4	4.5	5	5.5
4000B/74C (10 V)	30	60	5	1.2	1.3	0	V+/2	3	5 - 15	18
4000B/74C (5V)	50	90	2	3.3	0.5	0	V+/2	3	5 - 15	18
TTL										
74AS	2	4.5	105	8	20	0.5	1.5	4.5	5	5.5
74F	3.5	5	100	5.4	20	0.6	1.6	4.75	5	5.25
74ALS	4	11	34	1.3	8	0.1	1.4	4.5	5	5.5
74LS	10	15	25	2	8	0.4	1.1	4.75	5	5.25
ECL										
ECL III	1.0	1.5	500	60	—	—	−1.3	−5.19	−5.2	−5.21
ECL 100K	0.75	1.0	350	40	—	—	−1.32	−4.2	−4.5	−5.2
ECL100KH	1.0	1.5	250	25	—	—	−1.29	−4.9	−5.2	−5.5
ECL 10K	2.0	2.9	125	25	—	—	−1.3	−5.19	−5.2	−5.21
GaAs										
10G	0.3	0.32	2700	125	—	—	−1.3	−3.3	−3.4	−3.5
10G	0.3	0.32	2700	125	—	—	−1.3	−5.1	−5.2	−5.5

Source: Horowitz (W1HFA) and Hill, *The Art of Electronics—2nd edition,* page 570. © Cambridge University Press 1980, 1989. Reprinted with the permission of Cambridge University Press.

Table 22.38

Operational Amplifiers (Op Amps)

Listed by device number

Device	Type	Freq Comp	Max Supply* (V)	Min Input Resistance (MΩ)	Max Offset Voltage (mV)	Min dc Open-Loop Gain (dB)	Min Output Current (mA)	Min Small-Signal Bandwidth (MHz)	Min Slew Rate (V/µs)	Notes
101A	Bipolar	ext	44	1.5	3.0	79	15	1.0	0.5	General purpose
108	Bipolar	ext	40	30	2.0	100	5	1.0		
124	Bipolar	int	32		5.0	100	5	1.0		Quad op amp, low power
148	Bipolar	int	44	0.8	5.0	90	10	1.0	0.5	Quad 741
158	Bipolar	int	32		5.0	100	5	1.0		Dual op amp, low power
301	Bipolar	ext	36	0.5	7.5	88	5	1.0	10	Bandwidth extendable with external components
324	Bipolar	int	32		7.0	100	10	1.0		Quad op amp, single supply
347	BiFET	ext	36	106	5.0	100	30	4	13	Quad, high speed
351	BiFET	ext	36	106	5.0	100	20	4	13	
353	BiFET	ext	36	106	5.0	100	15	4	13	
355	BiFET	ext	44	106	10.0	100	25	2.5	5	
355B	BiFET	ext	44	106	5.0	100	25	2.5	5	
356A	BiFET	ext	36	106	2.0	100	25	4.5	12	
356B	BiFET	ext	44	106	5.0	100	25	5.0	12	
357	BiFET	ext	36	106	10.0	100	25	20.0	50	
357B	BiFET	ext	36	106	5.0	100	25	20.0	30	
358	Bipolar	int	32		7.0	100	10	1.0		Dual op amp, single supply
411	BiFET	ext	36	106	2.0	100	20	4.0	15	Low offset, low drift
709	Bipolar	ext	36	0.05	7.5	84	5	0.3	0.15	
741	Bipolar	int	36	0.3	6.0	88	5	0.4	0.2	
741S	Bipolar	int	36	0.3	6.0	86	5	1.0	3	Improved 741 for AF
1436	Bipolar	int	68	10	5.0	100	17	1.0	2.0	High-voltage
1437	Bipolar	ext	36	0.050	7.5	90		1.0	0.25	Matched, dual 1709
1439	Bipolar	ext	36	0.100	7.5	100		1.0	34	
1456	Bipolar	int	44	3.0	10.0	100	9.0	1.0	2.5	Dual 1741
1458	Bipolar	int	36	0.3	6.0	100	20.0	0.5	3.0	
1458S	Bipolar	int	36	0.3	6.0	86	5.0	0.5	3.0	Improved 1458 for AF
1709	Bipolar	ext	36	0.040	6.0	80	10.0	1.0		
1741	Bipolar	int	36	0.3	5.0	100	20.0	1.0	0.5	
1747	Bipolar	int	44	0.3	5.0	100	25.0	1.0	0.5	Dual 1741
1748	Bipolar	ext	44	0.3	6.0	100	25.0	1.0	0.8	Non-comp-ensated 1741
1776	Bipolar	int	36	50	5.0	110	5.0		0.35	Micro power, programmable
3140	BiFET	int	36	1.5×106	2.0	86	1	3.7	9	Strobable output
3403	Bipolar	int	36	0.3	10.0	80		1.0	0.6	Quad, low power
3405	Bipolar	ext	36		10.0	86	10	1.0	0.6	Dual op amp and dual comparator
3458	Bipolar	int	36	0.3	10.0	86	10	1.0	0.6	Dual, low power

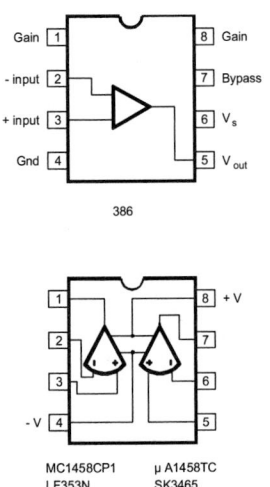

MC1458CP1 µ A1458TC
LF353N SK3465
N5558V ECG778
LM1458N LM358N

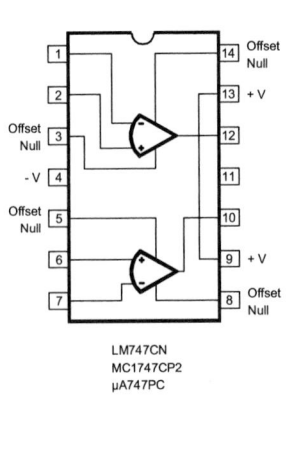

LM747CN
MC1747CP2
µA747PC

Top View

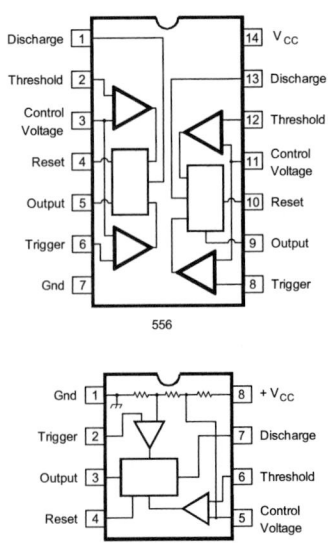

Device	Type	Freq Comp	Max Supply* (V)	Min Input Resistance (MΩ)	Max Offset Voltage (mV)	Min dc Open-Loop Gain (dB)	Min Output Current (mA)	Min Small-Signal Bandwidth (MHz)	Min Slew Rate (V/µs)	Notes
3476	Bipolar	int	36	5.0	6.0	92	12		0.8	
3900	Bipolar	int	32	1.0		65	0.5	4.0	0.5	Quad, Norton single supply
4558	Bipolar	int	44	0.3	5.0	88	10	2.5	1.0	Dual, wideband
4741	Bipolar	int	44	0.3	5.0	94	20	1.0	0.5	Quad 1741
5534	Bipolar	int	44	0.030	5.0	100	38	10.0	13	Low noise, can swing 20V P-P across 600
5556	Bipolar	int	36	1.0	12.0	88	5.0	0.5	1	Equivalent to 1456
5558	Bipolar	int	36	0.15	10.0	84	4.0	0.5	0.3	Dual, equivalent to 1458
34001	BiFET	int	44	106	2.0	94		4.0	13	JFET input
AD745	BiFET	int	±18	104	0.5	63	20	20	12.5	Ultra-low noise, high speed

LT1001   Precision op amp, low offset voltage (15 µV max), low drift (0.6 µV/°C max), low noise (0.3 µV p-p)
LT1007   Extremely low noise (0.06 µV p-p), very high gain (20×10^6 into 2 kΩ load)
LT1360   High speed, very high slew rate (800 V/µs), 50 MHz gain bandwidth, ±2.5 V to ±15 V supply range

Device	Type	Freq Comp	Max Supply* (V)	Min Input Resistance (MΩ)	Max Offset Voltage (mV)	Min dc Open-Loop Gain (dB)	Min Output Current (mA)	Min Small-Signal Bandwidth (MHz)	Min Slew Rate (V/µs)	Notes
NE5514	Bipolar	int	±16	100	1		10	3	0.6	
NE5532	Bipolar	int	±20	0.03	4	47	10	10	9	Low noise
OP-27A	Bipolar	ext	44	1.5	0.025	115		5.0	1.7	Ultra-low noise, high speed
OP-37A	Bipolar	ext	44	1.5	0.025	115		45.0	11.0	
TL-071	BiFET	int	36	10^6	6.0	91		4.0	13.0	Low noise
TL-081	BiFET	int	36	10^6	6.0	88		4.0	8.0	
TL-082	BiFET	int	36	10^6	15.0	99		4.0	8.0	Low noise
TL-084	BiFET	int	36	10^6	15.0	88		4.0	8.0	Quad, high-performance AF
TLC27M2	CMOS	int	18	10^6	10	44		0.6	0.6	Low noise
TLC27M4	CMOS	int	18	10^6	10	44		0.6	0.6	Low noise

*From −V to +V terminals

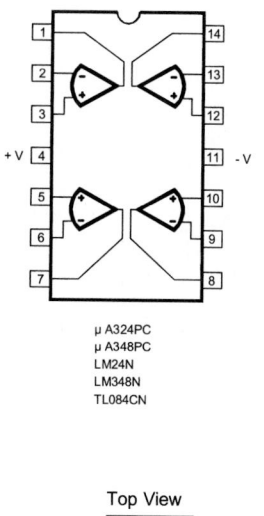

µ A324PC
µ A348PC
LM24N
LM348N
TL084CN

Top View

HBK05_07-13

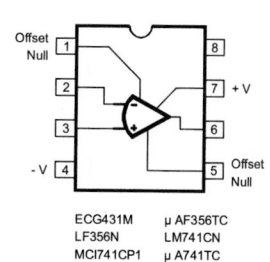

ECG431M µ AF356TC
LF356N   LM741CN
MCI741CP1 µ A741TC

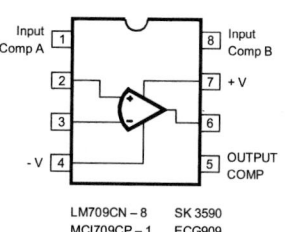

LM709CN – 8 SK 3590
MCI709CP – 1 ECG909

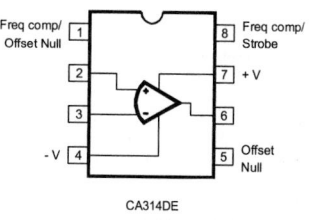

CA314DE

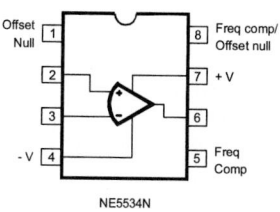

NE5534N

22.7 Tubes, Wire, Materials, Attenuators, Miscellaneous

Table 22.39
Triode Transmitting Tubes
The full 1988 *Handbook* table of power tube specifications and base diagrams can be viewed in pdf format on the *ARRLWeb* at **www.arrl.org/hf-tube-amplifiers.**

Type	Power Diss.(W)	Plate (V)	Plate (mA)	Grid dc (mA)	Freq (MHz)	Ampl Factor	Fil (V)	Fil (A)	C_{IN} (pF)	C_{GP} (pF)	C_{OUT} (pF)	Base Diagram	Service Class[1]	Plate (V)	Grid (V)	Plate (mA)	Grid dc (mA)	Input (W)	P-P (kΩ)	Output (W)
5675	5	165	30	8	3000	20	6.3	0.135	2.3	1.3	0.09	Fig 21	GG0	120	-8	25	4	¾	—	0.05
2C40	6.5	500	25	—	500	36	6.3	0.75	2.1	1.3	0.05	Fig 11	CTO	250	-5	20	0.3	¾	—	0.075
5893	8.0	400	40	13	1000	27	6.0	0.33	2.5	1.75	0.07	Fig 21	CT CP	350 300	-33 -45	35 30	13 12	2.4 2.0	— —	6.5 6.5
2C43	12	500	40	—	1250	48	6.3	0.9	2.9	1.7	0.05	Fig 11	CTO	470	—	38[7]	—	¾	—	9[2]
811-A	65	1000	175	50	60	160	6.3	4.0	5.9	5.6	0.7	3G	CT CP B/CG AB1	1500 1250 1250 1250	-70 -120 0 0	173 140 21/175 27/175	40 45 28 13	7.1 10.0 12 3.0	— — — —	200 135 165 155
812-A	65	1500	175	35	60	29	6.3	4.0	5.4	5.5	0.77	3G	CT CP B[2]	1500 1250 1500	-120 -115 -48	173 140 28/310	30 35 270[4]	6.5 7.6 5.0	— — 13.2	190 130 340
3CX100A5[6]	100 70	1000 600	125[5] 100[5]	50	2500	100	6.0	1.05	7.0	2.15	0.035	—	AGG CP	800 600	-20 -15	80 75	30 40	6 6	— —	27 18
2C39	100	1000	60	40	500	100	6.3	1.1	6.5	1.95	0.03	—	G1C CTO CP	600 900 600	-35 -40 -150	60 90 100[5]	40 30 50	5.0 ¾ ¾	— — —	20 40 ¾
AX9900, 5866	135	2500	200	40	150	25	6.3	5.4	5.8	5.5	0.1	Fig 3	CT CP B[2]	2500 2000 2500	-200 -225 -90	200 127 80/330	40 40 350[4]	16 16 14[3]	— — 15.68	390 204 560
572B T160L	160	2750	275	—	—	170	6.3	4.0	—	—	—	3G	CT B/GG[2]	1650 2400	-70 -2.0	165 90/500	32 —	6 100	— —	205 600
8873	200	2200	250	—	500	160	6.3	3.2	19.5	7.0	0.03	Fig 87	AB2	2000	—	22/500	98[3]	27[3]	—	505
8875	300	2200	250	—	500	160	6.3	3.2	19.5	7.0	0.03	—	AB2	2000	—	22/500	98[3]	27[3]	—	505
833A	350	3300	500	100	30	35	10	10	12.3	6.3	8.5	Fig 41	CTO CTO	2250 3000	-125 -160	445 335	85 70	23 20	— —	780 800
	450[6]	4000[6]	500	100	20[6]	35	10	10	12.3	6.3	8.5	Fig 41	CP CP B[2]	2500 3000 3000	-300 -240 -70	335 335 100/750	75 70 400[4]	30 26 20[4]	— — 9.5	635 800 1650
8874	400	2200	350	—	500	160	6.3	3.2	19.5	7.0	0.03	—	AB2	2000	—	22/500	98[3]	27[3]	—	505
3-400Z	400	3000	400	—	110	200	5	14.5	7.4	4.1	0.07	Fig 3	B/GG	3000	0	100/333	120	32	—	655
3-500Z	500	4000	400	—	110	160	5	14.5	7.4	4.1	0.07	Fig 3	B/GG	3000	—	370	115	30	5	750
3-600Z	600	4000	425	—	110	165	5	15.0	7.8	4.6	0.08	Fig 3	B/GG B/GG	3000 3500	— —	400 400	118 110	33 35	— —	810 950
3CX800A7	800	2250	600	60	350	200	13.5	1.5	26	—	6.1	Fig 87	AB2,GG[7]	2200	-8.2	500	36	16	—	750
3-1000Z	1000	3000	800	—	110	200	7.5	21.3	17	6.9	0.12	Fig 3	B/GG	3000	0	180/670	300	65	—	1360
3CX1200A7	1200	5000	800	—	110	200	7.5	21.0	20	12	0.2	Fig 3	AB2,GG	3600	-10	700	230	85	—	1500
8877	1500	4000	1000	—	250	200	5.0	10	42	10	0.1	—	AB2	2500	-8.2	1000	—	57	—	1520

Table 22-40
Tetrode Transmitting Tubes
Also see www.arrl.org/hf-tube-amplifiers.

[1] Service Class Abbreviations:
AB2GD = AB2 linear with 50-Ω passive grid circuit.
B = Class-B push-pull
CP = Class-C plate-modulated phone
CT = Class-C telegraph
GG = Grounded-grid (grid and screen connected together)
[2] Maximum signal value
[3] Peak grid-grid volts
[4] Forced-air cooling required.
[5] Two tubes triode-connected, G2 to G1 through 20kΩ to G2.
[6] Typical operation at 175 MHz.
[7] ±1.5 V.
[8] Values are for two tubes.
[9] Single tone.
[10] 24-Ω cathode resistance.
[11] Base same as 4CX250B. Socket is Russian SK2A.
[12] Socket is Russian SK1A.
[13] Socket is Russian SK3A.

Type	Max. Plate Diss. (W)	Max. Plate Volts (V)	Max. Screen Diss. (W)	Max. Screen Volts (V)	Max. Freq. (MHz)	Filament Volts (V)	Amps (A)	C_{IN} (pF)	C_{GP} (pF)	C_{OUT} (pF)	Base	Serv. Class[1]	Plate (V)	Screen (V)	Grid (V)	Plate (mA)	Screen (mA)	Grid (mA)	P_{IN} (W)	P-P (kΩ)	P_{OUT} (W)
6146/	25	750	3	250	60	6.3	1.25	13	0.24	8.5	7CK	CT	500	170	-66	135	9	2.5	0.2	—	48
6146A	25	750	3	250	60	6.3	1.25	13	0.24	8.5	7CK	CT	700	160	-62	120	11	3.1	0.2	—	70
8032	25	750	3	250	60	12.6	0.585	13	0.24	8.5	7CK	CT[6]	400	190	-54	150	10.4	2.2	3.0	—	35
6883												CP	400	150	-87	112	7.8	3.4	0.4	—	32
												CP	600	150	-87	112	7.8	3.4	0.4	—	52
6159B/	25	750	3	250	60	26.5	0.3	13	0.24	8.5	7CK	AB2[8]	600	190	-48	28/270	1.2/20	22	0.3	5	113
												AB2[8]	750	165	-46	22/240	0.3/20	2.6[2]	0.4	7.4	131
												AB1[8]	750	195	-50	23/220	1/26	100[3]	0	8	120
807, 807W	30	750	3.5	300	60	6.3	0.9	12	0.2	7	5AW	CT	750	250	-45	100	6	3.5	0.22	—	50
5933												CP	600	275	-90	100	6.5	4	0.4	—	42.5
												AB1	750	300	-35	15/70	3/8	75[3]	0	—	72
1625	30	750	3.5	300	60	12.6	0.45	12	0.2	7	5AZ	B[5]	750	—	0	15/240	—	55[3]	5.3[2]	6.65	120
6146B	35	750	3	250	60	6.3	1.125	13	0.22	8.5	7CK	CT	750	200	-77	160	10	2.7	0.3	—	85
8298A												CP	600	175	-92	140	9.5	3.4	0.5	—	62
												AB1	750	200	-48	24/125	6.3	—	—	3.5	61
813	125	2500	20	800	30	10.0	5.0	16.3	0.25	14.0	5BA	CTO	1250	300	-75	180	35	12	1.7	—	170
												CTO	2250	400	-155	220	40	15	4	—	375
												AB1	2000	750	-95	25/145	27[2]	0	0	—	245
												AB2[8]	2500	750	-90	40/315	1.5/58	230[3]	0.12[2]	16	455
												AB2[8]	2500	750	-95	35/260	1.2/55	235[3]	0.35[2]	17	650
4CX250B	250	2000	12	400	175	6.0	2.9	18.5	0.04	4.7	—	CTO	2000	250	-90	250	25	27	2.8	—	410
												CP	1500	250	-100	200	25	17	2.1	—	250
												AB1[8]	2000	350	-50	500	30	100	0	8.26	650
4-400A	400[4]	4000	35	600	110	5.0	14.5	12.5	0.12	4.7	5BK	CT/CP	4000	300	-170	270	22.5	10	10	—	720
												GG	2500	0	0	80/270[9]	55[9]	100[9]	39[9]	—	435
												AB1	2500	750	-130	95/317	0/14	0	0	4.0	425
4CX400A	400	2500	8	400	500	6.3	3.2	24	0.08	7	See[11]	AB2GD	2200	325	-30	100/270	22	2	9	—	405
												AB2GD	2500	400	-35	100/400	18	1	13	—	610
4CX800A	800	2500	15	350	150	12.6	3.6	51	0.9	11	See[12]	AB2GD	2200	350	-56	160/550	24	1	32	—	750
4-1000A	1000	6000	75	1000	—	7.5	21	27.2	0.24	7.6	—	CT	3000	500	-150	700	146	38	11	—	1430
8166												CP	3000	500	-200	600	145	36	12	—	1390
												AB2	4000	500	-60	300/1200	0/95	11	11	—	3000
												GG	3000	0	0	100/700[9]	105[9]	170[9]	130[9]	2.5	1475
4CX1000A	1000	3000	12	400	110	6.0	9.0	81.5	0.01	11.8		AB1[8]	2000	325	-55	500/2000	-4/60	—	—	2.5	2160
												AB1[8]	2500	325	-55	500/2000	-4/60	—	—	3.1	2920
												AB1[8]	3000	325	-55	500/1800	-4/60	—	—	3.85	3360
4CX1500B	1500	3000	12	400	110	6.0	10.0	81.5	0.02	11.8		AB1	2750	225	-34	300/755	-14/60	0.95	1.5	1.9	1100
4CX1600B	1600	3300	20	350	250	12.6	4.4	86	0.15	12	See[13]	AB2GD	2400	350	-53	500/1100	20	2	28	—	1600
												AB2GD	2400	350	-70	200/870	48	2	83[10]	—	1500
												AB2GD	3200	240	-57	200/740	21	1	33	—	1600

Table 22.41
EIA Vacuum-Tube Base Diagrams

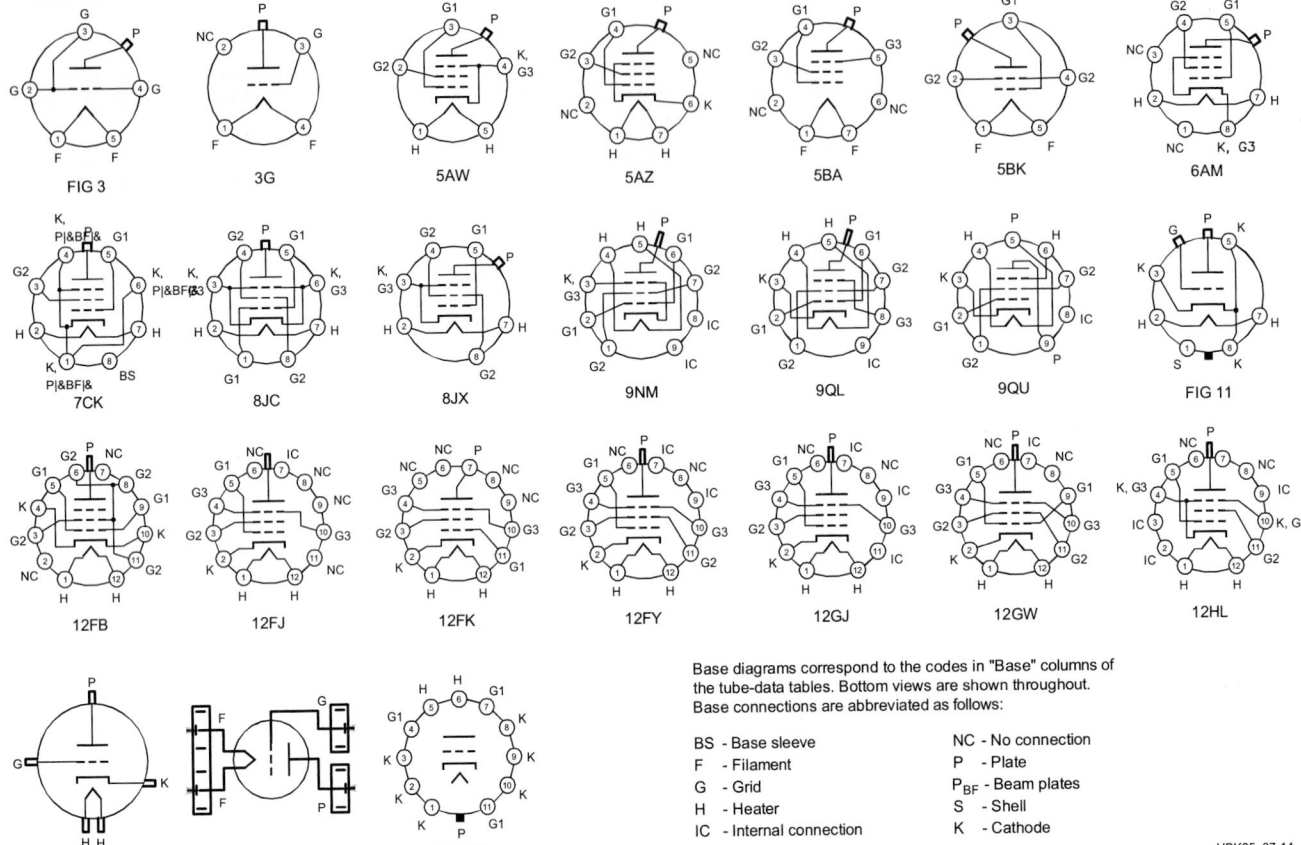

Base diagrams correspond to the codes in "Base" columns of the tube-data tables. Bottom views are shown throughout. Base connections are abbreviated as follows:

BS - Base sleeve
F - Filament
G - Grid
H - Heater
IC - Internal connection

NC - No connection
P - Plate
P_{BF} - Beam plates
S - Shell
K - Cathode

HBK05_07-14

Alphabetical subscripts (D = diode, P = pentode, T = triode and HX = hexode) indicate structures in multistructure tubes. Subscript CT indicates filament or heater center tap.

Generally, when pin 1 of a metal-envelope tube (except all triodes) is shown connected to the envelope, pin 1 of a glass-envelope counterpart (suffix G or GT) is connected to an internal shield.

Table 22.42

Metal-Oxide Varistor (MOV) Transient Suppressors

Listed by voltage

Type No.	ECG/NTE†† no.	V acRMS	Maximum Applied Voltage V acPeak	Maximum Energy (Joules)	Maximum Peak Current (A)	Maximum Power (W)	Maximum Varistor Voltage (V)
V180ZA1	1V115	115	163	1.5	500	0.2	285
V180ZA10	2V115	115	163	10.0	2000	0.45	290
V130PA10A		130	184	10.0	4000	8.0	350
V130PA20A		130	184	20.0	4000	15.0	350
V130LA1	1V130	130	184	1.0	400	0.24	360
V130LA2	1V130	130	184	2.0	400	0.24	360
V130LA10A	2V130	130	184	10.0	2000	0.5	340
V130LA20A	524V13	130	184	20.0	4000	0.85	340
V150PA10A		150	212	10.0	4000	8.0	410
V150PA20A		150	212	20.0	4000	15.0	410
V150LA1	1V150	150	212	1.0	400	0.24	420
V150LA2	1V150	150	212	2.0	400	0.24	420
V150LA10A	524V15	150	212	10.0	2000	0.5	390
V150LA20A	524V15	150	212	20.0	4000	0.85	390
V250PA10A		250	354	10.0	4000	0.85	670
V250PA20A		250	354	20.0	4000	7.0	670
V250PA40A		250	354	40.0	4000	13.0	670
V250LA2	1V250	250	354	2.0	400	0.28	690
V250LA4	1V250	250	354	4.0	400	0.28	690
V250LA15A	2V250	250	354	15.0	2000	0.6	640
V250LA20A	2V250	250	354	20.0	2000	0.6	640
V250LA40A	524V25	250	354	40.0	4000	0.9	640

††ECG and NTE numbers for these parts are identical, except for the prefix. Add the "ECG" or "NTE" prefix to the numbers shown for the complete part number.

Table 22.43

Crystal Holders

Note: Solder Seal, Cold Weld, and Resistance Weld sealing methods are commonly available. All dimensions are in inches

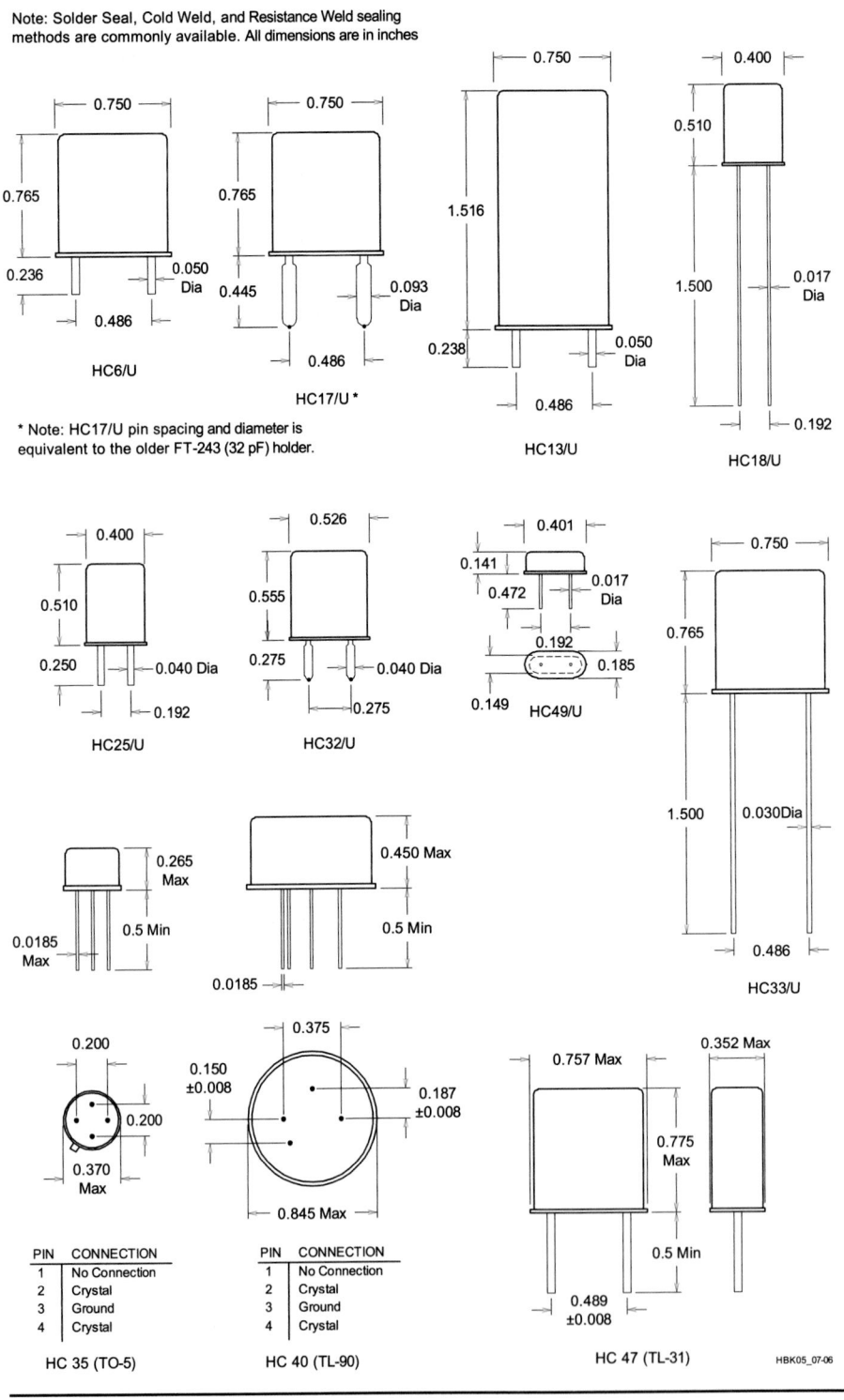

* Note: HC17/U pin spacing and diameter is equivalent to the older FT-243 (32 pF) holder.

PIN	CONNECTION
1	No Connection
2	Crystal
3	Ground
4	Crystal

HC 35 (TO-5)

PIN	CONNECTION
1	No Connection
2	Crystal
3	Ground
4	Crystal

HC 40 (TL-90)

HC 47 (TL-31)

HBK05_07-06

Table 22.44

Copper Wire Specifications

Bare and Enamel-Coated Wire

One mil = 0.001 inch

Wire Size (AWG)	Diam (Mils)	Area (CM[1])	Enamel Wire Coating Turns / Linear inch[2]			Feet per Pound Bare	Ohms per 1000 ft 25° C	Current Carrying Capacity Continuous Duty[3] at 700 CM per Amp[4]	Open air	Conduit or bundles	Nearest British SWG No.
			Single	Heavy	Triple						
1	289.3	83694.49				3.948	0.1239	119.564			1
2	257.6	66357.76				4.978	0.1563	94.797			2
3	229.4	52624.36				6.277	0.1971	75.178			4
4	204.3	41738.49				7.918	0.2485	59.626			5
5	181.9	33087.61				9.98	0.3134	47.268			6
6	162.0	26244.00				12.59	0.3952	37.491			7
7	144.3	20822.49				15.87	0.4981	29.746			8
8	128.5	16512.25				20.01	0.6281	23.589			9
9	114.4	13087.36				25.24	0.7925	18.696			11
10	101.9	10383.61				31.82	0.9987	14.834			12
11	90.7	8226.49				40.16	1.2610	11.752			13
12	80.8	6528.64				50.61	1.5880	9.327			13
13	72.0	5184.00				63.73	2.0010	7.406			15
14	64.1	4108.81	15.2	14.8	14.5	80.39	2.5240	5.870	32	17	15
15	57.1	3260.41	17.0	16.6	16.2	101.32	3.1810	4.658			16
16	50.8	2580.64	19.1	18.6	18.1	128	4.0180	3.687	22	13	17
17	45.3	2052.09	21.4	20.7	20.2	161	5.0540	2.932			18
18	40.3	1624.09	23.9	23.2	22.5	203.5	6.3860	2.320	16	10	19
19	35.9	1288.81	26.8	25.9	25.1	256.6	8.0460	1.841			20
20	32.0	1024.00	29.9	28.9	27.9	322.7	10.1280	1.463	11	7.5	21
21	28.5	812.25	33.6	32.4	31.3	406.7	12.7700	1.160			22
22	25.3	640.09	37.6	36.2	34.7	516.3	16.2000	0.914		5	22
23	22.6	510.76	42.0	40.3	38.6	646.8	20.3000	0.730			24
24	20.1	404.01	46.9	45.0	42.9	817.7	25.6700	0.577			24
25	17.9	320.41	52.6	50.3	47.8	1031	32.3700	0.458			26
26	15.9	252.81	58.8	56.2	53.2	1307	41.0200	0.361			27
27	14.2	201.64	65.8	62.5	59.2	1639	51.4400	0.288			28
28	12.6	158.76	73.5	69.4	65.8	2081	65.3100	0.227			29
29	11.3	127.69	82.0	76.9	72.5	2587	81.2100	0.182			31
30	10.0	100.00	91.7	86.2	80.6	3306	103.7100	0.143			33
31	8.9	79.21	103.1	95.2		4170	130.9000	0.113			34
32	8.0	64.00	113.6	105.3		5163	162.0000	0.091			35
33	7.1	50.41	128.2	117.6		6553	205.7000	0.072			36
34	6.3	39.69	142.9	133.3		8326	261.3000	0.057			37
35	5.6	31.36	161.3	149.3		10537	330.7000	0.045			38
36	5.0	25.00	178.6	166.7		13212	414.8000	0.036			39
37	4.5	20.25	200.0	181.8		16319	512.1000	0.029			40
38	4.0	16.00	222.2	204.1		20644	648.2000	0.023			
39	3.5	12.25	256.4	232.6		26969	846.6000	0.018			
40	3.1	9.61	285.7	263.2		34364	1079.2000	0.014			
41	2.8	7.84	322.6	294.1		42123	1323.0000	0.011			
42	2.5	6.25	357.1	333.3		52854	1659.0000	0.009			
43	2.2	4.84	400.0	370.4		68259	2143.0000	0.007			
44	2.0	4.00	454.5	400.0		82645	2593.0000	0.006			
45	1.8	3.10	526.3	465.1		106600	3348.0000	0.004			
46	1.6	2.46	588.2	512.8		134000	4207.0000	0.004			

Teflon Coated, Stranded Wire

(As supplied by Belden Wire and Cable)

Size	Strands[5]	Turns per Linear inch[2] UL Style No.		
		1180	1213	1371
16	19×29	11.2		
18	19×30	12.7		
20	7×28	14.7	17.2	
20	19×32	14.7	17.2	
22	19×34	16.7	20.0	23.8
22	7×30	16.7	20.0	23.8
24	19×36	18.5	22.7	27.8
24	7×32		22.7	27.8
26	7×34		25.6	32.3
28	7×36		28.6	37.0
30	7×38		31.3	41.7
32	7×40			47.6

Notes

[1] A circular mil (CM) is a unit of area equal to that of a one-mil-diameter circle ($\pi/4$ square mils). The CM area of a wire is the square of the mil diameter.

[2] Figures given are approximate only; insulation thickness varies with manufacturer.

[3] Maximum wire temperature of 212°F (100°C) with a maximum ambient temperature of 135°F (57°C) as specified by the manufacturer. The *National Electrical Code* or local building codes may differ.

[4] 700 CM per ampere is a satisfactory design figure for small transformers, but values from 500 to 1000 CM are commonly used. The *National Electrical Code* or local building codes may differ.

[5] Stranded wire construction is given as "count" × "strand size" (AWG).

Table 22.45
Standard vs American Wire Gauge

SWG	Diam (in.)	Nearest AWG
12	0.104	10
14	0.08	12
16	0.064	14
18	0.048	16
20	0.036	19
22	0.028	21
24	0.022	23
26	0.018	25
28	0.0148	27
30	0.0124	28
32	0.0108	29
34	0.0092	31
36	0.0076	32
38	0.006	34
40	0.0048	36
42	0.004	38
44	0.0032	40
46	0.0024	—

Table 22.46
Antenna Wire Strength

American Wire Gauge	Recommended Tension[1] (pounds)		Weight (pounds per 1000 feet)	
	Copper-clad steel[2]	Hard-drawn copper	Copper-clad steel[2]	Hard-drawn copper
4	495	214	115.8	126
6	310	130	72.9	79.5
8	195	84	45.5	50
10	120	52	28.8	31.4
12	75	32	18.1	19.8
14	50	20	11.4	12.4
16	31	13	7.1	7.8
18	19	8	4.5	4.9
20	12	5	2.8	3.1

[1]Approximately one-tenth the breaking load. Might be increased 50% if end supports are firm and there is no danger of ice loading.
[2]"Copperweld," 40% copper.

Table 22.47
Guy Wire Lengths to Avoid

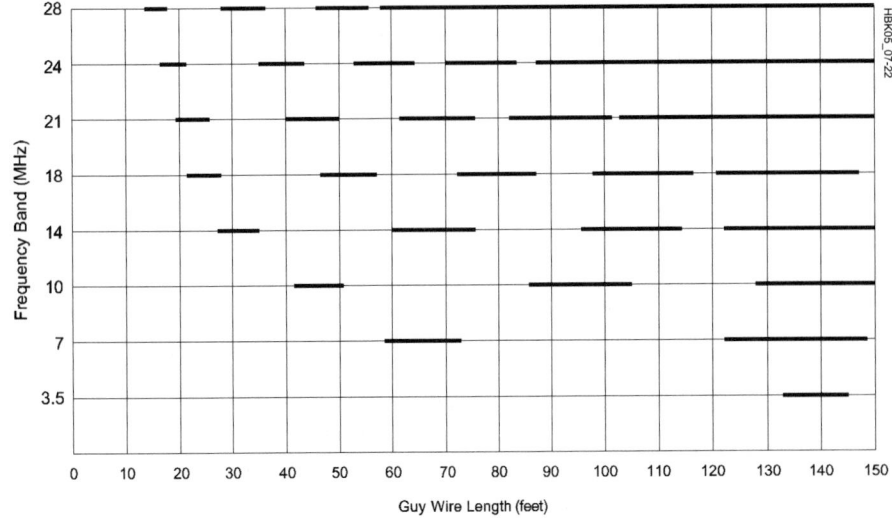

The black bars indicate ungrounded guy wire lengths to avoid for the eight HF amateur bands. This chart is based on resonance within 10% of any frequency in the band. Grounded wires will exhibit resonance at odd multiples of a quarter wavelength. *(Jerry Hall, K1TD)*

Table 22.48

Aluminum Alloy Specifications

Common Alloy Numbers

Type	Characteristic
2024	Good formability, high strength
5052	Excellent surface finish, excellent corrosion resistance, normally not heat treatable for high strength
6061	Good machinability, good weldability, can be brittle at high tempers
7075	Good formability, high strength

General Uses

Type	Uses
2024-T3	Chassis boxes, antennas, anything that will be bent or flexed repeatedly
7075-T3	
6061-T6	Mounting plates, welded assemblies or machined parts

Common Tempers

Type	Characteristics
T0	Special soft condition
T3	Hard
T6	Very hard, possibly brittle
TXXX	Three digit tempers—usually specialized high-strength heat treatments, similar to T6

Table 22.49

Impedance of Two-Conductor Twisted Pair Lines

Wire Size	Twists per Inch				
	2.5	5	7.5	10	12.5
#20	43	39	35		
#22	46	41	39	37	32
#24	60	45	44	43	41
#26	65	57	54	48	47
#28	74	53	51	49	47
#30			49	46	47

Measured in ohms at 14.0 MHz.

This illustrates the impedance of various two-conductor lines as a function of the wire size and number of twists per inch.

Table 22.50

Attenuation per Foot of Two-Conductor Twisted Pair Lines

Wire Size	Twists per Inch				
	2.5	5	7.5	10	12.5
#20	0.11	0.11	0.12		
#22	0.11	0.12	0.12	0.12	0.12
#24	0.11	0.12	0.12	0.13	0.13
#26	0.11	0.13	0.13	0.13	0.13
#28	0.11	0.13	0.13	0.16	0.16
#30			0.25	0.27	0.27

Measured in decibels at 14.0 MHz.

Attenuation in dB per foot for the same lines as shown above.

Table 22.51
Large Machine-Wound Coil Specifications

Coil Dia, Inches	Turns Per Inch	Inductance in μH Per Inch
1¼	4	2.75
	6	6.3
	8	11.2
	10	17.5
	16	42.5
1½	4	3.9
	6	8.8
	8	15.6
	10	24.5
	16	63
1¾	4	5.2
	6	11.8
	8	21
	10	33
	16	85
2	4	6.6
	6	15
	8	26.5
	10	42
	16	108
2½	4	10.2
	6	23
	8	41
	10	64
3	4	14
	6	31.5
	8	56
	10	89

Table 22.53
Small Machine-Wound Coil Specifications

Coil Dia, Inches	Turns Per Inch	Inductance in μH Per Inch
½ (A)	4	0.18
	6	0.40
	8	0.72
	10	1.12
	16	2.8
	32	12
⅝ (A)	4	0.28
	6	0.62
	8	1.1
	10	1.7
	16	4.4
	32	18
¾ (B)	4	0.6
	6	1.35
	8	2.4
	10	3.8
	16	9.9
	32	40
1 (B)	4	1.0
	6	2.3
	8	4.2
	10	6.6
	16	16.9
	32	68

Table 22.52
Inductance Factor for Large Machine-Wound Coils

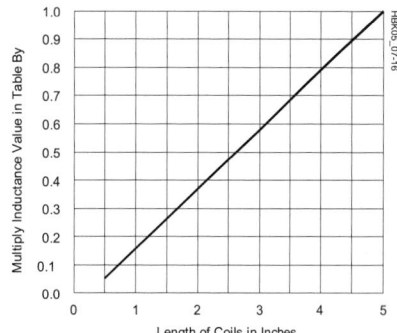

Factor to be applied to the inductance of large coils for coil lengths up to 5 inches.

Table 22.54
Inductance Factor for Small Machine-Wound Coils

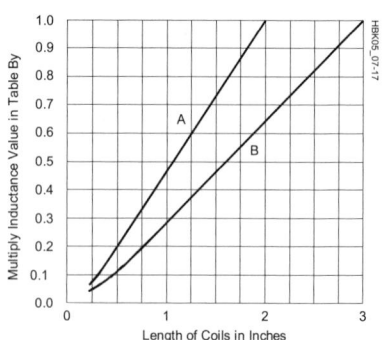

Factor to be applied to the inductance of small coils as a function of coil length. Use curve A for coils marked A, and curve B for coils marked B.

Table 22.55

Measured Inductance for #12 AWG Wire Windings

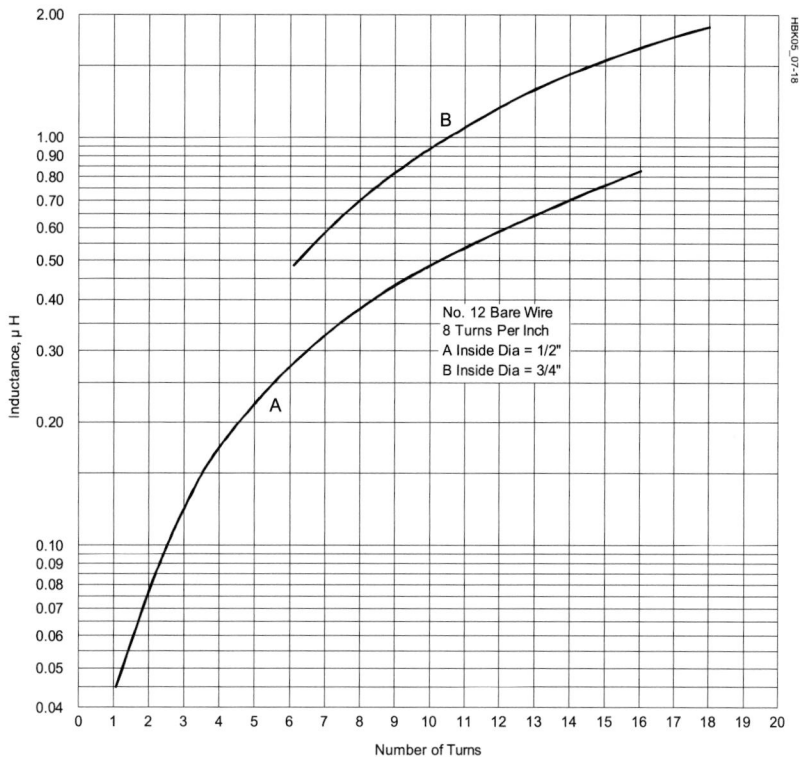

Values are for inductors with half-inch leads and wound with eight turns per inch.

Table 22.56

Relationship Between Noise Figure and Noise Temperature

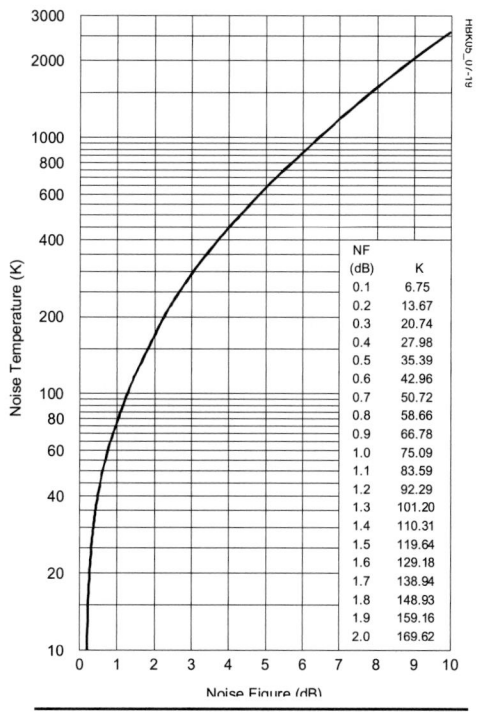

Table 22.57
Pi-Network Resistive Attenuators (50 Ω)

dB Atten.	R1 (Ohms)	R2 (Ohms)
1.0	870	5.77
2.0	436	11.6
3.0	292	17.6
4.0	221	23.8
5.0	178	30.4
6.0	150	37.4
7.0	131	44.8
8.0	116	52.8
9.0	105	61.6
10.0	96.2	71.2
11.0	89.2	81.7
12.0	83.5	93.2
13.0	78.8	106
14.0	74.9	120
15.0	71.6	136
16.0	68.8	154
17.0	66.4	173
18.0	64.4	195
19.0	62.6	220
20.0	61.1	248
21.0	59.8	278
22.0	58.6	313
23.0	57.6	352
24.0	56.7	395
25.0	56.0	443
30.0	53.2	790
35.0	51.8	1405
40.0	51.0	2500
45.0	50.5	4446
50.0	50.3	7906
55.0	50.2	14,058
60.0	50.1	25,000

Note: A PC board kit for the Low-Power Step Attenuator (Sep 1982 *QST*) is available from FAR Circuits. Project details are in the *Handbook* **template package STEP ATTENUATOR.**

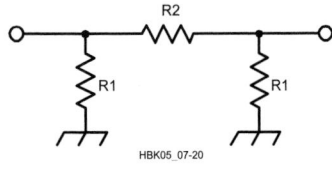

HBK05_07-20

Table 22.58
T-Network Resistive Attenuators (50 Ω)

dB Atten.	R1 (Ohms)	R2 (Ohms)
1.0	2.88	433
2.0	5.73	215
3.0	8.55	142
4.0	11.3	105
5.0	14.0	82.2
6.0	16.6	66.9
7.0	19.1	55.8
8.0	21.5	47.3
9.0	23.8	40.6
10.0	26.0	35.1
11.0	28.0	30.6
12.0	30.0	26.8
13.0	31.7	23.5
14.0	33.3	20.8
15.0	35.0	18.4
16.0	36.3	16.2
17.0	37.6	14.4
18.0	38.8	12.8
19.0	40.0	11.4
20.0	41.0	10.0
21.0	41.8	9.0
22.0	42.6	8.0
23.0	43.4	7.1
24.0	44.0	6.3
25.0	44.7	5.6
30.0	47.0	3.2
35.0	48.2	1.8
40.0	49.0	1.0
45.0	49.4	0.56
50.0	49.7	0.32
55.0	49.8	0.18
60.0	49.9	0.10

HBK05_07-21

22.8 Computer Connectors

Most connections between computers and their peripherals are made with some form of multi-conductor cable. Examples include shielded, unshielded and ribbon cable. **Table 22.59** shows a variety of computer connectors and pin outs, including some used for internal connections, such as power supplies and disk drives.

Table 22.59
ComputerConnector Pinouts

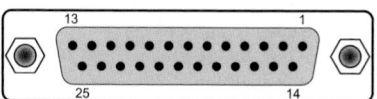

Parallel Port (DB 25 pin) Female

Pin	Signal	Pin	Signal
1	Strobe	10	Acknowledge
2	Data 0	11	Busy
3	Data 1	12	Paper Empty
4	Data 2	13	Select
5	Data 3	14	Auto Feed
6	Data 4	15	Error
7	Data 5	16	Initialize
8	Data 6	17	Select In
9	Data 7	18-25	GND

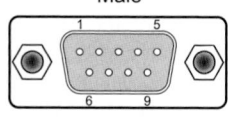

Serial Port (DB 9 pin) Male

Pin	Signal
1	DCD (Data Carrier Detect)
2	RxD (Receive Data)
3	TxD (Transmit Data)
4	DTR (Data Terminal Ready)
5	GND (Signal Ground)
6	DSR (Data Set Ready)
7	RTS (Request To Send)
8	CTS (Clear To Send)
9	RI (Ring Indicator)

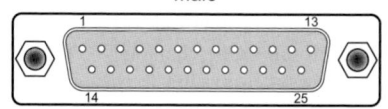

Serial Port (DB 25 pin) Male

Pin	Signal	Pin	Signal
1	N/C (not connected)	20	DTR (Data Terminal Ready)
2	TxD (Transmit Data)		
3	RxD (Receive Data)	21	N/C
4	RTS (Request To Send)	22	RI (Ring Indicator)
5	CTS (Clear To Send)	23	N/C
6	DSR (Data Set Ready)	24	N/C
7	GND (Signal Ground)	25	N/C
8	DCD (Data Carrier Detect)		
9-19	N/C		

USB Type A

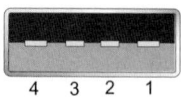
USB Type B

Pin	Signal
1	VBUS (+5 V)
2	D– (Data –)
3	D+ (Data +)
4	GND (Ground)

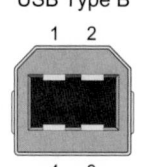

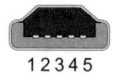
USB Mini-A   **USB Mini-B**

1 2 3 4 5   1 2 3 4 5

Pin	Signal
1	VBUS (+5 V)
2	D– (Data –)
3	D+ (Data +)
4	ID (host = GND; slave = N/C)
5	GND (Signal Ground)

USB Micro-A   **USB Micro-B**

1 2 3 4 5   1 2 3 4 5

Note: All figures not drawn to same scale.

HBK0723

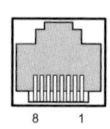

Ethernet Connector (RJ45-8 pin) Female

Pin	Signal
1	Output Transmit Data (+)
2	Output Transmit Data (–)
3	Input Receive Data (+)
4	N/C (not connected)
5	N/C
6	Input Receive Data (–)
7	N/C
8	N/C

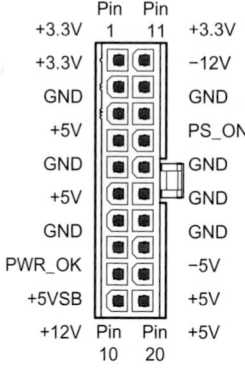

PC-ATX Type Power Connector Viewed from Connector End

	Pin 1	Pin 11	
+3.3V	1	11	+3.3V
+3.3V			–12V
GND			GND
+5V			PS_ON#
GND			GND
+5V			GND
GND			GND
PWR_OK			–5V
+5VSB			+5V
+12V	Pin 10	Pin 20	+5V

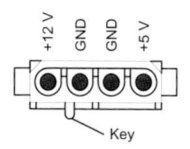

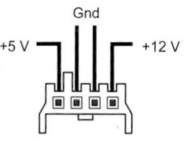

Disk Drive, CD and Other Device Power Connector Viewed from Connector End

+12 V   GND   GND   +5 V   +5 V   Gnd   +12 V

Key

22.9 RF Connectors and Transmission Lines

There are many different types of transmission lines and RF connectors for coaxial cable, but the three most common for amateur use are the UHF, Type N and BNC families. The type of connector used for a specific job depends on the size of the cable, the frequency of operation and the power levels involved. **Table 22.60** shows the characteristics of many popular transmission lines, while **Table 22.61** details coax connectors.

22.9.1 UHF Connectors

The so-called UHF connector (the series name is not related to frequency) is found on most HF and some VHF equipment. It is the only connector many hams will ever see on coaxial cable. PL-259 is another name for the UHF male, and the female is also known as the SO-239. These connectors are rated for full legal amateur power at HF. They are poor for UHF work because they do not present a constant impedance, so the UHF label is a misnomer. PL-259 connectors are designed to fit RG-8 and RG-11 size cable (0.405-inch OD). Adapters are available for use with smaller RG-58, RG-59 and RG-8X size cable. UHF connectors are not weatherproof.

Fig 22.19 shows how to install the solder type of PL-259 on RG-8 cable. Proper preparation of the cable end is the key to success. Follow these simple steps. Measure back about ¾-inch from the cable end and slightly score the outer jacket around its circumference. With a sharp knife, cut through the outer jacket, through the braid and through the dielectric — almost to the center conductor. Be careful not to score the center conductor. Cutting through all outer layers at once keeps the braid from separating. (Using a coax stripping tool with preset blade depth makes this and subsequent trimming steps much easier.)

Pull the severed outer jacket, braid and dielectric off the end of the cable as one piece. Inspect the area around the cut, looking for any strands of braid hanging loose and snip them off. There won't be any if your knife was sharp enough. Next, score the outer jacket about 5⁄16-inch back from the first cut. Cut through the jacket lightly; do not score the braid. This step takes practice. If you score the braid, start again. Remove the outer jacket.

Tin the exposed braid and center conductor, but apply the solder sparingly and avoid melting the dielectric. Slide the coupling ring onto the cable. Screw the connector body onto the cable. If you prepared the cable to the right dimensions, the center conductor will protrude through the center pin, the braid will show through the solder holes, and the body will actually thread onto the outer cable jacket. A very small amount of lubricant on the cable jacket will help the threading process.

Solder the braid through the solder holes. Solder through all four holes; poor connection

Fig 22.20 — Installing PL-259 plugs on RG-58 or RG-59 cable requires the use of UG-175 or UG-176 adapters, respectively. The adapter screws into the plug body using the threads of the connector that grip the jacket on larger cables. (*Courtesy Amphenol Electronic Components*)

83-1SP (PL-259) Plug with adapters (UG-176/U OR UG-175/U)

1. Cut end of cable even. Remove vinyl jacket 3/4" - don't nick braid. Slide coupling ring and adapter on cable.

2. Fan braid slightly and fold back over cable.

3. Position adapter to dimension shown. Press braid down over body of adapter and trim to 3/8". Bare 5/8" of conductor. Tin exposed center conductor.

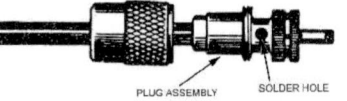

4. Screw the plug assembly on adapter. Solder braid to shell through solder holes. Solder conductor to contact sleeve.

5. Screw coupling ring on plug assembly.

HBK0460

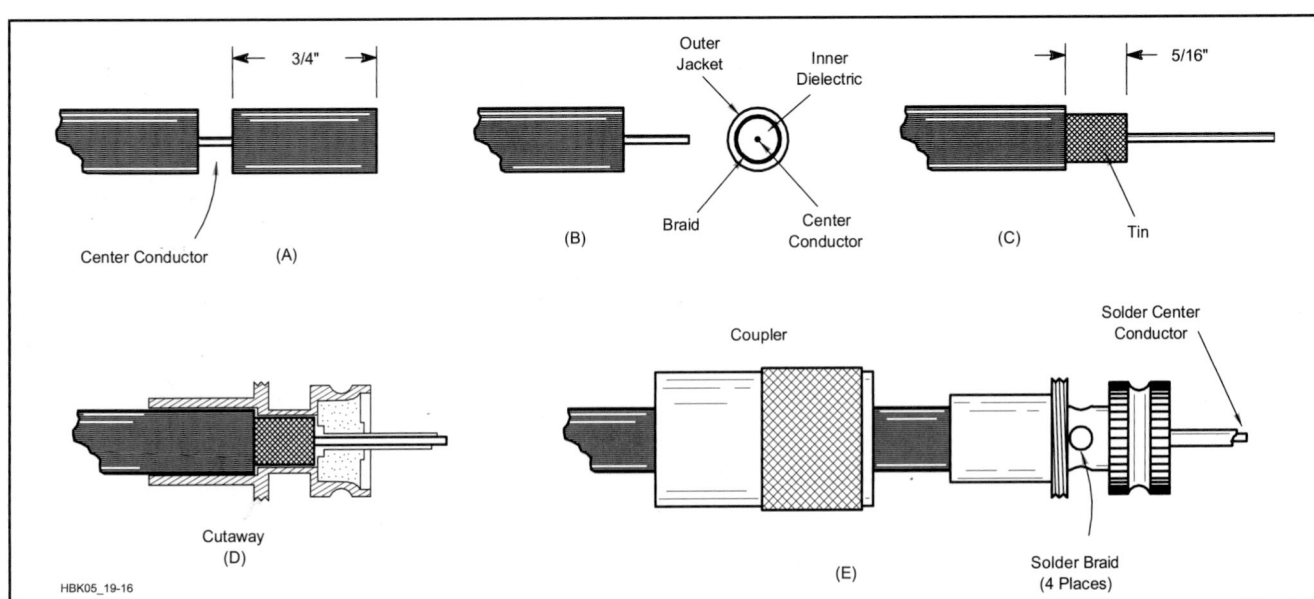

Fig 22.19 — The PL-259, or UHF, connector is almost universal for amateur HF work and is popular for equipment operating in the VHF range. Steps A through E are described in detail in the text.

to the braid is the most common form of PL-259 failure. A good connection between connector and braid is just as important as that between the center conductor and connector. Use a large soldering iron for this job. With practice, you'll learn how much heat to use. If you use too little heat, the solder will bead up, not really flowing onto the connector body. If you use too much heat, the dielectric will melt, letting the braid and center conductor touch. Most PL-259s are nickel plated, but silver-plated connectors are much easier to solder and only slightly more expensive.

Solder the center conductor to the center pin. The solder should flow on the inside, not the outside, of the center pin. If you wait until the connector body cools off from soldering the braid, you'll have less trouble with the dielectric melting. Trim the center conductor to be even with the end of the center pin. Use a small file to round the end, removing any solder that built up on the outer surface of the center pin. Use a sharp knife, very

fine sandpaper or steel wool to remove any solder flux from the outer surface of the center pin. Screw the coupling ring onto the body, and you're finished.

Fig 22.20 shows how to install a PL-259 connector on RG-58 or RG-59 cable. An adapter is used for the smaller cable with standard RG-8 size PL-259s. (UG-175 for RG-58 and UG-176 for RG-59.) Prepare the cable as shown. Once the braid is prepared, screw the adapter into the PL-259 shell and finish the job as you would a PL-259 on RG-8 cable.

Fig 22.21 shows the instructions and dimensions for crimp-on UHF connectors that fit all common sizes of coaxial cable. While amateurs have been reluctant to adopt crimp-on connectors, the availability of good quality connectors and inexpensive crimping tools make crimp technology a good choice, even for connectors used outside. Soldering the center conductor to the connector tip is optional.

UHF connectors are not waterproof and

must be waterproofed whether soldered or crimped as shown in the section of the **Safety** chapter on Antenna and Tower Safety.

22.9.2 BNC, N and F Connectors

The BNC connectors illustrated in **Fig 22.22** are popular for low power levels at VHF and UHF. They accept RG-58 and RG-59 cable, and are available for cable mounting in both male and female versions. Several different styles are available, so be sure to use the dimensions for the type you have. Follow the installation instructions carefully. If you prepare the cable to the wrong dimensions, the center pin will not seat properly with connectors of the opposite gender. Sharp scissors are a big help for trimming the braid evenly. Crimp-on BNC connectors are also available, with a large number of

(Text continues on page 22.51)

UHF Connectors
Braid Crimp - Solder Center Contact

Ferrule   Coupling Nut   Body assembly

Amphenol	Cable RG-/U	Cable Attachment		Hex Crimp Data			Stripping Dims, inches (mm)		
		Outer	Inner	Cavity for Outer Ferrule	Die Set Tool 227-994	CTL Series Tool No.	a	b	c
83-58SP	58, 141	Crimp	Solder	0.213(5.4)	227-1221-11	CTL-1	1.14 (29.0)	0.780 (19.9)	0.250 (6.4)
83-58SP-1002	400	Crimp	Solder	0.213(5.4)	227-1221-11	CTL-1	1.14 (29.0)	0.780 (19.9)	0.250 (6.4)
83-59DCP-RFX	59	Crimp	Solder	0255(6.5)	227-1221-13	CTL-1	1.22 (30.9)	0.574 (22.6)	0.543 (13.8)
83-58SCP-RFX	58	Crimp	Solder	0.213(5.4)	227-1221-11	CTL-1	1.22 (30.9)	0.574 (22.6)	0.543 (13.8)
83-59SP	59	Crimp	Solder	0.255(6.5)	227-1221-13	CTL-1	1.22 (30.9)	0.574 (22.6)	0.543 (13.8)
83-8SP-RFX	8	Crimp	Solder	0.429(10.9)	227-1221-25	CTL-3	1.22 (30.9)	0.574 (22.6)	0.543 (13.8)

See www.AmphenolRF.com for assembly instructions for all other connectore types. These dimensions only apply to Amphenol connectors and may not be correct for other manufacturers.

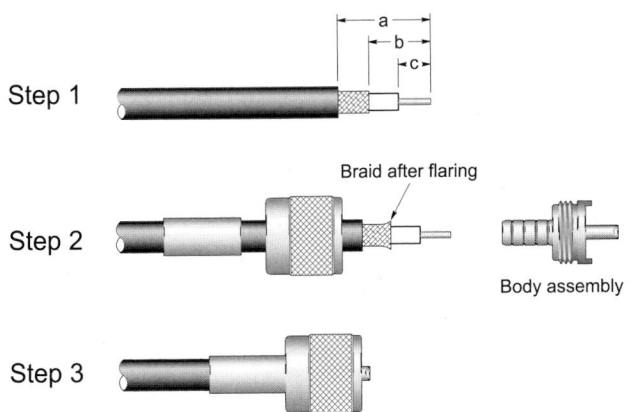

Step 1 Cut end of cable even. Strip cable to dimensions shown in table. All cuts are to be sharp and square. Do not nick braid, dielectric or center conductor. Tin center conductor avoiding excessive heat.

Step 2 Slide coupling nut and ferrule over cable jacket. Flair braid slightly as shown. Install cable into body assembly, so inner ferrule portion slides under braid, until braid butts shoulder. Slide outer ferrule over braid until it butts shoulder. Crimp ferrule with tool and die set indicated in table.

Step 3 Soft solder center conductor to contact. Avoid heating contact excessively to prevent damaging insulator. Slide/screw coupling nut over body.

HBK0475

Fig 22.21 — Crimp-on UHF connectors are available for all sizes of popular coaxial cable and save considerable time over soldered connectors. The performance and reliability of these connectors is equivalent to soldered connectors, if crimped properly. (*Courtesy Amphenol Electronic Components*)

Table 22.60

Nominal Characteristics of Commonly Used Transmission Lines

RG or Type	Part Number	Nom. Z_0 Ω	VF %	Cap. pF/ft	Cent. Cond. AWG	Diel. Type	Shield Type	Jacket Matl	OD inches	Max V (RMS)	Matched Loss (dB/100')			
											1 MHz	10	100	1000
RG-6	Belden 1694A	75	82	16.2	#18 Solid BC	FPE	FC	P1	0.275	300	0.3	.7	1.8	5.9
RG-6	Belden 8215	75	66	20.5	#21 Solid CCS	PE	D	PE	0.332	2700	0.4	0.8	2.7	9.8
RG-8	Belden 7810A	50	86	23.0	#10 Solid BC	FPE	FC	PE	0.405	300	0.1	0.4	1.2	4.0
RG-8	TMS LMR400	50	85	23.9	#10 Solid CCA	FPE	FC	PE	0.405	600	0.1	0.4	1.3	4.1
RG-8	Belden 9913	50	84	24.6	#10 Solid BC	ASPE	FC	P1	0.405	300	0.1	0.4	1.3	4.5
RG-8	CXP1318FX	50	84	24.0	#10 Flex BC	FPE	FC	P2N	0.405	600	0.1	0.4	1.3	4.5
RG-8	Belden 9913F	50	83	24.6	#11 Flex BC	FPE	FC	P1	0.405	300	0.2	0.6	1.5	4.8
RG-8	Belden 9914	50	82	24.8	#10 Solid BC	FPE	FC	P1	0.405	300	0.2	0.5	1.5	4.8
RG-8	TMS LMR400UF	50	85	23.9	#10 Flex BC	FPE	FC	PE	0.405	600	0.1	0.4	1.4	4.9
RG-8	DRF-BF	50	84	24.5	#9.5 Flex BC	FPE	FC	PE	0.405	600	0.1	0.5	1.6	5.2
RG-8	WM CQ106	50	84	24.5	#9.5 Flex BC	FPE	FC	P2N	0.405	600	0.2	0.6	1.8	5.3
RG-8	CXP008	50	78	26.0	#13 Flex BC	FPE	S	P1	0.405	600	0.1	0.5	1.8	7.1
RG-8	Belden 8237	52	66	29.5	#13 Flex BC	PE	S	P1	0.405	3700	0.2	0.6	1.9	7.4
RG-8X	Belden 7808A	50	86	23.5	#15 Solid BC	FPE	FC	PE	0.240	300	0.2	0.7	2.3	7.4
RG-8X	TMS LMR240	50	84	24.2	#15 Solid BC	FPE	FC	PE	0.242	300	0.2	0.8	2.5	8.0
RG-8X	WM CQ118	50	82	25.0	#16 Solid BC	FPE	FC	P2N	0.242	300	0.3	0.9	2.8	8.4
RG-8X	TMS LMR240UF	50	84	24.2	#15 Flex BC	FPE	FC	PE	0.242	300	0.2	0.8	2.8	9.6
RG-8X	Belden 9258	50	82	24.8	#16 Flex BC	FPE	S	P1	0.242	300	0.3	0.9	3.2	11.2
RG-8X	CXP08XB	50	80	25.3	#16 Flex BC	FPE	S	P1	0.242	300	0.3	1.0	3.1	14.0
RG-9	Belden 8242	51	66	30.0	#13 Flex SPC	PE	SCBC	P2N	0.420	5000	0.2	0.6	2.1	8.2
RG-11	Belden 8213	75	84	16.1	#14 Solid BC	FPE	S	PE	0.405	300	0.1	0.4	1.3	5.2
RG-11	Belden 8238	75	66	20.5	#18 Flex TC	PE	S	P1	0.405	300	0.2	0.7	2.0	7.1
RG-58	Belden 7807A	50	85	23.7	#18 Solid BC	FPE	FC	PE	0.195	300	0.3	1.0	3.0	9.7
RG-58	TMS LMR200	50	83	24.5	#17 Solid BC	FPE	FC	PE	0.195	300	0.3	1.0	3.2	10.5
RG-58	WM CQ124	52	66	28.5	#20 Solid BC	PE	S	PE	0.195	1400	0.4	1.3	4.3	14.3
RG-58	Belden 8240	52	66	29.9	#20 Solid BC	PE	S	P1	0.193	1400	0.3	1.1	3.8	14.5
RG-58A	Belden 8219	53	73	26.5	#20 Flex TC	FPE	S	P1	0.195	300	0.4	1.3	4.5	18.1
RG-58C	Belden 8262	50	66	30.8	#20 Flex TC	PE	S	P2N	0.195	1400	0.4	1.4	4.9	21.5
RG-58A	Belden 8259	50	66	30.8	#20 Flex TC	PE	S	P1	0.192	1400	0.5	1.5	5.4	22.8
RG-59	Belden 1426A	75	83	16.3	#20 Solid BC	FPE	S	P1	0.242	300	0.3	0.9	2.6	8.5
RG-59	CXP 0815	75	82	16.2	#20 Solid BC	FPE	S	P1	0.232	300	0.5	0.9	2.2	9.1
RG-59	Belden 8212	75	78	17.3	#20 Solid CCS	FPE	S	P1	0.242	300	0.2	1.0	3.0	10.9
RG-59	Belden 8241	75	66	20.4	#23 Solid CCS	PE	S	P1	0.242	1700	0.6	1.1	3.4	12.0
RG-62A	Belden 9269	93	84	13.5	#22 Solid CCS	ASPE	S	P1	0.240	750	0.3	0.9	2.7	8.7
RG-62B	Belden 8255	93	84	13.5	#24 Flex CCS	ASPE	S	P2N	0.242	750	0.3	0.9	2.9	11.0
RG-63B	Belden 9857	125	84	9.7	#22 Solid CCS	ASPE	S	P2N	0.405	750	0.2	0.5	1.5	5.8
RG-83	WM165	35	66	44.0	#10 Solid BC	PE	S	P2	0.405	2000	0.23	0.8	2.8	9.6
RG-142	CXP 183242	50	69.5	29.4	#19 Solid SCCS	TFE	D	FEP	0.195	1900	0.3	1.1	3.8	12.8
RG-142B	Belden 83242	50	69.5	29.0	#19 Solid SCCS	TFE	D	TFE	0.195	1400	0.3	1.1	3.9	13.5
RG-174	Belden 7805R	50	73.5	26.2	#25 Solid BC	FPE	FC	P1	0.110	300	0.6	2.0	6.5	21.3
RG-174	Belden 8216	50	66	30.8	#26 Flex CCS	PE	S	P1	0.110	1100	0.8	2.5	8.6	33.7
RG-213	Belden 8267	50	66	30.8	#13 Flex BC	PE	S	P2N	0.405	3700	0.2	0.6	2.1	8.0
RG-213	CXP213	50	66	30.8	#13 Flex BC	PE	S	P2N	0.405	600	0.2	0.6	2.0	8.2
RG-214	Belden 8268	50	66	30.8	#13 Flex SPC	PE	D	P2N	0.425	3700	0.2	0.7	2.2	8.0
RG-216	Belden 9850	75	66	20.5	#18 Flex TC	PE	D	P2N	0.425	3700	0.2	0.7	2.0	7.1
RG-217	WM CQ217F	50	66	30.8	#10 Flex BC	PE	D	PE	0.545	7000	0.1	0.4	1.4	5.2
RG-217	M17/78-RG217	50	66	30.8	#10 Solid BC	PE	D	P2N	0.545	7000	0.1	0.4	1.4	5.2
RG-218	M17/79-RG218	50	66	29.5	#4.5 Solid BC	PE	S	P2N	0.870	11000	0.1	0.2	0.8	3.4
RG-223	Belden 9273	50	66	30.8	#19 Solid SPC	PE	D	P2N	0.212	1400	0.4	1.2	4.1	14.5
RG-303	Belden 84303	50	69.5	29.0	#18 Solid SCCS	TFE	S	TFE	0.170	1400	0.3	1.1	3.9	13.5
RG-316	CXP TJ1316	50	69.5	29.4	#26 Flex BC	TFE	S	FEP	0.098	1200	1.2	2.7	8.0	26.1
RG-316	Belden 84316	50	69.5	29.0	#26 Flex SCCS	TFE	S	FEP	0.096	900	0.8	2.5	8.3	26.0
RG-393	M17/127-RG393	50	69.5	29.4	#12 Flex SPC	TFE	D	FEP	0.390	5000	0.2	0.5	1.7	6.1
RG-400	M17/128-RG400	50	69.5	29.4	#20 Flex SPC	TFE	D	FEP	0.195	1400	0.4	1.3	4.3	15.0
LMR500	TMS LMR500UF	50	85	23.9	#7 Flex BC	FPE	FC	PE	0.500	2500	0.1	0.4	1.2	4.0
LMR500	TMS LMR500	50	85	23.9	#7 Solid CCA	FPE	FC	PE	0.500	2500	0.1	0.3	0.9	3.3
LMR600	TMS LMR600	50	86	23.4	#5.5 Solid CCA	FPE	FC	PE	0.590	4000	0.1	0.2	0.8	2.7
LMR600	TMS LMR600UF	50	86	23.4	#5.5 Flex BC	FPE	FC	PE	0.590	4000	0.1	0.2	0.8	2.7
LMR1200	TMS LMR1200	50	88	23.1	#0 Copper Tube	FPE	FC	PE	1.200	4500	0.04	0.1	0.4	1.3

Hardline

RG or Type	Part Number	Nom. Z_0 Ω	VF %	Cap. pF/ft	Cent. Cond. AWG	Diel. Type	Shield Type	Jacket Matl	OD inches	Max V (RMS)	1 MHz	10	100	1000
1/2"	CATV Hardline	50	81	25.0	#5.5 BC	FPE	SM	none	0.500	2500	0.05	0.2	0.8	3.2
1/2"	CATV Hardline	75	81	16.7	#11.5 BC	FPE	SM	none	0.500	2500	0.1	0.2	0.8	3.2
7/8"	CATV Hardline	50	81	25.0	#1 BC	FPE	SM	none	0.875	4000	0.03	0.1	0.6	2.9
7/8"	CATV Hardline	75	81	16.7	#5.5 BC	FPE	SM	none	0.875	4000	0.03	0.1	0.6	2.9
LDF4-50A	Heliax – ½"	50	88	25.9	#5 Solid BC	FPE	CC	PE	0.630	1400	0.02	0.2	0.6	2.4
LDF5-50A	Heliax – ⅞"	50	88	25.9	0.355" BC	FPE	CC	PE	1.090	2100	0.03	0.10	0.4	1.3
LDF6-50A	Heliax – 1¼"	50	88	25.9	0.516" BC	FPE	CC	PE	1.550	3200	0.02	0.08	0.3	1.1

Parallel Lines

RG or Type	Part Number	Nom. Z_0 Ω	VF %	Cap. pF/ft	Cent. Cond. AWG	Diel. Type	Shield Type	Jacket Matl	OD inches	Max V (RMS)	1 MHz	10	100	1000
TV Twinlead (Belden 9085)		300	80	4.5	#22 Flex CCS	PE	none	P1	0.400	**	0.1	0.3	1.4	5.9
Twinlead (Belden 8225)		300	80	4.4	#20 Flex BC	PE	none	P1	0.400	8000	0.1	0.2	1.1	4.8
Generic Window Line		450	91	2.5	#18 Solid CCS	PE	none	P1	1.000	10000	0.02	0.08	0.3	1.1
WM CQ 554		440	91	2.7	#14 Flex CCS	PE	none	P1	1.000	10000	0.04	0.01	0.6	3.0
WM CQ 552		440	91	2.5	#16 Flex CCS	PE	none	P1	1.000	10000	0.05	0.2	0.6	2.6
WM CQ 553		450	91	2.5	#18 Flex CCS	PE	none	P1	1.000	10000	0.06	0.2	0.7	2.9
WM CQ 551		450	91	2.5	#18 Solid CCS	PE	none	P1	1.000	10000	0.05	0.02	0.6	2.8
Open-Wire Line		600	0.95-99***	1.7	#12 BC	none	none	none	**	12000	0.02	0.06	0.2	—

	1.8 MHz	7	14	30	50	150	220	450	1 GHz
RG-58 Style	1350	700	500	350	250	150	120	100	50
RG-59 Style	2300	1100	800	550	400	250	200	130	90
RG-8X Style	1830	840	560	360	270	145	115	80	50
RG-8/213 Style	5900	3000	2000	1500	1000	600	500	350	250
RG-217 Style	20000	9200	6100	3900	2900	1500	1200	800	500
LDF4-50A	38000	18000	13000	8200	6200	3400	2800	1900	1200
LDF5-50A	67000	32000	22000	14000	11000	5900	4800	3200	2100
LMR500	18000	9200	6500	4400	3400	1900	1600	1100	700
LMR1200	52000	26000	19000	13000	10000	5500	4500	3000	2000

Legend:

**	Not Available or varies	N	Non-Contaminating
***	Varies with spacer material and spacing	P1	PVC, Class 1
ASPE	Air Spaced Polyethylene	P2	PVC, Class 2
BC	Bare Copper	PE	Polyethylene
CC	Corrugated Copper	S	Single Braided Shield
CCA	Copper Cover Aluminum	SC	Silver Coated Braid
CCS	Copper Covered Steel	SCCS	Silver Plated Copper Coated Steel
CXP	Cable X-Perts, Inc.	SM	Smooth Aluminum
D	Double Copper Braids	SPC	Silver Plated Copper
DRF	Davis RF	TC	Tinned Copper
FC	Foil + Tinned Copper Braid	TFE	Teflon®
FEP	Teflon ® Type IX	TMS	Times Microwave Systems
Flex	Flexible Stranded Wire	UF	Ultra Flex
FPE	Foamed Polyethylene	WM	Wireman
Heliax	Andrew Corp Heliax		

Fig 22.22 (below) — BNC connectors are common on VHF and UHF equipment at low power levels. (*Courtesy Amphenol Electronic Components*)

BNC CONNECTORS

Standard Clamp

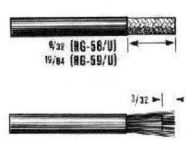

1. Cut cable even. Strip jacket. Fray braid and strip dielectric. **Don't nick braid or center conductor.** Tin center conductor.

2. Taper braid. Slide nut, washer, gasket and clamp over braid. Clamp inner shoulder should fit squarely against end of jacket.

3. With clamp in place, comb out braid, fold back smooth as shown. Trim center conductor.

4. Solder contact on conductor through solder hole. Contact should butt against dielectric. Remove excess solder from outside of contact. Avoid excess heat to prevent swollen dielectric which would interfere with connector body.

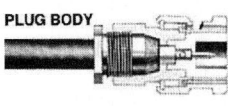

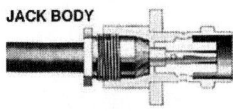

5. Push assembly into body. Screw nut into body with wrench until tight. **Don't rotate body on cable to tighten.**

HBK05_19-18

Improved Clamp

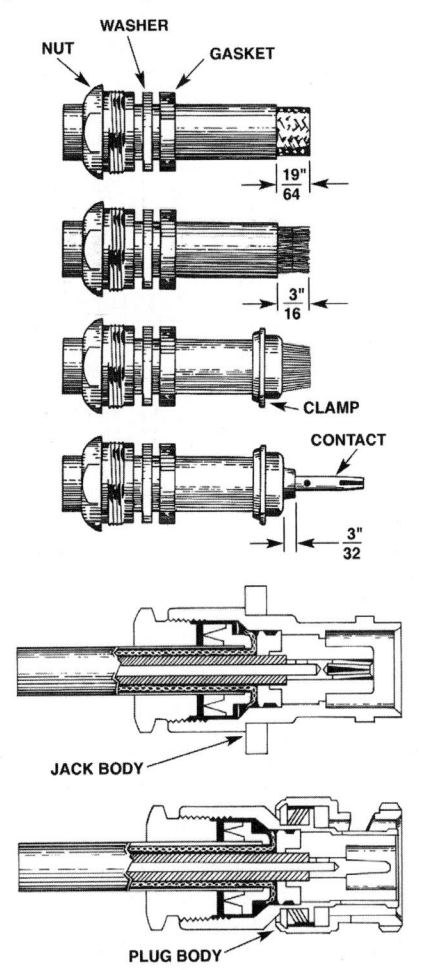

Follow 1, 2, 3 and 4 in BNC connectors (standard clamp) except as noted. Strip cable as shown. Slide gasket on cable *with groove facing clamp.* Slide clamp *with sharp edge facing gasket.* Clamp *should* cut gasket to seal properly.

C. C. Clamp

For Male Connectors (Plugs) (3/8" for Jacks)

1. Follow steps 1, 2, and 3 as outlined for the standard-clamp BNC connector.

2. Slide on bushing, rear insulator and contact. The parts must butt securely against each other, as shown.

3. Solder the center conductor to the contact. Remove flux and excess solder.

4. Slide the front insulator over the contact, making sure it butts against the contact shoulder.

5. Insert the prepared cable end into the connector body and tighten the nut. Make sure the sharp edge of the clamp seats properly in the gasket.

Table 22.61

Coaxial Cable Connectors

UHF Connectors

Military No.	Style	Cable RG- or Description
PL-259	Str (m)	8, 9, 11, 13, 63, 87, 149, 213, 214, 216, 225
UG-111	Str (m)	59, 62, 71, 140, 210
SO-239	Pnl (f)	Std, mica/phenolic insulation
UG-266	Blkhd (f)	Rear mount, pressurized, copolymer of styrene ins.

Adapters

PL-258	Str (f/f)	Polystyrene ins.
UG-224,363	Blkhd (f/f)	Polystyrene ins.
UG-646	Ang (f/m)	Polystyrene ins.
M-359A	Ang (m/f)	Polystyrene ins.
M-358	T (f/m/f)	Polystyrene ins.

Reducers

UG-175	55, 58, 141, 142 (except 55A)
UG-176	59, 62, 71, 140, 210

Family Characteristics:

All are nonweatherproof and have a nonconstant impedance. Frequency range: 0-500 MHz. Maximum voltage rating: 500 V (peak).

N Connectors

Military No.	Style	Cable RG-	Notes
UG-21	Str (m)	8, 9, 213, 214	50 Ω
UG-94A	Str (m)	11, 13, 149, 216	70 Ω
UG-536	Str (m)	58, 141, 142	50 Ω
UG-603	Str (m)	59, 62, 71, 140, 210	50 Ω
UG-23, B-E	Str (f)	8, 9, 87, 213, 214, 225	50 Ω
UG-602	Str (f)	59, 62, 71, 140, 210	—
UG-228B, D, E	Pnl (f)	8, 9, 87, 213, 214, 225	—
UG-1052	Pnl (f)	58, 141, 142	50 Ω
UG-593	Pnl (f)	59, 62, 71, 140, 210	50 Ω
UG-160A, B, D	Blkhd (f)	8, 9, 87, 213, 214, 225	50 Ω
UG-556	Blkhd (f)	58, 141, 142	50 Ω
UG-58, A	Pnl (f)		50 Ω
UG-997A	Ang (f)		50 Ω

Panel mount (f) with clearance above panel

M39012/04-	Blkhd (f)	Front mount hermetically sealed
UG-680	Blkhd (f)	Front mount pressurized

N Adapters

Military No.	Style	Notes
UG-29,A,B	Str (f/f)	50 Ω, TFE ins.
UG-57A.B	Str (m/m)	50 Ω, TFE ins.
UG-27A,B	Ang (f/m)	Mitre body
UG-212A	Ang (f/m)	Mitre body
UG-107A	T (f/m/f)	—
UG-28A	T (f/f/f)	—
UG-107B	T (f/m/f)	—

Family Characteristics:

N connectors with gaskets are weatherproof. RF leakage: –90 dB min @ 3 GHz. Temperature limits: TFE: –67° to 390°F (–55° to 199°C). Insertion loss 0.15 dB max @ 10 GHz. Copolymer of styrene: –67° to 185°F (–55° to 85°C). Frequency range: 0-11 GHz. Maximum voltage rating: 1500 V P-P. Dielectric withstanding voltage 2500 V RMS. SWR (MIL-C-39012 cable connectors) 1.3 max 0-11 GHz.

BNC Connectors

Military No.	Style	Cable RG-	Notes
UG-88C	Str (m)	55, 58, 141, 142, 223, 400	

Military No.	Style	Cable RG-	Notes
UG-959	Str (m)	8, 9	
UG-260,A	Str (m)	59, 62, 71, 140, 210	Rexolite ins.
UG-262	Pnl (f)	59, 62, 71, 140, 210	Rexolite ins.
UG-262A	Pnl (f)	59, 62, 71, 140, 210	nwx, Rexolite ins.
UG-291	Pnl (f)	55, 58, 141, 142, 223, 400	
UG-291A	Pnl (f)	55, 58, 141, 142, 223, 400	nwx
UG-624	Blkhd (f)	59, 62, 71, 140, 210	Front mount Rexolite ins.
UG-1094A	Blkhd		Standard
UG-625B	Receptacle		
UG-625			

BNC Adapters

Military No.	Style	Notes
UG-491,A	Str (m/m)	
UG-491B	Str (m/m)	Berylium, outer contact
UG-914	Str (f/f)	
UG-306	Ang (f/m)	
UG-306A,B	Ang (f/m)	Berylium outer contact
UG-414,A	Pnl (f/f)	# 3-56 tapped flange holes
UG-306	Ang (f/m)	
UG-306A,B	Ang (f/m)	Berylium outer contact
UG-274	T (f/m/f)	
UG-274A,B	T (f/m/f)	Berylium outer contact

Family Characteristics:

Z = 50 Ω. Frequency range: 0-4 GHz w/low reflection; usable to 11 GHz. Voltage rating: 500 V P-P. Dielectric withstanding voltage 500 V RMS. SWR: 1.3 max 0-4 GHz. RF leakage –55 dB min @ 3 GHz. Insertion loss: 0.2 dB max @ 3 GHz. Temperature limits: TFE: –67° to 390°F (–55° to 199°C); Rexolite insulators: –67° to 185°F (–55° to 85°C). "Nwx" = not weatherproof.

HN Connectors

Military No.	Style	Cable RG-	Notes
UG-59A	Str (m)	8, 9, 213, 214	
UG-1214	Str (f)	8, 9, 87, 213, 214, 225	Captivated contact
UG-60A	Str (f)	8, 9, 213, 214	Copolymer of styrene ins.
UG-1215	Pnl (f)	8, 9, 87, 213, 214, 225	Captivated contact
UG-560	Pnl (f)		
UG-496	Pnl (f)		
UG-212C	Ang (f/m)		Berylium outer contact

Family Characteristics:

Connector Styles: Str = straight; Pnl = panel; Ang = Angle; Blkhd = bulkhead. Z = 50 Ω. Frequency range = 0-4 GHz. Maximum voltage rating = 1500 V P-P. Dielectric withstanding voltage = 5000 V RMS SWR = 1.3. All HN series are weatherproof. Temperature limits: TFE: –67° to 390°F (–55° to 199°C); copolymer of styrene: –67° to 185°F (–55° to 85°C).

Cross-Family Adapters

Families	Description	Military No.
HN to BNC	HN-m/BNC-f	UG-309
N to BNC	N-m/BNC-f	UG-201,A
	N-f/BNC-m	UG-349,A
	N-m/BNC-m	UG-1034
N to UHF	N-m/UHF-f	UG-146
	N-f/UHF-m	UG-83,B
	N-m/UHF-m	UG-318
UHF to BNC	UHF-m/BNC-f	UG-273
	UHF-f/BNC-m	UG-255

Type N assembly instructions

CLAMP TYPES

Nut   Washer   Gasket   Clamp   Male Contact   Plug Body   Female Contact   Jack Body

Step 1

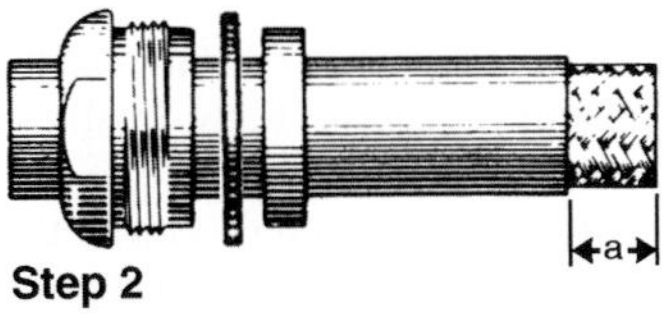

Step 2

Step 3

Step 4

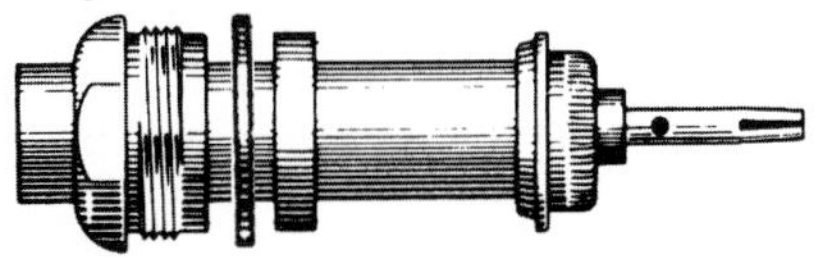

Step 5

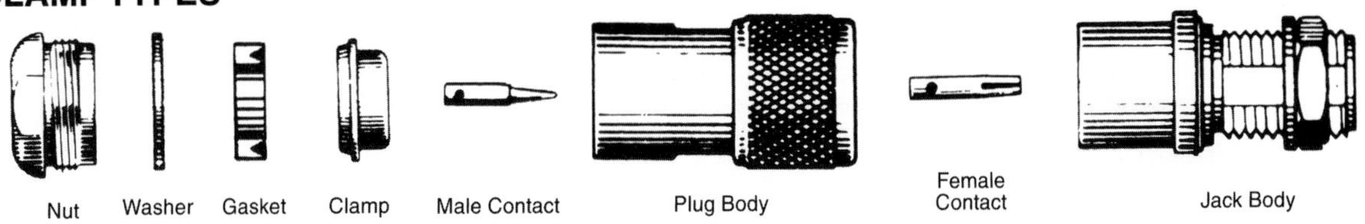

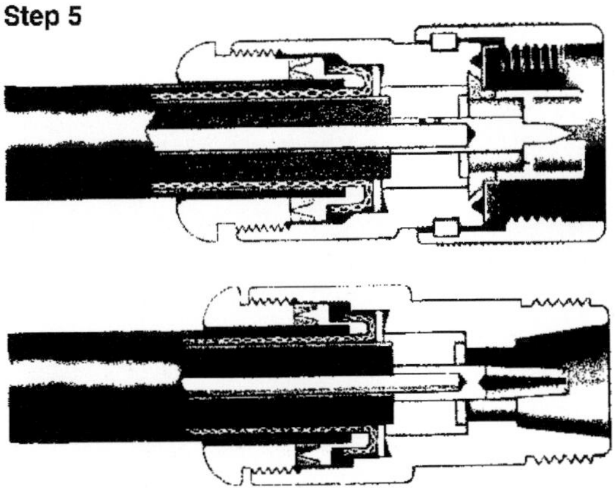

Amphenol Number	Connector Type	Cable RG-/U	Strip Dims., inches (mm)	
			a	c
82-61	N Plug	8, 9, 144, 165, 213, 214, 216, 225	0.359(9.1)	0.234(6.0)
82-62	N Panel Jack		0.312(7.9)	0.187(4.7)
82-63	N Jack	8, 9, 87A, 144, 165, 213, 214, 216, 225	0.281(7.1)	0.156(4.0)
82-67	N Bulkhead Jack			
82-202	N Plug	8, 9, 144, 165, 213, 214, 216, 225	0.359(9.1)	0.234(6.0)
82-202-RFX	N Plug	8, 213, 214	0.315(8.0)	0.177(4.5)
82-202-1006	N Plug	Belden 9913	0.359(9.1)	0.234(6.0)
82-835	N Angle Plug	8, 9, 87A, 144, 165, 213, 214, 216, 225	0.281(7.1)	0.156(4.0)
18750	N Angle Plug	58, 141, 142	0.484(12.3)	0.234(5.9)
34025	N Plug		0.390(9.9)	0.203(5.2)
34525	N Plug	59, 62, 71, 140, 210	0.410(10.4)	0.230(5.8)
35025	N Jack	58, 141, 142	0.375(9.5)	0.187(4.7)
36500	N Jack	59, 62, 71, 140, 210	0.484(12.3)	0.200(5.1)

See www.AmphenolRF.com for assembly instructions for all other connector types. These dimensions only apply to Amphenol connectors and may not be correct for other manufacturers.

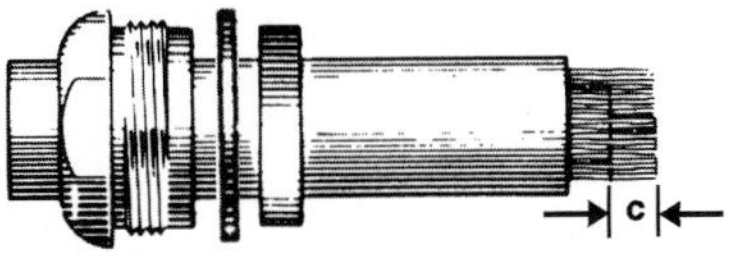

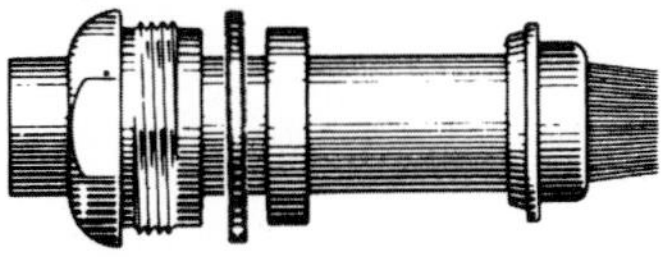

Step 1 Place nut, washer, and gasket, with "V" groove toward clamp, over cable and cut off jacket to dimension a.

Step 2 Comb out braid and fold out. Cut off cable dielectric to dim. c as shown.

Step 3 Pull braid wires forward and taper toward center conductor. Place clamp over braid and push back against cable jacket.

Step 4 Fold back braid wires as shown, trim braid to proper length and form over clamp as shown. Solder contact to center conductor.

Step 5 Insert cable and parts into connector body. Make sure sharp edge of clamp seats properly in gasket. Tighten nut.

Fig 22.23 — Type N connectors are a must for high-power VHF and UHF operation. (*Courtesy Amphenol Electronic Components*)

(Continued from page 22.47)

variations, including a twist-on version. A guide to installing these connectors is available on the CD-ROM accompanying this book.

The Type N connector, illustrated in **Fig 22.23**, is a must for high-power VHF and UHF operation. N connectors are available in male and female versions for cable mounting and are designed for RG-8 size cable. Unlike UHF connectors, they are designed to maintain a constant impedance at cable joints. Like BNC connectors, it is important to prepare the cable to the right dimensions. The center pin must be positioned correctly to mate with the center pin of connectors of the opposite gender. Use the right dimensions for the connector style you have. Crimp-on N connectors are also available, again with a large number of variations. A guide to installing these connectors is available on the CD-ROM accompanying this book.

Type F connectors, used primarily on cable TV connections, are also popular for receive-

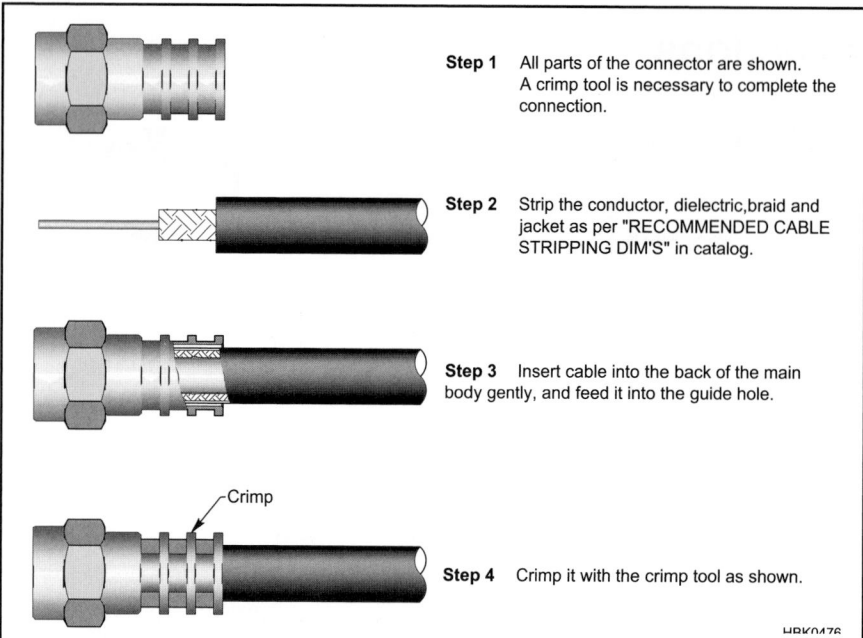

Step 1	All parts of the connector are shown. A crimp tool is necessary to complete the connection.
Step 2	Strip the conductor, dielectric, braid and jacket as per "RECOMMENDED CABLE STRIPPING DIM'S" in catalog.
Step 3	Insert cable into the back of the main body gently, and feed it into the guide hole.
Step 4	Crimp it with the crimp tool as shown.

HBK0476

Fig 22.24 — Type F connectors, commonly used for cable TV connections, can be used for receive-only antennas with inexpensive RG-59 and RG-6 cable. (*Courtesy Amphenol Electronic Components*)

only antennas and can be used with RG-59 or the increasingly popular RG-6 cable available at low cost. Crimp-on connectors are the only option for these connectors and **Fig 22.24** shows a general guide for installing them. The exact dimensions vary between connector styles and manufacturers — information on crimping is generally provided with the connectors. There are two styles of crimp; ferrule and compression. The ferrule crimp method is similar to that for UHF, BNC, and N connectors in which a metal ring is compressed around the exposed coax shield. The compression crimp forces a bushing into the back of the connector, clamping the shield against the connector body. In all cases, the exposed center conductor of the cable — a solid wire — must end flush with the end of the connector. A center conductor that is too short may not make a good connection.

22.10 Reference Tables

Table 22.62

US Customary Units and Conversion Factors

Linear Units

12 inches (in) = 1 foot (ft)
36 inches = 3 feet = 1 yard (yd)
1 rod = $5\frac{1}{2}$ yards = $16\frac{1}{2}$ feet
1 statute mile = 1760 yards = 5280 feet
1 nautical mile = 6076.11549 feet

Area

$1\ ft^2 = 144\ in^2$
$1\ yd^2 = 9\ ft^2 = 1296\ in^2$
$1\ rod^2 = 30\frac{1}{4}\ yd^2$
$1\ acre = 4840\ yd^2 = 43{,}560\ ft^2$
$1\ acre = 160\ rod^2$
$1\ mile^2 = 640\ acres$

Volume

$1\ ft^3 = 1728\ in^3$
$1\ yd^3 = 27\ ft^3$

Liquid Volume Measure

1 fluid ounce (fl oz) = 8 fluid drams = 1.804 in
1 pint (pt) = 16 fl oz
1 quart (qt) = 2 pt = 32 fl oz = $57\frac{3}{4}\ in^3$
1 gallon (gal) = 4 qt = 231 in^3
1 barrel = $31\frac{1}{2}$ gal

Dry Volume Measure

1 quart (qt) = 2 pints (pt) = 67.2 in^3
1 peck = 8 qt
1 bushel = 4 pecks = 2150.42 in^3

Avoirdupois Weight

1 dram (dr) = 27.343 grains (gr) or (gr a)
1 ounce (oz) = 437.5 gr
1 pound (lb) = 16 oz = 7000 gr
1 short ton = 2000 lb, 1 long ton = 2240 lb

Troy Weight

1 grain troy (gr t) = 1 grain avoirdupois
1 pennyweight (dwt) or (pwt) = 24 gr t
1 ounce troy (oz t) = 480 grains
1 lb t = 12 oz t = 5760 grains

Apothecaries' Weight

1 grain apothecaries' (gr ap)
    = 1 gr t = 1 gr
1 dram ap (dr ap) = 60 gr
1 oz ap = 1 oz t = 8 dr ap = 480 gr
1 lb ap = 1 lb t = 12 oz ap = 5760 gr

Conversion

Metric Unit = Metric Unit × US Unit

(Length)

mm	25.4	inch
cm	2.54	inch
cm	30.48	foot
m	0.3048	foot
m	0.9144	yard
km	1.609	mile
km	1.852	nautical mile

(Area)

mm^2	645.16	$inch^2$
cm^2	6.4516	in^2
cm^2	929.03	ft^2
m^2	0.0929	ft^2
cm^2	8361.3	yd^2
m^2	0.83613	yd^2
m^2	4047	acre
km^2	2.59	mi^2

(Mass) — **(Avoirdupois Weight)**

grams	0.0648	grains
g	28.349	oz
g	453.59	lb
kg	0.45359	lb
tonne	0.907	short ton
tonne	1.016	long ton

(Volume)

mm^3	16387.064	in^3
cm^3	16.387	in^3
m^3	0.028316	ft^3
m^3	0.764555	yd^3
ml	16.387	in^3
ml	29.57	fl oz
ml	473	pint
ml	946.333	quart
l	28.32	ft^3
l	0.9463	quart
l	3.785	gallon
l	1.101	dry quart
l	8.809	peck
l	35.238	bushel

(Mass) — **(Troy Weight)**

g	31.103	oz t
g	373.248	lb t

(Mass) — **(Apothecaries' Weight)**

g	3.387	dr ap
g	31.103	oz ap
g	373.248	lb ap

Multiply

Metric Unit = Conversion Factor × US Customary Unit

← Divide

Metric Unit ÷ Conversion Factor = US Customary Unit

Table 22.63

International System of Units (SI)—Metric Units

Prefix	Symbol			Multiplication Factor
exe	E	10^{18}	=	1,000,000 000,000,000,000
peta	P	10^{15}	=	1,000 000,000,000,000
tera	T	10^{12}	=	1,000,000,000,000
giga	G	10^{9}	=	1,000,000,000
mega	M	10^{6}	=	1,000,000
kilo	k	10^{3}	=	1,000
hecto	h	10^{2}	=	100
deca	da	10^{1}	=	10
		10^{0}	=	1
deci	d	10^{-1}	=	0.1
centi	c	10^{-2}	=	0.01
milli	m	10^{-3}	=	0.001
micro	μ	10^{-6}	=	0.000001
nano	n	10^{-9}	=	0.000000001
pico	p	10^{-12}	=	0.000000000001
femto	f	10^{-15}	=	0.000000000000001
atto	a	10^{-18}	=	0.000000000000000001

Linear

1 meter (m) = 100 centimeters (cm) = 1000 millimeters (mm)

Area

$1\ m^2 = 1 \times 10^4\ cm^2 = 1 \times 10^6\ mm^2$

Volume

$1\ m^3 = 1 \times 10^6\ cm^3 = 1 \times 10^9\ mm^3$
$1\ liter\ (l) = 1000\ cm^3 = 1 \times 10^6\ mm^3$

Mass

1 kilogram (kg) = 1000 grams (g)
  (Approximately the mass of 1 liter of water)
1 metric ton (or tonne) = 1000 kg

Table 22.64
Voltage-Power Conversion Table
Based on a 50-ohm system

RMS	Peak-to-Peak	dBmV	Watts	dBm
0.01 µV	0.0283 µV	−100	2×10^{-18}	−147.0
0.02 µV	0.0566 µV	−93.98	8×10^{-18}	−141.0
0.04 µV	0.113 µV	−87.96	32×10^{-18}	−134.9
0.08 µV	0.226 µV	−81.94	128×10^{-18}	−128.9
0.1 µV	0.283 µV	−80.0	200×10^{-18}	−127.0
0.2 µV	0.566 µV	−73.98	800×10^{-18}	−121.0
0.4 µV	1.131 µV	−67.96	3.2×10^{-15}	−114.9
0.8 µV	2.236 µV	−61.94	12.8×10^{-15}	−108.9
1.0 µV	2.828 µV	−60.0	20.0×10^{15}	−107.0
2.0 µV	5.657 µV	−53.98	80.0×10^{-15}	−101.0
4.0 µV	11.31 µV	−47.96	320.0×10^{-15}	−94.95
8.0 µV	22.63 µV	−41.94	1.28×10^{-12}	−88.93
10.0 µV	28.28 µV	−40.00	2.0×10^{-12}	− 86.99
20.0 µV	56.57 µV	−33.98	8.0×10^{-12}	−80.97
40.0 µV	113.1 µV	−27.96	32.0×10^{-12}	−74.95
80.0 µV	226.3 µV	−21.94	128.0×10^{-12}	−68.93
100.0 µV	282.8 µV	−20.0	200.0×10^{-12}	−66.99
200.0 µV	565.7 µV	−13.98	800.0×10^{-12}	−60.97
400.0 µV	1.131 mV	−7.959	3.2×10^{-9}	−54.95
800.0 µV	2.263 mV	−1.938	12.8×10^{-9}	−48.93
1.0 mV	2.828 mV	0.0	20.0×10^{-9}	−46.99
2.0 mV	5.657 mV	6.02	80.0×10^{-9}	−40.97
4.0 mV	11.31 mV	12.04	320×10^{-9}	−34.95
8.0 mV	22.63 mV	18.06	1.28 µW	−28.93
10.0 mV	28.28 mV	20.00	1 2.0 µW	−26.99
20.0 mV	56.57 mV	26.02	8.0 µW	−20.97
40.0 mV	113.1 mV	32.04	32.0 µW	−14.95
80.0 mV	226.3 mV	38.06	128.0 µW	−8.93
100.0 mV	282.8 mV	40.0	200.0 µW	−6.99
200.0 mV	565.7 mV	46.02	800.0 µW	−0.97
223.6 mV	632.4 mV	46.99	1.0 mW	0
400.0 mV	1.131 V	52.04	3.2 mW	5.05
800.0 mV	2.263 V	58.06	12.80 mW	11.07
1.0 V	2.828 V	60.0	20.0 mW	13.01
2.0 V	5.657 V	66.02	80.0 mW	19.03
4.0 V	11.31 V	72.04	320.0 mW	25.05
8.0 V	22.63 V	78.06	1.28 W	31.07
10.0 V	28.28 V	80.0	2.0 W	33.01
20.0 V	56.57 V	86.02	8.0 W	39.03
40.0 V	113.1 V	92.04	32.0 W	45.05
80.0 V	226.3 V	98.06	128.0 W	51.07
100.0 V	282.8 V	100.0	200.0 W	53.01
200.0 V	565.7 V	106.0	800.0 W	59.03
223.6 V	632.4 V	107.0	1,000.0 W	60.0
400.0 V	1,131.0 V	112.0	3,200.0 W	65.05
800.0 V	2,263.0 V	118.1	12,800.0 W	71.07
1000.0 V	2,828.0 V	120.0	20,000 W	73.01
2000.0 V	5,657.0 V	126.0	80,000 W	79.03
4000.0 V	11,310.0 V	132.0	320,000 W	85.05
8000.0 V	22,630.0 V	138.1	1.28 MW	91.07
10,000.0 V	28,280.0 V	140.0	2.0 MW	93.01

Table 22.65
Reflection Coefficient, Attenuation, SWR and Return Loss

Reflection Coefficient (%)	Attenuation (dB)	Max SWR	Return Loss, dB	Reflection Coefficient (%)	Attenuation (dB)	Max SWR	Return Loss, dB
1.000	0.000434	1.020	40.00	45.351	1.0000	2.660	6.87
1.517	0.001000	1.031	36.38	48.000	1.1374	2.846	6.38
2.000	0.001738	1.041	33.98	50.000	1.2494	3.000	6.02
3.000	0.003910	1.062	30.46	52.000	1.3692	3.167	5.68
4.000	0.006954	1.083	27.96	54.042	1.5000	3.352	5.35
4.796	0.01000	1.101	26.38	56.234	1.6509	3.570	5.00
5.000	0.01087	1.105	26.02	58.000	1.7809	3.762	4.73
6.000	0.01566	1.128	24.44	60.000	1.9382	4.000	4.44
7.000	0.02133	1.151	23.10	60.749	2.0000	4.095	4.33
7.576	0.02500	1.164	22.41	63.000	2.1961	4.405	4.01
8.000	0.02788	1.174	21.94	66.156	2.5000	4.909	3.59
9.000	0.03532	1.198	20.92	66.667	2.5528	5.000	3.52
10.000	0.04365	1.222	20.00	70.627	3.0000	5.809	3.02
10.699	0.05000	1.240	19.41	70.711	3.0103	5.829	3.01
11.000	0.05287	1.247	19.17				
12.000	0.06299	1.273	18.42				
13.085	0.07500	1.301	17.66				
14.000	0.08597	1.326	17.08				
15.000	0.09883	1.353	16.48				
15.087	0.10000	1.355	16.43				
16.000	0.1126	1.381	15.92				
17.783	0.1396	1.433	15.00				
18.000	0.1430	1.439	14.89				
19.000	0.1597	1.469	14.42				
20.000	0.1773	1.500	13.98				
22.000	0.2155	1.564	13.15				
23.652	0.2500	1.620	12.52				
24.000	0.2577	1.632	12.40				
25.000	0.2803	1.667	12.04				
26.000	0.3040	1.703	11.70				
27.000	0.3287	1.740	11.37				
28.000	0.3546	1.778	11.06				
30.000	0.4096	1.857	10.46				
31.623	0.4576	1.925	10.00				
32.977	0.5000	1.984	9.64				
33.333	0.5115	2.000	9.54				
34.000	0.5335	2.030	9.37				
35.000	0.5675	2.077	9.12				
36.000	0.6028	2.125	8.87				
37.000	0.6394	2.175	8.64				
38.000	0.6773	2.226	8.40				
39.825	0.75000	2.324	8.00				
40.000	0.7572	2.333	7.96				
42.000	0.8428	2.448	7.54				
42.857	0.8814	2.500	7.36				
44.000	0.9345	2.571	7.13				

$$\rho = \frac{SWR - 1}{SWR + 1}$$

where $\rho = 0.01 \times$ (reflection coefficient in %)

$$\rho = 10^{-RL/20}$$

where RL = return loss (dB)

$$\rho = \sqrt{1 - (0.1^X)}$$

where $X = A/10$ and A = attenuation (dB)

$$SWR = \frac{1 + \rho}{1 - \rho}$$

Return loss (dB) $= -8.68589 \ln (\rho) = -3.77155 \log (\rho)$
where ln is the natural log (log to the base e)

Attenuation (dB) $= -4.34295 \ln (1 - \rho^2) = 1.88578 \log (1 - \rho^2)$
where ln is the natural log (log to the base e)

Table 22.66

Abbreviations List

A

a—atto (prefix for 10^{-18})
A—ampere (unit of electrical current)
ac—alternating current
ACC—Affiliated Club Coordinator
ACSSB—amplitude-compandored single sideband
A/D—analog-to-digital
ADC—analog-to-digital converter
AF—audio frequency
AFC—automatic frequency control
AFSK—audio frequency-shift keying
AGC—automatic gain control
Ah—ampere hour
ALC—automatic level control
AM—amplitude modulation
AMRAD—Amateur Radio Research and Development Corporation
AMSAT—Radio Amateur Satellite Corporation
AMTOR—Amateur Teleprinting Over Radio
ANT—antenna
ARA—Amateur Radio Association
ARC—Amateur Radio Club
ARES—Amateur Radio Emergency Service
ARQ—Automatic repeat request
ARRL—American Radio Relay League
ARS—Amateur Radio Society (station)
ASCII—American National Standard Code for Information Interchange
ATV—amateur television
AVC—automatic volume control
AWG—American wire gauge
az-el—azimuth-elevation

B

B—bel; blower; susceptance; flux density, (inductors)
balun—balanced to unbalanced (transformer)
BC—broadcast
BCD—binary coded decimal
BCI—broadcast interference
Bd—baud (bids in single-channel binary data transmission)
BER—bit error rate
BFO—beat-frequency oscillator
bit—binary digit
bit/s—bits per second
BM—Bulletin Manager
BPF—band-pass filter
BPL—Brass Pounders League
BPL—Broadband over Power Line
BT—battery
BW—bandwidth
Bytes—Bytes

C

c—centi (prefix for 10^{-2})
C—coulomb (quantity of electric charge); capacitor
CAC—Contest Advisory Committee
CATVI—cable television interference

CB—Citizens Band (radio)
CBBS—computer bulletin-board service
CBMS—computer-based message system
CCITT—International Telegraph and Telephone Consultative Committee
CCTV—closed-circuit television
CCW—coherent CW
ccw—counterclockwise
CD—civil defense
cm—centimeter
CMOS—complementary-symmetry metal-oxide semiconductor
coax—coaxial cable
COR—carrier-operated relay
CP—code proficiency (award)
CPU—central processing unit
CRT—cathode ray tube
CT—center tap
CTCSS—continuous tone-coded squelch system
cw—clockwise
CW—continuous wave

D

d—deci (prefix for 10^{-1})
D—diode
da—deca (prefix for 10)
D/A—digital-to-analog
DAC—digital-to-analog converter
dB—decibel (0.1 bel)
dBi—decibels above (or below) isotropic antenna
dBm—decibels above (or below) 1 milliwatt
DBM—double balanced mixer
dBV—decibels above/below 1 V (in video, relative to 1 V P-P)
dBW—decibels above/below 1 W
dc—direct current
D-C—direct conversion
DDS—direct digital synthesis
DEC—District Emergency Coordinator
deg—degree
DET—detector
DF—direction finding; direction finder
DIP—dual in-line package
DMM—digital multimeter
DPDT—double-pole double-throw (switch)
DPSK—differential phase-shift keying
DPST—double-pole single-throw (switch)
DS—direct sequence (spread spectrum); display
DSB—double sideband
DSP—digital signal processing
DTMF—dual-tone multifrequency
DV—digital voice
DVM—digital voltmeter
DX—long distance; duplex
DXAC—DX Advisory Committee
DXCC—DX Century Club

E

e—base of natural logarithms (2.71828)
E—voltage

EA—ARRL Educational Advisor
EC—Emergency Coordinator
ECL—emitter-coupled logic
EHF—extremely high frequency (30-300 GHz)
EIA—Electronic Industries Alliance
EIRP—effective isotropic radiated power
ELF—extremely low frequency
ELT—emergency locator transmitter
EMC—electromagnetic compatibility
EME—earth-moon-earth (moonbounce)
EMF—electromotive force
EMI—electromagnetic interference
EMP—electromagnetic pulse
EOC—emergency operations center
EPROM—erasable programmable read only memory

F

f—femto (prefix for 10^{-15}); frequency
F—farad (capacitance unit); fuse
fax—facsimile
FCC—Federal Communications Commission
FD—Field Day
FEMA—Federal Emergency Management Agency
FET—field-effect transistor
FFT—fast Fourier transform
FL—filter
FM—frequency modulation
FMTV—frequency-modulated television
FSK—frequency-shift keying
FSTV—fast-scan (real-time) television
ft—foot (unit of length)

G

g—gram (unit of mass)
G—giga (prefix for 10^9); conductance
GaAs—gallium arsenide
GB—gigabytes
GDO—grid- or gate-dip oscillator
GHz—gigahertz (10^9 Hz)
GND—ground

H

h—hecto (prefix for 10^2)
H—henry (unit of inductance)
HF—high frequency (3-30 MHz)
HFO—high-frequency oscillator; heterodyne frequency oscillator
HPF—highest probable frequency; high-pass filter
Hz—hertz (unit of frequency, 1 cycle/s)

I

I—current, indicating lamp
IARU—International Amateur Radio Union
IC—integrated circuit
ID—identification; inside diameter
IEEE—Institute of Electrical and Electronics Engineers
IF—intermediate frequency

IMD—intermodulation distortion
in.—inch (unit of length)
in./s—inch per second (unit of velocity)
I/O—input/output
IRC—international reply coupon
ISB—independent sideband
ITF—Interference Task Force
ITU—International Telecommunication Union
ITU-T—ITU Telecommunication Standardization Bureau

J-K

j—operator for complex notation, as for reactive component of an impedance ($+j$ inductive; $-j$ capacitive)
J—joule (kg m^2/s^2) (energy or work unit); jack
JFET—junction field-effect transistor
k—kilo (prefix for 10^3); Boltzmann's constant (1.38x10^{-23} J/K)
K—kelvin (used without degree symbol) absolute temperature scale; relay
kB—kilobytes
kBd—1000 bauds
kbit—1024 bits
kbit/s—1024 bits per second
kbyte—1024 bytes
kg—kilogram
kHz—kilohertz
km—kilometer
kV—kilovolt
kW—kilowatt
kΩ—kilohm

L

l—liter (liquid volume)
L—lambert; inductor
lb—pound (force unit)
LC—inductance-capacitance
LCD—liquid crystal display
LED—light-emitting diode
LF—low frequency (30-300 kHz)
LHC—left-hand circular (polarization)
LO—local oscillator; Leadership Official
LP—log periodic
LS—loudspeaker
lsb—least significant bit
LSB—lower sideband
LSI—large-scale integration
LUF—lowest usable frequency

M

m—meter (length); milli (prefix for 10^{-3})
M—mega (prefix for 10^6); meter (instrument)
mA—milliampere
mAh—milliampere hour
MB—megabytes
MCP—multimode communications processor
MDS—Multipoint Distribution Service; minimum discernible (or detectable) signal
MF—medium frequency (300-3000 kHz)
mH—millihenry
MHz—megahertz

mi—mile, statute (unit of length)
mi/h (MPH)—mile per hour
mi/s—mile per second
mic—microphone
Mil—one-thousandth of an inch
min—minute (time)
MIX—mixer
mm—millimeter
MOD—modulator
modem—modulator/demodulator
MOS—metal-oxide semiconductor
MOSFET—metal-oxide semiconductor field-effect transistor
MS—meteor scatter
ms—millisecond
m/s—meters per second
msb—most-significant bit
MSI—medium-scale integration
MSK—minimum-shift keying
MSO—message storage operation
MUF—maximum usable frequency
mV—millivolt
mW—milliwatt
MΩ—megohm

N

n—nano (prefix for 10^{-9}); number of turns (inductors)
NBFM—narrow-band frequency modulation
NC—no connection; normally closed
NCS—net-control station; National Communications System
nF—nanofarad
NF—noise figure
nH—nanohenry
NiCd—nickel cadmium
NM—Net Manager
NMOS—N-channel metal-oxide silicon
NO—normally open
NPN—negative-positive-negative (transistor)
NPRM—Notice of Proposed Rule Making (FCC)
ns—nanosecond
NTIA—National Telecommunications and Information Administration
NTS—National Traffic System

O

OBS—Official Bulletin Station
OD—outside diameter
OES—Official Emergency Station
OO—Official Observer
op amp—operational amplifier
ORS—Official Relay Station
OSC—oscillator
OSCAR—Orbiting Satellite Carrying Amateur Radio
OTC—Old Timer's Club
oz—ounce (¹/₁₆ pound)

P

p—pico (prefix for 10^{-12})
P—power; plug
PA—power amplifier
PACTOR—digital mode combining aspects of packet and AMTOR
PAM—pulse-amplitude modulation
PBS—packet bulletin-board system

PC—printed circuit
PD—power dissipation
PEP—peak envelope power
PEV—peak envelope voltage
pF—picofarad
pH—picohenry
PIC—Public Information Coordinator
PIN—positive-intrinsic-negative (semiconductor)
PIO—Public Information Officer
PIV—peak inverse voltage
PLC—Power Line Carrier
PLL—phase-locked loop
PM—phase modulation
PMOS—P-channel (metal-oxide semiconductor)
PNP—positive negative positive (transistor)
pot—potentiometer
P-P—peak to peak
ppd—postpaid
PROM—programmable read-only memory
PSAC—Public Service Advisory Committee
PSHR—Public Service Honor Roll
PTO—permeability-tuned oscillator
PTT—push to talk

Q-R

Q—figure of merit (tuned circuit); transistor
QRP—low power (less than 5-W output)
R—resistor
RACES—Radio Amateur Civil Emergency Service
RAM—random-access memory
RC—resistance-capacitance
R/C—radio control
RCC—Rag Chewer's Club
RDF—radio direction finding
RF—radio frequency
RFC—radio-frequency choke
RFI—radio-frequency interference
RHC—right-hand circular (polarization)
RIT—receiver incremental tuning
RLC—resistance-inductance-capacitance
RM—rule making (number assigned to petition)
r/min (RPM)—revolutions per minute
rms—root mean square
ROM—read-only memory
r/s—revolutions per second
RS—Radio Sputnik (Russian ham satellite)
RST—readability-strength-tone (CW signal report)
RTTY—radioteletype
RX—receiver, receiving

S

s—second (time)
S—siemens (unit of conductance); switch
SASE—self-addressed stamped envelope
SCF—switched capacitor filter
SCR—silicon controlled rectifier
SEC—Section Emergency Coordinator

SET—Simulated Emergency Test
SGL—State Government Liaison
SHF—super-high frequency (3-30 GHz)
SM—Section Manager; silver mica (capacitor)
S/N—signal-to-noise ratio
SPDT—single-pole double-throw (switch)
SPST—single-pole single-throw (switch)
SS—ARRL Sweepstakes; spread spectrum
SSB—single sideband
SSC—Special Service Club
SSI—small-scale integration
SSTV—slow-scan television
STM—Section Traffic Manager
SX—simplex
sync—synchronous, synchronizing
SWL—shortwave listener
SWR—standing-wave ratio

T
T—tera (prefix for 10^{12}); transformer
TA—ARRL Technical Advisor
TC—Technical Coordinator
TCC—Transcontinental Corps (NTS)
TCP/IP—Transmission Control Protocol/Internet Protocol
tfc—traffic
TNC—terminal node controller (packet radio)
TR—transmit/receive
TS—Technical Specialist
TTL—transistor-transistor logic
TTY—teletypewriter
TU—terminal unit
TV—television
TVI—television interference
TX—transmitter, transmitting

U
U—integrated circuit
UHF—ultra-high frequency (300 MHz to 3 GHz)
USB—upper sideband
UTC—Coordinated Universal Time (also abbreviated Z)
UV—ultraviolet

V
V—volt; vacuum tube
VCO—voltage-controlled oscillator
VCR—video cassette recorder
VDT—video-display terminal
VE—Volunteer Examiner

VEC—Volunteer Examiner Coordinator
VFO—variable-frequency oscillator
VHF—very-high frequency (30-300 MHz)
VLF—very-low frequency (3-30 kHz)
VLSI—very-large-scale integration
VMOS—V-topology metal-oxide-semiconductor
VOM—volt-ohmmeter
VOX—voice-operated switch
VR—voltage regulator
VSWR—voltage standing-wave ratio
VTVM—vacuum-tube voltmeter
VUCC—VHF/UHF Century Club
VXO—variable-frequency crystal oscillator

W
W—watt (kg m^2s^{-3}), unit of power
WAC—Worked All Continents
WAS—Worked All States
WBFM—wide-band frequency modulation
WEFAX—weather facsimile
Wh—watthour
WPM—words per minute
WRC—World Radiocommunication Conference
WVDC—working voltage, direct current

X
X—reactance
XCVR—transceiver
XFMR—transformer
XIT—transmitter incremental tuning
XO—crystal oscillator
XTAL—crystal
XVTR—transverter

Y-Z
Y—crystal; admittance
YIG—yttrium iron garnet
Z—impedance; also see UTC

Numbers/Symbols
5BDXCC—Five-Band DXCC
5BWAC—Five-Band WAC
5BWAS—Five-Band WAS
6BWAC—Six-Band WAC
°—degree (plane angle)
°C—degree Celsius (temperature)
°F—degree Fahrenheit (temperature)
α—(alpha) angles; coefficients, attenuation constant, absorption factor, area, common-base forward current-transfer ratio of a bipolar transistor

β—(beta) angles; coefficients, phase constant, current gain of common-emitter transistor amplifiers
γ—(gamma) specific gravity, angles, electrical conductivity, propagation constant
Γ—(gamma) complex propagation constant
δ—(delta) increment or decrement; density; angles
Δ—(delta) increment or decrement determinant, permittivity
ε—(epsilon) dielectric constant; permittivity; electric intensity
ζ—(zeta) coordinates; coefficients
η—(eta) intrinsic impedance; efficiency; surface charge density; hysteresis; coordinate
θ—(theta) angular phase displacement; time constant; reluctance; angles
ι—(iota) unit vector
κ—(kappa) susceptibility; coupling coefficient
λ—(lambda) wavelength; attenuation constant
Λ—(lambda) permeance
μ—(mu) permeability; amplification factor; micro (prefix for 10^{-6})
μF—microfarad
μH—microhenry
μP—microprocessor
ξ—(xi) coordinates
π—(pi) ≈3.14159
ρ—(rho) resistivity; volume charge density; coordinates; reflection coefficient
σ—(sigma) surface charge density; complex propagation constant; electrical conductivity; leakage coefficient; deviation
Σ—(sigma) summation
τ—(tau) time constant; volume resistivity; time-phase displacement; transmission factor; density
ϕ—(phi) magnetic flux angles
Φ—(phi) angles
χ—(chi) electric susceptibility; angles
Ψ—(psi) dielectric flux; phase difference; coordinates; angles
ω—(omega) angular velocity 2πF
Ω—(omega) resistance in ohms; solid angle

Contents

Construction Techniques

Home construction of electronics projects and kits can be a fun part of Amateur Radio. Some say that hams don't build things nowadays; this just isn't so! A recent ARRL survey shows that more than half of active hams build some electronic projects. When you go to any ham flea market, you see row after row of dealers selling electronic components; people are leaving those tables with bags of parts. They must be doing something with them! Even experienced constructors will find valuable tips in this chapter. It discusses tools and their uses, electronic construction techniques, tells how to turn a schematic into a working circuit and then summarizes common mechanical construction practices. This chapter has been updated by Joe Eisenberg, KØNEB from material originally written and compiled by Ed Hare, W1RFI, and Jim Duffey, KK6MC.

Chapter 23 — CD-ROM Content

Supplemental Files

- "A No-Special-Tools SMD Desoldering Technique" by Wayne Yoshida, KH6WZ
- "Surface Mount Technology — You Can Work With It" by Sam Ulbing, N4UAU (Parts 1 - 4)
- "A Deluxe Soldering Station"
- Making Your Own Printed Circuit Boards
- "Reflow Soldering for the Radio Amateur" by Jim Koehler, VE5JP

23.1 Electronic Shop Safety

Building, repairing and modifying equipment in home workshops is a longstanding ham radio tradition. In fact, in the early days, building your own equipment was the only option available. While times and interests change, home construction of radio equipment and related accessories remains popular and enjoyable. Building your own gear need not be hazardous if you become familiar with the hazards, learn how to perform the necessary functions and follow some basic safe practices including the ones listed below and in the section on soldering. Let's start with our own abilities:

Consider your state of mind. Working on projects or troubleshooting (especially where high voltage is present) requires concentration. Don't work when you're tired or distracted. Be realistic about your ability to focus on the job at hand. Put another way, if we aren't able to be highly alert, we should put off doing hazardous work until we are able to focus on the hazards.

Think! Pay attention to what you are doing. No list of safety rules can cover all possibilities. Safety is always your responsibility. You must think about what you are doing, how it relates to the tools and the specific situation at hand. When working with tools, avoid creating situations in which you can be injured or the project damaged if things don't go "just right."

Take your time. If you hurry, not only will you make more mistakes and possibly spoil the appearance of your new equipment, you won't have time to think things through. Always plan ahead. Do not work with shop tools if you can't concentrate on what you are doing — it's not a race! Working when you are tired can also lead to problems, such as misidentification of parts or making poor or erroneous connections. Listening to the regular time notices on broadcast radio while you work will prompt you to take plenty of breaks so you do not work when you are too tired.

Protect yourself. Use of drills, saws, grinders and other wood- or metal-working equipment can release small fragments that could cause serious eye damage. Always wear safety glasses or goggles when doing work that might present a flying object hazard and that includes soldering, where small bits of molten solder can be flung a surprising distance. If you use hammers, wire-cutters, chisels and other hand tools, you will also need the protection that safety eyewear offers. Dress appropriately — loose clothing (or even hair) can be caught in exposed rotating equipment such as drill presses.

Don't work alone. Have someone nearby who can help if you get into trouble when working with dangerous equipment, chemicals or voltages.

Know what to do in an emergency. Despite your best efforts to be careful, accidents may still occur from time to time. Ensure that everyone in your household knows basic first aid procedures and understands how to summon help in an emergency. They should also know where to find and how to safely shut down electrical power in your shack and shop. Get medical help when necessary. Every workshop should contain a good first-aid kit. Keep an eye-wash kit near any dangerous chemicals or power tools that can create chips or splinters. If you become injured, apply first aid and then seek medical help if you are not sure that you are okay. Even a small burn or scratch on your eye can develop into a serious problem if not treated promptly.

What about the equipment and tools involved in shop work? Here are some basic safety considerations that apply to them, as well:

Table 23.1
Properties and Hazards of Chemicals often Used in the Shack or Workshop

Generic Chemical Name	Purpose or Use	Hazards	Ways to Minimize Risks
Lead-tin solder	Bonding electrical components	Lead exposure (mostly from hand contact) Flux exposure (inhalation)	Always wash hands after soldering or touching solder. Use good ventilation.
Isopropyl alcohol	Flux remover	Dermatitis (skin rash)	Wear molded gloves suitable for solvents.
		Vapor inhalation	Use good ventilation and avoid aerosol generation.
		Fire hazard	Use good ventilation, limit use to small amounts, keep ignition sources away, dispose of rags only in tightly sealed metal cans.
Freons	Circuit cooling and general solvent	Vapor inhalation Dermatitis	Use adequate ventilation. Wear molded gloves suitable for solvents.
Phenols and methylene chloride	Enameled wire/ paint stripper	Strong skin corrosive	Avoid skin contact; wear suitable molded gloves; use adequate ventilation.
Beryllium oxide	Ceramic insulator found in some power transistors and vacuum tubes that conducts heat well	Toxic when in fine dust form and inhaled	Avoid grinding, sawing or reducing to dust form.
Beryllium metal	Lightweight metal, alloyed with copper	Same as beryllium oxide	Avoid grinding, sawing, welding or reducing to dust. Contact supplier for special procedures.
Various paints	Finishing	Exposures to solvents	Adequate ventilation; use respirator when spraying.
		Exposures to sensitizers (especially urethane paint)	Adequate ventilation and use respirator.
		Exposure to toxic metals (lead, cadmium, chrome, and so on) in pigments	Adequate ventilation and use respirator.
		Fire hazard (especially when spray painting)	Adequate ventilation; control of residues; eliminate ignition sources.
Ferric chloride	Printed circuit board etchant	Skin and eye contact	Use suitable containers; wear splash goggles and molded gloves suitable for acids.
Ammonium persulphate and mercuric chloride	Printed circuit board etchants	Skin and eye contact	Use suitable containers; wear splash goggles and molded gloves suitable for acids.
Epoxy resins	General purpose cement or paint	Dermatitis and possible sensitizer	Avoid skin contact. Mix only amount needed.
Sulfuric acid	Electrolyte in lead-acid batteries	Strong corrosive when on skin or eyes. Will release hydrogen when charging (fire, explosion hazard).	Always wear splash goggles and molded plastic gloves (PVC) when handling. Keep ignition sources away from battery when charging. Use adequate ventilation.

Read instructions and manuals carefully... and follow them. The manufacturers of tools are the most knowledgeable about how to use their products safely. Tap their knowledge by carefully reading all operating instructions and warnings. Avoiding injuries with power tools requires safe tool design as well as proper operation by the user. Keep the instructions in a place where you can refer to them in the future.

Respect safety features of the equipment you work on and use. Never disable any safety feature of any tool. If you do, sooner or later you or someone else will make the mistake the safety feature was designed to prevent.

Keep your shop or work area neat and organized. A messy shop is a dangerous shop. A knife left laying in a drawer can cut someone looking for another tool; a hammer left on top of a shelf can fall down at the worst possible moment; a sharp tool left on a chair can be a dangerous surprise for the weary constructor who sits down.

Keep your tools in good condition. Always take care of your investment. Store tools in a way to prevent damage or use by untrained persons (young children, for example). Keep the cutting edges of saws, chisels and drill bits sharp. Protect metal surfaces from corrosion. Frequently inspect the cords and plugs of electrical equipment and make any necessary repairs. If you find that your power cord is becoming frayed, repair it right away. One solution is to buy a replacement cord with a molded connector already attached.

Make sure your shop is well ventilated. Paint, solvents, cleaners or other chemicals can create dangerous fumes. If you feel dizzy, get into fresh air immediately, and seek medical help if you do not recover quickly.

Respect power tools. Power tools are not forgiving. A drill can go through your hand a lot easier than metal. A power saw can remove a finger with ease. Keep away from the business end of power tools. Tuck in your shirt, roll up your sleeves and remove your tie before using any power tool. If you have long hair, tie it back so it can't become entangled in power equipment.

23.1.1 Chemicals

Chemicals such as cleaners, adhesives, construction materials, and coolants are used every day by amateurs without ill effects. Take the opportunity to become knowledgeable of the hazards associated with these materials and treat them with respect. **Table 23.1** summarizes the uses and hazards of chemicals and other materials used in the ham shack. It includes preventive measures that can minimize risk. For advanced infor-

---

Poison Control

If you think you have a chemical emergency, call your local poison control center immediately. Dial 1-800-222-1222 or use the map at **www.poison.org/otherPC** to find the center in your area or dial 911.

---

mation, the Centers for Disease Prevention and Control maintains an extensive database at **www.cdc.gov/niosh/topics/chemical-safety**. Meridian Engineering maintains a collection of materials safety information at **www.meridianeng.com/datasheets.html**. When in doubt, contact the manufacturer or distributor of the material for safety information or use an Internet search engine by entering the material name and "safety" into the search window.

Here are a few key suggestions for safely storing, handling, and using chemicals:

Read the information that accompanies the chemical and follow the manufacturer's recommended safety practices. If you would like more information than is printed on the label, ask for a material safety data sheet (MSDS). Manufacturers of brand-name chemical products usually post an MSDS on their product websites.

Store chemicals properly, away from sunlight and sources of heat. Secure their containers to prevent spills and so that children and untrained persons will not gain access. Always keep containers labeled so there is no confusion about the contents. It is best to use the container in which the chemical was purchased. If you transfer solvents to other

containers, such as wash bottles, label the new container with exactly what it contains.

Handle chemicals carefully to avoid spills. Clean up any spills or leaks promptly but don't overexpose yourself in the process. Never dispose of chemicals in household sinks or drains. Instead, contact your local waste plant operator, transfer station or fire department to determine the proper disposal procedures for your area. Many communities have household hazardous waste collection programs. Of course, the best solution is to only buy the amount of chemical that you will need, and use it all if possible. Always label any waste chemicals, especially if they are no longer in their original containers. Oil-filled capacitors and transformers were once commonly filled with oil containing PCBs. Never dispose of any such items that may contain PCBs in landfills – contact your county or city recycling office or local electric utility for information on proper disposal.

Always use recommended personal protective equipment (such as gloves, face shield, splash goggles and aprons). If corrosives (acids or caustics) are splashed on you *immediately* rinse with cold water for a minimum of 15 minutes to flush the skin thoroughly. If splashed in the eyes, direct a gentle stream of cold water into the eyes for at least 15 minutes. Gently lift the eyelids so trapped liquids can be flushed completely. Start flushing before removing contaminated clothing. Seek professional medical assistance. If using hazardous chemicals, it is unwise to work alone since people splashed with chemicals need the calm influence of another person.

Food and chemicals don't mix. Keep food, drinks and cigarettes *away* from areas where chemicals are used and don't bring chemicals to places where you eat.

23.2 Tools and Their Use

All electronic construction makes use of tools, from mechanical tools for chassis fabrication to the soldering tools used for circuit assembly. A good understanding of tools and their uses will enable you to perform most construction tasks.

While sophisticated and expensive tools often work better or more quickly than simple hand tools, with proper use, simple hand tools can turn out a fine piece of equipment. **Table 23.2** lists tools indispensable for construction of electronic equipment. These tools can be used to perform nearly any construction task. Add tools to your collection from time to time, as finances permit.

23.2.1 Sources of Tools

Electronic-supply houses, mail-order/web stores and most hardware stores carry the tools required to build or service Amateur Radio equipment. Bargains are available at ham flea markets or local neighborhood sales, but beware! Some flea-market bargains are really shoddy and won't work very well or last very long. Some used tools are offered for sale because the owner is not happy with their performance.

There is no substitute for quality! A high-quality tool, while a bit more expensive, will last a lifetime. Poor quality tools don't last long and often do a poor job even when brand

new. You don't need to buy machinist-grade tools, but stay away from cheap tools; they are not the bargains they might appear to be.

CARE OF TOOLS

The proper care of tools is more than a matter of pride. Tools that have not been cared for properly will not last long or work well. Dull or broken tools can be safety hazards. Tools that are in good condition do the work for you; tools that are misused or dull are difficult to use.

Store tools in a dry place. Tools do not fit in with most living room decors, so they are often relegated to the basement or garage.

Table 23.2
Recommended Tools and Materials

Simple Hand Tools
Screwdrivers

Slotted, 3-in, 1/8-in blade
Slotted, 8-in, 1/8-in blade
Slotted, 3-in, 3/16-in blade
Slotted, stubby, 1/4-in blade
Slotted, 4-in, 1/4-in blade
Slotted, 6-in, 5/16-in blade
Phillips, 2½-in, #0 (pocket clip)
Phillips, 3-in, #1
Phillips, stubby, #2
Phillips, 4-in, #2
Long-shank screwdriver with holding clip on
 blade
Jeweler's set
Right-angle, slotted and Phillips

Pliers, Sockets and Wrenches
Long-nose pliers, 6- and 4-in
Diagonal cutters, 6- and 4-in
Channel-lock pliers, 6-in
Slip-joint pliers
Locking pliers (Vise Grip or equivalent)
Socket nut-driver set, 1/16- to 1/2-in
Set of socket wrenches for hex nuts
Allen (hex) wrench set
Wrench set
Adjustable wrenches, 6- and 10-in
Tweezers, regular and reverse-action
Retrieval tool/parts holder, flexible claw
Retrieval tool, magnetic

Cutting and Grinding Tools
File set consisting of flat, round, half-round,
 and triangular. Large and miniature types
 recommended
Burnishing tool
Wire strippers
Wire crimper
Hemostat, straight
Scissors
Tin shears, 10-in
Hacksaw and blades
Hand nibbling tool (for chassis-hole
 cutting)

Scratch awl or scriber (for marking metal)
Heavy-duty jackknife
Knife blade set (X-ACTO or equivalent)
Machine-screw taps, #4-40 through
 #10-32 thread
Socket punches, 1/2 in, 5/8 in, 3/4 in, 1 1/8 in,
 1 1/4 in, and 1 1/2 in
Tapered reamer, T-handle, 1/2-in maximum
 width
Deburring tool

Miscellaneous Hand Tools
Combination square, 12-in, for layout work
Hammer, ball-peen, 12-oz head
Hammer, tack
Bench vise, 4-in jaws or larger
Center punch
Plastic alignment tools
Mirror, inspection
Flashlight, penlight and standard
Magnifying glass
Ruler or tape measure
Dental pick
Calipers
Brush, wire
Brush, soft
Small paintbrush
IC-puller tool

Hand-Powered Tools
Hand drill, 1/4-in chuck or larger
High-speed drill bits, #60 through 3/8-in
 diameter

Power Tools
Motor-driven emery wheel for grinding
Electric drill, hand-held
Drill press
Miniature electric motor tool (Dremel or
 equivalent) and accessory drill press

Soldering Tools and Supplies
Soldering station, adjustable temp,
 assorted tips
Soldering iron, 100 W or higher, 5/8-in tip

Soldering gun, 200 W or higher
Solder, 60/40, rosin core
Desoldering tool
Desoldering wick, 1/8-in and 1/4-in width
Liquid flux, pen or bottle
Isopropyl alcohol for flux removal

Safety
Safety glasses
Hearing protector, earphones or earplugs
Fire extinguisher
First-aid kit

Useful Materials
Medium-weight machine oil
Contact cleaner, liquid or spray can
RTV sealant or equivalent
Electrical tape, vinyl plastic
Sandpaper, assorted
Emery cloth
Steel wool, assorted
Cleaning pad, Scotchbrite or equivalent
Cleaners and degreasers
Contact lubricant
Sheet aluminum, solid and perforated,
 16- or 18-gauge, for brackets and shielding.
Aluminum angle stock, 1/2 × 1/2-in and 1/4-in
 diameter round brass or aluminum rod
 (for shaft extensions)
Machine screws: Round-head and flat head,
 with nuts and lockwashers to fit. Most useful
 sizes: 4-40, 6-32 and 8-32, in lengths from
 1/4-in to 1 1/2 in (Nickel-plated steel is satisfac-
 tory except in strong RF fields, where brass
 should be used.)
Bakelite, Lucite, polystyrene and copper-clad
 PC-board scraps.
Soldering lugs, panel bearings, rubber grommets, terminal-lug wiring strips, varnished-
 cambric insulating tubing, heat-shrinkable
 tubing
Shielded and unshielded wire
Tinned bare wire, #22, #14 and #12
Enameled wire, #20 through #30

Unfortunately, many basements or garages are not good places to store tools; dampness and dust are not good for tools. If your tools are stored in a damp place, use a dehumidifier. Sometimes you can minimize rust by keeping your tools lightly oiled, but this is a second-best solution. If you oil your tools, they may not rust, but you will end up covered in oil every time you use them. Wax or silicone spray is a better alternative.

Store tools neatly. A messy toolbox, with tools strewn about haphazardly, can be more than an inconvenience. You may waste a lot of time looking for the right tool and sharp edges can be dulled or nicked by tools banging into each other in the bottom of the box. As the old adage says, every tool should have a place, and every tool should be in its place. If you must search the workbench, garage, attic and car to find the right screwdriver,

you'll spend more time looking for tools than building projects.

SHARPENING

Many cutting tools can be sharpened. Send a tool that has been seriously dulled to a professional sharpening service. These services can sharpen saw blades, some files, drill bits and most cutting blades. Touch up the edge of cutting tools with a whetstone to extend the time between sharpening.

Sharpen drill bits frequently to minimize the amount of material that must be removed each time. Frequent sharpening also makes it easier to maintain the critical surface angles required for best cutting with least wear. Most inexpensive drill-bit sharpeners available for shop use do a poor job, either from the poor quality of the sharpening tool or inexperience of the operator. Also, drills should be sharp-

ened at different angles for different applications. Commercial sharpening services do a much better job.

INTENDED PURPOSE

Don't use tools for anything other than their intended purpose! If you use a pair of wire cutters to cut sheet metal, pliers as a vise or a screwdriver as a pry bar, you ruin a good tool and sometimes the work piece, as well. Although an experienced constructor can improvise with tools, most take pride in not abusing them. Having a wide variety of good tools at your disposal minimizes the problem of using the wrong tool for the job.

23.2.2 Tool Descriptions and Uses

Specific applications for tools are discussed throughout this chapter. Hand tools are used for so many different applications that they are discussed first, followed by some tips for proper use of power tools.

SCREWDRIVERS AND NUTDRIVERS

For construction or repair, you need to have an assortment of screwdrivers. Each blade size is designed to fit a specific range of screw head sizes. Using the wrong size blade usually damages the blade, the screw head or both. You may also need stubby sizes to fit into tight spaces. Right-angle screwdrivers are inexpensive and can get into tight spaces that can't otherwise be reached.

Electric screwdrivers are relatively inexpensive and very useful, particularly for repetitive tasks. If you have a lot of screws to fasten, they can save a lot of time and effort. They come with a wide assortment of screwdriver and nutdriver bits. An electric drill can also function as an electric screwdriver, although it may be heavy and over-powered for many applications.

Keep screwdriver blades in good condition. If a blade becomes broken or worn out, replace the screwdriver. A screwdriver only costs a few dollars; do not use one that is not in perfect condition. Save old screwdrivers to use as pry bars and levers, but use only good ones on screws. Filing a worn blade seldom gives good results.

Nutdrivers, the complement to screwdrivers, are often much easier to use than a wrench, particularly for nuts smaller than ⅜ inch. They are also less damaging to the nut than any type of pliers, with a better grip on the nut. Nutdrivers also minimize the chances of damage to front panels when tightening the nuts on control shafts. A set of interchangeable nutdrivers with a shared handle is a very handy addition to the toolbox.

PLIERS AND LOCKING-GRIP PLIERS

Pliers and locking-grip pliers are used to hold or bend things. They are not wrenches! If pliers are used to remove a nut or bolt, the nut or the pliers is usually damaged. To remove a nut, use a wrench or nutdriver. There is one exception to this rule of thumb: To remove a nut that is stripped too badly for a wrench, use a pair of pliers, locking-grip pliers, or a diagonal cutter to bite into the nut and start it turning. Reserve an old tool or one dedicated to just this purpose as it is not good for the tool.

Pliers are not intended for heavy-duty applications. Use a metal brake to bend heavy metal; use a vise to hold a heavy component. If the pliers' jaws or teeth become worn, replace the tool.

There are many different kinds of fine pliers, usually called "needle-nose" pliers or something similar, that are particularly useful

in electronics work. These are intended for light jobs, such as bending or holding wires or small work pieces. Two or three of these tools with different sizes of jaws will suffice for most jobs.

WIRE CUTTERS AND STRIPPERS

Wire cutters are primarily used to cut wires or component leads. The choice of blade style depends on the application. Diagonal blades or "dikes" are most often used to cut wire. Some delicate components can be damaged by cutting their leads with dikes because of the abrupt shock of the cut. Scissors or shears designed to cut wire should be used instead.

Specialized wire cutters are available to trim wires leads on circuit boards. These cutters are often called "flush cutters". Their cutting end is *not* designed to cut thicker wires. Use them *only* to clip smaller gauge wires, such as that on components used in circuits.

Wire strippers are available in manual and automatic styles. The manual strippers have a series of holes designed to remove insulation from a specific gauge of wire. Using the holes that are too big or too small will create nicks in the wire, which usually leads to the wire breaking at the nick. Automatic strippers grab and hold the wire for a consistent strip — some even judge the wire thickness automatically. If you strip a lot of wires, an automatic stripper may be worth the extra expense.

Wire strippers are handy, but with a little practice you can usually strip wires using a diagonal cutter or a knife. This is not the only use for a knife, so keep an assortment handy. Do not use wire cutters or strippers on anything other than wire! If you use a cutter to trim a protruding screw head or cut a hardened-steel spring, you will usually damage the blades.

FILES

Files are used for a wide range of tasks. In addition to enlarging holes and slots, they are used to remove burrs, shape metal, wood or plastic and clean some surfaces in preparation for soldering. Files are especially prone to damage from rust and moisture. Keep them in a dry place. The cutting edge of the blades can also become clogged with the material you are removing. Use file brushes (also called file cards) to keep files clean. Most files cannot be sharpened easily, so when the teeth become worn, the file must be replaced.

DRILL BITS

Drill bits are made from carbon steel, high-speed steel or carbide. Carbon steel is more common and is usually supplied unless a specific request is made for high-speed bits. Carbon steel drill bits cost less than high-speed or carbide types; they are sufficient for most

Table 23.3
Numbered Drill Sizes

No.	Diameter (Mils)	Will Clear Screw	Drilled for Tapping from Steel or Brass
1	228.0	12-24	—
2	221.0	—	—
3	213.0	—	14-24
4	209.0	12-20	—
5	205.0	—	—
6	204.0	—	—
7	201.0	—	—
8	199.0	—	—
9	196.0	—	—
10	193.5	—	—
11	**191.0**	**10-24**	—
		10-32	
12	189.0	—	—
13	185.0	—	—
14	182.0	—	—
15	180.0	—	—
16	177.0	—	12-24
17	173.0	—	—
18	169.5	—	—
19	166.0	8-32	12-20
20	161.0	—	—
21	**159.0**	—	**10-32**
22	157.0	—	—
23	154.0	—	—
24	152.0	—	—
25	**149.5**	—	**10-24**
26	147.0	—	—
27	144.0	—	—
28	**140.0**	**6-32**	—
29	**136.0**	—	**8-32**
30	128.5	—	—
31	120.0	—	—
32	116.0	—	—
33	**113.0**	**4-40**	—
34	111.0	—	—
35	110.0	—	—
36	**106.5**	—	**6-32**
37	104.0	—	—
38	101.5	—	—
39	**099.5**	**3-48**	—
40	098.0	—	—
41	096.0	—	—
42	093.5	—	—
43	**089.0**	—	**4-40**
44	**086.0**	**2-56**	—
45	082.0	—	—
46	081.0	—	—
47	**078.5**	—	**3-48**
48	076.0	—	—
49	073.0	—	—
50	**070.0**	—	**2-56**
51	067.0	—	—
52	063.5	—	—
53	059.5	—	—
54	055.0	—	—

equipment construction work. Carbide drill bits last much longer under heavy use. One disadvantage of carbide bits is that they are brittle and break easily, especially if you are using a hand-held power drill. When drilling abrasive material, such as fiberglass, the carbide bits last much longer than the steel bits.

Twist drills are available in a number of sizes listed in **Table 23.3**. Those listed in bold type are the most commonly used in construction of amateur equipment. You may not use all of the drills in a standard set, but it is nice to have a complete set on hand. You should also buy several spares of the more common sizes. Although Table 23.3 lists drills down to #54, the series extends to number #80. While the smaller sizes cannot usually be found in hardware stores or home improvement stores, they are commonly available through industrial tool suppliers and through various sources on the Internet.

A "step drill" consists of multiple drill diameters stacked as one bit, looking somewhat like a Christmas tree. These bits are very useful for drilling metal case material used in radio projects as the fluted edge of the bit in between sizes removes most if not all burrs from the hole you are drilling. Use them carefully and slowly to take advantage of their ability to remove burrs. A step drill also has the advantage of being able to drill a number of different standard size holes without having to change the bit.

SPECIALIZED TOOLS

Most constructors know how to use common tools, such as screwdrivers, wrenches and hammers. Although specialized tools usually do a job that can be done with other tools, once the specialty tool is used you will wonder how you ever did the job without it! Let's discuss other tools that are not so common.

A hand nibbling tool is shown in **Fig 23.1**. Use this tool to remove small "nibbles" of metal. It is easy to use; position the tool where you want to remove metal and squeeze the handle. The tool takes a small bite out of the metal. When you use a nibbler, be careful that you don't remove too much metal, clip the edge of a component mounted to the sheet metal or grab a wire that is routed near the edge of a chassis. Fixing a broken wire is easy, but something to avoid if possible. It is easy to remove metal but nearly impossible to put it back. Do it right the first time!

Deburring Tool

A deburring tool is just the thing to remove the sharp edges left on a hole after drilling or punching operations. See **Fig 23.2**. Position the tool over the hole and rotate it around the edge of the hole to remove burrs or rough edges. As an alternative, select a drill bit that is somewhat larger than the hole, position it over the hole, and spin it lightly to remove the burr. Be sure to deburr both sides of the hole.

Socket or Chassis Punches

Greenlee is the most widely known of the

socket-punch manufacturers. Most socket punches are round, but they do come in other shapes. To use one, drill a pilot hole large enough to clear the bolt that runs through the punch. Then, mount the punch as shown in **Fig 23.3**, with the cutter on one side of the sheet metal and the socket on the other.

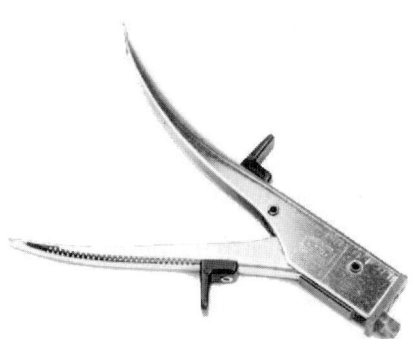
Fig 23.1 — A nibbling tool is used to remove small sections of sheet metal.

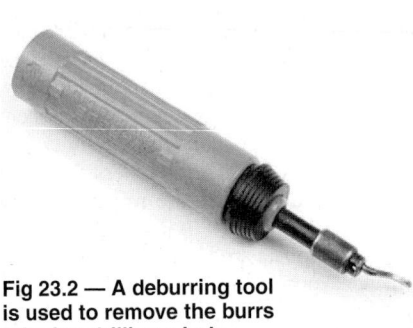
Fig 23.2 — A deburring tool is used to remove the burrs left after drilling a hole.

Fig 23.3 — A socket punch is used to punch a clean, round hole in sheet metal.

Tighten the nut with a wrench until the cutter cuts all the way through the sheet metal. These punches are often sold in sets at a significant discount to the same punches purchased separately. Hand-punches that operate by squeezing will also cut small holes by hand in light-gauge sheet metal, printed-circuit boards, and plastic.

Crimping Tools

The use of crimped connectors is common in the electronics industry. In many commercial and aerospace applications, soldered joints are no longer used. Hams have been reluctant to adapt crimped connections, largely due to mistrust of contacts that are not soldered, the use of cheap crimp connectors on consumer electronics, and the high cost of quality crimping tools or "crimpers." If high quality connectors and tools are used, the crimped connector will be as reliable a connection as a soldered one. The crimped connection is easier to make than a soldered one in most cases.

Crimped coaxial connectors are the most common crimped connector. MIL-spec or equivalent crimp connectors are available for the UHF, BNC, F and N-series MIL-spec connectors. Power connectors, such as the Anderson PowerPoles and Molex connectors are probably the second most commonly used crimped connections.

When purchasing a crimper, look for a ratcheting model with dies that are intended for the connectors you will be using. The common pliers-type crimper designed for household electrical terminals will have trouble crimping power connectors and is unsuitable for coaxial connectors. A good ratcheting crimper can be obtained for $50 to $100 with the necessary interchangeable dies. Large ratcheting crimpers suitable for the larger coaxial connectors can cost several hundred dollars. A good crimper and set of dies is an excellent investment for a club or group of like-minded hams.

Useful Shop Materials

Small stocks of various materials are used when constructing electronics equipment. Most of these are available from hardware or radio supply stores. A representative list is shown at the end of Table 23.2.

Small parts, such as machine screws, nuts, washers and soldering lugs can be economically purchased in large quantities (it doesn't pay to buy more than a lifetime supply). For items you don't use often, many radio supply stores or hardware stores sell small quantities and assortments. Stainless steel hardware can be kept on hand for outdoor use.

23.3 Soldering Tools and Techniques

Soldering is used in nearly every phase of electronic construction so you'll need soldering tools. This section discusses the tools and materials used in soldering.

23.3.1 Soldering Irons

A soldering tool must be hot enough to do the job and lightweight enough for agility and comfort. A heavy soldering gun useful for assembling wire antennas is too large for printed-circuit work, for example. A fine-tip soldering iron (sometimes called a "soldering pencil") works well for smaller jobs.

You may need an assortment of soldering irons to do a wide variety of soldering tasks. They range in size from a small 25 W iron for delicate printed-circuit work to larger 100 to 300 W soldering irons and guns. Small "pencil" butane-powered soldering irons and torches are also available, with a variety of soldering-iron tips. Most butane irons are not suited to use over long periods of time, but are ideal for small jobs and for performing tasks suited for a larger soldering iron, as they can be turned up to relatively high heating levels. Most small butane pencil irons can be set to provide as little as 10 W of equivalent heat, or up to 75-100 W. Butane powered irons also perform well outdoors, making them ideal for antenna work. Battery powered irons are available too, but are also not suited for use over long periods.

A 25 W pencil tip iron is adequate for printed circuit board work. A 40 W iron is necessary for larger jobs, such as soldering leads to panel connectors and making splices. These two irons or a temperature-controlled soldering station should handle most electronic homebrew requirements. A 100 W iron is good for bigger jobs, such as soldering antenna connections or soldering to power cables.

You should get several different sizes and shapes of tips when you purchase an iron. While most people prefer a conical tip, the chisel tip is also useful. The lower wattage irons will likely have a good selection of tip sizes and geometries. Irons 100 W and larger usually have a non-interchangeable chisel tip. For printed circuit board work, a good rule of thumb is to use a tip whose point is the same size as the component leads you are soldering.

If you buy an iron for use on circuits that contain electrostatic sensitive components, get one that has a grounded tip. Otherwise, you risk electrostatic damage to the components. Such irons are usually specified as having a grounded tip, and will have a three-prong plug. It is usually not necessary to have a grounded tip on the 100 W iron as it is not used on sensitive components.

Soldering guns are used for larger jobs and are too large for most electronics work. Where the soldering iron tip has an internal heater, the soldering gun tip is heated directly by current flowing in the tip. The nuts connecting the tip to the iron can loosen with each heat cycle, so they need tightening periodically. Soldering guns are available that have high and low heat levels controlled by an extra trigger position. Soldering gun tips are usually copper and do not last as long as the iron-clad tips of the smaller irons.

If you do a lot of building, a *soldering station* with a temperature-controlled tip is a better investment than a simple iron. The iron tip temperature can be precisely controlled and can be varied to the type of work being done. Soldering stations generally reach operating temperature more quickly than conventional irons and can be turned down to idle when you are not soldering. Some even have a digital tip temperature display. Soldering stations are available from numerous manufacturers. It is best to buy from an established manufacturer to insure future availability of tips and other components. Reconditioned soldering stations are often available and are good value.

If the only available soldering iron is a simple pencil type iron, an inexpensive means to vary the temperature of the iron is to use a lamp dimmer A simple control box can be made from a dimmer, wall socket, power cord, and outlet box available at any hardware store. Since a simple soldering iron is a resistive load, a lamp dimmer designed to work with incandescent lamps works well to cool down a soldering iron, and warm it up quickly once it is needed. This allows you to turn down the temperature of the iron while it is idle to prevent oxidation of the tip if left at full heat without being used.

RF-heated irons and hot-air soldering tools are primarily in use by industry and particularly useful for soldering surface-mount devices. The cost of these stations is high, although they are available as used and reconditioned.

MAINTAINING SOLDERING IRONS

Keep soldering tools in good condition by keeping the tips well-tinned with solder. Tinning is performed by melting solder directly onto the tip and letting it form a coating. Do not keep the tips at full temperature for long periods when not in use.

After each use, remove the tip and clean off any scale that may have accumulated. Clean an oxidized tip by dipping the hot tip in sal ammoniac (ammonium chloride) and then wiping it clean with a rag. Sal ammoniac is somewhat corrosive, so if you don't wipe the tip thoroughly, it can contaminate electronic soldering. There are proprietary "tip tinner" products available that can also be used for this function.

If a tip becomes oxidized during use, it can be restored to its shiny state by wiping the tip with a damp sponge or rag and then re-tinning. (Some tips are not supposed to be cleaned with water — check the manufacturer's recommendations.) A gentle scraping is also useful for stubborn cases. A copper or stainless-steel coil kitchen scrubber (do not use a scrubber that contains soap) works fine for this. Swipe the iron through the scrubber to clean it. A scrubber or sponge is most conveniently used by cramming it into a clean tuna can and placing it next to the solder station.

If a copper tip becomes pitted after repeated use, file it smooth and bright and then tin it immediately with solder. The solder prevents further oxidation of the tip. Modern soldering iron tips are nickel or iron clad and should not be filed, as the cladding protects the tip from pitting.

The secret of good soldering is to use the right amount of heat. Many people who have not soldered before use too little heat, dabbing at the joint to be soldered and making little solder blobs that cause unintended short circuits.

On a printed circuit board, examine your connections closely. If it looks like a rounded blob, reheat it with your soldering iron tip, drawing the tip away from the connection along the wire lead, forming a desired "Hershey's Kiss" type of appearance. A round blob will often hold the wire in place, but often makes little or no contact with the printed trace on the board's surface.

23.3.2 Solder

Solders have different melting points, depending on the ratio of tin to lead. Tin melts at 450 °F and lead at 621 °F. Solder made from 63% tin and 37% lead melts at 361 °F, the lowest melting point for a tin and lead mixture. Called 63-37 (or eutectic), this type of solder also provides the most rapid solid-to-liquid transition and the best stress resistance.

Solders made with different lead/tin ratios have a plastic state at some temperatures. If the solder is deformed while it is in the plastic state, the deformation remains when the solder freezes into the solid state. Any stress or motion applied to "plastic solder" causes a poor solder joint.

60-40 solder has the best wetting qualities. Wetting is the ability to spread rapidly, coat the surfaces to be joined, and bond materials uniformly. 60-40 solder also has a low melting point. These factors make it the most commonly used solder in electronics.

Some connections that carry high current can't be made with ordinary tin-lead solder because the heat generated by the current would melt the solder. Automotive starter brushes and transmitter tank circuits are two examples. Silver-bearing solders have higher melting points, and so prevent this problem. High-temperature silver alloys become liquid in the 1100 °F to 1200 °F range, and a silver-manganese (85-15) alloy requires almost 1800 °F.

Because silver dissolves easily in tin, tin-bearing solders can leach silver plating from components. This problem can be greatly reduced by partially saturating the tin in the solder with silver or by eliminating the tin. Commercial solders are available which incorporate these features. Tin-silver or tin-lead-silver alloys become liquid at temperatures from 430 °F for 96.5-3.5 (tin-silver), to 588 °F for 1.0-97.5-1.5 (tin-lead-silver). A 15.0-80.0-5.0 alloy of lead-indium-silver melts at 314 °F.

Rosin-core wire-type solder is formed into a tube with a flux compound inside. The resin (usually called "rosin" in solder) in a solder is a *flux*. Flux melts at a lower temperature than solder, so it flows out onto the joint before the solder melts to coat the joint surfaces. The solder used for surface-mount soldering (discussed later) is a cream or paste and flux, if used, must be added to the joint separately.

Flux removes oxide by suspending it in solution and floating it to the top. Flux is not a cleaning agent! Always clean the surfaces to be soldered before soldering. Flux is not a part of a soldered connection — it merely aids the soldering process.

After soldering, remove any remaining flux. Rosin flux can be removed with isopropyl or denatured alcohol. A cotton swab is a good tool for applying the alcohol and scrubbing the excess flux away. Commercial flux-removal sprays are available at most electronic-part distributors. Water-soluble fluxes are also available.

Never use acid flux or acid-core solder for electrical work. It should be used only for plumbing or chassis work. If used on electronics, the flux will corrode and damage the equipment. For circuit construction, only use fluxes or solder-flux combinations that are labeled for electronic soldering.

A basic tutorial on "Soldering 101" including a video demonstration is available from Sparkfun Electronics at **www.sparkfun. com/tutorials/213**.

LEAD-FREE SOLDER

In 2006, the European Union Restriction of Hazardous Substances Directive (RoHS) went into effect. This directive prohibits manufacture and import of consumer electronics which incorporate lead, including the common tin-lead solder used in electronic as-

sembly. California recently enacted a similar RoHS law. As a result of these directives there has been a move to lead-free solders in commercial use. They can contain two or more elements that are not as hazardous as lead, including tin, copper, silver, bismuth, indium, zinc, antimony and traces of other metals. Two lead-free solders commonly used for electronic use are SnAgCu alloy SAC305 and tin-copper alloy Sn100. SAC305 contains 96.5% tin, 3% silver, and 0.5% copper and melts at 217 °C. Sn100 contains 99.3% tin, 0.6% copper, as well as traces of nickel and silver and melts at 228 °C. Both of these melting points are higher than the 176 °C melting point of 60-40 and 63-37 lead-bearing solder, but conventional soldering stations will be able to reach the melting points of the new solders easily. Tin-lead solders are still available, but the move away from them by commercial manufacturers will probably lead to the day when they will be unavailable to hams who build their own gear. Be prepared.

The new RoHS solders can be used in much the same manner as conventional solders. The resulting solder joint appears somewhat duller than a conventional solder joint, and the lead-free solders tend to wick higher than the lead-tin solders. Due to the higher heat, it is important that the soldering iron tip be clean, shiny and freshly tinned so that heat is transferred to the joint to be soldered as quickly as possible to avoid excess heating of the parts being soldered. The soldering iron should be set to between 700 °F and 800 °F.

Solder and soldering equipment vendors provide numerous guides to hand soldering with lead-free solders. Weller's "Weller University" online presentation **www.elexp. com/tips/Weller_Coping_with_Lead_ Free.pdf** provides a great deal of detail about how soldering iron tips work with lead-free solder. More information is available from Kester (**www.kester.com**) in the Knowledge Base under the Hand Soldering link.

Most, if not all, RoHS-compliant components can be soldered with lead-tin solder. If the RoHS part has leads that are tinned or coated with an alloy to make soldering easier, it is necessary to use a hotter iron than would normally be required in lead-tin soldering. A soldering iron tip temperature of 315 °C (600 °F) or greater will be adequate for soldering RoHS parts with lead-tin solder. In contrast, a working tip temperature of 275 °C is generally adequate for working with conventional non-RoHS parts.

23.3.3 Soldering

The two key factors in quality soldering are time and temperature. Rapid heating is desired so that all parts of the joint are made hot enough for the solder to remain molten as it flows over the joint surfaces. Most unsuccess-

ful solder jobs fail because insufficient heat has been applied. To achieve rapid heating, the soldering iron tip should be hotter than the melting point of solder and large enough that transferring heat to the cooler joint materials occurs quickly. A tip temperature about 100 °F (60 °C) above the solder melting point is right for mounting components on PC boards.

Use solder that is sized appropriately for the job. As the cross section of the solder decreases, so does the amount of heat required to melt it. Diameters from 0.025 to 0.040 inch are good for nearly all circuit wiring. Sensitive and smaller components can be damaged or surfaces re-oxidized if heat is applied for too long a period.

If you are a beginner, you may want to start with one of the numerous "Learn to Solder" kits available from many electronics parts and kit vendors. The kits come with a printed-circuit board, a basic soldering iron, solder, and the components to complete a simple electronics project.

Here's how to make a good solder joint. This description assumes that solder with a flux core is used to solder a typical PC board connection such as an IC pin.

• Prepare the joint. Clean all conductors thoroughly with fine steel wool or a plastic scrubbing pad. Clean the circuit board at the beginning of assembly and individual parts such as resistors and capacitors immediately before soldering. Some parts (such as ICs and surface-mount components) cannot be cleaned easily; don't worry unless they're exceptionally dirty.

• Prepare the tool. It should be hot enough to melt solder applied to its tip quickly (half a second when dry, instantly when wet with solder). Apply a little solder directly to the tip so that the surface is shiny. This process is called "tinning" the tool. The solder coating helps conduct heat from the tip to the joint and prevents the tip from oxidizing.

• Place the tip in contact with one side of the joint. If you can place the tip on the underside of the joint, do so. With the tool below the joint, convection helps transfer heat to the joint.

• Place the solder against the joint directly opposite the soldering tool. It should melt within a second for normal PC connections, within two seconds for most other connections. If it takes longer to melt, there is not enough heat for the job at hand. If you have a variable heat soldering iron, adjust it so that the solder flows quickly for the size of wire and joints you are soldering. Much more heat can damage components and the board.

• Keep the tool against the joint until the solder flows freely throughout the joint. When it flows freely, solder tends to form concave joints called "fillets" between the conductors. With insufficient heat solder does not flow freely; it forms convex shapes — blobs. Once the solder shape changes from convex to concave, remove the tool from the joint. If a fillet won't form, the joint may need additional cleaning. Look for that "Hershey's Kiss" shape to know if it is done correctly.

• Let the joint cool without movement at room temperature. It usually takes no more than a few seconds. If the joint is moved before it is cool, it may take on a dull, satin or grainy appearance that is characteristic of a "cold" solder joint. Reheat cold joints until the solder flows freely and hold them still until cool.

• When the iron is set aside, or if it loses its shiny appearance, wipe away any dirt with a damp cloth or sponge. If it remains dull after cleaning, tin it again.

Overheating a transistor or diode while soldering can cause permanent damage, although as you get better at soldering, you'll be able to solder very quickly with little risk to the components. If the soldering iron will be applied for longer than a couple of seconds, use a small heat sink when you solder transistors, diodes or components with plastic parts that can melt. Grip the component lead with a pair of needle-nose pliers up close to the unit so that the heat is conducted away (be careful — it is easy to damage delicate component leads). A rubber-band wrapped around the pliers handles will hold the pliers on the wire. A small alligator clip or a flat spring type paper clip also makes a good heat sink.

Mechanical stress can damage components, too. Mount components so there is no appreciable mechanical strain on the leads. Be especially careful with small glass diodes and small disc capacitors as these components are easy to break when forming the leads.

Soldering to hollow pins, such as found on connectors, can be difficult, particularly if the connector has been used previously or has oxidized. Use a suitable small twist drill to clean the inside of the pin and then tin it. While the solder is still melted, clear the surplus solder from each pin with a whipping motion or by blowing through the pin from the inside of the connector. Watch out for flying hot solder — use safety goggles and protect the work surface and your arms and legs! A glass ashtray or small baking dish works great for catching the loose solder. Do not perform this operation near open electronic equipment as the loose solder can easily form short circuits. If the pin surface is plated, file the plating from the pin tip. Then insert the wire and solder it. After soldering, remove excess solder with a file, if necessary.

When soldering to the pins of plastic connectors or other assemblies, heat-sink the pin with needle-nose pliers at the base where it comes in contact with the plastic housing. Do not allow the pin to overheat; it will loosen and become misaligned.

23.3.4 Desoldering

There are times when soldered components need to be removed. The parts may be bad, they may be installed incorrectly, or you may want to remove them for use in another project.

There are several techniques for desoldering. The easiest way is to use a desoldering braid. Desoldering braid is simply fine copper braid, often containing flux. It is available under a wide variety of trade names wherever soldering supplies are sold. A good rule of thumb is to choose a width and thickness of desoldering braid that matches the size of the connection being desoldered.

The soldering braid is placed against the joint to be desoldered. A hot iron is pressed onto the braid. If you have a variable temperature soldering station, you might get better results by turning up the temperature, as the combination of the wick and the connection absorbs more heat. As the solder melts, it is wicked into the braid and away from the joint. Copper is an excellent conductor of heat, and the braid can get quite hot, so watch your fingers when using braid for desoldering. After all the leads have been treated in this manner the part can be removed. (A thin film of solder may remain, but is easily broken loose through the use of needle-nose pliers.) The part of the braid that wicked up the solder is clipped off.

Do not allow the used portion of the braid get too long, as it will absorb too much heat and not do as good of a job removing solder. When desoldering connections that have been made by using lead-free or "RoHS" solder, it can sometimes be difficult to achieve the high temperature needed to use solder wick, even with the iron heat turned all of the way up. A good tip in this case is to add a small amount of conventional leaded solder to the joint first, allowing it to mix. That lowers the melting point to a level that allows for much easier desoldering using solder wick, or other methods.

A desoldering vacuum pump can also be used. There are two types of desoldering vacuum tools, a simple rubber syringe bulb with a high temperature plastic tip and a desoldering pump. The desoldering pump is a simple manual vacuum pump consisting of a cylinder that contains a spring-loaded plunger attached to a metal rod inside a tip of high temperature plastic on the end of the pump. To desolder a joint, the plunger is pushed down, and locked in place. The tip is placed against the joint to be desoldered along with a soldering iron. When the solder melts, a button on the pump releases the plunger, which pulls the rod back, creating a vacuum that sucks the molten solder through the tip. The part being desoldered can then be removed. Pushing the plunger again ejects the solder from the desoldering tip.

The desoldering bulb employs a similar concept: heat the joint to be desoldered, squeeze the bulb, place the tip on the joint and release it to suck up the solder. Remove the part that was desoldered. If the first application of the desoldering pump doesn't suck up all the solder, reheat the joint and suck up the rest.

If the desoldering tool doesn't seem to be sucking up solder, the tip may be clogged with solder. The tip can be unclogged by pushing the solder through. You may have to clear the tip several times when doing a job that requires desoldering many joints. One can purchase small desoldering irons that contain a bulb on the handle that leads to a tip adjacent to the iron tip. This desoldering iron combination is somewhat easier to handle than the separate bulb and iron and does an effective job. Use a glass ashtray for a container to blow out the excess solder from this tool before using it again, as an ashtray is usually designed to handle heat.

Desoldering stations are also available. One type contains a vacuum pump in a console much like a soldering station. A vacuum line is connected to the tip of the soldering iron. There is a valve trigger on the iron that is used to open the tip to the vacuum when the solder is melted. The solder is sucked up the line into a receptacle in the station or in the hand piece. Another type of desoldering station that is commonly used in industry heats the joint with hot air so the component can be lifted off with pliers. These are particularly convenient to use and have the advantage that the hot air can be used for soldering surface mount devices.

Use surplus circuit boards to practice soldering and desoldering techniques. Old boards from all kinds of electronic items are often at very low cost or free at many flea markets. Use these surplus boards to practice different soldering and desoldering techniques without having to worry about damaging a project. Computer boards are often made with lead-free solder, and are ideal for practicing desoldering methods by adding leaded solder to the connections before desoldering.

23.3.5 Soldering Safety

Soldering requires a certain degree of practice and, of course, the right tools. What potential hazards are involved?

Since the solder used for virtually all electronic components is a lead-tin alloy, the first thing in most people's minds is lead, a well-known health hazard. There are two primary ways lead might enter our bodies when soldering: we could breathe lead fumes into our lungs or we could ingest (swallow) lead or lead-contaminated food. Inhalation of lead fumes is extremely unlikely because the temperatures ordinarily used in electronic soldering are far below those needed to vaporize lead. Nevertheless, since lead is soft and we may tend to handle it with our fingers, contaminating our food is a real possibility. For this reason, wash your hands carefully after any soldering (or touching of solder connections).

Using a small fan can keep the fumes away from your eyes and reduce your exposure to solder smoke. A small computer chipset fan, often only an inch or two wide, can be used. By reducing the voltage that feeds the fan, the speed and noise can be reduced, allowing the fumes to be blown away, yet not creating another problem with the airflow. Look in old computers for these little fans, often attached to video cards or to the bus chips on the motherboard. The CPU or case fans can be also used, but the voltage supplying them will definitely need to be reduced to create the desired level of airflow. There are also commercially available specialized fans with built-in filters designed for this purpose.

Soldering equipment gets *hot*! Be careful. Treat a soldering iron as you would any other hot object. A soldering iron stand is helpful, preferably one that has a cage that surrounds the hot tip of the iron. Here's a helpful tip — if the soldering iron gets knocked off the bench, train yourself not to grab for it because the chances are good that you'll grab the hot end!

When heated, the flux in solder gives off a vapor in the form of a light gray smoke-like plume. This flux vapor, which often contains aldehydes, is a strong irritant and can cause potentially serious problems to persons who suffer from respiratory sensitivity conditions such as asthma. In most cases it is relatively easy to use a small fan, like the small computer fans described previously to move the flux vapor away from your eyes and face. Opening a window provides additional air exchange.

Solvents are often used to remove excess flux after the parts have cooled to room temperature. Minimize skin contact with solvents by wearing molded gloves resistant to the solvent. If you use a solvent to remove flux, it is best to use the mildest one that does the job. Isopropyl alcohol, or rubbing alcohol, is often sufficient. You can purchase alcohol ranging from 70% to 92% concentration at local drug stores that works well in removing most types of fluxes. Some water-soluble solder fluxes can be removed with water.

Observe these precautions to protect yourself and others:

• Properly ventilate the work area. If you can smell fumes, you are breathing them.

• Wash your hands after soldering, especially before handling food.

• Minimize direct contact with flux and flux solvents. Wear disposable surgical gloves when handling solvents.

For more information about soldering hazards and the ways to make soldering safer, see "Making Soldering Safer," by Brian P. Bergeron, MD, NU1N (Mar 1991 *QST*, pp 28-30) and "More on Safer Soldering," by Gary E. Meyers, K9CZB (Aug 1991 *QST*, p 42).

23.4 Surface Mount Technology (SMT)

Today, nearly all consumer electronic devices are made with surface mount technology. Hams have lagged behind in adopting this technology largely due to the misconception that it is difficult, requires extensive practice and requires special equipment. In fact, surface mount devices can be soldered easily with commonly available equipment. There is no more practice required to become proficient enough to produce a circuit with surface-mount (SM) devices than there is with soldering through-hole (leaded) components.

There are several advantages to working with SMT:

• Projects are much more compact than if through-hole components are used

• SM parts are available that are not available in through-hole packages

• Fewer and fewer through-hole parts are being produced

• Equivalent SM components are often cheaper than through-hole parts, and

• SM parts have less self-inductance, less self-capacitance and better thermal properties.

There are several techniques that can be used effectively to work with SM components: conventional soldering iron, hot air reflow and hot plate/hot air reflow. This section describes the soldering iron technique. On-line descriptions of reflow techniques are available for the advanced builder that wants to try them. As is the case for through-hole soldering, kits are available to teach the beginner how to solder SMT components.

The following material on working with surface-mount technology contains excerpts from a series of *QST* articles, "Surface Mount Technology — You Can Work with It! Part 1" by Sam Ulbing, N4UAU, published in the April 1999 issue. (The entire series of four articles is available on the CD-ROM included with this book. Additional information on SMT is available at **www.arrl.org/surface-mount-technology**.) Additional information and illustrations were contributed by George Heron, N2APB.

23.4.1 Equipment Needed

You do not need lots of expensive equipment to work with SM devices.

• A fundamental piece of equipment for SM work is an illuminated magnifying glass. You can use an inexpensive one with a 5 inch diameter lens, and it's convenient to use the magnifier for all soldering work, not just for SM use. Such magnifiers are widely available from about $25. Most offer a 3× magnification and have a built-in circular light.

• A low-power, temperature-controlled soldering iron is necessary. Use a soldering iron with a grounded tip as most SM parts are CMOS devices and are subject to possible ESD (static) failure.

• Use of thin (0.020 inch diameter) rosin-core solder is preferred because the parts are so small that regular 0.031 inch diameter solder will flood a solder pad and cause bridging. Solder paste or cream can also be used.

• A flux pen comes in handy for applying just a little flux at a needed spot.

• Good desoldering braid is necessary to remove excess solder if you get too much on a pad. 0.100 inch wide braid works well.

• ESD protective devices such as wrist straps may be necessary if you live in a dry area and static is a problem.

• Tweezers help pick up parts and position them. Nonmagnetic, stainless-steel drafting dividers also work well. The nonmagnetic property of stainless steel means the chip doesn't get attracted to the dividers.

• Some hams prefer to hold components in place with a temporary adhesive such as DAP Blue-Stik while soldering rather than holding the part with tweezers.

23.4.2 Surface-Mount Parts

Fig 23.4 shows some common SM parts. (The **Component Data and References** chapter has more information on component packages.) Resistors and ceramic capacitors come in many different sizes, and it is important to know the part size for two reasons: Working with SM devices by hand is easier if you use the larger parts; and it is important that the PC-board pad size is larger than the part.

Discrete component packaging has shrunk to 0.12 × 0.06 inch, as shown in the "1206" capacitor in **Fig 23.5** compared to a penny. Even smaller packages are common today, requiring much less PC board area for the same equivalent circuits. Integrated circuit packaging has also been miniaturized to cre-

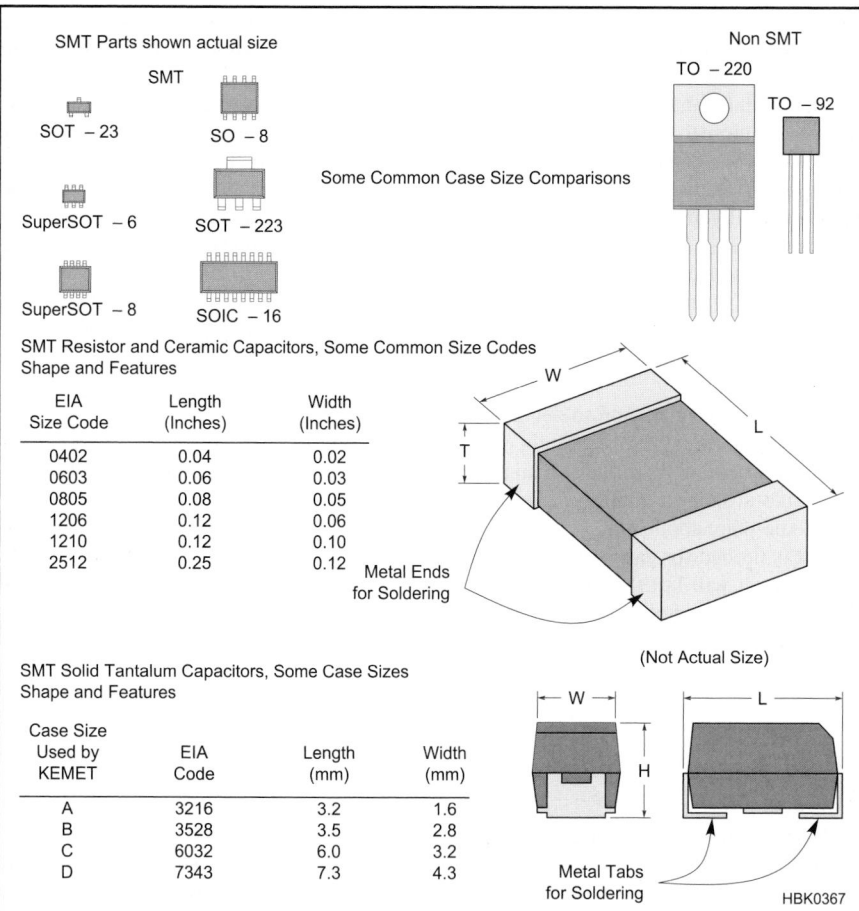

SMT Parts shown actual size

SMT

SOT – 23

SO – 8

SuperSOT – 6

SOT – 223

Some Common Case Size Comparisons

SuperSOT – 8

SOIC – 16

Non SMT

TO – 220

TO – 92

SMT Resistor and Ceramic Capacitors, Some Common Size Codes Shape and Features

EIA Size Code	Length (Inches)	Width (Inches)
0402	0.04	0.02
0603	0.06	0.03
0805	0.08	0.05
1206	0.12	0.06
1210	0.12	0.10
2512	0.25	0.12

Metal Ends for Soldering

(Not Actual Size)

SMT Solid Tantalum Capacitors, Some Case Sizes Shape and Features

Case Size Used by KEMET	EIA Code	Length (mm)	Width (mm)
A	3216	3.2	1.6
B	3528	3.5	2.8
C	6032	6.0	3.2
D	7343	7.3	4.3

Metal Tabs for Soldering

HBK0367

Fig 23.4 — Size comparisons of some surface-mount devices and their dimensions. See the Component Data and References chapter for more information about component packages and labeling.

Fig 23.5 — SMT components are small. Clockwise from left: MMIC RF amp, 1206 resistor, SOIC integrated circuit, 1206 capacitor and ferrite inductor.

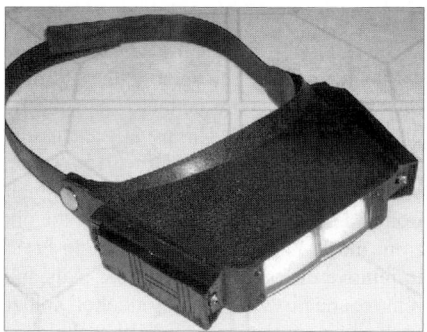

Fig 23.6 — A magnifying visor is great for close-up work on a circuit board. These headsets are often available for less than $10 at hamfests and some even come with superbright LEDs mounted on the side to illuminate the components being soldered.

ate 10 × 5 mm SOIC packages with lead separations of 0.025 inch.

Tantalum capacitors are one of the larger SM parts. Their case code, which is usually a letter, often varies from manufacturer to manufacturer because of different thicknesses. The EIA code for ceramic capacitors and resistors is a measurement of the length and width in inches, but for tantalum parts, those measurements are in millimeters times 10! Keep in mind that tantalum capacitors are polarized; the case usually has a mark or stripe to indicate the positive end. Nearly any part that is used in through-hole technology

is available in an SM package.

If you are a beginner, it is probably best to start with the larger sizes — 1206 for resistors and capacitors, SOT-23 for transistors, and SO-8 for ICs. When you get proficient with these parts you can move to smaller ones.

PREPARING FOR SMT WORK

The key to success with any construction project is selecting and using the proper tools. A magnifying lamp is essential for well-lighted, close-up work on the components. **Fig 23.6** shows a convenient magnifying visor. Tweezers or fine-tipped pliers allow you to grab the small chip components with dexterity.

A clean work surface is of paramount importance because SMT components have a tendency to slip from pliers or tweezers and fly off even when held with the utmost care. You'll have the best chance of recovering your wayward part if your table is clear. When the inevitable happens, you'll have lots of trouble finding it if the part falls onto a rug. It's best to have your work area in a room without carpeting, for this reason as well as to protect static-sensitive parts.

23.4.3 SMT Soldering Basics

If the project contains both SM parts and conventional through-hole parts and you intend to use a heat-gun or oven for reflow soldering, always mount the SM parts first, as through-hole components are not always designed to handle the higher reflow heat levels. Use junk PC boards to practice your soldering and desoldering techniques for SM parts until you are comfortable beginning your own project.

USING A SOLDERING IRON

Let's look at how to solder a surface-mount IC with a soldering iron. Use a little solder to pre-tin the PC board pads if the board is not already pre-tinned. The trick is to add just enough solder so that when you reheat it, it flows to the IC, but not so much that you wind up with a solder bridge. Putting a (very) little flux on the board and the IC legs makes for better solder flow, providing a smooth layer. You can tell if you have the proper soldering-iron tip temperature if the solder melts within 1.5 to 3.5 seconds.

Place the part on the board and then use dividers (or fingers) to push and prod the chip into position. Because the IC is so small and light, it tends to stick to the soldering iron and pull away from the PC board. To prevent this, use the dividers to hold the chip down while tack-soldering two IC legs at diagonally opposite corners. After each tack, check that the part is still aligned. With a dry and clean soldering iron, heat the PC board near the leg. If you do it right, when the solder melts, it will flow to the IC.

The legs of the IC must lie flat on the board. The legs bend easily, so don't press down too hard. Check each connection with a continuity checker placing one tip on the board the other on the IC leg. Check all adjacent pins to ensure there's no bridging. It is easier to correct errors early on, so perform this check often.

If you find that you did not have enough solder on the board for it to flow to the part, add a little solder. It's best to put a drop on the trace near the part, then heat the trace and slide the iron and melted solder toward the part. This reduces the chance of creating a bridge. Soldering resistors and capacitors is similar to soldering an IC's leads, except the resistors and capacitors don't have exposed leads. The reflow method works well for these parts, too.

Attaching wires that connect to points off the board can be a bit of a challenge because even #24 AWG stranded wire is large in comparison to the SM parts. First, make sure all the wire strands are close together, then pre-tin the wire. Pre-tin the pad, carefully place the wire on the pad, then heat it with the soldering iron until the solder melts.

USING REFLOW TECHNIQUES

While SMT projects can be built with conventional solder and a fine-point soldering iron, if you move on to reflow techniques you will need to use solder paste. Solder paste is a grayish looking paste made of a blend of flux and solder. A small dot of solder paste is put on the board at each location a component will need to be soldered. The components are then carefully placed on the board, with the paste loosely holding the parts in position. The whole board is then heated in an oven, or the area of the board being assembled is heated with a heat gun. With sufficient heat, the solder paste melts and flows onto the pad and component contacts, then the board is cooled leaving all of the components soldered in place.

A toaster oven can also be modified to perform reflow soldering as described in the article "Reflow Soldering for the Radio Amateur" by Jim Koehler, VE5FP, in the January 2011 issue of *QST* (this article is included on this book's CD-ROM). There are many on-line tutorials for adapting and using a toaster oven such as **http://hacknmod.com/hack/diy-reflow-surface-mount-soldering-smd-tutorial** and **www.youtube.com/watch?v=vduU4WWbpbM**. SparkFun also shows how to add a temperature control to a toaster oven at **www.sparkfun.com/tutorials/60**. If you plan on doing a lot of SMT assembly, learning reflow techniques is well worth the effort.

For occasional SMT use, a heat gun is a better choice. Many hobby/crafts stores sell a special heat tool that is used for melting embossing inks used in scrapbooking. Look for an embossing heat tool in the scrapbooking department of these stores. Do not get the heat gun too close to the board, as the airflow may move your components out of place. Hold the tool steady a couple of inches above the board until you see the solder paste turn silver and you see the component appear to be soldered in place, then gradually remove the heat gun. Never use a heat gun designed for paint stripping as the airflow is way too strong for this purpose and will blow the SM parts off the board.

Only a small amount of solder paste is required. Kester Easy Profile 256 is a good solder paste to start with. It is available at reasonable cost in a small syringe with a fine needle-point applicator. As only a small amount is needed for each solder joint, this small amount will last through several medium sized projects. Solder paste must be kept cold or it deteriorates. Kept in a household refrigerator or freezer it has a shelf life of at least a year.

23.4.4 Removing SMT Components

The surface-mount ICs used in commercial equipment are not easy for experimenters to replace. They have tiny pins designed for precision PC boards. Sooner or later, you may need to replace one, though. If you do, don't try to get the old IC out in one piece! This will damage the IC beyond use anyway, and will probably damage the PC board in the process.

Although it requires a delicate touch and small tools, it's possible to change a surface mount IC at home. To remove the old one, use small, sharp wire cutters to cut the IC pins flush with the IC. This usually leaves just enough of the old pin to grab with tweezers. Heat the soldered connection with a small iron and use the pliers to gently pull the pin from the PC board. Use desoldering braid to remove excess solder from the pads and remove and flux with rubbing alcohol. Solder in the new component using the techniques discussed above.

You can also use the embossing heat gun previously described to remove SM parts, especially ICs. Keep in mind that not only the desired component, but some adjacent parts as well may be loosened by this process, so be sure to not move the board during reheating to allow the other components to stay in place. Use a long-handled tweezers to remove a component once you see the solder become silvery again. Be careful to not disturb any adjacent components. Allow the board to thoroughly cool before moving it to prevent inadvertently allowing other components to shift before the solder solidifies again.

To remove individual components without a heat gun, first remove excess solder from the pads by using desoldering braid. Then the component will generally come loose from the pads if gently lifted with a hobby knife or dental tool. If the component remains attached to the pad, touch the pad with the soldering iron and lift the component off the pad. It may take one or two attempts to free the component from all pads. In extreme cases, it may be necessary to add solder to the pad to completely loosen the component.

23.4.5 SMT Construction Examples

The first project example is the DDS Daughtercard — a small module that generates precision RF signals for a variety of projects. This kit has become immensely popular in homebrew circles and is supplied with the chip components contained in color-coded packaging that makes and easy job of identifying the little parts, a nice touch by a kit supplier.

Fig 23.7A shows the DDS PC board, a typical layout for SMT components. All traces are on one side, since the component leads are not "through-hole." The little square pads are the places where the 1206 package-style chips will eventually be soldered. This project demonstrates the reflow technique using a soldering iron. **Fig 23.7B** illustrates how to use this technique.

(a) Pre-solder ("tin") one of the pads on the board where the component will ultimately go by placing a small blob of solder there.

(b) Carefully hold the component in place with small needle-nose pliers or sharp tweezers on the tinned pad.

(c) Reheat the tinned pad and component to reflow the solder onto the component lead, thus temporarily holding the component in place.

(d) Solder the other end of the component to its pad.

(e) Finally, check all connections to make sure there are no bridges or shorts.

Fig 23.7C shows the completed DDS board.

The second project example is the K8IQY Audio Amp — a discrete component audio amplifier that is homebrew-constructed "Manhattan-style" as described later in the chapter in which small pads are glued or soldered to the copper-clad base board wherever you need to attach component leads or wires. See **Fig 23.7D**.

Instead of using little squares or dots of PC board material for pads, you might decide to create isolated connection points by cutting an "island" in the copper using an end mill or pad cutter. No matter how the pads are created, SMT components may be easily soldered from pad-to-pad, or from pad-to-

Fig 23.7A — This DDS Daughtercard has all interconnections on the top side. Connections to the ground plane on the backside of the board are made by the use of "vias," wires through the PC board. Pin 28 of the SMT IC is shown being tack-soldered to hold it to the board, keeping all other pins carefully aligned on their pads. Then the other pins are carefully soldered, starting with pin 14 (opposite pin 28). Finally, pin 28 is reheated to ensure a good connection there. If you bridge solder across adjacent pads or pins, use solder wick or a vacuum solder sucker to draw off the excess solder.

Fig 23.7C — The fully-populated DDS Daughtercard PC board contains a mix of SMT and through-hole parts, showing how both packaging technologies can be used together.

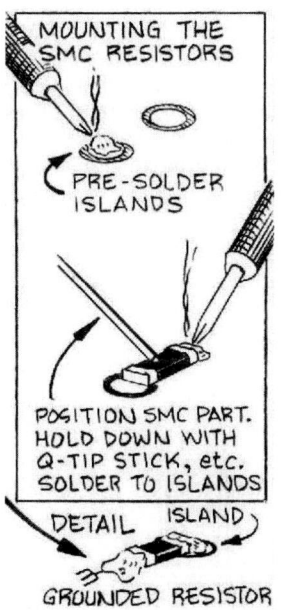

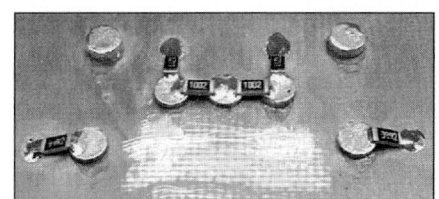

Fig 23.7D — SMT resistors soldered to base board of the Audio Amp in the beginning stages of assembly.

Fig 23.7F — Surface mount ICs can be mounted to general-purpose carrier boards, then attached as a submodule with wires to the base board of the homebrew project.

Fig 23.7B — Attaching an SMT part. It is a lot easier attaching capacitors, resistors and other discrete components compared to multi-pin ICs. Carefully hold the component in place and properly aligned using needle-nose pliers or tweezers and then solder one end of the component. Then reheat the joint while gently pushing down on the component with the pliers or a Q-tip stick to ensure it is lying flat on the board. Finally, solder the other side of the component.

Fig 23.7E — The completed homebrew Audio Amp assembly shows simple, effective use of SMT components used together with conventional leaded components when constructed "Manhattan-style."

ground plane to build up the circuit. **Fig 23.7E** shows the completed board, combining SMT and leaded components.

Homebrewing with SOIC-packaged integrated circuits is a little trickier and typically requires the use of an "SOIC carrier board" such as the one shown in **Fig 23.7F**, onto which you solder your surface-mount integrated circuit. You can then wire the carrier board onto your base board or whatever you're using to hold your other circuit components.

Full details on the DDS Daughtercard, the K8IQY Islander Audio Amp, and the Islander Pad Cutter may be found online at **www.njqrp.org**.

23.5 Constructing Electronic Circuits

Most of the construction projects undertaken by the average amateur involve electronic circuitry. The circuit is the "heart" of most amateur equipment. It might seem obvious, but in order for you to build it, the circuit must work! Don't always assume that a "cookbook" circuit that appears in an application note or electronics magazine is flawless. These are sometimes design examples that have not always been thoroughly debugged. Many home-construction projects are one-time deals; the author has put one together and it worked. In some cases, component tolerances or minor layout changes might make it difficult to get a second unit working. Using a solderless breadboard can make it easier to test this type of circuit design. For RF circuits above a few MHz, a solderless breadboard is not always practical due to long lead lengths that result.

Take steps to protect the electronic and mechanical components you use in circuit construction. Some components can be damaged by rough handling. Dropping a ¼ W resistor causes no harm, but dropping a vacuum tube or other delicate subassemblies usually causes damage.

Some components are easily damaged by heat. Some of the chemicals used to clean electronic components (such as flux removers, degreasers or control-lubrication sprays) can damage plastic. Check them for suitability before you use them.

23.5.1 Electrostatic Discharge (ESD)

Some components, especially high-impedance components such as FETs and CMOS gates, can be damaged by electrostatic discharge (ESD). Protect these parts from static charges. Most people are familiar with the static charge that builds up when one walks across a carpet then touches a metal object; the resultant spark can be quite lively. Walking across a carpet on a dry day can generate 35 kV! A worker sitting at a bench can generate voltages up to 6 kV, depending on conditions, such as when relative humidity is less than 20%.

You don't need this much voltage to damage a sensitive electronic component; damage can occur with as little as 30 V. The damage is not always catastrophic. A MOSFET can become noisy, or lose gain; an IC can suffer damage that causes early failure. To prevent this kind of damage, you need to take some precautions.

The energy from a spark can travel inside a piece of equipment to affect internal components. Protection of sensitive electronic components involves the prevention of static

build-up together with the removal of any existing charges by dissipating any energy that does build up.

MINIMIZING STATIC BUILD-UP

Several techniques can be used to minimize static build-up. Start by removing any carpet in your work areas. You can replace it with special antistatic carpet, but this is expensive. It's less expensive to treat the carpet with antistatic spray, which is available from electronics wholesalers. Adding humidity to the air can help reduce the presence of static charges as well.

Even the choice of clothing you wear can affect the amount of ESD. Polyester has a much greater ESD potential than cotton.

Many builders who have their workbench on a concrete floor use a rubber mat to minimize the risk of electric shocks from the ac line. Unfortunately, the rubber mat increases the risk of ESD. An antistatic rubber mat can serve both purposes.

Many components are shipped in antistatic packaging. Leave components in their conductive packaging. Other components, notably MOSFETs, are shipped with a small metal ring that temporarily shorts all of the leads together. Leave this ring in place until the device is fully installed in the circuit.

Use antistatic bags to transport susceptible components or equipment. Keep your workbench free of objects such as paper, plastic

and other static-generating items. Use conductive containers with a dissipative surface coating for equipment storage. Storing partially assembled projects in antistatic bags is also a good idea.

These precautions help reduce the build-up of electrostatic charges. Other techniques offer a slow discharge path for the charges or keep the components and the operator handling them at the same ground potential.

DISSIPATING STATIC

One of the best techniques is to connect the operator and the devices being handled to earth ground, or to a common reference point. It is not a good idea to directly ground an operator working on electronic equipment, though; the risk of shock is too great. If the operator is grounded through a high-value resistor such as 100 kΩ to 1 MΩ, ESD protection is still offered but there is no risk of shock.

The operator is usually grounded through a conductive wrist strap. This wrist band is equipped with a snap-on ground lead. A 1 MΩ resistor is built into the snap of the strap to protect the user should a live circuit be contacted. Build a similar resistor into any homemade ground strap.

The devices and equipment being handled are also grounded, by working on a charge-dissipating mat that is connected to ground. The mat should be an insulator that has been impregnated with a resistance mate-

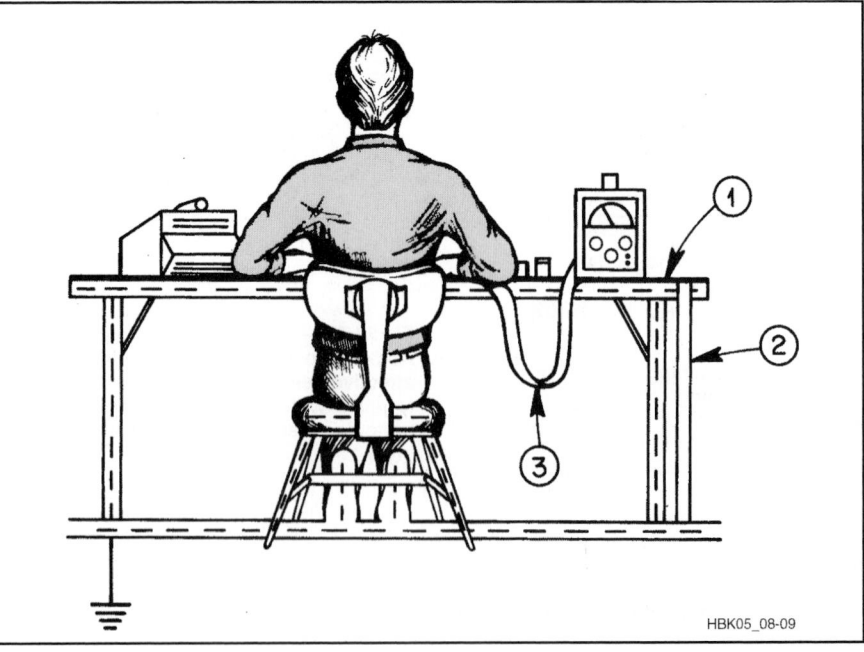

Fig 23.8 — A work station that has been set up to minimize ESD features (1) a grounded dissipative work mat and (2) a wrist strap that (3) grounds the worker through high resistance.

rial. Suitable mats and wrist straps are available from most electronics supply houses. **Fig 23.8** shows a typical ESD-safe work station.

The work area should also be grounded, directly or through a conductive mat. Use a soldering iron with a grounded tip to solder sensitive components. Most irons that have three-wire power cords are properly grounded. When soldering static-sensitive devices, use two or three jumpers: one to ground you, one to ground the work, and one to ground the iron. If the iron does not have a ground wire in the power cord, clip a jumper from the metal part of the iron near the handle to the metal box that houses the temperature control. Another jumper connects the box to the work. Finally, a jumper goes from the box to an elastic wrist band for static grounding.

23.5.2 Sorting Parts

When building a project, especially one packaged as a kit, finding the appropriate container to sort your components can be a problem. There are a number of things that make this task a lot easier and at a low cost.

Using a plastic egg carton or poking component leads into Styrofoam works quite well for sorting parts, but both methods can lead to ESD damage of sensitive components. Use this method for components such as resistors, capacitors and inductors that are relatively immune to ESD.

Metal cupcake trays are ideal, as the tray itself can be grounded through a 1 MΩ resistor to prevent static buildup. Cupcake trays typically come in three sizes: 6, 12 and 16 cups. The best sorting technique is to place the

---

How to Buy Parts for Electronics Projects

The number one question received by the ARRL Technical Information Service starts out "Where can I buy…" It seems that one of the most perplexing problems faced by the would-be constructor is where to get parts. Sometimes you are lucky — the project author has made a kit or a PC board or key parts available. But not every project has a ready-made kit. If you would like to expand your construction horizons, you can learn to be your own "purchasing manager." That means searching out parts sources and dealing with them yourself.

In reality, it is not all that difficult to find most parts. If you are lucky, you may have a local source of electronics parts. Look in the Yellow Pages under "Electronic Equipment Suppliers" to find the local outlets. RadioShack is one local source found nearly everywhere. They carry an assortment of the more common electronic parts.

Unfortunately, though, in many areas of the country it's just not possible to find a local electronic parts supplier that can provide everything for your next project. This is not surprising. Years ago an electronics supplier had to stock a relatively small number of electronic components — resistors, capacitors, tube sockets, a few relays and variable resistors. Technology has increased the number of components by a few orders of magnitude.

Nowadays, the number of integrated circuits alone is enough to fill a multi-volume book. No single electronics supplier could possibly stock them all. It has become a mail-order world; the electronics world is no exception. If you can't find what you need locally to fill your electronic needs, the good news is that an astonishing variety of components is literally a mouse click or phone call away.

Almost all of these suppliers have detailed catalogs on their websites, and some offer printed catalogs. Bookmark the websites and collect their catalogs, and you'll have a ready source of parts for your next project. It's good to have a variety of sources, as you'll probably need to order from more than one company. (It's almost a corollary to Murphy's Law: No matter how wide a selection you find at one supplier, there's always at least one part you must buy somewhere else!)

When planning for a project you want to build, you need to convert the parts list and/or schematic into a parts-order list that shows the catalog number and quantity required of each component. This may require a similar list for each vendor when one supplier does not have all the parts.

Check the type, tolerance, power rating and other key characteristics of the parts. Group the parts by those parameters before grouping them by value. If all of the circuit's components are already grouped by value on the parts list, you can just count the number of each value. Each time you add parts to the order list, check them off the published parts list. Sometimes the parts list does not include common components like resistors and capacitors. If this is the case, make a copy of the schematic and check off the parts as you build your shopping list.

It's generally a good idea to locate sources for all needed parts before ordering anything. Sometimes a key component can be hard to find in small quantities, prohibitively expensive, or obsolete. Be prepared to substitute similar parts if the exact ones needed are not available.

Although you'll probably be able to order exactly the right number of each part for a project, buy a few extras of some parts for your junk box. It's always good to have a few extra parts on hand; you may break a component lead during assembly, or damage a solid-state component with too much heat or by wiring it in backwards. If you don't have extras, you'll need to order another part. Even if you don't need the extras for this project, they may come in handy, and you'll be encouraged to build another project! Pick up an extra toroid or two as well.

Now's the time to decide whether you're going to build your project with ground-plane construction or PC board. If you need a PC board, FAR Circuits (**www.farcircuits.net**) and others have them for many ARRL book and *QST* projects. If one is available, that information is usually included in the article. If you're going to use ground-plane construction, buy a good-sized piece of single-sided copper-clad board — glass-epoxy board if you can. Phenolic board is inferior because it is brittle and deteriorates rapidly with soldering heat.

Don't forget an enclosure for your project. This is often overlooked in parts lists for most projects, because different builders like different enclosures. Make sure there's room in the box for all of the components used in your project. Some people like to cram projects in the smallest possible box, but miniaturization can be extremely frustrating if you're not good at it.

There are almost always a few items you can't get from one company and most have minimum orders. You may need to distribute your order between two or more companies to meet minimum-order requirements.

If you order enough parts, you'll soon find out which companies you like to deal with and which have slow service. It is frustrating to receive most of an order, then wait months for the parts that are on backorder. Fortunately, most of the larger suppliers indicate stock status on their websites. And most companies that cater to hams can tell you over the phone if a part is out of stock.

If you're ordering by mail and don't want the company to backorder your parts, write clearly on the order form, "Do not backorder parts." They will then ship the parts they have and leave you to order the rest from somewhere else.

If you are familiar with the websites, catalogs and policies of electronic-component suppliers, you will find that getting parts is not difficult. Concentrate on the fun part — building the circuit and getting it working. — *Bruce Hale, KB1MW*

most common or first-used components in the cups closest to the builder (usually resistors), with the least used parts (usually mounting hardware) in the farther cups.

Another great idea for parts sorting is to use a fishing tackle box with removable trays. Inexpensive tackle boxes are available at outdoor supply houses and department stores, and you can often find them on sale. The tackle box has a distinct advantage of allowing you to sort your parts into different compartments within each movable tray, and some trays have pieces that allow you to resize the compartments to better fit the parts for your project. Many tackle boxes have a larger open space in the top which allows you to store your partially finished boards, with the remaining components in the closeable trays below it. This arrangement is ideal to protect your project from damage and can be securely stored between work sessions.

23.5.3 Construction Order

When building a kit or DIY project, the question often arises as to what order the parts need to be mounted. In a kit, the manual often is very explicit, requiring the builder to construct the project stage by stage. Other kit manuals offer only a minimum of directions, leaving it up to the builder. Have the manual handy, either printed or on a laptop or tablet nearby for reference.

In stage by stage construction, each stage in the project is completed in order to allow the builder to test and troubleshoot that area of the project without having a more complex problem to solve with all components mounted. This method also allows the builder to learn the principles involved in the project and how each part of the circuit works from the power supply to the output. This is a great aid to future modification and repair.

When building a kit in stages, it is often better to sort the parts by stages as well, placing the parts from each stage in their own space. That way, when each stage is completed, there should be no extra parts left over in that stage's container. Number the stages, if they are not already numbered in the manual, and place a small piece of paper with that number in each compartment, indicating the stage that those parts belong to.

When given a minimum of assembly instructions, the best approach is to mount the resistors first, then the capacitors, and then the semiconductors, followed by the more unique components. This way, the majority of parts are mounted early in the process, so finding the remaining part locations is a lot easier. This technique also allows the builder to double check the usage of parts. Try to mount large parts after the smaller surrounding parts so as not to possibly block your ability to properly mount all of them.

Inventory the parts before commencing construction. In kits or DIY projects, if a change is introduced after a number of kits have been assembled, there is a chance of errors so that the parts list or board layouts do not reflect changes in the design of the circuit. Sometimes, a number of extra parts are supplied with a kit to facilitate different options, such as the choice of bands covered. Sometimes parts are eliminated or substituted with a change of other components in the circuit. Be sure to ask the kit supplier if you are not sure as to why you have an empty space or surplus parts. Always resolve any questions about component placement before powering up a completed project.

23.5.4 Component Mounting

When working with a large number of components, there are a few techniques that can be helpful should troubleshooting be required. Although resistors are not polarized, it is a good idea to mount them with the color codes reading the same direction to make it easier to spot a part that is not in its correct position. Polarity-sensitive parts, such as diodes, electrolytic capacitors and ICs, must be placed in their specified direction. In general, mount components so their values are readable without having to remove the component from the board or bending it, causing possible damage.

Axial-lead components such as resistors are mounted in one of two methods, upright and flat. To save space on a PC board, resistors are often mounted upright with one lead bent double in a manner resembling a hairpin. The best practice is to make this bend so that the color stripes denoting the resistor value begin at the top and the precision stripe (often silver or gold) is at the bottom, making it easier to read the values once mounted. Components with alphanumeric markings, such as diodes, should be mounted with the markings visible.

Non-polarized capacitors are best placed with markings facing in the same direction, unless the markings would be blocked by another component.

For polarized parts, always double-check its positioning before soldering. A commercially-prepared PC board often has stripes on the diode labels, indicating which end to place the cathode stripe on the diode. A "+" sign on the board inside or next to the circle for a capacitor denotes the positive lead which will often be longer. LEDs will have a flat spot on or a notch on a lead to identify the cathode. See the **Component Data and References** chapter and manufacturers' data sheets for more information on component body styles.

When mounting ICs, using a pin straightener helps align the leads for insertion. If one is not available, use a flat surface to align them at once. Be sure a pin does not get bent inward and that all pins go into the socket or PC board holes.

23.5.5 Electronics Construction Techniques

Several different point-to-point wiring techniques or printed-circuit boards (PC boards) can be used to construct electronic circuits. Most circuit projects use a combination of techniques. The selection of techniques depends on many different factors and builder preferences.

For one-time construction, PC boards are really not necessary. It takes time to lay out, drill and etch a PC board. Alterations are difficult to make if you change your ideas or make a mistake.

The simple audio amplifier shown in **Fig 23.9** will be built using various point-to-point or PC-board techniques. This shows how the different construction methods are applied to a typical circuit. (Surface-mount techniques are discussed in the previous section.)

POINT-TO-POINT TECHNIQUES

Point-to-point techniques include all circuit construction techniques that rely on tie points and wiring, or component leads, to build a circuit. This is the technique used in most home-brew construction projects. It is sometimes used in commercial construction, such as old vacuum-tube receivers and modern tube amplifiers.

Point-to-point wiring is also used to connect the "off-board" components used in a printed-circuit project. It can be used to interconnect the various modules and printed-circuit boards used in more complex electronic systems. Most pieces of electronic equipment have at least some point-to-point wiring.

GROUND-PLANE CONSTRUCTION

A point-to-point construction technique that uses the leads of the components as tie

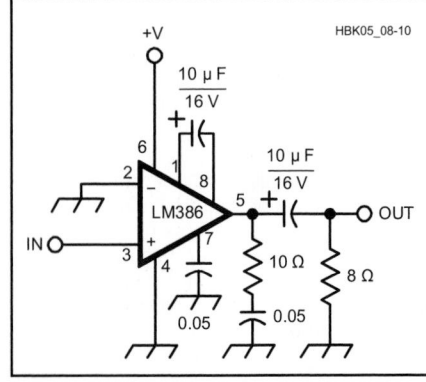

Fig 23.9 — Schematic diagram of the audio amplifier used as a design example of various construction techniques.

points for electrical connections is known as "ground-plane," "dead-bug" or "ugly" construction. "Dead-bug construction" gets its name from the appearance of an IC with its leads sticking up in the air. In most cases, this technique uses copper-clad circuit-board material as a foundation and ground plane on which to build a circuit using point-to-point wiring, so in this chapter it is called "ground-plane construction." An example is shown in **Fig 23.10**.

Ground-plane construction is quick and simple: You build the circuit on an unetched piece of copper-clad circuit board. Wherever a component connects to ground, you solder it to the copper board. Ungrounded connections between components are made point-to-point. Once you learn how to build with a ground-plane board, you can grab a piece of circuit board and start building any time you see an interesting circuit.

A PC board has strict size limits; the components must fit in the space allotted. Ground-plane construction is more flexible; it allows you to use the parts on hand. The circuit can be changed easily — a big help when you are experimenting. The greatest virtue of ground-plane construction is that it is fast.

Ground-plane construction is something like model building, connecting parts using solder almost — but not exactly — like glue. In ground-plane construction you build the circuit directly from the schematic, so it can help you get familiar with a circuit and how it works. You can build subsections of a large circuit on small ground-plane modules and string them together into a larger design.

Circuit connections are made directly, minimizing component lead length. Short lead lengths and a low-impedance ground conductor help prevent circuit instability. There is usually less inter-component capacitive coupling than would be found between PC-board traces, so it is often better than PC-board construction for RF, high-gain or sensitive circuits.

Use circuit components to support other circuit components. Start by mounting one component onto the ground plane, building from there. There is really only one two-handed technique to mount a component to the ground plane. Bend one of the component leads at a 90° angle, and then trim off the excess. Solder a blob of solder to the board surface, perhaps about 0.1 inch in diameter, leaving a small dome of solder. Using one hand, hold the component in place on top of the soldered spot and reheat the component and the solder. It should flow nicely, soldering the component securely. Remove the iron tip and hold the component perfectly still until the solder cools. You can then make connections to the first part.

Connections should be mechanically secure before soldering. Bend a small hook in the lead of a component, then "crimp" it to the next component(s). Do not rely only on the solder connections to provide mechanical strength; sooner or later one of these connections will fail, resulting in a dead circuit.

In most cases, each circuit has enough grounded components to support all of the components in the circuit. This is not always possible, however. In some circuits, high-value resistors can be used as standoff insulators. One resistor lead is soldered to the copper ground plane; the other lead is used as a circuit connection point. You can use ¼ or ½ W resistors in values from 1 to 10 MΩ. Such high-value resistors permit almost no current to flow, and in low-impedance circuits they act more like insulators than resistors. As a rule of thumb, resistors used as standoff insulators should have a value that is at least 10 times the circuit impedance at that point in the circuit.

Fig 23.11A shows how to use the standoff technique to wire the circuit shown at Fig 23.11C. Fig 23.11B shows how the resistor leads are bent before the standoff component is soldered to the ground plane. Components E1 through E5 are resistors that are used as standoff insulators. They do not appear in the schematic diagram. The base circuitry at Q1 of Fig 23.11A has been stretched out to reduce clutter in the drawing.

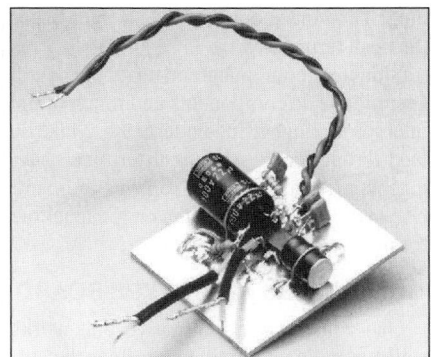

Fig 23.10 — The example audio amplifier of Fig 23.9 built using ground-plane construction.

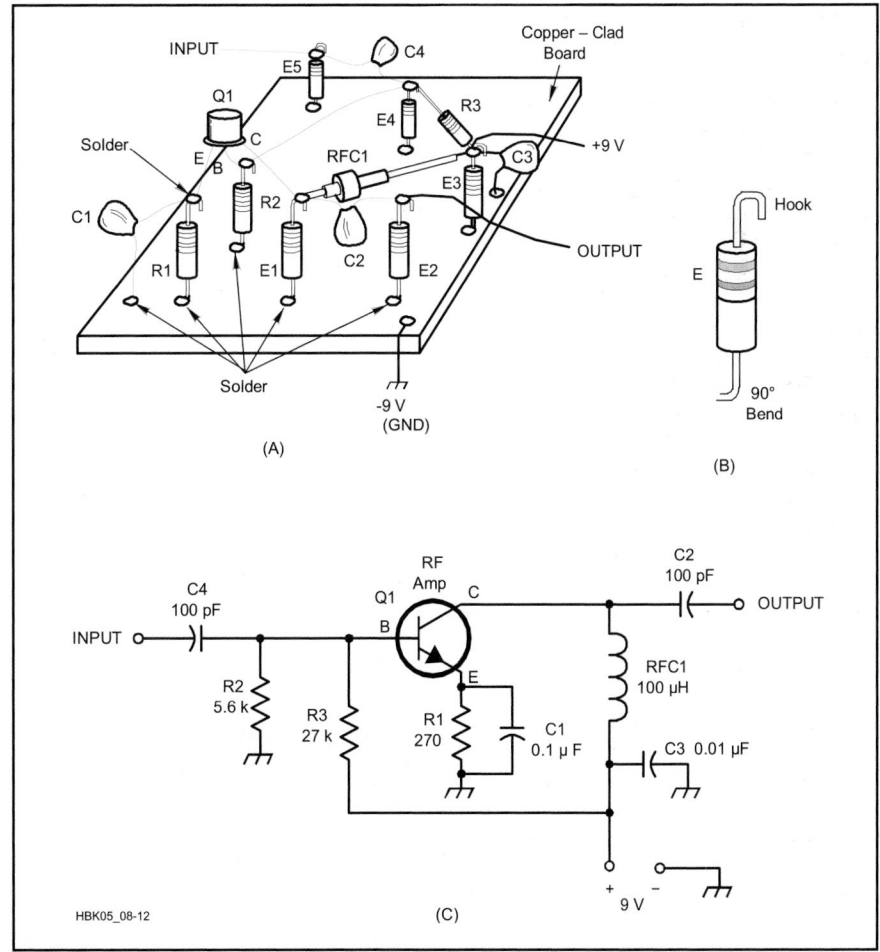

Fig 23.11 — Pictorial view of a circuit board that uses ground-plane construction is shown at A. A close-up view of one of the standoff resistors is shown at B. Note how the leads are bent. The schematic diagram at C shows the circuit displayed at A.

In a practical circuit, all of the signal leads should be kept as short as possible. E4 would, therefore, be placed much closer to Q1 than the drawing indicates.

No standoff posts are required near R1 and R2 of Fig 23.11. These two resistors serve two purposes: They are not only the normal circuit resistances, but function as standoff posts as well. Follow this practice wherever a capacitor or resistor can be employed in the dual role.

"MANHATTAN" CONSTRUCTION

Another solution to building up a circuit is called "Manhattan" construction, shown previously in the Surface-Mount Components section as Fig 23.7E. This method got its name from the appearance of the finished product, resembling the tall buildings in a city. Manhattan construction uses plain, unetched copper clad PC board material to make both the main board and the component connection points. The PC board material used may be single or double-sided.

After cutting the desired size and shape of the board required for the project, use the scraps left over to make the insulated contact pads. These pads can be made a number of ways, the most common being cutting the material into tiny squares about ¼ or ⅜ inch across. Another method to create the pads is to use a heavy-duty hole punch. This kind of punch often has changeable dies to create various sizes of round holes in materials such as sheet metal, and is available at many tool dealers. Once cut, the pads can be glued to the board in a pattern that accommodates the lead lengths of the parts to be connected.

Pads can be glued to the base board or soldered if the pads are double-sided PC board material. Use a tiny drop of instant glue, such as a cyanoacrylate "super glue" to mount the pads to the board. When soldering to the pads, use the minimum amount of heat required to avoid loosening the pads.

Component leads are soldered to the pads, and additional leads can be added to a pad by simply reheating the connection already there. The main board is used as the ground plane with all ground leads soldered to the board. This method of construction works well with RF circuits up to UHF.

WIRED TRACES — THE LAZY PC BOARD

If you already have a PC-board design, but don't want to copy the entire circuit — or you don't want to make a double-sided PC board — then the easiest construction technique is to use a bare board or perfboard and hard-wire the traces.

Drill the necessary holes in a piece of single-sided board, remove the copper ground plane from around the holes, and then wire up the back using component leads and bits

Fig 23.12 — The audio amplifier built using wired-traces construction.

of wire instead of etched traces (**Fig 23.12**).

To transfer an existing board layout, make a 1:1 photocopy and tape it to your piece of PC board. Prick through the holes with an automatic (one-handed) center punch or by firm pressure with a sharp scriber, remove the photocopy and drill all the holes. Holes for ground leads are optional — you generally get a better RF ground by bending the component lead flat to the board and soldering it down. Remove the copper around the rest of the holes by pressing a drill bit lightly against the hole and twisting it between your fingers. A drill press can also be used, but either way, don't remove too much board material. Then wire up the circuit beneath the board. The results look very neat and tidy — from the top, at least!

Circuits that contain components originally designed for PC-board mounting are good candidates for this technique. Wired traces would also be suitable for circuits involving multi-pin RF ICs, double-balanced mixers and similar components. To bypass the pins of these components to ground, connect a miniature ceramic capacitor on the bottom of the board directly from the bypassed pin to the ground plane.

A wired-trace board is fairly sturdy, even though many of the components are only held in by their bent leads and blobs of solder. A drop of cyanoacrylate "super glue" can hold down any larger components, components with fragile leads or any long leads or wires that might move.

PERFORATED CONSTRUCTION BOARD

A simple approach to circuit building uses a perforated phenolic or epoxy resin board known as perfboard. Perfboard is available with many different hole patterns. Choose the one that suits your needs. Perfboard is usually unclad, although it is made with pads that facilitate soldering.

Circuit construction on perforated board is easy. Start by placing the components loosely

on the board and moving them around until a satisfactory layout is obtained. Most of the construction techniques described in this chapter can be applied to perfboard. The audio amplifier of Fig 23.9 is shown constructed with this technique in **Fig 23.13**.

Perfboard and accessories are widely available. Accessories include mounting hardware and a variety of connection terminals for solder and solderless construction.

TERMINAL AND WIRE

A perfboard is usually used for this technique. Push terminals are inserted into the hole in a perfboard. Components can then be easily soldered to the terminals. As an alternative, drill holes into a bare or copper-clad board wherever they are needed (**Fig 23.14**). The components are usually mounted on one side of the board and wires are soldered to the bottom of the board, acting as wired PC-board "traces." If a component has a reasonably rigid lead to which you can attach other components, use that instead of a push terminal — a modification of the ground-plane construction technique.

If you are using a bare board to provide a ground plane, drill holes for your terminals with a high-speed PC-board drill and drill press. Mark the position of the hole with a center punch to prevent the drill from skidding. The hole should provide a snug fit for the push terminal.

Mount RF components on top of the board, keeping the dc components and much of the interconnecting wiring underneath. Make dc feed-through connections with terminals having bypass capacitors on top of the board. Use small solder-in feedthrough capacitors for more critical applications.

SOLDERLESS PROTOTYPE BOARD

One construction alternative that works well for audio and digital circuits is the solderless prototype board (protoboard), shown in **Fig 23.15**. It is usually not suitable for RF circuits above a few MHz.

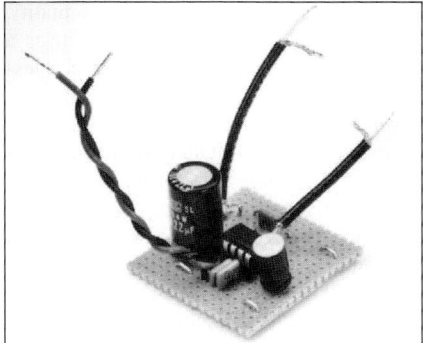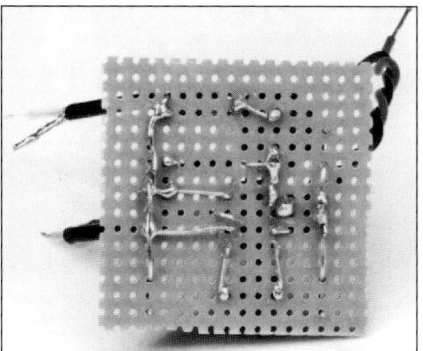

Fig 23.13 — The audio amplifier built on perforated board. Top view at left; bottom view at right.

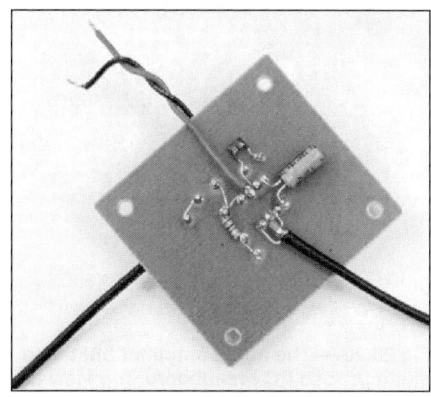

Fig 23.14 — The audio amplifier built using terminal-and-wire construction.

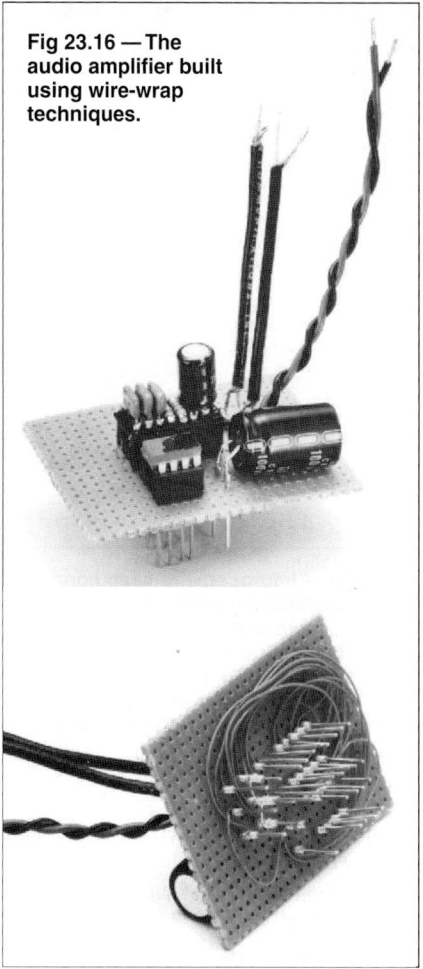

Fig 23.16 — The audio amplifier built using wire-wrap techniques.

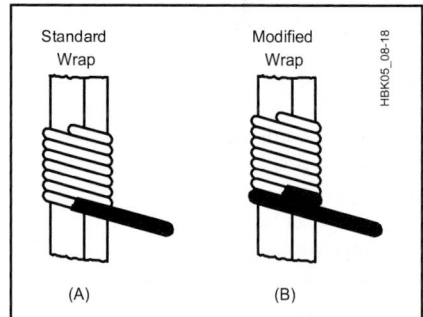

Fig 23.15 — The audio amplifier built on a solderless prototyping board.

A protoboard has rows of holes with spring-loaded metal strips inside the board. Circuit components and hookup wire are inserted into the holes, making contact with the metal strips. Components that are inserted into the same row are connected together. Component and interconnection changes are easy to make. Pre-made and color-coded jumpers make wiring these boards easier. (A length of phone system cable with four solid-conductor wires makes an excellent source of colored jumper wire.) Look for a protoboard that has power supply terminals already mounted on them for easier connection.

Protoboards have some minor disadvantages. The metal strips add stray capacitance to the circuit and jumper lengths can be long. Large-diameter component leads can deform the metal contacts of the strips — be sure to insert wire no larger than the manufacturer recommends.

WIRE-WRAP CONSTRUCTION

Wire-wrap techniques can be used to quickly construct a circuit without solder. Low- and medium-speed digital circuits are often assembled on a wire-wrap board. The technique is not limited to digital circuits, however. **Fig 23.16** shows the audio amplifier built using wire wrap. Circuit changes are easy to make, yet the method is suitable for permanent assemblies.

Wire wrap is done by wrapping a wire around a small square post to make each connection. A wrapping tool resembles a thick pencil. Electric wire-wrap guns are convenient when many connections must be made.

Fig 23.17 — Wire-wrap connections. Standard wrap is shown at A; modified wrap at B.

The wire is almost always #30 AWG wire with thin insulation. Two wire-wrap methods are used: the standard and the modified wrap (**Fig 23.17**). The modified wrap adds a turn of insulated wire which provides a bit of stress relief to the connection. The wrap-post terminals are square (wire wrap works only on posts with sharp corners). They should be long enough for at least two connections. Fig 23.17 and **Fig 23.18** show proper and improper wire-wrap techniques. Mount small components

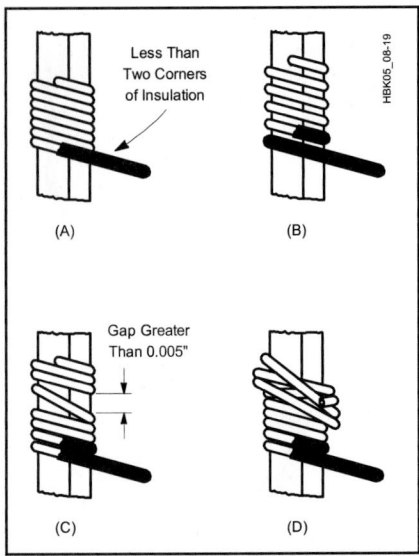

Fig 23.18 — Improper wire-wrap connections. Insufficient insulation for modified wrap is shown at A; a spiral wrap at B, where there is too much space between turns; an open wrap at C, where one or more turns are improperly spaced and an overwrap at D, where the turns overlap on one or more turns.

on an IC header plug. Insert the header into a wire-wrap IC socket as shown in Fig 23.16. The large capacitor in that figure has its leads soldered directly to wire-wrap posts.

"READY-MADE" UTILITY PC BOARDS

"Utility" PC boards are an alternative to custom-designed etched PC boards. They offer the flexibility of perforated board construction and the mechanical and electrical advantages of etched circuit connection pads. Utility PC boards can be used to build anything from simple passive filter circuits to computers.

Circuits can be built on boards on which the copper cladding has been divided into connection pads. Power supply voltages can

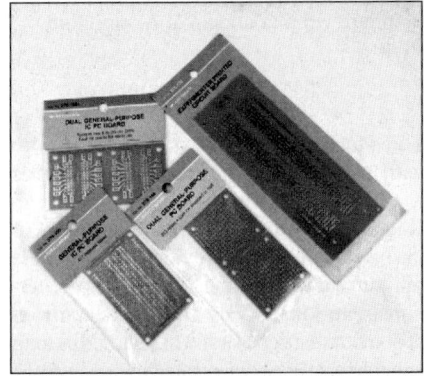

Fig 23.19 — Utility PC boards like these are available from many suppliers.

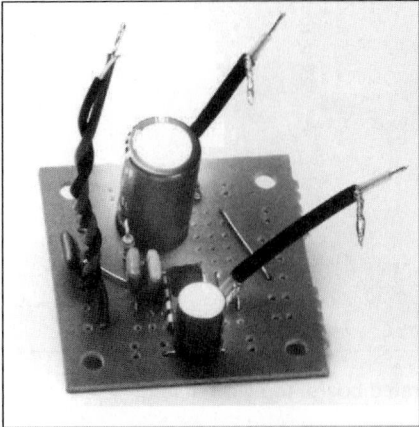

Fig 23.20 — The audio amplifier built on a multipurpose PC breadboard. Top view at A; bottom view at B.

be distributed on bus strips. Boards like those shown in **Fig 23.19** are commercially available.

An audio amplifier constructed on a utility PC board is shown in **Fig 23.20**. Component leads are inserted into the board and soldered to the etched pads. Wire jumpers connect the pads together to complete the circuit.

Utility boards with one or more etched plugs for use in computer-bus, interface and general purpose applications are widely available. Connectors, mounting hardware and other accessories are also available. Check with your parts supplier for details.

23.5.6 Printed-Circuit (PC) Boards

PC boards are everywhere — in all kinds of consumer electronics, in most of your Amateur Radio equipment. They are also used in most kits and construction projects. A newcomer to electronics might think that there is some unwritten law against building equipment in any other way!

The misconception that everything needs to be built on a printed-circuit board is often a stumbling block to easy project construction. In fact, a PC board is probably the worst

choice for a one-time project. In actuality, a moderately complex project (such as a QRP transmitter) can be built in much less time using other techniques such as those described in the preceding section. The additional design, layout and manufacturing is usually much more work than it would take to build the project by hand.

So why does everyone use PC boards? The most important reason is that they are reproducible. They allow many units to be mass-produced with exactly the same layout, reducing the time and work of conventional wiring and minimizing the possibilities of wiring errors. If you can buy a ready-made PC board or kit for your project, it can save a lot of construction time. This is true because someone else has done most of the real work involved — designing the PC board layout and fixing any "bugs" caused by inter-trace capacitive coupling, ground loops and similar problems. In most cases, if a ready-made board is not available, ground-plane construction is a lot less work than designing, debugging and then making a PC board.

Using a PC board usually makes project construction easier by minimizing the risk of wiring errors or other construction blunders. Inexperienced constructors usually feel more confident when construction has been simplified to the assembly of components onto a PC board. One of the best ways to get started with home construction (to some the best part of Amateur Radio) is to start by assembling a few kits using PC boards. A list of kit manufacturers can be found on the QRP ARCI website, **www.qrparci.org**, under "Links." Then click on "QRP Kits and Components".

ON-LINE PC BOARD FABRICATION SERVICES

In the past few years, on-line PC board fabrication services have become popular among hobbyists and professional designers. See the **Computer-Aided Circuit Design** chapter for a discussion of PCB design software and services. These services specialize in fast turnaround (two or three days, typically) of small boards in low quantities. Some accept artwork files in standard interchange formats and others have proprietary software packages. The cost per board is quite reasonable, considering the expense of maintaining the tools and techniques needed to construct boards in a home shop. The results are professional and high-quality.

PC-BOARD ASSEMBLY TECHNIQUES

Cleanliness

Make sure your PC board and component leads are clean. Clean the entire PC board before assembly; clean each component before you install it. Corrosion looks dark instead

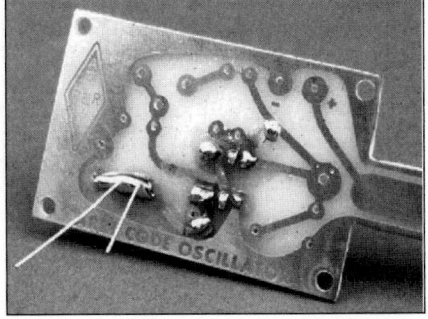

Fig 23.21 — The top photo shows how to solder a component to a PC board. Make sure that the component is flush with the board on the other side. Below is a solder bridge has formed a short circuit between PC board traces.

of bright and shiny. Don't use sandpaper to clean your board. Use a piece of fine steel wool or a Scotchbrite cleaning pad to clean component leads or PC board before you solder them together.

Installing Components

In a construction project that uses a PC board, most of the components are installed on the board. Installing components is easy — stick the components in the right board holes, solder the leads, and cut off the extra lead length. Most construction projects have a parts-placement diagram that shows you where each component is installed.

Getting the components in the right holes is called "stuffing" the circuit board. Inserting and soldering one component at a time takes too long. Some people like to put the components in all at once, and then turn the board over and solder all the leads. If you bend the leads a bit (about 20°) from the bottom side after you push them through the board, the components are not likely to fall out when you turn the board over.

Start with the shortest components — horizontally mounted diodes and resistors. Larger components sometimes cover smaller components, so these smaller parts must be installed first. If building a kit, follow the suggested order of mounting your parts if provided. Use adhesive tape to temporarily hold difficult components in place while you solder.

PC-Board Soldering

To solder components to a PC board, bend the leads at a slight angle; apply the soldering iron to one side of the lead, and flow the solder in from the other side of the lead. See **Fig 23.21A**. Too little heat causes a bad or "cold" solder joint; too much heat can damage the PC board. Practice a bit on some spare copper stock before you tackle your first PC board project. After the connection is soldered properly, clip the lead flush with the solder.

Make sure you have the components in the right holes before you solder them. Components that have polarity, such as diodes, ICs and some capacitors must be oriented as shown on the parts-placement diagram.

Inspect solder connections. A bad solder joint is much easier to find before the PC board is mounted to a chassis. Look for any damage caused to the PC board by soldering. Look for solder "bridges" between adjacent circuit-board traces. Solder bridges (Fig 23.21B) occur when solder accidentally connects two or more conductors that are supposed to be isolated. It is often difficult to distinguish a solder bridge from a conductive trace on a tin-plated board. If you find a bridge, re-melt it and the adjacent trace or traces to allow the solder's surface tension to absorb it. Double check that each component is installed in the proper holes on the board and that the orientation is correct. Make sure that no component leads or transistor tabs are touching other components or PC board connections. Check the circuit voltages before installing ICs in their sockets. Ensure that the ICs are oriented properly and installed in the correct sockets.

23.5.7 From Schematic to Working Circuit

Turning a schematic into a working circuit is more than just copying the schematic with components. One thing is usually true — you can't build it the way it looks on the schematic. The schematic describes the electrical connections, but it does not describe the mechanical layout of the circuit. Many design and layout considerations that apply in the real world of practical electronics don't appear on the schematic.

HOW TO DESIGN A GOOD CIRCUIT LAYOUT

A circuit diagram is a poor guide toward a proper layout. Circuit diagrams are drawn to be readable and to describe the electrical connections. They follow drafting conventions that have very little to do with the way the circuit works. On a schematic, ground and supply voltage symbols are scattered all over the place. The first rule of RF layout is — *do not lay out RF circuits as their schematics are drawn!* How a circuit works in practice depends on the layout. Poor layout can ruin the performance of even a well-designed circuit.

The easiest way to explain good layout practices is to take you through an example. **Fig 23.22** is the circuit diagram of a two-stage receiver IF amplifier using dual-gate MOSFETs. It is only a design example, so the values are only typical. To analyze which things are important to the layout of this circuit, ask these questions:

• Which are the RF components, and which are only involved with AF or dc?

• Which components are in the main RF signal path?

• Which components are in the ground return paths?

Use the answers to these questions to plan the layout. The RF components that are in the main RF signal path are usually the most critical. The AF or dc components can usually be placed anywhere. The components in the ground return path should be positioned so they are easily connected to the circuit ground. Answer the questions, apply the answers to the layout and then follow these guidelines:

• Avoid laying out circuits so their inputs and outputs are close together. If a stage's output is too near a previous stage's input, the output signal can feedback into the input and cause problems.

• Keep component leads as short as practical. This doesn't necessarily mean as short as possible, just consider lead length as part of your design.

• Remember that metal transistor cases conduct, and that a transistor's metal case is usually connected to one of its leads. Prevent cases from touching ground or other components, unless called for in the design.

In our design example, the RF components are shown in heavy lines, though not all of these components are in the main RF signal path. The RF signal path consists of T1/C1, Q1, T2/C4, C7, Q2, T3/C11. These need to be positioned in almost a straight line, to avoid feedback from output to input. They form the backbone of the layout, as shown in **Fig 23.23A**.

The question about ground paths requires some further thought — what is really meant by "ground" and "ground-return paths"? Some points in the circuit need to be kept at RF ground potential. The best RF ground potential on a PC board is a copper ground plane covering one entire side. Points in the circuit that cannot be connected directly to ground for dc reasons must be bypassed ("decoupled") to ground by capacitors that

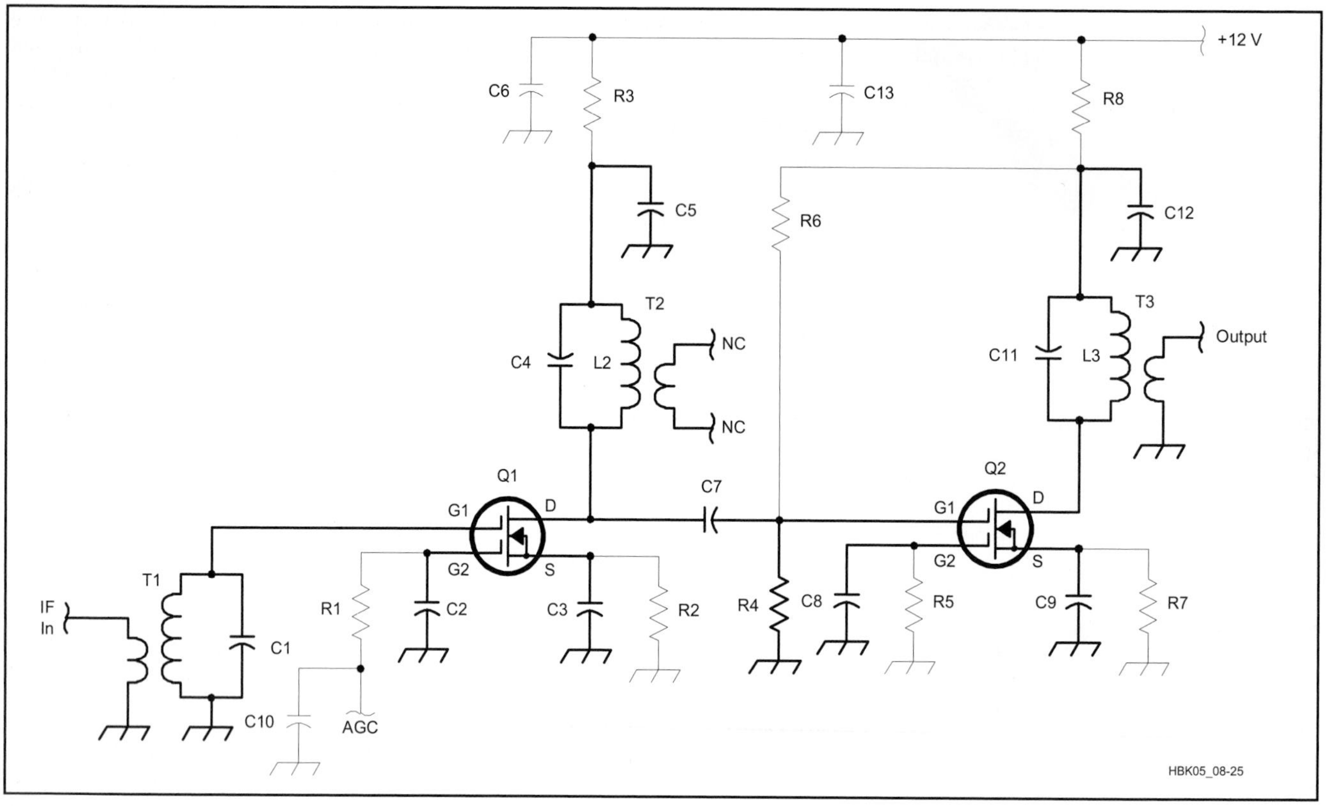

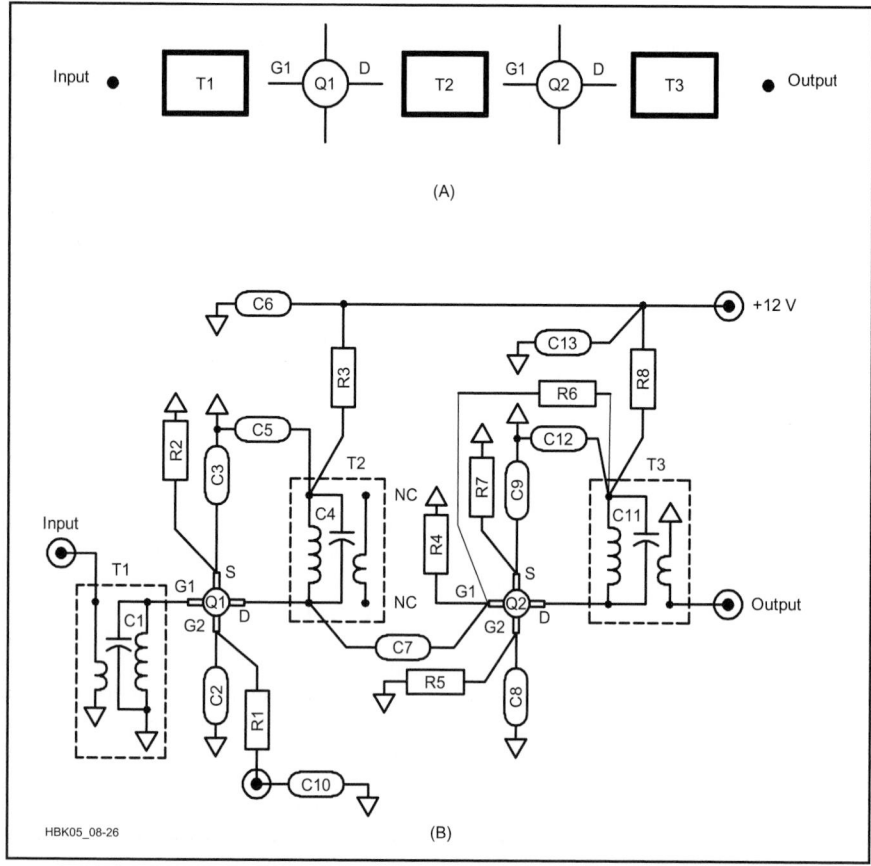

Fig 23.22 — The IF amplifier used in the design example. C1, C4 and C11 are not specified because they are internal to the IF transformers.

Fig 23.23 — Layout sketches. The preliminary line-up is shown in A; the final layout in B.

provide ground-return paths for RF.

In Fig 23.23, the components in the ground-return paths are the RF bypass capacitors C2, C3, C5, C8, C9 and C12. R4 is primarily a dc biasing component, but it is also a ground return for RF so its location is important. The values of RF bypass capacitors are chosen to have a low reactance at the frequency in use; typical values would be 0.1 µF at LF, 0.01 µF at HF and 0.001 µF or less at VHF. Not all capacitors are suitable for RF decoupling; the most common are disc ceramic capacitors. RF decoupling capacitors should always have short leads. Surface mount capacitors with no leads are ideal for bypassing.

Almost every RF circuit has an input, an output and a common ground connection. Many circuits also have additional ground connections, both at the input side and at the output side. Maintain a low-impedance path between input and output ground connections. The input ground connections for Q1 are the grounded ends of C1 and the two windings of T1. The two ends of an IF transformer winding are generally not interchangeable; one is designated as the "hot" end, and the other must be connected or bypassed to RF ground.) The capacitor that resonates with the adjustable coil is often mounted inside the can of the IF transformer, leaving only two component leads to be grounded as shown in Fig 23.22B.

The RF ground for Q1 is its source connection via C3. Since Q1 is in a plastic package that can be mounted in any orientation, you can make the common ground either above or below the signal path in Fig 23.23B, although the circuit diagram shows the source at the bottom. The practical circuit works much better with the source at the top, because of the connections to T2.

It's a good idea to locate the hot end of the main winding close to the drain lead of the transistor package, so the other end is toward the top of Fig 23.23B. If the source of Q1 is also toward the top of the layout, there is a common ground point for C3 (the source bypass capacitor) and the output bypass capacitor C5. Gate 2 of Q1 can safely be bypassed toward the bottom of the layout.

C7 couples the signal from the output of Q1 to the input of Q2. The source of Q2 should be bypassed toward the top of the layout, in exactly the same way as the source of Q1. R4 is not critical, but it should be connected on the same side as the other components. Note how the pinout of T3 has placed the output connection as far as possible from the input. With this layout for the signal path and the critical RF components, the circuit has an excellent chance of working properly.

DC Components

The rest of the components carry dc, so their layout is much less critical. Even so, try to keep everything well separated from the main RF signal path. One good choice is to put the 12 V connections along the top of the layout, and the AGC connection at the bottom. The source bias resistors R2 and R7 can be placed alongside C3 and C9. The gate-2 bias resistors for Q2, R5 and R6 are not RF components so their locations aren't too critical. R7 has to cross the signal path in order to reach C12, however, and the best way to avoid signal pickup would be to mount R7 on the opposite side of the copper ground plane from the signal wiring. Generally speaking, 1/8 W or 1/4 W metal-film or carbon-film resistors are best for low-level RF circuits.

Actually, it is not quite accurate to say that resistors such as R3 and R8 are not "RF" components. They provide a high impedance to RF in the positive supply lead. Because of R8, for example, the RF signal in T2 is conducted to ground through C5 rather than ending up on the 12 V line, possibly causing unwanted RF feedback. Just to be sure, C6 bypasses R3 and C13 serves the same function for R8. Note that the gate-1 bias resistor R6 is connected to C12 rather than directly to the 12 V supply, to take advantage of the extra decoupling provided by R8 and C13.

If you build something, you want it to work the first time, so don't cut corners! Some commercial PC boards take liberties with layout, bypassing and decoupling. Don't assume that you can do the same. Don't try to eliminate "extra" decoupling components such as R3, C6, R8 and C13, even though they might not all be absolutely necessary. If other people's designs have left them out, put them in again. In the long run it's far easier to take a little more time and use a few extra components, to build in some insurance that your circuit will work. For a one-time project, the few extra parts won't hurt your pocket too badly; they may save untold hours in debugging time.

A real capacitor does not work well over a large frequency range. A 10-µF electrolytic capacitor cannot be used to bypass or decouple RF signals. A 0.1-µF capacitor will not bypass UHF or microwave signals. Choose component values to fit the range. The upper frequency limit is limited by the series inductance, L_S. In fact, at frequencies higher than the frequency at which the capacitor and its series inductance form a resonant circuit, the capacitor actually functions as an inductor. This is why it is a common practice to use two capacitors in parallel for bypassing, as shown in **Fig 23.24**. At first glance, this might appear to be unnecessary. However, the self-resonant frequency of C1 is usually 1 MHz or less; it cannot supply any bypassing above that frequency. However, C2 is able to bypass signals up into the lower VHF range. (This technique should not be applied under all circumstances as discussed in the section on Bypassing in the **RF Techniques** chapter.)

Let's summarize how we got from Fig 23.22 to Fig 23.23B:
- Lay out the signal path in a straight line.
- By experimenting with the placement and orientation of the components in the RF signal path, group the RF ground connections for each stage close together, without mixing up the input and output grounds.
- Place the non-RF components well clear of the signal path, freely using decoupling components for extra measure.

Practical Construction Hints

Now it's time to actually construct a project. The layout concepts discussed earlier can be applied to nearly any construction technique. Although you'll eventually learn from your own experience, the following guidelines give a good start:
- Divide the unit into modules built into separate shielded enclosures — RF, IF, VFO, for example. Modular construction improves

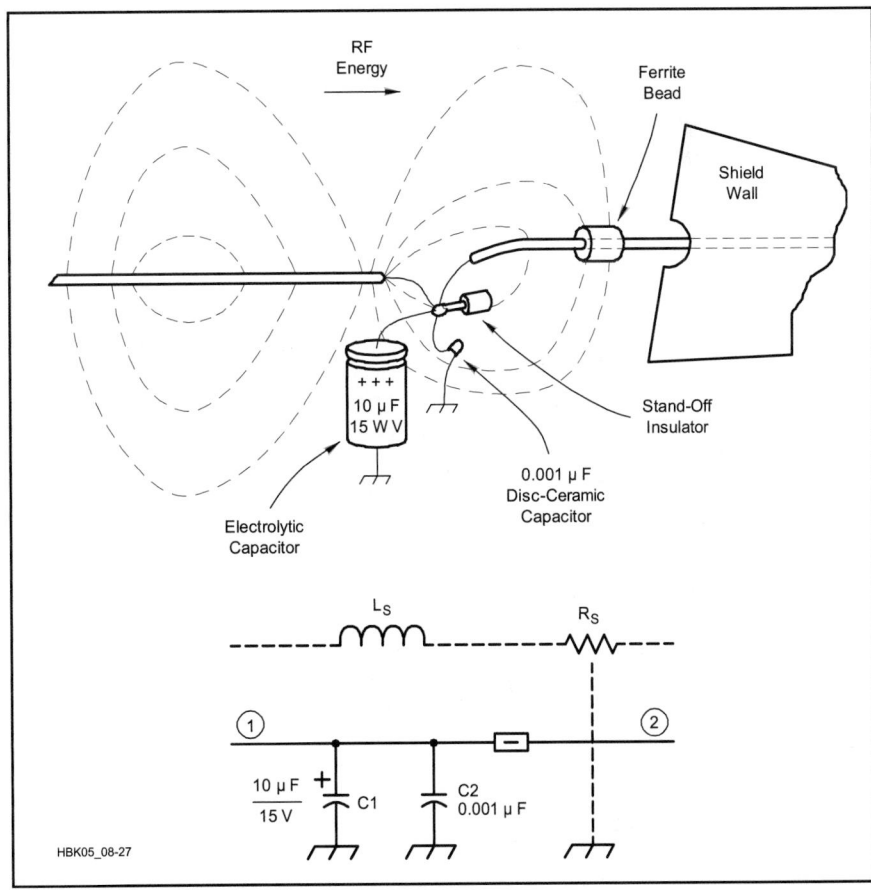

Fig 23.24 — Two capacitors in parallel afford better bypassing across a wide frequency range.

RF stability, and makes the individual modules easier to build and test. It also means that you can make major changes without rebuilding the whole unit. RF signals between the modules can usually be connected using small coaxial cable.

• Use a full copper ground plane. This is your largest single assurance of RF stability and good performance.

• Keep inputs and outputs well separated for each stage, and for the whole unit. If possible, lay out all stages in a straight line. If an RF signal path doubles back or re-crosses itself it usually results in instability.

• Keep the stages at different frequencies well-separated to minimize interstage coupling and spurious signals.

• Use interstage shields where necessary, but don't rely on them to cure a bad layout.

• Make all connections to the ground plane short and direct. Locate the common ground for each stage between the input and the output ground. Single-point grounding may work for a single stage, but it is rarely effective in a complex RF system.

• Locate frequency-determining components away from heat sources and mount them so as to maximize mechanical strength.

• Avoid unwanted coupling between tuned circuits. Use shielded inductors or toroids rather than open coils. Keep the RF high-voltage points close to the ground plane. Orient air-wound coils at right angles to minimize mutual coupling.

• Use lots of extra RF bypassing, especially on dc supply lines.

• Try to keep RF and dc wiring on opposite sides of the board, so the dc wiring is well away from RF fields.

• Compact designs are convenient, but don't overdo it! If the guidelines cited above mean that a unit needs to be bigger, make it bigger.

COMBINING TECHNIQUES

You can use a mixture of construction techniques on the same board and in most cases you probably should. Even though you choose one style for most of the wiring, there will probably be places where other techniques would be better. If so, do whatever is best for that part of the circuit. The resulting hybrid may not be pretty (these techniques aren't called "ugly construction" for nothing), but it will work!

Mount dual-in-line package (DIP) ICs in an array of drilled holes, then connect them using wired traces as described earlier. It is okay to mount some of the components using a ground-plane method, push pins or even wire wrap. On any one board, you may use a combination of these techniques, drilling holes for some ICs, or gluing others upside down, then surface mounting some of the pins, and other techniques to connect the

rest. These combination techniques are often found in a project that combines audio, RF and digital circuitry.

A Final Check

No matter what construction technique is chosen, do a final check before applying power to the circuit! Things do go wrong, and a careful inspection minimizes the risk of a project beginning and ending its life as a puff of smoke! Check wiring carefully. Make a photocopy of the schematic and mark each connection and lead on the schematic with a red X or use highlighter to mark the circuit when you've verified that it's connected properly.

23.5.8 Other Construction Techniques

WIRING

Select the wire used in connecting amateur equipment by considering: the maximum current it must carry, the voltage its insulation must withstand and its use.

To minimize leakage of RF that causes EMI, the power wiring and low-level signal wiring of all transmitters should use shielded wire or coaxial cable. Receiver and audio circuits may also require the use of shielded wire at some points for stability or the elimination of coupling to adjacent circuits. Coaxial cable is recommended for all 50 Ω circuits. It can also be used for *short* runs of high-impedance audio wiring.

When choosing wire, consider how much current it will carry. (See the Copper Wire Specifications table in the **Component Data and References** chapter for maximum current-carrying capability, called *ampacity*.) Stranded wire is usually preferred over solid wire because stranded wire better withstands the inevitable bending that is part of building and troubleshooting a circuit. Solid wire is more rigid than stranded wire; use it where mechanical rigidity is needed or desired.

Wire with typical plastic insulation is good for voltages up to about 500 V. Use Teflon-insulated or other high-voltage wire for higher voltages. Teflon insulation does not melt when a soldering iron is applied. This makes it particularly helpful in tight places or large wiring harnesses. Although Teflon-insulated wire is more expensive, it is often available from industrial surplus houses.

Solid wire is often used to wire RF circuits in both receivers and transmitters. Bare soft-drawn tinned wire, #22 to #12 AWG (depending on mechanical requirements) is suitable. Avoid kinks by stretching a piece 10 or 15 feet long and then cutting it into short, convenient lengths. Run RF wiring directly from point to point with a minimum of sharp bends and keep the wire well-spaced from the chassis or other grounded metal surfaces. Where the wiring must pass through a chassis wall or

shield, cut a clearance hole and line it with a rubber grommet. If insulation is necessary, slip spaghetti insulation or heat-shrink tubing over the wire. For power-supply leads, bring the wire through walls or barriers via a feedthrough capacitor.

In transmitters where the peak voltage does not exceed 500 V, shielded wire is satisfactory for power circuits. Shielded wire is not readily available for higher voltages — use point-to-point wiring instead. In the case of filament circuits carrying heavy current, it is necessary to use #10 or #12 AWG bare or enameled wire. Slip the bare wire through spaghetti then cover it with copper braid pulled tightly over the spaghetti. Slide the shielding back over the insulation and flow solder into the end of the braid; the braid will stay in place, making it unnecessary to cut it back or secure it in place. Clean the braid first so solder will flow into the braid with a minimum of heat.

ENAMELED WIRE

When connecting enameled wire leads, care must be taken to be sure a good connection is made. There are two methods that can be used to remove the insulation. With Thermaleze-type enamel, the heat from a soldering iron can be used to remove the insulation. You will first need to turn up the heat on your soldering iron for best results. After adding some melted solder to your soldering tip, move the drop slowly along the desired length of the wire lead. Moving it slowly gives the insulation time to melt and the solder a chance to tin the now-exposed wire. The tinned leads will then easily solder to a PC board or other mounting system.

If using other kinds of enameled wire, an emery board or small file can be used to remove the insulation. (Using a knife to scrape off the enamel usually nicks the wire which will eventually break at that point.) Be sure you have removed it completely from the desired lead. Follow that up with your soldering iron and solder to tin the wire using a thin coating of solder. Be sure to not make the tinned surface so thick that the lead cannot fit through the PC board holes. Preparing the leads in this manner will give the best possibility for a clean and secure connection.

HIGH-VOLTAGE TECHNIQUES

High-voltage wiring and construction requires special care. Read and follow the guidelines for high-voltage construction in the **Power Sources** chapter. You must use wire with insulation rated for the voltage it is carrying. Most standard hookup wire is inadequate above 300 or 600 V. High-voltage wire is usually insulated with Teflon or special multilayer plastic. Some coaxial cable is rated at 3700 V_{RMS} (or more) *internally* between the center conductor and shield but the outer jacket rating is usually considerably lower.

CABLE ROUTING

Where power or control leads run together for more than a few inches, they present a better appearance when bound together in a single cable. Plastic cable ties or tubing cut into a spiral are used to restrain and group wiring. Check with your local electronic parts supplier for items that are in stock.

To give a commercial look to the wiring of any unit, route any dc leads and shielded signal leads along the edge of the chassis. If this isn't possible, the cabled leads should then run parallel to an edge of the chassis. Further, the generous use of the tie points mounted parallel to an edge of the chassis, for the support of one or both ends of a resistor or fixed capacitor, adds to the appearance of the finished unit. In a similar manner, arrange the small components so that they are parallel to the panel or sides of the chassis.

Tie Points

When power leads have several branches in the chassis, it is convenient to use fiber-insulated terminal strips as anchors for junction points. Strips of this kind are also useful as insulated supports for resistors, RF chokes and capacitors. Hold exposed points of high-voltage wiring to a minimum; otherwise, make them inaccessible to accidental contact.

WINDING COILS

A detailed tutorial for winding coils by Robert Johns, W3JIP, titled "Homebrew Your Own Inductors!" from August 1997 *QST* can be found in the Radio Technology section of the ARRL TIS at **www.arrl.org/ radio-technology-topics** under Circuit Construction. Understanding these techniques greatly simplifies coil construction.

Close-wound coils are readily wound on the specified form by anchoring one end of the length of wire (in a vise or to a doorknob) and the other end to the coil form. Straighten any kinks in the wire and then pull to keep the wire under slight tension. Wind the coil to the required number of turns while walking toward the anchor, always maintaining a slight tension on the wire.

To space-wind the coil, wind the coil simultaneously with a suitable spacing medium (heavy thread, string or wire) in the manner described above. When the winding is complete, secure the end of the coil to the coil-form terminal and then carefully unwind the spacing material. If the coil is wound under suitable tension, the spacing material can be easily removed without disturbing the winding. Finish space-wound coils by judicious applications of RTV sealant or hot-melt glue to hold the turns in place.

The "cold" end of a coil is the end at (or close to) chassis or ground potential. Wind coupling links on the cold end of a coil to minimize capacitive coupling.

Winding Toroidal Inductors

Toroidal inductors and transformers are specified for many projects in this *Handbook*. The advantages of these cores include compactness and a self-shielding property. **Figs 23.25** and **23.26** illustrate the proper way to wind and count turns on a toroidal core.

The task of winding a toroidal core, when more than just a few turns are required, can be greatly simplified by the use of a homemade bobbin upon which the wire is first wound. A simple yet effective bobbin can be fashioned

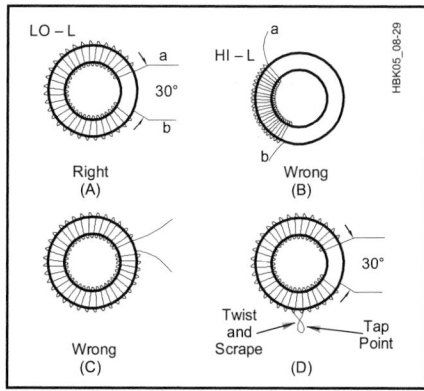

Fig 23.25 — The maximum-Q method for winding a single-layer toroid is shown at A. A 30° gap is best. Methods at B and C have greater distributed capacitance. D shows how to place a tap on a toroidal coil winding.

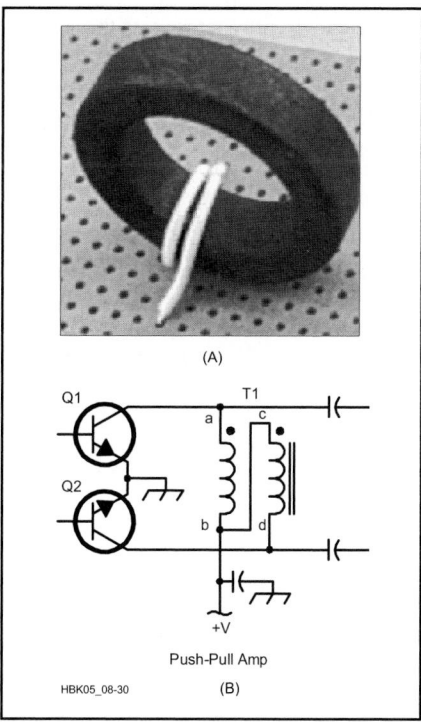

Fig 23.26 — A shows a toroidal core with two turns of wire (see text). Large black dots, like those at T1 in B, indicate winding polarity (see text).

from a wooden popsicle stick. Cut a "V" notch at each end and first wind the wire coil on the popsicle stick lengthwise through the notches. Once this is done, the wound bobbin can be easily passed through the toroid's inside diameter. While firmly grasping one of the wire ends against the toroidal core, the bobbin can be moved up, around, and through the toroidal core repeatedly until the wire has been completely transferred from the bobbin. The choice of bobbin used is somewhat dependent on the inside diameter of the toroid, the wire size, and the number of turns required.

Another method is to create a holder that allows you to grip the core, yet thread the wire easily around it. When winding toroids, be sure to seat the wire so it hugs the shape of the core and do not allow turns to overlap unless part of a twisted group of wires as in bifilar or trifilar windings. Do not pull the wire too tight as the thin wire used in some toroids can break if pulled too tightly.

When you wind a toroid inductor, count each pass of the wire through the toroid center as a turn. You can count the number of turns by counting the number of times the wire passes through the center of the core. See Fig 23.26A.

Multiwire Windings

A bifilar winding is one that has two identical lengths of wire, which when placed on the core result in the same number of turns for each wire. The two wires are wound on the core side by side at the same time, just as if a single winding were being applied. An easier and more popular method is to twist the two wires (8 to 15 turns per inch is adequate), then wind the twisted pair on the core. The wires can be twisted handily by placing one end of each in a bench vise. Tighten the remaining ends in the chuck of a small hand drill and turn the drill to twist the pair.

A trifilar winding has three wires, and a quadrifilar winding has four. The procedure

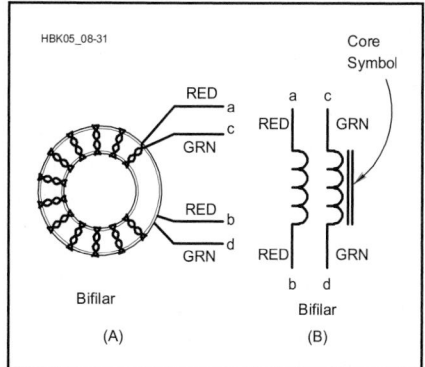

Fig 23.27 — Schematic and pictorial presentation of a bifilar-wound toroidal transformer.

for preparation and winding is otherwise the same as for a bifilar winding. **Fig 23.27** shows a bifilar toroid in schematic and pictorial form. The wires have been twisted together prior to placing them on the core. It is helpful, though by no means essential, to use wires of different color for multifilar windings. It is more difficult to identify multiple windings on a core after it has been wound. Various colors of enamel insulation are available, but it is not easy for amateurs to find this wire locally or in small-quantity lots. This problem can be solved by taking lengths of

wire (enameled magnet wire), cleaning the ends to remove dirt and grease, then spray painting them. Ordinary aerosol-can spray enamel works fine. Spray lacquer is not as satisfactory because it is brittle when dry and tends to flake off the wire.

You may also identify bifilar and trifilar toroid lead pairs of identical colors by using a continuity checker. It is a good idea to check all toroids with an ohmmeter or continuity tester to be sure there are no shorts between different windings and that there is a good connection on the ends of each winding.

Testing the leads after mounting can be difficult due to the circuit layouts, so be sure to test all toroids before mounting.

The winding sense of a multifilar toroidal transformer is important in most circuits. Fig 23.26B illustrates this principle. The black dots (called phasing dots) at the top of the T1 windings indicate polarity. That is, points a and c are both start or finish ends of their respective windings. In this example, points a and d are of opposite polarity to provide push-pull voltage output from Q1 and Q2.

23.6 Microwave Construction

Paul Wade, W1GHZ, updated this short tutorial to construction practices suitable for microwave frequency operation originally written by ARRL Laboratory Engineer, Zack Lau, W1VT. For more information on microwave equipment, read Paul's series of columns, "Microwavelengths," in *QST*.

Microwave construction is not only fun, but within the capabilities of most amateurs. To get on the air requires some degree of construction, since you can't buy a ready to go box. At a minimum, a few modules must be connected together with coax and power cables. At the other extreme, the whole station may be homebrewed from scratch, or from a kit.

The growth of mobile phones and wireless networking has caused a proliferation of microwave integrated circuits offering high performance at low cost. Some of these are also useful for ham applications, so that microwave construction has evolved from traditional waveguide "plumbing" to printed circuitry requiring surface-mount assembly of tiny components. Waveguide techniques are still valuable for some components, such as filters and antennas, so proficiency in both types of construction is valuable.

A problem we all face when building new microwave equipment is finding a nearby station to try it out. One solution is to convince a buddy to build a similar system, so that you will both have someone to work. If you have complementary skills, say one with metal and the other with surface-mount soldering, you can help each other. A larger group or club effort can be even better — someone with experience and test equipment can be an Elmer for the group.

In addition to the material here and in magazine columns, several excellent references for amateurs interested in working on microwave frequencies include:

• *Microwave Know How*, by Andy Barter, G8ATD (RSGB)
• *International Microwave Handbook*, by

Andy Barter, G8ATD (RSGB)
• *The ARRL UHF/Microwave Experimenter's Manual* (out of print but available used)
• *The ARRL UHF/Microwave Projects CD*

23.6.1 Lead Lengths

Microwave construction is becoming more popular, but at these frequencies the size of physical component leads and PC-board traces cannot be neglected. Microwave construction techniques either minimize these stray values or make them part of the circuit design.

The basic consideration in microwave circuitry is short lead lengths, particularly for ground returns. Current always flows in a complete loop with a return path through the "ground". (See **Fig 23.28**) The loop must be much smaller than a wavelength for a circuit to work properly — as the frequency gets higher, dimensions must get smaller. One area that requires particular care to ensure good ground return continuity is the transition from coax cable outside a cabinet to a PC board

inside the cabinet.

At microwave frequencies, the mechanical aspects and physical size of circuits become very much a part of the design. A few millimeters of conductor has significant reactance at these frequencies. This even affects VHF and HF designs in which the traces and conductors resonate on microwave frequencies. If a high-performance FET has lots of gain in this region, a VHF preamplifier might also function as a 10 GHz oscillator if the circuit stray reactances were just right (or wrong!). You can prevent this by using shields between the input and output or by adding microwave absorptive material to the lid of the shielded module. (SHF Microwave sells absorptive materials.)

23.6.2 Metalworking

Waveguide construction requires some basic metalworking skills, which are also useful for assembly and packaging of microwave systems. Some minimal tools would be a hacksaw, drills, files, and layout tools.

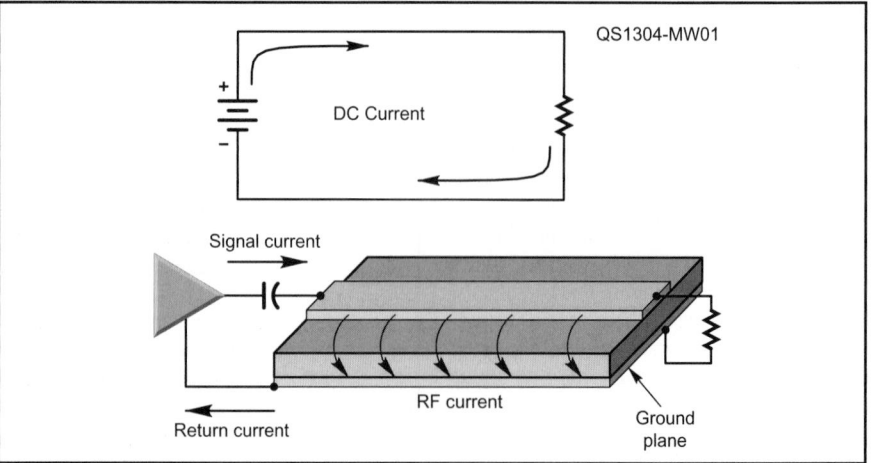

Fig 23.28 — Current must always flow in a complete loop. This is particularly important at microwave frequencies where wavelengths are very short.

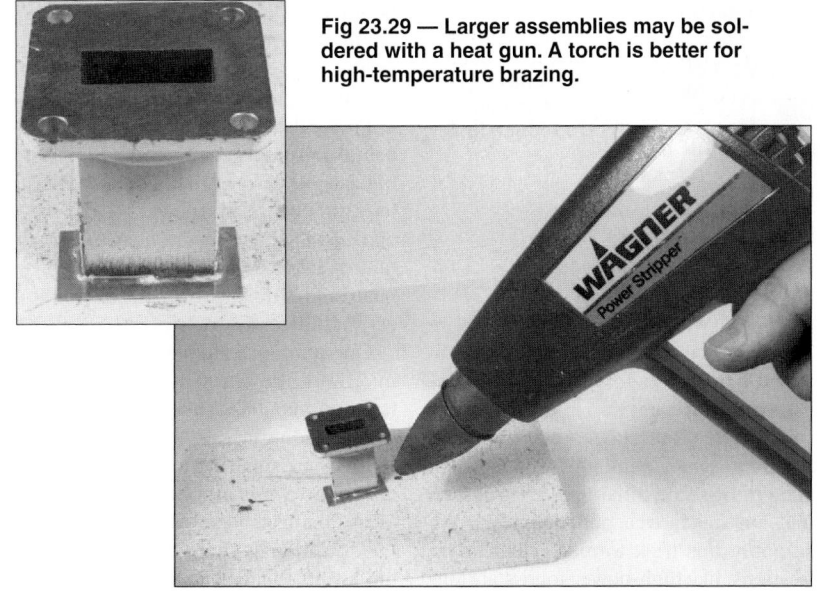

Fig 23.29 — Larger assemblies may be soldered with a heat gun. A torch is better for high-temperature brazing.

While a hand drill can suffice, an inexpensive drill press will make work more precise as well as safer. In metals, Dewalt Pilot Point drill bits make clean, accurate holes. Files and nibbling tools can make rectangular or odd-shaped holes.

Solder has a poor reputation for microwave losses — but, of course, it is essential. It does have higher resistance than copper or aluminum, but that only matters in locations of high current, for instance, the ground end of a high-Q quarter-wave resonator. Horn antennas and most waveguide structures have low Q so a clean solder joint, without excess blobs, has no effect. Traditional tin-lead solder works well; the silver solder with 1% silver has no performance advantage but looks pretty.

For things like waveguide and antennas that are too large for a soldering iron, a hot air gun does an excellent job. **Fig 23.29** shows the backshort of a waveguide transition as it is being soldered with a hot air gun. A small amount of paste flux (Kester SP-44) was applied to the waveguide end, and a ring of solder placed along the joint. Then heat is applied to the whole area until the solder melts and flows into the joint.

A torch can also be used for soldering, but tends to cause more oxidation of the metal. However, when a high-temperature brazed joint is desired, a torch is required. A MAPP gas torch and Silvaloy 15 (or Harris Staysilv 15) silver brazing material, which requires no flux, are suitable for small and medium-sized components.

Many hams seem to think that silver plating is desirable, but it rarely makes a difference — it just looks much prettier. The only application in which silver plating has been shown to make a significant difference is in UHF cavities with sliding contacts. In most other places, copper losses are low enough so that little improvement is possible, and aluminum is fine for most uses.

23.6.3 Circuit Construction

Modular construction is a useful technique for microwave circuits. Often, circuits are tested by connecting their inputs and output to known 50 Ω sources and loads. Modules are typically kept small to prevent the chassis and PC board from acting as a waveguide, providing a feedback path between the input and output of a circuit, resulting in instability.

PRINTED-CIRCUIT BOARDS AT MICROWAVES

A microwave printed circuit board (PCB) is typically double-sided, with transmission-line circuitry printed on one side and a ground plane on the far side. The ground return path is best provided by plated-thru holes (PTH) connecting the two sides with minimum length. Surface-mount components — integrated circuits, chip capacitors, and chip resistors — are soldered directly to printed pads and transmission lines, and to ground pads with embedded PTH. Many of the microwave integrated circuits come in really tiny packages, with lead pitch as small as 25 mils. See the section on Surface-Mount Technology earlier in this chapter for more information on this type of construction.

When using glass-epoxy PC board at microwave frequencies, the crucial board parameter is the thickness of the dielectric. It can vary quite a bit, in excess of 10%. This is not surprising; digital and lower-frequency analog circuits work just fine if the board is a little thinner or thicker than usual. Some of the board types used in microwave-circuit construction are a generic Teflon PC board, Duroid 5870 and 5880.

TRANSMISSION LINES AND CONNECTORS

From a mechanical accuracy point of view, the most tolerant type of construction is based on waveguide. Tuning is usually accomplished via one or more screws threaded into the waveguide. It becomes unwieldy to use waveguide on the amateur bands below 10 GHz because the dimensions get too large.

Proper connectors are a necessary expense at microwaves. At 10 GHz, the use of the proper connectors is essential for repeatable performance. Do not connect microwave circuits with coax and pigtails. It might work but it probably can't be duplicated. SMA connectors are common because they are small and work well. SMA jacks are sometimes soldered in place, although 2-56 hardware is more common.

At 24 GHz and above, even waveguide becomes small and difficult to work with. At these frequencies, most readily available coax connectors work unreliably, so these higher bands are really a challenge. Special SMA connectors are available for use at 24 GHz.

DEVICE SUBSTITUTION

It is important to copy microwave circuits exactly, unless you really know what you are doing. "Improvements," such as better shielding or grounding can sometimes cause poor performance. It isn't usually attractive to substitute components, particularly with the active devices. It may look possible to substitute different grades of the same wafer, such as the ATF13135 and the ATF13335, but these are really the same transistor with different performance measurements. While two transistors may have exactly the same gain and noise figure at the desired operating frequency, often the impedances needed to maintain stability at other frequencies can be different. Thus, the "substitute" may oscillate, while the proper transistor would work just fine.

You can often substitute MMICs (monolithic microwave integrated circuits) for one another because they are designed to be stable and operate with the same input and output impedances (50 Ω).

The size of components used at microwaves can be critical — in some cases, a chip resistor 80 mils across is not a good substitute for one 60 mils across. Hopefully, the author of a construction project tells you which dimensions are critical, but you can't always count on this; the author may not know. It's not unusual for a person to spend years building just one prototype, so it's not surprising that the author might not have built a dozen different samples to try possible substitutions.

Fig 23.30 — Printed comb-line filter uses two chip capacitors in parallel to reduce loss.

23.6.4 Capacitors for Microwave Construction

Ordinary ceramic chip capacitors cost a few cents while microwave chip capacitors are more than a dollar each. All have them have parasitic resistance and inductance, but the microwave versions use lower-loss materials which make a difference at higher microwave frequencies. Ordinary ceramic capacitors work fine in non-critical applications like blocking and bypass capacitors, even up to 10 GHz. In critical areas, like low-noise or power amplifiers, microwave capacitors are preferred.

For applications requiring high-Q for low loss, like the printed comb-line filter in **Fig 23.30**, two ordinary capacitors in parallel as shown have lower loss than one expensive microwave capacitor. By paralleling, the parasitic resistances are also paralleled, cutting the resistance in half. An additional advantage is that combinations can be chosen to yield non-standard capacitance values. We might apply this trick to really demanding applications, like high-power solid-state amplifiers, by paralleling microwave capacitors to reduce losses.

23.6.5 Tuning and "No-Tune" Designs

Microwave construction does not always require tight tolerances and precision construction. A fair amount of error can often be tolerated if you are willing to tune your circuits, as you do at MF/HF. This usually requires the use of variable components that can be expensive and tricky to adjust.

Proper design and construction techniques, using high precision, can result in a "no-tune" microwave design. To build one of these no-tune projects, all you need do is buy the parts and install them on the board. The circuit tuning has been precisely controlled by the board and component dimensions so the project should work.

One tuning technique you can use with a microwave design, if you have the suitable test equipment, is to use bits of copper foil or EMI shielding tape as "stubs" to tune circuits. Solder these small bits of conductor into place at various points in the circuit to make reactances that can actually tune a circuit. After their position has been determined as part of the design, tuning is accomplished by removing or adding small amounts of conductor, or slightly changing the placement of the tuning stub. The size of the foil needed depends on your ability to determine changes in circuit performance, as well as the frequency of operation and the circuit board parameters. A precision setup that lets you see tiny changes allows you to use very small pieces of foil to get the best tuning possible.

23.7 Mechanical Fabrication

Most projects end up in some sort of an enclosure, and most hams choose to purchase a ready-made chassis for small projects, but some projects require a custom enclosure. Even a ready-made chassis may require a fabricated sheet-metal shield or bracket, so it's good to learn something about sheet-metal and metal-fabrication techniques.

Most often, you can buy a suitable enclosure. These are sold by RadioShack and most electronics distributors. Select an enclosure that has plenty of room. A removable cover or front panel can make any future troubleshooting or modifications easy. A project enclosure should be strong enough to hold all of the components without bending or sagging; it should also be strong enough to stand up to expected use and abuse.

23.7.1 Cutting and Bending Sheet Metal

Enclosures, mounting brackets and shields are usually made of sheet metal. Most sheet metal is sold in large sheets, 4 to 8 feet or larger. It must be cut to the size needed.

Most sheet metal is thin enough to cut with metal shears or a hacksaw. A jigsaw or band saw makes the task easier. If you use any kind of saw, select a blade that has teeth fine enough so that at least two teeth are in contact with the metal at all times.

If a metal sheet is too large to cut conveniently with a hacksaw, it can be scored and broken. Make scratches as deep as possible along the line of the cut on both sides of the sheet. Then, clamp it in a vise and work it back and forth until the sheet breaks at the line. Do not bend it too far before the break begins to weaken, or the edge of the sheet might bend. A pair of flat bars, slightly longer than the sheet being bent, make it easier to hold a sheet firmly in a vise. Use "C" clamps to keep the bars from spreading at the ends.

Smooth rough edges with a file or by sanding with a large piece of emery cloth or sandpaper wrapped around a flat block.

23.7.2 Finishing Aluminum

Give aluminum chassis, panels and parts a sheen finish by treating them in a caustic bath. (See the information on chemical safety at the beginning of this chapter.) Use a plastic container to hold the solution and wear both safety goggles and protective clothing while treating aluminum. Ordinary household lye can be dissolved in water to make a bath solution. Follow the directions on the container. A strong solution will do the job more rapidly.

Stir the solution with a non-metal utensil until the lye crystals are completely dissolved. If the lye solution gets on your skin, wash with plenty of water. If you get any in your eyes, immediately rinse with plenty of clean, room-temperature water and seek medical help. It can also damage your clothing, so wear something old. Prepare sufficient solution to cover the piece completely. When the aluminum is immersed, a very pronounced bubbling takes place. Provide ventilation to disperse the escaping gas. A half hour to two hours in the bath is sufficient, depending on the strength of the solution and the desired surface characteristics.

23.7.3 Chassis Working

With a few essential tools and proper procedure, building radio gear on a metal chassis is a relatively simple matter. Aluminum is better than steel, not only because it is a superior shielding material, but also because it is much easier to work and provides good chassis contact when used with secure fasteners.

Spend sufficient time planning a project to save trouble and energy later. The actual construction is much simpler when all details are worked out beforehand. Here we discuss a large chassis-and-cabinet project, such as a high-power amplifier. The techniques are applicable to small projects as well.

Cover the top of the chassis with a piece of wrapping paper or graph paper. Fold the edges down over the sides of the chassis and fasten them with adhesive tape. Place the front panel against the chassis front and draw a line there to indicate the chassis top edge.

Assemble the parts to be mounted on the chassis top and move them about to find a satisfactory arrangement. Consider that some will be mounted underneath the chassis and ensure that the two groups of components won't interfere with each other.

Place controls with shafts that extend through the cabinet first, and arrange them so that the knobs will form the desired pattern on the panel. Position the shafts perpendicular to the front chassis edge. Locate any partition shields and panel brackets next, then sockets and any other parts. Mark the mounting-hole centers of each part accurately on the paper. Watch out for capacitors with off-center shafts that do not line up with the mounting holes. Do not forget to mark the centers of socket holes and holes for wiring leads. Make the large center hole for a socket *before* the small mounting holes. Then use the socket itself as a template to mark the centers of the mounting holes. With all chassis holes marked, center-punch and drill each hole.

Next, mount on the chassis the capacitors and any other parts with shafts extending to the panel. Fasten the front panel to the chassis temporarily. Use a machinist's square to extend the line (vertical axis) of any control shaft to the chassis front and mark the location on the front panel at the chassis line. If the layout is complex, label each mark with an identifier. Also mark the back of the front panel with the locations of any holes in the chassis front that must go through the front panel. Remove the front panel.

MAKING ENCLOSURES WITH PC BOARD MATERIAL

Much tedious sheet-metal work can be eliminated by fabricating chassis and enclosures from copper-clad printed-circuit board material. While it is manufactured in large sheets for industrial use, some hobby electronics stores and surplus outlets market usable scraps at reasonable prices. PC-board stock cuts easily with a small hacksaw. The nonmetallic base material isn't malleable, so it can't be bent. Corners are easily formed by holding two pieces at right angles and soldering the seam. This technique makes excellent RF-tight enclosures. If mechanical rigidity is required of a large copper-clad surface, solder stiffening ribs at right angles to the sheet.

Fig 23.31 shows the use of PC-board stock

Fig 23.31 — An enclosure made entirely from PC-board stock.

to make a project enclosure. This enclosure was made by cutting the pieces to size, then soldering them together. Start by laying the bottom piece on a workbench, then placing one of the sides in place at right angles. Tack-solder the second piece in two or three places, then start at one end and run a bead of solder down the entire seam. Use plenty of solder and plenty of heat. Continue with the rest of the pieces until all but the top cover is in place.

In most cases, it is better to drill all needed holes in advance. It can sometimes be difficult to drill holes after the enclosure is soldered together.

You can use this technique to build enclosures, subassemblies or shields. This technique is easy with practice; hone your skills on a few scrap pieces of PC-board stock.

23.7.4 Drilling Techniques

Before drilling holes in metal with a hand drill, indent the hole centers with a center punch. This prevents the drill bit from "walking" away from the center when starting the hole. Predrill holes greater than ½ inch in diameter with a smaller bit that is large enough to contain the flat spot at the large bit's tip. When the metal being drilled is thinner than the depth of the drill-bit tip, back up the metal with a wood block to smooth the drilling process.

The chuck on the common hand drill is limited to ⅜ inch bits. Some bits are much larger, with a ⅜ inch shank. If necessary, enlarge holes with a reamer or round file. For very large or odd-shaped holes, drill a series of closely spaced small holes just inside of the desired opening. Cut the metal remaining between the holes with a cold chisel and file or grind the hole to its finished shape. A nibbling tool also works well for such holes.

Use socket-hole punches to make socket holes and other large holes in an aluminum chassis. Drill a guide hole for the punch center bolt, assemble the punch with the bolt through the guide hole and tighten the bolt

to cut the desired hole. Oil the threads of the bolt occasionally.

Cut large circular holes in steel panels or chassis with an adjustable circle cutter ("fly cutter") in a drill press at low speed. Occasionally apply machine oil to the cutting groove to speed the job. Test the cutter's diameter setting by cutting a block of wood or scrap material first.

Remove burrs or rough edges that result from drilling or cutting with a burr-remover, round or half-round file, a sharp knife or chisel. Keep an old chisel sharpened and available for this purpose.

RECTANGULAR HOLES

Square or rectangular holes can be cut with a nibbling tool or a row of small holes as previously described. Large openings can be cut easily using socket-hole punches. A portable jig saw with the appropriate cutting blade can be used on thin sheet metal and plastic. The resulting edges will require filing.

23.7.5 Construction Notes

If a control shaft must be extended or insulated, a flexible shaft coupling with adequate insulation should be used. Satisfactory support for the shaft extension, as well as electrical contact for safety, can be provided by means of a metal panel bushing made for the purpose. These can be obtained singly for use with existing shafts, or they can be bought with a captive extension shaft included. In either case the panel bushing gives a solid feel to the control. The use of fiber washers between ceramic insulation and metal brackets, screws or nuts will prevent the ceramic parts from breaking.

PAINTING

Painting is an art, but, like most arts, successful techniques are based on skills that can be learned. The surfaces to be painted must be clean to ensure that the paint will adhere properly. In most cases, you can wash the item to be painted with soap, water and a mild scrub brush, then rinse thoroughly. When it is dry, it is ready for painting. Avoid touching it with your bare hands after it has been cleaned. Your skin oils will interfere with paint adhesion. Wear rubber or clean cotton gloves.

Sheet metal can be prepared for painting by abrading the surface with medium-grade sandpaper, making certain the strokes are applied in the same direction (not circular or random). This process will create tiny grooves on the otherwise smooth surface. As a result, paint or lacquer will adhere well. On aluminum, one or two coats of zinc chromate primer applied before the finish paint will ensure good adhesion.

Keep work areas clean and the air free

of dust. Any loose dirt or dust particles will probably find their way onto a freshly painted project. Even water-based paints produce some fumes, so properly ventilate work areas.

Select paint suitable to the task. Some paints are best for metal, others for wood and so on. Some dry quickly, with no fumes; others dry slowly and need to be thoroughly ventilated. You may want to select rust-preventative paint for metal surfaces that might be subjected to high moisture or salts.

Most metal surfaces are painted with some sort of spray, either from a spray gun or from spray cans of paint. Either way, follow the manufacturer's instructions for a high-quality job.

PANEL LAYOUT AND LABELING

There are many ways to layout and label a panel. Some builders don't label any controls or jacks, relying on memory to figure what does what. Others use a marking pen to label controls and inputs. Decals and dry transfers have long been a staple of home brewing. Label makers that print on clear or colored tape are used by many.

With modern computers and available software, it is not hard to lay out professional looking panels. One can use a standard drawing program for the layout. The grids available on these drawing programs are sufficient to make sure that everything is lined up squarely. If the panel label is laid out before the panel is drilled for controls, a copy of the label can be used as a drill template.

Computer-aided design (CAD) programs can also be used to lay out and label panels, although they can have a steep learning curve and may be overkill for many applications.

WB8RCR has written two software programs, *Dial* and *Panel*, that are specifically designed for laying out panels and dials. There are *Windows* versions and platform independent versions available for download at **www.qsl.net/wb8rcr**. These programs output a Postscript file.

These programs can be used in several ways. One can print out a mirror image of the layout on a transparency and then glue that to the front panel with the printing facing towards the panel. In this manner the transparency will protect the label. The panel layout can be printed out on card stock and affixed to the front panel with spray adhesive or self-adhesive contact paper can be used. If the printing is facing outward it can be sprayed with clear acrylic spray to protect it.

Surplus meters often find their way into projects. Unfortunately the meter faces usually do not have an appropriate scale for the project at hand. Relabeling meters has long been a mainstay to make home brew gear look professional. With the advent of computers this job has been made very easy. A software package, *MeterBasic*, by Jim Tonne, W4ENE, included on the *Handbook* CD-ROM, is very easy to use and results in professional looking meters that indicate exactly what you want them to indicate.

Contents

Chapter 24

Station Accessories

An Amateur Radio station is much more than the transceiver, amplifier and other major pieces of equipment found in a typical ham shack. In this chapter you will find a number of accessory items to make operation more convenient, or to tie other pieces of equipment together. Projects include antenna matching and switching solutions, transceiver/computer interfaces and other station accessories. Additional station accessory projects may be found on the *Handbook* CD-ROM.

Chapter 24 — CD-ROM Content

Projects

- "A Remote Power Controller" by Mike Bryce, WB8VGE
- "A Switched Attenuator" courtesy of RSGB
- "The ID-O-Matic Station Identification Timer" by Dale Botkin, NØXAS
- "The Tandem Match — An Accurate Directional Wattmeter" by John Grebenkemper, KI6WX
- "Two QSK Controllers for Amplifiers" by Jim Colville, W7RY, and Paul Christensen, W9AC

Support Files

- Code and support files for ID-O-Matic by Dale Botkin, NØXAS
- Support files for SWR Monitor by Larry Coyle, K1QW
- Support files for "Two QSK Controllers for Amplifiers" by Jim Colville, W7RY, and Paul Christensen, W9AC
- Trio of Computer Interfaces PCB template

24.1 A 100-W Compact Z-Match Antenna Tuner

A Z-Match tuner will match just about anything on all HF bands, and only uses two controls. They are well regarded, but acquiring the necessary air-wound inductors and variable capacitors can be difficult. In addition, air-wound inductors imply large tuners. The antenna tuner described here by Phil Salas, AD5X, is compact enough for portable HF operation, handles 100 W and uses readily available parts. The Z-Match circuit was originally described in "Multiband Tuning Circuits" by R.W. Johnson, W6MUR, in July 1954 *QST* and the article is included in this book's supplemental material.

The idea for this project came from an article by Charles Lofgren, W6JJZ, in the *ARRL Antenna Compendium Vol 5*. In that article, the author suggested using a toroid core-based inductor. This would solve the air-core inductor/size problem. Suitable variable capacitors are available at a good price. The result is the compact unit shown in **Fig 24.1**.

Fig 24.1 — The Z-Match Tuner handles 100 W and can be used from 80 through 10 meters.

TUNER CONSTRUCTION

The final circuit shown in **Fig 24.2** is based on W6JJZ's article. The only real change was to go from two switch-selected output links (10 turns and 4 turns) to a single 8-turn output link, and to increase the variable capacitor size to 400-pF. The tuner provides a good match from 80-10 meters!

The tuner is built into a 3 × 5¼ × 5⅞ inch (HWD) aluminum box. Toroid T1 is supported by its leads and some hot glue between the inductor and the frame of C2. Also put a little hot glue between T1 and the side of the case.

The variable capacitors must be insulated from ground, therefore mount both capacitors on a piece of single-sided circuit- or perf-board that is cut just wide enough to fit into the aluminum case. Then mount this capacitor/perf-board assembly in the case with stand-off screws. Capacitor shaft couplings are made from a ⅛-NPT brass nipple, available from the plumbing section of most hardware stores. These nipples have a ¼-inch inside diameter. Cut a 1-inch nipple in half to make two couplers. Then drill and tap holes for two #6 screws in each piece. See **Fig 24.3** for capacitor mounting and shaft coupler details. For the insulated shafts, use ¼-inch diameter nylon rods available from many hardware stores.

OPERATION

Tuning the Z-Match tuner is very easy. First adjust C2 for maximum receiver noise. Then apply some RF power and adjust C1 and C2 for minimum SWR. If you need more capacitance for matching, use S1 to switch in the extra section of C1, or switch in a fixed mica capacitor across C1. Balanced feed lines terminated in banana plugs can plug right into the center pin of the output SO-239 and the adjacent banana jack. To feed coax, ground one end of the output link with switch S2.

OPTICAL HF SWR METER FOR THE Z-MATCH TUNER

While you can use any external SWR meter with the Z-Match tuner — including the SWR meter built into most rigs — the author built an optical SWR meter into the same case for convenience. Refer to **Fig 24.4A**. It works well with high intensity LEDs rather than conventional meters. The SWR meter is built on a small piece of perf-board and mounted to the solder lug on the tuner's input SO-239. A little hot glue between the perf-board and the back of the chassis adds stability (**Fig 24.5**).

This broadband circuit works well at the 100-W power level through at least 30 MHz. With careful lead control, it should work up through 6 meters. The transformer is an FT37-43 ferrite core wound with 10 bifilar turns of #26 enameled wire. The primary is just the single wire passing through the center of the toroid. To calibrate the SWR bridge, connect the output to a resistive 50-Ω load. Apply RF power on any HF band and adjust the 20-pF variable capacitor until the REFL LED goes out.

The FWD LED gives an indication of transmitter forward power. You may want to increase the value of the 4.7 kΩ resistor in the FWD circuit if the green LED is too bright. Or you could eliminate this LED and just use the REFL LED.

A BAR-GRAPH DISPLAY FOR THE OPTICAL SWR METER

If you can supply dc power to your Z-Match Tuner and SWR meter, you may want to add the bar graph display shown in Fig 24.4B to display SWR reflected power. A version of the tuner with bar graph option is shown in **Fig 24.6** The nice thing about using the bar graph display is that it seems easier to null the reflected power because the display gives more of an analog meter "feel." Adjust the Z-Match tuner for minimum brightness of the REFL LED. When this occurs, the SWR should be less than 1.5:1.

This tuner addresses the issues of inductor size and finding reasonably priced multisection variable capacitors. The result is a wide-band, easily adjustable tuner for either portable or base station operation.

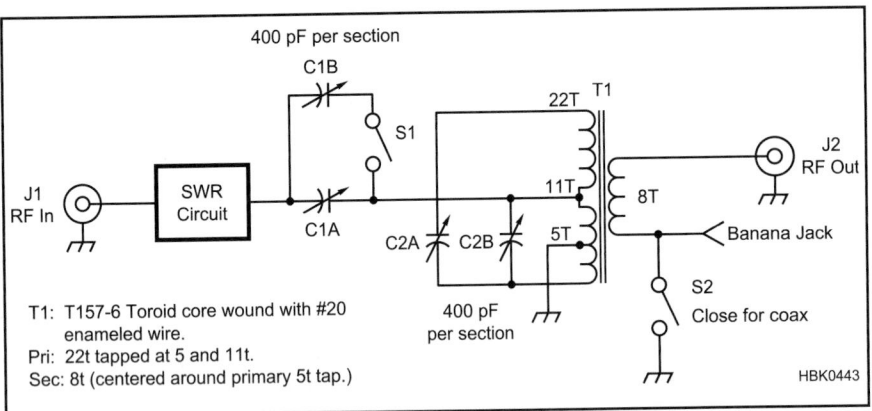

Fig 24.2 — Schematic of the 100-W Z-Match tuner.

C1, C2 — 400 pF per section, 2-section variable capacitor (Oren Elliott S2-399, www.orenelliottproducts.com). If you substitute another variable capacitor, use one with at least 350 pF per section, rated at 500 V RMS minimum.
J1, J2 — SO-239 connectors.
S1, S2 — SPDT mini-toggle switch.

T1 — Primary: 22 turns #20 AWG enam wire tapped at 5 and 11 turns. Secondary: 8 turns, with 4 turns either side of the primary 5 turn tap. Wound on T157-6 toroid (Amidon Associates, www.amidon-inductive.com).
Enclosure (Eagle, 40UB103 from www.mouser.com).

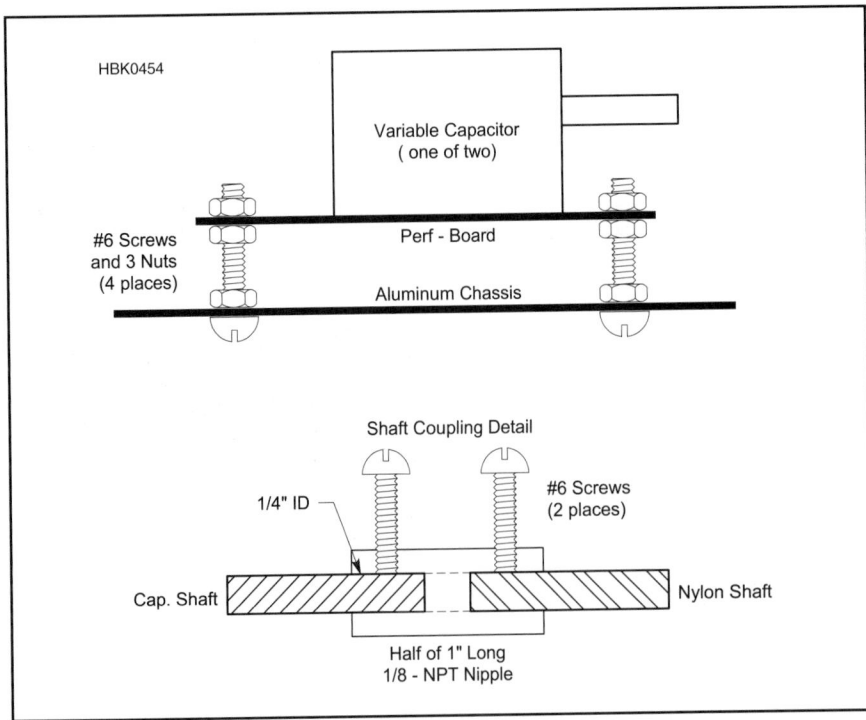

Fig 24.3 — The variable capacitors are insulated from the chassis by mounting them on perforated board. A 1/8-NPT nipple is made into a shaft coupling to couple the insulating shafts to the main capacitor shafts.

Fig 24.4 — Optical SWR meter schematic with the bar graph display modification (see text). The basic LED version is shown in A and the bar graph addition in B.

C3 — 2-20 pF variable capacitor (Mouser 24AA113 4-34 pF variable is suitable).

D1 — Red LED, high-intensity.

D2 — Green LED, high-intensity.

L1 — 10 bifilar turns of #26 enam wire on FT37-43 toroid core (Amidon Associates, www.amidon-inductive.com).

U1 — Display driver IC, LM3914 (All Electronics, www.allelectronics.com).

U2 — 10-LED bar graph display (Mouser 859-LTL-1000HR). Any 10-segment LED bar graph will work, but verify pinout.

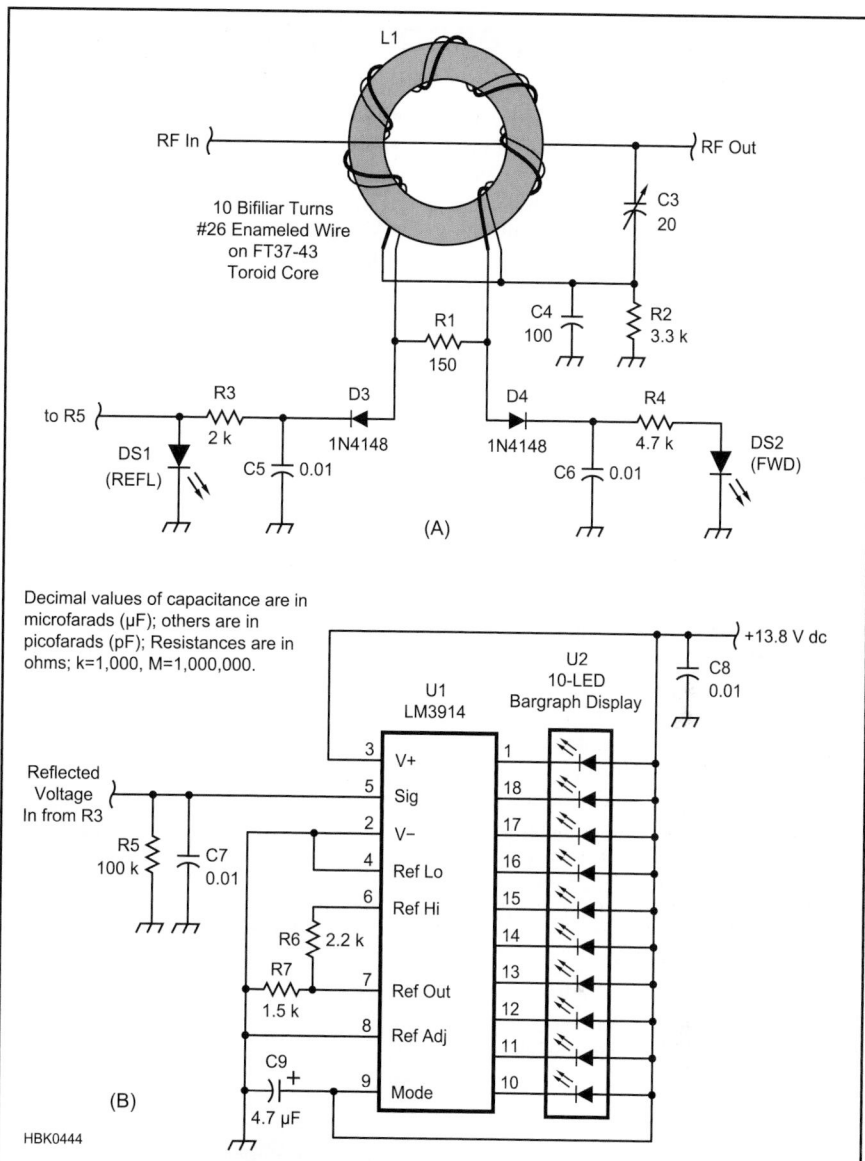

Fig 24.5 — The SWR meter circuit board is soldered directly to the RF input connector and attached to the back panel with hot glue.

Fig 24.6 — The bar graph display of reflected power makes tuning easier and more intuitive.

24.2 A Microprocessor Controlled SWR Monitor

This update of the basic SWR meter, designed by Larry Coyle, K1QW, uses a microprocessor to drive an analog panel meter, exploiting the best features of both analog and digital technology. The instrument also displays forward/reflected/net power and return loss (see the **Transmission Lines** chapter) with peak-hold capability. All display selections are selected by front panel switches. [*More information is available on K1QW's Web site, www.lcbsystems.com, and a complete package of information on this project with supporting files and photos is available on the CD-ROM accompanying this Handbook.*]

OVERVIEW AND BLOCK DIAGRAM

After a bit of work, the author came up with the SWR/Power/Return Loss monitor shown in block diagram form in **Fig 24.7**. There are three interacting assemblies: SWR sense head, front panel assembly and signal processor board. (The latter two make up the Display and Signal Processor unit.) The SWR sense head is a conventional Stockton dual-transformer bridge with Schottky diode detectors. This component converts the forward and reflected RF amplitudes to dc and connects to the Display and Signal Processor by an ordinary stereo audio cable.

The user interface to the SWR monitor is familiar and intuitive. As you can see in **Fig 24.8A**, a front-panel rotary switch selects the quantity to be measured and an analog meter displays the result.

•Forward power on a linear scale
•Reflected power as a percentage of forward power
•Net power delivered to the load

•SWR on a linear scale
•Return loss on a linear scale
Some additional features are:
•Peak-hold front panel switch
•A front panel switch to multiply the meter sensitivity by 3X
•Data logging to a computer using USB port.
•Transmitter Lockout (XLO) at high SWR
The rear panel, **Fig 24.8B**, shows the power switch and external connections

Transmitter Lockout (XLO) is a protective feature that lights a LED on the front panel and energizes a relay when the SWR exceeds a preset level. This opens a set of relay contacts which are connected to a pair of banana jacks on the rear panel. Once triggered, the relay remains energized for either three or nine seconds, depending on the configuration of the DIP switch located on the signal processor board. The same DIP switch also allows you to

select the SWR level at which the XLO relay triggers — 1.5, 2, 3 or 4:1. By connecting the relay contacts in series with your rig's PTT or keying line, transmitting into high SWR can be prevented.

The two dc signals from the sense head make up the analog input to the signal processor board. The analog output for driving the meter is a pulse width modulated (PWM) signal generated by the microprocessor and smoothed by a simple resistor-capacitor (RC) filter.

THE MICROPROCESSOR

The Parallax "Propeller" microprocessor (**www.parallax.com**) is unique in that it's really eight processors ("cogs") in one package, each one able to run its own separate program during its own time slot, controlled by a central "hub." This may sound challenging, but this actually makes programming the chip

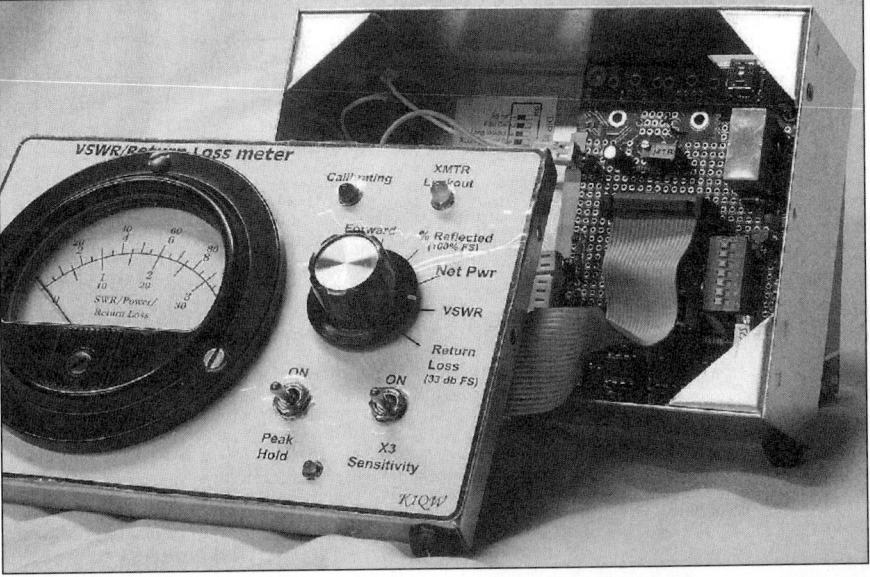

(A)

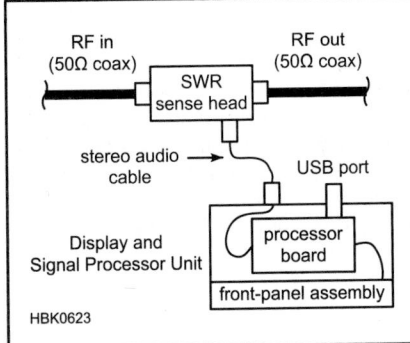

Fig 24.7 — This block diagram shows how the SWR sense head and the signal processor box are connected by an ordinary two-conductor shielded stereo audio cable. The signal processor box contains the processor board and the front panel assembly.

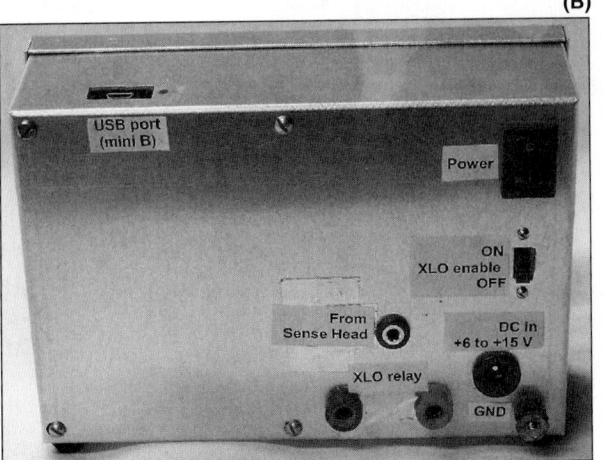

(B)

Fig 24.8 — (A) Front panel of the display and signal processor unit. The meter face and the front panel lettering were created using Microsoft *Visio*, printed on heavy paper stock and attached with contact cement. With the front cover moved aside, you can see the ribbon cable carrying signals between the front panel assembly and the processor board. (B) Rear panel of the display and signal processor. The mini-B USB connector is accessible through the opening on the top surface of the enclosure.

much easier. If you need to carry out several tasks at once — and most microprocessor applications do — just assign each job to its own individual cog and work on them separately. The central hub makes sure each cog gets its turn and keeps things synchronized.

The project uses the 3 × 4 inch Propeller USB Development Board from Parallax. It includes the Propeller chip, an Electrically-Erasable Programmable Read-Only Memory (EEPROM) for program storage, a 5 MHz crystal, two voltage regulators and, a USB mini-B connector with all the necessary USB interface circuitry.

The Propeller is programmed in a language called "Spin" which should make anyone who is familiar with the BASIC computer language feel at home. All the tools you need to write software for and to program the Propeller is available for free on the Parallax Web site, including an extensive library of pre-built code routines that you can just drop into your own programs.

Software is loaded onto the development board using the USB port from a host computer with a USB port. When not in use as a programming port, the same USB port is used as a serial data port for passing data between the SWR monitor and a computer running a terminal program such as HyperTerminal.

MICROPROCESSOR BOARD CIRCUITRY

The development board includes a proto-typing area large enough to hold all the parts of the SWR monitor except for front and rear panel components. Since most of the parts come assembled on the development board, construction of the digital board is quite easy. As you can see in **Fig 24.9A**, the electrical schematic, only two integrated circuits and a few analog components are required.

The dc signals representing forward and reflected power from the SWR sense head are received by dual op-amp U2. Each section of U2 is configured as a unity-gain voltage follower with a compensating diode (D1 and D2) to balance out the dc offset introduced by the Schottky rectifier diodes in the sensing head. (The technique of using a diode in the feedback path of an op amp to compensate for the non-linearity of an RF detector diode is described in detail in "A Compensated, Modular RF Voltmeter" in the **Test Equipment and Measurements** chapter of this book.) Each section also has an RC low-pass filter at its input to filter out unwanted RF — always a possibility around a ham shack.

The next stage in the signal path is U3, an analog multiplexer (MUX). This chip acts as a three-position switch under control of the microprocessor. (A fourth channel is unused.) The forward and reflected signals (on pins 14 and 15) are sampled alternately, 20 times per second, as long as the program is running. The

Table 24.1A
Digital Inputs

Digital Input	Originates at	What it does
X3SEL	front panel - toggle switch	increases meter sensitivity by 3X
PEAKENBL	front panel - toggle switch	enables peak hold function on the meter
RLOSSEL	front panel – rotary switch	displays return loss on the meter
SWRSEL	front panel – rotary switch	displays SWR on meter
NETSEL	front panel – rotary switch	displays net power on meter
REFSEL	front panel – rotary switch	displays percent reflected power on meter
FWDSEL	front panel – rotary switch	displays forward power on meter
HIREFSEL	proc board - DIP switch	drives meter to full scale for calibration (must be OFF for normal operation)
FASTCAL	proc board - DIP switch	speeds up meter response for calibration (must be OFF for normal operation)
XLOLONG	proc board - DIP switch	enables long xmtr lockout (about 9 seconds)
XLTHRESH0	proc board - DIP switch	XLO threshold bit 0
XLTHRESH1	proc board - DIP switch	XLO threshold bit 1
XLOCKENBL	rear panel switch	enables xmtr lockout function
DIAG	proc board - jumper JP1	displays raw input levels on terminal
W100	proc board - jumper JP2	changes terminal display to 0-100 watt range (required when using 100-watt sense head)

Table 24.1B
Digital Signals for SWR Threshold Select

XLTHRESH1	XLTHRESH0	Selects SWR threshold level
OFF	OFF	1.5:1
OFF	ON	2:1
ON	OFF	3:1
ON	ON	4:1

third input to the MUX (pin 12) is grounded, and is also sampled by the microprocessor from time to time. This gives a measure of any dc offsets introduced by the analog-to-digital conversion and is used to correct the forward and reflected readings in software. This automatic zero adjustment takes place every five seconds or so and is indicated by a brief flash from the front-panel "Calibrating" LED that also serves as a "digital heartbeat" to show that the software is running.

The MUX output (pin 13) is passed to a potentiometer voltage divider to allow the scale factor to be set during calibration. From there it goes to the Propeller microprocessor itself where a delta-sigma analog-to-digital conversion takes place. (See the **Analog Basics** chapter for information on analog-to-digital conversion.)

DIGITAL INPUT/OUTPUT

The microprocessor responds to the user's inputs by monitoring the state of 13 digital input pins. Most of these pins are controlled by the rotary and toggle switches on the front panel. The panel connects to the processor board by a short ribbon cable as shown in Fig 24.8A. See Fig 24.9B for a schematic of the separate front panel assembly. The ribbon cable also carries power for the LEDs. On the

processor board there are a multi-section dual in-line package (DIP) switch, S2, and two jumpers, JP1 and JP2, which select among the various operating modes. Lastly, the Xmtr Lockout (XLO) switch is mounted on the rear panel and visible in Fig 24.8B. These inputs to the microprocessor and the various modes of operation they control are summarized in **Table 24.1A**. The digital inputs are all active low; ie, the function is enabled when the signal line is at a low logic level. XLTHRESH1 and XLTHRESH0 form a two-bit code to select one of four SWR levels. When this SWR level is exceeded and if XLO is enabled, the transmitter lockout function is triggered.

There are only six digital *outputs*:

• Two address lines for the analog MUX to select among the three input channels
• Two lines to power the Calibrating and XLO LEDs located on the front panel
• One PWM signal to drive the front-panel analog meter
• One line to control the XLO relay.

Fig 24.8A shows the built-up processor board inside the 2 × 4 × 6 inch aluminum enclosure connected to the front panel with the ribbon cable. Here you can also see some of the wiring to the rear panel components. The USB connector is at the top edge of the circuit board and is accessible through a rectangular hole nibbled out of the top surface of the aluminum box.

THE RF SENSE HEAD

The SWR sense head is the business end of the SWR meter where the forward and reflected amplitude components of the RF present on the transmission line are separated and converted to dc. For this project I adopted a straightforward Stockton directional coupler circuit, with Schottky diode detectors and, unlike other designs, it requires no

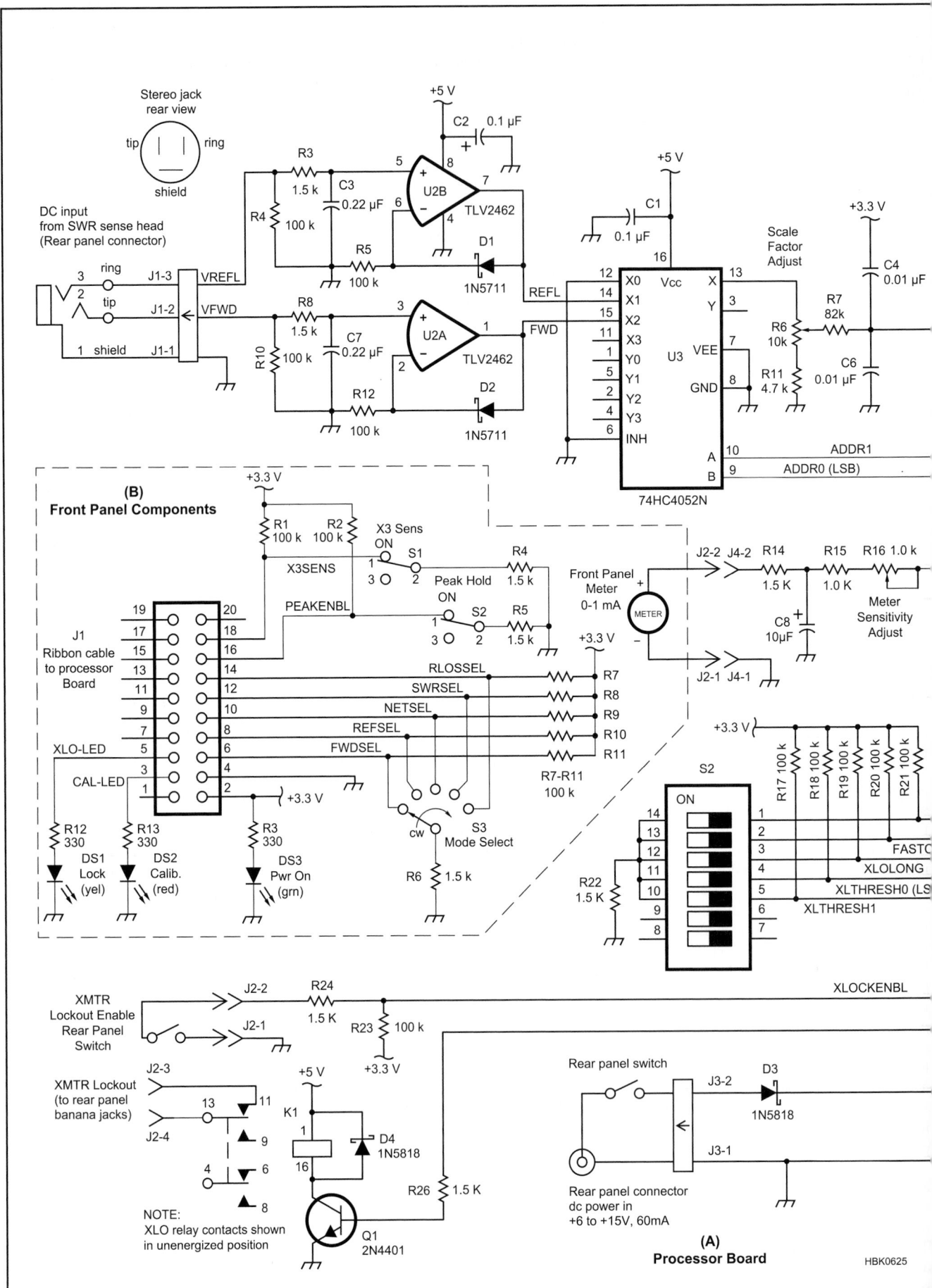

Fig 24.9 — (A) Processor board schematic. All of the parts shown here, except for the rear panel components, are mounted on the Parallax Propeller Development Board. The parts within the dotted line are already installed at the factory, and are shown here only for completeness. (B) Front panel schematic. Switches and LEDs connect to the processor board via a 20-pin dual-row header, J1, and a ribbon cable. For convenience in wiring, the analog meter connects to the processor board by a separate 2-pin connector. (See Parts List on page 24.8.)

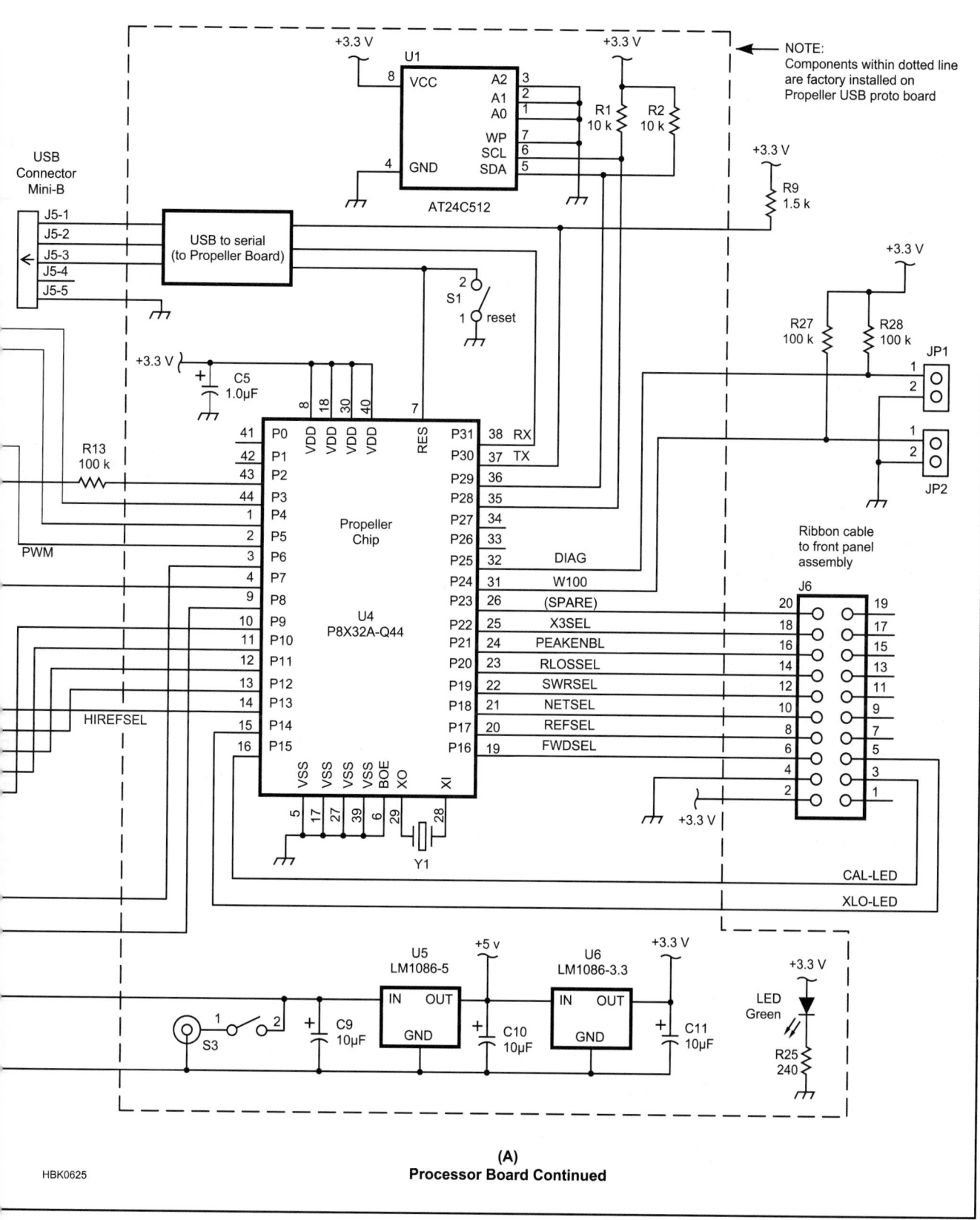

(A)
Processor Board Continued

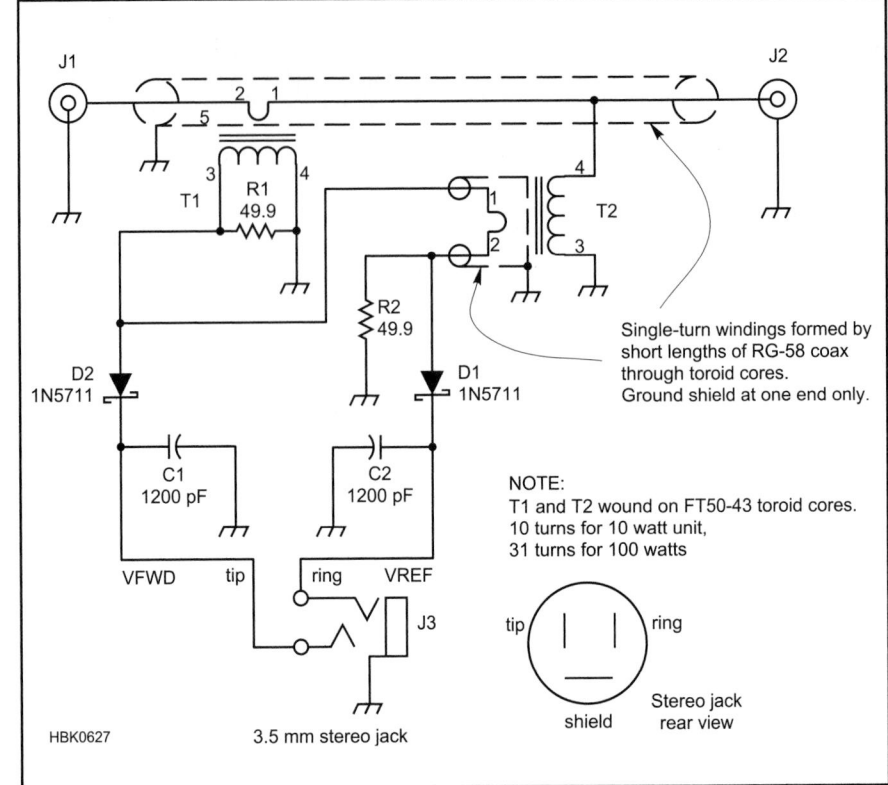

Fig 24.10 — SWR sense head schematic. All components except for the transformers are surface mount parts. R1 and R2 are size 1210, rated at ½ W. Transformers T1 and T2 are wound on FT50-43 ferrite toroid cores. For the 10-W sense head, use 10 turns of 24 AWG wire. For the 100-W head, use 31 turns.

balancing trimmers. The schematic is given in **Fig 24.10** and the circuit is described in the article by J. Grebenkemper, KA3BLO, "The Tandem Match — an Accurate Directional Wattmeter," *QST*, Jan 1987, pp 18-26 (also published in the 2009 and earlier editions of *The ARRL Handbook*).

The accuracy of this circuit can be greatly improved by taking care to match the diodes in the sensing head with the corresponding compensation diodes in the op amp input stages on the signal processor board (D1 and D2 in Fig 24.9A). The easiest way to do this is with the diode test function included on most utility digital multimeters which passes a small dc current through the diode being tested and displays the forward voltage drop. The high-frequency range can be extended by using surface-mount components in the sensing head.

There are two versions of the SWR sensing head. The QRP version has a maximum range of 10 W. The other can handle up to 100 W. The only difference between the two versions is the number of turns on the toroid transformers (see the parts list in Fig 24.10). The high frequency performance of the 100-W head at 50 MHz proved to be poorer than the 10-W unit. I attribute this to the added capacitance

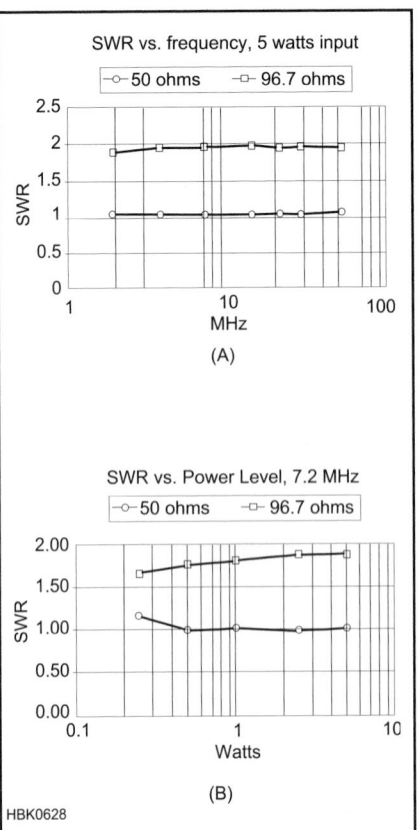

of the transformer windings (the higher power transformers have three times as many turns as the low-power ones).

The overall accuracy of the monitor compares fairly well to commercial devices. **Fig 24.11A** and **24.11B** show performance data I took using dummy loads of 50 and 96.7 Ω. At five watts input, using the 10-W sensing head, the SWR readings were within 3% over a range of 1.8 to 30 MHz, and within 5% at 50 MHz. Fig 24.11B shows good accuracy down to 500 mW input (great for QRP rigs!) and is even usable down to 250 mW.

24.4.6 Software

For those who want to duplicate this project, the Spin language software listing, ready to program into the Parallax USB Development Board, is available in the package of

Fig 24.11 — (A) SWR plotted against frequency, up to 50 MHz. (B) SWR as a function of input power level at a fixed frequency of 7.2 MHz. The data shown here were taken using the 10-W sense head, with dummy loads measured at 50 and 96.7 Ω.

supporting files on the CD-ROM. There are also details of the subroutines handled by each of the Propeller cogs plus additional text and graphic files covering operation and features of the instrument.

As explained above, the organization of the Propeller into several peripheral processors (cogs) makes it easy to follow the program flow. The main program object (MAIN) runs in its own cog and is the first to be executed when power is applied (or whenever the chip is reset). Its first order of business is to set individual microprocessor pins to be inputs or outputs, as required.

Next, MAIN starts up three separate cogs for (1) analog-to-digital conversion (the ADC3CH cog), (2) generating the pulse width modulated waveform to drive the analog panel meter (the PWM cog) and (3) communicating with the terminal via the USB port (the MONITOR cog). They run continuously, swapping data with the MAIN program by means of variables stored in memory. After these three cogs are started, the MAIN program object enters a loop, endlessly repeating the following steps:

1) Read forward and reflected amplitudes and SWR data generated by the ADC3CH object.

2) Calculate forward power, percent reflected power, net power, SWR and return loss in decibels.

3) Display the quantity selected by the front panel switch on the panel meter using the PWM object.

4) Convert power levels, SWR and return loss information to ASCII strings and send them to the MONITOR object, which writes to the USB port.

This is a shining example of building on the accomplishments of others who have made their work freely available. Without such an extensive library of pre-tested routines on hand, the author probably would never have even started this project.

24.3 A 160- and 80-Meter Matching Network for Your 43-Foot Vertical

Many amateurs are using 43-ft vertical antennas for multiband operation. These antennas have higher radiation resistance than shorter verticals, which minimizes ground losses. This is especially important when you have an electrically short antenna — a characteristic of even a 43-ft antenna on 160 and 80 meters.

When fed with a 1:4 unun (unbalanced-to-unbalanced transformer), a 43-ft antenna has a reasonable compromise SWR on 60-10 meters, which means that cable and unun losses are pretty much negligible on these bands. Performance on 160 meters, and to a lesser extent on 80 meters, is not as good unless you provide matching right at the antenna. The high capacitive reactance and low radiation resistance of a 43-ft antenna on 160 and 80 meters make the mismatch so bad that it is almost impossible to match from your shack if you are using low-loss coax. If you can match the antenna system from your shack, you will lose a lot of power in your coax and unun because of the very bad mismatch at the antenna. Phil Salas, AD5X, describes two impedance matching devices for use with 43-ft verticals designed to significantly reduce SWR-related losses and to "help out" inside tuners on 160 and 80 meters.

THE MATCHING REQUIREMENT

Antenna analyzer measurements of the author's 43-ft vertical antenna indicate a capacitive reactance of about 580 Ω on 160 meters. This value will vary with different kinds of 43-ft verticals, proximity to other objects and other factors unique to each installation but is usually in the 550-650 Ω range. With this amount of capacitive reactance, approximately 50 μH of inductance is

needed to resonate the antenna. On 80 meters, approximately 9 μH is needed. A 50 μH high-Q inductor is large, but a toroidal inductor will be more compact.

Fig 24.12 shows the circuit of a compact design that handles 1500 W SSB or CW on 80 and 160 meters. **Table 24.2** lists the necessary parts. The inductor consists of 35 turns of #14 AWG solid copper insulated house wiring wound on a T400A-2 toroid core, with the feed tapped 2 turns from the ground end for 80 meters, and 3 turns from the ground end for 160 meters. Start with 38 turns and then remove turns as necessary to get the network to resonate where you want it in the 160 meter band (more on this later).

The toroid assembly fits in a 6 × 6 × 4-inch NEMA enclosure. The assembly is secured with a 2.5-inch long #10 screw and associated hardware along with a 2 × 4 inch piece of unplated fiberglass PC board material (**Fig 24.13**). Before mounting the toroid, prepare it by scraping the insulation off the outside of the wire at turns 2, 3 and 11-13. Because of the high voltages possible with 1500 W (especially on 160 meters), wrap

the toroid with two layers of 3M #27 glass-cloth electrical tape for added insulation between the wire and the toroid core.

To select between 160- and 80-meter operation, the appropriate connections are made with external jumpers across binding

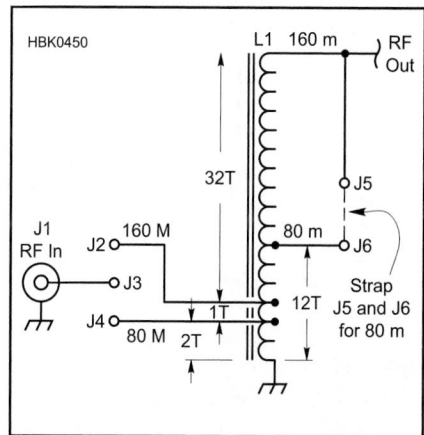

Fig 24.12 — 160/80-meter impedance matching network for use with a 43-ft vertical. See Table 24.2 for a parts list.

Table 24.2
Parts List: 160/80-Meter Impedance Matching Assembly

Qty	Description	Source/Part Number
1	6×6×4 inch NEMA enclosure	Hardware store/home center
1	T400A-2 powdered iron toroid (L1)	Amidon T400A-2
1	SO-239 connector (J1)	Mouser 601-25-7350
4	Black binding post (J2, J4-J6)	Mouser 164-R126B-EX
1	Red binding post (J3)	Mouser 164-R126R-EX
4	Banana plug	Mouser 174-R802-EX
1 roll	3M #27 glass tape (for L1)	Hardware store/home center

Misc: #14 AWG house wiring, stainless steel hardware.

posts as shown in **Fig 24.14**. Stainless steel #8 hardware (screws, washers, lockwashers, nuts) is used for the matching unit ground and RF output. Internal to the matching unit, a 2-inch-wide strip of aluminum duct repair tape makes a good low-impedance ground between the coax connector and the ground screw on the bottom of the case. Finally, use #14 AWG stranded insulated wire for all internal connections.

TUNING THE MATCHING NETWORK TO RESONANCE

Your particular installation will almost certainly require you to change the resonant frequency of the matching network. This is because there will be variations of the antenna impedance based on your particular installation and desired operating frequency range. The design is such that the overall inductance is too large for 160 meters, so the network should resonate at or below the lower band edge. Therefore, you will need to remove one or more of the upper inductor turns to resonate the network for the desired frequency on 160 meters. To do this, first solder wires from the turn 2 and 3 tap points on the coil to the two outer binding posts by the RF IN connector. The input tap points tend to be fairly non-critical and will probably always be the same for all installations.

Now solder a short wire from the RF IN center pin to the middle binding post. Next, externally jumper the middle binding post to the 160 meter binding post (turn 3). Connect the matching assembly to the base of your 43-ft vertical and use an antenna analyzer to find the minimum SWR point on 160 meters. If the resonant frequency is too low, remove a turn of wire and measure again. You should see a 50 kHz upward move in frequency per turn of wire removed.

When you have reached the desired resonant point on 160 meters, it is time to move to 80 meters. Externally jumper the input tap middle binding post to the 80-meter binding post (turn 2), and use a clip lead to short from the top of the coil to about turn 12. Again, find the frequency where minimum SWR occurs. Move the tap point up or down until you reach the desired resonance point. Solder a wire from this tap point to one of the binding posts. Solder another wire from the top of the coil to another binding post. Now you will be able to externally jumper these binding posts to go from 160 to 80 meters.

Fig 24.15 shows the author's final results for 160 and 80 meters as measured with an antenna analyzer connected directly to the matching network input at the base of the antenna. With resonance set for the CW end of these bands, the 2:1 SWR bandwidth on 160 meters is about 50 kHz and about 150 kHz on 80 meters. Even a 3:1 SWR on these bands results in negligible SWR-related cable losses and is easily matched with an in-shack tuner.

OPERATION

Using the matching unit is simple. Just disconnect your normal unun when you want to operate on 160 or 80 meters and connect this matching unit to the base of the antenna. Select either 160 or 80 meters with the external straps. You can connect both the unun and this matching unit to the antenna at the same time, and just leave off the ground wire from the unit that is not used. The matching unit connected to the base of a 43-ft vertical is shown in **Fig 24.16**.

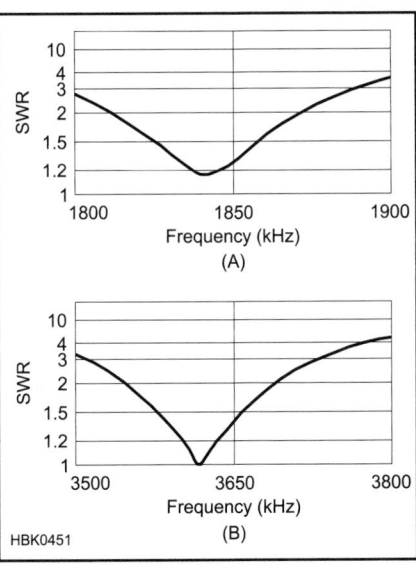

Fig 24.15 — Measured at the antenna base, the 2:1 SWR bandwidth is about 50 kHz for 160 meters and 150 kHz for 80 meters.

Fig 24.13 — Mounting details of the toroidal inductor for the 160/80-meters matching unit. The input side of the enclosure is on the left. Coax to the station connects to the RF IN SO-239, and a strap between the center binding post and one of the outer posts selects the input tap for operation on 160 or 80 meters. On the right side, a jumper between the two binding posts shorts a section of the coil for 80-meter operation. The RF OUT connection to the base of the antenna is made with via the #8 machine screw and hardware at the top, and the ground connection is on the bottom.

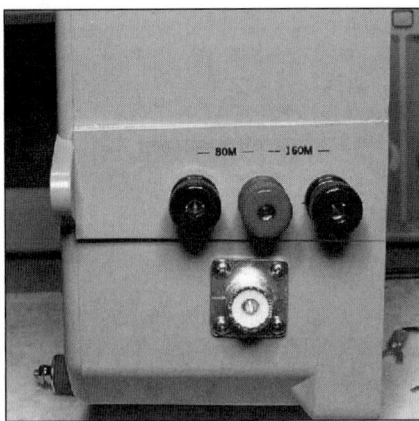

(A)

(B)

Fig 24.14 — Binding posts are used for jumpers to select 160 or 80 meter operation.

Fig 24.16 — The finished matching unit at the base of the author's 43-ft antenna (strapped for 80 meter operation).

24.4 Switching the Matching Network for Your 43-Foot Vertical

Another project in this chapter describes a simple 160 and 80 meter impedance matching network for installation at the base of a 43-ft vertical antenna. While that unit is very effective and inexpensive to build, it is also a little inconvenient in that you must connect it when it is needed, and you must also manually enable 160 or 80 meter operation using jumpers. Phil Salas, AD5X, enhances the original design with a more versatile matching assembly that can be controlled remotely for operation on all bands.

FIRST — A WORD ABOUT RF VOLTAGES

This unit uses relays for selecting the different bands, so a discussion of RF voltages is appropriate. RF voltages at the base of an untuned vertical can be quite high. As the antenna becomes shorter, the capacitive reactance becomes higher and so the resultant voltage drop across the combination of reactance and radiation resistance increases. With the 43-ft vertical, the worst-case situation will occur on 160 meters where the capacitive reactance is approximately 600 Ω. The radiation resistance is approximately 3 Ω.

With a perfect ground (no ground loss) and 1500 W of power properly matched to the antenna, peak RF voltage works out to about 19,000 V_{pk}:

$$I = \sqrt{1500\ W\ /\ 3\ \Omega} = 22.4\ A$$

$$|Z| = \sqrt{3^2 + 600^2} = 600\ \Omega$$

$$V_{RMS} = 22.4 \times 600 = 13,440\ V$$

$$V_{pk} = 13,440 \times 1.414 = 19,004\ V$$

Very few hams have a ground system that is lossless. With a ground loss of 10 Ω (better than most hams have), it works out to about 9100 V_{pk}. With the author's 600-W amplifier, assuming a ground loss of 10 Ω, it's about 5800 V_{pk}.

The author experimented with two different relays from Array Solutions. The RF-10 DPDT relay has 1.7 kV$_{pk}$ contact-to-contact and 3.1 kV$_{pk}$ contact-to-coil voltage breakdown ratings. This relay proved to be a good solution for up to about 500 W on 160 meters when the two sets of contacts are put in series. The second relay is the RF-3PDT-15 which has 3.1 kV$_{pk}$ contact-to-contact and 5.3 kV$_{pk}$ contact-to-coil voltage breakdown ratings. This relay has nearly twice the contact breakdown rating of the RF-10, and it's 3PDT so an additional set of contacts can be put in series to increase the breakdown voltage. This relay can be used in a full legal limit application if applied properly in the circuit (more on this later).

Depending on your power level and ground losses, the less expensive and smaller RF-10 may be all you need. Make the calculations to determine which relay is more appropriate for you. If you look for alternative relays, pay attention to the breakdown voltage rating; it's hard to find relays with suitable ratings for this application. The Array Solutions relays have 30 A contacts, but 15 A or more should be sufficient.

THE ALL-BAND MATCHING SOLUTION

As explained in the companion project describing a simple matching network, when fed with a 1:4 unun (unbalanced-to-unbalanced transformer), a 43-ft antenna has a reasonable compromise SWR on 60-10 meters. The antenna requires approximately 50 μH of inductance to resonate on 160 meters and 9 μH on 80 meters. For convenient all-band operation from the shack, the solution shown in **Fig 24.17** uses two relays to switch in the unun for 60-10 meter operation or the appropriate inductance for 160 and 80 meters. The unit will handle up to 1500 W from 160-10 meters.

Relay control requires inputs of 0 V, +12 V dc or –12 V dc. The matching unit operates as follows:

• With no control voltage applied, the 1:4 unun is connected to the antenna for 60-10 meter operation, and the inductor is disconnected, so the original antenna compromise SWR on these bands is preserved.

• With +12 V applied for 80-meter operation, K1 and K2 close. Contact K1C connects the unun secondary to L1 through K2C at the 200 Ω point (6 turns from the ground end). Contacts K2A and K2B short out the top part

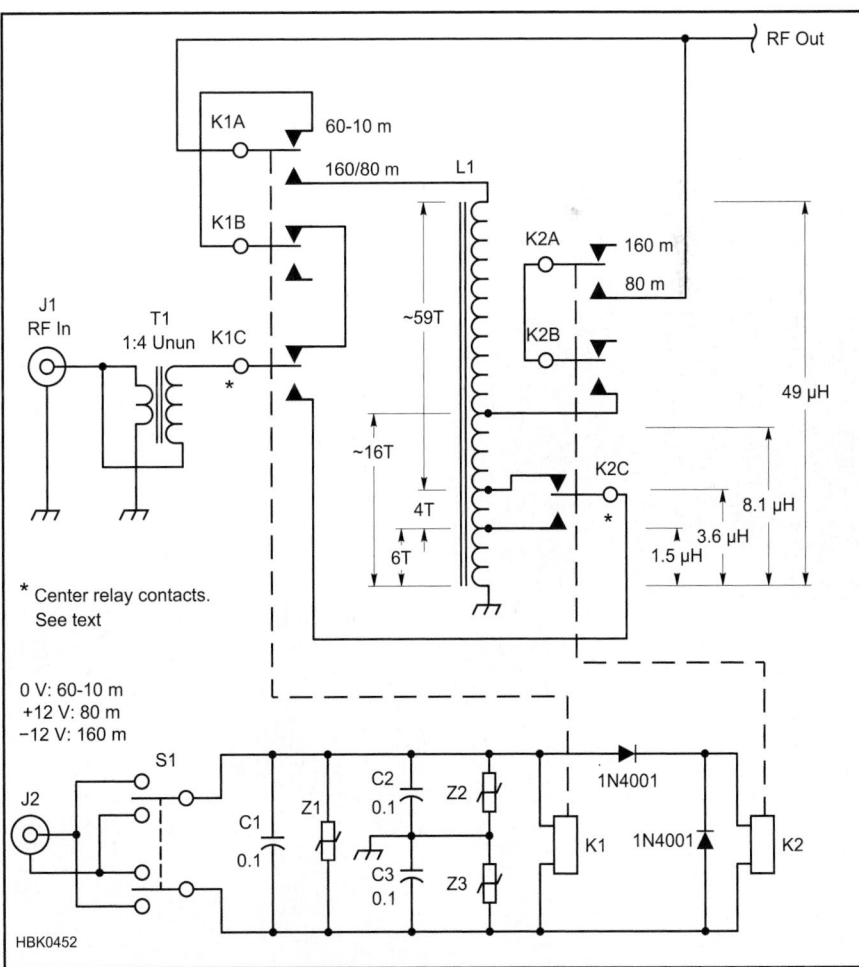

Fig 24.17 — This switched impedance matching network for use with a 43-ft vertical allows operation on 160-10 meters and is controlled from the station operating position. See Table 24.3 for a parts list.

of L1 (about 16 turns from the ground end), resonating the antenna on 80 meters.

• When the voltage is reversed (–12 V) for 160-meter operation, K1 stays closed but K2 opens. K2C now connects the unun secondary to L1 at the 200 Ω point for 160 meters (10 turns from the ground end). Contacts K2A and K2B open, removing the short and allowing the entire coil to be used to resonate the antenna on 160 meters.

Connecting the unun to L1 at the 200 Ω point keeps the unun secondary voltage reasonable, and the feed line and unun losses very low as well.

CONSTRUCTION

Fig 24.18 shows the layout. The matching unit is built in 8 × 8 × 4-inch NEMA weatherproof box available from most home centers. The author used an MFJ 404-0669 air-wound coil, and this box is too small to fit the entire inductor length needed. The inductor is split into two pieces; the long section consists of 61 turns, and the short section consists of 12 turns. As drawn in the schematic, the longer coil section is at the "bottom" (one end connects to ground), while the shorter coil is at the "top" (one end connects to RF OUT). If you have room for a larger box, a 12 × 12

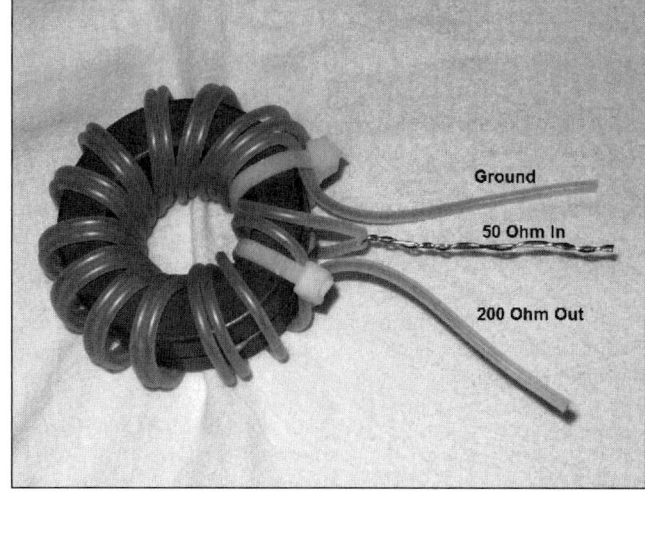

Fig 24.19 — Connections for a voltage unun (MFJ 10-10989D shown).

× 4-inch NEMA box fits the full inductor length without cutting.

Two Array Solutions RF-3PDT-15 relays are mounted to the case next to each other in the upper center. These are enclosed relays with all connections on one side, making for easy assembly. The 2.1 × 5.5 mm dc power jack and SO-239 RF IN connector are in the upper left corner. The MOVs and bypass capacitors for the control line are mounted on a terminal strip near the power connector.

The unun is mounted to the side of the enclosure using a machine screw and a piece of PC board material with the copper removed. The antenna is unbalanced, so a current balun or voltage unun (*not* a voltage balun) is typically used. A voltage unun should be wired as shown in **Fig 24.19**.

All internal wiring consists of insulated #14 AWG stranded wire. Attach the wires to the coil tap points using the MFJ coil clips called out in the parts list. (You can solder the wires directly to the coil, but this is difficult because of the size of the coil wire and the spacing of the turns.) Use #8 stainless steel screws, washers, lock-washers and nuts for coil mounting, and for the ground and RF OUT (antenna feed) terminals. **Table 24.3** lists the parts and part sources.

Relay Connections

As discussed earlier, the voltage across the full coil can be very high on 160 meters. Connecting the relay contacts in series as shown in the schematic increases the overall breakdown voltage. Also, the tap points on the inductor provide additional voltage-above-ground, which helps with the overall breakdown voltage.

The main concern is with the contact-to-

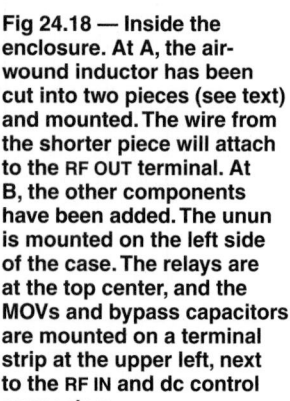

(A)

(B)

Fig 24.18 — Inside the enclosure. At A, the air-wound inductor has been cut into two pieces (see text) and mounted. The wire from the shorter piece will attach to the RF OUT terminal. At B, the other components have been added. The unun is mounted on the left side of the case. The relays are at the top center, and the MOVs and bypass capacitors are mounted on a terminal strip at the upper left, next to the RF IN and dc control connectors.

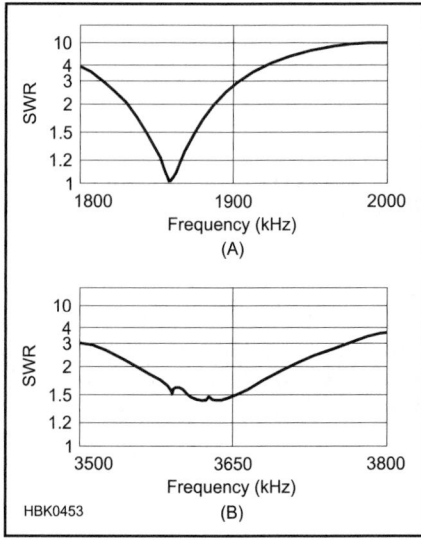

Fig 24.20 — Measured at the antenna base, the 2:1 SWR bandwidth is about 50 kHz for 160 meters and 150 kHz for 80 meters.

Table 24.3
Parts List: Switchable 160-10 Meter Impedance Matching Assembly

Qty	Description	Source/Part Number
1	8×8×4-inch NEMA box	Hardware store/home center
1	2-inch diam × 12-inch long coil, air wound with #12 AWG wire (79 µH, L1)	MFJ 404-0669
2	3PDT power relay (K1, K1)	Array Solutions RF-3PDT-15
4	Coil clips	MFJ 755-4001
1	SO-239 connector (J1)	MFJ-7721
1	1:4 unun, 1.5 kW rating (T1)	MFJ 10-10989D
1	2.1×5.5 mm dc jack (J2)	Mouser 163-1060-EX
3	18 V dc MOV (Z1-Z3)	Mouser 576-V22ZA2P
3	0.1 µf capacitor (C1-C3)	Mouser 581-SR215C104KAR
1	6-lug terminal strip	Mouser 158-1006
5	Micro-gator clips	Mouser 548-34
5	Test clips	Mouser 13AC130
1	2×1.38×0.8-inch plastic box (for control switch)	All Electronics 1551-GBK
1	DPDT center-off switch (S1)	All Electronics MTS-12
1	12-V dc, 1-A wall transformer	All Electronics DCTX-1218

Misc: #14 AWG house wiring, stainless steel hardware.

coil breakdown rating of 5.3 kV peak. The outer contacts are connected via insulated internal wires that are well separated from the coil by 0.1-0.2 inch. The problem is with the common wire for the center contacts, which is in contact with the coil. While the coil and common wire are both insulated, there is no air gap separation between them; this close proximity is what determines the relay breakdown voltage rating. Wired as shown in Fig 24.17, the center relay contacts are used for the lowest voltage points.

The switched control voltage input uses a 12-V, 1-A wall transformer supply. It's a good idea to keep the ±12 V control voltages separate and isolated from the regular station power supply to eliminate any possibility of shorting the main power supply when the control voltage polarity is flipped. The switching is easily done with a DPDT center-off switch (S1).

MATCHING NETWORK RESONANCE

Some initial adjustment is required to tune the resonant frequency of the matching network to account for variations in the antenna impedance based on your particular installation and desired operating frequency range. The overall starting inductance is too large for 160 meters, so the network will resonate at or below the lower band edge. Therefore, you can simply short one or more turns at the upper (RF OUT) end of the inductor to tune the network higher in frequency.

For the initial adjustments, don't connect the wire leads that attach between the relay contacts and the tap points on the coil. Instead, build short jumpers using the test clips and micro clips (perfect for the coil turn taps) called out in the parts list, and attach these between the relay contacts and coil using the suggested tap points shown on the schematic.

Next, connect the matching assembly to the base of your 43-ft vertical, enable 160 meter operation by applying –12 V dc and jumper turns from the RF OUT end of the inductor with a short clip lead until you find the desired 160 meter resonance point. Next, move the 160 meter relay tap point (from the unun secondary through K2C) until you get minimum SWR. (The starting point is 10 turns from the ground end.) Now permanently short the inductor turns by soldering a piece of #16 AWG buss wire across them. As shown in the photos, the author needed to short six turns (on the short coil).

Next apply +12 V to the relay assembly to enable 80-meter operation. Select the coil shorting point (from K2B — start 16 turns from the ground end) for your desired resonant frequency. Then adjust the tap point (from K2C — start 6 turns from ground) for best SWR. Finally remove the test clips leads, attach the coil clips, and solder wires between the coil clips and relay.

Fig 24.20 shows the results for 160 and 80 meters as measured with an antenna analyzer connected directly to the matching network input at the base of the author's antenna. With resonance set for the CW end of these bands, the 2:1 SWR bandwidth on 160 meters is about 50 kHz and about 150 kHz on 80 meters. Even a 3:1 SWR on these bands results in negligible SWR-related cable losses and is easily matched with an in-shack tuner.

OPERATION

Operation of this matching unit couldn't be simpler. When no control voltage is applied, the unun is connected and the antenna functions as it always has on 60-10 meters. For 80-meter operation, apply +12 V dc, and for 160-meter operation apply –12 V dc. The matching unit connected to the base of the author's 43-ft vertical is shown in **Fig 24.21**.

The matching network discussed in this article will permit very effective operation of your 43-ft vertical on all bands from 160-10 meters. Feel free to experiment a little. You might prefer to use the toroid from the basic matching network in the companion project instead of the air-core inductor. And if you are not running more than about 500 W, the less expensive RF-10 relays are all you need.

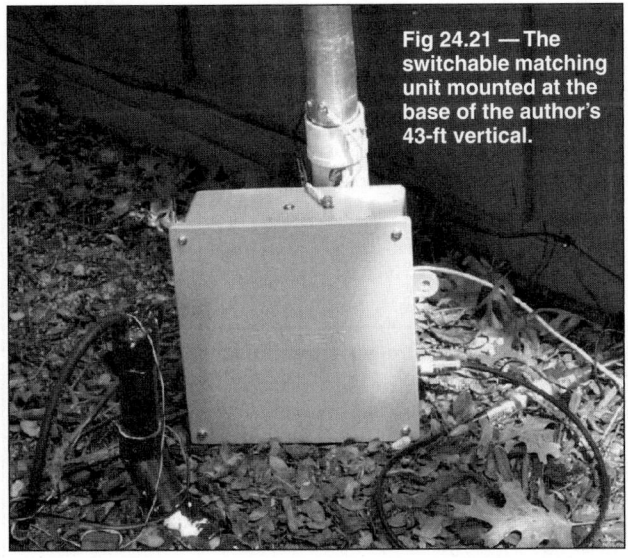

Fig 24.21 — The switchable matching unit mounted at the base of the author's 43-ft vertical.

24.5 An External Automatic Antenna Switch for Use With Yaesu or ICOM Radios

This antenna-switching-control project involves a combination of ideas from several earlier published articles.[1,2,3] This system was designed to mount the antenna relay box outside the shack, such as on a tower. With this arrangement, only a single antenna feed line needs to be brought into the shack. **Fig 24.22** shows the control unit and relay box, designed and built by Joe Carcia, NJ1Q. Either an ICOM or Yaesu HF radio will automatically select the proper antenna. In addition, a manual switch can override the ICOM automatic selection. That feature also provides a way to use the antenna with other radios. The antenna switch is not a two-radio switch, though. It will only work with one radio at a time.

Many builders may want to use only the ICOM or only the Yaesu portion of the interface circuitry, depending on the brand of radio they own. The project is a "hacker's dream." It can be built in a variety of forms, with the only limitations being the builder's imagination.

CIRCUIT DESCRIPTION

Fig 24.23 is a block diagram of the complete system. An ICOM or Yaesu HF radio connects to the appropriate decoder via the accessory connector on the back of the radio. Some other modern rigs have an accessory connector used for automatic bandswitching of amplifiers, tuners and other equipment. For example, Ten-Tec radios apply a 10 to 14-V dc signal to pins on the DB-25 interface connector for the various bands. Other radios use particular voltages on one of the accessory-connector pins to indicate the selected band. Check the owner's manual of your radio for specific information, or contact the manufacturer's service department for more details. You may be able to adapt the ideas presented in this project for use with other radios.

A single length of coax and a multiconductor control cable run from the rig and decoder/control box to the remotely located switch unit. The remote relay box is equipped with SO-239 connectors for the input as well as the output to each antenna. You can use any type of connectors, though.

ICOM radios use an 8-V reference and a voltage divider system to provide a stepped band-data output voltage. **Table 24.4** shows the output voltage at the accessory socket when the radio is switched to the various bands. Notice that seven voltage steps can be used to select different antennas. The ICOM accessory connector pin assignments needed for this project are:

Pin 1   +8 V reference
Pin 2   Ground
Pin 4   Band signal voltage
Pin 7   +12 (13.8) V supply

Yaesu radios provide the band information

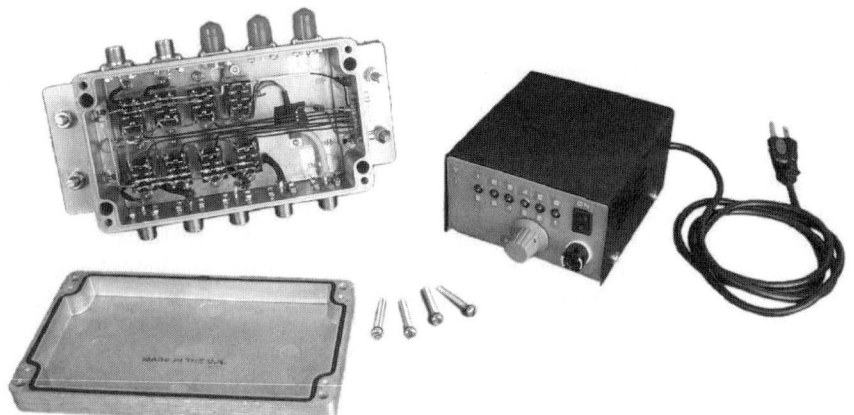

Fig 24.22 — Automatic antenna switch relay box (left) and control box (right).

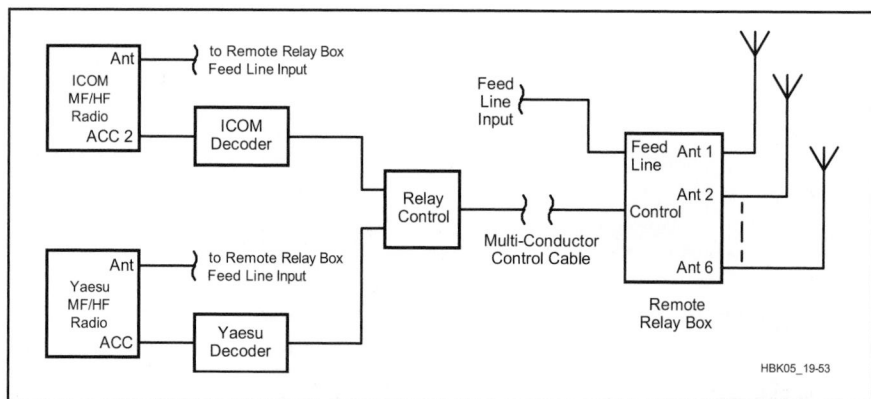

Fig 24.23 — Block diagram of the remotely controlled automatic antenna switch.

Table 24.4
ICOM Accessory Connector Output Voltages By Band

Band (MHz)	Output Voltage
1.8	7 – 8.0
3.5	6 – 6.5
7	5 – 5.5
14	4 – 4.5
18, 21	3 – 3.5
24, 28	2 – 2.5
10	0 – 1.2

Note: The voltage step between bands is not constant, but close to 1.0 V, and the 10-MHz band is not in sequence with the others.

Table 24.5
Yaesu Band Data Voltage Output (BCD)

Band	A (1)	B (2)	C (4)	D (8)	(BCD Equiv.)
1.8	5V	0V	0V	0V	1
3.5	0V	5V	0V	0V	2
7.0	5V	5V	0V	0V	3
10.1	0V	0V	5V	0V	4
14	5V	0V	5V	0V	5
18.068	0V	5V	5V	0V	6
21	5V	5V	5V	0V	7
24.89	0V	0V	0V	5V	8
28	5V	0V	0V	5V	9

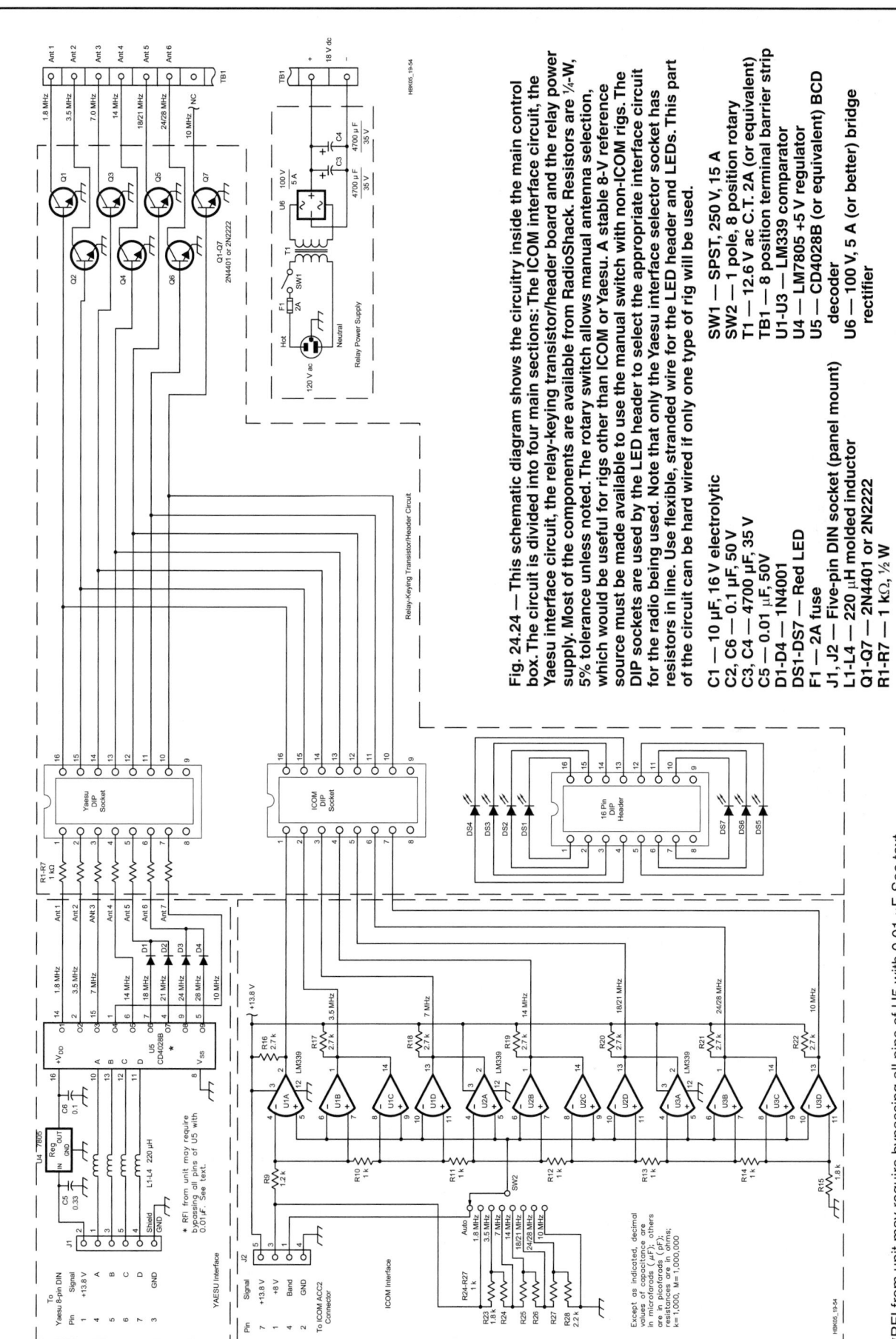

Fig. 24.24 — This schematic diagram shows the circuitry inside the main control box. The circuit is divided into four main sections: The ICOM interface circuit, the Yaesu interface circuit, the relay-keying transistor/header board and the relay power supply. Most of the components are available from RadioShack. Resistors are ¼-W, 5% tolerance unless noted. The rotary switch allows manual antenna selection, which would be useful for rigs other than ICOM or Yaesu. A stable 8-V reference source must be made available to use the manual switch with non-ICOM rigs. The DIP sockets are used by the LED header to select the appropriate interface circuit for the radio being used. Note that only the Yaesu interface selector socket has resistors in line. Use flexible, stranded wire for the LED header and LEDs. This part of the circuit can be hard wired if only one type of rig will be used.

C1 — 10 μF, 16 V electrolytic
C2, C6 — 0.1 μF, 50 V
C3, C4 — 4700 μF, 35 V
C5 — 0.01 μF, 50V
D1-D4 — 1N4001
DS1-DS7 — Red LED
F1 — 2A fuse
J1, J2 — Five-pin DIN socket (panel mount)
L1-L4 — 220 μH molded inductor
Q1-Q7 — 2N4401 or 2N2222
R1-R7 — 1 kΩ, ½ W

SW1 — SPST, 250 V, 15 A
SW2 — 1 pole, 8 position rotary
T1 — 12.6 V ac C.T. 2A (or equivalent)
TB1 — 8 position terminal barrier strip
U1-U3 — LM339 comparator
U4 — LM7805 +5V regulator
U5 — CD4028B (or equivalent) BCD
       decoder
U6 — 100 V, 5 A (or better) bridge
       rectifier

*RFI from unit may require bypassing all pins of U5 with 0.01 μF. See text.

as binary coded decimal (BCD) data on four lines. Nine different BCD values allow you to select a different antenna for each of the MF/HF bands. **Table 24.5** shows the BCD data from Yaesu radios for the various bands. The Yaesu 8-pin DIN accessory connector pin assignments needed for this project are:

Pin 1   +12 (13.8) V supply
Pin 3   Ground
Pin 4   Band Data A
Pin 5   Band Data B
Pin 6   Band Data C
Pin 7   Band Data D

Fig 24.24 is the schematic diagram for the control box. We will discuss each part of the control circuit later in this description. First, let's turn our attention to the external antenna box.

EXTERNAL ANTENNA BOX

Only the number of control lines going out to the relay box limits the number of antennas this relay box will switch. The unit shown in the photos has 10 SO-239 connectors, to switch the common feed line to any of nine antennas. Many hams will use an eight-conductor rotator cable (such as Belden 9405) to the relay box. Using eight wires, we can control seven relays (six for antennas and one to ground the feed line for lightning protection) plus the relay coil power supply and ground lead. The photos show eight relays, and the box can be expanded further if desired. There is also a connector for power and control lines. Use of a DB-15 allows for the addition of more relays and control lines later. A DB-9 connector would be suitable for use with the eight-conductor control cable, or you may wish to use a weatherproof connector. **Fig 24.25** shows the relay box schematic diagram.

Since the box will be located outside, use a weatherproof metal box — a Hammond Manufacturing, type 1590Z150, watertight aluminum box is shown. It's about 8.5×4.25×3.125 inches. This is a rather hefty box, meant to be exposed to years of various weather conditions. You can, however, use almost anything.

The coax connectors are mounted so each particular antenna connector is close to the relay, without too much crowding. For added weather protection (and conductivity), apply Penetrox to the connector flange mount, including the threads of the mounting screws. On the power/control line connector, use Coax Seal or other flexible cable sealant.

The aluminum angle stock on either side of the box is for mounting to a tower leg. The U-bolts should be of the proper size to fit the tower leg. They should also be galvanized or made of stainless steel.

ANTENNA RELAYS

One of the more difficult parts of this project was the modification of the relays (DPDT Omron LY2F-DC12). To improve isolation, the moveable contacts (armature) are wired in parallel and the connecting wire is routed through a hole in the relay case.

Remove the relay from its plastic case. Unsolder and remove the small wires from the armatures. Carefully solder a jumper across the armature lugs with #20 solid copper. Then solder a piece of very flexible wire (such as braid from RG-58 cable) to either armature lug. Obviously, the location of the wire depends on which side you wish to connect the SO-239. You will also need to make a hole in the plastic case that is large enough to accommodate the armature wire without placing any strain on the free movement of the armature. Slip a length of insulating tubing

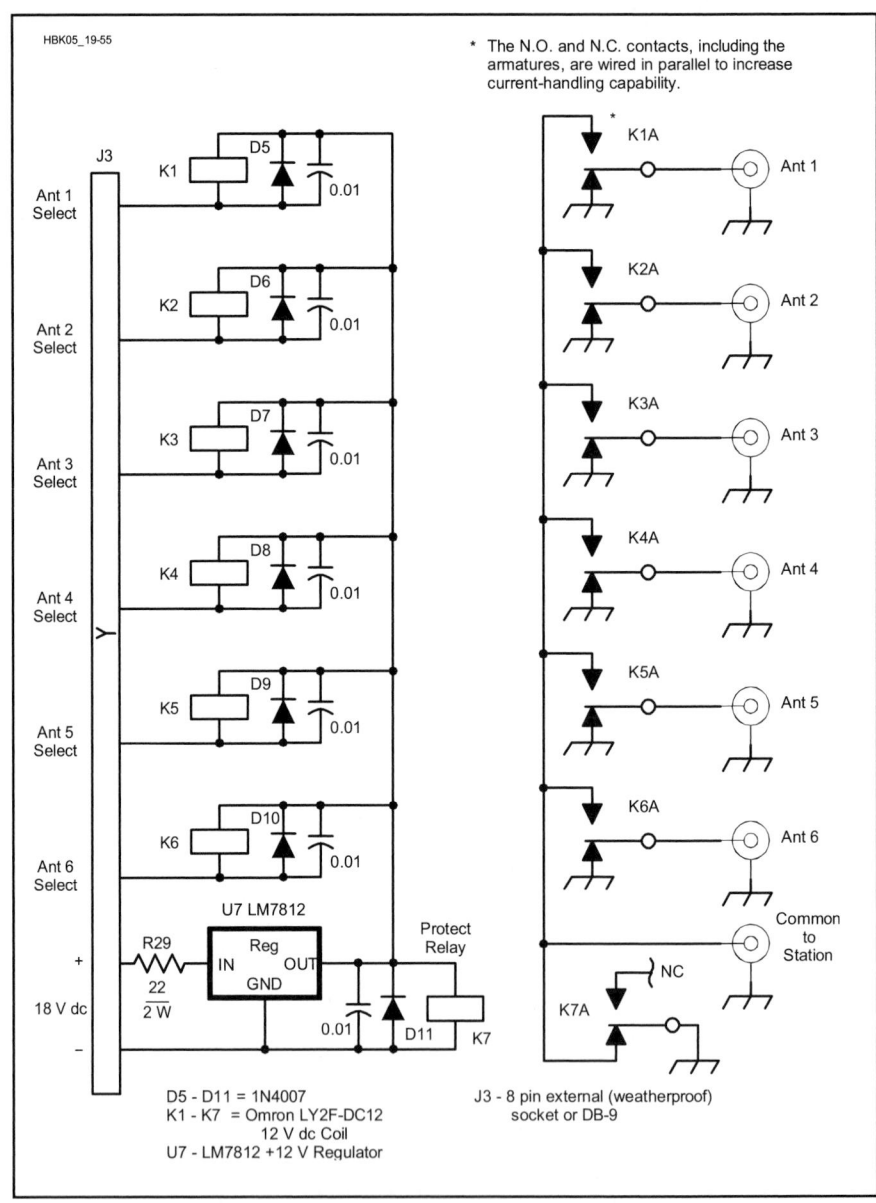

Fig 24.25— The schematic diagram for the external antenna relay box. All relays are DPDT, 250 V ac, 15 A contacts. R29 is used to limit the regulator current. Mount the regulator using TO-220 mounting hardware, with heatsink compound. With the exception of the normally closed and normally open contacts, all wiring is #22 solid copper wire.

D5-D11 — 1N4007.
J3 — 8 pin external weatherproof connector (or DB-9 with appropriate weather sealant).

K1-K7 — Omron LY2F-DC12 with 12 V coil.
U7 — LM7812 +12 V regulator.

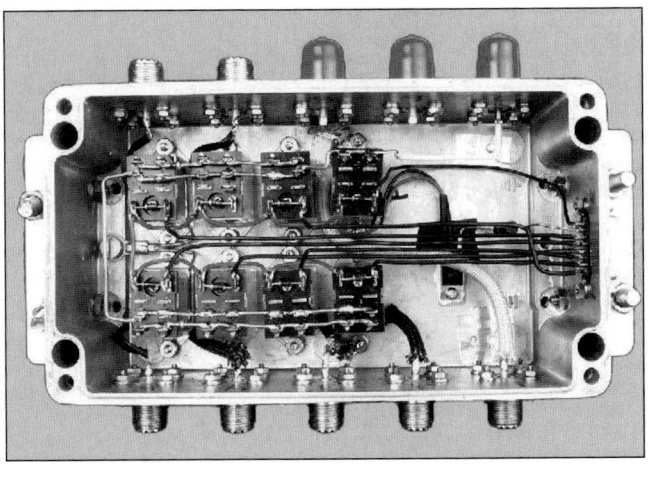

(A)   (B)

Fig 24.26 — At A, the external relay box. The LM7812 regulator is mounted to the bottom of the box. The relay normally open and normally closed contacts are wired in parallel using #12 solid copper. The two extra flange-mount SO-239 connectors at the upper right are for future expansion. At B, inside the control box. Components are mounted on RadioShack Universal Project Board. The bottom board in the enclosure, as well as the right half of the middle section, hold the ICOM circuit. The Yaesu interface is on the left side of the middle section. The top circuit board holds the DIP sockets and relay-selection transistors. All high voltage leads are insulated. The 7805 5-V regulator is mounted on the back panel using TO-220 mounting hardware, with heatsink compound.

over this wire to prevent it from shorting to the aluminum box.

The normally open and normally closed contacts are also wired in parallel. This can be done on the lugs themselves. For this, use #12 solid copper wire.

Mount the relays in the aluminum box, oriented so they can be wired together without difficulty. (See **Fig 24.26A**.) With the exception of the wire used for the relay coils (#22 solid wire), use #12 solid copper wire for the rest of the connections.

To eliminate the possibility of spikes or "back emf," a 1N4007 diode is soldered across the coil contacts of each relay. In addition, 0.01-μF capacitors across the diodes will reduce the possibility of stray RF causing problems with the relay operation.

Since the cable run from the shack to the tower can be quite long, consideration has to be given to the voltage drop that may occur. The relays require 12 V dc. The prototype used a 12 V dc regulator in the relay box, fed with 18 V dc (at 2 A) from the control box. If the cable run is not that long, however, you could just use a 12-V supply and skip the regulator.

One of the relays is used for lightning protection. When not in use, the relay grounds the line coming in from the shack. When the control box is activated, it applies power to this relay, thus removing the ground on the station feed line. All the antenna lines are grounded through the normally closed relay contacts. They remain grounded until the relay receives power from the control box.

CONTROL BOX

This is the heart of the system. The 18 V dc

power supply for the relays is located in this box, in addition to the Yaesu and ICOM decoder circuits and the relay-control circuitry. All connections to the relay box are made via an 8-position terminal barrier strip mounted on the back of the control box.

The front of the box has LEDs that indicate the selected antenna. A rotary switch can be used for manual antenna selection. The power switch and fuse are also located on the front panel.

The wiring schemes on the Yaesu and ICOM ACC sockets are so different that the unit shown in the photos has a 5-pin DIN connector for each rig on the control box. Since there is only one set of LEDs, use an 8-pin DIP header to select the appropriate control circuit for each radio. See Fig 24.26B. The unit is built with point-to-point wiring on Radio Shack Universal Project Boards.

ICOM Circuitry

This circuit originally appeared in April 1993 *QST* and is modified slightly for this application. The original circuit allowed for switching between seven antennas (from 160 to 10 meters). The Band Data signal from the ICOM radios goes to a string of LM339 comparators. Resistors R9 through R15 divide the 8-V reference signal from the rig to provide midpoint references between the band signal levels. The LM339 comparators decide which band the radio is on. A single comparator selects the 1.8 or 10 MHz band because those bands are at opposite ends of the range. The other bands each use two comparators. One determines if the band signal is above the band level and the other determines if it is below the band level. If

the signal is between those two levels, the appropriate LED and relay-selection transistor switch is turned on.

The ICOM circuit allows for manual antenna selection. The 8-V reference is normally taken directly from the ICOM ACC socket. If this circuit is to be used with other equipment, then a regulated 8-V source should be provided.

Yaesu Circuitry

The neat thing about Yaesu band data is that it's in a binary format. This means you can use a simple BCD decoder for band switching. The BCD output ranges from 1 to 9. In essence, you can switch among 9 antennas (or bands). Since the relay box switches just six antennas, steering diodes (D1 through D4 in Fig 24.24) allow the use of one antenna connection for multiple bands. One antenna connection is used for 17 and 15 meters, and another connection for 12 and 10 meters because the ICOM band data combines those bands. There is no control line or relay for a 30-meter antenna with this version of the project.

RFI has been reported to be caused by the CD4028 decoder IC (U5). Should this be the case for you, be sure the 7805 regulator is not oscillating and has the required bypass capacitors at the input and output. If the 7805 output is clean, add 0.01 μF, 50 V ceramic bypass capacitors to all pins of the CD4028 (except pin 8, the ground pin).

DIP Sockets and Header

A RadioShack Universal Project Board holds the DIP sockets along with the relay keying transistors. This board is shown as the top layer in Fig 24.26B. The Yaesu socket has 1-kΩ resistors wired in series with each input pin. The other header connects directly to the ICOM circuitry.

The DIP header is used to switch the keying transistors between the ICOM and Yaesu circuitry. The LEDs are used to indicate antenna number. Use stranded wire (for its flexibility) when connecting to the LEDs.

Relay Keying Transistors

Both circuits use the same transistor-keying scheme, so you need only one set of transistors. Each transistor collector connects to the terminal barrier strip. The emitters are grounded, and the bases are wired in parallel to the two 16-pin DIP sockets. The band data turns on one of the transistors, effectively grounding that relay-control lead. Current flows through the selected relay coil, switching that relay to the normally open position and connecting the station feed line to the proper antenna.

Power Supply

The power supply is used strictly for the relays. Other power requirements are taken from the rig used. There is room here for variations on the power supply theme; the parts used in this version were readily available.

Notes

[1]"An Antenna Switching System for Multi-Two and Single-Multi Contesting," by Tony Brock-Fisher, K1KP, January 1995 *NCJ*.
[2]"A Remotely Controlled Antenna Switch," by Nigel Thompson, April 1993 *QST*.
[3]*NA* Logging Program Section 11.

24.6 A Low-Cost Remote Antenna Switch

Getting multiple antenna feed lines into the shack is a common problem that can be easily solved with a suitable remote antenna switch. The project described here by Bill Smith, KO4NR, provides an easy and inexpensive solution. It originally appeared in April 2005 *QST*.

RELAY SELECTION

Interesting innovations in printed circuit board (PCB) power relay design have produced a number of compact units that exhibit high dielectric strength and can carry impressive amounts of current.

Although not factory tested for RF use, the American Zettler AZ755 series PCB relays work very well in this application. (See **www.americanzettler.com** for datasheets and information.) The AZ755 is rated for 480 W switched power with a resistive load and a maximum switched current of 20 A. Despite the relay's small physical size, the dielectric strength between the contacts and coil is 5 kV RMS, with an impressive 1 kV RMS between the open contacts. This means that

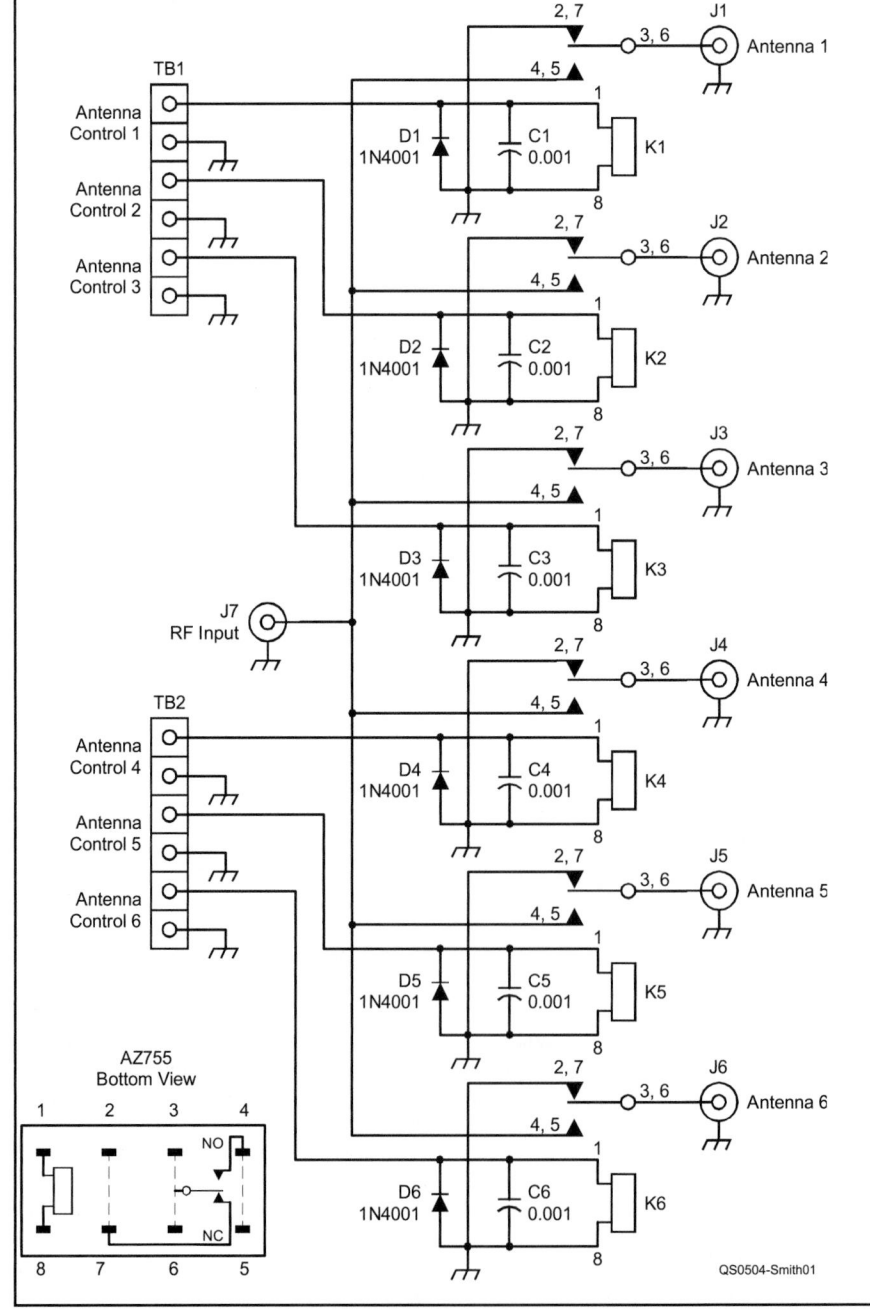

Fig 24.27 — Schematic and parts list for the remote antenna switch. Part numbers indicated with an M are available from Mouser Electronics (www.mouser.com). PC boards are available from FAR Circuits (www.farcircuits.net).

C1-C6 — 0.001 µF, 50 V disc ceramic (M 140-50P5-102K-TB).

D1-D6 — 1N4001 (M 512-1N4001).

J1-J7 — SO-239A chassis mount coaxial connector with silver-plated 4-hole square flange and Teflon insulation (Amphenol 83-798 available as Mouser 523-83-798).

K1-K6 — SPDT PC board power relay with 12 V dc coil (American Zettler AZ755-1C-12DE). Available from www.relaycenter.com. See text.

TB1, TB2 — PC board terminal block with 6 contacts (M 651-1729050).

Fig 24.28 — A completed remote antenna switch. The RF INPUT connector is in the center, with three ANTENNA connectors on each side. The control cable connects to the two terminal blocks.

Fig 24.29 — Connector flanges are soldered to the board's ground plane top and bottom. The ground ends of D1-D6 and C1-C6 are also soldered top and bottom. The relays have two pins each for the contact connections. Solder all pins and install eyelets supplied with the FAR Circuits board to ensure good connections for the relay common and normally closed pins.

the relay is resistant to a flashover that could damage the coil or pit the contacts.

The AZ755 series relays are offered in a wide variety of configurations. This project uses part number AZ755-1C-12DE. This model has a 12 V dc coil, but you can use any of the available coil voltages in the series. The contact style is Form C, which is SPDT. The E suffix indicates that the relay epoxy is sealed for better protection from dirt and moisture. For relays that are not epoxy sealed, drop the E from the part number.

CIRCUIT DESIGN

Fig 24.27 shows the circuit, which handles up to six antennas. The common contacts of the relays (K1-K6) are connected to SO-239 connectors (J1-J6) for the antenna feed lines. The normally open (NO) contacts are all connected to the RF INPUT connector, J7. The normally closed (NC) contacts are all connected to ground so that the antennas are grounded when not in use. To select an antenna, apply 12 V dc to the appropriate ANTENNA CONTROL terminal to energize the relay and connect the ANTENNA to the RF INPUT.

C1-C6 help to keep stray RF out. In addition, D1-D6 are installed across the coils to prevent voltage spikes when the power is removed from the coil. The finished board (available from FAR Circuits) with all components mounted is shown in **Fig 24.28**.

ASSEMBLY NOTES

The design is simple and assembly doesn't require special tools. In addition to the PC board and parts, you'll need a suitable enclosure to keep the board dry and pests away. The author used a plastic children's lunch box that hangs under his deck.

The most difficult part of the project is drilling the holes in the enclosure for the SO-239 connectors and getting everything to line up. You may find it easier to install the connectors in the enclosure first, and then solder them to the board. (Use the board to mark the center line and location of the connectors on the enclosure.) After the connectors are tacked in place, remove the screws holding the SO-239s to the enclosure and remove the total assembly. This ensures a perfect fit when reassembling.

Make sure the SO-239s are all the same type and brand to ensure a uniform fit. The board was designed around the Amphenol connectors recommended in the parts list. They have silver-plated center pins and flanges, and Teflon insulation. The silver plating makes the connectors easier to solder than nickel-plated connectors, and the Teflon insulation is much less prone to melting than the plastic often found on inexpensive connectors. Other SO-239 connectors may fit with possible modification for a good fit.

Use a *hot* soldering iron when soldering the flange of the SO-239 to the board's ground plane as shown in **Fig 24.29**. Be sure to solder top and bottom.

The PC board is double-sided, and FAR Circuits supplies eyelets with the board to use in the larger holes for the common and normally closed relay pins. They provide

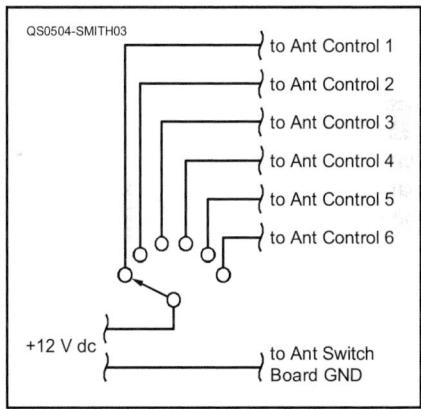

Fig 24.30 — Antennas are selected with a simple 12-V supply and rotary switch.

for a better connection to the component side of the board. In addition, soldering a short length of #14 bare copper wire into the center pin of each SO-239 will give you a better connection to the circuit trace on the board. Be sure to solder the ground ends of C1-C6 and D1-D6 on both the top and bottom sides of the board.

POWERING IT UP

The switch is controlled with a 12-V dc power supply and small ceramic rotary switch as shown in **Fig 24.30**. The switch simply sends 12 V to energize one of the relay coils and select the desired antenna. When 12 V is removed, the normally closed relay contacts connect all feed lines to ground. Connect the

PC board's ground plane to a ground rod for lightning protection and as a means of eliminating static.

The control cable can be the 8-conductor variety usually used for rotators. Long runs of wire may require large conductors to prevent unacceptable voltage drops at the relay coils, but the relays will work over a fairly wide range of coil voltages so it's not critical.

TESTING

Initial testing should be conducted on the bench with an antenna analyzer or a transceiver (internal antenna tuner off) and power/SWR meter connected to the RF INPUT jack.

Connect a 50 Ω load to one of the antenna jacks on the switch board. Apply a small of RF power. The SWR should be close to 1:1 (similar to readings without the switch in the line).

If everything looks okay, increase the power and check the SWR through the switch on all bands that you plan to use it on. Next, move the wattmeter to the switch output and verify that the power on that side is about the same as the power at the input. There should be little or no measurable loss through the switch. Repeat these tests for all of the switch positions.

If you run an amplifier, test the switch at high power. The author used the switch at 1-1.3 kW during normal intermittent operation (SSB and CW) with no problems.

Although they are not designed for RF, the relays perform well. The board exhibits low SWR, low insertion loss and good isolation over a wide frequency range. The ARRL Lab tested the completed antenna switch board. Insertion loss measured <0.1 dB for 2-50 MHz (for all ports to common). SWR measured 1.1:1 or less from 2-28 MHz, 1.2:1 or less on 50 MHz. Isolation was >60 dB for 2-28 MHz, except for the two innermost ports, which were 50 dB at 28 MHz. Worst-case isolation on 50 MHz was 45 dB.

24.7 Audible Antenna Bridge

The audible antenna bridge (AAB) shown in **Fig 24.31** was designed by Rod Kreuter, WA3ENK. This project would help a visually impaired ham, but sighted hams will also find it very useful. (For example, tuning a mobile screwdriver antenna while you're driving.) The AAB is rugged and easy to use, both in the harsh light of day and in the dark of night. It's also inexpensive to build, and a kit is available.[1]

An absorptive type of SWR meter, the AAB is used to indicate how close the match is between your transmitter and your antenna. Its output is a tone whose pitch is proportional to SWR. It also sends the SWR value in Morse code.

Unlike most SWR meters on the market, this one actually absorbs power from your transmitter while you're tuning, helping to protect your transmitter during tune-up. Many transmitters have SWR foldback capability and will decrease the output power if the antenna match is poor, but some do not.

By absorbing power in the AAB, no matter what you do at the antenna port, the transmitter never sees an SWR greater than 2:1. The trade-off is that 75% of your transmitter power ends up as heat in the bridge, so you need to switch out the AAB after you have tuned your antenna.

THEORY OF OPERATION

The AAB works on a principle taken right from the *ARRL Antenna Book*. The bridge consists of four impedances, or arms. See **Fig 24.32**. If three of the arms are 50 Ω resistors and the fourth is an antenna (or an antenna and tuner combination), then the bridge will be balanced when the antenna port is also 50 Ω. Diodes D1 and D2 sample the forward and reverse voltages, respectively. A periph-

Fig 24.31 — Front view of the audible antenna bridge.

eral interface controller (PIC) with a 10 bit analog to digital converter (A/D) digitizes these voltages and calculates the SWR.

The SWR is used to control a hardware pulse width modulator that produces the output tone. It's also used by the Morse code routine to send the actual value of SWR, at 15 WPM with 5 WPM spacing.

The power that the AAB can handle during tune-up is a function of the resistors used in the three reference arms of the bridge. The 30 W noninductive resistors, R1-R3, are made by a

number of companies but are not cheap. The bridge resistors must be noninductive. Don't use wire wound resistors!

Antenna tuning should be done with minimal power, and the AAB will work down to about 1 W. The 30 W resistors will handle tune-ups with about 60 W or perhaps 100 W for a short time. If you always remember to turn down your transmitter power, and you tune up quickly, the bridge resistors could be perhaps 5 W for normal rigs and 1 W for QRP rigs.

BUILDING THE AAB

A PC board is a nice way to build the AAB, but you could easily construct such a simple circuit using ugly construction. Keep the wires short in the RF section. See **Fig 24.33**.

Use a heat sink for the noninductive resistors. Like many power devices, the power rating depends on getting the heat out! These resistors get really hot running at 10 or 20 W. See **Fig 24.34**. Use a big heat sink with heat sink compound, give yourself plenty of room in the cabinet and drill some holes to let the heat escape.

Switch S1 chooses TUNE/OPERATE and S2, the INQUIRE button, is used to report on various conditions, such as the SWR value in Morse code. You can eliminate the TUNE/OPERATE switch, but remember to remove the AAB from your feed line after tuning up.

To use the AAB in a QRP station, you may want to reduce resistors R4 and R5. They are sized so that a 25 W transmitter will not saturate the analog to digital converter (ADC) on the PIC when using fresh batteries (more on this later). A good value for a 5 W transmitter would be about 20 kΩ. So long as the voltages at pin 12 and pin 13 of U1 are below the battery voltage, the system will work fine.

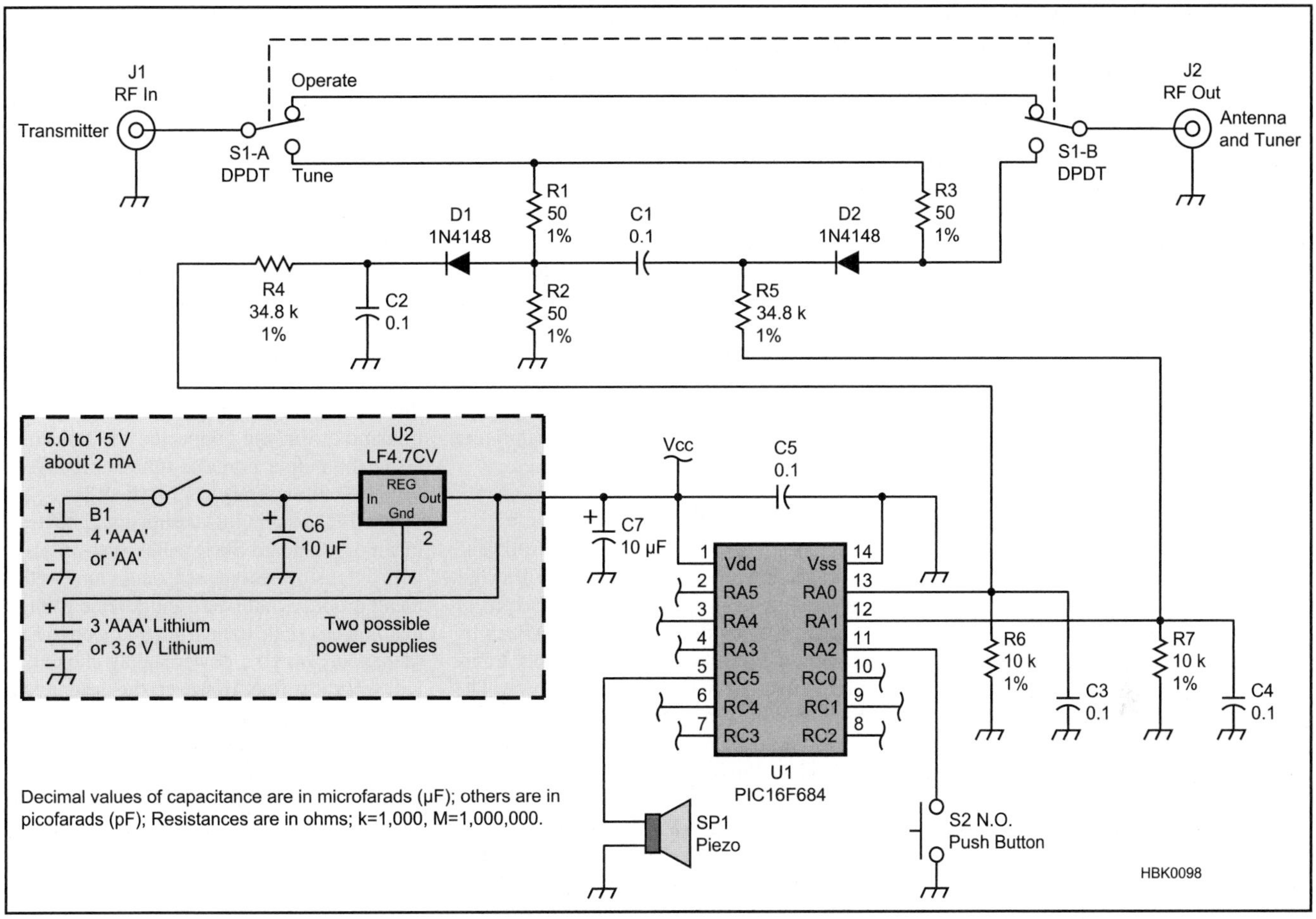

Fig 24.32 — Schematic and parts list of the AAB.

C1-C5 — 0.1 µF ceramic.
C6, C7 — 10 µF, 25 V electrolytic.
D1, D2 — 1N4148.
J1, J2 — SO-239 UHF coaxial jack.
R1 -R3 — Resistor, 50 Ω, 1%, 30 W
    Caddock noninductive
    (Mouser 684-MP930-50).

R4, R5 — Resistor, 34.8 kΩ, 1%.
R6, R7 — Resistor, 10 kΩ, 1%.
S1 — DPDT (Digi-Key SW333-ND).
S2 — Momentary contact
    (Mouser 612-TL1105LF250Q).
SP1 — Piezoelectric buzzer EFB-
    RD24C411 (Digi-Key P9924-ND).

U1 — PIC16F684 (Digi-Key PIC16F684-
    1/P-ND). Must be programmed before
    use. See Note 1.
U2 — LF4.7CV regulator
    (Mouser 511-LF47CV).

Fig 24.33 — Close-up view of three CaddockTO-220 style bridge resistors mounted on the heat sink.

Fig 24.34 — Interior view of the AAB, showing the battery pack at the bottom of the enclosure.

POWERING THE AAB

To save power the AAB uses a sleep mode. If you haven't pushed the INQUIRE button for about 5 minutes, the AAB will go to sleep after sending the letter N for "night night." During sleep the current draw is about 2 µA. Pushing the switch while it's sleeping will wake up the AAB and it will send the letter U ("up").

The first four prototypes used four AAA batteries and a low-dropout 4.7 V regulator. This made for a nice stiff supply and very good battery life, with only one problem. Even though the PIC uses less than 2 µA during sleep, the regulator uses 500 µA. Battery data says that alkaline AA batteries will last 6 months at this level and AAA batteries about half that long. Consider adding an ON/OFF switch if you go this route.

The other option is to use batteries without a regulator. Normally you couldn't get away with this without providing a separate voltage reference for the ADC in the PIC. In this case it doesn't matter as much because SWR is a ratio of two voltages.

What does matter is the battery voltage with respect to the voltage produced by the RF signal. Diodes D1 and D2 rectify the RF and produce a dc voltage proportional to RF power. This voltage is divided by resistors R4, R6 and R5, R7. The voltage at the ADC inputs of the PIC, pins 12 and 13, must be less than the power supply voltage. With a regulated supply this isn't much of a problem, and if you're careful it can also work with unregulated batteries.

Microchip recommends that these parts be run on 2.5 to 5.5 V if you need the 10 bit ADC to be accurate. Many battery combinations will provide this, but consider lithium batteries instead of alkaline. The voltage from an alkaline battery begins to fall very early in its life and continues at a downward slope until it is exhausted. A lithium battery, on the other hand, provides a nearly constant voltage throughout its useful life. A 3.6 V AA size lithium should last a few years, but with the lower output voltage you may want to reduce R4 and R5. The final prototype used three AAA lithium batteries (each about 1.7 V when new), with no regulator or ON/OFF switch, which should last at least two years.

U1's ADC converter will saturate at some level. With new batteries, let's say this level is 25 W when the power supply is 4.5 V. When the power supply has fallen to 3.5 V, the RF level that will saturate the converter will be 15 W. So if you find that the output of your rig, which used to be fine, is now saturating the ADC, it's time to change batteries.

USING THE AAB

Connect your transmitter to the INPUT and your antenna/tuner to the OUTPUT. Turn your transmit power down to 5 or 10 W and put S1 in the TUNE position. Wake the AAB up by pushing the INQUIRE button. Key your transmitter in a mode with some carrier power. The AAB should start to produce a steady tone. Adjust your tuner or antenna for the lowest tone pitch. Press the INQUIRE button for the numeric SWR value.

When the AAB is operating normally, it produces a tone proportional to SWR. The tone ranges from 250 Hz at an SWR of 1.0:1, rising to 4450 Hz at an SWR of about 15:1. Errors can occur and at times the AAB can't make a measurement. If an error occurs, either a high pitched tone or no tone is produced. Pushing the INQUIRE button will give you more information. Five different conditions can occur:

Normal operation. A tone proportional to SWR is produced. Press INQUIRE and a two digit SWR value is sent, separated by the letter R (Morse for the decimal point character).

Error 1. The forward power is too low for a reliable measurement. No tone is produced. INQUIRE button sends letter L (low).

Error 2. Forward power is too high (almost saturating the ADC). A pulsing tone proportional to SWR is produced. However, if the ADC is saturated this value might be meaningless. INQUIRE button sends letter S (saturated).

Error 3. SWR is greater than 9.9:1. A tone proportional to SWR is produced. INQUIRE button sends letter H (high).

Error 4. The reverse voltage is greater than the forward voltage. Two things can cause this error. You might be near a high power transmitter, or you connected the transmitter to the output port. No tone is produced. INQUIRE button sends letter F (fault).

The PIC used in this design could be used in a more standard SWR meter, such as those with a transformer type sensor. As long as the meter produces a forward and reverse voltage that you can scale to be from 0 to 3 up to 5 V (depending on your power supply choice), it will work just fine. However, the driving impedance of the voltage must be 10 kΩ or less.

The author thanks the Ham Radio Committee of the National Federation of the Blind for inspiration, and Tom Fowle, WA6IVG, and Bill Gerrey, WA6NPC, of the Smith-Kettlewell Research Institute for testing prototypes.

[1]A kit of parts including a printed circuit board is available from Q-Sat, 319 McBath St, State College, PA 16801. The price is currently $30, postpaid. PC boards are available for $7 postpaid and a programmed PIC is available for $6 postpaid. Source code for an unprogrammed PIC is available on the ARRL Web site. Go to **www.arrl.org/qst-in-depth** and look for Kreuter0607.zip in the 2007 section.

24.8 A Trio of Transceiver/Computer Interfaces

Virtually all modern Amateur Radio transceivers (and many general-coverage receivers) have provisions for external computer control. Most hams take advantage of this feature using software specifically developed for control, or primarily intended for some other purpose (such as contest logging), with rig control as a secondary function.

Unfortunately, the serial port on most radios cannot be directly connected to the serial port on most computers. The problem is that most radios use TTL signal levels while most computers use RS-232-D.

The interfaces described here simply convert the TTL levels used by the radio to the RS-232-D levels used by the computer, and vice versa. Interfaces of this type are often referred to as level shifters. Two basic designs, one having a couple of variations, cover the popular brands of radios. This article, by Wally Blackburn, AA8DX, first appeared in February 1993 *QST*.

TYPE ONE: ICOM CI-V

The simplest interface is the one used for the ICOM CI-V system. This interface works with newer ICOM and Ten-Tec rigs. **Fig 24.35** shows the two-wire bus system used in these radios.

This arrangement uses a CSMA/CD (carrier-sense multiple access/collision detect) bus. This refers to a bus that a number of stations share to transmit and receive data. In effect, the bus is a single wire and common ground that interconnect a number of radios and computers.

The single wire is used for transmitting and receiving data. Each device has its own unique digital address. Information is transferred on the bus in the form of packets that include the data and the address of the intended receiving device.

The schematic for the ICOM/Ten-Tec interface is shown in **Fig 24.36**. It is also the Yaesu interface. The only difference is that the transmit data (TxD) and receive data (RxD) are jumpered together for the ICOM/Ten-Tec version.

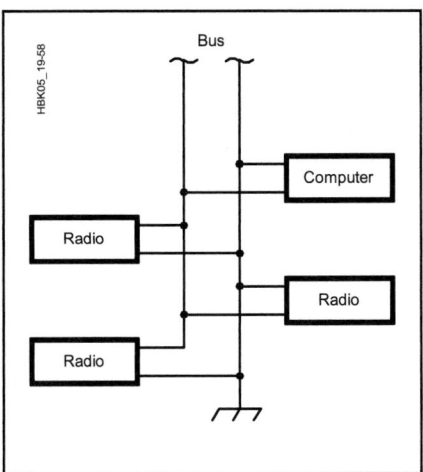

Fig 24.35 — The basic two-wire bus system that ICOM and newer Ten-Tec radios share among several radios and computers. In its simplest form, the bus would include only one radio and one computer.

The signal lines are active-high TTL. This means that a logical one is represented by a binary one (+5 V). To shift this to RS-232-D it must converted to –12 V while a binary zero (0 V) must be converted to +12 V. In the other direction, the opposites are needed: –12 V to +5 V and +12 V to 0 V.

U1 is used as a buffer to meet the interface specifications of the radio's circuitry and provide some isolation. U2 is a 5-V-powered RS-232-D transceiver chip that translates between TTL and RS-232-D levels. This chip uses charge pumps to obtain ±10 V from a single +5-V supply. This device is used in all three interfaces.

A DB25 female (DB25F) is typically used at the computer end. Refer to the discussion of RS-232-D earlier in the chapter for 9-pin connector information. The interface connects to the radio via a ⅛-inch phone plug. The sleeve is ground and the tip is the bus connection.

It is worth noting that the ICOM and Ten-Tec radios use identical basic command sets (although the Ten-Tec includes additional commands). Thus, driver software is compatible. The manufacturers are to be commended for working toward standardizing these interfaces somewhat. This allows Ten-Tec radios to be used with all popular software that supports the ICOM CI-V interface. When configuring the software, simply indicate that an ICOM radio (such as the IC-735) is connected.

TYPE TWO: YAESU INTERFACE

The interface used for Yaesu rigs is identical to the one described for the ICOM/Ten-Tec, except that RxD and TxD are not jumpered together. Refer to Fig 24.36. This arrangement uses only the RxD and TxD lines; no flow control is used.

The FT-990 and FT-1000/FT-1000D use an emitter follower as the TTL output. If the input

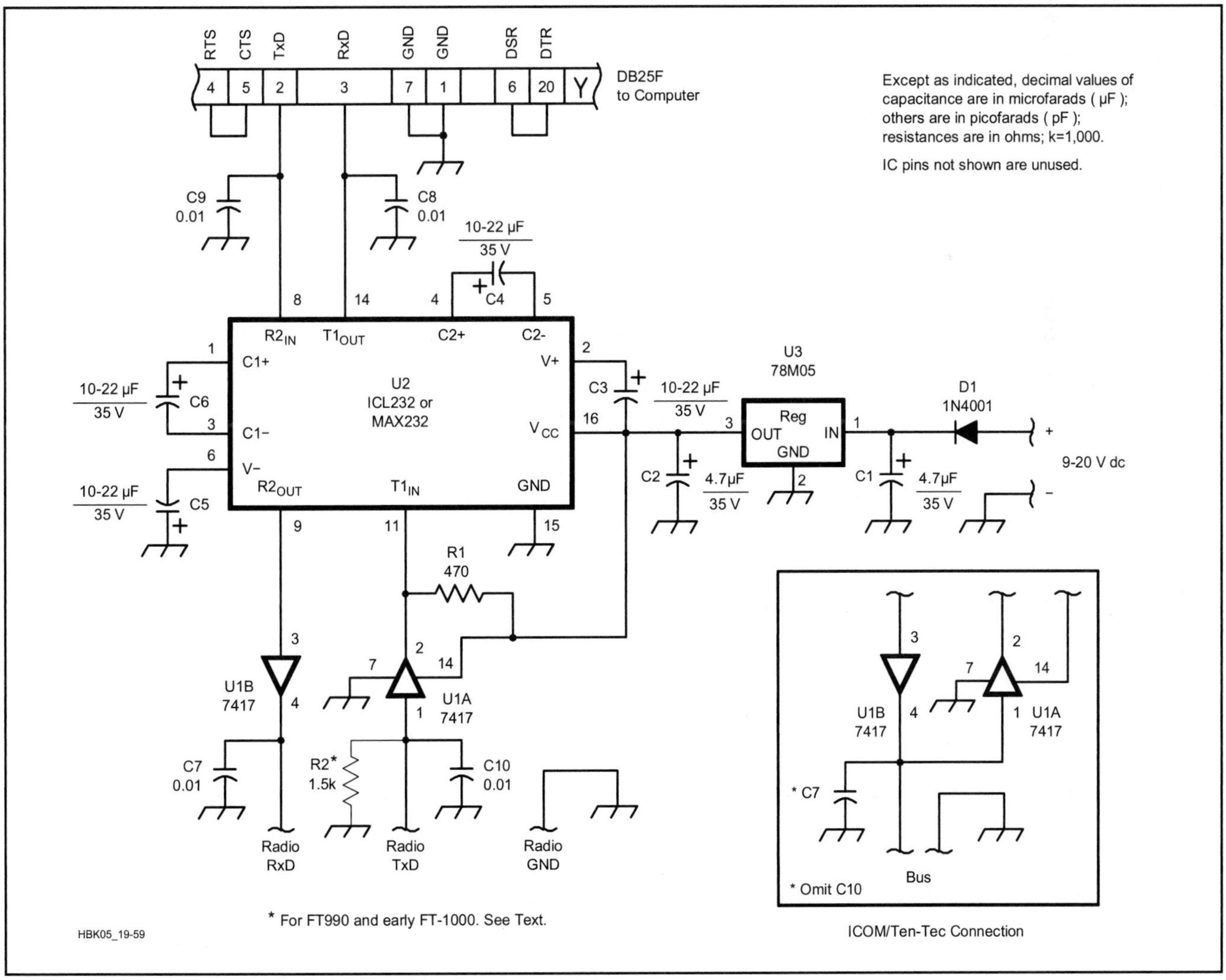

Fig 24.36 — ICOM/Ten-Tec/Yaesu interface schematic. The insert shows the ICOM/Ten-Tec bus connection, which simply involves tying two pins together and eliminating a bypass capacitor.

C7-C10 — 0.01-μF ceramic disc.   U1 — 7417 hex buffer/driver.   U2 — Harris ICL232 or Maxim MAX232.

impedance of the interface is not low enough to pull the input below threshold with nothing connected, the FT-990/FT-1000 (and perhaps other radios) will not work reliably. This can be corrected by installing a 1.5 kΩ resistor from the serial out line to ground (R2 on pin 1 of U1A in Fig 24.36.) Later FT-1000D units have a 1.5 kΩ pulldown resistor incorporated in the radio from the TxD line to ground and may work reliably without R2.

The same computer connector is used, but the radio connector varies with model. Refer to the manual for your particular rig to determine the connector type and pin arrangement.

TYPE THREE: KENWOOD

The interface setup used with Kenwood radios is different in two ways from the previous two: Request-to-Send (RTS) and Clear-to-Send (CTS) handshaking is implemented

and the polarity is reversed on the data lines. The signals used on the Kenwood system are active-low. This means that 0 V represents a logic one and +5 V represents a logic zero. This characteristic makes it easy to fully isolate the radio and the computer since a signal line only has to be grounded to assert it. Optoisolators can be used to simply switch the line to ground.

The schematic in **Fig 24.37** shows the Kenwood interface circuit. Note the different grounds for the computer and the radio. This, in conjunction with a separate power supply for the interface, provides excellent isolation.

The radio connector is a 6-pin DIN plug. The manual for the rig details this connector and the pin assignments.

Some of the earlier Kenwood radios require additional parts before their serial connection can be used. The TS-440S and R-5000

require installation of a chipset and others, such as the TS-940S require an internal circuit board.

CONSTRUCTION AND TESTING

The interfaces can be built using a PC board, breadboarding, or point-to-point wiring. PC boards and MAX232 ICs are available from FAR Circuits. The PC board template is available on the CD-ROM.

It is a good idea to enclose the interface in a metal case and ground it well. Use of a separate power supply is also a good idea. You may be tempted to take 13.8 V from your radio — and it works well in many cases: but you sacrifice some isolation and may have noise problems. Since these interfaces draw only 10 to 20 mA, a wall transformer is an easy option.

The interface can be tested using the data

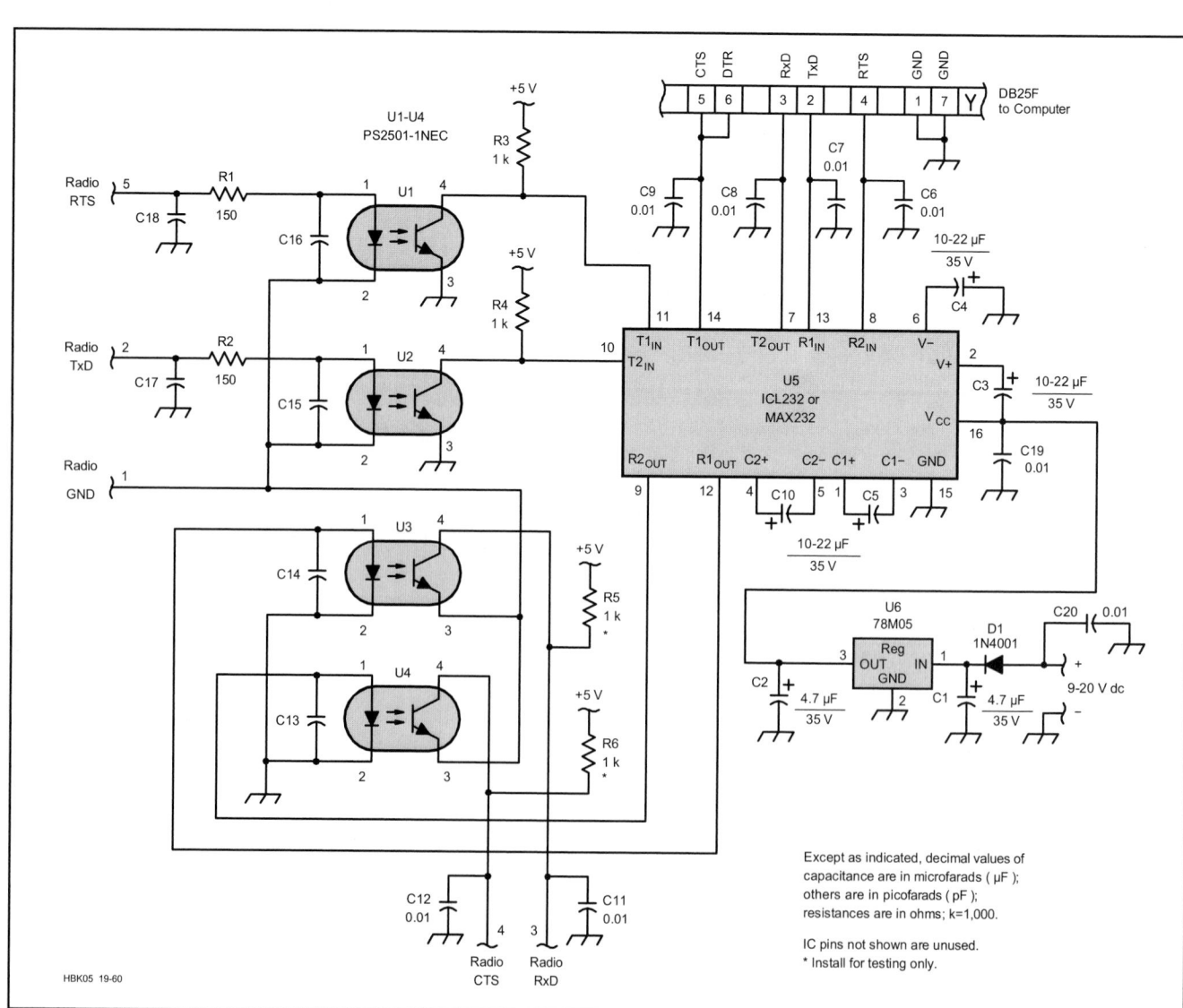

Fig 24.37 — Kenwood interface schematic.
C6-C9, C11, C12, C17, C18 — 0.01-µF ceramic disc.

C13-C16, C19-C21 — 0.01 µF ceramic disc.

U1-U4 — PS2501-1NEC (available from Digi-Key).
U5 — Harris ICL232 or Maxim MAX232.

Table 24.6 Kenwood Interface Testing	
Apply	*Result*
GND to Radio-5	−8 to −12 V at PC-5
+5 V to Radio-5	+8 to +12 V at PC-5
+9 V to PC-4	+5 V at Radio-4
−9 V to PC-4	0 V at Radio-4
GND to Radio-2	−8 to −12 at PC-3
+5 V to Radio-2	+8 to +12 V at PC-3
+9 V to PC-2	+5 V to Radio-3
−9 V to PC-2	0 V at Radio-3

Table 24.7 ICOM/Ten-Tec Interface Testing	
Apply	*Result*
GND to Bus	+8 to +12 V at PC-3
+5 V to Bus	−8 to −12 V at PC-3
−9 V to PC-2	+5 V on Bus
+9 V to PC-2	0 V on Bus

Table 24.8 Yaesu Interface Testing	
Apply	*Result*
GND to Radio TxD	+8 to +12 V at PC-3
+5 V to Radio TxD	−8 to −12 V at PC-3
+9 V to PC-2	0 V at Radio RxD
−9 V to PC-2	+5 V at Radio RxD

in **Tables 24.6, 24.7** and **24.8**. Remember, all you are doing is shifting voltage levels. You will need a 5-V supply, a 9-V battery and a voltmeter. Simply supply the voltages as described in the corresponding table for your interface and check for the correct voltage on the other side. When an input of –9 V is called for, simply connect the positive terminal of the battery to ground.

During normal operation, the input signals to the radio float to 5 V because of pullup resistors inside the radio. These include RxD on the Yaesu interface, the bus on the ICOM/Ten-Tec version, and RxD and CTS on the Kenwood interface. To simulate this during testing, these lines must be tied to a 5-V supply through 1-kΩ resistors. Connecting these to the supply without current-limiting resistors will damage the interface circuitry. R5 and R6 in the Kenwood schematic illustrate this. They are not shown (but are still needed) in the ICOM/Ten-Tec/Yaesu schematic. Also, be sure to note the separate grounds on the Kenwood interface during testing.

Another subject worth discussing is the radio's communication configuration. The serial ports of both the radio and the computer must be set to the same baud rate, parity, and number of start and stop bits. Check your radio's documentation and configure your software and the radio to have identical communications settings.

24.9 A Simple Serial Interface

A number of possibilities exist for building an RS-232-to-TTL interface level converter to connect a transceiver to a computer serial port. The interface described here by Dale Botkin, NØXAS, is extremely simple, requiring only a few common parts.

Most published serial interface circuits fall into one of the following categories:

• Single chip, MAX232 type. These are relatively expensive, require several capacitors, take up lots of board space and usually require ordering parts.

• Single chip, MAX233 type. This version doesn't need capacitors. Otherwise it shares most of the characteristics of the MAX232 type device, but is even more expensive.

• Transistors. Usually transistor designs use a mix of NPN and PNP devices, four to eight or more resistors, and usually capacitors and diodes. This is just too many parts!

• Other methods. Some designs use series resistors, or TTL logic inverters with resistors. The idea is that the voltages may be out of specification for the chip, but if the current is low enough it won't damage the device. This is not always safe for the TTL/CMOS device. With just a series resistor the serial data doesn't get inverted, so it requires the use of a software USART. It can work, but it's far from ideal.

A SIMPLER APPROACH

The EIA-232 specification specifies a

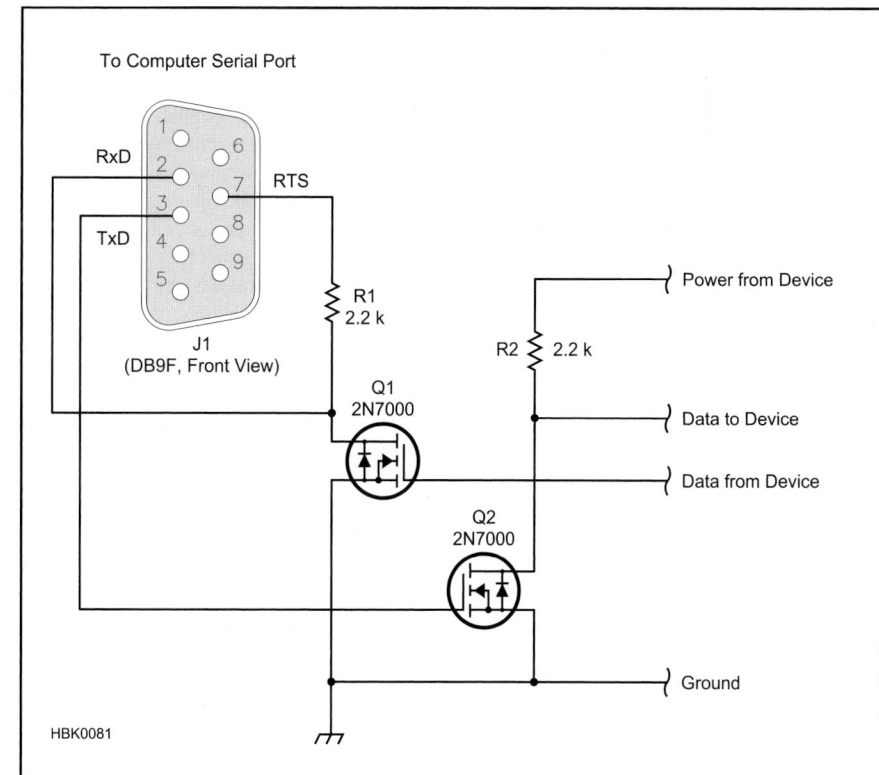

Fig 24.38 — Schematic for the simple serial interface.

J1 — DB9F or connector to fit device serial port. Note that pin assignments for RxD, TxD and RTS are different on DB25 connectors.

Q1, Q2 — Small signal MOSFET, 2N7000 or equiv.

signal level between ±3 V and ±12 V for data. The author looked at data sheets for devices ranging from the decades-old 1489 line receiver to the latest single-chip solutions. Not a single device found actually required a negative input voltage! In fact, the equivalent schematic shown for the old parts clearly shows a diode that holds the input to the first inverter stage at or near ground when a negative voltage is applied. The data sheets show that a reasonable TTL-level input will drive these devices perfectly well.

Relieved of the burden of exactly replicating the PC-side serial port voltages, the interface shown in **Fig 24.38** is designed for minimum parts count and low cost — under $1.

Why use MOSFETs for Q1 and Q2? In switching applications such as this, there are several advantages to using small signal MOSFETs over the usual bipolar choices like the 2N2222 or 2N3904. First, they need no base current limiting resistors — the MOSFET is a voltage operated device, and the gate is isolated from the source and drain. The gate on a 2N7000 is rated for ±20 V, so you don't need a clamping diode to prevent negative voltages from reaching the gate. They can also be had just as cheaply as their common bipolar cousins. (If sourcing 2N7000s is a challenge in Europe, the BS170 is basically identical except for the pins being reversed.) Now all that is needed is a drain limiting resistor, since the gate current is as close to zero as we are likely to care about.

Q1 gets its input from the TTL-level device. Q1 and R1 form a simple inverter circuit, with the inverted output going to pin 2 (RxD) of the serial connector. The pull-up voltage for RxD comes from pin 7 (RTS) of the serial connector, so the voltage will be correct for whatever device is connected — be it a PC or whatever else you have. Q2 and R2 are another inverter, driven by TxD from pin 3 of the serial connector. The pull-up voltage for R2 should be supplied by the low-voltage device — your radio or microcontroller project.

There you go. Two cheap little MOSFETs, two resistors. The 2.2 kΩ, ¼ W resistors seem to work with every serial port tried so far. The author has sold several hundred kits using this exact circuit for the serial interface without issues from any users. It's been tested with laptops, desktops, USB-to serial-converters, a serial LCD display, and even a Palm-III with an RS-232 serial cable. It's also been used with very slight modifications in computer interfaces for VX-7R and FT-817 transceivers. It's simple, cheap and works great — what more could you ask?

24.10 USB Interfaces For Your Ham Gear

Serial ports have disappeared from most new computers. The once-ubiquitous serial port, along with other interfaces that were once a basic part of every computer system, has largely been replaced by Universal Serial Bus (USB) ports. This presents problems to the amateur community because serial ports have been and continue to be the means by which we connect radios to computers for control, programming, backing up stored memory settings, contest logging and other applications. Radios equipped with USB ports are rare indeed, and many of the optional computer interface products sold by ham manufacturers are serial-only. The project described here by Dale Botkin, NØXAS, offers a solution.

THE UNIVERSAL SERIAL BUS

So what is USB, and how can it be used in the ham shack? The USB standard came about during the early 1990s as a replacement for the often inconvenient and frustrating array of serial, parallel, keyboard, mouse and various other interfaces commonly used in personal computer systems. The idea is to use a single interface that can handle multiple devices sending and receiving different types of data, at different rates, over a common bus. In practice, it works quite well because of the very well defined industry specification for signal levels, data formats, software drivers and so forth.

At least two and sometimes as many as six or more USB ports are now found as standard equipment on nearly all new computer systems. Support for USB ports and generic USB devices is built into most operating systems including *Windows, Linux, Solaris, Mac OS* and *BSD* to name a few. USB use has spread beyond computers, and the standard USB "B" or the tiny 4-conductor "Mini B" connectors are now found on all manner of electronic devices including cameras, cell phones, PDAs, TVs, CD and DVD recorders, external data storage drives and even digital picture frames. USB is indeed becoming *universal*, as its name implies — everywhere, it seems, except for Amateur Radio gear!

The EIA-232 serial interface specification we have used for decades is just that — an electrical interface specification that defines voltage levels used for signaling. The asynchronous serial format most commonly used with serial ports defines how data is sent over whatever interface is being used. Any two devices using the same protocol and format can talk to each other, whether they are both computers, both peripheral devices (like printers, terminals or TNCs) or any combination. All that is required is the proper cable and the selection of matching parameters (speed, data bits, handshaking) on each end of the connection.

This is not the case with USB devices. USB uses a tiered-star architecture, with well defined roles for the host controller (your computer) and devices. Devices cannot communicate directly with each other, only with a host.

See **Fig 24.39**. At the top of the stack is the *host controller device* (HCD), usually your computer, which controls all communication within the USB network. The computer may have several host controllers, each with one or more ports. The host controller discovers, identifies and talks to up to 127 devices using a complex protocol. No device can communicate directly with another device; all communication is initiated and controlled by the host controller. As a result, you cannot simply connect two USB devices together and expect them to communicate. (Some USB printers will talk directly to digital cameras because the printer also has a host controller built in.)

As you have probably already figured out, moving from serial to USB is not a simple

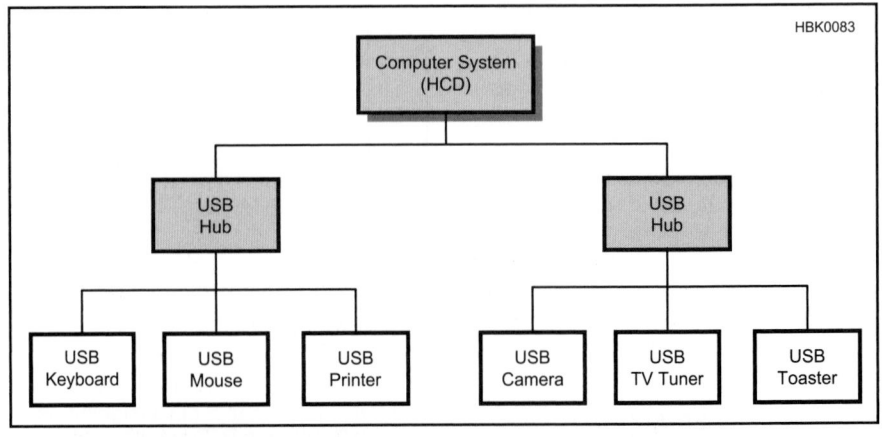

Fig 24.39 — Typical USB architecture.

matter of converting voltage levels or even data formats. USB devices all need some embedded "smarts," usually in the form of either a dedicated USB interface chip or a microcontroller using firmware to talk to the USB host. Where we could once use simple transistor circuits to adapt RS-232 serial to the TTL voltage levels used by many radios, we now must add the complete USB interface into the picture.

The USB specification is freely available and well documented, and there are books dealing with designing the software to use the interface. If you're starting from scratch it is a very daunting task to even understand exactly how USB works, let alone design and build your own USB-equipped device. If you're willing and dedicated enough, you can implement your own USB device using a fairly broad range of available microcontrollers. There are also dedicated function chips that implement a complete USB interface — or parts of it — in one or two devices.

Fortunately, several manufacturers have stepped up to make using USB relatively easy for the experimenter. One of the most commonly used solutions is from Future Technology Devices International (FTDI). FTDI manufactures a line of USB interface chips that plug into the Universal Serial Bus on one side and present a serial or parallel interface on the other side. The chips are quite versatile and are commonly used to add USB functionality to a wide range of devices ranging from cell phones to USB-to-serial cables. They are also available in several forms such as plug-in modules with a USB connector and the interface logic, all on a small PC board with pins that can be plugged into a standard IC socket.

CONSTRUCTION

This project uses an FT232R series chip from FTDI to bridge the USB interface from your computer to a TTL-level serial interface that can be adapted to use with many different ham rigs. The chip has built-in logic to handle the USB interface, respond to queries and commands from the computer, and send and receive serial data. By using special drivers available from FTDI (and built into *Windows* and many other systems), it can appear to your programs as a standard serial COM port.

Once the conversion to serial data is done, all that remains is to adapt the signals to the particular interface used by your equipment. In many cases, this serial data can simply be connected directly to the rig using the appropriate connector. In others it may require some extra circuitry to accommodate various interface schemes used by manufacturers. The most common is an open-collector bus type interface, which is easily accommodated using a few inexpensive parts.

Construction of this project can be handled in a couple of different ways. The FT232R series chips are available from various sup-

pliers at low cost. Unfortunately they are only available in SSOP and QFLP surface mount packages that can be difficult to work with unless you use a custom printed circuit board. (An additional letter is tacked onto the part number to identify the package. For example, the FT232RL uses a 28-pin SSOP package.) There are SSOP-to-through-hole adapter boards available, but they are relatively expensive and still require some fine soldering.

Fortunately FTDI also sells a solution in the form of the TTL-232R (**Fig 24.40**). It incorporates a USB "A" connector, an FT232RQ chip and the required power conditioning on a tiny PCB encapsulated in plastic, all about the size of a normal USB connector. A 6-ft long cable terminates in a 6-position inline header that provides ground, +5 V, and TTL level TxD, RxD, RTS and CTS signals.

You can connect these signals directly to your own microprocessor projects and have an "instant" USB connection. You can also build a variety of interface cables or boards for use with different radios. This arrangement is ideal for the experimenter and the person who might need several different interfaces for use with several different rigs — like many of us! The cost of the TTL-232R cable is quite reasonable, in fact lower than the cost of having a prototype PCB made to build your own equivalent device. Of course that option is still available to those who need more flexibility.

USING USB IN THE SHACK

With the USB-to-serial conversion handled so easily, the only remaining task is to adapt the TTL-level serial interface of the FT232R to the interface used by your particular radio. In the case of some rigs, such as the Yaesu FT-817, Kenwood TM-G707A and many others, it's simply a matter of wiring up the right connector, as shown in **Fig 24.41**.

For communicating with individual ICOM and Ten-Tec rigs, one simple solution is to simply connect the transmit and receive data signals together and wire up a ⅛-inch stereo plug as shown in **Fig 24.42**. This will work with a single device connected, but lacks the open-collector drivers needed to be compatible with attaching multiple devices to the CI-V bus.

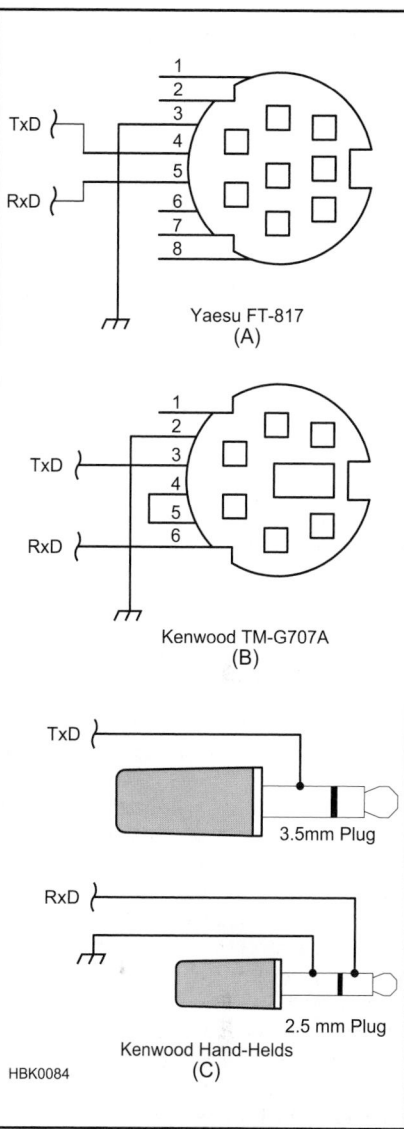

Fig 24.41 — Using the FTDI USB-to-serial interface with some radios is as simple as wiring up a connector. Connections for the Yaesu FT-817 are shown at A, the Kenwood TM-G707A (PG-4 type) at B, and Kenwood hand-helds (PG-5 type) at C. Ground, RxD and TxD connections are made to the 6-pin connector on the TTL-232R cable.

Fig 24.40 — The FTDI TTL-232R cable simplifies construction.

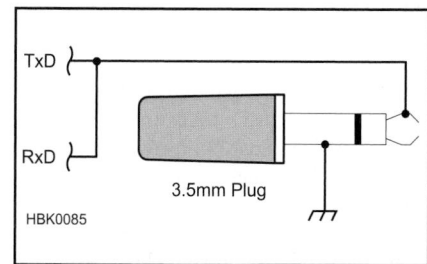

Fig 24.42 — The transmit and receive data signals can be connected together for use with a single Ten-Tec or ICOM CI-V device.

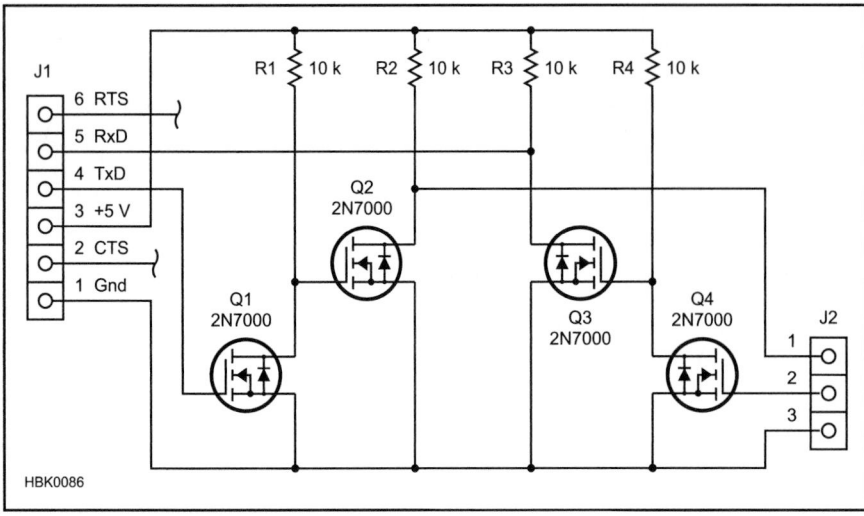

Fig 24.43 — Schematic of the basic interface buffer board.

J1 — 6-pin, 0.1-in pitch header to match TTL-232R cable.

J2 — 3-pin header to match radio cable.

Q1-Q4 — Small-signal MOSFET, 2N7000, BS-170 or equiv.

R1-R4 — 10 kΩ, ¼ W.

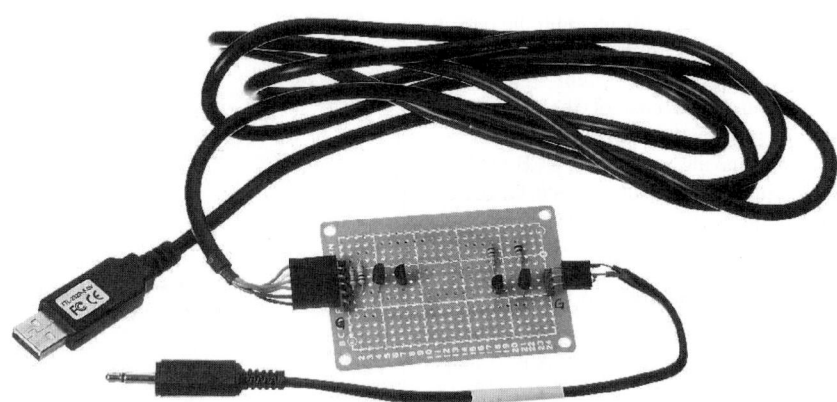

Fig 24.44 — The completed buffer board. Adapter cables for various radios plug into the header on the right edge of the board.

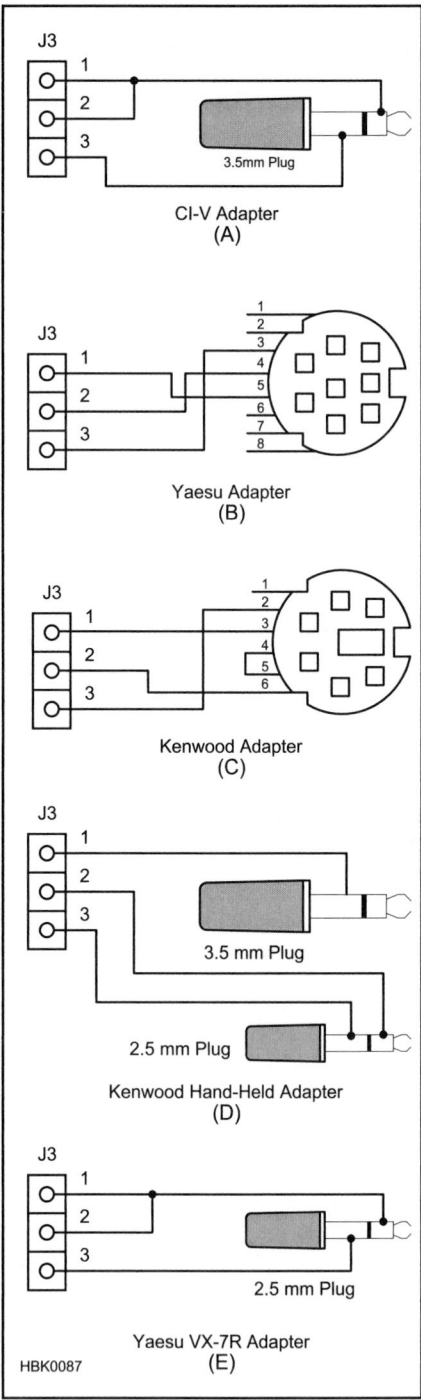

Fig 24.45 — Adapters for using the buffer board with various radios. J3 matches the adapter used on the buffer board.

A LITTLE MORE VERSATILITY

Many hand-held transceivers use a single data line for both transmit and receive. This method is very similar to CI-V, but requires an open-collector or open-drain configuration for a single device. Some additional buffering is required to provide an interface to the rig.

Fig 24.43 shows the design of the general-purpose buffer board for use with various radios. This circuit simply provides a noninverting, open-drain buffer for data in each direction. Adapters for various rigs can be plugged into J2 with no further changes needed. Use of MOSFETs instead of their bipolar counterparts simply reduces the parts count. No additional resistors are needed to limit base current, as you would need when using 2N2222 or similar transistors. Parts can be mounted on a small project board as shown in **Fig 24.44**.

This arrangement turns out to be extremely versatile and will work with CI-V, FT-817,

Yaesu VX-series handhelds and the other rigs the author was able to test as well. All that is required is the proper adapter cable from the base interface board to the rig. Some examples are shown in **Fig 24.45**.

ADDING PTT

Handling PTT is quite simple, and it can be done either with or without isolation as shown in **Fig 24.46**. For a non-isolated PTT line, we can simply drive the gate of a 2N7000 or similar MOSFET from the FT232RL's RTS output. If you prefer an optically isolated PTT output, all you need to add is one resistor and an optocoupler such as a 4N27 or PS-7142. Note that the TTL-232R provides a 100 Ω resistor in series with the RTS signal (the same is true for TxD), so a smaller value resistor (R1) can be used for the optocoupler's input than you would normally use.

While connecting the data lines directly to the rig is a suitable solution for many ap-

plications, some Kenwood HF rigs require a different approach. They require inverted TTL-level data, and may experience "hash" noise from the computer when not using an isolated interface. **Fig 24.47** shows an optically isolated method of connecting to a Kenwood rig such as the TS-850S. Note that this has not been tested but the design is similar to numerous examples of serial interfaces for the TS-850S and similar radios.

It's probably only a matter of time before

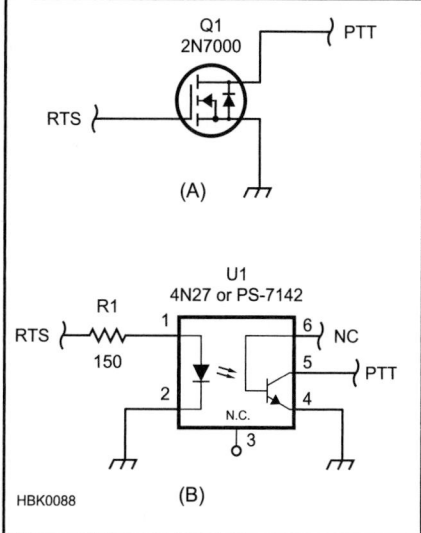

Fig 24.46 — PTT can be added either without isolation (A) or with an optoisolator (B).

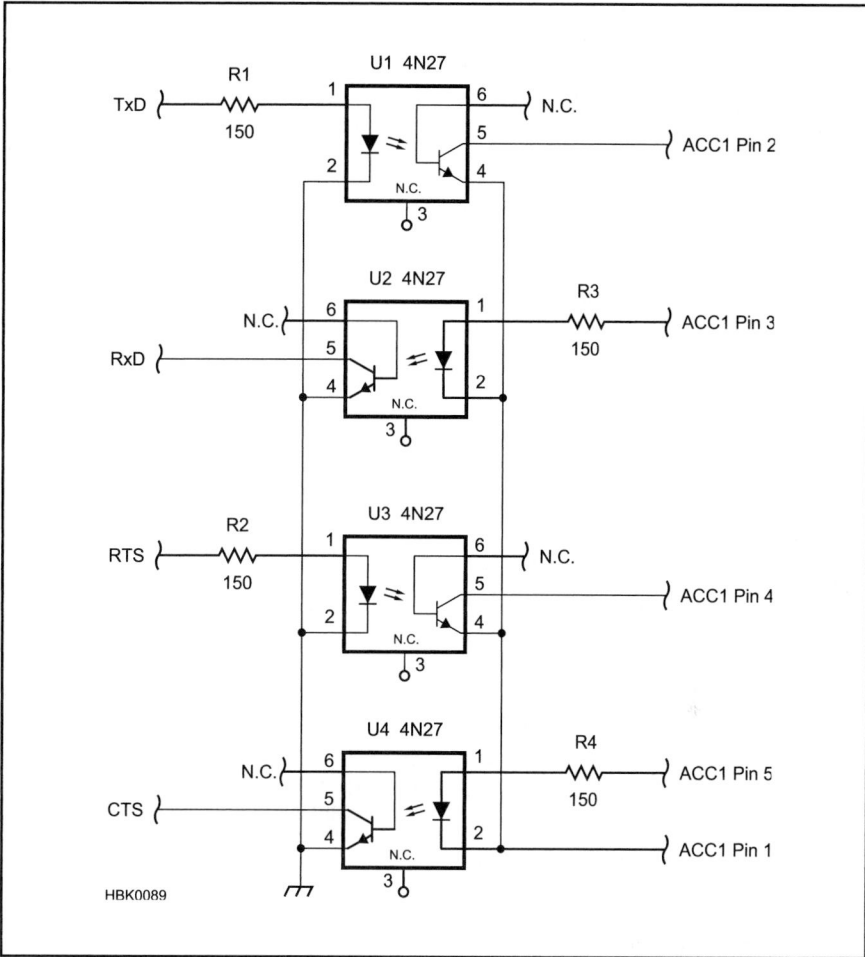

Fig 24.47 — Isolated Kenwood interface for the TS-850S and similar HF transceivers requiring inverted TTL-level data and isolated lines.

we start seeing the tiny USB "Mini B" connector appear on new ham gear. It has become inexpensive and easy to add USB to pretty much any product, and as serial ports become more and more rare the manufacturers will eventually face overwhelming consumer demand to switch. In the meantime, you can do a little homebrewing and use your new, serial port free computer to talk to all of your ham gear for very little cost.

24.11 The Universal Keying Adapter

When Dale Botkin, NØXAS, was about to start restoring his old Heathkit HW-16, he decided it was time to explore what would be needed to key this tube rig using an electronic keyer. He also remembered — not fondly — having been "bitten" a few times when he touched metal parts of the key. There was some substantial voltage on the key, and he wanted to avoid touching it again. The key jack of this rig has a negative voltage, something like –85 V or more. Even the 2N7000 output of a solid-state keyer, rated at 60 V, would be no match for that. The Universal Keying Adapter (UKA) shown in **Fig 24.48** bridges the gap between a modern keyer and a classic tube-type transmitter, and it can also be used to key older tube-type amplifier TR switching lines with solid-state transceivers that can't handle the voltage or current requirements.

KEYING SCHEMES

Older gear generally uses one of two keying

Fig 24.48 — This version of the Universal Keying Adapter 2 is available as a kit from www. hamgadgets.com. Parts count is low, and it can easily be built on a small project board.

schemes, grid-block or cathode keying. Grid-block keying requires that your key handle fairly high negative bias voltages, often in the 150 V range. Cathode keying is more demanding, with voltages up to +350 V or so. Most modern solid state keyers use a simple transistor output circuit suitable only for low-voltage, positive keying. Obviously, neither grid-block nor cathode keyed transmitters should be connected to such an output! At the very least it's going to damage the keyer.

Most grid-block keying adapters use one or two bipolar transistors and are designed specifically to key grid-block rigs. This is fine as far as it goes, but also requires that you have one setup for solid state rigs, another for grid-blocked rigs, and still another if you have a cathode keyed boat anchor as well. It's not an optimum solution. Of course one could use a relay to key just about anything, but relay contact noise would quickly get really irritating. Relays also eventually wear out, contacts get dirty and the coils can use quite a bit of current, meaning battery powered keyers will suffer from very limited battery life.

Solid-State Relays

Solid state relays exist, but they can be expensive. The author finally hit upon what seems to be an ideal solution. It's cheap, small, relatively low current, uses low voltage control and is capable of switching fairly high ac or dc voltages. Perfect! Crydom, NEC, Fairchild, Omron and other manufacturers make very similar small, inexpensive MOS-FET solid state relays. Suitable model numbers include the NEC PS7142 and PS7342, Fairchild HSR412, Omron G3VM-401B and others. Depending on the part selected they can handle up to a couple hundred milliamps and will switch up to 400 V ac or dc. All of this in a 6-pin DIP form factor, and for just a few bucks!

The MOSFET solid-state relay (SSR) is very similar to an optoisolator, but somewhat more versatile. Ac or dc loads can be switched, and the allowable load voltages are much higher than regular optoisolators. Driving the relay input involves supplying a few mA of current to turn on an LED inside the device. This requires about 1.4 V at a recommended forward current of 10 to 30 mA. The output is a pair of MOSFETs with common sources and a photo gate rather than a hard-wired gate. Turn on the LED and the MOSFETs conduct, turn off the LED and the MOSFETs shut off. Just like a relay, it's simple and elegant.

KEYING ADAPTER CIRCUITS

For a simple keying adapter arrangement, all that is needed is the MOSFET solid-state relay, one resistor, and a dc power source such as a battery. See **Fig 24.49**. The keying input from your hand key or electronic keyer completes the input circuit through the cur-

rent limiting resistor and the input side of the SSR, turning the output on. In this example, the value of R1 is determined using Ohm's Law to give between 10 and 30 mA of current at the desired input voltage. For example, for use with two AA alkaline batteries, a range of 100 to 300 Ω is okay; 150 or 220 Ω will work reliably with NiCd or NiMH cells as well. If you plan to use a 13.8 V dc power supply, a 1 kΩ resistor should be fine for R1.

It soon became apparent that other uses existed in addition to just isolating the key or keyer from a transmitter. The addition of a simple transistor inverter allows the use of

computer keying in parallel with input from a straight key, bug or keyer. This is handy, for example, to contesters using computers for keying who may need to send some information by hand from time to time. See **Fig 24.50**. The use of a 2N7000 MOSFET instead of the common NPN transistor makes it easier to accommodate both serial and parallel port use, since the gate will withstand up to 20 V positive or negative and does not need to be current limited. In this example, the value of R1 should be determined as mentioned earlier. R2's purpose is to keep the MOSFET gate from floating, so its value is not critical but

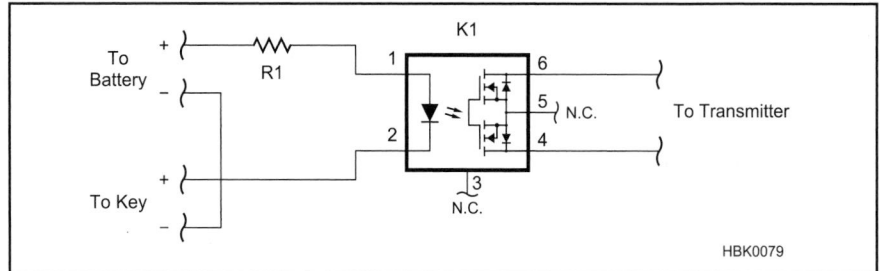

Fig 24.49 — The original Universal Keying Adapter circuit used a solid-state relay and a battery for power.

K1 — Optically coupled solid-state relay, Fairchild HSR412 or equivalent.
R1 — ¼ W resistor, varies with supply

voltage (see text). Use 150 or 220 Ω for two AA cells; 1 kΩ for 13.8 V dc power source.

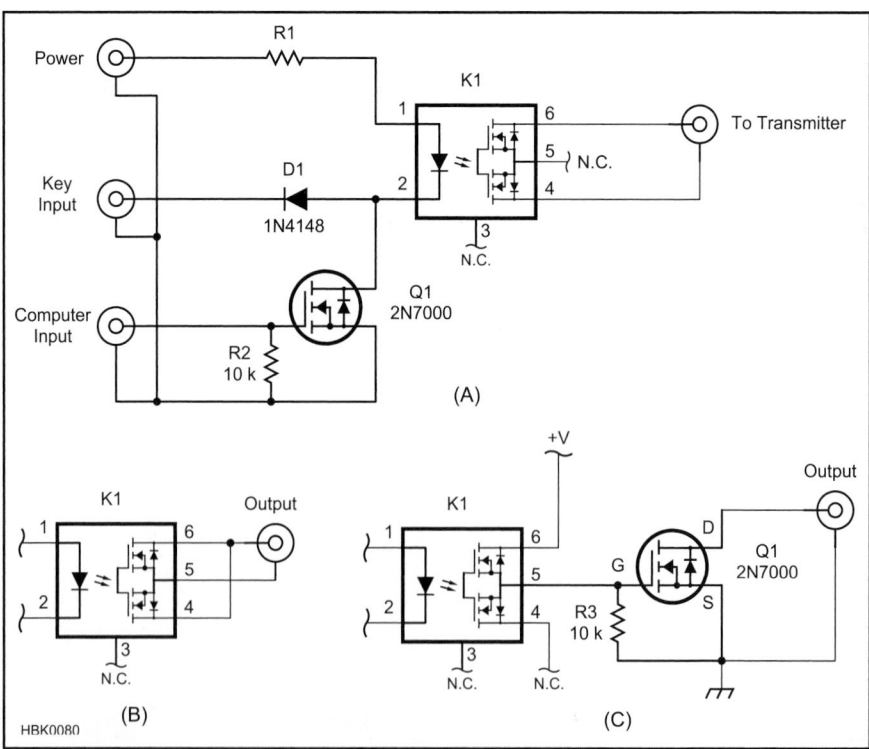

Fig 24.50 — The Universal Keying Adapter circuit with provisions for using computer keying in parallel with a hand key or keyer. Two methods of increasing the load current capacity are shown at B and C (see text for discussion).

K1 — Optically coupled solid-state relay, Fairchild HSR412 or equivalent.
R1 — ¼ W resistor, varies with supply

voltage (see text). Use 150 or 220 Ω for two AA cells; 1 kΩ for 13.8 V dc power source.

10 kΩ or 100 kΩ are good values.

The design eventually evolved to include a full-wave bridge rectifier allowing either ac or dc input, a Zener diode for voltage regulation and an LED keying indicator. The result was the Universal Keying Adapter 2, which is also available in kit form (see **www.hamgadgets.com**). The UKA-2 can accept any dc or ac power source up to about 30 V, and is adaptable to lower power battery operation by substituting or eliminating some of the power supply parts.

OTHER CONFIGURATIONS

It is worth noting that if only positive or only negative voltages will be used, the output current capacity can be greatly increased and the on-state resistance greatly decreased by connecting the two MOSFET drain outputs in parallel rather than serial. An example of this is shown in Fig 24.50B. For the HSR412, maximum load current increas-es from 140 mA for series output to 210 mA for parallel. An added feature of this arrangement is the presence of the intrinsic "body diodes" of the MOSFETS. If you are keying an amp that has an internal keying relay, this can serve to absorb some of the back EMF when the relay releases. Note that this arrangement requires that pin 5 of the IC be at a lower voltage than the other two pins — in other words, ground for positive keying, or –V for negative keying. Since the output is completely isolated from the input, the polarity of the output connector can be changed without the risk of exposing your other gear to dangerous voltages.

Although the IC solid-state relays are good for a wide range of uses, there are applications that present more of a load than the device is able to safely handle even with the outputs in parallel. For loads of more than the rated device current, an external MOSFET switch can be used. Again, the resulting output is optically isolated from the input. Resistor R3 shown in Fig 24.50C keeps the gate of the 2N7000 MOSFET low until the input is activated, so its value is not critical. The gate voltage +V needs only to be within the safe range for the device selected; in most cases 12 V will do fine.

The UKA design works well with both grid-block and cathode keyed rigs, as well as solid state. It's been tested with a TS-930SAT, Heath HW-16, FT-817, Rock-Mites, FT-480R and more. It has also been used with cathode keyed rigs such as the Heath DX-40, and many more examples of this circuit are in use keying various "boat anchor" rigs and amps. It works quite well to key older tube power amplifiers with solid-state transceivers that are not equipped with suitable keying relays or circuits. The input works well with any rig, key, electronic keyer, serial or parallel port tried so far.

24.12 The TiCK-4 — A Tiny CMOS Keyer

TiCK stands for "Tiny CMOS Keyer" and this is the fourth version of the chip. It is based on an 8-pin DIP microcontroller from Microchip Corporation, the PIC12CE674. This IC is a perfect candidate for all sorts of Amateur Radio applications because of its small size and high performance capabilities. This project originally used the TiCK-2 chip (based on the PIC12C509) and was described fully in Oct 1997 *QST* by Gary M. Diana, Sr, N2JGU, and Bradley S. Mitchell, WB8YGG. The keyer has the following features:

• Mode A and B iambic keying.
• Low current requirement to support portable use.
• Low parts count consistent with a goal for small physical size.
• Simple rig and user interfaces. The operator shouldn't need a manual and the rig interface should be paddles in, key line out.
• An audible sidetone for user-feedback functions and to support transceivers without a built-in sidetone.
• Paddle select, allowing the operator to swap the dot and dash paddles without having to rewire the keyer (or flip the paddles upside down!).
• Manual keying to permit interfacing a straight key (or external keyer) to the TiCK.

New for the TiCK-4 version are:
• Two 50-character message memories (up from one).
• Remembers operating parameters (but not message memories) when power is removed.

DESIGN

Fig 24.51 shows the schematic for the keyer. The PIC12CE674 has 3.6 kbytes of program read-only memory (ROM) and 128 bytes of random-access memory (RAM). This means that all the keyer *functions* have to fit within the ROM. The keyer *settings* such as speed, paddle selection, iambic mode and sidetone enable are stored in RAM. Unlike the previous TiCK keyers, this microcontroller also includes EEPROM memory that's used to store these operating parameters. So you can power the chip off, and upon power-up it will remember the last stored values for the settings.

A PIC12CE674 has eight pins. Two pins are needed for the dc input and ground connections. The IC requires a clock signal. Several clock-source options are available. You can use a crystal, RC (resistor and capacitor) circuit, resonator, or the IC's internal oscillator. The authors chose the internal 4-MHz oscillator to reduce the external parts count. Two I/O lines are used for the paddle input. One output feeds the key line, another output is required for the audio feedback (sidetone) and a third I/O line is assigned to a pushbutton.

USER INTERFACE

Using the two paddles and a pushbutton, you can access all of the TiCK's functions. Certain user-interface functions need to be more easily accessible than others; a prioritized list of functions (from most to least accessible) is presented in **Table 24.9**.

The TiCK employs a *single button interface* (SBI). This simplifies the TiCK PC board, minimizes the part count and makes for ease of use. Most other electronic keyers have multibutton user interfaces, which, if used infrequently, make it difficult to remember the commands. Here, a single button push takes you through the functions, one at a time, at a comfortable pace (based on the current speed of the keyer). Once the code for the desired function is heard, you simply let up on the button. The TiCK then executes the appropriate function, and/or waits for the appropriate input, either from the paddles or the pushbutton itself, depending on the function in question. Once the function is complete, the TiCK goes back into keyer mode, ready to send code through the key line.

The TiCK-4 IC generates a sidetone signal that can be connected to a piezoelectric element or fed to the audio chain of a transceiver. The latter option is rig-specific, but can be handled by more experienced builders.

THE TiCK LIKES TO SLEEP

To meet the low-current requirement, the authors took advantage of the PIC12CE674's ability to *sleep*. In sleep mode, the processor shuts down and waits for input from either of

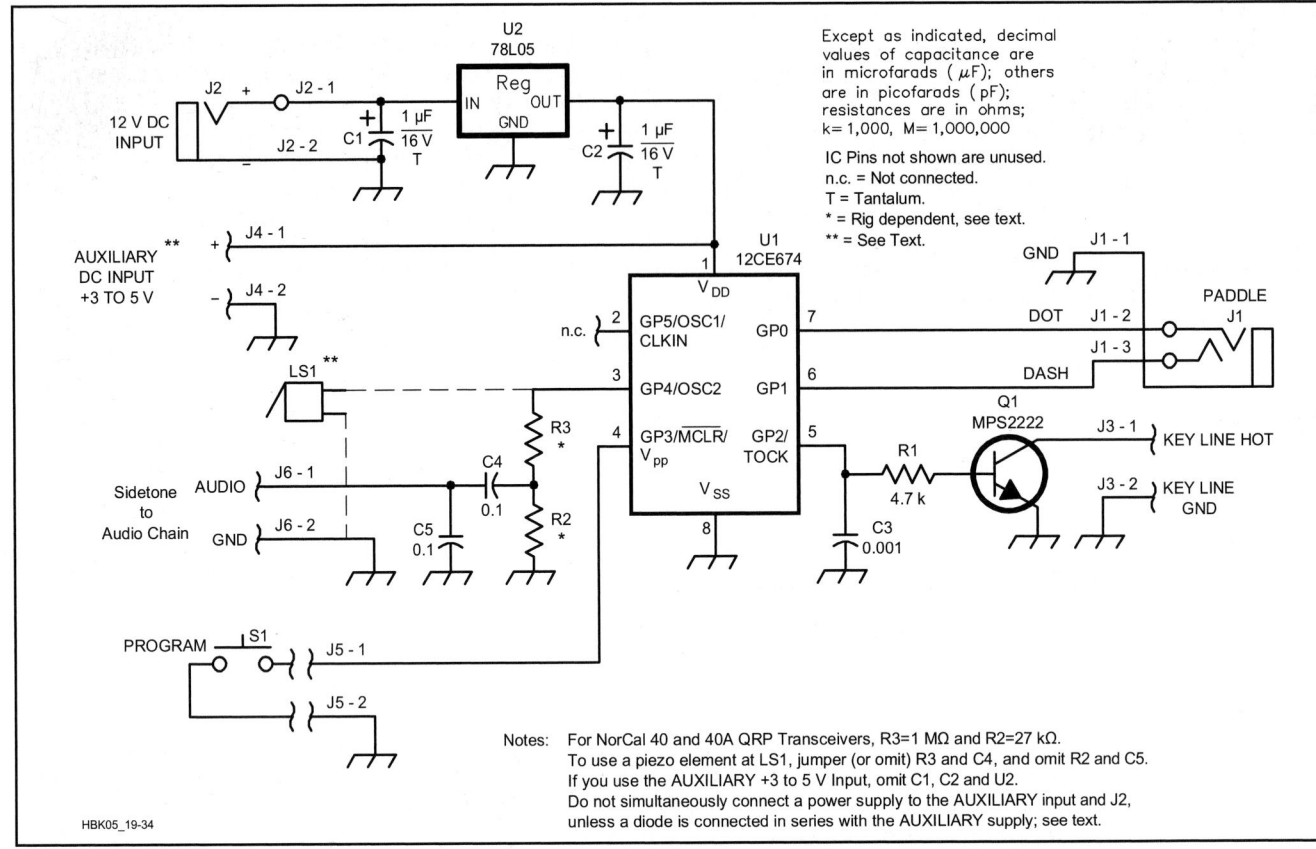

Fig 24.51 — Schematic of the TiCK-4 keyer. Equivalent parts can be substituted. Unless otherwise specified, resistors are ¼ W, 5%-tolerance carbon film units. The PIC12CE674 IC must be programmed before use; see the parts list for U1. RS part numbers in parentheses are RadioShack; M = Mouser (Mouser Electronics). Kits are available from Kanga US (www.kangaus.com).

C1, C2 — 1 µF, 16 V (or higher) tantalum
  (RS 272-1434; M 581-TAP105K035SCS).
J1 — 3-circuit jack (RS 274-249;
  M 161-3402).
J2 — 2-circuit jack (RS 274-251) or coaxial
  (RS 274-1563 or 274-1576).
J3 — 2-circuit jack (RS 274-251;
  M 16PJ135).

LS1 — Optional piezo element (RS 273-073).
Q1 — MPS2222A, 2N2222, PN2222, NPN
  (RS 276-2009; M 610-PN2222A).
S1 — Normally open momentary contact
  pushbutton (RS 275-1571).
U1 — Programmed PIC12CE674, available
  from Kanga US. Programmed ICs and

data sheets are available, as are kits
with the PC board and other parts.
Source code is not available.
U2 — 78L05 5 V, 100 mA regulator
  (M 512-LM78L05ACZ) or LP2950CZ-5.0
  low quiescent current regulator
  (Digi-Key LP2950CZ-5.0NS-ND).
Misc: PC board, 8-pin DIP socket, hard-
  ware, wire (use stranded #22 to #28).

Table 24.9
TiCK-4 User Interface Description

Action	TiCK-4 Response	Function
Press pushbutton briefly	none	Memory #1 playback.
Hold pushbutton down	E (dit)	Memory #2 playback.
Hold pushbutton down	S (dit-dit-dit)	Speed adjust: Press dit to decrease, dah to increase speed.
Hold pushbutton down	T (dah)	Tune: To unkey rig, press either paddle or pushbutton.
Hold pushbutton down	A (dit-dah)	ADMIN mode: Allows access to various TiCK IC setup parameters.
Press pushbutton briefly	I (dit-dit)	Input mode: Allows the user to enter message input mode.
Hold pushbutton down	1 (dit-dah-dah-dah-dah)	Message #1 input. Send message with paddle and push button momentarily when complete.
Hold pushbutton down	2 (dit-dit-dah-dah-dah)	Message #2 input. Send message with paddle and push button momentarily when complete.
Hold pushbutton down	P (dit-dah-dah-dit)	Paddle select: Press paddle desired to designate as dit paddle.
Hold pushbutton down	A (dit-dah)	Audio select: Press dit to enable sidetone, dah to disable. Default: enabled.
Hold pushbutton down	SK (dit-dit-dit dah-dit-dah)	Straight key select: Pressing either paddle toggles the TiCK to/from straight key/keyer mode. Default: keyer mode.
Hold pushbutton down	M (dah-dah)	Mode select: Pressing the dit paddle puts the TiCK into iambic mode A; dah selects iambic mode B (the default).
Hold pushbutton down	B (dah-dit-dit-dit)	Beacon mode: Pressing either paddle toggles the TiCK to/from Beacon/No-Beacon mode. Default: No-Beacon mode.
Hold pushbutton down	K (dah-dit-dah)	Keyer mode: If pushbutton is released, the keyer returns to normal operation.
Hold pushbutton down	S (dit-dit-dit)	Cycle repeats starting with Memory playback, Speed adjust, etc.

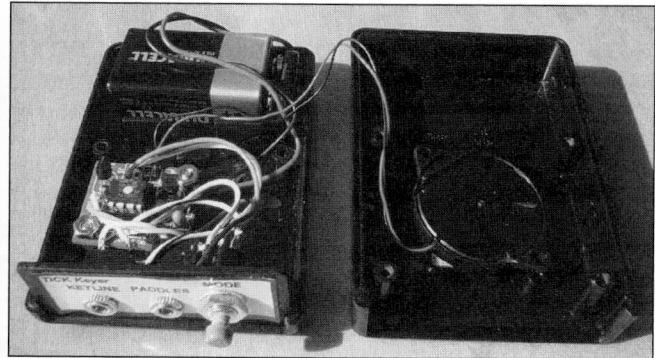

Fig 24.52 — Here's the completed TiCK-4 keyer complete with battery, mounted in a small case. (*Photos courtesy Bill Kelsey, N8ET*)

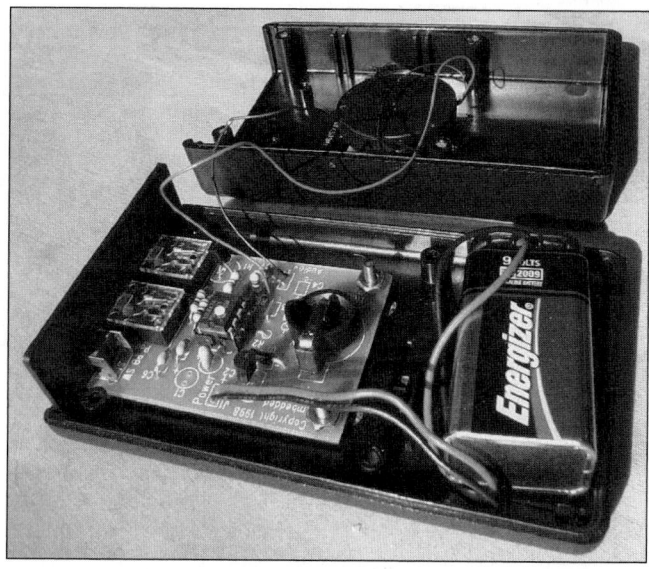

Fig 24.53 — The enhanced EMB version of the TiCK-4 includes a low-current regulator, battery backup and board-mounted jacks and switches.

the two paddles. While sleeping, the TiCK consumes about 1 μA. The TiCK doesn't wait long to go to sleep either: As soon as there is no input from the paddles, it's snoozing! This feature should be especially attractive to amateurs who want to use the TiCK in a portable station.

ASSEMBLING THE TICK

The TiCK-4's PC board size (1×1.2 inches) supports its use as an embedded and stand-alone keyer. See **Fig 24.52**. The PC board has two dc input ports, one at J2 for 7 to 25 V and another (J4, AUXILIARY) for 2.5 to 5.5 V. The input at J2 is routed to an on-board 5-V regulator (U2), while the AUXILIARY input feeds the TiCK directly. When making the dc connections, observe proper polarity: There is no built-in reverse-voltage protection at either dc input port.

The voltage regulator's bias current is quite high and will drain a 9 V battery quickly, even though the TiCK itself draws very little current. For this reason, the "most QRP way" to go may be to power the chip via the AUXILIARY power input and omit U2, C1 and C2. Another alternative is use of a low quiescent current 5 V regulator, such as the LP2950CZ-5.0, for U2.

When using the AUXILIARY dc input port, or if *both* dc inputs are used, connect a diode between the power source and the AUXILIARY power input pin, attaching the diode anode to the power source and the cathode to the AUXILIARY power input pin. This provides IC

and battery protection and can also be used to deliver battery backup for your keyer settings.

SIDETONE

A piezo audio transducer can be wired directly to the TiCK-4's audio output. Pads and board space are available for voltage-divider components. This eliminates the need to interface the TiCK with a transceiver's audio chain. Use a piezo *element*, not a piezo *buzzer*. A piezo buzzer contains an internal oscillator and requires only a dc voltage to generate the sound, whereas a piezo element requires an external oscillator signal (available at pin 3 of U1).

If you choose to embed the keyer in a rig and want to hear the keyer's sidetone instead of the rig's sidetone, you may choose to add R2, R3, C4 and C5. Typically R3 should be 1 MΩ to limit current. R2's value is dictated by the amount of drive required. A value of 27 kΩ is a good start. C4 and C5 values of 0.1 μF work quite well. C4 and C5 soften the square wave and capacitively couple J6-1 to the square-wave output of pin 3. Decreasing the value of R2 decreases the amount of drive voltage, especially below 5 kΩ. Use a 20 kΩ to 30 kΩ trimmer potentiometer at R2 when experimenting.

IN USE

To avoid RF pickup, keep all leads to and from the TiCK as short as possible. The authors tested the TiCK in a variety of RF environments and found it to be relatively

immune to RF. Make sure your radio gear is well grounded and avoid situations that cause an RF-hot shack.

The TiCK keys low-voltage positive lines, common in today's solid-state rigs. Don't try to directly key a tube rig because you will likely — at a minimum — ruin output transistor Q1.

To use the TiCK as a code-practice oscillator, connect a piezo element to the audio output at pin 3. If more volume is needed, use an audio amplifier, such as RadioShack's 277-1008.

The higher the power-supply voltage (within the specified limits), the greater the piezo element's volume. Use 5 V (as opposed to 3 V) if more volume is desired. Also, try experimenting with the location of the piezo element to determine the proper mounting for maximum volume.

SEVERAL VERSIONS

The TiCK keyer has been around for more than 10 years. The TiCK-4 is a direct plug-in replacement for the TiCK-1 and TiCK-2, so if you have built a keyer with an earlier version, you can upgrade to the TiCK-4 and gain the two memories and nonvolatile parameter memory features. The TiCK-3 is a surface-mount version with limited availability. An enhanced version, the TiCK-EMB-4, includes RF/static protection, low quiescent current regulator, battery backup and board-mounted jacks and switches. See **Fig 24.53**.

24.13 Adapting Aviation Headsets to Ham Radios

Aviation headsets are widely used for commercial and private aircraft. They have a frequency response well-suited for Amateur Radio and are comfortable when worn for long periods, even with glasses. The noise-canceling aviation microphone and the headphone earmuff's 23 dB isolation of acoustic noise are very helpful in noisy environments and contests. Adapting these headphones for amateur use requires impedance matching for the audio circuits and power for the standard electret microphone. Since aviation headsets are designed to a specific standard, other brands and models should have comparable performance.

ADAPTING THE HEADPHONE CIRCUIT

The headset earphones may be either 300 Ω stereo or 150 Ω monaural (some older sets are series connected 600 Ω). The plug fits a standard ¼ inch phone jack (two or three conductor to match). Stereo versions usually have a mono/stereo switch. If the plug is a mono version, a RadioShack 274-360 stereo-to-mono adapter can be used.

Most radios are designed for 8 Ω headphones. If your radio works with a Heil Pro Set type headset (200 Ω) no additional matching is required. To change the 8 Ω output to higher impedances, a RadioShack audio output transformer (273-1380) will do the job. Connect the low-impedance winding to the radio's headphone jack and the high-impedance winding to the headset earphones.

ADAPTING THE MICROPHONE

The microphone cannot be used with current radio models without an adapter. This project provides two different circuits — **Fig 24.54** uses a transformer-based circuit and **Fig 24.55** uses a resistive circuit with variable frequency response. **Fig 24.56** shows the relative size of a typical adapter circuit enclosure.

Aviation headsets use an electret circuit that mimics an old-style carbon microphone. The microphone amplifier requires an industry standard dc supply of 4 V at 15 mA of current. (Most headphones will tolerate higher voltages but the microphone circuit can be damaged if excess voltage is applied directly to the headset.) Current radio models can provide the needed bias voltage at the microphone jack (ICOM and Kenwood provide 8 V, Ten-Tec 9 V, Yaesu 5 V, and so on).

Nominal audio output is 400 mV into a 150-500 Ω load. A PL068 microphone plug is standard and requires a 0.210 inch three conductor jack. The tip contact is PTT (push-to-talk) and the ring contact carries the audio. A typical amateur transceiver specifies the mic input impedance at 200-10 kΩ, 600 Ω nominal.

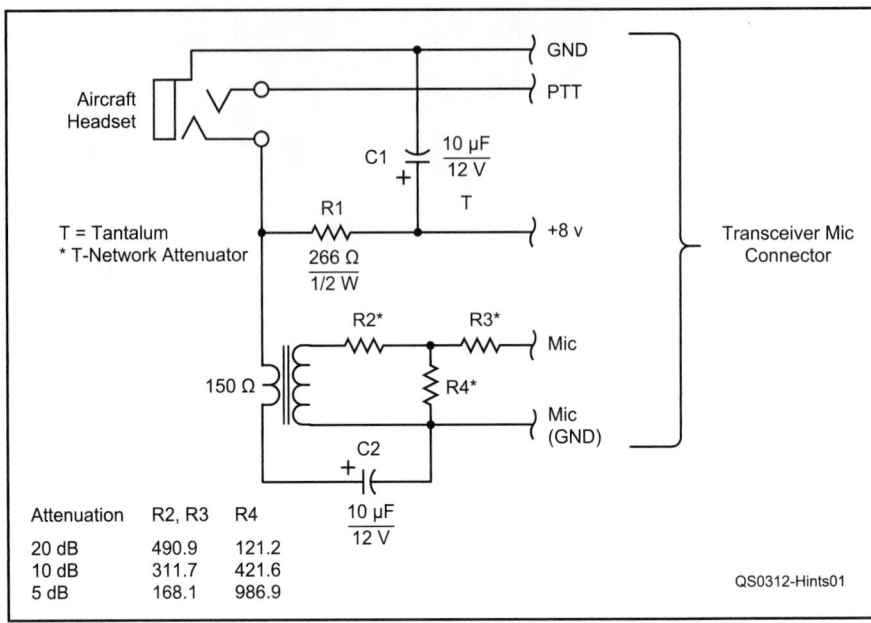

Attenuation	R2, R3	R4
20 dB	490.9	121.2
10 dB	311.7	421.6
5 dB	168.1	986.9

QS0312-Hints01

Fig 24.54 — K5ALQ's transformer-based aviation-headset interface circuit. C1 is optional, but its use is considered best practice. The headset requires a 150 Ω primary; the secondary must match the load presented by the radio and use appropriate attenuation.

TRANSFORMER-BASED ADAPTER

The adapter in Figure 24.54 provides proper bias to the headset microphone, removes dc from the primary of the transformer, matches the impedance of the headset output and provides appropriate attenuation for the necessary match of level and impedance to an amateur transceiver. The value of R1 is selected for a bias voltage of 8 V. To operate properly with 9 V bias from a radio or battery, R1 should be 330 Ω, ½ W. For 5 V bias, R1 should be 68 Ω.

A Digi-Key 237-1122ND transformer is used with 600 Ω/150 Ω primary and secondary windings. Connect the 150 Ω winding to the headset microphone and the 600 Ω secondary to the radio. The 600 Ω T-network 20 dB attenuator values for R2 and R3 are 490.9 Ω each, and R4 is 121.1 Ω. Since these are not standard values, use resistors with close tolerances, handpick the resistors, and combine parallel resistors to achieve these values. An alternative is to accept the attenuation yielded by standard-value resistors. Resistor values for other amounts of attenuation are given on the schematic.

RESISTIVE-ATTENUATOR ADAPTER

A circuit that uses resistive matching is shown in Figure 24.55 (the parts list is in **Table 24.10**). R1 provides the load for the microphone. R1 plus R2 supply the dc voltage from the radio (or battery). For Yaesu radios with 5 V, eliminate R2 and C1 and connect R1

Table 24.10

Aviation Headphone Adapter Parts List

C1	100 µF, 16 V (not needed for Yaesu)
C2, C3	10 µF, 16 V
C4	See Table 24.11
C5	0.001 µF, 50 V
J1	0.210 3-conductor jack
R1	470 Ω, ¼ W
R2	390 Ω, ¼ W (not needed for Yaesu)
R3	¼ W (see Table 24.11)
R4	620, ¼ W
R5	¼ W (see Table 24.11)
P1	Plug to match radio microphone jack
S1	SPDT miniature toggle switch
T1	1000 Ω center-tapped to 8 Ω audio transformer (RadioShack 273-1380 if needed, see text)

directly to the +5 V pin. C2 isolates the circuit from the microphone voltage and C3 isolates bias, if any, from the radio microphone pin (ie, ICOM radios). C5 reduces the chance of RFI via the bias supply. Aviation microphones are designed to be RFI resistant but a 0.001 µF capacitor across the circuit output to ground could be added if necessary.

R4 provides the load for the radio mic input circuit. The divider consisting of R3, R5, C4 and R4 reduces microphone audio output to an appropriate level. C4 creates a selectable "DX" rising frequency response. R5 reduces "normal" response by 2 dB to keep average

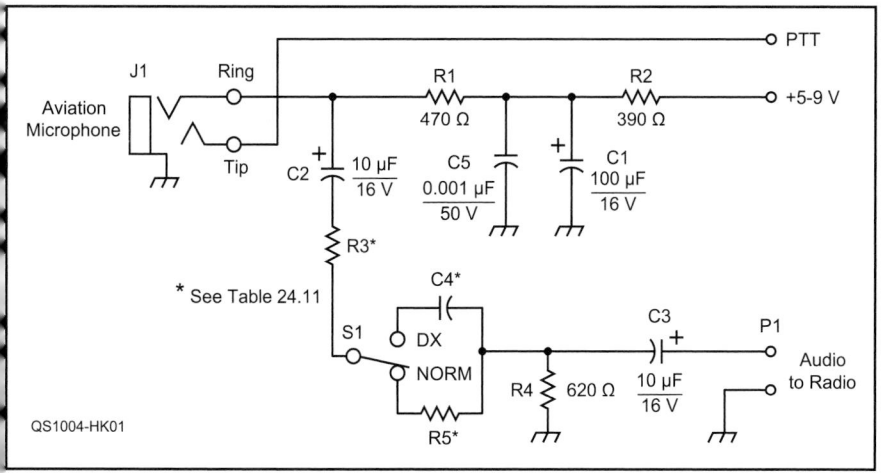

Fig 24.55 — Schematic of KØIZ's resistive-attenuator aviation headphones adapter. The parts list is given in Table 24.10.

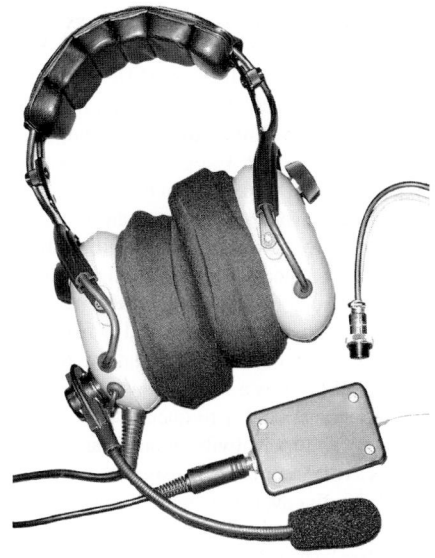

Fig 24.56 — A completed KØIZ headset/adapter. The black box contains the adapter circuit for the aviation headset.

Table 24.11

Aviation Headphone Adapter Microphone Equivalents

Microphone Equivalent Output	R3	R5	C4
Heil HC-5 (−68 dB)	18 k	3.5 k	0.01 µF
Dynamic microphone (−62 dB)	9.1 k	1.8 k	0.02 µF
Electret (−54 dB)	3.5 k	680	0.05 µF
Older ICOM radios (−46 dB)	1.5 k	300	0.12 µF

output similar to that of the "DX" response.

By selecting different values for R3 (and related R5 and C4) we can adjust the microphone output to match just about any microphone/radio combination. **Table 24.11** lists values to match a popular Heil cartridge, other dynamic and electret microphones (as used by ICOM, Yaesu, Kenwood, etc). Values are also

shown for older ICOM radios such as the IC-745/751/781 and early 706s, which require even higher output. Further adjustments to these values may be required to match radios that require different input levels.

The headset microphone jack is a Switchcraft C12B or S12B. The PTT connection can be ignored if not needed. You will also

need the appropriate plug for your radio's microphone jack. A RadioShack 1 × 2 × 3 inch plastic box (270-1801) will accommodate the components. If you also need to include a 9 V battery, a larger 270-1802 box will work.

A spectrum analysis of a SoftComm C-20 headset with this circuit shows the frequency response is flat through 3200 Hz, dropping off rapidly above that frequency. The "DX" frequency response is similar to Heil HC-4 cartridges.

24.14 An Audio Intelligibility Enhancer

Here's a simple audio processor (**Fig 24.57**) described by Hal Kennedy, N4GG, that can improve intelligibility, particularly for those with degraded high-frequency hearing. It might just give you an edge in that next DX pileup.

What we understand (intelligibility) of what we hear or detect is strongly affected by the end-to-end frequency response of a given communications circuit. Our equipment typically has a fixed audio bandwidth spanning about 300 Hz to 2700 Hz for SSB operation, but our hearing varies widely from operator to operator and it sometimes varies widely for an individual over the course of his or her life.

Speech intelligibility has been studied extensively since the 1940s. In brief, the "human speech signal" can be thought of as being made up of vowel and consonant sounds. The vowel sounds are low in frequency, with fundamental frequencies typically between

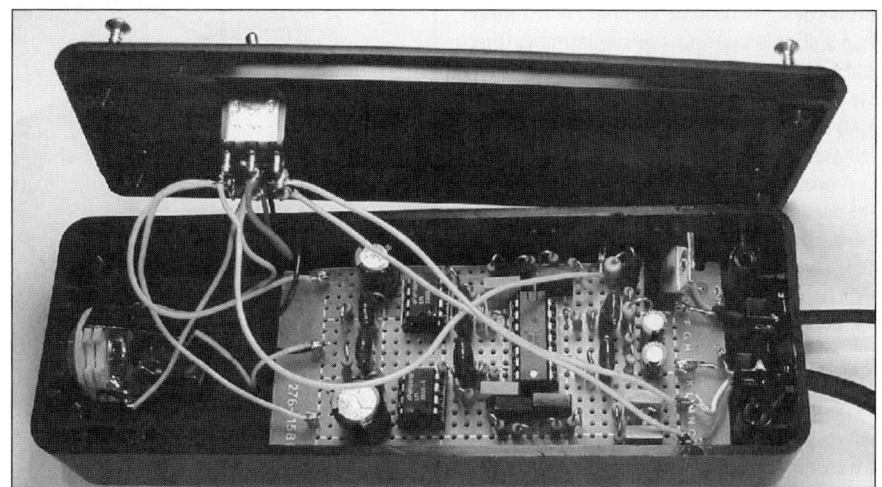

Fig 24.57 — The audio intelligibility enhancer is a compact audio processor that can help to compensate for degraded high-frequency hearing.

100 Hz and 400 Hz, and harmonics as high as 2 kHz. Consonant sounds occupy higher frequencies — typically from 2 kHz to as high as 9 kHz. (See "Speech Intelligibility Papers" at **www.meyersound.com/support/papers/speech/**). In spoken English, vowel sounds contain most of the energy in the speech signal, while consonant sounds are short, noise-like, and convey the majority of intelligibility. When consonants are not heard well, it sounds like the speaker is mumbling. It becomes difficult to distinguish *F* from *S* and *D* from *T*.

Frequently, hearing degradation manifests itself both as a loss of sensitivity over all frequencies and a frequency dependent loss where the sensitivity of our hearing degrades as the frequency increases. Sensitivity loss that is flat over frequency does not affect intelligibility, assuming that levels are high enough to be heard. A simple loss of sensitivity can be addressed by turning up the AF gain! Unfortunately, turning up the gain does not compensate for frequency-dependent loss, and frequency-dependent loss does degrade intelligibility by reducing the ability to hear the consonants in human speech. A quick Web search for "human hearing" will yield a multitude of sites describing the basic concepts.

The author's audiology test results are shown in **Fig 24.58**. Audiologists refer to this pattern as *age induced hearing loss* — a roll-off of hearing sensitivity as frequency increases. Fortunately, intelligibility can be improved significantly for those with this common condition, by simply adding amplification that increases as a function of frequency.

Hearing deficiencies can be much more complex than the typical high frequency loss associated with aging. One's hearing can include notches and/or peaking as a function of frequency, as well as significant mismatch between the left and right ears. Most of these can be addressed through use of audio circuits whose gain approximates the inverse function of the hearing problem(s). Analog circuit designs become complex when more than a simple high-pass or single-notch filter is required. Modern high-end hearing aids use DSP circuits that break the audio spectrum into narrow frequency bands, each with programmable gain.

A professional audiology exam and threshold sensitivity test results are recommended as the starting point for determining how much, if any, audio shaping might improve your ham radio activities.

A VARIETY OF PRACTICAL OPTIONS

The author's operating interests require stereo operation to make significant use of a transceiver's dual receiver function when chasing DX, and to operate two radios simultaneously in contests. In each case operation is in stereo, with each ear dedicated to one of

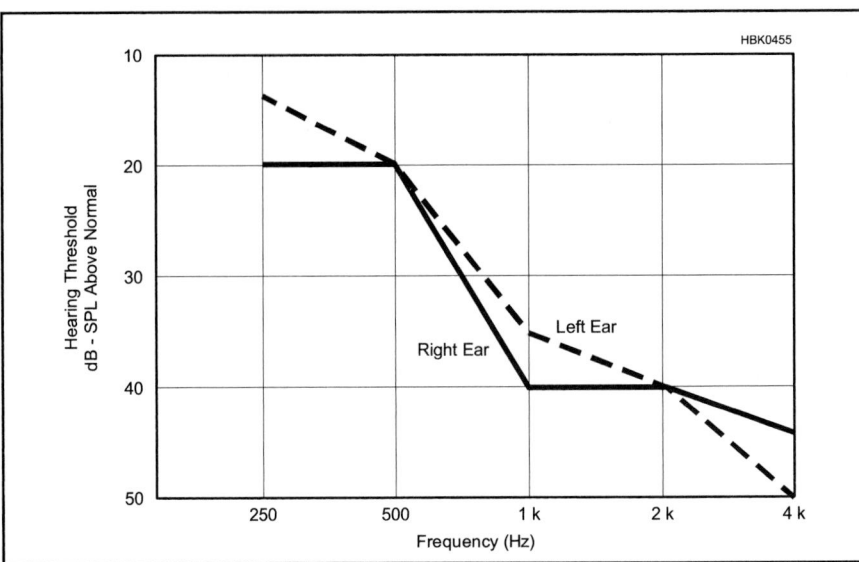

Fig 24.58 — The author's audiology exam results. Threshold sensitivity loss as a function of frequency is commonly associated with aging.

the two receivers. Monaural designs also have a potential drawback, in that the frequency response adjustments apply equally to both ears. In some cases the frequency response of each ear is different.

Stereo equalizers have been around for decades. The most basic are four-band graphic equalizers; they have adjustments to provide up to about 12 dB of gain or attenuation across frequency bands that are centered at four frequencies, each frequency typically one octave

above the next (an octave is a doubling of frequency). Seven, 10, 12 and 15 band equalizers are readily available, allowing for finer tailoring of amplification characteristics as the number of bands go up and the frequency range for each band gets smaller.

Parametric equalizers are also available — these differ from graphic equalizers in that the center frequency and bandwidth of each filter element is adjustable. Prices are very reasonable.

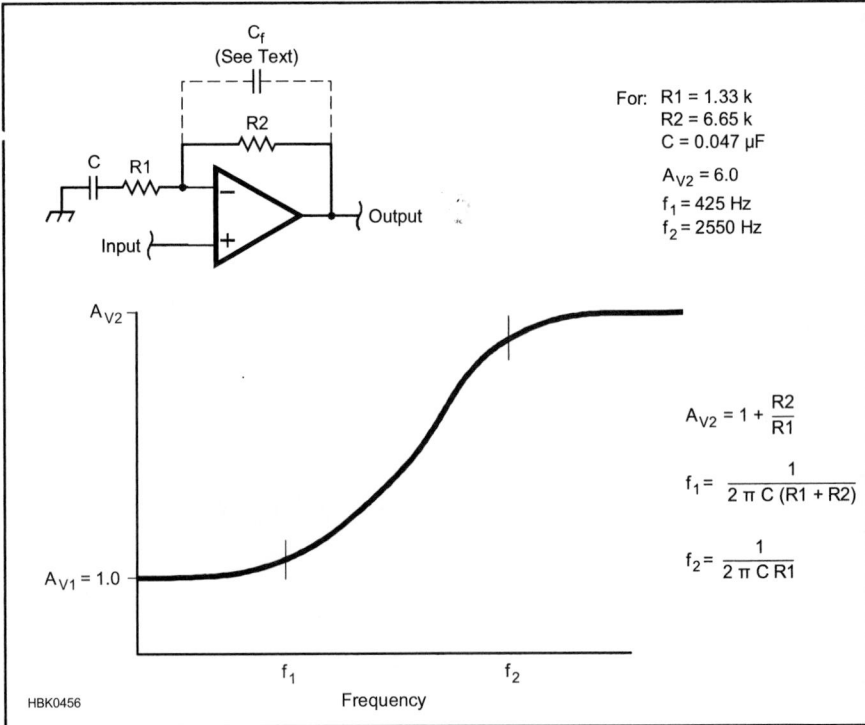

For: R1 = 1.33 k
R2 = 6.65 k
C = 0.047 µF

$A_{V2} = 6.0$
$f_1 = 425$ Hz
$f_2 = 2550$ Hz

$$A_{V2} = 1 + \frac{R2}{R1}$$

$$f_1 = \frac{1}{2\pi C (R1 + R2)}$$

$$f_2 = \frac{1}{2\pi C\, R1}$$

Fig 24.59 — Schematic and frequency response for a high-pass audio shelving filter.

An equalizer is a near-perfect solution to compensating for hearing deficiencies, and the ability to readily adjust gain at multiple frequencies allows optimizing via experimentation. Many equalizers are strictly low-level devices and lack sufficient output power to directly drive headphones, requiring an amplifier if the equalizer is to be inserted between rig and headphones.

Some transceivers include adjustable audio DSP circuits for receive as well as transmit audio. It may be possible to adjust the audio response of your transceiver to improve intelligibility without the use of an external device.

CIRCUIT DESCRIPTION

The heart of this audio intelligibility enhancer (AIE) is what audiophiles refer to as a *high-pass shelving filter*. Shelving filters are sometimes used in high-end audio systems to compensate for deficiencies in crossover networks or loudspeakers. The basic circuit and its frequency response are shown in **Fig 24.59**. Note that the shape of this response curve is the inverse of the hearing response of Fig 24.58. Referring to Fig 24.59, the circuit provides a gain, A_{V1}, equal to 1.0 for frequen-

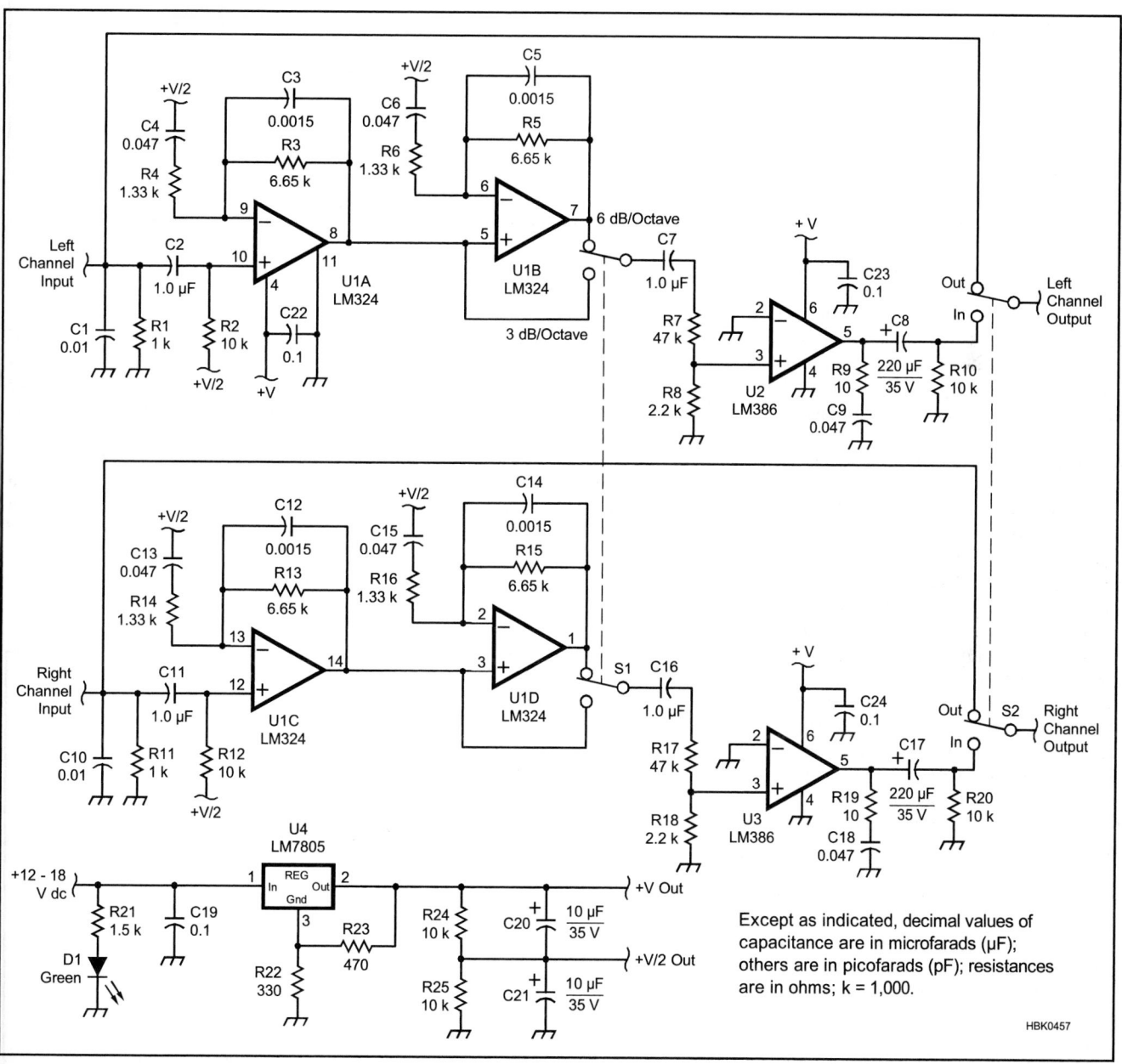

Fig 24.60 — The audio intelligibility enhancer schematic. Resistors are ¼ W. Parts are available from RadioShack, (www.radioshack.com) and Mouser Electronics (www.mouser.com) as well as many other suppliers. The part numbers shown in parentheses are RadioShack unless otherwise noted.

C1, C10 — 0.01 µF, 500 V ceramic (272-131).
C2, C7, C11, C16 — 1.0 µF, 250 V metal film (272-1055).
C3, C5, C12, C14 — 0.0015 µF, 100 V ceramic (Mouser 80-CK12BX152K).
C4, C6, C9, C13, C15, C18 — 0.047 µF, 50 V, metal film (272-1068).

C8, C17 — 220 µF, 35 V, electrolytic (272-1029).
C19, C22, C23, C24 — 0.1 µF, 50 V ceramic (272-109).
C20, C21 — 10 µF, 35 V, electrolytic (272-1025).
D1 — LED with holder, green (276-069).
S1, S2 — Miniature DPDT (275-626).

U1 — LM324 quad operational amplifier (276-1711).
U2, U3 — LM386 audio amplifier (276-1731).
U4 — LM7805, +5 V regulator (276-1770).
Project box, 6×2×1 in (270-1804) (see text).
Grid-style PC board, 0.1-in centers (276-158) (see text).

cies somewhat below f_1 and a higher gain, A_{V2}, that is given by $A_{V2} = 1 + R2/R1$ for frequencies somewhat above f_2. Frequency f_1 is given by $f_1 = 1/[2\pi C(R1+R2)]$, and frequency $f2 = 1/(2\pi CR1)$. Conveniently available component values of R1 = 1.33 kΩ, R2 = 6.65 kΩ, and C = 0.047 µF yield a high frequency gain, $A_{V2} = 6.0$, and cut-on and cut-off frequencies of $f_1 = 425$ Hz, and $f_2 = 2550$ Hz. With these values, the circuit gain rises at an average of 3 dB per octave from 300 Hz to 2700 Hz — the frequency range for SSB transmission. For those interested in modifying the design, or using a shelving filter for other purposes, it's important to note that rearranging the equations yields: $f_2/f_1 = A_{V2}$. The ratio of the two break frequencies always equals the maximum gain.

Fig 24.59 also shows a feedback capacitor, C_f, which, while not contributing to the shelving filter function, is included as good design practice. The LM324 op amps used to implement the filters have gain-bandwidth products in excess of 1 MHz. Operating them with gain at frequencies above the useful limits of our hearing (and all the way out to beyond 1 MHz) is an open invitation for RFI problems and unwanted high frequency noise. So, feedback capacitor C_f provides high frequency gain roll-off above the frequency $f = 1/(2\pi C_f R2)$ which for the value of C_f in this design (0.0015 µF) is 16.8 kHz.

Fig 24.60 is the complete schematic for the AIE. Two shelving filters, as described above, are used in series in each audio channel. Two filters in series provide sufficient gain change versus frequency to approximately match the author's high frequency hearing loss, which has a slope of about –7.5 dB/octave across the audiologist's measuring points from 250 Hz to 4 kHz. Two filters in series provide +6 dB/octave gain change from 300 Hz to 2700 Hz and 31 dB of total gain ($A_{Vtotal} = A_{V2} \times A_{V2} = 6 \times 6 = 36$ or about 31 dB). Some tips for tailoring the circuit to your hearing are included in the next section.

The shelving filters are followed by LM386 audio amplifiers (U2, U3). These provide sufficient output power to drive headphones or small speakers. LM386s have a built-in voltage gain of 20, so they are preceded by 20:1 voltage dividers (R7+R8, R17+R18) to maintain an end-to-end gain of 1 at low frequencies.

Voltage regulator U4 provides regulated and filtered dc voltage to the shelving filters and the output stages. Current draw for the AIE is approximately 35 mA from a +13.6 V dc source, if the LED pilot light (D1) is included. D1 draws 10 mA by itself, however, and R21 and D1 can be deleted if current consumption is a concern. Supply voltage is not critical — anything from 12 to 18 V dc will work.

Capacitors C1 and C10 are used to prevent RFI, by bypassing to ground RF that may

Fig 24.61 — A close-up of the AIE before mounting in the project box. It all fits neatly within a 6×2×1 inch enclosure if you have the patience to build within tight spaces.

enter at the inputs. The original AIE without these capacitors suffered interference from loud distorted sounds (unrectified SSB RF) during transmit. C1 and C10 eliminated RFI problems, but they could be insufficient in the presence of large RF fields. If necessary, additional RF filtering can be achieved by adding RF chokes in series with both audio input and output lines.

TAILORING THE DESIGN TO YOUR HEARING

The AIE design presented here is intended to address high frequency hearing loss, and not other hearing deficiencies. This AIE design does, however, have a reasonable amount of flexibility for tailoring to an individual's hearing.

The design as it is presented in Fig 24.60 includes two shelving filters per channel and switch S1. S1 allows selection of one (3 dB/octave amplification) or two (6 dB/octave amplification) filter stages. If you are the only operator of your station and you know your hearing deficiency is less than about 5 dB/octave, the switch and the second filter in each channel can simply be eliminated. If you are unsure of how much high frequency amplification you might need, or in the case where more than one operator may be sharing a station, the switch and the second stages will prove useful. S1 could also be replaced with two switches — one for each channel — if your high frequency hearing is not similar in both ears.

For gain slopes other than 3 dB/octave or 6 dB/octave, one or both filters in either channel can be modified using the equations given earlier. Taking this further, one or both of the op amps in either channel do not have to be shelving filters — they can be used to

implement a variety of other filter functions including notching, peaking, or multi-pole functions. Op-amp filter design is covered extensively in circuit design handbooks and on the Internet.

In addition to tailoring the gain change versus frequency, the overall gain of each channel can be adjusted to compensate for the sensitivity differences between ears. Overall gain is adjusted by changing the voltage dividers R7+R8 and R17+R18. R8 and R18 can also be replaced with 5-kΩ potentiometers if adjustable gain is desired.

When considering changing the design, it's important to know that human speech recognition is remarkably tolerant of deviations from an ideal amplitude versus frequency response curve. Also, personal preference comes into play — you may hear more comfortably with less than complete compensation for high frequency hearing loss. Although 6 dB/octave compensation closely matches the author's hearing loss, he usually has the AIE switched to 3 dB/octave compensation. At this setting, intelligibility is improved considerably.

CONSTRUCTION

With a few exceptions, the layout of the component parts is not critical. The leads on the IC power decoupling capacitors (C22, C23, C24) and all leads to the op amp inputs should be kept as short as possible. Be sure to make the U1 power connections at pins 4 and 11.

The completed AIE is shown in **Figs 24.61** and **24.62**. The prototype is built on a grid style PC board with 0.1 inch centers — these are available from RadioShack — cut to fit snugly in a 6×2×1 inch plastic project box. Parts numbers for the PC board and the proj-

Fig 24.62 — The AIE takes up little desk space. It plugs inline between the transceiver and headphones and requires 13.6 V dc. In the author's shack, the unit is powered from a transceiver accessory jack and is always switched on.

ect box are included in the parts list. A PC board is available from FAR Circuits (**www. farcircuits.net**).

Two cautions are in order. First, building an AIE as small as the prototype requires small tools, a large magnifying glass and patience. Since the circuit design is not particularly sensitive to layout, consider using a larger layout and a larger enclosure. The second caution is with regard to using a plastic enclosure. Susceptibility to RFI and 60 Hz hum pickup will be greatly reduced by using a metal enclosure. If you have lots of RF in your shack you may have no choice but to use a metal enclosure.

Several of the components are not available from RadioShack, but they are readily available from mail order suppliers and can be found in a well-stocked junk box. Several substitutions can be made to yield an "all RadioShack" design. The 1.33-kΩ resistors can be made from a 1-kΩ and a 330 Ω resistor in series; the 6.65-kΩ resistors can be made from two 3.3-kΩ resistors in series, and the 1.5-kΩ resistor can be made from a

1-kΩ and a 470 Ω resistor in series. The value of the 0.0015 μF feedback capacitors is not critical (C3, C5, C12, C14) — 0.001 μF disk ceramics can be substituted (RadioShack 272-126).

HOW WELL DOES IT WORK?

Intelligibility is absolutely crucial to maintaining high QSO rates and correct logging in contests, and in pulling the weak ones out of the noise when DXing or operating in less-than-ideal conditions. The AIE has provided the author with an immense improvement in both activities. It's just as helpful for reducing listening fatigue during ragchewing or general listening as well.

A real surprise was how much intelligibility improved on CW. This may seem counterintuitive at first, if you view "continuous wave" as comprising a single frequency. From a communications standpoint, however, the CW waveform is not continuous — it is comprised of elements — dots and dashes. The leading and trailing edges of each element are made up of frequencies higher than the carrier frequency,

and hearing these helps the listener separate one element from the next. When the rise and/or fall times of the elements are very fast compared to the carrier frequency, the high frequency content of the waveform becomes sufficiently large to cause key clicks. Very slow rise and/or fall times produce the opposite of clicks — a sound that experienced CW operators describe as "mushy." Mushy code is very difficult to copy, particularly at high speeds, because the ability to decipher when an element has started or stopped has been reduced. Degraded high frequency hearing has the same detrimental effect on CW intelligibility as slowing down the waveform edges — it makes a properly formed CW waveform sound mushy. The AIE restores perception of the waveform edges in CW, just as it does for speech.

How much an AIE or any other approach to compensating for your hearing deficiencies will help you depends, to a large degree, on the state of your hearing. The more you have lost, the more you have to gain back. Audio shaping may also offer some intelligibility improvement if your hearing is within normal limits. Typical receiver bandwidths are narrower than the audio range needed to fully support intelligibility — some of this can be gained back with high frequency amplification.

One cautionary note from Jim Collins, N9FAY, a designer in the hearing aid industry. At low volume levels, a device such as the AIE is helpful in improving intelligibility. At high volume levels, the audio will start to sound "tinny." This happens because as sound levels increase, the corrective frequency curve must flatten. At upper sound levels of comfortable hearing, it becomes almost a flat line with unity gain, In other words, the hearing response is close to normal at high sound levels. If a device like the AIE is used for prolonged periods at high volume levels, it may accelerate hearing loss at upper frequencies. For more discussion, see Technical Correspondence in May 2005 *QST*, p 70.

24.15 An Audio Interface Unit for Field Day and Contesting

For Field Day and multioperator contest operation, it's sometimes desirable for a logger or observer to listen to a transceiver in addition to the main operator. The project shown in **Fig 24.63**, by John Raydo, KØIZ, makes two-person operation more efficient. Dual stereo headphone jacks, volume controls and an external speaker jack are provided, as well as jacks for both dynamic and electret microphones and an intercom feature for operator and logger to more easily communicate via headset mics.

There are times when the operator and logger need to talk to each other, for example to resolve a call-sign ambiguity. With this unit the operator can initiate two-way intercom communication by closing an intercom switch. The logger, however, can only initiate one-way intercom to the operator. This is so that the logger can't inadvertently cut off the operator during a contact. To respond, the operator closes his intercom switch. In both instances, radio volume is reduced for "talk-over" during intercom use.

Both the operator and logger need headsets with microphones. The unit has two intercom switches plus jacks for optional foot switches for hands-free operation. Here's how it works:

Operator communicating with the logger: The operator presses and holds the OPERATOR TALK switch (S3 on the unit or by using a foot switch connected to J4). This disconnects the operator mic from radio. Both mics are connected via an internal amplifier to both headphones. Radio volume is reduced (adjustable using the RADIO VOL control). Operator and logger can talk to each other.

Logger communicating with the operator: The logger presses and holds the LOGGER TALK switch (S4 on the unit or by using a foot switch connected to J5). Logger mic is connected via an internal amplifier to both headphones. Operator mic remains connected to radio and can still transmit on the radio. Radio volume is reduced. The operator can respond to the logger by pressing the OPERATOR TALK switch.

CIRCUIT DESCRIPTION

The circuit diagram shown in **Fig 24.64** is designed for use with dynamic microphones such as the HC4 and HC5 elements used in various Heil headsets, or for Heil iC type headsets that use an electret element. The electret mic jacks supply the required dc for that type microphone. Mic bias current is limited to less than 1 mA to avoid damage if a dynamic mic is accidentally plugged into the wrong jack. The higher electret output is reduced by resistors R11 and R28.

Both mics have preamplifiers with 34 dB gain set by the ratio of the resistors R4/R3 and R9/R8. The INTERCOM volume controls, R14 and R15, allow operator and logger to individually adjust their intercom volumes. The preamps feed a single 2-W intercom amplifier.

Closing the OPERATOR TALK switch transfers the operator mic from radio to intercom preamp and amplifier via K1 and then to the headphones via K2. Pressing the LOGGER TALK switch will also activate K2 (but not K1). During intercom use, audio from the radio passes through control R20 (RADIO VOL) at reduced volume. S2 (L+R) allows selection of both left and right sides of the headphones or only the right side, for the intercom. The operator and logger each have individual headphone PHONE VOL controls R24 and R25 (the operator control also incorporates the POWER switch, S1).

CONSTRUCTION

The unit shown is built in a PacTec plastic box. The front panel has the control switches and volume controls described in the previous section. The back panel (**Fig 24.65**) has ¼-inch stereo jacks for two pairs of headphones, two 3.5-mm mono jacks for headset microphones, 3.5-mm stereo and mono jacks for connection to the radio's headphone and mic jacks, one stereo 3.5-mm jack for a speaker, two 3.5-mm mono jacks for foot switches and a jack for a 9-V power supply (a wall wart is okay). A 12-V dc supply can be substituted if K1 and K2 are changed to 12-V units and the LED series resistors R22 and R23 are changed to 1 kΩ.

Other than controls and jacks, most components are mounted on a RadioShack prototype circuit board supported by four ½ inch spacers and shown in the box in **Fig 24.66**. A suggested parts layout is shown in **Fig 24.67**. Point-to-point wiring is used between circuit board pads and is not particularly critical. A PC board for this project has been developed by FAR Circuits (**www.farcircuits.net**).

Connections to both mic jacks and the radio mic jack use shielded cable to reduce hum and risk of interference pickup. Connect the shields on the operator and radio mic cables to a common ground point on the circuit board near U1. Bypass capacitors C8, C9 and C16 and resistor R6 are mounted at their respective jacks. Rub-on lettering, followed by a couple light coats of clear plastic spray, helps give a professional appearance.

Connecting an external speaker in addition to two headphones might be too much of a load for some radios and cause reduced volume and/or distortion. Try an inexpensive amplified computer speaker plugged into the audio interface external speaker jack to reduce the load.

You will need two cables to connect the unit to your radio. A 3.5 mm mono plug-to-plug shielded cable for the mic circuit is compatible with a Heil headset mic. A 3.5 mm stereo plug-to-plug cable (RadioShack 42-2387 plus 274-367 adapter) works for the earphones.

Fig 24.63 — Front view of the audio interface box. Operator controls are on the left and logger controls are on the right.

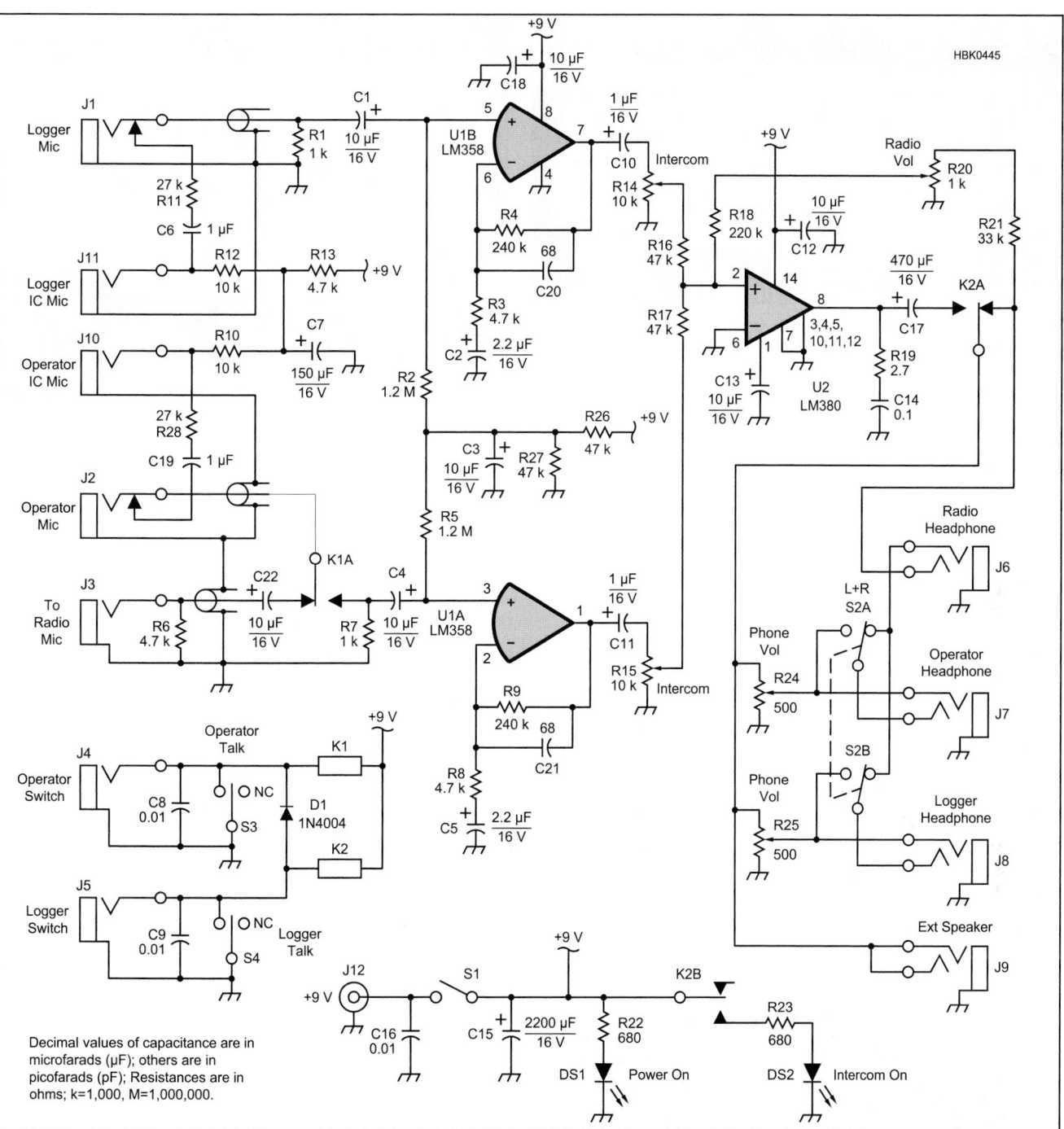

Fig 24.64 — Schematic diagram and parts list for the audio interface unit. Mouser parts are available at www.mouser.com, RadioShack parts at www.radioshack.com. Resistors are ⅛ W unless notes (¼ W are fine too.) A PC board is available from FAR Circuits (www.farcircuits.net).

C1, C3, C4, C12, C13, C18, C22 — 10 µF, 16 V electrolytic.
C2, C5 — 2.2 µF, 16 V electrolytic.
C6, C10, C11, C19 — 1 µF, 16 V electrolytic.
C7 — 150 µF, 16 V electrolytic.
C8, C9, C16 — 0.01 µF, 50 V ceramic.
C14 — 0.1 µF, 50 V ceramic.
C15 — 2200 pF, 16 V electrolytic.
C17 — 470 µF, 18 V electrolytic.
C20, C21—68 pF ceramic.
D1 — 1N4004.
DS1 — Red LED.
DS2 — Yellow LED.
J1-J5, J10, J11 — 3.5 mm mono jack, Mouser 16PJ135. Note: This is a

switched jack. The switch contact is not connected on J3-J5, J10 and J11.
J6, J9 — 3.5 mm stereo jack, Mouser 161-3402E.
J7, J8 — Three conductor ¼ inch jack, Mouser 502-12B.
J12 — 2.5 mm jack for dc input, Mouser 502-712A.
K1, K2 — DPDT relay, 9 V coil, Mouser 653-G6A-274P40-DC9.
R14, R15 — 10-kΩ potentiometer, Mouser 31JA401-F.
R20 — 1-kΩ potentiometer, Mouser 31JN301-F.
R24, S1 — 500-Ω potentiometer with SPST

switch, Mouser 31VM205-F.
R25 — 500-Ω, potentiometer, Mouser 31VA205-F.
S2 — DPDT toggle switch, Mouser 10TC260.
S3, S4 — SPDT momentary contact switch, center off, with long handle, Mouser 10TA445.
U1 — Dual op amp, LM358, Mouser 595-LM358P.
U2 — Audio amplifier, LM380 or NTE740A, Mouser 526-NTE740A.
Project box, 8 × 4.25 × 2.25 in, Mouser 616-72004-510-000.
Circuit board, RadioShack 276-158.

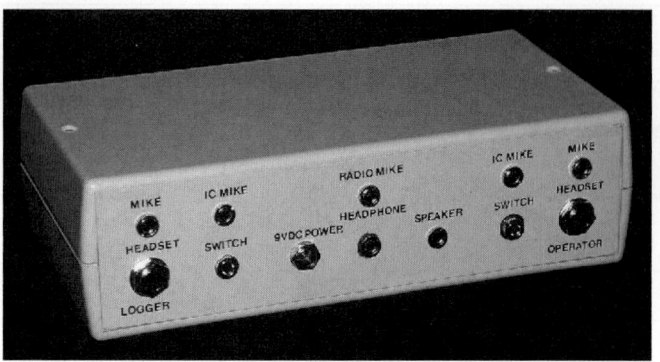

Fig 24.65 — Rear view of the audio interface box. Operator jacks are on the right, logger on the left. Connections to the radio are in the center.

Fig 24.66 — Inside view of the audio interface box. A RadioShack project board is used for most components.

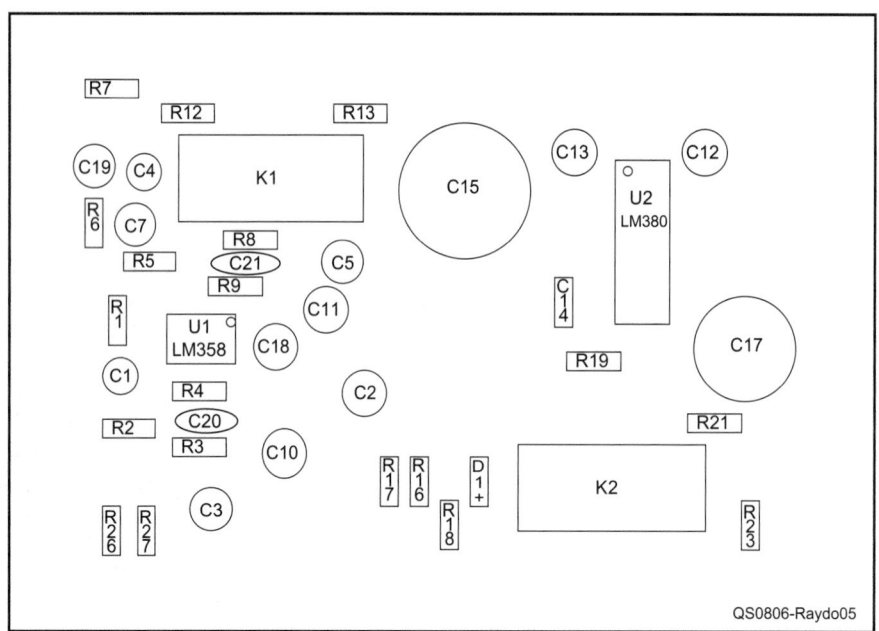

Fig 24.67 — Suggested parts layout for audio interface box.

24.16 Two QSK Controllers for Amplifiers

Many CW operators feel that QSK — or *full break-in* — operation is the most natural and conversational way to operate. "Full" break-in switches between transmit and receive so quickly that the operator can hear anything happening on the channel between individual code element dots and dashes. (*Semi break-in* uses a VOX-like circuit, remaining in transmit until a delay timer expires.) That sense of transparency enables the operator to respond immediately to interference or requests from other stations. Originally developed in support of high-speed CW traffic passing, QSK today is most often used by contest operators and for high-speed ragchews.

For low-power operation at 100 to 200 W, QSK circuits can use relatively small relays or solid-state PIN diode switches. However, when an amplifier is used, the requirements for voltage and current-handling go up dramatically. This leads, in turn, to the use of vacuum relays and high-power PIN diodes. *Hot-switching* — opening or closing mechanical contacts during current flow — greatly accelerates wear and can even destroy the relay contacts. Preventing hot-switching requires strict timing and interlock design to make sure the proper sequence of switching-then-transmitting occurs every time.

QSK controllers generally fall into one of two general types: hardware-only, in which the timing and logic of the sequential operations are built-in to a hard-wired circuit, and software, in which the switching circuits are under the control of a microprocessor. This pair of QSK controllers includes one from each camp: Jim Colville, W7RY, designed an analog controller and Paul Christensen, W9AC, programmed an Arduino processor for

his controller. You can choose to implement whichever project you feel most comfortable with. We'll begin with an overview of QSK operation by W7RY and then present both projects.

SAFETY

Should you decide to add QSK to your amplifier, you'll be working inside the amplifier. Safety is the top priority in working on any equipment that contains high voltage, such as a vacuum tube amplifier. The shorting straps and interlocks are there for your safety, please respect them and be very careful! Read and follow the recommendations of the High Voltage Techniques section of the **Power Sources** chapter and the **Safety** chapter of this book.

Also be sure to operate the amplifier with all shields in place. You don't want to exceed the MPE (maximum permissible exposure) of RF energy to your body. And you just never know when the cat is going to jump up on your workbench!

BASIC QSK PRINCIPLES

How is QSK different from PTT or semi-VOX operation in an amplifier? When switching is controlled by manual use of a PTT switch, the amplifier relays are usually energized long before the RF is output by the transmitter. With semi break-in, the receive/transmit transitions are usually between words or syllables rather than between dots and dashes.

Several things have to happen quite quickly with QSK, especially when you're sending CW at 30 WPM!

1. Switch the amplifying device bias from cutoff to the proper operating bias setting in less than 2 milliseconds.

2. Switch the RF relays from the bypass mode (in which the transceiver is connected directly to the antenna and the amplifier is biased OFF) to the transmit mode (in which the antenna is connected to the amplifier output and the transceiver is connected to the amplifier input). This must happen before RF starts flowing from the transceiver into the amplifier. (We are assuming the use of a modern transceiver for the remainder of this discussion. The same principles apply in switching separate receiver-transmitters.) **Fig 24.68** shows the basic relay wiring for a QSK system.

This design uses only a single-throw relay for routing transmitter output to the amplifier input. Using a double-throw relay with another contact for receive adds another opportunity for the contacts to become intermittent in the receive path. Henry Radio used the approach of Fig 24.68 for all of their later production runs of amplifiers. Relay contact materials

that are rated for transmitter output levels can become intermittent at the extremely low levels of received signals.

Basic Timing Considerations

When the key or keyer paddle is pressed, several things happen in the transceiver. First, the key or paddle state must be read by the rig's keyer or microprocessor. The transceiver then switches into transmit mode, which mutes the receiver, switches the amplifier keying circuit, and starts to output RF. Most modern transceivers have the option (usually a menu selection) for either a mechanical relay or FET amplifier keying circuit.

Proper QSK operation depends on the delay between when the transceiver's linear amplifier keying relay energizes (after contact bounce has stopped) and when the radio starts to transmit RF. This delay, which is easily measured, is the key element in control of linear amplifier antenna relay and bias switching. The author has measured many modern transceivers such as the

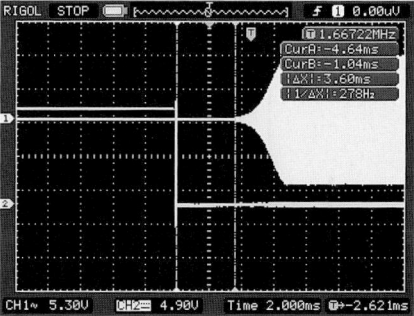

Fig 24.69 — The relay-close-to-RF timing for an ICOM IC-7200. RF output appears 3.6 ms after the amplifier relay contacts close. The relay-to-RF time can be increased 1.5-2.0 ms by substituting a small switching transistor for the relay.

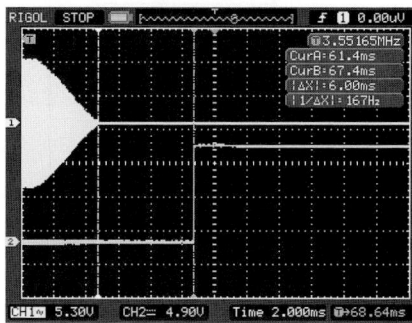

Fig 24.70 — The same ICOM IC-7200 timing from RF-stop to relay contacts open. There is 6 ms between the end of the RF output and the relays returning to receive.

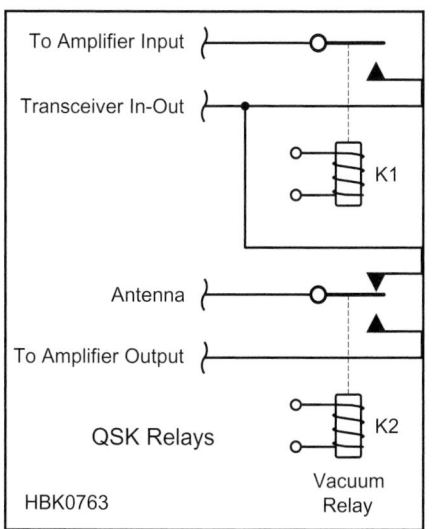

Fig 24.68 — Basic organization of relay switching for QSK.

HBK0763

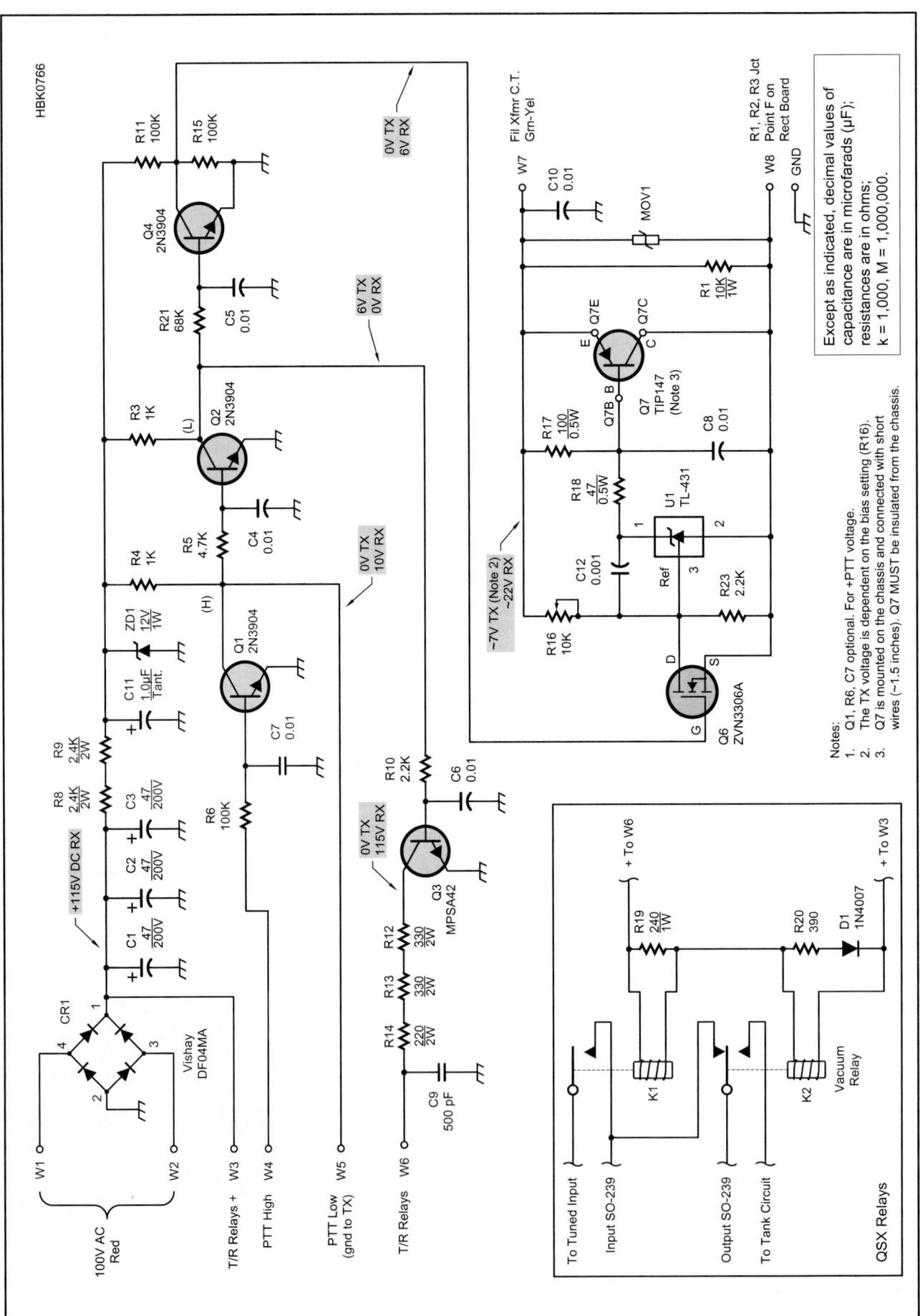

Fig 24.71 — The schematic for the SB-220 QSK controller. A PC board layout and parts list may be found on the *Handbook* CD-ROM.

ICOM IC-7600, IC-751A, IC-7200, IC-756 PRO-II, and IC-756 PRO-III. All of these transceivers typically have a delay from final contact closure to RF output of about 4 to 6 milliseconds, shown for the IC-7200 in **Fig 24.69**. (The IC-751A is actually adjustable with an internal control.) The timing in SSB and RTTY seems to follow the CW timing in most radios.

The same process works in reverse when going from transmit to receive. RF output stops, the amplifier switching circuit changes back to receive, and the receiver is un-muted. This delay is typically a bit longer and **Fig 24.70** shows the IC-7200 delay, measured at about 6 ms.

If PIN diodes are used for switching the RF, speed is generally not an issue as the diodes switch in microseconds. If electromechanical relays are used, however, the few milliseconds available when going from receive to transmit require a fast relay. Reed relays are typically used for low power, such as the output of the transceiver, and are fast enough to use for QSK. Higher power, high-speed vacuum relays must be used at the amplifier output. These relays open and close in around 1 ms and are rated for millions of open-close cycles.

All mechanical relays and switches have contact bounce because of flexing in the relay's moving armature. Contact bounce is clearly visible with an oscilloscope when measuring the time it takes for a relay to close or open. All of the measurements in this article include the contact bounce time until the contacts are ready for power. A more detailed discussion of modifications to speed up the relay circuits is available in the PDF article "W7RY QSK Relay Timing" also included on the *Handbook's* CD-ROM

Proper control of switching timing has benefits other than operational convenience. As W7RY relates, "After adding QSK to a Drake L-4, a Kenwood TL-922, and a Heathkit SB-200, the pesky arcing in the tuning capacitor stopped. The TL-922 was especially prone to it and the Heathkit SB-200 did it quite often when I was checking out the amplifier after I had purchased it. What happened?

"The arcing stopped because the amplifier was no longer hot-switching the relays and applying RF drive before the output of the amplifier was transferred to the antenna system! The Kenwood TL-922 has two huge, slow and loud antenna change-over and bias switching relays. With those s-l-o-w relays, the RF drive was getting to the tubes before the output relay contacts had switched! It was that simple. The Drake had arced in the tuning capacitor at least once. Now, it's smooth as silk. I used my TL-922 with QSK in the 2012 and 2013 ARRL RTTY Roundup contests and it worked perfectly! No arcs or sparks in the entire 30 hour contest on 10,

15 and 20 meters."

Note that properly designed QSK systems prevent hot-switching of the relays and possible arcing in the output tuning network whether full QSK is used or not. This will reduce wear of the relays and stress on the amplifier, leading to more reliable operation.

Tube Bias Control

In a vacuum tube linear amplifier, the QSK circuit must switch the tube's cathode-to-grid bias from cutoff in receive to a resting-current condition in preparation for transmit at the same rate as the antenna relays. The cutoff bias voltage must achieve the ZSAC (Zero Signal Anode Current, a.k.a. idling current) as stated by the tube manufacturer. For most 3-500Z amplifiers using ~3200 V, this voltage needs to be between 6 and 8 V.

Components and Construction for QSK

Placement of the relays in the amplifier isn't critical, but installing the relays near the input and output connectors is usually the most convenient and keeps leads short. Relay mounting isn't critical except for keeping them quiet. The author recommends mounting the relay on a rubber grommet through a hole or on a "bed" of silicone glue/sealant. Lead connection has a lot to do with keeping the relays quiet. The shield braid of RG-174 coax works well to absorb vibration. Just make sure you don't get the braid full of solder, or the extra stiffness defeats the purpose. Bend the braid into a fairly large horseshoe shape and solder to the relay terminals, and then solder

either to the RF connector directly or to the coax that connects the other circuitry.

Reed relays have glass-encapsulated contacts in a vacuum and a coil wound around the outside of that glass housing. The reed doesn't have to travel very far to make contact so the relay is extremely fast: closing is measured at less than 700 µs (0.7 ms) and opening slightly faster, on the order of 500 µs or less. A contact rating of 1 to 2 A is sufficient for 100 W or less of drive. Because of the speed, relay coil voltage is not critical. A resistor across the coil can compensate for higher or lower coil voltage and current. A 12 V coil reed relay is recommended.

Remember that when the amplifier is turned off or bypassed, the transmitter output RF doesn't pass through the reed relay. You can use a small mechanical relay, but they are slightly slower than a reed relay and are also noisier. Be cautious with the mounting and lead connections to keep the "sounding board" effect to a minimum.

The amplifier output vacuum relay must be of the smaller type such as the Jennings RJ1A, Kilovac HC1, Gigivac GH1, and the generic VHC-1. The Jennings RF1 style is just too slow, as are any of the glass-envelope vacuum relays. Do not use an HC3 or VHC-3 type of relay because they are rated only for dc, not RF.

The RJ1A relay's specified close time is 6 ms. Measurements show a typical closure time of about 3 ms. By using a higher initial voltage, the closure time can be reduced to as little as 1.5 ms, but the average time is around 2 ms.

Fig 24.72 — The S-QSK board showing the Arduino microprocessor and the LED error indicators. The RF SENSE input connector is at the lower left. [Paul Christensen, W9AC, photo]

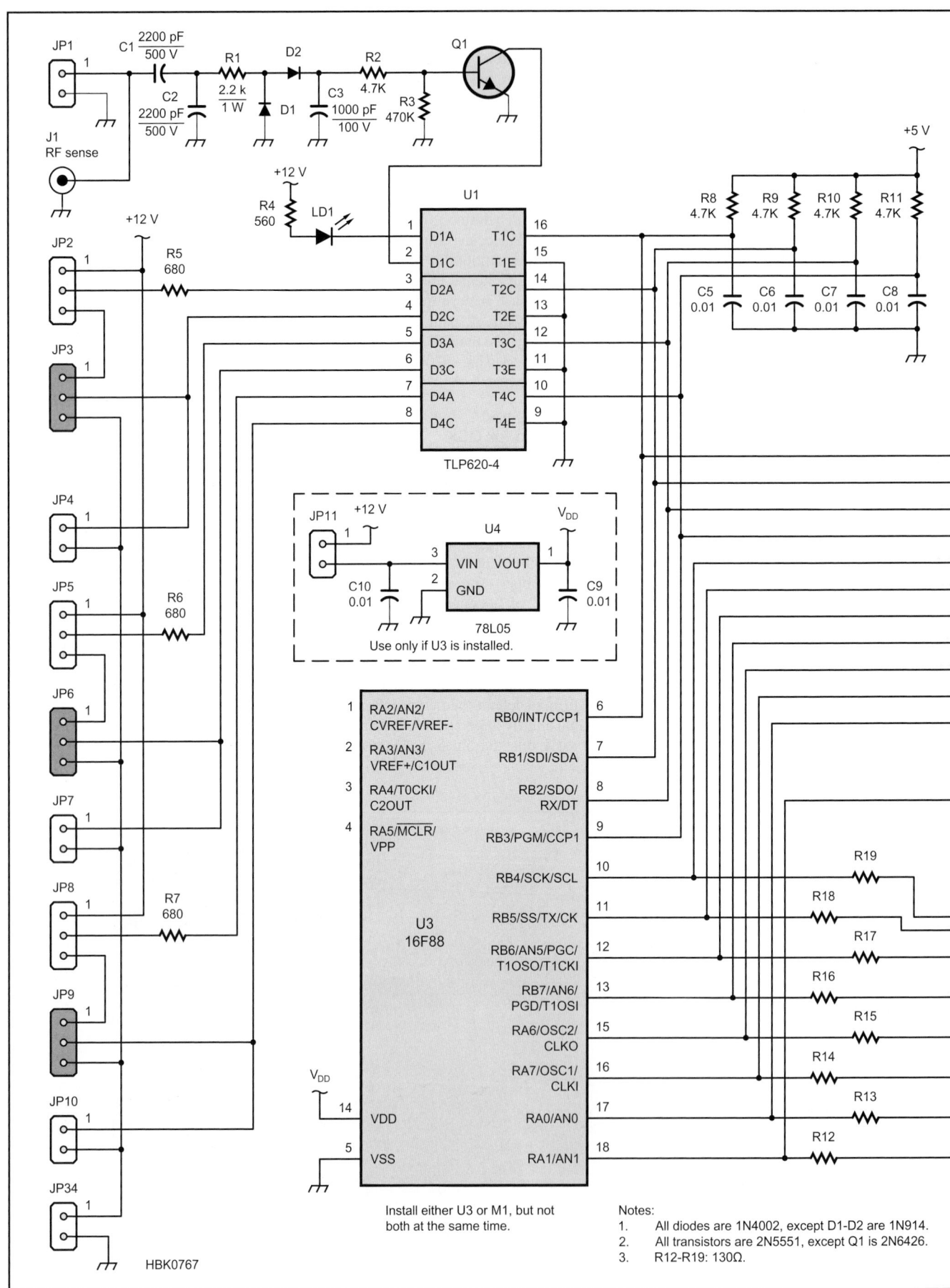

Fig 24.73 — The S-QSK schematic. A complete parts list is provided as a separate file on the book's CD-ROM.

Notes:
1. All diodes are 1N4002, except D1-D2 are 1N914.
2. All transistors are 2N5551, except Q1 is 2N6426.
3. R12-R19: 130Ω.

Install either U3 or M1, but not both at the same time.

Use only if U3 is installed.

HBK0767

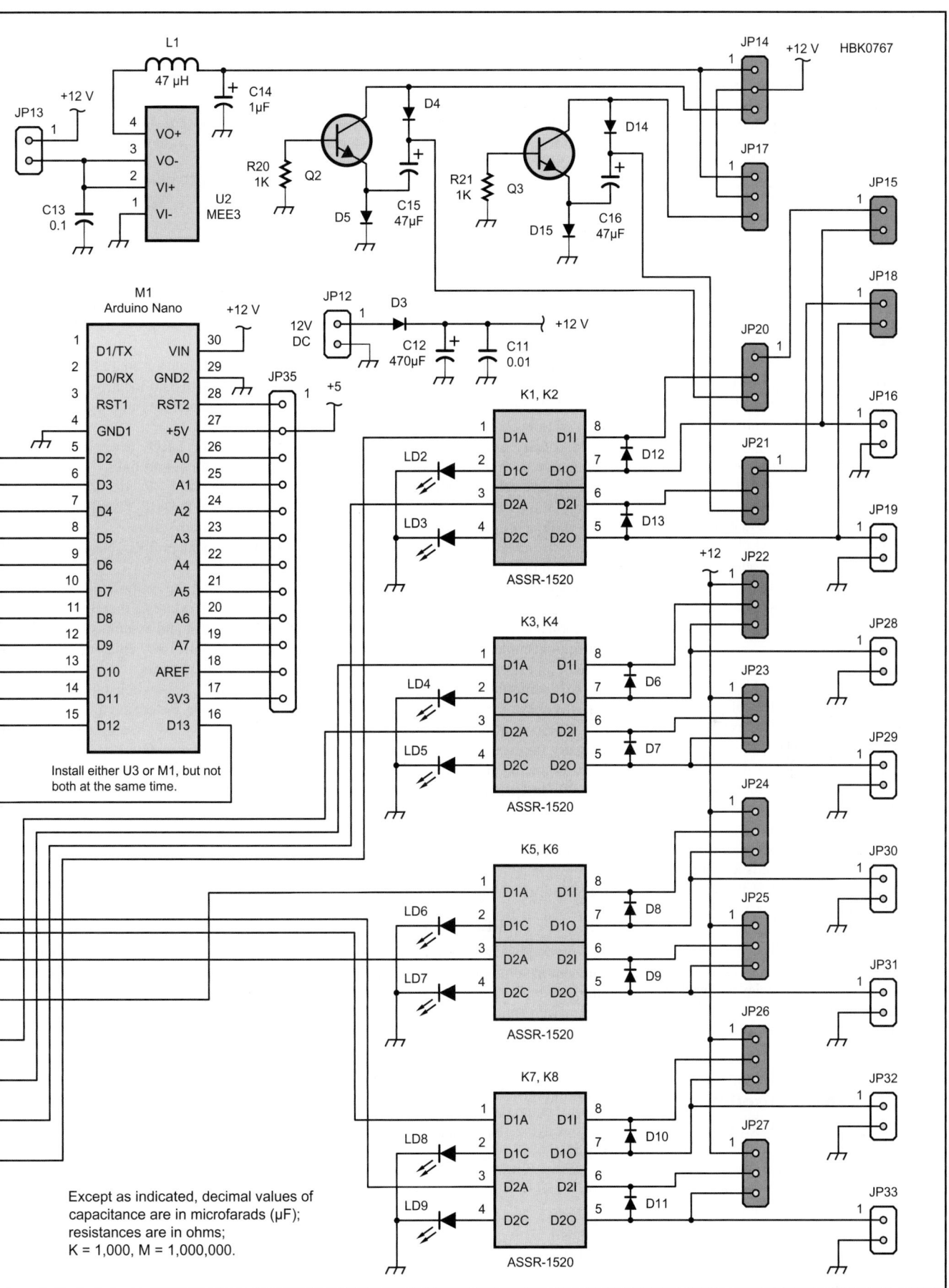

The relay's release time depends on the relay coil's kickback suppression circuit. The usual circuit is a reversed-biased diode connected across the coil. A resistor in series with the diode dissipates some of the energy in the coil and helps release the relay armature more quickly.

W7RY — Analog QSK Controllers

This project is really a series of controllers developed by W7RY for several popular non-QSK amplifier models. The SB-220 QSK controller shown in will be summarized as a design example — the schematic is shown in **Fig 24.71**. Additional construction details for the SB-220 controller are available on the CD-ROM accompanying this book. Design information for the Heathkit SB-200 and the Kenwood TL-922 QSK controllers are also included as separate packages of files.

The goal was to design a board that would be easy to install by anyone with electronics background with little to no support. The other requirement was little to no addition of major components to the amplifier. With the varieties of boards that the author designed, there is probably a board to fit most any amplifier. (See W7RY's web page **www. qrz.com/db/W7RY** for more information on these designs.)

As you may know, some older linear amplifiers used a fairly high voltage to switch the transmit-receive relay(s). Solid-state transceivers then had to use an external relay, which added even more delay to an already slow switching system, or install an aftermarket soft-key module into the amplifier.

The controllers use low voltage switching signals so that no additional power or interface signals are required. There is also a positive voltage input switching line if the available amplifier keying signal is positive-going. Some radios have a very fast positive-going keying signal that is even faster switching than the grounded amplifier relay switching signal. If the amplifier switching relay is a bit slow in your particular transceiver, you might be able to use a positive voltage to key the amplifier.

W9AC Universal Sequential QSK (S-QSK) Controller

The Universal S-QSK (S-QSK) board (**Fig 24.72**) is designed to control many RF switching applications, all controlled by an on-board Arduino Nano or PIC microcontroller. Through a choice of microcontroller programs, one S-QSK board can manage many different types of switching functions. Sample code, line-by-line documentation, and structured flowcharts are provided to assist the novice programmer with code customization. Structured logic is used for easy code modification. Copy and paste the source code into the Arduino software's edit window, then upload to the Nano board with a mini USB cable.

Presently, three programs have been written for S-QSK: an amplifier QSK controller; a TR switch for separate receiver-transmitter combinations, and a protection circuit for receiver inputs. (The latter two applications are discussed in the full version of this article on the CD-ROM.) When used as a QSK controller, S-QSK is designed to be used with the user's choice of external RF relays or a PIN diode switch. Circuit boards to support PIN diode TR switching and remote outputs that will interface with S-QSK's headers are being developed by the author.

The schematic of the S-QSK is shown in **Fig 24.73**. A complete parts list, PCB layout drawings, flow charts, and C++ source code are additionally provided on the CD-ROM accompanying this book. In addition, a double-sided professionally-made PCB is available from the author via his website at **www.qrz.com/db/w9ac**. Updates and revision information will also be made available on the same web page.

A powerful attribute of the Arduino microcontrollers is that they come pre-programmed with a bootloader, allowing the user to quickly upload new C++ code without the use of an external hardware programmer that is commonly needed for PIC and EEPROM chips. Connect a mini-USB cable to a PC USB port and you're ready to program. Optionally, the user can bypass the bootloader and program the microcontroller through the ICSP (In-Circuit Serial Programming) header.

The microcontroller and peripheral circuits are powered by the USB port during testing, and an external +7 to +12 V supply powers S-QSK during normal operation.

S-QSK FEATURE SET

A common hardware platform is used to implement many applications where precise timing is required between switching events. The hardware remains the same; only the source code changes for different applications.

• An Arduino Nano or PIC microcontroller plugs into the S-QSK motherboard. Programming a Nano requires no special hardware — code is uploaded via a USB cable from the host PC. For advanced programmers, a 16F88 PIC chip can be used in place of the Nano.

• Uses screw-down Phoenix-style I/O connectors. Each I/O connector can be unplugged from the motherboard for easy assembly and servicing.

• Four digital input channels; eight digital output channels. Header for access to optional analog channels.

• Optically isolated I/O for maximum RFI immunity. Photo-Darlington transistors on the input and solid-state PhotoMOS relays on the output.

• Each input channel can be selected for dry contact closure, fed from a solid-state open collector — or any other solid-state switching device.

• Each output channel can be selected to float or reference circuit ground. Each output can be jumper-selected to function as a current sink or current source.

• RF sampling via BNC connector or 2-pin header. RF is converted to a dc level and conditioned to drive photo-coupler. RF sensing activates with less than 100 mW of RF power.

• Remote RF sensor board can be used to sample RF at a distant location or where a T RF connection presents an unacceptable line mismatch.

• On-board LED diagnostic status indicators on all output lines.

• Uses a +12 V dc-dc converter, bootstrapped to the +12 V supply to provide +24 V for vacuum relays.

• Two optional, jumper-selected relay coil accelerators.

• Board can be populated with only the circuits of interest, thereby saving on construction cost and assembly time.

• Precise control and delay of all steps in 1 ms timing increments.

• Significant hot-switch fault protection built into the code. Before anything can switch between steps, RF is first sampled and judged with the state of the input key line.

• Small board size: S-QSK measures 4.5 × 3.5 inches. The remote RF sensor is 1.7 × 1.6 inches.

S-QSK OPERATION

S-QSK precisely sequences all critical timing elements between an RF power amplifier and transmitter/transceiver. The S-QSK board will work with QSK and non-QSK amplifiers. A sequenced electronic bias switching (EBS) system is created and supports both two-state and three-state bias control. Switching between types only requires a simple change to the microcontroller code.

The delay time between events is independently adjustable to accommodate various transmitter and amplifier timing characteristics. The input key and RF sensing lines are polled in a loop. Depending on the line states and the state of a flag bit, the S-QSK board's output key line, solid-state relays, and bias control are switched with time sequencing to avoid "hot-switch" effects. (See the flow chart files provided with the source code.)

S-QSK offers hot-switch RF protection by sampling the presence of the complete RF envelope. If RF excitation is present at the input to the S-QSK board before the input key line is active, a switch from receive to transmit is inhibited. Likewise, if the RF envelope has not decayed to zero after transmission, the S-QSK board will not switch back to receive. If either type of timing fault occurs, one of two LEDs will illuminate, showing a fault. The LEDs remain lighted until the timing fault clears. Each LED is pulsed to remain on for 0.5 second in the event of brief timing faults. Upon detection of a post-switch fault, bias lines are deactivated, providing further amplifier protection.

For three-stage EBS systems, "hang" bias is supported and is adjustable by the user from 0 ms to 255 ms in 1 ms increments.

In addition to establishing precise timing between amplifier switching elements, S-QSK can accelerate the relay activation time of frame-type output and input relays. The board contains two optional relay coil accelerators, each powered by a choice of +12 V or +24 V power buses. The accelerator circuits are engaged with header jumpers. For mechanical TR switching, Jennings RJ-1, Kilovac HC-1, and Gigavac GH-1 are all good choices for the RF output relay when they are powered at their rated supply voltage. For the input relay, consider the Aromat/Matsushita RSD-12V.

An RF input sample is capacitively coupled by C1 from the RF SENSE input connector J1 or JP1 header to a detector circuit consisting of R1-R3, C3, D1-D2 and Q1. RF sampling is optically isolated from the microcontroller digital inputs by a phototransistor in U1. A remote RF sampling board (under development as this edition was being prepared — see the author's website) can also be used in instances where a T RF connection presents an unacceptable line mismatch or where the sample point is at a long distance from the main S-QSK board. The detector circuit will indicate the presence of RF at less than 100 mW of power.

Contents

Chapter 25

Test Equipment and Measurements

This chapter, written by Alan Bloom, N1AL, covers test equipment and measurement techniques common to Amateur Radio. The initial sections discuss ac, dc and RF measurements and describe what to look for in commercially available test equipment. The concluding sections discuss transmitter and receiver performance testing, including the standard tests performed by the ARRL Lab. A selection of simple test equipment construction projects is presented as well.

Chapter 25 — CD-ROM Content

Supplemental Files
- Cathode ray tube theory
- Test and Measurement Bibliography

Project Files
- Gate Dip Oscillator articles and PCB artwork — Alan Bloom, N1AL
- Build a Return Loss Bridge — James Ford, N6JF
- Logic Probe — supporting photos and graphics — Alan Bloom, N1AL
- RF Power Meter — supporting files — William Kaune, W7IEQ
- Compensated RF Voltmeter articles — Sidney Cooper, K2QHE
- RF Sampler Construction details — Thomas Thompson, WØIVJ
- RF Step Attenuator — Denton Bramwell, K7OWJ
- Tandem Match articles — John Grebenkemper, KI6WX
- Transistor Tester — PCB artwork and layout graphics — Alan Bloom, N1AL
- Two-Tone Oscillator — PCB artwork and layout graphics — ARRL Lab

25.1 Introduction

According to the great physicist Sir William Thomson, Lord Kelvin, (1824-1907) "When you can measure what you are speaking about, and express it in numbers, you know something about it; but when you cannot measure it, when you cannot express it in numbers, your knowledge is of a meager and unsatisfactory kind." This is the purpose of electronic test and measurement, to express the characteristics and performance of electronic devices in numbers that can be observed, recorded and compared.

The amateur should undertake to master basic electronic and test instruments: the multimeter, oscilloscope, signal generator, RF power meter and SWR and impedance analyzers. This chapter, written by Alan Bloom, N1AL is the reader's guide to these instruments and the physical parameters they measure. The chapter goes beyond basic instrumentation to advanced instruments and measurements that are often encountered by amateurs. The material does not attempt to be complete in covering all available instruments and measurements. The goal of the chapter is to instruct and educate, present useful projects, and encourage the reader to understand more about these important facets of the radio art.

25.1.1 Measurement Standards

The measurement process involves evaluating the characteristic being tested using a *standard*, which is a rule for determining the proper numbers to assign to the measurement. In the past, instruments were calibrated to standards represented by physical objects. For example, the reference standard for length was a platinum bar, exactly 1.0 meters long, stored in an environmentally-controlled vault in Paris, France. In the latter part of the 20th century, most measurements were redefined based on fundamental physical constants. In 1983 the meter was defined as the distance traveled by light in free space in 1/299,792,458 second.

While most common everyday measurements in the United States still use the old Imperial system (feet, inches, pounds, gallons, and so on), electronic measurements are based on the international system of units, called *SI* after the French name *Système International d'Unités*, which is the modern, revised version of the metric system. The SI defines seven base units, which are length (meter, m), time (second, s), mass (kilogram, kg), temperature (kelvin, K), amount of substance (mole, mol), current (ampere, A), and light intensity (candela, cd). All other units are derived from those seven. For example, the unit of electric charge, the coulomb, is proportional to s × A and the volt has the dimensions $m^2 \times kg \times s^{-3} \times A^{-1}$.

In the United States, measurement standards are managed by the *National Institute of Standards and Technology* (NIST), a non-regulatory agency of the federal government which until 1988 was called the National Bureau of Standards (NBS).[1] (Notes may be found in the References section.) NIST's services include calibration of *transfer standards*. The transfer standards in turn are used to calibrate *working standards* which are used by companies to calibrate and measure their products. Such products are said to be *NIST traceable* if the rules and procedures specified by NIST have been followed. Most low-cost instruments used by hobbyists are another level down in accuracy, being calibrated by instruments that themselves may or may not be NIST traceable.

25.1.2 Measurement Accuracy

Accuracy and resolution are different specifications that are often confused. *Resolution* is the smallest distinguishable difference in a measured value. *Accuracy* is the maximum expected error in the measurement. For example an 8-digit frequency counter can measure the frequency of a 100 MHz signal to a resolution of 1 Hz, which is 0.01 ppm (parts per million). However, accuracy is determined by the time-base oscillator used as a reference, which typically would be significantly less accurate than 0.01 ppm in a low-cost instrument. Similarly, many digital voltmeters can display measurements to more digits of resolution than their accuracy permits. The extra resolution can be useful when comparing two or more values that differ only slightly because closely-spaced measurements tend to have nearly the same error. The difference can then be measured more accurately than the individual values.

An instrument's accuracy can be specified in absolute or relative terms or sometimes in both. An example of an absolute specification is an RF power meter with an accuracy of "5% of full scale." If the full-scale reading is 100 W, then the accuracy is plus or minus 5 W at all power levels. Theoretically a reading of 10 W could represent an actual power of 5 W to 15 W.

An example of a relative accuracy specification is an analog voltmeter with an accuracy of 3%. That means a voltage of 1.000 V can be measured to an accuracy of plus or minus 0.030 V. A voltage of 10.000 V can be measured to an accuracy of plus or minus 0.3 V. An example of a combined absolute and relative accuracy specification is a frequency counter with an accuracy of "1 ppm plus one count." When measuring a 100 kHz signal with 1 Hz resolution, one count is 10 ppm of the 100 kHz signal. The total accuracy is 1 + 10 = 11 ppm, or ±1.1 Hz.

An important point is that measurement error is not a mistake but rather a natural result of the imperfections inherent in any measurement. The sources of error can be sorted into several general classes. *Systematic error* is repeatable; it is always the same when the measurement is taken in the same way. An example is the inaccuracy of the voltage reference in a digital voltmeter that causes all measurements to be off by the same percentage in the same direction. *Random error* is caused by noise and results in a different measured value each time it is measured. Receiver sensitivity measurements involve measuring the signal-to-noise ratio of the audio output which varies due to the random fluctuations of the noise level.

Dynamic error results when the value being measured changes with time. The peak-envelope power (PEP) of a single-sideband transmitter varies with the modulation that is present at the particular time the measurement is taken. *Instrument insertion error* (also called *loading error*) is an often-overlooked factor. For example, a voltmeter must draw at least a little current from the circuit under test in order to perform the measurement, which can affect circuit operation. When performing high-frequency measurements, the capacitance of the measurement probe often can be significant.

25.2 DC Measurements

In discussions of instruments and measurements, the abbreviation *dc* generally refers to currents and voltages that remain stable during the measurement. If the dc current or voltage changes rapidly, it is said to have an *ac component* that may be measured separately.

25.2.1 Basic Vocabulary and Units

The **Electrical Fundamentals** chapter introduces basic electrical units of measurement and how they relate to circuit operation. The following material discusses additional points of particular relevance to test and measurement.

VOLTAGE

Another name for voltage is *potential difference*, because it is a measure of the difference in electrical potential between two points in a circuit. That is a very important point that is widely misunderstood among radio amateurs, technicians and even many engineers. It makes no sense to speak of the voltage "at" a particular point in a circuit unless you specify the reference point to which you are measuring the difference in potential. Normally the reference point is understood to be the chassis or circuit common. However, it is worth remembering that if the other lead of your voltmeter is connected to some other point in the circuit the readings will likely be different. Voltage is not a property of only a single circuit node, but rather is a measure of the *difference* in potential between two points. This is why voltmeters have two leads.

Unfortunately, the circuit common connection is conventionally called the "ground" which leads people to believe that it must be connected to the Earth for proper operation. In fact, the voltage with respect to Earth has no effect on the circuit as long as all the differences of potential within the circuit are correct. For example, a low-pass filter may need to be "grounded" to the chassis or circuit common connection for proper operation but it does not need to be "grounded" to the Earth.

The simplest way to calibrate an inexpensive meter is simply to use a more accurate meter. For example, when constructing a home-built power supply, the analog panel meter may be calibrated with a digital voltmeter (DVM) because most DVMs have better accuracy than an analog meter. Perhaps the most practical voltage reference for the home workshop is an integrated circuit voltage reference. Special circuit techniques in the IC are used to generate a very stable, low-temperature-coefficient reference based on the band-gap voltage of silicon, approximately 1.25 V, on the chip. Inexpensive devices with specified accuracy of 0.1% and better are available from companies such as Analog Devices, Linear Technology, Maxim and National Semiconductor.

CURRENT

Conventional current is the flow of positive charge and is considered to flow in the direction of positive to negative voltage, or from the higher voltage point in a circuit toward the lower voltage point. The flow of negatively-charged electrons in the opposite direction from negative to positive voltage is sometimes called *electronic current*. Electronic and conventional current are equivalent, but flow in opposite directions. Conventional current is used in most circuit design and is reflected in the direction of the arrow in the symbol for diodes and bipolar transistors. An *ammeter* is an instrument for measuring current. An ammeter may be calibrated by measuring the attractive force between two electromagnets carrying the current to be tested, but in practice it is easier to place a known resistor in series, measure the voltage drop across the resistor, and calculate the current using Ohm's law. Resistance and voltage can both be calibrated accurately, so that method gives good accuracy.

The most common analog ammeter type is the *D'Arsonval galvanometer* in which the pointer is attached to a rotating electromagnet mounted between the poles of a fixed permanent magnet. The modern form of this meter was invented by Edward Weston and uses two spiral springs to provide the restoring force for the pointer, providing good scale linearity and accuracy. Most commonly-available meter movements of this type have a full-scale deflection between about 50 µA and several mA.

Digital ammeters use an analog-to-digital converter to measure the voltage drop across a low-value resistor and scale the result to display as a value of current. They are generally more accurate than analog meters and are more rugged due to the lack of delicate moving parts.

RESISTANCE

Measuring resistance requires one or more precision resistors to use as a reference, either in a bridge circuit or as part of an ohmmeter as described below. Resistors with a 1% rating cost only a few pennies and are readily available in a wide variety of values from many suppliers. *Precision* resistors are generally considered to be those with a tolerance of 0.1% or less and are typically available for less than a dollar or so in small quantities. Expect to pay a few dollars each for parts with a 0.01% tolerance.

An *ohmmeter* is a meter designed for measuring resistance. Various circuits can be used, but most are variations of the simplified schematics in **Fig 25.1**. In the circuit at A, the battery is in series with the resistor under test. If the resistance is zero (the test leads are shorted) then the battery is in parallel with the voltmeter and it reads a maximum value. If

the resistance is infinite (test leads not connected) the meter reads zero. If the resistor equals R, the internal reference resistance, the meter reads half-scale.

In the circuit at B, the resistor under test is in parallel with the voltmeter. The meter indication is reversed from the series-connected circuit, that is, the meter reads zero for zero resistance, full-scale for infinite resistance, and mid-scale when the resistor equals R. In both circuits, an adjustment is normally provided to set the meter to full scale with the test leads shorted or open, as appropriate. In addition, a switch selects different values of the internal resistance R for measuring high or low-valued resistances.

A *Wheatstone bridge* is a method of measuring resistance that does not depend on the accuracy of the meter. Each arm of the bridge $(R_1 - R_S$ and $R_2 - R_X)$ forms a voltage divider. In **Fig 25.2**, the meter is a zero-center type so that it can read both positive and negative voltages between the center connections of the two voltage dividers. When variable resistor R_S in Fig 25.2B is adjusted for a zero reading on the meter, then the two arms of the bridge have the same ratio,

$$\frac{R_1}{R_S} = \frac{R_2}{R_X}$$

The resistor under test, R_X, can be calculated from

$$R_X = R_S \frac{R_2}{R_1}$$

Nowadays, the Wheatstone bridge is rarely used for measuring resistance since it is not as convenient as an ohmmeter, but the concept is important for several other types of measurement circuits that will be covered later.

25.2.2 Multimeters

A *multimeter* is probably the single most useful test equipment for an electronics experimenter. Besides measuring dc voltage and current, most also measure resistance and low-frequency ac voltage as well. Another common feature is a tone that sounds whenever the test leads are shorted, which is useful as a quick continuity tester. With most modern multimeters, the voltage they use for the resistance measurement is too low to forward-bias the junction of a silicon diode or transistor, which is a desirable characteristic for measuring in-circuit resistance. Many multimeters also have a special "diode" mode with a higher test voltage so that diodes and transistors can be tested as well. It is increasingly common for even low-cost digital multimeters to include functions such as frequency and capacitance measurement.

ANALOG MULTIMETERS

Nearly all analog multimeters use a D'Arsonval meter for the display. Compared to a digital display, analog meters have the advantage that it is easier to see whether the reading is increasing or decreasing as adjustments are made to the circuit under test. The experimenter may want to consider owning an inexpensive analog multimeter for that kind of testing as well as a higher-quality digital instrument for more precise measurements.

A *volt-ohm meter* (VOM) is a basic analog instrument that includes no electronic circuitry other than a switch and resistors to set the scale, a battery for the resistance-measuring circuit, and perhaps a diode to convert ac voltage to dc. Despite the name, most can also measure current as well as voltage and resistance.

A disadvantage of the VOM is that, when measuring voltage, the current to operate the meter must be drawn from the circuit under test. A figure of merit for a VOM it its *ohms-per-volt* (Ω/V) rating, which is just the reciprocal of the full-scale current of the meter movement. For example, if the VOM uses a meter that reads 50 µA full-scale then it has $1 / 50 \times 10^{-6} = 20,000$ ohms per volt. On the 1-V scale the meter has a resistance of 20 kΩ and on the 10-V scale, it is 200 kΩ.

Depending on the circuit being tested, drawing 50 µA may be enough to disrupt the measurement or the operation of the circuit. To solve that problem, some analog meters include a built-in amplifier with high input impedance. In the days of vacuum tubes, such meters were called *vacuum-tube volt-*

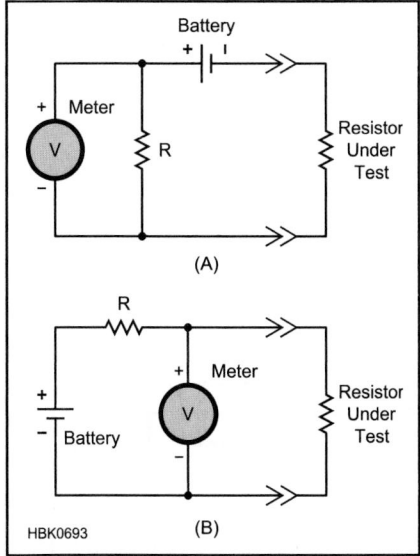

Fig 25.1 — Two ohmmeter circuits. At A, the meter reads full scale with a zero-ohm resistor and reads zero with no resistor connected. The circuit at B is the opposite; the meter reads full scale with no resistor connected (infinite resistance) and reads zero with zero resistance.

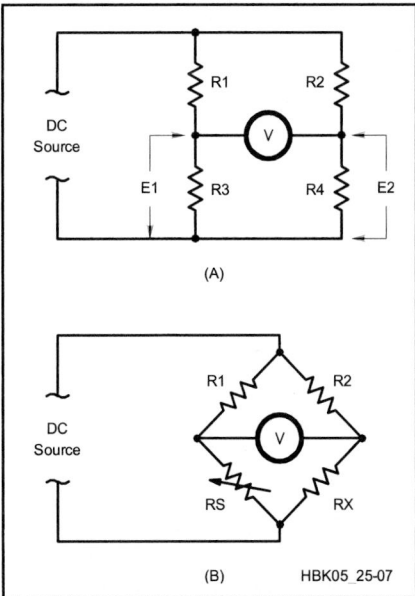

Fig 25.2 — A Wheatstone bridge circuit. A bridge circuit is actually a pair of voltage dividers (A). B shows how bridges are normally drawn.

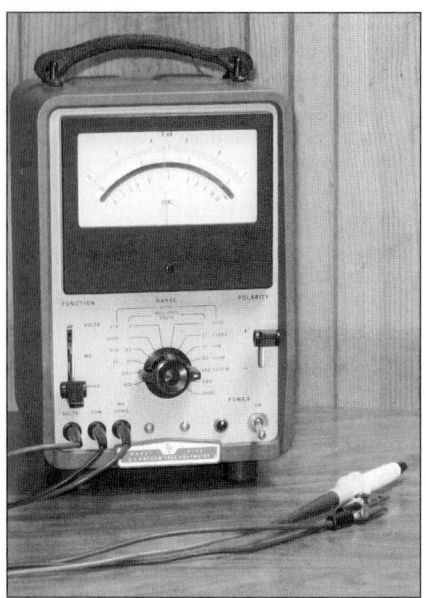

Fig 25.3 — This classic Hewlett-Packard HP412A vacuum-tube voltmeter (VTVM) has specifications that put many modern solid-state multimeters to shame.

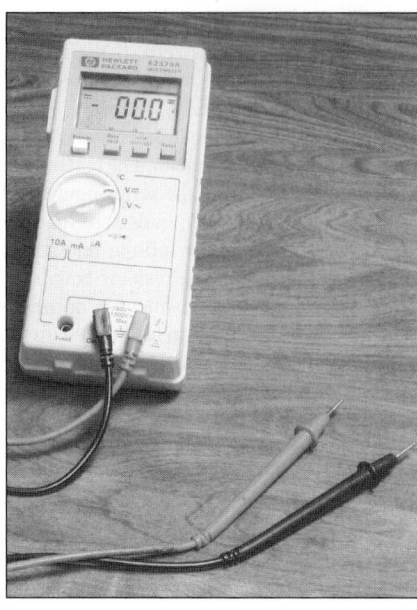

Fig 25.4 — A modern digital multimeter typically has a liquid crystal display readout.

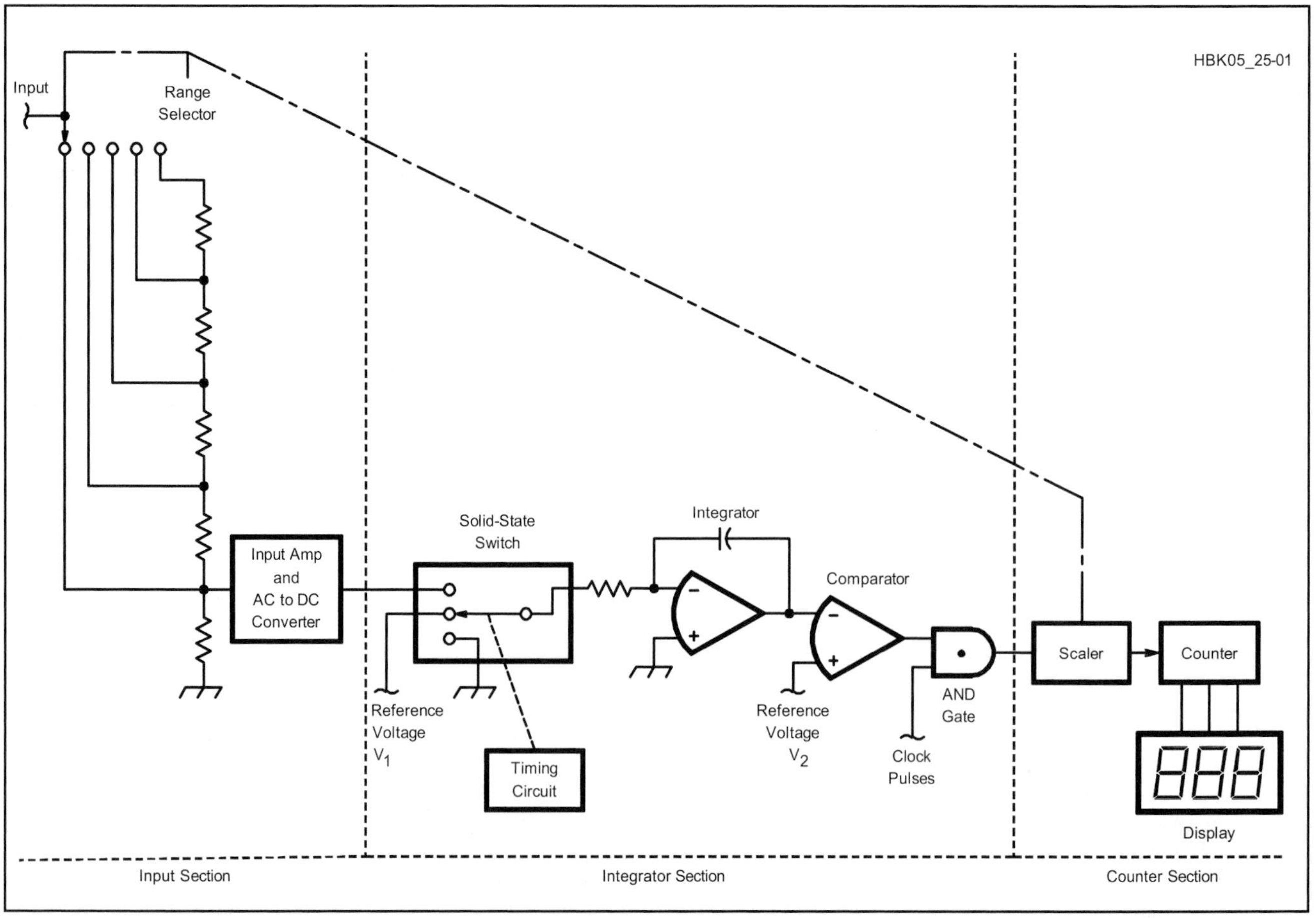

meters (VTVM). An example is shown in **Fig 25.3**. The modern equivalent is called an *electronic voltmeter* and generally uses an amplifier with field-effect transistors at the input. Again, despite the name, VTVMs and electronic voltmeters usually can measure current and resistance as well.

DIGITAL MULTIMETERS

Instead of using an analog meter to display its measurements, a *digital multimeter* (DMM) has a digital readout, usually an LCD display. A microprocessor controls the measurement process and the display. (See **Fig 25.4**.) To convert the analog voltage or current being measured to a digital number requires an analog-to-digital converter (ADC) controlled by a microprocessor. Most use a dual-slope type of ADC, which trades off a relatively slow measuring speed for excellent accuracy and low cost. (Analog-to-digital conversion is discussed in the **Analog Basics** chapter.)

A digital voltmeter is constructed as shown in **Fig 25.5**. The input section is the same as in an analog electronic voltmeter. A range

HBK05_25-01

Fig 25.5 — A typical digital voltmeter consists of three parts: an input section for scaling, an integrator to convert voltage to a pulse whose width varies with voltage, and a counter to measure the width of the pulse and display the measured voltage.

selector switch scales the input signal appropriately for amplification by an input preamplifier, which also includes a rectifier circuit that is used when in ac mode to convert the ac signal into dc, suitable for conversion by the ADC.

An additional feature that becomes possible with digital multimeters is *autoranging*. The selector switch is used only to choose between voltage, current, resistance and any other available functions. The scale or range is selected automatically based on the amplitude of the signal being measured, which is a nice convenience. If the signal is fluctuating such that it causes frequent range changes, there is usually a way to turn off auto-ranging and select the range manually.

Digital multimeters that feature a serial data interface can also act as a *data logger*, taking and storing measurements for use by a PC or under the control of a PC.[2] This is a very useful feature for experimenters.

HOW TO USE A MULTIMETER

When preparing to make a measurement, the first thing to do is to select the proper function on the multimeter — voltage, current, resistance or some other function. If the meter is not an auto-ranging type, set the range to the lowest value that does not over-range the meter. If you're not sure of the voltage or current being measured, start with a high range, connect the test probes then switch the range down until a good reading can be obtained. Choosing the correct function and range may involve connecting the test leads to specific connectors on the meter.

When measuring high voltage, special precautions must be taken. As little as 35 V should be considered dangerous because it can produce lethal current in the human body under some conditions. Grasp the test probes by the insulated handles, being careful to keep fingers away from the metal probe tips. Pay attention to the multimeter manufacturer's maximum voltage ratings. Special test probes are generally available for measuring high voltage. Do not exceed the meter's rated maximum voltage.

High current can be dangerous as well. If a probe accidentally shorts a power supply to ground, sparks can fly, damaging the equipment and endangering the operator. Be careful of metal jewelry such as rings and bracelets. If connected across a high-current circuit you could get a nasty burn. Most meters have a fuse to protect the instrument from an over-current condition in current-measuring mode. If the multimeter turns on but always reads zero, consult the manual on how to replace the fuse.

To measure a current, the meter must be inserted in series with the circuit. In a series-connected circuit, all components carry the same current, so it doesn't matter which component is disconnected to allow inserting the ammeter. Select the one that is most convenient, or the one that is at a low-voltage point if you're measuring a high-voltage circuit. For measuring ac power circuits without disconnecting them, clamp-on current probes are used.

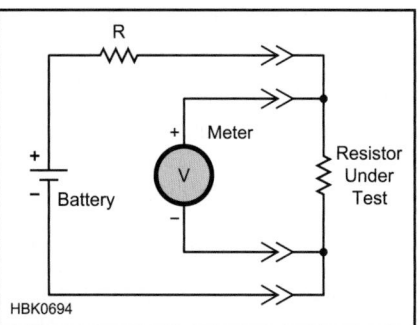

Fig 25.6 — The four-wire technique for measuring low-value resistors. By connecting the current source and the meter separately to the leads of the device under test, the error due to lead resistance is reduced.

It can be difficult to measure low-value resistors accurately because of the resistance of the meter leads, typically one or two tenths of an ohm. With the series-connected ohmmeter circuit described in the previous section, the calibration procedure compensates for that and some digital meters may have a means to compensate as well. However, accurate low-resistance measurement requires the *four-wire* technique, as illustrated in **Fig 25.6**. Two wires are connected to each end of the resistor, one to carry the test current and one to read the voltage with a high-input impedance voltmeter. In that way the resistance of the wires does not affect the measurement.

A similar technique can be used to measure high currents. Most inexpensive meters do not have a high-enough current range to measure the 20 A or so that is drawn from a 12-V power supply by a typical 100 W transceiver. The solution is to use an external *meter shunt*, which is a low value resistor placed in series with the current. See **Fig 25.7**. The multimeter reads the voltage, E, across the shunt, and then the current is calculated from Ohm's Law, $I = E/R_S$, where R_S is the resistance of the shunt. Resistors designed for this service may have four leads rather than two to allow a true four-wire measurement. The clamp-on current probes mentioned earlier may also be used.

There are several issues that apply specifically to analog multimeters. An analog meter movement is a rather delicate device. It can be damaged if the case is dropped or mechanically shocked. When transporting the instrument, it is a good idea to place a low resistance or short in parallel with the terminals of the meter movement, which increases damping and reduces the amount of pointer movement as the case is bounced around. Some multimeters have a special OFF position on the selector switch for that purpose. On others, switching to the highest current range accomplishes the same thing.

Most D'Arsonval meters have an adjustment screw located near the pointer's pivot point that may be accessed from the front of the meter. It should be adjusted so that the meter reads zero with no signal applied.

Some older VOMs include a high-voltage battery that is used on the highest resistance ranges. When testing solid-state devices, that voltage can be high enough to cause damage. If there is any doubt, use another high-impedance multimeter to test the voltage on the test leads of the VOM when it is set to the highest resistance ranges.

MULTIMETER CRITICAL SPECIFICATIONS

The first decision when selecting a multimeter is whether you want an analog or digital type. That is largely a matter of personal taste. Digital meters are generally more accurate but analog meters may make it easier to tune a circuit for a specific voltage or current. The next decision is between a hand-held or bench-type instrument. The latter tend to have more features and better specifications, but obviously are less portable and usually are more expensive.

The most obvious selection criteria are the features provided. Nearly all multimeters measure dc voltage, current and resistance and most also measure low-frequency ac voltage. Other common features on digital meters include autoranging and automatic turn-off to

Fig 25.7 — This 50-A, 50-mV current shunt has a resistance of 0.05 / 50 = 0.001 Ω. The two large terminals are for connecting to the circuit under test and the two small terminals are for connecting to a voltmeter.

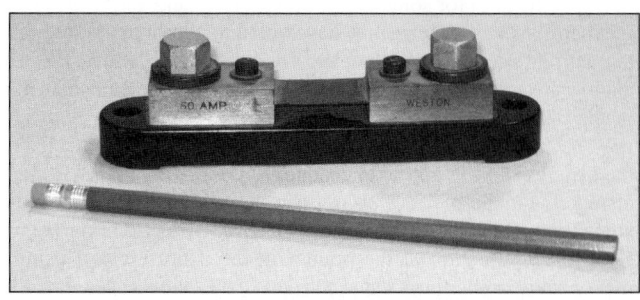

save the battery. More capable meters include features such as data hold, peak voltage, true RMS voltage, 4-wire resistance, capacitor tester, inductor tester, diode and transistor tester, logic tester, frequency counter, computer data logging and a graphical display.

One important feature that is often not specified is over-voltage and over-current protection. On analog meters, there may be internal back-to-back diodes across the meter movement to protect it from over-voltage. Most digital meters include autoranging which should prevent damage from any voltage below the specified maximum. The current-measuring input on both types is normally protected with a fuse.

The next most critical specification is the available measurement ranges of the voltage, current, resistance and any other functions provided. Analog meters tend to space the voltage and current ranges by a factor of three. For example, the full-scale voltage readings might be 3 V, 10 V, 30 V, 100 V, 300 V or perhaps 5 V, 15 V, 50 V, 150 V, 500 V. The latter is a particularly good choice since two voltages that are commonly measured, 12 V dc and 120 V ac, are near the top of a range where accuracy is best. The more range selections provided, the greater the span of values that can be measured. With auto-ranging instruments the measurement minimum and maximum range may not be obvious without consulting the manual.

The input impedance is important in minimizing the effect of the multimeter on the circuit under test. Impedance should be high for voltage measurements and low for current measurements. For VOMs the figure of merit for voltage measurements is the ohms-per-volt rating. Multiply the full-scale voltage range by ohms-per-volt to get the input resistance. Both digital and analog electronic voltmeters usually have the same input impedance on all voltage ranges, typically between 1 and 11 MΩ. The input resistance for current measurements is often specified by the *burden voltage*, which is the voltage drop across the test leads with a full-scale signal. Typical values vary widely, from a few millivolts to more than a volt. The voltage drop can often be reduced by switching to a higher current range, at the expense of measurement resolution.

Measurement accuracy can be an important specification for some applications. An inexpensive analog meter may have voltage and current accuracy of 3% or so. The best analog meters have accuracy specifications in the range of 1%. At that level, accuracy may be limited by *parallax*, which causes the meter reading to appear to change as you change the angle of view. To mitigate that, some high-end analog meters have a *mirror scale*. The reflection of the pointer in the mirror has a parallax error equal and opposite

to the unreflected pointer so that the correct reading is half way between the two.

You can't necessarily tell the accuracy of a digital meter from the number of digits in the display. Usually the accuracy is limited by the analog circuitry. However, the number of digits defines the resolution, which can be important when comparing two nearly-equal readings or when tuning a circuit for a peak or minimum value. A typical DMM might have a specified accuracy of 0.1% to 1% for dc measurements and perhaps 1% to 3% for ac voltage. (See the section on ac measurements.) Many inexpensive digital multimeters do not have published specifications and may not be very accurate.

Many bench-type multimeters and some hand-held units can be connected to a computer. That allows the computer to control the instrument to take automated readings and store the results in a computer file. Some older test equipment may have a GPIB (general purpose interface bus) interface, also known as IEEE-488 (HPIB on Hewlett-Packard equipment). GPIB-to-USB converters are available to allow connection to a PC.[3] Modern instruments typically have a USB or RS-232 interface.

USED AND SURPLUS MULTIMETERS

Fig 25.3 shows a typical multimeter that might show up at a ham flea market or on the Internet, a Hewlett-Packard model 412A VTVM. While obsolete, it is still a very useful instrument with capabilities that are rare in modern units. For example, it can measure dc voltage down to 1 mV (0.001 V) full scale and the burden voltage for current measurements is only 1 mV for currents up to 10 mA, rising to 100 mV at 1 A. The highest resistance range is 100 MΩ center scale, which allows reasonably-accurate resistance measurements up to about 1 GΩ. Don't overlook surplus equipment just because it is old. Some of it is a real bargain.

When buying a used analog multimeter, the most important thing to check is the meter movement itself. D'Arsonval meters are rather delicate and easy to damage. For a VOM or battery-operated multimeter, put the instrument in resistance-measuring mode then short and un-short the test leads. The needle should move smoothly between zero and full scale. If you can't do that test, at least make sure that the needle rests close to zero. Rotate the instrument back and forth to make the needle move and observe whether it appears to bind. It is difficult or impossible to repair a damaged meter movement.

Digital multimeters are less delicate than analog meters and many of their failures are electronic in nature so they sometimes can be repaired using normal troubleshooting techniques. Of course it is always a good

idea to test used equipment before buying, if possible. If at all possible, measure voltage and resistance with the meter to be sure it is fundamentally sound.

25.2.3 Panel Meters

Analog panel meters are quite expensive to buy new so many experimenters keep an eye open for flea-market bargains. You often can find old "boat anchor" equipment with good salvageable panel meters selling for less than the value of the meters.

The scale markings on surplus meters often represent what the meter was measuring in the equipment rather than the actual current flowing through the meter itself. Sometimes the full-scale current of the meter movement will be shown in small text at the bottom of the scale. However, that may not be the same as the current measured at the meter terminals because some meters include an internal shunt.

The only sure way to know the full-scale current and resistance of a surplus meter is to measure it. The resistance can be measured with the ohmmeter function of a multimeter, but be careful. On the lowest resistance ranges, the multimeter may output enough voltage and current to damage the meter under test. If the multimeter is an auto-ranging type you have no way to control test current unless you can turn auto-ranging off. With a non-autoranging ohmmeter, start the measurement at the highest resistance scale and then reduce the scale one step at a time until a valid reading is obtained, while keeping an eye on the meter under test to be sure it is not over-ranged.

A safer way to measure both the full-scale current and resistance of a panel meter is to place a high-value resistor in series with it and connect the combination to a dc power supply, perhaps a battery. A 1.5-V battery in series with a 100 kΩ resistor (15 μA of current) is a good starting point. Keep trying smaller and smaller resistance values until a good reading is obtained on the meter. Assuming the scale that is marked on the meter's scale face is linear, the full-scale current is

$$I_{FS} = I_{TEST} \frac{D_{FS}}{D_{TEST}}$$

where I_{FS} is the full-scale meter current, D_{FS} is the scale's full-scale marking, D_{TEST} is the needle indication measured with the test current, and I_{TEST} is the test current, which is equal to the voltage across the resistor divided by the resistance. The meter's resistance is the voltage across the meter divided by I_{TEST}.

USING PANEL METERS

Whether surplus or new, it is rare that a panel meter measures exactly what you need for a particular application. Usually you must change the current or voltage sensitivity.

To increase the full-scale current, place a *current shunt* in parallel with the meter. This is simply a resistor whose value is

$$R_{SHUNT} = R_M \frac{I_M}{I_{FS} - I_M}$$

where R_M is the meter resistance, I_M is the meter full-scale current and I_{FS} is the desired full-scale current reading. The shunt resistance is very small for high-current shunts, such that the resistance of the wires or circuit traces can cause a significant error. To reduce that error, connect the meter directly to the leads of the shunt, with no wires or circuit traces in common with the high-current path.

You can make a low-value shunt by wrapping a length of copper wire around a resistor or other component used as a form. Unfortunately, however, the resistance of copper has a poor temperature coefficient, typically around 0.4 percent per degree C. That means the meter reading can change more than 10% between a warm and a cold day. As the wire self-heats from the high current flowing through, the meter reading can easily be in error by 20% or more. Commercial shunts are made from a metal with a low temperature coefficient such as nichrome. Copper-wire shunts should only be used where accuracy is not important. (A table of wire resistance in ohms per foot (Ω/ft) is available in the **Component Data and References** chapter.)

If the panel meter is to be used to measure voltage, a *voltage multiplier* resistor is inserted in series with the meter. The value is

$$R_{MULT} = \frac{V_{FS}}{I_M} R_M$$

where V_{FS} is the desired full-scale voltage, I_M is the meter full-scale current and R_M is the meter resistance. If the meter has an internal current shunt, it should normally be removed to maximize the value of the multiplier resistor. For high-voltage applications, be aware that in addition to a power rating a resistor also has a working voltage specification, perhaps 200 to 250 V or so for a typical ¼ W, through-hole resistor. Applying voltages higher than the rating — even if the rated power dissipation is not exceeded — can result in arcing across the body of the resistor. If you need to measure a voltage higher than the voltage rating,

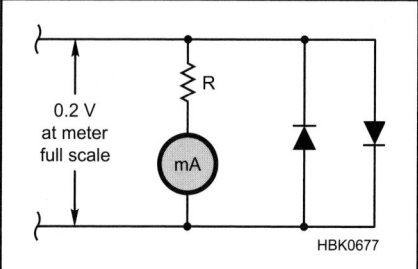

Fig 25.8 — Back-to-back silicon diodes protect the meter by limiting the maximum voltage. See the text for a discussion of how to select the value of R.

use several resistors in series. For example, to measure a 2000-V power supply, ten ¼ W resistors in series, each with a value of one-tenth the desired resistance, would be suitable.

If you intend to use the same panel meter for several different purposes in your project, be sure to use a *break-before-make* switch to make the selection. That protects the meter by making sure it is never connected to two circuits at the same time, even for an instant.

Analog D'Arsonval meters are easily damaged if subjected to excessive current. A standard technique to protect them is to wire back-to-back diodes in parallel with the meter. Silicon junction diodes have the property that they act like an open circuit for low voltages and start to conduct when the forward voltage reaches about 0.4 V. Most meters can withstand up to two times the full-scale current without damage. So choose resistor R in **Fig 25.8** such that the voltage across the diodes is about 0.2 V when the meter current is at full scale, that is

$$R = \frac{0.2\ V}{I_M} - R_M$$

where I_M is the full-scale meter current and R_M is the meter resistance. If that equation results in a negative value for R, replace each diode with two diodes in series and recalculate using 0.4 V instead of 0.2 V in the equation.

MAKING NEW PANEL SCALES

It seems like the scale that is printed on the meter scale is never what you need for your project. You can usually disassemble

the meter to remove the scale plate to modify the scale. Be sure you are working in a clean environment and temporarily reassemble the meter while you are working on the scale to prevent any dust or other contaminant from entering the delicate mechanism. Take particular care that magnetic metal debris cannot get into the meter movement where it is attracted by the permanent magnets.

There are several methods for modifying the scale markings for your own purposes. If only the labels are wrong and the tick marks are spaced appropriately, you can add new labels with a permanent marker or dry-transfer labels. If the old labels are not useful, sometimes they can be removed with a pencil eraser. If you need a completely new meter scale, one old trick is to turn the scale plate over and draw the new scale on the back. However, software is available to design and print custom scales. An Internet search will quickly find dozens of low-cost and free programs, such as *Meter Basic* by Jim Tonne, W4ENE, which is included on the CD-ROM that comes with this book.

DIGITAL PANEL METERS

Digital panel meters (DPMs) are available as preassembled modules that are almost as easy to use as analog meters. The displays are generally of the liquid crystal type, with or without a backlight, and typically have 3 to 4½ digits. A "½" digit is one that can display only a 1 or a blank. Displays with a half-digit usually have a full-scale input voltage of either 2 V or 200 mV minus one count, so that the full-scale voltage is 199.9 mV, for example. Most have programmable decimal points after each digit and some have indicators to indicate the units, such as µ, m, V, A and so forth. DPMs have a high input impedance so there is minimal loading on the circuit under test.

The required power supply voltage varies by model. Some require a floating supply, so if the power supply and the voltage being measured need a common ground connection, be sure the meter is capable of that.

Accuracy is typically 0.1% or better. The total accuracy is usually limited by the external circuitry that drives the meter, such as the amplifier, current shunt or attenuator that is required to get the signal within the input voltage range of the DPM.

25.3 AC Measurements

This section covers issues that affect all ac signals, while the following section on RF measurements concentrates on aspects of ac measurements that are particular to the higher frequencies.

25.3.1 Basic Vocabulary and Units

The **Electrical Fundamentals** chapter includes an introduction to basic electrical units of ac waveforms. The following material emphasizes some additional points that have particular relevance to test and measurement.

AC WAVEFORM VALUES

With ac signals, the voltage and current change periodically with time so, as you might expect, there are several ways to express their value. See **Fig 25.9**. The *average* voltage or current is the value averaged over one period and is equal to the *dc component* of the signal. For a symmetric, periodic ac signal with no dc component, the average is always zero. Non-periodic signals must be averaged over a long period of time to obtain a valid average.

The *peak* value of an ac signal is just as its name implies, the maximum value that the signal ever achieves. The *peak-to-peak* value is the difference between the positive and negative peaks. For a symmetrical-ac waveform such as a sine or square wave, the peak-to-peak is twice the peak.

The *RMS (root-mean-square)* value of voltage or current is that which would produce the same heating in a resistor as a dc voltage or current of the same value. For a sine wave, it is

$$V_{RMS} = \frac{1}{\sqrt{2}} V_{PK} \approx 0.707 \times V_{PK}$$

where V_{RMS} is the RMS voltage and V_{PK} is the peak voltage. A similar equation applies for RMS current. Don't forget that the equation only applies for sine waves. For example, the RMS voltage of a square wave is 1.0 times the peak. See **Fig 25.10**. Also be aware that the RMS value includes the effect of any dc component. If you wish to refer to the RMS value of the ac component only, be sure to state that explicitly.

The formulas for power apply for RMS voltage and current as well as for dc, $P = E_{RMS} I_{RMS}$, $P = E_{RMS}^2 / R$ and $P = I_{RMS}^2 R$, where P is power in watts, E_{RMS} is RMS potential difference in volts and I_{RMS} is RMS current in amperes. However, that assumes that the voltage and current are in phase, as in a resistor or an antenna at resonance. If they are not in phase, then the power in the above equations must be corrected by multiplying it by the *power factor*,

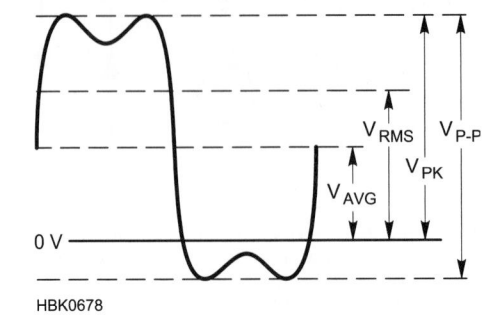

Fig 25.9 — Various methods to express the voltage or current of an ac signal. The average value, V_{AVG}, is the dc component. V_{RMS} is the root-mean-square, V_{PK} is the peak and V_{P-P} is the peak-to-peak value.

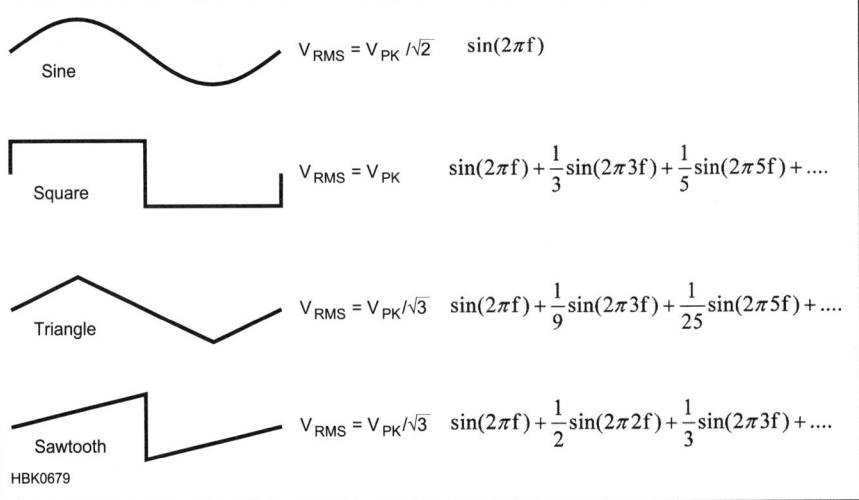

Fig 25.10 — Typical periodic ac waveforms seen in electronic circuitry. A single cycle of each example is shown. At the center are listed the relationships between V_{RMS}, the root-mean-square, and V_{PK}, the peak value, assuming the waveform has no dc component. Periodic waveforms are composed of sine waves at the fundamental and harmonic frequencies. At the right are the relative amplitudes of the various frequency components.

$$PF = \cos\theta$$

where θ = arctan (X / R), the phase angle caused by the reactance, X, and the resistance, R, of the circuit.

25.3.2 Using Multimeters for AC Measurements

Most multimeters can indicate the RMS value of ac voltage and current, but many do not measure the RMS value directly. Instead they measure the rectified average or peak voltage and then apply a correction factor so that the display reads RMS, assuming a sine wave. Unfortunately, that means that the RMS values are not accurate if the ac signal being measured does not have a sinusoidal waveform.

Some meters full-wave rectify the ac signal and then measure the average of that, internally correcting for the difference between

the average of a rectified sine wave

$$V_{AVG} = \frac{2}{\pi} V_{PK} = 0.637 \times V_{PK}$$

and the RMS value

$$V_{RMS} = \frac{1}{\sqrt{2}} V_{PK} \approx 0.707 \times V_{PK}$$

so that the reading is in RMS. Very inexpensive analog meters may only use a half-wave rectifier which causes RMS readings for asymmetric waveforms to vary with the orientation of the test connections.

Additional considerations may apply to RMS readings. For example, the accuracy of the RMS reading for most meters varies with frequency of the applied signal. Check the specifications of the multimeter for the frequency range over which it may be used to measure RMS values.

The only way to accurately measure RMS values of non-sinusoidal signals is with a meter that has *true RMS* capability. Such a meter uses circuitry or software to compute the RMS value of the signal. Note that the measurement bandwidth of the meter must include the significant harmonics of the signal as well as the fundamental in order to give accurate RMS readings.

An example of a measurement that requires a true RMS voltmeter is receiver sensitivity. For that, you need to measure signal and noise levels at the receiver audio output. Standard multimeters using a rectifier and averaging circuit give inaccurate results when measuring noise because noise and sine waves have different peak-to-RMS ratios. Another advantage of true RMS meters is that they tend to have better scale linearity, even for sinusoidal signals. A diode detector is nonlinear, especially at the low end of the scale.

Frequency response is another limitation when making ac measurements with a multimeter. Most are specified from below 50 or 60 Hz, to cover power-line frequencies, up to a few hundred Hz. Many receiver measurements use a 1 kHz test frequency, so a meter specified up to at least that frequency is especially useful.

For all of these and other reasons, the ac accuracy is usually significantly worse than the dc accuracy. Generally, an oscilloscope makes more accurate ac measurements than a multimeter. Modern digital oscilloscopes often have built-in capability to indicate peak, average and true RMS voltage.

One final issue with multimeters is the ac impedance of the probes. While the dc input resistance of a modern electronic multimeter is typically over 1 MΩ, the capacitive reactance can be a significant factor at radio frequencies. Even if all you care about is the dc voltage, if ac signals are present, reactance of the probe can affect the circuit's operation.

25.3.3 Measuring Frequency and Time

FREQUENCY COUNTERS

The basic instrument for measuring frequency is the *frequency counter*. A block diagram of a very basic design is shown in **Fig 25.11.** Three digits are shown, but typically there are more. The signal to be measured is routed through three cascaded decade counters. For 1 Hz frequency resolution, the counters count for 1 second. For 10 Hz resolution, they count for ¹⁄₁₀ second, and so on. The count time is determined by a high-stability crystal oscillator, whose frequency is divided down to 1 Hz, ¹⁄₁₀ Hz, or whatever resolution is desired. At the end of each count time, control circuitry stores the final count in latches that drive the digit displays and at the same time resets the counters for the next count period.

With this scheme, the displays are updated once per second when 1 second resolution is chosen, 10 times per second with 10 Hz resolution, and so on. One issue is that if the frequency is part-way between two adjacent displayed values, the least-significant digit will flicker back and forth between the two values on successive counts. Sometimes the designer chooses not to show the least-significant digit for that reason, even though that slows down the display update rate by a factor of 10 for any given display resolution.

Some frequency counters include the ability to measure time as well. In the block diagram, the connections to the first divide-by-10 stage input and the control circuit input are swapped. In that way the signal being measured controls the count time, and the reference oscillator provides the signal being counted. If the divided-down reference oscillator has a frequency of 1 kHz, for example, then the period can be measured to a resolution of 1 ms. This same technique can be used to measure low-frequency signals as well. For example, when measuring the frequency of a subaudible tone encoder, you need at least 0.1 Hz measurement resolution. Normally, that would require a 10-second count time which can be inconvenient. Some counters are able to measure the period, calculate the reciprocal, and display the resulting frequency. Since the count time is only one cycle of the measured signal, the display updates in real time.

Many frequency counters include a *prescaler*, a frequency divider between the input and the main part of the circuitry, to allow operation at higher frequencies. Usually the prescaler has a 50-Ω input. For low frequencies, there is a separate high-impedance input, typically 1 MΩ, that bypasses the prescaler.

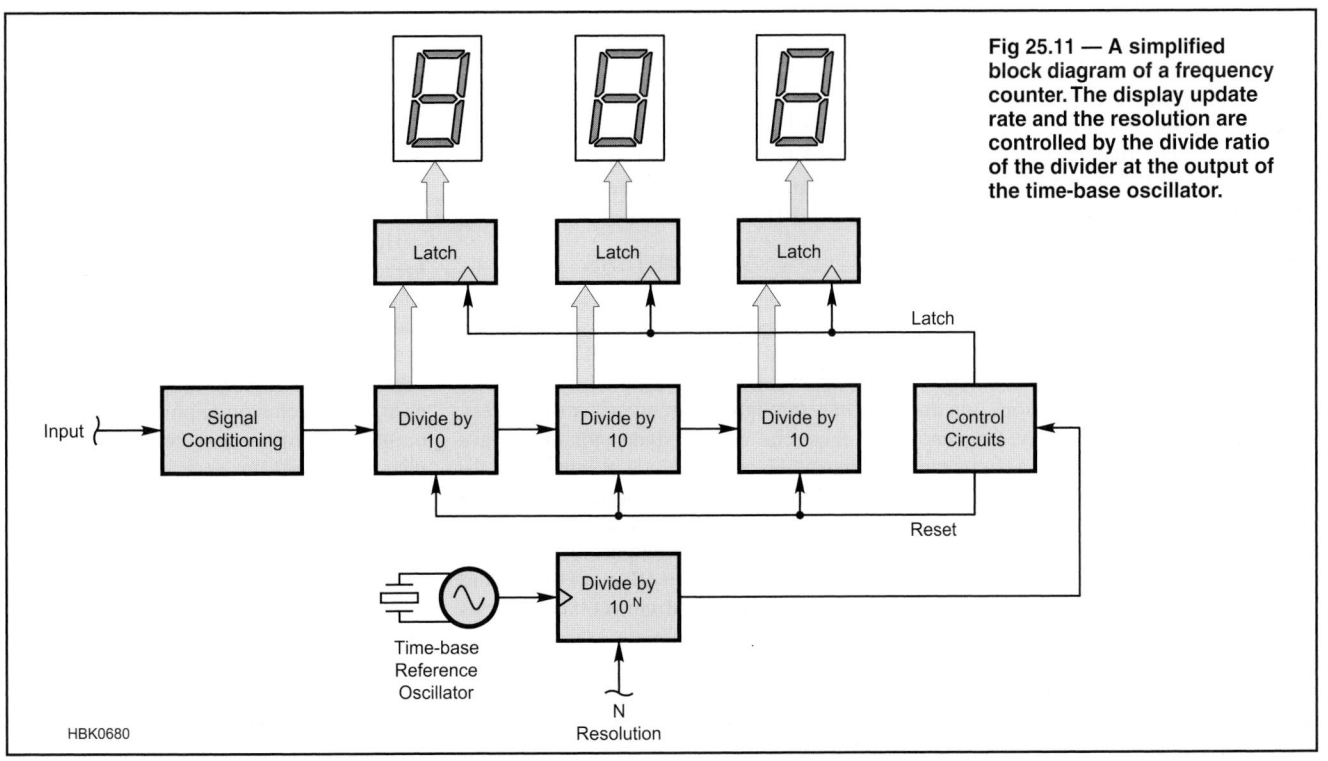

Fig 25.11 — A simplified block diagram of a frequency counter. The display update rate and the resolution are controlled by the divide ratio of the divider at the output of the time-base oscillator.

HBK0680

A switch selects between the two inputs.

It is important to realize that the so-called high-impedance input only has a high impedance at low frequencies. For example, if the stray input capacitance is 30 pF, then the impedance is only 177 Ω at 30 MHz. If you try to measure the frequency of an oscillator by connecting the frequency counter's input directly to the circuitry, it likely will alter the oscillator tuning enough to invalidate the measurement. If possible, connect the counter to the output of a buffer amplifier or at some other point in the circuitry that won't be adversely affected. If that isn't possible, another trick is to use a pickup coil placed near the oscillator. The coil could be a few turns of insulated wire soldered between the center conductor and shield of a coaxial cable that connects to the frequency counter input. Hold the coil just close enough to get a stable reading.

When connecting a frequency counter to a circuit, observe the maximum voltage and power ratings, both for dc and ac. An oscilloscope probe with a 10:1 attenuation ratio connected to the high-impedance input is a good method to reduce both the signal level as well as the capacitive loading.

Frequency counters tend to have sensitive inputs. Only a small fraction of a volt is typically enough for valid readings. It is quite practical to measure the frequency of a nearby transmitter off the air with a small whip antenna. The transmitter should not be modulated while measuring its frequency. SSB transmitters should be measured in CW mode.

The principle figure of merit for a frequency counter is its frequency accuracy, which is primarily determined by the reference oscillator, also known as the *time base*. The time base accuracy is affected by the temperature, power supply voltage, crystal aging and quality of calibration. For a temperature-compensated crystal oscillator (TCXO), the total accuracy is typically a few parts per million (ppm). If it is 10 ppm, for example, then the error at 144 MHz is 1.44 kHz.

Normally temperature is the factor with the greatest effect on the short-term stability. For best accuracy, calibrate the reference oscillator at the same temperature at which measurements will be taken.

Other important specifications are the number of digits in the display, frequency resolution, display update time, frequency range, input sensitivity, input impedance and, for portable units, power supply voltage and current. In choosing a frequency counter you'll need to decide if you want a desktop model or a portable handheld unit. Additional features to consider include the size and visibility of the display, the ability to measure period, a data hold feature, an external time base input, adjustable trigger level and polarity, input attenuator, switchable low-pass filter, and frequency ratio measurement.

Commercial frequency counters have become so common that it hardly pays to construct your own from scratch. Units with a wide range of prices, feature sets and performance levels are available both on-line and from local electronics distributors. Older used and surplus frequency counters tend to be less of a bargain than other types of test equipment because advances in solid-state electronics have made modern instruments inexpensive, lightweight and packed with features and performance.

FREQUENCY MARKER GENERATORS

Before the advent of digital frequency synthesizers, most shortwave amateur receivers and transceivers included a crystal calibrator, which is a low-frequency (typically 25 or 100 kHz) oscillator with strong harmonics throughout the HF spectrum, used to calibrate the analog dial on the radio. Early units used an actual 100 kHz crystal in the oscillator, but modern units use a higher-frequency crystal and a frequency divider to obtain the low-frequency signal.[4]

Most modern Amateur Radio equipment uses a crystal-controlled synthesizer or digital frequency display, so no crystal calibrator is needed. However, the idea can still be useful for testing homebrew gear. If the low-frequency signal consists of a series of narrow pulses rather than a square wave, then all the harmonics are of the same amplitude up to the point where the pulse width is a significant portion of a cycle. This can be approximated by placing a small-value capacitor in series with the output. The harmonics should have constant amplitude for frequencies below

$$f \ll \frac{1}{2\pi RC}$$

where f is the frequency in MHz, C is the capacitance in μF and R is the load resistance, usually 50 Ω.

WAVEMETERS

An *absorption wavemeter* is basically a tunable filter with some means of detecting the signal at the filter output. It allows crude spectrum analysis of a signal by manually tuning through the frequencies of interest. Commercial units designed for microwave frequencies often include a carefully-calibrated dial for reading the frequency with some precision. Typically a diode detector is used to indicate the output signal level. A wavemeter suitable for HF or VHF frequencies can be constructed with one or more coils and a variable capacitor.

DIP METERS

Old timers know this instrument as a grid-dip oscillator (GDO), so-called because the indicating "dip" was in the grid current of the vacuum-tube oscillator. Most dip meters these days are solid-state but the principle is the same. The oscillator coil is external, extending from the end of the instrument. A set of plug-in coils is provided to cover the unit's frequency range.

Along with resonant frequency measurements, the dip meter can also serve as a crude signal generator, capacitor and inductor meter, and antenna and transmission line tester, among other uses.[5] If you are purchasing a dip meter, look for one that is mechanically and electrically stable. On used units, the socket where the coils plug in is a common cause of intermittent operation. The coils should be in good condition. A headphone connection is helpful. Battery-operated models are convenient for antenna measurements.

If you hold the coil near a tuned circuit and adjust the dip meter tuning dial to the frequency of the tuned circuit, there is a dip in the meter reading as the resonant circuits interact with each other. To avoid detuning the circuit being tested, always use the minimum coupling that yields a noticeable indication.

Most dip meters can also serve as absorption wavemeters by turning off the oscillator and looking for a peak instead of a dip in the meter reading. Sometimes frequencies can be detected in this way that would be difficult to read on a frequency counter because of the presence of harmonics. Further, some dip meters have a connection for headphones. The operator can usually hear signals that do not register on the meter.

A dip meter may be coupled to a circuit either inductively or capacitively. Inductive coupling results from the magnetic field generated by current flow. Therefore, inductive coupling should be used when a coil or a conductor with relatively high current is convenient. Maximum inductive coupling results when the axis of the pick-up coil is placed perpendicular to the current path and the coil is adjacent to the wire.

High-impedance circuits have high voltage and low current. Use capacitive coupling when a point of relatively high voltage is convenient. An example might be the output of a 12-V powered RF amplifier. (For safety's sake, *do not* attempt dip-meter measurements on true high-voltage equipment such as vacuum-tube amplifiers or switching power supplies while they are energized.) Capacitive coupling is strongest when the end of the pick-up coil is near a point of high impedance. In either case, the circuit under test is affected by the presence of the dip meter.

To measure resonance, use the following procedure. First, bring the dip meter gradually closer to the circuit while slowly varying the dip-meter frequency. When a current dip occurs, hold the meter steady and tune for minimum current. Once the dip is found,

Table 25.1
Standard Frequency Stations
(Note: In recent years, frequent changes in these schedules have been common.)

Call Sign	Location	Frequency (MHz)
BPM	China	2.5, 5, 10, 15
BSF	Taiwan	5, 15
CHU	Ottawa, Canada	3.330, 7.850, 14.670
dcF	Germany	0.0775
HLA	South Korea	5.000
JJY	Japan	0.04, 0.06
MSF	Great Britain	0.06
RID	Irkutsk	5.004, 10.004, 15.004
RWM	Moscow	4.996, 9.996, 14.996
TDF	France	0.162
WWV	USA	2.5, 5, 10, 15, 20
WWVB	USA	0.06
WWVH	USA (Hawaii)	2.5, 5, 10, 15
ZSC	South Africa	4.291, 8.461, 12.724 (part time)

move the meter away from the circuit and confirm that the dip comes from the circuit under test (the depth of the dip should decrease with distance from the circuit until the dip is no longer noticeable). Finally, move the meter back toward the circuit until the dip just reappears. Retune the meter for minimum current and read the dip-meter frequency from the dial or with a calibrated receiver or frequency counter.

The current dip of a good measurement is smooth and symmetrical. An asymmetrical dip indicates that the dip-meter oscillator frequency is being significantly influenced by the test circuit, degrading the accuracy of the measurement. Increase the distance between the dip meter and the circuit until a shallow symmetrical dip is obtained.

FREQUENCY CALIBRATION

The best test equipment is of limited use if it is not well-calibrated. The traditional frequency calibration method is to zero-beat a crystal oscillator (or its harmonic) with a radio station of known frequency, preferably a standard frequency station such as WWV or WWVH. **Table 25.1** contains the locations and frequencies of some of those stations. A receiver is tuned to one of those frequencies and the oscillator is loosely coupled to the antenna. It may be necessary to use frequency multiplication or division to obtain a common frequency. The frequency difference between the two causes a *beat note*, a rapid variation in the strength of the tone received in the speaker that slows down as the frequencies are brought close together. Maximum beat-note modulation occurs when the off-the-air and oscillator signals are approximately equal in amplitude

While the transmitted frequencies from WWV and WWVH are highly accurate, better than 1 part in 10^{11}, after propagation via the ionosphere the received accuracy is significantly degraded by Doppler shift, typi-

cally to a few parts in 10^7. Also, due to fading of the received signal, it can be difficult to zero-beat the oscillator to better than about 1 Hz accuracy. Best results generally occur on the highest frequency that provides good reception.

VLF time standards and surplus rubidium standards can be used for frequency references.[6,7] The Global Positioning System (GPS) satellites offer further possibilities for very precise frequency calibration. Various companies sell *disciplined oscillator* units that correct the frequency using the cesium-clock-based signals from the GPS satellites. These can sometimes be found on the surplus market.[8] Amateur-level kits are also available or you can build one from scratch.[9]

25.3.4 Oscilloscopes

An *oscilloscope* ("scope" for short) is an instrument that displays voltage versus time on a screen, similar to the waveforms seen in electronics textbooks. Scopes are broken down into two major classifications: analog

and digital. This does not refer to the signals they measure, but rather to the methods used inside the instrument to process signals for display.

ANALOG OSCILLOSCOPES

Fig 25.12 shows a simplified diagram of a triggered-sweep oscilloscope. At the heart of all analog scopes is a cathode-ray tube (CRT) display. An electron beam inside the CRT strikes the phosphorescent screen causing a glowing spot. Unlike a television CRT, an oscilloscope uses electrostatic deflection rather than magnetic deflection. The exact location of the spot is a result of the voltage applied to the vertical and horizontal deflection plates. To trace how a signal travels through the oscilloscope circuitry, start by assuming that the trigger select switch is in the INTERNAL position.

The input signal is connected to the input COUPLING switch. The switch allows selection of either the ac part of an ac/dc signal or the total signal. If you wanted to measure, for example, the RF swing at the collector of an output stage including the dc level, you would use the *dc-coupling* mode. In the *ac-coupled* mode, dc is blocked from reaching the vertical amplifier chain so that you can measure a small ac signal superimposed on a much larger dc level. For example, you might want to measure a 25 mV 120-Hz ripple on a 13-V dc power supply. Note that you should not use ac coupling at frequencies below the low-frequency cutoff of the instrument in that mode, typically around 30 Hz, because the value of the blocking capacitor represents a high series impedance to very low-frequency signals.

After the coupling switch, the signal is connected to a calibrated attenuator. This is used to reduce the signal to a level within the range of the scope's vertical amplifier. The vertical amplifier boosts the signal to a level that can drive the CRT and also adds

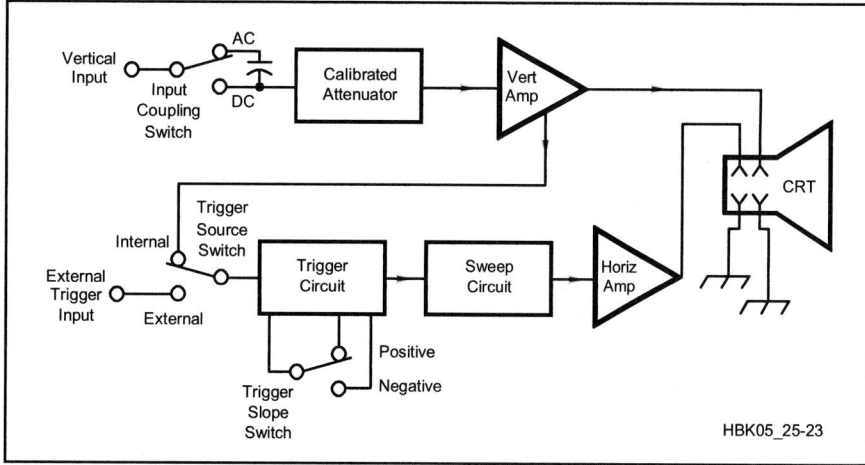

Fig 25.12 — Typical block diagram of a simple triggered-sweep oscilloscope.

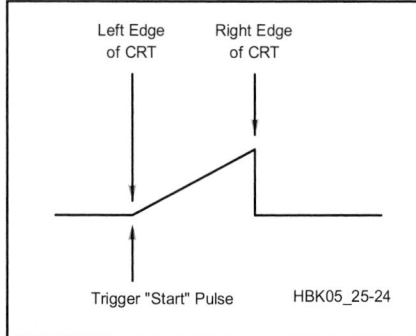

Fig 25.13 — The sweep trigger starts the ramp waveform that sweeps the CRT electron beam from side to side.

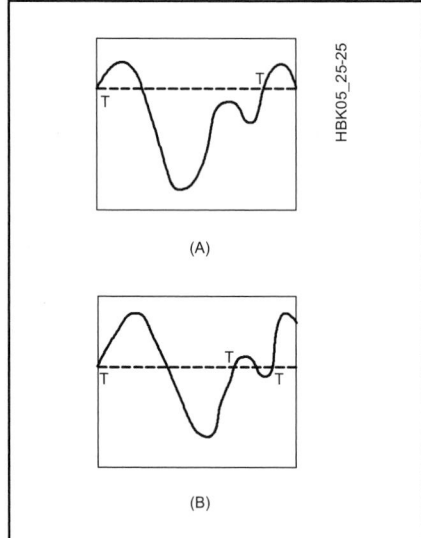

Fig 25.14 — In order to produce a stable display the selection of the trigger point is very important. Selecting the trigger point in A produces a stable display, but the trigger shown at B will produce a display that "jitters" from side to side.

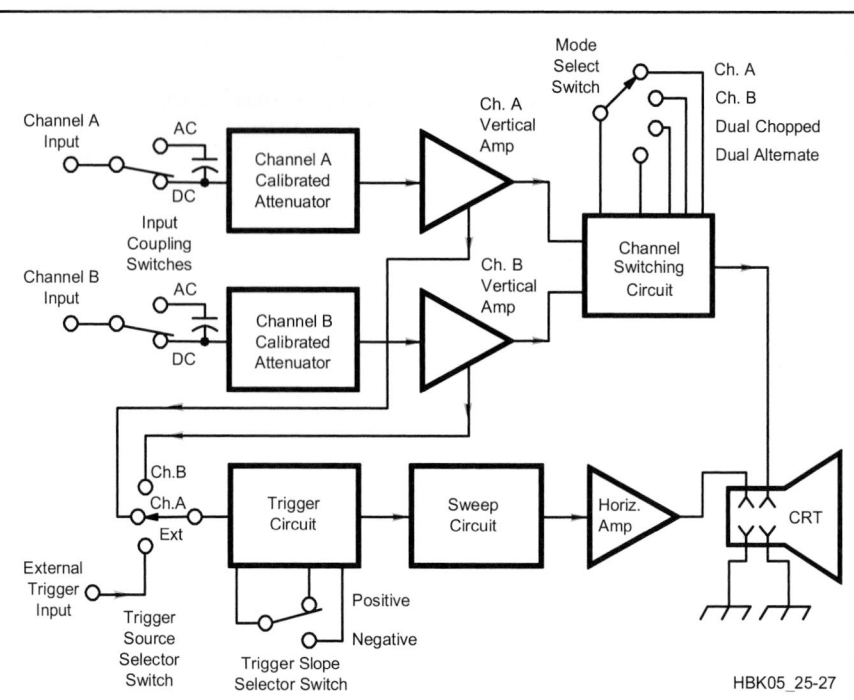

Fig 25.15 — Simplified dual-trace oscilloscope block diagram. Note the two identical input channels and amplifiers.

a bias component to position the waveform on the screen. The result is that the vertical position of the beam on the CRT represents input voltage.

A small sample of the signal from the vertical amplifier is sent to the trigger circuitry. The trigger circuit feeds a start pulse to the sweep generator when the input signal reaches a certain level (*level triggering*) or exhibits a positive- or negative-going edge (*edge triggering*). The sweep generator gives a precisely timed voltage ramp (see **Fig 25.13**). The rising edge of the ramp signal feeds the horizontal amplifier that, in turn, drives the CRT. This causes the scope trace to sweep from left to right, with the zero-voltage point representing the left side of the screen and the maximum voltage representing the right side of the screen. The result is that the horizontal

position of the beam on the CRT represents time. At the end of the ramp, the sharp edge of the ramp quickly moves the beam back to the left side of the screen.

The trigger circuit controls the horizontal sweep. It looks at the trigger source (internal or external) to find out if it is positive- or negative-going and to see if the signal has passed a particular level. **Fig 25.14A** shows a typical signal and the dotted line on the figure represents the trigger level. It is important to note that once a trigger circuit is "fired" it cannot fire again until the sweep has moved all the way across the screen from left to right. There normally is a TRIGGER LEVEL control to move the trigger level up and down until a stable display is seen. Some scopes have an AUTOMATIC position that chooses a level to lock the display in place without manual adjustment.

Fig 25.14B shows what happens when the level has not been properly selected. Because there are two points during a single cycle of the waveform that meet the triggering requirements, the trigger circuit will have a tendency to jump from one trigger point to another. This will make the waveform jitter from left to right. Adjustment of the TRIGGER LEVEL control will fix that problem.

It is also possible to trigger the sweep system from an external source (such as the system clock in a digital system). This is done by using the external input jack with the trigger select switch in the EXTERNAL position.

DUAL-TRACE OSCILLOSCOPES

Dual-trace oscilloscopes have two vertical input channels that can be displayed together on the screen. Although the best dual-trace scopes use a CRT with two electron beams, it is possible to trick the eye into seeing two traces simultaneously using a single-beam CRT. **Fig 25.15** shows a simplified block diagram of a dual-trace oscilloscope using this method. The only differences between this scope and the previous example are the additional vertical amplifier and the "channel switching circuit." This block determines whether we display channel A, channel B or both (simultaneously).

The dual display is not a true dual-beam display but the appearance of dual traces is created by the scope using one of two methods, referred to as *chopped mode* and *alternate mode*. In the chopped mode a small portion of the channel A waveform is written to the CRT, then a corresponding portion of the channel B waveform is written to the CRT. This procedure is continued until both waveforms are completely written on the CRT. The switching from one channel to the other is so fast that each trace looks as though it were continuous. The chopped mode is essential for *single-shot* signals (signals that do not repeat periodically). It is most useful at slow sweep speeds. At fast sweep speeds, the switching from channel to channel becomes visible, making each trace into a dotted line.

In the alternate mode, the complete channel A waveform is written to the CRT followed immediately by the complete channel B waveform. This happens so quickly that it appears that the waveforms are displayed at the same time. This mode of operation is not very useful at very slow sweep speeds since the two traces no longer appear simultaneous. It also does not work for single-shot events.

Most dual-trace oscilloscopes also have a feature called "X-Y" mode. This feature allows one channel to drive the horizontal amplifier of the scope (called the X channel) while the other channel (called Y in this mode of operation) drives the vertical amplifier. Some single-trace oscilloscopes support this mode as well. X-Y operation allows the scope to display *Lissajous patterns* for frequency and phase comparison and to use specialized test adapters such as curve tracers or spectrum analyzer front ends. Because of frequency limitations of most scope horizontal amplifiers the X channel is usually limited to a 5 or 10-MHz bandwidth.

DIGITAL OSCILLOSCOPES

In recent years, semiconductor and display technology have advanced to the point that much of the analog signal processing in an oscilloscope can be replaced with low-cost microprocessor-based digital circuitry. This results in dramatically improved accuracy for both amplitude and time measurements as well as enabling sophisticated features that would be difficult or impossible in an analog scope. For example, a trace can be displayed with infinite persistence and stored as a computer file if desired.

In a digital oscilloscope the vertical amplifiers are replaced with an analog-to-digital converter (ADC), which samples the signal at regular time intervals and stores the samples in digital memory. The samples are stored with an assigned time, determined by the trigger circuits and the microprocessor clock. The samples are then retrieved and displayed on the screen with the correct vertical and horizontal position.

Early digital oscilloscopes used a CRT with electrostatic deflection, similar to an analog scope, with the horizontal and vertical deflection signals generated by digital-to-analog converters (DACs). Modern instruments usually use either a raster-scan CRT similar to a television picture tube or a solid state (LCD) display. The microprocessor determines which pixels to light up to draw the traces on the screen. Even though a digital scope does not have the same internal circuitry as an analog scope, many of the same terms are used to control operation. For example, "Sweep speed" still describes the amount of time per horizontal division, even though there is no electron beam to be swept across the display in the original sense.

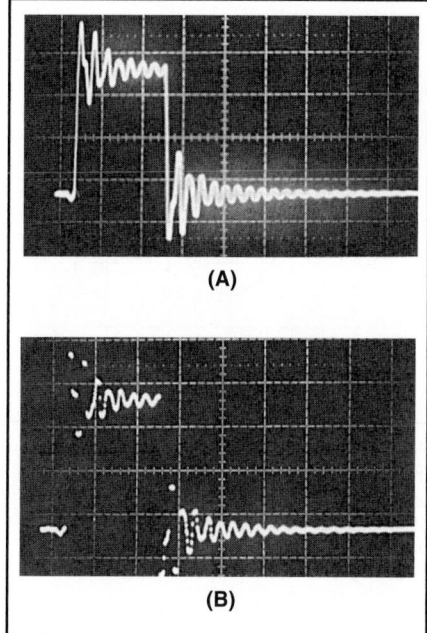

(A)

(B)

Fig 25.16 — Comparison of an analog scope waveform (A) and that produced by a digital oscilloscope (B). Notice that the digital samples in B are not continuous, which may leave the actual shape of the waveform in doubt for the fastest rise time displays the scope is capable of producing.

For the vertical signals you will see manufacturers refer to "8-bit digitizing," or perhaps "10-bit resolution." This is a measure of the number of digital levels that are shown along the vertical (voltage) axis. More bits give you better resolution and accuracy of measurement. An 8-bit vertical resolution means each vertical screen has 2^8 (or 256) discrete values; similarly, 10-bit resolution yields 2^{10} (or 1024) discrete values.

It is important to understand some of the limitations resulting from sampling the signal rather than taking a continuous, analog measurement. When you try to reconstruct a signal from individual discrete samples, you must take samples at least twice as fast as the highest frequency signal being measured. If you digitize a 100-MHz sine wave, you should take samples at a rate of 200 million samples a second (referred to as 200 Megasamples/second). Actually, you really would like to take samples even more often, usually at a rate at least five times higher than the input signal. (See the **DSP and Software Radio Design** chapter for more information on sampled signals.)

If the sample rate is not high enough, very fast signal changes between sampling points will not appear on the display. For example, **Fig 25.16** shows one signal measured using both analog and digital scopes. The large

spikes seen in the analog-scope display are not visible on the digital scope. The sampling frequency of the digital scope is not fast enough to store the higher frequency components of the waveform. If you take samples at a rate less than twice the input frequency, the reconstructed signal has a wrong apparent frequency; this is referred to as *aliasing*. In Fig 25.16 you can see that there is about one sample taken per cycle of the input waveform. This does not meet the 2:1 criteria established above. The result is that the scope reconstructs a waveform with a different apparent frequency.

Many older digital scopes had potential problems with *aliasing*. A simple manual check for aliasing is to use the highest practical sweep speed (shortest time per division) and then to change to other sweep speeds to verify that the apparent frequency doesn't change. Some modern oscilloscopes use a special technique to increase the effective sample rate for repetitive signals. The phase of the sample clock is adjusted slightly on each successive sweep, so that the new samples occur in between the previous ones. After several sweeps the missing data in the spaces between the original samples are filled in, producing a continuous trace. This only works with periodic signals that trigger at exactly the same point on each sweep.

OSCILLOSCOPE LIMITATIONS

Oscilloscopes have fundamental limits, primarily in frequency of operation and range of input voltages. For most purposes the voltage range can be expanded by the use of appropriate probes. The frequency response (also called the bandwidth) of a scope is usually the most important limiting factor. For example, a 100-MHz 1-V sine wave fed into an oscilloscope with a 100-MHz 3-dB bandwidth will read approximately 0.7 V on the display. The same instrument at frequencies below 30 MHz should be accurate to about 5%.

A parameter called *rise time* is related to bandwidth. This term describes a scope's ability to accurately display voltages that rise very quickly. For example, a very sharp and square waveform may appear to take some time in order to reach a specified fraction of the input voltage level. The rise time is usually defined as the time required for the display to show a change from the 10% to 90% points of the input waveform, as shown in **Fig 25.17**. Assuming the frequency response is primarily limited by a single-pole roll off in the amplifier circuitry, the mathematical definition of rise time is given by:

$$t_r = \frac{0.35}{BW}$$

where t_r = rise time in µs and BW = bandwidth in MHz of the amplifier.

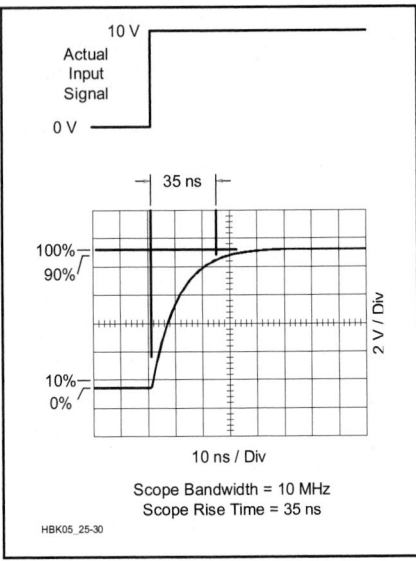

Fig 25.17 — The bandwidth of the oscilloscope vertical channel limits the rise time of the signals displayed on the scope.

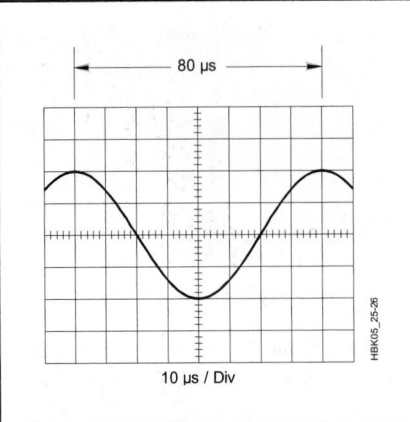

Fig 25.18 — An oscilloscope can measure frequency as well as amplitude. Here the waveform shown has a frequency of 80 microseconds (8 divisions × 10 μs per division) and therefore a frequency of 1/80 μs or 12.5 kHz.

It is also important to note that inexpensive analog oscilloscopes may not have better than 5% accuracy in these applications. Even moderately-priced oscilloscopes are still useful, however. The most important value of an oscilloscope is that it presents an image of what is going on in a circuit, which is very useful for troubleshooting waveforms and other time-varying phenomena. It can show modulation levels, relative gain between stages, clipping distortion, intermittent oscillations and other useful information.

USING AN OSCILLOSCOPE

An oscilloscope can measure a signal's shape, amplitude, frequency and whether it is dc, ac or a mixture of both. For example, in **Fig 25.18** it is clear from the shape that the signal is a sine wave. Assuming that the center horizontal line or axis represents zero volts, the signal has no dc component since there is as much above the axis as below it. If the vertical gain has been set to 1 V per division, then the peak value is 2 V and the peak-to-peak value is 4 V.

The horizontal travel of the trace is calibrated in units of time. If the sweep speed is known and we count the number of divisions (vertical bars) between peaks of the waveform (or any similar well-defined points that occur once per cycle) we can find the period of one cycle. The frequency is the reciprocal of the period. In Fig 25.18, for example, the distance between the peaks is 8 divisions. If the sweep speed is 10 μs/division then the period is 80 μs. That means that the frequency of the waveform is 1/80 μs, or 12,500 Hz. The accuracy of the measured frequency depends on the accuracy of the scope's ramp generator,

typically a few percent for an analog instrument. This accuracy cannot compete with even the least-expensive frequency counter, but the scope can still be used to determine whether a circuit is functioning properly.

Oscilloscopes are usually connected to a circuit under test with a short length of shielded cable and a probe. At low frequencies, a piece of small-diameter coax cable and some sort of insulated test probe might do.

However, at higher frequencies the capacitive reactance of the cable would be much less than the one-megohm input impedance of the oscilloscope. In addition each scope has a certain built-in capacitance at its input terminals (usually between 5 and 35 pF). The total capacitance causes problems when probing an RF circuit with relatively high impedance.

Most new oscilloscopes come with specially-designed *scope probes*, one for each vertical channel. They can also be purchased separately. The most common type is a *×10 probe* (called a "times ten" probe), which forms a 10:1 voltage divider using the built-in resistance of the probe and the input resistance of the scope. When using a ×10 probe, all voltage readings must be multiplied by 10. For example, if the scope is on the 1 V/division range and a ×10 probe were in use, the signals would be displayed on the scope face at 10 V/division. Some scopes can sense whether a ×10 probe is in use, and automatically change the scale of the scope's display.

Unfortunately a resistor alone in series with the scope input seriously degrades the scope's rise-time performance and bandwidth because of the low-pass filter formed by the series resistance along with the parallel capacitance of the cable and scope input. This may be corrected by using a compensating capacitor in parallel with the series resistor. If the capacitor value is chosen so that the R-C time constant is the same as the R-C network formed by the input resistance and capacitance of the scope, as

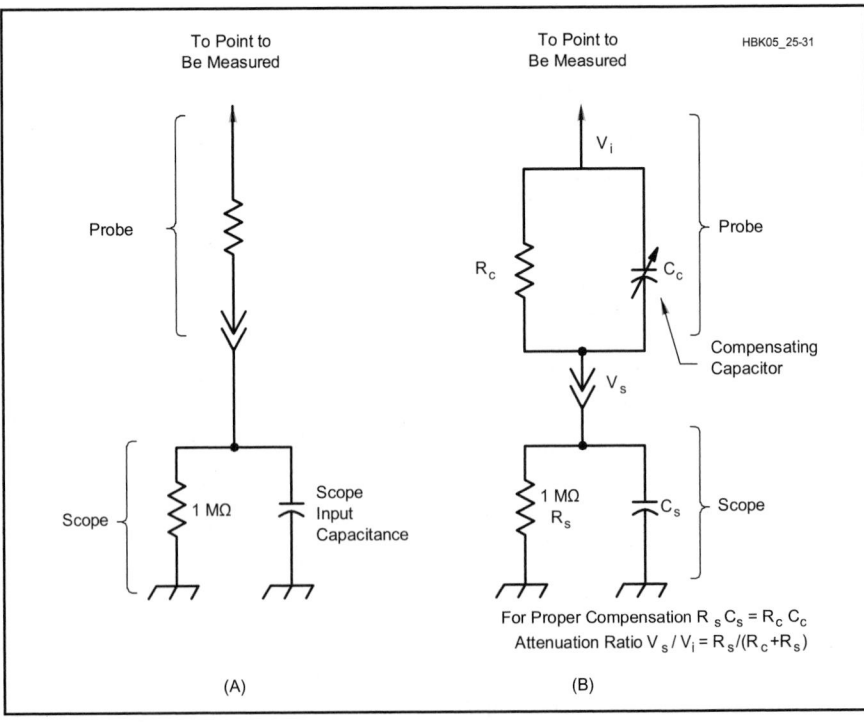

Fig 25.19 — Uncompensated probes such as the one at A are sufficient for low-frequency and slow-rise-time measurements. However, for accurate display of fast rise times with high-frequency components the compensated probe at B must be used. The variable capacitor is adjusted for proper compensation (see text for details).

shown in **Fig 25.19**, then the probe and scope should have a flat response curve throughout the whole bandwidth of the scope. A ×10 probe not only reduces the voltage by a factor of 10 but it increases the input resistance and reduces the capacitance as well, which reduces loading on the circuit under test.

To account for manufacturing tolerances in the scope and probe the compensating capacitor is made variable. Most scopes include a "calibrator" output that produces a fast-rise-time square wave for the purpose of adjusting the compensating capacitor in a probe. **Fig 25.20** shows possible responses when the probe is connected to the oscilloscope's calibrator jack. A misadjusted compensating capacitor can greatly affect the frequency response of the scope and create artifacts in signals that are not actually present.

If a probe cable is too short, do not attempt to extend the length of the cable by adding a piece of common coaxial cable. The compensating capacitor in the probe is chosen to compensate for the provided length of cable. It usually does not have enough range to compensate for extra lengths.

The shortest ground lead possible should be used from the probe to the circuit ground. Long ground leads act as inductors at high frequencies where they create ringing and other undesirable artifacts in the displayed signal.

For the best high-frequency performance, the scope probe can be eliminated entirely and the oscilloscope input converted to a 50-Ω impedance. Some scopes have a switch to choose between a high-impedance or 50-Ω input. For others, you can purchase a 50-Ω *through-line* termination, which is just a connector with a male BNC on one end, a female BNC on the other, and an internal 50-Ω termination resistor in parallel. Plug the male connector to the scope's high-impedance vertical input and connect the 50-Ω cable from the device under test to the female connector.

Some situations may require the use of a scope to measure a *current* waveform, rather than a voltage. Specialized current probes are available that make this task possible, with some capable of measuring both dc and ac.

Be wary of using a scope to judge harmonic content or distortion of a sine wave. A waveform that "looks good" to the eye may have significant distortion or high-frequency components unsuitable for on-the-air signals. A spectrum analyzer (described below) should be used for determining the spectral content of signals.

CHOOSING AN OSCILLOSCOPE

For many years a scope (even a so-called portable) was big and heavy. In recent years, microprocessors and other ICs have reduced the size and weight. Modern scopes can take other forms than the traditional large cabinet with built-in CRT. Nearly all digital oscilloscopes use an LCD display for true portability. (See **Fig 25.21**) You can use your personal computer as an oscilloscope with an external signal digitizer that connects to the PC via a high-speed USB or Firewire interface. This saves money by using the cabinet, power supply, processor and display of the PC. Even stand-alone scopes can attach to a PC and download their data for storage and analysis or transfer it to memory storage devices. Many high-end scopes now incorporate non-traditional functions, such as the Fast Fourier Transforms (FFT). This allows spectrum analysis or other advanced mathematical techniques to be applied to the displayed waveform.

Features

When choosing an oscilloscope, the first decision is analog versus digital. Used and surplus instruments are usually analog but more modern units are shifting to digital as the prices of solid-state displays and other components come down. Digital models are generally more accurate and tend to have lots

of features not found on analog oscilloscopes.

The next big decision is how many input channels you need. There are many situations in which having a second channel is extremely helpful and even more than two channels is often useful, especially when troubleshooting digital circuits. With two or more channels you typically also get X-Y mode.

Trace storage is essential for viewing very slow signals or single-shot transients. In the old days, high-end analog scopes used special CRTs that could store a trace in an analog fashion. They tended to be rather difficult to adjust properly and the CRTs are expensive and/or impossible to replace if they should ever fail. Trace storage is very easy to implement in a digital scope and nearly all of them have the feature. Ideally, one or more stored traces can be displayed simultaneously with the current trace for easy comparison.

Most scopes come with the ability to select internal, external or power-line synchronous triggering as well as an adjustable trigger level. Digital scopes may draw a horizontal line on the screen so you can see exactly where the trigger level is with respect to the signal, which is very handy. Automatic adjustment of trigger level is a common feature as well. Some scopes offer a noise-rejection feature that adds hysteresis to reduce false triggering. *Single sweep* is useful for looking at a non-repetitive signal. The single sweep can be *armed*, that is, reset and ready to start the next sweep, by pushing a button or sometimes with an external signal. *Trigger delay* allows horizontal centering of the display at a different place from the trigger point. *Trigger hold-off* inhibits the trigger for a selectable time after the sweep to prevent unwanted multiple triggers.

Some older tube-type scopes offered only

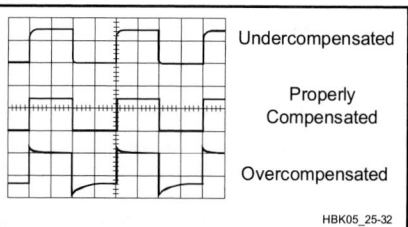

Undercompensated

Properly Compensated

Overcompensated

HBK05_25-32

Fig 25.20 — Displays of a square-wave input illustrating undercompensated, properly compensated and overcompensated probes.

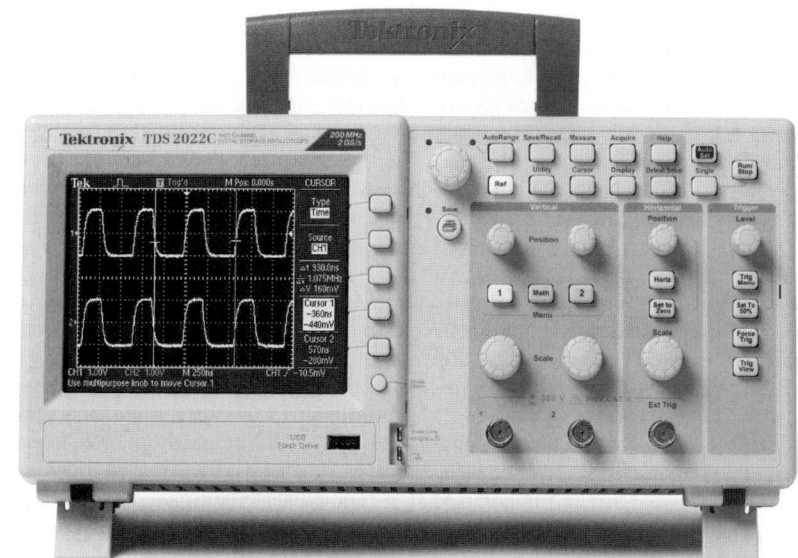

Fig 25.21 — The Tektronix TDS 2000C series is a typical example of digital oscilloscopes using an LCD display.

ac input coupling, but most nowadays have a switch to select between ac, dc and ground. As mentioned before, some include a switch to select 50-Ω or high input impedance. Some high-speed scopes include a switchable input low-pass filter to reduce wideband noise when measuring lower-frequency signals. A feature sometimes found on multiple-channel scopes is the ability to add or subtract two or more channels.

Digital oscilloscopes often include various data-analysis features. Averaging reduces noise and peak mode makes transient peaks visible. A common feature is the ability to calculate and display the peak, peak-to-peak and RMS values of a signal. For repetitive signals, the period and frequency can be calculated as well. Some scopes include amplitude and/or time markers that allow accurate measurements of specific points on a waveform. Sometimes the instrument state can be stored to one of several internal memories, which can be very handy on complicated instruments that take a long time to set up.

Computer connectivity used to be available only on high-end instruments but is now becoming more and more common. It allows storing traces for later reference and you can produce nice graphical screen shots for that magazine article or notebook entry about your latest creation. With the proper software it allows automated testing as well. Older instruments usually have a GPIB (general purpose interface bus) interface, also known as IEEE-488 or as HPIB on Hewlett-Packard equipment. GPIB-to-USB adapters are available from Prologix and a number of other companies. More modern scopes usually come with an RS-232 or USB interface.

Specifications

Bandwidth is the main "money spec" for an oscilloscope. The higher the frequency range, the more expensive it tends to be. Rise time is also sometimes specified. For digital scopes, the sample rate is just as important. Theoretically it must be at least twice the bandwidth but practically should be much higher than that to avoid aliasing. For repetitive waveforms, some scopes can do tricks with the sample phase to eliminate aliasing with a lower sample rate, as explained previously

The range of input signals is normally specified as the maximum and minimum volts per division. Amplitude accuracy is usually specified as a percent of the reading, typically 5-10% for inexpensive analog scopes and perhaps 1-2% for high-quality digital ones. Another important specification is the maximum input signal that can be accepted without damage. It is typically presented as a maximum dc plus peak ac voltage, that is, the maximum peak voltage of the complete signal. This is increased when using a ×10

probe, subject to the probe specifications.

For digital scopes, the resolution of the input analog-to-digital converter (ADC) is typically specified as a number of bits. Sometimes the lowest-voltage input ranges are obtained simply by using only the low end of the ADC range, which results in the displayed signal having a stepped response rather than a smooth curve.

The sweep speed is generally specified in seconds (milliseconds, microseconds) per division. The accuracy is typically a few percent for analog scopes and much better than that for digital scopes, often limited just by the screen resolution. The triggering system is a major factor that determines how useful a scope is in actual operation, but it can be hard to tell how well it works by studying the specifications. The trigger sensitivity is the main parameter to look for. It is specified in fractions of a division for internal trigger and in mV for external trigger.

BUYING A USED SCOPE

Many hams end up buying a used scope due to price. If you buy a scope and intend to service it yourself, be aware all scopes that use tubes or a CRT contain lethal voltages. Treat an oscilloscope with the same care you would use with a tube-type high-power amplifier. The CRT should be handled carefully because if dropped it will crack and implode, resulting in pieces of glass and other materials being sprayed everywhere in the immediate vicinity. You should wear a full-face safety shield and other appropriate safety equipment to protect yourself.

Another concern when servicing an older scope is the availability of parts. The CRTs in older units may no longer be available. Many scopes made since about 1985 used special ICs, LCDs and microprocessors. Some of these may not be available or may be prohibitive in cost. You should buy a used scope from a reputable vendor— even better yet, try it out before you buy it. Make sure you get the operator's manual also.

Older tube-type models are generally quite serviceable, often needing nothing more than a new tube or two. The massive lab-grade instruments from days of yore made by Tektronix and Hewlett-Packard can still give good service with a little care. They are so large and heavy that a special scope cart was often used to house them and allow easy movement from lab bench to lab bench.

Early generation digital scopes are now showing up on the surplus market at reasonable prices. Unfortunately, the user interface leaves much to be desired on some models, sometimes requiring the operator to use several layers of menus to access common functions. Look for one with an analog-like feel with separate buttons or knobs for most of the important functions.

25.3.5 Audio-Frequency Oscillators and Function Generators

There are a number of ways to generate an audio-frequency tone. If a square-wave output is acceptable, then a simple oscillator can be built with a 555 timer IC, two resistors and two capacitors. It is an inexpensive, time-tested design with good frequency stability.

If a sine-wave signal is needed, a *twin-T* oscillator made with a single bipolar transistor is about the simplest solution. The oscillator in **Fig 25.22** can be operated at any frequency in the audio range by varying the component values. R1, R2 and C1 form a low-pass network, while C2, C3 and R3 form a high-pass network. As the phase shifts are opposite, there is only one frequency at which the total phase shift from collector to base is 180°: Oscillation will occur at that frequency. C1 should be about twice the capacitance of C2 or C3. R3 should have a resistance about 0.1 that of R1 or R2 (C2 = C3 and R1 = R2). Output is taken across C1, where the harmonic distortion is least. Use a relatively high impedance load — 100 kΩ or more. Most small-signal AF transistors can be used for Q1. Either NPN or PNP types are satisfactory if the supply polarity is set correctly. R4, the collector load resistor may be changed a little to adjust the oscillator for best output waveform.

While the twin-T oscillator does give a

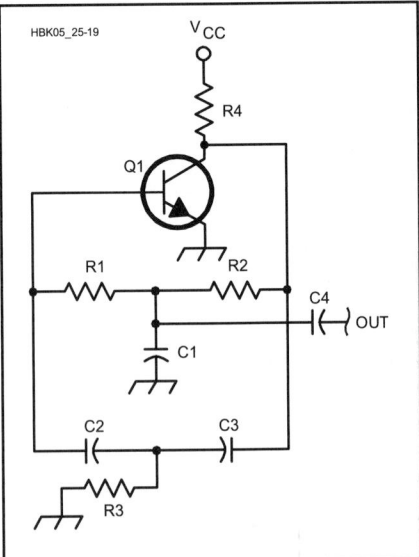

Fig 25.22 — Values for the twin-T audio oscillator circuit range from 18 kΩ for R1-R2 and 0.05 µF for C1 (750 Hz) to 15 kΩ and 0.02 µF for 1800 Hz. For the same frequency range, R3 and C2-C3 vary from 1800 Ω and 0.02 µF to 1500 Ω and 0.01 µF. R4 is 3300 Ω and C4, the output coupling capacitor, can be 0.05 µF for high-impedance loads.

roughly sinusoidal output, the distortion is rather high and the frequency is not easily tunable. Both of those problems are addressed in the *Wien bridge* oscillator illustrated in **Fig 25.23** The key to low distortion is to prevent the amplifier from going into saturation. If equal values of R and C are used in both sections of the positive feedback network, then the gain through that network is ⅓. If the negative feedback network also has a 3:1 ratio, then the total loop gain is exactly 1, the condition for oscillation. The light bulb in the bottom leg has a positive temperature coefficient. (Note that a true incandescent bulb must be used — an LED may not be substituted.) As the signal level increases, its resistance goes up, lowering the gain. In this way the amplitude is held steady so that the amplifier does not saturate. The frequency may be tuned with a single control by using a two-section potentiometer for the two resistors labeled R. Typically small fixed-value resistors are placed in series with each potentiometer section to give about a 10:1 tuning range. Additional ranges can be had by switching the capacitors, selected using the equation

$$C = \frac{1}{2\pi fR}$$

A suitable bulb is a number 327 or 1819. Under normal operation, it should be lit to a bit less than full brightness, which is determined by the value of the negative feedback resistor R_F, typically around 400 Ω.

The idea of using the positive temperature coefficient of resistance of a light bulb to stabilize the output of a tunable oscillator was used in the very first instrument produced by the *Hewlett-Packard* company (now Agilent Technologies). (The Model 200A audio oscillator circuit was first described in company founder Hewlett's 1939 college thesis.)

FUNCTION GENERATORS

A *function generator*, also known as a *waveform generator*, is a type of audio oscillator that can generate several waveforms. In addition to sine waves, most can also generate square waves, triangle waves and sawtooth ramp waveforms. A pulse output with variable duty factor is another common feature. Many models can also linearly sweep the frequency. Frequency coverage is typically from below audio frequencies up to a few MHz.

The heart of most analog function generators is a triangle wave oscillator. To create the sine wave, a diode shaping circuit rounds off the top and bottom of the triangles, resulting in a reasonably-accurate sine wave with distortion on the order of a percent or two. Circuitry that determines when the triangle wave is rising or falling is used to generate the square wave as well. The rise/fall duty factor of the triangle wave can be varied, which also varies the duty factor of the square wave. If the triangle wave duty factor is set near 100% or 0%, it becomes either a rising or falling sawtooth wave.

The easiest way to build your own analog function generator is to use a waveform generator IC that includes most of what you need in one package. The two most common such ICs available today are the MAX038 and XR2206. The ICL8038 is seen in many older circuits but is obsolete and no longer in production. The AD9833 direct digital synthesis (DDS) IC (see **Fig 25.24**) is also designed to be used as a low frequency function generator.[10] (DDS is discussed in the **Oscillators and Synthesizers** and **DSP and Software Radio Design** chapters.)

An *arbitrary waveform generator* (AWG or ARB) has capabilities that are a superset of a waveform generator. The sinusoidal waveform ROM look-up table in Fig 25.24 is replaced by RAM so that data which creates waveforms other than a sine wave can be used. Usually a large amount of memory is included so that long non-repetitive waveforms can be generated in addition to periodic signals.

ARBs generally include a microprocessor and provide means to generate complicated sequences that may combine repeating and nonrepeating segments.

CHOOSING A FUNCTION GENERATOR

For many applications, the basic analog sine/square wave generator based on a Wien bridge oscillator works fine. The sine wave is adequate for testing the gain, frequency response, and maximum signal level of an audio-frequency circuit. The square wave is useful for checking the transient response of circuits.

The additional wave shapes offered by a function generator are useful for more esoteric applications. Sawtooth and triangle waves can be used to sweep a voltage through a range to test the response of a circuit to various voltages. The pulse output is useful for testing digital circuitry, especially if the instrument has provision for adjusting the on and off voltage levels. Frequency sweep capability is very handy for testing the frequency response of an audio circuit, especially if a sweep ramp or sweep trigger output is provided for synchronizing an oscilloscope. Some instruments provide an input to control the frequency by means of an external voltage.

When selecting a function generator, the first specification to look at is the frequency range. Many units cover low radio frequencies as well as audio. The frequency accuracy may be important for many applications. Digitally-synthesized models use a quartz crystal clock oscillator so are much more accurate than analog models. Some analog models do have a built-in frequency counter, but the accuracy may still be limited by frequency drift.

The amplitude range and accuracy should also be considered. An amplitude of several volts is useful for testing power devices and the ability to accurately set the amplitude to a few millivolts may be needed for driving the microphone input of a transmitter. The output impedance is usually either 600 Ω or 50 Ω.

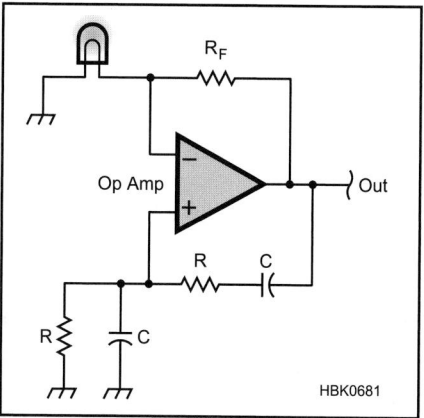

Fig 25.23 — A Wien bridge oscillator has lower distortion than a twin-T oscillator. The light bulb acts to stabilize the feedback to prevent distortion due to amplifier clipping.

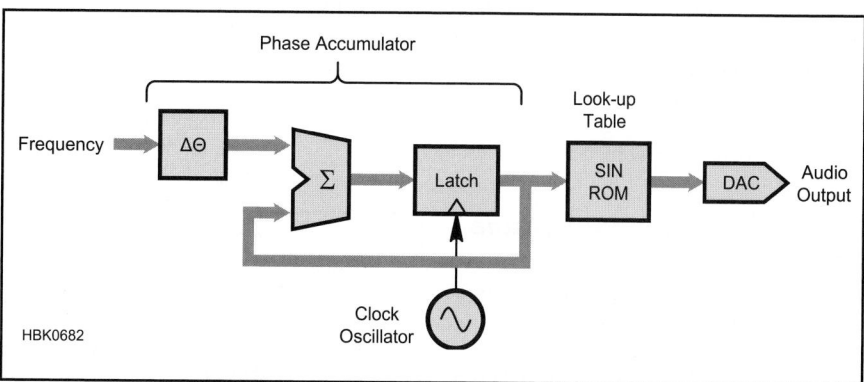

Fig 25.24 — A direct digital synthesizer (DDS) sine-wave generator. Not shown is a low-pass filter at the output that attenuates unwanted spurious frequencies above one-half the clock frequency.

Function generators are available for a wide range of prices, from inexpensive hand-held units to sophisticated bench instruments costing thousands of dollars. Surplus tube-type models can be good values, starting with the venerable Hewlett-Packard 200A, manufactured until the early 1970s. More-recent models can also be found on the used-equipment market made by HP, its successor company Agilent Technologies, Tektronix, B&K, Keithley, Wavetek, Leader, Fluke and others.

An even less-expensive solution is to use a computer sound card as a function generator. A variety of free function generator software can be found on the Internet. Since the sound card output is ac-coupled it is not possible to adjust the dc offset voltage as it is with most function generators. Also, the frequency range, the level range and accuracy, and the output drive capability are not as good as you would expect from a special-purpose instrument. However, quite sophisticated waveforms may be generated with the right software and the price is right.

25.3.6 Measuring Inductance and Capacitance

The traditional way to measure inductance (L), capacitance (C), or resistance (R) is with an *LCR bridge* consisting of a Wheatstone bridge driven by a sine-wave voltage and with an ac voltmeter for the null detector. In **Fig 25.25,** the box labeled Z is the reference component. It must be the same type as the device under test, an inductor, capacitor or resistor. The bridge is nulled when the ratio of the variable resistor R_V to R is the same as the

ratio of the impedance of the device under test to that of Z. The value of R_V is proportional to the resistance or inductance and inversely proportional to the capacitance. To make it proportional to C, simply swap the positions of R_V and R in the circuit. In all cases, when $R_V = R$, the null is achieved when the value of the device under test equals the value of Z.

Most commercial LCR bridges include a switch to select various values of Z. Typically they are in decade steps so that the dial calibration works on all scales. The frequency of the sine-wave source should be such that the reactance of the capacitor or inductor under test is not so low as to present too small a load impedance to the sine-wave source and not so high that the load presented by the ac voltmeter reduces the depth of the null. Often several frequencies may be selected, the lower frequencies being used for large-value

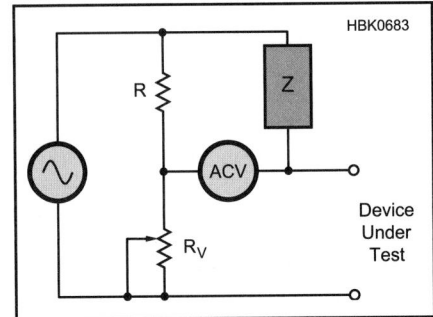

Fig 25.25 — An LCR bridge, which is a type of Wheatstone bridge used to measure inductors (L), capacitors (C) and resistors (R). The box labeled "Z" is the reference component to which the device under test is compared.

L and C and higher frequencies used for the smaller values.

A dip meter may be used for measuring either L or C as long as a component of the opposite type is available whose value is accurately known. The technique is to make a tuned circuit by connecting the inductor and capacitor in parallel and then measure the resonant frequency with a grid dip meter. The inductance or capacitance can be determined from the dip frequency, f, using the formula

$$L = \frac{1}{\left(2\pi f\right)^2 C} \quad \text{or}$$

$$C = \frac{1}{\left(2\pi f\right)^2 L}$$

Some multimeters have built-in capability to measure inductors and capacitors. It is also possible to build an adapter to allow measuring L and C with any multimeter. Circuit examples are given in the Construction Projects section at the end of this chapter.

The use of capacitors in high-frequency switchmode supplies (see the **Power Sources** chapter) makes it important to measure their *equivalent series resistance (ESR)* and *equivalent series inductance (ESL)*. ESR and ESL cause loss and affect the switching circuit's ability to regulate properly. An ESR meter measures a capacitor's ESR by using short pulses or ac signals. Some ESR meters can be used with the capacitor in-circuit although they should not be used with the capacitor charged or energized. ESL is typically measured with an inductance meter as descibed above.

25.4 RF Measurements

RF measurements are a special case of ac measurements. While everything in the previous section about measuring low-frequency ac applies, there are additional factors to take into consideration for high frequency measurements, such as parasitic values and operational bandwidth. This section concentrates on equipment and techniques especially suited to measuring at radio frequencies. The **RF Techniques** chapter has additional information on circuits at high frequencies.

25.4.1 Measuring RF Voltage and Current

An *RF probe* rectifies a radio-frequency signal so that its amplitude can be measured with a dc instrument such as a multimeter. The circuit is typically quite simple, consisting of a diode, a resistor and a couple capacitors, as in the example of **Fig 25.26.** The resistor value is chosen so that the rectified

output voltage is approximately equal to the RMS value of a sine wave input signal. The 390 kΩ value shown assumes the meter has a 1 MΩ input resistance. The diode can be a high-speed switching diode such as a 1N914 or 1N4148, or it can be a Schottky diode for greater sensitivity.

The detector is located as close to the measuring point as possible to minimize stray inductance and capacitance. The leads to the

dc voltmeter can be longer. The RF probe can be housed in any convenient enclosure that fits easily in the hand and the circuitry can be constructed on a scrap of perforated phenolic board, as long as the leads are kept short. A much more elaborate version with a coaxial input and an integrated compensated voltmeter is described on the CD-ROM that accompanies this book.

Measuring RF current is a little more dif-

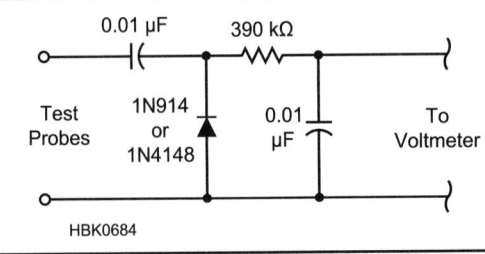

Fig 25.26 — A basic RF probe, used to convert an RF signal into a dc voltage that can be measured by a voltmeter.

ficult. When measuring the current on an antenna or feed line, it may not be practical to break the connection to insert an ammeter. Even when that can be done, the meter itself often upsets the measurement, either because it must be plugged into ac power or simply because of the instrument's size.

A time-honored technique is to wire a small incandescent lamp in series with the antenna or feed line to be measured. Although you can get a rough idea of the amount of RF current by comparing the brightness to the brightness with a known dc current, this method is only useful for relative measurements where the absolute value does not need to be known.

Another alternative is to construct an RF ammeter using a current transformer that clamps over the wire.[11] That method has the advantage that the wire does not need to be disconnected. It is important that the leads to the meter not couple to the wire or cable being measured.

Highly-accurate commercial RF ammeters made to measure the base current of AM broadcast station antennas can sometimes be found on the surplus market. If you can find one it would give much better accuracy than a homebrew device.

25.4.2 Measuring RF Power

Transmitter power is normally given in units of watts. Lower-power RF signals found in receiver and transmitter circuits may be specified in units of milliwatts (mW) or microwatts (μW) but it is perhaps more common to see units of *dBm*, decibels with respect to one milliwatt. For example, it is much easier to express the power of an S1 signal as –121 dBm than 0.00000000076 μW. The formula is dBm = 10 log (1000 P), where P is the power in watts. One milliwatt (P = 0.001) is 0 dBm, 10 mW is +10 dBm, 0.01 mW is –20 dBm and so on. **Table 25.2** lists common dBm and power equivalent.

Measuring RF power can be a little confusing because there are several ways to do it. We have already covered the difference between peak and RMS voltage and current. RF power is always based on RMS values. For example, if the RF voltage into a 50-Ω dummy load is 70.7 V RMS (100 V at the peak

of the RF sine waves), then the power is P = $E^2 / R = 70.7^2 / 50 = 100$ W. (See **www.eznec.com/Amateur/RMS_Power.pdf** for a more extensive explanation of power and RMS.)

Peak envelope power (PEP) has nothing to do with the difference between the peak and average voltage of a sine wave. It is a measure of the power of an RF signal at the modulation peak, averaged over one RF cycle. For a CW signal, the PEP is simply the power when the key is closed, as read on any wattmeter. However, for an SSB signal, the power is constantly changing as you speak. An average-reading wattmeter will read a value much lower than the PEP.

Measuring the peak envelope power is more difficult than measuring the average power. An oscilloscope can be a highly-accurate method within its bandwidth limitations if the load impedance is known accurately. Don't forget that the oscilloscope shows peak-to-peak rather than RMS voltage, so you must divide the maximum reading by $2\sqrt{2}$ or 2.828. Some wattmeters do have PEP-reading capability. Their circuitry must have very fast response to the detected RF signal to give an accurate reading of the peaks.

A *directional wattmeter* is a device that measures power flowing in each direction on a transmission line. The manner in which RF signals propagate on transmission lines is covered in the **Transmission Lines** chapter. Many amateurs keep a wattmeter permanently connected at their station to monitor the condition of their transmitter and antennas. See the **Station Accessories** chapter for a further discussion of directional wattmeters.

A *bolometer* is a device for measuring transmitter power by measuring the heat dissipated in a resistive load. A thermistor or other device measures the temperature. The device is calibrated with a dc voltage, since dc voltage and current can be measured very precisely. With careful construction and calibration, very high accuracy can be obtained. However, the response time of the measurement is very slow, so a bolometer is normally used to calibrate another wattmeter rather than being used directly for measurements. Bolometers are made commercially, however it is possible to homebrew one using a plastic picnic cooler.[12]

Commercial laboratory power meters are generally intended for measuring power levels in the milliwatt or microwatt range. There are two basic types, based on either diode or thermocouple detectors. Below a certain power level, the dc output from a diode detector is directly proportional to power, that is, the square of the RF voltage. With a suitable dc amplifier, the detector output can drive a meter with a linear scale to read power directly.

Thermocouple-type power meters feed the RF signal into a resistor that heats up in

proportion to the power level. The temperature is measured with a *thermocouple*, which consists of a pair of junctions of dissimilar metals. A voltage is generated based on the temperature difference of the two junctions, which is proportional to the RF power. The thermocouple method gives high accuracy and wide bandwidth, but the measurement time can be up to several seconds at low power levels, rather than being nearly instantaneous as with a diode detector. Older-model surplus diode and thermocouple-type meters can often be found for reasonable prices, but the proper sensors for the desired frequency range and power level often cost more than the power meter itself.

Analog Devices makes a series of integrated circuits that can detect RF signals and output a dc voltage proportional to the logarithm of the power level. That makes it easy to construct an RF power meter that reads directly in dBm. For example, the AD8307 covers dc to 500 MHz with 1-dB accuracy over an 88-dB (nearly 1 billion-to-one) power range.

25.4.3 Spectrum Analyzers

A spectrum analyzer is similar to an oscilloscope. Both present a graphical view of an electrical signal. The oscilloscope is used to observe electrical signals in the *time domain* (amplitude versus time). The time domain,

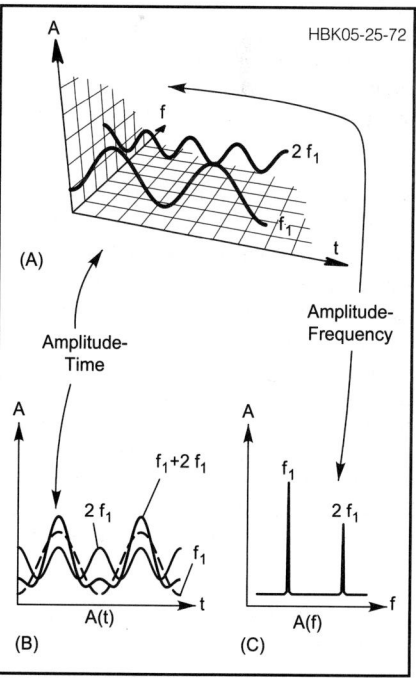

Fig 25.27 — A complex signal in the time and frequency domains. A is a three-dimensional display of amplitude, time and frequency. B is an oscilloscope display of time vs amplitude. C is spectrum analyzer display of the frequency domain and shows frequency vs amplitude.

Table 25.2
Power and dBm Equivalents

dBm	Power	dBm	Power
–60 dBm	1 nW	3 dBm	2 mW
–30 dBm	1 μW	6 dBm	4 mW
–20 dBm	10 μW	10 dBm	10 mW
–10 dBm	100 μW	20 dBm	100 mW
–6 dBm	¼ mW	30 dBm	1 W
–3 dBm	½ mW	60 dBm	1 kW
0 dBm	1 mW	61.7 dBm	1.5 kW

however, gives little information about the frequencies that make up complex signals, which are best characterized in terms of their frequency response. This information is obtained by viewing electrical signals in the *frequency domain* (amplitude versus frequency). One instrument that can display the frequency domain is the spectrum analyzer.

TIME AND FREQUENCY DOMAIN

To better understand the concepts of time and frequency domain, see **Fig 25.27**. The three-dimensional coordinates in Fig 25.27A show time as the line sloping toward the bottom right, frequency as the line rising toward the top right and amplitude as the vertical axis. The two discrete frequencies shown are harmonically related, so we'll refer to them as f1 and 2f1.

In the representation of time domain in Fig 25.27B, all frequency components of a signal are summed together. In fact, if the two discrete frequencies shown were applied to the input of an oscilloscope, we would see the solid line (which corresponds to the sum of f1 and 2f1) on the display.

In the frequency domain, complex signals (signals composed of more than one frequency) are separated into their individual frequency components. A spectrum analyzer measures and displays the power level at each discrete frequency; this display is shown at C.

The frequency domain contains information not apparent in the time domain and therefore the spectrum analyzer offers advantages over the oscilloscope for certain measurements, such as harmonic content or distortion as mentioned previously. For measurements that are best made in the time domain, the oscilloscope is the tool of choice.

SPECTRUM ANALYZER BASICS

There are several different types of spectrum analyzer, but the most common is nothing more than an electronically tuned superheterodyne receiver. The receiver is tuned by means of a ramp voltage. This ramp voltage performs two functions: First, it sweeps the frequency of the analyzer local oscillator; second, it deflects a beam across the horizontal axis of a CRT display, as shown in **Fig 25.28**. The vertical axis deflection of the CRT beam is determined by the strength of the received signal. In this way, the CRT displays frequency on the horizontal axis and signal strength on the vertical axis.

Most spectrum analyzers use an up-converting technique in which a wide-band input is converted to an IF higher than the highest input frequency. Up-conversion is used so that a fixed-tuned input filter can remove any image signals and only the first local oscillator needs to be tuned to tune the receiver. As with most up-converting communications receivers, it is not easy to achieve the desired

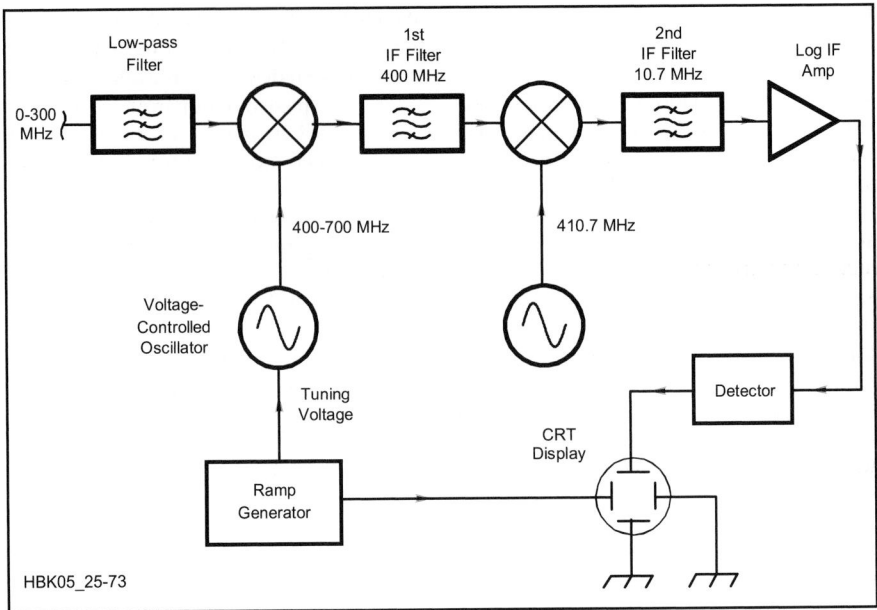

Fig 25.28 — A block diagram of a typical superheterodyne spectrum analyzer. Input frequencies of up to 300 MHz are up-converted by the local oscillator and mixer to a fixed frequency of 400 MHz.

ultimate selectivity at the first IF, because of the high frequency. For this reason, multiple conversions are used to generate an IF low enough so that the desired selectivity is practical. In the example shown, dual conversion is used: The first IF is at 400 MHz; the second at 10.7 MHz.

In the example spectrum analyzer, the first local oscillator is swept from 400 MHz to 700 MHz; this converts the input (from nearly 0 MHz to 300 MHz) to the first IF of 400 MHz. The usual rule of thumb for varactor-tuned oscillators is that the maximum practical tuning ratio (the ratio of the highest frequency to the lowest frequency) is an octave, a 2:1 ratio. In our example spectrum analyzer, the tuning ratio of the first local oscillator is 1.75:1, which meets this specification.

The range of image frequency extends from 800 MHz to 1100 MHz and is easily eliminated using a low-pass filter with a cut-off frequency around 300 MHz. The 400-MHz first IF is converted to 10.7 MHz where the ultimate selectivity of the analyzer is obtained. The image of the second conversion, (421.4 MHz), is eliminated by the first IF filter. The attenuation of the image should be large, on the order of 60 to 80 dB. This requires a first IF filter with a high Q, which is achieved by using helical resonators, SAW resonators or cavity filters. Another method of eliminating the image problem is to use triple conversion; converting first to an intermediate IF such as 50 MHz and then to 10.7 MHz. As with any receiver, an additional frequency conversion requires added circuit-

ry and adds potential spurious responses.

Most of the signal amplification takes place at the lowest IF, 10.7 MHz in this example. Here the communications receiver and the spectrum analyzer differ. A communications receiver demodulates the incoming signal so that the modulation can be heard or further demodulated for RTTY or packet or other mode of operation. In the spectrum analyzer, only the signal strength is needed.

In order for the spectrum analyzer to be most useful, it should display signals of widely different levels. As an example, consider two signals that differ by 60 dB, which is a thousand to one difference in voltage or a million to one in power. That means that if power were displayed, one signal would be one million times larger than the other. In the case of voltage one signal would be a thousand times larger. In either case it would be difficult to display both signals on a CRT. The solution to this problem is to use a logarithmic display that shows the relative signal levels in decibels. Using this technique, a 1000:1 ratio of voltage reduces to a 60-dB difference.

The conversion of the signal to a logarithm is usually performed in the IF amplifier or detector, resulting in an output voltage proportional to the logarithm of the input RF level. This output voltage is then used to drive the CRT display.

SPECTRUM ANALYZER PERFORMANCE

The performance parameters of a spectrum analyzer are specified in terms similar to those used for radio receivers, in spite of the fact that there are many differences between a receiver and a spectrum analyzer.

The sensitivity of a receiver is often specified as the *minimum discernible signal* (MDS), which means the smallest signal that can be heard. In the case of the spectrum analyzer, it is not the smallest signal that can be heard, but the smallest signal that can be seen. The *dynamic range* of the spectrum analyzer determines the largest and smallest signals that can be simultaneously viewed on the analyzer. As with a receiver, one factor that affects dynamic range is *second- and third-order intermodulation distortion* (IMD). IMD dynamic range is the maximum difference in signal level between the minimum detectable signal and the level of two signals of equal strength that generate an IMD product equal to the minimum detectable signal. (See the **Receivers** chapter for more information on IMD.)

Although the communications receiver is an excellent example to introduce the spectrum analyzer, there are several differences such as the previously explained lack of a demodulator. Unlike the communications receiver, the spectrum analyzer is not a sensitive radio receiver. To preserve a wide dynamic range, the spectrum analyzer often uses passive mixers for the first and second mixers. Therefore, referring to Fig 25.28, the noise figure of the analyzer is no better than the losses of the input low-pass filter plus the first mixer, the first IF filter, the second mixer and the loss of the second IF filter. This often results in a combined noise figure of more than 20 dB. With that kind of noise figure the spectrum analyzer is obviously not a communications receiver for extracting very weak signals from the noise but a measuring instrument for the analysis of frequency spectrum. For some applications, it may be useful to add an external wide-band, low-noise preamplifier to improve the noise figure.

The selectivity of the analyzer is called the *resolution bandwidth* (RBW). This term refers to the minimum frequency separation of two signals of equal level so that the signals are separated by a drop in amplitude of 3 dB between them. The IF filters used in a spectrum analyzer differ from a communications receiver in that they have very gentle skirts and rounded passbands, rather than the flat passband and very steep skirts of an IF filter in a high-quality communications receiver. The rounded passband is necessary because the signals pass through the filter passband as the spectrum analyzer sweeps the desired frequency range. If the signals suddenly pop into the passband (as they do if the filter has steep skirts), the filter tends to ring. A filter with gentle skirts has less ringing. Another effect, which occurs even with rounded-passband filters, is that the signal amplitudes are reduced at fast sweep rates, which distorts the display

and requires that the analyzer not sweep frequency too quickly. When adjusting the resolution bandwidth or the width of the *frequency span* (the range of frequencies being measured), the scan rate may need to be reduced so that the signal amplitude is not affected by fast sweeping.

The signal produced by the detector is known as the *video* signal. Most spectrum analyzers include a low-pass video filter to reduce the displayed noise level. The *video bandwidth* (VBW) is the bandwidth of this filter. Like the RBW, the VBW must also be taken into consideration when setting the sweep speed.

USING A SPECTRUM ANALYZER

Spectrum analyzers are used in situations where the signals to be analyzed are complex, for very low-level signals, or when the frequency of the signals to be analyzed is very high. Although high-performance oscilloscopes are capable of operation into the UHF region, moderately priced spectrum analyzers can be used well into the gigahertz region.

Unlike the oscilloscope which is a wide-bandwidth instrument, the spectrum analyzer measures the waveform using a narrow bandwidth; thus it is capable of reducing the noise power displayed.

Probably the most common Amateur Radio application of a spectrum analyzer is the measurement of the harmonic content and other spurious signals in the output of a radio transmitter. **Fig 25.29** shows two ways to connect the transmitter and spectrum analyzer. The method shown at A should not be used for wide-band measurements since most line-sampling devices do not exhibit a constant-amplitude output over a broad frequency range. Using a line sampler is fine for narrow-band measurements, however.

The method shown at B is used in the ARRL Lab. The attenuator must be capable

of dissipating the transmitter power. It must also have sufficient attenuation to protect the spectrum analyzer input. Many spectrum analyzer input mixers can be damaged by only a few milliwatts, so most analyzers have an adjustable input attenuator to provide a reasonable amount of attenuation for protection. The power limitation of the attenuator itself is usually on the order of a watt or so, however. This means that the power attenuator must have 20 dB of attenuation for a 100 W transmitter, 30 dB for a 1000 W transmitter and so on, to limit the input to the spectrum analyzer to 1 W. There are specialized attenuators that are made for transmitter testing; these attenuators provide the necessary power dissipation and attenuation in the 20 to 30-dB range.

When using a spectrum analyzer it is very important that the proper amount of attenuation be applied before a measurement is made. In addition, it is a good practice to start with maximum attenuation and view the entire spectrum of a signal before the attenuator is adjusted. The signal being viewed could appear to be at a safe level, but another spectral component, which is not visible, could be above the damage limit. It is also very important to limit the input power to the analyzer when pulse power is being measured. The average power may be small enough so the input attenuator is not damaged, but the peak pulse power, which may not be readily visible on the analyzer display, can destroy a mixer, literally in microseconds.

Spurious Responses in Spectrum Analyzers

When using a spectrum analyzer it is necessary to ensure that the analyzer does not generate additional spurious signals that are then attributed to the system under test. Some of the spurious signals that can be generated by

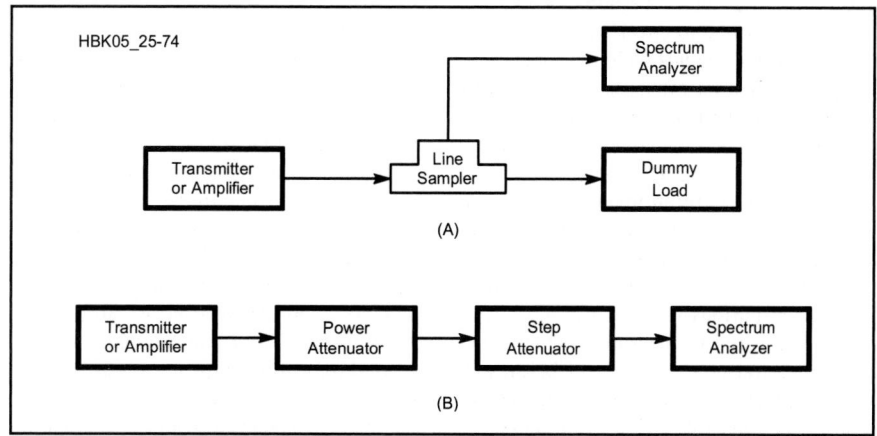

Fig 25.29 — Alternate bench setups for viewing the output of a high power transmitter or oscillator on a spectrum analyzer. A uses a line sampler to pick off a small amount of the transmitter or amplifier power. In B, most of the transmitter power is dissipated in the power attenuator.

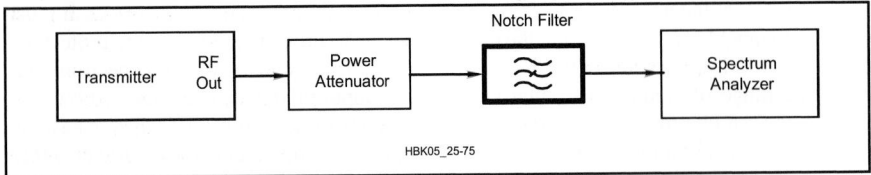

Fig 25.30 — A notch filter is another way to reduce the level of a transmitter's fundamental signal so that the fundamental does not generate harmonics within the analyzer. However, in order to know the amplitude relationship between the fundamental and the transmitter's actual harmonics and spurs, the attenuation of the fundamental in the notch filter must be known.

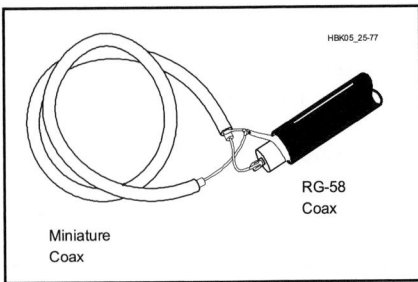

Fig 25.32 — A "sniffer" probe consisting of an inductive pick-up. It has the advantage of not directly contacting the circuit under test.

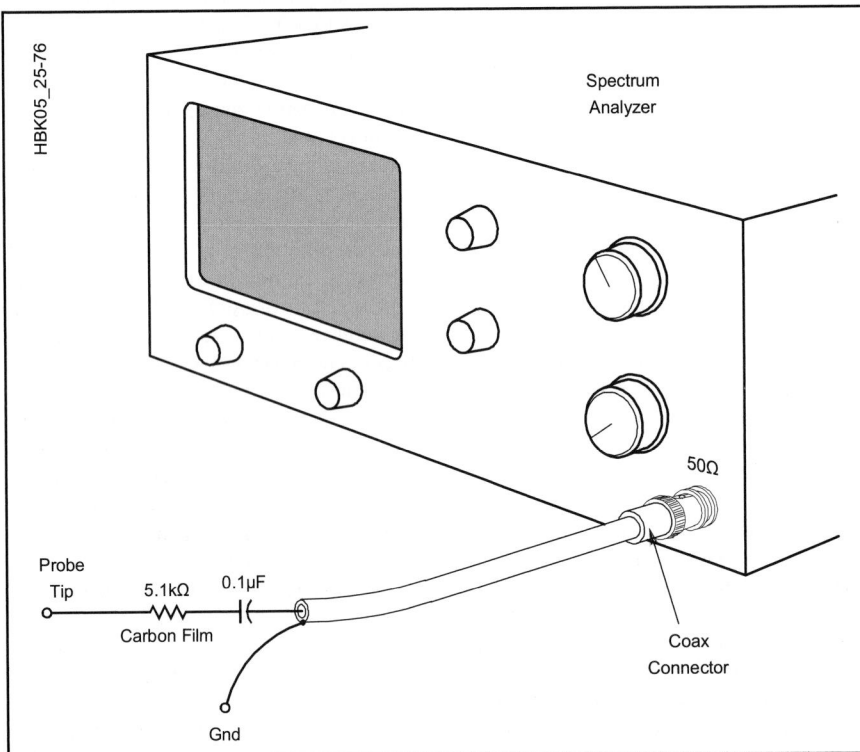

Fig 25.31 — A schematic representation of a voltage probe designed for use with a spectrum analyzer. Keep the probe tip (resistor and capacitor) and ground leads as short as possible.

a spectrum analyzer are harmonics and IMD. It is good practice to check for the generation of spurious signals within the spectrum analyzer. When an input signal causes the spectrum analyzer to generate a spurious signal, adding attenuation at the analyzer input will cause the internally generated spurious signals to decrease by an amount greater than the added attenuation. If attenuation added ahead of the analyzer causes all of the visible signals to decrease by the same amount, this indicates a spurious-free display.

If it is desired to measure the harmonic levels of a transmitter at a level below the spurious level of the analyzer itself, a notch filter can be inserted between the attenuator and the spectrum analyzer as shown in **Fig 25.30**. This reduces the level of the fundamental signal and prevents that signal from generating harmonics within the analyzer,

while still allowing the harmonics from the transmitter to pass through to the analyzer without attenuation. Use caution with this technique; detuning the notch filter or inadvertently changing the transmitter frequency will allow potentially high levels of power to enter the analyzer. In addition, use care when choosing filters; some filters (such as cavity filters) respond not only to the fundamental but notch out odd harmonics as well.

The input impedance for most RF spectrum analyzers is 50 Ω, however not all circuits have convenient 50-Ω connections that can be accessed for testing purposes. Using a probe such as the one shown in **Fig 25.31** allows the analyzer to be used as a troubleshooting tool. The probe can be used to track down signals within a transmitter or receiver, much like an oscilloscope is used. The probe shown offers a 100:1 voltage reduction and loads the

circuit with about 5000 Ω. A different type of probe is shown in **Fig 25.32**. This inductive pickup coil (sometimes called a "sniffer") is very handy for troubleshooting. The coil is used to couple signals from the radiated magnetic field of a circuit into the analyzer. A short length of miniature coax is wound into a pickup loop and soldered to a larger piece of coax that connects to the spectrum analyzer. The dimensions of the loop are not critical, but smaller loop dimensions enable the loop to more precisely locate the source of radiated RF. Coax is used for the loop to provide shielding from the electric field component (capacitive coupling). Connecting the coax shield on only one end provides a complete electrostatic shield without introducing a shorted turn.

The sniffer allows the spectrum analyzer to sense RF energy without contacting the circuit being analyzed. If the loop is brought near an oscillator coil, the oscillator can be tuned without directly contacting (and thus disturbing) the circuit. The oscillator can then be checked for reliable startup and the generation of spurious sidebands. With the coil brought near the tuned circuits of amplifiers or frequency multipliers, those stages can be tuned using a similar technique.

Even though the sniffer does not contact the circuit being evaluated, it does extract some energy from the circuit. For this reason, the loop should be placed as far from the tuned circuit as is practical. If the loop is placed too far from the circuit, the signal will be too weak or the pickup loop will pick up energy from other parts of the circuit and not give an accurate indication of the circuit under test.

The sniffer is very handy to locate sources of RF leakage. By probing the shields and cabinets of RF generating equipment (such as transmitters) egress and ingress points of RF energy can be identified by increased indications on the analyzer display.

Measuring Very Low-Level Signals

One very powerful characteristic of the spectrum analyzer is the instrument's capa-

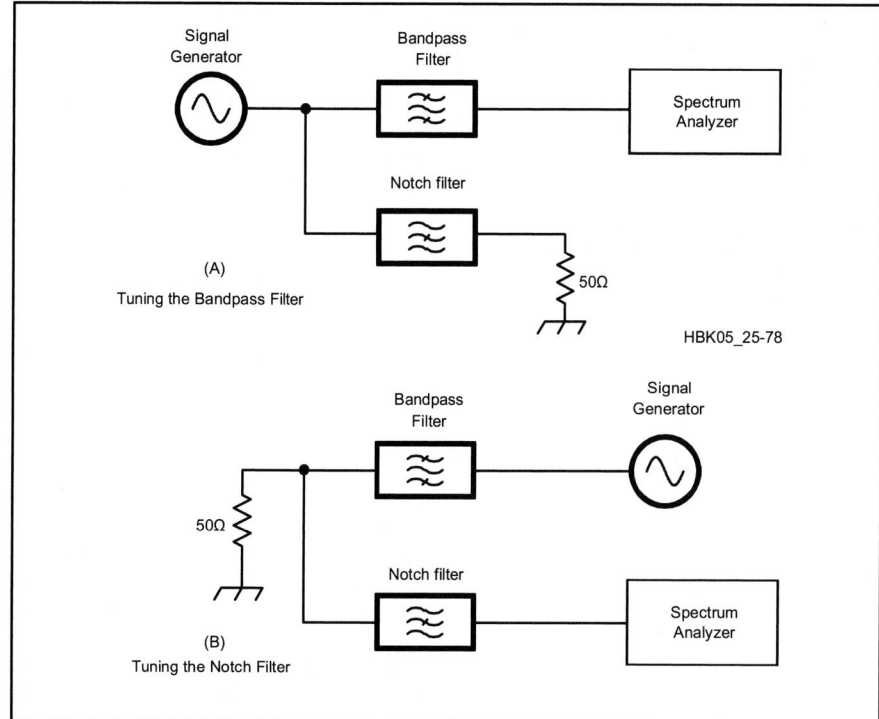

Fig 25.33 — Block diagram of a spectrum analyzer and signal generator being used to tune the band-pass and notch filters of a duplexer. All ports of the duplexer must be properly terminated and good quality coax with intact shielding used to reduce leakage.

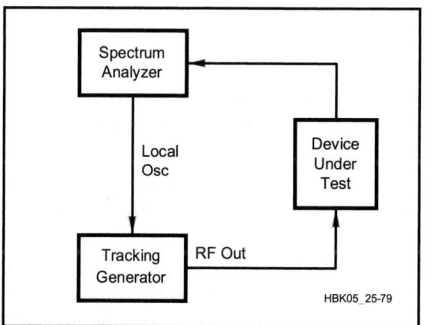

Fig 25.34 — A signal generator (shown in the figure as the "Tracking Generator") locked to the local oscillator of a spectrum analyzer can be used to determine filter response over a range of frequencies.

bility to measure very low-level signals. This characteristic is very advantageous when very high levels of attenuation are measured. **Fig 25.33** shows the setup for tuning the notch and passband of a VHF duplexer. The spectrum analyzer, being capable of viewing signals well into the low microvolt region, is capable of measuring the insertion loss of the notch cavity more than 100 dB below the signal generator output. Making a measurement of this sort requires care in the interconnection of the equipment and a well-designed spectrum analyzer and signal generator. RF

energy leaking from the signal generator cabinet, line cord or even the coax itself, can get into the spectrum analyzer through similar paths and corrupt the measurement. This leakage can make the measurement look either better or worse than the actual attenuation, depending on the phase relationship of the leaked signal.

EXTENSIONS OF SPECTRAL ANALYSIS

Many measurements involve a signal generator tuned to the same frequency as the spectrum analyzer. It would be a real convenience not to have to continually reset the signal generator to the desired frequency. It is, however, more than a convenience. A signal generator connected in this way is called a *tracking generator* because the output frequency tracks the spectrum analyzer input frequency. The tracking generator makes it possible to make swept frequency measurements of the attenuation characteristics of circuits, even when the attenuation involved is large.

Fig 25.34 shows the connection of a tracking generator to a circuit under test. In order for the tracking generator to create an output frequency exactly equal to the input frequency of the spectrum analyzer, the internal local oscillator frequencies of the spectrum analyzer must be known. This is the reason for the interconnections between the tracking

generator and the spectrum analyzer. The test setup shown will measure the gain or loss of the circuit under test.

Only the magnitude of the gain or loss is available; in some cases, the phase angle between the input and output would also be an important and necessary parameter. That is the function of a *vector network analyzer*, covered in a following section.

CHOOSING A SPECTRUM ANALYZER

The frequency range is perhaps the most important specification when choosing a spectrum analyzer. Of course it must cover the amateur bands you wish to test, but it must also cover harmonics of those frequencies if you wish to test transmitter harmonics. Many higher-frequency spectrum analyzers only cover down to 10 kHz, so they are not useful for audio spectrum analysis (spectrum analyzers for audio use are also available). Some instruments use a harmonic sampler on the higher microwave bands. You may need to use an external filter to remove any unwanted signals at harmonics or sub-harmonics of the signal you wish to examine.

Frequency stability is vital for narrow-band measurements such as looking at modulation spectra. Some older instruments include a frequency lock scheme that stabilizes the display after the analyzer is tuned to the desired signal.

The range of resolution bandwidths (RBW) available determines what kinds of measurements are possible. To measure two-tone intermodulation distortion of an SSB signal, the minimum RBW must be narrow enough to resolve the two tones, that is, no more than about 100 Hz. If you wish to view the demodulated time-domain video from a fast-scan television signal, the RBW must be wide enough to include the entire signal, perhaps 5 or 6 MHz.

Input sensitivity is normally not very important. An external preamplifier can always be added if needed for a particular measurement. However, dynamic range is a key specification, as previously discussed. Also, the screen display range is important. For example, if there are 8 divisions on the display and the maximum dB-per-division setting is 10 dB/div, then the maximum display range is 80 dB. Many spectrum analyzers can display signals in linear as well as logarithmic (dB) mode, which can be useful to look at modulation.

For accurate power measurements, the amplitude accuracy is important. For doing relative measurements, which are important for measuring spurious signals relative to the carrier, logarithmic linearity is the key specification.

There are several features that expand the spectrum analyzer's utility. *Markers* are pips that can be moved with a knob back and forth across the signals on the display. By placing a marker at the top of a signal you can read the frequency and amplitude directly. *Peak search* is a feature that automatically places a marker at the peak of the strongest signal. *Marker delta* measures the difference in frequency and amplitude of two markers.

Zero span means setting the sweep width to zero hertz to view the change in signal versus time. This is useful for looking at modulation. Many spectrum analyzers include a sweep triggering feature for this purpose, similar to the triggering circuit in an oscilloscope. Some include a speaker so you can hear what the modulating signal sounds like.

Modern spectrum analyzers have many of the features found in digital oscilloscopes, such as trace storage and retrieval. Most include computer connectivity, both to store measured data and to control the instrument with the proper software.

In addition to traditional swept-frequency spectrum analyzers, some modern units use the *Fast Fourier Transform* (FFT) either in place of, or in addition to, the swept-frequency architecture. The FFT allows for much faster screen updates when using narrow spans. Some RF and microwave instruments have a *digital IF* based on the FFT. For narrow spans, the local oscillator is held at a constant frequency and the "sweep" is performed mathematically by the FFT.

There are many used spectrum analyzers on the surplus market, at all different vintages and price points. As usual, try before you buy if at all possible and obtain the operator's and service manuals, if available. Another alternative is a spectrum analyzer built on a card that plugs into a personal computer or that consists of an external digitizing pod connected to the PC via a serial data link. As with PC-based oscilloscopes, they can represent excellent value because they save the cost of the display, cabinet, power supply, and microprocessor provided by the computer.

The least-expensive alternative is an audio spectrum analyzer based on a PC's sound card. Free spectrum analysis software is available on the Internet. The main limitation is the frequency coverage, which is typically about 40-45% of the sample rate of the sound card. The sound card that comes standard in a PC typically has a 48 kHz sample rate, so that the frequency response is limited to about 20 kHz. That is plenty for checking out the audio circuits in a communications receiver or transmitter.

25.4.4 Measuring RF Impedance

Most RF impedance-measuring devices are based on the Wheatstone bridge, as previously described in the section on LCR

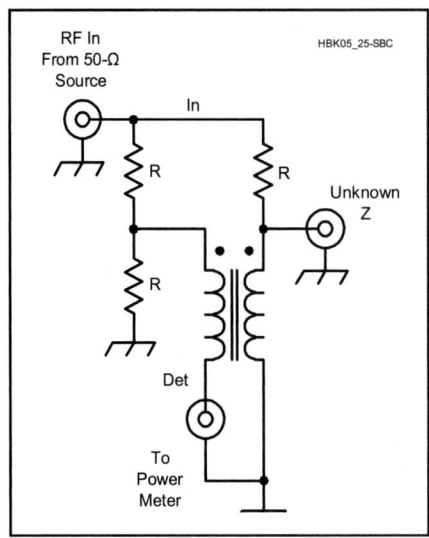

Fig 25.35 — A return-loss bridge for RF. See text for details.

bridges. The difference is that an impedance bridge does not require that the device under test be a pure inductance, capacitance or resistance but may be a complex impedance that includes both resistance and reactance. That means that two adjustments are required to balance the bridge, rather than just one as with an LCR bridge. Both the resistance and reactance must match.

In addition, since an impedance bridge is typically designed to operate at higher frequencies, circuit layout and the connection to the device under test are more important. Usually a coaxial connector is used to connect to the device and the impedance measurements are referenced to 50 Ω.

A *return-loss bridge* (RLB) is an RF bridge with fixed, usually 50-Ω, resistors in each leg except the one connected to the device under test. *Return loss* is the ratio of the reflected signal to the signal incident on a component, usually expressed in dB, which is always a positive number for a passive device.[13] (See the **RF Techniques** chapter for a discussion of return loss.)

The schematic of a simple return-loss bridge is shown in **Fig 25.35**. The circuit can also be used as a *hybrid combiner* (see the RF and Microwave Test Accessories section of this chapter), where the port labeled UNKNOWN is the common port and the other two are isolated from each other. For good results at high frequencies, it should be built in a small box with short leads to the coax connectors.

Apply the output of a signal generator or other signal source to the RF IN port of the RLB. The power level should be appropriate for driving the device connected to the UNKNOWN port, after accounting for the 6-dB

loss of the bridge. Connect the bridge POWER METER port to a power meter or other power-measuring device through a step attenuator and leave the UNKNOWN port of the bridge open circuited. Set the step attenuator for a relatively high level of attenuation and note the power meter indication.

Now connect the unknown impedance to the bridge. The power meter reading will decrease. Adjust the step attenuator to produce the same reading obtained when the UNKNOWN port was open circuited. The difference between the two settings of the attenuator is the return loss, measured in dB.

ANTENNA ANALYZERS

A return-loss bridge measures only the magnitude of the reflection coefficient. Among radio amateurs, an RF impedance analyzer capable of measuring both the magnitude and the phase is often called an *antenna analyzer*, since it is mostly used for measuring antennas.

To keep size and cost to a minimum, portable antenna analyzers usually use a narrow-band source (an internal oscillator) and wide-band detector (a diode). Some manufactured units include a microprocessor and can read out SWR, return loss, resistance, reactance and the magnitude and phase of the impedance on an LCD display. Antenna analyzers suitable for amateur use are available from a number of manufacturers — search for "antenna analyzers" on the Internet to find them. When shopping for an antenna analyzer, pay careful attention to the capabilities and limitations. Be aware that some units measure only the SWR or impedance magnitude while others do measure both the resistive and reactive parts of the impedance. Some units give the magnitude of the reactance but not the sign.

Some users have reported difficulties in obtaining accurate impedance measurements on low-band antennas when there is a nearby AM broadcast station. This is due to the wide-band detector responding to the incoming signal from the AM station.

Noise Bridges

Rather than use a narrow-band source and wide-band detector, another technique is to use a wide-band noise generator for the source and a receiver for the detector. Since most hams already have a receiver, this eliminates the need for an accurate, stable signal generator. You tune the receiver to the desired frequency and adjust the resistance and reactance controls for minimum noise in the receiver. If the receiver is fitted with a *panadapter* (a spectrum display for frequencies near the receive frequency using the receiver's IF circuits), you can locate the null as you adjust the controls, speeding the adjustment.

25.4.5 Network Analyzers

An electronic *port* is a pair of terminals with equal and opposite current flows. An antenna analyzer can measure the impedance of a single-port network such as an antenna, the input port of an antenna tuner, or a two-terminal component like a resistor or capacitor. An example of a two-port network is an amplifier. The input is one port, the output the other. A low-pass filter is another example. An example of a three-port network is a diplexer, used to connect one transceiver to two antennas. The theory behind two-port networks is covered in the **RF Techniques** chapter.

A *scalar network analyzer* (SNA) is an instrument that can measure, as a function of frequency, the magnitude of the gain or loss between the two ports of a two-port device as well as the magnitude of the return loss (equivalent to the SWR) of each port. The term "scalar" means that only magnitudes of those quantities are measured and not phase angles. In **Fig 25.36**, the signal labeled "REF" is proportional to the sweep oscillator output and the one labeled "A" is proportional to the signal that is reflected back from the device under test due to an imperfect 50-Ω match. The ratio of A to REF can be used to calculate the return loss and SWR of the device. Signal "B" is proportional to the signal that travels between the two ports of the device. The ratio of B to REF gives the device gain. To find the gain in the opposite direction and the return loss of the other port, simply turn the device around and swap the two ports. Some network analyzers come with a *test set* that includes RF switches to do that automatically.[14]

The box labeled "Signal processing and display" includes circuitry to take the logarithm of the signals so they can be displayed in dB. In modern instruments it normally includes a microprocessor, which does much of the signal processing as well as controlling the sweep oscillator and other circuitry.

A *vector network analyzer* (VNA) is an instrument that can measure both the magnitude and phase of the gain and return loss of a two-port network. The block diagram is identical to Fig 25.36 except that the detectors can measure not only the magnitude, but also the phase of each signal with respect to the reference, REF. The measured data are usually presented in terms of *scattering parameters*, or *S parameters*. S_{11} is the ratio of the incident to the reflected signal at port 1, S_{22} is the equivalent on port 2, S_{21} is the gain from port 1 to 2, and S_{12} is the gain from port 2 to 1. Each S parameter is a complex number, representing both magnitude and phase. Further discussions of S parameters can be found in the **Computer-Aided Circuit Design** and the **RF Techniques** chapters. Usually the VNA provides the capability to plot the

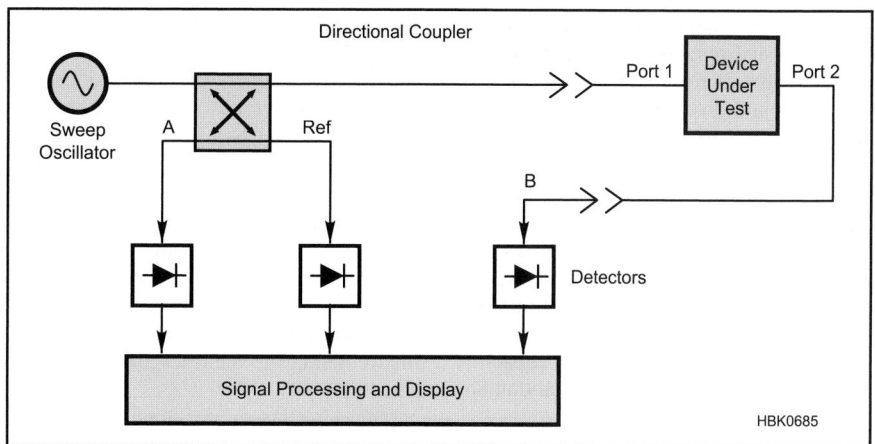

Fig 25.36 — Block diagram of a scalar network analyzer (SNA). A vector network analyzer has the same block diagram except that the detectors can measure phase as well as amplitude.

S parameters on a Smith chart overlaid on the display for easy evaluation. The Smith chart is discussed in the **Transmission Lines** chapter.

The accuracy of a VNA depends on the performance of the directional couplers, detectors, and other circuitry. The coupler directivity is a key contributor to measurement errors. For example, if the directivity is 20 dB, then even a perfect 50-Ω load will show 20 dB of return loss, equivalent to a 1.22:1 SWR. Fortunately it is possible to eliminate most errors by measuring several terminations of known impedance and calculating a series of correction factors. The mathematics involved is quite complex, but it is all handled by the instrument software and is invisible to the user. The most popular technique is called a Short-Open-Load-Through (SOLT) calibration because it uses four calibration standards, a short-circuit, an open-circuit, a 50-Ω load and a through connection between the two ports. One of the sources of error removed by the calibration is any error due to the length of the coaxial cables that connect to the device under test. For that reason, the calibration standards should always be connected at the end of the cables that are to be used for the test. A thorough discussion of VNA error-correction techniques can be found in an Agilent Technologies application note.[15]

CHOOSING A NETWORK ANALYZER

A spectrum analyzer with tracking generator can be the basis of a high-performance scalar network analyzer. Referring to Fig 25.36, the only pieces that need to be added are the directional coupler and a pair of 50-Ω loads. The tracking generator takes the place of the sweep oscillator. Connect the spectrum analyzer input to A, B or REF and terminate the other two with 50-Ω loads. Unlike a conventional scalar network analyzer, a spectrum analyzer has a tuned receiver so

it does not respond to spurious frequencies generated by the sweep oscillator or the device under test.

At the other end of the performance "spectrum" a simple SNA for measuring gain or attenuation can be made with any sweep oscillator, perhaps a function generator with sweep capability, plus a detector of some kind and an oscilloscope to display the detected voltage. If the oscilloscope bandwidth is great enough to cover the desired frequency range, no detector is needed and the RF signal can be displayed directly. Dynamic range is limited with this technique since the displayed signal is linear and not logarithmic. However it is sufficient for many purposes, such as measuring the passband ripple of a filter or the gain versus frequency of an amplifier.

Used scalar network analyzers are readily available on the surplus market. The sweep oscillator is normally a separate unit. Vector network analyzers also may be purchased surplus. Some require an external sweep oscillator while others include an internal signal source. The tests sets may be internal or external as well.

In recent years, several amateur VNA designs have appeared. These tend to be much smaller, lighter and lower-cost than commercial units. One by DG8SAQ down-converts the signals to be detected to frequencies in the audio range so they can be fed to the sound card on a personal computer.[16] It uses a pair of Analog Devices DDS chips for the sweep oscillator and the down-converter's local oscillator. All signal processing and control is done in the microprocessor in the PC. Paul Kiciak, N2PK, also developed a VNA using DDS chips for the source and a direct-conversion architecture which performs the detection function at dc. Like the DG8SAQ design, it relies on a PC for the processing, control, and display. His design has become available in kit form from several suppliers, and numerous third-parties have written software to support it. A number of VNA kits and products are available, most using a PC as a basis.[16,17,18]

25.4.6 RF Signal Generators

An *RF signal generator* is a test oscillator that generates a sine-wave signal that has an accurately-calibrated frequency and amplitude over a wide radio-frequency range. Usually AM and FM modulation capability is provided and sometimes other modulation types as well.

Some instruments also have built-in sweep capability. If not, narrow-band sweep can be obtained by feeding a sawtooth waveform into the FM input. Another feature that is very useful for transceiver testing is *reverse power protection*. It protects the sensitive output circuits in the event that the transceiver accidentally goes into transmit mode while connected to the signal generator.

Before the days of modern digital electronics, all signal generators used free-running oscillators. There is a wide variation in frequency stability and accuracy between the best and the worst. Nearly any modern synthesized signal generator has good enough accuracy for most amateur purposes. However that accuracy sometimes comes at the price of phase noise, a type of wideband noise caused by short-term fluctuations in phase that usually drops off gradually from the carrier frequency. The old tube-type Hewlett-Packard 608-series signal generators have better performance in this respect than some modern synthesized instruments. See **Fig 25.37**. Phase noise is discussed in the **Oscillators and Synthesizers** chapter.

Some signal generators, even professional laboratory-grade types, may not have enough output attenuation to accurately measure the minimum discernible signal (MDS) on a sensitive communications receiver. A 10-dB noise figure in a 500 Hz bandwidth implies an MDS of –137 dBm, which is well below the minimum amplitude level on many signal generators. The problem is easily resolved by adding an external fixed 20-dB attenuator. Just remember to subtract 20 dB from all amplitude readings.

Many signal generators include some way to connect them to a computer so they can be controlled for automated testing. That is useful for testing a receiver at many frequencies across a band to be able to plot sensitivity or dynamic range as a function of frequency, for example. Older instruments usually have a GPIB interface while modern ones are more likely to have RS-232, USB or Ethernet.

Probably the most common use for signal generators is for receiver testing. That is covered in detail in a later section. They are also useful for general-purpose test signals. For example, when developing a new receiver design, you could test the RF, IF and audio stages before the VFO is completed by using a signal generator as the local oscillator.

Critical specifications include the frequency range, frequency accuracy and stability, amplitude range, amplitude accuracy, and the spectral purity, including phase noise, harmonics and non-harmonic spurious emissions. Modulation accuracy and the capabilities of the internal modulation generator, if present, may be important as well, depending on the application.

Many inexpensive signal generators made for the hobbyist and consumer electronics service industry are not suitable for measuring receiver sensitivity because when the output attenuator is at maximum, there is more signal leaking out the cabinet and radiating from the power cord than is coming out the coax connector. Such instruments also tend to have poor frequency stability and the tuning dial tunes so fast that they are difficult to set to a precise frequency. They are useful for simple troubleshooting but are not capable of making accurate measurements.

Better-quality signal generators intended for servicing land-mobile and other communications equipment sometimes become available on the surplus market. They don't have the precision specifications of lab instruments but are generally rugged, reliable and easy to use.

Laboratory-grade signal generators are generally too expensive for the home hobbyist if purchased new but older models such as the HP8640B or its military version (AN/USM323) are widely available and have excellent specifications. Of course they may be less reliable than new instruments but in many cases the older technology is repairable without special equipment. Service manuals are sometimes available from the manufacturer on their website.

25.4.7 Using Noise Sources

Most of the time, noise is something to be avoided in electronic circuits. It can be quite useful for testing, however, precisely because the calibrated noise that you create on purpose has the same properties as the unwanted circuit noise that you are trying to minimize. Thus, calibrated sources of noise are used as test instruments for various measurements. (See the **RF Techniques** chapter for a discussion of noise and its associated terminology.)

MEASURING RECEIVER NOISE FIGURE

A noise source can be used to measure the noise figure of an SSB or CW receiver. The *excess noise ratio (ENR)* is the ratio of the noise added by the noise source to the thermal noise level, normally expressed in dB. An explanation of noise figure is given in the **RF Techniques** chapter. The basic idea is to connect the noise source to the receiver antenna input and then measure the difference in the noise level at the receiver speaker terminals when the noise source is turned on or off. When it is off, the measured noise contains only thermal noise and the noise contributed by the receiver. When it is on, the noise includes the excess noise from the noise source. The noise figure can then be

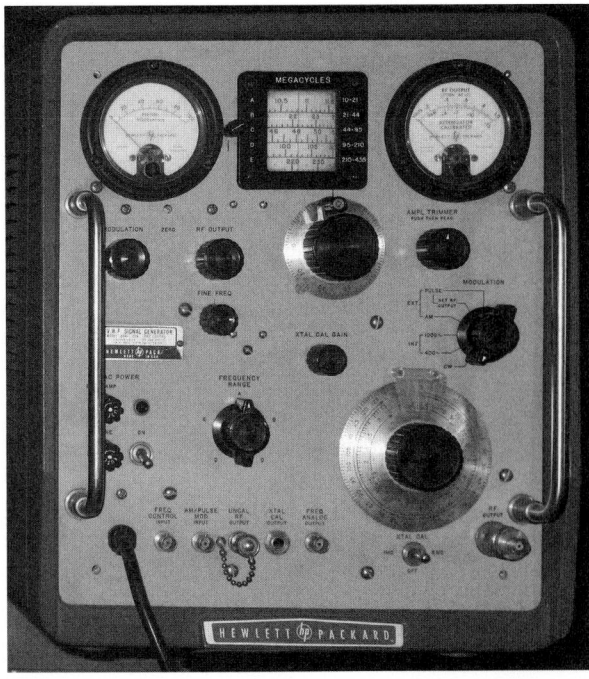

Fig 25.37 — A classic Hewlett-Packard 608F tube-type signal generator. While not as stable as modern synthesized signal generators, in other respects the performance is quite up to date.

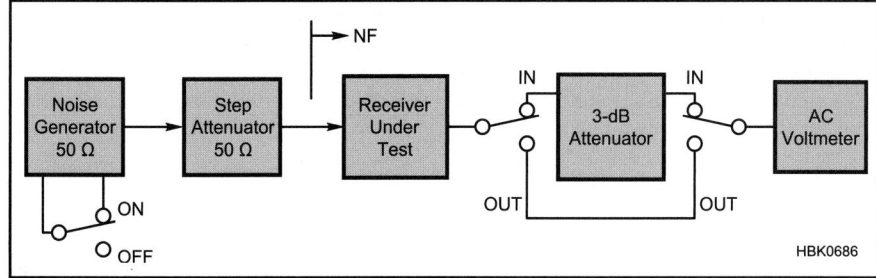

Fig 25.38 — Setup for measuring receiver noise figure.

calculated from the ratio of those two values,

$$NF_M = ENR - 10\log\left(\frac{E_{ON}^2}{E_{OFF}^2} - 1\right)$$

where NF_M is the measured noise figure in dB, ENR is the excess noise ratio in dB, E_{ON} is the RMS output noise voltage with the noise source on, and E_{OFF} is the RMS output noise voltage with the noise source off.

Rather than using the equation, another technique is to include a step attenuator with 1-dB steps at the noise source output. See **Fig 25.38**. Adjust the attenuator so that the noise increases by 3 dB (1.414 times the voltage) when the noise source is turned on. The noise figure is then simply the ENR of the noise source minus the attenuation.

The nice thing about measuring noise with noise is that the accuracy of the voltage reading is not important so long as the ratio of the two readings is accurate. A non-RMS ac voltmeter is not accurate when measuring noise, but so long as the inaccuracy is the same for both voltage levels the ratio is correct. When using a non-RMS instrument, follow the step attenuator technique described in the previous paragraph along with a 3-dB attenuator (1/1.414 voltage divider) in front of the ac voltmeter. First take a voltmeter reading with the noise source off and the 3-dB attenuator bypassed. Then turn on the 3-dB attenuator and the noise source and adjust the step attenuator for the same reading. The noise figure is the ENR of the noise source minus the attenuation.

There are some potential errors to watch out for. The receiver AGC must be turned off for this test so that the gain does not change when the noise source is turned on or off. In addition, the audio gain and perhaps the IF gain must be adjusted so that there is no clipping of the noise peaks under any condition. With the noise source turned on, the RMS value of the noise at the speaker output should be adjusted to no more than ⅓ the maximum output level to prevent the peaks from clipping.

OTHER USES FOR A NOISE SOURCE

You can take advantage of the flat spectrum of a noise source to measure the frequency re-

sponse of RF devices. It requires either a spectrum analyzer or a receiver that can be tuned manually across the frequency range of interest. With sufficient averaging, the output noise level at each frequency accurately reflects the response of the device at that frequency. For example it is quite easy to measure the total response of a receiver from antenna to speaker output using an RF noise source and free software running on a PC that turns the computer sound card into an audio spectrum analyzer. With the noise source connected to the receiver antenna and the audio output connected to the sound card input, the total response, including the IF crystal filter and all the audio stages, is shown by the average noise level in the spectrum analyzer window.

In addition to receivers, noise sources can also be used to measure the noise figure of other devices such as amplifiers. The technique is basically the same, except that the output signal to be measured is at a radio frequency instead of audio. Another difference is that the noise figure of the measuring instrument can affect the results. That is almost never a concern when measuring receivers since the gain from the antenna to the speaker output is normally so high as to swamp out the effect. The corrected noise figure is

$$NF = 10\log\left[10^{NF_M/10} - 10^{(NF_{NMI}-G)/10}\right]$$

where NF_M is the measured noise figure in

dB, NF_{NMI} is the noise figure of the noise-measuring instrument in dB, and G is the gain of the device under test in dB. For that equation to be valid, the bandwidth of the measuring instrument must be no greater than the bandwidth of the device under test.

Commercial noise figure meters combine everything you need to perform noise figure measurements of amplifiers and other devices into one box. While the noise source itself is usually a separate module, the switched power is provided by the noise figure meter. Calculations are performed internally and the noise figure reads out directly on a meter or display. The noise source is repeatedly switched on and off to produce a continuously-updated noise figure reading, which is handy for adjusting or tuning the device under test. Older units typically operate at a small number of fixed intermediate frequencies and it is up to the user to heterodyne the signal to the IF. Many modern units operate over a wide range of frequencies and some can sweep to provide a plot of noise figure versus frequency. Agilent Technologies has a good application note on noise figure measurement that is available for free download.[19]

26.4.8 RF and Microwave Test Accessories

The following section assumes that all coaxial devices are designed for 50-Ω characteristic impedance. While that is by far the most common value for Amateur Radio systems, most of the information applies as well for other impedances, such as 75 Ω which is common in the video and television industry.

ATTENUATORS AND TERMINATIONS

Most amateurs are familiar with the concept of a *dummy load*, which is nothing more than a high-power 50-Ω resistor with a coaxial connector. The term comes from the fact that it is often used as a dummy antenna to provide a low-SWR load for transmitter testing. The resistor must have low stray reactance throughout the desired frequency

Fig 25.39 — A 100-watt power attenuator rated for dc-3000 MHz. The transmitter is connected to the male type-N connector. The female connector is for the low-power connection.

range. Common wire-wound power resistors are not suitable for a dummy load because they have too much inductance. Commercial loads, such as the power attenuator in **Fig 25.39**, use special resistors encased in a finned enclosure for dissipating the heat generated by the power absorbed from the transmitter. One technique used in some amateur dummy loads is to enhance the power dissipation capability of less-expensive, lower-power resistors by submerging them in a bath of oil.

Low-power dummy loads, normally called *terminations*, are used whenever a device being tested needs to have one of its coaxial connectors terminated in a load that has good return loss over a wide band of frequencies. A *feed-through termination*, such as the one in **Fig 25.40**, is one that has two coaxial connectors with a straight-through connection between them as well as a 50-Ω load resistor connected between the center conductors and the shield. It is used to provide a 50-Ω load when using a measuring device with a high-impedance input, such as an oscilloscope. To avoid standing waves on the coaxial transmission line connected to the device under test, it is important to place the feed-through termination directly at the oscilloscope's input connector so that the transmission line is properly terminated.

Another technique to maintain a low SWR is to use a *T connector* at the oscilloscope input. The device under test is connected to one side of the T and some other device that provides a good 50-Ω load is connected to the other side. The high-impedance oscilloscope input does not excessively disturb the 50-Ω system. This is a good way to "tap into" a signal traveling between two devices on a coaxial line while maintaining the connection between the devices.

A *fixed attenuator* is useful both for reducing a signal level as well as for improving the 50-Ω match. For example, a 10-dB attenuator guarantees a minimum of 20 dB of return loss (1.22:1 SWR) even if the load SWR is infinite. A 20-dB attenuator makes a high-quality 50-Ω termination, even with nothing connected to its output.

A *step attenuator* is used when you need to adjust the signal level in fixed steps. The term "10-dB step attenuator" means that the step size is 10 dB. The maximum attenuation would typically be perhaps 70 to 120 dB. An attenuator with 1 dB steps and 10 dB of maximum attenuation would be called a "1 dB step attenuator" or a "0 to 10 dB step attenuator." Two attenuators, one with 10 dB steps and one with 1 dB steps, can be connected in series to obtain 1 dB resolution over a very wide range of attenuation. At microwave frequencies, continuously-variable attenuators are available for waveguide transmission lines. They can sometimes be obtained used at reasonable prices.

OTHER TEST ACCESSORIES

Other accessories available for waveguide include coax-to-waveguide adapters, detectors, directional couplers, isolators, absorption wavemeters, mixers and terminations of various kinds. Each size of waveguide covers about a 1.5:1 frequency range. Be aware that over the years there have been several systems for assigning letters to the various microwave frequency bands. For example, "X" band in the Agilent Technologies (formerly part of Hewlett-Packard) catalog is 8.2 to 12.4 GHz, but the old US Navy definition was 6.2 to 10.9 GHz, and the ITU assignment for X-band radar is 8.5 to 10.68 GHz. The IEEE standard definition is 8.0 to 12.0 GHz. When buying surplus waveguide accessories be sure they cover the frequency range you need.

A *coaxial detector* is just what the name implies, a diode detector in a coaxial package, usually with a 50-Ω RF load impedance. Silicon Schottky diodes are usually used to obtain good sensitivity although gallium arsenide devices are sometimes employed for the high microwave and millimeter-wave frequencies. Below a certain signal level, typically about −15 dBm, the detected output voltage is proportional to the square of the input voltage, that is, it is proportional to the input power. Sensitivity is typically specified in mV/mW, assuming a high-impedance load for the detected signal.

Wideband amplifiers can often be useful in test systems. They are available with various connector types in packages sized appropriately for the power level. A low-noise amplifier is useful as a preamplifier for a spectrum analyzer, for example. A higher-power amplifier can be used at the output of a signal generator either for receiver dynamic range testing or as an input signal for testing an RF power amplifier.

Directional couplers and bridges are useful network analyzer accessories as previously described but can also be used to acquire a small amount of signal for other test purposes. For example, a 30-dB coupler at the output of a 100-W transmitter produces a 100-mW

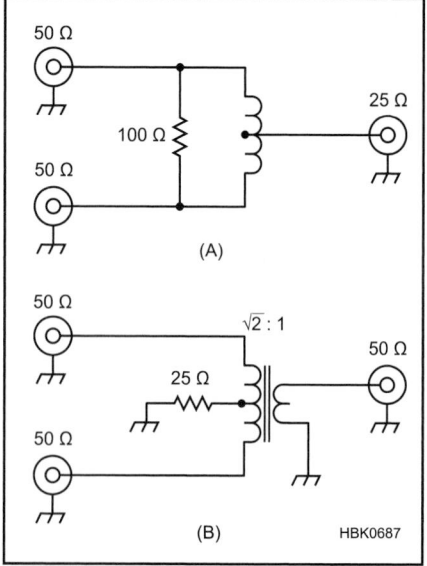

Fig 25.41 — Two forms of a lumped-element Wilkinson hybrid combiner. The one at B is 50 Ω on all three ports. In both circuits, the two input ports on the left are isolated from each other so long as the output port sees a low-SWR termination.

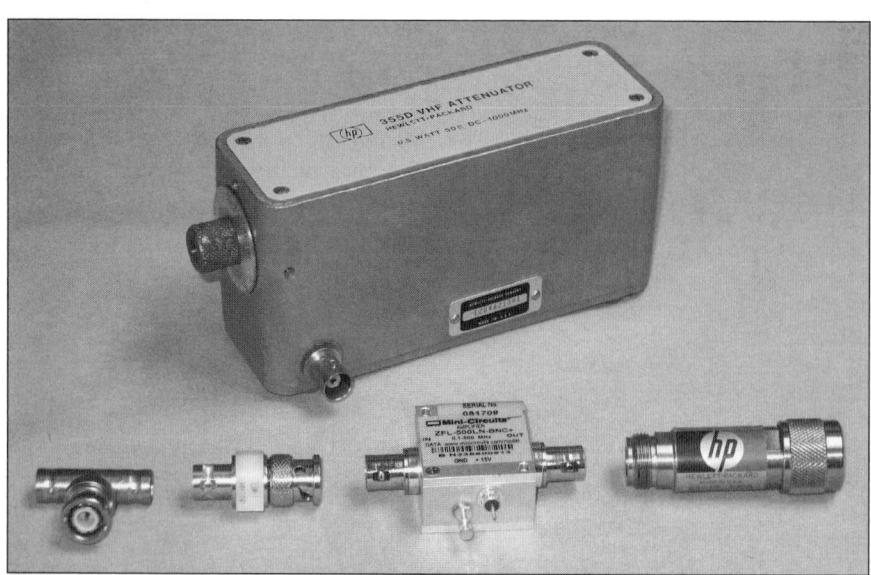

Fig 25.40 — Various RF test accessories. At the rear is a 10 dB step attenuator with a 0-120 dB range. In the foreground from left to right are a BNC "T" connector, a 50-Ω feed-through termination, a 0.1-500 MHz low-noise amplifier, and a 10-dB fixed attenuator.

signal that can be fed to a coaxial detector or a spectrum analyzer, while passing the rest of the 100-W signal through to the antenna or dummy load.

A *power divider*, also known as a *splitter*, is a device that divides an input signal equally among its output ports. A *combiner* is basically a power divider hooked up backward; the signals from the input ports are combined at the output port. Some power dividers consist only of a network of resistors. The insertion loss is typically at least 6 dB for 2-port splitters, meaning that a quarter of the power comes out each port and half the power is absorbed in the resistors.

A *hybrid combiner*, such as the *Wilkinson combiner* shown in **Fig 25.41**, includes a transformer as part of the network. It has the advantage that, so long as there is a good, low-SWR match on the output port, there is excellent isolation between the two input ports. That is very useful when combining the

Fig 25.42 — Coaxial connector adapters. From the left are: BNC male to type-N male, BNC female to UHF male, BNC female to SMA male, type-N male to SMA female, and type-N female to SMC male.

signals from two signal generators for high-level dynamic range testing of a receiver. The port-to-port isolation prevents the two signals from combining in the output amplifiers of the signal generators, where they could generate distortion products greater than the ones generated by the receiver that you are trying to measure. The insertion loss is nominally 3 dB.

The traditional Wilkinson combiner has a 2:1 ratio of input to output impedance. The variation in Fig 25.41B uses a transformer with a $\sqrt{2}$: 1 turns ratio (2:1 impedance ratio) to obtain 50 Ω on all ports.

Adapters such as those shown in **Fig 25.42** are useful for making connections among various pieces of test equipment.

25.5 Receiver Measurements

Receivers present a special measurement challenge because of the wide range of signal levels they must be able to handle. In a state-of-the-art receiver, the difference between the minimum discernible signal and the blocking level can be as high as 130 or 140 dB. It requires special care and high-quality test equipment to measure that level of performance.

25.5.1 Standard ARRL Lab Tests

The ARRL laboratory staff has standardized on a suite of tests for receivers, transmitters and several other types of RF equipment. The exact procedures for each test are meticulously specified in a 163-page document to ensure that tests performed at different times by different people are done in the same way. We won't attempt to repeat that level of detail here but rather describe the tests in general terms suitable for use by the home experimenter. (The full set of procedures is available online in the Product Review Testing area of the ARRL website at **www.arrl.org/product-review**.) Additional informa-

tion on receiver design and performance is available in the **Receivers** chapter.

RECEIVER SENSITIVITY

Several methods are used to determine receiver sensitivity. The modulation mode often determines the best choice. One of the most common sensitivity measurements is *minimum discernible signal* (MDS) or *noise floor*, which is suitable for CW and SSB receivers. The minimum discernible signal is defined as that which will produce the same audio-output power as the internally generated receiver noise. Hence, the term "noise floor."

To measure MDS, use a signal generator tuned to the same frequency as the receiver. Be certain that the receiver is peaked on the generator signal. In **Fig 25.43** a step attenuator is included at the receiver input since most signal generators cannot accurately generate the low signal levels required for MDS measurements. An audio-frequency ac voltmeter is connected to the receiver's speaker terminals. If no speaker is connected, then a resistor of the same resistance as the speaker impedance should be substituted. Set

the receiver to CW mode with a bandwidth of 500 Hz, or the nearest to that bandwidth that is available. Since the noise power is directly proportional to the bandwidth, always use identical filter bandwidths when comparing readings. Turn off the AGC, if possible. With the generator output turned off, you should hear nothing but noise in the speaker. Note the voltmeter reading at the receiver audio output. Next, turn on the generator and increase the output level until the voltmeter shows a 3-dB increase (1.414 times the voltage). The signal input at this point is the minimum discernible signal, which can be expressed in μV or dBm.

In the hypothetical example of Fig 25.43, the signal generator was adjusted to –133 dBm to cause the 3-dB increase in audio output power and the step attenuator is set to 4 dB. MDS is calculated with this equation:

$$MDS = -133 \text{ dBm} - 4 \text{ dB} = -137 \text{ dBm}$$

where the MDS is the minimum discernible signal and 4 dB is the loss through the attenuator.

Note that the voltmeter really should be a true-RMS type to accurately measure the RMS voltage of the noise. A typical average-reading multimeter is calibrated to indicate the RMS voltage of a sine wave but reads low on Gaussian noise. The error is small for the MDS test, however. To correct for the error, adjust the signal generator for a 3.2-dB increase (1.445 times the voltage) instead of 3.0 dB.

Another issue to watch out for is that the

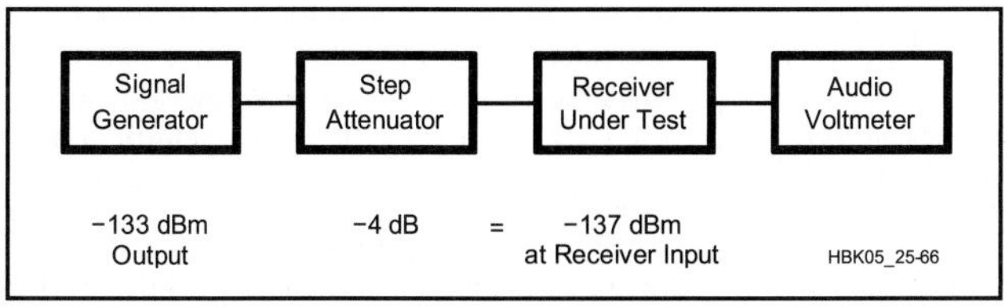

Fig 25.43 — A general test setup for measuring receiver MDS, or noise floor. Signal levels shown are for an example discussed in the text.

Signal Generator	Step Attenuator		Receiver Under Test	Audio Voltmeter
–133 dBm Output	–4 dB	=	–137 dBm at Receiver Input	HBK05_25-66

peak noise voltage is much greater than the RMS, typically by a factor of 4 or 5. The receiver volume should be adjusted so that the RMS output voltage is no more than about one-fifth the clipping level of the audio amplifier.

For AM modulation, receiver sensitivity is expressed as the RF signal level that results in a 10 dB signal plus noise to noise ratio at the audio output. The AM signal is 30% modulated with a 1000 Hz tone. Otherwise, the test setup is the same as for the MDS test. The signal generator output level is adjusted until there is a 10 dB (3.16 voltage ratio) increase in audio output voltage when the modulation is switched from off to on. Again, there is an error if a typical average-reading multimeter is used instead of a true-RMS voltmeter. For the AM sensitivity measurement, the error is 0.8 dB. Adjust the signal generator output to give a 10.8 dB (3.47 voltage ratio) increase in signal when the modulation is turned on. If an audio distortion meter is used in place of the RMS voltmeter, you can leave the modulation turned on and adjust the signal generator output level for 31.6% distortion.

For FM modulation, receiver sensitivity is expressed as the RF signal level that results in 12 dB SINAD. SINAD stands for "signal plus noise and distortion" and is calculated from

$$SINAD = 10 \log \left[\frac{signal + noise + distortion}{noise + distortion} \right] dB$$

where signal, noise and distortion are all entered in units of power (watts or milliwatts). It is very similar to the signal plus noise to noise ratio used for AM testing except that any distortion in the audio signal is added to the noise measurement. It means, however, that the signal, the noise and the distortion must all be measured at the same time, without turning off the modulation. For that, a distortion analyzer is needed as shown in **Fig 25.44**. A distortion analyzer includes a switchable band-reject filter to null out the 1 kHz tone for the noise-and-distortion measurement. The filter is bypassed for the signal-plus-noise-and-distortion measurement. For the test, the signal generator is set for FM modulation with a 1 kHz tone and the desired FM deviation, normally 3 kHz for VHF repeater operation. Adjust the signal generator output level until the distortion analyzer indicates 25% distortion, which is equivalent to 12 dB SINAD. Don't forget to subtract the attenuation of the step attenuator

from the signal generator output level, as was done for the MDS test.

Noise figure is a measure of receiver sensitivity that, unlike the other methods presented so far, is independent of the receiver bandwidth and operating mode. It can be calculated from the MDS so long as the noise bandwidth of the receiver filtering is known. *Noise bandwidth* is the bandwidth of a hypothetical perfect filter with a rectangular spectrum shape that would produce the same total noise power as the receiver filter. Accurate measurement of a filter's noise bandwidth requires integrating the spectral response using a swept signal source, but the filter's 3 dB bandwidth can be used as a reasonable approximation. The formula is

$$NF = MDS - \left(10 \log (BW) - 174\right) \text{ dB}$$

where MDS is the minimum discernible signal in dBm and BW is the noise bandwidth in Hz. Assuming the noise bandwidth is 500 Hz, we get NF = MDS + 147 dB. For example, if the MDS is –137 dBm then NF = –137 + 147 = 10 dB.

RECEIVER DYNAMIC RANGE

Dynamic range is a measure of a receiver's ability to receive weak signals without being overloaded by strong signals. It is easy to design a receiver with good sensitivity to weak signals. It is also easy to design a receiver that is not overloaded by strong signals. It is much more difficult to design a receiver that can do both at the same time.

Blocking dynamic range (BDR) is the dif-

ference between the noise floor and the signal level at which *blocking* occurs, that is, the level that causes a 1 dB reduction in gain for nearby weaker signals. The noise floor is just the MDS and can be measured using the technique described in the previous section.

The blocking level is measured using a test setup similar to the one used for measuring MDS except that two signal generators are connected to the input through a hybrid combiner, as shown in **Fig 25.45**. The receiver AGC should be turned off for this test. The mode is set to CW and the bandwidth to 500 Hz or the closest available. Two signal generators are used. One generates the weak signal that the receiver is tuned to. The ARRL standard specifies –110 dBm at the receiver input, which requires –97 dBm at the input to the hybrid combiner, assuming it has 3 dB loss. The other signal generator generates the strong interfering signal on a nearby frequency. Standard frequency separations are plus and minus 20, 5 and 2 kHz. The level of the strong signal is increased until the level of the weaker signal measured at the receiver audio output decreases by 1 dB.

Referring to Fig 25.45, the blocking level is the level from the signal generator minus the loss of the hybrid combiner and attenuator,

$$BL = -7 - 3 - 10 = -20 \text{ dBm}$$

The blocking dynamic range is given by

$$BDR = BL - MDS = -20 - (-137) = 117 \text{ dB}$$

assuming an MDS of –137 dBm as in the previous examples.

One complication is that it may be difficult to measure the amplitude of the audio tone because of the presence of noise caused by the phase noise of the signal generator and the receiver's local oscillator, especially at the 2 and 5 kHz frequency spacings. The solution is to use an audio-frequency spectrum analyzer to measure the change in tone amplitude. The

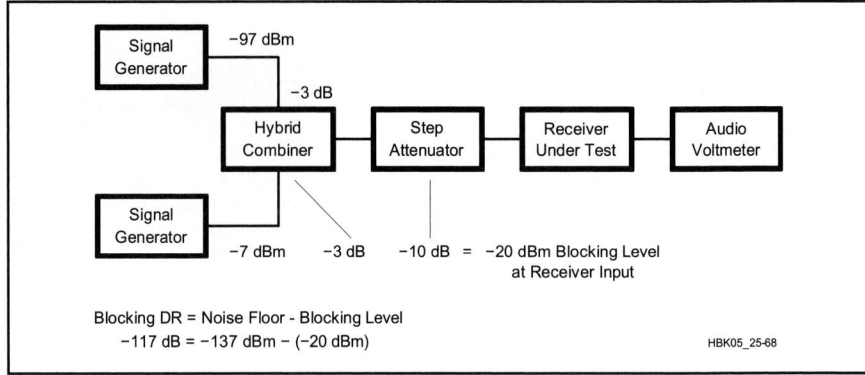

Fig 25.45 — Receiver blocking dynamic range is measured with this equipment and arrangement. Signal levels shown are for the example discussed in the text.

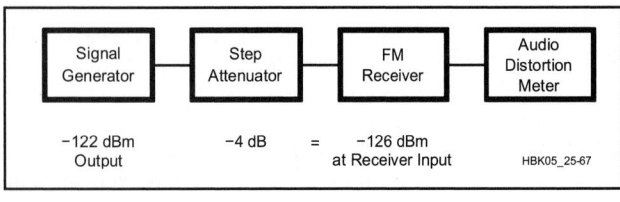

Fig 25.44 — FM SINAD test setup.

absolute accuracy of the spectrum analyzer is not important so long as it can accurately show a 1-dB change in signal level. An instrument based on a computer sound card and free spectrum analysis software should be adequate.

Reciprocal mixing is the name for the mixing of a nearby interfering signal with the phase noise of the receiver's local oscillator, which causes noise in the audio output. Although the ARRL BDR test eliminates this effect from the measurement, in actual on-the-air operation the phase noise is often the factor that limits the effective dynamic range. To address this issue, the ARRL test suite includes a separate measurement for reciprocal mixing. The test setup is the same as for MDS except that the signal generator in Fig 25.43 is replaced with a low-phase-noise crystal oscillator with an output power level of +15 dBm. The step attenuator block should include both a unit with 10-dB steps to be able to adjust the signal level over a wide range as well as a 1-dB step attenuator for fine adjustment. The receiver is tuned to 20, 5 or 2 kHz above or below the oscillator frequency. The output noise level is first measured with the oscillator turned off, then the oscillator is turned on and the step attenuator gradually reduced until the noise increases by 3 dB. Reciprocal mixing is expressed as a negative number:

$$\text{reciprocal mixing} = \text{MDS} - (+15 - A) \text{ dB}$$

where MDS is the noise floor, +15 is the signal level of the crystal oscillator in dBm, and A is the total attenuation in dB.

Intermodulation distortion (IMD) means the creation of unwanted signals at new frequencies because of two or more strong interfering signals modulating each other. If there are two interfering signals at frequencies f1 and f2, then *second-order IMD* products occur at f1 + f2. If f1 and f2 are close together, then the second-order products occur near the second harmonics. *Third-order IMD* products occur at 2f1 − f2 and 2f2 − f1. If f1 and f2 are close together, then the third-order products occur close by. For example, if f1 and f2 differ by 10 kHz, then the third-order IMD products are 10 kHz above the higher frequency and 10 kHz below the lower.

The two-tone IMD test setup shown in **Fig 25.46** is the same for second and third-order IMD. To obtain sufficient output power and isolation of the two signal generators, it may be necessary to follow each one with a wide-band power amplifier, not shown. The receiver under test is set to receive CW with the same bandwidth as for the MDS test and is tuned to the frequency of the distortion product to be measured. The two signal generators are always set to the same output amplitude level, which is increased until IMD

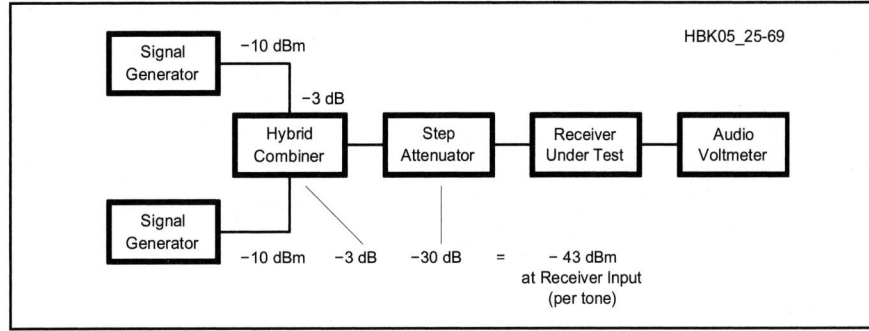

Fig 25.46 — The test setup for receiver intermodulation distortion dynamic range. Signal levels shown are for the example discussed in the text.

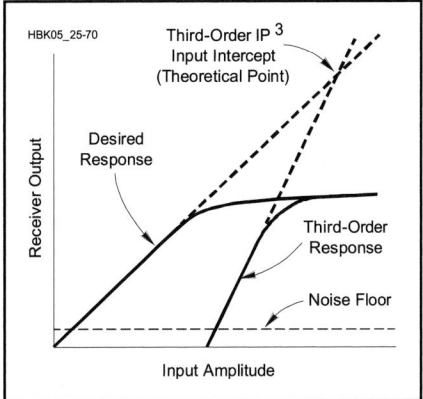

Fig 25.47 — The third-order intercept point can be determined by extending the lines representing the interfering signal level and the third-order intermodulation products on a plot of the signal levels in dB.

products equal in amplitude to the noise floor appear, resulting in a 3-dB increase in the audio voltmeter reading. The *IMD dynamic range* is then the difference in dB between the level of the interfering signals and the MDS.

For the second-order IMD measurement, the standard ARRL test sets the two signal generators to 6.000 MHz and 8.020 MHz and the receiver is tuned to 14.020 MHz. For the third-order IMD measurements, the two signal generators are set to frequencies 20, 5 and 2 kHz apart, separated from the receiver frequency by plus and minus 20, 5 and 2 kHz so that the lower or upper IMD product falls within the receiver passband.

For example, tune the receiver to 14.020 MHz and the two signal generators to 13.980 MHz and 14.000 MHz. With the signal generators turned off, measure the noise level with the audio voltmeter. Turn on the signal generators and increase their amplitudes until the voltmeter shows a 3-dB increase. If the signal generator amplitudes are −10 dBm, the loss in the hybrid combiner is 3 dB, and the step attenuator is set to 30 dB attenuation, then the third-order IMD level is

$$\text{IMD} = -10 - 3 - 30 = -43 \text{ dBm}$$

If the MDS is −137 dBm, then the third-order IMD dynamic range is

$$\text{IMD\_DR} = \text{IMD} - \text{MDS} =$$
$$-43 - (-137) = 94 \text{ dB}$$

As with the blocking dynamic range test, the phase noise of the signal generators and the receiver LO may obscure the IMD product being measured. Again, the solution is to use an audio spectrum analyzer to measure the tone amplitude. Calibrate the amplitude by temporarily tuning one of the signal generators to the receiver frequency and setting the amplitude level so that the signal level at the receiver is the MDS. Note the level on the spectrum analyzer. Then return the signal generator to the interfering frequency and adjust the signal generators' amplitudes until the IMD product is at the same level as the MDS signal. The ARRL test bench actually uses a third signal generator and a second hybrid combiner at the receiver input to generate the calibration signal so that it and the IMD product can be seen on the spectrum analyzer at the same time for a more-accurate comparison.

In most analog components such as mixers and amplifiers, the second-order products increase in amplitude by 2 dB for each 1 dB increase in the interfering signals and third-order products increase 3 dB per dB. If the output signal levels are plotted versus the input levels on a log-log chart (that is, in units of dB), the desired signal and the undesired IMD products theoretically trace out straight lines as shown for third-order products in **Fig 25.47**. Although the IMD products increase more rapidly than the desired signal, the lines never actually cross because blocking occurs before that level, however, the point where the extensions of those two lines cross is called the *third-order intercept point* (IP3).

Although the third-order intercept is an artificial point, it is a useful measure of the strong-signal-handling capability of a receiver. It can be calculated from the equation

$$IP3 = MDS + 1.5 \times IMD\_DR \text{ dBm}$$

where IP3 is the third-order intercept point, MDS is the minimum discernible signal in dBm and IMD_DR is the third-order IMD dynamic range in dB. Using the numbers from the previous example, $IP3 = -137 + 1.5 \times 94 = +4$ dBm.

The second-order intercept may be calculated in an analogous way.

$$IP2 = MDS + 2 \times IMD\_DR$$

where IP2 is the second-order intercept and IMD_DR is the second-order IMD dynamic range in this case.

An alternate method of measuring third-order IMD is to use S5 (–97 dBm) instead of the MDS as the reference level to which the IMD products are adjusted. That results in a higher IMD level but a smaller value of IMD dynamic range (the difference between the IMD level and the reference). It may be a more accurate method of determining the third-order intercept because the signal levels do not have to be measured in the presence of noise that is at the same level as the signal.

The third-order intercept is generally not a valid concept for software-defined receivers (SDRs) that do not use an analog front end. Some SDRs do not use a mixer but feed the signal from the antenna directly to an analog-to-digital converter (ADC). ADCs usually do not exhibit the 3 dB per dB relationship between signal level and third-order products, at least over major portions of their operating range. Comparing third-order dynamic range measurements of an SDR and a conventional analog radio may give misleading results.

Fig 25.48 shows the relationship between the various dynamic range values. The base line represents different power levels, with very small signals at the left and large signals at the right. The numbers listed are from the previous examples for a typical receiver. The thermal noise and the noise floor are referenced to a 500 Hz bandwidth. The third-order IMD dynamic range is less than the blocking dynamic range, which means that signals as low as –43 dBm may cause IMD interference while signals must exceed –20 dBm to cause blocking. However, the intermodulation distortion may actually cause fewer problems since the IMD products only appear at certain discrete frequencies. A signal that exceeds the blocking level can cause interference across the entire band.

Third-order IMD dynamic range may also be measured on an FM receiver. The test setup is the same as Fig 25.46 except that the audio voltmeter is replaced with an audio distortion meter, as in Fig 25.44. The frequencies involved are the same as in that example. For this

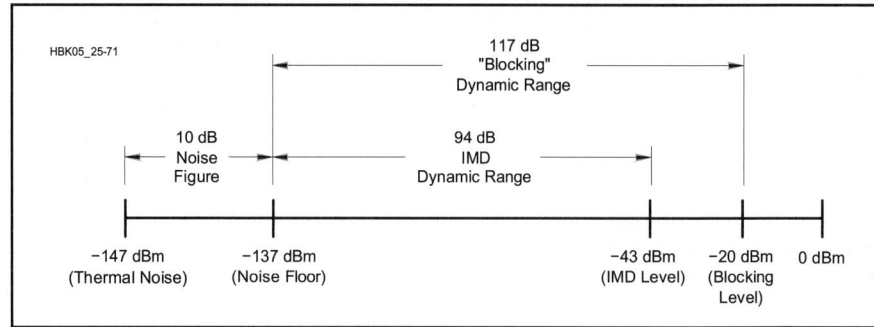

Fig 25.48 — Performance plot of the example receiver discussed in the text.

test, one of the signal generators (the one tuned to the frequency farthest from the receiver under test) is FM-modulated with a 1000 Hz tone at 3 kHz deviation. That causes the IMD products also to be FM-modulated, which can be measured with the distortion meter. The signal generator amplitudes are increased until the distortion product produces 12 dB SINAD (25% distortion) on the meter. The FM third-order IMD dynamic range is calculated using the same equation as for SSB and CW except that the SINAD sensitivity is substituted for the MDS. For example, if the 12 dB SINAD sensitivity is –120 dBm, the signal generator outputs are –10 dBm, the combiner loss is 3 dB, and the step attenuator is set to 30 dB, then

$$FM\_IMD\_DR = -10 - 3 - 30 - (-120) = 77 \text{ dB}$$

where FM_IMD_DR is the third-order FM IMD dynamic range.

OTHER ARRL RECEIVER TESTS

Most modern communications receivers are superheterodyne types. The first *IF rejection* and *image rejection* test measures the signal levels at the intermediate frequency and the image frequency that produces an audio output signal equivalent to the MDS, or noise floor. The test setup is the same as for the MDS test, as shown in Fig 25.43. The receiver is set to CW mode and 500 Hz bandwidth. The signal generator is tuned to the receiver first intermediate frequency or to the image frequency, which is the receiver frequency plus or minus two times the IF. The signal generator output level is gradually increased until there is a 3-dB increase in the audio voltmeter reading. As with the MDS test, if a multimeter is used instead of a true-RMS voltmeter, the signal generator should be adjusted for a 3.2-dB increase. The IF or image rejection is just the signal generator level at the receiver input less the MDS. For example, if the MDS is –137 dBm, the signal generator level is –40 dBm, and the attenuator is set for 10 dB, then the IF or image rejection is IR = –40 – 10 – (–137) = 87 dB.

FM adjacent-channel rejection is a measure of an FM receiver's ability to detect a weak signal in the presence of a strong interfering FM-modulated signal on an adjacent frequency channel. The test setup is the same as for the two-tone IMD dynamic range test shown in Fig 25.46 except that a distortion meter is substituted for the audio voltmeter, as was done for the FM SINAD test illustrated in Fig 25.44. The standard channel spacing is 20 kHz. The weak signal is modulated with a 1000 Hz tone and the interfering signal with a 400 Hz tone, both with 3 kHz deviation. The receiver under test is set for FM modulation and is tuned to the frequency of the weak signal. The weak signal is adjusted for 12 dB SINAD (25% distortion) on the distortion meter and then the interfering signal level is increased until the SINAD drops to 6 dB (50% distortion). The adjacent channel rejection is the difference between the power of the interfering signal and the 12 dB SINAD sensitivity, which is just the difference in level between the two signal generators.

Another test that applies especially to FM receivers is the squelch sensitivity test. As usual, the signal generator is set for FM modulation with a 1 kHz tone and 3 kHz deviation. If squelch is available for SSB, it can be tested in that mode as well. With the signal generator off, the squelch control on the receiver is adjusted just to the point where the noise is squelched. Then the signal generator is turned on and the level increased until the signal is heard.

Audio power output is tested with the same setup as the FM SINAD test illustrated in Fig 25.44. The receiver under test is set for SSB mode with the widest bandwidth available. A load resistor of the specified resistance, usually 8 Ω, is connected to the speaker output in place of the speaker. The signal generator level is set for an S9 level and the receiver is tuned for a 1 kHz output tone frequency. The receiver volume control is increased until the specified distortion level, usually 10%, is indicated on the distortion meter. The output power is given by the equation

$$P = V_{RMS}^2 / R$$

where P is the power in watts, V_{RMS} is the RMS output voltage, and R is the load resistance.

The audio and IF frequency response measures the audio frequencies at which receiver audio output drops by 6 dB from the peak. It includes the total response of the entire receiver, from the antenna connector to the speaker output. The test setup is the same as for the audio power output test except that some method must be included to measure the output audio frequency, such as a digital oscilloscope or frequency counter. The receiver is set for SSB mode at the bandwidth to be tested and the AGC is turned off. First tune the signal generator for a peak audio output signal and record the level of the audio signal. Then tune the signal generator downward until the signal drops 6 dB (½ the voltage) and record the audio frequency. Then tune the signal generator upward and record the high-end frequency at which the signal drops by 6 dB. The 6-dB bandwidth is the difference between the two frequencies.

There are several other miscellaneous ARRL receiver tests that are commonly reported in product reviews. The S meter test measures the signal level required to produce an S9 indication on the meter. The notch filter test uses a setup similar to the IMD dynamic range test in Fig 25.46 with an audio spectrum analyzer at the output. One signal is notched and the other is used as a level reference; the notch depth is the amplitude difference between the two tones. The DSP noise reduction test uses a similar setup except that one signal generator is replaced with a wideband noise generator. The signal generator is adjusted for S9 and the noise source for a 3-dB increase in the audio voltmeter so that the noise and signal are at the same level. The DSP noise reduction is then turned on and the reduction in noise level recorded.

25.5.2 Other Common Receiver Measurements

There are a few other useful receiver tests that are not covered in the standard ARRL test suite. For example, in addition to the IF and image response, it is possible for a receiver to have spurious responses at other frequencies as well. Testing for that can be a time-consuming process since it involves tuning the signal generator through a wide frequency range with the receiver tuned to each of a number of representative frequencies, typically at least one on each band. Start with the signal generator at maximum output power. When you find a response, reduce the level until the received signal level is at the noise floor (MDS). The spurious response amplitude is the difference between the signal generator level and the MDS.

Another time-consuming test is for internally-generated spurious signals, sometimes called *birdies* because they sometimes sound like a bird chirping as you tune through the signal. Birdies are typically caused by harmonics of the local oscillator(s) and BFO and their IMD products. For this test, connect a 50-Ω termination to the antenna connector and tune the receiver through its entire frequency range, writing down the frequency and S meter reading for each spurious signal found. You must tune very slowly because many birdies pass through the IF passband much faster than regular signals, sometimes in the opposite direction.

The ARRL S meter test only measures the response at the S9 level. If the S meter is accurate throughout its range it can be used to measure signal levels off the air. The standard definition is that, on the HF bands, S9 should correspond to –73 dBm (50 μV into 50 Ω) and each S unit corresponds to 6 dB, or a doubling of RF voltage. So S8 is –79 dBm or 25 μV, S7 is –85 dBm or 12.5 μV, and so on. The S unit calibration of most commercial equipment varies considerably from 6 dB per S unit, but the S meters accuracy can easily be measured with a calibrated signal generator.

The automatic gain control circuitry has a major effect on the operation of a receiver. The static AGC performance can be measured on a CW or SSB receiver with a signal generator and an audio voltmeter connected to the speaker output. You can plot a graph of audio output in dBV (decibels with respect to one volt) versus RF input level in dBm to see how good a job the receiver does in keeping the output level constant. Some receivers have menu settings to set the slope of the curve and the threshold, which is the small signal level at which the AGC circuitry starts to reduce the gain. The dynamic response is also important, although it is difficult to measure. A short attack time is important to reduce transients on sudden strong signals, but if it is too short then in-channel intermodulation distortion between strong signals can make signals sound "mushy."

The ARRL test suite includes a measurement of the 6 dB bandwidth at the audio output, but the sound is also affected by slope and ripple in the passband response. Using the same test setup as for the 6 dB measurement, you can measure the output voltage at a number of equally-spaced frequencies and plot the results on a graph. Output power is measured at the 10% distortion level, but it also is useful to measure the distortion at a volume level closer to what is used in actual practice. An output level of 1 V_{RMS} is commonly used for that test.

Frequency accuracy and stability are important performance criteria for a receiver. If a signal generator of sufficient frequency accuracy is not available, a standard time signal such as from radio station WWV can be used. The traditional method is to put the receiver in SSB mode and *zero-beat* the carrier, which means to tune the receiver until the audio tone (the beat note) is at zero Hz. Since most receivers' audio frequency response only extends down to a couple hundred Hertz, it is difficult to get good accuracy using that method. If a frequency counter or other means of measuring the audio output frequency is available, you can tune the receiver to obtain a 1000 Hz tone and add (for LSB) or subtract (for USB) 1000 Hz from the receiver's indicated frequency. The frequency drift from cold turn-on can be measured by plotting the audio output frequency versus time.

25.6 Transmitter Measurements

The signal levels found in a transmitter do not vary nearly as much as in a receiver. All the signals are generated internally, rather than being received off the air, so are much better controlled. Partly for that reason transmitter measurements tend to be simpler than receiver measurements and there are fewer of them.

25.6.1 Standard ARRL Lab Tests

For a transmitter, the RF power output is probably the first measurement that comes to mind. The test setup is straightforward: connect the transmitter output to the input of an RF wattmeter and connect the wattmeter output to a suitable dummy load. For CW, AM and FM modes, simply key the transmitter and measure the power on the wattmeter. For AM and FM the modulation should be turned off. For SSB, a two-tone audio oscillator should be connected to the microphone input to take the place of the voice signal. For the SSB test the wattmeter must be a peak-reading type to measure the PEP power

TRANSMITTER SPECTRAL PURITY

Ideally a transmitter confines its emissions to a narrow frequency range around the desired signal. Unwanted emissions can be divided into two categories, those that fall close to the desired signal and those that extend far away.

In the latter category are included harmonics and other discrete spurious frequencies. The measurement is done with the transmit-

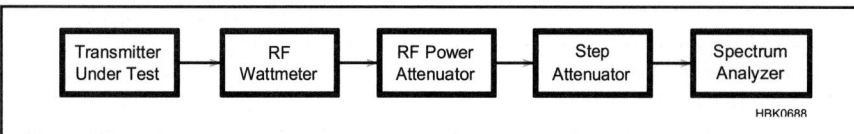

Fig 25.49 — The transmitter spectral purity test uses a spectrum analyzer to display the spurious frequencies.

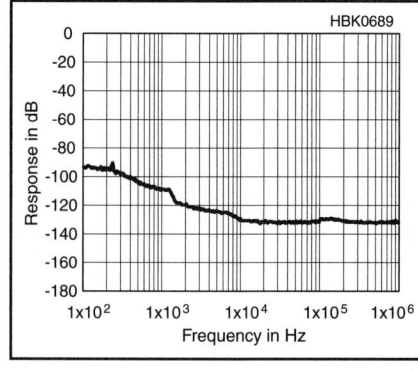

Fig 25.51 — Spectral display of an amateur transmitter output during composite-noise testing in the ARRL Lab. Power output was 200 W on the 14 MHz band. The carrier, off the left edge of the plot, is not shown. This plot shows composite transmitted noise 100 Hz to 1 MHz from the carrier on a logarithmic scale. The vertical scale is in dB with respect to the carrier.

ter in CW mode using the test setup shown in **Fig 25.49**. Tune up the transmitter as specified in the manual, set it for the desired power level, and adjust the frequency from one end of the band to the other while observing the spectrum analyzer. This should be done on each band. It may be necessary to retune the transmitter occasionally as the frequency is adjusted. The spurious-signal and harmonic suppression is the difference in dB between the carrier and the maximum spurious signal. It is important that the power level into the spectrum analyzer be maintained at a low enough level that spurious signals are not generated in the spectrum analyzer itself. To test for that, try changing the step attenuator setting; the desired carrier and all spurious signals should change by the same amount. If not, increase the attenuation until they do.

Representative spectrum analyzer plots are shown in **Fig 25.50**. The horizontal (frequency) scale is 5 MHz per division. The desired carrier frequency at 7 MHz appears 1.4 divisions from the left of the plot. Although not shown, a large apparent signal is often seen at the extreme left edge. This signal at zero Hz is caused by leakage of the spectrum analyzer local oscillator frequency and should be ignored.

In addition to the discrete spurious signals measured by the previous test, a transmitter may also generate broadband noise. It can be due both to the phase noise of the local oscillator as well as AM noise from all the devices in the amplifier chain. Generally the phase noise predominates, at least for frequencies close to the carrier. The composite noise test measures the total noise from both sources as well as any close-in spurious frequencies that don't show up in the wideband spurious signal and harmonic suppression test. Measuring phase noise requires high-performance test equipment and special measuring techniques.[20] The ARRL lab uses a special low-noise oscillator and a Hewlett-Packard (now Agilent Technologies) phase noise test set under computer control to perform this sophisticated measurement, which takes about 15 minutes to perform once it is set up. The result is a plot of the noise spectrum such as the one in **Fig 25.51**.

The most important unwanted emissions close to the carrier frequency are caused by distortion in the transmitter amplifier stages. In an SSB transmitter, this distortion creates

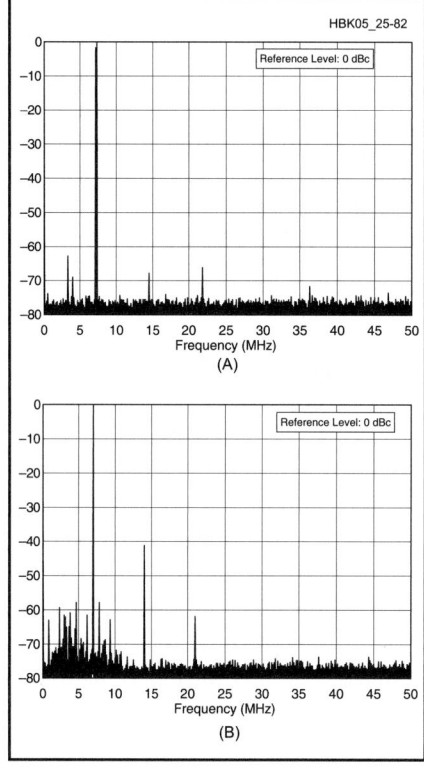

Fig 25.50 — Comparison of the spurious signal levels of two 100-watt transmitters, as shown on the spectrum analyzer display. The display on the top shows about 63 dB worst-case spurious signal suppression while the one on the bottom has a second harmonic suppressed approximately 42 dB. For transmitters below 30 MHz installed after January 1, 2003, the worst-case spurious emission must be at least 43 dB below the carrier power.

a signal that is wider than the bandwidth of the original modulation and causes interference to other stations. For this test, the same test setup is used as for the harmonics and spurious frequencies test shown in Fig 25.49 except that a two-tone audio generator is connected to the transmitter microphone input to simulate a voice signal, which contains many frequency components. The two tone frequencies must be non-harmonically related to prevent tone harmonics from being confused with IMD products. The ARRL Lab uses 700 Hz and 1900 Hz for these tests. Because many transmitters' modulation frequency response is not flat, the relative am-

plitudes of the two audio-frequency tones must be adjusted to obtain equal-amplitude RF tones on the spectrum analyzer. The test should be done for both lower and upper-sideband modes.

Two-tone IMD products are measured with respect to the transmitter peak-envelope power (PEP) which is 6 dB greater than the amplitude of either of the two tones. By adjusting the spectrum analyzer reference level to place the two tones 6 dB below the zero-dB reference line, as shown in **Fig 25.52**, the IMD distortion may be read out directly. For the signal shown, the third-order products are at −30 dB from PEP.

Carrier and unwanted sideband suppression may be measured with the same setup. In this case the 700 Hz tone of the two-tone generator is turned off and only the 1900 Hz tone is used. The single tone is set to the 0-dB reference line of the spectrum analyzer. For USB, the suppressed carrier shows as a small pip 1900 Hz below the desired signal and the unwanted sideband is 3800 Hz below. For LSB the unwanted signal frequencies are above the desired signal.

TESTS IN THE TIME DOMAIN

Oscilloscopes are used for transmitter testing in the time domain. Dual-trace instruments are best in most cases, providing easy to read time-delay measurements between keying input and RF- or audio-output signals. Common transmitter measurements performed with 'scopes include CW keying waveform and SSB/FM transmit-to-receive turnaround tests (important for many digital modes).

A typical setup for measuring CW keying waveform and time delay is shown in

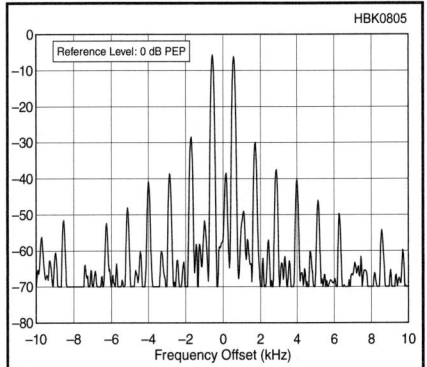

Fig 25.52 — An SSB transmitter two-tone test as seen on a spectrum analyzer. Each horizontal division represents 2 kHz and each vertical division is 10 dB. The third-order products are 30 dB below the PEP (top line), the fifth-order products are down 38 dB and seventh-order products are down 40 dB.

Fig 25.53. A keying test generator is used to key the transmitter at a controlled rate. The generator can be set to any reasonable speed, but ARRL tests are usually conducted at 20 ms on and 20 ms off (25 Hz, 50% duty cycle), which corresponds to a series of dits at 60 WPM. **Fig 25.54** shows a typical display. The first two dits at the beginning of the transmission are displayed in order to show

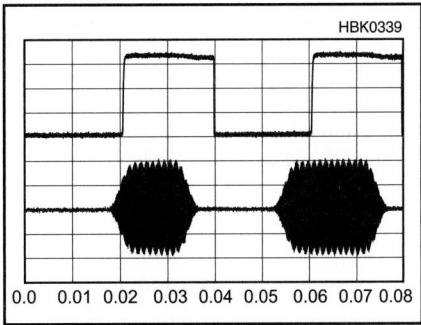

Fig 25.54 — Typical CW keying waveform for a modern Amateur Radio transceiver during testing in the ARRL Lab. This plot shows the first two dits in full break-in (QSK) mode using external keying. Equivalent keying speed is 60 WPM. The upper trace is the actual key closure; the lower trace is the RF envelope. (Note that the first key closure starts at the left edge of the figure.) Horizontal divisions are 10 ms. The transceiver was being operated at 100 W output on the 14 MHz band.

any transients or truncations that occur when the transceiver transitions from receive to the transmit state. The rise and fall times of the RF output pulse are measured between the 10% and 90% points on the leading and trailing edges, respectively. The delay times are measured between the 50% points of the

keying and RF output waveforms. Look at the **Receivers** and **Transmitters and Transceivers** chapters for further discussion of CW keying issues.

For voice modes, a PTT-to-RF output test is similar to CW keying tests. It measures rise and fall times, as well as the on- and off-delay times just as in the CW test. See **Fig 25.55** for the test setup. For SSB the transmitter is modulated with a single 700-Hz tone. For FM the transmitter is unmodulated. The keying generator is set to a speed that allows plenty of time for the transceiver to recover between dits. The ON or OFF delay times are measured from the 50% point of the falling or rising edge of the key out line to the 50% point of the RF waveform.

The transmit-receive turnaround time is the time it takes for a transceiver to switch from the 50% rise time of the key line to 50% rise of audio output. Turnaround time is an important consideration in some digital modes with required turnaround times of less than 50 ms in some cases. The test setup is shown in **Fig 25.56**. This test requires extreme care to prevent excessive transmitter power from reaching the signal generator and exceeding its specifications. The step attenuator is preset to maximum and gradually decreased until the receiver's S meter reads S9. Receiver AGC is usually on and set for the fastest response for this test but

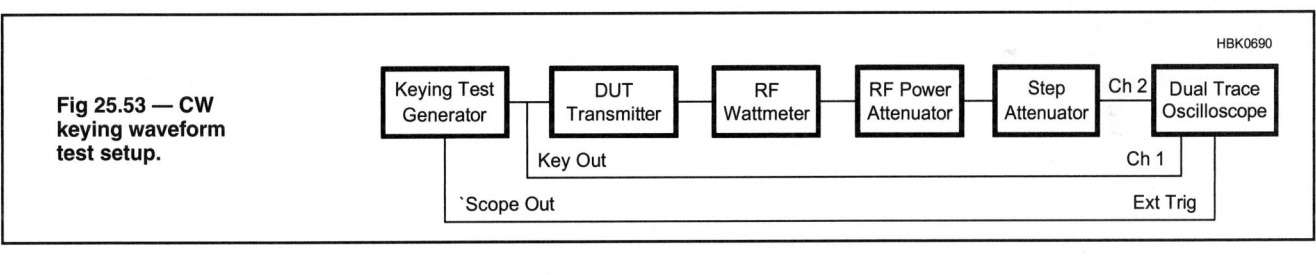

Fig 25.53 — CW keying waveform test setup.

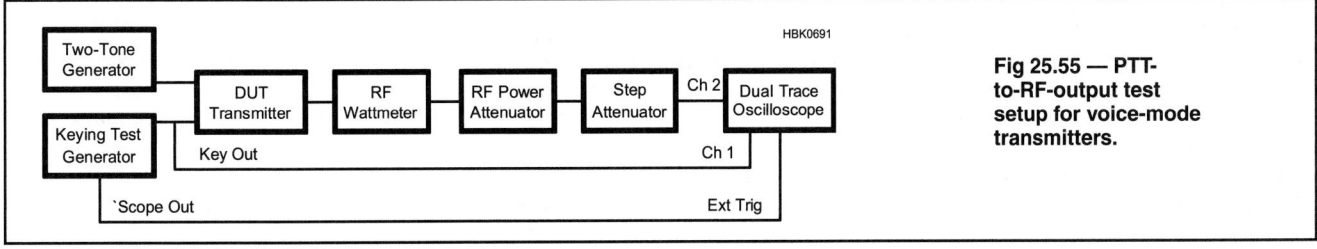

Fig 25.55 — PTT-to-RF-output test setup for voice-mode transmitters.

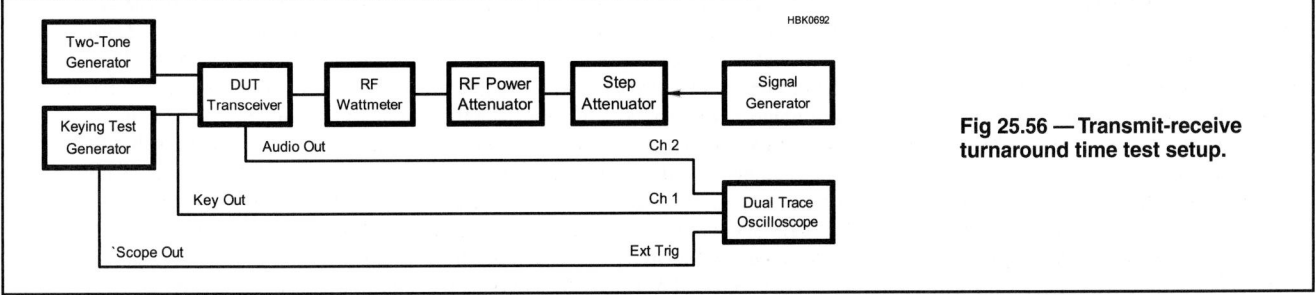

Fig 25.56 — Transmit-receive turnaround time test setup.

experimentation with AGC and signal input level can reveal surprising variations. As for the PTT-to-RF output test, the transmitter is tuned to full power output with a single 700 Hz tone. The keying rate must be considerably slower than the turnaround time; rates of 200 ms on/200 ms off or faster, have been used with success in Product Review tests at the ARRL Lab.

25.6.2 Other Common Transmitter Measurements

The peak envelope power (PEP) of a 100%-modulated AM signal is four times the average power. For that reason, the specified AM power level of most SSB transmitters is generally about 25% of the PEP SSB power rating.

The modulation percentage can most easily be measured with a wide-band oscilloscope connected to the 50-Ω dummy load. It is OK to exceed the specified bandwidth somewhat as long as a clean signal is dis-played since this is a relative measurement only. With 100% modulation, the negative modulation peaks just reach zero signal level and the positive peaks are twice the amplitude of the unmodulated carrier. The exact value can be calculated with the equation,

$$M = 100 \frac{max - min}{max + min}$$

where M is the modulation percentage, max is the signal amplitude at the peaks and min is the amplitude at the troughs. An alternative is to use an RF spectrum analyzer. With 100% sine-wave modulation, the two sidebands are each –6 dB with respect to the carrier. In this case, $M = 100 \times 10^{(S+6)/20}$, where S is the sideband level with respect to the carrier (a negative number) and M is the modulation percentage.

The equivalent measurement for an FM transmitter is the *deviation*, which is the amount the RF frequency deviates from the center carrier frequency. It is possible to purchase an instrument that measures FM deviation directly; the function is generally included in two-way radio test sets. Another way is to use slope detection with a standard analog spectrum analyzer. Start by choosing a resolution bandwidth (RBW) on the spectrum analyzer such that as you tune away from the unmodulated carrier the signal level changes approximately linearly (the same number of kHz per dB) over at least a 10 kHz range. An RBW of 10 kHz is usually about right. Record the kHz per dB sensitivity value. Then adjust the frequency for the middle of the linear range and set the spectrum analyzer for zero span. With modulation applied to the transmitter, the deviation is equal to one-half the peak-to-peak dB variation of the signal on the screen, times the kHz per dB value determined previously.

Yet another way to measure FM deviation with a spectrum analyzer is to use the fact that the carrier disappears for a modulation index of 2.405, as explained in the **Modulation** chapter. To set the deviation to 3 kHz, for example, apply sine wave modulation at a frequency of 3.0/2.405 = 1.25 kHz and adjust the modulation level to null the carrier.

25.7 Miscellaneous Measurements

While receiver and transmitter measurements are perhaps the type of testing most associated with Amateur Radio, the home experimenter will have occasion to do other measurements as well.

25.7.1 Testing Digital Circuitry

Virtually every electronic device these days includes digital circuitry. Even a simple QRP transceiver usually either includes or is used with a digital keyer and it may include other digital circuitry, such as a display, as well. A multimeter is generally of little use for testing digital circuits because it cannot respond fast enough to indicate the high and low transitions.

An oscilloscope is a very useful tool because it gives a visual display of the digital signal versus time. It can show impairments such as transient glitches, overshoot, slow rise times, and high or low logic levels that are out of spec. A scope with at least two channels is greatly desired because you often to need to see the time relationships between different signals. A separate external trigger input is also very useful for the same reason. The oscilloscope's bandwidth must be high enough to accurately display the signals to be tested. Bandwidths in the GHz range are needed for state-of-the-art digital circuitry, but 100 MHz should be adequate for most needs.

A hand-held *logic probe* is a device that indicates whether a circuit node is high, low or toggling in between the two states (even with very narrow pulses). While it does not give as much information as an oscilloscope, it is much smaller, cheaper and easier to use. In many cases it is all that is needed to troubleshoot a circuit of moderate complexity. For example, many circuit faults result in a "stuck bit" that remains high or low all the time, which is easy to detect with a logic probe. Typically the logic probe has two wires with clip leads on the ends that connect to the ground and power supply of the circuit under test. The operator then touches the probe tip to the point to be tested and LEDs light up to indicate the logic state.

At the other end of the complexity spectrum is the *logic analyzer*. This is typically the size of an oscilloscope with one or more external pods that can connect to dozens or hundreds of circuit nodes at the same time. Models that use an external digitizing pod and connect to a PC via USB are also available.

A graphics screen shows individual signals with an oscilloscope-like display or multiple signals can be treated as a bus, with the display reading out the values as a series of hexadecimal numbers versus time. Unlike an oscilloscope, a logic analyzer does not indicate actual voltage levels, but only whether a signal is high or low and the timing of the transitions. Sophisticated triggering modes allow synchronization to various clocks or data patterns. Usually a large on-board memory allows capturing long data traces for display or later analysis. Some models include a pattern generator to generate a long series of multi-bit test vectors. Logic analyzers are quite expensive to buy new but are commonly available on the surplus market.

25.7.2 Service Monitors

A *service monitor* is a "one-box tester" for transceivers. It includes a signal generator for testing the receiver and a spectrum analyzer for testing the transmitter, using the same RF connector so that only one connection to the transceiver's antenna jack is required. Other common features are an RF wattmeter and dummy load, a frequency counter, an FM deviation meter, audio tone generators to connect to the microphone, and an audio voltmeter and distortion analyzer/SINAD meter to connect to the speaker output. Some units contain additional features such as DTMF (touch-tone) and CTCSS (sub-audible tone) generators, an audio frequency counter and adjacent-channel power (ACP) measurement capability.

Older service monitors found on the surplus market were designed for testing analog two-way radios and repeaters. Many are portable for easy transportation to a mountaintop repeater site. Later units may be more oriented to testing cellular telephone base stations. Modern instruments cover the latest digital modes, with bit error rate (BER) testers for the receiver and various modulation quality tests for the transmitter.

While all the functions of a service monitor are available in separate instruments, having everything integrated in one box is more convenient and allows faster testing, which is something a commercial enterprise is willing to pay extra for. A brand-new service monitor is not inexpensive, but older used units made by such companies as Singer-Gertsch, Cushman and IFR that are suitable for testing analog radios can sometimes be found for reasonable prices.

25.7.3 Antenna Measurements

Traditionally, amateurs have measured coax-fed antennas using a standing wave ratio (SWR) meter. It is a good method for determining an antenna's resonant frequency and how well-matched it is to the characteristic impedance of the coaxial feed line. To obtain more detailed information about the complex impedance versus frequency, you can use an antenna analyzer to read out the magnitude and phase (or the resistance and reactance) of the antenna impedance. Standing wave ratio and SWR meters are covered in the **Transmission Lines** and **Station Accessories** chapters. Antenna analyzers are discussed in the Measuring RF Impedance section of this chapter.

Those instruments can tell you about the antenna's impedance but they say nothing about how well it actually radiates. For that, you need a way of detecting the signal level at some distance from the antenna. Probably the most common way to do that is to ask for signal reports on the air. Unfortunately, propagation is so variable on most amateur bands that a signal report gives only a very rough idea of how well your antenna is working, unless it can be compared with another antenna (perhaps at another local amateur's station) at the same time.

A *field-strength meter* is a device, generally with a built-in antenna, that picks up the radiated signal off the air and indicates the level on a meter or display. Professional field-strength meters use a carefully-calibrated antenna and circuitry that can read out the actual radiated signal strength in volts per meter or watts per square meter. Most amateur field-strength meters are not calibrated but give a relative indication only. They are useful for tuning an antenna, antenna tuner or transmitter for maximum signal as well as for comparing different antennas. A field strength meter is

a simple one-evening construction project.[21] If you have a reference antenna (such as a half-wave dipole) that has a known gain, then the gain of a second antenna can be calculated by alternately transmitting with each antenna and measuring the difference in signal level at a receiver whose antenna is located far enough away to be outside the *near field*, typically up to several wavelengths from the antenna under test. At first glance, it seems like this should be an easy measurement to make but in practice there are a number of devilish details that can ruin the measurement accuracy.

The biggest issue is reflections. If there is any significant reflector of RF signals between the transmitting and receiving antennas, the signal level will not be accurate. Even if there are no wires, fences or other conducting objects in the vicinity, the ground reflection can result in an apparent additional gain of 6 dB or conversely a loss of many dB if the receive antenna happens to be in a null. At microwave frequencies it is sometimes possible to use a high-gain, narrow-beamwidth receive antenna and mount the antenna under test high enough so that the ground reflection is outside the receive antenna's beamwidth.[22] At HF frequencies that is rarely possible. Commercial antenna companies use elaborate *antenna test ranges* that employ various techniques to assure measurement accuracy. Absent a proper test range, the best solution is probably to measure the reference antenna and antenna under test at various heights above ground to get an idea of how much ground reflections are affecting the measurement.

The subject of antenna measurements is addressed by Paul Wade, W1GHZ in his "Microwavelengths" *QST* columns for Oct 2012 (covering the antenna range) and Jan 2013 (discussing measurements and equipment).

25.7.4 Testing Digital Modulation

As digital modulation modes become more and more important in Amateur Radio, it is increasingly important to have ways of testing the performance. There are dozens of different digital formats in use, from traditional radio teleprinting (RTTY) using frequency-shift keying (FSK) to the latest systems that employ sophisticated error detection and correction along with various modulation types that pack multiple bits into each symbol. Despite the wide differences in modulation and coding, nearly all have in common a relatively-narrow bandwidth suitable for use with a voice transceiver using SSB or FM modulation.

If you're having trouble with reception or transmission of digital signals using a PC sound card, one straightforward trouble-

shooting technique is to install the software on two computers and see if you can transmit data from one computer to the other by connecting the sound card output of one to the input of the other and vice versa. If you don't get perfect reception, that indicates a problem with the computer software or hardware.

The next step is to transmit into a dummy load and receive the signal with a separate receiver located close by so it picks up the stray radiation from the dummy load. A piece of wire plugged into the receiver antenna connector can be moved around to adjust the signal level. Many software programs for receiving digital signals include a spectrum display, which can indicate faults in the transmitted signal such as distortion and *skew*, the amplitude imbalance among the tones of a multi-tone modulation signal. To see what the signal is supposed to look like, connect the two computers directly, as previously described. Then when you examine the RF signal transmitted into the dummy load, any additional bandwidth due to distortion or skew in the spectrum shape should be apparent.

If the demodulation software does not include a spectrum display, there are separate programs available that can display the spectrum of the signal at the sound-card input, as discussed in the Spectrum Analyzer section of this chapter. An RF spectrum analyzer measuring the RF output signal directly would give an even better idea of modulation quality because it is not affected by the filters and other circuitry of the receiver. A receiver panadapter as described earlier is a less-expensive substitute.

Comprehensive testing of a digital communications system is quite complicated because of all the variables involved. The *bit error ratio* (BER) is the number of single-bit errors divided by the number of bits sent in a certain time interval. It requires special test equipment to measure because the individual bits are typically decoded deep inside the demodulation software where they are difficult to access. The *packet error ratio* (PER) is easier to measure. In a packetized data system it is the number of incorrect packets divided by the number of packets sent. It can be measured either before or after error correction. In a non-packetized system like PSK31 the character error ratio is a useful figure of merit. BER is affected by the signal-to-noise ratio (SNR), interference, distortion, synchronization errors and multipath fading. PER is further affected by the effectiveness of the coding and error correction of the particular digital mode used.

It is interesting to measure BER or PER as a function of the SNR. For some digital systems with lots of error correction the errors are nearly zero down to a certain signal level and then degrade very sharply below that. However, in real-world operation the SNR is almost never constant. The signal is constantly changing, both in amplitude and phase, as propagation changes due to movement of the ionosphere (on HF) or of the vehicle (VHF and above), as explained in the **Propagation of Radio Signals** chapter. Measuring actual on-the-air performance is not a good way to compare systems because propagation varies so much at different times. For a repeatable test, you need a *channel simulator,* which is a device that intentionally degrades a test signal in a precise way as to simulate an over-the-air radio channel. Moe Wheatley, AE4JY offers a free software HF channel simulator which can be downloaded from the Internet.[23] A hardware HF channel simulator has been described by Johann B Forrer, KC7WW.[24]

25.7.5 Software-based Test Equipment

Most amateurs these days own a personal computer with a powerful microprocessor, tons of memory and mass data storage, a large color display and a sound card that provides stereo high-fidelity audio input/output. It doesn't take a great deal of imagination to realize that these resources can be harnessed to make low-cost measuring instruments of various types.

Audio-frequency instruments can be implemented directly using the sound card, which typically has a frequency response from perhaps 50 Hz up to about 20 kHz. (Professional and audiophile sound cards with higher performance are also available.) Free software is available on the Internet for instruments such as audio function generators, DTMF and CTCSS tone generators, DTMF and CTCSS decoders, two-tone generators for SSB transmitter testing, distortion/SINAD analyzers, oscilloscopes and audio spectrum analyzers. In addition to the frequency-response limitations of a typical sound card, another issue is that the device can be damaged by applying excessive voltage to the inputs or outputs. It is wise to add external buffer amplifiers that include overvoltage protection.

Radio-frequency test equipment can also use the sound card inputs by means of some type of frequency converter, consisting of a local oscillator and mixer. If the mixer is a quadrature type, the two outputs may be fed to the stereo sound card inputs so that software can treat the left and right channels as the in-phase and quadrature signals. A common application is a narrow-band RF spectrum analyzer. The RF bandwidth is typically limited to twice the sound card's audio

bandwidth. Low-cost hardware is available in kit form that can be used with free software downloaded from the Internet.[25]

One problem with using a sound card is that the signals may be susceptible to ground loops and radio-frequency interference (RFI). Since the computer and the device under test are grounded separately to the ac power system, hum and noise can be generated from currents flowing in the ground connection between the two. It is helpful to use a short, low-resistance ground connection between the sound card and the device under test. It is also possible to use isolation transformers or differential amplifiers to isolate the grounds and thus break the ground loop. Good quality cables and attention to proper shielding and grounding help prevent hum and noise pickup.

RFI may be an issue if you wish to make measurements while transmitting. Again, it may be helpful to use the shortest possible connections between the device under test and the sound card. For longer connections use shielded cable or twisted-pair wires such as found in Ethernet LAN cable. To prevent common-mode currents from entering the sound card, wrap several turns of the coax or twisted-pair wire through a ferrite toroid or clamp-on core as described in the **RF Techniques** and **RF Interference** chapters.

25.8 Construction Projects

25.8.1 Bipolar Transistor Tester

Here is a basic "good/bad" tester for bipolar transistors, designed by Alan Bloom, N1AL. This tester is small enough to carry in your pocket to a flea market (**Fig 25.57**). A printed-circuit board is available from FAR Circuits (**www.farcircuits.net**) but the simple circuit can easily be hand-wired on perfboard. (Printed-circuit board layout graphics are also available on this book's CD-ROM.)

To test an unknown NPN or PNP transistor, just remove the tester's working NPN or PNP transistor from its socket and replace it with the device to be tested. If you hear a tone in the headphones the transistor is good, otherwise it is bad. More elaborate instruments can measure various transistor parameters such as current gain, breakdown voltage and high-frequency performance, however this simple tester suffices in most situations. It

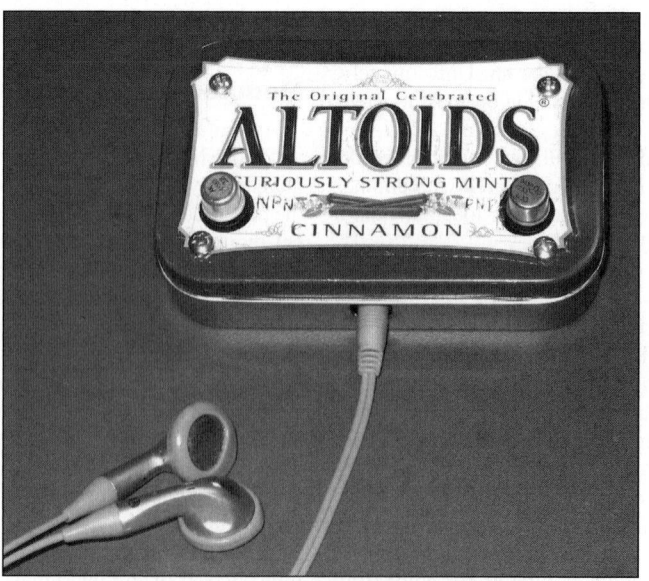

Fig 25.57 — A transistor tester built into an Altoids tin.

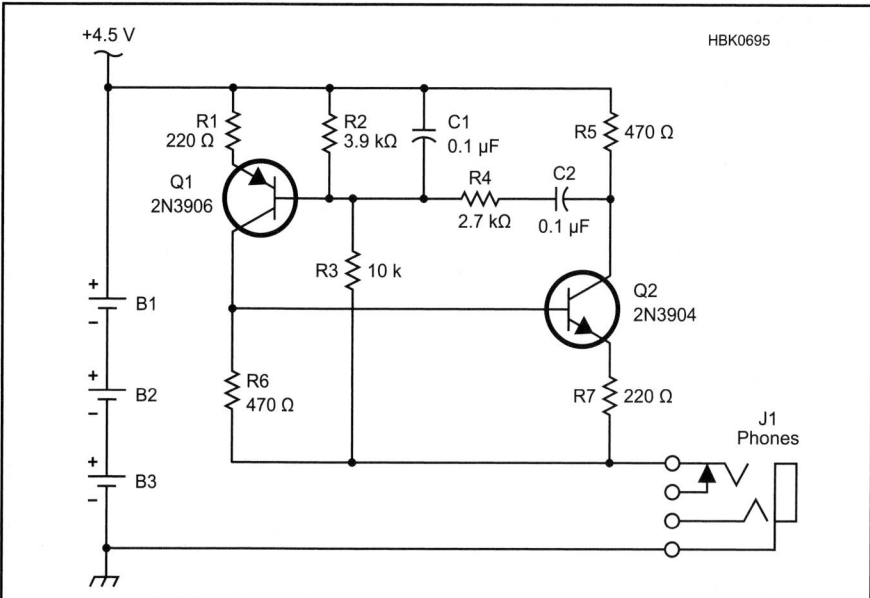

Fig 25.58 — Schematic diagram and parts list of the transistor tester. All parts can be obtained from Digi-Key at www.digikey.com except for the printed circuit board, available from FAR Circuits at www.farcircuits.net.

C1, C2 — 0.1 µF ceramic capacitor
  (Digi-Key 490-3873-ND)
J1 — Stereo 3.5 mm phone jack
  (Digi-Key CP1-3554NG-ND)
Q1 — 2N3906 PNP transistor
  (Digi-Key 2N3906FS-ND)
Q2 — 2N3904 NPN transistor
  (Digi-Key 2N3904FS-ND)
R1, R7 — 220 Ω, ¼ W, 5% resistor
  (Digi-Key 220QBK-ND)
R2 — 3.9 kΩ, ¼ W, 5% resistor (Digi-Key
  3.9KQBK-ND)

R3 — 10 kΩ, ¼ W, 5% resistor (Digi-Key
  10KQBK-ND)
R4 — 2.7 kΩ, ¼ W, 5% resistor (Digi-Key
  2.7KQBK-ND)
R5, R6 — 470 Ω, ¼ W, 5% resistor
  (Digi-Key 470QBK-ND)
Quantity 2 — 3-pin, TO-5 transistor socket
  (Digi-Key ED2150-ND)
Quantity 6 — AAA-size battery clips
  (Digi-Key 82K-ND)
PCB — Printed circuit board
  (FAR Circuits)

is rare for a transistor to be damaged in such a way that it still works but no longer meets its specifications.

When testing a batch of transistors of unknown condition you can use this tester to quickly sort them into a "good" and a "bad" pile and be fairly confident that the ones in the "good" pile are working correctly.

Metal-can TO-5 transistors are shown here, but small plastic transistors work just as well if you bend the leads a little. The TO-5 parts can be pressed down flat against the socket and so are less likely to fall out in a pocket.

The circuit in **Fig 25.58** is simply two transistors connected with positive feedback through a frequency-selective network, forming an oscillator at a frequency of approximately 500 Hz. Each transistor is configured for a voltage gain of about 2.0 and the feedback network has a gain of about ⅓, so that the total loop gain is a little greater than unity, the condition required for oscillation.

It should be nearly impossible to damage an unknown transistor by plugging it in wrong or into the wrong socket because the supply voltage is less than the base-emitter breakdown voltage of a bipolar transistor and the current is limited to a few milliamps. No on/off switch is included; simply unplug the headphones when you're done testing.

The prototype was built in an Altoids tin, as shown in **Fig 25.59**, but any handy enclosure would do. You'll need four mounting holes for the circuit board and two clearance holes for the transistor sockets. The headphone jack is best mounted on the side of the enclosure. If such a shallow enclosure is used, line the inside bottom with some insulating material such as electrical tape.

All components are mounted on the top side of the printed circuit board except the battery clips and the headphone jack, which go on the bottom. Leave a little extra lead length on the two 0.1 µF capacitors if they need to be bent over to clear the enclosure cover. The PC board from FAR Circuits does not have plated-through holes, so the leads of R5 and R6 must be soldered on both sides.

The transistor sockets are designed for TO-5 metal-can transistors but the smaller TO-18 or TO-92 plastic-cased devices can also be tested by bending the leads to fit. Nearly all TO-92 bipolar transistors have the base lead in the middle. Bend the center (base) lead toward the flat side of the transistor body, spread the three leads a little, and it should plug right in. Additional solder pads are provided for the base, emitter and collector of each transistor in case you wish to wire up additional sockets for other case types such as TO-220 or TO-3 power transistors. A transistor cross-reference guide is also handy to have to determine lead assignments.

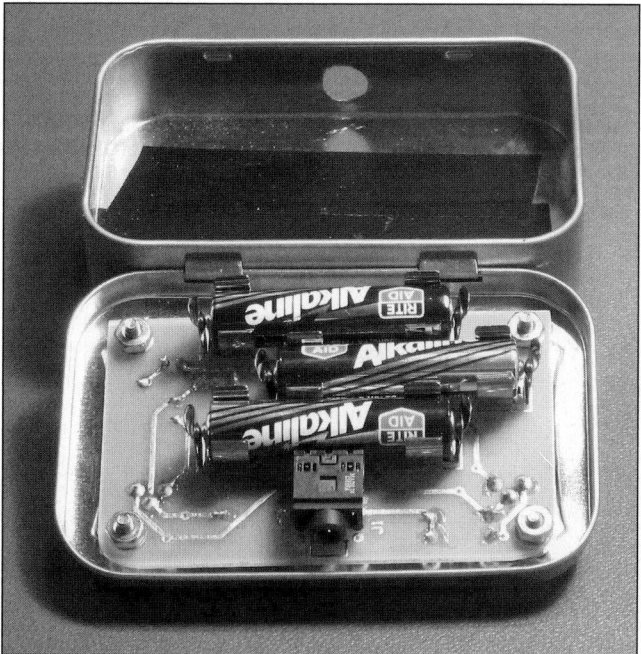

Fig 25.59 — Mounting of the circuit board inside the case.

25.8.2 Logic Probe

This simple logic probe (**Fig 25.60**) was designed by Alan Bloom, N1AL. It works with several different logic types, including TTL, 5 V CMOS and 3.3 V CMOS. A printed circuit board is available from FAR Circuits at **www.farcircuits.net** or the simple circuit may be hand-wired on perfboard. (Printed-circuit board layout graphics are also available on this book's CD-ROM.)

The purpose of a logic probe is to indicate whether signals are present at various circuit nodes. That's most of what troubleshooting a simple digital circuit requires. The probe has indicators to show whether the signal is high, low or toggling between the two states. It doesn't give as much information as an oscilloscope or logic analyzer but it is much smaller, cheaper and easier to use.

CIRCUIT OPERATION

This logic probe features a seven-segment LED display that forms letters to indicate the state of the signal at the probe's tip. A capital "L" is displayed if the signal is low and "H" if the signal is high. If it is toggling between low and high with roughly a 50% duty factor, the letter "B" is displayed to indicate that "Both" high and low are present. If the signal is mostly low with short-duration positive pulses, then a "C" is displayed. If the signal is mostly high with low-going pulses, the LED indicates an "A". "C" and "A" can be remembered as Cathode (mostly low) and Anode (mostly high), respectively.

The circuit is shown in **Fig 25.61**. Each of the common-anode seven-segment LED display segments is lit when its pin is low. The 74ACT04 inverters are arranged so that a continuous low input signal lights up the proper segments to form an "L" and a continuous high forms an "H". The 74LS122 retriggerable multivibrator outputs a 33-ms pulse whenever there is a positive-going tran-

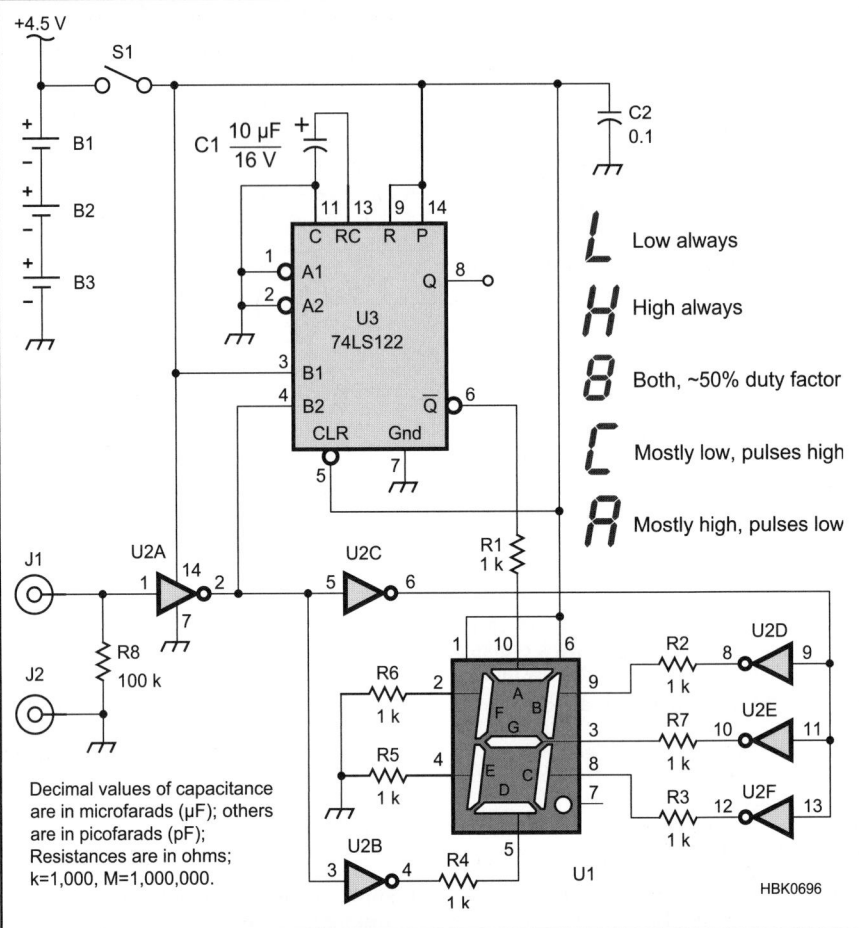

Fig 25.61 — Schematic diagram and parts list of the logic probe. All parts can be obtained from Digi-Key at www.digikey.com except for the printed circuit board, available from FAR Circuits at www.farcircuits.net.

C1 — 10 µF electrolytic capacitor (Digi-Key P807-ND)
C2 — 0.1 µF ceramic capacitor (Digi-Key 490-3873-ND)
J1 — Horizontal tip jack (Digi-Key J110-ND)
R1-R7 — 1 kΩ, ¼ W, 5% resistor (Digi-Key 1.0KQBK-ND)
R8 — 100 kΩ, ¼ W, 5% resistor (Digi-Key CF1/4100KJRCT-ND)

S1 — SPDT, right-angle slide switch (Digi-Key CKN9559-ND)
U1 — 7-segment common-anode LED (Digi-Key 160-1525-5-ND)
U2 — 74ACT04 hex inverter (Digi-Key 296-4351-5-ND)
U3 — 74LS122 monostable multivibrator (Digi-Key 296-3639-5-ND)
Qty 6 — Battery clips (Digi-Key 82K-ND)
Qty 2 — L-bracket (Digi-Key 621K-ND)
Printed circuit board (FAR Circuits)

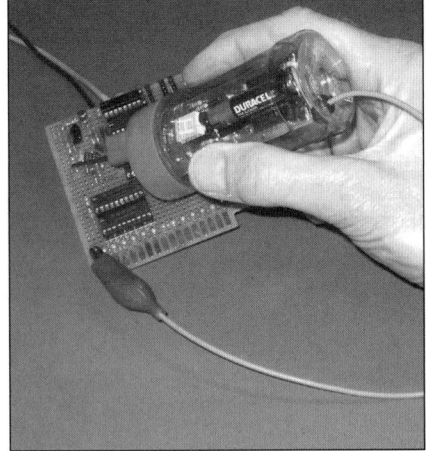

Fig 25.60 — A logic probe is small and easy to use.

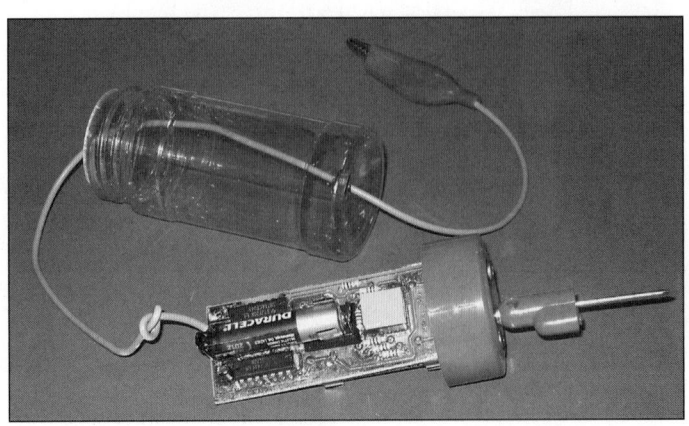

Fig 25.62 — The logic probe was built into a spice bottle.

sition on its B2 input. Repetitive transitions with a period less than about 33 ms (30 Hz or greater) assert the 74LS122 output continuously, which causes the top segment to light. If the signal is mostly low, the "L" turns into a "C" and if the signal is mostly high, the "H" turns into an "A".

CONSTRUCTION

The logic probe can be built in a plastic spice jar as in **Fig 25.62** but any clear plastic container big enough to accommodate the 1.2 × 3.5-inch printed-circuit board would also work. Batteries B1 and B2 and the test lead jack, J1, are mounted on the bottom side; all other components are on the top. Place a piece of tape under the top terminal of B3 to keep it from shorting to the ground plane. The PC board from FAR circuits does not have plated-through holes so at least the following must be soldered on both sides: R5, R6, R8, B1 and B3 (inner contacts), U2 pins 1 and 14, and U3 pins 1, 2, 3, 7 and 11. Also solder a wire to both sides of the via hole to the left of pin 1 of the display (U1). If you wish, you can use a socket for the LED display to raise it up higher, but don't use sockets for U2 and U3 since some of their leads must be soldered on both sides. Note that the L-brackets that attach the board to the bottle cap have a long and a short side. The long side is placed against the board.

The socket for the probe is a standard 0.08 inch diameter tip jack. The probe is just a nail soldered into the end of a mating tip plug. You can also use a standard test lead plugged into the jack. The component labeled "J2" on the schematic is just a solder pad for the ground lead. Tie a knot in the ground lead where it exits the bottle for strain relief.

USING THE LOGIC PROBE

The probe includes a self-contained, battery-operated power supply so only the ground lead needs to be connected to circuit ground.

It's amazing how much useful information you can get from such a simple device. One of the most common faults in a digital circuit is a node that is "stuck" low or high due to a short circuit or a faulty component. The logic probe is perfect for detecting that condition.

You can also get a rough idea of what percentage of time the signal is low or high by looking at the relative brightness of the segments. If the signal is low 100% of the time, an "L" is displayed. If there are narrow positive pulses, a "C" is shown. As the pulse width increases, the center and the two right segments start to glow dimly, finally forming a "B" when the duty factor is near 50%. As the duty factor increases further, the bottom segment dims, eventually turning the "B" into an "A". And finally, if the signal is high 100% of the time, the top segment goes out, leaving an "H".

25.8.3 Inductance Tester

Many inexpensive DVMs offer a capacitance measurement feature but measuring inductance is much less common. This project describes a simple test fixture for using a signal generator to measure inductance. It was originally published as "Mystery Inductor Box," by Robert J. Rogers, WA1PIO, in the January 2011 issue of *QST*.

MEASURING INDUCTANCE

One way to measure the value of an inductor is to connect it in parallel with a known capacitance and measure the resulting resonant frequency. In the past, grid-dip oscillators (GDO) have been used to determine the resonant frequency of such parallel tuned circuits. (See the project "Gate-Dip Oscillator" elsewhere in this section.) Using a dip oscillator, a dip, or drop in meter current was noted at the point of resonance.

This test fixture provides a convenient substitute for the dip meter method. A separate signal generator is used to provide the needed signal source and switched internal capacitors are used to resonate the inductors. An internal meter is provided as is a port for an external detector to indicate resonance. The schematic with parts list is shown in **Fig 25.63**.

BUILDING THE BOX

Fig 25.64 shows the layout of the internal components. Construction is straightforward with no critical dimensions or layout require-

ments. The 470-Ω resistor is used to provide a better peak in the voltage seen at resonance by the oscilloscope or meter movement. Most of the generator voltage will appear across this resistor until the point of resonance when the highest fraction of the applied signal voltage will then appear across the parallel circuit.

An oscilloscope is connected to the right BNC connector and its high input impedance, typically 1 MΩ, will not load the parallel tuned circuit at the relatively low frequencies used with this test fixture. Instead of the oscilloscope, the internal meter movement may be used to give a peak indication at resonance. Large capacitance values were used to minimize the effects of lead length, a concern while operating test equipment at higher frequencies.

USING THE TEST FIXTURE

The test fixture can be used to accurately determine the value of inductors in the mH and μH range. Selecting the 0.01 μF capacitor allows measurement of inductors in the μH range using a signal generator with a 159 kHz to 5 MHz output. Selecting the 0.22 μF capacitor measures larger mH-range inductors using signal generator frequencies in the 1 to 34 kHz region.

Fig 25.65 shows a typical test setup using the built-in meter as the peak indicator. A signal generator's sine wave output is connected to J1, the input BNC connector, and the unknown inductor to the terminal posts.

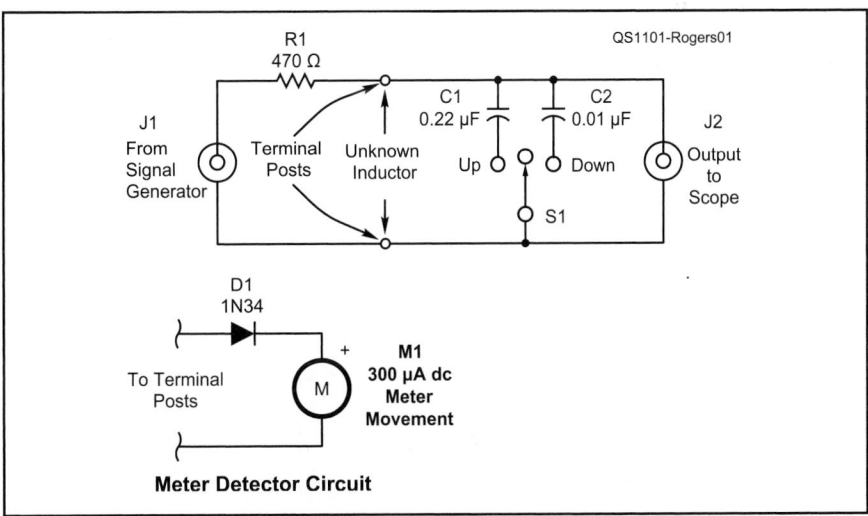

Fig 25.63 — Schematic diagram and parts list of the inductor test fixture. The switch's center-off position is used to disconnect the internal capacitors so that a desired value of capacitance can be placed in parallel with the inductor under test at the terminal posts. Measurement accuracy will depend on the actual value of C1 and C2 — determining the capacitor values with a capacitance meter will result in more accurate inductance values.

C1 — 0.22 μF ceramic capacitor.
C2 — 0.01 μF ceramic capacitor.
D1 — Germanium diode, 1N34 or 1N34A.
J1, J2 — BNC chassis-mount connectors.

M1 — 300 μA meter movement in a plastic case
R1 — 470 Ω, ¼ W resistor.
S1 — SPDT, center-off, toggle switch.

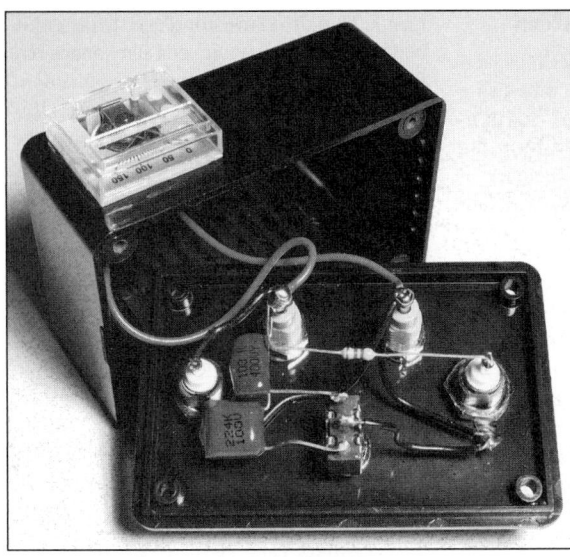

Fig 25.64 — Layout of the internal components.

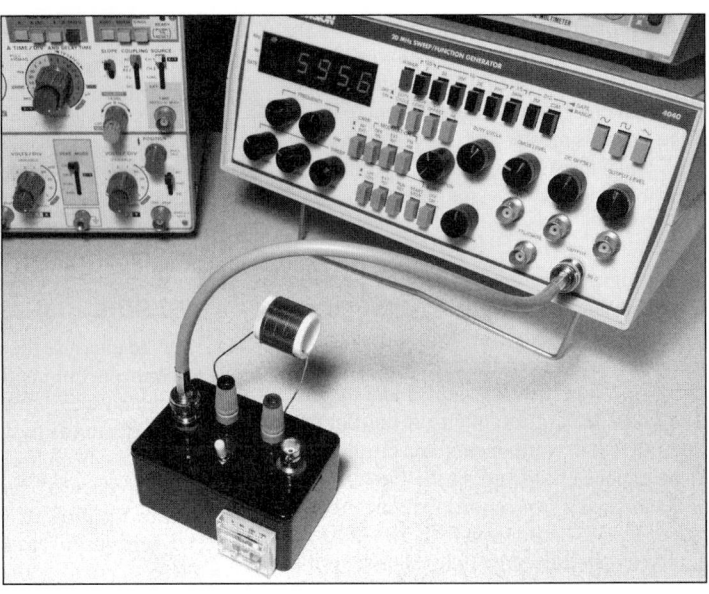

Fig 25.65 — The signal generator's output is 596 kHz (0.596 MHz) with the switch in the position selecting the 0.01 µF capacitor. Using the formula given results in an calculated inductance of 7.1 µH.

Select one of the capacitors with the switch. Connect an oscilloscope to J2, the output BNC connector, or use the built-in meter. Adjust the signal generator output to give some indication on the meter. Adjust the frequency of the signal generator until a peak is obtained on the oscilloscope or meter.

If the switch is in the wrong position, there will not be a sharp peak in the voltage seen by the oscilloscope or built-in meter. Avoid the band edges of the generator frequency range where output voltages tend to fall off from the mid-range values. Use the capacitor value selected, C, and the frequency of the signal generator at resonance, f, in the following formula:

$$L = \frac{1}{4\pi^2 f^2 C} = \frac{1}{39.5 f^2 C}$$

where f is in hertz, C is in farads and the resulting L is in henrys. A chart of frequency versus inductance for each of the two internal capacitors used to measure small and large inductors is shown in **Table 25.3**.

The SPDT switch also has a center-off position which disconnects the internal capacitors. This allows testing a particular capacitor-inductor combination connected to the terminal posts. The internal meter movement is always connected across the banana terminal posts.

25.8.4 Fixed-Frequency Audio Oscillator

An audio signal generator should provide a reasonably pure sine wave. The best oscillator circuits for this use are RC-coupled, the amplifiers operating as close to class A as possible. Variable frequencies covering the entire audio range are needed for determining frequency response of audio amplifiers.

A circuit of a simple RC oscillator that is useful for general testing is given in **Fig 25.66**. This Twin-T arrangement gives a waveform that is satisfactory for most purposes.

The oscillator can be operated at any frequency in the audio range by varying the component values. R1, R2 and C1 form a low-pass network, while C2, C3 and R3 form a high-pass network. As the phase shifts are opposite, there is only one frequency at which the total phase shift from collector to base is 180°: Oscillation will occur at this frequency. When C1 is about twice the capacitance of C2 or C3 the best operation results.

R3 should have a resistance about 0.1 that of R1 or R2 (C2 = C3 and R1 = R2). Output is taken across C1, where the harmonic distortion is least. Use a relatively high impedance load — 100 kΩ or more. Most small-signal AF transistors can be used for Q1. Either NPN or PNP types are satisfactory if the supply polarity is set correctly. R4, the collector load resistor may be changed a little to adjust the oscillator for best output waveform.

25.8.5 Wide-Range Audio Oscillator

A wide-range audio oscillator that will provide a moderate output level can be built

Table 25.3
Inductance Value at Resonant Frequency

0.22 µF Selected		0.01 µF Selected	
Inductance (mH)	Frequency (kHz)	Inductance (µH)	Frequency (MHz)
0.10	33.931	0.118	4.633
0.25	21.460	0.25	3.183
0.50	15.174	0.50	2.251
0.75	12.390	0.75	1.838
1.00	10.730	1.00	1.592
5.00	4.796	2.00	1.125
10.00	3.393	5.00	0.712
20.00	2.399	10.00	0.503
30.00	1.959	20.00	0.356
40.00	1.696	30.00	0.291
50.00	1.517	40.00	0.252
60.00	1.385	50.00	0.225
70.00	1.282	60.00	0.205
80.00	1.199	70.00	0.190
90.00	1.131	80.00	0.178
100.00	1.073	90.00	0.168
		100.00	0.159

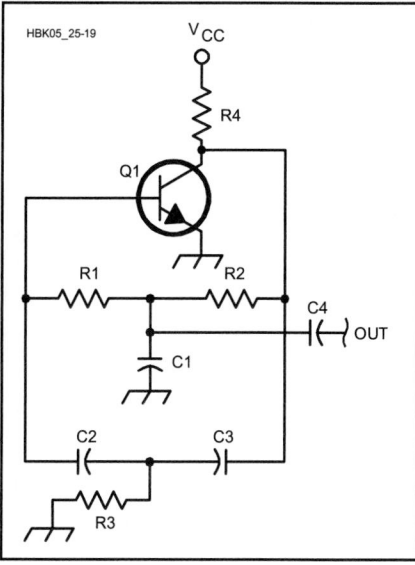

Fig 25.66 — Values for the twin-T audio oscillator circuit range from 18 kΩ for R1-R2 and 0.05 µF for C1 (750 Hz) to 15 kΩ and 0.02 µF for 1800 Hz. For the same frequency range, R3 and C2-C3 vary from 1800 Ω and 0.02 µF to 1500 Ω and 0.01 µF. R4 is 3300 Ω and C4, the output coupling capacitor, can be 0.05 µF for high-impedance loads.

from a single 741 operational amplifier (see **Fig 25.67**). Power is supplied by two 9-V batteries from which the circuit draws 4 mA. The frequency range is selectable from about 7 Hz to around 70 kHz. Distortion is approximately 1%. The output level under a light load (10 kΩ) is 4 to 5 V. This can be increased by using higher battery voltages, up to a maximum of plus and minus 18 V, with a corresponding adjustment of R_F.

Pin connections shown are for the eight-pin DIP package. Variable resistor R_F is trimmed for an output level of about 5% below clipping as seen on an oscilloscope. This should be done for the temperature at which the oscillator will normally operate, as the lamp is sensitive to ambient temperature. This unit was originally described by Shultz in November 1974 *QST*; it was later modified by Neben as reported in June 1983 *QST*.

25.8.6 Two-Tone Audio Generator

This generator is used in the ARRL Laboratory to test SSB transmitters for ARRL Product Reviews and makes a very convenient signal source for testing the linearity of a single-sideband transmitter. To be suitable for transmitter evaluation, a generator of this type must produce two non-harmonically related tones of equal amplitude. The level of harmonic and intermodulation distortion must be sufficiently low so as not to confuse the measurement. The frequencies used in this generator are 700 and 1900 Hz, both well inside the normal audio passband of an SSB transmitter. Spectral analysis and practical application with many different transmitters has shown this generator to meet all of the requirements mentioned above. While designed specifically for transmitter testing it is also useful any time a fixed-frequency, low-level audio tone is needed.

CIRCUIT DETAILS

Each of the two tones is generated by a separate Wien bridge oscillator, U1B and U2B. (see **Fig 25.68**) The oscillators are followed by RC active low-pass filters, U1A and U2A. Because the filters require nonstandard capacitor values, provisions have been made on the circuit board for placing two capacitors in parallel in those cases where standard values cannot be used. (The circuit board artwork for layout and part placement is available as graphics on the CD-ROM accompanying this book.) The oscillator and filter capacitors should be polystyrene or Mylar film types if available.

The two tones are combined by op amp U3A, a summing amplifier. This amplifier has a variable resistor, R4, in its feedback loop which serves as the output LEVEL control. While R4 varies the amplitude of both tones together, R3, the BALANCE control, allows the level of tone A to be changed without affecting the level of tone B. This is necessary because some transmitters do not have equal audio response at both frequencies. Multi-turn pots are recommended for both R3 and R4 so that fine adjustments can be made. Following the summing amplifier is a step attenuator; S3 controls the output level in 10-dB steps. The use of two output level controls, R4 and S3, allows the output to cover a wide range and still be easy to set to a specific level.

The remaining op amp, U3B is connected as a voltage follower and serves to buffer the output while providing a high-impedance load for the step attenuator. Either high or low output impedance can be selected by S4. The values shown are suitable for most transmitters using either high- or low- impedance microphones.

CONSTRUCTION AND ADJUSTMENT

Component layout and wiring are not critical, and any type of construction can be used with good results. Because the generator will

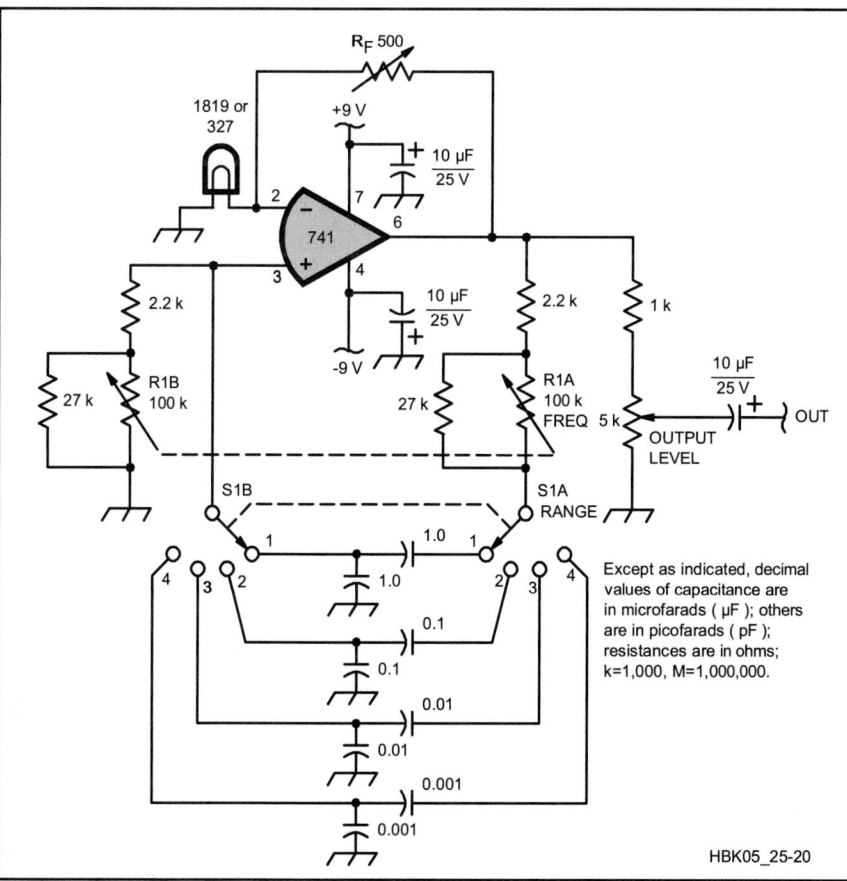

Fig 25.67 — An audio oscillator based on a single IC. The 741 op amp is shown but most op-amps will work in this application. The frequency range is set by switch S1.

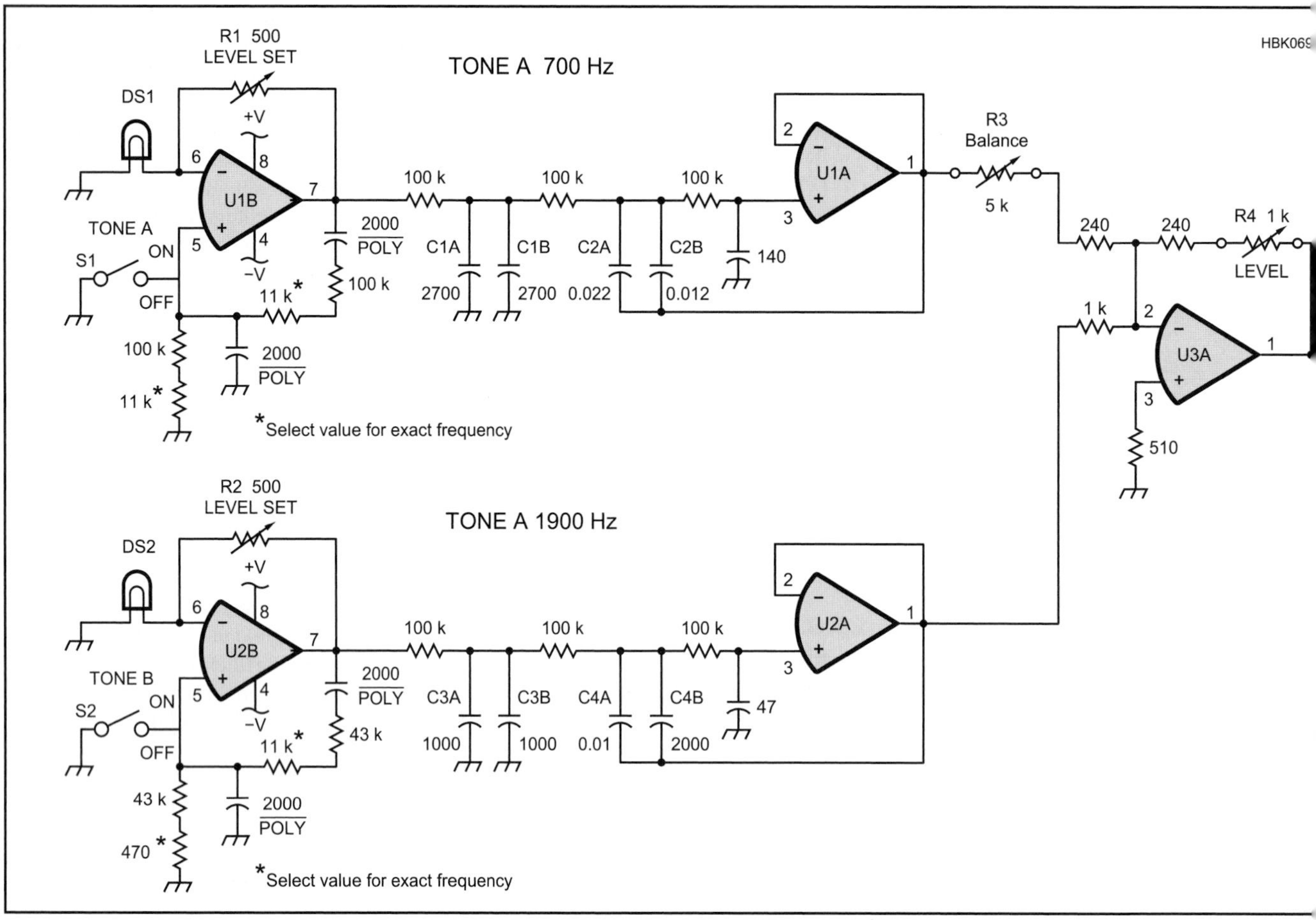

Fig 25.68 — Two-tone audio generator schematic. (the figure at top contains many labels)

Fig 25.68 schematic labels:

HBK069

R1 500 LEVEL SET
TONE A 700 Hz
DS1
+V
6 U1B 8
7
TONE A
S1 ON 5 4
OFF −V
100 k
11 k*
2000 POLY
11 k*
2000 POLY
100 k
*Select value for exact frequency
100 k 100 k 100 k
2000 POLY C1A C1B C2A C2B 140
2700 2700 0.022 0.012
2 U1A 1
+ 3
R3 Balance 5 k
240 240 R4 1 k LEVEL
1 k 2 U3A 1
+ 3
510

R2 500 LEVEL SET
TONE A 1900 Hz
DS2
+V
6 U2B 8
7
TONE B
S2 ON 5 4
OFF −V
11 k*
2000 POLY
43 k
43 k
2000 POLY
470*
*Select value for exact frequency
100 k 100 k 100 k
2000 POLY C3A C3B C4A C4B 47
1000 1000 0.01 2000
2 U2A 1
+ 3

Component list (Fig 25.68):

BT1, BT2 — 9 V alkaline.
C1A,B — Total capacitance of 0.0054 μF, ±5%.
C2A,B — Total capacitance of 0.034 μF, ±5%.
C3A,B — Total capacitance of 0.002 μF, ±5%.

C4A,B — Total capacitance of 0.012 μF, ±5%.
DS1, DS2 — 12 V, 25 mA lamp.
R1, R2 — 500 Ω, 10-turn trim potentiometer.
R3 — 500 Ω, panel mount potentiometer.
R4 — 1 kΩ, panel mount potentiometer.

S1, S2 — SPST toggle switch.
S3 — Single pole, 6-position rotary switch.
S4 — SPDT toggle switch.
S5 — DPDT toggle switch.
U1, U2, U3 — Dual JFET op amp, type LF353N or TL082.

normally be used near a transmitter, it should be enclosed in some type of metal case for shielding. Battery power was chosen to reduce the possibility of RF entering the unit through the ac line. With careful shielding and filtering, the builder should be able to use an ac power supply in place of the batteries.

The only adjustment required before use is the setting of the oscillator feedback trimmers, R1 and R2. These should be set so that the output of each oscillator, measured at pin 7 of U1 and U2, is about 0.5 volts RMS. A DVM or oscilloscope can be used for this measurement. If neither of these is available, the feedback should be adjusted to the minimum level that allows the oscillators to start reliably and stabilize quickly. When the oscillators are first turned on, they take a few seconds before they will have stable output amplitude. This is caused by the lamps, DS1 and DS2, used in the oscillator feedback circuit. This is normal and should cause no difficulty. The connection to the transmitter should be through a shielded cable.

25.8.7 RF Current Meter

The following project was designed by Tom Rauch, W8JI (**http://w8ji.com/building_a_ current_meter.htm**). The circuit of **Fig 25.69** is based on a current transformer (T1) consisting of a T157-2 powdered-iron toroid core with a 20-turn winding. The meter is used with the current-carrying wire or antenna inserted through the middle of the core as a one-turn primary

When 1 A is flowing in the single-turn primary, the secondary current will be 50 mA = primary current divided by the turns ratio of 20:1. R1 across the transformer flattens the frequency response and limits the output voltage. The RF voltage is then detected and filtered by the D1 (a low-threshold Schottky diode for minimum voltage drop) and C1. The adjustable sum of R2 and R3 allow for full-

scale (FS) calibration of the 100 μA meter. C2 provides additional filtering. The toroid core and all circuitry are glued to the back of the meter case with only R2 exposed — a screwdriver-adjustable calibration pot.

It is important to minimize stray capacitance by using a meter with all-plastic con-

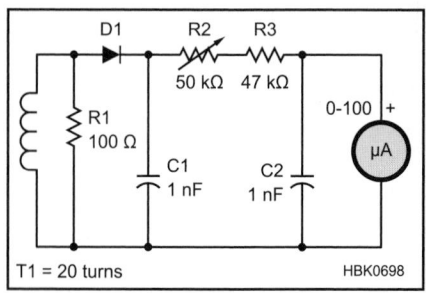

Fig 25.69 — The schematic of the RF current probe. See text for component information.

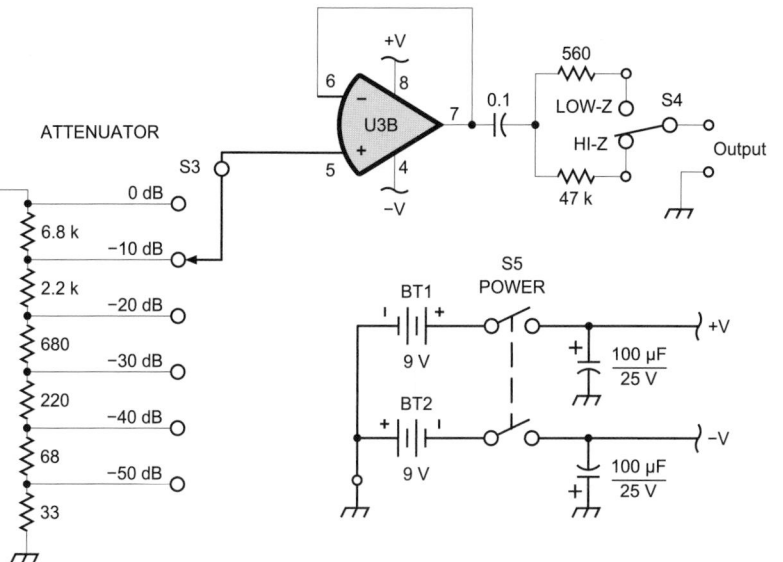

Decimal values of capacitance are in microfarads (µF); others are in picofarads (pF); Resistances are in ohms; k=1,000, M=1,000,000.

HBK0697

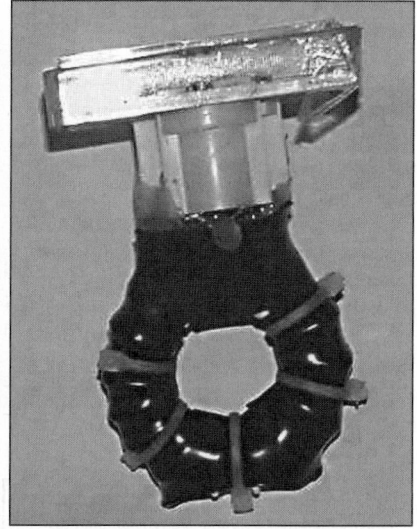

Fig 25.70 — Assembly of the RF current probe. Use an all-plastic meter and mount the circuits and toroid directly on the back of the meter case.

struction except for the electrical parts. The meter in **Fig 25.70** has an all-plastic case including the meter scale. The meter movement and all metallic areas are small. The lack of large metallic components minimizes stray

capacitance from the proximity of the meter. Low stray capacitance ensures the instrument has the least possible effect on the circuit being tested.

A value of 100 Ω for R1 gave the flattest response from 1.8 to 30 MHz. With 50 mA of secondary current, the voltage across R1 is $0.05 \times 100 = 5$ V$_{RMS}$. The peak voltage is then $1.414 \times 5 = 7.1$ V. At full current, power dissipation in R1 $= 50$ mA $\times 5$ V$_{RMS} = 0.25$ W so a ½-W or larger resistor should be used.

The meter used here was a 10,000 Ω/V model so for full-scale deflection from a primary current of 1 A producing a secondary voltage of ~7 V, the sum of R2 and R3 must be set to $7 \times 10,000 = 70$ kΩ. The low-current meter combined with high detected voltage improves detector linearity.

Calibration of the meter can be performed by using a calibrated power meter and a test fixture consisting of two RF connectors with a short piece of wire between them and through the transformer core. With 50 W applied to a 50-Ω load, the wire will be carrying 1 A of current. Full-scale accuracy is not required in comparison measurements, since the meter references against itself, but linearity within a few percent is important.

This transformer-based meter is much more reliable and linear than thermocouple

RF ammeters and perturbs systems much less. Stray capacitance added to the system being tested is very small because of the proximity of the meter and the compact wiring area. Compared to actually connecting a meter with its associated lead lengths and capacitance in line with the load, the advantages of a transformer-coupled meter become apparent.

25.8.8 RF Ammeters

When it comes to getting your own RF ammeter, there's good news and bad news as related by John Stanley, K4ERO. First, the bad news. New RF ammeters are expensive, and even surplus pricing can vary widely between $10 and $100 in today's market. AM radio stations are the main users of new units. The FCC defines the output power of AM stations based on the RF current in the antenna, so new RF ammeters are made mainly for that market. They are quite accurate, and their prices reflect that!

The good news is that used RF ammeters are often available. For example, Fair Radio Sales in Lima, Ohio has been a consistent source of RF ammeters. Ham flea markets are also worth trying. Some grubbing around in your nearest surplus store or some older ham's junk box may provide just the RF ammeter you need. Be sure you are really buying an RF ammeter as meters labeled "RF Amps" may just be regular current meters intended for use with an external RF current sensing unit.

RF AMMETER SUBSTITUTES

Don't despair if you can't find a used RF ammeter. It's possible to construct your own. Both hot-wire and thermocouple units can be homemade. Pilot lamps in series with antenna wires, or coupled to them in various ways, can indicate antenna current (F. Sutter, "What, No Meters?," *QST*, Oct 1938, p 49) or even forward and reflected power (C. Wright, "The Twin-Lamp," *QST*, Oct 1947, pp 22-23, 110 and 112).

Another approach is to use a small low-voltage lamp as the heat/light element and use a photo detector driving a meter as an indicator. (Your eyes and judgment can serve as the indicating part of the instrument.) A feed line balance checker could be as simple as a couple of lamps with the right current rating and the lowest voltage rating available. You should be able to tell fairly well by eye which bulb is brighter or if they are about equal. You can calibrate a lamp-based RF ammeter with 60-Hz or dc power.

As another alternative, you can build an RF ammeter that uses a dc meter to indicate rectified RF from a current transformer that you clamp over a transmission line wire (Z. Lau, "A Relative RF Ammeter for Open-Wire Lines," *QST*, Oct 1988, pp 15-17).

25.8.9 RF Step Attenuator

A good RF step attenuator is one of the key pieces of equipment that belongs on your workbench. The attenuator in this project offers good performance yet can be built with a few basic tools. The attenuator is designed for use in 50-Ω systems, provides a total attenuation of 71 dB in 1-dB steps, offers respectable accuracy and insertion loss through 225 MHz and can be used at 450 MHz as shown in **Table 25.4**. This material was originally published as "An RF Step Attenuator" by Denton Bramwell, K7OWJ, in the June 1995 *QST*.

The attenuator consists of 10 resistive π-attenuator sections such as the one in **Fig 25.71**. Each section consists of a DPDT slide switch and three ¼-W, 1 %-tolerance metal-film resistors. The complete unit contains single 1, 2, 3 and 5-dB sections, and six 10-dB sections. **Table 25.5** lists the resistor values required for each section.

The enclosure is made of brass sheet stock, readily available at hardware and hobby stores. By selecting the right stock, you can avoid having to bend the metal and need only perform a minimum of cutting.

CONSTRUCTION

The enclosure can be built using only a nibbling tool, drill press, metal shears, and a soldering gun or heavy soldering iron. (Use a regular soldering iron on the switches and resistors.) One method of cutting the small pieces of rectangular tubing to length is to use a drill press equipped with a small abrasive cutoff wheel.

Brass is easy to work and solder. For the enclosure, you'll need two precut 2 × 12 × 0.025-inch sheets and two 1 × 12 × 0.025-inch sheets. The 2-inch-wide stock is used for the front and back panels; the 1-inch-wide stock is used for the ends and sides. For the internal wiring, you need a piece of ⁵⁄₃₂ × ⁵⁄₁₆-inch rectangular tubing, a ¼ × 0.032-inch strip, and a few small pieces of 0.005-inch-thick stock to provide inter-stage shields and form the 50-Ω transmission lines that run from the BNC connectors to the switches at the ends of the step attenuator.

For the front panel, nibble or shear a piece of 2-inch-wide brass to a length of about 9½ inches. Space the switches from each other so that a piece of the rectangular brass tubing lies flat and snugly between them (see **Fig 25.72**). Drill holes for the #4-40 mounting screws and nibble or punch rectangular holes for the bodies of the slide switches.

Before mounting any parts, solder in place one of the 1-inch-wide chassis side pieces to make the assembly more rigid. Solder the side piece to the edge of the top plate that faces the "through" side of the switches; this makes later assembly easier (see **Fig 25.73**).

Table 25.4
Step Attenuator Performance at 148, 225 and 450 MHz

Table 25.4
Step Attenuator Performance at 148, 225 and 450 MHz
Measurements made in the ARRL Laboratory

Attenuator set for Maximum attenuation (71 dB)		Attenuator set for minimum attenuation (0 dB)	
Frequency (MHz)	Attenuation (dB)	Frequency (MHz)	Attenuation (dB)
148	72.33	148	0.4
225	73.17	225	0.4
450	75.83	450	0.84

Note: Laboratory-specified measurement tolerance of ±1 dB

Table 25.5
Closest 1%-Tolerance Resistor Values

Attenuation (dB)	R1 (Ω)	R2 (Ω)
1.00	866.00	5.60
2.00	436.00	11.50
3.00	294.00	17.40
5.00	178.00	30.10
10.00	94.30	71.50

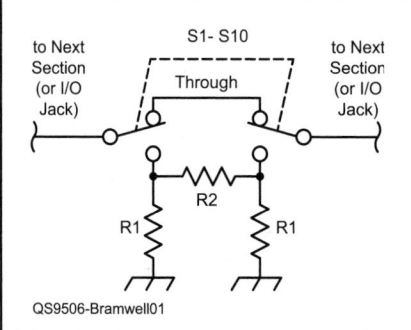

Fig 25.71 — Schematic of one section of the attenuator. All resistors are ¼-W, 1%-tolerance metal-film units. See Table 25.5 for the resistor values required for each attenuator section. There are six 10 dB sections and one each of 1, 2, 3 and 5 dB.

Although the BNC input and output connectors are shown mounted on the top (front) panel, better lead dress and high-frequency performance may result from mounting the connectors at the ends of the enclosure.

DPDT slide switches designed for sub-panel mounting often have mounting holes tapped for #4-40 screws. Enlarge the holes to allow a #4-40 screw to slide through. Before mounting the switches, make the "through" switch connection (see Fig 25.71) by bending the two lugs at one end of each switch toward each other and soldering the lugs together or solder a small strip of brass between the lugs and clip off the lug ends. Mount the switches above the front panel, using ⁵⁄₃₂-inch-high by ⁷⁄₃₂-inch-OD spacers. Use the same size spacer on the inside. On the inside, the spacer creates a small post that helps reduce capacitive

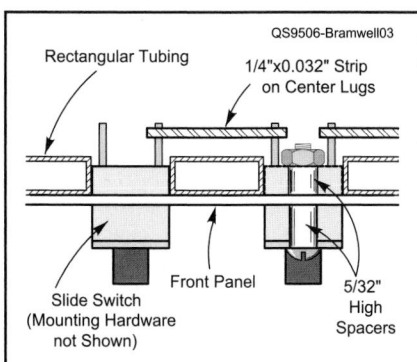

Fig 25.72 — Key to obtaining acceptable insertion loss in the "through" position is to make the whole device look as much as possible like 50-Ω coax. The rectangular tubing and the ¼ × 0.032-inch brass strip between the switch sections form a 50-Ω stripline.

coupling from one side of the attenuator to the other. The spacers position the switch so that the 50-Ω stripline can be formed later.

The trick to getting acceptable insertion loss in the "through" position is to make the attenuator look as much as possible like 50-Ω coax. That's where the rectangular tubing and the ¼ × 0.032-inch brass strip come into the picture (see Fig 25.72); they form a 50-Ω stripline. (See the **Transmission Lines** chapter for information on stripline.)

Cut pieces of the rectangular tubing about ¾-inch long, and sweat solder them to the front panel between each of the slide switches. Next, cut lengths of the ¼-inch strip long enough to conveniently reach from switch to switch, then cut one more piece. Drill ¹⁄₁₆-inch holes near both ends of all but one of the ¼-inch strips. The undrilled piece is used as a temporary spacer, so make sure it is flat and deburred.

Lay the temporary spacer on top of the rectangular tubing between the first two switches, then drop one of the drilled ¼-inch pieces over it, with the center switch lugs through the ¹⁄₁₆-inch holes. Before soldering, check the strip to make sure there's sufficient clearance between the ¼-inch strip and the switch lugs;

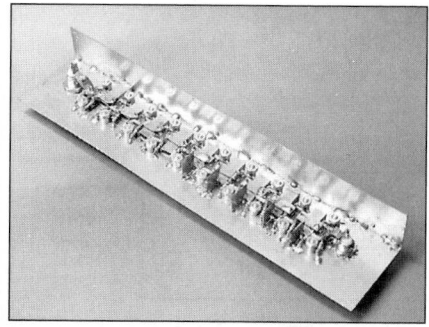

Fig 25.73 — Solder one of the 1-inch-wide chassis side pieces in place to make the assembly more rigid during construction. Solder the side piece to the edge of the top plate that faces the "through" side of the switches; this makes the rest of the assembly easier.

Fig 25.74 — The attenuator before final mechanical assembly. The ¼-inch strips are spaced 0.033 inch apart to form a 50-Ω connection from the BNC connector to the stripline. There are ½-inch square shields between 10-dB sections. The square shields have a notch in one corner to accommodate the end of the rectangular tubing.

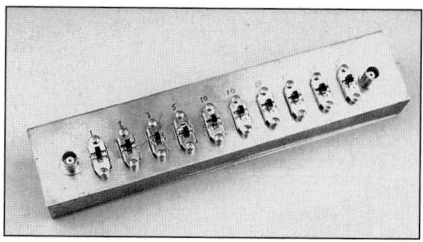

Fig 25.75 — The completed step attenuator in the enclosure of brass sheet. The BNC connectors may be mounted on the front panel at the end of the switches or on the end panels.

trim the corners if necessary. Use a screwdriver blade to hold the strip flat and solder the lugs to the strip. Remove the temporary spacer. Repeat this procedure for all switch sections. This creates a 50-Ω stripline running the length of the attenuator.

Next, solder in place the three 1%-tolerance resistors of each section, keeping the leads as short as possible. Use a generous blob of solder on ground leads to make the lead less inductive. Install a ½-inch-square brass shield between each 10-dB section to ensure that signals don't couple around the sections at higher frequencies.

Use parallel ¼-inch strips of 0.005-inch-thick brass spaced 0.033 inch apart to form 50-Ω feed lines from the BNC connectors to the switch contacts at each end of the stripline as shown in Fig 25.73. (Use the undrilled piece of 0.032-inch-thick brass to insure the proper line spacing.) The attenuator with all switches and shields in place is shown ready for final mechanical assembly in **Fig 25.74**

Finally, solder in place the remaining enclosure side, cut and solder the end pieces, and solder brass #4-40 nuts to the inside walls of the case to hold the rear (or bottom) panel. Drill and attach the rear panel and round off the sharp corners to prevent scratching or cutting anyone or anything. Add stick-on feet and labels and your step attenuator of **Fig 25.75** is ready for use.

Remember that the unit is built with ¼-W resistors, so it can't dissipate a lot of power. Remember, too, that for the attenuation to be accurate, the input to the attenuator must be a 50-Ω source and the output must be terminated in a 50-Ω load.

25.8.10 High-Power RF Samplers

If one wants to measure characteristics of a transmitter or high-powered amplifier, some

means of reducing the power of the device to 10 or 20 dBm must be used. The most straightforward way to do this is to use a 30 or 40 dB attenuator capable of handling the high power. A 30 dB attenuator will reduce a 100 W transmitter to 20 dBm. A 40 dB attenuator will reduce a 1 kW amplifier to 20 dBm. If further attenuation is needed, a simple precision attenuator may be used after the signal has been reduced to the 20 dBm level.

The problem with high-powered attenuators is that they are expensive to buy or build since the front end of the attenuator must handle the output power of the transmitter or amplifier. If one already has a dummy load, an RF sampler may be used to produce a replica of the signal at a reduced power level. The sampler described here was originally presented in *QST* Technical Correspondence for May 2011 by Tom Thompson, WØIVJ.

A transformer sampler passes a single conductor (usually the insulated center conductor from a piece of coaxial cable) from the transmitter or amplifier to the dummy load through a toroidal inductor forming a transformer with a single turn primary. The secondary of the transformer is connected to a resistor network and then to the test equipment as shown in **Fig 25.76**. Assume that the source, whether a transmitter or amplifier, is a pure voltage source in series with a 50-Ω resistor. This most likely is not exactly the case but is sufficient for analysis.

If a current, I, flows into the dummy load, then a current, I / N flows in the secondary of the transformer, where N is the number of turns on the secondary. Fig 25.76 also shows the equivalent circuit, substituting a current source for the transformer. The attenuation

is 40 dB and 15 turns for the secondary of the transformer. If $R_{SHUNT} = 15\ \Omega$, and $R_{SERIES} = 35\ \Omega$, then the voltage across a 50-Ω load resistor, R_{SAMPLE}, is 1/100 of the voltage across the dummy load, which is 40 dB of attenuation.

Reflecting this resistor combination back through the transformer yields 0.06 Ω in series with the 50-Ω dummy load impedance. This is an insignificant change. Furthermore, reflecting 100 Ω from the primary to the secondary places 22.5 kΩ in parallel with R_{SHUNT}, which does not significantly affect its value. The test equipment sees a 50-Ω load looking back into the sampler. Even at low frequencies, where the reactance of the secondary winding is lower than 15 Ω, the impedance looking back into the sample port remains close to 50 Ω.

The samplers described here use an FT37-61 ferrite core followed by two resistors as described above. The through-line SWR is good up to 200 MHz, the SWR is fair looking into the sampled port, and the useful bandwidth extends from 0.5 MHz to about 100 MHz. If you are interested in an accurate representation of the third harmonic of your HF transmitter or amplifier, it is important for the sampler to give accurate attenuation into the VHF range.

Fig 25.77 shows a photo of a sampler built into a 1.3 × 1.3 × 1 inch (inside dimensions) box constructed from single-sided circuit board material. The through-line connection is made with a short piece of UT-141 semi-rigid coax with the shield grounded only on one side to provide electrostatic shielding between the toroid and the center conductor of the coax. (Do not ground both ends of the shield or a shorted turn is created.) R_{SHUNT} is hidden under the toroid, and R_{SERIES} is shown connected to the sample port. This construction technique looks like a short piece of 200-Ω transmission line in the through-line which affects the SWR at higher frequencies. This can be corrected by compensating with two 3 pF capacitors connected to the through-line input and output connectors as shown

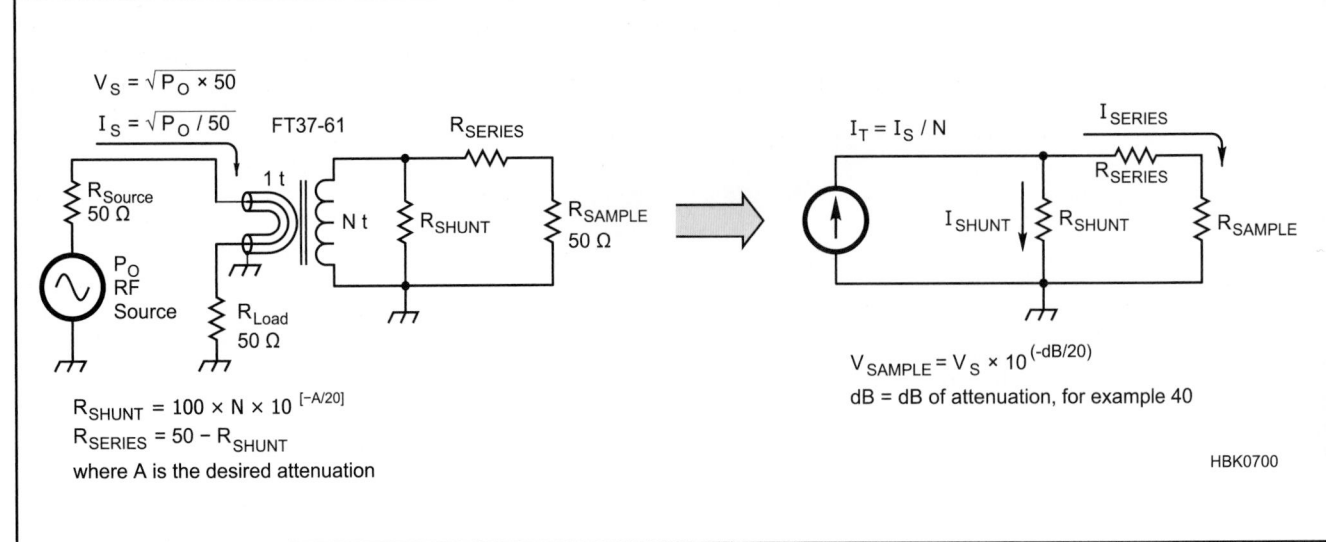

$V_S = \sqrt{P_O \times 50}$

$I_S = \sqrt{P_O / 50}$ FT37-61

$R_{SHUNT} = 100 \times N \times 10^{[-A/20]}$

$R_{SERIES} = 50 - R_{SHUNT}$

where A is the desired attenuation

$I_T = I_S / N$

$V_{SAMPLE} = V_S \times 10^{(-dB/20)}$

dB = dB of attenuation, for example 40

HBK0700

Fig 25.76 — RF sampler circuit diagram and equivalent circuit showing calculations.

in the photo. The through-line SWR was reduced from 1.43:1 to 1.09:1 at 180 MHz by adding the capacitors. This compensation, however, causes the attenuation to differ at high frequencies depending on the direction of the through-line connection. A sampler constructed using the box technique is useable from below 1 MHz through 30 MHz.

Fig 25.78 shows a different approach using ⁹⁄₁₆ inch diameter, 0.014 inch wall thickness, hobby brass tubing. This lowers the impedance of the through-line so that no compensation is needed. The through-line SWR for the tube sampler is 1.08:1 at 180 MHz which is as good as the box sampler and the sensitivity to through-line direction is reduced. Although the high frequency attenuation is not as good as the box sampler, the construction technique provides a more consistent result. A sampler constructed using the tube technique should be useable through 200 MHz.

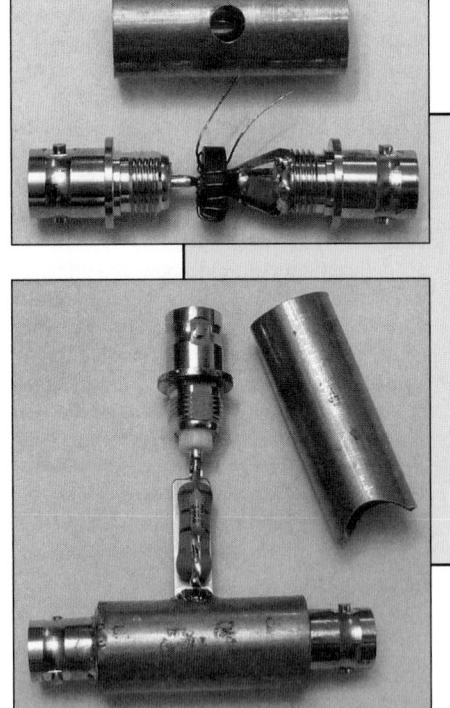

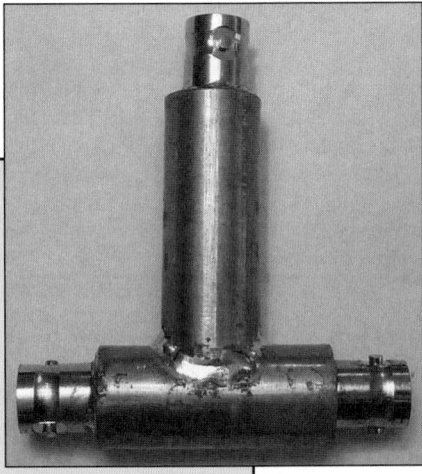

Fig 25.78 — RF sampler using tube construction.

Fig 25.77 — RF sampler using box construction.

CONSTRUCTION OF THE TUBE SAMPLER

Both samplers use 15 turns of #28 AWG wire on an FT37-61 core, which just fits over the UT-141 semi-rigid coax. R_{SHUNT} is a 15 Ω, 2 W, non-inductive metal oxide resistor and R_{SERIES} is a 34.8 Ω, ¼ W, 1% non-inductive metal film resistor. The power dissipation of the resistors and the flux handling capability of the ferrite core are adequate for sampling a 1500-W source. For those uncomfortable using BNC connectors at high power, an SO-239 version may be constructed using an FT50A-61 core and larger diameter tubing. Construction details are included on this book's CD-ROM.

25.8.11 RF Oscillators for Circuit Alignment

Receiver testing and alignment can make use of inexpensive RF signal generators which are available as complete units and in kit form. Any source of signal that is weak enough to avoid overloading the receiver usually will serve for alignment work and troubleshooting.

A crystal oscillator is often a satisfactory signal source for amplifier testing and receiver repair or alignment. Several example circuits can be found in the **Oscillators and Synthesizers** chapter. The output frequencies of crystal oscillators, while not adjustable, are quite precise and very stable. The Elecraft XG2 (**www.elecraft.com**) and NorCal S9 (**www.norcalqrp.org**) are good example of a simple fixed-frequency signal source kits. The harmonics of the output signals are on known frequencies and can also be used as low-level signal sources. The fundamental signals have known output amplitudes for calibrating S meters and other gain stages.

Variable frequency oscillators can be used as signal generators and there are several kits or assembled units available based on a direct digital synthesis (DDS) integrated circuit. (See the **DSP and Software Radio Design** chapter.) The Elecraft XG3 is a programmable signal source that operates from 1.5 to 200 MHz with four programmable output levels between −107 and 0 dBm. Another similar product is the DDSv4 by RMT Engineering (**www.rmt-tech.com**) with a 0-100 MHz range.

For receiver performance testing, precise frequency control, signal purity, noise, and low-level signal leakage become very important. A lab-quality instrument is required to make these measurements. Commercial and military-surplus units such as the HP608-series are big and stable, and they may be inexpensive. Recently, the HP8640-series of signal generators have become widely available at very attractive prices. When buying a used or inexpensive signal generator, look for these attributes: output level is calibrated, the output doesn't "ring" too badly when tapped, and doesn't drift too badly when warmed up.

25.8.12 Hybrid Combiners for Signal Generators

Many receiver performance measurements require two signal generators to be attached to a receiver simultaneously. This, in turn, requires a combiner that isolates the two signal generators (to keep one generator from being frequency or phase modulated by the other). Commercially made hybrid combiners are available from Mini-Circuits Labs (**www.minicircuits.com**).

Alternatively, a hybrid combiner is not difficult to construct. The combiners described here (see **Fig 25.79**) provide 40 to 50 dB of isolation between ports, assuming the common port is terminated in a 50-ohm load. Attenuation in the desired signal paths (each input to output) is 6 dB. Loads with low return loss typical of receiver inputs will reduce isolation.

The combiners are constructed in small boxes made from double-sided circuit-board material as shown in Fig 25.79A. Each piece is soldered to the next one along the entire length of the seam. This makes a good RF-tight enclosure. BNC coaxial fittings are used on the units shown. However, any type of coaxial connector can be used. Leads must be kept as short as possible and precision non-inductive resistors (or matched units from the junk box) should be used. The circuit diagram for the combiners is shown in Fig 25.79B.

The combiner may also be constructed and used as a *return loss bridge* as described in the *QST* article by Jim Ford, N6JF, "Build a Return Loss Bridge," from September 1997 and included on the CD-ROM included with this book. Return loss is discussed in the **RF Techniques** chapter.

25.8.13 Gate-Dip Oscillator (GDO)

The project is adapted from the May 2003 *QST* article, "A Modern GDO — the "Gate"

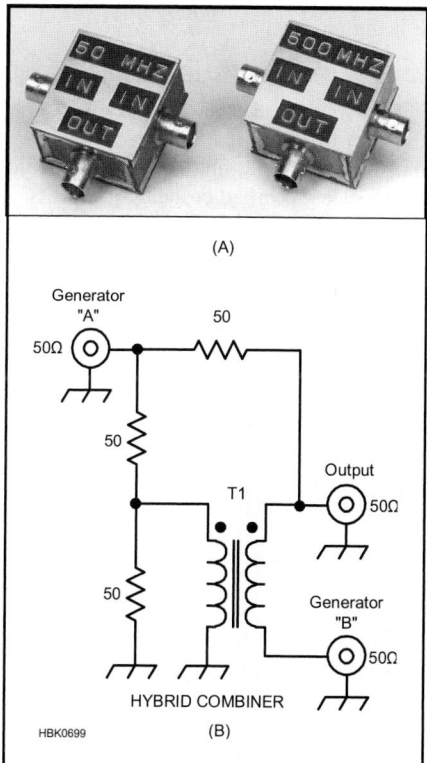

Fig 25.79 —The hybrid combiner on the left of A is designed to cover the 1 to 50-MHz range; the one on the right 50 to 500 MHz. B shows the circuit diagram of the hybrid combiner. Transformer T1 is wound with 10 bifilar turns of #30 AWG enameled wire. For the 1 to 50-MHz model, T1 is an FT-23-77 ferrite core. For the 50 to 500-MHz model, use an FT-23-67 ferrite core. Keep all leads as short as possible when constructing these units.

Dip Oscillator" by Alan Bloom, N1AL. A GDO is a tunable oscillator with the coil mounted outside of the chassis. The external coil allows you to measure the resonant frequency of a tuned circuit without any electrical connection to it. Just place the GDO coil near the tuned circuit and tune the GDO while watching for a dip in its meter reading.

This "no connection" measurement capability is handy in other applications. For example, you can measure the resonant frequency of a Yagi's parasitic elements that don't have a feed line connection or measure the resonant frequency of antenna traps. The GDO can "sniff out" spurious resonances in a linear amplifier's tank circuit with the amplifier powered down. (Be sure no high voltage is present!)

The GDO has other uses on the workbench. To measure inductance, temporarily connect a known-value capacitor in parallel with the unknown inductor and find the circuit's resonant frequency. Use the formula $L = 1/(2\pi f)^2C$ or the reactance vs frequency chart in the **Electrical Fundamentals** chapter to find the value of the inductor. (If C is measured in μF and f in MHz, the resulting value of L is in μH.) The same process works to find the value of a capacitor, such as that of an unmarked air-variable capacitor at a hamfest, by using the formula $C = 1/(2\pi f)^2L$.

A GDO can be used as a simple signal generator to test amplifiers, mixers, and filters. To troubleshoot a receiver, tune the GDO to each of the IF frequencies, starting with the last IF stage and hold the coil close to that part of the circuitry. If you can hear a signal at the receiver's output, then that IF stage and all circuitry after it are working. This is especially handy on densely packed, surface-mount boards that are difficult to probe.

With the oscillator turned off, the GDO functions as a tuned RF detector known as an *absorption frequency meter* or *wavemeter*. An obvious use is to determine if RF energy in a tuned circuit has the right frequency. Using the capacity probe as an antenna, it can be used as a frequency-selective field-strength meter. By holding the coil near a cable, you can detect RF current flowing on the outside of a shield and the GDO makes an excellent "sniffer" to detect RF leakage from a shielded transmitter at the fundamental and harmonic frequencies.

A headphone output is provided to listen for key clicks, hum or buzz, and low-frequency parasitic oscillations in a transmitted signal. By coupling the coil to an antenna, you can even use the GDO as a tunable "crystal radio"!

CIRCUIT DESIGN

Fig 25.80 shows the completed GDO. **Fig 25.81** is a view inside the case and **Fig 25.82** pictures the entire set of coils. The schematic and parts list are shown in **Fig 25.83**. A pair of source-coupled N-channel JFETs (Q1 and Q2) form the oscillator portion of the circuit. No RF chokes are required. This eliminates false resonances resulting from self-resonance of the chokes.

Q4, a bipolar 2N3904 transistor, serves a

Fig 25.80 — The "gate dipper" fits comfortably in the hand.

dual purpose. Its base-emitter junction acts as the RF detector. Further, it amplifies the rectified current flowing in the base and sends it to the emitter-follower Q5, a 2N2907. Transistor Q3 is a JFET source-follower amplifier for the output RF connector. The RF output may be used as a signal source or to drive a frequency counter for more accurate frequency display.

Determine the value of R5 using the formula $R5 = (1.5/I_m) - R_m$, where I_m is the full-scale meter current and R_m is the meter resistance. To measure your meter's resistance, be sure to use an ohmmeter that does not use more than the GDO meter's full-scale current. Most modern DVMs use test currents in the range off 50-200 µA for measuring resistance. The value of R5 in the parts list is for a 200 µA full-scale meter used by the author. Meters from 50 to 500 µA full-scale are suitable.

In detector mode, the power to the oscillator and RF buffer is turned off. Q4 detects and amplifies any signals picked up by the coil. The meter sensitivity control works in both oscillator and wavemeter mode and also controls the volume to the headphones. Battery current is 3-5 mA with the oscillator on and zero in wavemeter mode when no signal is being received.

CONSTRUCTION

Feel free to substitute parts on hand for those in the parts list. The exceptions are transistors Q1, Q2, and Q4, which should be the types specified or their equivalents.

The $2\frac{1}{4} \times 2\frac{1}{4} \times 5$ inch chassis is a compromise; it's large enough to allow all parts to fit easily and small enough to fit comfortably in the hand. The coil is mounted off-center, both to allow the shortest connection to the tuning capacitor and to afford easier coil coupling to an external circuit.

The prototype used a perforated test board for the RF portion of the circuitry and a solder terminal strip for the meter circuitry. If you prefer, a printed circuit board pattern and parts layout are available on this book's CD-ROM and circuit boards are available from FAR Circuits (**www.farcircuits.net**). Note that the battery's *positive* terminal is connected to ground which is backwards from the normal arrangement.

The coil forms are "⅜ inch" flexible plastic water tubing. The inside diameter is actually slightly less than ⅜ inch, which makes a nice force-fit onto the ⅜-inch threads of a chassis-mount BNC plug (not the more common chassis-mount receptacle). The lowest-frequency coil is wound on a pill bottle with two 1-inch diameter aluminum washers at the connector mount for added strength.

To construct the coils, start by running the wire through the center of each form and soldering it to the BNC connector's center pin connection. Then press the form onto the connector threads and cut a small notch in the form's opposite end to hold the wire in place for the winding. After winding the coil, cut and tin the wire end, then solder it to the ground lug mounted on the connector. Cover

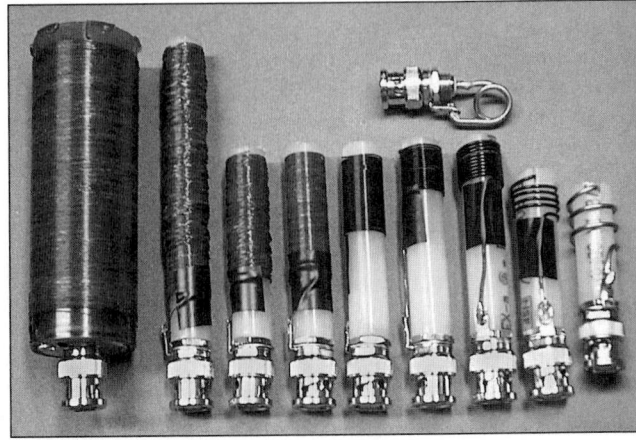

Fig 25.81 — Photograph of the GDO with the cover off. The tuning capacitor was oriented for the shortest possible connection to the coil connector. The prototype was constructed with two batteries but three result in better operation.

Fig 25.82 — This is how the coils look before covering them with heat-shrink tubing or tape. Small pieces of electrical tape are used to hold the turns in place during construction.

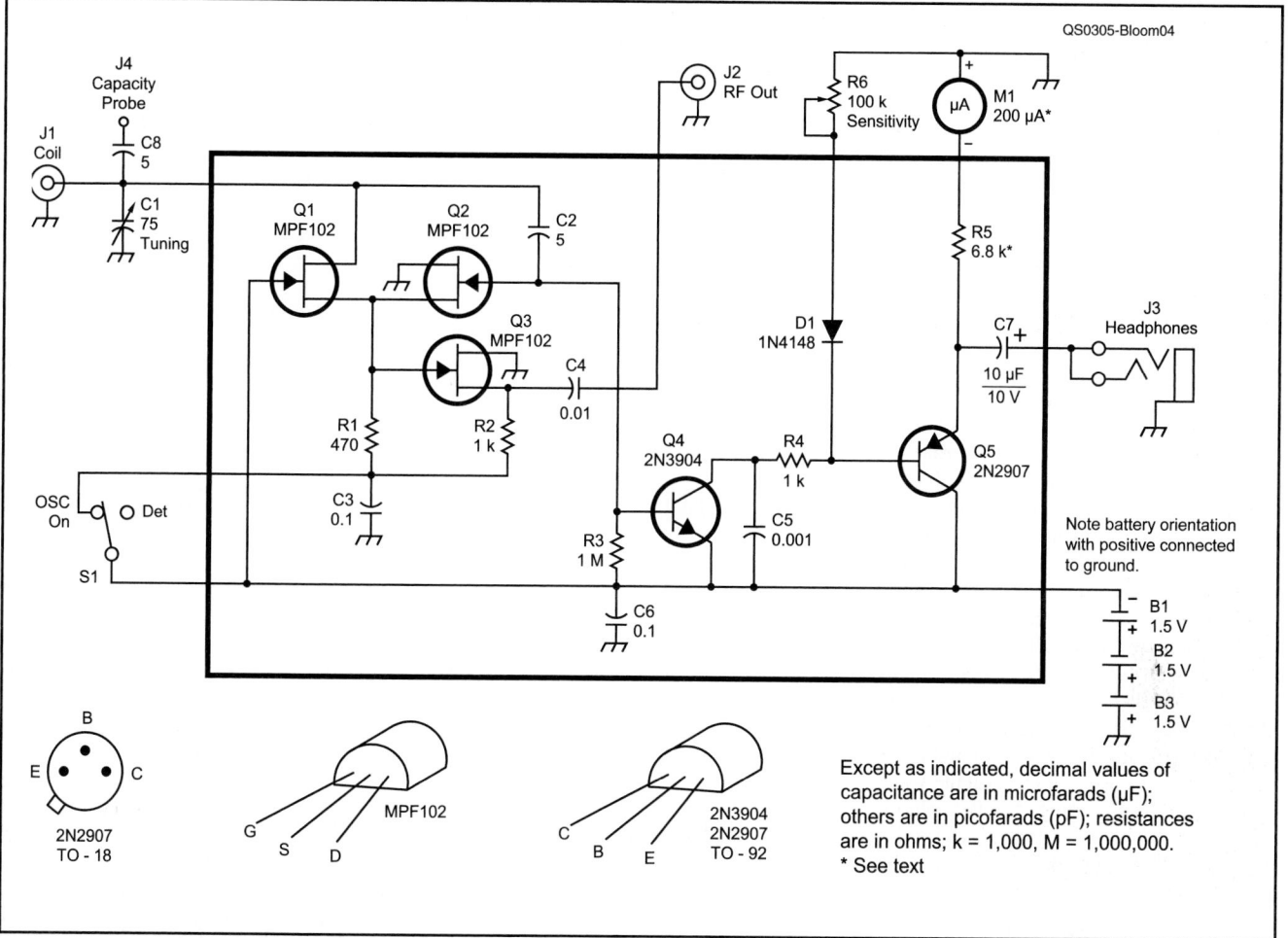

Fig 25.83 — Schematic diagram of the GDO.

B1, B2, B3 — 1.5 V AA or AAA cell.
C1 — 75 pF or 365 pF variable.
C2, C8 — 5 pF disc capacitor.
C3, C6 — 0.1 µF capacitor.
C4 — 0.01 µF capacitor.
C5 — 0.001 µF capacitor.
C7 — 10 µF, 10 V capacitor.
D1 — 1N4148 diode.
J1, J2 — BNC female chassis-mount
 connector.

J3 — 3.5 mm stereo phone jack.
J4 — Phone tip jack, phone tip plug.
M1 — 0-200µA, 1⅝ inch square panel
 meter, (see text).
Q1, Q2, Q3 — MPF102 transistor, JFET.
Q4 — 2N3904 transistor.
Q5 — 2N2907A transistor.
R1 — 470 Ω, ¼ W resistor.
R2, R4 — 1 kΩ, ¼ W resistor.

R3 — 1 MΩ, ¼ W resistor.
R5 — See text.
R6—100 kΩ potentiometer, linear taper.
S1 — SPST toggle switch.
**Misc: #18, #22 and #26 gauge enameled
wire; 10 BNC male chassis-mount
connector (for coils); dual AA or AAA
battery holder; 2¼ × 2¼ × 5 inch chassis;
3 ft of ⅜ inch PVC water tubing for coils.**

each coil with a layer of heat-shrink tubing or electrical tape to hold the turns in place and protect the wire. Clear tubing allows you to see the coil and a small label with the coil's frequency range can be placed under the tubing before shrinking. Fig 25.82 shows the coils before the heat shrink tubing is added.

Winding data is listed in **Table 25.6**. Unless you happen to duplicate the prototype exactly, the frequency range of each coil is likely to be different. However, this data can be used as a starting point for your own coil designs.

The heavy wire for the two highest-frequency coils is bare copper scrap from house wiring cable. The other coils are wound with enamel-insulated magnet wire. The exact wire gauge is not critical although it may

affect the number of turns required.

The three smallest coils (29.5-150 MHz) were space wound to the lengths listed. The remaining coils were close-wound. On the 0.9-1.5 and 1.5-2.8 MHz coils, there are too many turns to fit in a single layer. Wind these coils by overlapping turns a few at a time. Winding two complete overlapping layers increases the inter-winding capacitance which can cause spurious resonances and reduced tuning range.

The range of the lowest-frequency coil can be extended by connecting it to the GDO through a BNC "T" connector to which one or two 53 pF capacitors are attached using clip leads (see Table 25.6). Adding extra capacitance is a useful way to slow down

the tuning rate for measuring narrow-band devices such as crystals.

With the smallest coil, the GDO oscillates over only a small portion of the capacitor tuning range. Fortunately, this range covers the 2 meter amateur band. With the second-smallest coil, the oscillation amplitude drops off at lower frequencies but is useable down to about 62 MHz in the prototype. A lower-inductance tuning capacitor would improve performance at VHF.

The small 75-pF tuning capacitor results in a tuning range of only about 2:1. The advantage of the low ratio is that tuning is not so critical and the frequency dial is easier to read. The disadvantage is that it requires more coils to cover the overall frequency

Table 25.6
Coil Winding Data for the GDO

Wire Gauge	Form Diameter (in.)	Form Length (in.)	Coil Length (in.)	Number of Turns	Frequency Range
#12	0.375	—	0.125	2	130-150 MHz
#14	0.5	1.5	.5	3	62-108 MHz
#18	0.5	2.0	0.4	5	29.5-62 MHz
#18	0.5	2.5	0.5	10.5	16.5-35 MHz
#22	0.5	2.5	0.6	21	9.2-19 MHz
#26	0.5	2.5	1.0	46	5.1-10.5 MHz
#30	0.5	2.5	1.5	100	2.8-5.6 MHz
#30	0.5	2.5	1.6	180	1.5-2.8 MHz
#30	0.5	3.9	3.3	390	0.9-1.5 MHz
#30	1.25	3.9	3.4	230	620-980 kHz
				(+53 pF)	505-640 kHz
				(+106 pF)	440-525 kHz

range. Substituting a 365-pF AM broadcast radio tuning capacitor will increase the tuning range to nearly 3:1 with a consequent sharpening of adjustment.

The tuning dial in **Fig 25.84** was made from a circular piece of ¼-inch clear plastic glued to a knob. Making the diameter slightly larger than the chassis width allows one handed operation with a thumb adjusting the tuning as in Fig 25.80. The scale, cemented to the chassis under the dial, was drawn using a computer graphics program although measuring the frequency and creating the scales was very time-consuming. A better solution would be to include a small frequency counter with the GDO using the RF output. 4- or 5-digit accuracy is sufficient.

OPERATION

To measure the resonant frequency of a tuned circuit, switch to oscillator mode and adjust the meter sensitivity to about ¾ scale. Then orient the GDO coil close to and approximately parallel to the coil under test and tune the dial until you get a strong dip on the meter. Tune slowly or you may miss the dip. Overly-close coupling to the circuit under test causes such a strong dip that the oscillator is pulled far off-frequency. Once a dip is found, move the GDO coil farther from the circuit under test until the dip is barely visible — this results in the most accurate measurement.

Coupling to toroids or shielded coils can be difficult. One solution is to connect a wire from the capacitance probe to the "hot" end of the circuit to be tested. For looser coupling, just place the wire close to the circuit under test instead of connecting it to the circuit. When measuring the inductance of a toroid with a test capacitor, the capacitor leads often form enough of a loop to allow coupling to the GDO coil. Some authors recommend coupling the GDO to a one-turn loop through the toroid, but that creates a shorted secondary winding, changing the inductance of the toroid winding.

Antenna measurements are best made at the high-current point of the antenna conduc-

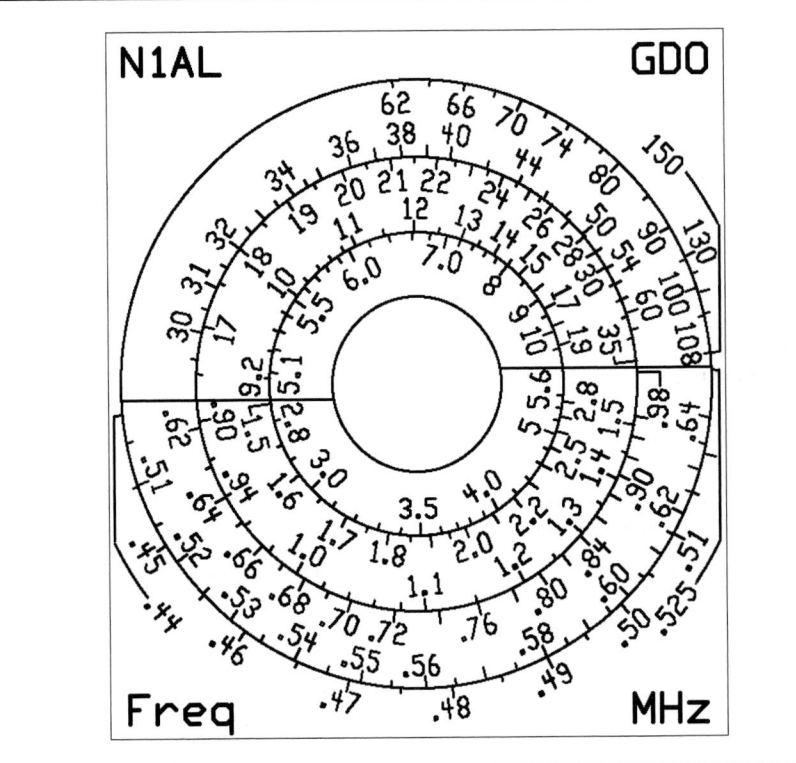

Fig 25.84 — The dial scale for the prototype GDO. Builders of this project should expect to create and calibrate a custom dial for the finished instrument or use a frequency counter as noted in the text.

tor. For a half-wave dipole, this is near the center. Orient the coil perpendicular to the conductor for maximum coupling. Be sure to short the feed point of the antenna before making the measurement. If you can't find the dip, make the shorting wire into a 1-turn loop for better coupling to the GDO coil You should also see resonance at the odd harmonic frequencies as well as the fundamental.

To measure the electrical length of a transmission line, couple the GDO to a small wire loop; connect it to one end of the line and leave the other end disconnected. (For best accuracy use the smallest loop that gives sufficient coupling.) The line is ¼ wavelength

long at the lowest resonant frequency, so the electrical length in meters is 75/f, where f is the resonant frequency in MHz. Again, you will also see resonance at the odd harmonics.

The voltage level at the RF output connector varies from coil to coil but typically runs about 250 mV$_{RMS}$ into an open circuit and 50 mV$_{RMS}$ into a 50-Ω load. That is sufficient for a typical frequency counter or to serve as a test signal for troubleshooting.

Additional uses of the GDO are presented in the following articles on this book's CD-ROM: "The Art of Dipping" in Jan 1974 *QST*, "Add-Ons for Greater Dipper Versatility" in Feb 1981 *QST*, and "What Can You

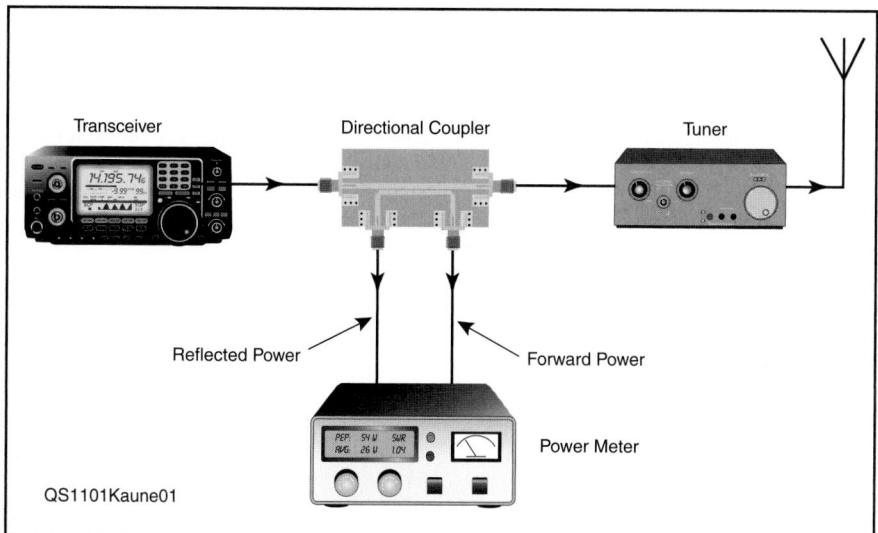

Fig 25.85 — W7IEQ station setup, including the power meter being described here.

coupled are reduced by a factor of $1/N^2$, where $N = 31$ is the number of turns of wire on each toroid. Thus the forward and reflected power samples are reduced by about 30 dB. For example, if a transceiver were delivering a power of 100 W to a pure 50 Ω load, the forward power sample from the directional coupler would be about 0.1 W (20 dBm).

The directivity of a directional coupler is defined as the ratio of the forward power sample divided by the reflected power sample when the coupler is terminated in 50 Ω. In this coupler, the directivity measured using an inexpensive network analyzer is at least 35 dB at 3.5 MHz and 28 dB at 30 MHz.

POWER/SWR METER — CIRCUIT DESCRIPTION

Fig 25.87 shows a front panel view of the power meter. An LCD displays the measured peak (PEP) and average (AEP) envelope powers as well as the standing wave ratio (SWR). The power meter calculates either the peak and average envelope power traveling from the transceiver to load (the forward power) or the peak and average envelope powers actually delivered to the load (the forward power minus reflected power). The average envelope power (AEP) represents an average of the forward or load powers over an averaging period of either 1.6 or 4.8 seconds.

A 1 mA-movement analog meter on the front panel facilitates antenna tuning. This meter continuously displays the quantity $1 - 1/SWR$, where SWR is the standing wave ratio on the line. Thus, an SWR of 1.0 corresponds to a meter reading of 0 — no deflection of the meter. An SWR of 2 results in a 50% deflection of the meter, while an SWR of 5 produces an 80% deflection of the meter.

The forward and reflected power samples from the directional coupler are applied to a pair of Analog Devices AD8307 logarithmic

Do with a Dip Meter?" in May 2002 *QST*.

25.8.14 RF Power Meter

The following section is an overview of the January 2011 *QST* article by Bill Kaune, W7IEQ, "A Modern Directional Power/SWR Meter". The complete article including firmware and printed circuit board artwork is available on the CD-ROM included with this book.

The primary use for this unit is to monitor the output power and tuning of a transceiver. The author's station configuration is shown in **Fig 25.85**. RF power generated by the transmitter is routed via RG-8 coaxial cable through a directional coupler to an antenna tuner, which is connected to the antenna with RG-8. The directional coupler contains circuits that sample the RF power flowing from

the transmitter to the tuner (the forward power) and the RF power reflected back from the tuner to the transmitter (the reflected power). These samples are sent via RG-58 cable to the two input channels of the power meter. This project includes the directional coupler and the power meter. Enough detail is provided in the full article so that an amateur can duplicate the device or modify the design.

DIRECTIONAL COUPLER

The directional coupler is based on the unit described in "The Tandem Match" by John Grebenkemper, KI6WX in the Jan 1987 issue of *QST* and also included on this book's CD-ROM. A pair of FT-82-67 toroids with 31 turns of #26 AWG magnet wire over lengths of RG-8 form the basis of the directional coupler shown in **Fig 25.86**.

The forward and reflected power samples

◄Fig 25.86 — Completed directional coupler.

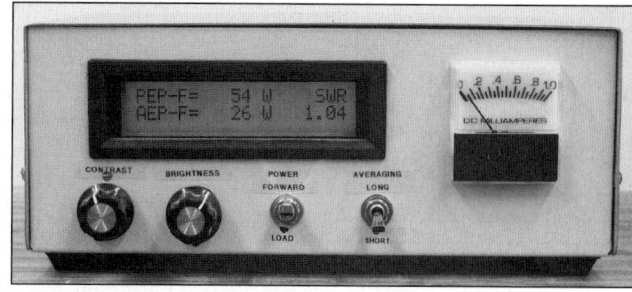

Fig 25.87 — Front panel of power meter. The LCD shows the peak envelope power (PEP), the average envelope power (AEP) and the SWR. The two knobs control the contrast and back lighting of the LCD. One toggle switch determines whether forward or load powers are displayed. A second switch sets the averaging time for the AEP calculation. The meter displays SWR and is used for tuning purposes.

detectors. External 20 dB attenuators (Mini-Circuits HAT-20) reduce the signals from the directional coupler to levels compatible with the AD8307. As noted earlier, the directional coupler has an internal attenuation of about 30 dB, so the total attenuation in each channel is about 50 dB. Thus, a rig operating at a power level of 1 kW (60 dBm) will result in an input to the forward power channel of about 10 dBm. (The schematic diagram and parts list of the power meter are provided on the CD-ROM version of the article.) The detectors are configured such that the time constant of their output follows the modulation envelope of the RF signal.

LF398 sample-and-hold ICs stabilize the voltages from the forward and reflected power logarithmic detectors. In this way both voltages can be sampled at the exact same time and held for subsequent analog-to-digital conversion and calculation of power and SWR by the PIC16F876A microprocessor (www.microchip.com). The processor also includes a pulse-width-modulated (PWM) output used to drive the analog SWR meter on the front panel.

25.9 References and Further Reading

25.9.1 References

1. National Institute of Standards and Technology, www.nist.gov
2. Hageman, Steve, "Build a Data Acquisition System for your Computer" (Tech Notes), *QEX*, Jan/Feb 2001, pp 52-57.
3. Prologix is one manufacturer of low-cost GPIB-to-USB converters. www.prologix.biz
4. Bryce, Mike, WB8VGE, "A Universal Frequency Calibrator," *QST*, Nov 2009, pp 35-37.
5. Bradley, Mark, K6TAF, "What Can You Do With a Dip Meter?," *QST*, May 2002, pp 65-68.
6. Miller, Bob, KE6F, "Atomic Frequency Reference for Your Shack," *QEX*, Sept/Oct 2009, pp 35-44.
7. Nash, Bob, KF6CDO, "An Event per Unit Time Measurement System for Rubidium Frequency Standards," *QEX*, Nov/Dec 2010, pp 3-17.
8. Jones, Bill, K8CU, "Using the HP Z3801A GPS Frequency Standard," *QEX*, Nov/Dec 2002, pp 49-53.
9. Shera, Brooks, W5OJM, "A GPS-Based Frequency Standard," *QST*, July 1998.
10. Richardson, Gary, AA7VM, "A Low Cost DDS Function Generator," *QST*, Nov 2005, pp 40-42.
11. Z. Lau, "A Relative RF Ammeter for Open-Wire Lines," *QST*, Oct 1988, pp 15-17.
12. Steinbaugh, Gary, AF8L, "An Inexpensive Laboratory-Quality RF Wattmeter," *QEX*, May/Jun 2010, pp 26-32
13. Bird, Trevor S., "Definition and Misuse of Return Loss," *QEX*, Sep/Oct 2010, pp 38-39.
14. Baier, Thomas C., DG8SAQ, "A Simple S-Parameter Test Set for the VNWA2 Vector Network Analyzer," *QEX*, May/Jun 2009, pp 29-32.

15. AN 1287-3, "Applying Error Correction to Network Analyzer Measurements," Agilent Technologies, www.agilent.com
16. Baier, Thomas C., DG8SAQ, "A Small, Simple USB-Powered Vector Network Analyzer Covering 1 kHz to 1.3 GHz," *QEX*, Jan/Feb 2009, pp 32-36. www.sdr-kits.net/VNWA/VNWA_Description.html
17. McDermott, Tom, N5EG, et al, "A Low-Cost 100 MHz Vector Network Analyzer with USB Interface," *QEX*, Jul/Aug 2004. www.tapr.org/kits_vna.html
18. Detailed instructions to build N2PK's vector network analyzer are at http://n2pk.com.
19. Application Note 57-1, "Fundamentals of RF and Microwave Noise Figure Measurements," Agilent Technologies, www.agilent.com. The older Hewlett-Packard application note AN57 is still on the Agilent website and may be more useful for legacy noise figure meters such as the HP340 series.
20. Pontius, Bruce E., NØADL, "Measurement of Signal-Source Phase Noise with Low-Cost Equipment," *QEX*, May/Jun 1998.
21. Noakes, John D., VE7NI, "The 'No Fibbin' RF Field Strength Meter, *QST*, Aug 2002, pp 28-29,
22. At microwave frequencies, the antenna and radio are sometimes tested as a system. Wade, Paul, W1GHZ, "Microwave System Test," Microwavelengths, *QST*, Aug 2010, pp 96-97.
23. PathSim, free HF channel simulation software. www.moetronix.com/ae4jy/pathsim.htm
24. Forrer, Johann B., KC7WW, "A Low-Cost HF Channel Simulator for Digital Systems," *QEX*, May/Jun 2000, pp 13-22.
25. Information on the Softrock I/Q downconverters is available at www.softrockradio.org.

25.9.2 Further Reading

Agilent Technologies, "Resistance; dc Current; ac Current; and Frequency and Period Measurement Errors in Digital Multimeters, Application Note AN 1389-2" www.agilent.com.

Agilent Technologies, "Test-System Development Guide Application Notes 1465-1 through 1465-8," www.agilent.com

Hayward, Wes, W7ZOI et al, *Experimental Methods in RF Design*, Chapter 7, "Measurement Equipment," ARRL, 2003.

Silver, H. Ward , NØAX, "Test Equipment for the Ham Shack." *QST*, May 2005, pp 36-39.

High-Frequency Measurements

Agilent Technologies, "Fundamentals of RF and Microwave Noise Figure Measurements, Application Note 57-1," www.agilent.com.

Carr, Joseph J., *Practical Radio Frequency Test and Measurement: a Technician's Handbook*, Boston, Newness Co., 1999.

Straw, R. Dean, N6BV, *The ARRL Antenna Book*, 21st edition, Chapter 27, "Antenna and Transmission-Line Measurements," ARRL, 2009.

Oscilloscopes

Agilent Technologies, "Agilent Technologies Oscilloscope Fundamentals, Application Note 1606," www.agilent.com.

R. vanErk, *Oscilloscopes, Functional Operation and Measuring Examples*, McGraw-Hill Book Co, New York, 1978.

V. Bunze, *Probing in Perspective — Application Note 152*, Hewlett-Packard Co, Colorado Springs, CO, 1972 (Pub No. 5952-1892).

The XYZs of Using a Scope, Tektronix, Inc, Portland, OR, 1981 (Pub No. 41AX-4758).

Basic Techniques of Waveform Measurement (Parts 1 and 2), *Hewlett-Packard Co, Colorado Springs, CO, 1980 (Pub No. 5953-3873).*

J. Millman, and H. Taub, *Pulse Digital and Switching Waveforms*, McGraw-Hill Book Co, New York, 1965, pp 50-54.

V. Martin, *ABCs of DMMs*, Fluke Corp, PO Box 9090, Everett, WA 98206.

Receivers

Rohde, Ulrich L., KA2WEU, "Theory of Intermodulation and Reciprocal Mixing: Practice, Definitions and Measurements in Devices and Systems, Part I," *QEX*, Nov/Dec 2002, pp 3-15.

Rohde, Ulrich L., KA2WEU, "Theory of Intermodulation and Reciprocal Mixing: Practice, Definitions and Measurements in Devices and Systems, Part II," *QEX*, Jan/Feb 2003, pp 21-31.

Wade, Paul, N1BWT, "Noise Measurement and Generation," *QEX*, Nov 1996, pp 3-12.

Spectrum Analysis

Agilent Technologies, "Eight Hints for Making Better Spectrum Analyzer Measurements, Application Note 1286-1", **www.agilent.com**

Stanley, John O., K4ERO, "The Beauty of Spectrum Analysis — Part 1", *QST*, June 2008, pp 35-38.

Stanley, John O., K4ERO, "The Beauty of Spectrum Analysis — Part 2", *QST*, July 2008, pp 33-35.

Network Analysis

Agilent Technologies, "Understanding the Fundamental Principles of Vector Network Analysis, Application Note AN 1287-1", **www.agilent.com**. Other Agilent application notes with further information on network analyzers are AN 1287-2 through AN 1287-10 and AN 1287-12.

Agilent Technologies, "S-Parameter Techniques for Faster, More Accurate Network Design, Application Note 95-1", **www.agilent.com**.

Hiebel, Michael, *Fundamentals of Vector Network Analysis*, Rohde & Schwarz, 2008.

See also "Test and Measurement Bibliography," an extensive listing of *QEX* and *QST* articles on the CD-ROM accompanying this book.

25.10 Test and Measurement Glossary

Accuracy — The maximum expected error in a measurement.

Ammeter — A device for measuring electrical current.

Antenna analyzer — A device to measure the RF impedance of a one-port network such as an antenna.

Antenna test range — An area designed to minimize the effect of RF reflections to permit accurate antenna gain and pattern testing.

Attenuator — A broadband device that reduces the amplitude of a signal by a specified, well-controlled amount.

Autoranging — The ability of a *multimeter* to set its range automatically based on the signal level.

AWG (arbitrary waveform generator, also ARB) — An instrument that can generate a complex signal based on waveforms stored in memory.

Birdies — Slang term for internally-generated spurious signals in a receiver that may be steady, warble, or pulsate.

Blocking dynamic range — The difference between the *blocking level* and the *MDS*.

Blocking level — The level of an interfering signal that causes weak signals to be reduced in amplitude by 1 dB.

Bolometer — A device for measuring RF power by measuring the heat dissipated in a dummy load.

Bridge — A circuit used to indicate the relative values of passive circuit elements by observing the null on a meter or other indicator. See this chapter's discussion of Wheatstone bridges.

Burden voltage — The full-scale voltage drop of an ammeter.

Coaxial detector — An RF detector with a coaxial connector.

Combiner — A device to combine signals from two sources. A *power splitter* in reverse. See *Hybrid combiner.*

D'Arsonval meter — The most common type of mechanical meter, consisting of a permanent magnet and a moving coil with pointer attached.

Dip meter (dip oscillator) — An instrument with an oscillator whose coil is external to the enclosure so it can be coupled to the circuit under test. The meter value drops (dips) when the oscillator frequency is tuned to the frequency of resonance of the circuit under test.

Directional coupler, directional bridge — A device that senses the RF power flowing in one direction on a transmission line.

DMM (digital multimeter) — A test instrument that measures voltage, current, resistance and possibly other quantities and displays the result on a numeric digit display, rather than on an analog meter.

DVM (digital volt meter) — See *DMM.*

Dynamic error — An error whose value depends on the time of measurement.

Dynamic range — The difference in dB between the strongest and weakest signals that a receiver can handle.

Electronic voltmeter — An amplified analog *multimeter.*

ENR (excess noise ratio) — The ratio of the noise added by a noise source to the thermal noise, normally expressed in dB.

Feed-through termination — See *Through-line termination.*

Field-strength meter — A device to indicate radiated RF field strength.

Four wire — A technique for measuring small resistances or large currents that uses four wires rather than two to reduce errors due to test lead resistance.

Frequency counter — A device that measures the frequency of a periodic signal and displays the result on a digital readout.

Frequency domain — Description of signals as a function of frequency, as opposed to as a function of time.

Function generator — An audio tone generator that generates tones in the shape of various functions, such as sine, square, triangle and ramp waveforms.

Frequency markers — Test signals generated at selected intervals (such as 25 kHz, 50 kHz, 100 kHz) for calibrating the dials of receivers and transmitters.

Gaussian distribution — The distribution of the probability of the instantaneous voltage of a noise signal as a function of the voltage. It forms a bell-shaped curve with a maximum at the dc value of the signal (typically zero volts).

GDO (grid-dip oscillator) — A *dip meter* that uses a vacuum tube.

Harmonic signal — A periodic signal. Its frequency spectrum consists only of the fundamental and harmonics of the repetition rate.

Hybrid combiner — A passive device used to combine two signals in such a way as to reduce the interaction between the signal sources.

IF and image rejection — The difference in level between an interfering signal at a receiver's IF or image frequency and the level of the desired signal that produces the same response.

IMD (intermodulation distortion) — The creation of unwanted frequencies because of two or more strong signals modulating each other.

IMD dynamic range — The difference in signal level between two signals that cause IMD products at the level of the MDS and the MDS.

LCR Bridge — A device for measuring inductors, capacitors or resistors using a Wheatstone bridge.

Lissajous Pattern — As used by amateurs, the combined display of two sine wave signals on an oscilloscope to determine the relationship between their frequencies. (Also called Lissajous figure.)

Logic analyzer — A sophisticated instrument for analyzing digital circuitry.

Logic probe — A simple device for sensing and displaying a digital signal's state.

Marker — An indicator on the screen of a *spectrum analyzer* that allows reading out the frequency and amplitude of a specific signal. See also *Frequency marker*.

Marker delta — A *spectrum analyzer* feature that reads out the difference in frequency and amplitude of two markers.

MDS (minimum discernible signal) — The level of the *noise floor* at the antenna connector of a receiver. Depends on the measurement bandwidth.

Multimeter — A device that measures several electrical quantities, such as voltage, current and resistance, and displays them on an analog meter or digital display.

Multiplier (voltage multiplier) — A resistor placed in series with a meter to increase the full-scale voltage reading.

Near field — The area close to an antenna where the electromagnetic wave is not completely formed. Antenna gain and pattern measurements are not valid in this region.

Network analyzer — An instrument to measure the return loss and transmission gain between the two ports of a two-port network.

NIST (National Institute of Standards and Technology) — A non-regulatory agency of the US federal government that manages measurement standards. It used to be called NBS, the National Bureau of Standards.

NIST traceable — Refers to a device whose accuracy is based on NIST standards using rules and procedures specified by NIST.

Noise bandwidth — The bandwidth of a rectangular-spectrum filter that would produce the same total noise power as the filter under consideration.

Noise density — Noise power per hertz.

Noise figure — A figure of merit for the sensitivity of receivers and other RF devices. It is the ratio of the effective noise level to the thermal noise level, usually expressed in dB.

Noise floor — The noise received by a receiver in a specified bandwidth, referenced to the antenna connector. It is the thermal noise in dBm plus the noise figure in dB.

Noise source — An instrument that generates well-calibrated white noise for test purposes.

OCXO (oven-controlled crystal oscillator) — A crystal oscillator mounted in a temperature-controlled oven to improve the frequency stability.

Ohmmeter — A meter that measures resistance. Usually part of a *multimeter*. See *VOM* and *DMM*.

Ohms per volt — A measure of the sensitivity of a voltmeter that has multiple scales. It is the reciprocal of the current drawn by the meter.

Oscilloscope — An instrument that displays signals in the *time domain*. Has a graphical display that shows amplitude on the vertical axis and time on the horizontal axis.

Peak search — A feature of a *spectrum analyzer* in which a marker is automatically placed on the strongest signal in the display.

Peak to peak value — The difference between the most positive and most negative signal values. For a symmetrical signal it is twice the peak value.

Peak value — The highest value of a signal during the measuring time.

Periodic — Refers to a signal that repeats exactly at a regular time interval, the period.

Phase noise — Wideband noise on an RF signal caused by random fluctuations of the phase.

Port — A pair of connections to a network, typically via a coaxial connector

Power divider (power splitter) — A device to divide an RF signal between two loads. A *combiner* in reverse.

Prescaler — A circuit used ahead of a counter to extend the frequency range. A counter capable of operating up to 50 MHz can count up to 500 MHz when used with a divide-by-10 prescaler.

Random error — A non-repeatable error caused by noise in the measurement system.

Reciprocal mixing — The mixing of a nearby interfering signal with the phase noise of the receiver local oscillator, which causes noise in the audio output.

Resolution — The smallest distinguishable difference in a measured value.

Resolution bandwidth — The smallest frequency separation between two RF signals that a *spectrum analyzer* can resolve. It is determined by the IF filters in the *spectrum analyzer*.

Return loss — The ratio, usually expressed in dB, between the incident and reflected RF signals.

Return-loss bridge (RLB) — A bridge used for measuring the return loss of an RF circuit, transmission line or antenna.

Reverse power protection — A *signal generator* feature to protect the instrument from accidental transmission by a transceiver under test.

RF probe — A hand-held probe with a detector to allow measuring RF signals with a dc voltmeter.

Scalar network analyzer (SNA) — A *network analyzer* that measures only the magnitude of the gain and return loss.

Scattering parameters (S parameters) — A set of four parameters to characterize the complex return loss and transmission gain in both directions of a two-port network.

Scope — Slang for *oscilloscope*.

Second-order IMD — IMD caused by strong signals at frequencies f1 and f2 that occurs at a frequency of f1 + f2.

Service monitor — An integrated package of test equipment packaged as a single instrument for testing receivers and transmitters.

Shunt (meter shunt) — A resistor connected in parallel with a meter to increase the full-scale current reading.

SI (Système International d'Unités) — The modern, revised version of the metric system.

Signal generator — An instrument that generates a calibrated variable-frequency RF signal, usually with adjustable amplitude and modulation capability.

SINAD (signal plus noise and distortion) — A measure of the relative level of signal compared to the noise and distortion at the audio output of an FM receiver.

Single-shot — Refers to a non-periodic signal that can be measured in a single time interval.

Spectrum analyzer — An instrument that displays signals in the *frequency domain*. Has a graphical display that shows amplitude (normally in logarithmic, or dB, form) on the vertical axis and frequency on the horizontal axis.

Spurious emissions, or spurs — Unwanted energy generated by a transmitter or other circuit. These emissions include, but are not limited to, harmonics.

Standard — A rule for the proper method to measure some quantity. A standard may involve a standard artifact that defines the unit.

Step attenuator — An attenuator that can be switched between different attenuation values.

Systematic error — A repeatable error due to some characteristic of the measurement system.

TCXO (temperature-compensated crystal oscillator) — A crystal oscillator that includes circuitry to compensate for frequency drift with temperature.

Termination — A resistor with a coaxial connector used to terminate a transmission line in its characteristic impedance.

Test set — A *network analyzer* accessory that automatically configures the connections to the device under test for various measurements.

Third-order IMD — IMD caused by strong signals at frequencies f1 and f2 that appears at frequencies 2f1 – f2 or 2f2 – f1.

Third-order IMD intercept point (IP3) — The power level representing the intersection of plots on a dBm scale of

the power level of the two strong signals and their IMD products.

Through-line termination — A termination with two coaxial connectors that connect straight through and a terminating resistor to ground.

Time base — A highly-accurate reference oscillator used in a frequency counter or other device that needs an accurate time or frequency reference.

Time domain — Refers to the variation in time of electronic signals, as opposed to their frequency spectrum.

Tracking generator — A *spectrum analyzer* accessory that consists of a signal generator whose frequency tracks the frequency of the analyzer.

True RMS — Refers to a meter that measures the actual RMS voltage or current instead of measuring the average or peak value and calculating the RMS value assuming a sinusoidal waveform.

VCXO (voltage-controlled crystal oscillator) — A crystal oscillator whose frequency can be adjusted slightly using an external applied voltage.

Vector network analyzer (VNA) — A *network analyzer* that measures both the magnitude and phase of the gain and return loss of a two-port network.

Video bandwidth — The bandwidth of the post-detection filter in a *spectrum analyzer*. It limits the maximum sweep speed.

VOM (volt-ohm meter) — A *multimeter* that does not include active circuitry such as an amplifier.

VTVM (vacuum-tube voltmeter) — A *multimeter* that uses one or more vacuum tubes.

Wheatstone bridge — See *Bridge circuit*.

Wilkinson combiner — A type of *hybrid combiner*.

Contents

Chapter 26 — CD-ROM Content

Supplemental Articles
- "Troubleshooting Radios" by Mel Eiselman, NC4L
- "Building a Modern Signal Tracer" by Curt Terwilliger, W6XJ
- "Hands-on Radio — Power Supply Analysis" by Ward Silver, NØAX
- "Amplifier Care and Maintenance" by Ward Silver, NØAX
- "Diode and Transistor Test Circuits" by Ed Hare, W1RFI

PC Board Templates
- Crystal controlled signal source template
- AF/RF signal injector template

Troubleshooting and Maintenance

The robust and self-reliant ethic of Amateur Radio is nowhere stronger than in the amateur's ability to maintain, troubleshoot and repair electronic equipment. Amateurs work with not just radios, but all sorts of equipment from computers and software to antennas and transmission lines. This flexibility and resilience are keys to fulfilling the Basis and Purpose for Amateur Radio.

The sections on troubleshooting approaches, tools and techniques build on earlier material written by Ed Hare, W1RFI. They will help you approach troubleshooting in an organized and effective manner, appropriate to your level of technical experience and tools at hand. This material shows how to get started and ask the right questions — often the most important part of troubleshooting.

Additional sections on troubleshooting power supplies, amplifiers, radios and antenna systems (contributed by Tom Schiller, N6BT, Ted Thrift, VK2ARA, and Ross Pittard, VK3CE) tackle the most common troubleshooting needs. Restoring and maintaining vintage equipment is a popular part of ham radio and so there are some sections by John Fitzsimmons, W3JN, and Pat Bunsold, WA6MHZ, on the special needs of this equipment.

This chapter is organized in three groups of sections to be consulted as required for any particular troubleshooting need. You will not need to read it from end-to-end in order to troubleshoot successfully. The first group of sections covers test equipment details, pertinent information about components, and safety practices. The second group presents general guidelines and techniques for effective troubleshooting. The third group presents specific advice and information on equipment that is commonly repaired by amateurs.

TROUBLESHOOTING — ART OR SCIENCE?

Although some say troubleshooting is as much art as it is science, the repair of electronic gear is not magic. It is more like detective work as you work carefully to uncover each clue. Knowledge of advanced math or electronics theory is not required. However, you must have, or develop, a good grasp of basic electronics and simple measurements, guided by the ability to read a schematic diagram and to visualize signal flow through the circuit. As with most skills, these abilities will develop with practice.

Not everyone is an electronics wizard; your gear may end up at the repair shop in spite of your best efforts. The theory you learned for the FCC examinations and the information in this *Handbook* can help you decide if you can fix it yourself. Even if the problem appears to be complex, most problems have simple causes. Why not give troubleshooting a try to the best of your abilities? Maybe you can avoid the effort and expense of shipping the radio to the manufacturer. It is gratifying to save time and money, but the experience and confidence you gain by fixing it yourself may prove even more valuable.

SAFETY FIRST! — SWITCH TO SAFETY

Always! Death is permanent. A review of safety must be the first thing discussed in a troubleshooting chapter. Some of the voltages found in amateur equipment can be fatal! Only 50 mA flowing through the body is painful; 100 to 500 mA is usually fatal. Under certain conditions, as little as 24 V can kill. RF exposure in a high-power amplifier can create severe burns very quickly. Batteries can deliver huge amounts of power that can melt tools and wires or create an explosion when short-circuited. Charging lead-acid cells can create a buildup of explosive hydrogen gas, as well.

Make sure you are 100% familiar with all safety rules and the dangerous conditions that might exist in the equipment you are servicing. A list of safety rules can be found in **Table 26.1**. You should also read the **Safety** chapter of this *Handbook* — all of it — before you begin to work on equipment.

Remember, if the equipment is not working properly, dangerous conditions may exist where you don't expect them. Treat every component as potentially "live." Some older equipment uses "ac/dc" circuitry. In this circuit, one side of the chassis is connected directly to the ac line, a condition unexpected by today's amateurs who are accustomed to modern safety standards and practices. This is an electric shock waiting to happen.

The maximum voltage rating of voltmeters and oscilloscopes is not often noted by the hobbyist but it is crucial to safety when working on voltages higher than the household ac line voltage. Test equipment designed to measure voltage always has a maximum safe voltage rating

Table 26.1
Safety Rules
1. Keep one hand in your pocket when working on live circuits or checking to see that capacitors are discharged.
2. Include a conveniently located ground-fault current interrupter (GFCI) circuit breaker in the workbench wiring.
3. Use only grounded plugs and receptacles.
4. Use a GFCI protected circuit when working outdoors, on a concrete or dirt floor, in wet areas, or near fixtures or appliances connected to water lines, or within six feet of any exposed grounded building feature.
5. Use a fused, power limiting isolation transformer when working on ac/dc devices.
6. Switch off the power, disconnect equipment from the power source, ground the output of the internal dc power supply, and discharge capacitors when making circuit changes.
7. Do not subject electrolytic capacitors to excessive voltage, ac voltage or reverse voltage.
8. Test leads should be well insulated and without cracks, fraying, or exposed conductors
9. Do not work alone!
10. Wear safety glasses for protection against sparks and metal or solder fragments.
11. Be careful with tools that may cause short circuits.
12. Replace fuses only with those having proper ratings.
13. Never use test equipment to measure voltages above its maximum rating.

between the circuit being measured and the equipment user — you! This is particularly important in handheld equipment in which there is no metal enclosure connected to an ac safety ground. Excessive voltage can result in a flashover to the user from the internal electronics, probes, or test leads, resulting in electric shock. Know and respect this rating.

If you are using an external high voltage probe, make sure it is in good condition with no cracks in the body. The test lead insulation should be in good condition — flexible and with no cracks or wire exposed. If practical, do not make measurements while holding the probe or meter. Attach the probe with the voltage discharged and then turn the power on. Turn power off and discharge the voltage before touching the probe again. Treat high voltage equipment with care and respect!

Soldering Safety

Remember that soldering tools and melted solder can be hot and dangerous! Wear protective goggles and clothing when soldering. A full course in first aid is beyond the scope of this chapter, but if you burn your skin, run the burn immediately under cold water and seek first aid or medical attention. Always seek medical attention if you burn your eyes; even a small burn can develop into serious trouble.

UNDERSTANDING THE BASICS

To fix electronic equipment, you need to understand the system and circuits you are troubleshooting. A working knowledge of electronic theory, circuitry and components is an important part of the process. When you are troubleshooting, you are looking for specific conditions that cause the symptoms

you are experiencing. Knowing how circuits are supposed to work will help you to notice things that are out of place or that indicate a problem.

To be an effective troubleshooter, review and understand the following topics discussed elsewhere in this book:
• Ohms law and basic resistor circuits (**Electrical Fundamentals**)
• Basic transistor and diode characteristics (**Analog Basics**)
• Fundamental digital logic and logic signals (**Digital Basics**)
• Voltage and current measurements (**Test Equipment and Measurements**)
• SWR and RF power measurement (**Transmission Lines**)
You would be surprised at how many problems — even problems that appear

complicated — turn out to have a simple root cause found by understanding the fundamentals and methods of one of these categories.

GETTING HELP

Other hams may be able to help you with your troubleshooting and repair problems, either with a manual or technical help. Check with your local club or repeater group. You may get lucky and find a troubleshooting wizard. (On the other hand, you may get some advice that is downright dangerous, so be selective.) Most clubs have one or two troubleshooting gurus who can provide guidance and advice, if not some on-the-workbench help.

There is a wealth of information available online, too. Many of the popular brands of equipment and even specific models have their own online communities or user's groups. The archives of these groups — almost universally free to join — contain much valuable troubleshooting, modification and operating information. If the problem doesn't appear to have been described, you can ask the group.

The Technology area of the ARRL's website also has an extensive section on Servicing Equipment (**www.arrl.org/servicing-equipment**). That page features articles and other resources, including links to schematic databases.

Your fellow hams in the ARRL Field organization may also help. Technical Coordinators (TC) and Technical Specialists (TS) are volunteers who are willing to help hams with technical questions. For the name and address of a local TC or TS, contact your Section Manager (listed in the front of any recent issue of *QST*).

Using Search Engines for Troubleshooting

The power of Internet search engines can save huge amounts of time when troubleshooting equipment. The key is in knowing how to construct the right list of words for them to find. Precision is your friend — be exact and use words others are likely to use if they had the same problem. Use the primary model number without suffixes to avoid being too specific. For example, when troubleshooting the well-known PLL potting compound problem exhibited by Kenwood TS-440 transceivers, entering the search string "TS-440 display dots" immediately finds many web pages dealing with the problem, while simply entering "Kenwood transceiver blank display" returns dozen of unrelated links.

Start with a very specific description of the problem and gradually use less exact terms if you don't find what you want. Learn how to use the "Advanced Search" functions of the search engine, too.

26.1 Test Equipment

Many of the steps involved in effiicient troubleshooting require the use of test equipment. We cannot see electricity directly, but we can measure its characteristics and effects. Our test equipment becomes our electrical senses.

The **Test Equipment and Measurements** chapter is where you can find out more about various common types of equipment, how to operate it, and even how to build some of your own. There are many articles in *QST* and in books and websites that explain test equipment and offer build-it-yourself projects, too. Surplus equipment of excellent quality is widely available at a fraction of its new cost.

You need not purchase or build every type of test equipment. Specialty equipment such as spectrum analyzers or UHF frequency counters can often be borrowed from a club member or friend — maybe one of those troubleshooting gurus mentioned earlier. If you own the basic instruments and know how to use them, you'll be able to do quite a bit of troubleshooting before you need the special instruments.

26.1.1 Senses

Although they are not test equipment in the classic sense, your own senses will tell you as much about the equipment you are trying to fix as the most-expensive spectrum analyzer. We each have some of these natural test instruments.

Eyes — Use them constantly. Look for evidence of heat and arcing, burned components, broken connections or wires, poor solder joints or other obvious visual problems.

Ears — Severe audio distortion can be detected by ear. The snaps and pops of arcing or the sizzling of a burning component may help you track down circuit faults. An experienced troubleshooter can diagnose some circuit problems by the sound they make. For example, a bad audio-output IC sounds slightly different from a defective speaker.

Nose — Your nose can tell you a lot. With experience, the smells of ozone, an overheating transformer, and a burned resistor or PC board trace each become unique and distinctive. Many troubleshooting sessions begin with "something smells hot!"

Finger — After using a voltmeter to ensure no hazardous voltages are present, you can use a fingertip to determine low heat levels— never do this in a high-voltage circuit. Use a temperature probe if using a finger is unsafe. Small-signal transistors can be fairly warm, but being very hot indicates a circuit problem. Warm or hot capacitors are always suspect. High-power devices and resistors can be quite hot during normal operation.

Brain — More troubleshooting problems have been solved with a multimeter and a brain than with the most expensive spectrum analyzer. You must use your brain to analyze data collected by other instruments.

26.1.2 Internal Equipment

Some test equipment is included in the equipment you repair. Nearly all receivers include a speaker. An S meter is usually connected ahead of the audio chain. If the S meter shows signals, that indicates that the RF and IF circuitry is probably functioning. Transmitters often have a power supply voltage and current meter, along with power output, SWR, ALC and speech compression readings that give valuable clues about what is happening inside the equipment.

The equipment also has visual indicator lights that provide additional information such as transmit status, high SWR, low voltage, squelch status, and so forth. These readings or indicators are often specifically referenced by the troubleshooting sections of manuals to help sort out problems.

Microprocessor-controlled equipment often provides error indications, either through a display or by indicator lights. In addition, faults detected by the control software are sometimes communicated through patterns of beeps or flashing of LEDs. Each sequence has a specific meaning that is described in the operating or service manual.

26.1.3 Bench Equipment

The following is a list of the most common and useful test instruments for troubleshooting. Some items serve several purposes and may substitute for others on the list. The theory and operation of most of this equipment is discussed in detail in the **Test Equipment and Measurements** chapter. Notes about the equipment's use for troubleshooting are listed here.

Multimeters —The most often used piece of test equipment, the digital multimeter or DMM, can often test capacitors of most values in addition to voltage, current and resistance. Most can test diodes and transistors on a go/no-go basis, while some can measure gain. Some can even measure frequency or use an external probe to measure temperature.

Some DMMs are affected by RF, so most technicians keep an old-style analog moving-needle VOM (volt-ohm-meter) on hand for use in strong RF fields. Some technicians prefer the moving needle for peaking or nulling adjustments.

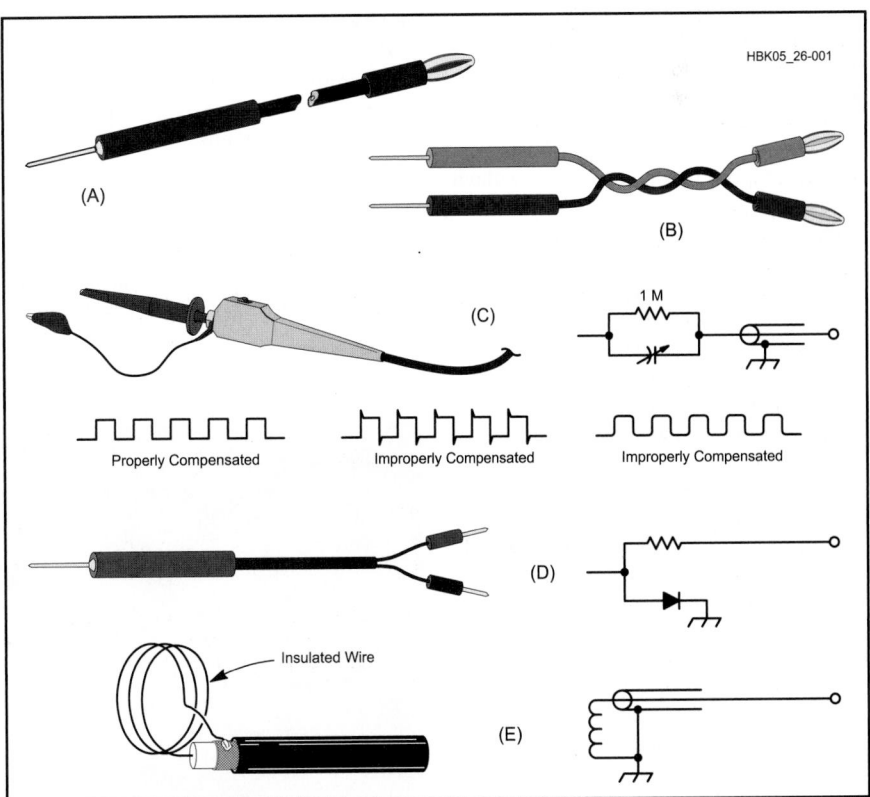

Fig 26.1 — **An array of test probes for use with various test instruments.**

Test or clip leads — Keep an assortment of these wires with insulated alligator clips. Commercially-made leads have a high failure rate because they use small wire that is not soldered to the clips, just crimped. You can slip off the clip jackets and solder the wire yourself for better reliability. Making a set of heavier-gauge leads is a good idea for currents above several hundred milliamps.

Individual wire leads (**Fig 26.1A**) are good for dc measurements, but they can pick up unwanted RF energy. This problem is reduced somewhat if the leads are twisted together (Fig 26.1B). Coaxial cable test leads can avoid RF pickup but also place a small capacitance across the circuit being measured. The added capacitance may affect performance.

Test probes — The most common probe is the low-capacitance (×10) oscilloscope probe shown in Fig 26.1C. This probe isolates the oscilloscope from the circuit under test, preventing the scope's input and test-probe capacitance from affecting the circuit and changing the reading. A network in the probe serves as a 10:1 divider and compensates for frequency distortion in the cable and test instrument.

Demodulator probes (see the **Test Equipment and Measurements** chapter and the schematic shown in Fig 26.1D) are used to demodulate or detect RF signals, converting modulated RF signals to audio that can be heard in a signal tracer or seen on a low-bandwidth scope.

You can make a probe for inductive coupling as shown in Fig 26.1E. Connect a two or three-turn loop across the center conductor and shield before sealing the end. The inductive pickup is useful for coupling to high-current points and can also be used as a sniffer probe to pick up RF signals without contacting a circuit directly.

Other common types of probes are the non-contact clamp-on probes shown in **Fig 26.2** that use magnetic fields to measure current. A high-voltage probe for use with DMMs or VOMs is shown in **Fig 26.3** and is discussed more in this chapter's section on power supply troubleshooting.

Thermocouple and active temperature sensor probes are also commonly available. These display temperature directly on the meter in °F or °C.

RF power and SWR meters — Simple meters indicate relative power SWR and are fine for adjusting matching networks and monitoring transmission line conditions for problems. However, if you want to make accurate measurements, a calibrated directional RF wattmeter with the proper sensing elements for the frequencies of signals being measured is required.

Dummy load — Do not put a signal on the air while repairing equipment. Defective equipment can generate signals that interfere

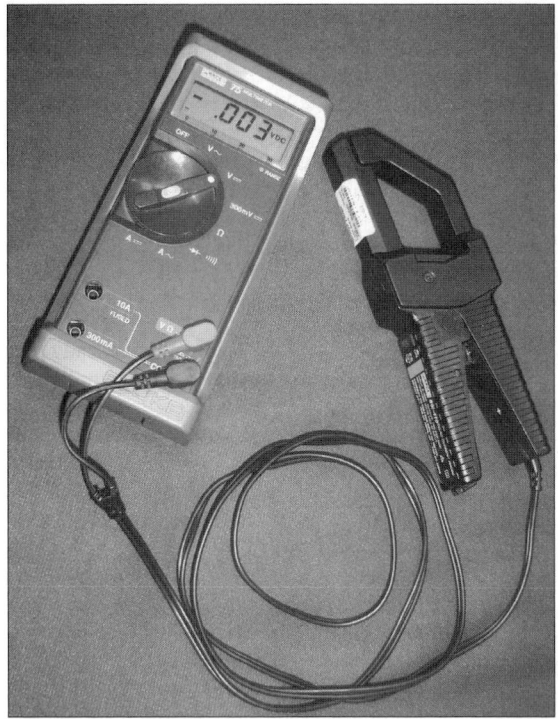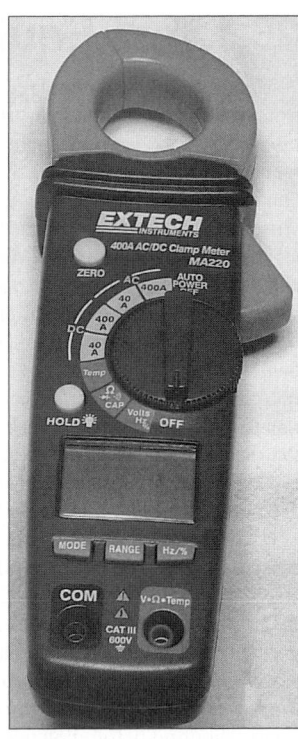

Fig 26.2 — A clamp-on meter probe is used with a digital multimeter for measuring ac current (left). Meters are also available integrated with the clamp-on probe (right).

with other hams or other radio services. A dummy load also provides a known, matched load (usually 50 Ω) for use during adjustments and test measurements. See the **Transmitters and Transceivers** chapter.

Dip meter — As described in the **Test Equipment and Measurements** chapter, dip meters are used to adjust and troubleshoot resonant circuits. Many can perform as an absorption frequency meter, as well. Dip meters can be used as low-power signal sources but are not very stable.

New dip meters are fairly rare. When purchasing a dip meter, look for one that is mechanically and electrically stable. All of the

coils should be present and in good condition. A headphone connection is helpful. Battery operated models are easier to use for antenna measurements. Dip meters are not nearly as common as they once were.

Oscilloscope — The oscilloscope, or scope, is the second most often used piece of test equipment, although a lot of repairs can be accomplished without one. The trace of a scope can give us a lot of information about a signal at a glance. For example, when signals from the input and output of a stage are displayed on a dual-trace scope, stage linearity and phase shift can be checked (see **Fig 26.4**).

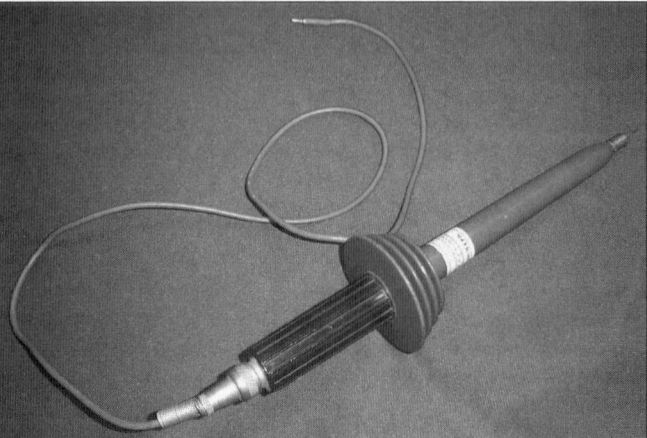

Fig 26.3 — A probe used for measuring high-voltage with a standard multimeter.

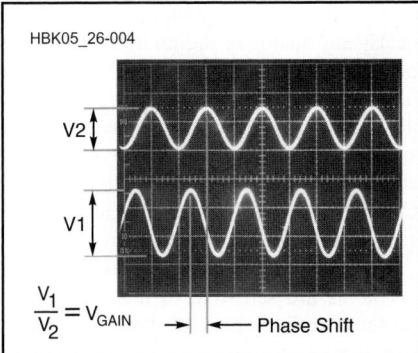

Fig 26.4 — A dual-trace oscilloscope display of amplifier input and output waveforms.

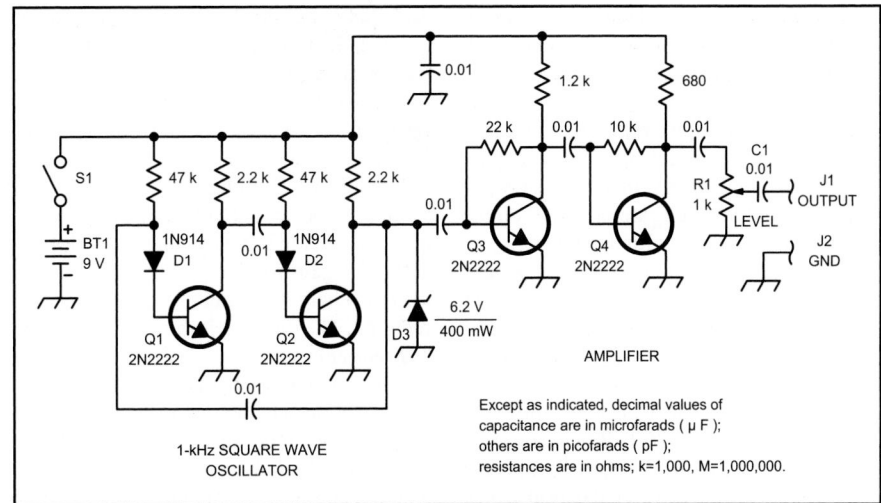

Fig 26.5 — Schematic of the AF/RF signal injector. All resistors are ¼ W, 5% carbon units, and all capacitors are disc ceramic. A full-size etching pattern and parts-placement diagram can be found in the Supplemental Files section of the *Handbook* CD-ROM.

BT1 — 9 V battery.
D1, D2 — Silicon switching diode, 1N914 or equiv.
D3 — 6.2 V, 400 mW Zener diode.
J1, J2 — Banana jack.

Q1-Q4 — General-purpose silicon NPN transistors, 2N2222 or similar.
R1 — 1 kΩ panel-mount control.
S1 — SPST toggle switch.

An oscilloscope will show gross distortions of audio and RF waveforms, but it cannot be used to verify that a transmitter meets FCC regulations for harmonics and spurious emissions. Harmonics that are down only 20 dB from the fundamental would be illegal in most cases, but they would not change the oscilloscope waveform enough to be seen.

When buying a scope, get the highest bandwidth you can afford. Old Hewlett-Packard or Tektronix instruments are usually quite good for amateur use.

Signal generator — Although signal generators have many uses, in troubleshooting they are most often used for signal injection (more about this later) and alignment of vintage equipment.

When buying a generator, look for one that can generate a sine wave signal. A good signal generator is double or triple shielded against leakage. Fixed-frequency audio should be available for modulation of the RF signal and for injection into audio stages. The most versatile generators can generate amplitude and frequency modulated signals. Used Hewlett-Packard and Tektronix units are typically available for reasonable prices but may not be repairable if they fail due to unavailable parts.

Good generators have stable frequency controls with no backlash. They also have multiposition switches to control signal level. A switch marked in dBm is a good indication that you have located a high-quality test instrument. The output jack should be a coaxial connector (usually a BNC or N), not the kind used for microphone connections.

In lieu of a fully tunable generator, you can build some simple equipment that generates a signal. For example, Elecraft makes the XG3 kit — a programmable signal source (**www.elecraft.com**) that generates 160 through 2 meter signals with four calibrated output levels. It's very useful for receiver calibration, sensitivity tests and signal tracing.

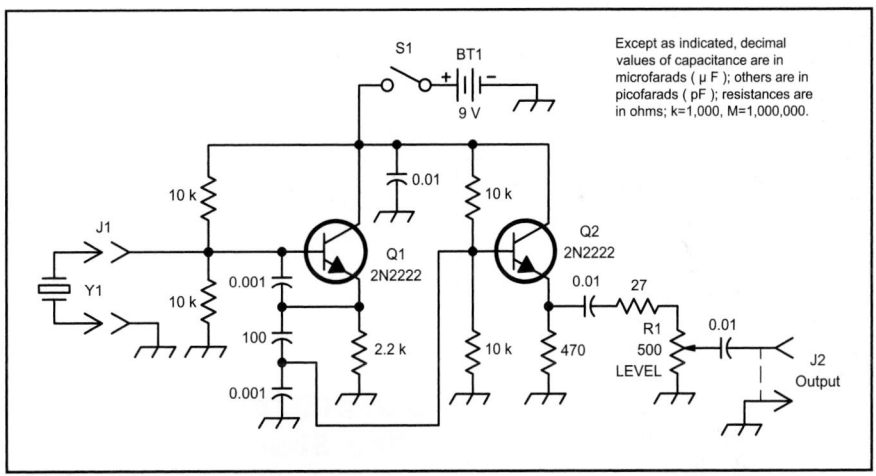

Fig 26.6 — Schematic of the crystal-controlled signal source. All resistors are ¼ W, 5% carbon units, and all capacitors are disc ceramic. A full-size etching pattern and parts-placement diagram can be found in the Supplemental Files section of the *Handbook* CD-ROM.

BT1 — 9 V transistor radio battery.
J1 — Crystal socket to match the crystal type used.
J2 — RCA phono jack or equivalent.
Q1, Q2 — General-purpose silicon NPN transistors, 2N2222 or similar.

R1 — 500 Ω panel-mount control.
S1 — SPST toggle switch.
Y1 — 1 to 15-MHz crystal.

Even simpler, you can homebrew the AF/RF signal-injector schematic as shown in **Fig 26.5**. If frequency accuracy is needed, the crystal-controlled signal source of **Fig 26.6** can be used. The AF/RF circuit provides usable harmonics up to 30 MHz, while the crystal controlled oscillator will function with crystals from 1 to 15 MHz. These two projects are not meant to replace standard signal generators for alignment and precision testing, but they are adequate for generating signals that can be used for general troubleshooting. (See the section on Signal Tracing and Signal Injection.)

Signal tracer — Signals can be traced with a voltmeter and an RF probe, a dip meter with headphones or an oscilloscope, but signal tracers combine these functions especially for signal tracing through a receiver or other RF signal processing circuit. Articles describing the use of signal tracers, including a project you can build yourself, are provided on the CD-ROM that comes with this book.

A general-coverage receiver can be also used to trace RF or IF signals, if the receiver covers the necessary frequency range. Most receivers, however, have a low-impedance input that severely loads the test circuit. To minimize loading, use a capacitive probe or loop pickup as in Fig 26.1. When the probe is held near the circuit, signals will be picked up and carried to the receiver. It may also pick up stray RF, so make sure you are listening to the correct signal by switching the circuit under test on and off while listening.

Transistor tester — Most transistor failures appear as either an open or shorted junction. Opens and shorts can be found easily with an ohmmeter or the diode junction checker of a standard DMM; a special tester is not required.

Transistor testers measure device current while the device is conducting or while an ac signal is applied at the control terminal. Transistor gain characteristics vary widely even between units with the same device number. Testers can be used to measure the gain of a transistor. DMM testers measure only transistor dc alpha and beta. Testers that apply an ac signal show the ac alpha or beta. Better testers also test for leakage.

In addition to telling you whether a transistor is good or bad, a transistor tester can help you decide if a particular transistor has sufficient gain for use as a replacement. It may also help when matched transistors are required. The final test is the repaired circuit.

Frequency counter — Most inexpensive frequency counters display frequency with 1 Hz resolution or better up to around low VHF frequencies. Some may include a prescaler that divides higher frequencies by 10 to extend the counter's range. Good quality used counters are widely available.

Power supplies — A well-equipped test bench should include a means of varying the ac-line voltage, a variable-voltage regulated dc supply and an isolation transformer.

AC-line voltage varies slightly with load. An autotransformer with a movable tap (also known by the trade name Variac) lets you boost or reduce the line voltage slightly. This is helpful to test circuit functions with supply-voltage variations.

An isolation transformer is required to work safely on vintage equipment that often ties one side of the ac line to the chassis. An isolation transformer is also required when working on any equipment or circuits that operate directly connected to the line. Note that your test equipment will also have to be powered through the isolation transformer in such cases!

A good multi-voltage supply will help with nearly any analog or digital troubleshooting project. Many electronics distributors stock bench power supplies. A variable-voltage dc supply may be used to power various small items under repair or provide a variable bias supply for testing active devices. Construction details for a laboratory power supply appear in the **Power Sources** chapter.

Heat and cold sources — Many circuit problems are sensitive to temperature. A piece of equipment may work well when first turned on (cold) but fail as it warms up. In this case, a cold source will help you find the intermittent connection. When you cool the bad component, the circuit will suddenly start working again (or stop working). Cooling sprays are available from most parts suppliers.

A heat source helps locate components that fail only when hot. A small incandescent lamp can be mounted in a large piece of sleeve insulation to produce localized heat for test purposes. The tip of a soldering iron set to low heat can also be used.

A heat source is usually used in conjunction with a cold source. If you have a circuit that stops working when it warms up, heat the circuit until it fails, then cool the components one by one. When the circuit starts working again, the last component sprayed was the bad one.

Stethoscope — A stethoscope (with the pickup removed — see **Fig 26.7**) or a long piece of sleeve insulation can be used to listen for arcing or sizzling in a circuit. Remove any metal parts at the end of the pickup tube before use for troubleshooting live equipment.

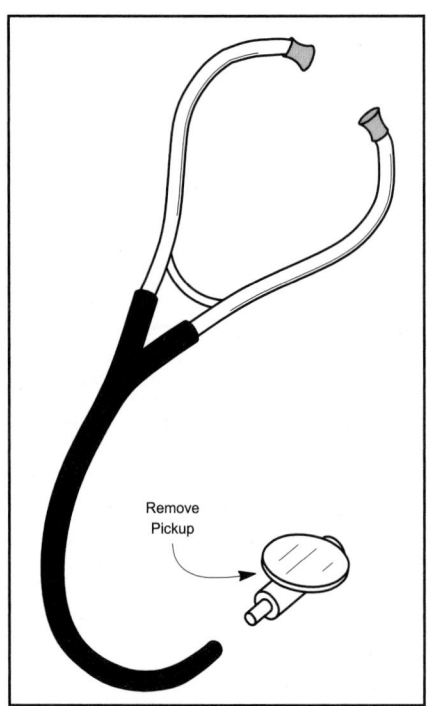

Remove
Pickup

Fig 26.7 — A stethoscope, with the pickup and all metal hardware removed from the listening tube, is used to listen for arcing in crowded circuits.

The Shack Notebook

A shack notebook is an excellent way to keep track of test results, wiring, assembly notes and so forth. If you haven't already started one, now is a good time. All it takes is an inexpensive composition book or spiral-bound notebook. The books filled with graph paper are especially good for drawing and making graphs.

The goal is to have one place where information is collected about how equipment was built, performs, or operates. The notebook is invaluable when trying to determine if performance has changed over time or what color code was used for a control cable, for example.

Before beginning a test session or when adding a new piece of equipment to your shack, get out the shack notebook first and have it available as you work. For new equipment, record serial numbers, when installed, whether it was modified or specially configured to work in your station, etc. For a new antenna or feed line, it's a good idea to make a few SWR measurements so you can refer to them later if something seems wrong in the antenna system. Be sure to include the date of any entries, as well.

You can also make good use of that digital camera, even the one in your mobile phone. Take pictures of equipment, inside and out, to document how it is assembled or configured. This can be very helpful when you have to maintain the equipment later.

26.2 Components

Once you locate a defective part, it is time to select a replacement. This is not always an easy task. Each electronic component has a function. This section acquaints you with the functions, failure modes and test procedures of resistors, capacitors, inductors and other components. Test the components implicated by symptoms and stage-level testing. In most cases, a particular faulty component will be located by these tests. If a faulty component is not indicated, check the circuit adjustments. As a last resort, use a "shotgun" approach — replace all parts in the problem area with components that are known to be good.

26.2.1 Check the Circuit

Before you install a replacement component of any type, you should be sure that another circuit defect didn't cause the failure. Check the circuit voltages carefully before installing any new component, especially on each trace or connection to the bad component. The old part may have died as a result of a lethal voltage. Measure twice — repair once! (With apologies to the old carpenter...) Of course, circuit performance is the final test of any substitution.

26.2.2 Fuses

Most of the time, when a fuse fails, it is for a reason — usually a short circuit in the load. A fuse that has failed because of a short circuit usually shows the evidence of high current: a blackened interior with little blobs of fuse element everywhere. Fuses can also fail by fracturing the element at either end. This kind of failure is not visible by looking at the fuse. Check even fuses thought to be good with an ohmmeter. You may save hours of troubleshooting.

For safety reasons, always use *exact* replacement fuses. Check the current and voltage ratings. The fuse timing (fast, normal or slow blow) must be the same as the original. Never attempt to force a fuse that is not the right size into a fuse holder. The substitution of a bar, wire or penny for a fuse invites a smoke party.

26.2.3 Wires and Cables

Wires seldom fail unless abused. Short circuits can be caused by physical damage to insulation or by conductive contamination. Damaged insulation is usually apparent during a close visual inspection of the conductor or connector. Look carefully where conductors come close to corners or sharp objects. Repair worn insulation by replacing the wire or securing an insulating sleeve (spaghetti) or heat-shrink tubing over the worn area.

When wires fail, the failure is usually caused by stress and flexing. Nearly everyone has broken a wire by bending it back and forth, and broken wires are usually easy to detect. Look for sharp bends or bulges in the insulation.

When replacing conductors, use the same material and size, if possible. Substitute only wire of greater cross-sectional area (smaller gauge number) or material of greater conductivity. Insulated wire should be rated at the same, or higher, temperature and voltage as the wire it replaces.

Cables used for audio, control signals, and feed lines sometimes fail from excessive flexing, being crimped or bent too abruptly, or getting pulled out of connectors. As with replacing wires, use the same cable type or one with higher ratings.

26.2.4 Connectors

Connection faults are one of the most common failures in electronic equipment. This can range from something as simple as the ac-line cord coming out of the wall, to a connector having been inserted into the wrong socket, to a defective IC socket. Connectors that are plugged and unplugged frequently can wear out, becoming intermittent or noisy. Inspect male connectors for bent pins, particular miniature connectors with very small pins. Check connectors carefully when troubleshooting.

Connector failure can be hard to detect. Most connectors maintain contact as a result of spring tension that forces two conductors together. As the parts age, they become brittle and lose tension. Any connection may deteriorate because of nonconductive corrosion at the contacts. Solder helps prevent this problem but even soldered joints suffer from corrosion when exposed to weather.

Signs of excess heat are sometimes seen near poor connections in circuits that carry moderate current. The increase in dissipated power at the poor connection heats the contacts, and this leads to more resistance and soon the connection fails. Check for short and open circuits with an ohmmeter or continuity tester. Clean those connections that fail as a result of contamination.

Occasionally, corroded connectors may be repaired by cleaning, but replacement of the conductor/connector is usually required, especially for battery holders supplying moderate currents. Solder all connections that may be subject to harsh environments and protect them with acrylic enamel, RTV compound or a similar coating. An anti-corrosion compound or grease is a good idea for connections located outside. See the entry on Weatherproofing RF Connectors in the Antenna and Tower Safety section of the **Safety** chapter.

Choose replacement connectors with consideration of voltage and current ratings. Use connectors with symmetrical pin arrangements only where correct insertion will not result in a safety hazard or circuit damage.

26.2.5 Resistors

Resistors usually fail by becoming an open circuit. More rarely they change value. Both failures are usually caused by excess heat. Such heat may come from external sources or from power dissipated within the resistor. Sufficient heat burns the resistor until it becomes an open circuit.

Resistors can also fracture and become an open circuit as a result of physical shock. Contamination on or around a high-value resistor ($100\,k\Omega$ or more) can cause a change in value by providing a leakage path for current around a resistor. This contamination can occur on the resistor body, mounts or printed-circuit board. Resistors that have changed value should be replaced. Leakage is cured by cleaning the resistor body and surrounding area.

In addition to the problems of fixed-value resistors, potentiometers and rheostats can develop noise problems, especially in dc circuits. Dirt often causes intermittent contact between the wiper and resistive element. To cure the problem, spray electronic contact cleaner into the control, through holes in the case, and rotate the shaft a few times.

The resistive element in wire-wound potentiometers eventually wears and breaks from the sliding action of the wiper. In this case, the control needs to be replaced.

Replacement resistors should be of the same value, tolerance, type and power rating as the original. The value should stay within tolerance. Replacement resistors may be of a different type than the original, if the characteristics of the replacement are consistent with circuit requirements. (See the **Electrical Fundamentals** chapter for more information on resistor types.)

Substitute resistors can usually have a greater power rating than the original, except in high-power emitter circuits where the resistor also acts as a fuse or in cases where the larger size presents a problem.

Variable resistors should be replaced with the same kind (carbon or wire wound) and taper (linear, log, reverse log and so on) as the original. Keep the same, or better, tolerance and pay attention to the power rating.

In all cases, mount high-temperature resistors away from heat-sensitive components. Keep carbon composition and film resistors away from heat sources. This will extend their life and ensure minimum resistance variations.

26.2.6 Capacitors

Capacitors usually fail by shorting, opening or becoming electrically (or physically) leaky. They rarely change value. Capacitor failure is usually caused by excess current, voltage, temperature or aging of the dielectric or materials making up the capacitor. Leakage can be external to the capacitor (contamination on the capacitor body or circuit) or internal to the capacitor.

TESTS

If you do not have a multimeter with a capacitor test function or a component tester, the easiest way to test capacitors is out of circuit with an ohmmeter. In this test, the resistance of the meter forms a timing circuit with the capacitor to be checked. Capacitors from 0.01 µF to a few hundred µF can be tested with common ohmmeters. Set the meter to its highest range and connect the test leads across the discharged capacitor. When the leads are connected, current begins to flow. Current is high when the capacitor is discharged, but drops as the capacitor voltage builds up. This shows on the meter as a low resistance that builds, over time, toward infinity.

The speed of the resistance build-up corresponds to capacitance. Small capacitance values approach infinite resistance almost instantly. A 0.01 µF capacitor checked with a meter having an 11 MΩ input impedance would increase from zero to a two-thirds scale reading in 0.11 second, while a 1 µF unit would require 11 seconds to reach the same reading. If the tested capacitor does not reach infinity within five times the period taken to reach the two-thirds point, it has excess leakage. If the meter reads infinite resistance immediately, the capacitor is open. (Aluminum electrolytics normally exhibit high-leakage readings.)

Capacitance can also be measured for approximate value with a dip meter by constructing a parallel-resonant circuit using an inductor of known value. The formula for resonance is discussed in the **Electrical Fundamentals** chapter of this book.

It is good practice to keep a collection of known components that have been measured on accurate L or C meters. Alternatively, a standard value can be obtained by ordering 1 or 2% components from an electronics supplier. A 10% tolerance component can be used as a standard; however, the results will be known only to within 10%. The accuracy of tests made with any of these alternatives depends on the accuracy of the standard value component. Further information on this technique appears in Bartlett's article, "Calculating Component Values," in Nov 1978 *QST*.

Older capacitors can also be checked and the dielectric reformed, if necessary, with a

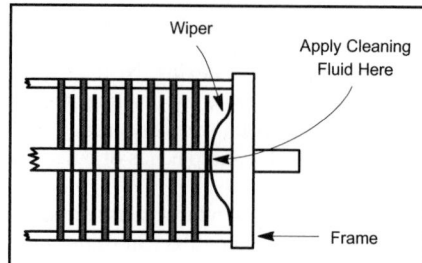

Fig 26.8 — Partial view of an air-dielectric variable capacitor. If the capacitor is noisy or erratic in operation, apply electronic cleaning fluid where the wiper contacts the rotor plates.

capacitor checker of similar vintage. See this chapter's section on Restoration and Repair of Vintage Radios for more information about this technique.

CLEANING

The only variety of common capacitor that can be repaired is the air-dielectric variable capacitor. Electrical connection to the moving plates is made through a spring-wiper arrangement (see **Fig 26.8**). Dirt normally builds on the contact area, and they need occasional cleaning. Before cleaning the wiper/contact, use gentle air pressure and a soft brush to remove all dust and dirt from the capacitor plates. Apply some electronic contact cleaning fluid. Do not lubricate the contact point. Rotate the shaft quickly several times to work in the fluid and establish contact. Use the cleaning fluid sparingly, and keep it off the plates except at the contact point.

Batteries and Tools

When working on equipment powered by a battery with a capacity of more than a few ampere-hours (Ah), take special care to avoid short circuits and always have a fuse or circuit-breaker in-line to the battery. If possible, fuse both the positive and negative connections. This is particularly important for vehicle batteries and mobile equipment. A short circuit can do thousands of dollars of damage to a vehicle's electrical power system in a matter of seconds! When working on mobile equipment, disconnect the battery's positive terminal or at least disconnect the circuit that powers the equipment by removing the fuse.

A tool that accidentally short-circuits the battery terminals can cause an instantaneous current flow of thousands of amps, often destroying both the battery and the tool and creating a significant fire and burn hazard.

REPLACEMENTS

Replacement capacitors should match the original in value, tolerance, dielectric, working voltage and temperature coefficient. Use only ac-rated capacitors for line service (capacitors connected directly to the ac line) to prevent fire hazards. If exact replacements are not available, substitutes may vary from the original part in the following respects: Bypass capacitors may vary from one to three times the capacitance of the original. Coupling capacitors may vary from one-half to twice the value of the original. Capacitance values in tuned circuits (especially filters) must be exact. (Even then, any replacement will probably require circuit realignment.)

If the same kind of capacitor is not available, use one with better dielectric characteristics. Do not substitute polarized capacitors for non-polarized parts. Capacitors with a higher working voltage may be used, although the capacitance of an electrolytic capacitor used significantly below its working voltage will usually increase with time.

The characteristics of each type of capacitor are discussed in the **Electrical Fundamentals** and **RF Techniques** chapters. Consider these characteristics if you're not using an exact replacement capacitor.

26.2.7 Inductors and Transformers

The most common inductor or transformer failure is a broken conductor. More rarely, a short circuit can occur across one or more turns of a coil. In an inductor, this changes the value. In a transformer, the turns ratio and resultant output voltage changes. In high-power circuits, excessive inductor current can generate enough heat to melt plastics used as coil forms.

Inductors may be checked for open circuit failure with an ohmmeter. In a good inductor, dc resistance rarely exceeds a few ohms. Shorted turns and other changes in inductance show only during alignment or inductance measurement.

The procedure for measurement of inductance with a dip meter is the same as that given for capacitance measurement, except that a capacitor of known value is used in the resonant circuit.

Replacement inductors must have the same inductance as the original, but that is only the first requirement. They must also carry the same current, withstand the same voltage and present nearly the same Q as the original part. Given the original as a pattern, the amateur can duplicate these qualities for many inductors. Note that inductors with ferrite or iron-powder cores are frequency sensitive, so the replacement must have the same core material.

If the coil is of simple construction, with the form and core undamaged, carefully count and write down the number of turns and their placement on the form. Also note how the coil leads are arranged and connected to the circuit. Then determine the wire size and insulation used. Wire diameter, insulation and turn spacing are critical to the current and voltage ratings of an inductor. (There is little hope of matching coil characteristics unless the wire is duplicated exactly in the new part.) Next, remove the old winding — be careful not to damage the form — and apply a new winding in its place. Be sure to dress all coil leads and connections in exactly the same manner as the original. Apply Q dope (a solution of polystyrene plastic) or a thin coating of plastic-based glue to hold the finished winding in place.

Follow the same procedure in cases where the form or core is damaged, except that a suitable replacement form or core (same dimensions and permeability) must be found.

Ready-made inductors may be used as replacements if the characteristics of the original and the replacement are known and compatible. Unfortunately, many inductors are poorly marked. If so, some comparisons, measurements and circuit analysis are usually necessary.

When selecting a replacement inductor, you can usually eliminate parts that bear no physical resemblance to the original part. This may seem odd, but the Q of an inductor depends on its physical dimensions and the permeability of the core material. Inductors of the same value, but of vastly different size or shape, will likely have a great difference in Q. The Q of the new inductor can be checked by installing it in the circuit, aligning the stage and performing the manufacturer's passband tests. Although this practice is all right in a pinch, it does not yield an accurate Q measurement. Methods to measure Q appear in the **Test Equipment and Measurements** chapter.

Once the replacement inductor is found, install it in the circuit. Duplicate the placement, orientation and wiring of the original. Ground-lead length and arrangement should not be changed. Isolation and magnetic shielding can be improved by replacing solenoid inductors with toroids. If you do, however, it is likely that many circuit adjustments will be needed to compensate for reduced coupling and mutual inductance. Alignment is usually required whenever a tuned-circuit component is replaced.

A transformer consists of two inductors that are magnetically coupled. Transformers are used to change voltage and current levels (this changes impedance also). Failure usually occurs as an open circuit or short circuit of one or more windings. Insulation failures can also occur that result in short circuits between windings or between windings and the core or case. It is also common for the insulated wire leads to develop cracks or abrasion where they come out of the case. This can be repaired easily by replacing the lead, or if the conducting wire strands are not burned or broken, by sliding an insulation sleeve over the wire to protect the insulation from further wear.

Amateur testing of power transformers is mostly limited to ohmmeter tests for open circuits and voltmeter checks of secondary voltage. Make sure that the power-line voltage is correct, then check the secondary voltage against that specified. There should be less than 10% difference between open-circuit and full-load secondary voltage. A test setup and procedure for evaluating power transformers is also provided in the **Power Sources** chapter.

Replacement transformers must match the original in voltage, volt-ampere (VA), duty cycle and operating-frequency ratings. They must also be compatible in size. (All transformer windings should be insulated for the full power supply voltage.)

When disconnecting a transformer for testing or repair, be sure to carefully record the color and connection for each of the transformer leads. In power transformers it is common for leads to be mostly one color but with a contrasting stripe or other pattern that can be overlooked. If in doubt, use tape or paper labels to note the connection for each lead. Recording transformer color codes and connections is a good use of the shack notebook.

26.2.8 Relays

Although relays have been replaced by semiconductor switching in low-power circuits, they are still used extensively in high-power Amateur Radio equipment for applications such as amplifier TR switching and in antenna systems or ac power control. Relay action may become sluggish. AC relays can buzz (with adjustment becoming impossible). A binding armature or weak springs can cause intermittent switching. Excessive use or hot switching ruins contacts and shortens relay life.

You can test relays with a voltmeter by measuring voltage across contacts (power on, in-circuit) or with an ohmmeter (out of circuit). Look for erratic readings across the contacts, open or short circuits at contacts or an open circuit at the coil. A visual inspection with a magnifying glass should show no oxidation or corrosion. Limited pitting is usually OK.

Most failures of simple relays can be repaired by a thorough cleaning. Clean the contacts and mechanical parts with a residue-free cleaner. Keep it away from the coil and plastic parts that may be damaged. Dry the contacts with lint-free paper, such as a business card; then burnish them with a smooth steel blade. Do not use a file to clean contacts because it will damage the contact surface.

Replacement relays should match or exceed the original specifications for voltage, current, switching time and stray impedance (impedance is significant in RF circuits only). Many relays used in transceivers are specially made for the manufacturer. Substitutes may not be available from any other source.

Before replacing a multi-contact relay, make a drawing of the relay, its position, the leads and their routings through the surrounding parts. This drawing allows you to complete the installation properly, even if you are distracted in the middle of the operation. (This is a good use of the shack notebook!)

26.2.9 Semiconductors

Testing diodes and transistors with the ohmmeter function of an analog VOM used to be the normal method. Today's inexpensive multimeters nearly always provide a forward and reverse junction voltage drop test function. This almost eliminates the need for resistance-based tests for functional troubleshooting with the attendant variability and dependence on meter circuits. However, it is occasionally useful to perform threshold and voltage-current testing to match components or troubleshoot a specialized circuit. In support of those tests, a short article "Diode and Transistor Test Circuits" containing test circuits and procedures for measuring leakage, gain, Zener point voltage and so forth is included on the CD-ROM accompanying this book. This section will focus on simple go/no-go testing.

DIODES

The primary function of a diode is to pass current in one direction only. They can be easily tested with an ohmmeter, and most multimeters have a diode junction test function built-in as well.

Signal or switching diodes — The most common diode in electronics equipment, signal diodes are used to convert ac to dc, to detect RF signals or to take the place of relays to switch ac or dc signals within a circuit. Signal diodes usually fail open, although shorted diodes are not rare.

Power rectifiers — Most equipment contains a power supply, so power rectifier diodes are the second-most common diodes in electronic circuitry. They usually fail shorted, blowing the power-supply fuse.

Other diodes — Zener diodes are made with a predictable reverse-breakdown voltage and are used as voltage regulators. Varactor diodes are specially made for use as voltage controlled variable capacitors. (Any semiconductor diode may be used as a voltage-variable capacitance, but the value will not be as predictable as that of a varactor.) A Diac is a special-purpose diode that passes only pulses of current in each direction.

Diode testing — There are several basic tests for most diodes. First, is it a diode? Does it conduct in one direction and block current flow in the other? A simple resistance measurement is suitable for this test in most cases.

Diodes should be tested out of circuit. Disconnect one lead of the diode from the circuit, then perform the test. We can also test diodes by measuring the voltage drop across the diode junction while the diode is conducting.

A functioning diode will show high resistance in one direction and low resistance in the other. A DMM with a diode-test function is the best instrument to use. If using an analog meter, make sure more than 0.7 V and less than 1.5 V is used to measure resistance. Use the highest resistance scale of the meter that gives a reading of less than full-scale. Check a known-good diode to determine the meter polarity if there is any question. Compare the forward and reverse resistance readings for a known-good diode to those of the diode being tested to determine whether the diode is good.

Diode junction forward voltage drops are measured by a multimeter's diode test function. Silicon junctions usually show about 0.6 V at typical test current levels, while germanium is typically 0.2 V. Junction voltage drop increases with current flow.

Multimeters measure the junction resistances at low voltage and are not useful for testing Zener diodes. A good Zener diode will not conduct in the reverse direction at voltages below its rating. See the CD-ROM article mentioned at the beginning of this section for procedures and a circuit to determine Zener diode performance.

Replacement diodes — When a diode fails, check associated components as well. Replacement rectifier diodes should have the same current and peak inverse voltage (PIV) as the original. Series diode combinations are often used in high-voltage rectifiers. (The resistor and capacitor networks used to distribute the voltage equally among the diodes are no longer required for new rectifiers but should be retained for older parts. See the **Power Sources** chapter for more information.)

Switching diodes may be replaced with diodes that have equal or greater current ratings and a PIV greater than twice the peak-to-peak voltage encountered in the circuit. Switching time requirements are not critical except in RF, logic and some keying circuits. Logic circuits may require exact replacements to assure compatible switching speeds and load characteristics. RF switching diodes used near resonant circuits must have exact replacements as the diode resistance

and capacitance will affect the tuned circuit.

Voltage and capacitance characteristics must be considered when replacing varactor diodes. Once again, exact replacements are best. Zener diodes should be replaced with parts having the same Zener voltage and equal or higher power rating, and equal or lower tolerance. Check the associated current-limiting resistor when replacing a Zener diode.

BIPOLAR TRANSISTORS

Transistor failures occur as an open junction, a shorted junction, excess leakage or a change in amplification performance. Most transistor failure is catastrophic. A transistor that has no leakage and amplifies at dc or audio frequencies will usually perform well over its design range. For this reason, transistor tests need not be performed at the planned operating frequency. Tests are made at dc or a low frequency (usually 1000 Hz). The circuit under repair is the best test of a potential replacement part. Swapping in a replacement transistor in a failed circuit will often result in a cure.

A simple and reliable test of bipolar transistors can be performed with the transistor in a circuit and the power on. It requires a test lead, a 10 kΩ resistor and a voltmeter. Connect the voltmeter across the emitter/collector leads and read the voltage. Then use the test lead to connect the base and emitter (**Fig 26.9A**). Under these conditions, conduction of a good transistor will be cut off and the meter should show nearly the entire supply voltage across the emitter/collector leads. Next, remove the clip lead and connect the 10 kΩ resistor from the base to the collector. This should bias the transistor into conduction and the emitter/collector voltage should drop (Fig 26.9B). (This test indicates transistor response to changes in bias voltage.)

Transistors can be tested (out of circuit) with an ohmmeter in the same manner as diodes or a multimeter with a transistor test function can be used. Before using the ohmmeter-transistor circuit, look up the device

characteristics before testing and consider possible consequences of testing the transistor in this way. Limit junction current to 1 to 5 mA for small-signal transistors. Transistor destruction or inaccurate measurements may result from careless testing.

The reverse-to-forward resistance ratio for good transistors may vary from 30:1 to better than 1000:1. Germanium transistors — still occasionally encountered — sometimes show high leakage when tested with an ohmmeter. Bipolar transistor leakage may be specified from the collector to the base, emitter to base or emitter to collector (with the junction reverse biased in all cases). The specification may be identified as I_{cbo}, I_{bo}, collector cutoff current or collector leakage for the base-collector junction, I_{ebo}, and so on for other junctions. Leakage current increases with junction temperature. (See the **Analog Basics** chapter for definitions of these and other transistor parameters.)

While these simple test circuits will identify most transistor problems, RF devices should be tested at RF. Most component manufacturers include a test-circuit schematic on the data sheet. The test circuit is usually an RF amplifier that operates near

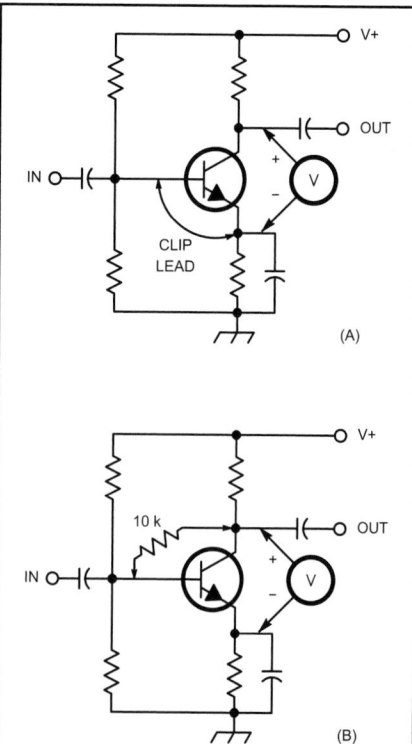

Fig 26.9 — An in-circuit semiconductor test with a clip lead, resistor and voltmeter. The meter should read V+ at (A). During test (B) the meter should show a decrease in voltage, ranging from a slight variation down to a few millivolts. It will typically cut the voltage to about half of its initial value.

the high end of the device frequency range. If testing at RF is not possible, substitution of a known-good device is required.

Semiconductor failure is sometimes the result of environmental conditions. Open junctions, excess leakage (except with germanium transistors) and changes in amplification performance result from overload or excessive current.

Shorted junctions (low resistance in both directions) are usually caused by voltage spikes. Electrostatic discharge (ESD) or transients from lightning can destroy a semiconductor in microseconds.

Transistors rarely fail without an external cause. Check the surrounding parts for the cause of the transistor's demise, and correct the problem before installing a replacement.

JFETs

Junction FETs can be tested with a multimeter's diode junction test function in much the same way as bipolar transistors (see text and **Fig 26.10**). Reverse leakage should be several megohms or more. Forward resistance should be 500 to 1000 Ω if measured with an analog meter.

MOSFETs

Small-signal MOS (metal-oxide semiconductor) layers are extremely fragile. Normal body static is enough to damage them. Even gate protected (a diode is placed across the MOS layer to clamp voltage) MOSFETs may be destroyed by a few volts of static electricity. MOSFETs used for power circuits and RF amplifiers are much more resistant to damage. The manufacturer's sheet will specify any special static-protection measures that are required. (See the **Construction Techniques** chapter for more information about managing static at the workbench.)

When testing small MOSFETs make sure the power is off, capacitors discharged and the leads are shorted together before installing or removing it from a circuit. Use a voltmeter to be sure the chassis is near ground potential, then touch the chassis before and during MOSFET installation and removal. This assures that there is no difference of potential between your body, the chassis and the MOSFET leads. Ground the soldering-iron tip with a clip lead when soldering MOS devices. The FET source should be the first lead connected to and the last disconnected from a circuit. The insulating layers in MOSFETs prevent testing with an ohmmeter. Substitution is the only practical means for amateur testing of MOSFETs.

FET CONSIDERATIONS

Replacement FETs should be of the same type as the original part: JFET or MOSFET, P-channel or N-channel, enhancement or

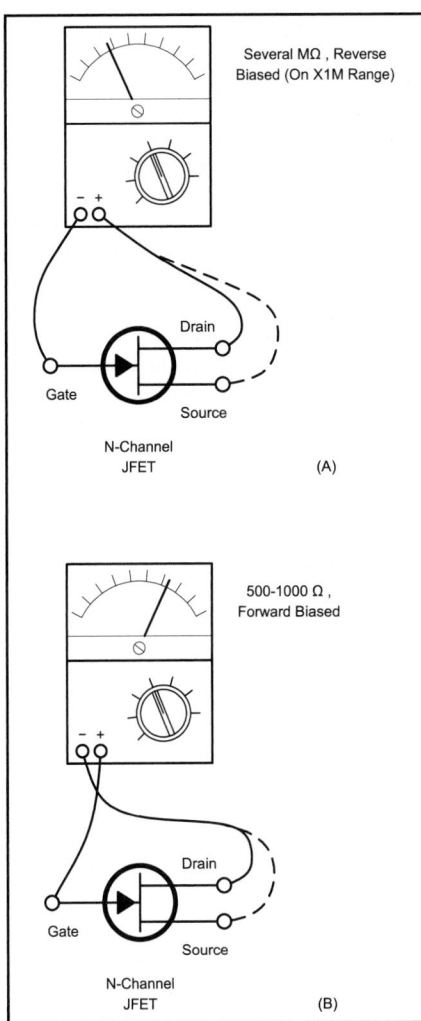

Fig 26.10 — Ohmmeter tests of a JFET. The junction is reverse biased at A and forward biased at B. (Analog meters are shown for convenience of illustration.)

depletion. Consider the breakdown voltage required by the circuit. The breakdown voltage should be at least two to four times the power-supply and signal voltages in amplifiers. Allow for transients of 10 times the line voltage in power supplies. Breakdown voltages are usually specified as $V_{(BR)GSS}$ or $V_{(BR)GDO}$.

The gate-voltage specification gives the gate voltage required to cut off or initiate channel current (depending on the mode of operation). Gate voltages are usually listed as $V_{GS(OFF)}$, V_p (pinch off), V_{TH} (threshold) or $I_{D(ON)}$ or I_{TH}.

Dual-gate MOSFET characteristics are more complicated because of the interaction of the two gates. Cutoff voltage, breakdown voltage and gate leakage are the important traits of each gate.

INTEGRATED CIRCUITS

The basics of integrated circuits are covered

in earlier chapters of this book. Amateurs seldom have the sophisticated equipment required to test ICs. Even a multi-channel oscilloscope can view only their simplest functions. We must be content to check every other possible cause, and only then assume that the problem lies with an IC. Experienced troubleshooters will tell you that — most of the time anyway — if a defective circuit uses an IC, it is the IC that is bad.

Linear ICs — There are two major classes of ICs: linear and digital. Linear ICs are best replaced with identical units. Original equipment manufacturers are the best source of a replacement; they are the only source with a reason to stockpile obsolete or custom-made items. If substitution of an IC is unavoidable, first try the cross-reference website of NTE mentioned in the sidebar. You can also look in manufacturers' websites and compare pin-outs and other specifications.

Digital ICs — It is usually not a good idea to substitute digital devices. While it may be okay to substitute an AB74LS00YZ from one manufacturer with a CD74LS00WX from a different manufacturer, you will usually not be able to replace an LS (low-power Schottky) device with an S (Schottky), C (CMOS) or any of a number of other families. The different families all have different speed, current-consumption, input and output characteristics. You would have to analyze the circuit to determine if you could substitute one type for another.

SEMICONDUCTOR SUBSTITUTION

In all cases try to obtain exact replacement semiconductors. Specifications vary slightly from one manufacturer to the next. Cross-reference equivalents such as NTE (**www. nteinc.com**) are useful, but not guaranteed to be an exact replacement. Before using an equivalent, check the specifications against those for the original part. When choosing a replacement, consider:

• Is it a PNP or an NPN?

• What are the operating frequency and input/output capacitance?

• How much power can it dissipate (often less than Vmax × Imax)?

• Will it fit the original socket or pad layout?

• Are there unusual circuit demands (low noise and so on)?

• What is the frequency of operation?

Remember that cross-reference equivalents are not guaranteed to work in every application. In cases where an absolutely exact replacement is required for an obsolete part, Rochester Electronics (**www.rocelec.com**) or 4 Star Electronics (**www.4starelectronics. com**) maintain extensive stocks, although the cost is likely to be rather high.

There may be cases where two dissimilar

devices have the same part number, so it pays to compare the listed replacement specifications with the intended use. If the book says to use a diode in place of an RF transistor, it isn't going to work! Derate power specifications, as recommended by the manufacturer, for high-temperature operation.

26.2.10 Tubes

The most common tube failures in amateur service are caused by cathode depletion and gas contamination. Whenever a tube is operated, the coating on the cathode loses some of its ability to produce electrons. It is time to replace the tube when electron production (cathode current, I_c) falls to 50 to 60% of that exhibited by a new tube.

Gas contamination in a tube can often be identified easily because there may be a greenish or whitish-purple glow between the elements during operation. (A faint deep-purple glow is normal in most tubes.) The gas reduces tube resistance and leads to runaway plate current evidenced by a red glow from the anode, interelectrode arcing or a blown power supply fuse. Less common tube failures include an open filament, broken envelope and inter-electrode shorts.

The best test of a tube is to substitute a new one. Another alternative is a tube tester; these are sometimes available at hamfests or through antique radio or audiophile groups. You can also do some limited tests with an ohmmeter. Tube tests should be made out of circuit so circuit resistance does not confuse the results.

Use an ohmmeter to check for an open filament (remove the tube from the circuit first). A broken envelope is visually obvious, although a cracked envelope may appear as a gassy tube. Interelectrode shorts are evident during voltage checks on the operating stage. Any two elements that show the same voltage are probably shorted. (Remember that some inter-electrode shorts, such as the cathode-suppressor grid, are normal.)

Generally, a tube may be replaced with another that has the same type number. Compare the data sheets of similar tubes to assess their compatibility. Consider the base configuration and pinout, inter-electrode capacitances (a small variation is okay except for tubes in oscillator service), dissipated power ratings of the plate and screen grid and current limitations (both peak and average). For example, the 6146A may usually be replaced with a 6146B (heavy duty), but not vice versa.

In some cases, minor type-number differences signify differences in filament voltages, or even base styles, so check all specifications before making a replacement. (Even tubes of the same model number, prefix and suffix vary slightly, in some respects, from one supplier to the next.)

26.3 Getting Started

INSTINCTIVE OR SYSTEMATIC

A systematic approach to troubleshooting uses a defined process to analyze and isolate the problem. An instinctive approach relies on troubleshooting experience to guide you in selecting which circuits to test and which tests to perform.

When instinct is based on experience, searching by instinct may be the fastest procedure. If your instinct is correct, repair time and effort may be reduced substantially. As experience and confidence grow, the merits of the instinctive approach grow with them. However, inexperienced technicians who choose this approach are at the mercy of chance.

A systematic approach is a disciplined procedure that allows us to tackle problems in unfamiliar equipment with a reasonable hope of success. The systematic approach is usually chosen by beginning troubleshooters.

26.3.1 The Systematic Approach

Armed with a collection of test equipment, you might be tempted to immediately dig in and start looking for the problem. While it is sometimes obvious what piece of equipment or subassembly inside equipment is at fault, the many connections that make up nearly all ham stations today make it far more effective to begin troubleshooting by looking at the problem from the system perspective. By *system*, we mean more than one piece of equipment or subassemblies connected together — nothing fancier than that.

Amateur stations are full of systems: a digital mode system is made up of a radio, power supply, and PC. An antenna system consists of the antenna tuner, feed line, and antenna. Inside a radio there is a system made up of the power circuits, receiver, transmitter, control panel, and transmit-receive switching circuits. Connections between parts of the system need not be cables; wireless data links can also be part of a system.

In general, it's best to approach any problem — even the supposedly obvious problems — from the perspective of the system it affects. The first step is to determine the system with the smallest number of components that exhibits the problem. Then you can start looking for the problem in one of those components or the connection between them. We'll start with an example to illustrate the process.

Problem — After a year of trouble-free operation, when you transmit with more than 50 W of output power using PSK31, other stations now report lots of distortion products around your signal on the waterfall display, even though the ALC is not active at all and no software level settings have been changed. This could be RF interference, a problem with the digital interface between the PC and the radio, settings in the PC, settings in the radio, a bad connection...there are lots of possibilities.

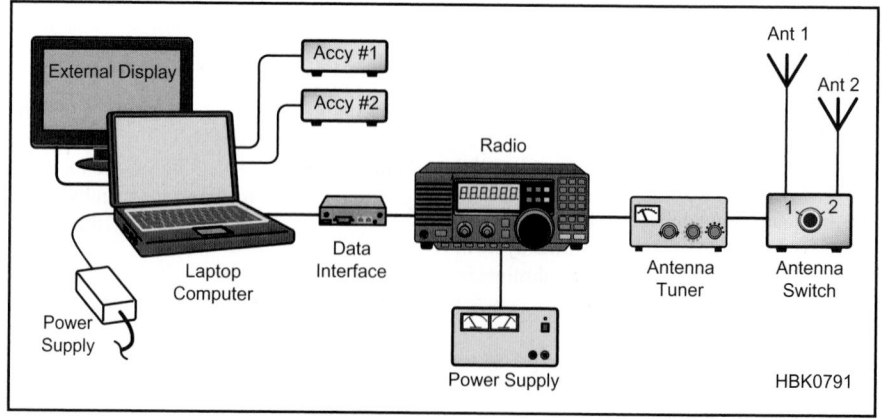

Fig 26.11 — The complete system of radio and power supply, data interface, laptop PC with a couple of accessories and power supply, antenna tuner, antenna switch, antenna 1 and 2, and interconnecting cables.

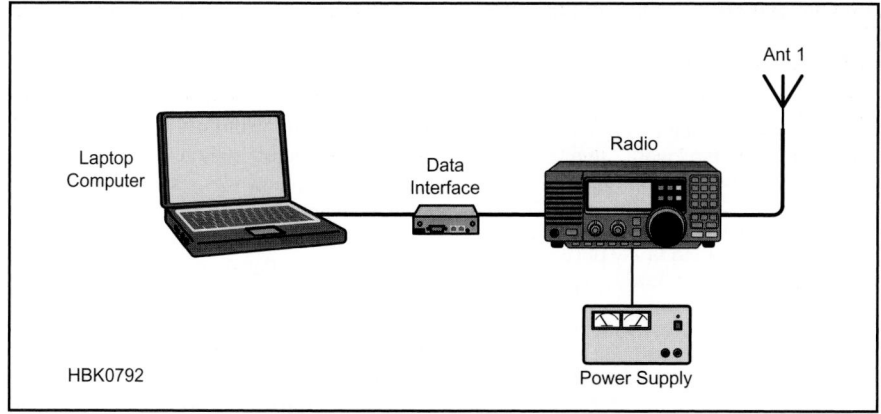

Fig 26.12 — The minimum system of radio and power supply, digital interface, laptop on battery power, antenna 1, and interconnecting cables.

Get a pad of paper and sketch out the system that is exhibiting the problem, such as in **Fig 26.11**. Write down what the symptom(s) is(are) and the conditions under which they occur. Fig 26.11 shows that you have a system that has a lot of parts involved, including the interconnecting cables. Let's simplify that system! For this type of problem, it would be very helpful to have a friend available to listen on the air as you troubleshoot. To find the minimum system, start by removing any accessories that aren't being used from the radio and PC and trying to reproduce the problem, writing down each step and noting whether the problem changed. Not only will you simplify the problem but in the process you will have an excellent opportunity to inspect and check connections and configurations — perhaps even discovering the problem!

Let's say that the distortion persists with the mouse and external display disconnected and the laptop running from its internal battery. Then you were able to remove the antenna tuner and antenna switch, manually attaching just one antenna to the radio at a time, and found that the problem only occurred when using antenna 1. Removing any other part of the system makes it non-functional, so you have found the minimum system with the problem as shown in **Fig 26.12**.

Now it's time to start making small changes in the minimum system to observe the effect, again taking notes as you go. Changing the microphone gain a little bit up and down has hardly any effect. Changing the PC sound card output audio level a little bit also has hardly any effect. Reducing RF output power a little bit has a *big* effect: as power is reduced, so are the distortion products. Below 35 W, the problem was completely gone! This is an important clue, dutifully recorded on the pad of paper. Replacing antenna 1 with a dummy load had a similar effect — the problem disappeared according to your friend, who lives close enough to hear the weak signal leaking

out of your greatly reduced antenna system. Putting antenna 1 back on the radio causes the problem to come back above 35 W, too.

The behavior of the problem points strongly to RF being picked up by a cable in your minimum system and interfering with the low-level digital audio. It's time for a close inspection of each remaining cable. Removing the radio's microphone connector shell and checking the wires inside didn't turn up anything — all of the connections were intact and only touching what they were supposed to touch. Looking at the cable between the digital interface and the sound card, though, you see that the shield of the cable at the PC end is badly frayed and barely making contact at all! In a few minutes, you've trimmed and re-soldered the cable, plugging it back into the PC. Testing even at full power output shows no distortion products on either antenna.

You fixed the problem by finding the minimum system and then inspecting one thing at a time until the root cause was found and eliminated.

What if the root cause didn't appear when you checked the cables? Then you would have to continue to test the system at the interfaces where the system components are connected together. You could measure signal levels with a sensitive voltmeter. You could add ferrite beads or cores on cables that might be picking up RF. You could check to see if the radio's power supply output is stable and clean as the transmitter power increases.

Another strategy would be to substitute known good components, one at a time, and see if the problem changes or disappears: You might swap in a different data interface. You could change the cables. Try a different power supply or radio or laptop.

Most systematic testing combines direct testing or inspection and a process of substitution. The goal is to either find the root cause or find the system component in which the

root cause is hiding. Once you have a system component identified as the culprit, treat the component as a system itself and start the process all over again. Eventually, you'll find the problem.

Here are some guidelines to systematic troubleshooting in the ham shack:

1. Take the time to define and understand the system exhibiting the problem.

2. Remove system components one at a time.

3. Verify that the problem still exists. If not, restore that component.

4. Continue to remove components and test until the minimum system exhibiting the problem has been reached.

5. Inspect what is happening at each interface between system components by testing or substitution. In the example, you inspected the cables connecting the radio and other system components.

6. If a problem is identified, correct it and re-test.

7. Otherwise, continue to test or substitute each system component until the root cause has been identified or isolated to one component.

8. Treat that component as a new system and return to step 1.

Systematically reducing the number of components and testing each interface between them is almost always the fastest way to determine where the problem's root cause really lies. You may be a lucky guesser and sometimes a little plume of smoke is a dead giveaway, but in the long run, the system approach will save you time and money.

26.3.2 Assessing the Symptoms

An important part of the troubleshooting process is a careful definition of just what the symptoms of the problem really are. It is important to note exactly how the problem manifests itself and the conditions under which the problem occurs. Avoid vague descriptions such as "broken" and "not working." Train yourself to use precise descriptions such as "relay fails to activate" or "no speaker output."

Ask yourself these questions:

1. What functions of the equipment do not work as they should; what does not work at all?

2. What kind of performance can you realistically expect?

3. Has the trouble occurred in the past? (Keep a record of troubles and maintenance in the owner's manual, shack notebook or log book.)

Write down the answers to the questions. The information will help with your work, and it may help service personnel if their advice or professional service is required.

Question your assumptions and verify what you think you know. Are you *sure* that power supply output is OK under all conditions? Did you actually confirm that there is continuity at every position of the antenna switch? If there is any doubt, make a confirming measurement or inspection. Countless hours have been wasted because of unjustified assumptions!

Intermittent problems are generally harder to track down, so try to note the conditions under which the problem occurs — those are often important clues.

Troubleshooting is a good reason to do regular maintenance. Not only will you have fewer problems, but when something is just a little off or out of place, you are much likelier to notice it. Learn to listen to the little voice in your head noting something out of the ordinary. Don't discount the wild cards that occasionally cause problems.

Occasionally step away from the workbench and relax. Take a walk. Your mind will continue to think about the problem and

Newly Constructed Equipment

What if you built a piece of equipment and it doesn't work? In most repair work, the troubleshooter is aided by the knowledge that the circuit once worked so that it is only necessary to find and replace the faulty part(s). This is not the case with newly constructed equipment.

Repair of equipment with no working history is a special, and difficult, case. You may be dealing with a defective component, construction error or even a faulty design. Carefully checking for these defects can save you hours. This is a good reason to test homebrew equipment at each possible step and on a section-by-section basis, if possible. In that way, you'll know more about what does work if the completed project has a problem.

you may surprise yourself when you sit back down to work!

If you can bounce ideas off a friend, try explaining the problem and letting the friend ask you questions. It's common for someone else with a different perspective to ask questions you haven't thought of.

26.3.3 External Inspection

Inspection is the easiest part of troubleshooting to do and careful, detailed inspection will often find the problem or a clue that leads to it. Make sure you have some paper to keep notes on as you go so when something occurs to you it can be recorded.

Try the easy things first. If you are able

to solve the problem by replacing a fuse or reconnecting a loose cable, you might be able to avoid a lot of effort. Many experienced technicians have spent hours troubleshooting a piece of equipment only to learn the hard way that the on/off switch was set to OFF or the squelch control was set too high, or that they were not using the equipment properly.

Next, make sure that equipment is plugged in, that the ac outlet does indeed have power, that the equipment is switched on and that all of the fuses are good. If the equipment uses batteries or an external power supply, make sure they supply the right voltage under load.

Check that all wires, cables and accessories are working and plugged into the right connectors or jacks. In a system of components it is often difficult to be sure which component or subsystem is bad. Your transmitter may not work on SSB because the transmitter is bad, but it could also be a bad microphone.

Connector faults or misconnections are common. Consider them prime suspects in your troubleshooting detective work. Do a thorough inspection of the connections. Is the antenna connected? How about the speaker, fuses and TR switch? Are transistors and ICs firmly seated in their sockets? Are all interconnection cables sound and securely connected? Are any pins bent or is a connector inserted improperly? Many of these problems are obvious to the eye, so look around carefully.

While you are performing your inspection, don't forget to use all of your senses. Do you smell anything burnt or overheated? Is something leaking electrolyte or oil? Perhaps a component or connector looks overheated and discolored. Is mounting hardware secure?

Once you're done with your inspection, retest the equipment to be sure the problem is still there or note if it has changed in some way.

26.4 Inside the Equipment

At this point, you've determined that a specific piece of equipment has a problem. A visual inspection of all the operating controls and connections hasn't turned up anything, but the problem is still there. It is time to really dig in, take it apart, and fix it!

26.4.1 Documentation

In order to test any piece of equipment, you'll probably need at least a user's manual. If at all possible, locate a schematic diagram and service manual. It is possible to troubleshoot without a service manual, but a schematic is almost indispensable.

The original equipment manufacturer is the best source of a manual or schematic. However, many old manufacturers have gone out of business. Several sources of equipment manuals can be located by a web search or from one of the *QST* vendors that sell equipment needs. In addition K4XL's Boat Anchor Manual Archive (**www2.faculty.sbc.edu/ kgrimm/boatanchor**) has hundreds of freely downloadable electronic copies of equipment manuals. If there is a user's group or email list associated with your equipment, a request to the group may turn up a manual and maybe even troubleshooting assistance.

If all else fails, you can sometimes reverse

engineer a simple circuit by tracing wiring paths and identifying components to draw your own schematic. By downloading datasheets for the active devices used in the circuit, the pin-out diagrams and applications notes will sometimes be enough to help you understand and troubleshoot the circuit.

THE BLOCK DIAGRAM

An important part of the documentation, the block diagram is a road map. It shows the signal paths for each circuit function. These paths may run together, cross occasionally or not at all. Those blocks that are not in the paths of faulty functions can be eliminated as suspects. Sometimes the symptoms point to a single block and no further search is necessary. In cases where more than one block is suspect, several approaches may be used. Each requires testing a block or stage.

26.4.2 Disassembly

This seemingly simple step can trap the unwary technician. Most experienced service technicians can tell you the tale of the equipment they took apart and were unable to easily put back together. Don't let it happen to you.

Take photos and lots of notes about the way you take it apart. Take notes about each component you remove. Take a photo or make a sketch of complicated mechanical assemblies before disassembly and then record how you disassembled them. It is particularly important to record the position of shields and ground straps.

Write down the order in which you do things, color codes, part placements, cable routings, hardware notes, and anything else you think you might need to be able to reassemble the equipment weeks from now when the back-ordered part comes in.

Put all of the screws and mounting hardware in one place. A plastic jar with a lid works well; if you drop it the plastic is not apt to break and the lid will keep all the parts from flying around the work area (you will never find them all). It may pay to have a separate labeled container for each subsystem. Paper envelopes and muffin pans also work well.

26.4.3 Internal Inspection

Many service problems are visible, if you look for them carefully. Many a technician has spent hours tracking down a failure, only to find a bad solder joint or burned component that would have been spotted in careful inspection of the printed-circuit board. Internal inspections are just as important as external inspections.

It is time consuming, but you really need to look at every connector, every wire, every solder joint and every component. A low power magnifying glass or head-mounted magnifier enables you to quickly scan the equipment to look for problems. A connector may have loosened, resulting in an open circuit. You may spot broken wires or see a bad solder joint. Flexing the printed-circuit board or tugging on components a bit while looking at their solder joints will often locate a defective solder job. Look for scorched components.

Make sure all of the screws securing the printed-circuit board are tight and making good electrical contact. Check for loose screws on chassis-mounted connectors. (Do not tighten any electrical or mechanical adjusting screws or tuning controls, however!) See if you can find evidence of previous repair jobs; these may not have been done properly. Make sure that ICs are firmly seated in sockets if they are used. Look for pins folded underneath the IC rather than inserted into the socket. If you are troubleshooting a newly constructed circuit, make sure each part is of the correct value or type number and is installed correctly.

POWER SUPPLIES

If your careful inspection doesn't reveal anything, it is time to apply power to the unit under test and continue the process. Observe all safety precautions while troubleshooting equipment. There are voltages inside some equipment that can kill you. If you are not qualified to work safely with the voltages and conditions inside of the equipment, do not proceed. See Table 26.1 and the **Safety** chapter.

You may be able to save quite a bit of time if you test the power supply right away. If the power supply is not working at all or not working properly, no other circuit in the equipment can be expected to work properly either. Once the power supply has either been determined to be OK or repaired, you can proceed to other parts of the equipment. Power supply diagnosis is discussed in detail later in this chapter.

With power applied to the equipment, listen for arcs and look and smell for smoke or overheated components. If no problems are apparent, you can move on to testing the various parts of the circuit. For tube equipment, you may want to begin with ac power applied at a reduced voltage as described in the sections on repair of vintage equipment.

26.4.4 Signal Tracing and Signal Injection

There are two common systematic approaches to troubleshooting radio equipment at the block level. The first is signal tracing; the second is signal injection. The two techniques are very similar. Differences in test equipment and the circuit under test determine which method is best in a given situation. They can often be combined.

Both of these approaches are used on equipment that is designed to operate on a

Block-Level Testing for DSP and SDR

More and more functions of today's radio equipment are implemented by microprocessor-based digital signal processing up to and including full SDR with direct digitization of RF signals very close to the antenna input. A look at the PC boards of such a radio show a few large, many-leaded ICs surrounded by control and interface components, power supply circuits, filtering, and transmitter power amplifiers. The individual stages of the classic super-heterodyne architecture are nowhere to be seen. How do you troubleshoot such a radio?

Start with the same basic approach as for an older radio — begin at the input or output and work your way towards the "other end" of the radio. In a DSP-based radio however, you'll rapidly encounter the point at which the signal "goes digital" and disappears into the microprocessor or an analog/digital converter. Jump to the point where the signal returns to analog form on the "other side" of the microprocessor (or PC in the case of most SDR equipment) and resume testing. In this way you can simply treat the microprocessor as one very large stage in the radio. If the problem turns out to be in the surrounding circuitry or in the interface between a PC and the RF circuits, you can troubleshoot it as you would any other piece of equipment.

If the problem turns out to be in the microprocessor or analog/digital converter, in all probability you will have to get a replacement board from the manufacturer. It is possible to replace the converter or processor (if you can obtain the pre-programmed part) but most manufacturers treat the entire circuit board as the lowest-level replaceable component. This makes repair more expensive (or impossible if no replacements are available) but the positive tradeoff is that the microprocessors rarely fail by themselves, resulting in fewer repairs being required in the first place.

Knowing When to Quit

It is common for experienced repair techs to be given "basket cases" — equipment that the original troubleshooter disassembled but then couldn't reassemble for whatever reason or couldn't find the problem. One of the most important decisions in the troubleshooting process is knowing when to quit. That is, realizing that you are about to go beyond your skills or understanding of the equipment.

If you proceed past this point, the chances of a successful outcome go down pretty quickly. It's far better to ask for help, work with a more knowledgeable friend, or carefully re-assemble the equipment and take it to a repair tech. Don't let your equipment wind up as a box full of partially connected pieces and mounting hardware under the table at the hamfest with a sign that says, "Couldn't fix — make offer!"

signal (RF or audio or data) in a sequence of steps. A transceiver based on the superheterodyne architecture is probably the best example of this type of equipment in the ham shack. Audio equipment is also built this way. More information about the signal injector and signal sources appears in "Some Basics of Equipment Servicing," from February 1982 *QST* (Feedback, May 1982).

Newer equipment that incorporates DSP and specialized or proprietary ICs is much less amenable to stage-by-stage testing techniques. (See the sidebar "Block-Level Testing for DSP and SDR") Nevertheless, the techniques of signal tracing and signal injection are still useful where the signal path is accessible to the troubleshooter.

SIGNAL TRACING

In signal tracing, start at the beginning of a circuit or system and follow the signal through to the end. When you find the signal at the input to a specific stage, but not at the output, you have located the defective stage. You can then measure voltages and perform other tests on that stage to locate the specific failure. This is much faster than testing every component in the unit to determine which is bad.

It is sometimes possible to use over-the-air signals in signal tracing, in a receiver for example. However, if a good signal generator is available, it is best to use it as the signal source. A modulated signal source is best.

Signal tracing is suitable for most types of troubleshooting of receivers and analog amplifiers. Signal tracing is the best way to check transmitters because all of the necessary signals are present in the transmitter by design. Most signal generators cannot supply the wide range of signal levels required to test a transmitter.

Equipment

A voltmeter with an RF probe is the most common instrument used for signal tracing. Low-level signals cannot be measured accurately with this instrument. Signals that do not exceed the junction drop of the diode in the probe will not register at all, but the presence, or absence, of larger signals can be observed.

A dedicated signal tracer can also be used. It is essentially an audio amplifier. (See this book's CD-ROM for a project to build your own signal tracer.) An experienced technician can usually judge the level and distortion of the signal by ear. You cannot use a dedicated signal tracer to follow a signal that is not amplitude modulated (single sideband is a form of AM). A signal tracer is not suitable for tracing CW signals, FM signals or oscillators. To trace these, you will have to use a voltmeter and RF probe or an oscilloscope.

An oscilloscope is the most versatile signal tracer. It offers high input impedance, variable sensitivity, and a constant display of the traced waveform. If the oscilloscope has sufficient bandwidth, RF signals can be observed directly. Alternatively, a demodulator probe can be used to show demodulated RF signals on a low-bandwidth oscilloscope. Dual-trace scopes can simultaneously display the waveforms, including their phase relationship, present at the input and output of a circuit.

Procedure

First, make sure that the circuit under test and test instruments are isolated from the ac line by internal transformers, an isolation transformer, or operate from battery power. Set the signal source to an appropriate level and frequency for the unit you are testing. For a receiver, a signal of about 100 µV should be

plenty. For other circuits, use the schematic, an analysis of circuit function and your own good judgment to set the signal level.

In signal tracing, start at the beginning and work toward the end of the signal path. Switch on power to the test circuit and connect the signal-source output to the test-circuit input. Place the tracer instrument at the circuit input and ensure that the test signal is present. Observe the characteristics of the signal if you are using a scope (see **Fig 26.13**). Compare the detected signal to the source signal during tracing.

Move the test instrument to the output of the next stage and observe the signal. Signal level should increase in amplifier stages and may decrease slightly in other stages. The signal will not be present at the output of a dead stage.

Low-impedance test points may not provide sufficient signal to drive a high-impedance

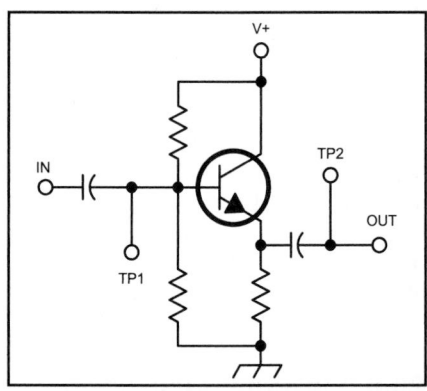

Fig 26.14 —Although the common-collector (emitter-follower) circuit functions as a current amplifier, the change in impedance from TP1 to TP2 results in the traces described in the text.

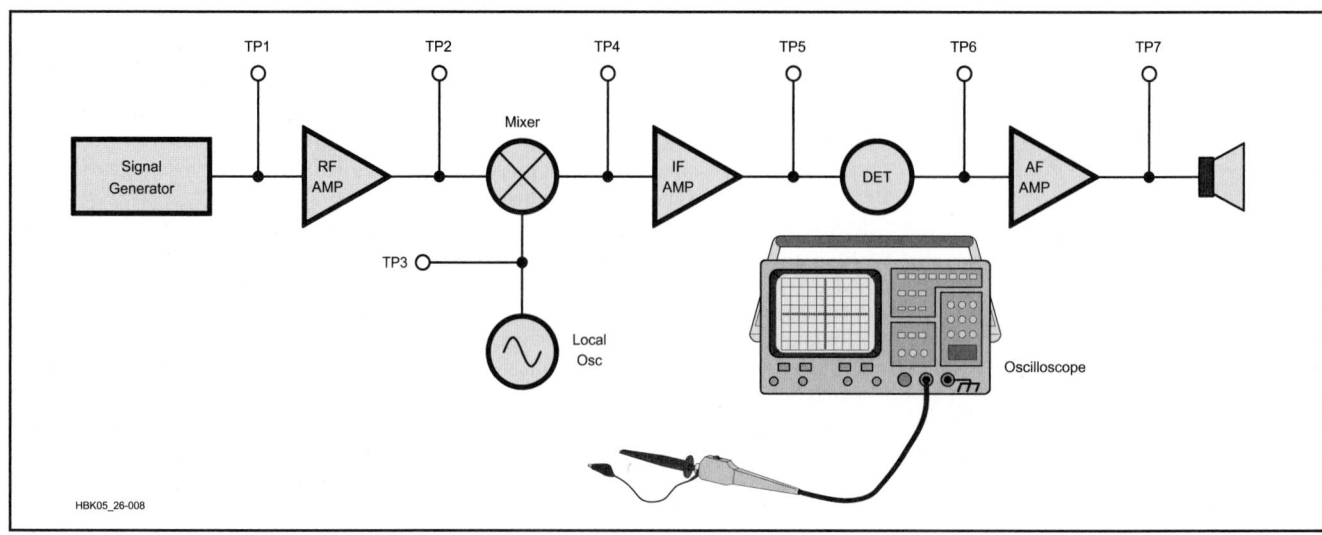

Fig 26.13 — Signal tracing in a simple superheterodyne receiver.

signal tracer, so tracer sensitivity is important. Also, in some circuits the output level appears low where there is an impedance change from input to output of a stage. For example, the circuit in **Fig 26.14** is a common-collector (emitter-follower) current amplifier with a high input impedance and low output impedance. The voltages at TP1 and TP2 are approximately equal and in phase.

There are two signals — the test signal and the local oscillator signal — present in a mixer stage. Loss of either one will result in no output from the mixer stage. Switch the signal source on and off repeatedly to make sure that the test instrument reading varies (it need not disappear) with source switching.

SIGNAL INJECTION

Signal injection is a good choice for receiver troubleshooting because the receiver already has a detector as part of the design. If the detector is working or a suitable detector is devised (see the signal tracing section), a test signal can be injected at different points in the equipment until the faulty stage is discovered.

Equipment

Most of the time, your signal injector will be a signal generator. There are other injectors available, some of which are square-wave audio oscillators rich in RF harmonics (see Fig 26.5). These simple injectors do have their limits because you can't vary their output level or determine their frequency. They are still useful, though, because most circuit failures are caused by a stage that is completely dead.

Consider the signal level at the test point when choosing an instrument. The signal source used for injection must be able to supply appropriate frequencies and levels for each stage to be tested. For example, a typical superheterodyne receiver requires AF, IF and RF signals that vary from 6 V at AF, to 200 µV at RF. Each conversion stage used in a receiver requires a different IF from the signal source. When testing the signal path of an AM radio, such as a broadcast receiver, you'll need a modulated RF signal before the detector stage. Use an unmodulated RF signal to simulate the local oscillator.

Procedure

If an external detector is required, set it to the proper level and connect it to the test circuit. Set the signal source for AF, and inject a signal directly into the signal detector to test operation of the injector and detector. Move the signal source to the input of the preceding stage, and observe the signal. Continue moving the signal source to the inputs of successive stages.

When you inject the signal source to the input of the defective stage, there will be no output. Prevent stage overload by reducing the level of the injected signal as testing progresses through the circuit. Use suitable frequencies for each tested stage.

Make a rough check of stage gain by injecting a signal at the input and output of an amplifier stage. You can then compare how much louder the signal is when injected at the input. This test may mislead you if there is a radical difference in impedance from stage input to output. Understand the circuit operation before testing.

Mixer stages present a special problem because they have two inputs, rather than one. A lack of output signal from a mixer can be caused by either a faulty mixer or a faulty local oscillator (LO). Check oscillator operation with an oscilloscope or by listening on another receiver. If none of these instruments are available, inject the frequency of the LO at the LO output. If a dead oscillator is the only problem, this should restore operation.

If the oscillator is operating, but off frequency, a multitude of spurious responses will appear. A simple signal injector that produces many frequencies simultaneously is not suitable for this test. Use a well-shielded signal generator set to an appropriate level at the LO frequency.

26.4.5 Microprocessor-Controlled Equipment

The majority of today's amateur equipment

and accessories have at least one microprocessor and sometimes several. While reliability of this equipment is greatly improved over the older analog designs, troubleshooting microprocessor-based circuitry takes a different approach. While a tutorial on microprocessor troubleshooting is well outside the scope of this *Handbook*, the following basic guidelines will help determine whether the problem is inside the microprocessor (or its firmware) or in the supporting circuitry. In addition, the **Digital Basics** chapter contains lots of information about the operation of digital circuits.

1) Start by obtaining the microprocessor datasheet and identifying all of the power and control pins if they are not identified on the equipment schematic. Determine which state the control pins should be in for the device to run.

2) Test all power and control pins for the proper state (voltage). Verify that the microprocessor clock signal is active. If not, determine the reason and repair before proceeding.

3) If there is an address and data bus for external memory and input-output (I/O) devices (less common in newer equipment), use a logic probe or scope to verify that they are all active (changing state). If not, there is a program or logic fault.

4) Determine which pins are digital inputs or outputs of the microprocessor and verify that a valid digital logic level exists at the pin. If not, check the external circuit to which the pin is attached.

5) Determine which pins are analog inputs or outputs of the microprocessor. Analyze the external circuit to determine what constitutes a proper voltage into or out of the processor. If the voltage is not valid, check the external circuit to which the pin is attached. If an external voltage reference is used, verify that it is working.

6) External circuits can often be checked by disconnecting them from the microprocessor and either driving them with a temporary voltage source or measuring the signal they are attempting to send to the microprocessor.

7) If all control and power signals and the external circuitry checks out OK, it is likely that a microprocessor or firmware fault has occurred.

If you determine that the microprocessor is faulty, you will have to contact the manufacturer in most cases, since firmware is most often contained within the processor which must be programmed before it is installed. Older equipment in which the program and data memory are external to the microprocessor can be very difficult to repair due to obsolescence of the parts themselves and the requirement to program EPROM or PROM devices.

26.5 Testing at the Circuit Level

Once you have followed all of the trouble-shooting procedures and have isolated your problem to a single defective stage or circuit, a few simple measurements and tests will usually pinpoint one or more specific components that need adjustment or replacement.

First, check the parts in the circuit against the schematic diagram to be sure that they are reasonably close to the design values, especially in a newly built circuit. Even in a commercial piece of equipment, someone may have incorrectly changed them during attempted repairs. A wrong-value part is quite likely in new construction, such as a home-brew or kit project.

26.5.1 Voltage Levels

Check the circuit voltages. If the voltage levels are printed on the schematic, this is easy. If not, analyze the circuit and make some calculations to see what the circuit voltages should be. Remember, however, that the printed or calculated voltages are nominal; measured voltages may vary from the calculations.

When making measurements, remember the following points:

• Make measurements at device leads, not at circuit-board traces or socket lugs.

• Use small test probes to prevent accidental shorts.

• Never connect or disconnect power to solid-state circuits with the switch on.

• Remember that voltmeters, particularly older analog meters may load down a high-impedance circuit and change the voltage, as will ×1 and low-impedance scope probes.

Voltages may give you a clue to what is wrong with the circuit. If not, check the active device. If you can check the active device in the circuit, do so. If not, remove it and test it, or substitute a known good device. After connections, most circuit failures are caused

directly or indirectly by a bad active device. The experienced troubleshooter usually tests or substitutes these first. Analyze the other components and determine the best way to test each as described earlier.

There are two voltage levels in most analog circuits (V+ and ground, for example). Most component failures (opens and shorts) will shift dc voltages near one of these levels. Typical failures that show up as incorrect dc voltages include: open coupling transformers; shorted capacitors; open, shorted or overheated resistors and open or shorted semiconductors.

Digital logic circuits require that signals be within specific voltage ranges to be treated as a valid logic-low or logic-high value.

26.5.2 Noise

A slight hiss is normal in all electronic circuits. This noise is produced whenever current flows through a conductor that is warmer than absolute zero. Noise is compounded and amplified by succeeding stages. Repair is necessary only when noise threatens to obscure normally clear signals.

Semiconductors can produce hiss in two ways. The first is normal — an even white noise that is much quieter than the desired signal. Faulty devices frequently produce excessive noise. The noise from a faulty device is often erratic, with pops and crashes that are sometimes louder than the desired signal. In an analog circuit, the end result of noise is usually sound. In a control or digital circuit, noise causes erratic operation: unexpected switching and so on.

Noise problems usually increase with temperature, so localized heat may help you find the source. Noise from any component may be sensitive to mechanical vibration. Tapping various components with an insulated screwdriver may quickly isolate a bad part. Noise can also be traced with an oscilloscope or signal tracer.

Nearly any component or connection can be a source of noise. Defective components are the most common cause of crackling noises. Defective connections are a common cause of loud, popping noises.

Check connections at cables, sockets and switches. Look for dirty variable capacitor wipers and potentiometers. An arcing mica trimmer capacitor can create static crashes in received or transmitted audio. Test them by installing a series 0.01 µF capacitor. If the noise disappears, replace the trimmer.

Potentiometers are particularly prone to noise problems when used in dc circuits. Clean them with a spray contact cleaner and rotate the shaft several times.

Rotary switches may be tested by jumpering the contacts with a clip lead. Loose

contacts may sometimes be repaired, either by cleaning, carefully rebending the switch contacts or gluing loose switch parts to the switch deck. Operate variable components through their range while observing the noise level at the circuit output.

26.5.3 Oscillations

Oscillations occur whenever there is sufficient positive feedback in a circuit that has gain. (This can even include digital devices.) Oscillation may occur at any frequency from a low-frequency "putt-putt" (often called *motorboating*) well up into the RF region.

Unwanted oscillations are usually the result of changes in the active device (increased junction or interelectrode capacitance), failure of an oscillation suppressing component (open decoupling or bypass capacitors or neutralizing components) or new feedback paths (improper lead dress or dirt on the chassis or components). It can also be caused by improper design, especially in home-brew circuits. A shift in bias or drive levels may aggravate oscillation problems.

Oscillations that occur in audio stages do not change as the radio is tuned because the operating frequency — and therefore the component impedances — do not change. RF and IF oscillations, however, usually vary in amplitude as operating frequency is changed.

Oscillation stops when the positive feedback is removed. Locating and replacing a defective (or missing) bypass capacitor may effect an improvement. The defective oscillating stage can be found most reliably with an oscilloscope.

26.5.4 Amplitude Distortion

Amplitude distortion is the product of nonlinear operation. The resultant waveform contains not only the input signal, but new signals at other frequencies as well. All of the frequencies combine to produce the distorted waveform. Distortion in a transmitter gives rise to splatter, harmonics and interference.

Fig 26.15 shows some typical cases of distortion. Clipping (also called flat-topping) is the consequence of excessive drive, a change in bias, or insufficient supply voltage to the circuit. The corners on the waveform show that harmonics are present. (A square wave contains the fundamental and all odd harmonics.) If this was a transmitter circuit, these odd harmonics would be heard well away from the operating frequency, possibly outside of amateur bands.

Harmonic distortion produces radiation at frequencies far removed from the fundamental; it is a major cause of electromagnetic interference (EMI). Harmonics are generated

---

Controlling Key Clicks

Key clicks are a special type of amplitude distortion caused by fast rising and falling edges of the output waveform or from abrupt disturbances in an otherwise smooth waveform. See the **Transmitters and Transceivers** chapter for more information about controlling key clicks. Most modern radios offer user configuration options to adjust keying rise and fall times. Older radios may require modification of a timing circuit that controls rise and fall time. Be a good neighbor to other operators and eliminate key clicks on your signal!

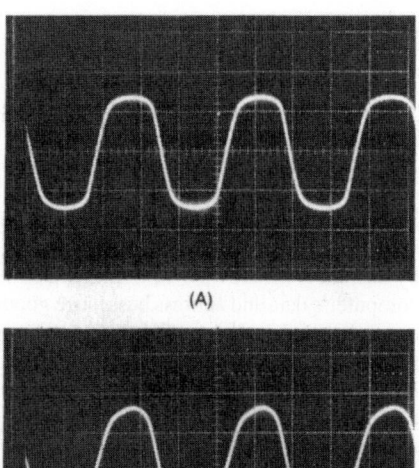

(A)

(B)

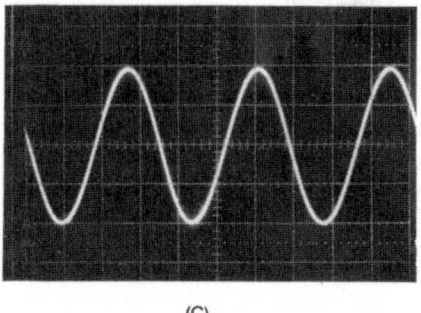

(C)

Fig 26.15 — Examples of distorted waveforms. The result of clipping is shown in A. Nonlinear amplification is shown in B. A pure sine wave is shown in C for comparison.

in nearly every amplifier. When they occur in a transmitter, they are usually caused by insufficient transmitter filtering (either by design, or because of filter component failure).

Anything that changes the proper bias of an amplifier can cause distortion. This includes failures in the bias components, leaky transistors or vacuum tubes with interelectrode shorts. In a receiver, these conditions may mimic AGC trouble. Improper bias of an analog circuit often results from a resistor that changed value or a leaky or shorted capacitor. RF feedback can also produce distortion by disturbing bias levels. Distortion is also caused by circuit imbalance in Class AB or B amplifiers.

Oscillations in an IF amplifier may produce distortion. They cause constant, full AGC action, or generate spurious signals that mix with the desired signal. IF oscillations

are usually evident on the S meter, which will show a strong signal even with the antenna disconnected.

26.5.5 Frequency Response

Every circuit, even a broadband circuit, has a desired frequency response. Audio amplifiers used in amateur SSB circuits, for example, typically are designed for signals between 300 and 3000 Hz, more or less. A tuned IF amplifier may have a bandwidth of 50 to 100 kHz around the stage's center frequency. Any change in the circuit's frequency response can alter its effect on the signals on which it operates.

Frequency response changes are almost always a consequence of a capacitor or inductor changing value. The easiest way to check frequency response is to either inject a signal

into the circuit and measure the output at several frequencies or use a spectrum analyzer or a signal generator's sweep function. In LC networks, it is relatively simple to lift one lead of the component and use a component checker to determine the value.

26.5.6 Distortion Measurement

A distortion meter is used to measure distortion of AF signals. A spectrum analyzer is the best piece of test gear to measure distortion of RF signals. If a distortion meter is not available, an estimation of AF distortion can sometimes be made with a function generator and an oscilloscope. Inject a square wave signal into the circuit with a fundamental frequency roughly in the middle of the expected frequency response.

Compare the input square wave to the output signal with an oscilloscope. **Fig 26.16** shows several effects on the square wave that are related to frequency response. These provide clues to what components or devices may be causing the problem. Severe distortion indicates some other problem besides frequency response changes.

26.5.7 Alignment

Alignment — the tuning or calibration of frequency sensitive circuits — is rarely the cause of an electronics problem with receiving or transmitting equipment, particularly modern equipment. Alignment does not shift suddenly and should be a last resort for treating sensitivity or frequency response problems. Do not attempt to adjust alignment without the proper equipment and alignment procedures. The process often requires steps to be performed exactly and in a specific order. Equipment misaligned in this way usually must be professionally repaired as a consequence.

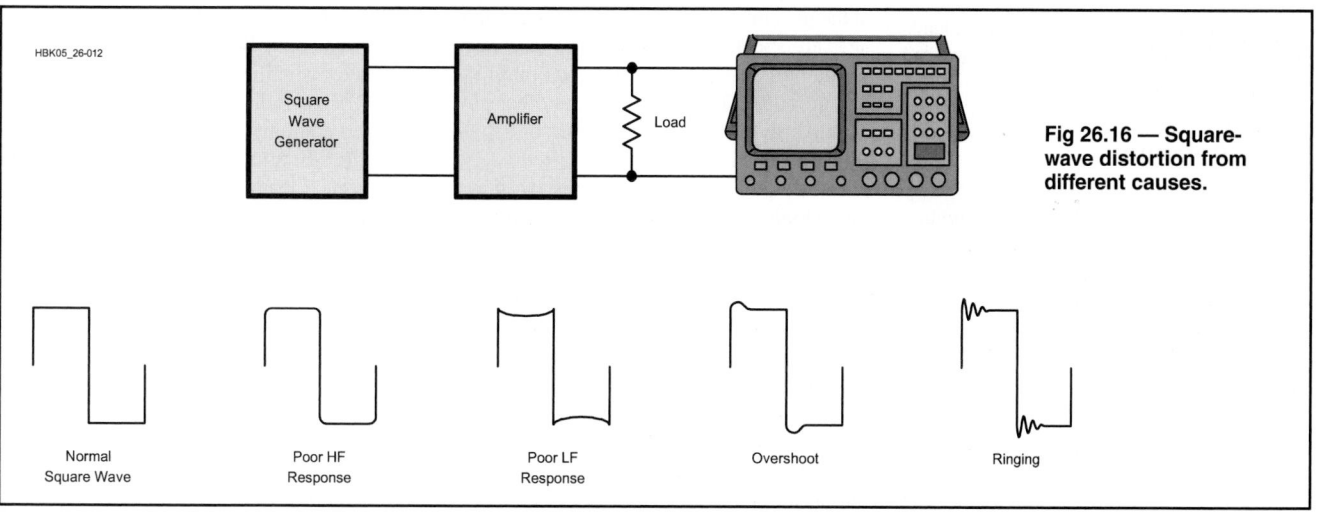

Fig 26.16 — Square-wave distortion from different causes.

26.5.8 Contamination

Contamination is another common service problem. Soda or coffee spilled into a piece of electronics is an extreme example (but one that does actually happen).

Conductive contaminants range from water to metal filings. Most can be removed by a thorough cleaning. Any of the residue-free cleaners can be used, but remember that the cleaner may also be conductive. Do not apply power to the circuit until the area is completely dry.

Keep cleaners away from variable-capacitor plates, transformers and parts that may be harmed by the chemical. The most common conductive contaminant is solder, either from a printed-circuit board solder bridge or a loose piece of solder deciding to surface at the most inconvenient time.

High-voltage circuits attract significant amounts of dust as described in the section on high-power amplifier maintenance later in this chapter. Cooling fans and ventilation holes also allow dust to accumulate, creating conductive paths between components. If not removed, the conductive paths gradually become lower in resistance until they begin to affect equipment performance. Vacuuming or blowing out the equipment is usually sufficient to clear away dust, although a carbon track may require a more thorough cleaning.

26.5.9 Solder Bridges

In a typical PC-board solder bridge, the solder that is used to solder one component has formed a short circuit to another PC-board trace or component. Unfortunately, they are common in both new construction and repair work. Look carefully for them after you have completed any soldering, especially on a PC-board. It is even possible that a solder bridge may exist in equipment you have owned for a long time, unnoticed until it suddenly decided to become a short circuit.

Related items are loose solder blobs, loose hardware or small pieces of component leads that can show up in the most awkward and troublesome places.

26.5.10 Arcing

Arcing is a serious sign of trouble. It may also be a real fire hazard. Arc sites are usually easy to find because an arc that generates visible light or noticeable sound also pits and discolors conductors.

Arcing is caused by component failure, dampness, dirt or lead dress. If the dampness is temporary, dry the area thoroughly and resume operation. Dirt may be cleaned from the chassis with a residue-free cleaner. Arrange leads so high-voltage conductors are isolated. Keep them away from sharp corners and screw points.

Arcing occurs in capacitors when the working voltage is exceeded. Air-dielectric variable capacitors can sustain occasional arcs without damage, but arcing indicates operation beyond circuit limits. Antenna tuners working beyond their ability may suffer from arcing. A failure or high SWR in an antenna circuit may also cause transmitter arcing. Prolonged, repeated, or high-power arcing can cause pits or deposits on capacitor plates that reduce capacitor voltage rating and lead to more arcing.

26.5.11 Digital Circuitry

Although every aspect of digital circuit operation may be resolved to a simple 1 or 0, or tristate (open circuit), the symptoms of their failure are far more complicated. The most common problems are false triggers or counts, digital inputs that do not respond to valid logic signals, and digital outputs stuck at ground or the supply voltage.

In most working digital circuitry the signals are constantly changing between low and high states, often at RF rates. Low-frequency or dc meters should not be used to check digital signal lines. A logic probe is often helpful in determining signal state and whether it is active or not. An oscilloscope or logic analyzer is usually needed to troubleshoot digital circuitry beyond simple go/no-go testing.

If you want to use an oscilloscope to give an accurate representation of digital signals, the scope bandwidth must be at least several times the highest clock frequency in the circuit in order to reproduce the fast rise and fall times of digital signals. Lower-bandwidth scopes can be useful in determining whether a signal is present or active but often miss short glitch signals (very fast transients) that are often associated with digital circuit malfunction.

LOGIC LEVELS

Begin by checking for the correct voltages at the pins of each IC. The correct logic voltages are specified in the device's datasheet, which will also identify the power pins (V_{cc} and ground). The voltages on the other pins should be a logic high, a logic low, or tristate (more on this later).

Most digital circuit failures are caused by a failed logic IC. IC failures are almost always catastrophic. It is unlikely that an AND gate will suddenly start functioning like an OR gate. It is more likely that the gate will have a signal at its input, and no signal at the output. In a failed device, the output pin will have a steady voltage. In some cases, the voltage is steady because one of the input signals is missing. Look carefully at what is going into a digital IC to determine what should be coming out. Manufacturers' datasheets describe the proper functioning of most digital devices.

TRISTATE DEVICES

Many digital devices are designed with a third logic state, commonly called tristate. In this state, the output of the device acts as an open circuit so that several device outputs can be connected to a common bus. The outputs that are active at any given time are selected by software or hardware control signals. A computer's data and address busses are good examples of this. If any device output connected to the bus fails by becoming locked or stuck in a 0 or 1 logic state, the entire bus becomes nonfunctional. Tristate devices can also be locked on by a failure of the signal that controls the tristate output status.

SIMPLE GATE TESTS

Most discrete logic ICs (collections of individual gates or other logic functions) are easily tested by in-circuit inspection of the input and output signals. The device's truth table or other behavior description specifies what the proper input and output signals should be. Testing of more complicated ICs requires the use of a logic analyzer, multitrace scope or a dedicated IC tester. If a simple logic IC is found to be questionable, it is usually easiest to simply substitute a new device for it.

CLOCK SIGNALS

In clocked circuits, check to see if the clock signal is active. If the signal is found at the clock chip, trace it to each of the other ICs to be sure that the clock system is intact. Clock frequencies are rarely wrong but clock signals derived from a master clock can be missing or erratic if the circuitry that creates them is defective.

RF INTERFERENCE

If digital circuitry interferes with other nearby equipment, it may be radiating spurious signals. These signals can interfere with your Amateur Radio operation or other services. Computer networking and microprocessor-controlled consumer equipment generate a significant amount of noise due to RF being radiated from cables and unshielded equipment.

Digital circuitry can also be subject to interference from strong RF fields. Erratic operation or a complete lock-up is often the result. Begin by removing the suspect equipment from RF fields. If the symptoms stop when there is no RF energy present, apply common-mode chokes as described in the **RF Interference** chapter..

The ARRL RFI Book has a chapter on computer and digital interference to and from digital devices and circuits. The subject is also covered in the **RF Interference** chapter of this book.

26.5.12 Replacing Parts

If you have located a defective component within a stage, you need to replace it. When replacing socket mounted components, be sure to align the replacement part correctly. Make sure that the pins of the device are properly inserted into the socket. See the **Construction Techniques** chapter for guidance on working with SMT components.

Some special tools can make it easier to remove soldered parts. A chisel-shaped soldering tip helps pry leads from printed-circuit boards or terminals. A desoldering iron or bulb forms a suction to remove excess solder, making it easier to remove the component. Spring-loaded desoldering pumps are more convenient than bulbs. Desoldering wick draws solder away from a joint when pressed against the joint with a hot soldering iron.

Removing soldered ICs is a lot simpler if you simply clip its leads next to the IC body using fine-point wire cutters, although it does destroy the IC. Then melt the solder and lift out the pin with tweezers or the wire cutter. Use the desoldering pumps or wick to remove the solder. This works well for both through-hole and SMT components.

26.6 After the Repairs

Once you have completed your troubleshooting and repairs, it is time to put the equipment back together. Take a little extra time to make sure you have done everything correctly.

26.6.1 All Units

Give the entire unit a complete visual inspection. Look for any loose ends left over from your troubleshooting procedures—you may have left a few components temporarily soldered in place, forgotten to reattach a wire or cable, or overlooked some other repair error. Look for cold solder joints and signs of damage incurred during the repair. Double-check the position, leads and polarity of components that were removed or replaced.

Make sure that all ICs and connectors are properly oriented and inserted in their sockets. Test fuse continuity with an ohmmeter and verify that the current rating matches the circuit specification.

Look at the position of all of the wires and components. Make sure that wires and cables will be clear of hot components, screw points and other sharp edges. Make certain that the wires and components will not be in the way and pinched or crimped when covers are installed and the unit is put back together.

Separate the leads that carry dc, RF, input and output as much as possible. Plug-in circuit boards should be firmly seated with screws tightened and lock washers installed if so specified. Shields and ground straps should be installed just as they were on the original.

26.6.2 Transmitter Checkout

Since the signal produced by an HF transmitter can be heard the world over, a thorough check is necessary after any service has been performed. Do not exceed the transmitter duty cycle while testing. Limit transmissions to 10 to 20 seconds unless otherwise specified by the owner's manual.

1. Set all controls as specified in the operation manual, or at midscale.
2. Connect a dummy load and a power meter to the transmitter output.
3. Set the drive or carrier control for low output.
4. Switch the power on.
5. Transmit and quickly set the final-amplifier bias to specifications if necessary
6. For vacuum tube final amplifiers, slowly tune the output network through resonance. The current dip should be smooth and repeatable. It should occur simultaneously with the maximum power output. Any sudden jumps or wiggles of the current meter indicate that the amplifier is unstable. Adjust the neutralization circuit (according to the manufacturer's instructions) if one is present or check for oscillation. An amplifier usually requires neutralization whenever active devices, components or lead dress (that affect the output/input capacitance) are changed.
7. Check to see that the output power is consistent with the amplifier class used in the PA (efficiency should be about 25% for Class A, 50 to 60% for Class AB or B, and 70 to 75% for Class C). Problems are indicated by the efficiency being significantly low.
8. Repeat steps 4 through 6 for each band of operation from lowest to highest frequency.
9. Check the carrier balance (in SSB transmitters only) and adjust for minimum power output with maximum RF drive and no microphone gain.
10. Adjust the VOX controls.
11. Measure the passband and distortion levels if equipment (wideband scope or spectrum analyzer) is available.

26.6.3 Other Repaired Circuits

After the preliminary checks, set the circuit controls per the manufacturer's specifications (or to midrange if specifications are not available) and switch the power on. Watch and smell for smoke, and listen for odd sounds such as arcing or hum. Operate the circuit for a few minutes, consistent with allowable duty cycle. Verify that all operating controls function properly.

Check for intermittent connections by subjecting the circuit to heat, cold and slight flexure. Also, tap or jiggle the chassis lightly with an alignment tool or other insulator.

If the equipment is meant for mobile or portable service, operate it through an appropriate temperature range. Many mobile radios do not work on cold mornings, or on hot afternoons, because a temperature-dependent intermittent was not found during repairs.

26.6.4 Button It Up

After you are convinced that you have repaired the circuit properly, put it all back together. If you followed the advice in this book, you have all the screws and assorted doodads in a secure container. Look at the notes you took while taking it apart; put it back together in the reverse order. Don't forget to reconnect all internal connections, such as ac power, speaker or antenna leads.

Once the case is closed, and all appears well, don't neglect the final, important step — make sure it still works. Many an experienced technician has forgotten this important step, only to discover that some minor error, such as a forgotten antenna cable, has left the equipment nonfunctional.

26.7 Professional Repairs

Repairs that deal with very complex and temperamental circuits, or that require sophisticated test equipment, should be passed on to a professional. Factory authorized service personnel have a lot of experience. What seems like a servicing nightmare to you is old hat to them. There is no one better qualified to service your equipment than the factory.

If the manufacturer is no longer in business, check with your local dealer or look through Amateur Radio magazines and websites. You can usually find one or more companies or repair services that handle all makes and models. Your local club or a user's group may also be able to make a recommendation.

If you are going to ship your equipment somewhere for repair, notify the repair center first. Get authorization for shipping and an identification name or number for the package.

26.7.1 Packing Equipment

You can always blame shipping damage on the shipper, but it is a lot easier for all concerned if you package your equipment properly for shipping in the first place.

• Take photos of the equipment before packing it to document its condition before shipping. Additional photos during the packing steps might also be useful to show that you took the proper care in packing the equipment.

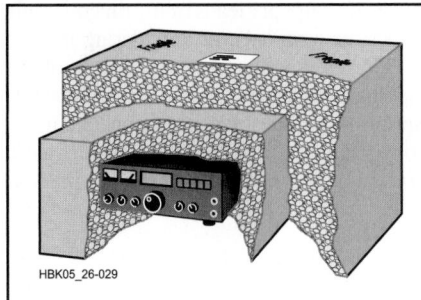

Fig 26.17 — Ship equipment packed securely in a box within a box.

• Firmly secure all heavy components, either by tying them down or blocking them off with shipping foam.
• Large transformers, such as for RF power amplifiers, should probably be removed and shipped separately.
• Large vacuum tubes should be wrapped in packing material or shipped separately.
• Make sure that all circuit boards and parts are firmly attached.

If you have the original shipping container, including all of the packing material, you should use that if the repair facility approves doing so. Otherwise use a box within a box for shipping. (See **Fig 26.17**). Place the equipment and some packing material inside a box and seal it with tape. Place that box inside

another that is at least six inches larger in each dimension.

Don't forget to enclose a statement of the trouble, a short history of operation and any test results that may help the service technician. Include a good description of the things you have tried. Be honest! At current repair rates you want to tell the technician everything to help ensure an efficient repair. Place the necessary correspondence, statement of symptoms, and your contact information in a mailing envelope and place it just inside the top covers of the outer box or tape it to the top of the inner box.

Fill any remaining gaps with packing material, seal, address and mark the outer box. Choose a good freight carrier and insure the package. If available, get a tracking number for the package so you can tell when it was delivered.

Even if you tried to fix it yourself but ended up sending it back to the factory, you can feel good about your experience. You learned a lot by trying, and you have sent it back knowing that it really did require the services of a pro. Each time you troubleshoot and repair a piece of electronic circuitry, you learn something new. The down side is that you may develop a reputation as a real electronics whiz. You may find yourself spending a lot of time at club meetings offering advice, or getting invited over to a lot of shacks for a late-evening pizza snack. There are worse fates.

26.8 Typical Symptoms and Faults

26.8.1 Power Supplies

Many equipment failures are caused by power supply trouble. Fortunately, most power supply problems are easy to find and repair. This section focuses on the common linear power supply. Some notes are also made about switchmode supplies. Both types are discussed in detail in the **Power Sources** chapter, including projects with typical schematics.

LINEAR POWER SUPPLIES

The block diagram for a linear power supply is shown in **Fig 26.18.** First, use a voltmeter to measure output. Complete loss of output voltage is usually caused by an open circuit. (A short circuit draws excessive current that opens the fuse, thus becoming an open circuit.) If output voltage appears normal, apply a small load (1/10th supply capacity or smaller) and test output voltage again. If the small load causes voltage to drop,

there is generally a problem in the regulator circuitry, often the pass transistors.

If the ac input circuit fuse is blown, that is usually caused by a shorted diode in the filter block, a failure of the output protection circuitry, or a short circuit in the device being powered by the supply. More rarely, one of the filter capacitors can short. If the fuse has opened, turn off the power, replace the fuse, and measure the load-circuit's dc resistance. The measured resistance should be consistent with the power-supply ratings. A short or open load circuit indicates a problem.

If the measured resistance is too low, troubleshoot the load circuit. (Nominal circuit resistances are included in some equipment manuals.) If the load circuit resistance is normal, suspect a defective regulator IC or problem in the rest of the unit.

IC regulators can oscillate, sometimes causing failure. The small-value capacitors on the input, output or adjustment pins of the regulator prevent oscillations. Check or replace these capacitors whenever a regulator has failed.

AC ripple (120 Hz buzz) is usually caused

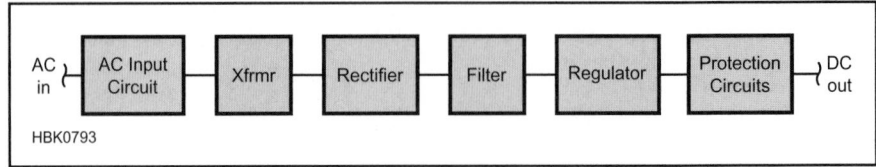

Fig 26.18 — Block diagram of a typical linear power supply.

by low-value filter capacitors in the power supply. Ripple can also become excessive due to overload or regulation problems. Look for a defective filter capacitor (usually open or low-value), defective regulator, or shorted filter choke if chokes are used (not common in modern equipment). In older equipment, the defective filter capacitor will often have visible leaking electrolyte: Look for corrosion residue at the capacitor leads. In new construction projects make sure RF energy is not getting into the power supply.

Here's an easy filter capacitor test: Temporarily connect a replacement capacitor (about the same value and working voltage) across the suspect capacitor. If the hum goes away, replace the bad component permanently.

Once the faulty component is found, inspect the surrounding circuit and consider what may have caused the problem. Sometimes one bad component can cause another to fail. For example, a shorted filter capacitor increases current flow and burns out a rectifier diode. While the defective diode is easy to find, the capacitor may show no visible damage.

If none of these initial checks find the problem, here are the usual systematic steps to locate the part of the supply with a problem:
- Check the input ac through switches and fuses to the transformer primary.
- Verify that the transformer secondary outputs ac of the right voltage. Disconnect the output leads, if necessary.
- Check the rectifiers for shorted diodes. Disconnect the rectifier output and test for output with a resistor load.
- Reconnect the filter and disconnect the regulator. Verify that the right dc voltage is present at the filter output.
- Reconnect the regulator and in the regulator IC or circuit, test every pin or signal, especially enable/disable/soft-start and the voltage reference.
- Disconnect any output protective circuitry and verify that the pass transistors are working with a resistor load.
- Reconnect and test the output protective circuitry.

SWITCHMODE POWER SUPPLIES

Switchmode or switching power supplies are quite different from conventional supplies. In a switcher, the regulator circuit is based on a switching transistor and energy storage inductor instead of a pass transistor to change power from one dc level to another that can be higher or lower than the input voltage. Switching frequencies range from 20-120 kHz for most supplies and up to the MHz range for miniature dc-dc converter modules.

Switchmode supplies operating from the ac line have similar input circuits to linear supplies. The transformer that supplies isolation is then located in the low-voltage, high frequency section.

Apply the same input and output block-level tests as for linear supplies. The regulator circuitry in a swichmode supply is more complex than for a linear supply, but it usually is implemented in a single regulator IC. Failure of the regulator IC, transistor switch, or feedback path usually results in a completely dead supply. While active device failure is still the number one suspect, it pays to carefully test all components in the regulator subsystem.

HIGH-VOLTAGE POWER SUPPLIES

Obviously, testing HV supplies requires extreme caution. See the safety discussion at the beginning of this chapter, the **Safety** chapter of this book, and the discussion of HV supplies in the **Power Sources** chapter. If you do not feel comfortable working on HV supplies, then don't. Ask for help or hire a professional repair service to do the job.

Most HV supplies used in amateur equipment are linear supplies with the same general structure as low voltage supplies. A typical supply is presented as a project in the **Power Sources** chapter and the same basic steps can be applied — just with a lot more caution.

Components in a string, such as rectifiers or filter capacitors, should all be tested if any are determined to have failed. This is particularly true for capacitors which can fail in sequence if one capacitor in the string shorts. Voltage equalizing resistors are not required for rectifier diodes available today such as the 1N5408. Consider replacing older rectifier strings with new rectifiers as a preventive maintenance step.

Interlocks rarely fail but verify that they are functioning properly before assuming they are in good working order.

26.8.2 Amplifier Circuits

Amplifiers are the most common circuits in electronics. The output of an ideal amplifier would match the input signal in every respect except magnitude: No distortion or noise would be added. Real amplifiers always add noise and distortion. Typical discrete and op-amp amplifier circuits are described in the **Analog Basics** chapter.

AMPLIFIER GAIN

Amplifier failure usually results in a loss of gain or excessive distortion at the amplifier output. In either case, check external connections first. Is there power to the stage? Has the fuse opened? Check the speaker and leads in audio output stages, the microphone and push-to-talk (PTT) line in transmitter audio sections. Excess voltage, excess current or thermal runaway can cause sudden failure of

semiconductors. The failure may appear as either a short circuit or open circuit of one or more PN junctions.

Thermal runaway occurs most often in bipolar transistor circuits. If degenerative feedback (the emitter resistor reduces base-emitter voltage as conduction increases) is insufficient, thermal runaway will allow excessive current flow and device failure. Check transistors by substitution, if possible. If not, voltage checks as described below usually turn up the problem.

Faulty coupling components can reduce amplifier output. Look for component failures that would increase series impedance, or decrease shunt impedance, in the coupling network. Coupling faults can be located by signal tracing or parts substitution. Other passive component defects reduce amplifier output by shifting bias or causing active-device failure. These failures are evident when the dc operating voltages are measured.

If an amplifier is used inside a feedback loop, faults in the feedback loop can force a transistor into cutoff or saturation or force an op amp's output to either power supply rail. In a receiver, the AGC subsystem is such a feedback loop. Open the AGC line to the device and substitute a variable voltage for the AGC signal. If amplifier action varies with voltage, suspect the AGC-circuit components; otherwise, suspect the amplifier.

In an operating amplifier, check carefully for oscillations or noise. Oscillations are most likely to start with maximum gain and the amplifier input shorted. Any noise that is due to 60 Hz sources can be heard, or seen with an oscilloscope triggered by the ac line.

Unwanted amplifier RF oscillations should be cured with changes of lead dress or circuit components. Separate input leads from

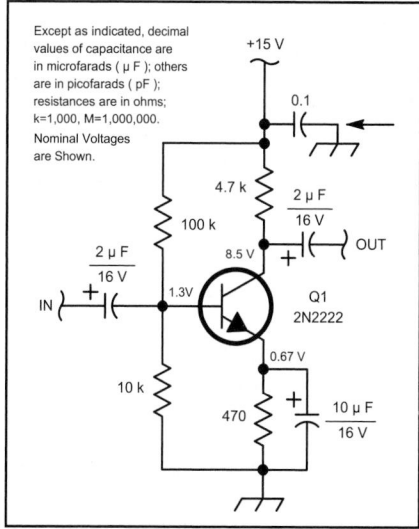

Fig 26.19 — The decoupling capacitor in this circuit is designated with an arrow.

output leads; use coaxial cable to carry RF between stages; neutralize inter-element or junction capacitance. Ferrite beads on the control element of the active device often stop unwanted oscillations.

Low-frequency oscillations (motorboating) indicate poor stage isolation or inadequate power supply filtering. Try a better lead-dress arrangement and/or check the capacitance of the decoupling network (see **Fig 26.19**). Use larger capacitors at the power supply leads; increase the number of capacitors or use separate decoupling capacitors at each stage. Coupling capacitors that are too low in value can also cause poor low-frequency response. Poor response to high frequencies is usually caused by circuit design.

COMMON-EMITTER AMPLIFIER

The common-emitter circuit (or common-source using an FET) is the most widely used configuration. It can be used as an amplifier or as a switch. Both are analyzed here as an example of how to troubleshoot transistor amplifier circuits. Other types of circuit can be analyzed similarly

Fig 26.20 is a schematic of a common-emitter transistor amplifier. The emitter, base and collector leads are labeled e, b and c, respectively. Important dc voltages are measured at the emitter (V_e), base (V_b) and collector (V_c) leads. V+ is the supply voltage.

First, analyze the voltages and signal levels in this circuit. The junction drop is the potential measured across a semiconductor junction that is conducting. It is typically 0.6 V for silicon and 0.2 V for germanium transistors.

This is a Class-A linear circuit. In Class-A circuits, the transistor is always conducting some current. R1 and R2 form a voltage divider that supplies dc bias (V_b) for the transistor. Normally, V_e is equal to V_b less the emitter-base junction drop. R4 provides degenerative dc bias, while C3 provides a low-impedance path for the signal. From this information, normal operating voltages can be estimated.

The bias and voltages will be set up so that the transistor collector voltage, V_c, is somewhere between V+ and ground potential. A good rule of thumb is that V_c should be about one-half of V+, although this can vary quite a bit, depending on component tolerances. The emitter voltage is usually a small percentage of V_c, say about 10%.

Any circuit failure that changes collector current, I_c, (ranging from a shorted transistor or a failure in the bias circuit) changes V_c and V_e as well. An increase of I_c lowers V_c and raises V_e. If the transistor shorts from collector to emitter, V_c drops to about 1.2 V, as determined by the voltage divider formed by R3 and R4.

You would see nearly the same effect if

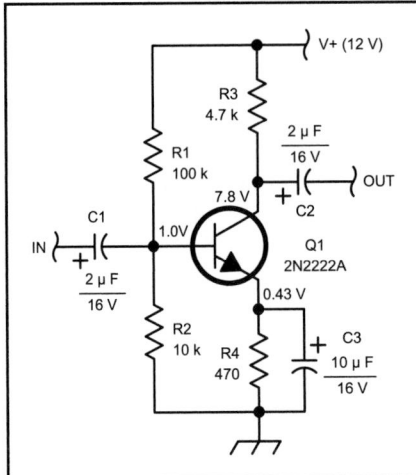

Fig 26.20 — A typical common-emitter audio amplifier.

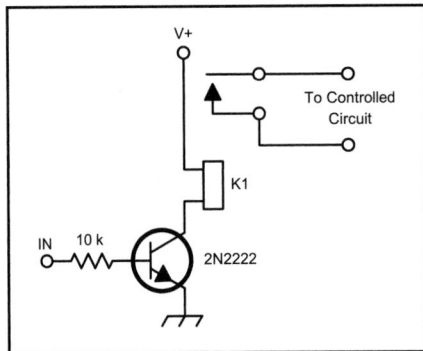

Fig 26.21 — A typical common-emitter switch or driver.

the transistor were biased into saturation by collector-to-base leakage, a reduction in R1's value or an increase in R2's value. All of these circuit failures have the same effect. In some cases, a short in C1 or C2 could cause the same symptoms.

To properly diagnose the specific cause of low V_c, consider and test all of these parts. It is even more complex; an increase in R3's value would also decrease V_c. There would be one valuable clue, however: if R3 increased in value, I_c would not increase; V_e would also be low.

Anything that decreases I_c increases V_c. If the transistor failed open, R1 increased in value, R2 were shorted to ground or R4 opened, then V_c would be high.

COMMON-EMITTER SWITCH

A common-emitter transistor switching circuit is shown in **Fig 26.21**. This circuit functions differently from the circuit shown in Fig 26.20. A linear amplifier is designed so that the output signal is a faithful reproduction of the input signal. Its input and output may have any value from V+ to ground.

The switching circuit of Fig 26.21, however, is similar to a digital circuit. The active device is either on or off, 1 or 0, just like digital logic. Its input signal level should either be 0 V or positive enough to switch the transistor on fully (saturate). Its output state should be either full off (with no current flowing through the relay), or full on (with the relay energized). A voltmeter placed on the collector will show either approximately +12 V or 0 V, depending on the input.

Understanding this difference in operation is crucial to troubleshooting the two circuits. If V_c were +12 V in the circuit in Fig 26.20, it would indicate a circuit failure. A V_c of +12 V in the switching circuit is normal when V_b is 0 V. (If V_b measured 0.8 V or higher, V_c should be low and the relay energized.)

DC-COUPLED AMPLIFIERS

In dc-coupled amplifiers, the transistors are directly connected together without coupling capacitors. They comprise a unique troubleshooting case. Most often, when one device fails, it destroys one or more other semiconductors in the circuit. If you don't find all of the bad parts, the remaining defective parts can cause the installed replacements to fail immediately. To reliably troubleshoot a dc coupled circuit, you must test every semiconductor in the circuit and replace them all at once.

26.8.3 Oscillators

In many circuits, a failure of the oscillator will result in complete circuit failure. A transmitter will not transmit, and a superheterodyne receiver will not receive if you have an internal oscillator failure. (These symptoms do not always mean oscillator failure, however.)

Whenever there is weakening or complete loss of signal from a radio, check oscillator operation and frequency. There are several methods:

• Use a receiver with a coaxial probe to listen for the oscillator signal.

• A dip meter can be used to check oscillators by tuning to within ±15 kHz of the oscillator, couple it to the circuit, and listen for a beat note in the dip-meter headphones.

• Look at the oscillator waveform on a scope. The operating frequency can't be determined with great accuracy, but you can see if the oscillator is working at all. Use a low capacitance (10×) probe for oscillator observations.

Many modern oscillators are phase-locked loops (PLLs). Read the **Oscillators and Synthesizers** chapter of this book in order to learn how PLLs operate.

To test for a failed LC oscillator, use a dip meter in the active mode. Set the dip meter to the oscillator frequency and couple it to the oscillator output circuit. If the oscillator is dead, the dip-meter signal will take its place and temporarily restore some semblance of normal operation.

STABILITY

Drift is caused by variations in the oscillator. Poor voltage regulation and heat are the most common culprits. Check regulation with a voltmeter (use one that is not affected by RF). Voltage regulators are usually part of the oscillator circuit. Check them by substitution.

Chirp is a form of rapid drift that is usually caused by excessive oscillator loading or poor power-supply regulation. The most common cause of chirp is poor design. If chirp appears suddenly in a working circuit, look for component or design defects in the oscillator or its buffer amplifiers. (For example, a shorted coupling capacitor increases loading drastically.) Also check for new feedback paths from changes in wiring or component placement (feedback defeats buffer action).

Frequency instability may also result from defects in feedback components. Too much feedback may produce spurious signals, while too little makes oscillator start-up unreliable.

Sudden frequency changes are frequently the result of physical variations. Loose components or connections are probable causes. Check for arcing or dirt on printed-circuit boards, trimmers and variable capacitors, loose switch contacts, bad solder joints or loose connectors.

FREQUENCY ACCURACY

In manually tuned LC oscillators, tracking at the high-frequency end of the range is

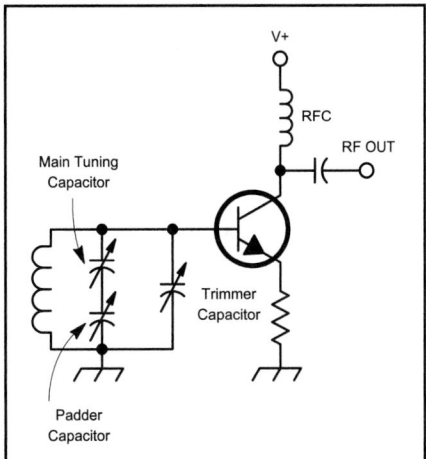

Fig 26.22 — A partial schematic of a simple oscillator showing the locations of the trimmer and padder capacitors.

controlled by trimmer capacitors. A trimmer is a variable capacitor connected in parallel with the main tuning capacitor (see **Fig 26.22**). The trimmer represents a higher percentage of the total capacitance at the high end of the tuning range. It has relatively little effect on tuning characteristics at the low-frequency end of the range.

Low-end range is adjusted by a series trimmer capacitor called a *padder*. A padder is a variable capacitor that is connected in series with the main tuning capacitor. Padder capacitance has a greater effect at the low-frequency end of the range. The padder capacitor is often eliminated to save money. In that case, the low-frequency range is set by adjusting the main tuning coil.

26.8.4 Transmit Amplifier Modules

Most VHF/UHF mobile radios and many small HF radios use commercial amplifier modules as the final amplifier instead of discrete transistors. These modules are quite reliable and can withstand various stresses such as disconnected antennas. However, replacement units are rarely available more than a few years after a particular radio model goes out of production. You may be able to find a damaged radio of the same model to scavenge for parts, but once the modules fail, you are usually out of luck. User's groups are often good sources of information about common failure modes of certain radios and possibly even sources of replacement parts.

The usual failure mode of an amplifier module is caused by thermal cycling that eventually leads to an internal connection developing a crack. The module becomes intermittent and then eventually fails completely. If you can open the module, you can sometimes identify and repair such a problem by soldering over the crack.

When reinstalling or replacing an amplifier module, be very careful to attach it to the heat sink as it was at the factory. Do not use excessive amounts of thermal compound or grease and make sure the mounting screws are secure. If you can find a datasheet for the module, check to see if there are recommendations for mounting screw torque and any other installation procedures. (See the **RF Power Amplifiers** chapter for guidelines on mounting power transistors.)

26.9 Radio Troubleshooting Hints

Tables 26.2, **26.3**, **26.4** and **26.5** list some common problems and possible cures for older radios that are likely to have developed problems with some later-model equivalents. These tables are not all-inclusive. They are a collection of hints and shortcuts that may save you some troubleshooting time. If you don't find your problem listed, continue with systematic troubleshooting.

Remember that many problems are caused by improper setting of a switch, a control, or a configuration menu item. Before beginning a troubleshooting session, be sure you've checked the operating manual for proper control settings and checked through the manual's troubleshooting guide. If possible, obtain a service manual with its detailed

procedures, measurements, and schematics.

26.9.1 Receivers

A receiver can be diagnosed using any of the methods described earlier, but if there is not even a faint sound from the speaker, signal injection is not a good technique. If you lack troubleshooting experience, avoid following instinctive hunches. Begin with power supply tests and proceed to signal tracing.

SELECTIVITY

Failure of control or switching circuits that determine the signal path and filters can cause selectivity problems. In older equipment, tuned transformers or the components

used in filter circuits may develop a shorted turn, capacitors can fail and alignment is required occasionally. Such defects are accompanied by a loss of sensitivity. Except in cases of catastrophic failure (where either the filter passes all signals, or none), it is difficult to spot a loss of selectivity. Bandwidth and insertion-loss measurements are necessary to judge filter performance.

SENSITIVITY

A gradual loss of sensitivity results from gradual degradation of an active device or long-term changes in component values. Sudden partial sensitivity changes are usually the result of a component failure, usually in the RF or IF stages. Excessive signal levels

Table 26.2
Symptoms and Their Causes for All Electronic Equipment

Symptom	Cause
Power Supplies	
No output voltage	Open circuit (usually a fuse, pass transistor, or transformer winding)
Hum or ripple	Faulty regulator, capacitor or rectifier, low-frequency oscillation
Amplifiers	
Low gain	Transistor, coupling capacitors, emitter-bypass capacitorAGC component
Noise	Transistors, coupling capacitors, resistors
Oscillations	Dirt on variable capacitor or chassis, shorted op-amp input
Oscillations, untuned (oscillations do not change with frequency)	Audio stages
Oscillations, tuned	RF, IF and mixer stages
Static-like crashes	Arcing trimmer capacitors, poor connections
Static in FM receiver	Faulty limiter stage, open capacitor in ratio detector, weak RF stage, weak incoming signal
Intermittent noise	All components and connections, band-switch contacts, potentiometers (especially in dc circuits), trimmer capacitors, poor antenna connections
Distortion (constant)	Oscillation, overload, faulty AGC, leaky transistor, open lead in tab-mount transistor, dirty potentiometer, leaky coupling capacitor, open bypass capacitors, imbalance in tuned FM detector, IF oscillations, RF feedback (cables)
Distortion (strong signals only)	Open AGC loop
Frequency change	Physical or electrical variations, dirty or faulty variable capacitor, broken switch, loose compartment parts, poor voltage regulation, oscillator tuning (trouble when switching bands)
No Signals	
All bands	Dead VFO or LO, PLL won't lock
One band only	Defective crystal, oscillator out of tune, band switch
No function control	Faulty switch or control, poor connection to front panel subassembly

Table 26.3
Receiver Problems

Symptom	Cause
Low sensitivity	Semiconductor degradation, circuit contamination, poor antenna connection
Signals and calibrator heard weakly	
(low S-meter readings)	RF chain
(strong S-meter readings)	AF chain, detector
No signals or calibrator heard, only hissing	RF oscillators
Distortion	
On strong signals only	AGC fault
AGC fault	Active device cut off or saturated
Difficult tuning	AGC fault
Inability to receive	Detector fault
AM weak and distorted	Poor detector, power or ground connection
CW/SSB unintelligible	BFO off frequency or dead
FM distorted	Open detector diode

or transients can damage input RF switching, amplifier, or mixing circuits. Complete and sudden loss of sensitivity is caused by an open circuit anywhere in the signal path or by a dead oscillator.

AGC

AGC failure usually causes distortion that affects only strong signals. All stages operate at maximum gain when the AGC influence is removed. An S meter can help diagnose AGC failure because it is operated by the AGC loop. If the S meter does not move at all or remains at full scale, the AGC system has a problem.

In DSP radios, the AGC function is often controlled by software which you cannot troubleshoot but inputs to the software such as signal level detectors may be causing a problem instead.

In analog receivers, an open bypass capacitor in the AGC amplifier causes feedback through the loop. This often results in a receiver squeal (oscillation). Changes in the loop time constant affect tuning. If stations consistently blast, or are too weak for a brief time when first tuned in, the time constant is too fast. An excessively slow time constant makes tuning difficult, and stations fade after tuning. If the AGC is functioning, but the timing seems wrong, check the large-value capacitors found in the AGC circuit — they usually set the AGC time constants. If the AGC is not functioning, check the AGC-detector circuit. There is often an AGC voltage that is used to control several stages. A failure in any one stage could affect the entire loop.

DETECTOR PROBLEMS

Detector trouble usually appears as complete loss or distortion of the received signal. AM, SSB and CW signals may be weak and unintelligible. FM signals will sound distorted. Look for an open circuit in the detector circuit. If tests of the detector parts indicate no trouble, look for a poor connection in the detector's power supply or ground connections. A BFO that is dead or off frequency prevents SSB and CW reception. In modern rigs, the BFO frequency is usually derived from the main VFO system.

26.9.2 Transmitters

Many potential transmitter faults are discussed in several different places in this chapter. There are, however, a few techniques used to ensure stable operation of RF amplifiers in transmitters that are not covered elsewhere.

RF final amplifiers often use parasitic chokes to prevent instability. Older parasitic chokes usually consist of a 51- to 100-Ω non-inductive resistor with a coil wound around the body and connected to the leads. It is

Table 26.4
Transmitter Problems

Symptom	Cause
Key clicks	Keying filter, distortion in stages after keying, ALC overshoot or instability

Modulation Problems

Loss of modulation	Broken cable (microphone, PTT, power), open circuit in audio chain, defective modulator
Distortion on transmit	Defective microphone, RF feedback, modulator imbalance, bypass capacitor, improper bias, excessive drive
Arcing	Dampness, dirt, improper lead dress
Low output	Incorrect control settings, improper carrier shift (CW signal outside of passband) audio oscillator failure, transistor or tube failure, SWR protection circuit

Antenna Problems

Poor SWR	Damaged antenna element, matching network, feed line, balun failure (see below), resonant conductor near antenna, poor connection at antenna
Balun failure	Excessive SWR, weather or cold-flow damage in coil choke, broken wire or connection
RFI	Arcing or poor connections anywhere in antenna system or nearby conductors

Table 26.5
Transceiver Problems

Symptom	Cause
Inoperative S meter	Faulty TR switching or relay
PA noise in receiver	Faulty TR switching or relay
Excessive current on receive	Faulty TR switching or relay
Arcing in PA	Faulty TR switching or relay
Reduced signal strength on transmit and receive	IF failure
Poor VOX operation	VOX amplifiers
Poor VOX timing	Adjustment, component failure in VOX timing circuits or amplifiers
VOX consistently tripped by receiver audio	AntiVOX circuits or adjustment

used to prevent VHF and UHF oscillations in a vacuum-tube amplifier. The suppressor is placed in the plate lead, close to the plate connection.

In recent years, problems with this style of suppressor have been discovered. See the **RF Power Amplifiers** chapter for information about suppressing parasitics. If parasitic suppressors are present in your transmitter, continue to use them as the exact layout and lead dress of the RF amplifier circuitry may require them to avoid oscillation. If they are not present, do not add them. When working on RF power amplifiers, take care to keep leads and components arranged just as they were when they left the factory.

Parasitic chokes often fail from excessive current flow. In these cases, the resistor is charred. Occasionally, physical shock or corrosion produces an open circuit in the coil. Test for continuity with an ohmmeter.

Transistor amplifiers are protected against parasitic oscillations by low-value resistors or ferrite beads in the base or collector leads. Resistors are used only at low power levels (about 0.5 W), and both methods work best when applied to the base lead. Negative feedback is used to prevent oscillations at lower frequencies. An open component in the feedback loop may cause low-frequency oscillation, especially in broadband amplifiers.

KEYING

The simplest form of modulation is on/off keying. Although it may seem that there cannot be much trouble with such an elementary form of modulation, two very important transmitter faults are the result of keying problems.

Key clicks are produced by fast rise and times of the keying waveform (see the previous sidebar, "Controlling Key Clicks"). Most transmitters include components in the keying circuitry to prevent clicks. When clicks are experienced, check the keying filter components first, then the succeeding stages. An improperly biased power amplifier, or a Class C amplifier that is not keyed, may produce key clicks even though the keying waveform earlier in the circuit is correct. Clicks caused by a linear amplifier may be a sign of low-frequency parasitic oscillations. If they occur in an amplifier, suspect insufficient power-supply decoupling. Check the power-supply filter capacitors and all bypass capacitors.

The other modulation problem associated with on/off keying is called backwave. Backwave is a condition in which the signal is heard, at a reduced level, even when the key is up. This occurs when the oscillator signal feeds through a keyed amplifier. This usually indicates a design flaw, although in some cases a component failure or improper keyed-stage neutralization may be to blame.

LOW OUTPUT POWER

Check the owner's manual to see if the condition is normal for some modes or bands, or if there is a menu item to set RF output power. Check the control settings. Solid-state transmitters require so little effort from the operator that control settings are seldom noticed. The CARRIER (or DRIVE) control may have been bumped. Remember to adjust tuned vacuum tube amplifiers after a significant change in operating frequency (usually 50 to 100 kHz). Most modern transmitters are also designed to reduce power if there is high (say 2:1) SWR. Check these obvious external problems before you tear apart your rig.

Power transistors may fail if the SWR protection circuit malfunctions. Such failures occur at the weak link in the amplifier chain: It is possible for the drivers to fail without damaging the finals. An open circuit in the reflected side of the sensing circuit leaves the transistors unprotected; a short shuts them down.

Low power output in a transmitter may also spring from a misadjusted carrier oscillator or a defective SWR protection circuit. If the carrier oscillator is set to a frequency well outside the transmitter passband, there may be no measurable output. Output power will increase steadily as the frequency is moved into the passband.

26.9.3 Transceivers

SWITCHING

Elaborate switching schemes are used in transceivers for signal control. Many transceiver malfunctions can be attributed to relay or switching problems. Suspect the switching controls when:

• The S meter is inoperative, but the unit otherwise functions. (This could also be a bad S meter or a consequence of a configuration menu item.)

• There is arcing in the tank circuit. (This could also be caused by a fault in the antenna system.)

• There is excessive broadband PA noise in the receiver.

Since transceiver circuits are shared, stage defects frequently affect both the transmit and receive modes, although the symptoms may change with mode. Oscillator problems usually affect both transmit and receive modes, but different oscillators, or frequencies, may be used for different emissions. Check the block diagram.

VOX

Voice operated transmit (VOX) controls are another potential trouble area. If there is difficulty in switching to transmit in the VOX mode, check the VOX-SENSITIVITY and ANTI-VOX control settings. Next, see if the PTT and manual (MOX) transmitter controls work. If the PTT and MOX controls function, examine the VOX control circuits. Test the switches, control lines and control voltage if the transmitter does not respond to other TR controls.

VOX SENSITIVITY and ANTI-VOX settings should also be checked if the transmitter switches on in response to received audio. Suspect the ANTI-VOX circuitry next. Unacceptable VOX timing results from a poor VOX-delay adjustment, or a bad resistor or capacitor in the timing circuit or VOX amplifiers.

26.9.4 Amplifiers

While this section focuses on vacuum-tube amplifiers using high-voltage (HV) supplies, it also applies to solid-state amplifiers that operate at lower voltages and generally have fewer points of failure. Amplifiers are simple, reliable pieces of equipment that respond well to basic care, regular maintenance and common sense. A well-maintained amplifier will provide reliable service and maximum tube lifetime. (The complete version of the original article is included on this book's CD-ROM.)

SAFETY FIRST

It is important to review good safety practices. (See the **Safety** and **Power Sources**

chapters for additional safety information.) Tube amplifiers use power supply voltages well in excess of 1 kV and the RF output can be hundreds of volts, as well. Almost every voltage in an amplifier can be lethal! Take care of yourself and use caution!

Power Control — Know and control the state of both ac line voltage and dc power supplies. Physically disconnect line cords and other power cables when you are not working on live equipment. Use a lockout on circuit breakers. Double-check visually and with a meter to be absolutely sure power has been removed.

Interlocks — Unless specifically instructed by the manufacturer's procedures to do so, never bypass an interlock. This is rarely required except in troubleshooting and should only be done when absolutely necessary. Interlocks are there to protect you.

The One-Hand Rule — Keep one hand in your pocket while making any measurements on live equipment. The hand in your pocket removes a path for current to flow through you. It's also a good idea to wear shoes with insulating soles and work on dry surfaces. Current can be lethal even at levels of a few mA — don't tempt the laws of physics.

Test Equipment Rating — Be sure your test equipment is adequately rated for the voltages and power levels encountered in amplifiers! This is particularly important in handheld equipment in which there is no metal enclosure connected to an ac safety ground. Excessive voltage can result in a flashover to the user from the internal electronics, probes, or test leads, resulting in electric shock. Know and respect this rating.

If you are using an external high voltage probe, make sure it is in good condition with no cracks in the body. The test lead insulation should be in good condition — flexible and with no cracks or wire exposed. If practical, do not make measurements while holding the probe or meter. Attach the probe with the voltage discharged and then turn the power on. Turn power off and discharge the voltage before touching the probe again. Treat high voltage with care and respect!

Patience — Repairing an amplifier isn't a race. Take your time. Don't work on equipment when you're tired or frustrated. Wait several minutes after turning the amplifier off to open the cabinet — capacitors can take several minutes to discharge through their bleeder resistors.

A Grounding Stick — Make the simple safety accessory shown in the High Voltage section of the **Power Sources** chapter and use it whenever you work on equipment in which hazardous voltages have been present. The ground wire should be heavy duty (#12 AWG or larger) due to the high peak currents (hundreds of amperes) present when

discharging a capacitor or tripping a circuit breaker. When equipment is opened, touch the tip of the stick to every exposed component and connection that you might come in contact with. Assume nothing — accidental shorts and component failures can put voltage in places it shouldn't be.

The Buddy System and CPR — Use the buddy system when working around any equipment that has the potential for causing serious injury. The buddy needn't be a ham, just anyone who will be nearby in case of trouble. Your buddy should know how to remove power and administer basic first aid or CPR.

CLEANLINESS

The first rule in taking good care of an amplifier is cleanliness. Amplifiers need not be kept sparkling new, but their worst enemy is heat. Excess heat accelerates component aging and increases stress during operation.

Outside the amplifier, prevent dust and obstructions from blocking the paths by which heat is removed. This means keeping all ventilation holes free of dust, pet hair and insects. Fan intakes are particularly susceptible to inhaling all sorts of debris. Use a vacuum cleaner to clean the amplifier and surrounding areas. Keep liquids well away from the amplifier.

Keep papers or magazines off the amplifier — even if the cover is solid metal. Paper acts as an insulator and keeps heat from being radiated through the cover. Amplifier heat sinks must have free air circulation to be effective. There should be at least a couple of inches of free space surrounding an amplifier on its sides and top. If the manufacturer recommends a certain clearance, mounting orientation or air flow, follow those recommendations.

Inside the amplifier, HV circuits attract dust that slows heat dissipation and will eventually build up to where it arcs or carbonizes. Use the vacuum cleaner to remove any dust or dirt. If you find insects (or worse) inside the amp, try to determine how they got in and plug that hole. Window screening works fine to allow airflow while keeping out insects. While you're cleaning the inside, perform a visual inspection as described in the next section.

Vacuuming works best with an attachment commonly known as a crevice cleaner. **Fig 26.23** shows a crevice cleaning attachment being used with a small paintbrush to dislodge and remove dust. The brush will root dust out of tight places and off components without damaging them or pulling on connecting wires. Don't use the vacuum cleaner brush attachment; they're designed for floors, not electronics. Some vacuums also have a blower mechanism, but these rarely

Fig 26.23 — A small paintbrush and a vacuum cleaner crevice attachment make dust removal easy.

have enough punch to clean as thoroughly as a brush. Blowing dust just pushes the dust around and into other equipment.

If you can't get a brush or attachment close enough, a spray can of compressed air will usually dislodge dust and dirt so you can vacuum it up. If you use a rag or towel to wipe down panels or large components, be sure not to leave threads or lint behind. Never use a solvent or spray cleaner to wash down components or flush out crevices unless the manufacturer advises doing so — it might leave behind a residue or damage the component.

VISUAL INSPECTION

Remove any internal covers or access panels and...stop! Get out the chicken stick (grounding stick), clip its ground lead securely to the chassis, and touch every exposed connection. Now, using a strong light and possibly a magnifier, look over the components and connections.

Amplifiers have far fewer components than transceivers, so look at every component and insulator. Look for cracks, signs of arcing, carbon traces (thin black lines), discoloration, loose connections, melting of plastic, and anything else that doesn't look right. This is a great time to be sure that mounting and grounding screws are tight. Does anything smell burnt? Learn the smells of overheated

components. Make a note of what you find, repair or replace — even if no action is required.

ELECTRICAL COMPONENTS

Let's start with the power supply. There are three basic parts to amplifier power supplies — the ac transformer and line devices, the rectifier/filter, and the metering/regulation circuitry. (See the **Power Sources** chapter for more information.) Transformers need little maintenance except to be kept cool and be mounted securely. Line components such as switches, circuit breakers and fuses, if mechanically sound and adequately rated, are usually electrically okay, as well.

Rectifiers and HV filter capacitors require occasional cleaning. Look for discoloration around components mounted on a printed circuit board (PCB) and make sure that all wire connections are secure. HV capacitors are generally electrolytic or oil and should show no signs of leakage, swelling or outgassing around terminals.

Components that perform metering and regulation of voltage and current can be affected by heat or heavy dust. If there has been a failure of some other component in the amplifier — such as a tube — these circuits can be stressed severely. Resistors may survive substantial temporary overloads, but may show signs of overload such as discoloration or swelling.

Amplifiers contain two types of relays — control and RF. Control relays switch ac and dc voltages and do not handle input or output RF energy. The usual problem encountered with control relays is oxidation or pitting of their contacts. A burnishing tool can be used to clean relay contacts. In a pinch a strip of ordinary paper can be pulled between contacts gently held closed. Avoid the temptation to over-clean silver-plated relay and switch contacts. It is easy to remove contact plating with excessive polishing and while silver-plated relay and switch contacts may appear to be dark in color, oxidized silver (black) is still a good conductor. Once the silver's gone, it's gone; contact erosion will then be pervasive. If visual inspection shows heavy pitting or discoloration or resistance measurements show the relay to have intermittent contact quality, it should be replaced.

RF relays are used to perform transmit-receive (TR) switching and routing of RF signals through or around the amplifier circuitry. Amplifiers designed for full break-in operation will usually use a high-speed vacuum TR relay. Vacuum relays are sealed and cannot be cleaned or maintained. When you replace RF relays, use a direct replacement part or one rated for RF service with the same characteristics as the original.

Cables and connectors are subjected to heavy heat and electrical loads in amplifiers. Plastics may become brittle and connections may oxidize. Cables should remain flexible and not be crimped or pinched if clamped or tied down. Gently wiggle cables while watching the connections at each end for looseness or bending. Connectors can be unplugged and reseated once or twice to clear oxide on contact surfaces.

Carefully inspect any connector that seems loose. Be especially careful with connectors and cables in amplifiers with power supplies in separate enclosures from the RF deck. Those interconnects are susceptible to both mechanical and electrical stress and you don't want an energized HV cable loose on the operating desk. Check the electrical integrity of those cables and make sure they are tightly fastened.

As with relays, switches found in amplifiers either perform control functions or route RF signals. Adequately rated control switches, if mechanically sound, are usually okay. Band switches are the most common RF switch — usually a rotary phenolic or ceramic type. A close visual inspection should show no pitting or oxidation on the wiper (the part of the switch that rotates between contacts) or the individual contacts. Arcing or overheating will quickly destroy rotary switches. **Fig 26.24** is a photo of a heavy-duty band switch that has suffered severe damage from arcing. Slight oxidation is acceptable

Fig 26.24 — The band switch section on the left clearly shows the signs of destructive arcing.

on silver-plated switches. Phosphor-bronze contacts can sometimes be cleaned with a light scrub from a pencil eraser, but plating can be easily removed, so use caution with this method and be sure to remove any eraser crumbs. Rotary switch contacts cannot be replaced easily although individual wafer sections may be replaced if an exact matching part can be obtained.

When replacing capacitors and resistors, be sure to use an adequately rated part. Voltage and power-handling ratings are particularly important, especially for components handling high RF currents. An RF tank capacitor replacement should be checked carefully for adequate RF voltage and current ratings, not just dc. HV resistors are generally long and thin to prevent arcing across their surfaces. Even if a smaller (and cheaper) resistor has an equivalent power rating, resist the temptation to substitute it. In a pinch, a series string of resistors of the appropriate combined value can be used to replace one HV unit. Don't use carbon resistors for metering circuits, use metal or carbon film types. The carbon composition types are too unstable.

If you are repairing or maintaining an old amplifier and manufacturer-specific parts are no longer available, the ham community has many sources for RF and HV components. Fair Radio Sales (**www.fairradio. com**) and Surplus Sales of Nebraska (**www. surplussales.com**) are familiar names. Hamfests and web sites often have amplifier components for sale. (See the **RF Power Amplifiers** chapter's sidebar on using surplus

or used parts for amplifiers.) You might consider buying a non-working amplifier of the same model for parts.

TUBES

Good maintenance of tubes starts with proper operation of the amplifier. Follow the manufacturer's instructions for input drive levels, duty cycles, tuning and output power level. Frequently check all metered voltages and current to be sure that the tubes are being operated properly and giving you maximum lifetime. Penta Labs' "Tube Maintenance & Education" (**www.pentalaboratories.com/ maintenance.asp**) is an excellent web page on maintaining power tubes.

The internal mechanical structures of tubes generally do not deal well with mechanical shock and vibration, so treat them gently. The manufacturer may also specify how the amplifier is to be mounted, so read the operating manual. Tubes generate a lot of heat, so it's important that whatever cooling mechanism employed is kept at peak efficiency. Airways should be clean, including between the fins on metal tubes. All seals and chimneys should fit securely and be kept clean. Wipe the envelope of glass tubes clean after handling them — fingerprints should be removed to prevent baking them into the surface. On metal tubes that use finger-stock contacts, be sure the contacts are clean and make good contact all the way around the tube. Partial contact or dirty finger stock can cause asymmetric current and heating inside the tube, resulting in warping of internal grids and possibly

cause harmonics or parasitics.

Plate cap connections and VHF parasitic suppressors should be secure and show no signs of heating. Overheated parasitic suppressors may indicate that the neutralization circuit is not adjusted properly. Inspect socket contacts and the tube pins to be sure all connections are secure, particularly high-current filament connections. Removing and inserting the tubes once or twice will clean the socket contacts.

Adjustments to the neutralizing network, which suppresses VHF oscillations by negative feedback from the plate to grid circuit, are rarely required except when you are replacing a tube or after you do major rewiring or repair of the RF components. The manufacturer will provide instructions on making these adjustments. If symptoms of VHF oscillations occur without changing a tube, then perhaps the tube characteristics or associated components have changed. Parasitic oscillations in high-power amplifiers can be strong enough to cause arcing damage. Perform a visual inspection prior to readjusting the neutralizing circuit.

Metering circuits rarely fail, but they play a key part in maintenance. By keeping a record of normal voltages and currents, you will have a valuable set of clues when things go wrong. Record tuning settings, drive levels, and tube voltages and currents on each band and with every antenna. When settings change, you can refer to the notebook instead of relying on memory.

MECHANICAL

Thermal cycling and heat-related stresses can result in mechanical connections loosening over time or material failures. Switch shafts, shaft couplings and panel bearings all need to be checked for tightness and proper alignment. All mounting hardware needs to be tight, particularly if it supplies a grounding path. Examine all panel-mounted components, particularly RF connectors, and be sure they're attached securely. BNC and UHF connectors mounted with a single nut in a round panel hole are notorious for loosening with repeated connect/disconnect cycles.

Rubber and plastic parts are particularly stressed by heat. If there are any belts, gears or pulleys, make sure they're clean and that dust and lint are kept out of their lubricant. Loose or slipping belts should be replaced. Check O-rings, grommets and sleeves to be sure they are not brittle or cracked. If insulation sleeves or sheets are used, check to be sure they are covering what they're supposed to. Never discard them or replace them with improperly sized or rated materials.

Enclosures and internal shields should all be fastened securely with every required screw in place. Watch out for loosely overlapping metal covers. If a sheet metal screw has stripped out, either drill a new hole or replace the screw with a larger size, taking care to maintain adequate clearance around and behind the new screw. Tip the amplifier from side to side while listening for loose hardware or metal fragments, all of which should be removed.

Clean the front and back panels to protect the finish. If the amplifier cabinet is missing a foot or an internal shock mount, replace it. A clean unit with a complete cabinet will have a significantly higher resale value, so it's in your interest to keep the equipment looking good.

SHIPPING

When you are traveling with an amplifier or shipping it, some care in packing will prevent damage. Improper packing can also result in difficulty in collecting on an insurance claim, should damage occur. The original shipping cartons are a good method of protecting the amplifier for storage and sale, but they were not made to hold up to frequent shipping. If you travel frequently, it is best to get a sturdy shipping case made for electronic equipment. Pelican (**www.pelican-shipping-cases.com**) and Anvil (**www.anvilsite.com**) make excellent shipping cases suitable for carrying amplifiers and radio equipment.

Some amplifiers require the power transformer to be removed before shipping. Check your owner's manual or contact the manufacturer to find out. Failure to remove it before shipping can cause major structural damage to the amplifier's chassis and case.

Tubes should also be removed from their sockets for shipment. It may not be necessary to ship them separately if they can be packed in the amplifier's enclosure with adequate plastic foam packing material. If the manufacturer of the tube or amplifier recommends separate shipment, however, do it!

CLEANING AND MAINTENANCE PLAN

For amateur use, there is little need for maintenance more frequently than once per year. Consider the maintenance requirements of the amplifier and what its manufacturer recommends. Review the amplifier's manuals and make up a checklist of what major steps and tools are required.

TROUBLESHOOTING

A benefit of regular maintenance will be familiarity with your amplifier should you ever need to repair it. Knowing what it looks (and smells) like inside will give you a head start on effecting a quick repair.

The following discussion is intended to illustrate the general flow of a troubleshooting effort, not be a step-by-step guide. Before starting on your own amplifier, review the

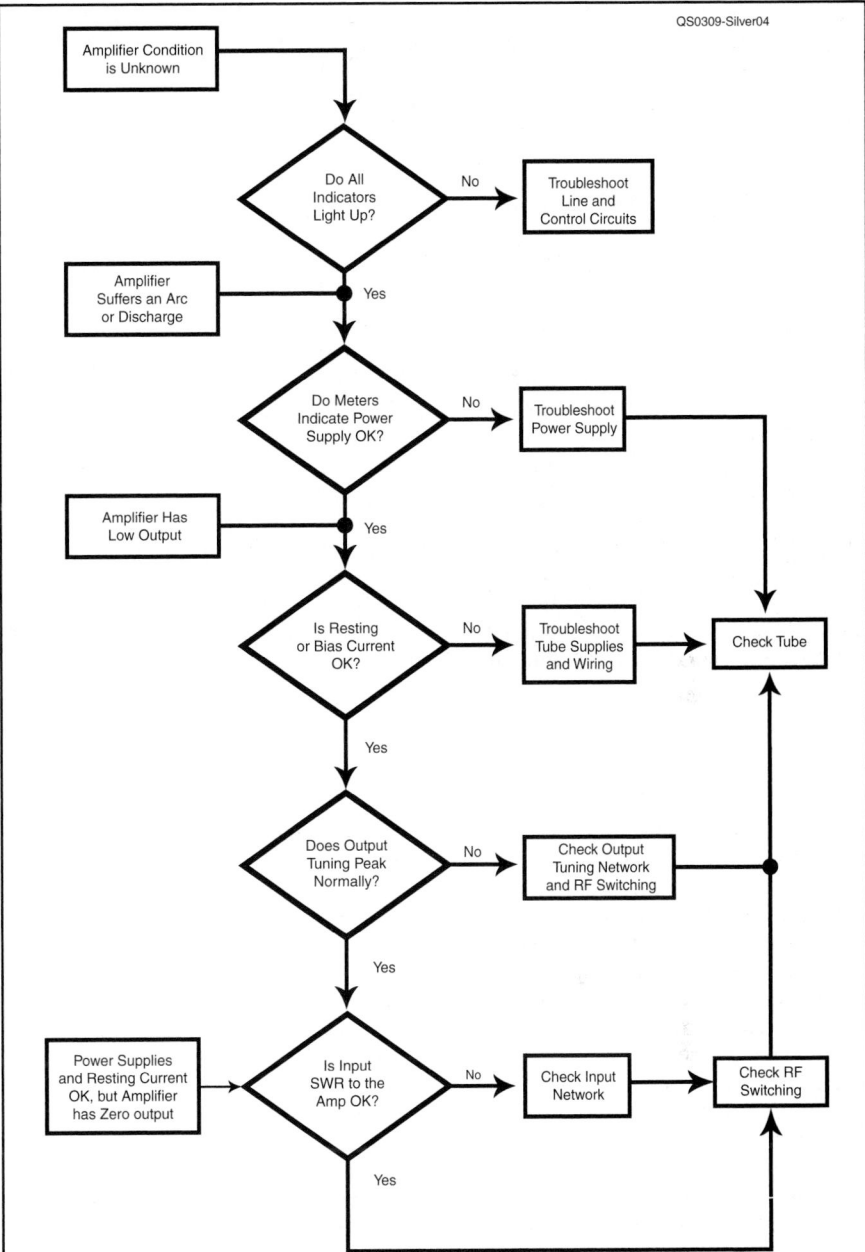

Fig 26.25 — This general-purpose flow chart will help identify amplifier problems. For solid-state units, substitute "Check Output Transistors" for "Check Tube."

amplifier manual's "Theory of Operation" section and familiarize yourself with the schematic. If there is a troubleshooting procedure in the manual, follow it. **Fig 26.25** shows a general-purpose troubleshooting flow chart. Do not swap in a known-good tube or tubes until you are sure that a tube is actually defective. Installing a good tube in an amplifier with circuit problems can damage a good tube.

Many "amplifier is dead" problems turn out to be simply a lack of ac power. Before even opening the cabinet of an unresponsive amplifier, be sure that ac is really present at the wall socket and that the fuse or circuit breaker is really closed. If ac power is present at the

wall socket, trace through any internal fuses, interlocks and relays all the way through to the transformer's primary terminals. If the amplifier operates from 220/240-V circuits, remember that the white wire in such wiring is hot!

Hard failures in an HV power supply are rarely subtle, so it's usually clear if there is a problem. When you repair a power supply, take the opportunity to check all related components. If all defective components are not replaced, the failures may be repeated when the circuit is re-energized.

Rectifiers may fail open or shorted — test them using a DMM diode checker. An open rectifier will result in a drop in the HV output of 50 percent or more but will probably not overheat or destroy itself. A shorted rectifier failure is usually more dramatic and may cause additional rectifiers or filter capacitors to fail. If one rectifier in a string has failed, it may be a good idea to replace the entire string as the remaining rectifiers have been subjected to a higher-than-normal voltage.

HV filter capacitors usually fail shorted, although they will occasionally lose capacitance and show a rise in ESR (equivalent series resistance). Check the rectifiers and any metering components — they may have been damaged by the current surge caused from a short circuit.

Power transformer failures are usually due to arcing in the windings, insulation failures, or overheating. HV transformers can be disassembled and rewound by a custom transformer manufacturer.

Along with the HV plate supply, tetrode screen supplies occasionally fail, too. The usual cause is the regulation circuit that drops the voltage from the plate level. Operating without a screen supply can be damaging to a tube, so be sure to check the tube carefully after repairs.

If the power supply checks out okay and the tube's filaments are lit, check the resting or bias current. If it is excessive or very low, check all bias voltages and dc current paths to the tube, such as the plate choke, screen supply (for tetrodes) and grid or cathode circuits.

If you do not find power supply and dc problems, check the RF components or RF deck. Check the input SWR to the amplifier. If it has changed then you likely have a problem in the input circuitry or one or more tubes have failed. Perform a visual check of the input circuitry and the band switch, followed by an ohmmeter check of all input components.

If input SWR is normal and applying drive does not result in any change in plate current, you may have a defective tube, tube socket or connection between the input circuits and the tube. Check the TR control circuits and relay. If plate current changes, but not as much as normal, try adjusting the output tuning circuitry. If this has little or no effect, the tube may be defective or a connection between the tube and output circuitry may have opened. If retuning has an effect, but at different settings than usual, the tube may be defective or there may be a problem in the tuning circuitry. A visual inspection and an ohmmeter check are in order.

The key to finding the trouble with your amplifier is to be careful and methodical, and to avoid jumping to false conclusions or making random tests. The manufacturer's customer service department will likely be helpful if you are considerate and have taken careful notes detailing the trouble symptoms and any differences from normal operation. There may be helpful guidelines on the manufacturer's web pages or from other Internet resources. Sometimes there is more than one problem — they work together to act like one very strange puzzle. Just remember that most problems can be isolated by careful, step-by-step tests.

26.10 Antenna Systems

This section is an abbreviated version of the Antenna System Troubleshooting chapter of the *ARRL Antenna Book* that was added to the 22nd edition. Because of the enormous variety of antenna systems, general guidelines must be presented, but the successful troubleshooting process usually follows a systematic approach just as for any other radio system.

26.10.1 Basic Antenna System Troubleshooting

Start with an inventory of the antenna system. Any of these can be the cause of your problem: supports, insulators, elements, feed point, balun (if any), feed line, grounding or transient protectors, impedance matching and switching equipment, RF jumper cables at any point. As with any troubleshooting process, be alert for mistaken or loose connections, loose or disconnected power and control cables, wires touching each other that shouldn't be, and so forth. Reduce the antenna system to the simplest system with the problem and it will likely look something like the system in **Fig 26.26**.

It is particularly important to remember that your station ground is often part of the antenna system. The length of the connection between the equipment ground bus and the ground rod is usually several feet at minimum and that can be an appreciable fraction of a wavelength on the higher HF bands. This can greatly affect tuning if there is common-mode current on the feed line or if a random-wire or end-fed type of antenna is being used. If touching equipment enclosures or the ground wire affects SWR or impedance readings, that will affect your antenna measurements as well.

DUMMY-LOAD TESTING

Begin by replacing components of your antenna system with a dummy load, starting at the output of the radio using a known-good jumper cable. Verify that the radio works properly into a 50 Ω load using a known-good directional wattmeter. Then move the dummy load to the output side of any antenna tuning or switching equipment, one component at a time until you have replaced the antenna feed line with the dummy load. If everything checks OK to this point, the problem is in the feed line or antenna. Don't forget to verify that the problem with the antenna is still present after each dummy load test. If the problem was a loose or intermittent connection, it is likely that swapping the dummy load in and out changed or eliminated the problem.

ANTENNA VISUAL INSPECTION

Now it is time to perform a visual inspection of the feed line and antenna itself. Start with the feed line connector at the last point where the dummy load was swapped in. Disassemble the connector and inspect it for damage from water or corrosion. If either are present, replace the connector and check the condition of the cable before proceeding. Note that if water can get into the cable at the antenna's elevated feed point, it is not unknown for it to flow downward through the braid both by gravity and by capillary action all the way to the shack! If the cable braid is wet at both ends, the cable must be replaced.

If you have a wire antenna, lower it and make a visual inspection of all the pieces.

• Cut away the waterproofing around the coax termination and inspect for water and/or corrosion.

• On the insulators at each end, there is no possibility of contact between the antenna wire and the supporting wire/ropes.

• If there are any splices in the wire elements, they are well crimped or soldered.

• At the center insulator, there is no possibility of contact between the element wires.

• At the balun or coax connection the element connections are soldered or firmly connected.

If you have a Yagi or vertical antenna, similar steps are required. Carefully check any feed point matching assembly, such as a gamma match, hairpin, or stub and make sure connections are clean and tight.

ANTENNA TEST

Assuming any mechanical problems have been rectified, proceed to retest the antenna and feed line. Replace the antenna with a dummy load and check the feed line loss with a wattmeter or antenna analyzer at the antenna end of the feed line. If the feed line checks out OK, the problem must be in the antenna. Reattach the feed line to the antenna and verify that the problem remains. Note that for wire antennas lowered to near ground level, the resonant point will change — this

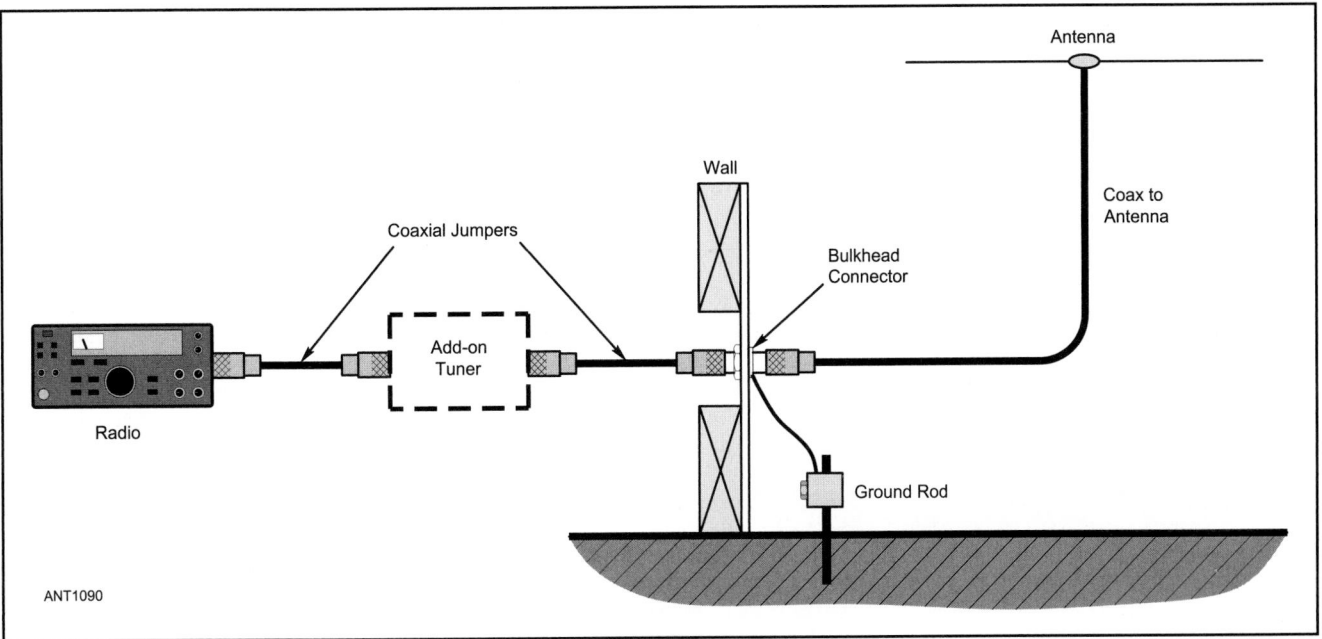

Fig 26.26 — A typical simple antenna system. An add-on antenna tuner is likely to be used if one is not built-in to the radio. More complex antenna systems have all of the same components plus some switching equipment.

is to be expected.

If the problem is still present, repeat the visual inspection at a closer level of detail. Check all dimensions and connections. Double-check any telescoping sections of tubing, transmission line stubs, in-line coax connectors, clamped connections between wires and between wires and tubing. If possible, give joints and connections a good shake while watching for intermittent readings on the wattmeter or antenna analyzer. If you cannot see inside a component or assembly, perform resistance tests for continuity. Remember to identify your signals since you are testing on the air at this point in the process.

The next step is to reinstall or raise the antenna at least ¼-wavelength off the ground and verify that the problem remains. If you have repaired the antenna, perform a re-check at this point to be sure everything is in good working order before returning the antenna to full height. Once you have re-installed the antenna, including full weatherproofing of any coaxial cable terminations or connectors, record in the shack notebook your measurements of the antenna along with what the problem was discovered to be and how you repaired it.

26.10.2 General Antenna System Troubleshooting

Think of the following topics as a kind of toolbox for troubleshooting. Many of them assume you are testing some type of Yagi or other beam antenna, but the general

guidelines apply to all types of antennas

It is important to remember this simple rule for adjustments and troubleshooting: Do the simplest and easiest adjustment or correction *first*, and only *one* at a time.

When making on-air comparisons, select signals that are at the margin and not pushing your receiver well over S9 where it can be difficult to measure differences of a few dB. Terrain has a lot to do with performance as well. If you are comparing with a large station, keep in mind that its location was probably selected carefully and the antennas were placed exactly where they should be for optimum performance on the property.

TEST MEASUREMENTS

A) Test the antenna at a minimum height of 15 to 20 feet. (See **Fig 26.27**.) This will move the antenna far enough away from the ground (which acts to add capacitance to the antenna)

and enable meaningful measurements. Use sawhorses *only* for construction purposes.

● 15 to 20 feet above ground does not mean 5 feet above a 10 to 15 foot high roof, it means above ground with nothing in between;

● Antenna resonant frequency will shift upward as it is raised;

● Feed point impedance will change with a change in height and this applies to both horizontal and vertical antennas;

● Some antennas are more sensitive to proximity to ground than others;

● Some antennas are more sensitive to nearby conductive objects (such as other antennas) than others.

B) Aiming the antenna upward with the reflector on the ground might coincide with some measurements on rare occasions, but there are no guarantees with this method. The reflector is literally touching a large capacitor (earth) and the driver element is very close,

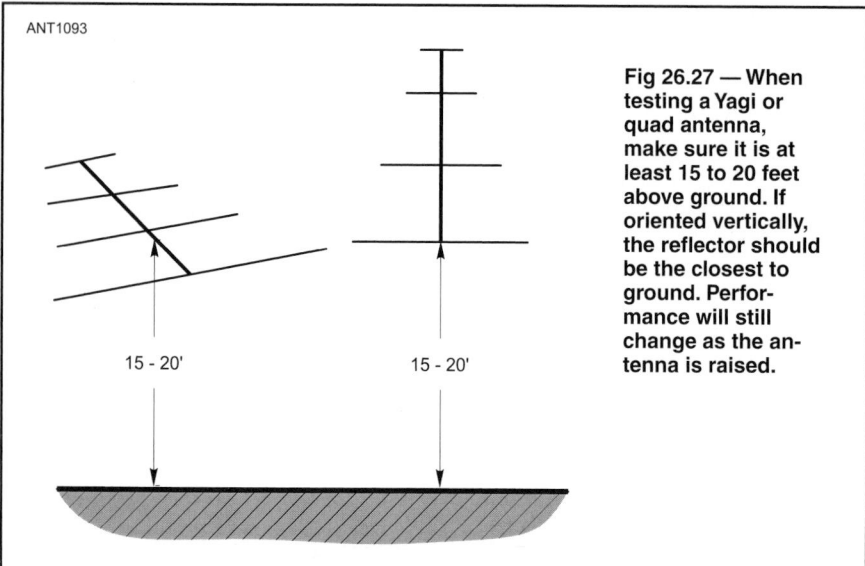

ANT1093

Fig 26.27 — When testing a Yagi or quad antenna, make sure it is at least 15 to 20 feet above ground. If oriented vertically, the reflector should be the closest to ground. Performance will still change as the antenna is raised.

15 - 20'   15 - 20'

too. Raise the antenna at least 15 to 20 feet off the ground.

C) When using a hand-held SWR analyzer you are looking for the dip in SWR, not where the impedance or resistance meter indicates 50 Ω. (Dip = frequency of lowest SWR value, or lowest swing on the meter.) On the MFJ-259/269 series, the left-hand meter (SWR) is the one you want to watch, not the right-hand meter (Impedance).

D) Does the SWR and frequency of lowest dip change when the coax length is changed? If so, the balun might be faulty, as in not isolating the load from the coax feed line.

Additionally, with an added length of coax and its associated small (hopefully small) amount of loss :

• The value of SWR is expected to be lower with the additional coax and,

• The width of the SWR curve is expected to be wider with the additional coax , when measured at the transmitter end of the coax.

E) Be sure you are watching for the right dip, as some antennas can have a secondary resonance (another dip). It is quite possible to see a Yagi reflector's resonant frequency, or some other dip caused by interaction with adjacent antennas.

MECHANICAL

A. Are the dimensions correct? Production units should match the documentation (within reason). When using tubing elements, measure each *exposed* element section during assembly and the element *half-length* (the total length of each half of the element) after assembly. Measuring the entire length is sometimes tricky, depending on the center attachment to the boom on Yagis, as the element can bow, or the tape might not lay flat along the tubing sections. Self-designed units might have a taper error.

B. Making the average taper diameter larger will make the equivalent electrical element longer. This makes the antenna act as if the physical element is too long.

C. Making the average taper diameter smaller will make the equivalent electrical element shorter. This makes the antenna act as if the physical element is too short.

D. If the element is a mono-taper (tubing element is the same for the entire length), larger diameter elements will be physically shorter than smaller diameter mono-taper elements to give the same electrical performance at the same frequency.

E. The type of mounting of the element to the boom affects the element length, whether it is attached directly to the boom, or insulated from the boom. Incorrect mounting/mounting plate allocation will upset the antenna tuning:

• A mounting plate 4 × 8 inches has an equivalent diameter of approximately 2.5 inches and 4 inches in length for each element half.

• A mounting plate that is 3 × 6 inches has an equivalent diameter of about 1.8 inches and a length of 3 inches for each element half.

• The mounting plate equivalent will be the first section in a model of the element half.

F. In a Yagi, if the elements are designed to be touching, are the elements touching the boom in the correct locations?

G. In a Yagi, if the elements are designed to be insulated, are the elements insulated from the boom in the correct locations?

H. The center of hairpin matching devices (i.e. on a Yagi) can be grounded to the boom .

I. The boom is neutral, but it is still a conductor! The center of a dipole element is also neutral and can be touched while tuning without affecting the reading. With a hairpin match , the center of the hairpin can also be touched while tuning and touching the

whole hairpin might not affect the readings much at all.

PROXIMITY

A. What else is near-by (roof, wires, guy lines, gutters)? If it can conduct at all, it can and probably will couple to the antenna!

B. Does the SWR change when the antenna is rotated? If so, this indicates interaction. Note that in some combinations of antennas, there can be destructive interaction even if the SWR does not change. Computer models can be useful here.

C. What is within ¼ wavelength of the antenna? Imagine a sphere (like a big ball) with the antenna in question at the center of the sphere, with the following as a radius, depending on frequency. Think in three dimensions like a sphere — up and down and front and rear. Any resonant conductor (antenna or not) with the following radii will couple to and probably affect the antenna you are testing or installing.

    160 meters = 140 feet
    80 meters = 70 feet
    40 meters = 35 feet
    20 meters = 18 feet
    15 meters = 12 feet
    10 meters = 9 feet

D. Interaction occurs whether or not you are transmitting on the adjacent antennas. When receiving, it simply is not as apparent as when transmitting.

E. Wire antennas under a Yagi can easily affect it. This includes inverted V dipoles for the low bands and multi-band dipoles. The wire antennas are typically for lower frequency band (s) and will not be affected by the Yagi(s), as the Yagis are for the higher bands.

FEED SYSTEM

The feed system includes:

• the feed line,

• switching mechanisms,

• pigtails from the feed point on the antenna to the main feed line or switch and,

• all feed lines inside the radio room.

The feed system is the *entire connection* between the radio and the feed point of the antenna.

A. Is the feed line (coax) known to be good? (Start with the easiest first.) Check the dc resistance across the cable with the far end open and shorted. Is there water in the coax? This can give strange readings, even frequency-dependent ones. If there is any question, swap the feed line for a known good one and re-test.

B. Are the connectors installed properly? Has a connector been stressed (pulled)? Is the rotation loop done properly to not stress the coax? Is it an old existing loop or a new one? Usually it's alright if new. Type N connectors (especially the older type) are prone to having the center conductor pull out due to the weight of the coax pulling down on the connector. Connectors are easy to do, using the right technique.

C. Is there a barrel connector (a PL-258 dual-SO-239 adapter) in the feed line anywhere? Has a new or different barrel been inserted? These are a common failure point, even with new barrels. The failures range from micro-bridges across the face of the barrel shorting out the center and shield, to resistance between the two ends of the barrel. Have the new barrels been tested in a known feed system? Always test them before installing. Use only quality RF adapters as these are common system failure points.

D. Is the coax intact and not frayed such that the shield can come into contact with anything? This can cause intermittent problems as the coax shield touches the tower, such as on rotation loops and coax on telescoping towers.

E. Is the tuner OFF on the radio? This is often overlooked when adding a new antenna.

F. Are there any new devices in the line? It might be a good idea to remove everything but the essential items when troubleshooting.

G. Is there a remote antenna switch? Swap to another port.

H. Is there a low pass filter in the line? The filter can be defective, especially on 10 meters, causing strange SWR readings

26.11 Repair and Restoration of Vintage Equipment

When purchasing a classic receiver or transmitter, unless you absolutely know otherwise, assume the radio will need work. Often you can get a top-of-the-line radio needing a bit of repair or clean-up inexpensively. Don't worry — these radios were designed to be repaired by their owners — and curiously, except for cosmetic parts such as cabinets and knobs, parts are much easier to find for 60-year-old radios than a 20-year-old imported transceiver!

Chances are the radio has gone for years without use. Even if it has been recently used, don't completely trust components that might be 60 or more years old. Don't start by plugging in your new acquisition! To do so might damage a hard-to-replace power transformer, or cause a fire.

Instead, if the radio didn't come with its owner's manual, get one. Several *QST* vendors sell old manuals in good condition. K4XL's Boat Anchor Manual Archive (**www2.faculty.sbc.edu/kgrimm/boatanchor**) is probably *the* best free resource for these manuals. Armed with the manual, remove the radio from its cabinet. You very likely will find evidence of unsightly repairs, modifications, or even dangling wires. While modifications aren't necessarily bad, they can certainly add some drama to any necessary subsequent troubleshooting. It's up to you to reverse or remove them.

Another option for working with vintage equipment is to refer to editions of the *ARRL Handbook* published around the time that the equipment was in common use. The circuit design and construction practices described in the *Handbook* are likely to be representative of those in the radio and may provide guidance for troubleshooting, repair, and adjustment. Similarly, the troubleshooting sections and chapters in previous editions provide valuable guidance for working with equipment of the same or earlier vintages.

26.11.1 Component Replacement

Correct any obvious problems such as dangling components. Replace the line cord with a three-wire, grounded plug, if not a transformerless "ac/dc" type as discussed below. If the radio is one with a live chassis, you should operate it from an isolation transformer for safety. If you don't have an isolation transformer, use a voltmeter to determine the orientation of the ac plug that places the chassis at ground potential. Avoid teaching touching the chassis and do not use knobs with set screws that contact metal control shafts. It's also a good idea to add a fuse, if the radio doesn't originally have one. Are we ready to give it the smoke test? Not so fast!

CAPACITOR RATINGS

Obviously, aged components deteriorate and capacitors are particularly prone to developing leakage or short-circuits with age. There are as many opinions on capacitor replacement as there are radio collectors, but *at the very least* you should replace the electrolytic filter capacitors. Here's why: they *will* short circuit sometime, and when they do, they'll probably take the rectifier tube and power transformer with them. Modern high voltage electrolytic capacitors are reliable and much smaller than their classic counterparts. Old paper-wax and black plastic tubular capacitors should also be replaced. Again, a short circuit in one of them could take out other components, too. Modern film capacitors of the appropriate voltage are great replacements. Opinions vary as to whether all should be replaced, but replacements are cheap and you have the radio apart now, so why not? If keeping the original components is important, follow the procedure for using a variable transformer to reform electrolytics as described below.

You can mount the new capacitors under the chassis by mounting a new terminal strip (do *not* just wire them to the old capacitor terminals), you can re-stuff an old electrolytic capacitor's can with new capacitors, or you can buy a new can from places such as **www.hayseedhamfest.com** or Antique Electronics Supply (**www.tubesandmore. com**). In any event, follow the manufacturer's schematic — don't assume that the – (minus) end of the capacitor goes to ground, as in some radios the ground path is through a resistor so as to develop bias for the audio output stage or RF gain circuit. Observe the polarity or you'll soon be cleaning up a stinky mess!

TESTING OLD CAPACITORS

All capacitors have a voltage rating written on the side of the cap unless it is a small disc. Surplus stores often have bins full of capacitors of unmarked voltage rating. Don't assume they are a high enough voltage to use — check them with a capacitor checker. There are many models out there by Knight Kit, Lafayette, Sencore, and Eico, but the best were the Heathkit IT-11 or IT-28. They are basically the same model but different colors. Both use a 6E5 Magic Eye tube to indicate the status of the capacitor. A selectable voltage from 3 to 600 V dc can be applied to check for leakage and operation. These are good for small disc or paper caps and large electrolytics.

Take the unknown voltage cap and place it in the test terminals. Advance the voltage control from minimum until the eye tube shows it breaking down. You now know what voltage it is good to.

If the capacitor tests good through the 600 V dc range, it probably is a 1000 V dc capacitor. It is best, though, to know for sure the rating of the cap. In tube equipment, most capacitors should be 500 or 1000 V rated. Mouser (**www.mouser.com**) and Digikey (**www.digikey.com**) do still sell caps for those voltage ranges, but they have become very expensive. You could also find new old stock (NOS) capacitors of the correct voltage rating at surplus stores.

REFORMING ELECTROLYTIC CAPACITORS

The best idea is to replace old electrolytic capacitors with a new unit. They are available cheaply in the voltage ranges required and more compact and reliable than the original electrolytic caps. If necessary, however, old electrolytics can often be revived by reforming the dielectric using a capacitor checker.

Disconnect the wires attached to the capacitor under test and connect it to the capacitor checker. Start at the lowest voltage rating and let it charge up the capacitor. You will know when it is charged by viewing the eye tube: if it is wide open, the cap is charged; if it is closed, the capacitor is either shorted or still charging. Advance gradually to the next voltage rating and wait until the eye fully opens—take plenty of time for the capacitor to stabilize. Continue on through the voltage ranges; each time it will take longer for the eye to open. The dielectric is being reformed. Finally, when you reach the voltage range of the capacitor and the eye is fully open, the cap has been fully revived and is ready for use.

The same process can be performed with a variable autotransformer (Variac) by advancing it a few volts at a time over several hours, but that is a coarse and unreliable process. A diode must be placed in series with the transformer to convert the ac voltage to dc. Monitor the voltage across the capacitor with a meter. If it suddenly jumps to zero, the capacitor has shorted and is now useless. Usually the capacitor can be revived successfully and will work just fine.

RESISTORS

Over time, carbon resistors in older radios can change value significantly, which can affect circuit operation. Disconnect one end of the questionable resistor and measure it with an accurate ohmmeter. If it is out of tolerance, replace it. Most resistors in the tube era were ½ W or greater. Most circuits today use ¼ W or smaller resistors, which will not dissipate the power tubes produce.

Carbon composition resistors are becoming rarer but there are still ample quantities of NOS in surplus stores. Be careful about power rating. Use metal oxide resistors if needed. Remember that wirewound resistors are very inductive and not good for RF circuits. They are excellent for power supply circuits and are usually found in the 1 to 25 W range.

REPLACING DIODES

Many old tube radios use rectifier tubes. It is not a good idea to replace these with solid-state rectifiers, as a shorted diode can take out the transformer. Selenium rectifiers, however, are good candidates to replace with a silicon diode. The 1N400x series of diode are usually fine for use and very cheap. For higher voltage supplies, be sure to use 1N4007 or higher rated diodes. This may result in higher ouput voltage from the supply. Add a series dropping resistor if necessary to reduce voltage.

TRANSFORMER REPLACMENT

The best bet is to find another radio of the same variety from which you can harvest the transformer. This is especially common in transmitters like the Heathkit DX-35 and DX-40 which frequently had transformers fail. Replacement transformers are generally no longer available. Some companies will rewind transformers, but that is usually prohibitively expensive. Output voltages are quite critical in the design of tube radios, so it is not a good idea to replace a 400 V ac transformer with a 600 V ac unit. It may be best to find a donor radio for a replacement transformer.

WIRE REPLACEMENT

Power cords should be replaced at the first sign of hardening and cracking. It is often a good idea to replace the two-wire power cords with three-wire cords, but this *cannot* be done on ac/dc transformerless radios in which one side of the ac line is connected directly to the chassis! Those must retain the two-wire cords. As noted previously, operating these radios with an isolation transformer is the safest option. If replacing the cord with a two-wire version, the neutral (larger blade) must be connected to the chassis.

Pre-WWII vintage radios often used a cotton covered power cord. To keep the radio as authentic as possible, cotton covered power cords and matching plugs can be found at suppliers such as Antique Radio Supply.

Many early radios had a two-pin power plug with fuses in the plug (the radio has no internal fusing). These are made by Elmeco and are still available (check eBay). Standard 3AG type fuses go in the power plug. Usually a 1 or 2 A fast-blow fuse will work fine except for a higher powered transmitter.

For using PVC-covered wire with terminal strips, solid wire is easier to attach before soldering than stranded wire, although stranded is very usable. You can also use Teflon covered wire that doesn't burn when the soldering iron hits the insulation.

TUBE REPLACEMENT

The sidebar "Using a Tube Tester" explains what a tube tester does. Watch swap meets and garage sales for tube substitution books. Many tubes are interchangeable or similar in purpose. Be sure to document any tube substitutions you make on a vintage radio. Remember that a new tube in the circuit may require re-peaking of the tuned circuits associated with it.

If a tube has a loose tube cap on top, you can easily repair it. Unsolder the cap and make sure that ample wire is still coming from the tube glass envelope. The tube cap should have a tiny hole in the center of the top which the wire will pass through. Mix a small amount of JB Weld epoxy (**www.jbweld.com**) and glue the tube cap back in place. Make sure the wire is sticking out of the hole. After the epoxy has hardened, solder the wire back onto the tube top. Don't let a loose tube cap break it off.

One might be anxious to wipe off the tube and clean it up. Be careful, as you might wipe off the tube number, and then you won't have any idea what the tube is. Many tubes have been lost because they have become unidentifiable. If you do decide to clean up the tube, make sure you steer well clear of the tube number.

Using a Tube Tester

Vacuum-tube testers are scarce but can be located through antique or vintage radio associations, audiophile groups, and sellers of tubes.

Most simple tube testers measure the cathode emission of a vacuum tube. Each grid is shorted to the plate through a switch and the current is observed while the tube operates as a diode. By opening the switches from each grid to the plate (one at a time), we can check for opens and shorts. If the plate current does not drop slightly as a switch is opened, the element connected to that switch is either open or shorted to another element. (We cannot tell an open from a short with this test.) The emission tester does not necessarily indicate the ability of a tube to amplify.

Other tube testers measure tube gain (transconductance). Some transconductance testers read plate current with a fixed bias network. Others use an ac signal to drive the tube while measuring plate current.

Most tube testers also check inter-element leakage. Contamination inside the tube envelope may result in current leakage and shorts between elements. The paths can have high resistance, and may be caused by gas or deposits inside the tube. Tube testers use a moderate voltage to check for leakage. Leakage can also be checked with an ohmmeter using the ×1M range, depending on the actual spacing of tube elements.

Tube sockets and tube pins easily become oxidized which result in radios not working or being intermittent. It is a good idea to pull each tube and spray the socket with DeoxIT or tuner cleaner. Re-insert the tube and wiggle it around in the socket to rub away any oxidation remaining.

REMOVING AND REPLACING COMPONENTS

Replacing capacitors and/or other components isn't difficult, unless they are buried under other components. The Hallicrafters SX-28 and SX-42 receivers are examples of receivers that have extremely difficult-to-reach components. There are different schools of thought on the proper component replacement method. You can use solder wick and/or a desoldering tool to remove the solder from a terminal, unwrap the wires, and install the new component by wrapping the lead around the terminal and soldering it securely. The proponents of this method point out that this is the preferred military and commercial method. I find it often will needlessly damage other components such as tube sockets and create solder droplets inside of the radio.

Back in the day, radio repairmen clipped out a component leaving a short stub of wire, made little coils in the new lead, then soldered the coiled lead to the old stub. This is a much faster, easier and neater method. Refer to books and websites on repairing vintage equipment for other useful tips and tricks.

26.11.2 Powering Up the Equipment

Get out your volt-ohm meter and measure the resistance from the B+ line to ground. Filter capacitors will cause an initial low-resistance reading that increases as the capacitors charge. If the resistance stays low or does not increase beyond tens of kΩ, find the short circuit before you proceed. Now it's time to plug in the radio. It's best to use a variable transformer such as a Variac and ramp up the voltage slowly, or use a "dim bulb tester" (a 100 W light-bulb wired in series with one leg of the ac power). Turn on the radio, and watch for any sparking, flashing or a red glow from the plates of the rectifier tube, or smoke. If any of these occur, immediately remove power and correct the problem. Observe that the tube filaments should light (although you won't see the glow from metal tubes, you should be able to feel them warm up). Again, any tubes that fail to light should be replaced before you continue.

Now hook up a speaker and antenna, and test the radio. With any luck you'll be greeted by a perfectly-performing radio. Seldom, however, is that the case. You may encounter any number of problems at this point. Dirty bandswitches and other controls manifest themselves by intermittently cutting out; they can be cleaned by DeoxIT contact cleaner applied with a cotton swab (don't spray the switch directly!). Scratchy volume or RF gain controls can be cleaned with some DeoxIT; in some cases you might need to remove the control and uncrimp the cover to reveal the carbon element inside.

If a receiver is totally dead at this point but the filaments and dial lights are lit, double-check to see that the Receive-Standby switch is in the receive position, and any battery plug or standby switch jumpers (as described in the manual) are in their correct place.

Although comprehensive troubleshooting is covered elsewhere in this chapter, the next step is comparing voltages with those stated in the user manual. If the manual doesn't have a voltage table denoting the expected voltage at each tube pin, expect between 200-350 V at the tube plate terminals, a few volts at the cathode (unless it's directly grounded), 70-200 V at the screen, and slightly negative voltage at the grid. If you're faced with this situation and a newcomer to troubleshooting vintage gear, help can be found at **http://amfone.net**, **www.antiqueradios.com**, or other forums that cater to boat-anchors and/or vintage radio repair and restoration.

26.11.3 Alignment

Over the years hams have been cautioned that alignment is usually the *last* thing that should be attempted to repair a radio. In general this is true — but it's also a certainty that a 50 year old radio *will* need alignment in order for it to perform at its best. In any case, replace the capacitors and any other faulty components before you attempt alignment — it'll never be right if it still has bad parts! You'll need a good signal generator and a volt-ohm-meter or oscilloscope. Follow the manufacturer's instructions, and with care you'll be rewarded with a radio that performs as good as it did when it was new.

RECEIVER ALIGNMENT

One last caution — alignment should not be attempted by the novice technician or if you do not have the proper equipment or experience. That said alignment may be justified under the following conditions:
• The set is very old and has not been adjusted in many years.
• The circuit has been subject to abusive treatment or environment.
• There is obvious misalignment from a previous repair.
• Tuned-circuit components or crystals have been replaced.
• An inexperienced technician attempted alignment without proper equipment.
• There is a malfunction, but all other circuit conditions are normal. (Faulty transformers can be located because they will not tune.)

Even if one of the above conditions is met, do not attempt alignment unless you have the proper equipment. Receiver alignment should progress from the detector to the antenna terminals. When working on an FM receiver, align the detector first, then the IF and limiter stages and finally the RF amplifier and local oscillator stages. For an AM receiver, align the IF stages first, then the RF amplifier and oscillator stages.

Both AM and FM receivers can be aligned in much the same manner. Always follow the manufacturer's recommended alignment procedure. If one is not available, follow these guidelines:

1. Set the receiver RF gain to maximum, BFO control to zero or center (if applicable to your receiver) and tune to the high end of the receiver passband.
2. Disable the AGC.
3. Set the signal source to the center of the IF passband, with no modulation and minimum signal level.
4. Connect the signal source to the input of the IF section.
5. Connect a voltmeter to the IF output.
6. Adjust the signal-source level for a slight indication on the voltmeter.
7. Peak each IF transformer in order, from the meter to the signal source. The adjustments interact; repeat steps 6 and 7 until adjustment brings no noticeable improvement.
8. Remove the signal source from the IF section input, reduce the level to minimum, set the frequency to that shown on the receiver dial and connect the source to the antenna terminals. If necessary, tune around for the signal — if the local oscillator is not tracking, it may be off.
9. Adjust the signal level to give a slight reading on the voltmeter.
10. Adjust the trimmer capacitor of the RF amplifier for a peak reading of the test signal. (Verify that you are reading the correct signal by switching the source on and off.)
11. Reset the signal source and the receiver tuning for the low end of the passband.
12. Adjust the local-oscillator padder for peak reading.
13. Steps 8 through 11 interact, so repeat them until the results are as good as you can get them.

26.11.4 Using Vintage Receivers

Connect a speaker, preferably the same impedance as the output impedance of the receiver. Some receivers have a 600 Ω and 3.2- or 4 Ω output. An 8 Ω speaker is fine — connect it to the low impedance output. Do not operate the receiver without a speaker, however, as the audio output transformer could be damaged by high voltage transients with no load. Alternatively, plug in a pair of headphones, keeping in mind that old receivers usually have high impedance headphone outputs and new headphones are usually low impedance. They'll work fine, but the volume may be considerably lower with the newer headphones.

If you're going to use the receiver in conjunction with a transmitter, you need to be able to mute the receiver while you're transmitting — otherwise, you'll end up with copious feedback from the receiver. Most receivers have mute terminals — some mute with a closed switch, others mute on open. Figure out which method your receiver and transmitter use. You'll need a relay if the receiver mute arrangement doesn't match that of the transmitter.

Some receivers — such as the older Hammarlund Super Pros and pre-WWII Hallicrafters models — use the mute terminals to open the B+ when putting the receiver in standby mode. This is extremely dangerous with 300 V or so on exposed terminals! An easy modification will save you from an almost certain shock. Open up the receiver and remove the wires from the standby terminals. Solder them together and insulate the connection with electrical tape or a wire nut. Better, solder them to an unused, ungrounded terminal if there's one handy. Next, examine the RF gain control and notice that one terminal is probably grounded. Cut this wire and solder a 47 kΩ resistor between ground and the RF gain control terminal. Connect a pair of wires from the terminals of the mute connection across the 47 kΩ resistor just installed. Now, with the mute terminals open the RF gain is all the way down so the receiver is essentially muted. Short the terminals to receive. The voltage here is low and not dangerous.

Next, connect an antenna and antenna relay in the same manner and tune the bands. You'll find that the best fidelity from the receiver occurs at its maximum bandwidth. The crystal filter, if fitted, can help notch out heterodynes as can tuning the receiver slightly higher or lower. The bandspread control can be used to fine tune. Now, just enjoy using your classic, vintage equipment!

26.11.5 Plastic Restoration

Sometimes an old radio has a meter lens or dial face that is badly damaged. If it is cracked, there isn't much you can do but find a replacement. If it is just scratched, you have a good chance at fully restoring it.

First, remove the meter lens from the meter movement if possible. Most just snap on. To avoid damaging the very delicate needle and meter movement, place it in a protected area and be sure metal filings cannot get to the movement's magnet.

Make sure the scratch hasn't gone all the way though the plastic to become a crack, although even a deep scratch can be buffed out. You will need various grades of wet/dry sandpaper. Obtain sheets of 320, 400 and 600 grit (600 is the finest). One sheet will last a long time. Cut a piece about ½ inch by 2 inches and fold it in half. Start with 320 and gently sand in *one* direction only, over the scratch. This is very tedious and will take a while before you sand away enough plastic to get through the scratch.

Once the scratch is removed, it is time to start reconditioning the plastic. Again sanding in one direction, use the 400 and finally the 600 to completely remove all traces of the earlier sanding scratches. The 600 should leave almost a powdered effect but the plastic will still be hazy and opaque. Sand a little more with the 600 just to be sure all traces of any sanding scratches are gone completely.

To remove the haze you will need a polishing compound called Novus #2 (**www.novuspolish.com** or hobby stores) and another compound called Novus #1, which is a plastic shiner and static remover. This will be important to remove the static on the meter cases when you finish. Static causes the needle to react strangely and erratically.

Start with just a drop of NOVUS #2 and a soft cloth such as the disposable shop towels found at auto parts stores for polishing. Cut a 2 inch by 3 inch piece and start polishing. Continue for a long time and bit by bit the haze will disappear — you will be left with a perfectly clear lens.

FRONT PANEL RESTORATION

The most important part of a radio restoration, cosmetically, is the front panel. The case should also look good, but the front panel is the highlight of the radio and needs the most attention. Usually, paint on a radio from the 50s and 60s is well oxidized and there may be some fine scratches as well.

Scratches can be touched up by using enamel model car paint. Buy a bottle of gloss white, gloss black, and the color closest to the panel you have. Into a small paper cup put a drop or two of the stock color. Using the white or the black, stirring in a small drop at a time, lighten or darken the color until it most closely matches the panel. Use a model paint brush with the finest tip you can find to fill just the scratch and not get it on the rest of the paint. Remove excess with a Q-tip. Let it dry completely.

Remove the panel completely, if possible, or at least remove all the knobs. You will work on them individually later. If the panel was originally a gloss panel, you are in luck. If it is a wrinkle finish or flat, this technique might not work for you — those will be addressed later.

Use the Novus #2 compound to remove a few microns of paint — just the oxidized layer. Place a drop on the panel, and with a soft cloth or shop towel, start working the Novus into the paint. You just want to remove the oxidation, so don't rub too hard. Be careful when working on paint that is a second layer above a base paint. It can be very thin and removed with the Novus. After a small amount of gentle polishing, get another cloth or towel and wipe off the panel which should appear just as it did when it came from the factory.

If your panel is a wrinkle finish, you cannot use the Novus. Use a gentle soapy type cleaner and carefully brush the ridges and peaks of the finish to remove layers of dirt and smoke. Sometimes cleaners like 409 and Simple Green will work very well, but be cautious that it doesn't take off the lettering. Be very carefully around the lettering. One way to work it in is with a plumber's acid brush with the bristles cut very short. The bristles will get down into the crevices of the wrinkle finish and clean it. Once the panel is clean, it can be shined up a bit with some WD-40. This also works very well on wrinkle finish cabinets. Just lightly brush it on and remove it with a rag. This can collect dust but gives a nice wet finish to the wrinkle finish paints.

KNOBS

Most knobs are plastic or have metal plates around the bottom. Use the Novus #2 to shine up the flat parts of the knob. Knobs with flutes on the sides are very tedious to clean, but look beautiful once restored. Take a fine pick and drag it down each flute to scrape out the accumulated dirt. Once cleaned, the Novus can be used to shine up the flutes as well. On knobs with metalized bases, the bases are also shined up with the Novus. Make sure you don't polish off any markings on the trim bases. Sometimes the knobs will have white or red lines in the tops. Those can be filled with model paint to restore them to full beauty.

HOLES IN FRONT PANELS

It may be most disconcerting to find someone has drilled a hole in the panel for one reason or another. Extra holes greatly devalue the rig because to properly restore it, you have to find a replacement panel. Panels with wrong holes can be salvaged with quite nice results by using JB Weld epoxy metal filler (**www.jbweld.com**).

First remove the panel. If the hole is a considerable distance from the other knobs, repairs can be done on the radio but it is not advised. You will need a special tape called Kapton tape. This is a high temperature polymide film tape that is widely available but not cheap. A little tape lasts a long time.

You will need a piece of tape slightly larger than the hole. Other types of tape may work but won't give as smooth a finish. If the panel is wrinkle finished, you may want a rough finish tape such as masking tape. The key is a tight fit to the front of the panel. Place a piece of tape across the hole on the front of the panel and seal it securely all around the hole.

Lay the panel on a flat surface tape-down so the tape will be held flat and not bulge at all. Mix a suitable amount of the JB Weld epoxy and flow it into the hole from the rear of the panel. Stir it while it is still very wet and pliable to make sure no air bubbles are in the hole. Let it cure securely overnight so the hole will be filled and rock hard.

Flip the panel over and remove the tape. If you have secured the tape well enough, none of the epoxy will have been drawn out onto the front surface. There should be a flat filled area exactly level with the rest of the panel.

Now you can touch up the repaired hole with an exact match of spray paint. Mask off the rest of the panel so you don't get any on the lettering. You can also use the model paint method, described for scratch repair Once painted, the offending hole should be virtually invisible.

BROKEN PLASTIC

Sometimes a microphone or other item will have a chunk of the plastic broken out of it. It is most unsightly and usually is the reason for discarding the item. But, using our repair technology, it can be saved and fully restored as in this example of repairing a damaged microphone case. This procedure will work well on broken Bakelite cases, too.

Once again, we will use JB Weld Epoxy and Kapton tape. Remove the microphone elements and switch from the case along with the cord and anything else not plastic. Clean the area around the broken part well. With the Kapton tape, make a backing area on the inside of the case to form a backing for the epoxy. The tape will hold it well and not deform.

Mix some JB Weld and pour it into the space around the break. Fill *higher* than the surrounding plastic. This will be difficult as the fill area will not be level. It may take a couple of fills to build up the area high enough to build up past the level of the surrounding plastic. Once cured and very hard (wait at least 24 hours), you now can begin to file the epoxy. File it down until it is nearly level with the plastic and then switch to sandpaper. Carefully sand the epoxy so it is exactly flush with the original plastic, and make sure the shape is correct. You can always add more JB Weld if too much is filed off.

Once sanded flush, finish the sanding with fine sandpaper (400 and 600 grade). The JB Weld will shine just like the original plastic, but will be the wrong color. Spray paint the entire case with a color close to the original. Once painted, the repair will be virtually undetectable!

26.12 References and Bibliography

J. Bartlett, "Calculating Component Values," *QST*, Nov 1978, pp 21-28.

J. Carr, *How to Troubleshoot and Repair Amateur Radio Equipment* (TAB Books, 1980).

D. DeMaw, "Understanding Coils and Measuring their Inductance," *QST*, Oct 1983, pp 23-26.

M. Eiselman, "Troubleshooting Radios," *QST*, May 2009, pp 30-33.

R. Goodman, *Practical Troubleshooting with the Modern Oscilloscope* (TAB Books, 1979).

A. Haas, *Oscilloscope Techniques* (Gernsback Library, 1958).

J. Lenk, *Handbook of Electronic Test Procedures* (Prentice-Hall, 1982).

J. Lenk, *Encyclopedia of Circuits and Troubleshooting Guide, Vol 1-4* (McGraw-Hill, 1993-1997).

G. Loveday, and A. Seidman, *Troubleshooting Solid-State Circuits* (John Wiley and Sons, 1981).

C. Smith, *Test Equipment for the Radio Amateur*, 4th edition (RSGB, 2011).

D. DeMaw, "Some Basics of Equipment Servicing — Part 1," *QST*, Dec 1981, pp 11-14.

G. Collins, "Some Basics of Equipment Servicing — Part 2," *QST*, Jan 1982, pp 38-41.

G. Collins, "Some Basics of Equipment Servicing — Part 3," *QST*, Feb 1982, pp 40-44, Feedback May 1982, p 43.

N. Bradshaw, "Some Basics of Equipment Servicing — Part 4," *QST*, Mar 1982, pp 40-44.

Contents

RF Interference

Chapter 27 — CD-ROM Content

Supplemental Files
- "What To Do if You Have an Electronic Interference Problem" — *CEA Handbook*
- TV Channel, Amateur Band and Harmonic Chart

Projects
- "A Home-made Ultrasonic Power Line Arc Detector and Project Update" by Jim Hanson, W1TRC (SK)
- "A Simple TRF Receiver for Tracking RFI" by Rick Littlefield, K1BQT
- "Active Attenuator for VHF-FM" by Fao Eenhoorn, PAØZR (article and template)
- "Simple Seeker" by Dave Geiser, W5IXM
- "Tape Measure Beam for Power Line Hunting" by Jim Hanson, W1TRC (SK)

Amateurs live in an increasingly crowded technological environment. As our lives become filled with technology, every lamp dimmer, garage-door opener or other new technical "toy" contributes to the electrical noise around us. Many of these devices also "listen" to that growing noise and may react to the presence of their electronic siblings. The more such devices there are, the higher the likelihood that the interactions will be undesirable.

What was once primarily a conversation about "interference" has expanded to include power systems, shielding, intentional and unintentional radiators, bonding and grounding, and many other related topics and phenomena. These are all grouped together under the general label of *electromagnetic compatibility (EMC)*. The scope of EMC includes all the ways in which electronic devices interact with each other and their environment.

The general term for interference caused by signals or fields is *electromagnetic interference* or *EMI*. This is the term you'll encounter in the professional literature and standards. The most common term for EMI involving amateur signals is *radio frequency interference (RFI)* and when a television or video display is involved, *television interference (TVI)*. RFI is the term used most commonly by amateurs. Whether it's called EMI, RFI or TVI, unwanted interaction between receivers and transmitters has stimulated vigorous growth in the field of electromagnetic compatibility! (This chapter will use the term RFI to refer to all types of interference to or from amateur signals, except where noted.)

This chapter begins with an overview of dealing with interference and includes relevant FCC regulations. This section is an excellent resource when you are confronted with an interference problem and are wondering "What do I do now?" The information here is based on the experiences of ARRL Lab staff in assisting amateurs with RFI problems.

The second part of this chapter is a discussion on identifying and locating RFI-related noise and signal sources then presents some effective ways of resolving the problem. A glossary of RFI terminology concludes the chapter.

The material in this chapter may provide enough information for you to solve your problem, but if not, the ARRL website offers extensive resources on RF interference at **www.arrl.org/radio-frequency-interference-rfi**. Many topics covered in this chapter are covered in more detail in the *ARRL RFI Book* from a practical amateur perspective.

Throughout this chapter you'll also find references to "Ott," meaning the book *Electromagnetic Compatibility Engineering* by EMC consultant Henry Ott, WA2IRQ. EMC topics are treated in far greater depth in Ott's book than is possible in this *Handbook*. Readers interested in the theory of EMC, analysis of EMC mechanisms, test methodology and EMC standards can purchase a copy through the ARRL Publication Sales department or the ARRL website.

27.1 Managing Radio Frequency Interference

Sooner or later, nearly every Amateur Radio operator will have a problem with RFI, but temper your dismay. Most cases of interference can be cured! Before diving into the technical aspects of interference resolution, consider the social aspects of the problem. A combination of "diplomacy" skills and standard technical solutions are the most effective way to manage the problem so that a solution can be found and applied. This section discusses the overall approach to solving RFI problems. Specific technical causes and solutions are described in subsequent sections.

27.1.1 Responsibility for Radio Frequency Interference

When an interference problem occurs, we may ask "Who is to blame?" The ham and the other party often have different opinions. It is natural (but unproductive) to assign blame instead of fixing the problem.

No amount of wishful thinking (or demands for the "other guy" to solve the problem) will result in a cure for interference. Each party has a unique perspective on the situation and a different degree of understanding of the personal and technical issues involved. On the other hand, each party has certain responsibilities and should be prepared to address them fairly. (Given the realities of amateur operation, one of the parties is likely to be a neighbor to the amateur and so the term "neighbor" is used in this chapter.)

Always remember that every interference problem has two components — the equipment that is involved and the people who use it. A solution requires that we deal effectively with both the equipment and the people. The ARRL recommends that the hams and their neighbors cooperate to find solutions. The FCC also shares this view. It is important therefore to define the term "interference" without emotion.

27.1.2 Proper Station Operation

A radio operator is responsible for the proper operation of the radio station. This responsibility is spelled out clearly in Part 97 of the FCC regulations. If interference is caused by a spurious emission from your station, you must correct the problem there.

Fortunately, most cases of interference are not the fault of the transmitting station. If an amateur signal is the source of interference, the problem is usually caused by fundamental overload — a general term referring to interference caused by the intended, fundamental signal from a transmitter. If the amateur station is affected by interference, electrical noise is most often the culprit. Typical sources include power lines and consumer devices.

27.1.3 Personal Diplomacy

Whether the interference is to your station or from your station, what happens when you first talk to your neighbor sets the tone for all that follows. Any technical solutions cannot help if you are not allowed in your neighbor's house to explain them! If the interference is not caused by spurious emissions from your station, however, you should be a locator of solutions, not a provider of solutions.

Your neighbor will probably not understand all of the technical issues — at least not at first. Understand that, regardless of fault, interference is annoying whether your signals are causing interference to the neighbor or a device owned by the neighbor is causing interference to you.

Let your neighbor know that you want to help find a solution and that you want to begin by talking things over. Talk about some of the more important technical issues, in non-technical terms. Explain that you must also follow technical rules for your signal, such as for spurious emissions, and that you will check your station, as well.

27.1.4 Interference To a Neighbor's Equipment

Your transmitted signals can be the source of interference to a neighbor's equipment. Assure your neighbor that you will check your station thoroughly and correct any problems. You should also discuss the possible susceptibility of consumer equipment. You may want to print a copy of the RFI information found on the ARRL website at **www. arrl.org/information-for-the-neighbors-of-hams**. (This document, "What To Do if You Have an Electronic Interference Problem," is also included on the CD-ROM accompanying this book.)

Your neighbor will probably feel much better if you explain that you will help find a solution, even if the interference is not your fault. This offer can change your image from neighborhood villain to hero, especially if the interference is not caused by your station. (This is often the case.)

Here is a good analogy: If you tune your TV to channel 3, and see channel 8 instead, you would likely decide that your TV set is broken. Now, if you tune your TV to channel 3, and see your local shortwave radio station (quite possibly Amateur Radio), you shouldn't blame the shortwave station without some investigation. In fact, many televisions respond to strong signals outside the television bands. They may be working as designed, but require added filters and/or shields to work properly near a strong, local RF signal.

Warning: Performing Repairs

You are the best judge of a local situation, but the ARRL strongly recommends that you do not work on your neighbor's equipment. The minute you open up a piece of equipment, you may become liable for problems. Internal modifications to your neighbor's equipment may cure the interference problem, but if the equipment later develops some other problem, you may be blamed, rightly or wrongly. In some states, it is even *illegal* for you to do *any* work on electronic equipment other than your own.

27.1.5 Interference From a Neighbor's Equipment

Your neighbor is probably completely unaware that his or her equipment can interfere with your operation. You will have to explain that some home electronics equipment can generate radio signals many times stronger than the weak signals from a distant transmitter. Also explain that there are a number of ways to prevent those signals from being radiated and causing interference. If the equipment causing the problem can be identified, the owner's manual or manufacturer of the device may provide information on the potential for RFI and for its elimination.

As with interference appearing to be caused by your station, explain that your intent is to help find a solution. Without further investigation it is premature to assume that the neighbor's equipment is at fault or that FCC regulations require the neighbor to perform any corrective action. Working together to find a mutually acceptable solution is the best strategy.

27.1.6 Being Prepared

In order to troubleshoot and cure RFI, you need to learn more than just the basics. This is especially important when dealing with your neighbor. If you visit your neighbor's house and try a few dozen things that don't work (or make things worse), your neighbor may lose confidence in your ability to help cure the problem. If that happens, you may be asked to leave.

Start by carefully studying the technical sources and cures for RFI in this book and in other references, such as the *ARRL RFI Book*. Review some of the ARRL Technical Information Service and *QST* articles about interference. If you are unfamiliar with any of the terms in this chapter, refer to the glossary.

LOCAL HELP

If you are not an expert (and even experts can use moral support), you should find some local help. Fortunately, such help is often available from your ARRL Section's Technical Coordinator (TC). The TC knows of any local RFI committees and may have valuable contacts in the local utility companies. Even an expert can benefit from a TC's help. The easiest way to find your TC is through your ARRL Section Manager (SM). There is a list of SMs on the ARRL website or in any recent *QST*. He or she can quickly put you in contact with the best source of local help.

Even if you can't secure the help of a local expert, a second ham can be a valuable asset. Often a second party can help defuse any hostility. When evaluating and solving RFI problems involving your station, it is very important for two hams to be part of the process. One can operate your station and the other can observe symptoms, and, when appropriate, try solutions.

PREPARE YOUR HOME AND STATION

The first step toward curing the problem is to make sure your own signal is clean and that devices in your home are not causing any problems. Eliminate all interference issues in your own home to be sure your station is operating properly and that your own electronic equipment is not being interfered with or causing interference to your station!

This is also a valuable troubleshooting tool for situations in which your station is suspected of being the source of interference: If you know your signals are "clean," you have cut the size of the problem in half! If the FCC ever gets involved, you can demonstrate that you are not interfering with your own electronics.

Apply RFI cures to your own consumer electronics and computer equipment. What you learn by identifying and eliminating interference in your own home will make you better prepared to do so in your neighbor's home. When your neighbor sees your equipment working well, it also demonstrates that filters work and cause no harm.

To help build a better relationship, you may want to show your station to your neighbor. A well-organized and neatly-wired station inspires confidence in your ability to solve the RFI problem. Clean up your station and clean up the mess! A rat's nest of cables, unsoldered connections and so on can contribute to RFI. Grounding is typically not a cure for RFI, but proper grounding will improve lightning safety and can greatly reduce hum and buzz from power systems. Make sure cable shields are connected properly and that RF current picked up from your transmitted signal by audio and power wiring is minimized.

Install a low-pass or band-pass transmit filter. (In the unlikely event that the FCC becomes involved, they will ask you about filtering.) Show your neighbor that you have installed the necessary filter on your transmitter. Explain that if there is still interference, it is necessary to try filters on the neighbor's equipment, too.

Operating practices and station-design considerations can cause interference to TV and FM receivers. Don't overdrive a transmitter or amplifier; that can increase its harmonic output.

Along with applying some of the interference-reducing solutions in this chapter, you can also consider steps to reduce the strength of your signal at the victim equipment. This includes raising, moving, or re-orienting the antenna, or reducing transmit power. The use of a balun and properly balanced feed line will minimize radiation from your station feed line. (See the **Transmission Lines** chapter.) Changing antenna polarization may help, such as if a horizontal dipole is coupling to a cable TV service drop. Using different modes, such as CW or FM, may also change the effects of the interference. Although the goal should be for you to operate as you wish with no interference, be flexible in applying possible solutions.

27.1.7 Contacting Your Neighbor

Now that you have learned more about RFI, located some local help (we'll assume it's the TC) and done all of your homework, make contact with your neighbor. First, arrange an appointment convenient for you, the TC and your neighbor. After you introduce the TC, allow him or her to explain the issues to your neighbor. Your TC will be able to answer most questions, but be prepared to assist with support and additional information as required.

Invite the neighbor to visit your station. Show your neighbor some of the things you do with your radio equipment. Point out any test equipment you use to keep your station in good working order. Of course, you want to show the filter you have installed on your transmitter's output.

Next, have the TC operate your station on several different bands while you show your neighbor that your home electronics equipment is working properly when your station is transmitting. Point out the filters or chokes you have installed to correct any RF susceptibility problems. If the interference is coming from the neighbor's home, show it to the neighbor and explain why it is a problem for you.

At this point, tell your neighbor that the next step is to try some of the cures seen in your home and station on his or her equipment. This is a good time to emphasize that the problem is probably not your fault, but that you and the TC will try to help find a solution anyway.

Study the section on Troubleshooting RFI before deciding what materials and techniques are likely to be required. You and the TC should now visit the neighbor's home and inspect the equipment installation.

AT YOUR NEIGHBOR'S HOME

Begin by asking when the interference occurs, what equipment is involved, and what frequencies or channels are affected, if appropriate. The answers are valuable clues. Next, the TC should operate your station while you observe the effects. Try all bands and modes that you use. Ask the neighbor to demonstrate the problem. Seeing your neighbor's interference firsthand will help all parties feel more comfortable with the outcome of the investigation.

If it appears that your station is involved, note all conditions that produce interference. If no transmissions produce the problem, your station *may* not be the source. (It's possible that some contributing factor may have been missing in the test.) Have your neighbor keep notes of when and how the interference appears: what time of day, what channels or device was being interfered with, what other equipment was in use, what was the weather? You should do the same whenever you operate. If you can readily reproduce the problem, you can start to troubleshoot the problem. This process can yield important clues about the nature of the problem.

The tests may show that your station isn't involved at all. A variety of equipment malfunctions or external noise can look like interference. Some other nearby transmitter or noise source may be causing the problem. You may recognize electrical noise or some kind of equipment malfunction. If so, explain your findings to the neighbor and suggest that he or she contact appropriate service personnel.

If the interference is to your station from equipment that may be in the neighbor's home, begin by attempting to identify which piece of equipment is causing the problem. Describe when you are receiving the interference and what pattern it seems to have (continuous, pulsed, intermittent, certain times of day, and so on).

Confirm that a specific piece of equipment is causing the problem. The easiest way to do this is to physically unplug the power source or by removing the batteries while you or the TC observe at your station. (Remember that turning a piece of equipment OFF with its ON/OFF switch may not cause the equipment to completely power down.) Removing cables from a powered-up piece of equipment may serve to further isolate the problem.

At this point, the action you take will depend on the nature of the problem and its source. Techniques for dealing with specific interference issues are discussed in the following sections of the chapter. If you are unable to determine the exact nature of the problem, develop a plan for continuing to work with the neighbor and continue to collect information about the behavior of the interference.

Product Review Testing and RFI

The ARRL Laboratory mostly considers RFI as an outside source interfering with the desired operation of radio reception. Power line noise, power supplies, motor controls and light dimmers are all *outside* a radio receiver and antenna system and can be a major annoyance that distracts from the pleasure of operating. Have you ever considered RFI generated *inside* your own receiver or by another legally operating Amateur Radio station?

Harmonics

RFI is just that: radio frequency interference. For instance, a harmonic from another radio amateur's transmitter could be interfering with the desired signal you're tuned to. One might think in this day and age, harmonics are minimal and do not cause interference. I ask you to reconsider.

The maximum spurious output of a modern amateur transmitter must be 43 dB lower than the output on the carrier frequency at frequencies below 30 MHz. While that figure may be "good enough" for an FCC standard, it's not good enough to eliminate the possibility of causing interference to other radio amateurs or possibly other services. Here at the ARRL Laboratory, the measurement of harmonic emission level is a measurement I make of RFI generated *outside* your receiver.

A radio amateur contacted me, concerned about a report that he was causing interference to operators on the 80 meter CW band while operating at full legal limit power during a 160 meter CW contest. Knowing FCC rule part 97.307, he made the effort to measure his 80 meter harmonic. Easily meeting FCC standards at 50 dB below carrier output on 160, he wondered how his transmitter could cause interference.

Here's a breakdown of power output and signal reduction:
0 dB down from 1500 W = 1500 W
10 dB down from 1500 W = 150 W
20 dB down from 1500 W = 15 W
30 dB down from 1500 W = 1.5 W
40 dB down from 1500 W = 150 mW
43 dB down from 1500 W = 75.2 mW (this is the FCC legal limit)
50 dB down from 1500 W = 15 mW

While 15 mW may seem too low of a power to cause interference, it wasn't in this case; the interfering signal was reported to be S9. QRP enthusiasts know that at 15 mW, signals can propagate well with the right conditions. Using a single band, resonant antenna will reduce interference caused by harmonics located on other bands, but today many stations employ antennas resonant on more than one ham band.[1]

Signals Generated in the Receiver

What about RFI generated *inside* the receiver you're operating? You're tuning across the 15 meter band when, all of a sudden, you hear what seems to be an AM broadcast station. Is it a jammer? It is definitely interference — radio frequency interference — caused by two strong shortwave stations! In this particular case, a Midwest radio amateur experienced interference on 15 meters and figured out what was happening. One station was transmitting near 6 MHz and another transmitting above 15 MHz. These two strong stations added up to created a second order IMD (intermodulation distortion) product at the 1st IF stage, and this unwanted signal was passed along to subsequent stages and to the speaker. The RFI in this case was caused by insufficient receiver performance (second order IMD dynamic range) where the frequencies of the two stations added up to exactly the frequency that the operator was tuned

The Lab uses these signal generators to test receivers for internally generated intermodulation distortion, as well as other key performance parameters.

to. Third-order IMD products from strong in-band signals are another form of RFI created within a radio receiver.

Nearby stations transmitting at or near the IF frequency will cause interference not because the transmitter is at fault, but because of a receiver's insufficient IF rejection. The same interference will be heard if a nearby transmitter is operating at an image frequency.

Power Supplies

RFI can also be created from another part of a radio system, such as an external power supply. In addition to transmitter and receiver testing, the Lab also measures the conducted emission levels of power supplies. This is an indication of the amount of RF at given frequencies conducted onto power lines from a power supply as described in this chapter.

Through our published Product Review test results in *QST* magazine, readers can compare the above figures of modern HF transceivers when considering the purchase of a new or used transceiver. Our published data tables spawn friendly competition between radio manufacturers who in turn, strive to perfect their circuit designs. The result is a better product for the manufacturer and a better product for you, the radio amateur. — *Bob Allison, WB1GCM, ARRL Test Engineer*

The ARRL Lab maintains a complete set of up-to-date equipment as well as an RF-tight screen room for Product Review testing. [Bob Allison, WB1GCM, photo]

[1]In this case, the use of a bandpass filter designed for 160 meters would significantly attenuate the harmonic on 80 meters, eliminating the interference.

27.2 FCC Rules and Regulations

In the United States most unlicensed electrical and electronic devices are regulated by Part 15 of the FCC's rules. These are referred to as "Part 15 devices." Most RFI issues reported to the ARRL involve a Part 15 device. Some consumer equipment, such as certain wireless and lighting devices, is covered under FCC Part 18 which pertains to ISM (Industrial, Scientific and Medical) devices.

The Amateur service is regulated by FCC Part 97. (Part 97 rules are available online at **www.arrl.org/part-97-amateur-radio.** See also the sidebar "RFI-related FCC Rules and Definitions.") To be legal, the amateur station's signal must meet all Part 97 technical requirements, such as for spectral purity and power output.

As a result, it isn't surprising that most interference complaints involve multiple parts of the FCC rules. (The FCC's jurisdiction does have limits, though — ending below 9 kHz.) It is also important to note that each of the three parts (15, 18 and 97) specifies different requirements with respect to interference, including absolute emissions limits and spectral purity requirements. The FCC does not specify *any* RFI immunity requirements. Most consumer devices therefore receive no FCC protection from a legally licensed transmitter, including an amateur transmitter operating legally according to Part 97.

Licensed services are protected from interference to their signals, even if the interference is generated by another licensed service transmitter. For example, consider TVI from an amateur transmitter's spurious emissions, such as harmonics, that meet the requirements of Part 97 but are still strong enough to be received by nearby TV receivers. The TV receiver itself is not protected from interference under the FCC rules. However, within its service area the licensed TV broadcast signal is protected from harmful interference caused by spurious emissions from other licensed transmitters. In this case, the amateur transmitter's interfering spurious emission would have to be eliminated or reduced to a level at which harmful interference has been eliminated.

27.2.1 FCC Part 97 Rules

While most interference to consumer devices may be caused by a problem associated with the consumer device as opposed to the signal source, all amateurs must still comply with Part 97 rules. Regardless of who is at fault, strict conformance to FCC requirements, coupled with a neat and orderly station appearance, will go far toward creating a good and positive impression in the event of an FCC field investigation. Make sure your station and signal exhibit good engineering and operating practices.

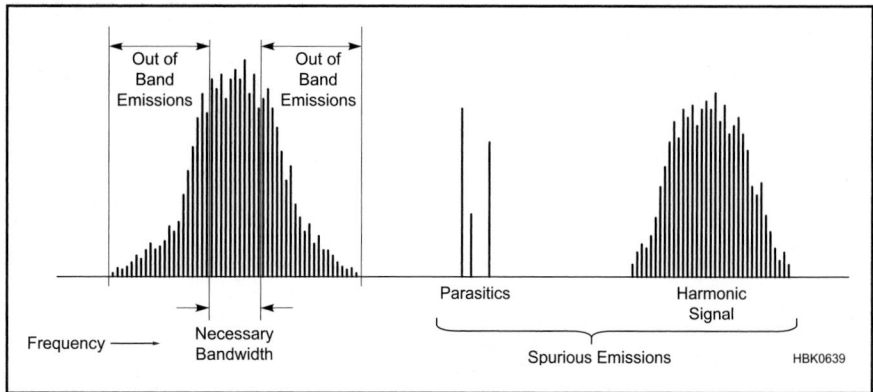

Fig 27.1 — An illustration of out-of-band versus spurious emissions. Some of the modulation sidebands are outside the necessary bandwidth. These are considered out-of-band emissions, not spurious emissions. The harmonic and parasitic emissions shown here are considered spurious emissions; these must be reduced to comply with §97.307.

What If the Police Are Called?

Many amateurs have had a similar experience. You are enjoying some time in front of the radio when the doorbell rings. When you answer the door, you find an irate neighbor has called the police about your transmissions interfering with their stereo (or cordless telephone or other home electronics). The officer tells you that you are interfering with your neighbor and orders you to stop transmissions immediately.

The bad news is you are in the middle of a bad situation. The good news is that most cases of interference can be cured! The proper use of "diplomacy" skills to communicate with a neighbor and standard technical cures will usually solve the problem. Even more good news is that if you are operating in accordance with your license and employing good engineering practices, the law and FCC rules are on your side.

Most RFI is caused by the unfortunate fact that most consumer equipment lacks the necessary filtering and shielding to allow it to work well near a radio transmitter. The FCC does not regulate the immunity of equipment, however, so when interference is caused by consumer-equipment fundamental overload, there is no FCC rules violation, and licensed stations have no regulatory responsibility to correct interference that may result.

Further, in 1982, Congress passed Public Law 97-259. This law is specific and reserves exclusive jurisdiction over RFI matters to the Federal Communications Commission. This national law preempts any state or local regulations or ordinances that attempt to regulate or resolve RFI matter. This is a victory for amateurs (and other services operating with the legal and technical provisions of their licenses).

Simply put, 97-259 says that cities and towns may not pass ordinances or regulations that would prohibit someone from making legal radio transmissions. But what do you do when your neighbor (or the police) confront you about RFI to their consumer electronics? First and foremost, remain calm. In all likelihood the officer or your neighbor has probably never heard of 97-259. Don't get defensive and get drawn into an argument. Don't make comments that the problem is with the neighbor's "cheap" equipment. While inexpensive radios are usually big culprits, any home electronics are potential problems due to inadequate technical designs.

Begin by listening to the complaint. Explain that while you understand, you are operating your equipment within its technical specifications. If your equipment doesn't interfere with your own home electronics, offer to demonstrate that to the officer. Also explain the basics of PL 97-259. If the officer (or neighbor) continues to insist that regardless of the law that you cease, consider temporarily complying with his or her request, with the understanding that you are doing so until the matter is resolved.

Work with your neighbors to understand that steps can be taken that should help resolve the problems (for example, placing toroids and filters on the consumer electronics). The ARRL website has lots of helpful information as you work to resolve the problems. If your club has a local RFI committee or ARRL Technical Specialist, get them involved — their expertise can really be helpful. But above all, remember that when you practice easy, level-headed "diplomacy" you can usually keep the situation from escalating. — *Dan Henderson, N1ND, ARRL Regulatory Information Manager*

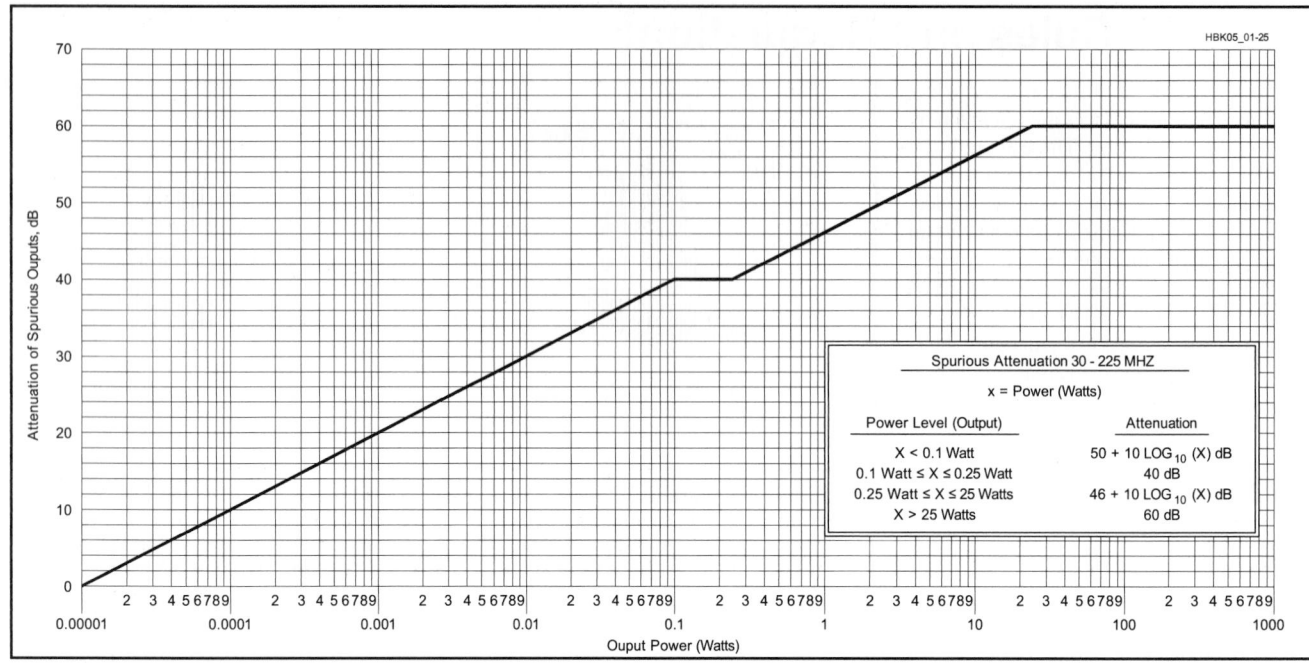

Fig 27.2 — Required attenuation of spurious outputs, 30-225 MHz. Below 30 MHz, spurious emissions must be suppressed by 43 dB for amateur transmitters installed after January 1, 2003.

The bandwidth of a signal is defined by §97.3(a)(8) while the paragraphs of §97.307 define the technical standards amateur transmissions must meet. Paragraph (c) defines the rules for interference caused by spurious emissions. As illustrated in **Fig 27.1**, modulation sidebands outside the necessary bandwidth are considered *out-of-band* emissions, while harmonics and parasitic signals are considered *spurious emissions*. Paragraphs (d) and (e) specify absolute limits on spurious emissions, illustrated in **Fig 27.2**. Spurious emissions must not exceed these levels, whether or not the emissions are causing interference. If spurious emissions from your transmitter are causing interference, it's your responsibility to clean them up.

Strict observance of these rules can not only help minimize interference to the amateur service, but other radio services and consumer devices as well.

27.2.2 FCC Part 15 Rules

In the United States, most unlicensed devices are regulated by Part 15 of the FCC's rules. While understanding these rules doesn't necessarily solve an RFI problem, they do provide some important insight and background on interference to and from a Part 15 device. (Part 18 devices and rules are similar in some respects) and will not be discussed separately — see the sidebar.)

There are literally thousands of Part 15 devices with the potential to be at the heart of an RFI problem. A Part 15 device can be almost anything not already covered in another

FCC Part 18, Consumer Devices

Some consumer devices are regulated by Part 18 of the FCC Rules which pertains to the Industrial, Scientific and Medical (ISM) bands. These devices convert RF energy directly into some other form of energy, such as heat, light or ultrasonic sound energy. Some common household Part 18 devices therefore include microwave ovens, electronic fluorescent light ballasts, CFLs, and ultrasonic jewelry cleaners. (Note that LED bulbs are covered under Part 15 because of the process by which they generate light.)

Consumer Part 18 devices are generators of RF — but not for communications purposes — and can cause interference in some cases. However, there are no rules that protect them from interference. The purpose of Part 18 is to permit those devices to operate and to establish rules prohibiting interference.

From the standpoint of an RFI problem, Part 18 rules aren't much different from Part 15. As with a Part 15 device, a Part 18 device is required to meet specified emissions limits. Furthermore, it must not cause harmful interference to a licensed radio service.

Part 18 Rules and the 33 cm Band

Part 18 specifies a number of bands for ISM (industrial, scientific, and medical) devices. The so-called ISM bands in some cases overlap amateur spectrum. For example, the entire 33 cm band from 902 to 928 MHz is both an amateur and ISM band. And, as the following rule from Part 97 indicates, the amateur service is not protected from ISM devices operating in this or any other ISM band:

§ 97.303 (e) Amateur stations receiving in the 33 cm band, *the 2400-2450 MHz segment, the 5.725-5.875 GHz segment, the 1.2 cm band, the 2.5 mm band, or the 244-246 GHz segment must accept interference from industrial, scientific, and medical (ISM) equipment.*

Additional restrictions apply in some areas of the country when using the 33 cm band.

• Amateurs located in some parts of Colorado and Wyoming may not transmit in some parts of this band.

• Amateurs located in some parts of Texas and New Mexico are prohibited from using this band.

• Amateurs located within 150 miles of White Sands Missile Range are limited to 150 W PEP.

Amateurs in these areas are responsible for knowing the boundaries of these areas and observing all applicable rules. See the latest edition of the *ARRL Repeater Directory* for more information and boundary details.

RFI-related FCC Rules and Definitions

Here are some of the most important rules and definitions pertaining to RFI and the Amateur Radio Service. Definitions in Part 2 are used in regulations that apply to all radio services.

§2.1 Definitions

Harmful Interference. Interference which endangers the functioning of a radionavigation service or of other safety services or seriously degrades, obstructs, or repeatedly interrupts a radiocommunication service operating in accordance with [the ITU] Radio Regulations.

Interference. The effect of unwanted energy due to one or a combination of emissions, radiations, or inductions upon reception in a radiocommunication system, manifested by any performance degradation, misinterpretation, or loss of information which could be extracted in the absence of such unwanted energy.

Out-of-band Emission. Emission on a frequency or frequencies immediately outside the necessary bandwidth which results from the modulation process, but excluding spurious emissions. Band does not mean "amateur band" here.

§97.3 Definitions

(a) The definitions of terms used in Part 97 are:

(8) Bandwidth. The width of a frequency band outside of which the mean power of the transmitted signal is attenuated at least 26 dB below the mean power of the transmitted signal within the band.

(23) Harmful interference. (see the previous Part 2 definition)

(42) Spurious emission. An emission, on frequencies outside the necessary bandwidth of a transmission, the level of which may be reduced without affecting the information being transmitted.

§97.307 Emission standards

(a) No amateur station transmission shall occupy more bandwidth than necessary for the information rate and emission type being transmitted, in accordance with good amateur practice.

(b) Emissions resulting from modulation must be confined to the band or segment available to the control operator. Emissions outside the necessary bandwidth must not cause splatter or keyclick interference to operations on adjacent frequencies.

(c) All spurious emissions from a station transmitter must be reduced to the greatest extent practicable. If any spurious emission, including chassis or power line radiation, causes harmful interference to the reception of another radio station, the licensee of the interfering amateur station is required to take steps to eliminate the interference, in accordance with good engineering practice.

(d) For transmitters installed after January 1, 2003, the mean power of any spurious emission from a station transmitter or external RF amplifier transmitting on a frequency below 30 MHz must be at least 43 dB below the mean power of the fundamental emission. For transmitters installed on or before January 1, 2003, the mean power of any spurious emission from a station transmitter or external RF power amplifier transmitting on a frequency below 30 MHz must not exceed 50 mW and must be at least 40 dB below the mean power of the fundamental emission. For a transmitter of mean power less than 5 W installed on or before January 1, 2003, the attenuation must be at least 30 dB. A transmitter built before April 15, 1977, or first marketed before January 1, 1978, is exempt from this requirement.

(e) The mean power of any spurious emission from a station transmitter or external RF power amplifier transmitting on a frequency between 30-225 MHz must be at least 60 dB below the mean power of the fundamental. For a transmitter having a mean power of 25 W or less, the mean power of any spurious emission supplied to the antenna transmission line must not exceed 25 µW and must be at least 40 dB below the mean power of the fundamental emission, but need not be reduced below the power of 10 µW. A transmitter built before April 15, 1977, or first marketed before January 1, 1978, is exempt from this requirement.

Part of the FCC rules. In fact, many Part 15 devices may not normally even be associated with electronics, RF or in some cases, electricity. While televisions, radios, telephones and even computers obviously constitute a Part 15 device, the rules extend to anything that is capable of generating RF, including electric motors and consumer devices such as baby monitors, wireless microphones and intercoms, RF remote controls, garage door openers, etc. With so many Part 15 consumer devices capable of generating and responding to RF, it isn't surprising therefore that most reported RFI problems involving Amateur Radio also involve a Part 15 device.

TYPES OF PART 15 DEVICES

Part 15 describes three different types of devices that typically might be associated with an RFI problem. A fourth type of device, called a *carrier current device*, uses power lines and wiring for communications purposes. As we'll see, the rules are different for each type.

Intentional Emitters — Intentionally generate RF energy and radiate it. Examples include garage door openers, cordless phones and baby monitors.

Unintentional Emitters — Intentionally generate RF energy internally, but do not intentionally radiate it. Examples include computers and network equipment, superheterodyne receivers, switchmode power supplies and TV receivers.

Incidental Emitters — Generate RF energy only as an incidental part of their normal operation. Examples include power lines, arcing electric fence, arcing switch contacts, dc motors and mechanical light switches.

Carrier Current Devices — Intentionally generate RF and conduct it on power lines and/or house wiring for communications purposes. Examples include Powerline and X.10 networks, Access or In-House Broadband-Over-Power-Line (BPL), campus radio-broadcast systems, and other power-line communications devices.

PART 15 SUMMARY

FCC's Part 15 rules pertain to unlicensed devices and cover a lot of territory. Although reading and understanding Part 15 can appear rather formidable — especially at first glance — the rules pertaining to RFI can be roughly summarized as follows:

- Part 15 devices operate under an unconditional requirement to not cause harmful interference to a licensed radio service, such as Amateur Radio. If such interference occurs, the operator of the Part 15 device is responsible for eliminating the interference.

- Part 15 devices receive no protection from interference from a licensed radio service. There are no FCC rules or limits with regard to Part 15 device RFI immunity.

When is the operator of a licensed transmitter responsible for interference to a Part 15 device?

- The rules hold the transmitter operator responsible if interference is caused by spurious emissions such as a harmonic. An example would be a harmonic from an amateur's transmitter interfering with a cordless telephone. In this case, the transmitter is generating harmful RF energy beyond its permitted bandwidth. A cure must be installed at the transmitter.

- The transmitter operator is not responsible when a Part 15 device is improperly responding to a legal and intentional output of the transmitter. An example of this case would be interference to a cordless telephone operation by the strong-but-legal signal from a nearby amateur transmitter. In this case, the Part 15 device is at fault and the cure must be installed there. It is important to note that this situation is typical of most interference to Part 15 devices.

Even though the causes and cures for these situations are different, the common element for all three situations is the need for personal diplomacy in resolving the problem.

PART 15 MANUFACTURER REQUIREMENTS

Under FCC rules, both device manufactures and operators of those devices share responsibility for addressing an RFI problem.

The rules for manufacturers are primarily designed to reduce the possibility of harmful interference. They do not however completely eliminate the possibility of an RFI problem. If and when interference does occur, the rules are designed to minimize and confine the scope of problems such that they can be addressed on a case by case basis. Responsibility

---

Part 15 Absolute Emissions Limits for Unintentional Emitters

§15.107 Conducted limits

(a) Except for Class A digital devices, for equipment that is designed to be connected to the public utility (ac) power line, the radio frequency voltage that is conducted back onto the ac power line on any frequency or frequencies within the band 150 kHz to 30 MHz shall not exceed the limits in the following table, as measured using a 50 µH/50 ohms line impedance stabilization network (LISN). Compliance with the provisions of this paragraph shall be based on the measurement of the radio frequency voltage between each power line and ground at the power terminal. The lower limit applies at the band edges.

Conducted Limits — Non Class-A Digital Devices

Frequency of emission (MHz)	Conducted limit (dBµV) Quasi-peak	Average
0.15–0.5	66 to 56*	56 to 46*
0.5–5	56	46
5–30	60	50

*Decreases with the logarithm of the frequency.

(b) For a Class A digital device that is designed to be connected to the public utility (ac) power line, the radio frequency voltage that is conducted back onto the ac power line on any frequency or frequencies within the band 150 kHz to 30 MHz shall not exceed the limits in the following table, as measured using a 50 µH/50 ohms LISN. Compliance with the provisions of this paragraph shall be based on the measurement of the radio frequency voltage between each power line and ground at the power terminal. The lower limit applies at the boundary between the frequency ranges.

Conducted Limits — Class-A Digital Devices

Frequency of emission (MHz)	Conducted limit (dBµV) Quasi-peak	Average
0.15–0.5	79	66
0.5–30	73	60

(c) The limits shown in paragraphs (a) and (b) of this section shall not apply to carrier current systems operating as unintentional radiators on frequencies below 30 MHz. In lieu thereof, these carrier current systems shall be subject to the following standards:

(1) For carrier current systems containing their fundamental emission within the frequency band 535-1705 kHz and intended to be received using a standard AM broadcast receiver: no limit on conducted emissions.

(2) For all other carrier current systems: 1000 µV within the frequency band 535–1705 kHz, as measured using a 50 µH/50 ohms LISN.

(3) Carrier current systems operating below 30 MHz are also subject to the radiated emission limits in §15.109(e).

(4) Measurements to demonstrate compliance with the conducted limits are not required for devices which only employ battery power for operation and which do not operate from the ac power lines or contain provisions for operation while connected to the ac power lines. Devices that include, or make provision for, the use of battery chargers which permit

operating while charging, ac adaptors or battery eliminators or that connect to the ac power lines indirectly, obtaining their power through another device which is connected to the ac power lines, shall be tested to demonstrate compliance with the conducted limits.

§ 15.109 Radiated emission limits

(a) Except for Class A digital devices, the field strength of radiated emissions from unintentional radiators at a distance of 3 meters shall not exceed the following values:

Radiated Limits — Non Class-A Digital Devices

Frequency of emission (MHz)	Field Strength (µV/meter)
30–88	100
88–216	150
216–960	200
Above 960	500

(b) The field strength of radiated emissions from a Class A digital device, as determined at a distance of 10 meters, shall not exceed the following:

Radiated Limits — Class-A Digital Devices

Frequency of emission (MHz)	Field Strength (µV/meter)
30–88	90
88–216	150
216–960	210
Above 960	300

(c) In the emission tables above, the tighter limit applies at the band edges. Sections 15.33 and 15.35 which specify the frequency range over which radiated emissions are to be measured and the detector functions and other measurement standards apply.

(d) For CB receivers, the field strength of radiated emissions within the frequency range of 25–30 MHz shall not exceed 40 microvolts/meter at a distance of 3 meters. The field strength of radiated emissions above 30 MHz from such devices shall comply with the limits in paragraph (a) of this section.

(e) Carrier current systems used as unintentional radiators or other unintentional radiators that are designed to conduct their radio frequency emissions via connecting wires or cables and that operate in the frequency range of 9 kHz to 30 MHz, including devices that deliver the radio frequency energy to transducers, such as ultrasonic devices not covered under part 18 of this chapter, shall comply with the radiated emission limits for intentional radiators provided in §15.209 for the frequency range of 9 kHz to 30 MHz. As an alternative, carrier current systems used as unintentional radiators and operating in the frequency range of 525 kHz to 1705 kHz may comply with the radiated emission limits provided in §15.221(a). At frequencies above 30 MHz, the limits in paragraph (a), (b), or (g) of this section, as appropriate, apply.

RFI-related Part 15 FCC Rules and Definitions

The FCC's Part 15 rules are found in Title 47 section of the Code of Federal Regulations (CFR). They pertain to unlicensed devices. Here are some of the more important Part 15 rules and definitions pertaining to RFI.

§15.3 Definitions.

(m) *Harmful interference.* Any emission, radiation or induction that endangers the functioning of a radionavigation service or of other safety services or seriously degrades, obstructs or repeatedly interrupts a radio communications service operating in accordance with this chapter.

(n) *Incidental radiator.* A device that generates radio frequency energy during the course of its operation although the device is not intentionally designed to generate or emit radio frequency energy.

(o) *Intentional radiator.* A device that intentionally generates and emits radio frequency energy by radiation or induction.

(z) *Unintentional radiator.* A device that intentionally generates radio frequency energy for use within the device, or that sends radio frequency signals by conduction to associated equipment via connecting wiring, but which is not intended to emit RF energy by radiation or induction.

(t) *Power line carrier systems.* An unintentional radiator employed as a carrier current system used by an electric power utility entity on transmission lines for protective relaying, telemetry, etc. for general supervision of the power system. The system operates by the transmission of radio frequency energy by conduction over the electric power transmission lines of the system. The system does not include those electric lines which connect the distribution substation to the customer or house wiring.

(ff) *Access Broadband over Power Line (Access BPL).* A carrier current system installed and operated on an electric utility service as an unintentional radiator that sends radio frequency energy on frequencies between 1.705 MHz and 80 MHz over medium voltage lines or over low voltage lines to provide broadband communications and is located on the supply side of the utility service's points of interconnection with customer premises. Access BPL does not include power line carrier systems as defined in §15.3(t) or In-House BPL as defined in §15.3(gg).

(gg) *In-House Broadband over Power Line (In-House BPL).* A carrier current system, operating as an unintentional radiator, that sends radio frequency energy by conduction over electric power lines that are not owned, operated or controlled by an electric service provider. The electric power lines may be aerial (overhead), underground, or inside the walls, floors or ceilings of user premises. In-House BPL devices may establish closed networks within a user's premises or provide connections to Access BPL networks, or both.

Some of the most important Part 15 rules pertaining to radio and television interference from unintentional and incidental radiators include:

§15.5 General conditions of operation.

(b) Operation of an intentional, unintentional, or incidental radiator is subject to the conditions that no harmful interference is caused and that interference must be accepted that may be caused by the operation of an authorized radio station, by another intentional or unintentional radiator, by industrial, scientific and medical (ISM) equipment, or by an incidental radiator.

(c) The operator of the radio frequency device shall be required to cease operating the device upon notification by a Commission representative that the device is causing harmful interference. Operation shall not resume until the condition causing the harmful interference has been corrected.

§15.13 Incidental radiators.

Manufacturers of these devices shall employ good engineering practices to minimize the risk of harmful interference.

§15.15 General technical requirements.

(c) Parties responsible for equipment compliance should note that the limits specified in this part will not prevent harmful interference under all circumstances. Since the operators of Part 15 devices are required to cease operation should harmful interference occur to authorized users of the radio frequency spectrum, the parties responsible for equipment compliance are encouraged to employ the minimum field strength necessary for communications, to provide greater attenuation of unwanted emissions than required by these regulations, and to advise the user as to how to resolve harmful interference problems (for example, see Sec. 15.105(b)).

§15.19 Labeling requirements.

(a) In addition to the requirements in part 2 of this chapter, a device subject to certification, notification, or verification shall be labeled as follows:

(3) All other devices shall bear the following statement in a conspicuous location on the device:

This device complies with part 15 of the FCC Rules. Operation is subject to the following two conditions: (1) This device may not cause harmful interference, and (2) this device must accept any interference received, including interference that may cause undesired operation.

And the following requirements apply for consumer and residential Class B digital devices. Different requirements apply for Class A digital devices which can only be used in industrial and similar environments:

§15.105 Information to the user.

(b) For a Class B digital device or peripheral, the instructions furnished the user shall include the following or similar statement, placed in a prominent location in the text of the manual:

This equipment has been tested and found to comply with the limits for a Class B digital device, pursuant to part 15 of the FCC Rules. These limits are designed to provide reasonable protection against harmful interference in a residential installation. This equipment generates, uses and can radiate radio frequency energy and, if not installed and used in accordance with the instructions, may cause harmful interference to radio communications. However, there is no guarantee that interference will not occur in a particular installation. If this equipment does cause harmful interference to radio or television reception, which can be determined by turning the equipment off and on, the user is encouraged to try to correct the interference by one or more of the following measures:

Reorient or relocate the receiving antenna.

Increase the separation between the equipment and receiver.

Connect the equipment into an outlet on a circuit different from that to which the receiver is connected.

Consult the dealer or an experienced radio/TV technician for help.

then falls on the device operator to correct the problem or cease using the device.

Manufacturers are subject to requirements that they use good engineering practice to help minimize the potential for interference. In addition, they must meet certain absolute conducted and radiated emissions limits for intentional and unintentional emitters. (See the sidebar for limits on conducted and radiated emissions.) These limits are high enough that S9+ interference levels can occur nearby, depending on frequency, distance and other factors. In fact, most reported Part 15 consumer products causing harmful interference to Amateur Radio are legal and meet these required limits. Therefore, the fact that a particular device is causing harmful interference is not in itself evidence or proof of a rules violation with regard to emissions limits.

With the exception of intentional emitters and carrier-current devices, there are no absolute radiated emissions limits below 30 MHz. The size of a Part 15 device is usually small relative to the wavelength at these frequencies. It is typically too small to be an effective antenna at these longer wavelengths. Therefore, under the FCC rules, only conducted emissions are specified below 30 MHz. (Note that cables and wiring connected to the devices are often effective at radiating signals and are frequent sources of radiated RFI.)

In general, radiated emissions limits are specified only at frequencies above 30 MHz. At the shorter wavelengths above 30 MHz, the device itself is large enough to be a radiator. Wiring connected to it can also be an effective antenna for radiating noise.

Although incidental radiators do not have any absolute emissions limits, as for all Part 15 devices, manufacturers must still employ good engineering practice to minimize the potential for interference.

The FCC also requires manufacturers to add information as a label to most Part 15 devices or as text in the device's operating manual. This information attests to the potential for interference and to the responsibility of the device operator. It must be placed in a conspicuous location on the device or in the manual and contain the following statement:

This device complies with part 15 of the FCC Rules. Operation is subject to the following two conditions: (1) This device may not cause harmful interference, and (2) this device must accept any interference received, including interference that may cause undesired operation.

Owners of Part 15 devices are frequently unaware of this information and surprised to find it on the device or in its manual. Reading this label can be an important step in resolving RFI issues. Additional details regarding labeling requirements can be found in the sidebar on Part 15 Rules.

EQUIPMENT AUTHORIZATION

FCC regulations do not require Part 15 devices to be tested by the FCC. In fact, very few devices must actually undergo FCC testing. In most cases, the requirements are met by the manufacturer testing the device and the test results either kept on file or sent to the FCC, depending on the type of device involved. Here is some general information concerning various FCC approval processes for RF devices:

• *Certification:* Requires submittal of an application that includes a complete technical description of the product and a measurement report showing compliance with the FCC technical standards. Certification procedures have now largely replaced the once familiar Type Acceptance, which is no longer used by the FCC. Devices subject to certification include: low-power transmitters such as cordless telephones, security alarm systems, scanning receivers, super-regenerative receivers, Amateur Radio external HF amplifiers and amplifier kits, and TV interface devices such as DVD players.

• *Declaration of Conformity (DoC)*: Is a declaration that the equipment complies with FCC requirements. A DoC is an alternative to certification since no application to FCC is required, but the applicant must have the device tested at an accredited laboratory. A Declaration of Conformity is the usual approval procedure for Class B personal computers and personal computer peripherals.

• *Notification*: Requires submittal to the FCC of an abbreviated application for equipment authorization which does not include a measurement report. However, a measurement report showing compliance of the product with the FCC technical standards must be retained by the applicant and must be submit-

---

RF Interference and the FCC

by Riley Hollingsworth, K4ZDH

Since 1999, the FCC has worked with the ARRL in a cooperative agreement whereby the staff at the ARRL Lab takes the first cut at resolving RFI. The lab works with the complainant to make sure that the noise is narrowed down to the most probable source. The success rate with this program has been very high, and in many cases — perhaps most — the ARRL and the complainant solve the problem without any FCC involvement.

The lab can help you with the proper testing you need to do in your shack and with the documentation you need in the event the matter is referred to the FCC. Just hearing the noise will often tip off the ARRL staff as to its source. I was always amazed at the number of situations in which noise that at first seemed to be power line related, was in fact (these are real examples) a nearby electric fence, a battery charger for a golf cart, an Ethernet adapter, a paper shredder or a circuit board in a brand new clothes washer in a room adjacent to the radio shack.

Don't assume anything. Test every possibility and document your testing. Not only will you learn a lot about the devices in your house and what causes noise and what doesn't, but good documentation will make your case stronger and easier to work. Follow to the letter the ARRL articles and website tutorials on tracking down noise in and around your shack. You can even hear noise samples on the website.

The documentation requirement is especially important if it is power line related and you have to start dealing with the power company. Take notes of every call you make, who you talked to and when. This helps not only the power company but also the FCC in the unfortunate event that FCC action is required. In many cases, power company staff has a lot of experience running down such noise and they take pride in locating it. In other cases, the power company has been bought, sold, merged or whatever and does not have staff with a lot of experience in these matters. Sometimes its staff has no experience.

The cost of the equipment required to track down power line noise is less than that of two employees and a bucket truck for a day. Often the source of power line noise is a simple piece of equipment that is loose or about to fail, so finding the source helps the power company maintain its system. Keep in mind, though, that some areas are just not suitable for Amateur Radio. If you live next to an old substation, or a conglomeration of old poles and transformers, or an industrial area, your situation is tenuous. Whether it's our roads or power grids, lots of the infrastructure in this country is just plain old and out of date.

You must test diligently and document thoroughly. Never go to the FCC with a situation when you have not already worked with the ARRL and the power company.

ted upon request by the Commission. Devices subject to notification include: point-to-point microwave transmitters, AM, FM and TV broadcast transmitters and other receivers (except as noted elsewhere).

• *Verification*: Verification is a self-approval process where the applicant performs the necessary tests and verifies that they have been done on the device to be authorized and that the device is in compliance with the technical standards. Verified equipment requires that a compliance label be affixed to the device as well as information included in the operating manual regarding the interference potential of the device. Devices subject to verification include: business computer equipment (Class A); TV and FM receivers; and non-consumer Industrial, Scientific and Medical Equipment.

PART 15 OPERATOR REQUIREMENTS

All Part 15 devices are prohibited from causing harmful interference to a licensed radio service — including the Amateur Radio Service. This is an absolute requirement without regard to the emitter type or a manufacturer's conformance to emissions limits or other FCC technical standards. It is important to note that the manufacturer's requirements are not sufficient to prevent harmful interference from occurring under all circumstances. If and when a Part 15 device generates harmful interference, it becomes the responsibility of the device operator to correct the problem. Upon notice from the FCC, the device operator may also be required to cease using the device until such time as the interference has been corrected.

27.3 Elements of RFI

27.3.1 Source-Path-Victim

All cases of RFI involve a *source* of radio frequency energy, a device that responds to the electromagnetic energy (*victim*), and a transmission *path* that allows energy to flow from the source to the victim. Sources include radio transmitters, receiver local oscillators, computing devices, electrical noise, lightning and other natural sources. Note that receiving unwanted electromagnetic energy does not necessarily cause the victim to function improperly.

A device is said to be *immune* to a specific source if it functions properly in the presence of electromagnetic energy from the source. In fact, designing devices for various levels of immunity is one aspect of electromagnetic compatibility engineering. Only when the victim experiences a *disturbance* in its function as a consequence of the received electromagnetic energy does RFI exist. In this case, the victim device is *susceptible* to RFI from that source.

There are several ways that RFI can travel from the source to the victim: *radiation*, *conduction*, *inductive coupling* and *capacitive coupling*. Radiated RFI propagates by electromagnetic radiation from the source through space to the victim. *Conducted RFI* travels over a physical conducting path between the source and the victim, such as wires, enclosures, ground planes, and so forth. Inductive coupling occurs when two circuits are magnetically coupled. Capacitive cou-

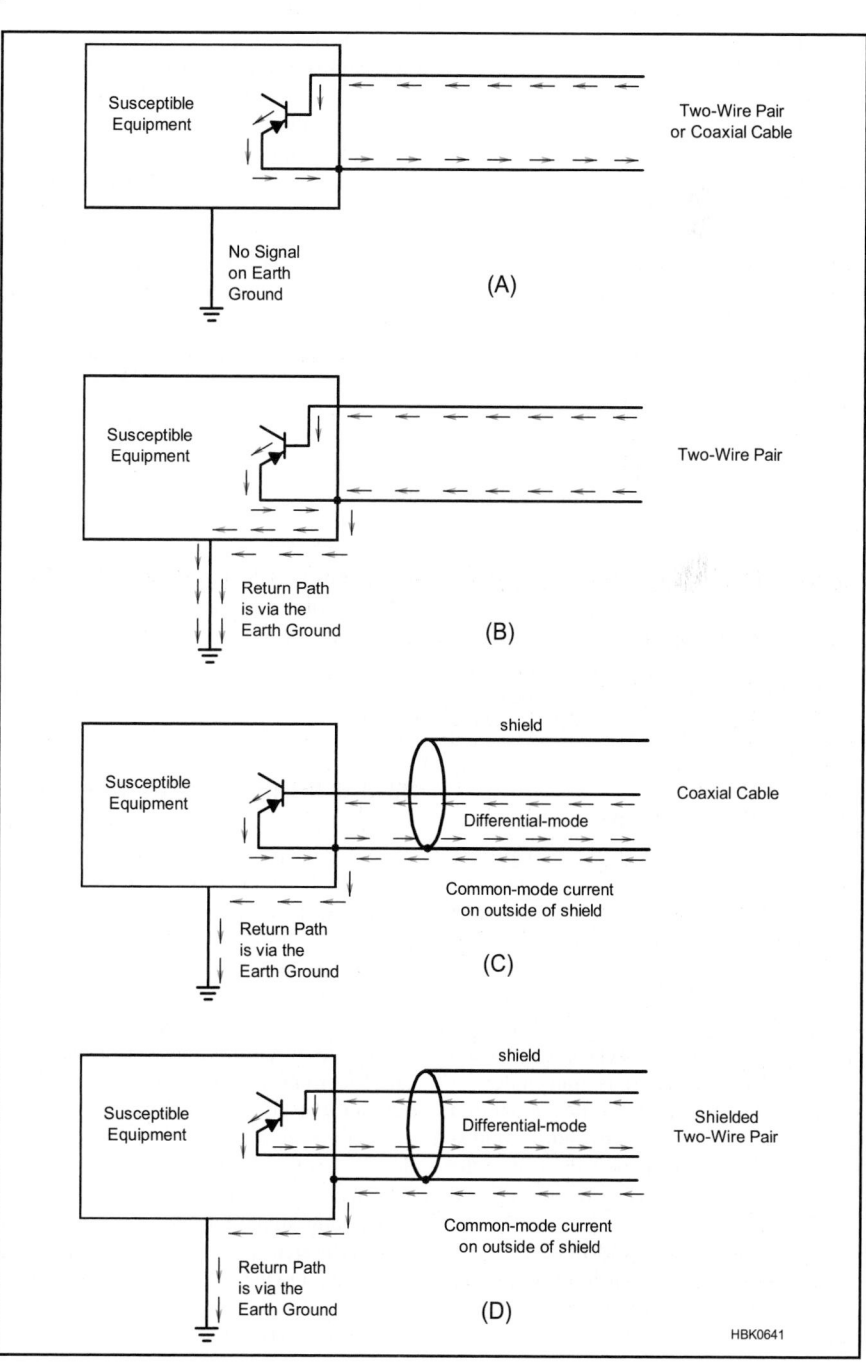

Fig 27.3 — Typical configurations of common-mode and differential-mode current. The drawing in A shows the currents of a differential-mode signal while B shows a common-mode signal with currents flowing equally on all of the source wires. In C, a common-mode signal flows on the outside of a coaxial cable shield with a differential-mode signal inside the cable. In D, the differential-mode signal flows on the internal wires while a common-mode signal flows on the outside of the cable shield.

pling occurs when two circuits are coupled electrically through capacitance. Typical RFI problems you are likely to encounter often include multiple paths, such as conduction and radiation. (See the section Shields and Filters, also Ott, sections 2.1-2.3.)

27.3.2 Differential-Mode vs Common-Mode Signals

The path from source to victim almost always includes some conducting portion, such as wires or cables. RF energy can be conducted directly from source to victim, be conducted onto a wire or cable that acts as an antenna where it is radiated, or be picked up by a conductor connected to the victim that acts like an antenna. When the noise signal is traveling along the conducted portion of the path, it is important to understand the differences between *differential-mode* and *common-mode* conducted signals (see **Fig 27.3**).

Differential-mode currents usually have two easily identified conductors. In a two-wire transmission line, for example, the signal leaves the generator on one wire and returns on the other. When the two conductors are in close proximity, they form a transmission line and the two signals have opposite polarities as shown in Fig 27.3A. Most desired signals, such as the TV signal inside a coaxial cable or an Ethernet signal carried on CAT5 network cable, are conducted as differential-mode signals.

A common-mode circuit consists of several wires in a multi-wire cable acting as if they were a single current path as in Fig 27.3B. Common-mode circuits also exist when the outside surface of a cable's shield acts as a conductor as in Figs 27.3C and 27.3D. (See the chapter on **Transmission Lines** for a discussion about isolation between the shield's inner and outer surfaces for RF signals.) The return path for a common-mode signal often involves earth ground.

27.3.3 Types of RFI

There are four basic types of RFI that apply to Amateur Radio. The first two occur in the following order of likelihood when the interfering source is an amateur transmitter intentionally generating a radio signal:

1) Fundamental Overload — Disruption or degradation of a device's function in the presence of a transmitter's fundamental signal (the intended signal from the transmitter).

2) Spurious Emissions — Reception of a radio signal interfered with by spurious emissions from the transmitter as defined in the previous section on Part 97 definitions and Fig 27.1.

The second two types of RFI occur, again in order of likelihood, when the reception of a desired signal is interfered with by RF energy received along with the desired signal.

3) External Noise Sources — Reception of a radio signal interfered with by RF energy transmitted incidentally or unintentionally by a device that is not a licensed transmitter

4) Intermodulation — Reception of a radio signal interfered with by intermodulation distortion (IMD) products generated inside or outside of the receiver

As an RFI troubleshooter, start by determining which of these is involved in your interference problem. Once you know the type of RFI, selecting the most appropriate cure for the problem becomes much easier.

27.3.4 Spurious Emissions

All transmitters generate some (hopefully few) RF signals that are outside their intended transmission bandwidth — out-of-band emissions and spurious emissions as illustrated in Fig 27.1. Out-of-band signals result from distortion in the modulation process or consist of broadband noise generated by the transmitter's oscillators that is added to the intended signal. Harmonics, the most common spurious emissions, are signals at integer multiples of the operating (or fundamental) frequency.

Transmitters may also produce broadband noise and/or parasitic oscillations as spurious emissions. (Parasitic oscillations are discussed in the **RF Power Amplifiers** chapter.) Overdriving an amplifier often creates spurious emissions. Amplifiers not meeting FCC certification standards but sold illegally are frequent sources of spurious emissions.

Regardless of how the unwanted signals are created, if they cause interference, FCC regulations require the operator of the transmitter to correct the problem. The usual cure is to adjust or repair the transmitter or use filters at the transmitter output to block the spurious emissions from being radiated by the antenna.

27.3.5 Fundamental Overload

Most cases of interference caused by an amateur transmission are due to *fundamental overload*. The world is filled with RF signals. Properly designed radio receivers of any sort should be able to select the desired signal, while rejecting all others. Unfortunately, because of design deficiencies such as inadequate shields or filters, some radio receivers are unable to reject strong out-of-band signals. Electronic equipment that is not a radio receiver can also suffer from fundamental overload from similar design shortcomings. Both types of fundamental overload are common in consumer electronics.

A strong signal can enter equipment in several different ways. Most commonly, it is conducted into the equipment by connecting wires and cables. Possible RFI conductors include antennas and feed lines, interconnect-

ing cables, power cords, and ground wires. TV antennas and feed lines, telephone or speaker wiring and ac power cords are the most common points of entry.

If the problem is a case of fundamental overload, significant improvement can often be observed just by moving the victim equipment and the signal source farther away from each other. The effect of an interfering signal is directly related to its strength, diminishing with the square of the distance from the source. If the distance from the source doubles, the strength of the electromagnetic field decreases to one-fourth of its power density at the original distance from the source. An attenuator is often a weapon of choice when encountering this type of problem. Reducing the level of the offending strong signal returns the receiver to normal operation and causes the undesirable effects to disappear. This characteristic can often be used to help identify an RFI problem as fundamental overload. If reducing the strength of the signal source causes the same effect that is also a signature of fundamental overload.

27.3.6 External Noise Sources

Most cases of interference to the Amateur Service reported to the FCC are eventually determined to involve some sort of external noise source, rather than signals from a radio transmitter. Noise in this sense means an RF signal that is not essential to the generating device's operation. The most common external noise sources are electrical, primarily power lines. Motors and switching equipment can also generate electrical noise.

External noise can also come from unlicensed Part 15 RF sources such as computers and networking equipment, video games, appliances, and other types of consumer electronics. Regardless of the source, if you determine the problem to be caused by external noise, elimination of the noise must take place at the source. As an alternative, several manufacturers also make noise canceling devices that can help in some circumstances.

27.3.7 Intermodulation Distortion

As discussed in the chapter on **Receivers**, intermodulation distortion (IMD) is caused by two signals combining in such a way as to create *intermodulation products* — signals at various combinations of the two original frequencies. The two original signals may be perfectly legal, but the resulting *intermodulation distortion products* may occur on the frequencies used by other signals and cause interference in the same way as a spurious signal from a transmitter. Depending on the nature of the generating signals, "intermod"

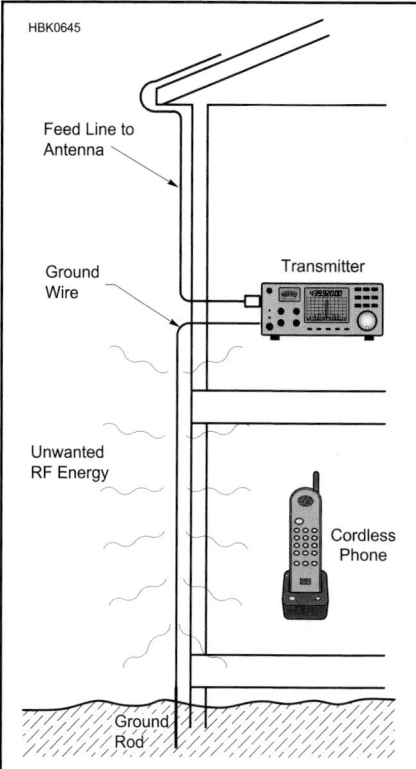

Fig 27.4 — An earth ground connection can radiate signals that might cause RFI to nearby equipment. This can happen if the ground connection is part of an antenna system or if it is connected to a coaxial feed line carrying RF current on the outside of the shield.

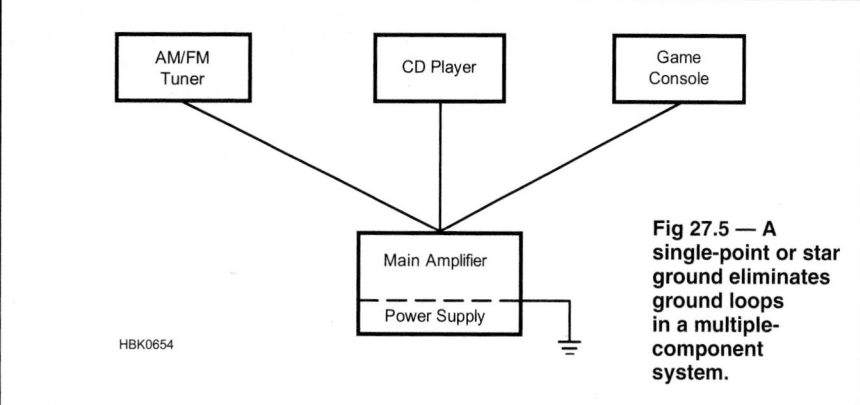

Fig 27.5 — A single-point or star ground eliminates ground loops in a multiple-component system.

can be intermittent or continuous. IMD can be generated inside a receiver by large signals or externally by signals mixing together in non-linear junctions or connections.

27.3.8 Ground Connections

An electrical ground is not a huge sink that somehow swallows noise and unwanted signals. Ground is a *circuit* concept, whether the circuit is small, like a radio receiver, or large, like the propagation path between a transmitter and cable-TV installation. Ground forms a universal reference point between circuits.

While grounding is not a cure-all for RFI problems, ground is an important safety component of any electronics installation. It is part of the lightning protection system in your station and a critical safety component of your house wiring. Any changes made to a grounding system must not compromise these important safety considerations. Refer to the **Safety** chapter for important information about grounding.

Many amateur stations have several connections referred to as "grounds"; the required safety ground that is part of the ac wiring system, another required connection to earth for lightning protection, and perhaps another shared connection between equipment for RFI control. These connections can interact with each other in ways that are difficult to predict.

Rearranging the station ground connections may cure some RFI problems in the station by changing the RF current distribution so that the affected equipment is at a low-impedance point and away from RF "hot spots." Creating a low-impedance connection between your station's equipment is easy to do and will help reduce voltage differences (and current flow) between pieces of equipment. However, a station ground is not always the cure-all that some literature has suggested.

LENGTH OF GROUND CONNECTIONS

The required ground connection for lightning protection between the station equipment and an outside ground rod is at least several feet long in most practical installa-

tions. (See the **Safety** chapter for safety and lightning protection ground requirements.)

In general, however, a long connection to earth should be considered as part of an RFI problem only to the extent that it is part of the antenna system. For example, should a long-wire HF antenna end in the station, a ground connection of *any* length is a necessary and useful part of that antenna and will radiate RF.

At VHF a ground wire can be several wavelengths long — a very effective antenna for any harmonics that could cause RFI! For example, in **Fig 27.4**, signals radiated from the required safety ground wire could very easily create an interference problem in the downstairs electronic equipment.

GROUND LOOPS

A *ground loop* is created by a continuous conductive path around a series of equipment enclosures. While this does create an opportunity for lightning and RFI susceptibility, the ground loop itself is rarely a cause of problems at RF. Ground loops are usually associated with problems of audio hum or buzz caused by coupling to power-frequency magnetic fields and currents. To avoid low-frequency ground-loop issues, use short, properly-shielded cables that are the minimum length required to connect the equipment and bundle them together to minimize the area of any enclosed loop. Ground-loop problems at RF are minimized by the use of a single-point or star ground system as shown in **Fig 27.5**.

27.4 Identifying the Type of RFI Source

It is useful to place an offending noise source into one of several broad categories at the early stages of any RFI investigation. Since locating and resolution techniques can vary somewhat for each type, the process of locating and resolving RFI problems should begin with identifying the general type of RFI source.

It is often impossible to identify the exact type of device generating the RFI from the sound of the interference. Because there are many potential sources of RFI, it is often more important to obtain and interpret clues from the general noise characteristics and the patterns in which it appears.

A source that exhibits a repeatable pattern during the course of a day or week, for example, suggests something associated with human activity. A sound that varies with or is affected by weather suggests an outdoor source. Noise that occurs in a regular and repeating pattern of peaks and nulls as you tune across the spectrum, every 50 kHz for example, is often associated with switchmode power supply or similar pulsed-current devices. A source that exhibits fading or other sky wave characteristics suggests something that is not local. A good ear and careful attention to detail will often turn up some important clues. A detailed RFI log can often help, especially if maintained over time.

Noise can be characterized as broadband or narrowband — another important clue. *Broadband noise* is defined as noise having a bandwidth much greater than the affected receiver's operating bandwidth and is reasonably uniform across a wide frequency range. Noise from arcs and sparks, such as power-line noise, tend to be broadband. *Narrowband noise* is defined as noise having a bandwidth less than the affected receiver's bandwidth. Narrowband noise is present on specific, discrete frequencies or groups of frequencies, with or without additional modulation. In other words, if you listened to the noise on an SSB receiver, tuning would cause its sound to vary, just like a regular signal. Narrowband noise often sounds like an unmodulated carrier with a frequency that may drift or suddenly change. Microprocessor clock harmonics, oscillators and transmitter harmonics are all examples of narrowband noise.

27.4.1 Identifying Noise from Part 15 Devices

The most common RFI problem reported to the ARRL comes from an unknown and unidentified source. Part 15 devices and other consumer equipment noise sources are ubiquitous. Although the absolute signal level

from an individual noise source may be small, their increasing numbers makes this type of noise a serious problem in many suburban and urban areas. The following paragraphs describe several common types of electronic noise sources.

Electronic devices containing oscillators, microprocessors, or digital circuitry produce RF signals as a byproduct of their operation. The RF noise they produce may be radiated from internal wiring as a result of poor shielding. The noise may also be conducted to external, unshielded or improperly shielded wiring as a common-mode signal where it radiates noise. Noise from these devices is usually narrowband that changes characteristics (frequency, modulation, on-off pattern) as the device is used in different ways.

Another major class of noise source is equipment or systems that control or switch large currents. Among them are variable-speed motors in products as diverse as washing machines, elevators, and heating and cooling systems. Charging regulators and control circuitry for battery and solar power systems are a prolific source of RF noise. So are switchmode power supplies for computers and low-voltage lighting. This type of noise is only present when the equipment is operating.

Switchmode supplies, solar controllers and inverters often produce noise signals every N kHz, with N typically being from 5 to 50 or more kHz, the frequency at which current is switched. This is different from noise produced by spark or arc sources that is uniform across a wide bandwidth. This pattern is often an important clue in distinguishing switching noise from power-line or electrical noise.

Wired computer networks radiate noise directly from their unshielded circuitry and from network and power supply cables. The noise takes two forms — broadband noise and modulated carriers at multiple frequencies within the amateur bands. As an example,

Ethernet network interfaces often radiate signals heard on a receiver in CW mode. 10.120, 14.030, 21.052 and 28.016 MHz have been reported as frequencies of RFI from Ethernet networks. Each network interface uses its own clock, so if you have neighbors with networks you'll hear a cluster of carriers around these frequencies, ±500 Hz or so.

In cable TV systems video signals are converted to RF across a wide spectrum and distributed by coaxial cable into the home. Some cable channels overlap with amateur bands, but the signals should be confined within the cable system. No system is perfect, and it is common for a defective coax connection to allow leakage to and from the cable. When this happens, a receiver outside the cable will hear RF from the cable and the TV receiver may experience interference from local transmissions. Interference to and from cable TV signals is discussed in detail later in this chapter.

27.4.2 Identifying Power-line and Electrical Noise

POWER-LINE NOISE

Next to external noise from an unknown source, the most frequent cause of an RFI problem reported to the ARRL involving a known source is power-line noise. (For more information on power-line noise, see the book *AC Power Interference Handbook*, by Marv Loftness KB7KK.) Virtually all power-line noise originating from utility equipment is caused by spark or arcing across some hardware connected to or near a power line. A breakdown and ionization of air occurs and current flows across a gap between two conductors, creating RF noise as shown in **Fig 27.6**. Such noise is often referred to as "gap noise" in the utility power industry. The gap may be caused by broken, improperly installed or loose hardware. Typical culprits include insufficient and inadequate hardware spacing such as a gap between a ground wire and a staple. Contrary to common misconception, corona discharge is rarely, if ever, a source of power-line noise.

While there may not be one single conclusive test for power-line noise, there are a number of important tell-tale signs. On an AM or SSB receiver, the characteristic raspy buzz or frying sound, sometimes changing in intensity as the arc or spark sputters a bit, is often the first and most obvious clue.

Power-line noise is typically a broadband type of interference, relatively constant across a wide spectrum. Since it is broadband noise, you simply can't change frequency to eliminate it. Power-line noise is usually, but not al-

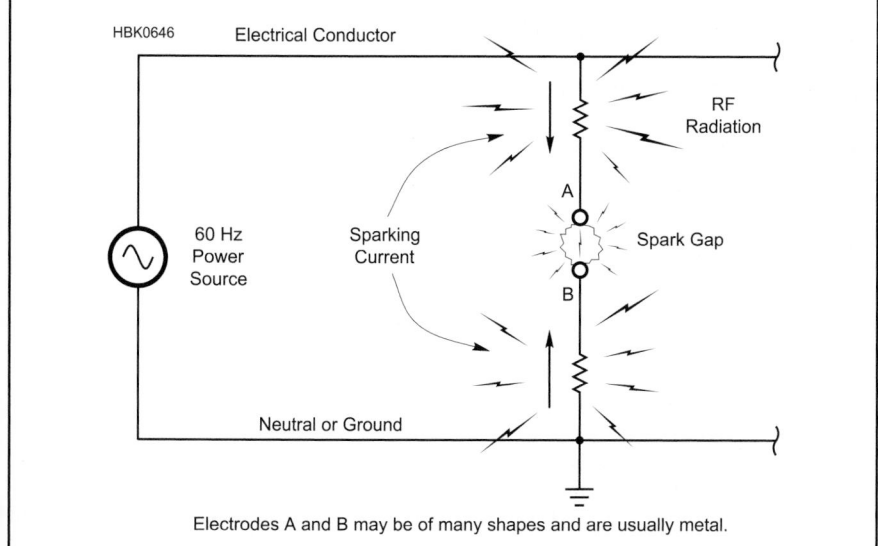

Fig 27.6 — The gap noise circuit on a power line — simplified. (From Loftness, *AC Power Interference Handbook*)

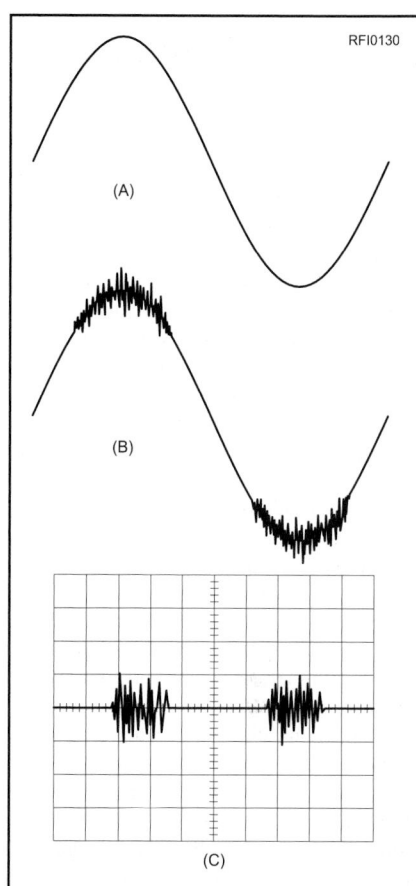

Fig 27.7 — The 60-Hz signal found on quiet power lines is almost a pure sine wave, as shown in A. If the line, or a device connected to it, is noisy, this will often add visible noise to the power-line signal, as shown in B. This noise is usually strongest at the positive and negative peaks of the sine wave where line voltage is highest. If the radiated noise is observed on an oscilloscope, the noise will be present during the peaks, as shown in C.

ways, stronger on lower frequencies. It occurs continuously across each band and up through the spectrum. It can cause interference from the AM broadcast band through UHF, gradually tapering off as frequency increases. If the noise is not continuous across all of an amateur band, exhibits a pattern of peaks and nulls at different frequencies, or repeats at some frequency interval, you probably do not have power-line noise.

The frequency at which power-line noise diminishes can also provide an important clue as to its proximity. The closer the source, the higher the frequency at which it can be received. If it affects VHF and UHF, the source is relatively close by. If it drops off just above or within the AM broadcast band, it may be located some distance away — up to several miles.

Power-line noise is often affected by weather if the source is outdoors. It frequently changes during rain or humid conditions, for example, either increasing or decreasing in response to moisture. Wind may also create fluctuations or interruptions as a result of line and hardware movement. Temperature effects can also result from thermal expansion and contraction.

Another good test for power-line noise requires an oscilloscope. Remember that power-line noise occurs in bursts most frequently at a rate of 120 bursts per second and sometimes at 60 bursts per second. Observe the suspect noise from your radio's audio output. (Note: The record output jack works best if available). Use the AM mode with wide filter settings and tune to a frequency without a station so the noise can be heard clearly. Use the LINE setting of the oscilloscope's trigger subsystem to synchronize the sweep to the line. Power-line noise bursts will remain stable on the display

and should repeat every 8.33 ms (a 120-Hz repetition rate) or less commonly, 16.67 ms (60 Hz), if the gap is only arcing once per cycle. (This assumes the North American power-line frequency of 60 Hz.) See **Fig 27.7** for an explanation. If a noise does not exhibit either of these characteristics, it is probably not power-line noise.

If a local TV station is transmitting analog TV signals on a lower VHF channel (very few remain as of early 2013), additional clues may be obtained by viewing the noise pattern on an analog TV set using an antenna (not a cable TV connection). Power-line noise usually appears as two horizontal bars that drift slowly upward on the screen through the picture. (This is due to the difference between the NTSC signal's 59.94 Hz field rate and the 60 Hz power-line frequency.) As one bar rolls off the screen at the top of the display, a new one simultaneously forms at the bottom. In cases where the noise is occurring at 60 bursts per second, there will be only one bar on the display. In addition, the power-line noise bursts may have slightly different characteristics at the positive and negative peaks. This can cause each half of the cycle to have a slightly different pattern on the screen.

ELECTRICAL NOISE

Electrical noise sounds like power-line noise, but is generally only present in short bursts or during periods when the generating equipment or machinery is in use. Noise that varies with the time of day, such as daytime-only or weekends-only, usually indicates some electrical device or appliance being used on a regular basis and not power-line noise. Unless it is associated with climate control or HVAC system, an indoor RFI source of electrical noise less likely to be

affected by weather than power-line noise.

ELECTRIC FENCES

A special type of electrical noise that is easy to identify is the "pop…pop…pop" of an electric fence. High voltage is applied to the fence about once a second by a charging unit. Arcs will occur at corroded connections in the fence, such as at a gate hook or splice. If brush or weeds touch the fence, the high-voltage will cause an arc at those points until the vegetation burns away (the arc will return when the vegetation re-grows). Each arc results in a short burst of broadband noise, received as a "tick" or "pop" in an HF receiver.

27.4.3 Identifying Intermodulation Distortion

Intermodulation distortion (IMD) often occurs within a receiver when two strong signals combine to produce intermodulation products that can interfere with a desired signal. Since the products are generated internally to the receiver, the strong signals must be filtered out or attenuated before they can enter the receiver circuits in which IMD products are generated.

Mixing of signals can also occur in any non-linear junction where the original signals are both strong enough to cause current to flow in the junction. This is a particular problem at multi-transmitter sites, such as broadcast facilities and industrial or commercial communications sites. Non-linear junctions can be formed by loose mechanical contacts in metal hardware, corroded metal junctions, and by semiconductor junctions connected to wires that act as antennas for the strong signals. Non-linear junctions also detect or demodulate RF signals to varying degrees, creating interfering audio or dc signals in some cases. IMD products generated externally cannot be filtered out at the receiver and must be eliminated where the original signals are being combined.

An intermodulation product generated externally to a receiver often appears as an intermittent transmission, similar to a spurious emission. (See the **Receivers** chapter for more information on intermodulation.) It is common for strong signals from commercial paging or dispatch transmitters sharing a common antenna installation to combine and generate short bursts of voice or data signals. AM broadcast transmissions can combine to produce AM signals with the modulation from both stations audible.

Like external intermodulation products,

those generated by a receiver acting non-linearly appear as combinations of two or more strong external signals. For example, intermodulation from two SSB or FM voice signals produces somewhat distorted signals with the modulating signals of both stations. Since the two signals are not synchronized, the intermodulation products come and go unpredictably, present only when both of the external signals are present.

Intermodulation within a receiver can often be detected simply by activating the receiver's incoming signal attenuator. If attenuating the incoming signals causes the intermodulation product to by reduced in strength by a greater amount than that of the applied attenuation, intermodulation in the receiver is a strong possibility. Since receivers vary in their IMD performance, differences in the interfering signal strength between receivers is also an indication of intermodulation.

The following simple "attenuator test" can be used to identify an IMD product, even in cases where it appears similarly in multiple receivers:

• If your receiver does not have one, install a step attenuator at its RF input. If you use an attenuator internal to your receiver, it must attenuate the RF at the receiver's input.

• Tune the receiver to the suspected intermod product with the step attenuator set to 0 dB. Note the signal level.

• Add a known amount of attenuation to the signal. Typically 10 or 20 dB makes a good starting point for this test.

• If the suspect signal drops by more than the amount of added attenuation, the suspect

signal is an IMD product. For example, if you add 10 dB of attenuation, and the signal drops by 30 dB, you have identified an IMD product.

• You can also compare the reduction in signal level between the suspect IMD product and a known genuine signal with and without the added attenuation. Use a known genuine signal that is about the same strength as the suspect signal with no added attenuation. If the suspect signal drops by the same as the added attenuation, it is not an IMD product.

Intermodulation distortion can be cured in a number of ways. The goal is to reduce the strength of the signals causing IMD so that the receiver circuits can process them linearly and without distortion.

If IMD is occurring in your receiver, filters that remove the strong unwanted signals causing the IMD while passing the desired signal are generally the best approach since they do not compromise receiver sensitivity. (The chapter on **Receivers** discusses how to add additional filtering to your receiver.) Turning off preamplifiers, adding or increasing the receiver's attenuation, and reducing its RF gain will reduce the signal strength in your receiver. Antenna tuners and external band-pass filters can also act as a filter to reduce IMD from out-of-band signals. Directional beams and antennas with a narrower bandwidth can also help, depending on the circumstances of your particular problem.

IMD can also be created in a broadcast FM or TV receiver or preamplifier. The solutions are the same — add suitable filters or reduce overall signal levels and gain so that the strong interfering signal can be processed linearly.

27.5 Locating Sources of RFI

Locating an offending device or noise source might sometimes seem like trying to find a needle in a haystack. With a little patience and know-how, it is often possible to find the source of a problem in relatively short order. RF detective work is often required and some cases require a little more perseverance than others. In any case, armed with some background and technique, it is often easier to find an offending source than the first-time RFI investigator might expect.

Whenever an unknown source of interference becomes an issue, begin the process of identifying the source by verifying that the problem is external to your radio. Start by removing the antenna connection. If the noise disappears, the source is external to your radio and you are ready to begin hunting for the noise source.

27.5.1 Noise Sources Inside Your Home

Professional RFI investigators and the experiences of the ARRL RFI desk confirm that most RFI sources are ultimately found to be in the complainant's home. Furthermore, locating an in-house source of RFI is so simple it that it makes sense to start an investigation by simply turning off your home's main circuit breaker while listening to the noise with a battery-powered portable radio. (Don't forget that battery-powered equipment may also be a noise source — remove the batteries from consumer devices, as well.) If the noise goes away, you know the source is in your residence. After resetting the breaker, you can further isolate the source by turning off individual breakers one at a time. Once you

know the source circuit, you can then unplug devices on that circuit to find it.

CAUTION: *Do not attempt to remove cartridge fuses or operate exposed or open-type disconnects if it is possible to make physical contact with exposed electrical circuits.*

27.5.2 Noise Sources Outside Your Home

It is often possible to locate a noise source outside your home with a minimum of equipment and effort. Because of Part 15's absolute emissions limits, most Part 15 noise sources are within a few hundred feet of the complainant's antenna. They are also often on the same power transformer secondary system as the complainant. This typically reduces the number of possible residences to relatively few.

If the noise source is not compliant with Part 15 limits, it may be blocks or even thousands of feet from your station.

Electrical noise sources in a home, such as an arcing thermostat or a noisy washing machine controller, can also be tracked down in the same way as noise from consumer electronic devices. Electrical noise from an incidental emitter, such as a power line, can propagate much farther than noise from an otherwise legal unintentional emitter. Some Part 15 devices, battery chargers for electric scooters and wheelchairs, for example, are notorious for exceeding Part 15 absolute emissions limits on conducted noise.

The following procedure can be used to trace a noise source to a private home, town house, apartment, or condominium. The number of homes that could be host to a source generating noise could make searching house by house impractical. In such cases, use noise tracking techniques discussed in the following sections to narrow the search to a more reasonable area.

1) Verify that the noise is active before attempting to locate it. Don't forget this all-important first step. You cannot find the source when it's not present.

2) If possible, use a beam to record bearings to the noise before leaving your residence. Walk or drive through the neighborhood with particular emphasis in the direction of the noise, if known. Try to determine the rough geographic area over which the noise can be heard. If the geographic area over which you can hear the noise is confined to a radius of several hundred feet or less, or it diminishes quickly as you leave your neighborhood, this confirms you are most likely dealing with a Part 15 consumer device.

3) Since the noise will be strongest at an electrical device connected to the residence containing the source, you want to measure the noise at a device common to the exterior of all the potential homes. Suitable devices include electric meters, main service breakers (whether outside or in a utility room), front porch lights, electric lamp posts, outside air conditioner units, or doorbell buttons. Whatever radiator you choose, it should be accessible at each home. The device you select to test as the noise radiator will be referred to in these instructions as the "radiator." Using the same type of device as a test point at each home helps obtain consistent results.

4) You are now ready to compare the relative signal strengths at the radiator on each of the potential source residences. Use a detector suitable to receive the noise, typically a battery-powered receiver. Preferably, the receiver should have a variable RF gain control. An external step attenuator will also work if the antenna is external to the radio. If the antenna can be removed, a probe can also be made from a small piece of wire or paper clip

to reduce the receiver's sensitivity. Start by holding the detector about two inches from the radiator at the residence where the noise source may be located. Turn the detector's RF gain control down to a point where you can just barely hear the noise. Alternately, increase the attenuation if using an external step attenuator. Record the RF gain or attenuator setting for each test.

5) Proceed to the next residence. Again hold the detector approximately two inches from the radiator. (The detector should be placed at the same location at each residence, as much as is practical.) Since you had previously set the detector to just barely hear the noise at the residence having the interference problem, you can move on to the next residence if you do not hear the noise. Remember, in order for your detector to hear the noise at then next house, the noise level will have to be the same or higher than the previous location. If you need to increase the detector's sensitivity to be able to hear the noise, you are moving away from the noise source.

6) When you reach the next residence, if the level is lower or not heard, you're moving further from the source. Continue your search to residences in other directions or across the street. If the level is higher, then you're headed in the right direction. Be sure to turn the gain control down to the point of just barely hearing the noise as its strength increases.

7) Continue on to the next house, repeating the previous steps as necessary. The residence with the source will be the one with the strongest noise at the radiator.

Depending on the circumstances of a particular situation, it may be possible to first isolate the power pole to which the source residence is connected. Walk or drive along the power lines in the affected area while listening to the noise with a battery-powered radio. Continue to decrease the receiver's RF gain as the noise gets louder, thus reducing the area over which you can hear it. Finally, isolate the loudest pole by reducing the RF gain to a point at which you can hear it at only one pole. Once the pole has been isolated, look to see which houses are connected to its transformer. Typically this will reduce the number of potential residences to a very small number.

CAUTION: *Always observe good safety practices! Only qualified people familiar with the hazards of working around energized electrical equipment should inspect power-line or other energized circuitry.*

When attempting to isolate the pole, it is often best to use the highest frequency at which you can hear the noise. Noise can exhibit peaks and nulls along a power line that are a function of its wavelength. Longer wavelengths can therefore make it difficult to pinpoint a particular point along a line. Furthermore, longer wavelength signals typi-

cally propagate further along power lines. You can often reduce your search area by simply increasing the frequency at which you look for the noise.

In some cases, tuning upward in frequency can also be used to attenuate noise. This can be especially helpful in cases where your receiver does not have an RF gain control. As mentioned previously, switchmode power supplies typically generate noise that exhibits a regular and repeating pattern of peaks and nulls across the spectrum. While a typical interval might be every 50 kHz or so, the noise will often start to diminish at the highest frequencies. The peaks in some cases might drift over time, but tuning to the highest frequency at which you can hear the noise will often attenuate it enough to help locate it. If the peak drifts however, be sure to keep your receiver set on the peak as you attempt to locate the source.

Under FCC rules, the involved utility is responsible for finding and correcting harmful interference that is being generated by its own equipment. In cases where a utility customer is using an appliance or device that generates noise, the operator of the device is responsible for fixing it — even if the noise is conducted and radiated by the power company's power lines.

27.5.3 Approaching Your Neighbor

Once you identify the source residence and approach your neighbor, the importance of personal diplomacy simply cannot be overstated. The first contact regarding an RFI problem between a ham and a neighbor is often the most important; it is the start of all future relations between the parties. The way you react and behave when you first discuss the problem can set the tone for everything that follows. It is important, therefore, to use a diplomatic path from the very start. A successful outcome can depend upon it!

A self-help guide for the consumer published jointly by the ARRL and the Consumer Electronics Association (CEA) often proves helpful when discussing an interference problem with a neighbor. Entitled *What To Do if You Have an Electronic Interference Problem*, it may be printed and distributed freely. It is available on the ARRL website at **www. arrl.org/information-for-the-neighbors-of-hams** and also on the CD-ROM accompanying this book. Be sure to download and print a copy for your neighbor before you approach him or her.

With the noise active and with a copy of the pamphlet handy, approach your neighbor with a radio in hand, preferably an ordinary AM broadcast or short-wave receiver. Let them hear it but not so loud that it will be offensive. Tell them this is the problem you are experiencing and you believe the source may be in their home. Don't suggest what you think the cause is. If you're wrong, it often makes matters worse. Give them the pamphlet and tell them it will only take a minute to determine whether the source is in their home. Most neighbors will agree to help find the source, and if they agree to turn off circuit breakers, it can be found very quickly. Start with the main breaker to verify you have the correct residence, then the individual breakers to find the circuit. The procedure then becomes the same as described for your own residence.

27.5.4 Radio Direction Finding

Radio direction finding (RDF) can be a highly effective method to locate an RFI source although it requires more specialized equipment than other methods. Professional interference investigators almost always use radio direction finding techniques to locate power-line noise sources. See the **Antennas** chapter for more information on direction finding antennas.

A good place to start, whenever possible, is at the affected station. Use an AM receiver, preferably one with a wide IF bandwidth. An RF gain control is particularly helpful but an outboard step attenuator can be a good substitute. If there is a directional beam capable of receiving the noise, use it on highest frequency band at which the noise can be heard using the antenna. If you can hear the noise

at VHF or UHF, you'll typically want to use those frequencies for RDF.

Select a frequency at which no other stations or signals are present and the antenna can discern a directional peak in the noise. Rotate the beam as required to get a bearing on the noise, keeping the RF gain at a minimum. Repeat with a complete 360° sweep using the minimum RF gain possible to hear the noise in its loudest direction. Try to decrease the RF gain to a point at which the noise clearly comes from one and only one direction. You can simultaneously increase the AF gain as desired to hear the noise.

Distant sources, including power-line noise, are generally easier to RDF at HF than nearby sources. Whenever possible, it's almost always better to use VHF or UHF when in close proximity to a source. Tracking a source to a specific residence by RDF at HF is sometimes possible. Such factors as balance and geometry of a home's internal wiring, open switch circuits and distance, may cause the residence to appear somewhat as a point source.

If the search is being conducted while mobile or portable, VHF and UHF are t ypically the easiest and most practical antennas. Small handheld Yagi antennas for 2 meters and 440 MHz are readily available and can serve double duty when operating portable. Many handheld receivers can be configured to receive AM on the VHF bands. Be sure to check your manual for this feature. VHF Aircraft band or "Air band" receivers

are also a popular choice since they receive AM signals.

Using RDF to locate an HF noise source while in motion presents significant challenges. Conducted emissions are typical from a consumer device or appliance. In this case, the emissions can be conducted outside the residence and on to the power line. The noise can then propagate along neighborhood distribution lines, which in turn acts as an antenna. The noise can often exhibit confusing peaks and nulls along the line, and if in the vicinity of a power line radiating it, RDF can be extremely difficult, if not impossible. Depending on the circumstances, you could literally be surrounded by the near field of an antenna! You would generally want to stay away from power lines and other potential radiators when searching at HF.

Antennas for HF RDF while walking typically include small loops and ferrite rod antennas. In some cases, a portable AM broadcast radio with a ferrite rod antenna can be used for direction finding. An HF dipole made from a pair of whip antennas may be able to be used to get an approximate bearing toward the noise. Mount the dipole about 12 feet above ground (remembering to watch out for overhead conductors!) and rotate to null out the noise. For all three types of antenna, there will be two nulls in opposite directions. Note the direction of the null. Repeat this procedure from another location then triangulate to determine the bearing to the noise.

27.6 Power-line Noise

This chapter's section "Identifying RFI Source Types" describes power-line noise, its causes, and methods to identify it. Power-line noise is a unique problem in several respects. First and foremost, the offending source is never under your direct control. You can't just simply "turn it off" or unplug the offending device. Nor will the source be under the direct control of a neighbor or someone you are likely to know. In the case of power-line noise, the source is usually operated by a company, municipality, or in some cases, a cooperative. Furthermore, shutting down a power line is obviously not a practical option.

Another unique aspect of power-line noise is that it almost always involves a defect of some sort. The cure for power-line noise is to fix the defect. This is almost always a utility implemented repair and one over which you do not have any direct control.

FCC rules specify that that the operator of a device causing interference is responsible for fixing it. Whenever encountering a power-line noise problem, you will be dealing with

a utility and won't have the option of applying a relatively simple technical solution to facilitate a cure, as you would if the device were located in your home. Utilities have a mixed record when it comes to dealing with power-line noise complaints. In some cases, a utility will have a budget, well-trained personnel, and equipment to quickly locate and address the problem. In other cases, however, the utility is simply unable to effectively deal with power-line noise complaints or even denies their equipment can cause RFI.

What does this mean for an amateur with a power-line noise complaint? Utilities can be of any size from large corporations to local co-operatives or city-owned systems. Regardless of the category in which your utility may fall, it must follow Part 15 of the FCC rules. Dealing with a company, coop or municipality, however, as opposed to a device in your home, or a nearby neighbor that you know personally, can present its own set of unique challenges. Multiple parties and individuals are often involved, including an RFI investi-

gator, a line crew and associated management. In some cases, the utility may never have received an RFI complaint before yours.

27.6.1 Before Filing a Complaint

Obviously, before filing a complaint with your local utility, it is important to verify the problem as power-line noise as best as possible and verify that it is not caused by a problem with electrical equipment in your home. Other sources, such as lighting devices and motors, can mimic power-line noise, especially to an untrained ear. Don't overlook these important steps. Attempting to engage your utility in the resolution of an RFI problem can not only waste time but can be embarrassing if the source is right in your own home!

Utilities are not responsible for noise generated by customer-operated devices — *even if the noise is being radiated by the power lines*. They are responsible for fixing only that noise which is being generated by their equipment.

27.6.2 Filing a Complaint

Once you have verified the problem to be power-line noise (see this chapter's section on Identifying Power-line and Electrical Noise) and that it is not coming from a source in your home or a nearby residence, contact your utility's customer service department. In addition to your local phone book, customer service phone numbers are included on most power company websites.

It is important to maintain a log during this part of the process. Be sure to record any "help ticket" numbers that may be assigned to your complaint as well as names, dates and a brief description of each conversation you have with electric company personnel. If you identify specific equipment or power poles as a possible noise source, record the address and any identifying numbers on it.

Hopefully, your complaint will be addressed in a timely and professional manner. Once a noise source has been identified, it is up to the utility to repair it within a reasonable period. You and the utility may not agree on what constitutes a reasonable period, but attempt to be patient. If no action is taken after repeated requests, reporting the complaint to the ARRL and requesting assistance may be in order. (Before contacting the ARRL review The Cooperative Agreement, a section of this chapter.)

It is also important to cooperate with utility personnel and treat them with respect. Hostile and inappropriate behavior is almost always counter-productive in these situations. Remember, you want utility and other related personnel to help you — not avoid you. Even if the utility personnel working on your case seem unqualified, hostile behavior has historically never been a particularly good motivator in these situations. In fact, most protracted power-line noise cases reported to the ARRL began with an altercation in the early stages of the resolution process. In no case did it help or expedite correction of the problem.

27.6.3 Techniques for Locating Power-line Noise Sources

Radio direction finding (RDF) techniques typically offer the best and most efficient approach to locating most power-line noise sources. It is the primary method of choice used by professionals. While RDF is usually the most effective method, it also requires some specialized equipment, such as a hand-held beam antenna. Although specialized professional equipment is available for RDF, hams can also use readily available amateur and homebrew equipment successfully. The CD-ROM accompanying this *Handbook* includes some power-line noise locating equipment projects you can build.

Although it is the utility's responsibility to locate a source of noise emanating from its equipment, many companies simply do not possess the necessary expertise or equipment to do so. As a practical matter, many hams have assisted their utility in locating noise sources. In some cases, this can help expedite a speedy resolution.

There is a significant caveat to this approach however. Should you mislead the power company into making unnecessary repairs, they will become frustrated. This expense and time will be added to their repair list. Do not make a guess or suggestions if you don't know what is causing the noise. While some power companies might know less about the locating process than the affected ham, indiscriminate replacement of hardware almost always makes the problem worse. Nonetheless, depending on your level of expertise and the specifics of your situation, you may be able to facilitate a speedy resolution by locating the RFI source for the utility.

27.6.4 Amateur Power-line Noise Locating Equipment

Much of the equipment that an amateur would use to locate power-line noise has previously been described in the "Radio Direction Finding" section. Before discussing how to locate power-line noise sources, here are a few additional equipment guidelines:

Receiver — You'll need a battery-operated portable radio capable of receiving VHF or UHF in the AM mode. Ideally, it should also be capable of receiving HF frequencies, especially if the interference is a problem at HF and not VHF. Some amateurs also use the aircraft band from 108 to 137 MHz. The lower frequencies of this band can sometimes enable an RFI investigator to hear the noise at greater distances than on 2 meters or 70 cm. An RF gain control is essential but an outboard step attenuator can be used as a substitute. A good S-meter is also required.

Attenuator — Even if your receiver has an RF gain control, an additional outboard step attenuator can often be helpful. It can not only minimize the area of a noise search but also provide added range for the RF gain control. As with other RFI sources — you'll need to add more and more attenuation as you approach the source.

VHF/UHF Antennas — You'll need a hand-held directional beam antenna. A popular professional noise-locating antenna is an eight-element Yagi tuned for 400 MHz. Since power-line noise is a broadband phenomenon, the exact frequency is not important. Either a 2 meter or 70 cm Yagi are capable of locating a power-line noise source on a specific power pole.

Although professional grade antennas can cost several hundred dollars, some hams can build their own for a lot less. See the CD-ROM accompanying this book for the article, "Adapting a Three-Element Tape Measure Beam for Power-line Noise Hunting," by Jim Hanson, W1TRC (SK). This low cost and easy to build antenna for locating power-line noise can be adapted for a variety of frequencies and receivers. Commercial 2 meter and 70 cm antennas for portable use are also suitable if a handle is added, such as a short length of PVC pipe.

Before using an antenna for power-line noise locating, determine its peak response frequency. Start by aiming the antenna at a known power-line noise source. Tune across its range and just beyond. Using minimum RF gain control, find its peak response. Label the antenna with this frequency using a piece of tape or marking pen. When using this antenna for noise locating, tune the receiver to this peak response frequency.

If you don't have a VHF or UHF receiver that can receive AM signals, see the CD-ROM accompanying this book for the article, "A Simple TRF Receiver for Tracking RFI," by Rick Littlefield, K1BQT. It describes the combination of a simple 136 MHz beam and receiver for portable RFI tracking.

HF Antennas — Depending on the circumstances of a particular case, a mobile HF whip such as a 7 or 14 MHz model can be helpful. Magnet-mount models are acceptable for temporary use. An RFI investigator can typically get within VHF range by observing the relative strength of the noise from different locations. Driving in a circle centered on the affected station will typically indicate the general direction in which the noise is strongest. As with beam antennas, determine the peak response frequency for best results.

Ultrasonic Pinpointer — Although an ultrasonic pinpointer is not necessary to locate the pole or structure containing the source, some hams prefer to go one more step by finding the offending noise source on that structure. Guidelines for the use of an ultrasonic device are described later in this section.

Professional-grade ultrasonic locators are often beyond the budget of the average ham. Home brewing options however, can make a practical ultrasonic locator affordable in most situations — and make a great weekend project too. See the CD-ROM accompanying this book for "A Home-made Ultrasonic Power Line Arc Detector" by Jim Hanson, W1TRC (SK).

Oscilloscope — A battery-powered portable oscilloscope is only required for signature analysis. See the next section, Signature or Fingerprint Method, for details.

Thermal/Infrared Detectors and Corona Cameras — This equipment is not recommended for the sole purpose of locating power-line noise sources. It is rare that an RFI source is even detectable using infrared techniques. Although these are not useful tools for locating noise sources, many utilities still use them for such purposes with minimal or no results. Not surprisingly, ARRL experience has shown that these utilities are typically unable to resolve interference complaints in a timely fashion.

27.6.5 Signature or Fingerprint Method

Each sparking interference source exhibits a unique pattern. By comparing the characteristics between the patterns taken at the affected station with those observed in the field, it becomes possible to conclusively identify the offending source or sources from the many that one might encounter. It therefore isn't surprising that a pattern's unique characteristic is often called its "fingerprint" or "signature." See **Fig 27.8** for an example.

This is a very powerful technique and a real money saver for the utility. Even though there may be several different noise sources in the field, this method helps identify only those sources that are actually causing the interference problem. The utility need only correct the problem(s) matching the pattern of noise affecting your equipment.

You as a ham can use the signature method by observing the noise from your radio's audio output with an oscilloscope. Record the pattern by drawing it on a notepad or taking a photograph of the screen. Take the sketch or photograph with you as you hunt for the source and compare it to signatures you might observe in the field.

Professional interference-locating receivers, such as the Radar Engineers Model 240A shown in Fig 27.8B, have a built in oscilloscope display and waveform memory. This is the preferred method used by professional interference investigators. These receivers provide the ability to switch between the patterns saved at the affected station and those from sources located in the field.

Once armed with the noise fingerprint taken at the affected station, you are ready to begin the hunt. If you have a directional beam, use it to obtain a bearing to the noise. If multiple sources are involved, you'll need to record the bearings to each one. Knowing how high in frequency a particular noise can be heard also provides a clue to its proximity. If the noise can be heard at 440 MHz, for example, the source will typically be within walking distance. If it diminishes beginning 75 or 40 meters, it can be up to several miles away.

Since each noise source will exhibit unique characteristics, you can now match this noise "signature" with one from the many sources you may encounter in the investigation. Compare such characteristics as the duration of each noise burst, pulse shape, and number of pulses.

If you have a non-portable oscilloscope, you may still be able to perform signature matching by using an audio recorder. Make a high quality recording of the noise source at your station and at each suspected noise source in the field, using the same receiver if possible. Replay the sounds for signature analysis.

27.6.6 Locating the Source's Power Pole or Structure

Start your search in front of the affected station. If you can hear the noise at VHF or UHF, begin with a handheld beam suitable for these frequencies. As discussed previously, the longer wavelengths associated with the AM broadcast band and even HF, can create misleading "hotspots" along a line when searching for a noise source as shown in **Fig 27.9**.

As a general rule, only use lower frequencies when you are too far away from the source to hear it at VHF or UHF. Generally work with the highest frequency at which the noise can be heard. As you approach the source, keep increasing the frequency to VHF or UHF, depending on your available antennas. Typically, 2 meters and 70 cm are both suitable for isolating a source down to the pole level.

If you do not have an initial bearing to the noise and are unable to hear it with your portable or mobile equipment, start traveling in a circular pattern around the affected station, block-by-block, street-by-street, until you find the noise pattern matching the one recorded at the affected station.

Once in range of the noise at VHF or UHF, start using a handheld beam. You're well on your way to locating the structure containing the source. In many cases, you can now continue your search on foot. *Again, maintain minimum RF gain to just barely hear the noise over a minimum area.* This is important step is crucial for success. If the RF gain is too high, it will be difficult to obtain accurate bearings with the beam.

Power-line noise will often be neither vertically nor horizontally polarized but somewhere in between. Be sure to rotate the beam's polarization for maximum noise response. Maintain this same polarization when comparing poles and other hardware.

27.6.7 Pinpointing the Source on a Pole or Structure

Once the source pole has been identified, the next step becomes pinpointing the offending hardware on that pole. A pair of binoculars on a dark night may reveal visible signs of arcing and in some cases you may be able to see other evidence of the problem from

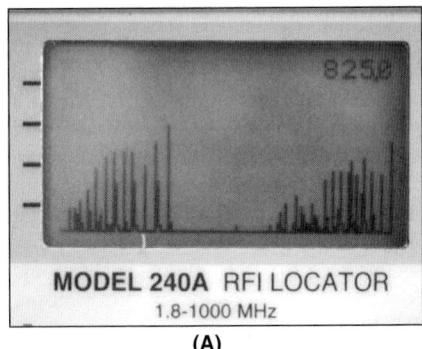

(A)

(B)

Fig 27.8 — An unknown noise source at the complainant's antenna is shown in A. During the RFI investigation, noise signatures not matching this pattern can be ignored, such as shown in B. Once the matching signature originally shown in A is found, the offending noise source has been located.

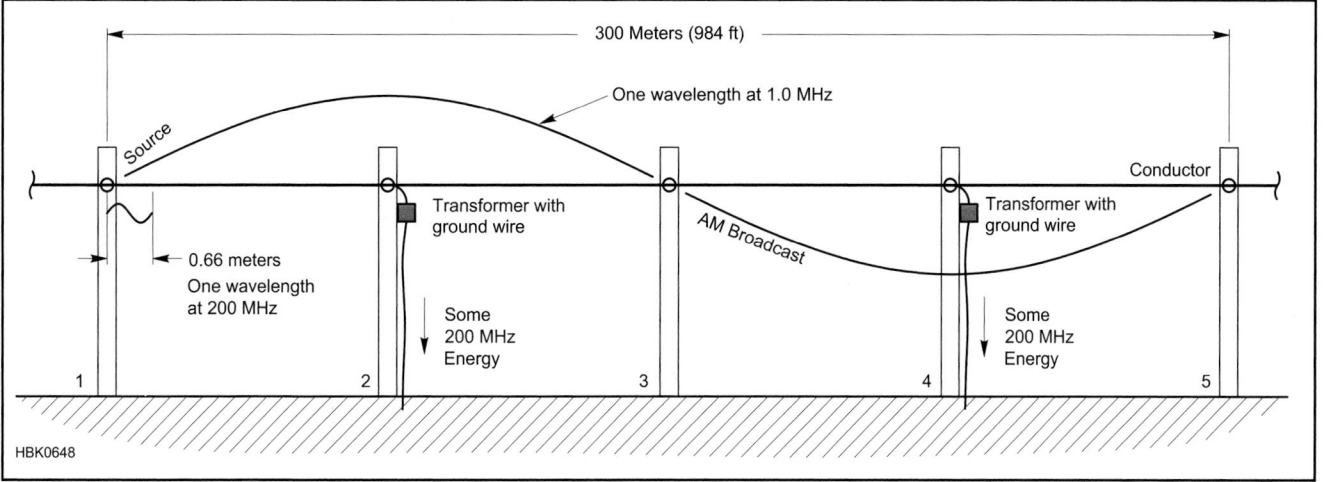

Fig 27.9 — Listening for noise signals on a power distribution line at 1 MHz vs 200 MHz can result in identification of the wrong power pole as the noise source. (From Loftness, *AC Power Interference Handbook*)

the ground. These cases are rare. More than likely, a better approach will be required. Professional and utility interference investigators typically have two types of specialized equipment for this purpose:

• An ultrasonic dish or pinpointer. The RFI investigator, even if not a lineman, can pinpoint sources on the structure down to a component level from the ground using this instrument. See **Fig 27.10**.

• An investigator can instruct the lineman on the use of a hot stick-mounted device used to find the source. This method is restricted to only qualified utility personnel, typically from a bucket truck. See **Fig 27.11**.

Both methods are similar but hams only have one option — the ultrasonic pinpointer.

Caution — Hot sticks and hot stick mounted devices are not for hams! Do not use them. Proper and safe use of a hot stick requires specialized training. In most localities, it is generally unlawful for anyone unqualified by a utility to come within 10 feet of an energized line or hardware. This includes hot sticks.

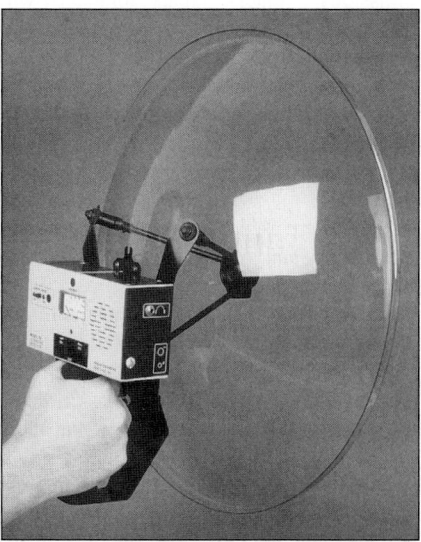

Fig 27.10 — The clear plastic parabolic dish is an "ear" connected to an ultrasonic detector that lets utility personnel listen for the sound of arcs.

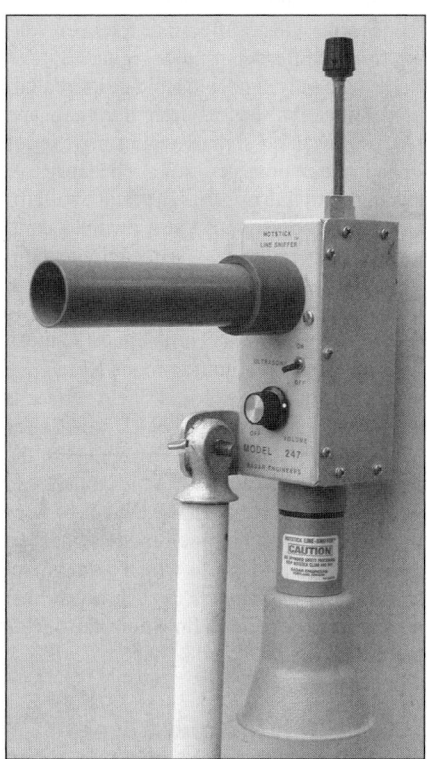

Fig 27.11 — The Radar Engineers Model 247 Hotstick Line Sniffer is an RF and ultrasonic locator. It is used by utility workers to pinpoint the exact piece of hardware causing a noise problem. Mounted on a hotstick, the sniffer is used from the pole or from a bucket.

ULTRASONIC PINPOINTER TIPS

An ultrasonic dish is the tool of choice for pinpointing the source of an arc from the ground. While no hot stick is required, an unobstructed direct line-of-sight path is required between the arc and the dish. This is not a suitable tool for locating the structure containing the source. It is only useful for pinpointing a source once its pole or structure has been determined.

Caveat: Corona discharge, while typically not a source of RF power-line noise, can and often is a significant source of ultrasonic sound. It can often be difficult to distinguish between the sound created by an arc and corona discharge. This can lead to mistakes when trying to pinpoint the source of an RFI problem with an ultrasonic device.

The key to success, just as with locating the structure, is using gain control effectively. Use minimum gain after initially detecting the noise. If the source appears to be at more than one location on the structure, reduce the gain. In part, this will eliminate any weaker noise signals from hardware not causing the problem.

27.6.8 Common Causes of Power-line Noise

The following are some of the more common power-line noise sources. They're listed in order from most common to least common. Note that some of the most common sources are not connected to a primary conductor. This in part is due to the care most utilities take to ensure sufficient primary conductor clearance from surrounding hardware. Note, too, that power transformers do not appear on this list:

• Loose staples on ground conductor
• Loose pole-top pin
• Ground conductor touching nearby hardware

- Corroded slack span insulators
- Guy wire touching neutral
- Loose hardware
- Bare tie wire used with insulated conductor
- Insulated tie wire on bare conductor
- Loose cross-arm braces
- Lightning arrestors

27.6.9 The Cooperative Agreement

While some cases of power-line noise are resolved in a timely fashion, the reality is that many cases can linger for an extended period of time. Many utilities simply do not have the expertise, equipment or motivation to properly address a power-line noise complaint. There are often no quick solutions. Patience can often be at a premium in these situations. Fortunately, the ARRL has a Cooperative Agreement with the FCC that can help. While the program is not a quick or easy solution, it does offer an opportunity and step-by-step course of action for relief. It emphasizes and provides for voluntary cooperation without FCC intervention.

Under the terms of the Cooperative Agreement, the ARRL provides technical help and information to utilities in order to help them resolve power-line noise complaints. It must be emphasized that the ARRL's role in this process is strictly a technical one — it is not in the enforcement business. In order to participate, complainants are required to treat utility personnel with respect, refrain from hostile behavior, and reasonably cooperate with any reasonable utility request. This includes making his or her station available for purposes of observing and recording noise signatures. The intent of the Cooperative Agreement is to solve as many cases as possible before they go to the FCC. In this way, the FCC's limited resources can be allocated where they are needed the most — enforcement.

As the first step in the process, the ARRL sends the utility a letter advising of pertinent Part 15 rules and offering assistance. The FCC then requires a 60-day waiting period before the next step. If by the end of 60 days the utility has failed to demonstrate a good faith effort to correct the problem, the FCC then issues an advisory letter. This letter allows the utility another 60-day window to correct the problem.

A second FCC advisory letter, if necessary, is the next step. Typically, this letter provides another 20- or 30-day window for the utility to respond. If the problem still persists, a field investigation would follow. At the discretion of the Field Investigator, he or she may issue an FCC Citation or Notice of Apparent Liability (NAL). In the case of an NAL, a forfeiture or fine can result.

It is important to emphasize that the ARRL Cooperative Agreement Program does not offer a quick fix. There are several built-in waiting periods and a number of requirements that a ham must follow precisely. It does however provide a step-by-step and systematic course of action under the auspices of the FCC in cases where a utility does not comply with Part 15. Look for complete details, including how to file a complaint, in the ARRL's Power-line Noise FAQ web page at **www.arrl.org/power-line-noise-faq-page**.

27.7 Elements of RFI Control

27.7.1 Differential- and Common-Mode Signal Control

The path from source to victim almost always includes some conducting portion, such as wires or cables. RF energy can be conducted directly from source to victim, be conducted onto a wire or cable that acts as an antenna where it is radiated, or be picked up by a conductor connected to the victim that acts like an antenna. When the energy is traveling along the conducting portion of its path, it is important to understand whether it is a differential- or common-mode signal.

Removing unwanted signals that cause RFI is different for each of these conduction modes. Differential-mode cures (a high-pass filter, low-pass filter, or a capacitor across the ac power line, for example) do not attenuate common-mode signals. Similarly, a common-mode choke will not affect interference resulting from a differential-mode signal.

It's relatively simple to build a differential-mode filter that passes desired signals and blocks unwanted signals with a high series impedance or presents a low-impedance to a signal return line or path. The return path for common-mode signals often involves earth ground, or even the chassis of equipment if it is large enough to form part of an antenna at the frequency of the RFI. A differential-mode filter is not part of this current path, so it can have no effect on common-mode RFI.

In either case, an exposed shield surface is a potential antenna for RFI, either radiating or receiving unwanted energy, regardless of the shield's quality. In this way, a coaxial cable can act as an antenna for RFI if the victim device is unable to reject common-mode signals on the cable's shield. This is why it is important to connect cable shields in such a way that common-mode currents flowing on the shields are not allowed to enter the victim device.

27.7.2 Shields and Filters

Breaking the path between source and victim is often an attractive option, especially if either is a consumer electronics device. Remember, the path will involve one or more of three possibilities — radiation, conduction, and inductive or capacitive coupling. Breaking the path of an RFI problem can require analysis and experimentation in some cases. Obviously you must know what the path is before you can break it. While the path may be readily apparent in some cases, more complex situations may not be so clear. Multiple attempts at finding a solution may be required.

SHIELDS

Shielding can be used to control radiated emissions — that is, signals radiated by wiring inside the device — or to prevent radiated signals from being picked up by signal leads in cables or inside a piece of equipment. Shielding can also be used to reduce inductive or capacitive coupling, usually by acting as an intervening conductor between the source and victim.

Shields are used to set boundaries for radiated energy and to contain electric and magnetic fields. Thin conductive films, copper braid and sheet metal are the most common shield materials for the electric field (capacitive coupling), and for electromagnetic fields (radio waves). At RF, the small skin depth allows thin shields to be effective at these frequencies. Thicker shielding material is needed for magnetic field (inductive coupling) to minimize the voltage caused by induced current. At audio frequencies and below, the higher skin depth of common shield materials is large enough (at 60 Hz, the skin depth for aluminum is 0.43 inches) that high-permeability materials such as steel or mu-metal (a nickel-iron alloy that exhibits high magnetic permeability) are required.

Maximum shield effectiveness usually requires solid sheet metal that completely encloses the source or susceptible circuitry or equipment. Small discontinuities, such as holes or seams, decrease shield effectiveness. In addition, mating surfaces between different parts of a shield must be conductive. To ensure conductivity, file or sand off paint or other nonconductive coatings on mating surfaces.

The effectiveness of a shield is determined by its ability to reflect or absorb the undesired energy. Reflection occurs at a shield's surface. In this case, the shield's effectiveness is independent of its thickness. Reflection is typically the dominant means of shielding for radio waves and capacitive coupling, but is ineffective against magnetic coupling. Most RFI shielding works, therefore, by reflection. Any good conductor will serve in this case, even thin plating.

Magnetic material is required when attempting to break a low-frequency inductive coupling path by shielding. A thick layer of high permeability material is ideal in this case. Low frequency magnetic fields are typically a very short range phenomenon. Simply increasing the distance between the source and victim may help avoid the expense and difficulty of implementing a shield.

Adding shielding may not be practical in many situations, especially with many consumer products, such as a television. Adding a shield to a cable can minimize capacitive coupling and RF pickup, but it has no effect on magnetic coupling. Replacing parallel-conductor cables (such as zip cord) with twisted-pair is quite effective against magnetic coupling and also reduces electromagnetic coupling.

Additional material on shielding may be found in Chapter 2 of Ott and in the **RF Technique**s chapter of this book.

FILTERS

Filters and chokes can be very effective in dealing with a conducted emissions problem. Fortunately, filters and chokes are simple, economical and easy to try. As we'll see, use of common-mode chokes alone can often solve many RFI problems, especially at HF when common-mode current is more likely to be the culprit.

A primary means of separating signals relies on their frequency difference. Filters offer little opposition to signals with certain frequencies while blocking or shunting others. Filters vary in attenuation characteristics, frequency characteristics and power-handling capabilities. The names given to various filters are based on their uses. (More information on filters may be found in the **RF and AF Filters** chapter.)

Low-pass filters pass frequencies below some cutoff frequency, while attenuating frequencies above that cutoff frequency. A typical low-pass filter curve is shown in **Fig 27.12**. A schematic is shown in **Fig 27.13**. These filters are difficult to construct properly so you should buy one. Many retail Amateur Radio stores that advertise in *QST* stock low-pass filters.

High-pass filters pass frequencies above some cutoff frequency while attenuating frequencies below that cutoff frequency. A typical high-pass filter curve is shown in **Fig 27.14**. **Fig 27.15** shows a schematic of a typical high-pass filter. Again, it is best to buy one of the commercially available filters.

Bypass capacitors can be used to cure dif-

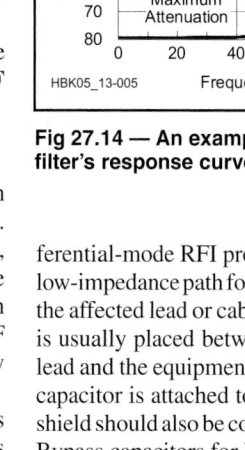

Fig 27.14 — An example of a high-pass filter's response curve.

ferential-mode RFI problems by providing a low-impedance path for RF signals away from the affected lead or cable. A bypass capacitor is usually placed between a signal or power lead and the equipment chassis. If the bypass capacitor is attached to a shielded cable, the shield should also be connected to the chassis. Bypass capacitors for HF signals are usually 0.01 µF, while VHF bypass capacitors are usually 0.001 µF. Leads of bypass capacitors should be kept short, particularly when dealing with VHF or UHF signals.

AC-line filters, sometimes called "brute-force" filters, are used to remove RF energy from ac power lines. Both common-mode and differential-mode noise can be attenuated by a commercially built line filter. A typical sche-matic is shown in **Fig 27.16**. Products from Corcom (**www.corcom.com**) and Delta Electronics (**www.delta.com.tw**) are widely available and well documented on their websites. Morgan Manufacturing (**www.morgan mfg.us**) sells stand-alone ac-line filters with ac plugs and sockets, such as the M-473 and M-475 series of AC Line Protectors.

AC-line filters come in a wide variety of sizes, current ratings, and attenuation. In general, a filter must be physically larger to handle higher currents at lower frequencies. The Corcom 1VB1, a compact filter small enough to fit in the junction box for many low

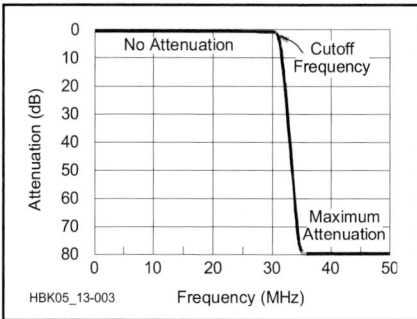

Fig 27.12 — An example of a low-pass filter's response curve.

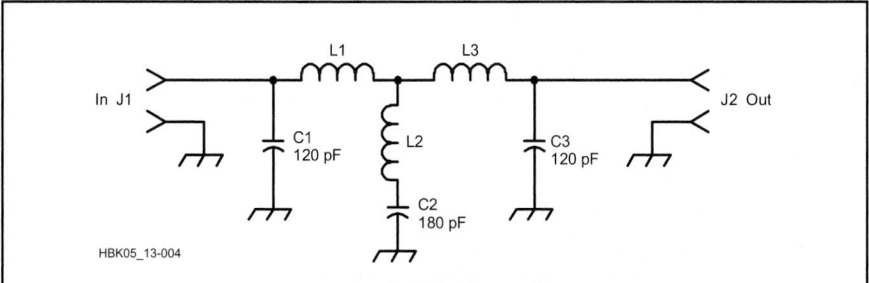

Fig 27.13 — A low-pass filter for amateur transmitting use. Complete construction information appears in the Transmitters chapter of *The ARRL RFI Book*. A high-performance 1.8-54 MHz filter project can be found in the RF and AF Filters chapter of this *Handbook*.

Warning: Bypassing Speaker Leads

Older amateur literature might suggest connecting a 0.01-µF capacitor *across an amplifier's speaker output terminals to cure RFI from common-mode signals on speaker cables. Don't do this!* Doing so can cause some modern solid-state amplifiers to break into a destructive, full-power, sometimes ultrasonic oscillation if they are connected to a highly capacitive load. Use common-mode chokes and twisted-pair speaker cables instead.

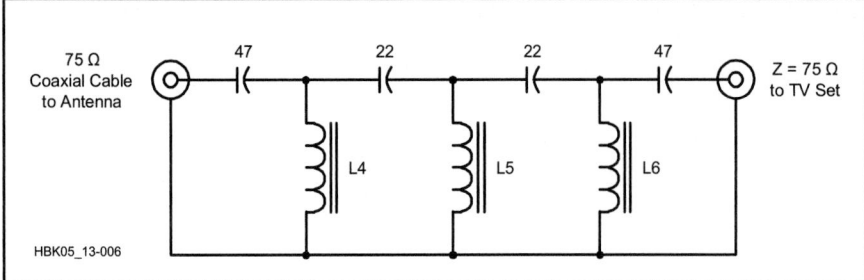

Fig 27.15 — A differential-mode high-pass filter for 75-Ω coaxial cable. It rejects HF signals picked up by a TV antenna or that leak into a cable-TV system. All capacitors are high-stability, low-loss, NP0 ceramic disc components. Values are in pF. The inductors are all #24 AWG enameled wire on T-44-0 toroid cores. L4 and L6 are each 12 turns (0.157 μH) and L5 is 11 turns (0.135 μH).

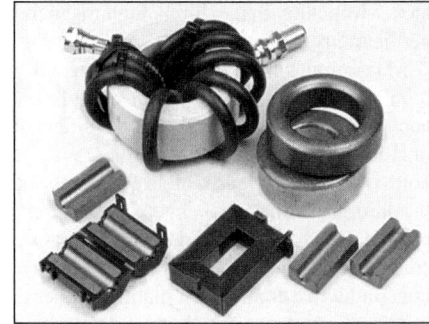

Fig 27.17 — A common-mode RF choke wound on a toroid core is shown at top left. Several styles of ferrite cores for common-mode chokes are also shown.

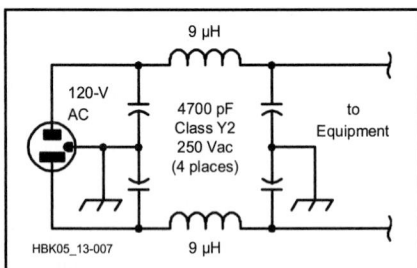

Fig 27.16 — A "brute-force" ac-line filter.

voltage lighting fixtures, provides good common mode attenuation at MF and HF and its 1 A at 250 V ac rating is sufficient for many LV lighting fixtures. In general, you will get more performance from a filter that is physically small if you choose the filter with the lowest current rating sufficient for your application. (Section 13.3 of Ott covers ac-line filters.)

Any wiring between a filter and the equipment being filtered acts as an antenna and forms an inductive loop that degrades the performance of the filter. All such wiring should be as short as possible, and should be twisted. Always bond the enclosure of the filter to the enclosure of the equipment by the shortest possible path. Some commercial filters are built with an integral IEC power socket, and can replace a standard IEC connector if there is sufficient space behind the panel. (IEC is the International Electrotechnical Commission, an international standards organization that has created specifications for power plugs and sockets.) Such a filter is bonded to the chassis and interconnecting leads are shielded by the chassis, optimizing its performance.

A capacitor between line and neutral or between line and ground at the noise source or at victim equipment can solve some RFI problems. ("Chassis ground" in this sense is not "earth," it is the power system equipment ground — the green wire — at the equipment.) Power lines, cords, and cables are often subjected to short-duration spikes of very high voltage (4 kV). Ordinary capacitors are likely

Feed Line Radiation — The Difference Between Balanced and Unbalanced Transmission Lines

The physical differences between balanced and unbalanced feed lines are obvious. Balanced lines are parallel-type transmission lines, such as ladder line or twin lead. The two conductors that make up a balanced line run side-by-side, separated by an insulating material (plastic, air, whatever). Unbalanced lines, on the other hand, are coaxial-type feed lines. One of the conductors (the shield) completely surrounds the other (the center).

In an ideal world, both types of transmission lines would deliver RF power to the load (typically your antenna) without radiating any energy along the way. It is important to understand, however, that both types of transmission lines require a balanced condition in order to accomplish this feat. That is, the currents in each conductor must be equal in magnitude, but opposite in polarity.

The classic definition of a balanced transmission line tells us that both conductors must be symmetrical (same length and separation distance) relative to a common reference point, usually ground. It's fairly easy to imagine the equal and opposite currents flowing through this type of feeder. When such a condition occurs, the fields generated by the currents cancel each other-hence, no radiation. An imbalance occurs when one of the conductors carries more current than the other. This additional "imbalance current" causes the feed line to radiate.

Things are a bit different when we consider a coaxial cable. Instead of its being a symmetrical line, one of its conductors (usually the shield), is grounded. In addition, the currents flowing in the coax are confined to the outside portion of the center conductor and the inside portion of the shield.

When a balanced load, such as a resonant dipole antenna, is connected to unbalanced coax, the outside of the shield can act as an electrical third conductor (see **Fig 27.A**). This phantom third conductor can provide an alternate path for the imbalance current to flow. Whether the small amount of stray radiation that occurs is important or not is subject to debate. In fact, one of the purposes of a balun (a contraction of balanced to unbalanced) is to reduce or eliminate imbalance current flowing on the outside of the shield. See the **Transmission Lines** chapter of this book for more information on baluns.

Fig 27.A — Various current paths are present at the feed point of a balanced dipole fed with unbalanced coaxial cable. The diameter of the coax is exaggerated to show the currents clearly.

to fail when subjected to these voltages, and the failure could cause a fire. Only Type X1, X2, Y1 and Y2 capacitors should be used on power wiring. AC-rated capacitors can safely handle being placed across an ac line along with the typical voltage surges that occur from time to time. Type X1 and X2 capacitors are rated for use between line and neutral, and are available in values between 0.1 μF and 1 μF. Type X2 capacitors are tested to withstand 2.5 kV, type X1 capacitors are tested to 4 kV. Type Y1 and Y2 capacitors are rated for use between line and ground; Y1 capacitors are impulse tested to 8 kV; Type Y2 to 5 kV. Note that 4700 pF is the largest value permitted to be used between line and ground — larger values can result in excessive leakage currents.

27.7.3 Common-Mode Chokes

Common-mode chokes on ferrite cores are the most effective answer to RFI from a common-mode signal. Differential-mode filters are *not* effective against common-mode signals. (AC-line filters often perform both common- and differential-mode filtering.) Common-mode chokes work differently, but equally well, with coaxial cable and paired conductors. (Additional material on common-mode chokes is found in sections 3.5 and 3.6 of Ott.)

The most common form of common-mode choke is multiple turns of cable wound on a magnetic toroid core, usually ferrite, as shown in **Fig 27.17**. The following explanation applies to chokes wound on rods as well as toroids, but avoid rod cores since they may couple to nearby circuits at HF.

Most of the time, a common-mode signal on a coaxial cable or a shielded, multi-wire cable is a current flowing on the outside of the cable's shield. By wrapping the cable around a magnetic core the current creates a flux in the core, creating a high impedance in series with the outside of the shield. (An impedance of a few hundred to several thousand ohms are required for an effective choke.) The impedance then blocks or reduces the common-mode current. Because equal-and-opposite fields are coupled to the core from each of the differential-mode currents, the common-mode choke has no effect on differential-mode signals inside the cable.

When the cable consists of two-wire, unshielded conductors such as zip cord or twisted-pair, the equal-and-opposite differential currents each create a magnetic flux in the core. The equal-and-opposite fluxes cancel each other and the differential-mode signal experiences zero net effect. To common-mode signals, however, the choke appears as a high impedance in series with the signal: the higher the impedance, the lower the common-mode current.

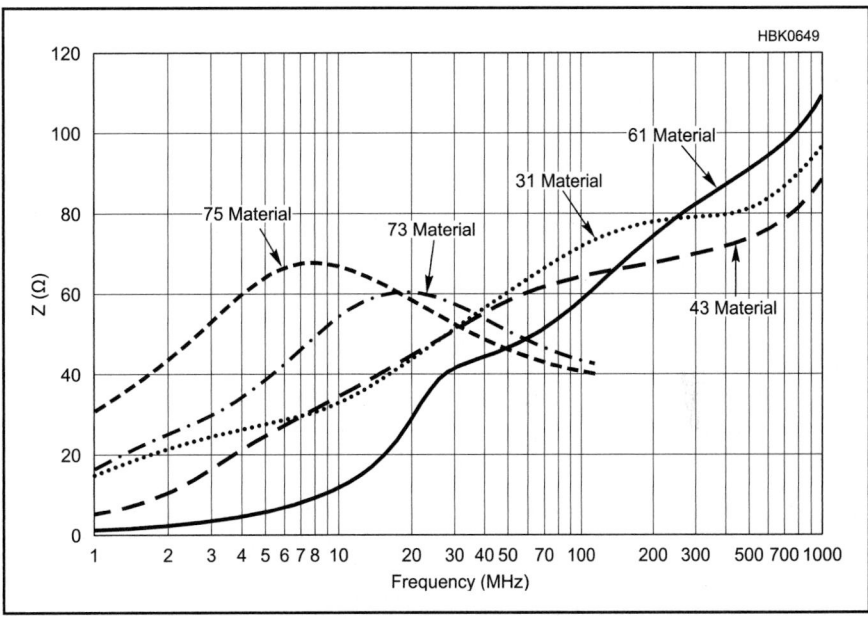

Fig 27.18 — Impedance vs. frequency plots for "101" size ferrite beads illustrate the effect of various ferrite materials across different frequency ranges. A 3.50 mm × 1.30 mm × 6.00 mm bead (Fair-Rite 301 size) was used for the above curve for material comparison, however all materials are not available in all shapes and sizes. Type 73 material is only available in smaller cores, type 31 is only available in larger cores, and type 75 is currently only available as a toroid core. (Graph provided courtesy of the Fair-Rite Corporation)

It is important to note that common-mode currents on a transmission line will result in radiation of a signal from the feed line. (See the sidebar for an explanation of balanced vs unbalanced transmission lines.) The radiated signal can then cause RFI in nearby circuits. This is most common when using coaxial cable as a feed line to a balanced load, such as the dipole in the sidebar. Using a common-mode choke to reduce common-mode feed line currents can reduce RFI caused by signals radiated from the feed line's shield.

The optimum core size and ferrite material is determined by the application and frequency. For example, an ac cord with a plug attached cannot be easily wrapped on a small ferrite core. The characteristics of ferrite materials vary with frequency, as shown by the graph in **Fig 27.18**. Type 31 material is a good all-purpose material for HF and low-VHF applications, especially at low HF frequencies. Type 43 is widely used for HF

through VHF and UHF. (See the **Component Data and References** chapter for a table of ferrite materials and characteristics.)

Ferrite beads and clamp-on split cores are also used for EMI control at VHF and UHF, both as common-mode chokes and low-pass filters. (These are essentially single-turn chokes as the cable passes just once through the bead or core.) Multi-turn chokes are required for effective suppression at HF. It is usually more effective to form a common-mode choke by wrapping multiple turns of wire or coaxial cable around a 1- to 2-inch OD core of the correct material.

Common-mode chokes can be used on single conductors unless the desired signal is in the RF spectrum. A common-mode choke creates a high resistive impedance at radio frequencies. At audio frequencies, however, it looks like a relatively small inductance, and is unlikely to have any effect on audio-frequency signals.

27.8 Troubleshooting RFI

Troubleshooting an RFI problem is a multi-step process, and all steps are equally important. First you must determine the type(s) of noise source(s) that are involved. Next, diagnose the problem by locating the noise source and the means by which it creates the noise. The final step is to identify the path by which the noise or signals reach the victim device. Only then can you cure the problem by breaking the path from source to victim.

Each step in troubleshooting an RFI problem involves asking and answering several questions: Is the problem caused by harmonics, fundamental overload, conducted emissions, radiated emissions or a combination of all of these factors? Should it be fixed with a low-pass filter, high-pass filter, common-mode chokes or ac-line filter? How about shielding, isolation transformers, a different ground or antenna configuration? By the time you finish with these questions, the possibilities could number in the millions. You probably will not see your exact problem and cure listed in this book or any other. You must not only diagnose the problem but find a cure as well!

Now that you have learned some RFI fundamentals, you can work on specific technical solutions. A systematic approach will identify the problem and suggest a cure. Most RFI problems can be solved in this way by the application of standard techniques. The following sections suggest specific approaches for different types of common interference problems. This advice is based on the experience of the ARRL RFI Desk, but is not guaranteed to provide a solution to your particular problem. Armed with your RFI knowledge, a kit of filters and tools, your local TC and a determination to solve the problem, it is time to begin. You should also get a copy of the *ARRL RFI Book*. It's comprehensive and picks up where this chapter leaves off.

27.8.1 General RFI Troubleshooting Guidelines

Before diving into the problem, take a step back and consider some of these "pre-troubleshooting steps."

Is It Really EMI? — Before trying to solve a suspected case of EMI, verify that the symptoms actually result from external causes. A variety of equipment malfunctions or external noise can look like interference.

Is It Your Station? — "Your" EMI problem might be caused by another ham or a radio transmitter of another radio service, such as a local CB or police transmitter. If it appears that your station is involved, operate your station on each band, mode and power level that you use. Note all conditions that produce interference. If no transmissions produce the problem, your station *may* not be the cause.

(Although some contributing factor may have been missing in the test.) Have your neighbor keep notes of when and how the interference appears: what time of day, what station, what other appliances were in use, what was the weather? You should do the same whenever you operate. If you can readily reproduce the problem with your station, you can start to troubleshoot the problem.

Take One Away — Can you remove the source or victim entirely? The best cure for an RFI problem is often removing the source of the noise. If the source is something broken, for example, the usual solution is to repair it. Power-line noise and an arcing electric fence usually fall into this category. If a switchmode power supply is radiating noise, replace it with a linear supply. Victim devices can sometimes be replaced with a more robust piece of equipment, as well.

Look Around — Aside from the brain, the eyes are a troubleshooter's best tool. Installation defects contribute to many RFI problems. Look for loose connections, shield breaks in a cable-TV installation or corroded contacts in a telephone installation. Have these fixed first. Look for wiring connected to the victim equipment that might be long enough to be resonant on one or more amateur bands. If so, a common-mode choke may be an easy cure. Ideally you'll generally want place the choke as close to the victim device as practical. If this placement proves too difficult or additional suppression is required, chokes placed at the middle of the wiring may help break up resonances. These are just a few of the possible deficiencies in a home installation.

At Your Station — Make sure that your own station and consumer equipment are clean. This cuts the size of a possible interference problem from your station in half! Once this is done, you won't need to diagnose or troubleshoot your station later. Also, any cures successful at your house may work at your neighbor's as well. If you do have problems in your own home, continue through the troubleshooting steps and specific cures and take care of your own problem first.

Simplify the Problem — Don't tackle a complex system — such as a telephone system in which there are two lines running to 14 rooms — all at once. You could spend the rest of your life running in circles and never find the true cause of the problem.

There's a better way. In our hypothetical telephone system, first locate the telephone jack closest to the telephone service entrance. Disconnect the lines to more remote jacks and connect one RFI-resistant telephone at the remaining jack. (Old-style rotary-dial phones are often quite immune to RF.) If the interference remains, try cures until the problem is

solved, then start adding lines and equipment back one at a time, fixing the problems as you go along. If you are lucky, you will solve all of the problems in one pass. If not, at least you can point to one piece of equipment as the source of the problem.

Multiple Causes — Many RFI problems have multiple causes. These are usually the ones that give new RFI troubleshooters the most trouble. For example, consider a TVI problem caused by the combination of harmonics from the transmitter due to an arc in the transmitting antenna, an overloaded TV preamplifier, fundamental overload generating harmonics in the TV tuner, induced and conducted RF on the ac-power connections, and a common-mode signal picked up on the shield of the TV's coaxial feed line. You would never find a cure for this multiple-cause problem by trying only one cure at a time.

In this case, the solution requires that all of the cures be present at the same time. When troubleshooting, if you try a cure, leave it in place even if it doesn't solve the problem. When you add a cure that finally solves the problem entirely, start removing the "temporary" attempts one at a time. If the interference returns, you know that there were multiple causes.

Take Notes — In the process of troubleshooting an RFI problem, it's easy to lose track of what remedies were applied, to what equipment, and in what order. Configurations of equipment can change rapidly when you're experimenting. To minimize the chances of going around in circles or getting confused, take lots of notes as you proceed. Sketches and drawings can be very useful. When you do find the cause of a problem and a cure for it, be sure to write all that down so you can refer to it in the future.

RFI Survival Kit — Table 27.1 is a list of the material needed to troubleshoot and solve most RFI problems. Having all of these materials in one container, such as a small tackle or craft box, makes the troubleshooting process go a lot smoother.

27.8.2 Transmitters

We start with transmitters not because most interference comes from transmitters, but because your station transmitter is under your direct control. Many of the troubleshooting steps in other parts of this chapter assume that your transmitter is "clean" (free of unwanted RF output).

Start by looking for patterns in the interference. Problems that occur only on harmonics of a fundamental signal usually indicate the transmitter is the source of the interference. Harmonics can also be generated in nearby semiconductors, such as an unpowered VHF

Table 27.1

RFI Survival Kit

Quantity	Item
(2)	300-Ω high-pass filter
(2)	75-Ω high-pass filter
(2)	Commercially available clamp-on ferrite cores: #31 and #43 material, 0.3" ID
(12)	Assorted ferrite cores: #31 and #43 material, FT-140 and FT-240 size
(3)	Telephone RFI filters
(2)	Brute-force ac line filters
(6)	0.01-μF ceramic capacitors
(6)	0.001-μF ceramic capacitors

Miscellaneous:

- Hand tools, assorted screwdrivers, wire cutters, pliers
- Hookup wire
- Electrical tape
- Soldering iron and solder (use with caution!)
- Assorted lengths 75-Ω coaxial cable with connectors
- Spare F connectors, male, and crimping tool
- F-connector female-female "barrel"
- Clip leads
- Notebook and pencil
- Portable multimeter

receiver left connected to an antenna, rectifiers in a rotator control box, or a corroded connection in a tower guy wire. Harmonics can also be generated in the front-end components of the TV or radio experiencing interference.

If HF transmitter spurs at VHF are causing interference, a low-pass filter at the transmitter output will usually cure the problem. Install the filter after the amplifier (if used) and before the antenna tuner. (A second filter between the transmitter and amplifier may occasionally help as well.) Install a low-pass filter as your first step in any interference problem that involves another radio service.

Interference from non-harmonic spurious emissions is extremely rare in commercial radios. Any such problem indicates a malfunction that should be repaired.

"Keeping It Simple"

Filters and chokes are the number one weapons of choice for many RFI problems whether the device is the source or the victim. They are relatively inexpensive, easy to install, and do not require permanently modifying the device.

Common-mode choke — Making a common-mode choke is simple. Select the type of core and ferrite material for the frequency range of the interference. (Type 31 is a good HF/low-VHF material, type #43 from 5 MHz through VHF) Wrap several turns of the cable or wire pairs around the toroid. Six to 8 turns is a good start at 10-30 MHz and 10 to 15 turns from 1.8 to 7 MHz. (Ten to 15 turns is probably the practical limit for most cables.) Ferrite clamp-on split cores and beads that slide over the cable or wire are not as effective as toroid-core chokes at HF but are the right solution at VHF and higher frequencies. For a clamp-on core, the cable doesn't even need to be disconnected from its end. Use type 31 or type 43 material at VHF, type 61 at UHF. At 50 MHz, use two turns through type 31 or 43 cores.

"Brute-Force" ac-line filters — RF signals often enter and exit a device via an ac power connection. "Brute-force" ac-line filters are simple and easy to install. Most ac filters provide both common- and differential-mode suppression. It is essential to use a filter rated to handle the device's required current.

General installation guidelines for using chokes and filters

1. If you have a brute-force ac-line filter, put one on the device or power cord. If RFI persists, add a common-mode choke to the power cord at the device.

2. Simplify the problem by removing cables one at a time until you no longer detect RFI. Start with cables longer than 1/10th-wavelength at the highest frequency of concern. If the equipment can't operate without a particular cable, add common-mode chokes at the affected or source device.

3. Add a common-mode choke to the last cable removed and verify its effect on the RFI. Some cables may require several chokes in difficult cases.

4. Begin reconnecting cables one at a time. If RFI reappears, add a common-mode choke to that cable. Repeat for each cable.

5. Once the RFI goes away, remove the common-mode chokes you added one at a time. If the RFI does not return, you do not need to reinstall the choke. If the RFI returns after removing a choke, reinstall it. Keep only those chokes required to fix the problem.

27.8.3 Television Interference (TVI)

An analog TV signal must have about a 45 to 50 dB signal-to-noise ratio to be classified as good-quality viewing. If interference is present due to a single, discrete signal (for example, a CW signal) the signal-to-interference ratio must be 57 dB or greater, depending on the frequency of the interference within the affected channel. Digital TV has somewhat better immunity but for both formats, clear reception requires a strong signal at the TV antenna-input connector so the receiver must be in what is known as a *strong-signal area.*

TVI to a TV receiver (or a video monitor) normally has one of the following causes:

- Spurious signals within the TV channel coming from your transmitter or station.
- The TV set may be overloaded by your transmitter's fundamental signal.
- Signals within the TV channel from some source other than your station, such as electrical noise, an overloaded mast-mounted TV preamplifier or a transmitter in another service.
- The TV set might be defective or misadjusted, making it look like there is an interference problem.

All of these problems are made potentially more severe because TV receiving equipment is hooked up to *two* antenna systems: (1) the incoming antenna or cable feed line and (2) the ac power line and interconnecting cables. These two antenna systems can couple significant levels of fundamental or harmonic energy into the TV set or video display! The *TVI Troubleshooting Flowchart* in **Fig 27.19** is a good starting point.

The problem could also be caused by direct pickup of the transmitted signal by an unshielded TV or device connected to the TV.

Certain types of television receivers and video monitors are reported to cause broadband RF interference *to* amateur signals — large-screen plasma display models seem to be the most frequent offender — and this may be difficult to cure due to the nature of the display technology. Fortunately, less-expensive, more power-efficient, and RF-quieter LCD technology seems to be displacing plasma technology.

The manufacturer of the TV or video equipment can sometimes help with an interference problem. The Consumer Electronics Association (CEA) can also help you contact equipment manufacturers. Contact them directly for assistance in locating help at **www.ce.org**.

COMMON SOURCES OF TVI

HF transmitters — A nearby HF transmitter is most likely to cause fundamental overload. This is usually indicated by interference to all channels, or at least all VHF channels.

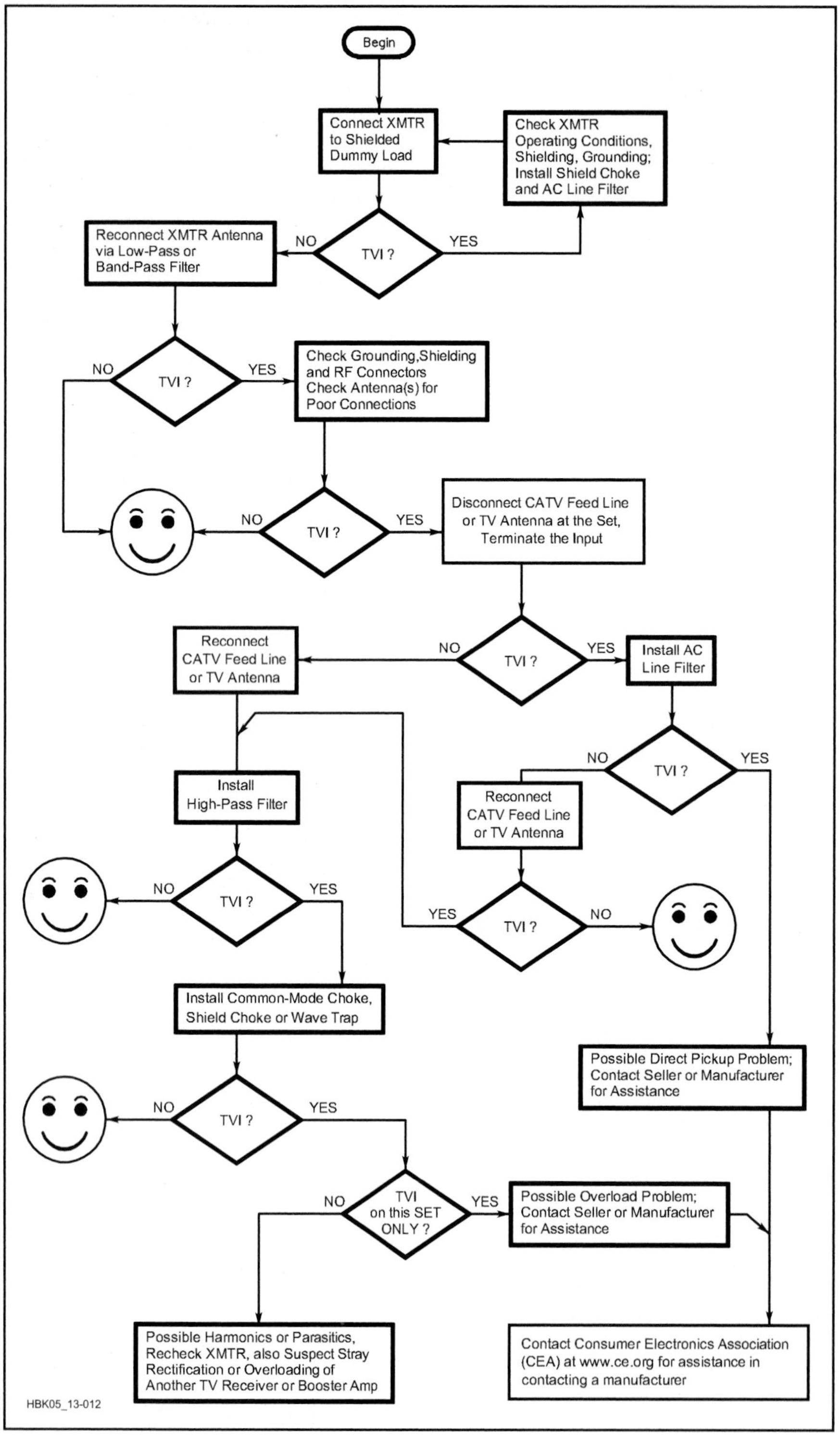

Flowchart (Fig 27.19)

Begin →

Connect XMTR to Shielded Dummy Load ← Check XMTR Operating Conditions, Shielding, Grounding; Install Shield Choke and AC Line Filter

TVI ? — YES → (back to Check XMTR Operating Conditions...)
TVI ? — NO → Reconnect XMTR Antenna via Low-Pass or Band-Pass Filter

Reconnect XMTR Antenna via Low-Pass or Band-Pass Filter →

TVI ? — NO → :)
TVI ? — YES → Check Grounding, Shielding and RF Connectors Check Antenna(s) for Poor Connections

Check Grounding, Shielding and RF Connectors Check Antenna(s) for Poor Connections →

TVI ? — NO → :)
TVI ? — YES → Disconnect CATV Feed Line or TV Antenna at the Set, Terminate the Input

Disconnect CATV Feed Line or TV Antenna at the Set, Terminate the Input →

TVI ? — YES → Install AC Line Filter
TVI ? — NO → Reconnect CATV Feed Line or TV Antenna

Install AC Line Filter →

TVI ? — YES → (to Possible Direct Pickup Problem)
TVI ? — NO → Reconnect CATV Feed Line or TV Antenna

Reconnect CATV Feed Line or TV Antenna →

TVI ? — NO → :)
TVI ? — YES → (to Install High-Pass Filter)

Reconnect CATV Feed Line or TV Antenna →

Install High-Pass Filter →

TVI ? — NO → :)
TVI ? — YES → Install Common-Mode Choke, Shield Choke or Wave Trap

Install Common-Mode Choke, Shield Choke or Wave Trap →

TVI ? — NO → :)
TVI ? — YES → TVI on this SET ONLY ?

TVI on this SET ONLY ? — YES → Possible Overload Problem; Contact Seller or Manufacturer for Assistance
TVI on this SET ONLY ? — NO → Possible Harmonics or Parasitics, Recheck XMTR, also Suspect Stray Rectification or Overloading of Another TV Receiver or Booster Amp

Possible Direct Pickup Problem; Contact Seller or Manufacturer for Assistance

Contact Consumer Electronics Association (CEA) at www.ce.org for assistance in contacting a manufacturer

HBK05_13-012

Fig 27.19 — TVI troubleshooting flowchart.

Fig 27.20 — To eliminate HF and VHF signals on the outside of a coaxial cable, use an 1- to 2-inch OD toroid core and wind as many turns of the cable on the core as practical.

pickup in which a signal is received by the TV set's circuitry without any conducting path being required. In that case, don't try to fix it yourself — it is a problem for the TV manufacturer.

High-pass filters *should not* be used in a cable TV feed line (Fig 27.21A) with two-way cable devices such as cable modems, set-top boxes, and newer two-way CableCARD-equipped TVs. The high-pass filter may prevent the device from communicating via the cable network's upstream signal path.

VHF Transmitters —Most TV tuners are not very selective and a strong VHF signal, including those from nearby FM and TV transmitters, can overload the tuner easily, particularly when receiving channels 2-13 over the air and not by cable TV. In this case, a VHF notch or stop-band filter at the TV can help by attenuating the VHF fundamental signal that gets to the TV tuner. Winegard (**www.winegard.com**) and Scannermaster (**www.scannermaster.com**) sell tunable notch filters. A common-mode choke may also be necessary if the TV is responding to the common-mode VHF signal present on the TV's feed line.

TV Preamplifiers — Preamplifiers are only needed in weak-signal areas and they often cause trouble, particularly when used unnecessarily in strong-signal areas. They are subject to the same overload problems as TVs and when located on the antenna mast it can be difficult to install the appropriate cures. You may need to install a high-pass or notch filter at the input of the preamplifier, as well as a common-mode choke on the input, output and power-supply wiring (if separate) to effect a complete cure. All filters, connections, and chokes must be weatherproofed. Secure the coax tightly to a metal mast to minimize common-mode current. (Do not secure twin-lead to a metal support.) Use two 1-inch long

To cure fundamental overload from an HF transmitter to an antenna-connected TV, install a high-pass filter directly at the TV set's antenna input. (Do not use a high-pass filter on a cable-TV input because the HF range is used for data and other system signals.)

A strong HF signal can also result in a strong common-mode signal on the TV's feed line. A common-mode choke will block signals on

the outside of the feed line shield, leaving the desired signals inside the feed line unaffected. **Fig 27.20** shows how a common-mode choke is constructed for a coaxial feed line.

These two filters can probably cure most cases of TVI! **Fig 27.21** shows a "bulletproof" installation for both over-the-air and cable TV receivers. If one of these methods doesn't cure the problem, the problem is likely direct

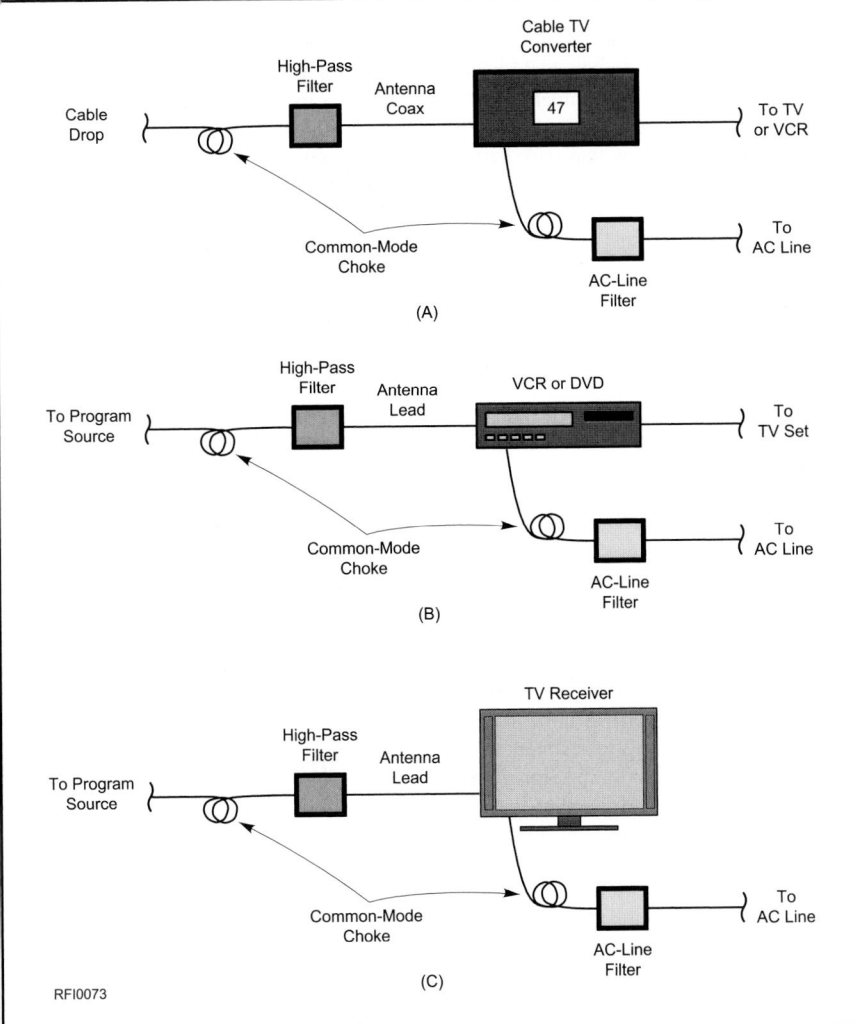

Fig 27.21 — Installing common-mode chokes and high-pass filters will cure most fundamental overload interference from HF sources. This technique does not address direct pickup or spurious emission problems.

type 43 clamp-on ferrite cores if VHF signals are causing the interference and type 61 material for UHF. HF choke design is discussed in the section on Common-Mode Chokes.

Spurious Emissions — You are responsible for spurious emissions produced by your station. If your station is generating any interfering spurious signals, the problem must be cured there. Start by analyzing which TV channels are affected. A TV Channel Chart showing the relationship of the amateur allocations and their harmonics to over-the-air and cable channels is provided on this book's CD-ROM. Each channel is 6 MHz wide. If the interference is only on channels that are multiples of your transmitting frequency, you probably have interference caused by harmonics of your transmitted signal.

It is not certain that these harmonics are coming from your station, however. Harmonics can be generated by overloaded preamplifiers or tuner input circuits. Harmonics can also be generated by non-linear junctions near your station transmitter or very near the TV receiver antenna. (See the section on Intermodulation Distortion.) If your transmitter and station check "clean" — check to see if you have interference on a TV set in your own home — then you must look elsewhere for the source of the harmonics.

Electrical Noise — Digital TV signals are fairly resistant to electrical noise, but in extreme cases can cause the picture to freeze or fail to be displayed as discussed in the following section on Digital TV Receivers. Electrical noise on an analog TV screen generally appears as horizontal bars. Because electrical noise is nearly synchronized with the ac line frequency, these artifacts drift vertically slowly on the screen. Noise from continuous arcing may appear as bright dots anywhere in the picture.

On an AM receiver (including SSB or CW receivers), electrical noise usually sounds like a buzz, sometimes changing in intensity as the arc or spark sputters a bit. If you have

a problem with electrical noise, refer to the section on Electrical Noise.

ANALOG TV RECEIVERS

Even though over-the-air TV broadcasting largely switched to a digital format in 2009, millions of analog TV receivers are still in use for cable TV, satellite TV, with converter boxes for digital broadcast signals, and for displaying video from DVDs and other video sources. Older VCR and DVD players may also include an analog TV tuner to receive analog TV signals.

Interference to video displays and monitors that do not receive RF signals from an antenna or RF modulator should be assumed to be common-mode interference or direct pickup. The same applies to interference to a TV set displaying video signals (not through the antenna input). Interference that is present only on the audio is probably a case of common-mode RFI. (See the Stereos and Home Entertainment Systems section of this chapter.)

DIGITAL TV (DTV) RECEIVERS

In 2009, nearly all over-the-air TV broadcasters in the US, with the exception of low-power TV stations and translators, switched from the older NTSC analog format to a new digital format called DTV. The FCC's Digital TV Transition web page (**www.dtv.gov**) has more information on this transition, including an FAQ page. Digital TV signals can operate with much lower signal-to-noise ratios, but are still susceptible to interference.

Interference to digital TV signals from amateur signals — narrowband interference, for instance, a CW carrier — to a 6 MHz-wide digital TV signal generally has two effects. If the interfering signal is strong enough, it will cause degraded *modulation error ratio* (MER) and degraded *bit error rate* (BER) in the digital video signal. If the amplitude of the interference is sufficient, the digital receiver's *forward error correction* (FEC) circuitry will be unable to fix the broken bits, and the digital video signal will "crash." (See the **Digital Modes** chapter for more information on coding and error correction in digital protocols.)

TV viewers watching any of the multiple video streams that may be contained within the digital video signal won't see any problems in the picture (or hear anything wrong in the sound), until the so-called "crash point" is reached. At that point, the picture will begin to show intermittent "tiling" (the picture breaking up into small squares) or blocking (freezing) in the image. As the amplitude of the interfering signal increases perhaps another 0.5 dB to 1 dB, the crash point or "digital cliff" is reached, and the picture and sound are gone! As you can see, there is a tiny window between receiving a perfect picture and receiving no picture. The same effect is produced by signal fading and may be difficult to distinguish from RFI.

Interference to the digital signal does not make its presence known through visual or audible artifacts such as streaks, lines, or tearing in the picture, or garbled audio. This means that a viewer experiencing interference may not be able to identify its source, but troubleshooting interference may also become more difficult. Nevertheless, the more robust digital modulation is often less susceptible to interference from narrowband amateur signals. A clue to the source of the interference is that interference caused by an amateur signal will occur in sync with the amateur's transmissions while other types of interference will have no correlation.

The techniques for curing interference between amateur and digital TV signals are largely the same as for analog TV. Fundamental overload generally responds well to filters in the antenna or RF inputs. Interference caused by spurious emissions from the amateur station can be eliminated by filtering at the amateur transmitter. Common-mode problems in which RF signals are conducted into the television receiver's circuitry by external audio, video, and power cables are no more or less likely than for analog TV sets and can be addressed as described elsewhere in this chapter.

27.8.4 Cable TV

Cable TV has generally benefited Amateur Radio with respect to TVI. The cable system delivers a strong, consistent signal to the TV receiver, reducing susceptibility to interference from amateur signals. It is also a shielded system so an external signal shouldn't be able to cause interference. Most cable companies are responsible about keeping signal leakage (*egress*) and *ingress* — the opposite of leakage — under control, but problems do happen. Cable companies are not responsible for direct pickup or common-mode interference problems, but are responsible for leakage, ingress, and any noise radiated by common-mode currents from their equipment.

Cable companies are able to take advantage of something known as frequency reuse. That is, all radio frequencies higher than 5 MHz are used to transmit TV signals. The latter is possible because the cables and components used to transport signals to and from paying subscribers comprise what is known as a closed network. In other words, a cable company can use frequencies inside of its cables that may be used for entirely different purposes in the over-the-air environment. As long as the shielding integrity of the cable network is maintained, the cable company's signals won't interfere with over-the-air services, and vice-versa.

The reality is that the shielding integrity of a cable network *is* sometimes compromised, perhaps because of a loose or damaged connector, a cracked cable shield, rodent damage, poorly shielded customer premises equipment (CPE) such as cable-ready TVs and VCRs, and problems that may happen when someone tries to steal cable service! §76.605(a)(12) of the FCC Rules defines the maximum allowable signal leakage (*egress*) field strength at specified measurement distances, and §76.613 covers harmful interference. FCC Rules also mandate that cable operators "…shall provide for a program of regular monitoring for signal leakage by substantially covering the plant every three months," and leaks greater than 20 microvolts per meter (μV/m) at a 10 ft. measurement distance repaired in a reasonable period of time. As well, an annual "snapshot" of leakage performance must be characterized via a flyover measurement of the cable system, or a ground-based measurement of 75% of the network.

CABLE TV FREQUENCY USAGE

A typical modern North American cable network is designed to use frequencies in the 5 to 1002 MHz spectrum. Signals that travel from the cable company to the subscriber occupy frequencies from just above 50 MHz to as high as 1002 MHz range (this is the downstream or forward spectrum), and signals that travel from the subscriber to the cable company are carried in the 5 to as high as 42 MHz range, known as the upstream or return spectrum. The downstream is divided into 6 MHz-wide channel slots, which may carry analog NTSC television signals or 64- or 256-QAM digitally modulated signals used for digital video, high-speed data, and telephone services. Upstream signals from cable modems and two-way set-top boxes are generally carried on specific frequencies chosen by the cable company. **Table 27.2**

summarizes cable downstream channel allocations that overlap Amateur Radio bands. The complete North American channel plan is shown in the TV Channel Chart on this book's CD-ROM.

Cable channels below about 550 MHz may carry analog NTSC signals or digital signals. Cable channels above 550 MHz generally carry digital signals, although there are exceptions. Cable channels above 870 MHz almost always carry only digital signals.

COMMON MECHANISMS FOR LEAKAGE AND INGRESS

As noted previously, cable TV leakage and ingress occur when the shielding integrity of the cable network is compromised. A large cable system that serves a major metropolitan area has literally millions of connectors, tens of thousands of miles of coaxial cable, thousands of amplifiers, hundreds of thousands of passives (splitters, directional couplers, and similar devices), and uncountable customer premises equipment connected to the cable network! Any of these may be a source of leakage and ingress.

DIGITAL SIGNAL LEAKAGE?

The digitally modulated signals carried in a cable TV network use 64-QAM or 256-QAM, the latter more common. (See the **Modulation** chapter for more information on Digital TV modulation.) If a QAM signal were to leak from a cable TV network, it is possible for interference to an over-the-air service to occur, but very unlikely to be identified as from a digital TV signal. The reason for this is that a QAM signal is noise-like, and sounds like normal background noise or hiss on a typical amateur receiver. The QAM signal's digital channel power — its average power over the entire occupied bandwidth — is typically 6 to 10 dB lower than what an analog TV signal's visual carrier peak envelope power

Table 27.2

Amateur Radio Bands Relative to Cable TV Downstream Channels

Amateur Band	Over-The-Air Frequency Range	Cable Channel	Cable Frequency Range
6 meters	50-54 MHz	Below Ch. 2	50-54 MHz, sometimes used for narrowband telemetry carriers
2 meters	144-148 MHz	Ch. 18	144-150 MHz
1.25 meters	222-225 MHz	Ch. 24	222-228 MHz
70 cm	420-450 MHz	Ch. 57	420-426 MHz
		Ch. 58	426-432 MHz
		Ch. 59	432-438 MHz
		Ch. 60	438-444 MHz
		Ch. 61	444-450 MHz
33 cm	902-928 MHz	Ch. 142	900-906 MHz
		Ch. 143	906-912 MHz
		Ch. 144	912-918 MHz
		Ch. 145	918-924 MHz
		Ch. 146	924-930 MHz

Table 27.3

VHF Midband Cable Channels

Cable Channel	Visual Carrier Frequency, MHz (STD)	Visual Carrier Frequency, MHz (IRC)	Visual Carrier Frequency, MHz (HRC)
98	109.2750	109.2750	108.0250*
99	115.2750	115.2750	114.0250*
14	121.2625	121.2625	120.0060
15	127.2625	127.2625	126.0063
16	133.2625	133.2625	132.0066
17	139.25	139.2625	138.0069
18	145.25	145.2625	144.0072
19	151.25	151.2625	150.0075
20	157.25	157.2625	156.0078
21	163.25	163.2625	162.0081
22	169.25	169.2625	168.0084

*Excluded from HRC channel set because of FCC frequency offset requirements

(PEP) would be on the same channel. As well, a QAM signal occupies most of the 6 MHz channel slot, and there are no carriers *per se* within that channel bandwidth. Note that over-the-air 8-VSB digital TV broadcast signals transmit a pilot carrier near the lower end of the digital "haystack," but the QAM format used by cable operators has no comparable pilot carrier.

What makes the likelihood of interference occurring (or not occurring) has in large part to do with the behavior of a receiver in the presence of broadband noise. While each downstream cable TV QAM signal occupies close to 6 MHz of RF bandwidth, the IF bandwidth of a typical amateur FM receiver might be approximately 20 kHz. Thus, the noise power in the receiver will be reduced by $10\log_{10}(6,000,000/20,000) = 24.77$ dB because of the receiver's much narrower IF bandwidth compared to the QAM signal's occupied bandwidth. In addition, there is the 6 to 10 dB reduction of the digital signal's average signal PEP.

Field tests during 2009 confirmed this behavior, finding that a leaking QAM signal would not budge the S meter of a Yaesu FT-736R at low to moderate field strength leaks, even when the receiver's antenna — a resonant half-wave dipole — was located just 10 feet from a calibrated leak. In contrast, a CW carrier that produced a 20 µV/m leak resulted in an S meter reading of S9 +15 dB, definitely harmful interference! When the CW carrier was replaced by a QAM signal whose digital channel power was equal to the CW carrier's PEP and which produced the same leakage field strength (the latter integrated over the full 6 MHz channel bandwidth), the S-meter read <S1. When the leakage field strength was increased to 100 µV/m, the CW carrier pegged the S meter at S9 +60 dB, while the QAM signal was S3 in FM mode and between S1 and S2 in USB mode. It wasn't until the leaking QAM

signal's field strength reached several hundred µV/m that the "noise" (and it literally sounded like typical white noise) could be construed to be harmful interference.

ANALOG TV CHANNEL LEAKAGE SYMPTOMS

When signal leakage does happen, interference to the amateur service may occur. And where there is leakage, there is probably ingress. That compounds the problem, because not only are you experiencing interference, but when you transmit you may interfere with cable service in the area. One of the most common signs of possible leakage is interference to the 2 meter amateur band, especially in the vicinity of standard (STD) cable channel 18's visual carrier on 145.25 MHz. If you suspect cable leakage, listen for the telltale buzz from the video signal on or near 145.25 MHz (sometimes buzz may not be heard), and check other STD, incrementally related carrier (IRC), and harmonically related carrier (HRC) visual carrier frequencies on nearby channels listed in **Table 27.3** using a wide range receiver or scanner. (Leakage of a digital TV signal on cable channel 18 sounds like broadband noise over the 144-150 MHz range.) Also listen for TV channel sound on the FM aural carriers 4.5 MHz above the visual carriers.

LOCATING LEAKAGE SOURCES

When a cable company technician troubleshoots signal leakage, the process is similar to Amateur Radio fox hunting. The technician uses radio direction finding techniques that may include equipment such as handheld dipole or Yagi antennas, Doppler antenna arrays on vehicles, near-field probes, and commercially manufactured signal leakage detectors. Many leakage detectors incorporate what is known as "tagging" technology to differentiate a leaking cable signal from an over-the-air signal or electrical noise that may exist on or

near the same frequency. Most leakage detection is done on a dedicated cable channel in the 108-138 MHz frequency range.

ELIMINATING LEAKAGE

A large percentage of leakage and ingress problems are not the result of a single shielding defect, although this does happen. For example, a squirrel might chew a hardline feeder cable, or a radial crack might develop in the shield as a consequence of environmental or mechanical damage. Most often, leakage and ingress are caused by several small shielding defects in an area: loose or corroded hardline connectors and splices, old copper braid subscriber drop cabling, improperly installed F connectors, subscriber-installed substandard "do-it-yourself" components, and the previously mentioned poorly shielded cable-ready TVs and other *customer premises equipment* (*CPE*).

After the cable technician locates the source(s) of the leakage, it is necessary to repair or replace the culprit components or cabling. In the case of poorly shielded TVs or VCRs, the cable technician cannot repair those devices, only recommend that they be fixed by a qualified service shop. Often the installation of a set-top box will take care of a cable-ready CPE problem because the subscriber drop cabling is no longer connected directly to the offending device. The latter usually resolves direct pickup problems, too, because the cable-ready TV must be tuned to channel 3 or 4 to receive the set-top signals.

VERIFYING AN RFI SOURCE TO BE LEAKAGE

Spurious signals, birdies, harmonics, intermodulation, electrical noise, and even interference from Part 15 devices are sometimes mistaken for cable signal leakage. One of the most common is emissions from Part 15 devices that become coupled to the cable TV coax shield in some way. Non-leakage noise or spurious signals may radiate from the cable TV lines or an amplifier location, but only because the outer surface of the cable shield is carrying the coupled interference as a common-mode current.

When interference from what sounds like a discrete carrier in the downstream spectrum is received, note the frequency and compare it to known analog TV channel visual and aural carrier frequencies. Consider the following example in which an interfering signal is being heard in the 70 cm amateur band and it falls within cable channel 60 (438-444 MHz). When a cable operator carries an analog TV channel that complies with the STD channel frequency plan (Table 27.3), the visual carrier is at 439.25 MHz. The aural carrier falls 4.5 MHz above the visual carrier, or 443.75 MHz. If the cable operator were using HRC channelization rather than STD channelization, channel 60's visual carrier would be at 438.0219 MHz and its aural carrier at 442.5219 MHz. Can you hear anything at either visual carrier frequency? What about the aural carrier frequencies? Since the visual carrier is the strongest part of an analog TV channel, that's the best place to listen for leakage.

The color subcarrier of STD channel 60 falls at 439.25 + 3.58 = 442.83 MHz, but this is a double sideband suppressed carrier. There are horizontal sidebands spaced every 15.734 kHz from the visual carrier (and also from the color subcarrier), but these are generally very low level. Analog TV channel visual carriers use AM, and the aural carriers use FM. If the interference is from the aural carrier of an analog TV channel, you may be able to hear the channel's audio. If the interfering carrier does not fall on expected cable frequencies, it may not be leakage.

A common non-leakage interference that may radiate from a cable network is broadband electrical interference or other noise in the MF and lower end of the HF spectrum. A common misconception is that since cable companies carry digital signals on frequencies that overlap portions of the over-the-air spectrum below 30 MHz, any "noise" that radiates from the cable plant must be leaking digital signals. This type of interference is almost always power-line electrical interference or other noise that is coupled to the cable network's shield as a common-mode signal.

Upstream digital signals from cable modems, which have channel bandwidths of 1.6, 3.2 or 6.4 MHz, are typically transmitted in the roughly 20 to 40 MHz range, and are bursty in nature rather than continuous like downstream digital signals. Set-top box upstream telemetry carriers are narrowband frequency shift keying (FSK) or quadrature phase shift keying (QPSK) carriers usually in the approximately 8 to 11 MHz range. This type of interference may respond to common-mode chokes. Radiated non-leakage noise-like interference from the outside of the hardline coaxial cables that happens outdoors, affecting MF and/or HF reception in the neighborhood, cannot be fixed using chokes. The source of the interference must be identified and repaired. A little RFI detective work may be necessary. Refer to the various RFI troubleshooting sections of this chapter.

If normal leakage troubleshooting techniques do not clearly identify the source of the interference, sometimes the cable company may temporarily shut off its network in the affected neighborhood. If the interference remains after the cable network is turned off, it is not leakage, and the cable company is not responsible for that type of interference. If the interference disappears when the cable network is turned off, then it most likely is leakage or something related to the cable net-

work. Turning off even a small portion of the cable network is a last resort and may not be practical because of the service disruption to subscribers. It may be easier for the cable company to temporarily shut off a suspect cable channel briefly. Here, too, if the interference remains after the channel is turned off, the interference is not leakage.

HOW TO REPORT LEAKAGE

If you suspect cable signal leakage is causing interference to your amateur station, *never attempt your own repairs to any part of the cable network, even the cabling in your own home*! Document what you have observed. For instance, note the frequency or frequencies involved, the nature of the interference, any changes to the interference with time of day, how long it has been occurring, and so forth. If you have fox-hunting skills and equipment, you might note the probable source(s) of the interference or at least the direction from which it appears to be originating.

Next, contact the cable company. You will most likely reach the cable company's customer service department, but ask to speak with the local cable system's Plant Manager (may also be called Chief Engineer, Director of Engineering, Chief Tech, VP of Engineering, or similar), and explain to him or her that you are experiencing what you believe to be signal leakage-related interference. If you cannot reach this individual, ask that a service ticket be created, and a technician familiar with leakage and ingress issues be dispatched. Share the information you have gathered about the interference. And as with all RFI issues, remember diplomacy!

In the vast majority of cases when cable leakage interference to Amateur Radio occurs, it is able to be taken care of by working with local cable system personnel. Every now and then for whatever reason, the affected ham is unable to get the interference resolved locally. Contact the ARRL for help in these cases.

27.8.5 DVD and Video Players

A DVD or similar video player usually contains a television tuner or has a TV channel output. It is also connected to an antenna or cable system and the ac line, so it is subject to all of the interference problems of a TV receiver. Start by proving that the video player is the susceptible device. Temporarily disconnect the device from the television or video monitor. If there is no interference to the TV, then the video player is the most likely culprit.

Next, find out how the interfering signal is getting into the video player. Temporarily disconnect the antenna or cable feed line from the video player. If the interference goes away, then the antenna line is involved. In this case,

you can probably fix the problem with a common-mode choke or high-pass filter.

Fig 27.21 shows a bulletproof video player installation. If you have tried all of the cures shown and still have a problem, the player is probably subject to direct pickup. In this case, contact the manufacturer through the CEA.

Some analog-type VCRs are quite susceptible to RFI from HF signals. The video baseband signal extends from 30 Hz to 3.5 MHz, with color information centered around 3.5 MHz and the FM sound subcarrier at 4.5 MHz. The entire video baseband is frequency modulated onto the tape at frequencies up to 10 MHz. Direct pickup of strong signals by VCRs is a common problem and may not be easily solved, short of replacing the VCR with a better-shielded model.

27.8.6 Non-radio Devices

Interference to non-radio devices is not the fault of the transmitter. (A portion of the *FCC Interference Handbook,* 1990 Edition, is shown in **Fig 27.22**. Although the FCC no longer offers this *Handbook,* an electronic version is available on this book's CD-ROM, from the ARRL at **www.arrl.org/fcc-rfi-information** or search the ARRL website for "cib interference handbook".) In essence, the FCC views non-radio devices that pick up nearby radio signals as improperly functioning; contact the manufacturer and return the equipment. The FCC does not require that non-radio devices include RFI protection and they don't offer legal protection to users of these devices that are susceptible to interference.

TELEPHONES

Telephones have probably become the number one, non-TVI interference problem of Amateur Radio. However, most cases of telephone interference can be cured by correcting any installation defects and installing telephone RFI filters where needed.

Telephones can improperly function as radio receivers. Semiconductor devices inside many telephones act like diodes. When such a telephone is connected to the telephone wiring (a large antenna) an AM radio receiver can be formed. When a nearby transmitter goes on the air, these telephones can be affected.

Troubleshooting techniques were discussed earlier in the chapter. The suggestion to simplify the problem applies especially to telephone interference. Disconnect all telephones except one, right at the service entrance if possible, and start troubleshooting the problem there.

If any single device or bad connection in the phone system detects RF and puts the detected signal back onto the phone line as audio, that audio cannot be removed with filters. Once the RF has been detected and turned into audio, it

PART II

INTERFERENCE TO OTHER EQUIPMENT

CHAPTER 6

TELEPHONES, ELECTRONIC ORGANS, AM/FM RADIOS, STEREO AND HI-FI EQUIPMENT

Telephones, stereos, computers, electronic organs and home intercom devices can receive interference from nearby radio transmitters. When this happens, the device improperly functions as a radio receiver. Proper shielding or filtering can eliminate such interference. The device receiving interference should be modified in your home while it is being affected by interference. This will enable the service technician to determine where the interfering signal is entering your device.

The device's response will vary according to the interference source. If, for example, your equipment is picking up the signal of a nearby two-way radio transmitter, you likely will hear the radio operator's voice. Electrical interference can cause sizzling, popping or humming sounds.

Fig 27.22 — Part of page 18 from the FCC *Interference Handbook* (1990 edition) explains the facts and places responsibility for interference to non-radio equipment.

Telephone Accessories — Answering machines and fax machines are also prone to interference problems. All of the troubleshooting techniques and cures that apply to telephones also apply to these telephone devices. In addition, many of these devices connect to the ac mains. Try a common-mode choke and/or ac-line filter on the power cord (which may be an ac cord set, a small transformer or power supply).

Cordless Telephones — A cordless telephone is an unlicensed *radio* device that is manufactured and used under Part 15 of the FCC regulations. The FCC does not intend Part 15 devices to be protected from interference. These devices usually have receivers with very wide front-end filters, which make them very susceptible to interference.

A likely path for interference to cordless phones is as common-mode current on the base unit's connecting cables that will respond to common-mode chokes. In addition, a telephone filter on the base unit and an ac line filter may help. The best source of help is the manufacturer but they may point out that the Part 15 device is not protected from interference.

Cordless phone systems operating at 900 MHz and higher frequencies are often less susceptible to interference than older models that use 49 MHz. It may be easiest to simply replace the system with a model less susceptible to interference.

AUDIO EQUIPMENT

Consumer and commercial audio equipment such as stereos, home entertainment systems, intercoms and public-address systems can also pick up and detect strong nearby transmitters. The FCC considers these non-radio devices and does not protect them from licensed radio transmitters that may interfere with their operation. The RFI can be caused by one of several things: pickup on speaker leads or interconnecting cables, pickup by the ac mains wiring or direct pickup. If the

cannot be filtered out because the interference is at the same frequency as the desired audio signal. To effect a cure, you must locate the detection point and correct the problem there.

Defective telephone company lightning arrestors can act like diodes, rectifying any nearby RF energy. Telephone-line amplifiers or other electronic equipment may also be at fault. Do not attempt to diagnose or repair any telephone company wiring or devices on the "telco" side of your service box or that were installed by the phone company. Request a service call from your phone company.

Inspect the telephone system installation. Years of exposure in damp basements, walls or crawl spaces may have caused deterioration. Be suspicious of anything that is corroded or discolored. In many cases, homeowners have installed their own telephone wiring, often using substandard wiring. If you find sections of telephone wiring made from nonstandard cable, replace it with standard twisted-pair telephone or CAT5 cable. If you do use telephone cable, be sure it is high-quality twisted-pair to minimize differential-mode pickup of RF signals.

Next, evaluate each of the telephone instruments. If you find a susceptible telephone, install a telephone RFI filter on that telephone, such as those sold by K-Y Filters. (**www. ky-filters.com**) If the home uses a DSL

broadband data service, be sure that the filters do not affect DSL performance by testing online data rates with and without a filter installed at the telephone instrument.

If you determine that you have interference only when you operate on one particular ham band, the telephone wiring is probably resonant on that band. Install common-mode chokes on the wiring to add a high impedance in series with the "antenna." A telephone RFI filter may also be needed. (See the section on DSL Equipment for filtering suggestions.)

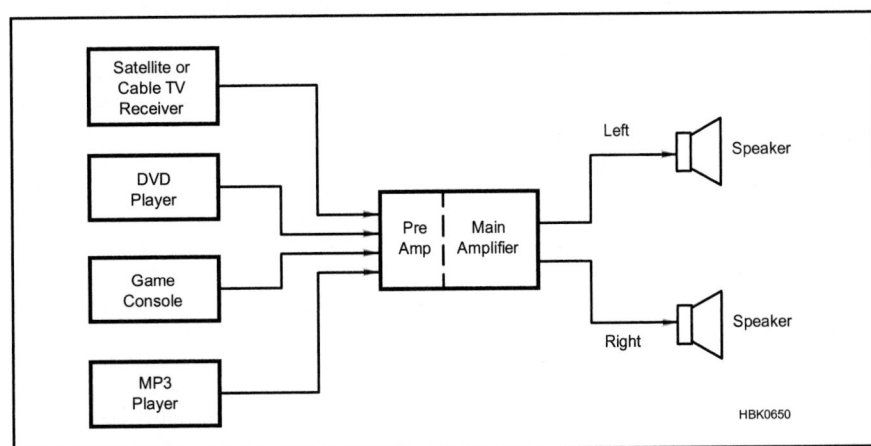

Fig 27.23 — A typical modern home-entertainment system.

interference involves wiring connected to the affected device, common-mode chokes are the most likely solution.

Use the standard troubleshooting techniques discussed earlier in this chapter to isolate problems. In a multi-component home entertainment system (as in **Fig 27.23**), for example, you must determine what combination of components is involved with the problem. First, disconnect all auxiliary components to determine if there is a problem with the main receiver/amplifier. Long speaker or interconnect cables are prime suspects.

Stereos and Home Entertainment Systems — If the problem remains with the main amplifier isolated, determine if the interference level is affected by the volume control. If so, the interference is getting into the circuit *before* the volume control, usually through accessory wiring. If the volume control has no effect on the level of the interfering sound, the interference is getting in *after* the control, usually through speaker wires.

Speaker wires are often effective antennas on HF and sometimes into VHF and above. The speaker terminals are often connected directly to the output amplifier transistors. Modern amplifier designs use a negative feedback loop to improve fidelity. This loop can conduct the detected RF signal back to the high-gain stages of the amplifier. The combination of all of these factors often makes the speaker cables the dominant receiving antenna for RFI.

There is a simple test that will help determine if the interfering signal is being coupled into the amplifier by the speaker leads.

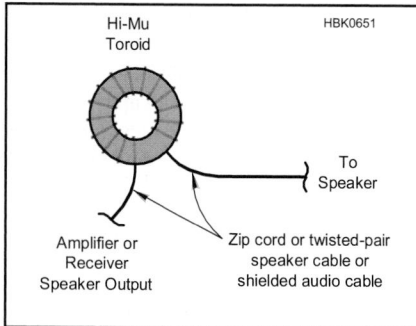

Fig 27.24 — Making a speaker-lead common-mode choke. Use ferrite material appropriate for the frequency of the RF interference.

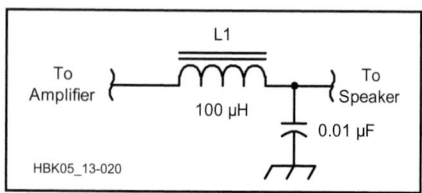

Fig 27.25 — A low-pass LC filter.

Temporarily disconnect the speaker leads from the amplifier, and plug in a test set of headphones with short leads. If there is no interference with the headphones, filtering the speaker leads will likely cure the problem.

Start by applying common-mode chokes. **Fig 27.24** shows how to wrap speaker wires around an large (2-inch O.D. or larger) ferrite core to cure speaker-lead RFI. Type 31 material is preferred at HF. (See the section on Common-Mode Chokes in this chapter.)

In some cases, the speaker wires may be picking up RF as a differential-mode signal. To reduce differential-mode pickup, replace the zip cord speaker wire with twisted-pair wire. (#16 AWG will work for most systems with higher-power systems requiring #12 AWG.)

Powered Speakers — A powered speaker is one that has its own built-in power amplifier. Powered subwoofers are common in home entertainment systems and small powered speakers are often used with computer and gaming systems. If a speaker runs on batteries and/or an external power supply, or is plugged into mains power, it is a powered loudspeaker. Powered loudspeakers are notoriously susceptibility to common-mode interference from internally misconnected cable shields and poor shielding. Apply suitable common-mode chokes to all wiring, including power wiring. If the RFI persists, try an RF filter at the input to the speaker, such as the LC low-pass filter in **Fig 27.25**. Unshielded speakers may not be curable, however.

Intercoms and Public-Address Systems — RFI to these systems is nearly always caused by common-mode current on interconnect wiring. Common-mode chokes are the most likely cure, but you may also need to contact the manufacturer to see if they have any additional, specific information. Wiring can often be complex, so any work on these systems should be done by a qualified sound contractor.

COMPUTERS AND OTHER UNLICENSED RF SOURCES

Computers and microprocessor-based devices such as video games or audio players can be sources or victims of interference. These devices contain oscillators that can and do radiate RF energy. In addition, the internal functions of a computer generate different frequencies, based on the various data signals. All of these signals are digital — with fast rise and fall times that are rich in harmonics.

Computing devices are covered under Part 15 of the FCC regulations as unintentional emitters. As for any other unintentional emitter, the FCC has set absolute radiation limits for these devices. As previously discussed in this chapter, FCC regulations state that the operator or owner of Part 15 devices must

take whatever steps are necessary to reduce or eliminate any interference they cause to a licensed radio service. This means that if your neighbor's video game interferes with your radio, the neighbor is responsible for correcting the problem. (Of course, your neighbor may appreciate your help in locating a solution!)

The FCC has set up two tiers of limits for computing devices. Class A is for computers used in a commercial environment. FCC Class B requirements are more stringent — for computers used in residential environments. If you buy a computer or peripheral, be sure that it is Class B certified or it will probably generate interference to your amateur station or home-electronics equipment.

If you find that your computer system is interfering with your radio (not uncommon in this digital-radio age), start by simplifying the problem. Temporarily remove power from as many peripheral devices as possible and disconnect their cables from the back of the computer. (It is necessary to physically remove the power cable from the device, since many devices remain in a low-power state when turned off from the front panel or by a software command.) If possible, use just the computer, keyboard and monitor. This test may identify specific peripherals as the source of the interference.

Ensure that all peripheral connecting cables are shielded. Replace any unshielded cables with shielded ones; this often significantly reduces RF noise from computer systems. The second line of defense is the common-mode choke. The choke should be installed as close to the computer and/or peripheral device as practical. **Fig 27.26** shows the location of common-mode chokes in a complete computer system where both the computer and peripherals are noisy.

Switchmode power supplies in computers are often sources of interference. A common-mode choke and/or ac-line filter may cure this problem. In extreme cases of computer interference you may need to improve the shielding of the computer. (Refer to the *ARRL RFI Book* for more information about this.) Don't forget that some peripherals (such as modems) are connected to the phone line, so you may need to treat them like telephones.

GROUND-FAULT AND ARC-FAULT CIRCUIT INTERRUPTORS (GFCI AND AFCI)

GFCIs are occasionally reported to "trip" (open the circuit) when a strong RF signal, such as an amateur's HF transmission, is present. GFCI circuit breakers operate by sensing unbalanced currents in the hot and neutral conductors of an ac circuit. In the absence of RF interference, such an imbalance indicates the presence of a fault somewhere in the circuit, creating a shock hazard. The breaker

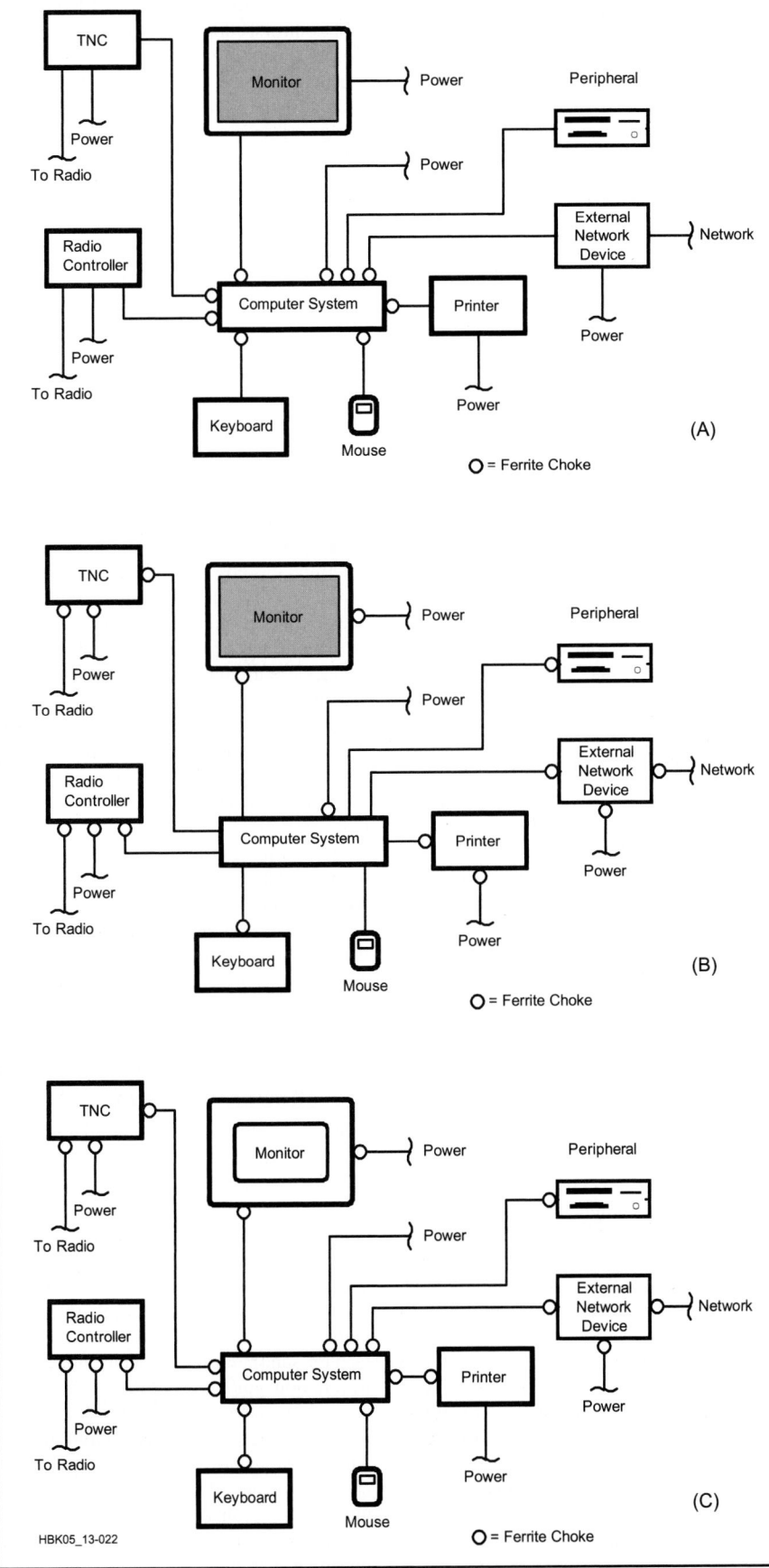

Fig 27.26 — Where to locate ferrite chokes in a computer system. At A, the computer is noisy, but the peripherals are quiet. At B, the computer is quiet, but external devices are noisy. At C, both the computer and externals are noisy.

then trips (opens) to remove the shock hazard.

An Arc Fault Circuit Interrupter (AFCI) circuit breaker is similar in that it monitors current to watch for a fault condition. Instead of current imbalances, the AFCI detects patterns of current that indicate an arc — one of the leading causes of home fires. The AFCI is not supposed to trip because of "normal" arcs that occur when a switch is opened or a plug is removed.

Under current codes, GFCI protection is required for all basement outlets, outdoor outlets, and for outlets in kitchens and bathrooms. AFCI protection is required for all circuits that supply bedrooms.

RF interference to GFCI breakers is caused by RF current or voltage upsetting normal operation of the imbalance detection circuit, resulting in the false detection of a fault. Similarly, RF current or voltage could upset the arc detection circuitry of an AFCI breaker. Some early GFCI breakers were susceptible to RFI, but as the technology has improved, fewer and fewer such reports have been received. While it is possible to add filtering or RF suppression to the breaker wiring, a simpler and less expensive solution is to replace the GFCI breaker with a new unit less susceptible to RF.

The ARRL Lab has received favorable reports on the following GFCI products:

• Leviton (**www.leviton.com**) GFCI outlets which are available in both 15 and 20 A versions for 120 V ac circuits as well as cord sets and user-attachable plugs and receptacles.

• Bryant (**www.bryant-electric.com**) ground fault receptacles which feature published 0.5 V immunity from 150 kHz to 230 MHz.

• Cooper (**www.cooperindustries.com**) GFCI products that are labeled "UL 943 compliant" on the package.

A web page on the ARRL website is maintained on GFCI/AFCI technology (**www.arrl.org/gfci-devices**). Reports have not been received on AFCI products as of early 2013. Since AFCI technology is newer than GFCI, immunity to RF may be less of a problem than with the initial GFCI products.

WALL TRANSFORMER SWITCHING SUPPLIES

While small, low-current linear power supplies known as "wall warts" have been widely used for many years, it is now becoming common for these devices to contain switchmode or "switcher" supplies. Because they must be manufactured very inexpensively, these supplies often have little or no RF filtering at either the ac input or dc output, frequently creating significant RFI to nearby amateurs.

The least expensive course of action may be to simply replace the switchmode supply with a linear model. If the system in which the supply is used is still under warranty, the distributor or manufacturer may be able to replace it. Otherwise, a third-party linear replacement may be available with the same voltage rating and current output equal to or higher than the original supply. Adapters may be available to convert output connector styles where necessary or new connectors installed.

If replacing the supply with a linear model is not an option, you will have to apply RF filters to the supply. These supplies are rarely serviceable, so filters much be installed externally. Noise is usually radiated from the output cable so winding the cable onto a ferrite core creates a common-mode RF choke

as described in this chapter's Elements of RFI Control section. Since the wall-wart style supply plugs directly into a wall-mounted receptacle, an ac filter must either be installed in the ac outlet box or a short ac extension cord can be made into a common-mode choke, as well.

27.9 Automotive RFI

Automobiles have evolved from a limited number of primitive electrical components to high technology, multi-computer systems on-wheels. Every new technology deployed can potentially interfere with amateur equipment.

Successful mobile operation depends on a multitude of factors such as choosing the right vehicle, following installation guidelines, troubleshooting and deploying the appropriate RFI fixes as needed.

A number of these factors will be covered in this section. Additionally, newly emerging electrical and hybrid-electric vehicles will be discussed, which pose unique challenges to amateur equipment installations and operation.

27.9.1 Before Purchasing a Vehicle

When shopping for a new vehicle intended for a mobile amateur installation, begin with research. A wealth of information is available on the Internet, and specifically at **www.arrl.org/automotive**, where the ARRL has compiled years of data from automotive manufacturers and other hams. Email reflectors and websites may provide information from hams willing to share their experiences concerning mobile communications in their own vehicles that may be the very make and model you were considering.

Armed with research, your next stop is your dealer. The manufacturer of each vehicle is the expert on how that vehicle will perform. The dealer should have good communication with the manufacturer and should be able to answer your questions. Ask about service bulletins and installation guidelines. You can also ask your dealer about fleet models of their vehicles. Some manufacturers offer special modifications for vehicles intended for sale to police, taxicabs and other users who will be installing radios (usually operating at VHF and UHF) in their cars.

When shopping for a vehicle, it is useful to take along some portable (preferably battery operated) receivers or scanners and have a

friend tune through your intended operating frequencies while you drive the vehicle. This will help identify any radiated noise issues associated with that model vehicle, which can be more difficult to resolve than conducted noise. If you intend to make a permanent transceiver installation, give some consideration to how you will mount the transceiver and route the power and/or antenna cables. While looking for ways to route the wiring, keep in mind that some newer cars have the battery located in the trunk or under the rear seat, and that may make power wire routing easier.

Test the car before you buy it. A dealer expects you to take the car for a test ride; a cooperative dealer may let you test it for radio operation, too. A fair amount of checking can easily be performed without digging too deeply into the car. Check the vehicle for noise with a portable receiver on VHF, where your handheld transceiver will do the job nicely. On HF, you can usually locate noise with a portable short-wave receiver, or operate your HF transceiver with a portable antenna and cigarette-lighter plug. With the engine running, tune across the bands of interest. You may hear some noise, or a few birdies, but if the birdies don't fall on your favorite frequency, this is an encouraging sign! Check with the vehicle completely off, with the key in the ignition, and with the vehicle running — electronic subsystems operate in different ways with the vehicle running or not running.

To test the vehicle for susceptibility to your transmitted signal, you must transmit. It is important to note that without a full and complete installation, you will not be able to fully assess the effects of full-power transmissions on a vehicle. Any testing done with temporary equipment installations cannot be considered an absolute guarantee because an installed transmitter connected directly to the vehicle's power source may cause the vehicle to act differently.

To perform transmit tests, bring your radio and a separate battery (if permitted by dealer) so you can transmit at full power while in mo-

tion without having to run cables to the vehicle battery. Use a magnet-mount antenna (several *QST* advertisers sell mounts suitable for HF) for temporary testing. (Use the magnet-mount carefully; it is possible to scratch paint if any particles of dirt get on the bottom of the magnets.) Transmit on each band you will use to see if the RF has any effect on the vehicle. Lack of response to your transmissions is a good sign, but does not mean the vehicle is immune to RF as a permanent installation will result in different (likely stronger) field strengths and distributions in and around the vehicle and a permanent antenna more effectively coupled to the vehicle.

On both transmit and receive, you may want to experiment with the placement of the antenna. Antenna placement plays an important role in operation, and you may be able to find an optimal location for the antenna that predicts good performance with a permanent installation.

27.9.2 Transceiver Installation Guidelines

While most amateurs are familiar with the process of installing a transceiver, there are preferred practices that will help minimize potential problems. These include support from the automotive dealer, typical "best practices" installations, and consideration of special situations.

The first step is to ensure that your installation complies with both the vehicle manufacturer's and radio transceiver manufacturer's installation guidelines. Links to domestic automotive manufacturer installation guidelines are found at **www.arrl.org/automotive**. Automotive manufacturers that import vehicles for sale here do not publish installation guidelines because their vehicles are not typically used in police, fire and taxicab applications within the US.

The installation guidelines of different manufacturers vary as to how to install a radio transceiver's power leads. Most manufacturers recommend that the positive and negative leads from the radio be run directly to the battery. This minimizes the potential for the interaction between the radio's negative lead currents and vehicle electronics. If the manufacturer recommends that both wires be connected to the battery, they will also require that both wires be fused. This is necessary because, in the unlikely event that the connection between the battery and the engine block were to fail, excessive current could be drawn on the radio's negative lead when the vehicle starter is engaged.

Some vehicles provide a "ground block" near the battery for a negative cable to be connected. On these vehicles, run the negative power lead, un-fused, to the "ground block." When this technique is recom -mended by the manufacturer, the interaction between the power return currents and vehicle electronics has been evaluated by the manufacturer. In all cases, the most important rule to remember is this: If you want the manufacturer to support your installation, do it exactly the way the installation guidelines tell you to do it!

If no installation guidelines are available for your vehicle, the practices outlined below will improve compatibility between in-vehicle transceivers and vehicle electronics:

1) Transceivers
• Transceivers should be mounted in a location that does not interfere with vehicle operator controls and visibility, provides transceiver ventilation, and be securely mounted.
• Ensure all equipment and accessories are removed from the deployment path of the airbag and safety harness systems.

2) Power Leads
• The power leads should be twisted together from the back of the rig all the way to the battery. This minimizes the area formed by the power leads, reducing susceptibility to transients and RFI.
• Do not use the vehicle chassis as a power return.
• The power leads should be routed along the body structure, away from vehicle wiring harnesses and electronics.
• Any wires connected to the battery should be fused at the battery using fuses appropriate for the required current.
• Use pass-through grommets when routing wiring between passenger and engine compartments.
• Route and secure all under hood wiring away from mechanical hazards.

3) Coaxial Feed Lines
• The coaxial feed line should have at least 95% braid coverage. The cable shield should be connected to every coaxial connector for the entire circumference (no "pigtails").
• Keep antenna feed lines as short as practical and avoid routing the cables parallel to vehicle wiring.

4) Antennas
• Antenna(s) should be mounted as far from the engine and the vehicle electronics as practical. Typical locations would be the rear deck lid or roof. Metal tape can be used to provide an antenna ground plane on non-metallic body panels.
• Care should be used in mounting antennas with magnetic bases, since magnets may affect the accuracy or operation of the compass in vehicles, if equipped.
• Since the small magnet surface results in low coupling to the vehicle at HF, it is likely that the feed line shield will carry substantial RF currents. A large (2-inch OD or larger toroid) common-mode choke at the antenna will help reduce this current, but will also reduce any radiation produced by that current.
• Adjust the antenna for a low SWR.

27.9.3 Diagnosing Automotive RFI

Most VHF/UHF radio installations should result in no problems to either the vehicle systems or the transceiver, while HF installations are more likely to experience problems. In those situations where issues do occur, the vast majority are interference to the receiver from vehicle on-board sources of energy that are creating emissions within the frequency bands used by the receiver. Interference to one of the on-board electronic systems can be trivial or it can cause major problems with an engine control system.

The dealer should be the first point of contact when a problem surfaces, because the dealer should have access to information and factory help that may solve your problem. The manufacturer may have already found a fix for your problem and may be able to save your mechanic a lot of time (saving you money in the process). If the process works properly, the dealer/customer-service network can be helpful. In the event the dealer is unable to solve your problem, the next section includes general troubleshooting techniques you can perform independently.

GENERAL TROUBLESHOOTING TECHNIQUES

An important aspect is to use the source-path-victim model presented earlier in this chapter. The path from the source to the receiver may be via radiation or conduction. If the path is radiation, the electric field strength (in V/m) received is reduced as a function of the distance from the source to the receiver. In most cases, susceptible vehicle electronics is in the near-field region of the radiating source, where the electric and magnetic fields can behave in complex ways. In general, however, the strength of radiated signals falls off with distance.

The best part of all this is that with a gen-

eral-coverage receiver or spectrum analyzer, a fuse puller and a shop manual, the vehicle component needing attention may be identified using a few basic techniques. The only equipment needed could be as simple as:
• A mobile rig, scanner or handheld transceiver, or
• Any other receiver with good stability and an accurate readout, and
• An oscilloscope for viewing interference waveforms

BROADBAND NOISE

Automotive broadband noise sources include:
• Electric motors such as those that operate fans, windows, sunroof, AM/FM antenna deployment, fuel pumps, etc.
• Ignition spark

If you suspect electric motor noise is the cause of the problem, obtain a portable AM or SSB receiver to check for this condition. Switch on the receiver and then activate the electric motors one at a time. When a noisy motor is switched on, the background noise increases. It may be necessary to rotate the radio, since portable AM radios use a directional ferrite rod antenna.

To check whether fuel pumps, cooling fans, and other vehicle-controlled motors are the source of noise, pull the appropriate fuse and see whether the noise disappears.

A note concerning fuel pumps: virtually every vehicle made since the 1980s has an electric fuel pump, powered by long wires. It may be located inside the fuel tank. Don't overlook this motor as a source of interference just because it may not be visible. Electric fuel-pump noise often exhibits a characteristic time pattern. When the vehicle ignition switch is first turned on, without engaging the starter, the fuel pump will run for a few seconds, and then shut off when the fuel system is pressurized. At idle, the noise will generally follow the pattern of being present for a few seconds before stopping, although in some vehicles the fuel pump will run almost continuously if the engine is running.

NARROWBAND NOISE

Automotive narrowband noise sources include:
• Microprocessor based engine control systems
• Instrument panel
• RADAR obstacle detection
• Remote keyless entry
• Key fob recognition systems
• Tire pressure monitoring systems
• Global positioning systems
• Pulse width modulating motor speed controls
• Fuel injectors
• Specialized electric traction systems found in newer hybrid/electric vehicles

Start by moving the antenna to different

locations. Antenna placement is often key to resolving narrowband RFI problems. However, if antenna location is not the solution, consider pulling fuses. Tune in and stabilize the noise, then find the vehicle fuse panels and pull one fuse at a time until the noise disappears. If more than one module is fed by one fuse, locate each module and unplug it separately. Some modules may have a "keep-alive" memory that is not disabled by pulling the fuse. These modules may need to be unplugged to determine whether they are the noise source. Consult the shop manual for fuse location, module location, and any information concerning special procedures for disconnecting power.

A listening test may verify alternator noise, but if an oscilloscope is available, monitor the power line feeding the affected radio. Alternator whine appears as full-wave rectified ac ripple and rectifier switching transients superimposed on the power system's dc power voltage (see **Fig 27.27**).

Alternators rely on the low impedance of the battery for filtering. Check the wiring from the alternator output to the battery for corroded contacts and loose connectors when alternator noise is a problem.

Receivers may allow conducted harness noise to enter the RF, IF or audio sections (usually through the power leads), and interfere with desired signals. Check whether the interference is still present with the receiver powered from a battery or power supply instead of from the vehicle. If the interference is no longer present when the receiver is operating from a battery or external supply, the interference is conducted via the radio power lead. Power line filters installed at the radio may resolve this problem.

27.9.4 Eliminating Automotive RFI

The next section includes various techniques to resolve the more common RFI problems. As a caveat, performing your own RFI work, in or out of warranty, you assume the same risks as you do when you perform any other type of automotive repair. Most state laws (and common sense) say that those who work on cars should be qualified to do so. In most cases, this means that work should be done either by a licensed dealer or automotive repair facility.

CONDUCTED INTERFERENCE

To reduce common-mode current, impedance can be inserted in series with the wiring in the form of common-mode chokes. (See this chapter's section on Common-Mode Chokes.) Wire bundles may also be wound around large toroids for the same effect.

Mechanical considerations are important in mobile installations. A motor vehicle is

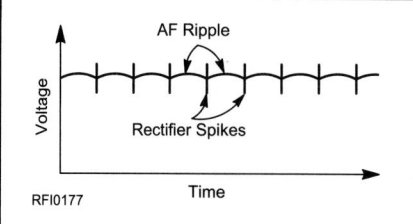

Fig 27.27 — Alternator whine consists of full-wave rectified ac, along with pulses from rectifier switching, superimposed on the vehicle's dc power voltage.

subject to a lot of vibration. If a choke is installed on a wire, this vibration may cause the choke to flex the wire, which may ultimately fail. It is critical that any additional shielding and/or chokes placed on wiring have been installed by qualified personnel who have considered these factors. These must be properly secured, and sometimes cable extenders are required to implement this fix.

RFI TO ON-BOARD CONTROL SYSTEMS

RFI to a vehicle's on-board control and electronic modules should be treated with common-mode chokes at the connection to the module. Some success has been reported by using braid or metal foil to cover a wire bundle as a shield and connecting the shield to the vehicle chassis near the affected module. Vehicle electronic units should not be modified except by trained service personnel according to the manufacturer's recommendations. The manufacturer may also have specific information available in the form of service bulletins.

FILTERS FOR DC MOTORS

If the motor is a conventional brush- or commutator-type dc motor, the following cures shown in **Fig 27.28** are those generally

used. As always, the mechanic should consult with the vehicle manufacturer. To diagnose motor noise, obtain an AM or SSB receiver to check the frequency or band of interest. Switch on the receiver, and then activate the electric motors one at a time. When a noisy motor is switched on, the background noise increases as well.

The pulses of current drawn by a brush-commutator motor generate broadband RFI that is similar to ignition noise. However, the receiver audio sounds more like bacon frying rather than popping. With an oscilloscope displaying receiver audio, the noise appears as a series of pulses with random space between the pulses. Such broadband noise generally has a more pronounced effect on AM receivers than on FM. Unfortunately, the pulses may affect FM receivers by increasing the "background noise level" and will reduce perceived receiver sensitivity because of the degraded signal-to-noise ratio.

ALTERNATOR AND GENERATOR NOISE

As mentioned previously, brush-type motors employ sliding contacts which can generate noise. The resulting spark is primarily responsible for the "hash" noise associated with these devices. Hash noise appears as overlapping pulses on an oscilloscope connected to the receiver audio output. An alternator also has brushes, but they do not interrupt current. They ride on slip rings and supply a modest current, typically 4 A to the field winding. Hence, the hash noise produced by alternators is relatively minimal.

Generators use a relay regulator to control field current and thus output voltage. The voltage regulator's continuous sparking creates broadband noise pulses that do not overlap in time. They are rarely found in modern automobiles.

Alternator or generator noise may be conducted through the vehicle wiring to the power

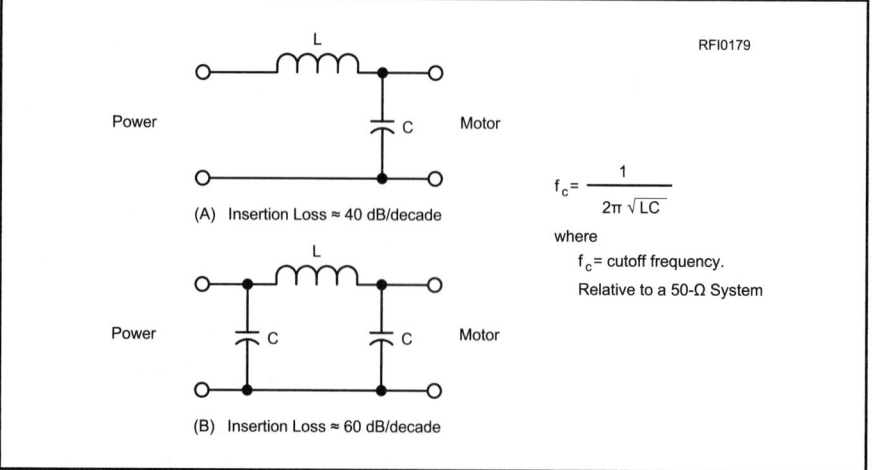

Fig 27.28 — Filters for reducing noise from dc motors

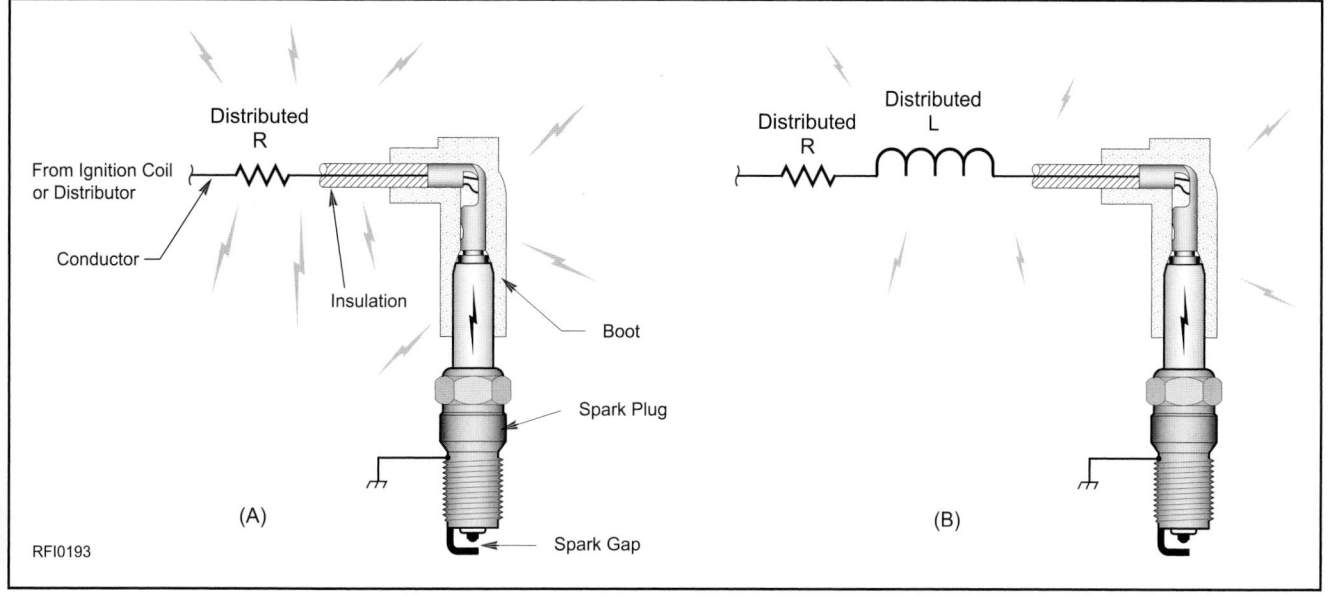

Fig 27.29 — Ignition noise suppression methods.

input of mobile receivers and transmitters and may then be heard in the audio output. If alternator or generator noise is suspected and an oscilloscope is not available, temporarily remove the alternator belt as a test. (This may not be possible in vehicles with a serpentine belt.)

IGNITION NOISE

Ignition noise is created by fast-rise-time pulses of coil current discharging across air gaps (distributor and spark plug). The theoretical models (zero rise time) of such pulses are called impulse functions in the time domain. When viewed in the frequency domain, the yield is a constant spectral energy level starting nearly at 0 Hz and theoretically extending up in frequency to infinity. In practice, real ignition pulses have a finite rise time, so the spectral-energy envelope decreases above some frequency.

It turns out that noise generated by ignition sparks and fuel injector activation manifest themselves as a regular, periodic "ticking" in the receiver audio output, which varies with engine RPM. If an oscilloscope were connected to the audio output, a series of distinct, separate pulses would appear. At higher speeds it sounds somewhat musical, like alternator whine, but with a harsher note (more harmonic content).

A distinguishing feature of ignition noise is that it increases in amplitude under acceleration. This results from the increase in the required firing voltage with higher cylinder pressure. (Noise at higher frequencies may also be reproduced better by the audio circuits.) Since ignition noise is usually radiated noise, it should disappear when the antenna element is disconnected from the antenna

mount. The radiation may be from either the secondary parts of the system or it may couple from the secondary to the primary of the coil and be conducted for some distance along the primary wiring to the ignition system, then radiated from the primary wiring.

Two main methods are employed to suppress this noise — one involves adding an inductance, and the other involves adding a resistance — both in the secondary (high voltage) wiring. This is shown in **Fig 27.29**. The addition of these elements does not have a measurable effect on the engine operation, because the time constants involved in the combustion process are much longer than those associated with the suppression components. (Note that modifying your vehicle's ignition system may be considered as tampering with your vehicle's emission control system and may affect your warranty coverage — work with your dealer or limit your efforts to changing spark plug wires or possibly shielding them.)

The resistance method suppresses RFI by dissipating energy that would have been radiated and/or conducted. Even though the amount of energy dissipated is small, it is still enough to cause interference to sensitive amateur installations. The other method uses inductance and even though the energy is not dissipated, suppression occurs because the inductor will store the pulse energy for a short time. It then releases it into the ignition burn event, which is a low impedance path, reducing the RFI.

For traditional "Kettering" inductive discharge ignition systems, a value of about 5 kΩ impedance (either real and/or reactive) in the spark plug circuit provides effective suppression and, with this value, there is no

detectable engine operation degradation. (Capacitor discharge systems, in comparison, are required to have very low impedance on the order of tens of ohms in order to not reduce spark energy, so they are not tolerant of series impedance). Most spark plug resistances are designed to operate with several kV across the plug gap, so a low-voltage ohmmeter may not give proper resistance measurement results.

The term "resistor wire" is somewhat misleading. High-voltage ignition wires usually contain both resistance and inductance. The resistance is usually built into suppressor spark plugs and wires, while there is some inductance and resistance in wires, rotors and connectors. The elements can be either distributed or lumped, depending on the brand, and each technique has its own merit. A side benefit of resistance in the spark plug is reduced electrode wear.

COIL-ON-PLUG IGNITION NOISE

Many newer spark-ignition systems incorporate a "coil on plug" (COP) or "coil near plug" (CNP) approach. There are advantages to this from an engine operation standpoint, and this approach may actually reduce some of the traditional sources of ignition system RFI. This is because of the very short secondary wires that are employed (or perhaps there are no wires — the coil is directly attached to the spark plug). This reduces the likelihood of coupling from the secondary circuit to other wires or vehicle/engine conductive structures.

There will always be some amount of energy from the spark event that will be conducted along the lowest impedance path. It may mean that the energy that would have been in the secondary circuit will be coupled back on the primary wiring harness attached to the coils. This means that the problem may go from a radiated to a conducted phenomenon.

The fix for this in some cases may actually be easier or harder than one might think. Two approaches that have been used with success are ferrite cores and bypass capacitors.

Ferrite cores are recommended as the first choice, since they require no electrical modification to the vehicle. Ferrite clamp-on split cores are added to the 12-V primary harness attached to the coils. Depending on the frequency of the noise and selection of the ferrite material, there can be significant improvement (as much as 10 dB). Key to optimizing the amount of suppression is to determine where the noise "peaks" and selecting the correct ferrite material for that frequency range (see this chapter's section Using Ferrite for RFI Suppression).

The second method is to add a bypass capacitor between the primary wire of the 12-V coil and ground in the harness near the coil assemblies (there may be 2, 3 or 4 coils). This must be done carefully because it could affect the functionality of the ignition system and — perhaps most importantly — may void the vehicle warranty. This "bypass" capacitor performs the same function that bypass capacitors in any other application perform — separating the noise from the intended signal/power.

27.9.5 Electric and Hybrid-Electric Vehicles

Electric vehicles (EV) and hybrid-electric vehicles (HEV) are quickly becoming a practical means of transportation. EV/HEVs are advanced vehicles that pose unique challenges for amateur equipment. While EV/HEVs provide improved emissions and fuel economy, EV/HEVs utilize switched high voltage and high current to control propulsion. The switching techniques used generate RFI within much of our frequency bands — a cause for concern, particularly for HF operation.

This section is designed to enlighten vehicle owners to the challenges and to make suggestions when installing mobile amateur equipments in an EV/HEV.

EV AND HEV ARCHITECTURE

Most EVs and HEVs have similar electrical traction system (ETS) architectures consisting of a high voltage battery supplying energy to an inverter which controls an electric motor within a transmission connected to the drive wheels. The main difference between the two is that an HEV includes an internal combustion engine to aid in propulsion and a pure

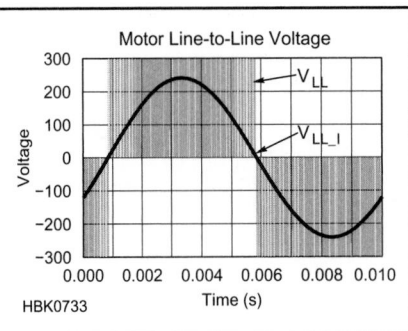

Fig 27.30 — Phase-to-phase motor terminal voltage.

EV is strictly electrically powered.

The heart of the ETS is a device called an inverter. It simply converts dc voltage from the high voltage battery (typical voltage range from 42 to 350 V dc) to an ac waveform supplying the electric motor. This dc-to-ac conversion is performed by a matrix of six transistor switches. The switches chop the dc voltage into systematically varying pulses called pulse-width-modulation (PWM) to form an adjustable frequency and RMS voltage suitable to power electric motor.

In most cases, the ac voltage from the inverter is a three-phase waveform similar to industrial applications because three-phase motors can be smaller, more efficient, and provide greater torque than single-phase motors.

IMPORTANT — Bright orange cables connect the battery pack to the inverter and the inverter to the drive motor, transferring voltage and current to and from the inverter. Because of the non-sinusoidal waveforms being transferred, these cables are shielded and terminated at each end. Under no condition should these cables be disconnected or modified, because the high voltage system depends on a delicate balance of sensors and safety mechanisms. Possible malfunction and damage to the ETS may occur if modified.

EV AND HEV RFI CONCERNS

The inverter uses PWM to convert dc battery voltage to an ac waveform. The phase-to-phase terminal voltage appears in **Fig 27.30** as rectangular blocks with positive and negative amplitude equal to the battery voltage. For example, a 300 V dc battery pack will provide 600 V peak-to-peak at the motor terminals. In Fig 27.30, the same terminal voltage signal is sent through a low pass filter to show how PWM forms a sinusoidal waveform. Each pulse is essentially a square wave. Harmonics from these pulses fall within most of our amateur HF bands, affecting radio performance. Because EV/HEV systems are evolving rap-

idly, check the ARRL's Automotive RFI web page (**www.arrl.org/automotive**) for more information.

EV AND HEV RFI REMEDIES

Troubleshooting techniques described earlier apply in diagnosing RFI from EV/HEV systems. Limited RFI remedies are available associated to components within the ETS. Work with your dealer when you suspect the ETS as the RFI source. Do not attempt to modify or repair your ETS; the dealer's service center is most qualified to inspect and repair your EV/HEV electrical traction system.

During installation, mobile equipment power cables and antenna coaxial cables should be routed as far as possible from the bright orange cables. Common-mode chokes can decrease noise on 12-V dc power cables. Additionally, antenna placement plays a critical role in mobile equipment performance. Areas such as the top of a roof or trunk sometimes provide additional shielding.

27.9.6 Automotive RFI Summary

Most radio installations should result in no problems to either the vehicle systems or any issues with the transceiver. However, manufacturer, make and models differ, thus introducing challenges during amateur equipment installations.

Begin by researching your vehicle of interest and visiting the dealer. Insist on transmitting and receiving your favorite frequencies as you test drive. Request information pertaining to the manufacturer's transceiver installation guidelines. If manufacturer information is not available, follow the guidelines described earlier.

After installation, RFI problems may appear. Report you problem to the dealer, because they have access to manufacturer service bulletins which may describe a repair solution. Additional troubleshooting and remedies are also described previously to assist in successful communication.

Limited RFI remedies are available associated to components within the ETS. Work with your dealer when you suspect the ETS is the RFI source. Do not attempt to modify or repair your ETS; the dealer's service center is most qualified to inspect and repair your EV/HEV electrical traction system.

Lastly, the latest version of the *ARRL RFI Book* contains additional information on RFI in automobiles. More details are given about noise sources, troubleshooting techniques, a troubleshooting flow chart, additional filtering techniques, and information on EV/HEVs.

27.10 RFI Projects

Note: Additional RFI projects appear on the CD accompanying this book.

27.10.1 *Project*: RF Sniffer

Every home is full of electrical equipment capable of emitting electromagnetic radiation to interfere with radio amateurs trying to listen to signals on the bands. This project detects the radiation that causes problems to the amateur, and the noise can be heard. This device will allow you to demonstrate the "noise" with which we have to contend.

CONSTRUCTION

The circuit (**Fig 27.31**) uses a telephone pick-up coil as a detector, the output of which is fed into a LM741 IC preamplifier, followed by a LM386 IC power amplifier. See **Table 27.4** for the complete component list.

The project is built on a perforated board (**Fig 27.32**), with the component leads pushed

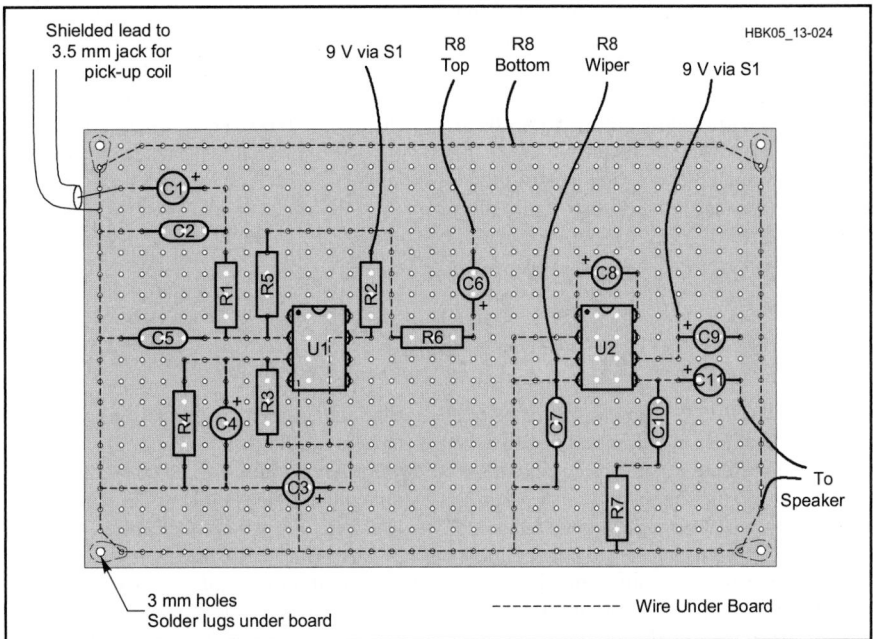

Fig 27.32 — The project is built on perforated board with point-to-point wiring underneath.

Table 27.4
Components List

Resistors	Value
R1	1 kΩ
R2, R6	100 Ω
R3, R4	47 kΩ
R5	100 kΩ
R7	10 Ω
R8	10 kΩ, with switch

Capacitors	Value
C1, C6	4.7 µF, 16 V electrolytic
C2, C5	0.01 µF
C3, C4	22 µF, 16 V electrolytic
C7	0.047 µF
C8	10 µF, 16 V electrolytic
C9, C11	330 µF, 16 V electrolytic
C10	0.1 µF

Semiconductors	
U1	LM741
U2	LM386

Additional Items	
LS1	Small 8-Ω loudspeaker
Perforated board	
9 V battery and clip	
3.5 mm mono-jack socket	
Case	
Telephone pick-up coil	

Table 27.5
Readings (pick-up coil near household items)

29-MHz oscilloscope	0.56 V
Old computer CRT monitor	0.86 V
Old computer with plastic case	1.53 V
New computer CRT monitor	0.45 V
New tower PC with metal case	0.15 V
Old TV	1.2 V
New TV	0.4 V
Plastic-cased hairdryer	4.6 V
Vacuum cleaner	3.6 V
Drill	4.9 V

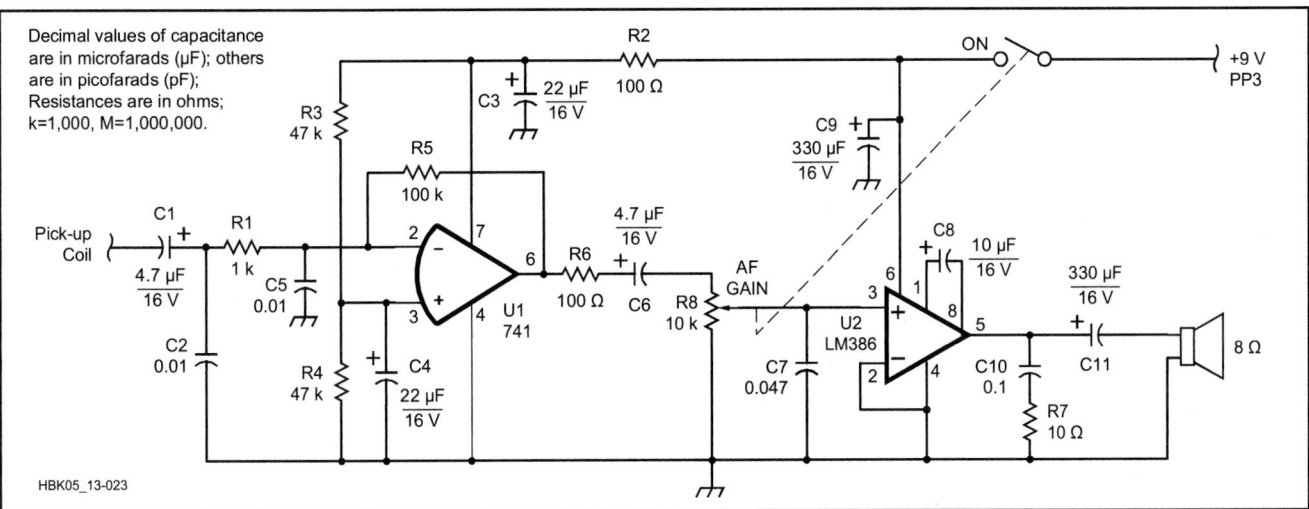

Fig 27.31 — The detector works by receiving stray radiation on a telephone pick-up coil and amplifying it to loudspeaker level.

through the holes and joined with hook-up wire underneath. There is a wire running around the perimeter of the board to form an earth bus.

Build from the loudspeaker backwards to R8, apply power and touch the wiper of R8. If everything is OK you should hear a loud buzz from the speaker. Too much gain may cause a feedback howl, in which case you will need to adjust R8 to reduce the gain. Complete the rest of the wiring and test with a finger on the input, which should produce a click and a buzz. The pick-up coil comes with a lead and 3.5 mm jack, so you will need a suitable socket.

RELATIVE NOISES

Place a high-impedance meter set to a low-ac-voltage range across the speaker leads to give a comparative readout between different items of equipment in the home. Sample readings are shown in **Table 27.5**.

27.11 RFI Glossary

Balanced circuit — A circuit whose two conductors have equal impedance to a common reference, such as a reference plane or circuit common.

Bond — (noun) A low-impedance, mechanically robust, electrical connection.

Common-mode — A voltage or current that is equal, in phase, and has the same polarity on all conductors of a cable. Common-mode current excites a cable as an antenna, and a cable acting as a receiving antenna produces common-mode current.

Conducted RFI — RFI received via a conducting path.

Coupled RFI — RFI received via inductive or capacitive coupling between conductors.

Differential mode — A signal that that is exists and is transmitted as a voltage *between* two conductors of a cable. At any instant, signal current on one conductor is equal to but of the opposite polarity to the current on the other conductor. Ordinary connections between equipment in systems are differential mode signals.

Disturbance — The improper operation of a device as a result of interference.

Electric field — The field present between two or more conductive objects as a result of potential difference (voltage) between those objects.

Electromagnetic field — The combination of a magnetic field and electric field in which the fields are directly related to each other, are at right angles to each other, and move through space as radio waves in a direction that is mutually perpendicular to both fields. An electrical conductor designed to produce electromagnetic fields when carrying an RF current is called an antenna.

Equipment ground — The connection of all exposed parts of electrical equipment to earth, or to a body that serves in place of earth.

Fundamental overload — 1. (Receiver Performance) Interference to a receiver caused by a signal at its input whose amplitude exceeds the maximum signal-handling capabilities of one or more receiver stages. 2. (RF interference) — Any disruption to the function of any RFI victim caused by the fundamental component of a transmitted signal or intended in-band output of a transmitter.

Ground — 1. A low impedance electrical connection to earth, or to a body that serves in place of earth. 2. A common signal connection in an electrical circuit.

Immunity — The ability of a device to function properly in the presence of unwanted electromagnetic energy. (After Ott, section 1.3)

Intentional radiator — A device that uses radio waves to transmit information by antenna action. A radio transmitter, with its associated antenna, is an intentional radiator.

Interference — 1. Disruption of a device's normal function as a result of an electromagnetic field, voltage, or current. 2. Disruption by a signal or noise of a receiver's ability to acquire and process a desired signal.

Magnetic field — The field produced by a permanent magnet or current flow through a conductor.

Path — The route by which electromagnetic energy is transferred from a transmitter to a receiver or from a source to a victim.

Radiated RFI — RFI received through radiation.

Shielding — A conductive barrier or enclosure interposed between two regions of space with the intent of preventing a field in one region from reaching the other region.

Source — A device that produces an electromagnetic, electric, or magnetic field, voltage, or current. If RFI is the result, the source is an *RFI source*.

Spurious emission — An emission outside the bandwidth needed for transmission of the mode being employed, the level of which may be reduced without reducing the quality of information being transmitted. Spurious emissions are most commonly the products of distortion (harmonics, intermodulation), of circuit instability (oscillation, including RF feedback), or of digital transmission with excessively fast rise times (including key clicks). Phase noise, such as that produced by a frequency synthesizer is also a spurious emission.

Susceptibility — The capability of a device to respond to unwanted electromagnetic energy. (After Ott, section 1.3)

System ground — A bond between one current-carrying conductor of the power system and earth.

Unintentional radiator — A device that produces RF as part of its normal operation but does not intentionally radiate it.

Victim — A device that receives interference from a *source*.

27.12 References and Bibliography

M. Gruber, ed, *The ARRL RFI Book*, Third Edition, ARRL, 2010

M. Loftness, *AC Power Interference*, Third Edition, Revised, distributed by the ARRL

H. Ott, *Electromagnetic Compatibility Engineering*, John Wiley and Sons, 2009

C. Paul, *Introduction to Electromagnetic Compatibility*, Wiley-Interscience, 1992

AES48-2005: AES standard on interconnections — Grounding and EMC practices — Shields of connectors in audio equipment containing active circuitry

Underwriter Labs Standard UL943, "Ground-Fault Circuit-Interrupters," **http://ulstandardsinfonet.ul.com/scopes/scopes.asp?fn=0943.html**

Contents

**Chapter 28 —
CD-ROM Content**

Supplemental Files

- *Electric Current Abroad* — U.S. Dept of Commerce
- "Shop Safety" by Don Daso, K4ZA
- "RF Safety at Field Day" by Greg Lapin, N9GL
- "Field Day Towers — Doing It Right" by Don Daso, K4ZA and Ward Silver, NØAX

Safety

This chapter focuses on how to avoid potential hazards as we explore Amateur Radio and its many facets. The first section, updated by Jim Lux, W6RMK, details electrical safety, grounding and other issues in the ham shack. The following section on antenna and tower safety was written by Steve Morris, K7LXC, a professional tower climber and antenna installer with many years of experience. Finally, the ARRL RF Safety Committee explains good amateur practices, standards and FCC regulations as they apply to RF safety.

Safety First — Always

We need to learn as much as possible about what could go wrong so we can avoid factors that might result in accidents. Amateur Radio activities are not inherently hazardous, but like many things in modern life, it pays to be informed. Stated another way, while we long to be creative and innovative, there is still the need to act responsibly. Safety begins with our attitude. Make it a habit to plan work carefully. Don't be the one to say, "I didn't think it could happen to me."

Having a good attitude about safety is not enough, however. We must be knowledgeable about common safety guidelines and follow them faithfully. Safety guidelines cannot possibly cover all situations, but if we approach each task with a measure of common sense, we should be able to work safely.

Involve your family in Amateur Radio. Having other people close by is always beneficial in the event that you need immediate assistance. Take the valuable step of showing family members how to turn off the electrical power to your equipment safely. Additionally, cardiopulmonary resuscitation (CPR) training can save lives in the event of electrical shock. Classes are offered in most communities. Take the time to plan with your family members exactly what action should be taken in the event of an emergency, such as electrical shock, equipment fire or power outage. Practice your plan!

28.1 Electrical Safety

The standard power available from commercial mains in the United States for residential service is 120/240-V ac. The "primary" voltages that feed transformers in our neighborhoods may range from 1300 to more than 10,000 V. Generally, the responsibility for maintaining the power distribution system belongs to a utility company, electric cooperative or city. The "ownership" of conductors usually transfers from the electric utility supplier to the homeowner where the power connects to the meter or weather head. If you are unsure of where the division of responsibility falls in your community, a call to your electrical utility will provide the answer. **Fig 28.1** shows the typical division of responsibility between the utility company and the homeowner. This section is concerned more with wiring practices in the shack, as opposed to within the equipment in the shack.

There are two facets to success with electrical power: safety and performance. Since we are not professionals, we need to pursue safety first and consult professionals for alternative solutions if performance is unacceptable. The ARRL's Volunteer Consulting Engineers program involves professional engineers who may be able to provide advice or direction on difficult problems.

28.1.1 Station Concerns

There never seem to be enough power outlets in your shack. A good solution for small scale power distribution is a switched power strip with multiple outlets. The strip should be listed by a nationally recognized testing laboratory (NRTL) such as Underwriters Lab, UL, and should incorporate a circuit breaker. See the sidebars "What Does UL Listing Mean?" and "How Safe are Outlet Strips?" for warnings about poor quality products. It is poor practice to "daisy-chain" several power strips

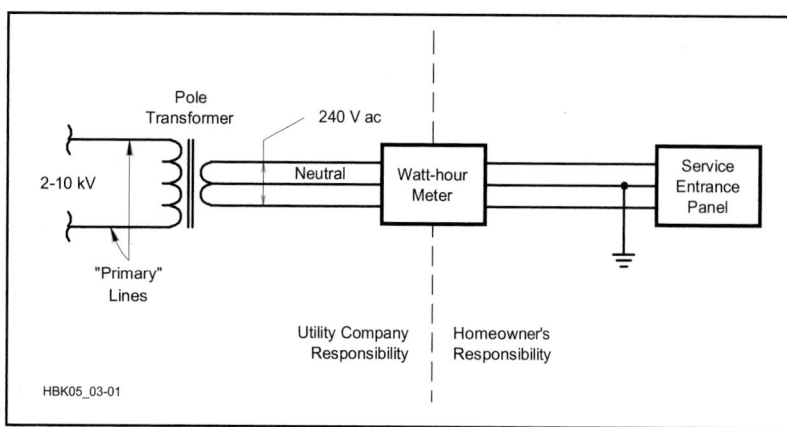

Fig 28.1 — Typical division of responsibility for maintenance of electrical power conductors and equipment. The meter is supplied by the utility company.

What Does UL Listing Mean?

UL is one of several nationally recognized testing laboratories (NRTLs), and probably the most well known. Listing *does not* mean what most consumers expect it to mean! More often than not the listing *does not* relate to the performance of the listed product. The listing simply indicates that a sample of the device meets certain manufacturers' construction criteria. Similar devices from the same or different manufacturers may differ significantly in overall construction and performance even though all are investigated and listed against the same UL product category. There is also a difference between a listed device and a listed component.

Many local laws and regulations, as well as the *NEC*, require that equipment and components used in electrical installations be listed by a NRTL. Some jurisdictions (Los Angeles County) require that any electrical equipment sold to consumers be listed.

The consumer must also be aware of the fine distinctions in advertising between a device or component that is advertised as "listed" or "designed to meet" or "meets." The latter two may not actually have been tested, or if tested, may have been tested by the manufacturer, and not an independent body.

It's also important to know that in some cases UL (and other standards organizations) only publish a standardized test procedure, but don't necessarily list or test the devices. Many standards also define varying levels of compliance, so knowing that your device meets some part of the standard may not be enough to know whether it meets *your* particular needs.

and may actually be a code violation. If you need more outlets than are available on a strip, have additional wall outlets installed.

Whether you add new outlets or use power strips, be sure not to overload the circuit. National and local codes set permissible branch capacities according to a rather complex process. Here's a safe rule of thumb: consider adding a new circuit if the total load is more than 80% of the circuit breaker or fuse rating. (This assumes that the fuse or breaker is correct. If you have any doubts, have an electrician check it.)

28.1.2. Do-It-Yourself Wiring

Amateurs sometimes "rewire" parts of their homes to accommodate their hobby. Most lo-

cal codes *do* allow for modification of wiring (by building owners), so long as the electrical codes are met. Before making changes to your wiring, it would be wise to determine what rules apply and what agency has the authority to enforce them. This is called the *authority having jurisdiction* (AHJ) and it varies from location to location. Also see the following section on the National Electrical Code.

Generally, the building owner must obtain an electrical permit before beginning changes or additions to permanent wiring. Some jobs may require drawings of planned work. Often the permit fee pays for an inspector to review the work. Considering the risk of injury or fire if critical mistakes are left uncorrected, a permit and inspection are well worth the effort. *Don't take chances* — seek assistance from the building officials or an experienced electrician if you have *any* questions or doubts about proper wiring techniques.

Ordinary 120-V circuits are the most common source of fatal electrical accidents. Line voltage wiring must use an approved cable, be properly installed in conduit and junction boxes, within a chassis with a cover or lid, or other means described in the electrical code. Remember that high-current, low-voltage power sources, such as car batteries and high-current power supplies, can be just as dangerous as high-voltage sources, from melting metal, sparks and short circuits.

Never work on electrical wiring with the conductors energized! Switch off the circuit breaker or remove the fuse and take positive steps to ensure that others do not restore the power while you are working, such as using a circuit-breaker lockout. (**Fig 28.2** illustrates one way to ensure that power will be off until you want it turned on.) Check the circuit with an ac voltmeter to be sure that it is "dead" *each time you begin work*.

Before restoring power, check the wiring with an ohmmeter: From the installed outlet, there should be good continuity between the neutral conductor (white wire, "silver" screw) and the grounding conductor (green or bare wire, green screw). An ohmmeter should indicate a closed circuit between the conductors. (In the power line, high voltage world, line workers apply a shorting jumper before starting work so if the power does get reapplied, the safety jumper takes the hit.)

There should be no continuity between the hot conductor (black wire, "brass" screw) and the grounding conductor or the neutral conductor. With all other loads removed from the circuit (by turning off or unplugging them), an ohmmeter should indicate an *open* circuit between the hot wire and either of the other two conductors.

Commercially available plug-in testers are a convenient way to test regular three-wire receptacles, but can't distinguish between the neutral and ground being reversed

How Safe Are Outlet Strips?

The switch in outlet strips is generally *not* rated for repetitive *load break* duty. Early failure and fire hazard may result from using these devices to switch loads. Misapplications are common (another bit of bad technique that has evolved from the use of personal computers), and manufacturers are all too willing to accommodate the market with marginal products that are "cheap."

Nonindicating and poorly designed surge protection also add to the safety hazard of using power strips. MOVs with too low a threshold voltage often fail in a manner that could cause a fire hazard, especially in outlet strips that have nonmetallic enclosures, because day-to-day voltage surges that are otherwise unexceptional degrade the MOV each time, until it fails shorted.

A lockable disconnect switch or circuit breaker is a better and safer station master switch.

Fig 28.2 — If the switch box feeding power to your shack is equipped with a lock-out hole, use it. With a lock through the hole on the box, the power cannot be accidentally turned back on. *(Photo courtesy of American ED-CO)*

28.1.3 National Electrical Code (NEC)

Fortunately, much has been learned about how to harness electrical energy safely. This collective experience has been codified into the *National Electrical Code*, or *NEC*, simply known as "the code." The code details safety requirements for many kinds of electrical installations. Compliance with the *NEC* provides an installation that is *essentially* free from hazard, but not necessarily efficient, convenient or adequate for good service (paraphrased from NEC Article 90-1a and b). While the *NEC* is national in nature and sees wide application, it is not universal.

Local building authorities set the codes for their area of jurisdiction. They often incorporate the *NEC* in some form, while considering local issues. For example, Washington State specifically exempts telephone, telegraph, radio and television wires and equipment from conformance to electrical codes, rules and regulations. However, some local jurisdictions (city, county and so on) do impose a higher level of installation criteria, including some of the requirements exempted by the state.

Code interpretation is a complex subject, and untrained individuals should steer clear of the *NEC* itself. The *NEC* is not written to be understood by do-it-yourselfers, and one typically has to look in several places to find *all* the requirements. (For instance, Articles 810, 250, *and* 100 all contain things applicable to typical Amateur Radio installations.) Therefore, the best sources of information about code compliance and acceptable practices are local building officials, engineers and practicing electricians.

The Internet has a lot of information about electrical safety, the electrical code, and wiring practices, but you need to be careful to make sure the information you are using is current and not out of date. The ARRL Volunteer Consulting Engineer (VCE) program can help you find a professional who understands the amateur radio world, as well as the regulatory environment. There are also a variety of websites with useful information (such as **www.mikeholt.com**), but you need to be aware that advice may be specific to a particular installation or jurisdiction and not applicable for yours. With that said, let's look at a few *NEC* requirements for radio installations.

HOMEBREW AND "THE CODE"

In many cases, there are now legal requirements that electrical equipment have been listed by an NRTL, such as Underwriter Laboratories. This raises an issue for hams and homebrew gear, since it's unlikely you would take your latest project down to a test lab and pay them to evaluate it for safety.

For equipment that is not permanently installed, there's not much of an issue with homebrew, as far as the code goes, because the code doesn't deal with what's inside the equipment. For a lot of low voltage equipment, the code rules are fairly easy to meet, as well, as long as the equipment is supplied by a listed power source of the appropriate type.

The problem arises with permanent installations, where the scope of the code and local regulations is ever increasing. Such things as solar panel installations, standby generators, personal computers and home LANs all have

About the National Electrical Code

Exactly how does the National Electrical Code become a requirement? How is it enforced?

Cities and other political subdivisions have the responsibility to act for the public safety and welfare. To address safety and fire hazards in buildings, regulations are adopted by local laws and ordinances usually including some form of permit and accompanying inspections. Because the technology for the development of general construction, mechanical and electrical codes is beyond most city building departments, model codes are incorporated by reference. There are several general building code models used in the US: Uniform, BOCA and Southern Building Codes are those most commonly adopted. For electrical issues, the *National Electrical Code* is in effect in virtually every community. City building officials will serve as "the authority having jurisdiction" (AHJ) and interpret the provisions of the *Code* as they apply it to specific cases.

Building codes differ from planning or zoning regulations: Building codes are directed only at safety, fire and health issues. Zoning regulations often are aimed at preservation of property values and aesthetics.

The *NEC* is part of a series of reference codes published by the National Fire Protection Association, a non-profit organization. Published codes are regularly kept up-to-date and are developed by a series of technical committees whose makeup represents a wide consensus of opinion. The *NEC* is updated every three years. It's important to know which version of the code your local jurisdiction uses, since it's not unusual to have the city require compliance to an older version of the code. Fortunately, the *NEC* is usually backward compatible: that is, if you're compliant to the 2008 code, you're probably also compliant to the 1999 code.

Do I have to update my electrical wiring as code requirements are updated or changed?

Generally, no. Codes are typically applied for new construction and for renovating existing structures. Room additions, for example, might not directly trigger upgrades in the existing service panel unless the panel was determined to be inadequate. However, the wiring of the new addition would be expected to meet current codes. Prudent homeowners, however, may want to add safety features for their own value. Many homeowners, for example, have added GFCI protection to bathroom and outdoor convenience outlets.

received increased attention in local codes.

28.1.4 Station Power

Amateur Radio stations generally require a 120-V ac power source, which is then converted to the proper ac or dc levels required for the station equipment. In residential systems voltages from 110 V through 125 V are treated equivalently, as are those from 220 V through 250 V. Amateurs setting up a station in a light industrial or office environment may encounter 208 V line voltage. Most power supplies operate over these ranges, but it's a good idea to measure the voltage range at your station. (The measured voltage usually varies by hour, day, season and location.) Power supply theory is covered in the **Power Sources** chapter.

Modern solid state rigs often operate from dc power, provided by a suitable dc power supply, perhaps including battery backups. Sometimes, the dc power supply is part of the rig (as in a 50-V power supply for a solid-state linear). Other times, your shack might have a 12-V (13.8 V) bus that supplies many devices. Just because it's low voltage doesn't mean that there aren't aspects of the system that raise safety concerns. A 15-A, 12-V power supply can start a fire as easily as a 15-A, 120-V branch circuit.

28.1.5 Connecting and Disconnecting Power

Something that is sometimes overlooked is that you need to have a way to safely disconnect all power to everything in the shack. This includes not only the ac power, but also battery banks, solar panels, and uninterruptible power supplies (UPS). Most hams won't have the luxury of a dedicated room with a dedicated power feed and the "big red switch" on the wall, so you'll have several switches and cords that would need to be disconnected.

The realities of today's shacks, with computers, multiple wall transformers ("wall-warts"), network interfaces and the radio equipment itself makes this tricky to do. One convenient means is a switched outlet strip, as used for computer equipment, if you have a limited number of devices. If you need more switched outlets, you can control multiple low-voltage controlled switched outlets from a common source. Or you can build or buy a portable power distribution box similar to those used on construction sites or stage sets; they are basically a portable subpanel with individual circuit breakers (or GFCIs, discussed later) for each receptacle, and fed by a suitable cord or extension cord. No matter what scheme you use, however, it's important that it be labeled so that someone else will know what to do to turn off the power.

AC LINE POWER

If your station is located in a room with electrical outlets, you're in luck. If your station is located in the basement, an attic or other area without a convenient 120-V source, you may need to have a new line run to your operating position.

Stations with high-power amplifiers should have a 240-V ac power source in addition to the 120-V supply. Some amplifiers may be powered from 120 V, but they require current levels that may exceed the limits of standard house wiring. To avoid overloading the circuit and to reduce household light dimming or blinking when the amplifier is in use, and for the best possible voltage regulation in the equipment, it is advisable to install a separate 240 or 120-V line with an appropriate current rating if you use an amplifier.

The usual circuits feeding household outlets are rated at 15 or 20 A., This may or may not be enough current to power your station. To determine how much current your station requires, check the VA (volt-amp) ratings for each piece of gear. (See the **Electrical Fundamentals** chapter for a discussion of VA.) Usually, the manufacturer will specify the required current at 120 V; if the power consumption is rated in watts, divide that rating by 120 V to get amperes. Modern switching power supplies draw more current as the line voltage drops, so if your line voltage is markedly lower than 120 V, you need to take that into account.

Note that the code requires you to use the "nameplate" current, even if you've measured the actual current, and it's less. If the total current required is near 80% of the circuit's rating (12 A on a 15-A circuit or 16 A on a 20-A circuit), you need to install another circuit. Keep in mind that other rooms may be powered from the same branch of the electrical system, so the power consumption of any equipment connected to other outlets on the branch must be taken into account. If you would like to measure just how much power your equipment consumes, the inexpensive Kill-A-Watt meters by P3 International (**www.p3international.com**) measure volts, amps, VA and power factor.

If you decide to install a separate 120-V line or a 240-V line, consult the local requirements as discussed earlier. In some areas, a licensed electrician must perform this work. Others may require a special building permit. Even if you are allowed to do the work yourself, it might need inspection by a licensed electrician. Go through the system and get the necessary permits and inspections! Faulty wiring can destroy your possessions and take away your loved ones. Many fire insurance policies are void if there is unapproved wiring in the structure.

If you decide to do the job yourself, work closely with local building officials. Most home-improvement centers sell books to guide do-it-yourself wiring projects. If you have any doubts about doing the work yourself, get a licensed electrician to do the installation.

THREE-WIRE 120-V POWER CORDS

Most metal-cased electrical tools and appliances are equipped with three-conductor power cords. Two of the conductors carry power to the device, while the third conductor is connected to the case or frame. **Fig 28.3** shows two commonly used connectors.

When both plug and receptacle are properly wired, the three-contact polarized plug bonds the equipment to the system ground. If an internal short from line to case occurs, the "ground" pin carries the fault current and hopefully has a low enough impedance to trip the branch circuit breaker or blow the fuse in the device. A second reason for grounding the case is to reduce the possibility of shock for a user simultaneously connected to ground and the device. In modern practice, however, shock prevention is often done with GFCI circuit breakers as described below. These devices trip at a much lower level and are more reliable. Most commercially manufactured test equipment and ac-operated amateur equipment is supplied with three-wire cords.

It's a good idea to check for continuity from case to ground pin, particularly on used equipment, where the ground connection might have been broken or modified by the previous owner. If there is no continuity, have the equipment repaired before use.

Use such equipment only with properly installed three-wire outlets. If your house does not have such outlets, either consult a local electrician to learn about safe alternatives or have a professional review information you might obtain from online or other sources.

Equipment with plastic cases is considered "double insulated" and fed with a two-wire cord. Such equipment is safe because both conductors are insulated from the user by two layers. Nonetheless, there is still a hazard if, say, a double insulated drill were used to drill an improperly grounded case of a transmit-

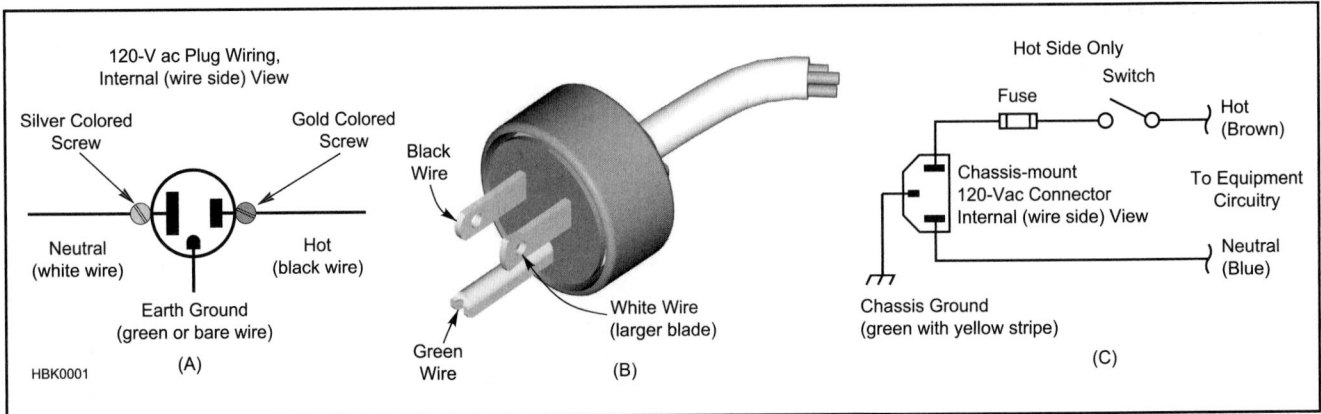

Fig 28.3 — 120 V ac plug wiring as viewed from the wire side (A) and viewed from the blade side (B). Wiring for an IEC type chassis connector is shown at C.

ter that was still plugged in. Remember, all insulation is prey to age, damage and wear that may erode its initial protection.

TRANSFER SWITCHES AND GENERATORS

More hams are adding standby generators and using alternate power sources such as solar panels or wind turbines, not as standalone systems like at Field Day, but interconnected with their home electrical system. These present some potential safety problems, such as preventing the local generator from "backfeeding" the utility's system during a power failure, and the fact that a solar panel puts out power whenever there is light falling on it.

For generators, the recommended approach is to use a *transfer switch*, which is a multipole switch that connects a selection of the house's circuits to the generator, rather than the utility power. The *NEC* and local regulations should be consulted for transfer switch selection and connection. The required wiring practices for permanently installed (stationary) generators are different from those for portable generators. Some issues that need to be considered are whether the neutral should be switched (many transfer switches do not switch the neutral, only the hot wire), and how the generator chassis is bonded to the building's grounding/bonding system. Most proper transfer switches are of the ON-OFF-ON configuration, with a mechanical interlock that prevents directly switching from one source to the other in a single operation.

The most dangerous thing to do with a generator is to use a so-called "suicide cord" with a male plug at each end: one end plugged into the generator's output receptacle and the other plugged into a convenient receptacle in the home. This is frequently illegal and at any rate should be avoided because of the inherent danger of having exposed, live contacts and the ease of overloading the circuit being fed.

Back-feeding your home's power panel should *never* be done unless the main breakers are in the OFF position or preferably removed. If your home's circuit-breaker panel does not have main breakers that can disconnect the external power line, *do not* use this technique to connect your generator to the home's wiring. Connect appliances to the generator directly with extension cords.

28.1.6 Ground-Fault and Arc-Fault Circuit Interrupters

GFCIs are devices that can be used with common 120 V household circuits to reduce the chance of electrocution when the path of current flow leaves the branch circuit (say, through a person's body to another branch or ground). An AFCI is similar in that it monitors current to watch for a fault condition. Instead of current imbalances,

Table 28.1
Traditional Divisions Among the Classes of Circuits

Class	Power	Notes
Class 1		
Power Limited	<30V, <1000VA	Transformer protected per Article 450. If not transformer, other overcurrent and fault protection requirements apply
Remote Control and Signaling	<600V	No limit on VA. Transformers protected as defined in Article 450
Class 2	Power supply <100VA Voltage <30V	
Class 3	Power supply <100VA Voltage <100V	

the AFCI detects patterns of current that indicate an arc — one of the leading causes of home fires. The AFCI is not supposed to trip because of "normal" arcs that occur when a switch is opened or a plug is removed.

The *NEC* requires GFCI outlets in all wet or potentially wet locations, such as bathrooms, kitchens, and any outdoor outlet with ground-level access, garages and unfinished basements. AFCI protection is required for all circuits that supply bedrooms. Any area with bare concrete floors or concrete masonry walls should be GFCI equipped. GFCIs are available as portable units, duplex outlets and as individual circuit breakers. Some early units may have been sensitive to RF radiation but this problem appears to have been solved. Ham radio shacks in potentially wet areas (basements, out buildings) should be GFCI equipped. **Figure 28.4** is a simplified diagram of a GFCI.

Older equipment with capacitors in the 0.01 µF to 0.1 µF range connected between line inputs and chassis as an EMI filter (or that has been modified with bypass capacitors) will often cause a GFCI to trip, because of the leakage current through the capacitor. The must-trip current is 5 mA, but many GFCIs trip at lower levels. At 60 Hz, a 0.01 µF capacitor has an impedance of about 265 kΩ, so there could be a leakage current of about 0.5 mA from the 120 V line. If you had several pieces of equipment with such capacitors, the leakage current will trip the GFCI.

Some early GFCI breakers were susceptible to RFI but as the technology has improved, fewer and fewer such reports have been received. While it is possible to add filtering or RF suppression to the breaker wiring, a simpler and less expensive solution is to simply replace the GFCI breaker with a new unit less susceptible to RF. Reports have not yet been received on AFCI products. For more information on RFI and GFCI/AFCI devices, check the ARRL web page **www.arrl.org/gfci-devices**.

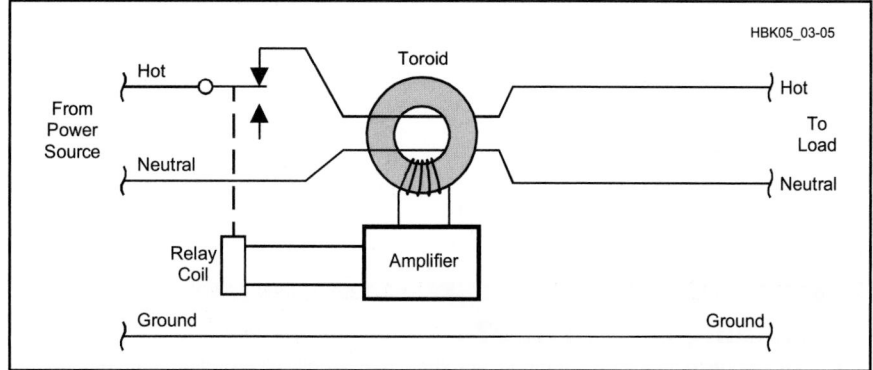

Fig 28.4 — Simplified diagram of a 120-V ac ground fault circuit interrupter (GFCI). When a stray current flows from the load (or outlet) side to ground, the toroidal current becomes unbalanced allowing detection, amplification and relay actuation to immediately cut off power to the load (and to the stray path!) GFCI units require a manual reset after tripping. GFCIs are required in wet locations (near kitchen sinks, in garages, in outdoor circuits and for construction work.) They are available as portable units or combined with over-current circuit breakers for installation in entrance panels.

28.1.7 Low-Voltage Wiring

Many ham shacks use low-voltage control wiring for rotators or antenna relays. The electrical code isn't consistent in what it calls low voltage, but a guideline is "less than 50 V." Article 725 of the code contains most of the rules for low voltage/low power remote control and signaling, which is what hams are typically doing. These circuits are divided into three classes, with Class 1 being further subdivided, as shown in **Table 28.1**. There used to be code rules defining the classes in terms of power and voltage, but these days, the code is written so that the class of the circuit is defined by the power source, which has to be listed and labeled with the class. That is, if you have something powered by a wall transformer that is listed and labeled as Class 2, the circuit is Class 2.

A typical example of a Class 1 Power Limited circuit that you might find in your home is 12 V low-voltage garden lighting or halogen lightning systems. A lot of amateur homebrew gear probably is also in this class, although because it's not made with "listed" components, it technically doesn't qualify. The other Class 1 would apply to a circuit using an isolation transformer of some sort.

Class 2 is very common: doorbells, network wiring, thermostats, and so on are almost all Class 2. To be Class 2, the circuit must be powered from a listed power supply that's marked as being Class 2 with a capacity less than 100 VA. For many applications that hams encounter, this will be the familiar "wall wart" power supply. If you have a bunch of equipment that runs from dc power, and you build a dc power distribution panel with regulators to supply them from a storage battery or a big dc power supply, you're most likely not Class 2 anymore, but logically Class 1. Since your homebrew panel isn't likely to be listed, you're really not even Class 1, but something that isn't covered by the code.

A common example of a Class 3 circuit that is greater than 30 V is the 70 V audio distribution systems used in paging systems and the like. Class 3 wiring must be done with appropriately rated cable.

WIRING PRACTICES

Low voltage cables must be separated from power circuits. Class 2 and 3 cannot be run with Class 1 low voltage cables. They can't share a cable tray or the same conduit. A more subtle point is that the 2005 code added a restriction [Article 725.56(F)] that audio cables (speakers, microphone, etc.) cannot be run in the same conduit with other Class 2 and Class 3 circuits (like network wiring).

Low voltage and remote control wiring should not be neglected from your transient suppression system. This includes putting appropriate protective devices where wiring enters and leaves a building, and consideration of the current paths to minimize loops which can pick up the field from transient (or RF from your antenna).

28.1.8 Grounds

As hams we are concerned with at least four kinds of things called "ground," even if they really aren't ground in the sense of connection to the Earth. These are easily confused because we call each of them "ground."

1) Electrical safety ground (bonding)
2) RF return (antenna ground)
3) Common reference potential (chassis ground)
4) Lightning and transient dissipation ground

IEEE Std 1100-2005 (also known as the "Emerald Book," see the Reference listing, section 28.1.13) provides detailed information from a theoretical and practical standpoint for grounding and powering electrical equipment, including lightning protection and RF EMI/EMC concerns. It's expensive to buy, but is available through libraries.

ELECTRICAL SAFETY GROUND (BONDING)

Power-line ground is required by building codes to ensure the safety of life and property surrounding electrical systems. The *NEC* requires that all grounds be *bonded* together; this is a very important safety feature as well as an *NEC* requirement.

The usual term one sees for the "third prong" or "green-wire ground" is the "electrical safety ground." The purpose of the third, non-load current carrying wire is to provide a path to insure that the overcurrent protection will trip in the event of a line-to-case short

circuit in a piece of equipment. This could either be the fuse or circuit breaker back at the main panel, or the fuse inside the equipment itself.

There is a secondary purpose for shock reduction: The conductive case of equipment is required to be connected to the bonding system, which is also connected to earth ground at the service entrance, so someone who is connected to "earth" (for example, standing in bare feet on a conductive floor) that touches the case won't get shocked.

An effective safety ground system is necessary for every amateur station, and the code requires that all the "grounds" be bonded together. If you have equipment at the base of the tower, generally, you need to provide a separate bonding conductor to connect the chassis and cases at the tower to the bonding system in the shack. The electrical safety ground provides a common reference potential for all parts of the ac system. Unfortunately, an effective bonding conductor at 60 Hz may present very high impedance at RF because of the inductance, or worse yet, wind up being an excellent antenna that picks up the signals radiated by your antenna.

RF GROUND

RF ground is the term usually used to refer to things like equipment enclosures. It stems from days gone by when the long-wire antenna was king. At low enough frequencies, a wire from the chassis or antenna tuner in the shack to a ground rod pounded in outside the window had low RF impedance. The RF voltage difference between the chassis and "Earth ground" was small. And even if there were small potentials, the surrounding circuitry was relatively insensitive to them.

Today, though, we have a lot of circuits that are sensitive to interfering signals at millivolt levels, such as audio signals to and from sound cards. The summary is that we shouldn't be using the equipment enclosures or shielding conductors as part of the RF circuit.

Instead, we design our systems to create a common reference potential, called the "reference plane," and we endeavor to keep equipment connected to the reference plane at a common potential. This minimizes RF current that would flow between pieces of equipment. (See the **RF Interference** chapter for more information.)

Some think that RF grounds should be isolated from the safety ground system — *that is not true! All grounds, including safety, RF, lightning protection and commercial communications, must be bonded together in order to protect life and property.* The electrical code still requires that antenna grounds be interconnected (bonded) to the other "grounds" in the system, although that connection can have an RF choke. Remember that the focus of the electrical code bonding requirement

---

Grounding or Bonding?

You may notice the term "bonding" is replacing "grounding" in many instances. A primary safety concern is for whatever carries fault currents to be mechanically rugged and reasonably conductive. It's also important that the fault-carrying conductor be connected to a ground rod, but that's a different consideration. Bonding is the term to use when contact between pieces of equipment or between conductors is the primary concern. Grounding is the term to use when the electrical potential of the conductor is the most important.

is safety in the event of a short to a power distribution line or other transient.

COMMON REFERENCE POTENTIAL (CHASSIS GROUND)

For decades, amateurs have been advised to bond all equipment cabinets to an RF ground located near the station. That's a good idea, but it's not easily achieved. Even a few meters of wire can have an impedance of hundreds of ohms (1 μH/meter = 88 Ω/meter at 14 MHz). So a better approach is to connect the chassis together in a well-organized fashion to ensure that the chassis-to-chassis connections don't carry any RF current at all as in **Fig 28.5**. (See the **RF Interference** chapter for more information.)

LIGHTNING DISSIPATION GROUND

Lightning dissipation ground is concerned with conducting currents to the surrounding earth. There are distinct similarities between lightning dissipation ground systems and a good ground system for a vertical antenna. Since the lightning impulse has RF components around 1 MHz, it is an RF signal, and low inductance is needed, as well as low resistance.

The difference is that an antenna ground plane may handle perhaps a few tens of amps, while the lightning ground needs to handle a peak current of tens of kiloamperes.

A typical lightning stroke is a pulse with a rise time of a few microseconds, a peak current of 20-30 kA, and a fall time of 50 μs. The average current is not all that high (a few hundred amps), so the conductor size needed to carry the current without melting is surprisingly small.

However, large conductors are used in lightning grounds for other reasons: to reduce inductance, to handle the mechanical forces from the magnetic fields, and for rug-gedness to prevent inadvertent breakage. A large diameter wire, or even better, a wide flat strap, has lower inductance. The voltage along a wire is proportional to the change in current and the inductance:

$$V = L\frac{di}{dt}$$

where

di/dt = rate of change in current, about 20kA/2μs for lightning, or 10^9 A/s, and
L = the inductance.

Consider a connection box on a tower that contains some circuitry terminating a control cable from the shack, appropriately protected internally with overvoltage protection. If the connection from the box to ground is high inductance, the lightning transient will raise the box potential (relative to the wiring coming from the shack), possibly beyond the point where the transient suppression in the box can handle it. Lowering the inductance of the connection to ground reduces the potential.

The other reason for large conductors on lightning grounds is to withstand the very high mechanical forces from the high currents. This is also the reason behind the recommendation that lightning conductors be run directly, with minimal bends, large radii for bends that are needed, and certainly no loops. A wire with 20,000 A has a powerful magnetic field surrounding it, and if current is flowing in multiple wires that are close to each other, the forces pushing the wires together or apart can actually break the conductors or deform them permanently.

The force between two conductors carrying 20,000 A, spaced a centimeter apart, is 8000 Newtons/meter of length (over 500 pounds/foot). Such forces can easily break cable strands or rip brackets and screws out.

This problem is aggravated if there are loops in the wire, since the interaction of the current and its magnetic field tends to make the loop get larger, to the point where the wire will actually fail from the tension stresses.

GROUNDING METHODS

Earth ground usually takes one of several forms, all identified in the *NEC* and NFPA 780. The preferred earth ground, both as required in the NEC, and verified with years of testing in the field, is a concrete encased grounding electrode (CEGR), also known as a *Ufer ground*, after Herb Ufer, who invented it as a way to provide grounding for military installations in dry areas where ground rods are ineffective. The CEGR can take many forms, but the essential aspect is that a suitable conductor at least 20 feet long is encased in concrete which is buried in the ground. The conductor can be a copper wire (#8 AWG at least 20 ft long) or the reinforcing bars (rebar) in the concrete, often the foundation footing for the building. The connection to the rebar is either with a stub of the rebar protruding through the concrete's top surface or the copper wire extending through the concrete. There are other variations of the CEGR described in the *NEC* and in the electrical literature, but they're all functionally the same: a long conductor embedded in a big piece of concrete.

The electrode works because the concrete has a huge contact area with the surrounding soil, providing very low impedance and, what's also important, a low current density, so that localized heating doesn't occur. Concrete tends to absorb water, so it is also less susceptible to problems with the soil drying out around a traditional ground rod.

Ground rods are a traditional approach to making a suitable ground connection and are appropriate as supplemental grounds, say at

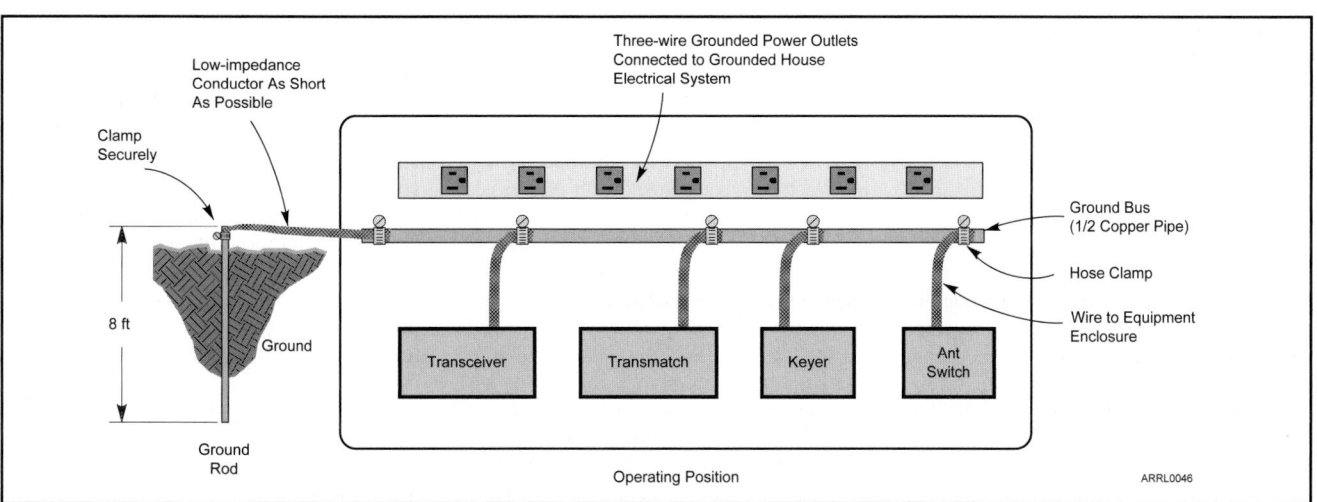

Fig 28.5 — An effective station safety ground bonds the chassis of all equipment together with low-impedance conductors and ties into a good earth ground.

the base of a tower, or as part of an overall grounding system. The best ground rods to use are those available from an electrical supply house. The code requires that at least 8 ft of the rod be in contact with the soil, so if the rod sticks out of the ground, it must be longer than 8 ft (10 ft is standard). The rod doesn't have to be vertical, and can be driven at an angle if there is a rock or hard layer, or even buried laying sideways in a suitable trench, although this is a compromise installation. Suitable rods are generally 10 ft long and made from steel with a heavy copper plating. Do not depend on shorter, thinly plated rods sold by some home electronics suppliers, as they can quickly rust and soon become worthless.

If multiple ground rods are installed, they should be spaced by at least half the length of the rod, or the effectiveness is compromised. IEEE Std 142 and IEEE Std 1100 (see the Reference listing) and other references have tables to give effective ground resistances for various configurations of multiple rods.

Once the ground rods are installed, they must be connected with either an exothermic weld (such as CadWeld) or with a listed pressure clamp. The exothermic weld is preferred, because it doesn't require annual inspection like a clamp does. Some installers use brazing to attach the wiring to the ground rods. Although this is not permitted for a primary ground, it is acceptable for secondary or redundant grounds. Soft solder (tin-lead, as used in plumbing or electrical work) should never be used for grounding conductors because it gets brittle with temperature cycling and can melt out if a current surge (as from a lightning strike) heats the conductor. Soft solder is specifically prohibited in the code.

Building cold water supply systems were used as station grounds in years past, but this is no longer recommended or even permitted in some jurisdictions, because of increased used of plastic plumbing both inside and outside houses and concerns about stray currents causing pipe corrosion.. If you do use the cold water line, perhaps because it is an existing grounding electrode, it must be bonded to the electrical system ground, typically at the service entrance panel.

28.1.9 Ground Conductors

The code is quite specific as to the types of conductors that can be used for bonding the various parts of the system together. Grounding conductors may be made from copper, aluminum, copper-clad steel, bronze or similar corrosion-resistant materials. Note that the sizes of the conductors required are based largely on mechanical strength considerations (to insure that the wire isn't broken accidentally) rather than electrical resistance. Insulation is not required. The "protective

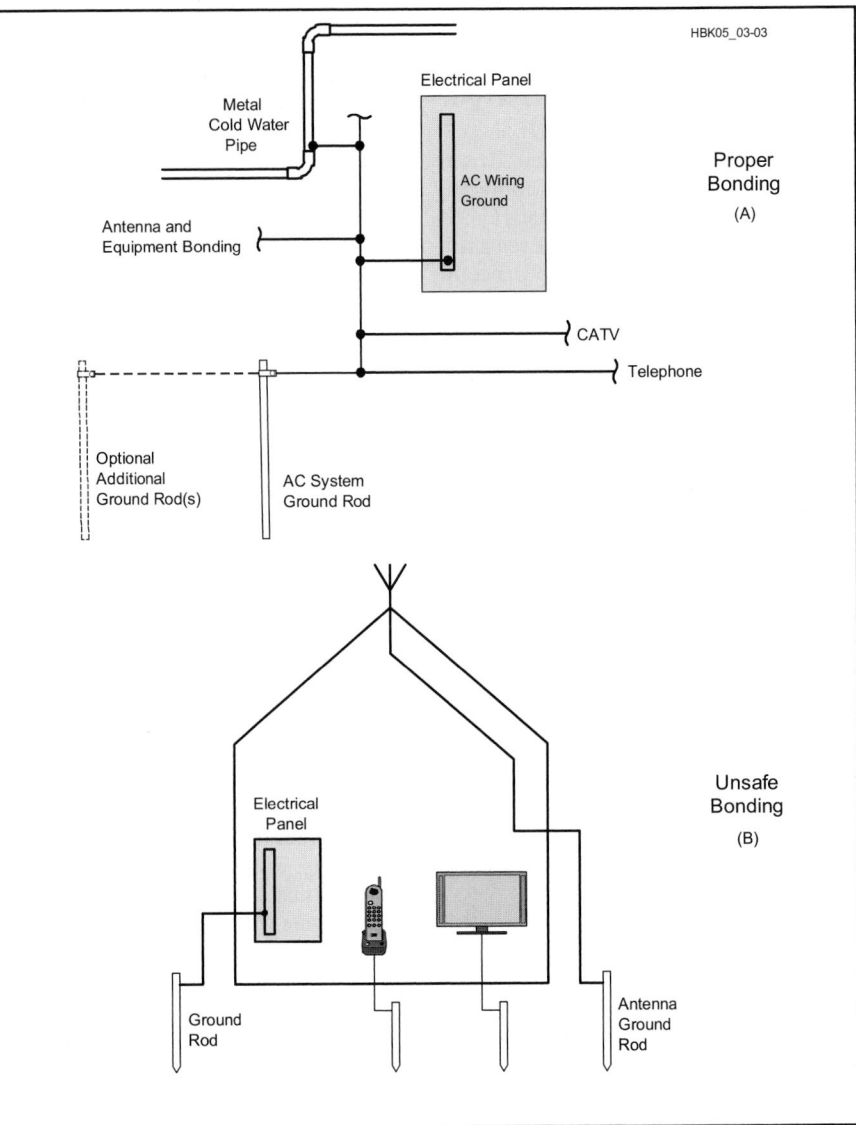

Fig 28.6 — At A, proper bonding of all grounds to electrical service panel. The installation shown at B is unsafe — the separate grounds are not bonded. This could result in a serious accident or electrical fire.

grounding conductor" (main conductor running to the ground rod) must be as large as the antenna lead-in, but not smaller than #10 AWG. The grounding conductor (used to bond equipment chassis together) must be at least #14 AWG. There is a "unified" grounding electrode requirement — it is necessary to bond *all* grounds to the electric service entrance ground. All utilities, antennas and any separate grounding rods used must be bonded together. **Fig 28.6** shows correct (A) and incorrect (B) ways to bond ground rods. **Fig 28.7** demonstrates the importance of correctly bonding ground rods. (Note: The *NEC* requirements do not address effective RF grounds. See the **RF Interference** chapter of this book for information about RF grounding practices, but keep in mind that RFI is not an acceptable reason to violate the *NEC*.) For additional information on good grounding

practices, the IEEE "Emerald Book" (IEEE STD 1100-2005) is a good reference. It is available through libraries.

Additionally, the *NEC* covers safety inside the station. All conductors inside the building must be at least 4 inches away from conductors of any lighting or signaling circuit except when they are separated from other conductors by conduit or insulator. Other code requirements include enclosing transmitters in metal cabinets that are bonded to the grounding system. Of course, conductive handles and knobs must be grounded as well.

28.1.10 Antennas

Article 810 of the *NEC* includes several requirements for wire antennas and feed lines that you should keep in mind when designing your antenna system. The single most

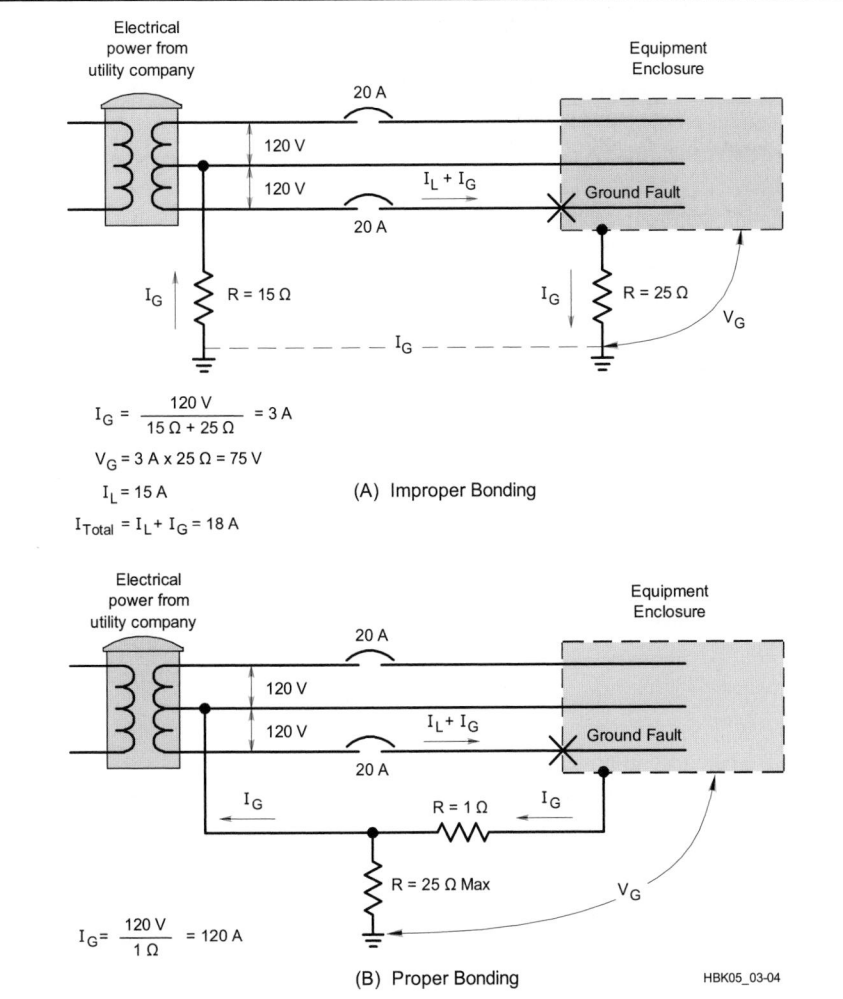

$$I_G = \frac{120\ V}{15\ \Omega + 25\ \Omega} = 3\ A$$

$$V_G = 3\ A \times 25\ \Omega = 75\ V$$

$$I_L = 15\ A \qquad \text{(A) Improper Bonding}$$

$$I_{Total} = I_L + I_G = 18\ A$$

$$I_G = \frac{120\ V}{1\ \Omega} = 120\ A$$

(B) Proper Bonding   HBK05_03-04

Fig 28.7 — These drawings show the importance of properly bonded ground rods. In the system shown in A, the 20-A breaker will not trip. In the system in B, the 20-A circuit breaker trips instantly. There is an equipment internal short to ground — the ground rod is properly bonded back to the power system ground. Of course, the main protection should be in a circuit ground wire in the equipment power cord itself!

Coax Shields as a Grounding Conductor

The importance of significant current-carrying capability in a grounding conductor is determined by your local circumstances. If lightning is a problem or contact with power-carrying conductors due to wind or ice is a possibility, the ability of the coax shield to carry a lot of current is important and a separate grounding conductor is a good idea. In areas where hazards are reduced, a smaller conductor may suffice. Contact a licensed electrician if you aren't sure about what is prudent for your area.

important thing to consider for safety is to address the potential for contact between the antenna system and power lines. As the code says, "One of the leading causes of electric shock and electrocution, according to statistical reports, is the accidental contact of radio, television, and amateur radio transmitting and receiving antennas and equipment with light or power conductors." (See Article 810.13, Fine Print Note.) The requirements in the code for wire sizes, bonding require-

ments, and installation practices are mostly aimed at preventing tragedy, by avoiding the contact in the first place, and by mitigating the effects of a contact if it occurs.

Article 820 of the *NEC* applies to Cable TV installations, which almost always use coaxial cable, and which require wiring practices different from Article 810 (for instance, the coax shield can serve as the grounding conductor). Your inspector may look to Article 820 for guidance on a safe installation

of coax, since there are many more satellite TV and cable TV installations than Amateur Radio. Ultimately, it is the inspector's call as to whether your installation is safe.

Article 830 applies to Network Powered Communication Systems, and as amateurs do things like install 802.11 wireless LAN equipment at the top of their tower, they'll have to pay attention to the requirements in this Article. The *NEC* requirements discussed in these sections are not adequate for lightning protection and high-voltage transient events. See the section "Lightning/Transient Protection" later in this chapter for more information.

ANTENNA CONDUCTORS

Transmitting antennas should use hard-drawn copper wire: #14 AWG for unsupported spans less than 150 feet, and #10 AWG for longer spans. Copper-clad steel, bronze or other high-strength conductors must be #14 AWG for spans less than 150 feet and #12 AWG for longer spans. Open-wire transmission line conductors must be at least as large as those specified for antennas. Stealth antennas made with light-gauge wire are not code-compliant.

LEAD-INS

There are several *NEC* requirements for antenna lead-in conductors (transmission lines are lead-in conductors). For transmitting stations, their size must be equal to or greater than that of the antenna. Lead-ins attached to buildings must be firmly mounted at least 3 inches clear of the surface of the building on nonabsorbent insulators. Lead-in conductors must enter through rigid, noncombustible, nonabsorbent insulating tubes or bushings, through an opening provided for the purpose that provides a clearance of at least two inches; or through a drilled windowpane. All lead-in conductors to transmitting equipment must be arranged so that accidental contact is difficult. As with stealth antennas, installations with feed lines smaller than RG-58 are likely not code compliant depending on how your local inspector interprets the code.

ANTENNA DISCHARGE UNITS (LIGHTNING ARRESTORS)

All antenna systems are required to have a means of draining static charges from the antenna system. A listed antenna discharge unit (lightning arrestor) must be installed on each lead-in conductor that is not protected by a permanently and effectively grounded metallic shield, unless the antenna itself is permanently and effectively grounded, such as for a shunt-fed vertical. Note that the usual transient protectors are *not* listed antenna discharge units.(The code exception for shielded lead-ins does *not* apply to coax, but to shields such as thin-wall conduit. Coaxial braid is neither "adequate" nor "effectively grounded" for lightning protection purposes.) An acceptable alternative to lightning arrestor installation is a switch (capable of withstanding many kilovolts) that connects the lead-in to ground when the transmitter is not in use.

ANTENNA BONDING (GROUNDING) CONDUCTORS

In general the code requires that the conductors used to bond the antenna system to ground be at least as big as the antenna conductors, but also at least #10 AWG in size. Note that the antenna grounding conductor rules are different from those for the regular electrical safety bonding, or lightning dissipation grounds, or even for CATV or telephone system grounds.

MOTORIZED CRANK-UP ANTENNA TOWERS

If you are using a motorized crank-up tower, the code has some requirements, particularly if there is a remote control. In general, there has to be a way to positively disconnect power to the motor that is within sight of the motorized device, so that someone working on it can be sure that it won't start moving unexpectedly. From a safety standpoint, as well, you should be able to see or monitor the antenna from the remote control point.

28.1.11 Lightning/Transient Protection

Nearly everyone recognizes the need to protect themselves from lightning. From miles away, the sight and sound of lightning boldly illustrates its destructive potential. Many people don't realize that destructive transients from lightning and other events can reach electronic equipment from many sources, such as outside antennas, power, telephone and cable TV installations. Many hams don't realize that the standard protection scheme of several decades, a ground rod and simple "lightning arrestor" is *not* adequate.

Lightning and transient high-voltage protection follows a familiar communications scenario: identify the unwanted signal, isolate it and dissipate it. The difference here is that the unwanted signal is many megavolts at possibly 200,000 A. What can we do?

Effective lightning protection system design is a complex topic. There are a variety of system tradeoffs which must be made and which determine the type and amount of protection needed. A amateur station in a home is a very different proposition from an air traffic control tower which must be available 24 hours a day, 7 days a week. Hams can easily follow some general guidelines that will protect their stations against high-voltage events that are induced by nearby lightning strikes or that arrive via utility lines. Let's

talk about where to find professionals first, and then consider construction guidelines.

PROFESSIONAL HELP

Start with your local government. Find out what building codes apply in your area and have someone explain the regulations about antenna installation and safety. For more help, look in your telephone yellow pages for professional engineers, lightning protection suppliers and contractors.

Companies that sell lightning-protection products may offer considerable help to apply their products to specific installations. One such source is PolyPhaser Corporation. Look under "References" later in this chapter for a partial list of PolyPhaser's publications.

CONSTRUCTION GUIDELINES
Bonding Conductors

Copper strip (or *flashing*) comes in a number of sizes. The *minimum* recommended grounding conductor for lightning protection is 1.5 inches wide and 0.051 inch thick or #6 AWG stranded wire. Do not use braided strap because the individual strands oxidize over time, greatly reducing the effectiveness of braid as an ac conductor. Bear in mind that copper strap has about the same inductance as a wire of the same length. While strap may be easier to install and provides a lower RF loss (if it's part of a vertical antenna grounding system, for instance), it doesn't provide significant improvement over round wire for power line frequencies (the NEC's concern) or lightning (where inductance dominates the effects).

Use bare copper for buried ground wires.

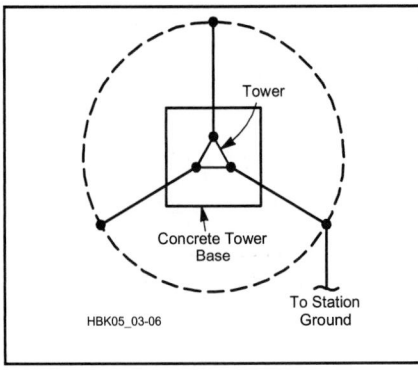

Fig 28.8 — Schematic of a properly grounded tower. A bonding conductor connects each tower leg to a ground rod and a buried (1 foot deep) bare, tinned copper ring (dashed line), which is also connected to the station ground and then to the ac safety ground. Locate ground rods on the ring, as close as possible to their respective tower legs. All connectors should be compatible with the tower and conductor materials to prevent corrosion. See text for conductor sizes and details of lightning and voltage transient protection.

(There are some exceptions; seek an expert's advice if your soil is corrosive.) Exposed runs above ground that are subject to physical damage may require additional protection (a conduit) to meet code requirements. Wire size depends on the application, but never use anything smaller than #6 AWG for bonding conductors. The NEC specifies conductors using wire gauge, and doesn't describe the use of flashing. Local lightning-protection experts or building inspectors can recommend sizes for each application.

Tower and Antennas

Because a tower is usually the highest metal object on the property, it is the most likely strike target. Proper tower grounding is essential to lightning protection. The goal is to establish short multiple paths to the Earth so that the strike energy is divided and dissipated.

Connect each tower leg and each fan of metal guy wires to a separate ground rod. Space the rods at least 6 ft apart. Bond the leg ground rods together with #6 AWG or larger copper bonding conductor (form a ring around the tower base, see **Fig 28.8**). Connect a continuous bonding conductor between the tower ring ground and the entrance panel. Make all connections with fittings approved for grounding applications. *Do not use solder for these connections.* Solder will be destroyed in the heat of a lightning strike.

Because galvanized steel (which has a zinc coating) reacts with copper when combined with moisture, use stainless steel hardware between the galvanized metal and the copper grounding materials.

To prevent strike energy from entering a shack via the feed line, ground the feed line *outside* the home. Ground the coax shield *to the tower* at the antenna and the base to keep the tower and line at the same potential. Several companies offer grounding blocks that make this job easy.

All grounding media at the home must be bonded together. This includes lightning-protection conductors, electrical service, telephone, antenna system grounds and underground metal pipes. Any ground rods used for lightning protection or entrance-panel grounding should be spaced at least 6 feet from each other and the electrical service or other utility grounds and then bonded to the ac system ground as required by the *NEC*.

A Cable Entrance Panel

The basic concept with transient protection is to make sure that all the radio and other equipment is tied together and "moves together" in the presence of a transient voltage. It's not so important that the shack be at "ground" potential, but, rather, that everything is at the *same* potential. For fast rise-time transients such as the individual strokes that make up a lightning strike, even a short wire has enough inductance that the voltage drop along the wire is significant, so whether you are on the ground floor, or the 10th floor of a building, your shack is "far" from Earth potential.

The easiest way to ensure that everything is at the same potential is to tie all the signals to a common reference. In large facilities, this reference would be provided by a grid of large diameter cables under the floor, or by wide copper bars, or even a solid metal floor. A more practical approach for smaller facilities like a ham shack is to have a single "tie point" for all the signals. This is often, but erroneously, called a "single-point ground", but what's really important is not only the shields (grounds) for the signals, but the signal wires as well are referenced to that common potential.

We want to control the flow of the energy

in a strike and eliminate any possible paths for surges to enter the building. This involves routing the feed lines, rotator control cables, and so on at least six feet away from other nearby grounded metal objects.

A commonly used approach to ensuring that all the connections are tied together is to route all the signals through a single "entrance panel" which will serve as the "single point ground" although it may not actually be at ground potential. A convenient approach is to use a standard electrical box installed in the exterior wall

Both balanced line and coax arrestors should be mounted to a secure ground connection on the *outside* of the building. The easiest way to do this is to install a large metal enclosure or a metal panel as a bulkhead and grounding block. The panel should be connected to the lightning dissipation ground with a short wide conductor (for minimum impedance), and, like all grounds, bonded to the electrical system's ground. Mount all protective devices, switches and relay disconnects on the outside facing wall of the bulkhead. The enclosure or panel should be installed in a way that if lightning currents cause a component to fail, the molten metal and flaming debris do not start a fire.

Every conductor that enters the structure, including antenna system control lines, should have its own surge suppressor on an entrance panel. Suppressors are available from a number of manufacturers, including Industrial Communication Engineers (ICE) and PolyPhaser, as well as the usual electrical equipment suppliers such as Square-D.

Lightning Arrestors

Feed line lightning arrestors are available for both coax cable and balanced line. Most of the balanced line arrestors use a simple spark gap arrangement, but a balanced line *impulse* suppressor is available from ICE.

DC blocking arrestors for coaxial cable have a fixed frequency range. They present a high-impedance to lightning (less than 1 MHz) while offering a low impedance to RF.

DC continuity arrestors (gas tubes and spark gaps) can be used over a wider frequency range than those that block dc. Where the coax carries supply voltages to remote devices (such as a mast-mounted preamp or remote coax switch), dc-continuous arrestors *must* be used.

28.1.12 Other Hazards in the Shack

UPS AND ALTERNATE ENERGY SOURCES

Many hams have alternate energy sources for their equipment, or an uninterruptible power supply (UPS), so that they can keep

operating during a utility power outage. This brings some additional safety concerns, because it means that the "turning off the breaker" approach to make sure that power is disconnected might not work.

In commercial installations, fire regulations or electrical codes often require that the emergency power off (EPO) system (the big red button next to the door) also disconnect the batteries of the UPS system, or at least, disable the ac output. This is so that firefighters who may be chopping holes with conductive tools or spraying conductive water don't face the risk of electrocution. (According to NEC, Articles 645-10 and 645-11, UPSs above 750 VA installed within information technology rooms must be provided with a means to disconnect power to all UPS supply and output circuits. This disconnecting means shall also disconnect the battery from its load. The code further requires that the control for these disconnecting means shall be grouped and identified and shall be readily accessible at the principal exit doors.)

A similar problem exists with solar panel installations. Just because the breaker is turned off doesn't mean that dangerous voltages don't exist on the solar panel. As long as light is falling on them, there is voltage present. With no load, even a relatively dim light falling on part of the panels might present a shock or equipment damage hazard. Modern grid-tie solar systems with no batteries often have the panels wired in series, so several hundred volts is not unusual.

Recent revisions of the *NEC* have addressed many of the aspects of photovoltaic (PV) installations that present problems with disconnects, bonding, and grounding. Consulting your local authorities is always wise, and there are several organizations such as the Southwest Technology Development Institute at New Mexico State University that have prepared useful information (see the references at the end of this section). In general, PV systems at 12 or 24 V aren't covered by the *NEC*.

ENERGIZED CIRCUITS

Working with energized circuits can be very hazardous since, without measuring devices, we can't tell which circuits are live. The first thing we should ask ourselves when faced with troubleshooting, aligning or other "live" procedures is, "Is there a way to reduce the hazard of electrical shock?" Here are some ways of doing just that.

1) If at all possible, troubleshoot with an ohmmeter. With a reliable schematic diagram and careful consideration of how various circuit conditions may reflect resistance readings, it will often be unnecessary to do live testing.

2) Keep a fair distance from energized circuits. What is considered "good practice" in

terms of distance? The *NEC* specifies minimum working space around electric equipment depending on the voltage level. The principle here is that a person doing live work needs adequate space so they are not forced to be dangerously close to energized equipment.

3) If you need to measure the voltage of a circuit, install the voltmeter with the power safely off, back up, and only then energize the circuit. Remove the power before disconnecting the meter.

4) If you are building equipment that has hinged or easily removable covers that could expose someone to an energized circuit, install interlock switches that safely remove power in the event that the enclosure is opened with the power still on. Interlock switches are generally not used if tools are required to open the enclosure.

5) Never assume that a circuit is at zero potential even if the power is switched off and the power cable disconnected. Capacitors can retain a charge for a considerable period of time and may even partially recharge after being discharged. Bleeder resistors should be installed, but don't assume they have discharged the voltage. Instead, after power is removed and disconnected use a "shorting stick" to ground all exposed conductors and terminals to ensure that voltage is not present. If you will be working with charged capacitors that store more than a few joules of energy, you should consider using a "discharging stick" with a high wattage, low value resistor in series to ground that limits the discharge current to around 5-10 A. A dead short across a large charged capacitor can damage the capacitor because of internal thermal and magnetic stress. Avoid using screwdrivers, as this brings the holder too close to the circuit and could ruin the screwdriver's blade. For

maximum protection against accidentally energizing equipment, install a shorting lead between high-voltage circuits and ground while you are working on the equipment.

6) Shorting a series string of capacitors does not ensure that the capacitors are discharged. Consider two 400 μF capacitors in series, one charged to +300 V and the other to –300 V with the midpoint at ground. The net voltage across the series string is zero, yet each has significant (and lethal) energy stored in it. The proper practice is to discharge each capacitor in turn, putting a shorting jumper on it after discharge, and then moving to the next one.

7) If you must hold a probe to take a measurement, always keep one hand in your pocket. As mentioned in the sidebar on high-voltage hazards, the worst path current could take through your body is from hand to hand since the flow would pass through the chest cavity.

8) Make sure someone is in the room with you and that they know how to remove the power safely. If they grab you with the power still on they will be shocked as well.

9) Test equipment probes and their leads must be in very good condition and rated for the conditions they will encounter.

10) Be wary of the hazards of "floating" (ungrounded) test equipment. A number of options are available to avoid this hazard.

11) Ground-fault circuit interrupters can offer additional protection for stray currents that flow through the ground on 120-V circuits. Know their limitations. They cannot offer protection for the plate supply voltages in linear amplifiers, for example.

12) Older radio equipment containing ac/dc power supplies have their own hazards. If you are working on these live, use an isolation transformer, as the chassis may be connected

directly to the hot or neutral power conductor.

13) Be aware of electrolytic capacitors that might fail if used outside their intended applications.

14) Replace fuses only with those having proper ratings. The rating is not just the current, but also takes into account the speed with which it opens, and whether it is rated for dc or ac. DC fuses are typically rated at lower voltages than those for ac, because the current in ac circuits goes through zero once every half cycle, giving an arc time to quench. Switches and fuses rated for 120 V ac duty are typically not appropriate for high-current dc applications (such as a main battery or solar panel disconnect).

28.1.13 Electrical Safety References

ARRL Technical Information Service Web page on electrical safety in the Technology area of the ARRL website.

Block, R.W., "Lightning Protection for the Amateur Radio Station," Parts 1-3 (Jun, Jul and Aug 2002 *QST*).

Block. R.W., "The "Grounds" for Lightning and EMP Portection," Polyphaser Corporation, 1993.

Federal Information Processing Standards (FIPS) publication 94: *Guideline on Electrical Power for ADP Installations*. FIPS are available from the National Technical Information Service.

IAEI: *Soares' Book on Grounding*, available from International Association of Electrical Inspectors (IAEI).

"IEEE Std 1100 - 2005 IEEE Recommended Practice for Powering and Grounding Electronic Equipment," *IEEE Std 1100-2005 (Revision of IEEE Std 1100-1999)*, pp 0_1-589, 2006. "This document pres-

Electrical Shock Hazards and Effects

What happens when someone receives an electrical shock?

Electrocutions (fatal electric shocks) usually are caused by the heart ceasing to beat in its normal rhythm. This condition, called ventricular fibrillation, causes the heart muscles to quiver and stop working in a coordinated pattern, in turn preventing the heart from pumping blood.

The current flow that results in ventricular fibrillation varies between individuals but may be in the range of 100 mA to 500 mA. At higher current levels the heart may have less tendency to fibrillate but serious damage would be expected. Studies have shown 60-Hz alternating current to be more hazardous than dc currents. Emphasis is placed on application of cardiopulmonary resuscitation (CPR), as this technique can provide mechanical flow of some blood until paramedics can "re-

start" the heart's normal beating pattern. Defibrillators actually apply a carefully controlled waveform to "shock" the heart back into a normal heartbeat. It doesn't always work but it's the best procedure available.

What are the most important factors associated with severe shocks?

You may have heard that the current that flows through the body is the most important factor, and this is generally true. The path that current takes through the body affects the outcome to a large degree. While simple application of Ohm's Law tells us that the higher the voltage applied with a fixed resistance, the greater the current that will flow. Most electrical shocks involve skin contact. Skin, with its layer of dead cells and often fatty tissues, is a fair insulator. Nonetheless, as voltage

increases the skin will reach a point where it breaks down. Then the lowered resistance of deeper tissues allows a greater current to flow. This is why electrical codes refer to the term "high voltage" as a voltage above 600 V.

How little voltage can be lethal?

This depends entirely on the resistance of the two contact points in the circuit, the internal resistance of the body, and the path the current travels through the body. Historically, reports of fatal shocks suggest that as little as 24 V *could* be fatal under extremely adverse conditions. To add some perspective, one standard used to prevent serious electrical shock in hospital operating rooms limits leakage flow from electronic instruments to only 50 μA due to the use of electrical devices and related conductors inside the patient's body.

ents recommended design, installation, and maintenance practices for electrical power and grounding (including both safety and noise control) and protection of electronic loads such as industrial controllers, computers, and other information technology equipment (ITE) used in commercial and industrial applications."

National Electrical Code, NFPA 70, National Fire Protection Association, Quincy, MA (**www.nfpa.org**).

Solar energy websites — **www.nmsu. edu/~tdi/PV=NEC_HTML/pv-nec/ pv-nec.html** and **www.solarabcs.org**

Standard for the Installation of Lightning Protection Systems, NFPA 780, National Fire Protection Association, Quincy, MA (**www.nfpa.org**).

See the Protection Group's list of white papers contained in the Knowledge Base **www.protectiongroup**.com.

28.2 Antenna and Tower Safety

By definition, all of the topics in this book are about radio telecommunications. For those communications, receive and transmit antennas are required and those antennas need to be up in the air to work effectively. While antenna design and construction are covered elsewhere, this section covers many of the topics associated with installing antennas, along with related safety issues.

A substantially more detailed treatment of techniques used to erect towers and antennas is available in these references:

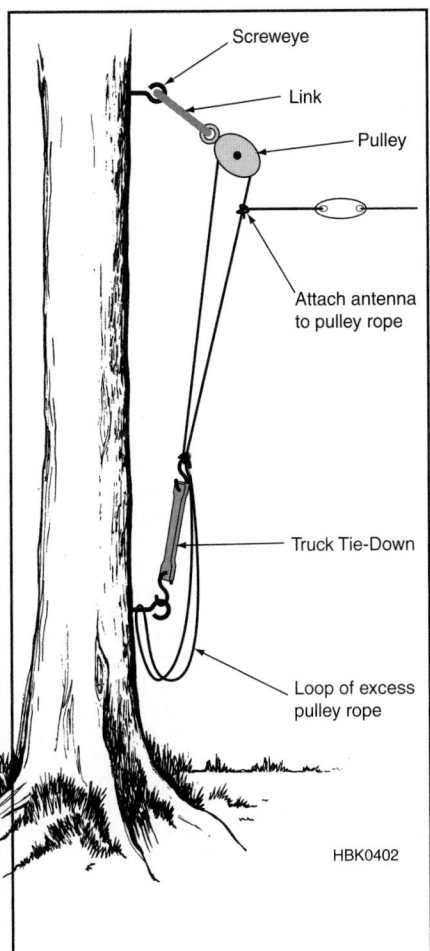

Fig 28.9 — Loop and halyard method of supporting wire antennas in trees. Should the antenna break, the continuous loop of rope allows antenna repair or replacement without climbing the tree.

The ARRL Antenna Book

Up the Tower: The Complete Guide to Tower Construction by Steve Morris, K7LXC

Antenna Towers for Radio Amateurs by Don Daso, K4ZA.

28.2.1 Legal Considerations

Some antenna support structures fall under local building regulations as well as neighborhood restrictions. Many housing developments have Homeowner's Associations (HOAs) as well as Covenants, Conditions and Restrictions (CC&Rs) that may have a direct bearing on whether a tower or similar structure can be erected at all. This is a broad topic with many pitfalls. Detailed background on these topics is provided in *Antenna Zoning for the Radio Amateur* by Fred Hopengarten, K1VR, an attorney with extensive experience in towers and zoning. You may also want to contact one of the ARRL Field Organization's Volunteer Counsels.

Even without neighborhood issues, a building permit is likely to be required. With the proliferation of cellular and other commercial wireless devices and their attendant RF sites, many local governments now require that the structures meet local building codes. Again, K1VR's book is extremely helpful in sorting all this out. Building permit applications may also require Professional Engineer (PE) calculations and stamp (certification). The ARRL Field Organization's Volunteer Consulting Engineer program may be useful with the engineering side of your project.

28.2.2 Antenna Mounting Structures

TREES AND POLES

The original antenna supports were trees: if you've got them, use them. They're free and unregulated, so it couldn't be easier. Single-trunked varieties such as fir and pine trees are easier to use than the multi-trunked varieties. Multi-trunked trees are not impossible to use — they just require a lot more work. For dipoles or other types of wire antennas, plan for the tree to support an end of the wire; trying to install an inverted V or similar configuration is almost impossible due to all of the intervening branches.

Install an eye-screw with a pulley at the desired height, trim away enough limbs to create a "window" for the antenna through the branches and then attach a rope halyard to the antenna insulator. Here's a useful tip: Make the *halyard* a continuous loop as shown in **Fig 28.9**. Since it's almost always the antenna wire that breaks, a continuous halyard makes it easy to reattach the wire and insulator. With just a single halyard, if the antenna breaks, the tree will have to be climbed to reach the pulley, then reinstall and attach the line. If you're unable to climb the support tree, contact a local tree service. Professional tree climbers are often willing to help out for a small fee.

Another way to get wires into trees is with some sort of launcher. Using a bow-and-arrow is a traditional method of shooting a fishing line over a tree to pull up a bigger line. There are now commercial products available that are easier to use and reach higher in the tree. For example, wrist-rocket slingshots and compressed-air launchers can reach heights of more than 200 feet!

Wooden utility poles offer a tree-related alternative but are not cheap, require special installation with a pole-setting truck, and there is no commercial antenna mounting hardware available for them. That makes them a poor choice for most installations.

TOWERS

The two most important parameters to consider when planning a tower installation are the maximum local *wind speed* and the proposed antenna *wind load*. Check with your local building department to find out what the maximum wind speed is for your area. Another source is a list of maximum wind speeds for all counties in the US from the TIA-222, *Structural Standard for Antenna Supporting Structures and Antennas*. This is an expensive professional publication so it's not for everyone, but the list is posted on the Champion Radio Products website under "Tech Notes." Tower capacities are generally specified in square feet of antenna load and antenna wind load specifications are provided by the antenna manufacturer.

Before beginning, learn and follow K7LXC's Prime Directive of tower construc-

tion — "DO what the manufacturer says." (And DON'T do what the manufacturer doesn't say to do.) Professional engineers have analyzed and calculated the proper specifications and conditions for tower structures and their environment. Taking any shortcuts or making different decisions will result in a less reliable installation.

Towers come in the two varieties shown in **Fig 28.10** — guyed and self-supporting. Guyed towers require a bigger footprint because the guys have to be anchored away from the tower — typically 80% of the tower height. Self-supporting towers need bigger bases to counteract the overturning moment and are more expensive than a guyed tower because there is more steel in them (the cost of a tower is largely determined by the cost of the steel).

The most popular guyed towers are the Rohn 25G and 45G. The 25G is a light-duty tower and the 45G is capable of carrying fairly big loads. The online Rohn catalog (see the References) has most of the information you'll need to plan an installation.

Self-supporting towers are made by several manufacturers and allow building a tower up to 100 feet or higher with a small footprint. Rohn, Trylon and AN Wireless are popular vendors. Another type of self-supporting tower is the *crank-up*, shown in **Fig 28.11**. Using a system of cables, pulleys and winches, crank-up towers can extend from 20 feet to over 100 feet high. These are moderately complex devices. The largest manufacturer of crank-up towers is US Towers.

Fig 28.10 — A guyed tower with a good-sized load of antennas at XE2DV-W7ZR in Baja California, is shown at the left. At the right, the Trylon Titan self-supporting tower of W7WVF and N7YYG in Bandon, Oregon. (*K7LXC photos*)

Fig 28.11 — The bottom of N6TV's crank-up tower is shown at left. The motor drive mechanism is on the left and a fishing net on the right catches and coils the feed lines and control cables as the tower is lowered. On the right, K6KR's fully loaded crank-up extended to its maximum height of 90 feet. (*K7LXC photos*)

Another simple and effective way to get an antenna up in the air is with a *roof-mounted tower*, seen in **Fig 28.12**. These are four-legged aluminum structures of heights from four to more than 20 feet. While they are designed to be lag-screwed directly into the roof trusses, it is often preferable to through-bolt them into a long 2×4 or 2×6 that acts a backing plate, straddling three to four roof trusses. In any case, if it is not clear how best to install the tower on the structure, have a roofing professional or engineer provide advice. Working on a roof-mounted tower also requires extra caution because of climbing on a roof while also working on a tower.

28.2.3 Tower Construction and Erection

THE BASE

Once all the necessary materials and the required approvals have been gathered, tower installation can begin. Let's assume you and your friends are going to install it. The first job is to construct the base. A base for a guyed tower can be hand-dug as can the guy anchors. For a self-supporting tower, renting an excavator of some sort will make it much easier to move the several cubic yards of dirt.

Next, some sort of rebar cage will be needed for the concrete. Guyed towers only require rudimentary rebar while a self-supporting tower will need a bigger, heavier and more elaborate cage. Consult the manufacturer's specifications for the exact materials and dimensions.

Typical tower concrete specs call for 3000 psi (minimum) compressive strength concrete and 28 days for a full cure. A local pre-mix concrete company can deliver it.

Pouring the concrete is easiest if the concrete truck can back up to the hole. If that's not possible, a truck- or trailer-mounted line pump can pump it up to 400 feet at minimal expense if using a wheelbarrow is not possible or practical. Packaged concrete from the hardware store mixed manually may also be used. Quikrete Mix #1101 is rated at 2500 psi after seven days and 4000 psi after 28 days.

TOOLS

Once the base and anchors are finished and the concrete has cured, the tower can be constructed. There are several tools that will make the job easier. If the tower is a guyed tower, it can be erected either with a crane or a *gin-pole*. The gin-pole, shown in **Fig 28.13**, is a pipe that attaches to the leg of the tower and has a pulley at the top for the haul rope. Use the gin-pole to pull up one section at a time (see below).

Another useful tool for rigging and hoisting is the *carabiner*. Shown in **Fig 28.14** (A and B), carabiners are oval steel or aluminum snap-links popularized by mountain climbers. They have spring-loaded gates and can be used for many tower tasks. For instance, there should be one at the end of the haul rope for easy and quick attachment to rotators, parts and tool buckets — virtually anything that needs to be raised or lowered. It can even act as a "third hand" on the tower.

Along with the carabineer, the *nylon loop sling* in Fig 28.14C can be wrapped around large or irregularly shaped objects such as antennas, masts or rotators and attached to ropes with carabiners. For a complex job, a professional will often climb with eight to ten slings and use every one!

A pulley or two will also make the job easier. At a minimum, one is needed for the haul rope at the top of the tower. A *snatch block* is also useful; this is a pulley whose top opens up to "snatch" (attach it to) the rope at any point. **Fig 28.15** shows two snatch-block pulleys used for tower work.

ROPES

Speaking of ropes, use a decent haul rope. Rope that is one-half inch diameter or larger affords a good grip for lifting and pulling. There are several choices of rope material. The best choice is a synthetic material such as nylon or Dacron. A typical twisted rope is fine for most applications. A synthetic rope with a braid over the twisted core is known

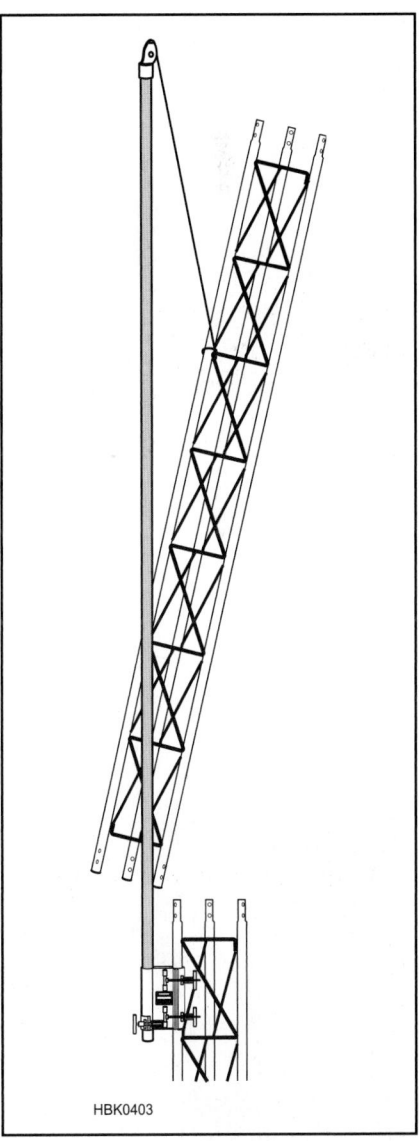

HBK0403

Fig 28.13 — A gin-pole consists of a leg clamp fixture, a section of aluminum mast and a pulley. It is used to lift the tower section high enough to be safely lowered into place and attached. (Based on Rohn EF2545.)

Fig 28.12 — The roof-mounted tower holding the AA2OW antenna system. (*AA2OW photo*)

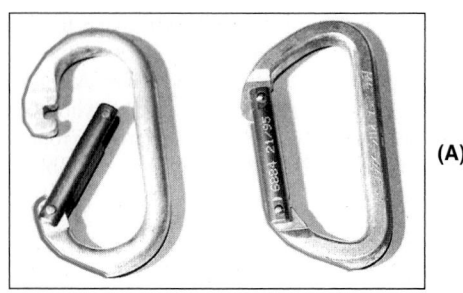

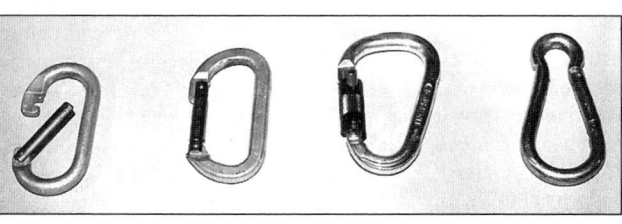

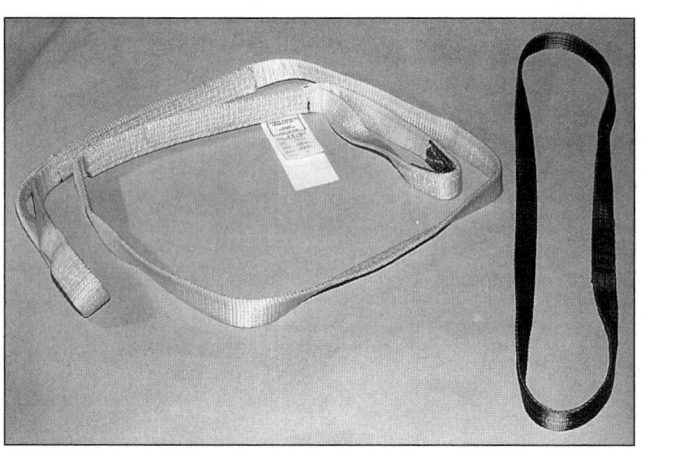

Fig 28.14 — (A) Oval mountain climbing type carabiners are ideal for tower workloads and attachments. The gates are spring loaded — the open gate is shown for illustration. (B) An open aluminum oval carabiner; a closed oval carabiner; an aluminum locking carabiner; a steel snaplink. (C) A heavy duty nylon sling on the left for big jobs and a lighter-duty loop sling on the right for everything else. (*K7LXC photos*)

(A)

(B)

(C)

as *braid-on-braid* or *kernmantle*. While it's more expensive than twisted ropes, the outer braid provides better abrasion resistance. The least expensive type of rope is polypropylene. It's a stiff rope that doesn't take a knot as well as other types but is reasonably durable and cheap. **Table 28.2** shows the safe working load ratings for common types of rope.

When doing tower work, being able to tie knots is required. Of all the knots, the *bowline* is the one to know for tower work. The old "rabbit comes up through the hole, around the tree and back down the hole" is the most familiar method of tying a bowline. Most amateurs are knot-challenged so it's a great advantage to know at least this one.

INSTALLING TOWER SECTIONS

The easiest way to erect a tower is to use a crane. It's fast and safe but more expensive than doing it in sections by hand. To erect a tower by sections, a gin-pole is needed (see Fig 28.13). It consists of two pieces — a clamp or some device to attach it to the tower leg and a pole with a pulley at the top. The pole is typically longer than the work piece being hoisted, allowing it to be held above the tower top while being attached or manipulated.

With the gin-pole mounted on the tower, the haul rope runs up from the ground, through the gin-pole mast and pulley at the top of the gin-pole, and back down the tower. The haul rope has a knot (preferably a bowline) on the end for attaching things to be hauled up or down. A carabiner hooked into the bight of the knot can be attached to objects quickly so that you don't have to untie and re-tie the bowline with every use.

It's a good idea to pass the haul rope through a snatch-block at the bottom of the tower. This changes the direction of the rope from vertical to horizontal, allowing the ground crew to stand away from the tower (and the fall zone for things dropped off the tower) to manipulate the haul rope while also watching what's going on up on the tower.

GUYS

For guyed towers, an important construction parameter is guy wire material and *guy tension*. Do *not* use rope or any other material not rated for use as guy cable as permanent tower guys. Guyed towers for amateurs typically use either ³⁄₁₆-inch or ¼-inch *EHS* (extra high strength) steel guy cable. The only other acceptable guy material is Phillystran — a lightweight cable made of Kevlar fibers. Phillystran is available witht he same breaking strength as EHS cable. The advantage of Phillystran is that it is non-conducting and does not create unwanted electrical interaction with antennas on the tower. It is an excellent choice for towers supporting multiple Yagi and wire antennas and does not have to be broken up into short lengths with insulators.

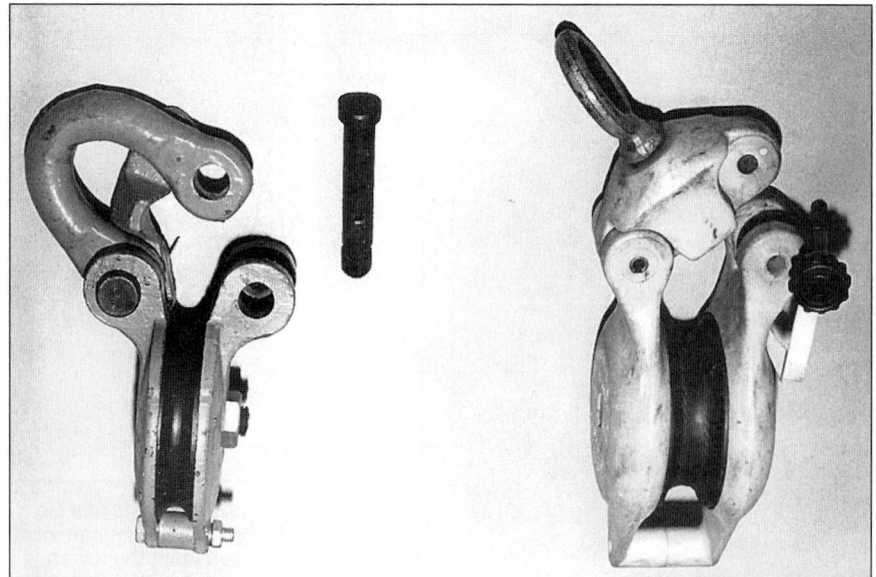

Fig 28.15 — Two snatch blocks; a steel version on the left and a lightweight high impact plastic one on the right in their open position. (*K7LXC photo*)

Table 28.2
Rope Sizes and Safe Working Load Ratings in Pounds

3 Strand Twisted Line

Diameter	Manila	Nylon	Dacron	Polypropylene
¼	120	180	180	210
⅜	215	405	405	455
½	420	700	700	710
⅝	700	1140	1100	1050

Double-Braided Line

Diameter	Nylon	Dacron
¼	420	350
⅜	960	750
½	1630	1400
⅝	2800	2400

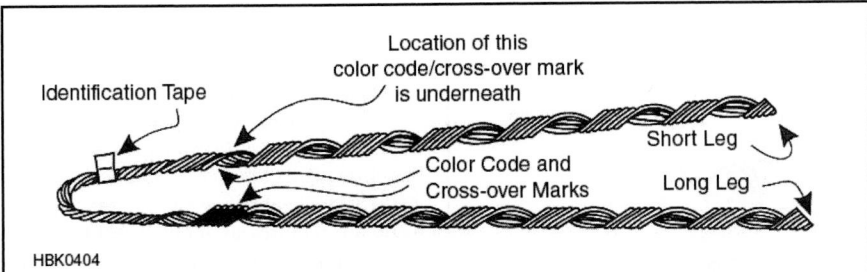

HBK0404

Fig 28.16 — A PreFormed Line Products Big Grip for guy wires.

EHS wire is very stiff — to cut it, use a hand-grinder with thin cutting blades or a circular saw with a pipe-cutting aggregate blade. Be sure to wear safety glasses and gloves when cutting since there will be lots of sparks of burning steel being thrown off. Phillystran can be cut with a utility knife or a hot knife for cutting plastic.

If the guys are too loose, the result will be wind-induced shock loading. Guys that are too tight exert extra compressive load on the tower legs, reducing the overall capacity and reliability of the tower. The proper tension of EHS or Phillystran guys is 10% of the material's *ultimate breaking strength*. For ³⁄₁₆-inch EHS the ultimate breaking strength is 4900 pounds and for ¼-inch it's 6000 pounds so the respective guy tension should be 490 pounds and 600 pounds. The easiest to use, most accurate, and least expensive way to measure guy tension is by using a Loos tension gauge. It was developed for sailboat rigging so it's available at some marine supply stores or from Champion Radio Products.

Guy wires used to be terminated in a loop with cable clamps but those have been largely replaced by pre-formed Big Grips, shown in **Fig 28.16**. These simply twist onto the guy wire and are very secure. They grip the guy cable by squeezing the cable as tension is applied. Be sure to use the right type of Big Grips for the thickness and material of the guy cable.

28.2.4 Antenna Installation

Now that the tower is up, install the antennas. VHF/UHF whips and wire antennas are pretty straightforward, but installing an HF Yagi is a more challenging proposition. With a self-supporting tower, there are no guy wires to contend with — generally, the antenna can just be hauled up the tower face. Sometimes it is that easy!

In most cases, short of hiring a crane, the easiest way to get a Yagi up and down a tower is to use the *tram* method. A single tramline is suspended from the tower to the ground and the load is suspended under the tramline. Another technique is the *trolley* method in which two lines are suspended from the tower to the ground and the antenna rides on top of the lines like a trolley car on tracks. Problems with the trolley technique include trying to get the lines to have the same tension, balancing the antenna so that it won't fall off of the lines, and the added friction of pulling the antenna up two lines. The tram method has none of these problems. **Fig 28.17** illustrates the tram method of raising antennas.

Tram and trolley lines are typically attached to the mast above the top of the tower. In the case of a big load, the lines may exert enough force to bend the mast. If in doubt, *back-guy* the mast with another line in the opposite direction for added support.

MASTS

A mast is a pipe that sticks out of the top of the tower and connects the rotator to the antenna. For small antenna loads and moderate wind speeds, any pipe will work. But as wind speed and wind load increase, more force will be exerted on the mast.

There are two materials used for masts — *pipe* and *tubing*. Pipe can be water pipe or

Table 28.3
Yield Strengths of Mast Materials

Material	Specification	Yield Strength (lb/in.²)
Drawn aluminum tube	6063-T5	15,000
	6063-T832	35,000
	6061-T6	35,000
	6063-T835	40,000
	2024-T3	42,000
Aluminum pipe	6063-T6	25,000
	6061-T6	35,000
Extruded alum. tube	7075-T6	70,000
Aluminum sheet and plate	3003-H14	17,000
	5052-H32	22,000
	6061-T6	35,000
Structural steel	A36	33,000
Carbon steel, cold drawn	1016	50,000
	1022	58,000
	1027	70,000
	1041	87,000
	1144	90,000
Alloy steel	2330 cold drawn	119,000
	4130 cold worked	75,000
	4340 1550 °F quench 1000 °F temper	162,000
Stainless steel	AISI 405 cold worked	70,000
	AISI 440C heat-treated	275,000

(From *Physical Design of Yagi Antennas* by David B. Leeson, W6NL)

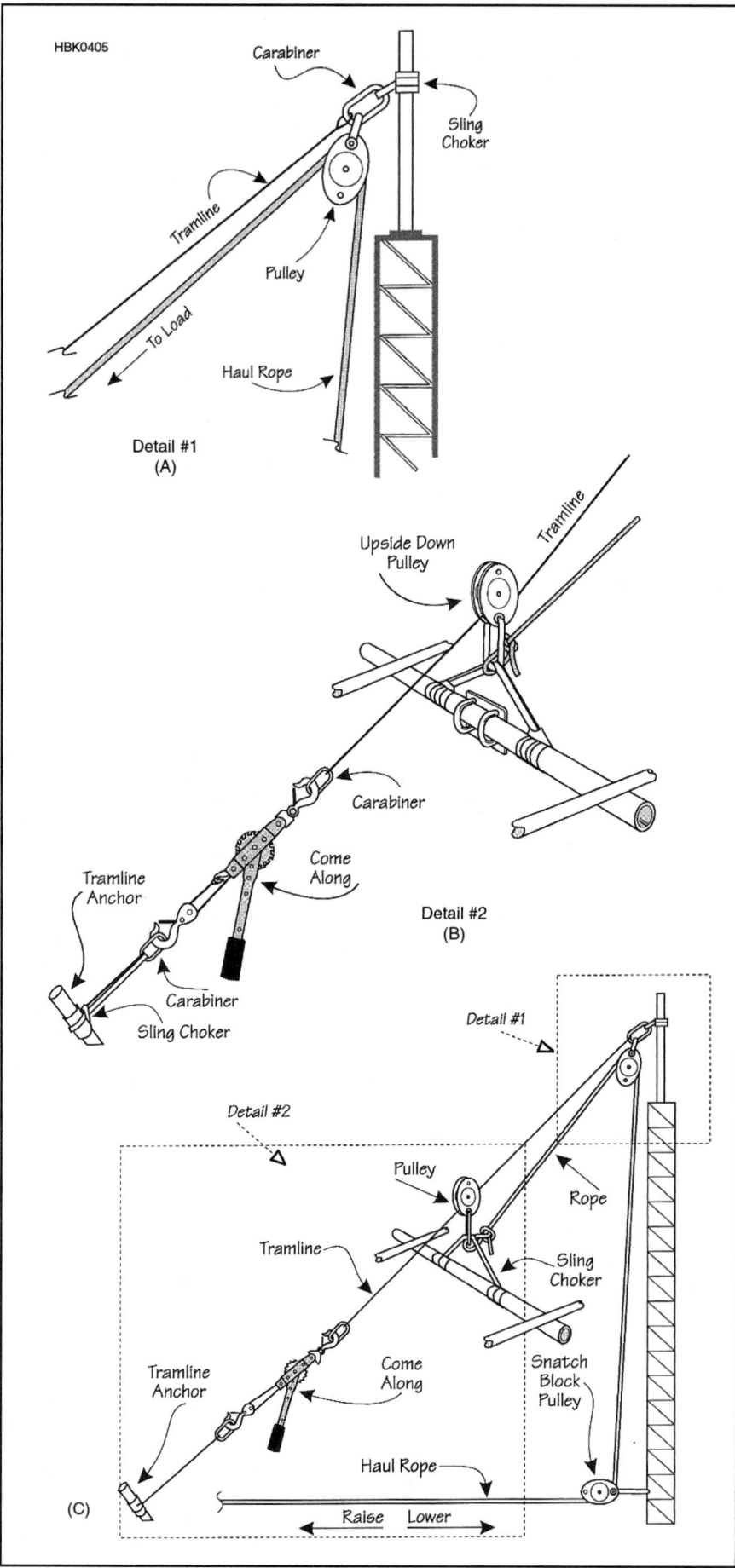

Detail #1
(A)

Detail #2
(B)

(C)

conduit (EMT). Pipe is a heavy material with not much strength since its job is just to carry water or wires. Pipe is acceptable as mast material for small loads only. Another problem is that 1.5-in. pipe (pipe is measured by its inside diameter or ID) is only 1.9-in. OD. Since most antenna boom-to-mast hardware is designed for a 2-in. mast, the less-than-perfect fit may lead to slippage.

For any larger load use carbon-alloy steel tubing rated for high strength. A moderate antenna installation in an 80 MPH wind might exert 40,000 to 50,000 pounds per square inch (psi) on the mast. Pipe has a yield strength of about 35,000 psi, so you can see that pipe is not adequately rated for this type of use. Chromoly steel tubing is available with yield strengths from 40,000 psi up to 115,000 psi but it is expensive. **Table 28.3** shows the ratings of several materials used as masts for amateur radio antennas.

Calculating the required mast strength can be done by using a software program such as the *Mast, Antenna and Rotator Calculator (MARC)* software. (See the References.) The software requires as inputs the local wind speed, antenna wind load, and placement on the mast. The software then calculates the mast bending moment and will recommend a suitable mast material.

28.2.5 Weatherproofing Cable and Connectors

The biggest mistake amateurs make with coaxial cable is improper weatherproofing. (Coax selection is covered in the chapter on **Transmission Lines**.) **Fig 28.18** shows how to do it properly. First, use high-quality electrical tape, such as 3M Scotch 33+ or Scotch 88. Avoid inexpensive utility tape. After tightening the connector (use pliers carefully to seat threaded connectors — hand-tight isn't good enough), apply two wraps of tape around the joint.

When you're done, sever the tape with a knife or tear it very carefully — *do not* stretch the tape until it breaks. This invariably leads to "flagging" in which the end of the tape loosens and blows around in the wind. Let the tape relax before finishing the wrap.

Next put a layer of butyl rubber *vapor wrap* over the joint. (This tape is also available in

Fig 28.17 — At A, rigging the top of the tower for tramming antennas. Note the use of a sling and carabiner. (B) Rigging the anchor of the tramline. A come-along is used to tension the tramline. (C) The tram system for getting antennas up and down. Run the antenna part way up the tramline for testing before installation. It just takes a couple of minutes to run an antenna up or down once the tramline is rigged.

the electrical section of the hardware store.) Finally, add two more layers of tape over the vapor wrap, creating a professional-quality joint that will never leak. Finally, if the coax is vertical, be sure to wrap the final layer so that the tape is going *up* the cable as shown in Fig 28.18. In that way, the layers will act like roofing shingles, shedding water off the connection. Wrapping it top to bottom will guide water between the layers of tape.

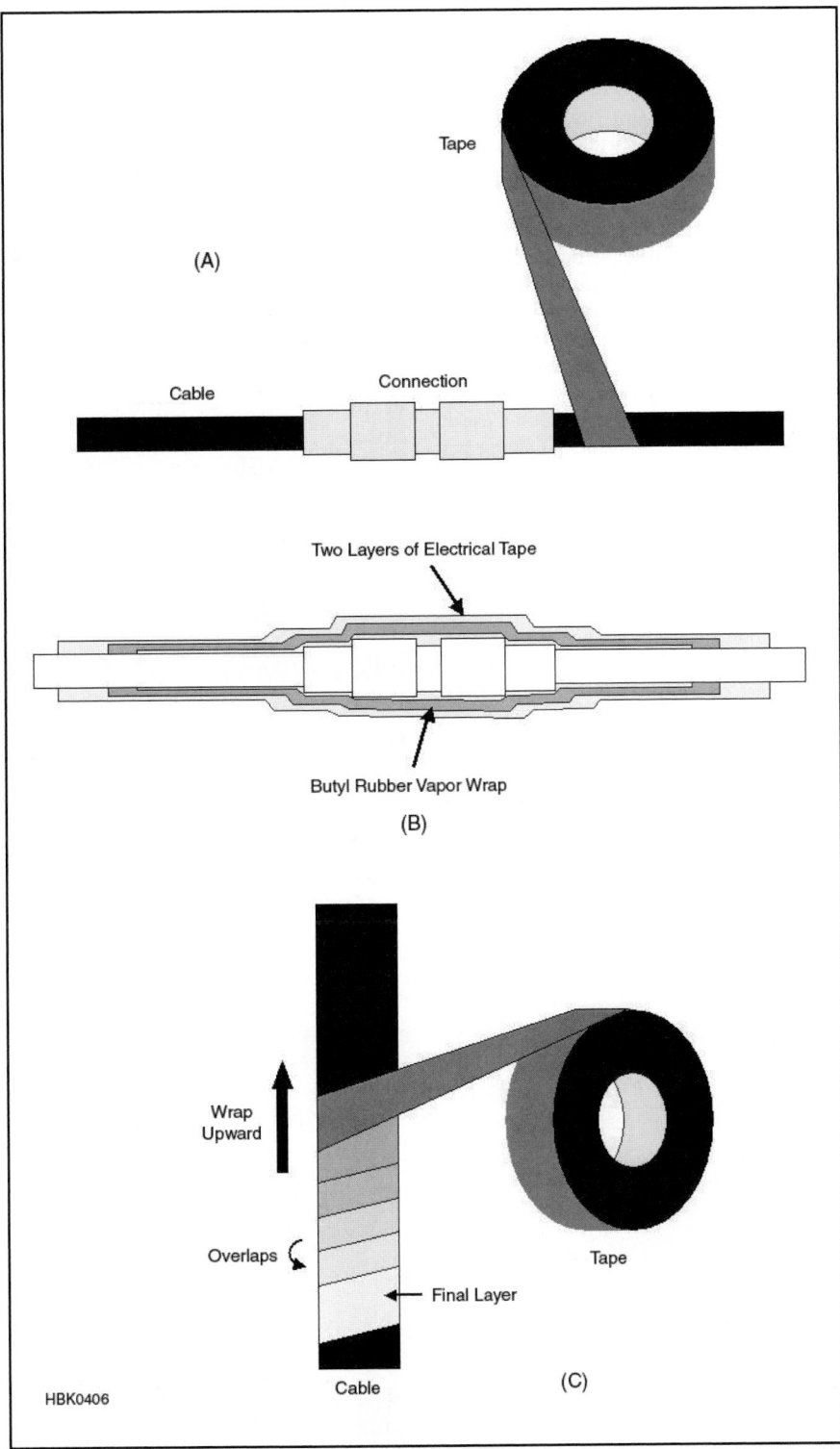

Fig 28.18 — Waterproofing a connector in three steps. At A, cover the connectors with a layer of good-quality electrical tape. B shows a layer of butyl rubber vapor wrap between the two layers of electrical tape. C shows how to wrap tape on a vertical cable so that the tape sheds water away from the connection. (Drawing (C) reprinted courtesy of *Circuitbuilding for Dummies*, Wiley Press)

HBK0406

28.2.6 Climbing Safety

Tower climbing is a potentially dangerous activity, so you'll need to use the proper safety equipment and techniques. OSHA, the Federal Occupational Safety and Health Administration, publishes rules for workplace safety. Although amateurs are not bound by those rules, you'll be much better off by following them. What equipment and techniques you use are up to you. As long as you've got the right safety equipment and follow the basic safety rules you won't have any problems.

SAFETY AWARENESS AND PREPARATION

One of the most important aspects of safety is having the knowledge and awareness to do a job safely and efficiently. You must have the mental ability to climb and work at altitude while constantly rethinking all connections, techniques and safety factors. Safely climbing and working on towers is 90% mental. Mental preparedness is something that must be learned. This is an occasion where there is no substitute for experience. The biggest obstacle for anyone is making the mental adjustment. Properly installed towers are inherently safe and accidents are relatively rare.

You should also check your safety equipment every time before you use it. Inspect it for any nicks or cuts to your belt and safety strap. Professional tower workers are required to check their safety equipment every day and you should check yours before each use.

One of the most important lessons for tower climbing is that you have four points of attachment and security — two hands and two feet. When climbing, move only one point at a time. That leaves you with three points of contact and a wide margin of safety if you ever need it. This is in addition to having your fall-arrest lanyard (see below) connected at all times.

Another recommended technique is to always do everything the same way every time. That is, always wear your positioning lanyard on the same D-ring and always connect it in the same way. Always look at your belt D-ring while clipping in with your safety strap. This way you'll always confirm that you're securely belted in. Always look!

PERSONAL SAFETY EQUIPMENT

The most important pieces of safety equipment are the *fall arrest harness* (FAH) you wear and the accompanying lanyards that attach to it (**Fig 28.19**). The FAH has leg loops and suspenders to help spread the fall forces over more of your body and has the ability to hold you in a natural position with your arms and legs hanging below you where you're able to breathe normally.

(A)

Fig 28.19 — (A) The well-dressed tower climber. Note the waist D-rings for positioning lanyard attachment as well as the suspenders and leg loops. At (B) is an adjustable positioning lanyard. The climber also has working boots, gloves, safety glasses and hardhat. (*K7LXC photos*)

Two or more lanyards are used. One is the positioning lanyard (**Fig 28.20**). That is, it holds you in working position and attaches to the D-rings at your waist. They can be adjustable or fixed and are made from different materials such as nylon rope, steel chain or special synthetic materials. An adjustable positioning lanyard will adjust to almost any situation whereas a fixed-length one is typically either too long or too short. A rope lanyard is the least expensive.

Leather safety equipment was outlawed some years ago by OSHA so don't use it. This includes the old-fashioned safety belt that was used for years but offers no fall arrest capability. If you drop down while wearing a safety belt, your body weight can cause it to rise up from your waist to your ribcage where it will immobilize your diaphragm and suffocate you. On the other hand you can use a safety belt for positioning when it's worn over an FAH. Just don't depend on it to catch you in case of a fall.

The other lanyard is the fall arrest lanyard, shown in **Fig 28.21**, which attaches to a D-ring between your shoulder blades. The other end attaches to the tower above the work position and catches you in case of a fall. The simplest is a 6 foot rope lanyard which is inexpensive but doesn't offer any

Fig 28.21 — The fall-arrest lanyard is above the climber so that the climber can climb up to it. The fall-arrest and positioning lanyards are then "leapfrogged" so that the climber remains attached to the tower 100 percent of the time. (*K7LXC photos*)

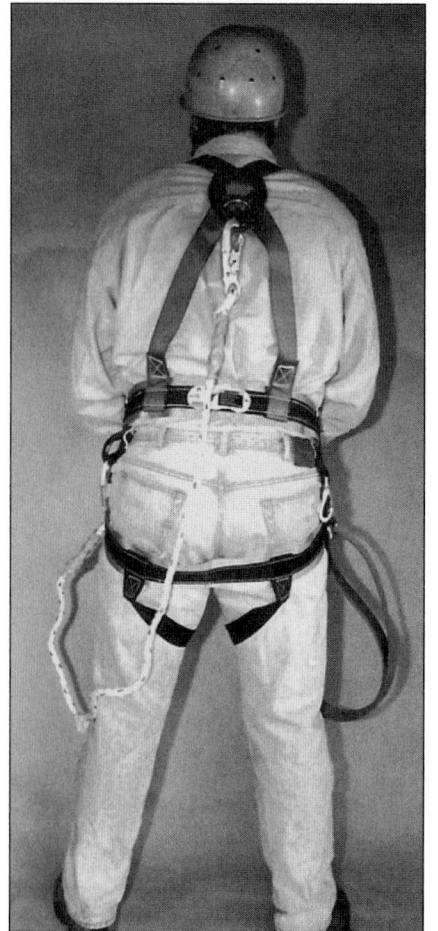

(B)

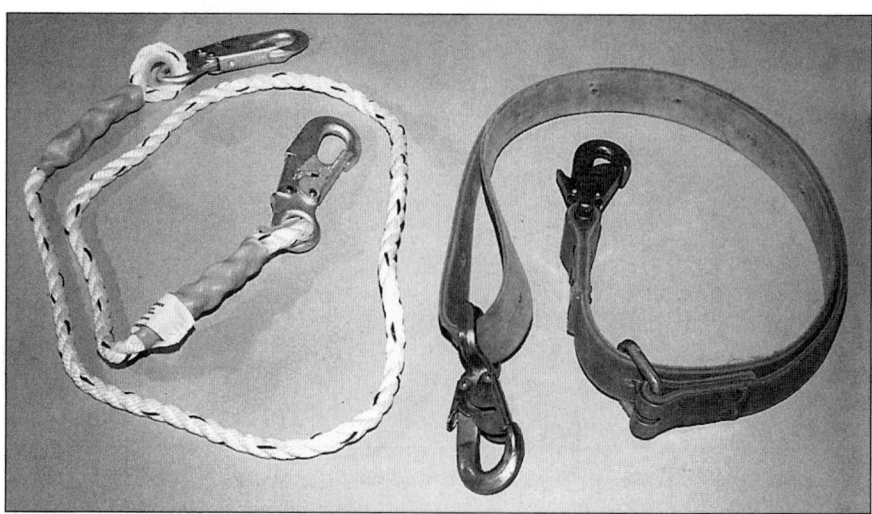

Fig 28.20 — A fixed-length rope positioning lanyard on the left and a versatile adjustable lanyard on the right. (*K7LXC photos*)

shock absorption. There are also shock absorbing varieties that typically have bar-tacked stitches that pull apart under force to decelerate you. Don't cut corners on buying or using safety equipment; you bet your life on it every time you use it!

OSHA rules and good common sense say you should be attached to the tower 100% of the time. You can do this several ways. One is to attach the fall arrest lanyard above you and climb up to it. Use your positioning lanyard to hold you while you detach it and move it up again. Repeat as necessary. Another option is to use a pair of fall arrest lanyards, attaching one then the other as you climb the tower.

Boots should be leather with a steel or fiberglass shank. Diagonal bracing on Rohn 25G is only 5/16 inch rod — spending all day standing on that small step will take a toll on your feet. The stiff shank will support your weight and protect your feet; tennis shoes will not. Leather boots are mandatory on towers like Rohn BX that have sharp X-cross braces. Your feet are always on a slant and that is hard on feet.

A hard hat and safety goggles is highly recommended. Just make sure they are ANSI or OSHA approved and that you and your crew wear them. As you'll be looking up and down a lot, a chin strap is essential to keep the hard hat from falling off.

If you do a lot of tower work, your hands will take a beating. Gloves are essential — keep several spare pair for ground crew members who show up without them. Cotton gloves are fine for gardening but not for tower work; they don't provide enough friction for climbing or working with a haul rope. Leather gloves are the only kind to use; either full leather or leather-palmed are fine. The softer the gloves the more useful they'll be. Stiff leather construction gloves are fine for the ground crew but pigskin and other soft leathers are better for the climber because you can thread a nut or do just about any other delicate job with these gloves on.

Safety Tips

1) Don't climb with anything in your hands; attach it to your safety belt if you must climb with it or have your ground crew send it up to you in a bucket.

2) Don't put any hardware in your mouth; it can easily be swallowed.

3) Remove any rings and/or neck chains; they can get hooked on things.

4) Be on the lookout for bees, wasps and their nests. If you do run into a hornet, wasp or other stinging insect, use Adolph's Meat Tenderizer on the sting — it contains the enzyme papain which neutralizes the venom. Have a small jar in your tool kit.

5) Don't climb when tired; that's when most accidents occur.

6) Don't try to lift anything by yourself; one person on a tower has very little leverage or strength. Let the ground crew use their strength; save yours for when you really need it.

7) If a procedure doesn't work, assess the situation and re-rig, if necessary, before trying again.

GROUND CREW SAFETY

The climber on the tower is the boss. Before tower work starts, have a safety meeting with the ground crew. Explain what is going to be done and how to do it as well as introducing them to any piece of hardware with which they may not be familiar (for example, carabiners, slings or come-along winches).

As part of the ground crew, there are a few rules to follow:

1) The climber on the tower is in charge.

2) Don't do anything unless directed by the climber in charge on the tower. This includes handling ropes, tidying up, moving hardware, and so on.

3) If not using radios to communicate, when talking to the climber on the tower, look up and talk directly to him or her in a loud voice. The ambient noise level is higher up on the tower because of traffic, wind and nearby equipment.

4) Communicate with the climber on the tower. Let him or her know when you're ready or if you're standing by or if there is a delay. Advise the climber when lunch is ready!

28.2.7 Antenna and Tower Safety References

AN Wireless — **www.anwireless.com**

ARRL Volunteer Counsel program — **www.arrl.org/volunteer-counsel-program**

ARRL Volunteer Consulting Engineer program — **www.arrl.org/volunteer-consulting-engineer-program**

D. Brede, W3AS, "The Care and Feeding of an Amateur's Favorite Antenna Support — the Tree," *QST*, Sep 1989, pp 26-28, 40.

Champion Radio Products — **www.championradio.com**

D. Daso, K4ZA, *Antenna Towers for Radio Amateurs* (ARRL, 2010)

D. Daso, K4ZA, "Workshop Chronicles" (column), *National Contest Journal*

F. Hopengarten, K1VR, *Antenna Zoning for the Radio Amateur* (ARRL, 2002)

Knot-tying website — **www.animatedknots.com**

Loos tension gauge — **www.championradio.com/rigging.html**

MARC software — **www.championradio.com/misc.html**

S. Morris, K7LXC, *Up the Tower: The Complete Guide to Tower Construction* (Champion Radio Products, 2009)

Rohn Tower — **www.rohnnet.com**

H.W. Silver, NØAX, ed., *The ARRL Antenna Book*, 22nd ed. (ARRL, 2012)

Trylon — **www.trylon.com**

US Towers — **www.ustower.com**

28.3 RF Safety

Amateur Radio is basically a safe activity. In recent years, however, there has been considerable discussion and concern about the possible hazards of electromagnetic fields (EMF), including both RF energy and power frequency (50-60 Hz) EMF. FCC regulations set limits on the maximum permissible exposure (MPE) allowed from the operation of radio transmitters. Following these regulations, along with the use of good RF practices, will make your station as safe as possible. This section, written by the ARRL RF Safety Committee (see sidebar), deals with the topic of electromagnetic safety.

28.3.1 How EMF Affects Mammalian Tissue

All life on Earth has adapted to live in an environment of weak, natural, low frequency electromagnetic fields, in addition to the Earth's static geomagnetic field. Natural low-frequency EM fields come from two main sources: the sun and thunderstorm activity. During the past 100 years, man-made fields at much higher intensities and with different spectral distributions have altered our EM background. Researchers continue to look at the effects of RF exposure over a wide range of frequencies and levels.

Both RF and power frequency fields are classified as *nonionizing radiation* because the frequency is too low for there to be enough photon energy to ionize atoms. *Ionizing radiation*, such as X-rays, gamma rays and some ultraviolet radiation, has enough energy to knock electrons loose from atoms. When this happens, positive and negative *ions* are formed. Still, at sufficiently high power densities, nonionizing EMF poses certain health hazards.

It has been known since the early days of

radio that RF energy can cause injuries by heating body tissue. Anyone who has ever touched an improperly grounded radio chassis or energized antenna and received an *RF burn* will agree that this type of injury can be quite painful. Excessive RF heating of the male reproductive organs can cause sterility by damaging sperm. Other health problems also can result from RF heating. These heat related health hazards are called *thermal effects*. A microwave oven is an application that puts thermal effects to practical use.

There also have been observations of changes in physiological function in the presence of RF energy levels that are too low to cause heating. These functions generally return to normal when the field is removed. Although research is ongoing, no harmful health consequences have been linked to these changes.

In addition to the ongoing research, much else has been done to address this issue. For example, FCC regulations set limits on exposure from radio transmitters. The Institute

of Electrical and Electronics Engineers, the American National Standards Institute and the National Council for Radiation Protection and Measurement, among others, have recommended voluntary guidelines to limit human exposure to RF energy. The ARRL maintains an RF Safety Committee, consisting of concerned scientists and medical doctors, who volunteer to serve the radio amateur community to monitor scientific research and to recommend safe practices.

The ARRL RF Safety Committee

Imagine you wake up one day and the newspaper headlines are screaming that scientists have discovered radio waves cause cancer. How would you react? How would your neighbor react? You may not have to imagine very hard because the news has been inundated with this type of story regularly over the past couple of decades. Clearly our society has not been decimated by epidemics of diseases since the vast increase in cellular telephone use. Some people deal with this discrepancy by ignoring all scientific reports. Others adopt a pessimistic attitude that technology is going to kill us all eventually, while still others treat every such story as "the truth" and militantly try to stop the transmission of RF energy. The reality is that while all scientific study is complex, the study of electromagnetic biological effects is even more so. Few newspaper reporters are capable of understanding the nuances of a scientific study and are even less able to properly report its results to the lay public. As a result many newspaper stories mislead the public into thinking that a scientific study has found something about which they need to be warned.

The ARRL has dealt with this dilemma by creating the RF Safety Committee, a group of experts in the facets of medical, scientific and engineering investigation needed to fully critique and understand the results of studies on electromagnetic biological effects. Experts in Dosimetry, Public Health, Epidemiology, Statistical Methods, General Medicine and specific diseases are well suited to reading and understanding published scientific reports and critiquing their validity.

It is not uncommon to examine how an experiment was performed only to realize that errors were made in the design of the experiment or the interpretation of its results. It takes a group of reviewers with a wide range of expertise to consider the implications of all aspects of the study to recognize the value of the results.

The field of biological effects of electromagnetic energy constitutes a complex combination of scientific disciplines. Many scientific studies in this field do not generate reliable results because they are not based on input from experts in the many fields that affect the interactions between electromagnetic energy and biological organisms. Even well designed scientific studies are subject to misinterpretation when the results are presented to a public that does not understand or appreciate the complex interactions that occur between the physical world and biological organisms and how these affect public health.

Since the 1960s there have been thousands of scientific studies that were intended to discover if electromagnetic energy had an adverse affect on biological tissue. A large number of these studies, designed and performed by biologists, did not accurately expose the subjects to known levels of electromagnetic energy. A field of expertise in RF engineering, called dosimetry, was developed to accurately determine the exact field strengths of both electrical and

magnetic fields to which subjects were exposed. It has been imperative that an expert in electromagnetic dosimetry be involved in study design, though even today this requirement is often ignored. The RF Safety committee contains expertise in dosimetry that often discovers experimental errors in published results due to misstatements of the amount of exposure that subjects experienced.

Epidemiological studies have the potential to recognize disease trends in populations. However, they can also develop misleading results. Epidemiology looks for health trends among people with similar types of exposures as compared to a similar group of people that does not have the same type of exposure. (This type of study has become difficult to perform with cellular telephones because it is hard to find people who do not use them). The great diversity of the population makes it difficult to know that there is not some other exposure that affects the study group. The RF Safety committee contains expertise in epidemiology to make sense of claims based on epidemiological evidence, and the review of the methods and results can reveal a lesser impact of the study than the author or the press had implied.

Some experimental studies correctly demonstrate biological changes due to exposure to electromagnetic fields. A change in a biological tissue that occurs because of the presence of some form of energy may be an interesting finding, but it does not imply that this change will lead to a public health problem. (An obvious example is contraction of the eye pupil in the presence of bright light, a form of electromagnetic energy). The RF Safety Committee contains expertise in Public Health that helps to determine if there may be a correlation between a laboratory finding and any potential concern for the health of people in our society.

The ARRL RF Safety Committee serves as a resource to the ARRL Board of Directors, providing advice that helps them formulate ARRL policy related to RF safety. The RFSC interacts with the ARRL HQ staff to ensure that RF safety is appropriately addressed in ARRL publications and on the ARRL website. The Amateur Radio community corresponds with the RFSC for help with RF safety-related questions and problems. RFSC members monitor and analyze relevant published research. Its members participate in standards coordinating committees and other expert committees related to RF safety. The RFSC is responsible for writing the RF safety text that is included in ARRL publications. The accuracy of RF safety-related issues in articles submitted to *QST* and *QEX* are confirmed by committee members. The RFSC also participates in developing the RF safety questions for FCC amateur question pools and works with the FCC in developing its environmental regulations. Radio amateurs with questions related to RF safety can contact the RFSC via its liaison, Ed Hare, W1RFI, **w1rfi@arrl.org**. The RFSC maintains a webpage at **www.arrl.org/arrl-rf-safety-committee**.

THERMAL EFFECTS OF RF ENERGY

Body tissues that are subjected to *very high* levels of RF energy may suffer serious heat damage. These effects depend on the frequency of the energy, the power density of the RF field that strikes the body and factors such as the polarization of the wave and the grounding of the body.

At frequencies near the body's natural resonances RF energy is absorbed more efficiently. In adults, the primary resonance frequency is usually about 35 MHz if the person is grounded, and about 70 MHz if insulated from the ground. Various body parts are resonant at different frequencies. Body size thus determines the frequency at which most RF energy is absorbed. As the frequency is moved farther from resonance, RF energy absorption becomes less efficient. *Specific absorption rate (SAR)* is a measure that takes variables such as resonance into account to describe the rate at which RF energy is absorbed in tissue, typically measured in watts per kilogram of tissue (W/kg).

Maximum permissible exposure (MPE) limits define the maximum electric and magnetic field strengths, and the plane-wave equivalent power densities associated with these fields, that a person may be exposed to without harmful effect, and are based on whole-body SAR safety levels. The safe exposure limits vary with frequency as the efficiency of absorption changes. The MPE limits Safety factors are included to insure that the MPE field strength will never result in an unsafe SAR.

Thermal effects of RF energy are usually not a major concern for most radio amateurs because the power levels normally used tend to be low and the intermittent nature of most amateur transmissions decreases total exposure. Amateurs spend more time listening than transmitting and many amateur transmissions such as CW and SSB use low-duty-cycle modes. With FM or RTTY, though, the RF is present continuously at its maximum level during each transmission. It is rare for radio amateurs to be subjected to RF fields strong enough to produce thermal effects, unless they are close to an energized antenna or unshielded power amplifier. Specific suggestions for avoiding excessive exposure are offered later in this chapter.

ATHERMAL EFFECTS OF EMF

Biological effects resulting from exposure to power levels of RF energy that do not generate measurable heat are called *athermal effects*. A number of athermal effects of EMF exposure on biological tissue have been seen in the laboratory. However, to date all athermal effects that have been discovered have had the same features: They are transitory, or go away when the EMF exposure is removed, and they have not been associated with any negative health effects.

28.3.2 Researching Biological Effects of EMF Exposure

The statistical basis of scientific research that confuses many non-scientists is the inability of science to state unequivocally that EMF is safe. Effects are studied by scientists using statistical inference where the "null hypothesis" assumes there is no effect and then tries to disprove this assumption by proving an "alternative hypothesis" that there is an effect. The alternative hypothesis can never be entirely disproved because a scientist cannot examine every possible case, so scientists only end up with a *probability* that the alternative hypothesis is *not* true. Thus, to be entirely truthful, a scientist can never say that something was proven; with respect to low-level EMF exposure, no scientist can guarantee that it is absolutely safe. At best, science can only state that there is a very low probability that it is unsafe. While scientists accept this truism, many members of the general public who are suspicious of EMF and its effects on humans see this as a reason to continue to be afraid.

There are two types of scientific study that are used to learn about the effects of EMF exposure on mammalian biology: laboratory and epidemiological.

LABORATORY STUDY

Scientists conduct laboratory research using animals to learn about biological mechanisms by which EMF may affect mammals. The main advantage of laboratory studies on the biological effects of EMF is that the exposures can be controlled very accurately.

Some major disadvantages of laboratory study also exist. EMF exposure may not affect the species of animals used in the investigations the same way that humans may respond. A common example of this misdirection occurred with eye research. Rabbits had been used for many years to determine that exposure of the eyes to high levels of EMF could cause cataracts. The extrapolation of these results to humans led to the fear that use of radio would harm one's vision. However, the rabbit's eye is on the surface of its skull while the human eye is buried deep within the bony orbit in the skull. Thus, the human eye receives much less exposure from EMF and is less likely to be damaged by the same exposures that had been used in the laboratory experiments on rabbits.

Some biological processes that affect tissue can take many years to occur and laboratory experiments on animals tend to be of shorter duration, in part because the life spans of most animals are much shorter than that of humans. For instance, a typical laboratory rat can be studied at most for two years, during which it progresses from youth to old age with all of the attendant physiological changes that come from normal aging. A disease process that takes multiple exposures over many years to occur is unlikely to be seen in a laboratory study with small animals.

EPIDEMIOLOGICAL RESEARCH

Epidemiologists look at the health patterns of large groups of people using statistical methods. In contrast to laboratory research, epidemiological research has very poor control of its subjects' exposures to EMF but it has the advantages of being able to analyze the effects of a lifetime of exposure and of being able to average out variations among large populations of subjects. By their basic design, epidemiological studies do not demonstrate cause and effect, nor do they postulate mechanisms of disease. Instead, epidemiologists look for associations between an environmental factor and an observed pattern of illness. Apparent associations are often seen in small preliminary studies that later are shown to have been incorrect. At best, such results are used to motivate more detailed epidemiological studies and laboratory studies that narrow down the search for cause-and-effect.

Some preliminary studies have suggested a weak association between exposure to EMF at home or at work and various malignant conditions including leukemia and brain cancer. A larger number of equally well-designed and performed studies, however, have found no association. Risk ratios as high as 2 have been observed in some studies. This means that the number of observed cases of disease in the test group is up to 2 times the "expected" number in the population. Epidemiologists generally regard a risk ratio of 4 or greater to be indicative of a strong association between the cause and effect under study. For example, men who smoke one pack of cigarettes per day increase their risk for lung cancer tenfold compared to nonsmokers and two packs per day increases the risk to more than 25 times the nonsmokers' risk.

Epidemiological research by itself is rarely conclusive, however. Epidemiology only identifies health patterns in groups — it does not ordinarily determine their cause. There are often confounding factors. Most of us are exposed to many different environmental hazards that may affect our health in various ways. Moreover, not all studies of persons likely to be exposed to high levels of EMF have yielded the same results (see sidebar on preliminary epidemiological studies).

28.3.3 Safe Exposure Levels

How much EMF energy is safe? Scientists and regulators have devoted a great deal of ef-

Preliminary Epidemiology

Just about every week you can pick up the newspaper and see a screaming banner headline such as: "Scientists Discover Link Between Radio Waves and Disease." So why are you still operating your ham radio? You've experienced the inconsistency in epidemiological study of diseases. This is something that every radio amateur should understand in order to know how to interpret the real meaning of the science behind the headlines and to help assuage the fears that these stories elicit in others.

Just knowing that someone who uses a radio gets a disease, such as cancer, doesn't tell us anything about the cause-and-effect of that disease. People came down with cancer, and most other diseases, long before radio existed. What epidemiologists try to identify is a group of people who all have a common exposure to something and all suffer from a particular disease in higher proportion than would be expected if they were not exposed. This technique has been highly effective in helping health officials notice excesses of disease due to things such as poisoning of water supplies by local industry and even massive exposures such as smoking. However, epidemiology rarely proves that an exposure causes a disease; rather it provides the evidence that leads to further study.

While the strength of epidemiology is that it helps scientists notice anomalies in entire populations, its weakness is that it is non-specific. An initial epidemiological study examines only two things: suspected exposures and rates of diseases. These studies are relatively simple and inexpensive to perform and may point to an apparent association that then bears further study. For instance, in one study of the causes of death of a selection of Amateur Radio operators, an excess of leukemia was suggested. The percentage of ham radio operators who died of leukemia in that study was higher than expected based on the percentage of the rest of the population that died of leukemia. By itself, this has little meaning and should not be a cause for concern, since the study did not consider anything else about the sample population except that they had ham licenses. Many other questions arise: Were the study subjects exposed to any unusual chemicals? Did any of the study subjects have a family history of leukemia? Did the licensed amateurs even operate radios, what kind and how often? To an epidemiologist, this result might provide enough impetus to raise the funds to gather more specific information about each subject and perform a more complete study that strengthens the apparent associations. However, a slight excess of disease in a preliminary study rarely leads to further study. Commonly, an epidemiologist does not consider a preliminary study to be worth pursuing unless the ratio of excess disease, also called the risk ratio, is 4:1 or greater. Unfortunately, most news reporters are not epidemiologists and do not understand this distinction. Rather, a slight excess of disease in a preliminary study can lead to banner headlines that raise fear in the society, causing unreasonable resistance to things like cell phones and ham radios.

Headlines that blow the results of preliminary epidemiological studies out of proportion are rarely followed by retractions that are as visible if the study is followed up by one that is more complete and shows no association with disease. In the case of the aforementioned epidemiological study of hams' licensing and death records, overblown publicity about the results has led to the urban legend that ham radio operators are likely to come down with leukemia. Not only is this an unfounded conclusion due to the preliminary nature of the original study, but a similar study was recently performed by the National Cancer Institute using a far larger number of subjects and no significant excess of any disease was found. Hams should be able to recognize when sensationalistic headlines are based on inconclusive science and should be prepared to explain to their families, friends and neighbors just how inconclusive such results are.

fort to deciding upon safe RF-exposure limits. This is a very complex problem, involving difficult public health and economic considerations. The recommended safe levels have been revised downward several times over the years — and not all scientific bodies agree on this question even today. The latest Institute of Electrical and Electronics Engineers (IEEE) C95.1 standard for recommended radio frequency exposure limits was published in 2006, updating one that had previously been published in 1991 and adopted by the American National Standards Institute (ANSI) in 1992. In the new standard changes were made

to better reflect the current research, especially related to the safety of cellular telephones. At some frequencies the new standard determined that higher levels of exposure than previously thought are safe (see sidebar, "Where Do RF Safety Standards Come From?").

The IEEE C95.1 standard recommends frequency-dependent and time-dependent maximum permissible exposure levels. Unlike earlier versions of the standard, the 1991 and 2006 standards set different RF exposure limits in *controlled environments* (where energy levels can be accurately determined and everyone on the premises is aware of the

presence of EM fields) and in *uncontrolled environments* (where energy levels are not known or where people may not be aware of the presence of EM fields). FCC regulations adopted these concepts to include controlled/occupational and uncontrolled/general population exposure limits.

The graph in **Fig 28.22** depicts the 1991 IEEE standard (which is still used as the basis of FCC regulation). It is necessarily a complex graph, because the standards differ not only for controlled and uncontrolled environments but also for electric (E) fields and magnetic (H) fields. Basically, the lowest E-field exposure limits occur at frequencies between 30 and 300 MHz. The lowest H-field exposure levels occur at 100-300 MHz. The ANSI standard sets the maximum E-field limits between 30 and 300 MHz at a power density of 1 mW/cm^2 (61.4 V/m) in controlled environments — but at one-fifth that level (0.2 mW/cm^2 or 27.5 V/m) in uncontrolled environments. The H-field limit drops to 1 mW/cm^2 (0.163 A/m) at 100-300 MHz in controlled environments and 0.2 mW/cm^2 (0.0728 A/m) in uncontrolled environments. Higher power densities are permitted at frequencies below 30 MHz (below 100 MHz for H fields) and above 300 MHz, based on the concept that the body will not be resonant at those frequencies and will therefore absorb less energy.

In general, the ANSI/IEEE standard requires averaging the power level over time periods ranging from 6 to 30 minutes for power-density calculations, depending on the frequency and other variables. The ANSI/IEEE exposure limits for uncontrolled environments are lower than those for controlled environments, but to compensate for that the standard allows exposure levels in those environments to be averaged over much longer time periods (generally 30 minutes). This long averaging time means that an intermittent RF source (such as an Amateur Radio transmitter) will result in a much lower exposure than a continuous-duty station, with all other parameter being equal. Time averaging is based on the concept that the human body can withstand a greater rate of body heating (and thus, a higher level of RF energy) for a short time.

Another national body in the United States, the National Council for Radiation Protection and Measurement (NCRP), also has adopted recommended exposure guidelines. NCRP urges a limit of 0.2 mW/cm^2 for nonoccupational exposure in the 30- 300 MHz range. The NCRP guideline differs from IEEE in that it takes into account the effects of modulation on an RF carrier.

The FCC MPE regulations are based on a combination of the 1992 ANSI/IEEE standard and 1986 NCRP recommendations. The MPE limits under the regulations are slightly

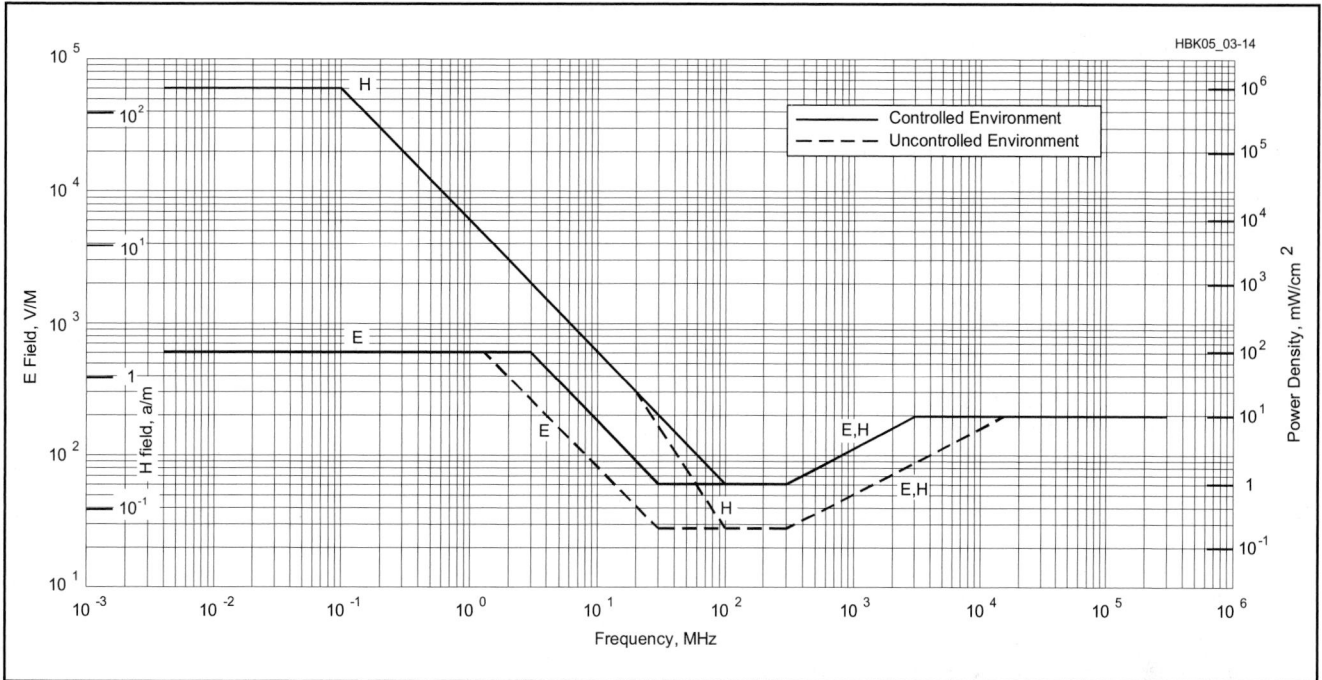

Fig 28.22 — 1991 RF protection guidelines for body exposure of humans. It is known officially as the "IEEE Standard for Safety Levels with Respect to Human Exposure to Radio Frequency Electromagnetic Fields, 3 kHz to 300 GHz."

Where Do RF Safety Standards Come From?

So much of the way we deal with RF Safety is based on "Safety Standards." The FCC environmental exposure regulations that every ham must follow are largely restatements of the conclusions reached by some of the major safety standards. How are these standards developed and why should we trust them?

The preeminent RF safety standard in the world was developed by the Institute of Electrical and Electronics Engineers (IEEE). The most recent edition is entitled *C95.1-2005: IEEE Standard for Safety Levels with Respect to Human Exposure to Radio Frequency Electromagnetic Fields, 3 kHz to 300 GHz.* The IEEE C95.1 Standard has a long history. The first C95.1 RF safety standard was released in 1966, was less than 2 pages long and listed no references. It essentially said that for frequencies between 10 MHz and 100 GHz people should not be exposed to a power density greater than 10 mW/cm². The C95.1 standard was revised in 1974, 1982, 1991 and 2005. The latest (2005) edition of the standard was published in 2006, is 250 pages long and has 1143 references to the scientific literature. Most of the editions of the IEEE C95.1

standard were adopted by the American National Standards Institute (ANSI) a year or two after they were published by IEEE. The 2005 edition was adopted by ANSI in 2006.

The committee at IEEE that developed the latest revision to C95.1 is called International Committee on Electromagnetic Safety Technical Committee 95 Subcommittee 4 and had a large base of participants. The subcommittee was co-chaired by C-K Chou, Ph.D., of Motorola Laboratories, and John D'Andrea, PhD, of the U.S. Naval Health Research Center. The committee had 132 members, 42% of whom were from 23 countries outside the United States. The members of the committee represented academia (27%), government (34%), industry (17%), consultants (20%) and the general public (2%).

Early editions of C95.1 were based on the concept that heat generated in the body should be limited to prevent damage to tissue. Over time the standard evolved to protect against *all known adverse biological effects* regardless of the amount of heat generated. The 2005 revision was based on

the principles that the standard should protect human health yet still be practical to implement, its conclusions should be based solely on scientific evidence and wherever scientifically defensible it should be harmonized with other international RF safety standards. It based its conclusions on 50 years of scientific study. From over 2500 studies on EMF performed during that time, 1300 were selected for their relevance to the health effects of RF exposure. The science in these studies was evaluated for its quality and methodology and 1143 studies were referenced in producing the latest standard.

Other major standards bodies have published similar standards. The National Council for Radiation Protection and Measurement (NCRP) published its safety standard entitled, *Report No. 86: Biological Effects and Exposure Criteria for Radiofrequency Electromagnetic Fields* in 1986. The International Commission on Non-Ionizing Radiation Protection (ICNIRP) published its safety standard entitled *Guidelines for Limiting Exposure to Time-Varying Electric, Magnetic, and Electromagnetic Fields (Up to 300 GHz)* in 1998.

FCC RF Exposure Regulations

FCC regulations control the amount of RF exposure that can result from your station's operation (§§97.13, 97.503, 1.1307 (b)(c)(d), 1.1310, 2.1091 and 2.1093). The regulations set limits on the maximum permissible exposure (MPE) allowed from operation of transmitters in all radio services. They also require that certain types of stations be evaluated to determine if they are in compliance with the MPEs specified in the rules. The FCC has also required that questions on RF environmental safety practices be added to Technician and General license examinations.

THE RULES

Maximum Permissible Exposure (MPE)

All radio stations regulated by the FCC must comply with the requirements for MPEs, even QRP stations running only a few watts or less. The MPEs vary with frequency, as shown in **Table A**. MPE limits are specified in maximum electric and magnetic fields for frequencies below 30 MHz, in power density for frequencies above 300 MHz and all three ways for frequencies from 30 to 300 MHz. For compliance purposes, all of these limits must be considered *separately*. If any one is exceeded, the station is not in compliance. In effect, this means that both electric and magnetic field must be determined below 300 MHz but at higher frequencies determining either the electric or magnetic field is normally sufficient.

The regulations control human exposure to RF fields, not the strength of RF fields in any space. There is no limit to how strong a field can be as long as no one is being exposed to it, although FCC regulations require that amateurs use the minimum necessary power at all times (§97.311 [a]).

Table A

(From §1.1310) Limits for Maximum Permissible Exposure (MPE)

(A) Limits for Occupational/Controlled Exposure

Frequency Range (MHz)	Electric Field Strength (V/m)	Magnetic Field Strength (A/m)	Power Density (mW/cm^2)	Averaging Time (minutes)
0.3-3.0	614	1.63	(100)*	6
3.0-30	1842/f	4.89/f	(900/f^2)*	6
30-300	61.4	0.163	1.0	6
300-1500	—	—	f/300	6
1500-100,000	—	—	5	6

f = frequency in MHz
* = Plane-wave equivalent power density (see Notes 1 and 2).

(B) Limits for General Population/Uncontrolled Exposure

Frequency Range (MHz)	Electric Field Strength (V/m)	Magnetic Field Strength (A/m)	Power Density (mW/cm^2)	Averaging Time (minutes)
0.3-1.34	614	1.63	(100)*	30
1.34-30	824/f	2.19/f	(180/f^2)*	30
30-300	27.5	0.073	0.2	30
300-1500	—	—	f/1500	30
1500-100,000	—	—	1.0	30

f = frequency in MHz
* = Plane-wave equivalent power density (see Notes 1 and 2).

Note 1: This means the equivalent far-field strength that would have the E or H-field component calculated or measured. It does not apply well in the near field of an antenna. The equivalent far-field power density can be found in the near or far field regions from the relationships: $P_d = |E_{total}|^2 / 3770$ mW/cm^2 or from $P_d = |H_{total}|^2 \times 37.7$ mW/cm^2.

Note 2: $|E_{total}|^2 = |E_x|^2 + |E_y|^2 + |E_z|^2$, and $|H_{total}|^2 = |H_x|^2 + |H_y|^2 + |H_z|^2$

Environments

The FCC has defined two tiers of exposure limits — *occupational/controlled limits* and *general population/uncontrolled limits*. Occupational/controlled limits apply when people are exposed as a condition of their employment and when they are aware of that exposure and can take steps to minimize it, if appropriate. General population/ uncontrolled limits apply to exposure of the general public or people who are not normally aware of the exposure or can-

not exercise control over it. The limits for general population/uncontrolled exposure are more stringent than the limits for occupational/controlled exposure. Specific definitions of the exposure categories can be found in Section 1.1310 of the FCC rules.

Although occupational/controlled limits are usually applicable in a workplace environment, the FCC has determined that they generally apply to amateur operators and members of their immediate households. In most cases, occupational/

---

different than the ANSI/IEEE limits and do not reflect all the assumptions and exclusions of the ANSI/IEEE standard.

28.3.4 Cardiac Pacemakers and RF Safety

It is a widely held belief that cardiac pacemakers may be adversely affected in their function by exposure to electromagnetic fields. Amateurs with pacemakers may ask whether their operating might endanger themselves or visitors to their shacks who have a pacemaker. Because of this, and similar concerns regarding other sources of EM fields, pacemaker manufacturers apply design methods that for the most part shield the pacemaker circuitry from

even relatively high EM field strengths.

It is recommended that any amateur who has a pacemaker, or is being considered for one, discuss this matter with his or her physician. The physician will probably put the amateur into contact with the technical representative of the pacemaker manufacturer. These representatives are generally excellent resources, and may have data from laboratory or "in the field" studies with specific model pacemakers.

One study examined the function of a modern (dual chamber) pacemaker in and around an Amateur Radio station. The pacemaker generator has circuits that receive and process electrical signals produced by the heart, and also generate electrical signals that stimulate (pace) the heart. In one series of experiments, the pacemaker was connected to a heart simu-

lator. The system was placed on top of the cabinet of a 1-kW HF linear amplifier during SSB and CW operation. In another test, the system was placed in close proximity to several 1 to 5-W 2-meter hand-held transceivers. The test pacemaker was connected to the heart simulator in a third test, and then placed on the ground 9 meters below and 5 meters in front of a three-element Yagi HF antenna. No interference with pacemaker function was observed in these experiments.

Although the possibility of interference cannot be entirely ruled out by these few observations, these tests represent more severe exposure to EM fields than would ordinarily be encountered by an amateur — with an average amount of common sense. Of course prudence dictates that amateurs with pace-

controlled limits can be applied to your home and property to which you can control physical access. The general population/uncontrolled limits are intended for areas that are accessible by the general public, such as your neighbors' properties.

The MPE levels are based on average exposure. An averaging time of 6 minutes is used for occupational/controlled exposure; an averaging period of 30 minutes is used for general population/ uncontrolled exposure.

Station Evaluations

The FCC requires that certain amateur stations be evaluated for compliance with the MPEs. Although an amateur can have someone else do the evaluation, it is not difficult for hams to evaluate their own stations. The ARRL book *RF Exposure and You* contains extensive information about the regulations and a large chapter of tables that show compliance distances for specific antennas and power levels. Generally, hams will use these tables to evaluate their stations. Some of these tables have been included in the FCC's information — *OET Bulletin 65* and its *Supplement B* (available for downloading at the FCC's RF Safety website). If hams choose, however, they can do more extensive calculations, use a computer to model their antenna and exposure, or make actual measurements.

Categorical Exemptions

Some types of amateur stations do not need to be evaluated, but these stations must still comply with the MPE limits. The station licensee remains responsible for ensuring that the station meets these requirements.

The FCC has exempted these stations from the evaluation requirement because their output power, operating mode and frequency are such that they are presumed to be in compliance with the rules.

Stations using power equal to or less than the levels in **Table B** do not have to be evaluated on a routine basis. For the 100-W HF ham station, for example, an evaluation would be required only on 12 and 10 meters.

Hand-held radios and vehicle-mounted mobile radios that operate using a push-to-talk (PTT) button are also categorically exempt from performing the routine evaluation.

Repeater stations that use less than 500 W ERP or those with antennas not mounted on buildings; if the antenna is at least 10 meters off the ground, also do not need to be evaluated.

Correcting Problems

Most hams are already in compliance with the MPE requirements. Some amateurs, especially those using indoor antennas or high-power, high-duty-cycle modes such as a RTTY bulletin station and specialized stations for moon bounce operations and the like may need to make adjustments to their station or operation to be in compliance.

The FCC permits amateurs considerable flexibility in complying with these regulations. As an example, hams can adjust their operating frequency, mode or power to comply with the MPE limits. They can also adjust their operating habits or control the direction their antenna is pointing.

More Information

This discussion offers only an overview of this topic; additional information can be found in *RF Exposure and You* and on the ARRL website at **www.arrl.org/rf-exposure**. The ARRL website has links to the FCC website, with OET Bulletin 65 and Supplement B and links to software that hams can use to evaluate their stations.

Table B
Power Thresholds for Routine Evaluation of Amateur Radio Stations

Wavelength Band	Evaluation Required if Power* (watts) Exceeds:
MF	
160 m	500
HF	
80 m	500
75 m	500
40 m	500
30 m	425
20 m	225
17 m	125
15 m	100
12 m	75
10 m	50
VHF (all bands)	50
UHF	
70 cm	70
33 cm	150
23 cm	200
13 cm	250
SHF (all bands)	250
EHF (all bands)	250
Repeater stations (all bands)	Non-building-mounted antennas: height above ground level to lowest point of antenna < 10 m *and* power > 500 W ERP
	Building-mounted antennas: power > 500 W ERP

*Transmitter power = Peak-envelope power input to antenna. For repeater stations only, power exclusion based on ERP (effective radiated power).

---

makers, who use handheld VHF transceivers, keep the antenna as far as possible from the site of the implanted pacemaker generator. They also should use the lowest transmitter output required for adequate communication. For high power HF transmission, the antenna should be as far as possible from the operating position, and all equipment should be properly grounded.

28.3.5 Low-Frequency Fields

There has been considerable laboratory research about the biological effects of power line EMF. For example, some separate studies have indicated that even fairly low levels of EMF exposure might alter the human body's circadian rhythms, affect the manner in which T lymphocytes function in the immune system and alter the nature of the electrical and chemical signals communicated through the cell membrane and between cells, among other things. Although these studies are intriguing, they do not demonstrate any effect of these low-level fields on the overall organism.

Much of this research has focused on low-frequency magnetic fields, or on RF fields that are keyed, pulsed or modulated at a low audio frequency (often below 100 Hz). Several studies suggested that humans and animals could adapt to the presence of a steady RF carrier more readily than to an intermittent, keyed or modulated energy source.

The results of studies in this area, plus speculations concerning the effect of various types of modulation, were and have remained somewhat controversial. None of the research to date has demonstrated that low-level EMF causes adverse health effects.

Given the fact that there is a great deal of ongoing research to examine the health consequences of exposure to EMF, the American Physical Society (a national group of highly respected scientists) issued a statement in May 1995 based on its review of available data pertaining to the possible connections of cancer to 60-Hz EMF exposure. Their report is exhaustive and should be reviewed by anyone with a serious interest in the field. Among its general conclusions are the following:

1. The scientific literature and the reports of reviews by other panels show no consistent, significant link between cancer and power line fields.

2. No plausible biophysical mechanisms for the systematic initiation or promotion of cancer by these extremely weak 60-Hz fields have been identified.

3. While it is impossible to prove that no deleterious health effects occur from exposure to any environmental factor, it is necessary to demonstrate a consistent, significant, and causal relationship before one can conclude that such effects do occur.

In a report dated October 31, 1996, a committee of the National Research Council of the National Academy of Sciences has concluded that no clear, convincing evidence exists to show that residential exposures to electric and magnetic fields (EMF) are a threat to human health.

A National Cancer Institute epidemiological study of residential exposure to magnetic fields and acute lymphoblastic leukemia in children was published in the *New England Journal of Medicine* in July 1997. The exhaustive, seven-year study concludes that if there is any link at all, it is far too weak to be of concern.

In 1998, the US National Institute on Environmental Health Sciences organized a working group of experts to summarize the research on power-line EMF. The committee used the classification rules of the International Agency for Research on Cancer (IARC) and performed a meta-analysis to combine all past results as if they had been performed in a single study. The NIEHS working group concluded that the research did not show this type of exposure to be a carcinogen but could not rule out the possibility either. Therefore, they defined power-line EMF to be a Class 2b carcinogen under the IARC classification. The definition, as stated by the IARC is: "Group 2B: The agent is possibly carcinogenic to humans. There is limited pidemiological evidence plus limited or inadequate animal evidence." Other IARC Class 2b carcinogens include automobile exhaust, chloroform, coffee, ceramic and glass fibers, gasoline and pickled vegetables.

Readers may want to follow this topic as further studies are reported. Amateurs should be aware that exposure to RF and ELF (60 Hz) electromagnetic fields at all power levels and frequencies has not been fully studied under all circumstances. "Prudent avoidance" of any avoidable EMF is always a good idea. Prudent avoidance doesn't mean that amateurs should be fearful of using their equipment. Most amateur operations are well within the MPE limits. If any risk does exist, it will almost surely fall well down on the list of causes that may be harmful to your health (on the other end of the list from your automobile). It does mean, however, that hams should be aware of the potential for exposure from their stations, and take whatever reasonable steps they can take to minimize their own exposure and the exposure of those around them.

Although the FCC doesn't regulate 60-Hz fields, some recent concern about EMF has focused on 60 Hz. Amateur Radio equipment can be a significant source of 60 Hz fields, although there are many other sources of this kind of energy in the typical home. Magnetic fields can be measured relatively accurately with inexpensive 60-Hz meters that are made by several manufacturers.

Table 28.4 shows typical magnetic field intensities of Amateur Radio equipment and various household items.

28.3.6 Determining RF Power Density

Unfortunately, determining the power density of the RF fields generated by an amateur station is not as simple as measuring low-frequency magnetic fields. Although sophisticated instruments can be used to measure RF power densities quite accurately, they are costly and require frequent recalibration. Most amateurs don't have access to such equipment, and the inexpensive field-strength meters that we do have are not suitable for measuring RF power density.

Table 28.5 shows a sampling of measurements made at Amateur Radio stations by the Federal Communications Commission and the Environmental Protection Agency in 1990. As this table indicates, a good antenna well removed from inhabited areas poses no hazard under any of the ANSI/IEEE guidelines. However, the FCC/EPA survey also indicates that amateurs must be careful about using indoor or attic-mounted antennas, mobile antennas, low directional arrays or any other antenna that is close to inhabited areas, especially when moderate to high power is used.

Ideally, before using any antenna that is in close proximity to an inhabited area, you should measure the RF power density. If that

Table 28.4
Typical 60-Hz Magnetic Fields Near Amateur Radio Equipment and AC-Powered Household Appliances

Values are in milligauss.

Item	Field	Distance
Electric blanket	30-90	Surface
Microwave oven	10-100	Surface
	1-10	12 in.
IBM personal computer	5-10	Atop monitor
	0-1	15 in. from screen
Electric drill	500-2000	At handle
Hair dryer	200-2000	At handle
HF transceiver	10-100	Atop cabinet
	1-5	15 in. from front
1-kW RF amplifier	80-1000	Atop cabinet
	1-25	15 in. from front

(Source: measurements made by members of the ARRL RF Safety Committee)

Table 28.5
Typical RF Field Strengths Near Amateur Radio Antennas

A sampling of values as measured by the Federal Communications Commission and Environmental Protection Agency, 1990

Antenna Type	Freq (MHz)	Power (W)	E Field (V/m)	Location
Dipole in attic	14.15	100	7-100	In home
Discone in attic	146.5	250	10-27	In home
Half sloper	21.5	1000	50	1 m from base
Dipole at 7-13 ft	7.14	120	8-150	1-2 m from earth
Vertical	3.8	800	180	0.5 m from base
5-element Yagi at 60 ft	21.2	1000	10-20	In shack
			14	12 m from base
3-element Yagi at 25 ft	28.5	425	8-12	12 m from base
Inverted V at 22-46 ft	7.23	1400	5-27	Below antenna
Vertical on roof	14.11	140	6-9	In house
			35-100	At antenna tuner
Whip on auto roof	146.5	100	22-75	2 m antenna
			15-30	In vehicle
			90	Rear seat
5-element Yagi at 20 ft	50.1	500	37-50	10 m antenna

is not feasible, the next best option is make the installation as safe as possible by observing the safety suggestions listed in **Table 28.6**.

It also is possible, of course, to calculate the probable power density near an antenna using simple equations. Such calculations have many pitfalls. For one, most of the situations where the power density would be high enough to be of concern are in the near field. In the near field, ground interactions and other variables produce power densities that cannot be determined by simple arithmetic. In the far field, conditions become easier to predict with simple calculations. (See the February 2013 *QST* article "Q and the Energy Stored Around Antennas" by Kai Siwiak, KE4PT and the **Antennas** chapter of this book for more information about stored energy density near antennas.)

The boundary between the near field and the far field depends on the wavelength of the transmitted signal and the physical size and configuration of the antenna. The boundary between the near field and the far field of an antenna can be as much as several wavelengths from the antenna.

Computer antenna-modeling programs are another approach you can use. *MININEC* or other codes derived from *NEC* (Numerical Electromagnetics Code) are suitable for estimating RF magnetic and electric fields around amateur antenna systems.

These models have limitations. Ground interactions must be considered in estimating near-field power densities, and the "correct ground" must be modeled. Computer modeling is generally not sophisticated enough to predict "hot spots" in the near field — places where the field intensity may be far higher than would be expected, due to reflections from nearby objects. In addition, "nearby objects" often change or vary with weather or the season, therefore the model so laboriously crafted may not be representative of the actual situation, by the time it is running on the computer.

Intensely elevated but localized fields often can be detected by professional measuring instruments. These "hot spots" are often found near wiring in the shack, and metal objects such as antenna masts or equipment cabinets. But even with the best instrumentation, these measurements also may be misleading in the near field. One need not make precise measurements or model the exact antenna system, however, to develop some idea of the relative fields around an antenna. Computer modeling using close approximations of the geometry and power input of the antenna will generally suffice. Those who are familiar with *MININEC* can estimate their power densities by computer modeling, and those who have access to professional power-density meters can make useful measurements.

While our primary concern is ordinarily the intensity of the signal radiated by an antenna, we also should remember that there are other potential energy sources to be considered. You also can be exposed to excessive RF fields directly from a power amplifier if it is operated without proper shielding. Transmission lines also may radiate a significant amount of energy under some conditions. Poor microwave waveguide joints or improperly assembled connectors are another source of incidental exposure.

28.3.7 Further RF Exposure Suggestions

Potential exposure situations should be taken seriously. Based on the FCC/EPA measurements and other data, the "RF awareness" guidelines of Table 28.6 were developed by the ARRL RF Safety Committee. A longer version of these guidelines, along with a complete list of references, appeared in a *QST* article by Ivan Shulman, MD, WC2S ("Is Amateur Radio Hazardous to Our Health?" *QST*, Oct 1989, pp 31-34).

In addition, the ARRL has published a book, *RF Exposure and You* that helps hams comply with the FCC's RF-exposure regulations. The ARRL also maintains an RF-exposure news page on its website. See **www.arrl.org/rf-exposure**. This site contains reprints of selected *QST* articles on RF exposure and links to the FCC and other useful sites.

SUMMARY

The ideas presented in this chapter are intended to reinforce the concept that ham radio, like many other activities in modern life, does have certain risks. But by understanding the hazards and how to deal effectively with them, the risk can be minimized. Common-sense measures can go a long way to help us prevent accidents. Traditionally, amateurs are inventors, and experimenting is a major part of our nature. But reckless chance-taking is never wise, especially when our health and well-being is involved. A healthy attitude toward doing things the right way will help us meet our goals and expectations.

Table 28.6
RF Awareness Guidelines

These guidelines were developed by the ARRL RF Safety Committee, based on the FCC/EPA measurements of Table 28.4 and other data.

• Although antennas on towers (well away from people) pose no exposure problem, make certain that the RF radiation is confined to the antennas' radiating elements themselves. Provide a single, good station ground (earth), and eliminate radiation from transmission lines. Use good coaxial cable or other feed line properly. Avoid serious imbalance in your antenna system and feed line. For high-powered installations, avoid end-fed antennas that come directly into the transmitter area near the operator.

• No person should ever be near any transmitting antenna while it is in use. This is especially true for mobile or ground-mounted vertical antennas. Avoid transmitting with more than 25 W in a VHF mobile installation unless it is possible to first measure the RF fields inside the vehicle. At the 1-kW level, both HF and VHF directional antennas should be at least 35 ft above inhabited areas. Avoid using indoor and attic-mounted antennas if at all possible. If open-wire feeders are used, ensure that it is not possible for people (or animals) to come into accidental contact with the feed line.

• Don't operate high-power amplifiers with the covers removed, especially at VHF/UHF.

• In the UHF/SHF region, never look into the open end of an activated length of waveguide or microwave feed-horn antenna or point it toward anyone. (If you do, you may be exposing your eyes to more than the maximum permissible exposure level of RF radiation.) Never point a high-gain, narrow-bandwidth antenna (a paraboloid, for instance) toward people. Use caution in aiming an EME (moonbounce) array toward the horizon; EME arrays may deliver an effective radiated power of 250,000 W or more.

• With hand-held transceivers, keep the antenna away from your head and use the lowest power possible to maintain communications. Use a separate microphone and hold the rig as far away from you as possible. This will reduce your exposure to the RF energy.

• Don't work on antennas that have RF power applied.

• Don't stand or sit close to a power supply or linear amplifier when the ac power is turned on. Stay at least 24 inches away from power transformers, electrical fans and other sources of high-level 60-Hz magnetic fields.

Contents

Assembling a Station

Although many hams never try to build a major project, such as a transmitter, receiver or amplifier, they do have to assemble the various components into a working station. There are many benefits to be derived from assembling a safe, comfortable, easy-to-operate collection of radio gear, whether the shack is at home, in the car or in a field. This chapter will detail some of the "how tos" of setting up a station for fixed, mobile and portable operation. Such topics as station location, finding adequate power sources, station layout and cable routing are covered. It includes contributions from Wally Blackburn, AA8DX, a section on mobile installations from Alan Applegate, KØBG. and information on gasoline generators from Kirk Kleinschmidt, NTØZ. Rick Hilding, K6VVA, contributed the section on remote stations.

Chapter 29 — CD-ROM Content

Portable Stations
- "A Look at Gasoline Powered Inverter Generators," by Bob Allison, WB1GCM
- "Field Day Towers — Doing it Right" by Ward Silver, NØAX and Don Daso, K4ZA

Remote Stations
- "Remote Station Resources by K6VVA - 2014 Edition" — a list of resources for remote station builders

29.1 Fixed Stations

Regardless of the type of installation you are attempting, good planning greatly increases your chances of success. Take the time to think the project all the way through, consider alternatives, and make rough measurements and sketches during your planning and along the way. You will save headaches and time by avoiding "shortcuts." What might seem to save time now may come back to haunt you with extra work when you could be enjoying your shack.

One of the first considerations should be to determine what type of operating you intend to do. While you do not want to strictly limit your options later, you need to consider what you want to do, how much you have to spend and what room you have to work with. There is a big difference between a casual operating position and a "big gun" contest station, for example.

29.1.1 Selecting a Location

Selecting the right location for your station is the first and perhaps the most important step in assembling a safe, comfortable, convenient station. The exact location will depend on the type of home you have and how much space can be devoted to your station. Fortunate amateurs will have a spare room to devote to housing the station; some may even have a separate building for their exclusive use. Most must make do with a spot in the cellar or attic, or a corner of the living room is pressed into service.

Examine the possibilities from several angles. A station should be comfortable; odds are good that you'll be spending a lot of time there over the years. Some unfinished basements are damp and drafty — not an ideal environment for several hours of leisurely hamming. Attics have their drawbacks, too; they can be stifling during warmer months. If possible, locate your station away from the heavy traffic areas of your home. Operation of your station should not interfere with family life. A night of chasing DX on 80 meters may be exciting to you, but the other members of your household may not share your enthusiasm.

Keep in mind that you must connect your station to the outside world. The location you choose should be convenient to a good power source and an adequate ground. If you use a computer, you may need access to the Internert. There should be a fairly direct route to the outside for running antenna feed lines, rotator control cables and the like.

Although most homes will not have an "ideal" space meeting all requirements, the right location for you will be obvious after you scout around. The amateurs whose stations are depicted in **Figs 29.1** through **29.3** all found the right spot for them. Weigh the trade-offs and decide which features you

Fig 29.1 — Admiral Scott Redd (ret), KØDQ, operates this station at WW1WW in pursuit of top ranking in DX contests. Notice that the equipment on the operating desk is laid out logically and comfortably for long periods "in the chair." [Woody Beckford, WW1WW, photo]

Fig 29.2 — Top RTTY contest operator Don Hill, AA5AU operates with low power from his effective station. RTTY operation emphasizes use of the computer mouse so Don's desk has lots of room for his "mouse hand". [Shay Hill, photo]

Fig 29.3 — Spreading out horizontally, John Sluymer, VE3EJ, has arranged his effective contest and DXing station to keep all of the controls at the same level for easy adjustment. [John Sluymer, VE3EJ, photo]

can do without and which are necessary for your style of operation. If possible pick an area large enough for future expansion.

29.1.2 Station Ground

Grounding is an important factor in overall station safety, as detailed in the **Safety** chapter. An effective ground system is necessary for every amateur station. The mission of the ground system is twofold. First, it reduces the possibility of electrical shock if something in a piece of equipment should fail and the chassis or cabinet becomes "hot." If connected to a properly grounded outlet, a three-wire electrical system grounds the chassis. Much amateur equipment still uses the ungrounded two-wire system, however. A ground system to prevent shock hazards is generally referred to as *dc ground*.

The second job the ground system must perform is to provide a low-impedance path to ground for any stray RF current inside the station. Stray RF can cause equipment to malfunction and contributes to RFI problems. This low-impedance path is usually called *RF ground*. In most stations, dc ground and RF ground are provided by the same system.

GROUND NOISE

Noise in ground systems can affect our sensitive radio equipment. It is usually related to one of three problems:
1) Insufficient ground conductor size
2) Loose ground connections
3) Ground loops

These matters are treated in precise scientific research equipment and certain industrial instruments by attention to certain rules. The ground conductor should be at least as large as the largest conductor in the primary power circuit. Ground conductors should provide a solid connection to both ground and to the equipment being grounded. Liberal use of lock washers and star washers is highly recommended. A loose ground connection is a tremendous source of noise, particularly in a sensitive receiving system.

Ground loops should be avoided at all costs. A short discussion here of what a ground loop is and how to avoid them may lead you down the proper path. A ground loop is formed when more than one ground current is flowing in a single conductor. This commonly occurs when grounds are "daisy-chained" (series linked). The correct way to ground equipment is to bring all ground conductors out radially from a common point to either a good driven earth ground or a cold-water system. If one or more earth grounds are used, they should be bonded back to the service entrance panel. Details appear in the **Safety** chapter.

Ground noise can affect transmitted and received signals. With the low audio levels required to drive amateur transmitters, and the ever-increasing sensitivity of receivers, correct grounding is critical.

29.1.3 Station Power

Amateur Radio stations generally require a 120-V ac power source. The 120-V ac is then converted to the proper ac or dc levels required for the station equipment. RF power amplifiers typically require 240 V ac for best operation.

Power supply theory is covered in the **Power Sources** chapter, and safety issues and station wiring are covered in the **Safety** chapter. If your station is located in a room with electrical outlets, you're in luck. If your station is located in the basement, an attic or another area without a convenient 120-V source, you will have to run a line to your operating position.

SURGE PROTECTION

Typically, the ac power lines provide an adequate, well-regulated source of electrical power for most uses. At the same time, these lines are fraught with power surges that, while harmless to most household equipment, may cause damage to more sensitive devices such as computers or test equipment. A common method of protecting these devices is through the use of surge protectors. More information on these devices and lightning protection is in the **Safety** chapter.

29.1.4 Station Layout

Station layout is largely a matter of personal taste and needs. It will depend mostly on the amount of space available, the equipment involved and the types of operating to be done. With these factors in mind, some basic design considerations apply to all stations.

THE OPERATING TABLE

The operating table may be an office or computer desk, a kitchen table or a custom-made bench. What you use will depend on space, materials at hand and cost. The two most important considerations are height and size of the top. Most commercial desks are about 29 inches above the floor. Computer tables are usually a couple inches lower for a more comfortable keyboard and mouse placement. This is a comfortable height for most adults. Heights much lower or higher than this may cause an awkward operating position.

The dimensions of the top are an important consideration. A deep (36 inches or more) top will allow plenty of room for equipment interconnections along the back, equipment about midway and room for writing or a keyboard and mouse toward the front. The length of the top will depend on the amount of equipment being used. An office or computer desk makes

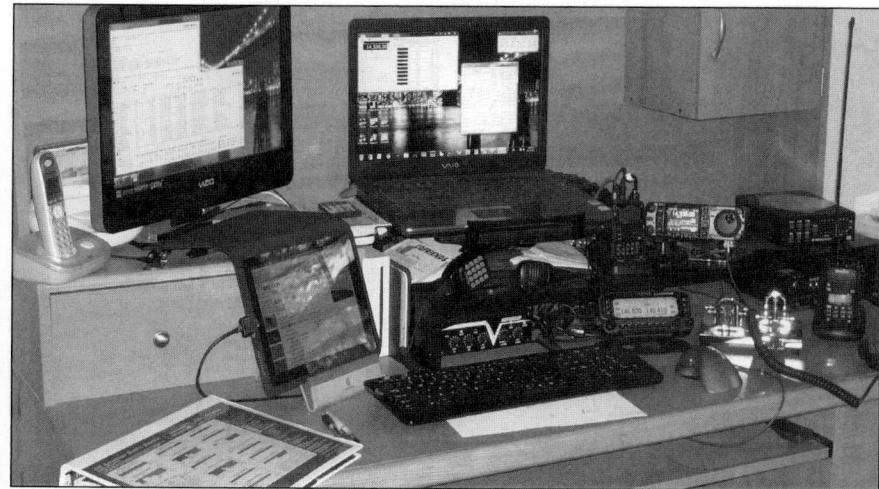

Fig 29.4 — Mike Adams, N1EN makes the most of his desktop to operate on the HF and VHF+ bands. His laptop and tablet computers are an alternative to the larger desktop systems. He uses a full-size keyboard with the laptop. [Mike Adams, N1EN, photo]

a good operating table. These are often about 36 inches deep and 60 inches wide. Drawers can be used for storage of logbooks, headphones, writing materials, and so on. Desks specifically designed for computer use often have built-in shelves that can be used for equipment stacking. Desks of this type are available ready-to-assemble at most discount and home improvement stores. The low price and adaptable design of these desks make them an attractive option for an operating position. An example is shown in **Fig 29.4**.

ARRANGING THE EQUIPMENT

No matter how large your operating table, some vertical stacking of equipment may be necessary to allow you to reach everything from your chair. Stacking pieces of equipment directly on top of one another is not a good idea because most amateur equipment needs airflow around it for cooling. A shelf like that shown in **Fig 29.5** can improve equipment layout in many situations. Dimensions of the shelf can be adjusted to fit the size of your operating table.

When you have acquired the operating table and shelving for your station, the next task is arranging the equipment in a convenient, orderly manner. The first step is to provide power outlets and a good ground as described in a previous section. Be conservative in estimating the number of power outlets

for your installation; radio equipment has a habit of multiplying with time, so plan for the future at the outset.

Fig 29.6 illustrates a sample station layout. The rear of the operating table is spaced about 1½ feet from the wall to allow easy access to the rear of the equipment. This installation incorporates two separate operating positions, one for HF and one for VHF. When the operator is seated at the HF operating position, the keyer and transceiver controls are within easy reach. The keyer, keyer paddle and transceiver are the most-often adjusted pieces of equipment in the station. The speaker is positioned right in front of the operator for the best possible reception. Accessory equipment not often adjusted, including the amplifier, antenna switch and rotator control box, is located on the shelf above the transceiver. The SWR/power meter and clock, often consulted but rarely touched, are located where the operator can view them without head movement. All HF-related equipment can be reached without moving the chair.

This layout assumes that the operator is right-handed. The keyer paddle is operated with the right hand, and the keyer speed and transceiver controls are operated with the left hand. This setup allows the operator to write or send with the right hand without having to cross hands to adjust the controls. If the operator is left-handed, some repositioning of equipment is necessary, but the idea is the same. For best results during CW operation, the paddle should be weighted to keep it from "walking" across the table. It should be oriented such that the operator's entire arm from wrist to elbow rests on the tabletop to prevent fatigue.

Some operators prefer to place the station transceiver on the shelf to leave the table top clear for writing or a computer keyboard and mouse. This arrangement leads to fatigue from having an unsupported arm in the air most of the time. If you rest your elbows on the tabletop, they will quickly become sore. If you rarely operate for prolonged periods, however, you may not be inconvenienced by having the transceiver on the shelf. The real secret to having a clear table top for logging, and so on, is to make the operating table deep enough that your entire arm from elbow to wrist rests on the table with the front panels of the equipment at your fingertips. This leaves plenty of room for paperwork, even with a microphone and keyer paddle on the table.

The VHF operating position in this station is similar to the HF position. The amplifier and power supply are located on the shelf. The station HF beam and VHF beam are on the same tower, so the rotator control box is located where it can be seen and reached from both operating positions. This operator is active on packet radio on a local VHF

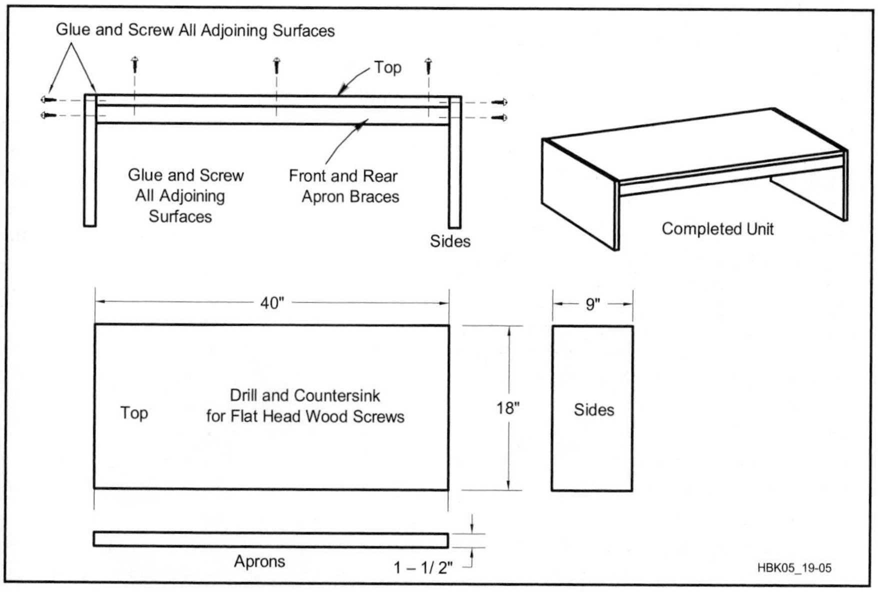

Glue and Screw All Adjoining Surfaces

Top

Glue and Screw All Adjoining Surfaces

Front and Rear Apron Braces

Sides

Completed Unit

40"

9"

Top

Drill and Countersink for Flat Head Wood Screws

18"

Sides

Aprons

1 – 1/ 2"

HBK05_19-05

Fig 29.5 — A simple but strong equipment shelf can be built from readily available materials. Use ¾-inch plywood along with glue and screws for the joints for adequate strength.

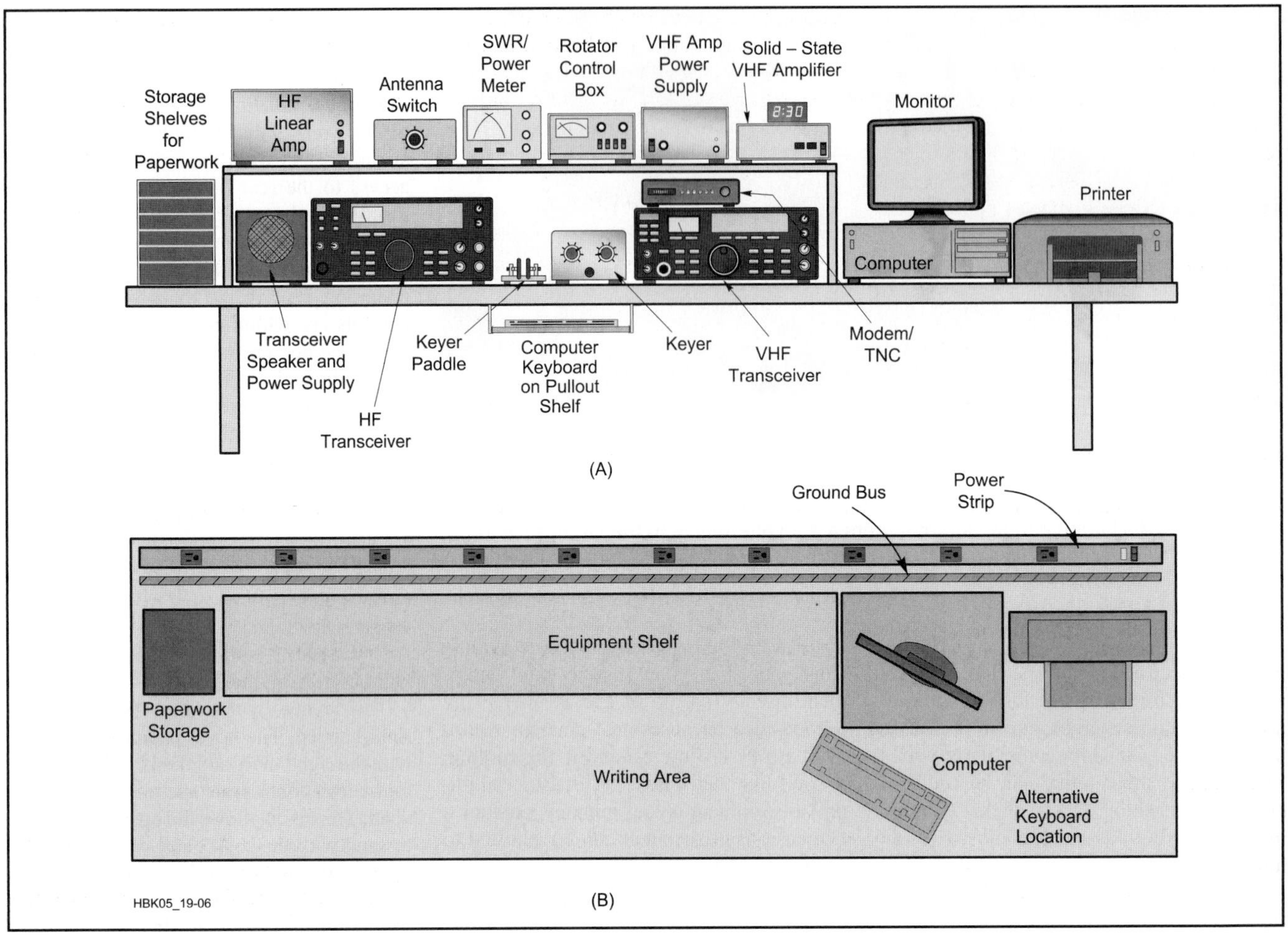

Fig 29.6 — Example station layout as seen from the front (A) and the top (B). The equipment is spaced far enough apart that air circulates on all sides of each cabinet.

repeater, so the computer, printer and terminal node controller (TNC) are all clustered within easy reach of the VHF transceiver.

This sample layout is intended to give you ideas for designing your own station. Study the photos of station layouts presented here, in other chapters of this *Handbook* and in *QST*. Visit the shacks of amateur friends to view their ideas. Station layout is always changing as you acquire new gear, dispose of old gear, change operating habits and interests or become active on different bands. Configure the station to suit your interests, and keep thinking of ways to refine the layout. **Figs 29.7** and **29.8** show station arrangements tailored for specific purposes.

ERGONOMICS

Ergonomics is a term that loosely means "fitting the work to the person." If tools and equipment are designed around what people can accommodate, the results will be much more satisfactory. For example, in the 1930s research was done in telephone equipment manufacturing plants because use of long-nosed pliers for wiring switchboards required

considerable force at the end of the hand's range of motion. A simple tool redesign resolved this issue.

Considerable attention has been focused on ergonomics in recent years because we

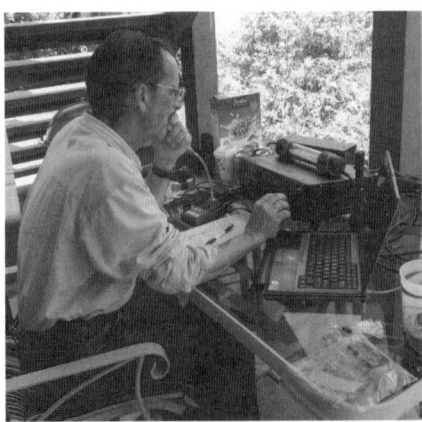

Fig 29.7 — Simple stations work best for portable operating, such as this 2013 Field Day setup for the two-operator NX9L team. [Photo courtesy of Andy Myers, NX9L]

have come to realize that long periods of time spent in unnatural positions can lead to repetitive-motion injuries. Much of this attention has been focused on people whose job tasks have required them to operate computers and other office equipment. While most Amateur Radio operators do not devote as much time to their hobby as they might in a full-time job, it does make sense to consider comfort and flexibility when choosing furniture and arranging it in the shack or workshop.

Fig 29.8 — Richard, WB5DGR, uses a homebrew 1.5-kW amplifier to seek EME contacts from this nicely laid out station.

Adjustable height chairs are available with air cylinders to serve as a shock absorber. Footrests might come in handy if the chair is so high that your feet cannot support your lower leg weight. The height of tables and keyboards often is not adjustable.

Placement of computer screens should take into consideration the reflected light coming from windows. It is always wise to build into your sitting sessions time to walk around and stimulate blood circulation. Your muscles are less likely to stiffen, while the flexibility in your joints can be enhanced by moving around.

Selection of hand tools is another area where there are choices to make that may affect how comfortable you will be while working in your shack. Look for screwdrivers with pliable grips. Take into account how heavy things are before picking them up — your back will thank you.

FIRE EXTINGUISHERS

Fires in well-designed electronic equipment are not common but are known to occur. Proper use of a suitable fire extinguisher can make the difference between a small fire with limited damage and loss of an entire home. Make sure you know the limitations of your extinguisher and the importance of reporting the fire to your local fire department immediately.

Several types of extinguishers are suitable for electrical fires. The multipurpose dry chemical or "ABC" type units are relatively inexpensive and contain a solid powder that is nonconductive. Avoid buying the smallest size; a 5-pound capacity will meet most requirements in the home. ABC extinguishers are also the best choice for kitchen fires (the most common location of home fires). One disadvantage of this type is the residue left behind that might cause corrosion in electrical connectors. Another type of fire extinguisher suitable for energized electrical equipment is the carbon dioxide unit. CO_2 extinguishers

require the user to be much closer to the fire, are heavy and difficult to handle, and are relatively expensive. For obvious reasons, water extinguishers are not suitable for fires in or near electronic equipment.

AIDS FOR HAMS WITH DISABILITIES

A station used by an amateur with physical disabilities or sensory impairments may require adapted equipment or particular layout considerations. The station may be highly customized to meet the operator's needs or just require a bit of "tweaking."

The myriad of individual needs makes describing all of the possible adaptive methods impractical. Each situation must be approached individually, with consideration to the operator's particular needs. However, many types of situations have already been encountered and worked through by others, eliminating the need to start from scratch in every case.

An excellent resource is the Courage Handi-Ham System. The Courage Handi-Ham System, a part of the Courage Center, provides a number of services to hams (and aspiring hams) with disabilities. These include study materials and a wealth of useful information on their comprehensive website. Visit **www.handiham.org** for more information.

29.1.5 Interconnecting Your Equipment

Once you have your equipment and get it arranged, you will have to interconnect it all. No matter how simple the station, you will at least have antenna, power and microphone or key connections. Equipment such as amplifiers, computers, TNCs and so on add complexity. By keeping your equipment interconnections well organized and of high quality, you will avoid problems later on.

Often, ready-made cables will be available. But in many cases you will have to make

your own cables. A big advantage of making your own cables is that you can customize the length. This allows more flexibility in arranging your equipment and avoids unsightly extra cable all over the place. Many manufacturers supply connectors with their equipment along with pinout information in the manual. This allows you to make the necessary cables in the lengths you need for your particular installation.

Always use high quality wire, cables and connectors in your shack. Take your time and make good mechanical and electrical connections on your cable ends. Sloppy cables are often a source of trouble. Often the problems they cause are intermittent and difficult to track down. You can bet that they will crop up right in the middle of a contest or during a rare DX QSO! Even worse, a poor quality connection could cause RFI or even create a fire hazard. A cable with a poor mechanical connection could come loose and short a power supply to ground or apply a voltage where it should not be. Wire and cables should have good quality insulation that is rated high enough to prevent shock hazards.

Interconnections should be neatly bundled and labeled. Wire ties, masking tape or paper labels with string work well. See **Fig 29.9**. Whatever method you use, proper labeling makes disconnecting and reconnecting equipment much easier. **Fig 29.10** illustrates the number of potential interconnections in a modern, full-featured transceiver.

WIRE AND CABLE

The type of wire or cable to use depends on the job at hand. The wire must be of sufficient size to carry the necessary current. Use the tables in the **Component Data and References** chapter to find this information. Never use underrated wire; it will be a fire hazard. Be sure to check the insulation too. For high-voltage applications, the insulation must be rated at least a bit higher than the intended voltage. A good rule of thumb is

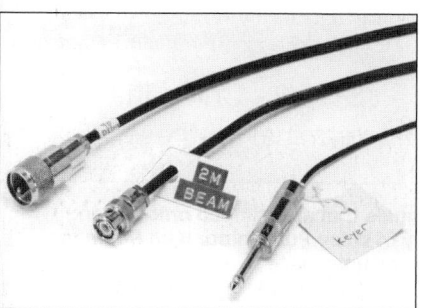

Fig 29.9 — Labels on the cables make it much easier to rearrange things in the station. Labeling ideas include masking tape, cardboard labels attached with string and labels attached to fasteners found on plastic bags (such as bread bags).

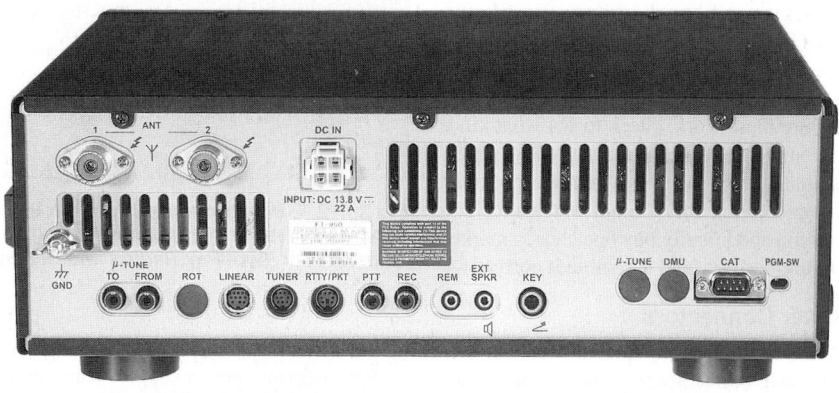

Fig 29.10 — The back of this Yaesu FT-950 transceiver shows some of the many types of connectors encountered in the amateur station. Note that this variety is found on a single piece of equipment.

to use a rating at least twice what is needed.

Use good quality coaxial cable of sufficient size for connecting transmitters, transceivers, antenna switches, antenna tuners and so on. RG-58 might be fine for a short patch between your transceiver and SWR bridge, but is too small to use between your legal-limit amplifier and antenna tuner. For more information, see the **Transmission Lines** chapter.

Hookup wire may be stranded or solid. Generally, stranded is a better choice since it is less prone to break under repeated flexing. Many applications require shielded wire to reduce the chances of RF getting into the equipment. RG-174 is a good choice for control, audio and some low-power RF applications. Shielded microphone or computer cable can be used where more conductors are necessary.

CONNECTORS

Connectors are a convenient way to make an electrical connection by using mating electrical contacts. There are quite a few connector styles, but common terms apply to all of them. Pins are contacts that extend out of the connector body, and connectors in which pins make the electrical contact are called "male" connectors. Sockets are hollow, recessed contacts, and connectors with sockets are called "female." Connectors designed to attach to each other are called "mating connectors." Connectors with specially shaped bodies or inserts that require a complementary shape on a mating connector are called "keyed connectors." Keyed connectors ensure that the connectors can only go together one way, reducing the possibility of damage from incorrect mating.

Plugs are connectors installed on the end of cables and *jacks* are installed on equipment. *Adapters* make connections between two different styles of connector, such as between two different families of RF connectors. Other adapters join connectors of the same family, such as double-male, double-female and gender changers. *Splitters* divide a signal between two connectors.

While the number of different types of connectors is mind-boggling, many manufacturers of amateur equipment use a few standard types. If you are involved in any group activities such as public service or emergency-preparedness work, check to see what kinds of connectors others in the group use and standardize connectors wherever possible. Assume connectors are not waterproof, unless you specifically buy one clearly marked for outdoor use (and assemble it correctly).

Power Connectors

Amateur Radio equipment uses a variety of power connectors. Some examples are shown in **Fig 29.11**. Most low power amateur equipment uses coaxial power connectors. These are the same type found on consumer

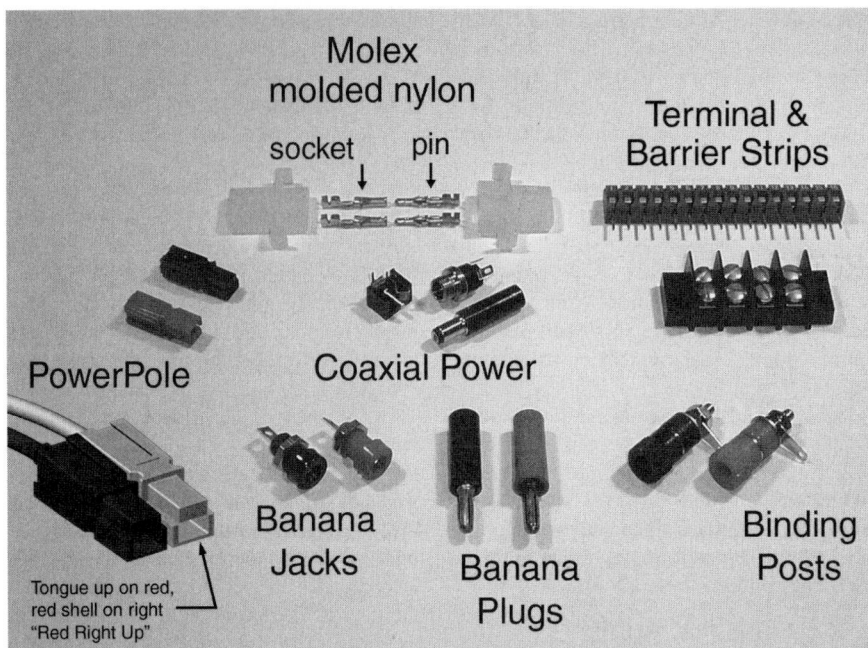

Fig 29.11 — These are the most common connectors used on amateur equipment to make power connections. The proper orientation for paired Powerpole connectors is with the red connector on the right and its tongue on top – "red-right-up". (Courtesy of Wiley Publishing, *Ham Radio for Dummies*, or *Two-Way Radios and Scanners for Dummies*)

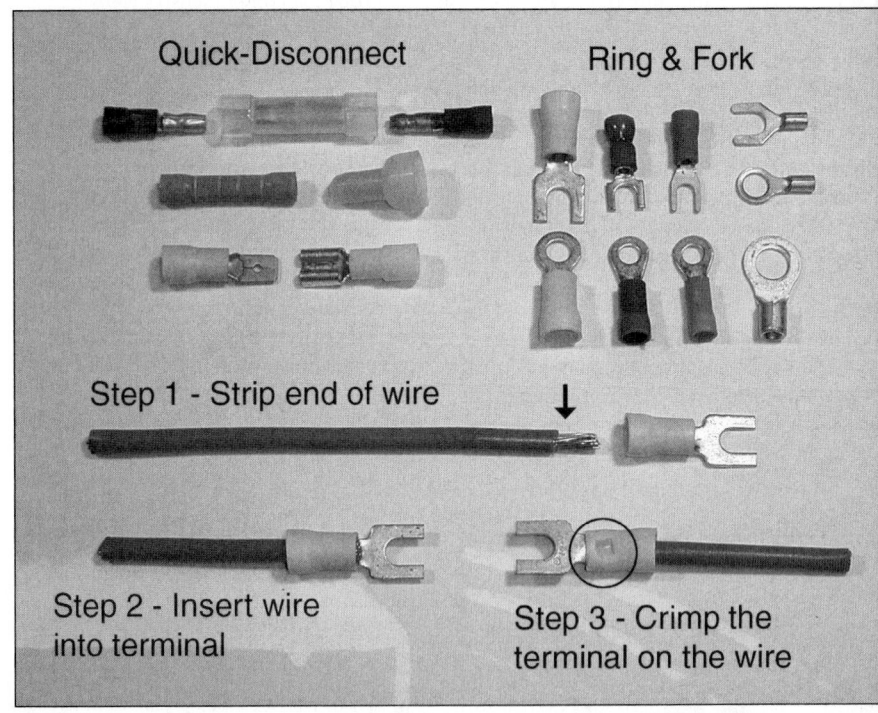

Fig 29.12 — Power connectors often use terminals that are crimped onto the end of wires with special crimping tools. (Courtesy of Wiley Publishing, *Ham Radio for Dummies*, or *Two-Way Radios and Scanners for Dummies*)

electronic equipment that is supplied by a wall transformer power supply. Transceivers and other equipment that requires high current in excess of a few amperes often use Molex connectors (**www.molex.com** — enter "MLX" in the search window) with a white, nylon body housing pins and sockets crimped on to the end of wires.

An emerging standard, particularly among ARES and other emergency communications

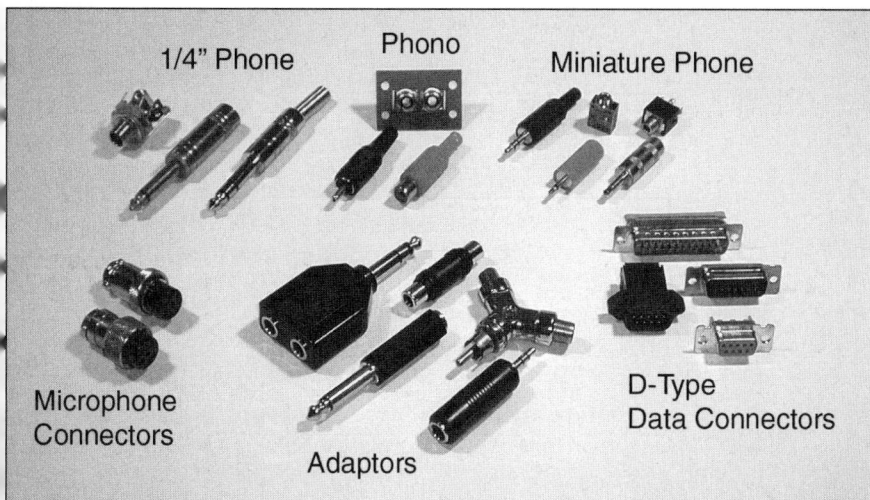

Fig 29.13 — Audio and data signals are carried by a variety of different connectors. Individual cable conductors are either crimped or soldered to the connector contacts. [Ward Silver, NØAX, photo]

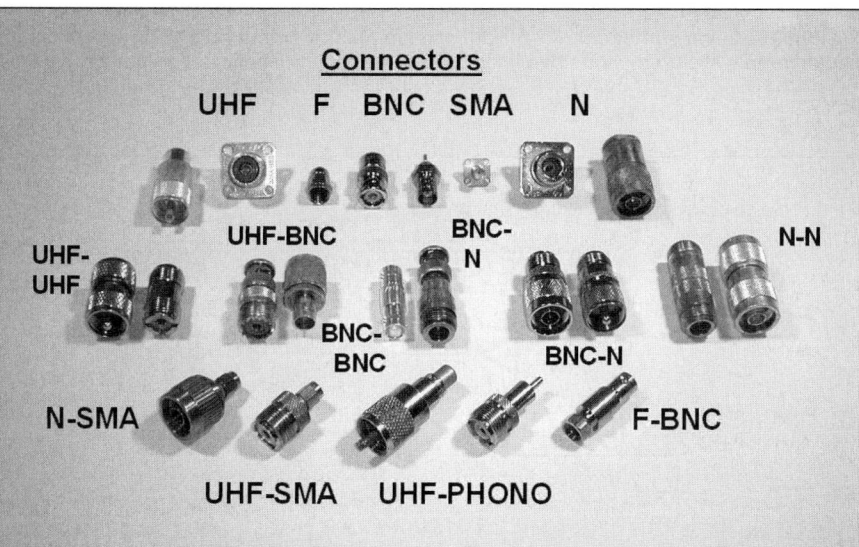

Fig 29.14 — Each type of RF connector is specially made to carry RF signals and preserve the shielding of coaxial cable. Adapters are available to connect one style of connector to another. [Ward Silver, NØAX, photo]

groups, is the use of Anderson Powerpole connectors (**www.andersonpower.com**). These connectors are "sexless" meaning that any two connectors of the same series can be mated — there are no male or female connectors. By standardizing on a single connector style, equipment can be shared and replaced easily in the field. The standard orientation for pairs of these connectors is shown in Fig 29.11. Using this orientation increases the compatibility of your wiring with that of other hams.

Molex and Powerpole connectors use crimp terminals (both male and female) installed on the end of wires. A special crimping tool is used to attach the wire to the terminal and the terminal is then inserted into the body of the connector. Making a solid connection requires the use of an appropriate tool — do not use pliers or some other tool to make a crimp connection.

Some equipment uses terminal strips for direct connection to wires or crimp terminals, often with screws. Other equipment uses spring-loaded terminals or binding posts to connect to bare wire ends. **Fig 29.12** shows some common crimp terminals that are installed on the ends of wires using special tools.

Audio and Control Connectors

Consumer audio equipment and Amateur Radio equipment share many of the same connectors for the same uses. Phone plugs and jacks are used for mono and stereo audio circuits. These connectors, shown in **Fig 29.13** come in ¼ inch, ⅛ inch (miniature) and sub-miniature varieties. The contact at the end of the plug is called the tip and the connector at the base of the plug is the sleeve. If there is a third contact between the tip and sleeve, it is the ring (these are "stereo" phone connectors).

Phono plugs and jacks (sometimes called RCA connectors since they were first used on RCA brand equipment) are used for audio, video and low-level RF signals. They are also widely used for control signals.

The most common microphone connector on mobile and base station equipment is an 8-pin round connector. On older transceivers you may see 4-pin round connectors used for microphones. RJ-45 modular connectors (see the section on telephone connectors below) are often used in mobile and smaller radios.

RF Connectors

Feed lines used for radio signals require special connectors for use at RF frequencies. The connectors must have approximately the same characteristic impedance as the feed line they are attached to or some of the RF signal will be reflected by the connector. Inexpensive audio and control connectors cannot meet that requirement, nor can they handle the high power levels often encountered in RF equipment. Occasionally, phono connectors are used for HF receiving and low-power transmitting equipment.

By far, the most common connector for RF in amateur equipment is the UHF family shown in **Fig 29.14**. (The UHF designator has nothing to do with frequency.) A PL-259 is the plug that goes on the end of feed lines, and the SO-239 is the jack mounted on equipment. A "barrel" (PL-258) is a double-female adapter that allows two feed lines to be connected together. UHF connectors are typically used up to 150 MHz and can handle legal-limit transmitter power at HF.

UHF connectors have several drawbacks including lack of weatherproofing, poor performance above the 2 meter band and limited power handling at higher frequencies. The Type-N series of RF connectors addresses all of those needs. Type-N connectors are somewhat more expensive than UHF connectors, but they require less soldering and perform better in outdoor use since they are moisture resistant. Type-N connectors can be used to 10 GHz.

For low-power uses, BNC connectors are often used. BNC connectors are the standard for laboratory equipment, as well, and they are often used for dc and audio connections. BNC connectors are common on handheld radios for antenna connections. The newest handheld transceivers often use small, screw-on SMA type connectors for their antennas, though.

The type of connector used for a specific job depends on the size of the cable, the frequency of operation and the power levels involved. More information on RF connectors may be found in the **Component Data and References** chapter.

Data Connectors

Digital data is exchanged between computers and pieces of radio equipment more than ever before in the amateur station. The connector styles follow those found on computer equipment.

D-type connectors are used for RS-232 (COM ports) and parallel (LPT port) interfaces. A typical D-type connector has a model number of "DB" followed by the number of connections and a "P" or "S" depending on whether the connector uses pins or sockets. For example, the DB-9P is used for PC COM1 serial ports.

USB connectors are becoming more popular in amateur equipment as the computer industry has eliminated the bulkier and slower RS-232 interface. A number of manufacturers make USB-to-serial converters that allow devices with RS-232 interfaces to be used with computers that only have USB interfaces.

Null modem or *crossover* adapters or cables have the same type of connector on each end. The internal connections between signal pins are swapped between ends so that inputs and outputs are connected together. This allows interfaces to be connected together directly without any intermediary equipment, such as an Ethernet switch or an RS-232 modem.

Pinouts for various common computer connectors are shown in the **Component Data and References** chapter. Several practical data interface projects are shown in the **Station Accessories** chapter and the **Digital Communications** supplement on the *Handbook* CD.

Telephone and Computer Network Connectors

Modular connectors are used for telephone and computer network connections. Connector part numbers begin with "RJ." The connectors are crimped on to multiconductor cables with special tools. The RJ11 connector is used for single- and double-line telephone system connection with 4 or 6 contacts. The RJ10 is a 4-contact connector for telephone handset connections. Ethernet computer network connections are made using RJ45 connectors with 8 contacts.

29.1.6 Documenting Your Station

An often neglected but very important part of putting together your station is properly documenting your work. Ideally, you should diagram your entire station from the ac power lines to the antenna on paper and keep the information in a special notebook with sections for the various facets of your installation.

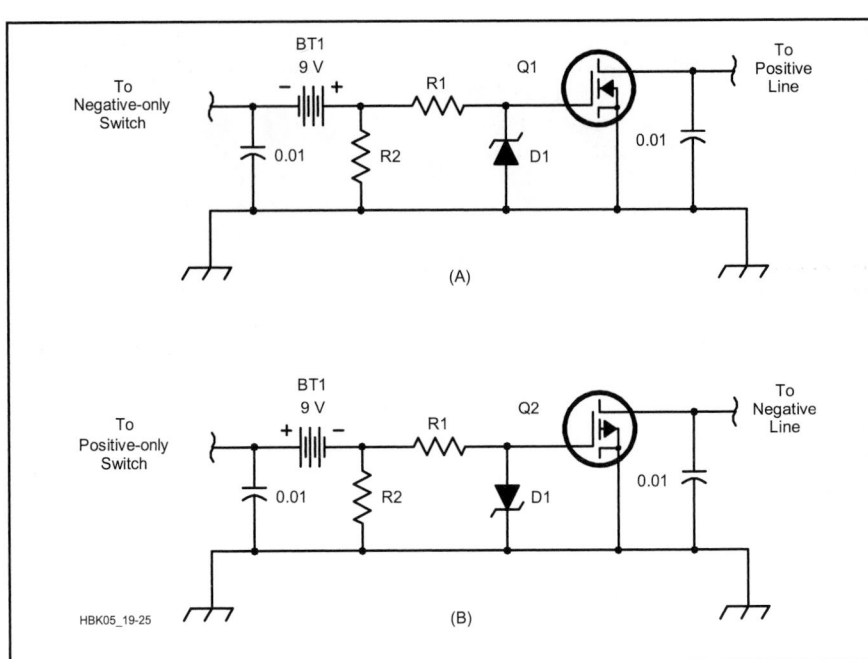

Fig 29.15 — Level-shifter circuits for opposite input and output polarities. At A, from a negative-only switch to a positive line; B, from a positive-only switch to a negative line.

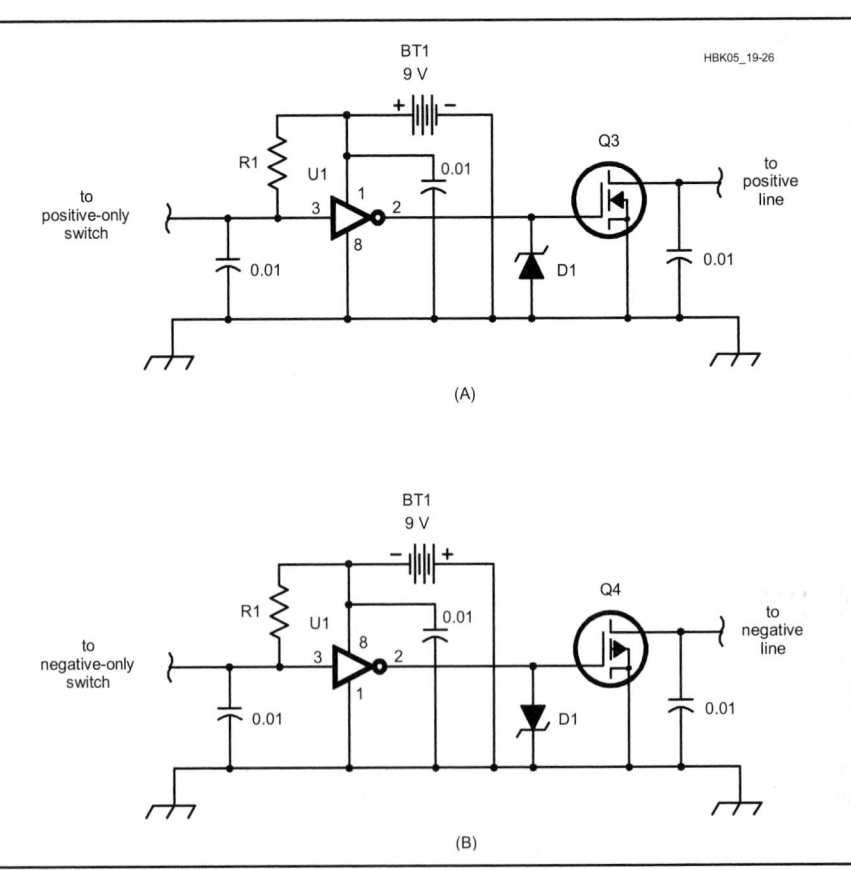

Fig 29.16 — Circuits for same-polarity level shifters. At A, for positive-only switches and lines; B, for negative-only switches and lines.

BT1 — 9-V transistor-radio battery.
D1 — 15-V, 1-W Zener diode (1N4744 or equiv).
Q3 — IRF620.
Q4 — IRF220 (see text).

R1 — 10 kΩ, 10%, ¼ W.
U1 — CD4049 CMOS inverting hex buffer, one section used (unused sections not shown; pins 5, 7, 9, 11 and 14 tied to ground).

Having the station well documented is an invaluable aid when tracking down a problem or planning a modification. Rather than having to search your memory for information on what you did a long time ago, you'll have the facts on hand.

Besides recording the interconnections and hardware around your station, you should also keep track of the performance of your equipment. Each time you install a new antenna, measure the SWR at different points in the band and make a table or plot a curve. Later, if you suspect a problem, you'll be able to look in your records and compare your SWR with the original performance.

In your shack, you can measure the power output from your transmitter(s) and amplifier(s) on each band. These measurements will be helpful if you later suspect you have a problem. If you have access to a signal generator, you can measure receiver performance for future reference.

29.1.7 Interfacing High-Voltage Equipment to Solid-State Accessories

Many amateurs use a variety of equipment manufactured or home brewed over a considerable time period. For example, a ham might be keying a '60s-era tube rig with a recently built microcontroller-based electronic keyer. Many hams have modern solid-state radios connected to high-power vacuum-tube amplifiers.

Often, there is more involved in connecting HV (high-voltage) vacuum-tube gear to solid-state accessories than a cable and the appropriate connectors. The solid-state switching devices used in some equipment will be destroyed if used to switch the HV load of vacuum tube gear. The polarity involved is important too. Even if the voltage is low enough, a key-line might bias a solid-state device in such a way as to cause it to fail. What is needed is another form of level converter.

MOSFET LEVEL CONVERTERS

While relays can often be rigged to interface the equipment, their noise, slow speed and external power requirement make them an unattractive solution in some cases. An alternative is to use power MOSFETs. Capable of handling substantial voltages and currents, power MOSFETs have become common design items. This has made them inexpensive and readily available.

Nearly all control signals use a common ground as one side of the control line. This leads to one of four basic level-conversion scenarios when equipment is interconnected:

1) A positive line must be actuated by a negative-only control switch.

2) A negative line must be actuated by a positive-only control switch.

3) A positive line must be actuated by a positive-only control switch.

4) A negative line must be actuated by a negative-only control switch.

In cases 3 and 4 the polarity is not the problem. These situations become important when the control-switching device is incapable of handling the required open-circuit voltage or closed-circuit current.

Case 1 can be handled by the circuit in **Fig 29.15A**. This circuit is ideal for interfacing keyers designed for grid-block keying to positive CW key lines. A circuit suitable for case 2 is shown in Fig 29.15B. This circuit is simply the mirror image of that in Fig 29.15A with respect to circuit polarity. Here, a P-channel device is used to actuate the negative line from a positive-only control switch.

Cases 3 and 4 require the addition of an inverter, as shown in **Fig 29.16**. The inverter provides the logic reversal needed to drive the gate of the MOSFET high, activating the control line, when the control switch shorts the input to ground.

Almost any power MOSFET can be used in the level converters, provided the voltage and current ratings are sufficient to handle the signal levels to be switched. A wide variety of suitable devices is available from most large mail-order supply houses.

29.2 Mobile Installations

Solid-state electronics and miniaturization have allowed mobile operators to equip their vehicles with stations rivaling base stations. Indeed, it is possible to operate from 160 meters through 70 cm with one compact transceiver. Adding versatility, most designed-for-mobile transceivers are set up so that the main body of the radio can be safely tucked under a seat, with the operating "head" conveniently placed for ease of use as shown in **Fig 29.17**.

Common power levels reach 100-150 W on HF, and 50-75 W on VHF. With proper antenna selection and placement (see the **Antennas** chapter), mobile stations can work the world, just like their base station counterparts. The only real difference between them is that you're trying to drive at the same time you are operating, and safe operating requires attention to the details.

For some of us living in antenna-restricted areas, mobile operating may offer the best solution for getting on the air. For others it is an enjoyable alternative to home-station operation. No matter which category you're in, you can enjoy success if you plan your installation with safety and convenience in mind.

29.2.1 Installation

Installing Amateur Radio equipment in modern vehicles can be quite challenging, yet rewarding, if basic safety rules are followed. All gear must be securely attached to the vehicle. Unsecured cup holder mounts, mounts wedged between cushions, elastic

Fig 29.17 — In this mobile installation, the transceiver control head is mounted in the center console, next to a box with switches for adjusting the antenna.

Table 29.1
Mobile Mount Sources

Gamber Johnson — **www.gamberjohnson.com**
Havis-Shields — **www.havis.com**
Jotto Desk — **www.jottodesk.com**
PanaVise Products — **www.panavise.com**
RAM Mounting Systems — **www.ram-mount.com**

(A)

Fig 29.18 — At A, the transceiver control head is attached to one of many available mounts designed for this purpose. Mounts are typically highly adjustable, allowing the control head or radio to be positioned close to the operator. An antenna controller is mounted below the microphone. At B, HF and VHF transceiver control heads and the microphone are all mounted to the dashboard, within easy reach.

(B)

cords, hook-and-loop tape, magnets, or any other temporary mounting scheme *must be avoided*! Remember, if it isn't bolted down, it will become a missile in the event of a crash. The radio mounting location must avoid SRS (airbag) deployment zones — virtually eliminating the top of the dash in most modern vehicles — as well as vehicle controls (see the sidebar "Air Bags and Mobile Installations).

If you're not into building a specific mount for your vehicle, there are many no-holes-needed mounts available from Amateur Radio dealers. Some mounts are even designed for a specific transceiver make and model. **Table 29.1** lists some suppliers.

Two other points to keep in mind when choosing a mounting location are convenience and lack of distraction. Microphone and power cabling should be placed out of the way and properly secured. The transceiver's controls should be convenient to use and to view. See **Fig 29.18** for examples.

Mounting radios inside unvented center consoles and overhead bins should also be avoided. Modern mobile transceivers designed for remote mounting allow the main body to be located under a seat, in the trunk, or in another out-of-the-way place (**Fig 29.19**) but be sure there is plenty of ventilation.

If you drive off-road or on rough roads, you may also wish to consider using shock mounts, also known as "Lord Mounts" for their original manufacturer. For more information, see the product information on "Plateform Mounts" from the Astrotex Company at **www.astrotex.com/lord.htm**.

Fig 29.19 — The main body of the radio may be mounted in the vehicle's trunk or other out-of-the-way spot. Allow for plenty of ventilation.

Air Bags and Mobile Installation

Since 1998 all passenger vehicles are required to have Supplemental Restraint Systems (SRS), better known as *airbags*. Side airbags and airbags for rear seat passengers have become commonplace. When used in conjunction with seat belts, they've become a great life saving device, but they do have a drawback — they literally explode when they deploy!

Airbags deploy within 200 ms, expanding at about 200 mph, driven by gas from a controlled explosion. **Fig 29.A** drawing shows a typical vehicle with several air bags deployed in the passenger compartment. Any radio gear within range of an airbag will be ripped free with great force and flung about the interior. This should eliminate from consideration any dash-top mounting scheme including windshield suction cup (mobile phone) mounts, so often employed.

Fig 29.B shows the passenger compartment of a vehicle with airbags deployed after a minor collision that caused less than $300 damage to the bumper. Note the loose piece of dashboard on top of the deflated air bag and the broken windshield. These are typical effects of a deploying airbag, whether from the top or center of the dash. Knowing how airbags deploy, avoid mounting radio gear anywhere near them.

It is always a difficult task finding a suitable mounting location for a transceiver and/or control head that is out of airbag range yet easily seen and operated. One workaround is a gooseneck mount (see **Table 29.1** for a list of suppliers). These attach via a seat bolt (no hole needed). They're a good alternative as long as they're placed away from the passenger airbag deployment area (the whole right side of the dashboard). The dealer for your make and model may have additional guidelines for mounting radios and control heads in the car.

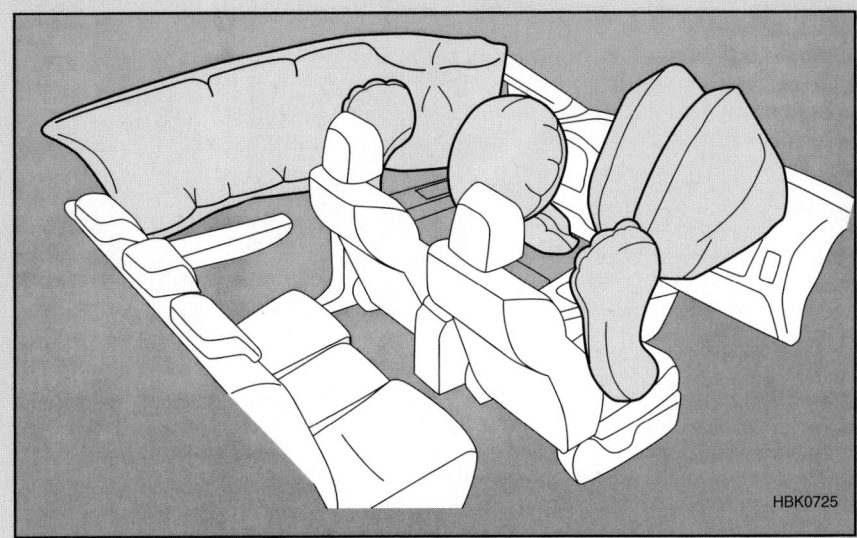

Fig 29.A — Airbag deployment zones in a modern vehicle.

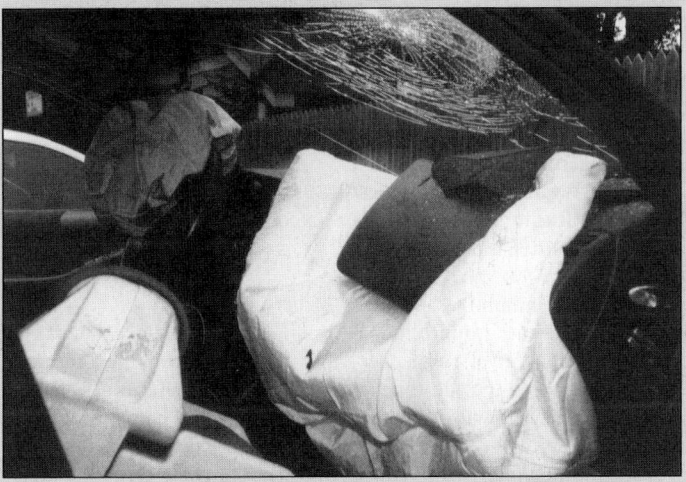

Fig 29.B — Airbag deployment following a minor collision caused significant disruption inside the passenger compartment.

29.2.2 Coaxial Cable

Cable lengths in mobile installations seldom exceed 15 feet, so coax losses are not a major factor except for 70 cm and above. Good quality RG-58A or RG-8X size coax is more than adequate for HF and VHF. While there is nothing wrong with using RG-8 size coax (0.405 inch), it is stiffer and has a larger bending radius, making it harder to work with in most mobile applications.

There are some caveats when selecting coax. Avoid solid center conductors such as in standard RG-58. It has a propensity to kink, is susceptible to failure from vibration and can be difficult to solder properly. Both RG-58A and RG-8X use foam dielectric, and care is needed when soldering PL-259 connectors — especially when reducers are being used. The **Component Data and References** chapter illustrates the correct installation procedure.

29.2.3 Wiring

Proper wiring is an essential part of any mobile installation. Consider the following points when selecting materials and planning the cable routing.

• Wire needs to be correctly sized and fused stranded wire.

• All cables need to be protected from abrasion, heat, and chemicals.

• Wiring needs to be shortened and/or bundled with appropriate wire ties to avoid interaction with passengers and mechanical devices.. **Fig 29.20** shows a typical vehicle wiring tray.

Power cables should be connected directly

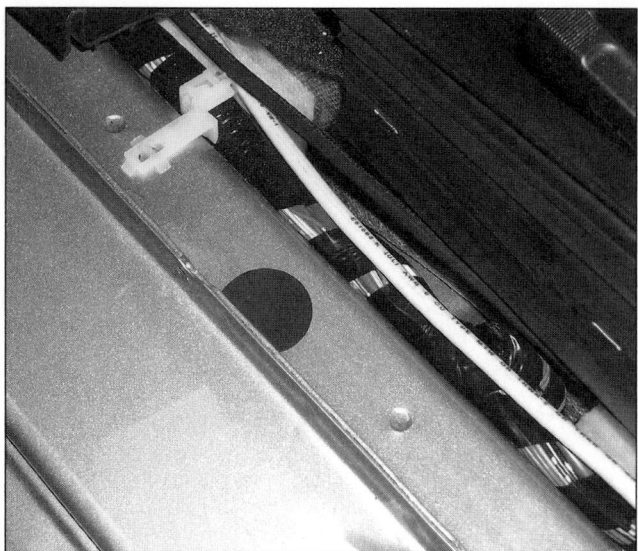

Fig 29.20 — Most vehicles have wiring troughs hidden behind interior body panels.

Fig 29.21 — Wiring attached to a fuse block.

to the battery following manufacturers' recommendations, with the requisite positive and negative lead fuses located close to the battery. **Fig 29.21** shows a typical fuse block. Accessory (cigarette lighter) sockets and power taps shouldn't be used except for very low current loads (<5 A), and then only with care. It pays to remember that a vehicle fire is both costly and dangerous! More information may be found at **www.fordemc.com/docs/download/Mobile_Radio_Guide.pdf** and **http://service.gm.com/techlineinfo/radio.html**.

Proper wiring also minimizes voltage drops and helps prevent ground loops. Modern solid-state transceivers will operate effectively down to 12.0 V dc (engine off). If the voltage drops below 11.6 V under load, some transceivers will reduce power, shut down or operate incorrectly. The vehicle chassis should not be used for ground returns; paint or other insulation can isolate different chassis sections and using a chassis return can create a ground loop.

Running cables through the engine firewall can be easy in some vehicles and nearly impossible in others. Using factory wiring grommets should be avoided unless they're not being used. In some cases, the only alternative is to drill your own hole. If you have any questions or concerns, have your local mobile sound shop or two-way radio dealer install the wiring for you.

Power for ancillary equipment (wattmeters, remotely tuned antennas and so on) should follow the same wiring rules. The use of a multiple outlet power distribution panel such as a RigRunner (**www.westmountain radio.com**) is also recommended. They're convenient, and offer a second level of protection.

WIRE SIZE

The **Component Data and References** chapter lists the current-handling capabilities of various gauges of wire and cable. The correct wiring size is one that provides a low voltage drop (less than 0.5 V under full load). Don't use wire at its maximum current-current carrying capacity.

Here's the formula for calculating the cable assembly voltage drop (V_d):

$$V_d = [(R_W \times 2 \, \ell \times 0.001) + 2 \, k] \times I$$

where
- R_W = resistance value (Ω per 1000 feet) from the **Component Data and References** chapter.
- ℓ = overall length of the cable assembly including connectors, in feet.
- k = nominal resistive value for one fuse and its holder. Note: Most power cables have two fuses. If yours doesn't, use 1 k in the formula. If you don't know the fuse and holder resistance, use a conservative value of 0.002 Ω.
- I = peak current draw in amperes for a SSB transceiver, or steady state for an FM radio.

For example, the peak current draw for a 100 W transceiver is about 22 A, and a typical power cable length is 10 feet. Using the resistive values for 1000 feet of #10 AWG wire (0.9987 Ω), and a conservative value for the fuse resistances (0.002 Ω each), the calculated drop will be 0.527 V.

It's important to reiterate that the wire size should be selected for minimum voltage drop, not maximum power handling capability. The voltage drop is often referred to as

"I-squared-R loss" — the current in amperes, squared, times the resistance — and should be held to a minimum whenever possible. In cabling, excessive I^2R losses can cause the wire to overheat with predictable results.

The insulation material of wire used in mobile installations should have a temperature rating of at least 90 °C, and preferably 105 °C. It should be protected with split-loom covering whenever possible, especially under-hood wiring.

Selecting the correct size fuse is also important. The average current draw for any given fuse should not exceed 60% of its rating. Thus, the correct fuse rating for a 22-A load is 30 A. That same 30-A fuse will handle a 40-A load for about 120 seconds, and a 100-A load for about 2 seconds. Therefore, it pays to be conservative when selecting the carrying capacities of both wire and fuses. **Fig 29.22** shows the characteristics of several sizes of automotive fuses.

29.2.4 Amplifiers

Mobile HF amplifiers have been around for many years, and with the advent of high-power solid state devices they are common. However, running high power in a mobile environment requires careful planning. Considerations include, but are not limited to:
- alternator current ratings and battery capacity
- wiring (in addition to safe current ratings, excessive voltage drop will create distortion of the output signal)
- antenna and feed line power ratings
- placement and secure mounting in the vehicle
- wiring and placing of remote controls

See **www.k0bg.com/amplifiers.html** for

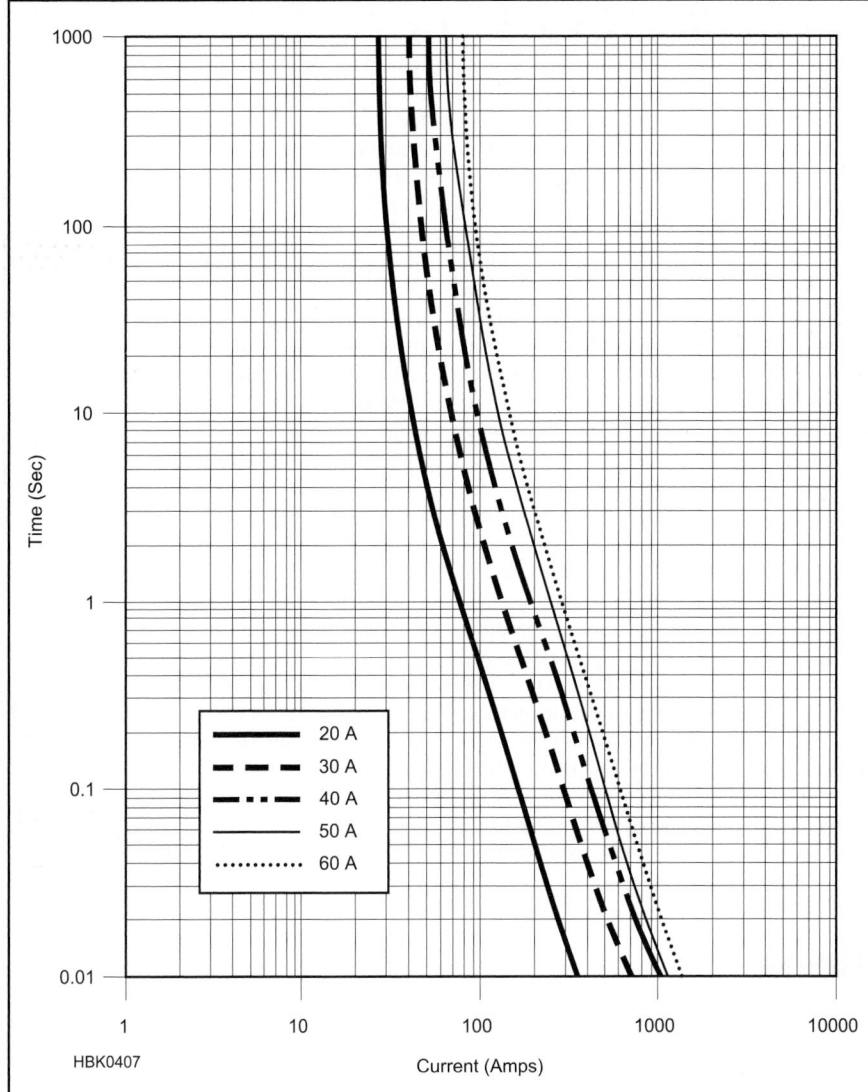

HBK0407

Fig 29.22 — Chart of opening delay versus current for five common sizes of Maxi fuses, the plastic-body high-current fuses common in vehicles. (*Based on a chart from Littelfuse Corp*)

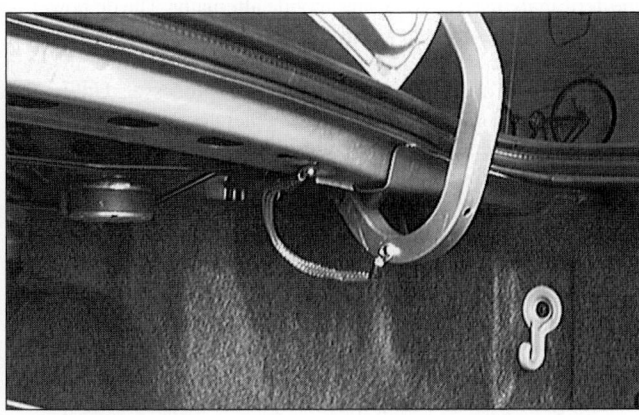

Fig 29.23 — Bonding vehicles parts — in this case, the trunk lid to the main body — can help reduce interference.

more information on these topics.

Before purchasing an amplifier, take a close look at your antenna installation and make sure it is operating efficiently. Using an amplifier with a poor antenna installation is counterproductive. Here's a rule of thumb applicable to any type of antenna: If the *unmatched* input SWR is less than 1.7:1 on 17 meters or any lower frequency band, then it isn't mounted correctly, and/or you need a better antenna. Whatever antenna you use, it must be capable of handling the amplifier power level — 500 W or more. More information on HF mobile antennas and installation techniques may be found in the **Antennas** chapter.

Mobile amplifiers for VHF/UHF operation are not as popular as they once were because most mobile transceivers have adequate output power (about 50 W). Boosting this to 150-300 W or more should be done with caution. Mobile VHF/UHF antennas for high power (>100 W) are rare, so check antenna ratings carefully. Those that are available need to be permanently mounted, and preferably on the roof to avoid inadvertent contact.

With any high-power mobile installation, pay careful attention to RF safety. More information on RF exposure can be found in the **Safety** chapter.

29.2.5 Interference Issues

In a mobile installation, radio frequency interference falls under two basic categories: *egress* (interference from the vehicle to your amateur station) and *ingress* (from your amateur gear to the vehicle). Most hams are familiar with ignition interference as it is the most common form of egress. RF interference to an auto sound system is a common form of ingress.

Both types of interference have unique solutions but they have at least one in common and that is *chassis bonding*. Chassis bonding refers to connecting accessory equipment or assemblies to the frame or chassis of the vehicle. For example, the exhaust system is isolated from the structure of the vehicle and acts like an antenna for the RFI generated by the ignition system. It should be bonded to the chassis in at least three places. **Fig 22.23** shows an example of bonding. More on these techniques is available at **www.k0bg.com/bonding.html**.

Other RFI egress problems are related to fuel pumps, HVAC and engine cooling fans, ABS sensors, data distribution systems and control system CPUs. These are best cured at the source by liberal use of snap-on ferrite cores on the wiring harnesses of the offending devices. Snap-on cores come is a variety of sizes and formulations called mixes. The best all-around ferrite core material for mobile RFI issues is Mix 31. Suitable cores are available from most Amateur Radio dealers. Unknown surplus units typically offer little HF attenuation and should be avoided. See the **RF Interference, RF Techniques,** and **Component Data and References** chapters for more information on ferrite cores.

Alternator whine can be another form of RFI egress. It is typically caused by an incorrectly mounted antenna resulting in a ground loop, rather than a defective alternator diode or inadequate dc power filtering as has been the traditional solution. Attempting to solve alternator whine with a dc filter can mask the problem and increases I^2R losses. Additional information on proper antenna mounting is in the **Antennas** chapter.

RFI ingress to the various on-board electronic devices is less common. The major causes are unchecked RF flowing on the control wires and common mode currents flowing on the coax cable of remotely-tuned HF antennas. Again, this points out the need to properly mount mobile antennas.

For more information on RFI issues, see the **RF Interference** chapter, *The ARRL RFI Handbook*, and the ARRL Technical Information Service (**www.arrl.org/tis**).

29.2.6 Operating

The most important consideration while operating mobile-in-motion is safety! Driver distraction is a familiar cause of vehicle crashes. While Amateur Radio use is far less distracting than mobile phones or texting, there are times when driving requires all of our attention. When bad weather, excessive traffic, or a construction zone require extra care, play it safe — hang up the microphone and turn off the radio!

In addition to properly installing gear, a few operating hints can make your journey less distracting. One of those is familiarization with your transceiver's menu functions and its microphone keys (if so equipped). Even then, complicated programming or adjustments are not something to do while underway.

Logging mobile contacts has always been difficult. Compact digital voice recorders have made that function easy and inexpensive. Units with up to 24 hours of recording time are available for less than $50.

For maximum intelligibility at the other end, avoid excessive speech processing and too much microphone gain. Don't shout into the microphone! It's human nature to increase your speaking level when excited or when the background level increases. In the closed cabin of a vehicle, your brain interprets the reflected sound from your own voice as an increase in background level. Add in a little traffic noise and by the end of your transmission you're in full shout mode! One solution is to use a headset and the transceiver's built-in monitor function. Doing so gives you direct feedback (not a time-delayed echo), and your brain won't get confused. Note that headset use is not legal in some jurisdictions and never legal if both ears are covered.

Overcoming vehicle ambient noise levels often requires the use of an external speaker and all too often it is an afterthought. Selecting a speaker that is too small accentuates high frequency noise which makes reception tiresome. Using adapters to interface with vehicle stereo systems isn't productive for the same reason. For speakers that are too large, mounting becomes a safety issue.

For best results, use at least a 4-inch speaker. Rather than mount it out in the open, mount it out of the way, under the driver's seat. This attenuates the high frequency noise, and enhances the mid-range response which increases intelligibility.

Most modern transceivers contain some form of DSP (digital signal processing) noise reduction as covered in the **DSP and Software Radio Design** chapter. Some are audio based and some are IF based, with the latter being preferred. But both types do a decent job of reducing high-frequency hiss, static spikes, and even ignition hash.

29.3 Portable Installations

Many amateurs experience the joys of portable operation once each year in the annual emergency exercise known as ARRL Field Day. Setting up an effective portable station requires organization, planning and some experience. For example, some knowledge of propagation is essential to picking the right band or bands for the intended communications link(s). Portable operation is difficult enough without dragging along excess equipment and antennas that will never be used.

Some problems encountered in portable operation that are not normally experienced in fixed-station operation include finding an appropriate power source and erecting an effective antenna. The equipment used should be as compact and lightweight as possible. A good portable setup is simple. Although you may bring gobs of gear to Field Day and set it up the day before, during a real emergency speed is of the essence. The less equipment to set up, the faster it will be operational.

29.3.1 Portable AC Power Sources

There are three popular sources of ac power for use in the field: batteries, dc-to-ac inverters, and gasoline powered generators.

Batteries and inverters are covered in detail in the **Power Sources** chapter. This section will focus on gasoline powered generators. The book *Emergency Power for Radio Communications* by Mike Bryce, WB8VGE is another good resource for generator application information.

Essentially, a generator is a motor that's operating "backward." When you apply electricity to a motor, it turns the motor's shaft (allowing it to do useful work). If you need more rotational power, add more electricity or wind a bigger motor. Take the same motor and physically rotate its shaft and it generates electricity across the same terminals used to supply power when using the motor as a motor. Turn the shaft faster and the voltage and frequency increase. Turn it slower and they decrease. To some degree, all motors are generators and all generators are motors. The differences are in the details and in the optimization for specific functions.

A "motor" that is optimized for generating electricity is an alternator — just like the one in your car. The most basic generators use a small gas engine to power an ac alternator, the voltage and frequency of which depends on rotational speed. Because the generator is directly coupled to the engine, the generator's rotational speed is determined by the speed of the engine. If the engine is running too fast or too slow, the voltage and frequency of the output will be off. If everything is running at or near the correct speed, the voltage and frequency of the output will be a close approximation of the power supplied by the ac mains — a 120 V ac sine wave with a frequency of 60 Hz. These are referred to as *constant-speed generators*. Most consumer models use two pole armatures that run at 3600 RPM to produce a 60 Hz sine wave.

Inverter generators produce high-voltage, multiphase ac that is rectified to dc — similar to an automobile alternator. The dc power is then converted back to very clean and consistent ac power by a solid state power inverter controlled by a microprocessor. Unlike the constant-speed generators, inverters can run at idle while still providing power, increasing speed to meet additional demand. This improves economy and reduces emissions. The most common models are available with capacities to approximately 2000 W output and some can be paralleled with special cables for higher capacity. The June 2012 *QST* Product Review "A Look at Gasoline Powered Inverter Generators" compares several popular models available at that time and is provided on the CD-ROM accompanying this book.

VOLTAGE REGULATION

There are several electronic and mechanical methods used to "regulate" the ac output — to keep the voltage and frequency values as stable as possible as generator and engine speeds vary because of current loads or other factors. Remember, a standard generator *must* turn at a specific speed to maintain output regulation, so when more power is drawn from the generator, the engine must supply more torque to overcome the increased physical/magnetic resistance in the generator's core — the generator *can't* simply spin faster to supply the extra oomph.

Most generators have engines that use mechanical or vacuum "governors" to keep the generator shaft turning at the correct speed. If the shaft slows down because of increasing generator demand, the governor "hits the gas" and draws energy stored in a heavy rotating flywheel, for example, to bring (or keep) the shaft speed up to par. The opposite happens if the generator is spinning too fast.

In addition to mechanical and vacuum speed regulating systems, generators that are a step up in sophistication additionally have electronic automatic voltage regulation (AVR) systems that use special windings in the generator core (and a microprocessor or circuit to monitor and control them) to help keep things steady near 120 V and 60 Hz. AVR systems can respond to short term load changes much more quickly than mechanical or vacuum governors alone. A decade ago AVR generators were the cream of the crop. Today, they're mostly used in medium to large units that can't practically employ inverters to maintain the best level of output regulation. You'll find them in higher quality 5 to 15 kW "home backup systems" and in many recreational vehicles.

Isolate Source and Load

Basic, inexpensive generators are intended to power lights, saws, drills, ac motors, electric frying pans and other devices that can reliably be run on "cruddy power." If you want the highest margin of safety when powering computers, transceivers and other sensitive electronics, a portable *inverter generator* is

the best way to go. Some popular examples are shown in **Fig 29.24**, and their key specifications are shown in **Table 29.2**. Available in outputs ranging from 1 to 5 kW, these generators use one or more of the mechanical regulation systems mentioned previously, but their ultimate benefit comes from the use of a built-in ac-dc-ac inverter system that produces beautiful — if not perfect — 60 Hz sine waves at 120 V ac, with a 1% to 2% tolerance, even under varying load conditions.

Instead of using two windings in the generator core, an inverter generator uses 24 or more windings to produce a high frequency ac waveform of up to 20 kHz. A solid state inverter module converts the high frequency ac to smooth dc, which is in turn converted to clean, tightly regulated 120 V ac power. And that's not all. Most inverter generators are compact, lightweight and quiet.

GENERATOR CONSIDERATIONS

In addition to capacity and output regulation, other factors such as engine type, noise level, fuel options, fuel capacity, run time, size, weight, cost or connector type, may factor in your decision. Consider additional uses for your new generator beyond Field Day or other portable operation.

Capacity

Your generator must be able to safely power all of the devices that will be attached to it. Simply add up the power requirements of *all* the devices, add a reasonable safety margin (25 to 30%) and choose a suitably powerful generator that meets your other requirements.

Some devices — especially electric motors

Fig 29.24 — Modern inverter generators from McCullough, Honda, Yamaha and Subaru/Robin.

See Table 29.2 for a partial list of specifications.

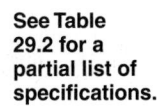

Table 29.2
Specifications of the Inverter Generators Shown in Fig 29.24

Make and Model	Output (W) (Surge/Cont)	Run Time (h) Full / 25% Load	Noise Range (dBA @21 feet)	Engine Type	Weight (Pounds)	Notes*
McCulloch FDD210M0	N/A / 1800	4 / N/A	60-70	105.6 cc, 3 HP	65 (shipping)	a,b
Honda EU2000i	2000 / 1600	4 /15	53-59	100 cc, OHC	46	a,b,c
Yamaha EF2400iS	2400 / 2000	N/A / 8.6	53-58	171 cc, OHV	70	a,b,c
Subaru/Robin R1700i	N/A / 1650	N/A / 8.5	53-59	2.4 HP, OHV	46	a,b,c

*a — has 12 V dc output; b — has "smart throttle" for better fuel economy; c — has low oil alert/shutdown

— take a lot more power to start up than they do to keep running. A motor that takes 1000 W to run may take 2000 to 3000 W to start. Many items don't require extra start-up power, but be sure to plan accordingly.

Always plan to have more capacity than you require or, conversely, plan to use less gear than you have capacity for. Running on the ragged edge is bad for your generator *and* your gear. Some generators are somewhat overrated, probably for marketing purposes. Give yourself a margin of safety and don't rely on built-in circuit breakers to save your gear during overloads. When operating at or beyond capacity, a generator's frequency and voltage can vary widely before the current breaker trips.

Size and Weight

Size and weight vary according to power output — low power units are lightweight and physically small, while beefier models are larger, weigh more and probably last longer. Watt for watt, however, most modern units are smaller and lighter than their predecessors. Models suitable for hamming typically weigh between 25 and 125 pounds.

Engines and Fuel

Low end generators are typically powered by low-tolerance, side valve engines of the type found in discount store lawnmowers. They're noisy, need frequent servicing and often die quickly. Better models have overhead valve (OHV) or overhead cam (OHC) engines, pressure lubrication, low oil shutdown, cast iron cylinder sleeves, oil filters, electronic ignition systems and even fuel injection. These features may be overkill for occasional use but desirable for more consistent power needs.

Run Time

Smaller generators usually have smaller gas tanks, but that doesn't necessarily mean they need more frequent refueling. Some small generators are significantly more efficient than their larger counterparts and may run for half a day while powering small loads. As with output power, run times for many units are somewhat exaggerated and are usually specified for 50% loads. If you're running closer to max capacity, your run times may be seriously degraded. The opposite is also true. Typical generators run from three to nine hours on a full tank of gas at a 50% load.

Noise

Except for ham-friendly inverter units — which are eerily quiet thanks to their high tech, sound dampening designs — standard generators are almost always too loud. Noise levels for many models are stated on the box, but try to test them yourself or talk to someone who owns the model you're interested

in before buying. Environmental conditions, distance to the generator and the unit's physical orientation can affect perceived noise levels.

Generators housed in special sound dampened compartments in large boats and RVs can be much quieter than typical "outside" models. However, they are expensive and heavy, use more fuel than compact models, and most don't have regulation specs comparable to inverter models.

Regulation

For hams, voltage and frequency regulation are the biggies. AVR units with electronic output regulation (at a minimum) and inverter generators are highly desirable and should be used exclusively, if only for peace of mind.

Unloaded standard generators can put out as much as 160 V ac at 64 Hz. As loads increase, frequency and voltage decrease. Under full load, output values may fall as low as 105 V at 56 Hz. Normal operating conditions are somewhere in between.

Some hams have tried inserting uninterruptible power supplies (UPSs) between the generator and their sensitive gear. These devices are often used to maintain steady, clean ac power for computers and telecommunication equipment. As the mains voltage moves up and down, the UPS's Automatic Voltage Regulation (AVR) system bucks or boosts accordingly. The unit's internal batteries provide power to the loads if the ac mains (or your generator) go down.

In practice, however, most UPSs can't handle the variation in frequency and voltage of a generator powered system. When fed by a standard generator, most UPSs constantly switch in and out of battery power mode — or don't *ever* switch back to ac power. When the UPS battery goes flat, the unit shuts off. Not *every* UPS and *every* generator lock horns like this, but an inverter generator is a better solution.

DC Output

Some generators have 12 V dc outputs for charging batteries. These range from 2 A trickle chargers to 100 A powerhouses. Typical outputs run about 10 to 15 A. As with the ac outputs, be sure to test the dc outputs for voltage stability (under load if possible) and ripple. Car batteries aren't too fussy about a little ripple in the charging circuit, but your radio might not like it at all!

Miscellaneous

Other considerations include outlets (120 V, 240 V and dc output), circuit breakers (standard or ground fault interrupter type), fuel level gauges, handles (one or two), favorite brands, warranties, starters (pull or electric), wheels, handles or whatever you require.

SETUP, SAFETY AND TESTING

Before starting the engine, read the user manual. Carefully follow the instructions regarding engine oil, throttle and choke settings (if any). Be sure you understand how the unit operates and how to use the receptacles, circuit breakers and connectors.

Make sure the area is clean, dry and unobstructed. Generators should *always* be set up *outdoors*. Do not operate gas powered engines in closed spaces, inside passenger vans, inside covered pickup beds, etc. If rain is a possibility, set up an appropriate canopy or other *outdoor protective structure*. Operating generators and electrical devices in the rain or snow can be dangerous. Keep the generator and any attached cords dry!

Exhaust systems can get hot enough to ignite certain materials. Keep the unit several feet away from buildings, and keep the gas can (and other flammable stuff) at a safe distance. Don't touch hot engines or mufflers!

When refueling, shut down the generator and let things cool off for a few minutes. Don't smoke, and don't spill gasoline onto hot engine parts. A flash fire or explosion may result. Keep a small fire extinguisher nearby. If you refuel at night, use a light source that isn't powered by the generator and can't ignite the gasoline.

Testing

Before starting (or restarting) the engine, *disconnect* all electrical loads. Starting the unit while loads are connected may not damage the generator, but your solid state devices may not be so lucky. After the engine has warmed up and stabilized, test the output voltage (and frequency), if possible, *before* connecting loads.

Because unloaded values may differ from loaded values, be sure to test your generator under load (using high wattage quartz lights or an electric heater as appropriate). Notice that when you turn on a hefty load, your generator will "hunt" a bit as the engine stabilizes. Measure ac voltage and frequency again to see what the power conditions will be like under load. See your unit's user manual or contact the manufacturer if adjustments are required.

Safety Grounds and Field Operation

Before we can connect *real* electrical loads in a Field Day situation we need to choose a grounding method — a real controversy among campers, RVers and home power enthusiasts.

To complicate matters, most generators have ac generator grounds that are connected to their metal frames, but some units do not bond the ac neutral wires to the ac ground wires (as in typical house wiring). Although they will probably safely power your ham station all day long, units with unbonded neutrals may appear defective if tested with a standard ac outlet polarity tester.

Some users religiously drive copper ground rods into the ground or connect the metal frames of their generators to suitable existing grounds, while others vigorously oppose this method and let their generators float with respect to earth ground (arguing that if the generator isn't connected to the earth, you can't complete the path to earth ground with your body should you encounter a bare wire powered by the generator; no path, no shock). Some user manuals insist on the ground connection, while others don't.

You can follow your unit's user manual, check your local electric code, choose a grounding method based on personal preference or expert advice, or do further research. Either method may offer better protection depending on exact circumstances.

Regardless of the grounding method you choose, a few electrical safety rules remain the same. Your extension cords *must* have intact, waterproof insulation, three "prongs" and three wires, and must be sized according to loads and cable runs. Use #14 to #16 AWG, three wire extension cords for low wattage runs of 100 feet or less. For high wattage loads, use heavier #12 AWG, three wire cords designed for air compressors, air conditioners or RV service feeds. If you use long extension cords to power heavy loads, you may damage your generator or your radio gear. When it comes to power cords, think *big*. Try to position extension cords so they won't be tripped over or run over by vehicles. And don't run electrical cords through standing water or over wet, sloppy terrain.

During portable operations, try to let all operators know when the generator will be shut down for refueling so radio and computer gear can be shut down in a civilized manner. Keep the loads disconnected at the generator until the generator has been refueled and restarted.

29.3.2 Portable Antennas

An effective antenna system is essential to all types of operation. Effective portable antennas, however, are more difficult to devise than their fixed-station counterparts. A portable antenna must be light, compact and easy to assemble. It is also important to remember that the portable antenna may be erected at a variety of sites, not all of which will offer ready-made supports. Strive for the best antenna system possible because operations in the field are often restricted to low power by power supply and equipment considerations. Some antennas suitable for portable operation are described in the **Antennas** chapter.

ANTENNA SUPPORTS

While some amateurs have access to a

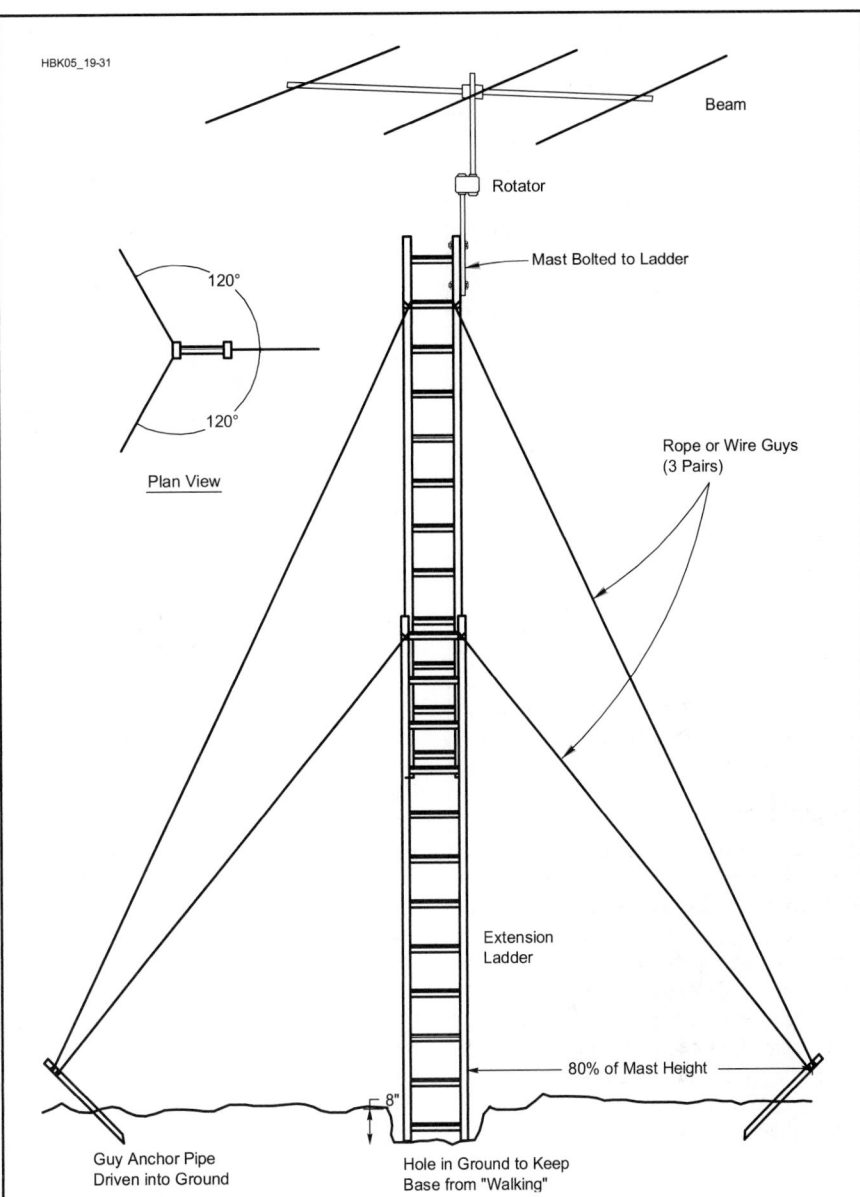

Fig 29.25 — An aluminum extension ladder makes a simple but sturdy portable antenna support. Attach the antenna and feed lines to the top ladder section while it is nested and lying on the ground. Push the ladder vertical, attach the bottom guys and extend the ladder. Attach the top guys. Do not attempt to climb this type of antenna support.

truck or trailer with a portable tower, most are limited to what nature supplies, along with simple push-up masts. Select a portable site that is as high and clear as possible. Elevation is especially important if your operation involves VHF. Trees, buildings, flagpoles, telephone poles and the like can be pressed into service to support wire antennas. Drooping dipoles are often chosen over horizontal dipoles because they require only one support.

An aluminum extension ladder makes an effective antenna support, as shown in **Fig 29.25**. In this installation, a mast, rotator and beam are attached to the top of the second ladder section with the ladder near the

ground. The ladder is then pushed vertical and the lower set of guy wires attached to the guy anchors. When the first set of guy wires is secured, the ladder may be extended and the top guy wires attached to the anchors. Do not attempt to climb a guyed ladder.

Telescoping fiberglass poles (**Figs 29.26** and **29.27**) are popular for supporting wire verticals, inverted Vs and small VHF/UHF antennas. These poles can extend up to 40 feet in length, yet retract to 4 to 8 feet for easy transport. They typically weigh less than 20 pounds, and some are much lighter.

Figs 29.28 and **29.29** illustrate two methods for mounting portable antennas described

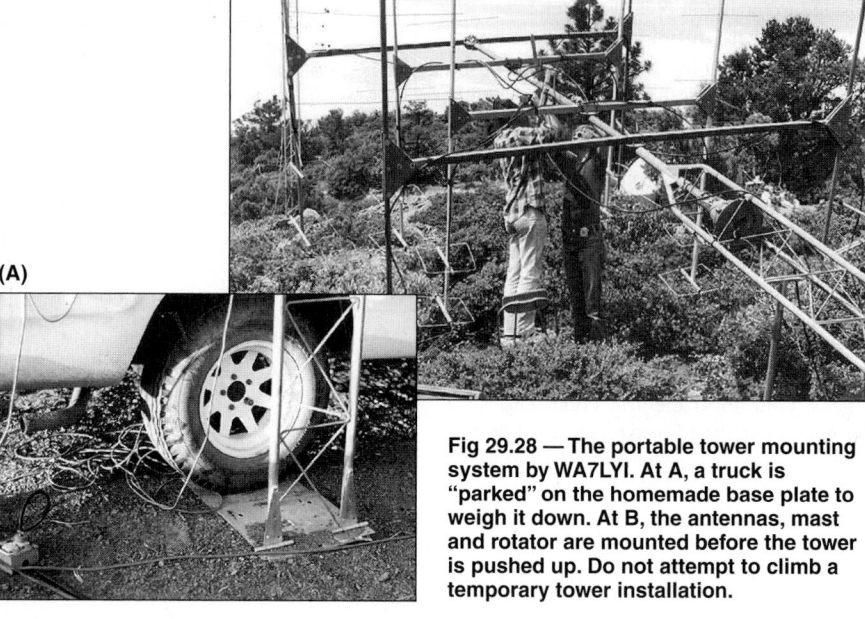

Fig 29.28 — The portable tower mounting system by WA7LYI. At A, a truck is "parked" on the homemade base plate to weigh it down. At B, the antennas, mast and rotator are mounted before the tower is pushed up. Do not attempt to climb a temporary tower installation.

Fig 29.27 — Close-up of a telescoping mast. Sections slide together and are secured by twisting a locking mechanism.

Fig 29.26 — Telescoping fiberglass poles can be used to support a variety of wire antennas or small VHF/UHF Yagis. This one is 40 feet long, yet collapses to 8 feet for storage.

by Terry Wilkinson, WA7LYI. Although the antennas shown are used for VHF work, the same principles can be applied to small HF beams as well.

In Fig 29.28A, a 3 foot section of Rohn 25 tower is welded to a pair of large hinges, which in turn are welded to a steel plate measuring approximately 18 × 30 inches. One of the rear wheels of a pickup truck is "parked" on the plate, ensuring that it will not move. In Fig 29.28B, quad array antennas

for 144 and 222 MHz are mounted on a Rohn 25 top section, complete with rotator and feed lines. The tower is then pushed up into place using the hinges, and guy ropes, anchored to heavy-duty stakes driven into the ground, complete the installation. This method of portable tower installation offers an exceptionally easy-to-erect, yet sturdy, antenna support. Towers installed in this manner may be 30 or 40 feet high; the limiting factor is the number of "pushers" and "rope pullers" needed to get into the air. A portable station located in the bed of the pickup truck completes the installation.

The second method of mounting portable beams described by WA7LYI is shown in Fig 29.29. This support is intended for use with small or medium-sized VHF and UHF arrays. The tripod is available from any dealer selling television antennas; tripods of this type are usually mounted on the roof of a house. Open the tripod to its full size and drive a pipe into the ground at each leg. Use a hose clamp or small U-bolt to anchor each leg to its pipe.

The rotator mount is made from a 6-inch-long section of 1.5-inch-diameter pipe welded to the center of an "X" made from two 2-foot-long pieces of concrete reinforcing rod (rebar). The rotator clamps onto the pipe, and the whole assembly is placed in the center of the tripod. Large rocks placed on the rebar hold the rotator in place, and the antennas are mounted on a 10 or 15 foot tall

mast section. This system is easy to make and set up.

TIPS FOR PORTABLE ANTENNAS

Any of the antennas described in the **Antennas** chapter or available from commercial manufacturers may be used for portable operation. Generally, though, big or heavy antennas should be passed over in favor of smaller arrays. The couple of decibels of gain a 5-element, 20 meter beam may have over a 3-element version is insignificant compared to the mechanical considerations. Stick with arrays of reasonable size that are easily assembled.

Wire antennas should be cut to size and tuned prior to their use in the field. Be careful when coiling these antennas for transport, or you may end up with a tangled mess when you need an antenna in a hurry. The coaxial cable should be attached to the center insulator with a connector for speed in assembly. Use RG-58 for the low bands and RG-8X for higher-band antennas. Although these cables exhibit higher loss than standard RG-8, they are far more compact and weigh much less for a given length.

Beam antennas should be assembled and tested before taking them afield. Break the beam into as few pieces as necessary for transportation and mark each joint for speed in reassembly. Hex nuts can be replaced with wing nuts to reduce the number of tools necessary.

| (A) | (B) | (C) |

Fig 29.29 — The portable mast and tripod by WA7LYI. At A, the tripod is clamped to stakes driven into the ground. The rotator is attached to a homemade pipe mount. At B, rocks piled on the rotator must keep the rotator from twisting and add weight to stabilize the mast. At C, a 10 foot mast is inserted into the tripod/rotator base assembly. Four 432 MHz Quagis are mounted at the top.

29.4 Remote Stations

The following section was contributed by Rick Hilding, K6VVA. Rick has constructed his own competitive remote HF contest station and presents issues for the remote station builder to consider. He also wishes to acknowledge the assistance and input of N6RK, K6TU, W6OOL and SM2O. Although this section focuses on "remoting" HF equipment, the same considerations apply to most remotely-operated stations, with the exception of repeater installations. For the purposes of this section, "remote station" applies to individually-operated stations, primarily using SSB, CW, and digital modes on the HF bands. Many of the same techniques and considerations apply to weak-signal remote operation on the VHF and UHF bands.

29.4.1 Introduction

Remote stations (stations operated by remote control) have been a part of Amateur Radio for decades, but usually in the form of VHF/UHF repeaters. In the past, a few remote stations operating on HF have been developed in impressive locations with significantly more land, equipment and expense than for a repeater installation. Prior to the expansion of Internet technologies, some of these remote pioneers utilized VHF/UHF links or commercial microwave equipment for connectivity between the remote and home stations. There has been a significant increase in the number of remote stations over the last five years and the trend is accelerating rapidly.

Since this section was first added in the 2011 edition, there have been several new developments and refinements of the technology available to remote station builders. As latency across the Internet has been reduced, operating in fast-paced contests over very long links is now commonplace — between the US and Japan, Finland and South Korea, and other similar spans. More than 4500 QSOs were made by OH2UA operating remotely as CQ8X in the Azores.

Local homeowner restrictions on antennas and interference issues from electrical power lines, plasma TVs, network routers and the like increasingly encumber HF operating. This is quite a turnabout from the days of amateur signals causing TVI! With today's technology, however, a remote HF station far from these problems may be as simple as a 100-W HF radio and all-band dipole or vertical with connectivity to the operator via the Internet. There is a flavor of remote station to meet almost every taste and budget.

This overview of remote station construction looks at three aspects of the project: planning the facilities, connectivity and remote station equipment. All three are equally important for a successful project and will be different for each installation. There is no cookbook approach to creating a remote station, so this discussion is intended to introduce topics that must be considered in the circumstances encountered by each station builder.

29.4.2 Evaluation and Planning

OPERATING REQUIREMENTS

Will your remote station be only for remote control from a home QTH or other single control point? Will it serve a dual purpose so you (and possibly others) can also operate at the remote site during favorable conditions? It is important to adequately plan physical space requirements for equipment and operator needs.

Is your remote station to be only for casual operating or will it be used for competitive contesting and DXing? If the latter, then the issue of *latency* (described below) will be much more significant.

Are you interested in operating on all HF bands including 60, 30, 17 and 12 meters? Will you operate on both CW and SSB?

Digital modes? Remember that on the low bands, you may need to deal with the SWR variations of an antenna across an entire band.

Will your remote station be low power with a single transceiver or do you want to also use an amplifier? If you're planning a competitive SO2R (single operator, two radio) HF contest station, your needs are going to be much more complex, involving band decoders, automatic dual-radio and antenna switching mechanisms and the like.

Do you plan to use a single multiband antenna, or maybe a tribander and wire antennas for the low bands, or are you planning to install the "Mother of all Antenna Farms"? Make sure you have enough physical space available at the intended site.

PURCHASE VS RENT OR LEASE

If you don't already own or have the funds to purchase the property for your remote site, there is still hope! Although having direct control over every aspect of what you want to do is most desirable, you may want to consider a rental or lease with a cooperative land owner. In either case, having a detailed agreement or contract prepared by a competent attorney is a must. Be sure to include every contingency you can think of, and especially what happens if the land owner decides to sell.

RELIABLE ACCESS BY ROAD

This is one of the most important things to consider in any remote site selection. Will you be able to get there safely, rain or shine? What happens if any problems develop that require immediate attention at the remote site? Pavement or a good all-weather, well-maintained gravel road should be available. Having more than one means of getting in or out should emergencies arise while you are at the site is also an important consideration. These concerns must be addressed before agreeing to purchase, rent or lease a remote site.

GEOLOGICAL AND OTHER HAZARDS

Evaluate the geological factors involved. In a hilltop or mountaintop site, is there evidence of soil movement that could affect access or damage your equipment shelter or tower installations? Does the property sit near an earthquake fault? Is there a significant risk of fire during fire season? Is the site exposed to high wind or storms? In flat-land sites, is there a risk of flooding or drainage problems? Remember, you won't be there to rescue your equipment! In consideration of both your initial construction activities and ongoing maintenance tasks, are there any wildlife hazards from animals or insects? What about allergies or reactions to plants such as poison ivy or poison oak? Diligently investigate all aspects

of a potential remote site *before* you end up with bad surprises down the line. For a reasonable fee, there are companies from which you can secure a Geological and Property Hazards report to aid in your assessment.

RAW SITE PREPARATION

It you are fortunate enough to locate suitable property, be sure to evaluate the site preparation needs. How much time and money will be necessary to develop the remote site to meet your specific needs? Is there brush and vegetation to be dispensed with? Will trees need to be cut down? Will bulldozers, excavators and trenchers need access to the site? Cement trucks? Service vehicles? Diligent evaluation of all costs involved will reduce unnecessary stress.

SITES WITH EXISTING FACILITIES

Finding a site with a self-supporting tower or two in place that can accommodate your needs, along with a building large enough for your equipment and operational needs can be like finding a pot of remote gold. However, unless you are experienced with towers and construction, hire professionals to evaluate everything. If you are considering co-locating at a remote site with existing antennas and equipment in use, consider hiring a professional to help you evaluate the likelihood of problems. Birdies and intermodulation could end up being a source of potential conflict and stress if problems can't be resolved.

TERRAIN CONSIDERATIONS

Evaluate the terrain on the bands you wish to operate. For HF, Dean Straw, N6BV, has developed a very useful program called *HFTA* (*High Frequency Terrain Assessment*) that is included with the *ARRL Antenna Book*, of which he is a former editor. Use *HFTA* to assist you in placement of your antennas. Don't expect excellent ground/soil conductivity if your remote site is on top of a rocky mountaintop.

If you plan to use vertical antennas, make sure there is sufficient room on the property to place the radials, whether on a mountaintop or flat area. Placing verticals near water, especially saltwater, is the most desirable way to achieve the low takeoff angles for long-haul DX. Remember that exposure to salt air or spray will create significant maintenance issues.

NOISE LEVELS

Since one of the reasons for building a remote station is to lower noise levels, you'll want to carefully evaluate existing and potential noise sources. Start by visiting the site with a receiver in several different types of weather and listening on all the bands you intend to use with a full-size dipole antenna. If noise levels are low, this is no guarantee that

problems won't crop up in different seasons or from hardware failures. You should avoid nearby substations or high-voltage and power distribution lines. If you will be sharing a site with other users, see the discussion below on noise and interference from and to their equipment.

TELEPHONE OR INTERNET ACCESS

If you are planning to operate your remote station via the Internet, having suitable telephone line access is required unless wireless Internet service is available. Be sure that broadband service such as DSL or cable TV can reach your site by asking the local service provider to do a site evaluation. Satellite Internet downlink speed may be fast, but uplink speed is often quite a bit slower. Be sure the minimum link speed is adequate in both directions.

If your control point is close enough to the remote site, another option is to implement your own private wireless "bridge" system. Even if multiple "hops" are involved, with the right equipment you can potentially have better results than using even broadband Internet.

The author has achieved a low 1 ms latency by using a dedicated point-to-point 5 GHz wireless link between his home QTH and the remote station site on a mountain ridge 4.4 miles away. Each remote link will have different results.

ELECTRICAL POWER

Having electrical power already available at your remote site can certainly save a lot of money. The most ideal situation is to have underground power to the remote site. Be sure to evaluate the full costs of installing ac power if it is not already present at the site or on the property. Truly remote sites may require wind, solar or generator power systems like that shown in **Fig 29.30**.

If your remote site will require solar/wind/battery power, it is critical that you properly assess sunlight path and duration, wind patterns, and provide sufficient battery storage to meet your intended operating needs during lengthy periods of inclement weather. The article "Designing Solar Power System for FM Translators" is a valuable resource for planning of remote HF station off-grid power.[1] You must thoroughly account for the power requirements of each and every piece of equipment.

There are various estimates for the duty cycle of transceivers. A conservative duty cycle of 50% for CW/SSB is recommended for power capacity evaluation. (RTTY/digital/AM/FM duty cycles will likely be higher.) Internet switches, routers, rotators, computers and anything electrical will also consume

Fig 29.30 — Solar panels can be used to charge a bank of batteries to supply power. Effective off-grid power requires an accurate and conservative estimate of power needs and available energy sources. [Rick Hilding, K6VVA, photo]

power. A supplementary diesel or propane remote-start generator to a basic off-grid system may be an appropriate option, particularly if you plan on using an amplifier. If your remote site is in a rugged terrain rural area, make sure a diesel or propane service provider vehicle can get to the location *before* you make final plans! It may be possible to use smaller propane tanks you can haul in a truck, but any manifold-type system necessitates insuring that the proper BTU requirements can be met based upon the propane generator manufacturer's ratings. Installing a solar and wind remote battery monitor system is advisable.

When you have made an accurate assessment of your complete system needs, add a generous additional contingency factor. It is also advisable to get quotes on your proposed system from at least two different "green power" equipment suppliers, and to make sure they fully understand your requirements and intended use *before* you purchase anything.

INSURANCE AND SECURITY

You've heard the expression "Stuff Happens." Unless you have blanket coverage in an existing homeowner's insurance policy that includes your remote site equipment, look into the ARRL "All-Risk" Ham Radio Equipment Insurance Plan. If you will

be hiring others to do work at your remote site, be sure that liability coverage issues are addressed. The same applies to any guests or visitors such as fellow hams. Have a competent attorney draw up a "Liability Release" agreement with strong "Hold Harmless" language.

If your remote site is yours alone and not a co-habitation arrangement with other services, be sure and evaluate all security needs (a necessity even if co-located). You might need one or more security gates, webcam surveillance and auto-remote notification of security breaches.

PERSONAL SAFETY

Regardless of your age, it is prudent to have someone with you at the site. You could severely cut yourself, or fall and break a leg. If there are wildlife hazards in the area, take appropriate steps to protect yourself.

Carry a first aid kit and have one readily available at your remote site. Keep some emergency supplies such as various sealed food items in a varmint-proof container and a case of bottled water in the event you find yourself stranded.

If your remote site is within the coverage area of a local repeater, take along a handheld radio (with batteries charged) as part of your SOP (Standard Operating Procedure). This is especially important for rural and remote operating sites.

29.4.3 Controlling Your Remote Site HF Station

Almost any modern all-mode transceiver can be controlled remotely by software. Several packages are listed in the Remote Station Resource document on the CD-ROM included with this book. (The resource listing also includes vendors of remote control interface equipment.) The manufacturer of your radio may also offer remote control software.

Fig 29.31 illustrates the design of a basic remote HF station system which uses Remoterig control interfaces.[2] ADSL is used to connect the interfaces via the Internet. A suitable radio, microphone and CW paddle are connected to the interface on each end. The Kenwood TS-480 radio in the figure is designed for simple connectivity to a network-style interface. Other radios may require an RS-232 interface as shown in **Fig 29.32**. The radio in this system is located entirely at the remote site and a computer running control software creates the operator's interface at the control point.

A popular alternative for contesters and DXers is to run the contest/logging software

Licenses and Remote Operation

Operating a remote station also carries with it an extra responsibility to be properly licensed and to identify your station correctly. Because the transmissions are made from the location of the remote station, you must follow the regulations that apply at the remote location. This is particularly important when the station is outside the jurisdiction of the regulatory authority that granted your license, such as a US ham operating a station in South America or a European ham operating a station in the US. You must be properly licensed to transmit from the remote station! This may require a separate license or reciprocal operating permission from the country where the remote station is located. Don't assume that because you have a US license, you automatically have a corresponding license to operate a remote station outside the US. For instance, CEPT agreements generally do not apply to operators who are not physically present at the transmitter. There may also be contest or award rules that apply to remote stations — the DXCC program does not accept remote station contacts, for example. Be sure to know and follow *all* rules that apply.

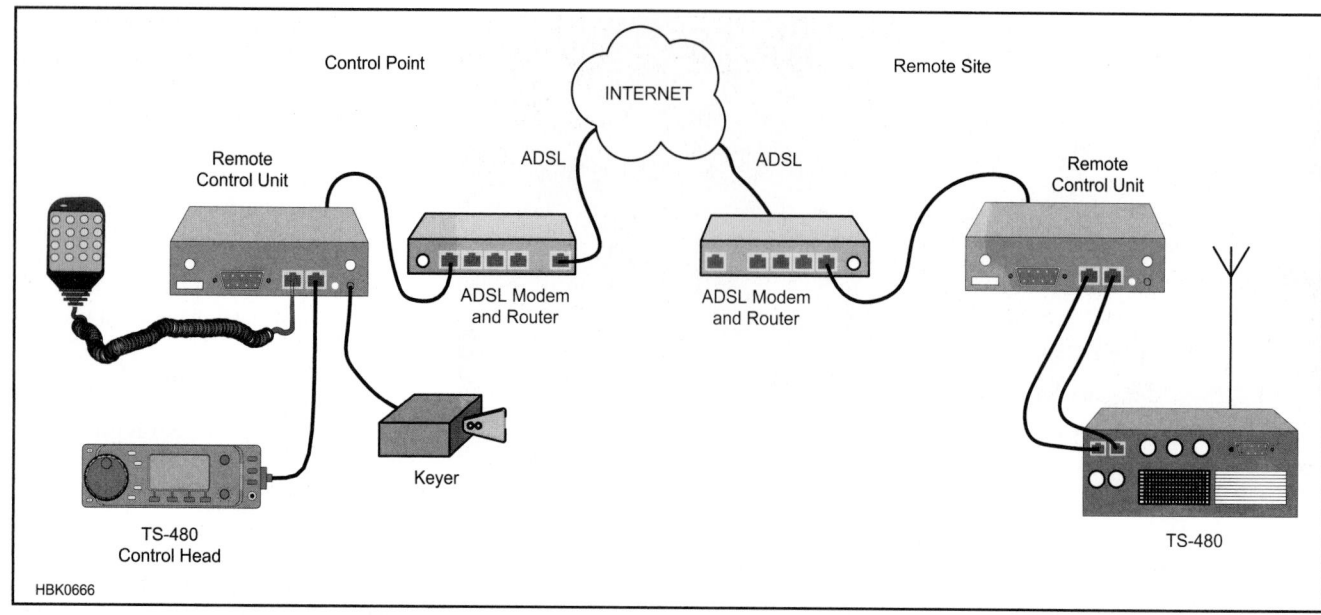

Fig 29.31 — Network-compatible interfaces and remote-ready radios such as the TS-480 can make assembling a remote HF station easier. The control head for the TS-480 can be used as the operator interface at the control point.

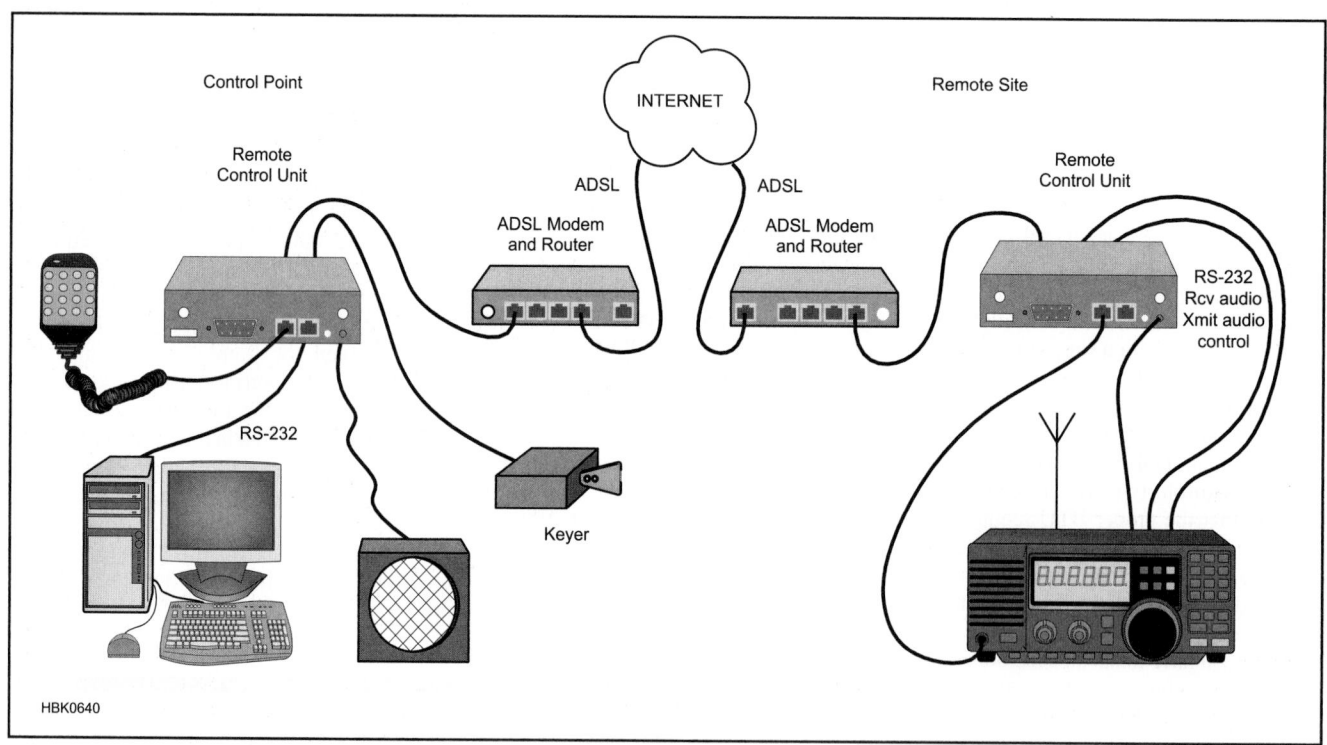

Fig 29.32 — Radios without network connectivity can be controlled by using an RS-232 interface at the remote site. A computer running control software at the control point forms the operator interface.

on a computer at the remote site with a Winkey CW keyer to generate CW under software control. *Windows XP Pro* has self-contained server capabilities for *Windows* Remote Desktop, and the WinRDP client can be used for access from the control point computer. For *Windows XP Home* operating systems, VNC (Virtual Network Computing — **http://en.wikipedia.org/**

wiki/Virtual_Network_Computing) utility software is available from several companies. For later versions of *Windows*, similar remote desktop user control software is available.

CONNECTING TO YOUR REMOTE SITE HF STATION

The most common way of connecting to a remote site station is via the Internet. You will

want to learn computer networking basics, find a ham friend with the necessary skills, or hire a computer networking consultant to help you. Terms like IP addresses, routers, port forwarding, DMZ, bandwidth, latency and others found in the Remote Station Glossary at the end of this chapter will become very familiar. A convenient list of other computer network terms can be found online at

Fig 29.33 — Solid-state disk drives in netbook-style computers and wireless point-to-point Ethernet bridges can be used to form a low-power, high-reliability computer-to-computer control link.

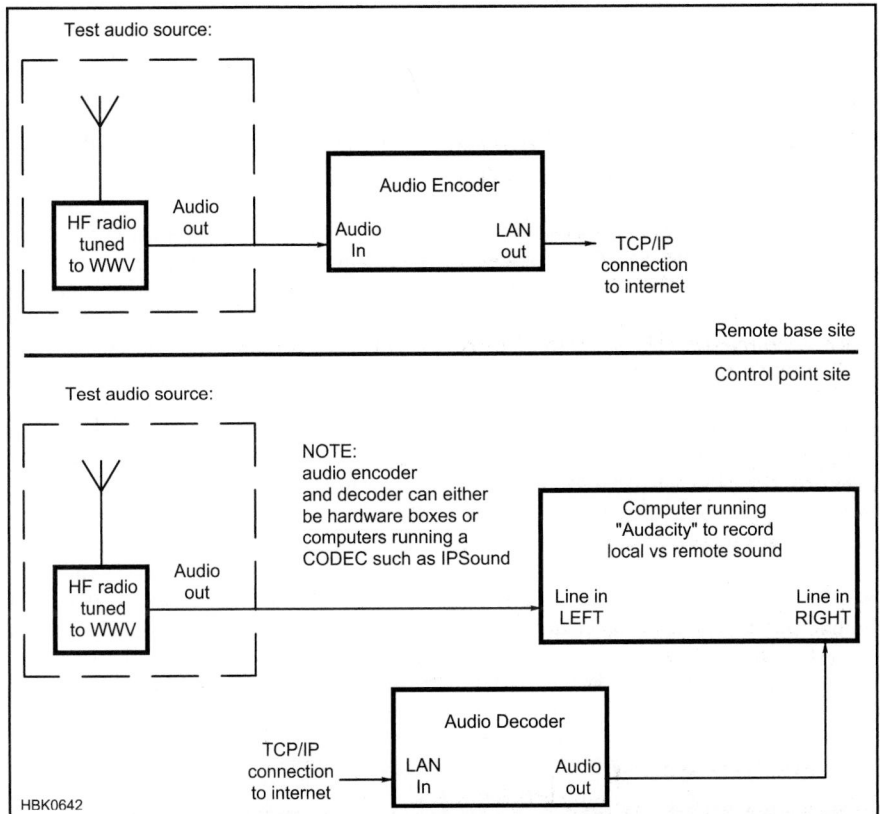

Fig 29.34 — By listening to a distant radio station at both the remote site and the control point, the delay in received audio from the remote site determines latency between the sites. (System design by N6RK.)

www.onlinecomputertips.com/networking/terminology.html.

If your remote site has no telephone or Internet service and you have a line-of-sight path between your home and the site, one logical choice for connectivity is to use a pair of wireless network bridge units and a pair of netbook computers with solid state drives (SSD) and low power consumption. **Fig 29.33**

shows an example of Airaya 5 GHz wireless bridges and ASUS Eee PC netbook computers. The NanoBridge M5 wireless system from Ubiquiti Networks (**www.ubnt.com**) is another option for this form of connectivity.

If a direct line of site doesn't exist, multiple bridge units can be used to overcome various terrain challenges. Of course, overall latency will increase somewhat and the additional

locations require a power source to operate the units. A word of caution about using unlicensed equipment and spectrum — you run the risk of potential interference from other users on the band.

UNDERSTANDING INTERNET LATENCY

High latency can seriously impair the usability of your remote station. For casual operating, you may be able to tolerate a latency of 250 ms or 300 ms. Serious contesters and DXers will most likely find this much delay frustrating, especially on CW. If you are a high-speed CW contest operator, a very low latency is required to be competitive.

ASSESSING LATENCY

You can easily evaluate basic Internet latency yourself. In a *Windows* (PC) environment, under Start | All Programs | Accessories you will find Command Prompt. From within this DOS-like window you can run the *ping*, *tracert* and or *pathping* commands to access a particular IP address and measure the latency. This is critical to know in advance or to troubleshoot connectivity issues once you are operational.

For the purposes of conducting Internet system audio latency tests, N6RK's method illustrated as a block diagram in **Fig 29.34** is very effective. Using a receiver at the remote site and one at the control point, the recorded latency can easily be determined. **Fig 29.35** shows a latency of 40 ms for the return audio over K6VVA's private control link using a Remoterig audio quality setting that requires a network speed of about 180 kbps. The audio is crystal clear and at 40 ms, little difference is noticed from having the TS-480 control head connected directly to the radio at home.

ESTIMATING INTERNET BANDWIDTH REQUIREMENTS

It is important to have a good idea of the actual bandwidth requirements for the effective operation of your remote HF station via the Internet. The higher the quality of the audio you need, the higher the bandwidth required. Unless you plan to use a private wireless bridge link, Internet speeds will determine your available bandwidth. Many remote station operators have found Skype (**www.skype.com**) a workable solution for remote audio routing as well as IP SOUND (**http://ip-sound.software.informer.com**). Another way to completely avoid Internet latency and bandwidth issues for audio is to use POTS (Plain Old Telephone System) connectivity with a dedicated analog phone line that might also be shared with a DSL data service.

Sharing an Internet connection can also cause reduction in bandwidth. If others in your home are online at the same time, their activity will reduce bandwidth available to

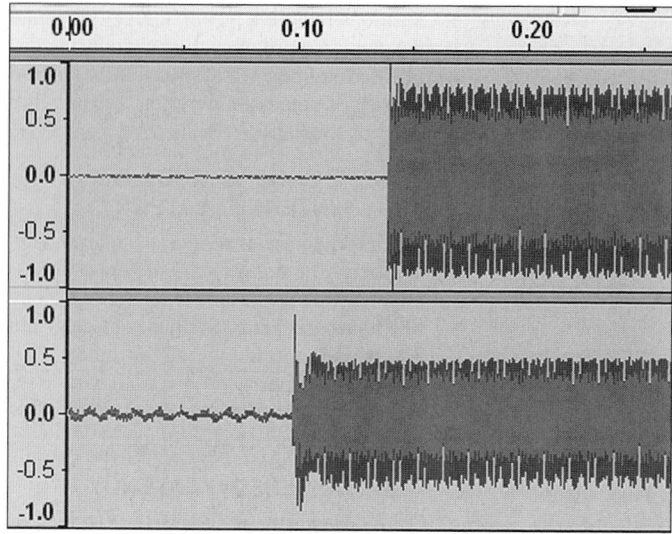

Fig 29.35 — The recordings of audio recorded from the remote site link and from the local receiver clearly show the estimated 40 ms of latency.

you. A dedicated line or service upgrade to support remote station operation may be required unless you can account for all the concurrent household bandwidth requirements and determine that your service is capable of handling it all. A router with "QoS" (quality of service) prioritizing capabilities might solve or at least minimize some of the multi-user problems on a single line Internet connection.

CW and SSB do not pose the same the bandwidth problems as RTTY, unless pan-adapter/waterfall type displays are used. If you plan to use a webcam or two for monitoring and security purposes, you must account for that bandwidth in your overall connectivity planning as well. Remember that screen changes of any type consume bandwidth.

29.4.4 A Basic Remote HF Station

There are several ways to approach implementing a remote HF station. One path is to start small with a single transceiver and a multi-band dipole, vertical and/or tribander. You can be up and running while working on your long-range plan. (A remote VHF/UHF station has a similar structure.)

K6VVA's SO2R (single-operator, two radio) remote station began with a single Kenwood TS-480 using W4MQ IRT system cables and hardware, LDG AT-200PRO tuner, a refurbished TH3, Jr. tribander on a 27-foot mast and a Carolina Windom 160. The ASUS Eee PC with *Windows XP Pro* enabled use of the *Windows* Remote Desktop server and also contained the *TR4W* contest logging software which generated CW via a Winkey keyer. The initial equipment placement was in a small used camper which also housed the solar-power storage batteries in plastic battery cases, and the Airaya wireless bridge with a small panel antenna mounted on the back of the camper. (You can see some of this equipment in **Fig 29.36**.)

The most significant improvement was made by adding Remoterig interfaces by SM2O (**www.remoterig.com**). While still located inside the small camper, the station was expanded to a functional SO2R operation using two Remoterig units and two TS-480 transceivers. The increase in audio quality at a low latency was significant.

Over time, the remote station grew to include a larger Tuff Shed mini-barn with interior alterations to provide for a separate battery and power system room, an interior operating "shack" with view window of a nearby reservoir and custom built-in operating desk, as well as an overhead sleeping loft and front entry area. The final build-out of this remote station has evolved into what seems like a never ending work-in-progress. More time and money has been expended on road repairs than in completing the final desired station. One thing to always keep in mind is that things "always take longer than expected, and always cost more than expected."

MURPHY'S LAW FOR REMOTE HF STATIONS

Murphy seems to visit remote stations more than home QTH stations. Plan on it. Don't be scared, but rather be prepared! A must for your remote site station is at least one fire extinguisher and regular checks that it is fully charged and ready for use. Remember to include adequate station grounding and lightning protection in your remote station planning!

Troubleshooting remote station problems will present more challenges in general. It would be nice to have backup redundancy for everything at the remote site (including radios, amplifiers, and the like), but for most this will not be practical. If you start out with quality equipment, new coax and cables, and invest with long term thinking in mind, you'll avoid creating the opportunity for problems.

When strange glitches occur while you are

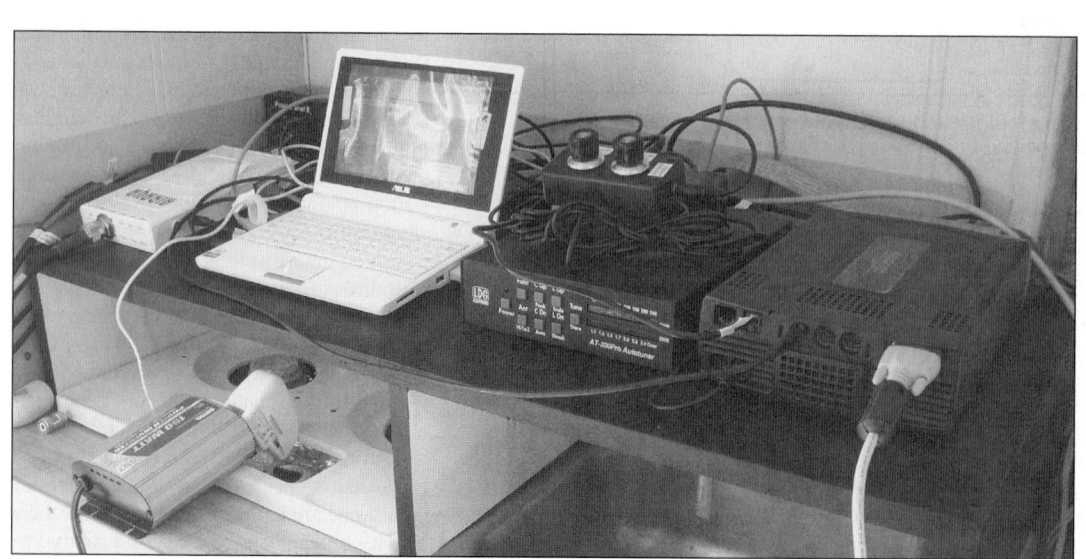

Fig 29.36 — The initial K6VVA remote station

on-air, first take a look at your latency and bandwidth to see if there are any issues which have developed in the main connectivity pipeline. In the author's case, trees on neighboring property had grown into the path of the microwave link. Fortunately, he was able to secure permission of that property owner to trim the foliage out of the way.

29.4.5 Remote HF Station Resources

There is no current "Bank of Remote HF" from which to make immediate withdrawals, however you may find the Remote Operating email reflector to be a helpful resource (**http://mailman.qth.net/mailman/listinfo/remoteoperating**). The ARRL has also published *Remote Operating for Amateur Radio*, a book on remote stations.[3]

For this edition there a few new items especially worthy of mentioning. One is a small serial device driver by Moxa (**www.moxa.com**). About the size of a packet of cigarettes, this compact unit has a built in web server, surge protection, and requires 1 W of power for operation which makes it very suitable for sites powered by solar panels and/or batteries. It is also available in multi-port configurations. **Fig 29.37** shows the single-port version.

For remote rotator control, the author recommends *PSTRotatorAZ* by Codrut, YO3DMU (**www.qsl.net/yo3dmu**). The software provides for a direct TCP/IP interface, which is used in conjunction with the Moxa serial device driver units for each remote rotator. Rotor-EZ boards from Idiom Press (**www.idiompress.com**) in the Hy-Gain rotator controllers are compatible with *PSTRotatorAZ*. A screenshot of the display taken prior to final configuration of the handy

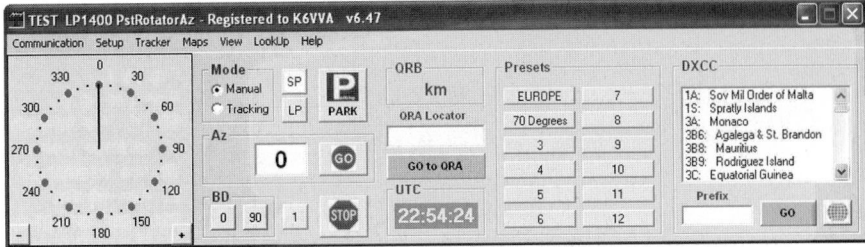

Fig 29.38 — Screen display of the *PSTRotatorAZ* remote rotator control panel.

Fig 29.39 — Elecraft K3 Twin Concept system utilizing Remoterig by SM2O.

Fig 29.40 — CommCat Mobile iPhone application for remote HF station control.

Fig 29.37 — The very small Moxa single-port interface is about the size of a pack of cigarettes.

PRESETS buttons is shown in **Fig 29.38**.

The Elecraft K3 100 W HF transceiver has also been adapted for remote stations. Utilizing the Remoterig interface, the latest K3 firmware supports a "K3 Twin Concept" that allows a regular K3 to be remotely controlled either by a complete K3 in "terminal" mode or a K3/0 (a lower cost K3 without RF components) at the operator control point with no PC required on either end of the link (see **www.elecraft.com/K3-Remote/k3_remote.htm**). **Fig 29.39** shows the combination of a K3 on the left and K3/0 on the right with the Remoterig interfaces. Flexradio's SDR transceivers (**www.flexradio.com**) are also well-suited to use in remote HF stations — see the **HF Transceiver Survey** on the

CD-ROM accompanying this book.

For those interested in controlling their remote HF stations while mobile, some interesting new options are now available or in development for Apple iOS devices. The Pignology/Etherpig products support control of transceivers and contact logging. CommCat has released a mobile application for iOS devices and their iPhone version is shown in **Fig 29.40**.

The most capable product for providing "hands on" rig control of a remote station remains the Remoterig by SM2O. The latest RRC-MKII version Control Point unit displayed at the top of **Fig 29.41** includes a built-in CW keyer with potentiometer for speed control. The unit shown at the bottom

Fig 29.41 — Remoterig by SM2O, the remote HF station control device assessed as having the best performance by the author.

of the figure is used at the remote station location. The author reports that his original Remoterig units have performed flawlessly for more than four years without incident.

It is predicted that the number of remote station installations will continue to increase dramatically. In particular, as more hams move to the antenna-restrictive environments of retirement and assisted living communities, remote stations offer continued activity, especially on the HF bands.

Although certainly not a complete list, the CD-ROM included with this book includes a listing for vendors of software and equipment mentioned in this section. Don't forget to do an Internet search for "Remote HF Stations," "Remote Base" and derivations thereof. You never know what might show up! By the time this edition appears in print, there will no doubt be a number of new items available for this rapidly expanding area of station design.

29.4.6 Remote Station Glossary

ADSL — Asymmetric Digital Subscriber Line. A common form of Internet connectivity which also allows for subscriber telephone use at the same time.

Band decoder — An interface device that reads frequency data from a modern transceiver and facilitates automatic control switching of other equipment such as an HF amplifier.

Bandwidth — Typically expressed in Mbps or kbps, this is used to represent both the capability of an Internet connection for data transfer as well as the amount of data to be transferred.

Charge controller — In solar and wind power generation, this device also serves as a regular to prevent overcharging of storage batteries.

Control point — Wherever the operator will be controlling the *remote site* station. This will normally be a home QTH, but could be anywhere connectivity to the remote site is available.

DHCP — Dynamic Host Configuration Protocol. This functionality is incorporated into most routers and provides automatic assignment of IP addresses to computer devices on a LAN where fixed or static IP addresses are not required.

Duty cycle — A device which is constantly "On" would normally be considered to have a 100% duty cycle. Morse code keying has spaces between the elements and the characters, therefore the duty cycle would be considerably less.

Firewall — Hardware or software capabilities in computers and routers which can be configured to minimize or completely block unauthorized users from access.

IP address — Every computer or device accessible via the Internet or a LAN has a unique numeric address, such as 192.168.1.1.

kbps — Kilobits per second (one thousand bits per second). Typically used as a reference to bandwidth data rate capability or actual usage.

LAN — Local Area Network. A home LAN enables multiple computers to connect to a single Internet service.

Latency — The delays involved in data routing from one point to another, but also a factor in A/D conversion processes.

Liability release — A written document to release one party or entity from getting sued in the event of injury.

Manifold — For propane generators, a distribution system which enables the use of multiple smaller tanks instead of one large tank.

Mbps — Megabits per second (one million bits per second). Typically used as a reference to bandwidth data rate capability or actual usage.

ms — Milliseconds. The unit of time typically used for measuring latency.

Netbook — A new generation of portable computers smaller than a traditional laptop or notebook variety. Many netbooks now use solid-state instead of rotating hard drives.

Off-grid — Generally refers to living off a main electrical power grid, but is also synonymous with extremely rugged, rural remote HF station locations at which solar/wind/battery and/or remote start gas/diesel/propane power must be used.

Packet — A formatted set of digital information. All information sent via the Internet is first converted into packets for transmission.

Pathping — A hybrid utility of the *path* and *tracert* functions which requires more time for analysis between two Internet connection points, but results in a more detailed analysis.

Ping — A utility used to ascertain the availability of another computer device over the Internet or LAN and also measure the round trip time required for the connection.

Port forwarding — Functionality most commonly used in routers to direct or re-direct incoming Internet traffic to specific destinations on a local network or LAN.

POTS — Plain Old Telephone Service. This generally refers to the use of a regular analog telephone system for purposes of remote audio and control signal routing in lieu of the Internet.

Remote site — The actual physical location of the transmitter, antennas and related equipment necessary to generate an RF signal on the HF bands. Some have used the term *Remote Base*.

Router — A device which allows traffic on a single Internet service line to be selectively distributed to multiple computer devices on a LAN. Many routers provide for assignment of local IP addresses as well as automatically via DHCP.

Smartphone — A generic term for the new generation of portable phones with integrated computer capabilities.

Switch — A device which allows multiple computer devices to share the same Internet or LAN connection.

Tracert — A utility for tracing the path taken by packets across the Internet and latency assessing analysis along the route.

Waterfall — A graphical display of digital information.

WiFi — A form of network connectivity without a physical wired connection, although with limited range. Also known by its controlling standard, IEEE 802.11.

Wireless bridge — Low powered transmitter-receiver devices for providing point-to-point digital communications, usually Ethernet, such as for interfacing Internet services to areas without traditional means.

WOL — Wake-on-LAN functionality enables a user to remotely turn on a computer.

29.5 References and Bibliography

APPENDIX

See this book's accompanying CD-ROM for a listing of vendors and other resources for remote station construction.

REFERENCES

[1]Sepmeier, RBR.com, **www.rbr.com/. radio/ENGINEERING/95/13603.html**

[2]Remoterig by SM2O, **www.remoterig. com**.

[3]S. Ford, WB8IMY, *Remote Operating for Amateur Radio*, ARRL, 2010.

BIBLIOGRAPHY

R. Hilding, K6VVA, "How Not to Build a Remote Station — Part 1," *National Contest Journal*, Jan/Feb 2010, pp 8-10.

R. Hilding, K6VVA, "How Not to Build a Remote Station — Part 2," *National Contest Journal*, Mar/Apr 2010, pp 19-23.

"Link and Remote Control" — **www.arrl. org/link-remote-control**

EchoLink® Ready

TM-V71A

With the Kenwood TM-V71A you have a choice of where you want your speaker, on the top or on the bottom of the radio. Simply remove the faceplate and flip the main body, then reattach the face, it's that simple! Yet another Kenwood **1st**, this dual band transceiver has ten dedicated EchoLink® memory channels as well as EchoLink sysop-mode operation. EchoLink connection to your PC via the optional PG-5H cable kit is easy with no expensive interface needed.

EchoLink® is a registered trademark of Synergenics, LLC. For more information please see: www.echolink.org.

Large dual color amber or green thirteen segment LCD.

Customer Support: (310) 639-4200
Fax: (310) 537-8235

Scan with your phone to
download TM-V71A brochure.

www.kenwoodusa.com

JQA-1205
ISO9001 Registered
Professional Systems Business Group
JVC KENWOOD Corporation

ADS#28413

Escape with the TM-281A

On or off the road, Kenwood's TM-281A is a mobile radio you can always count on.

As tough as nails, this MIL-STD-compliant transceiver delivers powerful performance, excellent audio clarity, and a host of advanced features. It offers superb operating ease day or night thanks to the large backlit LCD and illuminated keys. So the next time you take off, take the TM-281A.

KENWOOD

Customer Support: (310) 639-4200
Fax: (310) 537-8235

Scan with your phone to
download TM-281A brochure.

www.kenwoodusa.com

ADS#28513

Advertisers Index

Advertising Department Staff
Debra Jahnke, K1DAJ, *Sales Manager, Business Services*
Janet Rocco, W1JLR, *Account Executive*
Lisa Tardette, KB1MOI, *Account Executive*
Diane Szlachetka, KB1OKV, *Advertising Graphic Design*
Zoé Belliveau, W1ZOE, *Business Services Coordinator*

800-243-7768

Direct Line: 860-594-0207 ■ Fax: 860-594-4285 ■ e-mail: ads@arrl.org ■ Web: www.arrl.org/ads

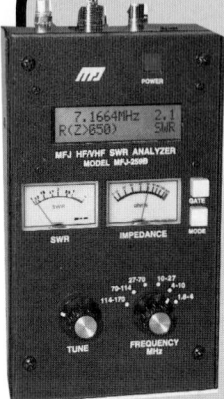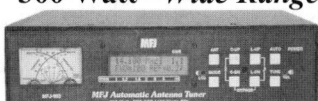

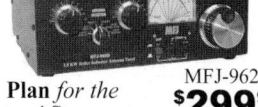

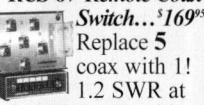

VHF/UHF Dual Band Transceiver
NEW ID–51A

Go more places than ever before.

Dualwatch with AM/FM Receive

Receives two bands simultaneously
with AM/FM broadcast station receiver

Slim, Compact, microSD Ready

Packed with multiple features in a compact size including
voice and data storage and memory data cloning*
*Optional microSD required

Integrated GPS

Provides position reporting for near repeater
and GPS logging* functions
*Optional microSD required

Information & Downloads

AMATEUR TOOL KIT | COMIC BOOKS | VIDEOS | WWW.ICOMAMERICA.COM

Electronic advertisements feature active links

D-STAR ready   IPX7

ICOM

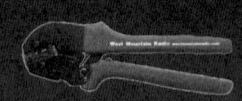

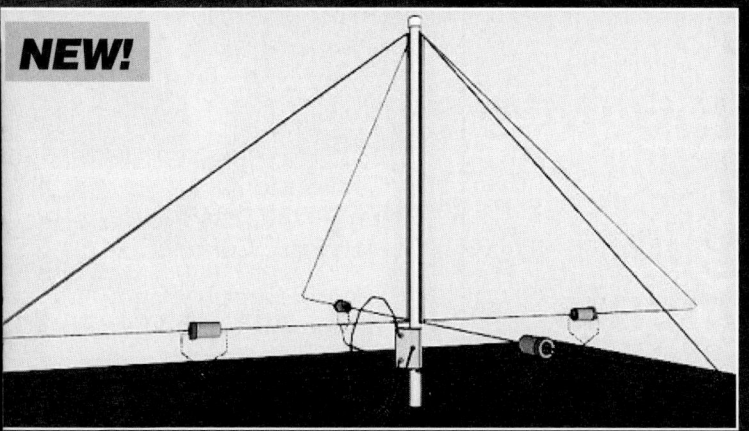

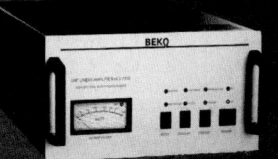

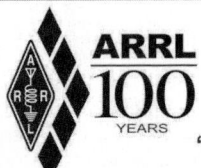

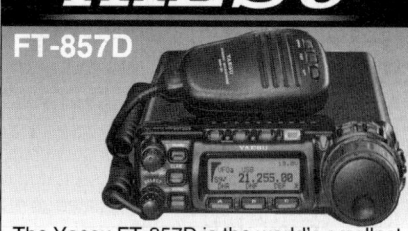

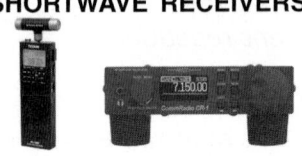

FLEX-6000™ Signature Series Direct Sampling HF/6m Transceivers with SmartSDR™

Imagine a transceiver that changes Ham Radio - forever...

FlexRadio Systems, a pioneer in software defined radios for the amateur radio community, proudly introduces the FLEX-6000 Signature Series with SmartSDR - A revolution in SDR technology. To learn more, visit www.flexradio.com

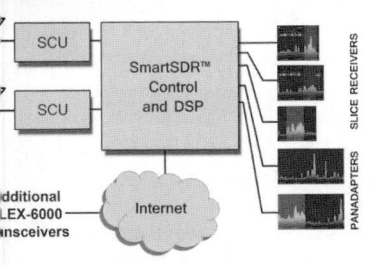

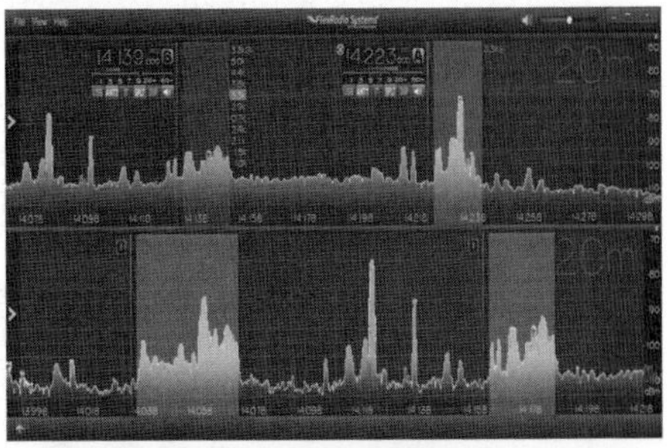

⬈ SIGNAL PROCESSING

The FLEX-6000 Signature Series packs more signal processing power into one radio than all other brands combined.

Up to 317 GMAC & 121 GFLOPS
ARM Cortex™ - A8/NEON™ CPU
Xilinx® Virtex®-6 FPGA
Reserve power for future apps

⬈ NETWORKING

Every radio is a network radio, whether across the shack or across the world.

• Ethernet interface
• Native remote capability
• Multi-user capable

⬈ SmartSDR

SmartSDR organizes all signal processing power into an advanced dynamically reconfigurable environment.

• Simple and elegant user interface
• Up to 8 Slice Receivers / Panadapters
• Hides complexity

4616 W. Howard Lane, Ste. 1-150, Austin, TX 78728

Tel 512-535-4713   Fax 512-233-5143

ail sales@flexradio.com   Website www.flexradio.com

FlexRadio Systems®
Software Defined Radios

Learn More, Hear More, Build More With Great New Books from W5YI!

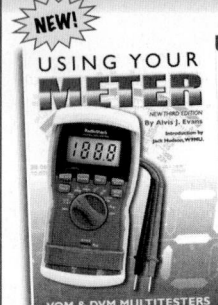

USING YOUR METER

Teach yourself the correct use of your multitester. Book explains fundamental concepts of electricity including conventional and electron current and series and parallel circuits. It teaches how analog and digital meters work and tells you what the voltage, current and, resistance measurements mean. Then it provides fully-illustrated, step-by-step instruction on using your meter in practical applications in the home, workshop, automotive and other settings. An excellent learning tool and reference for the hobbyist and ham. **METR $24.95**

WORLDWIDE LISTENING GUIDE

John Figliozzi's brand new, expanded 5th Edition *Worldwide Listening Guide* explains radio listening in all of today's formats – "live," on-demand, WiFi, podcast, terrestrial, satellite, internet, digital and, of course, AM, FM and SW. Includes a comprehensive program guide to what can be heard how, when and where. Link to WWLG website keeps you up-to-date on program schedule changes. Spiral bound to open in a flat, easy-to-use format. **WWLG $24.95**

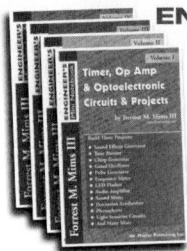

ENGINEER'S MINI NOTEBOOKS

These Mims classics teach you hands-on electronics! Study and build 100s of practical circuits and fun projects. Each volume contains several of his famous Mini Notebooks. Terrific ideas for science fair projects and a great way to learn about electronics! Useful reference guides for your workbench!

Vol. 1: *Timer, Op Amp, & Optoelectronic Circuits & Projects*   MINI-1 $12.95

Vol. 2: *Science & Communications Circuits & Projects*   MINI-2 $12.95

Vol. 3: *Electronic Sensor Circuits & Projects*   MINI-3 $12.95

Vol. 4: *Electronic Formulas, Symbols & Circuits*   MINI-4 $12.95

Order today from W5YI: 800-669-9594 *or on-line:* www.w5yi.org

The W5YI Group P.O. Box 200065 Arlington, TX 76006-0065   **Mention this ad for a free gift.**   QST

BIG BOY ROTATORS
The most powerfull antenna rotators available anywhere Amateur, Commercial, Government and Military pourposes

PST110D-PRO-HP

Model D digital solid state controller, the ultimate in easy to set up controller

- Built in RS232 and Preset
- Programmable soft start-stop
- Programmable North or South stop
- 500° rotation (- 70° + 70° overlap)
- 1° degree accuracy and more......

The ProSisTel rotators were designed to perform under tremendous stress with abnormally large antenna loads.

ProSisTel rotators give you: incredible starting and rotating torque, incredible braking resistence.

The rotators utilizing worm wheel technology far exceeds the specifications on any other amateur rotator made today.

Two years warranty

PRO.SIS.TEL.
Produzione Sistemi Telecomunicazioni

C.da Conghia 298
I-70043 Monopoli BA Italy
Phone ++39 0808876607
www.prosistel.net
Email: prosistel@prosistel.it

Ham Radio Books and CDs from ARRL!

www.arrl.org/shop

The NEW EZ HANG Square Shot Kit
www.ezhang.com

Suggestions from thousands of HAM'S and Cable installers around the world, led to a complete redesign of the **EZ Hang**. Custom designed for YOU, the user in mind. Now safer and easier to use, you will hit your mark every time, with less chance of misfires hitting the yoke.

OVER 10,200 SOLD AROUND THE WORLD!

$99.95 + $9.05 for shipping when paying by check

540-286-0176
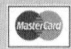 **www.ezhang.com**
EZ HANG
32 Princess Gillian Ct.
Fredericksburg, VA 22406

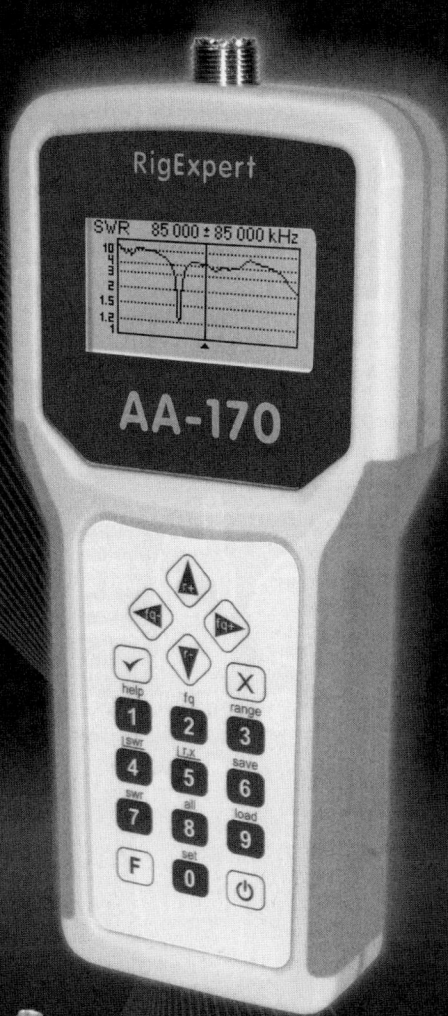
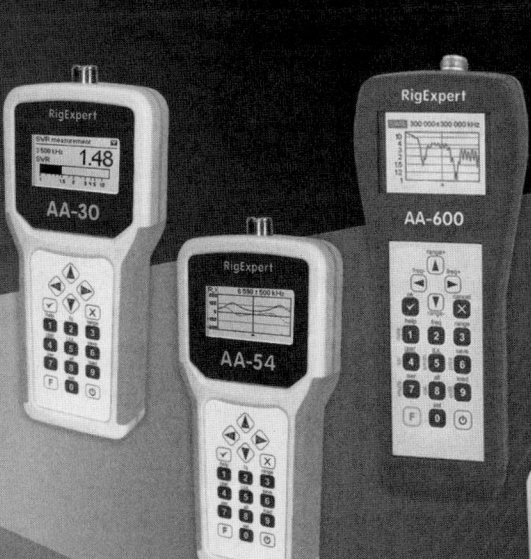
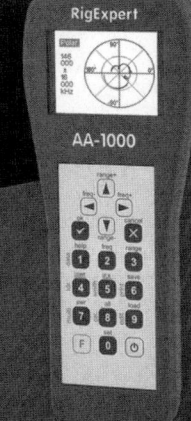
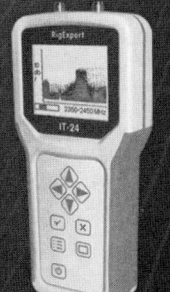

We've Got It ALL... at Great Prices

HamPROS

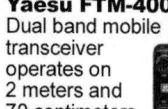

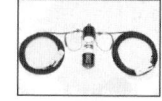
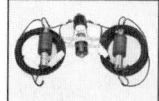

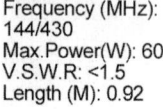
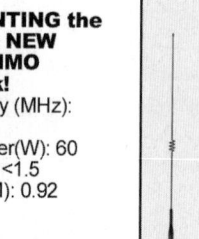
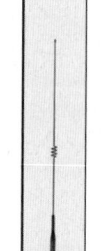

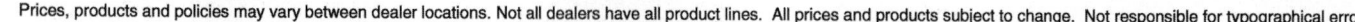

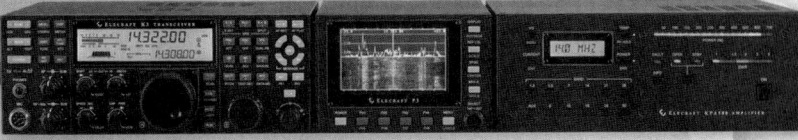
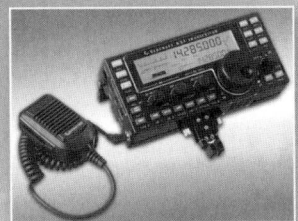

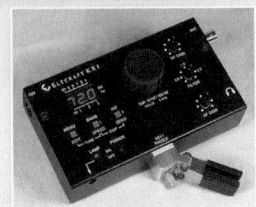

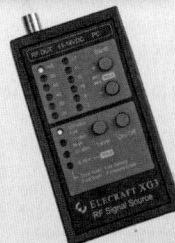

New Functions Enabled by the

The Choice of C4FM Digital

Compared to other digital modulation schemes within FDMA, C4FM has excellent communication quality (BER: Bit Error Rate characteristics). C4FM is the standard method for professional communication devices in FDMA, and is therefore considered to be the main stream digital communication mode in the future.

Automatic Mode Select(AMS) function detects the receive signal mode

The FT1DR/FTM-400DR operates in three digital modes and an analog mode.
Enjoy communication in the mode that suits each purpose.

1. V/D Mode (Simultaneous Voice/Data Communication Mode)
A high-speed data communication mode that uses the entire 12.5 kHz bandwidth for data communication. The FT1DR automatically switches to this mode when sending and receiving images, allowing a large amount of data to be transmitted quickly.

2. Voice FR Mode (Voice Full Rate Mode)
Half of the bandwidth is used for voice signal with error correction. The very effective error correction code provides benefits such as minimal interruption of communication.

3. Data FR Mode (High-speed Data Communication Mode)
This mode uses the entire 12.5 kHz bandwidth to transmit digital voice data. The larger voice data size allows voice communication with high sound quality.

4. Analog FM Mode
Analog FM is effective for communication with a weak signal that causes voices to break up in the digital modes. The analog mode allows communication even at distances where noise and weak signals make communication almost impossible.

V/D Mode — Voice FR Mode — Data FR Mode — Analog FM Mode

SILVER **BLACK**

C4FM FDMA / FM 144/430 MHz DUAL BAND 5W DIGITAL TRANSCEIVER
FT1DR

7.4V 1100 mAh Lithium Ion Battery FNB-101LI and battery charger PA-48 / SAD-11B(USA version), PC Connection Cable SCU-18 included

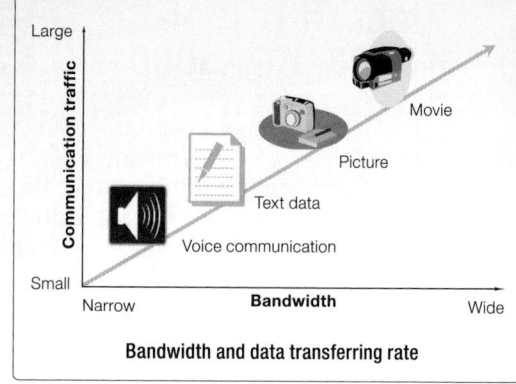

Bandwidth and data transferring rate

Digital Group Monitor (GM) Function

The digital GM function automatically checks whether members registered in a group are within communication range, and displays information such as distance and direction for each call sign on the screen. This convenient function makes it possible not only to see whether any friends are in communication range, but also to instantaneously determines the location and relationship between all members of the group. This function can also be used to send messages and data such as images between members of a group, permitting convenient and fun communication between friends when out for a drive or hike. Sent and received messages and images can be checked on the LOG List screen, with icons making them easy to distinguish.

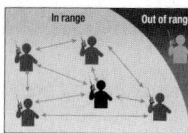
Group Monitor Function
In range   Direction

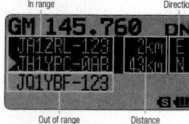
Out of range   Distance
Group Monitor screen

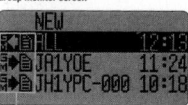
Icon display
LOG list screen

Smart Navigation Function

A real-time navigation function that records the location and direction of Group Monitor (GM) stations

Digital V/D Mode communicates information such as position data at the same time as the voice signal, allowing you to view the distance and direction of the other party in real time while communicating. This makes it possible to confirm your position and the other party's in situations such as hiking and driving where your positions are constantly changing, providing an easy way to meet up or join routes.

Display current position of other party (choose between other party's position and own position)
Displays direct distance to other party
Displays direction   Display own current position

Backtrack Function to Return to Departure Point

This function allows navigation back to the departure point, or a point previously added to the memory.
When hiking or camping, just register the starting point or the position of your tent and then you can constantly check the direction and distance from your current position. The arrow of the compass display constantly shows the direction to the registered point, making it extremely convenient in finding your way back to the registered place – just move in the direction so that the arrow in the heading-up display points straight upward.

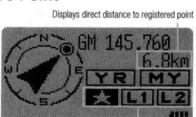
Displays direct distance to registered point
Registered points (★, L1 or L2)

C4FM FDMA Digital Transceiver

Digital Group Monitor

Smart Navigation Screen

APRS® Screen
* APRS® is a registered trademark of Bob Bruninga, WB4APR.

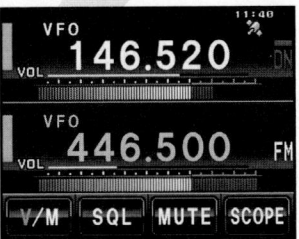

Dual Band Screen

3.5-inch full color
touch panel operation

The icon symbols, multi-function key display and pop-up messages are all displayed in high-resolution color thanks to the full-color, high luminance TFT liquid crystal screen. The settings and status of the wireless devices are displayed in an easy-to-understand format. You can perform various operations simply and easily by gently touching the screen.

Clock / Timer Screen

Message Screen

Band Scope Screen

Snapshot Function (Image Data Transmission)

Simply connect an MH-85A11U (option) microphone with camera.
Press the microphone shutter button to take snapshots, and then the image data can be displayed on the screen and easily sent to other C4FM FDMA digital transceivers.

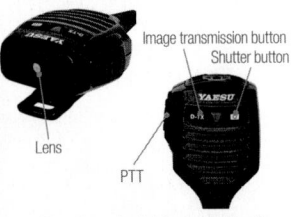

Image transmission button
Shutter button
Lens
PTT
* micro SD card is required by the snapshot function.

Image data which was sent from a group member is displayed on the full-color screen. This image data also retains a time record and the GPS location data of the snapshot. It is easy to navigate to that pictured location by using back track function.
In addition, you can observe on the screen, whether or not transmitted data was successfully received by the member station. The snapshot image or received data is stored in a high capacity micro SD card. You can recall and send that image data from the SD card anytime. The pictures and data files may be easily viewed and edited by using a personal computer.

C4FM FDMA / FM 144/430 MHz 50 W DUAL BAND TRANSCEIVER

FTM-400DR

DTMF Microphone MH-48A6JA, Mounting Bracket, Bracket for Controller, Control Cable 10ft (3m), PC Connection Cable SCU-20, and DC Power Cable included

YAESU
The radio

YAESU USA
6125 Phyllis Drive, Cypress, CA 90630 (714) 827-7600

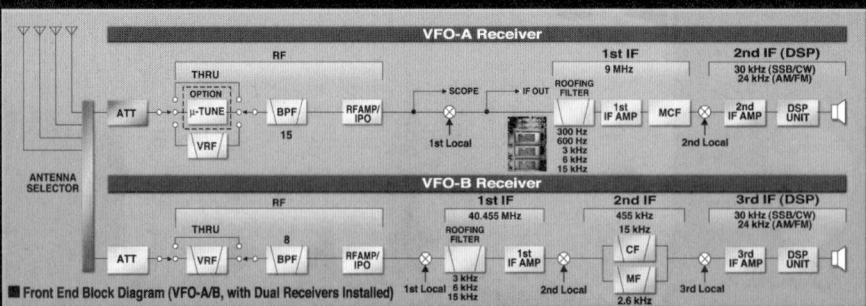
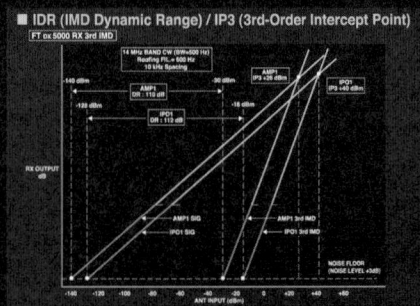

Index

Editor's Note: Except for commonly used phrases and abbreviations, topics are indexed by their noun names. Many topics are also cross-indexed. The names of software programs are italicized.

The letters "ff" after a page number indicate coverage of the indexed topic on succeeding pages.

A separate Project Index and Author Index follow the main index.

Project Index

Author Index

Author	Call	Topic	Section Ref	Page Ref
Halstead, Roger	K8RI	Amplifier tuning	17.9	17.26
		Surplus amplifier parts	17.11	17.44
Hansen, Markus	VE7CA	High-Performance HF Transceiver — HBR-2000	14.7.3	14.23
Harden, Paul	NA5N	Component Data and References	Chapter 22	22.1ff
Hare, Ed	W1RFI	Circuit Construction	Chapter 23	23.1ff
		RF Interference	Chapter 27	27.1ff
		Troubleshooting and Maintenance	Chapter 26	26.1ff
Hartnagel, Hans		The Dangers of Simple Usage of Microwave Software	Chapter 6	CD-ROM
Hayward, Roger	KA7EXM	JFET Hartley VFO	9.3.2	9.17
Hayward, Wes	W7ZOI	JFET Hartley VFO	9.3.2	9.17
		Oscillator temperature compensation	9.3.5	9.19
		RF semiconductor design	Chapter 5	5.1ff
		Segment-tuned VCO	9.7.7	9.50
Henderson, Dan	N1ND	RFI Management	27.2	27.4
Henderson, Randy	WI5W	Rock-bending Receiver for 7 MHz	12.2.1	12.5
Hilding, Rick	K6VVA	Remote Stations	29.4	29.18ff
Hollingsworth, Riley	K4ZDH	RF Interference and the FCC	27.2	27.10
Honnaker, Scott	N7SS	Digital Modes	Chapter 16	16.1ff
Hood, Mike	KD8JB	All-copper, 2 meter J-pole antenna	Chapter 21	CD-ROM
Hranac, Ron	NØIVN	Cable and digital television	27.8.3-27.8.4	27.29ff
		HDTV modulation	8.3	8.11
Humbertson, Ken	WØKAH	FLDIGI	16.1	16.5
Hutchinson, Chuck	K8CH	Antennas	Chapter 21	21.1ff
Hutchinson, Sylvia	K8SYL	75 and 10 meter dipole	Chapter 21	CD-ROM
Jones, Dave	KB4YZ	Slow-scan television	Image Comms	CD-ROM
Jones, Bill	K8CU	1.8 to 54 MHz Low-Pass Filter	11.11.6	11.49
Karlquist, Rick	N6RK	Audio latency measurement system	29.4.3	29.21
		Mixers, Modulators, and Demodulators	Chapter 10	10.1ff
Kay, Leonard	K1NU	Analog Basics	Chapter 3	3.1ff
		RF Techniques	Chapter 5	5.1ff
Keuken, Jack	KE2QJ	Base-loading system for whip antennas	21.8.3	21.54
Kleinschmidt, Kirk	NTØZ	Generators	29.3	29.13
Klitzing, James	W6PQL	All-Mode, 2 Meter, 80 W Linear Amp	17.11	17.38
Lakhe, Rucha		Mathematical stability problems in non-linear simulation programs	Chapter 6	CD-ROM
Lampereur, Steve	KB9MWR	10 GHz 2 W Amplifier	17.11	17.40
Lapin, Greg	N9GL	Analog Basics	Chapter 3	3.1ff
Larkin, Bob	W7PUA	RF semiconductor design	Chapter 5	5.1ff
Lau, Zack	W1VT	ARRL phase noise measurement	9.2.5	9.10
Lewellan, Roy	W7EL	JFET Hartley VFO	9.3.2	9.17
Lindquist, Rick	WW1ME	What Is Amateur Radio?	Chapter 1	1.1ff
Loveall, Pete	AE5PL	D-STAR repeaters	18.4	18.11
Luetzelschwab, Carl	K9LA	Propagation	Chapter 19	19.1ff
Lux, Jim	W6RMK	Electrical Safety	28.1	28.1ff
Martin, Mike	K3RFI	Power-line RF Noise	27.6	27.17ff
McClellan, Jim	N5MIJ	D-STAR repeaters	18.4	18.11
McCune, Earl	WA6SUH	Oscillators	Chapter 9	9.1ff
Miller, Russ	N7ART	2 meter RF power amplifier	Chapter 17	CD-ROM
Moell, Joe	KØOV	Radio Direction-finding antennas	21.12	21.72ff
		Radio Direction-finding techniques	Chapter 27	CD-ROM
Montgomery, Christine	KGØGN	Digital Basics	Chapter 4	4.1ff
Morris, Steve	K7LXC	Tower and antenna safety	28.2	28.12ff
Mullett, Chuck	KR6R	CAD for power supplies	17.11	17.28
Newkirk, David	W9VES	Computer-Aided Circuit Design	Chapter 6	6.1ff
		Mixers, Modulators, and Demodulators	Chapter 10	10.1ff
O'Hara, Tom	W6ORG	Amateur television	Image Comms	CD-ROM
Ott, Henry	WA2IRQ	Electromagnetic Compatibility	Chapter 27	27.1ff
Pearce, Gary	KN4AQ	FM voice repeaters	Chapter 18	18.1ff
Pittenger, Jerry	K8RA	3XC1500D7 RF Linear Amplifier	17.12.2	17.45
Pocock, Emil	W3EP	Propagation	Chapter 19	19.1ff
Portune, John	W6NBC	Coaxial Dipole for VHF or UHF	Chapter 21	21.64
Rado, John	KØIZ	Adapting Aviation Headsets to Amateur Radio	Chapter 24	24.34
Rauch, Tom	W8JI	RF Current Meter	25.8.7	25.44
Rogers, Robert	WA1PIO	Inductance Tester	25.8.3	25.41

Notes

Notes

Notes

Notes

Notes

Notes

Notes

Notes

FEEDBACK

Please use this form to give us your comments on this book and what you'd like to see in future editions, or e-mail us at **pubsfdbk@arrl.org** (publications feedback). If you use e-mail, please include your name, call, e-mail address and the book title, edition and printing in the body of your message. Also indicate whether or not you are an ARRL member.

Where did you purchase this book? ☐ From ARRL directly   ☐ From an ARRL dealer

Is there a dealer who carries ARRL publications within:

☐ 5 miles   ☐ 15 miles   ☐ 30 miles   of your location?   ☐ Not sure.

License class:

☐ Novice   ☐ Technician   ☐ Technician with code   ☐ General   ☐ Advanced   ☐ Amateur Extra

Name _____   ARRL member? ☐ Yes ☐ No

_____   Call Sign _____

Address _____

City, State/Province, ZIP/Postal Code _____

Daytime Phone   (   ) _____   Age _____

If licensed, how long? _____

Other hobbies _____   E-mail _____

_____

Occupation _____

For ARRL use only	2014 HBK
Edition	91 92 93 94
Printing	1 2 3 4 5 6 7 8 9 10 11 12

From _____

_____

_____

EDITOR, THE ARRL HANDBOOK
ARRL—THE NATIONAL ASSOCIATION FOR AMATEUR RADIO
225 MAIN STREET
NEWINGTON CT 06111-1494

please fold and tape

About the

Included CD-ROM

On the included CD-ROM you'll find this entire *Handbook*, including text, drawings, tables, illustrations and photographs — many in color. Using Adobe *Reader* (available from **Adobe.com**), you can view and print the text of the book, zoom in and out on pages, and copy selected parts of pages to the clipboard. A powerful search engine — Reader Search — helps you find topics of interest. Also included is supplemental information and articles, PC board template packages for many projects, and companion software mentioned throughout. See the README file on the CD-ROM for more information.

Adobe *Reader* must be installed on your computer to access the content and features of the ARRL 2014 Handbook. Adobe *Reader* is not installed with the ARRL Handbook but may be downloaded from **Adobe.com**.

INSTALLING YOUR ARRL HANDBOOK CD

Windows

1. Close any open applications and insert the CD-ROM into your CD-ROM drive. (Note: If your system supports "CD-ROM Insert Notification," the *ARRL Handbook CD* may automatically run the installation program when inserted.)

2. Select **Run** from the *Windows* **Start** menu.

3. Type **d:\setup** (where d: is the drive letter of your CD-ROM drive; if the CD-ROM is a different drive on your system, type the appropriate letter) and press **Enter**.

4. Follow the instructions to install the CD-ROM files on your system.

Macintosh

To install the CD files on the Mac, open the CD-ROM (ARRL 2014 Handbook) icon then click to go into the "Program Files" folder. Drag the "ARRL 2014 Handbook" folder icon to your hard drive.

To use the search capability in the *Handbook CD* you will need to have Adobe *Reader* installed. Mac Preview cannot access all of the features of the PDF files. If you need to install the Adobe *Reader* program, it may be downloaded from **Adobe.com**. After installation you will be then able to open the *Handbook* files in the *Reader* to view them.

USING YOUR ARRL HANDBOOK CD

Windows

Click on START, ALL PROGRAMS, then click ARRL SOFTWARE, and ARRL 2014 HANDBOOK CD to start the *Handbook CD* viewer. The ARRL 2014 HANDBOOK menu also has additional shortcuts for accessing the included Supplemental Files and Companion Software Files on the CD.

If you chose not to install the *Handbook* files on your hard drive, you will need to load the CD in the drive first, then go to MY COMPUTER, select your CD Drive, then click down into PROGRAM FILES, then ARRL 2014 HANDBOOK, and then click to start "intro.pdf".

Macintosh

Go to where you copied the ARRL 2014 Handbook folder onto your hard disk, then look for the "intro" file. Click to start this file in Adobe *Reader*.

If you chose not to install the *Handbook* files on your hard drive, you will need to load the CD in the drive first, then go to CD Drive icon (ARRL 2014 Handbook) on your desktop, then click down into PROGRAM FILES, then ARRL 2014 HANDBOOK, and then click to start "intro.pdf".

SUPPLEMENTAL FILES, TEMPLATES AND SOFTWARE

See the **Supplemental Files** and **Companion Software** sections in the *Handbook* CD-ROM Table of Contents for information about accessing the many supplemental files, template packages and other support features included on the CD.